XIX International Botanical Congress (IBC 2017)

Oral Presentations

Shenzhen, China
23 - 29 July 2017

Abstract Book I

ISBN: 978-1-5108-6169-5

Printed from e-media with permission by:

Curran Associates, Inc.
57 Morehouse Lane
Red Hook, NY 12571

Some format issues inherent in the e-media version may also appear in this print version.

Printed by Curran Associates, Inc. (2018)

IBC 2017
Lianhua Zhilu 1004
Futian District Shenzhen
P.R. China

Phone: (+86)755 8307 3632

office@ibc2017.cn

Additional copies of this publication are available from:

Curran Associates, Inc.
57 Morehouse Lane
Red Hook, NY 12571 USA
Phone: 845-758-0400
Fax: 845-758-2633
Email: curran@proceedings.com
Web: www.proceedings.com

ABSTRACT BOOK I

Oral Presentations

IBC 2017
XIX International Botanical Congress
Shenzhen China

XIX International Botanical Congress
Shenzhen Convention & Exhibition Center
July 23-29, 2017

Nomenclature Section
Peking University HSBC Business School
July 17-21, 2017

CONTENTS

PUBLIC LECTURES

Ginkgo: An evolutionary and cultural biography

Peter Crane

Oak Spring Garden Foundation, 1776 Loughborough Lane, Upperville, VA 20184, USA

School of Forestry and Environmental Studies, Yale University, 195 Prospect Street, New Haven, CT 06511, USA

Perhaps the world's most distinctive tree, living *Ginkgo biloba* is a botanical oddity that has remained little changed for at least 55 million years, and is also very similar to plants that flourished in the late Triassic and Jurassic over 200 million years ago. First brought to the attention of western science by Kaempfer, and named by Linnaeus only shortly before his death, *Ginkgo* was initially regarded as a conifer. It was later recognized as distinct in plant classifications of the early nineteenth century, a conclusion underlined subsequently by the remarkable discovery in *Ginkgo* (and also in cycads) of motile sperm cells. The precise phylogenetic position of *Ginkgo* among living and fossil seed plants remains uncertain, although most analyses using molecular data, as well as many morphological analyses that include fossils, associate *Ginkgo* with conifers. Classic work by Zhou Zhiyan has demonstrated that diverse *Ginkgo*-like plants existed during the Jurassic in China, and recent paleobotanical discoveries hint even greater diversity through the Jurassic and Early Cretaceous, as well as intriguing similarities to certain groups of Mesozoic "seed ferns". For much of the Mesozoic *Ginkgo*-like plants were widespread in both hemispheres, but their diversity declined during the mid-Cretaceous. They disappeared from the Southern Hemisphere early in the Cenozoic. In the Northern Hemisphere *Ginkgo* remained widely distributed through the Late Cretaceous and early Cenozoic, including at high paleolatitudes. Ultimately, however, *Ginkgo* barely survived the marked climatic changes of the later Cenozoic. It is last recorded in western North America about 15 million years before present. It disappeared still more recently from Europe, parts of Asia and Japan, and apparently survived the Pleistocene Ice Ages only in China. The only remaining putatively wild populations exist as relics in central-western and eastern China and recent research suggests that all living *Ginkgo* trees were ultimately derived from these two relictual populations. Despite its near extinction *Ginkgo* began its remarkable contemporary renewal and resurgence when people in China first found it useful as a "nut tree" about a thousand years ago. Subsequently *Ginkgo* was grown widely in China and appears to have been introduced into Korea and Japan in medieval times. In the mid-seventeenth century *Ginkgo* was introduced into Europe and by 1784 it was being grown outside of Philadelphia in North America. Today *Ginkgo* is an increasingly familiar living link to landscapes of the distant past. Its "nuts" are still enjoyed in the cuisines of eastern Asia, but it is also revered for its longevity, beloved for the elegance of its leaves, valued as a medicinal, and widely cultivated as one of the world's most popular street trees. In both the East and the West *Ginkgo* is also a source of artistic and spiritual inspiration. *Ginkgo* remains of great scientific importance and may provide our best window into the lives of diverse ancient seed plants that flourished between about 200 and 100 million years ago. The recent sequencing of the *Ginkgo* genome, and discoveries of new and exceptionally well-preserved fossil material, both offer exciting possibilities for a deeper insights into the life story of this remarkable tree, as well as its broader significance for understanding plant evolution.

People and plants – the unbreakable bond

Sandra Knapp

Life Sciences, Natural History Museum, Cromwell Road, London SW7 5BD, United Kingdom

Plants form the scaffold for Earth's green ecosystems, but they are also essential for human survival. Plants provide most of the food we eat (directly or indirectly), our medicines, clothes, buildings, and even the air we breathe; they also beautify our daily lives. So we need plants, but do they need us? In this time of increasing human impact on plants, animals and natural habitats, our actions can make a big difference in whether plants are a part of an ecological civilisation for the future. Plants do in fact need us - they need us to study and use them responsibly, both as scientists and as members of human societies. So what can plant science contribute to the survival of these species on which we depend? I will draw on my own experiences as a scientist working on a flowering plant family of enormous economic importance, the Solanaceae, to explore how basic science can become part of many different aspects of people's use of plants. Members of the Solanaceae have uses in horticulture (petunias), medicines (mandrakes, tobacco, and henbane), and they are among our most important foods (potatoes, tomatoes, and eggplants). I will look at how people have changed and modified these species, and explore the changes that have driven the use of some species and not others. The journeys taken by plants in the company of people involve many changes, on both sides. It could be said that rather than domesticating plants for our own use, plants are in fact domesticating people! So our relationship with plants is deep and complex – but what should it be in the future? Should we confine plant species to wild areas where we can see them, but not interact or use them? Or should we manage plant diversity strictly for human use and enjoyment by bringing more plants into cultivation and into close relationships with us? Or perhaps something in between? I will argue that to really develop the bonds that link plants to people and create a society in harmony with nature all of use, not just scientists, must work together in order to maintain these relationships now and for future generations.

Saving plants to save ourselves: The Shenzhen Declaration

Peter H. Raven

Missouri Botanical Garden, P.O. Box 299, St. Louis MO 63166, USA

Our planet consolidated into its present form about 4.5 billion years ago, with life appearing at least 3.9 billion years ago. The photosynthetic group of bacteria known as Cyanobacteria began producing oxygen about 2.7 billion years ago, with oxygen reaching present or even higher levels around the start of the Cambrian Period, 541 million years ago, about the same time that multicellular life first appeared. Animals, plants, and fungi colonized the land and began to diversify about 450 million years ago, with seed plants appearing about 370 million years ago and angiosperms some 135 million years ago. The ancestors of both dinosaurs and mammals appeared about 200 million years ago. The fourth great extinction occurred at about the time dinosaurs and mammals first appeared, with the last dinosaurs disappearing at the time of the fifth and last great extinction, at the end of the Cretaceous Period about 66 million years ago. Mammals then began to diversity, with primates, the mammalian order of which we are members, appearing about 55 million years ago. It is estimated that about 99% of all kinds of organisms that ever existed on earth are extinct, but when our ancestors diverged from the African apes about 6-8 million years ago, they did so in a world of biological richness, with perhaps 12 million species of eukaryotes and countless millions of species of prokaryotes supporting the functioning of the global ecosystem. The oldest fossils we know of our genus, *Homo*, date from about 2.8 million years ago, the oldest of our species, *Homo sapiens*, from about 200,000 years ago; these are both from Africa, with *Homo sapiens* migrating to Eurasia about 60,000 years ago. Therefore we are very new on Earth, but our activities have become devastating to our continued existence and that of other organisms as well.

The human development of agriculture as a main source of food about 10,000 years ago led to an explosion of our numbers from about 1 million people then to 7.4 billion today, projected to grow to 9.9 billion by mid-century, 33 years from now. This has made us an ecological force so great that we can be said to have entered the Anthropocene Period, one in which our actions will determined the fate of everything on Earth. For example, our bodies and those of our domestic animals now constitute 95% of all mammalian biomass on our planet. We are estimated to be consuming on a continuing basis about 164% of the planet's sustainable productivity, with some of us rich and many poor (www.footprintnetwork.org) . Like all organisms, we depend on the functioning of the global ecosystem for every aspect of our lives, and yet we are currently tearing it apart. Extinction is currently taking place at a rate 1,000 times or more greater than it has since the end of the Cretaceous Period 66 million years ago, so that as many as half of the estimated 12 million species of eukaryotes existing now are likely to have disappeared from nature by the end of the 21st Century, and we shall not even have recorded most of them.

Plants, and photosynthesis in general, do not receive the attention they deserve, and if we want to attain global sustainability, we shall have to work much harder to gain the knowledge that we need to save ourselves and to apply it to the maintenance of a sustainable world. We shall need to emphasize field studies in order to understand better the functioning of our globe. We have little or no knowledge about the majority of species of plants or other organisms on earth, despite our living during the last decades when many of them will exist. Nations and peoples have become starkly unequal, women and children often neglected or suppressed, and yet we are in danger of perishing together unless we do much better in these respects. We need to learn to cooperate with and love one another while there is still time to do so. We must attain a level population, equality between peoples, and better technologies while we still can do so. Plants, the subject of this Congress, and agriculture must receive greater attention since we depend so clearly and completely on them. Let us make this Congress a time of commitment to do better and resolutely seek a sound and sustainable future for all people.

Plants and people: History and future

Barbara A. Schaal
Washington University, USA

The relationship between humans and plants extends far back into hominid history. Plants have shaped our evolutionary past, adapting humans to a series of diet changes by alterations in morphological traits such as the structure of teeth, biochemical adaptations to novel foods and plant secondary compounds, and even alterations in our taste receptors. Humans early on have used plants as part of their culture, their medicine and their religion, forming a deep bond with native species. In turn, humans, with the advent of plant domestication and agriculture, have changed plants, altering them to increase yield, to simplify cultivation, and to modify their basic biochemistry for enhanced taste, and reduction in noxious compounds. This evolutionary dance between humans and plants has been central for the development and flourishing of agriculture. Agriculture, in turn, allowed humans to settle in towns and fostered the subsequent development of more complex civilizations. This legacy has persisted for at least 10,000 years with plants providing fiber and food for both people and animals. The complexity of our modern cities and cultures rests in part on the reliable production and transportation of food.

But, what is the future? How do we provide food security and proper nutrition to a growing global population? How do we grow more food on less land and reduce the environmental footprint of agriculture? Can plants be part of the new bio-economy, providing income to farmers in both developed and developing countries? And how do we affect all of these advances in the face of a rapidly changing climate? These are monumental changes facing the entire globe and they present both a challenge and an opportunity for plant science.

Green life, beautiful genes

Huanming Yang
BGI, Shenzhen, China

We have all the reasons to sing praises for plants: We survive by take-in plants everyday, and we, together with all other animals, breath with the oxygen made by plants every minute. We live in a world to which plants contribute so much to its sustainability and balanced ecological environment.

Plants do not only teach us the sense of harmony and colorfulness, but also help us open the doors to understanding the secrets of all life. It is the plant that provided the tools to discover the genetic rules of genes; it is the plant that helped us find out the genes are even able to "jump"; it is the plant, listed as one of the model organisms of the International Human Genome Project (HGP), helped understand ourselves, as well as providing the new strategy to understand plants.

The discovery of DNA double helix structure opened a new epoch of life sciences and biotechnologies, which revolutionized our understanding of the life and diversity of plants. The HGP, with its belief that "*Life is of Sequences, Life is Digital*", provides a now concept of genomics and all other omics, a new technology of sequencing, and a new culture of globalized collaboration under the HGP Spirit, "*needed by all, owned by all, done by all, and shared by all*".

The Earth BioGenome Project (EBP), which was announced in February this year, has the ambitious missions to sequence "all life on Earth", and plants accounting for a major part. We call again for international collaboration, especially all those working on plant sciences and omics-based breeding.

PLENARY LECTURES

Evolution of plant sexual diversity

Spencer C.H. Barrett
Department of Ecology & Evolutionary Biology, University of Toronto, Toronto, 25 Willcocks Street, Ontario, Canada M5S 3B2

Flowering plants possess striking sexual diversity despite the hermaphroditic condition of most species. Male and female gametes are deployed in a variety of spatial and temporal combinations at the flower, inflorescence and population level resulting in diverse sexual systems. As Darwin recognized, species with polymorphic sexual systems are especially valuable for evolutionary analysis, and studies of heterostyly and dioecy continue to provide key insights into evolutionary processes. Here, I provide an overview of our current work on the causes and genetic consequences of two of the most important evolutionary transitions in angiosperms: the pathway from outcrossing to predominant self-fertilization, and the origin of separate sexes (females and males) from hermaphroditism. The approaches used to address these problems include field studies of the ecology, demography and genetics of populations; comparative methods; genetic mapping of floral traits; and genomic and molecular evolutionary analyses.

The evolutionary breakdown of heterostyly to homostyly represents a classic system for investigating the shift from outcrossing to selfing. In *Eichhornia* this transition has occurred multiple times both within and among species. Evidence from *E. paniculata* indicates that the joint evolution of genetic drift and selection for reproductive assurance explains repeated transitions to selfing. Genotyping-by-sequencing and composite interval mapping of independently derived selfing forms that have colonized the Caribbean and Central America from Brazil, indicate that stamen modifications governing selfing occur on different linkage groups from the loci controlling tristyly. Each selfer has modifiers on a different linkage group consistent with their separate origins. This demonstrates that selfing can evolve via different genetic pathways within the same species. Transitions to selfing have resulted in a reduced efficacy of selection, the purging of deleterious mutations and changes in gene expression across the genome.

The transition from hermaphroditism to dioecy has occurred on multiple occasions in angiosperms. Although theory predicts 1:1 sex ratios in dioecious populations, biased ratios are common but the causes are not well understood. Our comparative analyses of angiosperms have revealed an association between heteromorphic sex chromosomes and female-biased sex ratios. In *Rumex*, a genus with heteromorphic sex chromosomes, natural populations often exhibit female-biased sex ratios and experimental studies have demonstrated that increased pollination intensity causes female bias. Moreover, theoretical work indicates that a balance between Fisherian sex-ratio selection and selection against deleterious mutations during the haploid phase can explain stable female biased sex ratios. These findings motivated us to investigate if the observed bias in populations is associated with Y-chromosome degeneration. Studies of sex-linked SNPs and transcriptome data have provided evidence of different degrees of Y-chromosome degeneration, depending on the time since recombination suppression, and a loss of neutral diversity consistent with selective interference among linked sites. Future comparative work on the genus will further investigate biased sex ratios, haploid selection and sex chromosome evolution across the phylogeny. Collectively, these studies of two polymorphic sexual systems both highlight the important genomic consequences of recombination suppression and also the insights that can be gained from intraspecific studies or natural variation.

Applications of genomics to crop plant improvement

Michael Bevan
John Innes Centre, Norwich, UK

Recent advances in genome sequencing technologies, coupled to radically improved assembly and annotation methods, have revolutionized genomics. These approaches are now being used to tackle crop plants with large and complex polyploid genomes. New methods for sequencing populations of crop plants and their wild relatives are also providing unprecedented access to a wide range of genetic variation. In my presentation I will use the wheat genome as an example of such progress. The question now is how can genomics guide and accelerate the breeding and engineering of crops with improved characteristics. I will provide examples of how our research teams at the John Innes Centre and the Earlham Institute in the UK are going about delivering the great promise of genomics.

On the exceptional value of model lineages in systematics and comparative biology

Michael J. Donoghue
Yale University, USA

With reference to a model portraying the scope of systematics and comparative biology, I will argue that comprehensive analyses of model lineages have a critical role to play in advancing our understanding of plants and their evolution. Such studies provide new avenues for analyzing how organisms are integrated and the evolutionary causes and consequences of this integration. They also provide a sound basis for the comparative analyses that are most likely to yield insights of lasting significance. As a concrete example I will highlight our ongoing studies of the angiosperm clade *Viburnum* (Adoxaceae). Specifically, I will develop connections between our analyses of leaf structure, function and evolution in *Viburnum* and global patterns in leaf form and leafing habit. These studies identify new traits of potential interest in studies of global change.

Fertilization mechanisms: A comparison of eudicots and grasses

Thomas Dresselhaus
Cell Biology and Plant Biochemistry, Biochemie-Zentrum Regensburg, University of Regensburg, 93053 Regensburg, Germany

After leaving their aquatic environment land plants adapted to various environmental challenges and evolved specialized reproductive structures. Originally, their sperm cells were motile and capable to swim to egg cells. This mode of fertilization is still preserved in most mosses and ferns preferring moist environments. However, flowering plants or angiosperms, which dominate in colder and dryer habitats, evolved embedded gametophytes to protect egg and immobile sperm cells. Male gametophytes or pollen grains develop in anthers, while female gametophytes or embryo sacs are produced deeply inside the maternal tissues of ovule and ovary, respectively. After release pollen grains of angiosperms are transported by wind or animals to female flower organs of the same species. They germinate, penetrate stigmatic tissue and grow through the style towards the ovule to deliver their passive sperm cell cargo. Inside the female gametophyte double fertilization is executed resulting in embryo and endosperm formation, respectively. To avoid fertilization failures and to promote outcrossing, angiosperms have developed various mechanisms to distinguish between own and alien pollen as well as pollen of the own species. Using *Arabidopsis* as a eudicot and maize as a grass model, I will compare morphological differences and focus on similar and distinct molecular mechanisms associated to fertilization. These include various hurdles along the pollen tube pathway, during sperm cell release and gamete activation. The analysis of genes involved in pollen adhesion, hydration and penetration, pollen tube growth and guidance as well as double fertilization and onset of embryo development will be presented. Some genes with key functions for fertilization and egg activation in maize lack homologs in *Arabidopsis*, while other genes are highly polymorphic. Molecular systems to discriminate own and foreign tubes will be presented and possibilities discussed to generate a new species.

Deconstructing the plant regulatory genome: Enabling technologies

Joseph R. Ecker
International Council Chair in Genetics
Investigator, Howard Hughes Medical Institute
The Salk Institute for Biological Studies, La Jolla, CA, USA

Genetic and epigenetic variation can impact gene transcription and phenotypic diversity and may thus contribute to adaptation. We have produced and characterized > 1000 high quality single-base resolution genomes, methylomes, and transcriptomes from a global collection of *Arabidopsis thaliana* ecotypes/accessions. Although the genetic basis of methylation variation is highly complex, geographic origin is a major predictor of global DNA methylation levels and altered gene expression caused by epialleles. Analysis of the cistromes and epicistromes of these 1001 accessions revealed significant association between transcription factor binding (TF) sites, methylation and nucleotide variation and co-expression modules. To more directly interrogate the effects of genetic and epigenetic variation on TF binding, we developed a high-throughput TF binding site discovery method and defined the *Arabidopsis* cistrome by resolving motifs and peaks for a large number of TFs. We have also developed a massively multiple yeast 2-hybrid based method for studies of protein:protein interactions and applied this approach to study of *Arabidopsis* TF:TF interactions. These approaches are advancing our understanding of how both genetic and epigenetic variation contribute to phenotype in natural populations.

1001 Genomes Consortium. 1,135 Genomes Reveal the Global Pattern of Polymorphism in *Arabidopsis thaliana*. *Cell*. 2016 Jul 14;166 (2):481-91. doi: 10.1016/j.cell.2016.05.063. Epub 2016 Jun 9. PubMed PMID: 27293186; PubMed Central PMCID: PMC4949382.

Kawakatsu T, Huang SS, Jupe F, Sasaki E, Schmitz RJ, Urich MA, Castanon R, Nery JR, Barragan C, He Y, Chen H, Dubin M, Lee CR, Wang C, Bemm F, Becker C, O'Neil R, O'Malley RC, Quarless DX; 1001 Genomes Consortium, Schork NJ, Weigel D, Nordborg M, Ecker JR. Epigenomic Diversity in a Global Collection of *Arabidopsis thaliana* Accessions. *Cell*. 2016 Jul 14;166 (2):492-505. doi:10.1016/j.cell.2016.06.044. PubMed PMID: 27419873.

O'Malley RC, Huang SS, Song L, Lewsey MG, Bartlett A, Nery JR, Galli M, Gallavotti A, Ecker JR. Cistrome and Epicistrome Features Shape the Regulatory DNA Landscape. *Cell*. 2016 May 19;165 (5):1280-92. doi: 10.1016/j.cell.2016.04.038. PubMed PMID: 27203113; PubMed Central PMCID: PMC4907330.

Structural characteristics of forests and diversity of woody plants in China

***Jingyun Fang**, Zhiyao Tang, Zhiheng Wang, Zehao Shen, Xiangping Wang, Chengyang Zheng*
Department of Ecology, College of Urban & Environmental Sciences, Peking University, Beijing 100871, China

China is rich in forest communities and woody plants. To systematically investigate species composition and structure of China's forest communities, we launched a long-term project consisting forest vegetation surveys across China's mountains in the mid 1990s. Over the study period, we have conducted vegetation surveys for more than 70 mountains and collected vegetation data from about 2000 forest plots, using consistent sampling protocols. Meanwhile, to explore geographic patterns and their environmental controls of woody species diversity, we established a huge database of distribution of China's woody plants that contains all 11405 woody species in the country. Our results indicated that as latitude increased, diameter at breast height (DBH) and height of trees increased, while individual density of trees and woody species richness decreased. Total basal area (TBA) of trees and species richness of herbs did not vary with latitude. Contemporary climate seems to drive these patterns: temperature was the leading factor for DBH, precipitation was most important for tree height and individual density, actual evapotranspiration (a surrogate of productivity) determined woody (trees and shrubs) species richness, and rainfall was the major controller of the herb species richness. The species–abundance relationship showed that species dominance (measured by the number of individuals per species) declined significantly from boreal forests to evergreen broadleaf forests from north to south. Further, using the database of distribution of woody plants, we compared the effects of habitat heterogeneity, human activities and climatic variables (particularly environmental energy, water–energy dynamics and winter frost) on the woody species diversity, and found that the contemporary climate explained the spatial variation of more than 80%, of which winter mean temperature of the coldest quarter was the strongest predictor, thus supporting the freezing-tolerance hypothesis.

Understanding the molecular basis of rice plant architecture

Jiayang Li

State Key Laboratory of Plant Genomics and National Center for Plant Gene Research (Beijing), Institute of Genetics and Developmental Biology, Chinese Academy of Sciences, Beijing, China

Rice is the most important staple crop in China and an ideal model plant. How to increase yield and improve quality has been the major challenge in rice research and breeding. However, to meet the security of food supply, the crop yield per unit area has to be increased by 50% before 2030. My lab has been focused on systematical dissection of key genes regulating rice plant architecture, elucidation of their molecular mechanisms and the genetic regulatory network, and application of this fundamental knowledge to breed new high-yield elite varieties.

Plant architecture is a primary determinant of crop yield and has profound adaptability in response to internal and external factors. Research on rice plant architecture and understanding of the molecular basis and regulatory mechanism are of great importance to both basic science and practical application. Plant architecture is mainly determined by plant height, tiller number, tiller angle and panicle morphology. My lab has molecularly characterized a series of rice mutants that show various altered plant tillering or panicle phenotypes, including *monoclum 1* (*moc1*), *tillering and dwarf 1* (*tad1*), *ideal plant architecture 1* (*ipa1*), *lazy1* (*la1*), *small panicle 1* (*sp1*), *dwarf 27* (*d27*), and *dwarf 53* (*d53*). We cloned these corresponding genes through map-based cloning approaches and in-depth characterization of their functions, demonstrating that the tillering phenotypes mainly result from either the deficiency in the branching plant hormone strigolactone biosynthetic and signaling pathways or the transcription factors that play key roles in plant architecture formation. The understanding of these gene functions enable us to elucidate a comprehensive regulatory network underlying the formation of plant architecture and greatly expanded our knowledge of plant development.

Based on the knowledge from the study on plant architecture of rice, our lab and collaborators have made enormous efforts in breeding new elite varieties through developing a molecular breeding system. These results demonstrate that this system could truly speed up the rice breeding course and generate new super rice with multiple advantages. However, to fully meet the challenge of food supply, more efforts need to be made by providing more genetic resources and designing different allele combinations to meet the unique demands in different regions in China.

Computational morphodynamic approaches to pattern formation in the shoot apical meristem of *Arabidopsis*

Elliot Meyerowitz

Division of Biology and Biological Engineering and Howard Hughes Medical Institute, California Institute of Technology, Pasadena, CA 91125, USA

The shoot apical meristem is the stem cell niche responsible for generating the above-ground parts of flowering plants, including stems, leaves and flowers. Though only a few hundred cells, it has a large number of overlapping but different domains of gene expression. These gene expression domains are dynamic – some genes change their expression relative to appearance of new leaf or flower primordia. Others are expressed in what appear to be stable domains in the meristem – but are not stable, as the cells that express the genes are dividing, and some progeny cells enter these domains as others depart, making them dynamic at the cellular level, even though stable in comparison to the meristem boundaries. How are these gene expression patterns controlled? By iteration between computational models of cell-cell interactions, and experiments to test the models, we have developed robust hypotheses for flower primordium-influenced gene expression that involve interactions between chemical (auxin) and mechanical signals in the meristem, and for meristem region-specific expression we have models where cell-cell communication is mediated by a combination of small-molecule (cytokinin) and peptide signals.

The origin of economically important species of *Cucumis*, *Citrullus*, and *Momordica* – combining Chinese genomics know-how and old European herbaria to find wild relatives

Susanne S. Renner

Department of Biology, University of Munich (LMU), Munich, 80628, Germany

China is World-leading in Cucurbitaceae genomics, having produced the draft genomes of cucumber, *Cucumis sativus* (Huang *et al.*, Nature Genetics 2009; also Qi *et al.*, Nature Genetics, 2013; Shang *et al.*, Science 2014; Zhou *et al.*, Nature Plants 2016) and watermelon, Citrullus 'lanatus' (Guo *et al.*, Nature Genetics 2013). Old-fashioned Sanger sequencing over the past 10 years, however, has revealed numerous misclassifications of material of Cucurbitaceae, from simple misidentification of germplasm to serious misplacements of species in unnatural genera, meaning that names do not reflect evolutionary relationships and cannot be trusted to guide sampling for genomic projects. The family Cucurbitaceae is underrepresented in herbaria because wild cucurbits often are climbers in disturbed habitats or desert plants that are hard to find at the right stage. Fruits of the family are difficult to preserve. Only with DNA barcoding has it become possible to figure out what is what (simply put). In the talk I will illustrate how molecular data are revolutionizing our knowledge about the wild relatives of *Cucumis sativus* (in India, Thailand, and tropical China), *Cucumis melo* (in India and Australia), *Citrullus lanatus* (in West or Northeast Africa), *Momordica charantia* (in Africa) and *Luffa* (in India and Australia), using examples of recent and ongoing work in my lab, in collaboration with genomics experts from China.

EXAMPLES:

Renner, S. S., H. Schaefer, and A. Kocyan. 2007. Phylogenetics of Cucumis (Cucurbitaceae): Cucumber (C. sativus) belongs in an Asian/Australian clade far from melon (C. melo). BMC Evolutionary Biology 7:58. DOI: 10.1186/1471-2148-7-58 ("highly accessed")

Schaefer, H., C. Heibl, and S. S. Renner. 2009. Gourds afloat: A dated phylogeny reveals an Asian origin of the gourd family (Cucurbitaceae) and numerous oversea dispersal events. Proceedings of the Royal Society B 276 (1658): 843-851.

Schaefer, H., and S. S. Renner. 2010. A three-genome phylogeny of Momordica (Cucurbitaceae) suggests seven returns from dioecy to monoecy and recent long-distance dispersal to Asia. Molecular Phylogenetics and Evolution 54 (2): 553-560.

Sebastian, P. M., H. Schaefer, I. R. H. Telford, and S. S. Renner. 2010. Cucumber and melon have their wild progenitors in India, and the sister species of Cucumis melo is from Australia. Proceedings of the National Academy of Sciences, USA 107 (32): 14269–14273.

Chomicki, G., and S. S. Renner. 2015. Watermelon origin solved with molecular phylogenetics including Linnaean material: Another example of museomics. New Phytologist 205 (2): 526-532.

Plant evolution in the Anthropocene

Loren H. Rieseberg

Department of Botany and Biodiversity Research Centre, University of British Columbia, Vancouver, Canada

Humans have emerged as the dominant selective force on our planet. This provides unparalleled opportunities for evolutionary biologists to study evolution in action. However, it also leads to the question of whether plants (and other organisms) can evolve fast enough to keep up with the rapid pace of human-caused environmental change. I will discuss what we have learned about plant adaptation and speciation from studies of contemporary evolution. Plant evolution in the Anthropocene is pervasive and can be rapid, but in many species it lags behind environmental change due to long generation times and limited dispersal. In the absence of intervention, extinction of such taxa seems likely. To maintain productive communities and ecosystems, plant biologists must develop strategies to aid adaptation when desired (e.g., evolving climate-adapted crops or forests) and hinder it when necessary (e.g., in weeds). More generally, our focus going forward should be on adaptation rather than preservation of the *status quo.*

Linking heterogeneous data in botanical research

Pamela Soltis

University of Florida, USA

Plant diversity is the product of complex interrelationships among genetics, ecology, and evolution, yet integration of these disciplines of botanical research remains elusive.

Emerging cyberinfrastructure and new data sources provide unparalleled opportunities for mobilizing and integrating massive amounts of information from organismal biology, ecology, genetics, phylogenetics, climatology, and other disciplines. Key among these data sources is the rapidly growing volume of digitized specimen records from natural history collections. With nearly 100 million specimen records currently publically available online, these data provide excellent information on species distributions and changes in phenology and distributions over time. Particularly powerful is the integration of phylogenies with specimen data, enabling analyses of phylogenetic diversity in a spatio-temporal context, the evolution of niche space, and more, while phylogenomic analyses that synthesize phylogenies and genome-scale data shed new light on the role of genomic change in plant diversification. The burgeoning field of phenomics offers still further data for integration with phylogenetic, genomic, and ecological data. However, a major challenge to synthetic research is the heterogeneous nature of complex data, and new methods are needed to link these divergent data types. Ongoing efforts to link and analyze diverse data are yielding new perspectives on a range of evolutionary and ecological problems. We will present multiple case studies that attempt to integrate heterogeneous data sources in analyses of plant diversity. Although many specific hypotheses may be addressed through integrated analyses of linked biodiversity and environmental data, additional value of such data-enabled science lies in the unanticipated patterns that emerge.

Decoding the epigenetic language of life

Jian-Kang Zhu

Shanghai Center for Plant Stress Biology, SIBS, CAS, China

Department of Horticulture and Landscape Architecture, Purdue University, West Lafayette, IN 47907, USA

Epigenetics refers to the study of heritable information that is not contained in DNA sequence. An important epigenetic mark conserved in mammals and plants is DNA methylation, a chemical modification of DNA that controls gene function. Proper DNA methylation patterns are critical for development, diseases and stress responses in humans as well as in plants. Plants are excellent biological systems to study how DNA methylation patterns are generated. DNA methyltransferase enzymes that deposit the DNA methylation mark are guided to specific DNA sequences, and DNA demethylase enzymes that remove the DNA methylation mark are also guided to distinctive sequences to erase unwanted DNA methylation. I will describe work in my lab that has shed light on how DNA methyltransferases and demethylases are guided to specific sequences, and how the antagonistic actions of the enzymes are coordinated to generate proper DNA methylation patterns. I will also describe some of our recent work on how DNA methylation influences transgenerational inheritance.

KEYNOTE LECTURES

Plant speciation across environmental gradients and the occurrence of hybrid zones

Richard Abbott

University of St Andrews, UK

The success of individual plant species to spread across environmental gradients is dependent on their ability to adapt to changed local conditions via adaptive phenotypic plasticity and/or adaptive genetic divergence. There are many examples of heritable adaptive changes in phenotype across environmental gradients and the underlying molecular genetic bases for some of these changes are beginning to be resolved. Such heritable change can occur in the presence or absence of gene flow and may be manifested as continuous change in phenotype (clinal variation) or the establishment of discrete ecotypes. In both cases, reproductive barriers may evolve, often gradually, between locally adapted forms and ultimately result in the formation of new species.

At an intermediate stage in this process, hybrid zones may originate as a result of primary intergradation (with continuous gene flow) or through secondary contact between locally adapted forms previously isolated by means of an eco-geographical barrier. The study of both types of hybrid zone can be highly informative of reproductive barriers that evolve between incipient species across an environmental gradient. Whereas in animal groups hybrid zones are recognised as useful systems for studying the nature and sequence of reproductive barriers that evolve across the speciation continuum, rather surprisingly, they have been little used for this purpose in plants, where they remain a relatively underutilised resource in studies of plant speciation. Nonetheless, a few recent studies of plant hybrid zones involving detailed genomic analysis have been conducted and begun to shed light on the nature of breeding barriers that may evolve during divergence across environmental gradients. These studies provide direction for the future use of hybrid zones in plant speciation research.

One possible reason for why studies of plant hybrid zones are relatively rare in the current literature is that though hybridization occurs frequently among plant species, the establishment of hybrid zones may be uncommon, particularly under relatively stable conditions, i.e. low levels of disturbance. Rarity of hybrid zones, if true, would be an interesting phenomenon in itself and raise questions as to why it is the case. However, more information is required on the occurrence of hybrid zones between plant species before reaching conclusions on their frequency.

Saving plants

Stephen Blackmore
Queens Botanist and Honorary Fellow, Royal Botanic Garden Edinburgh, UK

The international botanical community, meeting here in Shenzhen, is far more aware than most people in wider society of the fundamental role played by plants in shaping and maintaining the biosphere. However, a significant proportion of plant species are now threatened with extinction and plant biodiversity is being lost in other ways, so that even species that were once common are becoming rare. As botanists we also recognise that our current knowledge of plant diversity is far from complete: many plant species remain to be discovered and we are still in the early stages of exploring the genetic diversity. Maintaining plant diversity, at ecosystem, species and genetic levels, Is especially important in an era of global change and has direct relevance to future resilience and adaptation. The continuing erosion of plant diversity has major consequences for our planet and all of its inhabitants: threatening biodiversity in general and making the internationally agreed Sustainable Development Goals even more difficult to achieve.

I will explore some of the key actions humanity needs to undertake in order to conserve global plant diversity, considering conservation in the broadest sense to include any and all interventions that help to secure pant diversity and the ecosystem services delivered by plants. I will emphasise that many different sectors and stakeholders will need to work closely together if we are to succeed in passing on the Earth's rich, green inheritance to future generations. Direct action to conserve plants is carried out by agencies and disciplines as diverse as forestry, agriculture, conservation biology, urban and rural development planning, protected area management and other fields. An even wider set of institutions and organisations are involved in education and outreach, enabling people in wider society to understand the issues and to engage in addressing them.

My focus will be on the potential contribution of botanic gardens, arboreta, seed banks and other collections of living and preserved plants and on the strategic challenges they and other players need to overcome for global efforts in plant conservation to accelerate and expand to an appropriate degree.

Genes gone wild: Experimental evolution meets synthetic biology

Ralph Bock
Max Planck Institute of Molecular Plant Physiology, Am Mühlenberg 1, D-14476 Potsdam-Golm, Germany

In my talk, I will describe experimental approaches to study genome evolution in real time. I will discuss three fundamental processes in eukaryotic genome evolution and show how they can be reconstructed in laboratory experiments: (i) the transfer of organellar (plastid and mitochondrial) genes to the nuclear genome, (ii) the horizontal movement of organellar DNA between plants, and (iii) the movement of nuclear genetic material between plants by horizontal genome transfer. Finally, I will show how horizontal gene transfer can be employed as a versatile tool in plant biotechnology and synthetic biology.

Molecular mechanisms of histone demethylases in genome-wide targeting in *Arabidopsis*

Xiaofeng Cao
State Key Laboratory of Plant Genomics and National Center for Plant Gene Research, CAS Center for Excellence in Molecular Plant Sciences, Institute of Genetics and Developmental Biology, Chinese Academy of Sciences, Beijing 100101, China

Transcription activity of chromatin is regulated by covalent modifications on nucleosomes and DNA. Recent efforts have identified a lot of chromatin modifiers, which play important roles in various aspects of biological processes. These chromatin modifiers normally bind a subset of specific genomic loci. Discovering the mechanisms of recruiting these modifiers is essential for understanding the biological function of them. We previously identified *Arabidopsis* JMJ14 and JMJ12/REF6 as H3K4 and H3K27-specific histone demethylases, which use two distinct mechanisms to their genome-wide targets. The C-terminal FYR domain of JMJ14 interacts with a pair of NAC domain containing transcription factors, which bring JMJ14 to their common target genes; whereas REF6 recognizes its target loci by direct recognizing specific DNA sequence through its tandem C2H2-Zinc finger domains. In this talk, I will illustrate how other factors affect REF6 genome-wide targeting in *Arabidopsis*.

Pant sesquiterpenes: Biosynthesis and bio-interactions

Xiao-Ya Chen
National Key Laboratory of Plant Molecular Genetics, Shanghai Institute of Plant Physiology and Ecology, Chinese Academy of Sciences, Shanghai 200032, China
Plant Science Research Center, Shanghai Chenshan Botanical Garden, Shanghai 201602, China

Terpenoids constitute the largest group of plant secondary (or specialized) metabolites, and they play important roles in plant adaptation to environments. Sesquiterpenes are derived from the 15-carbon farnesyl diphosphate. We are interested in sesquiterpene biosynthesis in plant and the role of sesquiterpene phytoalexins in plant defense against insect pests.

The plant defense hormone jasmonate (JA) plays an important role in regulation of secondary metabolism pathways. The *Arabidopsis* MYC2, a bHLH transcription factor, acts as a key regulator of JA response. We found that in *Arabidopsis* inflorescence MYC2 integrates both gibberellin (GA) and JA signals in regulating the expression of terpene synthase genes. In addition, transcription factors of other families, such as WRKY and AP2/ERF, are also involved in induced formation of sesquiterpenes.

The level and accumulation of secondary metabolites change during plant development and growth. The miR156-targeted SPLs function as an age cue in plant phase transition and maturation, our recent study demonstrated that the SPL factors directly regulate sesquiterpene synthase genes in *Arabidopsis* and of *Pogostemon cablin*, linking the plant age cue to elevated accumulation. The SPLs also attenuate JA signaling through direct interaction with JAZ proteins, leading to age-dependent JA response decay. Conversely, phytoalexins may accumulate constitutively to a higher level in later stages, providing a strong insect resistance to old plants.

*I thank Ying-Bo Mao, Zong-Xia Yu, Gao-Jie Hong and others who contributed to this research.

Mechanisms of microRNA degradation

Xuemei Chen
University of California, Riverside, USA

The abundance of miRNAs is determined by both biogenesis and degradation. While the mechanisms of miRNA biogenesis are fairly well understood, how miRNAs are degraded in vivo, especially when miRNAs are associated with their effector protein ARGONAUTE 1 (AGO1), is poorly understood. In 2005, my group showed that miRNA degradation entails 3' truncation and 3' uridylation, but the enzymes responsible for these processes were unknown. In recent years, we identified the exonucleases and nucleotidyl transferases that are responsible for miRNA 3' truncation and 3' uridylation, respectively. We showed that these enzymes not only act on free miRNAs but also AGO1-bound miRNAs. While these enzymes are largely non-sequence-specific, the degradation of specific miRNAs in vivo can be achieved. One example of specificity determination is the promotion of miR165/6 degradation by AGO10. My talk will be focused on the molecular framework of miRNA degradation.

O-GlcNAc signaling mediates vernalization sensing in wheat

Shujuan Xu[1,2], *Lijing Xing*[1], *Yunyuan Xu*[1], ***Kang Chong***[1]*
1. Key Laboratory of Plant Molecular Physiology, Institute of Botany, Chinese Academy of Sciences, Beijing 100093, China
2 University of the Chinese Academy of Sciences, Beijing 100049, China

Wheat and rice show different preference to low temperature since their various origin and evolution. Cold can damage for living of rice, but be required for flowering of winter wheat. Sensitivity of rice for cold limits its planting only in certain climate-zones. Our previous study suggests COLD1 with a localization on plasmamembrane and ER, interacts with rice G-protein α subunit 1 (RGA1) to activate the Ca^{2+} channel for sensing chilling and to accelerate G-protein GTPase activity. Based on the characters of temperature sensor in mammalians and its Calcium signaling and electriphysiological response to temperature changes, COLD1 is a cold sensor in plants. Vernalization is the acquisition of a plant's ability to flower which requires a prolonged cold temperature. It is controlled by a genetic network such as *VRNs*. However, it's less known that sensing mechanism for prolonged cold, vernalization. Here reports that the *O*-GlcNAcylation and phosphorylation signaling were involved in sensing vernalization in wheat, and *O*-GlcNAc signaling mediated flowering through effecting epigenetic process in *Arabidopsis*.

The post-translational *O*-linked N-acetylglucosamine (GlcNAc) modification of proteins, catalyzed by *O*-GlcNAC transferase (OGT), is involved in multiple important biological processes in animals. However, whether protein *O*-GlcNAcylation affects plant development is largely unknown. Here, we show that loss of function of SECRET AGENT (SEC), an OGT in *Arabidopsis*, leads to weak early flowering time phenotype, which results from decreased transcription of FLC. And ChIP analysis indicates that H3K4me3 level is significantly reduced in sec-5 mutant comparing with that of wild type. We also found that SEC can interact with a protein and catalyzed *O*-GlcNAcylation in vivo to regulate H3K4 methylation impacting the transcription of *FLC*. So the *O*-GlcNAc signaling plays an important role in flowering by affecting the activity of specific protein and the epigenetic process in *Arabidopsis*. Meanwhile we studied the possible functions of *O*-GlcNAc signaling in wheat vernalization. The approaches of LWAC and iTRAQ-TiO2 were used to enrich *O*-GlcNAcylated peptides and phosphopeptides of the plumules from vernalization and non-vernalization wheat, respectively. As a result, about 200 *O*-GlcNAcylated proteins and 164 alternatively changed phosphoproteins were identified. Functional analysis showed that these proteins are mainly involved in metabolic processing, cellular processing and response to stimulus. The inhibitors, such as alloxan for OGT and PUGNAc for OGA were used to treat wheat plants with nonvernalization and vernalization. And the data showed that *O*-GlcNAc signaling in vernalization accelerated flowering transition in winter wheat. During vernalization, gradually increased *O*-GlcNAc modification was detected for TaGRP2, the RNA-binding protein directly binding to *TaVRN1* pre-mRNA to repress its activation. Phosphorylated VER2, a jacalin lectin, recognizes the *O*-GlcNAcylated TaGRP2 to repress its accumulation in the nucleus and attenuate its binding to *TaVRN1*, thereby releasing the repression of *TaVRN1*, ultimately promoting flowering. We speculated that *O*-GlcNAcylation and phosphorylation modification may act as signals to sense vernalization and regulate the network of *VRNs* for flowering in wheat.

Fitter with litter: Plant traits and their legacy for ecosystem services

J. Hans C. Cornelissen
VU University Amsterdam, Netherlands

Litter, i.e. the finer fraction of dead plant material, provides a wide variety of important ecosystem services. However, the role of litter in regulating ecosystem processes and services has mostly been studied in terms of litter presence or amount. Besides in biogeochemistry, we still do not know how litters from distinct plant species differ in their effects on other ecosystem processes and services including biodiversity support. I will propose a new conceptual model that will help us to investigate these differences based on species traits, by explicit reference to two highly distinct and possibly interacting plant 'trait spectra': the widely known (Plant or) Resource Economics Spectrum (RES), and the newly defined Size and Shape Spectrum (SSS). Different ecosystem services provided by plant litter are driven by either one of the trait spectra or by both. I will introduce this concept by drawing from examples from a range of contrasting ecosystem services, including regulatory, recreational, provisioning and biodiversity support services. I will encourage the audience to start empirical work to quantify the linkages between the multi-species RES-SSS trait space and these various services, thereby enhancing our understanding about how exactly dead plant matter matters not only to ecosystems but also to people in different parts of the world.

Seed germination and seedling emerging out of soil: Life and death decisions

*Hui Shi, Shangwei Zhong, **Xing Wang Deng***
State Key Laboratory of Protein and Plant Gene Research, Peking-Tsinghua Center for Life Sciences, School of Advanced Agricultural Sciences and School of Life Sciences, Peking University, Beijing 100871, China

In natural and agriculture systems, seeds often germinate and emerge from the soil to start new life cycles. How plants regulate seed germination and seedling development to emerge from soil is crucial for plants' survival and growth. In the past few years, our series of research greatly advance the knowledge of seed germination and seedling emergence out of soil. We identify the novel positive transcription factor and central repressor in mediating light-induced seed germination. Combining the genetics, molecular biology and mathematical modeling methods, we revealed that seeds use a multilevel regulatory circuit of triple feed-forward loops to sensitively and precisely control seed germination in various light conditions (Shi *et al.*, 2013; Shi *et al.*, 2015). In order to study the molecular mechanism underlying seedling emergence from the soil, we design a convenient and highly reproducible soil assay and show that the transcription factor in ethylene signaling pathway EIN3 is essential for the seedling emergence (Zhong *et al.*, 2014). We further demonstrate that the COP1-EBF1/2-EIN3 tandem E3 ligase module channels the light and mechanical pressure signals to elaborately adjust EIN3 protein levels to ensure seedling emergence out of soil (Shi *et al.*, 2016a). After growing out of the soil to reach the sunlight, dark-grown seedlings face dramatic environmental changes. Our recent studies show that photoreceptor phyB manipulates substrate-E3 ligase interactions in a light-dependent manner, thus directly controlling the stability of EIN3. Therefore, the plants adopt the tandem E3 ligase module to gradually adjust EIN3 protein accumulation in the soil, and elicit a rapid degradation of EIN3 to shut down the ethylene responses by a light-dependent "molecular glue" phyB upon emergence (Shi *et al.*, 2016b). Taken together, our findings demonstrate the molecular actions of key factors and regulatory mechanism during seed germination and seedling emergence out of soil, which helps us to understand the natural phenomena and provide the mechanistic basis for crop genetic improvement in future.

Shi, H., Zhong, S., Mo, X., Liu, N., Nezames, C.D., and Deng, X.W. (2013). HFR1 sequesters PIF1 to govern the transcriptional network underlying light-initiated seed germination in ***Arabidopsis***. Plant Cell 25,3770-3784.

Shi, H., Wang, X., Mo, X., Tang, C., Zhong, S., and Deng, X.W. (2015). ***Arabidopsis*** DET1 degrades HFR1 but stabilizes PIF1 to precisely regulate seed germination. Proc Natl Acad Sci U S A 112,3817-3822.

Shi, H., Liu, R., Xue, C., Shen, X., Wei, N., Deng, X.W., and Zhong, S. (2016a). Seedlings Transduce the Depth and Mechanical Pressure of Covering Soil Using COP1 and Ethylene to Regulate EBF1/EBF2 for Soil Emergence. Curr Biol 26,139-149.

Shi, H., Shen, X., Liu, R., Xue, C., Wei, N., Deng, X.W., and Zhong, S. (2016b). The Red Light Receptor Phytochrome B Directly Enhances Substrate-E3 Ligase Interactions to Attenuate Ethylene Responses. Dev Cell 39,597-610.

Zhong, S.W., Shi, H., Xue, C., Wei, N., Guo, H.W., and Deng, X.W. (2014). Ethylene-orchestrated circuitry coordinates a seedling's response to soil cover and etiolated growth. P Natl Acad Sci USA 111,3913-3920.

Distinct evolutionary dynamics of C4 and CAM photosynthesis

Erika Edwards
Brown University, USA

CAM and C4 photosynthesis are two carbon-concentrating mechanisms (CCMs) that increase photosynthetic efficiency, and together have evolved hundreds of times in flowering plants over the past 30 million years. Although C4 and CAM share a primary biochemical module, their anatomical specializations are quite distinct, and carry significant ecological repercussions. I'll present the latest ideas on the evolutionary trajectories of both C4 and CAM syndromes, highlighting areas of convergence and divergence. An emerging hypothesis is that anatomical modifications preceeded the emergence of CCM biochemistry during C4 evolution, while the reverse may be true for CAM. I'll discuss the evolutionary implications of this order of events, and suggest that structural innovations may act more generally as key evolutionary "bottlenecks": events that happen relatively rarely, and dramatically alter the adaptive landscape and subsequent evolutionary accessibility of novel phenotypes.

Early angiosperms: Key fossils and extinct diversity

Else Marie Friis[1], *Kaj Raunsgaard Pedersen*[2]

1. Department of Palaeobiology, Swedish Museum of Natural History, Stockholm, Sweden

2. Department of Geoscience, University of Aarhus, Aarhus, Denmark.

Fossil floral structures and other remains of angiosperms recovered from Cretaceous sediments have yielded a wealth of data on the organization and structure of early angiosperms. These discoveries are important in the search for character evolution and systematic diversification in early angiosperms. Results obtained so far from the study of Cretaceous angiosperms corroborates to a large extent patterns of diversification obtained from analyses of extant angiosperms using molecular data. However, the fossils also witness of extensive extinction among early angiosperms that cannot be detected studying extant plants alone. In several cases the fossils possess a suite of distinctive characters that are unknown among extant angiosperms, but that may help bridge morphological splits between extant lineages. In other cases the fossils cannot be place in any existing group. We illustrate extinctions and missing links among Cretaceous angiosperms using a number of exquisite preserved key fossils. Focus will be on fossil members of Chloranthaceae and Laurales, but other groups will also be discussed with examples from early diverging eudicots and core eudicots. It has long been suggested that the Chloranthaceae have an extensive geological history extending back to at least the mid-Early Cretaceous. Fossil flowers, fruits, seeds and stamens with pollen in situ particularly from the Early Cretaceous of Portugal have confirmed this suggestion and show considerable diversification very early in angiosperm history with several extinct forms, but also forms that are closely similar to extant taxa. A key fossil among extinct chloranthoids is *Canrightiopsis* that includes several different species. The affinity to extant Chloranthaceae is documented by unique seed coat features as well as details of the tiny embryo and nutritive tissue surrounding the embryo. *Canrightiopsis* is important as a key for understanding character evolution in the Chloranthaceae suggesting an evolutionary link between extinct *Canrightia* and extant *Chloranthus* and *Sarcandra* and a model for the derivation of the tripartite androecium of *Chloranthus*. *Canrightiopsis* co-occur with several other chloranthoid taxa, several of them not described earlier. A new fossil genus with three species under description from the Early Cretaceous is significant for understanding structural and systematic diversity among early members of the Laurales. The fossils show a combination of characters that is unknown among extant or fossil members of the group, but phylogenetic analyses clearly places the fossils in the Laurales. Several features are shared with extant core Laurales such as the presence of a single, pendulous seed and valvate anther dehiscence. In gynoecium features it is particularly similar to extant Gomortegaceae. However, other characters such as dicolpate pollen and continuous tectum are shared with extant Calycanthaceae, the earliest branching lineage of Laurales and the new genus thus provide a link between Calycanthaceae and core Laurales.

Another abominable mystery: Gene-body methylation

Brandon S. Gaut

University of California, Irvine, USA

Darwin thought the abrupt origin and rapid diversification of angiosperms to be an abominable mystery. Here I discuss another ongoing mystery within the genomes of land plants: the evolution and function of DNA methylation in coding genes. Plants methylate both genes and transposable elements (TEs). For the latter, the function is clear: DNA methylation suppresses TE activity. The methylation of TEs has corollary effects on the expression of nearby genes, and thus TE methylation has major consequences for the function and evolution of plant genomes. Just as TEs are methylated in plants, so is a subset of genes. Genic methylation was initially thought to be sporadic, but it is now clear that methylation is a conserved evolutionary property of individual genes. These genes tend to evolve slowly, they are expressed broadly, and their methylation is not a property of their GC content. The cause (s) of conservation remain unclear, as does the function of genic methylation, if indeed it has a function. I will present current information about evolutionary patterns of genic methylation, a recent idea about the mechanism of genic methylation, and varied evidence that genic DNA methylation correlates with gene expression. The function and evolution of genic methylation remains mysterious, however, because it has been difficult to separate cause from effect.

Genetic basis of complex traits and heterosis in rice

Bin Han

National Center for Gene Research, Shanghai Institute of Plant Physiology and Ecology, CAS-Center of Excellence for Molecular Plant Sciences, Chinese Academy of Sciences, 500 Caobao Road, Shanghai 200233, China

Most of agronomically important traits are called complex traits, which are usually controlled by multiple genes and affected by various environmental conditions. Although a lot of QTLs and genes related to rice complex traits have been cloned and functionally characterized, genetic basis and regulatory mechanisms underlying these complex traits are still unclear. We have implemented an integrated approach of genome-wide association study (GWAS) with functional analysis on agronomic traits in a diverse cultivated rice population, which is so called as a CAT-Gs (Catching Agronomic Trait Genes) approach. The associated loci with the agronomic traits such as panicle length, grain sizes, grain weight and grain filling rate can be further characterized through expressional profiling, in-depth genome analysis, transgenic study, genome editing, and population genetic analysis. We believe that allelic genetic variations responsible for the panicle and grain size complex traits can be effectively explored. The genetic basis of these complex traits will be characterized.

Exploitation of heterosis is one of the most important applications of genetics in agriculture. However, the genetic mechanisms of heterosis are only partly understood, and a global view of heterosis from a representative number of hybrid combinations is lacking. We have developed an integrated genomic approach to construct a genome map for elite hybrid rice varieties and their inbred parental lines. We therefore identified that the accumulation of numerous rare superior alleles with positive dominance is an important contributor to the heterotic phenomena. We have further done large-scale genomic mapping for yield related traits and heterotic effects by analyzing over 10,000 rice lines produced from 17 elite rice lines. The large data of genomics and phenomics from the well-designed populations enabled us, for the first time, to identify the genetic contributors comprehensively and find out the exact causes of heterosis. We found that modern rice varieties can be classified into three major types, reflecting the major breeding systems. Within each group a few genomic regions from female parents linked to heterosis effects for improved yields were identified, but these loci varied across the three groups. The key heterosis-related genes often controlled several yield-related components simultaneously, severing as the major contributors of heterosis. For the individual yield components, the heterozygous state of the heterosis-related genes generally acted through the way of dominance complementation. Taking all the components into account, the hybrids with yield heterosis resulted from an optimal combination of multiple yield-related components, meaning better performance of overall yield in crop productions. These results inform on the genomic architecture of heterosis for yield traits in rice, which will be useful information for crop improvement program.

Long-term field experiments and observation data in combination in Inner Mongolia grasslands of China successfully addressed many fundamental questions in ecology

Xingguo Han, *Jianhui Huang, Shiping Chen, Changhui Wang, Cunzheng Wei*
Institute of Botany, Chinese Academy of Sciences, China

The typical steppe ecosystem of China is an important part of the vast Eurasian grasslands in the world. With an area of 0.4 $*10^6$ km^2 in China, these ecosystems not only harbor rich biodiversity resources, but also serve as important bases of livestock production and ecological shelters for northern China. However, Inner Mongolia grassland is experiencing serious degradation resulting from overgrazing and climate change. During the last decades, we have conducted intensive studies focusing on the fundamental questions in ecology with collaborators from home and abroad. Through long-term field monitoring, transect surveys and controlled field experiments, we are trying to address the nature of fundamental concepts and principles in ecology includes 1) formation mechanisms of primary and secondary production, 2) biogeochemistry of major elements (C, N, S, P), 3) Coupled cycling of Carbon, nitrogen and H_2O, 4) stoichiometry of C: N: P ratio and its application. We seek to understand how organisms interact with each other and the abiotic environment, and also apply this knowledge to the management of grassland ecosystem, and the services it provides. Both the science of ecology and researchers on the Inner Mongolia grassland have come a long way over the last decades. Today, it is a dynamic international base for scientists with a focus on studying ecology in natural environments.

Ex situ cultivated flora and germplasm discovery for novel crop plants

Hongwen Huang
South China Botanical Garden, Chinese Academy of Sciences, International Association of Botanic Gardens, China

There are c. 3,000 botanical gardens in the world, distributed widely across different climatic zones and floristic regions. Most importantly they provide habitats for >120,000 vascular plants in *ex situ* cultivation, accounting 1/3 of known plant species in the world, including many economically important taxa or groups, such Aceraceae, Ericaceae, Fagaceae, Leguminosae, Magnoliaceae, as well as a large number of crop wild relatives, medicinal and aromatic, and horticultural plants. This *ex situ* cultivated flora contains an enormous amount of plant diversity in many well documented living collections in different botanical gardens and plays a critical role in maintaining the security of plant diversity, and in conservation, sustainable agriculture and other related bio-industries.

China has approximately 33,000 higher plants, of which more than half are endemic to China and encompasses a vast number of species of bryophytes (c. 2,200 species), pteridophytes (c. 2600 species), gymnosperms (c. 250 species), and angiosperms (>30,000 species), accounting for 9.1%, 22%, 26.7% and 10% of the world total, respectively. The Chinese flora also represents the living remnants of the early Miocene floras of the whole North Temperate regions and is the source of numerous crops and of medicinal and horticultural plants. There are c. 190 Chinese botanical gardens, harboring 22,375 species belonging to 3,927 genera and 382 families, of which 20,000 species, 2911 genera and 288 families are native to China, accounting for 91%, 86% and 60% of native families, genera and species, respectively. As an example of initiatives to utilize the ex situ cultivated flora to address plant diversity conservation and germplasm discovery for sustainable agriculture and the bio-industries, the *Ex situ Cultivated Flora of China* project aims to catalogue and document this mega diversity of plants that are cultivated in the Chinese botanical gardens.

The initiative has three main goals: 1) To enhance taxonomic research on common garden-based living collections: the morphological and biological data collected from these collections should provide adequate and accurate descriptions for the delimitation of difficult taxa in cases when traditional taxonomic revision has been based only on herbarium specimens; 2) To support comparative biology and frontier plant science research: with increasing awareness of environmental and habitat changes in the overall background of climate change on plant distributions *in situ*, the *Ex situ* cultivated flora program should provide detailed plant biological information from different gardens across a wide spectrum of different latitudes, climates and habitats for research on adaptive evolution of species, plant migration and distribution shifts and physiological or/and biochemical changes, *etc.*; 3) To strengthen germlasm discovery and sustainable uses of plant resources, which should enhance our current ongoing efforts in medicinal plants, industrial bio-energy plants, landscaping and ornamental plants, new functional fruits and vegetables, environment-ameliorating plants, *etc.*

The presentation will provide a comprehensive introduction and well documented case studies. The project of *Ex situ* Cultivated Flora is planned to be one of the most important initiatives of the plant diversity research platform for sustainable economic and social development in China.

Plant growth in a high CO_2 world

Christian Körner
University of Basel, Switzerland

Plant growth needs resources and physical environmental conditions that permit acquiring these resources. CO_2 is a central resource (close to 50% of plant biomass is carbon), hence, elevated CO_2 supply is widely held to enhance plant growth and mitigate growth constraints set by resources other than carbon and by environmental stress. I will present theory and evidence that questions such a simple rationale. It is a trivial fact that carbon can only become incorporated into new biomass to the extent soil nutrients permit, because plants are not made of carbon only. Hence, whenever soil nutrients are short, particularly when neighbouring plants compete for the same resources, there is little leeway for more carbon incorporation. Further, when abiotic limitations such as low temperature or low soil moisture cause soil nutrient shortage and a slowing of meristematic activity (tissue formation), it becomes a question of whether carbon uptake limitations or carbon investment limitations weigh more heavily. The two effects (source limitation *versus* sink limitation) are very hard to separate. At high CO_2, water shortage exerts less of a constraint, for stomatal CO_2 uptake, while at the same time, elevated CO_2 reduces stomatal conductance, thus, causing soil moisture to decline more slowly, also causing nutrients to remain available for a little longer. This non-photosynthetic effect of drought mitigation on growth can be simulated by similarly dosed water addition without any change in CO_2-supply. The effect of such water savings on future plant growth depends on future climate driven moisture regimes. Constraints of growth by low temperature do not seem to get less severe in a high CO_2 world. Alpine and arctic plant growth has been shown to be unresponsive to elevated CO_2. Similar constraints to CO_2 'fertilization' effects apply to ecosystems that have arrived at a natural steady state of the nutrient cycle and canopy density. In summary, the carbon limitation paradigm of plant growth is ill founded when long-term nutrient shortage, low temperature or drought come into play. The simple reason is that under those three conditions, it is not the photosynthetic machinery that sets the long-term limit to plant growth but sink activity (tissue stoichiometry, meristem functioning). Constraints acting upon sink activity are far more challenging to model than those on source activity. In most cases we simply lack appropriate algorithms and parameters to represent sink activity in a mechanistic way. The problem cannot be solved by tuning the output of source activity driven models, because the actual causality is the reverse: the ability to grow new tissue defines the demand for fresh carbon, a demand set by meristems and their direct stoichiometric and developmental controls, and these demands are communicated to foliage through the phloem pipelines. In other words, in all those cases, photosynthesis (source activity) is driven by the rate of growth (sink activity). Sink control over source activity (CO_2 uptake) can only be eliminated by ample nutrient and water supply, as happens under horticultural growth conditions, which do not represent appropriate test conditions for the likely responses of wild plants to elevated CO_2 concentrations in the atmosphere.

Tropical Plant-Animal Interactions: Coevolution in the Anthropocene

W. John Kress
National Museum of Natural History, Smithsonian Institution, USA

The complexity of life on earth is a product of the diversity of species multiplied by the diversity of their interactions. An understanding of species diversity, which requires correct species identification, is a prerequisite to accurately interpreting and understanding their interactions. Investigations of plant-herbivore and plant-pollinator interactions have provided classic examples of the interdependent relationships that exist among species within and across ecosystems. The charismatic tropical Zingiberales, including bananas, birds-of-paradise, heliconias, gingers, and prayer plants, offer illustrative and often surprising insights into the ecology of the animals with which they have coevolved. Specifically, the ecological interactions between colleopteran herbivores and hummingbird pollinators with their Zingiberalean hosts reveal intricate patterns of their evolutionary history and how they may respond to today's rapidly changing planet. The geographic mosaic of these relationships across tropical islands, fragmented landscapes, and elevational gradients suggests that human-caused habitat alterations, biological invasions, and climate change may significantly modify and disrupt through time and space the historical patterns of ecological interactions. The future of today's biological complexity in the Age of Humans, in the Anthropocene, remains to be determined.

A Ca^{2+} sensor switch for tolerance to elevated salt stress in *Arabidopsis*

Joerg Kudla
WWU Muenster, Institute of Plant Biology and Biotechnology, Schlossplatz 748149 Münster, Germany

Excessive sodium (Na^+) in soils inhibits plant growth and development. The Ca^{2+}-activated salt overly sensitive (SOS) pathway formed by the Ca^{2+} sensor SOS3/CBL4 interacting with the kinase SOS2/CIPK24 to activate the Na^+/H^+ antiporter SOS1, represents a core module for regulating Na^+ homeostasis and establishing salt tolerance in plants. In natural conditions, the actual soil Na^+ concentration can vary over a wide range. If and how the intensity of Na^+ stress is reflected by quantitative parameters of downstream signaling components and adjustment of the SOS remains to be investigated. Here, we identify a Ca^{2+} sensor switch that is specifically activated under highly elevated Na^+ stress conditions. We report that Na^+ triggers a primary Ca^{2+} signal specifically localized in root hair differentiation zone of the root from where it expands as a wave. The amplitude of this primary signal rises and the speed of wave propagation accelerates reflecting increasing external Na^+ concentration, thereby providing quantitative information about stress intensity to the plant. Moreover, by analyzing mutant and over-expression plants, we identified the Ca^{2+} sensor CBL8 as crucial for enhanced salt tolerance in *Arabidopsis*. Like SOS3/CBL4 also CBL8 interacted with SOS2/CIPK24 and activated SOS1. While the phosphorylation and conformation of SOS3/CBL4 remained largely unaffected by Ca^{2+}, CBL8 exhibited conformational changes and enhanced phosphorylation with increasing Ca^{2+} concentrations. Together our data identify a Ca^{2+} sensor switch to optimize plant salt tolerance over a range of stress conditions in which the SOS3-SOS2-SOS1 pathway conveys basal and essential salt tolerance, while in response to intensified Ca^{2+} signals, elicited by highly elevated Na^+, in addition CBL8-SOS2 complexes are activated to further enhance SOS1 activity for salt extrusion.

Understanding Kranz anatomy in maize with a view to achieving C_4 rice

Jane A. Langdale
Department of Plant Sciences, University of Oxford, UK

A multi-national effort aimed at introducing components of C_4 photosynthesis into the C_3 plant rice is underway. A fundamental requirement of the 'C_4 Rice Project' is the need to understand how Kranz anatomy develops in the leaves of C_4 plants so that leaf anatomy can be manipulated in rice. I will provide an overview of our understanding of Kranz development, discuss recent research aimed at identifying genetic regulators of Kranz anatomy and illustrate how this research fits into the wider C_4 Rice Project.

The quest for the origin and evolution of flowering plants using time-calibrated plastid phylogenomics

Hong-Tao Li[1 1], Ting-Shuang Yi[1*], Ting Zhang[1,3*], Lian-Ming Gao[2*], Jun-Bo Yang[1*], Peng-Fei Ma[1*], Matthew A. Gitzendanner[4,5*], Jie Cai[1], Yang Luo[2,3], Hong Wang[2], Shu-Dong Zhang[1], Douglas E. Soltis[4,5, 6], Pamela S. Soltis[4,6], Peter H. Hollingsworth[7], Peter W. Fritsch[8], Michelle Van Der Bank[9], Mark W. Chase[10],* ***De-Zhu Li[1,2, 3, 2z]***
1. Germplasm Bank of Wild Species, Kunming Institute of Botany, Chinese Academy of Sciences, Kunming, Yunnan 650201, China
2. Key Laboratory for Plant Diversity and Biogeography in East Asia, Kunming Institute of Botany, Chinese Academy of Sciences, Kunming, Yunnan 650201, China
3. Kunming College of Life Science, University of Chinese Academy of Sciences, Kunming, Yunnan 650201, China
4. Florida Museum of Natural History, University of Florida, Gainesville, FL 32611, USA
5. Department of Biology, University of Florida, Gainesville, FL 32611, USA
6. Genetics Institute, University of Florida, Gainesville, FL 32610, USA
7. Royal Botanic Garden Edinburgh, Edinburgh EH3 5LR, United Kingdom
8. Botanical Research Institute of Texas, 1700 University Drive, Fort Worth, TX 76017, USA
9. Department of Botany & Plant Biotechnology, University of Johannesburg, PO Box 524 Auckland Park, Johannesburg, Gauteng 2006, South Africa
10. Royal Botanic Gardens, Kew, Richmond, Surrey TW9 3DS, UK

The angiosperms, i.e., flowering plants, are the most species-rich plant lineage on Earth. Because angiosperms are critical components of nearly all terrestrial habitats, knowledge on their origin and evolution can provide the framework for understanding the history and composition of major terrestrial ecosystems and general patterns of biodiversity. Over the last two decades, molecular phylogenetic studies have greatly advanced our understanding of angiosperm relationships and evolution, as reflected in the latest update to the Angiosperm Phylogeny Group (APG) classification system (2016). However, the composition and relationships of some clades recognized as orders and families have remained weakly supported, with recent studies often based on either sparse gene sampling or low taxonomic coverage, or both. To address this problem, we assembled a large-scale plastid phylogenomic data set of angiosperms by sampling 80 genes from 2881 plastid genomes representing 2349 angiosperm species, 352 families (85% of the described families), and all 64 orders recognized in APG, with 163 species of gymnosperms comprising the outgroups. Our results yielded the diversification of eight major lineages in the early stages of angiosperm evolution. The ANA grade, i.e., Amborellales, Nymphaeales and Austrobaileyales, was confirmed to be the earliest-diverged set of lineages. However, the relationships of the mesangiosperms, including magnoliids, Chloranthales, monocots, Ceratophyllales and the eudicots, remain poorly resolved.

We estimated the divergence times of major angiosperm clades using 63 carefully verified fossils as calibration points. Our results support the origin of the crown angiosperms in the Rhaetian of the latest Triassic ca. 207 million years ago (mya). A rapid radiation of major lineages and orders of flowering plants occurred in the Middle to earlier Late Jurassic and Early Cretaceous. An early radiation may have occurred at the Callovian in the Middle Jurassic (ca. 165 mya) with the origin of three major lineages of mesangiosperms. Major radiations of the monocot lineages occurred in the Late Jurassic (ca. 157 mya) to the Valanginian in the earliest Cretaceous (ca. 133 mya). Core eudicots began to diverge at the Hauterivian (ca. 131 mya), mainly diversifying from the Aptian (ca. 121 mya) to the Turonian-Coniacian boundary (ca. 90 mya) in the Early Cretaceous, with the Paracryphiales and Dipsacales, the most recently diverged orders, at the mid-Campanian (ca. 79 mya) in the Late Cretaceous. Our divergence time estimates are generally congruent with others based on DNA sequence data and are corroborated by the phylogenomic-based estimate for the origin and diversification of major lineages and orders of holometabolous insects. However, they are significantly earlier than unequivocal fossil records of flowering plants. This may be attributed to the fact that even angiosperm-like pollen records from the Jurassic are spare and are thus difficult to find. The lack of resolution among mesangiosperms in our analysis may result from rapid radiation and extinction in the early stage of angiosperm evolution, particularly during the Jurassic.

Nuclear phylogenies of angiosperms at both deep and family levels: Implications on divergence times, morphological evolution and diversification rates

Hong Ma*, Chien-Hsun Huang, Liping Zeng, Ning Zhang, Yezi Xiang*
Ministry of Education of Key Laboratory of Biodiversity Sciences and Ecological Engineering, Institute of Plant Biology, School of Life Sciences, Fudan University, Shanghai, China

Angiosperms are the most successful plant group, with an estimated total of over 350,000 species. Angiosperm phylogeny is a

1 [*] These authors contributed equally to this abstract.
2 z Corresponding author.

foundation for a variety of other studies, including evolutionary biology and ecology, comparative genomics, and translational agriculture and has been examined using chloroplast sequences for many years. Rapid divergences of multiple lineages, frequent whole genome duplications and other factors have posed major challenges to the resolution of angiosperm relationship. Recent advances in DNA sequencing have availed large quantities of nuclear gene sequences, greatly facilitating the acquisition of needed marker information for addressing difficult questions in angiosperm phylogeny. We have used transcriptome sequencing and other approaches to obtain dozens, even hundreds, of nuclear gene sequences, and performed multiple phylogenetic analyses of major deep lineages. These investigations have yielded highly supported relationships that are largely in agreement with chloroplast sequence-based phylogenies; at the same time, the new hypotheses according to nuclear genes support well-resolved relationships for taxa whose relationships were unclear before, illustrating the effectiveness of nuclear genes. Well-resolved relationships include five deep lineages of the mesangiosperm clade representing nearly all angiosperms and several major branches of core eudicots. We have also performed phylogenetic analysis of several groups including Brassicaceae, Asteraceae, and Rosaceae, and will summarize recent results from these analyses, and discuss their implications on the divergence times of relevant groups, reconstruction of ancestral character of key morphological features, and shifts in rates of diversification during angiosperm evolution.

Mapping Asia plants

Keping Ma[1], *Bo Liu*[1], *Hongfeng Wang*[1], *Maofang Luo*[2], *Zheping Xu*[1], *Xuehong Xu*[2], *K. N. Ganeshaiah*[3]
1. Institute of Botany, Chinese Academy of Sciences, China
2. Biodiversity Committee, Chinese Academy of Sciences, China
3. University of Agricultural Sciences, GKVK Bangalore, India

Because of the huge population and rapid growth of economy, biodiversity including plants are being seriously threatened in Asia. Both conservation and sustainable use of biodiversity need species distribution data. Asia is the continent with highest diversity of plants. However, the information of plant taxonomy and distribution in Asia are less available and more scattered, compared with other continents such as Europe and America. So, a better infrastructure for plant distribution in Asia is important. In 2015, Asia Biodiversity Conservation and Databases Network (ABCDNet) started a project entitled Mapping Asia Plants for cataloguing species of plants and collecting distribution data from the following aspects: 1) specimens, 2) observations, 3) species checklists, 4) expert maps, 5) inventories, 6) vegetation plots, 7) literatures and 8) on line datasets. Major achievements so far are as follows. 1) for China, national higher plant checklists including 35,835 species, 9 million records of occurrence; 2) for Southeast Asia, 120 thousands species names, 1.3 million records of occurrence; 3) for north Asia, 6,900 species of higher plants and their distribution in 13 sub-regions. Some of the data collected are shared with the websites of ABCDNet (www.abcdn.org/).

Thirty clues to angiosperm exceptional evolutionary diversification

Susana Magallón[1], *Luna Luisa Sánchez-Reyes*[1], *Santiago Ramírez-Barahona*[1], *Tracy A. Heath*[2], *Tanja Stadler*[3]
1. Instituto de Biología, Universidad Nacional Autónoma de México, México City, México
2. Department of Ecology, Evolution and Organismal Biology, Iowa State University, Ames, Iowa, USA
3. Department of Biosystems, Science and Engineering, ETH Zürich, Switzerland

Angiosperms are the cornerstone of present-day terrestrial biomes, but little is known about the macroevolutionary processes underlying their diversification. A great controversy exists regarding the time when extant angiosperms lineages originated, and many questions remain regarding their evolutionary expansion: What is the phylogenetic position and temporal distribution of radiations leading to extant megadiversity? Are these radiations concentrated in particular phylogenetic regions or times? Are all angiosperm lineages radiating, or are some undergoing evolutionary depletions? Has a limit to angiosperm evolutionary expansion been reached? To investigate the absolute timing of angiosperm crown origin and lineage diversification, we integratively consider molecular sequence data of living species and information from the fossil record. Because sampled extant species with molecular data and fossils that provide absolute times are part of the same evolutionary process, we estimate divergence times and diversification parameters across angiosperms with the Fossilized Birth Death (FBD) Process, a model that estimates the rates of speciation, extinction, fossilization, and sampling probability of extant species, as a prior for divergence time dating. We contrast FBD age estimates for the origin of angiosperms with those derived from relaxed molecular clocks, and from two previous estimates derived from quantitative evaluations of the angiosperm fossil record. As the FBD directly incorporates information from the fossil record, it has a great potential for accurately estimating diversification and relative extinction parameters for angiosperms as a whole. To investigate radiations within angiosperms, we implement a Bayesian method (BAMM) that allows to identify significant diversification shifts across phylogenetic branches and through time. We conduct six alternative analyses that differ in the magnitude of the prior for the expected number of diversification shifts across the tree, and provide both technical and biological results. In all analyses, the posterior distribution of the number of shifts differs substantially from the prior. Moreover, a set of thirty diversification shifts highly supported by the data were congruently identified across the six analyses. Biologically, the placement of core diversification shifts suggests that extant angiosperm diversity is the result of several independent and partially overlapping radiations, interspersed with a few evolutionary depletions. Radiations are phylogenetically scattered and took place over a substantial length of angiosperm evolutionary history. Estimates of their long-term diversification trajectory suggest that angiosperm species accumulation has not reached a limit, but species turnover is increasing. The finding that several diversification shifts are distributed on two or more adjacent phylogenetic branches is congruent with the possibility that underlying causes of diversification may derive from synergic interactions among biological, ecological and physical factors.

How do plants age and die? – Some new insights

Hyojung Kim[1], *Junhyun Lim*[1], *Timilsina Rupak*[2], *Ji-Hwan Park*[3],

Sunghyun Hong[1], *Sukjoon Jung*[2], *Jin Hee Kim*[1], *Hye Ryun Woo*[2], *Pyung Ok Lim*[2], *Daehee Hwang*[1,2], ***Hong Gil Nam***[1,2]
1. Center for Plant Aging Research, Institute for Basic Science, Daegu, Republic of Korea
2. Department of New Biology, DGIST, Korea, Daegu, Republic of Korea
3. Department of Chemical Engineering, POSTECH, Pohang, Gyeongbuk, Republic of Korea

Aging followed by death is an inevitable fate in most organisms including human. Our group has been studying the molecular mechanisms of aging and death over two decades to solve the question 'How does living organisms know when and how to die?' We have been revealing that aging and death in plants are highly regulated at the chromatin, transcriptional, post-transcriptional, and epigenetic levels. In this presentation, we will report some new insights on the regulatory mechanisms of aging and death in plants with a comparative view on aging in animals.

Aging involves transitions of complex regulatory networks across lifespan. Yet, limited is understanding of causal relationship between age-associated networks and aging processes. Here, we built age-evolving causal regulatory networks of *Arabidopsis* NAC transcription factors across lifespan using mutants of 49 age-associated NACs. These age-evolving networks revealed a regulatory inversion at pre-senescent stage from activating to repressive regulatory modes, which was governed by three hub NACs that negatively regulated key pro-aging pathways, ROS and salicylic acid pathways. Knockout and overexpression of the three hub NACs (the NAC troika) delayed and advanced the regulatory inversion, respectively, rendering earlier and delayed aging. Thus, the network transition by the NAC troika at pre-senescent stage determines initiation and rate of the following aging process in *Arabidopsis*.

Meristems are small population of pluripotent stem cells located at shoot and root tips of plants. They have remarkable ability to maintain themselves for a long period and serve as the constant source of cells for newly developing tissue and organs, allowing the long lifespan often observed in many tree species. We are striving to reveal the mechanisms behind the long lifespan stemmed from the long-time maintenance of meristem cells, employing the root apical meristem (RAM) as a model. *Arabidopsis* RAM consists of mitotically inactive quiescent cells (QC) and surrounding active stem-cells. While active stem-cells are susceptible to various stresses, QC cells are highly stress resistant. We have found that *ORE15* has important role in maintenance of RAM. We will report the underlying mechanisms.

Light is one of the most important environmental factors which affect plants growth and development. The critical roles of red light through phytochrome-mediated light signaling has been reported recently. We examined the role of far-red light on leaf senescence and found that leaf senescence is modulated by red (R): far-red (FR) ratio (R/FR). We also found a strong correlation between FR-responsive leaf senescence of various ecotypes and latitudinal cline, suggesting that FR signal is used for latitudinal adaptation of *Arabidopsis* ecotypes. 'Climatome' analysis using climate information around ecotypes' habitat revealed that temperature in winter and spring is strongly associated with FR-responsive leaf senescence. We will report a possible mechanism on how *Arabidopsis* adapted to latitude-dependent environmental changes.

We are also comparing the regulatory mechanisms of aging between plants and animals. We will report the differential roles of tRNA fragments in regulating aging in *Arabidopsis*, *C. elegans* (worms), and mouse.

Phytochrome photosensory signaling and transcriptional networks

Peter H. Quail
Department of Plant and Microbial Biology, University of California, Berkeley, CA 94720, USA
USDA Plant Gene Expression Center, 800 Buchanan St., Albany, CA 94710, USA

Plants constantly monitor the ambient light environment for signals that enable them to adapt to the prevailing conditions. The phytochrome (phy) family of sensory photoreceptors plays a central role in this process. Light absorption induces conversion of the phy molecule to its active Pfr conformer which then migrates rapidly into the nucleus where it induces expression changes in target-genes within minutes. Photoactivated phys signal to their target transcriptional network by binding to, and inducing the phosphorylation and degradation of PIF transcription factors (phy-Interacting Factors), thereby evoking these changes. We have been examining the mechanisms underlying the signaling transactions at both the phy-PIF and PIF-genome interfaces in this process. Previously, we have shown that signaling requires direct binding of the phy Pfr conformer to the PIF protein, that this results in multisite phosphorylation of the bHLH factor, and that this in turn leads to polyubiquitination by Cullin3-BTB E3 ubiquitin ligases (called LRBs), and, consequently, proteasomal degradation. Recently, we have identified a small family of Photoregulatory Protein Kinases (PPKs) that are collectively necessary for normal phy-induced PIF3 phosphorylation in vivo. These kinases are recruited by photoactivated phyB to a nuclear-localized trimolecular complex with PIF3, where they catalyze multisite PIF3 phosphorylation. This discovery fills a longstanding mechanistic gap in our understanding of the short transduction process from the light-activated photoreceptor to transcriptional regulation. We have also identified a unique negative feedback signaling-attenuation mechanism, directly at the phy-PIF interface, whereby the phy-induced phosphorylation of, and LRB recruitment to, the PIF molecule, results concurrently in ubiquitination and degradation of the interacting phy molecule in a mutually-assured destruction (MAD) configuration. This coordinated PIF and phy degradation thus results concomitantly in both transcriptional regulation of target genes and immediate feedback attenuation of signaling intensity. Using genome-wide transcriptome analysis, we have identified PIF-regulated genes that respond rapidly to phy photoactivation, and using integrated ChIP-seq and RNA-seq analysis, we have identified a diverse network of these genes that are direct targets of PIF-quartet (PIF1,3, 4 and 5)-regulated transcription. Moreover, the evidence unveils an intriguing dual-layered mechanism of regulation, whereby both the level of promoter-binding-site occupancy, and in situ modulation of bound-transcription-factor intrinsic activity, together generate a complex matrix of shared, but quantitatively differential, gene expression patterns, under the control of the phy-PIF signaling pathway.

Endocytosis and the regulation of signaling in plants

Eugenia (Jenny) Russinova
VIB-UGent, Center for Plant Systems Biology, 9052 Gent, Belgium

Receptor-mediated endocytosis is an integral part of signal transduction, because, besides signal attenuation by removal of activated receptors and their bound ligands from the cell surface, it allows the spatial and temporal regulation of the signaling outputs from the endosomes. After receptors have been activated and internalized, they can be separated from their ligands and recycled back to the plasma membrane or transported for degradation. Crucial in the study of the interplay between endocytosis, recycling, and signaling of plant receptor kinases is the development of imaging tools (bioactive fluorescent probes) to visualize membrane-associated signaling events at a high spatiotemporal resolution. Recently, we visualized endocytosis of different receptor kinases in living *Arabidopsis thaliana* cells using fluorescent small-molecule and peptide ligands. The bioactive fluorescent probes together with genetic, biochemical and pharmacological analyses revealed distinct dynamics of endocytosis and differences in regulation of signaling outputs.

Mobilizing and integrating big data in analyses of phylogenetic and spatial patterns of biodiversity

Douglas Soltis
Department of Botany and the Genetics Institute, University of Florida, Gainesville, FL32611, USA

The pace of biological discovery is staggering in this era of "big data." We are at a crucial juncture when it is more important than ever to mobilize and integrate massive amounts of data to address fundamental issues in the study and preservation of biodiversity. The enormous challenges presently faced in biodiversity science require novel approaches that not only employ massive data sets, but also combine bioinformatics, phylogeny reconstruction at an unprecedented large scale, use enormous amounts of digitized specimen data, as well as large, highly complex, multi-faceted down stream phylogenetic analyses. These post-tree analyses include niche modeling, niche diversification, and ancestral niche reconstruction, as well as other ecological analyses. Recent tool development in multiple disciplines including phylogenetics, ecological niche modeling, combined with rapidly developing, powerful cyberinfrastructure coupled with the availability of multiple new data sources afford unique and powerful opportunities. Together these resources provide novel possibilities for mobilizing and integrating enormous quantities of biological, geological, climatic and other related biodiversity data, ultimately facilitating the discovery of complex patterns and new hypotheses for further study. These multifaceted and interdisciplinary developments are major advances--biodiversity data on the global scale now being collected and analyzed are inherently complex. This tool integration is transforming what we broadly term biodiversity science, enabling numerous new research opportunities that we refer to as "next-generation" investigations in biodiversity science that couple systematics, ecology, and evolution. The growth in tool building and use also demands new training for our students that integrates domain knowledge in biodiversity as well as data science skills –such connectivity and training is needed to accelerate biodiversity research. This next generation Integrative biodiversity science is of crucial importance--the future of global biodiversity will be greatly impacted by these efforts. We as researchers cannot continue to simply react to continued worldwide threats to biodiversity. In contrast, through the use of integrative, multifaceted research relying on big data, we now have the ability to make biodiversity projections—we can now provide crucial data to scientists, as well as to the public, land managers, policy makers, urban planners, and those in agriculture.

Thinking through the e- in e-Floras; or, Floras old, new, and not-yet

Kevin Thiele
School of Plant Biology, Faculty of Science, the University of Western Australia, Crawley, Western Australia 6009, Australia

When faced with a disruptive new technology such as the World Wide Web, it often takes a lot of work, and time, to escape conventional ways of thinking through a problem and to fully realise the implications of the new. So-called e-Floras are no exception. How much does the e- in the name really reflect a new way of approaching a Flora's core task (to communicate authoritative, synthesised taxonomic knowledge to as wide an audience as possible in the best way achievable)? If modern taxonomy and systematics were invented, or re-invented, now (in the age of the internet, social media, citizen science and the block chain), rather than in the 18th Century, would we do it all differently? And, what near-future disruptive technologies and paradigms do we need to be aware of when thinking about, or planning, an e-Flora project? Taking the new e-Floras of Australia and New Zealand as starting points, this talk will explore current and future thinking about e-Floras, and map out some ideas for, perhaps, the next wave.

Developing integrative systematics in the informatics and genomic era

Jun Wen
Department of Botany, MRC-166, Smithsonian Institution, PO Box 37012, Washington, DC 20013-7012, USA

Systematics is the science of discovering, organizing and interpreting the diversity of all living organisms on Earth. Recent developments in genomics and biodiversity informatics are transforming systematics and have opened up many new opportunities. Major digitization efforts have made available large amounts of biodiversity data. Developments in biodiversity informatics in the last decade have helped the systematics community explore ways to enhance the efficiency in organizing, publishing, and utilizing systematic information. We are now constructing the tree of life and classifications at various levels using information from hundreds to thousands of genes and sometimes the entire genomes. The new data and tools are bringing systematics into an exciting age of new discovery and interpretation of biodiversity and its assembly. Integrative systematics in the informatics and genomic era needs to be developed with biological collections continuing to serve as the foundation of the field. Discovering, documenting and analyzing the patterns and processes of biodiversity remain the core of systematics, yet the community needs to enhance the efficiency of systematic research and practices with user-friendly pipelines. We especially call for new efficient workflows for taxonomic revisions, monographs and floras, that utilize both the traditional strengths of synthesizing diverse collections-based taxonomic data and the capacity of online resources and biodiversity informatics tools. A new global community-wide cyberinfrastructure, Biodiversity Cyberbank, comparable to GenBank for genetic data, needs to be established as the repository of all biodiversity data, as well as the means to effectively organize and disseminate systematic data to biologists and the public, to ensure the long-term sustainability of the vast biodiversity data. Integrative systematics requires the training of the next-generation of systematists with integrative skills, to address grand questions about biodiversity and its assembly at different scales. Finally integrative systematics must proactively educate the public and policy makers on the importance of systematics and collections in the biodiversity crisis of the Anthropocene.

Establishing New Botanic Gardens in Indonesia: An ecoregion approach

Didik Widyatmoko

Center for Plant Conservation Bogor Botanic Gardens, Indonesian Institute of Sciences, Jl. Ir. H. Juanda No. 13 Bogor 16122 Indonesia

The development of new botanic gardens has been included in the 2015-2019 National Priority Programs and the Indonesian Biodiversity Strategy and Action Plan (2003-2020). Within international contexts, the development of new botanic gardens is in line with the global commitments such as the Convention on Biological Diversity and the target VIII of the Global Strategy for Plant Conservation. The ecoregion concept has been applied in the new botanic gardens development process in order to save and manage effectively the plant diversity occurred in different ecosystem types. This strategy should not only be valued as a protection effort of plant diversity per se, but also considering the need for empowering local communities. By conserving the plant diversity in their own habitat characteristics we can expect higher success, particularly for endemic, threatened species. This strategy is also less costly because the collected-selected plants are brought to and managed in the closest, most suitable gardens. Since Indonesia consists of 47 different ecoregions, at least one botanic garden should be located in each ecoregion in order to conserve the representatives of the Indonesian plant diversity. By the end of 2016, there have been 30 botanic gardens both developed and developing gardens, located in 17 (out of 47) different ecoregions. The implementation of the Indonesian Presidential Regulation No. 93 in 2011 has strengthened the development phases of the new botanic gardens by engaging the Ministry of Public Works and the Ministry of Internal Affairs. The establishment of new desirable botanic gardens and the relevant roles to conserve plant species ex situ will serve the public needs, provide fruitful visitation, as well as create social and economic benefits. Social welfare is indeed a key factor that must be taken into account if botanic gardens are to succeed in their goals of plant diversity conservation and the sustainable utilization.

International developments and responsibilities for the botanical community in plant conservation

Peter Wyse Jackson

Missouri Botanical Garden, P.O. Box 299, St Louis, Missouri 63166, USA

The development of the Global Strategy for Plant Conservation (GSPC) through the U.N. Convention on Biological Diversity can trace its origins back to a resolution of the International Botanical Congress held in St Louis, Missouri in 1999 and its 2002 adoption by the international community. Since then the GSPC has been of fundamental importance not only to provide a strategic framework for plant conservation undertaken by the world's governments, primarily within the context of the U.N.'s Strategic Plan for Biodiversity and its related Aichi targets, but also a guiding document used by numerous botanical institutions worldwide. The 16 targets of the GSPC are wide-ranging and highlight goals that are due to be achieved by 2020. These goals relate to understanding and documenting plant diversity, conserving and using plants sustainably, as well as public education and building capacity for plant conservation throughout the world. The strategy has had a transformative impact on many botanical institutions, particularly botanic gardens, helping to consolidate conservation as a primary purpose for many.

In this presentation, ways in which individual institutions have been contributing towards the achievement of the 2020 goals will be outlined. Examples from the Missouri Botanical Garden will be included where its most recent strategic plan has made a formal commitment to contributing to the GSPC through research in conservation biology, ecological restoration, community conservation and education programs, horticulture and ex situ conservation, as well as building conservation capacity with institutional partners worldwide.

A widely accessible Flora of all known plant species is a fundamental requirement for plant conservation. Without it, we have no framework to assess the diversity of plants or the threats they face on a worldwide scale. The creation of such a Flora was included in the GSPC as its first target. Subsequently, in response to this target, a World Flora Online (WFO) project was launched in 2012, which is now guided by an international WFO Consortium, involving almost 40 botanical institutions and organisations involved in plant systematics. The progress in the development of the project since 2012 will be outlined, including the preview of a new WFO portal that is planned to be available at the Congress.

The likely future of plant conservation beyond 2020 will be reviewed when greater integration between biodiversity conservation and sustainable development can be anticipated, within the context of the achievement of the U.N.'s Sustainable Development Goals for 2030.

Big data in plant genomics to connect nature-derived evolutionary innovations with future precision agriculture

Xun Xu

BGI, Shenzhen, China

DNA is the basic source code of life, connecting nearly all species in this planet during the past billions of years that produced magnificent biodiversity and remarkable innovations. To study the crucial role of plants in ecosystems, medicine development, and resources in food production and nutrition, big data of plant genomics by large-scale species sampling along the entire evolutionary line is an unprecedented but powerful way to understand the fundamental genetics and evolution. To achieve this goal, sampling and digitizing various non-model plants particularly by genome sequencing technology is the key, further to manipulate or edit/synthesize new life forms (or natural products) that are beneficial to human healthcare. From our pilot studies with careful scientific designs in algae, early land plants, and angiosperms that with innovative traits of interest, we have gained extensive novel insights in evolutionary forces that drive genome complexity and adaptation. However, all of these efforts are just the beginning, data is still far from enough, and our understanding to life is still much limited. Sequence most if not all of the species on this planet should be the final approach (every species/data is valuable indeed), just like what we have been working on the 'plant tree of life' project and even for the 'digitize the planet' project. All of which have been facilitated by the recent development in sequencing technologies (like BGISEQ-500/50 platform), through extensive global collaborations following a Data-Knowledge-Application mode in BGI. Through big data study of plant genomics, we have the opportunity to connect all valuable data/species together, explore biodiversity and gain inspirations from nature, and thus design 'super species' that are nutrition-rich and production-effective, and further guide future precision agriculture.

Diversity of tropical forests in South East Asia under the threat of rapid loss

Tetsukazu Yahara

Center for Asian Conservation Ecology, Kyushu University, Japan

In the world, 60,065 species of trees have been recognized (Beech *et al.*, 2017) and tree species richness is the highest in tropical rain forests of Brazil and SE Asia. Tree species richness in SE Asia is, however, largely underscored because taxonomic inventories of trees in SE Asia remain rather incomplete and thus many new species have been described every year. On the other hand, tropical rain forests of SE Asia are rapidly converted to plantations at the rate of 0.36% per year. Therefore, we need to speed up taxonomic inventories of tropical trees.

To obtain more accurate documentation of species richness in SE Asia, we have carried out a series of field surveys in 35 locations of SE Asia where we placed a total of 136 plots of 100 m x 5 m and collected 32,015 samples of the vascular plant species found in or near each plot even for sterile plants. To identify those collections, we used DNA sequences of three regions widely used for plant DNA barcoding: rbcL, mat K and ITS. However, those regions did not always give sufficient discrimination of closely related species and thus sequencing of three regions for more than 30,000 samples is not cost-effective.

A new DNA barcoding using MIG-seq, a recently developed method of genome-wide SNP genotyping with the next generation sequencing platform (Suyama and Matsuki 2015) enables us to discriminate closely related species more efficiently. We applied this method for some genera of Lauraceae, the most species-rich family in SE Asia. Consequently, we discovered at least the following 104 new species; 7 of *Actinodaphne*, 20 of *Cinnamomum*, 10 of *Cryptocarya*, 42 of *Machilus* and 25 of *Neolitsea*. There are more new species, possibly 100 more, in other genera including *Alseodaphne*, *Beilschmiedia*, *Dehaasia*, *Lindera*, *Litsea* and *Phebe* on which taxonomic studies are in progress.

A reliable estimate of the proportion of new species is available in Mt. Bokor, Cambodia where we published the tree flora as a Picture Guide and enumerated 770 species of 108 families. Among the 770 total, 65 species (8.4%) were undescribed; the family having the richest new species was Lauraceae in which 11 of 33 species (33.3%) was undescribed. Our 32,015 specimens from 35 locations of SE Asia include approximately 24,000 species among which ca. 2,000 species (24,000 x 0.084 =2016) would be undescribed. Because the cumulative number of species we collected are still linearly increasing, this is likely to be underestimated.

New species are often found as one or a few individuals in disturbed and fragmented stands of lowland forest. Considering the rapid loss of tropical forest in SE Asia, many species may be becoming extinct before those are collected and described. We need more intensive efforts to discover and describe species in tropical forest of SE Asia and the method we employed provide an efficient way of responding to this need.

Sequence assembly, evolution and functional application of the cotton genomes

Yu-Xian Zhu[1,2]

1. Institute for Advanced Studies and College of Life Sciences, Wuhan University, Wuhan 430072, China.

2. State Key Laboratory of Protein and Plant Gene Research, College of Life Sciences, Peking University, Beijing 100871, China.

We used the highly homozygous *G. hirsutum* (AtDt, $2n = 52$) genetic standard line TM-1 for DNA sequencing. We crossed a tetraploid *G. barbadense* cv. 3-79 with TM-1 to obtain 167 RIL individuals to construct the high-density genetic map. To successfully assemble the genome, we combined 180-fold whole-genome shotgun (WGS) sequences with a total of 100,187 BAC end sequences. Repeat sequences accounted for 57% of the D genome, 68.5% of the A genome and 67.2% of the AtDt genome. The AtDt genome displayed high degrees of conserved gene orders with the two diploids. Insertions of long terminal repeats (LTR), especially *Gypsy-type* retrotransposons, seems responsible for the two-fold size differences between the D (775.2 Mb) and A (1,694 Mb) genomes. Transposable elements (TEs) originated from Dt are more active than those from At. In-depth genome-wide analysis helped to resolve the puzzle with regard to gossypol biosynthesis and ethylene production in various cottons. The expression patterns of a large number of functional genes, especially both primary cell wall-specific and secondary cell wall-specific CesA or CesA-like (CSL) genes, are significantly different in the two subgenomes, suggesting that genome variation, caused likely by TE, especially by retrotransposon insertion, may play a crucial role during the concerted cotton genome evolution.

General Symposia

Theme	Symposium numbers
1: Biodiversity, Resources & Conservation	T1-01~T1-38
2: Taxonomy, Phylogenetics & Evolution	T2-01~T2-68
3: Ecology, Environment & Global Change	T3-01~T3-26
4: Development & Physiology	T4-01~T4-39
5: Genetics, Genomics & Bioinformatics	T5-01~T5-33
6: Plants & Society	T6-01~T6-15

THEME I: BIODIVERSITY, RESOURCES & CONSERVATION

T1-01: Developing priorities for the conservation of Karst flora (two sessions)

T1-01-01

Characteristics of carbon cycle and the effect of carbon sink in karst ecosystem

Jianhua Cao[1,2], *Hui Yang*[1,2], *Chunlai Zhang*[1,2], *Fen Huang*[1,2], *Azim Mallik*[3]

1. International Research Center on Karst under UNESCO

2. INstitute of Karst Geology, CAGS

3. Department of Biology, Lakehead University

Carbonate rock is the basic material of karst ecosystem, and its formation in geological era has an important role on the atmospheric CO_2 decreasing, about 99.5% of the carbon in the earth is sealed in carbonate rock; Karst ecosystem is characterized as rich calcium, alkaline, impacting carbon cycle: carbonate rock is the sedimentary rock deriving from clean ocean, its acidic insoluble matter is usually less than 10%, which lead shortage of the soil resources in karst area, therefore the total amount of limestone soil carbon in karst area is small. The high calcium makes more organic carbon stocked in shape of the slow and recalcitrant carbon pools, and improves the stability of limestone soil organic carbon. The proportion of underground biomass of vegetation in karst area may increased, because of shortage of soil resources, nutrients and water resources, the ration of underground/above vegetation biomass in the karst can be got to 30-50%, higher than non-karst area. The high content of HCO_3^- in karst water stimulates the photosynthesis of aquatic plants, inorganic carbon can be converted to organic carbon, this process improves the stability of the carbon in water flow. This paper try to clarify the state, migration and transformation of carbon in the interface of system of carbonate rock-soil-vegetation-atmosphere- water in karst ecosystem, meanwhile put forward the approaches to promote karst processes and carbon sequestration.

T1-01-02

Rooting depth determined with caves and DNA, and implicating water use under karst terrain

Yuqing Huang, *Guangping Xu, Danjuan Zen, Yingjie Sun, Longwu Zhou, Yefeng Yao, Fujing Pan, Chengxin He*

Guangxi Key lab of karst conservation and restoration ecology, Guangxi Institute of Botany

The solute carbonate rock of karst in south China develops an epikarst zone with a high fissure and rich spatial structure underground, which is favor for root growth and extend to deep karst layer. Belowground vertical community composition and maximum rooting depth of karst terrain in three sites of Guangxi, China were determined by using DNA sequence variation to identify roots from caves 0–40 m deep. 21 sample sites including 15 sites at Guilin secondly forest, 3 sites at Nonggang national park primary forest, and 3 at Mulun National Park primary forest, were investigated. Roots from caves were identified by comparing their DNA sequences for the internal transcribed spacer (ITS) region of the 18S–26S ribosomal DNA repeat against a reference ITS database and above cave ground species. The result presented root belong to 7 species extend to 10m, and two species to 29m. The result of stable isotope δD and δ18O analysis showed that water uptake by tree species *Cyclobalanopsis glauca* sourcing mainly shallow to depth relevance with wet to drought season, less source was from ground water. Even in the driest October, only 20% water use was from ground water. However, soil moisture did not change greatly except subsoil which was influenced by evaporation; meanwhile transpiration in March, July and October was fairly high. These result may implicate soil water must be supplemented by other source besides rain water. Consequently, combining root depth result, which rooting of *C. glauca* could be found at 5-10m and reaching to cave water, suggesting there would be the hydraulic lift by root which transport water from deep epkarst fissure water or ground water and deliver it at soil.

T1-01-03

Developing priorities for karst conservation

Alice Hughes

Xishuangbanna Tropical botanical garden, Chinese academy of sciences

Karsts represent a challenging ecosystem for conservation. With almost unparalleled levels of endemism, and an estimated 90% of cave fauna undesribed in some countries they represent priorities for further research. However growing demand for cement in many regions, and especially parts of Southeast Asia where accelerating cement demand caused China to use 6.6 gigatonnes in just a two year period, exceeding US historical construction and leading to an exponential increase in the demand for cement and consequently the increasing destruction of limestone karsts to meet that cement demand. Thus to ensure karst diversity is adequately protected, we require both an understanding of hotspots of biodiversity, and how karst landscapes can be effectively managed to maintain karst ecosystems. Here we review a recently initiated project which sets to do that. By initially developing methods to map karsts, then develop stratified inventories of karst biodiversity, hotspots of endemism and also evaluate how landscape configuration changes the microclimate and species present on karst ecosystems. We discuss how such stratified approaches can be implemented and scaled across the landscape to generate standardized information on karst biodiversity and understand how to manage karts and develop effective priorities to account for diversity and endemism. Given the almost unparalleled endemism in these systems it is essential that we rapidly develop priorities that most effectively protect biodiversity in these challenging systems.

T1-01-04

Genetic diversity and structure of *Primulina renifolia* (Gesneriaceae), an endangered Karst cave plants

Fang Wen[1,3], ***Xin Hong***[2,3], *Yi-Gang Wei*[1,3]

1. Guangxi Key Laboratory of Plant Conservation and Restoration Ecology in Karst Terrain, Guilin Botanical Garden, Guangxi Institute of Botany

2. School of Resources and Environmental Engineering, Anhui University

3. Gesneriad Conservation Center of China (GCCC)

We used 10 polymorphic microsatellite loci from transcriptome analyses data to assess the levels of genetic diversity and structure for 24 individuals from 3 populations in the type location. The result of genetic diversity showed that 63 alleles were identified across the ten microsatellites analysed while 3-10 alleles, as the mean value of 6 alleles, per locus. The mean number of alleles per locus (MNA) in all individuals was 6.3± 2.16; the values of observed heterozygosity (H_O) and expected heterozygosity (H_E) were 0.702 and 0.634 respectively, and *FIS* <0 showed few inbreeding. *Fst* result indicated a wide genetic variation between the No. 1 and No. 3 populations, however, no significant genetic differentiation were showed on No. 2 population, which were in the geographic middle position to other two populations; *Fst* analysis (0.0413) and AMOVA result supported small significant result as 76% of the total variance to the within-individuals component based on the microsatellite data, and the PCoA and structure analysis indicated that there just one clusters in total. Bottleneck analysis revealed that there was no historic bottleneck in these populations under TPM model, while Mode-shift test showed all the alleles form all populations are the L-shape.

T1-01-05

Conservation and biological implications of a diverse vascular plant flora in karst caves of SW China

Alexandre Monro[2], *Longfei Fu*[1], *Nadia Bystriakova*[3], *Fang Wen*[1], *Yigang Wei*[1]

1. Guangxi Institute of Botany

2. Royal Botanic Gardens, Kew

3. The Natural History Museum, London

Despite scientists long fascination with caves their diversity remains poorly documented, studied and understood with respect to both mineral and biological diversity. Alexander Von Humboldt was the first to document non-vascularized plants in caves in the 18th century. Surprisingly no vascular plant floras for caves have been documented since, despite several species having been documented as new to science and some investigation into adaptations of vascular plants to the low light environments in caves. Whilst documenting the karst flora of SW China field observations lead us to test the following hypotheses, 1) SW China caves contain a diverse flora, 2) which is a relic of a largely absent forest type lacking endemic species and, 3) that the light environment plants occupy is not distinct from that in forest understory. We surveyed 63 caves and used species accumulation curves (SAC) to estimate the total diversity of this flora. We evaluated the habitats and conservation value of the species documented using regional floras and existing species conservation assessments. We used a subsample of 14 caves to characterize the light environment. Four-hundred-and-eighteen vascular plant species were documented. SACs predict a total diversity of 529-846. Ninety-three percent are known from karst forest, 7% are restricted to caves. The proportion of threatened species is similar to that for the region. Forty-five percent of caves surveyed were disturbed. We documented a subset of species growing in the lowest light levels recorded for vascularised plants. We suggest that this flora is an extension of karst forest understory present prior to catastrophic deforestation and that in SW China caves serves as a refuge and germplasm source. We propose that caves represent a distinct plant biome, most similar to forest understory, differing in the absence of trees, leaf litter, root mats and low diurnal and annual variation in temperature and humidity, and elevated levels of CO_2. We highlight desertification, tourism and the absence of legislated protection of caves as threats and predict that plant-bearing caves are likely occur elsewhere in SE Asia.

T1-01-06

Patterns in karst colonisation by *Elatostema* (Urticaceae) inferred from a molecular phylogeny suggest roles for genome size, seed volume and range-size

Long-Fei Fu[1], *Alex Monro*[2], *Fang Wen*[1], *Yi-Gang Wei*[1]

1. Guangxi Key Laboratory of Plant Conservation and Restoration Ecology in Karst Terrain, Guangxi Institute of Botany

2. Identification & Naming Department, Royal Botanic Gardens

Elatostema is one of the most species-rich genera in the Urticaceae, encompassing ca 550 species that are distributed in Africa, Madagascar, Asia and Australasia. The centre of species-richness is in Asia. There are ca >200 species adapted to karst in SE Asia. In China, 184 species, 65% of the total diversity are associated with the SW China karst leading Wang (2014) to propose that the karst of this region is the centre of distribution and diversification for the genus. Karst is characterized by shallow soils deficient in N and P, but with excessive Ca and Mg and prone to seasonal droughts (Hao *et al.*, , 2015). Soil N is a limiting factor for genome size in karst (Kang *et al.,* 2015). This extreme habitat is associated with high species diversity and endemism for many groups of organism (Clements *et al.*, 2006; Schilthuizen & Clements, 2008). It is also very sensitive to deforestation which frequently results in desertification and the loss of this associated biodiversity. Adaptations which facilitate karst colonisation could include those which mitigate the impact of low N, high Ca, high Mg and water stress. In *Primulina* (Gesneriaceae), genes associated with calcium-permeable channels have been demonstrated to be under positive selection pressure in karst species of Tao *et al.*, (2106). Other possible adaptations to karst could include seed-volume (low N & P), small genome size (low N) and others associated with drought tolerance such as cyclic electron flow in photoprotection, pigments, osmotic adjustment and antioxidant enzymes (Liu *et al.*, 2011; Huang *et al.,* 2012). We use a phylogeny of 120 *Elatostema* species from SW China and SE Asia to explore patterns of karst colonisation within the genus. For all species we recorded the substrate on which they occur (karst versus non-karst) and their range size. For species in SW China we also measured genome size and ploidy level using flow cytometry and seed-size using SEM. Our results show contrasting patterns of speciation on karst between the basal clades and the most species-rich *'core Elatostema'*. By mapping a suite of characters on our tree we identified genome size, range-size and seed volume as being correlated with the colonisation of karst by *core Elatostema* but interestingly not for the basal clades of the genus. This has lead us to develop a series of hypotheses to explain karst colonisation and its consequences for range-size and species number that if confirmed will be important for the design of conservation strategies.

T1-01-07

Plant community retrogression leads to functional homogenization: Implications for Karst desertification

Azim Mallik[1,2], *Philippe St. Martin*[1]
1. Department of Biology, Lakehead University
2. Guangxi Institute of Botany

Effects of ecosystem retrogression on functional diversity are poorly studied. Retrogression is generally characterized by a loss of biomass, structural diversity and taxonomic diversity (TD). It has not been studied in the context of functional diversity (FD), which can affect community assembly and ecosystem function. Transformation of black forest into *Kalmia* dominated heath in eastern Canada has parallel to shrub-dominated karsts in southern China. We hypothesized that post-fire *Kalmia* heaths possess lower TD than forest communities and heath formation selects for a narrow range of species traits that can tolerate habitat stress resulting in biotic homogenization. We tested the hypothesis by i) assessing overall alpha and beta diversity of post-fire heath and forest communities and also individual life form groups to determine if ecosystem retrogression results in a loss of biodiversity or biotic homogenization and ii) comparing FD metrics and plant traits to null models to detect community assembly mechanisms. We measured TD and FD in nine heath and four forest chronosequences in Newfoundland, Canada. We detected significant loss of alpha and beta diversity for both TD and FD. A loss of functional alpha diversity for trees and herbs in heaths indicates a filtering or exclusion of particular traits. Functional beta diversity was lower in heaths than forests overall as well as for trees and herbs, indicating that retrogression through heath formation results in functional homogenization of the majority of life forms. Individual traits of three life forms either converged or diverged in both communities but to different extents. We conclude that heath formation leads to biotic homogenization in retrogressive succession through filtering of traits and that assessment of FD can better reflect ecosystem states of post-fire plant communities than just TD. Our results have implications for any long-lasting simplified vegetation resulting from ecosystem disturbance such as degraded karst vegetation after anthropogenic disturbance.

T1-01-08

Profiling elevational diversity patterns between microbes and plants in a typical karst Tiankeng of China

Gaozhong Pu
Guangxi Institue of Botany, Guangxi Zhuang Autonomous Region and Chinese Academy of Sciences

Karst Tiankeng, being discovered and named at the end of the 20th century in karst areas of China, has enjoyed increasing attention owing to its unusualgeologic processes, unique ecological communities and high-value for conservation and tourism in recent years. In such unique ecosystems, the distribution and composition of plants and animals in karst tiankengs are quite different from related external environments. Correspondingly, microorganism communities in karst tiankeng ecosystems may greatly differ from the related external environments since soil microorganisms and plants are dependent on each other. In this study, we examined the microbial abundance and composition in Shengmu Tainkeng, a typical mestone tiankeng of China. The results reveal that there are strong site differences in microbial community composition (including Bacterial and Fungi). However, diversity of the total microorganism community did not correlate with elevation. By contrast, as expected, plant diversity at the same sites declined along our elevational gradient. These results together suggest that elevational diversity patterns exhibited by eukaryotic microorganisms are fundamentally different from those of plants in Tiankeng ecosystems.

T1-01-09

Fossil records from Southwest China reveals a long existence of Karst vegetation

Jian Huang[1,3], *Lin-Bo Jia*[2,3], *Zhe-Kun Zhou*[1,2]
1. Xishuangbanna Tropical Botanical Garden, Chinese Academy of Sciences
2. Kunming Institute of Botany, Chinese Academy of Sciences
3. University of Chinese Academy of Sciences

Karst region occupys a large proportion of China's land area and is especially in in the Southwest China. Its vegetation ecology is very fragile and difficult to recover after over land-use, and even develops into "rocky desertification". The Wenshan and Maguan basins in southeastern Yunnan Province are Cenozoic sedimentary basins in the karst plateau, a large number of woody plant fossils were found in the Miocene strata of the basins. The flora seems represents an appearance of subtropical evergreen broad-leaved forest vegetation, but its composition are not as same as the typical modern evergreen broad-leaved forest. Combined with modern vegetation research we found that: many taxa in fossil flora also have a high importance value in today's limestone vegetation. For example, Calocedrus, Itea, Cladrastis, Dalbergia, Albizia, Rosa, Berchemia, Zelkova, Ulmus, Quercus miovariabilis, Carpinus, Coriaria, Mallotus, Cornus, Fraxinus, Pittosporum and *etc.* These plants above are main components of karst forest top-level community today. There are also some taxa bearing large amount fossils are sun-bearing and shrub-specific such as Pterolobium, Bauhinia, Paliurus ramosissimus, Ficus miochapaensis, Burretiodendron, Myrsine and *etc.* These findings suggest that in the early Miocene, due to soil and site constraints, there were karst forests in southeastern Yunnan distinct from typical evergreen broad-leaved forest, which was characterized by mixed of Calocedrus and many deciduous taxa. In addition, although the bare rock surface in the karst area can reach the forest community after a long succession, there was still a considerable area will have a long-term stability in the shrub status because of the environmental constraints. Since karst vegetation has been adapted to the environment from Miocene, this type of vegetation will still be available in this area if the human activities are reduced, the disturbed vegetation and biodiversity can also be restored gradually.

T1-01-10

Cycas fairylakea D.Y.Wang in danger and efforts made in its rescue

Qifeng Fu
Shenzhen Meilin Reservoir Management Office

Cycas fairylakea D.Y.Wang, of Cycadaceae and *Cycas*, is a kind of ancient relict plant and one of the rare and endangered plants in the world now. Under the "New International Grade of Endangered Species" by the International Union for Conservation of Nature (IUCN), *C. fairylakea* is involved in Endangered species. In China, it is listed as one of the National First-class Key Preserved Wild Plants. At present *C. fairylakea* only exists in small amount in Tanglang Mountain of Shenzhen City and Qujiang County of Shaoguan City, Heshan County of Jiangmen City in Guangdong Province, China. Due to development construction, recently the number of wild *C. fairylakea* is still dwindling, and till 2008, the total amount is even less than 2,000. Within the first-class drinking-water source reserve of Meilin Reservoir in Shenzhen City survive 1,013 wild *C. fairylakea.* But by 2007, most of the *C. fairylakea* in Meilin Reservoir was on the verge of death with the whole species in risk of extinction, because of their weak competitiveness, insect harm and other reasons. Since 2008, Shenzhen Meilin Reservoir Management Office has been constantly conducting the rescue and conservation of *C. fairylakea*, including the study of species properties, nurture technology, pandava prevention technology, the effect of light transmittance on the growth of *C. fairylakea*, pollination ecology, artificial pollination technology, seed transmission, seed seedling technology, *etc.*, exploring the threatening elements and protection methods for *C. fairylakea.* Through eight years of site protection and conservation, success has been achieved in saving *C. fairylakea.* Furthermore, with the growth status of *C. fairylakea* in Meilin Reservoir improving year by year, the vegetative growth and reproductive growth has fully restored, the species tends towards healthy from the brink of extinction, and the species size is rapidly expanding. To October 2016, in Meilin Reservoir, there are 1,107 *C. fairylakea* over 80cm in height, 4,126 seedlings less than 80cm, with species number totaling over 5,233 and species size expanding four times. At the meantime, more than 1,300 seedlings are planted in other placesprovided by us. To our pride, the rescue of endangered *C. fairylakea* is claimed as a success. In Shenzhen City, *C. fairylakea* is the unique species which belongs to the National First-class Key Preserved Wild Plants, and Meilin Reservoir is its largest habitat over the world. Thus Meilin Reservoir becomes the name card of ecological civilization of Shenzhen City for the rescue of *C. fairylakea. C. fairylakea* conservation breaks a new ground for the rescue of rare and endangered plants.

T1-01-11

Conflicts for plant conservation, *Delphinium caseyi* B.L. Burtt case.

Salih Gucel[1]*, Ozge Fuller*[2]
1. Near East University, Environmental Sciences Institute
2. Near East University, Department of Landscape Architecture, Faculty of Architecture

Biodiversity loss is one of the main concerns of the international scientific community. *Delphinium caseyi* is an endemic perennial in *Cyprus* and is distributed on two locations along Pentadactylos mountains. *D. caseyi* is a protected plant species both under local and international legislation and is categorized as "Endangered" by IUCN (1997) due to its' declining population on the island and is listed under EU Habitats Directive (92/43/EEC) as an Annex II plant species. The goals of this study are to determine the population growth areas and estimate the number of individuals. The next one is to update the IUCN category with the aid of the acquired data and to determine the reproductive biology. And the final goal is to develop specific conservation strategies.

T1-01-12

Tourism in Karst ecosystem: Economic development and sustainability of Lijiang river tourism

Yuefeng Yao[1]*, **Azim Mallik***[1,2]
1. Guangxi Institute of Botany
2. Department of Biology, Lakehead University

The spectacular karst landscape with abrupt eroded limestone hills and winding rivers of southern China attract tourists from all over the world. The four-hour cruise along the Lijiang river is one of the top five tourist attractions in China. With recent economic success there has been a big surge of tourists and concomitant increase in tour boats in the river despite a dramatic hike in ticket price. Rapid urban expansion and agricultural development has been drawing large amounts of water from the river. Hydro-electric dams constructed upstream interferes the natural flow of the river, which in turn interferes the navigation of tour boat. Extensive river engineering is employed to keep the river navigable at low water level. To determine the sustainability of Lijiang river tourism we examined i) mean monthly rainfall pattern, ii) stream flow, iii) number of tourists and tour boats per year, ticket price, and iv) river engineering expenditures from 1960-2010. The number of tour boat increased from 1 in 1963 to 500 in 2010. During this time the number of passengers increased from 400 to 2,210,000. Ticket price increased from < US$1 - $34 and river excavation costs increased from $< 1,000 to over $230 million. From 1990 to 2010 the number of irrigation pups increased from < 100 to 14,200 and water drawn for urban use increased exponentially. Seasonal water flow changed from pre-dam time with larger pick flow during mid-raining season and lower flow during late rainy season causing navigation difficulty. Since the GDP of greater Guilin, Guilin City and Yangshuo County all show a dramatic increase from 1990 to 2010 increase in tourism in the area will likely continue. We predict that tour boat traffic in Lijiang river at the current level is unsustainable. Furthermore, any significant climate change related shift in monthly rainfall pattern in the watershed will likely aggravate the situation even further.

T1-03: Crowdsourcing herbarium digitization: partnerships, platforms, and the worldwide engagement for digitizing biocollections (WeDigBio) event

T1-03-01

Four years of label transcription: What makes volunteers efficient on the web

Simon Chagnoux, Marc Pignal
Museum national d'Histoire Naturelle (MNHN)

The renovation of the Museum national d'Histoire Naturelle (MNHN)'s Paris Herbarium building was the opportunity to conduct a massive digitization of the P and PC collections. This resulted in almost 6 million images becoming available online in 2012. There is an important gap between the rich information available on the photographed specimen labels and the very few attributes indexed in the database. One of the ways to reduce this was to open the participation to the general public on the web. A first citizen science website, http://lesherbonautes.mnhn.fr/ was opened in late 2012 and the success of the operation has led to an improved and more flexible site in 2015. Both the quality and the extent of participation exceeded all initial expectations. The motivations of the volunteers to participate may vary, but a balance between gamification, scientific feedback and social interaction have structured a strong community who completed more than 60 transcribing missions, each with its own taxonomic, geographical or historical specifics. Quality was enforced by a progressive increase in newcomer's abilities, automated training via online quizzes, a strong redundancy in questions and some incentives to discuss special cases in a public forum to improve the collective knowledge. The site communication is more focused on finding answers than transcribing. The initial fixed set of questions on label content and georeferencing was extended in a more flexible way in a second version, thus enabling original workflows and the ability to transpose for use in zoology or paleontology. The heart of the community is a very active minority of power contributors. Their efforts provide the majority of the contributions. Despite some turnover, these volunteers have a long term involvement of several years. The Herbonautes community developed an original expertise about many aspects of the history of the Herbarium collections. Les Herbonautes also benefited from the patronage of the Fondation de la Maison de la Chimie, Paris and is now connected to hundreds of collections on the Recolnat infrastructure.

T1-03-02

Worldwide engagement for Digitizing Biocollections (WeDigBio)—The biocollections community's citizen science space on the Calendar

Elizabeth Ellwood[1], Paul Kimberly[2], Simon Chagnoux[3], Paul Flemons[4], Edward Gilbert[5], Robert Guralnick[6], Kevin Love[7], Austin Mast[1]
1. iDigBio; Florida State University
2. Smithsonian Institution
3. Muséum national d'Histoire naturelle
4. Australian Museum
5. Arizona State University
6. University of Florida
7. iDigBio

Digitization of biocollections, including herbarium specimens, is an ongoing and critical task that has been galvanized by technological advances and new resources, including innovations in crowdsourcing and citizen science. Involving citizen scientists in the digitization process increases their awareness of the number, kinds, and value of biodiversity specimens in collections, advances scientific literacy, increases support for biocollections, and builds sustainability for digitization activities. In turn, growing digital biocollections databases have direct implications for the global community who make use of those data for botanical research, education, and to inform policy. To build support for biocollections and their digitization activities and to increase digitization rates, we organized the annual Worldwide Engagement for Digitizing Biocollections (WeDigBio) Event. In the two years of the event, dozens of museums and classrooms have hosted onsite digitization events where participants transcribed specimen labels using one of five online platforms (DigiVol, Les Herbonautes, Notes from Nature, Smithsonian Institution's Transcription Center, and Symbiota). Thousands of additional citizen scientists also contributed online from more than one hundred countries, completing tens of thousands of transcription tasks. Planning and executing WeDigBio events required us to find efficient ways to integrate disparate transcription and participant data across platforms and projects. Here, we present information on the process of organizing an international citizen science event, an analysis of the event's effectiveness (e.g., transcription rates before, during, and after the event), lessons learned, and future directions.

T1-03-03

Building crowdsourcing campaigns for collections digitization using Biospex

Austin Mast, Greg Riccardi, Robert Bruhn, Elizabeth Ellwood
Florida State University

Biospex is an online basecamp for launching, advertising, and managing targeted efforts to digitize biodiversity specimens.

Biospex provides tools for curators, amateur enthusiast groups, and others to mint a webpage for a project, package specimens into one or a series of digitization "expeditions" associated with the project, launch the expeditions at crowdsourcing tools (e.g., Notes from Nature), widely recruit others to participate, and follow crowdsourcing progress. We will illustrate the functionality of Biospex using the WeDigFLPlants project, a collaboration between southeastern U.S. herbaria, the Florida Native Plant Society, and other Florida-focused amateur-enthusiast groups to transcribe > 50,000 herbarium specimens collected in Florida and held at > 50 herbaria. WeDigFLPlants is a WeDigBio Interest Group designed to maintain the momentum of the 4-day WeDigBio event throughout the year.

T1-03-04

WeDigBio: A two year tale: Unlocking over a century of natural history collecting and exploring; Accelerating increased accessibility to biodiversity data at The Field Museum, Chicago

Matthew Von Konrat[1], *Marilyn Alam*[3], *Megan Bradley*[1], *Laura Briscoe*[4], *Elana Carpinone*[1], *Lucille Carver*[1], *Robin Delapena*[1], *Christine Giannoni*[1], *Anni Glissman*[1], *Sharon Grant*[1], *Zoe Greenfield*[2], *Kate Golembiewski*[1], *Lauren Hancock*[1], *Kaccee Kellum*[1], *Jeff Konker*[1], *Crystal Maier*[1], *Christine Marsh*[1], *Joann Martinec*[1], *Paul Mayer*[1], *Susan Mochel*[1], *Angie Morrow*[1], *Caleb Mcmahan*[1], *Diana Ramirez*[2], *Yarency Rodriguez*[1], *Gretchen Rings*[1], *Pam Phelps*[5], *Jennifer Slate*[3], *Chloe Takala*[2], *Tom Thomas*[3], *Caitlin Vaughn*[6], *Kate Webbink*[1], *Barajas Sergio*[3], *Greif Matt*[3], *Alexandra Lopez*[7]

1. The Field Museum
2. Roosevelt University
3. Northeastern Illinois University
4. New York Botanical Garden
5. Oswego High School
6. DePaul University
7. Wilbur Wright College

Alarmingly, the world's biodiversity is diminishing rapidly and undergoing an extinction crisis. Natural history collections of museums and academic institutions, documenting the fossilized and living members of the world's ecosystems and their changes over time, are uniquely poised to inform the stewardship of life on Earth. The National Science Foundation (NSF) program Advancing Digitization of Biodiversity Collections (ADBC) is enhancing and expanding the national resource of digital data documenting biological and paleontological collections at a rapid pace. We provide a case study founded on these digitization efforts partnering with the annual "Worldwide Engagement for Digitizing Biocollections" (WeDigBio) Event to accelerate the pace of scientific discovery. Digitization of natural history collections is an ongoing and critical task that has been galvanized by technological advances and new resources, including innovations in crowdsourcing and citizen science. Involving citizen scientists in this process increases awareness of the number, kinds, and value of biodiversity specimens in collections, advances scientific literacy, increases public support for natural history collections, and builds sustainability for digitization activities. In turn, growing digital natural history collections databases has direct implications for the global community who make use of those data for research and education, including investigating climate and other environmental changes, conservation biology, population genetics, and even public health and safety. To build support for natural history collections, digitization activities and to increase digitization rates, the Gantz Family Collections Center of The Field Museum invited members of the public, including teens and young adults, to participate in a "transcription party" in 2015 and 2016. Partnering with college, university and high schools - especially those with a high proportion of students underrepresented in science - students received class credit for their work. Volunteers joined museums around the world as part of a global four day effort to digitize centuries of data about life on Earth organized by WeDigBio. The Field Museum was one of the most significant on-site event hosts, globally, with the support from over 400 citizen scientists working the equivalent of over 1,000 hours spanning the two events and producing over 12,000 collection records and pages of transcribed field notes. Projects spanned the entire collections including bryophytes, fungi, ferns, insects, fishes, fossil invertebrates, field notebooks and historical exhibit labels using a variety of online platforms. The majority of specimen and object data remain difficult to access, locked in the cabinets of museum and university collections in analog format, presenting the natural history collection community with many years of digitization work. Enlisting the public's help in transforming historic data labels into an open, globally accessible, digital resource is helping unlock valuable information that can help scientists accelerate the process of biodiversity discovery. We will discuss our experiences, assess performance and examine the balance of costs and benefit including quality of data and investment of staff time. We conclude that with careful training, planning, and targeting specific audiences these projects remain a cost-effective way of collecting data while also providing an excellent opportunity for people to become engaged with natural history collections.

T1-03-05

The notes from nature project: Evolutions and revolutions in enhancing engagement with data providers and citizen scientists

Julie Allen[1], *Michael Denslow*[2], *Austin Mast*[3], *Robert Guralnick*[1]

1. University of Florida
2. Southeast Regional Network of Expertise and Collections
3. Florida State University

Notes from Nature (http://www.notesfromnature.org; NFN) is a citizen science tool focused on public engagement and label transcription of natural history specimens. The project was developed collaboratively by biodiversity scientists, curators, and experts in citizen science, within the well-established Zooniverse platform. This project currently brings together digital images representing a wide variety of biodiversity records such as ledgers, herbarium sheets, pinned insects and fossils. Volunteer citizen scientists transcribe textual data contained in the specimen images. Since its launch in 2013, NFN has amassed > 1.2 million transcriptions from tens of thousands of volunteers worldwide. This talk focuses on innovations aimed to improve the platform, some incremental and others totally changing the way Notes from Nature works. We discuss improvements such as: (1) Tools to build consensus and quality metrics for the transcriptions to further enhance data quality for museum curators; (2) Better workflows to support image loading and repatriation of data to our providers; (3) Improved engagement about biodiversity by bringing in rich content from external sources such as Encyclopedia of Life and Map of Life. We also discuss engagement efforts and interoperability with other biodiversity informatics tools and databases that are important to our partners' curation needs. Such improvements help NFN take its place as a critical component of an ecosystem of tools needed to unlock the vast legacy biodiversity data for broad public good.

T1-03-06

Diversification in a transcription platform—costs and benefits

*Paul Flemons, **Rhiannon Stephens***
Australian Museum

Transcription of natural history collections through crowdsourcing has many benefits not just associated with the capture of data connected to the specimens of these collections. Benefits include broadening the understanding of the specific collections involved through participation; increasing scientific literacy particularly in relation to scientific collections; increasing the profile of the collection institutions and the specific collecting events involved; providing a true citizen science engagement for non professionals who have a passion for science. An ongoing challenge for transcription centres like DigiVol is the recruitment and maintenance of the transcription community that captures the data. DigiVol has pursued a strategy of engaging a broad range of institutions and types of data capture in an effort to provide a diverse engagement experience for transcribers. I will discuss the costs and benefits of this approach.

T1-04: Arboreta and botanical gardens as major centers for research and conservation of global trees and forests

T1-04-01

Plant diversity: What do we need to conserve?

Richard Corlett
Xishuangbanna Tropical Botanical Garden, Chinese Academy of Sciences

Botanical collections have traditionally included as many plant species as possible in order to provide both an attractive and interesting display, and a diversity of living material for botanical research. In recent decades, the increasing focus on plant conservation has resulted in repurposing of these living collections as *ex situ* conservation. A botanical garden that previously claimed to display 5,000 species now claims to conserve the same number. Seed banks make similar claims for their collections and protected areas are often assumed to provide *in situ* conservation for all the species recorded there. However, while even a single individual, perpetuated by cloning, may save a species from extinction, species are not the only level of diversity that we need to conserve. If we want to have more than a living museum specimen we need to save more than one individual. But, how many? And which ones? Modern high-resolution genomic techniques can detect differences between almost any pair of plants or plant populations, raising the question of when these differences are considered to be of conservation relevance. It is not practical to treat every genetically diagnosable difference as significant. Provenance trials with economic plants, show that local adaptation is widespread, suggesting that plant conservation needs to target the full range of adaptive variation within each species, rather than a single local population. In practice, information on adaptive genetic variation is rarely available and overall genetic variation is commonly used as a proxy for adaptive variation. However, restricted gene flow produces spatial genetic structure even if populations show little or no local adaptation, so this practice may exaggerate the scale of the conservation problem. New methods for the functional interpretation of genetic variation are being developed, but current best practice is still to use neutral markers and/or phenotypic information as well to ensure that adaptive variation has not been overlooked. These issues apply to both *in situ* and *ex situ* conservation, but the best solution will vary depending on the costs. *In situ* conservation is expensive, particularly if it requires land that that has alternative economic uses. The high costs limit the number of separate populations of each plant species that can practicably be protected. *Ex situ* conservation of orthodox (drying tolerant) seeds in seed banks is much cheaper, so it may be practical to conserve large samples from enough populations to allow for the uncertainties outlined above. For the minority of (mostly tropical) species with recalcitrant (drying intolerant) seeds, however, living collections are so far the most practical *ex situ* technique. As a result, conservation costs are again high and space constraints limit both the number of individual plants that can be grown and the number of populations that can be spaced far enough apart to maintain genetic isolation. High conservation costs mean that collections must be optimized, but there is currently no consensus on what this optimum is.

T1-04-02

Role of botanic gardens in conservation and management of rare and endemic Indian trees

Paramjit Singh
Botanical Survey of India

Botanic gardens play an important role in conservation of rare species, plant recovery programmes and phenological studies. Strengthening *ex situ* conservation is one of the essential components of India's National Biodiversity Strategy and Action Plan (NBSAP), which is based on the principles built in the National Environment Policy of the country. Being a signatory to Convention on Biological Diversity, India is committed towards conservation and sustainable utilization of plants of our country. Botanic gardens conserve the plant species in areas outside their natural habitats and serve as repositories of many threatened indigenous plant species. Botanical Survey of India (BSI), a premier research organization of this Ministry has been contributing immensely towards the *ex situ* conservation of threatened, rare and endemic plants. It is estimated that, Botanic gardens harbor about 30 per cent of country's indigenous threatened plants. Botanic gardens play an important role as centers for plant conservation, preservation of endangered species, seeds and propagules, cultivation of rare, vulnerable and endemic species. They are also centers for exchanging information and plant materials and act as a platform for creating awareness through education programme. Significant role of botanic gardens in developing the Global Strategy for Plant Conservation (GSPC) and its ultimate adoption by the Parties to the Convention on Biological Diversity (CBD) is well recognized. Indian Botanic Gardens plays a special and remarkable role in contributing to the targets of GSPC in conservation, education and public awareness. In India, for conservation of rare species in botanic gardens 'Assistance to Botanic Garden' programme is implemented through Botanical Survey of India (BSI) which is managing this programme through various initiatives. The presentation highlights recent initiatives undertaken to conserve endemic and rare trees in various botanic gardens in India. The success story of rehabilitation of some of the endemics in the wild like *Bentinckia nicobarica, Gymnocladus assamicus, Stapletonia arunachalense, Rhododendron* species is discussed in detail.

T1-04-03

Botanic gardens in developing countries: Challenges and opportunities in urban ecosystems

Mauricio Diazgranados

Natural Capital and Plant Health Department, Royal Botanic Gardens, Kew

Some of the most biodiverse countries have often serious social and environmental problems. A common element of developing nations is inequality between rural and urban communities, fuelling rapid urbanization. Growth of many the cities in these countries brings further social and environmental problems, and often they become massive overpopulated centres, acting as barriers for the surrounding ecosystems, and even holding introduced exotic species (sometimes invasive) in their green spaces. Under this context, botanic gardens have a great opportunity, beyond *ex situ* and *in situ* conservation: they can lead processes of conservation, propagation and research of native endangered trees and plants in general, using cities as experimental arenas. The challenge is obtaining the required funds in these countries. However, serving the city is a powerful formula, especially when botanic gardens become a vital part of cities administrations, with a more active involvement in improving livelihoods of citizens. Bogota, capital of Colombia, is a great example of analysis: a city of 8.5 million, that despite being in one of the most biodiverse countries, holds green spaces with exotic (and invasive) species. The city is home of the Bogota Botanic Garden (JBB), Colombia's oldest and most important botanic garden. In the past years, while I was JBB's director for science, a powerful strategy of service was developed, based on five large approaches for the city: 1. tree management (planting, pruning, monitoring); 2. restoration of strategic ecosystems within the city district (wetlands, rivers and creeks, Andean forest); 3. greening urban spaces (green roofs and walls, green public furniture, green bus stops); 4. leading urban agroecology (public urban orchards; orchards of medicinal plans); and 5. gardening with native (often threatened) species. When these approaches are combined with the use of native and endangered plant species, this model provides a great opportunity to contribute to their *ex situ* conservation, propagation and research. The JBB maintains an open access database with the trees from the city's public space (~1.7 million records), and over the last three years has planted more than 200,000 trees of 90 species, and has restored 16 areas planting more than 45,000 trees. Besides the success of the implementation, the strategy became an economically sustainable model, representing more than 80% of the income, and allowing a solid investment in the JBB's infrastructure: construction of new laboratories, research and public greenhouses, herbarium, conference centre, *etc.* While many other botanic gardens in developing countries are struggling to increase funding, the JBB's model not only secured funding for a historic development of its facilities, but also had a relevant role in transforming the green spaces of the city, with native and often endangered species of trees. May this be a model for botanic gardens in other developing countries?

T1-04-04

How to preserve native flora? A case from Chinese botanical gardens

Jin Chen

Xishuangbanna Tropical Botanical Garden, Chinese Academy of Sciences; Chinese Union of Botanical Gardens

Plant species extinction is among the most challenging threatens to the planet while effective actions, both globally and locally, appears to be limited. The information update for species distribution, the failure to convince decision makers and various conservation activities work separately without collective efforts, could be the challenging for solving the problem. To address biodiversity loss crisis, Xishuangbanna Tropical Botanical Garden (XTBG) has initiated a program termed 'Plant zero extinction program' in Xishuangbanna, a biodiversity hotspot in China's tropics. The concept for the program is trying to use the entire tools in conservation toolbox to achieve plant species to zero extinction in coming 5-10 years in the area. The preliminary results have shown it could be an effective and measurable approach to battle against local plant extinction. In 2014, by meaning of the network of the Chinese Union of Botanical Gardens (CUBG), this protocol has been expended to large areas throughout China, which is recognized as a key project of CUBG termed 'Complete coverage conservation program'. Totally, eight BGs (i.e., Xishuangbanna, Kunming, Wuhan, Tulupan, Guiling, Nanjing, Beijing and Shenyang), joined the program, targeting an area about 28% of total country territory. In this talk, I will present some of the progress. The weakness and challenging for achieving the safely conserve China's native flora will be discussed. As a whole, I wish to show a new and possibly workable model for botanical garden community acting as a leading role for collecting all efforts for effective conservation for native flora.

T1-04-06

The strengths and expertise botanical gardens can provide to the successful management and conservation of trees in the Anthropocene

Charles Cannon

Center for Tree Science, The Morton Arboretum

Human activity now dominates basic Earth processes, from the cycling of nutrients and energy in the environment to the fate of each species. As we make this transition into the Anthropocene, we must assess whether the traditional approaches to management and conservation are appropriate going forward. We need to anticipate the novelty and unpredictability of future climates and habitats, both of which pose unique challenges to long-lived sessile organisms like trees. In this talk, I will focus on the unique attributes and strengths that arboreta and gardens can bring to these challenges and how we can exploit the natural tendencies of tree reproduction and growth to potentially accelerate their adaptation and diversification, utilizing both traditional techniques like grafting and cutting-edge emerging technologies like genomics and metabolomics. I will also provide examples of how pro-active engagement can create opportunities for direct input into management and conservation of trees across the urban to wild spectrum.

T1-05: Climate change, conservation and the future of Asian pine species

T1-05-01
The influence of origin and climate upon tree growth and survival in a 100-year-old provenance trial in northern Arizona, USA: Lessons for managers

W. Keith Moser
U.S.D.A., Forest Service, Rocky Mountain Research Station

Early in the 20th century, a provenance trial of ponderosa pine (*Pinus ponderosa*) was established on what is now the Fort Valley Experimental Forest, 14.5 km north of Flagstaff, Arizona, USA. The purpose of the study was to assess genetic diversity and adaptation patterns with the goal of improving the management of ponderosa pine forests. Seed was collected from 18 National Forests throughout the western United States and planted on the study site from 1913 to 1917. The seeds were identified as belonging to one of two varieties, *Pinus ponderosa* var. *ponderosa* and *P. ponderosa* var. *scopulorum*. Only *P. ponderosa* var. *scopulorum* provenances have survived to this day. Annual survival was monitored until 1919, and then subsequently remeasured in 1928,1951,1964, and 1995. Tree height was measured in 1928, and height and diameter were measured 1995-1996. In 2014, we undertook a remeasurement of height, crown density, and diameter and, in 2016, we added surrounding tree density to the dataset. Selected trees were cored in 2014 to examine provenance × climate interactions with radial growth. We compared climate at the source location to that of the study site. We also looked at different variables of annual weather (precipitation, growing season, temperature) for clues as to their influence on individual tree performance. In this paper, we present results of our analyses, including the relationship between recent tree growth and climate and local weather patterns at the seed source site. We also evaluate the influence of irregular mortality and the residual effects of survivor density upon growth. Finally, we offer suggested field applications of these results and some plans for future studies.

T1-05-02
Adaptation and introgression between two conifers from the Qinghai-Tibet Plateau

***Kangshan Mao**, Yazhen Ma, Jianquan Liu*
College of Life Science, Sichuan University, Chengdu 610064, Sichuan, P.R. China

Climate change is the single largest threats to forest tree populations. To survive the rapidly changing environments, they may migrate to track favorable ecological niche spatially, persist through plasticity in shorter term, and adapt to new environment via generating of new genetic variation and mutation. Hybridization and introgression may also contribute to the adaptive responses of forest tree populations to changing environmental conditions by transferring standing genetic variations and mutations from one species (population) to another. Such cases have been reported for animals and model plants, yet very rare for non-model forest tree species.

Here, we examined the role of introgressive hybridization in transferring adaptive genetic variation between two cypresses in the Qinghai-Tibetan Plateau. We collected samples for 35 (11 populations) and 30 individuals (12 populations) of *Cupressus duclouxiana* and *C. gigantea*, respectively, and performed RNA-sequencing. The transcriptome of one *C. duclouxiana* invidivual were assembled and clean reads of the other individuals were then aligned to this reference transcriptome. After stringent quality filtering, 878,614 single nucleotide polymorphisms (SNPs) were identified in all 65 individuals, and 1,208,905 in *C. duclouxiana* and 1,204,335 in *C. gigantea*. Both principal component analysis and phylogeny analysis suggest that the two species were clustered into two monophyletic clades, respectively, and *C. duclouxiana* comprises of two groups which correspond to its northern and southern populations. Genetic clustrering indicates introgression occurred from *C. gigantea* to the northern population of *C. duclouxiana*. This hypothesis is also supported by the pairwise FST values as well as the identical-by-descent (IBD) analysis. FST between *C. gigantea* and *C. duclouxiana* southern population is higher than that between *C. gigantea* and *C. duclouxiana* northern population (0.3126 vs. 03060); 50 comparisons between *C. gigantea* and the *C. duclouxiana* northern population were found get shared haplotypes, while only six comparisons between *C. gigantea* and the *C. duclouxiana* southern population got the IBD haplotypes. Meanwhile, the best-fitting demographic model (out of 16 models) in our Fastsimcoal2 simulations suggests much higher gene flow from *C. gigantea* to *C. duclouxiana* northern population (1.42×10-6) than the opposite direction (5.02×10-8) or than that between *C. gigantea* and *C. duclouxiana* southern population (2.10×10-7,7.35×10-8). To further identify the introgressed loci in the *C. duclouxiana* northern population, we calculated the D statistic and the fdM-statistic, and finally identified a subset of 71 candidate introgression loci, 47 of which got annotations from Swiss-Prot database. Meanwhile, we used HKA (Hudson-Kreitman-Aguade) test to identify genes under natural selection, a total of 2831 candidate of positively selected unigenes were identified and 19 of them were potential introgressed loci. Differential expression analysis suggested that 700 and 1148 unigenes expressed at a significantly higher level *in C. gigantea* and *C. duclouxiana* southern population, respectively. Three candidate introgressive genes showing selection signature, ABA3, TR6055|c0_g1_i1 (best hit to At1g55270) and SPHK2, were detected to show significantly different expression levels between the two lineages ($P \leq 0.05$).

Putting together, multiple lines of evidence support introgression from *C. gigantea* to the northern population of *C. duclouxiana*, a set of 19 candidate introgressive loci under positive selection were identified and three of them show significant expression level difference between high and low altitude lineages. These candidate introgressive loci from *C. gigantea* to *C. duclouxiana* most likely have assisted the northward and probably upward expansion of *C. duclouxiana*. Our findings will shed light on the geographical and genetic consequences of forest tree population in the mountainous regions in the face of global climate change, and more importantly, emphasizes the conservation values and important roles of introgressive populations and hybrid zone in adaptation to changing climate.

T1-05-03

Phytogeography of Korean Pinaceae

Wooseok Kong
Kyung Hee University

Despite of ecological and landscape importance and public popularity of conifers, especially in case of Pinaceae in the Korean Peninsula, not much scientific information are known. Present work aims to understand the phytogeography of Korean native conifers, especially the family of Pinaceae or pine tree family, i.e., phylogeny, origin, life form, distribution, dispersal and migration. Korean native Pinaceae consists of five genera and fourteen species. Pinus or pine tree is systematically closely related to Picea or spruce and Larix or larch, but Abies or fir is related to Tsuga. Cold-tolerant boreal conifers which have migrated from NE Asia during the Pleistocene glacial epochs in search of glacial refugia and successfully survived, but now confined its ranges to the alpine and subalpine belts of the Korean Peninsula mainly due to climate amelioration since the Holocene period. Since the Last Glacial Maximum, species, such as Picea pungsanensis and Abies koreana have gradually adapted to local environment, and later became an endemic species of Korea. Disjunctive distribution of Pinus parviflora and Tsuga sieboldii on Ullung Island, which also occur in the Japanese Isles needs further study. Major dispersal agent of pine trees with winged seed is wind, but wingless pine tree seeds might dispersed by birds and rodents. Pine trees with bigger wings are easily dispersed by wind, and now show broader distribution. Species of Pinaceae with disjunctive distribution on the alpine and subalpine belts of the Korean Peninsula, which includes both ROK or South Korea and DPRK or North Korea seems to be more vulnerable to global warming.

T1-05-04

Phylogeography of *Abies koreana* and *A. nephrolepis* (Pinaceae) in Korea

Jongcheol Yang, *Insik Kim, Hyejin Kwon*
Korea National Arboretum

Genetic variations of *Abies koreana* and *A. nephrolepis*, taxonomically confusing taxa in *Abies*, were assessed using three mitochondrial DNA regions (*nad*5 intron 4, *nad*5 intron 1, and *cob*_pre), a chloroplast DNA region (trnV-UAC/trnT-GGU), and a nuclear DNA region (ITS) for 247 individuals from 19 natural populations to understand their phylogeographic relationships. In the mtDNA analysis, twelve polymorphic sites of the combined tree regions resulted in the resolution of five haplotypes (M1-M5). The average gene diversity within population (H_S) was 0.072, and total gene diversity (H_T) was 0.620. The populations were geographically divided into four groups (group1-4) by the SAMOVA analysis. In the cpDNA analysis, twelve polymorphic sites of a region resulted in the resolution of ten haplotypes (C1-C10). Average gene diversity within population (H_S) was 0.552, and total gene diversity (H_T) was 0.768. However, no variation was found in the nrDNA data. The populations of northern (group1) and southern (group2) areas were nearly structured for M1 and M2, respectively. The presence of the single mtDNA haplotype in the southern areas suggest that the current widespread populations could have expanded to central areas from a specific refugium following the last glacial periods (20,000 years ago). Based on the results from this study, the distribution of Abies koreana is postulated below Mt. Songni in Goesan-gun, Chungcheongbuk-do (group1 region). Additionally, the populations of M4 and M5 types (Mt. Worak in Jecheon-si, Chungcheongbuk-do and Mt. Yeongchuk in Yangsan-si, Gyeongsangnam-do) should be studied with closely related species of adjacent regions for specific taxonomic identify.

T1-05-05

How will subalpine conifer distributions be affected by climate change in Japan?

Ikutaro Tsuyama[1], *Motoki Higa*[2], *Katsuhiro Nakao*[3], *Tetsuya Matsui*[4], *Masahiro Horikawa*[5], *Nobuyuki Tanaka*[6]
1. Hokkaido Research Center, Forestry and Forest Products Research Institute
2. Faculty of Science, Kochi University
3. Kansai Research Center, Forestry and Forest Products Research Institute
4. Forestry and Forest Products Research Institute
5. Toyota Biotechnology and Afforestation Laboratory, Toyota Motor Corporation
6. Graduate School of Agriculture, Tokyo University of Agriculture

This study aimed to assess the climate change impacts on Japanese keystone subalpine conifers (*Tsuga diversifolia* and *Abies veitchii*) incorporating the uncertainties of general circulation models (GCMs) for spatial conservation planning. Potential habitats for *T. diversifolia* and *A. veitchii* under current and 20 GCMs in 2080-2100 were projected using generalized additive models. The number of GCMs under which current potential habitat was projected to lose was counted by a spatial unit and was used as indices of vulnerability and prediction uncertainty. We identified gaps between protected areas and potential habitats under future climates. Reasonably accurate models were developed. We revealed that *T. diversifolia* and *A. veitchii* require cool and wet summers with low winter precipitation. Less than 1% of potential habitats under current climate were projected to be sustainable with low uncertainties of GCMs. These certainly sustainable habitats were mostly designated as protected areas. About 88 and 97% of current potential habitats for *T. diversifolia* and *A. veitchii* were projected to be non-habitats under future climates with low GCMs-related uncertainties. The southern and low elevated northern range limits of the species were included in these certainly vulnerable habitats, of which 70% were designated as protected areas. Our results suggest that subalpine conifers in Japan are vulnerable to the climate change irrespective of GCMs variation. Strengthening of monitoring may be necessary at unprotected and certainly vulnerable habitats in lower subalpine areas where influences of the climate change will appear first. Active management, such as assisted regeneration, may be necessary for conserving the vulnerable and genetically endemic populations.

T1-05-06

Climatic control of zonal vegetation in Northern Asia and its phytogeographical background

Pavel Krestov
Botanical Garden-Institute FEB RAS

Large-scale vegetation studies are one of the fundamental aspects of ecology and biogeography and a key to the clarification of modern as well as past processes in vegetation cover. This study focuses at the problem of indication of climatic gradients by vegetation complexes at local and regional scales and aims to quantify the local and regional scale relations of vegetation units, their complexes and climatic parameters within boreal vegetation zone in Northeast Asia. We used phytosociological (over 5,000 releves), climatic (2,200 climatic stations) and floristic (distribution records for 22,000 species) databases. Identification of bioclimates was made in accordance with Rivas-Martínez *et al.*, (1999) approach using different climatic parameters that include Kira's warmth (WK) and coldness (CK) indices, continentality index (CI), ombro-evapotranspirational (OEI) index and winter precipitation (WP). Analysis of indices calculated with an aid of developed models showed the significant differences of vegetation units of the order rank in bioclimatic ranges. WK decreases from values over 75 in the middle temperate zone (Saso-Fagetalia, Aceri-Quercetalia) to 20 in subarctic zone. Among boreal vegetation units the orders of Betulo-Ranunculetea have the lowest warmth index in the circumboreal zone that can be explained by cool summer in conditions of oceanic climate. CK varies between values of -25 and -150 within boreal and temperate zones with prevalence of deciduous broadleaved, mixed and evergreen broadleaved forests. The boreal orders Lathyro-Laricetalia and Ledo-Laricetalia representing boreal deciduous coniferous forests are characterized by very low values of coldness index that, in this case, is comparable to that of subarctic orders. The ranges of orders along the continentality gradient are reflected by changes of vegetation types within a zone with proximity to the ocean. In boreal zone the lowest values of CI are characteristic to the forestless and not yet described class of Aleutian meadows, followed by *Alnus fruticosa* scrub and *Betula ermanii* forests of the class Betuletea ermanii. Highest values belong to deciduous coniferous orders Lathyro-Laricetalia and Ledo-Laricetalia. The snow affects vegetation strongly in the conditions of continental climate within northern subzone of temperate zone, where lack of snow in winter causes a shortage of moisture in spring and early summer, and in oceanic regions of boreal zone, where the strong accumulation of snow causes 2-3 week delay of its melting and a considerable shortening of growing season. The communities of *Betula ermanii*, *Alnus fruticosa* and tall-forb meadows are characteristic to the regions with slower-melting heavy snow deposits. Zonal subarctic vegetation is represented by communities of provisional class Betuletea glanduloso-divaricatae composed of species complexes of Pacific coastal mountainous regions. In boreal zone, vegetation of Asian ultracontinental to maritime sectors is represented by communities of Vaccinio-Piceetea, Asian suboceanic to hyperoceanic sectors—by communities of Betulo-Ranunculetea. The boreal classes were likely differentiated in Pre-Pleistocene time due to the well developed floristic centers under the different climatic situations.

This study was supported by Russian foundation for basic research (16-54-53057), Far Eastern Branch of Russian Academy of Sciences (the program "The Far East") and Korean National Arboretum.

T1-06: Reintroduction and management of rare plants in response to global change

T1-06-01

Conservation meets restoration - Rescuing threatened plant species by restoring their environments and restoring environments using threatened plant species

Sergei Volis
Kunming Institute of Botany

Plant conservation biology needs a new approach to cope with the rapid disappearance of species and ecosystems. I introduce a conceptual integration of conservation biology and restoration ecology in what can be called conservation-oriented restoration. The usage of this term is limited to the cases when restoration is applied to a still functioning ecosystem, excluding cases when the destroyed ecosystem must be recreated or altered to a desirable state. The underlining idea of this approach is the importance of habitat restoration for the majority of threatened species, and, on the other hand, usefulness of threatened plant species for restoration of natural habitats. It is proposed that threatened plant species should become an important part of many restoration projects and be introduced not only into locations where they currently grow or grew in the recent past, but also into suitable locations within their potential distribution range. Because the number of potentially suitable locations can be close to zero if we consider only untouched natural habitats as suitable, the re-introduction sites should include those that require restoration efforts. The general methodological guidelines are provided to show why and how ecological restoration using threatened plant species should become an integral part of the conservation biologist's armory.

T1-06-02

On the edge of a continent – An exploration of comprehensive rare plant conservation and restoration in Southern Florida, USA, over two decades

George D. Gann, *Craig Van Der Heiden*
The Institute for Regional Conservation

Southern Florida has undergone rapid transformation in just over a century. With barely 25,000 residents in 1900 the area now contains about 8 million people spread over an area of about 4 million hectares. The greater Everglades ecosystem, home to myriad tropical and subtropical species, was wracked by massive engineering works and expansion of the agriculture frontier. Coastal and upland ecosystems, containing many endemic species, were radically transformed to accommodate growing human populations. The entire region is now subject to many ecosystem and species-based restoration projects, including the Everglades Restoration, but population and other pressures continue to grow. Although there were some existing programs to conserve rare plants, in the mid-1990s The Institute for Regional Conservation (IRC) embarked on an ambitious and collaborative program to document and improve management of the region's native flora – a total of more than 1,400 species. Beginning with the cataloging of all existing information on plant distributions and rarity, intensive field surveys in dozens of protected areas, and the collection and databasing of herbarium specimens, IRC then assessed and ranked the entire native flora of southern Florida using a modification of the Heritage Program ranking system (now NatureServe). Results of this work were published online as the Floristic Inventory of South Florida (Gann *et al.,* 2001-2017) and as the book Rare Plants of South Florida: Their History, Conservation and Restoration (Gann *et al.,* 2002). Based on major findings in these publications, IRC then collaborated with multiple agencies at the local and national level to: 1) set up long-term monitoring programs for rare plants in fragmented ecosystems; 2) actively restore degraded habitat for several species listed by the U.S. Endangered Species Act and numerous other rare plants; 3) conduct baseline studies to evaluate the impact of the Everglades Restoration on rare plants in Everglades National Park (ENP); and, 4) conduct reintroduction and augmentation trials for several depleted species in ENP. This presentation will summarize the important results of IRC's work and how it has been used by the scientific and conservation communities to improve rare plant conservation and restoration efforts in southern Florida. It will also explore emerging concepts of plant conservation in the region given the reality of climate change (including sea level rise), fragmentation, invasive species and other factors. Finally, we will discuss how this work contributes to global initiatives, such as the Global Strategy for Plant Conservation, and serves as a model for other plant conservation initiatives in the Caribbean region.

T1-06-03

Conservation and introduction of *Camellia changii*, an endemic and rare plant in south China

Hai Ren
South China Botanical Garden, Chinese Academy of Sciences

Camellia changii is an endemic and rare species in south China with an extremely narrow range. It grew under the discontinuous canopy of secondary evergreen broad-leaved forest with well-drained acidic sandy loam. It is an insect-pollinated and self incompatible species. This species had characteristics of high blossoming rate but low seed setting rate. We found a total of 1039 wild individuals of this species during 2009-2013. The population structure was that of a declining population because no seedlings and few young plants were found. We confirmed the Critically Endangered status of this species, therefore, we established a nursery of plants generated by grafting of short cuttings (2-3 cm each) from all wild individuals to conserve the genetic diversity in the wild. We also conducted two types of conservation translocation experiments, augmentation and conservation introduction at two sites, Ehuangzhang and Tianxin, respectively, using 4-year-old grafted *C. changii* reproductive-size plants. The survival rate of transplanted plants was 100% after two years at both sites. The individuals at the augmentation site grew significantly faster than those at the introduction site ($P<0.05$). Transplanted plants at the augmentation site produced 26–28 flowers per plant per year, significantly less than did those at the introduction site (35–38 flowers) during the two-year (2013–2014) observation period ($P<0.05$). The content of anthocyanin, a main pigment in the petal of flowers, at the introduction site was less than that at the augmentation site. Compositions of the pollinating species of *C. changii* at both sites were similar, but the dominant pollinating species and their visiting frequency were different. Butterflies dominated the augmentation site and bees the introduction site. Nevertheless, the annual fruit and seed productions of both translocations in 2013 and 2014 were very similar. Our findings suggest that *C. changii* subject to the out-of-range conservation introduction did not suffer a decline in reproductive success when compared with augmentation. These findings suggest that certain rare and endangered plants may have the ability to cope with challenges that are brought on by conservation introduction in the face of climate change, at least in the short term.

T1-06-04

Orchid conservation translocation efforts in China and Florida: Which, why, who and future priorities

***Hong Liu*[1,2,3], *Haydee Borrero*[1,3], *Shicheng Shao*[4], *Jennifer Sanders*[5], *Kevin Burgess*[6]**
1. Florida International University, International Center for Tropical Botany
2. Guangxi University, College of Forestry
3. Fairchild Tropical Botanic Garden
4. Xishuangbanna Tropical Botanical Garden
5. Atlanta Botanical Garden
6. Columbus State University, College of Letters & Sciences, Department of Biology

Orchids are a highly diverse group of plants that are threatened due to habitat destruction and overharvesting. Globally, conservation translocation (sensu IUCN 2013) - the deliberate movement of organisms from one site for release in another with an intended conservation benefit – has been an important recovery strategy for orchids. In this study, we review current conservation translocation efforts of orchids in China and the United States to investigate variation in 1) the choice of plants for translocation; 2) the length of monitoring time; 3) the success of translocation efforts and 4) the type of agencies invoking translocation as a management action. We based our analyses on information gained from peer- and non-peer reviewed publications. We hope to unveil the motivation for orchid translocations between these two countries; both of which have notable contrasting socioeconomic and political systems. Based on our review of the literature we found 61 species (38 genera) targeted for conservation translocation across China (21 genera; 36 species) and the United States (17 genera; 25 species). We found that 2.6% (36) and 12.2% (25) of the national orchid species have been subjected to conservation translocation in China and the US, respectively. Five (13.9%) of the Chinese species and 11 (44%) of the US species were threatened on the global level while 22 (61%) of the Chinese species and the same number of US species (88%) were threatened on the national level. The type of species selected (habit) for translocation varied slightly among countries: for China, 42% (15) were epiphytes, 2.8% (1) were semi-terrestrial and 55.6% (20) were terrestrial while in the United States 40% (10) were epiphytes, 4% (1) were semi-terrestrial and 56% (14) were terrestrial. Furthermore the majority of plants used for reintroduction in China were adults (83%) while in the United States saplings were the most used growth stage (57%) in reintroduction efforts. The majority of the translocations had no or very short (1 year or less) monitoring periods. One year post transplanting survival was high in general, but survivorship declined with time since transplanting. Percentage survival was significantly higher for China (72.2%) compared to translocation efforts in the United States (35.2%) even though the mean number of propagules used in each reintroduction case did not differ significantly between China (mean = 267.5) and the United States (mean = 238.3). The majority of the translocation cases in China were carried out by state institutes. In contrast, the majority of the orchid translocation actions were carried out by private entities, such as nonprofits or universities, in the US. Overall, our results indicate that species are chosen for reintroduction more on the basis of threats at the national or local level than on a global level, and some were not chosen based on their biological rarity, threats or habit in both countries. We propose that social value (cultural and/or educational) be incorporated as a priority ranking along with more biological endangerment rankings of species.

T1-06-05

Domestication of endangered trees in economic forest plantation in Southeast Asia for restoring regional biodiversity

Taku Osoguchi

1. Environment Africa

2. Kasetsart University

The conversion of natural virgin forests into agriculture and forestland has continued to a severe extent in Southeast Asia due to the economic development and population increase. Tropical evergreen forest, one of hot spots for global diversity and climate change, is highly affected area where the natural forests are converted into para rubber, pulp, palm plantation and agricultural crops. *Aquilaria* species is one of critically endangered trees in Asia because the damage of *Aquilaria* trees by second-metabolism process might produce economically highly valuable resin-deposited tree called agrawood. People sometimes cut down the natural *Aquialria* trees to pursue the economic benefits without some adequate knowledge and practices in natural forests. Under these circumstances, the domestication of *Aquilaria* species in para rubber trees is investigated for restoring the forest lands. Since the late 1990s, smallholding farmers of Trat Province of eastern Thailand close to the Cambodian border covered by tropical evergreen forest has faced the problems that mono-cultural para rubber plantation make their livings and environments unsustainable due to loss of biodiversity, low price of para rubber sheet and soil degradation. We assess the life cycle of mono-cultural para rubber system and mixed forest system of *Aquilaria crassna* and para rubber based on field experiment data and modelling in terms of ecological, economic, social characteristics in Trat Province of eastern Thailand. It is realized that the aboveground species and nutrient richness and economic production of mixed forests will be significantly higher than that of the mono-culture in a life cycle. However the belowground component is not significantly different. The mixed forests system will be one of restoring regional biodiversity in mono-cultural para rubber plantation.

T1-06-06

The role of botanic gardens and partnership in plant conservation in the changing landscape of the southeastern United States of America

Jennifer Cruse-Sanders[1], *Carrie Radcliffe*[2], *Kylie Bucalo*[2], *Raya Pruner*[3], *Mincy Moffett*[4], *James Hamrick*[5]

1. State Botanical Garden of Georgia, University of Georgia

2. Atlanta Botanical Garden

3. Florida Park Service

4. Georgia Department of Natural Resources

5. University of Georgia

Increasingly, botanic gardens and arboreta are highlighted as effective partners for conserving plant species diversity and restoring natural communities at a time when the need for these activities is urgent. Capacity for restoration and conservation at botanic gardens comes directly from staff expertise in horticulture and research, but also from the role that gardens can take as good partners. Gardens are in a position to communicate information about rare plant species to owners and managers of public and private lands, and they can be instrumental in creating networks for effective conservation action. Several examples from the southeastern United States of America illustrate how this has been put into practice: 1) Mountain bog habitats have been reduced to less than 3% of their original distribution and they harbour a suite of rare species, including the mountain purple pitcher plants, *Sarracenia purpurea* var. *montana*. After more than two decades of restoration efforts, survival rates of augmented plants are more than 76% and population genetic diversity is not significantly different between restored and original populations. 2) Dwarf sumac, *Rhus michauxii*, is an endangered shrub that has become extinct in some states and critically rare in others. Through multiple partnerships, our on-the-ground management efforts have resulted in exponential increases in plant numbers and documented high levels of genetic diversity in the last population of this species at the southern most edge of its range. 3) Florida torreya, *Torreya taxifolia*, is considered one of the most endangered conifers in the world. Recent renewed efforts to survey, study and restore this species have resulted in significant new insights for understanding existing threats and available tools for preventing extinction of this species. Not only are botanic gardens achieving successes in plant conservation, through networked partnerships, they are setting priorities for conserving at-risk plants species. In November 2016, the first Southeastern Partners in Plant Conservation meeting convened at the Atlanta Botanical Garden. More than 160 people from 24 states and territories representing, federal, state and local government agencies, botanic gardens, universities and other partners met for three days to set priorities for 279 at-risk plant taxa across eight subregions. Priority actions were identified for nine areas of conservation need including, land management, *ex situ* conservation, and genetic and taxonomic research. What was unexpected was the identification of taxa that did not need conservation action or regulation due to new information or confirmation that they are stable on protected lands. Ultimately it will take continued coordinated efforts between agencies and botanical gardens to address priority needs for plants species and prevent loss of plant biodiversity.

T1-07: Urban parks: effects of management and planting schemes on insect diversity

T1-07-01

Using DNA metabarcoding to understand diet and pollination role of Cave Nectar Bat

Voon-Ching Lim[1,2], *Subha Bhassu*[1], *Rosli Ramli*[1], *John-James Wilson*[2]

1. Institute of Biological Sciences, Faculty of Science, University of Malaya

2. International College Beijing, China Agricultural University

Cave Nectar Bat, *Eonycteris spelaea*, is the largest nectarivorous bat in Southeast Asia. Due to its specialised diet of nectar and pollen, *E. spelaea* is the important pollinator of many plants including the monotypic *Oroxylum indicum*, the economically important *Durio zibethinus*, *Parkia speciose* and *P. timoriana*, and mangrove plants: *Soneratia alba* and *S. caseolaris*. The diet of *E. spelaea* was previously determined by identifying the pollens found (i) on the body and (ii) in the faeces of the bats based on the morphological characteristics of the pollen when examined under the microscope. However, this conventional method failed to identify many of the collected pollens to species level mainly due to the limitation of morphological identification which requires high level of plant taxonomic expertise. Therefore, we attempted to use DNA metabarcoding which utilises High-Throughput Sequencing at *rbcL* and *ITS2* regions to mass sequence the pollen and digested plant material in the faeces of *E. spelaea* roosting at Batu Caves, Kuala Lumpur city. We discovered plants including ferns that have not reported in diet studies of *E. spelaea*. Therefore, our study demonstrated the potential of DNA metabarcoding in dietary study of bats with specialised diet and provided insights into pollination role of *E. spelaea* in urban area.

T1-07-02

Butterfly diversity at a tourist hotspot; Sukhothai Historical Park, Sukhothai, Thailand

Narong Jaturas [1,2], *Kong-Wah Sing* [3], *Raxsina Polseela* [2], *Jedsada Sukantamala*[1], *John-James Wilson* [4]

1. Medical Science Academic Service Centre, Faculty of Medical Science, Naresuan University

2. Department of Microbiology and Parasitology, Faculty of Medical Science, Naresuan University

3. State Key Laboratory for Biology of Plant Diseases and Insect Pests, Institute of Plant Protection, Chinese Academy of Agricultural Sciences, Beijing, China

4. International College Beijing, China Agricultural University, Beijing, China

Sukhothai Historical Park in Thailand is a popular attraction and UNESCO World Heritage Site. The historical park is a 70 km^2 with a diversity of plants and offers the opportunity for the visitors to interact with wildlife besides the cultural sites. Butterflies play an important role in the natural food chain and act as nature's gardeners through their pollination of flowers. In addition butterflies are biodiversity indicators and ambassadors for insect conservation. However, to our knowledge no studies have been conducted to record butterfly communities in historical parks in Thailand. The aim of the present study was to survey the butterfly communities in historic town of Sukhothai. Such information is essential as the park serves as an important study area for many schools and park managers. We sampled butterflies using sweep nets at Sukhothai Historical Park on 23rd to 25th June 2016. A total 143 individuals from 25 species were recorded, comprised of the families Nymphalidae (35.7%), Pieridae (35.7%), Papilionidae (14.0%), Lycaenidae (8.4%) and Hesperiidae (6.3%). The results of this study could be useful for park managers planning to enhance butterfly diversity through their planting schemes and is expected to support the information on the conservation management and the ecological role of the butterfly's species in Sukhothai Historical Park.

T1-07-03

Butterfly assemblage in Kuala Lumpur's urban parks

Kong-Wah Sing[1], *John-James Wilson*[2]

1. Kunming Institute of Zoology, Chinese Academy of Sciences

2. International College Beijing, China Agricultural University

Kuala Lumpur is the capital city and urban centre of Malaysia. The city is experiencing rapid urban sprawl across the Klang Valley conurbation and continues to experience strong population growth. Given the location of Kuala Lumpur and the Klang Valley in a highly threatened biodiversity hotspot, understanding the biodiversity carrying potential of urban habitats and the associated influencing parameters is critical, but so far has received little attention. We examined the species diversity of butterflies in urban parks in the Federal Territory of Kuala Lumpur and investigated the relationships between butterfly species richness and three park variables: i) park size, ii) distance from the central business district and iii) park age. Standardised butterfly sampling was conducted across different microhabitat types at each park: i) groves, ii) hedges, iii) flowerbeds and iv) unmanaged areas. We sampled 572 butterflies belonging to 60 species from five families. In total, 58 (97%) of the species sampled were considered common species in peninsular Malaysia. The two rare species belonged to the genus *Taractrocera* (Hesperiidae); 4 species from *Taractrocera* are known from peninsular Malaysia and all are rare. Although species richness was positively correlated with park size and age and negatively correlated with distance from the central business district; the correlations were weak and not statistically significant. However, species richness of host-specialist species was significantly positively correlated with park size and age. The highest species richness was recorded in the unmanaged microhabitat. It is likely that both park planting scheme and the presence of early successional plants in unmanaged microhabitat led to highest butterfly species richness in parks that contained all four microhabitat types. Whether a diverse planting scheme and increased size and number of unmanaged areas in urban parks can improve the ability of parks to sustain populations of rare butterflies in the face of future urbanisation remains to be seen.

T1-07-04

Urbanization alters bee assemblages through species turnover rather than species loss

P.N.T Tsang, *T.C. Bonebrake*
The University of Hong Kong, Hong Kong, China

It is generally observed that species assemblages between urban and non-urban habitats are very different. However, it is unclear that if this pattern is mainly driven by species loss or species turnover. The evolutionary core hypothesis suggests that the more historical-persistent the stressor is, the more native species can respond to it. Compared with natural stressors such as altitudinal gradient, urbanization gradient is a relatively new stressor. Thus, dissimilarity between urban and non-urban habitats should be mainly driven by species loss rather than species turnover. In this study, we tested if the loss component is significantly higher than the turnover component in 13 non-urban and urban bee assemblages across Europe and North America. Surprisingly, we found that the turnover component is significantly higher than the loss component ($P = 0.01$), suggesting non-urban species are replaced by other species in urban areas. Removing exotic species from the analysis does not alter the conclusion. We also found that there is no significant difference between bee density and rarefied species richness between urban and non-urban habitats. Our results do not support the hypothesis that urbanization leads to beta diversity pattern through species loss. We speculate that while urban area is a relatively new habitat to species, its physical and biotic stress are still within the tolerance range of most native species.

T1-07-05

Structure and diversity of plants associated to *Medicago sativa* agro-ecosystem and its function as a refuge for butterfly and ant communities

Sombra Patricia Rivas-Arancibia, *Cielo Tais Silva-Rodríguez, Paola Fascinetto-Zago, Natalie Olmos-Santiago, Hortensia Carrillo-Ruiz, Agustina Rosa Andrés-Hernández*
Benemérita Universidad Autónoma de Puebla

Several agro-ecosystems can serve as alternative sites of refuges and conservation of species diversity. Many studies have reported that the structure and diversity of plants associated with farmland can support a high species richness of butterfly and ant species, which are important pollinators and soil fertilizers, respectively. In this study, we characterized the structure and diversity of vegetation found on the edges of *Medicago sativa* farms in central Mexico. We also monitored and identified the species of diurnals butterflies (Rhopalocera) and ants associated with this vegetation. We worked under the hypothesis that, even in farms under intensive cultivation, the diversity of plants on the field edges is enough to support high species richness of butterflies and a stable community of ants. Butterflies, ants and plants were counted along three transects at the edges of a cultivated field during 12 months. Twenty-two pitfall ant-traps and Van Someren Rydon butterfly traps were distributed across the length of each transect, which was 250 m. The plants surrounding each trap (area of 1 m x 1 m) were counted and identified. We found 2,700 individuals of plants, belonging to 48 different species from 24 families; 1,490 individuals of diurnal butterflies, belonging to 57 species from six families; and a total of 6,454 individual ants belonging to different 15 species. The most abundant plant species were Poaceae, Fabaceae, Asteraceae, Solanaceae and Malvaceae. Most than half of the plant species found (58%) were native flora of central Mexico, 91.66% of them with ethnobotanical use (as medicine, forage or food). Even though the Simpson's diversity index showed no significant differences between transects, the Jaccard index showed that the compositional similarity between species is very low. ANOSIM analysis showed no significant differences, confirming that three transects have slow compositional simililarity. In agreement with this, the beta diversity values (Wilson and Schmida) were high. The Canonical Correspondence Analysis showed that plant families explain 60% (first ordination axis) of the abundance of Rhopalocera species and 77% of the abundance of Formicidae species. The abundance of both butterfly and ant species can be divided into three groups associated with groups of plant families. Eight butterfly species were migratory, most of them associated with a group formed by Amaranthaceae, Euphorbiaceae, Annonaceae Lamiaceae Apiaceae and Fabaceae families. Most ant species were also found associated with these plant families. We have not finished conducting the necessary multiple regression analyses yet, but we hope to identify some key plant species that have made this particular agro-ecosystem a very important reservoir of diversity of butterflies and ants. This could provide useful information to implement eco-sustainable strategies for the management of similar agriculture lands.

T1-07-06

The role of urban plants in supporting diverse urban butterfly communities

Kwek Yan Chong
Department of Biological Sciences, National University of Singapore

Butterflies are among the most charismatic insect groups. Butterflies are closely dependent on plants. In the larval, or caterpillar, stage, most butterfly species are specialized for herbivory on a restricted group of related host plant species (although the degree of specialization varies). In the adult stage, most butterfly species feed on nectar and are less specialized, although proboscis length may restrict feeding to flowers of suitable corolla depth and nectary positions. These relatively predictable food-related dependencies suggest that plant selection in horticultural landscaping can be carefully tailored to support butterfly diversity and populations of targeted butterfly species in heavily human-modified areas, such as urban areas. In the urban parks of Singapore in Southeast Asia, we found evidence that flowering intensity and the diversity of native cultivated trees were positively associated with butterfly species richness. This suggests that phenology and the use of native species are important considerations when designing urban park plantings for butterflies. We also found that insectivorous bird species richness is associated with butterfly abundance, most likely due to the availability of caterpillar prey. Managing plant composition for butterflies could therefore also have knock-on effects for higher trophic levels. This could be generalized to other groups of insect herbivores, which tend to be specialized consumers of plant species. However, the landscape context of cultivated host plants also matters. Placement in unfavourable environments may result in the creation of population sinks and ecological traps that would reduce, rather than enhance, populations. I will report some preliminary results regarding this from an on-going study.

T1-08: Hot forests and cold storage

T1-08-01

The tropics: Hot, wet and recalcitrant

***Sershen**, Subbiah Ashley, Ramdhani Syd*
University of KwaZulu-Natal

The Tropics, parts of North and South America, Africa, Asia and Australia that lie between the latitudes of the Tropics of Cancer and Capricorn, account for only c. 36% of Earth's landmass yet host c. 80% of the world's terrestrial biodiversity. The tropical climate is best described as warm to hot and moist year-round (with only a limited dry season) and often supports lush vegetation. However, temperature and rain can vary greatly from one area to another within this zone, and so too can biodiversity. The meta-analysis conducted in the present study revealed that whilst most authors are non-committal regarding the exact biological richness of the Tropics, data available for plants suggest that they house the highest richness for numerous lineages; host most of the global centres of species richness; and contain regions that differ greatly in terms of species diversity. Importantly, they also host a number of species that produce recalcitrant seeds that exhibit desiccation sensitivity, a biological phenomenon that is antipodal to the desiccation tolerance observed in resurrection plants and the majority of plant seeds (labelled orthodox). Desiccation sensitivity has been extensively investigated in recalcitrant seeds largely because its consequences preclude their storage using conventional seed storage methods. Our analysis of primary and secondary data suggests that there are c. 600 recalcitrant-seeded taxa reported to date but this figure is ever-increasing. Of these reported taxa, 433 species (431 terrestrial and 2 aquatic) have been shown to be found in the Tropics. The 431 terrestrial tropical recalcitrant-seeded (TTR) species belong to 62 families (199 genera), 13 of which have $\geq$ 10 species with this seed storage physiology: Dipterocarpaceae (86), Meliaceae (30), Arecaceae (26), Lauraceae and Myrtaceae (21), Fabaceae (20), Moraceae and Saptoaceae (19), Fagaceae, Rhizophoraceae and Sapindaceae (15), Clusiaceae (11) and Anacardiaceae (11). Furthermore, six genera were shown to possess $\geq$ 10 TTR species: *Shorea* (40, Dipterocarpaceae), *Artocarpus* (17, Moraceae), *Dipterocarpus* (13, Dipterocarpaceae), *Hopea* (12, Dipterocarpaceae), *Quercus* (10, Fagaceae) and *Trichilia* (10, Meliaceae). Using these data as a backdrop, the current contribution identifies TTR species (as well as genera and families) that represent conservation priorities based on their International Union for Conservation of Nature (IUCN) status (e.g. certain Dipterocarpaceae species) and discusses some of practical challenges and successes associated with the short-, medium- and long- term seed germplasm conservation of TTR.

T1-08-02

Personalization of cryoprotection: A systematic approach to optimize chemical and osmotic effects

Haeng-Hoon Kim
Sunchon National University

Recent advances in developing cryopreservation methodology has underpinned the wider application of cryogenic storage for the long-term conservation of important agricultural crops and further has been applied for endangered wild species. Although an appropriate cryopreservation protocol is a prerequisite for basic studies, cryobanking and further cryopreservation studies, the process relies on trial and error. Among the vitrification-based cryopreservation techniques, droplet-vitrification produces high post-cryopreservation recovery. However, the protocol cannot universally solve all challenges in plant cryopreservation. Specifically, not all materials have the same (in) tolerance to desiccation and cytotoxicity to the most widely used cryoprotectant solutions. This presentation suggests alternative approaches to the development of cryopreservation protocols based on the initial characteristics of plant materials, by introducing specific steps in a droplet-vitrification method. Firstly, the osmotic tolerance of samples can be classified based on their response to overnight preculture at various final concentrations of sucrose. Secondly, some alternative vitrification solutions to the popular plant vitrification solutions 2 and 3 (PVS2, PVS3) can be considered based upon the cytotoxicity and size of the samples. Using this systematic approach we can identify whether the material is tolerant or sensitive to the osmotic stress and chemical toxicity of cryoprotection with vitrification solutions, thus revealing which is the main barrier to the successful application of vitrification methods. Based on the sensitivity of samples we can design a droplet-vitrification procedure, i.e. preculture, osmoprotection, cryoprotection, cooling and rewarming. This enables the development of appropriate, 'personalised' droplet-vitrification protocols that can be used to facilitate the preservation of plants, particularly endangered species for which there may be limited material available.

T1-08-03

Frozen architecture: Preserving cellular and tissue structure of explants for an optimal cryopreservation

Daniel Ballesteros Bargues
Royal Botanic Gardens, Kew

Recalcitrant seeds cannot be dried and cooled to the standard conditions of seed banks and cryopreservation of small explants is currently the most feasible option for their long term storage and conservation. But cryopreservation of recalcitrant materials (particularly the embryo axes) presents diverse challenges during, for example, the excision of the selected explants, partial drying, cryoprotection, and cooling. Aside from oxidative stress, ice crystal formation and growth, cell and tissue shrinkage during partial desiccation and/or cryoprotection, and uneven drying among diverse tissues of the explants are some of the major problems. They often involve the collapse of the cell and tissue architecture of the explants throughout the drying and cooling steps, which ultimately affects function (i.e., development and growth) during the *in vitro* recovery phase. Diverse tools have been employed in the last 25 years to investigate these processes, contributing to the understanding of the mechanisms that lead to the failure or success of the cryopreservation process from a structural perspective. With these specialized tools that combine comparative plant biology, biophysics, microscopy, and mechanics, researchers have been able to measure and observe the changes that happen while the explants are dried and cooled. This paper will review this topic, focusing on the most current research, and identifying future areas for investigation. By knowing the mechanisms that confer protection to partial desiccation, cryoprotection and exposure to cryogenic temperatures of diverse explants during the cryopreservation process, we should be able to better predict optimal cryopreservation protocols for the rich tropical (and temperate) flora that produce recalcitrant seeds.

T1-08-04

Cryobiotechnology for the long-term conservation of life

Maria Teresa Gonzalez Arnao
Univeridad Veracruzana

Life of many world's plant species is seriously threatened by different biotic, abiotic and anthropogenic reasons, which are provoking important losses at increasing rates. According to FAO, the highest rates of woody species loss are in the tropics, and forest species are the most affected by the anthropogenic impact. At present, the main challenge of plant cryobiotechnology is dealing with the preservation of highly endangered species, particularly as there is limited material to work with, little knowledge of their behavior, and the likelihood of material that is recalcitrant for storage and *in vitro* growth. The advances in plant cryobiotechnology associated with the use of *in vitro* culture techniques and cryogenic storage for different type of germplasm have provided new tools to face these difficulties and created new opportunities for multiplication, recovery, micrografting, and long-term *ex situ* conservation of such species at ultra-low temperature (-196°C). Current development of plant cryopreservation can be associated with the strategic adjustments in key cryoprotective steps. For vitrification-based protocols, these include: the exposure of material to highly concentrated formulations of osmoprotectants (loading and plant vitrification solutions); the encapsulation of specimens in calcium alginate; and the use of aluminum foil strips or cryo-plates instead of cryovials to achieve ultra-rapid cooling in liquid nitrogen. These procedures have significantly enhanced the effectiveness of cryopreservation of plant germplasm. Genetic transformation technologies can also be an innovative approach for improving cryostorage and longevity of less stress tolerant species. They may be used to induce intracellular accumulation of organic osmolytes increasing the tolerance and protection of cells against stress, and consequently improving cryopreservation outcomes.

T1-08-05

Cryopreservation of some Thai orchid species

***Kanchit Thammasiri*[1], *Nipawan Jitsopakul*[2], *Kathawut Sopalun*[3], *Upathum Meesawat*[4], *Supornchai Chaireok*[4]**
1. Department of Plant Science, Faculty of Science, Mahidol University, Rama VI Road, Phayathai, Bangkok 10400, Thailand
2. 2Department of Agro-Industry, Faculty of Agriculture and Technology, Rajamangkala University of Technology Isan, Surin Campus, Surin 32000, Thailand
3. Department of Microbiology, Faculty of Liberal Art Science, Kasetsart University, Kamphaeng Saen, Nakornpratom 73140, Thailand
4. Departmant of Biology, Faculty of Science, Prince of Songkla University, Hat-Yai 90110, Thailand

Thailand is the origin of about 1,300 tropical orchid species and 180-190 genera. Deforestation and over-collection of wild Thai orchids for trade has placed orchid species at a risk of extinction, such as the genus *Paphiopedilum* and *Dendrobium cruentum* which are listed in Appendix I of CITES. Therefore, the conservation, as well as sustainable use is urgently needed to conserve orchids by various means. Various methods of cryopreservation of Thai orchid species are implemented. Recent methods were used, namely vitrification [dehydration in PVS2 solution, consisted of 30% (w/v) glycerol, 15% (w/v) ethylene glycol, and 15% (w/v) dimethyl sulfoxide, prepared in modified Vacin and Went liquid medium], encapsulation-dehydration (encapsulation in calcium alginate beads followed by air-drying in a laminar air-flow cabinet), encapsulation-vitrification (encapsulation in calcium alginate beads followed by dehydration in PVS2 solution), droplet-vitrification (fast freezing from small drops of PVS2 solution on aluminium strip), and cryo-plate (a combination of encapsulation and droplet on very fast freezing aiuminium plate) dehydrated with silica gel and drying beads. Application of these methods in seeds was successful in *Dendrobium chrysotoxum* (99%, vitrification). *Dendrobium cruentum* (32%, virification), *Dendrobium draconis* (95%, vitrification), *Dendrobium hercoglossum* (80%, encapsulation-vitrification). *Doritis pulcherrima* (62%, vitrification). *Rhynchostylis coelestis* (85%, vitrification), *Vanda coerulea* (67%, vitrification), as well as in protocorms of *Dendrobium cruentum* (33%, vitrification; 27%, encapsulation-dehydration), *Dendrobium cariniferum* (15%, encapsulation-vitrification), *Grammaytophyllum speciosum* (14%, encapsulation-vitrification), *Rhynchostylis gigantea* (19%, vitrification), *Vanda coerulea* (40%, encapsulation-dehydration) and *Seidenfadenia mitrata* (67%, vitrification) and *Arundina graminifolia* (76% and 74%, cryo-plate dehydrated with drying beads and silica gel, respectively; 33% droplet-vitrification; 64% encapsulation-dehydration with drying beads or silica gel), and protocorm-like bodies of *Paphiopedilum niveum* (22%, vitrification). Cryopreserved seeds, protocorms, and protocorm-like bodies were able to develop into normal plantlets. These methods appear to be promising techniques for cryopreservation of some Thai orchid species.

T1-08-06

Current progress in understanding cryopreservation effects on oxidative stress and metabolism of Western Australian species: A combined biophysical and metabolomics approach

***Bryn Funnekotter*[1,2], *Eric Bunn*[2], *Ricardo L. Mancera*[1]**
1. Curtin University
2. Botanic Gardens and Parks Authority

The Australian continent hosts a unique assemblage of plants with many endemic species. Post-European settlement including agriculture, grazing and development of urban centres has resulted in drastically diminished native habitats with their attendant floras in both coastal and inland arable regions, hence many indigenous species are now in danger of extinction. Cryopreservation of somatic tissues of critically endangered species where there are few individuals (hence little or no seed), or seed axes of recalcitrant species provides the safest long-term *ex situ* conservation approach, however post-cryopreservation survival and regeneration varies substantially between plant species. In order to better understand this variability in cryo-capability, fundamental aspects affecting cryopreservation were assessed in a variety of Australian species, such as membrane stability and composition, oxidative stress and respiration rates. Membrane damage is a leading cause for a lack of cryogenic success, with the application of preconditioning treatments linked to improved membrane stability. The composition of membrane phospholipids, sterols and sugars was assessed before and after preconditioning to identify changes that may be beneficial for surviving the cryopreservation process. Oxidative stress during the cryopreservation process is another important factor that can alter cryopreservation success. Antioxidants, as the major defence against oxidative stress, were assessed throughout the cryopreservation process to determine if excessive oxidative stress was a factor in reducing post-cryogenic success. Mitochondria are the powerhouse of the cell, hence any damage done during cryopreservation will have a detrimental effect on ATP production, thus reducing synthesis of new compounds and repair of damaged components of the cells. Respiration rates in recovering shoot tips were analysed to determine if mitochondrial function correlates to cryogenic success.

T1-09: Exploration history on eastern Asian floras

T1-09-01

A hundred years of exploration on flora of China (1916-2016)

Jin-Shuang Ma
Shanghai Chenshan Plant Science Research Center

A hundred years has passed since the first modern taxonomic paper was published by a Chinese botanist in 1916. The general history of Chinese plant taxonomy can be roughly divided into the following periods: hard beginnings from 1910-1920s, wartime from 1930-1940s, restarting again from 1950-1960s, looking ahead from 1970-1980s, resplendent time from 1990-2000s, and finally flowering gloriously from 2000-2010s. The representative works of Plant Taxonomy in China are Flora Reipublicae Popularis Sinice (FRPS, Flora of China, Chinese edition, 1959-2004), up to 80 volumes, and Flora of China (FOC, Flora of China, English edition, 1993-2013), up to 25 volumes and 24 illustration respectively, plus more than 30 different local works both in provincial and higher levels, particularly from southwest China, such as Flora of Yunnan (20 volumes), Flora of Guizhou (10 volumes), and Flora of Xizang (Tibet, 5 volumes). There are about 300 families, more than 3,435 genera, and over 31,000 species of vascular plants from China, with more than fifty percent of them being endemic, which includes many relic elements, such as *Cathaya*, *Taiwania*, *Davidia*, as well as living fossils, such as *Ginkgo* and *Metasequoia*. Particularly, there are hundreds and thousands of ornamental plants either native or originating from China, such as *Rhododendron*, *Meconopsis*, *Primula*, *Gentiana*, *Camelia*, as well as Peony, Roses, Tea, which have been cultivated worldwide, which is why China has been called "Mother of Gardens" by Ernest H. Wilson (1929), a famous explorer from England who visited China during 1899-1911. This work is the first time that the general history of Plant Taxonomy in China has been summarized with various widely consulted publications both in China and abroad, including many archives which have never been reported before; and many facts have been confirmed for the first time. All in all, the first herbaria, gardens and arboreta, research institutions, journals and magazines, university along with their departments of Biology, Agricultural and Forestry, as well as the first graduates both from abroad and from China; all were established, founded, finished and completed during 1910-1930. However, since the wars as well as other issues, most of the taxonomic works, both research papers and monographic floras, have mainly been published in the last third of the century from 1980-2010s, including many young taxonomists as well as Ph.D. students have been trained, only very recently.

T1-09-02

Present data on Vietnam orchid inventory

***Leonid Averyanov**, Tatiana Maisak*
Komarov Botanical Institute of the Russian Avademy of Sciences

The orchid flora of Vietnam is the richest among all other regional floras of mainland Asia. Successive inventories show an increasing orchid number from 411 species in the year 1934, 718 spp. in 1992-1994, 897 spp. in 2003-2005, 1005 spp. in 2009 and 1090 species & 160 genera figured in 2011. Since last inventory, 116 species and 13 genera have been additionally discovered and documented in Vietnam. The present report summarizes published and new original data for the last 5 years. When the new data presented here were included, the orchid flora of Vietnam should comprise at least 1213 documented species from 174 genera. One genus discovered for this period in Vietnam has broad distribution in mainland subtropical and temperate Asia (*Cephalanthera*), 8 genera have broad Malesian distribution (*Abdominea*, *Hymenorchis*, *Cordiglottis*, *Grammatophyllum*, *Macropodanthus*, *Microtatorchis*, *Octarrhena*, *Tuberolabium*) and 4 genera are endemics of Indochina (*Bidoupia*, *Drymoda*, *Lockia*, *Vietorchis*). At least 116 species were discovered in Vietnam during last five years. Among them, 63 were discovered and documented as a new species for the flora of the country. These species form 4 groups: species with Mainland Asia – Himalayan distribution (19 species), species of South-Chinese floristic province (9 species), Malesian species (15 species), species of Indochinese Peninsula (20 species). Other 53 discovered species are local endemics of following floristic provinces: South-Chinese floristic province (6 species), North-Indochinese floristic province (15), Central-Annamese floristic province (4 species) and South-Annamese floristic province (28 species). All newly described endemics well illustrate unique character of local aboriginal floras of eastern Indochina and reveal some peculiar specific evolutionary threats in the orchid family. Brief illustrated data on floral morphology, distribution and ecology are provided for all newly discovered species. Expected IUCN Red List national status is identified for these species as well. Largest part of discovered species (67 species, or 58%) belongs to threatened categories: VU (20 species, or 17%), EN (25 species, or 22%) and CR (22 species, or 20%). Other discovered species (49 species, or 42%) belong to non-threatened categories: DD (32 species, or 28%) and LC or lover risk (17 species, or 15%). Totally for the last five years in studied area were discovered and described three new genera – *Bidoupia* Aver., *Lockia* Aver. and *Vietorchis* Aver., as well as at least 54 species new for science –

*Aerides phongii**, *Bidoupia phongii**, *B.khangii*, *Bulbophyllum bidoupense*, *B.blaoense*, *B.glabrichelia*, *B.phitamii*, *B.setilabium*, *B.striatulum*, *B.tipula*, *Cheirostylis glandulosa*, *Cleisostoma dorsisacculatum**, *C.lecongkietii**, *C.phitamii**, *C.tricornutum*, *Cordiglottis longipedicellata*, *Cymbidium repens**, *Dendrobium congianum*, *D.thinhii**, *Eria grandicaulis*, *E.praecox*, *Flickingeria xanthocheila*, *Gastrochilus kadooriei*, *Hymenorchis phitamii**, *Liparis caudata*, *L.honbaensis*, *L.longibracteata*, *L.longispica*, *L.ngoclinhensis*, *Lockia sonii**, *Luisia parviflora*, *Malaxis subtilis*, *Miguelia cruenta**, *Nervilia mekongensis*, *Oberonia dolichostachys*, *O.janae*, *O.ngoclinhensis*, *O.rhizoides*, *O.sonlaensis*, *O.tatianae*, *Octarrhena minuscula*, *Odontochilus aureus*, *Paphiopedilum tamphianum**, *Podochilus rotundipetala*, *Sarcoglyphis tichii**, *Schoenorchis hangianae**, *S.scolopendria*, *Taeniophyllum phitamii*, *Tainia acuminata*, *T.cornuta*, *Trichoglottis canhii**, *Vanda gracilis**, *Vanilla atropogon** and *Vietorchis furcata*. Almost all these taxa are strict endemics of Vietnam. Orchids, names of which are marked above by asterisk (*) stand presently on the verge of full extinction.

T1-09-03

The early exploration of the Nepal Himalaya: 1793-1850

Mark Watson

Royal Botanic Garden Edinburgh

The early scientific botanical studies of the flora of Nepal were undertaken by British natural historians, mostly in connection with the developing relationship between the British East India Company and Nepal. The first botanist to visit Nepal was the Scottish medic Adam Freer, who accompanied William Kirkpatrick on a short mission to Kathmandu in 1793. However, no botanical collections are known from this time and so the start of the scientific study of the Nepalese flora is credited to Francis Buchanan-Hamilton, who made extensive collections of herbarium specimens and living plants during his year in Kathmandu as surgeon to the diplomatic mission led by William Knox in 1802-3. Buchanan-Hamilton recorded over 1200 plant species, but most were never published at the time and so his achievements as the Father of Nepalese botany have not been fully appreciated. Botanical research resumed around 1818 when Nathaniel Wallich sent his local collectors to work with Edward Gardner, the first permanent Resident to the Nepalese Royal Court. Wallich sent specimens back to London, where David Don published the first floristic account of Nepal in 1825 - Prodromus Florae Nepalensis. Wallich later spent a year in Kathmandu and collected over 2000 species. These, together with many thousands of S Asian plants, were distributed by him around 1830 and documented in the 'Wallich Catalogue' (now online http://wallich.rbge.info). Joseph Hooker made extensive botanical collections in the Sikkim Himalaya in 1848, and from his base in Darjeeling he journeyed into eastern Nepal and made the first collections from this, the botanically richest part of the country. It would be another 100 years before foreign botanists ventured into Nepal, when this secretive Himalayan kingdom opened its boarders to foreigners. The achievements of these early botanical explorers will be discussed, along with a consideration their role in laying the foundation of botanical research in Nepal, the Himalayas and wider eastern Asia.

T1-09-04

Russian explorers of East Asian flora

Dmitry Geltman

Komarov Botanical Institute of the Russian Academy of Sciences

Russian botanists made a great contribution to the study of East Asian flora—both the territory of Russia itself as well as neighboring countries. Karl Maximowicz in 1853–1857 for the first time explored the south of the Russian Far East and also travelled along the rivers Amur (Hēilóng-jiāng) and Ussuri (Wūsūlǐ Jiāng). As a result the famous "Primitae Florae Amurensis" (1859) have been published. In 1859–1862 he continued his exploration of the Russian Far East and also visited Japan. Maximowicz became one of pioneers of the study of Japan plants, especially flora of Hokkaido where he successfully worked together with his assistant Sukawa Chonosuki. In 1854 Russian naval officer Alexander Schlippenbach made first small plant collection from the Korean Peninsula. Famous explorer Nikolay Przewalski in 1867 visited the Ussuri area and Manchuria. Later, in 1870–1886 he led several expeditions to China and Mongolia, mainly their desert parts, but also visited several areas with East Asian flora. His activity was continued by expeditions led by Grigory Potanin, Mikhail Pevtsov, Vsevolod Roborovski, Pyotr Kozlov, Veneamin Ladygin and others. Collections of mentioned above botanists and explorers have formed a special section in the Herbarium of the Imperial Botanical Garden in St. Petersburg (now—Komarov Botanical Institute). Since that time Russian botanists have the continued interest to the study of the flora of China, Mongolia, Japan and Korea. Vladimir Komarov during 1895–1897 made a great field exploration in Manchuria and also visited North Korea. The results of this expedition were published in 3-volume "Flora of Manchuria" (1909). Komarov not only presented a complete species description of the area but also made very important phytogeographical suggestions. This study can be assigned as a significant step in developing of his original approach to species definition (known as Komarov species concept) which later became basic for the "Flora of the USSR". Ivan Palibin prepared "Conspectus flora Coreanae" (1899-1901) — the first summary of the flora of Korea. During 1955–1957 a series of Soviet-Chinese biological expeditions visited badly explored areas of Yunnan and Sichuan. Some collections were made also in North Korea in 1957. The part of Russia with East Asian flora (current Primorsky and partly Khabarovsk Territories, South Sakhalin and part of Kuril Islands) was intensively studied during the whole XX century. Vladimir Komarov and Evgenia Klobukova-Alisova prepared and published "Manual to plants of Far East Territory" (1931-1932). Their activity was continued in 1950-1985 by Vladimir Voroshilov, Dmitry Vorobyev, Pyotr Gorovoi, Alexander Schreter and others: a number of field guides, articles on taxonomy and floristics were published. The greatest achievement in the study of this area is 8-volume series "Vascular Plants of the Soviet Far East" (1985–1996) led and edited by Sigizmund Kharkevich. The preparation of this series facilitated developing of teams of Russian botanists working on Far East flora in several centers (Vladivostok, St. Petersburg, Moscow, *etc.*).

T1-09-05

The forerunner of Chinese plant collecting—A study on the

collecting history of Mr. K.K. Tsoong

Shiyong Meng[1], *Huiyuan Liu*[2], ***Jin-Shuang Ma***[3]
1. Peking University, Beijing, China
2. Institute of Botany, Chinese Academy of Sciences, Beijing, China
3. Shanghai Chenshan Plant Science Research Center, Chinese Academy of Sciences, Shanghai, China

Through a carefully literature and specimen research, we studied the collecting history of K.K. Tsoong who was the Chinese plant collecting pioneer. After the study, we clarified that the earliest specimens collected by K.K. Tsoong were in 1903, Tianmu Mountain, Zhejiang Province, and are also the earliest plant specimens collected by Chinese scholars. Mr. K.K. Tsoong was the first Chinese scholar who large-scale collected plant specimens in China. From 1918 to 1921, he collected about 15,000 specimens from 11 provinces of China, e.g., Fujian, Guangdong, Yunnan, Zhejiang, *etc.* Specimens collected by K.K. Tsoong distributed in major herbariums from the United Kingdom, the United States and China. In particular, Museum of Biology, Peking University (PEY), Herbarium, Institute of Botany, CAS (PE), Beijing Normal University Herbarium (BNU) stored most of the specimens. His collections were the basement for the build of PEY, and play a great role in plant taxonomic review, the compiling of Flora of China, Flora of Hebei, and so on.

T1-09-06

New taxonomic bibliography of Chinese vascular plants

Cheng Du, *Wang Yuan, Shuai Liao, Jinshuang Ma*
Shanghai Chenshan Plant Science Research Center, Chinese Academy of Sciences, Shanghai Chenshan Botanical Garden, Songjiang, Shanghai, 201602, China

The Chinese and English version of the national flora completed in 2004 and 2013, respectively, cover more than 30,000 kinds of vascular plants, with most of the taxa published for decades ago. However, new taxonomic treatment, either new taxa or new nomenclatural data have been changed continuously. Therefore, there is an urgent need for the acquisition of the dynamic information of Chinese vascular plants. Due to the relatively loose requirements of ICN (International Code of Nomenclature for algae, fungi, and plants) on effective publication, many taxa published in regional journals and books, especially in a local or less important publications, are difficult to be included in the international retrieval mechanism (such as IPNI, Tropicos) timely. And incognizance of Chinese characters for foreign researchers also affects the efficiency for Chinese information retrieval. The project of Bibliography of Chinese Vascular Plants for New Taxonomic Data aims at to solve the above problems systematically. The objectives of this project is dealing with the taxonomic literature of new taxa, name changes, synonyms, records and typifications for all of the vascular plants distributed in China by steps, systematically collecting the information from the literature and digitally publishing the information. Specialized researchers will be able to retrieve the specific information after registration. These documents will be browsed, downloaded, borrowed, or accessed using the document delivery services according to the policy of the publisher company or database services. The database of Bibliography of Chinese Vascular Plants for New Taxonomic Data contains the following items: a, Publication information: the name of book or journal, publishing year, volume, number and the page number range; b, Article information: author (s), title and language; and c, Taxa information: name of taxon, category of taxonomic treatment, page of taxon, type specimen and the herbaria in where the specimens preserved, distribution in the world and China, and other pertinent information if possible. The database will be opened to professional scholars, and information retrieval of all items can be realized through a database entry. As a part of the renewal of A Bibliography of Eastern Asiatic Botany (1936) and A Bibliography of Eastern Asiatic Botany Supplement I (1958), the project of Bibliography of Chinese Vascular Plants for New Taxonomic Data only includes the work of vascular plants distributed in China, and will be completed in about five years. Two books in the series will be published before July 2017, covering the taxonomic literatures from 2000 to 2015. In the next five years, six books will also be added to fill gaps in the previous literatures, thus forming a complete database of plant taxonomic literature in China since 1958.

T1-10: Spatial phylogenetics as applied to floras around the world I and II (two sessions)

T1-10-01

Spatial patterns of phylogenetic diversity and endemism in the vascular flora of Chile

Rosa Scherson
University of Chile

Phylogenetically-based methods are increasingly important tools for conservation. The combination of ancient and more recent geological processes makes evolutionary history crucial to the understanding of distributional patterns. This study addresses questions about phylogenetic diversity and endemism across Chile, using Phylogenetic Diversity (PD), Phylogenetic Endemism (PE), Relative Phylogenetic Diversity (RPD), and Relative Phylogenetic Endemism (RPE) along with statistical tests based on spatial randomizations. Endemism is one of the main considerations when defining conservation priorities; an endemic territory is tightly correlated with the idea of irreplaceability and consequently with how vulnerable a place is. Chile is a highly endemic country, since it acts as a biogeographic island, concentrating unique evolutionary lineages with different spatial and temporal origins. When incorporating phylogenetic information, endemism can be seen as the geographic concentration of evolutionary history of lineages in any portion of a phylogenetic tree. Spatial patterns of phylogenetic diversity and endemism were studied for Chile using a genus level phylogeny of the native vascular plants of the entire country. Distribution for each genus was determined using georeferenced information from fieldwork, literature, and herbaria. While Central Chile (a recognized biodiversity hotspot) showed the highest concentration of taxon richness, PD and RPD were significantly higher than expected in the south of the country (below 35° latitude), while the north showed less PD and RPD than expected. Many areas of significant PE were found in the north of the country, with a concentration of neo-endemism. The south of the country showed areas of significant paleo-endemism. These patterns of endemism make sense from a geological point of view and considering the origins of the flora that inhabit the different areas. These analyses provide a new perspective on conservation since they highlight important areas that might not be conservation priorities. To study this, the areas of significant phylogenetic diversity and endemism were geographically compared to the country's protected areas (PAs) to see if and how much of the state system is currently protecting places that concentrate the most phylogenetic diversity and/or areas of significant paleo and neo-endemism.

T1-10-02

Spatial phylogenetics of the native California flora

Andrew Thornhill, Bruce Baldwin, ***William Freyman****, Sonia Nosratinia, Matthew Kling, Naia Morueta-Holme, Tom Madsen, David Ackerly, Brent Mishler*

University and Jepson Herbaria and Department of Integrative Biology, University of California, Berkeley

California is recognized as one of the world's top biodiversity hotspots and has a rich history of research on its vascular flora. Terms such as neo- and paleo-endemism were first used to describe significant elements of the California flora by Stebbins and Major and later adopted worldwide. Raven and Axelrod proposed various hypotheses as to how the diverse Californian flora has assembled over time, and much has been written on the biogeography and vegetation. All of these studies were carried-out using pre-cladistic taxonomy and species as the operational unit. A re-examination of the flora using newly available herbarium databases and novel phylogenetic approaches is critical given the need to prioritize conservation efforts in the face of rapid habitat loss and human-induced climate change across the state. We built a large spatial dataset comprised of specimen-based distributional data from the Consortium of California Herbaria and other collection databases, at a scale of 15 km x 15 km grid cells. Considerable effort was made to clean the spatial dataset to represent the natural range of the full California vascular flora; 1.38 million geo-referenced records remained after cleaning. We assembled a phylogenetic data matrix for 1,083 OTUs (i.e., terminal clades representing genera or monophyletic parts of genera) for 9 genes, beginning with data from GenBank and adding 1,182 new gene sequences targeted to fill in data gaps. A maximum likelihood phylogeny was constructed. These two datasets were used to find regions of significant phylogenetic diversity (PD) and phylogenetic endemism (PE) within California. We applied the recently developed metrics Relative Phylogenetic Diversity (RPD) and Relative Phylogenetic Endemism (RPE), as well as Categorical Analysis of Neo- And Paleo-Endemism (CANAPE). Observed biodiversity is not necessarily an indicator of significant biodiversity and therefore should not be used alone. We used a spatial randomization to develop null hypotheses used for statistical tests of the above indices. Using these tests we can determine whether more or less of the phylogeny occurs in an area than expected (PD), whether branch lengths in an area are longer or shorter than expected (RPD), and whether significant concentrations of long or short-branched endemism occur in an area CANAPE). We found: (1) extensive phylogenetic clustering in the southern coast ranges, southern Great Valley, and deserts of California; (2) significant concentrations of short branches in the Mojave and Great Basin Deserts and the south coastal ranges, and long branches in the northern Great Valley, Sierra foothills and the northwestern and southwestern parts of the state; (3) significant concentrations of paleoendemism in northwestern California, northern Great Valley, and western Sonoran Desert and neo-endemism in the White-Inyo Range, northern Mojave Desert, and southern Channel Islands. Multiple analyses were run to observe the effects on significance patterns of phylogenetic uncertainty, using uncalibrated trees (with raw molecular branch lengths) vs. time calibrated (ultrametric) trees, and artificially filling sampling gaps through species distribution modeling. We found that results are largely robust to phylogenetic uncertainty and time calibration, but ecologically modeled ranges tend to affect the ability to detect localized centers of endemism.

T1-10-03

A new spatial phylogenetic method for distinguishing neo-endemism, paleo-endemism, and meso-endemism: Range-weighted branch length difference (RWiBaLD)

Brent Mishler*[1], Nunzio Knerr[2], Joseph Miller[2,4], Shawn Laffan[3]*

1. University and Jepson Herbaria; University of California, Berkeley

2. Centre for Australian National Biodiversity Research, CSIRO

3. School of Biological, Earth and Environmental Sciences, University of New South Wales

4. Office of International Science and Engineering, National Science Foundation

It is important to understand spatial patterns of phylogenetic endemism (PE), both for academic studies of the ecological, biogeographic, and evolutionary causes of endemism, as well as for practical conservation applications. We propose a new method called range-weighted branch length difference (RWiBaLD) that extends a previously published method called categorical analysis of neo- and paleo-endemism (CANAPE). A range-weighted tree (RWT) starts with a phylogenetic tree with branch lengths and divides the length of each branch by its range size, thus the wider ranging branches are disproportionately shrunken down and most of the length of the tree is in range-restricted branches. The RWT is the basis for PE: PE is PD measured on a RWT. The RPE metric is a ratio comparing standard PE measured on the RW original tree to PE measured on a RW comparison tree (that retains the same topology but has all branch lengths adjusted to be of equal length before range-weighting). CANAPE is a two-step method that employs RPE and a spatial randomization to allow a clear, quantitative distinction between centers of neo- and paleo-endemism. Step one of CANAPE is to establish that a location is a center of high PE: it must be significantly high (one-tailed test) on either the RW original tree, the RW comparison tree or both. Step two is to examine the significance of the RPE ratio itself (two-tailed test). If the ratio is significantly high (greater than one), that means the range-restricted branches in that location are longer than expected (indicating a concentration of paleo-endemism). If it is significantly low (less than one), that means the range-restricted branches in that locality are shorter than expected (indicating a concentration of neo-endemism). When the ratio is not significantly high or low, there is some unknown mixture of lengths of range-restricted branches in that location that is not dominated by either neo-endemism or paleo-endemism. RWiBaLD delves into this mixture of branch lengths in a location by examining each branch of the phylogeny (both terminal and deeper branches), determining the amount and type of phylogenetic endemism (PE) it contributes, allowing a clear understanding of which branches on the tree are responsible for the broad, summary pattern given by CANAPE. The branches contributing the most PE are selected by taking those with the smallest ranges, and the type of PE is determined by the difference between the length of the branch on the range-weighted original tree and the length of the branch on the RW comparison tree. Where this difference is strongly positive, the branch is a paleo-endemic; where this difference is strongly negative, the branch is a neo-endemic. High PE branches that show only a small difference belong in an unrecognized category in the study of endemism: we here coin the term "meso-endemism" to refer to range-restricted lineages of intermediate length. We show how these categories of endemism can be mapped on the phylogeny as either continuous or quantitative traits for use in comparative methods, with an illustration using the genus *Acacia* in Australia.

T1-10-04

Spatial phylogenetics of the vascular plants of Florida: The effects of tree uncertainty and ultrametricity

Julie Allen[1], *Charlotte Germain-Aubrey*[1], *Narayani Barve*[1], *Kurt Neubig*[2], *Lucas Majure*[3], *Shawn Laffan*[4], *Brent Mishler*[5], *Hannah Owens*[1], *Robert Guralnick*[1], *Doug Soltis*[1], *Pam Soltis*[1]

1. University of Florida
2. Southern Illinois University
3. Desert Botanical Gardens
4. University of New South Wales
5. University of California, Berkeley

The recent data explosion in biodiversity and geophysical data, along with novel analytical methods and tools, has enabled an unprecedented capability to both model species distributions across space and time and integrate those results into broad-scale diversity assessments. As part of a larger project to understand plant biodiversity in Florida, U.S.A., we report on the development of phylogenetic trees explicitly built for the purpose of integration with spatial ecological and biodiversity knowledge. We chose Florida as a focus for study because it is home to a hyper-diverse biota, encompassing a broad range of terrestrial and aquatic habitats. Florida's ecosystems include over 4,000 species of vascular plants alone, and there is a unique floristic transition in Florida from temperate forests in the north to tropical elements in the south. We compare the results of phylogenetic diversity analyses based on a phylogenetic tree constructed specifically for this purpose with those derived from a common approach of pruning a subset tree from a pre-assembled supertrees using Phylomatic and the Open Tree of Life. Further, we compare the results from both phylograms and chronograms and highlight the differences in interpretations between these two types of trees. Finally, we take into account uncertainty in the phylogenetic process and how that may affect measures of spatial phylogenetics. We provide evidence that careful consideration of phylogenetic tree construction and use is essential for proper interpretation of spatial patterns.

T1-10-06

Spatial phylogenetics of the Australian flora

Joseph Miller[1,2], *Brent Mishler*[4], ***Shawn Laffan***[3], *Nunzio Knerr*[2], *Andrew Thornhill*[2,4], *Carlos Gonzalez-Orozco*[2,5]

1. National Science Foundation
2. Centre for Australian National Biodiversity Research
3. University of New South Wales
4. University of California - Berkeley
5. Corporación Colombiana de Investigación Agropecuaria, Corpoica

Biodiversity is usually measured by examining changes in the number of species across a region to identify areas of high species diversity, endemism and turnover. Our research group is applying a novel suite of spatial phylogenetic tools including two new metrics, Relative Phylogenetic Diversity and Relative Phylogenetic Endemism, and new methods called categorical analysis of neo- and paleo-endemism (CANAPE) to measure biodiversity. CANAPE searches for centers of endemism, and classifies them by the branch lengths of the rare taxa within them, allowing, for the first time, a clear, quantitative distinction between centers of neo- and paleo-endemism across an area. We apply these metrics on newly developed near species level phylogenies for *Acacia, Banksia,* and *Eucalyptus* and genus level phylogenies for the entire Australian flora (>1,900 genera). When we combine these phylogenies with the detailed spatial data from the Australian Virtual Herbarium available through the Atlas of Living Australia, we can understand patterns of endemism through time, space and clade. While patterns of phylogenetic diversity and endemism are similar to patterns of taxonomic diversity and endemism, in depth analysis of the distribution of long and short branch taxa provide compelling insight and significance tests of evolutionary patterns of endemism. Areas with an overrepresentation of short branches are primarily found in the arid interior of Australia while areas with an overrepresentation of long branches are found in southern and coastal areas. As expected the areas of significant endemism change depending on the phylogenetic scale of terminals and when different components of the flora are analyzed. For example the Southwest Australia is a significant area of phylogenetic endemism at the genus level in *Acacia, Banksia* and *Eucalyptus*. However the southwest is much less important than the Wet Tropics of Queensland in the analysis of the entire flora at the genus level. Observed PD and PE, as well as taxonomic richness and endemism, are inadequate to fully understand patterns of biodiversity. The combination of statistical tests applied here can be used to better explain biodiversity patterns and the evolutionary and ecological processes that have created them. The spatial phylogenetic methods used can also be used to identify conservation priorities at any geographic scale or taxonomic level. Understanding patterns of biodiversity on the landscape is important for conservation planning, given the need to prioritize efforts in the face of rapid habitat loss and human-induced climate change.

T1-10-07

High-resolution phylofloristics reveals evolutionary constraints on the assembly and future migration of a regional vascular flora

Daniel Spalink[1], *Ricardo Kriebel*[2], *Thomas Givnish*[2], *Mary Ann Feist*[2], *William Alverson*[2], *Donald Waller*[2], *Kenneth Cameron*[2], *Kenneth Sytsma*[2]
1. University of Utah
2. University of Wisconsin-Madison

The emerging discipline of phylofloristics analyzes the ecological and evolutionary processes driving past and future patterns of community assembly across landscapes. Here we integrate a phylogeny and ecological niche modeling and present a phylo floristic analysis for all members of the Wisconsin vascular flora to quantify fine-scale variation in potential species richness and phylogenetic diversity. We calculated the mean phylogenetic distance (MPD) of species in over 50,000 geographically structured assemblages throughout the state, and tested whether these assemblages are phylogenetically clustered or over-dispersed. A shift from phylogenetically conserved to over-dispersed assemblages occurs over only 4.5° of latitude, in a flora shaped by Holocene dispersal following deglaciation. Spatial phylogenetic turnover is most strongly related to changes in annual minimum temperature and secondarily to spatial distance. Species ranges and assemblage composition appear driven primarily by phylogenetic niche conservatism. Over the next 50 years, our models predict large-scale extirpation of northern species within Wisconsin and widespread increases in ranges of southern species. Substantial species turnover is expected throughout the state, with a net increase in regional diversity as Wisconsin becomes more climatically suitable for southern and eastern species. However, our models predict that the geographic structure of phylogenetic diversity will be largely lost as a result of this turnover. These analyses should provide a useful baseline for targeted conservation strategies, by identifying specific lineages and geographic assemblages that are most at risk. Our datasets and approaches should also be directly applicable to the analyses of similar patterns throughout North America.

T1-10-08

Delineate phytogeographical regions of China: A phylogenetic approach

Jianfei Ye[1,2], *Zhiduan Chen*[1], *Lingfeng Mao*[3], *Limin Lu*[1], *Bing Liu*[1], *Tuo Yang*[1]
1. Institute of Botany, Chinese Academy of Sciences, Beijing, China
2. University of Chinese Academy of Sciences, Beijing, China
3. University of Alberta, Alberta, Canada

Biogeographical regionalization provides an indispensable background for research of biodiversity and conservation. Previous regionalizations were proposed mainly based on taxonomic dissimilarity and/or endemism, while ignoring the evolutionary relationship among taxa. We used a mega-phylogeny of 2,591 genera of angiosperms and their geographical distributions to quantify both taxonomic and phylogenetic dissimilarity (or beta diversity) between grid cells. We applied a hierarchical cluster analysis on the phylogenetic beta diversity matrices to generate a phylogenetic regionalization of the Chinese flora. We also explored the differences of environmental variables among biogeographical regions and the correlations between environmental variable and phylogenetic turnover. An indicator genus analysis was used to characterize the genus composition of each of the biogeographical regions. Phylogenetic relatedness has significant correlation with taxonomic composition for floras in China. The lowest spatial turnover in phylogenetic relationships of the floristic assemblages occurred in Hengduan Mountains, Qin Mountains, Taihang Mountains and Yan Mountains, whereas the highest spatial turnover in both taxonomic and phylogenetic composition was inferred to be in the southernmost part of China. Two major groups (South Cluster and North Cluster) and five subgroups representing five floristic regions were recognized based on phylogenetic beta diversity matrices. The regions differ significantly in three environmental variables and indicator genera. Mean annual precipitation (MAP) is the most important environmental variable distinguishing between regions; Paleotropical region has the most indicator genera while Holarctic region has the least indicator genera. The five regions we recognized can help develop strategy to effectively conserve the taxa and floristic regions with different origins and evolutionary histories.

T1-10-09

Phylogenetic composition analysis of the world's tropical forests suggests that two major tropical forest types existed during Gondwanan breakup

Johan Willem Frederik Slik
Universiti Brunei Darussalam

Based on an extensive data set containing more than 4,000 plots, about a million trees and approximately 15,000 tree species, we compare the world's tropical forests based on the phylogenetic similarity of their angiosperm tree communities. We identify a basal division in the world's tropical forests that corresponds to the breakup of Gondwana in a western and eastern portion, with Africa containing elements of both. We also find a connection between Asia and America that likely represents remnants of the once extensive Boreo-tropical forests. Our analysis shows that Gondwana may already have contained two distinct tropical forest types prior to or shortly after breakup (ca. 100 million years ago), a time when angiosperms were still a relatively young group of plants. Our analysis shows the usefulness of including phylogenetic information in floristic analysis for biogeographic inference.

T1-10-10

The spatio-temporal patterns of diversification of global gymnosperms

Lyu Lisha[1,2], *Yong Yang*[3], *Xiaoting Xu*[1], *Dimitar Dimitrov*[4], *Nawal Shrestha*[1], *Hui Zeng*[2], *Zhiheng Wang*[1]
1. Department of Ecology, College of Urban and Environmental Sciences, and Key Laboratory of Earth Surface Processes of Ministry of Education, Peking University, Beijing 100871, China
2. School of Urban Planning and Design, Shenzhen Graduate School, Peking University, Shenzhen 518055, Shenzhen
3. Institute of Botany, Chinese Academy of Sciences, Beijing 100093, China
4. Natural History Museum, University of Oslo, PO Box 1172 Blindern, NO-0318 Oslo, Norway.

The evolutionary history of gymnosperms has long been a controversial issue for plant evolutionary biologists. Although this group is thought to be older than angiosperms, increasing number of recent studies have revealed that many gymnosperm species may be much younger. Here, we use 21 genes and 46 fossil calibrations to infer an dated phylogenetic tree including all 1,118 extant gymnosperms. We used this complete species level tree of the extant gymnosperms to study the historical patterns of their diversification. We also assess the effects of past diversification dynamics on the present day diversity patterns. We found that most extant gymnosperm species are rather young and have diverged in the Neogene. A strong increase in net diversification rate was observed from 10 Ma to present, which may be the results of rapid diversifications of cycads and conifers. Geographically, gymnosperm species diversity and evolutionary rates were found to vary between continents, which imply distinct evolutionary patterns across continents. We speculated that the paleoclimate and tectonic events rather than the modern environmental factors are the main determinants of gymnosperms' contemporary spatial pattern. Our study is the first to depict the global evolutionary history of gymnosperms in a spatial context and reveals the important role of Cenozoic climate and deep-time tectonic events in shaping the contemporary diversity pattern of gymnosperms.

T1-10-11

Quantitative geospatial data and area definition in modern hstorical biogeography: The continental Australian flora as a case study

Daniel Murphy[1], *Malte Ebach*[2], *Joe Miller*[3]
1. Royal Botanic Gardens Victoria
2. University of New South Wales
3. National Science Foundation

The study of historical biogeography requires the definition of biotic areas (areas of endemism). Areas are core to the entire research program of historical biogeography, whether for model-based or more traditional approaches, yet their definition remains problematic. With recent shifts towards model-based biogeographic analyses there is renewed interest in biotic areas, or rather methods to define barriers or turnover of biodiversity in a landscape – which are all related concepts. This is at its most challenging within a continental landmass, where there are few sharp barriers or boundaries to define biotic areas. For example, in Australia vast areas are arid or semi-arid environments, and these have traditionally been treated as a single huge area in many analyses. In addition, the definition of biotic areas has been inconsistently applied across Australia, with some regions comprised of many small biotic areas and others much larger in size, with no consistent methodology used to define scale or boundaries. Therefore, using Australian plants and the quantitative geospatial data now available through large-scale databasing of the Australian flora (through Australia's Virtual Herbarium and Atlas of Living Australia projects), we examine problems with the prevailing existing area definitions in Australia and discuss possible newer methods to define areas, include some recent results and a new bioregionalisation of Australia.

T1-10-12

The recent and rich high elevation flora of the southern south American at risk under climate change

Mary T.K Arroyo
Institute of Ecology and Biodiversity (IEB), University of Chile

Historical records around the globe show high altitude species can contend with global warming by migrating into cooler conditions. However, in seeking cooler conditions, the amount of suitable habitat available in the case of many species is expected to be reduced on account of the geometry of mountainous terrain, the orientation and degree of isolation of mountain ranges and their peaks, and regional changes in precipitation. High elevation, above-treeline or bioclimatically-equivalent habitats in South American Andes, found at progressively lower altitudes toward higher latitudes, support a rich alpine flora. Dated phylogenies and the nested positions of some genera within larger clades indicate a fairly recent origin for many clades. Distributional modelling employing Ensemble Forecasting (BIOMOD), georeferenced herbarium occurrence data, two climate models and two climate change scenarios on a large pool of high altitude species from the southern Andes (27°-56°S), predicts there will be far more losers than winners under climate change, even when unlimited dispersal capacity is assumed and a more conservative temperature increase is considered. Whether the severe habitat loss predicted for many species by these models becomes fully manifest will depend on the ability of high altitude species to hold on at the trailing edge and reach favorable thermal microsites within their present altitudinal ranges. Better knowledge of dispersal capacity and of establishment success at the leading edge of alpine species' altitudinal ranges is critical to our understanding of the impacts of global warming on high altitude floras, as is knowledge on physiological traits such as frost resistance and reproductive traits such as flower longevity.

T1-11: Biodiversity informatics and Mapping Asia Plants (MAP)

T1-11-01

Tri-trophic interactions among pubescent oak (*Quercus pubescens* Willd.), green oak leaf roller (*Tortrix viridana* L.) and their parasitoids

Onur Birol[1*], *Mehmet Faruk Gürbüz*[2]
1. Mehmet Akif Ersoy University
2. Süleyman Demirel University

In this study, tri-trophic interactions between green oak leaf roller (*Tortrix viridana* L.) which causes complete defoliation in oaks, heavily defoliated pubescent oak (*Quercus pubescens* Willd.) trees distributed in Akdağ Nature Park (Turkey) and parasitoids of *T. viridana* were investigated. The green oak leafroller (*T. viridana*) is a major defoliator of oak woodlands. Within the beneficial entomofauna accompanying the leaf rollers (Lepidoptera: *Tortricidae*) a big role is played by parasitic insects from Hymenoptera. They attack the larvae and pupae of tortricids, limiting the density of their population. As a result of this study many adult parasitoid individuals were emerged from *T. viridana* pupae. These parasitoids are members of Hymenoptera belonging to Ichneumonidae family *Itoplectis maculator*, Pteromalide family *Cyclogastrella simplex* and *Pteromalussemotus*, Torymidae family *Monodontomerus aer-*

eus and Chalcididae family *Brachymeria tibialis*. In Turkey, green oak leafroller pest mainly affects *Q. pubescens* L. Interactions between trees and their pathogenic, parasitic and herbivorous enemies are among the driving factors of co-evolutionary processes in forest ecosystems. Whether there is an interaction between the species that characterized with regard to morphological and biochemical was investigated on the basis of various statistical analyses. As a result of statistical analyses there is a significant relationship, which validated with observed species number from all stations, between parasitoid *I. maculator* habitat selection and amount of volatile 2 Hexenal, *T. viridana* and amount of trans Caryophyllene. This relationship is emerging as an S.O.S signal that released from *Q. pubescens* trees against *T. viridana* attacks. As a result environmental changes not only have an indirect importance in plant defense mechanism but also have an importance place in volatiles releasing mechanism in attacked plants by herbivore. If oak trees can take enough essential elements from soil they can activate signal mechanisms that are related with defense.

T1-11-02

An introduction of Southeast Asia Plant Biodiversity Information Infrastructure (SEADiv) from Asia Biodiversity Conservation and Database Network (ABCDNet)

Bo Liu, *Hongfeng Wang, Zheping Xu, Maofang Luo, Qinwen Lin, Keping Ma**

Institute of Botany, Chinese Academy of Sciences

Southeast Asia represents four global biodiversity hotspots, Sundaland, Wallacea. Philippines and Indo-Burma. Its complex and diverse topography, and varying climates, provide a partial explanation for the high biodiversity. So far, the rich flora is still largely unknown, and many taxonomic inventory work still needs to be done. As part of Asia Biodiversity Conservation and Database Network (ABCDNet), we aim to build a platform for sharing biodiversity information in Southeast Asia—Southeast Asia Plant Biodiversity Information Infrastructure (SEADiv), a pilot project. Our main aim is trying to integrate current available data based on flora books and herbaria collections, and then synonymize all scientific names to make a preliminary checklist for higher plants of Southeast Asia. First step, we made an inventory of 328 Southeast Asia flora books or book series and other taxonomic literatures, based on the collections of 32 libraries from Southeast Asia, and several other libraries from Netherlands, London, New York *etc.* where significant plant taxonomic work has been carried out in the colonial times. Based on literature analyse and key plant taxonomist interview, 35 flora books or book series are listed as key floras, the taxonomic names, and distributions are typed into a database. Second step, an inventory of 1,350,000 specimens from 105 herbaria of Southeast Asia and other countries were obtained through online resources such as GBIF, Tropicos, USDA, JSTOR, HerpNET, MaNIS, OBIS, ORNIS, REMIB or acquired based on directly inquiry. Third step, TNRS (Taxonomic Name Resolution Service v4.0) based on The Plant List program are used as the basis for synomyomize the data gathered from key flora books and specimen data. Final step, the online platform is being prepared to share information about the prelimanary higher plant checklist of Southeast Asia.

T1-11-03

Mining hidden biodiversity data which were collected in North Korea: Based on the herbarium collections before 2nd World War

Hui Kim[1], *Chin-Sung Chang*[2]

1. Mokpo National University

2. Seoul National University

The Korean Peninsula has been recognized as a biodiversity corridor that would connect in a consistent manner critical habitat in Northeast Asia. Six decades after the end of the Korean War, Korea remains trapped in mutually reinforcing security dilemmas, so called cold war. It has been hard to imagine a direct South-North scientific exchange as well as an indirect cooperation through the international platform such as the GBIF. As Peter Raven recommended, it is urgent to establish a regional biodiversity network which maintains all records of the occurrence of organisms that the organizers have been able to gather. The available data in North Korean region are inaccessible and not well managed and formatted. For publishing, harvesting and using biodiversity data from Korean Peninsula, which aims to document and synthesize knowledge on the botanical richness of Korea peninsula. The potential for improved knowledge in the form of well documented biodiversity information is now recognized as essential for a better understanding and more effective conservation of Korean plants. The Korean Peninsula Flora (KPF) database builds on earlier work that used data aggregated from heterogeneous sources before the end of the 2nd World War, such as herbarium collections from TI, KYO, TUS, E, A, and other diverse literatures. KPF database comprises ca. 65,000 accessions of vascular plants collected from Korean peninsula from 1850 to 1945. Georeferencing was to be accomplished on a largely individual place name and record basis, by using numerous maps, gazetteers, and spatial databases, and also through consulting collectors' itineraries. Among these, material from People's Republic of Korea (DPRK) is represented with ca. 33,000 accessions. The largest part of this material originates from Russsian botanist, Komarov, V.L and Japanese, Nakai, T. that are followed by other collectors, such as Imai, H., Mills, R. G., Furumi, M., Nomura, N., Saito, T., and Okuyama, S.

T1-11-04

Impact of future climate change on the distribution patterns of seed plants endemic to the Tibetan Plateau

Yujing Yan, ***Zhiyao Tang***

Department of Ecology, College of Urban and Environmental Sciences, Peking University

Evidences have shown that many species have already shifted their distributions as a response to climate change, which may greatly impact the function of local or even global ecosystems. However, few studies have discussed the species vulnerability on a regional basis. As one of the most sensitive regions to the global climate change and unique for its rich biodiversity, the Tibetan Plateau has undergone an earlier and faster warming process compared to the global mean along with changes in precipitation during last decades and may continue in the future at a faster pace. Here, we collected the distribution information for 1,672 endemic seed plants of Tibetan Plateau and assessed the effects of future climate change on the diversity and distribution of 1,095 species using an ensemble species distribution modeling framework. We projected range shifts of these species at year 2050 and 2070 under two RCP scenarios (RCP 2.6 and RCP 8.5) and further evaluated the threatened status for these species based on the International Union for Conservation of Nature criteria. We also evaluated the protection effectiveness of national and provincial nature reserves on these species. Main conclusions are summarized as follows: (1) Species distribution models performed well in modeling the current distribution of endemic seed plants on the Tibetan Plateau. Species richness is high in the mountainous area of south-eastern plateau and low in the north-western part of the Tibetan Plateau. Temperature seasonality is the main factor controlling the distribution of most species. (2) Average richness across the plateau would increase in the future if consider full dispersal. The center of richness will move upward and to the center of the plateau. Severe species loss may happen in the southern edge of the Tibetan Plateau, especially the southern part of the Hengduan Mountains. The turnover rate of species is high in the north-western part of the plateau, indicated the vulnerability of this region under future climate scenarios. (3) It was predicted that the range of about 30% species would decrease in the future even when considering full dispersal. Taken species dispersal into count may aggravate the risk of threat. National nature reserves on the Tibetan Plateau may not be so efficient in protecting these endemic species under current climate condition, but could better protect them in future climate change. Our study provided insights for biodiversity conservation management in this region under climate change.

T1-11-05

Moscow Digital Herbarium and the National Depository Bank of Live Systems initiative (Russia)

Alexey P. Seregin, *Tatiana P. Balandina, Natalya S. Gamova, Ksenia V. Dudova, Natalya K. Shvedchikova*
Lomonosov Moscow State University

Imaging of specimens is the modern trend in the herbarium management. Today, at least 62 herbaria have 1 M+ specimens and at least eight herbaria have estimated their collections as 1 M. Even in the larger herbaria mass digitisation is still not a common practice. In late 2014, the Moscow University Herbarium (MW) had received direct investment for digitisation within the grant #14-50-00029 from the Russian Science Foundation as part of the National Depository Bank of Live Systems initiative launched by the Moscow State University. As of November 2016, MW Herbarium holds 1,011,253 specimens being the 60th largest herbarium in the world and the second largest in Russia. An average annual increase in the last 12 years was 15,100 specimens. We employed new technical staff to facilitate further growth of the collections, intensified the collection management and 22,013 specimens were added in 2016 as a result. The MW Herbarium holds 37,100 species and subspecies of vascular plants and 2,223 species and subspecies of mosses and liverworts. Having the stable budget for the next four years, we decided to scan 1 M specimens at 300 dpi (TIFF + JPG) in 2015-2018, to digitise 4.6 K type specimens at 600 dpi, to scan 78 K labels from bryophyte capsules, and finally to database and georeference label data from as much specimens as possible until exhaustion of the budget. As of 6 January 2017, the Moscow Digital Herbarium consists of 785,887 specimens—712,925 images of vascular plants and 72,962 images of bryophyte labels. Vascular plant specimens are originated from eastern Europe—357,951 images; Asian Russia—159,663 images; Caucasus—97,400 images; the Crimea—31,790 images; Mongolia—27,323 images; South Asia—22,649 images; Africa—7,559 images; types at 600 dpi—4,630 images; *Herbarium Alchemillarum*—3,960 images; and labels from capsules of mosses—72,962 images. Collections from Middle Asia, western Europe, Americas, Australia and Oceania, liverwort labels, and historical collections are not yet digitised. At the moment, we operate the sixth largest digital herbarium in the world. The Moscow Digital Herbarium is the largest Russian biodiversity database. The Moscow Digital Herbarium publishes its data on the web portal of the National Depository Bank of Live Systems, a collaborative network of the Moscow University biological collections. Since October 2016, a fast public version of the botany portal with some functional limitations is available at http://plant.depo.msu.ru/, whereas an operational version is available at https://plant.mitotech.ru/. The National Depository online system is open for contributions from other Russian biological collections and soon will be visible via GBIF. We are nearly ready to start in 2017 data capturing from labels. To avoid mistakes and to make the first portion of label capturing easy, we will not enforce the operators to capture data from every single sheet. We will ask the operators to avoid capturing data from labels with at least one unclear word or to tick any uncertain entries. Also, we are ready to pay modest remuneration for professional botanists from small institutions and universities for capturing the data which might be interesting for them using the forms on our web portal.

T1-11-06

World vascular plant checklist in the Catalogue of Life

Yuri Roskov
Illinois Natural History Survey, University of Illinois at Urbana-Champaign

The Catalogue of Life (CoL, www.catalogueoflife.org/col) is a global taxonomic catalogue of valid species through all domains. It is built as a curated assembly of expert based Global Species Databases around the world. Today, the CoL contains basic information on 1.6 million valid species arranged in global classification, plus 1.5 million synonyms and over 400,000 common names. In 2015-2016, the CoL has aggregated world checklists for all families of vascular plants with 335,566 accepted species. This is about 99% of Tracheophyta species diversity recently recognised by taxonomists. Data from 18 plant databases has been assembled into the synonymic catalogue. Classes *Equisetopsida* (38 spp), *Lycopodiopsida* (1,393 spp), *Marattiopsida* (133 spp), *Polypodiopsida* (11,530 spp) and *Psilotopsida* (139 spp) have been provided by World Ferns database (Hassler M., 2017). Cycads (class *Cycadopsida*, 353 spp) has been taken from The World List of Cycads database (Calonje M., Stanberg L. & Stevenson D. (eds), 2015). Conifers (class *Pinopsida*, 615 spp) have been supplied by Conifer Database (Farjon A., Gardner M. & Thomas P., 2014). Classes *Ginkgoopsida* (1 spp) and *Gnetopsida* (112 spp) were taken from World Checklist of Selected Plant Families (Govaerts R. (ed), 2014). Flowering plants have been compiled from an array of 15 databases: AnnonBase (2,440 spp; Rainer H. & Chatrou L.W. (eds), 2014), Brassicaceae database (3,778 spp; Warwick S.I., Francis A. & Al-Shehbaz I.A., 2009), ChenoBase (2,007 spp; Walter J., 2014), Droseraceae Database (179 spp; Culham A. & Yesson C., 2008), Global Compositae Checklist (32,280 spp; Flann C. (ed), 2014), ILDIS World Database of Legumes (16,371 spp; Roskov Y., Zarucchi J., Novoselova M. & Bisby F. (eds), 2014), IOPI Global Plant Checklist (1,845 spp; Wilson K.L. & Berendsohn W.G. (eds), 2007), The Integrated Taxonomic Information System (Orrell T. & al., 2016), The Lecythidaceae Pages (311 spp; Mori S.A., Prance G.T. & Tulig M., 2006), Lacistemataceae Holistic Database (24 spp; Young F.E., 2009), Geranium Taxonomic Information System (344 spp; Aedo C., 2012), Solanaceae Source (1,082 spp; Knapp S., Walley L., Bohs L., Nee M. & Spooner D. (eds), 2008), World Checklist of Selected Plant Families (120,279 spp; Govaerts R. (ed), 2014), World Plants (133,957 spp; Hassler M., 2016) and World Wide Wattle database (1,469 spp; Maslin B., 2015). Global checklist of flowering plants in the CoL consists of 321,252 accepted species in total. CoL standard dataset includes accepted scientific names, synonyms, position in the classification, common names, references, distribution, ecological environment and credits to data providers and authors. We are updating CoL monthly. However, each database decides to have own schedule for updates depending on a progress. All data are available online as dynamic retrieval system, webservice, customised download service in DwC-A format and on annually published DVD (Roskov Y. & al., eds. (2017). Species 2000 & ITIS Catalogue of Life, 2017 Annual Checklist. Digital resource at www.catalogueoflife.org/annual-checklist/2017. Species 2000: Naturalis, Leiden, the Netherlands. ISSN 2405-884X).

T1-12: Contribution of Brazilian private natural heritage reserves to Atlantic forest biome flora conservation

T1-12-01

Are there more original species in remnants with greater phylogenetic diversity?

Vivian Almeida Assunção[1], *Mozart Sávio Pires Baptista*[1], ***Andressa Lima***[2], *Ângela Lúcia Bagnatori Sartori*[1], *Vidal De Freitas Mansano*[2]

1. Universidade Federal de Mato Grosso do Sul, CCBS. Laboratório de Botanica cidade Universitária.

2. Instituto de Pesquisas Jardim Botanico do Rio de Janeiro

Greater phylogenetic diversity in a community is related to more complementary resource use and application by different groups. The increase of complementarity is associated to the presence of species that present rare phylogenetic traits, with unique functions in the community. These species are known as 'original species', and can be identified through an originality index. This metric can describe the structure of the community and assist in the determination of priority conservation areas. In the Chaco, a seasonal domain, exclusive to South America (Argentina, Bolivia, Brazil and Paraguay), species composition varies throughout its territory, according to abiotic and biotic factors, resulting in different phytophysiognomies. The aim of this study was to analyze the phylogenetic diversity in two remnants of Chaco and to verify if the most diverse one presented a greater percentage of original species. Our experiment was carried out in two remnants, with an area of 1ha in phytophysiognomy of forest and in woodland chaco. For phylogenetic analysis we calculated the Net Relatedness Index (NRI) and Nearest Taxon Index (NTI) and to calculate originality the distance between each pair of species, by the sum of the branches, needed to bind the pairs in the phylogenetic dendrogram. The most original species were those with a sum of decreasingvalues around 50%. In both areas, phylogenetic diversity was lower than expected at random ($W = 1069$, $p = 0.44$), indicating aggregation, possibly due to environmental filters acting on a regional scale, such as climate. However, the lower phylogenetic diversity in the woodland chaco ($W = 1523$, $p = 0.01$), may be related to more intense environmental filtration due to the dystrophic soil that selects for the establishment of certain groups. In the forest chaco, the eutrophic soils favor the establishment of species belonging to different groups, also increasing species richness. We collected 75 species, of these 66 found in the forest chaco and 26 in the woodland one. Of these only 17 species (21.3%) found in both remnants represented 51.5% originality. Twelve of them in the forest chaco and five in the woodland, proportionally the woodland chaco presented more original species (19.2%) than the forest one (18.2%) considering the richness. Furthermore, the less restrictive environment had the capacity to maintain species belonging to different phylogenetic groups, while the other selected for related groups with less parity between species. However, the higher percentage of original species, where the environment was more restrictive, allowed unique functions to be performed in the community, minimizing local redundancy and contributing to the maintenance of its functioning. With this, we conclude that the Chaco community, under local filtering, contains species belonging to groups less related to each other than the community in which the environment is less restrictive and allows the establishment of several groups.

T1-12-02

Woody plant diversity in the seasonally dry tropical forests of the Urubamba valley, a threatened biodiversity hotspot in southern Peru

Isau Huamantupa-Chuquimaco[1,2], *Miguel A. Luza-Victorio*[2], *Reynaldo Linares-Palomino*[3], ***Andressa Lima***[1]

1. Programa de Pós-Graduação em Botânica, Escola Nacional de Botânica Tropical, Instituto de Pesquisas Jardim Botânico de Rio de Janeiro (ENBT/JBRJ)

2. Herbario CUZ, Facultad de Ciencias, Universidad Nacional San Antonio Abad del Cusco (UNSAAC)

3. Smithsonian Conservation Biology Institute

The woody flora of the seasonally dry tropical forests of the Middle Urubamba Basin, located in Cusco, southern Peru, was studied. Twenty plots of 50 m x 20 m (0.1 ha) were sampled from 700 to 1,300 Msnm; (DBH) $\geq$ 2.5 cm, representing all life forms (lianas, trees, shrubs, hemiepiphytes and cacti). 5,259 individuals have been recorded including 552 species, 303 genera and 82 families. The quantitative sampling reveals the presence of 92 species/0.1 ha, on average compared to the diversity already found in BTES of 150 species/0.1 ha with an alpha Fisher diversity index of 133.6 (Managua locality). The richest families were Fabaceae (69 species), Bignoniaceae (30), Moraceae (27), Apocynaceae (23) and Sapindaceae (22); the richest genera being *Inga* (13 species), *Aspidosperma* (11), *Ficus* (11), *Machaerium* (10) and *Eugenia* (9); the five most abundant species were *Allophylus punctatus* (142 individuals), *Pogonopus tubulosus* (122), *Warszewiczia coccinea* (111), *Annona neoulei* (106) and *Anadenanthera colubrina* (95). Half of all individuals (2,593,49.4%) were in the small DBH class (2.5-5 cm). The analysis of vegetation and abiotic variables show the existence of three types of BTES, the Amazonian pluviseasonal tropical dry forest, sub-Andean pluviseasonal tropical dry forest, and savanna-like pluviseasonal tropical dry forest. Initial data on phytogeographic relationships show that the BTES of the Urubamba are better related to the BTES nucleus of the Chiquitania of Bolivia and an almost null relation with other BTES nuclei Peruvian such as those of Tarapoto and Marañon. The state of conservation of these forests shows that much of these have been hugely deforested, and dry neotropical dry forest species such as *Amburana cearensis*, *Myroxylon balsamum* (Fabaceae), *Cedrela saltensis*, *Swietenia macrophylla* (Meliaceae), and other woody species have almost completely disappeared. Therefore, it is concluded that the high diversity of wood taxa, endemism's and particular phytogeographic relationships, indicate that the BTES of the Urubamba basin is part of the Andean-Amazon hotspots and forms an important nucleus of diversity for the BTESs. It has been concluded that these dry forests need urgent protection and conservation because of their rapid and alarming destruction.

T1-12-03

New species to science discovered at private nature reserves in the Brazilian Atlantic Forest Biome

Andressa Lima[2], *Maria Cristina Vieira*[1]

1. Associação de RPPN e outras Reservas Privadas de Minas Gerais

2. Instituto de Pesquisas do jardim Botânico do Rio de Janeiro

In Brazilian Private Natural Heritage Reserves located at the Atlantic Forest Biome and Hotspot some new species to science have been discovered. Publications in the Atlantic Forest Biosphere Reserves Booklets and Botanical Congresses have already mentioned some of these discoveries. In this presentation will be shown an updated compilation of the new species described in private nature reserves such as: 1. *Ficus lagoensis* C. C. Berg & Carauta – Figueira (Moraceae) - RPPN Fazenda Lagoa/Minas Gerais 2. *Neoregelia pernambucana* – Siqueira Filho & Leme - Bromélia – (Bromeliaceae) RPPN Frei Caneca/Pernambuco 3./ 4 *Couepia monteclarensis (Chrysobalanaceae),* e *Asterostigma lombardii.* (Araceae) - RPPN Feliciano Miguel Abdalla/Minas Gerais 5./ 6 - *Byrsonima pedunculata e Bunchosia pernambucana* RPPN Fazenda Bituri - Pernambuco.

T1-12-04

Conservation *in situ* of endangered species in private nature reserves in the Brazilian Atlantic Forest Biome

Maria Cristina Vieira, *Andressa Lima*

Associação de RPPN e outras Reservas Privadas de Minas Gerais

In this presentation the existing knowledge on flora conservation *in situ* in Private Natural Heritage Reserves in the Atlantic Forest Biome in Brazil shall be described. This information has already been partially published and presented in the Atlantic Forest Biosphere Reserve publications and also in several scientific events, such as Latin American and Brazilian Botanical Congresses. Several endangered species quoted in the Red Book of Brazilian Flora edited by the National Center for Plant Conservation of Rio de Janeiro Botanical Gardens (CNCFlora; cncflora.jbrj.gov.br) have been identified in private nature reserves in the Brazilian Atlantic Forest Biome. The following RPPN case studies will be presented: Fazenda Lagoa/MG; Frei Caneca/PE; Mata do Sossego/MG; Mira Serra/RS; Reserva Natural da Serra do Teimoso/BA; Santuário do Caraça/MG; Rizzieri/SP.

T1-12-05

Plant species richness in private nature reserves in the Brazilian Atlantic Forest Biome

Maria Cristina Vieira[1,2], *Andressa Lima*[2]

1. Brazilian Confederation of Private Natural Heritage Reserves - Technical and International Relations Director

2. Associação de RPPN e outras Reservas Privadas de Minas Gerais

The Atlas of Forest Remnnants of the Brazilian Atlantic Forest Biome published by the Foundation SOS Atlantic Forest points out that only 8.5% of the original forest cover has survived and considers this Biome as a Hotspot. Considering that 80% of the this Biome remaining forest ecosystems are in private land and that there is still a lack of information concerning the Flora Diversity the Floristic Surveys done in RPPNs are of utmost importance. In this presentation the results of floristic surveys developped at RPPNs and published in the Atlantic Forest Biosphere Reserve publications and presented in several scientific events, such as Latin American and Brazilian Botanical Congresses shall be described. The following case studies of RPPNs revealing high flora diversity shall be presented: RPPN Salto Morato/P ; RPPN Caraça/MG; RPPN Fazenda Lagoa/MG; RPPN Feliciano Miguel Abdalla/Caratinga, MG; RPPN Veracell/BA.

T1-13: The evolution redwoods of event in botany sciences

T1-13-01

Fossil histories of redwoods from China and their implications for plant conservation

Zhe-Kun Zhou[1], *Tao Su*[1], *Yong-Jiang Huang*[2]

1. Key Laboratory of Tropical Forest Ecology, Xishuangbanna Tropical Botanical Garden, Chinese Academy of Sciences

2. Key Laboratory for Plant Diversity and Biogeography of East Asia, Kunming Institute of Botany, Chinese Academy of Sciences

Redwoods is general name for three genera of *Metasequoia, Sequoia*, and *Sequoiadendron.* Fossil histories of *Metasequoia,* and *Sequoia* have fossil records from China are reported and their implication in conservation are discussed in the current presentation. Dawn redwood, *Metasequoia* was widely distributed in the Northern Hemisphere since the Late Cretaceous and more than 500 fossil records were reported but native living dawn redowood growing in center China only. In China, most Metasequoia fossils were reported from Paleogene strata of Northeastern China.

Recently, fossil leaves and corns were found from the middle Miocene of Zhengyuan, Yunnan province. This is a southernmost fossil record of *Metasequoia*. Coast redwood, *Sequoia,* is endemic to Pacific Coast Range, however it has widely distribution regions in the geological time. Most *Sequoia* fossils have been recognized from the Paleogene and early Neogene sediments of mid-northern latitudes. Fossils of *Sequoia* after the late Miocene are scarce and absent from Asia and Europe in the late Pliocene. New *Sequoia* fossil reported from the middle Miocene of Yunnan. These fossils are morphologically similar to *S. sempervirens*, indicating a close affinity to their modern relative. The genus probably disappeared from Yunnan some time after the late Miocene. Here may arise a question of why *Sequoia* have gone extinct in Yunnan but sustain in other regions. Investigations demonstrate that aridification of the dry season (winter and spring) associated with monsoon intensification might have played a role. Today, Yunnan has dry conditions from the late autumn to next early spring, whereas the Mediterranean regions and the western coat of America have wet winters and springs. As seeds of *Sequoia* are recalcitrant, they germinate immediately upon their maturity in autumn. Unfortunately, dry conditions after the autumn may prevent their seeds from germinating or their seedlings from developing new leaves. In contrast to the modern pattern, Yunnan was shown to have relatively humid conditions in the dry season during the Neogene. Humid and suitable habitats might explain the past existence of the two genera in the region. As the precipitation in the dry season decreased linked to the Asian monsoon intensification as is elaborated below, the dry winter and spring ultimately led to their extinction in Yunnan. The similar case can be observed in *Metasequoia* as well. So monsoon climate would cause extinction or relic of *Metasequoia* and *Sequoia* from China.

T1-13-02

The present status and future prospect of American coast redwood planting in China

Minghe Li

Central China Agricultural University

In 1972, the former US President Richard Nixon brought one small American coast-redwood tree to China as a gift to Premier Zhau Enlai. By rooting cuttings from that small gift tree, about 4,000 stecklings were propagated and planted in many localities in China. That Nixon redwood clone and other redwood trees from introduced seeds have grown very well in southern China, so far much better than local major planting species *Cunninghamia, Cryptomeria* and *Metasequoia*. It thus appears that coast redwood has great potential importance for China's and thus Earth's timber economy and for the enjoyment and well-being of future visitors to redwood parks in China as well. However, redwood has not yet been planted widely in China. This is mostly because China does not have reliable pedigreed redwood seeds, rooting redwood cuttings is not yet operationally satisfactory, redwood tissue-culture is complicated and costly, and redwood somatic embryogenesis is still in early stages of development. In 2014, the author tried an idea: to establish a redwood 'sprout orchard', in which many sprouts grow up from still-attached stems shallowly buried in soil. Those sprouts are cut underground and used as juvenile cuttings thus predisposed for rapid rooting. By 2016, average rooting percentages of over 90% have been obtained, the stecklings raised from them are apparently satisfactory, this process can be cycled 2 or 3 times in a year and the cost is less than 1/3 that of tissue-culture. China has land, labor, economic strength and technology, and therefore, may play an important role in Earth's future redwood planting.

T1-13-03

The basic situation of original dawn redwoods conservation

Haoran Xue[1], *Xiqun Wang*[2], *Momei Chen*[3,4], *Hua Wei*[5]

1. School of Life Sciences, Peking University, Beijing, China

2. Planning and Design Academy of Forest Products Industry, State Forestry Administration, Beijing, China

3. University of California, Berkeley

4. Beijing Normal University, Beijing, China

5. School of Humanties and Social Sciences, Beijing Forestry University, Beijing, China

As Hu Hsen-Hsu said in 1947, "the discovery of dawn redwoods is the most important and interesting new finding in Botany in China". In 1941, *Metasequoia* was described as a fossil genus by Shigeru Miki. In 1943, Wang Chan collected some specimens from an unidentified tree in Hubei, China. In 1946, Cheng Wan-Chun and Hu Hsen-Hsu connected these specimens to *Metasequoia*. In 1948, Cheng and Hu named the species as *Metasequoia glyptostroboides* and published a paper entitled "On the new family Metasequoiaceae and on *Metasequoia glyptostroboides*, a living species of the genus *Metasequoia* was found in Szechuan and Hupeh". This discovery shocked the world, as "living fossils of *Metasequoia* was found". However, after recent years of extensive introduction and cultivation, *M. glyptostroboides* has been distributed almost all over the world. Studies on *M. glyptostroboides* had provided significant information to paleobotany, paleoclimate, paleogeography, and the phylogeny of gymnosperms. Li Jianhua and Ban Jide classified dawn redwood-dominated communities into three types: (1) *Metasequoia glyptostroboides-Camellia cuspidata-Parathelypteris nipponica* community, (2) *Metasequoia glyptostroboides-Camellia cuspidata-Bigonia-Pilea* community, and (3) *Metasequoia glyptostroboides-Eurya hebeclados-Iris japonica-Parathelypteris nipponica* community. The largest *M. glyptostroboides* population is in Xiaohe, an open river valley in weastern Lichuan. Xiaohe is located at N30˚10', E108˚45', to the south of Fubao Mountain, east of Qiyue Mountain, west of Maqian Town, and north of Zhonglu Town, with altitudes range between 900 m and 1,350 m. Xiaohe has an area of around 60,000 ha, a length of about 30 km (from south to north), and a width of about 20 km (from east to west). It's an ideal place for dawn redwoods to grow, with short sunshine duration, low evaporation, high precipitation (1,500.9 mm per year), high relative humidity, and 230.9 growing days per year. Fubao Mountain and Qiyue Mountain block the coldness coming in winter and the heat in summer from the plain, which is to the southeast of Xiaohe. The temperature in Xiaohe fluctuates between -15.4°C and 35.4°C, with an annual average of 12.8°C. Dawn redwoods grow along both sides of the valley on alluvial soil and irrigated by foothill creek. There are approximately 5,000 dawn redwoods grow in this area with an average diameter of 20 cm. In 1972, a dawn redwood orchard was established in Hubei, China. In 1981, the Provincial Government of Hubei approved the establishment of Lichuan Xiaohe County Dawn Redwood Nature Reserve, aiming to protect the original dawn redwoods, especially "Lichuan Modaoxi dawn redwood type tree". In 2003, the State Council of P. R. China approved the establishment of Hubei Xingdoushan National Nature Reserve. The National Government of P. R. China has been directly investing the conservation of original dawn redwoods and their associated ecosystems. More than 5,700 dawn redwoods are included in this conservation. In 2012, Chinese Dawn Redwood Botanical Garden was established, with an area of more than 400 acres. A database for the original dawn redwoods has been established to facilitate the conservation.

T1-13-04

A whole-plant reconstruction of a Cretaceous redwood with novel morphology and ecology

Dori Contreras[1], *Garland Upchurch*[2]

1. University of California Berkeley and UC Museum of Paleontology

2. Texas State University, San Marcos

The three living redwoods (*Sequoia*, *Sequoiadendron*, and *Metasequoia*) are the sole remnants of a diverse group that was once widespread across both hemispheres. Redwoods (the Sequoid clade of the Cupressaceae) have a rich fossil record that leads back through the Cretaceous, yet there are relatively few whole-plant reconstructions of fossil species from which to understand their complex evolutionary history. We have previously reported an extinct Late Cretaceous redwood from the Jose Creek Member, McRae Formation, of New Mexico based on associated shoot impressions, pollen cones, and *in situ* permineralized stump. Additional field collections have recovered seed cones and pollen cones at various stages of maturity in organic attachment to foliage shoots. Here we reconstruct the entire plant, assess its relationships through phylogenetic analyses of living and fossil redwoods, and interpret its ecology based on morphological traits and our ongoing reconstruction of the fossil flora. The Jose Creek redwood is represented by over 500 specimens, collected primarily from one extensive ash-fall deposit that preserves a 74.7 million year old subtropical to paratropical forest. Shoot impressions preserving multiple orders of branching show heteromorphic leaves of two distinct morphologies, similar to *Sequoia*. The thickest axes bear highly reduced (scale-like) leaves with short pointed free tips. These axes give rise to up to 3 orders of planate branchlets bearing bifacially flattened leaves with narrowly elliptic shape and decurrent attachment to the stem (taxodioid leaf form). However, similar to living and fossil *Metasequoia*, the Jose Creek redwood exhibits opposite decussate leaf phyllotaxy, with leaves secondarily oriented into one plane by torsion along the basal decurrent portion of the leaves. Pollen cones are ellipsoid with spirally arranged microsporophylls and are shed individually. Similar to *Sequoia*, they are borne terminally on ultimate shoots, or on highly reduced axillary shoots that bear only a few scale-like leaves. Seed cones are ellipsoid in overall shape, borne singly at the tip of shoots with either scale-like leaves or taxodioid leaves. Mature bract-scale complexes are peltate, with centrally located tip that is recessed in a distal transverse groove with ridges that radiate from the groove. Overall, the Jose Creek redwood shares characteristics with both *Sequoia* and *Metasequoia*, but its combination of characteristics is unique compared with living and other known fossil redwoods. Phylogenetic analysis clearly places the Jose Creek redwood within the Sequoioid clade, but not in the distinct group formed by modern and fossil *Metasequoia*. The Jose Creek redwood was one of the dominant taxa in highly diverse sub- to paratropical forest that included palms, *Zingerberopsis*, a few other gymnosperms, and diverse magnoliids and eudicots. The absence of bud scales or scale-like leaves at the base of the branchlets and the absence of growth rings in the wood suggest absent or poorly developed dormancy mechanisms, unlike Cenozoic and modern relatives. This is consistent with other evidence suggesting a regional paleoclimate with low seasonality, and demonstrates that the ecology of the Jose Creek redwood is unlike that of other known redwoods.

T1-13-05

California coast redwood park and nature reserve: Protecting redwoods since 1918

Ivy Li[1], *Shikai Zheng*[2], *Momei Chen*[1,3]

1. University of California, Berkeley, USA

2. Chinese Academy of Forestry, Beijing, China

3. Beijing Normal University, Beijing, China

The first people to see the coast redwoods and the giant sequoias were the ancestors of today's native Americans-the nomadic hunters, fishermen and gatherers who migrated across North America about ten thousand years ago. These first discoverers were probably less astonished by the coast redwoods than were the first Europeans to see them, because they came to the redwoods through the ancient coniferous forests of the Pacific Northwest. Some of these trees were almost the equals of the coast redwoods for height, although none had ever been seen which rivaled the Giant Sequoia for sheer mass. The California Pacific Redwood Park and Nature Reserve has been protecting 36 very beautiful parks. Americans have long experienced the struggle to protect redwoods. Finally in 1918 an outstanding non-profit organization was set up: Save-the-Redwoods League. For nearly 100 years, the organization has educated generations for the benefit of mankind about protecting the excellent redwoods forest and guided California in the establishment of Redwood Park and its park protection guidelines. In 1928 the park and guidelines were consolidated into the California National Park System. Today, the California National Parks System is one of the best in the world, preserving vast natural resources and historic sites that provide education and entertainment for the benefit of the people and visitors from all over the world. Save-the-Redwoods League protects sequoias by purchasing sequoia forests and the surrounding land to nurture them. The principles of conservation biology guide our efforts to protect large, continuous mahogany land with forest diversity. With League's donors' support, Save-the-Redwoods League has protected over 200,000 acres (the size of 16 Manhattan Islands) and helped develop dozens of redwood parks and reserves for everybody to enjoy. League protection for redwood forests is unwavering and the vision for the future of redwoods is one in which the entire redwood forest ecosystem — public and private lands alike will continue to thrive and adapt to change. As our forests are increasingly threatened by unsustainable logging, development and climate change, it is essential for conservationists and foresters to work together.

This cooperative approach, in which active forests are managed alongside parks and reserves, allows for the permanent protection of more old-growth redwoods and wildlife habitat than would be possible otherwise. Science 1980's a Chinese American researcher at the University and Jepson Herbaria at UC Berkeley has initiated a new series of collaborations between Chinese and American botanists including Magnificent Chinese and American Redwoods, a bilingual book. Developments in the modern studies of molecular genetics, physiological ecology, and phylogeny -- underway in both the United States and China -- will continue to enhance this long tradition of collaboration and advance our ability to conserve these amazing trees and their forest ecosystem.

T1-13-06

The ancient redwoods: Where are the oldest, largest, and tallest redwoods?

Momei Chen

University of California, Berkeley

The three keystone species of redwoods have always been esteemed. (Taxonomy: The family Taxodiaceae traditionally comprises the redwoods (*Metasequoia*, *Sequoia*, and *Sequoiadendron*)). Modern molecular phylogenetic studies show that they are a monotypic genus of the family. The new family scientific name is Cupressaceae, which has nomenclatural priority over the traditional Taxodiaceae. The dawn redwood (*Metasequoia glyptostroboides* Hu et Cheng) was a "living fossil" widely distributed in high latitudes of the Northern Hemisphere in the early Cretaceous period (K2). During the Paleogene Period (E) and Neogene Period (N) of the Cenozoic, its distribution was extended and the genus included more than 10 species. In the Quaternary ice age, the genus became nearly extinct, with only one species surviving in central China within Shizhu (Chongqing), Lichuan (Hubei), and Longshan (Hunan). The range of the genus is broad at North Temperate Zone latitude 29°5′-31°32′, east longitude 108°-110°). Here, the sole surviving species, dawn redwood, occurs in extremely wonderful landscape areas. It persists on only one modest-sized stand and several scattered occurrences near the juncture of three provinces: Lichuan in West Hubei, Longshan in Hunan, and the deep valleys of Sizhu County of East Sichuan. The Cupressaceae includes two other related genera: the coast redwood (*Sequoia sempervirens* (D.Don) Endl) found in Pacific Coast Range. Redwood forests have been present in California for about 20 million years. They represent a unique and beautiful relic flora from the days of the dinosaurs. About a million years ago, advancing ice sheets confined the coast redwoods to their modern range. The glaciers left untouched a well-defined area, now known as the Redwood Belt, where the tall trees flourish, undisturbed. Giant sequoia (*Sequoiadendron giganteum* (Lindl.) J.Buchholz) are the tallest, most massive, and oldest trees in the world. They are found in the Sierra Nevada Range in California, which is 640 km long and 60–130 km wide, located mostly in E Calif. It rises to 4,418 m at Mt. Whitney, the highest peak in the United States outside Alaska. The range not only includes the renowned groves of massive giant sequoia, but the richness of the Sierran flora mirrors that of the state as a whole—of nearly 6,000 species of vascular plants known to occur in California, Calaveras Big Trees State Park, Yosemite National Park. The giant sequoia has the distinction of being the largest living thing in the world, past and present. The individual with the record size is the General Sherman Tree in Sequoia National Park. It has a volume of about 51,000 cubic feet (14,441,592 cubic meters) which, if made into a 2-by-4, would yield a board almost 175 miles (281,635 kilometers) long. To date, the bristlecone pines of the southwestern arid mountains are accepted as the oldest known living things; giant sequoias amongst the ancient ones. The oldest known giant sequoia is represented by a cut stump in the Converse Basin which, when the annual rings were counted, was determined to be at least 3,200 years of age.

T1-15: Building a global system for the conservation of all plant diversity

T1-15-01

The Millennium Seed Bank Partnership: A global seed conservation network

Jonas Muller

Royal Botanic Gardens Kew

Conserving plants *ex situ* as seeds in seed banks is a cost-effective conservation strategy intended as an insurance policy for the survival of plant species faced with threats in their natural habitats. The Millennium Seed Bank (MSB) Partnership is the world's largest and most ambitious *ex situ* conservation initiative of its type. The MSB Partnership is managed by the Royal Botanic Gardens, Kew (United Kingdom) and is a global network with currently 140 partner institutions in over 97 countries and territories covering all continents. More than half of those countries have active seed collecting projects running. The network is steadily growing, with more countries and institutions joining in, particularly from biodiverse countries. Over the last year, 14 partnership agreements have been renewed, and 13 new partner institutions have joined the MSB Partnership since 2015. Building on previous achievements and in line with the strategic outputs of Kew's Science Strategy, the MSB Partnership aims to collect 25% of the world's bankable flora by the end of 2020. Over 37,500 of the world's floral species are now secured in *ex situ* conservation, held in country of origin and in duplicate storage at Kew's Millennium Seed Bank. The majority of those seeds is available for restoration and research. Seeds held at the Millennium Seed Bank come from 187 countries worldwide, reaching beyond the geographical boundaries of the partner countries. The MSB Data Warehouse was established as a way of data sharing across the MSB Partnership. Currently, the MSB Data Warehouse holds data of 100,200 collections of 47,400 unique taxa (41,700 species) which are freely available to view for registered MSB Partnership users. 115 different users have consulted the MSB Data Warehouse in 2016. Training and capacity building is yet another important activity in MSB Partnership. During 2016 alone, Kew trained 245 individual partners from 27 countries. Our seed conservation experts provided 5,884 hours of training in 22 training courses.

T1-15-02

The Global Oak Conservation Partnership

Nicole Cavender[1], *Murphy Westwood*[1,2], *Kirsty Shaw*[2], *David Gill*[3], *Magin Georgina*[3]

1. The Morton Arboretum
2. Botanic Gardens Conservation International
3. Fauna & Flora International

Oaks (genus *Quercus*) are keystone species in a wide range of ecosystems around the world, such as oak-hickory forests and oak savannas in North America and subtropical broad-leaved evergreen forests in Southeast Asia. Globally, oaks are ecologically valuable for promoting terrestrial biodiversity, performing critical ecosystem functions, and providing food and habitat for countless species of animals. However, many oak species are threatened with extinction and oak-dominated ecosystems are declining worldwide. Furthermore, oak acorns cannot be seed banked through conventional methods, so long term *ex situ* conservation of threatened oak species must be through living collections. In order to prevent extinctions and ensure healthy oak populations for the future, conservation efforts must integrate tailored *in situ* protection and management activities in coordination with genetically diverse *ex situ* collections of living trees. In response to the global threats facing oak species, The Morton Arboretum, Botanic Gardens Conservation International, and Fauna & Flora International launched the Global Oak Conservation Partnership in 2015. The Partnership works with and supports organisations in regional oak diversity hotspots such as Mexico and Indochina to develop and implement conservation strategies and initiatives that directly support the vision and mission of the Global Trees Campaign the only global conservation program dedicated to saving all of the world's threatened tree species. The goals of the Partnership are: (1) Oak species of greatest conservation concern are identified and prioritised for action; (2) The world's most threatened oak species are secure and their wild populations are recovering through conservation action; (3) *Ex situ* conservation collections of threatened oak species are scientifically informed, efficiently managed, genetically diverse and representative, and climate change resilient; (4) Partners, practitioners and the wider community are empowered to act for oak conservation. The Partnership will work towards these goals by creating a multi-disciplinary program of conservation activities driven and implemented by a multi-national group of partners. These partners include local and international stakeholders from botanic gardens, academic institutions, government agencies, conservation NGOs, protected area managers, land owners, local communities, policy makers and the private sector thus leveraging the expertise, knowledge and networks of a variety of sectors. Through coordinated action at a global scale, we will ensure that oak species diversity is secure for the future.

T1-15-03

PSESP (Plant Species with Extremely Small Populations) conservation in China

Weibang Sun
Kunming Botanical Garden, Kunming Institute of Botany, Chinese Academy of Sciences

The new concept of PSESP (Plant Species with Extremely Small Populations) developed for rescuing most of the globally threatened plant species was firstly promulgated in Yunnan province of China in 2005, and it has being globally recognized and accepted. PSESP is characterized by small remaining populations (far lower than Minimum Viable Population, MVP), restricted habitat, extremely high risk of extinction, and exposure to serious human disturbance. Scientifically, there is no MVP threshold for plant species, and therefore 5000 mature individuals with less than 500 mature individuals for each isolated population has been proposed for qualification as a PSESP in China, which was based on both a review of literature addressing global MVP, and conservation practices in China. For such species with an extremely high risk of extinction, rescuing populations together with comprehensive study is undoubtedly urgent. During the past 12 years, especially from 2011, several national and regional-level conservation strategies for actions have been planned (120 species have been targeted by the State Forestry Administration and National Development & Reform Commission, 62 species have been approved by Yunnan provincial government, and in total some 143 species in Southwest China are targeted for action). Meanwhile, activities such as field surveys, creating the in-situ conservation sites, propagation for both *ex-situ* conservation and population restoration, and PSESP germplasm banking, have all made great advances in some parts of China. Recently, with particular support from the National Natural Science Foundation of China (NSFC) and Ministry of Science and Technology of China (MSTC), several major PSESP projects have been fruitful. This presentation will take an overview of PSESP conservation programs throughout China, and will highlight the most current outcomes of our work on typical species such as *Craigia yunnanensis*, *Acer yangbiense*, *Manglietia ventii* and *Manglietiastrum sinicum*.

T1-15-04

Capacity building for red listing of plants

Malin Rivers
Botanic Gardens Conservation International (BGCI)

There are nearly 400,000 plant species in the world, but less than 7% of these have been given a global conservation status on the IUCN Red List of Threatened Species. Yes, there are further species assessed in national, regional, non-IUCN conservation assessments-but despite these we are still missing conservation status information for nearly half of the world's species. We therefore need to step up red listing for plants in order for us to know which species are in need of conservation action and how to prioritise this action. Botanic gardens are ideally placed to be involved in red list assessments. They are hubs of information on plants in terms of the biodiversity collection they host: living collections, herbaria, libraries, seed collections, *etc.* They are also centres of the human expertise on plants. Botanic gardens are already very active in the production of red list assessments on large, medium and small scales, and many of their staff, volunteers and research associates are involved in red listing. However, we need to better organise the red list effort for plants to ensure that no plant is forgotten. We therefore need a coordinated network of plant red list experts with the goal to enhance plant red listing-a community of practice for plant red listing. This community of practise can share knowledge, share expertise, share data, share tools, share training and share outputs. This approach to share information and experiences will ensure that best-practises are established and continue to evolve. The approach also needs to be cost-effective and efficient. Through a coordinated effort, plant conservation prioritisation through red listing will work towards the prevention of plant species extinction.

T1-15-05

Saving our charismatic megaflora: Conservation of the world's most threatened tree species in their natural habitats

David Gill, *Georgina Magin, Victoria Price*
Fauna & Flora International

The world's estimated 60,000 tree species are one of humanity's most important natural resources. Trees provide a huge variety of timbers, medicines and foods for millions of people and enrich our daily lives by shaping many landscapes and cultures. However, extensive habitat loss and degradation, unsustainable harvesting and climate change threatens at least 9,600 different species with extinction. For many species, the long-term future of their wild populations now requires urgent and targeted conservation action. However such action is often limited in scope and the effectiveness of conservation interventions *in situ* is often constrained by limited understanding on how best to address the threats faced by individual species. Furthermore, many of the people responsible for managing the sites where threatened tree species occur have never had the opportunity to develop the technical skills required for tree conservation. The Global Trees Campaign, a partnership between Fauna & Flora International and Botanic Gardens Conservation International was established in 1999 to prevent tree species extinctions in the wild and ensure their benefits for people, wildlife and the wider environment. This presentation describes examples from FFI's field programme for threatened trees, as a contribution to the Global Trees Campaign. Various approaches used in the field to reduce threats, ensure sustainable use and reinforce and reintroduce some of the world's most threatened tree species will be described. Lessons learned from individual case studies will be shared, before opportunities to catalyse greater action for the world's threatened species are put forward for discussion.

T1-15-06

Sud Expert Plantes Développement Durable: A multilateral programme toward the development, the scientific cooperation, the support to collections, the capacity building and the avocacy for plant biodiversity in the Global South

Stéphanie Ardila-Chauvet[1], *Jean-Pierre Profizi*[1], *Anshuman Rana*[1], ***Maïté Delmas***[2]
1. Institut de Recherche pour le Développement (IRD)
2. Museum National d'Histoire Nuturelle

In keeping with the 2011-2020 Strategic Plan adopted by the Tenth Meeting of the Conference of the Parties to the CBD in Nagoya, Japan in October 2010, France scaled up its biodiversity action with a new National Biodiversity Strategy (NBS) for 2011-2020. Target 16 of the NBS aims at strengthening international solidarity and ensuring greater mainstreaming of biodiversity into French development assistance. Target 17 of this plan intents at reinforcing green diplomacy and international governance for biodiversity. In this context, France launched in 2015, the Sud Expert Plantes Développement Durable programme (SEP2D), a 5 years multilateral programme aiming at sustaining the scientific communities in 22 francophone countries in West and Central Africa, the Indian Ocean and South East Asia. Focussing on the sustainable management of plant biodiversity, SEP2D is supported by the French Development Agency (AFD), the French Global Environment Facility (GEF) and the French Ministry of Foreign Affairs and International Development (MAEDI) as well as other public and private partners among which the "Institut de Recherche pour le Développement" (IRD) and the "Muséum National d'Histoire Naturelle" (MNHN). The programme strategy is guided by a collegial expertise on the following themes: Forestry & Mines, the Valorisation of plants for medicinal, cosmetics & agro-biodiversity purposes, Research & collections and Training & Awareness. The scientific research on plant biodiversity is a cornerstone of the sustainable development in the Global South. It enables the identification and the building up of potential innovations derived from these countries' rich plant diversity but also provide the National policy makers the tools to preserve their sovereignty. Furthermore, a valorisation of the research on plant biodiversity is an essential pre-requisite to the development of a strong scientific expertise. A particular attention is needed on the appreciation of the enrichment of the collective knowledge, the individual expertise, the know-hows, the data collecting, their analysis and their valorisation to which the scientific community contributes. SEP2D is making every effort to bolster partnerships and collaboration in this regard, building bridges between research, education and societal demands. It encourages the scientific community to work on applied topics, such as the major challenges associated with tropical plant biodiversity assessment and conservation. It aims to reinforce the scientific research on plant biodiversity and the collections' and the botanical gardens' activities in these countries. With this approach, SEP2D not only aims to rely on natural resources markets as leverages for economic development, but also envisions to encourage long term mutually benefitting partnership inclusive of every stakeholders needs and aspirations. SEP2D seeks to participate to the development of a scientific research that would also be inclusive of national, sub-regional, regional and international public policy issues. A science, based on this South-North and South-South partnerships, that will advocate the National authorities, public stakeholders, private sector operators and associations in order to meet societal demands.

T1-16: Pollination and breeding systems of endangered plant species

T1-16-01

Fly pollination and rarity in the genus *Ceropegia* (Apocynaceae)

Jeff Ollerton
University of Northampton

Pollination by flies (Diptera) has been important to the diversification and ecology of the flowering plants, but is relatively neglected and poorly understood in contrast to pollination by (arguably more charismatic, better known and popular) groups such as bees, butterflies and birds. Species within the genus *Ceropegia* (Apocynaceae) employ a strategy of temporarily trapping their small (usually less than 4mm in length) fly pollinators, only releasing them following pollination and attachment of pollinaria.

The genus contains approximately 200 species, all of which were, until 2000, CITES listed due to the trade in wild collected plants. In the wild *Ceropegia* species tend to form very low-density populations and many are rare and endangered. Recent research on *Ceropegia* has resulted in some significant new findings in relation to the diversity of pollinators exploited by the genus, and the functional ecology of interactions with those pollinators. In this talk I will assess what we understand of these plant-pollinator interactions in relation to what is known about rarity and distribution of *Ceropegia* species. The analyses will be framed within the context of a more accurate understanding of phylogenetic relationships between the major groups than has previously been available. *Ceropegia* species have evolved a specialised trapping strategy to exploit atypical, "non-standard" pollinators. This strategy is ecologically (and evolutionarily) successful despite generally low population densities, raising the interesting question that rarity and pollination system may be causally linked in this fascinating group of plants.

T1-16-02

Reproductive ecology and conservation challenges of sub-alpine orchids in Yunnan after a decade of drought

Zong-Xin Ren[1], *Hong Wang*[1], *Peter Bernhardt*[2]

1. Key Laboratory for Plant Biodiversity and Biogeography of East Asia, Kunming Institute of Botany, Chinese Academy of Sciences

2. Department of Biology, Saint Louis University

Yunnan, southwest China, at the eastern edge of the Himalayas, has a far higher diversity of temperate, terrestrial orchids compared to their better studied, congeners distributed through Europe and North America. However, rapid increases in ambient temperature and changes in precipitation rates are recorded in the Himalayan Region as symptomatic of Global Warming. Consequently, during the past decade of vernal drought, most populations of montane orchid species in Yunnan have experienced a dramatic retreat. To make the status of these species worse there remains a large gap between efforts to protect remaining populations vs. a lack of information on their reproductive ecology. We monitored the reproductive ecology of populations of sub-alpine orchids on mountains in northern Yunnan using a series of field and lab techniques including fluorescence microscopy, insect pollen load analyses, flower-insect morphometrics, and floral scent analyses (GC-MS). This produced new data sets that determined breeding systems, flower-pollinator interactions and seed viability analyses to represent species in eight genera (*Amitostigma*, *Calanthe*, *Cypripedium*, *Epipactis*, *Habenaria*, *Herminium*, *Orchis* and *Spiranthes*). Results showed that population sizes decreased in most species over 3-5 years due to spring droughts at higher elevations with a decline the frequencies of those plants producing flowering stems. Some species, e.g. *Cypripedium margaritaceum*, remained at a higher risk of regional extirpation than others. While all species studied were self-compatible but pollinators were needed to produce the greatest numbers of viable seeds. Pollinating insect species differed between Yunnan species vs. their European/North American congeners and highly specialized pollination systems were found in some *Cypripedium* species. The fecundity of some species appears limited by their pollinator reward systems (nectar or food mimesis), pollinator guilds and the desynchronization of the flowering periods of the orchids, their sympatric co-blooming flora and the emergence of their primary pollinators. Severe Spring drought delayed the flowering time of some *Cypripedium* species, delayed fertilization and arrested fruit maturation by winter time. Seed set rates in some multiple-flowering species were also inhibited by a combination of inbreeding depression, low temperatures, herbivory and anthropogenic stress.

T1-16-03

The reproductive biology of Asteraceae on Oceanic Islands: Coping with rarity

Daniel Crawford[1], *Gregory Anderson*[2]

1. University of Kansas

2. University of Connecticut

Asteraceae are well represented on oceanic islands, often with the highest number of lineages and/or the most endemic/native species in an archipelago. This indicates that the members of the family are not only good dispersers and effective colonizers but also have the capacity to radiate and speciate in the island setting. Several reproductive attributes of the family contribute to their capacity to cope with rarity at different stages of the evolution of a lineage in the island setting. Initially, the colonizing ancestors will be very rare and face two major challenges: lack of compatible mates and/or pollen vectors; and insufficient genetic diversity for radiation and diversification. Two of these challenges can be met by self-compatible (SC), self-pollinating colonizers. However, SC colonizers will have been derived from highly selfing source populations, and thus will bring with them little genome-wide heterozygosity. As a consequence, the colonizers may well have little potential for diversification. Self-incompatible (SI) colonizers would have higher diversity but would lack both compatible mates (especially if derived from 1 or few disseminules) and likely lack pollinators. There is another alternative: some of the large insular lineages of Asteraceae arose from SI colonizers, but there is growing evidence that their colonizers may have been able to set some selfed seed (i.e., leaky, or pseudo-self-compatible, PSC). Thus, the SI-derived colonizer could establish a sexually reproducing population with higher diversity. In addition, dominance relationships among S-alleles in the sporophytic SI system of Asteraceae could increase the number of compatible mates in small populations. And, Asteraceae can be pollinated by a variety of vectors, making it feasible for them to utilize new pollinators in the islands. Despite high SI, low levels of PSC may result in populations with mixed mating systems, providing reproductive assurance in the colonization of disturbed areas on islands. We will discuss why the breakdown of the SI locus and associated origin of the selfing syndrome, both often associated with the colonization of new continental ranges, are rare occurrences in island lineages. Furthermore, habitats are fewer on young islands, and regular disturbances on islands may bring differently adapted species into contact, with resulting hybrids. Asteraceae are particularly amenable to hybridization because of their generalist pollinators. For rare species, hybridization could either increase genetic diversity, introduce new variations in reproductive biology, or drive them to extinction. Dioecy on islands has been considered advantageous because it promotes increased outcrossing. However, if populations decrease, skewed sex ratios may reduce effective population sizes and eventually drive species to extinction. Research on Asteraceae endemic to the Juan Fernández Archipelago and the genus *Tolpis* from the Macaronesian archipelagos will be used as examples. In addition, the reproductive biology of Asteraceae in the Juan Fernández and the Canary Islands will be contrasted, to show the possible roles of abiotic factors in shaping differences.

T1-16-04

Pollination syndromes in a biodiversity hotspot

Kingsley Dixon

1. Curtin University

2. Missouri Botanical Garden

3. Kings Park and Botanic Garden

The southwest Australian biodiversity hotspot contains approximately 8,000 species comprising 6 endemic families and one endemic order. Twenty million years of genetic isolation from eastern Australia and bounded by the Indian Ocean to the west combined with no glaciation for the past 250 million years has resulted in a region where the processes of evolution have played out to their fullest. This is seen in the diversity of life forms and ecological attributes from acute levels of fire resilience, the greatest diversity of carnivorous plants on earth to the remarkable attributes developed by plants in the region to grow, in fact thrive, on some of the most infertile landscapes on earth. But perhaps the most remarkable evolutionary force has been development of an array of pollination syndromes rarely matched by any other single biogeographic region. With 17% of plants pollinated by birds and the highest level of sexual deceptive pollination syndromes in orchids, pollination has driven levels of evolutionary specialisation across a wide variety of groups. My presentation will review some of the latest research into pollination including resolving one of the first cases of cryptic speciation in the Orchidaceae.

T1-16-05

How do population size and pollinator availability affect reproductive success, mating system and population genetics of a red-listed plant, *Polemonium caeruleum* L. (Polemoniaceae)?

Marcin Zych[1]*, Justyna Ryniewicz*[1]*, Katarzyna Roguz*[1]*, Emilia Brzosko*[2]*, Ada Wróblewska*[2]*, Izabela Tałałaj*[2]*, Beata Ostrowiecka*[2]*, Edyta Jermakowicz*[2]*, Paweł Mirski*[2]

1. Botanic Garden, Faculty of Biology, University of Warsaw

2. Department of Plant Ecology, Faculty of Biology and Chemistry, University of Białystok

Polemonium caeruleum L. (Polemoniaceae) is a boreal plant species producing dichogamous and predominantly bee-pollinated flowers. Literature data show that occasionally the plant may be pollen limited and suggest that the species may exhibit mixed-mating system. We surveyed 15 *P. caeruleum* populations and, working in the field, over two years conducted a five-variant pollination experiment and studied pollinator abundance and identity. Our study populations showed various degrees of compatibility: from clearly self-incompatible (SI) to fully self-compatible (SC) ones. In most SI populations, however, self-pollination occurred via geitonogamous pollen-transfer rather than autonomous selfing, since isolated flowers (those with restricted pollinator access) hardly produced any seeds. Only the smallest of the surveyed populations, composed of <20 flowering individuals, exhibited significant pollen limitation. This population was also clearly self-compatible, producing significant seed set following induced (geitonogamous) self-pollination. To relate this data to the plant's population genetics we performed AFLP analyses for all studied populations, which will be discussed in the paper.

T1-16-06

How general is the relationship between fitness and abundance in wind-pollinated plants?

James Rodger[1]*, Cang Hui*[2]*, John Pannell*[1]

1. University of Lausanne

2. Stellenbosch University

Pollinator visitation, pollen receipt and seed set were shown to be generally lower in smaller plant populations in a now decade-old meta-analysis. This is a threat to rare plant populations but may make it easier to manage invasive ones. However, the conclusion of the previous meta-analysis that fitness is generally positively related to population size in plants only applies to animal-pollinated species, as insufficient studies on wind-pollinated species were available to form conclusions for this group. We will present a meta-analysis of the relationship between abundance and fitness in wind-pollinated species, and a comparison of the effect of abundance on fitness between wind- and animal-pollinated species. We will also assess results for wind-pollinated species for consistency with models of transport of pollen by wind. In a preliminary dataset, positive relationships between abundance and fitness were found in 17 out of 18 wind-pollinated species.

T1-17: The World Flora Online – a response by the international plant taxonomic community to the global strategy for plant conservation (two sessions)

T1-17-01

Introduction and progress on developing the World Flora Online

Peter Wyse Jackson[1]*, Pierre-André Loizeau*[2]

1. Missouri Botanical Garden

2. Conservatoire et Jardin botaniques de la Ville de Genève

In 2010, the updated Global Strategy for Plant Conservation (GSPC) of the U.N. Convention on Biological Diversity (CBD) included as its first target (Target 1) the need for "An online flora of all known plants." Its aim was to ensure that a coherent, comprehensive, accessible and authoritative baseline of knowledge would become available to support efforts being made worldwide to conserve plant diversity. Therefore, in January 2012 in St. Louis, Missouri, USA, representatives from the Missouri Botanical Garden, the New York Botanical Garden, the Royal Botanic Garden Edinburgh, and the Royal Botanic Gardens, Kew — all members of the Global Partnership for Plant Conservation, took the initiative to organise a new global project to achieve this target. First steps included a proposed outline of the scope and content of a World Flora Online (WFO), as well as a decision to create an international Consortium of institutions and organizations to collaborate on providing its content. The WFO project was subsequently launched in October 2012 in India, at an event held during the 11th Conference of the Parties (COP) to the CBD, where the COP also adopted a decision welcoming the WFO initiative. In January 2013, a Memorandum of Understanding (MoU) on the WFO was opened for signature. To date, 35 institutions and other organizations have signed the MoU and are participating in the project. A range of other institutions and organizations worldwide is also invited to participate in the WFO Consortium. The project is managed by a WFO Council which meets twice a year, supported by several expert Working Groups. The WFO aims to be an open-access, web-based compendium of the world's plant species. It will be a collaborative, international project, building upon existing knowledge and incorporating information drawn from published Floras, checklists and taxonomic and other revisions. It will also require the collection and generation of new information on poorly know plant groups and plants in unexplored regions. At its core is a taxonomic 'backbone' including a consensus classification of known plant genera and species, including accepted names and synonyms. Originally based on 'The Plant List', (version 1.1) (www.theplantlist.org), the taxonomic core referential of the WFO is being actively reviewed, moderated and updated by a growing network of taxonomic experts worldwide. Other data included are data sources, geographical distributions and information on the current status of each species. Where possible, existing keys and other identification tools are being incorporated, as well as vernacular names. Work had progressed to deliver the WFO through a web portal, using software donated by RBG Kew and adapted by the Missouri Botanical Garden and other partners. The project aims to launch this the first public, albeit incomplete, version of a World Flora Online in mid-2017. The project represents a major step forward in developing a consolidated global information service on the world's flora. It also provides an important forum and mechanism where the world's plant taxonomists can ensure that the results of their work are acknowledged and presented for use by the global community.

T1-17-02

Transforming hard copy to digital for the World Flora Online

***William Thomas**, Melissa Tulig*
New York Botanical Garden

We estimate that there are 350,000-400,000 plant species and that descriptions for as many as half of these are only available in hard copy. Therefore, one of the greatest challenges in achieving the goals of the World Flora Online will be to make available the huge amount of botanical information that is not yet digitized. The New York Botanical Garden Press has been publishing floras and monographs for the past 125 years and most of the treatments are not freely available online. As part of our commitment to the WFO project, we used the Flora Neotropica Monograph series to test the conversion of a print series to marked up digital treatments of the individual taxa contained across all volumes. Optical character recognition software was used to convert images of the publications to text. Then project staff reviewed and edited the output and parsed treatments into fields that comply with DarwinCore standards. We then proceeded to do the same for the Memoirs of The New York Botanical Garden. Combined, these publications contain over 22,000 treatments that are now freely available online.

T1-17-03

The e-Flora of South Africa: An alternative method to compiling an online flora as a contribution towards the World Flora Online Initiative

***Margaretha Marianne Le Roux**[1,2], Ronell Klopper[1,3], Janine Victor[1]*
1. South African National Biodiversity Institute
2. University of Johannesburg
3. University of Pretoria

Floristic accounts are useful products that provide taxonomic information for all plant species within a defined geographic region. Traditionally, these accounts were compiled by one to several contributors and published in hard copy volumes. These projects often took long to complete and soon became outdated with difficulty of publishing updated versions. However, with the availability of electronic tools and a developing digital environment, Flora compilation has embraced new and innovative ways to overcome some of these challenges. In South Africa, the first floristic account was compiled by Thunberg in Flora Capensis in 1823. This was followed by Harvey and Sonder's Flora Capensis between 1859 and 1933, which was completed through contributions from multiple authors. In 1955 the Flora of southern Africa (FSA) project commenced with the aim to replace the outdated *Flora Capensis*. This project was never completed and as an alternative, a regional/provincial conspectus programme was initiated. To date, three volumes covering *ca.* 67% of South Africa's flora has been published (Flora of the northern provinces, 1996; the Greater Cape Floristic Region: core Cape Flora, 2012, and the extra Cape Flora, 2013), with the Flora of the Free State, Eastern Cape, KwaZulu-Natal and the Nama Karoo to follow. With the development of the updated Strategy for Plant Conservation (2011–2020), South Africa has committed to contribute a country-level Flora towards the World Flora Online (WFO) initiative (Target 1). After considering the remaining time in which to make such a contribution and the availability of published information (FSA volumes, published regional/provincial conspectuses and published taxonomic revisions), it was decided to compile an electronic Flora using existing information and following an aggregator portal approach. Available information for this endeavour covers *ca.* 85% of the South African flora. The remaining 15% of the flora will be dealt with in the conspectuses that are currently being compiled. The aggregator portal approach involves the use of published information by acquiring permission from the copyright holders, digitising the material if it does not exist in electronic format, mining the required data from the publication and aggregating it into a database from where it will be made available online in open access. Two disadvantages of this method are: (1) multiple descriptions for each taxon may be collected; (2) the format of descriptions will be inconsistent within and among species. These disadvantages may temporarily be overcome by presenting the most recent or appropriate description first and by displaying the data in a standard template. The main advantages of this method are: (1) existing information will be aggregated into one central portal and; (2) an e-Flora for South Africa will be completed by 2020. An overview of the history of Flora compilation in South Africa will be presented, along with the method currently followed to compile the e-Flora for the country. Advantages and disadvantages of this method and progress to date will be highlighted.

T1-17-04

The Flora of North America: Moving from hard copy to digital information sharing

Geoffrey Levin
Canadian Museum of Nature

The Flora of North America North of Mexico (FNA) is a 30-volume work that includes taxonomic treatments of all the native and naturalized plants growing in the region. All taxa are described and included in dichotomous keys, distributions of all species and infraspecific taxa are mapped, and about 20% of species are illustrated with line drawings prepared specifically for FNA. Producing FNA has created an international community of authors, reviewers and editors, integrating broad-scale taxonomic knowledge with regional floristic expertise, and fostering further collaboration. By providing keys and technical descriptions for numerous rare taxa and introduced species, FNA is a vital tool for botanists and others concerned with conserving the continent's flora. Soon after volumes are published, they are made available electronically through the efloras website and data are shared with JStor's Global Plants. As we near publication of the final volume in 2018, we are preparing to transition to a new website that moves beyond replication of the printed information. This new site will allow users to use either dichotomous or interactive keys and to search on any combination of characters, including morphology, phenology, habitat, and geographic distribution. We envision additional data sharing, including with the World Flora Online. The site will also include tools for moderated and versioned authoring and editing of treatments, allowing the Flora to remain up-to-date by continuously incorporating results of new research.

T1-17-05

World Flora Online—Technology & techniques to create a comprehensive data portal for all plants

Mark Watson[1], *Chuck Miller*[2], *William Ulate*[2]
1. Royal Botanic Garden Edinburgh
2. Missouri Botanical Garden

The World Flora Online (WFO) is an international collaborative project aiming to bring together floristic and monographic information on all the world's plants in fulfilment of Target 1 of the Global Strategy for Plant Conservation 2020: 'An online Flora of all known plants'. This builds on The Plant List (www.theplantlist.org) - a working list of all known species of vascular plants and bryophytes. The rationale of WFO is to meet the need for baseline information to support plant conservation and sustainable development, and so the main target audience is both those engaged in these activities and the taxonomic community providing authoritative information. The technical aspects of how the WFO will be achieved is handled by a Technical Working Group established by the WFO Council, and includes people with a range of skills from computer programmers to biodiversity informaticians to taxonomists. The requirements of the system were established through a Use Case study, where stakeholders and their needs were documented and prioritised. The data elements needed to fulfil the prioritised use cases were characterised and aligned with Darwin Core, Dublin Core and Plinian Core - new data elements with definitions were created only when absolutely necessary. The database and internet portal of the eMonocot system (http://e-monocot.org) was used as the basis for the WFO system. The WFO comprises of two logically separate components: the Taxonomic Backbone (nomenclature and classification) and the Descriptive Content of previously published floristic and monographic treatments linked to the Backbone. Contributors to either of these components send in their data conforming to the defined data standards and these are harvested into the WFO system. A crucial step in the ingestion is the resolution of and linage to the taxonomic names used in WFO and the contributor dataset, and for this WFO has established new identifiers for all names from subspecies to order and for all vascular plants and bryophytes. A demonstration portal (http://demo.worldfloraonline.org) is being used to test the system, and a production portal will be launched at IBC.

T1-17-06

Plants of the World Online Portal (POWOP), how the Royal Botanic Gardens, Kew will contribute to the World Flora Online

Abigail M Barker, *Daniel Borg, James Crowe, John Iacona, Don Kirkup, Nicola Kuhn, Nicky Nicolson, Isabelle Sims, Robert Turner*
Royal Botanic Gardens, Kew

In 2015, the Royal Botanic Gardens, Kew committed to delivering nine strategic outputs which aim to disseminate Kew's scientific knowledge of plants and fungi to maximize its impact in science, education, conservation policy and management. The Plants of the World Online Portal (POWOP), is one of those outputs. POWOP will enable users to access information on all the world's known seed-bearing plants by 2020. A key function is to ensure that Kew's floristic data can be easily harvested and ingested by the World Flora Online (WFO) portal enabling Kew to support the Global Strategy for Plant Conservation (GSPC) Target 1 2020 "An online Flora of all known plants". POWOP is a single point of access for authoritative plant species information, providing a multi-dimensional catalogue of plant life, including information on identification, distribution, traits, threat status, molecular phylogenies and uses. It uses Kew's extensive data resources including its Floras, alongside images from the digitisation of the collections. The portal has been designed to maximise accessibility and enables dissemination of plant information to users via mobile, tablet or desktop computer. The codebase is open source and available for use by partners in other countries to set up their own portals, which can be linked through to Kew and the WFO. Kew will help to enable this process through knowledge transfer and support, building on expertise in institutions across the globe to further develop a network of botanical data hubs. Ultimately, this will lead to a resource with global coverage which can empower and inform citizens, policy makers, conservationists and farmers everywhere, about the importance of plants and fungi to everyday life. POWOP will provide Kew's flora data to the WFO in Darwin Core Archive format linked by persistent identifiers from the International Plant Names Index (IPNI) and The Plant List. POWOP was launched in mid-November 2016 with a focus on the Floras of Tropical West and East Africa. It is these Floras which will form the basis of Kew's initial contribution to the WFO.

T1-17-07

The South African National Plant Checklist: History and utility

Ronell R Klopper, *M Marianne Le Roux, Brenda Daly, Janine E Victor*
South African National Biodiversity Institute

A national species checklist is an essential resource that forms the foundation for a range of research and biodiversity-related activities. South Africa has had the benefit of access to an updated national and regional plant checklist for many decades, which has underpinned botanical research in the country. The South African National Biodiversity Institute (SANBI) is mandated by the South African Government to co-ordinate taxonomy in South Africa, which includes maintenance, updating and dissemination of country-level checklists. Although at present the checklist only includes South African species, in the past it covered the entire Flora of Southern Africa (FSA) region (Botswana, Lesotho, Namibia, South Africa and Swaziland). Additionally, other southern African country-level plant checklists (for Angola, Botswana, Lesotho, Malawi, Mozambique, Namibia, Swaziland, Zimbabwe and Zambia) were compiled and published under the auspices of the Southern African Botanical Diversity Network (SABONET) lead by SANBI. Furthermore, SANBI was a key partner in compilation of the African Plant Checklist and Database. This monumental project was a collaboration between SANBI and Conservatoire et Jardin Botaniques de la Ville de Genèva, Switzerland, and later Tela Botanica and Missouri Botanical Garden. It initially combined two major sources of data: the Flora of Southern Africa dataset and the Lebrun & Stork Enumération des plantes à fleurs d'Afrique tropicale publications covering tropical Africa. This produced a global first inventory of the flowering plants of sub-Saharan Africa. Subsequent inclusion of names for angiosperms in North Africa and Madagascar in the online database has seen this resource grow to a continent-level catalogue. Plant checklists for South Africa and the FSA region were originally published in hardcopy and later made available in pdf format. However, the checklist is now available online as both a downloadable spreadsheet and in a searchable platform. Checklist and specimen data are housed in a BRAHMS database that enables the online platform to link the checklist to specimen records held in the SANBI herbaria [Compton Herbarium (NBG and SAM), Cape Town; KwaZulu-Natal Herbarium (NH), Durban; National Herbarium (PRE), Pretoria]. Future developments will expand coverage to include specimen level records from other South African herbaria. This is in line with global efforts to mobilise plant biodiversity data. Additional descriptive data, images and distribution maps will be linked to the checklist as part of the eFlora of South Africa project. This will enable country-level input to the World Flora Online Initiative that aims to fulfil Target 1 for the 2020 Global Strategy for Plant Conservation. This presentation will briefly outline the history of compiling and disseminating the South African National Plant Checklist. Information will be provided on the family-level classification followed, and the current processes and procedures applied in terms of data management, prioritisation, and quality control. The application and utility of the checklist as backbone for several foundational and applied biodiversity disciplines, such as herbarium specimen curation, conservation assessments, and biodiversity policy and planning activities, will be discussed. In addition, the importance of an updated national plant checklist during compilation of an eFlora for South Africa will be highlighted.

T1-17-08

Contribution to the WFO of completely digitized plants natural history collections: The case of Paris Muséum national d'histoire naturelle's vascular plants collection

Visotheary Ung, *Marc Pignal, Maïté Delmas, Serge Muller, Marc Jeanson, Germinal Rouhan, Véronique Schäfer, Thomas Haevermans*
1. Muséum national d'histoire naturelle. Institut de Systématique, Évolution, Biodiversité (ISYEB) UMR 7205 Centre national de la recherche scientifique - Muséum national d'histoire naturelle - École Pratique des Hautes Études – Université Pierre et Marie Curie, Sorbonne Universités, Paris Cedex 05, France
2. Muséum national d'histoire naturelle, Direction des Collections. Sorbonne Universités, Paris Cedex 05, France
3. Tela Botanica
4. Muséum national d'histoire naturelle, Direction des Relations Internationales. Sorbonne Universités, Paris Cedex 05, France

Large herbaria, i.e. global plant knowledge depositories constitute a suitable tool to address targets one and two of the Global Strategy for Plant Conservation: obtaining a complete list and conservation statuses for all known plant species. Paris Herbarium, one of the largest in the world has complete the digitization of its vascular plants specimens holdings, with more than 6,000,000 specimens, and 170,000 taxa. All images are freely available online. We have published a data paper in Scientific Data, where we described and explained precisely the whole digitization process. Besides, we have developed a citizen-science program: Les Herbonautes (lesherbonautes.mnhn.fr, Recolnat program - ANR11-INSB-0004) to gather additional data through its platform to complete our dataset as well as involving a large community of scientific volunteers. This ever growing amount of data could be integrated into the WFO portal to help accessing it while contributing to document plant biodiversity.

T1-17-09

Flora of North America: Nomenclature review

Kanchi Gandhi
Harvard University Herbaria

Since the beginning of the 20th Century almost all floras have relied upon nomenclatural and/or taxonomic indices/databases of various magnitudes serving either global (IPNI and Tropicos), or regional (e.g., *Fl. Eur.*), or individuals (e.g., *Selaginella* L.) for nomenclatural assessments. Each of the indices presents some problems, e.g., showing: a) misapplied names as later homonyms; b) prosynonyms as validly published. As the International Code of Nomenclature became stabilized since the 1930s, application of the Code enhanced nomenclatural accuracy asserting authorship, validity, legitimacy, and priority of names. For the 30-volume Flora of North America (FNA; 1993-2018?), we are making an effort to check the protologue of the included names to achieve nomenclatural accuracy, a worthy goal, which FNA has been seeking. Such verifications have led to corrections of even recently published names, e.g., "*Leptosiphon montanus* (Greene) J.M. Porter & L.A. Johnson (2000)" was found to be invalid because of the failure to provide a direct reference to the basionym. Occasionally, what have been perceived to be validly published names were found to be invalid making it time consuming to ascertain the first place of valid publication, e.g., "*Canotia* Torr. (1857)", originally proposed as a provisional name ("... there can be but little doubt of its constituting an undescribed genus, we may bestow upon it the provisional name of *Canotia*"), is believed to have been first validated by Asa Gray (1861). However, the boundary between a taxonomic doubt vs. a provisional name or a polite way (to express oneself that was all but customary in some periods/countries) is fluid, e.g., *Lespedeza angustifolia* var. *brevifolia* Britton 1893 ("... specimens of a peculiarly short-leaved form ... may claim recognition as var. *brevifolia*"); the preceding remark is considered to be polite. For apple, because of the uncertainty of the acceptance of the proposal to conserve the name *Malus domestica* Borkh. (1803, nom. illegit.) over *M. pumila* Mill. (1768), the latter name has been recognized in FNA. Citations of synonymy, if any, usually include legitimate names above the rank of forma; exceptions are basionyms at any rank and illegitimate homotypic replaced names. For well-established illegitimate names, if accepted for FNA, proposals were made for conservation, e.g., *Heteromeles* M. Roem., nom. cons., *Prunus serotina* Ehrh., nom. cons. If the rank of a name was uncertain in the original publication, the name was cited as unranked, e.g., *Zizia* [unranked] *Taenidia* Torr. & A. Gray (1840); the authors applied the sign § for both subgenus and section. Furthermore, for each genus name, etymology is provided; in this regard, pre-Linnaean, Linnaean, and post-Linnaean literature has been consulted; however, for a few names, the given or intended etymology is uncertain, e.g., *Causonis* Raf., *Nekemias* Raf., and *Oenothera* L. When nomenclatural inaccuracies were found in the published volumes, appropriate corrections have been made in the online versions.

T1-17-10

E-Flora India initiative and prospects

Paramjit Singh
Botanical Survey of India

Since 1990, Flora India has published its products as family and genus treatments in terms of 10 volumes and 27 Fascicles. An e-Flora will complement printed formats with electronic editions, offer unlimited access and instant updates, and increase the cost-effectiveness substantially by streamlining the production work flow. It will allow structured data entry, interactivity, multimedia, and enhanced accessibility. Moreover, it will also allow collaboration between several e-Flora projects (e.g., Australian eflora initiative and World Flora Online Initiative), users of taxonomic data, and other databases. This increases the possibilities for innovative scientific co-operation, also between other research fields and attracts worldwide contributors using services offered by the e-Flora. An overview is presented of available e-taxonomic products and ongoing projects contributing to Flora India. This is presented in the context of a strong plea to strengthen the implementation of state-of-the-art e-taxonomy tools to speed up the generation and publication of Flora India information. The increasing array of electronic taxonomy tools available for the elaboration and dissemination of floristic information has brought many advantages. It enables a shift from the traditional Flora concept as a static, printed account to a dynamic and interactive format, allowing for rapid updating and multiple uses of information. It provides taxonomists with means for instant interactive and remote co-operation, including continuous processes for evaluating preliminary results by peers and updating existing information. It also allows taxonomists to forge better links with their user-communities by making the products of their research more tailor made and accessible via internet. In order to strengthen and speed up Flora India related activities, that the Flora India community needed to adopt a more pragmatic and flexible attitude—flexible in terms of formats and publication strategy, and standards of robustness and confidence in the results. The Flora India project has recently gained new momentum by starting a new website (http://efloraindia.nic.in), by adopting e-taxonomy tools and by joining a broadly supported initiative that wants to promote the adoption of richly interactive and truly collaborative systems for the production and presentation of floras. Flora India work can be strengthened by increasing its global accessibility, facilitating efficient remote collaboration, making use of databases to safeguard data, changing the work flow in data preparation and presentation, and creating institutional commitment.

T1-18: Pollination by non-flying mammals

T1-18-01

When flowers smell cheesy: An investigation into functional floral traits of African and Australian mammal-pollinated proteas (Proteaceae)

Sandy-Lynn Steenhuisen
University of the Free State

Mammal-pollinated plants commonly produce inflorescences that emit very strong odours. Evolutionary shifts from bird to non-flying mammal pollination in the African plant genus *Protea* (Proteaceae) are most notably associated with modifications in inflorescence size and olfactory cues. Using headspace sampling and gas chromatography-mass spectrometry of inflorescences and dissected floral parts, we compared the emission rates and chemical composition of floral scents between several species from each of the mammal, bird and beetle pollination systems in this genus. Bird-pollinated species with larger tall colourful inflorescences emitted small amounts of "green-leaf" volatiles and benzenoid compounds, including benzaldehyde, anisole and benzyl alcohol. Beetle-pollinated species produce shorter and more open inflorescences that emit fruity scents that are more complex and on average 10-fold greater in the amount of scent emitted than odours sampled from bird pollinated species. These fruity scents are dominated by linalool, other monoterpenes and a variety of benzenoids. Floral scents of small dull-coloured non-flying mammal pollinated *Protea* species shared more similarities with those of beetle-pollinated species, with nectar emitting cheesy, buttery odours (diacetyl and acetoin in combination with dimethyl disulphide, several ketones and alcohols). Numerous choice tests with common rodent pollinators revealed that these mammals are significantly attracted to certain concentrations of ethanol and other fermentation volatiles, and preferred the sour-milk/cheesy odours of nectar from a rodent-pollinated *Protea* species over that of a solution of the same sugar composition, indicating that floral odours play a functional role in pollinator attraction. Remote cameras have revealed that these species are visited not only by a variety of rodents, but also elephant shrews and small carnivorous mammals such as mongooses and genets. As mammal-pollinated *Banksia* and *Grevillea* species in Australia also attract pollinators with incredibly strong floral odours, current collaborative work and preliminary results on the evolution of floral traits in the Proteaceae family across Africa and Australia will be discussed.

T1-18-02

A unique case of floral specialisation for pollination by elephant-shrews

Petra Wester[1,2], *Steven D. Johnson*[2]

1. Institute of Sensory Ecology, Heinrich-Heine-University, Düsseldorf

2. School of Life Sciences, University of KwaZulu-Natal, Pietermaritzburg

Pollination of plants by non-flying mammals (therophily) is a rare phenomenon. In South Africa, therophilous plant species investigated thus far are pollinated by mice (Rodentia) and sometimes secondarily by non-related elephant-shrews (Macroscelidea). In this study we investigated the pollination ecology of the holoparasitic South African endemic *Hyobanche atropurpurea* (Orobanchaceae). Its long-tubed and black flowers are visited most frequently by long-tongued elephant-shrews (*Elephantulus edwardii*) and to a lesser extent by short-tongued mice (mainly *Acomys subspinosus*). Because of the close mechanical fit between the long-tubed flowers and elephant-shrews' snout and tongue morphology, *H. atropurpurea* appears to be adapted to elephant-shrews as pollinators. Elephant-shrews carry much more pollen than mice in the fur around their snouts and in their scats. Extensive observations, video monitoring and pollen transfer experiments showed that elephant-shrews contact the plants' reproductive organs during almost every flower visit whereas flower visiting mice rarely make contact with the stamens and stigma. Compared to mice, the elephant-shrews transferred pollen more often and in significantly larger numbers to the stigma of the flowers. Moreover, flowers were often damaged or eaten by mice, but never by elephant-shrews. These results suggest that *H. atropurpurea* is specialised for pollination by elephant-shrews.

T1-18-03

Flying and non-flying mammalian pollinators of *Mucuna macrocarpa* (Fabaceae) and their effect on pollination

Shun Kobayashi, *Tetsuo Denda, Masako Izawa*

University of the Ryukyus

The genus *Mucuna* (Fabaceae) has a pantropical distribution and comprises approximately 100 species. Most species of *Mucuna* have inflorescences hanging down from vines. The stamens and pistil of the *Mucuna* flower are completely concealed within the keel-shaped petals (carina). Exposure of the reproductive organs within the carina is termed "explosive opening." It has been reported that the agents that trigger this step are specific bats or birds. Previous studies have suggested that the animals that open the flowers may also be the major pollinators of *Mucuna*, although there is no experimental evidence. This study targeted *Mucuna macrocarpa,* which is an evergreen vine distributed from Southeast Asia to Kyushu, Japan. The objectives of this study were (1) to confirm the necessity of the explosive opening step for successful pollination in *M. macrocarpa*, and (2) to identify the explosive openers in several habitats of *M. macrocarpa* with/without fruits bats. Bagging and crossing experiments indicated that *M. macrocarpa* was a self-compatible species but never sets fruit without explosive opening. These results demonstrate that explosive opening is indispensable for successful pollination in *M. macrocarpa*. A video camera trap survey showed that explosive openers were Ryukyu flying foxes (*Pteropus dasymallus*) on Okinawa (subtropical, Japan), whereas in the regions without fruit bats, Japanese macaques (*Macaca fuscata*) and Japanese martens (*Martes melampus*) in Kyushu (temperate, Japan), and red-bellied squirrels (*Callosciurus erythraeus*), Formosan striped squirrels (*Tamiops maritimus*), and masked palm civets (*Paguma larvata*) in Taiwan (subtropical) were openers. Among these, Ryukyu flying foxes, Japanese macaques, and red-bellied squirrels were effective openers in each region. When these mammals opened a flower, many pollen grains adhered to their body surface, suggesting that the explosive openers are the effective pollinators of *M. macrocarpa*. These explosive openers include both flying and non-flying mammals; however, with the exception of Japanese macaques, their explosive opening behaviors are common. The similar behaviors regardless of the difference in species may ensure the pollination success of this plant in each region. In terms of the flying ability of the opener and its effect on pollination, it is predicted that there are differences in flower visiting height and locomotory ability. Ryukyu flying foxes visited flowers at a height of 2 m or higher, whereas Japanese macaques and red-bellied squirrels visited flowers from ground level to the canopy. Consequently, fruits were observed only on the upper vines on Okinawa where the flying fox is the only opener. In contrast, fruits were observed from the ground to canopy in Kyushu and Taiwan where non-flying mammals were the openers. In addition, previous studies have shown that flying foxes have the largest home range size among the three effective openers, whereas that of squirrels is only one-fiftieth that of flying foxes. Furthermore, flying foxes can move more directly between feeding sites regardless of any terrestrial barriers. These characteristics may affect pollen transfer distances. Therefore, regional differences in the pollination of *M. macrocarpa* can be attributed to the flying ability of the explosive opener.

T1-18-04

Rodent responses to volatile compounds provides insights into the function of floral scent in mammal pollinated plants

Keeveshnee Govender, *Steve D Johnson*

School of Life Sciences, University of KwaZulu-Natal, Scottsville, Pietermaritzburg 3209, South Africa

Flowers pollinated by mammals such as bats and rodents have evolved in dozens of plant families. Volatiles are considered to play a key role in the attraction of mammals to these flowers. Volatiles that attract bats to flowers have been identified in previous studies, but the chemical ecology of rodent-plant interactions has scarcely been investigated. Volatile signals are likely important in rodent pollination systems because rodents are generally nocturnal and rely heavily on olfaction to locate food. The aim of this study was to identify compounds that may play a functional role in the attraction of rodents to flowers. Eighteen volatile compounds were identified in headspace samples taken from rodent-pollinated flowers in southern Africa and used in experiments that involved four native rodent species (*Aethomys namaquensis, Rhabdomys pumilio, Rhabdomys dilectus* and *Mastomys natalensis*). Compounds were tested individually in unrewarded y-maze experiments to determine their attractiveness to rodents. In addition, associative conditioning experiments using artificial nectar as a reward were conducted to determine the learning abilities of rodents when their food is associated with particular volatiles. Rodents generally responded positively to aliphatic compounds (ketones, esters and acids) and responded negatively to aromatic compounds and a sulphide compound that is known to be attractive to bats. Associative conditioning generally resulted in little change in the attractiveness of compounds. It was concluded that oxygenated aliphatic compounds, particularly ketones and esters, are key attractants of rodents to flowers adapted for pollination by these animals.

T1-18-05

Floral visitation and reproductive success in the rare *Grevillea obtecta* (Proteaceae)

Trevor Edwards, Scott Tunbridge, Rufaro George Kanjere, Kevin Farnier, Kylie Robert, Simon Watson, ***Susan Hoebee***
La Trobe University

In Australia vertebrates are ancient and important pollinators, birds occupy diurnal niches and small mammals prevail at night. While these pollinators seek similar rewards they rely on different cues. This study examined pollination of *Grevillea obtecta* which belongs to a clade of mostly ornithophilous species however, unlike most of its relatives, it produces a strong scent that also attracts nocturnal marsupials. Flowers produced unpredictable volumes of high quality nectar in dense, partially hidden inflorescences. Bird and mammal visitors were recorded using motion-triggered cameras. Antechinus and brush-tailed phascogale were the main nocturnal visitors while yellow-faced honeyeaters pollinated during the day. Excluding vertebrate pollinators reduced fruit set significantly, from 8.4% to 1.4%. The flowers of *G. obtecta* are yellow-pink to red-brown and mostly reflect long wavelengths of light, consistent with its close relatives. However, floral scent is unusual within the clade and GCMS revealed a mix of 31 compounds dominated by camphene (35.89%) and limonene (10.82%). We conclude that in *G. obtecta,* the separate pollinator suites (day and night) may be interpreted as double specialisation because the nocturnal and diurnal cues evolved independently and specialisation does not drive fitness trade-offs between these floral traits.

T1-18-06

Bees, birds, bats–and mice? Pollination syndromes and floral traits in Merianieae (Melastomataceae)

Agnes S. Dellinger[1]*, Silvia Artuso*[1]*, Diana Fernández Fernández*[5]*, Darin S. Penneys*[2]*, Fabián Michelangeli*[3]*, Marcela Alvear*[4]*, Frank Almeda*[4]*, Jürg Schönenberger*[1]
1. University of Vienna
2. University of North Carolina Wilmington
3. New York Botanical Garden
4. California Academy of Sciences
5. Instituto Nacional de Biodiversidad

Bee buzz-pollination constitutes the most widespread pollination system in the tropical plant family Melastomataceae. While bee pollinated species reward their pollinators only with pollen, approximately 3% of the ca. 5,500 Melastomataceae species have evolved additional reward systems (nectar, food bodies) and are pollinated by hummingbirds, bats and passerine birds. Furthermore, rodents have been observed visiting flowers in three cloudforest species of the Neotropical tribe Blakeae; their role as pollinators is under debate, however. Recent observations on one species *(Meriania sanguinea)* in the also Neotropical tribe Merianieae suggest another incidence of flower visitation and possible pollination by rodents in the family. Flowers are red, pendent, produce nectar and emit strong, sweet scent. Irregular diurnal visitors are hummingbirds while rodents have been detected in nocturnal video observations taking up nectar from the flowers. Following the concept of pollination syndromes, which assumes a convergence of floral phenotypes to the species' most important functional group of pollinator, rodent pollinated species are usually inconspicuous in colouration and emit yeasty or cheesy scent and thus stand in stark contrast to floral traits observed in *M. sanguinea*. Detailed floral morphological studies of 59 Merianieae species including representatives of all pollination systems place *M. sanguinea* within the hummingbird pollination syndrome. However, the species differs from other hummingbird pollinated species in inflorescence architecture, stamen structure and stamen arrangement, mechanism of nectar secretion and scent. Further analyses of pollen transfer, nectar and scent profile are planned to clarify if rodents constitute effective pollinators or rather are occasional secondary pollinator or mere nectar thieves.

T1-19: Ethnobotany of mountain regions

T1-19-01

Traditional knowledge in the Caucasus - changing knowledge patterns in Georgia after the end of soviet occupation.

Rainer W Bussmann[1]*, Narel Yaroslava Paniagua Zambrana*[2]*, Shalva Sikharulidze*[3]*, Zaal Kikvidze*[4]*, David Kikodze*[3]*, David Tchelidze*[3]*, Ketevan Batsatsashvili*[3]
1. Missouri Botanical Garden
2. Herbario Nacional de Bolivia
3. Institute of Botany and Bakuriani Alpine Botanical Garden, Ilia State University
4. 4-D Research Institute, Ilia State University

The territory of modern day Georgia has been continuously inhabited since the early Stone Age, and agriculture was developed during the early Neolithic era. In Georgian the name of the country is "Sakartvelo", and "Georgia" is semantically linked to Greek meaning "agriculture". Due to its long tradition, agriculture in Georgia is characterized by a great diversity of landraces, and endemic species of crops. These show a high level of adaptation to local climatic conditions and often high disease resistance. Fieldwork was conducted in Khevsureti, Khevi, Samtskhe, Javakheti, Tusheti, Svaneti, and Racha in July-August 2013, July-August 2014, and September-October 2015. Interviews using semi-structured questionnaires were conducted with over 175 participants after obtaining their oral prior informed consent. Our research indicates that while traditional crops like wheat, barley and rye have almost disappeared, a large number of species is traditionally also grown in home gardens, and a large part of the wild flora is used for food, medicine and cultural purposes. However, the related traditional knowledge is mainly held in the generation above 50, while younger people have started to loose traditional knowledge. No gender differences in knowledge could be observed. Overall, home gardens serve mostly as source for food, while wild plants are especially important fro medicine, and traditional pickles and jams. The main reason for genetic erosion of ancient crop varieties is the demographic decline in mountain regions due to harsh economic conditions and lack of modern infrastructure. The shift from ancient cultivars to modern high-yielding crops such as maize and potato, began in mountain villages after the end of Soviet occupation, when local inhabitants who had been forced to the lowlands, returned to their original villages.

T1-19-02

Changing markets – medicinal plant ethnobotany in the Andes of Bolivia, Peru and Colombia

Narel Yaroslava Paniagua-Zambrana[1], *Rainer W Bussmann*[2], *Robbie Hart*[2], *Laura Araceli Moya Huanca*[1], *Carolina Ronero*[2]

1. Herbario Nacional de Bolivia, Universidad Nacional de San Andres

2. William L. Brown Center, Missouri Botanical Garden

Given the importance of local markets as a source of medicinal plants for both healers and the population, literature on market flows and the value of the plant material traded is rather scarce. This stands in contrast to wealth of available information for other components of Andean ethnobotany. The present study attempts to remedy this situation by providing a detailed inventory of medicinal plant markets in the La Paz-El Alto (Bolivia); Lima, Trujillo and Chiclayo (Peru) and Bogota (Colombia). Both species composition, and medicinal applications, have changed considerably over time. From 2001-2015, semi–structured interviews were conducted with hundreds of plant vendors in order to elucidate more details on plant usage and provenance. The results of the present study were then compared to previous inventories of medicinal plants in the region, as much as available, to elucidate changes over time and impact of interview techniques. Over the years we we encountered and documented over 800 plant species. This indicates a great wealth of ethnobotanical knowledge in the Andean region of South America. All markets show a considerable change over the last few decades, with new species being introduced into the market chain, and other species being replaced. In course of the present study it became apparent that even well known species might often be replaced by other apparently similar but botanically unrelated species due to environmental and market forces. The present data indicate that, while the floristic composition of often remained relatively constant over the last decades, the number of indications for which certain species were used increased tremendously, and that profound differences exist even between markets in close proximity. The dramatic increase in previously not used species used per indication might pose serious risks for consumers. We found serious problems due to species replacements. Even plants that have a well established vernacular name, and are easily recognizable botanically, can be replaced by other species that can pose a serious health risk. Vendor education and stringent identification of the material sold in public markets are clearly needed.

T1-19-03

Ethnobotany of climate change in the greater Himalayan Region: a 3000 km transect

Robbie Hart[1], *Jan Salick*[1], *Zhendong Fang*[2], *Suresh K. Ghimire*[3], *Sangay Dema*[4], *Hassan Sher*[5]

1. William L. Brown Center, Missouri Botanical Garden

2. Shangri-La Alpine Botanical Garden

3. Central Department of Botany, Tribhuvan University

4. National Biodiversity Centre, Ministry of Agriculture and Forests

5. Center for Plant Sciences and Biodiversity, University of Swat

The Himalaya are experiencing rapid climate change: temperature increases, changes in precipitation, and changes in permanent snows and glaciers. The region is a center of temperate biodiversity and includes plants of utilitarian and cultural value to Himalayan peoples. Our Himalayan subnetwork of the Global Observation Research Initiative in Alpine Environments (GLORIA) monitors climate change, alpine vegetation, and ethnobotany across a 3000km transect. The subnetwork spans the greater Himalayan region from Swat Valley in Northern Pakistan, through the Manang, Langtang and Kanchenjunga areas of central and Western Nepal and the Jomolhari and Tampela areas of Bhutan to the Hengduan Mountains in southwest China. At each monitoring area, we sample four mountain summits on an elevational gradient from the treeline ecotone to the uppermost limit of plant life – in these areas 4000-5000 masl. To this basic methodology, which we share with other researchers in GLORIA, we add ethnobotanical interviews to determine alpine plant uses. Baseline data from all sites in our Himalayan transects show species diversity following the steep east-west precipitation gradient, with the easternmost sites, in "Medicine Mountains" of China, the most species-rich. These sites also exhibited the greatest percentage of medicinal and aromatic alpine plants of any Himalayan sites. At these sites – the three from which we've completed both baseline and resurvey data – useful plant diversity, species composition in general, and response to climate change over the seven-year period between surveys were all structured on a strong North-South aspect gradient. Understanding the importance of topography can inform our predictions for climate change effects on species composition in this biodiverse and culturally valuable flora. Across the entire subnetwork, mountain regions show high endemism. Many of the plants sampled, including indicator species, are ethnobotanically important. For these high alpine useful plants, we anticipate climate change threats both from increasing temperatures and from competition with lower-altitude plants extending their range. Indigenous Himalayan communities in each of the regions, already coping with the direct effects of climate change, also see traditionally used and marketed medicinal and aromatic plants becoming harder to find, alpine pastures declining as vegetation patterns change, and seasonal indicators falling out of synchrony. Adaptation and mitigation strategies vary in each location. In China and Nepal, mountain villages are experimenting with new crops to take advantage of warming temperatures. In Nepal and Pakistan, we work with local medicinal and aromatic plant gatherers to evaluate procedures for the sustainable collection of wild plants and the possibilities for their cultivation. These data that will continue to emerge from this subnetwork will deepen the understanding of climatic factors in shaping medicinal and aromatic alpine vegetation in the Himalayan ecosystems, and contribute to national sustainable development measures and international efforts to understand and address climate change and its ethnobotanical impacts.

T1-19-04

Landuse changes and plant biodiversity conservation in traditional mountainous agroecosystems

Chunlin Long[1,2], *Katsunori Tanaka*[3], *Bo Long*[4], *Zhe Chen*[5], *Kenji Kato*[6]
1. *Minzu University of China*
2. *Kunming Institute of Botany, Chinese Academy of Sciences*
3. *Hirosaki University*
4. *Yunnan University*
5. *Yunnan Institute of Forest Inventory and Planning*
6. *Okayama University*

People in Andes, Caucasus, Himalayas or other mountainous regions have domesticated local plants and animals for many centuries. In these traditional mountainous agroecosystems there is great biodiversity of plants from genetic level, species level to farming system level. In recent decades, however, the plant diversity in mountainous communities had been affected by various reasons, especially by landuse changes resulted from policy or market enforcements. To examine the loss of plant biodiversity in traditional mountainous agroecosystems in terms of landuse changed, villages from Asian countries including Bhutan, China, India, Laos, Malaysia, Myanmar, Nepal and Thailandhad been tentatively investigated. Xishuangbann, a tropical area in southwest China, had been selected as a case study site, taking its shifting cultivation as an example. The semi-structured interview, participatory observation, focus group discussion and ethnobotanical methods had been used in the investigations. Investigations had been conducted in Xishuangbanna to compare the changes of upland crop landraces. Data were collected from 1992, 2006 and 2016 in different ethnic groups (Dai, Jinuo, Hani, Yao, Miao, Bulang, Lahu & Yi) and Han Chinese. Crops investigated in swidden agroecosystems are upland rice (*Oryza sativa*), taro (*Colocasia esculenta*), Xishuangbanna cucumber (*Cucumis sativus* var. *xishuangbannaensis*), melon (*Cucumis melo*) and others. Shifting cultivation or swidden or slash-and-burn agriculture was the most important landuse type in agricultural history. It has been argued for many decades in the context of economic development and biodiversity conservation. Because of too many people and too little forest available, Chinahas issued policies of landuse to reduce and ban shifting cultivation since 1980s. Swidden agriculture had been stopped completely in the country by the end of 2010s. The results revealed the number of crops at both species and landrace level decreased rapidly in 1992-2006, at an average rate of 46.5% and 71.8%, respectively. Millet, flee herb (*Adenosma buchneroides*), sorghum, Xishuangbanna cucumber, and *Amorphophallus kachinensis* could not be found in the swidden systems. Almost all landraces in traditional swidden agroecosystems of Xishuangbanna had completely disappeared by the end of 2015. Only some individual households in a few remote villages of Yao, Hani, Miao or Bulang ethnic groups still maintain one or two landraces of these crops. Some new crops including passion fruit, rubber tree, macadamia and *Pennisetum sinese* were introduced into the local systems, which resulted in the loss of biodiversity. The results implied that more investigations should be conducted before landuse policy was issued, otherwise the biodiversity would be threatened or damaged by the unreasonable policies. In the Xishuangbanna case, the genetic resources should be collected and preserved in seed banks when the government issued the swidden ban. Lessons from our case study will be beneficial for other communities in traditional mountainous agroecosystems.

T1-19-05

Introduction of new analytical and quantitative approaches for the production and conservation of medicinal and aromatic plants for economic development of Pakistan

Hassan Sher[1], *Ali Imtiaz*[1], *Robbie E Hart*[2], *Rainer W Bussmann*[2]
1. *Centre for Plant Sciences and Biodiversity, University of Swat, Khyber Pakhtunkhwa*
2. *William L. Brown Center, Missouri Botanical Garden*

The project is being implemented by University of Swat in collaboration with Missouri Botanical Garden under the Pak-US Science and Technology Promotion Program, the program of HEC Pakistan. The current project has been evaluating opportunities to maximize farm income through introduction of high value medicinal and aromatic plants (MAPs) in the war-stricken district Swat of Pakistan, directly supporting new *Framework for Economic Growth* of the country. The hypotheses is the establishment of *ex-situ* experimental production plots, leading to the development of skills in horticultural production and marketing among people in the valley and help rebuild commercial connections between this area and the rest of Pakistan. District Swat host about 90% of its population depend directly or indirectly on agricultural and forestry products. This translate the importance of the present project for maximize farm income and other livelihood to uplift the socio-economic conditions of the farmers through introduction of high value MAPs and make them self sufficient to achieve the goal of poverty alleviation and to move the country towards economic stability. This project has covered a range of interventions such as local awareness campaigns, capacity-building training, market survey, and MAPs production in the farms land of community. Local and indigenous communities have evolved traditional wisdom about the cultivation of conventional crops with very low economic return. This paper focuses on the cultivation of ten high value MAPs with farmers in three different locations/villages. This study also focus on the economic analysis and profitability of selected MAPs production and to explore the prospects of MAPs cultivation as a potential economic venture in the region and a way of ensuring the long-term conservation of these plants in the wild. The study has also incorporate evaluation of its performance in introducing standardized production technology and appropriate post-harvest management, which represent the prime 'engines of growth' for the local economy. These strategic economic development areas are entirely based upon, and closely interlinked, with the management and conservation practices of high value MAPs, and intact landscapes. It was suggested that the cultivation of important and threatened MAPs is very effective means to satisfy market demand, to provide income to communities dependent on the wild natural resource, and to reduce pressure on the wild population of plants species. Moreover, the available marginal, sub-marginal and cultivable wasteland can be developed and planted with selected MAPs species. It was also recommended that the farmers should be further guided through extension visits in order to educate them regarding the adoption of new technologies related to MAPs production and conservation.

T1-19-06

Ecological calendars: Enabling the participation of indigenous and civil society to cope with climate change in the Asian highlands

Jianchu Xu

1. Kunming Institute of Botany, Chinese Academy of Sciences

2. World Agroforestry Center

One of the greatest challenges of climate change is the difficulty in predicting its effects at the local community level due to erratic and unprecedented weather patterns. It is therefore an urgent need to develop anticipatory capacity in indigenous and civil society, as this capacity will provide these peoples with the ability to envision possible futures and develop a dynamic plan to deal with these uncertainties. We argue that development of a local adaptation strategy to climate change that is culturally and ecologically grounded will require re-conceptualization of the notion of time within an ecological framework. This will require the characterization of ecological calendars that are place-based and integrated with indigenous knowledge that has grown over generations and is sensitive to the biophysical signals. Such ecological calendars need however to be tested, validated and possibly recalibrated due to shifts in ecological and climatic regimes. These recalibrated calendars, together with climate scenario will enable us to anticipate and plan activities for future conditions and outcomes. While recognizing that some local groups are more resilient than others, the adaptive capacity of indigenous people and local groups can be strengthened through the revival, validation and recalibration of ecological calendars when appropriately assisted in partnerships with universities, research academies, government and non-government organizations. Intercultural dialogue must be entrenched in these partnerships, and is best achieved by interaction between both local and scientific knowledge systems. Thus, we call for establishment of the Ecological Calendar Network in Asian Highlands as a seed for change, as they develop innovative solutions to livelihood and climate challenges in the world's mountainous regions. We endorse the IPCC's recognition of the importance of local communities in climate change adaptation, and the necessity of respecting cultural and spiritual values in efforts to achieve sustainable development in mountain regions where local communities live and work.

T1-20: Biodiversity and conservation of aquatic plants

T1-20-01

Development of a global conservation strategy for freshwater plants

Richard Lansdown

IUCN SSC Freshwater Plant Specialist Group

Freshwater plants are low on the global conservation agenda. To counter this, the IUCN SSC Freshwater Plant Specialist Group (FPSG) was established with the aim to promote and further the conservation of freshwater wetland-dependent plants. The remit of the FPSG includes vascular plants, bryophytes, algae and some lichens which could not survive if the wetlands in which they occur were drained and probably includes more than 30,000 species worldwide. The FPSG takes a mainly species-focussed approach, using the IUCN Red List of Threatened Species to derive lists of taxa which are threatened, then developing conservation strategies for individual species, genera or families. One of our aims is to use Red List Assessments to establish a global Red List Index of freshwater plants which will enable monitoring of trends both in the conservation condition of individual species and in that of the wetlands which they support. Another aim of the FPSG is to raise the profile of freshwater wetland species, habitats and sites which are overlooked by the conservation industry, such as seasonal wetlands, flushes and river or stream headwaters. In addition we are working to identify taxa which are particularly vulnerable to climate change and to develop protocols by which such species may be used to monitor climate change in wetlands. However, we still must concentrate on fundamental issues which together form the foundation of all conservation. The taxonomy of many groups is still so poorly elucidated that we cannot start to address their conservation needs, while in some regions the taxonomic expertise necessary is lacking and field botanists use working names for species yet to be formally described. Similarly, we still lack the basic ecological information necessary to understand the conservation requirements of many wetland plant species and further to this, we have only a very limited experience of application of field conservation methods to wetland-dependent plants. A major emphasis within the FPSG is to support and promote research to acquire the information needed to provide credible, informed conservation. Through IUCN and other avenues, such as the publication of Species Conservation Strategies, the FPSG disseminates information on freshwater wetland-dependent plants to a wider audience from policy makers to the general public.

T1-20-02

Overlooked and undervalued: Diversity and conservation of Australia's seasonal freshwater wetlands

Adam Cross

1. Curtin University

2. Kings Park and Botanic Garden

3. University of Western Australia

Australia's dynamic climate and ancient landscape have produced one of the world's largest suites of seasonal aquatic habitats. These ecosystems harbour an immense assemblage of species ranging from specialist to generalist, and from widespread to highly restricted. Seasonal freshwater wetlands in Australia display unusual features in a global context, with a high proportion of species occurring in rain-fed, rapidly desiccating habitats experiencing high ecological stress. Plant endemism in these challenging habitats is remarkable, driven by geological antiquity, insularity, and their comparative long-term stability compared with other aquatic ecosystems, and their floras display distinctive ecological passport features associated with bet hedging, reproductive resilience and niche exploitation. However, data relating to the ecology and biogeography of Australia's seasonal aquatic flora are sparse, particularly in the remote tropical north where many wetland plants are actively growing when conditions are least conducive to accessibility. Seasonal freshwater wetlands have been poorly surveyed, face increasing pressure from a number of pressing threats, and our understanding of their true biodiversity is hampered by high levels of cryptic and undescribed diversity across many taxonomic groups. Historically, conservation initiatives and land management projects in Australia have focused predominantly on large wetland systems such as lakes and river deltas, and high-profile communities such as mound springs and swamp-based rainforest. As the frequency of range-restriction and ecological specialisation appear to increase with a decrease in habitat scale, the strong emphasis placed on these ecosystems only partially captures wetland biodiversity, and an increased attention to rain-fed wetland habitats of reduced size and local catchment must become an integral part of future conservation planning in Australia.

T1-20-03

Air, water, soil and soil pore water, a review of the four dimensional challenge for wetland plant translocation

Thomas Abeli[1], *Graziano Rossi*[1], *Simone Orsenigo*[2]
1. University of Pavia
2. University of Milan

Translocations (*sensu* IUCN) can be effective conservation tools only if they are undertaken with underpinning science to support the actions and outcomes. Translocations are rarely a simple exercise of moving species and/or populations for conservation purposes, the so-called 'gardening approach', rather, behind each translocation there should be a level of pragmatic and focused science to support the actions and to interpret the long-term viability of the action. Various authors have shown that the success of reintroductions can be at risk with even minor gaps in knowledge of the target species biology, ecology, interactions with other species, threats, *etc.* leading to failure. In the last decade important improvements in translocation techniques have increased their general rate of success. However, there is limited information available on the rate of success of translocation in wetland plants. Wetland plants are often grown *ex situ* but, unlike terrestrial plants, the interplay of different ecological factors such as the climate, the water quality, the soil and the water circulating in the soil, make acquiring a real understanding of the target species ecology, the identification of threats and their removal, the selection of suitable release site, *etc.* more challenging. A worldwide review of the rate of success in wetland plants translocation, including strictly aquatic plants and other wetland-dependent species is presented. The review takes into account how translocations were carried out, the types and aims of ecological studies performed, how threats were identified and removed, reasons for success or failure and very importantly, how the post-translocation success has been evaluated. Specific examples of successes and failures are also presented. In particular, failures usually remain unpublished, but understanding why a translocation failed can be as useful as a successful story.

T1-20-04

The road ahead for the *ex situ* conservation of aquatic plants: A project to restore the extinct species *Eriocaulon heleocharioides* to the wild

Norio Tanaka[1], *Yuju Horiuchi*[2], *Kohei Suzuki*[3], *Shoh Nagata*[4], *Takashi Kamijo*[5]
1. Tsukuba Botanical Garden, National Museum of Nature and Science
2. Graduate School of Life and Environmental Science, University of Tsukuba
3. Graduate School of Environmental Studies, Nagoya University
4. Specified Nonprofit Corporation AQUA CAMP
5. Faculty of Life and Environmental Sciences, University of Tsukuba

The IUCN Red List of threatened species includes 61 species classed as extinct in the wild. An organism is a component of the ecosystem and can be maintained as a species only in the wild; therefore, returning extinct species to the wild is important. Stable and long-term *ex situ* conservation, conservation biological research, voluntary conservation activities, building of co-operative relations with concerned parties and social understanding are essential for the successful return of a species to the wild. However, *ex situ* conservation generally faces some problems in terms of supplying populations to return to the wild. Decline in the genetic diversity of ex-situ populations is a particularly serious challenge. As the cultivation period increases and/or as the cultivation condition differs from the wild, the genetic diversity and fitness may decline, preventing a successful return of the species to the wild. To solve these problems, in 2008, we pioneered a conservation project to return the extinct species *Eriocaulon heleocharioides* (Eriocaulaceae) to the wild. This species had disappeared from its last habitat, the Sanuma Lake, Japan, after a change in the lake's water management plan in 1994 and has only been preserved *ex situ*, such as the Tsukuba Botanical Garden. In 2008, the project team first persuaded the concerned parties to return to the previous method of water management in Sanuma Lake. Subsequently, optimal growth environment conditions, pollinators, reproductive traits and seed storage and germination conditions were studied. In 2009, the plant flowered and bore fruit in the wild–after a period of 14 years. Nevertheless, the conservation team still cannot maintain the population without repeated introduction. Moreover, our preliminary results suggested that there has been genetic degradation of the species in *ex situ* conservation. Here, we describe the project and discuss problems and solutions regarding the *ex situ* conservation of aquatic plants with a view to reintroduction to the wild.

T1-20-05

Challenges and risks of conservation of an aquatic macrophyte extinct in the wild

Simone Orsenigo[1], *Ilda Vagge*[1], *Graziano Rossi*[2], *Alfons J P*

Smolders[3], Thomas Abeli[2]
1. Department of Agricultural and Environmental Sciences - Production, Landscape, Agroenergy, University of Milan, Milan, Italy
2. Department of Earth and Environmental Sciences; University of Pavia
3. Department of Aquatic Ecology and Environmental Biology, Radboud University

The reintroduction of a plant species regionally extinct in the wild poses a stimulating conservation challenge. However, when the issue concerns an aquatic dioecious macrophyte, with vigorous asexual reproduction forming very dense floating mats that can alter aquatic communities (through competition and allelopathy), each step must be very well planned. This is the case of *Stratiotes aloides* L. (water soldier), a keystone species that is declining in Europe, but is recognized as invasive in North America. The species, once abundant in the wetlands of North-Eastern Italy, is currently extinct in the wild in Italy. Only female Italian plants have been saved from extinction and are cultivated *ex situ*. To assess the feasibility of the de-extinction of *Stratiotes aloides* in Italy we investigated the water quality, the genetic diversity, and the potential spread of the species in its European range. The high concentrations of surface water inorganic nitrogen (especially nitrates) were identified as the likely cause of decline and extinction of *S. aloides*. The analyses of genetic diversity revealed the presence of different genetic patterns across Eurasia, that could be explained through long-distance bird-mediated dispersal events. Moreover, chromosome counts revealed the presence of diploid ($2n = 24$) and tetraploid ($2n = 48$) populations. According to our results the reintroduction of vital populations of *Stratiotes aloides* in Italy could be carried out using male individuals from the same genetic cluster of the Italian females, to avoid outbreeding depression. High-density stands must be reintroduced to permit self-facilitation of *S. aloides* under high nitrogen loads. However, considering the presence of high nitrogen loads in historical growing sites of *S. aloides*, there is a high risk of failure. Though in case of successful reintroduction, the risk to spread vegetatively downstream and the long-distance bird-mediated dispersal of seeds with unknown consequences on already threatened freshwater ecosystems cannot be ruled out. Therefore, considering the risks and benefits of the de-extinction of *Stratiotes aloides,* chaperoned managed relocation of female individuals in isolated water bodies basin (like gravel pits) could be the best solution.

T1-20-06

Degradation of aquatic vegetation in lakes in the Middle and Lower Reaches of Yangtze River in the past 60 years: Implications for the conservation of aquatic ecosystems

***Zhichun Lan**[1,2], Yasong Chen[1], Wei Shi[1], Yingzheng Wan[1], Hua Xing[1], Liyan Jiao[1], Jiakuan Chen[1,2]*
1. Institute of Watershed Ecology, Ministry of Education Key Laboratory of Lake Poyang Environment and Resource Utilization, Nanchang University
2. Coastal Ecosystems Research Station of the Yangtze River Estuary, Ministry of Education Key Laboratory for Biodiversity Science and Ecological Engineering, School of Life Sciences, Fudan University

There is widespread degradation in aquatic vegetation by a mixture of human activities. But it's difficult to evaluate the relative contribution of each factor to the vegetation degradation, because of data lacking. Such knowledge gap will limit our ability to make effective policies to conserve aquatic vegetation. To address these uncertainties, we compiled published data of aquatic vegetation in lakes along the middle and lower Yangtze River, China, and data of environmental properties and indexes of human disturbance during 1950s-2010s. (1) Aquatic vegetation of almost all lakes suffered significant degradation and succession in the past 60 years. For example, species richness had decreased by 16%, vegetation area decreased by 88%, and productivity decreased by 79% for aquatic vegetation. (2) Plant functional groups explain most of inter-species variations in the temporal trend, with submerged plants most sensitive to human activities. (3) For changes in aquatic vegetation, extensive aquaculture was the dominant factor, and water pollution and hydraulic projects were less important, while reclamation showed much inferior effect. We also found that disconnection between lakes and Yangtze river triggered community succession, with the dominant species shifting from submerged species to floating-leaved species or algae. To mitigate the degradation of aquatic vegetation, we should minimize the scale of aquaculture in natural lakes and keep the natural interconnected river-lake networks.

T1-21: Green digitization: online botanical collections data answering real-world questions

T1-21-01

From field collections to digital data: A workflow and digitization pipeline for reconstruction of a fossil flora

Dori Contreras
University of California Museum of Paleontology

The integration of curation and digitization with project-focused data collection is a key component to performing time- and resource-efficient studies from new fossil collections. Currently, standard workflows for processing fossil specimens starting from initial field collection and continuing through digitization are not widely established. Field work, data collection, curation, and digitization are often widely separated in time, involving different teams of personnel and separate funding. In many cases this follows the necessity of bringing older museum collections into the digital age. For new collections, however, more efficient use of personnel and monetary resources can be made by integrating these processes into a streamlined workflow carried out by researchers during the course of their study. Such workflows simultaneously meet the short term needs of a scientific study and the long term needs of the museum for archival of specimens and enhancement of public access to collections. Here I present my workflow for reconstruction of a diverse Late Cretaceous flora from plant macrofossils that are preserved in an extensive recrystallized volcanic ashfall deposit. This project ultimately aims to address fundamental questions about Late Cretaceous forests during the rise of angiosperms, including taxonomic and function diversity and spatial structure. Over the course of four years, I have made extensive new collections, establishing 26 quarries that span the >1.2 km length of exposure. Field censuses of leaf specimens have been performed at each quarry, and over 2,000 fossiliferous slabs with voucher specimens have been collected for further study and curation at the University of California Museum of Paleontology. Basic tasks involved in the reconstruction of the flora include differentiation and description of unique leaf types, identification and census counts of leaf specimens at each quarry, and measurement of leaf characteristics from digitized specimens. I will discuss how these tasks have been worked into a pipeline to streamline data collection and curation, including both specimen imaging and generation of geo-referenced individual specimen records as part of the digitization process. I will delve into parts of the workflow that have caused processing bottlenecks, and what measures have been taken to resolve them, especially how physical re-organization of space and both people and physical resources have been key to developing the workflow. Lastly, I will show how the digitized specimen data is being used in the context of this study.

T1-21-02

Developing standards for scoring phenology from herbarium specimens

Jenn Yost[1], *Gil Nelson*[2]
1. California Polytechnic State University
2. Florida State University

The digitization of herbarium specimens has advanced our ability to understand complex and changing biological systems. However, when a digital record consists only of a taxon name, date, and location, we are limited in the types of research questions we can address. While basic data such as these can be used to address changes in species distributions, herbarium specimens are rich with additional information regarding plant health, reproductive condition, and morphology that is generally not captured in digitization workflows. Phenology—the timing of life-cycle events such as flowering—in particular, has cascading effects on multiple levels of biological organization from individuals to ecosystems but data about these events are rarely interpreted, captured, or systematically scored in digitization efforts. Community wide data standards for scoring phenology from herbarium specimens are currently lacking. This talk will address the current efforts being taken by the Plant Phenology Ontology and the Phenology Standards Working Group. We will present the draft standards developed by the working group. The goal is to incorporate these standards into the Apple Core and potentially the Darwin Core, and develop the training tools necessary for phenological digitization to be included in all future digitization workflows.

T1-21-03

Virtual Herbaria tracking usage and benefits for biological collections an example from Australasia

David Cantrill
Royal Botanic Gardens Victoria

The Australasian Virtual Herbarium (AVH) makes available nearly 6 million records from all major collecting institutions in Australia and New Zealand. Traditionally loan records and research visitor information have provided data on the usage and importance of the collection. However, having released herbarium specimen data freely into the digital world the challenge for collecting institutions is to understand how their records are being used. Tools developed by the Atlas of Living Australia provide important insights to overall usage that can be used to inform funding bodies about the importance of these collections. Individual case studies also provide insights about how the sum of the data is more valuable than the parts.

T1-21-04

Current status and the applications of online botanical collection data in China

Zheping Xu, *Keping Ma, Cui Xiao*
Institute of Botany, Chinese Academy of Sciences, Beijing, China

In recent years, the government of China pays more and more attention to the conservation and application of the biodiversity. However, one of the most important works to achieve the goal of some national biodiversity action plan is to obtain the vast high-quality biodiversity data. Luckily, people can more easily get these data from different sources in China. For specimen data, NSII (China National Specimen Information Infrastructure) is the largest online specimen data portal in current China, which is funded by Ministry of Science and Technology (MOST). It has more than 14 million records (about 12.64 million records can be accessed via the internet) from more than 120 institutes or universities. From the view of Kingdom, 9.24 million records (73.1%) are plant specimens and 3.27 million records (25.87%) are animal specimens. For field data of botanical collection, we can get them from CERN (Chinese National Ecosystem Research Network), SinoBON (BON is short for Biodiversity Observation Network) and CFH (China Field Museum). Besides, you can also get other botanical collection data from other government projects, internal cooperation projects, citizen science platforms, digital literatures and social media. In China, there are also many applications after the integration of the botanical collection data from different data sources. People can find these applications in research articles, APPs (like for flower identification with Baidu and Microsoft), national/regional/global projects (for example, Southeast Asia Plant Biodiversity Information Infrastructure, SEADiv) and national policy reports. From the applications, we can see that the botanical collection data has played a very important role in the activities of research work and government work. Furthermore, we also need more digitization and put them online.

T1-21-05

Rebranding botanical collections: Global societal and biodiversity data needs for novel research

Shelley James[1], *Pamela Soltis*[1], *Gil Nelson*[2], *Deborah Paul*[2], *Matthew Collins*[3]
1. iDigBio, Florida Museum of Natural History
2. iDigBio, Florida State University
3. ACIS, University of Florida

Building on centuries of research based on herbarium specimens and their digitized data collected across time and space, a new era of discovery, synthesis and prediction has begun. Aggregated, open access botanical and associated biological, environmental and ecological data sets, from genes to ecosystem, can be used to document the impacts of global change on communities, organisms and society, predict future impacts, and help to drive the remediation of change. Within the collections and research community there is a need for advocacy for botanical collections, their expansion, continued and sustainable digitization (specimen data, associated media and information), including non-traditional digitized data fields and annotations, and born-digital field data to enable the rapid access of rich, digitally available, datasets for research, education, informed decision-making, and other scholarly and creative activities. Researchers are currently getting enormous benefits from data type-specific aggregators like iDigBio and Biodiversity Heritage Library (BHL) but assistance is still needed when working with large and disparate data sets. One example of infrastructure development is the Global Unified Open Data Architecture (GUODA) collaboration, which approaches this problem with processes to automatically retrieve and format data from providers of large open access datasets for easy processing by researchers. Tools for data discovery, visualization, representation and analysis, along with skills training, are also becoming increasingly important for inspiring novel research, in turn increasing the intrinsic value of the physical collections.

T1-21-06

Invasive or Not? A collection-based investigation of a historically unseen, persistent green algal bloom on Pacific coral reefs

Tom Schils[1], *Christian Boedeker*[2], *Frederik Leliaert*[3]
1. University of Guam Marine Laboratory
2. Victoria University of Wellington
3. Botanic Garden Meise

In 2012, a coarsely-filamented, unattached, large-celled, tropical *Chaetomorpha* (hereafter, CULT *Chaetomorpha*) suddenly appeared on reef flats in Guam, Mariana Islands, and caused ephemeral, localized outbreaks. By 2014, the bloom turned into a perennial phenomenon and occupied most of the southeastern shoreline of Guam (approximately 22% of the island's coastline). A further expansion into the west coast became apparent in late 2014 and in 2016 similar blooms were documented for Saipan in the northern Mariana Islands. This study investigates the invasion history and origin of the CULT *Chaetomorpha* in the Mariana Islands by screening online botanical collections. DNA sequence analysis was used to identify the bloom-forming alga and to determine the probable origin of different strains of CULT *Chaetomorpha*s from the Mariana Islands, the Western Pacific, and online aquarium retailers.

T1-22: Microbiomes of bryophtyes

T1-22-02

Cyanobacteria-bryophyte associations

Silvia Pressel[1], *Katie Field*[2], *Martin Bidartondo*[3], *Sebastian Schornack*[4], *Jeffrey Duckett*[1]
1. Natural History Museum, London
2. Leeds University
3. Imperial College and Royal Botanic Gardens, Kew
4. University of Cambridge

Land plants form associations with a great diversity of microorganisms. Partnerships with fungi, mycorrhizas or mycorrhizas-like, occur in over 80% of extant plant species, spanning the land plant phylogeny and including the earliest diverging bryophyte clades (liverworts and hornworts). The best understood and most actively researched interactions between plants and bacteria are rhizobia—the endosymbiotic nitrogen-fixing association of the Gram-negative soil bacterial genus *Rhizobium* with roots of legumes, not least because of their very important role in agriculture. Less well understood are the partnerships between nitrogen-fixing cyanobacteria (mostly in the genus *Nostoc*) and their host plants. These range from epiphytic associations, such as those reported in a few mosses, most notably *Sphagnum* and the feather-moss *Pleurozium schreberi*, which have been shown to contribute significantly to the N economy in the Arctic, the Antarctic, and boreal forest regions, to endophytic ones. Endophytic plant-cyanobacteria symbioses are rare in liverworts, restricted to two basal genera in the Blasiales (*Blasia* and *Cavicularia*), but ubiquitous in the hornworts. They are found sporadically in vascular plants: in one fern, *Azolla*, in all Cycads and in only one flowering plant family, the Gunneraceae. While liverworts harbouring cyanobacteria are notably fungus-free, hornworts are much more promiscuous, variably also forming associations with a broad diversity of Glomeromycotina and Mucoromycotina fungi, sometimes simultaneously. The significance of these tripartite associations remains unexplored, with key questions on possible nutritional and other interactions between fungi and cyanobacteria inhabiting the same host hornwort hitherto unanswered. We will present our novel insights into these tripartite partnerships and discuss these in the context of plant-microbe symbiosis evolution.

T1-22-03

Symbiotic options for the conquest of land

Katie Field[1], *Martin Bidartondo*[2,3], *William Rimington*[2,3,4], *Jeff Duckett*[4], *Silvia Pressel*[4]
1. University of Leeds
2. Imperial College London
3. Royal Botanic Gardens, Kew
4. Natural History Museum, London

The colonisation of Earth's continental land masses by plants > 475 Ma marked a turning point in the development of the terrestrial biosphere, with widespread consequences for the future of terrestrial life. It has long been hypothesised that the terrestrialisation of plants was facilitated by nutritional symbioses with Glomeromycota fungi to form arbuscular mycorrhiza-like associations. The establishment of this symbiosis would likely have enhanced access to mineral nutrients for early, rootless plants in exchange for carbon fixed through photosynthesis. Our recent discoveries that the earliest diverging clade of land plants (Haplomitriopsida liverworts) forms mutualistic symbioses with partially saprotrophic Mucoromycotina fungi challenge this long-standing paradigm. Using a combination of microscopic, molecular and isotope tracer techniques, we have shown that other lineages of early-diverging liverworts form simultaneous mutualistic associations with both Glomeromycota and Mucoromycotina fungi. Together with recent fossil evidence, these findings bring into question the identity, biology and functioning of the fungal symbionts of the earliest land plants. Here, we suggest that there were symbiotic options available to the earliest land plants and that such relationships may have been far more varied and transient than hitherto assumed.

T1-22-05

Notes on microbiomes of mosses from some active tufa deposits, China

Zhaohui Zhang[1,2], *Zhihui Wang*[1]
1. Guizhou Normal University
2. Guizhou Key Laboratory for Mountainous Environmental Information and Ecological Protection

Tufa depositing sites are widely distributed in the southwestern China, and contain a number of rare and unusual microbiomes communities of mosses such as cyanobacteria, protozoa, diatoms, fungi and algae. Particularly, calcicolous mosses and cyanobacteria are often abundant on tufa deposits. Supported by the National Natural Science Foundation of China (NSFC. No.31360043; No.31360035; No. 31160042), the biodiversity of microbiomes of mosses from some active tufa deposits sites were investigated in central Guizhou, China.After our preliminary identification and analysis in laboratory, 39 biota taxa belonging to five groups have been found from karst deposits in research area. Among them, there are 3 species for cyanobacteria while 15 for protozoan, two for green algae, 10 for diatoms and 9 for mosses.The relationship between microbiomes communities, mosses and tufa deposition have been discussed in this paper, including (1) the surfaces of biota acting as sites for nucleation and crystal growth of calcium carbonate from fresh water; (2) the process of photosynthesis removes carbon dioxide from the water, favoring carbonate precipitation; (3) living microbiomes communities providing a framework for carbonate deposition (4)the morphology of karst deposition may be affected by the growth of living microbiomes communities. Biodiversity of microbiomes and mosses may play an important ecology and biogeological role in karst deposition process at active tufa sites.

T1-23: Plant biodiversity data mobilization and management in the 21st century

T1-23-01

The digitization of a large herbarium, new opportunities & new responsibilities

Quentin Groom
Botanic Garden Meise

Globally a large number of herbaria are digitizing their collections. Researchers are now able to access these images and data of specimens through web portals and aggregators. This process promises many new opportunities for new and improved botanical research. It will also improve access to collections and reduce the historical inequalities between scientists. Herbarium digitization should reduce the cost of taxonomic research, but also expand the use of herbaria in ecology, biogeography and evolution. Indeed, we have only began to imagine what will be possible once the world's herbaria are digital. In 2015 the Botanic Garden Meise, Belgium, received a large grant from the Flemish government to fund the digitization of the Garden's African and Belgian collections. This will result in the digitization of more than a million specimens that will be accessible on a new web portal. We also received funding to modernize our imaging equipment so that the complete digitization of our collections no longer remains a distant possibility. Nevertheless, digitization is not a mere adjunct onto traditional curatorial activity. It will change the way collections are managed and biodiversity informatics will be increasingly central to the functioning of a modern herbarium. Digitization opens up many questions on how data are managed. For example, who should have access to those data? What data standards should be used? How can data be referenced permanently and how can the data publishers get recognition for their work? This means we need to create new policies, plans and procedures to maintain the quality of our output. Though the possibilities of digitization are exciting, we must be attentive to the new responsibilities it presents, so that we can ensure the long term sustainability of this research infrastructure.

T1-23-02

Everyday operational issues associated with managing the digital collections of one of the largest herbaria in the World

*Sarah Phillips, **Alan Paton**, Marie-Hélène Weech, Laura Green*
Royal Botanic Gardens Kew

We will present challenges, related to prioritisation, data quality and data curation in the context of the daily operation of digital collections management at Kew and discuss possible solutions. Over the last decade, digitisation rates have increased dramatically with ca.13% of the herbarium now databased. However, these specimens are physically dispersed throughout the collection, having been selected for digitisation based on specific scientific research and digitisation projects or through image requests from external researchers. The priorities of these stakeholders have not always matched those of the institution, and has led to a bias for certain taxonomic or geographical units not necessarily representative of the whole collection. As any one cupboard can be searched through multiple times for different specimens, this digitisation approach is not the most efficient. However, the recent emergence of industrial scale imaging and processing means it is now a reality that the 7 million specimens could be imaged within a 5 year time frame, working sequentially through the cupboards. Securing funding for mass digitisation is challenging, the sheer volume of specimens making it expensive and funding has traditionally been easier to obtain for geographically scoped projects or projects targeting particular species to answer individual research questions. Currently digitised and un-digitised material are intermingled, which is not operationally efficient or strategic, providing a bias to digitally accessible records and limiting value to users if this bias is not understood. The increasing volume of digitised collections brings new challenges to curators of these specimens and corresponding digital records. Un-digitised material should not be stored amongst sections of already digitised material and the digital collection needs to stay aligned with the physical collection. As specimens are moved and filed under different taxonomic names their digital records need to be updated. This takes additional resource often using software not designed efficiently for these tasks. Feedback received from image and data end users through the herbarium catalogue and other aggregator portals are sent to curators for assessment and records updated accordingly.

Feedback is encouraged to enhance data quality but takes time to manage, the efficiency of which could be improved through provision of easy to use automated annotation tools and linking of duplicates in other herbaria. In the current digitisation workflow, full label transcription is completed from the images rather than from the physical specimens. Different methods have been employed to capture label data from images including outsourcing, crowdsourcing and working in collaboration with project partners and to a smaller scale individual researchers. These approaches all have their own challenges with differing depths of data capture, quality and format of data returned. However all require decisions on level of data cleaning necessary before records are imported into the collection management system and made publically available. Capturing label data post imaging lends itself to user prioritisation of transcription where data is enhanced for specimens when there is an identified need. Future options for consideration include sharing the transcription and associated data validation, working as a global community rather than at an institutional level.

T1-23-03

Integrated digitization workflow and management system for herbarium plant specimens

Kun-Chan Tsai[1], *Jenn-Che Wang*[2]

1. Microtek International Inc.

2. National Taiwan Normal University

Large amounts of herbaria specimens represent as invaluable assets in academic research and policy-making, attributing spatial-temporal distribution pattern and phenotypic traits of species against emergent problems of natural resource management, global food malnutrition and climate-related dynamics. Accordingly, less than one third of herbaria specimens were digitally transcribed, while the estimated cost of sheet photographing and label recognition was between 5 and 7 USD dollars per specimen. The current efficiency and maintenance of digitization workflow, despite various modules and strategies were already proposed, is mainly limited to the duration and scale of funding, time, and staff. In addition, lack of integration of distinct digitization steps from specimen bar-coding, followed by label recognition and manual typing, to final data upload and website management, into a total solution package further hinders the establishment of a cost-effective coordinated digitization system. Based on the primary scanning result at TNU, we present an advanced digitization system, whose functions combine specimen barcode reading, on-top scanning platform, Optical Character Recognition (OCR) of specimen label, and registration-verification-based system for specimen image archive and intuitive database management, specifically for the digitization of herbarium plant specimens. Compared to conventional convey-belt and camera-based photographing system, this system design can significantly cooperate with recruited volunteers with limited botany background and experience, economically maximizing image capture efficiency at an ideal rate of roughly 200 records of complete label information per hour. The intuitive interface of database management system offers registration-based parallel verification options on specimen images and corresponding label data for the convenience of internal file management and external control over public-access website. The default file format and database structure follow the JSTOR standard, facilitating the import/export of database information among GPI members.

T1-23-04

Data mobilization for the Global Tree Assessment

Malin Rivers

Botanic Gardens Conservation International

The Global Tree Assessment is an initiative to provide conservation assessments of all tree species by the year 2020. This will be the first comprehensive assessments to effectively protect and safeguard the world's trees. The Global Tree Assessment is utilising a wide range biodiversity data sources, tools and platforms to achieve its goal. The Global Tree Assessment is led by Botanic Gardens Conservation International (BGCI) and the IUCN Species Survival Commission (SSC) Global Tree Specialist Group (GTSG), and to date we have already completed a complete list of the world's tree species, developed a robust procedure for identifying those that are not threatened according to international criteria and identified those tree species that are known to be threatened at national, regional or international level.

A global tree checklist with about 60,000 accepted tree species and their country distributions - GlobalTreeSearch - forms the backbone of the Global Tree Assessment. The assembly of this list of the world's trees relied on over 700 data sources, including various digitized biodiversity information, such as online floras, checklists, and specimen databases, as well as not yet digitized information. The expansion of the availability of floras online, and more detailed distribution information of species (including species status/specimen references/date of observation) would facilitate and improve this data source. Equally, much of this data is available, but is not open access. Increasing the amount of data freely available would be beneficial.

The Global Tree Assessment has also focussed on developing a methodology to identify widespread, not threatened species. An initial workshop determined data sources, tools and methodologies to identify not threatened trees. Open sourced data and data manipulation tools will be used for rapid identification and documentation of these species, this in turn will allow us to concentrate on collecting information on the species most at risk.

Another database, ThreatSearch, in which all known plant conservation assessments are collated, including national, regional and global levels, has also been used to identify tree species that are already known to be threatened.

The Global Tree Assessment has been made possible by the progress in plant biodiversity data mobilization. However, we still need to continue to develop an extensive global collaborative partnership for the completion of this project. This partnership will rely on biodiversity informatics tools and platforms in order to work efficiently across political, geographical and linguistic boundaries.

T1-23-05

Rapid response to invasive species in the digital era

Baudewijn Odé, *Laurens Sparrius*

FLORON Dutch Botanical Research Foundation

Rapid availability of observation data of invasive species is an important means to be able to prevent establishment and spread of these species and to help management for already widespread invasive species. Yet, these data have to be trustworthy and also have to be made available to management organisations fast to be able to adequately diminish the influence of invasive species in the ecosystem. In The Netherlands a vast majority of botanical observation data is collected in the National Database Flora and Fauna. In this database many private, corporate, scientific and governmental parties are working together to collect data on the distribution of plants and other organisms and recently comprises 100 million observations. These data are not only collected by professionals, but mainly through a network of volunteers. Data from different apps and web portals are send to the National Database. Data quality is assured through a validation protocol. This first automatically checks data integrity and the probability of a species' determination, based upon earlier and nearby observations of the species. Remaining doubtful observations are then manually checked by species group experts. In this way, more than 90% of the data becomes available to different types of end users within a few days after being collected. In the case of invasive species, local managers can rapidly act after receiving an e-mail alert or accessing the data portal.

T1-23-06

The Global Biodiversity Information Facility: A review of progress and challenges in plant data mobilization and integration

*Siro Masinde, **Hobern Donald***

Global Biodiversity Information Facility (GBIF)

GBIF—the Global Biodiversity Information Facility (http://www.gbif.org/) is the leading aggregator of species occurrence data, including botanical and fungal data. As of January 2017, GBIF had more than 700 million records harvested from over 32,000 data resources published by more than 800 data providers. Since its establishment in 2002, GBIF has provided leadership in delivering tools, standards and best practices for mobilizing biodiversity data for free and open use by the research community and to support policy applications. Established tools and processes now support the aggregation and integration of species distribution data from natural history collections, field research, environmental genomics, citizen science and literature into a single, unified resource. Modern approaches to linked-open data management point the way to this data resource becoming the focal point for connecting the full spectrum of recorded information about species and communities. GBIF strives to provide well integrated data of high quality and in large quantity to enable users to query and summarize data, with answers that are as complete and accurate as possible. This can only be achieved if data providers worldwide mobilize large, high quality data across a wide spatial, temporal and taxonomic spectrum and publish them openly using interoperable standard data formats. We review GBIF progress in plant data mobilization and integration, and provide insights on future directions. We also highlight the challenges encountered and invite a discussion on future expectations of the botanical community and the role the community can play in developing and curating the GBIF database to meet the needs and expectations of users.

T1-24: Systematics, population biology, morphology and conservation of the grape family Vitaceae: overview and recent advances

T1-24-01

Phylogeny and diversification of *Tetrastigma* L. (Vitaceae)

***Li-Min Lu**[1], Sadaf Habib[1,2], Viet Cuong Dang[1,2], Jun Wen[3], Zhi-Duan Chen[1]*

1. Institute of Botany, Chinese Academy of Sciences

2. University of Chinese Academy of Sciences

3. Smithsonian Institution

Tetrastigma is one of the most species rich genera of the economically and agronomically important family Vitaceae. It encompasses ca. 95 species that are primarily distributed in the tropical and subtropical regions of Asia and Australia, and species of the genus exhibit great diversity in both vegetative and reproductive characters. Species of *Tetrastigma* can be distinguished from other taxa of Vitaceae by their 4-lobed stigmas. The genus has attracted a great deal of attention due to its unique host-parasite relationship with Rafflesiaceae, especially *Rafflesia* (the genus with the largest flower on the world). However, relationships among major clades of *Tetrastigma* remain contentious probably because of insufficient phylogenetic signals. With 10 chloroplast markers and expanded taxon sampling (73/95 species), we recognized six major clades within *Tetrastigma* and relationships among these six major clades were well resolved. With extensive field collections and herbarium work, we systematically documented seed morphology throughout *Tetrastigma* and traced the evolution of six key seed characters. Morphological synapomorphies for each major clade of *Tetrastigma* were detected based on character optimizations and traditional morphology-based infrageneric classifications of *Tetrastigma* were evaluated. Ancestral area reconstruction and molecular dating suggest that *Tetrastigma* might have originated from the Sino-Himalaya and the Indo-China region during the Eocene and it then reached Australia and Malesia several times independently. A series of diversification rate tests of *Tetrastigma* were conducted to identify potential diversification rate shifts. The relationships between diversification rate shifts, adaptive traits, and habitats of *Tetrastigma* were assessed. To provide a comprehensive infrageneric classification of *Tetrastigma*, further field investigations are needed especially in southeastern Asia, where many new species or forms of Vitaceae have been discovered in the recent decade. Nuclear phylogeny of *Tetrastigma* is in progress to test potential phylogenetic incongruence among plastid and nuclear genomes and investigate the underlying mechanisms concerning the adaptation and diversification of *Tetrastigma.*

T1-24-02

Genomic analyses reveal deep phylogenetic relationship, biogeographic history and adaptations to different ecological environments of Eurasian grapes

***Zhi-Yao Ma**[1], Jun Wen[2], Stefanie M. Ickert-Bond[3], Ze-Long Nie[4], Long-Qing Chen[1], Xiu-Qun Liu[1]*

1. Key Laboratory of Horticultural Plant Biology (Ministry of Education), College of Horticulture and Forestry Science, Huazhong Agricultural University

2. *Department of Botany, National Museum of Natural History, MRC166, Smithsonian Institution*
3. *UA Museum of the North Herbarium and Department of Biology and Wildlife, University of Alaska Fairbanks*
4. *Key Laboratory of Plant Resources Conservation and Utilization, College of Biology and Environmental Sciences, Jishou University*

The application of whole-genome resequencing based on next-generation sequencing (NGS) provides an unprecedented opportunity for researchers to resolve long-standing evolutionary questions. The taxa of grape genus (*Vitis* L.), especially the Eurasian grapes, are widely recognized as important genetic resources for improvement of cultivated grapes. However, resolving the deep phylogenetic relationships within *Vitis* is a significant challenge, which limits our understanding of the evolutionary history of *Vitis* and impedes the utilization of valuable wild grape resources. Meanwhile, the biogeographic history and the adaptive evolution of Eurasian species of subgenus *Vitis* are still poorly understood. Identifying the genetic basis of adaptive variations may be the key to understanding the genetic diversity and the speciation of Eurasian species of subgenus *Vitis*. In this study, we carry out whole-genome resequencing of 38 accessions (representing 37 taxa) in the genus *Vitis* based on the *V. vinifera* L. reference genome. A neighbor-joining (NJ) phylogenetic tree within the genus *Vitis* based on 2,068 single-copy orthologous genes are reconstructed with high resolution and bootstrap values of almost all branches are 100%. The phylogenetic relationship clearly reflects the continental distribution of taxa of subgenus *Vitis* (*i.e.* America, Europe and East Asia) and the sister-clade relationship of European and East Asian *Vitis* species is robustly supported for the first time. Four subclades from East Asia have been identified, in which three subclades can be recognized as early diverging in the East Asian clade: subclade I, subclade II and subclade III, while, subclade IV is the most derived subclade. The taxa of subclade I are mainly distributed in northern China and the Far East, and subclade II and subclade III are widely distributed from northern China/the Far East to the southern China. In contrast, the taxa of subclade IV are mainly distributed in southern China and Southeast Asia. We infer that the ancestral taxa of East Asian *Vitis* are distributed in northern China and the Far East and then migrated to the southern China and Southeast Asia. There are preferences to different ecologic environments within Eurasian grapes, i.e. the bushy, sprawling to low climbing species with small leaves generally occur in shrub habitats and the high climbing species with big leaves generally occur in forest habitats. Further analyses comparing the two groups based on selective sweeps identify 6,370,000 bp and 6,870,000 bp of selection regions in the high climbing species group and the low climbing species group, respectively. Among the genes embedded in selected regions of high climbing *Vitis* species group for the forest habitats adaptation, several selected genes involved in the regulation of stem elongation growth and increased apical dominance may play a significant role for high climbing species to get light at the forest canopy. In contrast, some selected genes of low climbing *Vitis* species group associate with reduced apical dominance with shortened primary stems and increased number of branches. We provide insights into both phenotypic- and genetic-level adaption to environments of Eurasian grapes.

T1-24-03

Evolutionary history of *Vitis* based on genotyping-by-sequencing data

Laura Klein[1], *Claudia Ciotir*[1], *Allison Miller*[1], *Jason Londo*[2]
1. *Saint Louis University*
2. *United States Department of Agriculture - Agricultural Research Service*

As concerns mount about sustainable agriculture in a changing climate, attention is focusing on understanding diversity and diversification in crop wild relatives. The domesticated European grapevine (*Vitis vinifera*) is among the most economically important berry crops in the world and is becoming an increasingly valuable component of rural agricultural development. In many areas of North America and around the world, grapevine cultivation includes not only *V. vinifera* but also abiotic and biotic stress resistant native North American *Vitis* species, which, aside from being grown outright, contribute to hybrid scions and rootstocks. Despite this, relatively little is known about geographic patterns of variation among natural *Vitis* populations or the evolutionary processes shaping them. The evolutionary history of the grapevine genus has been notoriously challenging, in part due to likely rampant hybridization of sympatric species. Within subgenus *Vitis*, a large North Temperate group that includes ~50 *Vitis* taxa (~18 in North America), it appears that most species are interfertile. In North America it is common for two, three, or up to five *Vitis* species to grow in sympatry, further increasing the likelihood of interspecific hybridization. In this study, we leveraged living germplasm collections maintained by the United States Department of Agriculture Agricultural Research Service (USDA-ARS) Grape Genetics Research Unit (Geneva, NY) and National Clonal Germplasm Repository (Davis, CA) to investigate relationships among *Vitis* species. Single nucleotide polymorphisms (SNP) were identified using genotyping-by-sequencing. After sequencing and filtering we generated a dataset of 358 individuals representing 17 *Vitis* species (12 from North America, the European *V. vinifera*, and four Asian species) and three *Ampelopsis* species. Individuals were genotyped at approximately 11,000 sites with a minimum mean depth of coverage of 10. To infer phylogeny, we used the Geneious Tree Builder software and the RaxML plugin with the GTRCAT model and rapid bootstrapping. Preliminary results provide support for previously recognized groups including 1) monophyletic *Vitis*; 2) monophyletic subgenus *Vitis*; 3) within subgenus *Vitis*, a monophyletic group of Eurasian species. In addition, these data suggest close relationships between *V. acerifolia, V. arizonica, V. riparia* and *V. rupestris*, and close relationships between *V. cinerea, V. palmata,* and *V. vulpina*. In some cases, sampled individuals from a single species did not cluster together in these analyses. This could represent misidentification within the USDA system; alternatively (or in addition), it is likely that interspecific hybridization is contributing to reticulate patterns of evolution within the genus *Vitis*. Future areas of work should include detailed population level sampling, and morphological and genetic analyses of *Vitis* populations where multiple species occur in sympatry as well as in regions where each of the species occur in isolation.

T1-24-04

Leaf anatomy and micromorphology of the New World *Vitis*: Implications for taxonomy and evolutionary shifts to new habitats

S.M. Ickert-Bond[1], *Aj Harris*[2], *S. Lutz*[2], *J. Wen*[2]
1. Herbarium (ALA), UA Museum of the North and Department of Biology and Wildlife, University of Alaska Fairbanks
2. Department of Botany, MRC-166, Smithsonian Institution

The economically important grapes of *Vitis* subgenus *Vitis* occupy diverse habitats in North America and the Neotropics, but they are especially diverse in the mesic forest areas in the southeastern U.S. and the xeric scrublands in southwestern North America (e.g., the Edwards Plateau of Central Texas and adjacent areas). Species in subgenus *Vitis* comprise a clade (except for *V. californica*) according to ongoing molecular phylogenetic studies. Nevertheless, species delimitation within North American subgenus *Vitis* remains difficult due to limited variation in traditionally applied macromorphological characters, especially tendril frequencies along branches, tendril branching, floral sexuality, stipule size, leaf shape, inflorescence size, berry size and coating of berries. Therefore, we examined the c. 18 species of *Vitis* subgenus *Vitis* in the New World for leaf anatomical and micromorphological characters using light and scanning electron microscopy, and we analyzed our data using principal component analysis (PCA). We observed that micormoprhological characters of taxonomic importance include trichome type (prostrate, ribbon-like, unicellular; upright, multicellular); stomata morphology; mesophyll organization and midrib vascularization. Our PCA revealed seven clusters of species according to their morphological variation: (1) *Vitis labrusca* and *V. aestivalis*, (2) *V. acerifolia*, *V. riparia* and *V. rupestris*, (3) *V. cinerea* var. *cinerea*, *V. cinerea* var. *baileyana*, *V. cinerea* var. *helleri*, *V. biformis* and *V. tiliifolia*, (4) *V. palmata* and *V. vulpina*, (5) *V. mustangensis* and *V. shuttleworthii*, (6) *V. arizonica* and *V. monticola*, and (7) *V. californica* and *V. girdiana*. The PCA showed high loadings for the compactness or airiness of the spongy mesophyll on PCA axis one and of the height of the palisade layer on axis two. Leaf anatomical and micromorphological characters are often strongly related to the environmental preferences of species. However, when we represented the distributions of species within biomes in ordination space, we found little correlation with the clusters revealed in our PCA. According to micromorphological features, *Vitis labrusca* is more closely related to *V. aestivalis* than to *V. mustangensis* and *V. shuttleworthii* with the former two lacking a ridge above the midvein, and having four distinct bundles in the midvein, and raised stomates. *Vitis mustangensis* and *V. shuttleworthii* share a protruding stomatal apparatus abaxially and highly papillose epidermal cells adaxially. *Vitis cinerea* var. *baileyana* and var. *floridana* have similar trichome morphology and leaf anatomy. *Vitis biformis* from Mexico and the *V. cinerea* complex show simple trichomes with a multicellular base. *Vitis californica*, *V. arizonica*, and *V. girdiana* all have an airy spongy layer, two rows of cells in the parenchyma bridge of tertiary veins, and striate subsidiary cells. We hypothesize that adaptations to new habitats or biomes are reflected by the micromoprhological character states that we observed with xeric species possessing a spongy layer with large air spaces and well-developed, large vascular bundles, and the tropical rainforest species *V. tiliifolia* having thin leaves with a poorly-differentiated mesophyll and hyathodes adaxially.

T1-24-05

Taxonomic diversity and phylogenetic classification of Vitaceae

Jun Wen[1], *Li-Min Lu*[2], *Stefanie Ickert-Bond*[3], *Ze-Long Nie*[4], *Zhiduan Chen*[2]
1. Smithsonian Institution
2. Institute of Botany, Chinese Academy of Sciences, Beijing, China
3. University of Alaska
4. Jishou University, Jishou, China

Vitaceae (the grape family) consists of about 15 genera and 900 species primarily distributed in tropical regions. The family is most well-known for the economic importance of grapes and is also ecologically significant with many species as dominant climbers in tropical and temperate forests. Recent phylogenetic and phylogenomic analyses of sequence data from all three genomes have supported five major clades within Vitaceae: (1) *Cayratia* Juss., *Cyphostemma* (Planch.) Alston and *Tetrastigma* (Miq.) Planch., with the latter two nested within *Cayratia*; (2) the *Cissus* L. clade; (3) the grape genus *Vitis* L. and its close tropical relatives *Ampelocissus* Planch., *Pterisanthes* Blume and *Nothocissus* (Miq.) Latiff, with *Nothocissus* and *Pterisanthes* nested within *Ampelocissus*; (4) *Parthenocissus* Planch. and *Yua* C.L. Li; and (5) *Ampelopsis* Michx., *Nekemias* Raf., the African *Rhoicissus* Planch., and *Clematicissus* Planch. Based on the phylogenetic and morphological evidence, the new classification places the 900 species of the family into 15 genera: *Acareosperma* Gagnep. (1 sp., poorly known from Laos; never sampled in any of the recent phylogenetic studies), *Afrocayratia* gen. nov. (10 spp.), *Ampelocissus* (110 spp., including 20 species of *Pterisanthes*, and the type species of *Nothocissus*), *Ampelopsis* (15 spp.), *Causonis* Raf. (25 spp.), *Cayratia* (25 spp.), *Cissus* (350 spp.), *Clematicissus* (7 spp.), *Cyphostemma* (200 spp.), *Nekemias* (9 spp.), *Parthenocissus* (c. 15 spp.), *Rhoicissus* (12 spp.), *Tetrastigma* (80 spp.), *Vitis* (70 spp.), and *Yua* (2 spp.). *Causonis* and *Afrocayratia* are segregate genera from *Cayratia* s.l. *Nekemias* is a segregate of *Ampelopsis* s.l. *Clematicissus* is expanded to include five neotropical species in the *Cissus striata* Ruiz & Pavon complex, and represents a genus with an intercontinental disjunction between Australia and the Neotropics. The phylogenetic classification further recognizes three tribes in Vitaceae: (1) tribe Ampelopsieae trib. nov. (43 species in 4 genera: *Ampelopsis*, *Nekemias*, *Rhoicissus* and *Clematicissus*); (2) tribe Cissieae trib. nov. (665 species in 7 genera: *Cissus*, *Cayratia*, *Causonis*, *Afrocayratia*, *Acareosperma*, *Cyphostemma* and *Tetrastigma*); and (3) tribe Vitieae (207 species in 4 genera: *Ampelocissus*, *Vitis*, *Parthenocissus* and *Yua*).

T1-24-06

The fossol record of the native grape family (Vitaceae) from the Cenozoic of Australia

Andrew Rozefelds[1], *Marcelo Pace*[2]
1. Queensland Museum
2. Smithsonian Museum of Natural History

The fossil record of the Vitaceae family from Australia includes pollen, seeds and wood. The oldest fossil pollen referred to the family from Australia is Late Eocene in age from the Bass and Gippsland Basin in south eastern Australia and Vitaceae pollen is also recorded from the Oligo-Miocene coal floras of south eastern Australia. In Australia, two generalised types of seeds are recognised in the Vitaceae, with pyriform and oval-shaped seeds present since the Eocene and this indicates that the family had already diversified by the Eocene in Australia. Re-examinaton of *Cissocarpus jackesieae* from the Oligocene Capella locality in central Queensland shows that the seed are oval in shape, have a linear chalazal scar, paired ventral parallel grooves (infolds) that are parallel but diverge slightly towards the apex, with a columnar, moderately thick sclerotesta. A phylogenetic position near the *Cissus*, *Cayratia*, *Tetrastigma* clade seems likely. The second undescribed taxon from this locality has a pyrifrom seed with short, ventral infold slits that are apically divergent. Fossil wood from the Capella locality is also recognized as belonging to the family and the bark and wood anatomy of a new genus *Austrovideira dettmannae* sp. nov. is also described. The species shares with the other species of Vitaceae described to date the presence of secondary xylem with two diameter classes of vessels, wide vessels usually solitary, narrow vessels forming radial chains, very wide and tall rays, scanty paratracheal parenchyma and septate fibres. The secondary phloem has a clear stratified structure, with fibre bands alternating with the other axial elements and rays rapidilly dilating towards the periderm. A new genus *Austrovideira* is proposed because the fossil wood from Capella differs from *Vitaceoxylon,* the only other genus of Vitaceae wood known, in having scalariform intervessel pits and homocellular rays composed of exclusively procument cells, which are characters not previously seen in fossil Vitaceae wood. The only known extant species of Vitaceae with these combination of features is *Vitis* and *Parthenocissus*. *Parthenocissus* is present in the Australian flora, while *Vitis* is a Northern Hemisphere genus occurring in temperate regions of North America, Europe and Asia. *Austrovideira* differs from *Parthenocissus* in lacking vessels in tangential groups. This new discovery is the first fossil Vitaceae wood from the Australian flora and the Southern Hemisphere. The fossil wood record indicates that the lianescent habit in the Vitaceae was established by the Eocene in the Northern Hemisphere, and by the Oligocene in the Southern Hemisphere. The fossil record, based upon pollen and seeds, shows that the Vitaceae were present in Australia from the Eocene and based upon variation in seed morphology the family had already diversified by this time. The family, based upon seeds and wood, is well represented in the Capella flora with at least two, possibly three taxa. The modern North Queensland Vitaceae Flora includes sympatric genera *Clematocissus*, *Cissus*, *Tetrastigma, Ampellocissus* and *Cayratia*, so the co-occurrence of a number of Vitaceae species in the same fossil flora could be expected.

T1-25: Science of the Jade Dragon Mountain

T1-25-01

Phenological progressions of Mt. Yulong Rhododendron flowering over season and elevation

Robbie Hart[1], *Jan Salick*[1], *Sailesh Ranjitkar*[2,3], *Elizabeth Georgian*[4], *Jianchu Xu*[2,3]

1. William L. Brown Center, Missouri Botanical Garden

2. Key Laboratory of Plant Diversity and Biogeography of East Asia, Kunming Institute of Botany, Chinese Academy of Sciences

3. World Agroforestry Centre, East and Central Asia, Kunming

4. Department of Botany and Genetics, Vilnius University, Vilnius, Lithuania

Phenology-the seasonal timing of life-history events-is a critical dimension of natural history. In plants, reproductive phenology is particularly important, affecting gene flow, population persistence and species boundaries. Phenology is also one of the earliest and most noticeable traits by which organisms respond to climate change. However, these responses are complex, and only beginning to be understood, especially in the montane and alpine environments that are among the ecosystems most vulnerable to climate change. Drawing from diverse data sets and employing multiple methodologies, we examine how climate affects phenology in Rhododendron spp. of Mt. Yulong, Yunnan, China. Mt. Yulong is a center of diversity for Rhododendron, home to a suite of species that are diverse, dominant, and ecologically and culturally salient. They are also extraordinarily well represented by a unique set of historical plant collections. We reassemble these historical collections to build a proxy record of Rhododendron phenology, and use this record to show that although increased annual temperatures are associated with earlier flowering, increased fall temperatures are associated with delayed flowering. These contrasting effects have resulted in opposing changes in flowering time, even during rapid recent anthropogenic warming. In addition, we comprehensively monitor flowering phenology in the most common Rhododendron species on Mt. Yulong over gradients of season and elevation. We measure individual response to inter-annual temperature change, contextualize the results within the herbarium record, and characterize the effects of phenology on reproductive success. Mt. Yulong Rhododendron spp. show an elaborate sequenced progression of flowering over season and elevation. Phenological response to temperature varies among species, elevation, and season. Both phenology and response to temperature directly impact reproductive success, making this progression vulnerable to future climate change. A community phylogeny shows how phylogeny affects this phenological progression. Uniquely among phenological traits, last flowering day shows a phylogenetic signal, and is associated with flower size and fruit size. We suggest that this pattern may be driven by the limited time for fruit development before the onset of cold temperatures in autumn and highlight the unique and asymmetric pressures on phenology at seasonal extremes. Finally, we explore the dynamics of ecological knowledge of Rhododendron phenology around Mt. Yulong in two ethnic groups: the indigenous Naxi and immigrant Nuosu Yi. Local knowledge is rich and local understandings of changes and drivers parallels results from herbarium and ecological methods. While these knowledge systems are dynamic and able to adapt to change, they are also threatened by urbanization and changing lifeways. In combination, these studies reveal the complexity of phenological response to climate change. Multiple methodologies allow a deeper exploration than simple 'earlier spring' models of phenological response, while similar patterns seen among the diverse data sources increases our confidence in each. Mt. Yulong Rhododendron spp. flower in a progression over season and elevation, constrained to some degree by phylogeny. Both phenology and response to temperature directly impact reproductive success. Their responses to temperature vary among species and across elevations, and the effects of warming in different seasons drive contrasting responses.

T1-25-02

Quaternary palaeoecology and climate change in the Jade Dragon Snow Mountain Region, Yunnan, SW China

Yi-Feng Yao[1], *Xiao-Yan Song*[2], *Xia Wang*[1], *Anjali Trivedi*[3], *Alexander H. Wortley*[4], *Yu-Fei Wang*[1], *Stephen Blackmore*[4], *Cheng-Sen Li*[1]
1. State Key Laboratory of Systematic and Evolutionary Botany, Institute of Botany, Chinese Academy of Sciences
2. Shanxi Agricultural University
3. Birbal Sahni Institute of Palaeobotany
4. Royal Botanic Garden Edinburgh

The Hengduan Mountains, with a distinct altitudinal differentiation and strong vertical vegetation zonation, occupy an important position in Southwest China as a global hotspot of biodiversity. Pollen analysis of lake sediments sampled along an altitudinal gradient in this region, enables us to understand how the vegetation zonation arose and how it has responded to climate change and human impacts through time. Here we present new findings on Late Quaternary palaeoecology and climate change in the Jade Dragon Snow Mountain, one of the highest peaks of the Hengduan Mountains region, based on pollen records from Haligu, Wenhai and Shuanghaizi Lakes. Our results show: 1) From 22.6 to 19.4 ka BP, vegetation in the southern Jade Dragon Snow Mountain was dominated by broad-leaved forest, accompanied by needle-leaved forest. The period 19.4 to 17.9 ka BP was a transition stage from broad-leaved forest to needle-leaved forest. Between 17.9 and 9.3 ka BP, needle-leaved forest declined and broad-leaved forest began to increase at first, while needle-leaved forest expanded and broad-leaved forest shrank towards the end of the period. From 9.3 ka BP to the present, the vegetation changed to evergreen broad-leaved forest, mixed needle- and evergreen broad-leaved forest, needle-leaved forest. Vegetation on the eastern side of the mountain was of coniferous and broad-leaved mixed forest between 11.2 and 1.3 ka BP, followed by needle-leaved forest dominated by *Pinus* and *Abies* since 1.3 ka BP. 2) Climate in the period of 22.6–17.9 ka BP was characterized by cold and dry conditions, possibly corresponding to the Last Glacial Maximum (LGM, the start of which perhaps occurred prior to the basal age). Between 17.9 and 9.3 ka BP, temperature and humidity increased at first, followed by a colder and drier climate at the end (YD: 12.9–11.2 ka BP). After 9.3 ka BP, climate showed unstable with three significant phases of rapid climate change (at 9.3–8.7 ka, 7.0–6.6 ka, 3.7–2.4 ka BP), accompanied by a significant decline in humidity. 3) Plant diversity variation indicated by Simpson index, Shannon-Wiener index, Pielou's evenness and rarefied species richness was congruent with the temperature changes over the Holocene. 4) The Jade Dragon Snow Mountain region was probably under cultivation and experiencing other anthropogenic activities since the past 10 ka BP, particularly human settlement and human impact increased in intensity over the last 2.4 ka BP. These new findings are helpful not only for us to understand the history of biodiversity in the Jade Dragon Snow Mountain Region, but also useful for taking measurements on biodiversity conservation in the future.

T1-25-03

Origins and objectives of the Jade Dragon Research Station

Stephen Blackmore[1], ***De-Zhu Li***[2]
1. Botanic Gardens Conservation International
2. Kunming Institute of Botany

Around the world, Research Stations occupy a special role in the infrastructure of science. They support long-term, site-specific and multidisciplinary studies that are critically important in an era of global change. They also play an important role in the training and development of new generations of scientists, providing practical experience of living and working in what are generally remote and challenging conditions. Beyond this Research Stations often make a direct contribution to the protection and conservation of biodiversity and the natural environment. With this in mind, the Kunming Institute of Botany and the Royal Botanic Garden Edinburgh began in 2001 to develop the Jade Dragon Research Station set in a surrounding landscape designated as the Lijiang Alpine Botanic Garden. The concept was to make the Jade Dragon Snow Mountain a focal point and centre for long term interdisciplinary research and conservation biology within alpine zone of the Hengduan Mountain Biodiversity Hotspot. From the beginning the project had an additional dimension of relationship building between China and the UK. In 2004 the Research Station was formally recognised by CAS and the UK Government as the First China UK Joint Scientific Laboratory and it has since been the focus of a wide diversity of international investigations. This introduction to the origin and objectives of the Jade Dragon Research Station draws parallels with other Research Stations around the world. It will also review recent developments and investments in scientific facilities at the research station and nearby, including the establishment and enumeration of a 25 hectare permanent monitoring plot following the methodology of the Tropical Forest Science and Forest Global Earth Observatory (CTFS-ForestGEO). The symposium will also discuss and review the prospects for future research and conservation activities on the Yulong Mountain and the wider Hengduan mountain region.

T1-25-04

Phylogeny and ecological processes on pollination networks across an elevation gradient in the Jade Dragon Mountain, SW China

Yan-Hui Zhao, *Hong Wang, Zong-Xin Ren, Wei Zhou, Hai-Dong Li, Zhi-Bin Tao, De-Zhu Li*
Kunming Institute of Botany, Chinese Academy of Sciences

Ecological interactions are often phylogenetically conserved. However, species adapt to ecological conditions (i.e. competition and facilitation) could diminish or eliminate this phylogenetic constraints. Until now, changes in the strength of the effect of phylogenetic relationships on interspecific interactions (phylogenetic signal) across ecological gradients are poorly known. In addition, the adaptive mechanisms to promote species coexistence (i.e. shift partners, temporal segregation) are rarely studied at the network level. Here, we constructed 36 quantitative phenological plant-pollen vector networks by combining field observations and insect pollen load analyses at four sites along an extreme elevation gradient (2,725–3,910 m) in nine temporally distinct sampling periods. For each site, we also constructed eight temporal cumulative networks by pooling the phenological networks sequentially along flowering periods. To estimate the phylogenetic diversity and phylogenetic signal in the interaction networks, we reconstructed the phylogenetic trees of plant and pollen vectors using barcoding markers. We first test the effects of pollen vector to plant ratio (A/P ratio), resources density and phylogenetic diversity on phylogenetic signal in interactions of phenological plant-pollen vector networks. Then, we examined the role of temporal niche segregation in coexisting of plant and pollen vector species by comparing the phylogenetic signal between temporal cumulative networks and their corresponding phenological networks. For the first time, we found that plant phylogenetic signal in interactions tended to decrease with elevation, whereas pollen vector phylogenetic signal did not show systematically variation with elevation. The low strength of plant phylogenetic signal was related to low A/P ratio, low flower density and low plant phylogenetic diversity. However, pollen vector phylogenetic signal did not influenced by A/P ratio, pollen vector density and pollen vector phylogenetic diversity. In addition, results showed that the strength of plant and pollen vector phylogenetic signal tended to increase with the time accumulation of interactions in some communities. The phylogenetic signal of temporal cumulative networks pooled interactions of all the nine censuses higher than their corresponding phenological networks for both plants and pollen vectors across the four sites. Our results highlight that phylogenetic conservatism in plant interactions is low, and decrease with the increase in environmental stress. However, pollen vector interactions are mainly phylogenetic constrained and are not sensitive to variation in ecological processes. This study also indicates that temporal segregation among sympatric closely related species interacting with similar partners is an important mechanism to promote species coexistence.

T1-25-05

Survey and germplasm conservation of PSESP (Plant Species with Extremely Small Populations) in Southwest China: Role of China's Botanical Gardens of Kunming, Lijiang and Shangri-La

Weibang Sun

Kunming Botanical Garden, Kunming Institute of Botany, Chinese Academy of Sciences

PSESP (Plant Species with Extremely Small Populations) is a conservation concept developed for rescuing most of the globally threatened plant species, and it was firstly promulgated in Yunnan province of China in 2005. PSESP are characterized by small remaining populations (far lower than Minimum Viable Population, MVP), restricted habitat, extremely high risk of extinction and exposure to serious human disturbance. Survey and Germplasm Conservation of PSESP in Southwest China (2016-2021) is a national key project supported by the Ministry of Sciences and Technology of China in 2016. A total of 231 PSESP are targeted in this project which covers an area of 1459300 square kilometers in some 490 counties in Yunnan, Guizhou, Sichuan, Chongqing, west Guangxi and southeast Xizang. Based on the survey and investigation of the targeted species, at least 70 of them are required to be propagated and conserved *ex-situ* in Botanical Gardens, while the germplasm of over 100 PSESP will be stored in national germplasm banks. Survey and germplasm collection of 82 PSESP from 100 counties with 456100 square kilometers in the "Yunnan-Xizang" region of Yunnan plateau, north west Yunnan and southeast Xizang, has been planned as a sub-project. Undoubtedly, the resources of the botanical garden of Kunming, Lijiang and Shangri La located in this region are rather important for implementing the project. This presentation will provide an overview introduction of the project goals and its current progress, and emphasize the importance of the three botanical gardens for the success of the project. Special mention will be made of the Lijiang Alpine Botanic Garden, associated with the Jade Dragon Research Station, and its future development.

T1-25-06

Forest biodiversity monitoring in North-western Yunnan at the 25 hectare Yulongxueshan Forest Dynamic Plot and associated satellite plots

Kun Xu

Kunming Institute of Botany, Chinese Academy of Sciences

One of the main goals of the 25 ha-Yulongxueshan Forest Dynamic Plot is to explain the high species diversity of subalpine forests. Spruce-fir forest is the best protected forest vegetation on the eastern slope of the Yulong Mountains. To better understand its species composition and community structure, a 25 ha permanent plot was established in the Yulongxueshan Provincial Nature Reserve by Kunming Institute of Botany, Chinese Academy of Sciences and Yulongxueshan Administration of Provincial Nature Reserves from 2012 to 2014 following the standards of the Center for Tropical Forest Science (CTFS). We analyzed the flora, community composition, structure, size class and spatial distribution based on data collected from all woody plants (except bamboo) with DBH$\geqslant$1 cm. A total of 47,751 free-standing individuals with DBH $\geqslant$ 1 cm belonging to 62 species, 41 genera and 26 families were recorded with northern temperate plants being the main floristic elements. The Importance Values (IV) of 19 species were 1, and these species contributed the IV of 91.23% of the community. The forest community was divided into four layers, the dominant canopy species being *Abies forrestii, Picea likiangensis* and *Quercus guajavifolia. Gamblea ciliata.* var. *evodiifolia, Acer pectinatum* and *Padus brachypoda* occupied the sub-canopy where the absence of small size-class trees suggests their populations may decline in the future. *Sorbus prattii, Viburnum betulifolium* and *Rhododendron yunnanense* dominated the sub-tree layer, while the shrub layer was strongly dominated by *Berberis fallax.* 30 satellite plots, containing three samples of 1000 square meters, were established in different mountains in the northwest Yunnan in the area upstream of three great rivers of Asia, recognized for its importance for biodiversity conservation, and as a security barrier for regional ecology. The aim of the satellite plots is to describe the differences in plant community structures in the many different vegetation types found in the area and to explore the mechanisms which maintain the plant communities. The dominant tree species include *Pinus armandii, Pinus yunnanensis, Pinus densata, Quercus pannosa, Larix gmelinii, Picea likiangensis, Picea brachytyla var.complanata, Abies georgei, Abies delavayi, Abies forrestii,*

Taxus yunnanensis and *Alnus nepalensis*. The coordinates, DBH and height of trees were recorded together with surveys of shrubs, herbs, litter, deadwood and biomass. We found that random distribution of species was common in the mature forest, while in the young forest clumped distribution and uniform distribution were more typical. The reserves of carbon in alpine and subalpine coniferous forests and soils were richer than for forests at lower altitude, and had slower regeneration rates. Once the alpine and subalpine forest has been damaged it may face a long period of recovery, and more carbon from the forest and soil released into the atmosphere with important consequences for global change.

T1-26: Plants on inselbergs: speciation patterns, ecology and conservation

T1-26-01

New vistas on plant evolution: Lessons from terrestrial islands

Luiza De Paula, *Stefan Porembski*
Universität Rostock, Institut für Biowissenschaften

The study of oceanic islands has provided interesting hypotheses for our understanding of ecological and evolutionary processes that affect the biodiversity of ecosystems. Inselbergs are ancient monolithic outcrops mainly in tropical and subtropical regions, which rise abruptly above their surroundings, thus forming naturally occurring terrestrial islands. Nevertheless, these isolated environments have been so far overlooked concerning evolution and speciation. Due to harsh environmental conditions, both non-vascular and vascular plants of inselbergs are clearly distinguished from the surrounding matrix, represented by specialized species adapted to drought, heat and high irradiation. In southeastern Brazil, particularly in the *Mata Atlântica* domain, they occur in large numbers and in various degrees of isolation. This core area, which harbors a highly diverse inselberg flora, has been recently baptized as Sugar Loaf Land (SLL). Different plant communities are found on these outcrops, and mats are the most prominent habitat, consisting especially of long-lived monocots occurring on open rocky slopes, literally glued by dense wiry roots. In most cases mats are formed by angiosperms (e.g. within Bromeliaceae, Cactaceae, Cyperaceae, Orchidaceae, Poaceae, Velloziaceae) and ferns (e.g. *Selaginella*). Based on our newly generated data, Brazilian monocot-mats are among the richest in species in the world, and it can be expected that SLL is the hotspot of inselberg plant diversity concerning numbers of species and endemics. Particularly outstanding is the high beta diversity, i.e. the high species turnover over between neighboring rock outcrops in the above mentioned region. It can be assumed that the high alpha and beta diversity are due to the concentration of large numbers of inselbergs in the core area, the existence of phytogeographic links to other rock outcrop types (e.g. *campos rupestres*, ironstone outcrops) and a high degree of demarcation against the surrounding matrix. Also remarkable within the mat-formers is the widespread occurrence of species displaying a large spectrum of morphological variation, causing extreme difficulties in their taxonomic treatment. Highly variable (polymorphic) species, whose variation though partly correlated with ecology and geography, are called ochlospecies. In our study, broad intraspecific variation was found especially in populations of monocot taxa. The consideration of these species-complexes is important not only for solving taxonomic problems, but also for understanding processes of species evolution in terrestrial islands such as inselbergs.

T1-26-02

Patterns of plant species richness on inselbergs: A global overview

Stefan Porembski
Universität Rostock, Institut für Biowissenschaften, Allgemeine und Spezielle Botanik

In many parts of the world rock outcrops such as granitic and gneissic inselbergs form important landscape elements that contain numerous plant and animal species and provide important ecosystem services. They are millions of years old and rise often abruptly above their surroundings. Typical are extreme environmental conditions (e.g. lack of soil and water, high temperatures and insolation). Consequently their flora and vegetation clearly deviates from those of the surroundings. Throughout the tropics a typical set of plant communities and habitat types is present on inselbergs. Prominent are cryptogamic crusts, monocotyledonous mats and ephemeral flush communities. Many vascular plant species have developed particular adaptive traits in order to cope with life on inselbergs. In particular, desiccation-tolerant vascular plants are dominant mat-formers on tropical inselbergs with members of Cyperaceae and Velloziaceae being of particular importance which show a Gondwanian distribution. We estimate that globally 15,000-20,000 species of vascular plants occur preferentially on inselbergs. Particularly rich in inselberg specific species are southeastern Brazil ("Sugar Loaf Land"), Madagascar and southwestern Australia. Moreover, rock outcrops in these regions are characterized by high numbers of endemics and high rates of beta diversity. For certain countries and regions (e.g. East Africa, India, China) almost no data are available concerning the floristic composition of granitoid inselbergs. Due to negative human impacts (e.g. mining, fire, tourism, invasive weeds) inselbergs are threatened on a global level what could result in the loss of many unique and endemic taxa.

T1-26-03

Malagasy inselbergs: Floristic diversity and aspects of conservation

Marina Rabarimanarivo[1], *Stefan Porembski*[2], *Noromalala Ravololomanana*[1], *Roger Andriamiarisoa*[1], *Peter Phillipson*[1,3]
1. Missouri Botanical Garden
2. University of Rostock
3. Muséum national d'Histoire naturelle, Département Systématique et évolution

Globally, inselbergs which are isolated rock outcrops, mostly consisting of granite or gneiss, constitute old terrestrial habitat islands which are widespread throughout the tropics. Madagascar is known to be rich and diverse in terms of species and ecosystems and is renowned for its high levels of endemism, unparalleled in the world. Specific habitats such as inselbergs which have been largely neglected hitherto contribute to this status and also provide various ecosystem services such as important sources of mineral water. They form prominent landscape features, mainly in the central highland regions and have clearly a global importance with regard to their high species richness of certain plant groups such as succulents, terrestrial orchids, resurrection and carnivorous plants. Most of them are still unexplored and biologically unknown but recent extensive fieldworks conducted on more than 40 Malagasy inselbergs have contributed to a better knowledge about their floristic diversity and habitats. Up to now, more than 800 species were recorded to occur on Malagasy inselbergs. The most important families with largest numbers of species are Orchidaceae, Cyperaceae, Rubiaceae, Asteraceae, Apocynaceae, Poaceae, Fabaceae, Asparagaceae, Euphorbiaceae and Lamiaceae whereas *Cynorkis, Cyperus, Euphorbia, Kalanchoe, Aloe, Cynanchum, Exacum, Angraecum, Senecio, Xerophyta, Utricularia* and *Jumellea* represent the most speciose genera. Inselbergs are physiognomically identical throughout the island, but differ significantly in respect of their floristic composition as well as the degree of endemism. With southeastern Brazil and southwestern Australia, Malagasy inselbergs have been identified as one of the three hot spots of inselberg plant diversity. Effectively, Malagasy inselbergs are currently threatened by habitat degradation due to various human pressures (mining, quarrying and fire, the common illegal collection of plants for commercial purposes) resulting in loss of floristic diversity and the ecosystem services they are providing. Many of their species are highly endangered now, for ca. 20% of the species which have been assessed for the IUCN Red List, ca. 50% have a threatened status. Moreover, inselbergs are under-represented within the Protected Area network of Madagascar. Conservation of this unique ecosystem is urgently needed at regional, national as well as international levels and the most efficient way could be the creation of additional protected areas. In addition, creating awareness of their importance for the local population is also urgent.

T1-26-04

Midland lateritic hills of South India: Cradle of neoendemism?

C. Pramod [1], *A.K. Pradeep* [2]

1. Government Brennen College, Thalassery, Kerala

2. University of Calicut, Kerala

The Western Ghats, which is one of the major global biodiversity hotspots, has been recently declared by the UNESCO as a World Heritage Site and the Western Ghats Ecology Expert Panel (WGEEP) constituted by the Ministry of Environment and Forests (MoEF), Government of India, submitted their report in August 2011, to address the issue of conservation of biodiversity of the Western Ghats. Unfortunately there has been never an effort to recognize the rare biodiversity of coastal lateritic plateaus and hills, parallel to the Western Ghats, which bore a large proportion of rare and endemic plants. Northern Kerala of South India, which is the type locality of laterite, is characterized by wide lateritic plateaus which provide extremely harsh physical environment for life leading to the development of specialized plant communities with large number of endemic and habitat specific species. Madayippara, a midland lateritic plateau located in the Kannur District of Northern Kerala, South India, is an icon of the rich floristic diversity and endemism shown by South Indian lateritic hills. The plateau forms a number of microhabitats due to the difference in geographic terrain and soil cover, thus forming a complex of peculiar habitats with diverse forms of plants, mainly ephemeral herbs. Out of the 665 flowering plant taxa recorded from the microhabitats of the plateau, within a limited area of 3.65 square kilometers, 161 (24.21%) are endemics. Most of the endemic species occur in specialized microhabitats. The plateau is the type locality of nine taxa, *viz., Rotala malabarica* (Pradeep *et al.,* 1990), *Nymphoides krishnakesara* (Joseph & Sivarajan, 1990), *Justicia ekakusuma* (Pradeep & Sivarajan, 1991), *Lepidagathis keralensis* (Madhusoodanan & Singh, 1992), *Coelachne madayensis* (Pramod & Pradeep, 2012), *Eriocaulon madayiparense* (Swapna *et al.,* 2012), *Lindernia madayipparensis* (Ratheesh Narayanan *et al.,* 2012), *Parasopubia hofmannii* (Pradeep & Pramod, 2013) and *Parasopubia hofmannii* var. *albiflora* (Pradeep & Pramod, 2013). Two recently described species from South India, viz., *Eriocaulon gopalakrishnanum* (Rashmi & Krishnakumar, 2013) and *Lindernia tamilnadensis* (Prasad & Sunojkumar, 2014) also occur in this plateau. Lateritic plateaus of Northern Kerala, which are associated with characteristic and rich biodiversity, are now under varied types of anthropogenic threats such as large scale mining for bricks and clay. Any slight disturbance in the micro ecosystems can easily take away a number of short-lived herbaceous species which cannot be easily conserved outside its natural habitat. For the conservation of the rich diversity and microhabitats of the plateau, *in situ* conservation of the entire habitat is the only solution as *ex situ* conservation measures cannot provide complex microclimatic requirements artificially.

T1-26-05

Conservation and exploration of Rio de Janeiro's inselberg flora, Brazil

Luana Paula Mauad [1], *Leandro Jorge Telles Cardoso* [1], *Stefan Porembski* [2], *João Marcelo Alvarenga Braga* [1]

1. National School of Tropical Botany, Rio de Janeiro Botanical Garden Research Institute

2. University of Rostock, Institute of Biosciences, Department of Botany

Inselbergs worldwide host a very particular vegetation, being a refuge for rupicolous plant communities which do not exist in the surrounding matrix. Inselbergs occur both in the tropics (e.g. southeastern Brazil, Madagascar) and in temperate zones (e.g. southwestern Australia). In many regions their unique biodiversity is threatened by human activities, such as quarrying, fires, illegal collection of ornamental plants, tourism and weed invasion. In the metropolitan area of Rio de Janeiro these negative effects have caused the degradation of many inselbergs. The region forms a biodiversity hotspot where inselbergs harbour many local endemics and threatened species. In the city of Rio de Janeiro more than 200 inselbergs occur in an urban agglomeration with c. 6.000 inhabitants/ km^2. Southeastern Brazil (incl. Rio de Janeiro) forms together with Madagascar and southwestern Australia one of the most important regions globally, with regard to the richness and degree of endemism of rupicolous species. Remarkably, little is known about the flora, vegetation and conservation status of inselbergs of the megadiversity country Brazil. Based on own fieldwork and data available from other sources we present an overview on the inselberg flora of Rio de Janeiro. In total 193 species of vascular plants occur on inselbergs in the study area. More than 80 species (> 40%) are restricted to the Atlantic Rain Forest domain. Richest families are Bromeliaceae (17 spp.), Orchidaceae (11 spp.) and Cactaceae and Araceae (6 spp. each). In total 25 species (13%) are endemic to the federal state of Rio de Janeiro with Bromeliaceae (10 spp.), Orchidaceae (3 spp.) and Gesneriaceae (2 spp.) comprising species that are often restricted to a few (< 10) inselbergs. In addition, 32 species (17%) are listed under particular threat categories. Based on the major risk criteria of the four lists checked (IUCN, CNCFlora, MMA, SMAC), 6 species are Critically Endangered (CR), 14 Endangered (EN), 11 Vulnerable (VU), and one Near Threatened (NT). The families with high numbers of threatened species are Bromeliaceae (12 spp.), Cactaceae and Begoniaceae (4 spp. each), and Orchidaceae (2 spp.), which also usually figure among the richest families on inselbergs in Brazil. Certainly many other species will have to be included in these lists, once their conservation status has been evaluated. Furthermore, as new studies on inselberg vegetation in southeastern Brazil are underway now, new species have been discovered, as well as new distribution records for the state. It becomes clear that inselbergs, even in the middle of one of the biggest urban areas of Brazil, still host important scientific novelties, since there is still a lack of knowledge regarding their vegetation, mainly in the steepest parts. Thus, the information provided by us can contribute to conservation actions, in order to prevent the extinction of the most vulnerable species. Seeking to guarantee the conservation and diversity of the rupicolous community as a whole, might help to guide public policies towards *in situ* conservation of species on inselbergs in Rio de Janeiro.

T1-26-06

Development of SSR markers for species of *Chresta* (Asteraceae, Vernonieae) from rock outcrops of the Brazilian Caatinga using genomic data

Carolina Siniscalchi[1,3], *Benoit Loeuille*[2], *José Rubens Pirani*[1], *Jennifer Mandel*[3]

1. *Universidade de São Paulo*
2. *Universidade Federal de Pernambuco*
3. *University of Memphis*

Chresta is a genus of Asteraceae almost endemic to Brazilian *Cerrado* and *Caatinga*. Despite the relatively small number of species, the genus presents high morphological variation, which raises interesting evolutionary questions. From the 18 species of the genus, seven occur in the semi-arid *Caatinga* region, usually growing directly over quartzitic or granitic rock outcrops, forming distinct populations that are isolated from each other and spread apart in a dense semi-deciduous forest matrix. Five of these species present similar morphological features, which sometimes overlap. Usually populations in the extremes of the distributions present different morphologies from those found in the "core" populations. This variation led to the acknowledgment of some populations as different species, such as *C. hatschbachii* and two new species yet to be published, which were all segregated from *C. harleyi.* The morphology of *C. filicifolia,* which occurs in the Atlantic Forest, is closely related to those from *Caatinga* species, what may indicate a *Caatinga*-Atlantic Forest link that has been observed in other plant groups. In the present work, we used genomic data previously obtained by the sequencing of approximately 1000 nuclear markers from four different *Chresta* species and the whole chloroplast sequence from seven samples to design SSR markers that present variation among five *Caatinga* species and *C. filicifolia* and among individuals from the same species. The markers will be further used to evaluate genetic diversity and gene flow among populations from the same species and different species, in order to investigate processes of speciation and evolution in this species group and in the *Caatinga* as a whole. The chloroplast sequences of seven *Chresta* species were aligned using MAFFT, and the SSR finder Phobos was used to locate possible repeat regions, which were then selected based on putative differences in size among the seven sequences. The set of nuclear markers was assembled in HybPiper and then alignments for each locus were individually searched with Phobos. Also, contigs assembled with Spades in the HybPiper pipeline were searched with Phobos and contigs with possible candidates were then blasted in GenBank to assure they were not chloroplast sequences. Nineteen pairs of primers were developed, twelve from nuclear markers and seven from chloroplast; most of them amplified with all species, using the same PCR recipe and program. Fragment analysis was carried out in 56 individuals from different populations of six species (*C. filicifolia*: 4 individuals, *C. harleyi*: 28, *C. hatschbachii*: 4, *C. martii*: 10, *C.* sp1: 12, *C.* sp3: 2) with 12 pairs of primers, four pairs being chloroplast markers and eight nuclear. From the 12 tested primers, only 4 nuclear markers showed variation among all species. Three of the chloroplast primers presented variation among individuals of *C. martii*, but were not variable in the other species.

T1-27: The evolution of tropical biodiversity: a global comparative perspective (two sessions)

T1-27-01

Global signals of extinction and adaptation in megafaunal-fruited plants: Insights from a keystone tropical plant family (Palms, Arecaceae)

Renske E. Onstein[1], *Søren Faurby*[2], *William J. Baker*[3], *Thomas L.P. Couvreur*[4], *Jens-Christian Svenning*[5], *W. Daniel Kissling*[1]

1. *University of Amsterdam*
2. *University of Gothenburg*
3. *Royal Botanic Gardens Kew*
4. *Institut de Recherche pour le Développement UMR-DIADE*
5. *Aarhus University*

Frugivory (fruit-consumption and subsequent seed dispersal by animals) is a key plant-animal interaction in tropical rainforests. Of particular importance for this mutualism are large-bodied frugivores because they disperse the largest seeds and contribute disproportionally to long-distance dispersal events. However, the consequences of global change on the diversification and adaptation of small- (< 4 cm) and megafaunal-fruited (> 4 cm) plants at large spatial and temporal scales remain largely unexplored. Here, we combine an all-evidence species-level phylogeny for all palms (Arecaceae, ca. 2600 species), global species distributions and a novel fruit size dataset to study the consequences of dispersal by megafaunal frugivores (mammals 44 kg) on palm diversification.

We show that palms (a keystone resource for frugivores) with megafaunal fruits have lower speciation rates than smaller-fruited palms, possibly due to frequent long-distance dispersal events that increase gene flow among populations and thereby limit allopatric speciation. Furthermore, a time-window analysis demonstrates that since the onset of the Quaternary (2.6 Mya), a period a dramatic global change, extinction rates of megafaunal-fruited New World palm lineages have increased whereas Old World palms show an increasing transition rate (evolving smaller fruits from megafaunal fruits). We suggest that high extinction rates in the New World could be triggered by climate-driven habitat fragmentation potentially coupled with a temporary decrease in megafauna diversity in the initial stages of the Great American Biotic Interchange. Increased transition rates in the Old World could reflect adaptation to dispersal by smaller-bodied, volant frugivores against a background of sea level fluctuations in island-dominated environments of Southeast Asia. Our results suggest that plant lineages with megafaunal fruits are disappearing, either due to increased extinction rates or rapid evolution to smaller fruit sizes.

T1-27-02

Exploring the evolutionary history of the Australian rainforest flora

Samantha Yap
National Herbarium of NSW

Despite representing <1% of the Australian land surface, rainforests have high biodiversity and biogeographic significance. Rainforest vegetation was once widespread throughout the supercontinent Gondwana. As the Australian continent drifted northwards on the Sahul shelf, its rainforests became isolated. The increasing continental aridification associated with these geological processes, also caused major distributional contractions and lineage losses. This loss of diversity was somewhat reversed as the Sahul continental plate approached and eventually collided with the Sunda shelf. The increased proximity between landmasses promoted biotic exchange between the two floras, with extensive immigration of Indo-Malesian rainforest lineages (i.e. with Sunda ancestry) to Australia. An increasing number of phylogenetic studies is revealing the contrasting biogeographical histories of these two floras within the Australian continent. Yet, so far, little is known about the impact that distinct ancestries have on the geographic distribution and landscape-level dynamics of rainforest species, as well as on the relative prevalence of specific functional attributes. I will present continent-wide data showing how these two biogeographic groups can be broadly distinguished based on a range of functional attributes, and on their geographic distribution. I will then discuss contrasting population dynamics patterns based on multi-species investigations using next-generation sequencing (NGS) technologies (*in silicio* chloroplast DNA assembly). These analyses detected significant differences in functional trait variation, and in continental dynamics between species with Sunda and Sahul ancestry. We also confirmed the location and relative importance of long-term refugias.

T1-27-03

Comparative approaches to studying plant diversification in tropical rain forests

Wolf L. Eiserhardt[1], *Thomas L. P. Couvreur*[2], *William J. Baker*[1]
1. Royal Botanic Gardens, Kew
2. Institut de Recherche pour le Developpement

Warm and wet environments have played a key role in the diversification of plants. The vegetation of these environments, known as tropical rain forest (TRF), has been estimated to harbour 43-45% of extant vascular plant species despite only covering ~7% of the Earth's land surface. In addition, many plant lineages now occurring outside TRF trace back their origins to this biome. Although the history of TRF has been well studied, many fundamental questions are insufficiently answered. How important is the old age of TRF for their extraordinary diversity? Do plant species originate more readily, and/or disappear less frequently, in TRF compared to other biomes? Which factors promote speciation and regulate extinction in TRF? How much plant diversity did TRF receive from other biomes during their history? Is there a limit to the number of species that can exist in a biome at any time, and is this limit higher in TRF than elsewhere? Multispecies comparative studies can provide crucial answers to these questions, but rely on phylogenetic trees as an explicit evolutionary framework. Phylogenies link fossils to extant taxa, allow process-based modelling of evolutionary and biogeographic change, and facilitate the estimation of speciation and extinction rates. Critical insights can be gained at two levels: globally, diversification in TRF can be systematically compared to diversification in other biomes, modelling ancestral biome associations on a phylogeny and identifying the frequency and drivers of speciation, extinction, and biome shifts. Meanwhile, TRF is not a homogeneous entity, and differences in diversification dynamics within the biome can yield important insights, using a similar approach but focusing on environmental gradients and historical differences inside TRF. We illustrate these possibilities using case studies, mainly from the palm family (Arecaceae), which has been extensively used as a model for plant diversification and ecology in TRF. Based on this we point to areas where more research is needed, emphasizing the need for broad comparative studies across taxa and regions. The feasibility of broad comparative studies is, in spite of the "big data" revolution, still limited by the availability of basic biodiversity data across the dazzling diversity of TRF plants. Not least, the pace at which TRF plant taxa are included in reliable phylogenetic frameworks will determine how soon the comparative approach to TRF plant diversification comes to full fruition. Given the rampant destruction of TRF across its range, time to collect such data is pressing.

T1-27-04

Exploring the evolutionary and ecological drivers behind intracontinental geographic disjunctions: The African Rand Flora

Isabel Sanmartin[1], *Tamara Villaverde*[1], *Lisa Pokorny*[2], *Mario Mairal*[1], *Sanna Olsson*[3], *Ricarda Riina*[1]
1. Real Jardin Botanico, CSIC
2. Royal Botanic Gardens, Kew
3. INIA-CIFOR

Compared to other biogeographic regions located at tropical latitudes, North and Central Africa remain surprisingly poor in species numbers. Sometimes termed as the "odd-man out" in the global Latitudinal Diversity Gradient, a species-poor Africa is not universal across all groups of organisms, but in angiosperms examples of lineages with low diversity in Africa and high diversity in other tropical regions (the Neotropics, South East Asia) abound. This pattern is often associated to striking intra-continental disjunctions, in which closely associated species are distributed at opposite margins of the African continent, and in most cases separated by geographic distances of thousands of kilometres. Many of these taxa are seemingly confined to their restricted distributions by environmental barriers, i.e., the presence of intermediate, inhospitable land that lies outside the set of environmental conditions in which species can reproduce and maintain viable populations. One example is the Rand Flora, a continental-scale pattern in which different angiosperm families share a similar disjunct distribution, with related species confined to the margins of Africa and adjacent islands such as Macaronesia and Socotra. The barriers in this case are the deserts of Sahara in the north and Kalahari and Namibia in the south, or the tropical lowlands of Central Africa. Whether such patterns were formed by dispersal in recent times and ecological constraints (niche-conservatism) or through the fragmentation of a wider distribution in the past by environmental barriers ("ecological vicariance") is unclear. For the Rand Flora, the interest lies in the possibility to study the effects of large-scale extinction associated to the ongoing aridification process that started in the Miocene and climaxed in North Africa c. 7 million years ago with the formation of the Sahara Desert. Rand Flora relicts would be in this case survivors of climatic extinction, confined to their current distribution ("climatic refugia") on the edges of the continent by the disappearance of populations in the intermediate regions. Unlike background extinction rates, intrinsic to each lineage and dependent on biological traits, large-scale climatic extinction is a process concurrently affecting multiple lineages that inhabit the same region. The difficulty to study this type of extinction lies in the need to integrate different evolutionary processes: Red Queen (biological, intrinsic) versus Court Jester (abiotic, extrinsic), and – since aridification in Africa is a long-term, ongoing process – contrasting evolutionary levels: from across species (macroevolutionary) to within-populations (microevolutionary). The latter has become feasible with the recent development of next-generation sequencing techniques that allow scaling up the number of genes, individuals, populations and species sampled from non-model organisms. We will present evidence for the use of such approach to unravel the origins of the Rand Flora pattern. Using molecular (genomic), palaeobotanical and palaeoclimatic data, combined with species-distribution models and novel cross-lineage Bayesian biogeographic methods, we demonstrate that climatic extinction associated with ecological vicariance is the most plausible explanation for the Rand Flora pattern.

T1-27-05

Amazonia is the evolutionary heart of Neotropical biodiversity

Christine Bacon, *Fernanda Carvalho, Ruud Scharn, Alexander Zizka, Daniele Silvestro, Fabien Condamine, Alexandre Antonelli*
University of Gothenburg

The American tropics (the Neotropics) harbor the highest number of species on Earth and for centuries scientists have attempted to understand the origins and evolution of this biodiversity. It is now clear that different regions and clades have responded differently to climatic and geological changes, and that the Neotropical diversity is largely due to this historical complexity in addition to fine-scale ecological processes. Despite this, there is a lack of basic understanding of how Neotropical biodiversity was assembled over evolutionary timescales. Here we infer the timing and origin of the biota in all major Neotropical regions by performing a cross-taxonomic biogeographic analysis based on species from six major clades (angiosperms, anurans, birds, ferns, mammals, and squamates), species occurrence data, and dated molecular phylogenies. We show that Amazonia is at the heart of Neotropical diversity, serving as a source for thousands of lineages to other regions since the Late Cretaceous. Most biotic interchange has taken place between Amazonia and western South America, closely followed by exchange with Central America. The West Indies and the northern South American savannas are primarily sinks of diversity. In several cases, congruent patterns in the timing of biotic interchange across clades suggest that abiotic factors played a key role in shaping major patterns of biodiversity. These results demonstrate the complex origin of tropical ecosystems, and highlight the need of establishing biotic corridors across biomes and elevational gradients to safeguard future biotic interchange and diversification.

T1-27-06

Historical biogeography and diversification in the pantropical Bombacoideae

Alexander Zizka[1,2], *Jefferson Carvalho-Sobrinho*[3], *Suzana Alcantara*[4], *Luciano Queiroz*[5], *David Baum*[6], *Toby Pennington*[7], *Alexandre Antonelli*[1,2,8], *Christine Bacon*[1,2]
1. *University of Gothenburg*
2. *Gothenburg Global Biodiversity Centre*
3. *Universidade Federal do Vale do São Francisco*
4. *Universidade Federal de Santa Catarina*
5. *Universidade Estadual de Feira de Santana Feira de Santana*
6. *University of Wisconsin*
7. *Royal Botanic Garden Edinburgh*
8. *Gothenburg Botanical Garden*

The species of the pantropical clade Bombacoideae (Malvaceae) are characteristic elements of the vegetation in both Africa and South America and include such charismatic tree species as the baobabs (*Adansonia* spp.) and kapok (*Ceiba* spp.). Most species of the clade occur in South America, congruent with the commonly observed pattern of high species richness in South American lineages. Here we use fossil calibration from explicit morphological analyses to date a molecular phylogeny of 120 species of Bombacoideae based on nuclear and plastid data. We combine this phylogeny with large scale species distribution and morphological data to reconstruct the biogeographic history and trait evolution within the clade. Specifically, we date the dispersal events out of America and test the relationship of dispersal-related traits and climatic niche breadth with range size and diversification rates. Finally, we compare the results to analyses from other taxonomic groups and discuss the results in the context of recent studies suggesting high diversification rates and species turnover as causes for the extraordinary plant species richness in tropical South America, and proposing South America as global 'species pump'.

T1-27-07

Recent origin and diversification of the Andean orchid flora in the world's richest plant biodiversity hotspot

Oscar Alejandro Pérez-Escobar[1], *Guillaume Chomicki*[2], *Adam P. Karremans*[3,4], *Diego Bogarín*[3,4], *Alexandre Antonelli*[5,6,7]

1. Royal Botanic Gardens, Richmond, Surrey, TW9 3DS, UK

2. University of Munich (LMU), 67 Menzinger Str., Munich 80638, Germany

3. Lankester Botanical Garden, University of Costa Rica, P.O. Box 302-7050 Cartago, Costa Rica

4. Naturalis Biodiversity Center, Leiden, The Netherlands

5. University of Gothenburg, 41319 Gothenburg, Sweden

6. Gothenburg Botanical Garden, Carl Skottsbergs gata 22A, 41319, Gothenburg, Sweden

7. Gothenburg Global Biodiversity Centre, Box 461, SE-405 30 Göteborg, Sweden

Species richness is highly uneven across the globe, with both species-poor and mega-diverse areas. 'Biodiversity hotspots' could be areas of environmental stability where species accumulate through time or alternatively areas of major landscape change that directly or indirectly promote rapid species diversification. The Andean Mountains in South America host *ca* 15% of the world's plant species in only 1% of the world's land surface, hence it is the most species-rich biodiversity hotspot on Earth. One of the most prominent and diverse components of the Andean and neotropical flora are the orchids, yet the mode and tempo of evolution of their remarkable diversity are still poorly understood. We address these issues by inferring the biogeographical history and evolutionary dynamics of the two largest neotropical orchid groups (Cymbidieae and Pleurothallidinae, with *ca* 7,000 species). Using comparative phylogenetic methods, geological and biological datasets, and novel molecular sequences, we find that the majority of Andean orchid lineages only originated in the last 15 million years. Most Andean lineages are derived from lowland Amazonian ancestors, with additional contributions from Central America and the Antilles. This contrasts with previously reported patterns of mountain flora origins, where highland lineages were mostly derived from high-altitude, already pre-adapted immigrants. Species diversification is correlated with Andean orogeny, and multiple migrations and re-colonisations across the Andes indicate that mountains do not constrain orchid dispersal over long timescales. Our study suggests that landscape changes can greatly participate in the formation of anomalously rich areas – biodiversity hotspots.

T1-27-08

A community phylogenomics approach to the study of the origin and evolution of hyper-diverse Southeast Asian forests

Lisa Pokorny, *Steven Dodsworth, Timothy M. A. Utteridge, William J. Baker, Ilia J. Leitch, Félix Forest*

Royal Botanic Gardens, Kew

Most tree species are concentrated in tropical latitudes, with the Indo-Pacific region being as rich as the neotropics and each region estimated to harbor *c.* 22,000 tree species. Although tree species diversity is similar in these two regions, the level of threat they face is in sharp contrast. SE Asian forests have the highest deforestation rate of all tropical regions and, given that they are currently in a refugial state as a result of Quaternary Glaciations, these forests are particularly vulnerable to disturbance. As part of the ongoing research programme on "Global Tree Conservation through Seed Science" funded by the Garfield Weston Foundation, we present an overview of our project investigating angiosperm tree community assembly through time across SE Asia and neighboring regions. Our overall aim is to investigate how the drastic geological and climate shifts experienced throughout Earth's history have affected speciation, extinction, and migration processes in the tropical forest trees of SE Asia. This will be achieved by integrating new analytical methods and high-throughput sequencing techniques, with climate and niche modeling, to create a community phylogenomics framework of SE Asian forests at the genus level. We will then use this framework to explore a range of factors which may have contributed to the hyper-diverse SE Asian forests. For example, we aim to explore whether the SE Asian tropical tree flora has arisen predominantly by the immigration of species from neighboring areas or whether it is the product of *in situ* speciation/extinction processes modulated by climate shifts and geological processes (e.g. plate tectonics, volcanism). Since much of the lowland forests in SE Asia have experienced periodic changes in sea level which in turn have led to changes in niche availability, we will also examine what impact these changes have had on diversification rates (i.e., speciation, extinction, migration). This will be done by looking for evidence that changes in diversification rates in the lowland forests are linked to past sea level shifts. Similarly, the impact of changes in precipitation regimes (e.g., monsoons) will also be factored in. Finally, we wish to address the impact of past geological dynamics since we hypothesize that mountain formation as a result of plate tectonics will lead to opening of new niches and therefore to a biodiversity increase. Such a hypothesis would be supported if we were to observe increases in speciation or immigration across highland communities in relation to orogenic events in the region. Overall, the data generated by this study will highlight areas within the SE Asian region comprising the highest levels of unique diversity (e.g. phylogenetic diversity, phylogenetic endemism) and hence of most pressing conservation concern.

T1-27-09

The origin and phylogeography of mangroves *Xylocarpus* (Meliaceae) From the Indo-West Pacific

Zixiao Guo, *Haidan Wu, Yelin Huang, Suhua Shi*

State Key Laboratory of Biocontrol and Guangdong Provincial Key Laboratory of Plant Resources, School of Life Sciences, Sun Yat-sen University

The *Xylocarpus* genus is the only mangrove genus in family Meliaceae. Besides the ecological importance as mangrove species, plants of *Xylocarpus* are also valuable in medicinal usage that biological activities extracted from these plants show potent in antimalarial, antiallergic, antifeedant, antifilarial, antibacterial, antidiarrhoeal and antineoplastic properties. Recent molecular data placed it within the subfamily Swietenioideae, but its generic boundary between sister genera is unclear. By sequencing five chloroplast and five nuclear fragments, we aimed at resolving phylogenetic problems on both species and population levels. Phylogenetic trees indicated that *Xylocarpus* is closest to *Carapa* genus and diverged with *Carapa* in Africa within 16.39-27.52 million years ago during Miocene, and later diverged into three species, *X. granatum*, *X. moluccensis* and *X. rumphii*. Furthermore, *Xylocarpus* may have underwent a "out-of-Africa" dispersal, in which long distance dispersal (LDD) played important role. Phylogeographic analysis on the population level revealed that populations of *X. granatum* have split into three groups, Indian Ocean (IDO), Southeast Asia (SEA) and Australasia (AUA), whereas those of *X. moluccensis* split into West and East group with boundary along the Malay Peninsula. The chronological order of haplotypes suggested a west to east dispersal both in *X. granatum* and *X. rumphii*. Though closely related, barriers of land mass Sunda shelf and Ocean currents in Wallacea have different influences on *X. granatum* and *X. moluccensis*, which may attributed to the difference in their fruits' ability of LDD. Low level of genetic diversity was found on population level in all the three species, that reminds of conservative concern.

T1-27-10

Australasian (non-Orchidaceae) asparagales biodiversity

Bee Gunn[1], *Daniel Murphy*[1], *Neville Walsh*[1], *John Conran*[2], *Chris Pires*[3], *Terry Macfarlane*[4], *Joanne Birch*[5]
1. Royal Botanic Gardens Victoria
2. University of Adelaide
3. University of Missouri, Columbia
4. Western Australian Herbarium
5. University of Melbourne

Asparagales comprise about one third of all monocots in 12 - 14 families, predominant in the tropics (1,122 gen.; 36,205 spp.). The order includes Orchidaceae and economically important taxa (e.g. asparagus, onions and vanilla). Despite a high diversity within Australia (12 families; 48 genera and over 300 species), phylogenetic studies to-date have typically neglected sampling Australasian Asparagales taxa. This has led to difficulties in classifying Australian lineages and, as a result, evolutionary relationships among lineages remain uncertain. Additionally, extreme morphological heterogeneity in many lineages (e.g. Asparagaceae subf. Lomandroideae) makes accurate estimation of taxonomic boundaries difficult. In this newly initiated study we are using phylogenomic approaches and high throughput sequencing techniques, such as shotgun whole genome sequencing and exon capture, to generate a dated phylogeny, with dense sampling of Australian taxa. This study will investigate cryptic diversity at a fine evolutionary scale within *Lomandra*, one of the most complex genera of Australasian Asparagales, with c. 51 spp. found across a wide range of biomes. Phylogeographic analyses will enable investigation of the environmental processes that have shaped patterns of diversification in this genus, which includes widespread species and narrow endemics. The phylogenetic framework provided will be used to ask these questions: i) what are the biogeographical patterns of sister taxa among biomes? and ii) is there congruence in the patterns consistent with historical events? We will present a preliminary high throughput dataset, which comprises several hundred loci from chloroplast genome sequences from representative Asparagales families including lineages present in Australian mesic, arid and monsoonal tropics biomes. We will discuss trends in diversification of the Australasian Asparagales in the context of the expansion and contraction of Australian biomes. This study will provide insights into the influences of macroevolution and macroecology on Asparagales lineages by understanding diversification across multiple clades in historical climatic and geologic contexts.

T1-27-11

A new way to evaluate the evolutionary tempo of community assembly

Marlien Van Der Merwe[1], *Charles Foster*[2], *Maurizio Rossetto*[1], *Simon Ho*[2], *Hannah Mcpherson*[1]
1. Royal Botanic Garden, Sydney, Australia
2. University of Sydney, Sydney, Australia

Understanding the age of plant community assemblies can greatly improve our understanding of landscape genetics and will improve conservation management plans. Adding a temporal component to community turn-over can greatly increase our understanding of the environmental change and fluctuations that a community has experienced, which components of the assembly has experienced recent rapid turn-over and which components may have formed part of the more distant past of the assembly. In this paper we will discuss a simple method to obtain assembly age profiles using full chloroplast SNP data. The example data set uses whole genome Illumina libraries obtained from over 70 rainforest tree species with samples collected from multiple important rainforest refugial regions along the east coast of Australia. For each species and each library Chloroplast DNA was extracted bioinformatically and for each species a chloroplast reference genome was assembled. We could then estimate a species specific substitution rate for each species by using the sequence data (along with other publicly available chloroplast genomes) in a fossil calibrated phylogenetic analysis using an MCMC approach. The estimated substitution rates for each terminal branch was then used to estimate the TMRCA for each species at each of the three regions using a coalescent approach. The estimates were used to obtain age profiles for each region and each species. The results from this exercise highlights that a simple method can be applied to a relatively simple data set in order to obtain complex background information. The results support the more recent findings that the rainforests of Australia are dynamic assemblages and indicate how each region has undergone unique species-turnover. As expected from a dynamic system the results indicate that for each region a different set of species forms the older and younger components of the region. This method can be applied to any vegetation type and can be applied to landscape level ecological studies as in the given example of the Australian east coast rainforests.

T1-27-12

Floristics of the "Andean *tepuis*" of Ecuador and Peru: Diversity, endemism and phytogeographical links to the Guiana Shield

David Neill, *Mercedes Asanza*
Universidad Estatal Amazónica

The ancient sandstone mountains of the Guiana Shield, the *tepuis* of southern Venezuela and adjacent countries, are relatively well-known floristically with numerous endemic genera and species that have been described since the early 19th century. Less well-known are the "Andean *tepuis*, a term we apply to the sandstone mountain ranges with a north-south orientatation located east of the main cordillera of the Andes, including the Galeras and Cutucú ranges in Ecuador, the Cordillera del Cóndor along the Ecuador-Peru border, and the Escalera, Azul, and Yanachaga cordilleras in Peru. The Andean tepuis are much younger geologically than those of the Guiana Shield, but share similar edaphic conditions with soils derived from the nutrient-poor quartzitic sandstone substrate. Whereas the Precambrian Roraima Formation that constitutes the tepuis of the Guiana Shield has been available for plant colonization at its present location and elevation since the beginning of the rise of the angiosperms, the Cretaceous sandstone of the Hollín Formation that constitutes the Cordillera del Cóndor and other Andean tepuis has been uplifted with the recent tectonic movements of the Andes and has been exposed at its present elevation only within the last 5 million years. Disjunct distributions between the Guiana highland tepuis and the Andean tepuis have been recorded for a number of plant families, notably within genera harboring 30-40 species in the Guiana Shield and one or two outliers on the Andean tepuis; examples are *Bonnetia* (Bonnetiaceae) and the *Crepinella* group of *Schefflera* (Araliaceae). Similar disjunctions between the Andean tepuis and the white-sand areas of lowland Amazonia include *Pagamea* (Rubiaceae) and *Guatteria* (Annonaceae). Numerous locally endemic taxa have been described from the Andean *tepuis* in recent decades; notable among these are species within species-rich genera which are mostly mid-canopy trees in Andean cloud forests such as *Clethra* (Clethraceae), *Weinmannia* (Cunoniaceae) and *Miconia* (Melastomataceae), for which the local endemics on the highest summits of the Andean tepuis are low shrubs growing in open shrublands on the oligotrophic sandstone outcrops. Many of the Andean tepui endemics share a characteristic of small, thick sclerophyllous leaves, evidently an adaptation to the nutrient-poor soils. Much more floristic and taxonomic work is required before the floras of the Andean tepuis can be adequately characterized and enumerated, and efforts are currently underway by botanical researchers in Ecuador and Peru to meet this challenge in coming years. Phylogenetic and phylogeographic analyses of the Guiana Shield-Andean tepui disjunctions are lacking at present but are an important goal for future research.

T1-28: The biology of mycoheterotrophic plants (two sessions)

T1-28-01

Comparative plastome genomics in Ericaceae: Plastid gene losses and rearrangements across all trophic levels

Thomas Braukmann[1,2], *Michael Broe*[3], *John Freudenstein*[3], ***Sasa Stefanovic***[1]
1. University of Toronto Mississauga
2. Biodiversity Institute of Ontario
3. Ohio State University

Non-photosynthetic, heterotrophic plants often have a highly reduced plastome associated with the transition from an autotrophic to mixotrophic to fully heterotrophic nutritional mode. They are either parasitic via direct connection with other plants (i.e. haustorial parasites) or are epiparasitic on mycorrhizal fungi that are themselves in symbiotic relationships with autotrophic plants (i.e. mycoheterotrophs; MHT). There are at least 21 independent origins of heterotrophy among eudicots. Ericaceae, the heather family, is a large and diverse group of plants that forms elaborate symbiotic relationships with mycorrhizal fungi, and includes autotrophs as well as fully and partially mycoheterotrophic plants. Based on input from a broad Southern hybridization survey, and using a next generation sequencing approach, we sequenced plastomes of 14 Ericaceae species across the trophic spectrum. This includes plastomes of four autotrophic (*Enkianthus, Gaultheria, Epigaea, Arctostaphylos*), three mixotrophic (*Chimaphila, Moneses, Pyrola*), and seven (*Allotropa, Hemitomes, Hypopitys, Monotropa, Monotropsis, Pityopus, Pterospora*) genera. To investigate potential intraspecific variation, the latter group saw two individuals sequenced for *Allotropa virgata*, *Monotropa uniflora*, and *Pityopus californica*. Previously sequenced plastomes of autotrophic Ericaceae (*Arbutus, Vaccinium*) were also included in analyses as well as those of *Actinidia* (Actinidiaceae) and *Camellia* (Theaceae), as progressively more distant outgroups. Our results indicate that the rearrangements of the large single copy region of the plastome are commonplace for most member of Ericaceae. Associated with these extensive structural rearrangements of the plastome across the family is an increase in forward and palindromic repeats, including in fully mycoheterotrophic species. Additionally, most Ericaceae share the loss of the plastid *ycf1* and *ycf2* genes, which is rare even among other fully heterotrophic plant lineages. *clpP* is either absent or present as a highly divergent ORF. Mixotrophic plants retain most genes relating to photosynthesis but are variable for the plastid *ndh* genes. The plastomes of fully heterotrophic Ericaceae are highly reduced in both size (~33-41 kbp) and content, having lost a copy of the inverted repeat and all photosynthesis related genes; they are reduced to encoding housekeeping genes and an unusual plastid acetyl-CoA carboxylase subunit (*accD*)-like open reading frame. Intraspecific variation within MHT Ericaceae ranges from a few differences (*Allotropa*) to extensive population divergence (*Monotropa*). Despite an increase in the rate of their nucleotide substitutions, we found that the plastomes of Ericaceae are generally under strong purifying selection, including those of partially and fully mycoheterotrophic plants in their remaining protein-coding genes. However, several classes of genes are under positive selection, including the *ndh* genes. Furthermore, positive selection is acting on the branch leading to subfamilies Vaccinoideae and Ericoideae for the *rps*, *rpl*, and *rpo* genes. We also identify a number of ribosomal proteins under relaxed or neutral selection. Finally, we demonstrate and discuss the limitations of large plastid data sets in resolving the number of independent origins of mycoheterotrophy in Ericaceae.

T1-28-02

Partial mycoheterotrophy is more widespread among orchids than previously assumed: A multi-element stable isotope natural abundance approach

Gerhard Gebauer, *Julienne Schiebold*
University of Bayreuth, BayCEER - Laboratory of Isotope Biogeochemistry

Based on their isotopic positioning between autotrophic non-orchids and fully mycoheterotrophic orchids some green-leaved orchids have been identified during the last decade as gaining simultaneously carbon from two different sources, own photosynthesis and fungal supply. This mode of mixotrophic carbon gain is called partial mycoheterotrophy. All of these ^{13}C- (and ^{15}N-) enriched partially mycoheterotrophic orchids are mycorrhizal with fungi simultaneously forming ectomycorrhizas with forest trees, while the majority of green orchids worldwide are mycorrhizal with fungi of the phylogenetic heterogeneous assemblage of rhizoctonias. Recent work has documented that carbon and nitrogen isotope abundance is not suited to unequivocally identify partially mycoheterotrophic species among rhizoctonia-mycorrhizal green orchids. Here we provide an essentially new approach to identify species with partially mycoheterotrophic nutrition among rhizoctonia-mycorrhizal orchids. In addition to carbon and nitrogen stable isotope abundance we analysed hydrogen isotope abundance in dry tissues of a fully MH orchid, in green orchids associated either with ectomycorrhizal fungi or with rhizoctonias, in autotrophic reference plants and in sporocarps of ectomycorrhizal and saprotrophic fungi growing together in the understorey of a dark beech forest and of a broad range of orchids and autotrophic reference plants from sunny montane meadows. It turned out that almost all orchids were significantly enriched in heavy hydrogen isotopes relative to autotrophic reference species irrespective whether they were enriched in heavy carbon isotopes or not. In addition to carbon, nitrogen and hydrogen stable isotope natural abundance, we also analysed oxygen stable isotope abundance in order to rule out different transpiration rates of orchids compared to reference plants as responsible for the differences found for the hydrogen isotope abundances. Thus, partial mycoheterotrophy appears to be much more widely distributed among green orchids than previously assumed.

T1-28-03

Arbuscular mycorrhizal interactions of mycoheterotrophic plants

Vincent Merckx[1], *Sofia Gomes*[1,2]
1. Understanding Evolution, Naturalis Biodiversity Center
2. Institute of Environmental Sciences, CML, Leiden University

More than half of all species of fully mycoheterotrophic plants—plants that live on fungi during their entire lifecycle—are associated with mycorrhizal fungi of the phylum Glomeromycota. These fungi form arbuscular mycorrhizas with over 80% of all land plant species, which makes this the most common and widespread mycorrhizal interaction on earth. Unlike the mutually beneficial interactions between photosynthetic plants and arbuscular mycorrhizal fungi, which are generally characterized by low specificity, mycoheterotrophic plants often show high specificity towards particular lineages of arbuscular mycorrhizal fungi. Here we investigate these mycoheterotrophic interactions in an evolutionary and ecological context. We demonstrate that not all mycoheterotrophic flowering plant species grow on an equally narrow range of arbuscular mycorrhizal fungi. In two mycoheterotrophic plant lineages with extremely specific interactions we measure a high level of phylogenetic niche conservatism, indicating that specificity is retained during evolution and diversification. Within one of these mycoheterotrophic lineages a comparison of the mycorrhizal diversity between roots of mycoheterotrophs and surrounding plants shows that mycoheterotrophic plants consistently target more narrow lineages of fungi than photosynthetic plants, despite the larger fungal pool available in the soil. These data indicate that host specialization is a common consequence of the evolution of mycoheterotrophy in the arbuscular mycorrhizal symbiosis. The tendency of mycoheterotrophic plants to target specific lineages of arbuscular mycorrhizal fungi supports the idea that different fungal lineages have different ecological functions.

T1-28-04

Global distribution of mycoheterotrophic plants

Sofia Gomes[1,2], *Nadia Soudzilovskaia*[2], *Peter Van Bodegom*[2], *Vincent Merckx*[1]
1. Understanding Evolution, Naturalis Biodiversity Center
2. Institute of Environmental Sciences, CML, Leiden University

Mycoheterotrophic plants are achlorophyllous, non-photosynthetic plants, with the exceptional mode of life of obtaining carbon from the mycorrhizal fungi in their roots. This ability of cheating mycorrhizal networks is relatively widespread in land plants and evolved multiple times independently. The diversity of fungi involved in this interaction is quite large, including members of the phylum Glomeromycota, Basidiomycota and Ascomycota. Mycoheterotrophic plants associated with arbuscular mycorrhizal fungi (Glomeromycota) generally occur in tropical rainforests, while the mycoheterotrophs associated with ectomycorrhizal fungi, and more rarely saprotrophic fungi, (Basidiomycota and Ascomycota) are restricted to temperate areas, with some exceptions in the tropics for the latter. While many mycoheterotrophic plants show a high specificity towards narrow lineages of fungi, the distribution of mycoheterotrophic plants is clearly more limited than that of their fungal partners. Thus, the occurrence of mycoheterotrophy is not explained by the presence of suitable host fungi alone. There is an insufficient knowledge on the ecological drivers that limit mycoheterotrophic plants' distribution. Here, we attempt to infer global ecological niches of these plants by using species distribution modelling. We focus on fully mycoheterotrophic plants in angiosperms, and include representatives of all lineages growing on arbuscular, ectomycorrhizal and saprotrophic fungi. Our results contribute to the understanding of ecological conditions where mycorrhizal networks are cheated at a global scale.

T1-28-05

Organellar phylogenomics and molecular evolution in mycoheterotrophic plants

Vivienne Lam[1], *Hayley Darby*[1], *Qianshi Lin*[1], *Wesley Gerelle*[1], *Vincene Merckx*[2], ***Sean Graham***[1]
1. University of British Columbia
2. Naturalis Biodiversity Center

About 50 independent lineages of land plants have completely lost photosynthetic ability as a consequence of mycoheterotrophy —they obtain their carbon and other nutrients from mycorrhizal or saprophytic fungi housed in specialized root systems. These plants likely evolved from photosynthetic plants that complemented photosynthesis for all or part of their life cycle in this manner. Examples of the latter (partial and initial mycoheterotrophs) are known in several families, including orchids. Full-blown mycoheterotrophy can have profound effects on genome evolution, particularly the plastid ('chloroplast') genome, which is thought to degrade in a relatively ordered manner after loss of photosynthesis, according to recent evolutionary models of gene loss and retention. Long-term persistence of the plastid genome in non-photosynthetic plants is likely a consequence of delayed losses of multiple non-bioenergetic genes. Here we report on plastid genomes retrieved from one conifer family and nearly all of the 10-11 angiosperm families that include mycoheterotrophic plants. We document instances of gene loss and retention, changes in selective regime, and patterns of genome structural change and rate elevation across recent and ancient examples of photosynthesis loss, and several partially mycoheterotrophic lineages. Although often idiosyncratic, some general evolutionary patterns emerge. We also address the molecular evolution of the mitochondrial genomes of mycoheterotrophs, and apply these new organellar data in a phylogenetic context to address higher-order placements and relationships among mycoheterotrophic lineages of plants.

T1-28-06

Plastid genome evolution in mycoheterotrophic orchids

Craig Barrett[1], *John Freudenstein*[2], *Aaron Kennedy*[3], *Chodon Sass*[4]
1. West Virginia University
2. Ohio State University
3. US Department of Agriculture - APHIS
4. University of California, Berkeley

Heterotrophic plants present numerous examples of extreme morphological and genomic reduction. Models of plastid genome (plastome) degradation have been put forth recently for these plants, focusing initially on gene loss/retention and pseudogene formation. The most recent models have evolved to incorporate additional genomic characteristics including base frequency changes, indel accumulation, genomic rearrangement, repeat accumulation, substitution rates (both synonymous and non-synonymous), and structural dynamics of the inverted repeat. Nearly half of the approximately 400 described holomycotrophs (HM) belong to Orchidaceae. The number of transitions from the partially mycoheterotrophic (PM) to HM lifestyle in Orchidaceae is over 30, and this number may increase as fine-scale phylogenetic and physiological studies are completed. Several researchers have published complete plastomes of holomycotrophic orchids, which display a range in the degree of degradation due to relaxed selective constraints of photosynthesis. However, most have focused on single plastomes, allowing only comparisons to (relatively) distantly related orchids. Aside from plastome evolution, very little is known about differences in gene expression or nuclear genome evolution across the spectrum of autotrophy, partial mycoheterotrophy, and holomycotrophy. We have developed the leafless, North American genus *Corallorhiza* as a non-standard model for studying morphological, physiological, ecological, and genomic consequences of obligate parasitism upon fungi. *Corallorhiza* is a promising system because the amount of green, putatively photosynthetic tissue varies significantly among its 12 species. Whole plastome sequencing has demonstrated that *Corallorhiza* is early in the process of plastome degradation, compared to some other parasites and mycoheterotrophs. The genus originated ~2.6 million years ago, and the two holomycotrophic lineages likely arose ~1.9 (*C. striata* complex) and ~1.3 (*C. maculata* complex) million years ago. Expanded sampling of 48 plastomes across the geographic range of the *C. striata* complex revealed substantial infraspecific variation in plastome size and gene content, with plastomes ranging from 141 kb in *C. striata* var. *striata* to 124 kb in the *C. bentleyi-C. involuta* clade. Furthermore, a phylogeographic approach to assessing plastome degradation in this species complex revealed two putatively independent transitions to HM in the complex, in addition to that in the *C. maculata* complex. Another North American genus containing HM species is *Hexalectris*. This clade of approximately 10 species represents a powerful system for comparison with *Corallorhiza* and members of Neottieae, the only orchid taxa for which both green, photosynthetic members and non-green, nonphotosynthetic members have been sequenced in a phylo-comparative fashion. Thirty plastomes were sequenced from *Hexalectris* and the related, photosynthetic *Bletia* and *Basiphyllaea*. Results will be compared with those from *Corallorhiza* and Neottieae, to test predictions within the context of recent models of plastome evolution, and to refine these models. Additionally, data comparing plastid and nuclear gene expression patterns between PM and HM *Corallorhiza* will be discussed. The ultimate goal of this research program is to take a comparative approach across all of Orchidaceae, in order to characterize patterns and processes of plastome degradation, and to similarly test hypotheses based on nuclear genome evolution and gene expression.

T1-28-07

Phylogenetic analysis of the mycoheterotrophic genus *Thismia* (Thismiaceae, Dioscoreales) based on molecular and morphological data

E. Shepeleva[1], *M. Nuraliev*[1,2], *M. Logacheva*[1], *S. Chantanaorrapint*[3], *M. Hroneš*[4], *S.S. Mar*[5]
1. M.V. Lomonosov Moscow State University, Moscow
2. Joint Russian-Vietnamese Tropical Scientific and Technological Center, Hanoi
3. Prince of Songkla University, Hat Yai
4. Palacky University, Olomouc
5. Ying Wa College

Thismia Griff., comprising more than 60 species, is the largest genus of the family Thismiaceae. It is widely distributed in tropical regions with a few species in subtropical and temperate zones. Members of the genus *Thismia* are small mycoheterotrophic herbs with highly reduced vegetative structure and intricate floral morphology. Current views on phylogenetic affinities within this genus are controversial. Recent analyses have revealed paraphyly of the genus *Thismia*; however, they were based on limited sampling, and the genes used in these studies had high substitution rates. Thus, these results remain provisional and are still to be verified; besides, most of the species of *Thismia* have never been studied in terms of molecular phylogeny. We have examined the relationships between members of the genus *Thismia* on the basis of extended sampling which included originally collected specimens. A total of 21 species were used for phylogenetic studies, which were carried out by analyzing molecular and morphological data. The morphological dataset was compiled basing on careful consideration of available literature and original observations. The molecular phylogenetic analysis involved markers from the nuclear (18S, 5.8S rRNA genes and the internal transcribed spacers, ITS1 and ITS2) and mitochondrial (*atp1* gene) genomes. We have found that internal transcribed spacers show high level of interspecific divergence, which makes it impossible to use them for assessing phylogenetic relationships at the generic level. Nevertheless, ITS regions are revealed to be significant in resolving the phylogenetic affinities within groups of closely related species. Analysis of the other molecular markers allowed reconstructing phylogenetic relationships at the level of the whole genus *Thismia*. As in the previous studies, in our trees *T. panamensis* (Standl.) Jonker from Central America is distantly related to the Old World taxa. Then, several monophyletic groups were determined. One of them is a strongly supported group comprising five species: *T. annamensis* K.Larsen & Aver. fromVietnam, *T. filiformis* Chantanaorr. and *T. alba* Holttum ex Jonker fromThailand, *T. aseroe* Becc. and *T. inconspicua* ined. from Borneo. Another one consists of the Vietnamese *T. mucronata* Nuraliev and *T. puberula* Nuraliev and Thai *T. nigricans* Chantanaorr. & Sridith and *T. mirabilis* K.Larsen. Evolutionary pathways of morphological characters were studied by mapping their conditions onto our combined molecular phylogenetic tree. The following characters were examined: floral symmetry, presence of a mitre, leaves, appendages of tepals, a wing-like appendage of a connective, number of stamen appendages and some others. Evolution of some of them, including the presence of transverse bars inside the hypanthium and the number of stamen appendages, was found to be homoplastic. Our investigation provides important data for establishing a new taxonomic system of the genus *Thismia*. As they suggest, certain morphological characters traditionally employed in its taxonomy show more complicated evolution than it was believed before. We propose a list of characters which can be used for morphological description of the monophyletic groups supported by molecular data.

T1-28-08

Mycorrhizal diversity and nutritional strategies in the fully mycoheterotrophic orchid *Epipogium roseum*

Lorenzo Pecoraro[1,2], *Lei Cai*[1], *Gerhard Gebauer*[3], *Guo-Qiang Zhang*[2], *Yong-Zhao Diao*[1], *Mubashar Raza*[1], *Zhong-Jian Liu*[2]

1. State Key Laboratory of Mycology, Institute of Microbiology, Chinese Academy of Sciences

2. Shenzhen Key Laboratory for Orchid Conservation and Utilization, The National Orchid Conservation Center of China and The Orchid Conservation and Research Center of Shenzhen

3. BayCEER - Laboratory of Isotope Biogeochemistry, University of Bayreuth

All orchid species completely rely on their mycorrhizal fungi for the provision of nutrients, at least during the non-photosynthetic protocorm stage. At the adult stage, most orchids develop a photoassimilating apparatus and become autotrophic, while fully mycoheterotrophic orchids are achlorophyllous and remain dependent on fungal sugars even at maturity. Recent advances in studying of orchid mycorrhizas are linked to the development of molecular techniques that enable identification of fungal symbionts, whereas the recent application of stable isotope analyses has represented a powerful tool for understanding nutrient exchange between orchids and fungi. In spite of the increasing number of studies describing fungal diversity in orchids, there is still more to be learnt about the identity and functional role of fungal partners and specificity in orchid mycorrhizal associations. We investigated the diversity of root fungal associates in the fully mycoheterotrophic forest orchid *Epipogium roseum* Lindl., using morphological and molecular techniques. We also applied stable nitrogen and carbon isotope abundance analyses in order to test whether nutrients are gained from the fungal symbiont. Orchid underground organs (stolons and tubers) were collected from fifteen *E. roseum* adult individuals, in three geographically distinct protected areas in Yunnan Province (South China). Total underground organ DNA was extracted and fungal ITS regions were PCR amplified using ITS1F/ITS4, ITS1/ITS4-Tul, and ITS1-OF/ITS4-OF primer pairs and sequenced. Microscopic observations of *E. roseum* hypogeous system sections revealed an extremely low level of fungal colonization. Molecular analysis uncovered a range of fungi, belonging to *Basidiomycota* and *Ascomycota*, in the rhizomatous system of the investigated orchid species. We did not find any dominant associated fungal taxon. On the contrary, fungal species belonging to different genera, such as *Tulasnella*, *Coprinellus*, *Ilyonectria*, and *Fusarium* were sporadically detected in the analysed orchid hypogeous organs. Carbon and nitrogen stable isotope abundances were measured from flower stalk samples of *E. roseum* and leaves of three non-orchid species collected around each orchid individual, maximum 1 m apart. All orchid plants were significantly enriched in both ^{13}C and ^{15}N compared with neighbouring reference species, showing that the investigated orchid species depends on mycorrhizal partners for nutrition. The analysis of mycorrhizal strategies in *E. roseum*, performed in our work, provides valuable information on the biology of this threatened orchid species that could be of great value for its conservation.

T1-28-09

Phylotranscriptomic analysis of mycoheterotrophic and non-mycoheterotrophic lineages in the Pandanales and Dioscoreales

Claude Depamphilis[1], *Prakash Raj Timilsina*[1], *Alma Piñeyro Nelson*[2], *Joel Mcneal*[3], *Saravanaraj Ayyampalayam*[4], *Eric Wafula*[1], *James Leebens-Mack*[4], *Elena Alvarez-Buylla*[2]

1. Pennsylvania State University

2. Universidad Nacional Autonoma de Mexico

3. Kennesaw State University

4. University of Georgia

Mycoheterotrophy has evolved independently within the sister monocot orders Pandanales and Dioscoreales. Transcriptome data were generated for exemplars representing all APG III families for these two orders including mycoheterotrophic species *Gymnosiphon sp.* and *Burmannia biflora* within the Burmanniaceae (Dioscoreales), and *Lacandonia schismatica* and *Triuris* sp. within the Triuridaceae (Pandanales). Phylogenomic analyses of these data are resolving relationships among families and elucidating the timing of gene and genome duplication events in the Pandanales/Dioscoreales clade. Further, divergence of the molecular and biological functions of genes expressed in mycoheterotrophic and non-mycoheterotrophic lineages is being characterized. Results from these analyses are forming a firm foundation for investigations of the origin of mycoheterotrophy in the Pandanales and Dioscoreales.

T1-29: Ethnobotany and ethnopharmacy

T1-29-01

The mission and development of molecular pharmacognosy

Lu-Qi Huang
China Academy of Chinese Medical Sciences

Traditional Chinese medicine (TCM) resources which is a kind of special resources, has a great value in utilization and development. It is related to the matter of people's livelihood, social stability, ecological environment protection and development of the emerging strategic industries. As the embodiment of national advantages in global competition, TCM resources has important strategic significance. With the development of society, the demand of medicinal materials keeps increasing. However, the awareness on how to develop and use TCM resources rationally is far from enough, which leads to that the sustainable development and utilization on TCM resources faces enormous pressure. With the continuous development of molecular biology, more and more technologies have been applied in sustainable development and utilization on medicinal plants. The biosynthetic pathway of secondary metabolites from medicinal plants has gradually been discovered. Using synthetic biology approaches by heterologous expression of the entire biosynthetic pathway in microorganism and channeling metabolic flux, could produce plant natural products with a high productivity. This strategy could be used for production of rare active ingredients, and provide an alternative route for the production of single-component from Chinese herbal medicine and its extract. This biosynthetic pathway analysis and the strategy of synthetic biology production have played a more and more important role in the production of plants' natural products, such as the famous Artemisinin, Opioid Analgesics, Paclitaxel, *etc.* Since 1995, when the molecular pharmacognosy was put forward for the first time, molecular pharmacognosy has become a mature inter-discipline which has a stable research direction, leading technology level, innovative theory, widespread academic influence and reasonable discipline team during 20 years of development. At present, more than 20 universities have opened the course of molecular pharmacognosy for undergraduate and graduate students. In 2012, molecular pharmacognosy (English version) which was published by Springer press, marked that molecular pharmacognosy has gained the international recognition. Currently, TCM synthetic biology has become the key research area in molecular pharmacognosy. Moreover, getting the medicinal active component by using the strategy of TCM synthetic biology is becoming the one of the most potential way for achieving resources recognized by international authority in this research are. It will become an important approach on TCM resources sustainable utilization.

T1-29-02

The specific use of wild species as medicinal plants of small ethnic group of the North

Kirill Tkachenko
Komarov Botanical Institute of RAS, St. Petersburg

Li Shizhen ideas set out in of the Compendium of Materia Medica are the basis for the development of Ethnobotany and ethnopharmacy. Currently, it is very important to have time to collect endangered folk details about the features of the use of small nations species of local flora as useful species (medicinal, food, hygiene). Collection and analysis ethnobotany data about using of wild species of local flora as medicinal plants people from small ethnic population on the North of Russia. Flora of North-West part Russia includes approximately 2700–3000 species of vascular plants. Small ethnic groups look like as Veps, Karelians, Izhora, Vodes, Seto-Estonians, Komi-Zyrian, Saami-they using as helpful (medicinal and edible) plants only some species from local flora (not more than 1-2% from the total number of species from flora of North-West of Russia). Modern information technologies penetrate deeper into the lives of different segments of the population. This leads to a rapid loss of traditional knowledge accumulated by many centuries. Everything is quickly erased the memory of the use of different kinds of plants to treat diseases, using them as amulets, and their role in many ceremonies. Only in small remote villages can still find people who preserve old knowledge about the use of plants as useful. It is important to have time to collect and store grain of people's knowledge and skills in the use of wild plant species as useful (food, medicinal, ceremonial, construction, ritual). Fewer young people who pass native folk knowledge about the use of plants. Particular attention should be paid to informants (who to interview and how). Many healers, who keep native people's knowledge, often do not go to the contact, and do not share their knowledge to use as a medicinal plant. In many countries, especially in the last decade, a lot of attention to the collection of ethnobotanical and ethnopharmacy data. These works are important to and among minorities of the North, whose lives are still largely dependent on the connection with Mother Nature, the ability to use the plant as useful. In the data collection process it has been revealed that a single national group, but residing remotely from each other, one and same species are not equally used. Different people often use various types of plant for the treatment of diseases or as edible. Currently requires the organization of a wide collection of information on the use of wild species of local flora of the local population as useful, especially as food and medicines. The most complete knowledge about potential new sources from local floras for the development of new drugs of plant origin, derived from data collected ethnobotanical, will identify promising new kinds of complex medical and biological research and development of new highly effective drugs.

T1-29-03

Using a bioassay-guided approach to investigate the medicinal properties of plants routinely utilized in traditional Chinese medicine (TCM)

***Elliot Altman**, Ying Gao*
Middle Tennessee State University

The Tennessee Center for Botanical Medicine Research (TCBMR) at Middle Tennessee State University (MTSU), Murfreesboro, Tennessee, USA, in collaboration with the Guangxi Botanical Garden of Medicinal Plants (GBGMP), Nanning, China and the Institute of Medicinal Plant Development (IMPLAD), Beijing, China, has employed a bioassay-guided approach to investigate the medicinal properties of 40 plants that have been utilized routinely in traditional Chinese medicine (TCM). The study served two purposes; (1) to evaluate the effectiveness of bioassay-guided screens and (2) to evaluate the correlation between the properties of the plants that were identified in the bioassays versus what was predicted according to the Chinese pharmacopoeia. The TCBMR is an interdisciplinary center that has expertise in using rapid *in vitro* cellular screening approaches to develop antibacterial, anticancer, antifungal, antiprotozoal, antiviral, or immunosuppressant agents and in the purification and identification of compounds. In this study a series of extracts were prepared from each of the 40 plants and then the extracts were tested for antibacterial, anticancer, antifungal, antiprotozoal, antiviral, or immunosuppressant activities. A number of plants with antibacterial, anticancer, antifungal, antiprotozoal, antiviral, or immunosuppressant activities were identified and while there was good correlation with the bioactivities that were identified versus what is predicted by the Chinese pharmacopoeia, a number of new yet unknown bioactivities were discovered as well. Pure compounds were isolated from some of the more promising potent plant extracts using bioassay-guided fractionation. Some of the compounds that were isolated, such as apigenin, an antiviral agent, hederagenin, an anticancer agent, and pandanusphenol, an anticancer agent, have been identified before, but some of the compounds that were identified were novel, such as *cis*- and *trans*- gnetin H, two anticancer agents and iso-iridogermanal, an antiprotozoal agent. This study further demonstrates the importance of known botanicals with medicinal properties as sources for new pharmaceutics.

T1-29-04

From comprehensive analysis to quality standards of Chinese herbal medicines

Dean Guo
Shanghai Institute of Materia Medica, CAS

Chinese herbal medicine is an extremely complex system, which contains hundreds or even thousands of chemical components in one herb, not to say compound formulas composed of several or tens of herbs. This generates great difficulty to clarify its chemical composition, which is the fundamental basis for the research of quality control, mechanism, pharmacological and toxicological actions, PK/PD, clinical trial, new drug discovery, *etc.* Thanks to the new analytical technique advancement, comprehensive analysis for traditional Chinese medicines and other herbal medicines becomes feasible. In the past several years, we dedicated ourselves to the herbal analysis by using some of the new techniques or instruments, such as UPLC, 2DLC, HRLCMS, UPC2, *etc.* to facilitate the elaboration of holistic quality control standards of herbal complex systems. In this lecture, several typical herbs were exemplified by using this approach. For examples, similar herbs of Panax species named *Panax ginseng*, *Panax quinquefolium, and Panax notoginseng* were analyzed by LCQ Advantage ion-trap mass spectrometer to profile their ginsenosides. To fully understand the chemical diversity of Tumeric (the rhizomes of *Curcuma longa*), multiple neutral loss/precursor ion scanning combined with substructure recognition and statistical analysis was utilized. Using this approach, 846 terpecurcumins (terpene-conjugated curcuminoids) were discovered from turmeric, including a number of potentially novel compounds. In addition, some new developed techniques as 2DLC and UPL2 were also applied in the herbal analysis. Based on the comprehensive analysis results, holistic quality standard of TCM complex systems was elaborated. The strategy and key factors of holistic standard were established and the quality monographs of a number of TCM herbs were elaborated and adopted by several mainstream pharmacopoeias including Chinese Pharmacopoeia, United States Pharmacopoeia and European Pharmacopoeia. It can be concluded that with the new development of analytical techniques, more and more chemical components in the complex herbal system will be clarified to facilitate their quality control and clarification of therapeutic substances.

T1-29-05

Beyond GAP -- Ecological agriculture considering the entirety of an ecosystem in the cultivation of medicinal plants

Thomas Garran
China Academy of Chinese Medical Sciences

Cultivation of medicinal plants is an important issue across the globe. In countries such as China, India, and some other nations, large tracks of land are used for this increasingly important and economically valuable resource. However, in our rush to serve this need, we often do not use sustainable agricultural practices in the management of this land. Meanwhile throughout the world the harvest of wild medicinal plant resources is putting enormous pressure on biodiversity, both of the harvested species and the entire ecosystem that depend on these species. The vast majority of medicinal plants used globally still come from the wild, while most of those in cultivation have only been domesticated for a relatively short period of time. Finding ways to improve the yield of those already in cultivation, as well as understanding how to bring wild plants into an agricultural settings is a serious challenge. Putting aside the exceptions of weedy species used in medicine, i.e. *Taraxicum* sp., *Plantago* sp., *etc.,* the vast majority of medicinal plants are found in relatively specific ecosystems and under specific bio-regional conditions. However, experience has taught us that simply growing a plant within that bio-region may be insufficient for long-term sustainable cultivation, and, in fact, such methods often yield very poor results. China has a long history of a concept known as *Daodi* herbs, which is to say that a medicinal plant produced in a specific region/area (within its native habitat) is considered the best clinically therapeutic material, this concept was largely based on the wild material, but it has been transferred to modern agricultural needs. Furthermore, a review of the material on cultivation of medicinal plants, while extensive in many areas, generally overlooks the concept of the soil as an ecosystem, and thus compares data on plant available nutrients in the soil (N, P, K, *etc.*), pH, *etc.* as well as the above-ground ecosystem, i.e. temperature range, annual rainfall, average number of days without frost, elevation, *etc.* This has created a clash between archaic agricultural concepts used for growing food with a developing field of medicinal plant agriculture. However, food production has, largely, failed to consider the soil as an ecosystem consisting of bacterial, fungi, protozoan, nematodes, earthworms, and other micro- and macroarthropods that plants depend on in a multitude of ways. This dependence is especially true for plants being moved from wild habitats to cultivated habitats, because these wild plants have long developed a relationship with the soil ecosystem consisting of these organisms, and are essential for their healthy growth. Looking at past work that has considered these issues for the cultivation of various plants, we will discuss current research showing how to apply this concept to what can be truly called "ecological agriculture" of medicinal plants.

T1-29-06

Therapeutic Principles and Methods guided identification of renoprotective substances from Chinese medicines

Yong-Xian Cheng

Shenzhen University Health Science Center

Chronic kidney disease (CKD) has become a global public health problem that accounts for more than 10% of the population in the developed countries. Renal fibrosis is known to be the inevitable consequence of CKD which could in turn lead to a progressive and irreversible loss of renal function, a process known as end-stage renal disease (ESRD) needing dialysis or organ transplantation. Thus finding new approaches aimed at attenuating renal fibrosis is of great importance. In the recent years, search for renoprotective substances from Chinese medicines has become our research focus. Inspired by the traditional uses of Chinese medicine, in particular, guided by unique therapeutic principles and methods of traditional Chinese medicine (TCM), we selected medicinal resins and *Ganoderma* fungi respectively with blood circulation promoting, tonic effects as our research materials, and structurally diverse renoprotective substances were successfully identified targeting TGF-β1/Smads, Wnt/β-catenin and Nrf2/Keap1 signaling pathways. Our practice justifies that therapeutic principles and methods based exploration of Chinese medicines might aid gaining new insights into the scientific connotations of Chinese medicines.

T1-30: TM and drug discovery

T1-30-01

Discovery of the effective components of Chinese herbal medicines

Min Ye

Peking University

Chinese herbal medicines are gaining increasing attention worldwide due to their significant therapeutic effects. However, the effective components, i.e. chemical constituents responsible for the therapeutic effects, are usually not well defined. This is challenging due to the complicated chemical composition. A comprehensive understanding on the chemical constituents and *in vivo* metabolism is critical to dissect the effective components of herbal medicines. We have developed a series of analytical methods, including LC/MS, 2DLC and SFC to globally profile the chemical constituents of herbal medicines, and to discover novel compounds with significant bioactivities. We have also proposed a strategy to monitor the multi-component pharmacokinetics of herbal medicines. Combined with biological evaluations, these new techniques have been used to elucidate the effective components of licorice (Gan-Cao).

T1-30-02

Discovery of novel chemical structures from some traditional medicinal plants and their bioactives

Ming-Hua Qiu

Key Laboratory of Phytochemistry and Plant Resources in West China, Kunming Institute of Botany, Chinese Academy of Sciences

In the Foreword of famous monograph "Natural Pruducts — Their Chemistry and Biological Significance", its authors, some famous professors said: Naturally occurring organic chemical compounds (natural products) have always fascinated chemists. Interesting and intriguing chemistry is involved in their *in vivo* production and their laboratory utilization, and their importance as structural material and biologically active molecules (substrates for life processes, toxins, hormones, drugs, *etc.*) is of unparalleled importance. It was very difficult that we got similar felling with them, because they are of great scholars, but we already were subdued by the natural product structures of fascinating and diversity. Additionally as well-known, new molecules from traditional medicinal plants are highly signi? Cant sources of new discovery and development process. Therefore, our research group have been focused on the discovery of novel chemical structures from traditional medicinal plants and their bioactives. In the presentation, I would like to share the three features of our results, which finished in investigation of chemistry and bioactives of some TM plants in a couple of years, as following: 1. discovery of novel cycloartane triterpenoids from Cimicifuga spp. and their bioactives as well as the mechanism of bioactive molecules; 2. discovery of novel structures of triterpenoids and meroterpenoids from Ganoderma spp. and their bioactives; 3. discovery of novel structures of triterpenoids and meroterpenoids from Meliaceae and their physiological effects.

T1-30-03

Chemical and bioactive constituents of wolfberry

***Hao Gao**, Zheng-Qun Zhou, Wan-Yang Sun, Hong-Xia Fan, Xin-Sheng Yao*

Institute of Traditional Chinese Medicine & Natural Products, College of Pharmacy, Jinan University

The fruit of *Lycium barbarum*, also called wolfberry and goji berry, has been used as a traditional Chinese medicine (TCM) to nourish liver and kidney and brighten eye. Wolfberry has also been marketed as food and dietary supplements in North America, Caribbean countries, European Union, Oceania, and Southeast Asia. A systematic study on wolfberry chemical constituents including their bioactivities was carried out. 87 chemical constituents were obtained and identified including 19 dicaffeoylspermidine derivatives, 15 phenylpropionoyl phenylethylamine derivatives, and 53 polyphenols, in which 31 constituents are new compounds. Their anti- Alzheimer's disease (AD), antioxidant, and hepatoprotective activities were assayed. The discovery of chemical constituents and their bioactivities is beneficial for understanding the scientific basis of the effects of wolfberry. Dicaffeoylspermidine derivatives, a rare kind of plant secondary metabolites, are primarily distributed in the family of Solanaceae. All dicaffeoylspermidine derivatives discovered in wolfberry are new compounds and named lycibarbarspermidines A-S. Lycibarbarspermidines A-S are the first glycosidic products of dicaffeoylspermidine derivatives, and lycibarbarspermidines N-O are the first cyclization products of dicaffeoylspermidine. Moreover, dicaffeoylspermidine derivatives were first isolated and identified from wolfberry and they exhibited anti-AD and antioxidant activities which are related to the effects of wolfberry. The content of lycibarbarspermidines in wolfberry is high and estimated to be beyond 2.1 g/kg, which are potential natural anti-AD drug candidates.

T1-30-04

Research progress on *Panax notoginseng* for medicinal purposes: Effects of *Panax notoginseng* in neurodegenerative diseases

Wensheng Zhang, *Dina Zhu*
Beijing Normal University, Beijing, China

Panax notoginseng (sanqi), the root of *Panax notoginseng* (Burk.) F.H. Chen, is a well-known traditional Chinese herbal medicine which has been widely used for diseases related to the circulatory system, such as cardio- and cerebro-vascular disorders. Now it has been accepted as a functional food and is used globally as a natural medicine. Ginsenosides are responsible for various pharmacological properties in notoginseng, and have been shown to produce many beneficial effects in the nervous system. Our work reveals novel mechanisms involving in reducing Aß accumulation by activating PPARγ and up-regulating insulin degrading enzyme (IDE), increasing neuronal excitability and modulating cell strength, which contributes to the protective effects of notoginseng or ginsenosides against neurodegeneration and synaptic dysfunction. Our study suggests that notoginseng may potentially be useful as a therapeutic agent for Alzheimer's disease.

T1-30-05

Mapping the polysaccharides in rhizome and root of *Panax ginseng* by HPGPC-UPLC- CAD and UPLC-QQQ- MS/MS determination and chemometric analysis

Qi-Lei Chen, *Jun Xu*, *Hu-Biao Chen*
Hong Kong Baptist University, Hong Kong, China

Ginseng, a wildly used herbal medicine in China and many other countries, is derived from the rhizome and root of *Panax ginseng* C.A. Mey. It has been well studied that ginsenosides are the main secondary metabolites and important bioactive ingredients of *P. ginseng*, however, the carbohydrates are often overlooked. In fact, phytosaccharidesare as wellmajor active components in herbal medicines and carry much weight in the total tonic effect of herbal medicines. It is therefore essential to investigate polysaccharides in *P. ginseng*, where chemical profiling and structure determination of the polysaccharides is fundamental. The objectives was to investigate distribution and accumulation of polysaccharides in *P. ginseng*; to characterize primary structures of tissue-specific polysaccharides in the medicine. Ten batches of *P. ginseng* were separated by tissues using laser microdissection (LMD). Polysaccharides were extracted and purified from the cut sections respectively. High-performance gel permeation chromatography on an ultra-high performance liquid chromatography with a charged aerosol detector (HPGPC-UPLC-CAD) and UPLC coupled to triple quadrupole mass spectrometry (UPLC-QQQ-MS/MS) was performed to analyze the polysaccharides both qualitatively and quantitatively. Principle component analysis (PCA) was conducted to give clear comparison of polysaccharides from different sections of the medicine. Three macro-sectioned parts, namely, rhizome, main root, and branch root, and five micro-dissected tissues, namely, cork, cortex, resin canal, phloem, and xylem, of ginseng were separated. The different parts can be distinguished according to their respective polysaccharide profiles and sugar compositions. The skin, especially the cork, contained polysaccharides with smaller molecular weights and higher compositional percentages of galactose and galacturonic acid; the flesh, including xylem, phloem, and resin canal, contained larger carbohydrates consisting glucose as a dominant sugar residue. The PCA pattern showed distinctiveseparation of the five microdissected tissues and unified tissue construction in both main and branch roots. It is therefore conjectured that the flesh of ginseng contained mostly neutral polysaccharides, while the skin had a higher compositional proportion of acidic pectins. The present work paves the way for further structural determination and pharmacological studies of ginseng polysaccharides. Combining the present results and previously investigated ginsenoside distribution, the ancient literature record and traditional usages of *P. ginseng* can be explained, inspiring more economical and efficient application of the medicine.

T1-30-06

Chemical constituents and their anti-AD activities of the medicinally important plants: *Hypericum perforatum* (St John's Wort)

Yongbo Xue*, *Yi Guo*, *Na Zhang*, *Shuangshuang Xie Xie*, *Yonghui Zhang**
School of Pharmacy, Tongji Medical College, Huazhong University of Science and Technology

The genus *Hypericum* is one of the largest genera in Clusiaceae, consisting of 484 species throughout the world. *Hypericum perforatum*, also known as "St John's Wort", is widely distributed in Europe, North Africa, Asia, and North America. Since the early 1800s, extracts of *H. perforatum* (EHP) have been utilized externally for wound healing and internally to relieve the symptoms of neurological disorders. In addition to the use of EHP for alleviating mild to moderate depression that is mainly attributed to the inhibition of various neurotransmitter receptors, further studies indicated that EHP also exerts versatile biological properties, including antibacterial, antiviral, anti-inflammatory, cytotoxic, and neuroprotective properties. To date, types of organic substances such as flavonoids, proanthocyanidins, naphthodianthrone, prenylated acylphloroglucinols and xanthones have been previously isolated from this plant. Among them, the polycyclic prenylated acylphloroglucinols (PPAPs) typified by hyperforin, characterized by a highly oxygenated and densely substituted bicyclo[3.3.1] nonane-2,4, 9-trione or other related core structures, have attracted great interest from the scientific community due to their intriguing molecular structures and interesting pharmacological effects. Aimed to discover structurally unique and bioactive secondary metabolites from the genus of *Hypericum* collected in central China regions, our group reported the isolation of almost one hundred new PPAPs from *H. sampsonii*, *H. attenuatum*, *H. japonicum* and *H. ascyron*. Although an abundance of reports exist on the biological activities and many organic substances including 26 PPAPs were previously reported to be derived from *H. perforatum* by other research groups, the present chemical study by our research group on this species has led to the isolation of 40 new PPAPs, along with 15 known ones. Some of the newly isolated PPAPs exhibited potent inhibitory activities against acetylcholinesterase and beta-secretase (BACE1) enzymes in Alzheimer's disease. The finding will provide new insights into the skeletal diversity of PPAPs and more scientific evidences in favor of sustainable exploitation and utilization of the important plant resources of *Hypericum* for natural anti-AD agents.

T1-31: Plants in human health and well being

T1-31-01

An ethnobotanical perspective of traditional medicinal plants from Fairy Meadows, Western Himalaya, Pakistan

Zia-Ur-Rehman Mashwani, *Rhmat Wali*
PMAS Arid Agriculture University, Rawalpindi

Ethno medicinal traditional knowledge about the use of plants for various human ailments is known by the elders of the community members. This explorative ethnobotanical survey was attempted to document, build up and publicize the gigantic local knowledge of century's practiced restorative use of medicinal plants, their mode of preparation and administration by the inhabitants of this secluded territory. A total of 146 informants (101 males and 45 females) were interviewed through questionnaire method. Ethnomedicinal data was quantitatively analyzed through use value (UVi), frequency of citation (FC), relative frequency of citation (RFC) along with a Pearson correlation coefficient (PCC). All the plants were checked with previously published articles for their novel uses. The demographic inventory of the participants was also recorded. Collected specimen were preserved, mounted, labeled on herbarium sheets, cataloged and deposited with voucher numbers in Arid Agriculture University Rawalpindi (AAUR). 90 plants species belonging to 75 genera and 49 different families have been documented in the current study. These medicinal plants were used against 55 pathological disorders in the area. Among these, most commonly treated ailments, stomachic (23.3%), cough (17.7%), asthma and fever (16.6%) *etc.* Most of the collected plants were wild (86%) and (14%) were cultivated. Herbs were the dominant life forms contributing 63% followed by trees 26% and shrubs 11%. Asteraceae was the leading family contributing 10% of the total families. Leaves are the most widely used parts (27%) followed by fruit (19%), root (16%) and stem (12%) for curing different ailments. Most common method of herbal preparation was powder (27%), decoction (16%) and direct use (13%). Among these, 81.7% preparations were administered orally. The medicinal plant with highest use values were *Berberis lyceum* (5.47), *Thymus palmata* (5.07) and *Rhododendron anthopogon* (5) while the plants with greater relative frequency of citation were *Berberis lyceum* (0.97), *Thymus serphyllum* (0.89) and *Rhodendron anthopogon* (0.75). The pearson correlation coefficient was found 0.836 between RFC and UV showing high positive association. This is the first ethnobotanical study in Diamir district and comparative results with other districts of Gilgit Baltistan revealed that 25.5% of the plants have been documented ethnomedicinally for the first time in Gilgit Baltistan. In this study, we have reported plentiful indigenous knowledge about local communities.

T1-31-02

Uptake of heavy metals in the peel and flesh of potatoes irrigated with untreated wastewater in the presence of plantain peel biochar

Christopher Nzediegwu[1,2], *Shiv Prasher*[1], *Eman Elsayed*[1], *Jaskaran Dhiman*[1], *Ali Mawof*[1]
1. Bioresource Engineering Department, McGill University
2. Federal University of Technology Owerri

Consuming crops heavily loaded with heavy metals can lead to serious health damage in human body. Heavy metal loading in crops, in most cases, results from uptake from contaminated soils. Just like the essential nutrient that are taken up by plants during their growing season, these heavy metals, when present in irrigation water, can be available for uptake in the soil solution. Biochar, produced in the pyrolysis of organic waste (s), has been reported to reduce the uptake of heavy metals to crops grown on contaminated soils. The duration of this effectiveness of biochar as a biosorbent is uncertain. Some researchers have found biochar effective for just one year/season, while others have found it effective for more than one year/season. A study was conducted in six outdoor PVC lysimeters (1.0 X 0.45 m) following a randomized complete block design with 3 replicates. For two consecutive years (2015 and 2016), we investigated the effectiveness of plantain peel biochar applied to sandy soil and irrigated with untreated wastewater to study the uptake of six heavy metals (Cd, Pb, Cr, Zn, Cu and Fe) by the flesh and peel of Russet Burbank Potatoes. Biochar (1% w/w) was mixed in the top 10 cm with sandy soil only in the first year before planting the potatoes, while in the second year, no new biochar was applied. For both years, the potatoes were irrigated with untreated wastewater applied at the just ponding rate. Upon harvesting, the flesh and peel was sampled, washed (trice) with deionized water, oven-dried (60°C for 48 hrs), digested with nitric acid, and analyzed for the heavy metals with an Inductively Coupled Plasma Mass Spectrometry (ICP-MS). In the first year, there was significant reduction ($P<0.05$) in only Cd and Zn, while in the second year there was significant reduction ($P<0.05$) in all six heavy metals in both the flesh and peel of the biochar amended treatments, compared to the control (without biochar). This suggests that biochar can reduce heavy metal uptake by the edible parts of Russet Burbank Potatoes. Likewise, it suggests that plantain peel biochar became more effective in the second year.

T1-31-03

Biosynthesis of major secondary metabolites in the well-known anti-aging wolfberry

Ying Wang, *Shaohua Zeng*, *Yongliang Liu*
Key Laboratory of South China Agricultural Plant Molecular Analysis and Genetic Improvement, Provincial Key Laboratory of Applied Botany, South China Botanical Garden, Chinese Academy of Sciences, Guangzhou, 510650, China

Lycium L. is a genus of Solanaceae containing about 80 species distributed in the temperate and subtropical zones. *Lycium* species are mostly found in dry, semi-saline environments. Chinese Pharmacopoeia (2010) recorded lycii fructus (*Gouqizi*, wolfberry, dry red fruit [RF] of *L. barbarum*), and lycii cortex (*Digupi*, dry root bark of *L. chinense* and *L. barbarum*). Black fruits (BF) of *L. ruthenicum* have also been used as folk medicine, especially in Tibetan and Mongolian medicine. Therefore, wolfberry (or Goji, Gouqi) nowadays in China refers to the products prepared from *L. chinense*, *L. barbarum*, and *L. ruthenicum*, which is one of the most famous anti-aging herbs. Genomic sequencing, transcriptome sequencing and chloroplast development have been investigated for identification of key genes involved in the biosynthesis and accumulation of carotenoids, especially zeaxanthin dipalmitate, and betaine. The failure of the chromoplast development in BF causes low carotenoid biosynthesis levels and continuous carotenoid degradation, which ultimately leads to undetectable carotenoid levels in ripe BF. In contrast, the successful chromoplast biogenesis in RF furnishes the sink necessary for carotenoid storage. The abundant zeaxanthin accumulation in RF is primarily determined via the high levels of carotenoid biosynthesis, transportation, and storage, as well as the lack of carotenoid degradation, which are regulated at the transcriptional level. Function of several key genes will be characterized and reported. Glycine betaine (GB) plays significant role in salinity tolerance in many plants. The last biosynthetic step of GB is catalyzed by aminoaldehyde dehydrogenase (AMADH) serving as betaine aldehyde dehydrogenase (BADH) activities. *LrAMADH1* is characterized as the true Lr-BADH which also responses to salt stress in *L. ruthenicum*.

T1-31-04

Ethnobotanical studies on medicinal plants traded in local periodic markets in rural Guizhou, China

Chunlin Long[1,2], *Liya Hong*[1], *Jiangju Zhou*[3], *Qiyi Lei*[1,3], *Binsheng Luo*[1], *Hang Shu*[1], *Yuanyuan Ji*[1]

1. College of Life and Environmental Sciences, Minzu University of China

2. Kunming Institute of Botany, Chinese Academy of Sciences

3. Kaili University

Guizhou, a province in Southwest China is well-known for its biological and cultural diversity. Miao, Dong, Buyi, Yao, Shui, Maonan and other ethic groups living in this region depend on local medicinal markets for medical treatment and family healthcare. However, information on the market traded medicinal plants is not well documented from traditional medicinal markets in southern Guizhou. The study aimed to document medicinal plants traded by local people in the periodic markets. Ethnobotanical market survey had been adopted for documenting herbal plants and associated traditional knowledge. A total of 387 medicinal plant species, belong to 309 genera and 123 families, were investigated and collected together with their medicinal uses by the local people. The plants were used to treat 256 human diseases. Of the 387 species, 234 (60.47%) were obtained from the wild whereas 71 (18.35%) were from home gardens, and 82 (21.19%) species were from both home gardens and wild habitats. Herbaceous species (247) were the most utilized plants, accounting for 63.82% of the species, followed by shrubs (59,15.25%). Fabaceae came out as a leading family with 25 medicinal species (6.46%) while Asteraceae (5.94%) followed with 23 species. The most frequently used plant parts were the whole plants (148), followed by roots (84). Decoction was a widely used method to prepare traditional herbal medicines. Our research results indicated that the study area might be probably one of the richest diversity centers in both species and cultural level of the country's ethnomedicines, to compare with other parts of China. The medicinal plants which spontaneously sold in the open air markets are collected by the local people from wild and their living environment, even the toxic plants. There is at risk of ingesting toxic plants sold in popular markets, because there is no hard distinction made between medicinal and toxic plants. The local government should regulate the traditional markets for misusing of the herbal medicine and protecting the local biodiversity and associated traditional knowledge.

T1-31-05

A friendly relationship between endophytic fungi and host medicinal plants

Lu-Ping Qin

Dept. of Phaemacognosy, School of Pharmacy, Second Military Medical University

Endophytic fungi or endophytes exist widely inside the healthy tissues of living plants, and are important components of plant micro-ecosystems. Over the long period of evolution, some co-existing endophytes and their host plants have established a special relationship with one and another, which can significantly influence the formation of metabolic products in plants, then affect quality and quantity of crude drugs derived from medicinal plants. Our research group focused on the endophytic fungi and their correlations with the host medicinal plants. The following contents were carried out: (1) Research on endophytic fungi diversity and population structure; (2) Chemical diversity and bioactive secondary metabolites of endophytic fungi; (3) The effects of endophytic fungi on host plants and the correlation between the qualities of TCMs and endophytic fungi. More than 5,000 endophytic fungal strains were isolated from more than 70 medicinal plants. In addition, fermented products of 32 strains of endophytic fungi with bio-activities were separated. More than 200 compounds were identified. Many endophytic fungi could produce secondary metabolites originally-found from host plants, some could promote the growth of host plant. For example the fungus ZJ44 was able to increase the growth of the host plant-adlay *Coix lacryma-jobi* var. *mayuen* and Arobidopis. Interestingly, one endophytic fungus D16 (*Trichoderma atroviride*) isolated from the root of *Salvia miltiorrhiza,* was found out to produce tanshinone and tanshinone A. Moreover, it was discovered that D16 could significantly enhance the biomass of *S. miltiorrhiza* hairy roots and promote the biosynthesis of tanshinones notably, and protein polysaccharide fraction (PSF) from D16 was the main active fraction to promote tanshinone biosynthesis. Furthermore, technologies of metabolomics, transcriptomics and proteomics were used to investigate the molecular mechanism of PSF on the growth and secondary metabolism of *S. miltiorrhiza* hairy roots. The results indicate that the distribution and population structure of endophytes can be considerably affected by factors, such as the genetic background, age, and environmental conditions of their hosts. On the other hand, the endophytic fungi can also confer profound impacts on their host plants by enhancing their growth, increasing their fitness, strengthening their tolerances to abiotic and biotic stresses, and promoting their accumulation of secondary metabolites.

T1-31-06

Phylogenomics and authentication of the resource medicinal plants using complete chloroplast and nuclear genomes simultaneously obtained from low coverage NGS data

Tae-Jin Yang
Seoul National University

Cytoplasmic chloroplast (cp) genome and 45S nuclear ribosomal DNA gene cluster (nrDNA) regions are the most popular targets for research of plant genetics diversity, barcoding and evolution. We developed a high throughput method to obtain complete whole cp genome and a major unit for 45S nrDNA tandem array as well as characterization of major repeats simultaneously by de novo sequence assembly using low coverage whole genome shotgun sequence (dnaLCW method). The dnaLCW method can be applied for any plant species without advanced genome data and provide a new era for elucidation of genetic diversity and fundamental understanding the tree of life and evolution. Moreover, the genomic information can be utilized for development of practical barcoding markers not only at inter species level but also at intra species level. The genome information produced by dnaLCW method can be applied to improve breeding of the target species and also to develop the practical barcode markers which can be applied to authenticate the target species or the target cultivars from the related. The species-unique or cultivar-specific markers can be applied to prohibit the illegal trade of economically motivated adulteration instead of the target plants in the production of functional foods or traditional oriental medicinal drugs.

This work was supported by the grant funded by Next-Generation BioGreen21 Program (No. PJ01103001, PJ01100801), Rural Development Administration, the Bio & Medical Technology Development Program (NRF-2015M3A9A5030733) of the NRF funded by the Korean government, and "16172MFDS229" from Ministry of Food and Drug Safety in 2016, Republic of Korea.

T1-32: Traditional medicinal plants

T1-32-01

Herbal remedies used in the treatment of scorpion stings in Saudi Arabia

Amal Fakhry[1], *Ghalia Elgadani*[2], ***Masarrat Migahid***[1]
1. Alexandria University
2. King Abdulaziz University

Traditional medicine has a long history of serving people all over the world. In recent years, the use of traditional medicine information on plant research has again received considerable interest. Medicinal plants are an important element of indigenous medical systems in Saudi Arabia as well as elsewhere. Ethno-botanical survey of the plant based folk practices used for treatment of scorpion stings was carried out in two regions in Saudi Arabia; Tabuk in the northwestern region and Jeddah in southwestern region. The people in the study area still use plant therapy to considerable extent although the modern medical facilities are available. From the survey, a total of a total of twelve plant species belonging to eleven families of angiospermae are used for treatment of scorpion stings in the study area. Six plant species are recorded in Tabuk and other six plant species in southern Jeddah. Alliaceae family is common in the two studied regions. *Astragalus spinosus* (Fabaceae) and *Heliotropium bacciferum* (Boraginaceae) are the most commonly used plant species for treatment of scorpion stings in Tabuk region, while *Allium stivum* (Alliaceae) and *Linum usitatissimum* (Linaceae) in southern Jeddah. It is to be noted that wild plant species are the most commonly used in the northwestern region of Saudi Arabia where five species out of six (about 83%) are wild desert plants, while cultivated and imported plant species are the most commonly used (about 67%) in the southwestern region of Saudi Arabia. Generally leaves, bulbs and seeds are the commonest parts of plants used, while decoctions, infusions, pulverization and grinding are the main methods of preparations.

T1-32-02

Studies on large scale tissue culture of medicinal plants

Wenyuan Gao, *Juan Wang*
School of Pharmaceutical Science and Technology, Tianjin University

Large scale tissue culture of medicinal plants is one important way to solve the resource problem of TCM in China. We have carried out experiments in *Panax ginseng, Panax quinquefolium, Panax notoginseng* and *Glycyrrhiza uralensis*. We have optimized culture condition and realized large scale bioreactor culture, constituting 500 l bioreactor scale. In terms of ginseng adventitious roots, we selected three fungi strains (*Aspergillus niger*, *Aspergillus flavus* and *Aspergillus oryzae*) as elicitors prepared from mycelium or fermentation broth to improve ginsenosides production in adventitious roots culture. The results showed that ginsenosides production (29.90 ± 4.67 mg g^{-1}) was significantly enhanced upon elicitation with 200 mg L^{-1} *A. niger* elicitor prepared from mycelium, which was 3.52-fold of untreated group. Furthermore, we found that *A. niger* significantly enhanced accumulation of Nitric oxide (NO), salicylic acid (SA) and jasmonic acid (JA) involved in plant defense response, and significantly up-regulated the expression of the geranyl diphosphate synthase (GPS), farnesyl diphosphate synthase (FPS), squalene synthase (SS), squalene epoxidase (SE), dammarenediol synthase (DS), Two cytochrome P450 (CYP) genes (CYP716A47 and CYP716A53v2) and three UDPglycosyltransferases (UGT) genes (UGTAE2, UGT94Q2 and UGTpg100). In terms of *Panax quinquefolium* adventitious roots, we combine elicitors and transcriptomics to investigate the inducible biosynthesis of the ginsenoside from the *Panax quinquefolium*. Treatment of *P. quinquefolium* adventitious root with methyl jasmonate (MJ) results in an increase in ginsenoside content (43.66 mg/g compared to 8.32 mg/g in control group). Therefore, we sequenced the transcriptome of native and MJ treated adventitious root in order to elucidate the key differentially expressed genes (DEGs) in the ginsenoside biosynthetic pathway. Through DEG analysis, we found that 5,759 unigenes were up-regulated and 6,389 unigenes down-regulated in response to MJ treatment. Several defense-related genes (48) were identified, participating in salicylic acid (SA), jasmonic acid (JA), nitric oxide (NO) and abscisic acid (ABA) signal pathway. Additionally, we mapped 72 unigenes to the ginsenoside biosynthetic pathway. Four cytochrome P450s (CYP450) were likely to catalyze hydroxylation at C-16 (c15743_g1, c39772_g1, c55422_g1) and C-30 (c52011_g1) of the triterpene backbone. UDP-xylose synthases (c52571_g3) was selected as the candidate, which was likely to involve in ginsenoside Rb_3 biosynthesis. In terms of *Glycyrrhiza uralensis* adventitious roots, the protein fragment more than 10 kDa of *Escherichia coli* significantly increased the metabolites contents in *Glycyrrhiza uralensis* adventitious roots. The results showed the highest accumulation of total flavonoids (7.59 $mg \cdot g^{-1}$), glycyrrhizic acid (0.29 $mg \cdot g^{-1}$), glycyrrhetinic acid (0.27 $mg \cdot g^{-1}$) and polysaccharide (93.11 $mg \cdot g^{-1}$) by up to 2.27-fold, 2.64-fold, 2.70-fold and 2.32-fold than control roots, respectively. Besides, the protein fragment more than 10 kDa significantly activated the defense signaling and extremely up-regulated the expression of defense-related genes and functional genes in glycyrrhizic acid and flavonoids biosynthesis.

T1-32-03

Comparative transcriptome of *Gastrodia elata* in response to fungus symbiosis and genomic divergence in saprophytic *Gastrodia*

Yu-Chung Chiang[1], *Chi-Chu Tsai*[2], *Tsai-Wen Hsu*[3], *Tzen-Yuh Chiang*[4]
1. Department of Biological Science, National Sun Yat-Sen University, Kaohsiung
2. Kaohsiung District Agricultural Research and Extension Station, Pingtung
3. Endemic Species Research Institute, Nantou
4. Department of Life Sciences, National Cheng Kung University, Tainan

Gastrodia elata is a traditional Chinese herb used for treating neural diseases. *Gastrodia*, with 50 to 60 species mostly distributed in the tropics and subtropics, is one of the saprophytic and achlorophyllous genera in the Orchidaceae. Taiwan, with a high species diversity (19 taxa), represents a hot spot of *Gastrodia*. All *Gastrodia* species are mycoheterotrophic, dwelling underground of forests with rich litter and obtaining nutrients from the symbiotic mycorrhizal fungi. Such a unique growth form makes *Gastrodia* species hard to be found in the field except the flowering and fruiting seasons, generally one month after budding. Seed and pollen disperse to a limited distance with the aides of insects, likely resulting in strong genetic differentiation between populations and divergence among species. Using next generation sequencing techniques, transcriptomes of *G. flavilabella* and *G. elata* rhizome were constructed. We investigated 9 *Gastrodia* species, specifically belonging to three subgenera with separate flowering times, for testing the genetic divergence and the mode of speciation. Compared with other angiosperms, *Gastrodia* attained the interspecific reproductive isolation much sooner due to temporal isolation. We also conducted a comparative analysis of transcriptomes obtained from vegetative propagation corms and tubers of *G. elata* and *Armillaria mellea*. We detected two unigenes that encode monooxygenase and glycosyltransferase, with upregulating gene expression from vegetative propagation corms to tubers, providing insights into the molecular basis of gastrodin biosynthesis in *G. elata*.

T1-32-04

How the West was lost: Exploring the disconnect between traditional Chinese medicine and Western medical science

Yilin Qiao, *Xue Li*, ***Robert Borris***
School of Pharmaceutical Science and Technology, Tianjin University

The World Health Organization (WHO) has stated that the majority of people worldwide rely on some form of traditional medicine to address their medical needs. Some of these traditional practices, including Traditional Chinese Medicine (TCM) for example, have been used, documented and refined over the space of thousands of years. Efficacy is well documented and often undeniable. Nonetheless, many published and unpublished investigations in Western laboratories have failed to detect and/or isolate the chemical principles that are responsible for the observed biological activities in man. The inability of Western science to 'validate' the efficacy of TCM in the laboratory has contributed to the skepticism about traditional medicine that is prevalent throughout much of the "Developed World", severely limiting the acceptance of TCM outside of China. The present study explores some potential reasons for the apparent disconnect between the observed clinical efficacy of TCM and the disappointing results in laboratory studies.

T1-32-05

The quality comparison between saffron from China and Iran by multiple component analysis

Jiang Di Liu[1,2], ***Bin Yang***[1]
1. Institute of Chinese Materia Medica, China Academy of Chinese Medical Sciences
2. College of Pharmacy, Jiangsu University

Saffron is the dried stigma of *Crocus sativus* L. As a traditional spice, it is widely used among the world. Meanwhile, saffron is also a traditional herb, which has abilities like reducing bruises, promoting blood circulation, calming the mind and body, providing anti-tumorous effects and reducing the sugar concentration in blood, *etc.* According to the reference, crocetin, safranal and picrocrocin are the material basis of the color, scent and bitterness of saffron respectively, and they are used as indices to estimate the quality of saffron. Saffron primarily distributed in the Mediterranean regions around Iran and Greece. In Tang Dynasty, it was introduced into China from India through Tibet, and now it forms scale plantings in places like Shanghai, Zhejiang, Jiangsu and Henan. As well known, the growing environment is one of the important factors in the quality of medicinal herbs. Will there be any difference between the saffron grown in China and Iran? In the present study, about 100 kinds of saffron samples were purchased or harvested from Iran and China, and they were divided into different quality levels according to the methods specified in ISO-3632. An HPLC fingerprint combined with multiple components analysis method was established to qualitatively and quantitatively analyze the saffron samples. Also, the difference between the quality of saffron samples grown in China and Iran was explored using chemometrics method. It showed that no significant difference was found between saffron samples from China and Iran on the content of crocetin I, but significant differences were found on the contents of crocetin II and safranal. And the samples could be divided and grouped based on whether it's from China or from Iran according to the data from HPLC fingerprint.

T1-32-06

Impacts of cultivation on genetic diversity pattern of medici-

nal plants in China

***Qing-Jun Yuan**, Lu-Qi Huang*
China Academy of Chinese Medical Sciences

There are about 500 species of commonly used in traditional Chinese medicine, among which 250 species are primarily cultivated. The cultivation of medicinal plants began since 2,600 years ago, but large-scale cultivation (about 200 species) was initiated in 1958. Different medicinal plants may have distinct cultivation histories, which could have different genetic consequences for the cultivated populations. However, few studies have been compared the genetic diversity patterns of medicinal plants with different cultivation histories. *Scutellaria baicalensis* Georgi (Huangqin or Chinese Skullcap) is a newly-cultivated species (ca. 60 years of cultivation), but *Angelica sinensis* (Oliv.) Diels (Danggui) was cultivated since 1,000 ago. The two medicinal species may represent an ideal model to evaluate the impacts of different cultivation histories on the genetic diversity pattern of medicinal plants. Investigations of three polymorphic chloroplast fragments and fifteen nuclear microsatelite loci in 28 wild and 22 cultivated populations of *S. baicalensis* found that there was a slight genetic bottleneck during the cultivation and cultivation populations derived from multiple origins. However, short-term cultivation has lead to extensive gene flow among cultivated populations compared to wild populations. In contrast, through surveying the genetic diversity of 57 cultivated populations and 6 wild populations on eighteen microsatellite loci, a severe cultivation/domestication bottleneck was found in *A. sinensis*. Its cultivated populations formed an independent group in STRUCTURE analysis, although their ancestor remains unclear. This study indicates that different cultivation histories have distinct impacts on the genetic diversity and genetic structure of cultivated medicinal plants, which provides powerful insights into the conservation of genetic resources of traditional medicinal plants in China.

T1-33: The megadiverse island of New Guinea: diversity, evolution, and conservation of a tropical wilderness

T1-33-01

Forest dynamics and conservation in New Guinea's lowland forests

***Timothy Whitfeld**[1,4], George Weiblen[2,4], Vojtech Novotny[3,4], Francesca Dem[4], Kenneth Molem[4], Lucie Houdková[7,4], Dominic Rinan[4], Byron Siki[4], Owen Lewis[6,4], Yves Basset[5,4]*
1. Brown University
2. University of Minnesota
3. Czech Academy of Sciences
4. New Guinea Binatang Research Center
5. Smithsonian Tropical Research Institute
6. University of Oxford
7. University of South Bohemia

The Center for Tropical Forest Science-Forest Global Earth Observatory includes a network of large-scale research plots across the globe's tropical and temperate forests. Just one of these 50-hectare (ha) plots is located in the Melanesian tropical region, east of Wallace's line. The plot is in the lowland rainforest of northern New Guinea, near the village of Wanang, Madang Province, Papua New Guinea. The New Guinea plot is in a contiguous area of mature, mixed evergreen forest at 100-200 m a.s.l. and encompasses the full elevation range of the region. The initial plot census and full topographic survey were completed in 2012. All 265,771 individual trees with diameter at breast height ⩾ 1 cm were mapped, measured, tagged, and, where possible, identified. The plot includes 584 species with a further 65 taxa that remain either unnamed or undescribed. The first re-census of the New Guinea plot is currently underway. This effort represents a singular opportunity to examine forest dynamics (tree mortality, recruitment, and growth rates) in this unusual biogeographic region and helps place New Guinea's forest into a global context. Many previous studies have investigated forests in the Neotropics and Southeast Asia but these studies do not necessarily allow us to make predictions about New Guinea forests where there is only a minute overlap in species composition with other tropical areas and significant differences in soils. In fact, preliminary analysis of re-census data from the New Guinea plot suggests the island's recently uplifted, steep slopes, and unstable soils combined with high seismic activity make its forests particularly dynamic and prone to natural disturbance. This has lead to higher than average tree mortality and lower carbon stocks per hectare compared to other tropical forests. There is also lower than expected species turnover across environmental gradients in the plot compared to other tropical areas. In addition to the censuses of trees and saplings, 800 small seedling plots spread throughout the 50-ha plot provide an opportunity to investigate dynamics during the critical transition phase between germination and establishment. These plots track nearly 3,000 seedlings representing at least 270 species. There are also fruit traps across the plot to monitor seed production as well as ongoing monitoring of seed eating insects and experiments to document seed distribution. All research in the New Guinea 50-ha plot is overseen and managed by local landowners in collaboration with researchers from the New Guinea Binatang Research Center. This unique relationship has had a direct impact on forest conservation efforts by strengthening the community-lead development of a 10,000 ha conservation area in the center of an active logging concession. Ongoing visits from international scientists provide employment to village assistants and visitor fees support community development including the only elementary school in the area and improved access to medical care. Additionally, the partnership is enriched through the sharing of indigenous forest knowledge and scientific thinking. The relationship has become a role model for indigenous conservation internationally and was awarded the UNDP Equator Prize for innovative conservation in 2015.

T1-33-02

Morphological variation in Australian and Papua New Guinean *Polyosma* Blume (Escalloniaceae)

Oliver Paul
PNG Forest Research Institute

The morphological variation within and between species of *Polyosma* Blume from Australia and Papua New Guinea (PNG) was investigated. The taxonomic usefulness of morphological characters was determined. A total of 161 Operational Taxonomic Units (OTU) of *Polyosma*, comprising 35 known species, three undescribed taxa and nine un-identified herbarium specimens were assessed by the use of 30 characters. Of these characters, 20 were qualitative (non-overlapping) and 10 were quantitative (overlapping). The variability within these morphological characters proved to be useful for circumscribing the currently recognised species. In the analysis, two main groups were obtained based on whether the characters were hairy or glabrous, which were further down had resolved into thirteen groups. Twenty four species were recognised from Papua New Guinea. Three un-published names were discovered. Of these, two species grouped well with published names where as one was significantly different to the published named species it was seen grouped with. Australia has six species, with a possible new species from Mt. Spurgeon, Queensland. More materials are required to analysis the Mt. Spurgeon species. The characters and character-states used were consistent within the entire taxon. In future, few more characters will be included and a molecular phylogeny to compliment this morphology study of the taxon in the two study areas.

T1-33-03

A suryey of the origin, diversity, and conservation status of New Guinean tree ferns (Cyatheaceae, Dicksoniaceae)

Marcus Lehnert[1], *Rodrigo Cámara-Leret*[2], *Tim Utteridge*[2]
1. Nees-Institute for Plant Biodiversity, University of Bonn
2. Identification & Naming Department, Royal Botanic Gardens, Kew, Richmond

Tree ferns are conspicuous elements of tropical and southern temperate vegetation and most diverse in wet mountain rainforests of South America, Madagascar and southeast Asia. New Guinea is assumed to be one of the epicenters of diversity, with an estimated 130 species of scaly tree ferns (Cyatheaceae) and hairy tree ferns (Dicksoniaceae). However, this estimate is based on a comparatively low specimen number from rather sporadic collection efforts. Our understanding of the ecological adaptations and morphological variation of the species is still in its infancy. Field observations indicate that high local and regional diversity of tree ferns may be due to adaptation to different life strategies (pioneers vs. climax species; understory vs. canopy species) and substrate preferences, meaning that some species are more indicative of undisturbed habitats than others. We present the results of the latest taxonomical assessment of New Guinean tree ferns, which forms the basis for modeling the potential distributions and ecological niches of the species. We use these results to assess the degree of threat for each species according to IUCN guidelines and to identify areas with concentration of rare species, which should be prioritized for protection.

T1-33-04

Trees of New Guinea

Timothy Utteridge, *Laura Jennings, Marie Briggs*
Royal Botanic Gardens, Kew

The tropical island of New Guinea is one of the least explored regions and largest remaining wilderness areas on the planet. There are many gaps in our knowledge of the island's biodiversity as there is no complete flora or species checklist for the island, no island-wide vegetation map, and no conservation strategy. The Trees of New Guinea project (TONG) at Kew will publish a generic guide to the trees on the island of New Guinea. The island includes 698 arborescent genera (smaller taxa have been excluded using a height of 4 m as a cut-off) which account for approx. 5,600 species, or which *c*. 5,000 are arborescent. This presentation will outline the guide, some of the key taxa, and detail the final stages and planned production.

T1-33-05

Morphological phylogenetic analysis of the Astronieae (Melastomataceae)

Jeffrey Mancera[1], *Frank Almeda*[1], *Darin Penneys*[2], *Peter Fritsch*[3]
1. California Academy of Sciences
2. University of North Carolina Wilmington
3. Botanical Research Institute of Texas

The Astronieae (Decne.) Triana represent one of the earliest diverging tribes of the pantropical family Melastomataceae Juss. and are endemic to the rainforests of Southeast Asia and the Pacific islands. They comprise about 150 species of trees and woody shrubs in four currently accepted genera: *Astrocalyx* Merr., *Astronia* Blume, *Astronidium* A. Gray, and *Beccarianthus* Cogn. To date, only one published phylogenetic analysis has focused on the Astronieae, but sampling was insufficient to draw conclusions about generic circumscriptions or species-level relationships. Moreover, an ongoing molecular phylogenetic reexamination of the family reveals that the enigmatic *Tessmannianthus* Markgr. from tropical America has closer affinities with the Astronieae than the neotropical Merianieae Triana, where it was originally placed. This study presents the first phylogenetic analysis of the tribe based on morphological data. Maximum parsimony analyses of 57 ingroup taxa using 205 morphological characters were conducted to test support for the monophyly of each constituent genus and to verify the placement of *Tessmannianthus* within the Astronieae. Characters were partitioned into two sets to identify potential areas of incongruence: discrete and continuous, the latter was coded using the differential gap-weighting method. The global analysis yielded two equally most parsimonious trees (L = 4,800, CI = 0.207, RI = 0.484). The strict consensus tree shows that *Astrocalyx*, *Astronia*, and *Astronidium* are each monophyletic. *Beccarianthus* is biphyletic, consisting of one clade with oblong anthers and another with linear anthers, the latter being restricted to eastern New Guinea. Likewise, *Astronia* section Astigmata Maxw. & Veldk. was found to be polyphyletic. Intergeneric relationships within the tribe are well-resolved but these need further confirmation and support. Finally, *Tessmannianthus* forms monophyletic genus that groups with the Merianieae. However, it is noteworthy that analysis of only discrete characters suggests the monophyly of *Astronia* sect. Astigmata with a single reversal and the sister relationship of *Tessmannianthus* and the Astronieae. The phylogeny of the Astronieae sensu stricto supports the evolution of axile placentae into ascending and bilobed basal-axile, and ultimately, into collapsed ones. Pending molecular analyses, the resurrection of *Bamlera* K. Schum. & Lauterb. to accommodate the deviant Papuasian *Beccarianthus* clade is proposed.

T1-34: DNA metabarcoding biodiversity

T1-34-01

Plant distribution patterns and the factors influencing its diversity in the desert region of Northwest China

Yanlei Liu[1,2], *Wenpan Dong*[1,3], *Chao Xu*[1,2], *Wen Zhang*[1,2], *Xun Chen*[1,4], *Yufei Wang*[1], *Shiliang Zhou*[1]
1. Institute of Botany, Chinese Academy of Sciences, Beijing, China
2. University of Chinese Academy of Sciences, Beijing, China
3. Peking University, Beijing, China
4. Northeast Forestry University, Haerbin, China

Spatial plant diversity patterns underpins our understanding of ecosystem functioning and guides our environmental management decisions on plant resource conservation priority and utilization. Most researches have been focusing on biodiversity rich areas (hotspots) and ignoring biodiversity "poor" areas (coldspots). Most biodiversity coldspots locate on environmentally sensitive and economically undeveloped areas. Although the visible biodiversity is poor in such areas, the invisible biodiversity may not so poor and some species may even have great ecological importance. In order to understand the spatial plant diversity patterns in coldspots, we barcoded the plant species diversity in different parts of deserts in Northwest China to disclose factors that shape the patterns. We sampled 145 plots over the whole regions of deserts in Northwest China. Surface soils were collected and the ecological data were recorded. DNA were extracted from the soil samples usig Powersoil DNA isolation kit. Four chloroplast regions, *rbc*Lb, *mat*K, and *trn*L-F, were used DNA barcodes. These DNA barcodes were sequenced by metabarcoding method and plant species were determined by blast our data base. Different barcodes performed variably. The *rbc*Lb revealed 574 species belonging to 443 genera. The *trn*L-F identified 737 species belonging to 503 genera. The *mat*K uncovered 94 species belonging to 73 genera. A total number of 1,315 plant species was identified, they belong to 863 genus of plant. It is a common notion that a biodiversity coldspot such as the deserts in Northwest China is a plant diversity poor area. Our results showed that the plant species is not so poor as imagined. The species diversity each plot is underway and then association between plant species diversity and ecological factor could be analyzed.

T1-34-02

Unveiling the mask of microbiome in *Panax notoginseng* via DNA metabarcoding

Chao Xu[1], *Yanlei Liu*[1], *Zhongjian Chen*[2], *Shiliang Zhou*[1]
1. Institute of Botany, Chinese Academy of Sciences, Beijing, China
2. Institute of Sanqi Research, Wenshan College, Wenshan, China

Microorganisms dominate the whole world as key decomposers. Many microorganic species coexist together and some are fatal pathogens. Understanding the species of pathogens is the first step toward controlling them and curing diseases. *Panax notoginseng*, a very popular traditional Chinese medicinal plant in resolving blood problems, suffers from microorganism-associated diseases which cause severe economic loss. Our current knowledge of the pathogenic microorganisms is still limited depending on identification of cultured strains. High-throughput sequencing approaches have a potential application in pathogen identification by DNA barcoding, we therefore conducted a study for characterizing microbiomes in *P. notoginseng* by DNA metabarcoding. DNA were extracted from both healthy and infected rhizomes, leaves and seeds and nuclear gene 16S (V1-V4) and ITS were used to catalogue prokaryotes and eukaryotes separately. Both genes of *P. notoginseng* were blocked from amplification. In total, eighteen orders of prokaryotic microbes were detected. For eukaryotes, Dikarya taxa of seventeen orders probably account for fungal induced diseases while eukaryotic animals from Metazoa of fourteen families were characterized. DNA metabarcoding shows extremely high sensitivity in recovering microorganism like *Alternaria*, *Botrytis* and so on which were previously reported to be responsible for a series of plant diseases in *P. notoginseng*. This study found both culturable and unculturable microbiomes in *P. notoginseng*, which would illuminate the way to its disease control, even find the obstacles in continuous cropping.

T1-34-03

Using meta-barcoding to investigate bamboo diversity and plant/animal interactions

Linda Neaves, *Peter Hollingsworth*
Royal Botanic Garden Edinburgh

Next Generation Sequencing (NGS) methods have facilitated species identification from environmental samples, via meta-barcoding. This not only expands our ability to rapidly generate DNA-based inventories for plants and obtain information on levels of phylogenetic and species diversity in a region but can also to provide insights into the interactions between animals and plants. However, in some taxonomic groups the standard 'universal' barcodes provide only limited resolution and multiple group specific barcodes are required. Capture-based approaches provide a means for assessing multiple barcodes and mini-barcodes as well as allowing specific gene regions, which may be of interest or relevant to species identification and/or phylogenetic diversity. These methods can also be used to identify and validate regions for more targeted assays (e.g. qPCR). The relationship between specialist herbivores and the plants they feed on can often be complex than anticipated, as the full diversity of species consumed is overlooked. The giant panda is a key example of this. Whilst feeding almost exclusively on bamboo, there have been over 60 species reportedly consumed. Difficulties in distinguishing bamboo species, due to the infrequent flowering complicates out understanding of this relationship and research typically focus on a much smaller number of species. Standard 'universal' plant barcodes have limited species resolution due to low rates of molecular evolution, and even whole chloroplast sequencing shows limited differences. Using existing genomic data from bamboos and other grasses, a capture based approach will allow identification of regions/SNPs useful for distinguishing bamboo species as well as information of specific gene regions of interest (e.g. flowering genes and lignin biosynthesis). This information can subsequently be used to develop assays to provide robust and reliable identifications and hence detailed information of the distribution of bamboo species. In addition, the level of genomics data can provide insights into the bamboo (and non-bamboo) species giant panda predominantly feed on, and why they may be selected.

T1-34-04

Monitoring the Chinese herbal medicine market using metabarcoding method

Shiliang Zhou, *Yanlei Liu, Yuzhe Sun*
Institute of Botany, CAS

China is rich in plant species diversity. About 12800 species are used as medicinal plants. However, only nearly 800 medicinal plants species could probably be found in the medicine markets in China. Medicines made of these species are consumed not only by Chinese, but also by people all over the world, especially those in Korea, Japan, European Union and USA. In 2015,1.79 million tons of herbal medicines were exported, a drop of 8.84% from 2014. The governments committed a general check-up on major local medicinal markets and famous companies to find out fakes, adulteration, involvement of endangered species and many other aspects of herbal medicines. As an example, we scrutinized the Bozhou medicine market at Bozhou City, Anhui Province, China. In total 676 samples were collected from several companies. DNA was extracted, and four DNA barcodes were amplifies. PCR fragments of each sample were labeled by a unique DNA sequence adhered to primers. All PCR products were then mixed together to build a sequencing library for a PE250 run on Illumina HiSeq X platform. The resulting reads were sorted into genes according to the primer sequences and into samples according to the unique label sequences. Reads of different alleles of the same gene of a sample were separated with Usearch by limiting a similarity at 99%. Consensus sequences were created from the aligned contigs. The consensus sequences were identified by blasting them against the reference database. If unexpected species were found in a sample, the sample is considered adulterated; if no expected species was found, the sample is considered fake. The 676 samples belong to 607 medicines, which are supposed to be from 477 species, including six poisonous plants and 24 endangered medicinal plant species under cultivation. Although some samples have been processed, PCR is proved to be successful. The percentages of sequencing successes for ITS, *matK*, *rbcL* and *trnL-F* are 92.9%, 95.6%, 95.9 and 94.8%, respectively. Identification by DNA barcoding indicated that misidentification is not a big problem. However, contamination is common, which is likely to be due to adulteration on purpose. The Bozhou herbal medicine market is generally safe in terms of herbal medicine materials.

T1-34-05

Increasing efficiencies in the inventory of tropical rainforest floras: A case study from the Brazilian Amazon

Benjamin M. Torke[1], *Thaís Elias Almeida*[2], *Thiago André*[2], *Leandro L. Giacomin*[2], *Jonathan D. Kuziel*[1], *Vidal De Freitas Mansano*[3]
1. New York Botanical Garden
2. Universidade Federal do Pará-Santarém
3. Jardim Botânico do Rio de Janeiro

The under-documentation of tropical rainforest floras continues to inhibit research on global patterns of plant diversity and endemism and impedes efforts to assess the conservation status of numerous plant species. Carrying out floristic inventories in tropical rainforests presents many challenges, among the most notable are the relatively intensive efforts needed to reach exhaustive sampling of species in hyper-diverse rainforests, and the paucity of resources that are available to identify the plant species that occur in many tropical areas. We are in the midst of assembling floristic lists of the vascular plants of two large protected areas in an under-sampled part of the Brazilian Amazon. We present preliminary data comparing the relative efficiency of various sampling techniques as measured by the number of woody species encountered per time and sampling effort. We also assess the efficacy of DNA barcoding as a tool for expediting identifications of taxa in a study area that lacks a local barcode reference database. Sampling in ten contiguous 2 m x 50 m (= 0.1 ha) transects was found to have higher sampling efficiency than other widely used methods, such as general collecting and quantitative sampling in square plots. Barcoding employing matK sequences obtained using universal primers and a non-local reference database yielded accurate generic identifications for approximately 75% of species. We conclude that the integration of transect-based sampling and identification of sterile voucher specimens using a combination of DNA barcoding and traditional approaches to identification holds promise for increasing the rate and accuracy of plant species inventory in tropical rainforests.

T1-34-06

Use of DNA barcoding to validate identity of living *Angiopteris* Hoffman species in a university fernarium

Reggie Y. Dela Cruz, *Clene Rose Palle*
Central Mindanao University

The study was focused on the validation of the species of living *Angiopteris* in a university fernarium through DNA barcoding using *rbc*L gene. Extraction of genomic DNA was carried out using modified cetyltrimethylammonium bromide (CTAB) DNA extraction method. A set of primers, *rbc*LaF (5′ATGTCACCA-CAAACAGAGACTA AAGC-3′; Levin *et al.*, 2003) and *rbc*LaR (5′GTAAAATCAAGTCCACCRCG-3′; Kress and Erickson, 2007) was used in this study for the amplification of *rbc*L gene of the chloroplast. A genomic sequence of 539 base pairs was obtained for 14 collected DNA samples of *Angiopteris*. BLASTn analysis revealed that the unknown species of a previous unpublished work matched at 99% similarity with *Angiopteris evecta*. Previously identified *A. elmeriana* also matched with *A. palmiformis*. All *A. palmiformis* samples were confirmed as the same species through BLASTn. Neighbor joining tree further showed that the unidentified species, *Angiopteris* sp. 1 and 2, were found in the same cluster with bootstrap value of 70%. The previously identified *A. elmeriana,* clustered together with *A. palmiformis* with bootstrap support of 69% and 66%. Five potential barcoding regions were identified through haplotype construction. *Angiopteris* sp. 1 and 2 exhibit a distinguishable haplotype from the other *A. palmiformis* species. *Angiopteris elmeriana* however had same haplotype as *A. palmiformis*. Thus, the two previously unidentified *Angiopteris* species were identified as *A. evecta* and the identified *A. elmeriana* species was re-identified as *A. palmiformis* using the *rbc*L sequence. The *A. palmiformis* samples were however confirmed as the same species. It could be seen that *rbc*L as DNA barcode could be used to discriminate the *Angiopteris* species in this study.

T1-35: Plant conservation, learning from the past!

T1-35-01

The evolution of monsoonal climate and its influence on plant diversity in Southwestern China

Tao Su[1], *Yong-Jiang Huang*[2], *Shu-Feng Li*[1], *Yao-Wu Xing*[1], *Zhe-Kun Zhou*[1]

1. Xishuangbanna Tropical Botanical Garden, Chinese Academy of Sciences

2. Kunming Institute of Botany, Chinese Academy of Sciences

The paleoclimate history played an important role in the distribution and evolution of plants. Southeastern China is dominated by monsoonal climate at the present day, which is characterized by wet summers and dry winters. Quantitatively paleoclimate reconstructions by fossil floras indicate that the monsoon gradually intensified since the Neogene, which means the seasonality of precipitation in the past was less significant the present day. Fossil records indicate that, some plants disappeared in this region because of the intensification of monsoonal climate, such as *Cedrus*. Combined with information on *Cedrus* fossil records and its seed physiology, we concluded that the intensification of a drier climate after the late Pliocene may have prevented the survival of *Cedrus* seedlings, leading to the eventual disappearance of *Cedrus* in western Yunnan. Meanwhile, many other plants, e.g., *Pinus*, and *Quercus* subgenus *Cyclobalanopsis*, survived and diversified successfully here by the morphological plasticity under paleoclimatic changes. These findings suggest that the evolution of monsoonal climate is crucial in shaping the patterns of plant diversity and floristic structures in southwestern China.

T1-35-02

The Miocene forest in western Amazonia and its response to abiotic changes, lessons from the past

Carina Hoorn[1], *Raymond Van Der Ham*[2], *Laura Lagomarsino*[3], *Felipe De La Parra*[4], *Sonia Salamanca*[5], *Hans Ter Steege*[2], *Hannah Banks*[6], *Wim Star*[2], *Bertie-Joan Van Heuven*[2], *Rob Langelaan*[2], *Fernanda Antunes*[7], *Guillermo Rodriguez-Forero*[8], *Andres Pardo*[9]

1. University of Amsterdam

2. Naturalis

3. University of Missouri

4. University of Oxford

5. no affiliation

6. Royal Botanic Gardens at Kew

7. Universidade Federal do Rio Grande do Norte

8. Instituto Colombiano del Petroleo

9. Universidad de Caldas

The Amazonian forest is a global treasure and its preservation concerns us all. Climate change and destruction by mining, logging and farming constitutes a real threat. If current rates of destruction and global warming persist within 100 years most of the forest might be lost. It is thus of great importance to properly understand the response of the forest to past environmental change and climatic optima. Based on palynological records from western Amazonia we evaluated two extinct, key components of the Miocene forest that occurred there around 16-17 million years ago. We established the nearest living relatives, past distribution, phylogeny, and closest associates of these taxa. Our data suggests that these fossils represent Malvaceae and that they occurred during the mid Miocene global climatic optimum (MMCO), when regional climatic patterns were quite different from the Present. At the time the Eastern Cordillera (Andes) was not yet fully formed as orographic barrier, and the Orinoco and Amazon drainage basins were still connected. In the course of the middle Miocene this situation changed with the development of a large wetland in western Amazonia and an influx of water and sediments from the Andes. We hypothesize that climate and sedimentary environments changed to such a degree that our Malvaceae perished, while new taxa (e.g. Arecaceae) evolved, as they were better suited to the new environmental conditions. At least one of these extinct Malvaceae was related to a taxon of the Gondwanan flora, which had its past distribution in southern South America, Antarctica, Australia and New Zealand. The other extinct Malvaceae taxon possibly was an immigrant from northern latitudes. Our study suggests that during the MMCO, when climate was warmer and more humid than Present, the forest did not perish. Instead, it was rather diverse albeit of different composition than at Present. Perhaps a better analogue for future scenarios of climate change in Amazonia might be found in late Miocene records, when global climatic cooling (albeit still warmer than at Present) and lower global sea level created generally dryer conditions, and coincides with the development of the savanna biome.

T1-35-03

First fossil flowers from the Danian (earliest Paleocene) of South America reveal complex biogeographic history of Gondwanan flora

Nathan Jud[1], *Maria Gandolfo*[1], *Ari Iglesias*[2], *Peter Wilf*[3]

1. Cornell University

2. Universidad Nacional del Comahue

3. Pennsylvania State University

The early Danian (earliest Paleocene) Salamanca Formation in southern Chubut, Patagonian Argentina yields well-preserved plant fossils from within 2.5 million years of the end-Cretaceous extinction event. Flowers attributable to Rhamnaceae and Cunoniaceae were found among the Salamanca fossils. Herein, we describe these flowers, and we discuss the implications of their presence just above the K-Pg boundary of Patagonia. The rhamnaceous flowers are pentamerous with an obconic hypanthium, keeled sepals, stamens opposite cucullate petals, and a floral disk. Today, Rhamnaceae are cosmopolitan with nearly 1,000 species, and the combination of molecular phylogenetic and paleobotanical studies suggests the family originated in tropical forests during the Cretaceous. The presence of Rhamnaceae in the Salamanca Formation indicates that the family spread to subtropical climates at mid-latitudes from the tropical forests by the earliest Paleocene. By contrast, the cunoniaceous flowers are 8-10 merous with notched petals, a pubescent and superior ovary, and two styles. This combination of features indicates an affinity with the tribe Schizomereae, a primarily southern-temperate clade with 36 species previously known only from Australasia and South Africa. The Salamanca fossils are the oldest evidence of the Schizomereae, and they indicate that the evolution of this group was not constrained to Australasia, strengthening the Gondwanan connection between Patagonia, Australia, and South Africa during the early Paleogene.

T1-35-04

Environmental impact on differentiation development of Late Cenozoic floras between East China and Japan

Wei-Ming Wang

Nanjing Institute of Geology and Palaeontology, Chinese Academy of Sciences

Floristically, East China and Japan belong to the East China-Japan Forest Floral Subregion. There are 12 seed plant families in the Subregion which do not occur in Japan. Some well represented Tertiary plants in Japan, such as *Keteleeria*, *Pseudolarix*, *Cunninghamia*, *Glyptostrobus*, *Carya, Hemiptelea, Sassafras, Liquidambar* and *Sycopsis*, are currently only growing in China. To explore these floral changes in the geological past, some representative Late Cenozoic floras in Japan are evaluated in comparing with those in North China. Evidences show that some major environment events have exerted a great influence on the development of local floras. Pollen floras from Neogene strata of Northern China and Japan show a detailed changing process prior and after some major Neogene environmental events. Climate reflected by the Neogene pollen floras of both places are mostly similar at the early stage, showing a gradual climatic resuming process in the Early Miocene, followed by a period of a period of climatic optimum in the late Early-early Middle Miocene. Since the Middle Miocene (ca. 14 Ma), the flora in North China was under the control of ever intensified continental condition, while the one in Japan had undergone the impact of prevailing marine climate. The floras in Japan are indicated by high occurrence of conifer and amentiferous elements at the early Early Miocene, followed by a distinctive floral flourishing period at the late Early Miocene-early Middle Miocene. Pollen floras in the Middle-Late Miocene Japan display a monotonous composition with Pinaceae predominated in the early Middle Miocene, and some broad-leaf and Tertiary elements, such as evergreen *Quercus, Carya, etc.* in the late Middle Miocene. A rise in boreal components and a decline in the broad-leaf and Tertiary elements are recognized again in the Late Miocene (since ca. 7 Ma) flora. Early Pliocene flora shows some Tertiary and warm-temperate broad-leaved types partly re-expanded, while in the Late Pliocene flora conifers are prominent again, and broad-leaf trees are minor with the exception of *Fagus*. Cause of the differentiation development of Neogene floras between East China and Japan is suggested to be raised mostly by the opening of the Sea of Japan started at the early stage of the Miocene. It has already been revealed that the initial stage of the opening of the Sea of Japan began 16 Ma ago, and the main stage of the opening was 15 Ma ago. By 14 Ma ago, the opening of the Sea of Japan had been almost completed. Since then, the Japanese Islands has been surrounded by the prevailed marine environment, while the mainland of China dominated by the continental condition, resulting in the differential development of their floras. Stratigraphic distribution of plant macrofossil shows a gradually extinction of Tertiary plant types in the Pliocene-Pleistocene of Central Japan. It is deduced that these extinct events are closely related with the global climate cooling, racial crustal movement, formation of the modern mountain system, and frequent volcano activities, *etc.*

T1-35-05

Impact of the Plio-Pleistocene climate and geomorphological changes on floral diversification and plant extinction in Japan

Arata Momohara

Graduate School of Horticulture, Chiba University

Plio-Pleistocene floral changes in central Japan are characterized by extinction of major elements in the Neogene flora, including *Metasequoia* and *Glyptostrobus*, and expansion of plants dominant in modern cool-temperate and subalpine forests. The floral change was stepwise, and the events were concentrated in transition periods of climatic fluctuation and/or in a downward shift of the glacial climate in the late Pliocene (ca. 3.3 – 2.5 Ma) and late Quaternary since 1.4 Ma. The last occurrences are generally recorded in warm stages followed by significantly colder stages. Their extinction is thus attributed to extirpation of refugia during the cold stages in and around the sedimentary basin. Increased dryness and coldness provided by winter monsoon that developed in a stepwise manner would have altered the climate regime and the ability of plants to grow and establish populations. Mountain uplifts became active after the late Pliocene and may have promoted plant extinction by constituting barriers of migration. It is because the decreasing winter temperatures in higher altitudes would have hindered migration to the north and inland over mountain ranges, even as exotic taxa surviving in southern sedimentary basins had the potential to expand their distribution under improving climatic conditions. However, *Metasequoia* and *Glyptostrobus* occurred during the Early Pleistocene glacial stages and it indicates their tolerance for cool-temperate climate condition. Along with mountainous areas surrounding the basins, the landform of the lowland plains in the sedimentary basins was also changed by tectonic movement and sea level changes, which increased in magnitude during the later stages of the Quaternary. This change influenced directly on habitats of wetland plants such as *Metasequoia* and *Glyptostrobus* which were distributed in the lowermost altitude in basins. Pronounced sea level changes after the late Early Pleistocene also transformed alluvial plains. During interglacial stages, basins and coastal plains were submerged during marine transgressions that reached near mountain bases and diminished habitat of wetland plants. Furthermore, alluvial plains that lost water after sea levels decreased became drier uplands with steep and narrow valleys at the lowest sea level stands in glacial stages. Such drastic environmental changes in the lowlands possibly influenced the regeneration and growth of *Metasequoia* and *Glyptostrobus*, which dominated wetland vegetation before the late Early Pleistocene.

T1-35-06

Learning from the past: Relic plants in SE and E Asia and paleoecology of their fossil relatives in Europe

Lutz Kunzmann[1], *Zlatko Kvacek*[2], *Karolin Moraweck*[1], *Tao Su*[3], *Zhe-Kun Zhou*[3]

1. Senckenberg Natural History Collections Dresden

2. Charles University Prague

3. Xishuangbanna Tropical Botanical Garden, CAS

Relatives of E and SE Asian plants were major constituents in the Paleogene and Neogene European vegetation and thus tend to be important indicators for deep time climate and environments. These extant genera and species are often relics and endemic to SE Asia. Relics have very specialized demands on habitat conditions such as soil nutrients, water availability and temperatures which evoke relatively narrow ecological amplitudes for those plants. This constraint is important for conservation purposes of endemics and relics. In contrast, their fossil relatives in Europe often dominated characteristic plant associations from particular depositional settings. Combined knowledge from phytosociological composition of fossil assemblages, from burial sediments and accompanying fauna provide often evidence that ecological demands of the fossil-taxa remarkably differ from those of the extant relative species and/or genera. This could be caused either by ecological shifts within the respective taxon, or, alternatively, fossil-species had a distinct autecology in comparison to the extant relic species, with or without overlap in the amplitudes. Thus extinction of fossil-species within a certain genus refers to cuts in the ecological amplitude of the genus ending with the narrow ecological amplitude of the extant relic. This contribution gives several examples from the European fossil record and compares their ecological demands to those from their extant relatives in E and SE Asia. Selected fossil examples are *Cathaya bergeri*, *Cyclocarya cyclocarpa*, *Diplopanax limnophilum*, *Eotrigonobalanus furcinervis*, *Eurya stigmosa*, *Fortunearia europaea*, *Gironniera pyriformis*, *Glyptostrobus europaeus*, *Keteleeria hoehnei*, *Platanus neptuni*, *Rhederodendron wiesaense*, *Trigonobalanopsis rhamnoides* and others. This could potentially give valuable hints for conservation policy and protection strategies of relics. Mismatch in ecological demands between fossil and extant taxonomical relatives should be paid particular attention because the extant relative could not serve as a proxy for paleoenvironmental and paleoclimatic reconstructions.

T1-36: Plant diversity and conservation in Southeast Asia

T1-36-01

Patterns of species distribution and diversity along elevational gradients in sparsely documented areas in Northern Myanmar

Phyo Kay Khine, *Georg Miehe, Jürgen Kluge*
Philipps University of Marburg

The high landscape heterogeneity along gradients of latitude, elevation and geological formations shapes the physical environment and vegetation structure and favours the diversity and richness in Myanmar flora. The northern part of Myanmar is located between eastern Himalaya and the southwestern China Biodiversity Hotspot, but unfortunately the area is highly under-sampled. Most of the floristic surveys were engaged in the late 19th and the early 20th century. The taxonomic knowledge and distribution patterns are thus very poorly known. The assessment of the elevational gradients of species richness is a major issue of biogeography worldwide, but plot-based complete floristic gradients in species rich tropical mountain areas are not yet available. We undertook the task to close the gap of knowledge in this biodiversity hotspot especially for spermatophytes and pteridophytes. In order to make inventories and patterns comparable to respective studies worldwide, we used a standard protocol of plot-based complete floristic sampling. Our floral inventories were conducted in Natma Taung (Mt. Victoria; 400-3062 m) and Hponyin Razi (400-4057 m). Four plots of 400 m^2 were exhaustively sampled at every elevational step of 200 m starting from 400 m above sea level, making up a total of 132 plots. Here we will report on species composition and richness of pteridophytes and compare the results to adjacent mountain regions from Nepal Yunnan (China) and Japan. Species richness showed hump-shaped patterns for both transects, however with different elevations of highest richness. The highest number of species was found at 2200 m in Natma Taung and at 1200 m in Hponyin Razi. The patterns were mainly driven by epiphyte species. Since most of the ferns species are perennial, the rainfall and relative humidity is crucial for their survival. The occurrence of epiphyte species limited with treeline ecotone. Climate is therefore thought to be the most prevailing factor that influence species richness pattern. This finding coincides with results from regional studies of central Himalayas in Nepal and local studies of islands in Japan. Epiphytes are susceptible to the environmental changes, so the region with high epiphytic species number needs to prioritize for documentation and conservation. To have an overview about the flora of Myanmar and understand more about the species distribution pattern with elevation, the transect study should be extended to more north in the Kachin Sate, northeast to the Yunnan border and southwest to Naga hill such as Mt. Saramati.

T1-36-02

Bryology in Borneo: A review from 1996 to 2016

Monica Suleiman
Universiti Malaysia Sabah

This paper reviews the progress and challenges of bryology in Borneo for the past 20 years (1996-2016). Data presented is based on all research carried out by the author, research collaborators and students since 1996. The number of mosses recorded in Borneo has increased from 649 to 755 taxa, with several new species. However, little progress is seen in the other two groups of bryophytes, the liverworts and hornworts. The current number of liverworts and hornworts in Borneo has not changed much, which is 623 taxa. Furthermore, most studies in Borneo are biased to Sabah (East Malaysia), with less progress in Sarawak (East Malaysia), Brunei and Kalimantan (Indonesia). Sadly, only a few local students are interested to pursue postgraduate studies in bryology, especially in taxonomy and ecology, due to job prospects. However, interest in applied studies such as on pythochemical contents, antioxidants, antiproliferative activities and novel chemical compounds of Bornean bryophytes have been increasing in recent years.

T1-36-03

Plant community homogenization due to forest fragmentation in South China, Xishuangbanna

Daniele Cicuzza
University Brunei Darussalam

Habitat fragmentation causes a rapid change, in tropical forest, for species composition, biomass, phylogenetic diversity and increase of disturbance-adapted species. Xishuangbanna Prefecture, south of west China, has experienced a severe change in forest cover in the last decades due to the expansion of banana, tea and mainly rubber plantations. We conducted a study over the entire Xishuangbanna Prefecture (~20.000 km^2) assessing the tree and herbs diversity related to the forest structure and patches sizes. The inclusion of different forest patches allowed us to gather species diversity and community composition over a large area and a variety of forest structure. The results shows the importance of the study scale and the different ecosystems. The northern part has large and continuous forest surrounded by numerous small patches, whereas the south has few large forest patches surrounded by extensive cultivation. General conclusion underline that plant community changes with fewer and disturbance-adapted species in small patches and slightly richer community in bigger patches with more forest species. Tree diversity, forest structure (number of trees and DBH per plot) varies, as expected, according to the forest patch size, with simpler diversity and structured forest in small patches and more articulated in species and structure in mature forest. These methods allowed us to obtain important field-based data on plant community diversity and forest structure over a large human modified landscape. We underline how rare species, community composition and forest structure highlight the threat for important areas such as lowland forest or karst ecosystems. These methods can be used and expanded in other south China tropical areas where landscape with high level of biodiversity have been recently heavily modified.

T1-36-04

Inventory and assessment of herbaceous flowering plants in Long Term Ecological Research sites (LTER) of Mindanao, Philippines

Florfe Acma, *Dave Buenavista, Jerome Gaas, Fulgent Coritico, Victor Amoroso*
Central Mindanao University

The Philippines harbor floral diversity including the herbaceous angiosperms which are prone to threats brought by human and natural causes. It is therefore essential to inventory and assess these herbaceous angiosperms. One-hectare permanent plots with twenty-five 20 m x 20 m subplots were established in the five Mindanao LTER sites, Philippines *viz.*, Mt. Apo of North Cotabato, Mt. Hamiguitan of Davao Oriental, Mt. Kitanglad of Bukidnon, Mt. Malindang of Misamis Occidental and Mt. Musuan of Bukidnon. Species richness, diversity and assessment of herbaceous angiosperms were determined. Among the five LTER sites, Mt. Malindang has the highest species richness with 57 species, followed by Mt. Musuan (56), Mt. Kitanglad (53), Mt. Apo (51) and Mt. Hamiguitan (43). In terms of diversity, Mt. Malindang (H' = 1.58) was recorded to be most diverse ecosystem. This is followed by Mt. Apo with H' = 1.39, Mt. Hamiguitan and Mt. Kitanglad both have H' = 1.29 and the least diversity value is Mt. Musuan with H' = 1.18. Further, Mt. Malindang has the greatest frequency (23.92) of species with *Alpinia alpina* as the most frequent species. Mt. Kitanglad revealed to have the highest density value with *Sarcandra glabra* as the densest species. The high frequency and density of the said species could be related to dispersal since the aforementioned species are prone to herbivory by animals that serve as pollinators and dispersers of their pollen and seeds. Mt. Musuan has the lowest diversity value since it is prone to anthropogenic disturbances because the site is situated at the altitude of 388 m asl. which is near to a residential community hence, vulnerable to exploitation. Mt. Malindang is observed in the mid-montane vegetation which serves as a transition zone between lower montane and upper montane forest where species are observed to be rich and diverse. Species with the highest species importance value (SIV) in Mt. Apo and Mt. Hamiguitan is *Freycinetia* sp., in Mt. Kitanglad is *Sarcandra glabra,* in Mt. Malindang is *Impatiens platypetala* and in Mt. Musuan is *Donax canniformis* sp. These herbaceous angiosperms dominate the understory of each site where these species exert a great influence to the forest ecosystem. Assessment of the herbaceous angiosperms in the five sites recorded 2 endangered, 1 vulnerable, 9 economically important species, 1 rare and 1 other wildlife species. These endangered species are *Nepenthes copelandii* and *Paphiopedilum ciliolare* which are recorded only in Mt. Apo and Mt. Hamiguitan respectively. Most of the recorded economically important species belong to the Orchidaceae family which are usually highly priced ornamentals. These species of herbaceous angiosperms should be conserved and protected to prevent the loss of the country's valuable biodiversity.

T1-36-05

History of floristic exploration in Myanmar including modern collections towards a Flora of Myanmar

Kazumi Fujikawa[1], *Moe Aung Zaw*[2], *Mark Watson*[3]
1. The Kochi Prefectural Makino Botanical Garden
2. Forest Research Institute
3. Royal Botanic Garden Edinburgh

Myanmar, located at the western edge of Southeast Asia and the eastern edge of the Indian sub-continent, is a unique place where four floristic regions intersect: the Indian, Sino-Japanese, South-East Asian and Malesiana floristic regions. Myanmar is known for its rich plant diversity due to the combination and interaction of geography, topography, climate, pattern of seasonal rainfall, presence of high mountains and major rivers. The first scientific collections were made by Francis Buchanan-Hamilton in 1795, and many herbarium specimens were gathered by British explorers in the 150 years that followed. However, very little inventory-based botanical fieldwork has been undertaken since then, whilst in surrounding countries field investigations and also preparations and publications of their floras have been actively progressing. Myanmar is a major part of a global biodiversity hotspot, yet remains a knowledge gap in the floristic knowledge of Asia. During the last 15 years several botanical explorations have been undertaken, led by botanists from Japan and Myanmar. Over 26,500 specimens of vascular plants have been collected, housed in the herbarium of Myanmar's Forest Department and in Makino Botanical Garden. The ongoing project aims to publish a Taxonomic Enumeration of Natma Taung, Chin State, and we, with international collaborative team of botanists, have so far identified 1,843 species representing 169 families. The level of new knowledge generated for the country is remarkable, for example, the Asteraceae consists of 138 species, of which 27 are new records for Myanmar and one species of each *Ainsliaea*, *Himalaiella* and *Taraxacum* appear to be new to science based on these recent collections. We continue on field collections and taxonomic research collaborated with international team of botanists, which are the vital first steps towards a Flora of Myanmar.

T1-36-06

Regional patterns of local processes based on the '101° E' forest transect from Southwest China to the Central-Southern Peninsula

Luxiang Lin
Xishuangbanna Tropical Botanical Garden, CAS

One large forest plot (16-50 ha) can be used as a platform to explore ecological processes underlying tree species coexistence at local scales (<1 km^2). However, little is known about how these local processes vary at regional scale. The '101° E' forest transect from Southwest China to the Central-Southern Peninsula (here after '101 transect') can acts as a platform to explore how ecological processes occurring at local scale change along an environmental gradient at larger scale. This is so-call 'regional patterns of local processes'. The '101 transect' consists of ten large forest plots, from five of which we have collected key functional traits of more than 1,000 tree species. We use trait difference among species pairs which co-occur in the same plot to quantify relative importance of environmental filtering and similarity limitation underlying species coexistence, especially we compare the difference between common and rare species in trait variation to distinguish how common and rare species co-exist at local tree assemblages. Based on above, we want to test a hypothesis at regional scale, namely the strength of environmental filtering increases with habitat harshness, while the strength of similarity limitation decreases with habitat harshness. I will report the advance on construction of the '101 transect' and research based on it in my oral presentation.

T1-37: New insights on the assembly and biodiversity of the flora of North America (two sessions)

T1-37-01

The roles of molecular evolutionary rates and functional traits in the colonization success of maples in North America

A.J. Harris[1], *Jeiyao Feng*[2], *Jun Wen*[1]
1. Smithsonian Institution, Department of Botany
2. T.C. Williams High School

We are using *Acer* L. (maples, Sapindaceae) to investigate the roles of molecular evolutionary rates and functional traits in successful colonization of new habitats by plants in North America. Colonization success is an understudied historical aspect of the North American flora and has important implications for patterns in the spatial organization of biodiversity. Colonization of an available area is dependent upon the regional species pool, and the species pool hypothesis asserts that colonization is logically more probable for larger areas that have persisted for longer times. Thus, area processes drive colonization success. However, colonization from regional species pools is undoubtedly also strongly affected by biological processes such as life history and evolutionary rate, including molecular substitution rate, speciation, and extinction. Our study is one of the first to seek correlations between colonization success and evolutionary rate and life history at a continental scale. To obtain evolutionary rates for all eight North American lineages of *Acer*, we are reconstructing a dated phylogeny from existing molecular sequences and fossils. We are using the dated phylogeny to estimate substitution rates as well as speciation and extinction rates while controlling for time. We are using functional traits as a proxy for life history strategies, and we are obtaining functional trait data from herbarium specimens and the literature. Our method of measuring colonization success is to compare the size of the distributional range of each lineage of *Acer* with the size of its climatically suitable range in North America according to species distribution models (i.e., ecological niche models); that is, how much area it does occupy compared to how much area it could occupy. We are evaluating the relationships between colonization success and evolutionary rate and the states of functional traits using standard statistical methods. Our study provides insights into the biological influence on colonization success of North American maples and has relevance for other woody plant clades. Moreover, the study offers a methodological framework for integrating biological processes into future work on colonization.

T1-37-02

Assembly and biodiversity of the Flora of North America: Areas of endemism of Mexican vascular plants

Victoria Sosa[1], *J. Arturo De-Nova*[2]
1. Instituto de Ecologia, Xalapa
2. Instituto de Investigación de Zonas Desérticas, Universidad Autónoma de San Luis Potosí

From the approximately 25,000 species of native vascular plants of Mexico, nearly 50% are endemic. Moreover a number of lineages are endemic to Mexico, including one family, several genera and clades in different plant families comprising many species. The most remarkable are the monotypic family Setchellanthaceae, and several clades/genera in Asteraceae and Cactaceae, as well as nine monocot genera. With regard to their distribution, the majority of the regions with the highest indices of endemism have dry climates, however areas of endemism have been recorded in tropical humid climates as well. Among objectives of this study are: 1) identify areas of neo- and palaeoendemism, 2) examine variation of phylogenetic diversity in areas of endemism, 3) identify whether areas of endemism are associated with current climate conditions and historic climate stability. Occurrence points for almost 400 species belonging to angiosperm endemic lineages of Mexico were obtained from literature and databases and they were scored taking into account biogeographic provinces. Phylogenies for the endemic groups in Cactaceae and Asteraceae were constructed and molecular dating analyses were used for estimating divergence times. In addition, a megatree was obtained to estimate phylogenetic diversity and different metrics related to this diversity. Current and past ecological niche were modeled. Results suggest that most of the areas are identified by neoendemisms congruent with hypotheses of the assembly of the Mexican flora. The biogeographic regions, the Trans-Mexican Volcanic Belt and the Tehuacán Valley are outstanding for their endemism. Areas of endemism are explained by historic climate rather than current climate.

T1-37-03

Adaptation, competition, and the distributions of North American woody plants

Robert Ricklefs
University of Missouri-St Louis

Distributions of species within regions (the "regional community") are influenced by adaptations to the physical environment and by interactions—particularly competition—among species. These factors can be revealed indirectly by analyzing the distributions of related species with respect to characteristics of the environment and to each other. Comparisons of tree species richness among temperate forests on different continents indicate significant regional influences, in addition to the effects of present-day climate, presumably reflecting the unique history and geography of each region (Qian and Ricklefs, 2000; Ricklefs and He, 2016). Within North America, climate variables explained about 70 per cent of the variance in the ordinated positions of 340 forest plots based on tree species composition. Considering the ordinated positions of species based on their distributions across plots, related (congeneric or confamilial) species tended to be marginally closer than randomly-selected species, indicating a stronger role for phylogenetically shared adaptation compared to competition in shaping distributions (Ricklefs, 2015). Moreover, the abundance of individual species within plots was not related to the number of co-occurring closely related (congeneric) species, further suggesting that competition is not stronger among closely related compared to distantly related species. Comparisons of the distributions of related (i.e., congeneric) herbaceous (but not woody) taxa in eastern Asia and North America suggest a general conservatism of latitudinal distribution and geographic range size over long evolutionary time scales (Ricklefs and Latham, 1992; Qian and Ricklefs, 2004). In contrast, a taxonomically nested ANOVA of tree distributions in eastern North America showed little evolutionary conservatism in population traits (range position, distributional extent, and abundance); nor were local abundance and geographic (ecological) dispersion related (Ricklefs, 2015). These present-day patterns can be compared to the postglacial fossil pollen record of plants in North America, which suggests that the distributions of individual species changed relatively independently during the past 21,000 years, such that the positions of pollen localities in ordination space based on species became uncorrelated over time. Finally, beta diversity (turnover of species with distance) is greater with respect to both geographic distance and climate in the more diverse eastern Asian flora than in the North American flora (Qian *et al.*, 2005), suggesting that diffuse competition constrains ecological distributions more strongly within eastern Asia. These analyses suggest that the distribution and abundance of species in temperate deciduous forests are largely determined by labile adaptations to the physical and biological environments and are independent of competition among closely related species, although diffuse competition within the entire local flora apparently has an important impact. Nonetheless, factors that influence the relative abundance of species and their distributions within large regions, which vary over several orders of magnitude, remain poorly understood.

Thanks for enjoyable and productive collaborations to Hong Qian, Qinfeng Guo, Fangliang He, Roger Latham, and Jessica Blois; financial support from the U.S. National Science Foundation and the Curators of the University of Missouri.

T1-37-04

Floristic changes from Paleocene through Miocene time across midlatitude western North America

Steven Manchester[1], *Kathleen Pigg*[2], *Melanie Devore*[3]
1. University of Florida
2. Arizona State University
3. Georgia College & State University

Floristic inventories of Tertiary localities in the middle latitudes of western North America document patterns that reflect climatic shifts and biotic interchange with other regions. Most Paleocene floras are dominated by broadleaved deciduous taxa, including Platanaceae (*Macginitiea*, *Platanites*, *Platanus*), Coryloideae (*Palaeocarpinus*), Juglandaceae (*Polyptera*, *Cyclocarya*, *Juglandicarya*), Cercidiphyllaceae, Nyssaceae (*Davidia, Amersinia, Browniea*), plus *Aesculus*, *Cornus* and *Celtis*. *Glyptostrobus*, *Metasequoia*, Taxaceae, *Ginkgo*, and occasional cycads were present, as were palms and Zingiberaceae. Dominant dicots had serrate leaves with long petioles. Ferns included *Onoclea* and *Woodwardia*. Although richness per locality is typically low (< 20 species), many taxa were shared with the Paleocene of Greenland, Scotland and Asia. The Paleocene-Eocene transition, well documented in north central Wyoming, coincides with a peak in mean annual temperature. Most of the usual Paleocene elements are absent, and new taxa appear, including abundant legumes, Sapindales and Hernandiaceae. Early Eocene floras of the same region are relatively low in megafossil diversity (< 10 taxa per locality), commonly including *Alnus*, *Platycarya*, Malvaceae sl., Simaroubaceae (*Ailanthus*, *Chaneya*), and *Salvinia*. Possibly endemic, extinct genera of uncertain affinity, are common in these floras. Early middle Eocene sites of the Rocky Mountain region are dominated by Platanaceae (*Macginitiea*), Salicaceae (*Populus, Pseudosalix*), and Ulmaceae (*Cedrelospermum*), with abundant legumes and Sapindales (Anacardiaceae, Sapindaceae). Climbers included *Lygodium*, Vitaceae, Icacinaceae and Menispermaceae. Long-petioled taxa no longer dominated. Similarly aged floras near the west coast shared some of these taxa (e.g. *Macginitiea*, *Cedrelospermum*, Icacinaceae, *Lygodium*), but had fewer Salicaceae and Leguminosae and a greater diversity of mesic, thermophilic taxa (diverse ferns, Icacinaceae, Lauraceae, Mastixiaceae, Menispermaceae, Sabiaceae, Torricelliaceae, Musaceae, and cycads, along with *Juglans* and *Ulmus*. Many genera are shared with coeval European floras. Rosaceae, Betulaceae, Sapindaceae, Trochodendraceae, and diverse conifers proliferated in parts of Washington and British Columbia, but remained relatively rare in other western floras. *Cedrelospermum* remained common through the late Eocene, as new elements arrived. Late Eocene Rocky Mountain floras show increased occurrence of *Koelreuteria*, *Carya*, *Cercis*, Rosaceae (*Cercocarpus, Crataegus, Rosa*) as well as some extinct genera (*Asterocarpinus*, *Fagopsis*, *Florissantia*). Overall, this inland late Eocene flora has markedly smaller leaves than those of the west coastal Eocene and Oligocene floras. Early Oligocene floras contain *Pinus*, Cupressaceae (*Metasequoia, Se-*

quoia, Tetraclinis), many deciduous angiosperms, e.g., Betulaceae (*Alnus, Asterocarpinus, Betula, Ostrya*), Aceroideae, Fagaceae, Hamamelidaceae, Juglandaceae, Tilioideae, Ulmaceae, *Cedrela* and only few broadleaved evergreens, e.g. *Mahonia*. Rare Late Oligocene floras indicate lower diversity of arborescent taxa. They included *Juniperus*, *Pinus* and a few angiosperms e.g., *Salix*, *Mahonia*, with smaller leaf size. Phytoliths indicate increasing presence of grasses coinciding with cooling and drying. Middle Miocene floras in Washington, western Idaho and adjacent areas indicate floristically diverse mixed mesophytic assemblages with leaves of moderate size. Many taxa are shared with Miocene floras of Europe and Asia, including those extirpated from western North America and Europe, but surviving today in eastern North America and/or eastern Asia, e.g., diverse conifers (*Abies, Cunninghamia, Calocedrus, Metasequoia, Taxodium*), Magnoliaceae (*Liriodendron, Magnolia*), *Liquidambar*, Fagaceae (*Castanea, Quercus, Lithocarpus*), Betulaceae, Juglandaceae, and *Cercidiphyllum*.

T1-37-05

Spatial phylogenetics of the North American flora

Brent Mishler, *Bruce Baldwin, William Freyman, Andrew Thornhill*
University and Jepson Herbaria; University of California, Berkeley

Biodiversity has usually been measured by examining changes in the number of species across a region to identify areas of particularly high diversity and endemism. Beta-diversity, or turn-over on the landscape, is likewise typically measured by comparing proportions of species shared among subareas. However, investigations based on species distributions alone miss the full richness of analyses that can result from taking a phylogenetic approach. Fortunately, advances in digitization of natural history collections, broad-scale DNA sequencing of many taxa increasingly represented in public databases, and rapid scaling-up of methods for building phylogenies have made it possible to apply a "big data" evolutionary approach to assessment of biodiversity and endemism that can be termed spatial phylogenetics. Our research group has been developing a novel suite of phylogenetic tools including two new metrics, relative phylogenetic diversity and relative phylogenetic endemism, and three new methods, categorical analysis of neo- and paleo-endemism (CANAPE), range weighted branch length difference (RWiBaLD), and phylogenetic range-weighted turnover (PhyloRWT). These methods are rank-free since it does not matter what taxonomic levels the terminals represent, as long as they are monophyletic and their geographic distribution can be characterized, and are thus relatively robust to lumping and splitting decisions by taxonomists. CANAPE searches for centers of endemism, and classifies them by the branch lengths of the rare taxa within them, allowing, for the first time, a clear, quantitative distinction between centers of neo- and paleo-endemism across an area. RWiBaLD complements CANAPE and makes a close-up examination of the distribution of branch lengths within a single subarea, allowing a clear understanding of which branches on the tree are responsible for the patterns seen in CANAPE. Both CANAPE and RWiBaLD are both tested statistically using a spatial randomization of terminal taxa on the map. PhyloRWT examines turnover in amount of the phylogeny shared among subareas, while emphasizing the branches that are range-restricted. It serves as a particularly useful measure of phylobetadiversity for purposes of understanding changes in phylogenetic assemblages across the landscape. PhyloRWT can be applied for a variety of purposes including bioregionalization, ecological studies of causes for beta-diversity, and complementarity analyses for applied conservation studies, which are important given the need to prioritize efforts in the face of rapid habitat loss and human-induced climate change. These new phylogenetic methods allow assessments of protected lands and can identify complementary areas of biodiversity that have unique evolutionary histories in need of conservation. The prospects for applying these methods across North America are rapidly improving given advances in herbarium digitization; the time is ripe for planning a continent-wide analysis. The first step is identifying gaps in spatial data and taxon sampling for genetic data. Regional studies have been completed for California and Florida; these will be presented as springboards for the continental study. In addition we will present a very preliminary analysis for North America using existing (albeit incomplete) data in order to look at preliminary patterns as well as to highlight data gaps.

T1-37-06

The role of land bridges, ancient environments, and migrations in the assembly of the North American flora

Alan Graham
Missouri Botanical Garden

North America was connected and separated from adjacent and more distant regions during the Late Cretaceous and Cenozoic (100 Ma to the Present) by five land bridges and through significant alterations in climate. Beringia (connecting NW North America to NE Eurasia) was mostly continuous for much of the time with warm-temperate to tropical climates in the Paleogene trending toward cool to cold temperate in the Neogene, and becoming discontinuous with each of the 18-20 interglacial cycles of the Quaternary. The North Atlantic (Greenland, the Arctic Islands, Iceland, NW Europe) was essentially continuous land until about the early Eocene with disruption (northward promulgation of the mid-Atlantic Ridge) corresponding to a change in climate from hot house (through the Middle Eocene Climatic Optimum) to ice house conditions in the late Cenozoic. The Antilles (NE South America, SE U.S. via the Bahamas, Yucatán region of Mexico) afforded discontinuous stepping stones throughout their history (present consensus, with an alternate interpretation) and muted mean annual temperatures varying by ca. 3-5°C. Central America (connecting South America with Mexico into the southern U.S.) had an open seaway through the Darian region until about 3.5 Ma (present consensus) after which began the Great American Biotic Interchange; climates were comparable to those of the Antilles. The more distant Magellan Land Bridge (southern South America, Antarctica, Australasia and regions beyond) was physically and climatically disrupted by isolation of Antarctica (separation from South America) at about 34 Ma (early Oligocene) but *Nipa* and *Nothfagus* are reported in earlier Eocene deposits all the way to Texas with *Nothofagus* extending further northward. *Podocarpus* also likely moved into North America from the south; it is known from the Eocene/Oligocene of the Mississippi Embayment becoming extinct there with later Cenozoic cooling. Its present northern distribution is represented by *P. matudae* in the cloud forests of Tamaulipas, Mexico. The message emerging from a more precise knowledge of the physical and climatic history of the New World land bridges, and the taxonomic relationships and phylogeny of the organisms utilizing them, is that over time they provided opportunity for migration of a diverse assemblage of organisms and ecosystems ranging from subtropical and warm temperate to cool and cold temperate accounting for the diversity of the North American flora, the vast extent and widespread similarity of the boreal forest in the harsh environments of the far north (ca. nine principle genera of trees), as well as disruptions that better explain disjunct biogeographical affinities like between the dry regions of SE U.S./northern Mexico and the Mediterranean (ca. 15 genera; e.g., *Ephedra*, *Juniperus*, Pinus, Quercus; Madrean-Tethyan vegetation), temperate eastern U.S. and eastern Mexico (ca. 20 genera; e.g., *Acer*, *Alnus*, *Carya*, *Fagus*, *Fraxinus*, *Juglans*, *liquidambar*, *Magnolia*, *Nyssa*, *Tilia*, *Ulmus*), and the ca. 50-60 trees of temperate eastern North America and central China.

T1-37-07

Diversification and assembly of the North American sedges (Cyperaceae): Evolution of geographical place and niche space

Daniel Spalink[1], *Bryan Drew*[2], *Matthew Pace*[3], *John Zaborsky*[4], *Pan Li*[5], *Kenneth Cameron*[4], *Thomas Givnish*[4], *Kenneth Sytsma*[4]

1. University of Utah
2. University of Nebraska
3. New York Botanical Garden
4. University of Wisconsin-Madison
5. Zhejiang University, Hangzhou, China

The North America sedge (Cyperaceae) flora contains over 840 species, making it the third most diverse family on the continent. Here, we analyze the evolution of these species in the context of geographical and ecological diversification. Utilizing over 200,000 georeferenced specimens, we characterized the geographical distribution and climatic and edaphic niche space occupied by each species in a continuous manner, using a modified Parsimony Analysis of Endemicity approach in conjunction with ordination statistics. We integrate these data with a supermatrix phylogeny of the North American sedges, and use a multifaceted approach to assess the role of geographical and ecological divergence on lineage diversification. In addition to measuring phylogenetic signal for these traits, we measured pairwise phylogenetic distance of species within floristic regions, calculated rates of speciation, and tested for correlations of speciation rate to tempo of geographical and ecological evolution. Our analyses consistently show that evolutionarily related species tend to be geographically unrelated. Rates of geographical and ecological diversification are closely linked to tempo of speciation, and exploration of geographical place coincides with divergence in ecological niche space. We highlight the benefits of treating geography in a continuous manner, and stress the importance of employing a diverse suite of analytical approaches in testing hypotheses regarding the evolution of range and niche. We then analyze broad patterns of community assembly throughout the continent, and demonstrate a latitudinal shift from phylogenetic overdispersion to phylogenetic conservatism as lineages become filtered at high latitudes.

T1-37-08

Richness and geographic distribution of selected angiosperm groups in Mexico

Aarón Rodríguez[1,2], *Guadalupe Munguía-Lino*[3], *Ofelia Vargas-Ponce*[1,2], *Georgina Vargas-Amado*[1,2], *Arturo Castro-Castro*[4], *Pilar Zamora-Tavares*[1,2], *Jesús González-Gallegos*[4], *Pablo Carrillo-Reyes*[1,2], *Marco Anguiano-Constante*[1], *Marco Carrasco-Ortiz*[1], *Miguel García-Martínez*[1], *Brandon Gutiérrez-Rodríguez*[1], *Juvenal Aragón-Parada*[1], *Christian Valdes-Ibarra*[1]

1. Herbario Luz María Villarreal de Puga, Departamento nica y Zoología, Centro Universitario de Ciencias Biológicas y Agropecuarias, Universidad de Guadalajara
2. Laboratorio Nacional de Identificacióny Caracterización Vegetal (LaniVeg), Consejo Nacional de Ciencia y Tecnología
3. Catedrático CONACYT-Universidad de Guadalajara
4. Catedrático CONACYT-Instituto Politécnico Nacional, Centro Interdisciplinario de Investigación para el Desarrollo Integral Regional, Unidad Durango (CIIDIR-Durango)

México is a megadiverse country. 22,126 species of angiosperms have been registered on its territory and 11,001 are endemic. However, their geographical distribution is far from homogeneous. In addition, Mexico is the center of origin and diversification of several groups. Our analysis focused on such groups. The aims were to identify areas of species richness and endemism. A data matrix with 25,509 geographical records of Tigridieae (Iridaceae); *Solanum*, *Lycianthes* and Physalinae (Solanaceae); *Cosmos*, *Dahlia* (Asteraceae); Poliantheae, *Echeandia* (Asparagaceae); *Calochortus* (Liliaceae); *Bletia* (Orchidaceae); *Hylocereus* (Cactaceae); *Salvia* subgenus Calosphace (Lamiaceae); Crassulaceae and Amaryllidaceae, was analyzed. Using Geographic Information Systems (GIS), we determined the richness and distribution on a 60 km × 60 km cell grid. The areas of emdemism were estimated using the endemicity analysis (EA). The Transmexican Volcanic Belt (TMVB) concentrates the highest number of taxa and endemics. The TMVB has been recognized as a province on the basis of geologic, tectonic, geomorphologic, physiographic and biogeographical observations. It is a 1,000 km long volcanic arc between 80 and 230 km wide, located in central Mexico between 17°30' and 20°25' N and between 96°20' and 105°20' W. The radiation of several plant groups coincides with the formation of the TMVB, which began 19 mya, during the Middle Miocene.

T1-37-09

Origins and diversity of the fern flora of the Pacific Northwest

Melanie Link-Perez

Oregon State University

The Pacific Northwest (PNW) is a large and geographically complex region in western North America comprising the Yukon Territory and British Columbia of Canada, and the states of Alaska, Washington, Oregon, Idaho, and western Montana in the United States. The region has been developing over the past 200 million years beginning with the western movement of the North American plate following the breakup of the supercontinent Pangea. The subduction of the Pacific Ocean floor beneath the leading edge of the North American plate has added landmass to the Pacific Northwest through volcanism and the accretion of volcanic island chains and ocean-floor rocks through plate collision. Much of the region has been repeatedly covered by glaciers, which may have impacted the ability of the flora to assemble and diversify. Among vascular plants, the ferns (Polypodiopsida) are well represented, with taxa from seven of the 10 orders currently recognized by the Pteridophyte Phylogeny Group occurring within the region and spanning the diversity of the fern tree of life. Fifteen families and 27 genera have been recorded. We may wonder, however, why there isn't more fern diversity in the PNW, given the high degree of topographic complexity and the range of Köppen climate types that are present. The assembly of a fern flora is of particular interest considering that fern spores are able to disperse thousands of kilometers and aren't reliant on other organisms for their migration. However, for ferns to persist both the gametophytes and sporophytes must be able to survive the environment in which the dispersed spore germinates. The gametophyte and sporophyte stages of ferns are independently exposed to the selective pressures of their environment and their fundamental niche may be smaller than we might expect given their reproductive capacity and the spore vagility. The current research is motivated by the following questions: 1) what diversity of ferns is present using various measures (for example, species and genus richness, phylogenetic diversity), 2) what environmental factors correlate with distributional patterns of selected species, 3) what ancestral areas gave rise to the extant fern flora, and 4) how may we expect the fern flora to respond to predicted climate change? To address these questions, we incorporated data from phylogenetics and ecological modeling. Presence-only distribution data were obtained from the Consortium of Pacific Northwest Herbaria and the Oregon Flora Project. Chloroplast DNA sequence data were obtained from GenBank and used in phylogenetic reconstruction to infer ancestral areas and assess phylogenetic diversity. Bioclimatic data were downloaded from WorldClim.org and maximum-entropy approaches were used to determine environmental predictors for select taxa, including *Equisetum*, *Botrychium*, and *Polystichum*. We will share results of our ancestral area reconstructions and present hypotheses regarding the factors that have shaped the fern assemblage of the Pacific Northwest.

T1-37-10

The role of Pleistocene climate fluctuations in the assembly of boreal steppe communities of Northwest North America

Mary Edwards[1,4], *Andrea Lloyd*[2], *Carl Roland*[3], *Scott Armbruster*[4]
1. University of Southampton
2. Middlebury College
3. US National Park Service
4. University of Alaska Fairbanks

Boreal or subarctic steppe is an azonal vegetation type dominated by perennial grasses and forbs, plus sagebrush, which occurs as disjunct communities on south-facing slopes within the boreal zone. It is restricted to areas of continental climate in Northwest North America, as well as Northeast Siberia and the Russian Far East. In order to gain insight into the timing and routes of dispersal of species occupying islands of steppe in boreal western North America, we examined elevational distributions of species on the modern landscape. Our study areas were open, south-facing mountain slopes in Alaska, Yukon, and the Russian Far East. We assumed that Asian migrants crossed the Bering Land Bridge under climates cooler and drier than today, and we therefore predicted that these species should have broad elevational ranges extending upward to tundra. In contrast, we assumed North American species migrated during interglaciations or interstadial periods, including the Holocene, as during cold periods they would have been excluded by ice sheets. We predicted they would be restricted to warmer sites. We found that steppe species that are Asian disjuncts indeed tended to have large altitudinal ranges, reaching cold steppe and tundra. North American disjuncts were largely restricted to the warmest, lower-slope locations. Fossil evidence nevertheless indicates at least one North American species survived full-glacial conditions in Beringia. Invertebrates associated with the boreal steppe environment show similar patterns, supporting the concept that the Bering Land Bridge acted a cold filter to species migrating from Asia. The boreal steppe flora of Northwest North America likely comprises species that have migrated during different Pleistocene climate cycles, reflecting assembly over an extended period of time from multiple source regions.

T1-37-11

Isoetes diversity in Eastern North America

Elizabeth Zimmer[1], *Peter Schafran*[1,2], *Lytton Musselman*[2], *Gabriel Johnson*[1], *W. Carl Taylor*[1]
1. Smithsonian National Museum of Natural History, Department of Botany
2. Old Dominion University, Department of Biology

The early lineage of vascular plants, the genus *Isöetes* (quillworts), is found throughout North America, with the highest diversity in the eastern United States. Species of *Isöetes* are all morphologically very similar, causing considerable confusion for both taxonomists and field biologists. Additionally, taxonomy of these closely related species of *Isoëtes,* likely the result of a relatively recent radiation, has been difficult to infer, and is further complicated by the occurrence of many, and often complex, polyploid species. With the recent developments in molecular methods, DNA sequencing is currently the predominant tool for estimating species boundaries and phylogeny in *Isöetes*. To date, we have used both a nuclear gene intron and whole chloroplast genome (plastome) sequences to infer a phylogeny of over fifteen diploid taxa and to investigate the origins of several polyploid taxa. Among our results, we have found that the phylogeny of the entire plastome for diploids was well supported across various models of evolution, with a deep divergence representing a geographic split between the Atlantic seaboard and the Gulf Coast/Midwest being inferred. In an initial study of polyploid *Isöetes*, we have identified different diploids as nuclear genome donors to species originally identified as monophyletic, and used the plastome sequences to identify their chloroplast donor (likely maternal) species. The molecular data also has the potential to generate biogeographic hypotheses regarding range shifts of species of *Isöetes* in the recent geologic past. Finally, mapping spore characters and phenology to our phylogeny indicates white, tuberculate megaspores and white, papillose microspores that mature in the spring are plesiomorphic to this eastern American clade, with some other character states subsequently evolving several times.

T1-37-12

Origins and diversification of conifers in North America and implications for the assembly of the conifer flora of the Northern Hemisphere: An example using *Abies* (Pinaceae)*

Qiaoping Xiang[1], *Ran Wei*[1], *Cunfeng Zhao*[1], *Xianchun Zhang*[1], *Harris AJ*[2]

1. State Key Laboratory of Systematics and Evolutionary Botany, Institute of Botany, Chinese Academy of Sciences, Beijing, China

2. Department of Botany, National Museum of Natural History, Smithsonian Institution, Washington, D.C., USA

The relationships and diversification histories of the floras of North America and eastern Asia are widely studied. Studies using phylogeny and historical biogeographic reconstructions, often infer that Eastern Asia is the origin of angiosperm groups. In contrast, these studies frequently find that conifers originated and migrated out of North America. The 'out-of-North America' hypothesis in conifers is also congruent with the fossil record of many angiosperm lineages. However, it remains challenging to integrate fossils into historical biogeographic reconstructions across phylogenies. Thus, insights into conifer origins and geographic radiation may provide key penetration into patterns and processes in angiosperms. In the conifer genus, *Abies*, two East Asia-North American disjunct sections were formerly recognized based on morphological characters: sect. *Balsamea* and sect. *Amabilis*. In order to examine the geographic origins of these two sections, we inferred a dated phylogeny using nearly complete species sampling of the genus, fossil evidence, and DNA data from three genomes: nrITS, low copy nuclear gene, chloroplast DNA intergenic spacers, and mitochondrial intergenic spacers. We used the dated phylogeny to perform historical biogeographic reconstruction. Our biogeographic reconstruction supported 'out-of-North America' dispersal in section *Amabilis* via the Bering Land Bridge from western North America to Japan during the middle Miocene. The situation in sect. *Balsamea* is unclear because the hybridization, with parent-ship from eastern Asia and North America, was probably involved in its origination. Our present provides insight on the assembly of the conifer flora in the Northern Hemisphere, and we expect that future work may improve understanding of how diverse, early conifer lineages in North America impacted the assembly of the conifer flora on other continents. In the future, we intend to use next generation sequencing methods, including RAD-Seq, to better resolve the phylogeny of *Abies*. Moreover, we expect to use resolved phylogenies of *Abies* and increasingly available, stable phylogenies of other conifer genera to further test the out-of-North America biogeographic hypothesis.

T1-38: Biogeographic patterns and adaptive radiations in the Pacific

T1-38-01

The biogeographic and evolutionary origins of the Hawaiian peppers (Piperaceae: Peperomia)

Jun Ying Lim[1], *Elizabeth Zimmer*[2], *Warren Wagner*[2], *Rosemary Gillespie*[1]

1. University of California, Berkeley

2. National Museum of Natural History, Smithsonian Institution

Peperomia represents one of the most species-rich genera globally (over 1,000 species), and is widely distributed across the world's tropics. In addition, owing to their high dispersal ability, *Peperomia* have also colonized and diversified across the islands of the Pacific, which collectively harbor a tenth of its global diversity. On Hawaii, *Peperomia* represents one of the most species-rich genera among the flowering plants of Hawaii, comprising 25 native species, 22 of which are unique to the archipelago. Taxa of *Peperomia* on Hawaii are also highly diverse morphologically and exhibit a range of ecologies. However, it remains unclear whether the endemic taxa on Hawaii originated through diversification from a single successful colonist, or if they were derived from multiple colonization events and thus, parallel radiations. Furthermore, the islands of Hawaii may also represent a "stepping-stone" for dispersal and hence act as a biogeographic source of colonists for the rest of the Pacific. Thus, a greater understanding of patterns of dispersal and diversification on Hawaii will provide insights to their evolutionary success across the wider Pacific region. Here, we constructed a multi-loci molecular phylogeny using a multiplex high-throughput sequencing approach for 45 taxa of *Peperomia* across the Pacific, with particular focus on Hawaiian taxa. Using this phylogeny, we then inferred patterns of dispersal in this group across the Pacific and within the Hawaiian archipelago using probabilistic historical biogeographic models. Our results shed light on the evolutionary history and extraordinary radiation of this diverse group across the Pacific, and provide insight into the evolutionary assembly of island floras.

T1-38-02

Applying RADseq to resolve the phylogeny of the Hawaiian radiation of *Myrsine* (Primulaceae, Myrsinoideae)

Warren Wagner[1], *Claudia Paetzold*[2], *Kenneth Wood*[3], *Marc Appelhans*[1,2]

1. Smithsonian Institution

2. University of Goettingen

3. National Tropical Botanical Garden

Myrsine ranks among the largest genera of the primrose family (Primulaceae s.l.) and its +/- 200 species are distributed in subtropical and tropical areas worldwide except Europe. A particularly high number of species are found in the Pacific area. The largest of these pacific radiations is that of the Hawaiian Archipelago which consists of 20 endemic species. *Myrsine* has colonized all major Hawaiian Islands except Kaho'olawe and Ni'ihau and is most diverse on the oldest islands Kaua'i (14 species of which 10 are endemic to the island) and O'ahu (8 species, 2 endemic). Hawaiian *Myrsine* are woody plants and range from small and many-branched shrubs to trees of up to 8 m in height. While many species have a very narrow distribution and show little morphological variability, the species *M. lessertiana* is extremely variable. Several subspecies and varieties have been described for this species, and populations that show intermediate characters with *M. emarginata*, *M. pukooensis* and *M. sandwicensis* have been observed. The genus *Myrsine* and especially the Hawaiian taxa have been largely neglected by systematists and virtually nothing is known about their relationships among each other. Not a single DNA sequence of the Hawaiian species has been deposited at Genbank so far. Our study aims at resolving the phylogenetic history and the inter-island colonization pattern of Hawaiian *Myrsine*. Since other examples of radiations from the Hawaiian Islands have shown that Sanger sequencing does mostly not deliver enough information to untangle phylogenetic relationships at the species level, we decided to use the NGS method RADseq (restriction-site associated DNA sequencing) in our study. Our taxon sampling includes all but one species of Hawaiian *Myrsine* as well as *M. africana* and two species of the closely related genus *Ardisia* as outgroups. Seven specimens of the variable *M. lessertiana* were sampled from all major islands.

T1-38-03

Assembling a colonisation timeline for the endemic tree flora of New Zealand

Benjamin C. M. Potter[1], *Alexei J. Drummond*[2], *Richard D. Newcomb*[1,3], *William G. Lee*[1,4]
1. School of Biological Sciences, University of Auckland, Auckland, New Zealand
2. Department of Computer Science, University of Auckland, Auckland, New Zealand
3. Plant & Food Research, Auckland, New Zealand
4. Landcare Research, Dunedin, New Zealand

The extant flora of New Zealand contains 271 species of endemic forest trees, spread across a broad spectrum of plant diversity (147 asterids, 83 rosids, 20 conifers, 10 magnoliids, 6 ferns, and 5 monocots). Here, we set out to characterise the colonisation history of this endemicity. Bayesian relaxed molecular clocks were used to date the NZ vs. overseas floristic disjunctions, and place the NZ immigrations along an evolutionary timeline. The primary aims were: (1) to contrast the historical arrivals with significant geological and climatological events, (2) build a time-calibrated framework for tree diversification and biogeography research, and (3) assemble a database useful for community assembly questions, focused on the role of deep-time priority effects in shaping plant community structure. Towards this an assessment of the molecular dating literature was carried out (focusing exclusively on studies employing relaxed clock models in BEAST), resulting in immigration ages for 39 tree genera. For the remaining 46 genera that were undated in the literature, molecular dating analyses were conducted using a combination of publically available and newly generated DNA sequence data. The results suggest that although the Zealandia continental fragment has been geographically isolated for *ca.* 80 million years, over 85% of the endemic tree colonisations can be attributed to dispersal events in the Early Miocene or later (mean ages). Broadly, most of these dates are in line with the 'fly-paper of the Pacific' hypothesis of NZ biogeography, although some lineages represent possible examples of continuous post-Gondwanan presence.

T1-38-04

Biogeography and evolution of the largest adaptive radiation of woody plants (Melicope, Rutaceae) on the Hawaiian Islands

Marc Appelhans[1,2], *Claudia Paetzold*[1], *Jun Wen*[2], *Kenneth Wood*[3], *Warren Wagner*[2]
1. University of Goettingen
2. Smithsonian Institution
3. National Tropical Botanical Garden

The Hawaiian Islands are a perfect natural laboratory to study recent and species-rich radiations. With an age of approximately 5 millions of years Kauai and Niihau are the oldest of the current main islands and the archipelago ranks among the most isolated landmasses on the planet. Due to their recent age and their isolated geography, only relatively few organismal lineages successfully colonized the archipelago and some of them underwent adaptive radiations that lead to great numbers of species. There are often vast morphological differences between those species, which is usually contrasted by very low genetic variation. The Hawaiian Melicope lineage is a very good example for this phenomenon as it shows a wide array of morphological variation (e.g., growth form, connation of carpels, leaf shape, indumentum), but low genetic variation. Melicope constitutes one of the most species-rich plant lineages on the Hawaiian archipelago and, with a total number of 54 species, it accounts for the largest lineage of woody plants. The bird-dispersed genus Melicope also occurs on other Pacific Islands, Australasia, Malesia, Southern Asia, Madagascar and the Mascarene Islands. Altogether, with about 230 species, Melicope is the largest genus of Rutaceae. Phylogenetic reconstructions of Melicope based on Sanger sequencing datasets were highly supported and well resolved for most branches, but the Hawaiian clade showed relatively poor resolution. Similar results have been reported for other Hawaiian plant lineages. We here present the first attempt to investigate the phylogeny and biogeography of Hawaiian Melicope species using Next-Generation Sequencing. We chose RADseq (restriction site associated DNA sequencing) since it is perfectly suited for evolutionary young groups and does not require a reference genome. Out of the 54 known species, five are believed to be extinct and 39 species from all major islands have been sampled in our study. Our results clearly demonstrate the suitability of the RADseq method in resolving rapid radiations. More than 70,000 RAD loci proved to be sufficient to reconstruct a well-resolved and supported phylogeny. Our results show that the traditional division of Hawaiian Melicope into four sections is artificial since only one of the sections proved to be monophyletic. Our molecular dating results suggest that the diversification of Hawaiian Melicope (crown group) began in the Late Miocene or Early Pliocene (7.5Ma; 4.7-11.7Ma), so that the lineage might be older than the current main islands. Other islands such as Necker and Nihoa, which are now shallow islands without forests, might therefore have been the first habitat of the ancestor of the Hawaiian lineage. From these islands, Melicope colonized the other islands mainly following the rule of progression (i.e. directional from the older to the emerging islands).

T1-38-05

Integrating biogeography, threat, and evolutionary data to explore extinction crisis in the taxonomic group of cycads.

Kowiyou Yessoufou
University of Johannesburg

Will the ongoing extinction crisis cause a severe loss of evolutionary information accumulated over millions of years on the tree of life? This question has been largely explored, particularly for vertebrate and angiosperms. However no equivalent effort has been devoted to gymnosperm. Here we address this question focusing on cycads, the gymnosperm group exhibiting the highest proportion of threatened species in the plant kingdom. We assembled the first complete phylogeny of cycads, and assessed how species loss under three scenarios would impact the cycad tree of life. These scenarios are: i) all top 50% of Evolutionarily Distinct (ED) species are lost; ii) all threatened species are lost, and iii) only all threatened species in each IUCN category are lost. Finally we analysed the biogeographical pattern of cycad diversity hotspots and tested for gaps in the current global conservation network. First, we showed that threatened species are not significantly clustered on the cycad tree of life. Second, we showed that the loss of all vulnerable or endangered species do not depart significantly from random loss. In contrast the loss of all top 50% ED, all threatened or all critically endangered species would result in a greater loss of PD than expected. To inform conservation decisions, we defined five hotpots of diversity, and depending on the diversity metric used, these hotspots are located in southern Africa, Australia, Indo-Pacific and Mexico and all are found within Protected Areas. We conclude that the phylogenetic diversity accumulated over million years in the cycad tree of life would not survive the current extinction crisis. As such, prioritizing efforts based on ED or concentrating efforts on critically endangered species particularly in southern Africa, Australia, Indo-Pacific and Mexico are required to safeguarding the evolutionary diversity in the cycad tree of life.

T1-38-06

Revisiting *Astrocalyx* Merr. (Astronieae: Melastomataceae): New morphological observations on the Philippine endemic, enigmatic, endangered, and monotypic genus

Jeffrey Mancera[1], *Darin Penneys*[2], *Fulgent Coritico*[3]

1. Department of Botany, California Academy of Sciences

2. Department of Biology and Marine Biology, University of North Carolina Wilmington

3. Center for Biodiversity Research and Extension in Mindanao (CEBREM), Central Mindanao University

The monotypic *Astrocalyx* Merr. (Astronieae: Melastomataceae) was collected in 2015 from the Mount Kitanglad Range Natural Park (Mindanao Island, the Philippines) during a joint expedition conducted by the Central Mindanao University (CMUH) and the California Academy of Sciences (CAS). Since the original publication of the genus, no other detailed account of its enigmatic morphology has been published. Examination of herbarium vouchers revealed discrete variations in morphology, which may be indicative of reproductive isolation or presence of emerging selection pressures among populations of the species across the archipelago. This documentation aims to rekindle interest and initiate approaches toward conservation of this poorly known endemic taxon, already listed as endangered in the Philippine Islands.

THEME II: TAXONOMY, PHYLOGENETICS & EVOLUTION

T2-01: Evolution, ecology, and physiology of carnivorous plants - Part I and Part II (two sessions)

T2-01-01

Evolution of carnivory in angiosperms and diversification of trap types

Andreas Fleischmann

Botanische Staatssammlung München

Approximately 800 species of carnivorous plants are currently recognized, which have evolved at least ten times from non-carnivorous ancestors in five angiosperm orders (Poales: 4 species in 3 genera, Caryophyllales: ca. 370 spp./6 genera, Oxalidales: 1 monotypic genus, Ericales: 33 spp./4 genera, Lamiales: ca. 390 spp./5 genera). In the 19 genera of carnivorous plants, a variety of trap types have evolved to attract, catch and digest small animals – ranging from simple pitfall traps made up by the entire plant body in the tank-forming carnivorous bromeliads, leaf-derived epiascidiate pitcher leaves of various pitcher plant genera, active and passive adhesive (flypaper) traps, eel traps as well as sophisticated, fast-moving snap traps and suction traps. Putative steps towards the evolution of the carnivorous syndrome in flowering plants, and the diversification of trap types within certain lineages are discussed in the light of dated phylogenetic reconstructions.

T2-01-02

What conditions favor carnivorous plants? Cost/benefit analysis and the context-specific advantages of botanical carnivory

Thomas Givnish[1], *Kenneth Sparks*[1], *Steven Hunter*[1], *Andrej Pavlovic*[1]

1. University of Wisconsin-Madison

2. Palacky University

The cost/benefit model for the evolution of carnivorous plants advanced in 1984 addressed the ecological value of carnivory in detail for the first time. It provided a potential explanation for why carnivorous plants are common in habitats that are not only nutrient-poor but sunny and moist as well, and for how plants should adjust their allocation to carnivory as a function of environmental conditions. Over the past 32 years, this model has stimulated a large amount of research, and provided a conceptual framework for many studies of photosynthesis, respiration, carbon balance, and relative growth rates in carnivorous plants, of the causes of their ecological distributions and variation in allocation to carnivory under different conditions, and of the evolutionary pathways to carnivory. Here we re-assess this model 32 years after its proposal, discuss its predictions and how the model and its predictions can be extended in a number of important ways, and evaluate the extent to which the model's assumptions and predictions have proven correct. We briefly review what is known about photosynthesis, respiration, relative growth rate, and resource allocation in carnivorous vs. non-carnivorous plants, the degree to which various resources are known to limit their growth and stoichiometry, adaptation to different kinds of prey, and what this means about the model's predictions. We conclude by assessing how future studies might test the cost/benefit model for carnivory in more rigorous fashion.

T2-01-03

Floral features, gland distribution and architecture in flowers of Lentibulariaceae with different pollination syndromes

Bartosz Jan Płachno[1], *Małgorzata Stpiczyńska*[2], *Richard W. Jobson*[3], *Hans Lambers*[4]
1. *Jagiellonian University in Kraków*
2. *University of Warsaw, Botanic Garden*
3. *National Herbarium of New South Wales, Royal Botanic Gardens and Domain Trust*
4. *The University of Western Australia*

The aim of our study is to compare glands and other floral features in diverse species from the genera *Pinguicula*, *Genlisea*, and *Utricularia* that are known to be pollinated by different vectors. We analysed four groups of species including potential ornithophilous species such as *Utricularia menziesii, Utricularia quelchii* and several Mexican *Pinguicula* species, entomophilous species, suspected sexually deceptive species (*Utricularia capilliflora* complex), and also sterile species (e.g. *Utricularia bremii*, *Utricularia australis*), that do not produce seeds and therefore don't need pollinators. In entomophilous species from genera *Genlisea* and *Utricularia* the palate possesses diverse micro-morphology comprising unicellular, papillae (conical to villiform) and/or multicellular, uniseriate glandular trichomes. In some entomophilous species, the palate functions as an unguentarius. Nectariferous trichomes are observed in floral spurs. In the sexually deceptive species the nectary spur is reduced, and pollinators are probably attracted largely by the insectiform configuration of the flower and volatilisation of fragrance produced by glandular trichomes distributed upon the modified floral appendages. The flowers of ornithophilous species possess colours that may attract birds, along with well developed spurs that reward pollinators with nectar. Acknowledgements: This study was funded by the National Science Centre, Poland. Contract grant number: DEC-2013/09/B/NZ8/03308.

T2-01-04

Sarracenia purpurea microecosystem as a model system for understanding and predicting aquatic ecosystem dynamics and tipping points

Amanda C. Northrop[1], *Rachel Brooks*[1], *Jéssica Duarte Sousa*[2], *Aaron M. Ellison*[3], *Bryan A. Ballif*[1], *Nicholas J. Gotelli*[1]
1. *University of Vermont*
2. *Saint Michael's College*
3. *Harvard Forst, Harvard University*

The northern pitcher plant, *Sarracenia purpurea*, has long been used as a model system for understanding ecological processes such as community assembly and food web interactions. Recently, we have investigated the utility of the *S. purpurea* system to serve as a model system for understanding state change dynamics and identifying early warning signals of tipping points in aquatic ecosystems. Using field and greenhouse experiments, we have shown that a tipping point in the *S. purpurea* pitcher ecosystem can be induced by enriching pitchers with organic matter. The microecosystem undergoes a change from an oxic state to a hypoxic state, characterized by an increase in microbial productivity and resulting rapid decline in dissolved oxygen. This change is similar to changes in larger aquatic ecosystems in which enrichment leads to a shift from a green food web to that of a brown food web, driven by microbial decomposition of organic matter, ultimately causing anoxia or hypoxia in bottom waters. Using metagenomics and metaproteomics, we have identified differences in the composition and function of functionally active bacteria in organic-mater enriched and non-enriched pitchers. Functionally active bacteria in enriched pitchers were dominated by Betaproteobacteria in the family Neisseriaceae and were predominantly facultative anaerobes. In non-enriched pitchers, aerobic bacteria, especially those associated with the family Comamonadaceae, comprised the majority of bacteria contributing to protein expression. Species evenness of functionally active bacteria was higher in non-enriched pitchers. There was more variation among non-enriched replicates in taxa contributing to protein expression, protein identifications, and functional pathway representation than in enriched pitchers. A time series metaproteomic study identified 50S ribosomal proteins (L2 and L5) and molecular chaperone GroEL as potential early warning protein biomarkers of an impending change to a hypoxic state in the *S. purpurea* pitcher microecosystem. The relationship between organic matter concentration and dissolved oxygen concentration in enriched pitchers was nonlinear, displaying hysteretic behavior similar to that between driver and response variables in larger aquatic ecosystems. To date, our results imply that the *S. purpurea* microecosystem can be experimentally manipulated to undergo a state change, that such states are distinguishable from one another using environmental proteomics methods, and that results from experiments using the *S. purpurea* model microecosystem may be scalable to larger aquatic ecosystems.

T2-01-05

Carnivorous plants are ideal model systems for experimental research

Aaron Ellison
Harvard University

The more than 600 carnivorous plants have evolved in at least six independent radiations within five angiosperm orders. Despite these independent origins, there is a remarkable morphological convergence of carnivorous plant traps, physiological convergence of mechanisms for digesting and assimilating prey, and functional convergence in interspecific interactions between the plants and their associated microbial and macrobial fauna. These convergent traits have made carnivorous plants model systems for addressing questions in plant molecular genetics, physiology, autecology and community ecology, and evolutionary biology. For example, the observation that carnivorous plant genera with morphologically complex traps have higher relative rates of gene substitutions than do those with simple sticky traps has suggested two alternative mechanisms for the evolution and diversification of carnivorous plant lineages. Comparative studies of carnivorous plant physiology and morphology have revealed new types of sticky and wettable surfaces that facilitate prey capture and have suggested new types of biomaterials. Experimental studies of pitcher-plants and their associated infauna have provided compelling examples of convergent interactions that are being used to identify new symbioses in previously unexplored habitats. These microecosystems also have been developed as scalable models for identifying and manipulating rapid state changes in ecological systems. Finally, the widespread popularity of carnivorous plants among people of all ages and nationalities has led to the development of educational curricula and citizen-science initiatives that are advancing both botanical literacy and scientific discovery. Nearly 150 years after Charles Darwin first experimentally demonstrated that plants could attract, trap, eat, and digest insects, we continue to be surprised and fascinated by the knowledge gained from studying these "most wonderful plants in the world."

T2-01-06

Ecological and evolutionary genomics of carnivorous *Nepenthes* pitcher plants

Mathias Scharmann
ETH Zürich

Carnivorous plants of the genus *Nepenthes* are widely known for their bizarrely modified pitcher leaves that can trap and digest animals. While recently much has been learned about the physiology, mechanics and ecology of these enigmatic organs, little is known about the vast variation in the carnivorous syndrome among the 150+ species of the genus. Is divergence within *Nepenthes* dominated by adaptive or neutral processes? Could it be that the carnivorous syndrome is directly involved in adaptative divergence, or even play a key role in speciation? Population genomics (ddRAD-seq, coalescent modelling) of natural sympatric *Nepenthes* communities in Southeast Asia reveal that reproductive isolation is far from complete, and multiple species engage in divergence with gene flow simultaneously. In theory, gene flow between lineages should lead to the erosion of neutral divergence. Maintenance of genetic and phenotypic divergence in the presence of gene flow therefore indicates adaptive processes, presumably through ecological selection against certain hybrids. Thus, many of the differences that are observed among long-term introgressing sympatric species may be adaptive, such as trap morphologies. A transcriptomic approach elucidates for the first time the molecular reaction of *Nepenthes* traps to feeding, and in combination with proteomics identifies genes underlying the digestive process. In a pair of recently diverged sister species, expression divergence across all expressed genes is higher in pitchers than in leaves. Branch-site tests over the transcriptomes of multiple species at phylogenetic and phenotypic extremes reveal that candidate carnivory genes are more commonly under positive (diversifying) selection than expected by chance. Overall, it appears that aspects of the carnivorous syndrome may be under divergent selection in introgressing sympatrics, and are among the most important targets of selection within the *Nepenthes* radiation, at both shallow and deep time scales.

T2-01-07

From prey to predator – Nutrient flows from animals to temperate carnivorous plants (terrestrial Pinguicula, aquatic Utricularia) elucidated with a stable isotope natural abundance approach (^{15}N, ^{13}C)

***Saskia Klink**, Gerhard Gebauer*
University of Bayreuth, BayCEER Laboratory of Isotope Biogeochemistry

Carnivorous plants fascinate humans all over the world due to their habit of trapping animals by creating diverse, extraordinary trap structures. Many studies were performed about carnivorous plants of hotspot tropical and subtropical regions. With exception of *Drosera* L. species knowledge on nutrient-translocation from prey to plant in temperate occasions is rather limited. The utilization of stable isotope natural abundance approaches is an elegant minimal destructive method to analyse nutrient flows from prey to plant as well as within the plant species. Thereby, benefits of a carnivorous life strategy can be elucidated using the stable nitrogen isotope (^{15}N), while the stable carbon isotope (^{13}C) mirrors cost and physiological adaptions to environmental conditions. Terrestrial carnivorous butterworts (*Pinguicula alpina* L. and *P. vulgaris* L.) as well as aquatic bladderworts (*Utricularia australis* R.Br. and *U. minor* L.) from Germany and Austria were analysed for their efficiency in gaining prey-derived nutrients. A stable isotope natural abundance approach with ^{13}C and ^{15}N was applied. (^{13}C) and benefits (^{15}N) of the carnivorous life strategy of the temperate plants (plant-family Lentibulariaceae) were investigated. To calculate the nutrient-uptake efficiency non-carnivorous references (terrestrial autotrophic plants, aquatic phytoplankton) and prey animals (insects, spiders, zooplankton) were used as end-members for the linear mixing model. Prey animals were graded by trophic level to identify trophic-level-dependent influences of plants' enrichment. The temperate carnivorous plants were successful in translocating prey-derived nitrogen to their tissue (nitrogen-uptake-efficiency: *Pinguicula alpina* 63%, *Utricularia australis* 100% and *U. minor* 21%). Main prey-source for *P. alpina* and *U. minor* was phytophagous prey, while zooplankton and prey of higher trophic levels were main sources for *U. australis*. While *P. vulgaris* was shown to be successful in nitrogen gaining from prey in a nitrogen-poor alpine habitat (54%), efficiency-calculation for *P. vulgaris* inhabiting a lowland forest side failed due to an unsuited data structure. Apart from nitrogen the carnivorous *Pinguicula* species utilize a proportion of carbon. The efficiency of terrestrial *Pinguicula* species in gaining prey-derived nutrients is similar to other sticky leaf plants (e.g. *Drosera*). The calculation of the efficiency using the linear mixing model depends on suitable data (difference in isotopic values between references, target plant and prey), which can be influenced by habitat properties (e.g. N-deposition). A certain proportion of carbon is taken up from prey, but habitat properties seem to influence carnivores' ^{13}C values additionally (e.g. stomatal behaviour). Aquatic *Utricularia* species turned out to be equally effective as terrestrial carnivorous plants. The utilization of plankton as end-members of the linear mixing model states itself as elegant method to circumvent issues with a lack of suitable aquatic reference plants. The position of carnivorous plants as indicators for climate change and environmental pollution makes these plants key-species for further studies to better understand influences of anthropogenic hazards and find solutions to preserve unique habitats.

T2-01-08

Phylogenetic analysis of carnivorous plants *Nepenthes* with their phytogeography in southeast Asia and evolutionary trends of the peristom

Firman Alamsyah[1], *Motomi Ito*[2]
1. Indonesia International Institute for Life Sciences
2. Graduate School of Arts and Sciences, The University of Tokyo

Nepenthes is a genus of carnivorous plant that belongs to the monotypic family Nepenthaceae. The genus currently contains 120 species, which are primarily distributed across Southeast Asia. Molecular phylogenetic analysis of this genus had been performed based on the DNA sequence of the plastid *trn*K intron. But further study revealed that the phylogeny of the *trn*K intron did not represent the true phylogeny of *Nepenthes*. Therefore, further phylogenetic study of this genus is required to clarify its evolutionary trends and to establish its infra-generic taxonomy. In total, 57 ITS sequences from 56 *Nepenthes* species were analyzed. Phylogenetic analysis were conducted using 4 different methods: Maximum parsimony, Bayesian analysis, Neighbor Joining, and Maximum Likelihood. Bootstrap values were calculated from 1000 replicates. Character states for distribution areas and the peristome of upper pitchers were mapped onto 1 of the MPTs by using the MacClade program (v4.06), with ACCTRAN optimization, to study the phytogeography of *Nepenthes* in Southeast Asia and the evolutionary trends of their morphological characteristics. Using *A. robertsoniorum* and *D. muscipula* as outgroups, 2 basal branches and 8 subclades were recognized in the phylogenetic tree. The 2 basal taxa were *N. pervillei* from Seychelles and *N. madagascariensis* from Madagascar. Subclade I comprised species distributed in New Guinea and outlying areas; and a widely distributed species. Subclade II contained a species from New Caledonia, a species from Misool Island, and 2 widely distributed species. Subclade III consisted of a species endemic to Sulawesi and a species distributed in Sulawesi and Borneo. Subclade IV and V, both contained species that were restricted to the Philippines and a species endemic to Borneo at the basal position. Subclade VI contained only species from Borneo. Subclade VIIA comprised exclusively 4 species from Peninsular Malaysia and Indochina, a species endemic to Borneo, and a species distributed in Borneo and Sumatra at the basal position. Subclade VIIB comprised 13 *Nepenthes* species from Sumatra with 1 species that was also distributed in Java. Judging from the distribution patterns and phylogenetic relationships, it is possible that *Nepenthes* have expanded from the eastern region to the western region of Southeast Asia. The tree topology suggested that *Nepenthes* had diverged in Borneo, resulting in the evolution of many local species and could have also migrated to adjacent areas. Therefore, Borneo could be a secondary center of diversification for *Nepenthes* in the Sunda Shelf of Southeast Asia. The evolutionary trends of the peristomes in this genus with 3 character states: narrow (<10 mm), intermediate (10-20 mm) and broad (>20 mm), suggested that the narrow peristome is the plesiomorphic state in this genus. The intermediate and broad peristomes likely evolved at least 7 and 8 times, respectively, in this genus, whereas the narrow peristome evolved again at least 4 times after the widening of the peristome. These results suggested that repeated evolution of these 3 peristome characteristics occurred in this genus. Judging from the tree topology, the evolutionary trends of the peristomes revealed the limitedness of Danser's system of classification for *Nepenthes*.

T2-01-09

Convergent evolution of two Mediterranean-climate carnivorous plants: investigating the relationship between investment and reward for *Byblis gigantea* and *Drosophyllum lusitanicum*

Laura Skates[1,2], *Maria Paniw*[3], *Fernando Ojeda*[4], *Kingsley Dixon*[5], *Gerhard Gebauer*[6], *Jason Stevens*[2], *Adam Cross*[2,5]
1. School of Biological Sciences, University of Western Australia, Perth
2. Biodiversity Conservation Centre, Kings Park and Botanic Gardens, Perth
3. Department of Evolutionary Biology and Environmental Studies, University of Zurich, Zurich
4. Departamento de Biología-IVAGRO, Universidad de Cádiz, Campus Río San Pedro, 11510-Puerto Real
5. Department of Environment and Agriculture, Curtin University, Perth
6. BAYCEER Laboratory of Isotope Biogeochemistry, University of Bayreuth, Bayreuth

Convergent trait evolution is apparent amongst many of the world's recognised carnivorous plant groups. Adhesive traps, for example, have evolved independently within Caryophyllales (*Drosera* from Droseraceae, *Drosophyllum* from Drosophyllaceae, and *Triphyophyllum* from Dioncophyllaceae), Ericales (*Roridula* from Roridulaceae), and Lamiales (*Byblis* from Byblidaceae, and *Pinguicula* from Lentibulariaceae). With some structural differences, all of these groups possess species with mucilage-producing, glandular leaves capable of insect capture. In this study, we compare two morphologically and ecologically similar yet phylogenetically and geographically distinct carnivorous plant species with adhesive traps: *Byblis gigantea* Lindl. from southwest Australia, and *Drosophyllum lusitanicum* (L) Link. from the western end of the Mediterranean Basin. Both species occur in nutrient-poor and fire-prone environments, within globally significant biodiversity hotspots with Mediterranean climates. Despite their similarities, these two species seem to exist at opposite ends of the spectrum of investment in carnivory: *D. lusitanicum* actively attracts and captures prey with fully vascularised sticky-glands, while *B. gigantea* passively captures prey with simple sticky-glands. To better understand the ecological relationship between degree of investment in carnivory and the degree of success in gaining prey-derived nutrients, we applied stable isotope methodology. Samples of each carnivorous plant, along with neighbouring non-carnivorous plants, prey insects, and soil were collected *in situ*. Leaf and insect tissue samples were analysed to determine $d^{15}N$, using an elemental analyser coupled to an isotope ratio mass spectrometer. Percentage of prey-derived nitrogen was calculated using a two-source linear mixing model. This study provides empirical evidence for the different reliance on carnivory of *B. gigantea* and *D. lusitanicum*. These differences in nutritional gain link with the differences in gland morphology, providing a framework for the positive relationship between investment in carnivorous traits and success in gaining prey-derived nutrients. Determining how other adhesive-trap carnivorous plants fit into this model will enhance our understanding of convergent trait evolution amongst carnivorous plants globally.

T2-01-10

A possible new carnivorous plant in the monocots, and the molecular evolution of its plastid *ndh genes*

Qianshi Lin, *Gregory Ross, Fushi Ke, Sean Graham*
University of British Columbia

At least nine independent origins of carnivory are known in five angiosperms orders. Most of these are in the eudicots: there are only four species known to be carnivorous in monocots, all in the order Poales. Here we report a possible new instance of carnivory in the monocot order Alismatales (*Triantha*, Tofieldiaceae), the first report in this order. *Triantha* has four recognized species, three of which occur in North America (*T. glutinosa, T. occidentalis and T.racemosa*), with a fourth in Japan (*T. japonica*). In Canada we found plants in this genus growing in bogs with other carnivorous plants, such as sundew and butterwort. Plants appear to capture insects using sticky glandular hairs on infloresence stalks. Enzymatic tests of phosphatase activity are positive for their trichomes, supporting the possibility that they are carnivorous. We sequenced the plastid genomes of two species (*T. glutinosa* and *T. occidentalis*) and found that one or more of 11 plastid *ndh* genes (which code for subunits of the plastid NADH dehydrogenase-like complex, NDH-1) have been lost or pseudogenized. A similar phenomenon has been observed in some members of carnivorous Lentibulariaceae. We also performed a population-level survey of *ndh* genes in *Triantha* across North America, and find that the degree of gene loss varies among different populations in a manner that is inconsistent with current species boundaries.

T2-01-11

Molecular evolution of the key digestive enzyme nepenthesin II in the genus *Nepenthes*

Firman Alamsyah[1], *Ito Motomi*[2]
1. Indonesia International Institute for Life Sciences
2. Graduate School of Arts and Sciences, The University of Tokyo

Nepenthes is a genus of carnivorous plant that specifically evolved to inhabit marginal environments with nutrient deficiency. *Nepenthes* have adapted the trapping process to live and grow in rainy tropical habitats. They augment their nutrient uptake by carnivory with their pitchers. Morphological adaptations, including wetness-dependent peristome, encourage the prey to fall into the pitcher. Then the body of the prey is drowned by the pitcher fluid and subsequently digested by enzymes within the pitcher fluid. Nepenthesins are major constituents of *Nepenthes* digestive enzymes. Nepenthesin I and II had been purified from *N. distillatoria*, and their partial internal sequences had been determined. However, the molecular genomic study of nepenthesin-encoded genes, in addition to their structure-function relationships, have never been conducted. The structure of nepenthesin-encoded genes might be related to the functioning of genes in response to environmental stress, as the habitat characteristics where *Nepenthes* live. Therefore, the objectives of this study are to examine the molecular evolution of the *Nep2* gene, which expresses the nepethesin II digestive enzyme in pitcher fluid. In the present study, 29 DNA sequences of *Nep2* gene have been succeeded amplified for the first time. The *Nep2* genes from 29 *Nepenthes* species varied in length between 1314-1317 bps, but mostly 1317 bp. All of the DNA sequences of the *Nep2* genes show no introns. Intron-less *Nep2* gene will produce its protein product rapidly for digesting the trapped prey. This rapid production of nepenthesin II enzyme may help to avoid putrefaction of trapped prey, which resulted in an accumulation of ammonium that may harm the pitcher to die. Thus, *Nep2* gene is supposed to have adapted specifically to produce nepenthesin II digestive enzymes rapidly by removing its introns during the course of molecular evolution. All of the enzyme of nepenthesin II examined contained 12 cysteine residues which would form 6 disulphide bonds expected to contribute greatly to the stability of the enzyme and allows the protein to be resistant to protease degradation, thus can remains in the pitcher fluid without digestion. In this study, *Nep2* gene suggestively experienced significant positive selection on *N. sumatrana*, a species inhabits the lowest altitude habitats (0-800 m) amongst Sumatran endemic species, which most of them are highland species. Adaptation to lowland habitats in Sumatra would be influenced by multiple physiological factors. For instance, at the physiological level, an obvious adaptive phenotype is the development of unusual 2 types of lower pitchers of *N. sumatrana*. The functions of both types of lower pitchers are to trap creeping insects. Therefore, the 2 distinct types of lower pitchers of *N. sumatrana* are mirrored by the strategy employed to obtain nutrients from the trapped prey in the lowland habitats. *Nepenthes* are highly adaptable plants and actively respond to the environmental conditions and availability of the prey in their habitats, by changes in phenotypic and genotypic characteristics. These changes are correlated to the fitness of the plants and are supposed under strong selective pressure of nutrient and environment that characterize the habitats of *Nepenthes*.

T2-02: The moss tree of life: phylogenomic approaches to reconstruct moss evolution, diversification, biogeography and biotic interactions

T2-02-01

Phylotranscriptomics reveals widespread gene duplication associated with the diversification of pleurocarpous mosses

Matthew Johnson[1], *Yang Liu*[2], *Rafael Medina*[4], *Bernard Goffinet*[2], *Jonathan Shaw*[3], *Norman Wickett*[1,5]
1. Chicago Botanic Garden
2. University of Connecticut
3. Duke University
4. Augustana College
5. Northwestern University

The pleurocarpous mosses (Hypnanae) are the largest group of extant mosses, and share a common ancestor less than 100 million years ago. The most speciose families within the Hypnanae experienced a rapid early burst of diversification, associated with a change in basic growth form (pleurocarpy). In addition to making phylogenetic inference among pleurocarpous moss groups difficult, the radiation likely left signatures of genome evolution observed in other plant radiations, including evidence of whole genome duplication, gene family expansion, and changing evolutionary pressures. Using transcriptomes assembled from over 90 moss species, including proto-pleurocarps and acrocarpous mosses, we constructed more than 27,000 orthologous gene families. Using a species phylogeny constructed from over 600 single-copy loci, we mapped the expansion and contraction of gene families. Gene family expansions greatly outnumbered gene family contractions, and one-sixth of all gene families expanded in size in the ancestors of pleurocarpous mosses. We test whether the proliferation of gene families is associated with paleopolyploidy, discuss the enrichment of functional annotations in expanded gene families, and examine the possible origin of pleurocarpous moss gene families with no homology in currently published genomes.

T2-02-02

Do moss host phylogenetics shape microbiome composition and function?

Lily Lewis[1], *Hannah Holland-Moritz*[2], *Noah Fierer*[2], *Jose-Miguel Ponciano*[1], *Michelle Mack*[3], *Stuart Mcdaniel*[1]
1. University of Florida
2. University of Colorado
3. Northern Arizona University

Bryophytes and their associated microbes are critical components of carbon (C) and nitrogen (N) cycling in arctic tundra and boreal forest ecosystems. Mosses account for approximately half of the net primary productivity in these biomes, and moss-associated microbes are the major source of biologically fixed N into these nitrogen-limited ecosystems. The extent to which climate change will perturb the ecosystem C and N dynamics critically depends upon how the moss-microbiome assemblages respond to new temperature and precipitation regimes. Here we evaluate the roles of environmental factors as opposed to moss host genotype in shaping moss-microbiome community assembly and function. We collected ten moss species from multiple sites near Fairbanks and Toolik Lake, Alaska. We used stable isotope incubations to estimate the rates of nitrogen and methane fixation in individual samples, 16S amplicon sequencing and shotgun metagenomic analyses to survey the community of bacteria hosted on each sample, and sequence capture to generate a moss phylogeny. Microbiomes from different moss species and sites fixed variable quantities of N and C. We found high relative abundances of members of novel bacterial phyla as well as a high diversity of putative N_2-fixing bacteria. Furthermore, we identified large numbers of nifH sequences from these metagenomes, including many from poorly described lineages. We are currently developing approaches to link biogeochemical processes to specific bacterial communities to understand the role of moss phylogeny in shaping microbiome function, and the potential for climate change to influence the function of the moss-microbiome.

T2-02-03

The Carboniferous set the stage for the major moss lineages which later experienced multiple independent rate shifts

Michael Krug[1], *Philip Testroet*[1], *Sarah Noben*[1], *Yang Liu*[2], *Bernard Goffinet*[2], *Kai Müller*[3], *Dietmar Quandt*[1]
1. Nees Institute for Biodiversity of Plants, University of Bonn
2. University of Connecticut, Department of Ecology & Evolutionary Biology
3. Institute for Evolution and Biodiversity, University of Muenster

Uncovering the branching order of major moss lineages is central to reconstruct the transformation of morphological traits during their evolution. This is particularly critical for understanding the order in which major peristome types arose. Current reconstructions mainly based on multi gene approaches either using chondrome data and/or coding regions from the plastome are inconclusive, as the short deep backbone nodes splitting the major clades receive inconsistent support. We generated a multi loci data set using mainly fast evolving regions from both organellar compartments. This data set comprised 15,759 characters (plastome: 8,576 char; chondrome 7,183 char.) plus 1,719 indels generated by simple indel coding. Phylogenetic analyses (Bayes & RAxML) yielded a highly supported backbone phylogeny of mosses that revealed the nematodontous lineages (Polytrichopsida & Tetraphidopsida) as a grade leading towards the arthrodontous mosses (Bryopsida). Among arthrodontous mosses the two first branching lineages contain mosses characterized by an unique endostome architecture (Buxbaumiidae & Diphysciidae) followed by the gymnostomous or cleistocarpous Gigaspermidae. Gigaspermidae are followed by the Funariidae indicating the diplolepideous-opposite peristome as the ancestral peristome type of which the haplolepideous (Dicranidae) and diplolepideous-alternate (Bryidae) peristome as well as the *Timmia*-type was derived. According to the BEAST analyses these major peristomes types were already established 300-350 MYA, indicating that the Carboniferous set the stage for the split of the major moss lineages (subclass) known today. Overall four independent rate shifts (BAMM) were detected, of which two affect the two biggest moss lineages, i.e. the haplolepideous (~ 4000 species) and the pleurocarpous mosses (~ 5000 species), that experience rate shifts at different stages of earth history. Both lineages are not only characterized by different peristome types (haplolepideous versus diplolepideous-alternate) but also by their different growth form (i.e. acrocarpy versus pleurocapy). The BEAST analyses based on an extended data set including a representative sampling of all other major land plant lineages indicate that the rate shift at the pleurocarpous branch in the Cretaceous can be associated with the radiation of angiosperms, i.e. new habitats, especially epiphytic habitats. Most likely the exploitation of the new habitats was aided by the invention of pleurocarpy. In contrast, the haplolepideous lineages represent a much older radiation experiencing a rate shift already in the middle/upper Jurassic that leads to a diversification in the Cretaceous. This finding goes in line with the increase in origination of spore bearing plants in the recovery phase of the late Permian mass extinction. Whether the evolution of the haplolepideous peristome that facilitates spore distribution enabled a higher colonization and origination rate needs to be shown.

T2-02-04

Sex-specific gene expression in the pleurocarpous moss species *Brachythecium rivulare*

Sanna Olsson[1,2], ***Sanna Huttunen***[2], *Matt Johnson*[3], *Fernando Puente-Sánchez*[4], *Norm Wicket*[3], *Tiina Sävilammi*[2], *Erica Leder*[2]
1. INIA Forest Research Centre (INIA-CIFOR)
2. Univeristy of Turku
3. Chicago Botanic Garden

4. *Centro Nacional de Biotecnología (CSIC)*

Molecular data on structure and evolution on sex chromosomes on early land plant lineages including mosses, liverworts and hornworts with ca. 20 000 species is currently based only on five species. Among mosses presence of sex chromosomes has been confirmed only in one species, but otherwise genetic regulation of sexuality remains unstudied. Based on scattered phylogenetic position of dioecious species, shifts to separate sexes have taken place several times independently among mosses and even within families, such as Brachytheciaceae. Evolution of characters that are directly linked to reproduction, such as position of female gametangia and sporophytes on moss shoot have also had important consequences in diversification of mosses. Evolution of pleurocarpy, a condition where female gametangia develop on reduced lateral branches may have resulted in diversification one of the most specious group of bryophytes, pleurocarps. The aims of the project were to 1) Identify differentially expressed (DE) genes between female and male samples in *B. rivulare*. 2) Find out whether DE genes are unique to pleurocarpous and/or dioicous mosses. We identified DE genes that are present in pleurocarps but are absent in Physcomitrella patens, the hermaphrodite acrocarpous model species. 3) Investigate if the DE genes are expressed both in autoicous and dioicous species by comparison to several Brachytheciaceae transcriptomes. We sequenced the transcriptomes of male and female samples of the dioicous, pleurocarpous moss species *Brachythecium rivulare* (Brachytheciaceae) using Illumina short read sequencing platform. The high-throughput sequencing was followed by *de novo* transcriptome assembly. The transcriptome data was compared with the transcriptomes of six Brachytheciaceae species, one other pleurocarp, Anomodon rostratus, and Physcomitrella patens. Differential expression between the sexes was detected in 820 expressed transcripts, representing 770 open reading frames. A search for orthologous sequences revealed 102 of them to be present in at least four other pleurocarpous mosses but not in Physcomitrella. Most of them were present in both autoicous and dioucous mosses. These genes will be analyzed in detail to learn more about their function and evolutionary history. BLAST searches against land plant, bacteria, fungi and other eukaryote data bases will show if they are ancient but lost in Physcomitrella, or synapomorphous in pleurocarps.

T2-02-05

The origin of the model taxon *Physcomitrella patens*: Resolving a rapid diversification using a phylogenomic approach based on enriched genomic libraries

Rafael Medina[1], *Matthew Johnson*[3], *Yang Liu*[2], *Jessica M. Budke*[4], *Nicholas Wilding*[5], *Bernard Goffinet*[2]

1. *Augustana College*
2. *University of Connecticut*
3. *Chicago Botanic Garden*
4. *University of Tennessee Knoxville*
5. *University of La Réunion*

The Funariaceae comprise about 250 species of mosses, mostly specialist of ephemeral habitats, including *Physcomitrella patens* and *Funaria hygrometrica*, two taxa frequently used as model organisms. From the morphological point of view, the Funariaceae combine a rather uniform gametophytic generation with a sporophyte that greatly varies across taxa, displaying a diverse range of architectures. The evolutionary history underlying this morphological diversity is obscure. Historically, relationships among species were drawn from shared sporophytic traits but inferences from DNA data revealed that the transformations of the sporophyte were homoplasious, and hence poor indicators of shared ancestry. Furthermore, the reconstruction of ancestral traits, and hence of the polarity of the transformations was hampered by the much of the diversity from a rapid diversification. To resolve the relationships among lineages within the Funariaceae, we combined high throughput sequencing and enriched genomic libraries to obtain the full organellar exome plus a selection of about 600 single copy nuclear genes from a representative selection of 91 Funariaceae. The analysis of these datasets provides a new timeframe for this diversification plus an increased resolution among taxa. As a result, we confirm the impact of sporophytic homoplasy across the family and making possible a comparison among morphological and genomic divergences. Our results also provide a genetic framework for the model organism *P. patens* concluding that extreme sporophytic reductions are not linked to significant genetic divergence in the studied datasets.

T2-02-06

Resolving the backbone phylogeny of mosses, using targeted NGS data from plastid, mitochondrial and nuclear genomes

Yang Liu[1,10], *Matt Johnson*[2], *Rafael Medina*[3], *Nicolas Devos*[4], *James Shevock*[5], *Lars Hedenäs*[6], *Alain Vanderpoorten*[7], *Neil Bell*[8], *Blanka Shaw*[4], *Dietmar Quandt*[9], *Norm Wickett*[2], *Jon Shaw*[4], ***Bernard Goffinet***[1]

1. *University of Connecticut*
2. *Chicago Botanic Garden*
3. *Augustana College*
4. *Duke University*
5. *California Academy of Sciences*
6. *Swedish Museum of Natural History*
7. *University of Liège*
8. *Royal Botanic Garden Edinburgh*
9. *University of Bonn*
10. *Shenzhen Botanic Garden*

Mosses compose a diverse lineage of land plants that arose soon after the conquest of land. The evolution of mosses is marked by transformations of the sporophyte and gametophyte generation, with ontogenetic changes of the spore-bearing generation defining the major lineages (i.e., classes) of mosses. However, homology among characters and their states remains ambiguous, and hence the relationships among these lineages are unresolved. Further, inferences from DNA sequences highlighted several lineages lacking specific apomorphies (e.g., Oedipodiales, Gigaspermales) in likely critical position along the backbone of moss tree of life, adding to the difficulty of reconstructing the transformation of the sporophyte. Finally, phylogenetic inferences have been drawn from few discrete loci that often lack the power to resolve the relationships along the tree. Here we present the first inferences based on variation in all organellar and close to 150 nuclear protein-coding genes sampled based on a liquid phase enrichment strategy for 134 exemplars of all orders of mosses. The conserved nature of the targeted loci avoids, following critical bioinformatics processing and manual inspection, uncertainty or conflict arising from alignment issues and hence the exclusion of data. Furthermore, the gene sequences can be translated and the signal of the nucleotides and amino acids compared, to assess artifacts due to saturation. We explored the homogeneity of the signal across loci and genomes, the congruence among topologies inferred from each of the three genomic compartments, and finally the distribution of the support among nuclear loci for individual nodes along the backbone. The moss tree of life is well supported at most nodes, but ambiguity remains in particular regarding the monophyly of mosses with nematodontous peristomes (i.e., Tetraphidales and Polytrichales), and their shared ancestry with the Oedipodiales, the sister-group to the Funariales, and the lineages emerging from the diversification of mosses with a diplolepideous alternate peristome. The monophyly of the Hypnanae (pleurocarpous mosses) is robust and their sister-group consistently consists of the Aulacomniaceae only. The Takakiales are resolved as sister to the Sphagnales based on chloroplast data but mitochondrial and nuclear data support them being sister to all mosses. The origin and evolution of the arthrodontous peristome and its derivation from a nematodontous one remain uncertain and depend on the resolution of relationships among the Nematodonteae, but also further ontogenetic studies to establish homology. However, with the arthrodontous mosses, those with an opposite arrangement of exostomial teeth and endostomial segments compose a grade leading to the lineages wherein teeth and segments alternate. Our study highlights the benefits of a broad genomic sampling and the advantages of using coding regions for resolving ancient evolutionary events (e.g., older than 150 my) but also draws attention to the need to investigate the nature of the support, as robust support can be drawn from few, or at least a minority of loci, and that the majority of the loci may not be congruent in the signal that they carry for a particular node.

T2-03: Ericaceae – phylogeny, biogeography and evolution

T2-03-01

Evolution and diversification of Ericaceae - Insights from whole plastome sequences

Kathleen Kron[1], *Benjamin Potter*[2], *Emily Gillespie*[3], *Darren Crayn*[4]

1. Wake Forest University

2. School of Biological Sciences, University of Auckland, Auckland

3. Marshall University, Huntington, WV

4. Australian Tropical Herbarium, James Cook University, McGregor Road, Smithfield, QLD

The Ericaceae clade is remarkable for its diversity of growth form, physiology, anatomy and morphology and includes a diversity of mycorrhizal associations greater than any other angiosperm family. To date, most of the major clades within Ericaceae have been identified and the relationships among these groups has been resolved to some extent from phylogenetic analyses of selected regions within the chloroplast and nuclear genomes. The results of these analyses agree with most of the previous studies, but the relationships at the deeper nodes of Ericaceae are still not strongly supported. Previously published complete chloroplast genomes of *Arbutus unedo* (strawberry tree) and *Vaccinium macrocarpon* (cranberry) showed unexpected levels of diversity in gene order and content and suggest that plastome structure and content within the Ericaceae may vary significantly reflecting the already documented variation in habit, mycorrhizal association and morphology. We sequenced the plastome of more than 40 representatives from the eight currently recognized subfamilies (Enkianthoideae, Arbutoideae, Monotropoideae, Cassiopoideae, Ericoideae, Harrimanelloideae, Epacridoideae, Vaccinioideae) and 20 named tribes. Among these we obtained plastomes that included representatives of different growth forms and functionalities: e.g., green and non-green (mycoheterophic); different mycorrhizal associations (vesicular-arbuscular, arbutoid, ericoid, monotropoid, cavendishioid); tropical cloud forest, open heathlands, boreal habitats; woody, herbaceous. Preliminary results indicate that ericaceous plastomes are much more variable in gene order, content and size than expected. In addition, some plastomes showed organizational shifts/changes that were not obviously correlated with, e.g., physiology, but were coincident with previously identified major clades within Ericaceae. Further analyses of plastome variation across the ericads indicates that even within closely related groups, e.g., Vaccinieae (blueberries), the plastome can significantly vary in size and gene content. These new findings provide important insight into the evolution and diversification of the Ericaceae and its biogeography, while also indicating that there is likely a more diverse and dynamic pattern of plastome evolution in general than previously thought.

T2-03-02

Historical biogeography of the Australasian Ericaceae: Evaluating competing vicariance and dispersal scenarios

Caroline Puente-Lelievre[1,2], *Darren M. Crayn*[2], *Benjamin C. M. Potter*[3]

1. US Food and Drug Administration

2. Australian Tropical Herbarium

3. School of Biological Sciences, University of Auckland

The subfamily Epacridoideae, Ericaceae (commonly known as the epacrids) comprises 37 genera and ca. 640 species. This clade includes mostly perennial woody shrubs that occur in heathlands, sandplains, and tropical forests. Their distribution is mostly restricted to Gondwanan fragments (Australia, New Zealand, New Caledonia, and South America) with outliers in SE Asia and Hawaii. The epacrids are however most diverse in South-Western (WA) and South-Eastern Australia (EA). The Epacridoideae also typify many of the controversies in the historical biogeography of Australasia and the contributions of long distance dispersal versus Gondwanan vicariance. Such hypotheses can be tested with molecular dating and ancestral area reconstruction analyses. While a vicariance scenario predicts patterns of near-simultaneous divergences in multiple lineages, long-distance dispersal predicts independent, asynchronous divergences. We present the most densely sampled phylogeny to date that includes representatives from all genera and ca. 65% of the species in the Epacridoideae. Based on a 5-gene dataset (the nuclear ribosomal ITS region, and the chloroplast regions *rbc*L, *mat*K, *psb*A-*trn*H, *atp*B-*rbc*L) we estimated the phylogenetic relationships and tempo of evolution within the subfamily with a focus on the relationships within the Australian clades, and their relationships with the New Zealand and New Caledonian clades. Furthermore, we used the resulting phylogeny to evaluate multiple calibration schemes, molecular clock models, and speciation priors, and estimated the ages of the disjunctions between these clades. Our results under different schemes consistently suggest that the multiple disjunctions between the Australian, New Zealand, and New Caledonian clades are independent events and not older than ca. 10 Ma. They also indicate that the WA and EA clades diverged approximately over the same period of time during the Miocene (10 – 20 Ma), after the emergence of the Nullarbor Plain. We conclude that the current distributional patterns of the epacrids can be explained by both vicariance and long distance dispersal events.

T2-03-03

Ericaceae phylogenetics revisited: Developments and prospects

Benjamin C. M. Potter[1], *Caroline Puente-Lelievre*[2], *Andrew H. Thornhill*[3], *William A. Freyman*[3], *Kathleen A. Kron*[4], *Darren M. Crayn*[5],

1. School of Biological Sciences, University of Auckland, Auckland, New Zealand

2. US Food and Drug Administration, 5001 Campus Drive, College Park, MD, USA

3. Department of Integrative Biology, University and Jepson Herbaria, University of California, Berkeley, CA, USA

4. Department of Biology, Wake Forest University, 1834 Wake Forest Road, Winston-Salem, NC, USA

5. Australian Tropical Herbarium, James Cook University, McGregor Road, Smithfield, QLD, Australia

The Ericaceae, with ca. 4200 species and 125 genera, has a near-global distribution and many taxa of great economic (e.g. blueberries, cranberries, and rhododendrons) and ecological (e.g. northern heathlands, montane-heath shrubberies, and Cape heaths) importance. Here, we review the current state of systematics of the family, highlight where major problems exist, and evaluate how close we are to a well-resolved comprehensive phylogenetic estimate for all taxa. Towards this we briefly discuss the recent literature, and present new phylogenies based on a DNA sequence matrix sampling 22 regions and 2027 species (ca. 50% species-level coverage, with 5-96% coverage per marker). The resulting phylogenies are then compared to the prevailing classification system of Kron *et al.*, (2002). Additional studies on the systematics and historical biogeography of the family will be outlined and discussed, as well as directions for future research.

T2-03-04

Biogeography, classification, breeding system evolution, and new species discovery in the Gaultherieae (Ericaceae)

Peter W Fritsch[1], *Lu Lu*[2], *Kathleen A Kron*[3], *Catherine M Bush*[4], *De-Zhu Li*[2], *Hong Wang*[2]

1. Botanical Research Institute of Texas

2. Kunming Institute of Botany

3. Wake Forest University

4. Elon University

The tribe Gaultherieae (Ericaceae: Vaccinioideae) contains ca. 280 species with a circum-Pacific distribution that encompasses a wide range of bioclimatic areas. We employed an eight-gene phylogeny based on three nuclear and five plastid gene regions to reconstruct the biogeographic history of the tribe, reassess subdivisional classification, and trace breeding systems evolution. Results of combined Bayesian divergence time and ancestral area analyses recovered an origin for crown clade Gaultherieae of ca. 33 Ma in temperate North America. Of the 15 dispersals inferred from the analysis, five occur across the Northern Hemisphere, and seven occur from north to south. Two of the latter are from temperate North America to tropical America, and three from temperate East Asia to tropical East Asia, consistent with global drying and cooling during this time. In addition, one dispersal is inferred from tropical America to temperate South America, one from temperate South America to New Zealand, and one from New Zealand to Australia. All 15 dispersals occurred later than 23 Ma, and so the dispersal from South America to Australasia must have been over-water, likely by birds because an analysis of fruit type and color reconstructed the ancestor of the Australasian lineage as fleshy and red. *Gaultheria* exhibits variation in breeding systems, with hermaphroditic, gynodioecious, and dioecious species. Tracing the evolution of breeding system onto the phylogeny recovered the evolution of gynodioecy from hermaphroditism in two major clades, i.e., Himalaya-Hengduan Shan species and the clade of temperate South American and Australasian species. In the latter clade, full gynodioecy evolved in temperate South America. Gynodioecy had already evolved in the ancestral lineage that migrated from South America to Australasia, and then descendants largely retained gynodioecy in New Zealand, confirming the allochthonous origin of dicliny in New Zealand for the genus as previously hypothesized. Because *Diplycosia* and *Tepuia* are nested within *Gaultheria*, these three genera are best treated either as a single genus (*Gaultheria*) or as six genera, i.e., with the former two genera retained as is and *Gaultheria* divided into four genera. We prefer the former because one of the large clades that would require its recognition as a genus is not diagnosable with morphology. The infrageneric classification of *Gaultheria* requires revision at the sectional level. Through the combination of field, herbarium, and molecular studies we have discovered 14 new species in the tribe from China and SE Asia. Because most species grow in high elevation zones with difficult access, further exploration in these areas are likely to uncover many more.

T2-03-05

Fruit color variation and diversification of a circum-Pacific clade of plants, the tribe Gaultherieae (Ericaceae)

Lu Lu[1], *Peter Fritsch*[2], *Hong Wang*[1], *Catherine Bush*[3], *Kathleen Kron*[4], *De-Zhu Li*[1]

1. Kunming Institute of Botany, Chinese Academy of Sciences

2. Botanical Research Institute of Texas

3. Elon University

4. Wake Forest University

Fleshy fruit is generally thought to have evolved for effective seed dispersal by frugivores. However, the variation of fleshy fruit colors and its adaptive significance in plant diversity has received little attention in an evolutionary context. Here we investigate fruit evolution in the circum-Pacific ericoid tribe Gaultherieae, a clade with a wide range of fruit types and colors, with diversification and correlated evolution tests in a phylogenetic context. We found evidence for independent diversification rate accelerations facilitated by a switch from red to the colors in the violet part of the spectrum (blue, purple, or black), suggesting that violet fruit is a key innovation in the tribe. We show that violet fruits predominate in either low-latitude or high-elevation environments. With phytochemical analysis, we isolated cyanidin 3-O-β-D-xylopyranosyl- (1→2)-β-D-glucoside and delphinidin 3-O-β-D-xylopyranosyl- (1→2)-β-D–xylopyranoside from red and violet fruits of the Gaultherieae, respectively, and detected no pigments in white-fruited species. Dephinidin and its derivatives have been found to have greater absorption in the UVB portion of the spectrum than cyanidin. Because UV intensity increases with elevation and decreases with latitude, the spatial distribution of fruit color suggests that the the delphinidin derivative in violet fruits functions as a stronger UV protectant for progeny (seeds) than does the cyanidin derivative of red fruits. This can be tested more directly by experiments involving the insertion of UV-detecting microprobes into Gaultherieae fruits of various colors.

T2-03-06

Optimizing RADseq methods and inferring histories of diversification of rhododendrons in mountain ranges of Asia

Valerie Soza[1], *George Argent*[2], *Dale Lindsley*[1], *Elizabeth Ramage*[1], *Michelle Stitzer*[4], *Alex Twyford*[3], *Benjamin Hall*[1]
1. University of Washington
2. Royal Botanic Garden Edinburgh
3. University of Edinburgh
4. University of California Davis

The genus *Rhododendron* consists of about 1000 species with centers of diversification in the Himalayas and the Malesian islands. The Himalayan diversification is predominantly represented by subgenus *Hymenanthes* with 225 species historically classified into 24 subsections (Chamberlain 1982). The Malesian diversification is primarily represented by subgenus *Rhododendron* section *Schistanthe* (formerly sect. *Vireya*) with 320 species currently classified into 4 subsections (Craven *et al.,* 2011). Previous molecular phylogenetic studies on these two groups using several loci showed conflicting data or lack of resolution in phylogenetic reconstructions. Therefore, in this study, we employed genomic sampling of loci using single-digest restriction site associated DNA sequencing (RADseq) with *PstI* to infer biogeographic histories for subg. *Hymenanthes* and sect. *Schistanthe*. We sampled 90 species from all subsections of subgenus *Hymenanthes* and 108 species from all subsections of sect. *Schistanthe*, as well as 31 outgroups sampled from all subgenera of *Rhododendron* and several closely related genera. We compared results from denovo versus reference guided assemblies of our RADseq data. We also compared results using single-end RAD sequences (100 bp) versus paired-end RAD contigs (mean 450 bp). Resulting phylogenetic analyses based on entire RAD loci were calibrated with several primary and secondary calibration points using penalized likelihood for ancestral area reconstructions. Our preliminary results infer an origin of subg. *Hymenanthes* in northeast China during the Oligocene with subsequent radiation into the Himalayas during the Miocene. For sect. *Schistanthe*, we infer an origin in Indochina during the Miocene with subsequent radiation into the mountains of Borneo beginning in the Miocene and New Guinea beginning in the Pleistocene.

T2-04: Phylogenomic perspectives on flagellate plant evolution

T2-04-01

Can ancestral state reconstructions of dioecy in mosses be Independently confirmed using coalescence of sex linked genes?

Sarah Carey, *Adam Payton, Stuart Mcdaniel*
University of Florida

A major goal of many phylogenetic analyses is to reconstruct the evolution of key organismal features. Although sophisticated methods exist to reconstruct ancestral states, in general, there is much uncertainty in the reconstruction of evolutionarily labile traits, even in groups with a rich fossil history. Here we take advantage of unusual features of the moss sex determining system to evaluate our previous reconstruction of sexuality across the mosses. Sex in dioecious mosses is often determined by a UV sex chromosomal system. In this system, each sex has a non-recombining chromosome (U for females, and V for males) that pair at meiosis in the monomorphic sporophyte, and segregate to the male and female haploid gametophytes. Because the sex chromosomes are transcriptionally active in the haploid stage, and therefore subject to purifying selection, we expect many homologous genes will be retained between the U and V chromosomes. Thus, we expect that the most ancient sex-linked loci should coalesce at or below nodes where the common ancestor was reconstructed as dioecious. To find sex-linked genes, we first sequenced transcriptomes from ten male-female sibling pairs of the moss *Ceratodon purpureus*. Next we assembled the transcriptomes using standard genome-guided RNAseq approaches. Since the reference genome was female, we de novo assembled the unmapped male reads and used BLAST to identify female homologs for these transcripts. We used a genetic map to confirm the sex linkage of many of these loci. We estimated Ks between the male and female homologs, and retained those with $Ks > 0.1$ for further study. We identified additional homologs using the 1KP data, and we built phylogenetic trees using RAxML to estimate the placement of the coalescence between the *C. purpureus* male and female homologs. We are currently increasing taxon sampling to test the accuracy of multiple ancestral reconstruction methods.

T2-04-02

GoFlag: A next generation sequencing perpective on flagellate plant phylogeny

Gordon Burleigh, *Christine Davis, Stuart Mcdaniel, Emily Sessa*
University of Florida

GoFlag is an NSF-funded project to construct a species-level phylogenetic tree of thousands of flagellate plant species, including bryophytes, lycophytes, ferns, and gymnosperms, and link these trees to relevant phenotypic, environmental, and geographic data. We report on a pilot study to use a sequence capture approach to generate a common set of ~500 low-copy nuclear loci and well as chloroplast data for phylogenetic inference across all flagellate plant groups. The pilot study is based on 192 phylogenetically diverse flagellate plant species. We discuss the methodological challenges of our approach as well as analytical issues of analyzing this next-generation sequence data. We also discuss new approaches for linking metadata to the phylogenetic tree for evolutionary and ecological analyses. Although constructing a well-resolved phylogenetic tree for flagellate plants is a difficult and complicated endeavor, new sequencing technologies and analytical methods provide much promise for revealing the rich history of flagellate plants, and we describe how the GoFlag project will expand upon this pilot study to build a comprehensive flagellate plant tree.

T2-04-03

A phylogenomic analysis of the cycad genus *Cycas* using RADseq

Nathalie Nagalingum[1], *Patrick Griffith*[2], *Shugaung Jian*[3], *Jiban Khuraijam*[4], *Anders Lindstrom*[5], *Rita Singh*[6]

1. *California Academy of Sciences*
2. *Montgomery Botanic Center*
3. *South China Botanical Garden*
4. *CSIR-National Botanical Research Institute*
5. *Nong Nooch Tropical Botanical garden*
6. *Indraprastha University*

Relationships among cycad species have long been difficult to infer, principally due to their recent rapid radiation and the reliance on data from few, relatively slowly evolving, uninformative markers. Next Generation Sequencing (NGS) data offer promise for resolving these relationships through the sheer vast number of rapidly evolving nuclear markers. Since cycads have large genomes (~14,500 Mb), we used a reduced-representation sequencing approach in the form of RADseq (Restriction-site Associated DNA sequencing). Here we focus on *Cycas*, the largest cycad genus, which has over 100 species and is the sole member of Cycadaceae. We generate a phylogeny for three-quarters of species, covering the range of *Cycas* in Asia, Australia, Madagascar, India and the southwestern Pacific Ocean islands. Using the RAD phylogeny, we infer species relationships, assess biogeographic patterns, and evaluate the current classification of the genus.

T2-04-04

A phylogenetic approach to assessing the evolution of genomic repetitive elements across the fern family pteridaceae

Erin M. Sigel[1], *Amanda L. Grusz*[2], *Eric Schuettpelz*[3]

1. *University of Louisiana at Lafayette*
2. *University of Minnesota Duluth*
3. *Smithsonian Institution*

Due to their large genome sizes and high chromosome numbers, ferns have been sidelined from most phylogenomic and comparative genomic analyses. While published fern genomes are still forthcoming, low-coverage genome skimming has provided initial glimpses into the broad-scale genomic composition of these plants, particularly with respect to their abundance of non-coding repetitive elements relative to seed plants. That said, the relative contributions of transposable elements and other repeat types within major fern lineages remain unexplored. Here, we present the results of a comparative study of repeat content, in a phylogenetic context, by analyzing genome skimming data for 32 species across the polypod fern family Pteridaceace. This family accounts for approximately 10% of extant fern diversity and comprises five monophyletic clades spanning diverse ecologies and morphologies. We assess whether the abundance of various repetitive elements are correlated with phylogenetic distance and genome size. This work lays the groundwork for exploring the role of repetitive elements in shaping genome structure, content, and size across fern family, and greatly expands genomic resources for these plants.

T2-04-05

Evolutionary divergence in the coralloid root microbiome of cycads

Angelica Cibrian-Jaramillo, *Francisco Barona-Gomez, Pablo Cruz-Morales, Antonio Corona-Gomez, Karina Gutierrez-Garcia*

National Laboratory for Genomics of Biodiversity, Langebio, CINVESTAV

Cycads are the only gymnosperms and seed plants that have evolved a specialized coralloid root to host endophytic bacteria. There is no precedent of exploring the taxonomic, phylogenetic or functional diversity of the microbiome of this million-year old association. We provide a genomic characterization of the cycad coralloid root microbiome of the Mexican *Dioon* species collected from natural populations, and yield insights on its evolutionary history. We explored the microbiome of the coralloid root employing a co-culture-based experimental strategy to unveil rare bacteria, in par with phylogenomic analyses to capture the biological diversity and the presence of Bacterial Genetic Clusters (BGCs) associated to specialized metabolites. Most taxa were identified as diazotroph plant endophytes and included known and previously undescribed Cyanobacteria OTUs. We sequenced three of their genomes, which formed a monophyletic group that suggests a level of specialization typical of obligate or closely co-evolved symbiotic relationships. This contrasted with our finding of their large genome sizes, typical of facultative endosymbionts of complex alternative lifestyles. Nine out of 23 novel BGCs are specific to these coralloid root endophytes. Testing of vertical transfer of the root endophytes via the seeds, gene expression profiles of the coralloid roots, and population genetics of the genus is ongoing. Combined, our results thus far show that the highly diverse taxonomic, phylogenetic composition of the coralloid root, and a specialized biosynthetic repertoire, correlate more with specificity to the *Dioon* plant host than other closely related plant endosymbionts, or to the environment. We add to the growing notion that symbiotic relations occur under heavy influence of chemical and genomic interactions, providing a rich source of novelty for drug discovery and understanding of the evolution of gymnosperm-bacteria symbiosis.

T2-04-06

Analyses of structure-function relationships in the evolution of embryophyte shoot meristems: A macroevolutionary transcriptomic approach

Margaret Frank[1], *Michael Scanlon*[2]

1. Donald Danforth Plant Science Center
2. Cornell University, Department of Plant Biology

Shoot apical meristems (SAMs) have increased in complexity over the course of embryophyte evolution from single apical initials that generate the gametophytic body plans in bryophytes, to the histologically stratified, multicellular SAMs that give rise to the entire sporophytic shoot in angiosperms. Despite this diversity in shoot apical structure, our understanding of the genes involved in SAM function is largely limited to the angiosperms. We used laser microdissection paired with RNA-sequencing (LM-RNA-seq) to generate shoot meristem molecular profiles for species from anciently evolved land plant lineages. This data was used to identify genetic networks that correlate with three major advances in embryophyte SAM structure: (1) the transition to multidimensionally growth; (2) the rise of indeterminate apical growth in the sporophyte generation; and (3) structural diversification of the sporophytic SAM in seedless vascular plant lineages. Our comparative transcriptomic analyses contribute to new molecular models for these pivotal events in embryophyte evolution. Specifically, we propose that the combination of asymmetric cell division and shoot meristem maintenance programs is associated with multi-dimensional shoot meristem function in the moss *Physcomitrella patens*. In addition, we suggest that pleiotropic gene expression in both the gametophytic meristem and the non-meristematic sporophyte of gametophyte-dominant plants contributed to the rise of indeterminate SAM growth in the sporophyte generation of early polysporangiophytes. Finally, we present transcriptomic evidence for the convergent evolution of meristem structures in the seedless vascular plant lineages containing *Selaginella* and *Equisetum*. The comprehensive nature of the data leads to testable, functional models for developmental genetic programs that contributed to fundamental strategies in embryophyte evolution.

T2-05: Hybridization and phylogenetic reconstruction

T2-05-01

Phylogenetics and the discovery of plant hybrid lineages

Gonzalo Nieto Feliner
Real Jardín Botánico, CSIC. Madrid

Reticulation is a pervasive issue in phylogenetic inference. Because hybridization is not normally something you look for but something you run into, you may face it under two perspectives. By far the first, and most common one, is trying to account for hybridization avoiding the disturbing effects that it introduces in the reconstruction of phylogenetic relationships of group of organisms in which we assume a predominance of divergence processes. Under this focus, hybridization events are hurdles that one has to overcome mainly by telling apart such events from incomplete lineage sorting (ILS). Different approaches have been developed for this task, which is particularly challenging when tracing both old events and those lying in the so-called 'coalescent stochasticity zone' in which attributing gene conflict to hybridization is hampered by the high frequency of non-coalesced genes. The second perspective is to focus on the process itself, i.e., on hybridization events associated to 'discordant' lineages. The main contribution of phylogenetic studies to this aim is unveiling unsuspected hybrid lineages. However, compared to studies of speciation that usually focus on reproductive isolation, phylogenetic studies also provide a systematic and evolutionary frame, including a temporal component, that represent a basis for understanding the evolutionary consequences of the hybridization event (s). Whether the merging of two genomes contributes positively to evolution depends on a number of factors that need to be explored in each case. Phylogenetic approaches may provide assessments of factors such as how closely related are the hybridizing species, how old are their speciation events, how old is the hybridization event, or how close is the hybrid lineage from the progenitors from a genetic standpoint. A quick look at cases of reticulation in phylogenetically oriented studies suggests a wide range of situations in which hybridization is involved and also shows how such studies can throw light on the circumstances surrounding hybridization events, making phylogenetics a relevant complementary approach to the field of hybrid speciation.

T2-05-02

Genomic relationships of recent species radiations

Alex Twyford
University of Edinburgh

Phylogenetic reconstructions with next generation sequencing data may resolve challenging species relationships where Sanger Sequence data have failed. Restriction site associated DNA sequencing (RAD-seq) and related approaches are particularly promising for phylogenies of non-model taxa due to their scalable recovery of many homologous regions. However, this wealth of sequence data may still fail to resolve relationships in recent species radiations where hybridisation and incomplete lineage sorting are common. I present genome-scale phylogenies of diverse plant taxa that demonstrate how geography rather than current taxonomy frequently explain genome-wide clustering in closely related species complexes. In contrast, taxonomy may be supported by a subset of causal loci underlying traits of interest. These results caution against using concatenated data matrices to delimit taxa in recent radiations. This also highlights the need for careful analysis of gene genealogies for regions that may promote species divergence.

T2-05-03

Inferring species trees given coalescence and reticulation

Michael Pirie
Johannes Gutenberg-Universität, Mainz

Hybridisation is an important process in plant evolution, its impact apparent at all levels from the genome duplications shared by all angiosperms through to recent speciation events. Patterns of inheritance caused by hybridisation (and other reticulate processes) appear to be incompatible with one of the fundamental tools of evolutionary biologists and ecologists seeking to understand the evolution of biological diversity: the phylogenetic tree. The current paradigm for inferring relatedness of organisms – the 'species tree' – can be summed up in one word: coalescence. Current analytical approaches serve to bias against inferring reticulate processes, even though they may be common and of direct importance both for the evolutionary process itself and for the performance of methods used to infer it. I present a brief account of current methodologies and draw on examples from both plant and virus datasets to illustrate the importance of reticulate processes in evolutionary inference. An example from danthonioid grasses shows how our inference of the direction and frequency of long distance dispersal events can be impacted by post-dispersal hybridisation; one from the northern heathers (genus *Erica*) shows the impact on interpretations of morphological evolution. In viruses, recombination can lead to pathogenic strains, and our ability to infer this process, even on human timescales, is dependent on correctly interpreting differences between gene trees. I illustrate an easily implemented approach that may be used to infer the sequence and timing of gene and genome divergences given conflict between individual gene trees, even when the processes underlying that conflict cannot be distinguished. It can in principle be applied to any group of organisms and can be extended to explicitly model both reticulation and coalescence without prior knowledge of the species tree topology.

T2-05-04

Assessing ancestral niche suitability and geographic range dynamics as drivers of hybridization in *Heuchera* (Saxifragaceae)

Ryan Folk, *Clayton Visger, Robert Guralnick, Pamela Soltis, Douglas Soltis*
Florida Museum of Natural History

Hybridization has received a resurgence of attention from evolutionary biologists, owing to powerful approaches using phylogenomics and the coalescent model that enable their robust detection with molecular data. Sometimes, ancient hybridization has been detected among lineages that are not (currently) extant, consistent with our knowledge of the historically dynamic geographic ranges of taxa. Recent developments in the exploration of niche space using phylogenies have enabled explicit tests of these kinds of hypotheses. As an empirical case study, we demonstrate how the integration of a recently developed ancestral reconstruction approach, using ecological niche models of extant species and a robust phylogeny, can test whether ancestral co-occurrence could have enabled historical hybridization among widely disjunct taxa. Extremely high incongruence between nuclear and chloroplast data has previously been demonstrated in *Heuchera*; coalescent simulations favored a hypothesis of hybridization over incomplete lineage sorting to explain this incongruence. Taxa inferred to have hybridized are as much as hundreds of kilometers disjunct. We show that ancestral parapatry under a Pleistocene glacial scenario likely explains gene flow, and how novel Bayesian applications of ancestral niche reconstruction allow projection of ancestral suitability into past climate models to produce geographic maps of suitable range.

T2-05-05

Population genetics, phylogenomics and hybrid speciation of Chinese *Juglans* (Juglandaceae) determined via whole chloroplast genomes, transcriptomes, and Genotyping-By-Sequencing (GBS)

Peng Zhao[1], *Yi-Heng Hu*[1], *Hui-Juan Zhou*[1], *Daniel Potter*[3], *Meng Dang*[1], *Xiao-Jia Feng*[1], *Gui-Fang Zhao*[1], *Wen-Zhe Liu*[1], *Meng Yue*[1], *Zhan-Lin Liu*[1], *Keith Woeste*[2]
1. Northwest University
2. Purdue University
3. University of California

Genomic data provide powerful tools for elucidating the processes involved in the evolution and divergence of species, the fundamental units in the fields of ecology, population genetics, and biogeography. We used phylogenomic and population genetic analyses of data from transcriptomics, Genotyping-By-Sequencing (GBS), and whole chloroplast genomes (Cpgs) to infer processes of lineage formation among the five native Chinese species of the walnut genus (*Juglans,* Juglandaceae), a widespread, economically important group. We also used ecological niche modeling and morphological analysis to verify ecological speciation. Our analyses supported the genetic distinctiveness of *Juglans regia*, *J. cathayensis*, *J. mandshurica*, and *J. sigillata*. Comparisons among whole chloroplast genomes and nuclear transcriptome analyses provided strong evidence that *J. hopeiensis*, a species considered threatened, arose through relatively recent hybridization between *J. mandshurica* (female parent) and *J. regia* (male parent). The genetic structure of *J. cathayensis* and *J. mandshurica*, two closely related taxa, indicate they are poorly differentiated. GBS data revealed clear introgression/gene flow between *J. regia* and *J. sigillata*, while population genetic analysis detected previously unrecognized deep lineages within *J. regia*, and the presence of individuals with highly differentiated genomes within otherwise geographically delimited populations. Using trees based on chloroplast and GBS markers, we reconstructed the evolutionary history and estimated divergence times of the five Chinese *Juglans*. The divergence time between sect. *Juglans/Dioscaryon* and sect. *Cardiocaryon* was estimated at 62.93 Myr, while *J. regia* and *J. sigillata* diverged much more recently (0.01 Myr). The current spatial distribution of genetic variability among the five Chinese walnut species likely results from ecological speciation based on local adaptation resulting in morphological and floral differentiation, even in the presence of gene flow and introgression. Together our results highlight the outstanding capacity of NGS sequencing data (GBS, transcriptomes, and Cpgs) for elucidating the complex processes of phylogeny, biogeography, and hybrid speciation, particularly among relatively closely related species.

T2-05-06

Natural hybridization and repeated hybrid speciation in Actinidia through genome resampling from rapidly radiated backbone lineages

Yifei Liu
South China Botanical Garden, CAS

Reticulate speciation caused by interspecific hybridization is now recognized as an important mechanism in creating biological diversification. However, depicting the pattern of phylogenetic network for lineages that have undergone inter-specific gene flow is challenging, and the difficulty can be further compounded by the presence of ancestral polymorphisms owing to incomplete lineage sorting (ILS). We sequenced 25 taxa representing the high diversity deposited in the genus *Actinidia* with an average mapping depth of 26× on reference to reconstruct their reticulate history. Using genome-wide data, we found evidence, including significant gene tree discordance, cytonuclear conflicts, and changes in genome-wide heterozygosity across taxa, collectively supporting extensive reticulation in the genus. Further examinations indicated that at least two separately parental species pairs were involved in the repeated origin of the hybrid lineages, in which some triggered a further phase of syngameon. On the basis of the elucidated hybridization relationships, we finally obtained a highly resolved backbone phylogeny constructed by taxa exhibiting no evidence of hybrid origin. The backbone taxa have distinct demographic histories and are the product of recent rounds of adaptive radiations by sorting of ancestral variation under variable climatic and ecological conditions. They commonly serve as a genomic base contributing raw parental material for hybrid speciation of other *Actinidia* taxa. Our results suggest a potentially universal model for consecutive plant diversification through two layers of adaptive radiations, including the rapid evolution of backbone lineages and further hybrid swarm derived from them.

T2-06: Causes and consequences of the Cretaceous Terrestrial Revolution to the tree of life with focus on seed-free land plants and trophic interactions

T2-06-01

Angiosperms as engineers of ecology and climate

C. Kevin Boyce
Stanford University

Ecosystems changed dramatically over the Cretaceous and on into the Cenozoic. Flowering plant evolution is an obvious consequence, but can also have been a key driver of these changes. Angiosperms have dramatically higher transpiration capacities than other plants. The recycling of precipitation via transpiration is important for the maintenance and expansion of everwet environments. Flowering plants themselves are diverse in the everwet tropics, but seedless plants have also been beneficiaries of this climate engineering. The elevated transpiration capacities of angiosperms correspond to higher productivity potential as well and the combination of increasing productivity with the environmental stability afforded by angiosperm buffering of climate may also have led to increased trophic complexity.

T2-06-02

Burmese amber provide insights into the phylogenetic composition of terrestrial habitats in the KTR

Harald Schneider[1], Schmidt Alexander[2], Jochen Heinrichs[3]
1. Natural History Museum, London
2. Georg August University, Goettingen
3. Ludwig Maximilian University, Munich

The Cretaceous Terrestrial Revolution is a period of profound changes in the ecology of terrestrial environment which are marked not only by the rise of angiosperms. This ecological revolution is not only marked by diversification of some animal lineages dominated extent terrestrial forest habitats such as various lineages of insects such as ants, rise of birds and mammals but also by the the diversification of non-angiosperm land plant lineages such as liverworts and ferns. In turn, some lineages of ecological importance in the pre-KTR forests, declined or were replaced by phylogenetically related lineages. Burmese amber besides other Cretaceous amber lagerstaetten, provide unique insights in the taxonomic and phylogenetic composition of forests habitats during the KTR. The amber inclusions of these localities include often the first appearances of lineages of great ecological importance to extant forest habits or the latest record of lineages in decline. These fossils provide also information on the phylogenetic composition and in turn provide insights to the phylogenetic context of the trophic networks during the KTR. As a result, the Burmese amber fossils are not only providing evidence to improve efforts to date phylogenetic hypotheses but also provide a unique window to study the dynamics of changes in trophic networks during major ecological transitions in the history of land plants. This patterns are explore here by placing recently describe Burmse amber fossils of bryophytes, ferns, and insects into their phylogenetic context integrating both diversification time estimates and niche reconstructions. This innovative approach is discussed and outlined here for the first time.

T2-06-03

Mesozoic fern floras from China: anatomy, diversity and evolutionary implications

Yongdong Wang[1,2], Ning Tian[3,4]
1. Nanjing Institute of Geology and Palaeontology, Chinese Academy of Sciences, Nanjing 210008
2. State Key Laboratory of Palaeobiology and Stratigraphy, Chinese Academy of Sciences, Nanjing 210008
3. College of Paleontology, Shenyang Normal University, Shenyang 110034
4. Key Laboratory for Evolution of Past Life in Northeast Asia, Ministry of Land and Resources, Shenyang 110034

Ferns are very abundant and well developed in the Mesozoic floras in both northern and southern China, mainly ranging from Late Triassic to Early Cretaceous intervals. During last decade, more than 45 species referred to 16 genera of Early Mesozoic fern fossils have been investigated in China by our research group, including six families such as Marattiaceae, Osmundaceae, Gleicheniaceae, Matoniaceae, Dipteridaceae and Dicksoniaceae. These fossils are mainly recovered from the Late Triassic, Early and Middle Jurassic deposits (a few taxa are from the Late Permian) in a variety of plant fossil localities, including Hubei, Sichuan, Anhui and Jiangsu of southern China phyoprovince as well as Hebei, Liaoning, Nei Mongol and Xinjiang regions in northern China phytoprovince. In our series studies, we mainly focus on the anatomy of reproductive structures, such as sori or synangia, sporangia, annuli, spore output and spores in situ using both LM and SEM approaches, such as *Marattia*, *Asterotheca* and *Daeneopsis* of Marattiaceae, *Oligocarpia* of Gleicheniaceae, *Phlebopteris* of Matoniaceae, *Dictyophyllum*, *Clathropteris* and *Hausmannia* of Dipteridaceae, as well as *Coniopteris* and *Eboracia* of Dicksoniaceae. In particular, we have investigated fossil spore ultrastructurs of several exceptionally well preserved fern sporangia, including *Marattia asiatica* (Marattiaceae), *Oligocarpia kepingensis* (Gleicheniaceae) and *Dictyophyllum nilssonii* (Dipteridaceae) from Late Permian of Xinjiang and Early Jurassic of western Hubei using both SEM and TEM approaches. These results demonstrate that fossil and living spore ultrastructures are a stable character that endured a long history, providing unequivocal evidence for the evolutionary relationships of fern lineages. The anatomy of fossil osmundaceous rhizomes have been studied from the Jurassic deposits in Northern Hebei and western Liaoning provinces. There are about 20 species ascribed to two genera such as *Ashicaulis* and *Millercaulis* in these localities, including about 12 new species. The development and phylogenetic evolution of these plants are analyzed based on both morphology and anatomical features. Their forming palaeonenvironmental background are discussed based on sedimentary and palaeoecological analyses of the Middle Jurassic deposits in western Liaoning. The fossil record, diversity variation and tempo-spatial distribution pattern of some fern groups, including three families, such as Marattiaceae, Gleicheniaceae, Dipteridaceae and Osmundaceae are investigated with emphasizes on evolutionary implidations.

T2-06-04

Re-evaluation on the systematic position of the Mesozoic fern genus *Coniopteris*

Xin-Yuan Miao[1], *Li-Bing Zhang*[2], *Chun-Xiang Li*[1]
1. Nanjing Institute of Geology and Palaeonotology, Chinese Academy of Sciences
2. Missouri Botanical Garden

The extinct fern genus *Coniopteris* was a typical component in the Mesozoic flora with a wide distribution in both Northern and Southern Hemispheres from Early Jurassic to Early Cretaceous, and played a very important role in indicating strata. At the very beginning of the discovery of *Coniopteris,* it was placed in the family Dicksoniaceae based on its cup-shaped indusia (bivalvate only in *C. venusta*), resembling that of *Thyrsopteris elegans* in Dicksoniaceae. However, based on molecular phylogeny *T. elegans* has been moved from Dicksoniaceae to its own family Thyrsopteridaceae, leaving the relationships of *Coniopteris* more unresolved. In this study, we analyzed all phylogenetic and morphological data to re-evaluate the systematic position of *Coniopteris.* Our major results include: 1) All extant genera in Dicksoniaceae have bivalvate indusia, while species of *Coniopteris* have cup-shaped indusia; 2) Most Dicksoniaceae occur in the tropical zone, different from the temperate habitats we thought for *Coniopteris*; 3) *Coniopteris* is herbaceous, very small and slender, less than 20 cm tall, whereas extant Dicksoniaceae are mostly tree-like ferns; 4) Dicksoniaceae and Thyrsopteridaceae all have oblique annuli, while *Coniopteris* has vertical ones. Based on these differences, *Coniopteris* should not be kept in Dicksoniaceae. Instead we found that *Coniopteris* is most similar to Dennstaedtiaceae in having bowl- or cup-shaped indusia, herbaceous habit, and vertical annulus, and living in temperate habitats). On the other hand, molecular dating indicates that the stem groups of Dicksoniaceae, Thyrsopteridaceae, and Dennstaedtiaceae all originated from Lower Jurassic when *Coniopteris* made its first appearance. Our study leads us to propose that *Coniopteris* may be more closely related with Dennstaedtiaceae (Polypodiales) than with Dicksoniaceae.

T2-07: Mimicry in plants

T2-07-01

Key challenges for understanding floral mimicry

Steven Johnson
University of KwaZulu-Natal

Mimicry is considered a classical example of adaptation through natural selection, yet many aspects of the evolution of mimicry are still shrouded in uncertainty. The purpose of this presentation is to identify a broad conceptual framework for mimicry research across the kingdoms of living organisms and then to identify particular challenges that are encountered in the study of floral mimicry. Floral mimics are usually non-rewarding, but rewarding plants can also use mimicry to attract pollinators, and not all non-rewarding plants are mimics. Floral mimics imitate the cues that animals use to locate their food-sources, mating partners and oviposition sites, and thereby exploit the behaviour of these animals for the purposes of pollination. The evolution of mimicry therefore involves complex ecological interactions among multiple species in communities. Mimicry theory was originally developed around ideas of conditioning in signal receivers, but pre-existing sensory bias by pollinators may explain most cases of floral mimicry and also the evolution of imperfect mimicry. Selection for traits that confer mimicry has driven phenotypic diversification in many different plant lineages and this highlights the general importance of community interactions for explaining evolutionary processes and patterns.

T2-07-02

Cryptic coloration in alpine plants: Function and divergence

Yang Niu, *Hang Sun*
Kunming Institute of Botany, Chinese Academy of Sciences

Camouflage is a common anti-predator strategy utilized by animals, which may also occurs in plants, given that they face a world with herbivores. Although there is anecdotal evidence of camouflage in plants, it receives few rigorously experimental studies, possibly because a lack of ideal study system. In the alpine area of Qinghai-Tibet Plateau and the nearby region, several plants from far related taxa exhibit typical cryptic coloration. Studying on a host plant - herbivore system (*Corydalis – Parnassius*) in the alpine region of SW China, we investigated the evolution of camouflage in plants. Take advantage of the leaf color dimorphism (green and gray) in *Corydalis*, we first tested the premise that whether herbivores (*Parnassius* butterflies) can distinguish gray leaves from their scree backgrounds, using a butterfly's color vision model. Then we investigated the adaptive significance of cryptic leaf color under the context of cost-benefit analysis. Potential costs of inconspicuousness were estimated by comparing the photosynthetic performance and visual attractiveness to flower visitors of the two color morphs. Potential benefits were examined by comparing damage, survivorship and female reproductive success. Given that the efficacy of camouflage through background matching depends on the specific visual environment that organism lives, and the background color of plants varies between populations, we tested the hypothesis that *Corydalis* may exhibit population-specific leaf color to get optimal local camouflage in different visual environments. We show that it's difficult for butterfly herbivore to distinguish gray leaves against the background. This cryptic leaf color originates in a specific combination of anthocyanins and chlorophylls. The two color morphs had similar photosynthetic performance, visual attractiveness and female reproductive success. However, gray morphs had significantly lower herbivore damage and higher survivorship. And as we expected, there is a remarkable divergence in leaf colors of *Corydalis*. Leaves are better matched against their native backgrounds than against foreign ones, providing local camouflage. In conclusion, the cryptic leaf color benefit alpine *Corydalis* by reducing herbivores' attack with limited cost. The divergence of cryptic leaf color may have evolved in response to local herbivore selection in various visual environments.

T2-07-03

How flower colours and floral colour patterns trigger pollinator behaviour

Klaus Lunau
Heinrich-Heine-University Dusseldorf

The innate colour preferences of bees reflect the evolution of flower colours and floral colour pattern. To discuss this hypothesis it is argued that the yellow and UV-absorbing colour of pollen and anthers, caused by protective flavonoid pigments, was the primary visual signal of flowers predating the evolution of corollas and other secondarily attractive structures of insect-pollinated angiosperm flowers. Many recent bee-pollinated flowers still visually display pollen, anthers or pollen-mimicking floral guides contrasting against the corolla. Hoverflies and bees actively eat and/or collect pollen grains and respond to visual pollen signals. Naïve hoverflies innately respond to this colour pattern by extending their proboscis exclusively towards yellow and UV-absorbing colours. In contrast, bees target at the yellow and UV-absorbing colours of pollen and pollen-mimicking floral guides for the first physical contact with a flower by means of their antennae preceding landing. Of note, not only the behavioural reaction differs among bees and flies, also the colour parameter triggering the response is different. Hoverflies respond to the colour hue, whereas bees respond to the colour saturation of these signalling structures. Since pollen that is eaten or collected by flower visitors is no longer available for pollen transfer to conspecific flowers bee-and hoverfly-pollinated plants benefit both from hiding pollen from their pollinators and from displaying pollen-mimicking structures replacing signalling real pollen. Bee-pollinated angiosperms that either visually display yellow-UV-absorbing pollen or anthers or pollen-mimicking floral guides and their pollinators constitute the most speciose mimicry system of the world. It is discussed that, besides mimicry, sensory exploitation of innate responses and signal standardization are important features of the most common floral signalling system. In addition it will be demonstrated that even hummingbird-pollinated flowers are affected by this signalling system in order to avoid the attraction of nectar- and pollen-robbing bees. In this respect it is important to state, that even experienced bees and hoverflies use their innate preferences when visiting flowers. Hoverflies cannot be conditioned to exhibit the proboscis reaction towards other than UV-absorbing yellow colours. Bees can easily be trained to colours of slightly deviant wavelength, but continue to prefer more saturated colour over trained colours in learning experiments.

T2-07-04

When flowers play dead: Linking the flower microbiome to brood-site deceptive pollination in the common pawpaw *Asimina triloba*

Kyle Martin, *Spencer Debenport, Daniel Buckley, Robert Raguso*
Cornell University

Nearly all eukaryotes form close relationships and interactions with microbes that are essential for their fitness. The presence of microbes on and within plant tissues is ubiquitous and accordingly, the phenotype of most plants is a combination of both plant and microbial-expressed genes. In this respect, the plant-microbe entity can be considered a holobiontic organism expressing plant derived traits interspersed with microbial derived traits that can be overlooked due to the limits and biases of human perception and investigation. In particular, some aspects of floral mimicry, including brood-site deceptive floral scents that mimic fermenting fruit, carrion, or feces, show the chemical signatures of microbial metabolism but have not been investigated to determine whether they are of plant or microbial origin. A particularly striking example is the common pawpaw *Asimina triloba* (Annonaceae), a small understory tree native to the deciduous forests of eastern North America. The clonal patches of trees produce an early spring bloom of small (3cm), maroon colored flowers with a floral scent that is strongly reminiscent of fermenting fruit or bread dough. Not surprisingly, the floral scent of *Asimina triloba* is dominated by aliphatic fermentation products including ethanol, ethyl acetate, 3-methyl-1 butanol, acetoin, and acetic acid and consequently attracts sap beetles (Coleoptera; Nitidulidae) and Drosophilid flies that are deceived while searching for oviposition/mating sites. It remains unclear how these floral traits have evolved, or whether the flowers have outsourced this function to third-party agents (microbes) to circumvent the long process of evolution by natural selection. In this study we investigated the microbiome of *Asimina triloba* whole flowers, as well as the change in microbial community during protogynous floral ontogeny and between distinct spatial units of the flower, using both culture independent and culture dependent methods. In addition, we performed manipulative experiments including flower sterilization, ^{13}C tracing, and the exclusion of biotic microbial vectors to test the hypothesis of microbial signaling within the flowers. Our results document unexpected diversity and abundance of yeasts and bacteria living on the flowers, as well as spatio-temporal patterns of microbial community composition during the flowers' short lifespan. Further, our manipulative experiments have revealed candidate microbially produced volatiles and suggest opportunities for further investigation with bioassays that promise to shed light on the potential for microbial-mediated plant-pollinator interactions.

T2-07-05

Floral mimicry of fruit within Annonaceae

Katherine Goodrich
Widener University

Pollination strategies involving floral mimicry of non-floral resources have long fascinated researchers. These pollination strategies are scattered across many angiosperm families, with documented floral mimicry of models including carrion, dung, insects, fungi, and fruit. Floral mimicry of ripe or rotting fruit is infrequently documented in existing literature, yet may represent a common mimicry type in some plant groups. Within the family

Annonaceae, many species release floral scents similar to ripe or rotting fruits and have fleshy, dull-colored petals. These floral features attract beetles thought to be in search of fruit substrates as food or brood sites for oviposition, yet the similarity of olfactory or other cues between putative mimics (flowers) and models (locally-available fruits) has yet to be tested. The North American species *Asimina triloba* and *Asimina parviflora* (Annonaceae) produce yeasty floral scents similar to fermenting fruit and are known to be pollinated by flies and beetles. I hypothesize that their flowers may mimic locally available fruits. I have analyzed the floral scents of these two species along with volatiles generated by fruit pulp from several species inoculated with *Saccharomyces cerevisiae*, and from fruits of *Asimina triloba* at different stages of ripeness and decay within the understory of a natural habitat in order to distinguish between "general food mimicry" and a match to an ecologically-relevant model. While the odor of fresh fruit pulp for *A. triloba* is dominated by methyl and ethyl esters of hexanoic and octanoic acids, fermentation volatiles including ethyl acetate, ethanol, acetoin, and 2,3-butanediol begin to appear several days after the fruit drops to the ground. Floral scents of *A. triloba* and *A. parviflora* share these fermentation volatiles but lack the esters seen in fresh fruit of *A. triloba*. Floral scent of *A. parviflora* also contains isoamyl acetate, isobutyl acetate, and 2-heptanone, all compounds associated with some ripe fruits but not seen in the fruit odor of *A. triloba*. I use non-metric multidimensional scaling to compare the scent blends from these different sources. The comparison of floral scent chemistry and other floral cues to ecologically-relevant potential models is an important first step to distinguishing Batesian floral mimicry from generalized food mimicry. Furthermore, knowledge of the ecology and behavior of pollinating taxa and the evolutionary history of floral mimics are important to establishing the degree and efficacy of floral mimicry. Future work in these areas is underway for this particular system.

T2-07-06

Inaccurate color discrimination by pollinators promotes evolution of discrete color polymorphism in food-deceptive flowers

Kotaro Kagawa[1], *Gaku Takimoto*[2]
1. The Institute of Statistical Mathematics
2. The University of Tokyo

Many plant species employing food-deceptive pollination strategy show discrete or continuous floral polymorphism within their populations. Previous studies have suggested that negative frequency-dependent selection (NFDS) caused by learning behavior of pollinators is responsible for maintenance of floral polymorphism. However, NFDS alone does not explain why and under what conditions discrete or continuous polymorphism evolves. In this study, we use an evolutionary simulation model to propose that inaccurate discrimination of flower colors by pollinators leads to evolution of discrete flower color polymorphism. Simulation results showed that associative learning based on inaccurate color discrimination by pollinators caused disruptive selection on flower colors where intermediate color variants in the flower color distribution suffered fitness disadvantage. This was because intermediate colors between other common color variants received twofold collateral unpopularity from pollinators who had learnt to avoid these neighboring colors. The degree of inaccuracy determined the number of discrete flower colors that evolved. Our results suggest that animal behavior based on inaccurate discrimination may be a general cause of disruptive selection that promotes discrete trait polymorphism.

T2-08: Cyperaceae I: early diverging lineages

T2-08-01

A new infrageneric classification for the genus *Scleria* (Cyperaceae) based on molecular phylogenetic and morphological data

Kenneth Bauters[1], *Pieter Asselman*[1], *David A. Simpson*[2], *A. Muthama Muasya*[3], *Paul Goetghebeur*[1], ***Isabel Larridon***[2]
1. Ghent University
2. Royal Botanic Gardens Kew
3. University of Cape Town

The genus *Scleria* is placed in the monotypic tribe Sclerieae. Its distribution is primarily pantropical and currently ca. 250 species are recognized. Previously, infrageneric classifications have been proposed but none has been based on molecular phylogenetic evidence and most studies have treated the genus from only one geographical region. DNA sequence data from more than 100 species of *Scleria* from across its distributional range and all infrageneric groups was generated for three molecular markers. Phylogenetic hypotheses were constructed using Bayesian inference and maximum likelihood approaches. Additionally, ancestral states of four morphological characters were reconstructed at important nodes using the program BayesTraits. A new infrageneric classification based on molecular evidence and supported by morphology is presented. *Scleria* is now confirmed as monophyletic and sister to tribe Bisboeckelereae. A subdivision of *Scleria* into four subgenera (*Scleria* subgenus *Browniae*, subgenus *Hypoporum*, subgenus *Scleria* and subgenus *Trachylomia*) is supported in our results. Twelve major clades are recovered within *Scleria* subgenus *Scleria*. Clear morphological diagnostic features match these clades. Ancestral state reconstruction indicates that androgynous spikelets, a zoniform hypogynium, a paniculate inflorescence and normal linear leaves are ancestral in *Scleria*. A more detailed molecular study of *Scleria* subgenus *Hypoporum* revealed several new species and resolved some of the more complex synonymy questions in this group. Species from this subgenus have few identifying morphological characters, and this led to many erroneous interpretations of species relations in the past. In our study, we clarify those relationships, the synonymy and typification of species belonging to *Scleria* subgenus *Hypoporum*.

T2-08-02

Biogeography and diversification in the megadiverse genus *Carex* (Cyperaceae)

Pedro Jiménez-Mejías[2,4], *Marlene Hahn*[1], ***Santiago Martín-Bravo***[2,3], *Tamara Villaverde*[1,3,5], *Modesto Luceño*[3], *Marcia J. Waterway*[6], *Eric H. Roalson*[2], *Andrew L. Hipp*[1]
1. Morton Arboretum
2. Washington State University
3. Universidad Pablo de Olavide

4. *New York Botanical Garden*
5. *Real Jardin Botanico, Madrid*
6. *McGill University*

Carex, with about 2,000 species is one of the largest angiosperm genera in the World, the second largest in the temperate zones, and the largest among monocots. It has a cosmopolitan distribution, with high species richness in temperate and cold areas of the northern hemispheres and a presumed origin in SE Asia. *Carex* may be found in a wide range of habitats, from Arctic swamps to semi-desert areas, and is sometimes ecologically important. As a result of a 4-years project we have compiled a 3-markers mega-phylogeny of *Carex* based on an extensive sampling of about 70% of the species diversity of the genus and representing all the main lineages. This increased amount of data has allowed us to perform the first studies on the evolution of the genus as a whole. Using a time-calibrated tree, with several *Carex* fossils as calibration points, we performed biogeographic and diversification analyses on a spatial and temporal framework. We performed an ancestral area reconstruction to test the hypotheses of the different centers of origin and to assess colonization, dispersal and extinction processes underlying possible range shifts. We also studied the patterns of diversification rates between different lineages, looking for shifts in speciation rates and its possible correlation with different morphological and biogeographic traits. Overall, our objective was to investigate the different evolutionary processes that have shaped the present day distribution of the genus, its impressive species diversity, and the contrasting patterns in species richness between regions.

T2-08-03

Tribe Schoeneae (Cyperaceae): The taxonomic iceberg of sedges

Russell L. Barrett[1], ***Jeremy J. Bruhl***[2], *Paul Goetghebeur*[3], *Isabel Larridon*[4], *Kenneth R. Mills*[2], *Muthama Muasya*[5], *George T. Plunkett*[2], *Karen L. Wilson*[6]
1. *Australian Biological Resources Study*
2. *University of New England*
3. *Ghent University Research Group Spermatophytes & Botanical Garden*
4. *Royal Botanic Gardens, Kew*
5. *University of Cape Town*
6. *National Herbarium of New South Wales*

The tribe Schoeneae is not monophyletic, yet even in a strict sense, it will include more genera than other tribes in the Cyperaceae. We present a synthesis of recent phylogenetic studies and an outline of proposed taxonomic changes that are needed at the tribal and generic levels. We use several genera including *Costularia*, *Lepidosperma* and *Tetraria* to demonstrate recent findings and continuing challenges to understanding the systematics, biology (including Honey Bees collecting resins from sedges) and ecology (including myrmecochory) of these fascinating plants.

T2-08-04

The *Schoenus* clade of the tribe Schoeneae (Cyperaceae)—generic circumscriptions, taxonomic complexity and polyploidy

Tammy Elliott[1], *Petr Bures*[2], *A. Muthama Muasya*[1]
1. *Department of Biological Sciences, University of Cape Town*
2. *Department of Botany and Zoology, Masaryk University*

The tribe Schoeneae is unique among Cyperaceae, as it is predominantly restricted to nutrient-poor sandstone habitats in the temperate continents of the southern hemisphere. Recent evidence suggests that the tribe originated in Australia during the Paleogene and dispersed to southern Africa in five independent events. Two of these lineages are currently included in *Tetraria* P. Beauv., which has an estimated 46 of its 54 species in the Cape Floristic Region (CFR) of South Africa and is considered a 'Cape floral clade' because of its evolutionary history. Molecular and morphological evidence suggests this morphologically-complex genus is polyphyletic with two distinct clades, one of which includes *Epischoenus* C.B. Clarke—a small genus of eight species endemic to the CFR. The clade of *Tetraria* that includes *Epischoenus* is currently referred to as the *Schoenus* clade of *Tetraria*, since recent phylogenetic evidences places the generic type of *Schoenus* (*Schoenus nigricans* L.) within the clade. Relatively little taxonomic attention has been given to the *Schoenus* clade of *Tetraria* compared to other 'Cape floral clades', most likely because of character similarities in taxa, possible hybridization and a paucity of collections, which has led to confusion in species boundaries. In addition, preliminary data indicate that genome sizes within the *Schoenus* clade are substantially larger than the genomes of other Schoeneae species, suggesting that polyploidy and possibly hybridization could play an important role in speciation within the clade. We will present results from recent taxonomic work within the *Schoenus* clade of *Tetraria*, including a taxonomic realignment of the genus, changes in species boundaries and the description of several new species. In addition, we will discuss diversification in the *Schoenus* clade in the context of speciation patterns among the Cyperaceae and Cape lineages and present future research questions related to genomes sizes and evolutionary relationships within the clade.

T2-08-05

Phylogenomics of Cyperaceae subfam. Mapanioideae

Christina J. Prychid[1], *Felix Forest*[1], *David A. Simpson*[1], *Shu-Biao Wu*[2], *Isabel Larridon*[1], *Jeremy J. Bruhl*[2]
1. *Royal Botanic Gardens, Kew*
2. *University of New England, Armidale*

Previous phylogenetic studies, combining *trnL-F*, *rps16* and pollen data, more-or-less resolved the Mapanioideae into two distinct well supported lineages, the Hypolytreae and the Chrysitricheae, providing a backbone for further taxonomic reorganisation and evolutionary study of this group. Although bootstrap values were mostly high, *Hypolytrum*, *Mapania* and *Scirpodendron* were not recovered as monophyletic. In the present study fresh, silica dried and herbarium samples, including collections from New Caledonia, have been used to expand on the existing analyses. *Hypolytrum* in particular was densely surveyed, with 70 additional samples. This has resulted in a unique mapaniid dataset of 140 terminals, consisting of sequences for six plastid and ribosomal gene regions (*trnL-F*, *rps16*, *rbcL*, *matK*, ETS, ITS), yielding a highly resolved dated phylogenetic tree. The talk provides a summary of the molecular data, as well as comparative morphological and evolutionary analyses towards the goal of unravelling the mechanisms involved in the emergence and evolution of mapaniids.

T2-08-06

Geographical and ecological drivers of diversification in the Mediterranean basin: The evolutionary history of the *Carex panormitana-C. reuteriana* complex (Cyperaceae)

Santiago Martín-Bravo[1], *Carmen Benítez-Benítez*[1], *Marcial Escudero*[2], *Pedro Jiménez-Mejías*[3]
1. Universidad Pablo de Olavide
2. Universidad de Sevilla
3. New York Botanical Garden

The Mediterranean basin is one of the most important hotspots of plant biodiversity, characterised by the high species richness and endemicity (ca. 50%) of its flora. *Carex*, with ca. 150 species, is one of the richest genera in this region. Within the mainly circumboreal *Carex* section *Phacocystis*, *C. reuteriana* and *C. panormitana* are two sister, disjunctly distributed species inhabiting margins of Mediterranean creeks and brooks. *Carex reuteriana* is endemic to the Iberian Peninsula and North West Africa, while *C. panormitana* is found in Sardinia and in two single isolated populations in Sicily and Tunisia. A molecular study based on DNA sequencing and fingerprinting data, with population sampling covering both species' range, as well as bioclimatic niche modeling, were carried out to elucidate the geographical and ecological drivers of diversification in this group. Phylogenetic and divergence time analyses revealed that the two species form a monophyletic group originated in the late Pliocene-early Pleistocene. A clear genetic differentiation was detected between both species, suggesting long term reproductive isolation. Moreover, the populations of each taxon formed genetic clusters according to disjunct areas, indicating restricted gene flow among them and the role of geographic barriers in shaping the observed genetic variation. Bioclimatic modeling detected an important niche overlapping of between the two species, although also certain degree of ecological differentiation. The projection to the Pliocene showed that Central Europe was a suitable area for *C. panormitana-C. reuteriana*, which was validated by the occurrence of fossils assignable to this group in this area. This suggests bioclimatic niche conservation through time rather than adaptation to new habitats as the ecological strategy of these plants. In conclusion, allopatric divergence and niche conservatism seem to be the main evolutionary features of the *C. panormitana-C. reuteriana* group. Thus, the observed phylogeographic structure appears to be consequence of the splitting of a larger range into allopatric sets of populations established in ecologically suitable areas.

T2-09: Electronic Floras–integrating the best of the past with the possibilities of the future

T2-09-01

The New Zealand eFlora - delivering what's needed in the electronic age to the people of New Zealand

Peter De Lange
Department of Conservation

The Flora of New Zealand has been traditionally served to the tax payer through a series of technical hard copy publications. The most recent of these, the New Zealand Flora Series, was initiated in the 1940s with the first volume published in 1961 and the last (as a second edition) in 2010. Although excellent for their time they served as static accounts, often dated when they went to print, and because of this sometimes causing confusion, or in the case of biosecurity matters a false impression of priorities. The move to an eFlora model for New Zealand has been relatively recent, the first treatment - The Hypericaceae was published in 2010. Since then 49 treatments have been published online covering 30 Moss, 16 Pteridophyte and nine flowering plant families (values as of 9 November 2016). In this presentation I explore the eFlora of New Zealand as seen by a major end user - the New Zealand Department of Conservation.

T2-09-02

eFlora of New Zealand – integrating the best of the past with the possibilities of the future

Rob Smissen, *Aaron Wilton, Ilse Breitwieser*
Allan Herbarium, Landcare Research

Floras are an authoritative summation of taxonomic knowledge for a given region at a given time. Traditionally Floras are books, but because of the considerable time and resources required to compile and produce a Flora, these books are often out of date as soon as the Flora is printed. Developments in biodiversity informatics make it now possible to use different approaches to create Floras, and worldwide there are numerous "next-generation Flora" projects underway. These range from digitisation of existing Floras to development of novel information systems that replace or extend traditional Floras. Such systems variously include tools to address different parts of the Flora writing process from accumulating and synthesising data through to generating products to meet specific audience needs. These new approaches substantially change the way in which we can both construct and manage Floras, and make our knowledge widely available and accessed. It may also, at least potentially, substantially change the way we do our work, and think about our information, its management, and the ways in which we can help our users. This talk is an introduction to the symposium on electronic Floras that will explore some of the implications of the eFlora platform for taxonomy, systematics, and the world by using the eFlora of New Zealand as an example. We have started to deliver the next-generation of New Zealand Floras as a dynamic resource: a resource that is both up-to-date and engages a wide range of users. We will present our experiences in developing this next-generation of New Zealand Floras, our planning processes, our work processes, some of the difficulties we encountered, the lessons we learned, what we have achieved so far, and some of our plans.

T2-09-03

Building the Australasian e-flora collaboratively - It's more than science and technology

Judy West[1], ***Zoe Knapp***[2]
1. Australian Government Department of the Environment and Energy
2. Australian Biological Resources Study

Creating descriptive content for a flora, whether it is a book series or a web based e-flora platform, follows a familiar pathway. The Australian Biological Resources Study has been working with Australian botanists to develop standardised taxonomic treatments within its publications for over forty years. The shift of the *Flora of Australia* from a hard-copy book series to a truly interactive e-flora has not only involved the careful transfer, review and update of existing floral profiles, it has also seen the need to take a much more challenging journey. This presentation will outline some of the key governance considerations and cultural changes that need to be taken into account when building a national or regional scale e-flora. The *Australasian e-flora* platform, built by the Atlas of Living Australia and project managed by the ABRS and the Council of Heads of Australasian Herbaria can, after addressing many technical challenges, effectively accommodate a continental scale flora. As significant as this technical achievement has been, project partners have also needed to work closely with the Australian and New Zealand botanical communities - to build a consensus view of accepted taxonomic concepts for all forms of the kingdom Plantae - as part of the developing [Australian] National Species List and the established New Zealand Organisms Register. Developing accepted taxonomies across multiple-jurisdictions has proven to be one of the more challenging but critically important elements in building a regional e-flora, its taxonomic hierarchy and related diagnostic keys. The mix of science, technology and ability for people to work together are all important factors in successfully creating an e-flora.

T2-09-04

VicFlora: Turning hardcopy digital

David Cantrill

Royal Botanic Gardens Victoria

VicFlora is a innovative digital system that presents up to date knowledge on the Flora of Victoria (https://vicflora.rbg.vic.gov.au/). The system was developed through the ingestion of a hardcopy (printed) flora and then has been designed so that species profiles can be easily added (or deleted) including adding taxa to the identification tools (keys). Distribution information is not only generated from MEL herbarium records but harvested from web resources. Checklists for regions can be generated by users is geographic polygons that are overlaid on the distribution data. This talk focuses on the key lessons learnt in developing a digital flora from hardcopy.

T2-09-06

Thinking about the e- in e-Floras; or, Floras old, new, and not-yet

Kevin Thiele

University of Western Australia

When faced with a disruptive new technology such as the World Wide Web, it often takes a lot of effort, time and creativity, to escape conventional ways of thinking through a problem and to fully realise the implications of the new. So-called e-Floras are no exception. How much does the e- in the name really reflect a new way of approaching a Flora's core task (to communicate authoritative, synthesised taxonomic knowledge to as wide an audience as possible in the best way achievable)? If modern taxonomy and systematics were invented, or re-invented, now (in the age of the internet, social media, citizen science and the block chain), rather than in the 18th Century, would we do it all differently? And, what near-future disruptive technologies and paradigms do we need to be aware of when thinking about, or planning, an e-Flora project? Taking the new e-Floras of Australia and New Zealand as starting points, this talk will explore current and future thinking about e-Floras, and map out some ideas for, perhaps, the next wave.

T2-10: Polyploidy: evolution, function, and diversification

T2-10-01

Of dups and dinos: Evolution at the K/Pg boundary

Yves Van De Peer[1,2,3], *Rolf Lohaus*[1,2]

1. Ghent University (UGent)

2. VIB

3. University of Pretoria

Fifteen years into sequencing entire plant genomes, more than 30 paleopolyploidy events could be mapped on the tree of flowering plants (and many more when also transcriptome data sets are considered). While some genome duplications are very old and have occurred early in the evolution of dicots and monocots, or even before, others are more recent and seem to have occurred independently in many different plant lineages. Strikingly, a majority of these duplications date somewhere between 55 and 75 million years ago (mya), and thus likely correlate with the K/Pg boundary. If true, this would suggest that plants that had their genome duplicated at that time, had an increased chance to survive the most recent mass extinction event, at 66 mya, which wiped out a majority of plant and animal life, including all non-avian dinosaurs. I will review several processes, both neutral and adaptive, that might explain the establishment of polyploid plants, following the K/Pg mass extinction.

T2-10-02

Higher frequency of polyploidy associated with the colonization of land

Zheng Li, *Thomas I. Kidder, Chris R. Reardon, Stacy A. Jorgensen, Anthony E. Baniaga, Hannah E. Marx, Xinshuai Qi, Michael S. Barker*

Dept. of Ecology & Evolutionary Biology, University of Arizona

Increased rounds of whole genome duplication (WGD) in the ancestry of the most derived lineages of green plants. Here we provide the first summary of paleopolyploidy across the entire evolutionary history of the viridiplantae. Our analyses of genomic data from more than 1600 species revealed that the incidence of ancient polyploidy increased nearly 500% from the algae to flowering plants. We identified 235 WGDs in the history of the viridiplantae. Nearly 150 WGDs were newly discovered in this study. We also found that nearly every lineage of land plants has experienced at least one round of ancient polyploidy. In contrast, most algae have no evidence of paleopolyploidy in their genomes. Intriguingly, the green algal clade with a land plant like frequency of paleopolyploidy was the Zygnematophyceae. This clade has relatively recently been recognized as the most likely sister lineage to land plants and was confirmed by our present phylogenomic analyses. These results suggests that perhaps the proclivity for polyploidy found among the land plants, and especially the flowering plants, may have its roots deep in the history of the green plant phylogeny. Overall, our study demonstrates that paleopolyploidy has been an increasingly important component of green plant evolution, rising from zero rounds among the algae to more than five rounds of WGD within the flowering plants.

T2-10-03

Subgenome dominance in an interspecific hybrid, synthetic allopolyploid, and a 140 year old naturally established neo-allopolyploid monkeyflower.

Patrick Edger
Michigan State University

The importance and applications of polyploidy have long been recognized, from shaping the evolutionary success of flowering plants to improving agricultural productivity. Recent studies have shown that one of the parental subgenomes in ancient polyploids is generally more dominant - having both retained more genes and being more highly expressed - a phenomenon termed subgenome dominance. How quickly one subgenome dominates within a newly formed polyploid, if immediate or after millions of years, and the genomic features that determine which genome dominates remain poorly understood. To investigate the rate of subgenome dominance emergence, we examined gene expression, gene methylation, and transposable element (TE) methylation in a natural less than 140 year old allopolyploid *(Mimulus peregrinus)*, a resynthesized interspecies triploid hybrid *(M. robertsii)*, a resynthesized allopolyploid *(M. peregrinus)*, and diploid progenitors *(M. guttatus* and *M. luteus)*. We show that subgenome expression dominance occurs instantly following the hybridization of two divergent genomes and that subgenome expression dominance significantly increases over generations. Additionally, CHH methylation levels are significantly reduced in regions near genes and within transposons in the first generation hybrid, intermediate in the resynthesized allopolyploid, and are repatterned differently between the dominant and submissive subgenomes in the natural allopolyploid. Our analyses reveal that the subgenome differences in levels of TE methylation mirror the increase in expression bias observed over the generations following the hybridization. These findings not only provide important insights into genomic and epigenomic shock that occurs following hybridization and polyploid events, but may also contribute to uncovering the mechanistic basis of heterosis and subgenomic dominance.

T2-10-04

The allotetraploid grass *Brachypodium hybridum*: An evolutionary dead-end or a newly emerging lineage?

Pilar Catalan[1], *Bruno Contreras-Moreira*[2], *Ruben Sancho*[1], *Sean P. Gordon*[3], *John Vogel*[3]
1. Universidad de Zaragoza
2. Estación Experimental de Aula Dei-Consejo Superior de Investigaciones Científicas
3. DOE Joint Genome Institute

Hybrid allopolyploids have been alternatively viewed as drivers of biodiversity or evolutionary dead-ends; the 'dead-end' case reflecting either a non-significant contribution to evolution or a higher probability of extinction compared to those of parental diploids. Defendants of the drivers-of-diversity-and-evolution hypothesis argue that common recurrent polyploidization has led to polyploid species showing higher genetic variation from multiple populations of diploid progenitors and highly dynamic genomes generating novel genetic variation. By contrast, oponents interpret polyploids as blind alleys unable to accumulate new variation or to give rise to newly divergent lineages. We compare the two competing hypotheses in *Brachypodium hybridum*, a model species for polyploid speciation in grasses [allotetraploid *B. hybridum* (2n=4x=30) derived from the cross of diploids *B. distachyon* (2n=2x=10) and *B. stacei* (2n=2x=20), followed by WGD]. Comparative genomics between the reference genomes of parental *B. distachyon* (Bd21 line) and *B. stacei* (ABR114) species, and the respective reference subgenomes of *B. hybridum* (ABR113; D=distachyon-type, S=stacei-type) show both a high colinearity between the *B. distachyon/B. hybridum*-D and *B.stacei/ B.hybridum*-S genomes, and a similar syntheny between the two progenitor genomes (*B. distachyon/ B. stacei*) and the two *B. hybridum* subgenomes (D/S), suggesting that few genomic variation has been incorporated into the *B. hybridum* subgenomes since the hybridization took place. However, phylogenomic analysis of a genome-mapped SNP data set (>1800 kbp nDNA) including several samples of *B. hybridum*, *B. distachyon* and *B. stacei* reconstructs distinct evolutionary placements for the differently inherited D and S subgenomes. Phylogenomics of plastome data confirm the bidirectional crossing origin of *B. hybridum*. Barcoding data from nuclear genes and SSR data further demonstrate the existence of recombinant or new allelic variants in *B. hybridum*. Metabolomic and phenotypic data also detect new characters and significant differences in other traits in *B. hybridum* with respect to its progenitors. All these evidences suggest that *B. hybridum* is not an evolutionary dead-end species but a newly emerging lineage. It still maintains relatively intact parental genomes, probably due to the short time elapsed since its estimated first origin (~1 Mya), though it is reproductively isolated and genomically divergent from its parents. *B. hybridum* ocuppies a native circum-Mediterranean distribution area broader than those of its progenitor species and has also successfully colonized other continents. *B. hybridum* might also have potential evolutionary future as other high-ploidy level polyploid *Brachypodium* species have apparently originated from low-ploidy level polyploids.

T2-10-05

Polyploidy and the origins of novelty: Impact of duplication on genomes and network evolution

J. Chris Pires
University of Missouri

Ancient whole genome duplications (WGDs) are ubiquitous throughout the evolutionary history of higher eukaryotic lineages. These events have been hypothesized to be the basis for major evolutionary transitions, including the origin of novel traits in large species radiations across plants, fungi, protozoa, and animals. Repeated rounds of WGDs, or polyploid events, have been best documented among the flowering plants. We have analyzed the impact of WGDs in *Arabidopsis*, *Brassica* and other members of the mustard family (Brassicaceae). Phylogenomic analyses of these WGDs show striking correlations of duplicate gene retention and novel traits. For example, the origin of two novel classes of chemical defenses, indole and met- derived glucosinolates (i.e. mustard oils), are associated with duplicated regulatory and biosynthetic pathways that arose via WGDs. Our analyses suggest that the origin of these novel defense compounds spurred an evolutionary arms-race with insect herbivores, resulting in massive co-radiations of both the host plant and predatory insects. The use of new systems biology and network approaches to discover coevolved modules will be discussed.

T2-10-06

Asymmetrical evolution of polyploid genomes: The role of recombination linking genomic variation to traits in *Brassica napus*

Chaobo Tong[1], *Yemao Liu*[2], *Minqiang Tang*[1], *Yuanyuan Zhang*[1], *Rod Snowdon*[3], *Junyan Huang*[1], *Yanqiu Zhou*[1], *Linli Yang*[1], *Ming Hu*[1], *Qingyong Yang*[2], ***Shengyi Liu***[1]

1. Oil Crops Research Institute of CAAS

2. Huazhong Agricultural University

3. Justus Liebig University

The process of genome polyploidization and subsequent rapid diploidization has provided significant sources of genetic variation for plant adaptive evolution and new species formation. However, the way in which molecular evolution of polyploid genomes builds up genetic architecture underlying speciation is unclear and its impact on trait formation related to duplicate genes is unknown. Based on our previous findings that polyploid subgenomes have undertaken an asymmetrical evolution, we have further been using *Brassica napus* (AACC) as a model to present the role of recombination in the process. First we demonstrated recombination of the subgenome A is dominant over C by using diverse segregation populations. Further we re-sequenced 420 core germplasm collections and confirmed asymmetrical recombination between A and C. Recombination links to selection and further link DNA and epigenetic variation to traits. By analysis of SNP and DNA deletion and their association with traits, we showed that there are high correlations between recombination rate and either of abundance of SNP, DNA deletion and QTLs. By genome-wide analysis of gene expression, small RNA and DNA methylation in representative collections, we showed that these three genomic parameters are highly correlated (positively or negatively) with recombination rate and all other parameters of genome characteristics mentioned above. The above relationships also exist in many blocks of meso-subgenomes (each A or C contains three subgenomes originated from meso-polyploidization ca.15MYA). These results suggested a mode of molecular evolution of plant polyploid genomes.

T2-11: Next-generation botany: Outcrossing evidence, integrating knowledge

T2-11-01

Integrating morphology, molecules, and fossil flowers: First results from the Paleo-eFLOWER initiative

Jürg Schönenberger[1], *Maria Von Balthazar*[1], *Susana Magallón*[2], *Béatrice Albert*[3], *Charlotte Prieu*[3], *Hervé Sauquet*[3]

1. Department of Botany and Biodiversity Research, University of Vienna

2. Instituto de Biología, Universidad Nacional Autónoma de México

3. Laboratoire Écologie, Systématique, Évolution, Université Paris-Sud

The fossil record of angiosperms has increased substantially both in size and quality in recent decades. This is mainly due to the discovery of numerous Cretaceous localities yielding 3-dimensionally preserved, charcoalified mesofossils comprising a great diversity of angiosperm reproductive structures. In addition, novel technical applications such as high resolution computed tomography have recently boosted our abilities to retrieve high-quality structural data from fossil specimens. This rapid increase in knowledge of the fossil record has provided important information about the antiquity of different floral morphologies and of different phylogenetic lineages of angiosperms. However, what has not kept pace with these recent developments, are our abilities to test the systematic placement of fossils based on explicit phylogenetic analyses across the tree of angiosperms, mainly due to the lack of adequate morphological data sets for extant taxa. Here, we propose a new approach, centered on a multi-user database (PROTEUS) and an angiosperm-wide floral data set for 792 extant species (exemplar approach) representing all orders and 367 families (87%) of angiosperms. In the framework of the eFLOWER initiative, we have recently used this data set to infer ancestral flowers across the angiosperm tree. For the present study, we have supplemented the original data set with additional floral and pollen traits, now adding up to a total of 30 characters. To demonstrate the potential of the present paleo-eFLOWER initiative, we scored 30 published Cretaceous flowers for the same set of characters. Our main long term goals with the inclusion of fossil taxa into our data set are: (i) to develop an online tool for the paleobotanical community to systematically test the phylogenetic position of fossil flowers in an angiosperm-wide context, (ii) to test the impact of fossils on phylogenetic relationships (of extant taxa), and (iii) to incorporate the fossil record into the reconstruction of trait evolution. For this presentation, we focus on our first goal. We use a molecular backbone approach and apply parsimony as well as model-based analyses to evaluate the phylogenetic position of selected fossils. We find that some are placed congruently with earlier suggestions, while others are found in various equivocal positions, including the stem branches of very large clades. Possible reasons for equivocal results include the low number of characters for some fossils, or the lack of distinct combinations of apomorphic characters for others as well as the still limited sampling of extant species in our current data set. We also report on various sensitivity analyses (e.g., random resampling of taxa), which inform us about the consistency of our approach. Although further work is needed, including total evidence analyses, these first results are promising. While such angiosperm-wide analyses will in many cases only provide a coarse estimate of the phylogenetic position of a fossil taxon and will need to be followed by finer-scale analyses focusing on a particular clade with more and different characters, we envision to develop future automated 'phylo-scanning' community tools so that such broad-scale analyses become commonplace for the initial, exploratory testing of angiosperm fossil relationships.

T2-11-02

Fitting models to the real world: The use of external validation of spatial and temporal parameters in phylogeographic studies on oceanic islands

Javier Fuertes-Aguilar[1], *Alejandro González Fernandez De Castro*[1], *Juan Carlos Moreno Saiz*[2]
1. Real Jardín Botánico, CSIC
2. Universidad Autónoma de Madrid

The accumulation of phylogeographic studies on certain geographic regions has stirred the interest for linking the estimated genetic patterns to the species' natural history traits, its interactions with other species within ecological networks, or the species evolutionary history. Trying to fit fine-scale phylogeographic studies and associated genetic data in plants to historical events and the dynamic structure of ecosystems involves a series of methodological hurdles that often makes this task frustrating and leads to inconclusive results if only reliant on genetic data. Oceanic island species offer the advantage of providing a finite spatial framework for ecological and ecological evolutionary studies. We present a study on the bird-pollinated Macaronesian paleoendemic *Navaea phoenicea* (Malvaceae), a monotypic genus exclusive to Tenerife (Canary Islands) with a disjunct distribution in two geographically isolated areas. To understand the origin of this phylogeographic pattern, we estimated population genetic parameters and compared them with the geological history of the island and present biological data. The latter were used as external validators of evolutionary traits. First, we examined the influence of the complex geological history of Tenerife in the disagreement/discordance between the potential *versus* the realized niche of the species. In plants, the bioclimatic niche is an important predictor of the species spatial distribution. We constructed ecological niche models (ENMs) under high-resolution bioclimatic layers to examine whether the species potential niche corresponds to the realized niche, and how present populations distribute across them. The gaps observed in the potential ENM prediction corresponded to areas in the island that have undergone a series of catastrophic events. Second, we used demographic coalescent models (IMa2) to test whether the temporal scale of these events matches the divergence time estimation for the two lineages present at Tenerife. To determine empirically the generation times for the species (a critical parameter in coalescence models), we performed a demographic study of the natural populations. Using these generation times in IMa2, we obtained absolute time estimates for the genetic isolation between populations, which compared with the geological dating of the large-scale landslides and eruptions in Tenerife helped to set a temporal window for the influence of these events in the observed disjunct pattern. Finally, we carried out a third comparative analysis to try to explain the role of pollinators in the present genetic structure of populations. Estimates of intra and inter-population paternal gene flow obtained from nuclear hypervariable genetic markers were compared with empirical values of maximum distance in pollen dispersal recorded for three bird species, which are here confirmed as natural pollinators. Interestingly, the observed spatial genetic structure matched the one expected from flower visitation data by present pollinator. A second validation test via paternity analysis of plant individuals confirmed these results and identified the limits of current inter-population gene flow. In sum, our results showcase the potential of using multiple sources of information as external validators in phylogeographic studies, such as species natural history traits (reproductive biology), demographic studies, and a region's geological history to obtain reliable inferences on species evolutionary responses and distribution patterns.

T2-11-03

Bayesian biogeographic models and their application in macroevolution and macroecology: Integration and partitioning of multiple sources of evidence

Isabel Sanmartin
Real Jardin Botanico, CSIC

Bayesian hierarchical biogeographic (BHB) models allow joint estimation of the posterior distribution of tree topology, divergence times, and ancestral ranges given molecular data and the geographic location of sequences. Because of their relatively simple underlying biogeographic model, based on a continuous-time Markov chain (CTMC) process with single discrete states, and the use of Bayesian inference to ease computational tractability, these models have become very popular in phylogeography to answer a wide range of questions, from the transmission of diseases to routes of viral spread or historical patterns of gene flow across populations. In 2008, we first proposed the use of these models in biogeography to estimate dispersal rates and area carrying capacities (equilibrium frequencies) from DNA sequences and species geographical distributions. Here, I focus on the use of these models to test general hypotheses in ecology and evolution by using datasets of multiple lineages that inhabit the same geographic region. An advantage of BHB models is their ability to integrate external sources of evidence (geological connectivity, paleoclimatic data, fossil record, species biological traits..) in the form of new variables, scaling factors, or prior distributions informing the posterior estimates of parameters. The use of a Bayesian hierarchical approach allows estimation of global biogeographic parameters over multiple clades – thus decreasing the uncertainty associated to a single character set – while integrating over clade-specific biological traits such as the rate of molecular evolution, age, or ability to disperse, as nuisance parameters. These models have been used to partition the contribution of different abiotic (geographic distance, area) and biotic factors (habitat specialization) in migration rates and area carrying capacities in animals versus plant lineages. The original CTMC process implemented in BHB was a time-homogenous model, assuming constancy of rates over time. I describe here extensions of the BHB model to relax the time homogeneity of the Markov process by allowing dispersal rates to vary over time and to test for the existence of temporal barriers to colonization. Non-equilibrium models can be employed to detect the signal of mass extinction events, evidenced as a sudden decrease in area carrying capacities, due to past catastrophic (geologic, climatic) events that partly wipe out the biota of a region. By allowing the intensity of these events to vary depending on clade-specific traits, these models could be used to "predict" the fate of lineages, for example, did some island guilds react more severely to catastrophic events (environmental perturbation) than others? Further work focuses on: a) extending these models to their use with phylogenomic "big" data (hundreds of sequences and genes) and to account for intraspecific processes (gene flow, incomplete lineage sorting, introgression); b) modeling the interaction among species (competition, facilitation) and among processes (trait- and geographic-dependent models). I will illustrate the use of these models with empirical studies on island and continental floras.

T2-11-04

Recent advances in biogeography: Integrating genomic and ecological data

Elena Conti[1], *Spyros Theodoridis*[1], *Gabriele Casazza*[2], *Florian Boucher*[1,3]

1. University of Zurich

2. University of Genova

3. Stellenbosch University

Some of the key questions that have fascinated naturalists and evolutionary biologists ever since Darwin and Wallace include: Why do species occur where they do? How did species attain their current distribution? Why do some species have very broad ranges of distribution, and others very narrow ones? How do ecological preferences, geologic/climatic events, and evolutionary processes shape species distributions? What are the relative roles of geographic barriers and adaptation in driving speciation? The recent advent of new methodological and analytical approaches opens unprecedented opportunities to gain novel insights into these classic biogeographic questions. I will illustrate how the integration of high-resolution molecular, distributional, geologic and climatic data with powerful analytical tools in phylogenomics and population genomics, ancestral area inference, species distribution modeling, and demographic modeling enabled us to answer long-standing questions on the biogeography of alpine, Mediterranean, and island plants, with emphasis on the first. By employing the integrative approach described above, we tested two competing hypotheses on the distributional responses of cold-adapted, alpine plants to past climatic oscillations. Specifically, we asked whether *Primula farinosa* contracted or expanded its range during interglacials of the Late Quaternary. At variance with prevailing notions that explain the currently fragmented distribution of cold-adapted species in terms of postglacial contraction, we found that *P. farinosa* expanded its range during interglacials. Using a combination of high-throughput genomic data and forecasting models of species distribution, we predict that lineages within the same species harboring sufficient ecological and genomic diversity are resilient to climate warming, while less diverse infra-specific lineages are threatened with extinction. Comparisons of species distribution models also allowed us to test the hypothesis that the ecological preferences of newly arisen polyploids should be both distinct and broader than those of their diploid relatives. As expected, we discovered that the ecological niches of species in *Primula* sect. *Aleuritia* at different ploidy levels differ significantly in both climatic and geographic space. However, contrary to expectations, we found that the ecological niches of polyploid species are narrower than those of their diploid relatives. The integration of ecological niche modeling with quantitative analyses of floral traits allowed us to ask whether niche and/or floral shifts favor the establishment and persistence of newly arisen polyploid populations within *Primula marginata*. In this case, we found that neither the ecological niches nor the floral traits of populations at different ploidy levels within the same species differ significantly, hence the disjunct distribution of dodecaploids in *P. marginata* is probably maintained by dispersal limitation and minority-cytotype exclusion. Expanding the focus of our analyses to the entire family Primulaceae, we used sister-species comparisons of distributional ranges and ecological preferences to investigate the relative roles of geographic vs. ecological divergence as major drivers of speciation. Our results confirmed predominant models of allopatric speciation as a main engine of biodiversity in plant evolution, especially in alpine/arctic environments.

T2-11-05

Estimating divergence times and ancestral breeding systems in *Ficus* and Moraceae

Qian Zhang[1], *Renske Onstein*[1,2], *Stefan Little*[1], *Hervé Sauquet*[1]

1. Ecologie Systématique Evolution, Univ. Paris-Sud, CNRS, AgroParisTech, Université Paris-Saclay, Orsay, France

2. Institute for Biodiversity and Ecosystems Dynamics (IBED), University of Amsterdam, Sciencepark 904,1098 XH Amsterdam, the Netherlands

Flowers are the highly variable reproductive structures of angiosperms. Although most flowers are bisexual (hermaphrodite), many species are characterized by other breeding systems (including monoecy, dioecy, polygamodioecy). The relationship between breeding system and diversity is still not clear. Dioecy was previously thought to be an evolutionary dead end, while in the latest decade, this perspective has gradually changed. In some clades, dioecy has been found to evolve into other breeding systems frequently and these transitions could be linked to habitat, pollination strategies and other environmental or ecological factors. Family Moraceae (40 genera, ca. 1100 spp.) harbors four different breeding systems: monoecy, androdioecy, gynodioecy and dioecy. *Ficus* (ca. 750 spp.), which is the largest genus in Moraceae, has two breeding systems: monoecy and gynodioecy. Previous studies using parsimony found monoecy and dioecy to be ancestral in *Ficus* and Moraceae, respectively. Here, we reconstruct ancestral breeding systems of *Ficus* and Moraceae by using both parsimony and model-based approaches and we explore the sensitivity of our results to various parameters of the inference (including phylogeny, age, model, and taxon sampling). To do so, we first estimated a new timescale for diversification of *Ficus* and Moraceae by combining a revised set of 12 fossil calibrations and a densely sampled molecular data set of eight loci and 320 species (representing all families of Rosales and 36 out of 40 genera in Moraceae). We estimated the crown group-ages of *Ficus* and Moraceae to 40.8-56.0 Ma and 73.3-84.9 Ma, respectively. Using this new timescale and both maximum likelihood and Bayesian approaches to reconstruct breeding system evolution, we found the ancestral state of *Ficus* to be particularly sensitive to model selection. Gynodioecy was inferred as ancestral in *Ficus* with strong support with the unequal-rate maximum likelihood model, whereas, the equal-rate maximum likelihood model and parsimony supported ancestral monoecy in *Ficus*. The Bayesian approach instead suggests high uncertainty in this reconstruction. Conversely, all approaches consistently supported dioecy as the ancestral state in Moraceae. Our results show that monoecy and gynodioecy evolved from dioecy in Moraceae, although uncertainty remains on whether gynodioecy was an intermediate step in evolution to monoecy in *Ficus*.

T2-11-06

A phylogenetic framework for the carpel development regulation: Mixing and matching old with new

Kai Pfannebecker, ***Annette Becker***
Justus-Liebig-University Gießen

A major evolutionary innovation in the plant lineage is the angiosperm carpel, their unifying character and most complex plant organ, composed of many clearly distinct tissue types to ensure reproductive success. However, the origin of the carpel is unknown, but many components of the gene regulatory network (GRN) governing carpel development and their genetic interactions are described in *Arabidopsis thaliana*. To unravel the evolution of the carpel GRN and to discriminate between "early" and "late" steps in carpel evolution we calculated thorough phylogeny reconstructions based on sequenced genomes such that orthologs of the major *A. thaliana* carpel GRN are now placed in their phylogenetic context. We find that the carpel GRN components are of various ages, and identify especially high retention rates for carpel development genes in Brassicaceae leading to Brassicaceae-specific interactions of carpel GRN members. Further, our data indicate that developmental processes present already in the most recent common ancestor of seed plants, such as reproductive meristem termination or adaxial/abaxial polarity specification requires few interacting transcription factors, which are not retained in duplicates after whole genome duplications (WGD). In contrast, developmental processes associated with derived carpel characters, such as the transmitting tract require larger numbers of interacting transcription factors which were retained as duplicates after WGD.

T2-12: Pattern, function, and evolution of biodiversity across the ancient EA-ENA floristic disjunction (two sessions)

T2-12-01

A genomics analysis of the eastern Asia - eastern North America floristic disjunction

Anthony E Melton[1], *Gane Ks Wong*[2], *Douglas E Soltis*[1], *Pamela S Soltis*[1], *Matthew A Gitzendanner*[1]
1. University of Florida
2. The University of Alberta

Large disjunctions in taxon distributions provide excellent opportunities to study biogeographic patterns and the processes that shape them. One of the most notable disjunctions is that of eastern Asia and eastern North America (EA-ENA). This disjunction is thought to have been formed when a large temperate forest circling the Northern Hemisphere was fragmented due to by global cooling and the loss of important habitat and land bridges during the Miocene, followed by further fragmentation during the Pleistocene. Sixty-five angiosperm genera exhibiting this disjunct distribution have been identified. In many of these genera, species richness is greater in EA than ENA, and this pattern has been attributed to higher rates of molecular evolution and speciation in EA. We used transcriptomes from species in 10 genera to gain new insights into how this disjunction may have affected gene and genome evolution and to evaluate whether rates of molecular evolution are elevated in EA relative to ENA. Single to low-copy nuclear genes were identified using MarkerMiner and classified into GO categories. We have focused on seven genera (*Campsis, Cotinus, Gelsemium, Nelumbo, Penthorum, Phrymus,* and *Saururus*) for which two species exhibit the disjunction, and three genera with three or more taxa (*Calycanthus*, *Cornus*, and *Hamamelis*). If rates of molecular evolution are indeed elevated in EA, then genetic divergence from the common ancestor of EA-ENA species to EA species would be greater than that to ENA species, leading to greater species richness in EA, as identified previously in numerous genera. We determined rates of evolution for many genes as a means of testing this hypothesis and determined which GO categories of genes have experienced strong natural selection. No statistically significant differences were identified between EA and ENA species pairs, suggesting equal rates of evolution for both species within pairs, despite long isolation. For larger genera, no pattern suggesting EA taxa evolve at faster rates was identified. Our results suggest that many genes across multiple GO categories are under strong purifying selection, with few exceptions, and indicate that greater species richness in EA may be due to factors other than rates of molecular evolution.

T2-12-02

Exploring intercontinental disjunctions in a rare complex system of sumac-gall aphids, host plants and symbiotic bacteria

Zhumei Ren[1], *Rebecca Dikow*[2], *Aj Harris*[3], *Tingshuang Yi*[4], *Yang Zhong*[5], *Jun Wen*[6]
1. Shanxi University
2. Smithsonian Institution
3. Smithsonian Institution
4. Kunming Institute of Botany, Chinese Academy of Sciences
5. Fudan University
6. Smithsonian Institution

We investigated the co-evolutionary and biogeographic relationships among three associated taxa representing three kingdoms of life: sumac-gall aphids, sumac plants, and bacteria. Sumac-gall aphids (Aphididae: Eriosomatinae: Fordini) live obligately on species of sumac plants (Anacardiaceae: *Rhus* subgenus *Rhus*) for part of their life cycle, and show the classical eastern Asian (EA, 5 genera and 11 species) – eastern North American (ENA, 1 monotypic genus) biogeographic disjunction, a common pattern in many plant groups. While sumac-gall aphids are restricted to *Rhus* hosts in EA and ENA, *Rhus* subgenus *Rhus* has a slightly broader intercontinental disjunction, with two species extending to southern Europe and Hawaii. Furthermore, the sumac-gall aphids depend on the highly variable endosymbiont *Buchnera aphidicola* (Gammaproteobacteria: Enterobacteriales), a bacterium, which supplies the animal host with certain essential amino acids while the endosymbionts receive nutrients from the aphids feeding on the host plant's sap. To explore the diversification of this unique complex co-evolutionary system of sumac aphids, sumac summer hosts and the aphid's endosymbiont *Buchnera* across their geographic range and through time, we obtained complete mitochondrial genome sequences for the sumac-gall aphids and complete genome sequences of *Buchnera* bacteria via the shotgun genome skimming of total aphid genomic DNA. For the host plants *Rhus*, we utilized seven plastid and two nuclear markers. We conducted extensive phylogenetic analyses for all three major associated lineages, inferred their biogeoraphic range evolution, and dated the co-evolution and major disjunctions. We tested the hypotheses that (1) such an aphid/sumac host plant relationship dates back to the early Eocene; (2) the coevolution between the aphids and the sumacs has driven species diversification in sumac aphids; and (3) there is biogeographic parallel evolution of the aphids and the endosymbiont *Buchnera aphidicola*.

T2-12-03

Species delimitation, phylogenetics and biogeography of *Stewartia* s.l. (Theaceae) based on plastomic and morphological evidence

Hanyang Lin[1,2], *Pingping Yin*[1,2], *Yunpeng Zhao*[1,2], *Chengxin Fu*[1,2]
1. The Key Laboratory of Conservation Biology for Endangered Wildlife of the Ministry of Education, College of Life Sciences, Zhejiang University
2. Laboratory of Systematic & Evolutionary Botany and Biodiversity, Institute of Ecology and Conservation Center for Gene Resources of Endangered Wildlife, Zhejiang University

Stewartineae (Airy-Shaw) Chang is the basal-most clade of Theaceae, and comprises *Stewartia s.l.* which includes *Stewartia* L. and *Hartia* Dunn. The former genus exhibits typical eastern Asian and North American disjunction, while the latter ranges from Southern China to Indo-China. The circumscription of the two genera and delimitation of some taxa had been controversial and the phylogenetic relationship among the clades within *Stewartia s.l.* remained unclear. To examine species delimitation, phylogenetics and biogeography of *Stewartia s.l.*, we collected complete plastid genome sequences for 19 of the 24 taxa and morphological traits from both specimens and field observations. Phylogenetic reconstruction suggested that the evergreen *Haria* Dunn was monophyletic and nested in the deciduous *Stewartia* L. *S. malacodendron* L. represented the earliest diverged species. Hartia was sister to the clade consisted of the other North American species, *S. ovata* (Cav.) Weath., and the sub-clade composed of Asian deciduous taxa. Thus, both genera were supported to be merged. The phylogenetic relationships of all studied species were robustly resolved. The earliest split within the Asian *Stewartia* clade resulted in speciation of *S. koreana* Nakai ex Rehder. The ancestral distribution reconstruction and dating suggested that *Stewartia s.l.* was originated in eastern North America in the mid-Miocene (14~20 Mya). The evergreen and deciduous taxa in Asia resulted from two independent biogeographic events in 8~15 Mya and 7~13 Mya, respectively. The within-clade divergence was initiated earlier in *Hartia* Dunn than that in Asian *Stewartia* L. (4.7~9.2 Mya vs. 1.4~3.5 Mya). The latter clade experienced rapid speciation events started in the Mid-Pliocene. The speciation of the China-endemic *Stewartia* L. involved three independent events. Morphological analyses lent further supports for the species delimitation. Both molecular and morphological evidence supported the controversial taxa as species, e.g., *S. shensiensis* Hung T. Chang, *S. acutisepala* P. L. Chiu & G. R. Zhong and *S. koreana* Nakai ex Rehder.

T2-12-04

Phylogeny and biogeography of Podophylloideae (Berberidaceae) with an emphasis on the diversification of *Dysosma endemic* to China

Wen-Qing Ye, *Yi-Han Wang, Rui Guo, Yun-Rui Mao, Pan Li, Ying-Xiong Qiu (Corresponding Author)*
College of Life Sciences, Zhejiang University

Podophylloideae Eaton (Berberidaceae) is a small subfamily of 12 species, which contains *Sinopodophyllum* (Royle) Ying (1 sp.), *Dysosma* Woodson (7 spp.), *Podophyllum* L. (1 sp.), and *Diphylleia* Michaux (3 spp.). This subfamily has a disjunct distribution in eastern Asia (EA) and eastern North America (ENA) with ten species in temperate region of EA. Despite great medicinal pharmacogenetic importance, the phylogenetic relationships of the four genera within this subfamily are still controversial due to limited taxon sampling and few genetic markers employed in previous studies. In addition, the monophyly of *Dysosma* has not yet been tested. Here we reconstructed the phylogeny with extensive taxon sampling (including all 12 species and the close allies) using chloroplast genome sequencing and restriction site associated DNA sequencing (RAD-seq) data. We also examined the biogeography and diversification of this subfamily as well as the spatio-temporal processes of divergence within *Dysosma*. *Our* phylogenetic analyses, for the first time, strongly support *the monophyly* of *Dysosma* and a sister-group relationship between *Diphylleia sinensis* from China and *Diphylleia cymosa* (ENA)-*Diphylleia grayi* (Japan). We inferred that the maximum crown age of Podophylloideae is approximately 19 million years ago (Mya) and that the Himalayas is likely to be its ancestral area, with both migration events across Beringian land bridge (BLB) and vicariant events leading to its distribution in ENA. Climatic changes since the Early *Miocene* and the uplift of the Qinghai-Tibetan Plateau (QTP) probably drove the diversification of major lineages of Podophylloideae and speciation of three high-elevation *Dysosma* species from Southwest China and the south-eastern slopes of the QTP. For the *Dysosma versipellis-pleiantha* complex, our phylogeographic results support that the early divergence of *D. versipellis* and *D. pleiantha* proceeded through allo-peripatric speciation, possibly triggered by Early Pleistocene climate change, while subsequent climate-induced cycles of range contractions/expansions enhanced the eco-geographical isolation of both taxa. Furthermore, modelling of population-genetic data indicated that major lineage divergences within *D. versipellis* likely resulted from long-term allopatric population isolation in multiple localized refugia over the last glacial/interglacial periods, and which in turn fostered endemic species formation (*D. difformis*, *D. majoensis*) from within *D. versipellis* in Southwest China.

T2-12-05

Ecological traits and recent population dynamics of eastern Asian-eastern North American disjunct tree species in North America

Jeremy Lichstein[1], *Tao Zhang*[1], *Miao Sun*[1], *Michelle Mack*[2], *Sarah Graves*[1], *Mark Whitten*[1], *Johanna Jantzen*[1], *John Park*[1], *Stephanie*

Bohlman[1], Matthew Gitzendanner[1], Douglas Soltis[1], Pamela Soltis[1]
1. University of Florida
2. Northern Arizona University

The biogeography of plant species with disjunct distributions between eastern Asian (EA) and eastern North American (ENA) suggests evolutionary stasis (because sister taxa on the two continents tend to occur in similar climates) and a tendency towards drought-intolerance (e.g., disappearance from western North America since the late Tertiary due to increased aridity). These observations suggest that the ecological traits and dynamics (e.g., population trends) of EA-ENA disjuncts may be distinct from co-occurring non-disjunct species. We compared the ecological traits and recent population trends of EA-ENA disjunct and non-disjunct tree species in ENA using a database of ecological traits and systematically sampled forest inventory data collected by the U.S. Forest Service in the 1980s and 2000s. Disjuncts tended to be less drought-tolerant and more shade-tolerant than non-disjuncts. A majority of disjunct species were classified as being pyrophobic (negative population response to fire), whereas a majority of non-disjuncts were classified as being pyrophilic (positive population response to fire). These and other ecological/functional traits displayed a moderate degree of phylogenetic conservatism when examined in the context of a phylogeny of ENA tree species. Despite ecological trait differences between disjunct and non-disjunct species, the two groups had similar population trends from the 1980s to 2000s in the north-central, northeastern, and southeastern USA. We suggest that trait differences had counter-acting effects on disjunct and non-disjunct species, so that the two groups had similar population trends despite being ecologically distinct. For example, drying trends over recent decades in the north-central and southeastern USA would be expected to adversely affect drought-intolerant disjunct species, but the expected population declines may have been counter-balanced by fire-suppression, which favors pyrophobic disjuncts over pyrophilic non-disjuncts.

T2-12-06

Macroecology of leaf morphological traits of Chinese woody plants

***Zhiheng Wang**, Yaoqi Li*
College of Urban & Environmental Sciences, Peking University

Diverse leaf morphologies reflect the adaptation of plants to their habitats, and can directly influence carbon uptake and water loss of plants. The relationship between patterns in leaf morphology and climate has long intrigued botanists. Among all leaf morphological traits, leaf size and shape have been widely observed to vary along climatic gradients, but quantitative description of the variations and the underlying mechanisms are still largely lacking. Here, using distribution maps and leaf morphological data (including leaf margin states, leaf length, leaf width, and length-width product/ratio) of 10480 Chinese woody dicots and dated family-level phylogenies, we evaluated the phylogenetic signals of leaf size and shape, and demonstrated the geographical patterns in leaf size and shape and their relationships with climate across different life-forms (evergreen and deciduous; trees, shrubs and lianas) and species with different family ages. Significant phylogenetic signals were found for leaf margin state of families, but not for mean leaf size and shape. Our results indicated that 1) proportions of species with entire leaves in regional floras were significantly correlated with mean winter temperature. However, their relationship was substantially influenced by precipitation regimes, plant life-form and evolutionary history, which suggests that caution is needed in the usage of the widely-used method (leaf margin analysis, LMA) for paleo-temperature reconstruction. 2) Leaf size (including leaf length, leaf width and length-width product) significantly decreased from the tropics and subtropics towards the high latitudes. The geographical variations in leaf size were most strongly correlated with annual actual evapotranspiration (AET), net primary productivity and gross primary productivity (GPP) among all environmental variables tested and therefore could be used to reconstruct paleo-AET and ecosystem productivity.

T2-12-07

Evolutionary history and Environment drive Global Diversity Patterns of *Quercus* L.

***Xiaoting Xu**[1], Zhiheng Wang[2], Carsten Rahbek[3], Jean-Philippe Lessard[4], Jingyun Fang[2]*
1. College of Life Science, Sichuan University
2. College of Urban and Environmental Sciences, Peking University
3. Center for Macroecology, Evolution and Climate, Natural History Museum of Denmark, University of Copenhagen
4. Department of Biology, Concordia Univers

Evolutionary process, such as speciation, extinction and dispersal, can directly change species diversity. Climate could have influence on species diversity through these evolutionary process. Therefore the species diversity-climate relationship may be originated from evolution history. Early studies mainly focused on species diversity-climate relationship and proposed many hypothesis to explain the spatial variation of species diversity. Recent researches on spatial variation of species richness proposed new hypothesis based on evolutionary history, such as evolutionary rate hypothesis, time-for-speciation hypothesis and niche conservatism hypothesis *etc.* However, what drives the species diversity-climate relationship and the link between evolutionary history and species diversity-climate relationship is still poor studied. By incorporating a phylogenetic perspective, this question may be best studied. Here we will use a phylogenetic approach to address the causes of species diversity-climate relationship in Oaks (*Quercus* L.) by combining data on distribution, leaf habitat, climate, Quaternary climate and phylogeny. Specifically, we are aiming to answer the following three key questions to address the origin of oak species diversity-climate relationship.1) How the historical event or historical climate change influences the evolutionary rate, niche evolution, leaf habitat evolution and species diversification of oaks? Dose the historical event influence species diversification by speed the evolutionary rate, hence force promote the niche and leaf habitat evolution? 2) Does evolution history drive the species diversity-climate relationship? If it is true, which hypothesis plays a key role in the origin of species diversity-climate relationship, the evolutionary rate hypothesis, time-for-speciation hypothesis or niche conservatism? 3) Dose evolutionary history have influence on the relationship between species diversity and Quaternary climate change? Using the clades with different evolutionary history, we will compare the effects of Quaternary climate change on species diversity among clades and identify which clade in which area is easily to be affected by climate change. This project aims to enhance the understanding of mechanisms of the species diversity patterns; especially the influence of evolutionary history on the species diversity-climate relationship. Taking into evolutionary perspective, the achievements of this project could be used in the conservation policy plan under global climate change.

T2-12-08

Microbial biodiversity in the disjunct forests of Eastern Asia

*Teng Yang, **Haiyan Chu**, Yingying Ni, Yuying Ma*
Institute of Soil Science, Chinese Academy of Sciences

The mixed deciduous forest in Eastern China represents important proportions of relics of once widespread Tertiary mesophytic forests that spanned the Northern Hemisphere. This forest type shared dozens of genera of seed plants that have closely related species occurring in China and America, which were also referred to as disjunct taxa. In the context of floristics, we sampled soils and leaves strictly corresponding to 347 disjunct tree individuals in five mountain forest across Eastern China, and explored the diversity and community composition of plant-associated fungi by using Illumina sequencing of the ITS1 and 18S region. Preliminary analyses indicated the strong site effect on soil fungal assemblages, and that pH, nutrient contents were the best predictors for soil fungal communities. A significant distance-decay pattern was observed for fungal community across whole area and individual mountains. Partial Mantel test showed the relative effect of geographic distance (Mantel r=0.491) was slightly stronger than that of environmental factors (Mantel r=0.447). We found a very strong effect of tree species identity on fungal assemblages and significant interactive effect with site. A large of unexplained variations may could be explained by the plant phylogenetic distance and functional traits.

T2-12-09

Phylogenetic, ecological, and functional diversity in the disjunct forests of eastern Asia and eastern North America

Pamela Soltis
University of Florida

Biodiversity is multidimensional, comprising genetic and phenotypic variation above and below the species level and diverse functional roles played by genotypes across space and time. In this project, co-funded by the US and Chinese National Science Foundations, we are addressing phylogenetic, genetic, and functional diversity in the context of plant and microbial communities that span an ancient floristic disjunction to discover new components of biodiversity that arise at the intersection of these three dimensions. The deciduous forests of eastern Asia (EA) and eastern North America (ENA) represent relics of once-widespread mixed mesophytic forests of the Northern Hemisphere. Despite over a century of observation and study, many questions remain regarding the origin, evolution, and assembly of these forests. Disjunct EA and ENA forests provide an excellent natural system to examine the link between species diversity and ecological function within the context of a shared phylogenetic history; our project is enabling novel analyses of plant community diversity, structure, and function in forests that are linked both ecologically and evolutionarily. Moreover, this biogeographic disjunction provides opportunities to assess the relative roles of historical constraint, local adaptation, and time in shaping extant diversity and function of co-distributed plant and microbial communities. Collaborations among plant phylogeneticists, plant and ecosystem ecologists, environmental microbiologists, and microbial geneticists in the USA and China are exploring biodiversity across spatial and temporal scales to discover those factors that shape biodiversity through space and time.

T2-12-10

Microbial biodiversity in the disjunct forests of eastern North America

***Christopher Marshall**[2], Haitao Wang[1], Jack Gilbert[1]*
1. University of Chciago
2. University of Pittsburgh

Biodiversity is multidimensional, comprising genetic and phenotypic variation above and below the species level and diverse functional roles played by genotypes across space and time. In this project, we are addressing phylogenetic, genetic, and functional diversity in the context of plant and microbial communities that span an ancient floristic disjunction to discover new components of biodiversity that arise at the intersection of these three dimensions. The deciduous forests of eastern Asia (EA) and eastern North America (ENA) represent relics of once-widespread mixed mesophytic forests of the Northern Hemisphere. Despite over a century of observation and study, many questions remain regarding the origin, evolution, and assembly of these forests. Using 16S rRNA and ITS rRNA sequencing of the bacterial and fungal communities associated with soil and leaves from trees in different forests from EA and ENA we investigated the associations in the microbiome that defined each region and the different host tree phylogeny. This revealed differences in the microbial community composition associated with the different disjunct genera. Interestingly, the microbial communities associated with *Acer* sp. and *Nyssa* sp. cluster together in Ordway sites. Archaea are likely driving these similarities (and relative dissimilarities to the other tree genera) with Thermoplasmata (Euryarchaeota phylum) and NRP-J (Crenarchaeota) orders the predominant members of the archaeal fraction. The *Nyssa* and *Acer* associated taxa in Thermoplasmata are similar to methylotrophic methanogens, which could be explained by the presence of methylated root exudates. A site associated difference was observed between Ordway and Talladega sites. However, no tree-species effect was found on the soil microbial communities. We will discuss these details and highlight ongoing results that are currently being generated.

T2-12-11

New insights into phylogenetic and phylogeographic relationships among intercontinental disjunct members in subfamily Orontioideae (Araceae)

***Seon-Hee Kim**[1], Sangryong Lee[2], Masayuki Maki[2], Joon Seon Lee[3], Jun Wen[4], Seung-Chul Kim[1]*
1. Sungkyunkwan Universtiy
2. Tohoku University

3. University of British Columbia
4. Smithsonian Institution

Subfamily Orontioideae includes three north temperate genera (*Symplocarpus*, *Lysichiton*, and *Orontium*) of the primarily tropical Araceae. Orontioideae and its monotypic sister subfamily Gymnostachydoideae are referred to as the proto Araceae and fossil evidence suggests Orontioideae dated back to the late Cretaceous in the temperate Northern Hemisphere. The monotypic genus *Orontium* is restricted to eastern North America (ENA) and *Lysichiton* with two species occurs in eastern Asia (EA; *L. camtschatcensis*) and western North America (WNA; *L. americanus*). Genus *Symplocapus* includes five species and is disjunctly distributed in EA (4 spp.) and ENA (1 sp.). Based on extensive sampling, we newly evaluated phylogenetic relationships among genera and species within Orontioideae and estimated the timing of intercontinental disjunct events in northern hemisphere. Overall phylogenetic relationships among three genera are congruent with the earlier study. However, we found a deep divergence between two major lineages within *Symplocarpus*, which correspond to their ploidy levels: a diploid lineage includes *S. nipponicus*, *S. egorovii*, and *S. renifolius* (Korea), while a tetraploid lineage consists of *S. foetidus*, *S. nabekuraensis*, and *S. renifolius* (Japan and Russia). This lineage divergence event within *Symplocarpus* was estimated to be 34 mya (late Eocene). Two distinct lineages of *S. renifolius* in Korea and Japan show marked cytological and morphological differences. The origin and diversification of these species were estimated to occur during mid Miocene. Two intercontinental disjunct events in *Lysichiton* (between EA and WNA) and *Symplocarpus* (between EA and ENA) are estimated to be 27.7 mya (late Oligocene) and 17.8 mya (early Miocene), respectively. Based on the phylogenetic framework of Orontioideae, we carried out three phylogeographic studies based on the same four highly variable chloroplast noncoding regions (*trn*Q-*rps*16, *psb*J-*pet*A, *trn*S-*trn*G, and *rpl*32-*trn*L). First, for the phylogeographic study of *S. nipponicus* in EA, we sampled a total of 457 individuals (15 populations with 224 individuals in Korea and 21 populations with 233 individuals in Japan). We found no haplotype sharing between populations in Korea and Japan, higher genetic diversity in populations from Korea, two highly divergent haplotype groups from Japan, and no geographical structuring within each country. Second, we assessed phylogeographic patterns of tetraploid intercontinental disjunct species pair, *S. foetidus* in ENA (32 populations; 485 individuals) and *S. renifolius* in EA (Japan) (22 populations; 233 individuals). We found higher haplotype diversity in *S. renifolius*, no haplotype sharing between two taxa, and two divergent haplotype groups in *S. renifolius* in EA. The detailed phylogeographic patterns of ENA *S. foetidus* only will also be presented. Lastly, we conducted phylogeographic study of intercontinental genus *Lysichiton* (EA-WNA disjunct) (*L. camtschatcensis*, 20 populations with 191 individuals and *L. americanus*, 20 populations with 225 individuals) and found that WNA species had twice the haplotype diversity of EA species. We also found diverse haplotypes present within populations in WNA species. Comparing phylogeographic patterns in *Lysichiton* and *Symplocarpus*, we suggest different timing and direction in two intercontinental disjunct genera.

T2-12-12

Divergent morphological and ecological evolution in the bulrushes *(Scirpus)* and cotton-grasses *(Eriophorum)* explains EA-ENA and circumboreal distributions

Daniel Spalink[1], *Étienne Léveillé-Bourett*[2], *Julian Starr*[2], *Kenneth Sytsma*[3]
1. University of Utah
2. Canadian Museum of Nature
3. University of Wisconsin-Madison

The *Scirpus-Eriophorum* (Cyperaceae) complex is an ideal clade to test hypotheses regarding the timing and directionality of migration throughout the Holarctic, the historical processes resulting in either circumboreal (*Eriophorum*) or disjunct, eastern Asian-eastern North American distributions (*Scirpus*), and the evolution of morphologcal and physiological traits associated with arctic climates. Here, we present the most densely sampled phylogeny of *Scirpus* and *Eriophorum* to date, utilizing chloroplast, nuclear ribosomal, and low copy nuclear genes in conjunction with over 60,000 SNPs recovered from nearly 10 billion bases of Next Generation Sequencing data. Our analyses recover unprecedented resolution to the relationships among the bulrushes and cotton grasses. We complement our extensive molecular dataset with both traditional and geometric morphometric analyses. We reconstruct the ancestral biogeography and ancestral climatic tolerances of *Scirpus* and *Eriophorum*, demonstrating that the relictual eastern North America – eastern Asia disjunct pattern characterizing *Scirpus* is the result of both "out of America" and "out of Asia" migrations across the North Atlantic and Beringian land bridges. We highlight the evolutionary shifts and accompanying morphological and ecological adaptations in *Eriophorum* potentially associated with its persistence in boreal and arctic climates throughout the northern hemisphere.

T2-13: Zingiberales: taxonomy, phylogeny and evolution

T2-13-01

Historical introgressions among *Roscoea* species of different elevations within the south slope of the Himalayas

Jian-Li Zhao[1,2], *Qing-Jun Li*[1,2]
1. Yunnan University
2. Xishuangbanna Tropical Botanical Garden of the Chinese Academy of Sciences

The Himalayas locates in the southern margin of the Qinghai-Tibetan Plateau (QTP), it is one of the well-known biodiversity hotspots in the world. Although numerous studies have been conducted to reveal the temporal history of organisms on/around the QTP region, mechanism of diversification and *in-situ* speciation within the Himalayas is still sparsely investigated. *Roscoea* shows a disjunct distribution between the Himalayas and the Northern Indochina. Species distribution of the genus within the Himalayas exhibits distinctive altitudinal differences. There is no habitat overlapping among all species. In elevation, *R. bhutanica*, *R. auriculata*, *R. nepalensis* overlap with *R. alpina*, *R. capitata* overlaps with *R. purpurea*. *R. tumjensis* occurs within the elevational range of *R. purpurea*. We collected samples across the entire distribution range of each species. 12587 SNPs were generated using RAD-seq approach. Hierarchical *D*-statistics analyses were employed to test the introgression directions among different species and clades. Ecological niche modeling with estimation of migration vectors were used to simulate habitat shift of *Roscoea* species during the ice cycles. Clear genetic boundaries among species were reconstructed based on a maximum likelihood method. Hierarchical *D*-statistics analyses estimated significant bidirectional introgression between species with close elevational range but no introgression between *R. alpina* and *R. purpurea* and between *R. alpina* and *R. capitata* (distant elevation between species). Ecological niche modeling predicted that high-elevation *Roscoea* within the southern slope of Himalayas moved to low elevation but low-elevation *Roscoea* seems to be stable during ice age. The habitat rose up back to high elevation during warm periods. These results suggested Quaternary climatic fluctuations likely triggered introgression among *Roscoea* species by secondary contact. Divergences among *R. alpina*, *R. purpurea* and *R. capitata* may be caused by pre-Quaternary uplift of the Himalayas. This is likely the first study to explore *in-situ* speciation processes of for a plant lineage within the Himalayas. It also provides a new insight to understand the origin of biodiversity within the Himalayas.

T2-13-02

Floral development reveals the existence of a sixth staminode on the labellum of basal Globbeae

Limin Cao[1], *Shiuri Ishigooka*[2], *Akitoshi Iwamoto*[2], *Louis Ronse De Craene*[3]

1. Hengyang Normal University, Hengyang, China

2. Tokyo Gakugei University, Tokyo, Japan

3. Royal Botanic Garden Edinburgh, United Kingdom

The flowers of Zingiberaceae are characterized by a complex perianth consisting of sepals, petals, and petaloid staminodes. Compared to other Monocots and more basal Zingiberales with a full androecium of two trimerous whorls, only a single adaxial stamen of the inner whorl is fertile, and the other stamens are variously transformed in a colourful labellum. The labellum is traditionally considered to consist of two staminodes belonging to the inner whorl, in addition to two smaller lateral lobes representing two staminodes of the outer whorl. The original abaxial staminode is thought to be lost. The tribe Globbeae consists of two smaller basal genera *Hemiorchis* and *Gagnepainia*, and the large genus *Globba*, consisting of about 100 species in Tropical Asia. While *Hemiorchis* and *Gagnepainia* have a well developed labellum with a central appendage, flowers of *Globba* are characterized by a small bilobed labellum. To understand the nature of the labellum in the tribe, the floral development was carried out in 10 *Globba* species representing the major subsections, the two species of *Gagnepainia*, and one species of *Hemiorchis*. In most *Globba* species there is no trace of a third staminode on the labellum, even in early development. The only exception is *G. geoffrayi*, which showed the development of a small third central lobe, that was consecutively lost in older buds. *Hemiorchis* also lacked a third staminode as most *Globba* did, but the evidence of a third staminode was clearly observed in the floral development of *Gagnepainia*, where it persists as the small appendage visible on the matured labellum. These observations indicate that the missing staminode is not lost independently of labellum formation, but is integral part of the labellum. The presence of the lost staminode in basal genera of Globbeae indicates that it has become progressively lost within the clade. Further studies in the other tribes of Zingiberaceae could reveal more cases of this ancestrally retained character, and is a clear link to other members of Zingiberales.

T2-13-03

Specific adaxial-abaxial polarity participates in the evolutionary stamen-to-petal conversion in *Canna indica*

Xueyi Tian, *Jingping Liao*

South China Botanical Garden, Chinese Academy of Sciences

The Zingiberales is an order of tropical monocots comprising eight families (Musaceae, Lowiaceae, Strelitziaceae, Heliconiaceae, Zingiberaceae, Costaceae, Marantaceae and Cannaceae). The main line of the evolution of Zingiberales is the number of fertile stamens decreasing by (6)5 1 1/2, and the number of petaloid organs increasing contrarily. Bearing only a half fertile stamen in the flower, *Canna indica* (Cannaceae) is one of the representative species to investigate the developmental evolution of Zingiberales. In the androecium of *Canna indica*, 3~4 petaloid staminodes develop in place of stamens, and the only fertile stamen consists a one-theca anther and a petaloid appendage. Current evidences concerning ABC homeotic genes are insufficient to explain the stamen-to-petal conversion since C-function gene is expressed in the sterile petaloid androecial members. The one-theca fertile stamen can be seen as an intermediate state organ from a two-thecae stamen to a completely petaloid staminode, thus we examined the differences of molecular basis for the differentiation of anther and petaloid appendage. Class III HD-ZIP family members, which specify adaxial identity, were mostly expressed lower in petaloid appendage than in anther; while KANADIs and YABBYs, which promote abaxial identity, exhibited opposite expression patterns. Additionally, asymmetric expressions of marker genes for adaxial-abaxial polarity were dynamically detected during the bisymmetric-to-asymmetric change process of stamen primordium; at last, the adaxial marker gene was mainly expressed in the region between the two protrusions of the anther, while the abaxial marker was mainly expressed in petaloid appendage. We hypothesize that the fertile theca represents the adaxial part of the stamen, while the sterile petaloid appendage represents the abaxial part, and that the adaxial-abaxial polarity participates in the distinctive anther-petaloid appendage patterning within the stamen of *Canna indica*. These results can be applied in further study of the evolution of Zingiberales.

T2-13-04

Ovary structure and evolution in the Zingiberaceae and Costaceae

Bruce Kirchoff

University of North Carolina at Greensboro

Recent work on visual character description has established methods for extracting character data from otherwise intractable sources, such as developmental sequences. The methods are based on standardized photographs of homologous developmental stages, or homologous positions in complex structures. The photographs are then sorted into hierarchical similarity clusters called "character cladograms." Each photograph represents a character state, while the cladogram as a whole is the character. A character cladogram is a graphical representation of a single character, based on photographs not verbal descriptions. When used to reconstruct phylogeny each character cladogram serves as a hypothesis of phylogeny, and each dichotomy in the character cladogram represents a character state. The similarity groups in character cladograms may be hierarchically arranged, allowing for multiple levels of data extraction from the same set of photographs. This visual character description method avoids many of the problems that arise from the application of verbal labels, problems that include difficulty in representing variation in character states, and the inability to capture information that can be easily seen but that is hard to describe. In this study two methods of morphological character description were applied to the ovaries of the Zingiberaceae and Costaceae. The traditional method of character description, based on the analysis and verbal description of features, was contrasted with visually-based character cladograms. Taxa in the Zingiberaceae and Costaceae were selected based on published phylogenies of these taxa. To describe the characters, we identified similar positions in in the ovary and took standardized photographs of cross-sections. The levels selected for photography were below the locule, ¼ of the way from the bottom to the top of the locule, in the middle of the locule, ¾ of the way from the bottom to the top of the locule, and above the locule in the middle of the tissue that closes the ovary. Four photographs were taken of the following positions at each level: full cross section, one locule, one septa, the epidermal region opposite a locule. The photographs were used to describe 12 traditional characters and create 16 character cladograms: four character cladograms at each of the four levels, corresponding to the four positions described above. Once created, both the traditional characters and the character cladograms were converted to numerical values and plotted on a combined phylogeny of the Zingiberaceae and Costaceae constructed from published phylogenies of these families. Calculation of individual and ensemble consistency indices allowed comparison of the two methods and an assessment of their strengths and weaknesses.

T2-13-05

Pollen transfer and explosive style movement in Marantaceae

Regine Classen-Bockhoff

Johannes Gutenberg-University Mainz

Marantaceae (29/550) have a unique pollen transfer mechanism including secondary pollen presentation and a rapid explosive style movement. As the flowers have only a single chance to get pollinated, floral structures need to be highly synorganised. The fertile part of the androecium is reduced to a single bisporangiate theca; all other elements are sterile, showy (outer staminodes) and functionally specialised. The pollination complex consists of the style and the hooded staminode which holds the style. The style is a compact highly turgescent structure with a head differentiated into a pollen plate at its back, a rather large stigmatic orifice at the upper side and a sticky gland separating the two parts. The hooded staminode is a thin structure bearing a trigger appendage and a basal plate. Already in the bud, the pollen plate is loaded with own pollen. Shortly before flowering, the style elongates more than the hooded staminode enveloping the style. Thereby, tension is set up. The tension is hold by the hooded staminode and released when a pollinator touches the trigger appendage. The mechanical stimulus is transferred via the basal plate to the contact point at the base of the style head where the style presses against the hood of the staminode. Within a spit of a second, the style springs forward. While rapidly rolling in, foreign pollen is scraped from the pollinator into the stigmatic orifice and own pollen is glued onto the same position of the pollinator. The pollination mechanism in Marantaceae is highly derived and much more complex than the pollen transfer process in the sister group Cannaceae (1/10). The high number of species suggests that the explosive style mechanism might be a key innovation for speciation. However, in view of the low number of seed set throughout the family and the vigorous clonal growth of the plants, the question raises how far the pollination mechanism indeed contributes to speciation. Reconstructing reproductive isolation mechanisms throughout the family reveals that geographic isolation and adaptation to different pollinators may have influenced radiation more than the pollination mechanism. The existence of highly diverse and elaborate floral structures should thus be interpreted as inventions optimizing the reproductive system by pollen saving or reducing investment costs.

T2-13-06

The evolutionary history of the Zingiberales: The final Word on a Charismatic Lineage of tropical plants using targeted gene enrichment

W. John Kress[1], *Monica M. Carlsen*[1], *Tomá Fér*[2], *Roswitha Schmickl*[2], *Jana Leong-Skornickova*[3], *Mark Newman*[4]

1. National Musuem of Natural History, Smithsonian Institution

2. Charles University

3. Singapore Botanic Gardens

4. Royal Botanic Gardens Edinburgh

The evolutionary relationships of the eight families of the tropical order Zingiberales has been debated by taxonomists for decades. Nearly every possible topological arrangement for these eight lineages has been proposed and defended with anatomical, morphological, DNA sequence, and genomic data. Well-supported reconstructions have been especially problematic among the basal branches (the "Banana Families") due to an early rapid evolutionary radiation in the lineage. Previous investigations have suffered in part from the lack of adequate generic sampling within families and the use of a limited number of genetic markers, mostly plastid loci. We have employed hybridization-based targeted enrichment to obtain sequences from 4,618 exons (all longer than 120 bp) within 1,180 orthologous nuclear loci (all longer than 960 bp), and 78 coding regions from the plastome. Concatenated data matrices were analyzed using ExaML and supported individual gene trees were merged using a coalescent framework in ASTRAL and a supertree method (MRL). Only one fully supported topology was recovered in all three multi-locus analyses (concatenated, coalescent, and supertree) with no significant conflicts among data sets and/or analytic procedures. The single topology is composed of two primary clades, which diverged early in the evolution of the order. One clade, the "Ginger Families, " which includes (Costaceae + Zingiberaceae) (Marantaceae + Cannaceae), has been well-resolved and well-supported in all previous studies. The other primary clade, the "Banana Families, " has until now never been strongly supported as a monophyletic group. The topology among these four families, Musaceae (Heliconiaceae (Lowiaceae + Strelitziaceae)), is finally fully and unambiguously resolved.

T2-14: Integrated research in Magnoliales

T2-14-01

Which fleshy fruit traits allow for long-distance dispersal? A deep-time perspective from the custard apple family (Annonaceae)

Renske E. Onstein[1,2], *W. Daniel Kissling*[1], *Thomas L.P. Couvreur*[3], *Lars W. Chatrou*[4], *Hélène Morlon*[5], *Hervé Sauquet*[2]
1. University of Amsterdam
2. Université Paris-Sud
3. Institut de Recherche pour le Développement UMR-DIADE
4. Wageningen University and Research
5. école Normale Supérieure

Tropical rain forests are the most species-rich ecosystems on Earth and this diversity may have resulted from *in-situ* diversification as well as colonization by (megathermal) lineages through long-distance dispersal (LDD). Frugivory (fruit-eating and seed dispersal by vertebrates) is the primary dispersal mode for the majority of tropical plants, but it remains unclear which fleshy fruit traits have allowed these lineages to colonize new continents historically. Here, we test whether certain fruit functional traits are associated with long distance dispersal events during the Cenozoic in the custard apple family (Annonaceae, ca. 2400 species). Annonaceae has a pantropical distribution and shows a wide-range of vertebrate-dispersed fruit types, which differ in size (0.3 to > 40 cm), color and display (e.g. cauliflory). We use a phylogenetic framework to reconstruct the biogeographic history of Annonaceae in relation to the evolution of fifteen fruit functional traits. We test whether ancestral fruit traits during LDD events are significantly distinct from the average ancestral Annonaceae fruit type, and we classify them according to several Cenozoic dispersal routes. We find that LDD events during the Eocene climatic optimum, when Annonaceae may have taken the boreotropical route of dispersal, are associated with fruits consisting of a few large monocarps (the dispersal unit, ca. 3 cm), growing on relatively short stipes (ca. 0.7 cm) and bearing several seeds (ca. 4). These fruits are typically appealing to large-bodied mammals that are expected to contribute disproportionally to long-distance dispersal events. Several oceanic dispersal events may have been associated with typical bird-dispersed fruits, such as moniliform or dehiscent fruits with bright (red, black, yellow) colors. These results suggest that the evolution of fruit syndromes linked to large-bodied mammals and strong-flying birds may have allowed fleshy-fruited rain forest taxa to disperse and colonize new continents during the Cenozoic when climatic conditions were suitable. This may in part explain the disjunct distribution of megathermal plant taxa across the tropics.

T2-14-02

Subito allegro – the pace of diversification in the Annonaceae

Lars W. Chatrou[1], *Eva Drukker*[1], *Paul H. Hoekstra*[1,2], *Joeri S. Strijk*[3]
1. Wageningen University and Research, Wageningen
2. Naturalis Biodiversity Center, Leiden
3. College of Forestry, Guangxi University, Nanning

Despite considerable efforts in recent years to elucidate phylogenetic relationships within the Annonaceae, we only have a faint understanding of diversification patterns. The reasons for this are twofold: first, our efforts to infer the ages of Annonaceae are seriously hampered by lineage-specific heterogeneity of substitution rates. The main effect of the disregard of these differences is the underestimation of node ages in the subfamily Malmeoideae, which in turn results in the overestimation of diversification rates. The second cause is the incompleteness of taxon sampling in most family-wide analyses so far. Tree shape of the two main clades of the Annonaceae, the subfamilies Annonoideae and Malmeoideae, is very different. Phylogenetic signal in the Malmeoideae is fairly evenly distributed over the branches from the crown node to the tips, whereas the topological patterns in the Annonoideae are more radiation-like, with long branches subtending abrupt lineage diversifications. A commonly used approach of sampling a small of number of species per genus is likely to leave many patterns of diversification unrevealed. To explore node ages and diversification patterns in the light of these observations, we gathered a data set containing over 1000 species, for which sequence data of multiple plastid markers are available. In our analyses we accommodate for substitution rate heterogeneity among the major lineages, and apply three undescribed and previously unexploited fossils as calibrations. Using Bayesian approaches to the inference of node ages and diversification rates, we demonstrate the evolutionary buildup of species diversity in the family.

T2-14-03

Phylogenetic analyses and the position of Cretaceous fossils in Magnoliales: Implications for character evolution, age of clades, and past diversity

James A. Doyle[1], *Peter K. Endress*[2]
1. University of California Davis
2. University of Zurich

Phylogenetic analyses have confirmed and refined relationships of several Cretaceous fossil taxa to Magnoliales, providing minimum ages for nodes, insights on the course of character evolution, and surprises concerning the past morphological and ecological diversity of the order. Archaeanthus, based on bilobed leaves and flowers with an elongate receptacle from near the Albian-Cenomanian (Early-Late Cretaceous) boundary in Kansas, is probably a stem relative of Magnoliaceae; a recent suggestion that it was nested in the crown group was based on use of inappropriate outgroups. Our earlier analysis associated an older fossil, Endressinia, known as leafy shoots and flowers from late Aptian lake deposits in northeastern Brazil, with the living clade consisting of Degeneria, Galbulimima, Eupomatia, and Annonaceae, which is united by inner staminodes (retained in Anaxagorea but lost in the remaining Annonaceae). However, a more recent analysis suggests that Endressinia and Schenkeriphyllum, another genus from the same deposits, may also be stem relatives of Magnoliaceae, attached below Archaeanthus. This would imply that inner staminodes were originally present in the ancestors of Magnoliaceae but later lost.

Based on their leaf morphology and associated paleoclimatic indicators, these fossils suggest that some early Magnoliales were able to adapt to more arid tropical conditions than might be inferred from living members. Futabanthus, a flower from the Coniacian (Late Cretaceous) of Japan, may be a near-basal member of the clade comprising all Annonaceae other than Anaxagorea, based on its lack of inner staminodes. This clade is also represented by seeds with lamelliform endosperm ruminations from near the end of the Cretaceous. Cecilanthus, a bisexual flower with numerous whorled parts from the Cenomanian of Maryland, has conflicting suites of characters, some suggesting relationships with Magnoliales, others with Nymphaeales. However, consideration of the most likely states of uncertain characters favors a position in Magnoliales, which would mean that the early radiation of the order produced now-extinct lines with floral morphology quite different from that of any living members. None of these fossils have associated pollen; in some cases, information on their pollen morphology or seed anatomy could resolve uncertainties on their systematic position.

T2-14-04

Gene tree discordance and coalescent methods support ancient intergeneric hybridisation between *Dasymaschalon* and *Friesodielsia* (Annonaceae)

Xing Guo[1], *Daniel C. Thomas*[1,2], *Richard M.K. Saunders*[1]
1. School of Biological Sciences, The University of Hong Kong, Hong Kong, P. R. China.
2. Singapore Botanic Gardens, Singapore

Although hybridisation between closely related species is common and known to be important in plant evolution, hybridisation at the generic level or above is comparatively rare. We address ancient intergeneric hybridisation in the early-divergent angiosperm family Annonaceae by reconstructing the phylogenetic and divergence history of the genus *Dasymaschalon* using a multi-locus approach based on molecular data from five chloroplast (*matK*, *psbA-trnH*, *ndhF*, *rbcL*, and *trnL-F*) and five nuclear (ITS, ETS, AP3, *PhyA*, and *PhyC*) DNA markers. We demonstrate incongruence among different gene trees: *Dasymaschalon* is retrieved as monophyletic in the nuclear ribosomal tree (based on ITS and ETS), but is non-monophyletic in the chloroplast and Phy-gene trees, with the majority of species assigned to a strongly supported clade but three species (*D. filipes*, *D. longiflorum* and *D. tibetense*) more closely related to the sister genus *Friesodielsia*. Three contrasting approaches—a coalescent method based on molecular dating, incongruence pattern comparison, and a multi-accession phylogenetic reconstruction—are used to assess the patterns of this gene tree incongruence and test hypotheses of ancient hybridisation and incomplete lineage sorting. Our results support a late Miocene intergeneric hybridisation between members of the *Dasymaschalon* and *Friesodielsia* lineages in continental Asia-west Malesia.

T2-14-05

Next generation Annonaceae phylogenetics: The way forward in Annonaceae evolutionary biology

Thomas Couvreur[1], *Erik Koenen*[3], *Rita Brandao*[4], *Stefan Little*[2], *Roy Erkens*[4], *Herve Sauquet*[2]
1. Institut de Recherche pour le Développent (IRD)
2. Université Paris-Sud
3. Institut für Systematische und Evolution?re Botanik
4. University of Maastricht

Annonaceae has been an important model family to address questions about tropical rain forest evolution and biodiversity. The phylogenetic relationships of the family have been inferred mainly at the generic level and based on few plastid markers. Lots of unresolved relationships remain, especially in subfamily Miliuseae. However, with the arrival of Next Generation Sequencing we now have the potential to generate a well resolved species level phylogeny based on hundreds of nuclear genes. Here, we used transcriptomes sequenced for five species from across the family to generate over 100 nuclear markers amplifiable across Annonaceae for inter and intra species evolutionary studies. The markers are generated with MyBaits and can thus be used to sequence hundreds of species at a time as well as degraded herbarium material. We use these new markers and sequence most species of the enigmatic African Piptostigmateae tribe (Miliuseae sub family). We show how useful these markers are at resolving generic, species and infra species relationships. We hope that the generation of these markers will mark a new age of Annonaceae evolutionary biology studies.

T2-14-06

Phylogenomics and historical biogeography of the recalcitrant annonaceae subfamily Malmeoideae

Daniel C. Thomas[1], *Tatiana Arias*[2], *Magnus Lundberg*[3], *Thomas L.P. Couvreur*[4], *Hervé Sauquet*[5], *Laetitia Carrive*[5], *Roy Erkens*[6], *Rita D. Brandao*[6], *Lars W. Chatrou*[7], *Jenifer De Carvalho Lopes*[8], *Bine Xue*[9], *Tanawat Chaowasku*[10], *Richard M.K. Saunders*[3]
1. Singapore Botanic Gardens
2. Corporacion para Investigaciones Biologicas
3. University of Hong Kong
4. Institut de Recherche pour le Développement
5. Université Paris Sud
6. Maastricht University
7. Wageningen University
8. University of São Paulo
9. South China Botanical Garden
10. Chiang Mai University

Annonaceae (ca. 2400 species of trees, shrubs and lianas) are characteristic and ecologically important elements in tropical lowland forests of both the Neotropics and the Old World. There have been substantial advances in resolving the backbone phylogeny of the family in recent years, but there are notable exceptions. Intertribal relationships in subfamily Malmeoideae (7 tribes, 47 genera, >770 species) and intergeneric relationships in its largest tribe, Miliuseae (26 genera, ca. 540 species), remain very poorly understood. This has impeded downstream analyses such as investigations of the historical biogeography and character evolution in the family. We used a phylogenomic approach to resolve the relationships in subfamily Malmeoideae. Data from sequencing of total genomic DNA on the Illumina HiSeq2000 platform was used to assemble near-complete plastome DNA sequences. The dataset comprised 55 newly sequenced accessions including samples of all seven tribes and representatives of 45 of 47 currently accepted genera in the subfamily. These data were analysed separately and by integrating a supermatrix of previously published DNA data including accessions of all currently accepted genera in Malmeoideae using likelihood and Bayesian phylogenetic reconstruction methods. The resulting phylogenetic reconstructions largely resolve intertribal relationship and intergeneric relationships in tribe Malmeeae, and provide various new insights into intergeneric relationships within tribe Miliuseae. Parts of the Miliuseae backbone, however, remain unresolved. Likelihood ancestral range estimations indicate a wide ancestral distribution for the most recent common ancestor of the Malmeoideae crown group including the Neotropics, Africa and Asia. This is inferred as the result of a range expansion from Africa between the stem and the crown node sometime during a time interval ranging roughly from the mid-Cretaceous to the mid-Eocene. In combination with the fossil record, the results indicate that the 'boreotropics', a frost-free and humid climate corridor in the Northern mid-latitudes in the Paleocene-Eocene, likely played an important role facilitating intercontinental dispersal of this group.

T2-15: Phylogeny and evolution of the lamiids

T2-15-01

Early lamiid evolution: Basal relationships, ancestral characters, and ancient polyploidy

Gregory Stull[1], *Clayton Visger*[2], *Douglas Soltis*[2], *Pamela Soltis*[2], *Stephen Smith*[1]

1. University of Michigan

2. University of Florida

Although our understanding of angiosperm phylogeny has improved tremendously over the past ca. 25 years, several major clades have eluded resolution, despite two decades of study and increasingly large molecular data sets. *Lamiidae*, an asterid clade with ca. 40,000 species (or 15 percent of angiosperm species richness), is a salient example. The clade includes four families previously unplaced at the ordinal level and an unresolved basal topology, hindering efforts to reconstruct ancestral states and early patterns of morphological evolution. However, recent work—employing plastome sequencing across key under-sampled lineages, namely genera of Icacinaceae s.l., Metteniusaceae, and Oncothecaceae—has greatly clarified our understanding of basal lamiid relationships. We summarize recent progress in inferring lamiid phylogeny, focusing on the positions and circumscriptions of previously unplaced taxa (e.g., Icacinaceae and Metteniusaceae). With this improved phylogenetic framework, we reconstruct the evolution of multiple morphological characters considered important in earlier classifications of asterids (e.g., unitegmic vs. bitegmic ovules, crassinucellate vs. tenuinucellate ovules, presence vs. absence of iridoids) as well as additional characters such as habit and fruit type. Finally, we explore instances of ancient polyploidy (or whole-genome duplication, WGD) in *Lamiidae* based on genetree reconstructions across a broad sampling of lamiid transcriptomes. Ancient WGDs have been identified in numerous major angiosperm lineages, underscoring the importance of polyploidy in flowering plant evolution. Preliminary data (KS-plots derived from lamiid transcriptomes) suggest the possibility of one or several WGDs early in the evolution of lamiids. Our results will provide important initial insight on the role of ancient polyploidy in the evolution of this major angiosperm clade.

T2-15-02

Phylogenies and proposed classifications of Hydrophyllaceae and Namaceae

Genevieve Walden

University of Washington Herbarium (WTU) at the Burke Museum

Recent research focusing on the systematics of Boraginales (Lamiidae) has resulted in revised infrafamilial, generic, and species circumscriptions. The clade comprising members of Namaceae have long been treated as a tribe (Namaeae Choisy) within Hydrophyllaceae, but has now been recognized as a distinct lineage, closely related to an emended Hydrophyllaceae. Namaceae includes 75 spp. in North and Central America, western South America, the Caribbean, and Hawaii. Hydrophyllaceae includes 240–260 spp. of North, Central, and western South America. Results of molecular phylogenetic analyses of Hydrophyllaceae and Namaceae found support for clades corresponding to some previously recognized infrafamilial tribes. However, reconciliation of previous classifications is complicated and requires revision to better reflect evolutionary relationships. Results were used to examine evolution of non-molecular characters, and compare diversification rates and patterns across clades. Proposed classifications are discussed.

T2-15-03

Progress in Gentianales systematics: Molecular phylogeny, morphological evolution and divergence times of a cosmopolitan Asterid order

Lena Struwe[1], *Tatyana Livshultz*[2], *Hannes Hettling*[3], *Ruud Scharn*[4], *Alexandre Antonelli*[5]

1. Department of Ecology, Evolution, and Natural Resources, Rutgers University, New Brunswick, NJ, USA

2. Department of Biodiversity Earth and Environmental Sciences, Academy of Natural Sciences of Drexel University, Philadelphia, PA, USA

3. Naturalis Biodiversity Center, Leiden, The Netherlands

4. Department of Biological and Environmental Sciences, University of Gothenburg, Göteborg, Sweden

5. Department of Biological and Environmental Sciences, University of Gothenburg & Gothenburg Botanical Garden, Göteborg, Sweden

There has been substantial progress in our understanding of the phylogeny and biogeography of the Gentianales (Lamiidae, Angiospermae) in the last decades. Its five families are well-circumscribed and well-defined with morphological synapomorphies and key characteristics. Rubiaceae is the largest family (ca. 13550 species, 617 genera), Apocynaceae a medium-sized family (ca. 4700 species, 370 genera), the recently reclassified Gentianaceae has ca. 1700 species and 101 genera, and the remaining species belong to the two small families Gelsemiaceae (11 species, three genera), and Loganiaceae (ca. 500 species, 16 genera). Most order-wide phylogenies show Rubiaceae as sister to the other families, but relationships among the remaining four families have been in conflict. Rubiaceae comprise three subfamilies, Cinchonoideae, Ixoroideae, and Rubioideae, with an uncertain position for the genus *Luculia*. Apocynaceae are divided into five subfamilies (three monophyletic) and 23 tribes. Generic delimitation is a work in progress, with many genera, e.g. *Ceropegia, Vincetoxicum,* and *Marsdenia*, still being para- or polyphyletic as historically circumscribed. A new family classification of Gentianaceae will be published in 2017 recognizing seven tribes, including the resurrection of tribe Voyrieae and the acceptance of several new genera segregated from *Gentiana* and *Fagraea*. Gelsemiaceae has been expanded from two to three genera to include the Southeast Asian genus *Pteleocarpa*, one of the few Gentianales taxa with alternate leaves. Within Loganiaceae, the paraphyletic Australian genera Logania and Mitrasacme have been redefined, resulting in several new genera. In terms of distribution and biogeography, the Gentianales is cosmopolitan with most species in the tropics or subtropics, but certain lineages are more temperate or even alpine, particularly in Rubiaceae and Gentianaceae. Pantropical distributions are common in many clades/tribes. The area of origin is currently uncertain, but appears to be the New World tropics for Gentianaceae and the paleotropics for Rubiaceae. In an effort to provide a well-sampled, large-scale phylogeny for classification and a dated, biogeographic analysis of the whole order, we use automated methods implemented in the SUPERSMART pipeline for phylogenetic inference (www.supersmart-project.org). The time calibration is provided by fossils from the order Gentianales. To this end, we compiled a curated set of molecular sequences from all Gentianales species available in GenBank that are suitable for the analysis. Phylogenetic inference is accomplished in a two-step procedure, in which a genus-level backbone phylogeny is combined with species-level trees estimated under the multi-species multi-locus coalescent model. Our method uses Bayesian inference to ensure that uncertainty in the molecular data is reflected in the posterior node probabilities in the final phylogeny. We also outline key prospects for future research in the phylogenetics and biogeography of this diverse plant order.

T2-15-04

Nuclear phylogeny of Lamiids and related taxa: Implications on divergence and character evolution

Hong Ma[1], *Caifei Zhang*[1], *Yezi Xiang*[1], *Chien-Hsun Huang*[1], *Federico Luebert*[2], *Maximilian Weigend*[2], *Michael Frohlich*[3], *Mathew Rees*[3], *Yi Hu*[4]

1. Ministry of Education of Key Laboratory of Biodiversity Sciences and Ecological Engineering, Institute of Plant Biology, School of Life Sciences, Fudan University

2. University of Bonn

3. The Royal Botanical Garden, Kew

4. Department of Biology, The Pennsylvania State University

Lamiids form a large group of angiosperms and one of two major subdivisions of Asterids, which also include Campanulids (the other major subdivision, with sunflowers, lettuce, carrot and celery) and the orders Cornales and Ericales. Well-known Lamiid species include tomato, potato and tobacco in the Solanaceae, mint, basil, sage, thyme in Lamiaceae, olive, and many ornamental flowers. Current systematics according the APG IV divides Lamiids into eight orders: four of them (Boraginales, Gentianales, Lamiales, Solanales) are relatively large with multiple families; the other four are much smaller (Garryales, Icacinales, Metteniusales, and Vahliales), each with one or two small families, totaling about 50 families. Molecular phylogenetic analyses in the recent past have largely resolved well-supported relationships amongst many of the families and orders, but some relationships and placements remain uncertain. Similarly, in the larger context of Asterids, there are also many unresolved relationships at the level of families and orders. The present study has obtained hundreds of nuclear gene sequences from newly sequenced transcriptomes and other sequences using a set of criteria including length of aligned regions, species coverage, and putative orthology, and performed multiple phylogenetic analyses of Lamiid lineages and their Asterid relatives. The phylogenetic studies resolve strongly supported relationships, most of which are in agreement with previous results; in addition some relationships that were previously uncertain are now resolved. The nuclear gene sequences were used to estimate divergence times of orders and families, and the newly established phylogeny provides a framework for ancestral character reconstruction.

T2-15-05

Dispersal routes in the southern hemisphere and Neotropics: Examples from Lamiales

Richard Olmstead

University of Washington

Detailed phylogenetic study of Lamiales has permitted the enumeration and directionality of migration paths between southern hemisphere continents and throughout the Neotropics in several families. Dated trees, biogeographic inference, and habitat preferences of contemporary species, suggest that most intercontinental dispersal events involved long distance, over-water dispersal. Families of South American origin with contrasting dispersal patterns include Bignoniaceae with dispersal primarily to Australasia, and Verbenaceae with dispersal primarily to Africa. Scrophulariaceae originated in southern Africa and reached South America at least two times. Within the Neotropics, amphitropical disjuncts pose some of the same questions about dispersal timing and routes with long-distance vs. terrestrial migration routes possible. Several genera in Verbenaceae exhibit a pattern of disjunction between the arid zones of North and South America. With an origin in South America, the direction is clearly south to north. However, evidence seems to suggest that both dispersal paths occur in the history of the family.

T2-15-06

Evolutionary trends in Boraginales

Maximilian Weigend, *Julius Jeiter, Hans-Jürgen Ensikat, Adeel Mustafa, Federico Luebert, Juliana Chacón-Pinilla*

Nees-Institut für Biodiversität der Pflanzen, Universität Bonn

The phylogeny and systematics of Boraginales (ca. 125 genera and 2700 spp.) have been the subject of several recent, large-scale studies, largely resolving evolutionary relationships within the group. These have been translated into a revised classification both at the order level and at the infrafamilial level for the largest family (Boraginaceae, >80 genera, ca. 1600 spp.). The order is resolved with two major clades, Boraginales I comprising the families Wellstediaceae, Codonaceae and Boraginaceae, Boraginales II comprising the families Cordiaceae, Coldeniaceae, Ehretiaceae, Heliotropiaceae, Hoplestigmataceae, Hydrophyllaceae, Namaceae, and Lennoaceae. The well-resolved relationships now permit the study of character evolution across this order. A range of different morphological traits appears to show interesting evolutionary trends. Fruit morphology, the classical character complex used to classify Boraginales, can be shown to have a relatively complex character evolution, with the one-seeded mericarpids in Boraginales I and Boraginales II derived independently from capsule-fruited ancestors. Across the order, there has been a trend towards the reduction of seed number and segregation of the seeds into one-seeded mericarpids. Seed dispersal (wind-dispersal) seems to be the ancestral condition in Boraginales I and II, more specialized mechanisms of fruit or mericarp dispersal have then arisen in the more derived groups of both clades. In Boraginaceae the ancestral diaspore dispersal mode was epizoochory, probably associated with the presence of glochidiate ornamented nutlets or calyces, and that shifts to other dispersal modes occurred at least twelve times during the evolution of the clade. In Boraginales II there is an overwhelming trend towards endo-zoochorous dispersal, with several transitions to abiotic mechanisms. Detailed studies on floral morphology and function, especially in Hydrophyllaceae, Codonaceae and Boraginaceae, reveal a wealth of diversifications, especially with regard to flower organization and access to nectar, but also with regard to reward and breeding systems. Morphological trends mainly centre around the formation of specialized corolla outgrowths, which are well-developed in Boraginales I and Hydrophyllaceae. Additionally, a trend towards specific modifications of the nectary, gynoecium, style and stigma in the context of increasingly hypocrateriform corollas and flower compartmentalization can be observed in several lineages. There were multiple inventions of polymery and specialized functional syndromes, including dioecy, heterostyly, buzz-pollination and obligate autogamy across the order. Vegetative morphology and growth habits are diversified, as would be expected from any plant group of this size. It has also long been known that many representatives of Boraginales are characterized by a distinct, rough indument. However, detailed studies on the biomineralization of these trichomes have not been carried out in the past, and we present some data revealing complex mineralization patterns and striking trends, including individual, one-celled trichomes with up to three different biominerals in their cell wall. Our data show that trichomes with three different biominerals (calcium carbonate, calcium phosphate and silica) are found in both Boraginales I and Boraginales II. The basally branching families Codonaceae, Namaceae and Hydrophyllaceae are predominantly mineralized with calcium carbonate and phosphate, whereas the more derived Cordiaceae and Boraginaceae appear to be predominantly mineralized with calcium carbonate and silica.

T2-16: Orchid phylogenomics: Diversification, evolution and biogeography

T2-16-01

Orchid phylogenomics, multiple drivers of extraordinary diversification, and historical biogeography

***Thomas Givnish*[1], *Daniel Spalink*[2], *Mercedes Ames*[1], *Stephanie Lyon*[3], *Steven Hunter*[1], *Alejandro Zuluaga*[4], *William Iles*[5], *Alfonso Doucette*[1], *Giovanny Giraldo Caro*[1], *James Mcdaniel*[1], *Mark Clements*[6], *Mary Arroyo*[7], *James Leebens-Mack*[8], *Lorena Endara*[9], *Ricardo Kriebel*[1], *Kurt Neubig*[10], *Mark Whitten*[9], *Norris Williams*[9], *Kenneth Cameron*[1]

1. University of Wisconsin-Madison
2. University of Utah
3. University of Wisconsin-Stevens Point
4. Universidad del Valle
5. University of California at Berkeley
6. Centre for Australian National Biodiversity Research
7. Universidad de Chile
8. University of Georgia
9. University of Florida
10. Southern Illinois University

Orchidaceae is the most diverse family of angiosperms, with over 25,000 species – more than mammals, birds and reptiles combined – and one of the most broadly distributed families as well. Tests of hypotheses to account for orchid diversity have been stymied by the lack of a fully resolved broad-scale phylogeny. Here, my colleagues and I provide such a phylogeny, based on 75 chloroplast genes for 39 species representing all orchid subfamilies and 16 of 17 tribes, time-calibrated against 17 angiosperm fossils. A supermatrix analysis places an additional 144 species based on three plastid genes, including representatives of almost all orchid subtribes. Orchids appear to have arisen roughly 112 million years ago (Mya); subfamilies Orchidoideae and Epidendroideae diverged from each other at the end of the Cretaceous; and the eight tribes and three previously unplaced subtribes of the upper epidendroids diverged rapidly from each other between 37.9 and 30.8 Mya. Orchids appear to have undergone one significant acceleration of net species diversification in the orchidoids, and two accelerations and one deceleration in the upper epidendroids. Consistent with theory, such accelerations were correlated with the evolution of pollinia, the epiphytic habit, CAM photosynthesis, distribution in extensive tropical cordilleras, and pollination via Lepidoptera, euglossine bees, or deceit. Deceit pollination appears to have elevated the number of orchid species by one-half. The highest rate of net species diversification within the orchids (0.382 sp sp^{-1} My^{-1}) is 6.8 times that at the Asparagales crown. We reconstructed orchid historical biogeography using BioGeoBears. Orchids appear to have arisen on the Australian Plate and then spread to the Neotropics overland via Antarctica by 90 Mya, when all three continents were in close contact and Antarctica had a subtropical climate. Ancestors of the vanilloids, cypripedioids, and orchidoids + epidendroids appear to have arisen in the Neotropics 84 to 64 Mya. Repeated long- and short-distance dispersal occurred throughout orchid history. Stochastic mapping identified a mean of 74 long-distance dispersal events, for an average rate of 0.8 long-distance dispersal events every million years. Of the 33 orchid genera having disjunct distributions on continents separated by tropical oceans, Dressler argued that only five might be old enough to reflect vicariance driven by continental drift. Our data, however, indicate that all of these groups are too recent to reflect continental drift and are instead products of long-distance dispersal. We estimate that 15 transoceanic long-distance dispersal events resulted in the origin of orchid tribes or smaller lineages restricted to individual continents, and that all occurred long after Gondwana had begun to break up. The importance of modest amounts of long-distance dispersal for orchid diversification is highlighted by the fact that 97% of all orchid species are restricted to individual continents or subcontinents. Across orchid history, Southeast Asia was the most important source and maximally accelerated net diversification; across epidendroids, the Neotropics maximally accelerate diversification.

T2-16-02

Next generation species delimitation and phylogeography of a recently radiated clade of Australian *Corybas*

Stephanie Lyon[1], *Thomas Givnish*[2], *Mark Clements*[3]
1. University of Wisconsin Stevens Point
2. University of Wisconsin Madison
3. CSIRO

Problems with elucidating species boundaries and evolutionary history abound in groups that have high rates of speciation, such as orchids. In order to reconstruct the phylogenetic and phylogeographic history in a recently radiated, Australian endemic clade (the Corysanthes clade) in the terrestrial orchid genus *Corybas* (Acianthinae, Diurideae), we used both sequencing of standard molecular markers (plastid spacers, nrITS, and the single-copy nuclear gene phytochrome C) as well as a form of reduced representation library sequencing, genotyping-by-sequencing (GBS). Phylogenetic analyses of the GBS data provided an unprecedented degree of resolution within this group, while remaining highly congruent with the results from the standard molecular regions. Based on our findings, we recommend reducing the number of species in this clade from 10-12, to seven monophyletic, strongly supported, morphologically distinct species (*Corybas fimbriatus*, *C. hispidus*, *C. pruinosus*, *C. diemenicus*, *C. incurvus*, *C. recurvus*, and *C. despectans*). *Corybas dentatus* appears to be a hybrid of *C. diemenicus* and *C. incurvus*. *Corybas limpidus* forms a basal grade relative to *C. despectans*, and *C. expansus* appears embedded within *C. despectans*. On the other hand, several morphological forms recognized by local botanists do appear to represent distinct evolutionary lineages and should be protected to preserve infraspecific genetic diversity and evolutionary potential.

T2-16-03

Development of a phylogeny for Australian *Caladenia* orchids to explore the evolution of specialised pollination

Ryan Phillips, Monica Ruibal, Darren Wong, Celeste Linde, ***Rod Peakall***
Research School of Biology, The Australian National University

Floral traits associated with the evolution of specialised pollination systems have been hypothesised to represent key innovations that lead to increased rates of speciation. However, the evolution of specialised floral traits may also constrain subsequent floral evolution. In orchids, the evolution of sexual deception from the more generalised food deception strategies may have important consequences for both diversification and long-term evolutionary persistence. Pollination by sexual deception is predicted to increase diversification rates, particularly when reproductive isolation is potentially readily generated by mutations that control the chemistry of pollinator attraction. The evolution of sexual deception may also lead to a specialised pollination system incapable of undergoing reversals to more generalised strategies. Until now, tests of these hypotheses have been impossible due to a lack of well-resolved phylogenies. *Caladenia* is a diverse genus of Australian terrestrial orchids comprising over 360 species. The genus is unusual among sexually deceptive genera in that it also contains species pollinated by food-reward and food-deception, providing a unique opportunity to investigate the evolution of specialised pollination systems. However, like many other orchid groups, existing phylogenies based on just a few chloroplast and nuclear DNA loci are poorly resolved at both the tips and the deep branches. Therefore, we are presently developing and applying an Exome Capture method with Next Generation Sequencing (NGS) for the phylogenetic analysis of *Caladenia* and other Australian terrestrial orchids. The probe design was based on cDNA sequences from the leaf transcriptomes of 11 species spanning the Australia Diurid orchids and outgroups, including two *Caladenia*, plus additional floral transcriptomes of four other *Caladenia* species. We are also targeting a subset of the 315 putative single locus orchid wide loci identified by Deng et al (2015), which will allow sequence sharing with other studies. In a key distinction from typical methods where the capture probes are designed to target short segments (<200 bp) across several thousand coding loci, we are simultaneously targeting longer regions (>1500 bp) consisting of both protein coding sequences and introns for up to 1500 loci. Modifications during the library preparation phase are designed to maximise the sequencing coverage of the predicted, but unknown introns. The downstream bioinformatics analysis will include *de novo* transcript assembly, alignment with the reference sequences, and extraction of coding only, coding and intron, and intron only sequence subsets for the phylogenetic and population genetic level analyses. At the time of writing the abstract, the laboratory work is still in progress. In the talk itself we will reveal the outcomes of our phylogenetic analysis, and discuss the insights revealed by our overlays of chemical and morphological traits onto the phylogeny. We will also provide a critical analysis of the strength and limitations of this NGS based technique and explore its applicability beyond our study system.

T2-16-04

Phylogenomic insights into the evolution of Australia's rich endemic orchid flora

Katharina Schulte[1,2], *Natascha Wagner*[3,4], *Kamolphat Atsawawaranunt*[5], *Allison Mertin*[6], *Lars Nauheimer*[1], *Leah Kaslar*[1], *Mark Clements*[7]
1. Australian Tropical Herbarium
2. CSIRO National Collections and Marine Infrastructure
3. University of Kassel
4. Georg-August-Universität Göttingen
5. Imperial College London
6. University of Wollongong
7. Center for Australian National Biodiversity Research

The Australian flora is characterized by high levels of endemicity, with 92% of its ca. 19,324 native vascular plant species not occurring anywhere else. The three orchid tribes Diurideae, Cranichideae, and Dendrobieae account for nearly 90% of the species diversity of the rich and highly endemic Australian orchid flora, which harbours many threatened species. Several lineages within these tribes underwent extensive diversification on the Australian continent, such as in the subtribes Prasophyllinae, Caladeniinae and Thelymitrinae (Diurideae), Pterostylidinae (Cranichideae), and sections of the epiphytic mega genus *Dendrobium* (Dendrobiinae). Previous phylogenetic studies in these groups were based on few molecular markers resulting in partially unresolved tree topologies with low support, and had gaps in the sampling of critical taxonomic groups. Hence, they provided only limited insights into phylogenetic relationships, morphological character evolution and spatio-temporal evolution. To improve our understanding of the evolution and biogeographic history of Australia's exceptional orchid flora we carried out phylogenomic studies in the Diurideae (nine subtribes), Pterostylidinae, and Dendrobiinae. Altogether, plastome data for 520 samples was generated via shotgun high-throughput sequencing, representing all genera and major intrageneric lineages within the eleven subtribes. 246 samples were included for Diurideae, 107 samples for Pterostylidinae, 143 samples for Dendrobiinae, plus the respective outgroup samples. Maximum likelihood analyses and Bayesian tree inference were carried out with RAxML and MrBayes based on concatenated alignments of ~ 46,000 bp length representing 64 plastid genes. Divergence time estimation was conducted based on fossil and secondary calibration points in BEAST. Key morphological and ecological traits were coded and character state changes traced along the phylogenies using maximum parsimony and maximum likelihood approaches. Distributions were coded based on major disjunctions, and ancestral areas were reconstructed based on the BEAST chronogram using model based inference methods using BioGeoBears. The phylogenetic analyses of our taxonomically comprehensive data sets for these eleven subtribes yielded highly resolved and well supported tree topologies. This is greatly improving our understanding of phylogenetic relationships within Diurideae, Pterostylidinae, and Dendrobiinae, e.g. the relationships between the nine Diurideae subtribes are now fully resolved and well supported. This also allowed for the reconstruction of key morphological and ecological characters, such as important floral traits or pollination syndromes, for these iconic orchid groups. The ancestral area analyses provided novel insights into the spatio-temporal evolution of Australasian orchid lineages and into the assembly of the Australian orchid flora.

T2-16-05

Phylogenomic Insights into the spatio-temporal evolution of Asian and Australasian *Bulbophyllum* (Orchidaceae)

Lalita Simpson[1,2], *Mark Clements*[3], *Darren Crayn*[1,2], *Katharina Schulte*[1,2]

1. Australian Tropical Herbarium

2. Centre for Tropical Biodiversity and Climate Change, James Cook University

3. Centre for Australian National Biodiversity Research, CSIRO Plant Industry

The mainly epiphytic orchid genus *Bulbophyllum* Thouars (Dendrobieae, Bulbophyllinae) is the second largest genus of flowering plants with about 2,200 species in 97 sections. The genus has a pantropical distribution with main centers of diversity in Australasia (New Guinea) and Asia, which harbour 67 sections and ~ 1,700 species. Previous molecular phylogenetic studies identified four main lineages within the genus: a large Asian/Australasian clade, an African and a Malagasy clade, and a South American clade. Intrageneric relationships within the Asian/Australasian clade are still poorly understood. This talk will present results from three phylogenetic studies with a focus on the Asian and Australasian *Bulbophyllum*. The first study examines evolutionary relationships within the Asian/Australasian clade based on a representative sampling of the sectional diversity within this clade and plastome data of 70 genes. Bayesian and Maximum likelihood analyses were carried out and yielded highly resolved and well-supported phylogenies providing new insights into intersectional relationships. BEAST analysis based on fossil and secondary calibrations and a broader outgroup sampling were run to estimate divergence times within the tribe Bulbophyllinae. Based on the BEAST chronogram ancestral area reconstructions were carried out in BioGeoBears providing insights into the spatio-temporal evolution of the genus. The second study focuses on the Australasian *Bulbophyllum* section *Adelopetalum*, which diversified mainly in tropical Australia. Phylogenetic analyses were conducted on a combined plastid dataset sampling up to three representatives of the every species of the section plus representatives of closely related *Bulbophyllum* sections as identified in our broader phylogenomic study. Reconstruction of phylogenetic relationships based on three genes (ITS, *mat*K and *ycf*1) identified main clades and resolved interspecific relationships within the majority of these clades. Nevertheless, relationships between major clades remained largely unclear. Therefore, plastome data of 70 genes was generated for each major lineage within the *Adelopetalum* alliance and combined with the previous plastid data set into a supermatrix which was then analysed using ML and Bayesian inference. Our results reveal section *Adelopetalum* is comprised of three main lineages and is paraphyletic with Australasian representatives of section *Minutissima* nested within it. Divergence age estimations using BEAST and ancestral area analyses using BioGeoBears were carried out to examine the spatio-temporal evolution of the *Adelopetalum* alliance in Australasia. The third study investigates patterns of intraspecific genetic diversity of two *Bulbophyllum* species of section *Adelopetalum* endemic to the mountain tops of Australia's Wet Tropics (*B. lageniforme* and *B. newportii*) to assess the likely impact of future climate change on the genetic diversity of Australia's mountain top orchids. Restriction site associated DNA (ddRAD) genomic data was generated for 8-15 individuals per population and SNP and multi-locus alignments produced using the pyRAD software pipeline. Our phylogeographic analysis reveal phylogeographic structure is present in both species corresponding to major climate refugia of the Plio- and Pleistocene.

T2-16-06

Genome size and phylogenetics of Vanilloideae (Orchidaceae) inferred from nextgen anchored phylogenomics

Kenneth Cameron, *Giovanny Giraldo*

University of Wisconsn, Madison

Phylogenetic relationships among genera in tribe Vanilleae (Vanilloideae) remain elusive among various analyses that have addressed this tribe specifically or focused at the family level.

One of the reasons why phylogenetic reconstruction of this group has been so difficult is that five of the nine genera of the tribe are achlorophyllous mycoheterotrophs (e.g., unplaced *Lecanorchis*), and hence are poorly or not at all represented in molecular studies that relied heavily on plastid loci. Interestingly, even the relationships among some photosynthetic genera remain unresolved (e.g., the relationship between South American *Epistephium* and the clade of New Caledonian endemic genera *Clematepistephium+Eriaxis*) Through the use of Anchored Phylogenomics via Next Generation DNA Sequencing, we have inferred the phylogenetic relationships of most genera in the Vanilleae tribe using data from 23 low copy nuclear markers. Our results indicate that *Epistephium* forms a clade with *Clematepistephium + Eriaxis,* as well as *Lecanorchis*. Most earlier studies showed *Epistephium* to be sister to the entire tribe. Although the relationship of *Epistephium* to the New Caledonian taxa has been suggested before, this is the first time such a relationship is recovered with high support. The placement of *Lecanorchis* in the group reflects what would be expected based on the morphology of the flower, as it has a calyculus, an overgrowth of the top of the ovary that is also present in the other three genera of this clade. In addition to phylogenetic reconstruction, we also estimated genome size for most genera of Vanilloideae, since published reports for the subfamily to date have been entirely from Vanilla. Among angiosperms, orchids have the most variable genome size, ranging from 0.33 pg in *Trichocentrum maduroi* to 55.4 pg in *Pogonia ophioglossoides*, a member of Vanilloideae. Not only does the subfamily have the largest known genome, but it also contains the largest chromosome number reported for Orchidaceae: *Epistephium lucidum* (2n= ca. 170). Here we report the genome size for several additional members of Vanilloideae: several species of *Vanilla, Clematepistephium smilacifolium*, *Eriaxis rigida*, *Epistephium duckei*, mycoheterothroph *Erythrorchis cassythoides*, both *Isotria* species, tropical *Cleistes*, and others. Our data shows that there is at least an 18-fold range of genome size in Vanilloideae, from 2.985 pg for *Eriaxis rigida* to 55.4 pg in *Pogonia ophioglossoides*. The implications of these findings will be discussed in the context of the family's phylogeny.

T2-17: Biodiversity and phylogeography of bryophytes

T2-17-01

Combining ecological niche models and dispersal simulations to predict bryophytes dynamic response to climate changes in Europe

Florian Zanatta[1], *Robin Engler*[2], *Katul Gabriel*[3], *Jairo Patiño*[4], *Antoine Guisan*[2], *Alain Vanderpoorten*[1]

1. University of Liège

2. University of Lausanne

3. Duke University

4. Instituto de Productos Naturales y Agrobiologia

Dispersal is a key evolutionary force that determines the survival, growth and reproduction of individuals, cycles of colonization and extinction of populations and globally drives species dynamic responses to their environment. In the context of extent climate changes, these dynamics take on even greater importance as the survivability of species depends on their ability to shift their distribution or adapt to changes in local climatic conditions. We focus on bryophytes, whose documented long distance dispersal ability, together with their lack of vascular system, position as likely indicator species to monitor the effects of climate changes. We use an approach combining Ecological Niche Models (ENM) and Dispersal Simulations (MigClim) to model the actual niche and simulate the future potential response of bryophytes to environmental changes in Europe. For this purpose we computed specific dispersal kernels using WALD analytical models that integrate both experimentally derived spores settling velocity and continental wind and forest canopy data to infer potential species dispersal capacities. MigClim cellular automaton was then used to simulate atmospheric dispersal over maps of habitat suitability generated from ENMs and enabled us to question European bryophytes ability to colonize suitable areas under actual, and predicted future climatic conditions.

T2-17-02

Biogeographic patterns of Southern South American bryophytes: New evidence based in recent exploration

Juan Larraín

Pontificia Universidad Católica de Valparaíso

The southern Andes are isolated from the rest of the South American continent, and can not be classified as part of the Neotropics. The Atacama desert by the north, the high elevation of the central Andes reaching 7,000 m at Mount Aconcagua, and the dry Patagonian steppes by the southeast, constitute a strong barrier that only a few species can overcome. Remarkable bryogeographic patterns are found in the flora of this region, with the southernmost part of the Andes bearing about 50% of endemic taxa, and a large proportion of taxa shared with distant Austral land masses like New Zealand, SE Australia, Tasmania, the Subantarctic islands, and even with Antarctica. There are four times more species that have this latter distribution pattern than the species that can reach the Tropical Andes. By the other hand, central Chile has a very different bryophyte flora, with a rich endemic component and a large number of Mediterranean elements that links it with the bryophyte floras found in Mediterranean Europe, western North America, South Africa and western Australia, with many ephemeral taxa and a remarkable diversity of Pottiaceae, Bryaceae, and thalloid liverworts. The land between central Chile and the southernmost part of the continent show a combination of these two patterns resulting in a very rich and interesting bryophyte flora. Recent intense bryophyte exploration in different parts of Chile and Argentina (Valparaíso Region, Bío-Bío Region, Chiloé, Aisén, south Argentinian Andes, and Cape Horn) have yielded many new records and new species described for the area, improving the knowledge we previously had about both the diversity and the biogeographic relationships of the bryophyte flora of this part of the world. There is still little information about the conservation status of many rare and/or endemic species, and at the same time the human impact on ecosystems increases year by year, especially in the central Chilean regions. This new information gathered in the last years will help addressing these questions, as well as providing a rich source of specimens of all groups of bryophytes available for bryologists world wide to help understanding the diversity and phylogeographic patterns of many taxa.

T2-17-03

Taxonomy and diversity of Lejeuneaceae (Marchantiophyta): Past, present and future

Rui-Liang Zhu, *Lei Shu*
East China Normal University

Lejeuneaaceae, a predominantly pantropical family with richest epiphyllous species, is usually considered the most difficult group of bryophytes owing to the extensive morphological homoplasy and great morphological plasticity, minute plants, and rich species diversity. It is the largest and most diverse family of bryophytes. Since 1820, over 100 genera have been described and 81,95,68, and 71 genera were accepted in 1983,1997,2013, and 2016, respectively. At present, the Lejeuneaceae contains 1873 species in 70 genera, comprising over 1/4 of the extant liverwort and hornwort species. Although recent morphological and molecular phylogenetic studies have substantially altered our understanding of this notoriously difficult family, the classification and phylogenetic relationships within the Lejeuneaceae remain incompletely known owing to the limited taxon sampling and little monographic work. In the present presentation, we review the taxonomic history of this family and report our recent taxonomic and molecular phylogenetic progresses based on a considerable extension of the taxon sampling. Five new genera and a number of new species were found. To reveal the morphological evolution, diversity, and phytogeography of the Lejeuneaceae, the future work should focus on intensive surveys in diversity-rich areas and comprehensive monographs derived from integrative evidence.

T2-17-04

How molecular evidence transformed our understanding of bryophyte diversity and phylogeography in Europe, the bryologically best known continent

Lars Hedenäs
Swedish Museum of Natural History

Bryophyte diversity and species' distributions in Europe have been studied intensively at least since the age of Hedwig at the end of the 18th Century. The bryoflora of Europe is therefore likely better known than that of any other continent. However, during the last few decades investigations of intraspecific molecular variation and phylogeographic patterns have revolutionized our understanding of European bryophyte diversity and its distribution. Earlier, bryophyte species were viewed as less variable than vascular plants, and it was thought that they encountered no dispersal limitations due to their small spores (often < 20 μm). Widespread species were assumed to be genetically more or less homogeneous over their entire distribution areas. This is clearly contradicted by the molecular variation unveiled below to just above the traditional morphological species level, as well as by the clear phylogeographic patterns discovered in many species. We now have evidence on how some European species evolved before the Pleistocene and on how other species recolonized Scandinavia after the latest glacial period. We have uncovered many earlier hidden species thanks to integrative taxonomy, have revealed (so far) cryptic species, and have disclosed paraphyletic species and other complicated molecular relationships among closely related species. I will present how our perception of the European bryophyte diversity has gradually changed from the relatively straight-forward species-centered interpretation before the molecular era, to a more complex and far more interesting modern and comprehensive view. Specifically, I will outline the implications of these new data (i) on our understanding of the extant European bryophyte diversity, its distribution, and evolution; (ii) on how to manage bryophytes in order to safeguard their diversity not only at the (morphological) species level, but their total diversity; and (iii) for bryophyte barcoding, in Europe and elsewhere.

T2-17-05

Diversity, biogeography and conservation of the thalloid liverworts and hornworts of Sri Lanka

N. C. Sumudu Ruklani[1], *Sumudu C. K. Rubasinghe*[1,2], *Laura L. Forrest*[3], *David G. Long*[3]
1. Postgraduate Institute of Science, University of Peradeniya, Sri Lanka
2. Faculty of Science, University of Peradeniya, Sri Lanka
3. Royal Botanic Garden Edinburgh

Sri Lanka is a tropical island in the Indian Ocean, with a total land area of about 65,610 km^2. The island harbors a diversity of ecosystems, with a varied climate and topography. Together with the Western Ghats of India, Sri Lanka forms a global biodiversity hotspot. The bryophyte flora of Sri Lanka remains poorly researched, and therefore does not contribute much towards national biodiversity statistics. According to literature-based checklists, there are 560 species of mosses (Bryophyta), 327 species of liverworts (Marchantiophyta) and 5 species of hornworts (Anthocerotophyta). Thalloid liverworts recorded include 8 families, 9 genera and 20 species of complex thalloids (Marchantiidae) and 6 families, 8 genera and 19 species of simple thalloids (Pelliidae and Metzgeriidae). There is no Flora of Sri Lankan bryophytes; endemic and threatened taxa have not been identified, a major impediment to carrying out research on these important lineages. Of the three bryophyte phyla, Marchantiophyta and Anthocerotophyta are the least studied. The present study aims to investigate the biodiversity and biogeography of thalloid liverworts and hornworts of Sri Lanka. Field explorations were carried out throughout the country, from the coastline and lowlands to the central highlands, to collect fresh thalli of liverworts and hornworts. Morphological and anatomical characteristics of the collected samples were observed using light and stereo microscopes. Spore morphological characteristics were observed using scanning electron microscopy (SEM). Specimens were identified to generic and specific levels using taxonomic keys and monographs. Taxa identified include 8 families, 9 genera, 13 species of complex thalloid liverworts (Marchantiidae), 4 families, 6 genera, 6 species of simple thalloid liverworts (Pelliidae and Metzgeriidae) and 3 families, 5 genera, 6 species of hornworts. This includes two new records of hornworts for Sri Lanka. The study presents the first phylogenetic framework of Sri Lankan thalloid liverworts and hornworts, based on DNA sequence data from two plastid genes (*psb*A and *rbc*L) and one nuclear gene region (ITS), sequenced for 100 accessions, representing samples collected throughout the geographic range of the island. Phylogeographic patterns of all identified species were traced and mapped using GIS. The highest diversity of thalloid liverworts and hornworts was encountered in the central highlands (>1000 m) of the country, under a cool temperate climate. Their microhabitats include wet soil, rock crevices, rotten wood, mountain slopes and roadside verges. Many of their habitats are under severe threat due to construction and cultivation. Air pollution by vehicle emissions causes changes in the microclimatic conditions and may force out some sensitive bryophytes from their natural habitats. This study emphasizes the importance of protecting these habitats, and thereby conserving the existing diversity of the bryophyte flora of Sri Lanka.

T2-17-06

Inferring species delimitation of the problematic *Riccia fluitans* species complex (Ricciaceae, Marchantiidae) with integrative taxonomy

Lucile Rabeau, *Mathilde Ruche, Catherine Reeb*
Muséum National d'Histoire Naturelle

Understanding processes by which species diversify is a fundamental problem in ecological and evolutionary genetics. Among thalloid liverworts, thecosmopolitan *Riccia*genus contains a large number of species and paradoxically a great endemism rate, common pattern among angiosperms but quite rare among bryophytes. Unlike other Marchantiidae, capsules containing the spores are developing inside the thallus itself.*Riccia*'s spores are amongst the largest in liverworts and bear some of the most discriminant characters. Furthermore, their dispersion mode remains unknown. This set of particularities makes*Riccia*a key genus to understand the evolution of Marchantiidae and how lineages diversify through time and space. Substrate's nature has a strong impact on their installation, implying micro-adaptations likely to ease endemism and radiation events. *Riccia fluitans* species complex contains about ten species including *Riccia fluitans*, *R.rhenana, R. canaliculata, R. duplex* and *R. stricta*. Their high phenotypic plasticity hence the lack of vegetative reliable characters and the rare fertile state induce a challenging taxonomy, uneasy characterization and identification. We selected this species complex as a model to answer the following questions: how do taxa isolate and diversify at a large spatiotemporal scale? How to reconcile a great dispersion and endemism inside the same complex? We used integrative taxonomy to determine the fine scale relationship of 70 samples of the*Riccia fluitans* species complex. Species were sampled from a wide range of geographical locations and environmental niches. Both terrestrial and aquatic habitats are represented. The aims are to delimit species and then to produce a phylogenetic frame to study diversification modes, presence/absence of radiations and the origin of endemism. Study of morphological and ecological data will also allow us to test the relationship between diversification, morphological innovations, habitats and therefore adaptative characteristics related to the diversification. Molecular species delimitation was conducted from different analyses based of chloroplastic and nuclear markers: barcode, genetic distances, coalescence and haplowebs. Several species seem to be confirmed by our study when other are not discriminated. Analysis of morpho-anatomical characters allowed for the reconstruction of evolutionary history. Incongruence across some results suggests a complex evolutionary history for this group. Insights on evolution of spore's ornamentation were drawn. Challenges and future directions are discussed; even though morphological characters are inefficient to delineate species, the power of spores' morphometry approach seems promising. The coming study of type specimens will allow to truly clarify the taxonomy of this complex.

T2-18: Plant speciation in the 21st century (two sessions)

T2-18-01

Molecular evolutionary genetics of hybrid lethality between sympatric *Mimulus* species

Matthew Zuellig, ***Andrea Sweigart***
University of Georgia

As a common cause of reproductive isolation in diverse taxa, hybrid incompatibilities are fundamentally important to speciation. Recent work has provided insight into the molecular functions of genes involved in hybrid dysfunction, but we still know little about how such genes initially evolve within species. Here we describe a simple genetic incompatibility that causes lethality in hybrids between two closely related species of yellow monkeyflower (*Mimulus guttatus* and *M. nasutus*). This hybrid incompatibility, which causes a fraction of F_2 hybrid seedlings to lack chlorophyll and die shortly after germination, occurs between sympatric populations that are connected by ongoing interspecific gene flow. In a series of genetic crosses between plants collected from throughout the species' ranges, we have discovered that both hybrid lethality alleles are widely distributed and rarely fixed within populations. Moreover, polymorphism at one of the loci appears to be driven by introgression between the two *Mimulus* species. We have now genetically mapped hybrid lethality to two small regions that contain duplicated copies of a promising candidate gene. This polymorphic hybrid lethality system offers a unique chance to directly investigate the evolutionary dynamics of incompatibilities in nature.

T2-18-02

Phylogenomics reveals common features of recent plant radiations in the Solanaceae

Leonie Moyle[1], *Meng Wu*[1], *Jamie Kostyun*[1], *James Pease*[3], *David Haak*[2], *Matthew Hahn*[1]
1. Indiana University, Bloomington
2. Virginia Tech
3. Wake Forest University

For decades, studies of rapid species radiations have provided insight into the causes and consequences of diversification, including the factors that can accelerate speciation. Recently, these analyses have been revolutionized by the availability of next-generation sequencing technologies which enable a detailed characterization of clade-wide genetic variation across hundreds to millions of genomic sites. In addition to illuminating the phylogenetic relatedness of population and species, these data have begun to reveal complex patterns of genetic differentiation and sharing that appear to characterize radiations in a diverse range of organisms. Using clade-wide comparative transcriptomic data in two closely related, ecologically diverse, genera in the Solanaceae—*Solanum sect. Lycopersicum* (wild tomatoes) and *Jaltomata*—we describe several features of these genome-wide analyses of diversification. First, we find evidence for abundant shared ancestral variation across whole clades. Second, we show that phylogenetic reconstruction is characterized by rampant gene tree discordance due to incomplete lineage sorting (ILS) that arises from rapid successive lineage splitting events. And, third, we detect discrete post-speciation introgression events between lineages, that also contribute to phylogenetic discordance. Despite these complexities, we show that (when carefully handled) phylogenomic data can still be used to identify candidate loci contributing to rapid diversification. This includes loci where lineage-specific adaptive evolution is associated with ecological differentiation in *Solanum*, and rapid floral evolution in *Jaltomata*.

T2-18-03

What can the genomic composition of the *Helianthus* hybrids tell us about hybrid speciation?

Gregory Owens, *Rieseberg Loren*
University of British Columbia

How do the genomes of different species merge to form a new homoploid hybrid species? How quickly do new hybrid genomes become stabilized? Do different parts of the genome stabilize at different rates? Is the process largely stochastic or is governed by natural selection? To address these questions, we employed transcriptome sequence data and a newly developed maximum likelihood approach to analyze the genomic composition of three homoploid hybrid species in the sunflower genus, *Helianthus*. These species are particularly suitable for this research because they are known to have arisen from the same parental species. We find that the hybrid species are older than previous estimates based on microsatellites. Genome composition is highly consistent within hybrid species and even has appreciable consistency between species, suggesting common selection during speciation. Parental fragments are overwhelmingly tiny, but larger (>10cM) fragments exist, highlighting regions potentially targeted by selection. Despite the ages of the hybrid species, we find that the genomes are not fully stabilized and residual interspecific heterozygosity remains. Surprisingly, we find evidence that two of the hybrid species, *H. anomalus* and *H. deserticola*, may have originated from a single hybrid speciation event and then subsequently speciated.

T2-18-04

The evolution of reproductive isolation in orchids

Giovanni Scopece, *Salvatore Cozzolino*
University Federico II of Naples

The type of reproductive barrier predominant in the initial stages of species divergence can affect the nature and rate of emergence of additional barriers that subsequently strengthen reproductive isolation between species. Different groups of Mediterranean deceptive orchids are characterized by different levels of pollinator specificity. Whereas food-deceptive orchids show weak pollinator isolation, the sexually deceptive species display a highly specialized pollination. Comparative analyses revealed that orchids with high pollinator specificity mostly rely on few premating reproductive barriers and have very little postmating isolation. By contrast, orchids with weak pollinator isolation, such as food-deceptive orchids, show strong evidence for multiple, postmating reproductive barriers. These different types of isolating barriers (and related genetic background) are likely to affect not only the first phase of species divergence but also the evolution of other barriers that emerge when species meet in secondary contact zones. Indeed, secondary contact zones between closely related species are the natural arenas where reproductive isolation is exposed to natural selection and additional reproductive barriers may arise between former independent species as a consequence of ecological interactions or due to selection against hybridization. This process acts on prezygotic mechanisms but it is likely to be affected by different levels of pollinator isolation and by the contribution of other pre-existing barriers. In a case study of sexually deceptive orchids, two species coexist despite the common pollinator and the virtual absence of later barriers. Here, sympatric plants overlap in phenology but have more divergent floral traits than allopatric ones in this way shaping a mechanical barrier that impede or strongly limit the exchange of interspecific pollen. In food-deceptive orchids with weak pollinator isolation natural selection on reproductive barriers usually do not affect pollinator interactions but phenological overlap and postmating prezygotic barriers as pollen-stigma incompatibility. Indeed, in a closely related species pair, allopatric individuals do not show the pollen-stigma incompatibility that characterize sympatric individuals. Also, in several cases, this postmating barrier is highly asymmetric among sympatric species and is generally stronger in the crossing direction that is more likely to occur in nature (i.e. from the early flowering to the late flowering species) thus suggesting that it may have evolved as a consequence of a local selective pressure. These findings highlight that the subsequent evolution of later prezygotic barriers may take different avenues depending on the initial level of pollinator isolation evolved in the early stage of species divergence.

T2-18-05

Genome-wide investigation on parallel speciation in *Oryza nivara*

Zhe Cai[1,2], *Ning-Ning Ren*[1,2], *Lian Zhou*[1,2], ***Song Ge***[1,2]
1. State Key Laboratory of Systematic and Evolutionary Botany, Institute of Botany, Chinese Academy of Sciences, Beijing 100093, China
2. University of Chinese Academy of Sciences, Beijing 100049, China

Parallel speciation is a typical form of parallel evolution in which similar traits associated with reproductive isolation evolve repeatedly in different populations within species and strongly implicates the action of natural selection. Despite numerous documentations in animals, example of parallel speciation with unequivocal evidence has been lacking in plants. Particularly, the genomic basis of parallel speciation is largely unknown. Two wild rice species, the perennial *Oryza rufipogon* and annual *O. nivara*, are evolutionarily young and distinct in ecology and morphology and thus offer a unique opportunity for studying speciation. Here we take the advantage of high-throughput sequencing, in combination with traditional common garden experiment, to investigate the origin model of *O. nivara* as well as the causes and consequences of genomic divergence between the two species. Based on the re-sequencing data for 24 natural populations, in conjunction with population genomics approach and statistic modeling, we tested for alternative models of *O. nivara* origin and revealed that *O. nivara* originated multiple times under similar habitats from different *O. rufipogon* populations. Common garden experiment on eight sympatric pairs of *O. rufipogon* and *O. nivara* populations showed significantly phenotypic divergence between species but not within species. We showed evidence that the shared phenotypes in different *O. nivara* populations have evolved under natural selection. Importantly, we found an effective premating isolation between species but not within species. In addition, genome-wide screening detected both a large proportion of genomic regions with significant divergence between species, including regions with parallel genomic divergence across multiple population pairs and those unique to individual population pairs. These divergent regions contain many genes associated with functional terms involving phenotypic and ecological differentiation between species. Our study suggests that parallel speciation as a consequence of adaptation to alternative environments might be widespread in plants and provides new insights into the genomic basis underlying parallel speciation in plants.

T2-18-06

Does speciation drive biodiversity? A systematic analysis of multiple mangrove species on the Indo-Malayan coasts and a novel mechanism of speciation

Ziwen He[1], *Xinian Li*[1], *Ming Yang*[1], *Chung-I Wu*[1,2], ***Suhua Shi***[1]

1. State Key Laboratory of Biocontrol and Guangdong Provincial Key Laboratory of Plant Resources, School of Life Sciences, Sun Yat-sen University

2. Department of Ecology and Evolution, University of Chicago

High biodiversity may be a modern "mystery of mysteries"; in particular, speciation mechanisms have not been one of an ensemble of explanations. Indo-Malayan coasts, having continuous geological records and possessing a high level of biodiversity that includes 70% of the living mangrove species, may offer new insights into the possible connections between biodiversity, geography and speciation. Here, we analyze the pattern of species and population divergence on these coasts by collecting DNA sequences from ~1,700 plants of 5 common mangrove species. Good species and geographical populations provide the upper and lower bound estimates of speciation time, respectively. This large dataset, superimposed on biogeographical observations, reveals that mangrove speciation is driven mainly by the isolation at the Strait of Malacca. However, the isolation is periodically punctuated by migration due to fluctuations in sea level leading to the opening/closing of the Strait. As a result, speciation has generally proceeded through cycles of isolation and migration, with each cycle averaging 100,000 years. Theoretical analyses indicated that while gene flow is permitted during divergence, isolation phase is still indispensable for genetic incompatibility to evolve. This mechanism, referred to as the mixing-isolation-mixing (MIM) cycles, extends several proposals that aim at integrating gene flow and geographical isolation in a single framework of speciation and provides the basis for further investigations of adaptive evolution in mangroves. After n cycles, the mechanism could potentially yield m^n (m>1) species and may contribute to the maintenance of high biodiversity.

T2-18-07

Genomic insights into the combination of differentiated alleles that promote diploid hybrid speciation

Jianquan Liu

Sichuan University

Homoploid hybrid speciation is increasingly viewed as a contributor to current species diversity. In addition to the recombination of chromosomal structural differences, the recombination of alleles that differentiate parent species across their genomes may promote reproductive isolation (RI) of a new homoploid hybrid species. We tested this hypothesis by conducting transplant studies and examining allele combinations at multiple loci in the diploid hybrid species, *Ostryopsis intermedia* (Betulaceae), and its two parents, *O. davidiana* and *O. nobilis,* through whole genome sequencing. First, we examined RI between the hybrid and its parents by transplanting *O. davidiana* into sites occupied by one or other of its two parents. We found that RI between *O. intermedia* and *O. nobilis* resulted from both a pre-pollination flowering time difference and a post-zygotic reduction in fruit set. In contrast, RI between *O. intermedia* and *O. davidiana* appeared to involve habitat selection with *O. davidiana* less able to tolerate high levels of soil iron content. Second, our genomic analyses indicated that more than 70% of genes of *O. intermedia* were closely related to those of *O. davidiana*, including those likely to affect flowering and fruiting, while the remainder grouped with those of *O. nobilis* and included key genes responsible for absorption and transport of iron ions in plants. Further functional tests of parental alleles for one of the latter genes confirmed that they greatly affected iron absorption. We conclude that population genomic data support three independently evolving lineages of *Ostryopsis* with one, *O. intermedia*, sharing alleles that distinguish its two parent species. Our findings highlight the importance of combining differentiated alleles at multiple loci in the promotion of RI of a homoploid hybrid species from its parents.

T2-18-08

Parallel speciation during adaptive radiation in an Australian wildflower

Daniel Ortiz-Barrientos

The University of Queensland

Deterministic forces can create order in nature. For instance, natural selection can lead to organic discontinuity by reducing the homogenizing effects of sexual reproduction between populations, thus freeing traits to evolve along different trajectories. Equally, if numerous environments occur repeatedly across landscapes, evolution of form can echo over time and space, and traits with similar expression can arise independently in multiple populations. This form of repeated evolution is not uncommon in nature, and seems to define the evolution of costal and alpine forms in the species complex *Senecio lautus*. Here, through as series of ecological and genetic experiments we show that environmental dependency is crucial to the origins of organic discontinuity, but also show that natural selection can lead to the irreversible breakdown of sexual reproduction between populations occupying contrasting environments. Repeated trait evolution seems to originate from abundant levels of standing genetic variation, but only a few mutations seem to break down sexual reproduction between populations. Whether the genetic architectures of trait evolution and speciation evolve in step with one another remains unknown, but it is clear that natural selection can create discontinuity at different levels of divergence. These results have implications for our understanding of the origin and also the nature of new species.

T2-18-09

The role of retrotransposons in reproductive isolation among wild wheats

Christian Parisod
University of Neuchâtel

Interspecific hybridization leads to new interactions among divergent genomes, revealing the nature of genetic incompatibilities having accumulated after the origin of species. Conflicts associated with misregulation of transposable elements in hybrids expectedly results in their activation and genome-wide changes that may be a key issue to species boundaries. Here, we first used experimental reciprocal F1 hybrids between three *Aegilops* species to assess restructuring and epigenetic repatterning in relation to the strength of reproductive isolation. Asymmetrical reorganization of long terminal repeat retrotransposons (LTR-RTs) predicted to cause conflicting interactions matched differential survival of F1 hybrids. In particular, non-random sequence losses and methylation appeared coherent with the necessary repression of incompatible TE loci in sustaining hybrid viability. We then used natural hybrid zones between *Aegilops geniculata* and *Aegilops triuncialis* to further tackle the consequences of such conflicts for maintaining species boundaries in face of recurrent gene flow. Coherent with insights from experimental populations, conflicting LTR-RTs presented asymmetrical introgression and specific reorganization among natural hybrids. Genomic conflicts involving LTR-RTs are thus highlighted as key incompatibilities supporting reproductive isolation and shaping species boundaries in the long term.

T2-18-10

Repeated speciation and long-term evolution of allopolyploids in *Achillea* (Asteraceae)

Jun-Nan Wan[1], *Yan-Ping Guo*[2], *Guang-Yuan Rao*[1]
1. School of Life Sciences, Peking University, Beijing
2. College of Life Sciences, Beijing Normal University, Beijing

Allopolyploidy is considered to be an important mechanism of speciation in plants. Many polyploidy species, including a number of allopolyploid species, have been known of recurrent formations with strong evidence. Due to the complexity of polyploidy data, the long-term evolution after recurrent polyploid formation, as well as its influence on speciation remain to be less understood and need more studies on polyploid lineages with relatively long evolutionary histories. Previous work demonstrated that two East Asia *Achillea* allotetraploid species, *A. alpina* and *A. wilsoniana*, share diploid progenitors, one *A. acuminata*-like and the other *A. asiatica*-like. Here, we described a new progenitor-specific amplicon sequencing approach that allows us to separately estimate the divergence of two allotetraploid subgenomes from their progenitors within a framework of population demographic theory. We found dramatic differences between *A. alpina* and *A. wilsoniana* based on both progenitor-specific subsets of nuclear genetic markers, suggesting that the two tetraploid species were formed independently from at least two allopolyploid events with the same diploid species but different populations as progenitors. Furthermore, plastid sequence data indicated at least two formations of *A. alpina alone*. In sum, our analyses show recurrent allopolyploid formations, some of which are consistent with the species delimitation or differentiation on morphology, and some are infra-species. Moreover, with this working system, we were able to reconstructed long-term evolution of allopolyploids after their establishment using Approximate Bayesian Computations (ABC). The analytical result suggested potential competitions between new formed tetraploids and their diploid progenitors. Our method with progenitor-specific genetic markers could also be applied to other allopolyploid species with little genomic information, to uncover the allopolyploid speciation process.

T2-18-11

Evolution of the nuclear and plastid pangenome: Divergence vs introgression in the model grass *Brachypodium distachyon*

Pilar Catalan[1], *Bruno Contreras-Moreira*[2], *Ruben Sancho*[1], *Daniel Woods*[3], *Richard Amasino*[3], *Sean P. Gordon*[4], *John Vogel*[4]
1. Universidad de Zaragoza
2. Estación Experimental de Aula Dei-Consejo Superior de Investigaciones Científicas
3. University of Wisconsin, Madison
4. DOE Joint Genome Institute

A pan-genome is the non-redundant collection of all DNA sequences present in the entire population of a species. It is composed of a "core genome", containing sequences present in all individuals, and a "conditionally dispensable genome", that includes sequences not present in all lines. Though "dispensable", these genes may be of high value under some conditions as they are enriched for functions associated with potential adaptive traits (e.g., stress tolerance, disease resistance and development). We created a robust nuclear pan-genome of *Brachypodium distachyon*, a wild annual grass endemic to the circum-Mediterranean region and a powerful model system for the grasses. We analyzed 54 whole nuclear genome sequences, including both genes and repeatitive elements, for *B. distachyon* lines chosen to represent a large portion of the genomic and phenotypic diversity of this species. We characterized the interplay between core/dispensable genomes, population structure, gene expression, predicted gene function and non-coding sequences. We constructed a phylogenomic tree based on >4 million SNPs, largely congruent with groupings based on median amino acid similarity in pan-genome clusters among lines, as well as copy number of dispensable pan-genome clusters. There was strong population structure in the tree, which showed a main split into an extremely delayed flowering (EDF) clade and a second clade subdivided into two groups defined by geography (East, West). Many flowering time traits and putative flowering time genes showed strong evidence of correlated evolution with flowering class and weak correlation with latitude. Genetic structure analysis showed that some lines were admixed, indicating recent, though infrequent, interbreeding among groups. This was confirmed through the phylogenomic analysis of their plastomes, that reconstructed a plastome tree highly congruent with the nuclear pan-genome tree showing nine additional chloroplast capture and two ptDNA introgression and micro-recombination events.

Estimations from nested dating analysis suggested that the *B. distachyon* ancestor split in Mid-Pleistocene times (~0.9 Mya), and that intra-specific divergences occurred very recently, during the last half million years. Flowering time variation is a main factor driving rapid intra-specific divergence (microspeciation) in *B. distachyon*, though it is counterbalanced by repeated introgression between previously isolated lineages. Swapping of plastomes between the three different genomic groups (EDF, West, East) likely resulted from random backcrossing followed by stabilization through selection pressure.

T2-18-12

Eucalypt systematics: NGS data do not support the Monophyly of *Corymbia* and indicate a complex evolutionary history

Tanja Schuster[1,2], *Sabrina Setaro*[3], *Michael Bayly*[1]

1. The University of Melbourne

2. The Royal Botanic Gardens Victoria

3. Wake Forest University

We used Next Generation Sequencing (NGS) to investigate the phylogenetic relationships of *Angophora* Cav., *Corymbia* K.D.Hill & L.A.S.Johnson, and *Eucalyptus* L'Hér. In particular, we addressed the question of monophyly of *Corymbia* (the bloodwoods and ghost gums). Chloroplast and nuclear data were skimmed from NGS data of genomic DNA. The dataset includes 121,017 base pairs for nine species of *Angophora*, 57 species of *Corymbia* (some with two accessions) and 31 species of *Eucalyptus*, with *Allosyncarpia* and *Stockwellia* as outgroups. Maximum Likelihood and Maximum Parsimony methods show that *Corymbia* and *Angophora* are related as one lineage. *Corymbia* is not monophyletic because *Angophora* is nested within it. Relationships among species of *Corymbia* are not all congruent with current infrageneric classification, some previous phylogenies based on nrITS data, or morphological characters. However, our results confirm outcomes from some previous analyses using fewer chloroplast and nuclear markers. Analyses of our chloroplast data show evidence of chloroplast capture within *Corymbia* and a complex evolutionary history of this group. Our results frame a new working hypothesis for the evolutionary relationships of *Angophora*, *Corymbia*, and *Eucalyptus*. Taxonomic changes are likely after the future re-evaluation of morphological characters and analysis of additional nuclear data.

T2-19: Evolution of the mega-diverse genus *Begonia*

T2-19-01

Species-level functional variation in *Begonia*

Catherine Kidner[1,2], *Mobina Shaukat Ali*[1,2], *Katie Emelianova*[1,2]

1. Royal Botanic Gardens, Edinburgh

2. University of Edinburgh

What distinguishes the many species of *Begonia* from each other? We have used QTL analysis and Differential Expression analysis to determine the genetic differences between two closely related species of *Begonia*. *B. conchifolia* and *B. plebeja* are Central American species found in contrasting environments. With the aid of a genetic map we have identified 31 QTLS controlling vegetative and reproductive form, physiology, growth and micro morphology. The degree of variance associated with these QTLs suggests that the genetic differences between these two species are not great, despite their very different morphologies and habitats. To identify candidate genes for these QTLs and to investigate interspecific variation in traits not scorable in our mapping populations we generated transcriptomes from 6 tissues for each species for differential gene expression analysis. This has enabled us to identify the key genetic pathways, enzyme classes and classes of genes which differ in gene copy number, sequence and expression between the species. Our data suggest that the different fate of duplicated genes in isolated populationsis a key factor in the evolution of so many *Begonia* species.

T2-19-02

Insights of the genome architecture for the species *Begonia conchifolia*

Aureliano Bombarely[1], *Katie Emelianova*[2,3], *Catherine Kidner*[2,3]

1. Virginia Polytechnic Institute and State University

2. University of Edinburgh

3. Royal Botanic Gardens, Edinburgh

Begonia conchifolia A. Dietr. (Begonia section Gireoudia) is a neotropical herb native from Central America that can be found in disturbed areas from Costa Rica, El Salvador and Panama. Although *B. conchifolia* is not a popular ornamental plant, it is an important model to study leaf morphological variation and sex flower ratio across the Begonia genus. Flow cytometry experiments have estimated its genome size in 587 Mb. We have sequenced the *B. conchifolia* genome using a combination of Illumina and PacBio sequencing technologies. The estimation of the genome size based in Illumina read Kmer distribution was comprised between 831 and 910 Mb with a heterozygosity of 0.48%. We have obtained a draft genome assembly of 25,304 scaffolds for a total size of 736 Mb (N50 = 61.3 Kb). The genome annotation reported 31,935 genes. The pairwise analysis of Ks distribution revealed a Whole Genome Duplication (WGD) dated between 13 and 33 MY ago. The terpenoid cyclases gene family showed a notable expansion (78 genes) compared with other plants species such as *Arabidopsis thaliana* (26), *Glycine max* (23), *Populus trichocarpa* (49) and *Vitis vinifera* (45). The development of a reference genome for the Begonia genus will help to elucidate the complex evolutionary history of one of the more diverse genus in the plant kingdom.

T2-19-03

Species radiation by a random walk through niche space in the mega-diverse genus *Begonia* (Begoniaceae)

Peter Moonlight[1], *Tiina Sarkinen*[1], *Mark Tebbitt*[2], *Mark Hughes*[1]

1. Royal Botanic Garden Edinburgh

2. Department of Biological and Environmental Sciences, California University of Pennsylvania

Begonia is among the world's largest genera and has radiated rapidly throughout its pantropical distribution. It is widely assumed that the majority of large, rapid radiations are adaptive. Central to the theory of an adaptive radiation is that the majority of speciation events are adaptive, implying speciation driven by adaptation to different environments and resulting in niche shifts. We evaluate whether observed ecological disparity among species in the mega-diverse genus *Begonia*, calculated through species distribution models, best fits the predictions of models of adaptive or non-adaptive radiations. Patterns of ecological diversification in three clades of *Begonia* strongly diverge from those predicted by an adaptive model, and we conclude *Begonia* is not an adaptive radiation with respects to climatic niche. Ecological disparity within *Begonia* clades best fits the predictions of a non-adaptive, geographic model of speciation and we suggest the remarkable diversity of the genus has developed through geographic speciation, and adaptation to local environments. This phenomenon may be widespread among plant radiations in topographically- and environmentally-heterogeneous areas, and a re-evaluation of putative adaptive radiations throughout the world's montane regions may be necessary.

T2-19-04

Systematic and biogeography of the section *Casparya* (*Begonia*, Begoniaceae)

Orlando Adolfo Jara-Muñoz[1], *James E. Richardson*[2,3], *Santiago Madriñán*[1,4]

1. Universidad de los Andes
2. Universidad del Rosario
3. Royal Botanical Garden Edinburgh
4. Jardín Botánico de Cartagena Guillermo Piñeres

Section *Casparya* include around 41 species, being one of the most species rich groups of *Begonia* in the North Andean region. This is a monophyletic and morphologically well defined section, characterized by a unique type of fruit among the American begonias. The mostly North Andean distribution of *Casparya* make to this section an appropriate system to study the patterns and process of diversification in this region. We generate a multilocus and well sampled phylogeny of *Casparya*, dated through Bayesian inference to estimate the ancestral ranges of distribution and reconstruct his historical biogeography using the model DEC (dispersal-extinction-cladogenesis). In order to understand the dominant modes of speciation and the changes in climatic niche, we performed niche modeling and polygons of distribution for each species and then we measure the age-range and age-niche correlation using independent comparisons. We show that allopatric speciation has been the dominant mode of speciation in *Casparya*, and that climatic niche overlap has not varied from the null expectation, suggesting niche conservatism. The ancestral area reconstructions shows that *Casparya* originated in the Cordillera Oriental of Colombia and colonized during the Pliocene all the major montane systems of the northern Andes, but was during the Pleistocene when most of the extant species were originated. We concluded that *Casparya* have responded to the climatic variation mostly moving around the optimum conditions in the mesic cloud forest, generating species allopatrically, but also have been able to cross occasionally the climatic tolerance adapting to extreme habitats as the páramos.

T2-19-05

New discoveries of *Begonia* from Sangkulirang Nature Reserve, borneo, indonesia

Deden Girmansyah
Research Center for Biology, Indonesian Institute of Sciences

An expedition to Sangkulirang Nature Reserve, East Kalimantan, Borneo was organised in order to visit previously uncollected areas of karst limestone, of potentially high conservation importance. We documented and collected seven native *Begonia* species. Of these, six represent new species, highlighting both the previous lack of knowledge of the area and its high endemism. Living plants and seeds were deposited at Bogor Botanic Gardens for cultivation, and to provide fertile material for description. All species belong the highly species rich section Petermannia, and phylogenetic evidence will be presented as to their biogeographic origin. An overview of the distinctive characters of the species and their distribution will be presented. The study highlights the role of living collections in supporting taxonomic research and conservation, and the high conservation value of the Sankulirang Karst.

T2-19-06

Five new species of *Begonia* (section *Baryandra*, Begoniaceae) from Palawan Island, Philippines

Rosario Rubite[1], *Ching-I Peng*[2], *Che-Wei Lin*[3], *Mark Hughes*[4]

1. University of the Philippines Manila
2. Biodiversity Researach Center, Academia Sinica
3. Taiwan Forestry Research Institute
4. Royal Botanic Garden Edinburgh

In a joint expedition to the island province of Palawan, botanists from Taiwan, United Kingdom and the Philippines found 5 unknown *Begonia* and compared with potentially allied species. The unknown begonias are similar in their rhizomatous stem, persistent stipules, large olivaceous, asymmetric leaves, four-tepaled flowers, inferior three-locule ovary with bilamellate placentas and pendulous, three-winged capsule. The 5 are clearly assignable to section *Baryandra*. Thorough studies of literature, herbarium specimens, and living plants support the recognition of the five new species: *Begonia elnidoensis, Begonia gironellae, Begonia quinquealata, Begonia tabonensis* and *Begonia tenuibracteata.* This brings the total of *Begonia* species in Palawan to 22, and represents a 40% increase in the number of species in section *Baryandra* from Palawan. Phylogenetic evidence shows the species to have originated during the Pliocene-Pleistocene, and sharing of chloroplast haplotypes between distantly related species according to nuclear DNA data shows evidence of recent hybridisation.

T2-20: The Brassicaceae family – a model system to study evolutionary processes with high resolution in space and time

T2-20-01

Resolution of Brassicaceae phylogeny using nuclear genes uncovers nested radiations and supports convergent morphological evolution

Chien-Hsun Huang[1], *Renran Sun*[1], *Yi Hu*[2], *Liping Zeng*[1], *Ning Zhang*[3], *Liming Cai*[1], *Qiang Zhang*[4], *Marcus A. Koch*[5], *Ihsan Al-Shehbaz*[6], *Patrick P. Edger*[7], *J. Chris Pires*[8], *Dun-Yan Tan*[9], *Yang Zhong*[1], *Hong Ma*[1]

1. Institute of Plant Biology, School of Life Sciences, Fudan Uni-

versity
2. Department of Biology, the Huck Institute of the Life Sciences, the Pennsylvania State University
3. Department of Botany, National Museum of Natural History
4. Guangxi Key Laboratory of Plant Conservation and Restoration Ecology in Karst Terrain, Guangxi Institute of Botany, Guangxi Zhuang Autonomous Region and Chinese Academy of Sciences
5. Biodiversity and Plant Systematics, Centre for Organismal Studies (COS) Heidelberg, Heidelberg University
6. Missouri Botanical Garden
7. Department of Horticulture, Michigan State University
8. Division of Biological Sciences, University of Missouri
9. Xinjiang Key Laboratory of Grassland Resources and Ecology, College of Grassland and Environment Sciences, Xinjiang Agricultural University

Brassicaceae is one of the most diverse and economically valuable angiosperm families with widely cultivated vegetable crops and scientifically important model plants, such as *Arabidopsis thaliana*. The evolutionary history, ecological, morphological and genetic diversity and abundant resources and knowledge of Brassicaceae make it an excellent model family for evolutionary studies. Recent phylogenetic analyses of the family revealed three major lineages (I, II, and III), but relationships among and within these lineages remain largely unclear. Here, we present a highly supported phylogeny with six major clades using nuclear markers from newly sequenced transcriptomes of 32 Brassicaceae species and large datasets from additional taxa for a total of 55 species spanning 29 out of 51 tribes. Clade A consisting of Lineage I and *Macropodium nivale* is sister to combined Clade B (with Lineage II and others) and a new Clade C. The ABC clade is sister to Clade D with species previously weakly associated with Lineage II and Clade E (Lineage III) is sister to the ABCD clade. Clade F (the tribe Aethionemeae) is sister to the remainder of the entire family. Molecular clock estimation reveals an early radiation of major clades near or shortly after the Eocene-Oligocene boundary and subsequent nested divergences of several tribes of the previously polytomous Expanded Lineage II. Reconstruction of ancestral morphological states during the Brassicaceae evolution indicates prevalent parallel (convergent) evolution of several traits over deep times across the entire family. Deep analyses with intense taxa samplings covering all tribes and lage genera are ongoing to reveal the complete relationships in this family, which may form a foundation for future evolutionary analyses of structures and functions across Brassicaceae.

T2-20-02

Towards a backbone phylogeny of Brassicaceae and principles of crucifer evolution – unravelling evolutionary patterns and processes

***Marcus Koch**, Multiple Authors Brassibase Consortium*
Heidelberg Universitiy, Centre for Organismal Studies (COS) Heidelberg

The Brassicaceae family (mustards or crucifers) comprises nearly 4000 species based on our newly established species check-list for the entire family and includes *Arabidopsis thaliana* as one of the most important model species in plant biology and a number of important crop plants as well, such as the various *Brassica* species (e.g. cabbage, canola and mustard). Moreover, the family comprises an increasing number of species that serve as study systems in many fields of plant science and evolutionary research such as *Arabidopsis*, *Brassica*, *Boechera*, *Capsella*, *Cardamine*, *Thellungiella*, and *Arabis* species. As any evolutionary hypothesis needs to be placed in a phylogenetic and temporal context, reliable and highly resolved phylogenetic hypotheses of the Brassicaceae are essential. However, those are still missing. Herein we aim to unravel a comprehensive and highly-resolving phylogeny for the entire Brassicaceae family based on whole genome sequences of two organellar genomes: namely the mitochondrial and plastid genomes. Family-wide genomic data are analysed within the framework of the entire Superrosid clade, thereby allowing elaboration on a comprehensive time-calibrated framework with divergence time estimates allowing stem and crown group age calculation for any lineage and tribe of Brassicacae. These data will allow testing the hypothesis of generally occurring diversification-lag-phase after ancient polyploidization (mesopolyploidization) giving rise to many of these tribes and lineages and with more than 43% neopolyploid taxa. Furthermore we aim to elaborate on the idea that spatio-temporal diversification patterns may correlate among the various evolutionary lineages, thereby allowing us to draw conclusions about major past-climate/environmental transitions contributing to the radiation of the entire family in concert with extensive polyploidization, which might either predate or follow these environmental transitions throughout the last 35 million years.

T2-20-03

New insights into crucifer genome evolution

***Karol Marhold**[1,2], Martin A. Lysak[3]*
1. Institute of Botany, Slovak Academy of Sciences, Bratislava
2. Department of Botany, Charles University, Praha
3. Central European Institute of Technology (CEITEC), Masaryk University, Brno

The Brassicaceae family comprises 320 genera and over 3,600 species classified into some 50 bona fide monophyletic tribes clustered in three or five major phylogenetic lineages. The whole-genome sequence projects along with comparative cytomolecular maps for roughly two dozens of crucifer species (e.g., *Arabidopsis* spp., *Arabis alpina*, *Brassica* spp., *Camelina sativa*, *Cardamine* spp., *Schrenkiella parvula*) allowed us to reconstruct ancestral genomes and infer prevalent trends of genome evolution in the *Brassicaceae*. While two major ancestral genomes, ACK (n=8) and PCK (n=7), remained conserved in a number of extant species, in other groups they were reshuffled by inversions and translocations with or without centromere elimination/inactivation. Multiple independent descending dysploidy events dominate the genome evolution in the Brassicaceae. In tribes which experienced recent whole-genome duplication events, the extent and tempo of genome reshuffling towards quasi-diploid genomes exceed that observed in ?true diploid" genomes. While we are revealing the mechanisms of descending dysploidy, we are less clear on the triggers and evolutionary significance of this process. The importance and consequences of the gene linkage, reduction of recombination, and telomere/centromere reduction will be discussed. More detailed examples are provided from the tribes *Alysseae*, *Camelinae* and *Cardamineae* illustrating different approaches in the search for parental species of polyploids, origin of dysploidy and genetic differentiation of polyploid lineages.

T2-20-04

The role of gene and genome duplication in network evolution

Patrick Edger[1], *Chris Pires*[2]
1. Michigan State University
2. University of Missouri

Gene duplications, derived from polyploidy and smaller scale events, are prevalent and recurring throughout the history of higher eukaryotes including angiosperms. Retained duplicate genes, often that functionally diverge, has helped shape the ultimate success of flowering plants. This presentation will focus on the impact of gene duplications on gene networks, and provide examples of both the origin of novel pathways and network rewiring that were and continue to be selected for (naturally and artificially) that encode plant adaptive traits.

T2-20-05

Recruitment of ecologically relevant duplicates in alpine Mustards

Christian Parisod
University of Neuchâtel

Polyploid taxa represent excellent models to address the underpinnings of genome evolution and the building up of new species in heterogeneous environments. Here, I present an overview of recent works in the alpine *Biscutella laevigata* autopolyploid complex (Brassicaceae). Transcriptomics inferred recurrent whole genome duplication (WGD) events specific to clade of species and that were used to infer processes fostering genome evolution across different timescales: (i) After a 7-8 million years old WGD event, intense chromosomal repatterning selected for clusters of retained duplicates enriched in functions associated with responses to abiotic stresses. Low coverage genome sequencing unraveled the dynamics of several retrotransposons, supporting interplay between genome reorganization and environmental opportunities in shaping the evolution of paleopolyploids. (ii) Retrotransposons in autotetraploids having recolonized the Alps after the ice ages showed considerable dynamics going along with ecological radiation following this recent WGD. Ecological genomics involving transplant experiment indeed supported distinct autopolyploid gene pools firmly associated with contrasted habitats despite gene flow. These ecotypes demonstrated adaptive differentiation at loci whose functions match habitat requirements. WGDs thus recurrently fostered genome reorganization and adaptive recruitment of genes responding to environmental factors, indicating that similar proximate and ultimate factors of genome dynamics may consistently act through time.

T2-20-06

Increased genetic variation from mesopolyploidy preferentially selected during domestication of *Brassica* rapa crops

Xinshuai Qi
University of Arizona

Many crops are polyploid or have a polyploid ancestry. Whole genome duplication could provide raw genetic material for later rapid adaptation and diversification of crop species under strong artificial selection. Despite long interest in the connection between polyploid and domestication, this hypothesis has not been empirically tested. Crops in the genus *Brassica* provide a unique system to explore this question because the genus experienced multiple rounds of polyploidization and these crops show extraordinary morphological variation within a few thousand years under artificial selection. Here we analyzed transcriptome data of 102 representative *B. rapa* cultivated accessions and asked to what extent the *Brassica* mesohexaploidy contributes to the diversification and domestication of *B. rapa* crops. We first categorized *Brassica rapa* genes as derived and not derived from the mesohexaploidy (paleologs and non-paleologs). Using three different approaches, we then identified *B. rapa* genes experiencing recent positive selection or significant differential gene expression. Approximately 68% of the *B. rapa* genes were directly derived from the mesohexaploidy. Genes derived from the ancient hexaploidy have four times higher mean nucleotide diversity than other genes. Candidate genes associated with domestication were significantly enriched with these paleologs (10% - 15% higher than expected). Our results for the first time directly demonstrated in what extent polyploidization contributes to the crop diversification at the microevolutionary scale. The observed pattern should also apply to other crops and wild plants which rapidly diversified after polyploidy.

T2-21: Cyperaceae II: later diverging lineages of the Cyperoideae

T2-21-01

Sumatroscirpus: Taxonomic revision and the discovery of a new monotypic Southeast Asian sedge tribe (Cariceae-Dulichieae-Scirpeae Clade, Cyperaceae)

Étienne Léveillé-Bourret[1], *Julian R. Starr*[1], *Bruce A. Ford*[2], *Thị Kim Thanh Nguyễn*[3], *Anh Tài Vũ*[4]
1. University of Ottawa
2. University of Manitoba
3. Vietnam National University
4. Vietnam Academy of Science and Technology

Our knowledge of phylogenetic relationships within CDS (Cariceae-Dulichieae-Scirpeae Clade, Cyperaceae) has grown dramatically in recent years due to increased efforts to identify the sister-group to *Carex* and to clarify the taxonomy of the paraphyletic tribe Scirpeae. Approximately half of all CDS species (total ~2,050 spp.) have now been included in phylogenetic studies, but a single genus has escaped attention due to its extreme rarity: the putatively monotypic *Sumatroscirpus*, which is believed to be endemic to remote mountains in North and West Sumatra. As part of ongoing tribal- and generic-level revisionary work on CDS, we have reexamined the taxonomy and phylogenetic position of this genus. Its current placement in tribe Dulichieae appears at odds with several unique characteristics such as sheathing spikelet prophylls and inflated tubercles on fruits. Molecular phylogenetic analyses support these observations in showing its isolated position within CDS, and justify the recognition of *Sumatroscirpus* as the sole member of a new Cyperaceae tribe. In addition, field studies combined with the examination of all extant herbarium material indicate that the taxonomic diversity of *Sumatroscirpus* has been dramatically underestimated, with a total of four morphologically-distinct species now recognized, and with important new records extending its range into continental Southeast Asia.

T2-21-02

Ficinia clade (Cypereae, Cyperaceae): Evolving taxonomy, diversification patterns and biogeography

A. Muthama Muasya
University of Cape Town

Cypereae is the second largest tribe in the Cyperaceae with over 1200 species which have a predominantly tropical and subtropical distribution. The tribe has a crown age in late Eocene (36 mya) and comprises two lineages, the Cyperus and Ficinia clades, with bulk of the diversity in the predominantly tropical Cyperus clade. The Cyperus clade has recently been a focus of taxonomic and phylogenetic studies, showing rapid diversification in the C_4 clade and complex taxonomy where genera were previously described based on autapomophic (and frequently homoplasious) characters. This presentation will focus on the Ficinia clade, comprising about 160 species in eight genera (*Afroscirpoides, Dracoscirpoides, Erioscirpus, Ficinia, Hellmuthia, Isolepis, Karinia, Scirpoides*) and distributed in cool tropics and temperate areas. The Ficinia clade has a crown age in early Miocene, with earlier emerging lineages now restricted to Mediterranean regions in Central America (Mexico, *Karinia*), Eurasia (*Scirpoides, Erioscirpus*) and Southern Africa (*Afroscirpoides, Dracoscirpoides, Hellmuthia, Scirpoides*), whereas *Isolepis* and *Ficinia* have a centre of diversity in Mediterranean South Africa and distributed into Australasia and elsewhere. Unlike the Cyperus clade that has experienced taxonomic mergers of genera, the Ficinia clade has newly described genera (*Afroscirpoides, Dracoscirpoides*) and now includes two genera previously thought to belong to Scirpeae (*Erioscirpus, Karinia*). Furthermore, *Isolepis* and *Ficinia* show Miocene diversification into emerging niches in the South African Fynbos biome and multiple dispersal events out of South Africa to Australia as well as towards emerging temperate island (volcanic mountains) habitats in tropical Africa.

T2-21-03

Sedges of Vietnam: Species discoveries and insights into the evolution of *Carex* (Cyperaceae)

Bruce A. Ford[1], *Julian R. Starr*[2], *Francesco H. Janzen*[2], *Étienne Léveillé-Bourret*[2], *Nguyễn Kim Thanh*[3], *Vũ Anh Tài*[4]
1. Department of Biological Sciences, University of Manitoba
2. Department of Biology, University of Ottawa
3. Department of Botany, Faculty of Biology, VNU University of Science, Vietnam National University
4. Institute of Geography, Vietnam Academy of Science and Technology

Although Vietnamese *Carex* species represent a small fraction of global sedge diversity (ca. 85 of 1966 spp.), their inclusion in systematic studies has led to unprecedented insights into the evolution and diversification of this genus. Most recently our investigations have shown that *Carex* is comprised of two major evolutionary lineages, with over 98% of all species (1936 spp.) falling within the Major *Carex* Alliance. The remaining 30 species are restricted to East and Southeast Asia and comprise the Minor *Carex* Alliance. Our studies have also revealed the presence of three other previously unrecognised, early diverging, East and Southeast Asian lineages: a "Hypolytroides Clade" (sect. Hypolytroides, 2 spp.), which along with the Siderostictae Clade comprise the Minor *Carex* Alliance; a "Dissitiflora Lineage" (sect. Mundae, 3 spp.); and a morphologically diverse "Small Core *Carex* Clade" (sects. Graciles, Decorae, Mapaniifoliae, Euprepes, Indicae, ca. 70 spp.), which is sister to the temperate and species rich Large Core *Carex* Clade (> 1,300 spp.). Despite the relatively small number of species in these newly recognized lineages, its members exhibit an extraordinary range of morphologies that are rare or unique in *Carex* (e.g., densely capitate or cymose inflorescences, white stigmas and anthers, grass-like vegetative structures, wide pseudopetiolate leaves). Overall, our findings suggest that morphological diversification may have occurred in clades dominated by East and Southeast Asian species followed by canalization of a narrower range of phenotypes in species-rich, cosmopolitan lineages. Our results also suggest a Southeast Asian origin for the genus. Vietnamese *Carex* have not only afforded unprecedented insights into deep patterns of relationship within the genus, they have also upon closer examination revealed species groups that are hidden under a single name. One example of this unrecognized diversity is seen in *Carex kucyniakii* s.l. (Siderostictae Clade, Minor Carex Alliance), a taxon that represents a complex of three distinct species. These taxa are endemic to the southern slopes of the Hoang Lien Mountains and exhibit some of the most extraordinary morphological features found in the Cyperaceae including pseudopetiolate leaves up to 12 cm wide, and in some instances inflorescences suffused with pink or red. Each taxon is characterized by a suite of morphological and molecular traits that allows them to be consistently distinguished from one another and justifies their recognition as separate species. At the same time, these taxa share a number of attributes, including molecular synapomorphies, geographic range, and distinctive morphological features that show that they are closely related and constitute a monophyletic group. Despite being one of the largest genera in the Vietnamese flora, remarkably few *Carex* species have been described in recent years from this country, with the most recent discoveries dating back to the late 1970s. This contrasts with the unprecedented rate of taxonomic discovery that has taken place in China over the past 30 years. We are aware of other undescribed *Carex* species and propose that the sedge flora of Vietnam could reveal a comparable number of taxonomic novelties given a more focused attention by cariciologists.

T2-21-04

Relationships within the CDS Clade (Cariceae-Dulichieae-Scirpeae): New tribes, new genera and new species

Julian Starr[1], *Étienne Léveillé-Bourret*[1], *Bruce Ford*[2], *Claire Gilmour*[1], *Robert Naczi*[3], *Sabina Donadío*[4], *Nguyễn Thị Kim Thanh*[5], *Vũ Anh Tài*[6], *Chen Binghua*[7], *Daniel Spalink*[8], *Kenneth Sytsma*[9], *Emily Moriarty Lemmon*[10], *Alan Lemmon*[10]

1. University of Ottawa
2. University of Manitoba
3. New York Botanical Garden
4. Instituto de Botánica Darwinion
5. VNU University of Science
6. Vietnam Academy of Science and Technology
7. Fujian Normal University
8. University of Utah
9. University of Wisconsin
10. Florida State Uni

With approximately 40% of all sedge species (>2,050), the Cariceae-Dulichieae-Scirpeae (CDS) clade is a cosmopolitan lineage that is found in habitats ranging from deserts to tundra and from temperate wetlands to tropical rain forests. Although comprised of three tribes, the CDS clade is dominated by the Cariceae, a monotypic tribe for *Carex*. Long recognised as one of the world's largest plant genera, this taxon's diverse biogeography (e.g. amphiatlantic, bipolar, Gondwanan), intriguing cytology (holocentric chromosomes: n=6 to 56), and morphological singularity have made it an ideal model for studying speciation, ecology and biogeography. Despite many significant advances in our knowledge of relationships within *Carex*, our understanding of its relationships within the CDS clade remain unclear, a fact that affects the conclusions drawn in all *Carex* studies since outgroup relationships have a significant impact on homology assessment, character polarisation and biogeographical analyses. Over the past decade, we have focused our research on the relationships of Cariceae to the approximately 86 species and infraspecific taxa in tribes Scirpeae and Dulichieae that make up the remaining members of the CDS clade. Using representatives for all 18 genera currently recognised in CDS and ca. two thirds of the species that comprise tribes Scirpeae and Dulichieae, our five marker DNA sequence dataset (*matK*, *ndhF*, *rps16*, ETS 1f, ITS), in conjunction with morphological, anatomical and embryological data, suggest there are seven well supported major lineages within the CDS clade with Dulichieae and *Khaosokia* successively sister to a clade containing the previously recognised Cariceae, *Trichophorum*, *Calliscirpus*, *Zameioscirpus* and *Scirpus+Eriophorum* Clades. Although the backbone of our previous phylogenetic analyses has often been weak or unresolved due to a rapid radiation among CDS lineages (10 million years), we have since employed universal flowering plant enrichment probes for a subset of key taxa to generate a matrix of 461 nuclear genes using next generation sequencing technology. The results of these phylogenomic analyses are fully congruent with our Sanger-sequence datasets and morphology, but we now have unprecedented support for backbone nodes (100% BS) that were previously unsupported (<60% BS). Given these results, this talk will propose a new tribal and generic classification for the CDS clade, including the naming of four new tribes, a new subtribe, and the discovery of several new genera and species.

T2-21-05

Multiple approaches to unveil the phylogeny, biogeography and biodiversity of Himalayan sedges (Cyperaceae) of Pakistan

Uzma Uzma[1], *Shafique Atif*[1], *Fakhar Imran* [1], *Bashir Tanzeela* [1], *Shah Iqra*[1], *Amir Rabia*[1], *Hayat Muhammad Qasim*[1], ***Hipp Andrew***[2]
1. National University of Sciences and Technology, H-12 Islamabad, Pakistan
2. University of Chicago/The Morton Arboretum, USA

Cyperaceae, the botanical family of sedges, consists of 5,500 species and 109 genera globally. In Himalayan Pakistan it is represented by 22 genera and 200 Species. Little information is available about the diversity, phylogeny and biogeography of Himalayan sedges. In this study, we are reporting: new field collection to herbarium specimens; morphology to character evolution; GIS mapping to biogeography; DNA bar coding to molecular phylogeny; Phytochemicals to chemotaxonomy. Further, by using the above data, we try to answer the big questions about the western Himalayan sedges diversity. Such as, "Is the western Himalayan sedges assemblage consist of immigrants from various other regions of the world?" Are they subjected to convergent evolution? This study also tested the hypothesis, whether diversity in western Himalayan Cyperaceae is primarily due to speciation during the uplift of the Himalaya (in case of endemic species). These basic questions are of prime interest—understanding the shape of plant biodiversity in this floristically rich region of the world—as well as applied interest: our basic understanding of species origins and distribution as a crucial first step to conserving sedge biodiversity in the region. This project will be a valuable addition to taxonomy of Cyperaceae worldwide and more broadly interpret Himalayan plant biodiversity globally.

T2-21-06

Taxonomic revision of genus *Eleocharis* R. Br. (Cyperaceae) in India

Arun N. Chandore[1], *Devias B. Borude*[2], *Shrirang R. Yadav*[3]
1. Rayat Shikashan Sansth's, Department of Botany, Abasaheb Marathe Arts and New Commerce, Science College, Rajapur, District- Ratnagiri- 416 702, (MS), India
2. Rayat Shikashan Sansth's, Department of Botany, Abasaheb Marathe Arts and New Commerce, Science College, Rajapur, District- Ratnagiri- 416 702, (MS), India
3. Department of Botany, Shivaji University, Kolhapur-416 004, (MS), India

Genus *Eleocharis* R. Br. (*Heleos* = a marsh, wet and *charis* = grace, beauty) is generally called as a beauty of marsh. It is characterized by the leaves reduced to the bladeless sheaths, inflorescence reduced to a single terminal spike, spikelets with few to many bisexual flowers, achenes usually with hypogynous bristles, style swollen at base, achene with a beak of persistent style base. The genus with about 300 species is widely distributed from tropical to temperate regions of both hemispheres. In India, it is represented by about 31 taxa. The genus was neglected in general floristic studies being an aquatic plants growing in muddy marshy areas difficult to collect and also due to their minute flowers and microscopic taxonomic characters and unavailability of standard literatures. Thus, number of specimens of *Eleocharis* were found to be wrongly identified in various herbaria of the country. Therefore an attempt has been made to revise the genus for India. The revision resulted into discovery of four novelties in the genus viz. *Eleocharis wadoodii* S.R. Yadav, Lekhak & Chandore, *E. khandwaensis* Mujaffar, Chandore & S.R. Yadav, *E. konkanensis*

Chandore, Borude, Kambale & S.R. Yadav and *Eleocharis* sp. nov. (in press). Similarly, *Eleocharis setifolia* (A. Rich) Raynal, *E. wichurae* Boeckeler and *E. yunnanensis* Svenson have been added to the flora of India. Some eight species of *Eleocharis* viz. *E. mitracarpa* Steud., *E. parishii* Britton, *E. philippinensis* Svenson, *E. quinqueflora* (Hartmann) O. Schwarz, *E. tuberculosa* (Michx.) Roem. & Schult., *E. uniglumis* (Link) Schult., *E. variegata* (Poir.) C. Presl, *E. yokoscensis* (Franch. & Sav.) Tang & F.T. Wang are reported from India on authority but their distribution in India is doubtful as there are no specimens of the species available in Indian herbaria (AHMA, ARUN, ASSAM, BSA, BSD, BSHC, BSI, BSJO, CAL, CALI, DRC, LWG, MH, NBU, PBL, PUN, RHT, SUK) of India. Therefore these species are excluded from the study. Presently, the genus is reported to have 31 taxa (25 species, 3 subspecies and 3 varieties) in India. The revision of the genus resulted into an addition of 10 taxa to the flora of India. Six taxa viz. *E. andamanensis* Govind., *E. atropurpurea* var. *vengurlaensis* Shaikh R.I & R.D. Tour, *E. khandwaensis* Mujaffar, Chandore & S.R. Yadav, *E. konkanensis* Chandore, Borude, Kambale & S.R. Yadav, *E. lankana* subsp. *mohamadii* Wadood Khan and *E. wadoodii* S.R. Yadav, Lekhak & Chandore are so far known to be endemic to India. This revisionary work on genus *Eleocharis* in India provides presentation of important morphological characters through photographs, line drawing and details of SEM characters of achenes useful in identification of Indian species. Botanical keys based on important taxonomic characters are also provided for easy identification.

T2-22: Mechanical forces in flowers: An evolutionary perspective

T2-22-01

Forces in flowers and their evolutionary significance

Regine Classen-Bockhoff
Institut für Spezielle Botanik, Johannes Gutenberg-Universität Mainz

Though many flowers demand physical force from their pollinators little is known about the functional and evolutionary significance of barriers in flowers. Among many open questions, two are of particular importance: Are weak insects excluded from pollen transfer? And do low forces have a selective significance? To get an insight into force dimensions, we measured floral forces with a highly sensitive force measurement device. We selected flowers with a force demanding pollination mechanism from six angiosperm families: Lamiaceae: *Salvia* (> 30 ssp.), Zingiberaceae: *Roscoea* x *beesiana*, Fabaceae: 11 genera (13 ssp.), Polygalaceae: *Polygala myrtifolia*, Calceolariaceae: *Calceolaria* (4 ssp.) and Asclepiadaceae: *Asclepias* (3 ssp.). Results fall into two groups: In Fabaceae, forces range from very low values up to more than 200mN. Compared to published data on bee forces, it is evident, that weak insects are excluded by some keel flowers. Surprisingly, in all other flowers, forces are rather weak not exceeding an average of 15 mN. They do not exclude bees from pollen transfer completely, but may have a functional significance during food plant selection. To test the functional significance of low forces, we constructed artificial '*Salvia*' flowers, i.e. a flower model with a lever-like barrier and a control model without any barrier. We trained honeybees (*Apis mellifera*) and bumblebees (*Bombus terrestris*) to the models and found, that both bee species were able to discriminate low forces. They preferred the model with no barrier as long as nectar rewards were equal. However, choice experiments indicated that the bees preferred the model with barrier when it offered a higher concentrated sucrose solution. Thereby, bumblebees were more successful on models with barriers than honeybees. We preliminarily conclude that flowers demanding low forces may play a selective role relative to the environmental conditions of the plant species. In the presence of competing bee species, they may contribute to niche formation by offering stronger bees a nectar source with reduced competition.

T2-22-02

The impact of floral developmental constraints on the evolution of flowers.

Louis Ronse De Craene
Royal Botanic Garden Edinburgh

Flower development and shape is regulated by the interaction of a genetic program and environmental factors, such as influence of pollination systems and biophysical forces acting on developing flowers. The initiation of floral meristems and floral organs is mainly regulated by spatial constraints. Three major factors are responsible for creating these spatial constraints and influence the morphology of flowers: time, size, and pressure. Changes in morphology are mainly induced by shifts in time of initiation of organs, pressure of previously initiated organs, and alterations of the size of the floral meristem, and these operate independently of genetic factors. A number of examples demonstrate this interaction and its importance in the establishment of different floral forms.

Delays in the timing of initiation of organs lead to the establishment of novel morphologies by absorption of tissue, such as in stamen-petal primordia, or by inducing shifts of organ positions, such as the phenomenon of obdiplostemony.

Floral meristem expansion will provide more space and will lead to an increase of stamen and carpel number, the development of hypanthia and accessory structures, and changes in merism. Merism is mainly established by a balance between the size of organs versus the size of the floral apical meristem.

Pressures can act centripetally by the action of external organs, or centrifugally by the inner meristem size directing the number and arrangement of organs.

It is demonstrated that changes in the physical environment of apical meristems has an important influence on ontogenetic processes leading to changes in floral configurations and major shifts in floral form, which is reflected in the phylogeny.

T2-22-03

Spatial constraints influence the shape of spikelets and flowers in Cyperoideae (Cyperaceae)

Alexander Vrijdaghs[1], ***Erik Smets***[2]
1. Plant Conservation and Population Biology, KU Leuven, Leuven, Belgium
2. Botanic Garden Meise, Belgium
3. Naturalis Biodiversity Center, Leiden, The Netherlands

Cyperaceae is a large grass-like family in Poales with two subfamilies, the small Mapanioideae and the species-rich Cyperoideae, each subfamily corresponding to a well-supported sister clade. Based on scanning electron microscopic observations we studied the development of the flowers and inflorescences of representatives in a wide range of genera of Cyperoideae.

Cyperoideae are characterized by spikelets composed of of wind-pollinated, typically trimerous, monocotyledonous flowers. The spikelets are the units of compound inflorescences, namely panicles of spikelets. The panicles are often compacted and contracted into dense head-like inflorescences. In such inflorescences, spatial constraints are obviously influencing the development of the spikelets as well as the individual flowers.

Reduction of the number of flowers in the spikelet, loss of the spikelet prophyll, metatopic displacements of the bracts subtending a flower, and reduction of the floral parts of the flowers are the main consequences of spatial constraints.

Within the flowers, spatial constraints may result in the reduction of the number of stamens to three instead of six, delaying or totally reducing the development of the abaxial floral parts, the loss of the perianth and modifications of the gynoecium. In the most derived subtaxa of Cyperoideae, the compacted inflorescence tends to take over the floral function, forming dense heads consisting of single-flowered spikelets. The question remains whether these evolutionary trends were driven by mechanical forces or by pollination strategies, or a combination of both.

T2-22-04

Development of actively morphing structures in plants

Naomi Nakayama

Institute of Molecular Plant Science/SynthSys/CSEC, School of Biological Sciences, University of Edinburgh

Throughout the phylogeny of seed plants, flowers and fruits contain actively moving structures to facilitate pollination or seed dispersal. Examples include pollinator-induced movement of stamens and/or carpels, diurnal flower opening, and moisture-controlled dispersal of pinecones and wheat awns. These functional structures require unique cellular and tissue-level features that are precisely patterned and generated during development. In this talk I will present an overview of the biomechanics behind the moving parts in flowers and fruits and highlight the commonly found morphological and anatomical strategies. By understanding better about the hydrodynamic and tissue material features underlying the actively controled organ movement, we can identify the candidate key developmental events that are also likely to have been targetted by the evolutionary innovation of such morphing structures. I will then report our effort to establish an experimental platform to untangle the physico-chemical feedback loops regulating plant organ development, which can be used to test hypotheses generated above.

T2-22-05

Forces in flower ontogeny

Kester Bull-Hereñu[1,2], *Patricia Dos Santos*[3], *João Felipe Ginefra Toni*[4], *Akitoshi Iwamoto*[5], *Juliana Hanna Leite El Ottra*[6], *Louis Ronse De Craene*[7]

1. Museo Nacional de Historia Natural

2. Fundación Flores

3. Centre for Ecology, Evolution and Environmental Changes, University of Lisbon

4. University of Basel

5. Department of Biology, Tokyo Gakugei University

6. Instituto de Biociências, Universidade de São Paulo

7. Royal Botanic Garden Edinburgh

From an evolutive perspective, the flower can be understood as the historical outcome that serves reproduction and maintenance of the plant's lineage. At the same time, but from an organismic point of view, the flower can be seen as the outcome of the ontogenetic process in each individuum. Questions regarding the ontogenetic process of the flower involve multiple aspects including cell proliferation and growth, hormone flux, gene expression patterns, as many others. One of these elements that has received undeservedly little attention are mechanical forces itself. In this work we present a revision of a number of examples where mechanical forces have been shown to play a fundamental role in shaping the flower at different moments throughout ontogeny. The palette of action of these forces go from the early ontogeny at the microscopic level, - as the organ initiation patterns on the flower meristem in relation to the pressure exerted by inflorescence axes or neighboring primordia -, to the the macroscopic level in the late ontogeny, as the twisting of a pedicel of a resupinating flower. The premise here conceived is that without these forces ocurring during ontogeny, the recalled flower phenotype wouldn't be expressed as it is actually known, implying that the mentioned force plays a causal-generative role on the flower's phenotype, unaware of further genetic or molecular response coocurrence.

T2-22-06

Pollinator-driven floral evolution with and without pollinator shifts in *Impatiens* (Balsaminaceae)

Saroj Ruchisansakun[1,2], *Steven B. Janssens*[4], *Erik F. Smets*[1,2], *Timotheüs Van Der Niet*[3]

1. Naturalis Biodiversity Center

2. Leiden University

3. School of Life Sciences, University of KwaZulu-Natal

4. Botanic Garden Meise

Floral diversity may evolve as a result of plant-pollinator interactions by two distinct processes: through the use of different functional pollinator groups, or divergent use of the same pollinator. However, the relative importance of these processes is largely unknown. Here we use the recently radiated, large plant genus *Impatiens*, characterized by variable, zygomorphic or asymmetric flowers, to understand the association between floral diversity and pollination ecology, and to study its evolution. A comparative field study of co-flowering *Impatiens* species in northern Thailand revealed that out of seven species, six are dependent on animal visits for pollination. Several different types of animals visited the flowers, but not all were effective pollinators. Four *Impatiens* species are pollinated by bees, whereas two species are pollinated by bees and butterflies. The main features that distinguish pollination systems included the shape and dimensions of the floral chamber and length of the (nectar) spur. Co-flowering species that shared bee pollinators deposited their pollen on different parts of the bee body, including through an unusual form of the asymmetric perianth. Based on our field-observations we predicted pollinators at the level of functional groups for a large number of *Impatiens* species. The majority of species was inferred to be bee-pollinated, and floral asymmetry was significantly more common among bee-pollinated species than those with other pollination systems. According to the phylogenetic analysis, shifts between functional pollinator groups occurred at least 17 times. Floral asymmetry evolved at least 11 times, never reversed, and was not associated with large clades. Shifts to floral asymmetry were uncoupled from shifts between functional pollinator groups, and always happened in lineages that were ancestrally bee-pollinated. These results are consistent with a scenario according to which floral diversity either evolves by means of pollinator shifts, or by divergent use of the same, frequently exploited pollinators. Once floral asymmetry evolves it does not appear to function as key-innovation that allows species to invade a new niche leading to clade proliferation. Further analyses, which take the co-occurrence of *Impatiens* species into account, could reveal whether competition between co-flowering species is an important driver of floral evolution.

T2-23: New data on early Cretaceous seed plants (two sessions)

T2-23-01

Diversity of corystosperms from the Early Cretaceous of Mongolia

Gongle Shi[1], *Fabiany Herrera*[2], *Patrick S. Herendeen*[2], *Niiden Ichinnorov*[3], *Masamichi Takahashi*[4], *Peter R. Crane*[5,6]

1. Nanjing Institute of Geology and Palaeontology, Chinese Academy of Sciences, 39 East Beijing Road, Nanjing, 210008, P. R. China

2. Chicago Botanic Garden, 1000 Lake Cook Road, Glencoe, Illinois 60022, USA

3. Paleontological Center, Mongolian Academy of Sciences, Ulaanbaatar-51, P.O. Box 260, Mongolia

4. Department of Environmental Sciences, Faculty of Sciences, Niigata University, 8050, 2-cho, Ikarashi, Nishi-ku, Niigata, 950-2181, Japan

5. Oak Spring Garden Foundation, 1776 Loughborough Lane, Upperville, VA 20184

6. School of Forestry and Environmental Studies, Yale University, 195 Prospect Street, New Haven, CT 06511, USA

The corystosperms (=Umkomasiaceae) are a key extinct group for understanding the phylogeny of seed plants and the evolution of their reproductive structures, including the carpel and ovule of angiosperms. Corystosperms first occur in the Late Permian and appear to have attained their highest diversity and widest distribution in Gondwana during the Middle and Late Triassic. The group appears to have been extinct by the early Paleogene. Corystosperms have generally been thought to be distributed primarily in Gondwana, with only few scattered occurrences in the Northern Hemisphere. Recently however, study of a diverse, excellently preserved, conifer-dominated, plant assemblage from the Early Cretaceous Tevshiin Govi Formation at Tevshiin Govi, in central Mongolia, has revealed diverse seed-bearing organs of corystosperms, which can be recognized as three distinct species, all of which can be assigned to *Umkomasia*. All three *Umkomasia* species consist of a bract subtending an axis that branches distally, with each fork bearing a reflexed cupule and each cupule bearing a partially enclosed, single, triangular seed with the micropyle oriented towards the cupule stalk. In these respects all three species closely resemble the original material of *Umkomasia* from the Triassic of South Africa. However, the Mongolian *Umkomasia* species are associated with *Pseudotorellia*, a leaf with *Ginkgo*-like characteristics, rather than the pinnately compound *Dicroidium* leaves that are typical of Triassic Gondwanan Corystospermales. The three species of *Umkomasia* from Mongolia are distinguished by the number of branches of the main axis, the prominence of the extension of cupule stalk, as well as the number of cupule lobes. Together with the nearly contemporary North American permineralized seed cone *Doylea*, these species suggest that corystosperms were more diverse in the Early Cretaceous of Northern Hemisphere than previously supposed. They appear to exhibit an unusual combination of features seen in *Ginkgo* and conifers as well as other Mesozoic seed plants – such as *Caytonia* and *Petriella* – and as such raise important questions about our current understanding of seed plant phylogeny.

T2-23-02

Floristic composition of an Early Cretaceous swamp from Mongolia and the implications for seed plant evolution

Fabiany Herrera[1], *Gongle Shi*[2], *Niiden Ichinnorov*[3], *Masamichi Takahashi*[4], *Eugenia Budgaeva*[5], *Peter Crane*[6], *Patrick Herendeen*[1]

1. Chicago Botanic Garden

2. Nanjing Institute of Geology and Palaeontology, Chinese Academy of Sciences

3. Institute of Paleontology and Geology, Mongolian Academy of Sciences

4. Department of Environmental Sciences, Faculty of Sciences, Niigata University

5. Institute of Biology and Soil Science, Far East Branch, Russian Academy of Sciences

6. Oak Spring Garden Foundation

Fossils and molecular evidence suggest that all extant and crown conifer families were established by the Early Cretaceous. In addition ginkgophytes were an important element across Asia in the Early Mesozoic. However, recognition of the lineages that lead to extant conifers groups (e.g., Cupressaceae s. l., Pinaceae) and ginkgophytes remains incomplete, mostly due to an incomplete and inadequately understood fossil record. Exquisitely preserved lignite fossils (seed and pollen cones, foliage, and wood) from the Aptian-Albian (~100-125 Mya) of central Mongolia (Tevshiin Govi and Tugrug localities) provide taxonomic diversity and morphological and anatomical data that contribute to a more complete understanding of the complex pattern of conifer and ginkgophyte evolution. New Mongolian conifer fossils include at least seven taxa of seed cones, three types of pollen cones, at least five leaf morphospecies, and four wood types. *Krassilovia*, one of the seed cones, bears multilobed ovuliferous scales that suggest an affinity with extinct members of the voltzian clade of the Voltziales ("transitional conifers"); *Krassilovia* is associated with *Podozamites* foliage and also shows that voltzian seed cones with complex bract-scale arrangement and morphology persisted into the later Mesozoic. Three other seed cones show affinity with stem and crown Pinaceae, including the stem lineage taxon *Schizolepidopsis*; other Pinaceae cones include *Pityostrobus* and one the oldest records of *Picea* (spruce). Three additional taxa show affinities with the subfamilies Cunninghamioideae (*Elatides*, *Pentakonos*) and Taiwanioideae (*Stutzeliastrobus*) of the Cupressaceae s.l. The new Mongolian fossils also reveal that the enigmatic ginkgophyte *Umaltolepis* has a peltate, cupulate seed-bearing organ, unlike any other ginkgoalean fossil and extant *Ginkgo biloba. Umaltolepis* bore simple, strap-shaped *Pseudotorellia* leaves on well-developed short shoots. The new interpretation of *Umaltolepis* reveals that early ginkgophytes showed considerable morphological diversity as well as adaptations for wind pollination and ovule protection during the Early Mesozoic. Other major plant groups from Mongolia include corystosperms (*Umkomasia*), filmy ferns and bryophytes. However, conifers dominated the lignite floras. The deposition and taphonomic conditions of the lignite deposits suggest that these plants inhabited permanently flooded systems (e.g., forest-moor swamps), however, the presence of charcoal in the same sedimentary sequences suggest that fire events were also part of the Early Cretaceous landscape in Mongolia.

T2-23-03

Cretaceous/Paleogene angiosperm floras from Patagonia, Argentina

Maria Alejandra Gandolfo[1], *Maria Del Carmen Zamaloa*[2], *N. Rubén Cúneo*[3]

1. Plant Biology Section, SIPS, Cornell University

2. Facultad de Ciencias Exactas y Naturales, Universidad de Buenos Aires.

3. Museo Paleontológico Egidio Feruglio

The southernmost area of South America, Patagonia, is considered to be critical for understanding South American modern floras as it had a fundamental role in the origin and diversification of seed plants in the Southern Hemisphere. In this contribution, we will discuss the evolution of angiosperms in Patagonia from their first appearance during the Barremian (Early Cretaceous) to the Cretaceous/Paleogene boundary. Data was gathered from three major Patagonian basins: San Jorge Gulf, Deseado Massif, and Austral, and from the Chubut Extra Andean Region, representing 26 formations that were deposited during the earlier Cretaceous (Late Barremian, 130 MA) to the Cretaceous/Paleocene (Danian, 61.6 MA) which offer a wealth of fossil angiosperm pollen and leaves. We will report on the taxonomical components of the floras, dominating groups, the development of plant communities, and the environments in which they lived. As expected, during the Cretaceous the angiosperms went from being a minor component of the paleoecosystems that were dominated by gymnosperms and pteridophytes to be the dominating clade at the end of the Cretaceous (Campanian-Maastrichtian)/Paleocene (Danian). The fossil record confirms the presence of members of the ANITA grade, Chloranthales and several monocotyledons during the Aptian-Albian, while the eudicots and the rosids become more diverse and the ones dominating the paleoecosystems at the end of the Cretaceous. The first appearance of angiosperms in Patagonia dates from the Barremian Springhill Fm (Austral Basin), and they were represented by pollen grains assigned to the *Clavatipollenites* complex. Angiosperm leaves appeared later at the Early Aptian Anfiteatro de Ticó Fm (Deseado Massif Basin) as three morphotypes show clearly angiosperm characteristics. At end of the Cretaceous, the angiosperms from La Colonia (Campanian-Maastrichtian) and Lefipán (latest Maastrichtian) formations were undoubtedly the dominant components of the paleocommunities confirming the declining of the gymnosperms and pteridophytes and the rise of the angiosperms at the Cretaceous/Paleogene boundary in Patagonia.

T2-23-04

The early angiosperms of the Primorye region, Russian Far East

Elena Volynets[1], *Eugenia Bugdaeva*[1], *Valentina Markevich*[1], *Ge Sun* [2]

1. Federal Scientific Center of the East Asia Terrestrial Biodiversity, FEB RAS, Vladivostok, Russia

2. Shenyang Normal University, Shenyang, China

Recently we discovered new localities of Aptian–Albian age with angiosperms in the Razdolnaya River, Partizansk, and Alchan River basins in Primorye region, Russia. The plant-bearing deposits have alluvial, sometimes volcanogenic origin. Hence it appears that the environments inhabited by these first angiosperms were high-stress and unstable. For the first time we have revealed in the Aptian coaliferous deposits and the Cenomanian red beds of the Razdolnaya River Basin the remains of the flowering plants. They are represented by extremely rare leaf fragments of unidentifiable monocots and dicots, as well as by dispersed cuticle of probable Platanaceae. The angiosperm pollen *Tricolpites* sp., *T. vulgaris*, *T. micromunus*, *Clavatipollenites hughesii*, *Quercites sparsus, Retitricolpites georgiensis* were obtained from these Aptian deposits. The Albian angiosperms are represented by *Sapindopsis variabilis*, *Laurophyllum* sp., *Cercidiphyllum sujfunense*, *Menispermites* sp., and unidentifiable dicots. The palynological assemblage has increasing numbers of angiosperm pollen including the new taxa *Asteropollis asteroides*, *Fraxiniopollenites variabilis*, *Rouseia delicipollis*, *Polyporites clarus*. The abundant leaves of platanoids are recognizable in the Cenomanian. The angiosperm pollen *Tricolpites micromunus*, *Fraxiniopollenites variabilis*, *Rouseia delicipollis*, *Polyporites clarus* were found. In the Partizansk Basin the early angiosperms were collected in the Albian deposits. These are rare isolated leaf fragments of *Laurophyllum* sp., *Sassafras ussuriensis*, *Sapindopsis* cf. *angusta*, *"Aralia" lucifera*, *Cissites* sp., *Sapindopsis* sp., *Artocarpidium* sp., and unidentifiable dicots. Recently in the south-western and central parts of the Partizansk Basin we have found imprints of early flowering plants, which are represented by entire plants, including rhizomatous stems with tubers and adventitious roots, branches terminating in fruits, and pinnatisect leaves, intact solitary flowers and heads of achenes or follicles. These angiosperms were described as *Achaenocarpites capitellatus* and *Ternaricarpites floribundus*. The Albian angiosperm pollen is represented by rare *Asteropollis asteroides, Cyclusphaera psilata*, *Clavatipollenites incisus*, *Fraxiniopollenites variabilis, Tricolpites micromunus*. In the Cenomanian of this basin the following flowering plants were found: *Araliaephyllum* spp., *Cercidiphyllum* sp., *Magnoliaephyllum* sp., *Platanophyllum* spp., *Menispermites* sp. In the Alchan River Basin the early angiosperms appear in the middle Albian. These are very rare fragments of small leaves of *Laurophyllum* sp., *Sapindopsis* sp., and unidentifiable dicots. During the late Albian the diversity and abundance of the flowering plants increase. They are represented by *Magnoliaephyllum* sp., *Sapindopsis variabilis, S. brevifolia, Cinnamomoides* ex gr. *ievlevii, Cissites* sp., *Sassafras* aff. *ussuriensis, Quercophyllum* sp., *Celastrophyllum* ex gr. *oppositifolius, C.* cf. *serrulatus, Lindera jarmoljukii, Menispermites* sp., *Kenella harrisiana, K. filatovii.* The extremely rare tricolpates *Tricolpites* sp. and monocolpates *Clavatipollenites incisus* occur in the Albian deposits. In the Cenomanian the platanoids mainly occur along with aquatic angiosperms. Among angiosperm pollen, including *Tricolpites* sp., *Clavatipollenites* sp., the new taxa *Tricolpites micromunus, Tricolporopollenites* sp., *Polyporites clarus*, and *Triporopollenites* sp. appear.

T2-23-05

Early Cretaceous vegetation and the emergence of angiosperms in Europe and North America

***Else Marie Friis*[1], *Kaj Raunsgaard Pedersen*[2], *Peter R. Crane*[3]**

1. Swedish Museum of Natural History, Stockholm

2. Department of Geoscience, University of Aarhus, Aarhus

3. Oak Spring Garden Foundation, Upperville, Virginia and Yale School of Forestry & Environmental Studies, New Haven, Connecticut

The Early Cretaceous was marked by the initial rise of angiosperms, and also by the diversification of modern lineages of conifers (e.g. Pinaceae) and ferns (e.g.water ferns) that became important in later plant communities. The Early Cretaceous also saw the decline of ancient conifers, ferns, ginkgophytes and cycads, that were characteristic of earlier Mesozoic vegetation, as well Bennettitales, Erdtmanithecales and other seed plant lineages that are now extinct. This large-scale floristic transition is seen worldwide through the Early Cretaceous, although the details differ regionally as a consequence of palaeopalaeogeography and palaeoclimate. Most research on this transition has focused on the emergence of angiosperms, but changes in other elements of the vegetation through the Early Cretaceous are also important. Here we focus on the fossil flora collected at a large clay pit near Torres Vedras, northeast of Forte de Forca, Portugal, and illustrate the vegetational changes that took place in Europe and North America during the Early Cretaceous by comparisons with other mesofossil floras that range in age from Berriasian to Albian (ca. 145 to 100 myr BP). The Torres Vedras flora is likely the oldest of the Portuguese mesofossil flora so far described that contains angiosperms remains. There are many earlier mesofossil floras, all of which are devoid of angiosperms, and there are many later floras in which angiosperms are more common. Fossil assemblages from Torres Vedras are dominated by twigs and seeds of extinct conifers and fragments of fern foliage, some of which are fertile. Megaspores of Isoetaceae and Selaginellaceae are also common, as are chlamydospermous seeds (assigned to the extinct genera *Tomcatia, Quadrispermum* and *Ephedrispermum*). Angiosperms are not common in the Torres Vedras assemblages compared to remains of other groups of plants, suggesting that they did not dominate the local vegetation. However, the diversity of angiosperms is relatively high, with about 50 different species, represented mainly by forms with no clear modern relatives, or by forms related to early diverging lineages, such as ANA-grade angiosperms and Chloranthaceae. Other than pollen grains of probable Araceae there is no evidence of monocots at Torres Vedras and there are only few taxa of possible magnoliid affinity. Eudicots are extremely rare and are represented in the mesoflora only by two types of tricolpate pollen grain, one known from a single coprolite and the other (with irregular apertures) known from a single pollen sac. There is also one (perhaps two) kind (s) of pantoporate grains that could be of eudicot affinity. There is no evidence of any extant group of early diverging eudicot, or of later branching lineages of core eudicots, such as rosids or asterids. In mesofossil floras from later in the Early Cretaceous, from both Portugal and North America, angiosperms are more common, taxonomic diversity is higher, and there are several fossil species that can be assigned to various groups of extant magnoliids and early diverging eudicots. Unequivocal core eudicots are know only from about the latest Albian-earliest Cenomanian onwards (ca. 100 myr BP).

T2-23-06

The Early Cretaceous seed plants from Crato Flora, South America Brazil

***Flaviana Jorge Lima*[1], *Else Marie Friis*[2], *Renan Alfredo Machado Bantim*[1], *Antônio Álamo Feitosa Saraiva*[1], *Juliana Manso Sayão*[3]**

1. Department of Biology, Universidade Regional do Cariri

2. Department of Palaebiology, Swedish Museum of Natural History

3. Department of Biology, Universidade Federal de Pernambuco

The Early Cretaceous (Aptian/Albian) Crato Formation from the Araripe Basin (Brazil) contains one of the best known paleofloras from low palaeolatitudes from this important time interval in the history of angiosperms. The flora is diverse characterized primarily by pteridophytes and non-angiospermous seed plants including many taxa related to the Gnetales. The gnetalean component includes stems and twigs with opposite branching, reduced or absent leaves and compact cone-like reproductive structures. Angiosperm macrofossils are also diverse in the Crato Formation, sometimes preserved as whole plants with roots, stems, leaves and reproductive organs organically connected. Many of the angiosperms were clearly herbaceous and several were interpreted as aquatics. However, the non-angiospermous component of the Crato flora shows structural features consistent with arid environments (fibrous leaves, sunken stomata, well-developed indumentum, possible deciduous branches). The Crato Formation (lacustrine-carbonatic) together with the upper part of the underlying Barbalha Formation (deltaic) constitutes the lacustrine Aptian/Albian sequence of the post rift phase of the Araripe Basin. The plant fossils occur in laminated limestones that were developed under tropical, arid and semi-arid climatic conditions, with long intervals of dry climate and periodic precipitation. Some of the fossils possess sufficient characters for a systematic assignment at the family or higher level, and many taxa have already been described including taxa assigned to Nymphaeales and Magnoliales as well as other eumagnoliid angiosperms and several types of monocots. Other plant fossils are less informative systematically, but provide useful information on habit and diversity of early angiosperms in the palaeoequatorial region. New data on angiosperms from Crato Formation will be presented together with a floristic analyses showing the diversity and taxonomic composition of the flora.

Among the new angiosperm fossils is a tiny floral structures borne in small spikelets, subtended at the base by a pair of bracts. The inflorescence has 16 small flowers, with stamens (the number is uncertain), preserved three dimensionally.

T2-23-07

New findings of the gymnosperms female reproductive structures and seeds (*Nagrenia, Ginkgo, Allicospermum*) from the Middle Jurassic of Angren, Uzbekistan

Natalya Nosova
Komarov Botanical Institute of the Russian Academy of Sciences

The material studied comes from the Middle Jurassic Angren Formation in in the Angren coalfield (Uzbekistan). The first findings of ginkgoalean female reproductive structures in Angren were described as *Ginkgo gomolitzkyana* Nosova (2012). The fossil remains are represented by numerous detached seeds and some ovules with minor pedicel fragments. Those specimens lacked hairs and papillae on the ordinary cells of the outer integument cuticle. The investigation of the new specimens of seeds of *G. gomolitzkyana* suggests that this feature is not constant: hairs and papillae may occur irregularly on the surface of the outer integument epidermis varying locally from numerous to absent. Based on this I provided the emended diagnosis for *G. gomolitzkyana.* One more finding of a ginkgoalean female reproductive structure (collar complexes) from Angren was described as *Nagrenia samylinae* Nosova (2013), the only species of the genus *Nagrenia.* The specimens of *N. samylinae* are associated with the leaves of *Sphenobaiera angrenica* Nosova (2013) and share a similar epidermal pattern. The investigation of the morphological and epidermal features of the new findings of collar complexes from Angren permitted to describe two new species: *N. pilosa* and *N. leviana. N. pilosa* differs from *N. samylinae* in cuticle thickness and in having hairs and hollow papillae on the cuticle surface. Unlike these two species, remains of *N. leviana* are characterized by entirely preserved inner surface of the collars with a conical protrusion in the center that is similar to *Ginkgo biloba*. The co-occurrence of *N. pilosa* with leaves of *Ginkgo asiatica* Nosova and seeds of *G. gomolitzkyana* in one sample suggests that these remains may have belonged to the same plant. The sediments of the Angren Formation have yielded numerous remains of detached seeds. Besides *Allicospermum angrenicum* Nosova (2013) and *A. budantsevii* Gordenko (2015), I distinguish three other types of *Allicospermum* (unnamed species, preparing paper). Seeds of these species differ in sizes as well as in the structure of the outer epidermis of the seed coat and of the megaspore membrane. In shape and dimensions, the seeds of *Allicospermum* sp. 1 are comparable to *A. angrenicum* and *A. budantsevii*, but have a megaspore membrane that is 2.5–3 times thicker. Also they differ by the very thin outer integument cuticle lacking stomata. Seeds of *Allicospermum* sp. 2 are smaller than other seeds of *Allicospermum* from Angren. They have a thick outer integument cuticle, which also lacks stomata. The main difference of these seeds is unbranched sinuous bacula of the pattern layer of the megaspore membrane, arranged parallel to the foot layer. The seed of *Allicospermum* sp. 3 differs considerably from others by long resin bodies and the structure of the outer integument epidermis. There are numerous *Ginkgo*-type stomata, oriented along the axis of the seed. Flattened papillae occur in the most of the ordinary cells of the outer epidermis, and mycropilar cells have long papillae.

T2-23-08

The presumed ginkgophyte *Umaltolepis* has seed-bearing structures resembling those of Peltaspermales and Umkomasiales

Fabiany Herrera[1], Gongle Shi[2], Niiden Ichinnorov[3], Masamichi Takahashi[4], Eugenia Bugdaeva[5], ***Patrick Herendeen[1]****, Peter Crane[6]*
1. Chicago Botanic Garden
2. bNanjing Institute of Geology and Palaeontology, Chinese Academy of Sciences
3. Institute of Paleontology and Geology of Mongolian Academy of Sciences
4. Niigata University
5. Federal Scientific Center of the East Asia Terrestrial Biodiversity, Far Eastern Branch of the Russian Academy of Sciences
6. Oak Spring Garden Foundation

The origins and phylogenetic relationships of the five groups of living seed plants, including the single relictual species *Ginkgo biloba*, are poorly understood, in large part because of very imperfect knowledge of extinct seed plant diversity from the late Paleozoic and Mesozoic. Here we describe abundant and exceptionally well-preserved material from the Early Cretaceous of Mongolia of the previously enigmatic Mesozoic seed plant reproductive structure *Umaltolepis*, which has been presumed to be a ginkgophyte. The new material shows that *Umaltolepis* is a seed-bearing cupule, which was borne on a stalk at the tip of a well-developed short shoot. Each cupule is an umbrella-like organ with a central column that bears a thick, resinous, four-lobed outer covering, which opens from the base. Each cupule encloses four, pendulous, winged seeds that are attached to the upper part of the column. Evidence from morphology, anatomy and field association suggests that the short shoots bore simple, elongate *Pseudotorellia* leaves that have similar venation and resin bodies to leaves of living *Ginkgo. Umaltolepis* seed-bearing structures were previously poorly understood, but the Mongolian material shows that while very different from those of *Ginkgo* they are very similar to fossils described previously as *Vladimaria. Umaltolepis* and *Vladimaria* do not closely resemble the seed-bearing structures of any living or extinct plant, but are comparable in some respects to those of certain Peltaspermales (peltasperms) and Umkomasiales (corystosperms). Vegetative similarities of the *Umaltolepis* plant to *Ginkgo*, and the reproductive similarities to extinct peltasperms and corystosperms, support previous ideas that *Ginkgo* may be the last surviving representative of a once much more diverse clade, which included many kinds of extinct plants, several of which exhibited various degrees of ovule enclosure.

T2-23-09

Two oldest fossils of Altingiaceae from the Upper Cretaceous of New Jersey

Yangjun Lai[1]*, Maria A. Gandolfo[2], William L. Crepet[2], Kevin C. Nixon[2]*
1. Nanjing Institute of Geology and Palaeontology, Chinese Academy of Sciences
2. L. H. Bailey Hortorium, Section of Plant Biology, Cornell Uni-

versity

We report two new fossil materials from the Turonian (Upper Cretaceous, ~90-94 million years B.P.) Raritan Formation of New Jersey. The fossils are charcoalifed and reserved detail with tiny pistillate inflorescences and fruits. The multi-features composition—solitary capitate infructescence, syncarpous bicarpellate, 2-locules, unique sterile phyllomes structure—indicate that the fossil taxa close affinities to modern Altingiaceae. Also, it was supported by the evidence of phylogenetic analyses of molecular sequences combined with morphological data. Compared with *Microaltingia* which found in the same formation, the significant difference is the morphology and position of phyllomes. *Microaltingia*'s gynoecium is surrounded by two-wheeled numerous short, sterile phyllomes, the new fossil species have enough open space between carpel and involucres for phyllomes. One with incompactly florets, granular phyllomes, haired at the base of style, the other is density florets, one circle wide triangular phyllomes longer than the style, and suspected the remaining perianth. The primitive types indicate the possibility of insect pollination and the latter supposed a bisexual flower though we did not find the *in situ* pollen grain in the stamen-like phyllomes. The tiny fruit possibility suggests wind-dispersed. Considering the importance of phyllomes to classification in Altingiaceae and the matrix tree structure, we erected a new genus *Paleoaltingia* and two new species.

T2-23-10

Radiography of Cretaceous ovuliferous cones – case study

Jiri Kvacek
National Museum Prague

X-ray and neutron radiography are becoming widely used non-invasive methods to investigate preserved fossil plants. Various modes of preservation require different approaches. X-ray and neutron radiography are both used to document 3D internal structures of preserved fossil plants. Charcoalified ovuliferous cones of *Pityostrobus andraei* (Coemans) Seward from the type locality Baume, from the Belgian Wealden were studied, using X-ray radiography. Bracts are short, but separate from the scale base. Distally, the ovuliferous scales broaden and thicken to form a rhomboidal apophysis, with a dorsal umbo. Their seeds are always two per cone scale. It represents one of the basal members of Pinaceae. Charcoalified ovuliferous cones attached to and associated with twigs of *Cyparisidium bohemicum* of the Bohemian Cenomanian were studied with X-ray radiography. The ovuliferous cone scales are flat, typically bearing two seeds per ovuliferous scale, characteristic of Pinaceae. However, the twigs are typically cupressoid. Flat ovuliferous scales with wingless seeds are similar to basal members of the Cupressaceae Taiwania and Cunninghamia. A silicified ovuliferous cone of unknown affinity from the Bohemian Cenomanian was documented using neutron radiography. The ovuliferous cone is characterized by a slender axis and densely, spirally arranged ovuliferous complexes, which are long and curved. They make a wide rhomboidal pattern on the cone surface, formed by the distal ends of the woody ovuliferous complexes. Seeds are large and wing-less. Botanical affinity of this plant is not known. It may be associated with dwarf stems of Dammarites albens Presl in Sternberg.

T2-23-11

Early cretaceous seed plant diversity in Burmese amber

Chao Shi[1], ***Shuo Wang***[2,3]
1. Kunming Institute of Botany, Chinese Academy of Sciences
2. Institute of Zoology, Chinese Academy of Sciences
3. Kunming University of Science and Technology

Several groups of seed plants had their rapid diversification in Early Cretaceous. In the ~100 mya Burmese amber, we found plenty evidence for these fossil floras, including conifers, yews, Gnetales, Bennettitales, and early angiosperms. Some well conserved flowers, fruits, and leaves indicated that specific groups of flowering plants, ranging from basal angiosperms to higher taxa of monocots and eudicots, may have reached great diversity at that time, and some extant clades may have originated earlier. Their emergence brought major changes into plant reproductive biology, and affected dinosaurs, insects, mammals, and other organisms in their evolvement and evolution. Traces of insect activities were also witnessed on the fossil floras, which provides new insights into the co-diversification of seed plants and their associated fauna. We'd like to share our findings in Burmese amber concretely in the following presentation, and collaborate on studying the ecological landscape of Early Cretaceous with paleontologist and botanists allover the world.

T2-24: The Origin of Plants: rocks, genomes and geochemistry

T2-24-01

Resolving the phylogeny of early land plants by integrating phylgenomics and fossil evidence using morphological evidence

Harald Schneider[1], *Mark Puttick*[1], *Jennifer Morris*[2], *Philipp Donoghue*[2], *Davide Pisani*[2], *Cymon Cox*[3], *Silvia Pressel*[1], *Paul Kenrick*[1], *Charles Wellman*[4], *Dianne Edwards*[5], *Ziheng Yang*[6]
1. Natural History Museum, London
2. University of Bristol
3. Centro de Ciencias do Mar, Universidade do Algarve, Faro
4. University of Sheffield
5. University of Cardiff
6. University College London

The relationships among the four main lineages of land plants, namely hornworts, liverworts, mosses, and vascular plants, are still controversial. The land plant grade hypothesis with the liverworts basal have been preferred for several years over alternative hypotheses such as monophyletic bryophytes sister to vascular plans or hornworts sister to all other land plants. However, studies using phylogenomic approaches challenged this hypothesis. In this study, we integrate phylogenomic evidence and morphological evidence using a total evidence approach to explore the relationships among the four lineages not only based on extant plant diversity, but also by integrating the early land plant fossil record. Using this combined approach, we test not only the support for alternative hypotheses in the context of consistency with transcriptome sequence data but also in the context of morphological evolution and divergence time estimates. To achieve this, we carry out several phylogenetic analyses to evaluate the phylogenetic signal in support to alternative hypotheses with focus on four hypotheses namely: "land plant grade with liverworts basal", "monophyly of bryophytes", " land plant grade with hornworts basal and mosses and liverworts sister", and "hornworts sister to vascular plants and mosses liverworts sister to mosses". Finally, we highlight the consequences of the alternative hypotheses on studies aiming to reconstruct the original of land plants and the ecological consequences of the diversification of land plants to the formation of terrestrial environments.

T2-24-02

Latest approaches in establishing early land plant phylogeny and a timescale for plant evolution

Jennifer Morris[1], *Mark Puttick*[1], *Silvia Pressel*[2], *Paul Kenrick*[2], *Charles Wellman*[3], *Ziheng Yang*[4], *Sandra Arndt*[1], *Dianne Edwards*[5], *Phil Donoghue*[1], *Harald Schneider*[6]

1. University of Bristol
2. Natural History Museum, UK
3. University of Sheffield
4. University College London
5. Cardiff University
6. Sun Yat-sen University

The colonisation of plant life on land was one of the most significant evolutionary events to transform the planet. Major anatomical, biochemical and morphological innovations in the early evolution of land plants (e.g. complex body plans, leaves, rooting systems, symbioses with fungi) led not only to the evolutionary success of the lineage, but also to fundamental changes to global biogeochemical cycles, such as increased rates of continental weathering and sequestration of carbon over geological timescales. However, the nature and timing of the early evolutionary history of land plants and their interactions with the Earth is far from substantiated. There is currently no consensus on the fundamental phylogenetic relationships between living bryophytes (liverworts, mosses and hornworts) and tracheophytes, with conflicting results from numerous analyses using either molecular or morphological data. Support can be found for most hypotheses, including monophyletic bryophytes, each solution implying very different scenarios of the sequence and tempo of the origin of land plant characters. These conflicts are a consequence of a dramatic increase in the availability of molecular data and the development of more realistic models of molecular evolution. However, these models have yet to be applied to genome-scale data. Secondly, living bryophytes possess a mosaic of anatomical characters that are difficult to determine if derived from convergent or common descent. Our project aims to test the competing hypotheses of living plant phylogeny using a combination of phenotypic data with genome-scale sequence alignments with the latest computational models of molecular evolution. Genomic data is obtained from published sequences. Phenotypic data is in the form of a morphological character matrix of representative taxa across the plant kingdom, but with particular emphasis on bryophytes and 'lower' land plants. However, living taxa only represent a fraction of the diversity and disparity of early land plants that we known from the fossil record, thus fossil taxa are included in the phenotypic character matrix to understand their phylogenetic positions among their living relatives. One assemblage of particular importance in understanding early embryophyte evolution is a group of charcoalified fossils from rocks of Early Devonian age (~415 Ma), Shropshire, UK, that possess both bryophytic and tracheophytic characteristics. These rare fossils are preserved in three-dimensions to subcellular level by charcoalification. Traditionally they have been sectioned and characterised using Scanning Electron Microscopy. However, by using the latest techniques in synchrotron radiation X-ray Tomographic Microscopy, they can be characterised in a non-invasive and non-destructive method. Finally, after establishing the evolutionary relationships of living bryophytes and tracheophytes and the phylogenetic position of fossil taxa, we will determine the timescale of early land plant evolution. This will be achieved by applying the latest developments in molecular clock methodology, using molecular sequence data, models of evolution and fossil age constraints, with consideration of facies biases that can severely impact upon the fossil record. By determining the nature and timing of early land plant evolution it will then be possible to test hypotheses on the co-evolution of plants and the planet, particularly the far-reaching impacts on global biogeochemical cycles.

T2-24-03

The rise of land plants: Ecological context and geochemical implications

Paul Kenrick, *To Be Confirmed Others*
The Natural History Museum, London

The origin and early evolution of plants was part of the development of complex terrestrial ecosystems and, through their interactions with the physical environment, plants and associated microorganisms are thought to have had very broad impact on major geochemical cycles and on climate globally. Understanding how terrestrial ecosystems assembled and their broader impact requires the integration of evidence from several sources, including the fossil record, phylogenetics of living species and also the geochemical record. Furthermore, gauging the broader impact on Earth Systems depends to a large extent on identifying and measuring effects in modern analogues and analogue systems. A growing body of fossil evidence indicates that early terrestrial ecosystems most closely resembled modern biological soil crusts and cryptogamic ground covers (CGCs), which are communities that comprise bacteria, fungi, algae, lichens, bryophytes and arthropods. Critical comparisons between the biota of the earliest well-preserved terrestrial ecosystem (407Ma Rhynie Chert) and plausible modern CGC analogues demonstrates many points of similarity but also highlights some key differences in the nature of the growth forms of the plants, the absence of certain key animal groups (e.g., Oligochaeta, Formicidae, Isoptera) and the presence of important enigmatic organisms (e.g., *Prototaxites*). Symbioses between fungi and various autotrophic organisms including cyanobacteria, green algae and plants were present at the outset and were key elements. The evolution of lichens and mycorrhizal-like symbioses in plants is thought to have had a major effect on the geochemical carbon cycle principally through increasing the rate of calcium-magnesium silicate weathering. Likewise, the later evolution of roots amplified this process through physical disruption of regolith and by enhancing the development of soils, but the magnitude of these effects is poorly constrained. Furthermore, once vascular plants evolved and began to diversify recalcitrant complex organic polymers (e.g., lignin) became common in the environment, but in the absence of modern lignin decomposing fungi (e.g., Agaricomycetes) it is unclear how such matter was recycled and to what extent it contributed to soil organic carbon. Calibrated molecular phylogenetic trees provide a means of understanding when key organ systems (e.g., roots, wood) and relations (e.g., mycorrhizae) evolved enabling one to place constraints on when their effects would be expected to manifest in the rock record. Recent analyses however give highly divergent ranges and there is a need to develop better constrained calibrations. A second promising approach that we are developing is to characterise the micro-dissolution features in soil minerals. This could provide a general but indirect means of evaluating the nature of biotic weathering in sediments and palaeosols and thereby its broader effects on the environment. Finally, incorporating Ordovician-Devonian vegetation parameters inferred from fossils and calibrated phylogenetic trees into global vegetation models will provide insights into their likely effects on climate and carbon cycle.

T2-24-04

On the monophyly of bryophytes; Evidence based on organellar and nuclear protein-coding gene data

Cymon J. Cox, *Filipe De Sousa*
Centro de Ciencias do Mar (CCMAR/UALG)

Determining the pattern and timing of morphological changes that accompanied the origin and early diversification of plants on land is fundamental to our understanding the evolution of terrestrial ecosystems. The often cited consensus opinion is that the bryophytes diverged first, with either liverworts or hornworts identified as being the earliest-diverging lineage of land plants, and that therefore tracheophytes are derived from bryophyte ancestors. The primary molecular evidence for this conclusion comes from analyses that include chloroplast and nuclear protein-coding genes which can be shown to possess the greater-part of the supporting phylogenetic signal even when included in multi-genome data sets. Strikingly, when the protein translations of the same chloroplast genes are analysed, they predict a very different tree; one in which the liverworts and mosses are strongly supported as a monophyletic group, and together they are often grouped with the hornworts. Here we diagnose the main source of the conflict between analyses of chloroplast genes and their implied proteins as being due to the presence of synonymous substitutions in the gene sequences: non-synonymous substitutions and proteins, when modelled and analysed independently, give largely congruent results. Our analyses suggest that conventional substitution models are unable to adequately account for synonymous substitutions in protein-coding gene data and demonstrate that conflicting results can be attributed to the influence of a directional mutation pressure among synonymous substitutions that drives a composition bias. We find that currently the best supported hypothesis based on the evaluation of data from all three plant genomes is that bryophytes are monophyletic. If indeed the first divergence of land plants is a split between bryophytes and tracheophytes then the ancestral traits of the land plant ancestor could conceivably, and most likely would, have included some tracheophyte characters that were hitherto thought to have evolved subsequent to the diversification of bryophytes.

T2-24-05

Molecular dating of angiosperms in the genomic era

Charles Foster[1], *Hervé Sauquet*[2], *Marlien Van Der Merwe*[3], *Hannah Mcpherson*[3], *Maurizio Rossetto*[3], *Simon Ho*[1]
1. University of Sydney
2. Université Paris-Sud
3. National Herbarium of New South Wales

The evolutionary timescale of angiosperms has long been a key question in biology. The oldest crown-group fossils date to the early Cretaceous (~140 Ma), yet fossils attributed to the angiosperm stem lineage date back to 247.2–242.0 Ma. Molecular estimates of the angiosperm evolutionary timescale have shown considerable variation, being influenced by differences in taxon sampling, gene sampling, fossil calibrations, evolutionary models, and choices of priors. However, a common theme in modern molecular dating studies is that the angiosperm crown age far predates the oldest fossils by a non-trivial amount of time, even by up to ~100 Ma. Did crown-group angiosperms really arise in the Triassic? Or are the substantially older molecular dating estimates a product of methodological biases? Here, we report on two of our studies in which we analyse plastid-level data sets to (1) evaluate the impact of genomic data and priors on Bayesian estimates of the angiosperm evolutionary timescale, and (2) compare different partitioning methods to account for rate heterogeneity across lineages, and investigate the impacts on the precision of molecular dating estimates. In our main analysis, based on a data set comprising 76 protein-coding genes from the chloroplast genomes of 195 taxa spanning 86 families, we find that crown angiosperms arose 221 (251–192) million years ago during the Triassic. Based on a range of additional sensitivity and subsampling analyses, we found that our date estimates were generally robust to large changes in the parameters of the birth-death tree prior and of the model of rate variation across branches. In our reduced analysis of 79 protein-coding genes from 54 taxa, we find that all genes are best considered to have been evolving under a single pattern of among-lineage rate variation. Despite this, increasing the number of relaxed clock models to two led to a dramatic increase in precision of posterior age estimates. For example, the uncertainty surrounding the age of crown angiosperms halfved. Additionally, further increases to as many as 20 relaxed clock models always led to increased precision of posterior time estimates, but the precision appeared to be close to the limit imposed by the uncertainty in fossil calibrations. Overall, our results suggest that increases in gene sampling are unlikely to produce substantial changes in estimates of the angiosperm evolutionary timescale. Instead, our understanding of this important question is likely to improve through increased taxon sampling, significant methodological changes, and/or new information from the fossil record. Additionally, increasing the number of molecular clock models may be an effective way to increase the precision of posterior age estimates if computationally feasible, but whether this results in spurious precision remains to be evaluated.

T2-25: Poaceae: Systematics and phylogeny of major Lineages

T2-25-01

A molecular phylogeny and classification of the Cynodonteae (Poaceae: Chloridoideae)

Paul M. Peterson[1], *Konstantin Romaschenko*[1], *Yolanda Herrera Arrieta*[2]
1. Smithsonian Institution
2. Instituto Politecnico Nacional

Morphologically, the tribe Cynodonteae is a diverse group containing about 840 species in 100 genera and 21 subtribes, found primarily in Africa, Asia, Australia, and the Americas. However, there are still seven unplaced genera within the tribe. We have genetic evidence that rapid radiation has occurred within the grass tribe Cynodonteae and we want to identify the ecological and genetic factors that characterize the early evolutionary history of these grasses. The Cynodonteae are C4 grasses, i.e., with an extra system to fix carbon dioxide, that can outcompete other plants in warm, arid climates but we know very little about their early evolutionary history and how they became dominant in short grass and savannah grasslands. A modern classification of the Cynodonteae based on the study of molecular characters has resulted in major changes in the circumscription and alignment of genera. During the last 10 years we have conducted several phylogenetic analyses on nearly 600 species, subspecies and varieties (67% of the Cynodonteae species) using sequence data from eight plastid regions (*rps16-trnK* spacer, *rps16* intron, *rpoC2*, *rpl32-trnL* spacer, *rps3*, *ndhF*, *ndhA* intron, *ccsA*) and the nuclear ribosomal internal transcribed spacer regions (ITS 1 & 2) to infer evolutionary relationships and refine the current classification. *Allolepis, Jouvea, Kalinia, Lepturidium*, and *Sohnsia* do not align with other genera in our trees but occur as isolated branches in a New World clade along with other members of the Boutelouinae, Hilariinae, Monanthochloinae, Muhlenbergiinae, and Scleropogoninae. However, *Decaryella, Kampochloa*, and *Vietnamochloa* are still incertae sedis within the Cynodonteae and have not been included in any modern molecular study. In our new classification we recognize three new subtribes: Dactylocteninae that includes *Acrachne, Brachychloa, Dactyloctenium, and Neobouteloua*; Orininae with *Cleistogenes* and *Orinus*; and Zaqiqahinae with a single genus, *Zaqiqah*; resurrect Hubbardochloinae with seven genera; and describe six new genera: *Eleusinella* (Eleusininae) with two species, *Hyalopterus* (Eleusininae) with a single genus, *Orthacanthus* (Traginae) with a single species, *Triplasiella* (Gouiniinae) with a single species, *Tripogonella* (Tripogoninae) with three species, and *Zaqiqah*. Subgeneric classifications have been completed on *Bouteloua* (60 spp.), *Distichlis* (11 spp.), and *Muhlenbergia* (182 spp.). In the future we hope to clarify the branching pattern of the early diverging lineages (crown nodes) within the Cynodonteae and continue to refine our classification.

T2-25-02

Phylogeography and classification of *Stipa* (Poaceae: Pooideae: Stipeae)

Konstantin Romaschenko, *Paul M. Peterson, Robert J. Soreng*
Smithsonian Institution

The genus *Stipa* includes approximately 144 species and is a dominant member of the Eurasian steppe grasslands. Earlier molecular phylogenies included a few species with up to 10 DNA markers or only two DNA markers and 52 species. Consequently, relationships among the species are not well resolved and the phylogeography has not been thoroughly explored. The subgeneric structure among species of *Stipa* is complex and the number of the natural groups is unknown. Our earlier molecular phylogeny using 10 DNA markers revealed two major clades within *Stipa,* Asian and European. The European clade includes more than 115 species (80%) and originated in the late Pliocene (2.5 – 3.6 mya) via rapid radiation. As a result, the lineage is genetically homogeneous and shows little polymorphism in nucleotide sequences when using conventional markers (cpDNA regions, nrDNA ITS). In our study we found low copy nDNA genes of the Conserved Ortholog Set (COS) - *AroB, Agt1*, and *At103* to be variable enough to resolve phylogenetic relationships among the major lineages and detect multiple origins. We clarify subgeneric relationships among the species of *Stipa* and detect the following distributional patterns: 1) broad diagonal distribution from W (SW) to NE in ser. *Pulcherrimae* (from West Atlas to the mountain ranges of the Central Asia) and ser. *Penicilliferae* (from the Alpine ranges to the Siberian grasslands); 2) eastern European longitudinal distribution of ser. *Rubentes* associated with the periglacial steppe zone; 3) another diagonal pattern SW?NE (Irano-Turanian to Mediterranean) is seen in sect. *Barbatae* which is most speciose in Central Asia and not present in Central and Eastern Europe; and 4) sect. *Smirnovia* that is restricted to Central Asia. We identify ancestral areas for all major lineages within *Stipa* and reconstruct a dispersal pattern for separate groups and for separate species. We exploit dating and location of rapid radiation to establish essential patterns of evolution within *Stipa* which was under the pressure of gradual aridification during the late Miocene-Pleistocene. We juxtapose the biogeographic patterns for the major clades with phylogeny, and investigate whether or not colonization and radiation events were triggered by climate shifts. Our study supports the premise that rapid radiation is a key factor in evolution and speciation within a lineage.

T2-25-03

Phylogenetic classification of the Poaceae (Gramineae): A comparison with the past

Robert J. Soreng[1], *Paul M. Peterson*[1], *Konstantin Romaschenko*[1], *Patricia Barberá Sánchez*[2], *Lynn J. Gillespie*[3]
1. Smithsonian Institution
2. Real Jardín Botánico de Madrid, Consejo Superior de Investigaciones Científicas (CSIC)
3. Canadian Museum of Nature

Three modern worldwide classifications were produced just prior to or at the beginning of the molecular phylogenetic revolution [Clayton & Renvoize, Genera graminum, Kew Bull. Addit. Ser. 13: 1–389. 1986; Tzvelev, The system of grasses (Poaceae) and their evolution, Bot. Rev. 55: 141–203. 1989; Watson & Dallwitz, The grass genera of the world, CAB International, 1038 p. 1992]. Linnaeus published 40 grass genera in 1753 and this number has increased by an average of 8.5 per year up to 1986, when Clayton & Renvoize accepted 651. Since Clayton & Renvoize's publication (30 yrs) there have been 5.7 genera published per year (170 total). Currently there are ca.1500 legitimate generic names (typified by different species), roughly half of which are treated as heterotypic synonyms, the other half accepted. Based on molecular data two recent phylogenetic classifications were presented [Soreng, Peterson, Romaschenko, Davidse, Zuloaga, Judziewicz, Filgueiras, Davis, Morrone, A Worldwide phylogenetic classification of the Poaceae (Gramineae), J. Syst. Evol. 53: 117–137. 2015; Kellogg, XIII Flowering Plants, Monocots, Poaceae in Kubitzki, The Families and Genera of Vascular Plants, Springer International Publishing, 416 p. 2015] accepting 763 and 711 genera, respectively. These two classifications share 674 accepted genera, but diverge by 89 and 37 genera, respectively. Soreng *et al.,* listed 783 accepted and synonymous names as having been investigated using molecular methods while only 87 accepted genera were lacking DNA data. As of May 2016 only 50 genera were lacking DNA data entered in GenBank or in our own unpublished sequences. The number of accepted subfamilies, tribes, and subtribes reported in Soreng *et al.,* versus Kellogg are, respectively: 12 & 12,52 & 30, and 80 & 53. Obviously Kellogg used a broader concept and did not recognize tribes or subtribes when classifying monotypic lineages or those with few genera. In comparison with Soreng *et al.,* only *Alloeochaete*, *Danthonidium*, and *Phaenanthoecium* were placed in different subfamilies, and *Streptogyna* was left as incertae sedis by Kellogg, versus placement in Oryzoideae by us. At the rank of tribe Soreng *et al.,* and Kellogg placed 11 genera differently, mostly as incertae sedis in a closely related tribe or in a narrower tribal delimitation. Subtribal placements of genera differed more between the two classifications, primarily because of our narrower subtribal concept. Kellogg placed more genera in synonymy, i.e., lumped, especially in *Trisetaria* (7), *Elymus* (6), *Eragrostis* (4), but also in *Chinonachne*, *Ehrharta*, *Saccharum*, *Tristachya* (all 3 each), and in 30 other cases (all 1 each). However, the two classifications were remarkably similar, especially when compared to the discrepancies at all ranks among the three earlier classifications. The power of molecular phylogenetics to interpret evolutionary relationships to stabilize the classification is apparent. We are continually updating our current classification by expanding synonymy, adjusting tribe and subtribe membership, including new genera and new molecular results for existing genera that have been overlooked. Updated arrangements for tribes, subtribes, and generic alignments are noted across the family.

T2-25-04

A new phylogenetic classification of Poeae subtribes Poinae, Coleanthineae, and Miliinae (Poaceae)

Lynn J. Gillespie[1], *Robert J. Soreng*[2]

1. Canadian Museum of Nature

2. Smithsonian Institution, National Museum of Natural History

Subtribes Poinae, the largest subtribe in the grass family with over 700 species in 29 genera, Miliinae (1 genus, 5 spp.), and Coleanthineae (10 genera, 163 spp.) together form one of three clades in tribe Poeae, the largest tribe in Poaceae (118 genera, 2776 spp.). Here we present a revised classification of this clade based on molecular phylogenetic data (ITS, ETS, *trnT-trnL-trnF*, *MatK*, and *rpoB-trnC*). Our analysis includes 39 of the 40 genera (data lacking for *Agrostopoa*), with broad sampling within all of the large genera and all or most of the species in the small genera (>800 samples). We compare our classification with previous classifications based on morphology and/or molecular data, and discuss the evolution of several morphological characters. Subtribes Miliinae and Coleanthineae resolved as monophyletic with strong support [Soreng, Gillespie, Koba, Boudko, Bull. J. Syst. Evol. 53: 138-162. 2015]. However, Poinae sensu Soreng *et al.,* 2015 [Soreng, Peterson, Romaschenko, Davidse, Zuloaga, Judziewicz, Filgueiras, Davis, Morrone, A worldwide phylogenetic classification of the Poaceae (Gramineae), J. Syst. Evol. 53: 117–137. 2015], where Alopecurinae, Beckmanniinae, Cinninae, Phleinae were included in synonomy, is paraphyletic because Miliinae is embedded within it. The genus *Poa* is more closely related to *Milium*, and possibly *Phleum*, than it is to other genera traditionally placed within Poinae s.s. The latter genera, together with genera of Alopecurinae, Beckmanniinae and Cinninae, form a well supported clade. Rather than synonymyzing Miliinae in Poinae, thereby making that subtribe even larger and more heterogeneous, we prefer to recognize strongly supported clades within Poinae s.l. as subtribes. Here we redefine subtribe Poinae to include only the genus *Poa*, recognize both Miliinae and Phleinae as monogeneric subtribes, and subdivide the remaining 26 genera of Poinae s.l. into several well-defined subtribes, some of which are reticulate in origin. Morphological and habit characters traditionally used in the classification of the group, such as annual versus perennial, one-flowered versus multi-flowered spikelets, glume length relative to floret length, spike-like panicles, and awn development and insertion, were determined to be highly homoplasious among and within the subtribes.

T2-25-05

A molecular phylogenetic study of temperate grasses in Poeae chloroplast group 1 (Poaceae: Pooideae: Poeae) based on nuclear ribosomal and plastid DNA sequence data

Jeffery M. Saarela[1], *Roger D. Bull*[1], *Michel J. Paradis*[1], *Sharon N. Ebata*[1], *Paul M. Peterson*[2], *Robert J. Soreng*[2], *Beata Paszko*[3]

1. Canadian Museum of Nature

2. Smithsonian Institution

3. W. Szafer Institute of Botany, Polish Academy of Sciences

We conducted phylogenetic analyses of >2500 new DNA sequences from nuclear ribosomal (ITS and ETS) and five plastid regions to clarify relationships within and among genera and subtribes that are part of the grass tribe Poeae chloroplast group 1 and allies. A large clade comprises the subtribes Agrostidinae and Calothecinae, neither of which is monophyletic. The type species of *Deyeuxia* is allied with species of *Calamagrostis* s.s., thus *Deyeuxia* is a synonym of *Calamagrostis*, and relationships among species of *Calamagrostis* s.s. and related genera are mostly poorly resolved. *Agrostis* and *Polypogon* are distinct from *Calamagrostis* s.s. but are not reciprocally monophyletic. Species of *Agrostis* and *Polypogon* are intermixed in a strongly supported clade in plastid trees, including *Lachnagrostis*, and there is incongruence among data partitions in a subset of species. *Ammophila* and *Calamagrostis* s.s. are not distinguished as separate genera, and *Calamagrostis coarctata*, *Dichelachne*, *Echinopogon* and *Relchela* are closely related. Taxa of Sesleriinae are allied with Aveninae s.s. (*Arrhenatherum*, *Avena*, *Helictotrichon*, *Tricholemma*) in nuclear ribosomal trees, and with taxa of chloroplast group 2 in plastid trees. Koeleriinae is strongly supported and includes two major subclades: clade A comprises *Avellinia*, *Gaudinia*, *Graphephorum* and *Trisetum* sect. *Trisetum* p.p. (nuclear trees), *Koeleria*, *Rostraria*, *Trisetaria*, *Trisetum* sect. *Trisetum* p.p. and *Trisetum* sect. *Trisetaera*, and *Koeleriinae* clade B comprises *Calamagrostis*/*Deyeuxia* p.p., *Graphephorum* and *Trisetum* sect. *Trisetum* p.p. (plastid trees), *Peyritschia*, *Leptophyllochloa*, *Sphenopholis*, *Trisetopsis* and *Trisetum* sect. *Deschampsioides*. Within each clade most aspects of relationship are unresolved, and conflict between data partitions is identified. Affinities of *Lagurus ovatus* with respect to Aveninae s.s. and Koeleriinae are incongruent in nuclear ribosomal and plastid trees. All but one of the eight sampled species of *Deyeuxia* sect. *Stylagrostis* are part of the *Deschampsia* lineage, and these are transferred to *Deschampsia*. Substantial changes to the current classification will be needed to produce generic circumscriptions consistent with phylogenetic evidence.

T2-25-06

Reticulate evolution of Triticeae revisited: New insights based on multiple chloroplast and nuclear genes

Jia Liu[1,2], *Ruiwu Yang*[2], *Xiuying Kong*[3], ***Shiliang Zhou***[1]
1. State Key Laboratory of Systematic and Evolutionary Botany, Institute of Botany, Chinese Academy of Sciences, Beijing 100093
2. College of Life Science, Sichuan Agricultural University, Yaan 625014, Sichuan
3. The National Key Facility for Crop Gene Resources and Genetic Improvement, Institute of Crop Sciences, Chinese Academy of Agricultural Sciences, Beijing, China

Triticeae is one of the economically important tribe in Poaceae. Due to complex reticulate evolution in the tribe, the genetic relationships among genera and among genomes are unclear, and the circumscription of genera remains questionable. In this study we analysed 26 of 27 genera in the tribe using two nuclear genes and 18 hypervariable chloroplast regions. Our well resolved phylogenies provide new insights into the evolution in Triticeae. (1) There are eight lineages in Triticeae, plus *Bromus*. (2) Genome Xe is a variant of genome P, genome Y is a variant of genome V, and genome Xm is closely related to genome Ns. (3) The origins of nine polyploid genera was proposed and *Elymus*, *Kengyilia*, and *Leymus* are of multiple origins involving several genomes. (4) The divergence time of Triticeae was dated to the Oligocene, but most genera originated in the Miocene and the speciation of polyploid species was in the Pliocene. (5) The cradle of Triticeae is deduced to be in Asia. (6) Taxonomic treatments of the genera are provided. This study clarifies generic relationships, genome relationships and the origins of hybrid genera in Triticeae, which is helpful for understanding the evolution in the tribe.

T2-26: Lamiales: Phylogenetic insight into the patterns and mechanisms of phenotypic diversity (two sessions)

T2-26-01

Phylogeny and systematics of Lamiales: Historical perspective and present concept

Richard Olmstead
University of Washington

We have come a long ways since the concept of Lamiales in its present form was first introduced in 1992. At the St. Louis IBC in 1999 a symposium on the phylogeny and systematics of Lamiales presented the first summary of what we know about this large clade. We now recognize over 1000 genera assigned to 26 families and including more than 22,000 species. With more than one-third of all species of Lamiales included in molecular phylogenetic studies, we now have a detailed understanding of the phylogeny of the clade. I will present the history of phylogenetic studies within Lamiales, an overview of the diversity at the family level, and conclude with a new dated phylogeny of the clade, which suggest that the age of crown-group Lamiales is approximately 92 million years (+/-12 MY).

T2-26-02

Probe into floral symmetry and A-class genes in a new model plant *Chirita pumila*

Xia Yang, *Jing Liu*, *Jie Wu*, *Yin-Zheng Wang*
State Key Laboratory of Systematic and Evolutionary Botany, Institute of Botany, Chinese Academy of Sciences

The development and evolution of the flowers has always attracted much attention of botanists. In spite of extensive investigation in two classical model species *Antirrhinum majus* and *Arabidopsis thaliana*, there are still many unanswered questions in the fields of floral organ identity and floral symmetry. For example, the functions of A-class floral organ identity genes, so far limited to *Arabidopsis* and its close relatives, have not yet been testified in other lineages, and whether the BC model (lacking the A-function) is general in eudicots is still an open question. In the field of floral symmetry, it is still a challenge to decipher how related floral symmetry genes interplay to generate zygomorphic flowers in other Lamiales species, how they regulate their downstream targets, and how their own activities are controlled. *Chirita pumila* D. Don belongs to the Gesneriaceae family, one of the most basal groups in Lamiales, is an ideal model system to address these questions. Here we select *C. pumila* as a model to deeply address the function and regulatory networks of floral symmetry related genes, involving *CYCLOIDEA*-, *RADIALIS*- and *DIVARICATA*-like genes, and the role of A-class floral organ identity genes, mainly *APETALA1* and *APETALA2* genes. Real-time qPCR and *in situ* hybridization results hinted that *CpCYC1* and *CpCYC2* might have partially redundant function. However, knock-out of either one via a strong repressor SRDX motif was sufficient to generate the fully ventralized actinomorphic flowers due to concurrent down-regulation of another one, confirming the positive auto- and cross-regulation between them, as previously reported in *Primulina heterotricha*. Furthermore, knock-out of either *CpCYC1* or *CpCYC2* led to the up-regulation of *CpRAD* with *CpDIV*'s expression pattern unaffected. Further transgenic experiments involving *CpRAD* and *CpDIV* and transcriptome analyses are ongoing. In addition to function studies, we also focused on the regulatory mechanisms underlying the dorsal-specific expression patterns of *CpCYC1* and *CpCYC2* that play a pivotal role in establishing floral zygomorphy in *C. pumila*. We have identified a series of key elements in the proximal promoters of the two genes by sequential truncation transgenic experiments, and further function verification experiments are ongoing. As for the two A-class genes, we have performed real-time qPCR and *in situ* hybridization experiments. The results showed that the expression patterns of both *CpAP1* and *CpAP2* were different from the corresponding orthologs in *A. thaliana* and *A. majus*. Whether *CpAP1* and *CpAP2* have A-function and if they activate B- and C-function genes deserve further function studies.

T2-26-03

Phylogenetic pattern and underlying mechanisms of hummingbird-adapted flowers in the genus *Penstemon*

Carolyn Wessinger[1], *Amanda Katzer*[1], *Mark Rausher*[2], *Lena Hileman*[1]
1. *University of Kansas*
2. *Duke University*

Penstemon is ancestrally adapted to pollination by bees, and flowers adapted to hummingbird pollination have evolved through coordinated evolutionary transitions in a complex of physiologically and developmentally distinct floral traits. Using a phylogenomic RADseq approach on a subset of *Penstemon* species (sampled primarily from herbarium material), we identified an asymmetric pattern of pollination syndrome evolution: transitions from bee to hummingbird syndrome have repeatedly occurred with little evidence for evolutionary reversal. Results from QTL analyses and comparative morphological and gene expression studies suggest that loci of major effect and loci that pleiotropically influence multiple floral traits underlie adaptations to hummingbird pollination. This relatively efficient genetic architecture may help explain the frequency of evolutionary transitions to hummingbird pollination in this genus.

T2-26-04

Speciation patterns and processes in Gesneriaceae: New world *Columnea*, old world *Primulina*

James Smith[1], *Yin-Zheng Wong*[2], *Peng-Wei Li*[2], *John Clark*[3], *Oscar Marín-Gómez*[5], *Marisol Amaya-Márquez*[4]
1. *Boise State University*
2. *State Key Laboratory of Systematic and Evolutionary Botany, Institute of Botany, Chinese Academy of Sciences*
3. *The Lawrenceville School*
4. *Universidad Nacional de Colombia*
5. *Instituto de Ecología, INECOL A. C*

Phylogenetic analyses based on molecular data have revolutionized systematics and with well-supported phylogenetic hypotheses we are now able to examine processes that affect speciation as well as the patterns. These can resolve which morphological, biogeographic region, or environmental parameters are ancestral and how changes have occured across the evolutionary history of the group being examined. Likewise, methodologies also exist to infer which, if any of the above alter the rate of diversification. Here we examine two genera of Gesneriaceae from different parts of the world. *Columnea* is restricted to the New World tropics and is the largest genus in the nearly exclusively Neotropical subfamily, Gesnerioideae. Species are mostly epiphytic or epipetric and most are presumed to be hummingbird pollinated and the seeds dispersed by birds consuming the fleshy fruits. In contrast, *Primulina* is restricted to the Old World and is one of the larger genera in the Paleotropical subfamily Didymocarpoideae. Species in this genus are mostly epipetric and are found on calcareous outcroups or in caves where moisture levels are high. Pollination is presumably by bees and seeds are gravity or wind dispersed from dry dehiscent fruits. We examine biogeography and ancestral character state evolution in both of these lineages and investigate the parameters that are driving diversification.

T2-26-05

Say it with monkeyflowers *(Mimulus)*: Genetics, development, and evolution of phenotypic diversity and novelty

Yaowu Yuan
University of Connecticut

The monkeyflower genus *Mimulus* (now *Erythranthe*) has long been recognized as a classic ecological and evolutionary model system that is particularly suitable for the studies of local adaptation, speciation, plant-pollinator interaction, mating system evolution, and species range limitation. Yet its potential as an excellent genetic and developmental model system has only been realized in the past few years. Instead of focusing on one specific story, in this talk I will present an overview of the sophisticated functional tools and genetic and genomic resources developed for this system, and highlight some recent progresses in using *Mimulus* to solve problems on the genetic control of ecologically important floral traits (that cannot be studied using *Arabidopsis*), developmental mechanism underlying the formation of periodic patterns in biological objects, molecular basis of inter-species phenotypic variation, and the origin and evolution of phenotypic novelty.

T2-26-06

Evolution of floral morphology in Bignonieae (Bignoniaceae): Implications for Community Assembly

Suzana Alcantara[1], ***Lucia Lohmann***[2]
1. *Universidade Federal de Santa Catarina*
2. *Universidade de São Paulo*

Flowering plants represent one of the most diverse groups of organisms in the Planet. A large portion of this diversity results from the multitude of floral forms encountered nowadays. An improved understanding of the patterns of floral variation and of the historical and ecological factors that led to the evolution of such diversity in floral forms is critical for a better understanding of the processes that led to the diversification of angiosperms. The tribe Bignonieae (Bignoniaceae) is the most diverse clade of Neotropical lianas and represents an excellent model for the study of floral evolution due to its high diversity of floral forms. In this study we characterize the: (i) pattern of evolution of discrete floral traits and floral morphologies in Bignonieae, and its potential associations with pollinators; (ii) phylogenetic signal and rates of evolution of continuous floral traits; (iii) pattern and magnitude of phenotypic integration among floral traits across the phylogenetic history of tribe Bignonieae; and, (iv) influence of phylogeny, floral morphology and abiotic factors for the patterns of species co-occurrence and the structure of communities of Bignonieae in the Neotropics. Our findings indicate high lability in the evolution of discrete floral traits and floral forms, contrasting the significant phylogenetic signal encountered in 16 continuous floral traits examined. However, the phylogenetic signal differs between traits of different floral whorls. Rates of evolution also varied among different characters, suggesting the action of different selective pressures or differential responses to selection in different floral parts. Overall, the patterns of phenotypic integration are constant during the history of Bignonieae, despite the homoplastic evolution of the magnitude of correlation among characters. This apparent evolutionary complexity, leading to different patterns in different traits, is not reflected in the ecological structure of communities given that neither floral morphology nor phylogeny influences species co-occurrence. On the other hand, we found species specialization to abiotic factors, suggesting that environmental filtering played a key role in the structure of communities of Bignonieae. These results reject the hypothesis that saturation caused by competition for pollinators is the main factor determining intra-community structure in Bignonieae (Funding: FAPESP, CAPES & CNPq, Brazil).

T2-26-07

Global Biogeographic patterns of diversification in Lamiales

David Tank[1], *Richard Olmstead*[2]
1. University of Idaho, Moscow, Idaho, USA
2. University of Washington, Seattle, Washington, USA

With over 20,000 species and a cosmopolitan distribution, Lamiales represent one of the major asterid radiations. Minimum estimates of the crown (~65-75 my) and stem ages (~89-113 my) suggest an origin and initial diversification at the time of the initial split of Gondwana (~100 my), thus suggesting that while vicariance may have played a role in their early diversification, the continental distributions within clades were likely the result of intercontinental dispersal. Using a dated Lamiales megaphylogeny and geographic data obtained from GBIF, we quantify the rates of intercontinental migrations, as well as transition rates between global biomes, to address how intercontinental migration events and rates of global biome evolution have led to present distributions in Lamiales, thereby obtaining a first pass at the dynamics of biogeographic history on a global scale for this large complex group.

T2-26-08

Evolution of floral symmetry and orientation as functional integration promotes speciation and diversification in Lamiales

Yin-Zheng Wang
State Key Laboratory of Systematic and Evolutionary Botany, Institute of Botany, Chinese Academy of Sciences, 20 Nanxincun, Xiangshan, Beijing 100093, China

Flowers have provided an important model for studying the evolution of integration relevant to prezygotic mechanism in speciation and diversification. The association between floral orientation and symmetry was recognized as a floral complex trait very early the 19th century. However, the floral orientation to date has received little attention even though the floral symmetry has been of great interest to evolutionary biologists because the evolution of zygomorphy is considered a major trend in angiosperm radiation. In fact, the horizontal orientation, possibly the first step toward floral specialization, is widely linked to and co-varied with zygomorphy in angiosperms, especially in Lamiales. Recent study shows that the floral horizontal orientation and floral zygomorphy acting as a functional integration are controlled by a single CYC-like TCP gene via repressing the basal-dorsal floral tube, dorsal petals and stamens. This finding shed significant light on our understanding of the complex mechanisms behind floral adaptive evolution at molecular level in angiosperms. It has becoming increasingly apparent that floral horizontal orientation and zygomorphy play a key role in effecting specific-pollinator attraction, approach and behavior, on which scent and pigmentation traits develop further enhancing prepollination isolation. Given frequent variation in zygomorphic lineages, CYC-like genes could be hotspots of evolution in both floral orientation and symmetry, and dominate in the evolution of the complex prezygotic-isolation mechanism that underlies the rapid diversification of angiosperms.

T2-26-09

Inflorescence evolution in the Scrophulariaceae, with focus on the genus *Buddleja*

John Chau[1], *Nataly Oleary*[2], *Wei-Bang Sun*[3], *Richard Olmstead*[1]
1. University of Washington
2. Instituto de Botánica Darwinion
3. Kunming Institute of Botany

Inflorescence architecture is important in the fitness and reproductive success of many angiosperm groups. Breeding system, pollinator attraction and behavior, ovary competition, and fruit dispersal and predation are all affected by inflorescence structure. As such, evolution in inflorescence architecture may be an important driver of plant adaptation and speciation. A diversity of inflorescence architectures has evolved in flowering plants, which may be correlated with different habitat types, pollinator communities, and life histories. In the Lamiales, including the family Scrophulariaceae, determinate cymose inflorescence units are common, but these cymes are arranged in compound inflorescences in various ways. *Buddleja* displays a diversity of inflorescence architectures across its 108 species, including racemose, thrysoid, verticillate, capitate, and paniculate forms. Using a recently inferred molecular phylogeny as a framework, inflorescence characters are mapped and analyzed to investigate evolutionary patterns in inflorescence architecture and correlations with environmental factors and other traits. We also survey inflorescence form across Scrophulariaceae to place inflorescence evolution in *Buddleja* within a broader context.

T2-26-10

Phylogeny and floral trait evolution in *Eremophila* (Scrophulariaceae, Lamiales)

Rachael Fowler[1,2], *Michael Bayly*[1], *Daniel Murphy*[2]
1. The University of Melbourne
2. Royal Botanic Gardens Victoria

Tribe Myoporeae is a member of the large and cosmopolitan plant family Scrophulariaceae (Lamiales). Previously considered a family in its own right, Myoporeae was recently recircumscribed as a tribe within Scrophulariaceae sensu stricto based upon sequences of plastid genes. Myoporeae consists of seven genera, predominantly distributed throughout Australia. By far the largest of these genera is *Eremophila*, with ~220 species, ~60 subspecies and a large number of putative taxa still awaiting formal description. Commonly known as emu bushes, *Eremophila* species are significant, often dominant, components of the Australian arid and semi-arid zone flora, where they tolerate harsh, dry and saline conditions and provide food and shelter for a wide range of insect and bird species. Morphologically, *Eremophila* species are incredibly diverse particularly with regards to flower shape, colour and structure. This diversity is largely attributed to variation in modes of pollination. Eremophilas can be broadly divided into two groups, those that are bird pollinated (~20%) displaying brightly coloured, tubular flowers, and those that are insect pollinated (~80%) displaying purple, mauve or white flowers and patterns of spotting and banding. Species relationships within *Eremophila* have been unclear and the classification of this group to date has been based on morphology alone. With high throughput sequencing methods, we now have the opportunity to study the evolution of this group using substantial genomic datasets. The focus of our research has been to generate a comprehensive molecular phylogeny of *Eremophila* and related genera within tribe Myoporeae using plastid and nuclear data. This phylogenetic reconstruction using entire chloroplast genomes and high copy sections of nuclear DNA has provided greater understanding of evolutionary relationships both in *Eremophila* and Myoporeae, as well as biogeographic inferences for arid Australia. This molecular phylogeny provides a framework for inferring of patterns of floral evolution in *Eremophila*, with multiple floral trait transitions likely to have occurred during evolution of the group.

T2-26-11

Phylogeny, systematics and evolution of the largest hemiparasitic genus *Pedicularis* (Orobanchaceae)

Hong Wang[1], ***Wen-Bin Yu***[2], *Min-Lu Liu*[1,3], *De-Zhu Li*[1]

1. Kunming Institute of Botany, Chinese Academy of Sciences

2. Xishuangbanna Tropical Botanical Garden, Chinese Academy of Sciences

3. Third Institute of Oceanography, State Oceanic Administration

Pedicularis, with more than 500 species, is the largest genus in Orobanchaceae. Species of *Pedicularis* are biennial or perennial hemiparasitic herbs, and are primarily distributed in mountain ranges throughout the north temperate zone. Striking interspecific variation in floral traits in *Pedicularis* has given rise to controversy concerning species delimitation and infra-generic classifications. Evaluating DNA barcodes showed that nuclear ribosomal internal transcribed spacer (nrITS) or nrITS + *rbc*L can discriminate at least 78% of species in the genus. Furthermore, molecular methods are effective in helping to identify and discover cryptic species. Based on extensive sampling of *Pedicularis* (~ 257 species), with an emphasis on species from the Himalaya-Hengduan Mountains, phylogenetic analyses strongly recovered 13 major clades. There is little consensus between the phylogenetic tree and Tsoong's classification of *Pedicularis*. Only two of the 13 groups (15.4%), and 19 of the 56 series (33.9%) with more than one sampled species were found to be strictly monophyletic. Most opposite-/whorled-leaved species fell into a single clade, i.e. clade 1, while alternate leaved species were placed in the remaining 12 clades. Excluding the widespread *P. verticillata* in clade 1, species from Europe and North America fall into clades 6–8. Our results suggested that combinations of morphological and geographic characters associated with strongly supported clades are needed to elucidate a comprehensive global phylogeny of *Pedicularis*. Alternate leaves were inferred to be plesiomorphic in *Pedicularis*, with multiple transitions to opposite/whorled phyllotaxy. Anatomical observation on 39 *Pedicularis* species revealed two nectary forms [bulged (N = 27) or elongated (N = 5)] or the absence of nectaries (N = 7). Evolutionary inferences indicated that the bulged nectary should be the ancestral state, nectaries were independently lost in some beaked species, and elongated nectaries evolved independently in some clades of beakless species. Phylogenetic path analysis showed that nectary presence was indirectly correlated with beak length/pollinator behavior through an intermediate factor, nectar production. No significant correlation was found between nectary type and nectar production, beak length or pollinator behavior. The nectary in beaked species may be a vestigial structure retained during a recent rapid radiation of *Pedicularis*, especially in the Himalaya– Hengduan Mountains of south-western China.

T2-26-12

Phylogenetic relationships, character evolution, and biogeography of *Salvia* (Lamiaceae), with special focus on Iranian taxa

Fatemeh Moein[1], *Ziba Jamzad*[3], *Mohammadreza Rahiminejad*[1], *Evgeny Mavrodiev*[2], *Mattew Gitzendanner*[2], *Douglas Soltis*[2], *Pamela Soltis*[2]

1. University of Isfahan

2. University of Florida

3. Research institute of Forest and Rangelands

Salvia L. is the largest genus in Lamiaceae (Mentheae-Salviinae), with approximately 1000 species that are mostly found in five major centers of diversity: Central and South America (500 spp.), Western Asia (200 spp.), Eastern Asia (100 spp.), Africa (60 spp.), and Europe (36 spp.). *Salvia* is characterized by a unique staminal structure of two theca separated by an elongated connective, which is attributed to a special pollination system referred to as a lever mechanism. This character evolved in parallel between the Old World and New World *Salvia* species. A high percentage of endemism and high morphological diversity are of special interest in understanding the evolutionary patterns within *Salvia*. Compared to New and Old World *Salvia*, taxa from Southwest Asia have not been well sampled in previous phylogenetic studies. Iran contains 62 species of *Salvia*, of which 19 are endemic, and the area is considered a center of diversity for *Salvia* in Southwest Asia. In this study, the phylogenetic relationships of Iranian species of *Salvia* are reported for the first time. Two nuclear regions, ITS/5.8S and ETS, and the plastid locus *ycf1* were amplified and sequenced for 58 of the 62 species of Iranian *Salvia*. Maximum likelihood analyses for 274 species including Genbank data suggest that Iranian *Salvia* species are not monophyletic; instead, they are placed in three main clades. These clades are congruent with morphological characters. To assess the evolutionary patterns and diversification of *Salvia*, key morphological characters and habitat for 245 species were reconstructed across the phylogeny. In addition, biogeographic patterns of *Salvia* in the Old World were analyzed and will be discussed.

T2-27: Asian and Chinese bryology (two sessions)

T2-27-01

Systematic study of the genus *Yakushimabryum* and related genera in the Pylaisiadelphaceae (Musci)

Hiroyuki Akiyama
Museum of Nature and Human Activities, Hyogo

Phylogenetic relationships among the members of an epiphytic genus *Yakushimabryum* H. Akiyama, Y. Chang, T. Yamag. & B. C. Tan and related genera of the Pylaisiadelphaceae were studied based on phylogenetic analyses using cpDNA (*rbc*L, *rps*4, and *trn*L-F) and mtDNA (*nad*5) gene sequences as well as morphological features of type and authentic specimens. Our present phylogenetic analysis did not support the monophyly of the family. While, it suggested that *Yakushimabryum* to be monophyletic and should be treated as a distinctive genera in the family. The genus was revealed to be closely related to *Taxithelium* and *Pylaisiadelpha* in the family. It was revealed to include three species, all of which had been reported in East Asia. Systematic position of *Clastobryum*, which had been classified in the Pylaisiadelphaceae or Sematophyllaceae, was left ambiguous. One new genus *Orientobryum* H. Akiyama and a new species *Y. brevigemmium* H. Akiyama were recognized and described. Low genetic differentiation among geographically distantly separated populations were found in *Yakushimabryum spp.*, *O. oligonema*, *Pylaisiadelpha tenuirostris*, *Clastobryum scalare*, *Isocladiella surcularis*, *Gammiella ceylonensis*, all of which are epiphytes and reproduced by means of light and small propagules produced in leaf axils.

T2-27-02

Phytogeographic affinities of the moss flora of China relative to those of North America, Japan, and mainland Southeast Asia

Si He
Missouri Botanical Garden

A phytogeographic affinity of one region with another refers to similarities of their current floras and is an outcome of their common evolutionary histories and past and contemporary connectivity. The existence of taxa (families, genera and species) is often expressed by their phytogeographic classifications. The level of similarities between two floras is commonly explained by their similarity coefficients. The moss flora of China contains a total of 2031 species and subspecific taxa belonging to 434 genera and 87 families. This study was a first attempt to analyze floristic composition and phytogeographical connections of the Chinese mosses based on distributional patterns of taxa. The distributional data corresponding to 2031 Chinese taxa were classified into northern, southern, southwestern, and widely spread elements according to their distribution records by provinces. The phytogeographic affinities of the Chinese mosses relative to the mosses of North America, Japan, and the mainland Southeast Asia for the first time were explored in this study, respectively. The Sørensen-Dice coefficients were used to measure the similarities between the two floras in the comparisons. The analysis shows that the moss flora of China not only holds Holarctic elements in the north of Yangtze River (northern flora), but also holds many tropical elements in the south of Yangtze River (southern flora). The northern flora shares more similarity to that of North America which has predominantly Holarctic temperate elements and a few tropical elements in the southern end. Their similarities are 100% at the familial level, 82.97% at the generic level, and 48.96% at the species level. The study also shows that the moss flora of northern China in contrast to the southern flora is more similar to that of Japan with 92.50% similarity at the family level, 81.23% at the generic level, and 61.48% at the species level. The southern flora and particularly the flora of southwestern (SW) China are more closely related to that of mainland Southeast Asia. The SW region alone represents nearly all the similarity between southern China and the mainland Southeast Asia with coefficients of 84.29% at the familial level, 68.42% at the generic level, and 39.57% at the species level. The data indicate that SW China is largely tropical in nature, although it also contains a significant number of temperate elements. Nearly all Chinese records of the following larger genera occur in SW China: *Bartramia*, *Brachymenium*, *Brachythecium*, *Bryoerythrophyllum*, *Bryum*, *Campylopus*, *Dicranodontium*, *Didymodon*, *Ditrichum*, *Entodon*, *Gollania*, *Hypnum*, *Leucobryum*, *Mnium*, *Neckeropsis*, *Orthomnionm*, *Philonotis*, *Pinnatella*, *Plagiomnium*, *Plagiothecium*, *Pogonatum*, *Pohlia*, *Rhizomnium*, and *Trichostomum*. This phenomenon suggests that the SW region occupies an important transitional position between the Paleotropical flora and the Holarctic flora, possibly serving as a significant corridor for the migration of moss species.

T2-27-03

All is about *Herbertus aduncus* (Dicks.) Gray, Species Delimitation in *Herbertus* (Marchantiophyta)

Xiaolan He, *Yu Sun*
Finnish Museum of Natural History, University of Helsinki

No consensus has been settled on the taxonomy of the genus *Herbertus* since it was established in 1821. Much of the taxonomy of the genus, has in the past, been inaccurate and in a confused state in considerable measure, but has unfortunately been followed subsequently. The taxonomic uncertainties within the genus reside in the lack of global-scaled taxonomic studies and in the presence of a high degree of morphological variations. Because the genus is extremely variable from one local population to another as well as over wide geographic ranges, it has been suggested that the problem of delimiting species is unlikely to be solved without the study of large quantities of material. In the present study, an attempt on the species delimitation of *Herbertus* of the Northern Hemisphere was made based on morphology and multigene sequence data. Over six hundreds herbarium specimens were studied. Our results show that morphology alone offers limited useful taxonomical characters and that there are little genetic differentiations among currently recognized species *Herbertus aduncus*, *H. dicranus*, *H. hutchinsiae*, *H. longifissus*, *H. stramineus*, and *H. delavayi*, and among *H. sendtneri*, *H. armitanus* and *H. circinatus*. Our results do not support the current morphology-based species concept of the genus. Most of the species included in the present study are expected to be synonymized.

T2-27-04

Phytogeography of East Asian mosses with special emphasis on the East Asian element

Masanobu Higuchi
National Museum of Nature and Science

A survey is given of the distribution patterns of East Asian element of mosses. East Asia is one of the richest floristic regions in the world. This region roughly overlaps the Eastern Asiatic region, one of the phytogeographic regions which have been historically recognized on the basis of the distribution of flowering plants. This region was not significantly glaciated in the Pleistocene, which is one of the reason why this region has richer flora than that of Europe. For example East Asiatic *Nipponolejeunea* is only recorded by fossils in Europe. As compared with Europe, however, exploration of the moss flora of the East Asia has been ongoing, yet substantial gaps remain in our floristic understanding of this region. In the system of Takhtajan the Eastern Asiatic region belongs to the Holarctic Kingdom, is bordered by the Circumboreal region in the north, the Irano-Turanian region in the west and the Indian, Indochinese and Malesian regions of the Paleotropical Kingdom in the south. As the result the moss flora of the Eastern Asiatic region consists of several floristic elements such as circumboreal element, East Asian element, tropical element, cosmopolitan, *etc.*, but it is most characterized by East Asian element. The range of mosses is usually much wider than that of the flowering plants. Analysis of the range of the mosses of the East Asian element indicates that there are many taxa which show their limits of the distribution around the border of the Eastern Asiatic region recognized by the flowering plants. The causes leading to the accordance is discussed.

T2-27-05

Dr. Benito C. Tan 1946 to 2016 — The pillar and legacy of bryology in Asia

Boon-Chuan Ho
Singapore Botanic Gardens

The symposium on Asian and Chinese Bryology would be incomplete without a tribute to Dr. Benito C. Tan who has sadly passed away on 23rd December 2016. The research focus of Benito was in the taxonomy, systematics, and evolution of Asian mosses, as well as in the conservation and biogeography of the bryophyte Floras in Asia. He was undoubtedly the foremost expert on Asian mosses, specialising in one of the most difficult moss families, the Sematophyllaceae. His 40-year long career in Bryology has resulted in over 300 publications and significantly improve the knowledge of Asian bryophyte Floras. In 2004, he had been awarded the Richard Spruce Award by the International Association of Bryologists, for his outstanding research contribution in Bryology. A joyful and generous man, he has trained and guided numerous students into the study of bryophytes in Asia.

T2-27-06

Moss Flora of Russia project

Michael Ignatov[1,2], *Elena Ignatova*[2], *Vladimir Fedosov*[2]
1. Tsitsin Main Botanical Garden of Russian Academy of Sciences
2. Moscow State University, Faculty of Biology

Recent studies considerably expanded knowledge of the moss flora of Russia, with especially great additions to the flora of its Asian part. A number of families, such as Pseudoditrichaceae and Andreaeobryaceae, previosly considered as endemic of North America were found in Siberia. More southern Erpodiaceae and Ortodontiaceae were discovered in Asian Russia, as well as Oedipodiaceae, a family with hyperoceanic distribution. In the last 25 years a number of moss species known in Russia raised from 1050 to 1280, and new findingds continue. This addition in fact is not 230, but over 300 species, as more that 70 species were excluded as erroneous or non-confirmed records in the course of revisions. In average 10 newly recorded species per year and two species new for science per year were being discovered in Russia during last twenty-five years. In 2017 the first out of six volumes of the Moss Flora of Russia was printed, while other volumes are planned to be accomplished in the next five to ten years. The flora comprises taxonomic revisions based on molecular and morphological data. Among others, it turned out that some species which were considered as circum-Holarctic, e.g., *Rhynchostegium riparioides, Schistidium apocarpus, Hedwigia ciliata, Polytrichum formosum, Brachythecium glareosum* are absent in Asian Russia, and all their previous records in Russia east of Urals belong to other species. At the same time, recently described from North America *Coscinodon yukonensis, Pylaisia steerei* and *Sphagnum alaskanum* appeared to be widespread in Asia; *Ditrichopsis clausii* described from Yunnan was found in Transbaicalia, and *Struckia enevis* described from Sichuan appeared to be widespread in Siberia, reaching eastern Yakitia, the coldest area in the Northern Hemisphere. Further taxonomic revisions, with the special attention to cross-comparison of East Asian, Siberian and North American specimens may reveal more similarity between these areas than it was previously thought.

T2-27-07

Taxonomy and phylogeny of *Leptolejeunea* (Spruce) Schiffn.

Lei Shu, *Rui-Liang Zhu*
East China Normal University

Leptolejeunea is a pantropical, epiphyllous leafy liverwort genus of Lejeuneaceae, with approximately 50 extant species. Due to their small and easily degraded ocelli, it is hard to distinguish species of *Leptolejeunea* from one another. This genus has not been revised around the world until now. The definitions of species, subgenus and even the circumscription of genera are still unclear. Previous taxonomic schemes for the *Leptolejeunea* were re-evaluated in the light of more representative species derived from a Bayesian, maximum-likelihood and parsimony reconstructions with two plastid regions (*rbc*L, *trn*L-F) and nuclear ribosomal ITS region. The reconstruction revealed that *Leptolejeunea* as currently defined was not monophyletic, with *L. spinistipula* belonging to another genus. Some previous taxonomic proposals for subgenera of *Leptolejeunea* that do not conform to the strongly supported monophyletic groups here were recovered and four subgenera were proposed with two new subgenera *Neoleptolejeunea* and *Subasticta*. The consequence of molecular dating and ancestral area reconstructing indicate that *Leptolejeunea* may go through extinction events during the initial evolution period, which leads to two ancestral areas. Three molecular markers (chloroplast *trn*L-*trn*F and *trn*G, nuclear ITS region) were used to explore the circumscription of some ill-defined species. The accessions of the widespread species *L. elliptica* is discovered to be not a monophyletic group but nested with the species from the same origin regions respectively. Several new species discovered in Southeast Asia suggests more exploration and investigation needed in other tropics areas.

T2-27-08

Generic delimitation and biogeography of *Meteorium* and *Papillaria*, focusing on Asian species

Thien-Tam Luong[1], *Sanna Huttunen*[1], *Juliana R.P.M. Oliveira*[2], *Dietmar Quandt*[3]

1. University of Turku

2. Universidade Federal de Espirito Santo, Alegre

3. Nees Institute for Biodversity of Plants, University of Bonn

The genera *Meteorium* Dozy & Molk. and *Papillaria* Lorentz comprise two major clades within the family Meteoriaceae with a great number of species occurring mainly in the humid tropical and warm temperate forests. The identity of these two genera has gone through several changes during the last twenty years, especially when the traditional key characters to separate the two genera have been questioned. For example, Buck (1994) synonymized *Papillaria* with *Meteorium* with the argument that the number of papillae over the lumen of each laminal cell is an unstable character depending on the environment. Huttunen *et al.*, (2008) and Quandt *et al.*, (2004) favored recognizing the two genera using the structure of axillary hairs as a distinguishing morphological character. The generic delimitation by the latter authors is well-supported by molecular phylogenetic analysis but required some complex nomenclatural changes. However, the tangled taxonomical classification of *Meteorium* species still exists and species delimitation of these two genera remains unclear and needs further examinations, especially for South East Asian taxon. Within the genera taxonomy is hampered especially by plasticity of characters such as leaf shape and leaf laminal cell papillosity. This study aims to reconstruct a molecular phylogeny, to discuss the evolution of characters, to investigate morphological synapomorphies, and to refine the species delimitation of *Papillaria* and *Meteorium*, focusing on tropical Asian species. The biogeography of the two genera will also be reviewed.

T2-27-09

A taxonomic study of family bryaceae *s. l.* (Musci) in China

Jian-Cheng Zhao[1], *Yong-Ying Liu*[1,2], *Yu-Lu Niu*[1,3], *Xiao-Rui Wang*[1], *Lin Li*[1], *Shuo Shi*[1], *Min Li*[1]

1. Hebei Normal University

2. Jiaozuo Teachers College

3. Hengshui University

The Bryaceae (*s. l.*) is one of the important moss groups with many species characterized by diverse sporophytic and gametophytic morphologies. However, the taxonomic system is confused, genera and species in it are considerable difficult to identifying with a global distribution. Recent researches based on molecular data have greatly changed understanding of relationships in the Bryaceae (*s. l.*). Based on observation in the field and specimen study at the laboratory, 85 morphological characters, including 55 gametophytic characters, 26 sporophytic characters and four habitat conditions, were used for cladistics analysis. Four chloroplast DNA sequences, *rps4*, *atpB-rbcL*, *trnL-trnF* and *trnG*, from 46 species, 65 specimen of Bryaceae (*s. l.*) and its related groups were studied. Combining morphological and molecular data in the analyses showed that the traditional Bryaceae should be divided into four families. Members of re-circumscribe Bryaceae include five genera, *Anomobryum*, *Brachymenium*, *Bryum*, *Plagiobryum* and *Rhodobryum* in China, total 77 species and two varieties. The genera *Orthodontium* and *Leptobryum* were respectively transferred from Bryaceae (*s. l.*) to family Orthodontiaceae and Meesiaceae. Although the molecular analysis showed that the group with the central genus *Pohlia* has more close relationships with genera of the Mniaceae than the Bryaceae, genera of Mniaceae differ from *Pohlia* in both gametophytic and sporophytic morphology. So we adopted views that the Mielichhoferiaceae was circumscribed as a single family. There are five genera, *Mielichhoferia*, *Pohli*a, *Pseudopohlia*, *Epipterygium* and *Synthetodontium*, in the Mielichhoferiaceae in China, total 36 species and two varieties. *Synthetodontium kunlunense* J. C. Zhao et Y. Y. Liu and *Pohlia mamtiminii* Y. Y. Liu & J. C. Zhao were named as new species. A new combination, *Synthetodontium gossypinum* (X. J. Li & M. Zang) J. C. Zhao & Y. L. Niu, was proposed. *Bryum blandum* ssp *handelii* (Broth.) Ochi was treated as *Bryum handelii* Broth. *Bryum petelotii* Thér. & Henr. and *Bryum yuennanense* Broth. were treated as independent species of *Bryum*. Six species of sensu lato Bryaceae were reported for the first time from China, e.g., *Bryum weigelii* Spreng. The genus *Synthetodontium* Card. was newly recorded in China. *Pohlia marchica* Osterw. and *Anomobryum concinnatum* (Spruce) Lindb. were reinstated at species level. *Pohlia minor* Schleicher ex Scheaegr. was reduced as *Pohlia elongata* var. *greenii* (Brid.) Shaw. Three propaguliferous Pohlious species, *Pohlia flexuosa* Harv., *Pohlia leucostoma* (Bosch & Sande Lac.) M. Fleisch. and *Pohlia proligera* (Lindb. ex Breidl.) Lindb. ex Arn. were correct identified in China. Determining how climate change may impact multiple ecosystem levels and lead to various species adaptations is necessary for both biodiversity conservation and sustainable biological resource utilization. We employed Maxent to project the potential distributions of species of Bryaceae (*s. l.*) under current and future climate change scenarios and determined the dominant environmental variables that affect changes in the distribution.

Our work was funded by the National Natural Science Foundation of China (No. 31370237 & No. 31070184), Specialized Research Fund for the Doctoral Program of Higher Education of China (No. 20131303l2005) and Research Foundation of Hebei Normal University (No. L2015B13).

T2-27-10

Molecular phylogeny and subfamilial classification of the family Pottiaceae (Bryophyta)

Yuya Inoue[1,2], *Hiromi Tsubota*[1]

1. Miyajima Natural Botanical Garden, Graduate School of Science, Hiroshima University

2. Hattori Botanical Laboratory

The family *Pottiaceae* is the most generic and species rich family of mosses (*Bryophyta*), with around 1,400 species in 83 genera, comprising more than 10% of the extant moss species (Frey & Stech 2009). Widely distributed in the world, its species adapt to a wide range of habitat types including xeric, mesic and hydric, grow on various substrata including saxicolous, tericolous and corticolous, and possess a variety of life strategies including perennial, annual and ephemeral. The family exhibits a great variety of apparent morphological, physiological and genecological adaptations to their particular environments (Zander 1993). Among the various types of habitat where they are found, most species exhibit a great tolerance of hot and dry environments. Adaptation to such harsh selective pressures often leads to the presence of parallel or convergent characters which develop in response to the same environmental stimuli. In addition the difficulty in effecting fertilization or allowing for the maturation of sporophytes under xeric conditions leads to the lack of mature sporophytes which provide taxonomically important characters in mosses. These environmental features have complicated the phyletic assessment and classification of the family based on morphological criteria. The species concepts are often not well understood, and the family has been variously classified without an understanding of its phylogenetic relationships, leaving many ambiguous or poorly understood taxa unresolved (Satio 1975). A source of independent taxonomic criterion is therefore needed for the sound classification of this family, and during the past 20 years a number of molecular phylogenetic analyses have been conducted in an attempt to resolve relationships within the family. From the early 19th century, variously attempts at a subfamilial classification of the family have been proposed based on morphological criteria, while the limited number of morpho-molecular systematics have been proposed (Werner *et al.*, 2004, Zander 2006, Stech & Frey 2009). In this talk we review the current state of knowledge on phylogenetic relationships and classification at subfamilial level within the *Pottiaceae*, and reassess the subfamilial relationships within the family based on molecular phylogenetic inference using concatenated sequences of chloroplast *rbcL* and *rps4* genes, with codon substitution model which is a statistically higher precision model than nucleotide and amino acid substitution models for the evolutionary analysis of protein-coding sequences. The supposed ingroup species represent all the subfamilies of *Pottiaceae* recognized by Werner *et al.*, (2004): *Merceyoideae*, *Pottioideae* and *Trichostomoideae*. We also include the genus *Streblotrichum* whose phylogenetic position has remained ambiguous (Kucera *et al.*, 2013). The resultant tree confirms the monophyly of the *Pottiaceae* with four major clades within the family, corresponding to *Merceyoideae*, *Pottioideae*, *Trichostomoideae*, and the newly proposed *Streblotrichoideae*. Merceyoideae is resolved as the most basal clade within the family. *Pottioideae* comprise the sister-group to *Trichostomoideae* and this clade is sister to *Streblotrichoideae*. Based on the tree we recognize four subfamilies in *Pottiaceae*.

T2-27-11

Regeneration and vegetative proliferation of peat moss (*Sphagnum squarrosum*)

*Wenqian Zhao, Shirui Zheng, Zeling Li, Qiuping Li, Youfang Wang, Ruiliang Zhu, **Yue Sun***
Life Science School of East China Normal University

Sphagnum (peat moss) are the major components of peat lands which are very important ecosystems for world water conservation and air carbon uptake. They has great financial potential for their utilization as pre-prepared 'growing media' that allow uniform high-quality plants to be grown at very high productivity levels. Sphagnum have unique morphological and developmental characteristics also. During their vegetative development process, there were 4 different tissues including filamental and unistratose thallose protonema, stem and mature leaf with the slender chlorophyllose cells and diamond shape hyaline cells. In Sphagnum, there are one-dimensional filament protonema, two-dimensional plate protonema, three-dimension leafy shoot and strong differentiated leaf cells. Besides the importance on earth ecologic system and potency on horticulture and other applications, Sphagnum are good material for plant cell develop and differentiation study. Sphagnum gametophore tissue culture system have been set up already while there is no report about stable protonema culture yet. The detailed developmental process of peat mosses vegetative growth were still unclear.Through continuously observation, we summed up the regeneration principle of Sphagnum gametophore fragments. The thallose protonema is the most interesting structure of Sphagnum. How does it look like? How does it developed from linear protonema? How does gametophore out growth? How phytohormones regulate Sphagnum development? They are the major questions that we endeavored to find answers. There are multidirectional cell development and regeneration relationship among the 4 vegetative tissues in Sphagnum vegetative tissues. A relative stable peat moss protonema culture method was set up based on that.

T2-27-12

Stress imprint of two moss species

***Weiqiu Liu**, Binyang Liu, Shan Li, Yiyun Guan*
Guangdong Key Laboratory of Plant Resources, School of Life Sciences, Sun Yat-sen University

Plants are often exposed to stress environments, which generally induce morphological, physiological or genetic modifications in the plants. And, these modifications always improve the resistance of the plants to future stresses. This feature is described as 'stress memory' or 'stress imprint'. However, the evidences of moss's stress imprint are still lack. We studied the responses of two moss species widely distributed in South China, *Hypnum plumaeforme* Wils. and *Pogonatum cirratum* Bird. subsp. *fuscatum (*Mitt.) Hyv?nen, to high N, alone or combined with drought or low temperature stress, as well as their responses after a short term recovery or after a recurring stress. We found that after a 10-day recovery, oxidative stress induced by N application, drought or low temperature was significant decreased in both species, but it was still higher than non-stressed controls. At the same time, SOD activity and ABA content in the mosses after a short-term recovery were also higher than non-stressed controls, and we speculated that this might enable the mosses to better resist the future stress. To verify our speculation, we exposed mosses to recurring stress or cross stress of drought and low temperature with the time intervals of 7 days, 45 days and one year respectively. Our results showed that single drought or low temperature stress caused an increase in MDA content and inhibited the activity of antioxidant enzymes, however the level of MDA in *Hypnum plu-*

maeforme subjected to recurring stress or cross stress after a 7-day interval, was lower than corresponding single-stressed plants. The activity of SOD and CAT in *Hypnum plumaeforme*, exposed to recurring stress or cross stress were higher than corresponding single-stressed plants, showing the evidence of 'stress imprint'. However, those in *Hypnum plumaeforme* showed such rules only for recurring low temperature stress. If the second stress happened after a time interval of 45 days, MDA content in the recurring low temperature stressed or drought-low temperature stress treated samples was still lower than those of single low-temperature stressed samples, but for the samples subjected to recurring drought stress or low temperature-drought stress, no significant differences were found comparing with the single drought stress samples. The activity of SOD in *Hypnum plumaeforme* exposed to recurring or cross stress was still higher than corresponding single-stress plants, but this did not occur for CAT, suggesting that imprint of the first stress weakened gradually. Our results also indicated that after a time interval of one year, the memory of the first stress in the mosses disappeared. In summary, our results provide the first evidence for moss's stress imprint, confirming our speculation that mosses can 'remember' the stress and show higher resistance when subjected to recurring or cross stresses. However, this 'memory' is time limited.

T2-28: Evolutionary genomics and adaptation of *Arabidopsis* relatives

T2-28-01

What forces maintain genetic variation for large-effect trans eQTLs?

Emily Josephs[1], *Stephen Wright*[2], *John Stinchcombe*[2]

1. University of California, Davis

2. University of Toronto

The evolutionary forces that maintain genetic variation in quantitative traits within populations remain poorly understood. Specifically, how much of the variation present within populations is deleterious? Does balancing selection make an important contribution to within-population variation? We use gene expression as a model trait to address these questions by mapping the genetic basis of leaf gene expression ('eQTLs) within a single population of the outcrossing plant, Capsella grandiflora. First, we map local eQTLs and show that alleles at these loci are rarer than expected and exhibit a negative correlation between phenotypic effect size and frequency, consistent with the action of purifying selection. Second, we identify a number of eQTLs affecting large groups of coexpressed genes in trans and investigate the role of balancing selection in maintaining variation for these loci and the effect of these QTLs on traits. In total, our findings illuminate the role that both purifying and balancing selection play in shaping genetic variation for phenotype within populations.

T2-28-02

Population genomics of transposable elements in *Capsella*

Jasmina Uzunovic1, Emily Josephs[2], *John Stinchcombe*[1], ***Stephen Wright***[1]

1. University of Toronto

2. University of California, Davis

Transposable elements (TEs) comprise a significant proportion of eukaryotic genomes, but their importance as contributors to standing genetic variation in natural plant populations remains unclear. Given the action of host silencing mechanisms and the ability of TEs to move regulatory sequences around the genome, there is strong potential for TEs to have major contributions to gene expression variation. Here, we examined the population genomic variation for transposable elements and their contribution to gene expression variation in a large population of the outcrossing plant *Capsella grandiflora*. Most TE insertions were found at low population frequencies, particularly near highly expressed genes, consitent with the action of strong purifying selection. We observe a genome-wide pattern suggesting that rare TE insertions tend to drive gene expression towards extreme values, consistent with their widespread importance in maintaining natural variation in gene expression. We discuss the relative contribution of TEs and single nucleotide polymorphism in driving variation and evolution of gene expression.

T2-28-03

Evolutionary divergence between populations of *Arabidopsis lyrata* and its consequences for reproductive isolation

Outi Savolainen[1], *Tuomas Hämälä* [1], *Tiina Mattila*[1], *Yongfeng Zhou*[1], *Jaromir Guzinski*[1,2], *Helmi Kuittinen*[1]

1. University of Oulu, Department of Ecology and Genetics

2. CNRS, UPMC, UMR7144, Adaptation et diversité en Milieu Marin, Équipe Diversité et Connectivité, dans le paysage marin côtier, Station Biologique de Roscoff

Arabidopsis lyrata consists of disjunct populations belonging to two subspecies, ssp. *petraea* in Eurasia and ssp. *lyrata* in North America, which have diverged 180 KYA based on whole genome data analysis. The Central European populations have diverged about 135 KYA from those currently found in Scandinavia, after post-glacial colonization. A population in eastern Karelia (western Russia) likely is derived from an independent post-glacial colonization, as it is highly differentiated from other European populations. During the migration and colonization, populations have experienced bottlenecks, become locally adapted to their current climatic conditions, and diverged phenotypically in many traits. The divergence has also resulted in incipient reproductive isolation, found earlier as cytoplasmic male sterility in hybrids between populations of North America and Europe. More recent studies have also demonstrated that between Russian and North American populations there exists a low level of incompatibility at the seed stage, which is associated with high levels of transmission ratio distortion in F2 populations. The consequences of the various individual incipient barriers were also examined in an artificial hybrid zone comprising Russian and Swedish populations. We analyzed both the mating patterns and the fate of chromosomal segments in this experiment.

T2-28-04

Long-term balancing selection contributes to the adaptation of *Arabidopsis* relatives

Qiong Wu[1,2], *Ting-Shen Han*[1,2], *Xi Chen*[1,2], *Zi-Wen Li*[1,2], *Jia-Fu*

Chen[1,2], Yu-Pan Zou[1,2], Yong-Chao Xu[1,2], **Ya-Long Guo**[1,2]
1. State Key Laboratory of Systematic and Evolutionary Botany, Institute of Botany, Chinese Academy of Sciences
2. University of Chinese Academy of Sciences

Balancing selection maintains genetic variations within a population, which plays important roles in the adaptation of organisms. What genes, genome-wide, are under balancing selection and to what extent these genes are involved in adaptation are largely unknown. To address these questions, we performed a genome-wide scan of shared polymorphisms, and then finally, trans-species polymorphisms - a robust signature of balancing selection, across two plant species, the model plant *Arabidopsis thaliana* and its close relative species *Capsella rubella*, which diverged about 8 million generations ago. Interestingly, we identified hundreds of genes with candidate sites, from which five of them, were confirmed to be under long-term balancing selection, involved in fundamental biochemical function or response to biotic or abiotic stress. Particularly, each of the five genes exhibited significant ecological diversification between its two haplotype groups, suggesting balancing selection must have been very important to the adaptation during a long time span. Our results highlight that in plants, beyond S-locus genes and resistance genes, other genes related with resistance to stress and fundamental functions are under balancing selection, which must have contributed to the adaption to diverse habitats.

T2-28-05

Gene expression and sequence variation in the genus *Boechera*: Implications for the evolution and maintenance of apomixis

Ana Marcela Florez-Rueda, *Ueli Grossniklaus*
University of Zurich

The genus *Boechera* arose through extensive and complex hybridization events that occurred multiple times among many diploid species in the genus, with current species being allopolyploid as well as diploid hybrids. Hybridization has been proposed to be the basis for gametophytic apomixis (diplospory), which is widespread in the genus. Apomictic *Boechera* species appear to be the only well-documented example of diploid apomixis in angiosperms, making the study of its genome evolution and transcriptional landscape of high relevance for the identification of genes relevant to the engineering of apomixis in crops, which would allow the genetic fixation of complex traits, including heterosis. Genomic differences between apomictic and sexual species have been reported with apomictic individuals being highly heterozygous, whereas sexual diploids having low levels of heterozygosity and high inbreeding. Furthermore, several genetic and cytological studies point to a particular genomic region of low recombination, and therefore high heterozygosity to be the diplospory-determining region in *Boechera*. We performed laser-capture microdissection of egg cells coupled with RNAseq and analyzed patterns of gene expression and sequence variation. We will present the results of our transcriptomic analysis of four sexual *Boechera* species and will compare them to four apomictic *Boechera* species. We will describe the enriched gene ontology terms and discuss their relevance in the establishment and maintenance of parthenogenesis. We have used the transcriptomic data to inquire the nucleotide variation between apomict and sexual species using SNPs from thousands of genes. This rich data set allow us to perform a comparative genome scan that describes the differences in nucleotide diversity between species of different reproductive modes, particularly validating the long standing observation that apomictic species have higher levels of heterozygozity and aiding in the delimitation the apomixis determining region. Moreover we have performed molecular evolution analyses and detected signatures of selection in the *Boechera* genome and will discuss the implications of this selective evolutionary force in shaping the maintenance of the different reproductive modes and hybrid networks in the genus.

T2-28-06

Population genomics of heavy metal adaptation of *Noccaea caerulescens*

Mark Aarts
Laboratory of Genetics, Wageningen University

Noccaea caerulescens (Brassicaceae) is one of the rare metal hyperaccumulator plant species that evolved adaptation to multiple heavy metals, i.e. exposure to high levels of Zn, Ni, Cd or Pb. There are several accessions of this species, each with different metal accumulation and tolerance characteristics. All accessions hyperaccumulate Zn, preferentially in their leaves, and sometimes Ni, Cd or Pb. The species is widely distributed over North, West and Central Europe, though often found in separate local populations, especially the metallicolous ones. We use *N. caerulescens* as a model to study the evolution of plant adaptation to extreme environments. An inbred line of accession 'Ganges', originating from the south of France (Les Avinières, St.-Laurent-le-Minier), has been used for whole genome sequence analysis. This accession is extremely tolerant to high Zn and Cd exposure and hyperaccumulates these metals in its leaves. The whole genome sequence was assembled based on Illumina Hi-Seq and PacBio generated sequences. This assembly comprises around 30,000 predicted, annotated gene models, suggesting nearly all of the genic regions to be included in the assembly. When compared to *Eutrema salsugineum*, its closest relative of which a whole genome sequence is available, some 1000-1500 genes are found in more copies in N. caerulescens than in *E. salsugineum*. Many of these multiplied gene copies were previously found to be higher expressed in *N. caerulescens* than in related species, including several orthologues of genes known to be involved in mineral homeostasis of *Arabidopsis thaliana*. Quite a few of these genes show copy number variation (CNV) when comparing different metallicolous and non-metallicolous populations of *N. caerulescens*. Thus, CNV appears to have played an important role in the adaptation of *N. caerulescens* to heavy metals. Our progress in understanding the functional importance of this CNV will be discussed. Next to CNV, population genomics has revealed some signatures of selection associated with adaptation to metallicolous conditions. The function of the genes involved will be discussed.

T2-29: Apocynaceae: phylogeny, biogeography and diversity

T2-29-01

Phylogenomics of Apocynaceae: From tools to trees

Shannon Straub[1], *Tatyana Livshultz*[2], *Julien Boutte*[1], *Kevin Weitemier*[3], *Aaron Liston*[3], *André Simões*[4], *Mark Fishbein*[5]
1. Hobart and William Smith Colleges
2. Drexel University
3. Oregon State University
4. Universidade Estadual de Campinas
5. Oklahoma State University

The advent of next generation sequencing and methods for targeted sequencing of nuclear genomes have created the opportunity for development of tools for phylogenomics in Apocynaceae. Sequencing of the genome and transcriptome of the common milkweed (*Asclepias syriaca*) have provided the basis for design of probes to specifically target putatively single-copy nuclear genes, which is combined with genome skimming for assembly of high copy targets, such as the plastome, in the Hyb-Seq approach. These tools were utilized to understand the evolutionary history of North American milkweeds (*Asclepias*), a recent rapid radiation of ca. 115 species that has been difficult to investigate using traditional molecular systematics approaches due to low levels of DNA sequence variation and problematic levels of incomplete lineage sorting. Phylogenetic relationships in *Asclepias* were estimated using DNA sequence data from ca. 768 nuclear genes and whole plastomes. The relationships of major clades in the genus are supported by both analyses, but produce conflicting results concerning species relationships within each clade. Nuclear loci exhibit low levels of phylogenetic signal and what signal exists is highly discordant among loci, providing challenges to accurately estimating the species tree with confidence. Building upon the *Asclepias* work, Hyb-seq tools were developed for wider use in the study of evolution of Apocynaceae. Genome skimming has produced the data to infer a plastome phylogeny of the family with sampling to the subtribal level and facilitated estimation of divergence times of major clades. Probes for targeted sequencing of ca. 800 single-copy nuclear genes that work across the family were also developed and have allowed a fresh look at both phylogeny and character evolution.

T2-29-02

The rapid radiation of Asclepiadeae (Apocynaceae) in Africa with special reference to the subtribe Asclepiadinae

Stoffel Petrus Bester
1. National Herbarium (PRE), South African National Biodiversity Institute, Private Bag X101, Pretoria, 0001.
2. Unit for Environmental Sciences and Management, North-West University, Private Bag X6001, Potchefstroom, 2520.

The family Apocynaceae consists of approximately ± 366 genera with at least 5 745 accepted names. Since the last XVIII IBC meeting in Melbourne increased resolved clade structure has lowered the number of genera slightly, a tendency that seems to be on-going. On the African continent there are some 1,679 taxa (including 13 naturalized taxa) and 174 genera (including 10 naturalized genera). With almost a third of the known species in the family found on the African continent, it is well-represented including members of all the subfamilies: Apocynoideae (± 29% of the genera); Rauvolfioideae (± 47% of the genera); Periplocoideae (± 69% of the genera); Secamonoideae (± 13% of the genera); and Asclepiadoideae (± 57% of the genera). Globally the Asclepiadoideae has three main centres of diversity: a Tropical Asian centre (± 38 genera); a tropical Central and South America centre (± 47 genera); and the Eastern and Southern Africa centre that is the most diverse (± 93 genera). Due to the recent radiation of the Asclepiadoideae, members are well-represented in Africa with ± 1,266 taxa mainly within the two large tribes: Asclepiadeae (± 36% of the taxa) and Ceropegieae (± 59% of the taxa). In an analysis of African members of the Asclepiadeae the largest six genera are: *Asclepias* (± 80 taxa); *Aspidoglossum* (± 39 taxa); *Cynanchum* (± 49 taxa); *Pachycarpus* (± 41 taxa); *Schizoglossum* (± 45 taxa) and *Xysmalobium* (± 44 taxa). These genera account for approximately 65% of the taxa and 71% of the genera in the tribe has less than 10 species. In this presentation an overview of the genera and species in the Asclepiadeae in Africa with specific focus to the centre of diversity in eastern and southern Africa, emphasis on the subtribe Asclepiadinae, and a synthesis of the genera and species of Africa will be presented. The presentation will also highlight some of the outcomes of the revision based on the morphological study and phylogenetic relationships of a selection of genera of southern African members of the subtribe Asclepiadinae. As a specific example, reference to members of the *Schizoglossum*-complex in the subtribe Asclepiadinae will be discussed.

T2-29-03

Pollination of *Ceropegia* pitfall flowers: Chemical mimicry and pollinator specificity

Annemarie Heiduk[1,2], *Andreas Jürgens*[3], *Sigrid Liede-Schumann*[1], *Ulrich Meve*[1], *Stefan Dötterl*[2]
1. University of Bayreuth
2. University of Salzburg
3. Technische Universität Darmstadt

Ceropegia L. (Apocynaceae, Asclepiadoideae) comprises more than 200 species that occur mainly in subtropical and tropical habitats of the Old World. Species in this genus are characterized by complex and non-rewarding deceptive pitfall flowers. Pollination is mediated by flies of different families, such as Ceratopogonidae, Chloropidae, Drosophilidae, Milichiidae, and Sciaridae. It has been suggested that floral scents mimic food sources, sex pheromones or oviposition sites of the flies and are key players leading to a highly functional specialization in terms of pollinator specificity. We identified flower visiting/pollinating flies of 14 *Ceropegia* species and analyzed their floral scent composition using dynamic headspace and gas chromatography coupled to mass spectrometry (GC/MS). We further tested for relationships between floral scent and visitor/pollinator patterns. Based on genetic similarities of the plants, we also checked for phylogenetic signals in scent chemistry and pollinator spectra. Gas chromatography coupled to electroantennographic detection (GC/EAD) and behavioral studies were applied to identify scent components responsible for pollinator attraction. We found that most *Ceropegia* species are associated with only one or two pollinating fly families or genera with strong differences in pollinator spectra among species. Floral scent chemistry was highly variable among species but did not correlate with flower visitor/pollinator patterns. We did not find phylogenetic signal in overall scent chemistry but in fly pollinator assemblages. Although scent patterns did not correlate with pollinator spectra, the highly specific pollination systems in *Ceropegia* still can be explained by floral scent chemistry and the resultant mimicry strategies. This is supported by more detailed chemical ecological studies on *C. sandersonii* for which we revealed the model mimicked by the flowers, i.e. western honey bees under attack.

T2-29-04

Distribution patterns in large Asclepiadoid genera

Sigrid Liede-Schumann
University of Bayreuth

Asclepiadeae, possessing pendent pollinia, constitute the largest tribe of subfamily Asclepiadoideae (Apocynaceae). From an African origin, an early split separates an exclusively New World lineage from a lineage centered in the Old World. The latter consists of three subtribes, Asclepiadinae, Cynanchinae and Tylophorinae, each of which comprising one large genus and few small relict genera. For all three subtribes, reconstruction suggests an East African origin. In Asclepiadinae, a New World and a southern African radiation contribute most to species diversity. In the New World, a niche shift to temperate climates, followed by a speciose radiation, has taken place. In Cynanchinae, the most speciose radiation is centered in Madagascar, and the New World has been reached only late in the history of the subtribe. Only single species extend beyond tropical and subtropical zones. Tylophorinae has centers of diversity in Africa and Asia, it is naturally absent from the New World. It comprises two lineages that have undergone a shift to temperate climates and subsequently radiated. One of these lineages comprises members that are neophytic to the New World and behave as noxious weeds in suitable niches. The phylogenetic history and possible adaptations enabling niche shifts will be discussed.

T2-29-05

The Apocynaceae, subfamilies Rauvolfioideae and Apocynoideae, in Cambodia, Laos and Vietnam

David Middleton
Singapore Botanic Gardens

In Cambodia, Laos and Vietnam there are 119 known species in 38 genera of Apocynaceae, subfamilies Rauvolfioideae and Apocynoideae (or the Rauvolfioid and Apocynoid informal groups). The diversity in Vietnam is much higher than in Cambodia and Laos which is a reflection of both the greater diversity of habitats in Vietnam but also the very much lower collection densities in Cambodia and Laos. The Rauvolfioids and Apocynoids show similar patterns of diversity in the Cambodia, Laos and Vietnam region compared to neighbouring Thailand. Based on their distributions in Thailand, and habitat availability across Southeast Asia, there are only very few species known from Thailand that would be expected to be found Cambodia, Laos and Vietnam that have not yet been found there, despite the lower collection densities. There are 17 known endemic species in Cambodia, Laos and Vietnam of which 16 are in Vietnam and one in Laos. There are, however, only two known endemic species of *Alstonia* and no known endemic species of *Wrightia* even though in neighbouring Thailand there are several narrowly endemic species of each genus in limestone habitats that are also present in Cambodia, Laos and Vietnam. New species of both of these genera, and possibly also of *Kopsia*, are expected with better exploration in the region. There is one endemic genus in the FCLV region, *Ixodonerium*. This genus has, unfortunately, never been investigated using molecular sequencing techniques. It shows affinities to *Anodendron* but does occur in a region of Vietnam known for its distinct plant diversity and high level of generic endemism in a range of families. Seven species of Rauvolfioid and Apocynoid Apocynaceae have been assessed under IUCN criteria as being threatened but many species are assessed as data deficient as no collections have been made for decades.

T2-29-06

Diversity of *Ceropegia* L. (Ceropegieae: Apocynaceae) in India

Sharad S. Kambale[1], *Siddharthan Surveswaran*[2], *Shrirang R. Yadav*[3]
1. Department of Botany, Maratha Vidya Prasarak Samaj's Arts, Commerce & Science College, India.
2. School of Biological Sciences, The University of Hong Kong, Hong Kong.
3. Department of Botany, Shivaji University, India

Ceropegia L. (*Ceropegieae*: Apocynaceae) with about 244 species is widely distributed in the Old World ranging from the Spanish Canary Islands in the west, through central, southern, and northern Africa, Madagascar, Arabia, India, South-East Asia to northern Australia in the east. The genus is characterized by tuberous, fleshy or fibrous roots, solitary flower or flowers in cymose inflorescences, tubular corolla, uniseriate (*Ceropegia santapaui*) or biseriate corona, pair of follicles and comose seeds. Presently, in India *Ceropegia* is represented by 60 taxa (57 species, two varieties and one forma) of which 50 species are distributed in the peninsular India, nine in North-east India and one in Andaman Islands. Ceropegias are also known for endemism and 44 species are endemic to the country. Northern Western Ghats has 17 endemic species while Southern Western Ghats has only eight. Most of them are narrow endemics. Out of 60 taxa, nine are known only from type localities. Northern Western Ghats seem to be important region of diversification for the Ceropegias in India and probably most of the species are of recent origin. There are 24 non-tuberous species while 36 are with tuberous rootstock. Non-tuberous species are confined to tropical high rainfall areas with evergreen to semi-evergreen forests while species adapted to semi-dry or dry regions are always tuberous, a survival strategy during dry period of a year. 20 taxa are erect while 40 are twiners. *Ceropegia fimbriifera* Bedd., C. *mahabalei*, *C. nampyana*, *C. noorjahaniae* and *C. spiralis* become twiners under captivity. *Ceropegia* species show considerable variations in vegetative as well as floral characters especially with reference to colour, shape and size. *Ceropegia juncea* and *C. bulbosa* show Crassulacean Acid Meetabolism (CAM) pathway. Propagation and cultivation of Ceropegias is herculean task; however *Ceropegia candelabrum, C. elegans, C. gardneri, C. intermedia, C. juncea and C. maculata* can be propagated through stem cuttings. In the present study chromosome numbers for nine species are reported. All species show $2n = 22$. New cytotype ($2n$=22) for *C. elegans* has been reported. *Ceropegia concanensis, C. ravikumariana* and *C. nampyana* are the new species described during the study. *Ceropegia gardneri* was earlier reported from Sri Lanka which has been now collected from Karnataka (India). In present study lectotypes for 17 binomials have been designated based on original material. Preliminary studies have revealed that the Indian species of *Brachystelma* nest in Indian clade of *Ceropegia* and African *Brachystelma* into clade of African Ceropegias. Even morphologically *Brachystelma* is very close to Ceropegias but differs only in its corolla divided almost upto the base. Taxonomically, inclusion of *Brachystelma* (ca. 120 species) in *Ceropegia* (240+ species) would solve part of the problem; however, the morphological concept of *Ceropegia* would be demolished, and the many name changes would be unpleasant. In future there is every possibility of merging *Brachystelma* into *Ceropegia*. Present revisionary study concludes that the further explorations in Indian subcontinent may result into novelties in Ceropegias.

T2-30: Evolution and development of the flower: from genes to development

T2-30-01

The underlying mechanisms of pollen wall development in *Larix decidua* and *Polemonium caeruleum*. experimental modeling of exine

Nina Gabarayeva, *Valentina Grigorjeva*
Komarov Botanical Institute of Russian Academy of Sciences

The sporopollenin-containing part of the pollen wall - the exine - is one of the most complex cell wall in plants. By tracing each stage of *Larix decidua* (Pinaceae) and *Polemonium caeruleum* (Polemoniaceae) microspore development at a high level of TEM resolution, we aim to understand the establishment of pollen wall in connection with organelle activities and to unravel the suggested self-assembly underlying mechanisms. In *Larix*, the sequence of events observed in the microspore periplasmic space, leading to exine emergence is as follows: the appearance in the periplasmic space of spherical units; their accumulation and arrangement to radial columns; initial sporopollenin accumulation on spherical units and their columns; the appearance of the primordial endexine lamella; multiplication of the white-lined endexine lamellae; the appearance of granulate ectexine after callose dissolution and final sporopollenin accumulation; the formation of intine. In *Polemonium*, in spite of different kind of the exine structure, the main steps of exine ontogeny are the same: spherical units, gradually transforming into columns, then to rod-like units (procolumellae); the appearance of the initial tectum; the appearance of the endexine lamellae; the appearance of the intine in aperture sites. The distinctions in *Polemonium* sequence is the emergence of a sponge-like layer and the foot layer on the base of it late in the microspore development. The sequences of the developmental events in both species fit well to the sequence of self-assembling micellar mesophases in colloidal solution with increasing concentration of surfactants: spherical micelles; columns of spherical micelles, in places rearranged to cylindrical micelles; laminate micelles, separated by stratum of water and seen with TEM as white-lined lamellae. The same fitness was observed in our ontogenetic previous studies on other species. A number of patterns were obtained in our modeling experiments in colloidal solutions in vitro conditions, when the influence of the genome is absent, in the course of condensation (a kind of self-assembly process), simulating structures at several stages of exine development in *Larix*. Therefore, pure physico-chemical processes as self-assembly, which are not under direct genetic control, play an important role in exine development and share control with the genome. It is highly probable that self-assembly is an intrinsic instrument of evolution.

T2-30-02

Perianth structure across Eudicotyledoneae and its relationship with diversification

Adriana Benitez-Villaseñor[1], *Susana Magallón*[2], *Hervé Sauquet*[3]
1. Posgrado en Ciencias Biológicas, Universidad Nacional Autónoma de México
2. Instituto de Biología, Universidad Nacional Autónoma de México
3. Laboratoire Ecologie Systématique et Evolution, Université Paris-Sud

Flowering plants display an extraordinary diversity of floral structural configuration and morphological variation. Structural stabilization in the flower often allows architectural elaboration associated with specialized reproductive syndromes, including complex associations with pollinators. The eudicot clade (Eudicotyledoneae) contains over 70% of angiosperm species. It consists of an early-diverging grade of mostly depauperate lineages and a very large clade, Pentapetalae, which contains the vast majority of the species of the group. The early-diverging eudicot lineages encompass a wide variety of floral structural configurations, some of which lack a fixed floral pattern, and have little or no differentiation between perianth cycles. Within Pentapetalae the most common floral groundplan is characterized by (4- or) 5-parted flowers and well differentiated calyx and corolla. This basic Pentapetalae floral pattern has been greatly elaborated along many different lineages to constitute complex floral architectural configurations. However, it is not clear if this structurally stable Pentapetalae floral groundplan evolved one or more times within the clade; in which branch (es) did it originate; and if there is correlation between the calyx and corolla and the meristic pattern organization. In this study, we investigate the distribution of the bipartite perianth and meristic pattern across eudicots, and evaluate its relationship with increased rate of species diversification. We apply an exemplar approach on a 622 terminal dated phylogeny representing 295 eudicot families to document the distribution of three floral structural characters. Floral data was obtained from different electronic resources and combining direct observations of fresh and herbarium flowers. The evolutionary history and ancestral morphological characters were estimated with maximum likelihood using Mk model in BayesTraits. Subsequently, we conducted state-dependent diversification analyses to evaluate the potential link between these morphological attributes and the rate of diversification using the Hidden-State Speciation and Extinction (HiSSE) model. HiSSE also takes into account the possibility of unmeasured (hidden) states that also affect the dynamics of diversification. Finally, we analyze the correlation between the presence of a bipartite perianth and meristic pentamerous pattern. We found that the evolution of floral pentamery is decoupled from perianth differentiation in eudicots. Pentamerous merosity may have originated on the stem lineage of Pentapetalae, or after the diversification of this clade in subsequent branches. Our results provide an estimation of the number and phylogenetic placement of shifts towards a bipartite perianth, and reconstructions of the ancestral perianth structure in major eudicot lineages, including eudicots as a whole. These results provide a framework for a more detailed evaluation of the floral evolution in the Eudicotyledoneae, in particular, and the correlation between the different structural characters.

T2-30-03

Evolution, development and genetic basis of floral symmetry in *Anacyclus* L. (Anthemideae, Asteraceae)

María A. Bello[1], *Inés Álvarez*[1], *Pilar Cubas*[2], ***Javier Fuertes-Aguilar***[1]
1. Real Jardín Botánico, CSIC
2. Centro Nacional de Biotecnología, UAM-CSIC

Most of the diversity in the pseudanthia of Asteraceae is based on the differential symmetry and sexuality of its flowers. Several studies have been In *Anacyclus* (tribe Anthemideae), where there are (1) homogamous capitula, with bisexual, mainly actinomorphic and pentamerous flowers; and (2) heterogamous capitula, with peripheral zygomorphic, trimerous and long-/short-rayed female flowers, we investigated the floral ontogeny and the evolution and expression of *CYC2* to infer their origin. The floral development in the capitulum of *Anacyclus* found separate sequence for disc and ray-flowers. The disc flowers, subtended by paleae, initiate acropetally. Their perianth and androecium initiation is unidirectional and simultaneous, and a late zygomorphy occurs by enlargement of the adaxial perianth lobes. In contrast, ray flowers, subtended by involucral bracts, initiate after the proximal disc buds, breaking the inflorescence acropetal pattern. Early zygomorphy is manifested through the fusion of the lateral and abaxial perianth lobes and the arrest of the adaxials. We found and interpreted atypical phenotypes with peripheral 'trumpet' flowers from natural populations. The peripheral 'trumpet' buds initiate after disc flowers, but maintain an actinomorphic perianth. Once established de developmental sequence, we investigated the genetic basis of the morphological diversity by using a candidate genes approach. We know that homologs of the *CYC/TB1* gene family have been independently recruited many times across the eudicots to control aspects of floral symmetry. The family Asteraceae exhibits the largest known diversification in this gene paralog family which is accompanied by a parallel morphological floral richness in its specialized capitulum inflorescence. In Asteraceae, whether or not *CYC/TB1* gene floral symmetry function is preserved along different tribes and gene lineages was unknown. In this part of the study, we used phylogenetic, structural and expression analyses with an emphasis in the highly derived genus *Anacyclus* to address this question. Phylogenetic reconstruction recovered eight main gene lineages present in Asteraceae: two from *CYC1*, four from *CYC2* and two from *CYC3*-like genes. The species phylogeny was recovered in most of the gene lineages, allowing the delimitation of orthologous sets of *CYC/TB1* genes in Asteraceae. The gene expression study included RNA qPCR and RNA in-situ hybridization. The quantitative real-time PCR results found that in *Anacyclus* three of the four isolated *CYC2* genes are more highly expressed in ray flowers. The expression of the four *AcCYC2* genes overlaps in several organs including the ligule of ray flowers, as well as in anthers and ovules throughout development. We concluded that despite the high diversity observed in the capitulum morphology of Anthemideae, the set of genes associated to floral symmetry detected in other tribes has conserved its function.

T2-30-04

Evolution of Darwin's peloric gloxinia (*Sinningia speciosa*) under domestication: Single gene speciation by pleiotropy

Yang Dong[1,2], *Yin-Zheng Wang*[2]
1. Crop Genetics Department, John Innes Centre
2. State Key Laboratory of Systematic and Evolutionary Botany, Institute of Botany, Chinese Academy of Sciences

Unlike most crop domestication that occurred through long periods of unconscious selection by ancient humans, horticultural plants were mostly domesticated with intentional selection over short time periods. The origin and spread of novel traits in the domestication process has remained unexplored in horticultural plants. The gloxinia (*Sinningia speciosa*) with attractive brilliant *peloric* flowers that influenced the thinking of Darwin has been cultivated since the early 19th century, but its origin and genetic basis has been unknown. By employing multiple experimental approaches including artificial hybridization, genotype-phenotype association, gene expression and functional repression, we determined that a single gene *SsCYC* encoding TCP protein controls the floral shape transition under domestication in gloxinia. In addition to the significant expression in the dorsal region of the second and third floral organs, a novel expression domain of *SsCYC* was observed in the dorsal basal floral tube that pre-pattern the development of the gibbous structure that force the floral tube orientate horizontally. We revealed that the loss-of-function of SsCYC protein caused by a *de novo* 10-bp deletion was targeted by intensive cross-based artificial selection, which underlies the evolution of the *peloric* flowers in gloxinia. Based on literature review and genetic reference combined with the findings herein, we further traced the origin of modern *peloric* gloxinias to a single individual with the mutant allele that originated in domestication 200 years ago. This study provides the first clear example for the evolutionary dynamics of horticultural plants. We also find that the floral orientation is causally associated with floral symmetry, acting as a key combinatorial complex floral structure relating to pre-pollination isolation in angiosperms, which is controlled by a single pleiotropic TCP gene. Our results suggest that pleiotropic changes with selective advantage could promote coordinated evolution of the complex integrated floral organs with subsequent speciation and diversification.

T2-30-05

Floral evolution and new hypotheses on perianth function in Magnoliidae

Julien Massoni[1,2,3], *Maria Von Balthazar*[4], *Thomas L.P. Couvreur*[5], *Yannick Staedler*[4], *Jürg Schönenberger*[4], *Hervé Sauquet*[1]

1. Laboratoire Ecologie, Systématique, Evolution, Université Paris-Sud
2. Institute für Mikrobiologie, ETH Zurich
3. Center for Adaptation to a Changing Environment, ETH Zurich
4. Department of Botany and Biodiversity Research, University of Vienna
5. Institut de Recherche pour le Développement (IRD), UMR-DIADE

With about 10,000 species in four orders (Cannellales, Laurales, Magnoliales, Piperales), the Magnoliidae are the largest clade of angiosperms outside Monocotyledonae and Eucotyledonae. Their floral biology and diversity of floral architectures greatly differ from those of other lineages. The magnoliid perianth is highly variable and includes a broad range of patterns present in other angiosperm lineages. Furthermore, traits such as perianth phyllotaxis and whorl number are more labile than in other flowering plants lineages. In this study, we investigate the diversification of the magnoliid perianth using a newly generated data set of seven floral characters scored for 198 exemplar species. This data represent more than 75% of the existing genera in the group and match the most recently published dated phylogeny of Magnoliidae. Ancestral states were reconstructed using maximum parsimony, maximum likelihood, and Bayesian reversible-jump approaches exploring multiple models of morphological evolution. The perianth morphology of the most recent common ancestor of all extant Magnoliidae is reconstructed as having at least two trimerous whorls of tepals. This ancestral morphology subsequently evolved further via gains (e.g. Winteraceae, Annonaceae, Magnoliaceae) or losses (e.g. Myristicaceae) of whorls, increases in the number of organs per whorl (e.g. Atherospermataceae, Hernandiaceae), or transitions from whorled to spiral phyllotaxis of the perianth (e.g. Calycanthaceae, Gomortegaceae, Degeneriaceae). Based on our current knowledge of magnoliid flower biology and these new results, we discuss functional roles of the perianth in floral evolution. For instance, we hypothesize that spiral perianth phyllotaxis has evolved independently in several magnoliid lineages as a response to floral bud predation.

T2-30-06

Variation on a theme: evolutionary-developmental insights into the Asteraceae flower head

Oriane Hidalgo[1], *Maarten Christenhusz*[1,2], *Benjamin Coquillas*[1], *Lin Fu*[1,3], *D. J. Nicholas Hind*[1], *Luis Palazzesi*[4], *Luca Pegoraro*[1], *Jaume Pellicer*[1], *Paula Rudall*[1], *Thais Vasconcelos*[1], *Ilia J. Leitch*[1]
1. Royal Botanic Gardens Kew
2. Plant Gateway
3. South China Botanical Garden
4. Museo Argentino de Ciencias Naturales – CONICET

Plant reproduction relies on flowers and their arrangement into an inflorescence –two interrelated traits that exhibit impressive variation between plant lineages and constrain crop yields and horticultural forms. A key aim in evolutionary biology is to unravel how the modification of genetic networks has led to the diversification of reproductive plant architecture and floral form. We address this aim by seeking to understand how two fundamental biological processes, floral architecture and symmetry, have evolved in the daisy family Asteraceae (Compositae), with unique combinations of these traits. Much of the research on Asteraceae has concentrated on model plant systems with simple heterogamous capitula (e.g. *gerbera, sunflower*). We propose an original and novel evolutionary-developmental (evo-devo) approach to floral evolution, by bringing to this field the species of Asteraceae with secondary heads (=syncephalia). This unique inflorescence architecture has arisen from two rounds of capitulescence (i.e. inflorescence condensation followed by simplification) during their evolution. Syncephalous species which comprise ca. 70 genera from 11 tribes provide an invaluable source of material for deciphering the genetic basis of capitulum evolution since their genomes bear the signature of two inflorescence shifts toward capitulescence. Our approach, integrating evo-devo and genomics, provides a unique opportunity to ask fundamental questions relating to coordinated trait evolution, and to increase our understanding of how genetic pathways have been altered or co-opted during the evolutionary diversification of flowers.

T2-31: Key questions on angiosperm macroevolution: where do we stand, and where are we heading now? (two sessions)

T2-31-01

Measuring how the probability of long-distance dispersal depends on geographic distance, environmental distance, and co-evolving seed dispersal traits: BioGeoBEARS on a mega-phylogeny of angiosperms

Nicholas Matzke
The Australian National University

Organism traits must be important in historical biogeography. In particular, rates of dispersal (both range-expansion dispersal, and jump dispersal leading to founder-event cladogenesis) must depend to some degree on traits such as seed traits and the vectors that transport seeds (e.g. animals, wind, water). However, to date no full-likelihood historical biogeographical models have been available that allow geographic range and traits to co-evolve on the phylogeny, with traits influencing dispersal ability. I present an addition to the R package BioGeoBEARS that enables an evolving discrete trait to influence both anagenetic and cladogenetic dispersal. This model can be freely combined with models adding parameters for jump dispersal (e.g., DEC+J), distance as a predictor of dispersal (+x models, with dispersal rate multiplied by distance^x), and other variants (e.g. +n, where the multiplier is environmental distance^n). I use Maximum Likelihood to apply the model to a phylogeny of over 14,000 angiosperms with areas coded for over 50 biome/continent combinations, and show that each predictor causes massive improvements in model fit, as measured by change in log-likelihood (delta lnL). Parameter x adds 15231 lnL, n adds another 618, and 3 parameters for seed traits add 2344. The model infers that plants with seeds classified as wind-dispersed have the lowest average rate of dispersal at this global scale (muliplier 0.001), animal-dispersed intermediate (multiplier 1) and water-dispersed the highest (8.18). Seed traits are classified very crudely in this analysis, so there must be care in interpretation (winged maple seeds and dandelion seeds are both "wind dispersed", and sampling of species and traits is incomplete), but overall the approach shows great promise. The study of long-distance dispersal need not rely on "dumb luck" and near-miraculous events to explain dispersal events, but rather, in the main, the causes of long-distance dispersal can be inferred statistically and understood mechanistically. BioGeoBEARS updates are at http://phylo.wikidot.com/biogeobears.

T2-31-02

How (much) do flowers vary? –Unbalanced disparity among flower functional modules and a mosaic pattern of morphospace occupation.

Marion Chartier[1], *Stefan Löfstrand*[1], *Maria Von Balthazar*[1], *Sylvain Gerber*[2], *Florian Jabbour*[2], *Hervé Sauquet*[3], *Jürg Schönenberger*[1]
1. University of Vienna
2. Muséum National d'Histoire Naturelle
3. Université Paris Sud

Angiosperm evolution has given rise to an overwhelming diversity of floral morphology. Applying morphospace analyses to a large dataset including extant and fossil species from the order Ericales, we quantify and explore phylogenetic and temporal patterns of floral morphological diversity (disparity). In addition, we compare disparity levels among the three functional modules of a flower, i.e., perianth, androecium, and gynoecium. We find that the floral morphospace of Ericales is organized as a continuous cloud, in which most clades occupy distinct regions in a mosaic pattern, indicating that different phylogenetic lineages have explored separate evolutionary trajectories of morphological diversification. Further, we show that disparity increases with clade size rather than age, likely because speciation is narrowly linked with floral diversity in angiosperms. The fossil taxa, although not closely related to each other, all fall in a narrow portion of the morphospace, suggesting that they share many plesiomorphic characters. Finally, our analyses clearly show that the androecium contributes more to total floral disparity than either the perianth or the gynoecium. Likely, this is due to particularly flexible ontogenetic patterns that allow for high levels of structural and functional diversity in the androecium of Ericales. The lower level of disparity in the gynoecium likely reflects architectural constraints exerted by the ubiquitous syncarpy in Ericales. The perianth with its simple organs displays the lowest disparity. Taken together, our analyses demonstrate that a morphospace approach is a powerful tool to disentangle phylogenetic and functional components of overall floral disparity in a major group of angiosperms.

T2-31-03

Combining macroevolutionary approaches and ecophylogenetics to uncover the role of pollinators in plant diversification

Ezgi Ogutcen[1], *Jana Vamosi*[1], *Yan-Bing Gong*[2]
1. University of Calgary
2. Wuhan University

The functional relationships between plants and their pollinators are thought to be important contributors to plant biodiversity. Phylogenetic approaches can elucidate the rate at which transitions to specialization occur during lineage diversification and how trait evolution affects plant-pollinator interactions within communities. On a more ecological timescale, studies on phylogenetic betadiversity in plants and pollinators have brought important developments to our understanding of the role pollinators play in angiosperm diversification. Here, we summarize the diversity and complementarity of approaches, focusing on i) how pollinator composition heterogeneity alters selection pressures on floral traits; ii) whether switches between pollinator functional groups affect diversification rates in angiosperm clades with diverse pollination modes, iii) the roles of functional traits versus phylogeny in assembling plant communities, and iv) the limitations and potential solutions to further improve the phylogenetic methods to elucidate the relationship between plants and pollinators. We put forth several lines of inquiry that we feel are especially promising for providing a more holistic view of plant-pollinator interactions that incorporates space and time. Using analyses on Antirrhineae (Plantaginaceae) as a case study, we show that shifts in pollinator preference can be observed upon dispersing into new regions, but these transitions do not necessarily confer increases in diversification rates in a clade. We also find a lack of correlation between floral colour and pollinator preference in the New World Antirrhineae, suggesting pollinator inferences based on pollination syndromes should be approached cautiously. Our multi-year studies of floral colour and phylogenetic structures of alpine communities in Canada and China show that the colour overdispersion with communities is not due to phylogenetic history, but due to the avoidance of interspecific pollinator moves and heterospecific pollen deposition, which suggests competition for pollinators plays an important role in plant diversification. The key to understanding the macroevolutionary history of plant-pollinator interactions is generating reliable phylogenetic reconstructions. Phylogenetic analyses heavily depend on the explanatory power of the phylogeny used as backbone. Therefore, the use of informative priors and assumptions in these analyses greatly affect the downstream analyses. Potential limitations to any phylogenetic analysis also include, but are not limited to, model selection, taxon sampling, missing data, and time calibration, and these should be addressed carefully to make reliable inferences from the phylogenetic data. Obtaining estimates of turnover in plants and pollinators over space and time are critical steps to understanding the contribution of pollinator shifts and pollinator breadth changes in plant macroevolution. Phylogenetic studies provide invaluable data to better explain plant-pollinator interactions in detail, and to elucidate the factors affecting plant biodiversity.

T2-31-04

Are floral nectar spurs a key evolutionary innovation? Phylogenetic, ecological and developmental insights from the Antirrhineae (Plantaginaceae)

Mario Fernández-Mazuecos[1,2], *Pablo Vargas*[2], *Beverley J. Glover*[1]
1. University of Cambridge
2. Real Jardín Botánico, CSIC

Floral nectar spurs are tubular outgrowths containing nectar that have evolved in multiple angiosperm families. Evolutionary variation in spur length may promote pollinator specialization by restricting the access of different pollinators to the nectar reward. Nectar spurs can thus potentially mediate reproductive isolation and eventually speciation. For this reason, they have been hypothesized as a key evolutionary innovation promoting diversification. Snapdragons and relatives (tribe Antirrhineae, Plantaginaceae) constitute an ideal study system to test this hypothesis at multiple evolutionary scales. The Antirrhineae include c. 30 genera of varying species diversity, including spurless and spurred ones. A comprehensive, time-calibrated phylogeny confirmed the recurrent acquisition of nectar spurs in the course of Antirrhineae diversification, and allowed us to test the key innovation hypothesis using state-of-the-art comparative methods. To understand the mechanisms underlying spur evolution at shallower evolutionary scales, we are investigating the role of spur length variation in speciation within a recently-diversified Western Mediterranean clade of the genus *Linaria* using an eco-evo-devo approach. Phylogenetic relationships were resolved using coalescent analyses of genome-wide sequence data generated by genotyping-by-sequencing. The sister relationship between the two species with the longest (*L. salzmannii*) and shortest (*L. clementei*) spurs suggested a role of spur length change in a recent speciation event. Nectar quantifications and pollinator observations indicated that this change was driven by a loss of nectar production in *L. clementei*, linked to a shift in the main type of reward offered to pollinators. A comparison of spur development in *L. salzmannii* and *L. clementei* indicated that contrasting spur lengths in *Linaria* can be achieved by changes in the initial phase of cell division generating the spur primordium.

T2-31-05

Macroevolutionary analyses questioned the role of angiosperms on insect diversification

Fabien Condamine
CNRS

Phytophagous insects represent more than 25% of the global biodiversity. A remarkable diversity of lifestyles parallels this great diversity including mouthparts, gut morphological adaptations to the host-plant preference, and behavioural adaptations for use of particular plants. Interactions between insects and plants are considered as an important driver of diversification and evolutionary changes, yet we still know little regarding the macroevolutionary role and effect of angiosperms on insects. Insects appeared about 500 million years ago (Ma), while angiosperms are estimated to appear about 200-140 Ma. During the Cretaceous terrestrial revolution (between 125–80 Ma), the fossil record and phylogenetic trees revealed that angiosperms experienced an explosive radiation, which culminated in about 80% of the floral composition at the end of the Cretaceous. This angiosperm revolution likely provided new ecological and evolutionary opportunities for insects. However, there are contrasting supports for the role of angiosperms on insect diversification. Although angiosperms may have been a driver of insect diversification, other factors may have been at play such as climate change, or tectonic evolution. In this study, the aim is to investigate whether angiosperms had a role in the insect diversification, and to compare this impact to other time-variable determinants (temperature changes, gymnosperm diversity, and number of continents through time). Both family-level fossil and phylogenetic data were used to assess the role of each factor using macroevolutionary statistical frameworks. Very interestingly, both fossil-based and phylogeny-based analyses inferred the same results for each of the determinants of diversification. The rise of angiosperms correlated negatively with both origination and extinction, suggesting that origination and extinction decreased when the angiosperm diversity increased in the Cretaceous. The opposite role is inferred with the diversity changes of gymnosperms, which had a positive correlation with origination and extinction, thus indicating that insect origination and extinction increased with gymnosperm diversity. These results indicate that temporal changes in plant composition led to important Cretaceous turnover in insect diversification. Regarding abiotic variables, the temperature variations correlated positively with origination and extinction rates, signifying that insect diversification dynamics tracked global climate change. Strangely, the number of continents through time correlated negatively with origination and positively with extinction such that origination decreased and extinction increased when the break-up of Pangea and Gondwana occurred (vicariance events did not foster higher diversity). More importantly, the model with angiosperm diversity significantly outperformed the other models for both fossil-based and phylogeny-based analyses. The results question the positive role of angiosperms on the insect diversification. The rise of angiosperms was not a trigger of insect diversity, but instead a turnover, and may be explained because insects first diversified on gymnosperms before shifting to generalist-angiosperm associations. As the gymnosperms were replaced by angiosperms during the Late Cretaceous and Paleogene, the adaptations to gymnosperm were likely not advantageous as compared to adaptations on angiosperms. The concomitant decline of gymnosperms and rise of angiosperms led to the respective loss and evolution of feeding strategies and host associations (e.g. spore feeders were replaced by pollen feeders, sap suckers were replaced by nectar suckers).

T2-31-06

Testing the morphological clock using an angiosperm-wide floral dataset

Hervé Sauquet[1], *Von Balthazar Von Balthazar*[2], *Susana Magallón*[3], *Isabel Sanmartín*[4], *Jürg Schönenberger*[2]
1. Université Paris-Sud
2. University of Vienna
3. Universidad Nacional Autónoma de México
4. Real Jardín Botánico, CSIC

Probabilistic models of evolution are now widely used in all aspects of macroevolution, including molecular sequence evolution, branching processes (diversification), biogeography, and morphological evolution. Some of these models have been tested and improved considerably over the last few decades. However, models for morphological evolution, especially those for discrete traits, have remained little explored and are considered too simplistic by many biologists, despite their attractive statistical properties. In particular, most of these models make the assumption that morphological traits evolve at a constant rate through time and across lineages. It remains uncertain whether this assumption is reasonable for morphological rates, while many studies have now demonstrated that molecular and diversification rates of evolution vary in many groups across the Tree of Life. Further, most model-based inferences of discrete morphological evolution are still being conducted using the equal-rates (ER, Mk1) model, which assumes equal transition rates among all states of a character, despite the availability of more realistic, but also parameter-rich models such as the all-rates-different (ARD) model. Thus, accounting for heterogeneity and asymmetry in rates of trait evolution is considered to be a key challenge for macroevolution. The angiosperm-wide dataset of 27 floral traits scored in 792 species (367 families) compiled for the eFLOWER project provides an unprecedented opportunity to explore these questions at a very large taxonomic scale. Specifically, we ask: (1) Is floral evolution time-reversible (i.e., symmetrical rates)? (2) Do rates of floral evolution vary through time and/or across lineages? In other words, is there a floral morphological clock? (3) Last, what is the impact of model selection on ancestral state reconstruction? We address these questions by fitting different models of morphological evolution to the floral dataset using a maximum likelihood and a Bayesian approach. We find strong evidence for among-state rate asymmetry. For instance, out of 27 floral characters, 20 were best fit by an ARD model and 2 by unidirectional models, both supporting asymmetry, while 4 were best fit by a symmetrical (SYM) model and only one by an ER model. We also find strong signal for among-lineage rate heterogeneity, suggesting the absence of a strict floral morphological clock, in other words implying that floral character evolution has not been a homogeneous process throughout angiosperm diversification. However, importantly, and perhaps surprisingly, we find ancestral state reconstruction of these floral traits in key nodes of the angiosperm phylogeny to be remarkably robust to model choice and different assumptions on rate heterogeneity, with a few notable exceptions. Although further work on these and other traits and clades is necessary, and new models are required, we anticipate that relaxing the morphological clocks will represent an increasingly important development in the study of angiosperm macroevolution in the years to come.

T2-31-07

Drivers of origination, extinction, and dispersal of flowering plants in deep time

Daniele Silvestro, *Christine D Bacon, Alexandre Antonelli*

Dep. of Biological and Environmental Sciences, University of Gothenburg

Angiosperms originated at least 145 millions of years ago in a world dominated by non-flowering plants and survived through drastic geological and climatic changes and at least one mass extinction event. Since their origination they have undergone an unprecedented taxonomic, phenotypic, and functional diversification, which allowed angiosperms to become the most dominant taxon in several terrestrial ecosystems. However, which factors have influenced their diversification and rise to dominance remain debatable. Here we compile the largest fossil data set of flowering plants to date and analyze it to understand how origination, extinction and dispersal processes have contributed to shaping their diversity through time and across space. We use birth-death-dispersal models to assess the relative importance of different factors on the macroevolutionary history of flowering plants. In particular, we estimate whether changes in the angiosperm origination–extinction balance through time is affected by biotic interactions with other organisms (non-flowering plants and insects) on a macroevolutionary scale. We compare these effects with the impact of abiotic factors (e.g. changes in the configuration of land masses, climate change, and atmospheric composition) using a novel Bayesian analytical framework. Our results indicate that origination, dispersal, and extinction processes in angiosperms are heterogeneous in time and space and respond to multiple biotic and abiotic factors.

T2-31-08

Detecting hidden rate shifts, Type I errors, and revising our concept of angiosperm innovations using trait-dependent models of diversification

Jeremy Beaulieu[1], *Brian Omeara*[2]

1. University of Arkansas

2. University of Tennessee

The distribution of diversity within plants can vary considerably from clade to clade. Attempts to understand these patterns often employ state-dependent speciation and extinction models to determine whether the evolution of a particular novel trait has increased speciation rates and/or decreased extinction rates. It is still unclear, however, whether these models are uncovering important drivers of diversification in plants, or whether they are simply pointing to more complex patterns involving many unmeasured and co-distributed factors. Here we describe a new way forward by extending the popular state-dependent speciation and extinction models to account for the presence of unmeasured factors that could impact diversification rates estimated for the states of any observed trait, addressing at least one major criticism of BiSSE methods. Specifically, our model, which we refer to as HiSSE (Hidden-State Speciation and Extinction), assumes that related to each observed state in the model are "hidden" states that exhibit potentially distinct diversification dynamics and transition rates than the observed states in isolation. We also demonstrate how our model can be used as character-independent diversification models that allow for a complex diversification process that is independent of the evolution of a character, and how it can be used to potentially solve the rather serious issues related to Type I errors and "phylogenetic pseudoreplication". We will focus on the evolution of fruit type in campanulid angiosperms, and discuss the important ways in which HiSSE provides a more nuanced understanding of trait-dependent diversification.

T2-31-09

Radiation of the grassland biome during a falling CO_2 world

Luis Palazzesi[1], *Oriane Hidalgo*[2], *Viviana Barreda*[1], *Felix Forest*[2], *Sebastian Hoehna*[3]

1. Museo Argentino de Ciencias Naturales, CONICET

2. Jodrell Laboratory, Royal Botanic Gardens, Kew

3. Department of Integrative Biology, University of California

Grasslands occur extensively around the globe and their productivity provides important services to humanity. Global scenarios for the year 2100 predicted that grassland ecosystems will experience a major biodiversity change. A better understanding of how a grasslands responded to past environmental variations will help us to comprehend present scenarios, inform conservation decisions, and predict the effects of future environmental change. Here, we explore the relationship between past atmospheric CO_2 fluctuations and the shifts in diversification rates of grasses (Poaceae) and daisies (Asteraceae), two exceptionally species-rich families (~11,000 species and ~23,000 species, respectively) dominant in grasslands. We introduce a novel species-diversification Bayesian model that simultaneously allows, for the first time, 1) the estimation of diversification-rates through time using phylogenetic trees with non-uniform incomplete sampling, and 2) the evaluation of correlations between speciation/extinction rates and environmental variables. The newly proposed model runs in the open-source package 'RevBayes' and provides a novel pathway to explore speciation and extinction rate shifts in hyper-diverse lineages, taking into account putative environmental drivers of biotic change. We found strong evidence that support a simultaneous increase in diversification rates both for daisies and grasses after the Cenozoic most significant reduction of atmospheric CO_2 (~34 Mya). Our estimations support earlier predictions that link periods of CO_2 starvation with the expansion of grasslands.

T2-31-10

The phloem of lianas: a comparative study across the tracheophytes

Marcelo Pace[1,2], *Veronica Angyalossy*[2]

1. Smithsonian Institution

2. Universidade de São Paulo

The lianescent habit has independently evolved many times in the history of tracheophytes, converging into a number of morphoanatomical features related to the ability to climb and sustain a large canopy with a generally slender stem.The anatomy of lianas is very unique, combining structures related to an increased flexibility and a more efficient hydraulic system, such as cambial variants, very wide and long vessels associated with narrow ones, tall and heterocellular rays, fewer fibres and more abundant parenchyma. These convergences are so strong that even distantly related lineages look extremely alike in their wood anatomies, making them discernible mainly by the different cambial variants present. It is very likely that convergences of this type are also present in the phloem; however, that has never been explored. Therefore, the present work aimed to conduct a thorough analysis to determine any correlation between phloem characters and the lianescent habit within a phylogenetic framework. We used phylogenetically controlled pairs of lianas and self-supporting species across the entire tracheophyte phylogeny, comparing phloem 1) qualitatively by delimiting character and character states and checking whether they evolved in correlation with habit, or other aspects, using Pagel's 94 correlation test, and 2) quantitatively by means of paired *t*-tests.Our results showed that lianas, as a whole, have wider sieve elements, with wider sieve pores, and taller rays than their self-supporting relatives; aspects related, respectively, with a presumably more efficient conduction in the phloem and an increased flexibility for the stem. In self-supporting species derived from lianas, the sieve tubes lost the large width, but the rays and cambial variants tended to retain lianescent features, a phenomenon interpreted as phylogenetic inertia. Overall, however, modifications in the phloem of lianas are solely related to the width of sieve elements and ray height. All other anatomical aspects have remained equal in lianas and their self-supporting relatives, in contrast to what occur in the xylem anatomy.This study demonstrates that the phloem of lianas has undergone convergent evolution, generating, in the process, an efficient conductive system across all the tracheophytes, while retaining a strong phylogenetic signal for its other cell types.

T2-31-11

The time tree of order Malvales with the fossilized birth-death process

Rebeca Hernández-Gutiérrez[1], *Susana Magallón*[2]

1. Posgrado en Ciencias Biológicas, Universidad Nacional Autónoma de México

2. Instituto de Biología, Universidad Nacional Autónoma de México

The fossil record is an important source of information about the evolution of organisms through time. One way to analyze the evolutionary history of groups or organisms is by combining the information of the fossil record with molecular-based phylogenetic trees. The Fossilized Birth-Death Process is a method that estimates lineage divergence times along with speciation and extinction rates by taking into account the age and phylogenetic position of fossils. The order Malvales comprises 10 families and approximately six thousand species, most of which are important components of tropical forests around the world. There is a substantial number of fossils described for most of the families of the order, which makes this group an excellent system to study macroevolutionary processes by integratively combining the fossil record and molecular-based phylogenies. This study aimed to illustrate the evolutionary history of the order Malvales by applying the Fossilized Birth-Death Process (FBD). We generated a phylogeny derived from eight molecular markers for all the available malvalean genera in GenBank, extracted with PHLAWD. We reviewed the fossil record of each of the families and built a dataset containing the age and phylogenetic position of 126 fossils. We applied the FBD model to the phylogeny and the fossil dataset and compared the results against those with the "node dating" method (calibration points with a prior distribution). We obtained divergence time for all major clades within order Malvales as well as speciation and extinction rates of the order. The FBD model allowed us to use all the described fossils even if they are not circumscribed into a genus.

T2-31-12

Temporal patterns of extant species diversification of the families of flowering plants

Santiago Ramírez-Barahona[1], *Jürg Schönenberger*[2], *Hervé Sauquet*[3], *Susana Magallón*[1]

1. Universidad Nacional Autónoma de México

2. Universität Wien

3. Université Paris-Sud

Flowering plants represent the structural and energetic basis of nearly all terrestrial ecosystems and their phylogenetic and ecological radiation has promoted the diversification of several biological lineages, including other plants, animals and fungi. Since their first appearance in the fossil record, angiosperms have gone through an exceptional species proliferation and an astounding morphological, functional and ecological diversification. More than 290,000 extant species of angiosperms have been described and the total number is estimated at ca. 350,000 species. Despite the many advances in the fields of comparative phylogenetics and paleontology, the question still remains whether the acquisition of extant species richness is dissociated from the proliferation of major phylogenetic branches. More specifically, present-day species richness in a particular family may have accumulated only recently (i.e., within the crown group), long after the time of origin of the family (i.e., stem group). The number of extant flowering plant species does not allow for a complete (or even nearly complete) sampling of species in phylogenetic studies. Thus, the most recent phylogenies have aimed at sampling representaive taxa of every flowering plant family (ca. 90% of the 434 families currently recognized in the Angiosperm Phylogeny Website). However, these phylogenetic trees include crown node age estimates for less than 30% of families. In addition, crown-group age estimates for several families are probably underestimated due to poor taxonomic representation in the phylogenies. Thus, the temporal pattern of acquisition of extant species diversity across families remains largely unresolved. In this study we address this question by estimating crown-group ages for over 90% of the families of flowering plants, which together represent 99.3% of the total present-day species diversity. Based on the most current version of the Angiosperm Phylogeny Website, we compiled a list of families and reviewed available phylogenetic studies for each family in order to select specific taxa representing the family's crown node. We then reconstructed a time-calibrated phylogeny using four cpDNA genes (i.e., rbcL, atpB, matK, ndhF) and two nuclear markers (i.e., 18S, 26S), with over 120 fossil-based calibration points. The resulting phylogeny represents a significant contribution to flowering plant systematics, allowing for a better description of the mode and tempo of evolution of extant species. This phylogeny will serve as a starting point for the more precise investigation of morphological evolution and lineage diversification in this key clade.

T2-32: The Juan Fernandez Archipelago: A model system for studying plant evolution and biogeography in oceanic islands

T2-32-01

Speciation in the endemic flora of the Juan Fernández Archipelago

Daniel Crawford[1], *Eduardo Ruiz*[2]

1. University of Kansas

2. University of Concepción

The process of speciation may be studied in two general ways. One approach is to study divergence before speciation is complete while the other infers process from comparing traits in distinct species. The latter approach will be employed in this presentation. Some 40% of the genera (lineages) have single species endemic to Robinson Crusoe Island (RC), which is nearer to continental Chile and older than the younger and more distant Alejandro Selkirk Island (AS), which has 16% endemics. Thus, geographical (peripatric) speciation accounts for a large percentage of the endemics in the archipelago. In some cases, the closest continental relatives are reasonably certain and changes associated with speciation are known. In other instances, the relatives are obscure. The largest lineage is the endemic genus *Dendroseris* (Asteraceae: Cichorieae: Sonchinae) with 11 species, eight on RC and three on AS. Sister species on RC are ecologically or altitudinally divergent; the three species on AS are the result of independent colonizations, with divergence subsequent to disperal. The species on RC do not appear to have postzygotic isolating factors, as hybrids between various species are known when they occur together in cultivation. The genus *Robinsonia* (Asteraceae: Senecioneae: Senecioninae) is dioecious, wind-pollinated, and consists of six extant species, five on RC, one on AS, and two extinct species. Two or more species occurring in close proximity sometimes grow in different microhabitats or have different flowering phenologies. In other cases, the factors isolating the species are not obscure. Molecular markers indicate that all species of *Robinsonia* are distinct genetically. The genus *Erigeron* (Asteraceae: Astereae: Conyzinae) is represented by six endemic species (apparently from a single colonization), five restricted to AS and one species on both islands. Species are rare and in some cases not easily distinguished morphologically. Molecular markers do not invariably group all individuals/populations assigned to the same species based on morphology. The species/populations of *Erigeron* on the younger AS are ostensibly in earlier stages of divergence than are species of *Dendroseris* and *Robinsonia* on RC. Several species of *Erigeron* may be isolated by habitat and altitude. Overall in Juan Fernández, differences in flowering time, autonomous self-pollination and pollinator specificity have not been significant in divergence and speciation. Other than hummingbirds, there are essentially no known animal pollinators in the archipelago. Wind, either directly or indirectly, is the most important mechanism for pollen distribution. Very few examples of interspecific hybrids have been documented, and there are insufficient data to judge whether any represent stabilized lines distinct from their parents, i.e., are homoploid hybrid species. The potential limitations of inferring modes of speciation for the flora of the Juan Fernández from comparative studies of extant species will be discussed in the context of the ontogeny of the islands, especially the loss of area and habitat diversity.

T2-32-02

Chromosome diversity and evolution in the flora of the Juan Fernández Islands

Koji Takayama[1], *Carlos M. Baeza*[2]

1. Museum of Natural and Environmental History, Shizuoka

2. University of Concepción

Evolutionary changes in chromosome number, both polyploidy and dysploidy, have long been recognized as a common phenomenon among angiosperms. Biologists have been interested in explaining such cytological variations, and correlations have been sought between polyploidy and distribution. It is estimated that more than 60% of angiosperms exist at the polyploid level, in part due to reticulate evolution via allopolyploid hybridization. Chromosome number surveys of endemic plants in oceanic islands, including the Juan Fernández Islands, have revealed very little variation in chromosome number within genera despite high levels of endemism within these floras. Attempts to understand this chromosomal stasis have invoked suggestions of low levels of hybridization, short periods of geological time, and selection against chromosomal change in divergent habitats during evolution in the islands. Smaller karyological changes between sibling species at the same chromosomal level, however, can lead to reproductive isolation and subsequent genetic and morphological differentiation. These data, therefore, can be helpful for understanding taxonomic affinities within the endemic flora. As an example, the genus *Taraxacum* Wigg. (dandelion) is quite common in the temperate zones of the Northern and Southern Hemispheres. The basic chromosome number in *Taraxacum* is eight, and the genus reveals broad variation of chromosome numbers from $2n = 2x = 16$ to $2n = 12x = 96$. During an expedition to the Juan Fernández Archipelago, Chile, in February 2011, seeds of the endemic *Taraxacum fernandezianum* Dahlst. from Robinson Crusoe and Alejandro Selkirk Islands were collected. A karyological analysis of the two populations revealed both as triploid with $2n = 3x = 24$, with a formula 18m + 3sm + 3sm-sat (i.e., 18 metacentric, 3 submetacentric, and 3 submetacentric with satellite) chromosomes. These populations are karyologically identical to those reported in Japan and India from *Taraxacum officinale*. On this basis, we suggest that the previously designated endemic species, *Taraxacum fernandezianum*, is most probably *Taraxacum officinale*, a weed widely distributed throughout the world.

T2-32-03

Reproductive biology in the endemic plants of the Juan Fernández Archipelago (Robinson Crusoe Islands)

Gabriel Luis Mario Bernardello[1], *Gregory J. Anderson*[2]

1. Museo Botánico, Facultad de Ciencias Exactas, Físicas y Naturales, Universidad Nacional de Córdoba-CONICET

2. Department of Ecology and Evolutionary Biology, University of Connecticut

We surveyed the reproductive features of the angiosperms of the archipelago including the pollination, breeding systems and sexual systems. The collection of original data was based on our team's hundreds of hours of field and lab experiments and observations combined with data from the limited existing literature. This is the first and most comprehensive analysis of an entire island flora. The Juan Fernández flora is not large. It is notable in terms of the proportion of endemics per unit area (the greatest for any island archipelago), and for unusual species (e.g., *Lactoris* and many Asteraceae, e.g., *Dendroseris* and *Robinsonia*). The reproductive biology is notable for several reasons. Although, the level of dioecy is not particularly high (9%), it is higher than that on the Canaries, and significantly less than in Hawaii. As studies move beyond the necessary alpha-ecological first steps to careful in-depth analyses of various taxa, our understanding of breeding systems, and of the incidence of dioecy, will change (i.e., more reports of cryptic dioecy as we discovered with *Pernettya rigida*). The pollination is notable for two reasons. First, because bird pollination, hummingbirds in this case, is prominent, serving some 10% of the flora; one of the hummingbirds is the only hummingbird endemic to an oceanic island. Secondly, because other than the hummingbirds, there are virtually no animal pollinators. The impressive pioneering work by Skottsberg cited a number of likely zoophilous flowers in the flora, but, given that there are no virtually no bees (virtually, though the just discovered new species might become more important), or flower-visiting Lepidoptera or Diptera, animal pollination is restricted to birds. Thus, we have concluded that the island-ubiquitous autogamy or geitonogamy, and anemophily, are the key breeding systems and pollination syndrome. That lack of potential animal pollinators, coupled with a relatively young flora, has led to another particularly notable feature: many more of the island endemics and natives have retained features that are similar to their presumed ancestors than expected. To be sure, a significant portion of the flora fits the usual island model that Carlquist (1974) recognized, where many of the flowers are small, green and, in terms of pollination, are not showy. However, a number of the features would seem to manifest retention of features of the colonizing forms (e.g., larger brightly colored corollas, the expected pollen to ovule ratios for biotically pollinated taxa, and the presence of nectar – all characteristic of zoophilous pollination) – presumably as a result of the lack of selection among species that are autogamous or geitonogamous and/or are wind pollinated. The flora of these little islands is indeed interesting in many ways. And, it seems, at least for the reproductive biology, to manifest an ability to "make do", to improvise success and sustainability, characteristics shared with its alternative namesake – the Robinson Crusoe Islands – from the literature.

T2-32-04

Modeling species diversity in the Juan Fernandez Archipelago

Tod Stuessy[1], *Ulf Swenson*[2]

1. Ohio State University

2. Natural History Museum, Stockholm

Oceanic islands have long been investigated for insights on patterns and processes of evolution. One principal interest has been on understanding factors that regulate species diversity. The Equilibrium Theory of Island Biogeography, presented by MacArthur and Wilson in the 1960s, stressed the importance of size of island and distance from source area as the major factors controlling diversity. These dimensions constrain rates of immigration and extinction, which lead to an equilibrium number of species within a particular island. Despite stimulating aspects of this theory, it did not deal with many other aspects that impact species diversity, such as ecological diversity, inter-island dispersal, modes of speciation, and island subsidence and erosion. In the past decade, new dynamic biogeographic models have appeared that attempt to explain species diversity during the ontogeny of oceanic islands. After volcanic origin, islands subside on submarine tectonic plates and they are slowly eroded, such that after several million years the area and ecological diversity is much less than early in their formation. Eventually the islands disappear under the surface of the sea. The Juan Fernández Archipelago is a simple island system that contains only two main islands, each approximately 50 sq. kms.: Robinson Crusoe Island, 667 kms west of continental Chile, and 4 million years old; and Alejandro Selkirk Island, 181 kms further west in the Pacific Ocean, and 1-2 million years old. The main source area of the vascular flora is western South America. Because of their similar size, different ages, and direct East-West orientation from the major source area, the archipelago serves as an excellent model to test factors that might be contributing to species diversity. Since the flora of the archipelago is well known, we can begin with the species number on Robinson Crusoe Island

and attempt to identify factors that may have contributed to reaching this number. Using these same factors, we can then test their efficacy by applying them to the younger island and checking how accurately they predict the actual number of species. Important in these calculations are the distances from the major source area, the positions of the islands at their formation, their changing surface areas (size) over their ontogenies, intra-island cladogenetic speciation, and inter-island anagenetic speciation. By application of the earlier MacArthur and Wilson theory, the species diversity on the younger island is predicted at 18% higher than currently exists. By taking more dynamic factors into consideration, particularly island ontogeny, the actual number of species of Alejandro Selkirk Island can be closely approximated.

T2-32-05

Invasive plants and threats to the native flora of the Juan Fernández Islands and other archipelagos

Josef Greimler[1], *Christian Schulze*[1], *Patricio Lopez-S*[2], *Patricio Novoa*[3], *Alejandro Gatica*[4], *Karl Reiter*[1], *Carlos Baeza*[2], *Patricio Penailillo*[5], *Eduardo Ruiz*[2], *Tod Stuessy*[6]

1. University of Vienna, Department of Botany and Biodiversity Research

2. Universidad de Concepción, Departamento de Botánica

3. Corporación Nacional Forestal, Jardin Botánico, Viña del Mar

4. Instituto de Ecología y Biodiversidad (IEB), Santiago

5. Universidad de Talca, Departamento de Biología Vegetal y Biotecnología

6. The Ohio State University, Herbarium, Museum of Biological Diversity

Biological invasion is a hot topic in conservation. Alien biota have had and continue to have a severe impact on biological diversity on oceanic islands whereby alien invasive plants such as the Strawberry Guava (*Psidium cattleianum*) on the Hawai´i Islands often benefit from habitat destruction by alien animals such as feral pigs. Lightheaded introduction of ornamental plants and poor understanding of their biology often leads to massive invasion. This is e.g. the case with the ornamental tree *Miconia calvescens* on Tahiti, where it replaces high proportions of the former native forest and drives the endemic forest understorey plants towards extinction. Alien plant invasion is also a hot topic on Robinson Crusoe Islands (Juan Fernández archipelago). The flora of the archipelago sums up to 475 taxa (excluding those behind garden fences) of which 207 (44%) are native and 268 (56%) are alien (introduced) taxa. Among the native plants 133 (64%) are endemics and 74 (36%) are also occurring on the continent. The high proportion (>55%) of alien plants on the Robinson Crusoe Islands is only surpassed by Easter Island (75%) among the Chilean islands and is fairly high compared to other oceanic archipelagos such as the Hawaiian Islands (44%) and the Galapagos Islands (29%) based on numbers published in the new millennium. Alien plants on the Robinson Crusoe Islands have increased continuously in number since the first records. A few of them such as the shrubs *Aristotelia chilensis, Rubus ulmifolius* and *Ugni moline* as well as the herbs and grasses *Acaena argentea*, *Rumex acetosella*, *Hypochaeris radicata*, and *Anthoxanthum odoratum* have become invasive. On Robinson Crusoe Island the invasive shrubs replace now a high proportion of native forest which is accompanied by a notable decline of endemic taxa. Additionally introduced animals cause several problems: Livestock (cattle) on Robinson Crusoe Island enters the native forests and the forest-grassland ecotone foraging the tree seedlings. Rabbits (*Oryctolagus cuniculus*) have increased in their number to an immense quantity on this island. Goats (*Capra hircus*) were introduced on both islands and are estimated at present to a population size of more than 2000 individuals on Alejandro Selkirk. On this island we found also a significant decline of endemic plants compared to data collected 100 years ago, while alien plants appeared with significantly higher frequencies in 2011. The most common aliens (*Anthoxanthum odoratum* and *Rumex acetosella*) were found at lower frequencies in 1917 and one (*Hypochaeris radicata*) was entirely missing in all old records. This may be due to a more recent arrival of *H. radicata* on Alejandro Selkirk and subsequent invasion of nearly all plant assemblages in less than 100 years. Goats certainly played a substantial role in this change. Other introduced animlas such as rats (*Rattus spp.*) and coatis (*Nasua nasua*) are dangerous for the native bird fauna. Besides the direct impact on endemic plants of which some show very small and declining populations herbivores such as goats and rabbits are effective dispersal agents for the seeds of alien plants. The native Austral Thrush (*Turdus falcklandii*) produces a dilemma for conservation decisions because the fleshy fruits of the invasive shrubs *Aristotelia chilensis*, *Rubus ulmifolius*, and *Ugni molinae* are mainly dispersed by this bird. Conservationists argue that control or even eradication of the bird would limit the spread of the invasive shrubs significantly.

T2-32-06

Genetic diversity and conservation in endemic species of the Juan Fernández Archipelago

Patricio López-Sepúlveda, *Glenda Fuentes A.*

Department of Botany, Faculty of Natural Sciences and Oceanography, University of Concepción

Located at 667 km from continental Chile, the Juan Fernández Archipelago is composed of two major islands: Robinson Crusoe (= Masatierra) and Alejandro Selkirk (= Masafuera), separated by 181 km, plus a much smaller island, Santa Clara. A fundamental difference between the major islands is their estimated geological age, approximately 4 my for Robinson Crusoe and 1-2 my for Alejandro Selkirk. The high level of endemic plant species (66%) makes the Archipelago a hot spot of biodiversity and a unique location for understanding the evolutionary processes that have occurred there. The effects of invasive plant species (*Aristotelia chilensis, Rubus ulmifolius* and *Ugni molinae*) and animals (*Capra hircus*, *Oryctolagus cuniculus* and *Turdus falcklandii*) on the endemic and native vegetation, as well as the human impact and natural processes on the islands such as erosion, have resulted in the delimitation of areas of protection based mainly on the diversity of endemic species in the different zones of Robinson Crusoe Island. The use of modern molecular tools such as AFLPs and SSRs, as well as traditional data based on isoenzymes, allows generation of information on genetic variation within many endemic species of the Juan Fernández Archipelago. This contribution identifies areas with the greatest genetic richness in each of the islands, using selected species and different molecular markers. Genetic diversity datasets from the following 13 species of endemic vascular plants of the Juan Fernández Archipelago were analyzed: *Drimys confertifolia* Phil. (Winteraceae); *Dysopsis hirsuta* (Müll. Arg.) Skottsb. (Euphorbiaceae); *Erigeron fernandezia* (Colla) Solbrig, *E. rupicola* Phil. (Asteraceae); *Gunnera bracteata* Phil., *G. masafuerae* Skottsb. and *G. peltata* Phil. (Gunneraceae); *Myrceugenia schulzei* Johow (Myrtaceae); *Rhaphithamnus venustus* (Phil.) B.L. Rob. (Verbenaceae); *Robinsonia evenia* Phil., *R. gayana* Decne., *R. gracilis* Decne. and *R. masafuerae* Skottsb. (Asteraceae). Neutral and outlier matrices were obtained, which allowed comparing levels of genetic diversity between each of the populations for each of the species of Robinson Crusoe and Alejandro Selkirk Islands, as well as geographic partitioning. Based on these values, the genetically most diverse regions are the southeast areas of Robinson Crusoe Island and southwest regions of Alejandro Selkirk Island. This study was financed by projects FONDECYT N° 1160794 and FWF N° P21723-B16.

T2-33: Evolution of mangroves: Phylogeny, conservation genetics and genomics

T2-33-01

Genetic and functional trait variation along three expanding black mangrove *(Avicennia germinans)* distributional ranges in the United States

John Paul Kennedy[1], *Ilka C. Feller*[2]

1. Smithsonian Marine Station

2. Smithsonian Environmental Research Center

Climate-driven shifts in species' distributional patterns can have profound ecological consequences, particularly when these shifts involve ecosystem foundation species. Mangroves are (sub)tropical intertidal foundation species that influence both ecosystem function and associated community structure. With a warming climate, mangroves have experienced poleward range expansion worldwide and greater encroachment into areas historically dominated by temperate salt marsh. In the United States, mangrove expansion has been documented along three coastlines: Texas, West Florida, and East Florida. Northernmost range limits along these coastlines are dominated by black mangrove (*Avicennia germinans*), the most cold-tolerant mangrove species in the United States. We present findings from our assessment of both genetic and morphological variation among 41 *A. germinans* populations (a total of 1,103 individuals) that encompass this species' entire distributional range in the United States. We use both microsatellite molecular markers and functional leaf traits to outline population differences with a combination of classic population genetics, Bayesian, and multivariate approaches. The results of this study provide baseline information on *A. germinans* population structure in the United States and contribute to our understanding of the evolutionary and environmental forces shaping these populations. Continued research will focus on empirical approaches to test for patterns of local adaptation at mangrove range limits in the United States and the potential ecological consequences a shift from salt marsh to mangrove dominance may have on these coastal ecosystems.

T2-33-02

Global phylogeography of the mangrove genus Rhizophora: Separation and reunion across the ocean

Koji Takayama[1], *Yoichi Tateishi*[2], *Tadashi Kajita*[2]

1. Museum of Natural and Environmental History, Shizuoka

2. University of the Ryukyus

The global distribution of mangroves is attributed to interactions between geographical barriers and long-distance seed dispersal via ocean currents, which are manifest in species distribution, genetic structuring and evolutionary history. Extreme long-distance dispersal, especially, is an important process in mangrove biogeography, because such events can lead to changes in distribution ranges of mangrove species, to rapid diversification due to founder effects, and to secondary contact with sibling species. A mangrove genus *Rhizophora* (Rhizophoraceae) has been considered as a key taxon to reveal the global and disjunctive distribution patterns of mangroves because the plants dominate in most of the mangrove forests in both Indo-West Pacific (IWP) and Atlantic-East Pacific (AEP). Furthermore, the fossil records of the genus and family are relatively abundant due to the distinct shape of the propagules, which is useful features to trace historical distribution. Due to its wide distribution and presence of fossil records, however, comprehensive phylogeography on a global scale that includes all species have not been clarified. To investigate the phylogeography of the genus, we conducted chloroplast and nuclear DNA sequences, and microsatellites analyses in *Rhizophora* species collected worldwide. All molecular markers strongly supported the genetic disjunction between IWP and AEP, except for *R. samoensis* in the South Pacific Islands. The most recent common ancestor between IWP and AEP *Rhizophora* species were at 11 million years ago estimated by calibration on fossil records, indicating that the continental drift involving the closure of the Tethys seaways and the reduction of mangrove distribution followed by Mid-Miocene cooling could be key factors to promote their disjunctive distribution. DNA sequences and microsatellite supported that both lineages met in the South Pacific Island through the immigration of *R. samoensis* by an extreme long-distance dispersal and experienced gene exchange after long-term isolation. Global genetic disjunctions between IWP and AEP and its collapse by long-distance dispersal are key factors to explain current patterns of diversification of *Rhizophora*.

T2-33-03

Mangrove floristics and biogeography revisited: Corresponding biodiversity hot spots and ancestral discontinuities formed by common evolutionary processes

Norman Duke

James Cook University, TropWATER Centre

This treatment provides a novel re-assessment of common biodiversity patterns and the evolution of mangrove plants based on the ancestral biogeography, extant floristics and current distribution records of all species entities. It is generally acknowledged that mangrove plants occur where they do in the world because past and current factors have influenced their dispersal, diversification and establishment. The premise here is that, while all 80 extant mangrove taxa from 18 family lineages with mangrove representatives responded individually, they did so under the overwhelming common influences of large-scale geophysical factors associated with continental drift and periods of glacial maxima. Accordingly, the combined distribution and phylogeny of all mangrove genotypes are likely to have shared common pathways from their genesis and dispersal to their current day distributions. In this treatment, I present the hypothesis where evolutionary processes of diversification, divergence and speciation have been driven largely by similar geophysical circumstances and events of isolation and reunion. As such, it is proposed that all species and their phylogenies may have been influenced by the same dominant factors and events, including: land barriers, water separation and climate conditions; albeit reflected differently in each genetic lineage depending on the dates and location of the genesis of each, and regards individual capacities and tolerances affecting mutation rates, dispersal capacities, propagule buoyancy, longevity and other species-specific factors.

T2-33-04

Mangrove genomes are extensively modified and point to novel mechanisms of adaptation

Ziwen He, *Shaohua Xu, Zhang Zhang, Wuxia Guo, Suhua Shi*
State Key Laboratory of Biocontrol, Guangdong Provincial Key Laboratory of Plant Resources, School of Life Sciences, Sun Yat-sen University

Fewer than 100 species of woody plants, collectively known as mangroves, inhabit the interface between terrestrial and marine environments. Since mangroves have been suggested to be critically endangered at present (Duke *et al.*, 2007), the future could be uncertain when the sea level rises. Understanding how mangroves adapted to these habitats, which have experienced frequent sea level changes, may inform about their future. Here, we report the sequencing of the whole genomes of species in three mangrove genera (*Avicennia*, *Rhizophora* and *Sonneratia*) and the transcriptomes of 14 others, using the PacBio single-molecule real-time (SMRT) sequencing and Illumina paired-end/mate-paired short reads sequencing. High quality sequences of scaffolds are assembled (N50 = 2.3,5.4 and 5.5 Mb, respectively). The data show that the three species invaded new habitats independently within a relatively small window of time (43-54 million years before present), suggesting that climates favorable for invasion have occurred infrequently. In adapting to new habitats, mangroves, instead of expanding their genomes as might have been expected, have undergone extensive convergent reductions in both repetitive sequences and gene families. They have also convergently reconstituted their amino acid compositions across the genome. Amino acid changes between dissimilar residues, rarely found in other species, are much more common in mangroves. Their genetic diversity is extremely low as a consequence of small populations occupying shifting habitats under continual sea level changes. The genomes of mangroves suggest a precarious existence throughout the history. Hence, their ability to withstand the double assaults of rising sea levels and increasing human perturbations is hardly assured. Taken together, the genomic features suggest extensive modifications of mangrove biology for the adaptation to the intertidal habitats and open up new research avenues in adaptation, speciation and conservation.

T2-33-05

The integration and application of genetics in mangrove conservation

Alison K.S. Wee[1], *Gustavo M. Mori*[2], *Aidan W. Short*[1], *Tianhui Yu*[1], *Catarina F. Lira-Medeiros*[3], *Juan Núñez-Farfán*[4], *Koji Takayama*[5], *Leanne Faulks*[6], *Suhua Shi*[7], *Yoshiaki Tsuda*[6], *Yoshihisa Suyama*[8], *Takashi Yamamoto*[9], *Takaya Iwasaki*[10], *Yukio Nagano*[11], *Zhengzhen Wang*[7], *Shin Watanabe*[12], *Tadashi Kajita*[12]
1. Guangxi University
2. São Paulo State University
3. Jardim Botânico do Rio de Janeiro
4. National University of Mexico (UNAM)
5. Museum of Natural and Environmental History
6. University of Tsukuba
7. Sun Yat-sen University
8. Tohoku University
9. Kagoshima University
10. Kyoto Uni
11. Saga University
12. University of the Ryukyus

Mangroves are a highly valuable ecosystem under severe anthropogenic threats. With the current rate of degradation, it is possible that the ecosystem services provided by mangroves will be lost or impaired within the next century. Here, we present an article prepared as a joint effort by mangrove conservation geneticists worldwide, advocating for a renewed focus to integrate and apply findings from mangrove phylogeny, biogeography and adaptive evolution to conservation planning. We highlighted three major areas of importance: (1) the determination of species identity and conservation priorities, (2) the identification of evolutionary significant units, and (3) the understanding of adaptation in the face of climate change. For each of these areas, we put forth suggestions on how state-of-the-art molecular data on mangrove evolution and ecology can inform and facilitate its conservation. Specifically, we present our work on the cold tolerance of mangroves as an example of how genetic data is essential for the effective conservation of mangroves. In this study, we conduct transcriptome analysis on individuals of *Kandelia obovata* and *Bruguiera gymnorhiza* under different temperature treatments to understand the molecular and physiological regulatory mechanism behind cold tolerance. Our findings have important implications on the distribution of mangroves, especially their response to climate change. Ultimately, we call for conservation efforts to aim toward "managing evolution", whereby the adaptive and evolutionary potential of mangroves is understood and preserved. This article serves as a mission statement for the field of conservation genetics in mangroves and provides a common ground for establishing international collaboration.

T2-33-06

A multiple disciplinary approach to unveil the evolutionary history of New World *Rhizophora*

Gustavo Maruyama Mori[1], *Takaya Iwasaki*[2], *Mariana Vargas*[3], *Stephanie Karenina Bajay*[3], *Koji Takayama*[4], *Yu Matsuki*[5], *Rafael Silva Oliveira*[6], *Yoshihisa Suyama*[5], *Anete Pereira De Souza*[3,6], *Suhua Shi*[7], *Maria Imaculada Zucchi*[8], *Tadashi Kajita*[9]
1. Biosciences Institute – Coastal Campus - São Paulo State Uni-

versity, São Vicente
2. Center for Ecological Research, Kyoto University
3. Centro de Biologia Molecular e Engenharia Genética – Universidade Estadual de Campinas
4. Museum of Natural and Environmental History
5. Kawatabi Field Science Center, Graduate School of Agricultural Science, Tohoku University
6. Departamento de Biologia Vegetal – Universidade Estadual de Campinas
7. State Key Laboratory of Biocontrol and Guangdong Provincial Key Laboratory of Plant Resources, Sun Yat-sen University
8. Agência Paulista de Tecnologia dos Agronegócios
9. Iriomote Station, Tropical Biosphere Research Center, University of the Ryukyus

Humans' impact on Earth is undeniable such that people have become a geological force. Expectedly, living beings also are influenced by human activities. One of the phenomena that have severe impact on organisms are the current global climate changes (GCC), human-driven alterations on the climate properties, whose biological effects are unquestionable. Despite its global influence, different species may differently respond to these changes. Mangrove forests, for instance, are projected to be specially impacted such that distributional shifts are currently being reported in different regions of the world. Because retrospective information is a valuable input to make better predictions of the responses to a current changing world, our objective is to describe the evolutionary history of Western hemisphere *Rhizophora* species. Due to the complex nature of the ecological and evolutionary drivers of organisms' distribution, we are using a multiple disciplinary approach based on climatic niche, genetic diversity and transcriptome analyses of *Rhizophora mangle* and *R. racemosa* besides their putative hybrid *R. harrisonii* from the Atlantic-Caribbean-West Pacific biogeographic region. As a whole, the climatic niche pairwise comparison indicated that *R. racemosa* and *R. mangle* presented a similar (but not equivalent) climatic niche. Climatic niche models projected in three past moments showed a substatial regional heterogeneity in distributional response such that, for instance, whereas in North America there was a poleward expansion, in the South American Atlantic coast, we did not observe this latitudinal shift. Moreover, for *R. racemosa,* the Last Glaciation has reduced its climatic niche to few areas. It suggests that this species has been under cycles of population expansion-retraction since the Last Interglacial Period. These changes in species distribution may also provide cues to insterspecific hybridization, a relevant biological process for mangrove species, specially for Western World *Rhizophora.* Using genome-wide genetic markers, we observed that individuals identified as *R. racemosa* from the Pacific Coast of Central America are more closely related to *R. mangle* individuals from the same coastline than *R. racemosa* individuals from the Central American Atlantic littoral. Within species, we observed previously described genetic structure patterns indicating, for example, that the American continent is a major barrier to *R. mangle* gene flow whereas the Atlantic and Pacific oceans do not impede this species' migration. Moreover, in the South American Atlantic coast, for *R. mangle,* we found a clear subdivision of two populations (from North and South coasts of Brazil). We are currently analyzing the transcriptomes of *R. mangle* individuals from both North and South populations. These preliminary results provide cues from different perspectives that may shed light on the mangrove trees' response historical changes that may support better predictions of how they may respond to the current GCC.

T2-34: Evolution and ecology of bamboos I: Phylogenetics and evolution of woody and herbaceous bamboos, II: Ecology and biogeography of temperate and tropical woody bamboos (two sessions)

T2-34-01

Negative correlation between rates of molecular evolution and flowering cycles in temperate woody bamboos revealed by plastid phylogenomics

Peng-Fei Ma[1], *Maria S. Vorontsova*[2], *Zhen-Hua Guo*[1], *Thomas Haevermans*[3], *De-Zhu Li*[1]
1. Kunming Institute of Botany
2. Royal Botanic Gardens
3. Muséum national d'histoire naturelle

Heterogeneous rates of molecular evolution are universal across the tree of life, posing challenges for phylogenetic inference. Here we present 15 newly sequenced plastid genomes in the temperate bamboo tribe Arundinarieae, including the first genomes ever sequenced from Madagascar representatives. The woody bamboos are noted for its extremely slow evolutionary rates within the rapidly evolving family Poaceae, supposedly caused by their mysterious monocarpic reproduction. However, the correlation between evolutionary rates and flowering cycles has not been formally tested. A data matrix of 46 plastid genomes representing all 12 lineages of the tribe was assembled for phylogenetic and molecular evolutionary analyses. We conducted phylogenetic analyses using different sequences (e.g., coding and noncoding) combined with different data partitioning schemes as well as using site-heterogeneous models, revealing conflicting relationships involving internodes among several lineages. A great difference in branch lengths were observed among the major lineages, and topological inconsistency could be attributed to the long-branch attraction (LBA). Using clock model-fitting by maximum likelihood and Bayesian approaches, we furthermore demonstrated extensive rate variation among these major lineages. Rate accelerations mainly occurred for the isolated lineages with limited species diversification, totaling 11 rate shifts during the tribe's evolution. As an example of recently radiated tribe, we further illuminate the universality of rate heterogeneity in plants. More importantly, we find a strong negative correlation between rates of molecular evolution and flowering cycles for Arundinarieae, in accordance with a putative generation time effect. The correlation identified will be helpful for studying the unpredictable flowering in these bamboos.

T2-34-02

Progresses in Bamboo (Poaceae – Bambusoideae) macroecology and diversification studies

Thomas Haevermans[1], *Peng-Fei Ma*[2], *Meng-Yuan Zhou*[2], *Xian-Zhi Zhang*[3], *De-Zhu Li*[2]
1. Muséum national d'histoire naturelle. Institut de Systématique, Évolution, Biodiversité (ISYEB) UMR 7205 Centre national de la recherche scientifique - Muséum national d'histoire naturelle

- *École Pratique des Hautes Études – Université Pierre et Marie Curie, Sorbonne Universités, Paris Cedex 05, France*
2. Plant Germplasm and Genomics Center, Germplasm Bank of Wild Species, Kunming Institute of Botany, Chinese Academy of Sciences, Kunming, Yunnan 650201, China
3. College of Forestry, Northwest A&F University, Yangling, Shaanxi, China

Bamboos are a seemingly well-known flagship taxon whatever the region, be it tropical or temperate. Its classification and phylogeny have been the subject of numerous studies, but much work is still needed, especially in notoriously difficult groups such as the *Dendrocalamus/Bambusa/Gigantochloa* complex or for addressing still debated questions such as obtaining supported phylogenetic relationships for the *Olyrea* (the herbaceous bamboos). The aim of this communication is to show the progresses made by our team in understanding bamboos evolution and ecology at a global scale, without focusing on a too specific group, as so much knowledge is still lacking to get enough resolution outside of the macroscale. Our study is thus focusing on macroscale patterns of selected morphological traits within a macro-ecological framework. Our results, exposed during this presentation, are heavily relying on distributional data and especially phylogenetic backbones produced through projects held in collaboration with Kunming Institute of Botany, RPC China, and Paris Muséum national d'histoire naturelle, in France.

T2-34-03

The phylogeny of temperate woody bamboos (Arundinarieae) based on ddRAD data

Zhenhua Guo, *Cen Guo, Xiaying Ye, Ying Guo, Dezhu Li*
Kunming Institute of Botany

The tribe Arundinarieae (the temperate woody bamboos) has long been recognized as a monophyletic group. However, the phylogenetic relationship within the tribe are still intricate due to the difficulty of detecting sufficient markers and their reticulate evolution history. Here, 422 species (417 temperate woody bamboos, and five tropical woody bamboos as outgroups) were sampled and candidate SNP markers were identified using the double digest restriction-site associated DNA sequencing (ddRAD-seq) for our phylogeny analysis. To improve data quality and obtain more robust phylogenetic topologies, the grouping method of Stacks and Bowtie2-Stacks mapping strategy were finally adopted in generic and higher level, respectively. Our results revealed 14 highly supported lineages within Arundinarieae using maximum likelihood inference. In this scheme, *Ampelocalamus calcareus* was the most basal branch to diverge from Arundinarieae, with the ADH clade as the second branch, thereafter followed by *Gaoligongshania megalothyrsa* and the core Arundinarieae which can be further divided into the monopodium clade and the sympodium clade according to their rhizome features. The monophyly of Arundinarieae and core Arundinarieae were well supported. Overall, compared with previous limited data sets consisting of either a single or a few genes, ddRAD data sets have been significantly advanced our ability to resolve historically difficult phylogenies. This well-resolved phylogenetic framework of Arundinarieae reflects a major step forward in our understanding of bamboo biodiversity and evolutionary history.

T2-34-04

Phylogeny of *Fargesia* (Poaceae: Bambusoideae) inferred from multiple DNA sequence data and chloroplast genome

Yuqu Zhang, ***Yun Zhou****, Juqing Kang, Jianqiang Zhang, Yi Ren*
Shaanxi Normal University

The majority of the woody bamboo species have been descried frequently without knowledge of the flowers so that the infrageneric taxonomy was mainly based on the subtle and variable vegetative features. But it has not been proven that either the vegetative-feature-based infrageneric taxonomy or the relationships between genera are cladistically sound. The infrageneric classification of *Fargesia* based on vegetative characteristics has not yet been tested by molecular phylogeny. To prove the infrageneric classification, we explored phylogenetic relationships within *Fagesia* based on three cpDNA (*mat*K, *psb*A-*trn*H and *trn*L-*trn*F), nrITS and *rpb*2 sequences. A representative sample of 50 bamboo species, including 37 species of *Fargesia* and nine species of *Yushania* was used to estimate the phylogeny of *Fargesia*. To estimate deeper-level phylogenetic relationships for *Fargesia*, we assembled a data set of whole genomes from 26 bambooos including 17 *Fargesia* species and 4 *Yushania* species, two newly added *Drepanostachyum* species (namely *D. melicoideum* and *D.saxatile*,) one *Bashania* species, one *Ampelocalamus* species and one *Pleioblastus* species by using next-generation sequencing. The phylogenetic trees of 5 DNA fragements showed that *Fargesia* (*sensu* Yi, 1988) is polyphyletic, with *F. crassinoda* and *F. damuniu* in *Thamnocalamus*, *F. ampullaris*, *F. semiorbiculata*, *F. gyirongensis* and *F. collaris* in *Drepanostachyum,* and the rest of the species of *Fargesia* and all species of *Yushania* in a *Fargesia* + *Yushania* clade. We suggest that *Yushania* should be merged into *Fargesia* and that the *Fargesia* + *Yushania* clade identified in this study should form the new *Fargesia* (s. l.). The phylogeny of genus *Fargesia* (s. l.) diverges into Clade A and Clade B, consistent with the changes in the morphology of flowers and inflorescences. Species of Clade A have compressed and unilateral inflorescences subtended by a series of spathe-like leaf sheaths; the flowers have three stigmas, blooming in spring, and Clade A tends to occupy the northern part of the distribution of *Fargesia* (s. l.), where it is colder in winter and spring. Species of Clade B have open inflorescences subtended by variably expanded but never spathe-like leaf sheaths; most have two stigmas, bloom mainly from June to August, and occupy the southern part of the distribution where it is warm. Reconstruction of ancestral states indicated that the derived "spathe-like leaf sheaths" type likely evolved as a result of an adaptive advantage of compressed inflorescences in colder temperature. There were several differences based on plastid phylogenomics compared with the previous results of 5 DNA fragments: 1) Clade A, two newly added species of *Drepanostachyum*, genus *Bashania* and *Phyllostachys* formed a strong supported clade. 2) Clade B was polyphyletic clade. Genus *Yushania* and some species of *Fargesia* in clade B constructed a strong support clade, which is sister to the clade of "*Drepanostachyum* + *Bashania* + *Phyllostachys* + Clade A". And the remaining species of clade B plus *F. crassinoda* and *Fargesia* sp. formed another strong support clade.

T2-34-05

Speciation of BDG Complex (*Bambusa, Dendrocalamus & Gigantochloa*): the molecular evidence of hybridization

Nianhe Xia, *Caohan Zheng*
South China Botanical Garden, Chinese Academy of Sciences

BDG complex (*Bambusa-Dendrocalamus-Gigantochloa*) have numerous species and are widely distributed in tropical and subtropical aeras. Most species with high economic value and extensive utilization belong to these three genera. The generic and infrageneric boundaries between/within these three genera are problematic, and the taxanomic status of many taxa are arguable. This study was conducted to uncover the potential hybrid origins in the woody BDG complex (*Bambusa-Dendrocalamus-Gigantochloa*) and the phylogenetic relationship and the taxonomic positions of Bambusoideae (Poaceae) by molecular analyses, both separated and combined, of nuclear (GBSS I partial regions) and chloroplast (*psdA-trnH*, *rpl16* intron, and *rps16-trnQ*) sequences from the BDG complex, including 113 species, varieties, forms and hybrids from the three genera. All the samples were classified into eight basic taxa: hard thorn group of *Bambusa* subg. *Bambusa* (BBh), soft thorn group of *Bambusa* subg. *Bambusa* (BBS), *Bambusa* subg. *Leleba* (BLe), *Bambusa* subg.*Dendrocalamopsis* (BD), *Bambusa* subg. *Lingnania* (BLi), *Dendrocalamus* subg. *Dendrocalamus* (DD), *Dendrocalamus* subg. *Sinocalamus* (DS) and *Gigantochloa* (Gi). The taxa of hybrid origins in the BDG complex are verified and exclusively studied. The main results are as follows. 1. The taxa with soft thorns in *Bambusa* subg. *Bambusa* (BBs) are initially confirmed to be natural hybrid taxa. Its parental taxa are from the hard thorn taxa in *Bambusa* subg. *Bambusa* (BBh) and *Bambusa* subg. *Leleba* (BLe). Most BBs species' maternal origins are from BLe, except *B. lapidea*, *B.subaequalis* and a few species have maternal taxa from BBh. 2. Some species in *Bambusa* subg. *Lingnania* (BLi) are confirmed to be natural hybrids of *Bambusa cerosissima* or its closely related, as the maternal origin, and BLe, *Dendrocalamus* or *Gigantochloa*, as the paternal origins. 3. The hypothesis of *Bambusa* subg. *Dendrocalamopsis* (BD) as a natural hybrid taxa is further proved. Moreover, BD species have multiple parental origins. The result shows that natural hybridization are common in BDG Complex; it has significant reference for demarcation of the generic scope and taxonomic studies on BDG complex.

T2-34-06

Endemism, ecology and generic circumscription in woody bamboos in India

Paramjit Singh, *P. Kumari*
Botanical Survey of India

Bamboos form a major component of the secondary forests and as pure strands in India mainly in North eastern states, Western Ghats and Andaman and Nicobar Islands. According to the present estimate more than 140 species are recorded in India of which 46 species are endemic to the country. Most of these are neoendemics. Two genera are described from and are endemic to India. The endemic genus *Stapletonia* with 2 species is reported from Arunachal Pradesh and Munrochloa, is reported from Penninsular India. Generic delimitation in the bamboos has stabilized in the subcontinent. Among the major dominant genera with maximum number of endemic species, *Bambusa* is represented with 10 species in NE region and *Ochlandra* with 9 species in Western Ghats. 2 species of *Dendrocalamus* are from NE region and 1 each from North and Penninsular India. 3 species of *Cephalostachyum* is endemic to North eastern states Arunachal Pradesh, Meghalaya, Nagaland and Sikkim. *Melocalamus* and *Sinarundinaria* are having 2 endemic species to India found in NE region. Species of *Schizostachyum* are found in Andaman and Nicobar islands and out of three species 2 are endemic to the country. Also genus *Dinochloa* is restricted to the Andaman and Nicobar islands of which 1 species *Dinochloa nicobariana* is an endemic species to the Nicobar Islands. Genera like *Chimonobambusa, Chimonocalamus, Drepanostachyum, Melocanna* and *Thamnocalamus* are having 1 species each endemic to NE India whereas *Pseudoxytenanthera* and *Teinostachyum* are having 1 endemic species from Kerala. 1 endemic species under *Gigantochloa* has been recently reported from Bastar district of Chhattisgarh state. This presentation is meant for comprehensive information on the endemic bamboo species found only in Indian territories. With a brief description of identifying characters of the species, accompanied with their illustrations or the photographs, local names, habitat and occurrence, information of flowering and new shoots, conservation status as per IUCN guidelines, distribution with map and specific notes.

T2-34-07

Madagascar's grasses and bamboos

Maria Sergeevna Vorontsova[1], *Olinirina Prisca Nanjarisoa*[2], *Jan Hackel*[3], *Guillaume Besnard*[3], *Peter Linder*[4]
1. Royal Botanic Gardens, Kew
2. Kew Madagascar Conservation Centre
3. Universite Paul Sabatier
4. Institute of Systematic Botany

The dominant narrative regarding human arrival in Madagascar has long stated that people arrived 2,500 years ago to a largely forested island, with anthropogenic destruction responsible for the 65% of Madagascar's land area now covered in grasslands and savannas. Scarcity of lake bed deposits have made paleoenvironmental reconstruction difficult. One piece of the puzzle regarding Madagascar's environmental history remains missing: the Poaceae flora has never been fully documented or considered in an ecological perspective. Herbarium specimens of Malagasy grasses (family Poaceae) were studied, taxonomic revisions produced, and new species described during the six years of this project. Regular field work has targeted locations of poorly known endemic grasses and species knowledge has been built up throughout the island. Phylogenetic trees were reconstructed and dating was carried out in order to document all instances of Poaceae diversifications into Madagascar. Plot data recording species composition, phylogenetic diversity, and disturbance levels has been gathered at 80 sites. 217 of Madagascar's 541 Poaceae species are endemic, a level of endemicity in line with the grass floras of other subtropical islands. We have identified 97 endemic lineages (including 4 lineages of bamboos), which have colonised Madagascar primarily from Africa, with a mean age of 3.5 million years. Half of the grass dispersals into Madagascar were C4 lineages pre-adapted to open habitats. The earliest endemic lineage *Lecomtella mada-*

gascariensis arrived ca 22 million years ago. The largest endemic clade recovered corresponds to the bamboo subtribe Hickeliinae. Madagascar's High Plateau is home to a unique and endemic grass flora. A checklist if the Itremo Protected Area has revealed 99 species of Poaceae, with 20 restricted to the High Plateau. Phylogenetic diversity within grassy ecosystems of Madagascar decreases in areas with strong physical disturbance, indicating ecosystems dynamics typical of natural assemblages. We demonstrate that Madagascar is home to an ancient and diverse grass flora, with local species assemblages functioning similarly to natural ecosystems. This indicates that pre-human Madagascar was home to at least some tropical grassy biomes. Work continues to determine where these were located, and how can they be distinguished from anthropogenic landscapes resulting from forest destruction.

T2-35: Insights into the evolution and biogeography of the Compositae: Beyond the book (two sessions)

T2-35-01

The complicated phylogeny of the Gnaphalieae: Perspectives, outstanding issues and the path forward

Rob Smissen, *Ilse Breitwieser*
Landcareresearch.co.nz

The publication of *Systematics, Evolution, and Biogeography of Compositae* coincided with the consolidation of new phylogenetic hypotheses that explained much of the diversity of the Gnaphalieae. A basal grade suggests a southern African origin of the tribe with multiple lineages spreading out around the globe. However, at the time significant discrepancies were emerging between nuclear ribosomal and plastid DNA trees. The Australasian Gnaphalieae and *Helichrysum* and its allied genera were proving recalcitrant to phylogenetic analysis. Few of the diverse South American Gnaphalieae had been included in molecular phylogenetic studies. Several large genera remained polyphyletic and the subtribal classification problematic. Since then progress has been patchy, leaving many opportunities. Important progress has been made with South American Gnaphalieae. Several African and Australian genera have been examined and refined. A major study of *Helichrysum* has been published as well as a separate detailed study of its satellite genus *Anaphalis*. Three informally named non-African clades; the FLAG, HAP and Australasian clade have proved useful hypotheses accounting for c. 140 genera (of c. 180) and c. 1,700 (of c. 2,100) species of the tribe. Some important genera are still probably polyphyletic (e.g. *Syncarpha*). Relationships among major clades, including those in southern Africa and the HAP, FLAG and Australasian clades are not robustly resolved. Hypotheses of allopolyploid origins for the Australasian and FLAG clade have been advanced, but lack independent tests. Within the HAP clade, clades with allopolyploid origin are also hypothesised, although these also require testing. Although these hypotheses of reticulate evolution have explanatory power, the devil in the detail is clear and conflict among DNA sequence phylogenies of different loci suggest those so far advanced may be just the tip of the iceberg. While many systematists consider monophyly to be a necessary characteristic of taxa this can be problematic when relationships are reticulate. While a number of species have been confidently excluded from some genera, accommodating them without erecting small or monotypic genera, or wholesale lumping, remains an issue. A corollary of an allopolyploid origin for the Australian and FLAG clades is that reduction in chromosome number and genome size has progressed in several genera of Australian Gnaphalieae and in the FLAG clade at least in *Leontopodium*. The Australian Gnaphalieae provide ample opportunities to study genome evolution after polyploidy, since many genera retain the putative ancestral complement of chromosomes while others have undergone varying degrees of reduction. Developments in sequencing technology provide the means to advance our understanding of Gnaphalieae phylogeny in two immediate areas. Firstly, whole plastid genome sequencing will increase the detection of robust incongruence between plastid and nuclear phylogenies and force more nuanced interpretation of both. More excitingly, the capacity to unpick the phylogeny of many nuclear gene families at once should enable tests of existing hypotheses of reticulate evolution and others that will emerge. Finally we will argue for a conservative new subtribal classification consistent with all the available phylogenetic evidence as an interim step until progress is made on the outstanding issues of phylogeny within the tribe.

T2-35-02

Multiple polyploidization events across Asteraceae with two nested events in the early history revealed by nuclear phylogenomics

Chien-Hsun Huang[1], *Caifei Zhang*[1], *Main Liu*[1], *Yi Hu*[2], *Tiangang Gao*[3], *Ji Qi*[1], *Hong Ma*[1]
1. Institute of Plant Biology, Institute of Biodiversity Sciences, Center for Evolutionary Biology, School of Life Sciences, Fudan University
2. Department of Biology, the Huck Institutes of the Life Sciences, the Pennsylvania State University
3. State Key Laboratory of Evolutionary and Systematic Botany, Institute of Botany, the Chinese Academy of Sciences

Biodiversity results from multiple evolutionary mechanisms, including genetic variation and natural selection. Whole-genome duplications (WGDs), or polyploidizations, provide opportunities for large-scale genetic modifications. Many evolutionarily successful lineages, including angiosperms and vertebrates, are ancient polyploids, suggesting that WGDs are a driving force in evolution. However, this hypothesis is challenged by the observed lower speciation and higher extinction rates of recently formed polyploids than diploids. Asteraceae includes about 10% of angiosperm species, is thus undoubtedly one of the most successful lineages and paleopolyploidization was suggested early in this family using a small number of datasets. Here, we used genes from 64 new transcriptome datasets and others to reconstruct a robust Asteraceae phylogeny, covering 73 species from 18 tribes in six subfamilies. We estimated their divergence times and further identified multiple potential ancient WGDs within several tribes and shared by the Heliantheae alliance, and also observed nested duplications in the early history of this family at core Asteraceae (Asteroideae–Mutisioideae) and shared by the sister family Calyceraceae. Our analysis of the history of Asteraceae reveals two rounds of WGDs that each coincide with an extinction event in the Earth history and are followed by two kinds of great increase of biodiversity in the family; the older one proceeded the divergence of at least 10 subfamilies within 10 My, with great variation in morphology and physiology, whereas the other was followed by extremely high species richness in the clade named Heliantheae alliance. Our results provide different evidence for several WGDs in Asteraceae and reveal potential association among WGD events, dramatic changes in environment and species radiations, providing a possible scenario for polyploids to overcome the disadvantages of WGDs and to evolve into lineages with high biodiversity.

T2-35-03

Big data for a big family: Phylogenomics of the compositae (Asteraceae)

Jennifer Mandel[1], ***Linda Watson***[2], *Rebecca Dikow*[3], *Vicki Funk*[3]
1. University of Memphis
2. Oklahoma State University
3. Smithsonian Institution

Phylogenomics, made possible by advances in Next Generation Sequencing, holds great promise for elucidating complex relationships in plant lineages. Such approaches make use of large-scale taxon and gene sampling to reconstruct phylogeny to address questions of evolutionary relationships. Despite enormous effort put forward to resolve the Compositae's evolutionary relationships through traditional molecular systematics, many areas of the phylogeny remain poorly understood, particularly along the backbone of the tree. Comprehensive and well-supported phylogenetic reconstructions at the family-wide level have been difficult to obtain due to the paucity of phylogenetically-informative universal markers for a family of this magnitude. Moreover numerous gene and genome duplication events coupled with large-scale and rapid radiations have further complicated phylogeny reconstruction. To date, molecular phylogenies for the Compositae have largely been based on molecular markers derived from Sanger sequencing focused at the tribal level and below with few exceptions. In addition, family-wide level phylogenies have largely been reconstructed using a metatree approach due to the difficulties associated with alignment issues for a common set of phylogenetically-informative markers across this large and diverse family. To overcome these issues, we have developed and implemented a targeted capture approach for low copy number nuclear markers for the Compositae using Next Generation Sequencing on the Illumina platform. Through the HybSeq method, we are able to sequence ~1000 nuclear loci from the COS II marker set along with chloroplast genomes as off-target capture reads. These data enable the comparison of nuclear and chloroplast phylogenies. We have employed both gene and species tree estimation methods, as well as maximum likelihood and Bayesian analyses. Much work remains going forward, but some key findings from our work thus far are that some subfamilies of Compositae are not recovered as monophyletic, and the backbone relationships surrounding these subfamilies in question remain tenuous. We hypothesize that a major cause of this discordance and low resolution in key areas is related to gene and genome duplication due to the polyploid nature of the family. We have begun exploring the use of transcriptomes to more precisely place genome duplication events on the tree, and on resolving issues of orthology and paralogy for the nuclear markers, all of which affects phylogenetic reconstruction for many major groups of plants that possess a rich history of polyploidy.

T2-35-04

Advances in our understanding of the evolution and biogeography of the compositae

Vicki Funk
Smithsonian Institution

The Compositae (Asteraceae) are the largest and most diverse flowering plant family (ca. 24,000 spp/ 1,700 genera). Members occur worldwide in all but the most extreme habitats, reaching their greatest numbers in arid and temperate regions and on tropical and subtropical mountains. While many species have restricted ranges in areas that are threatened with high extinction rates (i.e., Pacific Islands, Cape Floristic Region, Tibetan Plateau), the family also includes some of the world's most noxious weeds (i.e., dandelion, ragweed, thistle) some of which cost the billions (in USD) annually. Numerous species produce novel secondary compounds that have many industrial and biomedical uses and others have been domesticated for food. In 2009 a comprehensive book was published that resolved some issues but left many questions unanswered, however, it triggered substantial new research based on new fossil and morpholgoical evidence including phylogenies and character evolution studies based on next generation methods such as gene target enrichment. It has been hypothesized that Compositae originated in South America and experienced a subsequent explosion in Africa afterward moving across the globe. But the placement of enigmatic taxa has cast some doubt on parts of the story. New data from traditional and next generation sequencing have provided additional insights on the radiation of the basal grade and the Cichorioideae bringing into question the monophyly of the both the Carduoideae and the Cichorioideae. In both cases one clade moves in and out of the respective subfamily and may disrupt the current hypotheses. Potential causes of this may be the type of data used in each analysis but it may also reflect a past history of hybridization. If two valid hypotheses exist then it brings up the fundamental question of which one should be used for studies in evolution and classification.

T2-35-05

Global biogeographic divergences and chromosome evolution of the paper daisy tribe Gnaphalieae

Ze-Long Nie[1], *Ying Meng*[1], *Hang Sun*[2], *Vicki Funk*[3], *Jun Wen*[3]
1. Jishou University
2. Kunming Institute of Botany, Chinese Academy of Sciences
3. National Museum of Natural History, Smithsonian Institution

The paper daisy tribe Gnaphalieae (Asteraceae) has complex biogeographic distributions characterized by disjunctions as well as radiations worldwide. Variation of chromosome numbers in the tribe is also extremely complicated with basic chromosome numbers ranging from 3-14. We herein explored the possible link between the global radiation and the complex chromosomal variation in this species-rich herbaceous tribe within a phylogenetic framework based on the hitherto largest taxon sampling, a total of 835 terminal accessions representing 80% of the genera, with nuclear internal transcribed spacer (ITS) and external transcribed spacer (ETS) sequences. Our broadly sampled study produces a robustly supported phylogenetic hypothesis for the backbone relationships of the tribe. The tribe experienced several dynamic early diversifications in southern Africa, resulting in at least three near-endemic groups (the *Relhania* clade, the *Metalasia* clade, and the *Lasiopogon* clade). Analyses using ChromEvol suggests that the ancestral base chromosome number for the tribe is most likely seven, and that the three basal lineages have experienced several chromosome gain or loss events. Four well-supported lineages are recognized in the crown group (i.e., HAP, *Gnaphalium* s.s., FLAG, and the Australasian group); and each of the four lineages has a world-wide distribution or independent large radiations. The HAP clade has a South African and diploid origin, followed by a radiation in southern Africa and several migrations towards the north of the African continent, the Mediterranean basin and Asia. These expansions and diversifications have been accompanied by a large number of genome duplications, as shown by several independent events of tetraploidy, hexaploidy and octoploidy. The FLAG clade is inferred to have a European diversification with a southern African and diploid origin, supporting a single dispersal/migration event in the early Miocene associated with both numerous duplications and dysploidizations. The Australasian group was likely founded by a single dispersal event from southern Africa that resulted in a radiation of more than 550 species, with numerous genome duplications as well as gain or loss events, as indicated by all chromosome base numbers presented in this group. Our data strongly suggested that Gnaphalieae originated in southern Africa during the Oligocene, followed by repeated migrations into the rest of Africa and the Mediterranean region, with subsequent entries into other continents during various periods starting in the Miocene. Expansions in the late Miocene to Pliocene appear to have been the driving force that shaped the global distribution of the tribe as forests were progressively broken up by the midcontinent aridification and savannas and grasslands expanded into the interior of the major continents. Both polyploidy and dysploidy are indicated as the most significant evolutionary trend in chromosome number within the tribe and may have partly or even mainly contributed to the recent colonizations and rapid radiations of the Gnaphalieae worldwide.

T2-35-06

Island-like mountain radiations in Asia: The case study of the genera *Saussurea* and *Jurinea*

Alfonso Susanna
Botanic Institute of Barcelona (CSIC-Ajuntament de Barcelona)

Evolutionary radiations represent events in which many species or lineages evolved from a common ancestor in a short period of time. Many plant radiations have been triggered by island-like ecological opportunities following mountain uplift; the mountain ranges with the steepest and widest environmental gradients, such as the Andes, are home of exceptional regional species pools (and also exceptional endemism rates), mainly derived from evolutionary radiations. The Himalayan-Qinghai-Tibetan Plateau (HQTP) and the adjacent Hengduan Mountains (HM) are considered one of the main biodiversity hotspots of the world thanks to its richness in species and endemics. Both regions show extreme altitudinal ranges compressed in short distances as a consequence of the collision of the Indian and Eurasian plates, and thus constitute ideal evolutionary scenarios to study diversification processes in mountain regions. We have identified two possible cases of alpine radiations in the *Saussurea-Jurinea* complex (Compositae-Cardueae), involving some 550 species in total. *Saussurea* shows an amazing number of species (more than 300) in the HQTP and Hengduan mountains, although a considerable number of species are also found on the west side of the mountains of Middle Asia (Tian Shan and Pamir-Alay). *Jurinea*, in contrast, has the highest number of species (150 sp.) in the Tian Shan and Pamir-Alay. Our general objectives are the following: a) To carry out an extensive sampling of *Saussurea* and *Jurinea*, especially centred in the two main radiation areas in the HQTP and Tian Shan mountains; b) To generate well-resolved phylogenies of both genera using a multi-loci approach through next-generation-sequencing (NGS) analyzed by Bayesian inference and parsimony, and explore coalescent-based species tree estimation with our NGS data set; c) On the basis of the new phylogenies, conduct phylogenetic comparative analyses and multi-model biogeographical inference to address the following questions: Do the alpine species of both genera in Asian mountains constitute clades with clearly higher rates of diversification than their lowland relatives? How many independent radiations took place in the complex? If several, did they occur at the same period, and are they comparable in terms of speciation rates? Which factors (intrinsic or extrinsic) shape species radiations, i.e., do the radiations follow a geographical model, an adaptive model, or a mixed model? Our results will be a major contribution to the study of alpine radiations especially in the HQTP, and will open a methodological pathway for the analysis of very large radiations in other genera.

T2-35-07

Phylogenetics of *Baccharis* (Astereae), a mega-diverse genus of compositae

Gustavo Heiden[1], *José Rubens Pirani*[2], *Alexandre Antonelli*[3]
1. Embrapa Clima Temperado
2. Universidade de São Paulo
3. Gothenburg Global Biodiversity Centre, University of Gothenburg

Baccharis L. is the fourth largest genus in the Compositae and the most speciose within tribe Astereae. With species number estimates ranging from 350 up to 500, *Baccharis* is among the largest plant genera in the Neotropics, and has long been considered a taxonomic nightmare. A comprehensive phylogenetic hypothesis for this American genus is necessary to test its monophyly, clarify the infrageneric relationships and shed light on its evolutionary and biogeographical history. Given its great morphological diversity and wide geographic distribution, phylogenetic studies based on molecular data to test the monophyly of the genus and investigate the relationships of its infrageneric groups are required, allowing the recognition of clades with evolutionary and biogeographical significance. Therefore, we provide the hitherto most comprehensive phylogenetic hypothesis for *Baccharis*. We performed Bayesian phylogenetic analyses based on four molecular regions from the nuclear and plastid genomes (ETS, ITS, *trn*H-*psb*A, *trn*L-F; in total 5041 aligned nucleotides) and including 248 species of *Baccharis*, from all previously segregated genera and all subgenera previously recognized, and 44 outgroups belonging to tribe Astereae, totalizing 1080 new sequences. Our results confirm subtribe Baccharidinae to be a monogeneric group, and support a wide circumscription of *Baccharis* as monophyletic, including the genera previously segregated from it (*Baccharidastrum*, *Baccharidiopsis*, *Heterothalamulopsis*, *Heterothalamus* and *Lanugothamnus*). Seven main lineages are recognized and treated as subgenera, all of which re-circumscribed as monophyletic lineages. *Baccharis* subgen. *Baccharis*, *B.* subgen. *Molina* and *B.* subgen. *Tarchonanthoides* roughly match previous subgeneric concepts, while *B.* subgen. *Pteronioides* and *B.* subgen. *Stephananthus* are synonyms. Two previously segregated genera (*Heterothalamulopsis* and *Heterothalamus*) and two formerly existing sections (*Coridifoliae* and *Oblongifoliae*) are placed at subgeneric rank with new circumscriptions. Four new sections are recognized, and three new statuses at the rank of section are proposed to accommodate taxa not corresponding to any previously described section, while 22 sections are reduced to synonymy. In total, we propose that *Baccharis* should comprise 433 species classified into 7 subgenera and 47 sections. This study represents a major advance in elucidating the relationships among species in such a large predominantly Neotropical plant genus, and thus constitutes the basis for further exploration of the evolutionary processes that have allowed the group to attain its wide geographic distribution, morphological variation, varied sexual reproductive strategies, and complex chemical contents.

T2-35-08

Analysis of genetic diversity and marker-trait associations in the medicinal non-model plant Chamomile (*Matricaria recutita l.*) using genome-wide snp-markers

Lars-Gernot Otto[1], *Prodyut Mondal*[2], *Jonathan Brassac*[1], *Sang He*[1], *Susanne Preiss*[2], *Jörg Degenhardt*[2], *Jochen Christoph Reif*[1], *Timothy Francis Sharbel*[1,3]

1. Leibniz Institute of Plant Genetics and Crop Plant Research (IPK) Gatersleben

2. Martin-Luther University Halle-Wittenberg

3. Global Institute for Food Security (GIFS) Saskatoon

Matricaria recutita L. (syn. *Chamomilla recutita* (L.) Rauschert, commonly known as German chamomile, is a traditional medicinal plant already mentioned by Hippocrates, whose works can be considered as the basis of modern medicine, in the 5th century BC. In the last decades, rather limited research has been done on chamomile, often focusing on biochemical pathways analyses for medicinally-active compounds like alpha-bisabolol. NGS approaches are frequently used to exploit genetic resources and to improve breeding in major crops. Here, we applied genotyping-by-sequencing (GBS) to unravel the genetic structure of cultivated medicinal chamomile and create a first genomic resource for future applications to improve and facilitate breeding. A high number of SNP-markers were mined to characterize the genetic diversity in the panel of 91 plants from 33 predominantly cultivated varieties, populations and accessions, including diploids and artificially generated tetraploids, and from different geographic regions. 18 compounds (terpenoids) were measured by GC-MS from flower heads. Analyses of flowering time demonstrated that diploids generally flowered earlier than tetraploids. As revealed by STRUCTURE, PCoA and neighbor-joining analysis using 6495 high-quality SNPs, the outgroup *M. discoidea* was clearly distinct from *M. recutita*. The STRUCTURE analysis further showed that tetraploids tend to display lower genetic diversity than diploids, potentially reflecting their origin by artificial polyploidisation from only a part of the chamomile genetic background. No connection between geographic origin and genetic structure was found, but varieties with genotypes possessing high alpha-bisabolol content (e.g. 'Manzana') could be genetically discriminated. GWAS enabled us to resolve a number of SNPs significantly associated with flowering time and high alpha-bisabolol content. Four sequences characterized by these alpha-bisabolol-associated SNPs are involved in plant biotic and abiotic stress response in different plants species, whereas six of these sequences are described to play a role in universal metabolic pathways. One sequence associated with flowering time is described to be involved with self-pollination in *Arabidopsis thaliana*. Further research could lead to the development of reliable and cost-effective markers applicable for marker assisted selection in chamomile breeding.

T2-35-09

High altitude conservation biogeography: richness, turnover, endemism and regional hotspots of Andean Compositae

Andres Moreira-Munoz[1], *Mauricio Diazgranados*[2]

1. Instituto de Geografia, Pontificia Universidad Catolica de Valparaiso

2. Natural Capital and Plant Health department, Millennium Seed Bank, Royal Botanic Gardens Kew

The Andes is the largest, widest and highest mountain range in the world. It covers all possible climatic ranges, from the driest to the humid, and from the warmest to the coldest. From the biological point of view, it includes the Tropical Andes Biodiversity hotspot, considered the world's most biodiverse hotspot. It has been estimated that along the Andes there are about 50,000 plant species (~ 50% endemic). However, the patterns that shape Andean biodiversity show differences between plant groups and are still not well understood. Being the most diverse South American angiosperm family, and showing high levels of endemism, the Compositae is an excellent model for the study of Andean biogeography. There are some 3,000-3,500 Andean species of the family, with both ancestral and recent origins. This work aims to understand the patterns of richness, species turnover and endemism of the Compositae family throughout the Andes, to help understand the mechanisms that have facilitated its radiation in these high-altitude environments, and to identify regional areas of importance for conservation (regional Compositae hotspots). As a first step in this long-term study program, all georeferenced Asteraceae records are being compiled and analyzed in the different databases of biological collections, and analyses are being carried out on the basis of richness estimators, diversity indexes, Beta-diversity models and sampling efforts. The results show three to four richness centres mainly in the northern and Central Andes of Colombia, Ecuador and Bolivia. The southern Andes show comparatively reduced collection efforts, but levels of endemism appear remarkable. These results disaggregated at the tribe level, allow to test hypothesis regarding the radiation of the different lineages within the family, identifying areas with sampling gaps, and regional priority areas for conservation efforts, eventually related to the ancestral uses of the species.

T2-35-10

Systematics of Chinese compositae

***Tiangang Gao**, Zhixi Fu, Guojin Zhang, Bao Nie, Bohan Jiao*
Institute of Botany, Chinese Academy of Sciences

The sunflower family (Compositae) is the largest and the most diverse flowering plant family, comprising
24,000–30,000 species and 1,600–1,700 genera. In China, Compositae are also the largest family, with approximately 2,336 indigenous species in 248 genera. In the past two decades, molecular phylogenetic analyses has contributed greatly to our understanding of the systematics of Compositae. Nevertheless, the large-scale analyses and knowledge about the relationships of Chinese Compositae at the generic level as a whole are far from complete due to difficulties in sampling. In this study, we presented a three-marker (*rbcL*, *ndhF*, and *matK*) phylogeny of Compositae, including 506 genera (i.e., approximately one-third of Compositae genera). The study sampled 200 Chinese genera (i.e., approximately 80% of Chinese Compositae genera). The backbones of the new phylogeny were largely congruent with earlier studies, with 13 subfamilies and 45 tribes recognized. Chinese Compositae were distributed in 7 subfamilies (Mutisioideae, Wunderlichioideae, Carduoideae, Pertyoideae, Gymnarrhenoideae, Cichorioideae, and Asteroideae) and 22 tribes (Mutiseae, Hyalideae, Cardueae, Pertyeae, Gymnarrheneae, Vernonieae, Cichorieae, Doroniceae, Senecioneae, Astereae, Anthemideae, Gnaphalieae, Calenduleae, Inuleae, Athroismeae, Helenieae, Coreopsideae, Neurolaeneae, Tageteae, Millieae, Eupatorieae, and Heliantheae). Chinese Compositae lacked 6 basal subfamilies and 23 tribes. Several previously ambiguous relationships were clarified. Our analyses also resolved some unplaced genera within Chinese Compositae. Finally, our phylogenetic tree was used to revise the classification for all genera of Chinese Compositae. In total, 255 genera, 22 tribes, and 7 subfamilies in China are recognized.

T2-35-11

An updated phylogeny of *Moquiniastrum* (Gochnatieae, Compositae) revealed by a novel molecular marker set using low cost phylogenomic methods

***Morgan Gostel,** Vicki Funk*
Department of Botany, Smithsonian Institution

The genus *Moquiniastrum* (Compostiae) includes 21 species and belongs to the Gochnatieae, a tribe located near the base of the Compositae phylogeny. The genus is unique in Gochnatieae based on the combination of dioecious flowers, 2–5 armed pubescence, and paniculiform capitulescences. Moreover, species of *Moquiniastrum* have a widespread distribution throughout South America in varied and sometimes disjunct habitats. We have designed and validated a 576-locus marker dataset for microfluidic PCR-based target enrichment in *Moquiniastrum*. Using this dataset we anticipate improved phylogenetic resolution not only in this genus, but more broadly within the tribe Gochnatieae. Ongoing research is focused on improving species-level resolution in each of the seven clades in Gochnatieae as well as resolving infrageneric relationships in the tribe, which have proven historically intractable. Our results will help to understand the biogeography and evolution of species of *Moquiniastrum* as well as traits that may be associated with transition from mesic and dry habitats throughout South America. The tribe Gochnatieae is the last branch of the South American base of the Compositae and followed by significant radiations into North America and Africa; as such, the findings in this study are critical to understanding broad scale biogeography and evolution in the family. Samples from more distantly related taxa in the Gochnatieae and Compositae suggest this phylogenomic marker set can be effective for phylogeny reconstruction at various scales of evolutionary distance in the family.

T2-35-12

Is the vernonieae (compositae) truly the "Evil Tribe?" Clues to the answer from phylogeny, biogeography and niche conservatism

***Sterling C Keeley** [1], Jason Cantley[2], Timothy Gallaher[3]*
1. University of Hawaii, Manoa
2. Bucknell University
3. University of Washington

The Vernonieae (aka the Evil Tribe) includes 1500+ species distributed throughout the subtropics in both the New and Old World. It is unusual in that it has a nearly unbroken distribution in both hemispheres, running from Argentina through to northernmost Mexico with a similarly continuous distribution from South Africa through Ethiopia. Species are also found in eastern North America and the Caribbean in the NW, and in the OW in the Arabian Peninsula, SE Asia, Australia, and the Hawaiian Islands. The relationships among taxa within and between hemispheres and the biogeographical history of the tribe have been difficult to unravel, however. Twenty-four subtribes have been described based on DNA sequence data and morphology providing a framework to test putative relationships and clarify the tribe's biogeography. Chloroplast and nuclear DNA sequences were obtained from 400+ taxa from across the tribe's geographical range and including all subtribes. A dated phylogeny was produced using a BI tree in BEAST. This confirmed the origin of the Vernonieae in SE Africa/ Madagascar, however, its origin was dated at ~45my, twice as old as previous estimates. Additionally, within 10 my of its origin the Vernonieae were established in all major geographic regions in which they are found today. Long-distance dispersal occurred directly from Africa to South America, Australia, Hawaii, and SE Asia. As previously suggested, the NW and OW lineages evolved independently once separated. There were, however, several cases of back dispersal from the NW to the OW (but not in the reverse direction). The nearly continuous distribution of the tribe in both hemispheres is associated with the nearly continuous creation of volcanic mountain chains from south to north in both Africa and South America (40-5 my). Coincident with the new habitat created by volcanic orogeny the world's climate gradually became drier and more seasonal, shrinking the area occupied by tropical wet forests and opening up large expanses of sunny shrub/grassland habitats. The latter favor the family Compositae; most taxa prefer open sunny seasonal habitats. While retaining these family preferences, the Vernonieae are distinct in specializing on metal rich (especially iron containing volcanic) soils, and many taxa are equally well suited to limestone, bauxite, serpentine, and dolomite exposed by weathering and upthrust. The Vernonieae also grow from sea level to over 4,200 m. Additionally, the high mineral content of soils underlying savannahs and open areas of Africa and South America greatly expanded the possibilities for colonization in non-montane areas. And one other piece of family phylogenetic conservatism favoring the Vernonieae's ability to reach and colonize new areas is the production of vast numbers of wind-dispersed seeds, year after year. Rather than being called the "Evil Tribe" perhaps the Vernonieae should be renamed "The Lucky Devil tribe." Its extensive distribution and species radiations are due in large part to a fortuitous combination of phylogenetic/niche conservatism at the family level, edaphic specialization within the tribe, and the continuous formation of new areas open to colonization due to geo/tectonic events and climate change over the past 40 my.

T2-36: Systematics Agenda 2050: call for strategies for the future of systematic biology

T2-36-01

Systematics in latin America: current status and future challenges

Mauricio Diazgranados

Natural Capital and Plant Health Department, Royal Botanic Gardens, Kew

Latin America is a region united by culture, history, geography and diversity. Mainly tropical and subtropical, it contains the most diverse terrestrial ecosystems, the wettest and the driest areas on Earth, seven of the 25 Global Biodiversity Hotspots, six megadiverse countries, twelve major biomes, three of the world´s largest wetlands, and some of the largest and most pristine forests. Its immense plant diversity contains 90,000-110,000 plant species (30-35% of global plant diversity), in a land area of 9.6% of total land area worldwide. However, Latin American biodiversity is being lost or seriously threatened by human activities at all levels, being scenario for vast environmental problems, including the highest rates of deforestation (overall 4.6 million ha/year). The region includes seven of the 20 countries whose plant species are the most threatened. Although 21% of the land area is protected, the present configuration of protected areas is not always optimal, with strong imbalances when comparing ecosystem types. Numerous areas are still unexplored, there is no good knowledge of species distributions, or no checklists at all. Georeferenced databases are not always made available, and overall there is still poor knowledge on the majority of the species. In addition, there is a striking inequality in the knowledge of plant systematics within Latin American countries, and compared to developed countries, with no positive correlation between plant diversity and national researchers working on it. In some countries the community of taxonomists is down to a minimum, with no taxonomists at all for several groups of organisms. A bibliometric analysis of conferences at regional and worldwide level shows a relatively low collaboration index in Latin America, and a notable difference in the main research topics. Some of the 50 most common words in abstracts at the IBC or BSA meetings, e.g. phylogeny, phylogenetics, phylogenomics, genome and gene, are replaced at the CLB by words such as species, diversity, flora, and distribution. In the light of increased threats for plant species, the systematic agenda 2050 for Latin America must consider to: 1) carry out explorations and inventories guided by gap analyses, to allow a better understanding of local/regional plant diversity, and the implementation of important areas of conservation; 2) develop georeferenced databases, available to all the scientific community; 3) complete and monograph Latin American flora; 4) investigate and compile knowledge associated to plants; 5) promote research on micro- and macroevolution evolution and biogeography of Latin American plants; and 6) support capacity building, to incorporate new technologies, improve local research facilities and train Latin American taxonomists and systematists.

T2-36-02

Collections in 21st century science: More important than ever

Vicki Funk

Smithsonian Institution

The 19th Century ushered in a new age as naturalists undertook large-scale collecting expeditions leading to field observations and preserved specimens in the short term, and to major scientific advances in the long term. Notable among these were the founding of Physical Geography, Meteorology, Ecology (Humboldt), Biogeography (Hooker), and the theory of Evolution (Darwin, Wallace). In the 20th Century collections were central to paradigm shifts, including theories of Continental Drift (Eigenmann) and Phylogenetic Systematics (Hennig, Brundin). Past expeditions provided tissues for all the cladograms as the era of Phylogenetics took over biological thought. Will this tide of collections-based scientific advancement continue? In the first 15 years of the 21st Century we have seen tree-thinking pervade the life sciences, leading to the emergence of Evolutionary Ecology, Evolutionary Medicine, and new Food Safety methods, and collections data increasingly are used for climate change studies. Collections are a gold mine of information and are now leading the way to advances in three main areas: collections contain vast quantities of genomic data accessible through Next-Generation Sequencing techniques and phylogenomics, allowing us to address big evolutionary questions such as the frequency of genome duplication and its role in species diversification; Open access to specimen data, allowing us to model changes in diversity through time; and Estimating extinction risk and conservation priorities, by linking collections and climate data with phylogenies. Our 'grand challenge' is to determine where we want to be with collections-based research in 2050 and plan a strategy to get there.

T2-36-03

The coming age of systematic biology: Global collaboration from big data to big impact

Xiao-Quan Wang

State Key Laboratory of Systematic and Evolutionary Botany, Institute of Botany, Chinese Academy of Sciences

Systematic biology is a discipline that aims to discover and classify all species on the earth, to reconstruct the species phylogeny and answer how species evolved, and to educate the public in biodiversity knowledge and biological conservation. Recent years have witnessed great progress in systematic biology, from big data collection to our understanding of phylogeny and genome evolution of living organisms. To further develop systematic biology and promote its impact on social services, I propose to strengthen global collaborations in two areas. One is to build shared platforms (DNA banks and photo banks) that aim to collect DNA and photos of all plant species. The shared DNA banks can greatly accelerate the reconstruction of tree of life and are very helpful to develop DNA barcodes for monitoring invasive species, while the shared photo banks can provide massive information for developing programs that identify species automatically using mobiles. The other is to sequence transcriptomes of all families and genera of land plants, which will provide unprecedented opportunities to interpret how plants adapt to different environments and which genes are important for plant adaptation.

T2-36-04

Strong, connected, demanded and alive: Four crucial characteristics for biological systematics in 2050

Gonzalo Nieto Feliner
Real Jardín Botánico, CSIC. Madrid

Designing strategies for developing biological systematics (BS) in the next three decades requires integrating retrospect and prospect perspectives. Two elements will be crucial for BS and how to respond to people's demands towards this field: the concerns about the global-change-driven biodiversity threat and the changes in the human life, determined on the one hand by our aims, interests and priorities, and on the other, by the technologies incorporated into the common daily life. Imagining the interaction between these two aspects will be the theme of my talk. In the best possible scenario, the demands of society for BS in three decades time will not only be directly related to the so-called services, which plants have provided since the dawn of civilization, but also to a more basic and intellectual request for knowledge. In a world in which technology will facilitate better analyzing and interpreting complex systems, it is conceivable that citizens will aim at understanding biological systems, which unlike other natural sciences subjects, involve multiple organization levels, are spatio-temporally constrained and non-ideal for reductionist interpretations. In all, I envisage—or would like to imagine—a 2050 scenario in which BS is strong, efficiently integrating all relevant data from different sources, using the ultimate technologies to better achieve its goals; connected through truly distributed taxonomic information systems allowing full accessibility of information including collections; demanded by society, so that knowledge will be provided not only by the research community but also by the society in collaboration with the scientific community through the expansion of citizen science initiatives; and certainly alive, so that BS will continue to have the funding, the intellectual appeal and the human resources necessary for projecting its live well through the second half of the 21s Century.

T2-36-05

Global-scale spatial patterns of diversity in the Compositae family

Alexander Schmidt-Lebuhn[1], *Christiana Mcdonald-Spicer*[2], *Nunzio Knerr*[1], *Marcel Cardillo*[2]
1. CSIRO
2. Australian National University

With an estimated 20-30,000 species, the Compositae account for one in ten flowering plants. They occur on all continents except Antarctica and in a wide range of habitats. Semi-arid to arid environments and mountain ranges are generally considered to be their centres of diversity, but no formal, quantitative spatial study has been conducted to date. We examined patterns of spatial diversity of the Compositae at the global scale, combining occurrence data from the expertly curated Global Compositae Checklist (GCC) and a genus level phylogeny derived from a supermatrix of Genbank and BOLD sequences. Spatial diversity metrics were weighted by land area of the regions used by the GCC. Species richness was highest in the South African Cape Province, Peru, Turkey and Iran. Oceanic Islands, the Cape Province, Madagascar, Western Australia, and the New Zealand South Island showed the highest Corrected Weighted Endemism. Phylogenetic diversity was highest in the Andes and Mexico. We also conducted a regression analysis of diversity scores against environmental variables to examine the correlates of high diversity in the family. Finally, we used the phylogeny for the estimation of ancestral areas and biogeographic events under different models, to compare the results against an earlier study by Funk *et al.,* that was based on a supertree and ancestral character reconstruction.

T2-37: Evolution of floral traits

T2-37-01

Spatio-temporal stability of an island endemic plant-pollinator interaction involving floral colour change

Jeff Ollerton
University of Northampton

Endemic oceanic island plant-pollinator relationships have generated particular interest because there is a theoretical mismatch in the expectations of the outcomes of dispersal and establishment of biotically pollinated plants to islands. The ancestors of endemic plants should, in theory, be more likely to establish on an island if they are generalists capable being pollinated by a wide range of taxa. At the same time, island pollinator faunas tend to be of rather low diversity and abundance, and plant reproduction is often pollinator limited, making pollination niche segregation more likely via specialisation on a single pollinator or functional group of pollinators. The relative importance of specialist versus generalist strategies for biotic pollination are therefore unclear for island plants. At the moment, however, too few detailed studies of oceanic island plant-pollinator interactions have been conducted to properly assess the generality of any patterns and further case studies are required, particularly over a longer time scales and across multiple populations to assess spatiotemporal variation in the interactions between plants and their pollinators. Since 2003 I have been studying the pollination ecology of a Canary Island endemic plant species, the Canary Wallflower *Erysimum scoparium* (Brassicaceae), on Tenerife. *E. scoparium* employs a flower colour change strategy to manipulate its principal pollinators, which comprise a guild of medium sized bees, dominated by an endemic solitary bee *Anthophora alluaudi* (Apidae). The interaction between *E. scoparium* and *A. alluaudi* is stable over that time period and in multiple populations over an altitudinal range of almost 1,000 m (range = 1,300 to 2,250 m.a.s.l). The period of study includes one of the driest winters recorded on Tenerife and I discuss the effect of changes in community composition on the ecology of this endemic plant-pollinator interaction, and the reproductive success of the plant.

T2-37-02

Flower beak in bumblebee-pollinated *Pedicularis*: Combined functions in controlling pollen removal

Ze-Yu Tong, *Shuang-Quan Huang*
Institute of Evolution and Ecology, School of Life Sciences, Central China Normal University

Pollen release could be governed by floral structures that regulate pollen removal in the form of packaging or dispensing pollen grains, limiting pollen removal in a single visit of pollinators. As the pollen presentation theory (PPT) assumed, staggered pollen presentation corresponds to frequent visitation which reduces available pollen for each visit, increasing total pollen dispersal, while simultaneous pollen presentation could be more advantageous in a scarcity of pollinators. Flower shape is greatly diverse in a species-rich genus *Pedicularis* which species are exclusively pollinated by bumblebees. The upper lip forms a narrow-tunnel shaped galea enclosing the anthers in many species, and is extended into a long, slender beak. It has been hypothesized that the featured various beaks may govern pollen removal, however an empirical study remains unknown. We conducted series of experiments to test whether *Pedicularis* species match PPT predictions, and whether the beak structure could affect pollen removal by pollinators. We investigated anther dehiscence pattern in 10 *Pedicularis* species in a field station in Shangri-La, southwest China. Pollinator visitation, pollen removal and deposition by the pollinator in a single visit to 4 beaked species and 1 beakless species, as well as the time spent per visit were quantified. The differences of pollen removal/deposition between species with different beak lengths were also compared. We observed that five of the 10 *Pedicularis* species dehisced two of the four anthers in the first day and dehisced the remaining two anthers in the next day. Gradually dehiscing species could be either beaked or beakless, and received more than 2.5 times of pollinator visits than those simultaneously dehiscing species on average. Bumblebee pollinators removed 35-61% available pollen and deposited 17-102 pollen grains in a single visit, and the bees removed less proportion of available pollen in *Pedicularis* species with longer beaks. Although the beaked nectarless species provided only pollen as reward, their pollen production or P/O were even much lower than beakless nectariferous species, in conflict to the prediction of compensative sex allocation, but in accordance with a competition driven specialization model which based on the cost to male fitness. Our results confirmed the functions of *Pedicularis* beak structures as constraining pollen release to pollinators. Our results also suggested that, due to different pollination environments, the control of pollen removal by *Pedicularis* flowers could be fulfilled by non-conflict multiple mechanisms, and further supported the assumption that higher visitation of pollinators could drive plants allocating more to male function.

T2-37-03

Non-equilibrium dynamics and floral trait interactions shape extant angiosperm diversity

Pamela Diggle[1], *Brian C Omeara*[2], *Stacey D. Smith*[3], *Scott Armbruster*[4], *Lawrence D Harder*[5], *Christopher Hardy*[6], *Lena C Hileman*[7], *Larry Hufford*[8], *Stephen Smith*[9], *Susana Magallón*[10], *Peter Stevens*[11], *Charles Fenster*[12]
1. University of Connecticut
2. University of Tennesee
3. University of Colorado
4. University of Portsmouth
5. University of Calgary
6. Millersville University
7. University of Kansas
8. Washington State University
9. University of Michigan
10. Universidad Nacional Autonoma de Mexico
11. University of Missouri
12. South Dakota State University

Why are some traits and trait combinations exceptionally common across the tree of life, whereas others are vanishingly rare? The distribution of trait diversity across a clade at any time depends on the ancestral state of the clade, the rate at which new phenotypes evolve, the differences in speciation and extinction rates across lineages, and whether an equilibrium has been reached. Here we examine the role of transition rates, differential diversification (speciation minus extinction), and non-equilibrium dynamics on the evolutionary history of angiosperms, a clade well known for the abundance of some trait combinations and the rarity of others. Our analysis reveals that three character states (corolla present, bilateral symmetry, reduced stamen number) act synergistically as a key innovation, doubling diversification rates for lineages in which this combination occurs. However, this combination is currently less common than predicted at equilibrium because the individual characters evolve infrequently. Simulations suggest that angiosperms will remain far from the equilibrium frequencies of character states well into the future. Such non-equilibrium dynamics may be common when major innovations evolve rarely, allowing lineages with ancestral forms to persist, and even outnumber those with diversification enhancing states, for tens of millions of years.

T2-37-04

Testing the Grant-Stebbins Hypothesis: Experimental sympatry between pollination ecotypes

Kathleen Kay
University of California, Santa Cruz

Under the classic Grant-Stebbins model, variation in pollinator assemblages across a geographic range leads to divergent specialization on the most effective local pollinator, resulting in reproductive isolation among populations. Alternatively, adaptation to a one pollinator may come at little cost to the effectiveness of another and therefore be unlikely to reduce gene flow among populations. I test this model in two species in the genus *Clarkia* with abutting ranges in California that are pollinated by a variety of diurnal pollinators, including bees, flies and hummingbirds versus by nocturnal hawkmoths. I ask to what extent geographical divergence in floral adaptation contributes to reproductive isolation upon experimental secondary sympatry. Using multiple populations of each species, I quantify ethological and mechanical floral isolation by exposing flower-naive hawkmoths to both homo- and heterospecific floral arrays. Although hawkmoths are more motivated to visit the hawkmoth-adapted species in single species arrays, they show no visitation preference in mixed arrays, and thus contribute no ethological isolation. For mechanical floral isolation, the likelihood of heterospecific pollen transfer depends on the ecotypic origin of the diurnally-pollinated species used in the trial. Populations with small flowers and narrow floral tubes are strongly isolated from the hawkmoth-adapted species, whereas larger-flowered populations experience no isolation. Counter to the Grant-Stebbins model, adaptation to a novel pollinator in one part of the geographic range has not automatically resulted in reproductive isolation; rather, the outcome of sympatric contact depends on the specific pairings of ecotypes. Whereas most floral isolation studies involve young plant species that are currently sympatric, my work shows weak ethological and floral mechanical isolation prior to widespread secondary contact, but also that genetic variation for floral isolation is abundant and could itself respond to selection.

T2-37-05

Functional differentiation and trait convergence in accordance with pollinator shifts in Merianieae (Melastomataceae)

Agnes S. Dellinger[1], *Silvia Artuso*[1], *Diana Fernández Fernández*[5], *Darin S. Penneys*[2], *Fabián Michelangeli*[3], *Marcela Alvear*[4], *Frank Almeda*[4], *Jürg Schönenberger*[1]

1. University of Vienna
2. University of North Carolina Wilmington
3. New York Botanical Garden
4. California Academy of Sciences
5. Instituto Nacional de Biodiversidad

Pollinator shifts, i.e. shifts from one functional group of pollinators to another (e.g., bee to bird), are known to affect floral traits relevant for pollinator attraction, pollen deposition on the pollinator, and pollen pick-up by the stigma. The heavily debated concept of pollination syndromes summarizes these trait combinations under the assumption of convergent evolution of floral traits in response to similar selective pressures exerted by a given functional group of pollinators. Our study focuses on the Neotropical tribe Merianieae (Melastomataceae) as a model system for the study of pollinator mediated selection on floral traits. The Merianieae are characterized by repeated independent shifts from bee to hummingbird/bat, and passerine pollination, respectively. We combine qualitative trait mapping of 40 floral traits relevant for pollination with a modern micro CT 3D-geometric morphometric assessment of flowers across 60 Merianieae species. Our goal is (i) to pinpoint functional trait differentiation in pollen transfer mechanisms between the different pollination systems and (ii) to challenge the concept of pollination syndromes and evaluate its applicability in Merianieae. Our analyses indicate a clear functional differentiation in pollen expulsion and deposition mechanisms including elaborate vibratile buzz-pollination in bee pollinated systems, saltshaker-like pollen release upon stamen contact in hummingbird/bat pollinated species, and pollen release by a specialized "bellows"-mechanism in passerine pollinated species. Particularly strongly affected floral traits associated with the different functional groups of pollinators include reward type, stamen appendage shape, thecal wall structure and attachment as well as size of the staminal pore. Geometric morphometric analyses of 3D landmark data further support this pattern and reveal convergence in floral shape and function in relation to the functional group of pollinators. Thus, our preliminary results provide tentative support for the concept of pollination syndromes in the Merianieae.

T2-37-06

The macroevolution of floral hydraulic traits and strategies for maintaining turgor

Adam Roddy[1], *Matt Guilliams*[2], *Todd Dawson*[3], *Craig Brodersen*[1]

1. Yale University
2. Santa Barbara Botanic Garden
3. University of California, Berkeley

During the Cretaceous, early angiosperm leaf evolution was dominated by decreases in cell size that facilitated increases in vein and stomatal densities. Together these traits enabled leaves to more efficiently transport water and thereby increase photosynthetic assimilation rates. At the same time, flowers were rapidly diversifying morphologically in association with their animal pollinators. Due to their novel developmental pathways and unique functions, floral physiological traits are predicted to have evolved independently from those in leaves. Because flowers are heterotrophic, selection may have favored reduced water transport capacity in flowers and reliance on other physiological strategies to maintain turgor. Yet, how flowers remain turgid despite often being placed in the hottest, driest parts of the plant canopy is essentially unknown. Furthermore, with increasing aridity threatening plants globally, determining the physiological function of flowers is critically important to our understanding and predicting biotic responses to climate change. Here we present data from a series of studies employing a water balance framework to characterize the macroevolutionary patterns in floral hydraulic traits and physiological function. Variation in hydraulic traits among extant species suggests that monocot and eudicot flowers have few, if any, stomata, less leaky epidermises, lower vein densities, and lower hydraulic conductances compared to basal angiosperm and magnoliid flowers. The lack of stomata on the corollas of many monocot and eudicot species may minimize transpiration rates under cool, humid conditions but comes with the tradeoff of not being able to regulate transpiration rates under hot, dry conditions. As a result, selection in dry habitats favors species with low epidermal conductances. Reducing stomatal density is predicted to have relaxed constraints on the floral vascular system of supplying high fluxes of water. Indeed, the hydraulic structure-function relationships differ among major angiosperm clades, with basal angiosperm flowers having relationships more typical of leaves and monocot and eudicot flowers having lower vein densities than would be needed to optimally meet transpirational demands. While the upper limit of leaf vein density increased dramatically during the Cretaceous, the upper limit of floral vein density did not increase, but its lower limit declined. These results suggest that the hydraulic strategies used by flowers to maintain turgor underwent substantial transitions during early floral evolution. While magnoliid flowers rely predominantly on high investments in water (high hydraulic conductances and hydraulic capacitances), eudicot flowers rely more on osmotic gradients to prevent turgor loss and compensate for underbuilt vascular systems, leading to lower carbon costs in monocot and eudicot flowers. Untethered by high physiological costs, other morphological traits of flowers may more rapidly and more directly track pollinator preference.

T2-38: Plant DNA barcoding in evolution, ecology, and conservation (two sessions)

T2-38-01

Analyses of Andean and Amazonian Tree Communities in Ecuador using DNA barcoding

Samantha Worthy[1], *Rosa Jiménez*[2], *Alvaro Pérez*[2], *Alex Reynolds*[3], *Jennifer Cruse-Sanders*[4], *John Barone*[1], *Renato Valencia*[2], ***Kevin Burgess***[1]

1. Columbus State University

2. Pontificia Universidad Católica del Ecuador

3. The Lovett School

4. The Atlanta Botanical Garden

The Amazonian and Andean rain forests of Ecuador are richly populated with tropical tree species that have rarely been studied in a community phylogenetic context. Studies that incorporate DNA barcoding can be used to test a broad set of environmental and evolutionary factors that may be affecting tree community structure. For example, to test abiotic and biotic factors that potentialy determine forest community structure along elevational gradients in montane forest communities, traditional diversity metrics (Shannon's and Simpson's indices) can be coupled with DNA barcode phylogenies. In lowland habitats, traits such as phytochemical content can be combined with DNA barcodes to better understand the distribution of medicinal plants. The goal of this presentation is to highlight two applications of DNA barcoding that (1) evaluate community phylogenetic structure and correlate phylogenetic analyses with diversity metrics among Andean tree species along an elevational gradient at the Siempre Verde Reserve, Ecuador and (2) construct a tropical tree community phylogeny using DNA barcodes to test for phylogenetic signal in the occurrence of phytochemicals among tree species within Yasuní National Park, Ecuador. In the montane forest at Siempre Verde, 595 individuals were tagged, collected and identified, comprising 36 families, 53 genera, and 88 species. Of these individuals, 152 were sequenced for the *rbc*L and *mat*K gene regions. Preliminary results indicate that species richness decreases with elevation but the number of stems of common species increases. Furthermore, significant clumping was found at the highest two plots within the transect for all three metrics (PD, MPD, and MNTD) tested. This result correlates well with findings based on Shannon's and Simpson's diversity indices; there are more closely related species at higher elevations. Habitat filtering may be the cause of diversity peaks at mid-elevations where clouds inundate the forest causing extreme differences in habitat above and below this elevation. Given that only four species span the transect, abiotic stressors could be limiting species distribution and central drivers of community structure. In the upland Amazonian forest of Yasuní, 337 common tree species were barcoded and trait distribution of phytochemical presence was determined for 181 genera distributed across 56 families. Of these individuals, 247 species were successfully sequenced for the *rbc*L and/or *mat*K gene regions with 67 of these classified as medicinal. Preliminary findings show that the mean phylogenetic distance (MPD), mean nearest taxa distance (MNTD), and Fritz and Purvis' D statistic support a random distribution of the medicinal trait within the phylogeny. Based on these findings it would appear that evolutionary processes controlling phytochemical distribution among tree species are weak. Overall our preliminary findings indicate that our use of DNA barcoding in a community phylogenetic context shows that plant communities are constructed in non-random ways along elevational gradients in Andean tree communities and traits such as phytochemicals are randomly distributed in lowland Amazonian tree communities. Future studies that include taxonomic, functional, and phylogenetic diversity along with more complete floristic sampling will improve statistical power and the ability to detect fine scale community structure patterns in both tree communities.

T2-38-02

Plant DNA barcoding: Advances in forest community studies in China

Nancai Pei[1], *Bufeng Chen*[1], *John Kress*[2]

1. Research Institute of Tropical Forestry, Chinese Academy of Forestry

2. National Museum of Natural History, Smithsonian Institution

DNA barcoding, a bio-technology characterized by standardization, universality and efficiency, is widely used in multiple disciplines including biology, environmental science, forensics and inspection, *etc.* Generally, these applications can be divided into the clade and community levels. Forest dynamic plots (FDPs) provide a unique opportunity to study large-scale research and multidisciplinary researches on forests including DNA barcoding. Presently, China had set up four, nine and six FDPs in tropics, subtropics and temperate zone, respectively. Amongst, at least seven FDPs (i.e., Dinghushan, Gutianshan, Changbaishan, Xishuangbanna and three FDPs in Taiwan) are available with plant DNA barcoding sequences (mostly *rbc*L, *mat*K, *trn*H-*psb*A and occasionally ITS2). The first community-level research article on plant DNA barcoding in China was published in 2011, utilizing a well-resolved DNA barcode phylogeny to explore tree-habitat associations in Dinghushan FDP. Subsequently, approximately a dozen of international and domestic literatures closely-related or comparative to this topic were published. The work is important and novel for since it demonstrates four previous progresses (e.g., species discrimination, community phylogenetic reconstruction, phylogenetic community structure exploration and biodiversity index evaluation), three major challenges (e.g., absent of impetus to sequence all woody species at forest community levels, lack of novel scientific hypotheses, and scarce of big-data analysis ability) and several possible directions. These attempts might be valuable to explore large-scale biodiversity patterns, preservation and utilization of biological resources, origin and evolution of life.

T2-38-03

Assessment of phylogenetic community structure for a mixed dipterocarp forest in Brunei Darussalam using DNA barcoding

Jacqueline Heckenhauer[1], *Mark W. Chase*[2,3], *Kamariah Abu Salim*[4], *Toby R. Pennington*[5], *David F. R. P. Burslem*[6], *Maria Ellen*

Kaye[6], *Michael H. J. Barfuss*[1], *Rosabelle Samuel*[1]
1. *Faculty of Life Sciences, University of Vienna, Rennweg 14,1030 Vienna*
2. *Jodrell Laboratory, Royal Botanic Gardens, Kew, Richmond TW9 3DS*
3. *School of Plant Biology, The University of Western Australia, Crawley, Western Australia, 6009*
4. *Environmental and Life Sciences, Faculty of Science, University of Brunei Darussalam, Tungku Link Road, BE 1410*
5. *Royal Botanic Garden, Edinburgh EH3 5LR*
6. *School of Biological Sciences, University of Aberdeen, Aberdeen*

DNA barcoding is a fast and reliable tool to assess and monitor biodiversity and has a role to play in investigating processes that are responsible for species interactions and thus the community structure of forests. Here, we use DNA barcoding to assess phylogenetic community structure of a 25 ha forest dynamics plot in a mixed dipterocarp forest in Brunei Darussalam which is part of the Center for Tropical Forest Science - Forest Global Earth Observatory (CTFS-ForestGEO) network of plots. Leaves and bark of shrubs and trees were sampled from 70 subplots (10x10 m) that varied in relative elevation, convexity and slope. We generated DNA sequence data for the two widely used plastid coding regions *rbcL* (3,118 sequences) and *matK* (2,598 sequences) for 599 morphotaxa identified in the 70 subplots. These represented 24 orders, 69 families and 189 genera with Dipterocarpaceae (16%) and Euphorbiaceae (9%) as the dominant families. Based on the barcode sequences, we reconstructed phylogenetic relationships among the taxa. To compare phylogenetic resolution and to test if it has an influence on phylogenetic diversity (PD) metrics, we generated a second phylogeny using the program Phylomatic which is widely used to infer phylogenetic community structure. It estimates the phylogenetic hypothesis for the taxa in the study plot on the basis of the Angiosperm Phylogeny Group IV. The phylogenetic trees were largely congruent, but our community scale phylogeny obtained from barcode sequences showed higher resolution than the Phylomatic tree, which resolved only at family or generic level. Both phylogenetic trees will be used to calculate two widely used metrics of PD, Mean Phylogenetic Distance (MPD) and Mean Nearest Taxon Distance (MNTD). Further, Webb's Net Relatedness index (NRI) and Nearest Taxon Index (NTI) will be calculated. The metrics will be compared to a null model of community assembly to determine whether species are assembled into local communities at random or if they are more closely (phylogenetic clustering) or more distantly (overdispersion) related than expected by chance. Further, besides biotic interactions habitat filtering often plays a role in species interactions. We will investigate the influence of environmental factors on community structure.

T2-38-04

Plant DNA barcoding and conservation

Alison Shapcott[1], *W. John Kress*[2], *Dan Faith*[3], *Paul Forster*[4], *Bill Mcdonald*[4]
1. *University of the Sunshine Coast*
2. *Smithsonian Institution National Museum of Natural History*
3. *Australian Museum*
4. *Queensland Herbarium*

Currently conservation planning primarily rely's on identification of different community types, species richness or species endemism as measures of diversity and or uniqueness. None of these measures captures the genetic diversity among taxa and all communities and all species are treated as essentially of equal importance and of making equal contributions to diversity. Tallying the numbers of species restricted to particular geographic locations as a measure of uniqueness again does not incorporate the genetic distinctiveness of the taxa. DNA barcodes provide a standardised measure of genetic distance among unrelated taxa and with the development of comprehensive reference libraries for geographic regions and or vegetation community types are enabling assessments to be made for conservation planning that incorporate standardised measures of phylogenetic distance among taxa and hence better assessments for biodiversity conservation. We illustrate with examples of this application to the South East Queensland rainforest estate of Australia. While phylogenetic distance and species richness are correlated in rainforest species richness tends to underestimate the phylogenetic diversity. Here we have compared the diversity of different regional rainforest ecosystem types (RE) and identified limitations in the representativeness of the protected area network. We assessed how well rainforest plant biodiversity is conserved within biogeographic subregions and where to strategically target new protected areas to maximise biodiversity conservation. We looked at fine scale patterns and utilised this approach to identify the significance of the rainforests in the Fraser Island World Heritage area despite their relatively low species richness and to identify high diversity refugial areas on the Sunshine Coast that are under the greatest threats due to urbanisation and agriculture.

T2-38-05

Ten years of barcoding at the African Centre for DNA barcoding

Michelle Van Der Bank[1], *Simeon Bezeng*[1], *Barnabas Daru*[2], *Ronny Kabongo*[1], *Olivier Maurin*[3], *Jonathan Davies*[4]
1. *University of Johannesburg*
2. *Harvard University*
3. *Royal Botanic Gardens KEW*
4. *McGill University*

The African Centre for DNA Barcoding (ACDB) was established in 2005 as part of a global initiative to accurately and rapidly document biodiversity using short DNA sequences. Over the past decade, the ACDB has generated a DNA barcode library for African flora, contributing more than 21,000 DNA barcodes to the Barcode of Life Database. Although the ACDB was initially established to generate molecular data for systematics studies, its characteristics suggest that it has major implications for understanding the fundamental processes by which species diversity is recognized, distributed and maintained. Thus, research at the ACDB in the past decade has been a major step in that direction. Here, we present several case studies utilizing DNA barcode data generated at the ACDB, to explore questions relating to the evolutionary processes underlying the formation, distribution and maintenance of biological diversity in sub-Saharan Africa based on five major focus areas: (1) plant community dynamics in response to large mammal disturbances, (2) phylogenetically informed biogeographical regionalization, (3) species conservation, (4) species adaptive responses to environmental change e.g. responses of African trees to fire regimes, and (5) biological invasion. Together, our research at the ACDB demonstrates that DNA barcode data, integrated with systematics, ecology, evolutionary biology and conservation, can help to address questions about the mechanisms governing regional diversity patterns, especially in the face of climate change.

T2-38-06

DNA barcoding, genomics and plant identification

Peter Hollingsworth
Royal Botanic Garden Edinburgh

DNA barcoding is based on sequencing a few standard DNA markers to tell species apart. The approach was developed using Sanger sequencing technologies and workflows. DNA barcoding is now at something of a junction, between (1) Sanger sequencing of standard markers, (2) the use of new sequencing platforms to improve efficiency and recovery of standard barcode markers, and (3) extending the concept of DNA barcoding to give greater depth of coverage of more DNA regions per individual. In this talk I will provide an overview of the strengths, weakness and insights from standard DNA barcoding approaches, and explore the opportunities for improving the power of DNA barcoding with new sequencing technologies.

T2-38-07

DNA barcoding performs better in understory herbs than trees for species identification in a 25 ha subalpine forest dynamics plot of Southwestern China

Shaolin Tan[1,2], *Yahuang Luo*[1], *Jie Liu*[1], *Kun Xu*[4], *Dezhu Li*[3], *Lianming Gao*[1]
1. Kunming Institute of Botany, Chinese Academy of Sciences
2. University of Chinese Academy of Sciences, Beijing 100049, China
3. Germplasm Bank of Wild Species, Kunming Institute of Botany, Chinese Academy of Sciences, Kunming 650201, Yunnan, China
4. Lijiang Forest Ecosystem Research Station, Kunming Institute of Botany, Chinese Academy of Sciences, Lijiang 674100, China

Abstract: Understory herbs contribute substantially to the species diversity of both temperate and tropical forests. DNA barcoding has been widely used for the identification of forest plants, but most of the applications were focusing on trees. In this study, a total of 880 specimens of both woody species and herbaceous species were collected in a 25 ha subalpine temperate forest dynamics plot, which represented 216 species (73 woody species and 143 herbaceous species) belonging to 141 genera and 64 families of seed plants. The performance of four DNA barcodes (*rbc*L, *mat*K, *trn*H-*psb*A, and ITS) were evaluated for the 216 species, which were represented by a minimum of two individuals for each species where possible. We quantified the rates of species discrimination (RSD) by the barcodes singly and in combinations using the tree-based method. Amplification and sequencing success were highest for *rbc*L (with 99% amplification success rate and 97% sequencing success rate, 99%/97%), and *trn*H-*psb*A (96%/92%) ranked the second, performed better than ITS (95%/91%) and *mat*K (93%/90%). The highest RSD of single barcode was shown for *trn*H-*psb*A (81%) and ITS (81%), followed by *mat*K (77%) and *rbc*L (75%). In combinations, the RSD of *rbc*L+*mat*K, *rbc*L+*mat*K+*trn*H-*psb*A, *rbc*L+*mat*K+ITS, and rbcL+*mat*K+*trn*H-*psb*A+ITS were 78%, 81%, 82%, and 86%, respectively. The four DNA barcodes performed pretty well in most of the taxa, but the species level resolution of some taxonomical complex taxa, such as *Acer, Litsea, Rhododendron, Salix, Clinopodium*, and *Poa* were unresolved. The RSD of the barcodes, either singly or in combinations, were much higher for herbaceous plants than that of woody plants. For example, the RSD of *matK* for herbaceous species was 83%, but only 68% for woody species.The RSD of the four barcodes in combination for herbaceous species and woody species were 88% and 82%, respectively. Our DNA barcode database would be a valuable resources for further community phylogeny studies of both trees and herbs in the forest dynamics plot.

T2-38-08

DNA barcoding reveals cryptic lineages of *Schistidium* at the interface between the natural and the anthropogenic environment

Wolfgang Karl Hofbauer[1], ***Laura Lowe Forrest***[2]
1. Fraunhofer Institute for Building Physics (IBP)
2. The Royal Botanic Garden Edinburgh (RBGE)

Previous studies have shown that in the moss *Schistidium*, a narrow species concept better fits the observed diversity; this is in concordance with molecular genetic results. In our recent work (SYNTHESYS grant GB-TAF-3881) genetic barcoding assisted in identification of morphologically poorly-developed pioneer forms occurring on anthropogenic structures like masonry, that otherwise are only assignable to genus. We found unexpectedly high diversity, and were able to identify certain taxa; however, we could not identify all the recovered lineages. Similar taxa occur on man-made structures in geographically distinct regions throughout Europe. In order to increase their identification, sampling of high-quality determined material is needed for reference. Our new project (SYNTHESYS grant GB-TAF-5965) extends the previous study. Further samples and specimens from the extensive RBGE herbarium, which holds >600 *Schistidium* specimens from around the world, will be examined according to the current taxonomy: a subset shall be characterized morphologically with light and scanning electron microscopy (leaf tip structure, leaf and nerve surface, spore and capsule peristome structure), and DNA-barcoded for up to 4 loci (comprising *rbc*L, *mat*K, *psb*A-*trn*H and ITS2) to elucidate the complex taxonomy in *Schistidium*. Using this reference DNA barcode library for Western European species, we will not only increase our understanding of pioneer *Schistidium* species on artificial surfaces, but also target selected lineages for deliberate cultivation. Our secondary aim is to establish a sound technique for extensive moss cultivation on building structures. An effective way to enhance homogenous moss growth on urban surfaces offers several benefits: Moss cover on buildings can help retain water, reduce noise, and in building acclimatisation. Further benefits include reduction of air pollutants, and carbon sequestration. On the other hand, detailed knowledge of pioneer moss species on man-made structures can also allow specially tailored countermeasures where the occurrence of moss growth is deemed undesirable.

T2-38-09

Revisiting Darwin's naturalization conundrum: Explaining invasion success of non-native trees and shrubs in southern Africa

Simeon Bezeng
University of Johannesburg

Invasive species are detrimental ecologically and economically. Their negative impacts in Africa are extensive and call for a renewed commitment to better understand the correlates of invasion success. Here, we explored several putative drivers of species invasion among woody non-native trees and shrubs in southern Africa, a region of high floristic diversity. We tested for differences in functional traits between plant categories using a combination of phylogenetic independent contrasts and a simulation-based phylogenetic ANOVA. We found that non-native species generally have longer flowering duration compared with native species and are generally hermaphroditic, and their dispersal is mostly abiotically mediated. We also revealed that non-native trees and shrubs that have become invasive are less closely related to native trees and shrubs than their non-invasive non-native counterparts. Non-natives that are more closely related to the native species pool may be more likely to possess traits suited to the new environment in which they find themselves and thus have greater chance of establishment. However, successful invaders are less closely related to the native pool, indicating evidence for competitive release or support for the vacant niche theory. Non-native trees and shrubs in southern Africa are characterized by a suite of traits, including long flowering times, a hermaphroditic sexual system and abiotic dispersal, which may represent important adaptations promoting establishment. We suggest that differences in the evolutionary distances separating the native species pool from invasive and non-invasive species might help resolve Darwin's naturalization conundrum.

T2-38-10

Can DNA barcoding provide any help for conservation of the Turkish geophytes?

***Ali A. DÖnmez**, Zübeyde UĞurlu Aydin, Aslı DoĞru Koca*
Hacettepe University, Molecular Plant Systematic Laboratory (MOBIS), Faculty of Science, Department of Biology

Turkey is one of the most important plant biodiversity centre comprising more than 700 geophytes within approximately 10,000 vascular plant species. Many species of tulip, lily, snowdrop, hyacinth, orchid, cyclamen, crocus, and of the other bulbous plants naturally grow in Turkey. It is well known that DNA-based methods can be applied quickly and efficiently. Recently, studies have showed that the use of DNA barcodes might be a promising tool for identification and conservation of biodiversity. We aimed to investigate utility of eleven loci for barcoding in 36 geophytes, using both single region and multi region approaches. The study has been carried out on the taxa belong to more than a hundred samples from 36 geophyte species. The sequence data of each species belonging to the *atp*B, *atp*F-*atp*H, *rbc*L, *rpo*B, *rpo*C1, *trn*L-*trn*F and ITS (ITS1+ITS2) markers have been obtained. After basic studies on the row alignments, p-distance values of the sequences have been evaluated for their reliability. Consequently, the data have been analyzed with the Neighbor Joining and Maximum Composite Likelihood analyses applied for discrimination value. Based on the results, in consideration of sequence quality and PCR success as a basic requirement of being a barcode region, *rpo*B is one of the most distinguishing genes among them. Intraspecific variation of the species of the genera *Cyclamen* L., *Adonis* L., *Tulipa* L., *Fritillaria* L., *Narcissus* L., *Galanthus* L., *Muscari* Mill. *Arum* L. and *Orchis* L. are very low (<0.03) or absent according to *rpo*B sequences. Hence, this gene region can be used for barcode gene for the mentioned genera. However, the other markers have low discrimination value for the studied geophyte genera. Moreover, these markers have additional problems for being barcode as low PCR success, multiple bands and primer universality. Geophytes comprise many endemic and economically important plants in Turkey. Moreover, new geophytes are still discovered from the nature by botanical explorations. Hence, as a new tool for discovering biodiversity, barcoding of plant is not sufficient for identify and characterize of these plant species alone.

T2-38-11

Evaluation of DNA barcoding markers for Thai Epiphyllous Bryophytes

***Ekaphan Kraichak**[1], Sorrasak Yodphaka[1], Thomas Gleason[2], Kansri Boonpragob[3], H. Thorsten Lumbsch[2]*
1. Kasetsart University
2. Field Museum of Natural History
3. Ramkhamheang University

Leaf-colonizing (epiphyllous) bryophytes are ubiquitous in tropical ecosystems and can be found on a number of vascular plant leaves, many of which are economically important plants. However, the studies on diversity of these bryophytes have been limited, because of minute and incomplete specimens, as well as lack of taxonomical expertise. The recent development of the DNA barcoding approach has allowed the taxon identification and species discovery of many obscure groups of organisms. We compared the efficiencies of six DNA markers (*rbc*L, *mat*K, *trn*L-F, *trn*H-*psb*A, nrITS1 and nrITS2) in their ability to amplify and discriminate among Thai epiphyllous species. The amplification success was relatively high (70-90%) in all of the markers, except for *matK* which yielded no success. The barcoding gap, as calculated from the difference between inter- and intraspecific genetic distances, was the highest in nrITS2, while the highest number of monophyletic groups were found in nrITS2 and *rbc*L. Consistent with the recommendation by the Consortium of Barcoding of Life Working Group, *rbcL* should be used a a main barcoding marker for epiphyllous species. The development of DNA barcoding as a tool for discovering currently poorly known species diversity will provide a rapid and reliable identification tool for epiphyllous bryophytes in Thailand.

T2-38-12

A mini barcode signature for the identification of Rhodiolae Crenulatae Radix et Rhizoma and its products

*Jianqiang Zhang, Dalv Zhong, **Ruowei Zhu***
Shaanxi Normal University

The roots and rhizomes of *Rhodiola crenulata* (J. D. Hooker &

Thomson) H. Ohba have been included in the Pharmacopoeia of China as the authentic Rhodiolae Crenulatae Radix et Rhizoma, to enhance inner spiritual power, concentration and physical endurance. The accelerated and uncontrolled use of *R. crenulata* in China has led to deforestation, and rapid increasing commercial utility of the raw material has made other species of *Rhodiola* been sold as Rhodiolae Crenulatae Radix et Rhizoma in the market. DNA barcoding methods provide us an effective and quick way to identify *R. crenulata* and its product. However, the successful identification relies on a suitable barcode marker and a comprehensive survey of plants all over its distribution area. As the proposed internal transcribed spacer 2 (ITS2) is not sufficient to resolve heavily processed materials, we surveyed 308 individuals of *R. crenulata* from 16 populations from Yunnan, Sichuan, and Xizang provinces covering all its natural distribution area, to search a short barcode for the identification of processed materials. All ITS2 sequences were obtained for the 308 individuals, ranging from 209 to 223 bp in length. We detected six haplotypes in all individuals. Through comparing sequences from 185 other *Rhodiola* species, we found a 32 bp (5'-TACGAGGCCTTGCG-CCTTTGAGCTTT GCG CG T-3') nucleotide signature, which is unique in *R. crenulata*. Then, we designed primers to amplify the nucleotide signature region from processed materials. Out of the 23 samples we purchased online, 12 were from other *Rhodiola* species, and 3 were not *Rhodiola*. Thus, using nucleotide signature provides us an effective alternative for identify heavily processed traditional Chinese plant materials. We also evaluated the use of secondary structure of ITS2 for identification of *R. crenulata*. The results revealed that all of the ITS2 sequences had a common four-helix folding, but the structure of *R. crenulata* is different from that of its closest relatives. We thus propose that ITS2 sequence structure information could be used as an alternative mini-barcode when a full-length ITS is not present. Our results demonstrated that both a mini barcode from ITS2 and the structure of ITS2 are effective markers for identification of *R. crenulata*, after surveying representatives from all its distribution area. This study greatly broaden the application of DNA barcoding in identifying Chinese patent medicines and other products with degraded DNA.

T2-39: Ecology and evolution of parasitic plants

T2-39-01

Driving forces for rapid spreading of *Pedicularis kansuensis* in Western China

Ai-Rong Li[1], *Xiao-Lin Sui*[1], *Wen-Jun Li*[2], *Yan-Yan Liu*[2], *Rui-Juan Xue*[1], *Kai-Yun Guan*[1,2]

1. Key Laboratory of Economic Plants and Biotechnology, Kunming Institute of Botany, Chinese Academy of Sciences

2. Key Laboratory of Biogeography and Bioresource in Arid Land, Xinjiang Institute of Ecology and Geography, Chinese Academy of Sciences

Pedicularis L. (Orobanchaceae) is a large linkage of root hemiparasitic species widely distributed in the temperate zone of Northern Hemisphere, with about 800 species worldwide (2/3 of the species distributed in China). In most cases, they show rather patchy distribution patterns and do not cause significant reduction in plant productivity at a large scale. However, in the past two decades, dense occurrence in large scales and rapid expansion of *P. kansuensis* in the west of China have been severely reducing forage crop productivity in grasslands, hence threatening the local livestock industry. To better understand the spreading of *P. kansuensis* for a more effective and sustainable control against it, we investigated the driving forces for the spreading. Both lab-based studies and field trials have been conducted to understand the internal and external driving forces. Our results showed that high genetic diversity, high morphological as well as physiological plasticity conferred *P. kansuensis* with strong adaptation to various habitats. Climate changes with increased rainfall and temperature facilitated seed germination and dispersal of this root hemiparasite. Monoculture of forage grasses that function as preferred but vulnerable hosts of *P. kansuensis*, lack of quality control for forage and restoration plant seeds, and lack of quarantine control of seed transportation contributed the most to rapid spreading of this species over a large scale in a short period of time.

T2-39-02

Evolution of the organellar genomes in the Piperales with a focus on the holoparasitic Hydnoraceae

Julia Naumann[1], *Matthias Jost*[2], *Erik K. Wafula*[1], *Guo Chunce*[1], *Kong Hongzhi*[3], *Christoph Neinhuis*[2], *Claude Depamphilis*[1], ***Stefan Wanke***[2]

1. Department of Biology and Institute of Molecular Evolutionary Genetics, The Pennsylvania State University

2. Institut für Botanik, Technische Universität Dresden

3. State Key Laboratory of Systematic and Evolutionary Botany, CAS

Our knowledge of plastome evolution has greatly increased in recent years while comparatively little is known about plant mitochondrial genomes. In contrast to plant mitochondrial genomes, plastid genomes of photosynthetic flowering plants are usually highly conserved in size, structure and gene content. However, the plastomes of parasitic and mycoheterotrophic plants may be released from selective constraint due to the reduction or loss of photosynthetic ability resulting in size and gene reduction. The purge of plastome genes is well understood and appears to follow a very similar pattern. Piperales have both photosynthetic and nonphotosynthetic lineages and the holoparasitic genera *Hydnora* and *Prosopanche* have split from their photosynthetic ancestor presumably 100 MY ago. That makes the Piperales a particularly interesting system for studying organellar evolution. Genomic sequence data of the three species *Hydnora visseri*, *Prosopanche americana* and *Aristolochia fimbriata* allowed us to identify, characterize, and compare their plastomes as well as to obtain first insights into their mitochondrial genomes. Following the publication of the *H. visseri* plastome, it was comparatively easy to identify and characterize the plastome of *P. americana*. Despite the extreme reduction and sequence divergence of the plastomes to photosynthetic plants, they likely remain functional. Surprisingly, the assembly and characterization of the plastome of the photosynthesic *A. fimbriata* has proven to be more challenging. In contrast to the plastomes, the mitochondrial genomes of the two parasites are among the larger ones across the angiosperms and the gene set is surprisingly conservative, representing a fairly complete and functional set of common plant mitochondrial genes. While horizontal gene transfer (HGT) is fairly common in parasitic plants due to the physical connection to the host plant, there is no evidence for HGT based on phylogenetic efforts so far. Ongoing research in Piperales, including several more lineages has the potential to greatly complete our understanding of organellar genome evolution in nonphotosynthetic plants that diverged long time ago from their photosynthetic ancestor.

T2-39-03

Integrating nuclear gene data sheds further light on plastid genome degradation in parasitic plants

Katharina Schäper, Michelle Thönnes, Julian Koch, Kai F. Müller, ***Susann Wicke***
Institute for Evolution and Biodiversity, University of Muenster

Parasitism represents the most extreme interaction between plants, where the parasitic plant steals water and nutrients from a host plant by connecting directly to its vascular tissue. The secondary transition to a nonphotosynthetic way of life in plants has brought about massive reconfigurations at the molecular level. Particularly plastid genomes (plastomes) of nonphotosynthetic plants have experienced rampant gene losses and an acceleration of molecular evolutionary rates. Although general trends of reductive plastome evolution are known, we still have no understanding of the genomic reconfigurations pertaining to the parasites' nuclear genomes or of the coevolutionary forces acting on nuclear and plastid genes under relaxed selective constraints in parasitic plants. Here, we investigate the fate of nuclear-encoded photosynthesis-associated genes along the transition from an autotrophic to a nonphotosynthetic lifestyle. Our comparative genomic and molecular evolutionary analyses of 15 nonparasitic and parasitic species from the Broomrape family (Orobanchaceae), some of which suffering from extensive plastid gene losses, reveal a surprisingly long (physical) retention of genes directly involved in photo-autotrophic pathways in holoparasites. However, transcriptome data indicate that many of those genes play no more role in molecular and cellular pathways. Analysis of selectional regimes indicate that purifying selection is relaxed in most - but not all - nuclear photosynthesis elements in the various holoparasitic lineages. Our analyses also show that shifts in selection intensity have occurred already in the Orobanchaceae ancestor or set in as parasitism evolved within the family, that is, long before photosynthesis was lost. These results imply that already the transition to parasitism relaxes selectional constraints on photosynthesis genes. We also observe that the degradation of photosynthesis-associated genome fragments appears to proceed much faster in the plastome than in the nuclear genome. Our findings highlight the importance of integrating data from various genomic compartments for understanding mechanisms of molecular evolution in subgenomes and reconfigurations of biological pathways along lifestyle transitions.

T2-39-04

Plastome reduction in parasitic plants: Independent evolutionary events?

Wen-Bin Yu[1], *Christopher Randle*[2], *Hong Wang*[3], *Jun-Bo Yang*[4], *Yu Song*[1], *Chao-Nan Fu*[3], *Claude Depamphils*[5], *De-Zhu Li*[4]
1. Xishuangbanna Tropical Botanical Garden, Chinese Academy of Sciences
2. Department Biological Sciences, Sam Houston State University
3. Key Laboratory for Plant Diversity and Biogeography of East Asia, Kunming Institute of Botany, Chinese Academy of Sciences
4. Plant Germplasm and Genomics Center, Germplasm Bank of Wild Species, Kunming Institute of Botany, Chinese Academy of Sciences
5. Department of Biology, Pennsylvania State University

Many parasitic plants are characterized specialized organs (i.e. haustoria) by which they obtain water and/or nutrients from host plants. There are around 4,000 parasitic flowering species placed in 12 orders and 19 families. Of them, about 90% of parasitic species (in four orders) retain photosynthetic capability (hemiparasites), while 10% of parasitic species (in 10 orders) have lost photosynthetic capability (holoparasites). In free-living plants, the plastome is very conserved in size (120 – 170 kb), gene contents (110 –130 genes) and a quadripartite structure (LSC, IRA, SSC and IRB). Loss of photosynthetic capability in holoparasites is often correlated with degradation or loss of photosynthesis-related and other non-housekeeping genes. To date, plastomes of 11 parasitic orders have been published or examined (except Ehretiaceae, Boraginales). In hemiparasitic taxa such as *Cassytha* (Lauraceae, Laurales), Krameriaceae (Zygophyllales), Orobanchaceae (Lamiales) and Santalaceae s.l. (Santalales), plastome size and/or gene contents showed high similarity to its non-parasitic relatives. Some NADH-polymerase (*ndh*) genes were found to be degraded or lost in some species, *Cassytha* ssp. had no inverted repeat region, and gene rearrangements were found in hemiparastic Orobanchachece. Genome reductions were found in all holoparastic plants. Genome size in holoparasitic plants varied from 11,348bp (*Pilostyles aethiopica*, Apodanthaceae, Cucurbitales) to 150,504bp (*Lathraea squamaria*, Orobanchaceae). Photosynthesis-related genes were (partitially) retained in *Cuscuta* spp. (Convolvulaceae, Solanales) and some holoparastic Orbanchachaceae (e.g. *Harveya* spp., *Lathraea* spp., *Orobanche californica*). Nevertheless, the photosynthesis-related genes, as well as some ribosomal protein genes, were lost in other holoparasitic lineages, i.e., *Cynomorium coccineum* (Cynomoriaceae, Saxifragales), *Cytinus hypocistis* (Cytinaceae, Malvales), *Hydnora visseri* (Hydnoraceae, Piperales), *Mitrastemon yamamotoi* (Mitrastemonaceae, Ericales) and *Pilostyles* spp., as well as in some Orobancheceae, e.g. *Christisonia hookeri*, *Cistanche phelypaea*, *Conopholis americana*, *Epifagus virginiana*, and *Orobanche* spp. Moreover, a plastome was not detected in Rafflesiaceae (Malpighiales) and Balanophoraceae (Santalales). Phylogenetic analyses have demonstrated that parasitism has evolved independently at least 12 times. When considering the evolutionary ages of parasitic lineages, plastome degradation follows a pattern: degradation of *ndh* genes in hemiparasites, followed by photosynthesis-related genes in newly derived holoparasites, loss of protein synthesis and other functional genes, and finally retention of ribosomal RNAs in ancient holoparasites, or complete elimination or with possible incorporation of plastome remnants into mitochondria and nuclear genomes. Seven ancient holoparastic lineages, i.e. Ericales, Cucurbitales, Malvales, Malpighiales, Piperales, Santalales, Saxifragales, have extremely reduced (or no) plastomes. By contrary, young holoparasitic lineages in *Cuscuta* and Orobanchaceae showed a gradual reduction pattern of plastomes.

T2-39-05

Parasitic plants: Genome scale analyses reveal core parasite genes and extensive horizontal gene transfer

Claude Depamphilis[1], *Zhenzhen Yang*[1], *Eric Wafula*[1], *Yeting Zhang*[1], *Loren Honaas*[1], *Gunjune Kim*[2], *Joel Mcneal*[3], *Huiting Zhang*[1], *Elizabeth Kelly*[1], *Christopher Clarke*[2], *Michael Timko*[4], *John Yoder*[5], *James Westwood*[2]

1. The Pennsylvania State University

2. Virginia Polytechnic Institute and State University

3. Kennesaw State University

4. University of Virginia

5. University of California at Davis

Parasitic plants, which obtain water and nutrients directly from a photosynthetic host plant, comprise 1.5% of all angiosperms. Phylogenetic analyses have resolved 12 independent origins of direct stem and root parasites, with a similar number of lineages having evolved the ability to parasitize mycorrhizal relationships (mycoheterotrophs). Parasitic plants have long challenged and enlightened our understanding of organelle genome evolution, the function and evolution of plant-plant interactions, and the dynamics of RNA and DNA transfer between species. Here, I will describe recent efforts by the Parasitic Plant Genome Project (PPGP) to leverage comparative transcriptomics and genomic-scale analyses to identify and study a core set of genes involved in parasitism in Orobanchaceae, the family of parasitic plants with the widest range of parasitic abilities. Current efforts to understand the function of parasite genes promises to lead to control measures for parasitic weeds, especially in parasites where RNA movement from host to parasite allows host-induced gene silencing (HIGS). The discovery of bidirectional movement of RNA and DNA between parasites and hosts (resulting in horizontal gene transfer, HGT) has opened up new dimensions in our understanding of parasite-host interactions and their long-term impact. Genome-scale survey of expressed genes has identified more than 50 high confidence functional HGTs in Orobanchaceae and more than twice that number in *Cuscuta*, a stem parasite with strong parasitic dependence. Results support the thesis that parasitic plants have stolen genes from their host plants via HGT and are using them in parasitic processes.

T2-39-06

Genomic basis of taxonomic complexity in hemiparasitic *Euphrasia*

Alex Twyford

University of Edinburgh

Hemiparasitism is a remarkably flexible life-history strategy whereby a given parasite can survive to flower without a host, but performs much better after host attachment. This strategy occurs in thousands of plant species, and is of special importance as it may represent an intermediate stage in the transition from free-living autotrophism to obligate holoparasitism. Here, I present recent work on the facultative hemiparasitic plant genus *Euphrasia* (Orobanchaceae). The genus includes incredible taxonomic complexity, which may in part be due to plasticity in response to host quality. Using large-scale population genomic analyses, in conjunction with a common garden study, I show surprising species relationships that have been obscured by plant performance in the field. Genotyping by sequencing data reveal deep divergence corresponding to ploidy, but limited support for species within a ploidy level. Plants grown on clover flower many months earlier than those without a host, while other morphological characters used in species delimitation show little plasticity. These results reveal the potential for common garden studies of non-specialised parasitic plants like Euphrasia for clarifying taxonomic boundaries and host preference, while also emphasizing the need to perform genomic studies of complex recent species radiations.

T2-40: Systematics and evolution of palms - a model group for the tropics

T2-40-01

Time-calibrated phylogeny of the American palms (Arecaceae) based on NGS sequence-capture methods

Ángela Cano[1], *Mathieu Perret*[1], *Tobias Hofmann*[2], *Isabel M. Liberal*[2], *Christine D. Bacon*[2], *Fred D. Stauffer*[1], *Mats Töpel*[2], *Alexandre Antonelli*[2]

1. Conservatoire et Jardin botaniques de la Ville de Genève and University of Geneva

2. Department of Biological and Environmental Sciences, University of Gothenburg

During the last decade, resolution of the phylogenetic relationships within the palm family has been largely improved. Thanks to these efforts palms became a model group to investigate the origin of tropical rain forest and the evolution of plant diversity at a global scale. However, the genus-level phylogeny available today and the difficulty to resolve interspecific relationships within genera using Sanger DNA sequencing, strongly limit our abilities to reveal evolutionary processes occurring at a finer geographical scale. To overcome this difficulty, palm scientists, now organized in the Palm Phylogeny Working Group, recognize the need of generating new data towards a complete species-level phylogeny of palms. Next Generation Sequencing (NGS) and sequence-capture in particular, have been considered as promising techniques to achieve this goal. However, so far, no studies have evaluated the feasibility and advantages of NGS to estimate large species-level phylogenies on palms. To address this issue, we aimed to reconstruct the species-level phylogeny of the American palms, which represent about a third of all palm species. Sequence data were obtained using the sequence-capture procedure, with available probes designed for cross-taxonomic sequencing within Arecaceae. Nearly 30% of all American palms, including all subclades and 59% of the genera were sampled from private collections (Fairchild Tropical Botanical Garden, *Instituto Plantarum*, *Jardín Botánico del Quindío*, Montgomery Botanical Center) and during intensive fieldwork in Colombia and Panama. Emphasis was given to the richest clades: Bactridinae (55 species sampled/232 accepted species), Chamaedoreeae (80/123) and Geonomateae (60/84). Sequences data was processed using the bioinformatics seqcap_processor pipeline (https://github.com/AntonelliLab/seqcap_processor). Alignments were analyzed using Maximum Likelihood (RAxML) and Bayesian Inference (BEAST) methods. Divergence times were estimated using fossil data available for palms. Illumina sequencing of our gene libraries produced a mean of ca. 650,000 reads per sample that allowed the alignment of 326 targeted exons in Arecoideae. The phylogenetic analysis of the 50 alignments (totaling 21,380 bp) with the highest taxonomic coverage confirms the current palm classification and the monophyly of most of the subclades and genera. We will discuss the advantages and limitations of the sequence-capture method and notably the performance of the palm-probes, with the challenges imposed by differential loci capture among clades and loci variation. Our study is a promising first step towards the reconstruction of a complete species-level phylogeny of the American palms. This phylogeny will soon be complemented with the inclusion of all the remaining genera and 259 additional species that we have already sampled. Our result will contribute to better understand the causes and processes of plant diversification in the Neotropics. Preliminary results concerning the origin of palm diversity in Central America will be presented.

T2-40-02

Using a NGS approach to explore the macroevolutionary history of a Neotropical palm genera

Oriane Loiseau[1], *Margot Paris*[2], *Jonathan Rolland*[1], *Mary Laure De La Harpe*[2], *Ingrid Olivares*[3], *Anna Weigand*[3], *Michael Kessler*[3], *Christian Lexer*[2], *Nicolas Salamin* [1]
1. University of Lausanne
2. University of Fribourg
3. University of Zurich

Understanding why the Neotropics host the greatest plant diversity on Earth is a question that has puzzled evolutionary biologists for a long time and is still a vibrant field of research. Recent phylogenetic comparative studies have brought new insight not only into the general patterns of diversity but also into the mechanisms that promoted the diversification of plant lineages. Our study focuses on *Geonoma*, a genus of small understory palms distributed from Southern Mexico to Southern Brazil. With 68 species it is one of the most diverse palm genus in the Neotropics and it occurs from sea level to above 3,000 m in Central and South American rain forests. Species delimitation has proved challenging because of high intra-specific morphological variation and 20% of *Geonoma* species are in fact considered as species complexes. *Geonoma* originated and diversified in the Neogene and this recent origin, combined with a slow rate of molecular evolution has led to poor resolution in current phylogenetic trees based on few genetic markers. We selected a target sequencing approach to obtain thousands of single copy nuclear DNA regions, for about 300 individuals and a taxonomic sampling covering 95% of the described species, in order to reconstruct a robust phylogenetic hypothesis and better understand the evolutionary history of this group. Using the phylogeny we investigated the biogeographical history as well as the drivers of diversification of *Geonoma* by exploring several intrinsic and extrinsic factors that could have played a role in the evolution of the group. Our results contribute to broadening the general knowledge about the emergence of the current outstanding neotropical plant diversity.

T2-40-03

Towards a species-level phylogeny of Indo-Pacific arecoid palms (Arecaceae): First insights from New Caledonian endemics

Sidonie Bellot, *Steven Dodsworth*, *Wolf Eiserhardt*, *Felix Forest*, *Ilia Leitch*, *William Baker*
Royal Botanic Gardens, Kew

Palms are a prominent plant family of rainforests, and humans daily benefit from hundreds of palm species as food, medicine or timber. Many palm species are endemics on small islands in the Pacific and Indian Oceans, which makes them a good model for comparative studies of the mechanisms of speciation in archipelagos. Such analyses were so far impaired by the lack of a well-resolved species-level phylogeny of the family, leading palm specialists to initiate an international cooperation to fill this gap. Our contribution to this effort will be: (i) to generate a genus-level phylogeny of all palms (180 genera) using a combined genome-skimming and targeted sequence capture approach (Hyb-Seq, Weitemier *et al.*, 2014, Applications in Plant Sciences 2: 1400042), (ii) to generate a species-level phylogeny of tribe Areceae (ca. 600 species) using the same approach adapted to palms (Heyduk *et al.*, 2016, Biological Journal of the Linnean Society 117: 106-120), and (iii) to combine those new phylogenies with ecological and morphological data in order to infer the biogeography of Areceae and the role of seed traits in their adaptation to changing climates and sea levels. The potential of this approach will be illustrated by the results of a pilot study bringing new insights to the evolution of endemic New Caledonian Areceae.

T2-40-04

Genome evolution and diversification in Palms

Craig Barrett[1], *Christine Bacon*[2], *Alexandre Antonelli*[3], *Michael Mckain*[4]
1. West Virginia University
2. University of Gothenburg
3. University of Gothenburg
4. Donald Danforth Plant Science Center

Estimates of genome size variation and sequenced genomes have revealed no clear correlation between genome size and organismal complexity; i.e., the "C-value paradox." Processes contributing to genome size variation include polyploidy (genome doubling), tandem duplication, expansion/contraction of repetitive regions such as transposable elements, illegitimate/intramolecular recombination, fractionation/diploidization, and selection for smaller genomes. The palms (family Arecaceae) are diverse, widespread, and ecologically successful components of tropical ecosystems, with >2,500 species. Based on available data across the family, genome size and chromosome number vary immensely. Modal ploidy levels are 16 and 18 within the species-rich subfamilies Arecoideae and Coryphoideae, while patterns of genome sizes based on C-values are less clear. Recent studies demonstrate strong correlations between genome size and species diversity, suggesting polyploidy potentially confers selective advantages due to the introduction of genetic novelty. Here we aim to test that hypothesis in palms and their close relatives, the commelinid monocots (grasses, gingers, dasypogonoids, *etc.*). We address potential correlations between ploidy level and C-value across the family, accounting for phylogeny, and reconstruct ancestral chromosome numbers and genome sizes. We use available genomic and transcriptomic data to investigate evidence for polyploidy among the commelinids and within the palms (at the subfamily level) by comparing the age distribution of paralogous sequences. Lastly, we investigate the mechanisms of genome size variation by exploring repeat type and content based on 'genome skim' data. Disentangling the causes of genome size variation in palms will help us understand the genetic conditions facilitating adaptive radiation and ecological success in other important clades.

T2-40-05

A molecular systematic study of the palm tribe Euterpeae

Fritz Pichardo, Nicolle Macdonald, ***Julissa Roncal***
Memorial University of Newfoundland

The palm tribe Euterpeae, one of the 13 tribes in the subfamily Arecoideae, is distributed in the Neotropical region of Central and South America including the Caribbean. The Euterpeae is of economic importance for local people in the tropics, as it contains palms useful for harvesting palm hearts, acai berries, and oil for exportation. The ecological importance of the tribe is reflected in the fact that three Euterpeae species are among the 20 most abundant tree species in the Amazon. In this study, a molecular phylogenetic analysis was conducted using one plastid (*trnD-trnT*) and four low-copy nuclear DNA regions (*CISP4*, *PHYB*, *RPB2*, and *WRKY6*). We sequenced 34 individuals within 26 of the 31 species in the tribe, and 42 outgroup species to calibrate the tree using the fossil record of palms. Phylogenetic trees were reconstructed using maximum parsimony, maximum likelihood, and Bayesian inference methods. The Euterpeae and all its five genera were monophyletic. *Hyospathe* was basal to the rest of the genera, *Prestoea* and *Neonicholsonia* were sister, *Euterpe* was sister to the *Prestoea/Neonicholsonia* clade, and *Oenocarpus* was sister to the *Prestoea/Neonicholsonia/Euterpe* clade. This tree topology rejected Balick's hypothesis that *Oenocarpus*' hippuriform (horsetail shape) inflorescence was derived from *Euterpe*'s inflorescence by loss of the adaxial rachillae because *Euterpe* is a more recent genus. Furthermore, the hypothesis that *Euterpe catinga* is an ancestral species because it inhabits a unique and geologically old substrate type in Amazonia (white sand) was also rejected because it was found to be one of the most recent taxa within the genus and the tribe. Resolution at the species level was particularly challenging for *Prestoea* and *Euterpe* for which we should sequence many more DNA regions, and complete the taxon sampling by including the narrow endemic species unavailable for this study. Our phylogenetic analysis recovered quite different intergeneric relationships than those recovered in a previous maximum parsimony analysis based on 54 morphological and anatomical characters. The biogeographic history of the group will be analyzed in the future using the dated molecular phylogenetic tree.

T2-40-06

The de-evolution of diversity in Amazonia

Christine Bacon[1]*, Rosane Collevatti*[2]
1. University of Gothenburg
2. Universidade Federal de Goias

The Neotropics is recognized as harboring biodiversity hotspots, such as Amazonia, but there are some areas where diversity may have decreased over time, particularly in oligarchic forests. Understanding of the evolutionary origins and maintenance of oligarchic forests in the context of geologic change and historical climate can improve our predictions of how organisms, habitats, and Neotropical biomes may respond to comparable changes in the face of climate change. The genus *Mauritia* (Arecaceae) presents an ideal system for investigating the evolution of oligarchic forests, or the de-evolution of Amazonian biodiversity. The genus has a rich pollen record and next-generation sequencing techniques, gene capture probes, have been developed specifically for palms that we show are appropriate for investigating demographic structure on both ecological and evolutionary time-scales. Furthermore, *Mauritia* is the most widely used and economically important native South American palm. We hypothesize that historical extinctions in *Mauritia* correlate with increased diversification rates in eudicotyledonous plants, the expansion of South American savannas, and/or the constriction of lowland rainforests during the Neogene. The inference of biogeography and diversification is an integral window into the past that enables the investigation of how geographic regions, biomes, and communities assembled through time and how they may evolve in the future. By integrating data across fields, for example from geology, genetics, fossils, species distribution and abundance, and morphology, we can reach a more comprehensive framework for the understanding of biome evolution and extinction.

T2-41: Practical importance of polythetic approaches to the taxonomy of plant-associated Ascomycota

T2-41-01

Systematics and biodiversity of Didymellaceae, one of the largest families in the fungal kingdom

Qian Chen, ***Lei Cai***
Institute of Microbiology, Chinese Academy of Sciences

The Didymellaceae is one of the largest families in the fungal kingdom, and most species in this family are known as plant pathogens causing leaf and stem lesions on a very broad range of hosts. We have recently assessed the Didymellaceae employing a polyphasic approach using morphological examination and multi-locus (ITS, *28S*, *tub2*, *rpb2*) analyses, with the family delimited to 17 well-supported monophyletic genera. As a result nine new genera and 84 new combinations were introduced respectively. Subsequently several more genera were added to this family, which are discussed in this paper. During our most recent survey in China and six other countries, 100 Didymellaceae isolates were obtained, of which 64 isolates represent previously undescribed species. Twenty-six novel species were described from 40 different host plants or other substrates, such as air, soil, water and faeces from several karst caves in China. For the fungal-host association in Didymellaceae, only a few genera exhibit certain level of host-specificity, for example, *Ascochyta* to Fabaceae, *Neoascochyta* to Poaceae, and *Phomatodes* to Cruciferae. Fabeceae, Poaceae, Asteraceae, Ranunculaceae, Rosaceae and Solanaceae are the six most common host families related to Didymellaceae species.

T2-41-02

The evolution of fungal taxonomy over two centuries: A case

study in the Ophiostomatales

Z. Wilhelm De Beer, *Tuan A. Duong, Miranda Erasmus, Brenda D. Wingfield, Michael J. Wingfield*

Forestry and Agricultural Biotechnology Institute (FABI), University of Pretoria

The Ascomycete order *Ophiostomatales* accommodates more than 320 species of mostly arthropod-associated species. These include some serious tree pathogens such as the causal agents of Dutch Elm Disease, as well as the human pathogens causing sporotrichosis. The taxonomic history of this order largely reflects events that shaped Ascomycete systematics over the past two centuries. The type species of the order, *Ophiostoma piliferum*, was described in 1823 by Fries in a cryptic, 11-word Latin paragraph without illustration. This style persisted during the 19th century in all the major attempts to describe and classify fungi. By the turn of the century, new taxa were often described as part of long, detailed studies on the biology, ecology or pathogenicity of fungi. By that time, the connections between sexual and asexual states were also understood, and in 1932 Buisman was the first to successfully cross two asexual isolates of an Ascomycete, *Ophiostoma ulmi*, in the laboratory to produce its sexual state. During the 21st century detailed morphological studies also led to species concepts being refined based on microscopy, ultrastructure, biochemical and physiological traits. However, throughout the century several monographs on genera in the order were hampered by their focus on either sexual or asexual state morphology, with generic concepts often contradicting each other. In 1992 ribosomal DNA sequences showed that the asexual human pathogen *Sporothrix schenckii* grouped among sexual, wood-infesting species in *Ophiostoma*. Soon protein coding DNA sequences were used together with rDNA sequences to distinguish between cryptic species, reflecting phylogenetic relatedness, rather than phenotypic traits that were often the result of convergent evolution, leading to taxonomic confusion. The ability to link sexually and asexually reproducing isolates provided the impetus for the One Fungus One Name movement that culminated in the abolishment of the dual nomenclature system, which allowed for one fungal species to have two or more names, at IBC in Melbourne in 2011. During the past decade population genetics has also changed the way taxonomists look at species, and hybridization and horizontal gene transfer often explain incongruence within phylogenies. Full genome sequences have now been produced for almost 30 Ophiostomatalean species, and phylogenies based on more than 2700 gene regions extracted from these genomes, formed the basis for the selection of five genes that were amplified and sequenced for more than 200 species in the order. The resulting phylogenies enabled the delineation of genera in the *Ophiostomatales* that clarified confusing and overlapping morphological traits that plagued the group's taxonomy for many decades. Robust phylogenies such as these also reveal and explain evolutionary processes that went undetected in morphology-based taxonomy. The current challenge for fungal taxonomists and nomenclaturalists is the incorporation of environmental nucleic acid sequences (ENAS) into taxonomic studies. One species, *Hawksworthiomyces sequentia*, known only from ITS sequences has recently been described in the *Ophiostomatales*, although the Melbourne Code does not provide for the valid description of such taxa for which no specimen exists.

T2-41-03

New Canadian lineages of ergot fungi revealed by multi-gene phylogeny

Parivash Shoukouhi[1], *Jim G. Menzies*[2], *Zlatko Popovic*[2], *Keith A. Seifert*[1], *Rafik Assabgui*[1], ***Miao Liu***[1]

1. Agriculture and Agri-Food Canada, Ottawa Research and Development Centre

2. Agriculture and Agri-Food Canada, Cereal Research Centre

Claviceps purpurea, the type species of the genus, was first described by Tulasne 1853 based on a specimen from rye. Other cereal crops (i.e. *Triticum* and *Avena*) and grasses (*Alopercuri*, *Dactylis*, *Lolium*, and *Poa*) were noted as additional hosts. To date, the reported host range has increased to more than 400 species in at least 19 genera belonging to 3 subfamilies of grasses (Poaceae). Unlike other *Claviceps* spp. that infect only a few genera in the same tribe, *C. purpurea* is unique in having a wide host range, leading to suspicion that it might be a complex of multiple species. Studies have confirmed cryptic speciation correlated with ecological niches. The taxonomy of *Claviceps* species occurring in Canada has been little explored. Among 600 specimens of *Claviceps* in the National Mycology Herbarium (DAOM), 80% are labeled as *C. purpurea*. To better understand species composition and genetic diversity of ergot fungi in Canada, we conducted multi-gene phylogenetic studies using recently collected samples from different parts of the country, along with historical specimens housed in DAOM. The loci sequenced include rDNA ITS, house-keeping genes (*EF1-alpha*, *BT* and *RPB2*) and ergot alkaloid synthesis genes (*easE*, *cloA*). Based on individual locus and concatenated alignments, the analyses recognized six lineages within the *C. pupurea* complex, among which three appear to be new. Morphological and phenotypic studies were undertaken to characterize these lineages.

T2-41-04

Using molecular dating in all cases of fungal classification or only in higher taxa?

Sinang Hongsanan[1], *Santiago Sánchez-Ramírez*[2,3], *Rui-Lin Zhao*[4], *Kevin D. Hyde*[1]

1. Center of Excellence in Fungal Research, Chiang Rai 57100

2. Department of Natural History, Royal Ontario Museum, Toronto, ON M5S 2C6

3. Department of Ecology and Evolutionary Biology, University of Toronto, Toronto, ON M5S 3B2

4. State Key Laboratory of Mycology, Institute of Microbiology, Chinese Academy of Sciences, Beijing 100101

The classification and taxa delimitation in the Fungi are often subjective, being based on morphological differences and therefore other studies (e.g. phylogenetics) are needed to clarify unresolved taxa. The ranking of fungal species especially at the higher levels needs sets of evidence for their delimitation. Morphology and molecular data have been used for fungal identification and classification for many years. However, there are many problematic groups that need further collections and evidence and new techniques for their clarification. Many species also need to be restudied since they have been introduced repeatedly from different parts of the world. Thus, giving wrong names can cause many problems in identification and classification. The molecular dating technique is a key tool for interpreting the history of life of any organisms and it presently has become a common tool for many biologists in research areas. Molecular clock analysis can use information from anatomy embryology, biogeography, fossils and biological molecules. Fungal evolutionary study has used fossil evidence and molecular data to estimate the evolutionary events. Many thousands of papers have used molecular dating for estimating species divergences times. The evolutionary events can explain the rate of mutation of the interested organisms and present their history of life. The major factor that can cause mutation in each evolutionary can be estimated. Presently, researchers have increasingly used larger datasets that can estimate a much greater diversity of species. However, do we need to use the molecular dating in all cases? and do we presently have enough evident for molecular clock analysis?

T2-41-05

Secondary metabolites as chemotaxonomc and phylogenetic markers in an important order of Ascomycota

Marc Stadler[1], *Wendt Lucile*[1], *Eric Kuhnert*[1], *Esteban B. Sir*[2]
1. Helmholtz Centre for Infection Research, Dept. Microbial Drugs
2. Fundación Miguel Lillo, CONICET, Institute of Mycology, Miguel Lillo 251, San Miguel de Tucumán 4000, Tucumán

The Xylariaceae (Xylariales, Sordariomycetes) are a large, hyperdiverse family of the Ascomycota, comprising macrofungi that produce conspicuous stromata on wood or dung, as well as microfungi that belong to the predominant endophytic mycobiota of seed plants. On the other hand, xylariaceous fungi are well-known as producers of bioactive metabolites with selective activities against human, animal and plant pathogens, as well as industrial enzymes, or as biocontrol agents. However, little is currently known about their taxonomy, phylogeny, biogeography, and ecology. Our recent work involves taxonomic and chorological studies on the tropical members of the family, to provide a basis for concise identification of the genus *Hypoxylon* and allied taxa. In addition, we have studied the stromata and cultures of these fungi for the occurrence of biologically active secondary metabolites. While some of these compounds constitute valuable chemotaxonomic marker molecules that helped to resolve species complexes, others show strong and selective activities in biological systems and may serve as lead compounds for development of anticancer agents, antibiotics or agrochemical fungicides. The results of our taxonomic work, with particular emphasis of a study of *Annulohypoxylon* and *Hypoxylon* in Argentina will be presented. A new multi-gene genealogy that resulted in the rearrangement of the Xylariales and the genera of Xylariaceae will be illustrated. Furthermore, some of our recent work on the discovery of novel secondary metabolites will be highlighted.

T2-41-06

Fungal biodiversity and functions in the pine barrens ecosystem

Emily Walsh, Stephen Miller, Jing Luo, Stacy Bonos, Zane Helsel, John Dighton, Debashish Bhattacharya, ***Ning Zhang***
Rutgers University

Pine barrens ecosystem has acidic, sandy and nutrient-poor soils. Pines and oaks are the dominant trees, and other common plants in pine barrens include switchgrass and other grasses (Poaceae), sedges (Cyperaceae), orchids (Orchidaceae), blueberries and other heath family members (Ericaceae). Scarce attention has been paid to studies of fungi in the pine barrens, and much remains unknown about fungal biodiversity and functions in this ecosystem. The largest and most uniform area of pine barrens in the United States is the 1.4 million acre (57,000 km^2) pine barrens of New Jersey (NJ) located in the southern region of the state. The NJ Pine Barrens represents one of a series of barrens ecosystems along the eastern seaboard of USA and one of a series of similar ecosystems around the world. The podzolic soil in the NJ Pine Barrens is highly acidic (pH ~4.0 with very low cation exchange capacity), sandy, dry (low moisture holding capacity), nutrient poor (e.g. low in P, K, N) and containing elevated levels of soluble aluminum. During the 1600's and 1700's when settlers first came to this area they discovered most of the region's soils would not support the growing of grain crops from traditional European agriculture. Therefore they named the region "Barrens". But why and how do wild grasses thrive in the pine barrens acid and nutrient poor soil? Do fungi play a role in aiding plant establishment and survival in such stressed environments? In order to address these questions, we have been studying root-associated fungi in the pine barrens using both culture method and Illumina metabarcoding method. Our data showed that with regard to fungal diversity, the pine barrens is anything but barren. We found that grass (Poaceae) roots in the pine barrens are one of the major reservoirs of novel fungi with 47% being undescribed species. Importantly, we observed that *Acidomelania panicicola* (Leotiomycetes), a new genus and species we described from switchgrass in the NJ Pine Barrens increases root hair growth of switchgrass and rice in low nutrient conditions. We also described another new genus *Pseudophialophora* (Magnaporthales, Sordariomycetes) that contains several pathogenic species. Eleven new species from the pine barrens have been published. Naming and description of a number of other new fungal taxa are in progress. We compared the root fungal community between the wild switchgrass from the NJ Pine Barrens and cultivated switchgrass in managed farms using both culture and metabarcoding methods. Both methods suggest that Leotiomycetes are dominant fungi in the switchgrass roots from pine barrens soils, while Sordariomycetes are dominant in the roots growing in the rich farm soils. Metabarcoding method uncovered a more diverse microbiome from the plant roots. More experiments are being conducted to test our hypothesis that *Acidomelania* and other Leotiomycetes play a role in increasing plant tolerance to abiotic stresses (e.g., low pH, low nutrients, drought) and contribute to improved establishment in poor soils. Results from this work will facilitate ecological and evolutionary studies on root-associated fungi.

T2-42: Evolution of land plants and their interaction with the environment

T2-42-01

Evolution of MIKC-type MADS-domain proteins and the origin of novelties in land plants

Guenter Theissen[1], *Florian Ruempler*[1], *Marcus Blohs*[1], *Rainer Melzer*[2]
1. Department of Genetics, Friedrich Schiller University Jena
2. School of Biology and Environmental Science, University College Dublin

MIKC-type MADS-domain transcription factors (MIKC TFs) represent a remarkable synapomorphy of streptophytes (charophyte green algae + land plants). Based on the presence of a keratin-like K domain they have the capacity to constitute different dimeric and tetrameric complexes in a combinatorial way. By binding to specific *cis*-regulatory elements on DNA they constitute floral quartet-like complexes (FQCs) that control diverse sets of target genes and hence developmental processes in land plants. The expansion and diversification of MIKC TFs is strongly correlated and probably causally linked to the origin of major novelties, such as ovules/seeds, flowers and fruits, during the evolution of land plants. These novelties have significantly shaped the way in which land plants interact with their environment. Since the constituents of FQCs are all paralogous MIKC TFs, events such as gene duplication, sub- and neofunctionalisation, paralog interference and escape form adaptive conflict all may have played important roles during the evolution of MIKC TFs in land plants. To better understand the proximate causes that generated major novelties in land plants, we study the interaction of MIKC TFs in a phylogenetic framework throughout the streptophytes. We will report about recent insights concerning the origin of FQCs during evolution and the molecular basis of FQC formation and function. We hypothesize that the unique phylogenetic trajectories of paralogous MIKC TFs interacting in FQCs facilitated the diversification of efficient developmental switches and hence the origin of novelties in land plants.

T2-42-02

A temporal evolutionary framework of land plants from multiple experimental molecular clock analyses

Yin-Long Qiu, *Alexander Taylor*
University of Michigan, Department of Ecology and Evolutionary Biology

Time-related information is essential to understand the history of life on earth. In this study, four aspects of data and methods were explored in a series of Bayesian statistical analyses of divergence time implemented in BEAUTi and BEAST, to develop a temporal evolutionary framework of land plants. First, four different matrices were constructed: a 5-gene (cp-*atp*B, cp-*rbc*L, mt-*atp*1, mt-*nad5*, and nu-*18S*) and 616-species matrix, and three phylogenomic matrices: chloroplast, mitochondrial, and combined chloroplast-mitochondrial. The 5-gene matrix sampled >70% families and the phylogenomic matrices included most orders of land plants. Second, one full- and two sub-sets of fossil calibration points were assembled, which spanned across the entire history of land plants and green algae. Third, different topologies of relationships among angiosperm-gymnosperm lineages and bryophyte-tracheophyte lineages were experimented, in addition to a series of phylogenetic constraints used in the analyses. Fourth, various combinations of rate, substitution, and speciation models were used in the analyses. These schemes of data assembly and analytic experimentations allowed comprehensive evaluation of data sufficiency and model adequacy that had not been done in most molecular clock analyses, and they offered quantitative assessment of robustness of the molecular clock analysis results. When fossils are still being discovered for many important groups of land plants, e.g., bryophytes and basal angiosperms, and molecular clock analyses seem to produce unreasonably old and a wide range of estimates for ages of some lineages, e.g., crown group angiosperms, it is imperative that molecular analyses are conducted rigorously and internally consistent and precise results are obtained.

T2-42-03

The importance of physiology for understanding the ecology and evolution of Paleozoic plants

C. Kevin Boyce
Stanford University

Paleoecology is necessarily founded on modern ecology, but that presents a distinct challenge: the modern world is utterly dominated by one highly unique lineage, the flowering plants, that have only been around for a quarter of land plant history. Thus, even when differences in form and structure are expressly recognized in the fossil record, implicit assumptions based on modern angiosperm ecophysiology can lead to very misleading expectations about the Paleozoic origins of vascular plant vegetation. I will present arguments that productivity of non-flowering plants has always been consistently low relative to what is seen among modern flowering plants. The implications of low productivity will then put important constraints on the potential biology of Paleozoic fossil plants. With this context, the early evolution of plant architecture and ecology across the vascular plants will be broadly considered.

T2-42-04

Synchronized radiations of fungi and plants linked to symbiosis

François Lutzoni[1], *Michael D. Nowak*[2], *Michael E. Alfaro*[3], *Valérie Reeb*[4], *Jolanta Miadlikowska*[1], *David Swofford*[1], *A. Elizabeth Arnold*[5], *David Hibbett*[6], *Khidir Hilu*[7], *Timothy Y. James*[8], *Dietmar Quandt*[9], *Susana Magallón*[10]
1. Duke University
2. University of Oslo
3. University of California, Los Angeles
4. University of Iowa
5. University of Arizona
6. Clark University
7. Virginia Tech
8. University of Michigan
9. Reinische Friedrich-Wilhelms-Universität
10. Universidad Nacional Autónoma de México

Ecological interactions between fungi and plants, ranging from parasitism to mutualism and including plant decomposition by saprotrophic fungi, have been invoked as a central mechanism underlying their respective macroevolutionary success. Accordingly, their putative synchronized radiations might be detectable phylogenetically. Here we compare the origins of plant-fungal symbioses and saprotrophy, detecting linked and drastic shifts in diversification rates of both kingdoms in a comprehensive time-calibrated phylogenetic framework. The largest radiation during the evolution of land plants (tracheophytes) and fungi (Leotiomyceta) occurred simultaneously. However, fungi colonized land before embryophytes, likely facilitating their terrestriality through endomycorrhizal symbioses and possibly through endophytic symbioses aboveground. In contrast, the diversification of saprotrophic mushrooms occurred in the Late Paleozoic as terrestrial forests of seed plants (spermatophytes) started to dominate the landscape. The subsequent diversification and explosive radiation of Agaricomycetes, and eventually of ectomycorrhizal mushrooms, were associated with the evolution of Pinaceae in the Mesozoic, and the establishment of angiosperm-dominated biomes in the Cretaceous, respectively.

T2-42-05

Before angiosperms: Insect pollination during the Mid Mesozoic

Conrad Labandeira
Smithsonian Institution

Gymnosperm-dominated floras of the mid-Mesozoic world were significantly different from the angiosperm-dominated floras that succeeded them. Diverse assemblages of Late Jurassic to Early Cretaceous gymnosperm groups were morphologically disparate at the highest taxonomic levels and included cycads, ginkgophytes, corystosperms, caytonialeans, bennettitaleans and gnetophytes that largely were replaced by the angiosperms that originated during the Early Cretaceous and continued to expand throughout the mid Cretaceous. Similarly, insect faunas occurring on gymnosperms previous to the advent of angiosperms were taxonomically distinct from those later associated with angiosperms. Much of the ecologic structure of these two evolutionary biotas can be gleaned from the myriad of plant-insect interactions present as pollination, herbivory and mimicry found in gymnosperm-rich deposits that essentially lack angiosperms. Such gymnosperm-rich compression-impression deposits include the Jiulongshan Formation, of latest Middle Jurassic age (about 165 million years old) in Inner Mongolia, and the Yixian Formation, of mid Earlly Cretaceous age (about 125 million years old) in Liaoning, both from Northeastern China. Other examples of well-preserved deposits include three, spectacularly preserved amber deposits of earlier Cretaceous age from Lebanon, Myanmar and Spain that range in age from 125 to 99 million years, spanning much of the Early Cretaceous. These compression-impression and amber deposits harbor many important plant-insect associations of the pre-angiosperm Mesozoic and provide fundamental insights into a lost world of plant and insect evolutionary and ecological convergences that parallel the analogs of today. The only difference is that the plant and insect interactors are taxonomically distinct while the associations--involving plant and insect morphologies, nectar and pollen rewards, mouthpart structure and functional feeding groups--are virtually identical. Four major insect pollination modes have been documented on gymnosperms during the pre-angiosperm Mesozoic. These documented gymnosperm-insect pollination modes include the four pollination modes of: (i) the thrips pollination mode consisting of mouthcone mouthparts engaged in punch-and-sucking feeding, pollen rewards and occurrences on ginkgoalean hosts; (ii) the beetle pollination mode bearing mandibulate mouthparts, a chewing feeding style, pollen rewards and hosted by cycads; (iii) the long-proboscid pollination mode found in scorpionflies, lacewings, flies and moths that possess a siphonate feeding style of imbibing pollination-drop fluids from deep-throated or tubular ovulate fructifications; and (iv) the labellate pollination mode made possible by the expanded labellum of brachyceran flies that fed by sponging from specialized ovulate cones of cheirolepidiaceous conifers. These pollination modes also exhibited four evolutionary-ecological patterns during the Aptian-Albian Gap in which the gymnosperm-to-angiosperm transition occurred. These changes included: early pollinator lineages associated with gymnosperms that became extinct; other gymnosperm-associated lineages that remained on gymnosperm hosts but significantly decreased in diversity, some surviving to the present day; still other gymnosperm-associated lineages that transitioned onto angiosperms via plant-host shifts; and later evolving pollinator lineages that originated with angiosperms, lacking any gymnosperm affiliation in their histories. This combination of several major pollinator modes and patterns of evolutionary-ecological change essentially resulted in the modern insect pollinator fauna.

T2-42-06

Embryophyte evolution as the main driver of global climate change and mass extinctions during the mid-Paleozoic Era

Thomas Algeo
1. University of Cincinnati
2. China University of Geosciences-Wuhan

The evolution of embryophytes over an extended, ~100-Myr-long interval of the mid-Paleozoic Era may have been the most important factor in driving contemporaneous climate change and triggering global biotic crises. The role of early vascular plants in elevating rates of burial of organic carbon, drawing down atmospheric CO_2 levels, and triggering a major glaciation during the Devonian Period was recognized twenty years ago (*Algeo, T.J. and Scheckler, S.E., 1998. Terrestrial-marine teleconnections in the Devonian: links between the evolution of land plants, weathering processes, and marine anoxic events. Philosophical Transactions of the Royal Society of London B: Biological Sciences, 353:113-130*). More recently, attention has shifted toward the possible role of early pre-vascular (bryophyte-grade) land plants in drawing down atmospheric CO_2 levels and triggering global cooling during the mid- to late Ordovician. As the record of fossil discoveries of the earliest land plants improves, evidence for such connections is accumulating. Both the Ordovician and Devonian events had profound effects on contemporaneous marine faunas, although of differing nature. The Ordovician cooling trend initially promoted increased animal diversification, probably due to invigorated oceanic circulation, improved ventilation, and enhanced nutrient upwelling and productivity. As marine productivity and organic carbon burial peaked in the late Ordovician, however, 'climatic overshoot' resulted in a hypercooling event (the Hirnantian glaciation at ~445 Ma) that led to a global mass extinction (the first of the so-called 'Big Five' mass extinctions of the Phanerozoic). The Devonian cooling trend occurred in a stepwise manner over an interval of ~30 Myr that resulted in repeated second-order biotic crises, culminating in the end-Devonian Hangenberg mass extinction at ~360 Ma (*Algeo, T.J., Berner, R.A., Maynard, J.B. and Scheckler, S.E., 1995. Late Devonian oceanic anoxic events and biotic crises: "rooted" in the evolution of vascular land plants. GSA Today, 5 (3): 45,64-66*). In this presentation, I will review the record of embryophyte evolution from the Ordovician through Devonian periods as well as contemporaneous changes in global climate, biodiversity, and marine and terrestrial environmental conditions. I will propose that embryophyte evolution was the key driver of global changes during a ~100-Myr-long interval of the mid-Paleozoic, and that long-term climatic cooling driven by expansion of terrestrial floras was ultimately responsible for setting the stage for the Permo-Carboniferous Ice Age, the most severe and most protracted (~350-290 Ma) glacial event of the Phanerozoic. The influence of embryophytes on global events was exceptional during the mid-Paleozoic because it represented a transition from a largely unvegetated to a fully vegetated world. Since the Devonian, global plant biomass has probably varied little (despite major changes in its composition, such as the Cretaceous advent of angiosperms), and, thus, land plants have not continued to have an outsize influence on global events.

T2-43: Natural history of *Ginkgo biloba* and its interaction with human

T2-43-01

Incongruent range dynamics between co-occurring Asian temperate tree species facilitated by life history traits - *Ginkgo biloba* as an example with new insights from plastid and mitochondrial genomes

Marcus Koch[1], *Eva Wolf*[1], *Nora Hohmann*[1], *Philippe Rigault*[1], *Markus Kiefer*[1], *Junpeng Zhao*[2], *Cheng-Xin Fu*[2]
1. Heidelberg Universitiy, Centre for Organismal Studies (COS) Heidelberg
2. The Key Laboratory of Conservation Biology for Endangered Wildlife of the Ministry of Education, College of Life Sciences, Zhejiang University

Postglacial expansion to former range limits varies substantially among species of temperate deciduous forests in eastern Asia. Isolation hypotheses (with or without gene flow) have been proposed to explain this variance, but they ignore detailed population dynamics spanning geological time and neglect the role of life history traits. Using population genetics to uncover these dynamics across their Asian range, we infer processes that formed the disjunct distributions of Ginkgo biloba and the co-occurring Cercidiphyllum japonicum (published data). Phylogenetic, coalescent, and comparative data suggest that Ginkgo population structure is regional, dichotomous (to west–east refugia), and formed ~51 kya, resulting from random genetic drift during the last glaciation. This split is far younger than the north–south population structure of Cercidiphyllum (~1.89 Mya). Significant (recent) unidirectional gene flow has not homogenized the two Ginkgo refugia, despite 2 Nm > 1. Prior to this split, gene flow was potentially higher, resulting in conflicting support for a priori hypotheses that view isolation as an explanation for the variation in postglacial range limits. Isolation hypotheses (with or without gene flow) are thus not necessarily mutually exclusive due to temporal variation of gene flow and genetic drift. In comparison with Cercidiphyllum, the restricted range of Ginkgo has been facilitated by uncompetitive life history traits associated with seed ecology, highlighting the importance of both demography and lifetime reproductive success when interpreting range shifts. Additional evidence comes from plastid and mitochondrial genome sequence data highlighting diversification during the Late Pleistocene.

T2-43-02

Demographic history of *Ginkgo bioloba* revealed by population genomics

Yun-Peng Zhao[1], *He Zhang*[2], *Guang-Yi Fan*[2], *Ping-Ping Yin*[1], *Xin Liu*[2], *Xiao-Ning Hong*[2], *Fu-Min Zhang*[3], *Xun Xu*[2], *Jian Wang*[2], *Huan-Ming Yang*[2], *Cheng-Xin Fu*[1], *Song Ge*[3], *Wen-Bin Chen*[2]
1. The Key Laboratory of Conservation Biology for Endangered Wildlife of the Ministry of Education, College of Life Sciences, Zhejiang University
2. BGI-Shenzhen
3. State Key Laboratory of Systematic and Evolutionary Botany, Institute of Botany, Chinese Academy of Science

Ginkgo biloba L. (Ginkgoaceae) is one of the best-known and most distinctive trees worldwide, representing one of the four lineages of extant gymnosperm without living relatives. It is a living fossil that has been essentially unchanged in gross morphology for more than 200 million years. It survived glaciations as a relic in China followed by a human-aided global redistribution, and thus provides an inspiring example of how human can help a species survive and renew. Ginkgo's resilience has enabled it to become widespread and popular across the world. This resilient tree demonstrates an outstanding resistance or tolerance to both herbivores and pathogens, accounting in part for the longevity of the individual tree and also, in turn, for the longevity of this species. Despite substantial advances in recent two decades, knowledge gaps remains for ginkgo's natural history followed by human-assisted introduction. As a follow-up project of *de novo* sequencing of ginkgo genome (10.61 Gb), we resequenced the genomes of 300 big ginkgo trees selected from the evidenced refugial populations and introduced populations to 5× raw data coverage using BGISEQ-550 platform. We will investigate genome-wide variation patterns in ginkgo populations and will obtain a tremendous volume of high-quality single nucleotide polymorphisms (SNPs) after excluding sites with missing data in any accession. Using these population SNP data, we will be able to reconstruct the comprehensive picture of demographic and introduction history of ginkgo populations in China and to unveil the relevant mechanisms. Genes with significantly differential diversity can be identified in diverse populations, which might represent candidate regions undergoing selection either naturally or during introduction/domestication. These genes will greatly facilitate the identification of functional variations and agronomically important genes. Dense variation data will also be useful for marker-assisted resource evaluation, breeding and gene mapping of ginkgo. The present study will provide a milestone case study of population genomics for non-model organisms in particular with large genomes.

T2-43-03

Strategies and approaches for population genome resequencing of non-model plants: Implication of genome rese-

quencing of ginkgo populations

Wenbin Chen[1], *Yunpeng Zhao*[2,3], *Chengxin Fu*[2,3], *Song Ge*[4], *Guangyi Fan*[1], *Xin Liu*[1], *He Zhang*[1], *Rui Guan*[1,5]
1. BGI-shenzhen
2. The Key Laboratory of Conservation Biology for Endangered Wildlife of the Ministry of Education, College of Life Sciences, Zhejiang University, Hangzhou 310058
3. Laboratory of Systematic & Evolutionary Botany and Biodiversity, Institute of Ecology and Conservation Center for Gene Resources of Endangered Wildlife, Zhejiang University, Hangzhou 310058
4. State Key Laboratory of Systematic and Evolutionary Botany, Institute of Botany, Chinese Academy of Sciences, Beijing 100093
5. State Key Laboratory of Bioelectronics, School of Biological Science and Medical Engineering, Southeast University, Nanjing 210096

The "living fossil"*Ginkgo biloba* L. (Ginkgoaceae) is the sole extant species representing one of the four lineages of gymnosperm with a large genome (~10.61 Gb) composing of ~70% repeat sequences. Resequencing of ginkgo populations will be another milestone in illustrating population structure, demographic history, adaptation, secondary metabolism and origin of important traits (seed, pollen tube, carpel, sexuality and *etc.*) of ginkgo. Here, we elaborately selected ginkgo accessions from all over the world to perform whole-genome re-sequencing using BGISEQ-500 sequencing platform. To our knowledge, this is the first case study of genome resequencing on non-model plants with such large and complicated genome, which challenging our ability of handling enormous sequencing data. By using BGISEQ-500 along with efficient alignment and variants calling algorithm, we dramatically reduced time consuming and resource dependency. High-quality population-level genetic markers (SNP, INDEL, SVs) were generated for identification of genetic structure, demographic history and local adaptation. Combining phenotype information and genome-wide association study (GWAS), we further identified candidate region controlling important traits, especially dioecious reproduction. This study highlights the effectiveness and efficiency of the strategies and approaches which were successfully applied in population re-sequencing of ginkgo genomes, shedding new light on re-sequencing large and complex genomes of non-model plants.

T2-43-04

Regional characteristics in genetic variation and introduction history of *Ginkgo biloba*, inferred using giant old trees in Japan

Keiko Katakura, *Tomohiro Kawakami*, *Watanabe Yoichi*, *Eijiro Fujii*, *Koichi Uehara*
Chiba University

Ginkgo biloba, which is originated in China, is one of common planted tree species in Japan. In Japan, the species is spread by people and mainly planted in temples and shrines, it means that the species was introduced by association of buddhists. It is considered that initial introduction of the species into Japan had occurred during Asuka (ca. 600-700 A.D.), Kamakura (ca. 1190-1330 A.D.) or Muromachi (ca. 1400-1570 A.D.) period in the history. In the present, 70 giant old trees (over 10m in girth at breast height, GBH) live in Japan. Because *Ginkgo biloba* is long-lived, these giant-old trees are considered to be residuals in initial introduction stage into Japan. Therefore, these trees are considered to be representatives of genetic variations in the initial introduction. Here, we collected 272 samples from giant old trees, which are over 8 m in GBH, from Kyushu to Tohoku regions in Japan and analyzed using eight nuclear microsatellites previously developed to understand characteristics and history of the introduction for *Ginkgo biloba* into Japan.

T2-43-05

Gingo biloba, its close relationship with people

Ye Liu
Kunming institute of botany, Chinese academy of sciences

Environment degradation has grown in recent decades because increased human population density, which has influenced the biological evolution in a way. The famous relic plant, *Ginkgo biloba*, is a representative example for this case. However, in another way, *Ginkgo biloba* influences human life in many aspects, such as, culture, foods, and medicine. In this study, based on the evidences from population demography, biological geography, and genetic information, we explored how *Ginkgo biloba* dispersed into the world from its wild population in eastern China and how it was used by human being. The results from this study will be useful to preserve the green earth and ourselves. Also, the methods used in this study will be extended in other fields.

T2-43-06

Application of Gydle bioinformatics tools to *Ginkgo* organellar studies

Philippe Rigault[1,2], *Eva Wolf*[2], *Nora Hohmann*[2], *Marcus Koch*[2]
1. Gydle
2. Heidelberg University

The bioinformatics methods and software developed by Gydle allow the analysis of complex genomes using NGS data. These tools enable the genome assembly of genetically diverse organisms, and the study of genomic elements with different evolution dynamics across populations and species. We present a Ginkgo plastid genomes study demonstrating high-throughput tools for the assembly of complete chloroplast genomes from total DNA sequence data in plants with large nuclear genomes. The finished-level quality of assemblies provides unprecedented resolution for phylogeography analyses. We show that Ginkgo plastid genomes exhibit extremely low levels of sequence diversity, and present methods to identify often-overlooked sequencing and assembly artifacts in order to increase phylogenetic signals. The assembly of high-quality and complete plant mitochondrial genomes from total DNA sequence data is a more challenging task, owing to large variations in size (typically 300 kbp to 10 Mbp), the potential for recombination, repeat expansion and exchange with the plastid genome, and the different evolution dynamics of core genes with regard to other mitochondrial genome regions. Our results on Ginkgo mitochondrial genomes will be shown as well as comparisons between mitochondrial and plastid phylogenetic analyses. For large genomes such as Ginkgo, we present novel methods enabling the study of the gene space by combined assembly of RNA-Seq and DNA-

Seq data, and the application of comparative genomics based on simultaneous analysis of allelic diversity and copy number.

T2-44: Building and exploring the green plant tree of life (two sessions)

T2-44-01

Consensus and conflicts in genome and gene histories across the green tree of life

Jim Leebens-Mack
University of Georgia

Since their origin approximately one billion years ago, green plants (Viridiplantae) have formed the foundation for energy acquisition across the vast majority of ecosystems on our planet including marine, fresh water, terrestrial and agricultural systems. Great progress is being made on understanding ecological, molecular, biochemical and taxonomic diversity in green plants. Here we present a comprehensive phylogenomic estimation of relationships and patterns of diversification across the Viridiplantae. Analyses of nearly 400 universally single copy nuclear genes sampled from transcriptomes and available genomes for nearly 1,200 species representing all extant green plant diversity has yielded a robust phylogenetic inference for the evolutionary history of the Viridiplantae including repeated bouts of rapid diversification associated with developmental and ecological innovations. These diversification events are marked by a high degree of discordance between gene tree and estimated species trees. The implications of our phylogenetic inferences and gene tree - species tree discordance will be discussed.

T2-44-02

Plastome phylogenomics and evolution of the monocots

Thomas Givnish[1], *Alejandro Zuluaga*[2], *Sean Graham*[3], *James Leebens-Mack*[4], *J. Chris Pires*[5], *Jerrold Davis*[6], *Dennis Stevenson*[7], *Wendy Zomlefer*[4], *Craig Barrett*[8]
1. University of Wisconsin-Madison
2. Universidad del Valle
3. University of British Columbia
4. University of Georgia
5. University of Missouri-Columbia
6. Cornell University
7. New York Botanical Garden
8. University of West Virginia

Phylogenetic analyses of just a handful of genes provided an initial division of ca. 83 monocot families into 12 orders, but failed to provide decisive evidence regarding the relationships of several of these orders relative to each other (e.g., commelinids, Asparagales, Liliales) and the relationships of many families to each other within most of the orders. Here we present a phylogenomic analysis of relationships among monocot families and orders based on aligned sequences of 75 genes from the plastid genome and of entire plastomes for more than 350 placeholders. These analyses resolve relationships among all monocot families and orders with strong support for the first time and provide a backbone phylogeny on which to base many future studies of character-state evolution, ecological radiation, historical biogeography, and determinants of net rates of species diversification. Remarkably, however, even entire plastomes lack sufficient information to resolve a few deep but short branches (e.g., within Zingiberales) with 100% bootstrap support. Plastome-scale data also provide strong support for the placement of many mycoheterotrophic taxa; these have proven highly problematic in the past, due to loss of several plastid genes and greatly accelerated evolution of many others. A supermatrix approach extends the plastome backbone to produce a monocot phylogenetic hypothesis including placeholders for most monocot genera. Supermatrix phylogenies provide a means for testing several hypotheses regarding the drivers of extraordinary diversification in orchids, and for reconstructing historical biogeography in orchids and the order Liliales. In several orders, phylogenomic analyses point to biogeographic origins in Gondwana, with inferred overland dispersal via Antarctica when that continent had a subtropical climate and was connected to the Australian Plate and South America.

T2-44-03

From cacti to carnivores: A genome-wide perspective on Caryophyllales phylogeny

Michael Moore[1], *Ya Yang*[2], *Joseph Walker*[3], *Samuel Brockington*[4], *Stephen Smith*[3]
1. Oberlin College
2. University of Minnesota
3. University of Michigan
4. University of Cambridge

The Caryophyllales comprise approximately 6% of angiosperm species diversity (approximately 12,500 spp. in 40 families) and are found on all continents and in all terrestrial ecosystems, from tundra to tropical rainforests, but are particularly diverse and abundant in semiarid, saline, and temperate environments. In conjunction with this ecological variation, the clade exhibits extreme life history diversity, ranging from tropical trees, temperate annual herbs, and long-lived succulent cacti to a diverse array of carnivorous plants, including sundews and epiphytic pitcher plants. This remarkable diversity has fostered a long history of research, making Caryophyllales one of the best-characterized major angiosperm clades with respect to morphology, ecophysiology, and life history. For example, the Caryophyllales exhibit repeated origins of such fundamental morphological and ecophysiological traits as a differentiated perianth, betalain and anthocyanin pigmentation, C4 and CAM photosynthesis, and leaf and stem succulence. Hence, resolving the backbone phylogeny of Caryophyllales and its constituent families with confidence would provide the necessary framework for future comparative evolutionary analyses of these and other key traits. Recent multigene phylogenetic analyses have helped to clarify the backbone relationships among many Caryophyllales lineages, but the core Caryophyllales in particular contain several relatively ancient and apparently rapid radiations that have so far proved difficult to resolve (e.g. early core Caryophyllales, early Portulacineae, and early phytolaccoids). To resolve these relationships, we employed transcriptome sequencing to generate a large data set of over 150 transcriptomes for key Caryophyllales taxa. Phylogenetic analyses of these data have helped to resolve the backbone relationships of Caryophyllales while simultaneously revealing high levels of gene tree incongruence at certain key nodes.

T2-44-04

Deciphering the genetic and genomic basis of extreme plant adaptation in the Caryophyllales

Ya Yang[1], *Michael J. Moore*[2], *Joseph F Walker*[3], *Samuel F Brockington*[3], *Stephen A. Smith*[3]

1. University of Minnesota-Twin Cities
2. Oberlin College
3. University of Michigan, Ann Arbor

Modern transcriptomic and genomic techniques offer the ability to explore the patterns and drivers of plant diversification on a scale never before possible. The plant order Caryophyllales contains ~12,500 species in 40 families with an estimated crown age of approximately 67–121 Ma. Species of the Caryophyllales are found on every continent including Antarctica, and in all terrestrial ecosystems as well as aquatic systems. Familiar members of the group include cacti, diverse array of carnivorous plants (e.g., the sundews *Drosera* and pitcher plants Nepenthes), and a number of crop plants (e.g. beet, spinach, and quinoa). Using a data set of genomes and transcriptomes from over 300 species across the Caryophyllales we found evidence for at least 21 ancient genome duplication events, which is more than previously known in all of angiosperms. From the data set of >15,000 genes per species, we discovered that many gene duplication events were associated with key adaptive trait changes, such as evolution of pigmentation and plant carnivory. Our work integrates not only new molecular phylogenomic and computational methods, including several developed for this study, but also field- and collection-based research to connect rapid phenotypic shifts to their genetic and genomic bases.

T2-44-05

Complete plastid genome sequences and the green plant tree of life

Matthew Gitzendanner[1], *Gane Wong*[2], *Pamela Soltis*[1], *Douglas Soltis*[1]

1. University of Florida
2. University of Alberta

The plastid genome has a long history of use in plant molecular systematics. From the initial single gene *rcb*L studies in the early 1990s through today, the plastid genome has served as the workhorse of plant molecular systematics. With the advent of second generation sequencing technologies, sequencing of entire plastid genomes for large numbers of taxa has become feasible. We report on the phylogeny of 1,899 green plant taxa using complete plastid gene sequences. Overall support of the phylogeny, as assessed by bootstrap analysis, is high with many historically problematic nodes receiving high support, and the relationships supported by the tree consistent with previous hypotheses based on plastid data. Our analysis supports previous findings of the Zygnematales as sister to the land plants. We recover a monophyletic bryophte clade, with mosses, liverworts and hornworts sister to the vascular plants. Within the gymnosperms, we recover support for the Gne-Cup hypothesis. *Amborella* alone is recovered as sister to the remaining angiosperms, followed by Nymphaeales, Austrobaileyales, and Chloranthales and Magnoliales. In addition to reviewing the topology of the plastid phylogeny, we will make comparisons to a comparable nuclear gene-based phylogeny and discuss the challenges of filtering transcriptome data to generate reliable sequences.

T2-44-06

Rosaceae nuclear phylogeny and fruit type evolution in the context of geological times and genome duplication

Hong Ma[1], *Yezi Xiang*[1], *Chien-Hsun Huang*[1], *Yi Hu*[2], *Jun Wen*[3], *Jun Xiang*[4], *Tingshuang Yi*[5]

1. Ministry of Education Key Laboratory of Biodiversity and Ecological Engineering, Institute of Plant Biology, Center of Evolutionary Biology, School of Life Sciences, Fudan University
2. Department of Biology, the Huck Institutes of Life Sciences, the Pennsylvania State University
3. Smithsonian Institution,
4. Hubei Key Laboratory of Economic Forest Germplasm Improvement and Resources Comprehensive Utilization, School of Life Sciences, Huanggang Normal College
5. Plant Germplasm and Genomics Center, Germplasm Bank of Wild Species, Kunming Institute of Botany, Chinese Academy of Sciences

Fruits are the defining feature of angiosperms, likely have contributed to angiosperm successes by protecting and dispersing seeds, and provide foods to humans and other animals, with many morphological types and important ecological and agricultural implications. Rosaceae is a family with ~3,000 species and an extraordinary spectrum of distinct fruits, including fleshy peach, apple, and strawberry prized by their consumers, as well as dry achenetum and follicetum with features facilitating seed dispersal, excellent for studying fruit evolution. To address Rosaceae fruit evolution and other questions, we generated 125 new transcriptome and genomic datasets and identified hundreds of nuclear genes to reconstruct a well-resolved Rosaceae phylogeny with highly supported monophyly of all subfamilies and tribes. Molecular clock analysis revealed an estimated age of ~101.6 Mya for crown Rosaceae and divergence times of tribes and genera, providing a geological and climate context for fruit evolution. Phylogenomic analysis yielded strong evidence for numerous whole genome duplications (WGDs), supporting the hypothesis that the apple tribe had a WGD and revealing another one shared by fleshy fruit-bearing members of this tribe, with moderate support for WGDs in the peach tribe and other groups. Ancestral character reconstruction for fruit types supports independent origins of fleshy fruits from dry-fruit ancestors, including the evolution of drupes (e.g., peach) and pomes (e.g., apple) from follicetum, and drupetum (raspberry and blackberry) from achenetum. We propose that WGDs and environmental factors, including animals, contributed to the evolution of the many fruits in Rosaceae, which provide a foundation for understanding fruit evolution.

T2-44-07

Phylogenomics of angiosperms from the 1KP Consortium

Pamela Soltis

University of Florida

The 1KP initiative provides transcriptome resources and phylogenetic inference across the approximately half a billion species of *Viridiplantae*. Angiosperms represent ~3/4 of *Viridiplantae* and account for ~2/3 of the transcriptomes assembled by 1KP. Phylogenomic analyses using a supermatrix comprising ~800 species of angiosperms and over 400 nuclear genes provide a comprehensive view of evolutionary history from the perspective of the nuclear genome. Notable phylogenetic results include: *Amborella* sister to all other extant angiosperms, with subsequent divergence of Nymphaeales and Austrobaileyales prior to the radiation of *Mesangiospermae*, within which relationships remain poorly resolved and/or supported. Ceratophyllaceae are sister to all other mesangiosperms, with monocots sister to a clade of magnoliids + (Chloranthales + eudicots), but with weak support. Within magnoliids, there is strong support for Magnoliales + Laurales sister to Piperales + Canellales. Relationships among basal eudicots are not strongly supported, but Trochodendraceae are sister to Buxaceae + core eudicots. Within core eudicots, there is general support for *Superrosidae* and *Superasteridae*, although the placement of Santalales within *Superrosidae* requires further attention. As found previously with nuclear data, the COM clade of *Rosidae* is sister to malvids, whereas nearly complete plastomes place it with fabids. Other key points of angiosperm phylogeny will be discussed and comparisons made between nuclear and plastid trees. The transcriptome resources of 1kp can be used in a variety of ways, at both deep and shallow evolutionary scales, and provide for discovery of putatively single-copy nuclear genes for use in phylogeny reconstruction of other groups.

T2-44-08

Phylogenomic analysis of plastid genomes confirmed the position of Mitrastemonaceae and revealed rapid radiation of Ericales

Chaonan Fu, *Dezhu Li, Lianming Gao*
Kunming Institute of Botany, Chinese Academy of Sciences

Ericales is a highly morphologically heterogeneous order sister to Euasterids in the largest Asterids clade in angiosperms. It comprises 22 families including a parasite family Mitrastemonaceae. Previous molecular studies have circumscribed some major clades of the order, but deeper nodes are largely unresolved, and Mitrastemonaceae has never been incorporated in phylogenetic analyses of Ericales. A total of 105 plastid genomes covering all families of Ericales were obtained here to explore the interfamilial relationships in the order and ensure the systematic place of Mitrastemonaceae. Our results showed that the plastid genome of Mitrastemonaceae was highly reduced at a size of ~26 kb and lack of inverted repeat region compared to other members of Ericales. Fourteen conserved protein coding genes and four rRNA genes found in plastome of Mitrastemonaceae were selected in all plastomes of Ericales and other 32 orders of Eudicots to verify the position of Mitrastemonaceae. Results revealed that the parasite family was placed in Ericales, but the interfamily relationship was uncertain. Phylogenetic relationships within Ericales were reconstructed using Maximum likelihood method base on 83 genes of 105 Ericales species with outgroups of 29 Cornales and three Caryophyllales. Some strongly supported clades were resolved: (1) Marcgraviaceae- (Tetrameristaceae-Balsaminaceae) was sister to the remaining families of Ericales; (2) Polemoniaceae-Fouquieriaceae subsequently formed a clade sister to the remainder of the order; (3) (Primulaceae-Ebenaceae)-Sapotaceae; (4) Pentaphylacaceae-Sladeniaceae; (5) ((Actinidiaceae-Roridulaceae)-Sarraceniaceae)- ((Ericaceae-Cyrillaceae)-Clethraceae); (6) (Styracaceae-Diapensiaceae)-Symplocaceae. Phylogenetic relationships/position of some families and clades were unresolved. The systematic position of Mitrastemonaceae and Lecythidaceae within Ericales was also not fully confirmed in this study. Very short internal branches of families in the phylogenetic tree may indicate a rapid radiation diversification of Ericales, which posing a particularly difficult challenge for phylogenetic inreference of Ericales.

T2-44-09

Revisiting the relationships among bryophyte lineages with increased taxon and gene sampling, and phylogenomic approaches

Norman Wickett, *Matthew Johnson*
Chicago Botanic Garden

Bryophytes comprise three monophyletic lineages: the mosses, liverworts, and hornworts. The extant members of one of these lineages, or some combination of the three, represent the sister group to the vascular plants and, as such, play a key role in understanding the evolution of terrestrial plants approximately 480 million years ago. Several molecular phylogenetic studies using chloroplast genomes and a limited set of nuclear genes in the 1990s and early 2000s led to the consensus view that bryophytes comprise a paraphyletic grade with liverworts and mosses successively sister to a clade composed of hornworts and vascular plants. However, alternative topologies have been proposed and supported by both molecular and morphological data sets. More recently, phylogenomic studies employing hundreds of nuclear markers have suggested that the consensus view of successive sister groups might not represent the true relationships, with growing support for the monophyly of mosses and liverworts, with the hornworts remaining somewhat unstable in their placement. Here, we discuss past, current, and ongoing efforts to reconstruct bryophyte relationships, with an emphasis on understanding conflicting phylogenetic signal among genes and datasets.

T2-44-10

Completing the plant tree of life

William Baker, *Abigail Barker, Steven Dodsworth, Wolf Eiserhardt, Ester Gaya, Jan Kim, Olivier Maurin, Joe Parker, Lisa Pokorny, Felix Forest*
Royal Botanic Gardens, Kew, Richmond, Surrey, TW9 3AE

Evolutionary trees are powerful tools for prediction, species discovery, monitoring and conservation. Through comparative analysis of DNA sequence data, the backbones of the plant and fungal trees of life are relatively well understood, and many subcomponents have been studied in great detail. However, DNA data are still lacking for numerous genera and the vast majority of species of plants and fungi, preventing their accurate placement within this evolutionary framework and hindering downstream science. To better understand how the world's plants and fungi are related to each other and how they have evolved, we have initiated a project at the Royal Botanic Gardens, Kew to complete the Plant and Fungal Trees of Life (PAFTOL). We will utilise our collections and work with our collaborative networks to produce extensive new DNA sequence data (whole plastid genomes, 100s of nuclear loci) for a representative species from each genus of plant and fungus using high-throughput sequencing technologies. This comprehensive investigation of phylogenetic relationships will be a rich resource enabling the discovery and study of evolutionary patterns in the plant and fungal kingdoms, and will provide a unifying framework for comparative research. The project is an essential step towards the compilation of genomic data for all known species. This talk will focus on the plant tree of life component of the project (focusing on angiosperms), highlighting current progress in sampling, phylogenomic methodologies, bioinformatic approaches and collaborative opportunities.

T2-44-11

Tree of Life: China Project

Zhi-Duan Chen[1], *Li-Min Lu*[1], *Tuo Yang*[1], *Jian-Fei Ye*[1,2], *Bing Liu*[1], *Hai-Hua Hu*[1,2], *Dan-Xiao Peng*[1,2], *Yan-Ting Niu*[1,2], *Min Chen*[1]
1. State Key Laboratory of Systematic and Evolutionary Botany, Institute of Botany, the Chinese Academy of Sciences
2. University of Chinese Academy of Sciences

China is ranked among the top six megadiverse countries of the world and has 31,362 species of vascular plants belonging to 3,328 genera and 312 families. We herein reconstructed the Tree of Life of the Chinese vascular plants at the generic level based on five molecular markers by sampling 6,098 species (representing 93% vascular genera from China). The comprehensive large phylogeny supports most relationships among and within families recognized by recent molecular phylogenetic studies for lycophytes, ferns (monilophytes), gymnosperms, and angiosperms. For angiosperms, most families in Angiosperm Phylogeny Group IV are supported as monophyletic, except for a paraphyletic Dipterocarpaceae and Santalaceae. The infrafamilial relationships of several large families and monophyly of some large genera are well supported by our dense taxonomic sampling. To facilitate further application of such large-scale phylogeny to other biology fields, the SoTree (http://www.darwintree.cn/flora/index.shtml) software was introduced to enable the efficient generation of the phylogenetic trees by providing sub-datasets with interested species lists for studies concerning the origin, ecology, and biogeography of the local flora in China. On the basis of a robust phylogenetic framework, we further investigate the temporal divergence patterns of the Chinese angiosperm flora using a dated mega-phylogeny. Our results showed that the majority of Chinese generic flora (61%) originated during the Neogene and Quaternary, and the remaining genera diverged in the Paleogene (32%) and the Cretaceous (7%). We found evidence for an older divergence history in East China with higher phylogenetic diversity (PD), while a younger divergence history in West China with lower PD. Our data suggest that West China represents an evolutionary cradle while East China is a museum for Chinese angiosperm genera.

T2-44-12

Evolution of genus *Ludwigia* L. (Onagraceae) and polyploid evolution of section Macrocarpon

Shih-Hui Liu[1,2], *Peter C. Hoch*[2], *Peter H. Raven*[2], *Janet C. Barber*[1]
1. Saint Louis University
2. Missouri Botanical Garden

Ludwigia L. (Onagraceae) is a worldwide wetland genus of plants that includes 82 species in 23 sections. This genus has a wide distribution, with endemic groups of species in North and South America, Africa, and Asia. Several species are of medicinal or horticultural interest; some are invasive weeds in wetlands worldwide. With a comprehensive multilocus phylogenetic analysis representing 73% of species in *Ludwigia*, we have tested the current classification of the genus and its validity as a reflection of the phylogenetic relationships of the species. Our results suggest that the North Temperate Haplostemonous (NTH) group, the *Microcarpium* complex, and sections *Ludwigia*, *Jussiaea*, and *Macrocarpon* are monophyletic clades, while others, including the largest section, *Myrtocarpus*, are paraphyletic and/or polyphyletic. Our results indicate that hybridization and polyloidy have each played important roles in the evolutionary history of the genus. We discuss the origins of several polyploid taxa. Moreover, based on our multilocus phylogeny, the divergence time estimations suggest that sect. *Ludwigia* and the *Microcarpium* complex split at ca. 26.7 mya during the late Oligocene Period of the Paleogene Era, sect. *Jussiaea* diverged ca. 41.0 mya during the middle Eocene, and the crown age of sect. *Macrocarpon* is ca. 11.5 mya during late Miocene. Furthermore, ancestral range estimations show that the NTH clade likely originated in North America, while other taxa originated in Central or South America, followed by several independent dispersal events to the Old World. The ancient state reconstruction analyses support the following character state pairs as plesiomorphic/derived: diplostemony/haplostemony, seeds free/seeds embedded, pluriseriate seeds/other patterns, diploid/polyploid, and tetramery/pentamery or more. We explored polyploid evolution of section *Macrocarpon* as a model for the genus, using both genome skimming and target enrichment methods. This section contains two diploids (*L. bonariensis*, *L. lagunae*) and one tetraploid (*L. neograndiflora*), all with restricted distributions; a fourth polyploid, *L. octovalvis*, distributed throughout the warmer regions of the world. Phylogenies based on organelle genomes together with 32 enriched putative nuclear single copy genes suggest that *L. lagunae* is closely related to *L. bonariensis,* and *L. bonariensis* may have contributed a genome to the tetraploid *L. neograndiflora*. A series of Mexican diploid populations grouped within *L. octovalvis* deserve taxonomic recognition. Hexaploid *L. octovalvis* likely arose from tetraploid strains through the functioning of an unreduced gamete. The methodology, involving organelle genomes and nuclear single copy genes, that we have used in this study should be useful to further studies of *Ludwigia* and other polyploid taxa.

T2-45: Plant genome size diversity and evolution: From genomics to ecology

T2-45-01

Plant genome size diversity: Towards a comprehensive understanding across angiosperms

Jaume Pellicer[1], *Oriane Hidalgo*[1], *Maarten Christenhusz*[1,2], *Ilia J*

Leitch[1]
1. Royal Botanic Gardens, Kew
2. Plant Gateway

In recent years, research into genome size evolution in land plants has increased significantly, uncovering a huge diversity across different taxonomic lineages. Despite such interest and the vast amount of data now available (> 9,500 species collated in the Plant DNA C-values database (http://data.kew.org/cvalues/), many plant families are still lacking a single estimate. Certainly, such taxonomic gaps are particularly notably in angiosperms, where nuclear DNA contents are only available for c. 2,095 genera out of the c. 13,000 genera currently recognised and > 9000 species (c. 2.5% of the c. 352,000 species recognised). At the family level, this translates into estimates for about 262 families out of 416 (*sensu* the Angiosperm Phylogeney Group - APG IV*), i.e. nearly 62% of the currently accepted diversity. With the currently available data, we have been able to unveil a staggering diversity of nuclear DNA contents in angiosperms that spans up to 2400-fold. Notwithstanding, the question still remains as to (i) what is the full extent of diversity of genome sizes in angiosperms? and (ii) what is yet to be uncovered? In order to bring further robustness to our current understanding of angiosperm genome size diversity, we have set up a project aimed at filling most –if not all- the taxonomic gaps, starting at the family level. In the age of high throughput sequencing technologies, C-values provide important baseline information not only for genomic sequencing projects, but also to shed light on why such genome size diversity exists and what impact does it have on plant evolution. Our goal is to provide at least one estimate per family recognised by APG IV, focusing our efforts on those lineages largely unstudied. So far we have obtained first C-values for 216 genera, which represent new additions to 82 families that previously lacked any genome size information. These data have already highlighted the existence of plant lineages with distinctive genome size profiles and contribute towards our knowledge of how ups and downs of genome size evolution in plants, may be linked to and impact on the diversity of intrinsic and extrinsic processes which are shaping plant genome diversity.

T2-45-02

Ecological consequences of angiosperm genome size under macronutrient limitation and herbivory stress.

Maité Guignard[1,2], *Ilia Leitch*[1], *Michael Crawley*[3], *Richard Nichols*[1], *Rob Knell*[1], *Andy Macdonald*[4], *Mark Trimmer*[1], *Andrew Leitch*[1]
1. Royal Botanic Gardens, Kew
2. Queen Mary University of London
3. Imperial College
4. Rothamsted Research

Genome size (GS) is a fundamental trait influencing cellular, developmental and ecological parameters in angiosperms. There is huge variation in GS among species, yet most genomes are small despite the prevalence of polyploidy in many angiosperm lineages. Nucleic acids are amongst the most nitrogen (N) and phosphate (P) demanding molecules, and GS may thus influence a plant's nutrient demands and its ability to grow in environments where N and P are limited. We first test the hypothesis that large genome sizes are costly to build and to maintain by analysing the impact of N and P treatment on the abundance of angiosperm species with a range of genome sizes, ploidy levels, and competition strategies in natural plant communities at the Park Grass Experiment (Rothamsted, UK), the world's longest running ecological experiment. We further test whether GS influences how species and plant communities respond to herbivore pressure in long-term exclusion experiments at Silwood Park (Imperial, UK). Support for macronutrient limitation of plants with larger GS was found in the Park Grass Experiment, where plants with large genomes (e.g. 1C-value > 5 pg) are under greater macronutrient limitation in comparison to plants with small genomes. Polyploidy further influences the abundances of plants with large GS. In terms of herbivory, the key finding is a highly significant negative association between GS and rabbit herbivory. Our results suggest that ploidy and genome size are functional traits which contribute to how plants respond to nutrient and predation stress, thus influencing plant abundances, community composition, and hence, potentially food web and nutrient cycling dynamics.

T2-45-03

Population perspectives on genome size evolution through polyploidy: Dynamics of unreduced gametes and phenotypic divergence within species

Brian Husband
University of Guelph

Variation in genome size, achieved through multiplication of genome copy number (i.e. polyploidy), is widespread in plants and represents a unique source of phenotypic and genomic novelty with important implications for adaptation and speciation. Biologists have long sought to explain the prevalence of polyploidy by studying its genetic, genomic, and phenotypic consequences. Ultimately, however, the influence of these mechanisms on genome size evolution must be interpreted in terms of their impact on key population processes (formation, establishment, persistence). This perspective is lacking for all but a few species despite the abundance of intraspecific variation. Here, I present insights into two aspects of the population biology and evolutionary dynamics of polyploidy: 1) polyploid formation via unreduced gametes, and 2) polyploid establishment via rapid phenotypic and ecological divergence. Polyploid formation is governed by the frequency and distribution of unreduced gametes, and mating patterns within populations, neither of which are well understood in natural populations. Using new techniques in flow cytometry and high throughput screening of gametes across multiple species, I identify factors governing the persistence of 2n gametes and enhancing the likelihood of polyploid formation. In contrast to unreduced gametes, the role of phenotypic divergence in polyploid establishment is well recognized; however, the immediate effects of genome duplication on phenotype, and the factors determining the magnitude of divergence are limited to only a few cases. I will present recent evidence from two species (*Chamerion angustifolium, Arabidopsis thaliana*) that suggest the effects of genome doubling are highly variable within species but dependent on progenitor genotype and phenotype. These results suggest that the phenotypic impact of genome multiplication and therefore the potential for polyploid establishment may be non-uniform within species and across the geographic range. Collectively, this research highlights the importance of a population perspective when explaining the prevalence and uneven distribution of genome size via genome multiplication. Through further work, we hope to better understand why some taxa are predisposed to polyploid evolution and others not; and under what conditions new polyploids will arise, establish and evolve.

T2-45-04

Retrotransposons and whole genome duplications shape genome (size) variation

Christian Parisod
University of Neuchâtel

The dynamics of retrotransposons (RT) and whole genome duplications (WGD) are major molecular drivers of genome size, but their causes and consequences among plant populations deserves further attention. As reported in both autopolyploid Buckler Mustards and allopolyploid wild wheats, hybridization coupled with WGD induces conflicting interactions among RT families and promotes their proliferation. The prevalence of gene flow as a proximal cause of RT dynamics and genome size variation among plant populations can now be addressed with adequate tools. Available evidence further indicates that RT proliferation supports genome fractionation and species diversification, but the relationship with genome size variation and to what extent this is non-neutral has to be elucidated. Here, I highlight recent insights on interactions between RTs and duplicated genes during the colonization of stressing environments, and discuss how pseudogenisation and innovation may impact on genome trajectories towards either downsizing or obesity.

T2-45-05

Repeat and genome size dynamics in *Nicotiana* polyploids of different ages

Steven Dodsworth
Royal Botanic Gardens, Kew

Genome size in most angiosperms is dictated by changes in repetitive element (repeat) abundance, that often can constitute a majority of nuclear genomic DNA (gDNA). The other main mechanism by which plants increase their total gDNA is through rounds of polyploidy, or whole genome duplication. Post-polyploidisation, however, changes also occur to the repeat content of genomes, and this can be dramatic. In the genus *Nicotiana* (Solanaceae) polyploidy is pervasive, with allotetraploids being formed recurrently over different timescales and with alternative parental lineages. This sets *Nicotiana* as an excellent model system in which to look at changes in genome size and structure post-polyploidisation. Here I present the results of high-throughput sequencing across the genus in order to compare repeat profiles and genome size dynamics. In young allopolyploids formed less than 1 Mya (e.g. *N. rustica* or *N. tabacum*) the polyploid genome is almost additive for its two putative parental subgenomes, and this is also reflected in repeat dynamics. However, for much older allopolyploids (c. 4-7 Mya) genome downsizing is the norm, and concurrent with this is much restructuring of the repetitive fraction of the genome. In all cases there also seems to be significant bias as to gDNA loss, in favour of retaining maternal DNA, i.e. as predicted by the cytonuclear discordance hypothesis. In summary this presents a timescale for the diploidisation process that occurs in most angiosperms at some point following rounds of polyploidy, where genomes are restructured and streamlined to resemble their diploid ancestors.

T2-45-06

Influence of genome size on stomatal size and photosynthetic efficiency in *Fritillaria* (Liliaceae)

Sarah Seco
1. Queen Mary University of London
2. Royal Botanical Gardens, Kew
3. Petersham Lodge

The genus *Fritillaria* (Liliaceae), which comprises ca 140 species of bulbous perennial plants, includes species with the largest diploid genomes so far reported (1C-values ranging from 30.8pg to 100.4pg). This contrasts strikingly with the majority of angiosperms which are characterized by much smaller genomes (data for >10,000 species show the modal 1C-value is 0.6 pg /1C, and mean = 5.1 pg/1C). While there is ongoing research into the molecular dynamics underpinning the origin and evolution of such large genomes, our understanding of how this impacts various physiological processes is more limited. This talk will focus on how photosynthesis is affected by genome size via its impact on stomatal size. Previous studies across a diverse range of plants have shown that there is a broad correlation between genome size and stomatal guard-cell size, and that this may, in part be driven by changes in atmospheric CO_2 concentration over geological time (Franks *et al.*, 2012). Given that guard-cell size is likely to influence, for example, gas and water exchange dynamics needed for photosynthesis and hence potentially photosynthetic efficiency, we have probed this subject further by analyzing three different factors that contribute to photosynthetic efficiency in ten species of *Fritillaria* that differ in genome size (37.41 - 89.2 pg/1C). The approaches used were (i) infra-red gas analyser (IRGA) to investigate the rates of CO_2 uptake and Rubisco activity, (ii) pulse amplitude modulation (PAM) fluorometry to gain insights into the efficiency of light energy harvesting for photosynthesis, and (iii) stomatal measurements (size and density). The results shed light on how genome size interacts with other plant traits and abiotic factors to influence photosynthesis dynamics in these genome giants.

T2-46: Plant phylogeography in East Asia

T2-46-01

A review and meta-analysis of plant phylogeographical studies in the Ryukyus

Koh Nakamura
Hokkaido University Botanic Garden

Phylogeographical studies on continental archipelagos generally seek explanations for geographical patterns of genetic differentiations in the history of island connections and separations. Old straits, thereby, have been considered to be the most likely barriers in the determination of genetic boundaries among island populations. However, caution should be taken when discussing phylogeographical patterns based on geological history. Biogeographic patterns have several possible explanations, of which historical and ecological biogeography are contrasting approaches. Historical biogeography seeks explanations in geological processes such as sea level change, crustal deformation and plate tectonics, which acted thousands to millions of years in the past. Ecological biogeography, in contrast, attempts to explain biogeographical patterns by the effects of more recent and current environments. The theories of historical and ecological biogeography have developed separately, but any biological phenomena are not the result of a single factor, either historical or current environmental. It is therefore necessary to analyze and explain biogeographical patterns from ecological as well as historical biogeographical viewpoints. The islands of the Ryukyus provide interesting settings for biogeographic studies that are rarely available elsewhere on the Earth: chain-like arrangement of the islands stretching over a thousand kilometers, landbridge formation/split during the past few million years, and latitudinal climatic gradient. Preceding phylogeographical studies in the Ryukyus chiefly considered the oldest straits that split the landbridge, the Tokara and Kerama gaps, as the most effective barriers that caused phylogeographic differentiation. Of the phylogeographical surveys so far conducted on terrestrial plants from the Ryukyus, some revealed large genetic differentiations across the Tokara and/or Kerama gaps, while others found a lack of differentiation across either gap. At this stage, there is no clear trend in the dispersal strategies/modes of studied plants on whether they showed genetic differentiation across the gaps. I therefore presume that ecological factors had (and still have) an additional role in the origin and maintenance of lineage divergence in the Ryukyus. It is feasible, for example, that the spatial configuration of the islands, including their isolated distribution, small size, and low elevation, imposes obstacles to the dispersal through limiting the availability of habitats for some plant species. In addition, adaptive processes should not be excluded from consideration. If an environmental selection gradient and related ecophysiological differentiation exist among populations, migration would be limited and, as a result, populations would be genetically differentiated even at neutral genetic loci. Examination of ecological factors (e.g. latitudinal low-temperature gradient) and related physiological differentiations among populations (e.g. cold-tolerance differentiation along the latitude) can offer explanations for the different phylogeographical patterns without assuming stochastic events. Integration of geohistorical and ecological explanations is essential to advance our understanding of plant phylogeography in the Ryukyus.

T2-46-02

Spatial and temporal evolution of the Arid Central Asia inferred from molecular phylogeography of desert plants

Yan Yan[1,2], *Chaoju Qian*[1], *Xing-Ke Fan*[1], *Xiao-Yue Yin*[1], *Hengxia Yin*[1], *Yong Shi*[1], ***Xiao-Fei Ma***[1]

1. Key Laboratory of Stress Physiology and Ecology in Cold and Arid Regions, Gansu Province, Northwest Institute of Eco-Environment and Resources, Chinese Academy of Sciences

2. Heihe Key Laboratory of Ecohydrology and Integrated River Basin Science, Northwest Institute of Eco-Environment and Resources, Chinese Academy of Sciences, Lanzhou 730000, Gansu

The formation and evolution of the Arid Central Asia (ACA) was associated with the uplift of the Qinghai-Tibet Plateau (QTP) and the development of East Asian Monsoon system (EAMS). However, most of the evidences came from loess sediment records. To fully understand the spatial and temporal evolution of the ACA, here, we thoroughly investigated molecular phylogeography of one zonal constructive shrub *Reaumuria soongarica* and one pioneer annual plant species *Agriophyllum squarrosum*, both of which were widely distributed across all the desert ecosystems from the ACA. Based on cpDNA, nrITS and 13 nuclear loci, we found a complete divergence between the western and eastern populations could be traced to 2.96 ~ 0.8 Ma in *R. soongarica*. However, ecological niche modeling supported that the frequent historical global climate changes could comprehensively fragmented the habitat of this desert shrub. Thus, the present distribution of the desert ecosystems were shaped by both the long term habitat fragmentation driven by the uplift of northeast QTP and Pleistocene climate oscillations. Both cpDNA and nrDNA supported that *R. soongarica* could bidirectionally colonize into the Gurbantunggut Desert (the north desert) from the eastern and western since ~ 0.37 Ma, inferring to the asynchronous aridification of the regional ACA. Approximate Bayesian Computation (ABC) simulation strongly suggested scenario that the North group could originate with two hybridizations between the East and West groups, which could be facilitated by the multiple aridification events in the late of Pleistocene. As the desertification processes of the ACA were much more influenced by development of East Asian Winter Monsoon, which strongly supported by wide stratifigraphic records from the marginal lakes and loess around ACA. Our molecular data on cpDNA and nrITS indicated that *A. squarrosum* could originate from the north Gurbantunggut Desert or the neighbor area since ~1.6 Ma, and subsequently underwent three waves of colonisation into other deserts and sandy lands corresponding to several glaciations during late Pleistocene. The rapid population expansion and distribution range shifts of *A. squarrosum* from monsoonal climate zones suggested that the development of the monsoonal climate significantly enhanced the population growth, and unbalanced gene flow of *A. squarrosum* which could be an indicator for the historical movements of sand dunes in response to global climate change. Therefore, more efforts should be focused on profiling the gene flow of these desert plants with genomic data. These studies on psammophytes exemplified how the desert ecosystems of the ACA responded to the geological and paleo-climate changes, for the first time from the molecular phylogeographic view.

T2-46-03

Secondary contact, hybridization and polyploidization add to the biodiversity in the Hengduan Mountains, exemplified by the widespread *Corallodiscus lanuginosus* (Gesneriaceae)

Ping Zhou[1,2], *Jie Li*[2], ***Michael Moeller***[3]

1. Yunnan University of Traditional Chinese Medicine

2. Xishuangbanna Tropical Botanical Garden, Chinese Academy of Sciences

3. Royal Botanic Garden Edinburgh

Corallodiscus lanuginosus is distributed 12,000 km across its northeast-southwest boundaries in Central Asia, showing its highest morphological diversity in the Hengduan Mountains, in Southwest China. We investigated the mechanisms behind the high morphological diversity exhibited by this species. We acquired molecular sequence data across the distribution range, reconstructed a dated phylogeny, projected the resulting clades onto geographic maps and linked the results with inferred ploidy levels of the populations. We found strong geographic patterns for the ITS clades but with extensively overlapping ranges, particularly in the Hengduan Mountains region. We used the presence of ITS polymorphisms to infer a high level of hybridization events mainly in areas of clade overlap, where we also inferred repeated tetraploidization events to have taken place. We deduced an origin of *C. lanuginosus* in the mid Oligocene. Historical pre- and post-Quaternary climate oscillation-driven expansions and contractions of distribution ranges have apparently allowed secondary contacts, hybridization and introgression to have occurred that shaped the current patchy distribution of morphological diversity in the species. Polyploidization is seen as a reinforcing genetic barrier added genetic diversity in space and time particularly in the Hengduan Mountains.

T2-46-04

Range expansions of plant species in subtropical China during the Quaternary

Zhiyong Zhang, *Dengmei Fan, Yixuan Kou*
Laboratory of Subtropical Biodiversity, Jiangxi Agricultural University

Range expansions have occurred repeatedly in the history of most species and they are presently occurring at an increasing rate owing to rapid climatic changes. Because subtropical China was not glaciated during the Quaternary, most phylogeographic studies revealed that multiple refugia and limited range expansion from refugia seem to be a paradigm for plant species in this region. However, this conclusion could be incomplete because early phylogeographic studies in the region biased toward endangered species with narrow distributions. Recently, rapid advances have be achieved concerning range expansions in subtropical China during the Quaternary climatic oscillations. 1) Extensive range expansions may be the rule rather than exceptions for plant species, as the region underwent profound climate changes throughout the Quaternary; 2) Molecular dating suggested that the time for range expansion events varied among different plant species, most species occurring in the Holocene or the last interglacial period. Glacial expansions have also been implicated in several studies. 3) Due to the relatively mild Quaternary climate in subtropical China, the genetic signature of multiple range expansions across different glacial-interglacial cycles was likely to be preserved. Phylogeographic inference in subtropical China is prone to be complicated by multiple range expansions; 4) With increasing cases of range expansion, several recolonization routes, which largely coincide with major mountains in subtropical China, have been identified, especially through the integration of phylogeographic and geospatial data. These new findings provide important insights into evolutionary histories of plant species in subtropical China and have profound implications for biodiversity conservation in the face of recent climate change.

T2-46-05

Biogeographic history of alpine flora in the Japanese Archipelago

Hajime Ikeda
Okayama University

High mountains in the Japanese Archipelago are one of the southernmost areas in the enormous ranges of arctic-alpine plants as well as alpine plants distributed around Beringia. One can hypothesize that alpine flora in this region may represent ephemeral occurrence of cold adapted species that dispersed southward during the last glacial maximum (LGM) and have survived the postglacial warm climate in relict populations on high mountains. On the other hand, the origin of alpine flora may have predated the LGM and populations of alpine plants may have persisted in Japan for prolonged periods in the Pleistocene. Recent phylogeographic studies on various alpine plants consistently revealed genetic divergence between populations in northern and central Japan, which implies prolonged persistence of alpine plants in central Japan. In this talk, I will show recent progresses on phylogeographic studies and provide insight into the biogeographic history of alpine flora in Japan. Analyses based on coalescent simulations demonstrated that genetic divergence of alpine plants between northern and central Japan predated the LGM. In contrast to the genetic uniqueness in central Japan, genetic similarity as well as recent divergent history was revealed between northern Japan and the Kamchatka Peninsula. Accordingly, these findings suggest that alpine flora at least in central Japan is unlikely the ephemeral occurrence of cold adapted species in the postglacial warm period but has persisted for several cool-warm climate cycles in the Pleistocene. On the contrary, alpine flora in northern Japan would originate from alpine and/or arctic-alpine plants that spread widely in northern East Asia during the LGM, while its persistence in the Pleistocene remains to be examined.

T2-46-06

Phylogeographic history of lithophytic plant genera disjunct in geologically active regions in East Asia

Shota Sakaguchi[1], *Ying-Xiong Qiu*[2], *Hiroaki Setoguchi*[1], *Yuji Isagi*[1]
1. Kyoto University
2. Zhejiang University

East Asian warm-temperate zone includes peripheral regions characterized by rapid geological movements and earthquakes due to active faults. The geological movements have continued to produce landslides, forming complex landscapes full of rocky outcrops, steep slopes and waterfalls. These structures harbor the habitats for lithophytic plant groups, which prefer to survive on rocks or rock-associated environments. Among these groups, two genera of *Shortia* (Diapensiaceae) and *Tanakaea* (Saxifragaceae) show remarkable disjunct distributions between the Japanese Archipelago-Ryukyu Archipelago-Taiwan and Yunnan-Sichuan regions. Contrary to most lithophytic plants, they have shallow root systems with fine roots extending on the surfaces of rock-associated habitats, which requires year-round wet microenvironment to maintain populations. The specific ecological requirements and their conservatism would have led to similar disjunct distribution in *Shortia* and *Tanakaea*. Therefore, these genera can be suitable model systems to investigate the relative impacts of geological activity and historical fluctuations of rainfall on phylogeographic structures and population demographies in East Asia, where most plant phylogeographic studies focused on temperature changes in late Pleistocene. This study will present the results of comparative analyses of *Shortia* and *Tanakaea* using chloroplast genome sequencing and ddRAD-sequencing/nuclear microsatellite markers to discuss the roles of environmental drivers that led to present disjunct distribution patterns.

T2-47: Taxonomy and conservation of ferns and Lycopods (two sessions)

T2-47-01

Phylogenomics analysis resolve the backbone of tree of life in extant ferns

Hui Shen[1], *Dongmei Jin*[1], *Xile Zhou*[1], *Xianchun Zhang*[2], ***Yue-Hong Yan***[1]
1. *Shanghai Chenshan Botanical garden/Shanghai Chenshan Plant Science Research Center, Chinese Academy of Sciences*
2. *Institute of Botany, Chinese Academy of Science*

Exant ferns are monophyletic group and the closest living relatives to seed plants and originated about 380 million years agao. However, the evolutionary relationships of extant ferns remain controversial based on different molecular evidences with different morophological interpretation. Here we aimed to resolved the early evolutionary relationships in ferns using high-quality RNA sequencing data obtained by high-throughput sequencing for 69 species representing all extant 11 orders and 30 families. We compared and analysed two supermatrices of homologous proteins encoded by 2,392 genes and 1,334 genes. From this analysis, we obtained a new topolgical structure with strong support by optimisation of the ML methods.These results allowed us for the resolution of key nodes that had been unclear in the past and revealed the evolutionary relationship of vertical annulate sporangia from exannulate sporangia. This well-resolved tree of ferns will constitute a framework for further studies of evolution and development in land plants.

T2-47-02

Diversity of ferns and lycophytes in Brazil

Jefferson Prado[1], *Regina Yoshie Hirai*[1], *Eric Schuettpelz*[2]
1. *Instituto de Botanica*
2. *Smithsonian Institution*

This paper presents the diversity of ferns and lycophytes in Brazil, after the previous compilation published in 2015. The methodology to estimate the diversity consisted in collecting data from regional checklists, taxonomic revisions, selected databases, and literature in general. The results show 1,310 species: 1,143 of ferns and 167 of lycophytes. This number is 4.35% higher than the previous one (1,253 spp.), reported in 2015. Were recognized 39 families and 150 genera (vs. 36 families, 133 genera in 2015). The 10 most diverse families are Pteridaceae (196 spp.), Dryopteridaceae (184), Polypodiaceae (164), Thelypteridaceae (97), Hymenophyllaceae (89), Selaginellaceae (80), Aspleniaceae (78), Lycopodiaceae (64), Anemiaceae (61), and Cyatheaceae (45). The three most diverse genera are *Elaphoglossum* (87 spp.), *Selaginella* (80), and *Asplenium* (74). The richest phytogeographic domain continues to be in the Atlantic Rainforest with 895 species which also has the largest number of endemic and threatened species, followed by the Amazon Rainforest (516), Cerrado (263), Pantanal (34), Caatinga (29), and Pampa (10). Minas Gerais remains as the richest state (666 spp. vs. 657 in 2015).

T2-47-03

Community phylogenetics of Florida ferns: Insights from integrated analyses of trees, traits, and climate data

Emily Sessa, *Sally Chambers*, *Benjamin Baiser*
University of Florida

Community phylogenetic methods enable us to examine the dynamics of plant community and ecosystem assembly in the context of species' evolutionary histories. Combined with ecological, functional, and climatic data, we can take an integrated approach to understanding the various biotic, abiotic, and evolutionary components driving patterns of diversity in communities at different temporal and spatial scales. The state of Florida in the United States has the richest fern flora of any state in the continental U.S. It is home to 149 species of ferns, including ~120 that are thought to be native. We are using community phylogenetic methods to investigate community structure and the ways in which trait associations and climatic factors shape the assembly of fern species at the county level. We have built a phylogeny for all species of Florida ferns using chloroplast markers, and constructed a matrix of functional trait data for all species using morphological descriptions in floras. We find that phylogenetic diversity and species richness are strongly correlated, as expected, across the state of Florida, and several traits show strong phylogenetic signal. We are also using species distribution models, which can be used to examine species' contemporary geographic ranges and identify the climatic factors that may have shaped these distribution patterns. These models can also be used to forecast anticipated shifts in distribution in response to changing climates. We have quantified the climatic niches of all fern species in Florida, and are examining the specific factors that drive each species' distribution, and how these relate to functional traits and species' evolutionary histories. Preliminary results suggest that species diverge in their tolerance for warmer temperatures and greater amounts of precipitation, with phylogenetic signal appearing at the family level and generic level. Species from the same genus and family tend to occupy similar temperature and/or precipitation regimes. Projected distributions for the year 2070 indicate that a few species may be able to expand their distributions in the future, if they are able to disperse to areas with newly appropriate climates, but the vast majority of Florida fern species are likely to experience severe reductions in their geographic distributions.

T2-47-04

Nuclear phylogenetics of ferns and its implications for taxonomy

Carl Rothfels
University of California, Berkeley

Much of our understanding of fern phylogeny, and thus the foundations of classifications that seek to recognize monophyletic groups, derives from inferences from plastid sequence data. This reliance on plastid data has potential weaknesses: plastid data come from a single linkage group so plastid-only studies can't corroborate their inferences with independent data; approaches that involve multiple gene trees (such as "species tree" inference) are unavailable; and being uniparentally inherited, plastid data are ill-suited for studies of hybridization or allopolyploidy. Fortunately, genomic resources for ferns have become much more available in the past five years, and there has been a corresponding increase in the use of nuclear markers for inferring fern phylogeny. Here, I'll discuss some of the taxonomic implications of these studies, ranging from those focused on the deep relationships among fern lineages to those involving more fine-scaled patterns, such as the prevalence of allopolyploid species. New opportunities bring with them new challenges, and nuclear data are no exception; I'll close with a discussion of some of the difficulties associated with using nuclear data for fern phylogenetics, and ways that the pteridological community may be able to move forward to an increasing robust classification.

T2-47-05

Phylogeny and biogeography of *Isoetes*

Eva Larsén, *Niklas Wikström, Catarina Rydin*
Stockholm University

The lycopods of the genus *Isoetes* have long held a fascination because of their fossil relatives within Isoetales, including some of the first trees. But the extant species also hold some intriguing mysteries. *Isoetes* are a mostly semi-aquatic genus with species all over the world. They are heterosporous and ligulate just like their closest living relative *Selaginella*. But the sister groups diverged long time ago, in the vicinity of 380 million years before present, and in their adaptation to aquatic environments the ancestors of present-day *Isoetes* likely lost many similarities to the ancestors of *Selaginella*. Because they are also substantially different on a molecular level it has been difficult to use outgroups in phylogenetic analyses of *Isoetes*. Their very uniqueness makes it difficult to place them within a framework. The phylogeny presents some intriguing insights into the history of the genus. Despite the high similarity between species there is well supported evidence for such old patterns as a Gondwana clade and a corresponding Laurasia clade. In my presentation I will disclose new details on the phylogeny and biogeography of *Isoetes*.

T2-47-06

Current condition of diversity information on Asian ferns and lycopods: An overview through the Asian Red List compilation

Atsushi Ebihara
National Museum of Nature and Science

Integrating diversity information on ferns and lycopods on an Asia-wide scale is an urgent and important task for taxonomists possessing massive knowledge about species of the region since it is not always possible to obtain sufficient information about Asian species from worldwide database and name index projects (e.g. The Plant List, GBIF, IUCN Red List *etc.*). The number of species (including subspecies and varieties, excluding hybrids) in Asia (excluding New Guinea) is estimated to be 4,650 according to our integrated checklist based on the latest flora and monographs. For screening globally threatened species, it is essential to extract candidate species on unified criteria throughout Asia, but it is quite difficult mainly due to difference in the degree of progress of local check-listings and red-listings. We therefore extracted candidate species by excluding certain LC (Least Concern) species from the total check list. As a result, 2,185 candidate species remained after exclusion of 2,465 LC species, and a 1-degree gridded preliminary distribution map showing the species density was produced based on the information from presently available sources (online specimen databases, floras, monographs, original descriptions *etc.*). It would be most important to refine the list by adding more information on taxon status, up-to-date local conservation status and distribution records in collaboration with as many specialists as possible.

T2-47-07

A worldwide molecular phylogeny of the morphologically cryptic genus *Botrychium* (Ophioglossaceae)

Benjamin Dauphin, *Jason Grant, Donald Farrar, Alessio Maccagni*
University of Neuchâtel

Background — The description of biodiversity is an essential starting point for any study related to evolutionary biology and conservation. Among these investigations, molecular phylogenies have led systematists to characterize more taxa, sometimes having similar morphologies despite their deep phylogenetic divergence, and among them, moonworts (*Botrychium* s.s.) are an emblematic example of morphologically cryptic species. Previous molecular studies on this genus did not include a complete taxonomic sampling of the taxa, nor multiple specimens from throughout the known geographical range of each taxon, and therefore to investigate the evolutionary relationships of its major clades, we increased both the taxonomic representativeness as well as the phylogenetic resolution by including additional chloroplast markers. *Emphasizing main questions and goals* — Our study aims to better understand basic phylogenetic relationships within *Botrychium*, then retrace the maternal lineages of allopolyploid taxa, and ultimately identify species diversity centers in the northern hemisphere. *Materials and methods* — Out of 35 published and thirteen putatively new taxa, we sampled from 5 to 15 populations per taxon throughout their geographical range (Asia, Europe, North America, Oceania, and South America) for a total of 365 specimens, most of which were genotyped with 20 nuclear enzyme loci to confirm taxa identification and provide evidence from both maternal and paternal parentage of allopolyploids. We then analyzed four chloroplast regions (*matK* intron, $trnH^{GUG}$–*psbA* and $trnL^{UAA}$-$trnF^{GAA}$ intergenic spacers, and *rpL16* intron region) on those specimens and we conducted a phylogenetic analysis based on Maximum Likelihood (ML) and Bayesian Inference (BI) methods to explore the genetic diversity and phylogenetic relationships between taxa. *Results* — We found strong support for the monophyly of the three major clades: Lanceolatum, Lunaria, and Simplex-Campestre, and resolved 17 subclades among those. Our results suggest multiple origins for at least four polyploid taxa. The Simplex-Campestre clade had the largest number of species despite having a similar total number of haplotypes to the Lunaria clade (62 and 59 respectively), which has the broadest worldwide distribution. In total, 47 taxa are considered in this new molecular phylogeny based on multiple accessions per taxon. *Conclusions and perspectives* — Our results support the taxonomic delineation of nearly all moonwort taxa previously described on the basis of morphology, chloroplast, and nuclear data. These results have also revealed several new morphologically cryptic taxa as well as substantial genetic diversity hidden within common species. Our study also strongly corroborates previously hypothesized maternal parentage of allopolyploid taxa, and provides support for multiple independent hybridization events generating the same allopolyploid taxon. Strong asymmetry in maternal parentage of allopolyploids by certain diploid species raises questions and provides clues about the understanding of how the process of polyploidy takes place within natural populations. Finally, our large sampling also provides evidence of long-distance migratory ability of *Botrychium* taxa on intra– and inter–continental scales, as well as probable directions of migration. Vectors of migration undoubtedly involve wind dispersal of spores and possibly bird migrations from the Northern to Southern Hemisphere, and further studies are needed to retrace the phylogeography history of the genus, especially in Asia where the taxonomic diversity remains poorly known.

T2-47-08

Toward a community-derived classification for extant lycophytes and ferns

Eric Schuettpelz
Smithsonian Institution

Phylogeny has long informed pteridophyte classification and, as our ability to infer evolutionary trees has improved, classifications aimed at recognizing natural groups have become increasingly predictive and stable. The ordinal and familial scheme by Smith *et al.*, (2006, *Taxon* 55: 705–731) was the first higher-level pteridophyte classification published in the molecular era. Founded on the principle of monophyly, while also recognizing the importance of maintaining well-established names and circumscriptions, this classification established a new standard in fern taxonomy. Over the past decade, many important advances have been made in our understanding of relationships, but only some of these were reflected in subsequent higher-level schemes. The Pteridophyte Phylogeny Group (PPG) was established to provide a modern, comprehensive classification for lycophytes and ferns, down to the genus level, utilizing a community-based approach. A global assemblage of 94 pteridologists ultimately came together to produce a classification treating an estimated 11,916 species in 337 genera, 51 families, 14 orders, and two classes. This first installment of the PPG classification is not intended as the final word on lycophyte and fern taxonomy, but rather a summary statement of current hypotheses, derived from the best available data and shaped by those most familiar with the plants in question. The hope is that it will serve as a resource for those wanting references to the recent literature on pteridophyte phylogeny and classification, a framework for guiding future investigations, and a stimulus to further discourse.

T2-47-09

Plastid phylogenomics resolve the deep relationships of Polypodiales

***Jin-Mei Lu**, Xin-Yu Du, De-Zhu Li*
Plant Germplasm and Genomics Center, Germplasm Bank of Wild Species, Kunming Institute of Botany, Chinese Academy of Sciences, Kunming 650201

A series of molecular phylogenetic studies in recent years have reconstructed framework of ferns. However, the circumscriptions of some families remain controversial and their relationships are not thoroughly resolved. Polypodiales represents 80~90% of the species diversity of ferns, and the relationships among the basic groups of Polypodiales remain unresolved or poorly supported, and phylogenetic positions of some families in Eupolypods are not completely resolved. The primary objective of this study is to explore the application of plastid phylogenomics in resolving difficult backbone relationships of Polypodiales. We sequenced plastomes from 37 species representing 19 families of polypod ferns, and constructed the phylogeny of Polypodiales based on all sequenced plastomes of ferns. Plastid phylogenomics can provide large chloroplast gene data to resolve difficult deep nodes in Polypodiales. Our results supported the sister relationships between Dennstaedtiaceae and Pteridaceae (BP > 90), and the Eupolypod clade was sister to the Dennstaedtiaceae-Pteridaceae clade (BP = 100). The Didymochlaenaceae-Hypodematiaceae clade was the firstly diverged lineage in eupolypods I (BP = 100), and Didymochlaenaceae was sister to Hypodematiaceae (BP = 100). The plastid genomic features, such as gene order, intron gain/loss, may also offer independent tests of phylogenetic hypotheses. The pseudogene is universal in ferns, and the pseudogene of *rps16* was found in all nine Polypodiaceae species with sequenced plastomes, while it only was found in *Odontosoria* (Dennstaedtiaceae) except for Polypodiaceae. Furthermore, *rps16* of *Elaphoglossum* (Dryopteridaceae) and *Lomariopsis* (Lomariopsidaceae) contains 2-3 terminators.

T2-47-10

Asplenium sect. *Thamnopteris* (Aspleniaceae): The morphological delimitation of species and molecular phylogeny based on chloroplast DNA sequences

***Shi-Yong Dong**[1], Zheng-Yu Zuo[1,2], Hui-Guo Zhao[1,2], Hui-Hui Ding[1,2], Ling-Ling Wei[1,2], Yi-Shan Chao[3]*
1. Key Laboratory of Plant Resources Conservation and Sustainable Utilization, South China BotanicalGarden, Chinese Academy of Sciences, Guangzhou 510650
2. University of Chinese Academy of Sciences, Beijing 100093
3. Department of Biomedical Science and Environmental Biology,Kaohsiung Medical University, Kaohsiung

The section *Thamnopteris* in *Asplenium* (Aspleniaceae), or called *A. nidus* complex, is unique in morphology (fronds simple, entire, with free veins connected at margin) and in cytology (only tetraploids and a hexaploid reported). Some 20 species are estimated within this group and the morphological limits between species are confusing. To determine the morphological delimitation of species we conducted comprehensive comparisons of morphology as well as phylogenetic analyses based on two datasets of chloroplast DNA sequences. Our results showed that most species in the complex can be clearly circumscribed by the combination of four morphological characters, i.e., type of stipe scales, shape of fronds' midrib, density of sori, and pattern of perispores. Phylogenetic analyses indicated that *A.* sect. *Thamnopteris* is not monophyletic. In *A.* sect. *Thamnopteris* the species sampled in our analyses were well resolved into two clades, *A. nidus* clade and *A. phyllitidis* clade. The clade *A. nidus*, morphologically supported by narrow scales, comprises *A. australasicum*, *A. goudeyi*, *A. harmanii*, *A. musifolium*, *A. nidiforme*, *A. nidus*, *A. oblanceolatum*, *A. setoi*, and *A. wagauensis* (ined.). A species outside *A.* sect. *Thamnopteris*, *A. amboinense*, was resolved to be sister to *A. nidus* clade. In the clade *A. phyllitidis*, which is supported by broad scales, clustered seven species: *A. antiquum*, *A. antrophyoides*, *A. cymbifolium*, *A. humbertii*, *A. palawanense* (ined.), *A. phyllitidis*, and *A. riswanii*. Some species with fronds finely dissected, such as *A. beccarianum*, *A. prolongatum*, *A. sampsonii*, and other species outside the complex, were suggested to be sister to *A. phyllitidis* clade. The re-circumscribed *A. nidus* is still a complex which contains at least three cryptic species and currently we fail to find stable morphological characters to distinguish them.

T2-47-11

Taxonomy and evolutionary history of the neotropical fern genus *Salpichlaena* (Blechnaceae)

Glenda Gabriela Cárdenas Ramírez, *Samuli Lehtonen, Hanna Tuomisto*

University of Turku

The neotropical fern genus *Salpichlaena* (Blechnaceae) is interesting both because it has an unusual climbing habit and because there has been much debate during the past three decades about how many species the genus contains. Some researchers recognize 2 or 3 species whereas others consider there to be just one. The habit of these plants has caused some difficulty for taxonomical work. They have twice pinnate twining leaves that can become more than 15 m long, and the pinnae vary considerably in both size and shape according to their distance from the rhizome. We have now systematically observed 62 morphological traits on 183 herbarium specimens of *Salpichlaena*. E We also sequenced three plastid genes (*rbcL*, *rpoC1* and *rps4*), two plastid intergenic spacers (*trnH-psbA* and *trnG-trnR*) and a nuclear gene (*pgiC*) from 46 *Salpichlaena* and nine out-group specimens to run phylogenetic analyses. We found that pinna texture, shape of pinna apex, type of pinna margin, shape of the scales on the costa and presence of buds on pinna axils are the most informative traits for distinguishing the species. Our preliminary results indicate the presence of at least three species in the genus *Salpichlaena*, and that their distribution is related to geographical and ecological factors.

T2-47-12

Lycophytes and ferns of Natmataung National Park in Myanmar

Byung-Yun Sun[1], *Myung-Ok Moon*[3], ***Yunwe Thet***[2], *Narae Yoon*[1], *Xian-Chun Zhang*[4]

1. Chonbuk National University

2. Yangon University

3. Jeju National University

4. Institute of Botany, Chinese Academy of Science

The flora of Myanmar was extensively documented by British botanists and plant hunters in the early nineteenth century. Dickason (1946) reported 460 species of lycophytes and ferns as first records for Myanmar. More recently, Thet YuNwe (2009) studied the fern diversity of Kyaik-Hti-Yoe Wildlife Sanctuary, reporting 68 species. Because of lack of man powers and available funds, the flora, especially lycophytes and ferns are pooly known. As a part of compilation of lycophytes and ferns of Myanmar, field works were conducted in Natma Taung National park, Chin State, Myanmar from 2015 to 2017 to make a checklist of lycophytes and ferns. Natma Taung National Park in southern Chin State, western Myanmar, was established in 1994 and is located close to the border with India and Bangladesh. It covers around 723 km^2 of Chin Hills and contains the highest peak in southern and central Myanmar: Mt. Natma Taung (Mt. Victoria, elevation 3,051 m). As a result, we found some lycophytes and ferns which include first recorded taxa from the flora of Myanmar.

T2-48: Plant diversity and evolution in Africa: Challenges and opportunities in a changing world

T2-48-01

How threatened is the flora of continental tropical Africa? A response to reaching Targets 1 & 2 of the Global Strategy for Plant Conservation (GSPC)

Thomas Couvreur[1], *Tariq Stévart*[2], *Gilles Dauby*[3]

1. Institut de Recherche pour le Développent (IRD)

2. Missouri Botanical Garden

3. Université Libre de Bruxelles, Evolutionary biology and Ecology unit

Two of the GSPC Targets for 2020 are: 1) an online world flora, and 2) an assessment of the conservation status of all known plant species. Progress toward achieving both targets is far from sufficient. Tropical African biodiversity face a high level of threat linked to explosive population growth and resulting expansion of industrial activities such as agriculture and mining. Knowing distribution of plants across the continent and estimate extinction risks is thus of major conservation importance. However, distributional data for plants remains insufficient and lack a general standardization. Estimates of conservation status are available for only a limited portion of species (around 11%). To speed up this process, we have developed 1) a mega database of tropical African plants called RAINBIO and 2) a new approach to undertake fast-track preliminary conservation assessments based on the IUCN categories under criteria A and/or B. RAINBIO represents the first-ever near-comprehensive database of vascular plant occurrences in sub-Saharan tropical Africa, currently comprising distribution information for over 22,000 native species. This allowed us to generate a check list of known species at regional and country levels contributing to Target 1. Our cost-effective and efficient conservation assessment approach enabled the preparation of a preliminary working set of potentially threatened species at multiple scales contributing to Target 2. We used this approach on RAINBIO and processed over 18,000 species in less than 24 hours. Our results suggested that around half of tropical African plants are potentially or likely threatened by extinction. We identified areas with high concentrations of threatened species. Synthesizing databases and applying new fast-track approaches to entire floras provides a powerful example of how we could achieve

GPCS Targets by 2020.

T2-48-02

The grasses and grasslands of Madagascar: A multidisciplinary investigation

Maria Sergeevna Vorontsova[1], *Olinirina Prisca Nanjarisoa*[2], *Jan Hackel*[3], *Guillaume Besnard*[3], *Peter Linder*[4]
1. Royal Botanic Gardens, Kew
2. Kew Madagascar Conservation Centre
3. Universite Paul Sabatier
4. Institute of Systematic Botany

The dominant narrative regarding human arrival in Madagascar has long stated that people arrived 2,500 years ago to a largely forested island, with anthropogenic destruction responsible for the 65% of Madagascar's land area now covered in grasslands and savannas. Scarcity of lake bed deposits have made paleoenvironmental reconstruction difficult. One piece of the puzzle regarding Madagascar's grasslands remains missing: the grasses themselves have never been fully documented or considered in an ecological perspective. Herbarium specimens of Malagasy grasses (family Poaceae) were studied, taxonomic revisions produced, and new species described during the six years of this project. Regular field work has targeted locations of poorly known endemic grasses and species knowledge has been built up throughout the island. Phylogenetic trees were reconstructed and dating was carried out in order to document all instances of Poaceae diversifications into Madagascar. Plot data recording species composition, phylogenetic diversity, and disturbance levels has been gathered at 80 sites. 217 of Madagascar's 541 grass species are endemic, a level of endemicity in line with the grass floras of other subtropical islands. We have identified 97 endemic lineages of grasses, which have colonised Madagascar primarily from Africa, with a mean age of 3.5 million years. Half of the grass dispersals into Madagascar were C4 lineages pre-adapted to open habitats. The earliest endemic lineage *Lecomtella madagascariensis* arrived ca 22 million years ago. Madagascar's High Plateau is home to a unique and endemic grass flora. A checklist if the Itremo Protected Area has revealed 99 species of Poaceae, with 20 restricted to the High Plateau. Phylogenetic diversity within grassy ecosystems of Madagascar decreases in areas with strong physical disturbance, indicating ecosystems dynamics typical of natural assemblages. We demonstrate that Madagascar is home to an ancient and diverse grass flora, with local species assemblages functioning similarly to natural ecosystems. This indicates that pre-human Madagascar was home to at least some tropical grassy biomes. Work continues to determine where these were located, and how can they be distinguished from anthropogenic landscapes resulting from forest destruction.

T2-48-03

Contrasting dispersal histories of broad- and fine-leaved temperate Loliinae grasses: Range expansion, founder events, and the roles of distance and barrier

Pilar Catalan[1], *Miguel Minaya*[1], *Jan Hackel*[2], *Mary Namaganda*[3], *Christian Brochmann*[4], *Maria S. Vorontsova*[5], *Guillaume Besnard*[2]
1. Universidad de Zaragoza
2. University of Toulouse III, CNRS-IRD-UPS-ENSFEA
3. Makerere University
4. Natural History Museum, University of Oslo
5. Royal Botanical Gardens, Kew

Successful colonisation after long-distance dispersal (LDD) depends on the availability of suitable habitats and competitive ability. Here we address the hypothesis that two widely distributed sister grass lineages (broad- and fine-leaved Loliinae; hereafter BL and FL) that differ in their habitat requirements and competitive ability also differ in their biogeographic history, with particular reference to LDD. Sampling was designed to represent all genera, subgenera and sections, and the geographical distribution of the Loliinae, particularly in the less explored Southern Hemisphere (sub-Saharan African taxa were studied for the first time). We reconstructed a comprehensive phylogeny of the Loliinae, based on nuclear and plastid markers and estimated divergence times using fossil calibrations. Biogeographical events were estimated using analysis of range evolution, comparing different models. Numbers and rates of dispersals were estimated for BL and FL using stochastic mapping with best-performing and baseline biogeographical models, and examined for correlation with distance, disjunction type, and phenotypic syndrome. The most recent common ancestor of the Loliinae likely split at the Oligocene-Eocene transition (22.50 ± 3.95 Ma), pre-dating previous estimates, whereas the ancestors of the BL and FL Loliinae likely began to diversify in the Early Miocene (18.91 ± 4.15 and 17.50 ± 3.50 Ma, respectively). A model of range evolution integrating founder events and scaling of dispersal by shortest distance between areas performed best among a set of alternative models and recovered a mean of 83 dispersal events in Loliinae. Overall dispersal rates were significantly higher in BL than in FL. Per-route dispersal rates showed a significant negative exponential relationship to shortest distance but were not affected by phenotypic syndrome or disjunction type. Loliinae originated in the Northern Hemisphere and evolved through recurrent LDDs. Sub-Saharan Africa was colonised four times. A continental colonisation of Tropical Africa from the West Palaearctic was inferred to have occurred in the BL lineage in the Early-Mid Miocene, and from there to S. Africa and Madagascar in Late Miocene to Quaternary, resulting in a Tropical African-South African clade (including *Pseudobromus*). By contrast, a transoceanic colonisation of Tropical Africa from S. America was inferred to have occurred in Late Miocene (Messinian) – Early Pliocene leading to the origin of the FL Afroalpine clade; it was followed by two Late Pliocene – Quaternary dispersals from there to Madagascar and to the Mascarenes, thus expanding the distribution of the Tropical African 'sky islands' flora to the high peaks of the Indian Ocean islands. Higher competitive ability, potentially related to the broad-leaved syndrome (i.e., tall strong-rhizomatous habit, long-living individuals, occupancy of more stable habitats), may explain higher dispersal rates in BL compared to FL Loliinae. However, the dominant factor influencing dispersal in both BL and FL Loliinae is the distance between suitable areas.

T2-48-04

Multiple origins and a narrow genepool characterise African tea germplasm: Concordant patterns revealed by nuclear and plastid DNA markers

Moses Wambulwa[1], *Kasun Meegahakumbura*[3], *De-Zhu Li*[2], *Lian-*

Ming Gao[2]
1. *World Agroforestry Centre*
2. *Kunming Institute of Botany*
3. *Coconut Research Institute*

Despite the highly economic value of tea in Africa, its genetic and geographic origins remained largely unknown. We collected 439 samples across 11 countries in Asia and Africa to investigate the origin and genepool composition of African tea based on 23 nuclear microsatellites loci (nSSRs) and three cpDNA intergenic spacer regions. Our results concluded that the African tea represents a potpourri originating from multiple introductions over time. The nSSR analysis revealed that the majority (79%) of tea accessions collected in Africa belong to Indian Assam tea which have likely originated from India and/or Sri Lanka. China tea was initially introduced from China directly to Africa. Patterns of nSSR variation also showed that Chinese Assam tea, genetically distinct from Indian Assam tea, has rarely been used in African tea breeding efforts, as only 4% of the African tea accessions possessed this genotype. We found a total of 22 cpDNA haplotypes, which grouped into three main geographic clades that were concordant with the distribution of microsatellite genotypes. Several private cpDNA haplotypes were identified in Chinese Assam tea in Southern Yunnan province of China. Therefore Chinese Assam tea may be important for the enrichment of African tea gene pools. Our results will be a useful guide in future breeding programmes and allow a strategic approach to the conservation of tea germplasm in Africa.

T2-48-05

Connectivity in the extremely fragmented African 'Sky Island' flora: Historical migration corridors or long-distance dispersal?

Abel Gizaw[1], *Christian Brochmann*[1], *Desalegn Chala*[1], *Magnus Popp*[1], *Niklaus Zimmermann*[2], *Dorothee Ehrich*[3], *Vegar Bakkestuen*[4], *Gerald Eilu*[5], *Pantaleo Munishi*[6], *Geoffrey Mwachala*[7], *Sileshi Nemomissa*[8], *Afroalp Team*[1,5,6,7,8]
1. *Natural History Museum, University of Oslo*
2. *Swiss Federal Research Institute WSL, Birmensdorf*
3. *UiT-The Arctic University of Norway, Tromsø*
4. *Norwegian Institute for Nature Research*
5. *Makerere University, Kampala*
6. *Sokoine University of Agriculture*
7. *National Museums of Kenya, Nairobi*
8. *Addis Ababa University, Addis Ababa*

The flora on the isolated high mountains of eastern Africa is renowned in biogeography. These mountains were mainly colonized by lineages pre-adapted to frost in remote areas of the world, some possibly before the formation of the current high mountains, some at the time of their final uplift phase, and some long after their formation. By combining ecological niche modeling and genetic analysis, we here address the relative importance of long-distance dispersal (LDD) and migration corridors potentially formed during past climates in shaping the isolated floras on different mountains. We are currently performing comparative ecological and genetic analyses of 29 plant species (AFLP genotypes of ~4000 plants from ~800 populations). A preliminary analysis revealed relatively high median genetic divergence between mountains (F_{ST} = 0.24-0.47), but we found large variation both among different species and among different mountains. In some species, there was virtually no differentiation among mountains, whereas populations of other species were extremely divergent. The phylogeographic histories thus seem to differ considerably among species. Our ecological niche modeling work suggests that most alpine populations on different mountains remained isolated from one another even during the Last Glacial Maximum, when the alpine habitat extended 1000 m further downslope and was eight times larger than today. Intermountain gene flow may therefore to a large degree depend on LDD, a highly stochastic process, at least for species that exclusively occur in the alpine zone of the mountains.

T2-48-06

Exploring the phylogenetic composition of pteridophyte flora of Eastern Afromontane Biodiversity Hotspot

Hongmei Liu[1], *Harald Schneider*[2]
1. *Fairy Lake Botanical Garden, Shenzhen & CAS*
2. *Natural History Museum*

Studies on the fern and lycophytes established a reliable taxonomic and geographical framework to identify the factors that shaped the extant diversity of these plants in East African Biodiversity Hotspot (EABH). The formation of geological young mountain chains, such as the Andeans and the Himalaya, has been recognized as major factors contributing to the assembly of extant pteridophyte diversity hotspots in South America and Asia. In contrast, relatively little attention has been given so far to the impact the rise of the East African Mountains on the local and global diversity of ferns and lycophytes. Here, we present the first comprehensive analyses on the phylogenetic composition of the fern and lycophyte diversity of this region. This analysis enables us to explore the support for previously introduced hypotheses such as the existence of geographically differentiated pteridophyte floras in Africa such as the East African pteridophyte flora and the Central-West African lowland forest pteridophyte flora. The analyses include also the reconstruction the role of niche conservatisms, long distance dispersal, and local adaptive radiations on the accumulation of ferns and lycophyte diversity across Africa but specifically to the East African Biodiversity Hotspot. Given the distinct geographic and geological structure of this region, it is expected that the evolutionary processes shaping the extant plant diversity of this region are distinct from those shaping the accumulation of species diversity of the main mountain chains in Asia and South America. Instead, some evidence suggested that the accumulation of the plant diversity of the mountains is more comparable to those on Oceanic islands

T2-49: Origin and diversification of the East Asian flora

T2-49-01

The diverse origins of *Lobelia* in China

Chunjiao Li[1,2], *Eric Knox*[2]
1. *Northeast Normal University*
2. *Indiana University*

Lobelia originated in southern Africa, but has since colonized all habitable continents and many oceanic islands. The 2011 revision by Thomas Lammers recognized 18 sections of *Lobelia*, five of which are represented in the flora of China. Section *Delostemon* is centered in Africa, but one lineage colonized southern and eastern Asia, with three species present in China. Section *Stenotium* also has an African origin, and independently reached China, where it is represented by the widespread *L. heyneana* and the rare *L. hainanensis*. Section *Hypsela* originated in Australia, with *L. nummularia* and *L. chinensis* derived from separate dispersal events. Section *Speirema* is endemic to southeastern Asia, with three species that range north into China. Section *Rhynchopetalum* originated in eastern Asia (following dispersal from Madagascar) and subsequently dispersed to Africa and South America, but also gave rise to section *Speirema* and the Hawaiian endemic sections *Revolutella* and *Galeatella*. In total, the Chinese species of *Lobelia* are a diverse assemblage derived from five independent, intercontinental dispersal events.

T2-49-02

The East Asian origin of the giant Lobelias

Eric Knox[1], *Chunjiao Li*[1,2]
1. Indiana University
2. Northeast Normal University

The cosmopolitan radiation of *Lobelia* that began about 20 million years ago (mya) has two lineages derived from African ancestry. Precise reconstruction of the early biogeographic details of the first lineage is not possible because it colonized the Mediterranean, North America, South America, and Australia in such rapid succession. The second lineage had a slower tempo, with initial colonization of Madagascar before an extended period of diversification in eastern Asia. The four distinct East Asian lineages provide compelling evidence of the evolutionary center of origin for the giant lobelias. However, the giant lobelias comprise two lineages that dispersed in different directions about 10 mya. One lineage colonized the ancient Hawaiian archipelago and also reached isolated, tall islands in the South Pacific. The other lineage dispersed to East Africa, where it colonized ancient crystalline block mountains and the tall volcanos as they arose, but also reached Ethiopia (across the desert of northern Kenya), West Africa (across the Congo basin), and Brazil (across the South Atlantic Ocean).

T2-49-03

Origin of East Asian Campanuloideae (Campanulaceae) diversity

Nico Cellinese[1], *Andrew Crowl*[1], *Kyung-Ah Kim*[2], *Kyeong-Sik Cheon*[2], *Ki-Oug Yoo*[2]
1. University of Florida, Florida Museum of Natural History
2. Division of Life Sciences, Kangwon National University, Chuncheon, Korea

In Eastern Asia, the Campanuloideae (Campanulaceae) clade is represented by 14 genera and over 190 species. Almost half of these are endemic to this region. Results from phylogenetic and biogeographic analyses, using both plastid and nuclear loci, will be presented for this group, with special emphasis on East Asian taxa. Evolution of the Eastern Asian campanuloids appears to be the result of numerous, independent movements from Africa, Eurasia, and North America. Movement of the largest East Asian clade (platycodonoids) occurred approximately 37 mya from an African ancestor. More recently, we found evidence for 13 additional movements into this region, primarily from Eurasia and a single dispersal event from western North America. Results from diversification analysis will also be presented in the context of genome and character evolution.

T2-49-04

Evolution of Musaceae: Diversification of bananas was driven by the geological and climatic history of Southeast Asia

Steven Janssens[1], *Filip Vandelook*[1], *Rony Swennen*[2]
1. Botanic Garden Meise
2. KU Leuven

Tropical and subtropical Southeast Asia, which harbors most of the Musaceae biodiversity, is one of the most species-rich regions in the world. Its high degree of endemism is shaped by the region's tectonic and climatic history, with large differences between northern Indo-Burma and the Malayan Archipelago. Here, we aim to demonstrate a link between the diversification and biogeography of Musaceae and the geological history of the Southeast Asian subcontinent emphasizing northern Indo-Burma. Age estimates of the Musaceae family are inferred using a phylogenetic framework encompassing 163 species from all Zingiberales families. Evolutionary patterns within Musaceae were investigated using ancestral area reconstruction and diversification rate analyses. All three Musaceae genera - *Ensete*, *Musa* and *Musella* - originated in northern Indo-Burma during the early Eocene. *Musa* species dispersed from 'northwest to southeast' into Southeast Asia with only few back-dispersals towards northern Indo-Burma. Musaceae colonization events of the Malayan Archipelago subcontinent are clearly linked to the geological and climatic history of the region. *Musa* species were only able to colonize the region east of the Wallace line from the late Miocene onwards after the availability of emergent land.

T2-49-05

Uplift-driven diversification in the Hengduan Mountains, a temperate hotspot of biodiversity

Yaowu Xing[1,2], *Richard Ree*[1]
1. Integrative Research Center, The Field Museum
2. Xishuangbanna Tropical Botanical Garden, Chinese Academy of Sciencese

A common hypothesis for the rich biodiversity found in mountains is uplift-driven diversification—that orogeny creates conditions favoring rapid in situ speciation of resident lineages. We tested this hypothesis in the context of the Qinghai-Tibetan Plateau (QTP) and adjoining mountain ranges, using the phylogenetic and geographic histories of 18 groups of plants to infer the tempo (rate) and mode (colonization vs. *in situ* diversification) of biotic assembly through time and across regions. We focused on the Hengduan Mountains region, which in comparison to the QTP and Himalayas was uplifted more recently (since the late Miocene), is smaller in area, and richer in species. The time-calibrated phylogenetic analyses, which made no prior assumptions about when any region was uplifted, showed that about 8 million years ago, the rate of *in situ* diversification increased in the Hengduan

Mountains, significantly exceeding that in the geologically older QTP and Himalayas, and marked the point at which cumulative speciation overtook colonization. By contrast, in the QTP and Himalayas during the same period, the rate of *in situ* diversification remained relatively flat, with colonization dominating lineage accumulation. This indicates that the Hengduan Mountains flora has been assembled disproportionately by recent *in situ* diversification that coincides temporally with independent estimates of orogeny, and is the first quantitative evidence to support the uplift-driven diversification hypothesis.

T2-49-06

Plant radiations in hotspots of biodiversity surrounding the Qinghai-Tibet Plateau

Alexandra Muellner-Riehl, *Jana Ebersbach, Adrien Favre*
Leipzig University

Orogenesis is often considered the main driver for the radiation of plants in mountains. Yet, the role of climate oscillations and key opportunities as well as key innovations as contributors to the establishment of high levels of biodiversity around the Qinghai-Tibet Plateau (QTP) is often overlooked. In order to understand the contribution of orogenesis to plant radiations and the importance of the QTP to the diversification of temperate plants, we investigated the evolutionary history of two taxa with high species numbers in the QTP region, Gentianinae and *Saxifraga*. We performed molecular dating, historical biogeography, diversification rate and climatic niche evolutionary analyses. Contrary to the "out of Tibet" hypothesis, not all lineages that radiated in the QTP region originated there, and immigration contributed to the high species richness of this region. Similar to other studies, we found a significant delay between the start of orogenesis and radiations. In fact, our studies underline the complexity of processes and factors underpinning plant radiations in mountain systems. For example, rather than geomorphological processes, evolution of key innovations such as berry-like fruits and the cushion life form as well as habitat shifts and niche evolution were found to be important processes connected to these radiations. Interestingly, even in closely related lineages occupying the same life zone, radiations are not necessarily governed by the same factors. In conclusion, the underlying causes for plant radiations in mountains can be multi-faceted and do not necessarily derive from orogenesis, but rather result from an intricate interaction of lineage-specific traits and complex geological settings (a prerequisite) and/or climatic modifications providing key opportunities.

T2-50: Systematics of Orchidaceae

T2-50-01

Relationships of Liparis species in Russia based on nuclear and plastid sequences phylogenetic analysis

Dmitry Lyskov, *Elena Terenteva, Tahir Samigullin, Galina Degtjareva, Tatyana Varlygina, Sergey Efimov*
M.V. Lomonosov Moscow State University

Liparis is an orchid genus containing over 400 terrestrial and epiphytic species widely distributed in tropical and temperate regions of the world. Six species of *Liparis* are known in Russia: *L. loeselii* (L.) Rich. is found in European part of Russiaand Siberia; the other five species (*L. japonica* (Miq.) Maxim., *L. krameri* Franch. et Savat., *L. kumokiri* F. Maek., *L. makinoana* Schlechter and *L. sachalinensis* Nakai) are distributed inRussia's Far East. The last one probably is a narrow endemic to Sakhalin island. *Liparis* is one of the most complicated genera in Orchidaceae family, in spite the fact that numerous researches were dedicated to the systematics of this genus. Numerous morphological features – such as size of perianth parts, shape of lip and amount of its twisting, length of inflorescence, pedicle and petiole, shape of leaf blade – are usually used in *Liparis* species delimitation. However, intraspecific variability of these features makes *Liparis* species difficult for delimitation. Taking this fact into consideration, the number of *Liparis* species in flora ofRussiaturns out to be an open-ended question. We analyzed sequences of nuclear (*nrITS*) and chloroplast (*trnL-F* spacer; *psbA-trnH* spacer; partial *ycf1b* sequence) DNA in order to discover additional molecular features applicable for delimitation of *Liparis* species. 37 accessions of 6 species of *Liparis* distributed inRussiawere involved in analysis, including critical specimen from Kunashir (*Liparis sp. nov.*?). The final set contained 52 accessions. The results of *nrITS* analysis show that representatives of Far East populations form 3 well supported clades. Critical specimen from Kunashir occupies its own isolated position and have specific nucleotide substitutions. Furthermore the results of analysis indicate paraphyly of genus *Liparis*. Results of chloroplast *trnL-F* and *psbA-trnH* spacer analises are coherent with results of *nrITS* analysis. The results of *ycf1b* analysis did not revealed specific substitutions or indels for the studied species. This fact eliminate *ycf1b* marker from the group of potential molecular-diagnostic features.

T2-50-02

RADseq elucidates relationships within phylogenetically challenging groups of European Orchids

Gabor Sramko[1,3], *Ovidiu Paun*[2], *Marie Kristine Brandrud*[2], *Levente Laczko*[3], *Attila Molnar V.*[3], *Paula J. Rudall*[4], *Richard M. Bateman*[4]
1. MTA-DE
2. Plant Ecological Genomics Research Group, University of Vienna
3. Department of Botany, University of Debrecen
4. Royal Botanical Gardens, Kew

Intensive molecular and morphological research has led to robust phylogenies of most orchid genera that occupy Europe and Asia Minor. However, at least three genera apparently underwent recent radiations that have generated extensive species complexes wherein morphological and especially molecular divergence are at best subtle. Consequently, the delimitation, rank and relationships of species remain highly controversial. The three clades – Bee Orchids (*Ophrys*), Tongue Orchids (*Serapias*) and Helleborines (*Epipactis* subgenus *Helleborine*) – are each characterised by unusual pollination mechanisms: pseudocopulation in *Ophrys*, shelter provision in *Serapias*, and near-obligate autogamy in many of the *Epipactis* taxa. We have exploited recent technical advances in applying next generation sequencing (NGS) approaches to non-model organisms in order to re-evaluate phylogenetic relationships among putative species within *Ophrys* and *Epipactis* subgenus *Helleborine*. Detailed, long-term field sampling across the full geographic range of both genera has allowed us to apply the RADseq NGS technique to comprehensive suites of taxa. The resulting matrices of single-nucleotide polymorphisms each contain several thousand phylogenetically informative characters, inevitably generating phylogenetic trees that incur stronger statistical support than any previous trees. The RADseq trees broadly corroborate the topologies the more rigorous of the previous candidate-gene trees but differ in many details, as well as challenging some previous assertions of both species status and species circumscription. In addition to elucidating relationships within these much-discussed clades of orchids, the presentation will outline the advantages and disadvantages of RADseq for reconstructing phylogenies and inferring evolutionary mechanisms, as viewed through the lenses of these two novel case-studies.

T2-50-03

Phylogenetic analysis and morphological revision of the orchid genus *Holothrix*

Timothée Le Péchon [1,2], *Steve Johnson* [1], *Benny Bytebier*[1]
1. School of Life Science, University of KwaZulu-Natal
2. Université Sorbonne Universités, Université Pierre et Marie Curie, Institut de Systématique, Évolution, Biodiversité, ISYEB - UMR 7205 – CNRS- MNHN-UPMC-EPHE

The African orchid genus *Holothrix* includes between 46 and 55 species, 23 of which occur in South Africa. A treatment of the Southern Africa taxa based on morphology was published by Immelman in 1996, but a comprehensive revision is lacking. Furthermore, no phylogenetic analysis has ever been performed. There is also uncertainty at the generic level as the genus *Bartholina* shares a similar ecology, as well as some key morphological characters with *Holothrix*. Therefore, the separation of these two genera might be artificial and needs reassessment. The main aim of our study was to reconstruct the first phylogeny for *Holothrix* and its allied genus *Bartholina*. We included 32 species of the *Holothrix* alliance as well as 19 outgroup taxa and gathered 3461 molecular characters from three different DNA regions: ITS (nuclear), matk and *trn*L-F (plastid). Bayesian inferences, maximum parsimony and maximum likelihood were used to examine the phylogenetic relationships. The resultant phylogeny is generally very well-resolved and most of the clades are strongly supported. *Bartholina* is deeply nested within *Holothrix* but the two together form a monophyletic clade, strongly supported by molecular characters as well as by floral morphology. The three different sections defined by Schlechter (i.e. *Euholothrix*, *Tryphia* and *Scopularia*) appear polyphyletic. In the light of the molecular phylogeny, we identify the taxonomic significance of several morphological characters and lay the foundation for a new classification of the *Holothrix* lineage.

T2-50-04

Phylogenetic affinities and diversity of New Zealand potato orchids (*Gastrodia*)

Carlos Lehnebach[1], *Jeremy Rolfe*[2], *Kenji Suetsugu*[3], *Yu-Chung Chiang*[4]
1. Museum of New Zealand Te Papa Tongarewa
2. Department of Conservation
3. Kobe University
4. National Sun Yat-sen University

Gastrodia is one of the largest genera of non-photosynthetic myco-heterotrophic orchids. It comprises about 70 species and although the highest diversity of species and endemism occurs in Asia, *Gastrodia* is well represented across Africa, Oceania and islands of the Indian and Pacific Ocean. The southernmost limit of distribution of *Gastrodia* is in New Zealand (NZ) where it reaches ca. 47°S. Five species of *Gastrodia* occur in NZ. Two of them are self-pollinating and the other three out-crossers, depending fully on insect pollinators for fertilisation of their flowers. Only one of these species, *G. sesamoides*, does also occur in other nearby landmasses. The remaining species are endemic to NZ, and have a contrasting extent of occurrence and conservation status. In this talk we will present results from phylogenetic analyses of nuclear and mitochondrial DNA sequences that suggest long distance dispersal of *Gastrodia* to NZ, followed by local and recent species diversification. Diversification seemed to have been accompanied by changes in floral structures and, consequently, breeding systems. This study represent one of the first outcomes of collaboration between orchid researchers from multiple international institutions that aims to understand biogeographic patterns in *Gastrodia* and phylogenetic affinities among species.

T2-50-05

The power of movement in orchids, a kinematics study of *Porroglossum* (Pleurothallidinae) evaluated in the context of their phylogeny reconstructed using nextgen GBS data

Kenneth Cameron, ***James Mcdaniel***
University of Wisconsin, Madison

Charles Darwin was fascinated by the power of movement in plants. In particular, he had a strong interest in the various types of climbing plants, leaves exhibiting sleep movements, and insectivorous plants exhibiting fast-action snap-traps. Had Darwin known about the orchid genus *Porroglossum*, he surely would have considered these small plants of subtribe Pleurothallidinae the most remarkable of orchids because physical stimulation of the flower's labellum causes it to actively snap inward thrusting pollinators against the column. *Porroglossum* is composed of 53 described species, most of them endemic to Ecuador, that are distributed throughout the Andean cloud forests of South America. Recently, we reconstructed the evolutionary history of *Porroglossum* using a combination of nuclear (*XDH*), ribosomal (nr*ITS*), and plastid (*matK*, *ycf1*, *rpl32-trnL*) markers, but the backbone of the phylogeny and interspecific relationships remained unresolved. Additionally, a majority of the species were characterized by relatively short branch lengths indicating the possibility of a recent origin for the genus. To compensate, we turned to genotyping by sequencing (GBS) for phylogenetic inference at the species-level. By utilizing high-throughput computing, we implemented the Stacks pipeline to generate thousands of SNPs from the raw sequence data for phylogenetic inference. Support values for the clades recovered using traditional Sanger sequencing were bolstered (100% bs) and the tree was fully resolved. For our kinematics study we recorded high-speed videos (60fps) of the active, floral snap-trap of all species at the orchid nursery Ecuagenera in Gualaceo, Ecuador. For each video, we included a scale bar (e.g., a ruler), applied a single stimulus to the callus of the labellum, and did not cease recording until the labellum of the flower was fully closed. As a result, we obtained >900 videos spanning 30 species of *Porroglossum* (>20 samples per species). By utilizing the program Tracker, we were able to calculate the average time to snap-trap closure for each species as well as the average velocity and acceleration of the snap-trap for each species. Furthermore, we implemented statistical methods in R to map the aforementioned continuous traits for each species onto the GBS phylogeny to calculate phylogenetic signal. By doing so, we documented strong phylogenetic signal when using both Pagel's λ (λ = 1.0840, p = 0.0014) and Blomberg's K statistic (K= 1.7001, p = 0.0027) indicating that closely related species resemble each other more than expected by chance in relation to time, velocity, and acceleration.

T2-50-06

Biogeography and diversification of Zygopetalinae (Cymbidieae, Orchidaceae)

Janice Valencia-D.[1], *Kurt Neubig*[1], *W. Mark Whitten*[2]
1. Southern Illinois University
2. University of Florida

Determining historical biogeography is fundamental to understanding diversification in Neotropical Orchidaceae and the mechanisms underlying spatial patterns. Widely distributed throughout tropical America, subtribe Zygopetalinae comprises 36 genera and more than 430 species adapted to live as epiphytes in temperate premontane and humid lowlands forests below 1500 m. This habitat's specificity is reflected in the high level of endemism exhibited in the group, with almost 70% of the species distributed in small areas within any one country. To assess the historical biogeography of these orchids, we inferred phylogenetic relationships from three plastid DNA sequences, *matK*, *ycf1*, and *trnL*-F intergenic spacer and the nuclear ribosomal ITS region. Bayesian, maximum likelihood, and parsimony analysis of combined datasets revealed phylogenetic results, which were congruent with previous studies. We also calculated divergence dates, an ultrametric analysis using BEAST. Based on a crown clade age of 11.8 My, Zygopetalinae orchids appeared by middle Miocene, and their diversification happened in the last 5 My, coinciding with major geologic events, such as the uplift of the Andes and the final emergence of the Panama Isthmus. Biogeographic reconstruction for some genera of these orchids indicated a clear origin in South America and recent dispersal events to Central America, like in *Benzingia, Chaubardiella, Chondroscaphe* and *Warczewiczella;* meanwhile for others, *Dichaea*, *Cryptarrhena*, *Huntleya* and *Kefersteinia,* the pattern is unclear. The northwest of South America was identified as the center of diversification and the richest area in endemic species; two other regions were also found, one in Central America and the other in the Southeast of Brazil. Given the high habitat specificity in the group and the increasing alteration of the natural forest in some of these rich areas, the protection of the Andean montane and humid lowland forest is a priority for the conservation of these Neotropical orchids. This study provides information that highlights the vulnerability of Neotropical Orchidaceae.

T2-51: Systematics and evolution of Liliales and Asparagales

T2-51-01

The Monocotyledonous underground: Excavating the evolutionary history of geophytes

Cody Coyotee Howard[1,2], *Jeremy Beaulieu*[3], *Nico Cellinese*[1,2]
1. Florida Museum of Natural History
2. University of Florida
3. University of Arkansas

Geophytes, plants with resting buds located beneath the soil surface, are able to survive periods of unfavorable growing conditions by lying dormant below the soil surface. Examples of this alternative lifestyle include rhizomes, corms, tubers and bulbs. These adaptations are hypothesized to have evolved in response to an increase in seasonality, and are renowned landscape beautifiers in areas such as the Mediterranean Basin or the Cape Floristic Region. In addition to these geographic hotspots, geophytic lineages can be found across the globe and are distributed across the plant tree of life. However, geophytic taxa are most prevalent within the monocot clade and include members such as ginger, taro, arums and tulips. Interestingly, some monocot clades appear to favor a certain underground adaptation whereas other groups contain a concoction of different traits. This breadth of diversity has led many of these groups to become highly prized for both their horticultural and agricultural value, which has warranted researchers' attention for centuries, yet a majority of studies on the groups possessing these adaptations have primarily focused on select clades. Broad phylogenetic inferences of these traits have yet to be carried out. Here we attempt to answer several questions related to these fascinating adaptations in order to uproot their evolutionary complexity. We dig into questions such as: (1) Do certain climatic variables (e.g., precipitation vs temperature) favor certain geophytes? (2) Does climate or shared ancestry play a larger role in geophyte evolution? (3) How has biogeography affected the evolution of these traits? (4) Are there selective advantages for any of the traits? Using a transcriptomic dataset, in addition to a large, comprehensive phylogeny we investigate these questions in the monocot clade. By digging deeper into the complex evolution of geophytic lineages we can bring to the surface clades in need of more research in order to fully tease apart the complex evolution of geophytism, highlight geographic areas that favor geophytism and may need conservation, and predict the future response of these groups to climate change.

T2-51-02

Overall view on genus *Dipcadi Medik.* (Asparagaceae) in India

Usha S. Yadav
Department of Botany, Willingdon College, India.

Genus *Dipcadi* Medik. of about 55 species with main center of diversity in Africa is represented by about 9 species in India of which 5 species are endemic to the country. It largely remained under explored in India from the point of view of taxonomy, distribution, phenology, ecology, reproductive biology, cytology and understanding of diversification of the genus in the region.

Taxonomically the genus is difficult because of unvarying external morphology of species and there are no good qualitative characters to differentiate and delimits of the species. The species of the genus are primarily distinguished on the basis of taxonomically bad quantitative characters. Distribution, phenology, odor of flower, chromosome number, karyotype and overall morphology are collectively of considerable importance in the taxonomy of Indian species of the genus and their reproductive isolation and diversification in India. Therefore, concentrated efforts were made to collect populations of *Dipcadi* from various localities from all over India and to grow them in garden under similar ecological conditions. Continuous and critical observations were made on their morphology, phenology, floral characters including odor of flowers and blooming time and stigma receptivity during last decade to assess Indian species of *Dipcadi*. Different diagnostic features of taxonomic value have been identified to delimit the species and probable mechanisms operating in speciation and diversification of the genus have been identified. Cytological data combined with morphology is found to be useful in distinguishing and delimiting *Dipcadi* species in India. On the basis of chromosome number, the Indian species can be grouped into three groups viz. *Dipcadi* species with 2n = 12,20 and 22. Flowering period, somatic chromosome number (2n), length of flower, extent of union of perianth lobes, length of bract, width of leaf and odor of flower are of diagnostic value. Odor of flower has played significant role in reproductive isolation of populations and speciation of the genus in India. The paper gives systematic account of the genus in India, a botanical key for determination of the Indian species and discusses speciation in the genus with reference to India.

T2-51-03

Comparative genomics and phylogenomics of East Asian tulips (Amana, Liliaceae)

Pan Li[1], *Ruisen Lu*[1], *Chengxin Fu*[1], *Kenneth Cameron*[2]
1. Zhejiang University
2. University of Wisconsin-Madison

The genus *Amana Honda* (Liliaceae), when it is treated as separate from Tulipa, comprises six perennial herbaceous species that are restricted to China, Japan and the Korean Peninsula. Although all six Amana species have important medicinal and horticultural uses, studies focused on species identification and molecular phylogenetics are few. Here we report the nucleotide sequences of six complete Amana chloroplast (cp) genomes. The cp genomes of Amana range from 150,613 bp to 151,136 bp in length, all including a pair of inverted repeats (25,629–25,859 bp) separated by the large single-copy (81,482–82,218 bp) and small single-copy (17,366–17,465 bp) regions. Each cp genome equivalently contains 112 unique genes consisting of 30 transfer RNA genes, four ribosomal RNA genes, and 78 protein coding genes. Gene content, gene order, AT content, and IR/SC boundary structure are nearly identical among all Amana cp genomes. However, the relative contraction and expansion of the IR/SC borders among the six Amana cp genomes results in length variation among them. Simple sequence repeat (SSR) analyses of these Amana cp genomes indicate that the richest SSRs are A/T mononucleotides. The number of repeats among the six Amana species varies from 54 (*A. anhuiensis*) to 69 (*A. kuocangshanica*) with palindromic (28–35) and forward repeats (23–30) as the most common types. Phylogenomic analyses using likelihood inference and Bayesian analysis based on these complete cp genomes and 74 common protein-coding genes strongly support the monophyly of the genus, and a sister relationship between Amana and Erythronium, rather than a shared common ancestor with Tulipa. Nine DNA markers (*rps15–ycf1, accD–psaI, petA–psbJ, rpl32–trnL, atpH–atpI, petD–rpoA, trnS–trnG, psbM–trnD* and *ycf4–cemA*) with number of variable sites greater than 0.9% were identified, and these may be useful for future population genetic or phylogeographic studies of Amana species.

T2-51-04

Accelerated rates of climatic-niche evolution in Old World after migrating from the New World: Evidence from spatiotermporal diversification history of the Greenbrier family (Liliales: Smilacaceae)

Zhe-Chen Qi[1,2,3], *Pan Li*[1,2], *Yun-Peng Zhao*[1,2], *Kenneth M. Cameron*[4], *Cheng-Xin Fu*[1,2]
1. Laboratory of Systematic & Evolutionary Botany and Biodiversity, College of Life Sciences, Zhejiang University, Hangzhou, 310058
2. The Key Laboratory of Conservation Biology for Endangered Wildlife of the Ministry of Education, Hangzhou, 310018
3. Zhejiang Province Key Laboratory of Plant Secondary Metabolism and Regulation, and College of Life Sciences, Zhejiang Sci-Tech University, Hangzhou 310018
4. Department of Botany, University of Wisconsin, Madison, WI 53706

Explain differences in species richness between regions and among clades is an important topic in biogeography. The diversification rate of clades is a key parameter for understanding both patterns. Many studies have showed examples in animal and plant groups with regional and cladistical difference in speciation rate. These disparities might correlate to climatic shifts and topographic heterogeneity. Smilacaceae is a widespread monocot group with both Old World and New World distribution. We combined phylogenetic, fossil records, and climatic data for 150 species of greenbrier family (Smilacaceae) to assess the spatiotermporal diversification focused on Old World and New World clades. Old World clade of Smilacaceae underwent a distinct increase in speciation rate around Late Eocene to Middle Miocene (34 Ma-15 Ma) compare to their New World allies. Accelerated diversification in the Old World Smilax is associated with climatic-niche evolution among species triggered by intensive topographic change and dramatic climatic fluctuation during the period of accretion of the Indian subcontinent to Eurasia.

T2-51-05

Phylogenomics and historical biogeography of Asparagales

Changkyun Kim, *Joo-Hwan Kim*
Gachon University

Phylogenomics refers to the use of phylogenetic trees to explain genome evolution and to use of genome-scale data to reconstruct phylogenetic trees. As more complete genomes are sequenced thanks to the now widespread availability of next generation sequencing technologies, phylogenomic has advanced rapidly in the past decade and recent studies have demonstrated the power of this approach. Asparagales is the largest order of monocots (14 families, 1122 genera, and 26,070 species; 50% of monocots), and it is economically important as ornamentals and crop plants. Although a number of studies have generally clarified the relationships among the big families as like those of Asparagaceae s.l., Alliaceae s.l., and Xanthoreaceae in Asparagales, the exact relationships of families, especially in lower asparagoids distributed in the Southern Hemisphere, are unclear. Here, we present the phylogenomic analyses of relationships among all 14 families of Asparagales based on 79 coding genes of chloroplast genome. Based on the phylogenetic framework, fossil-calibrated chronogram, we also provide the historical biogeography of Asparagales taxa with a focus on their main routes for the migration between the Northern and the Sourthern Hemispheres.

T2-51-06

A hidden evolutionary story in Liliales from organelle genome data

Hoang Dang Khoa Do, *Joo-Hwan Kim*
Gachon University

The order Liliales include potentially medicinal and economically ornamental species which are either worldwide or natively distributed. Exploring organelle genome data of this order will improve the understanding of genomic evolution and contribute to further research on conservation and biodiversity. Therefore, we sequenced chloroplast genome of 45 representative species which cover all 10 families of Liliales and did a comprehensively comparative analysis. The results showed highly conserved chloroplast genomes in the photosynthetic group and a significantly structural reorganization in Corsiaceae which contain mycoheterotrophic taxa. Among Liliales families, Melanthiaceae possess a higher frequency of genomic events including gene duplication, deletion, and inversion. Further investigation revealed that deletion of gene and inversion in chloroplast genome was occurred in accordant with the diversification of species (ie. *Veratrum, Trillium,* and *Colchicum*), suggesting that genomic events reflected the phylogenetic relationship in Liliales. Furthermore, by comparing junction among large single copy (LSC), small single copy (SSC) and inverted repeat (IR) regions of chloroplast genomes from basal angiosperm to most of monocotyledon orders, we proposed that the border of LSC/IR could be categorized into 5 types which expanded from *rpl2* to *rps3* genes. In contrast, the SSC/IR junction was commonly located within ycf1 coding region which was supposed to be a potential region for developing marker among angiosperms. Additionally, the first data of mitochondrial genomes in Liliales demonstrated a contrasting trend between photosynthetic groups and mycoheterotrophic species in which a higher rate of gene transferred from chloroplast to mitochondrial genome was found in the former. Overall, organelle genomes of Liliales improve the understanding of plastid and mitochondrial genome evolution of angiosperms in general and of monocotyledons in particular.

T2-52: The evolution of apomixis in natural plant populations

T2-52-01

The role of ploidy and niche shifts, apomixis, residual sexuality and mutational load in the formation, dynamics and evolution of agamic complexes

Diego Hojsgaard
University of Göttingen

Polyploidy is one of the most relevant factors promoting plant diversity. Polyploid plants frequently associate with diploid parents to form interbreeding complexes in which each cytotypes gene pool retain genetic cohesivity whereas allowing a certain level of recurrent genetic exchange. In some cases, the genomic shock resulting from a change in ploidy is accompanied by a reproductive shift from sexual to asexual seed formation, or apomixis. Because apomixis skips meiosis and syngamy, it highly restricts gene flow and modifies the genetic structure of natural populations. Additionally, in absence of recombination deleterious mutations can't be purged by selection, and stochastic loss of individuals with the least-loaded mutational class would lead to an irreversible increase of mutational load in asexuals (a phenomenon called *Muller's Ratchet*). Thus, deleterious mutation accumulation is expected to drive clonal lineages to a genomic decay and extinction. However, 1) the ratchet mechanism is efficient in small diploid and obligates asexual populations, while polyploids mask recessive mutations and buffer the effect of mutation accumulation, 2) apomictic plants in polyploid complexes usually are widespread over extended geographic areas (depicting a pattern of *geographical parthenogenesis*), and 3) apomixis in plants is hardly ever obligated and variable levels of residual sexuality are normally observed among genotypes. Using frequencies of meiotic recombination events observed in progenies of apomictic plants, a mathematical modeling for predicting the dynamics of mutant frequencies as a function of time show that low levels of meiotic recombination may counteract the ratchet mechanism. Under various selection coefficients and plant-specific mutation rates, low recombination rates suffice to prevent facultative apomictic plants from the accumulation of slightly deleterious mutations. Uniparental reproduction, cytotype niche differentiation, fixation of adapted genotypes and small proportions of new recombinants in facultative apomicts facilitate range expansion by coping and occupying new ecological niches, and help to explain geographical parthenogenesis in sexual-asexual plant complexes. High levels of intragenomic (allelic) diversity, epigenetic variability, and physiological plasticity further enhance fast ecological flexibility, adaptation, and dispersal. Finally, occasional re-stabilization of meiotic pathways, reversals to sexuality in apomictic lineages and establishment of new sexual populations will be constrained by *Allee* effects, but when possible, the occurrence of new sexuals in areas far from the parental species can promote divergent evolution and speciation. Phylogenetic patterns of apomixis and biodiversity parameters support this hypothesis and place a potential role of facultative apomixis as a promoter of plant biodiversity.

T2-52-02

Next-generation sequencing of apomictic fern (*Woodwardia*

orientalis) and non-vivipary fern (_Woodwardia japonica_) reveals candidate genes for vivipary character

Baskaran Xavier Ravi[1,2], *Zhang Shouzhou*[1], *Liao Wenbo*[2]
1. Fairy Lake Botanical Garden, Shenzhen & Chinese Academy of Sciences, Shenzhen– 518004
2. Sun Yat-sen University, Guangzhou

Vivipary is an asexual reproduction (apomixis) process which occurs in plants diversity due to environmental stress and lack of phyto-hormone function. Pseudo-vivipary describes that plants produce apomictic or asexual propagules like bulbils or plantlets in place of sexual reproductive structure. The objective of our present study is to compare Chinese pseudo-viviparous medicinal fern, *Woodwardia orientalis* (W.O) and non-viviparous fern, *W. japonica* (W.J) in order to reveal their functional viviparous mutant genes for viviparous character and spore dormancy. The transcriptomics data was acquired in our study using newly developed Illumina Hiseq™ 2000 RNA sequencing. Among differentially expressed genes (DEGs), 6997 up-regulated genes and 39912 down-regulated genes were identified in mature leaf of *W.orientalis* and their vivipary buds, respectively. Of them, 75 up-regulations and 39 down-regulations of transcription factors were also exhibited. All of these DEGs were associated with 112 metabolic pathways. In addition, all of them are involved in starch and sucrose metabolism, phenylpropanoid, secondary metabolites, and hormones biosynthesis pathways. Moreover, all the genes in fatty acid, brassinosteroid biosynthesis and basal transcription factors were up-regulated. Totally, 12936 simple sequence repeats (SSRs) in *W. orientalis* and 9920 SSRs in *W. japonica* were known as possible molecular markers, which may helpful for genetic diversity studies. All the unigenes were annotated to numerous databases (Nr, Swiss-Prot, GO, COG, KEGG) to understand their functional classification. In pseudo-vivipary fern (W.O), we identified the vivipary mutant genes (*NOT2/NOT3/NOT5* family isoform 2), ABA content reducing gene (*Abscisic acid deficient 4*) and spore dormancy reducing gene (transcription factor viviparous 1). The pseudo-vivipary fern (W.O), exhibited a special type of protein, dehydrin which performs cellular protection against osmotic stress while the same gene was not present in non-vivipary fern (W.J). Moreover, DELLA protein DWARF8 reduces or represses gibberellic acid (GA) inducible genes and also present in viviparous fern (W.O). There are no genes which are responsible for reducing spore dormancy, ABA content, gibberellic acid (GA), and vivipary mutant genes in non-viviparous fern, *W.japonica*. In addition, greater expression level of ABA (Abscisic acid) and GA (Gibberellic acid) inducing genes in non-vivipary fern (W.J) and lower expression of these same genes found in vivipary fern (W.O) transcriptomic sequencing data. Furthermore, significant functional genes were identified to confirm their presence in mature leaves, vivipary buds of *W. orientalis* and mature leaves of *W. japonica* using RTPCR technique. However, our present study opens the entry to understand the molecular mechanism of pseudo-vivipary fern for the first time in pteridophyte's scientific research.

T2-52-03

Progress towards understanding the causes of geographical parthenogenesis in diploid sexual and autopolyploid apomictic easter daisies (_Townsendia hookeri_: Asteraceae).

Jeannette Whitton
University of British Columbia

Ecological divergence between sexual polyploids and their diploid progenitors permits new polyploids to avoid minority cytotype interactions that would limit their establishment. These dynamics are potentially different when polyploids are apomictic, because apomixis can protect nascent polyploids from hybridization. Further, apomicts that retain male function can reduce recruitment of sexual diploids by siring interploidy hybrids with sexual diploid mothers, suggesting that sexual diploids are threatened by the presence of apomicts, even when apomictic polyploids are the minority cytotype. *Townsendia hookeri* includes autopolyploid (triploid and tetraploid) apomicts and sexual diploids. The ranges of the two cytotypes are distinct, but overlap in southern Wyoming and Northern Colorado (USA), with the sexuals extending to southern Colorado, and the apomicts extending north to southern Alberta and British Columbia (Canada). Species distribution models reveal evidence of differentiation in climatic niches of the cytotypes, but suggest that diploids may have the capacity to spread further into the polyploid range. Crossing experiments in the field indicate that pollen from apomicts can readily fertilize ovules of sexual diploids, producing viable offspring of varying DNA content, mostly above diploid levels. Along with the results of niche modeling, our crossing results suggest that sexual expansion could be limited by reproductive interference rather than ecological factors. In order to test the role of ecological divergence in limiting the distribution of apomicts and sexuals, we set up a large- scale experiment that includes common gardens that span allopatric and sympatric parts of the range of each cytotype. Eight transplant gardens were established in the summer of 2014, and are being monitored for survival, growth and reproductive traits, with a separate experiment to look at early establishment success. Our predictions, that apomictic polyploids would have low fitness in sexual diploid sites, is not supported to date, and apomicts show advantages in early life stages. The study of ecological and evolutionary interactions between cytotypes in an apomict such as Townsendia hookeri, emphasizes the need to consider the selective pressures that act on diploids, during and after the establishment of asexual polyploids.

T2-52-04

Effects of light and temperature stress on the expression of apomixis in _Ranunculus_

Simone Klatt, *Franz Hadacek, Christoph Cf Schinkel, Elvira Hörandl*
Georg-August-University of Goettingen, Albrecht-von-Haller-Institute for Plant Sciences

Apomixis, the asexual reproduction from seed, occurs in natural plant populations in many angiosperms lineages. Most apomictic species retain the ability for facultative sexuality and thus reproduce with variable frequencies of sexual and apomictic seed formation. The factors determining the frequencies of sex and apomixis are still not completely understood. Abiotic factors, like temperature and length of the photoperiod affect growth and development of plants. Unfavorable and stressful environmental conditions are supposed to play an important role in the expression of apomixis and might influence the apomictic or sexual pathway directly. One of our model system is the hexaploid (allopolyploid) natural hybrid *R. carpaticola* x *cassubicifolius* (*R. auricomus* complex), a forest understory plant native to the western Carpathians, Slovakia. We hypothesized that light stress would enhance the production of reactive oxygen species and oxidative DNA damage in the flowering buds. We consider meiosis as a repair mechanism of DNA damage, and hence we expect under stress a higher frequency of sexuality. We tested if variation in facultative apomictic (aposporous) reproduction in this species was affected by light stress (16.5 h photoperiod, compared to 10 h in the control). We analyzed two developmental stages: the ovule directly after megasporogenesis (via microscopy) and the seed stage (via flow cytometric seed screening, FCSS). The prolonged photoperiod in *R. auricomus* resulted in higher frequencies of sexual megaspores versus aposporous initial cells, while the frequencies of sexual versus asexual seeds were not affected, suggesting that only meiosis and sporogenesis are sensitive to light stress. A concomitant change of secondary metabolite profiles at the early developmental stage supports the hypothesis that oxidative stress could have affected megasporogenesis. Our second model plant, *R. kuepferi*, occurs with diploid and polyploid cytotypes in the European Alps. A correlation between mode of reproduction and temperature was observed in natural populations. While the predominantly obligate sexual diploids grow in lower altitudes, the facultative tetraploid cytotype is more widespread at higher elevations with lower mean annual temperature. We tested experimentally if low temperature and repeated frost treatments triggered the production of apomictic seeds in diploid and tetraploid *R. kuepferi* plants (FCSS). We observed a slight but significant increase of apomictic seed production in cold treated diploid *R. kuepferi* plants, while the likewise slight increase in tetraploids was not significant. Light stress and low temperature can affect the mode of reproduction in *Ranunculus*. However, in natural populations and environments a combination of diverse abiotic and temporary stressful growth conditions is likely. Facultative apomixis might be an adaptive strategy for plants to reproduce under unfavorable environmental conditions. The actual impact of stress most likely depends not only on the severity or combination of parameters, but also on the reproductive stage and on ploidy level of the plant.

T2-52-05

Adaptive advantages of partial clonality

Katja Reichel[1], *Jean-Pierre Masson*[2], *Solenn Stoeckel*[2]

1. Freie Universität Berlin

2. Institut National de la Recherche Agronomique

Plants are notorious for the kaleidoscopic diversity of their reproductive systems. For example, many plants undergo both sexual and clonal (apomictic, asexual) reproduction simultaneously. Thus, in contrast to the either/or choice often evoked to discuss the evolution of sex, plant sexual/clonal reproductive strategy may be tuned on a much finer scale. Which factors act to calibrate it? Partial clonality might improve the potential for adaptation to new selective pressures. Using sexual reproduction to create new and clonal reproduction to maintain well-adapted genotypes, populations of partially clonal plants could be able to combine the "best of both worlds". We tested this hypothesis with a mathematical model, comparing different selection schemes across the whole range from completely sexual to completely clonal reproduction. For each scheme, we investigated which rates of clonality make adaptation at the population level most probable, fastest and most efficient. We found selection for a heterozygous genotype (heterosis) to be the only selection scheme with clear adaptive advantages for partial clonality. This ties in well with the frequent coincidence of apomixis and hybridization in plants, and adds another angle to our understanding of the evolutionary maintenance of apomictic systems. We discuss further questions raised by our result, as well as its implications for the frequencies of sexual reproduction expected to be found in natural plant populations.

T2-52-06

Genome evolution in *Taraxacum agamospecies* and comparisons with triploids

Js (Pat) Heslop-Harrison[1], *Rubar Hm Salih*[1], *Ľuboš Majeský*[2], *Trude Schwarzacher*[1]

1. University of Leicester

2. Palacky University Olomouc

Taraxacum officinale agg. (dandelion) includes triploid (2n=3x=24) agamospecies showing extensive and characteristic morphological variation. A key question asks how the apomictic species gain variation without sexual recombination? We looked at three apomictic microspecies using molecular markers and next generation sequencing to reconstruct the chloroplast genomes and other genome regions. In particular, we will report on the nature, organization and abundance of the repetitive DNA present in the genomes. We have identified these elements in unselected genomic raw-read DNA sequence data from three agamospecies. Most of the repetitive DNA can be identified as being retrotransposon-related, although a significant proportion of the sequence reads are repetitive but unrelated to known retrotransposon sequences. Few satellite repeats were found. Most retrotransposons show similar abundance in the three microspecies. A few classes show substantially different copy numbers in the accessions, which is also seen in the contrasting chromosomal distributions, indicating amplication, homogenization and loss of sequences during separation of the taxa. IRAP markers (Inter-retroelement amplified polymorphisms) can be used to group and classify the microspecies, showing that retroelements are active and related to evolution. The results from nuclear repetitive DNA will be compared with chloroplast sequences (http://dx.doi.org/10.1371/journal.pone.0168008) and the genomic organization of repetitive DNA in other triploid species associated with humans, saffron crocus and banana. The contribution of repetitive sequences to genome evolution and variation in *Taraxacum* will be considered in relation to phylogeny, selection and generation of variation.

T2-53: Role of plant-pollinator interactions in generating and maintaining biodiversity

T2-53-01

Plant-pollinator interactions, reproductive isolation, and the diversification of flowering plants

Kathleen Kay

University of California, Santa Cruz

Flowering plants are spectacularly diverse, and flowers themselves have been implicated in species diversification because of the effects they have on plant mating patterns. In comparative studies across the angiosperms, we typically see higher diversification rates in plant lineages with animal pollination and floral traits indicative of specialized plant-pollinator relationships. Although most coevolutionary relationships between plants and pollinators are diffuse and do not involve cospeciation, divergence in plant-pollinator interactions can contribute substantially to plant reproductive isolation, act synergistically with other forms of ecological divergence, and allow closely related species to coexist without reproductive interference. Floral divergence can involve adaptation to different pollinators or more subtle changes in the way a plant uses the same type of pollinator, and importantly, floral isolation can evolve as an indirect byproduct of floral adaptation or through natural selection to reinforce reproductive isolation *per se.* I explore both types of floral divergence and modes of selection primarily using the Neotropical spiral ginger genus *Costus*, which has rapidly diversified into dozens of species pollinated by either orchid bees or hummingbirds. I ask how strikingly novel flowers that fit a new type of pollinating animal evolve, when often the inferred transitional states appear maladaptive, and to what extent an evolutionary shift in pollinators contributes to reproductive isolation and coexistence. Traits that deter ancestral pollinator types are crucial to providing reproductive isolation, but may evolve late in the process of a pollinator shift. I further investigate how divergence in flower shape can allow two closely related spiral ginger species to use the same hummingbird pollinators while risking little cross contamination of pollen. In this case, I ask whether natural selection to reduce hybridization in sympatric areas has driven the divergence, and how selection interacts with the genetic architecture underlying the flower morphology. Overall, we see that floral divergence proceeds along pathways facilitated by existing traits and genetic correlations, takes advantage of pollinator assemblages that are already in place, and can be the target of direct natural selection upon secondary contact. This work helps elucidate the myriad ways plant-pollinator relationships can contribute to reproductive isolation and promote plant diversification.

T2-53-02

The diversity and evolution of pollination systems in the ANITA grade of earliest-diverging angiosperms: New insights

Shixiao Luo[1], *Susanne Renner*[2]

1. Key Laboratory of Plant Resources Conservation and Sustainable Utilization, South China Botanical Garden, The Chinese Academy of Sciences

2. Systematic Botany and Mycology, University of Munich (LMU)

Pollination in phylogenetically early-diverging flowering plants has been thought to mostly be generalized, but this hypothesis is based on field observations of less than 10% in the ANITA grade families. The most species-rich family Schisandraceae (*Illicium*, 40 species; *Kadsura*, 16 species; and *Schisandra*, 22 species) holds great potential for understanding the origin of traits that are characteristic of more derived evolutionary lineages. *Schisandra* and *Kadsura* species bear separate staminate and pistillate flowers. *Illicium* was historically segregated in the Illiciaceae, and bears bisexual flowers, but is now known to be unambiguously sister to the *Schisandra-Kadsura* clade. We try to use an integration of pollination biology, molecular phylogeny, and character-evolution inferring methods to analyze coevolutionary/pollinator shift relationships between Schisandraceae plants and their pollinators. Our field work finds that, in Asia, gall midges are effective pollinators of Schisandraceae. Trees of Subgen *Cymbostemon* bear red bisexual flowers which are pollinated by gall midges, but intriguingly exhibit post-anthetic floral thermogenesis as a pollinator reward; this suggests that floral heating-possibly an ancestral trait in the family-has been co-opted for post-anthetic nurturing of gall midge larvae. While in Subgen *Illicium* often bear white flowers and was pollinated beetles. Our field observation and previous studies also found that species in *Schisandra* and *Kadsura* pollinated by gall midges (Cecidomyiidae), which have a special visiting behavior. In this presentation, I will show a highly diversified species-specific pollination mutualism between the base angiosperm family, Schisandraceae, and a resin midge genus, *Resselillea*. In this mutualism association, the female resin midge visit the flowers and lay eggs in the flower (male and female or bisexual) at night, and the larvae feed on the post-andthetic secretion of flower for growth. These results also suggest that the diversification of insect-pollinator mutualism in early angiosperms may have been facilitated by the exploitation of preexisting sensory biases of herbivores.

T2-53-03

Quantifying integration of native and alien pollinators on the Galápagos Islands

Kristian Trøjelsgaard[1], *Ruben Heleno*[2], *Anna Traveset*[3]

1. Aalborg University, Denmark

2. University of Coimbra, Portugal

3. Institut Mediterrani d'Estudis Avancats (CSIC-UIB), Spain

Natural ecosystems are increasingly threatened by biotic homogenization due to the expansion of some "winning species" and the concomitant extinction of "looser species" - a situation persistently fuelled by anthropogenic disturbances and species introductions. One key question arising under such a scenario is to what extent these new winning species will fully replace the functions lost with the extinction of the indigenous biota. The functional role of species is to a large extent rooted in the way that species engage in complex interaction networks with other co-occurring species. Indeed, ecological networks have proved a most valuable way of analysing an entire ecological community as a coherent unit. Although species and interaction turnover can be high between spatially or temporally separated networks, recent studies also suggest that species roles can be conserved and that species do not interact with a random assemblage of the potential partners. Phenological and morphological matches likely drive such non-random choice of partners. If species differ in their interaction behaviour it is therefore unlikely that 'winning' alien species will replace the roles of 'loosing' indigenous species. Here, we investigate this by focusing on alien and native insect pollinators on the Galápagos Islands. Using network analysis as a tool for understanding functional roles, we quantified whether alien and native pollinators are integrated differently into pollination networks on these islands. Surveys were undertaken at five major islands within the Galápagos archipelago *viz.* San Cristobal, Santa Cruz, Santiago, Pinta and Fernandina. In predefined areas on each island, plant species were surveyed for pollinators during the peak flowering season for two consecutive years, allowing the construction of quantitative pollination networks. If aliens and natives were functionally equivalent, the community response to sequential removal of either natives or aliens would be similar. Performing 200 iterations, we sequentially and randomly removed all alien pollinators from a given community and at each step we re-calculated structural network descriptors (e.g. weighted nestedness and specialization) to quantify how these removals affected the network structure. Subsequently, we repeated the process but instead randomly removed an equal number of native pollinators. This allowed us to compare how the two different removal scenarios affected the network topology, and essentially to quantify whether alien and native species are integrated similarly. For several islands, but especially the two most invaded ones (San Cristobal and Santiago), there were substantial differences between the native and alien removal simulations, suggesting that native and alien pollinators are indeed integrated differently. Tentatively, we infer that small individual differences in the integration of native and alien pollinators on the Galápagos can produce large accumulated differences at the community level. These findings provide a piece to the puzzle of understanding how the threatening biotic homogenization may influence natural communities, and cautiously suggest that an exact replacement of lost functions is highly dubious.

T2-53-04

Ecological speciation through floral scent/pollinator divergence in Asian Mitella

Yudai Okuyama[1], *Tomoko Okamoto*[2]
1. Tsukuba Botanical Garden, National Museum of Nature and Sciences
2. Laboratory of Insect Ecology, Faculty of Applied Biological Sciences, Gifu University

Floral scents are among the key signals used by pollinators to navigate to specific flowers. Therefore, evolutionary changes in floral scents should have strong impacts on plant speciation through pollinator shifts. Nevertheless, few cases of ecological speciation initiated by floral scent changes are reported, and thus the presumed divergent selection acting on the floral scents has rarely been demonstrated. Here we present an ongoing research project that aims to extensively isolate and analyze genes involved in floral scent production in Asian *Mitella* (Saxifragaceae), where floral scent composition repeatedly switched between species adapted to pollination by long-tongued and short-tongued fungus gnats. With the combined use of RNA-seq, recombinant protein assay, and linkage analysis with interspecific cross, we successfully identified several genes involved in the production of the key floral scent compounds for pollinator specificity in Asian *Mitella*. Similar approach would be effective for any plant lineages that utilize scents as the major signals for communicating with their mutualists and/or antagonists.

T2-53-05

Patterns of flowering phenology and floral Trait diversification in two Neotropical Gesneriaceae Genera

Marten-Rodríguez Silvana[1], *Quesada Mauricio*[1], *Fenster Charles B.*[2], *Ramírez Aguirre Erandi*[1]
1. Universidad Nacional Autónoma de México
2. South Dakota State University

Pollinators have long been recognized as important selective agents in floral trait evolution. Evidence for pollinator-mediated selection on traits associated with pollinator attraction and pollen transfer effectiveness is compelling in many plant taxonomic groups. However, the role of biotic interactions on the diversification of flowering phenology patterns is not yet well understood. Here, we use a comparative framework to assess patterns of diversification of morphological and phenological traits in the subtribe Gesneriinae and the genus *Achimenes* in the sister subtribe Gloxiniinae. Pollinator shifts in both genera are associated with floral trait diversification, but not with phenological patterns. In the *Achimenes* in particular, phenology appears to be constrained by a geophytic growth form, a condition that may have triggered evolutionary changes in floral traits to reduce interspecific hybridization or competition for pollinators. This study highlights the importance of studying phenology along with other morphology and other floral traits for a more comprehensive understanding of the factors underlying floral diversification in different plant lineages.

T2-53-06

Pollinators determine assembly of flower colour patterns in communities of mass flowering annual daisies

Jurene E. Kemp, *Allan G. Ellis*
Stellenbosch University

Communities are thought to be assembled either neutrally or non-randomly (through ecological sorting, or evolutionary interactions). In addition to abiotic processes structuring communities, biotic interactions can also play a key role in how communities are assembled. For example, the pattern of assembly of floral phenotypes (or pollination syndromes) within plant communities likely reflects ecological filtering or floral evolution in response to the community of available pollinators and the nature of interactions between plants for pollination services. Overdispersion of floral phenotypes within communities would be expected if competitive interactions between plants favour partitioning of the pollinator resource. Clustering of floral phenotypes would be expected when plant-plant interactions are facilitative, or when filtering or floral divergence occur in response to differences in the available community of effective pollinators across communties. Here we assess the landscape-level patterns of community assembly of flower colour patterns in annual mass daisy flowering displays, in the winter rainfall deserts of South Africa. While the open shape of daisy inflorescences does not preclude visits from any insects, annual daisy species in these communities exhibit a wide array of bullseye colour patterns which may represent a syndrome trait associated with particular pollinators. Using a null modelling approach we show that colour patterns are clustered within communities, with the dominant colour pattern varying across the landscape. This suggests different communities are dominated by pollinators with dissimilar preferences, and pollinators either facilitate the co-occurrence of plant species with similar floral phenotypes, or plant species converge in phenotype in response to effective pollinators. Analysis of community visitation networks demonstrates that the dominant flower colour pattern within communities is associated with particular fly pollinator species. If different fly species prefer particular floral phenotypes, we expect pollinators to exert divergent selection on floral phenotype. We find turnover of dominant colour patterns across the landscape corresponds to turnover in the fly pollinator community, and experimental evidence suggests that fly species exhibit divergent preferences for colour patterns. Together these lines of evidence suggest non-random assembly of flower colour patterns resulting from convergence of multiple species within communities on traits favoured by dominant fly pollinators.

T2-54: Ranunculaceae and the Evo-Devo of the flower

T2-54-01

Perianth evolution in Ranunculales

Sophie Nadot[1], *Hervé Sauquet*[1], *Florian Jabbour*[2], *Boris Domenech*[3], *Catherine Damerval*[4]

1. University Paris-Sud
2. Museum national d'Histoire naturelle
3. Biodiversity Centre
4. CNRS

Angiosperms are characterised by one of the most striking evolutionary novelties of land plants, the flower, which presents an extraordinary diversity in shape, size and colour in relation with pollination, a key step in the life cycle of flowering plants. This evolutionary novelty accounts for the fact that angiosperms are by far the most successful clade of land plants. In most angiosperm species flowers possess a differentiated perianth composed of sepals and petals, but a significant number of species displays an undifferentiated perianth composed of a single category of organs, called tepals. Ranunculales are one of the few orders that includes species with either a differentiated or an undifferentiated perianth. This order is the first diverging order within eudicots (the largest clade of angiosperms), it is therefore a key group to understand the evolution of the flower. Progress has been made towards the elucidation of phylogenetic relationships in the Ranunculales, providing now a relatively robust framework to address the key issue of the ancestral vs. derived nature of a differentiated perianth within the order as a whole. Here, we present ancestral state reconstructions for several perianth characters including perianth differentiation, symmetry, shape of petals and presence/absence of nectaries, with a focus on Ranunculaceae.Our results challange the traditional hypothesis of petals having evolved repeatedly in the family.

T2-54-02

The ABC model of flower development in non-core eudicots: A functional synthesis from studies in the ranunculid *Thalictrum thalictroides*

Verónica Di Stilio, *Kelsey Galimba, Valerie Soza, Jesús Martínez-Gómez*

University of Washington, Department of Biology, Seattle WA 98195

The elucidation of the "ABC model" of flower development in the 1990's promised a mechanistic basis for J.W. Goethe's intuitive idea that all floral organs must have arisen from a leaf *bauplan*. Yet the first attempts to induce floral organs from leaves by overexpression of B- or C-class genes failed to produce full conversion. The discovery of a fourth gene class, the E-class, filled this gap by elegantly demonstrating the recapitulation of floral organs in seedling leaves via overexpression of all four gene classes. The "extended" ABC model was complete and the ensuing floral quartet model clarified how these transcription factors, belonging mostly to the *MADS*-box gene family, interacted as tetramers while binding and looping DNA to turn on flower development cascades. As conserved as the ABC model appeared, first among the distinct model systems *Arabidopsis thaliana* and *Antirrhinum majus*, later in other eudicots and in monocots, much remained to be elucidated in light of mounting evidence for the importance of gene duplication in the evolution of development. Multiple duplication events affecting the *MADS*-box gene family and flower morphologies that diverged drastically from those described originally required modifications to the canonical model. As part of this evolutionary journey into the tinkering of the ABC model, be it via gene duplication, changing protein-protein interactions or changes in binding affinities and downstream targets, functional studies in a variety of plant groups began to emerge. In the past few years, my research group has functionally characterized all B, C and E class genes in *Thalictrum thalictroides*, a representative of the ranunculids (the sister lineage to the rest of the eudicots). Our work contributes to a broader effort to fill the gap in functional knowledge outside of the core eudicots in order to better understand the evolution of this important lineage of floral *MADS*-box genes. To that end we used Virus Induced gene Silencing, gene expression, protein-protein interactions and mutant analyses. Here, I attempt to summarize these efforts, highlighting particular cases of sub- and neo-functionalization, as well as redundancy, and ending with a proposed ABC model for this plant and its modifications in floral homeotic mutants.

T2-54-03

Ranunculales as new model order to gain insight into the molecular base of floral trait evolution

Annette Becker[1], *Yafei Zhao*[2], *Paula Elomaa*[2]

1. Justus-Liebig-University Gießen
2. University of Helsinki

Ranunculales, the sister group to all other eudicots, is an amazingly diverse plant order. Many species with diverse floral architectures are now amenable to genetic manipulation with VIGS enabling the functional characterization of key gene regulating floral development. Comparison between Ranunculales flower architecture and the respective genes regulating this architecture allows to address questions concerning the evolution of floral traits. We summarize recent advances towards understanding the evolution of floral variation and present recent results on the *CYCLOIDEA* gene function in Ranunculales.

T2-54-04

Identifying the genetic basis of a key innovation: How Aquilegia got (& lost) their spurs

***Scott Hodges*[1], *Evangeline Ballerini*[1], *Elena Kramer*[2]**
1. University of California, Santa Barbara
2. Harvard University

As this symposium is testament to, the Ranunculaceae are well known for their diversity of floral morphology. Much of this variation is due to the striking diversity of petal shape, from their complete absence to elaborate modifications such as tubular nectar spurs. Such nectar spurs are especially interesting as they have independently evolved within the Ranunculaceae as well as across the Angiosperms and these origins are strongly associated with bursts of species diversification. While we now know that many aspects of petal development appear to be highly conserved, little is known about the genetic basis underlying the great diversity of petal shape. Ideally, we would like to know how cell division and expansion interact to sculpt petal shape and what genes control these processes. In addition, we would like to know how novel organ shapes initially evolve and the genetic basis of morphological variation between taxa. We are using the dramatic development of the Aquilegia nectar spur to address these questions. In particular, we are investigating the genetic basis of the very early initiation of petal nectar-spur development. Key advantages to using Aquilegia is the great diversity among species in spur morphology from completely lacking to 180 mm in length, the fact that fertile hybrids can be formed between most species and a high quality reference genome sequence. To identify candidate genes underlying spur development we are using a QTL approach to first identify genomic region (s) associated with spur development. We are then using comparative gene expression profiles of stages of petal development across multiple species to narrow the list of candidate genes. These studies, combined with functional analyses of candidate genes will be discussed.

T2-54-05

Morphological development of petals in Ranunculaceae

Yi Ren
College of Life Sciences, Shaanxi Normal University

The petals, or the honey-leaves, are of great divergence in morphology in Ranunculaceae, i. e., tubular, bilabial, cup-shaped, flat, concaved or scaled at the base, with or without spur or succate. The previous observations showed that although the petals differ in mature morphology, they showed great similarity in the early development stage. The petal primordia are all hemispherical, rounded and much smaller than the sepal primordia, a relatively long plastochron exists between the last sepal and the first petal and differentiate into a blade and a short stalk. Thus, we assumed that the different morphology of the mature petals might be due to the morphological repatterning of petals in the development. To prove the hypothesis, the morphological development of the petals from 22 species from 20 genera, recovering all ten petalous clades and the major morphological types, in Ranunculaceae was observed by scanning electron microscope (SEM). The young petal undergoes the following developmental stages to the mature petal after it differentiates into blade and stalk. In the first stage, a depression appears at the base of the blade and the nectary tissue will appear in the depression in the later development. In the second stage, two bulges appear at the base of the depression that makes the petal bilabial and the bulges will be the upper lip of the petal and thus the blade will be the lower lip. In the third stage, two bulges become larger and fuse with one another at first and then fuse with the margins of the blade in each side, or each of the bulges fuses with the margin of the blade at first and then fuses with one another, or the bulges stop further growth and the depression deepened to form the succate or the spur. In the fourth stage, the lips, the two fused sides and the stalk growth in different speed. The divergence of development of different petals happens mainly in the third and the fourth stages and less divergence in the second and then the first stages. For example, a tubular petal of *Helleborus thibetanus* undergoes the following developmental stages: a depression appear at the base of the blade, then two bulges appear at the base of the depression, and then the bulges fuse with one another to form the upper lip, the upper lip fuses with two margins of the blade that makes the petal oblique cup-shaped, the growth speed of the upper lip is faster than the two fused sides and that of the later is faster than the lower lip that makes the petal to be tubular; a spurred petal of *Aquilegia yabeana* undergoes the following developmental stages: a depression appears at the base of the blade, then two bulges appears at the base of the depression, the depression deepens to form the spur and the bulges stop further growth. According to the molecular systematic results, the genera in two basal most clades, *Glaudidium* and *Hydrastis*, are apetalous. *Coptis* and *Xanthorhiza* are in the next basal most clade, the petal in this clade only has, or sometimes not clear, the depression at the base of the blade and the nectary tissue appears in the depression and the petals in these genera are not bilabial. The rest of the genera form the core Ranunculaceae in which two bulges appear at the base of the depression. The appearance of the bulges can be considered as a key innovation and let the petals get the diverged potential in morphology and finally resulted in the divergence of the genera. The development processes of the petals in different petalous clades, or even in different petalous genera in the same clade, are different. The bulges might be lost in *Consolida* clade, in this case, the spur is well developed, or only one bulge appears in *Ranunculus* and *Nigella* clades, or both the bulges and the depression might be lost in *Adonis* which has the flat petals without nectary tissue. The growth of the bulges which becomes the upper lip of the petal or the depression which becomes the succate or the spur can result in the hidden of the honey from the nectary tissue and thus can increase the difficulty for the nectar foraging pollinators and avoiding the non nectar foraging insects and then increase the pollination efficiency and increase the adaptability to different nectar foraging insects.

T2-54-06

Ranunculaceae and the Evo-Devo of the flower

Hongzhi Kong
Institute of Botany, Chinese Academy of Sciences

Flowers show tremendous diversity in size, color, structure, symmetry, display, and function, *etc.* The mechanisms that underlie the origin and diversification of the flower, however, remain largely unclear. One of the obstacles that prevent us from knowing more about the underlying mechanisms of flower evolution is the lack of suitable model systems. To solve this problem, we have developed the basal eudicotyledonous family Ranunculaceae into a model system to address important evolutionary developmental questions that cannot be easily addressed by using all the existing models. In this talk, I will first introduce the advantages of using the Ranunculaceae as a model and then present our new results on three topics: 1) the mechanisms that underlie parallel petal losses within the Ranunculaceae; 2) the molecular basis and its flexibility of the floral organ identity determination program in *Nigella damascena*, a species with spiral flower; and 3) the tempo, mode, and mechanisms of character evolution during petal elaboration within the genus *Nigella*.

T2-55: Neuro-pollination Biology

T2-55-01

The specific neural circuit of pollinating flies responses to volatile compound of orchid flower.

Binyan Lu[1], *Wufan Zhang*[1], *Liyuan Wang*[2], *Yi Liu*[1], *Yibo Luo*[1]
1. Institution of Botany, Chinese Academy Of SCience
2. School of Life Science, Tsinghua University

Pollinators display innate attractions to a range of volatile compounds of flowers, and this vital plant-insect interaction shapes the ecological and evolutionary fates of flowering plants. However, the neuronal and behavioral basis of these interactions remain poorly understood. In our study of the deceptive pollination system of *Cypripedium bardolphianum*, we first found out the precise *Drosophila* species of pollinators which mostly are female *D. immigrans*, and identified ethyl tiglate, one of the floral volatile compounds as the key "attractive cue" in this interaction. Behavioral experiments revealed that Ethyl tiglate intrigues female flies in the role of aggregation pheromone, feeding signal and ovipositing signal, which should promote the probability and latency of landing of these pollinators. Furthermore we found out that the specific neural circuit, V glomerulus in antennal lobe of flies, mediated the olfactory signal of ethyl tiglate, which referred as "pollination glomerulus". And the volume of this "pollination glomerulus" of pollination flies was enlarged compared to that of the non-pollination flies. Taken together, our study demonstrated that vital floral volatile compound ethyl tiglate exhibits innate multifunctional attraction of specific *Drosophila* species to accomplish the pollination task and the specialized expanded neural circuit mediating this olfactory signal evolved to serve the dramatic preference of ethyl tiglate in pollinating *Drosophila* species.

T2-55-02

The sweet scent of pollination success

Alexander Haverkamp[1], *Felipe Yon*[2], *Ian W. Keesey*[1], *Christine Mißbach*[1], *Christopher Koenig*[1], *Ian T. Baldwin*[2], *Danny Kessler*[2], *Bill S. Hansson*[1], *Markus Knaden*[1]
1. Max Planck Institute for Chemical Ecology, Department of Evolutionary Neuroethology
2. Max Planck Institute for Chemical Ecology, Departmant of Molecular Ecology

In order to forage efficiently pollinators need to make informed choices about individual flowers. In a natural environment however, floral odors are quickly intermixed with other volatile compounds from the environment such as green leaves or neighboring flowers. Due to this the hawkmoth *Manduca sexta* might need to additionally evaluate individual flowers at a close range before investing energetically costly hovering time. When we tested hawkmoths in a semi-natural free flight tent assay we did find that the moths approached wild-type *Nicotiana attenuata* flowers just as often as neighboring flowers in which the emissions of the main flower compound had been genetically silenced. In spite of this, nectar uptake was drastically reduced in scentless flowers, indicating a low degree of foraging success by the moth. Further tests in a wind tunnel revealed that also pollen take-up and pollen delivery on scentless flowers was strongly reduced due to a lower time investment of the hawkmoth into odor-free flowers. However, as these tests indicated that the hawkmoth detected volatiles of individual flowers mainly in the close vicinity, we hypothesized an additional mechanism for close-range odor detection in the moth. Using morphological, molecular and electrophysiological approaches we demonstrated the presents of additional olfactory neurons on the moth proboscis, which responded strongly to floral volatiles. To further explore the behavior significance of these neurons, a costume build two-choice assay was developed, in which in a freely flying moth only the proboscis was exposed to the flower odors. Through this we could demonstrate that the proboscis neurons were indeed sufficient to mediate the increased time investment into scented flowers and could dues also help to explain the higher foraging and pollination success of the moth on these flowers. Taken together, our results highlight the importance of volatile signals for plant-pollinator interactions and thus also for the co-evolution of these two mutualistic partners.

T2-55-03

Linking brain activity to odour-guided behaviour in the pollination model *Manduca sexta*

Sonja Bisch-Knaden, *Ajinkya Dahake*, *Markus Knaden*, *Silke Sachse*, *Bill Hansson*
Max Planck Institute for Chemical Ecology

Most hawkmoths feed on floral nectar, and are therefore important pollinators of night-blooming plants. To locate these nectar sources, moths use mainly olfactory cues. The hawkmoth *Manduca sexta* is predominately nectaring at flowers of the desert plants *Datura wrightii*, and *Agave palmeri*, which emit very dissimilar floral bouquets. While *Datura* has a strong, sweet smell, *Agave* has a weaker scent, reminiscent of rotten fruits. Odours are coded as spatial activation patterns across the antennal lobe, the first olfactory neuropil of the insect´s brain. We investigated if the valence of an odour is represented already at this early processing level. Using *in vivo* calcium imaging, we first established a functional atlas of the antennal lobe by stimulating the moths with a panel of 80 monomolecular compounds belonging to different chemical classes (terpenes, aromatics, aldehydes, alcohols, esters, ketones, acids, and nitrogen-bearing molecules). A subset of these compounds was used as diagnostic odours that enabled us to identify 23 activity spots, so-called olfactory glomeruli. These functional units of the antennal lobe had distinct response spectra but were mainly broadly tuned, indicating a combinatorial olfactory coding strategy. In order to study the innate significance of the tested odours, i.e. which chemicals might act as feeding stimuli, we next tested the compounds in two-choice experiments (odour *versus* solvent) in a wind tunnel. We found that foraging moths were highly attracted by some of the odours, especially by aromatics and esters, which are characteristic components of *Datura* (aromatics), and *Agave* (esters) flowers. Linking our behavioural results with the obtained functional activity patterns in the brain revealed that innate olfactory valence is encoded in the moth's antennal lobe by a group of glomeruli that is preferentially activated by attractive odours.

T2-55-04

Floral odor signals guide optimal flower selection in a hawkmoth-pollination system

Markus Knaden, *Alexander Haverkamp, Brynn Cook, Julia Bing, Bill Hansson*
Max Planck Society

In order to forage efficiently, pollinators need to make informed choices about individual flowers. This is particularly important for pollinators such as hummingbirds or hawkmoths, which feed during hovering flight - making their decisions particularly costly. If possible these pollinators should therefore recognize rewarding flowers already at a distance or at least before engaging into energetically demanding hovering maneuvers. Darwin already predicted that the best flower choice for a hawkmoth pollinator should be those flowers, which match its proboscis in length and that this foraging advantage would drive the co-evolution between hawkmoths and the flowers they pollinate. However, it has been unclear whether hawkmoths innately recognize and prefer matching flowers. Here, by measuring both the energetic costs and gains of the hawkmoth foraging in a free-flight respiration chamber, we demonstrate first that hawkmoths do indeed obtain the highest net reward from matching flowers. Moreover, we reconstructed the moth flight behavior from 3D video recordings and show that the hawkmoth preferred matching flowers most and did respond to the best matching flower already at the first encounter with the floral odor plume.

T2-55-05

Experimental test of pollination syndromes using hawkmoths and tobacco flowers

Rainee Kaczorowski[1,2], *Karen Chang*[1], *Jose Manuel Sevenello*[1,3], ***Robert Raguso***[1]
1. Cornell University
2. University of Pittsburgh
3. Benemérita Universidad Autónoma de Puebla

The pollination syndrome concept continues to be debated as an organizing principle in floral ecology and evolution. A primary assumption undergirding syndromes is that pollinators show pre-existing sensory bias for specific floral traits, such as colors and scents, which translates into innate preference. Another assumption is that certain floral morphologies or shapes are easier for certain pollinators to learn to manipulate as they seek floral rewards, making them more profitable. When such flowers physically match the dimensions of their pollinators, efficient pollen transfer should result. In combination, these two features provide the ethological (attraction and preference) and mechanical (handling and fit) aspects, respectively, of Verne Grant's concept of floral isolation. We used a lineage of South American tobacco plants (*Nicotiana*; Solanaceae) and a laboratory colony of flower-naïve, inexperienced hawkmoths (*Manduca sexta*; Sphingidae) to test aspects of innate preference and handling efficiency. Individual hawkmoths were flown in no-choice flight cage assays with blooming *Nicotiana* plants whose flowers represented bee-, hummingbird-, settling moth- and hawkmoth-pollinated phenotypes. To assess pre-existing bias, we measured the latency (time to response) and the overall proportion of test moths responding to each *Nicotiana* species. To assess handling, we measured discovery time and success rate for moths to probe the nectary of each *Nicotiana* species with their extended proboscis. All bioassays were filmed and data were scored blind by student volunteers on video. As expected, *M. sexta* moths showed the shortest median latency (25 sec) and highest proportion (100%) of responses to *N. alata*, a hawkmoth-syndrome species, at rates that were twice as responsive as the next most attractive species. Conversely, moths showed lowest responses and highest latencies in response to *N. forgetiana* and *N. rastroensis*, two red-colored, poorly scented hummingbird-syndrome species. Median nectary discovery time (2 sec) did not differ appreciably between species, with slightly lower values for the small, white, upright flowers of *N. bonariensis*, a settling-moth syndrome flower. However, median success rate (40%) was at least four-fold higher at flowers of *N. alata* than at any other tobacco species. These results are largely consistent with the expectations of the syndrome concept, and reinforce recent studies showing that mechanical fit between *M. sexta* tongue length and *Nicotiana* floral tube length predicts energetic profitability in this same system. We will discuss the relative contributions of scent, color and floral posture to these results and those of other studies using similar model systems.

T2-55-06

Perfume-collecting male oil bees? Evidence for a novel pollination system involving *Anthurium acutifolium* (Araceae) and *Paratetrapedia chocoensis* (Apidae, Tapinotaspidini)

Florian Etl[1], *Anna Franschitz*[2], *Antonio J. C. Aguiar*[3], *Stefan Dötterl*[4], *Jürg Schönenberger*[1]
1. University of Vienna
2. Institute of Science and Technology Austria
3. Universidade de Brasília
4. University of Salzburg

It has been known since Stefan Vogel's observations in 1969 that solitary female oil bees collect fatty floral oils and thereby pollinate the flowers of various specialized oil-secreting plant taxa. The bees collect the oil with the help of distinct hair tufts on either their legs or their abdomen. The female bees use this oil-reward as food for their larvae and/or to line their brood cells. Similar adaptations are also known from male oil bees, although the purpose of their oil-collecting behavior has not yet been clarified. Based on an investigation made in the tropical lowland of Costa Rica, we describe a novel pollination system involving male *Paratetrapedia* oil bees and the tropical herb *Anthurium acutifolium*. We present ultrastructural morphological details of bee and plant structures involved in this interaction and the composition of floral scents likely mediating pollinator attraction. Inflorescences of *A. acutifolium* are visited almost exclusively by male *P. chocoensis* oil bees. During their visits on anthetic *A. acutifolium* inflorescences, the bees press a patch of hairs located on their abdominal sterna 3 onto the inflorescence surface. Due to this behaviour, both on staminate and pistillate stage inflorescences, the bees' abdomens and legs become loaded with pollen and also get in contact with receptive stigmas. In contrast to what has been observed in other angiosperms visited for the collection of fatty floral oils, our anatomical investigation shows that the inflorescences/flowers of *A. acutifolium* do not have specialized structures for oil secretion, i.e., they do not have any elaiophores. However, these inflorescences are strongly scented during the time they are visited by the bees. Gas chromatography/mass spectrometry (GC/MS) analyses of dynamic headspace floral samples revealed that inflorescences of both anthetic phases emitted scent bouquets consisting mainly of aliphatic esters, indole and uncommmon terpenoids (megastigmanes). Interestingly enough, our data suggest that the unusual floral scent of *A. acutifolium* is a perfume reward collected by male *P. chocoensis*. This pollination system thus bears a remarkable resemblence with the interactions between perfume-collecting male euglossine bees and their preferred flowers, discovered by Stefan Vogel half a century ago.

T2-56: Polyploidy: early effects and their evolutionary implications

T2-56-01

Effects of abiotic stress on duplicated gene expression and alternative splicing in polyploid *Brassica napus*

Keith Adams

University of British Columbia

Alternative splicing is a fundamental aspect of gene expression because differential AS patterns can alter the protein products produced and/or level of transcripts, which can have phenotypic consequences. In this study we investigated the effects of abiotic stress treatments on expression levels and alternative splicing of homoeologous genes in *Brassica napus*, which is an allotetraploid derived from *Brassica rapa* and *Brassica oleracea*. We conducted a comprehensive transcriptome analysis by strand-specific transcriptome sequencing to examine expression levels and alternative splicing of homoeologous genes in *B. napus* under heat stress, cold stress, and drought stress compared with untreated plants. Cold stress resulted in the greatest changes in gene expression compared to the other stress conditions. More genes had decreased expression levels under the stress treatments than increased expression levels. There were more gene pairs with A_t subgenome biased gene expression than C_t genome biased expression among the homoeolog pairs (27,360) analyzed. Overall the C_t subgenome showed a bias in alternative splicing patterns (quantitatively and qualitatively) in all conditions. The plants under cold stress and drought stress shared more changes in alternative splicing patterns than either with heat stress. These results provide insights into the effects of abiotic stress treatments on homoeologous gene expression and alternative splicing patterns. Some of the changes in expression and alternative splicing may help the plants cope with abiotic stresses.

T2-56-02

The effect of polyploidy on transcriptome size and dosage response in natural and synthetic polyploids

Jeremy Coate[1], ***Jeff Doyle***[2]

1. Reed College

2. Cornell University

"Genome downsizing" is a commonly observed phenomenon in polyploid evolution, in which the size of the polyploid's genome is smaller than would be expected given the genome size of its diploid progenitor (s). Much less is known, however, about the response of the overall transcriptome to polyploidy. We have shown previously that *Glycine dolichocarpa*, an allopolyploid ($2n = 80$) relative of soybean (*G. max*, $2n = 40$) shows "transcriptome downsizing", having a leaf transcriptome that is around 1.4x the size of its two diploid progenitors (i.e., 0.7x per genome), both of which have similar transcriptome sizes. We have extended this observation to the leaf transcriptomes of other *Glycine* allopolyploid species, and also to the root transcriptome of *G. dolichocarpa*, all of which show an approximately 1.5x larger size in polyploids. Knowledge of transcriptome size permits the dosage responses of each individual gene to be estimated, and we have found that the distribution of responses is unimodal, with the mode at around 1.5x per cell, which equates to ca. 0.75x per gene copy. The biological relevance for this response remains unknown. Because these natural *Glycine* allopolyploid species are all several hundred thousands of years old, it is also unclear whether this downsizing is an immediate effect of allopolyploidy, or a function of parallel divergence in these independently formed polyploids. To explore the earliest stages of polyploidy, we have measured transcriptome size in a synthetic *Arabidopsis thaliana* polyploid. As in *Glycine*, the *Arabidopsis* C24 autotetraploid exhibits leaf transcriptome downsizing (0.8-0.9x expression per genome), as well as a unimodal distribution of gene dosage responses centered near 0.8x per gene copy. Estimates of transcriptome size per cell are complicated by the fact that, unlike in *Glycine*, *Arabidopsis* leaf tissue includes endopolyploid cells, and the degree of endopolyploidy varies between accessions. Allowing for endopolyploidy, average transcriptome size on a per cell basis is only about 1.1-1.2x larger in the tetraploids than in their diploid parents. Approximately 20% of genes are dosage compensated (no change in expression following a doubling of copy number), and these are enriched for primary metabolic functions, including translation and photosynthesis. Approximately 34% of genes show a 1:1 dosage effect (a doubling of transcription following a doubling of copy number), and these are enriched for GO terms relating to signal transduction and transport, similar to responses in *Glycine* allopolyploids. The fact that a doubling of gene content does not produce a doubling of transcription, even in the first generations after genome duplication, indicates that less than doubled gene expression is an immediate effect of genome duplication, and perhaps an emergent effect of polyploidy.

T2-56-03

Allotetraploidy generates extensive homeologous expression rewiring and phenotypic diversity at the population level in rice

Yue Sun[1], *Ying Wu*[1], *Jonathan Wendel*[2], *Lei Gong*[1], *Chunwu Yang*[1],

Bao Liu[1]
1. Key Laboratory of Molecular Epigenetics of the Ministry of Education (MOE), Northeast Normal University, China
2. Department of Ecology, Evolution, and Organismal Biology, Iowa State University, USA

Allopolyploidization, i.e., concomitant merging and doubling of two or more divergent genomes in a common nucleus/cytoplasm, is known to instantly alter transcriptome dynamics, a phenomenon referred to as "transcriptomic shock". However, the immediate effects of transcriptomic alteration in generating phenotypic diversity at the population level remain under-investigated. Here, Mass ARRAY-based Sequenom platform was employed to assess and compare orthologous, allelic, and homeologous gene expression in leaf and root of randomly-chosen individuals from populations of parental subspecies (*indica* and *japonica*), *in vitro* "hybrids" (parental mixes), reciprocal F1 hybrids and tetraploids. The data revealed that hybridization and whole genome duplication (WGD) have opposing effects on allelic and homeologous expression in the F1 hybrids and tetraploids, respectively. Whereas hybridization exerts strong attenuating effects on allelic expression differences in hybrids, WGD augments these differences and generates extensive and variable homeologous content and expression patterning among tetraploid plants. Coupled with the extensive phenotypic diversity among tetraploid individuals, our results provide experimental evidence in support of the notion that allopolyploidy catalyzes rapid phenotypic diversification in higher plants. Our data further suggest that homeolog content reshuffling rather than alteration in total expression level may be an important feature of evolution in young allopolyploids.

T2-56-04

Genomic and gene expression divergence between two subgenomes in tetraploid *Oryza minuta*

Xin-Hui Zou[1], *Ming-Sheng Chen*[2], *Yu-Su Du*[1], *Song Ge*[1]
1. Institute of Botany, Chinese Academy of Sciences
2. Institute of Genetics and Developmental Biology, Chinese Academy of Sciences

Polyploidization have long been recognized as a common phenomenon and played important roles in the evolution of plants. Recent studies show that polyploidization is usually accompanied by extensive genome modifications, and the fractionation process and gene expression divergence occurred preferentially toward one subgenome, such as in *Zea mays* and *Brassica rapa*, etc, although the precise mechanisms remain largely unknown. Of about 20 wild species in the rice genus, approximately one half of them are tetraploids that originated through interspecific hybridization and genome doubling, representing a unique system for polyploid research. In our previous studies, we determined the origins of BBCC tetraploids and found that all the BBCC species were formed through polyploidization within the last one million years. In this study, we used the tetraploid *O. minuta* and its diploid progenitors as a study system. We sequenced the genomic regions surrounding *DEP1* locus by isolating orthologous BACs from OMAP BAC libraries, in conjunction with published sequences from three other genomic region, and selected 36 low-copy genes that harbor diagnostic single-nucleotide polymorphisms (SNPs) between the two subgenomes for cDNA pyrosequencing analysis. We compared the homoeologous genome sequence of BBCC tetraploid with its diploid progenitors and investigated the differentiation of two subgenomes at genetic and expression level by DNA sequences analysis and cDNA pyrosequencing. The results indicated various divergent patterns of duplicated genes and intergenic regions in the two subgenomes of *Oryza* tetraploid, including the biased gene structure variation and different expression patterns among three tissues. These findings provide several novel insights into our understanding of the consequences of genome merger and doubling in plants.

T2-56-05

The genomic drivers of ecological divergence in sibling *Dactylorhiza* allopolyploids (Orchidaceae)

Thomas Wolfe[1], *Emiliano Trucchi*[1], *Marie Brandrud*[1], *Francisco Balao*[2], *Juliane Baar*[1], *Maria Teresa Lorenzo*[2], *Terezie Mandakova*[3], *Mikael Hedren*[4], ***Ovidiu Paun***[1]
1. University of Vienna
2. University of Seville
3. Masaryk University
4. Lund University

Early-generation allopolyploids need to quickly accommodate divergent genomes into one nucleus by adjusting organization and function, thereby altering ecological properties and adaptive success. With powerful genomic approaches we are screening the genome-wide natural diversity among ecologically-divergent, sibling alllopolyploids in *Dactylorhiza* (Orchidaceae) in order to identify the mechanisms that, after iterative genome doubling, drive adaptation to different environments leading to isolation. The array of *Dactylorhiza* polyploids are of different ages, 15,000 years or younger. With restriction-site associated DNA sequencing (RADseq) we document a genome-wide absence of genetic differentiation between the allopolyploids. In addition, we bring evidence of frequent gene flow between the polyploids in sympatry, which points toward a strong divergent selection required in order to maintain the observed phenotypic divergence. With RNAseq and smRNAseq we document the progression through time of alterations in gene expression and regulation after allopolyploidization, and their importance to the ecological properties of the allopolyploids. Our experiment includes 31 transcriptomes isolated from wild individuals collected on a European-wide scale, after they have been grown in uniform conditions. We observe a general trend of increased overexpression of genes in the youngest polyploid lineages in comparison to the oldest *D. majalis*, whose transcriptome generally resembles more closely those of the diploid parents. We estimate the diploid parental species to have diverged 10 MYA; they show today highly divergent transcriptomes, indicating a significant genomic shock associated with the origin of the allopolyploids. The phenotypic divergence between the polyploids is mediated by a general parental dominance in opposite directions in the sister polyploids, a patterns retained partly also at the level of transgressively expressed transcripts. The differential expression between the polyploids affects several genes related to metabolic processes and responses to stimuli, with a putative ecological function. For example, several genes in the nitrate metabolic pathway appear to have been significantly down-regulated in *D. traunsteineri*, which became adapted to nitrogen poor environments. The differential posttranscriptional regulation via smRNAs is less significant, and affects a different population of transcripts, processes and functions. We conclude that the major transcriptomic divergence among the diploid parental species of these polyploids became reconciled in different ways in the sibling *Dactylorhiza* polyploids, most probably as a result of an interplay between stochastic genomic alterations and distinct selection pressures specific for divergent environments. We finally discuss the importance of qualitative versus quantitative expression alterations in allopolyploids, and their role for diversification in general.

T2-56-06

Insights into reticulate evolution and hybridization in the phylogeny of willows (*Salix* L.) using RAD sequencing

*Natascha Dorothea Wagner, **Susanne Gramlich**, Elvira Hörandl*
Georg-August-Universität Göttingen, Department of Systematics, Biodiversity and Evolution of Plants

Hybridization is a frequent phenomenon in flowering plants, nevertheless, the evolutionary impact is still controversial discussed and in detail poorly understood. The genus *Salix* is an interesting model system for studying ancient and recent hybridization. Willows have the highest frequencies of hybrid formation among all flowering plants and polyploidy occurs frequently, ranging from diploid to octoploid species. Crosses between distantly related species from different morphological sections appear regularly in natural populations. Ecological and geographical crossing barriers can break down easily after secondary contact hybridization. Information on morphological patterns and presence of polyploidy suggest that the evolutionary history of the genus has been shaped by reticulate evolution and ancient hybrid speciation. However, traditional DNA Sanger sequencing markers could not resolve interspecific relationships within Salix. Especially the big and species-rich clade of northern hemisphere shrub willows (subg. *Chamaetia* and subg. *Vetrix*) was so far unresolved. It remained unclear whether polytomies in phylogenetic trees were the result of conflicting phylogenetic signals or of low genetic divergence. In deeper time levels the evolutionary history might have been further influenced by incomplete lineage sorting. We used the next generation sequencing approach RAD sequencing (Restriction site Associated DNA sequencing) to analyse the phylogenetic relationships of representatives of the 16 European sections of the *Vetrix-Chamaetia* clade. By mining a large number of single-nucleotide polymorphisms (SNPs) we could reconstruct the phylogenetic history and the temporal framework for reticulate events within the genus. We will discuss methods of discriminating incomplete lineage sorting from hybridization. We will further present experience on the applicability of this genomic marker system on high polyploids to understand evolutionary origin, parentage and diversification of polyploid willow species.

T2-57: Evolutionary and biogeographic studies of gymnosperms in the genomic era

T2-57-01

Gnetales is sister to the pine family but experienced convergent evolution with angiosperms: evidence from a comparative transcriptomic analysis

*Jin-Hua Ran, Ting-Ting Shen, **Xiao-Quan Wang***
State Key Laboratory of Systematic and Evolutionary Botany, Institute of Botany, Chinese Academy of Sciences

After decades of molecular phylogenetic studies, evolutionary relationships among major clades of seed plants are still controversial, especially the phylogenetic placement of Gnetales. What factors are responsible for the phylogenetic conflicts? To address this question, and to resolve the phylogeny of gymnosperms, we conducted a comparative transcriptomic analysis with a sampling of 22 genera representing all 13 families of gymnosperms and 13 species representing main lineages of angiosperms. Multiple data sets containing up to 1,135,848 sites across 1,114 loci were analyzed, using partitioned and unpartitioned concatenations as well as coalescence approaches. Our study generated a consistent and well-resolved phylogeny of seed plants, which strongly supports Gnetales as sister to Pinaceae (the Gnepine hypothesis), and cycads plus *Ginkgo* as sister to the remaining extant gymnosperms. It is interesting that the values of d_N, d_S, ω, R_N and R_S are very similar between angiosperms and Gnetales, implying that these two groups might have experienced similar selective pressures in their evolutionary histories. Convergent molecular evolution, rather than compositional heterogeneity and heterotachy, could be responsible for the topological discordance between gene trees of seed plants.

T2-57-02

Advances in using Hyb-Seq for molecular evolutionary studies of Pinus and Pinaceae

***David S. Gernandt**[1], Ann Willyard[2], Xitlali Aguirre Dugua[1], Alejandra Moreno Letelier[1], Alejandra Vázquez-Lobo[3], Rubén Montes Montiel[1], Aaron Liston[4]*
1. Universidad Nacional Autónoma de México
2. Hendrix College
3. Universidad Autónoma del Estado de Morelos
4. Oregon State University

Large multilocus datasets offer an outstanding opportunity for evolutionary studies of closely related species, but there remains a great need to evaluate the effect of alignment length, number of loci, evolutionary rate, and clocklike behavior on improving the accuracy of phylogenetic results. Early molecular studies of Pinaceae based on the nuclear, plastid, and mitochondrial genomic compartments mainly found agreement with morphology-based phylogenetic hypotheses among genera in the family and within *Pinus*, its largest genus. However, single locus studies of closely related species of *Pinus* have given results that are at odds with each other or with other sources of evidence such as morphology and anatomy. We adopted the Hyb-Seq method, target capture of nuclear loci combined with high copy number plastid genomes and nuclear ribosomal DNA, to characterize representative lineages of Pinaceae and *Pinus*. RNA probes for target enrichment of 1,000 low-copy number nuclear genes were based on published studies of *Pinus taeda*. Fewer than 25% of these genes have been mapped to chromosomes. High-copy number plastome and nuclear ribosomal sequences were also obtained. Using stratified taxonomic sampling (within *Pinus* subsections and among principal lineages of Pinaceae) we inferred phylogenetic relationships using multispecies coalescent methods that account for both incomplete lineage sorting and introgression. We tested the hypothesis that loci with clocklike substitution rates give more accurate phylogenetic inferences than those with unequal rates.

T2-57-04

The intergeneric plastid phylogenomics of Gymnosperms

Xiao-Jian Qu[1,2], *De-Zhu Li*[1,2], *Ting-Shuang Yi*[1]
1. Kunming Institute of Botany, Chinese Academy of Sciences
2. Kunming College of Life Sciences, University of Chinese Academy of Sciences

The phylogenetic relationships among major clades of gymnosperms have been addressed by multiple studies. However, relationships among some genera are still not fully resolved. We carried out plastid phylogenomic study including all 86 gymnosperm genera. Relationships among most genera in gymnosperms were fully resolved with strong support except those in one Podocarpaceae clade. Different lineages show different degree of substitution rate variation and plastomic rearrangements. Correlation between nucleotide substitution rates and plastomic rearrangements was revealed and should be explained by the improper mismatch repair system. This comprehensive plastid phylogenomics offers important insights into deep phylogenetic relationships and diversification of plastome structures of gymnosperms. The robust phylogenetic backbone we provide establish a framework for future comparative studies on gymnosperm evolution.

T2-57-05

Evolution and speciation of the Chinese white pine, *Pinus armandii*

Yan-Yan Liu, ***Xiao-Xin Wei***, *Xiao-Quan Wang*
Institute of Botany, Chinese Academy of Sciences

Speciation is a fundamental issue in evolutionary biology and it is responsible for the world's biodiversity. The process of species diversification is strongly linked with historical and ecological factors. The Hengduan Mountains, in the southeastern fringe of the Qinghai-Tibetan Plateau (QTP) in southwestern China, is one of the world's biodiversity hotspots. It is characterized with a series of parallel, roughly north-south running high ridges, separated by deep valleys. These massive mountains provide an exceptional range of climates, and thus would be an excellent system to investigate how genetic, geographical, ecological and evolutionary factors interact to create species. *Pinus armandii* Franch is an endemic and dominant species of coniferous forest in central and southwest China. Two distinct genetic groups or ecotypes, i.e., the northern and southern provenances, which roughly divided by Gongga Mountains, the highest mountain in the eastern QTP, have been recognized by previous provenance test and cytoplasmic DNA markers. In this study, transcriptome data was used to reconstruct the evolutionary history of *P. armandii* complex. Population genomics analysis further confirmed the significant genetic differentiation between the north and south populations of *P. armandii*. Many lines of evidence suggested that speciation has occurred in *P. armandii*. In the process of speciation, the massive Gongga Mountain played a key role as a physical barrier in the initial isolation of the north and south populations, and the contrasting environments resulted from the formation of Gongga Mts. subsequently accelerated their differentiation. This study sheds novel insights into the extremely high species diversity of conifers in the Hengduan mountains region.

T2-57-06

Biogeography and ecological niches of Asian *Gnetum* (Gnetales)

Chen Hou[1], *Catarina Rydin*[1,2]
1. Stockholm University
2. Bergius Foundation, The Royal Swedish Academy of Sciences

Macroevolutionary patterns are the result of complicated processes, including speciation and extinction events as well as dispersal and vicariance, which may operate over long time periods. It can therefore be particularly difficult to infer the evolutionary history of a group if a fossil record is deficient. This is true for *Gnetum*, a gymnospermous genus that comprises 30-40 tropical species with the life forms trees, shrubs and lianas. Asia hosts the majority of the species and is the hotspot diversity region of the genus. Time-calibrated phylogenies have indicated the Gondwana continent as the ancestral region of *Gnetum*, and its subsequent breakup appears to have facilitated early divergence of the genus. However, no investigations that explicitly elucidate historical processes behind current biogeographic patterns have so far been performed, and the biogeography of Asian *Gnetum* is particularly poorly understood. To address the latter question, we will first address the delimitation of the sister-species of the remaining Asian clade, *G. gnemon*, an arborescent species with a broad distributional range across tropical regions of the Southeast Asia. Second, ancestral areas of Asian clades of *Gnetum* will be inferred using an extensive sampling of species. We will also investigate whether geological or climatic variables exert considerable influences on the diversification of Asian *Gnetum*. Modelling of environmental niches of Asian *Gnetum* species will be conducted in the context of a new phylogeny. The study enriches the knowledge of species delimitation, biogeography and ecology of *Gnetum*. Furthermore, it attempts to uncover mechanisms that drive diversification in extant gymnosperms.

T2-58: Orchid pollination

T2-58-01

Pollination of *Specklinia* (Orchidaceae: Pleurothallidinae) by nectar-feeding *Drosophila*

Adam Karremans[1,2], *Franco Pupulin*[1], *Barbara Gravendeel*[2]
1. Lankester Botanical Garden, University of Costa Rica
2. Naturalis Biodiversity Center

Background: The orange flowered species of *Specklinia* (Orchidaceae: Pleurothallidinae) form a monophyletic but morphologically diverse group. Fruit flies, *Drosophila* spp., have been reported to be pollinators of *Specklinia endotrachys*, one of the most morphologically aberrant and variable species in the group. Based on morphological, geographical, ecological and molecular data, we have found that *S. endotrachys* represents a species complex. Objective: Our aim was to discover which strategy of pollinator attraction is employed by orange flowered *Specklinia*. Methods: Pollinators and visitors of species of the complex, that includes *S. endotrachys*, *S. dunstervillei*, *S. pfavii*, *S. remotiflora* and *S. spectabilis*, were observed and caught in the wild and in a green house in Costa Rica. Using DNA barcoding more than 10 closely related species of *Drosophila* were identified, including *D. hydei* and *D. repleta*. The flies are arrested by the aggregation pheromones ethyl tiglate, methyl tiglate and isopropyl tiglate, to which at least *D. hydei* was shown to be very sensitive in ElectroAntenography (EAG) tests. Using SEM of the sepals we were able to detect glands on the abaxial surface and stomata and nectar on the adaxial surface. Results: TEM photographs of the glands were identified as possible osmophores and the stomata as nectar secreting stuctures. Visible nectar drops on the adaxial surface of sepals are secreted by those nectar-secreting stomata, encouraging male and female *Drosophila* to linger on the flowers for several hours at a time. The flies frequently show courtship behavior, occasionally copulating. Several different *Drosophila* species can be found on a single *Specklinia* species. Conclusions: Species of the *S. endotrachys* group share a similar pollination syndrome based on aggregation pheromones and there seems to be no species-specific relationship between the orchids and the flies. However, it is not expected that *Specklinia* species will hybridize naturally as their populations do not overlap geographically.

T2-58-02

Exploring the evolutionary origin of floral organs of the emerging orchid model *Erycina pusilla*

Anita Dirks-Mulder
University of Applied Sciences Leiden

Background: thousands of flowering plant species attract pollinators without offering rewards, but the evolution of this deceit is poorly understood. Rewardless flowers of the orchid *Erycina pusilla* have an enlarged median sepal and incised median petal ('lip') to attract oil-collecting bees. These bees also forage on similar looking but rewarding Malpighiaceae flowers that have five unequally sized petals and gland-carrying sepals. The lip of *E. pusilla* has a 'callus' that, together with winged 'stelidia', mimics these glands. Different hypotheses exist about the evolutionary origin of the median sepal, callus and stelidia of orchid flowers. Results: the evolutionary origin of these organs was investigated using a combination of morphological, molecular and phylogenetic techniques to a developmental series of floral buds of *E. pusilla*. The vascular bundle of the median sepal indicates it is a first whorl organ but the morphology of the epidermal cells reflects convergence of petaloid features. Expression of some MADS-box gene lineages is low in the median sepal, possible correlating with its petaloid appearance. A vascular bundle indicating second whorl derivation leads to the lip. Some *MADS*-box gene lineages are highest expressed in lip and callus, consistent with current models for lip identity. Six vascular bundles, indicating a stamen-derived origin, lead to the callus, stelidia and stamen. Some *MADS*-box gene lineages are highest expressed in the stamen. Other *MADS*-box gene lineages are highest expressed in the stelidia, suggesting these may be required for fertile stamens. Conclusions: the median sepal, callus and stelidia of *E. pusilla* appear to be derived from sepals, petals, and stamens, respectively. Duplications, diversifying selection and changes in spatial expression of different *MADS*-box genes shaped these organs, enabling the rewardless flowers of *E. pusilla* to mimic an unrelated rewarding flower for pollinator attraction. These genetic changes are not incorporated in current models and urge for a rethinking of the evolution of deceptive flowers.

T2-58-03

Deceptive pollination of New Zealand *Corybas* (Orchidaceae)

***Carlos Lehnebach*[1]**, *Andreas Zeller*[2], *Philip Garnock-Jones*[2]
1. Museum of New Zealand Te Papa Tongarewa
2. Victoria University of Wellington

There are over 100 species of orchids in New Zealand (NZ). Although most genera are also found in neighbouring landmasses over 70% of NZ orchid species are endemic. *Corybas*, a genus of terrestrial orchids locally known as spider and helmet orchids, comprises 22 species and ca. five entities which taxonomic status requires further study. Pollination of *Corybas* has puzzled botanist and orchid enthusiasts for decades. Fungus gnats (*Mycetophila*, Diptera) have been reported as occasional floral visitors, and based on these observations, it has been suggested pollination by brood-site deception occurs in these orchids. How does this mechanism exactly operate and the degree of specificity of such interaction has remained largely unstudied. Similarly, whether this deceptive mechanism has promoted diversification of *Corybas* in NZ is unknown. To answer these questions we will present evidence from floral scent analyses, scanning electron microscopy and observations under UV light of the floral structures as well as DNA analyses of floral visitors, eggs and larvae collected from flowers and mushrooms. Also, preliminary data from *ex situ* bio-assays using virgin flowers and naïve pollinators reared in captivity will be presented.

T2-58-04

Pollination and reproductive isolation in *Cryptostylis* orchids

***Alyssa M Weinstein*[1]**, *Ryan D Phillips*[1], *Björn Bohman*[2,1], *Gavin R Flematti*[2], *Celeste C Linde*[1], *Rod Peakall*[1,2]
1. The Australian National University
2. The University of Western Australia

Sexually deceptive orchids provide a unique opportunity to study plant speciation due to their extreme pollinator specificity, which can rapidly drive diversification via pollinator-mediated speciation. The genus *Cryptostylis* presents an exception to the norm of a different insect species pollinating each sexually deceptive orchid species, with all five Australian species of *Cryptostylis* being exclusively pollinated by the ichneumonid wasp *Lissopimpla excelsa*. Despite frequently occurring in sympatry, the species do not hybridise. We aimed to identify mechanisms that could be contributing to reproductive isolation, and to compare traits relevant to pollinator attraction across species. In a mark-recapture experiment we found that 20% of individual wasp pollinators visited multiple *Cryptostylis* species, and displayed the necessary behaviour for inter-species pollen transfer to occur. Different ploidy levels have previously been reported for two Australian species, leading us to investigate this further across all five species. Flow cytometry revealed inter-species differences in orchid genome size, which may suggest that ploidy differences are contributing to the maintenance of reproductive isolation between the species. Hand cross-pollinations have been conducted to compare the percentage of viable seed resulting from intra- and inter- species pollination events. To determine whether an inter-species difference in requisite fungal symbionts could be preventing the establishment of hybrids, fungal isolations and sequencing of peloton rich orchid sections have been undertaken. Floral orientation (which varies among species) was artificially manipulated, yet pollinator sexual attraction and positioning behaviour in relation to the floral parts remained constant. Floral shielding experiments, where orchids are placed inside a black shield to remove visual cues, indicated long-range pollinator attraction to be chemically mediated. One semiochemical that attracted the pollinator was isolated using bioassay-guided fractionation, with extracts being fractionated via column chromatography and high-pressure liquid chromatography. This semiochemical was found to be present across Australian *Cryptostylis* species. Investigation of one of 18 Asiatic *Cryptostylis* species, *Cryptostylis arachnites*, revealed that this species does not contain the semiochemical. To provide further insight into the evolution and diversification of this genus, it is aimed to determine the pollination strategy of Asiatic *Cryptostylis*. Further, a comprehensive phylogeny of the genus will be created, which will yield insights into the phylogenetic relationships between the Australian and Asian taxa.

T2-58-05

Floral modifications for pollination by Microdiptera (Chloropidae) in the genus *Corunastylis* (Orchidaceae)

Peter Bernhardt[1], *Zong-Xin Ren*[2], *Wendy Grimm*[3], *Brian Towle*[4], *Qi Qiao*[5], *Retha Meier*[6]

1. Department of Biology, Saint Louis University

2. Kunming Institute of Botany, Chinese Academy of Sciences

3. Ku-ring-gai wildflower garden

4. Ecological Solutions Australia

5. College of Agriculture Henan University of Science, Technology

6. Department of Education, Saint Louis University

Flies 1.5 - 2.5 mms pollinate Corunastylis species in eastern Australia. The orchids' high conversion ratio of ovaries into fruit (75-95%) is explained based on the evolution of modifications that exploit Microdiptera including secretory staminodia, trembling labella and summer-autumn flowering.

T2-58-06

Wasp pollination in two sympatric orchids: *Coelogye fimbriata* and *Panisea cavaleriei* (Orchidaceae)

Jin Cheng

Beijing Forestry Univeisity

Orchidaceae is one of the most species-rich and highly evolved plant families and is well-known for its special pollination mechanism.Pollination by a single wasp species is rare in orchids except in species with sexually deceptive flowers that are pollinated exclusively by male insects. Consequently, finding and interpreting examples and attraction mechanism of wasp-pollinated orchid species that are pollinated by deception, but lack sexually attractive characters, becomes a key point in understanding the evolutionary ecology of wasp-orchid interactions and even pollinator-orchid interactions. Moreover, it is also important in discussing the effect of deceptive strategies in speciation, diversification and evolution of orchids and even angiosperms. In this study, we focused on the wasp-pollination system in two sympatric and co-flowering orchids in Yachang Orchid National Nature Reserve: *Coelogye fimbriata* and *Panisea cavaleriei* with nectarless flowers which were pollinated by females of the same *Vespula* species (Vespidae, Hymenoptera). Quantitative study of floral scent and color were employed in order to learn the deceptive strategies to wasps. The floral odor was considered to be the important cue to attract the female wasps in both two orchids. Reproductive isolation is one of the key steps in speciation and the foundation of species independence, especially in sympatric and co-flowering species. Mechanical isolation mediated by *Vespula* sp. was the prezygotic isolation mechanism between *Coelogye fimbriata* and *Panisea cavaleriei.*

T2-59: Phylogenomics and Evolution of Legumes

T2-59-01

Plastid phylogenomics reveals deep relationships of Fabaceae

Ting-Shuang Yi, *Jian-Jun Jin, Yin-Huan Wang, Rong Zhang, De-Zhu Li*

Kunming Institute of Botany, CAS

The family Fabaceae (Leguminosae) is the third largest angiosperm family in terms of species richness with c. 770 genera and over 19,500 species. In ecological and economic terms the family is also one of the most important plant groups of the world and it has been recently the focus of numerous taxonomic, phylogenetic and evolutionary studies. However, the classical and long standing three subfamilies classification has been questioned, and categories of many tribes, subtribes and clades need to be clarified. Despite remarkable advances over the last 20 years, some deep phylogenetic relationships within the family remain unresolved or weakly supported. Plastid phylogenomic analyses were conducted on 350 species representing 300 genera of Fabaceae. Trees obtained from different data sets and/or tree reconstruction methods are largely consistent. We found better support for some short nodes when noisy sites or saturated genes are excluded from the data matrix. The major phylogenetic relationships of Fabaceae were fully resolved with high support, and most deep relationships are clarified.

T2-59-02

A phylogenomic approach to study the evolutionary origin of Detarioideae, a clade of ecologically dominant tropical African trees.

Manuel De La Estrella[1], *Dario I. Ojeda*[2], *Erik Koenen*[3], *Sandra Cervantes*[2], *Jeremy Migliore*[2], *Anne Bruneau*[4], *Olivier Hardy*[2], *Felix Forest*[1]

1. Royal Botanic Gardens, Kew

2. Universite Libre de Bruxelles

3. University of Zurich

4. Universite de Montreal

Tropical rain forests are one of the most biologically diverse habitats on the planet. Yet their description and the understanding of their evolutionary history are far from complete. West Central Africa represents the area of greatest biodiversity richness in tropical Africa and has the highest percentage of untouched pristine forest in all of Africa and Madagascar. Our study is focused on the recently established subfamily Detarioideae (Leguminosae), which are the dominant tree components of West Central African forests. Detarioideae includes 81 genera (and approximately 750 species), 58% of which occur in Africa and Madagascar (with 329 species). Our aim is to generate the first fully sampled phylogeny, including genera never sequenced before, using a phylogenomics approach (sequence capture). Our bait design includes 283 nuclear genes (359,222 bp) and we will obtain plastid genome sequences as well. We use this new phylogeny to study the biogeographical history of the group and investigate biome shifts within the subfamily. Detarioideae originated in Africa-South America and are retained as a Primary Forest clade in its origin according to our analysis. Implications for the origin of the group and its role in shaping African diversity are discussed.

T2-59-03

Solving difficult phylogenetic problems in Leguminosae using multiple genome-scale sequence data sets

Erik Koenen[1,7], *Jens Ringelberg*[1], *Catherine Kidner*[2,4], *James Nicholls*[3,4], *Toby Pennington*[4], *Madhugiri Nageswara-Rao*[5], *Donovan Bailey*[5], *Jan Wieringa*[6], *Freek Bakker*[7], *Royce Steeves*[8], *Dario Ojeda-Alayon*[9], *Jérémy Migliore*[9], *Olivier Hardy*[9], *Luciano De Queiroz*[10], *Gillian Brown*[11], *Gwilym Lewis*[12], *Anne Bruneau*[13], *Colin Hughes*[1]

1. Department of Systematic & Evolutionary Botany, University of Zurich

2. Institute of Molecular Plant Sciences, School of Biological Sciences, University of Edinburgh

3. Institute of Evolutionary Biology, School of Biological Sciences, University of Edinburgh

4. Royal Botanic Garden Edinburgh

5. Department of Biology, New Mexico State University

6. Naturalis Biodiversity Center

7. Biosystematics department, Wageningen University

8. Fisheries and Oceans Canada

9. Université Libre de Bruxelles

10. Universidade Estadual de Feira de Santana

11. Queensland herbarium, DSITI

12. Comparative Plant and Fungal Biology Department, Royal Botanic Gardens, Kew

13. Institut de recherche en biologie végétale, Université de Montreal

Several parts of the legume phylogeny suffer from lack of resolution. Here we investigate whether genome-scale data can resolve these problems, with a focus on three specific parts of the legume phylogeny: (1) the root of the Leguminosae and relationships among the six subfamilies, (2) relationships among the major lineages in the recently recircumscribed Caesalpinioideae and (3) the large polytomy formed by the clade that includes the c. 37 genera in the non-monophyletic tribes Ingeae and Acacieae p.p. and c. 1,850 of the c. 3,300 species in the mimosoid clade (the former Mimosoideae). We harness the power of large datasets including fully sequenced genomes, transcriptomes, complete chloroplast exomes/proteomes and hybrid capture sequence data for 967 low copy nuclear genes for the recircumscribed Caesalpinioideae. We explore different phylogenetic reconstruction methods using heterogeneous models of molecular evolution on both DNA and protein sequence data and gene-tree species-tree reconciliation methods. Our results show that the root of the legumes remains problematic to resolve, even with large volumes of data. After dissecting the phylogenetic signal among hundreds of nuclear genes, we find that significant numbers of genes support each of the different possible rootings. We suggest this might be caused by incomplete lineage sorting, rapid diversification of the early legume lineages, extinction of early legume diversity, whole genome duplication, or most likely, a combination of these factors. Within the recircumscribed Caesalpinioideae, we can resolve the most probable branching order of the major lineages in the subfamily. Previously, lack of resolution across this part of the tree has hindered progress towards a phylogenetic classification of the legume family, even though it was clear that former subfamily Mimosoideae needed to be included in a recircumscribed Caesalpinioideae to avoid having to recognize a large number of additional, small subfamilies. Our results provide the basis for a new tribal classification, and recognition of additional clades with evolutionary significance (e.g. a petiolar nectary clade and an aggregated pollen clade). Within the large clade combining former tribes Ingeae and Acacieae p.p., large scale plastid and nuclear gene data resolve several major subclades for the first time, supporting some previously hypothesized informal generic alliances based on morphology while rejecting others. Furthermore, substantial geographic and ecological structure is apparent across this large clade, but this requires further investigation to understand the relative contributions of dispersal limitation and niche conservatism in shaping the mimosoid phylogeny. Future research will focus on mimosoids, and aims to infer a robust backbone phylogeny for the clade with complete sampling of genera and dense sampling of larger genera using the same set of genes via further hybrid capture work. Our results have implications for revising the tribal classification of the legume family and our study has wider significance in terms of methods for inferring phylogenies of large (plant) clades combining different types of (genome-scale) molecular data.

T2-59-04

Polyploidy, nodulation, and Legume genome evolution

Susan Strickler[1], *David Lyon*[1,2], *Ivana Rodriguez*[1,3], *Adrian Powell*[4], ***Jeff Doyle***[4]

1. Boyce Thompson Institute

2. Franklin & Marshall College

3. New Mexico State University

4. Cornell University

Polyploidy has played a major role in shaping the genomes of species in the legume family (Leguminosae, Fabaceae). It is now known that independent whole genome duplications took place in the ancestors of several of the major lineages of the family; only one of these, in the papilionoid subfamily (Papilionoideae), has been well characterized. Symbiotic nitrogen fixation involving soil bacteria (rhizobia) housed in plant-produced nodules occurs in most members of the Papilionoideae, and more sporadically in the Caesalpinioideae (within which the predominantly nodulating mimosoid clade is embedded), but is absent outside of these two sister lineages. The scattered distribution of nodulating lineages in Caesalpinioideae has raised the question of the number of origins of the phenomenon in legumes. Assessment of nodule homology using morphological, anatomical, biochemical, or molecular phylogenomic approaches has produced equivocal results. In an effort to determine whether non-nodulating legumes have lost nodulation or never had it, we have identified conserved noncoding sequences (CNS) in the genome sequences of papilionoid legumes, with over 100,000 CNS shared between soybean and *Medicago truncatula*; we are assessing the association of CNS with genes involved in nodulation. In addition to characterizing an important component of legume genomes, understanding the evolution of CNS could help elucidate how genes were recruited for nodulation from pre-existing functions such as mycorrhizal symbiosis, and how polyploidy could have played a role in the evolution of nodulation. The assumed (though still uncertain) correlation of the distribution of polyploidy and nodulation in papilionoids led to competing hypotheses: 1) polyploidy enabled or facilitated the origin of nodulation; vs. 2) polyploidy refined or "enhanced" pre-existing nodulation. The first hypothesis is untestable, but we are testing the second by studying the effects of polyploidy on nodulation in recently formed (<500 K years) allopolyploids in wild perennial members of *Glycine*, the genus that includes soybean. Although the effects of genome merger and duplication are not always predictable, we do see a general enhancement of nodulation in allopolyploid *Glycine* species. Specifically, *G. dolichocarpa* appears to have enhanced nodulation responses when compared to its diploid progenitors. We also have evidence that the allopolyploid exudes reduced levels of formononetin, an isoflavone that inhibits symbiotic signaling, compared to one of its diploid progenitors, and that it has an enhanced nodulation-related hormone signaling response, and reduced defense response, compared to the other progenitor, when treated with rhizobia.

T2-59-05

Phylogenetic and biogeographical implication of *Caragana, Calophaca* and *Chesneya* to the relation of Central Asia and East Asia flora

Mingli Zhang

Xinjiang Institute of Ecology and Geography, CAS, Institute of Botnay, CAS

Caragana, Calophaca and *Chesneya,* have a distribution concentration pattern of Central Asia and East Asia, especially occur in Himalayas-Tibet and Central Asian Mountains and desert. Based on our recent molecular phylogeny and biogeography of these legume genera, the perspectives of relation of Central Asian and East Asian flora are concluded as following floristic aspects. Most taxa at generic level, have the temporal origin and diversification in late Oligocene to early Miocene. Mountains chairs, such as Himalayas, Tianshan, Pamir-Alai, Hindu Kush, are hypothesized as the origin, diversification and evolutionary center, as well as the dispersal and migration routes. Geological and climate events occurring late Tertiary, such as QTP uplift, Tethys retreat, Central Asian aridification, and East Asian monsoon, are thought as evolutionary driving force.

T2-59-06

The legume genus dussia as a model to understand the complexity of taxonomic tree diversity, biogeography and origin of the neotropical flora

*Toby Pennington[1], Cathy Winterton[2], **Nelson Zamora**[3]*

1. Royal Botanic Garden Edinburgh

2. Royal Botanic Garden Edinburgh

3. Museo Nacional de Costa Rica

Dussia is a genus of trees, described in 1892 from Martinique Island, confined to tropical rainforests and distributed from S. Mexico and Antilles to Bolivia and Amazon basin of Brazil. *Dussia* is poorly known, both taxonomically and biologically because its species are rare trees that have been seldom collected due to their large size. The most recent monographic revision (1963), stated that delimitation of taxa was "tentative" because of inadequate material and recognized 10 species; this revision involved no field study. Due to variation patterns in *Dussia* are highly complex and not amenable to simple delimitation of species. It appeared clear that to resolve problems of species delimitation in *Dussia* required extensive field study to gather more specimens, and to evaluate whether any "field" characters could provide extra feature for species delimitation. A forthcoming taxonomic revision based upon extensive fieldwork and upon many more specimens is likely to recognize up to eight new species, the highest diversity for the genus is found in Costa Rica. The phylogeny of *Dussia* clearly demonstrates stronger affinities of the Chocó with Central America than with Amazonia because the Pacific Ecuador accessions are nested in a Central American clade. In contrast, the phylogeny suggests that there have been no trans-Andean migrations between Amazonia and the Chocó region. The Chocó clade is nested in the Central American clade. This suggests that subsequent to the south-north migration event between South and Central America, a north-south migration event must have occurred ~7 Ma. The earliest migration between South America and Central America on the *Dussia* tree is dated at ~15 Ma, which substantially pre-dates the widely accepted date (ca. 3 Ma) of the formation of the Panamanian land bridge, so over-water dispersal seems possible. This dated phylogeny for *Dussia* shows there were multiple migrations between South and Central America. Three main biogeographical influences were found, the North Pacific Ocean, the Chocó region and the Andes.

T2-60: Use of target capture and high-throughput sequencing as a tool for plant systematics

T2-60-01

Data assembly and post-processing in aTRAM for museum phylogenomics

***Ryan Folk**, Julie Allen, Pamela Soltis, Douglas Soltis, Robert Guralnick*

Florida Museum of Natural History

NGS methods have quickly become a central resource for phylogenomics, enabling robust phylogenetic inference in a wide variety of taxa and greatly facilitating the use of historical DNAs. Methods for data collection and multilocus phylogenetic analysis are now well-trodden territory, but in the middle ground there is a conspicuous literature gap of identifying appropriate methodologies for data assembly. Much has been proposed in the way of alternative methods, but little exists in the way of comparative methodological studies. Several pipelines have appeared, among which aTRAM is particularly promising. aTRAM is a hybrid reference-based and reduced *de novo* method (that is, a *reference-guided* method). Reads matching a reference by BLAST are pulled and assembled as small *de novo* subproblems; this process is iterative in the sense that assembly results are the new reference for each successive round of BLASTing and assembly. aTRAM is highly parallelizable within each assembly problem and among loci and taxa, and is being actively optimized for performance and ported to python; the *de novo* portion integrates multiple methods with highly flexible parameters. Here, we report aTRAM results for (1) a deep divergence, comprising exon-capture data collected in the Saxifragales. This late Cretaceous radiation presents an assembly challenge due to deep nucleotide divergence and our reliance on primarily museum DNAs. We also show (2) results for a shallow divergence in *Heuchera*, a ~4 my divergence where low nucleotide variation and the acute need for gene-tree based analyses present the primary challenges. We compare these to *reference-based* short read aligning approaches and present new work in post-processing exonic data in a protein- and paralog-aware manner, complementary to alternative consensus-based approaches. We advocate a pluralistic approach to NGS assembly; much like *de novo* assembly, researchers should begin by testing several methods, and perhaps report comparative results among assemblies, which could shed light on methodological or biological issues that would be hidden in isolation.

T2-60-02

Using target-enriched high-throughput sequencing to gain robust nuclear genetic evidence to reconstruct the phylogeography of the Ponderosae, a recalcitrant clade of conifers

Ann Willyard
Hendrix College

Species relationships (and even species delineations) have been elusive in subsection *Ponderosae* (Pinaceae; *Pinus*; subgenus *Pinus*; section *Trifoliae*). The identification of morphological synapomorphies in this group was hindered by plasticity, within-population variability, and convergent evolution as populations migrated into similar ecological niches. The relatively recent origin of the clade, incomplete lineage sorting, and some instances of plastid transfer have complicated previous attempts at creating molecular phylogenies. The *Pinus* (pine) nuclear genome is huge, even compared to other plant species. At 22 billion base pairs, the recently published *P. taeda* genome is 163 times the size of *Arabidopsis* and the *P. ponderosa* genome is estimated to be 24 billion base pairs. As retrotransposon activity repeatedly copied genomic regions, many pseudogenes and larger gene families were created, making isolation of orthologous low-copy nuclear genes for phylogenetic evidence a challenging endeavor even though pines are all diploids. Nonetheless, the economic and ecological importance of these forest trees has inspired a substantial amount of molecular genetic research. Over 14 thousand unigenes were published for *P. taeda* that were successful targets for solution-based enrichment (retrieving complementary biotin-labeled oligonucleotides with streptavidin beads) prior to high-throughput short-read sequencing (Illumina). We selected 700 of these genes for an array of bait probes (MYcroarray MYbaits) and successfully multiplexed 48 target-enriched *Ponderosae* samples per lane. After a series of filtering steps to exclude putative paralogs, we aligned over 300 putative nuclear orthologs from 75 individuals. By including some raw DNA along with the target-enriched DNA in the sequencing lane, we were able to assemble nearly-complete plastomes as well. Coalescence-based species trees inferred from this dramatically larger data set provided a substantial step forward in our understanding of these species relationships.

T2-60-03

Dealing with phylogenetic noise in phylogenomic data: A pilot study applying the plant anchored hybrid enrichment method to new world sages (*Salvia* Subgenus *Calosphace*; Lamiaceae)

Itzi Fragoso-Martínez[1], *Gerardo A. Salazar*[1], *Martha Martínez-Gordillo*[1], *Susana Magallón*[1], *Alejandra Moreno-Letelier*[1], *Cássio Van Den Verg*[2], *David S. Gernandt*[1], *Emily Moriarty Lemmon*[3], *Alan Lemmon*[3], *Carolina Granados Mendoza*[4]
1. National Autonomous University of Mexico
2. Univsidade Estadual de Feira de Santana
3. Florida State University
4. Instituto Potosino de Investigación Científica y Tecnológica A.C.

Large-scale phylogenomic data has substantially increased our understanding about relationships within several lineages across the tree of life. These data can be obtained through varied genomic partitioning strategies (GPS), one of them is the Anchored Hybrid Enrichment (AHE) method. AHE was originally developed for vertebrates and only recently applied to flowering plants. This method uses the Angiosperm V. 1 kit, that contains short RNA probes, designed from 25 angiosperm genomes, to target up to 517 low-copy nuclear genes, potentially informative at a range of phylogenetic levels. We conducted a pilot study using AHE to resolve relationships among a mostly Neotropical sage lineage that may have undergone a recent evolutionary radiation. *Calosphace* is one of the two monophyletic subgenera of *Salvia*. It includes ca. 600 species, most of them endemic to the Neotropics, being the most diverse lineage in the genus. Four diversity centers have been postulated for *Calosphace*: Mexico, the Andean Region, Eastern Brazil and the Antilles. Conventional markers (ITS, *trnL-trnF* and *trnH-psbA*) have not been able to resolve the relationships among species nor along portions of the backbone of the clade. The aims of this study were to test: 1) the effectiveness of gene capture and massive sequencing in obtain quality data for this group; 2) the performance of the captured genes in resolving and supporting internal relationships; 3) the effect of noise reduction strategies on phylogenetic estimation; and 4) the potential of the enrichment by-catch of the two GPS in recovering plastid genomes for the study group. For the AHE method we sampled 12 representative species of subgenus *Calosphace* and included one species of *Salvia* s.l.'s closest relative, *Lepechinia*, as outgroup. Gene capture and sequencing yielded 448 alignments of individual loci with an average length of 704 bp, 64% of them containing all the taxa sampled. The performance of the AHE data in phylogenetic estimation was superior to that of conventional markers, increasing both support and resolution. Captured loci vary in the amount of net phylogenetic informativeness at different phylogenetic depths, making these data promising in phylogenetic estimation for this group, and possibly for other lineages within Lamiales. Noise reduction through the exclusion of sites with unusually high substitution rates resulted in increased support and resolution for shallow nodes in maximum likelihood phylogenetic trees resulting from concatenated analyses of all the loci. Additionally, it is expected that an increase in sampling (preferably taxa) will aid in resolving weakly supported, short internal branches at deep phylogenetic levels. From the 25 reference genomes used to design the probes from the Angiosperm v. 1 kit, the closest relative to *Salvia* is *Mimulus* (Phrymaceae), also from Lamiales. The genetic distance between these lineages allowed the hybridization of some of the probes with chloroplast DNA, resulting in the recovery of complete to nearly complete chloroplast genomes using by-catch from off-target reads.

T2-60-04

Phylogenetic analyses at different evolutionary levels using hybridization-based target enrichment

Tomas Fer[1], *Roswitha Schmickl*[2], *W John Kress*[3], *Monica Carlsen*[3], *Monika Pospisilova*[1], *Lenka Flaskova*[1], *Jana Leong-Skornickova*[4]

1. Dept. of Botany, Charles University in Prague

2. Institute of Botany, The Czech Academy of Sciences, Pruhonice

3. Botany Department, National Museum of Natural History, Smithsonian Institution, Washington, D.C.

4. Research & Conservation Branch, Singapore Botanic Gardens

Using a high number of independent and putatively orthologous low-copy loci is a large benefit for phylogenetics. Species trees reconstructed from big phylogenomic datasets are often completely resolved and well supported. Hybridization-based target enrichment (Hyb-Seq) is one of the methods for effective sampling of thousands of exons from hundreds of genes. Because hybridization enrichment is not 100% efficient nearly complete chloroplast genomes can be obtained from off-target reads. Huge amount of data generated by the Hyb-Seq NGS method need an appropriate bioinformatic pipeline for processing raw reads up to the reconstruction of species tree. We developed HybPhyloMaker, a modular set of BASH scripts for UNIX-like environment that is designed for compact and easy-to-use processing of raw Illumina paired-end reads, selecting suitable loci and constructing gene and species trees using different methods. Probe set covering 1,180 genes was designed for the genus *Curcuma* (Zingiberaceae) comparing *C. longa* transcriptome and *C. ecomata* genome-skimming data. The probes were successfully applied to enrich sequencing libraries not only within the genus *Curcuma* but also at the level of the whole family Zingiberaceae and the order Zingiberales. More than 1,000 loci were usable for the genus *Curcuma* and around 700 loci were used to establish the phylogeny of Zingiberaceae and Zingiberales. With our work we demonstrated really broad evolutionary applicability of the set of orthologous genes initially designed for generic level phylogeny.

T2-60-05

The potential of target enrichment and hybrid capture for improving our understanding of complex diversity patterns within the lettuce alliance (Asteraceae)

Katy Elizabeth Jones[1], *Jennifer Mandel*[2], *Paul Johnston*[3,4,5], *Schilling Edward*[6], *Norbert Kilian*[1]

1. Botanischer Garten und Botanisches Museum, Freie Universität Königin-Luise-Straße 6-8, 14195 Berlin, Germany

2. University of Memphis, Ellington Hall 301, Memphis, TN 38152

3. Freie Universität Berlin, Evolutionary Biology, Königin-Luise-Straße 1-3, 14195 Berlin, Germany

4. Berlin Center for Genomics in Biodiversity Research, Königin-Luise-Straße 6-8, 14195 Berlin, Germany

5. Leibniz-Institute of Freshwater Ecology and Inland Fisheries (IGB), Müggelseedamm 301, 12587 Berlin, Germany

6. University of Tennessee, Herbarium TENN, Department of Ecology & Evolutionary Biology, University of Tennessee, Knoxville, Tennessee 37996

The lettuce alliance (*Lactucinae*, *Cichorieae*, Asteraceae) contains ca. 200 species of varying ploidy level, distributed in Europe, Africa, Asia and North America. Phylogenetic studies to date have relied mostly on ITS and chloroplast markers that have highlighted the non-monophyletic status of currently recognized taxonomic groups within the alliance and revealed complex patterns of reticulation and incongruence between nuclear and chloroplast regions. In order to gain greater phylogenetic resolution and to unravel such complex patterns of diversity, we aimed to capture a greater portion of the nuclear genome compared to previous studies, via target enrichment and hybrid capture. We made use of already available probes that previous studies had shown to capture highly informative and potentially low copy nuclear genes for a number of Asteraceae taxa. The resulting dataset consisted of >800 nuclear loci for a range of taxa from across the lettuce alliance. The bioinformatics for preparation and analyses of such large datasets is continually being developed. We present the results of different phylogenomic analyses of our hybrid capture data including PHYLUCE, HybPiper and clustering. We discuss the application and potential of these different analytical approaches for understanding diversity within complex flowering plant groups such as the lettuce alliance.

T2-60-06

An angiosperm-wide bait kit for targeted sequencing of 354 low-copy nuclear genes across flowering plants

Steven Dodsworth[1], *Lisa Pokorny*[1], *Matt G. Johnson*[2], *Alison Devault*[3], *Niroshini Epitawalage*[1], *Olivier Maurin*[1], *Ilia J. Leitch*[1], *Félix Forest*[1], *Norman J. Wickett*[2], *William J. Baker*[1]

1. Royal Botanic Gardens, Kew

2. Chicago Botanic Garden

3. MYcroarray

Utilizing high-throughput sequencing efficiently for phylogenomics requires methods that reduce genomic complexity whilst maximizing the number of taxa/samples. The use of RNA baits to pull out genomic regions of interest (a.k.a. bait hybridization, sequence capture, targeted enrichment) is now becoming widespread as one of the best methodologies to achieve this goal.We have developed a new angiosperm-wide probe kit from existing transcriptomes (1 KP) and genomes (Phytozome), in collaboration with the Chicago Botanic Garden and MYcroarray, and in the context of

Kew's Plant and Fungal Trees of Life (PAFTOL) project. Our angiosperm-wide kit targets 354 low-copy nuclear genes that are shared across land plants and sister algal lineages. Our bait design included up to 15 variants for each of the 354 genes, such that we could capture 95% of sequence diversity across angiosperms with no target sequences being more than 25% divergent from the bait sequence. This resulted in 75,151,120-mer probes, with 3x tiling density. To check the breadth of our kit we have tested the baits on all angiosperm orders and all angiosperm families that were not included in 1KP. Additionally, we tested the potential of these loci for shallower level systematics, by using several small sets of ~10 species within genera belonging to major flowering plant lineages. We have also tested the kit on a variety of samples, including herbarium specimens of varied quality, which were dried in different ways (e.g. air vs. ethanol), and those that are over a century old. Here we present results of the efficiency of capture across angiosperms and sample types, and at different phylogenetic levels. We also present the most up-to-date angiosperm phylogenetic tree at the family level, by combining our initial data with 1KP alignments.

T2-61: Diversification, evolution, biogeography and conservation of the family Magnoliaceae (two sessions)

T2-61-01

Magnolia genome: An additional reference for understanding angiosperm diversification

Jongsun Park[1], Jong Bhak[2], Sangtae Kim[3]
1. InfoBoss
2. UNIST
3. Sungshin Women's University

Magnolia has received keen interest from botanists because it is one of the early-diverging angiosperms in the evolutionary history of angiosperms. Recent molecular phylogenetic studies show that Magnoliaceae is placed at the "Magnoliids" clade after the *Amborella*, Nympheaceae, and Austrobailales. Genomic study of magnolia, a member of basal angiosperm grade, highlights the importance of understanding the early diversification of angiosperms. We obtained more than 190 Gbp of filtered NGS sequences from *Magnolia kobus*, a popular garden tree in Magnoliaceae and an endangered species in Korea, with six different libraries (2 pair-end and 4 mate-pair libraries). The current draft of *M. kobus* genome assembly (ver. 0.4) consists of 1.91 Gbp with 26,553 bp and 9,852 bp of N20 and N50, respectively, which is slightly smaller than that from k-mer analysis (2.09 Gbp) but larger than that estimated from flow cytometry (1.44 Gbp). Gene prediction by AUGUSTUS showed 371,292 putative genes, of which 84,663 (22.80%) genes have known functional domains defined by InterProScan. *Magnolia* genome shares 5,282 functional domains (79.05%) with the *Amborella* genome, and 1,088 domains (16.28%) are *Magnolia*-specific. In the gene family analyses, 39 MADS-box genes were identified in the *M. kobus* genome, showing a similar number as in the *Amborella* genome (34 genes). However, members of AP2 (95 genes) and cytochrome P450 (567 genes) gene families identified in the *Magnolia* genome were almost two-fold higher in number than that that from the *Amborella* genome. For the improvement of gene annotation as well as for understanding the floral development of *M. Kobus*, we sequenced transcriptomes from nine different tissues. To find lineage-specific traits in Magnoliaceae, five additional species (*M. coco*, *M. sieboldii*, *M. grandiflora*, *M. portorensis*, and *Liriodendron chinense*) were sequenced (8.0~9.1 Gbp, respectively). We detected 120,334 simple sequence repeats (SSRs) throughout the genome of *M. kobus* and 171,317 SSRs from five other species of Magnoliaceae. Sequences and analyses results of magnolia species are accessible in the Magnolia Genome Database (http://www.magnoliagenome.net/), which is a web-based integrated platform for comparative genomics of Magnoliaceae.

T2-61-02

Genome-wide approach on the phylogeny of Magnoliaceae using a target-enrichment NGS sequencing

Sangtae Kim[1], Suhyeon Park[1], Jongsun Park[2]
1. Sungshin Women's University
2. InfoBoss

Reconstruction of the phylogeny of Magnoliaceae, a member of early-diverging angiosperms, plays a key role in understanding the diversification of angiosperms. With the development of new molecular techniques, various phylogenetic studies on Magnoliaceae have been conducted such as studies based on chloroplast RFLP, and sequencing of single gene, multiple genes, and chloroplast whole genome. However, the evolution of Magnoliaceae remains unclear in many parts, especially the relationships among major sublineages in the family because overall base substitution rate in Magnoliaceae is very low, as compared to other angiosperm groups. In this study, we used a genome-wide approach on the phylogeny of Magnoliaceae with high throughput target-enrichment NGS sequencing. We identified 504 putative single/low copy gene regions (443 kbp) as targets for captured NGS sequencing. These regions are 1) expressed single-copy genes extracted from a preliminary genome assembly of *Magnolia kobus* (ver. 0.4; see talk 1401) using gene clustering (Tribe-Mcl) with its transcriptomes and 2) shared single/low copy genes among *Arabidopsis, Polulus, Oryza, Vitis*, and *M. kobus* . Probes of these regions were designed and hybridized with 130 taxa representing all previously reported sublineages (sections and tribus) of Magnoliaceae. Index sequences were added in the captured fragments and sequenced altogether using the Illumina HiSeq platform. Phylogenetic trees based on exons of targeted nuclear genes as well as adjacent intergenic sequences (including introns) were compared with those based on chloroplast genome data. The results provide a basis for the evolutionary diversification in the family as well as new classification system of Magnoliaceae.

T2-61-03

From investigation to implementation: BGCI's integrated conservation action to save threatened plants in China – the example of Magnoliaceae

Xiangying Wen[1], Joachim Gratzfeld[2], Shouzhou Zhang[3], Daoping Yu[4]
1. Botanic Gardens Conservation International (BGCI) China Programme Office, Guangzhou, China.
2. Botanic Gardens Conservation International (BGCI), Richmond, United Kingdom

3. Fairylake Botanical Garden, Shenzhen & Chinese Academy of Sciences
4. Emeishan Botanical Garden, Sichuan Provincial Institute of Natural Resource Sciences

The family Magnoliaceae is among the most ancient groups of angiosperms. Comprising over 300 species in 2 genera – *Magnolia* L. and *Liriodendron* L., Magnoliaceae are found in the temperate and tropical regions of East and Southeast Asia and the Americas. Some 109 taxa are native to China of which more than half are endemic. Southwest and South China including Yunnan, Guangxi, Guangdong, Hainan and Guizhou are the centre of extant Magnoliaceae diversity of the Old World flora. 50% of China's magnolias are threatened with extinction in the wild (*The Red List of Magnoliaceae,* 2016). As in other areas of high magnolia diversity, forest habitat conversion, logging, over-exploitation and limited, natural reproduction, present major threat factors. Under the umbrella of the Global Trees Campaign (GTC) and working with Chinese partners including botanic gardens, academia and local authorities and communities, BGCI has been developing and implementing integrated conservation projects for various species of Magnoliaceae over the past 10 years. Activities range from establishing distribution maps to plant material sampling for propagation at botanic gardens aimed at *ex situ* conservation collections as well as population reinforcement and reintroduction in the wild. Public outreach and educational activities form an integral part of all the projects. Over the years, more than 10 rare and/or threatened species including *M. sinostellata, M. omeiensis, M. longipedunculata, M. coriacea, M. phanerophlebia, M. angustioblonga, M. hebecarpa, M. ingrata, M. odoratissima* and *M. aromatica* were included in practical conservation work. As an immediate insurance policy against their extinction in the wild, *ex situ* collections for a number of these species have been established at Chinese botanic gardens and local forestry research stations. BGCI's most recent magnolia conservation work focusses on the Critically Endangered *Magnolia omeiensis* as well as the Endangered *Magnolia sinostellata.* Known from only two locations in temperate forest on Mount Emei, southern Sichuan, less than 75 individuals of *M. omeiensis* have been recorded. Trees display low seed production and germination rates. Intensive logging is the single-most important threat to the species and no special protection is yet in place to ensure that the population remains intact. *Magnolia sinostellata* is known from 5 locations in southern Zhejiang province with some 500 individuals remaining in the wild. As with *M. omeiensis*, all populations are exhibiting limited reproductive performance. As a high-value ornamental plant, over-collection in the wild as well as ongoing deforestation are responsible for the continuing decline of *M. sinostellata* populations. BGCI is working with Emeishan Botanical Garden and Shenzhen Fairy Lake Botanical Garden to develop *ex situ* conservation collections and to reinforce populations of the species in the wild. Over the past two years, saplings of *M. omeiensis* and *M. sinostellata* have been established, generated from seed, cuttings and grafts. Overall survival rates of saplings planted *in situ* in 2016 to reinforce existing populations are at 95.4% for *M. omeiensis.* Reinforcement for *M. sinostellata* was initiated in March 2017. Conservation work continues in 2017 and beyond to consolidate and scale up the propagation and species recovery work. While grafting has been employed as a last resort to establish sizable stocks of plants in the short-term, major emphasis will be given to generating seedlings for use in future population reinforcement programmes. Representatives from local communities and authorities will be offered advanced training in propagation techniques, whilst a series of public outreach campaigns aims to enhance awareness of the needs and values to conserve China's unique magnolia diversity as a matter of national pride.

T2-61-04

Abundance, rarity and geographical range size in neotropical species of *Magnolia*: implications for conservation efforts in Mexico, Colombia and Ecuador

David Neill[1], *Mercedes Asanza*[1], *Álvaro Cogollo*[2], *Joachim Gratzfeld*[3], *Lou Jost*[4], *J. Antonio Vázquez-Garcia*[5]
1. Universidad Estatal Amazónica
2. Jardín Botánico de Medellín
3. Botanic Gardens Conservation International
4. Fundación EcoMinga
5. Universidad de Guadalajara

The species abundance curves of trees in neotropical forests, at local, regional and continental scales, typically exhibit a "reverse-J" distribution with a few very abundant species, a larger number of species that are moderately abundant to uncommon, and a long "right-hand tail" of rare to very rare species. For instance, a recent floristic analysis of 1,170 one-hectare tree inventory plots from throughout the 6 million km2 landscape of lowland Amazonia, estimated 16,000 tree species for the region, but just 227 "hyper-dominant" species account for half of all the trees in Amazonia, and the most abundant species have total population sizes upwards of 10^9 individuals. In contrast, this study estimated 6,000 very rare tree species in the Amazon with a total population < 1,000 individuals – a very long "tail" of "hyper-rare" species. Although the quantitative data is scanty, evidence suggests that many *Magnolia* species in neotropical forests are within that rightward tail of species that exhibit the most restricted of the "seven forms of rarity" – small local population size, small geographic range, and narrow habitat breadth. For example, *Magnolia canandeana* is known from just two adult trees, each over 30 m with trunks over 100 cm DBH, within a 1300-hectare private nature reserve in the Pacific coastal region of Ecuador, in a region of active logging and conversion of the forest to African oil plantations. *Magnolia yantzazana* is known from just 13 adults in a 5 km^2 of forest in the Cordillera del Cóndor region of southeast Ecuador, and the entire population is within an area of a gold mining concession; in this case the mining company, within its environmental management plan, is taking steps for *in situ* and *ex situ* conservation of this and other locally endemic plants. The extreme rarity of many neotropical *Magnolia* species poses considerable challenges for conservation efforts. At least 115 of a total of 160 neotropical species of *Magnolia* are considered threatened according to the IUCN Red List Criteria. Mexico, Colombia and Ecuador harbor the greatest number of Neotropical Magnolias, with 40, 34 and 25 species, respectively, reported for these three countries. Conservation programs for threatened *Magnolia* species including *in situ* and *ex situ* components, are currently in progress in these three countries. Methods to overcome the challenges of *ex situ* conservation, including difficulties of obtaining seed, and difficulties in propagation of recalcitrant seeds, are being addressed through various methods including micro-propagaion *in vitro* of vegetative meristems. Reintroduction of *Magnolia* plants into appropriate native habitats is in initial stages in several areas of these three countries.

T2-61-05

The Lirianthe Spach (Gwillimia + Blumiana clade) as the basal lineage of Magnoliaceae

Ernesto De Castro-Arce, Gerardo Hernández-Vera, Laura Guzmán-Dávalos, ***J. Antonio Vázquez-García***
Universidad de Guadalajara

Evolutionary relationships among members of the family Magnoliaceae have been difficult to clarify due to the lack of informative morphological and molecular characters together with unbalanced sampling. Thus, several hypotheses have been proposed regarding the position and taxonomic delimitation of the groups that conform the family. We aimed to clarify the evolutionary relationships within Magnoliaceae using DNA sequence data from three chloroplast (*mat*K, *psb*A, ORF 350) and two nuclear (GAI, PHYA) genomic regions obtained from GenBank and 15 newly sequenced Neotropical species for a total of 80 taxa, from both Asia and America. Phylogenetic analyses were conducted for a concatenated alignment of chloroplast and nuclear sequences using Bayesian inference (BI) and Maximum likelihood (ML). Substitution models for each genomic region were selected with the Akaike Information Criterion. The nomenclature used mostly followed Figlar and Nooteboom´s classification. Results from the BI and ML analyses of the dataset were largely in agreement, except for some differences in the tree topology and bootstrap support values for the inferred clades. Except for subgenera and the basal lineage of the family, several sections or subsections were recovered as monophyletic with high support values. Phylogenetic relationships among clades were recovered in general agreement with previous analyses, yet improved in terms of resolution. Results from the BI show that the base of subfamily Magnolioideae is composed of the lineage that originated extant *Gwillimia* + *Blumiana* species, recognized as the genus *Lirianthe* Spach by some authors. This basal configuration differs from those of previous phylogenetic hypotheses resulting in either a polytomy or placing *Macrophylla* Figlar & Noot. or *Talauma* Juss. as the basal clade. Besides, our results provide a well-supported hypothesis with improved resolution of the phylogenetic relationships within Magnoliaceae. Many subsections were recovered as monophyletic with high support values except for the clades *Gwillimia* + *Blumiana*, *Dugandiodendron*, *Manglietiastrum* + *Gymnopodium* and the complex *Aromadendron* + *Maingola* + *Michelia* + *Ermerillia*. Subsect. *Auriculata* was placed as a sister clade to *Oyama* + *Rhytidospermum* + *Manglietia*), sharing a common ancestor; its similarity with *Macrophylla* is a convergence, given their placement in independent clades; the latter more related to the *Magnolia* clade. The largest and well supported clade consisted of the subgenera *Yulania* + *Gynopodium* + section *Kmeria* of subgenus Magnolia. In relation to Figlar & Nooteboon´s Classification of Magnoliaceae, 12 taxa were supported in our results: six sections (*Gwillimia*, *Auriculata* Figlar & Noot., *Manglietia* Blume, *Macrophylla*, *Magnolia* L., and *Kmeria* (Pierre) Figlar & Noot.); and six subsections (*Talauma* Baill., *Cubenses* Imkhan., *Oyama* (Nakai) Figlar & Noot., *Rhytidospermum* Spach, *Tulipastrum* Spach) Figlar & Noot. and *Yulania* Spach.). Alternatively, ten genera could here be recognized as monophyletic in this study: *Houpoëa* N.H.Xia & C.Y.Wu, *Kmeria* (Pierre) Dandy, *Lirianthe*, *Magnolia*, *Manglietia* Blume, *Metamagnolia* Sima & S.G.Lu, *Oyama* (Nakai) N.H.Xia & C.Y.Wu, *Paramagnolia* Sima & S.G.Lu, and *Talauma*. In order to facilitate our comprehension of the evolution of biogeography of Magnoliaceae, an updated classification system is urgently needed, one that integrates the evolutionary relationships presented here with those reported in previous efforts.

T2-61-06

Tempo and mode of diversification in Magnoliaceae

Ernesto De Castro-Arce*, Gerardo Hernández-Vera, Laura Guzmán-Dávalos, J. Antonio Vázquez-García*
Universidad de Guadalajara

Diversification, the interplay between speciation and extinction, has recently rise a profound interest among researchers due to the advent of new analytical methods developed since the seminal paper of Nee *et al.,* (1994). These macroevolutionary metrics have been increasingly used in trying to explain why some lineages are more diverse than others and shed light on the factors and processes that drive such diversity imbalance. In flowering plants, many species-rich groups of relatively recent divergence have been subjected to these kind of analyses, while groups near to the base of the angiosperm phylogenetic tree have been overlooked, thus missing the opportunity to study interesting plant lineages with long evolutionary histories.The family Magnoliaceae represents a basal group of angiosperms composed of c. 400 species and a temporal range of evolution spanning c. 100 My. The family exhibits a strong disparity in species richness among its constituent lineages with almost half of the total diversity contained in two groups; the Neotropical basal lineage *Talauma* and the Paleotropical *Michelia*. Using a phylogenetic approach, the temporal pattern of diversification in Magnoliaceae was investigated with lineage through time (LTT) plots and the constant-rates test (gamma statistic). The likelihood-based method MEDUSA was employed to identify shits in the diversification rate across lineages in the phylogeny of Magnoliaceae. The LTT plot and the gamma statistic failed to reject a model of variable rate of diversification in time while the MEDUSA analyses favored a model with three diversification rate shifts. Our results indicate that the mode and tempo of diversification in Magnoliaceae started at a slow steady pace that has been maintained until the present with a high turnover, probably due to extinction. The relatively high diversity of *Michelia* can be explained by a shift in diversification rates recovered at the crown node of the group while no shift in rates was suggested for *Talauma* nor for any other lineage within the family. This result suggests that other mechanisms of origination and maintenance of diversity should operate for *Talauma* while the emergence of key innovations may be argued for *Michelia*.

T2-61-07

Historical biogeography and ancestral areas of Neotropical Magnoliaceae

Viacheslav Shalisko, J. Antonio Vázquez-García, ***Gerardo Hernández-Vera****, Alma Rosa Villalobos-Arámbula, Miguel Angel Muñiz-Castro*

University of Guadalajara

The improved understanding of patterns in species diversity and relationships within Neotropical clades of *Magnolia* could provide valuable information for understanding the global and regional biogeography of Magnoliaceae. In this study, we analyze the historical biogeography of Magnoliaceae with the complex Dispersal–Extinction–Cladogenesis model, using newly available information on Neotropical taxa and molecular phylogenetic reconstruction based on cpDNA loci from 74 species. The availability of fossil record for Magnoliaceae, as well as known phylogenetic positions of fossil taxa allowed us to test and interpret the results of the Dispersal–Extinction–Cladogenesis model. The probable ancestral area of the whole Magnoliaceae has been identified as a combination of temperate and subtropical regions in Eurasia and North America, as the connected mid-latitude zone of the Northern Hemisphere during the early Eocene. This ancestral area for Magnoliaceae is consistent with previously proposed hypothesis of fragmentation of Eocene 'Arcto-Tertiary flora' or more recent 'Boreo-Tropical flora'. During the early Tertiary, Magnoliaceae included two lineages that persisted until modern time: *Liriodendron* ancestors and species of *Magnolia* clade. In *Liriodendron*, the integrity of the wide distribution range shared between Eurasia and North America persisted until the Miocene, while in *Magnolia* the geographical separation between America and Eurasia started the middle Eocene with evolving of geographically independent sections *Gwillimia*, *Manglietia*, *Kmeria*, *Michelia* in Eurasia, and *Magnolia*, *Macrophylla*, *Auriculata*, *Talauma* in North America. However, in the case of the subsections *Yulania* and *Tulipastrum*, the interchange between Eurasia and North America persisted until the Oligocene, and in the section *Rytidospermum* until the middle Miocene. We revealed a possible general migration pathway of a Neotropical *Magnolia* ancestor from eastern North America through Greater Antilles that resulted in the establishment of the modern *Talauma* distribution. *Talauma* as one of the first modern sections to evolve in the middle Eocene within *Magnolia*, has its ancestral area shared between the subtropical zone of North America and the subtropical zone of Eurasia, but evidently went extinct in the latter one. We interpreted the historical habitat availability data in relation to the actual and inferred ancestral distribution of *Talauma* to reveal the possible historical migration pattern. The ancestor group for modern Neotropical section *Talauma* as well as related *Dugandiodendron* and *Splendentens* taxa had started migration southwards from the southeastern North America during middle Eocene, and experimented diversification initially in Greater Antilles, later in Oligocene penetrated to northern South America, northern Andes and Central America. The section *Macrophylla* actually known form Mexico and USA had the ancestral area in subtropical and temperate North America in middle or late Eocene, unrelated to distribution of *Talauma* clade. The detailed biogeographical scenario for neotropical Magnolia will require extended representative sampling within *Talauma*, *Dugandiodendron* and *Splendentens*.

T2-61-08

Genetic population structure and diversity of *Magnolia pacifica A.*Vázquez species complex in southwestern Mexico Inferred from ISSR Markers

Miguel Angel Muñiz-Castro[1], Patricia Castro-Félix[2], Ahtziri Socorro Carranza-Aranda[2], Anne Santerre Lucas[2], J. Antonio Vázquez-García[1]

1. Universidad de Guadalajara, Centro Universitario de Ciencias Biológicas y Agropecuarias, Departamento de Botánica y Zoología, Instituto de Botánica, Laboratorio de Ecosistemática, Herbario IBUG.

2. Universidad de Guadalajara, Centro Universitario de Ciencias Biológicas y Agropecuarias, Departamento de Biología Celular y Molecular, Centro de Genética, Lab. de Biomarcadores y Genética Molecular.

The delimitation of species, the fundamental units of evolutionary biology, is an essential task of current systematic studies. Magnoliaceae is still considered a relatively primitive group of angiosperms, comprising ca. 350 species of trees and shrubs native to Asia and America. Although several molecular phylogenetic studies based on DNA sequence data have been conducted on the Magnoliaceae family, several major and minor clades have not been completely resolved. *Magnolia pacifica s.s.*, *M. pugana* and *M. vallartensis* are recently described species grouped in the *Magnolia pacifica* A. Vázquez species complex. These species are endemic, endangered, and distributed in isolated fragments of tropical montane cloud forest and gallery forest in southwestern Mexico. In this study, we used inter-simple sequence repeats (ISSR) in order to: (1) to analyze species boundaries within the M. pacifica complex under a population genetics approach, (2) evaluate genetic diversity and (3) provide information for species conservation. Six ISSR primers amplified a total of 76 loci from 278 individuals sampled in four localities of *M. pugana*, three of *M. pacifica s.s.* and three of *M. vallartensis*. Sixty five percent of loci were polymorphic among all samples. Bayesian cluster analysis, the Exact Test for population differentiation and Unweighted Pair-Group Mean Average (UPGMA) cluster analysis were consistent with the morphological recognition of *M. pugana* and *M. pacifica* at the species level. Two subpopulations of M. pugana were identified, they are separated mainly by the Santiago river canyon (500-700 m depth and 3-15 km width). ISSR data suggests that *M. vallartensis* is in a process of incipient speciation with a recent partial divergence from *M. pacifica s.s.* Positive and significant correlation between geographic and genetic distances (Mantel Test $r = 0.8$, $p = 0.001$) indicates isolation by geographic distance. The Analysis of Molecular Variance (AMOVA) showed that most of the genetic variation was within populations (83%), lower proportion among taxa (8%) and among populations of the same taxon (8%). G_{ST}, θ and D_{EST} indices revealed significant differentiation among localities within each taxon, but differentiation was higher in *M. pugana* ($G_{ST} = 0.168$, $\theta = 0.2$, $D_{EST} = 0.028$) than in *M. pacifica* ($G_{ST} = 0.102$, $\theta = 0.131$, $D_{EST} = 0.018$) and *M. vallartensis* ($GST = 0.127$, $\theta = 0.159$, $DEST = 0.021$). Diversity indices were lower for M. pugana HT = 0.158, HS = 0.134 than for *M. pacifica* and *M. vallartensis* ($H_T = 0.175$, $H_S = 0.159$ and $H_T = 0.171$, $H_S = 0.153$, respectively). Findings of the present study bring out the usefulness of molecular data and a population genetics approach to help delineate species boundaries within species complexes. The two species, *M. pacifica* and *M. vallartensis*, and the two subpopulations of *M. pugana*, should be considered as separate units of conservation. Concerted efforts must be made to protect these three taxa, but a greater focus on *M. pugana* is needed.

T2-61-09

Taxonomic revision of Sect.*Yulania* of *Magnolia*

Yaling Wang[1], *Shouzhou Zhang*[2]
1. Xi'an Botancial Garden of Shaanxi Province
2. Laboratory of Southern Subtropical Plant Diversity, Fairylake Botanical Garden

Magnolia Sect. *Yulania* Spach. is an important taxa in Magnoliaceae which with comparatively evolved features such as deciduous leaves, pollen lateral cracking. This section contains 16 general recognized species (including two varieties) which are mainly distributed in north subtropical and warm temperate zone. Among these species, *M. acuminata* L. and *M. acuminata* var. *subcordata* L. distributed in southeast of America, *M. kobus* (DC.) Spach, *M. stellata* M. and M. *salicifolia* (Sieb. & Zucc.) Maxim.distributed in Japan, the other 11 species are all occurred in China. The systematic position and internal classification of this group has been controversial. More than 28 new species were published recently in China. In this study, 267 individuals from 45 populations representing 17 taxa in sect.*Yulania* were studied for the taxonomic revision of sect.*Yulania*. The morphology variation and geographical distribution, as well as molecule data analysis on the combine data of three universal DNA regions (*mat*K, *rbc*L, *trn*H-*psb*A) and three self-screening chloroplast gene spacer sequences (*pet*A-*psb*J, *psb*E-*pet*L, *trn*T-*psb*D), were conducted to detect the phylogenetic relationships. The results revealed that sect. *Yulania* was divided into two subsections (Subsection *Yulania* and subsection *Tulipastrum*) with four gourp, 11 speceis 8 varieties: Accuminata group (*M. accuminata* L., *M. acuminata* var. *subcordata* (Spach) Dandy), Binodii group (*M. biondii* Pampan., *M. biondii* Pamp. var. *purpurascens* Y.L.Wang & S.Z.Zhang, *M. biondii* var. *zenii* (Cheng) Y.L.Wang & S.Z.Zhang, *M. kobus* DC, *M. salicifolia* (Sieb. & Zucc.) Maxim., *M. sinostellata* P.L.Chiu et Z.H.Chen, *M. stellata* (Sieb. & Zucc.) Maxim.), Sprengeri group (*M.campbellii* Hook f. &Thous, *M. sprengeri* Pampan., *M. sprengeri* var. *sargentiana* (Rehder & Wilson) Y.L.Wang & S.Z.Zhang), Denudata group (*M. cylindrica* Wils., *M. cylindrica* E.H.Wilson var. *purpurascens* Y.L.Wang & S.Z.Zhang,, *M. denudata* Desr., *M. denudata* var. *glabrata* Y. L. Wang et S. Z. Zhang, *M. denudata* var. *pilocarpa* (Z. Z. Zhao & Z. W. Xie) Y. L. Wang et S. Z. Zhang, *M. liliiflora* Desr., *M. liliiflora* Desr. var. *polytepala* (Law, R.Z. Zhou & R.J. Zhang) Y. L. Wang et S. Z. Zhang).

T2-61-10

An overview of species richness, conservation and ecology of Neotropical *Magnolia*

Marcela Serna Gonzalez[1,3], *Eduardo Calderon*[2], *Ligia Estela Urrego*[3], *Alejandro Palmarola*[4], *Luis Roberto Gonzalez*[5], *Ramona Oviedo*[6], *Yenia Molina*[7], *Ramon Elias Castillo*[8], *Reyna Dominguez*[9], *Gerardo Hernández-Vera*[10], *Frank Arroyo*[11], *Alvaro Perez*[12], *Jose Luis Gomez*[6]
1. Tecnológico de Antioquia - Institución Universitaria
2. Reserva Natural El Refugio
3. Universidad Nacional de Colombia sede Medellin
4. Jardin Botanico de Cuba
5. Plant Life Conservation Society
6. Instituto de Ecología y sistemática CITMA
7. Instituto de Investigaciones F
8. Reserva Ebano Verde
9. Universidad de La Sierra de Juarez
10. Universidad de Guadalajara
11. Universidad Nacional Agraria La Molina
12. Pontificia Universidad Catolica del Ecuador

The classification of Magnoliaceae has been controversial, especially at the genus level including infrageneric categories. In the Neotropics, there are approximately 130 species of Magnoliaceae belonging to subgenus *Magnolia*, sections *Macrophylla, Magnolia* and *Talauma*. This last section comprises four subsections: *Cubensis* (West Indian species), *Dugandiodendron* (Species from North of South America including Colombia and Venezuela), *Talauma* (from Mexico to Bolivia) and a new one recently described based on morphological characters: *Chocotalauma* (species distributed throughout the Choco Biogeographical Region). Mexico, with 40 species, is the country with the highest diversity of *Magnolia* species in the Neotropics, followed by Colombia with 36 species reported up to date. Several Neotropical species of *Magnolia* have been recently described, especially from Mexico, Ecuador and Peru. In the Neotropics, *Magnolia* species are characterized by a restricted geographical distribution in highly disturbed forests, populations with few individuals and in some cases, their trees have been overharvested. Consequently, most species are in some risk of extinction and are currently considered as a priority for conservation. The very low seedling recruitment observed in natural populations suggests that forest fragmentation could affect tree reproduction by altering pollination interactions, increasing seed predation and herbivory on juveniles. In some species, fruit formation and seed production are extremely low despite the availability of flowers. This aspect may be associated to specific plant-pollinator interactions that could be affected by high levels of forest fragmentation. Frugivory could also affect the reproductive success of these species. However, little is known about their ecology. The goal of this work is to show general aspects on species richness, conservation and ecology of Magnolias from the Neotropics, specifically those related to reproductive biology as a key factor to be considered for successful conservation initiatives and future research programs.

T2-61-11

Early evolution and dispersal of the magnoliaceae—an alternate hypothesis

David Dilcher
Indiana University

The Magnoliaceae is an ancient family that dates from the early radiation of flowering plants. The Magnoliales and related taxa are among the Basal Angiosperms that were diversifying 127 mya, even before the emergence of the Basal Eudicotyledons. During the mid-Cretaceous and into the Paleogene, the Tethys Seaway served as a tropical corridor connecting Asia with the islands of Europe to the north, Africa to the south and the Americas to the far west. Welwitschia, restricted to southern Africa today, was present in the Lower Cretaceous of Brazil as well as in some angiosperm remains found in lower Cretaceous sediments in North Africa and Portugal, which demonstrates potential ancient land connections and plant migration patterns. This low latitude east-west movement of tropical plants can also be seen in the relationships of some floristic elements occurring in Paleocene and Eocene plant fossils of southeastern North America to plants currently living in tropical Asia, tropical South America and Africa today. Common elements are also shared in the Paleocene and Eocene pollen floras of these regions. Also found in the fossil record of Cenozoic sediments in southeastern North America are remains of *Dugetia* in the Annoniaceae, now found in South America and Africa; *Gordonia* in the Theaceae, now found in South America and tropical Asia; and *Nypa*, a palm now found only in tropical southern Asia. These are some tropical floristic elements shared between Asia and the Americas and were components of floras that migrated along the low latitude lands bordering the Tethys Seaway rather than moving over a high latitude route. The extant taxa of the Magnolialiales are predominately distributed today in the Asian and American tropics and to some extent in the African tropics. These plants could have easily traversed the low latitude tropical islands of the Tethys Seaway extending between Asia and North America. This could explain the disjunct distributions of many tropical American and Asian plants today. Here I propose the hypothesis of a *Low Latitude Migration Pattern* for many of the disjunct truly tropical genera including those of the Magnoliaceae found in the American and Asian tropics today rather than by the migrations of these taxa over a temperate to cool temperate Bering Sea or cooler North Atlantic land bridges. Currently much of the published research seems to support of high latitude migrations of both temperate and tropical plant taxa. This should be reconsidered in relation to the ecological tolerances of the various taxa involved and the various climates of the migration routes proposed through time. Because of the extreme tropical nature of many of these plants, it is questionable that so many tropical taxa would have migrated between the tropical Americas and tropical Asia even during the most extreme Paleocene Eocene Thermal Maximum (PETM). During the Late Cretaceous the major taxa of many angiosperms, including the Magnoliaceae, had already migrated and were similar to extant taxa.

T2-61-12

Molecular phylogeny of Magnoliaceae based on cpDNA sequence data

Hiroshi Suzuki-Azuma
Kyoto University

A molecular phylogenetic study of Magnoliaceae was conducted using sequences of coding and noncoding chloroplast DNA regions (total ca. 9,000 bp) to reveal the phylogenetic positions of about 100 Magnolia taxa covering all subgenera (Magnolia, Yulania, and Gynopodium), sections (Magnolia, Gwillimia, Talauma, Manglietia, Kmeria, Rhtidospermum, Auriculata, Macrophylla, Yulania, Michelia, Gynododium, and Manglietiastrum), and subsections (total 20). In this study, several tropical Asian Neotropical species are newly sequenced and added to the sequence data matrix previously conducted. Phylogenetic analysis of the data matrix indicates that subsect. Dugandiodendron is sister to subsect. Cubenses, and the clade is sister to subsect. Talauma. That is, the three Neotropical groups (Dugandiodendron, Cubenses, and Talauma) are closely related and formed a monophyletic clade. However, close relationship between the Neotropical clade and southeast Asian groups (subsects. Gwillimia and Blumiana) is still controversial. Phylogenetic relationships among these tropical Asian species are also still unclear. Although I am still obtaining additional sequence data from Asian and American species, the tree will be helpful for understanding the taxonomy and the evolution of morphological characters in Magnoliaceae.

T2-62: Evolutionary and developmental morphology of basal monocots

T2-62-01

Implications and importance of morphological and developmental diversity of Aponogentanaceae in Indian subcontinent with special emphasis on embryo diversity

Shrirang R. Yadav
Angiosperm Taxonomy Laboratory, Department of Botany, Shivaji University, India

Family Aponogetonaceae with a single genus *Aponogeton* L.f. of about 58 aquatic tuberiferous species is distributed in tropical and subtropical regions of the world. The genus is well studied from phylogeny and biogeography point of view. Three new species have been added to Aponogetons of India and presently there are about nine species viz. *A. appendiculatus, A. bruggenii, A. crispus, Aponogeton* species, *A. lakhonensis, A. natans, A. nateshii, A. satarensis* and *A.undulatus* in India. Among the species, *Aponogeton appendiculatus, A. bruggeni, A. nateshii* and *A. satarensis* endemic to India are of special significance either from their phytogeography, reproductive biology or embryo morphology point of veiw. Similarly, *Aponogeton* species similar and close to *A. jacobsenii* from Sri Lanka is reported to occur in India. In present investigation implications of newly described species of Aponogetons, interrelationships among Indian species of the genus and unique embryo morphology have been presented and discussed. *Aponogeton satarensis* is the only Indian dioecious relict endemic species with biforked spike whose close relative viz. *A. decaryi* occurs in Madagascar. This has great phytogeographical significance. *Aponogeton bruggenii* is narrow endemic with barriers of unknown nature in sexual reproduction and seldom sets seeds. It propagates through formation of daughter tubers on mother tuber. *Aponogeton nateshii* is a unique in its embryo morphology. The embryo is globular and possesses 15-22 appendages all around the embryo. Such embryo is so far not reported in any other angiosperm except *A. appendiculatus* which has 3-5 tortuous appendages. Three population of *A. nateshii* showed variation in number of appendages per embryo and their fusion with tuberous part of embryo. All Indian species are closely allied to each other except for *A. satarensis*. The size, shape, appendages (cotyledons), position of plumule and emergence of radical root varies with species. Attempts have been made to understand morphological and developmental diversity of Aponogetonaceae in Indian subcontinent with special reference to embryo diversity and diversification of the genus.

T2-62-02

Pollen diversity and parallel evolution in Alismatales

Elena Severova[1], *Terry D. Macfarlane*[2], *Olga Volkova*[1]
1. Moscow State University

2. Western Australian Herbarium

The order Alismatales comprises Tofieldiaceae, Araceae and core Alismatales (including 'tepaloid' and 'petaloid' clades). Some core Alismatales demonstrate adaptations for hydrophilous pollination accompanied with severe modifications and reductions in sporoderm structure (Furness & Banks, 2010). The early-divergent member of the tepaloid clade, Aponogetonaceae, is characterized by monosulcate pollen that may represent a plesiomorphic feature in Alismatales and monocots in general. Many other Alismatales have inaperturate pollen, but pantoporate (Alismataceae, petaloid clade) pollen is also known. Diaperturate pollen is not common but remarkably diverse in Alismatales. In *Tofieldia* (Tofieldiaceae), one aperture is distal and another is proximal, which is confirmed by our developmental data. In contrast, *Calla* (Araceae) has two equatorial sulci or a ring-like aperture (Ulrich *et al.,* 2013). Thus the increase in number of apertures has been realized in different ways in different lineages. *Scheuchzeria* occupies a key position in the tepaloid clade. Scheuchzeriaceae, Juncaginaceae and Maundiaceae all possess inaperturate pollen and are wind-pollinated. These families are the three successive members of the grade leading to a more derived monophyletic group of tepaloid clade that includes several seagrass lineages with filiform pollen, but also some members with aperturate or tenuitate pollen (*Ruppia* and *Althenia* spp.). This pattern suggests evolutionary loss and re-gains of apertures in the 'tepaloid clade'. In the light of this scenario, a question arises whether any rudimentary traces of an aperture can be seen at ultrastructural and developmental levels in the inaperturate members of the tepaloid clade. Our data on pollen development in *Scheuchzeria* does not reveal any possible aperture sites in the process of microsporogenesis that favours the hypothesis of the complete evolutionary loss and subsequent re-gain of the apertures (Volkova *et al.,* 2016). The unusual permanent pollen dyads of *Scheuchzeria* can be considered as adaptation to fertilization of both ovules typically present in carpels. *Althenia* s.l. (Potamogetonaceae) includes two dissimilar species with in-aperturate (omniaperturate) and several species with diaperturate pollen. Species with diaperturate pollen fall in two groups based on structure of sporoderm revealed by scanning and transmission electron microscopy. Fine structure of sporoderm in the apertural region in each group is similar to the sporoderm along the entire surface of pollen grain in one of the two omniaperturate species. Along with molecular phylogenetic evidence (Ito *et al.,* 2016), this suggests that the omniaperturate condition appeared twice in course of evolution of *Althenia*, each time by expansion of the apertural area onto entire surface of the pollen grain. We speculate that the transition from inaperturate to diaperturate pollen in Potamogetonaceae should be viewed as an evolutionary step towards a complete loss of interapertural type of sporoderm. Thus the evolution of spherical omniaperturate pollen was homoplastic, like the origin of the filiform omniaperturate pollen in different families of tepaloid Alismatales. More data on pollination biology are needed to understand the biological significance of different types of pollen grains with reduced sporoderm. The work on pollen morphology and development was performed by EES and OAV with funding from the Russian Science Foundation (project 14-14-00250).

T2-62-03

Morphological diversity of core Alismatales

Dmitry Sokoloff*, Anna Platonova, Ingrid Lock, Margarita Remizowa*

M.V. Lomonosov Moscow State University

The core Alismatales includes all families of the order except Tofieldiaceae and Araceae. This is a rare example of a group of angiosperm families commonly recognized in pre-molecular era (as subclass Alismatidae) and then fully supported by molecular data in the same circumscription. Aquatic or semi-aquatic habit is of course the most remarkable feature of the group (hence the historical name Helobiae), but in terms of diagnostic characters, the presence of intravaginal squamules and the large storage embryo are the most remarkable. These two features are present in all members of core Alismatales (the squamules of Scheuchzeria are transformed in hairs) and are otherwise very rare among monocots. Both features are present in some Araceae being homoplastic within the family, and the squamules are present in Acorus. Morphological nature of intravaginal squamules is enigmatic. Arber (1923) highlighted that the squamules are attached above the leaf to the base of stem internode. We hypothesize that the squamules are homologous to collateral buds that are characteristic to monocots in general but not to the core Alismatales. The number and morphology of intravaginal squamules are of taxonomic significance in the core Alismatales, for example, distinguishing *Potamogeton* and former Zannichelliaceae. Leaves of the core Alismatales are diverse in terms of external morphology and anatomy. It is unclear whether they can be considered unifacial in any of the core Alismatales other than *Tetroncium* (Juncaginaceae). Leaves or leaf petioles have three-dimensional venation in several families, normally with a system of small peripheral vascular bundles. As in angiosperm succulents and halophytes (Ogburn & Edwards, 2013), the three-dimensional leaf venation in thick aquatic and helophyte leaves of Alismatales serves to reduce transport distances between veins and photosynthetic cells (Platonova *et al.,* 2016). Like in the eudicot order Caryophyllales (e.g., Melo-de-Pinna *et al.,* 2014), the patterns of orientation of peripheral collateral bundles (with inverted adaxial or abaxial bundles) are unstable in Alismatales. These slender bundles cannot be used for the identification of unifacial leaves. *Maundia* (Maundiaceae) is remarkable in having similar anatomical structure of foliage leaves and inflorescence peduncles, both with xylem of small peripheral bundles oriented towards periphery of the organ. The similarity is due to their shared function as photosynthetic organs, though functional significance of particular types of bundle orientation remains unclear (Platonova *et al.,* 2016). Compared to many other monocot lineages, the core Alismatales are remarkably diverse in terms of number and position of floral organs and degree and mode of intercarpellary fusion. Floral characters are homoplastic in Alismatales and inferring directions of their transformation is problematic. Patterns of floral diversity in Alismatales can be only partially explained by adaptations to aquatic habitats. In most cases, evolutionary reductions in number of floral organs left no rudiments in the core Alismatales, with notable exceptions in the family Juncaginaceae. The study is supported by the Russian Science Foundation (project 14-14-00250).

T2-62-04

Floral development of petaloid Alismatales as a clue to the origin of the trimerous pentacyclic flower of the Monocots

Akitoshi Iwamoto[1], *Ayaka Nakamura*[1], *Ayumi Otani*[1], *Shinichi Kurihara*[1], ***Louis Ronse De Craene***[2]
1. Tokyo Gakugei University
2. Royal Botanic Garden Edinburgh

The monocots are remarkably homogeneous in sharing a common trimerous pentacyclic floral Bauplan. A major factor affecting monocot evolution is the unique origin of the clade as an offshoot of the basal angiosperms. The origin of the floral Bauplan of monocots is still controversial, as no immediate sister groups with similar structure can be identified among basal angiosperms, and there are several possibilities for an ancestral floral structure. Either more complex flowers with a higher stamen and carpel number, or strongly reduced flowers could be ancestral, while a stable Bauplan is only established beyond the divergence of Alismatales. We revisited the floral development of members of the three 'petaloid' Alismatales families Butomaceae, Hydrocharitaceae, and Alismataceae. Outer stamen pairs are present in Alismataceae and Butomaceae, It is demonstrated that paired stamens always arise independently, and are either shifted opposite the sepals or close to the petals. The position of stamen pairs is a matter of development of the petals. In Butomaceae the perianth is undifferentiated and not delayed; the larger petals push the stamens in antesepalous pairs. Alismataceae have delayed petals and stamens are shifted in close proximity to the petals, leading to a close association of pairs with the petals in so-called stamen-petal complexes. In the studied Hydrocharitaceae paired stamens are replaced by larger single stamens and petals are not delayed. Despite the fact that the petaloid Alismatales are not at the base of monocot divergence, the floral Bauplan of Butomaceae and some members of Tofieldiaceae (e.g. Pleea) are probably close to the ancestral flowers of monocots. Among basal angiosperms, the comparative Bauplan is only retrieved in members of Cabombaceae, such as *Brasenia*, and may indicate that Monocots evolved as an offshoot from aquatic basal angiosperms, with variable stamen and carpel number.

T2-62-05

Taxonomy and evolutionary morphology of Australian *Althenia* (Potamogetonaceae, Alismatales)

Terry Macfarlane[1], *Dmitry Sokoloff*[2]
1. Western Australian Herbarium
2. Moscow State University

Traditionally *Althenia*, with two species in Eurasia and Africa, was regarded as one of four genera of submerged aquatics comprising family Zannichelliaceae, related to Cymodoceaceae. Molecular phylogenetic data confirmed monophyly of Zannichelliaceae but placed the clade with *Potamogeton* and its segregates, so that current classifications include Zannichelliaceae in Potamogetonaceae. A recent study (Ito *et al.*, 2016) using several molecular markers demonstrated that *Althenia* is nested within the Australian and New Zealand *Lepilaena*, most closely related to *L. preissii*. Consequently the two genera were combined under the name *Althenia*. According to Ito *et al.*, (2016), the genus comprises seven species (five in Australia). However, many aspects of morphological diversity, diagnostic characters, taxonomic composition and geographical distribution of *Althenia* remain unresolved. Our current data shows *Althenia* comprises at least eight species in Australia. This first comprehensive study of Australian herbarium collections showed up to 30 percent were incorrectly identified. Re-identification of all available material resulted in much better understanding of the ecological requirements and distribution patterns of Australian species of *Althenia*. One species is restricted to marine habitats or estuaries. *Althenia* is of interest with respect to several aspects of morphological evolution. The genus comprises both dioecious and monoecious species (one species probably possessing monoecious and dioecious populations). Monoecy is likely plesiomorphic in *A. australis*, which was shown to be sister to the rest of the genus (Ito *et al.*, 2016). In other cases, monoecy is possibly derived from dioecy as a reversal. In some dioecious species, foliage leaves associated with male inflorescences are pronouncedly ligulate while those associated with female inflorescences lack ligules. The sheathing leaf bases and ligules are apparently involved in the mechanism of release of pollen grains because they may enclose anthetic stamens. Traditionally, *Althenia* s.str. was distinguished from *Lepilaena* mainly by male flowers with 2-sporangiate (vs. 4-12-sporangiate) anthers and funnel-shaped, circular, polysymmetric (vs. monosymmetric) stigmas. The discovery of a new dioecious species from Western Australia challenges these criteria and supports molecular phylogenetic evidence for lumping the two genera. The species has polysymmetric circular stigmas like *Althenia* s.str. but 12-sporangiate anthers like *Lepilaena*. Like some other Australian species, the new taxon has strongly dimorphic male and female inflorescences. Scanning electron microscopic study of developmental series of male and female inflorescences of the new species showed that in spite of habit differences, major structural and developmental features of floral units are identical in both sexes. Among interesting features is the possession in male floral units of very long filamentous (distally flattened) terminal structures that resemble tubular and filamentous terminal structures that sporadically develop in inflorescences of *Potamogeton*. However in the case of our *Althenia* species their regular occurrence in only one of the two sexes indicates that they cannot be treated as a developmental abnormality and must have a certain function, unknown as yet. Phylogenetic studies of *Althenia* using molecular data are being expanded with additional geographic sampling and new species. This work was supported with funding from the Russian Science Foundation (project 14-14-00250).

T2-62-06

Secondary carpel margin in ascidiate carpels of Alismatales

Margarita Remizowa
Lomonosov Moscow State University

Completely ascidiate carpels (often with funnel-shaped stigmas) are especially common among tepaloid alismatids (basal monocot order Alismatales). Along all their length, these carpels represent a tubular structure with ventrally attached single pendent ovule. There is a considerable piece of tissue between ovule and carpel margin. Developmentally, the ovule originates on the ventral placenta which is located in the cross-zone. Usually, ovule initiation happens early in the development when the carpel is represented by low meristematic rim. As the carpel walls grow, the ovule attachment becomes lifted above the receptacle. Thus, the ovary is formed by circular zonal growth below the level of placenta. The most intriguing process is style development which proceeds via circular growth above the level of placenta. We speculate that directly above the cross-zone a new, secondary carpel margin appears. The cross-zone itself is a marginal meristem to build up the ventral carpel wall and to produce the ovule. Technically, the cross-zone becomes exhausted by ovule initiation and the growth of carpel wall ceases ventrally after ovule initiation in many angiosperms. To continue with tubular style, a new meristematic activity is needed above the ovule. The secondary ventral carpel margin as a result of such an activity develops more slowly than lateral flanks. This leads to slit appearance of carpel orifice which faces ventrally. During further development, the carpel orifice becomes more or less horizontal. The work was performed with funding from the Russian Science Foundation (project 14-14-00250).

T2-63: Ecological and biogeographic implications of Asian Oligocene and Neogene fossil floras

T2-63-01

Cenozoic plant diversity of Yunnan: A review

Yong-Jiang Huang[1], *Lin-Bo Jia*[1], *Qiong Wang*[2], *Volker Mosbrugger*[3], *Torsten Utescher*[3,4], *Tao Su*[2], *Zhe-Kun Zhou*[1,2]

1. Key Laboratory for Plant Diversity and Biogeography of East Asia, Kunming Institute of Botany, Chinese Academy of Sciences

2. Key Laboratory of Tropical Forest Ecology, Xishuangbanna Tropical Botanical Garden, Chinese Academy of Sciences

3. Senckenberg Research

4. Steinmann Institute, University of Bonn

Yunnan in southwestern China is renowned for its high plant diversity. To understand how this modern botanical richness formed, it is critical to investigate the past biodiversity throughout the geological time. In this review, we present a summary on plant diversity, floristics and climates in the Cenozoic of Yunnan and document their changes, by compiling published palaeobotanical sources. Our review demonstrates that thus far a total of 386 fossil species of ferns, gymnosperms and angiosperms belonging to 170 genera within 66 families have been reported from the Cenozoic, particularly the Neogene, of Yunnan. Angiosperms display the highest richness represented by 353 species grouped into 155 genera within 60 families, with Fagaceae, Fabaceae, Lauraceae and Juglandaceae being the most diversified. Most of the families and genera recorded as fossils still occur in Yunnan, but seven genera have disappeared, including *Berryophyllum*, *Cedrelospermum*, *Cedrus*, *Palaeocarya*, *Podocarpium*, *Sequoia* and *Wataria*. The regional extinction of these genera is commonly referred to an aridification of the dry season associated with Asian monsoon development. Floristic analyses indicate that in the late Miocene, Yunnan had three floristic regions: a northern subtropical floristic region in the northeast, a subtropical floristic region in the east, and a tropical floristic region in the southwest. In the late Pliocene, Yunnan saw two kinds of floristic regions: a subalpine floristic region in the northwest, and two subtropical floristic regions separately in the southwest and the eastern center. These floristic concepts are verified by results from our areal type analyses which suggest that in the Miocene southwestern Yunnan supported the most Pantropic elements, while in the Pliocene southwestern Yunnan had abundant Tropical Asia (Indo-Malaysia) type and East Asia and North America disjunct type that were absent from northwestern Yunnan. From the late Miocene to late Pliocene through to the present, floristic composition and vegetation types changed markedly, presumably in response to altitude changes and coeval global cooling. An integration of palaeoclimate data suggests that during the Neogene Yunnan was warmer and wetter than today. Moreover, northern Yunnan witnessed a pronounced temperature decline, while southern Yunnan experienced only moderate temperature changes. Summer precipitation was consistently higher than winter precipitation, suggesting a rainfall seasonality. This summary on palaeoclimates helps us to understand under what conditions plant diversity occurred and evolved in Yunnan throughout the Cenozoic.

T2-63-02

Cenozoic floras in the Qinghai-Tibetan Plateau: Were they isolated?

Tao Su[1], *Yong-Jiang Huang*[2], *Shu-Feng Li*[1], *Zhe-Kun Zhou*[1]

1. Key Laboratory of Tropical Forest Ecology, Xishuangbanna Tropical Botanical Garden, Chinese Academy of Sciences

2. Key Laboratory for Plant Diversity and Biogeography of East Asia, Kunming Institute of Botany, Chinese Academy of Sciences

The Qinghai-Tibetan Plateau (QTP) and adjacent areas are well known for the high biodiversity worldwide at the present day. Plant fossils are pivotal to understand their histories of biodiversity and biogeography under dramatically paleoenvironmental changes in this large and fascinating region; however, fossil floras there have been far from fully investigated. During recent five years, we found several paleofloras ranging from Eocene to Pliocene in high altitudes of the QTP. For the early Eocene Liuxiang flora (~ 55 Ma) in southern Tibet, Moraceae are the most abundant; the late Eocene Kajun flora (~34.6 Ma) in eastern Tibet is dominated by *Cyclobalanopsis* and *Betula*, representing evergreen-deciduous broadleaf forest; For the early Miocene Lunpola flora (~23 Ma) in central Tibet, grasses firstly occurred together with Arecaceae and *Koelreuteria*. The Pliocene Zhada flora (~ 4 Ma) in western Tibet is represented by small-sized leaves. The floristic history proves dramatically environmental changes throughout geological time in the QTP. Meanwhile, it suggests that the floristic modernization in the QTP occurred no later than the late Eocene. Some extinct taxa, e.g., *Hemitrapa*, *Limbophyllum*, and *Cedrelospermum*, as well as many taxa being still widely distributed in the QTP, e.g., *Betula*, *Cyclobalanopsis*, *Excentrodendron*, *Koelreuteria*, and *Rosa*, indicate that there were frequently floristic exchanges between the QTP and other parts of the North Hemisphere. This work is supported by National Natural Science Foundation of China (41661134049,31470325, U1502231), Excellent Young Scientists Foundation-CAS (QYZDB-SSW-SMC016), and Youth Innovation Promotion Association, CAS (2017439).

T2-63-03

Oligocene and Neogene vegetation change in Siberia and the northeast of Russia reconstructed from palaeocarpological

data

Svetlana Popova[1], *Torsten Utescher*[2], *Dmitry Gromyko*[1], *Volker Mosbrugger*[3], *Louis François*[4]
1. Komarov Botanical Institute
2. Steinmann Institute
3. Senckenberg Research Institute and Natural History Museum
4. Unité de Modélisation du Climat et des Cycles Biogéochimiques (UMCCB) Université de Liège

The Cenozoic continental deposits of Siberia are reknowned for their carpological record. P. Nikitin, V. Nikitin and P. Dorofeev carried out comprehensive studies there throughout the 20th century. Based on the flora-bearing horizons four main evolutionary stages were outlined by V. Nikitin. In the first phase, the Pre-Turgayan, a subtropical flora existed (Late Cretaceous – Eocene). The second, Turgayan phase is characterized by the expansion of a boreal, warm temperate flora. This flora evolved during the Early Oligocene and was replaced by diverse mesophilous mixed coniferous-broad-leaved forests during the Late Oligocene to Early Miocene. The last phase, the Post-Turgayan (Middle Miocene to Early Pliocene), mainly shows the dominance of forested steppe. Steppe landscapes evolving since the end of the Pliocene represent the modern stage of the vegetation in the southern part of Western Siberia (WS). The analysis of the diversity of plant functional types (PFTs) is proven to be very useful for quantifying spatial vegetation patterns and gradients and their evolution during the Cenozoic. Here we use the diversity of 27 herbaceous to woody PFTs to classify 148 carpofloras from WS and the northeast of Russia (NER) covering the time-span from the OLigocene to the late Pliocene. Our novel PFT classification scheme, designed for use in biome modeling, comprises 26 herbaceous to arboreal PFTs based on physiognomic characters and bioclimatic tolerances of plants, completed by an aquatic PFT. Using multivariate statistics, localities with similar PFT spectra are grouped and interpreted in terms of biomes. The results are visualized on palaeovegetation maps. The PFT technique facilitates interregional comparisons and unravels differentiations in biome type that are hardly quantifiable in classical vegetation reconstruction from the palaeobotanical record. Moreover, the use of diversity of single plant functional types instead of specimen abundance minimizes taphonomic bias, which may be considerable when carpofloras are regarded. In a number of cases it can be shown that PFT diversity gradients in WS and NER follow climatic patterns. Our study reveals a longitudinal gradient between WS and NER existing in Eurasia already in the early Miocene but getting more significant along with the Neogene continental cooling. In WS, late Cenozoic vegetation change relates to a decline of arboreal components, together with a strong diversity increase in herbaceous plants. This process already began in the late Palaeogene, with the onset of a global step-wise cooling, and intensified from the middle Miocene on. In the later Neogene, increasing diversity of xerophytes in the southern part of WS indicates drying of the continental interior. At the same time, the declining diversity of thermophilous PFTs in the northern part of WS testifies the cooling of the higher latitudes. According to our findings, NER was persistently covered by forests until the Pliocene, including the High Arctic. The humid and thermophilous aspect of late Miocene to Pliocene palaeofloras from the western part of NER contravenes the global cooling trend and could mirror a later Neogene intensification of the East Asian Monsoon System.

T2-63-04

Oligocene-Miocene floral changes in the circum-Japan Sea areas- climate changes and floristic separation

Atsushi Yabe[1], *Eunkyoung Jeong*[2], *Boris Pavlyutkin*[3], *Kyungsik Kim*[2], *Kazuhiko Uemura*[1]
1. Department of Geology and Paleontology, National Museum of Nature and Science
2. Department of Biological Sciences, Chonbuk National University
3. Far East Geological Institute, Far East Branch, Russian Academy of Sciences

Oligocene–Miocene floral changes in the circum-Japan Sea areas were reviewed to understand the floristic response to global and local climates as well as to the effects of the opening of Sea of Japan during the early–middle Miocene. We recognized two floral stages in the Oligocene and four in the Miocene, based on detailed stratigraphic correlations. The early Oligocene, here referred to as Stage O-1, is represented by the Wakamatsuzawa flora in Hokkaido and the Kobe flora in Central Honshu, Japan. Some assemblages in western Primorye (Kraskino, Rettikhovka) and North Korea (Kogeonwon, Kungshim) can be correlated to this stage. Floristic composition of each assemblage differs one to the other possibly due to latitudinal temperature changes; however, most of them typically contain lobed-oaks, Palaeocarya and show similar physiognomic characters. The next Stage O-2 of late Oligocene is recorded only in western Japan. Two assemblages represent temperate forest vegetation with few evergreen species. Stage M-1 in the earliest early Miocene in Japan is called Aniai-type flora. Comparable assemblages have been recorded in the Korean Peninsula (Janggi, Yongdong) and western Primorye (Sinij Utes). All these assemblages contain deciduous temperate elements like Fagus and various species of Betulaceae, Ulmaceae, and Sapindaceae with no or few evergreen angiosperms. Stages M-2 and 3, denoted here as late early and middle Miocene, are comparable with the Daijima-type floras of Japan, which are distributed throughout the Japanese Islands. They are composed of a mixture of evergreen and deciduous species of dicotyledons in Honshu, but represented more by deciduous elements in Hokkaido. Sinjeongri (South Korea) and Tongcheon (North Korea) floras are compared with M-2, while Yeonil and Hamjindong floras of South and North Korea are compared with M-3 assemblages. Marine organisms that indicate tropical–subtropical oceanic conditions have been recorded from the strata dated between Stages M-2 and 3, in which pollen grains of mangrove plants have been reported from over ten sites in central to northern Honshu. Assemblages from the Nezhino and Novokachalinsk horizons in Primorye are considered contemporaneous with Stages M-2 and 3. However, floristic composition of the Russian assemblages differ considerably from those in Honshu and southern Korean Peninsula. Stage M-4 assemblage, denoted as the late Miocene, is known as the Mitoku-type floras in Japan. The Ust Suifun flora in western Primorye is compared stratigraphically. In Japan, a number of inland basins started to develop at this stage. There is a strong difference between those from inland basins and coastal lowlands. The former is represented by deciduous temperate species, while the latter consists of mixture of evergreen and deciduous species. Quantitative climate analysis of these assemblages from lowland vegetation indicates a trend roughly parallel to the global climate pattern. However, after the separation of Japanese Islands from the continent, the extent of temperature increase from stages M-1 and 3 evidently differed between assemblages on the continent and those on the island, which is probably due to the influence of cold and warm water currents both toward the north and south of the Sea of Japan.

T2-63-05

Elaeocarpus L. (Elaeocarpaceae J.) fossil fruits from Oligocene-Miocene of Guangxi, South China

Xiaoyan Liu[1], *Cheng Quan*[2], *Jianhua Jin*[1]

1. State Key Laboratory of Biocontrol, Guangdong Provincial Key Laboratory of Plant Resources, and School of Life Sciences, Sun Yat-sen University, China

2. Research Center of Paleontology and Stratigraphy, and MOE Key-Lab for Evolution of Past Life and Environment in Northeast Asia, Jilin University, Changchun 130026, China

Elaeocarpus as a genus of tropical and subtropical evergreen trees and shrubs is widespread in the Indo-Pacific region from India, Southeast Asia, southern China, and Japan, Malaysia, New Guinea, Australia to the outliers including Madagascar, New Caledonia and New Zealand. Centers of high diversity include northeastern Australia, New Caledonia and the Malesiana region. In China, there are approximately 39 extant species in the genus, which are mainly distributed in the South and Southwest China. The fossil record of *Elaeocarpus* is rare, including some pollen, leaves and fruits from several Oligocene and younger deposits in Australia and India. However, the fossil pollen and leaves attributed to the genus are questionable. *Elaeocarpus* fruits, by contrast, are recognizable from other genera by their unique woody and highly variable morphology and ornamentation types of mesocarps, which are extremely important in understanding systematic relationships within the genus. Therefore, the most reliable and the earliest fossil record of *Elaeocarpus* is the fruit stones of *E. spackmaniorum* Rozefelds, *E. cunningii* Rozefelds and *E. rozefeldsii* Dettmann and Clifford from the early-late Oligocene of central Queensland, Australia. In this article, well-preserved *Elaeocarpus* fruit stones recovered from the Oligocene of the Nanning Basin and Miocene of the Guiping and Nanliu Basin, Guangxi Province, South China were studied. Micro-CT was used in reconstructing three dimensions and sections of the fossil and extant specimens, six species types are confirmed based the highly variable morphology and ornamentation of mesocarps. They are similar to extant species of *Elaeocarpus glabripetalus* Merrill, *E. rugosus* Roxburgh, *E. lacunosus* Wallich ex Kurz, *E. serratus* L., *E. prunifolioides* Hu, and *E. sikkimensis* Masters. The origin for the genus was inferred as southern hemisphere origination and migrated to the northern hemisphere via birds, based on the fossil records and modern distribution pattern of the genus. This is the first fossil record for the occurrence of the genus in the lowest latitude of Asia, indicating that the genus has been spread into South China as early as Oligocene and diversified in the Miocene, and providing new and significance evidence for understanding the origin and evolution of the genus and family.

T2-63-06

Earliest fossil fruit record of *Canarium* (Burseraceae) in eastern asia and its implications for phytogeographical history

Meng Han[1], *Steven R. Manchester*[2], *Yan Wu*[1], *Cheng Quan*[3], *Jianhua Jin*[1]

1. Sun Yat-sen University

2. University of Florida

3. Jilin University

As a distinctive genus within the Burseraceae, *Canarium* contains approximately 78 species of predominantly medium to large sized rainforest trees that are naturally distributed in low to middle altitudes of tropical or subtropical regions. It is characterized by distinctive drupaceous fruit with a trilocular endocarp derived from three fused pyrenes. Recently some new *Canarium* fossil fruits were collected from the late Oligocene of Yongning Formation, the Miocene of Erzitang Formation and the late Miocene of Foluo Formation in Guangxi Province, South China, providing the first confirmed fossil occurrences of *Canarium* in eastern Asia. The fruits of *Canarium guangxiensis* Han, Manchester, Wu, Jin et Quan sp. nov. are ovoid to spindle shaped, 22.8-34.3 mm long, 10.7-14.6 mm wide. Computed tomography (CT) scan was used to study the morphological and anatomical characters of fossil and modern *Canarium*, which facilitates identification of the fossil fruits. This new occurrence supplements other megafossil records of *Canarium* fruits from the Early Eocene in North America, the Middle Eocene to Oligocene in Europe, the Oligocene to Miocene in Asia and from the Pleistocene in Australia and the Pacific islands. The fossil records indicate a wide dispersal of *Canarium* over the Northern Hemisphere during the Eocene and Oligocene, followed by a phytogeographical contraction during the Miocene as the result of its extinction from North America and Europe. The origin and migratory routes of this genus are not clearly resolved, but based on the fossil records known so far, we hypothesize the *Canarium* may have had a North American Eocene origin, with subsequent spread to Eurasia and followed by dispersal to the Southern Hemisphere continents.

T2-64: Floral adaptation and mating efficiency

T2-64-01

Quantum dots—a new tool for studying components of male fitness in plants

Corneile Minnaar, *Bruce Anderson*

Stellenbosch University

Apart from the advent of molecular tools, pollination biology has been a "bucket and spade science", and the basic toolkit has remained largely unchanged for more than a century. Perhaps the reason why pollination biology has had such a long and distinguished history, is that with only a few simple and inexpensive tools (rulers, mesh bags, sticky labels, and permanent markers), biologists have made spectacular advances in understanding natural selection and evolution-primarily through counting seeds. However, easy as it is to count seeds, every pollination biologist at some point must confront an uncomfortable truth-seed counts only quantify the female aspect of fitness. What about pollen export and components of male fitness? The responses of male and female fitness components to selection may not always be similar, and many floral traits are thought to have evolved primarily through male components of fitness. Despite this, male fitness is very rarely quantified. This is because our tool box is missing an important tool: one which allows us to track pollen grains and quantify different avenues of pollen export from individual plants. Here, in the grand tradition of buckets and spades, I introduce a long overdue addition to the pollination biologist's basic collection of rulers, bags and sticky labels: a simple method to directly tag individual pollen grains and track their subsequent fates. The method, developed in our lab over three years, attaches different coloured fluorescent nanocrystals (q-dots) directly to pollen grains. This allows us to uniquely identify their origins upon recovery from pollinators, stigmas, or any other surface they may end up on and enables us to quantify all potential avenues of pollen movement. Unique q-dot labels can be applied directly to dehisced anthers in the field and attached to pollen grains within minutes, allowing for rapid assessment of pollen movement *in situ*. I will demonstrate details of the method with examples of systems in which we have employed q-dots for pollen tracking. Our method makes tracking pollen grains and quantifying male fitness nearly as easy as quantifying female fitness. We hope that this method will usher in a new era in pollination biology, where tracking pollen movement, and easily quantifying male fitness, is finally possible.

T2-64-02

The impact of pollinator behaviour on the adaptive function of heterostyly exposed by quantum dots

Marinus De Jager, *Corneile Minnaar, Bruce Anderson*
Stellenbosch University

Intraspecific variation in stigma placement, or heterostyly, has been a staple of evolutionary inquiry ever since Darwin first considered it a century and a half ago. Tristyly is a subset of heterostyly characterised by reciprocal herkogamy of anthers and stigmas at three levels and self/intra morph incompatibility. Traditional theory posits that tristyly evolved in order to promote disassortative mating between floral morphs exhibiting anthers and stigmas at the same level. Other explanations hold that it evolved to limit pollen wastage and opportunities for selfing. These theories have been largely untested, mostly due to an inability to track pollen from anther to stigma. The genus *Oxalis* represents an excellent system to elucidate the evolutionary mechanisms behind tristyly as all South African species exhibit two anther rings and one stigma ring contained within a central column inside actinomorphic flowers. Here we investigate pollen fates of self-incompatible *Oxalis purpurea* visited by either nectar or pollen foraging pollinators by employing a novel technique for tracking individual pollen grains. Using fluorescent quantum dots, we label pollen grains from each anther level in each stylar morph with distinct colours before allowing honeybees to forage in a controlled laboratory setup. Our results reveal that opportunities for selfing are rampant despite a well-developed tristylous syndrome. Of pollen exported, disassortative mating was more frequent relative to random expectations for all three morphs. Of pollen imported, long styled flowers in particular received predominantly legitimate pollen. We further discuss the influence of different pollinator behaviours on male and female plant fitness in heterostylous species.

T2-64-03

Comparative biomechanics of pollen dispersal in wind- and animal-pollinated plants

David Timerman, *Spencer C.H. Barrett*
University of Toronto

Aerodynamic forces have shaped the structure and function of many plant organs, including flowers and inflorescences. Reproductive organs are frequently exposed to some amount of airflow which can dislodge pollen from anthers. Approximately 10% of flowering plant species rely on wind for pollination, and aerodynamic pollen release is the first critical step in successful reproduction. Conversely, releasing pollen into airflows is costly to plants which rely solely on animal pollinators. The contrasting demands of these two pollination modes suggest that wind- and animal-pollinated species will differ with respect to their adaptation to aerodynamic forces. Convergent evolution of floral traits in wind-pollinated species supports this hypothesis as it is thought to result from biophysical adaptation for fluid dynamic processes. However, some species do not exhibit the classical wind pollination syndrome and yet may still rely on wind pollination or a mix of animal and wind pollination (ambophily). Biophysical analysis at the wind-flower interface is needed to interpret the structural characteristics of flowers controlling pollen release, and yet such analyses are rare. We investigated pollen release mechanisms in wind- and animal-pollinated species of *Thalictrum* (Ranuculaceae), a moderately large genus with multiple independent transitions between animal- and wind-pollination syndromes. Using high speed video of pollen release in a wind tunnel, a laser particle counter and computer vision algorithms, we compared the vibration response of stamens to mechanical perturbation and measured pollen release probability. We characterized stamen vibrations because turbulence-induced shaking of reproductive structures has been shown in species of angiosperms, gymnosperms, bryophytes and fungi to initiate the dispersal of pollen or spores by wind. We also investigated the effects of stamen vibration on pollen release rate and export in field arrays of the putatively ambophilous *T. pubescens*. Arrays exhibiting variation in stamen vibration were either open pollinated or placed in insect exclusion cages to determine the relation between stamen vibration, aerodynamic pollen release rate and the amount of pollen exported by pollinators. Our studies identified novel biomechanical floral traits that are related to pollen release probability including the fundamental frequency of vibration and stamen stiffness. In many cases these traits confound conventional descriptions of discrete pollination syndromes. Consequently, pollen release may be more common among species exhibiting an animal-pollinated syndrome than previously thought and factors other than aerodynamic efficiency may contribute to shaping the wind-pollination syndrome. Moreover, aerodynamic pollen release may play contrasting roles in the evolution of specific pollination strategies in environments that differ with respect to atmospheric conditions (wind speed and turbulence intensity) and pollinator availability. We discuss the potential costs and benefits of atmospheric pollen release and the likely evolutionary outcomes for floral structures.

T2-64-04

The function and adaptive significance of morphological

polymorphisms characterizing the heterostyly syndrome

Joana Costa[1], *Sílvia Castro*[1,2], *João Loureiro*[1], *Spencer C. H. Barrett*[3]

1. CFE, Centre for Functional Ecology and Department of Life Sciences, Faculty of Sciences and Technology, University of Coimbra, Calçada Martim de Freitas, 3000-456, Coimbra, Portugal

2. Botanical Garden of the University of Coimbra, Calçada Martim de Freitas, 3000-456 Coimbra, Portugal

3. Department of Ecology and Evolutionary Biology, University of Toronto, 25 Willcocks Street, Toronto, Ontario, Canada M5S 3B2

Animal-pollinated plants display a remarkable diversity of structural adaptations that function to promote cross-pollen dispersal and reduce pollen wastage. Heterostyly is a convergent floral syndrome that has evolved in ~28 angiosperm families and is one of the most striking examples of the functional fit between flowers and pollinators. Most heterostylous species possess a reciprocal arrangement of stigmas and anthers (reciprocal herkogamy), diallelic self-incompatibility and ancillary polymorphisms of pollen and stigma. Here, we investigate the function of the morphological polymorphisms that characterize the heterostyly syndrome in two unrelated taxa of heteromorphic species through experimental approaches. We used tristylous *Lythrum salicaria* (Lythraceae) to (1) evaluate Darwin's hypothesis that reciprocal herkogamy promotes disassortative pollination, and to (2) examine the extent to which floral design, specifically the location of sex-organs within a flower, influences disassortative pollination. Using experimental arrays we detected varying degrees of disassortative pollination among the three morphs with sex-organ height playing an important role in determining compatible and incompatible pollen transfer and capture. Our studies provide support for the Darwinian hypothesis that heterostyly functions to promote disassortative pollination, with floral design playing a critical role in determining its intensity. To investigate the functional significance of ancillary polymorphisms, we conducted field experiments on three species of Plumbaginaceae to determine the extent to which pollen and stigma polymorphisms influence the composition of pollen loads. Specifically, we evaluated Dulberger's topographical complementarity hypothesis, which proposes that ancillary polymorphisms function in the rejection of incompatible pollen thus promoting disassortative pollination. To test this hypothesis (1) we investigated patterns of pollen transfer and capture in natural populations of two pollen-stigma dimorphic species lacking reciprocal herkogamy (*Armeria maritima* and *A. pubigera*) and a distylous species with pollen stigma dimorphism (*Limonium vulgare*) and (2) we examined pollen adherence and germination patterns in *A. maritima* following controlled hand-pollinations. We detected disassortative pollination in populations of each species, and no difference in compatible pollen capture between mating types was observed, although 'cob stigmas' captured more incompatible pollen. Our controlled hand-pollination experiment revealed the failure of incompatible pollen to adhere and germinate on stigmas. These results demonstrate that pollen-stigma dimorphisms serve to promote disassortative pollination by limiting incompatible pollen deposition through an intimate combination of structural and physiological mechanisms. Our studies provide novel insights into the functioning of key components of the heterostylous syndrome, and demonstrate that both reciprocal herkogamy and ancillary polymorphisms promote disassortative pollination and reduce pollen wastage.

T2-64-05

Quantum dots reveal the mechanics behind handedness in *Wachendorfia* flowers

Bruce Anderson, *Corneile Minnaar*

Stellenbosch University

Stylar polymorphisms such as enantiostyly (handedness) in plants are underpinned by a classic body of compelling theory, much of which remains poorly tested. A key part of that theory is that the spatial separation of male and female reproductive parts (herkogamy) to reduce selfing, presents a problem to plants: The spatial separation also reduces the efficiency of pollen transfer. In theory, this problem was elegantly solved by the evolution of stylar polymorphisms like enantiostyly and heterostyly because they retain the spatial separation of the sexes but facilitate pollen transfer between reciprocally matching stylar morphs. Biologists have been unable to test whether this solution holds up to scrutiny because until recently it has been almost impossible to evaluate the efficiency of pollen transfer due to the difficulty in marking and tracking granular pollen grains. Using fluorescent nanocrystals (quantum dots) which attach to the surface of pollen, we were able to identify the origins of pollen grains on the bodies of pollinators and on the stigmas of flowers. We show that pollen from left versus right facing anthers separate perfectly on either side of pollinating insects. This results in exceptionally strong assortative pollen movement, where most pollen moves between morphs rather than within morphs. This suggests that a shift from herkogamy to enantiostyly would indeed be accompanied by more efficient pollen transfer.

T2-64-06

Effect of floral phenotype on mating patterns and male and female fitness in experimental populations of *Narcissus papyraceus*

Violeta I. Simón-Porcar, ***Juan Arroyo***

University of Seville

The floral phenotype and its variation is the result of natural selection affecting the pollination success of flowering plants. In single hermaphroditic plants, reproductive success is divided into its female and male components; while the measurement of the former is straightforward through seed production and quality, a direct measure of the latter requires paternity analyses in the population to determine siring success of particular plants. The floral phenotype may also determine affinities between particular phenotypes, which might be disposed to cross with each other more likely, as it occurs in heterostylous and other reciprocal polymorphic plants. *Narcissus papyraceus* (Amaryllidaceae) is a self-incompatible style-dimorphic species with individuals presenting either long-styled or short-styled flowers in a single population, a condition which is similar and evolutionarily ancestral to heterostyly. Dimorphic and long-styled monomorphic populations are associated with different pollinator faunas in the central and north regions of its range, respectively. Previous work on experimental populations of this species (Simón-Porcar, Meagher & Arroyo 2015) demonstrated the effect of the style morph in the mating success of individuals, but nothing is known about the possible influence of other floral traits. Here we explore the effect of the floral phenotype in mating and reproductive success of individuals of *Narcissus papyraceus* under different natural conditions. We present new results from the experimental populations of Simón-Porcar *et al.*, (2015), which were exposed to ambient pollinators in the central and marginal regions of the range of the species and subjected to genetic paternity analysis. We characterize the whole floral phenotypes of the individuals and relate them to their maternal and paternal reproductive success, and investigate their influence on mating patterns. Our comprehensive approach allows to prove the existence of natural selection on the floral phenotype, and to correlate male and female components of fitness. Thus, we advocate that experiments and observations set in natural conditions should always account for both maternal and paternal fitness when aiming to assess similar questions.

T2-65: Using fossil evidence to explore the plant evolution, diversity, and their response to global changes

T2-65-01

Paleobotanical evidence of environmental constrains of early hominid expansion into southern Europe

Angela A. Bruch
ROCEEH Reserch Centre, Senckenberg Research Institute

The earliest hominid expansions from Africa into Eurasia date to about 2 Ma based on the Dmanisi record of *Homo erectus* in Georgia. In South Europe however records (notably from Spain and Italy) are much younger and extend back to 1.2 and possibly 1.4 Ma. The scarce records may represent a single or several out-of-Africa events, and questions remain as to what extent climate and environmental change facilitated or hindered the occupation of Europe by early *Homo*. Climate change is considered a major driving factor of hominin evolution and dispersal patterns. In fact directly or indirectly by its severe influence on vegetation, physiography of landscape, and animal distribution, climate modulates the availability of resources. Several researchers suggest that orbital cycles (especially the obliquity 41 ka cycle) were driving regional climate providing short time windows of favorable landscape and vegetation conditions for hominine expansions in Europe. Early Pleistocene vegetation successions with more and less favorable conditions for hominid occupations at the end of each glacial period (warm, but still dry enough to enable open landscapes) have been documented from Southern Europe. However, this is not the case throughout the possible expansion routes. Based on own results from the Caucasus region, we extrapolate the maximum extend of forests and mosaic landscapes for different climatic phases during Early Pleistocene as a prerequisite for the reconstruction of early human environments. In comparison with existing paleobotanical records from Southern Europe possible climatic constraints of early hominid expansions into Europe become evident and imply restricted expansion corridors during some very warm and humid climatic phases.

T2-65-02

A review of three Eocene floras from southeast China and southeast United States: Implications for understanding the extant floristic disjunction between eastern Asia and eastern north America

Hongshan Wang[1], *Jianhua Jin*[2], *Xiaoyan Liu*[2], *Jane Blanchard*[1], *David Dilcher*[3]
1. Florida Museum of Natural History
2. Sun Yat-sen University
3. Indiana University

The floristic similarity between temperate eastern Asia (EA) and eastern North America (ENA) was first recognized by Linnaeus in 1750. This disjunct distribution of plants has intrigued both neobotanists and paleobotanists for more than 260 years. An early interpretation, often referred to as Gray's hypothesis, is that the floras of eastern North America (ENA) and eastern Asia (EA) were more similar to each other than to the flora of western North America (WNA). Different approaches, focusing on disjunct groups of plants at generic, familial, or community levels, have been employed to interpret this relationship and to hypothesize the processes that shaped this distribution pattern. The geological records indicate that Eurasia and North American were geographically connected throughout most of the Cenozoic or during the Paleogene by two major land bridges, i.e., the Bering land bridge and the North Atlantic land bridge. It is widely accepted that these land bridges significantly contributed to the floristic similarities between eastern Asia and North America. Paleoclimatic and molecular analyses of some groups of angiosperms from the two areas seem to support the hypothesis that most taxa with this disjunct pattern are the result of vicariance. With the application of new techniques such as molecular phylogenetics and the molecular clock, the origin, time of disjunctions, and relationships of some lineages are now better understood. Recent molecular phylogenetic analyses have yielded two important results. One is that, contradictory to the traditional view of the floristic similarity between eastern Asia (EA) and eastern North America (ENA), the examination of this disjunct pattern in the broader context of Northern Hemisphere biogeography suggests a closer relationship between eastern North America (ENA) and western North America (WNA) than between eastern North America (ENA) and eastern Asia (EA). The disjunction between eastern Asia (EA) and eastern North America (ENA) mostly represents a relic of a wider distribution in the Cenozoic and is resulted from complex processes of dispersal/migration, extinction, speciation, vicariance, and morphological convergence and stasis. The second result of the use of new techniques is that most intercontinental species pairs studied are not sister species and that some taxa, traditionally treated as disjunct, are now considered paraphyletic and polyphyletic. We present our preliminary results based on observations of thousands of fossil specimens of reproductive material from three Eocene floras, the Claiborne Flora from southeastern United States and the Changchang and Maoming floras from southeastern China. After comparing the floristic components of these Eocene floras at the generic and familial levels, we discuss the floristic patterns and their possible causes. We then compare the distribution of the Eocene geofloras with that of the extant floras of eastern Asia and eastern North America. The result will provide some insights for interpreting the modern floristic relationship between these two regions.

T2-65-03

Impact of Mid-Miocene to Late Holocene climate changes on the development of the eastern Himalayan flora: evidence from plant mega and microfossils

Subir Bera
University of Calcutta

Indian monsoon system characterised by seasonally reversing winds and precipitation changes associated with asymmetric heating of land and sea is a substantial component of the global atmospheric circulation that brings about 80% rainfall to the Indian subcontinent. In order to understand the evolution, dynamics and response of the plant communities to these changes of Indian monsoon, long term palaeoclimate records are essential. A synthesis of the quantitative palaeoclimate records to throw light on the evolution of the Indian monsoon since the mid-Miocene time and climate driven vegetation dynamics in the Darjeeling Himalaya is focused here. CLAMP (Climate Leaf Analysis Multivariate Program) analysis of a mid-Miocene fossil flora from the Darjeeling Himalayan Siwalik suggests a warm (tropical to subtropical) humid climate with a distinctive monsoon signature. A moderately high rainfall with over 2000 mm mean annual precipitation (MAP) and an average annual humidity of 80% might have harboured an evergreen to moist deciduous forest in the mid- Miocene Darjeeling Himalaya. Darjeeling fossil flora estimating a ratio of wet and drier months as 3.8:1 is indicative of a more even distribution of rainfall throughout the year, apparently due to a wetter dry season. A mean annual temperature (MAT) of 25.4 ± 2.8 °C (all uncertainties ±2 sigma) with warm month mean temperature (WMMT) of 28.4 ± 3.39 °C and cold month mean temperature (CMMT) of 17.9 ± 4 °C also further suggests a conducive environment for sustenance of a dense evergreen to moist deciduous forest. A further analysis of palynofossils through Coexistence approach (CA) and other proxy data portray palaeomonsoon oscillations in the Darjeeling Himalaya over the last ca. 50 ka. CA estimates coupled with other proxies indicate significant oscillations in rainfall during 46.4 and 25.9 ka, 25.9–15.6 ka, and 5.4 to 3.5 ka. Span between ca 46.4–31 ka is characterized by a comparatively low monsoonal activity and slightly higher temperature than that during ca 31 ka onwards. Simultaneous expansion of deciduous trees and C4 grasses also imply a drier and warmer phase. Between 31 and 22.3 ka, higher precipitation and a slightly cooler temperature led to an increase in evergreen elements over deciduous taxa and wet-loving mixed C3-C4 grasses over dry-loving C4 grasses than earlier. After ca 22.3 ka, shrinking of forest cover, expansion of C4 grasses and dry-loving heliophytic elements in the vegetation with lowering of temperature and precipitation characterized the onset of the LGM that persisted till 18.3 ka. A restoration in the forest cover and in the temperature and precipitation regime marks the end of the LGM. Later, a strong monsoonal activity during 5.4 to 4.3 ka supported a dense moist evergreen forest cover that subsequently showed a thinning during 4.3 to 3.5 ka with an increase in deciduous elements and non-arboreals. A reduction in rainfall with rise in temperature might be responsible for a shrink in the forest cover during this phase. A multiproxy comparison of the global data of the ~50 ka implies that these fluctuations in plant succession were mainly driven by monsoonal variations.

T2-65-04

Early cretaceous angiosperm reproductive diversity

David Dilcher[1], *Hongshan Wang*[2], *Ge Sun*[3], *Bernard Gomez*[4]
1. Indiana University
2. University of Florida
3. Shenyang Normal University
4. University of Lyon

Angiosperm origins and earliest radiations can be traced back most consistently to the early Lower Cretaceous with some reports of Jurassic and even Triassic remains. Angiosperm evolution is centered upon a wide variety of reproductive strategies to encourage and/or require outcrossing. Evidences of Angiosperms are commonly found in Early Cretaceous sediments from which their diversity rapidly radiates. The early records include reproductive axes that contain both ovule-bearing organs subtended by pollen-bearing organs. The homologies of these organs may be interpreted as representing individual female and male flowers on an axis or alternatively as one unit as a whole, similar to a compound inflorescence. In either case, the *Archaefructus* axis appears to represent an early attempt to achieve a successful outcrossing system in the evolution of the bisexual flower. The active engagement of insects, co-evolving with early angiosperms, during the Late Jurassic and Early Cretaceous, shaped the evolution of both the insects and the angiosperms. The co-evolution of insect pollinators had and continues to have a profound influence on angiosperm evolution. This is joined by the evolution of wind pollination that became a common feature of genetic exchange in many angiosperm lineages. At the same time, some lines of angiosperm evolution were using other systems to manage their reproductive biology. Fossils similar to the extant *Ceratophyllum* were widespread and diverse. These angiosperms managed outcrossing in aquatic environments. Water was the abiotic agent that moved genetic material within and between populations. This is a syndrome used by diverse land plants since the Paleozoic to manage outcrossing. This syndrome seems to have persisted in the earliest angiosperms as well. While the first flower in the world remains a mythical concept, we need to focus our research upon the evidence of the steps of change preserved in the fossil record. Then perhaps we can understand the complex reproductive biology of these plants as they functioned in their biotic and abiotic environments.

T2-65-05

The fossil history of *Mesocyparis* (Cupressaceae) and its response to the global cooling across Cretaceous-Paleogene boundary

Yi-Ming Cui[1,2], *Wei Wang*[1], *Yu-Fei Wang*[1], *Jian Yang*[1], *David Ferguson*[1,3]
1. Institute of Botany, Chinese Academy of Sciences
2. University of Chinese Academy of Sciences
3. University of Vienna

The fifth largest mass extinction on the earth was occurred across the Cretaceous-Paleogene Boundary (K-Pg Boundary, ca. 66.0 million years ago), which accompanied by a dramatic global climate cooling in both oceans and landmasses. There are some zoological cases to show how animal taxa responded to climate changes across the K-Pg boundary in richness and morphological changes, but few in plant taxa. *Mesocyparis*, an extinct genus, was distributed from Eastern Asia to Western North America across this time interval. Here we report a new species of *Mesocyparis* with seed cones and foliage in the early Paleocene Wuyun Coalmine from NE China, which has ovate seed cones, 4-decussate cone bracts and flattened foliage. Phylogenetic analysis show that *Mesocyparis* is a member of Cupressoideae and is sister to the clade containing *Juniperus*, *Cupressus*, *Hesperocyparis*, and *Xanthocyparis*. Within *Mesocyparis*, two Western North American species formed a clade, whereas three Asian species formed anther. The paleogeographic distribution of this genus, together with phylogenetic analysis reveal it originated at high latitudes in the North Hemisphere, and divided into an Eastern Asiatic clade and Western North American clade. Morphological comparison shows the genus changed markedly across K-Pg boundary, i.e. the seed cones became larger, the umbo was erect (vs. reflexed) and the position of the umbo varied from middle part of cone scale to the apex. The northern limit of *Mesocyparis*, with those of two other arboreal taxa *Metasequoia* (gymnosperm) and *Nordenskioeldia* (angiosperm), migrated 4 - 5° paleolatitude southward, which might be driven by the global mean surface cooling across K-Pg boundary. These findings give a case on how a specific plant taxon responded to the K-Pg climate change.

T2-65-06

Mesozoic fossil wood diversity variations and paleoclimate implications in China

Yongdong Wang[1,2], *Ning Tian*[3,4], *Zikun Jiang*[5], *Xiaoju Yang*[1], *Qiuhong Ding*[6]

1. Nanjing Institute of Geology and Palaeontology, Chinese Academy of Sciences, Nanjing 210008

2. State Key Laboratory of Palaeobiology and Stratigraphy, Chinese Academy of Sciences, Nanjing 210008

3. College of Paleontology, Shenyang Normal University, Shenyang 1

4. Key Laboratory for Evolution of Past Life in Northeast Asia, Ministry of Land and Resources, Shenyang 110034

5. Chinese Academy of Geological Sciences, Beijing, 100037

6. Shenyang Geological Center, China Geological Survey, Shenyang 110034

As an important component of plant remains, fossil wood plays a significant role in understanding the floral composition and evolution of plants in the geological past. Fossil wood is also one of the significant proxies for terrestrial palaeoclimate and palaeogeographical reconstruction in earth history. Diversified fossil wood has been documented from the Mesozoic deposits in China after a long time of investigations. During the past few years, many new fossil wood materials were reported from a variety of horizons in some fossil localities, including Sichuan, Chongqing, Yunnan, Xinjiang and Liaoning provinces, ranging from Triassic to the Cretaceous, providing significant references for reconstructing the Mesozoic palaeoclimate. In this paper, we summarize the recent advances in Mesozoic fossil wood studies in China. Particularly, the new discoveries of fossil wood from the Upper Triassic in Sichuan of southern China, the Jurassic Yanliao Biota and the Early Cretaceous Jehol Biota in western Liaoning are reviewed with emphasis on the diversity variations, palaeoclimate perturbations and tempo-spatial distribution of the Mesozoic wood in China. The future research directions of Mesozoic fossil wood in China are further discussed.

T2-66: Evolution of intercontinental disjunctions in ferns and lycophytes

T2-66-01

The biogeographic history of hairy tree ferns (Dicksoniaceae-Cyatheales): A parting with different pathways and paces

Marcus Lehnert[1], *Sarah Noben*[2], *Michael Kessler*[2], *Anna Weigand*[2], *Michael Krug*[1], *Susann Wicke*[3], *Dietmar Quandt*[1]

1. Nees-Institute for Plant Biodiversity, University of Bonn

2. Institute for Systematic and Evolutionary Botany, University of Zurich

3. Institute for Evolution and Biodiversity, University of Münster

The tree fern family Dicksoniaceae, encompassing the genera *Calochlaena* (5 spp. paleotropical), *Lophosoria* (3–4 spp. neotropical), and *Dicksonia* (26 spp., widespread), dates back to the early Cretaceous (ca. 125–140 Mya) and is considered a typical Gondwana element. A phylogeny of cp marker (*rpl16, trnL-F, trnG-R* and *matK*) confirmed the monopyhly of the genera and found three monophyletic groups within *Dicksonia*, whose stem ages are younger than the separation of the tectonic units of Gondwana to which they seem restricted today. Other geographical disjunctions, e.g. between Australia and Malesia, are even more recent and thus more likely to be the result of long-distance dispersal. The different scenarios are discussed in the context of paleoclimate and geologic events.

T2-66-02

Biogeography of the brake fern genus *Pteris* (Pteridaceae)

Liang Zhang[1], *Hang Sun*[1], *Li-Bing Zhang*[2]

1. Kunming Institute of Botany, CAS

2. Missouri Botanical Garden

The brake fern genus *Pteris* contains about 200–250 species, some of which are widely used as ornamental plants or arsenic hyperaccumulators. The global distribution of the genus makes it an ideal system to test biogeographical hypotheses. Earlier analyses were largely based on unresolved phylogenies using chloroplast markers. In this study, analyses of five chloroplast markers and one nuclear marker of 160 species (60–80% of the all species in the genus) resulted in a robust phylogeny and confirmed the monophyly of *Pteris* s.l. Our main results include: (i) The genus might have originated from tropical Asia-Australasia; (ii) The early diverging lineages of *Pteris* underwent accent range expansions and diversified in Africa, Eurasia, and Neotropics; (iii) The Southern Hemisphere endemics are closely related to one another than to the Northern Hemisphere endemics; (iv) The Neotropical endemics form two independent clades suggesting that they originated independently; (v) The island endemics are all resolved within the 15 clades defined by our previous study and long-distance dispersals contributed to the successful colonization from Mainland Asia to Taiwan, Japan, and Hawaii, and from Mainland Africa to the near archipelagos.

T2-66-03

From Asia to Africa: Biogeography of the fern genus *Pyrrosia* (Polypodiaceae)

Chun-Xiang Li[1], *Xin-Mao Zhou*[2], *Li-Bing Zhang*[2]

1. State Key Laboratory of Palaeobiology and Stratigraphy, Nanjing Institute of Geology and Palaeontology, Chinese Academy of Sciences

2. Missouri Botanical Garden

The Old World fern genus *Pyrrosia* (Polypodiaceae) has representatives in Africa and Asia, but its highest species diversity is clearly found in southern Asia and Malesia. Earlier phylogenetic analyses based on small sampling did not confirm the monophyly of the genus and the the relationships within the genus were not well resolved, and therefore it was not possible to understand the biogeography of the genus. In the present study, DNA sequences of five plastid markers of 220 accessions of Polypodiaceae representing two species of *Drymoglossum*, 14 species of *Platycerium*, 47 species of *Pyrrosia*, and the only species of *Saxiglossum* (subfamily Platycerioideae), and 12 species of other Polypodiaceae representing the remaining four subfamilies are used to infer a phylogeny of *Pyrrosia*. Our major results include: (1) *Pyrrosia* is supported as a monophyletic group; (2) The Himalayan *P. costata* and the Southeast Asian *P. stigmosa* together are sister to the remaining species in the genus; (3) The genus apprears to have originated in Asia, while the African species appear to have arrived via long-distance dispersal from southern China/Southeast Asia; and (4) Divergence time estimation suggests that the first occurrence of *Pyrrosia* species was at ca. 25 Ma in the Himalaya-Southeast Asia.

T2-66-04

Climatic niche evolution in polyploid and diploid vascular plants

Anthony Baniaga[1], *Hannah Marx*[1], *Nils Arrigo*[1,2], *Marc Bogonovich*[3], *Michael Barker*[1]

1. Department of Ecology & Evolutionary Biology, University of Arizona

2. Department of Ecology & Evolution, University of Lausanne

3. Department of Biology, Indiana University

The dynamics of ecological niche evolution has long been hypothesized to be a critical determinant of polyploid success and establishment. In general, newly formed polyploids are reproductively isolated from their parental taxa, yet initially remain in the same geographic area and/or niche as their parents. Polyploids also typically begin with a small population size and face a number of short term and long-term ecological and evolutionary challenges. Previously, it was thought that niche divergence was a prerequisite for polyploid success and comparative studies from microhabitat associations and regional floras show mixed support for this hypothesis. Here, we developed a broad comparison of polyploid and diploid niche evolution to ask three key questions: 1) Is climatic niche divergence necessary for polyploid success and establishment, 2) Is the rate of climatic niche evolution different for polyploids than congeneric diploids, and 3) Are polyploids with divergent ecological niches associated with higher net diversification rates? Using a manually curated dataset of known polyploids, their diversification rates, and their parental taxa, we analyze data across vascular plants and compare and contrast patterns and processes between pteridophytes and seed plants.

T2-66-05

Understanding the odd man out pattern in African plant: New insights from ferns

Emily Sessa

University of Florida

Africa has long been recognized as the "odd man out" in terms of global plant distribution patterns. The continent has a whole has between three and nine times fewer species per square kilometer than neighboring continent-scale areas to the west and east (South America and the Indo-Malaysian region, respectively). This pattern has been documented across plant groups, including in palms, orchids, and ferns and lycophytes. Differences in species richness can be caused by different rates of speciation, extinction, and migration, which in turn are driven by myriad factors specific to a given place and time, such as climate, proximity to other areas, geologic history, and time available for speciation, among other things. In ferns, data collected to date for three genera (*Dryopteris*, *Marsilea*, and *Cheilanthes*) suggest that ferns may only have undergone significant radiation in Africa over the last 20 million years, despite the fact that Africa has existed in largely its current location and orientation for far longer than that (at least 300 million years). In addition, the last 20 million years have largely corresponded with a period of aridification in sub-Saharan Africa, and if ferns in general have indeed diversified in Africa during this time, it would be striking given that we tend to think of ferns as moisture-loving organisms dependent on water for their survival. I am using a combination of phylogenetic, biogeographic, and diversification analyses to try and understand patterns of fern diversity and species richness in Africa, and to shed light on the "odd man out" phenomenon in African plants.

T2-66-06

Biogeography of Tectariaceae

Li-Bing Zhang[1], *Liang Zhang*[2]

1. Missouri Botanical Garden

2. Kunming Institute of Botany, Chinese Academy of Sciences

Tectariaceae are typically characterized by the presence of multicellular and uniseriate (ctenitoid) hairs and costae that are non-sulcate adaxially. They contain seven genera and have pantropical distribution with *Arthropteris*, *Tectaria*, and *Triplophyllum* being in the Old and New Worlds, *Draconopteris* and *Hypoderris* in the New World, and *Malaifilix* and *Pteridrys* in the Old World. Earlier pre-molecular works or molecular works with small sampling were not able to understand how the family acquired today's intercontinental distribution The current study included the largest taxon and character (plastid and nuclear markers) sampling so far. Based on the phylogeny reconstructed and the extant geographical distributions of taxa in the family, it is postulated that: (1) the family originated in the Old World; (2) none of the genera were involved in more than one intercontinental dispersal between the Old and New Worlds; (3) there might have been five independent dispersals from the Old World to the New World: single dispersal of *Draconopteris* and *Tectaria* each from Asia to the New World, single dispersal of *Arthropteris* and *Triplophyllum* each from Africa to the New World, and single dispersal of *Hypoderris* from the Old World to the New World; (4) within the Old World there were multiple dispersals among Africa, Asia, and the Pacific islands within *Tectaria*; and (5) African species of *Tectaria* evolved from Asian ancestors.

T2-68: The evolution of Plant Diversity in the "third pole", Qinghai-Tibetan Plateau

T2-68-01

Relationships through time between the orography of Tibet, the Himalaya, Asian monsoon systems, and the vegetation of southern Asia

Robert Spicer
The Open University

The mechanisms underlying Asian monsoon dynamics are poorly understood but the Tibetan plateau and the Himalaya modify climate systems associated with seasonal latitudinal migrations of the Inter-Tropical Convergence Zone (ITCZ). ITCZ migrations must have generated monsoonal precipitation regimes at low latitudes (~10°- ~30°) throughout Earth history so much of southern Asia must have experienced monsoonal regimes since at least the early Cretaceous. Abundant evidence shows that a high (>3 km) proto-Tibetan edifice existed prior to the onset of the India - Eurasia collision at around 55 ± 10 Ma and if an elevated Tibet in any way amplifies monsoon systems over southern Asia the rich biota of the region must have evolved under strong monsoonal influences. However, these Paleogene monsoons were different from those of today. A monsoon climate is more than just strong seasonal variations in precipitation. Charting the evolutionary history of Asian monsoon systems has been hampered by the lack of deep time proxies compatible with meteorological monsoon definitions. However, the leaves of woody dicots have to be particularly well adapted to the seasonal extremes they experience across all climate variables (thermal and hydrological) simultaneously. These adaptations give rise to unique monsoon 'fingerprints' encoded in fossil leaf architecture. Multivariate analysis reveals early Eocene (~55 Ma) to latest Oligocene (~23 Ma) leaf form across southern Asia was adapted to climates characteristic of ITCZ type monsoons and not those of the modern South Asia Monsoon (SAM), which is influenced by a high Himalaya. Recent analysis of Tibetan plant fossils suggests that the Himalaya formed against a pre-existing Gangdese Highland (palaeoelevation ~4.5 km) soon after the India-Eurasia collision began. By the end of the Oligocene the proto-Himalaya had reached an elevation of ~2.3 km, and then rose to ~5 km by 15 Ma. Only later did the Himalaya project above the average elevation of the Tibetan plateau to generate the modern SAM circulation.

T2-68-02

Out-of-Tibet: Taxonomy and spatio-temporal evolution of Gentiana (Gentianaceae)

Adrien Favre[1], *James S. Pringle*[2], *Sabine Matuszak*[3], *Hang Sun*[4], *Yong-Ming Yuan*[5], *Alexandra N. Muellner-Riehl*[1,6]
1. Leipzig University
2. Royal Botanic Gardens, Hamilton
3. Goethe University and Biodiversity and Climate Research Centre (BiK-F), Frankfurt am Main
4. Kunming Institute of Botany, Chinese Academy of Sciences
5. Research & Development Centre, Firmenich Aromatics (China) Co. Ltd., Shanghai
6. German Centre for Integrative Biodiversity Research (iDiv) Halle-Jena-Leipzig

The origin and evolution of species richness in Asian mountainous hotspots of biodiversity is complex and probably caused by non-mutually exclusive factors such as explosive radiations, lineage persistence and biological interchange. Whereas a plethora of studies has claimed that orogenesis alone triggered the radiation of alpine plants around the world, the role of climate oscillations, niche evolution, and the immigration of pre-adapted lineages are usually overlooked (or underestimated) as contributors to local biodiversity. After revising the taxonomy within Gentianinae and *Gentiana*, we investigated the evolutionary history of *Gentiana* with regard to the uplift of the Qinghai-Tibetan Plateau (QTP) with phylogenetic reconstructions, molecular dating, ancestral area reconstructions, and diversification rate analyses. The common ancestor of *Gentiana* and subtribe Gentianinae lived in the QTP region at around 34 (25–45) million years ago (Ma), and 40 (29–52) Ma, respectively. From the surroundings of the QTP, *Gentiana* lineages dispersed to eastern China, Taiwan, Europe, North and South America, Australia and New Guinea, from mid-Miocene onward (c. 15 Ma–present), with only one older dispersal event to Europe (c. 37–21 Ma). Diversification rates gradually increased over time, and one switch of diversification rates (c. 7 Ma) was identified in the lineage encompassing *Gentiana* sections *Pneumonanthe* and *Cruciata*, following a niche shift to lower elevations. Our results confirm the general role of the QTP in promoting the adaptation of tropical lineages to temperate habitats, and underline its status as a source area for alpine plants worldwide. However, because a slow and steady increase in diversification rates coincides with the extension of the QTP, we argue that the museum theory rather than the explosive radiation theory prevails for gentians in this region, and that the sharp increase of diversification rates is unrelated to the uplift, but rather corresponds to niche shifts and key innovations.

T2-68-03

Diversity of *Arisaema* (Araceae) in the Himalayas, Hengduan Mountains and Tibet inferred from an updated molecular phylogeny

Jin Murata
Botanical Gardens, University of Tokyo

In 2016, we presented an updated genus-wide phylogenetic tree of *Arisaema* based on a large data set of four plastid non-coding regions (a total 2,858 bp after alignment) from 138 taxa representing the diversity of the genus. It shows eight major clades and one branch with unique sequence variation, although the relationships were unclear due to a polytomy and weak support. These nine clades mostly correspond to the sections recognized in recent classifications. Four sections are distributed mainly in the Himalayan region. Three of them, sections *Dochafa*, *Tortuosa* and *Tenuipistillata* are included in the first clade. Most of the species are distributed in and around Himalayan region but *Arisaema flavum* subsp. *flavum* is subendemic to Tibet and *A. wardii* is endemic to Tibetan-Qinghai plateau. Section *Arisaema* corresponding to the second clade is diversified both in Himalayan region and Hengduan Mounatains. It have been well known that *A. costatum* of this section is sister to an African group. Section *Sinarisaema* is distributed widely throughout China to South India and Sri Lanka. This section is included in the final large clade together with other four sections. Of them sect. *Sinarisaema* and sect. *Pistillata* share a spilodistichous leaf arrangement and are considered to be more recent origin than the sections which are distributed in Himalayan region mentioned above.

T2-68-04

Phylogeographic perspectives on plant evolution in the Qinghai-Tibetan Plateau (QTP) region and adjacent areas

Hans Peter Comes[1], *Hang Sun*[2], *Ying-Xiong Qiu*[3]

1. Department of Ecology and Evolution, Salzburg University

2. Kunming Institute of Botany, Chinese Academy of Sciences,

3. College of Life Sciences, Zhejiang University

Over the last decade, molecular phylogeographic studies have led to an improved understanding of the origins of exceptionally high plant diversity and endemicity of the Qinghai-Tibetan Plateau (QTP) itself and the forested southern and eastern margins of the Plateau, i.e. the Himalaya-Hengduan Mountains (HHM). Here we review recent spatial genetic and palaeo-modelling analyses of QTP–HHM plant diversity related to the importance of these mountain ranges and deep valley systems as floristic hotspots, in which high topographic and climatic complexity not only promoted strong population structure, multiple and independent refugial isolation, and (incipient) allopatric species formation, but probably also buffered against lineage extinction. Moreover, molecular dating studies spanning the flora of the QTP region – and especially within the southwest – continue to implicate a major influence of palaeo-geological and/or -climatic events on intraspecific lineage divergence and demographic histories from the Late Pliocene/Early Pleistocene onwards, thus being largely consistent with the most recent and rapid episodes of uplift of the eastern part of the QTP. Finally, by using the perennial understory herb genus *Dysosma* Woodson (Berberidaceae, Podophylloideae; *c.* 7 spp.) as a model system, we demonstrate how the establishment of (sub)alpine plant diversity in the QTP region might have contributed to the origin of high levels of species diversity and endemism in the warm-temperate deciduous and evergreen forest biomes of subtropical China.

T2-68-05

Nectar guide change and it mechanism in *Arnebia szechenyi* Kanitz (Boraginaceae)

Yang Yongping

Kunming Institute of Botany, Chinese Academy of Sciences

Communications between plants and their pollinators involve a complex interplay of different sensory modalities. Nectar guides are the contrasting patterns on the corolla, and considered to be advertising the presence and indicating the location of floral rewards (nectar or pollen) to pollinators. However, the changes of nectar guides across a flower's lifetime are poorly understood, and the mechanisms of nectar guide changes remain unclear. In *Arnebia szechenyi* Kanitz (Boraginaceae), a montane perennial herb native to China, flowers typically display nectar guides on the first day of anthesis until they start fading out and then disappear altogether on the second day, with no apparent link to the longevity of individual flowers (ca. 4 days). We explore the functions of the nectar guides and their changes by designing different treatments to the nectar guides and by performing field observations, and examine the different gene expressions by next generation sequencing. Our results illustrate that nectar guides disappear in netted flowers, suggesting that these subtle color changes are intrinsic, although pollination could facilitate fading of nectar guide in *A. szechenyi*. Control experiments indicated that nectar guides increase male but not female reproductive fitness. Plants with more flowers attracted more pollinators and received more visits, but the visitation rates to single flower with nectar guides did not increase with enlarged floral display size. Transcriptome analysis show that 154,021 genes are differentially expressed in different stages of anthesis, and among these transcripts, low expression levels of key genes in the anthocyanin biosynthesis during flowering process are *CHS*, *F3H*, *UF3GT*, *ANS* and *DFR*. Notably, except for these vital genes, there is one gene encoding subgroup 6 R2R3-MYB transcription factor greatly show down-regulated expression, which are known to activate the structural genes in anthocyanin biosynthesis. Conclusively, we propose that nectar guides and their changes have developed via selection on the male function of *A. szechenyi*, and older, nectar-depleted flowers increase the floral display size and are therefore likely to be maintained by selection through the enhanced pollen export rates to flowers with nectar guides on the same plant. The set of down-regulated genes indicates the crucial role of anthocyanin regulation in the nectar guides disappearance, and the modulation of R2R3-MYB transcription factor may play an important role in regulating the nectar guide changes in *A. szechenyi*.

T2-68-06

Tempo and mode of floristic assembly in the Hengduan Mountains

Richard Ree[1], *Yaowu Xing*[2]

1. The Field Museum

2. Xishuangbanna Tropical Botanical Garden

The Hengduan Mountains region is one of the world's richest hotspots of plant diversity, and is unusual in being neither tropical nor Mediterranean in climate. How was its flora assembled, in terms of tempo (rate of species accumulation) and mode (*in situ* diversification vs colonization) through time? A recent biogeographic analysis of 19 plant clades shows a remarkable increase in the rate of in situ diversification around 8 Ma, corresponding to independent estimates of when Hengduan Mountains were tectonically uplifted. This supports the idea of uplift-driven diversification. However, many Hengduan clades show signatures of older radiations that may have coincided with uplift of the Himalayas.

Many outstanding questions about assembly of the Hengduan flora remain, such as the conspicuous lack of polyploidy and the role of biotic interactions.

THEME III: ECOLOGY, ENVIRONMENT & GLOBAL CHANGE

T3-01: Plant climatic adaptation in the new genomic era

T3-01-01

Metabolomic changes of stomatal responses to CO_2 in the changing climate

Sixue Chen[1], *Sisi Geng*[1], *Ning Zhu*[2], *Craig Dufresne*[3], *Evaldo De Armas*[3], *Lloyd Sumner*[4], *Hans Alborn*[5]

1. Department of Biology, University of Florida, Gainesville, Florida

2. Proteomics and Mass Spectrometry, Interdisciplinary Center for Biotechnology Research, University of Florida, Gainesville, Florida

3. Thermo Fisher Scientific, West Palm Beach, Florida

4. Samuel Roberts Noble Foundation, Ardmore, OK

5. Chemistry Research Unit, Agricultural Research Service, United States Department of Agriculture, Gainesville, Florida

Stomatal movement in respond to various stimuli is an important process that balances plant carbon dioxide (CO_2) uptake and water transpiration. Elevated CO_2 induces stomatal closure, while low CO_2 induces stomatal opening. Here we report the metabolomic responses of *Brassica napus* guard cells to elevated and low CO_2 using three hyphenated metabolomics platforms: gas chromatography (GC)-mass spectrometry (MS), liquid chromatography (LC)-multiple reaction monitoring (MRM)-MS, and ultra high-performance LC (UHPLC)-quadrupole time-of-flight (QToF)-MS. A total of 358 and 411 metabolites were quantified in a time-course response of guard cells to elevated CO_2 and low CO_2, respectively. Major alterations in fatty acid, flavonoid, organic acid, sugar, phenylpropanoid, and amino acid metabolic pathways indicated changes in both primary and specialized metabolic pathways in guard cells. Most interestingly, multiple hormones including jasmonic acid (JA) were altered in the course of the CO_2 treatments, indicating hormone involvement in guard cell CO_2 responses.

T3-01-02

Omics in wild soybean (*Glycine soja*) – insights into the plant climate adaptation and stress management

Bao-Hua Song, *Hengyou Zhang, Larry Leamy, Neha Mittal, Janice Kofsky*

University of North Carolina at Charlotte

Climate change has generated various critical challenges to agriculture susceptibility and food security. These challenges may be met by the development of novel crop varieties with increased biotic or abiotic resistance that enable them to thrive in changing environments. Crop wild relatives (CWRs) harbor a much higher level of genetic diversity than cultivated crops and have the potential to meet these challenges. The wild soybean (*Glycine soja*), from which domesticated soybeans (*Glycine max*) were derived, is widely distributed throughout diverse habitats in East Asia. We use *G. soja* as our study system to investigate its genomic diversity, population structure, climate adaptation, and biotic stress resistance, integrating genomic, transcriptomic, and metabolomic approaches. Using ~42,000 genome-wide single nucleotide polymorphism (SNP) data, we estimated that environmental factors contributed 23.6% to population differentiation, much more than that for geographic factors (6.6%). Precipitation variables largely explained the divergence of the groups along longitudinal axes whereas temperature variables contributed more to latitudinal divergence. We also used a genome scan of outlier loci to identify candidate loci involved in climate adaptation. This scan uncovered a total of 32 known genes with diverse functions, seven of which are involved in abiotic stress responses. We also integrated genome-wide association studies, RNA-seq comparisons, and metabolomic comparisons to dissect the genetic basis of wild soybean resistance to soybean cyst nematode (SCN), the most devastating pest of soybean. We identified some candidate genes involved in SCN type-specific resistance, as well as others exhibiting broad-spectrum resistance to multiple types of SCN. These candidate genes will not only shed light on the molecular mechanisms of plant and SCN interactions, but also will facilitate the molecular breeding and development of soybean varieties with improved SCN resistance.

T3-01-03

How plants adapt to harsh environments on the Qinghai-Tibet Plateau at genomic level

Ticao Zhang[1], *Qin Qiao*[2], *Qia Wang*[1], *Miao An*[3], *Jinling Huang*[4], *Hang Sun*[1], *Yang Zhong*[3]

1. Key Laboratory for Plant Diversity and Biogeography of East Asia, Kunming Institute of Botany, Chinese Academy of Sciences

2. School of Agriculture, Yunnan University

3. Key Laboratory for Biodiversity Science and Ecological Engineering, School of Life Sciences, Fudan University

4. Department of Biology, East Carolina University

Interpretation of genetic variation at the whole genome level can contribute to understanding how organisms adapt to changing environments. The Qinghai-Tibet Plateau (QTP) has the highest biodiversity for an extreme environment worldwide, and provides an ideal natural laboratory to study adaptive evolution. Little genome/transcriptome based study has been conducted on how plants adapt to the environments of QTP compared to numerous studies on vertebrates. 1) We generated a draft genome sequence of cyanobacteria *Trichormus* sp. NMC-1 in the QTP and performed whole transcriptome sequencing under low temperature to investigate the genetic mechanism by which *T.* sp. NMC-1 adapted to the extreme environment. The genome sequence of *T.* sp. NMC-1 was 5.9 Mb with a G + C content of 39.2% and encompassed a total of 5362 CDS. Genome comparison between *T.* sp. NMC-1 and six closely related species showed that the functions of unique, significant positively selected, expanded orthogroups, and differentially expressed genes were related to specific adaptation traits, such as signal transduction mechanisms, cell wall/membrane biogenesis, secondary metabolite biosynthesis, defense mechanisms, and energy production and conversion. 2) *Crucihimalaya himalaica* is a close relative of *Arabidopsis* with typical QTP distribution, and hoped to be a new model system to study speciation and ecological adaptation in extreme environment. We de novo generated a transcriptome sequence of *C. himalaica*, with a total of 49,438 unigenes. Compared to five relatives, we have identified 487 extremely significantly positively selected genes (PSGs) in *C. himalaica* transcriptome. Theses PSGs were enriched in functions related to specific adaptation traits, such as response to radiation, DNA repair, nitrogen metabolism, and stabilization of membrane. These functions are responsible for the adaptation of *C. himalaica* to the high radiation, soil depletion and low temperature environments on QTP. 3) *Thlaspi arvense* is an emerging model wild plant distributed across an elevation range of nearly 4,500 m, for which a complete genome sequence has been released. We generated ~150 Gb of data using a Pool-seq genome resequencing approach in two high and two low altitude populations, and identified more than three million single nucleotide polymorphisms. Using sliding-window ZFst (rms) approach, we identified 135 candidate naturally selected genes associated with altitude variation. We also measured the transcriptome expression level of different altitude *T. arvense* seedlings. Combining these analyses, we identified several genes related to cell wall biogenesis and polysaccharide metabolism that might be responsible for adaptation to distinct environments. In conclusion, differ from previous studies on animals, our results provide novel insights into genetic mechanisms of highland adaptation in plants.

T3-01-04

Quaternary niche dynamics and range dynamics in alpine plants

Da Pan, *Gerald Schneeweiss*
Department of Botany and Biodiversity Research, University of Vienna

The ecological niche is a central issue to understand ecology, evolution and biogeography of species. In this context, two hypotheses have been proposed to explain range dynamics, niche conservatism and niche evolution. Under the first hypothesis, species are tracking their ancestral climatic niche, whereas under the second hypothesis range shifts are achieved by adaptation to new niches. The recurrent (re-)colonization history of alpine plants due to Quaternary climate fluctuations renders the European Alps a suitable system to study range dynamics in the context of climate change. In this study, we employed RAD sequencing and climate modelling to compare intraspecific niche dynamics and phylogeographic patterns of eight alpine species. We address (i) whether (postglacial) range shifts in alpine plants are characterised by niche evolution or by niche conservatism and (ii) whether ecologically similar species have similar biogeographic histories.

T3-01-05

Local adaptation in the context of climate change: Insights from field studies of the subalpine mustard plant, *Boechera stricta*

Jill Anderson
University of Georgia

Abiotic and biotic environments vary across the landscape. Over evolutionary time, divergent selection regimes can favor the evolution of local adaptation, wherein ecotypes have elevated fitness in their home environment and depressed fitness in contrasting environments. Human activities are simultaneously modifying multiple abiotic and biotic agents of selection, likely leading to growing discrepancies between current and optimal phenotypes. We hypothesize that novel selection imposed by climate change shifts fitness landscapes, disrupting local adaptation. As a consequence, local populations could experience diminished growth rates. Climatic variation across mountains affords the opportunity to test hypotheses about the evolution local adaptation to continuous environmental variation. We predict that in future climates, local ecotypes will have reduced fitness in their home sites relative to low elevation families in that same site, whose modern climates are similar to projections for higher elevations. We test this prediction in the mustard *Boechera stricta,* which is native to the Rocky Mountains and inhabits elevations as low as 1,500 m and as high as 3,500 m. In this region, warming winter temperatures reduce snowpack and warming spring temperatures cause the remaining snow to melt early. We simulated these climate change dynamics via snow removal experiments in five common gardens (elevations: 2,530 m, 2,710 m, 2,890 m, 3,133 m, 3,340 m) near the Rocky Mountain Biological Lab including transplants of 28,490 seeds and 17,832 juveniles (N = 270 families from 68 populations). We have found clear evidence for adaptation to local environments over short geographic distances in *Boechera stricta*. We have also found consistent genetic clines in heritable life history traits: high elevation families flower early, at a small size, and for a short duration, likely representing an evolutionary response to short growing seasons. Foliar herbivory increases with provenance elevation in all gardens, revealing limited herbivore resistance of high elevation families. Short cool summers could limit herbivore numbers and reduce selection for defense at high elevations. The rapid and favorable responses of herbivores to climate change could disproportionately diminish the future fitness of high elevation plants. Poor performance of high elevation families at lower sites foreshadows future maladaptation as temperatures warm. Across sites in 2014, fecundity declined by 15% for every 100 m difference between source and garden elevations ($p < 0.0001$) and snow removal depressed fecundity by 10% relative to ambient snowmelt ($p = 0.0026$), as expected. Furthermore, lower elevation families appear to outperform local families even under control conditions, suggesting that climate warming may have already disrupted local adaptation.

T3-01-06

Evolution of wild *Arabidopsis thaliana* populations over the last 150 years

Lua Lopez[1], *Stephanie Marciniak*[1], *Heritiana Randrianatoandro*[1,2], *George Perry*[1], ***Jesse Lasky***[1]
1. Pennsylvania State University
2. University of Antananarivo

While major phenotypic changes, especially in phenology, over the course of anthropogenic climate change have been observed, the role of evolution in this is unclear. We resequenced whole genomes of ~100 herbarium specimens of *Arabidopsis thaliana*, with densest sampling of three populations. We tested for evidence of polygenic adaptation, asking whether changes in allele frequency over time corresponded to alleles associated with particular phenotypes or environments. We also tested whether alleles that changed frequency in association with temporal climate fluctuations' were also associated with certain phenotypes. We included multiple phenotypes likely important to climate adaptation: dormancy, flowering time, $d^{13}C$, and stomatal and trichome density.

T3-02: Patterns and processes of the subtropical/warm-temperate vegetation of East Asia

T3-02-01

Flora of subtropical China: Ancient, rich, and threatened

Jordi López-Pujol

Botanic Institute of Barcelona (IBB-CSIC-ICUB)

China is one of the richest countries in the world in terms of plant diversity. The estimated number of vascular plant species is about 33,000 (30,000 angiosperms, 250 gymnosperms, and 2,600 pteridophytes). In agreement with this, four of the world biodiversity hotspots are located totally or partially within China. The endemic component of the Chinese flora is considerable: ~240 genera and over 17,000 plant species are probably endemic to China, a substantial part of them being relict elements. Most of these relict taxa belong to lineages that were widely distributed along the Northern Hemisphere at the Miocene but that at present are restricted to small and isolated ranges along the mountains of central and southern China; conspicuous examples include *Cathaya argyrophylla*, *Eucommia ulmoides*, *Ginkgo biloba*, *Glyptostrobus pensilis*, *Metasequoia glyptostroboides*, and *Taiwania cryptomerioides*. Although multiple reasons are responsible for this relict, highly endemic, and rich flora, reduced extinction rates during the late Cenozoic global cooling—thanks to the existence of large refugia—is certainly the main driving factor. Since early 2000s, a growing body of research (including studies aimed to identify centres of endemism, studies on phylogeography/genetic diversity, and ecological niche modelling) has revealed the existence of large refuge areas in China, which were mostly concentrated in the current subtropical region. Instead of a single refugium, the mountains of subtropical China followed a pattern of "refugia-within-refugia" (that is, plant species often had multiple refugia, sometimes even within the same mountain range). Unfortunately, subtropical China, like the rest of the country, is facing enormous threats directly related to the global change and China's burgeoning economic and population growth (some of the major concentrations of population of China are within the subtropical area, such as the Sichuan Basin, Wuhan's region and the Yangtze Delta region). At national level, nearly 4,000 plant species have been listed as endangered, and at least 200 taxa having already become extinct since the 1950s. To avoid further losses, a series of conservation measures should be urgently implemented, which may include: (1) increasing the research in any aspect of plant biology, especially for rare and threatened species, (2) expanding the lists of protected plants to all the endangered species of China, (3) improving the efficiency of the already established protected areas, (4) setting up new protected areas in the floristically richest regions (hotspots), (5) ensuring that all endemic and threatened plants are well represented in botanical gardens and germplasm banks, (6) stressing upon enforcement of environmental laws, (7) promoting the sustainable use of wild plant resources, (9) strengthening the environmental awareness and education of Chinese society.

T3-02-02

Patterns and processes of the subtropical vegetation of Southwestern China

Cindy Q. Tang

Institute of Ecology and Geobotany, Yunnan University

Southwestern China (SW China, 21°08′-34°20′N, 97°30′-110°10′E, including Yunnan, Sichuan, Guizhou provinces and Chongqing Municipality) falls into latitudes classified as subtropical, except for the small tropical area of southernmost Yunnan. With that exception, altitudinal climates range from subtropical, to temperate, subalpine, alpine and frigid. I propose to elucidate the patterns and processes of the subtropical vegetation, in the context of historical events (e.g., geology and evolution), and with reference to ecological factors and human activities. Fossil evidence indicates that in the Early Tertiary period, the areas now including eastern Yunnan, Sichuan, Chongqing and Guizhou bore subtropical dry and semi-dry sparse woodlands, while most areas now comprising Yunnan had a warm-rainy winter and hot-dry summer climate supporting tropical and subtropical evergreen broad-leaved forests. In the Late Tertiary period, the vegetation of SW China became a subtropical mixed evergreen and deciduous broad-leaved forest, except in southern Yunnan, home to tropical and subtropical evergreen broad-leaved forests. In the Quaternary, vegetation changes were taking place at about 10,000 years BP (years before the present), and by 7,000 years BP the pattern of the mixed evergreen and deciduous forest had become more complex. Some parts of SW China had not been covered by the ice sheet during the Quaternary. In the region, the present subtropical vegetation, including the semi-humid, the monsoon, the mid-montane moist and the typical humid evergreen broad-leaved forests and several other forest types and minor plant communities, reflect variable environmental conditions and levels of disturbance. The monsoon and the mid-montane moist evergreen broad-leaved forests achieve maximum floristic diversity and structural complexity. Subtropical primary evergreen broad-leaved forests, dominated by species of *Castanopsis*, *Lithocarpus*, *Cyclobalanopsis*, *Machilus*, *Phoebe*, *Schima*, *etc.* are now restricted to less accessible steep slopes, remote mountains, temple grounds and nature reserves. Also within this subtropical zone, there are rare and precious Tertiary relict forests dominated by gymnosperms such as *Ginkgo*, *Metasequoia*, *Taiwania*, *Thuja*, or broad-leaved deciduous *Davidia*, *Tetracentron*, *Liriodendron* in specific habitats, mainly scree slopes, limestone areas and stream banks. In the past few centuries human activities have increasingly changed the subtropical vegetation ecosystem. Secondary stands now prevail over large areas on foothills and lower mountain slopes. In some cases bamboo communities appear after the original evergreen broad-leaved forests have been destroyed. Dramatic expansions of coniferous *Pinus*, *Cunninghamia*, *Cryptomeria* and exotic broad-leaved *Eucalyptus* appear. Complex mosaics of stands of varying ages, composition and structure intermingle with agricultural fields or the mono-species of fast-growing plantations. SW China is a land of plant diversity, vegetation dynamics and succession in time and space.

T3-02-03

Altitudinal zonation of tree species in Philippine mountain forests, 1996-2016

Inocencio Jr Buot[1], *Lita Sopsop*[2], *Geofe Cadiz*[3], *Ivy Amor Lambio*[1]
1. University of the Philippines (UPLB/UPOU)
2. Western Philippines University
3. University of the Philippines Cebu

The paper discusses the pattern of vegetation structure and dynamics along altitudinal gradient in selected Philippine mountain forests from 1996-2016. It presents synthesis, insights and suggest future research directions and conservation measures. Many mountain forests are beset with anthropogenic disturbances affecting the vegetation structure and dynamics and consequently the ecosystem services for the surrounding human communities. Field investigations for more than 20 years revealed anomalous zonation patterns of vegetation, quite different from the normal trend in the tropics. In these mountain forests, lower elevation dominants encroach in upper elevations, replacing original dominants destroyed by farming. *Pinus kesiya* of northern Philippines (Mount Pulag and Mount Akiki), replaces the *Lithocarpus* and other oaks. The *Diplodiscus* of Mount Makiling in southern Luzon colonized the nich of *Dipterocarpus* destroyed during the extensive logging in the 1940s. The *Astronia* of the volcanic Mount Mayon in southern Luzon and the *Artocarpus* of Mount Tabunan on Cebu island, central Philippines and that of the Aborlan Guba System in Palawan island are encroaching higher altitudes. Multivariate regression analysis shows that elevation, temperature, pH, moisture, water holding capacity and human disturbance, significantly influenced the distribution of tree species. In northern Luzon island, *Pinus* becomes a topo-edaphic climax displacing the oaks. The case of *Diplodiscus*, *Artocarpus* and *Astronia* could be a temporary seral stage of a long succession process initiated by logging and swidden agriculture, drastically changing the edaphic and microclimatic conditions, favorable for the lower dominants. There should be a strong will in protecting core forest zones and in pursuing rehabilitation efforts such as establishment of biodiversity corridors. Additionally, however, there should be sustained community biodiversity education for kids, youth and adults.

T3-02-04

Climate change impact on the potential habitats of dominant evergreen broad-leaved tree species in the Taiwan-Japan archipelago

Katsuhiro Nakao[1], *Motoki Higa*[2], *Ikutaro Tsuyama*[1], *Cheng-Tao Lin*[6], *Chyi-Rong Chiou*[3], *Tzu-Ying Chen*[4], ***Tetsuya Matsui***[1], *Nobuyuki Tanaka*[5]
1. Forestry and Forest Products Research Institute
2. Kochi University
3. National Taiwan University
4. National Ilan University
5. Tokyo Agriculture University
6. National Chiayi University

Ecosystem vulnerability to climate change remains elusive in the species-rich Taiwan-Japan archipelago. We predicted potential habitats (PHs) of ten dominant evergreen broad-leaved tree species by using the current and twenty potential climate change scenarios using Generalised Additive Models. The presence/absence records of each species, extracted from vegetation database, were used as response variables. Four climatic and one spatial variable were used as explanatory variables. The results showed that the interaction terms of spatial variable, indicating historical range shifts or species interactions, restricted the distribution of all the target species as much as that by the each climatic variable. The PHs of all the target species were predicted to consistently increase, and in particular, to expand northward and upward to the cool temperate zone. However, the PHs were predicted to decrease within the range of 23.6–38.1% in the Ryukyu Islands for *Castanopsis sieboldii* and *Elaeocarpus japonica*, respectively, and within the range of 32.4–42.3% in Taiwan for *Camellia japonica* and *Distylium racemosum*, respectively. These findings suggest that the four species will be vulnerable at the southern range limits; however, the remaining six species will potentially increase within the PH areas in the future at all regions.

T3-02-05

Vegetation patterns of abandoned farms in Penghu southern islands

Cheng-Tao Lin
Department of Biological Resources, National Chiayi University

Abandoned farms have a long history linking to the human cultivation and activities. The vegetation patterns on these sites are influenced by cultivation legacy such as biomass alternation, fertilization, *etc.* To understand the vegetation patterns influenced by abandoned farms and stocking goat, a succession scheme are applied to elucidate the ecological processess. The target area is the South Penghu National Park among 119°30'E-119°41'E and 23°14'N-23°16'N, which is located at the southern regions of Penghu archipelagos. These islands include four largest islands: Dong-Ji, Si-Ji, Dong-Yuping and Si-Yuping. We established 81 vegetation plots of standard size with 25 m^2 and collect environmental factors, such as soil properties, wind speed, wind directions and topographical factors for exploratory analysis. Succession index was used to construct the succession scheme to explain the process and seral stage among different vegetation types. The results showed that the vegetation was classified into seven types: I. *Tetragonia tetragonoides-Leucaena leucocephala* type, II. *Epaltes australis-Fimbristylis cymosa* type, III. *Portulaca pilosa-Evolvulus alsinoides* type, IV. *Imperata cylindrica* var. *major-Vitex rotundifolia* type, V. *Wedelia prostrata-Cassia sophora* var. *penghuana* type, VI. *Bothriochloa glabra-Sorghum nitidum* fo. *aristatum* type, VII. *Sporobolus virginicus-Peucedanum japonicum* type. The succession degree suggests that the vegetation type III in plateau would be the early stage of succession seres, and it would be transitioned into vegetation types IV, V and VI. The succession seres in coastal region are early seral stages (types II and VII). If the continuous expansion of *Leucaena lecocephala* proceeds, these two types would be transitioned into type I. For the broad view of the southern Penghu islands, major vegetation belongs to the early seral stage and disturbed by wild goats continuously. We also found that the species diversity would be affected by the wild goats, but it needs to be confirmed by further experiment.

T3-02-06

Climatic and human impact on Holocene vegetation change inferred from a pollen record in the northern edge of Wuyi Mountains, eastern subtropical China

Ting Ma[1,2], *Zhuo Zheng*[1], *Pavel Tarasov*[2]
1. Sun Yat-Sen University
2. Free University of Berlin

The study core was taken from a mountain peat bog at 900 m elevations in the northern edge of Wuyi Mountains, eastern subtropical China. Pollen and charcoal analyses were conducted to examine the vegetation changes caused by both climatic variation and anthropogenic activities over the past 8,200 years. Our results show a humid climatic phase between c. 8,200 and 5,600 cal. yr BP, marked by high proportion of pollen from evergreen broad-leaved forests (e.g., *Quercus* and *Castanopsis*) and hygrophilous alder (*Alnus trabeculosa*), which is consistent with the Holocene moisture maximum in eastern China. After 5,600 cal. yr BP, the spread of the wooded swamp taxon, *Glyptostrobus*, suggests the formation of local swamp caused by gradual drying. The drying trend generally corresponds with the speleothem isotope records from this region, revealing a weakening East Asian summer monsoon due to a decrease in Northern Hemisphere summer insolation. Peaks in the abundance and concentration of *Glyptostrobus* pollen at c. 4,600–4,400 cal. yr BP and c. 3,300–3,000 cal. yr BP suggest two periods of swamp expansions, which coincide with the drought intervals revealed by the speleothem records. The pollen and charcoal record demonstrates that human-induced land cover change was negligible before 3,600 cal. yr BP. This may be resulted from the geographical factor that huge mountain ranges inhibit cultural expansion of agriculture civilization from lower Yangtze plain and Hangzhou Bay toward mountainous southern China. We consider the first signal of intensive human activity and landscape clearing to be the noticeable increase in charcoal particles at around 3,600 cal. yr BP, followed by dramatic decrease in arboreal pollen and sharp increase in Poaceae (including rice pollen) and pioneer fern (*Dicranopteris*) and trees (such as *Pinus*), likely reflecting a transition to rice-paddy agriculture in the study area (Grant No. 41230101).

T3-03: Nectaries and nectars: from molecules to ecosystems

T3-03-01

Floral nectar of the obligate outcrossing *Canavalia gladiata* (Jacq.) DC. (Fabaceae) contains only one predominant protein, a class III acidic chitinase

Hong-Xia Ma[1*], *Richard I Milne*[2,3], *Hong-Xia Zhou*[1], *Jiang-Yu Fang*[1], ***Hong-Guang Zha***[1]
1. College of Life and Environment Sciences, Huangshan University, Anhui, China
2. Institute of Molecular Plant Sciences, University of Edinburgh, Edinburgh, UK
3. Royal Botanic Garden, Edinburgh, UK

Floral nectar can affect the fitness of insect-pollinated plants, through both attraction and manipulation of pollinators. Self-incompatible insect-pollinated plants receive more insect visits than their self-compatible relatives, and the nectar of such species might face increased risk of infestation by pathogens carried by pollinators than self-compatible plants. Proteins in nectar (nectarins) play an important role in protecting the nectar, but little is known regarding nectarins in self-incompatible species. The nectarins from a self-incompatible and insect-pollinated leguminous crop, *Canavalia gladiata*, were separated using two-dimensional electrophoresis and analysed using mass spectrometry. The predominant nectarin gene was cloned and the gene expression pattern investigated using quantitative real-time PCR. Chitinolytic activity in the nectar was tested with different substrates. The *C. gladiata* nectar proteome only has one predominant nectarin, an acidic class III chitinase (CaChi3). The full-length CaChi3 gene was cloned, coding for a protein of 298 amino acids with a predicted signal peptide. CaChi3 is very similar to members of the class III chitinase family, whose evolution is dominated by purifying selection. CaChi3 was expressed in both nectary and leaves. CaChi3 has thermostable chitinolytic activity according to glycol-chitin zymography or a fluorogenic substratem but has no lysozyme activity. Chitinase might be a critical protein component in nectar. The extremely simple nectar proteome in *C. gladiata* disproves the hypothesis that self-incompatible species always have more complex nectar proteomes. Accessibility of nectar might be a significant determinant of the evolutionary pressure to develop nectar defence mechanisms.

T3-03-02

Nectaries and nectar secretion in Cleomaceae

Nuria Koteyeva, *Elena Voznesenskaya, Alexandra Ivanova, Alexey Shavarda*
Komarov Botanical Institute of Russian Academy of Sciences

The complex study of floral nectary development and functioning in 15 *Cleome* s. l. species and two closely related (outgroup) species was carried out. Three groups were distinguished according to the secretion circadian dynamics, duration of secretion and yield of nectar. The diversity of nectary morphology and anatomy alongside with ultrastructure of secretory cells was demonstrated. The genus was characterized by high diversity of the nectar composition which is not related to the nectary morphology and ultrastructure. Data on nectary vascular system and stomata formation, as well as starch storage dynamics confirm the passive role of phloem and stomata in nectar secretion. It was shown that secretion of sugars from secretory cells into the apoplasm is an active ATPase driven process regardless of the nectar composition, morphology of the nectary and ultrastructure of secretory cells. The data obtained give an evident support for the universal hypothesis of nectar secretion as a pressure-driven continuous mass flow. Analysis of nectary-related evolutionary traits in Cleomaceae lineages shows that the morphology of nectaries is one of the important characters that could be used for systematics of the genus *Cleome*. However neither the nectar composition nor the ultrastructure of secretory tissue can be used as systematic features. Results indicate the multiple transitions between different types of nectaries in Cleomaceae evolution.

T3-03-03

The diversity of substances produced by floral nectaries in Anacardiaceae and their functional and systematic significance

Elisabeth Dantas Tölke[1], *Elimar Lima*[1], *Leonardo Galetto*[2], *Diego Demarco*[3], *Sandra Carmello-Guerreiro*[1]
1. Universidade Estadual de Campinas
2. Universidad Nacional de Córdoba
3. Universidade de São Paulo

Anacardiaceae includes approximately 800 species grouped in two subfamilies: Anacardioideae and Spondioideae. Most species present a floral nectariferous disk. However, its structure, nectar sugar composition and chemical nature of other secreted substances are poorly understood. In a previous study, the production of lipids, phenolic compounds and nectar by the nectariferous floral disk of *Tapirira guianensis* Aubl (Spondioideae) were detected. The ultrastructural study of *T. guianensis* showed that the cellular structure of the disk is more directed towards for the production of oil than sugars. At the same time, nectar sugar composition is also poorly explored in the family, showing some variations between sucrose and hexose dominant types. Based on this background, we examined the anatomy and histochemistry of the floral nectary in 13 different species belonged to both subfamilies of Anacardiaceae, but also the proportion of secreted sugars. Results include ultrastructural data to gain insights on the production and route of secretion of the different substances. The morpho-anatomical study was carried out through transmission, scanning electron and light microscopy. Histochemical tests were performed to detect different compounds. Nectar sugars were identified via spectrophotometry. The floral nectary in Anacardiaceae is a nectariferous disk or it can be trichomatous. The nectariferous disk comprises a single-layered epidermis, a nectariferous and subnectariferous parenchyma, with or without phloem supply. Our results showed evidence of a mixed secretion in all the studied species, composed by sugars, phenols, hydrophilic substances and lipids. The production of lipids by the floral nectary seems to be a synapomorphy of Anacardiaceae, evidenced mainly by the ultrastructural data. This set of species showed a great variability in the nectar sugar composition, with sucrose-dominant, sucrose-rich, hexose-rich and hexose-dominant species. This variability on nectar sugar composition, even among species belonging to the same genera, suggests the absence of a clear phylogenetic signal. The nectariferous cells showed dense cytoplasm, large nucleus, mitochondria, plastids with osmiophilic inclusion, oil drops scattered in the cytosol, dictyosome, and rough endoplasmic reticulum segments. Some species also presented abundant amyloplasts in pre-secretory phase, indicating that they function as a resource in the formation of nectar. Subcellular evidence suggests that the abundance of mitochondria contribute to eccrine mechanism of secretion, although dictyosomal vesicles may contribute to a granulocrine process. A symplastic pathway, via plasmodesmata, is the most likely to be occurring. The manner of nectar exudation varies according to the species. Some species that present nectarostomata, and it would be the most likely route; however, the cuticle may also play a role in this process. In the other species nectar is exuded exclusively thought the cuticle, without disruption. The lipids are exuded through the permeability of the cuticle and phenolics seem not to be secreted and remain stored within the vacuoles. We conclude the floral nectary of Anacardiaceae is responsible for the production of a mixed secretion, with lipids and sugars always present. This may be related with the attraction of pollinators, since lipids are substances that provide more energy than nectar.

T3-03-04

Comparative functional genomics of nectaries and nectars in the dicots

Clay Carter
University of Minnesota

Plants attract mutualistic animals by offering a reward of nectar. Specifically, floral nectar (FN) is produced to attract pollinators, whereas extrafloral nectar (EFN) mediates indirect defenses through the attraction of mutualist predatory insects to limit herbivory. Nearly 90% of all plant species, including 75% of domesticated crops, benefit from animal-mediated pollination, which is largely facilitated by FN. Moreover, EFN represents one of the few defense mechanisms for which stable effects on plant health and fitness have been demonstrated in multiple systems, and thus plays a crucial role in the resistance phenotype of plants producing it. In spite of its central role in plant-animal interactions, the molecular events involved in the development of both floral and extrafloral nectaries (the glands that produce nectar), as well as the synthesis and secretion of the nectar itself, have been poorly understood until recently. To date, a holistic and coordinated characterization of nectar secretion from a genomic and molecular perspective has been lacking. Toward this end, we have evaluated the transcriptomes of floral and extrafloral nectaries throughout development across twelve dicotyledonous species and identified core sets of genes involved in the synthesis and secretion of nectar across species, as well as its regulation. Similarly, metabolite profiling coupled with transcriptomic data has identified specific loci potentially responsible for different nectar characteristics that influence mutualist visitation.

T3-03-05

Nectar secretion dynamics in different flowering plants and its effect on honey bee foraging and proboscis extension responses

Hussain Ali
1. Agriculture Research Institute Tarnab, Peshawar, Pakistan
2. Visiting Research Fellow at School of Life Sciences, Arizona State University Tempe, USA

Plants produce nectar and pollen, food of honey bees. The experiment was carried out to check nectar secretion dynamics and its role in attracting different pollinators. Nectar secretion dynamics were tested at different times of the day during flowering season. The honey bee foraging was also recorded in relation to nectar secretion. The nectar was collected from flowers and different races/subspecies of honey bees were tested against these nectars for Proboscis extension responses (PER). A significant positive correlation was recorded between nectar secretion and temperature in *Acacia* and *Zizuphus* species. The nectar secretion was peaked at noon time, whereas in morning and evening the secretion was significantly less. In some flowering plants the nectar secretion remained similar with temperature. A was positive correlation between nectar secretion and honey bee foraging was also observed. The nectar was then tested for Proboscis extension responses (PER) in three races/subspecies of *Apis mellifera.* The PER was tested at two daytimes, morning and afternoon. A significant difference was recorded in PER among the three races during morning and evening. PER is used to examine the quality and attractiveness of the bee forages, and to understand how the attractiveness of a specific bee forage varied according to bee genotype and weather.

T3-03-06

Genome editing tools meet plant chemical ecology

Sang-Gyu Kim[1], *Sang-Tae Kim*[1], *Je Wook Woo*[1], *Hyeran Kim*[1], *Hyomin Ahn*[1], *Ian Baldwin*[2], *Jin-Soo Kim*[1]

1. Center for Genome Engineering, Institute for Basic Science

2. Max Planck Institute for Chemical Ecology

Genome editing tools have been developed to manipulate genomic DNA in living organisms; in particular, the recent genome editing tool, CRISPR/Cas9, allows us to induce targeted mutagenesis in several plant species. CRISPR/Cas9 also allows the genetic bases of ecological traits of interest to be explored in plants and the qualities of crop plants to be improved. In this talk, I will talk about how we can apply CRISPR/Cas9 system for plant genome editing. Especially, I will introduce a DNA-free plant genome-editing tool with Cas9 protein mixed with a single guide RNA (ribonucleoprotein, RNP) and a simple vector system for *Agrobacterium*-mediated plant transformation. We are now generating targeted knock plants to understand nectary development and nectar chemistry.

T3-04: Rising atmospheric CO_2 and higher plants: physiology, genetics and biological diversity of plant responses

T3-04-01

The impact of elevated [CO_2] on the hydraulic traits and growth of seedlings of four evergreen and four deciduous trees in SW China

Pei-Li Fu[1], *Ya Zhang*[2], *Ze-Xin Fan*[1], *Jin-Hua Qi*[1]

1. Xishuangbanna Tropical Botanical Garden, Chinese Academy of Sciences

2. Institute for Systematic Botany and Ecology, Ulm University

Elevated [CO_2] can reduce plants stomatal conductance, and could further change the plants hydraulic traits. However, there was no agreement on the effect of elevated [CO_2] on plants hydraulics, some studies found elevated decreased plant hydraulic conductance, while other studies found the opposite results. We used four close top growth chambers, two chambers were ambient [CO_2] and two chambers were elevated [CO_2] (700 ppm), to investigate the impact of elevated [CO_2] on the leaf gas exchange, leaf and branch hydraulic traits and biomass of seedlings of four evergreen and four deciduous trees in a subtropical forest in SW China. We found that elevated [CO_2] increased leaf area-based photosynthesis rate for all the eight species, and all the tree species excerpt *Populus davidiana* decreased stomata conductance. However, there was no significant change of leaf hydraulic conductance for all the tree species excerpt for *Lithorcarpus hancei.* The branch hydraulic conductance increased significantly for *Populous davidiana,* while there was no significant change in the other tree species. The biomass increased for all plant tree species excerpt *Eriobotrya bengalensis*. The increase of biomass in the three evergreen tree species ranged from 38% to 128%, however, the increase of biomass in deciduous trees ranged from 21% to 51%. Our results indicated that elevated [CO_2] reduced plant stomatal conductance, but had little impact on either leaf or branch xylem water transport capacity. Evergreen trees had great variations than those of deciduous trees in their response to elevated [CO_2].

T3-04-02

Phosphorus limits photosynthesis in tropical trees on four continents

David S. Ellsworth[1], *Martin De Kauwe*[2,3], *Kristine Y. Crous*[1], *Ian J. Wright*[2]

1. Hawkesbury Institute for the Environment, Western Sydney University

2. Dept. of Biological Sciences, Macquarie University

3. ARC Centre of Excellence for Climate Extremes, The University of New South Wales

Over a third of the world's soils have low plant-available phosphorus. Phosphorus might strongly constrain forest productivity and growth on these soils, with tropical forest growth processes particularly limited by phosphorus. If this is the case, then global modellers will need relationships between plant metabolism and leaf phosphorus to reproduce productivity in their models, but relatively few such relationships have been examined across a broad geographical swath and a wide number of plant families. With key data contributors, we compiled a large global dataset of leaf nutrient concentrations and photosynthetic function, comprising over 500 species from six continents from over 250 plant families to examine whether there were broad relationships between leaf photosynthesis and leaf phosphorus and explore the physiological phenomena underlying these relationships. We used the biochemical photosynthesis model of Farquhar *et al.* (1980) and subsequent revisions to parameterise relationships between photosynthetic function and leaf phosphorus, considering soils and site phosphorus status, leaf nitrogen and leaf mass per area as variables that could modulate this relationship. These factors were considered in direct analysis as well as a multiple regression framework designed to consider a principal driving variable with other explanatory variables held constant. Our dataset contains plants with some of the lowest leaf phosphorus concentrations measured (<1 mg/g), as well as plants from habitats where phosphorus was not considered limiting or where leaf N:P ratio was low. We found relatively strong overall relationships between both leaf photosynthesis and leaf phosphorus, and leaf RuBP regeneration capacity and leaf phosphorus across the four continents in the tropics. An Australian and Amazonian site showed some of the lowest leaf P concentrations (0.6 mg/g) and highest N:P ratios, yet maintained moderate light-saturated net photosynthesis (11 ± 3 μmol CO_2 m^{-2} s^{-1}) and RuBP regeneration capacity. The overall relationships we observed suggest that global plant productivity models should consider soil or plant phosphorus as a potential constraint to future climate change responses, and employ triose-phosphate limitations in simulating photosynthesis across the globe.

T3-04-03

Transgenerational exposure to elevated CO_2 attenuates the impact of drought on wheat

Xiangnan Li[1], *Yafei Li*[1,2], ***Fulai Liu***[1,2]
1. Department of Plant and Environmental Sciences, University of Copenhagen
2. Sino-Danish Center for Education and Research (SDC)

Climate change scenarios predict more frequent drought spells along with an elevation in atmospheric CO_2 concentration ($e[CO_2]$). Despite the effects of drought or a single generation exposure to $e[CO_2]$ on the physiological and yield responses of winter wheat (*Triticum aestivum* L.) have been intensively studied, the transgenerational effect of $e[CO_2]$ on plant water use relations and nitrogen nutrition in response to drought stress remains largely unknown. Seeds harvested from plants after two generations (2014-2015) continuously grown in ambient CO_2 ($a[CO_2]$, 400 mmol l^{-1}) and $e[CO_2]$ (800 mmol l^{-1}) were sown in 4 L pots, and the plants were grown separately in greenhouse cells with either $a[CO_2]$ or $e[CO_2]$. At stem elongation stage, in each of the cells half of the plants were subjected to progressive drought stress until all the plant available soil water was depleted, and the other half were well-watered and served as controls. The results showed that transgenerational exposure of the winter wheat plants to $e[CO_2]$ could attenuate the negative impact of drought stress on dry biomass accumulation, nitrogen uptake, and water use efficiency. The modulations of multi-generational exposure to $e[CO_2]$ on stomatal control over plant water use and nitrogen uptake could have contributed to the improved water use efficiency and nitrogen nutrition. These findings provide new insights into the response of wheat plants to a future drier and CO_2-enriched environment.

T3-04-04

Simulation of carbon dynamics of *Tectona grandis* forest in Western Ghats of Kerala, India using Century model

Manjunatha Munishamappa
Environmental Management and Policy Research Institute

Century model can be used to simulate carbon dynamics and the effects of other components in ecosystems. Many studies have measured the carbon stocks of Teak (*Tectona grandis)* but those studies only captured a static view without taking into account the role of other components of the ecosystem. This research has simulated the carbon dynamics of *Tectona grandis* in Western Ghats of Kerala, India using Century model. The research stages included model parameterization, validation, and analysis of the pattern of carbon accumulation in a *Tectona grandis* stand grown in a reforestation setting. The parameterization was done by adjusting the model parameters to the characteristics of *Tectona grandis* and the environment condition of the study area. The validation was conducted by comparing the simulation results to analysed statistical data from the field measurements of carbon stocks in *Tectona grandis* stands 0-5, 06-10, 11-20, 21-30 and above 30 years old. The validation process demonstrated that the output of simulation approaches analysed statistical data. Pattern of the simulated dynamics in 50 years shows that the carbon accumulated in the forest system, *Tectona grandis* biomass, and necromass increase as the age of stand increases. However, the accumulation of soil carbon initially decreases until it reaches a relatively constant value.

T3-04-05

Physiological integration ameliorates effects of UV-B radiation in clonal herb *Duchesnea indica*

Qian Li[1,2], *Xiao Liu*[2], *Yong-Fu Chai*[2], *Ming Yue*[2]
1. Xi'an Botanical Garden of Shanxi Province
2. Key Laboratory of Resource Biology and Biotechnology in Western China, Ministry of Education, Northwest University

Ultraviolet-B (UV-B) radiation can result in deleterious effects on many plant growth processes. In clonal plants, resources and hormones are often shared between connected ramets to deal with various environment. However, little comprehensive study is known about whether and how clonal integration influences the ability of clonal plants to respond to UV-B radiation. In this paper, the change in physiological integration intensity of water transportation rate and photosynthetic ability and survival and growth of *Duchesnea indica* under homogeneous and heterogeneous ultraviolet-B radiation (280–315 nm) were evaluated. Ramet pairs of *Duchesnea indica* were subjected to four treatments: (1) connected ramet pairs under a homogeneous condition, i.e. pairs of ramets were kept under no additional UV-B conditions and the stolon connection between the ramets remained intact; (2) disconnected ramet pairs under the same UV-B condition as in treatment (1); (3) connected ramet pairs under a heterogeneous condition, i.e. one of the ramets was exposed under additional UV-B radiation and the stolon connection between the ramets remained intact; (4) disconnected ramet pairs under the same UV-B condition as in treatment (3). When ramets were connected and exposed to heterogeneous UV-B radiation, the velocity of water transportation from the UV-B treated ramet to its connected sister ramet was markedly lower. Moreover, stolon connection resulted in an increase of biomass and ramet number in the UV-B stressed part of the clonal fragments at the expense of that in UV-B unstressed part, which meant that full compensatory growth effect was existed in *Duchesnea indica* under heterogeneous UV-B radiation. In addition, chlorophyll contents and UV-B absorbing compounds were shared between connected ramets under heterogeneous UV-B radiation to benefit for the whole clonal fragment. This inference extends the concept of physiological integration in clonal plants to include defensive responses. Net photosynthesis rate (P_N), stomatal conductance (g_s), intercellular CO_2 concentration (C_i) were lower and transpiration rate (E) were higher in UV-B stressed part, and C_i in UV-B unstressed part were reversed. Moreover, in UV-B unstressed part, maximum photochemical efficiency of PSII (F_v/F_m) and photochemical quenching coefficient (q_P) had no significant difference, the electron transport rate (ETR) were lower and non-photochemical quenching (NPQ) were higher, and in UV-B stressed part, additive UV-B radiation resulted in a notable decrease of F_v/F_m, ETR and q_P and an increase of NPQ, stolon connection made F_v/F_m, ETR, q_P and NPQ changed more in the connected group. Clonal integration significantly reduced the stressful impact imposed by enhanced UV-B radiation with a cost, and the whole clone fragment will benefit, from survival and growth, chlorophyll and UV-B-absorbing compounds concentrations, photosynthetic efficiency in heterogeneous UV-B micro-habitat.

T3-04-06

Dendrochronology of *Fagus grandifolia* subsp. *mexicana* in the state of Hidalgo, Mexico

Adriana Beatriz Ortiz Quijano[1], ***Árturo Sánchez González***[1], *Lauro López Mata*[2], *Jose Villanueva Díaz*[3], *Consuelo Cuevas Cardona*[1]
1. Centro de Investigaciones Biológicas, Universidad Autónoma del Estado de Hidalgo
2. Colegio de Postgraduados Campus Montecillo
3. Centro Nacional de Investigación Disciplinaria en Relación Agua, Suelo, Planta, Atmosfera, Instituto Nacional de Investigaciones Forestales Agrícolas y Pecuarias

The analysis of tree rings growth in environmental history is a valuable source of information for the reconstruction of past events. In this research we used the information from historical archives, interviews and chronology of the growth rings of the species *Fagus grandifolia* subsp. *mexicana* to identify the main causes of disturbance. Sampling of the nuclei of growth was carried out in three localities where the species mentioned dominates the canopy, forming the beech forests. A total of 3,355 tree rings were measured and the standardization of the time series was elaborated. The annual averages of ring width were similar between localities, ranging from 0.98 to 1.08. The analysis of the percentage of change of growth allowed to recognize a multiple pattern of suppressions and releases, associated with local events and the natural development of the species. Although the correlation values between chronologies were not statistically significant, similar trends of variation were observed in the localities El Gosco and La Mojonera. The linear correlation values between the ring width index and the climatic variables (annual total precipitation, minimum, maximum and average temperature) of each locality were not statistically significant. In the Hidalgo state, there have been no major natural or human disturbances of at least 150 years, with the exception of El Gosco, where anthropogenic disturbance has increased dramatically in the last decade.

T3-05: Niche evolution and biome assembly through time (two sessions)

T3-05-01

Despite phylogenetic effects, C_3-C_4 lineages bridge the ecological gap to C_4 photosynthesis

Marjorie Lundgren, *Pascal-Antoine Christin*
University of Sheffield

C_4 photosynthesis is a physiological innovation involving several anatomical and biochemical components that emerged recurrently in flowering plants. This complex trait evolved via a series of physiological intermediates, broadly termed 'C_3-C_4', which have been widely studied to understand C_4 origins. While this research program focused on biochemistry, physiology, and anatomy, the ecology of these intermediates remains largely unexplored. Here, we use global occurrence data and local habitat descriptions to characterize the niche of multiple C_3-C_4 lineages, as well as their close C_3 and C_4 relatives. While C_3-C_4 taxa tend to occur in warm climates, their abiotic niches are spread along other dimensions, making it impossible to define a universal C_3-C_4 niche. Phylogeny-based comparisons suggest that, despite shifts associated with photosynthetic types, the precipitation component of the C_3-C_4 niche is particularly lineage specific, being highly correlated with that of closely related C_3 and C_4 taxa. Our large-scale analyses suggest that C_3-C_4 lineages converged toward warm habitats, which may have facilitated the transition to C_4 photosynthesis, effectively bridging the ecological gap between C_3 and C_4 plants. The intermediates retained some precipitation aspects of their C_3 ancestor's habitat and likely transmitted them to their C_4 descendants, contributing to the diversity among C_4 lineages seen today.

T3-05-02

Australia: A dry continent with small terrestrial succulents, but not big ones

Joseph Holtum[1], *Klaus Winter*[2]
1. James Cook University
2. Smithsonian Tropical Research Institute

Why are there no big terrestrial succulent plants native to Australia, a continent within which ca. 75% of the land area is arid or semi-arid? Why, in such a water-limited landscape, are only seven terrestrial Australian species known to exhibit crassulacean acid metabolism (CAM), a water-use efficient mode of photosynthesis that is globally associated with succulent plants? On the basis of the frequency of CAM in the planetary angiosperm flora one might expect ca. 1,300 CAM species in Australia. It has been suggested that the unpredictable rainfall in arid Australia cannot support massive water-storing succulents. Yet large succulents can do well in Australia. In the 1920s exotic *Opuntia* occupied 24 million hectares. A recent resurgence has seen opuntias infest ca. 2 million hectares. Although depauperate in large succulents, Australia does support a spatially-widespread terrestrial flora of endemic small leaf-succulents. Low level of CAM is known in species of *Anacampseros* (Anacampserotaceae), *Calandrinia* (Montiaceae), *Carpobrotus* (Aizoaceae), *Crassula* (Crassulaceae), *Disphyma* (Aizoaceae) and *Portulaca* (Portulacaceae). Substantial CAM is known only from *Cynanchum* (Apocynaceae), Australia's largest terrestrial succulent. A surprising observation is the preponderance in the small terrestrial succulents of drought-induced facultative (i.e. optional) CAM, a form of CAM considered uncommon, and previously known from ca. 20 species worldwide. Facultative CAM in Australian succulents is present in species in which the principal photosynthetic phenotype in well-watered plants is C_3 but is also present in Portulaca, in which the principal photosynthetic phenotype in well-watered plants is C_4. It is not yet clear whether the high incidence of facultative CAM seen in the succulents is solely an Australian phenomenon or a more widespread one that has been overlooked so far. As the Australian genera with facultative CAM have close relatives in the New World and Africa, more thorough surveys of small succulents in those regions may be instructive. It is not difficult to imagine how small annual succulents with a facultative water-use efficient mode of photosynthesis might be selected in warm, low-nutrient landscapes subject to low and unpredictable rainfall. However, the absence of large stem succulents in remains enigmatic. Environments that support large succulents may be uncommon in contemporary Australia, but they exist. There must be more to the absence of large stem-succulents than just the lack of dry environments with a predictable rainy season. Perhaps taxa with the evolutionary capacity to develop stem-succulence simply did not make it to Australia, or perhaps there was insufficient niche space in existing vegetation assemblages with their associated fire regimes. Deeper exploration of the Australian extant succulents will surely be informative.

T3-05-03

General features of the evolutionary assembly of biomes

Erika Edwards[2], ***Michael J. Donoghue***[1]
1. Yale University
2. Brown University

Major changes in climate over geological time have been responsible for the emergence and spread of modern biomes. Prolonged directional selection has resulted in the massive convergence of climate-related traits, perhaps through gradual in situ evolutionary change. The extent and the phylogenetic distribution of this convergence is also a function of the relative accessibility of particular adaptive "solutions" in different lineages. Consequently, we observe distinctly clumped patterns of convergence, with multiple shifts occurring within particular lineages. An additional factor in explaining the shifting of lineages among biomes is the extent (in both space and time) to which biome boundaries have existed. Evolutionary shifts between adjoining biomes are far more common than they are between disjunct biomes. Our research has focused on direct shifts between warm tropical forests and cold temperate forests. We argue that many key evolutionary changes that have accompanied this shift have been driven by changes in the rhythm of growth with the occupation of cold climates. This, we think, has driven the evolution of both the deciduous leaf habit and aspects of leaf form, including broadly ovate leaf shapes and toothed or lobed leaf margins. Although the specific drivers of evolutionary change will certainly differ with the occupation of different biomes, we suspect that changes in the rhythm of growth underlie many such changes, irrespective of biome. An evaluation of this proposition (and other such generalities about biome assembly) will require multiple integrative studies of well-sampled clades whose members straddle biome boundaries.

T3-05-04

Biogeography of Berberidaceae and biome shifts

Chih-Chieh Yu[1], *Erika Edwards*[2], ***Kuo-Fang Chung***[1,3]
1. School of Forestry and Resource Conservation, National Taiwan University
2. Ecology and Evolutionary Biology, Brown University
3. Biodiversity Research Center, Academia Sinica

Berberidaceae is one of the most characteristic components of northern temperate environments, comprising about 17 genera within which mostly are herbaceous perennials. However, at the species level, more than two-thirds of species diversity are found in *Berberis* L., a shrubby group almost entirely restricted to montane to alpine habitats globally. Why is *Berberis* the most successful group among its perennial sisters and what are the evolutionary processes in shaping diversification rates and current distribution of the genus? Recent phylogenetic and biogeographic studies have reconstructed the evolutionary trajectory for the genus, unraveling an uncommon arid-adapted origin in North America with a key niche shift into global high-mountain systems. While molecular dating and diversification models indicate an increase of diversification rate from Oligocene onward, the successful transition into a cold temperate biome of *Berberis* may be triggered and influenced by the paleoclimate deterioration during Neogene. Here we further identified two more recent mountain radiations (Andes vs Himalayas-southwestern mountains of China) by using BAMM and BiSSE analyses. Preliminary investigation on leaf area (LA) from herbarium collections among *Berberis* and sister genera also suggested the habitat and LA had a positive interactive effect on the diversification. Together, *Berberis* may entail a unique case of biome shifts in the assembly of global temperate biota, and its high diversity is likely a result of adaptive radiation facilitated by leafing traits evolution

T3-05-05

Strong regional effects in Northern Hemisphere tree physiology and phenology

Susanne S. Renner, *Constantin Zohner*
University of Munich (LMU)

Donoghue and Edwards (AREES 2014) stress the expected, but under-researched link between *biomes*, which are defined by climate and eco-physiology, and *biogeographic realms*, which reflect isolated centers of diversification for multiple groups of organisms. But into which conceptual framework do we fit the following experimental finding: trees and shrubs native to Eastern North America (ENA) on average have a three-week shorter vegetative season than woody species in East Asia and Europe because the leaf-out later and break down their chlorophyll 11 ± 4 days earlier (Zohner and Renner, in review)? Modern average temperatures cannot explain this difference because mean annual temperature and temperature seasonality experienced in species' natives ranges today have small effect sizes on experimentally investigated leaf-out and leaf senescence in ca. 400 woody species from throughout the Northern hemisphere. Nor can biogeography explain these regional effects because closely related species in different biogeographic realms exhibit the local strategy. Instead, we propose that the physiology and phenology of the modern ENA woody flora is the product of multiple pre-Pleistocene climate disruptions that selected for conservative growth strategies. These results highlight climate as a major driver of plant traits, such as the timing chlorophyll breakdown, and also underscore that physiological traits will remain 'invisible' unless we combine indirect comparative approaches with experiments.

T3-05-06

Geographic divergence and functional evolution of Magnoliaceae

Hui Liu[1], *Marjorie Lundgren*[2], *Robert Freckleton*[2], *Louis Santiago*[3], *Qiuyuan Xu*[1], *Qing Ye*[1]
1. South China Botanical Garden
2. University of Sheffield
3. University of California

The early diverged Magnoliaceae shows a historical temperate-tropical distribution among lineages, indicating a divergent evolution. Analyse the relationship between phylogeny, environmental constraints and associated trait variances could reveal how species distributions were shaped. Phylogenetic comparative models can be used to disentangle the relative roles of spatial and phylogenetic distances in explaining trait variances, reveal phylogenetic signals in ecophysiological traits, and test whether these traits are involved in correlated evolution. We analyzed variances in 15 plant traits and 12 environmental variables of 183 Magnoliaceae species in China using four comparative models. We also measured 20 ecophysiological traits of 27 species of the four largest Magnoliaceae genera. Species richness gradually decreased from south to north, with phylogenetic clustering along the southern coast, and in north and central China. Variances of plant diameter, length and width of leaves and tepals, and fruit length were strongly associated with phylogenetic relatedness. In contrast, variances of plant height, flowering and fruiting time, and all of the 12 environmental variables were associated with both spatial proximity and phylogenetic distance among species. Strong phylogenetic signals in water-conducting and nutrient-use related traits were identified, and correlated evolution of several key functional traits was demonstrated. Among the three evergreen genera of tropical origin, *Lirianthe* had the lowest hydraulic-photosynthetic capacity and the highest drought tolerance compared with *Manglietia* and *Michelia*. Contrastingly, the temperate centred deciduous *Yulania*, showed high rates of hydraulic conductivity and photosynthesis at the cost of drought tolerance. Overall, the present tropical-temperate distribution of Magnoliaceae was associated with the divergent evolution of plant functional traits under both geographical separations and environmental constraints. Specifically, the regulation of hydraulic and photosynthetic processes during the adaptations to local environments, was the ecophysiological basis for phylogenetic niche conservatism in functional traits among divergent lineages of Magnoliaceae.

T3-05-07

The past is the key to the present: Using fossil plant to uncover the continent-specific assembly of grassland ecosystems

Caroline A. E. Stromberg
University of Washington

Reconstructing how terrestrial ecosystems were assembled through time is vital for understanding the processes that affect them today—and in the future. The grassland biome currently covers nearly 40% of Earth land surface, and has a major influence on the climate system, the faunas that inhabit it, as well as global silica and carbon cycling. Documenting the Cenozoic rise of this important biome has preoccupied paleontologists and evolutionary biologists for nearly 140 years. The rich Cenozoic terrestrial record on many continents has allowed detailed study of functional morphology of animals, fossilized soils, and stable isotopes, documenting broad patterns of grassland evolution. In contrast, paleobotanical data have historically been scarce; however, in recent years, plant silica (phytoliths) has emerged as a highly suitable tool for tracking diversification of Poaceae lineages and the spread of grassland vegetation. Unlike pollen and most macrofossils, grass phytoliths are often diagnostic of ecologically distinct grass subclades, providing vital information about grass community composition. In addition, plant silica can be found in sediment types representing depositional environments that do not preserve other kinds of plant remains, extending the range of sampled habitats to include, for example, dry "uplands". In combination with other lines of evidence, phytolith data collected so far in North and South America, Europe, Asia, and Africa have shown that the assembly of the grassland biome was substantially more complex than previously assumed. For instance, it now appears that open-habitat grasses (pooids, PACMADs) diversified long before becoming ecologically dominant in both North and South America, indicating that the traits associated with tolerance of open, dry habitats may have evolved in response to a different set of environmental triggers than those that promoted the expansion of open, grass-dominated habitats. It is also clear that the earliest Miocene grasslands were dominated by C_3, pooid grasses that were only later replaced by C_4 grasslands in at low-mid latitudes (e.g., North America, Asia). Furthermore, comparison between faunal and phytolith assemblage data shows that grazing herbivores did not evolve in tight coevolution with grasslands in North America and Eurasia, but instead appeared several millions years after vegetation changed. In contrast, in southern South America, grazer-like morphology originated independently of grass-dominated habitats, in the context of palm-dominated shrublands that seemed to have occupied this region for much of the Cenozoic. This open, dry-adapted, but nearly grass-free vegetation type, which lacks clear modern analogs, likely expanded as a result of increasingly arid regional climates during the middle-late Eocene. The exact ecology of the pooid and PACMAD grasses that formed a minor component of this vegetation remains unclear. Taken together these results suggest that non-analog conditions have fundamentally shaped the history of grasslands in many regions and point to essential differences in ecosystem assembly among continents.

T3-05-08

Insights into the historical construction of East Asian subtropical evergreen broadleaved forests revealed by the temporal history of Theaceae

Xiang-Qin Yu[1], *Lian-Ming Gao*[1], *Douglas E. Soltis*[2], *Pamela S. Soltis*[2], *Jun-Bo Yang*[1], *Liang Fang*[3], *Shi-Xiong Yang*[1], *De-Zhu Li*[1]
1. Kunming Institute of Botany, Chinese Academy of Sciences
2. Florida Museum of Natural History, University of Florida
3. Jiujiang University

Subtropical evergreen broadleaved forests (SEBLFs) are among the most important vegetation types in East Asia, contributing fundamentally to both the biodiversity function and ecosystem service of this region. The East Asian SEBLFs were proposed as unique zonal vegetation that might have developed in concert with the Asian monsoon system, particularly the East Asian summer monsoon (EASM). Although paleovegetation reconstructions revealed that the SEBLFs have existed in Southwest China during the Miocene, the historical construction of the SEBLFs and its relationship with the dynamics of the EASM in East Asia are poorly known. We used the tea family (Theaceae), a characteristic component of the SEBLFs in East Asia, to gain insights into the assembly of this important biome. Fifty-three representatives from all three tribes and nine genera of Theaceae were represented, covering major distributional range of the family. Eighteen species from seven families within Ericales were used as outgroups. The phylogenetic relationships within Theaceae were reconstructed using plastome and nuclear ribosomal DNA sequence data. The temporal history of the tea family was estimated using eight fossils from Theaceae and related families in Ericales. Ages of other six structural genera of the SEBLFs from Fagaceae, Lauraceae and Magnoliaceae were also integrated based on previous studies. Analyses of speciation rate and potential rate shifts within Theaceae were conducted using a Bayesian approach. The ancestral states of vegetation and habit of Theaceae were inferred based on the maximum likelihood method. The recovered phylogenetic framework of Theaceae in our study is robust, providing well-resolved relationships among tribes and genera in the family. The phylogenetic backbone of Theaceae is in agreement with the evolution of the endosperm. The early-diverged Stewartieae contain the most copious endosperm, and the other two sister tribes (Theeae and Gordonieae) retain only a thin layer of or perhaps no endosperm. The reconstructed temporal history of the tea family suggests that the major lineages of the SEBLFs possibly originated around the Oligocene-Miocene (O-M) boundary. A consistent pattern was also found in four genera of other three characteristic families within the SEBLFs. In addition, the accelerated speciation rate and transition from deciduous to evergreen leaves within Theaceae were detected close to the O-M boundary. It seems that a rapid assembly of the SEBLFs in East Asia might have occurred around the O-M boundary. Moreover, two structural genera from Fagaceae and Lauraceae date back to the Eocene, suggesting that the establishment of the SEBLFs could be possibly traced back to the Eocene epoch. The initiation of the EASM in the Eocene and its subsequent intensification around the O-M boundary may have triggered the establishment and rapid assembly of the SEBLFs in East Asia, respectively. Our results indicate that the historical construction of the SEBLFs in East Asia was closely related to the dynamics of the EASM, which brought abundant rainfall and enhanced humidity within the region.

T3-05-09

Into the tropics: High tropical diversification not time-for-speciation explains latitudinal gradient in *Rhododendron* species richness

Nawal Shrestha, *Zhiheng Wang*

Department of Ecology and Key Laboratory for Earth Surface Processes of the Ministry of Education, College of Urban and Environmental Sciences, Peking University

The latitudinal gradient in species richness is a well-known biological phenomenon which has puzzled ecologists since the early 19th century. The mechanisms underlying this pattern have remained one of the most controversial issues in ecology and biogeography. Several hypotheses have been proposed to explain this diversity gradient, many of which can be broadly grouped into two categories; ones that regard contemporary environmental factors (e.g., water, energy, habitat heterogeneity, *etc.*) and others that consider evolutionary history as the major determinant. Although species diversity shows strong relation with the environment, large scale pattern of species diversity must ultimately result from the process of speciation, extinction, and dispersal that occur over evolutionary time scales. The time-for-speciation hypothesis (TSH) and diversification rate hypothesis combine both these ecological and evolutionary perspectives to provide an explanation for large-scale richness patterns. Here, we evaluate the relative effects of contemporary environment and evolutionary history on the diversity pattern of a species-rich plant genus *Rhododendron*, which comprises over 1,000 species globally. Using global climate, a complete distribution data of global species, and a comprehensive phylogeny of *Rhododendron* constructed herein, we reconstructed ancestral area, ancestral climate and traced biogeographical history of this group for the first time. Specifically we aim to 1) compare the relative contribution of contemporary environment, time-for-speciation and diversification rate on species diversity, and 2) evaluate the relationship between net diversification rates and environmental variables. Our findings suggest that *Rhododendron* likely originated in the temperate region of Northeast Asia in the early Paleocene, from where it dispersed southward to the Himalayas and south-western China. This dispersal history reflects a rare into-the-tropics phenomenon. Interestingly, despite being a temperate group, *Rhododendron* diversity is higher and the species are younger in the tropics, showing evidence inconsistence with the prediction of TSE. The accumulation of species diversity away from their origin centre suggests that longer time does not necessarily yield more species. Instead, the differences in net diversification rates across tropical and temperate latitudes provide a better explanation for uneven species richness. We conclude that high *Rhododendron* diversity in the tropics is, particularly, due to increased allopatric speciation and elevated net diversification rates facilitated by the formation of complex topography in the south-eastern fringe of the Tibet-Himalayan region and the Malay Archipelago.

T3-05-10

Niche divergence driven by range overlap and time in *Mimulus*

Qin Li, *Amy L. Angert*

Departments of Botany, University of British Columbia, Vancouver, British Columbia

Species' niches and distributions interplay across multiple spatial scales through evolutionary time. In particular, the ways that niches diverge between close relatives can reveal determinants of co-occurrence, while the degree of range overlap can determine when selection favours niche convergence or divergence. Quantitative analysis of niche divergence and range overlap in a phylogenetic framework can provide important insights into niche evolution and speciation modes. In areas of range overlap, competitive interactions may drive greater niche divergence. Alternatively, shared selective pressures due to environmental filtering may lead to greater niche conservatism. We tested these hypotheses in 16 pairs of western North American monkeyflowers (*Mimulus*), using data on various niche axes, geographic ranges and a robust phylogeny. We found that coarse-scaled environmental niche divergence decreased with increasing range overlap and phylogenetic distance, consistent with environmental filtering that operates as range overlap increases following allopatric speciation. No such relationship was detected for local-scaled microhabitat niche divergence, except a positive association with phylogenetic distance along one microhabitat niche axis (related to vegetation cover). Furthermore, greater microhabitat niche divergence was found in sympatry than in allopatry along at least one niche axis for each species pair with partially overlapping ranges. Together, these results reveal that different niche axes show different patterns of evolutionarily lability, perhaps because the relative importance of environmental filtering versus biotic interactions depends on spatial scale.

T3-05-11

Evolution along the C_3-CAM photosynthetic continuum in *Rumicastrum* (Montiaceae)

Lillian Hancock, *Joseph Holtum, Erika Edwards*
Brown University

Crassulacean Acid Metabolism (CAM) photosynthesis is a complex trait that has evolved many times from the more common C_3 photosynthesis and improves water use efficiency in plants under drought and temperature stress. In addition to strong CAM, there are recognized C_3-CAM intermediate phenotypes (constitutive low-level CAM and facultative CAM), which are hypothesized to represent transitional states along the CAM evolutionary trajectory; however there has been no explicit demonstration of how these potentially intermediate types relate to one another, or to the evolution of the full CAM syndrome. *Rumicastrum* (Montiaceae) is an ideal lineage in which to investigate the evolutionary assembly of CAM photosynthesis. It comprises ~70 species of annual and perennial succulent herbs endemic to Australia, and species within this group are known to span the C_3 to CAM phenotypic spectrum. Using a targeted gene enrichment approach, we sequenced 100s of loci from multiple gene families of interest across *Rumicastrum* and other Montiaceae. Here we present the first, complete phylogeny for *Rumicastrum*, its relationship to the rest of Montiaceae, and how the lineage likely radiated across the Australian continent. In addition, we explore the evolution of CAM within this group, integrating physiological, isotopic, and anatomical measurements to more fully understand the expression, evolution, and reversibility of the CAM photosynthetic syndrome.

T3-05-12

Plant mating system predicts island biogeography and range size

Dena Grossenbacher[1], *Jeremiah Busch*[2], *Yaniv Brandvain*[3], *Emma Goldberg*[3]
1. California Polytechnic State University, San Luis Obispo
2. Washington State University, Pullman
3. University of Minnesota

Key life history traits, such as shifts from outcrossing to selfing, can have major impacts on the geographic properties of species. The reproductive assurance provided by selfing may allow selfers to successfully establish after long distance dispersal events (e.g., to oceanic islands), and to rapidly expand their ranges. Using phylogenies, geographic occurrence and mating system data, we conduct a large-scale test of this hypothesis in flowering plants. We find that 1) self-compatible species are significantly enriched on oceanic islands relative to self-incompatible species and 2) automatically selfing species tend to have larger geographic ranges than their outcrossing relatives, and that this tendency increases with time since a selfing species diverged from its outcrossing relative. Thus, traits that improve mating success may be at least as important as other traits, such as dispersal ability and environmental tolerance, in helping plants to establish new populations and increase their range size.

T3-06: Functional traits explaining plant responses to past and future climate changes

T3-06-01

Phenology constraints of plant growth under climatic change

Christian Körner
University of Basel

It is widely held that climatic warming will stimulate metabolism, growth, length of growing season, and thus, productivity of plants and ecosystems. This rationale does not account for the basic role of phenology in plant life. Plant phenology, that is, the progression of seasonal development, is controlled by plant-internal and external drivers. Phenology responses are under strong evolutionary selection because they (1) insure plants against biological activity at the wrong time (e.g., escaping from periods with risk of freezing damage in the temperate zone), and (2) they ensure gene transfer in sexual reproduction (synchrony of flowering). Both of these functions of phenology are so fundamental for successful plant life, that evolution selected against dominant influences of the curiosities of weather, that is temperature or moisture only. Yet, if controls were that simple, selection would have driven phenology to strictly follow an astronomic calendar (photoperiod) that minimizes the statistical risk of exposure to dangerous conditions. Since the photoperiod is the same in spring and autumn, it requires at least a second control, a certain chill-dose, signalling the passing of winter. Photoperiod thresholds are selected for by the regional long-term probability of freezing in the case of a pronounced temperature seasonality or other regularly occurring adverse events (drought season). These 'escape' mechanisms come into play in long-lived plants that cannot rely on a long-term seed bank to survive 'misbehaviour' (e.g., flushing or flowering at the wrong time). In long-lived temperate zone taxa (trees) thermal forcing within a safe window set by photoperiod comes in as a third driver. Because, what commonly is observed, is largely this temperature response, once other signals have 'opened a window of perceptivity', temperature has widely been held as the master-driver of phenology, which clearly is not the case in late successional, native tree species. It can safely be expected that evolution does not play lottery, but prevents plants from being active at the wrong time. Across the entire plant world and different climatic zones an immense number of combinations of these three phenology controls can be expected. There is also space for opportunistic behaviour. Hence responses of phenology to climatic change are species and region specific, which is severely constraining predictive models. As a rule of thumb, the control of phenology by temperature alone declines with (1) a species' longevity, (2) the annual amplitude of natural habitat temperature, (3) unpredictability of temperature (including topography driven snow cover), and, (4) as a combination of (2) and (3), with latitude. Since most soil organisms exert immediate responses to climatic conditions, rapid climatic change will create a mismatch between soil activity (provision of nutrients) and plant activity (demand in nutrients) until micro-evolutionary selection has tracked those climatic changes, which may take centuries in the case of trees.

T3-06-02

Functional traits are good predictors of global change impacts on tree species abundance dynamics in a subtropical forest

Ronghua Li, ***Qing Ye***
South China Botanical Garden, Chinese Academy of Sciences

Significant changes in the composition of tree species have been observed in various forests worldwide. We hypothesized that these changes might result from variable sensitivities of species to global change, and species sensitivities might be quantified using functional traits. Employing long term (1978-2010) species abundance data of 48 tree species from a permanent subtropical forest plot, where multiple global change factors have been observed, including soil drying, we examined the relationships between temporal trends in abundance and suits of functional traits. We found that species with high photosynthesis rates, leaf phosphorus and nitrogen concentrations, specific leaf area, hydraulic conductivity, turgor loss point, and predawn leaf water potential had increased in abundance, while species with opposite trait patterns had decreased. Our results demonstrate that functional traits underlie tree species abundance dynamics in response to drought stress, thus linking traits to compositional shifts in this subtropical forest under global changes.

T3-06-03

Clonality-latitude relationship and the underlying mechanisms in Australia

Hongxiang Zhang[1,2], *Stephen Bonser*[2], *Sichong Chen*[2], *Daowei Zhou*[1], *Angela Moles*[2]
1. Northeast Institute of Geography and Agroecology, CAS
2. the University of New South Wales

We provide a large-scale quantification of the relationship between latitude and the proportion of species with clonal reproduction. The parasite pressure is thought to be higher at lower latitudes, favoring sexual reproduction. The abiotic conditions is relatively harsh at high latitudes, which may harm sexual reproduction. Therefore, we predicte that there would be a higher proportion of clonal species at high latitudes than at low latitudes. We collected data of 4,386 native seed plant species from 446 genera and 99 families present in ABRS *Flora of Australia*. Species' occurrence records were taken from the *Atlas of Living Australia*, including 817,450 species-site combinations spanning 34.5° of latitude. Logistic regression showed that the proportion of clonal species significantly increased with latitude, rising from just 3.3% clonal species at 9.25°S to 26.7% clonal species at 43.75°S. The overall average proportion of clonal species in Australian seed plants was 9.4%. It reveals Australian vegetation contains a relatively low proportion of clonally reproducing species. In order to test which factor forcing the latitudinal gradient in clonality, we collected temperature, precipitation and net primary productivity (NPP) data from CRUTEM dataset, GPCC dataset and the MOD173A annual product, respectively. Growth form and plant height data were also collected from the species descriptions of ABRS, reliable websites (e.g., FloraBase, PlantNET) or from previous publications. We aim to investigate the main environmental factors that drive plant reproductive strategy, determine whether plant traits are related to reproductive strategy and detect which is more important to shape reproduction strategy selection, plant own traits or environmental factors. The results showed that annual average temperature, maximum temperature in hottest month, and the seasonal variation in precipitation were significantly negatively related to the proportion of species with clonal reproduction. The average monthly precipitation in driest month and NPP were significantly positively related to clonality. Among the environmental factors, annual average temperature explained the largest variation. Plant growth form and height significantly affect reproductive strategy. Herbaceous and short species tend to have clonal reproduction. This study indicated a positive relationship between latitude and the proportion of species with clonal reproduction and illustrates the forcing factors, which adds to our growing understanding of dramatic latitudinal gradients in the way plants grow and reproduce. This project was funded by China Scholarship Council, the National Natural Science Foundation of China (41571055), and the State Key Basic Research Development Program (2015CB150800) to HX Zhang and by an ARC discovery grant to SPB and ATM (DP110100210).

T3-06-04

Water relations are coupled with photosynthetic light- and water-response functions across 12 mosses

Zhe Wang, *Maaike Bader*
University of Marburg

Trade-offs between water conservation and carbon gain are well-studied in vascular plants, were stomata regulate gas-exchange rates. In bryophytes, water loss and access to CO_2 are not regulated but depend on moss morphology and moisture conditions. Although water is essential for photosynthetic activity, excess water hinders CO_2 diffusion to the chloroplasts and thus hinders photosynthesis. Bryophytes thus need to balance between being too wet and too dry. We hypothesize that species can follow two main strategies: 1) staying wet as long as possible, with a low photosynthetic capacity needed because CO_2 supply will be limiting, or 2) avoiding external water, with high photosynthetic rates needed to be very productive before drying out completely. Consequently, we hypothesis that the water relations and photosynthetic light- and water-response functions of bryophytes should be inter-dependent. In the current study, we aimed to find out 1. How does the maximum water content (WC_{max}) of moss species relate to their water loss rate? 2. How are the position and the 90%-range of the optimal water content for photosynthesis (WC_{opt} and $WC_{90\%\text{-}range}$) related? 3. How are water relations and photosynthetic light- and water-response traits of mosses related? We determined the light-compensation and -saturation points, dark respiration rates, mass-based light saturated assimilation rates (A_{max}), WC_{max}, 50% water loss time (50%time), water-loss decay constants (DC), WC_{opt} and $WC_{90\%\text{-}range}$ of 12 forest-floor mosses from central Germany. Principal component analysis and bivariate correlations were used to explore the associations between these traits and their distributions among the studied species. The first PCA axis was dominated by the extreme trait values of the two *Sphagnum* species, which had a very high WC_{max} and 50%time, a low DC, a high WC_{opt} and large $WC_{90\%\text{-range}}$, and a high A_{max}. *Sphagnum* species are known for having specialized water-holding cells, thereby avoiding external water while still remaining wet as long as possible. WC_{max}, 50%time, WC_{opt} and $WC_{90\%\text{-range}}$ were generally positively related, and negatively related to DC. A_{max} was positively related to WC_{max}, WC_{opt} and $WC_{90\%\text{-range}}$, but this relationship was strongly dominated by the two *Sphagnum* species. Excluding these species, A_{max} was negatively related to WC_{max} and had no signifiant relationships to the photosynthetic water-response traits. When also excluding the two Polytrichaceae, species with internal water-conducting tissues, no relationships between water relations and photosynthetic capacities could be detected. This was due to the lower variation in trait values and to the low remaining sample size (n = 8). Our study shows that water relations and photosynthetic water-response traits are inter-related across moss species, but these relationships depend strongly on the species selected for analysis. Thereby *Sphagnum* and Polytrichaceae having over-proportionally large effects due to their special water-related adaptations, allowing alternatives to the two hypothesized strategies. Further studies should test the generality of these relationships for larger sets of species, including other habitats and climate zones and at the level of plasticity within species. Modelling could help to understand the suggested constraints of the trait space by simulating moss carbon gain in different wetting and drying scenarios.

T3-06-05

Extremely thick cell walls and low mesophyll conductance: Welcome to the world of ancient living!

Linda-Liisa Veromann-Jürgenson[1], *Tiina Tosens*[1], *Lauri Laanisto*[1], *Ülo Niinemets*[1,2]

1. Estonian University of Life Sciences

2. Estonian Academy of Sciences

Mesophyll conductance is thought to be an important photosynthetic limitation in gymnosperms, but they currently constitute the most understudied plant group in the extent to which photosynthesis and intrinsic water use efficiency (WUE_i) are limited by mesophyll conductance. However, it is necessary to study a variety of taxa to predict vegetation responses and the carbon-cycle in the changing climate. A comprehensive analysis of leaf gas-exchange, photosynthetic limitations, mesophyll conductance (g_m, calculated by three methods previously used for across-species comparisons), and the underlying ultra-anatomical, morphological and chemical traits in 11 gymnosperm species in varying evolutionary history was performed to gain insight into the evolution of structural and physiological controls on photosynthesis at the lower return end of the leaf economics spectrum. Two primitive herbaceous species were included in order to enhance evolutionary context. Low mesophyll conductance was the main limiting factor of photosynthesis in the majority of species. The strongest sources of limitation were extremely thick mesophyll cell walls (T_{cwm}), high chloroplast thickness and variation in its shape and size, and the exposed surface area of chloroplasts per unit leaf area. In gymnosperms, the negative relationship between net assimilation per mass and leaf mass per area (LMA) reflected an increased T_{cwm}, whereas the easy-to-measure integrative trait, LMA failed to predict the underlying ultrastructural traits limiting g_m. These adaptations may be relics from the atmospheric concentrations in which these genera evolved, still present due to evolutionary constraints, but may give evergreen gymnosperms an advantage over angiosperms in the rising CO_2 levels of this century.

T3-06-06

Intraspecific plasticity of leaf physiology in widely distributed eucalypts to combined water and heat stress varies under current and future climates

Guomin Huang[1], *Paul Rymer*[2], *Zhonghua Chen*[2], *Michael Aspinwall*[2], *John Drake*[2], *Sebastian Pfautsch*[2], *Mark Tjoelker*[2], *David Tissue*[2]

1. Nanchang Institute of Technology

2. University of Western Sydney

The capacity of plant species to cope with rapid climate change will depend upon phenotypic plasticity. For woody species, populations originating from different environments usually show differentiation in response to warming or drought. However, few studies have examined intraspecific variation in woody plant response to simultaneously occurring climate extremes. Here, we investigated the effects of warming and co-varying climate extremes (i.e., drought and/or heat waves) on growth and physiology of two widely distributed *Eucalyptus* species (*E. grandis* and *E. tereticornis*), each of which contained a temperate (drier and cooler) and a tropical (wetter and warmer) provenances. The two species in general did not show interspecific differentiation in response to the same environmental changes, but we observed significant differences between the temperate and tropical provenances in plant growth response to warming and in photosynthetic response to heat waves, both of which depended on taxon temperature of origin. However, provenances of both species responded similarly in growth and physiology to drought. Heat stress alone generally had little effect on plant growth rates and photosynthesis, but the synergism between drought and heat imposed significantly greater impact on plants than each applied separately. Furthermore, two distinct strategies (senescence of older mature leaves *vs.* complete closure of stomata) were observed, and both proved to be effective, in coping with combined drought and heat stress. Taken together, our results suggest that plant populations of widespread tree species may differ in their response to climate warming and heat waves depending on the climate of origin, but may not necessarily show difference in response to drought. Drought is likely to be the dominant stressor during heat waves. Widespread tree species may possess different strategies to cope with climatic extremes and show interspecific or even intraspecific variation.

T3-07: Interspecific pollen transfer: ecological circumstances, behavioral mechanisms, and evolutionary consequences

T3-07-01

Temporal variation in pollinator network and pollen transfer network: How common is misplaced pollen and its causes?

Shuang-Quan Huang
Central China Normal University, School of Life Sciences

Studies of plant-pollinator interaction in diverse communities have shown that one pollinator species may visit multiple plant species in one foraging bout. Such interspecific visits could deposit heterospecific pollen on stigmas, causing deleterious effects on sexual reproduction in flowering plants. Heterospecific pollen transfer (HPT) is deleterious to both pollen donor and receipt species, suggesting that co-flowering species may have evolved adaptive strategies to reduce the effects of HPT. For example, heterospecific pollen carried by pollinators may not be picked up by stigmas of the next visiting plant species. To a better understanding pollen transfer network in natural communities, we need quantify the relationship between the three stages of pollination process, i.e., plant-pollinator interaction, pollen placement on the pollinator body and pollen receipt by stigmas. Our recent survey in an alpine meadow in the eastern Himalaya, southwest China, showed that plant species with more outgoing pollinator moves tended to disperse more of their own pollen to others, but plant species which received more pollinator interspecific moves received less heterospecific pollen. In the natural community, HPT varied from 0 to over 50% that misplaced pollen on stigmas was more than conspecific pollen in some species. We still know little about what factors affect the degree of HPT in natural communities and whether generalized and specialized flowers differ in donors and receipts of heterospecific pollen. We have quantified temporal variation of HPT in the alpine meadow involving over 30 co-flowering species for three years, to explore why some species could experience relatively higher HPT but other species do not. Pollen-transfer network was constructed by investigating pollen placement of the bumblebee body, which bees acted as the predominant pollinators. I will present recent results of pollination ecology at the community level from a field station in Shangri-La. These studies highlight the importance of understanding how co-flowering plants affect each other from multiple scales of pollination, improving our understanding of plant-plant interactions, interspecifc pollen flow, reproductive isolation and coexistence of diverse species.

T3-07-02

Tradeoffs of pollinator visitation rate and pollination quality

James Thomson
University of Toronto

I will review older literature that considered heterospecific pollen transfer as a component of the broader issue of plant competition for pollination service. In that view, the negative effects of competition would have a quantity component (visitation rate) and a quality component (including effects of heterospecific transfer). Both components are expected to vary spatially, depending on how much two plant species overlap and intermingle in space. However, much of the new literature on HPT has not explicitly considered spatial effects. Because it is hard to examine such effects in the field, I have modeled the process using artificial flowers that dispense and receive particles of food dye as analogues for pollen grains. Arrays of flowers are visited by bumble bees in flight cages. I will present data from experiments that simulate a species boundary between a nectar-rich plant species and a nectar-poor one. The nectar-poor species receives more visitation in the vicinity of the nectar-rich species, indicating a positive "magnet species effect, " but the increased visitation entails a cost of greater HPT.

T3-07-03

Ecological character displacement: Style length variation in *Pelargonium* driven by geographic mosaics of plant community structure.

Ethan Newman, *Bruce Anderson*
Stellenbosch University

Geographically heterogeneous patterns of selection are thought to be important in generating intraspecific floral variation. Usually, geographic mosaics in the pollinator landscape are inferred as the primary agents of divergent selection on floral traits. However, it is also possible that geographic heterogeneity in plant community structure can give rise to divergent selection and consequent intraspecific floral variation. Here we investigate ecological character displacement as the possible driving force behind floral variation in *Pelargonium* communities. First, we demonstrate that the co-occurrence of *Pelargonium* species may have negative consequences in terms of interspecific pollen transfer. In particular, we show that the precise area of pollen placement on the body of the pollinator is determined by the amount of style exertion. Using visitation experiments, we demonstrate that individuals from different species experience high interspecific pollen transfer if they have similar style lengths. In contrast, individuals from different species experience less interspecific pollen transfer if they have very different style lengths. Using crossing experiments, we also show that interspecific pollen transfer has negative consequences on seed set, suggesting that co-occurring species should evolve style length differences in sympatry to reduce interspecific pollen transfer. Patterns of style length variation over 22 different *Pelargonium* communities indicate that style lengths are always very different when species co-occur, however style lengths can be very similar when the same species occur in allopatric populations. Given the apparently strong selection pressure for competition to drive style length divergence in sympatry, we suggest that character displacement is the most parsimonious explanation for the observed patterns of style length variation in a guild of long proboscid fly pollinated *Pelargonium*.

T3-07-04

Costs of interspecific pollen transfer for Neotropical bat-pollinated flowers

Nathan Muchhala
University of Missouri - St. Louis

Although competition for pollination is often invoked as a driver of broad-scale evolutionary and ecological patterns, we still lack a clear understanding of the mechanics of such competition. When flower visitors alternate between two species of flower, heterospecific pollen transfer takes place. The impact of these mixed loads on the female reproductive success of a recipient has received considerable attention, but the concomitant loss of male reproductive success—due to pollens grains being lost to foreign stigmas—has received less. Furthermore, pollen losses are not limited to grains that land on stigmas, but can also include deposition on non-stigmatic surfaces of the intervening flowers, or loss from the animal's body through passive detachment or active grooming. Here I quantify pollen transferred by nectar bats between focal flowers (*Aphelandra acanthus*) with and without intervening visits to one of two competitor species. One competitor (*Centropogon nigricans*) places its pollen in the same region of bats' heads as the focal species, while the other (*Burmeistera sodiroana*) places its pollen farther forward. I found that 1) any intervening visit caused some reduction in the number of pollen grains transferred, 2) competitor flowers with similar pollen-placement locations caused greater reductions in pollen transfer, and 3) of these competitors, those in male-phase (dispensing pollen) caused greater pollen loss than those in female-phase (without pollen). This study provides rare empirical support for the detrimental effects of competition for pollination on male fitness via pollen misplacement, and is the first to show an added cost imposed by male-phase competitors. Although this competition is especially strong when competitors overlap in pollen placement, diverging in pollen placement will not completely eliminate pollen loss during visits to foreign flowers simply because pollen sheds or is groomed from pollinator's bodies at some background rate over time. This suggests that any angiosperms that share pollinators face a pervasive selective pressure through male fitness to diverge in floral traits, alleviating competition by attracting different pollinators, altering floral phenology, or encouraging floral constancy.

T3-07-05

Heterospecific pollen transfer alters distribution of closely related species through reproductive interference

Sachiko Nishida[1], *Koh-Ichi Takakura*[2], *Takayoshi Nishida*[2]

1. Nagoya University Museum

2. School of Environmental Science, The University of Shiga Prefecture

Heterospecific pollen transfer may reduce fitness of a recipient species, and this phenomenon is known as Reproductive Interference (RI). The most conspicuous property of RI is its positive frequency dependence, which stimulates its self-reinforcing impact via positive feedback. For example, when pollen of one species exerts adverse effect on seed set of another species, it reduces the number of descendants of the latter. We have focused on RI from an alien species on a native congener, and have studied how the alien species has altered the distribution of native species in *Taraxacum*. *Taraxacum officinale* was introduced to Japan in the early 1900s and it is now widely distributed throughout the country. Most of the species populations, including hybrids between the species and native species, are triploid and reproduce asexually, although they produce pollen. Some of the native *Taraxacum* species in Japan, on the other hand, reproduce sexually and are self-incompatible. The alien and native species share flower visitors, allowing heterospecific pollen transfer. Through a field survey and a hand-pollination experiment, we showed that RI from the alien to the native species was moderate to strong in regions where the alien species outnumbered the native species, and marginal where it did not. A simulation model using the field data revealed that the native species could be eliminated much faster if the alien species exert RI, compared to the situation without the RI. We also observed pollen tube behaviour following the heterospecific pollination on the native species, and concluded that RI from the alien dandelion might occur through pre-emption of the native ovules by the heterospecific pollen. RI should be considered when devising strategies to protect native biodiversity from invasive species.

T3-07-06

Keep the old, attract the new: Floral color change by plants for a full exploitation of site-faithful pollinators

Kazuharu Ohashi[1], *Miki Suzuki*[1], *Takashi Makino*[2], *Kentaro Arikawa*[3]

1. University of Tsukuba

2. Yamagata University

3. Sokendai-Hayama

Floral color change—the retention of old, nonreproductive, rewardless, but unwilted flowers in an altered color—has been suggested as an adaptive strategy for plants to attract pollinators from a distance with an increased display while minimizing pollinator visits to nonreproductive, rewardless flowers using distinct coloration. However, considering that most visually oriented pollinators would respond similarly to these aspects of floral display, the reason for low prevalence of floral color change in nature is unclear. Here we briefly review our recent studies aiming to provide evolutionary explanations for the sporadic distribution of floral color change in angiosperms. First, based on spectral data for 219 flower species, we show that pollinators with UV-sensitive eyes can see color changes of flowers more often than humans. We then present the results from phylogenetically independent analyses suggesting that the occurrence of floral color change is consistently correlated with bee pollination. A plausible explanation for the association between floral color change and bee pollination may lie in the behavioral characteristics of bees—their tendency to forage within small areas and learn to make frequent returns to profitable plants or patches. As our laboratory experiments with bumble bees illustrate, plants with discolored old flowers receive persistent return visits from resident foragers by presenting a useful color guide to rewarding flowers, as well as they could attract the passersby by increasing the size of an entire display. Thus, floral color change can be considered as a strategy to fully exploit site-faithful pollinators such as bumble bees, by maintaining experienced foragers while attracting inexperienced ones. Moreover, our field data on color-changing and non-color-changing species in *Weigela* suggest that these advantages of floral color change could be highly context dependent. In natural habitats of the non-color-changing species, most insects visit flowers in an opportunistic way, making little or no return visits to the same plants or patches. This pattern holds even for bumble bees because the non-color-changing species often grows in large monoculture stands. In contrast, bumble bees are more site-faithful in habitats of the color-changing species, where small number of plants are intermingled with other co-flowering species that offer different levels of reward; in such situations, bumble bees focus on small areas containing more rewarding plants to maintain a high foraging performance. Based on these observations, we suggest that the color-changing species gains advantage from site-faithful bumble bees only in such multi-species plant communities, where interspecific competition for pollinators is intense as well as heterospecific pollen receipt could impose substantial fitness costs. More studies are needed to clarify whether this hypothesis could be extended to other angiosperm species with color-changing flowers.

T3-08: Ecology and evolution of plant mating systems (two sessions)

T3-08-01

Mating system and density-dependent effects on fitness in *Noccaea caerulescens*

Céline Devaux[1], Russell Lande[2], Agnès Mignot[1], Emmanuelle Porcher[3]
1. Université de Montpellier - Institut des Sciences de l'Evolution
2. Norwegian University of Science and Technology - Department of Biology
3. Muséum National d'Histoire Naturelle

We use *Noccaea caerulescens*, a heavy metal tolerant Brassicaceae that relies in part on pollinators to set seeds, to test theories we developed of the evolution of flowering phenologies and selfing rates under pollinator constrains. These theories predict intermediate selfing rates will emerge in *N. caerulescens* via geitonogamy generated by pollinator foraging behaviour among flowers of the same plant. A corollary predicts that different pollinators select for different floral displays and selfing rates. We also expect individual variance of flowering time to amount to that at the population level due to expected large stochasticity in pollinator visitation rates. At the community level and because of competition among plant species for shared pollinators, we predict staggered species flowering phenologies. Using common garden experiments in spring 2016 and a factorial design crossing ecotypes (from normal versus metal-polluted soils) and pollination modes, we excluded autonomous selfing in *N. caerulescens* and showed that its fruit set is pollen limited, results almost equally from pollinators and wind, and is affected by neighbouring plant density (~1m). Using weekly observations during the springs of 2015 and 2016 of four natural populations growing on either normal or metal-polluted soil in Southern France (Cévennes), we tested for the effect of pollinator communities on mean population selfing rates, and the effect of pollinator visitation rates on both temporal variation in fitness and individual selfing rates. We estimated in each population (1) *N. caerulescens* plant densities at several spatial scales, (2) individual flowering phenologies of 12 *N. caerulescens* random focal plants and of all their neighbouring conspecifics in a 0.8 m radius, (3) temporal variation of seed and fruit sets of the 12 focal plants across the seasons, and (4) the composition and flowering phenologies at the species level of the plant and pollinator communities. Two consecutive years of bad weather prevented the estimation of direct effects of pollinator visitation rates. Mean population selfing rates ranged from 10 to 50%, and contrary to previous results, ecotypes showed divergent changes between years. Fruit set was less variable among plants in denser populations found on metal-polluted soils. We are currently estimating the temporal variation of selfing rates at the plant level, genotyping seeds along the main inflorescence of each of the 12 focal plants in each population and each year, to correlate it to variation in fitness components, flowering phenology, plant density and plant/pollinator communities.

T3-08-02

Sexual mimicry promotes outcrossing and multiple paternity in sympatric orchid species

Michael Whitehead[1,2], Rod Peakall[2]
1. School of BioSciences, The University of Melbourne
2. Research School of Biology, The Australian National University

Most flowering plants engage animals to carry out the essential service of pollination. The behaviour of animals therefore exerts a profound influence on important microevolutionary processes such as mating and gene flow. However, our knowledge of exactly how different kinds of pollinators control plant gene flow patterns remains poor. Specialist pollination systems offer the most tractable systems to explore this relationship because observed patterns of plant mating can safely be ascribed to a single species or functional group of pollinator. This multidisciplinary study examines the consequences of highly specific sexually deceptive pollination in the mating and pollen dispersal patterns of two sympatric Australian orchids. First I describe a detailed and innovative study of behaviour and movement ecology for a species of wasp pollinator. Genetic studies of mating and pollen flow then provide evidence that pollination via sexual deception results in near exclusive outcrossing despite clonality, as well as multiple paternity–a rarity for orchids. In summary, multiple lines of evidence support sexually deceptive pollination as an adaptive solution to the problem flowers face of simultaneously attracting pollinators while persuading them to leave quickly.

T3-08-03

Pollinator-mediated constraints on the evolution of selfing

Rachel Spigler[1], **Susan Kalisz**[2]
1. Temple University
2. University of Tennessee, Knoxville

Hermaphroditic plants can produce both selfed and outcrossed progeny, termed mixed mating. Evolutionary models of mating systems classically predict two outcomes, either complete selfing or complete outcrossing, depending on the balance of two genetic factors: the automatic gene transition advantage of selfing and inbreeding depression. If inbreeding depression is less than 0.5, complete selfing is predicted to evolve, otherwise complete outcrossing should be favored. The abundance of species with mixed-mating systems has thus presented something of a paradox. A longstanding question posed by plant scientists who study mating system evolution is: Why do high levels of selfing persist within populations despite high levels of inbreeding depression? In general, the evolution of autonomous self-pollination and its role in providing reproductive assurance when mates or pollinators are scarce have been raised as answers to this question and to explain the maintenance of mixed mating systems more generally. Instead, we turn the original question around to ask: Why do high levels of outcrossing persist in some populations despite low levels of inbreeding depression? We propose a new functional hypothesis: the evolution of autonomous self-pollination effectively results in an asymmetry in the benefit of plant-pollinator mutualism such that plants may not be able to escape visitation by pollinators. That is, while autonomously selfing plants are less reliant on their pollinator mutualists for seed production, pollinators may still strongly rely on specific species for pollen and nectar. This results in continued outcross pollen delivery, halting or hindering the progression of the evolution of high selfing. We will discuss evidence supporting our hypothesis, including data from a recent analysis of inbreeding depression from mixed mating species and their inbreeding depression levels (Winn *et al.*, 2011) and the constraints that plants may face in circumventing pollination mutualism relative to other types of mutualisms. We conclude by offering future directions for testing predictions from this "persistent pollinators" hypothesis. This perspective can provide novel insight on the evolutionary transitions between outcrossing and selfing and the maintenance of mixed mating despite low inbreeding depression.

T3-08-04

Selfing as an evolutionary dead-end? Insights from a model of inbreeding depression due to stabilizing selection on quantitative characters

Emmanuelle Porcher[1], *Russell Lande*[2]
1. Muséum national d'Histoire naturelle
2. Imperial College, London

Phylogenetic analyses suggest that self-fertilization is an evolutionary dead-end, characterized by higher rates of extinction than in predominantly outcrossing taxa, but maintained by frequent evolutionary transitions from outcrossing to selfing. Higher rates of extinction of highly selfing taxa are thought to be related to lower rates of adaptation, but the details of the mechanisms are still largely unknown. Here we explore an alternative explanation in a constant environment, by analyzing the maintenance of quantitative genetic variance in a mixed-mating system of self-fertilization and outcrossing. We consider two models, in which purely additive genetic variance is maintained by mutation and recombination under stabilizing selection on the phenotype of one or more quantitative characters. The two models have contrasting architectures of quantitative characters. Our approach differs from previous theory by distinguishing lineages according to the number of consecutive selfing events in their ancestry since their last outcross. We examine how equilibrium genetic variance, and the resulting inbreeding depression, varies across selfing rates. In both models, at low selfing rates the equilibrium genetic variance of characters remains almost the same regardless of the selfing rates, maintained by compensatory effects of (1) an increase in genic variance with increased selfing rates caused by increased homozygosity and (2) stabilizing selection, creating negative covariance among loci. Above a threshold selfing rate, the value of which depends on the number of characters under selection, the equilibrium genetic variance declines and reaches a "purged equilibrium". In one model, the purged equilibrium also exists at low selfing rates, and the purging of the genetic variance and inbreeding depression is associated with an accumulation of high genic variance, hidden by large negative covariance. This hidden variance is released in outcrossing events, which prevents a transition from selfing to outcrossing. The implications for evolution of selfing rates, and for adaptive evolution and persistence of predominantly selfing species, provide a theoretical basis for the classical view of Stebbins that predominant selfing constitutes an "evolutionary dead end."

T3-08-05

Population-genetic expectations for trait filtering of self-incompatibility on islands

Jeremiah Busch, *Nathan Layman*
Washington State University

Environmental and genetic factors play a critical role in the ecology and evolution of reproductive traits in nature. A major goal of modern biology is to provide mechanistic explanations for non-random patterns. The most common evolutionary trend in flowering plants is the transition from obligate outcrossing to selfing. Indeed, about half of all angiosperm species have self-incompatible (SI) breeding systems but this condition is frequently and irreversibly lost, producing self-compatible (SC) lineages. The hypothesis of reproductive assurance attributes this pattern to the advantages of selfing when mates or pollinators are difficult to find. The colonization of oceanic islands has long been heralded as an iconic exemplar of this idea, since island plant communities maintain more SC plants. Baker was impressed by this pattern and suggested, "With self-compatible individuals a single propagule is sufficient to start a sexually reproducing colony.." (Baker 1955). However, it is theoretically unclear whether island breeding systems reflect extinction of colonists or adaptive evolution after arrival. To address these hypotheses, it is necessary to model island colonization and explicitly study the costs of self-incompatibility: extinction and breeding system evolution. We consider an island-mainland model with evolution at the S-locus and dynamic inbreeding depression at many loci throughout the genome. First, we treat the probability of extinction by reproductive failure in purely SI island colonists, assuming they originate from mainland populations at equilibrium. Analytical results show that unless the number of colonists and S-alleles are both very low, immediate extinction of SI colonists is unlikely given adequate pollinator availability. However, with continued sampling in recently founded SI island populations, extinction rates increase when population growth is slow. Second, we consider polymorphism at the S-locus and show that SC mutations are often maintained at low frequencies in mainland populations. While the presence of SC mutations at the S-locus slightly reduces the chances of reproductive failure immediately after colonization, this advantage is offset by the low probability of their inclusion during founding. These results highlight several important points. With few island colonists, SC mutations are often not sampled, and extinction eliminates SI populations in short order. With more colonists, extinction rates decline but the spread of SC mutations is favored unless population growth is rapid and sufficient diversity at the S-locus is present. This work clarifies the spectrum of costs that self-incompatible lineages face on islands and quantifies their importance beyond the verbal appeal of Baker's rule.

T3-08-06

Interactions between drift and selection in the evolution of plant mating systems

Laura Galloway[1], *Matthew Koski*[1], *Jeremiah Busch*[2], *Dena Grossenbacher*[2]

1. *University of Virginia*
2. *Washington State University*

Typically, mating system variation is explained by hypotheses focusing on the benefits of selfing, such as reproductive assurance, that may vary with ecological context. However, mating system evolution may also be driven by historical changes in genetic load. For example, colonization from glacial refugia to current distributions is expected to shape population genetic structure. Increases in homozygosity will reduce cost of inbreeding and create conditions permissive for the evolution of selfing. Here we bring together studies of the contemporary pollination environment as well as an evaluation of genetic structure due historical colonization events to explain range-wide variation in mating system in the insect pollinated herb, *Campanula americana*. 24 populations sampled from throughout the range from putative southern refugia north to the range edge were grown in a pollinator-free environment. Autonomous selfing increased in populations that were further north and west in the range. Decreased duration of male phase (dichogamy) in this protandrous species was associated with increased selfing while herkogamy was not. The same 24 populations were evaluated for pollen limitation and reproductive assurance in each of two years. Pollen limitation was greater for populations located in northern regions of the range and these populations had fewer bumblebee pollinators. Despite substantial levels of autonomous fruit set under pollinator-free conditions, no reproductive assurance was found in natural populations. These results will be combined with ongoing work in the same focal populations estimating selfing rates in nature, inbreeding depression following one generation of self fertilization, and patterns of genetic diversity and genetic load across the range. Data to date indicate variation in mating system among *C. americana* populations is more tightly associated with geography and genetic structure due to post-glacial migration than pollinator abundance and pollen limitation. Such historic processes are likely to be common across many temperate taxa, and create broad expectations for range-wide patterns in mating system.

T3-08-07

Reinforcement and speciation between outcrossing and selfing *Clarkia* species

***David Moeller**, Ryan Briscoe Runquist*
University of Minnesota

Studies of reproductive isolation often find that prezygotic barriers evolve more rapidly than postzygotic barriers between incipient species. However, it has been challenging to determine whether selection has directly caused elevated isolation (reinforcement) or whether is has occurred as a by-product of adaptation to alternative environments. In plants, pollinator isolation has commonly been found to influence reproductive isolation. However, little is known about how mating system divergence may contribute to the speciation process. In both cases, few studies have established whether natural selection can directly enhance reproductive barriers via reinforcement. In *Clarkia xantiana*, there are two subspecies that differ in mating system despite both being self-compatible. One subspecies, *xantiana*, is primarily outcrossing and bee pollinated, with pronounced dichogamy and herkogamy. The other, *parviflora*, primarily autonomously self-fertilizes, with minimal dichogamy and protandry. The self-fertilizing taxon is monophyletic and recently derived from the primarily outcrossing taxon (~65,000 yrs) and phylogeographic studies show that they have come into secondary sympatry in a narrow zone. Field experiments assessing male and female mating success under pollen limitation have shown evidence of disruptive selection where divergent mating systems are favored by selection, whereas intermediate phenotypes suffer reduced fitness. Moreover, there is a partial postzygotic barrier to the production of F1 hybrids. Across the range of the two subspecies, there is a pronounced pattern of reproductive character displacement (RCD) where floral traits are most strongly diverged in sympatry and only minimally diverged between the most allopatric populations. This is a key signature of reinforcement. We used field experiments pairing sympatric and allopatric genotypes to test whether reinforcement selection has directly caused the evolution of RCD in floral traits. Our results indicate that hybridization between incipient species is strongly reduced between sympatric genotypes but is approximately twice as likely when allopatric genotypes of the two taxa are paired, consistent with the reinforcement hypothesis. By contrast, we found no evidence that floral evolution has occurred as by-product of adaptation to contrasting pollination environments in the allopatric and sympatric regions. These results are novel in demonstrating a contribution of both reinforcement selection and mating system divergence to the speciation process.

T3-08-08

Why do plant mating systems often vary widely among populations?

***Jeffrey Karron**[1], Randall Mitchell[2], Michael Whitehead[1,3]*
1. *University of Wisconsin-Milwaukee*
2. *University of Akron*
3. *University of Melbourne*

Classical genetic models predict that disruptive selection on outcrossing rate will cause mixed-mating to be rare and transitory. To more fully characterize the extent of among-population variation in outcrossing, we surveyed all published microsatellite and allozyme studies reporting multilocus outcrossing rates (t_m) for three or more natural populations of hermaphroditic Angiosperms. Our dataset includes t_m values for 753 populations from 105 species. These taxa are distributed across 43 families and 80 genera. Many of the species in our survey exhibit substantial among-population variation in outcrossing rate. For 32/105 species the range between the lowest and highest t_m was > 0.3. Nearly two-thirds of all species have at least one mixed-mating population. Given this marked variation among populations, there is a critical need for studies that explore how both ecological context and heritable differences in floral traits influence mating system phenotypes. For mating systems to evolve, there must be heritable variation for mating system traits within natural populations. Nearly all studies estimate outcrossing rates for entire populations, rather than for individuals, because estimates for individuals often have large error due to violations of assumptions about the local pollen pool and statistical limitations of the maximum likelihood procedure. However, if there is sufficient genetic variation at marker loci to facilitate unambiguous paternity assignment to all sampled offspring, it is possible to quantify individual outcrossing rates, and to explore relationships between outcrossing rate, siring success, and pollen discounting. We are using this approach to characterize individual variation in outcrossing rates within and among populations of *Mimulus ringens*, a wetland perennial native to eastern North America. By clonally replicating genets and exposing them to different ecological contexts, we can explore how changes in the pollination environment influence the pattern of selection on the mating system. We are complementing these experimental studies with studies of natural populations that address whether rates of selfing and levels of inbreeding depression correlate with natural variation in ecological context and floral display.

T3-08-09

Consequences of clonal growth on plant mating systems

Wan-Jin Liao[1], *Da-Yong Zhang*[1], *Lawrence D. Harder*[2], *Spencer C.H. Barrett*[3]

1. Beijing Normal University

2. University of Calgary

3. University of Toronto

Clonal growth is widespread among angiosperm species, so many plants have a mixed reproductive strategy of both sexual and asexual reproduction. Clonality has been suggested to negatively impact aspects of sexual reproduction, especially mating patterns. Importantly clonality could reduce female fitness, owing to among-ramet self-pollination (geitonogamy), especially in species with clumped ramets that display many flowers simultaneously. However, most previous studies compared the effects of variation in clone size, rather than between clonal and non-clonal individuals with the same resource investment in floral display. We investigated this issue experimentally with bumble bees foraging on artificial flowers and through computer simulations. Individual bees visited fewer flowers on plants with multiple inflorescences than on plants with a single inflorescence with the same total number of flowers. In simulations with all flowers receiving and donating pollen, this visitation behavior caused less geitonogamy for plants with clonal growth. Thus, when flowers simultaneously receive and donate pollen clonality can promote pollination quality without increasing geitonogamy. However, a single large inflorescence was more advantageous if segregation of sex roles among flowers (dichogamy or monoecy) reduced geitonogamy. To examine the influence of clonality on pollinator foraging and mating we investigated natural populations of *Aconitum kusnezoffii* and *Delphinium glaucum*, both of which produce clumped clones with protandrous flowers that are visited by bumble bees. On *Aconitum kusnezoffii*, bumble bees primarily moved between flowers within ramets. As a result, autogamy, within-ramet, and between-ramet geitonogamy accounted for 12%, 68%, and 20% of the selfing rate, respectively. In *Delphinium glaucum*, unmanipulated genets with two ramets outcrossed less than those for which we tied the ramets together to produce a single large inflorescence. The results from natural population of *Aconitum kusnezoffii* and *Delphinium glaucum* were consistent with the results of our simulations and demonstrated that a single large inflorescence can be more advantageous than multiple flowering ramets for dichogamous plants. Future research on the mating consequences of clonality should address the combined effects on female and male function and the influence of clonal architecture. Comparison between clonal and non-clonal sister taxa and manipulative experiments of display characteristics would be of particular value in identifying the influence of the distribution of floral resources within and among ramets on plant mating patterns.

T3-08-10

The evolutionary breakdown of heterostyly to homostyly in *Primula chungensis* (Primulaceae)

Wei Zhou[1], *Spencer C.H. Barrett*[2], *De-Zhu Li*[1], *Hong Wang*[1]

1. Kunming Institute of Botany, Chinese Academy of Sciences

2. University of Toronto

The evolutionary pathway from predominant outcrossing to high levels of self-fertilization has been followed in numerous taxa of herbaceous flowering plants. A classic example of this mating system transition involves the evolutionary breakdown of the sexual polymorphism heterostyly to a monomorphic condition. This evolutionary change is of particular value for studies of mating system transitions because the origin of selfing in heterostylous groups generally has a simple genetic basis and is usually followed by a characteristic sequence of changes to the floral morphology and morph structure of populations. Here we address three questions on this topic involving the population genetics and floral biology of the breakdown of distyly to selfing in *Primula chungensis* (Primulaceae). This species of primrose is native to SW China and is unusual among heterostylous plants in possessing both rare distylous and common homostylous populations. The specific questions addressed in our study are: 1) Is there evidence that homostylous populations are derived from distylous populations, and if so has the transition occurred more than once in different parts of the geographical range? 2) What are the patterns of floral variation following transitions from outcrossing (distyly) to selfing (homostyly) and how are these likely to influence mating patterns? We genotyped plants from 20 populations throughout the entire range, using the chloroplast intergenic spacer region (*trn*L-*trn*F), nuclear ribosomal ITS, and 10 nuclear microsatellite loci and determined genetic relationships among populations and variation in floral traits associated with homostyle evolution. The marker data consistently identified two multi-population lineages (Tibet and Sichuan) and one single-population lineage (Yunnan), a pattern consistent with at least two independent origins of homostyly. Evidence from flower and pollen size variation indicated that transitions to selfing have likely arisen by the same genetic mechanism, although the two main lineages have followed divergent evolutionary pathways resulting in populations with both approach and reverse herkogamy. The evolution of herkogamy may limit opportunities for predominant selfing and small flowers to evolve in homostylous populations. Our results provide several new insights into the historical processes governing the breakdown of distyly to selfing and on the trajectories of floral evolution in selfing populations.

T3-08-11

Variation in mating system, life history and floral traits of *Incarvillea sinensis*, a species with annual and perennial populations

Wei-Ning Bai[1], *Yue Ma*[1], *Spencer C. H. Barrett*[2]

1. Beijing Normal University

2. *University of Toronto*

Flowering plants display considerable variation in mating system, specifically the frequency of cross- and self-fertilization. This diversity is distributed non-randomly with respect to life history. Annual species commonly exhibit high selfing rates whereas perennial species are more often outcrossing. Few studies have explicitly investigated the correlated evolution of life history, floral biology and mating system because species that maintain variation among populations in these characteristics are rare. Here, we present results of investigations on variation in mating system and floral traits in *Incarvillea sinensis* (Bignoniaceae), a bee-pollinated species with annual and perennial populations. We also examined seasonal variation in the mating system of *I. sinensis* to investigate the influence of abiotic and biotic environmental factors on day-to-day variation. *Incarvillea sinensis* has a broad geographical range in China and we surveyed 16 populations across seven provinces spanning 2,000 km of the range. In each population, we measured five floral (corolla length and width, corolla tube depth, distance between stigmas and anthers, daily floral display) and two life-history traits (plant height and branch number) and collected open-pollinated seed families and data on population size and density. Of the 16 populations seven were annual, eight were perennial, and one contained annual and perennial plants. Annual populations were distributed in more northern localities and perennial populations were distributed in southern locations. In the mixed population, daily floral display, corolla tube length, stigma-anther separation and corolla length and width were greater in perennial than annual plants. Among remaining populations, corolla length, corolla tube length, stigma-anther separation and branch number in perennial populations were greater than annual populations, whereas plant height and daily flower number were greater in annual populations. Our studies of mating patterns are in progress but we predict, based on theory and empirical evidence from other species, that annual populations are likely to exhibit higher selfing rates than perennial populations. Our investigation of seasonal variation in mating patterns in an annual desert population from Mu Us Sand Land, Inner Mongolia, revealed the dynamic nature of mating. Flower density increased significantly over most of the 50-day flowering season, but was associated with a decline in pollinator service by bees, particularly on windy days. Multilocus estimates of outcrossing rate indicated high selfing rates of ~80%. There was evidence for a significant increase in selfing as the flowering season progressed and pollinator visitation declined. Biparental inbreeding also declined significantly as the flowering season progressed. Delayed selfing by corolla dragging largely explains the occurrence of mixed mating in this population of *I. sinensis,* and this mode of self-fertilization probably functions to promote reproductive assurance when pollinator service is limited by windy environmental conditions and competition from co-occurring flowering plants. The striking geographical variation in life history and floral traits in *I. sinensis* provides outstanding opportunities to examine the demographic and environmental factors causing mating system variation and provides a suitable model system for investigating the correlated evolution of life history and mating patterns.

T3-08-12

Pollen dispersal distance within a population is influenced by the frequency of females in the gynodioecious geranium, *Geranium maculatum*

Shu-Mei Chang, *Dorothy Christopher*
University of Georgia

Breeding systems in the flowering plants are highly diverse – species ranges from being entirely hermaphroditic with perfect flowers as the only floral form to being dioecious such that every plant has either pistillate or staminate flowers only with many different combinations of floral forms within or between individuals existing between these two extremes. A key evolutionary question is, thus, to understand how such vast diversity in the flower sexual expression (breeding systems) has evolved and been maintained in nature. A powerful approach to address this question is through studying systems that are polymorphic in floral sexual expression - such as gynodioecy, a system considered to be the most important intermediate evolutionary stage from hermaphroditism to dioecy. In gynodioecious species, male-sterile (i.e. Female; with pistillate flowers only) genotypes are found to coexist with hermaphrodites (with perfect flowers) in some natural populations. We chose to study the interactions between the gynodioecious wild geraniums, *Geranium maculatum,* and their pollinators. Specifically, we are testing the hypothesis that because pistillate flowers do not produce pollen rewards and are much smaller than hermaphroditic flowers, their presence in a population will influence the flight pattern of the pollinators and, hence, influence the pollen dispersal distance. To test this hypothesis, we sampled seeds of 10-20 maternal plants from 10 populations that range from 0-35% females. We genotyped both seeds and maternal plants for 10 microsatellite loci in each population. Using the software POLDISP, we estimated the average pollen dispersal distance in each population to range from 1.2 m to 1.7 m. In addition, we found that the presence of females in a population significantly reduces average pollen dispersal distance. We will discuss this result in the context of how the presence of female individuals may change not only pollinator behavior but also the patterns of population genetic structures – both at the within-population fine-scale and across a larger geographical region.

T3-09: Seeds and climate change

T3-09-01

Climate warming and regeneration from seeds in cold adapted plants

Andrea Mondoni, *Maria Tudela Isanta*
University of Pavia, Department of Earth and Environmental Science

Climate has a large influence on plant recruitment. Understanding the impacts of climate change on regeneration from seedlings is therefore urgently needed to predict future plant population dynamics and migration. Moreover, because of the substantial variation in both genetic and phenotypic plasticity of seed dormancy, germination and longevity regeneration from seeds is expecting to play a crucial role in plant adaptation to climate change. In this context, arctic and alpine ecosystems are considered highly threatened by climate change because they are predicted to experience above-average warming and because they are characterized by species adapted to low temperatures. Consequently, an increasing number of studies have been addressing the effects of climate warming on different life-history traits and functions of plants in these ecosystems, including seed germination and longevity responses. Here we present the results of several studies investigating the effects of pre- and post-dispersal climate warming on seed dormancy release, germination requirements and seed longevity across alpine plants. We show that climate warming could change the germination phenology and the longevity of seeds of alpine plants both directly (i.e. after dispersal) and indirectly (i.e. via parental effects), by shifting the timing of emergence from spring to autumn, by increasing the proportion of emerged seedlings and by extending the seed resistance to heat stress. Strictly cold-adapted chorotypes were more affected by warming, the extent to which was linked to seed dormancy traits: with deeply dormant species showing major changes in response to germination temperature and less dormant species in response to winter vernalization. The most significant aspect relating to warming on alpine ecosystems dealt with the extension of the snow-free period in spring that enhanced recruitment success, at least in the short term. We conclude that range shift and/or changes of local dominance might occur in response to changes in recruitment success. The overall success will be dependent on interactions with range shifts in new species (i.e. more thermophilous plants). In this context, transgenerational plastic adjustments of seed traits might play a crucial role in future plant adaptation to climate change.

T3-09-02

Thermal sensitivity of seed germination to a changing climate

Charlotte E. Seal, *Elena Castillo Lorenzo, Hugh W. Pritchard*
Royal Botanic Gardens, Kew

Seed germination is a critical step in the natural regeneration of the majority of the world's flora. Germination is highly regulated by environmental temperature, yet the sensitivity of germination to climate change has been quantified for relatively few species. Thermal time models are one tool to quantify the response of species to temperature. These seed population based models use germination progress over a range of thermal environments to establish thresholds for germination. This approach enables comparative studies of germination responses between species and habitats, and when related to the growing environment, are fundamental to understanding germination niche in the context of climate change. Examples are given where thermal time approaches have provided an overview of plasticity in the thermal germination phenotype, including for the Cactaceae family, and for a range of other species including crop wild relatives. The limitations of seed population based models are also discussed. Revealing the extent of sensitivity or resilience to environmental change has identified descriptors of germination plasticity that are critical for assessing how well species may cope in the future. This information is currently being used to identify species most at risk and to predict the proportion of species that will be "winners" and "losers" as a result of climate change.

T3-09-03

Climatic sensitivity of early life stages: Plant traits as indicators of environmental tolerance range

Jessamine Finch[1,2], *Ridma Bandara*[3], *Jeffrey Walck*[3], *Siti Hidayati*[3], *Kayri Havens-Young*[2]
1. Northwestern University
2. Chicago Botanic Garden
3. Middle Tennessee State University

Plant sensitivity to climate change will be dictacted in part by the range of conditions a group of organisms can tolerate for any given environmental variable (e.g., temperature, water availability). Those with a broad environmental tolerance range (TR) are expected to be buffered against changing climate–at least in the short term–while those with a narrow TR are predicted to be more sensitive. However, TR varies across life stages, species, and presumably even populations. Recently, early plant life stages (e.g., seed dormancy break, germination, and seedling establishment) were identified as more sensitive to climate change than adult stages. However, traditional approaches to forecasting plant responses to climate change are largely based on the geographic distribution of a species, which does not explicitly represent the TR of early life stages, potentially overlooking a large bottleneck to plant recruitment. I will present results from two germination studies investigating variation in TR both among species and, importantly, populations. All study species possess physiologically dormant seeds, but differ in key traits expected to influence genetic neighborhood size, and thus environmental TR. One study examined three common forb species (an *Asclepias*, *Physalis*, and *Penstemon*) with differing seed dispersal mechanisms (wind, animal, gravity). The second study examined three congeners (*Asclepias* spp.) with wind-dispersed seed that vary in multiple other traits relevant to gene flow, including pollination syndrome, breeding system, habitat specificity, and degree of clonality. Multiple populations of each study species were collected along a latitudinal gradient throughout the Midwest United States. A subset of populations included in the first study were collected over two subsequent years. Tolerance ranges were calculated based on germination proportion across a range of treatments manipulating cold stratification length, cold stratification temperature, and incubation temperature. Consistent with our hypotheses, both studies found average TR breadth reflected presumed genetic neighborhood size for each species based on known plant traits; however, there was relatively little difference in TR between wind and animal dispersed species. In the first study, TR shifted, but not consistently between collection years: TR of *Asclepias* populations decreased from Year 1 to Year 2, while TR of *Physalis* populations increased. In the second study, two of the three *Asclepias* taxa exhibited a negative linear relationship between TR breadth and collection latitude, while one showed no relationship. Interestingly, this trend was not present in the *Asclepias* populations included in the first study. Taken together, our findings suggest that known plant traits associated with gene flow via seed or pollen dispersal may serve as effective indicators of germination tolerance range; however, TR is readily influenced by local climate, interannual climatic variation, and presumably other factors. Results from this research have direct implications for forecasting plant responses to climate change, substantially improving current predictive, species distribution models by incorporating environmental limitations of early life stages, as well as population-level variation.

T3-09-04

Are seeds better physically defended towards the equator?

Si-Chong Chen[1,2], *Angela Moles*[2]
1. Jacob Blaustein Institutes for Desert Research, Ben-Gurion University of the Negev
2. Evolution and Ecology Research Centre, School of Biological, Earth and Environmental Sciences, UNSW Australia

There have been heated debates on the existence and generality of the latitudinal gradient in biotic interactions. Empirical studies and data syntheses have shown that seed predation is not more intense towards the tropics. One possible explanation for this finding is that seeds may be better defended at low latitudes than at high latitudes. Our goal was to provide the first broad-scale, quantitative analysis of the latitudinal gradient in physical investment in tissues surrounding seed reserve. We measured the biomass ratio of protective tissue to seed reserve for 250 species-site combinations ranging from 15°30'S to 43°35'S along the east coast of Australia. Contrary to expectations, we found that the proportional investment in tissues surrounding seed reserve was greater towards high latitudes in seeds exposed to pre-dispersal seed predators, while seeds exposed to post-dispersal seed predators did not show a latitudinal gradient in the proportional investment in tissues surrounding seed reserve. The physical investment in tissues surrounding seed reserve was not correlated with the level of seed predation, at either the pre-dispersal or the post-dispersal stage. Fleshy-fruited species invested proportionally less biomass in tissues surrounding seed reserve while suffering less pre-dispersal seed predation. Our findings indicate that natural seeds in their natural habitats are not better defended at lower latitudes. Together with recent studies of latitudinal gradients in seed predation, leaf herbivory and defence, this study adds to the growing evidences that challenge the idea of a declined latitudinal gradient in biotic interactions, and therefore cast doubt on our current understanding of the factors as plant-pest interactions in driving the latitudinal gradient in plant diversity. Our findings also demonstrate a defensive role of fleshy fruits in deterring pre-dispersal seed predators.

T3-09-05

Climate change on seedling germination and growth of *Larix principis-rupprechtii* and *Pinus tabulaeformis*

Runmei Gao
College of Forestry, Shanxi Agricultural University

Early development stages of trees are expected to be more sensitive to climate change, and represent a major bottleneck to plant community recruitment. In North China, drought index had increased dramatically from 1990s, understanding on seedling's response to anticipated climate changes is urgently needed. *Larix principis-rupprechtii* and *Pinus tabulaeformis* are two dominant coniferous tree species in North China, and play significant ecological roles in water conservation. We collected seeds of these two species from Guandi Mountain, Shanxi Province, and conducted a controlled greenhouse experiment to investigate how climate changes (warming and precipitation change) would affect their seedling emergence and growth. Three levels were set for both temperature and water supply. Temperature levels were monthly mean temperature (T_0) over the past 15 years (1999-2013), increased by 2 °C (T_1) and increased by 6 °C (T_2). Water supply levels were monthly mean precipitation (W) over past the 15 years (1999-2013), reduction by 20% (W^-) and addition by 20% (W^+). A two-way ANOVA was used to compare seedling growth (including seedling germination percentage, needle length, height, length of main root, number of lateral root, and length of lateral root) between the two treatments to the species. If the interaction was tested to be significant ($P < 0.05$), one-way ANOVA was used to test the effects of water treatments on seedlings indexes, and multiple comparison was used otherwise. The least significant difference method (*LSD*) was used to test the significant differences of all data at level of 0.05. Seed germination of two species was affected by temperature and precipitation changes. Seedling stable germination time reduced and seedling germination percentage improved by increased temperature of 2°C and water addition. However in the condition of increased temperature of 6 °C or water reduction, seedling stable germination time delayed and seedling germination percentage decreased. Seedlings of the two species presented different responses to the climate change. *L. principis-rupprechtii* was affected more greatly by warming and drying climate than *P. tabulaeformis*. Seedling morality of *L. principis-rupprechtii* increased sharply and seedling height decreased by higher temperature and water reduction. The changes from increased temperature or precipitation showed no significant effects on seedling survival and aboveground-growth of *P. tabulaeformis*. Underground growth of both *L. principis-rupprechtii* and *P. tabulaeformis* increased, indicating by longer primary and lateral roots. Soil drought was induced by dramatically increased temperature and reduced precipitation, which resulted in adaptive strategy of the two species by reducing aboveground growth and giving priority to root growth. Seedlings showed an interspecific variation of response to water deficit. *L. principis-rupprechtii* was more sensitive to climate change than *P. tabulaeformis*. The regeneration of *L. principis-rupprechtii* was inhibited by warming and drying climate. The seedling growth was negatively affected by highly increased temperature in growing period, and it could not be improved by water addition.

T3-09-06

Project Baseline: A time capsule of seeds to study evolution

Julie R. Etterson[1], *Steven J. Franks*[2], *Susan J. Mazer*[3]
1. Department of Biology, University of Minnesota Duluth, Duluth, Minnesota, USA
2. Department of Biological Sciences, Fordham University, Bronx, New York, USA
3. Department of Ecology, Evolution & Marine Biology, University of California, Santa Barbara, California, USA

Project Baseline is a new research-grade seed bank established in 2011-2015 that offers an unprecedented opportunity to study spatial and temporal dimensions of microevolution during an era of rapid environmental change. This seed bank is specifically designed to be used over the upcoming 50 years for research that employs the "resurrection approach". This kind of research entails withdrawing ancestral seed from the seed bank, recollecting contemporary seed from the same locations, and then growing antecedent and successor plants side-by-side to directly observe evolutionary change. Project Baseline includes seeds from 65 native and nonnative species collected from an average of 15 populations across their respective species' ranges in the continental USA in sites that are likely to be preserved into the future (see http://baselineseedbank.org/ for more detail). More than half of the species include collections from sister taxa. Seeds were collected from multiple species at 100 sites, providing rich resource for testing evolutionary hypotheses. The seeds are stored by maternal line (100-200 lines per population) under conditions that maximize seed longevity at the ARS USDA National Center for Genetic Resources Preservation in Ft. Collins, CO and are accessioned with site and population data. This open-access resource will be available to researchers at regular intervals to evaluate contemporary evolution with the first call for proposals anticipated in 2019. Our expectation is that research using the Project Baseline collection will vastly improve our understanding of temporal and spatial changes in phenotype and the evolutionary processes underlying it.

T3-10: Plant traits under global change: from traits-environment-ecosystem function relationships to trait-based vegetation modelling (two sessions)

T3-10-01

Towards a trait-based universal model of terrestrial primary production

Han Wang[1,2], *Colin Prentice*[1,2,3], *Ian Wright*[2], *Trevor Keenan*[2,4], *Tyler Davis*[3,5], *William Cornwell*[6], *Changhui Peng*[1]
1. Northwest A&F University
2. Macquarie Univeristy
3. Imperial College
4. Lawrence Berkeley National Laboratory
5. United States Department of Agriculture-Agricultural Research Service
6. University of New South Wales

Based on well-established ecophysiological principles and consideration of the adaptive significance of traits, we propose universal relationships between photosynthetic traits (carbon fixation capacity, and the ratio of electron transport capacity to carbon fixation capacity) and primary environmental variables, which capture observed trait variations both within and between plant functional types. Moreover, incorporating these traits into the standard model of C_3 photosynthesis allows gross primary production (GPP) of natural vegetation to be predicted by a single equation with just two free parameters, which can be estimated from independent observations. The resulting model performs as well as much more complex models. Our results provide a fresh perspective with potentially high reward: the possibility of a deeper understanding of the relationships between plant traits and environment, simpler and more robust and reliable representation of land processes in Earth system models, and thus improved predictability for biosphere-atmosphere interactions and climate feedbacks.

T3-10-02

Sources and ecological consequences of variation in bark traits

J. Hans C. Cornelissen
Systems Ecology, Department of Ecological Science, Vrije University

The study of variation in plant traits, as a tool for understanding and predicting vegetation composition and ecosystem functions, is still hot. However, compared to other plant organs, bark traits have not received the research attention they deserve based on the important functions of bark for woody plants. I will first focus my talk on key aspects of variation in bark traits between and within species, including the role of ontogeny. Then I will discuss the various pathways by which variation in bark traits affects carbon cycling, from the role of bark in carbon gain and transport, in protecting carbon-rich tissues in the xylem against biotic attack and abiotic stress, and from its roles in carbon release. I will give empirical examples of how bark trait variation, through afterlife effects on bark litter, can be important for the two main carbon release pathways, i.e. decomposition and fire regimes. These bark trait afterlife effects can be direct, via the bark decomposability and flammability of different species, and indirect, by influencing the decomposition and flammability of the wood inside the bark. I hope to convince the audience that the study of bark traits should be high on the agenda for various woody ecosystems in the world.

T3-10-03

Quantifying and modelling the responses of leaf gas exchange to drought: From traits-drought function relationships to trait-based vegetation modelling

Shuang-Xi Zhou[1], *I. Colin Prentice*[2], *Belinda Medlyn*[3]
1. CSIRO
2. Imperial College London
3. University of Western Sydney

Global climate change is expected to increase drought duration and intensity in certain regions while increasing rainfall in others. The quantitative consequences of increased drought for ecosystems are not easy to predict. Process-based models must be informed by experiments to determine the resilience of plants and ecosystems from different climates. This study provides quantitative information aimed at improving land surface models (LSMs). It includes four papers. (1) Responses of leaf-atmosphere gas exchange to short-term drought were analysed, across plant functional types and climates, based on a synthesis of previous experiments. Explicit and consistent definitions of stomatal *versus* non-stomatal responses were adopted. Both types of response were shown to be important, and plants adapted to arid climates responded very differently from others. (2) Parallel responses of stomatal conductance, mesophyll conductance, and photosynthetic capacity were found in two glasshouse experiments with tree species from Australia and Europe, revealing a common, co-ordinated pattern of increasing tolerance in plants from drier environments. (3) Xeric and riparian species of *Eucalyptus* were subjected to short- and long-term drought. The species were found to differ not only in their tolerance for short-term drought, but also in the extent to which they could acclimate to long-term drought through morphological adjustments. (4) Experimentally based drought responses were used to define new, plant type-dependent relationships of stomatal sensitivity and photosynthetic capacity to soil water potential in the Community Atmosphere Biosphere Land Exchange (CABLE) LSM. Comparison with CO_2 and latent heat flux measurements from eddy covariance flux measurement sites in Europe during the 'heatwave' year of 2003 showed that discrepancies between model results and observations were not substantially improved by the inclusion of the more realistic functions, due to a persistent positive bias in the model's simulation of evapotranspiration which overshadowed the differences between different representations of drought response functions.

T3-10-04

TRY – a global database of plant traits

Jens Kattge[1,2], *Sandra Diaz*[3], *Sandra Lavorel*[4], *Colin Prentice*[5], *Paul Leadley*[6], *Gerhard Bönisch*[1], *Wirth Wirth*[7,2]

1. Max Planck Institute for Biogeochemistry

2. German Centre for Integrative Biodiversity Research (iDiv) Halle-Jena-Leipzig

3. CONICET - Universidad Nacional de Córdoba, Instituto Multidisciplinario de Biología Vegetal

4. Laboratoire d'Ecologie Alpine, Université Joseph Fourier

5. Imperial College

6. Laboratoire Ecologie Systématique Evolution, Université Paris Sud

7. Special Botany and Functional Biodiversity, University of Leipzig

The global database initiative named TRY (www.try-db.org) was established in 2007 and since then has united a wide range of the plant trait research community worldwide. This talk will briefly introduce the TRY initiative and then present recent key developments based on the unprecedented buy-in of plant trait data, i.e. ongoing efforts to characterize the global spectrum of plant form and function, to better understand trait-environment relationships and to provide global maps of traits relevant for ecosystem function.

T3-10-05

Trait-based reconstructions of palaeoclimate

Sandy P. Harrison[1], *Han Wang*[2], *Zhiyong Zhang*[3], *I. Colin Prentice*[4]

1. University of Reading

2. Northwest A & F University

3. Chinese Academy of Sciences

4. Imperial College London

Leaf size and margin dissection have been used to reconstruct palaeotemperature from fossil leaf assemblages from many regions of the world, but reconstruction uncertainties are very large and show the confounding effects of other aspects of climate. We have used logistic regression to investigate the relationship between a suite of 25 leaf morphometric traits and summer temperature, plant-available moisture and temperature seasonality and the frequency of 25 leaf morphometric traits, based on data from 96 sites from the China Plant Trait Database sampling the range of climate and vegetation types found in China. Variations in the frequency of leaf morphometric traits can be explained as a response to gradients in summer temperature, plant-available moisture and temperature seasonality. These analyses identify distinct syndromes of morphometric traits associated with the carbon cost of leaf formation are recognized: warm, wet climates with low seasonality favour broad leaves that are evergreen, large, tough, thick, and dark in colour; dry, warm climates with high seasonality favour small, fleshy, glaucous leaves of intermediate thickness; cold, wet climates of high seasonality favour needle leaves that are evergreen. Leaf size and margin dissection are influenced by both temperature and moisture gradients. However, some individual traits (e.g., surface patterning, rugosity) show strong relationships with temperature but not moisture gradients, while others (e.g., hairs, drip tips) are related to moisture gradients but not temperature. Many specific relationships between traits and bioclimate variables are conservative across all woody life forms, and often between woody and non-woody life forms. Given that the observed variations in morphometric traits along climate gradients are both predictable and physically understandable, we then use the climate-leaf trait relationships established for the present day to build predictive models for estimating the temperature seasonality and moisture availability. We apply these models to predict the climate using fossil leaf floras from the Fushun Basin (Eocene), from Shandong Basin (early Miocene), from Xiaolongtan Basin (Miocene), and from Shengzhou/Zheijiang (Pliocene). These palaeoclimate reconstructions are well-constrained and produce estimates that are consistent with our physical understanding of climate changes during these past intervals when CO_2 was higher than today and regional climates warmer. These analyses suggest that exploiting known climate-trait relationships for a large set of morphometric traits simultaneously could offer a way to make robust reconstructions of pre-Quaternary climates.

T3-10-06

Plant traits, leaf economics and next-generation ecosystem models

Iain Colin Prentice

Imperial College London

It is now widely recognized that the "plant functional type" (PFT) concept, which was a valuable simplification in the early stages of global vegetation model development, has largely outlived its usefulness. This insight has come mainly from empirical studies of plant functional traits, which have revealed (a) that variation in most traits is continuous, (b) that most traits vary as much or more within conventionally defined PFTs as they do between PFTs, and (c) that a substantial component of this within-PFT variation is adaptive and systematically related to environmental gradients. Many traits also show (d) a component of similar magnitude that is related to phylogeny rather than environment, accounting for considerable variation that is expressed within the plant community. The data analysis techniques of generalized linear modelling, trait gradient analysis and constrained ordination have been deployed to quantify each of these aspects of trait variation and covariation and their relationships to environmental predictors. The leaf economics spectrum (LES) has provided an organizing principle for understanding a key leaf-level trade-off, between leaf longevity and specific leaf area. However, both leaf area – a trait showing strong, adaptive relationships to macro- and microclimate – and photosynthetic capacity, expressed on a per unit area basis, appear to show variation orthogonal to the LES. This finding suggests that models to account properly for functional trait diversity need to take account of at least three dimensions of trait variation. Variation in photosynthetic capacity, moreover, can be understood in terms of an optimality principle: the return on investment in photosynthetic capacity falls off abruptly at the point where light availability becomes limiting to photosynthesis, so the realized photosynthetic capacity is expected to be close to this point. Its actual value is also dependent on temperature, ambient CO_2 concentration, and the ratio of leaf-internal to ambient CO_2 – itself a key trait (indexed by the stable carbon isotope ratio) that depends in a predictable manner on temperature, vapour pressure deficit and atmospheric pressure, and appears to be perfectly plastic along environmental gradients. A growing body of empirical evidence supports the idea that ecophysiological traits, including leaf nitrogen content (through its relationship to photosynthetic capacity) and respiration rate, are tightly regulated at the leaf level following well-defined optimality principles. Indeed, these principles can be said to be the ultimate source of predictability in plant and ecosystem function. Next-generation ecosystem models will have to account for continuous variation in plant functional traits, both between and within local environments; and the huge expansion in the collection and availability of trait data can provide a sound empirical basis for such models. But data alone, however advanced the statistical analysis, cannot provide the key relationships and parameter values that will be required by models that are supposed to remain valid under altered environmental conditions. The necessary theoretical underpinning for models will be provided by optimality hypotheses that are explicitly testable with data. Achievement of this goal will require the adoption of a hypothesis-driven "strong inference" paradigm in ecosystem model development.

T3-10-07

How tightly tree form and function are integrated within a local forest community? A perspective from plant architectural, economics, hydraulic, anatomical and physiological traits

En-Rong Yan, *Yan-Jun Song*

East China Normal University/ School of Ecological and Environmental Sciences

The global plant economic spectrum provides a useful framework for improving models that predict future vegetation based on continuous variation in plant form and function. However, whether the global spectrum of plant form and function still holds in the local tree communities remains largely unknown. We measured five tree architectural, four economic, six physiological, eight hydraulic, three anatomical, and one stem thermal exchange traits for 56 individual trees (18 species) in an evergreen forest and 33 individual trees (8 species) in a deciduous forest in subtropical China. Principal component analysis (PCA) was employed to construct the local spectrum of tree form and function in each community types. We find that architectural, economic, physiological, hydraulic, anatomical, and stem thermal traits are highly intra- and inter correlated, thus formulating a local spectrum of plant form and function in two communities. Multi-trait variations in PCAs reveal that the structure of the local spectrum of plant form and function depends on community types. In evergreen forest, three notable dimensions of trait variation stand out. The first dimension is defined mainly by tree architecture and water transportation rate, i.e., short trees with large proportion of dispersed leaves and specific leaf area, but small and shallow tree crown, low light saturation point, stomatal density and vessel diameter, are capable of moving water and assimilating light and CO_2 slowly while exchanging stem temperature rapidly. The reverse is true for tall trees with larger proportion of clumped leaves. The second dimension is defined mainly by water transpiration rate and light capture ability, i.e., running from tall trees with high transpiration rate, stomatal conductance, light compensation point, leaf nitrogen concentration and proportion of vertical stretch branch, to short trees with high dark respiration rate, the apparent quantum efficiency, and proportion of horizontal stretch branch. The third dimension is defined mainly by hydraulic properties, i.e., trees with high wood density and Huber value of petiole tend to have low leaf water potential, vessel density, and hydraulic conductivity in both leaf and branch. In deciduous forest, there are two independent dimensions of trait variation. The first dimension is defined by tree architecture, water transportation and economics traits, almost showing the similar pattern compared to the first dimension in evergreen forest. The second dimension is structured by the coupling relationships between physiological, hydraulic, and architectural traits. Relative to short trees, tall trees with larger proportion of dispersed leaves and wider vessel diameter have low Huber value of petiole, dark respiration rate in leaves and vessel density, whereas higher hydraulic conductivity, transpiration rate, stomatal conductance, the maximum photosynthesis rate, light saturation point, and leaf nitrogen concentration. In conclusion, our results suggests that tree height induced variations in architectural, economics, hydraulic, anatomical and physiological traits that relate to water, carbon and nutrients are highly integrated at the local tree communities. The finding of this local spectrum of plant form and function would be beneficial for modelling carbon and water fluxes in the changing climate.

T3-10-08

Eco-evolutionary dynamics of plant community structure and function in Tibetan alpine grasslands

Kechang Niu[1], *Shiting Zhang*[2], *Jinsheng He*[3], *Martin J Lechowicz*[4]

1. Department of Biology, Nanjing University

2. Department of Ecology, Lanzhou University

3. Department of Ecology, Peking University

4. Department of Biology, McGill University

The functional composition of plant communities responds to environmental change and disturbance through a combination of ecological and evolutionary adaptions involving the balance between intraspecific trait variability (ITV, resulting from both phenotypic plasticity of individuals and the adaptive evolution of populations) and changes in community composition (CCC, resulting from both species turnover and shifts in species abundance). Working in this conceptual framework, we analyze the eco-evolutionary dynamics of community structure and function in Tibetan alpine meadows distributed along soil resource gradients and in response to grazing. For every species within each plot in alpine meadows both at differing elevations and aspects on each of three Tibetan mountains and at grazed and ungrazed plots in five sites across the Qinghai–Tibetan Plateau, we measured two chemical traits (leaf nitrogen and leaf phosphorus concentrations) and two morphological traits (specific leaf area and leaf dry matter content) that characterize fundamental tradeoffs in resource uptake strategies associated with growth rate. We distinguished the relative importance of ITV and CCC in affecting shifts in the community-weighted means of these foliar traits in response to soil resource availability and to grazing. In resource limited environments, alpine plant communities on Tibetan mountains increasingly tend to be dominated by species with a conservative resource strategy, while subordinate and infrequent species have an exploitive resource strategy and fast growth rates that increase their relative abundance and buffer decline in trait diversity at the community level. In contrast, less resource limited environments favor dominance by fast-growing species with exploitive resource use, while subordinate and infrequent species have a conservative resource use strategy. In short, the eco-evolutionary interplay between CCC and ITV effects acts to stabilize community function along soil resource gradients, promoting variation in species composition among plots at alpine meadow sites on Tibetan mountains. This interplay between CCC and ITV could provide an important buffering mechanism maintaining the structure and function of alpine communities under global change. In response to continuous grazing, plants in Tibetan alpine meadows increase specific leaf area and foliar nutrients but tend to have lower leaf dry matter content, a response consistent with faster growth and regrowth under grazing. This ITV mediated response drives a shift in community function from conservative, slow-growing resource use in ungrazed meadows to exploitative resource use under grazing. This community-wide functional response enhances forage quality, in turn favoring the secondary productivity of small herbivorous mammal communities, but also contributes to accelerated depletion of soil available phosphorus.

T3-10-09

Assessing the spatiotemporal dynamics of vascular epiphytes on the forest canopy via a mechanistic forest-epiphyte simulation model

Gunnar Petter[1,2], *Gerhard Zotz*[3,4], *Holger Kreft*[2], ***Juliano Sarmento Cabral***[2,5]

1. Forest Ecology, ETH Zurich

2. Biodiversity, Macroecology & Biogeography, University of Göttingen

3. Functional Ecology, University of Oldenburg

4. Smithsonian Tropical Research Institute

5. Ecosystem Modeling, Center for Computational and Theoretical Biology (CCTB), University of Würzburg

Vascular epiphytes live in a three-dimensional, highly dynamic, and heterogeneous habitat that is the forest canopy. Knowledge about how forest structure and dynamics influence the spatio-temporal distribution of epiphytes is still limited owing to difficulties in accessing the forest canopy and a lack of long-term observations. In this study, we present a coupled forest-epiphyte model that is three-dimensional, spatially-explicit, niche-, trait-, and individual-based. The model considers species-specific light requirements and simulates individual dispersal, recruitment, growth, and mortality. By coupling spatiotemporal distribution of trunks and branches obtained from a functional-structural forest model with our epiphyte model, we first assessed whether the model results satisfactorily match real-world data. Thereafter, we addressed how differences in long-term natural forest dynamics, selective logging, and the size of forest fragment influence epiphyte assemblages. The model generated vertical distributions of individuals and species and rank-abundance distributions that were comparable with field data from lowland forests in Panama and Ecuador. Higher forest dynamics had complex effects on epiphyte assemblages, with stable forests having generally high abundances, low extinction rates and highest saturation. Selective logging of larger trees resulted in lower abundances, species numbers, and saturation levels of epiphytes. Strikingly, a slight reduction in minimum harvested tree size from 45 to 40 cm DBH had a catastrophic effect on the epiphyte assemblage and led to a near complete extinction of epiphytes. Increasing fragment size decreased species extinction rates. Our model demonstrated that the average abundance and biomass of epiphytes are strongly influenced by forest dynamics and human activities. We concluded that modeling approaches that explicitly account for forest dynamics and epiphyte demography provide valuable insights on epiphyte assemblages, with potential applications for assessing long-term impacts of human-induced changes.

T3-10-10

Macroscale patterns on the functional diversity of pollination modes among conserved, restored and secondary areas of a tropical forest

Paula Maria Montoya Pfeiffer

University of Sao Paulo

Questions: (1) How does the functional diversity (FD) of tree pollination modes vary in relation to the species turnover within semi-deciduous forests at a regional scale? (2) In comparison with conserved forest areas, are there any significant difference in the FD among restored and secondary vegetation areas? Methods: an extensive literature review was carried out about the pollination modes of tree species obtained from 40 floristic surveys at 18 remnants, 11 restored and 11 secondary vegetation areas. Species and functional dissimilarities, species and abundance proportions of pollination modes and functional richness and evenness were calculated and compared within and among forest types. Results: (1) Species dissimilarity was much higher than functional dissimilarity within forest areas (mean Morisita-Horn distances > 80%, < 35% respectively). (2) Conserved tree communities were characterized by the predominance of generalized insect pollinated species (49%), followed by secondary proportions of bee (11%), wind (10%) and moth (8%) pollination. Other pollination modes were under-represented (< 4%). FD indices among restored and secondary forest were not significantly different from remnants. However, significant changes were found in the species and abundance proportions of several pollination modes. On restored forests, reductions were found in generalized insect, moth, wind, fly, pollen flowers and very small insect pollination, while the species pollinated by bees and bats were more than duplicated. On secondary vegetation areas, reductions included moth, fly and fig-wasp pollination, whereas bee, beetle, big animal and small insect pollination were increased. Such changes could not be related to the successional time on restored and secondary forests. Conclusions: Our results indicate a rather stable functional assemblage of pollination modes and also high ecological redundancy among trees regardless of the species replacement at a regional scale. Thus, major changes in FD between forest types would primarily rely on human disturbance. Tree communities showed a relatively low FD of pollination modes; hence, greater FD among plants with different growth forms is expected. Efforts in recovering such patterns on FD should be encouraged in order to avoid possible negative effects in plant-pollinator interactions.

T3-10-11

Facilitation as a driver of plant assemblages in Caatinga

Meira-Neto Joao, *Juan Rodríguez, Nayara Mota, Markus Gastauer*

Universidade Federal de Viçosa

Facilitation is a positive interaction in which a nurse plant reduces the environmental severity experienced by neighboring plants by providing shade, enabling nutrient accumulation or repelling herbivores within patches of vegetation. Nurse plants within facilitated patches should preferentially promote the coexistence of ecologically dissimilar species with little niche overlap, and if ecological traits are conserved within evolutionary lineages (i.e., niche conservatism), this should result in phylogenetic overdispersion. In contrast to competition, facilitation is expected to increase species richness. Therefore, to examine the role of facilitation as a driver of plant assemblages in Caatinga, we quantified the functional traits of nurse species and compared species richness, phylogenetic diversity, phylogenetic structure of the tree layer and of the herbaceous layer between patchy Caatinga and Caatinga with segregated plants. Furthermore, we determined the phylogenetic community structure (whole community) of Caatinga patches. Results show that functional traits related to resilience (resprouting) and resistance against herbivory (spines, thorns, urticancy and toxicity) seem to be crucial for facilitation in Caatinga. Autochory occurs at a higher frequency in nurse plants than in Caatinga in general. The herbaceous layer of patchy Caatinga is richer in species than Caatinga with segregated plants and facilitation is the suggested cause. As the whole community of the 196 patches is phylogenetically overdispersed compared to the null expectations for the metacommunity, facilitation seems to predominantly promote the coexistence of dissimilar species with little niche overlap. Therefore, niche diversification promoted by facilitation in patchy Caatinga likely caused its increased herbaceous richness, its increased phylogenetic diversity and its phylogenetic overdispersion.

T3-10-12

Transpiration is more important for leaf temperature regulation of the plants from a hot dry habitat than the plants from a hot wet habitat

Hua Lin[1], *Yajun Chen*[1], *Houlei Zhang*[2], *Peili Fu*[1], *Zexin Fan*[1]

1. Xishuangbanna Tropical Botanical Garden

2. Nanjing University of Science & Technology

Leaf temperature has important impact on the micro-environment of plants and physiological processes in the leaves. Plants from different habitats may show different leaf temperatures even under the same environment. However, the mechanism that causes temperature differences among plant groups is unclear because of the coupling of transpiration and morphological traits. We investigated 21 morphological leaf traits and thermal properties of 38 species in the common garden. Among them, 18 species were dominant species from the hot wet habitat (HW) and 20 species were dominant species from the hot dry habitat (HD). To separate the impact of transpiration and morphological traits on leaf temperature regulation, we measured the diurnal course of leaf temperature with and without transpiration. The temperature of a reference leaf beside each individual was measured simultaneously to exclude the diurnal thermal variance of environment. Generally, the species from HD showed lower leaf temperatures than the species from HW under the same condition. Plants from HD dissipate heat more efficiently by developing higher transpiration capacity and synergic leaf morphological traits. Moreover, leaf temperature was more dependent on transpiration than leaf morphological traits for the plants from HD. Among the leaf morphological traits, leaf length and the thickness of the abaxial epidermis were significantly correlated with the temperature of the leaf without transpiration. Transpiration is efficient to cool leaves, which provides the possibility of fast growth when water was available. Leaf morphological traits can damp heat stress to some extent, but they are not enough to provide a safe temperature for photosynthesis for the species in present study. Our results complemented the strategies of plants adaptation to hot environment with the regard to thermal regulation.

T3-11: Interaction between plants and insects

T3-11-01

Accumulation, spillover and spillback: An introduced thistle as a reservoir for insect and pathogen enemies of native plants

Peter M. Kotanen, *Krystal A. Nunes, Jason D. Verbeek*

University of Toronto

Invading plants often escape their native-range natural enemies, as described by the Enemy Release Hypothesis. However, once in a new region, exotic invaders acquire new enemies recruited from among invaded-range species. As well, enemies from the invader's original range may be accidentally or deliberately co-introduced; these species also may attack local natives. As a result, many invaders soon accumulate a new community of natural enemies, many of which are shared with co-occurring natives. Such shared enemies may participate in feedbacks on natives through spillover (spread of introduced enemies from exotic hosts to natives) and spillback (spread of native enemies to exotics followed by spread back onto their original hosts). Here, we present results from ongoing research into the potential of a common non-native thistle, *Cirsium arvense*, to act as a reservoir for insects and diseases attacking co-occurring native plants in its invaded range in Ontario, Canada. We documented the potential of *C. arvense* to host herbivores and pathogens by sampling plants over a 800 km transect from agricultural southern to boreal northern Ontario. We also monitored colonization of artificial populations, and investigated soil feedback using inocula sampled at both local (inside/outside existing populations) and geographic scales. Results indicate that thistle populations host a diverse set of herbivores and pathogens, but their impacts depend upon location. Leaf damage declined strongly with latitude. Different seed predators showed contrasting latitudinal patterns, while damage by the bacterial pathogen *Pseudomonas syringae* varied irregularly. Plants inoculated with northern soils outperformed those inoculated with southern soils, suggesting an accumulation of pathogens at southern sites. Artificial populations were colonized rapidly by herbivores, but even short (< 100 m) separation from other populations significantly slowed this process. Similarly, plants grown in soils inoculated with samples from outside thistle populations performed better than plants inoculated with soil from invaded areas only 50-100 metres away. A second set of experiments was then established to assess impacts of these enemies on co-occurring native plants. First, we compared insect damage and populations of shared herbivores on native plants in proximity to *vs*. isolated from populations of *C. arvense*. Second, we conducted an experiment in which we compared the effects of fungicide on survival of seeds of native competitors buried inside *vs*. outside thistle populations. Third, we inoculated root systems of co-occurring natives with soil biota sampled from inside *vs*. outside thistle populations. Finally, we inoculated native competitors with a bacterium, *Pseudomonas syringae*, sampled from *C. arvense*. Results indicate that spillover and spillback of some, but not all, of the insects and microbial enemies commonly attacking *C. arvense* might increase the ecological impacts of this invader on co-occurring native species. However, the magnitude of these interactions likely depends on the location and isolation of the host thistle population, the host-specificity of the enemy in question, and the phylogenetic relationship between *C. arvense* and co-occurring natives.

T3-11-02

Impact of pollen grain properties on pollen collectability by *Bombus terrestris* (Apidae)

Klaus Lunau
Heinrich-Heine-University Duesseldorf

The pollination of plants via bees faces a dilemma due to the fact that bees do not only passively transport pollen grains between conspecific flowers, but instead actively collect pollen to provision their larvae. Therefore, several kinds of protection against pollen collection by bees have evolved in bee-pollinated plants, e.g., toxic pollen, pollen hidden in floral tubes. Here we report about a new mechanism of protection against collection: Spines of large pollen grains impede the storage of pollen grains in the corbiculas of bees. Micrographs show that the spines of pollen grains of the the Malvaceae family are not covered with pollenkitt. Nevertheless, the spiny pollen grains adhere very well to the bristles of the bees' body. We analysed the influence of pollen grain properties – diameter, spine length, and spine density – on the collectability of echinate pollen by bumblebees. Buff-tailed bumblebees (*Bombus terrestris*, Apidae) were kept in a flight cage to forage on plants of the families Malvaceae, Cucurbitaceae, Dipsacaceae, and Campanulaceae which feature pollen with an echinate exine structure. The bumblebees foraged pollen significantly less on the Malvaceae and the Dipsacaceae species. Both the number of visited flowers and the weight of collected pollen are positively correlated with the handling time of the flowers, which in turn is negatively correlated with the pollen grains' spine length. To further test the influence of pollen structure on the collection behaviour of bumblebees, five plant species (*Alcea ficifolia*, *Lavatera thuringiaca*, *Cucurbita pepo*, *Centaurea jacea*, and *Syringa* spec.) were offered progressively over the course of five days. The results show that bumblebees stopped collection of pollen following the failure to groom pollen grains into their corbiculas. Melittophilous plants benefit from exine structures that render pollen uncollectable for polylectic bees which are seldom the most efficient pollinators. This does not deter nectar foragers from visiting the flowers and contributing to pollination. Consequently, mechanical defence of pollen grains against collection by non-specialist bees proves to be an effective solution to the pollen dilemma.

T3-11-03

Extrafloral nectaries: Helps in behavioural defences for Plant-Herbivore interaction between ant-plant communities

Ghosh Debasree
Department of Botany and Forestry, Vidyasagar University, Midnapore, West Bengal

Extrafloral nectar (EFN) (or extranuptial nectar) is secreted on the vegetative and less commonly the reproductive parts of a plant and does not contribute to its pollination. Indeed, ants attracted to nectar that is secreted around flowers can even repel flower visitors, which can have a significant ecological cost for the plant. Most, if not all, extrafloral nectary bearing plants secrete EFN to attract ants, wasps, and parasitoids as the enemies of their enemies to enhance top-down control of the herbivores and thereby their defense against herbivores. Recent study was conducted on some members of Malvaceae family to see the ant-plant interactions and that's depend on extrafloral nectary secretion. In the present study we observed and recorded EFN visiting ants and other insect activity in *Gossypium herbaccum, Hibiscus mutabilis, Thespesia populnea, Urena lobata, Urena rigida* over a period of one year and compiled all the data on a large statistical database SPSS version 11 using contingency tables and Kruskal-Wallis test. Besides this we also monitored the taxonomic composition of EFN, abundance, frequency and aggressiveness of the EFN associated ant species. To study the protective role of the observed ant assemblage, we analyzed the data by Spearman rank correlation coefficients, and analyzed the relationships between both EFN presence and abundance and ant defense behaviors against (1) stimulated herbivores and (2) natural nectar thieving butterflies.

T3-11-04

A survey of Queen Anne's Lace on Nantucket Island, Massachusetts, USA and an assessment of its effect on native pollination

Adam Ramsey, *Steven Ballou, Jennifer Mandel*
University of Memphis

Queen Anne's Lace (*Daucus carota*, QAL) is a non-native plant species that was introduced to North America from Europe during the early colonial period, and its presence on Nantucket Island, Massachusetts has been documented since at least the nineteenth century. Research in other plant systems has investigated whether native pollination was affected by the presence of a non-native, and the results have been mixed. Depending on the system studied a native may be negatively affected, may not be affected or may be positively affected by the presence of the non-native. As yet, there has been no survey of QAL on Nantucket, and no study known to the authors has investigated the effect of QAL on a native species. In summer 2015 and 2016 a survey of QAL populations was performed on properties owned and managed by the Nantucket Conservation Foundation, Linda Loring Foundation and Nantucket Islands Land Bank, as well as other, public properties. Following this, a pollinator observation study was performed on QAL and a native species, Toothed White-top Aster (Sericocarpus asteroides, TWTA). Pollinators were observed for three hours per day and five days each for the 2 species individually and both species cohabitating together. Additionally, in summer 2016 a removal treatment was performed on sites where both species cohabitate. The sites were observed as described for one day. After observation was over QAL umbels were removed from the site and pollinators were observed on TWTA as described the following day. Finally, TWTA populations from a range of distances from QAL were assayed for the presence and magnitude of heterospecific pollen. We had four hypotheses: 1) QAL would be present across Nantucket Island; 2) The presence of QAL would increase pollinator visitation rates and diversity indices on TWTA; 3) The removal of QAL would restore pollinator visitation rates and diversity indices to those found on sites with only TWTA; 4) Higher levels of heterospecific pollination would be found on TWTA in populations located close to QAL. We found QAL to be abundant across the island, with an average density of 12.2 plants/m^2 and a high density of 109.4 plants/m^2. While both plant species are generalist pollinated, QAL attracted 15 insect families and TWTA attracted only 8, the majority of each were flies and bees. Each removal treatment received at least 50% fewer pollinator visits after QAL was removed.

T3-11-05

Islands in the trees: A biogeographic exploration of epiphytes and the spider communities inhabiting them.

Francisco Emmanuel Méndez-Castro[1], *Maaike Bader*[1], *Glenda Mendieta-Leiva*[1], *Dinesh Rao*[2]
1. Philipps-Universität Marburg
2. Instituto de Biotecnología y Ecología Aplicada INBIOTECA

According with the equilibrium theory of island biogeography, within oceanic and habitat islands, the species number of a given taxa is determined by the interaction of two antagonist factors: islands size and isolation. It has been proposed that epiphyte plants in the tropics function like habitat islands for canopy arthropods. The base for this statement is the positive relationship found between the epiphytes' size and the species richness of the arthropod communities inhabiting them. Despite this observation, it stills remains unknown if there is an isolation effect produced by the spatial-arrangement of epiphyte plants. Epiphytes spatial distribution reflects the conjugation of intrinsic (plant microclimatic and physiological traits) and extrinsic factors (host tree phenology and morphology), but the isolation effect is related to the dispersal and locomotion abilities of each singular arthropod taxa. In the nature epiphyte islands are occurring in two different ways, like single plants or like plant mats. The abundance and diversity of epiphyte plants conforming an epiphyte-island is an approximation to habitat heterogeneity of oceanic islands, and recently this has been proposed like a third factor driving island's species richness. In this work we used spider communities inhabiting epiphytes to test if there is an isolation effect produced by the plants' spatial distribution. We collected all epiphytes and their companion spiders from tree threes in a shade coffee plantation of Veracruz, Mexico. Epiphytes tridimensional position was recorded at one by one centimeter scale, then we calculated the distance between all epiphytes within each tree. Spiders are capable of navigate through their environment by means of both cursorial and aerial way, therefore we used two different distance systems (Euclidian and Topological) in order to better understand the implications of this divergence. We found that spiders' diversity increased with epiphyte-islands' size and that this effect was negatively related to isolation. In the other hand, the effect of epiphyte-islands' heterogeneity was at most marginal and inconsistent. We also found that crown position and isolation across epiphyte islands affected spider species similarity. Regarding spiders' locomotion, walking distances (Topological) explained more of the diversity patterns than flying distances (Euclidean). This is the first study presenting evidence of species-isolation processes driving the diversity of epiphyte-dwelling spider communities. The three-dimensional structure, abundance and high biodiversity of epiphyte communities present an excellent opportunity for further testing the generality of island-biogeographic theory predictions.

T3-11-06

Diaspore dispersal by carnivorous wasps

Dylan Burge
University of California, Los Angeles

The California-endemic spicebush (*Calycanthus occidentalis*; Calycanthaceae) produces achenes with an auxiliary organ that is attractive to carnivorous wasps of the genus *Vespula* (Vespidae). Studies on the presentation, dispersal, and chemistry of the achenes suggest that these wasps may be a major agent of dispersal of spicebush diaspores, and that a protein reward is involved (the auxiliary organ is more than 22% protein by dry weight). Foraging worker wasps appear to be strongly attracted to an unidentified volatile compound that probably mimics the high-protein items (e.g., carrion and other insects) that are the typical targets of these wasps. Studies are currently underway to determine if a similar dispersal system exists in the other members of the Calycanthaceae, found in the southwestern United States and China. In addition, there is preliminary evidence that seeds of the California endemic pipevine (*Aristolochia californica*) are also dispersed by these wasps.

T3-12: Ecological Genetics

T3-12-01

Clonal extent of a temperate canopy tree species (*Tilia cordata* Mill.) across the UK geographic range

Paul Ashton, *Carl Barker*
Edge Hill University

Vegetative asexual propagation (clonality) is a common mode of plant reproduction. However the extent of clonality, how it is combined with sexual reproduction and the factors influencing the relative balance of the two is rarely explored. The genus *Tilia* is a group of woody tree species that have mixed mating systems. Within the UK, *T. cordata* is a native species of ancient woodlands that often grows gregariously. This study uses molecular markers to identify the extent of clonal growth within the species across its UK range. Results from an examination of the proportion of clonality, the size and spatial arrangement of clonal groups, fine-scale spatial genetic structure and the potential relationship of asexual reproduction to location/climate in several populations are considered.

T3-12-02

Uplifting of the Qinghai –Tibetan Plateau led to speciation of shrubs *Spiraea alpina* and *S. mongolica* (Rosaceae)

Faqi Zhang[1], *Khan Gulzar*[1], *Qingbo Gao*[1], *Zunzhe Tian*[1,2], *Shilong Chen*[1]

1. Key Laboratory of Adaptation and Evolution of Plateau Biota, Northwest Institute of Plateau Biology, Chinese Academy of Sciences

2. University of Chinese Academy of Sciences

The phylogeographic history of two closely related species, *Spiraea alpina* and *S. mongolica*, were analysed across the Qinghai–Tibetan Plateau (QTP) before, during, and after the most recent glaciation or orogenic events using chloroplast DNA (cpDNA) and microsatellite data sets. Data were obtained by generating and analysing data sets for 851 individuals from 63 natural populations. Three cpDNA markers were used to infer phylogenetic relationships using Network and PAUP*. Microsatellite data were analysed in STRUCTURE 2.2.3. Patterns of variation among and within populations were analysed in Arlequin 3.5. Divergence time was calibrated in BEAST 1.7.5 with *Sorbaria kirilowii* fossils. Phylogenetic relationships of the two species revealed two main groups and three subgroups. Alternatively, the microsatellite data revealed three main clusters. Demographic analysis based on chlorotypes revealed population recolonization in the late Pleistocene after the most recent glaciation and high genetic diversity in both species. Our results indicate that the high genetic diversity of these shrubs is attributable to Pliocene and Pleistocene climatic changes, combined with orogenic events of the QTP. The distribution of these spiroides species on the QTP resulted from the combine effects of climatic instability during quaternary glacial–interglacial periods and uplift of the QTP.

T3-12-03

Introduction of Sahara mustard (*Brassica tournefortii*) in North America: Evolutionary and environmental mechanisms promoting invasive species success

Daniel Winkler[1], *Kenneth Chapin*[2], *Travis Huxman*[1]

1. University of California, Irvine

2. University of California, Davis

Sahara mustard (*Brassica tournefortii*) has successfully colonized semi-arid regions eastward from California to Texas and northward into Nevada and Utah since its introduction in California in the early 20th century. The dispersal and genetic mechanisms that have enabled its success remain unexplored and are a major hurdle in the successful control of the species. The recent, rapid invasion of this species provides an opportunity to address hotly debated questions in invasion biology and natural selection. We used a next-gen genotyping by sequencing technique to generate reduced representation genomics for 943 individuals from 55 locations and six states across the species' invaded range to identify population structure and invasion history. Further, we hypothesized that human-mediated dispersal is driving colonization in the US. To address this, we tested for roadways as dispersal corridors as an alternative to isolation by distance models. We inferred discrete populations and calculated the hierarchical genetic structure within and among locations. Further, we identified a unique pattern of private alleles that helps explain structure. Last, we identified introduction, expansion, and admixture events in the species' history. We followed our genetic analyses with a multi-generational common garden experiment to test for trait variation and identify adaptive traits across the range. We identified 1,164 single nucleotide polymorphisms across the 55 locations sampled. We found no evidence of isolation by distance among locations, revealed a significant amount of genetic variation across and within locations, and found evidence for distinct genetic clusters of Sahara mustard in the US. Overall genetic diversity was low across sites, perhaps due to the species dispersal modality and natural history. Surprisingly high heterozygosity across the species range likely indicates admixture of multiple introduction sites after initial invasions. Overall, we found moderate to low levels of genetic structure supporting human-mediated migration along roadways. Additionally, we found genetically-controlled trait differences including phenological, morphological, and physiological traits that may explain the species success in local environments. We present Sahara mustard as an eminently-suited study system to understand invasion evolutionary biology, and discuss our next steps aimed at genomic analyses in the species native range.

T3-12-04

The battle between historical bottleneck and gene flow during population expansion in natural forest of *Cyclobalanopsis neglecta* (Fagaceae)

Xueqin Zeng, *Gunter Fischer*

Kadoorie Farm and Botanic Garden, Hong Kong

The anthropogenic disturbance causes massive forest decline, and the ensuing forest conservation would make forest expanding rapidly. However, it is still uncertainty how populations genetically respond to the demographic change. Hong Kong has current expanding forest which was mostly developed after severe habitat disturbance in 1940's during the wars. We took an oak species *Cyclobalanopsis neglecta* as an example to study the demographic genetics of young natural forest in Hong Kong. Genetically diverse compositions were detected in old generations and increasing genetic compositions in young generations which indicated the genetically diverse founder populations. Historical bottleneck effect was shown by the low population allelic diversity, high among-population genetic differentiation and high within-population spatial genetic structure (SGS) in old generations compared to young generations. Gene flow could erase the negative consequence of bottleneck effect. In conclusion, both historical bottleneck effect and current gene flow determined the demographic genetics, and rapid population expansion would speed up the process of genetic recovery. Understanding the dynamics of genetic structure and genetic diversity could determine the relative role of stochastic and deterministic ecological and evolutionary forces, which can guide the practice of forest restoration and conservation.

T3-12-05

The immediate capacity for adaptation and its realization in natural plant populations

Mason Kulbaba[1], *Seema Sheth*[2], *Rachel Pain*[1], *Vincent Eckhart*[3], *Ruth Shaw*[1]

1. University of Minnesota

2. University of California, Berkeley

3. Grinnell College

Adaptation results from evolutionary response to natural selection operating across all life-history stages of an organism's existence. The rate of this adaptation, as predicted by the Fundamental Theorem of Natural Selection, depends on the additive genetic variance for fitness, σ_A^2 (*w*). Explicit quantitative predictions regarding the rate of adaptation can be made from estimates of σ_A^2 (*w*) and compared with measured, realized adaptation in order to fully evaluate the dynamics of the evolutionary response to selection. Whereas this prospective, quantitative approach to estimating the capacity for adaptation has been available for over 80 years, studies have tended to focus retrospectively on qualitative changes in fitness. Such qualitative assessments of the changes in fitness demonstrate adaptation, but do not directly illuminate the adaptive process. In contrast, prediction of the change in mean fitness via σ_A^2 (*w*), together with evaluation of the realized change in mean fitness, elucidates ongoing adaptation from standing additive genetic variation in natural populations. This talk describes results from a multi-generation study of the annual legume *Chamaecrista fasciculata* to evaluate the immediate capacity for adaptation, and the degree to which that capacity is realized under contemporary natural selection. Pedigreed seeds generated from greenhouse crosses within three populations (1,737-3,593 seeds) were planted into restored prairie near the seed source for each population (Minnesota and Iowa, USA). Each individual was followed throughout its complete lifespan with fitness outcomes recorded at each life-cycle stage used cumulatively to estimate σ_A^2 (*w*). For the following year, progeny resulting from the reproduction of the initial pedigreed populations were planted into each respective site. In addition, individuals from the initial pedigreed greenhouse population were planted again so that mean lifetime fitness of the two generations could be evaluated in common, contemporaneous environments. The use of aster modeling permitted inclusion of all components of fitness (e.g., seeds that did not germinate) that traditional statistical treatments omit to accommodate statistical assumptions. Further, explicit modeling of the dependence of components of fitness measures on those expressed at earlier life-history stages (e.g., number of seeds produced given total number of fruits set) produces more precise assessments of fitness. This talk will report on a large-scale experiment to empirically predict the rate of adaptation, and evaluate the accuracy of this prediction with realized measures of adaptation over two generations in three natural populations. These results will lay the foundation for the interpretation of natural selection and adaptive evolutionary responses. Finally, these results also directly relate to assessing the potential for evolutionary rescue in the face of ongoing climate change.

T3-12-06

Genomic, epigenomic and metagenomic analyses reveal parallel ecological divergence in *Heliosperma pusillum* (Caryophyllaceae)

Emiliano Trucchi[1], *Aglaia Szukala*[1], *Juliane Baar*[1], *Bozo Frajman*[2], *Clara Bertel*[2], *Peter Schoenswetter*[2], ***Ovidiu Paun***[1]

1. University of Vienna

2. University of Innsbruck

Mosaic distribution of interbreeding taxa with contrasting ecology and morphology constitute an ideal opportunity to study the pace and mode of neutral and adaptive processes during repeated evolution of ecotypes. Applying a combination of population genetic, epigenomic, transcriptomic, ecologic and metagenomic tools we investigate here the evolutionary history of an alpine and a montane ecotype of *Heliosperma pusillum* (Caryophyllaceae) in the south-eastern Alps. From six pairs of geographically close populations of the two ecotypes we obtained a high-coverage RADseq dataset that was further used for demographic inference to test the hypothesis of parallel evolution of the two ecotypes. The data demonstrate repeated ecological divergence in *H. pusillum*, uncovering up to five polytopic origins of one ecotype from the other. Coalescent-based demographic simulations evidence a complex evolutionary history, with local isolation-with-migration in two population pairs and intra-ecotype migration in two others. In all cases, the time of divergence or secondary contact was inferred as postglacial. A metagenomic analysis on exogenous contaminant RAD sequences demonstrates divergent microbial communities between the ecotypes, whereas measured abiotic conditions are significantly divergent between the two types of localities, but consistent within each ecotype. The similar set of biotic interactions is involved, together with abiotic factors, in shaping common selective regimes at different growing sites of each species. However, a lack of shared genomic regions of high divergence across population pairs illustrates the power of drift and/or local selection in shaping genetic divergence across repeated cases of ecological divergence, together with the complexity of molecular basis of ecologically-relevant traits. Epigenetic alterations can be expected to be particularly important in the early phases of adaptation and divergence. Employing a novel approach (bisulfite-converted RAD sequencing–bsRADseq), we screened DNA methylation patterns across more than 200,000 nucleotide positions in the six population pairs of the two ecotypes. Comparing by sequencing DNA methylation state and context (CpG, CHG, CHH) in 120 individuals, we demonstrate that epigenetic patterns are much more conserved across representatives of the two ecotypes than predictions given the distinct native environments and that several candidate loci for convergent epigenetic modification during parallel adaptation to a similar environment are present. From progeny screens across three generations we infer high heritability of DNA methylation, that is however, context dependent. Altogether, our work supports the hypothesis that repeated ecological divergence, observed here at an early stage, may be a common process of species diversification.

T3-13: Halophyte physiology

T3-13-01

How do recretohalophytes secrete salt? From structures to genes

Fang Yuan, Zhongtao Feng, Yunquan Deng, Bingying Leng, ***Baoshan Wang***
Shandong Normal University

To survive in a saline environment, halophytes have evolved many strategies to resist salt stress. The salt glands of recretohalophytes are unique features for directly secreting salt out of a plant. Knowledge of the pathways of salt secretion in relation to the function of salt glands may help us to change the salt-tolerance of crops and to cultivate the extensive saline lands that are available. Recently, ultrastructural studies of salt glands and the mechanism of salt secretion, particularly the candidate genes involved in salt secretion, have been illustrated in detail. In this report, we summarize current researches on salt gland structure, salt secretion mechanism and candidate genes involved, and provide an overview of the salt secretion pathway and the asymmetric ion transport of the salt gland. (1) Salt gland structure. To date, eleven families (65 species) have been discovered to have salt gland structures. The number of component cells has been used to separate multi-cellular salt gland and bi-cellular salt gland. Typical characteristics were identified in salt gland: the cuticles surrounding the salt gland, plasmodesmata between the mesophyll cells and the salt gland, and accumulated vesicles, which were considered to play a significant role in salt secretion combined with both apoplastic and symplastic transport. (2) Salt secretion mechanism. Three hypotheses of salt secretion were also supported by the recent studies: the role of the osmotic potential in salt secretion, a transfer system similar to liquid flow in animals and salt solution excretion by vesicles in the plasma membrane (exocytosis). In the last few years, more and more evidence has supported the third one. In addition, the membrane-bound translocating proteins are likely to be involved in salt secretion and these transporters mainly distributed asymmetrically in the other side of plasma membranes of salt gland cells adjacent to the collecting chamber. (3) As a typical recretohalophyte possessing salt glands, *Limonium bicolor* is a potential model plant for investigating the salt secretion mechanism by salt glands in the future. 102 genes high likely to be involved in salt secretion of *L. bicolor* have been screened by high-throughput RNA sequencing. The roles of these genes are being identified in our laboratory. For a decade, various ion transporters have been shown to participate in salt resistance as verified in *Arabidopsis*. However, the in-depth mechanism of halophyte salt tolerance, in particular the molecular mechanism of salt gland secretion, is largely unknown. More and more scientists have begun to use halophytes as materials to study the salt tolerance mechanism.
Acknowledgements: This work has been supported by the NSFC (National Natural Science Research Foundation of China, project No. 31570251), Programs Foundation of Ministry of Education of China (20123704130001) and Natural Science Research Foundation of Shandong Province (ZR2014CZ002).

T3-13-02

ZxAKT1 is essential for K^+ uptake and K^+/Na^+ homeostasis in the succulent xerophyte *Zygophyllum xanthoxylum*

Qing Ma[1], Jing Hu[1], Xiang-Rui Zhou[1], Hui-Jun Yuan[1], Tanweer Kumar[1], Sheng Luan[2,3], ***Suo-Min Wang***[1]
1. State Key Laboratory of Grassland Agro-ecosystems, College of Pastoral Agriculture Science and Technology, Lanzhou University
2. Department of Plant and Microbial Biology, University of California at Berkeley
3. NJU-NJFU Joint Institute for Plant Molecular biology

The inward rectifying K^+ channel AKT1 constitutes an important pathway for K^+ acquisition in plant roots. In glycophytes, excessive accumulation of Na^+ is accompanied by K^+ deficiency under salt stress. However, in the succulent xerophyte *Zygophyllum xanthoxylum*, which exhibits excellent adaptability to adverse environments, K^+ concentration remains at a relatively constant level despite increased levels of Na^+ under salinity and drought conditions. In this study, the contribution of *ZxAKT1* to maintaining K^+ and Na^+ homeostasis in *Z. xanthoxylum* was investigated. Expression of *ZxAKT1* rescued the K^+-uptake-defective phenotype of yeast strain CY162, suppressed the salt-sensitive phenotype of yeast strain G19, and complemented the low-K^+-sensitive phenotype of *Arabidopsis akt1* mutant, indicating that *ZxAKT1* functions as an inward-rectifying K^+ channel. *ZxAKT1* was predominantly expressed in roots, and was induced under high concentrations of either KCl or NaCl. By using RNA interference technique, we found that *ZxAKT1*-silenced plants exhibited stunted growth compared to wild-type *Z. xanthoxylum*. Further experiments showed that *ZxAKT1*-silenced plants exhibited a significant decline in net uptake of K^+ and Na^+, resulting in decreased concentrations of K^+ and Na^+, as compared to wild-type *Z. xanthoxylum* grown under 50 mM NaCl. Compared with wild-type, the expression levels of genes encoding several transporters/channels related to K^+/Na^+ homeostasis, including *ZxSKOR*, *ZxNHX*, *ZxSOS1* and *ZxHKT1;1*, were reduced in various tissues of a *ZxAKT1*-silenced line. These findings suggest that *ZxAKT1* not only plays a crucial role in K^+ uptake but also functions in modulating Na^+ uptake and transport systems in *Z. xanthoxylum*, thereby affecting its normal growth.

T3-13-03

Short and long term effects of salinity on photosynthesis and biomass of *Phragmites karka*: A biofuel candidate

Erum Shoukat, Zainul Abideen, ***Muhammad Zaheer Ahmed****, Salman Gulzar, Muhammad Ajmal Khan*
Institute of Sustainable Halophyte Utilization, University of Karachi

Phragmites karka (tall reed) considered a potential biofuel crop due to high productivity and ligno-cellulosic contents. Whereas, photosynthesis is a growth controlling mechanism and affected by salinity particularly in C_3 grasses. However, plant exhibit different strategies to deals with the time-course effects of salinity. Knowledge about plant response to short and long term exposure of salinity could help in better manage our crop in saline areas. Therefore, the aim of this study was to investigate the short (0-7 days) and long (15-30 days) term effects of salinity on growth and photosynthetic responses of *Phragmites karka.* Plant growth, gas exchange, chlorophyll fluorescence, soluble sugar (SS), stomatal stoichiometry, photosynthetic pigments, hydrogen peroxide (H_2O_2) and malondialdehyde (MDA) content were measured after different time periods (7,15 and 30 days) in 0 (control), 100 (moderate) and 300 (high) mM NaCl treated plants. In short term exposure of moderate salinity, plant maintain biomass and photosynthetic performance but in high salinity reduced growth link with the decline in whole plant photosynthesis (A_W). Reduction in net photosynthetic rate (A_{NET}) at 300 mM NaCl could be accounted by lower stomatal conductance (gs), transpiration (E) and internal carbon dioxide concentration (Ci). During long term salinity treatment, *P. karka* reduced growth about 35% and 80% at 100 and 300 mM NaCl, respectively. Similarly, inhibition of A_W was more severe under high salinity. In moderate salinity plant maintained A_{NET} similar to control by improve water use efficiency (WUE) and instantaneous carboxylation efficiency (COE), with no change in relative excess of light energy (EXC), maximum and effective quantum yield of PS II (Fv/Fm, YII). Moderate salinity reduced the stomatal size but increase number of stomata, photosynthetic pigments, dark respiration (Rd) and sugar content. In long term exposure with 300 mM NaCl, A_{NET} decreased along with gs, E, stomatal size and density while Ci was maintained. High salinity reduced COE and YII, whereas Fv/Fm declined at noon but not at predawn time. Moreover, higher H_2O_2 and MDA content in plant of 300 mM NaCl were linked with the increase in photosynthetic pigments and EXC. In conclusion, *P. karka* maintained photosynthesis via alter stomatal stoichiometry and increase the efficiency of CO_2 fixation while reduce growth for long term survival in moderately saline areas. Whereas high salinity severely decreased biomass production and photosynthesis due to stomatal limitation (in short term effect) and lower efficiency of PSII and CO_2 fixation (in long term effect). Our results indicate that *P. karka* could cultivate under moderately saline areas or with brackish water irrigation for source of energy.

T3-13-04

Coping with extremes: Molecular and physiological basis of salinity stress tolerance in halophytes

Sergey Shabala
University of Tasmania

Sustainability of world agriculture is critically dependent on the development of new generation crops that will be more resilient to drought and capable of coping with increasing concentration of salt in the soil, without yield penalties. Recently we have argued (Shabala *et al* (2014) Trends Plant Sci) that salinity tolerance of traditional crops may be improved by incorporating some specific traits found in halophytes. In this talk, I will focus on two major mechanisms conferring salinity tolerance in halophytes that can be targeted in crop breeding programs. The first one is effective Na^+ sequestration in either internal (vacuoles) or external (epidermal bladder cells; EBC) structures. The second mechanism is plant's ability to maintain efficient photosynthetic gas exchange via stomata under hyperosmotic stress conditions. Working along the first lines, we show that internal (vacuolar) Na^+ sequestration in halophytes is achieved by the orchestrated regulation of at least four major traits: (1) higher expression levels and operation of tonoplast NHX Na^+/H^+ exchangers; (2) higher tonoplast H^+-pump activity; (3) better cytosolic K^+ retention originating from intrinsically higher plasma membrane H^+ pumping ability; and (4) efficient control of SV and FV tonoplast channels to prevent futile Na^+ cycling between the cytosol and vacuole. I then analyse mechanism underlying sequestration of toxic Na^+ and Cl^- species in external structures such as epidermal bladder cells. I describe the essentiality of the tissue-specific expression of specific transporters and show experimental evidence for the functional expression and operation of HKT, SOS1, and HAK transporters in epidermis-stalk cell- EBC continuum. In the final part of my presentation I discuss peculiarities of stomata operation in halophytes. The questions asked are: (1) is there a requirement for Na^+ as an osmoticum for stomatal movements in halophytes? (2) is there any difference in guard cell sensitivity to abscicic acid (ABA) and hydrogen peroxide (H_2O_2) between glycophyte and halophyte species? (3) what is the molecular identity and regulation of key membrane transporters mediating stomatal movement in halophytes? Once molecular identity of key genes and control modes of the above processes are understood, plant breeders would be able to devise breeding strategies to incorporate desirable features into traditional crops via molecular or traditional breeding tools thus expanding the sustainable limits for productive use of soil and water resources.

T3-13-05

Comparative study of salt tolerance mechanisms in *Oryza coarctata* and *Spartina anglica*

Nuria Koteyeva[1], *Elena Voznesenskaya*[1], *Anna Berim*[2], *David Gang*[2], *Gerald Edwards*[2]
1. Komarov Botanical Institute of Russian Academy of Sciences
2. Washington State University

Oryza (=Porteresia) coarctata is a wild rice species which has high tolerance to salt with traits which are of interest for improving tolerance of rice growing under saline conditions; while, *Spartina anglica* is a serious invasive species in natural saltmarsh ecosystems. In this study mechanisms of tolerance to salinity were analyzed. The leaf anatomy and ultrastructure of salt glands, pattern of salt excretion, features of gas exchange, accumulation of key photosynthetic enzymes, leaf water content, Na^+ and K^+ accumulation, and osmolality along with levels of some osmolytes and betaine aldehyde dehydrogenase expression, were compared in the two species grown without salt, with 200 mM NaCl, and with 200 mM KCl. Measurements of gas exchange showed photosynthesis was not affected by salt treatments. Despite differences in the type of photosynthesis (*S. anglica* is C_4 while *O. coarctata* is a C_3 plant), the two species have a very similar leaf anatomy; in both the abaxial side is flat while the adaxial side has distinctive ridges. Both species have salt glands which are located deep in the leaf ridges; however they are very different in their anatomy and ultrastructure. *S. anglica* has bicellular glands with basal and cap cells. However, *O. coarctata* has unicellular glands with wall labyrinths near the apical part of the cell which are more developed than in the basal gland cell in *S. anglica*. In contrast to a previous report, our results show that the multiple unicellular hairs which occur in *O. coarctata* are not responsible for salt secretion. Both species accumulate compatible solutes with increasing osmolality but they differ in osmolyte composition. In summary, *Spartina anglica* and *Oryza coarctata* show high tolerance to salt in function of photosynthesis, maintenance of water content, and in excretion of salt from leaves; but, they have distinct differences in the structure of salt glands and accumulation of osmolytes.

T3-13-06

Genetic improvement of alfalfa by transferring genes from xero-halophyte *Zygophyllum xanthoxylum*

Ai-Ke Bao, *Peng Kang, Fei Wang, Zhulatai Bao, Xiao-Wen Zhang, Tanweer Kumar, Suo-Min Wang*
State Key Laboratory of Grassland Agro-ecosystems, College of Pastoral Agriculture Science and Technology, Lanzhou University

Alfalfa is an important legume forage containing a variety of vitamins and high protein content. In recent years, the demand for alfalfa has been shown a dramatic increase in China, since it is the vital foundation for Chinese dairy industry. Most of the existing alfalfa cultivars show high water consumption, weak tolerance to salinity and nutrient (such as P and N) deficiency. Therefore, under the background of limited arable land, developing alfalfa cultivars suitable for growing on the marginal lands is necessary for alfalfa production in northern China. However, it is hard to significantly enhance the stress tolerance of alfalfa only relying on the existing alfalfa resources and using intraspecific hybridization. *Zygophyllum xanthoxylum* is a xero-halophyte distributing in desert regions of northern China with excellent adaptability to adverse environments. Our previous studies showed that the transporters/channels involved in uptake, transport and compartmentation of Na^+, K^+ and other important nutrient elements play vital roles for *Z. xanthoxylum* to adapt to arid and saline environments. Tonoplast Na^+/H^+ antiporter (ZxNHX) is essential for controlling Na^+ and K^+ uptake, long-distance transport and their homeostasis at whole-plant level through feedback regulating the expression of genes involved in Na^+ and K^+ transport, thereby determining the salt tolerance of *Z. xanthoxylum*. On the other hand, the surface of *Z. xanthoxylum* is covered by thick cuticle, which can prevent the water loss via leaf in dry and heat conditions. In the present study, we co-expressed *Z. xanthoxylum ZxNHX* and *ZxVP1-1* genes in alfalfa. The transgenic alfalfa grew faster than wild-type under either greenhouse or field conditions by developing a larger root system, and showed enhanced salt and drought tolerance with higher leaf relative water content and greater photosynthesis capacity, which was associated with more Na^+, K^+ and Ca^{2+} accumulation in plants. As a consequence of increased vacuolar compartmentation, the addition of 50 mM NaCl significantly alleviated the deleterious impact of osmotic stress on the growth of transgenic alfalfa. Interestingly, transgenic alfalfa showed enhanced tolerance to P deficiency, which was consistent with more P accumulation, and improved adaptation to low nitrogen condition with a higher nodule formation ability after inoculated with *Sinorhizobium meliloti*. Compared with wild-type, transgenic alfalfa developed more and bigger multiple-branched nodules. Finally, the assessment of forage quality indicated that transgenic plants accumulated more crude protein, crude fat, and have higher relative feeding value (RFV) than wild-type, especially under stress conditions and in the field. To further fortify the stress tolerance of transgenic alfalfa, we also introduced another gene *ZxABCG11* of *Z. xanthoxylum* into alfalfa and are developing the transgenic lines with thick cuticle on leaf surface, since ABCG11 was considered as playing a vital role in cuticular lipids secretion. This study provides a solid foundation for creating the alfalfa cultivars with high yield, good quality and wide adaptability on saline, dry and nutrient-deprived marginal lands of northern China.

T3-14: Harnessing new genetic approaches to improve restoration success

T3-14-01

Restore & renew: Providing evolutionary, environmental and ecological information on a large scale to support restoration practices

Maurizio Rossetto
National Herbarium of NSW

The flagship research project Restore & Renew, is a world-first initiative that will create a comprehensive and easy to use website readily accessible by the community and supporting effective and sustainable land restoration. Utilising novel whole-genome sequencing and environmental modelling techniques, the project will uncover an unparalleled level of information to understand how genetic diversity is partitioned across the landscape, and if genetic provenances are associated with climatic and environmental variables. The project will collect, analyse and share genetic, adaptive, environmental, and ecological data for over 200 plant species commonly used in restoration projects across Australia's Eastern seaboard and representing the regional floristic, ecological and phylogenetic diversity. No other project on this scale and scope exists worldwide and I will present the concept and preliminary outcomes of this large-scale study. Restore & Renew will deliver a readily accessible resource of comprehensive restoration and management guidelines on an unprecedented scale. This enduring community resource will improve the success and long-term viability of land restoration projects, as well as improve predictive capacity to respond to climate change. Beside supporting restoration practices, the project will also provide significant information from species to landscape levels, discover regions of high genetic diversity, identify commonalities among taxonomic and functional groups that will improve our ability to generalise beyond the 200 species, and enable us to explore how species and assemblages are likely to respond through time. Understanding and predicting how plants will grow and interact in changing ecosystems is vital to conserving, or restoring, resilient natural habitats.

T3-14-03

Revegetation rewilds the soil bacterial microbiome of an old field

Martin Breed
University of Adelaide

Ecological restoration is a globally important and well-financed management intervention used to combat biodiversity declines and land degradation. Most restoration aims to increase biodiversity towards a reference state, but there are concerns that intended outcomes are not reached due to unsuccessful interventions and land use legacy issues. Monitoring biodiversity recovery is essential to measure success, however most projects remain insufficiently monitored. Current field-based methods are hard to standardise and are limited in their ability to assess important components of ecosystems, such as bacteria. High-throughput amplicon sequencing of environmental DNA (metabarcoding of eDNA) has been proposed as a cost-effective, scalable and uniform ecological monitoring solution, but its application in restoration remains largely untested. Here we show that metabarcoding of soil eDNA is effective at demonstrating the return of the native bacterial and fungal communities in an old field following native plant revegetation. Microbiome composition shifted significantly after 8 years of revegetation, where younger sites were more similar to cleared sites and older sites were more similar to remnant stands. Revegetation of the native plant community strongly impacted on the belowground microbiome, despite the revegetated sites having a long and dramatically altered land use history (i.e. > 100 years grazing). We demonstrate that metabarcoding of eDNA provides an effective way of monitoring changes in microbiomes that would otherwise go unchecked with conventional monitoring of restoration projects. With further development, awareness of microbial diversity in restoration has significant scope for improving the efficacy of restoration interventions more broadly.

T3-14-04

Metabarcoding reveals soil microbial community succession patterns over a 30-year chronosequence of post-mining forest ecosystem restoration

Siegfried Krauss[1], *Martin Breed*[2], *Andrew Bissett*[3]
1. Kings Park and Botanic Garden
2. Adelaide University
3. CSIRO

Advances in next generation sequencing (NGS) methods are promising to revolutionise the practical contribution of genetics to ecological restoration. Novel applications include an enhanced capacity to measure adaptive genetic variation for seed sourcing, gene expression analysis as a measure of genetic resilience of restored populations, and high-throughput assessment and monitoring of natural and restored biological communities aboveground and belowground. Soil microbial community characterization through high throughput meta-barcoding provides a powerful tool to assess responses and trajectories of restoration site soils post-disturbance. This study assesses changes in bacterial and fungal communities at 36 sites through a 30-year-old chronosequence of highly diverse forest-community rehabilitation after bauxite mining in south-west Australia. Bacterial and fungal communities were analysed using Illumina MISEQ sequencing of the 16S rRNA and ITS1 gene regions respectively. These were assessed against changing restoration practises, above ground biodiversity, adjacent intact remnant communities, and physical/chemical properties of the soil. Measuring below-ground richness using NGS considerably alters perceptions of biodiversity in restored and natural landscapes, and helps to optimise reconstructed substrates for aboveground and belowground biodiversity.

T3-14-05

A test of climate-informed restoration in forests adjacent to the north shore of Lake Superior, MN, USA

Julie R. Etterson[1], *Mark A. White*[2], *Laura C. Kavajecz*[1], *Jordan S. Mead*[1], *Meredith W. Cornett*[2]
1. University of Minnesota Duluth
2. Minnesota - North Dakota - South Dakota, The Nature Conservancy, Duluth, MN

Great Lakes forests will likely experience some of the most dramatic effects of climate change in the continental U.S.A. Simulation modeling indicates that temperate tree species that currently occur in the region at low abundance levels (e.g., *Quercus rubra*, *Q. macrocarpa* and *Pinus strobus*) will be well-adapted to future climate conditions, but also demonstrates that the natural colonization rates of these species cannot keep pace with the rapidly changing climate leaving many suitable niches unrealized. In the first climate-informed restoration project in this region, we tested the efficacy of within-range planting of climate adapted species sampled from more than one seed transfer zone. In 2013 and 2014 we planted a total of 4,400 seedlings in sixteen sites using both local and seed sources from further south or west where climate conditions are warmer. For three years (2014-2016), we assessed survival, growth, and spring and fall phenology. We compared the fit of the data using four models that featured contrasting factors: 1) seed transfer zones, 2) forest types, 3) geographical position, and 4) climate variables. Overall, the geographical model fit the data best for most traits according to DAICc. Differences between plants sampled from contrasting seed zones was found for both oak species and in some cases for the pine. Interestingly, the plant ecotype that performed best in northern conditions differed between species. In sum, early results suggest climate-informed restoration within the geographic range of species may be an effective strategy for maintaining key forest functions in a changing climate.

T3-14-06

The genomic basis of climate adaptation in lodgepole pine and interior spruce

Kathryn Hodgins[1], *Sam Yeaman*[2], *Katie Lotterhos*[3], *Kristin Nurkowski*[4], *Loren Rieseberg*[4], *Whitlock Mike*[4], *Tongli Wang*[4], *Sally Aitken*[4]
1. Monash University
2. University of Calgary
3. Northeastern University
4. University of British Columbia

Local adaptation is common in widespread conifer species and current reforestation policy reflects this through local seed sourcing and breeding programs. However, as the climate changes local tree populations may become mismatched to their local environments. Our goal was to identify the genes responsible for climatic adaptation in western Canada's two most economically important conifers, lodgepole pine (*Pinus contorta*) and interior spruce (*Picea glauca*, *P. engelmannnii*, and their hybrids). To do this we conducted a population genomics study of local adaptation for both these species across similar climatic gradients. Using a sequence capture approach that targeted the exome, we used environment allele associations, and phenotype allele associations to identify candidate regions of the genome important for temperature adaptation. Our comparative analysis of these regions between the two species revealed that adaptation to temperature has a polygenic signature of convergence at the genomic level. This is despite the fact that these species have been diverged for more than 140 million years. This suggests that adaptation to climate is somewhat genetically constrained, with key genes, particularly transcription factors, playing non-redundant roles. Our results will be important for designing reforestation policies that consider future climates, and for understanding the genetic capacity of natural populations to adapt to new climatic conditions.

T3-15: Effects of thermophilisation on alpine vegetation

T3-15-01

No more late bloomers: Warming induces late-season flower deficits in bumble bee host plants

Nicole Miller-Struttmann[1,2], *James D. Franklin*[3], *Peter G. Kevan*[4], *Candace Galen*[3]

1. Webster University

2. Missouri Botanical Garden

3. University of Missouri

4. University of Guelph

Climate-induced changes contribute to recent pollinator declines by disrupting pollination mutualisms through phenological, functional or behavioral mismatches. Social insects, such as bumble bees, have complex life histories that require flower resources during three key stages: queen emergence, worker accrual, and mating. Bumble bees are dominant pollinators in climate-sensitive high altitude environments which forecast the effects of climate change in other ecosystems and where abundance of bumble bee host flowers has declined with warming temperatures. The seasonal spacing of these declines will shed light on the implications for bumble bee life cycles. In one montane habitat, a mid-season flower deficit has emerged with recent climate change, potentially restricting colony growth but not queen overwintering provisions. Here, we determine the impact of temperature and snowpack on the onset, peak and duration of flowering for historical bumble bee host plants on three mountains in Colorado: Pennsylvania Mountain, Niwot Ridge Long-Term Ecological Research station, and Mt. Evans Wilderness. Using phenology data collected across a 300 m altitudinal gradient on Pennsylvania Mountain from 1977-1980 and at all three sites from 2012-2014, we tested for climate-induced shifts in flowering onset, peak and duration. We asked whether temperature and snowpack variation among sites gives rise to geographical variation in flowering phenology, congruent with historical trends on Pennsylvania Mountain. Taken together, these data elucidate whether flower availability aligns with key bumble bee life history stages at the scale of bumble bee foraging. In comparison to historical norms on Pennsylvania Mountain, bumble bee host plants flowered for 11 fewer days, on average, in 2012-2014. Late flowering plants have advanced more in both onset and peak flowering with warming than early flowering species, resulting in a compression of the flowering season. Similarly, warmer temperatures were associated with compressed flowering on three geographically distinct mountains with flowering duration declining on average by 3.15 days per 1°C warming. While the flowering season was compressed, flowering overlap did not increase. Indeed, co-blooming species richness decreased by 0.81 species per 1°C warming. Plots with higher species richness bloomed for longer, indicating that fewer species bloom in each plot when it is warmer. Overall, these results suggest that warming-induced flower deficits will be most severe at end of the colony cycle when new reproductive bees are developing, mating, and provisioning for over-winter torpor. Contrary to mid-season deficits seen at lower altitude, the late-season phenological mismatch between flowering and colony needs could weaken colony production and queen overwintering success in high-altitude bumble bees.

T3-15-02

The genus *Saxifraga* as a model to study drivers of alpine diversification and global climate change risks

Michelangelo Moerland[1,2], *Jurriaan De Vos*[1], *William Baker*[1], *Alexander Papadopulos*[1], *Julie Hawkins*[2], *Wolf Eiserhardt*[1]

1. Royal Botanic Gardens, Kew

2. School of Biological Sciences, University of Reading

Mountains can harbour high and unique plant diversity, but in spite of long standing scientific interest the fundamental principles responsible for the diverse mountain biota are not fully understood. The species-rich genus *Saxifraga* L. (ca. 440 spp.) has a predominantly alpine distribution and is emerging as a model system to study the eco-evolutionary assembly of mountain floras. We are assembling a high-resolution phylogenetic tree of *Saxifraga* and to this end use next generation sequencing (HybSeq) to obtain genome-wide data to overcome problems due to a lack of genetic variation among recently diverged species. In addition, a global database will be built of *Saxifraga* distribution, habitat and functional traits. Combined with the species-level phylogeny this database can be used to infer historic dispersal events and radiations, and the role of habitat- and trait evolution therein. Furthermore, many *Saxifraga* taxa are narrow endemics in sensitive alpine habitats. Our study also aims to unveil to what extent species diversity and functional diversity are at risk due to global climate change.

T3-15-03

Range shifts in response to climate change of *Ophiocordyceps sinensis*, a fungus endemic to the Tibetan Plateau

Yujing Yan[1], ***Yi Li***[2,3], *Wen-Jing Wang*[2], *Jin-Sheng He*[1], *Rui-Heng Yang*[2], *Hai-Jun Wu*[2], *Xiao-Liang Wang*[2], *Lei Jiao*[2], *Zhiyao Tang*[1], *Yi-Jian Yao*[2]

1. College of Urban and Environmental Sciences, Peking University

2. Institute of Microbiology, Chinese Academy of Sciences

3. Institute of Plant Virology, Fujian Agricultural and Forestry University

Recent climate change has been widely recognized to influence the distribution of many plants and animals, while its impacts on the distribution of fungi remain largely unknown. Here, we used *Ophiocordyceps sinensis*, an entomopathogenic fungus and important traditional Chinese medicine whose distribution range was reported as decreased on the Tibetan Plateau in recent decades, as an example to predict the current potential distribution and the possible range shifts in response to climate change of a fungus by using extensive field records and an ensemble species distribution modeling method. It is demonstrated that the distribution range of the fungus wound decrease significantly, shifting upward in altitude and toward the central part of the Plateau. In an unlimited dispersal scenario, net habitat losses of 19% for both years 2050 and 2070 under representative concentration pathways (RCP) 2.6 and of 8% and 4% under RCP8.5 for the years 2050 and 2070, respectively, were predicted. If a non-dispersal scenario was considered, 36–39% of the current habitats would be lost in the future. The results presented here will not only provide useful information for the conservation of *O. sinensis*, but also provide a representative case of evaluating impacts of climate change on fungal distribution using species distribution modeling method.

T3-15-04

Plant geographic range size and climate stability in China: Growth form matters

Wubing Xu[1], *Jens-Christian Svenning*[2], *Guoke Chen*[1], *Bin Chen*[3], *Jihong Huang*[4], *Keping Ma*[1]
1. Institute of Botany, Chinese Academy of Sciences
2. Section for Ecoinformatics and Biodiversity, Aarhus University
3. Shanghai Chenshan Botanical Garden
4. Institute of Forest Ecology, Environment and Protection, Chinese Academy of Forestry

Geographic variation of range size is often proposed to be linked to climate seasonality and Quaternary glacial-interglacial oscillations. In addition, the importance of long-term climate stability is thought to be modulated by species' dispersal abilities. For plants, growth forms integrate numerous traits and likely also dispersal ability. Here, we assessed the relative importance of short- and long-term climate stability for range size patterns and whether the importance of the latter is greater for more poorly dispersed growth forms. Location: China. A comprehensive botanical dataset including over 30,000 vascular plant species was used to quantify the geographic patterns of range size across China for all species and eight growth forms separately. Then, spatial and non-spatial regressions with information-theoretic multi-models selection were performed to estimate explanatory importance of short- and long-term climate stability as well as land area and contemporary average climate conditions. The proportion of endemic species for each growth forms was also calculated. Geographic range sizes were significantly associated with both short- and long-term climate stability, with large range sizes occurred in areas with strong climate seasonality and historical climate change. The association between range size and climate change velocity was stronger for perennial herbs and shrubs than ferns, annual herbs, climbers and trees and stronger for small trees than large trees, i.e., more important for the growth forms with the lowest general dispersal capacities. In line with this interpretation, growth forms strongly influenced by velocity also tend have high proportions of endemism, which further confirm their generally low dispersal abilities. Our results suggest that range size patterns of vascular plants in China are shaped by both climate seasonality and long-term climate stability, despite that the region was relatively mildly influenced by the Pleistocene glaciations, in supplement of other factors, with a stronger influence of paleoclimatic stability on more poorly dispersed groups. The differential influences of paleoclimatic stability across growth forms suggest that biodiversity conservation needs consider plants with multiple growth forms to respond further climate changes.

T3-15-05

From the great plain to the Carpathians – How the global climate change affects the roses and hawthorns

Viktor Kerényi-Nagy[1,2], *Bence Borus*[3], *Szilárd Czóbel*[1], *Orsolya Szirmai*[1], *Károly Penksza*[1]
1. Szent István University, Faculty of Agricultural and Environmental Sciences
2. The Museum and Library of Hungarian Agriculture
3. REX Doghome Foundation

The affects of the global climate change can be felt specially in the Carpathian-Pannon region due to its geological conditions. The problem is complex since the behaviour and reactions of the species are determined by human activities, incorrect habitat management, forest and hunting management and also by extreme weather conditions. The efforts done locally to protect the sepcies might fail if the negative global trends would not change. The *Crataegus nigra* Waldst. et Kit. grows in the flood plains of the River Danube and Dráva (which closest relative is the *Crataegus chlorosarca* Maxim - which can be found in the Far-East, so in China also). It has has lost 50% of its habitat in the last century. One of the reasons might be the global warming, the drying of the wetlands beside the river and forest regulations. The *Curvisepala*-group: *Crataegus lindmanii* Hrab.-Uhr., *C. ovalis* Kit., *C. rosaeformis* Janka are prety tight tolerance (steoecicus) species. They grow from mid-mountains till high mountains and have long sepals. Beside the incorrect forest regulations the other reason of their desappearing is the negative change in the climate of the forests (extremely rainy weather in May and extremely hot and dry weather during the summer). As a first step they can only bloom (produce pollen), but are unable to produce fruit, as a result of this the populations can not renew. As a second step, through producing pollen they hibridize with euroecicus species. These hibrids are more viable. Its effect is easy to notice: these species are desappearing from the Hungarian mid-mountains but the hibrids can be found in great amounts. From these processes we can conclude the following: due to the global climate change the same will happen in the high mountains of the Carpathians. The Mediterranean *C. brevispina* Kunze was very rare in the Carpathian Basin hundred years ago, but it grows now in the submediterranean part of midmountains of Carpathian Basin. The most sensitive roses for the global warming are the *Villosa*-group (*R. villosa* L., *R. ciliato-petala* Besser, *R. velebitica*, *R. coziae*) and the *Tomentosa*-group (*R. tomentosa*, *R. kitaibelii*, *R. sherardii*, *R. floccida*, *R. umbelliflora*). These species can be found only in mid- and high mountains. These two groups have been retracting since the Ice age, the populations in Northern-Europe are disjunct, towards the south they become desperse. The split off-populations are on a unique evolution way, local small species have formed. The herbarium revision of the *R. villosa* showed that the number of speciments which were collected during the last cca 200 years has decreased, in spite of looking for the known-habitats the population was not found because the vast majority become extinct even from the high mountains. The Hungarian populations of the *R. ciliato-petala* partly have extincted, they remained only on the dolomit vegetation in the north side. The *Rosa velebitica* and *R. coziae* grows on the top of a few high mountains, the global climatechange risks mainly these populations. The *R. tomentosa* és *R. sherardii* behaves like the *R. villosa* but its populations only grow thinner, big populations can be found only in the high mountains. Several taxa (*R. kitaibelii*, *R. floccida*, *R. umbelliflora*) show termophil nature due to the global warming. They are trapped on the "islands" of the mid- and highmountains southern, warmer parts. It is not known how these micropopulations will deal with any further climate change. They might survive or disappear. The *Rosa pendulina* L. grows in the Transdanubian mountains in the XVIII. century, but it is extinct from there now and this species grows in the North midmontains and the Carpathians. The *R. glauca* Pourr. grow in the highmountains in Europe, it is common species in the Pyrenees, Alps and Velebit mountains, but this species is extinct from the Carpathian Basin during in the last century.

T3-15-06

Effects of climate change on Eurasian and North American alpine vegetation – a general overview based on field studies

Patrícia Pósa, *András Demeter, Csilla Ficsor, Kata Kurucz, Szilárd Czóbel*

Szent István University

Despite broad evidence that climate change continuously effects alpine vegetation at different levels, the mechanistic understanding of these changes is still missing. A rapid warming observed in Eurasian and North American mountains is predicted to increase plant upper distributional limits, vegetation cover and abundance of species adapted to warmer climate. Plant community composition analysis throughout arctic and alpine North American and European sites show, that increases in the relative abundance of species with a warmer thermal niche were observed in response to warmer summer temperatures. However, effect sizes were greater over broad-scale spatial gradients relative to either temporal variability in summer temperature within a site or summer temperature increases induced by experimental warming. A pan-European survey showed that thermal differences caused by solar radiation determine plant species diversity on temperate mountains. The field studies indicate that the direction and magnitude of the changes differed considerably among the studied communities and were dependent on initial species composition, vegetation structure and geographical context of individual vegetation belts. It seems that plant species responses to ongoing climate change will not be unidirectional upward range shifts but rather multi-dimensional, species-specific and spatially variable.

T3-16: Floral chemistry and plant-pollinator diversification

T3-16-01

Pollinator-mediated hybrid and introgression under the diffuse coevolution may improve *Ficus* diversification greatly

Gang Wang, *Jin Chen*

Xishuangbanna Tropical Botanical Garden, Chinese Academy of Sciences

Whether and how coevolution between plant and insect pollinator improved plant diversification is one of the central question for both ecology and evolution researching. The extreme diversity of figs (*Ficus*) and their obligate pollinators, the fig wasps (Agaonidae), provides a nice opportunity to target that question. Here, following the diffuse coevolution mode, in which, a group of figs species coevolution with a group of pollinating fig wasps and the hybridization and introgression among fig species mediated with pollinator sharing or pollinator shifts are allowed, we try to test whether the diffuse coevolution is common in this obligate pollination system and how it contribute to the species diversification of the system. Using five sympatric dioecious fig taxa and their pollinators, we examine the degree of pollinator sharing and inter-taxa gene flow. We experimentally test pollinator preference for floral volatiles, the main host recognition signal, from different figs. All five fig taxa shared pollinators with other taxa and gene flow occurred between fig taxa within and between sections. Floral volatiles of each taxon attracted more than one pollinator species. Floral volatiles were more similar between closely related figs, which experienced higher levels of pollinator sharing and inter-taxa gene flow. This study demonstrates that pollinator sharing and inter-taxa gene flow occurs among closely related sympatric dioecious fig taxa and that pollinators choose the floral volatiles of multiple fig taxa. The diffuse coevolution should be common in this system. We further tested how many "good" fig species may speciation with hybridization and introgression via pollinator sharing or shifts as above diffuse coevolution. A chloroplast *Ficus* phylogeny with de novo assembled full cp genome sequences of 27 species of 16 sections of 6 subgenus was established and compared with recent nuclear *Ficus* phylogeny. Primary result shows chloroplast genome *Ficus* phylogeny get very high support in almost all deep and tip nodes. However, the cyto-nuclear discordance was repeatedly detected in both phylogeny of specie and section level, which may suggest the role of hybridization speciation via above diffuse coevolution should be one of important way for *Ficus* diversification.

T3-16-02

Floral chemistry and plant-pollinator diversification: Introduction to the symposium

Robert Raguso

Cornell University

Plant evolutionary biologists have long appreciated the role of key innovations, such as nectar spurs, in the diversification of flowering plants. Recent evidence suggests that chemical features of flowers – their scent, nectar, pollen and oils – play important roles in the full spectrum of generalized to specialized, and mutualistic to deceptive relationships with pollinators. These roles include the structuring of plant-pollinator community networks through pollinator attraction and repellence, innate pollinator preference in specialized reward-based systems, partner choice in obligate mutualism and sensory exploitation in deceptive pollination. Improved methods in chemical analysis combined with new statistical, genomic and phylogenetic approaches make it possible to combine chemical ecological methods with the study of diversification in species-rich angiosperm lineages (e.g., orchids, figs, aroids). In this symposium, we showcase the state of the art of these combined chemical ecological and comparative approaches to explore the potential for floral chemical traits to function as key innovations in the diversification of angiosperm lineages. Southern China is an inspiring place to discuss such a topic, with botanical diversity across its biomes and emerging local research on chemical aspects of pollination biology.

T3-16-03

The chemistry and evolution of pollination by sexual deception in Australian orchids

Rod Peakall[1], *Bjorn Bohman*[2], *Ryan Phillips*[1], *Darren Wong*[1], *Gavin Flematti*[2], *Eran Pichersky*[3]

1. Research School of Biology, The Australian National University

2. School of Chemistry and Biochemistry, The University of Western Australia

3. Department of Molecular, Cellular, and Developmental Biology, University of Michigan

In highly specialised pollination systems, floral volatiles, floral morphology, or a combination of both likely underpin pollinator specificity. Floral volatiles in particular frequently play a key role in long-distance attraction, yet we often know very little about the chemistry involved. When pollinator specificity is strongly controlled by chemistry, we expect to find one of two main patterns: Either specificity is achieved using unusual blends of commonly occurring volatile compounds, or alternatively via novel or unusual compounds that are rare in nature. The former is predicted to be the most likely evolutionary scenario, since the semiochemicals can be readily produced by the existing biosynthetic machinery. However, testing this prediction is challenging since elucidation of functional floral volatiles is likely be strongly biased towards known and common compounds. Semiochemicals are of paramount importance in sexually deceptive orchids as they are used to lure male insects through mimicry of the female's sex pheromone. The strategy has evolved repeatedly in orchids, with a wide diversity of insect groups exploited. In Australian terrestrial orchids that attract male thynnine wasps by sexual deception, several different chemically unrelated sex pheromone systems are now known. These include: chiloglottones, representing a new class of floral volatiles when first discovered, new (hydroxymethyl) pyrazines, and blends of novel (methylthio) phenols not seen before in plants. Critically, confirmation of these compounds as semiochemicals is fully backed by bioassays in the field. We have also confirmed an unusual dual semiochemical system requiring a specific 1:4 blend of the widespread monoterpene plant volatile, citronellol, with an unusual disubstituted acetophenone not previously known as a floral volatile. Further work on the biosynthesis of these two compounds in the orchid has fully elucidated the biochemical pathway for citronellol production for the first time in any plant, and points strongly to a polyketide based biosynthesis of the acetophenone. Pollinator attraction in this case thus requires the coordination of not one, but two unrelated biosynthetic pathways for semiochemical production. An overlay of these semiochemicals onto the phylogeny of Australian terrestrial orchids reveals that there have been multiple independent origins of sexual deception in Australian orchids, with major switches in chemistry occurring even within orchid clades. Contrary to expectation, a diversity of floral volatiles new to science, and unusual combinations of known volatiles, underpin specific pollinator attraction in Australian sexually deceptive orchid systems. Furthermore, it is evident that we have barely scratched the surface of our understanding of the chemistry of sexual deception in Australian orchids, with our focus so far limited to male thynnine wasp pollinated systems, to the neglect of other groups of sexually exploited pollinators involving other groups of Hymenoptera, and Diptera. Our finding that both novel compounds, and unusual combinations of compounds, are frequent in our study system lead us to predict that similar discoveries await both other sexually deceptive systems, and other specialised pollination systems. Indeed, the chemistry of pollinator specificity may yet prove to be more often based on novel and unusual compounds, than on unusual blends of common floral volatiles.

T3-16-04

Herbivory, pollination and the diversification of floral scent in evening primroses (Onagraceae)

Tania Jogesh[1], *Rick P. Overson*[1], *Geoffrey T. Broadhead*[2], *Robert A. Raguso*[2], *Krissa Skogen*[1]

1. Chicago Botanic Garden

2. Cornell University

Floral trait evolution is frequently attributed to pollinator-mediated selection but herbivores can play a key role in shaping plant reproductive biology, especially if the same floral traits mediate both mutualistic and antagonistic interactions. Due to their opposing effects on plant fitness, herbivores and pollinators have the potential to influence the evolution of floral phenotypes across geographic landscapes with implications for speciation and diversification at higher taxonomic scales. Here, we examine the role of herbivory in the diversification of floral scent chemistry in evening primroses, Onagraceae. Evening primroses represents the most diverse group of hawkmoth-pollinated plants in North America with remarkable divergence in floral morphology and chemistry. Preliminary analyses suggest that inter- and intraspecific variation in floral scent is pervasive, especially for two enantiomers of the monoterpene alcohol, linalool. In *Oenothera harringtonii*, a geographic (R) (-) linalool polymorphism is maintained in spite of high levels of gene flow, suggesting strong selection on floral scent. In many *Oenothera* species, (R) (-) linalool is negatively correlated with the presence of *Mompha*, a specialist flower and fruit-feeding moth genus. These data suggest that (R) (-) linalool might function as an herbivore deterrent, and that geographic variation in the strength of plant-herbivore interactions might diversify floral scent phenotypes with implications for plant-pollinator interactions. We examined the role of floral chemistry in mediating pollinator and herbivore interactions and in driving the diversification of evening primroses at multiple evolutionary timescales. We performed a manipulative field experiment, adding (R) (-) linalool to flowers of *Oenothera harringtonii* in a population lacking linalool, and measured pollinator visitation, herbivory and seed fitness. Controlled laboratory assays were used to determine whether hawkmoth pollinators preferentially oviposit on plants that vary in linalool presence/chirality. Finally, we surveyed the presence (and enantiomeric composition) of linalool and other floral volatiles in 61 populations of 16 species of Onagraceae (*Oeno-*

thera and *Clarkia*) across western North America, in conjunction with measurements of pollinator visitation and herbivore damage.

T3-16-05

Pollination of sexually deceptive genus *Lepanthes* (Orchidaceae: Pleurothallidinae)

Diego Bogarín[1,2,3], *Franco Pupulin*[1,4,5], *Erik Smets*[3,7,8], *Barbara Gravendeel*[3,6,7,8]
1. University of Costa Rica
2. Herbarium UCH, Universidad Autónoma de Chiriquí, Chiriquí, Panama
3. Naturalis Biodiversity Center, Leiden, The Netherlands
4. Harvard University Herbaria, 22 Divinity Avenue, Cambridge, Massachusetts, U.S.A.
5. Marie Selby Botanical Garde, Sarasota, Florida, U.S.A.
6. University of Applied Sciences Leiden, Leiden, The Netherlands
7. Institute Biology Leiden, Leiden University, Leiden, The Netherlands
8. KU Leuven, Ecology, Evolution and Biodiversity Conservation, Leuven, Belgium

Pleurothallidinae is the largest Neotropical orchid subtribe encompassing > 5,000 species that are mainly dipteran pollinated. Of the few pollination systems documented in detail for the subtribe, the most interesting one occurs in the genus *Lepanthes*. Species of this genus are pollinated by sexual deception through genitalic pseudocopulation involving male fungus gnats (*Bradysia*, Sciaridae), probably attracted by a female pheromone-mimicking strategy. The pseudocopulation strategy of *Lepanthes* has not yet been fully investigated and remains largely hypothetical. Our objective was to investigate the pheromone-mimicking strategy in *Lepanthes*. We studied the ultrastructure of *Lepanthes* flowers focusing on their adaptations to pollinators through the integration of various microscopy methods (LM, SEM, TEM and micro-CT), histochemistry and chemical analyses of floral fragrances by GC/MS. The sepals do not react with any histochemical tests but show secretory trichomes likely involved in fragrance synthesis. The petals and lip are papillose and react with various staining tests for carbohydrates and lipids. Micro-CT scans show a close 3D fit between flowers and pollinators. Our results suggest that *Lepanthes* flowers do not produce obvious rewards. We further found that fragrances are composed of long chain hydrocarbons (alkanes, alkenes, esters and acetates). Some of these compounds act as pheromone precursors or are pheromone mimicing compounds. For example, Z-13-Octadecen-1-yl acetate, detected in *Lepanthes turialvae*, was reported as the pheromone mimic compound of the rice leafroller moth, *Cnaphalocrocis medinalis*, an economic pest of rice in China and Korea. Our data support the theory of a sex pheromone-mimicking strategy in *Lepanthes*. Floral morphology of *Lepanthes* varies astonishingly but is always based on similar (micro)morphology and chemistry. Thus, it is likely that a female pheromone-mimicking strategy is widespread in the genus. Variation in micromorphology and fragrance composition among species might act as isolation barrier thus promoting speciation.

T3-16-06

Landscape perspectives on pollinator-driven evolution of floral scent signals

Steven Johnson
University of KwaZulu-Natal

There is a general consensus that most phenotypic diversification of plants occurs in geographically separated populations and that this eventually translates into character differences between species. It is therefore important to understand the basis of phenotypic divergence among populations, particularly the roles of neutral versus adaptive processes. Floral volatile emissions have been shown in many cases to be correlated with pollination systems at the macro-evolutionary level and to be functional for pollinator attraction, but can we also detect a population level process of adaptation to local pollinator assemblages? The study of geographical variation in floral scent emissions is still in its infancy given the serious analytical challenges involved in collecting and analysing such data. However, available evidence shows that volatile emissions are often highly variable among populations, presenting an example of phenotypic variation which is usually cryptic to the human observer. Some recent studies based on reciprocal translocation experiments and laboratory choice tests have provided some tantalizing evidence for local adaptation of plants in terms of their volatile emissions.

T3-17: Increasing the sustainability of agricultural ecosystems

T3-17-01

Applying plant ecological knowledge to increase agricultural sustainability

Jacob Weiner
University of Copenhagen

Plant ecological knowledge accumulated over the past 150 years has enormous implications for agriculture, but most of these implications have not been appreciated by ecologists or agriculturalists. Here, I present several of the most salient examples: Agriculturalists refer to “improvements”, but plant ecologists know that “trade-offs” represent a better conceptual framework for agricultural production. There is much evidence for trade-offs between yield and resource use efficiency, and between individual fitness and populations yield. I argue that there is also a “limiting trade-off” between short term yield and sustainability, and it is important to take this into consideration if we are serious about increasing sustainability. At the local level, agricultural sustainability is about maintaining or improving soil fertility, but this is not a priority in most agricultural systems worldwide. Natural plant communities with the highest productivity do not resemble the very earliest stages of succession with extremely low standing biomass, suggesting that it is possible to develop highly productive, more sustainable, plant productions systems, which share many properties of somewhat later successional stages. I present such a vision, which I call “High Biomass Cropping Systems”. Plant ecology tells us that rotation of crops with very different nutritional needs and very different pests, diseases and weeds can make a major contribution to sustainability. The very limited crop rotations practiced in most modern plant production systems is a clear indication that farming practices are based on short-term economic and regulatory factors, without much if any consideration for sustainability. The modern scientific method tells us how we should test hypotheses, but it says nothing about how hypotheses are generated. We need to address the agricultural research agenda if it is to serve the interests of farmers, consumers and society as a whole, rather than narrow but powerful economic interests.

T3-17-02

Investigating past domestication to improve sustainable agriculture in future croplands

Rubén Milla, *Pablo García-Palacios, Silvia Matesanz*
Universidad Rey Juan Carlos

Securing sustainable food provision in response to current global change and demographic growth poses challenges that must be addressed from diverse perspectives. Bringing better agronomic practices contributes to that endeavour. However, not only management affects food provision. Crop traits also play significant roles in modulating the ecosystem services (ESs) that secure sustainable yields in the face of global change. Here we advocate that understanding how crop traits evolved under domestication and impacted ESs delivery should help breed future crops. Thus, we address the effects of crop evolution (from initial domestication to current times) on relevant ESs, including crop yield and its stability over time, soil carbon (C) sequestration, and soil nitrogen (N) retention. There is initial evidence that crop domestication affects the delivery of ESs. However, most of those evolutionary effects are understudied. Therefore we propose research gaps that merit attention, including (1) assessing whether performance in polyculture is lower for crop mixtures than for mixtures of wild plants, and breeding for more efficient crop polycultures; and (2) investigating how crop evolution impacted plant effects on C storage and N cycling, and how such effects contribute to the stability of yields. The sustainable provision of cropland ESs, and thus food security, will benefit from exploring those avenues

T3-17-03

Using positive interactions between species to help sustain agricultural ecosystem

Xin Chen, *Jianjun Tang*
Zhejiang University

Facilitative interactions between species (one species makes conditions more favorable for another) are receiving increasing attention in Ecology. It is also been recognizing that exploiting facilitative interactions between species play an important in the restoration of degraded ecoysytems and in the development of sustainable agricuture with biodiversity. In agro-ecosystem, although multiple-species cropping has showed over-yielding, however, additional species might not always increase agricultural productivity of the main crop due to competition (for nutrients, light, *etc.*). Thus in the utilization of species diversity in agriculture, it is critical to understand the characters of species interactions and to exploit the positive interaction (facilitation) between species. Here, we synthesize the current advances in our understanding of how agriculture systems with co-cultured species can ensure sustainable food production while minimizing environmental damage by a meta-analysis and a field survey. Crop yield in legume-cereal intercropping or rice-fish co-culture or agroforestry, and the corresponding monoculture was used in the meta-analysis. Papers published in the past twenty-years (1990-2012) were search in the public database using key words legume-cereal intercropping, or rice-fish, or agroforestry. A data set of 276 paired-field survey (rice-fish co-culture vs monoculture) was also used to do the meta-analysis. The effect size was calculated as the natural log of the response ratio (R, the mean of intercropping, or rice-fish, or agroforestry divided by the mean of monoculture). A positive effect size with confidential intervals (CIs) that do not overlap zero would indicate that the crop yield in intercropping, or rice-fish, co-culture or agroforestry than with monoculture. Our results from meta-analysis indicated that legume–cereal intercropping significantly increased cereal yield but did not affect legume yield. The positive effect size on cereal yield was greater with pesticide use but was not influenced by the other five factors (climate, planting pattern, irrigation, fertilization, or weed control). For legume yield, the effect size was not influenced by any of the six factors. For rice-fish co-culture system, meta-analysis indicates a positive effect size of rice–fish co-culture on rice yield regardless of fish culture type (single fish species or mixed fish species) in the co-culture system. For intercropped agroforestry systems, however, crop yields significantly increased only in some cases (41 studies among 115 studies). For most cases (74 studies among 115 studies), crop yields significantly decreased compared to monoculture. These yield reductions were mainly due to shading and root competition. Results from field survey on rice-fish co-culture showed that utilizing species diversity can sustain rice yield with less chemical fertilizer and pesticide inputs when the coexisting-species generate facilitative effects. This study indicates that co-cultured species in agriculture can increase crop yield, but these enhancements are sensitive to environmental conditions, planting techniques, and management.

T3-17-04

Crop diversity enhances agroecosystem sustainability via improving soil fertility in north-western China

Xiao-Fei Li[1], *Ping Wang*[2], *Chengbao Wang*[3], *Sicun Yang*[3], *Lingguo Gui*[2], *Weiping Zhang*[1], ***Long Li***[1]
1. College of Resources and Environmental Sciences, China Agricultural University
2. Institute of Agricultural Resources and Environment Sciences, Ningxia Academy of Agriculture and Forestry Sciences, Yinchuan, China
3. Institute of Soils, Fertilizers, Wate-saving Agriculture, Gansu Academy of Agricultural Science

We face a major challenge feeding an expanding world population with highly degraded soils. Reversing the trend of soil degradation and ensuring food security and a sustainable future have therefore become essential. Although multispecies systems often are productive, the sustainability has been questioned in longer temporal and wider spatial variations as continuous overyielding generally removes more nutrients from soils. In order to address these issues, two field experiments were conducted initially in 2009 at the soils which were reclaimed for 3 years at Hongsipu in Ningxia (Exp. 1) and for 20 years at Jinyuan in Gansu (Exp. 2), China. The Exp. 1 was split-split-plot design, where main plot treatments included inoculation with rhizobium or not, and the sub-plot treatments included five N-application rates (0, 75, 150, 225 and 300 kg N ha^{-1}), and the sub-sub-plot treatments were sole faba bean, maize, and faba bean/maize intercropping. The Exp. 2 was two-factorial design, where main factor was P application rates (0, 40, 80 kg ha^{-1}) and the second factor was cropping systems, including canola/maize, chickpea/maize, faba bean/maize, and soybean/maize intercropping, and their corresponding monocropping. Each treatment has three replications in both field experiments. Productivity, soil physical and chemical properties (bulk density, water-stable aggregates, infiltraction, compactivity, soil organic matter, total N, Olsen-P, Exchangable K, CEC, and pH), and some enzyme activities were measured in 2012 and 2013 in both experiments. Results showed that overyielding were observed in faba bean/maize and chickpea/maize intercropping, whose productivity stability was maintained or increased compared with corresponding monocropping. The percentage of water-stable macro-aggregate (> 2.0 mm) was increased by intercropping compared to monocropped faba bean (67.9% vs 56.0%). Soil organic matter and total N were increased significantly by intercropping compared to monocropped maize 6.86 g/kg and 0.27 g/kg vs 6.03 g/kg and 0.22 g/kg) in the Exp. 1. In the Exp. 2, similarly, percentages of soil water-stable macro-aggregate were significantly increased by faba bean/maize, chickpea/maize, canola/maize intercropping, but not by soybean/maize intercropping. Soil organic and total N presented similar trends to Exp. 1. In summary, soil fertility was significantly improved after 4-5 years of intercropping compared with monocropping, especially in rewly-reclaimed desert soil. In other words, crop diversity can enhance agroecosystem sustainability via increasing crop productivity as well as inproving soil fertility in a long-term scale.

T3-17-05

Soil organic carbon accumulation, climate variability and crop production in Tanzania's semi-arid agro-ecological zone: A case study of the Kongwa District

Msafiri Y. Mkonda[1,2], *Xinhua He*[1,3]

1. Centre of Excellence for Soil Biology, College of Resources and Environment, Southwest University, Chongqing 400715, China

2. Department of Physical Sciences, Faculty of Science, Sokoine University of Agriculture, Morogoro 3038, Tanzania

3. School of Plant Biology, University of Western Australia, Crawley 6009, Australia

Various ecological and environmental indicators including climate change, soil fertility, water availability and proper agronomic practices that form optimal agricultural systems are needed to be integrated for increasing agricultural productivity in the Tanzanian semi-arid agro-ecological zone. Among these indicators, climate change and soil fertility are the major limiting factors to affect crop yields in this semi-arid agro-ecological zone. To improve crop productivity, this study assessed the accumulation of soil organic carbon (SOC), the trend of climate variability and crop production in the Kongwa District, a semi-arid zone in central Tanzania. In doing so, climate data and soil samples were collected from two representative villages of Mnyakongo and Ugogoni. Household surveys, informative interviews, physical observations, group discussions and literature reviews were employed for data collection and crosschecking. The Walkley-Black Method was used for SOC analyses while the SPSS v. 20, Mann-Kendall and Sen's Slope Test for climate, crop yields and climate-smart agriculture analyses. Qualitative data were analyzed using the theme content analysis and Community-based Risk Screening Tool-Adaptation and Livelihoods (CRiSTAL). Results showed that the accumulation of SOC was significantly greater in soils under organic fertilization (1.15 and 0.80 MgC ha^{-1} at soil 0-20 cm and 20-30 cm depth) than under no-fertilization (0.35 and 0.30 MgC ha^{-1} at 0-20 cm and 20-30 cm) and decreased with increasing soil depths. So did the crops yields (2 tn ha^{-1} under organic fertilization vs. 0.87 tn ha^{-1} under no-fertilization). The mean annual rainfall or temperature (1980–2015) fluctuated at a decreasing ($R^2 = 0.21$) or an increasing trend ($R^2 = 0.30$). Meanwhile, crop yields of maize, sorghum or millet fluctuated at a decreasing trend at $R^2 = 0.07$, 0.05 or 0.85, respectively. In whole, rainfall variability positively correlated with SOC or crop yields. Thus, to increase crop production, there is an immediate need to apply manure, irrigation and drought-tolerant crop seed. These results confirm that the studied semi-arid areas are among the most vulnerable regions to climate change impacts. Given to this vulnerability, this study proposes a district adaptation plan to increase the resilience of smallholder farmers. Similarly, it calls for more proactive practices to intervene the authentic and potential consequences in the country. A serious action to improve agronomic practices, mitigate and adapt to climate change impacts should concurrently be a priority in semi-arid areas to limit the level of vulnerability.

T3-17-06

Sustainability and management of water resources in *Hordeum vulgare* agro-ecosystems in relation to the allelopathic effect of *Retama raetam*

Kholod Khatab[1], ***Salama El-Darier***[2], *El-Nage Abdel Hady*[3]

1. university of Damnhour, faculty of science

2. university of Alexandria, faculty of science

3. Faculty of Arts and Sciences, Benghazi University, Libya

The main objective of the present study was to manage water as a natural resource in *Hordeum vulgare* agro-ecosystems depending on the allelopathic potential of *Retama raetam* shoots aqueous extract (RRSAE). Germination percentage (GP), coleoptile (CL) and radicle lengths (RL) of the recipient species significantly increased under the allelopathic effect of RRSAE relate to control. The highest values (99%, 12.8 cm and 12.4 cm) were achieved at 3% concentrations level relative to the three parameters respectively compared to control. Furthermore, the effect of *R. raetam* shoots crude powder (RRSCP) on some growth parameters (number of tillers, spikes, grains $spike^{-1}$ and grains $plant^{-1}$) as well as shoot dry weight, grains yield and grain harvest index of *H. vulgare* was evaluated. Results showed that the highest values were achieved at 3% concentration levels of RRSCP and 25% field capacity (low water field capacity) compared to the control. Phenolic derivatives (mg/kg DW) (especially catechin) of *H. vulgare* seedling was increased under the lowest water field capacity (25%) at 3% concentration level of RRSCP. Such increase initiatives stimulation in all studied growth parameters with adequate amount of water that help in conservation and management of water as a natural resource in agro-ecosystems.

T3-18: Plant invasion risk, control and ecosystem security

T3-18-01

Predicting potential benefits and risks of the ragweed leaf beetle as a biological control agent: From ecological to evolutionary studies

Heinz Mueller-Schaerer[1], *Yan Sun*[2], *Suzanne Te Lommen*[1], *Benno A Augustinus*[3], *Urs Schaffner*[3]
1. Department of Biology, University of Fribourg, CH-1700 Fribourg, Switzerland
2. Institute of Evolution and Ecology, University of Tübingen, Germany
3. CABI CH-1700 Delémont, Switzerland

An important but largely unknown outcome of biological control involves evolution, e.g., will a biocontrol herbivore over time be able to feed on a new non-target host, will the target host populations become tolerant, and what will be the outcome of the plant-antagonist interaction under climate change. Common ragweed *Ambrosia artemisiifolia* is native to North America and was accidentally introduced to many parts of the world, including Europe, Asia, South America, Australia and New Zealand. Like no other plant common ragweed has raised the awareness of invasive plants in Europe, causing great damage to our society due to its highly allergenic pollen, and as an important and hard-to-control crop weed. Both its spread and impact are likely to increase with changing climate, posing a significant risk even in countries presently not yet affected. The recently and accidentally introduced ragweed leaf beetle *Ophraella communa* (Col.: Chrysomelidae), is native to North America and presently used as a successful biological control agent in Asia and Australia. This beetle has the potential to reduce the population density and pollen production of ragweed in Europe and to halt its further spread, but it also bears the risk to attack taxonomically closely related sunflower and native endangered species. Following its first observation in Europe in 2013, we conducted extensive field surveys in Northern Italy to monitor the spread and target and non-target impact of *O. communa*, as well as experimental host specificity experiments under quarantine and field conditions. *Ophraella communa* expanded from early sites in S-Switzerland and N-Italy in 2013 by 500km both west- and eastwards up to 2016. Regression models showed that the exceptionally low amounts of airborne *Ambrosia* pollen observed in the Milan area since 2013 is most likely related to the presence of large numbers of *O. communa,* leading to yearly savings in human health costs of millions of Euros. Field survey and experimental biosafety studies indicate a certain risk to sunflower, particularly later in the season when most ragweed plants are defoliated by the beetle. We also initiated long-term studies to explore the evolvability of both the target ragweed populations to selection by *O. communa*, as well as of the *Ophraella* populations when kept on the sunflower as compared to ragweed. We present first results of a novel evolutionary approach to assessing the beetle's potential for genetically-based changes to novel biotic (host plants) and abiotic (future temperature) conditions, using next generation sequencing and bioassay approaches. Assessing evolvability of a biological control agent will greatly improve predicting its future benefits and risks.

T3-18-02

Adaptive genetic divergence and increased competitive ability in invasive *Ambrosia artemisiifolia*

Yan Sun[1,2], *George Roderick*[1]
1. Department of Environmental Science, Policy & Management, University of California, Berkeley
2. Plant Evolutionary Ecology, Institute of Evolution & Ecology, University of Tübingen

Invasive alien plant species face different abiotic and biotic conditions in their native and introduced range; furthermore, biological control programs will create an additional mosaic of species interaction in the introduced range. Together, these make them ideal study systems to address the basic question on rapid evolutionary changes of plant population. Differences in the selection and genetic variation in relevant traits can make adaptation difficult to predict. We used a common-garden experiment to assess patterns of adaptation in the invasive plant, *Ambrosia artemisiifolia*, sampled across 11 populations in its native (USA) and 10 invasive (China) ranges, with the 10 Chinese populations having a different history of biological control management. We tested for adaptive trait evolution by comparing differentiation at neutral genetic loci (F_{ST}) with differentiation at quantitative traits (Q_{ST}). We estimated multivariate Q_{ST}-F_{ST} tests to determine if phenotypic differences between countries are due to adaptation. Between countries, we found that two growth traits are distinguishable from genetic drift ($Q_{ST} > F_{ST}$), and interestingly those two traits are correlated with climatic variation in the USA, but not in China. Within countries, we found more traits under selection in the USA than that in China. Moreover, we found increased adaptive potential in the Chinese populations without biological control history ($Q_{ST} >> F_{ST}$) than with biological control history. Additionally, a pairwise greenhouse competition experiment using 10 invasive Chinese populations in competition with a native USA grass showed a reduced competitive ability for populations with than without biological control history, which might be due to a tradeoff between herbivore resistance and plant growth. Our results indicate that adaptation in invasive Chinese populations does not parallel patterns in native US populations due to differences in abiotic and biotic selection, genetic constraints, or both.

T3-18-03

Impact of plant invasions on biodiversity

Bo Li
Fudan University

Plant invasions are one of the five major driving forces that lead to biodiversity loss worldwide. In this presentation, I will first explore how invasive plants affect native biodiversity through hybridization, disruption of ecological relations between species (e.g., native plants and pollinators), and alterations of ecosystem processes. I will then use exotic smooth cordgrass (*Spartina alterniflora*) as a case to show the effects of its invasion in the east coast of Chinaon biodiversity in saltmarshes.

T3-18-04

Higher resource sharing in invasive than in native clonal plants

Feihai Yu[1], ***Yongjian Wang***[2]
1. Taizhou University
2. Huazhong Agricultural University

What confers invasive plants with competitive advantages over native ones remains an unresovlved question in invasion ecology. Many of the invasive plant species and many of the world's worst invasive plant species are capable of clonal growth. Clonal growth confers clonal plants with distinct traits such as clonal integration (resource sharing between connected ramets). While clonal integration has been shown to benefit the growth of some clonal invasive species, it is still unknown whether invasive clonal plants benefit more from clonal integration than native ones. We selected five pairs of invasive and native clonal plant species in China, and grew their disconnected (preventing clonal integration) or connected (allowing clonal integration) ramet pairs under heterogeneous light, soil nutrient and water conditions that are commonly encountered by alien plants during their invasion into new areas. To account for the phylogenetic relatedness, the invasive and native species of a pair were always from the same genus. Invasive clonal plant species grew better than native ones, and clonal integration increased the growth of both invasive and native clonal plants in all three types of heterogeneous environments. More interestingly, the benefits of clonal integration were significantly higher in invasive than in native clonal plants. Thus, clonal integration can confer invasive clonal plants with competitive advantages over native ones, and may partly explain the invasive success of alien clonal plants when invading new areas where light, soil water and/or nutrients are commonly heterogeneously distributed.

T3-18-05

A biocontrol agent may benefit a native plant in competing with an invasive plant under warming climate

Xinmin Lu[1,2], *Evan Siemann*[3], *Minyan He*[1], *Hui Wei*[1], *Xu Shao*[1], *Jianqing Ding*[1,4]
1. Wuhan Botanical Garden, Chinese Academy of Sciences
2. Central China Normal University
3. Rice University
4. Henan University

Climate warming may affect competition between native and non-native invasive plants directly or indirectly via herbivory. Here, we show that an introduced biocontrol beetle shifted competition between a native plant and an introduced plant under elevated temperature. In a 2-yr field experiment at 30.5°N, we found that experiment warming (+ 1.8°C) increased overwintering ability of the introduced beetle which prefers the invasive plant over the native plant, as a result, the beetle established sustainable population in the presence of the invasive plant throughout the trial. But under ambient temperature, the beetle failed to emerge in the second year. Warming also shifted life history of the native plant from annual to perennial. At the end of the second year, beetle suppressed the invasive plant and shifted the community from invasive plant dominated to native plant dominated under elevated temperature. Under ambient temperature, the invasive plant dominated the community under differing herbivore treatments. In addition, in a field survey from 21°N to 30.5°N where the beetle was present, the native populations were perennial < 25.8°N, but only annual populations were observed > 26.5°N, and the invasive plant only dominated the community > 26.5°N. Our results highlight that herbivore can play an important role in determining plant community structure and non-native plant invasions under warming climate. Thus, understanding how biotic interactions, herbivory in particular, shape responses of plant communities to climate change is crucial for predicting the risk of plant invasions under climate change scenarios.

T3-18-06

Changes in competitive ability, defense and allelopathic effect between invasive plants and conspecifics in their native ranges

Yu-Long Feng
College of Bioscience and Biotechnology, Shenyang Agricultural University

The Evolution of Increased Competitive Ability (EICA) hypothesis predicts that after liberation from herbivores, invasive plants evolve increased vigor and decreased defense by reallocating resources from costly defensive traits to growth or reproduction. However, invasive plants may not escape from generalists and may even experience increased damage in introduced ranges, although they generally release from specialists. Shifting Defense Hypothesis (SDH) predicts that invasive plants may evolve to decrease defense against specialist herbivores but maintain or even increase defense (increase toxic chemicals) against generalist herbivores in introduced ranges. It has been documented that plant toxins may also function as allelopathic agents. If so, the increased toxins may also contribute to the increased competitive ability of invasive plants by increasing allelopathic inhibition on co-occurring native plants. The Novel Weapons Hypothesis (NWH) predicts that unique chemicals of invasive plants may provide disproportionately allelopathic, defense, or anti-microbial advantages in their nonnative ranges. With respect to allelopathy, novel chemicals may have stronger effects on naïve native plants from nonnative ranges of the invaders, which have not adapted to the unique chemicals and therefore are more vulnerable to them than adapted plants from native ranges of the invaders. If novel plant chemicals conferred competitive advantages in the nonnative ranges of invaders, then selection pressure for the chemicals may be greater on the genotypes in the invaded regions than on the conspecific genotypes remaining at home. If so, the evolution of competitive ability might derive in some cases from selection for increased production of allelopathic compounds. The EICA hypothesis and the NWH are not mutually exclusive; plants can evolve larger size, greater production of novel allelochemicals, or changes in quantitative defense. However, little effort has been made to explore how these processes might be integrated in invasion success. In this presentation I will compare biogeographical differences in competitive ability, defense, and allelopathic effects of noxious invasive plants in China, and discuss how increased vigor and production of novel allelochemicals contribute to increased competitive ability of the invasive plants.

T3-19: Species interaction and community structure across environmental conditions ecosystem (two sessions)

T3-19-01

Life in the canopy: Comparative and community phylogenetic analyses of epiphytic growth in ferns

Joel H. Nitta[1], *Ravahere Taputuarai*[2], *James E. Watkins*[3], *Charles C. Davis*[4]
1. Department of Botany, National Museum of Nature and Science
2. Association Te Rau Ati Ati a Tau a Hiti Noa Tu
3. Department of Biology, Colgate University
4. Department of Organismic and Evolutionary Biology and Harvard University Herbaria, Harvard University

Epiphytic growth is an important and recurring theme in fern evolution: transitions from terrestrial to epiphytic growth have occurred in at least five large fern clades, and may be associated with increased rates of diversification coinciding with the widespread appearance of angiosperm-dominated forests during the late Cretaceous and early Cenozoic. Despite the significance of epiphytic growth for fern evolution, however, few studies have investigated traits related to epiphytism in ferns while accounting for phylogeny. Here, we use the ferns of Moorea, French Polynesia, which include representatives of all five Cretaceous epiphytic fern radiations, as a study system to investigate evolutionary patterns in traits related to epiphytic growth in ferns. We analyze traits at both the species and community level, and test for degree of phylogenetic conservatism as well as phenotypic clustering. Our results are consistent with the canopy environment acting as a strong filter to restrict both phylogenetic and functional diversity of epiphytic fern communities. Our study provides an important phylogenetic perspective on the role of functional traits in ferns, and paves the way for more detailed studies examining how niche space is filled within terrestrial and epiphytic habitats.

T3-19-02

Dune stabilization and shrub size together regulate shrub-understorey herbs interactions in a semi-arid sandy ecosystem

Yuxuan Bai[1], *Yuqing Zhang*[1,2], *Richard Michalet*[3], *Weiwei She*[1], *Xin Jia*[1,2], *Shugao Qin*[1,2]
1. Yanchi Research Station, School of Soil and Water Conservation, Beijing Forestry University
2. Key Laboratory of State Forestry Administration on Soil and Water Conservation, Beijing Forestry University
3. UMR CNRS 5805 EPOC, University of Bordeaux

Questions: (i) do the stabilization stage of dunes and individual size of shrubs influence the direction and intensity of interactions between shrubs and their understorey species? (ii) do understorey species with different life-histories respond differently to dune stabilization and shrub size? (iii) does facilitation vary depending on the understorey species performance (i.e. growth or richness)? Location: Sand-fixing communities of the Mu Us Desert, China. Methods: We randomly selected shrubs of different sizes (large and small) below shrubs and in adjacent open areas along a dune stabilization gradient. We measured shrub crown size, environmental characteristics and the height, abundance, biomass and richness of subordinate species. Relative interaction index (RII) was calculated at the community-level and for annuals and perennials separately in all paired plots and analysed with ANOVAs. Understorey species composition was analyzed with Principal Component Analysis (PCA). Results: *A. ordosica* had dominant positive effects on understorey species in this dune system. However, we found contrasting variations in community-level facilitative effects across the gradient of dune stabilization and with shrub size depending on understorey species performances. Overall, facilitation for growth and in particular biomass decreased with increasing disturbance and shrub size, whereas the converse was observed for species richness. Notably, dune's stabilization and shrub's size were not only functioned independently but interacted to regulate the shrub-grass interactions. In addition, perennials were strongly affected by the shrub-grass interactions than annuals on both growth and recruitment. Species composition was changed due to the interactions, causing the invasion of perennials and reduction of annuals under the process of dune stabilization. Conclusions: Our observational study showed that community-level increase in facilitation for growth with increasing dune stability and shrub size were mostly explained by the turnover of abundance and biomass of perennials vs. annuals and the higher sensitivity of perennials to the effects of the shrubs, which were positive until a threshold of size. In contrast, the community-level increase in facilitation for richness with increasing disturbance and shrub size was mostly explained by the high recruitment strain of perennials in the disturbed open patches of the semi-fixed dunes, which was more likely to be ameliorated below large than small shrubs. Our study provides additional support to the theory arguing that communities are set of hidden interaction groups which might have contrasting responses to dominant neighbours in a single community depending on their functional strategies.

T3-19-03

Species associations in a subtropical forest in South China

Deyi Yin
South China Botanical Garden, Chinese Academy of Sciences

The spatial pattern of plants may provide significant insights into processes and mechanisms that maintain species richness. We applied data from a fully mapped 50-ha subtropical forest plot in Heishiding (HSD) Nature Reserve, China to conduct a community-wide assessment of the type and frequency of spatial association of species pairs. By analyzing the spatial association and attributes of the species pairs, we found that 55% of all species pairs showed negative bivariate association, which occurred less frequently among species sharing the same habitat association. In contrast, positive associations occurred more frequently among species with similar habitat requirements. The observation that species with similar habitat preference tend to aggregate suggests the vital role of environmental filtering in shaping community structure of this subtropical forest.

T3-19-04

Phenotypic effects of the nurse *Thylacospermum caespitosum* on dependent plant species along regional climate stress gradients

Sa Xiao[1], *Xingpei Jiang*[1], *Richard Michalet*

1. State Key Laboratory of Grassland and Agro-ecosystems, School of Life Sciences, Lanzhou University, Lanzhou, P. R. China.
2. University of Bordeaux, Allée Geoffroy Saint-Hilaire Pessac, France.

Contrasting phenotypes of alpine cushion species have been recurrently described in several mountain ranges along small-scale topography gradients, with tight competitive phenotypes in stressful convex topography and loose facilitative phenotypes in sheltered concave topography. The consistency of phenotypic effects along large-scale climate stress gradients have been proposed as a test of the likely genetic bases of the differences observed at small-scale. Inversely, plastic phenotypic effects are more likely to vanish at some points along climate stress gradients. We tested this hypothesis for two phenotypes of the alpine cushion species *Thylacospermum caespitosum* at four points along regional gradients of cold and drought stress in northwest China. We measured the traits of the two cushion phenotypes and quantified their associated plant communities and environmental variables along the regional temperature and aridity gradients. Cushion height, convexity and stem density overall showed significant effect of phenotypes. Difference in tightness of cushions between phenotypes was consistent across climate conditions, whereas differences in cushion convexity and height between phenotypes increased with increasing cold stress. Phenotypic effects on species richness and abundance were consistent along both climate gradients but not effects on species composition, while there were no phenotypic effects on environmental variables. Additionally, RII curves were linear along the drought gradient but unimodal along the temperature gradients, likely due to the occurrence of contrasting species pools at the different sites. We conclude that the consistency of phenotypic effects of *T. caespitosum* was high for species richness and abundance and mainly explained by differences in interference mediated by likely heritable differences in cushion tightness. Additionally, our study shows that the shapes of the relationship between plant responses to neighbours and environmental stresses are not necessarily driven by niche-based deterministic factors.

T3-19-05

Changes in species diversity and community structure of the steppes along climate gradients on the Mongolian Plateau: Implications for conservation and adaptive management

Frank Yonghong Li, *Cunzhu Liang, Wenhong Ma, Liqing Zhao, Luomeng Chao, Tiejun Bao, Lixin Wang, Lu Wen, Jinghui Zhang*
School of Ecology and Environment, Inner Mongolia University

Understanding vegetation patterns along climate gradients are useful to predict vegetation dynamics and develop strategies for vegetation conservation and adaptive management under the global change. We investigated the steppe vegetation patterns on four transects along climate gradients radiating from the desert region in the central towards the relative humid areas in the north, northeast, east and south parts of the Mongolian Plateau. We found that (i) plant species richness and evenness in steppe vegetation declined, whereas the richness and dominance of C_4-species increased, with increasing climate aridity; (ii) the species-area curves showed a plateau at a lower level in relatively arid than humid regions, whereas the curve of species saturation (ratio of the number of species in a certain area to total number of species at a site) showed no difference on climate gradients; and (iii) the gradational replacement of the species of the same genera (ecological vicariance) were common along the climate gradients. These findings suggest that C_4-speceis are more adaptable to the environment of high temperature and low humidity, and that it is necessary to conserve plant communities in different sections on the climate gradient for biodiversity conservation. The results also suggest it appropriate to use the quadrat of the same size to survey plant diversity in different types of steppe vegetation.

T3-19-06

The extent to which mutualists affect the geographical range of plants in a changing environment

Karl Duffy
KU Leuven

Understanding the factors that underlie species distributions is a fundamental goal of ecology. Geographically disparate populations of species are likely to adapt to local biotic and abiotic conditions, potentially forming ecologically adapted populations (ecotypes), which may determine the range limits of species. However, we still do not fully understand the relative importance of environmental factors that lead to population differentiation. Despite having the potential to tolerate a broad range of environmental conditions, populations of species often only exploit a narrow range of resources. As their ranges are determined by abiotic factors (e.g., temperature, precipitation), and also by biotic factors (e.g., pollinators, mycorrhizal fungi), orchids are particularly sensitive to the local availability of environmental resources. As orchid seeds rely entirely on fungi to provide nutrients in order to germinate and establish, mycorrhizal fungi may determine the population occurrence of orchids. Hence, availability of mycorrhizas may lead to population differentiation according to the composition of the local mycorrhizal community. Sampling 50 populations over a 2,500 km range of the broadly distributed European orchid *Spiranthes spiralis*, I identified mycorrhizal communities using Illumina Miseq sequencing with the mycorrhizal specific primer pair, ITS3/ITS4OF. In addition, niche models were constructed to quantify the extent to which key abiotic environmental variables determine the distribution of *S. spiralis* and its mycorrhizas under current conditions and future climate scenarios. Results indicate that *S. spiralis* associates with a broad range of mycorrhizas, but that there is turnover in mycorrhizal communities. Temperature and precipitation play important roles in determining the current geographical range of both *S. spiralis* and its mycorrhizas, but its range will contract and shift under climate change scenarios. Results will be discussed in light of how such broad scale studies can help inform us to ameliorate plant declines due to anthropogenic manipulation of the environment, and shifting distributions due to climate change.

T3-19-07

Experience of inundation or drought alters the responses of plants to subsequent water conditions

Shu Wang[1,2], *Ragan M. Callaway*[2], *Daowei Zhou*[3], *Jacob Weiner*[4]
1. Northeast Normal University
2. University of Montana

3. Northeast Institute of Geography and Aroecology, Chinese Academy of Sciences
4. University of Copenhagen

The availability of water is often highly variable over the life of a plant in nature, and most plants experience episodic extremes in water scarcity and abundance. The importance of plant plasticity in coping with such experiences is widely recognized, but little is known about how plastic responses to current conditions are affected by prior environmental experiences. Our objectives were to investigate the effects of early inundation or drought on the subsequent responses of plant species to the same, opposite or more favourable conditions. To address these questions, we subjected four invasive and four native herbaceous perennial species from different habitats (xeric, mesic, hydric) to two rounds of hydrological treatments (drought, moderate water, inundation) and analysed the effects of the early treatments on survival and performance (total biomass and relative growth) of individuals in the later treatments. In general, (i) early drought reduced the performance of more species than did early inundation, and decreased the final total mass of all species; (ii) early inundation and early drought did not lead to lower survival immediately or later, but improved the relative growth of survivors across all late conditions; (iii) late drought resulted in the highest mortality and lowest performance after any early treatment. With respect to habitat of origin: (i) early inundation was more beneficial for species from wet habitats than for other species; (ii) species from xeric habitats had the strongest drought tolerance after early drought; (iii) mesic species were more likely to suffer reduced later growth after either inundation or drought experience. Invasive species benefitted more from early inundation than did native species, but native species grew better after experiencing early drought. Results indicate that early exposure to inundation or drought conditions alters how plants respond to later conditions and suggest that exposure to extreme events can induce physiological or morphological changes that improve tolerance for either extreme conditions later. This increased tolerance can be at the cost of poorer performance under more benign conditions. Early inundation or drought experience may be harmful immediately, but can be beneficial for the later growth of plants. The ability of species to utilize early hydrological experiences was associated with the water range of their habitats and whether the species is invasive or native. The ability to modulate future plastic responses may be as important as short-term plasticity in adapting to temporal environmental heterogeneity. Such 'metaplasticity' can optimize current performance, while avoiding the potential costs of maintaining a high degree of plasticity throughout life.

T3-19-08

Trait-based adaptability of *Phragmites australis* to effects of soil water and salinity in the Yellow River Delta

Xiaona Yu, *Weihua Guo*
Institute of Ecology and Biodiversity, College of Life Science, Shandong University

Ecological responses to soil properties of the *Phragmites australis* community were experimentally investigated through 96 field studies along the coastal-inland regions in the Yellow River Delta of China. The aim was to evaluate the relationship between phenotypic plasticity and environmental variables and we asked whether the selected functional traits could better respond to changes in electrical conductivity and soil water content. Within the range of soil water content (9.39–36.92%) and electrical conductivity (0.14–13.29 ms/cm), strong phenotypic plasticity of morphological and physiological based functional traits were observed. The effects of soil water content and salinity were not equally important for characterization of the morphological and physiological variability. When compared with soil water content, the response of salinity on plant traits, leaf traits and stem traits are more remarkable. The relationship between functional traits and electrical conductivity were generally linear and logarithmic, indicating that suitable electrical conductivity permits a wider range of functional variations in *P. australis*. Based on these results, salinity leads to reduced average height, specific leaf area, leaf area and base stem diameter, but increased leaf water content and leaf thickness. Overall, lower relative growth rate, longer leaf longevity, and preservation of internal resources were found to be the responses to salt stress to enable highly efficient utilization of resources. However, further studies, including underground traits and genetics, are also needed to clearly understand different responses to related ecotypes.

T3-19-09

Progresses and challenges on algal inhibition potential of hydrophyte allelopathy in shallow eutrophic Chaohu Lake

Shengpeng Zuo, *Shoubiao Zhou, Jianchun Xie, Liangtao Ye, Xiaofeng Jiang, Wenxiu Hong*
College of Environmental Science and Engineering, Anhui Normal University

Besides purifying and restoring the polluted water body by metabolizing, stabilizing, and transforming pollutants, aquatic macrophytes will kill harmful algae by producing active allelochemicals including phenols, terpenoids, fatty acids, esters and heterocycle group. Moreover, bio-diversity can promote algal suppression of typical hydrophytes by some formats such as positive interaction of many allelochemicals, the reasonable mixing culture of different ecotypes of macrophytes, and the efficient combination of hydrophytes and predatory zooplankton or benthic fauna. We have verified synergistic interactions among five allelochemicals [coumarin, ρ-hydroxybenzoic acid, protocatechuic acid, stearic acid, ρ-aminobenzene-sulfonic acid] with antialgal effects on bloom-forming *Microcystis aeruginosa*. Meanwhile, in the watershed of Chaohu Lake, through both field survey and lab experiment, three typical aquatic macrophytes (*Alternanthera philoxeroides*, *Ranunculus sceleratus* and *Trapa incise*) showed combined enhanced-effect on algal suppression. Interestingly, four predatory zooplankton, i.e. *Stentor polymorphus* Müller, *Brachionus calyciflorus* Pallas, *Acartia spinicauda*, and *Moina macrocopa* will help the macrophytes fighting with the noxious algae. Similar findings also occurred for some dominant zoobenthos such as *Limnodrilus hoffmeisteri* (oligochaete), *Chironomus plumosus* (chironomid larva), *Palaemonetes sinensis* (crustacean), and *Corbicula fluminea* (mollusk). Especially, the pollution conditions will influence algal inhibition potential by representative hydrophytes. Low level pollution reduced the allelopathic inhibition of microalgae by allelochemicals. However, many challenges need be solved. For example, how shall the exotic invasion of aquatic macrophytes like *A.*

philoxeroides (Mart.) Griseb, *Eichhornia crassipes* (Mart.) Solms, and, *Spartina anglica, etc.* affect the interaction between hydrophytes and algae, and ecosystem health and safety. Of course, the development of local travel, wading birds, and human interfering activities such as mowing and agricultural chemicals application also deserve more attentions.

T3-19-10

Effect of frugivore *Ocycerous gingalensis* on seed dispersal and seed germination of four threatened plant species in Mihintale sanctuary

Iresha Wijerathne, *Sriyani Wickramasinghe*
Department of Biological Sciences, Faculty of Applied Sciences, Rajarata University

Frugivory is a common interaction mechanism between plants and animals. Sri Lankan grey hornbill is one of the frugivore as well as considered as omnivore which plays considerable dispersal task in dry zone forest ecology as old word endemic bird to Sri Lanka. Main goal was to determine the contribution to the dispersal of threatened plants by Sri Lankan Grey Hornbill. Study was conducted within Mihintale sanctuary (Latitude. 8.3333°, Longitude. 80.5167°) at Anuradhapura district. To determine whether SLGH could be legitimate disperser of threatened plants, fecal samples were collected during breeding season and focal sample analysis was done during non-breeding within 2014-2016. It was concluded that SLGH swallow the whole fruit and regurgitate the intact seeds. In final experiment germination trials were conducted for germination percentages and germination success of the seeds. The study shows the great inter specific variation in the effect of frugivore and seed germination percentage. As threatened plant species *Diospyros ebenum, Stychnos nux vomica, Strychnos potatorum* and *Manilkara hexandra* are considerably endangered and vulnerable plants which SLGH contribute the dispersal and enhance the germination among other dominant dry zone fruit bearing plant species such as *Locxoccocus* sp, *Azarichta indica, Drypetes separia, Filicium dicipiens* and *Ficus* sp. However pulp removal by frugivore was an important factor for germination success specially the enhancing ability to conserve threatened plants.

T3-19-11

Seed rain pattern in a subtropical evergreen broad-leaved forest in southwestern China

Liang Zhao[1,2,3], *Yang Yongchuan*[1,2,3], *Qian Shenhua*[1,2,3], *Lin Dunmei*[1,2,3], *Xiao Jing*[1,2,3]
1. Key Laboratory of Three Gorges Reservoir Region's Eco-Environment, Ministry of Education, Chongqing University, Chongqing 400045, China
2. National Centre for International Research of Low-carbon and Green Buildings, Chongqing University, Chongqing 400045, China
3. Joint International Research Laboratory of Green Building and Built Environment, Ministry of Education, Chongqing University, Chongqing 400045, China

Seed-rain pattern is a crucial process of natural forest regeneration. We conducted a three years field study on seed rain in subtropical evergreen broad-leaved forest in Chongqing, southwestern China. 172 seed traps (0.5 m^2) were erected throughout a 1ha permanent plot in Oct .2013. Seeds were collected monthly. More than 34,000 seeds from 26 species was found in the study period. There were highly differences in seed production between mass (147 seeds seeds m^{-2} yr^{-1} and 202 seeds seeds m^{-2} yr^{-1} in 2014 and 2016, respectively) and few (53 seeds seeds m^{-2} yr^{-1} in 2015) on community level. Seven species represent more than 90% of total seed rain in the community. The seed rain of dominant species *Castanopsis fargesii* represents different pattern compared to community seed rain (54 seeds seeds m^{-2} yr^{-1} and 45 seeds seeds m^{-2} yr^{-1} in 2014 and 2015, respectively, 22 seeds seeds m^{-2} yr^{-1} in 2016). The seed rain of sub dominant species *Elaeocarpus japonicas, Adenandra bockiana, Syraplocos setchuanensis, Engelhardtia roxburghiana* represent same pattern with community level. Our result suggested that longer time study should be conducted to better understand the seed rain in evergreen broad-leaved forest.

T3-19-12

Heritability and plasticity of competitive ability of invasive species

Daniel Montesinos
Centre for Functional Ecology - University of Coimbra

Introduced exotic species can rapidly adapt to conditions in non-native ranges, including changes in size and competitive ability, which result in different invasive success for each individual species. Increased competitive ability generally increases invasive success, but the role of plasticity of competitive effects and responses on invasive success is unclear. Additionally, little is known about the heritability of such changes, and on the importance of hypothetical reintroductions in the maintenance of locally adapted traits. To test for this, we experimentally produced cohorts of the invasive *Centaurea solstitialis*, and the non-invasive *C. calcitrapa* and *C. sulphurea*, of ancestry from either the native range in Spain, the invasive range in California, or (for the invasive *C. solstitialis* only) inter-regional hybrids from both ranges. We then set up one-to-one competition experiments with the grass *Bromus hordeaceus*. The invasive *C. solstitialis* was the only species presenting significantly larger competitive effects and responses in the non-native range of California, but also a significantly lower plasticity for this trait. Additionally, while descendants of two Californian parents had almost twice the competitive effect on *B. hordeaceus* than descendants of two Spanish parents, inter-regional hybrids between Spain and California showed intermediate values. These results suggest that, at this stage of invasion, homogenizing gene flow resulting from hypothetical re-introductions from the native range could reduce their competitive effects on invaded plant communities.

T3-20: Interactions of plants and herbivores under global change

T3-20-01

Quantifying coexistence mechanisms contribution's to diversity maintenance in a species-rich semiarid grassland

Verónica Zepeda, *Carlos Martorell*
Universidad Nacional Autónoma de México

Despite the remarkable advances in ecological theory during the last two decades, the evidence for the mechanisms that maintain species diversity is still lacking. Coexisting species that compete for the same set of resources challenge classic theory, as the number of species in natural communities seems to exceed the limits imposed by competition. Modern coexistence theory recognizes two different mechanisms for coexistence. Equalizing mechanisms arise when the fitness differences between species are small, so no one species is capable of excluding rapidly the others in the community, leading to long-term coexistence. Such mechanisms are unstable as nothing prevents extinctions. In contrast, stabilizing mechanisms grant indefinite coexistence as the species' population growth rates increase when they are rare, averting extinction. Stabilizing mechanisms can be fluctuation-independent or fluctuation-dependent. The storage effect and the relative nonlinearity are fluctuation-dependent stabilizing mechanisms because they depend critically on fluctuations in population densities and environmental factors. Studies of the maintenance of species diversity have traditionally examined individual coexistence mechanisms. However, in nature, multiple mechanisms of coexistence are likely to operate simultaneously. Quantitative measurements are required to assess the relative importance of each mechanism, how they relate to each other, and if they depend on species traits such as their life history. Our aim was to determine the roles of different coexistence mechanisms in maintaining species diversity in a species-rich semiarid grassland. We quantify the contributions of equalizing and stabilizing mechanisms to the coexistence of 19 species, and, because of the large interannual variations in climate and species abundances in the study site, we place special emphasis in the two fluctuation-dependent mechanisms: relative nonlinearity and storage effect. Our analysis showed that 17 of the 19 species have the potential to coexist stably with the others in community. We found strong evidence that species diversity of this semiarid grassland is maintained by stabilizing mechanisms. Our results confirm recent findings that fitness differences between species are very small, and that niche differences are much larger than needed to stabilize the community. The extremely small fitness differences points toward a strong effect of equalizing mechanisms in maintaining diversity. Fluctuation-independent mechanisms were most important for species coexistence as a result of strong intraspecific competition compared to interspecific competition. Relative non-linearity has been usually considered to be an unlikely mechanism for coexistence in temporally fluctuating environments, but our results indicate that it may be far more important than storage effect, although secondary to fluctuation-independent mechanisms.

T3-20-02

Effects of climate change in Mediterranean and semiarid rangelands: Known and unknown in time of uncertainty

Marcelo Sternberg[1], *Zalmen Henkin*[2]

1. Department of Molecular Biology and Ecology of Plants, Faculty of Life Sciences, Tel-Aviv University

2. Agricultural Research Organization, Newe Ya'ar Research Center, Ramat Yishai 30095

Climate change is expected to reduce annual precipitation and increase its standard deviation in eastern Mediterranean rangelands. We examined how these expected changes may affect herbaceous aboveground net primary production (ANPP), and its inter-annual coefficient of variation (CV), in natural rangelands along a desert–Mediterranean precipitation gradient. We examined sites representing Arid, Semi-Arid and Mediterranean ecosystems, which exhibited a positive linear relationship between herbaceous ANPP and annual precipitation. Scenarios of reduced annual precipitation and increased inter-annual precipitation variability were defined by manipulating mean annual precipitation (MAP) and its standard deviation. For each scenario we simulated precipitation and calculated ANPP using current ANPP–precipitation relationships. Our model predicts that reduced precipitation will strongly reduce ANPP in arid and semi-arid sites. Moreover, the effect of reduced precipitation on CV of ANPP along the entire gradient may be modified by a change in inter-annual variability in MAP. In accordance with previous research in the region, reduced precipitation combined with decreased precipitation variability had greater relevance at the arid end of the gradient due to the strong effect on mean ANPP. In contrast, soil seed banks showed high resistance to rainfall reduction and increasing variability along the aridity gradient. The results will be discussed under different climate change scenarios combining known and unknown facts of these particular ecosystems under increasing uncertainty.

T3-20-03

Global change effects on ecosystem functions and services provided by African dryland rangelands

Anja Linstaedter

University of Cologne

In Africa's drylands, plant growth is mainly limited by low and variable rainfall, which constrains agricultural activities to livestock production. Thus, livelihood security relies heavily on forage provision from rangeland vegetation. A sustainable rangeland management requires a profound understanding of how ecosystem structure and functioning will respond to changing climate and land-use. However, this ecological understanding is still surprisingly poor. In my talk I will give an overview on my various research activities in Africa's savannas and grasslands that aim at improving this critical understanding. I will also discuss how my research contributes to help farmers cope with future climatic conditions.

T3-20-04

Overgrazing and carrying capacity: A new insight into rangeland management

Avi Perevolotsky

Agricultural Research Organization

A prudent ecological management of rangeland resources is strongly related to proper determination of the carrying capacity (CC) of the pasture and avoidance from overgrazing (OG). Even though these concepts have been extensively discussed in the scientific and professional literature we believe that they require further and more realistic analysis. Hereby we discuss these issues in relation to different vegetation types: herbaceous vs. woody plants and annual vs. perennial species. We urge to interpret CC and OG according to the specific vegetation formation which is embedded in a specific a-biotic envelop. We suggest to distinguishing between overgrazing and overutilization, each requiring a different managerial approach. The outcome of long-term field experiments in which grazing treatments are imposed on Mediterranean ecosystem helps to demonstrate our claims.

T3-20-05

Effect of endophyte infection and clipping treatment on growth and locust resistance of *Achnatherum sibiricum*

*Junhua Qin, **Anzhi Ren**, Yubao Gao*
Nankai University

Endophytes live asymptomatically within many cool season grasses for at least a portion of their life cycle. Thus far, the best-studied endophytes are *Neotyphodium lolii* and *Neotyphodium coenophialum*, which colonize perennial ryegrass (*Lolium perenne*) and tall fescue (*Lolium arundinaceum*), respectively. It is well documented that endophytes are able to enhance the resistance of host grasses to herbivory, and the anti-herbivore properties of endophyte infection are largely attributable to the production of a variety of alkaloids. Up to now, four classes of alkaloids associated with infected host grasses have been detected, including saturated aminopyrrolizidines (lolines), pyrrolopyrazines (peramines), ergot alkaloids and indolditerpenes (lolitrems). Endophytes not only exist in cultivated grasses such as tall fescue and perennial ryegrass but are also widely distributed in native grasses. Tall fescue and perennial ryegrass may produce three to four different classes of alkaloids. In contrast, endophytes in native grasses usually produce fewer classes and a lower concentration of alkaloids, and some endophytes do not even produce alkaloids. Can endophyte infection in native grasses, which do not produce alkaloids or produce a low concentration of alkaloids, enhance the herbivore resistance of host plants? The limited interactions reported are highly variable. *Achnatherum sibiricum* (L.) Keng is a caespitose perennial grass that is widely distributed in northern China and commonly infected by *Epichloë* endophyte. In this study, endophyte-infected (EI) and endophyte-free (EF) *A. sibiricum* were adopted as plant material. *Locusta migratoria*, a common herbivore of grasses, was adopted as the feeding herbivore. Clipping is a common practice in our sampling grassland; thus, we use clipping as the interference. We wondered whether endophyte infection has a positive effect on the insect resistance of *A. sibiricum* and whether clipping can influence insect resistance of infected *A. sibiricum*. We found that locusts preferred EF plants to EI plants, and the weight of locusts fed on EI plants was significantly lower than those fed on EF plants. Hence, endophyte infection significantly enhanced the resistance of the host to *L. migratoria*. Endophyte infection significantly decreased the concentration of soluble sugar and amino acids while significantly increased the concentration of total phenolic content, and these metabolites may contribute to herbivore resistance of the host. The clipping treatment further strengthened the locust resistance advantage of EI over EF plants. After clipping, the weight of the locusts fed on EI plants significantly decreased compared with those fed on unclipped plants, whereas the weight of the locusts fed on EF plants increased significantly. The results suggested that endophyte infection could increase herbivore resistance while decreasing the tolerance of the host grass by mechanisms apart from endophyte-conferred alkaloid defense.

T3-20-06

The effect of warming on root dynamics and soil microbial community in alpine meadow

Wang Changting
Southwest University for Nationalities

Studies have substantially explored the effects of global change on above-ground vegetation in grassland ecosystems. However, few studies have investigated the responses of root dynamics and soil microbes to simulated warming in an alpine meadow of the Qinghai-Tibet Plateau using a field OTC experiment. Root production, mortality, and turnover were observed in situ using minirhizotron tubes, while Soil microbial community composition and diversity were assessed by using Phospholipid-derived fatty acids (PLFA) and Biolog-Eco plates. The results showed clearly that 1) warming shifted the root standing crop distribution to deeper soil while increased the accumulated root mortality and root production at 0–10 cm soil depth; 2) warming increased the abundance of fungal PLFAs while decreased the abundance of other functional groups; 3) warming decreased the soil microbial functional diversity at the different soil layer; 4) bacterial and fungal PLFA contents were positively correlated with root mortality and turnover but was negatively correlated with root standing crop; 5) root dynamics significantly affect carbon utilization of soil microbes community by warming; and warming induced changes in root dynamics could associate with soil microbial community structure and function. Overall, these findings indicate that warming alters root production, mortality, and turnover and the relative abundance of different microorganisms in different soil layers. As a result, warming tended to shift microbial communities from bacteria towards fungi, leading to changes in the microbial community structure in different soil layers.

T3-21: Ecological processes of grassland ecosystems in a changing world (two sessions)

T3-21-01

Diversity patterns in and historical biogeography of Pannonian and western Pontic steppe grasslands

Matthias Kropf[1]*, Kristina Plenk*[1]*, Olga N. Demina*[2]*, Maria Höhn*[3]*, Wolfgang Willner*[4]
1. University of Natural Resources and Life Sciences, Institute for Integrative Nature Conservation Research
2. Karachay-Circassian State University
3. Szent István University, Department of Botany and Botanical Garden of Soroksár
4. University of Vienna, Department

Biodiversity and biodiversity loss are important topics in grassland ecosystem conservation. Beside recent (negative) impacts on biodiversity due to anthropogenic activities and global environmental change, diversity patterns are also significantly influenced by the biogeographical history of species and communities. Eurasian steppe grassland ecosystems are the largest on earth; they extend westwards to Central Europe forming there one of the most species-rich vegetation types. Their historical establishment and recent decline occurred both probably under (major) human impact. However, in our research we focus on Pannonian and western Pontic dry grasslands covering the steppe zone within the European part of Russia towards Eastern and Central Europe. We are aiming at analysing diversity patterns and biogeographical history in time and space using different sources of data: one the one hand, we use phytosociological relevés not only for classification of vegetation types, but also for extracting species richness within different communities. On the other hand, we exemplarily investigate characteristic plant species of different grassland communities, intraspecifically, using genetic data. Both data types will jointly allow reconstruction of possible historical biogeography of dry and semi-dry grasslands. More specifically, we ask whether and to what extent patterns and processes are similar across different grassland communities. Therefore, we study plant species representing differing, but typical plant communities in parallel, i.e. diagnostic species from the the grassland orders *Brometalia erecti, Festucetalia valesiacae* and *Stipo-Festucetalia pallentis*, respectively. Phylogeographical molecular analyses will focus on cpDNA sequence variation within the different diagnostic species sampled across the Pannonian area and adjacent regions, including potential long-term refugial regions. Aligning different biodiversity patterns across the large geographical range of Pannonian and western Pontic steppe grasslands will be achieved by a rigorous sampling design using grid cells for both vegetation (i.e. phytosociological community) data and intraspecific genetic diversity measures. This will not only provide insights into the biogeographical history of species and vegetation, but will also help defining important diversity regions in grassland ecosystems.

T3-21-02

Correlation between genome size and altitude in purple false brome

Signem Oney-Birol[1], *Selma Tabur*[2], *Metin Tuna*[3]
1. Department of Molecular Biology and Genetics, Faculty of Arts and Sciences, Mehmet Akif Ersoy University, Burdur
2. Department of Biology, Faculty of Arts and Sciences, Suleyman Demirel University, Isparta
3. Department of Field Crops, Namik Kemal

Over many years there have been challenging questions in plant evolutionary biology that study correlations between genome size (GS) and changing ecological parameters or climatic parameters such as latitude, longitude, or altitude. It is still a contradictory question to give an idea about genome size and altitude above sea level or geographical latitude. Altitudinal gradients are excellent models for ecological and evolutionary studies to test the responses of climatic factors such as temperature, precipitation, wind, and sunshine on plant growth and development. Purple false brome (*Brachypodium distachyon* L.) is a model crop for temperature grasses. It belongs to the Poaceae family and is related to the cereals such as wheat and barley.We have We used 541 individuals belonging to 91 accessions distributed throughout Turkey ranging from sea level up to 1,219 m to answer how much correlation is there in genome size under various altitudinal habitats? We have used flow cytometry to determine DNA content (pg). We analysed data using statistical procedures for ANOVA, regression analysis, and principal component analysis, factor analysis, discriminant analysis in Minitab 8 to understand how change genome size among different altitudes. GS values varied widely among individuals, from 0.54 to 0.92 pg. The mean 2C DNA content of *B. distachyon* was determined to be 0.736 ± 0.02 pg, ($P < 0.001$). The scatter plot of all individuals shows that there was no correlation between altitude and genome size. However, altitudinal gradients have been grouped from 0 to 600 meters and 601 to 1,219 meters for testing the level of altitude to see differences in genome size. With these two group altitudinal levels we found a positive correlation between altitude and genome size. In the first altitudinal gradient (from 0 to 600 m) belonging to 409 individuals, the mean 2C DNA content was 0.734 ± 0.001 pg, ($P < 0.005$) and the second group (from 601 to1,219 m) belonging to 132 individuals, the mean 2C DNA content was 0,742 ± 0.002, ($P < 0.005$) pg genome size. Consequently, although gradual elevation gradient (every 100 meters) changes have not caused a correlation between genome size and altitude for individuals of the species, in parallel to higher groups of altitude (0 to 600 m and 601 to 1,219 m) the genetic answer is changed.

T3-21-03

Degradation of Tibetan grasslands: Consequences for soil organic carbon and nutrients losses

Shibin Liu, *Per-Marten Schleuss, Yakov Kuzyakov*
Göttingen University

The *Kobresia* pastures, commonly known as "alpine meadow", cover the southeastern quarter of the Tibetan Highlands (~450,000 km^2). They host important grazing ground for livestock (i.e. yaks, sheep and goats) and thus ensure the livelihood of the Tibetan herders. The *Kobresia* pastures also store huge amount of soil organic carbon (SOC) and nutrients (e.g., nitrogen (N) and phosphorus (P)), which are required for sufficient forage production. In recent decades, the *Kobresia* pastures have experienced severe degradation due to anthropogenic activities and climate change, which has initiated high losses of SOC and nutrients and threatened the functioning of this ecosystem. Plenty studies have been implemented showing the response of degradation on SOC and nutrients levels on local scale. They classify these alpine pastures into various degradation stages that are mainly based on vegetation characteristics (e.g., vegetation coverage, proportion of edible plants). Within this study we synthesized their results in a review for a better understanding of SOC and nutrients losses following pasture degradation across the whole ecosystem. We aggregated the degraded *Kobresia* pastures into five degradation stages: Non-degraded, Light degradation, Moderate degradation, Heavy degradation and Extreme degradation. Results show that degradation from light to extreme stages has lost on average 42 ± 2% SOC, 33 ± 6% N and 17 ± 4% P as compared to the non-degraded pastures. This implies strong reduction of soil fertility and an exacerbation prevailing N and P limitations. Concurrently, degradation has decreased aboveground and belowground biomass by ~42 ± 3% and ~45 ± 6%, which reflects (a) decreasing photosynthetic C input and (b) less available forage for livestock. Besides, the declining vegetation promotes wind and water erosion. In conclusion, our results provide an overview and a quantification of degradation impacts on plant characteristics and soil properties that improve estimations regarding SOC and nutrients losses across the whole ecosystem. This highly matters because large amounts of SOC have been lost due to erosion and mineralization. Most likely this has polluted the Tibetan headwaters and contributed to climate change, respectively. Further, the decreasing N and P losses have reduced soil fertility lowering forage production. Therefore, it endangers the livelihood of the Tibetan herders, which highly rely on forage to feed their livestock. Despite plenty of ameliorations (e.g., fertilization, grazing enclosure, reseeding) have been proposed and implemented at many locations, their impacts on pasture ecosystems (especially on soil fertility) are still subtle and thus require further investigations.

T3-21-04

Wind erosion caused by land use changes significantly reduces ecosystem carbon storage and carbon sequestration potentials in grassland in Northern China

Ping Li[1,2], *Yonggang Chi*[1], *Jing Wang*[1,2], *Lingli Liu**[1,2]

1. State Key Laboratory of Vegetation and Environmental Change, Institute of Botany, Chinese Academy of Sciences, Beijing, China

2. University of Chinese Academy of Sciences, Beijing, China

Wind erosion and deposition of soil could greatly alter soil C pool and ecosystem C balance in arid and semi-arid ecosystems. Land use change is one of the most important driving forces influencing the intensity of wind erosion in those regions. However, how wind erosion under different land use scenarios will affect ecosystem C balance and its capacity for future C sequestration are poorly quantified. Here, we established an experiment in Xinlinhot, Inner Mongolia, and simulated different intensity of land use in grassland: control, 50% of aboveground vegetation removal (50R), 100% vegetation removal (100R) and tillage (TI). We monitored lateral and vertical carbon flux components from 2013 to 2016. We found that wind deposition resulted in net C gain during growing seasons, whereas wind erosion induced net C loss. Adding up the lateral C flux across the whole year, ecosystems under control behaved as a C sink of 44.85 g C m^{-2} yr^{-1}. However, under relative degradation ecosystems, the disturbance managements resulted in a net C loss, and the loss strength increased from 3.35 g C m^{-2} yr^{-1} under 50R to 135.86 g C m^{-2} yr^{-1} under TI and the erosion intensity increased rapidly with experimental duration. Land use also significantly altered the biological vertical carbon flux. The net ecosystem exchange (NEE) shifted from a net uptake of 86.85 g C m^{-2} yr^{-1} under control to a net emission of 35.12 g C m^{-2} yr^{-1} under TI treatment. With the increase in land use intensity, the contribution of lateral C flux to ecosystem C balance increased from 34% under control to 79% under TI. Wind erosion caused by land use changes not only result in dramatically surface soil C loss, but also significantly decrease soil C sequestration potential by altering soil texture. Permanent losses of organic carbon sequestration potential were 0.10 kg C m^{-2},0.12 kg C m^{-2} and 0.31 kg C m^{-2} in 50R, 100R, and TI, respectively. Overall, our study demonstrated that wind erosion could result in irreversible soil degradation and shape the landscapes of arid and semi-arid regions in the long term. Appropriate land use is critical to protect grassland ecosystems from being crashed.

T3-21-05

The underlying basis for the tradeoff between leaf size and leafing intensity

Yingxin Huang

Northeast Institute of Geography and Agroecology, Chinese Academy of Sciences

Recent studies have reported a consistent tradeoff between leaf size (mass) and leafing intensity (the number of leaves produced per unit of supporting stem tissue volume), however, a theoretical basis for this tradeoff has not been described. We explore the mechanistic basis for this tradeoff and assess the relationship in light of other prominent theories for allometric biomass partitioning. We show algebraically how the allocation of mass to leaves versus stems and the density of stem tissue can potentially influence this tradeoff. To assess these possible effects, we compared the relationship between leaf size and leafing intensity, expressed on both mass and volume basis, at the level of a single branch as well as the entire above-ground plant in 61 forbs over a 3-year period. Our results support the idea that the tradeoff between leaf size and volume-based leafing intensity depends on both biomass investment (leaves vs. stems) and stem bulk density (mass vs. volume), whereas the tradeoff between leaf size and mass-based leafing intensity only depends on the biomass investment. Similar exponents in the scaling of leaf mass vs. stem mass and stem mass vs. stem volume at branch- and whole plant-level lead to similar tradeoffs. An isometric tradeoff between leaf size and volume-based leafing intensity is consistent with a constant biomass partitioning between leaves and stems as well as constant stem tissue density. Conversely, an allometric tradeoff between leaf size and volume-based leafing intensity arises when biomass allocation is allometric and stem bulk density varies with plant size.

T3-21-06

Long-term dynamics of flowering shoots in alpine lichen heath and *Festuca varia* grassland (Teberda Reserve, the North-western Caucasus)

Tatiana Elumeeva

Lomonosov Moscow State University

In alpine environment, flowering strongly depends on weather conditions of the previous growing season, when the buds of the most of the perennial species develop. The dynamics of reproductive shoots shows higher variability comparing with vegetative shoots. Under current climate changes it may influence not only plant species regeneration, but also populations of other organisms, such as pollinators and herbivorous insects feeding in flowers. I analysed long-term flowering shoot dynamics in alpine lichen heath (ALH, 2,800 m a.s.l.) and *Festuca varia* grassland (FVG, 2,100 m a.s.l.) at the Malaya Khatipara Mountain, the Teberda reserve, the North-western Caucasus, Russia. These communities are typical in the alpine belt of the Caucasus and contain a number of species common in both communities. ALH occupy ridges and FVG south-eastern slopes with low snow accumulation. I tested (1) whether there are trends in interannual flowering shoot number dynamics, (2) which weather conditions are the most important in determining of their dynamics. The number of flowering shoots was counted on the permanent plots every year, from 1981 to 2016 in ALH, and from 1987 to 2016 in FVG. The total area of observations was 62 m^2 in ALH and 50 m^2 in FVG. Temperatures and precipitation during the growing season were derived from the nearby Teberda weather station. To reveal trends in flowering, I run Detrended Correspondence Analysis (DCA), where years were used as "plots" and number of shoots was a criterion of species abundance. Rare species were downweighted in the course of the analysis. Spearmen's correlation coefficients were calculated between axes scores and weather conditions of the previous and the current year. For both communities, the first DCA axis was significantly ($p < 0.001$) correlated with ordinal number of year and with the mean August temperatures of the previous year ($p < 0.001$). DCA also revealed groups of decreasing species (including, for example, *Carex umbrosa, Carex sempervirens, Anemone speciosa* in ALH and *Cruciata laevipes, Galium verum, Leontodon hispidus* in FVG) and increasing species (*Euphrasia ossica, Plantago saxatilis, Aster alpinus, Alchemilla caucasica, Ranunculus oreophilus* in ALH and *Campanula collina, Euphrasia ossica, Hedysarum caucasicum* in FVG). The observed patterns in the dynamics of flowering shoots are similar with those in the dynamics of the total shoot number in these communities. Despite the fluctuations, when the species composition and abundance of their flowering shoots are close to those observed in any previous years, our data suggest trends in the total dynamics of the studied communities. The field works were supported by RFBR (grants № 11-04-01215,14-04-00214). Data analysis was supported by RSF (grant № 14-50-00029).

T3-21-07

Climate modulates the effects of grazing on plant functional group diversity in *Stipa krylovii* grassland in Inner Mongolia, China

Yantao Song, *Xiaohong Zhang, Guangwei Huo, Yunna Wu*
Dalian Minzu University

Grazing is one of the prevalent human activities effects on grassland ecosystems. There has been a rapidly increasing interest in the effect of grazing on species diversity or functional group diversity in grasslands. In this study we analysed the effects of grazing on plant species diversity, plant functional group (PFG) diversity of *Stipa krylovii* grassland in Inner Mongolia, China. We analysed plant diversity data from 60 sampling plots by three estimated classes of low, medium and high grazing intensity in 2014 (wet year) and 2015 (dry year). The sample site located in Kerulen basin in New Barag Right Banner. We also classified a priori all plant species in four PFGs: perennial grasses, perennial forbs, shrubs/subshrubs, and annual/biennial herbs. The results showed that the species richness was highest in medium grazing condition in both 2014 and 2015. The species richness in high grazing intensity was higher than in low grazing intensity. There was no significant difference in PFG richness with grazing intensity. However, each PFG showed different response to grazing intensity. The perennial forbs was the dominated PFG in community composition, and it contained more than 50% species richness in each grazing intensity. The species richness in PFG of annual/biennial herbs increased as grazing intensity increased, especially in wet year of 2014. There was little changed of the species richness in PFG of shrubs/subshrubs as grazing intensity increased. Both of the species richness and PFG richness in 2014 were significant higher than those in 2015 in each grazing intensity because of the dry weather in 2015. In 2014, the degree of dominance of perennial grasses was higher than other PFGs in each grazing intensity. In 2015, the degree of dominance of perennial grasses decreased as grazing intensity increased, and the degree of dominance of perennial forbs was increased as grazing intensity increased. This finding may imply that medium grazing promotes the species richness, it could be predicted by an intermediate disturbance hypothesis. The perennial forbs have strong resistance with high grazing intensity in dry year. The annual/biennial herbs was more sensitive to precipitation than grazing. So, climate modulates the effects of grazing on plant functional group diversity in *Stipa krylovii* grassland in Inner Mongolia, China.

T3-21-08

Vegetation pattern of grassland habitats on local and landscape-level an its changes in changing geographycal environment in the Danube-Tisza Interfluve's Sand-ridge, Hungary

Áron József Deák, *Zsuzsanna Ladányi, János Rakonczai*
University of Szeged, Department of Physical Geography and Geoinformatics

The Danube-Tisza's Interfluve's Sand-ridge, a Pleistocene alluvial cone of the river Danube is situated in the middle of the Carpathian-basin on the western edge of the forest steppe zone. The average annual precipitation income is very rare, rather significantly under average precipitation-income favoring for dry grasslands or above average precipitation-income favoring for wet fens, wet salt communities and forests are typical helping - together with soil, hydrogeographical, geomorphological features - the maintain of forest-grassland mosaics. Wet years are typical after dry years within 2-3 years compensating somewhat the affect of dry years. This climatic feature with temporary changes and shifts to more boreal, atantic, sub-mediterranean and continental conditions was typical here during the Holocene resulting high climatic adaptive capacity of plants and plant communities, special vegetation pattern, transitional and endemic communities (see Pannonian sand grasslands). The local and landscape-level vegetation pattern and its gradients were studied with habitat-mapping using the General National Habitat Classification System (ÁNÉR). In consecutive wet and dry years the shift of zonation, formation and change of transitional communities, their species composition and the coverage of species were studied with coenological plots alongside hydrogeographical gradients and with repeated habitat-mappings on the western edge of this mesoregion (Turján-land). In this sample area (around Kunpeszér, Kiskunság National Park) the natural grasslands remained on significant areas, where zonation-shifts, transitional communities with fast annual changing species-composition and coverage are typical in years with different precipitation income. The vegetation pattern is highly influenced by the geomorphology, the groundwater-level and its Na-salt content in this landscape. On dunes open sand grasslands with scattered patches of polar or oak forests are typical due to the thin, nutrient-poor soils and deeper groundwater-level, whereas on flat sand sheets and residual ridges closed sand steppes are typical with formerly more distributed closed poplar and oak forests. Uniquely *Molinia* fens, rich fens, tall sedge-beds, mesotrophic wet meadows and Pannonian, more continental salt vegetation (salt meadows, *Puccinellia* and annual salt pioneer swards, salt marshes) appear in the blow-out depressions next to each other. The precipitation is not enough for fens, sedge-beds and wet meadows, so they appear in the groundwater up-welling zones, while further from these zones salt meadows, even further *Puccinellia* and annual salt pioneer swards are dominant in the deepest parts as soil solutions become concentrated due to evapotranspiration and salt-accumulation. In the Dorozsma-Majsa Sand-ridge this pattern shows a local and landscape level northwest-southeast gradient covering soil-type, pH, salt and organic material gradients as well. The observed groundwater-decrease due to precipitation-decrease, drainage and groundwater take-out resulting the out-leaching of salt habitats as well threatens the communities of the depressions. Sand steppe grasslands can occupy their places, but landscape-use decreased their extension significantly making this shift more difficult. So conservation of the full zonation is important for the better adaptation to climate change. Decreasing groundwater-level, desalinization, increasing differences and uncertainties in precipitation, accumulation of organic materials and nutrients due to the lack of cow-grazing, mowing result the decrease of plant diversity, spread of reeds making wet communities more characterless.

T3-21-09

Facilitation by leguminous shrubs increases along a precipitation gradient

Hai-Yang Zhang[1,2*], *Xiao-Tao Lü*[1], *Alan K. Knapp*[3], *Henrik Hartmann*[2], *Edith Bai*[1], *Xiao-Bo Wang*[1], *Zheng-Wen Wang*[1], *Xiao-Guang Wang*[4], *Qiang Yu*[5], *Han Han*[6]

1. Erguna Forest-Steppe Ecotone Research Station, Institute of Applied Ecology, Chinese Academy of Sciences, Shenyang, China

2. Max Planck Institute for Biogeochemistry, Jena, Germany

3. Department of Biology, Graduate Degree Program in Ecology, Colorado State University, Fort Collins, Colorado, USA

4. College of Environment and Resources, Dalian Minzu University, Dalian, China

5. National Hulunber Grassland Ecosystem Observation and Research Station/Institute of Agricultural Resources and Regional Planning, Chinese Academy of Agricultural Sciences, Beijing, China

6. State Key Laboratory of Vegetation and Environmental Change, Institute of Botany, Chinese Academy of Sciences, Beijing, China

Combining nutrient dynamics (plant nutrient uptake and soil fertility) can help uncover mechanisms of shrub-grass interactions and assess the validity of the stress-gradient hypothesis, which predicts that facilitation between plants increases in stressful environments. However, how facilitation via shrub-mediated nutrient increases varies with precipitation is poorly resolved. We first synthesized a global dataset from 66 studies and evaluated how shrubs affected soil organic carbon (C) and nitrogen (N) in grasslands along a precipitation gradient. We then made new measurements in a single grassland type encroached by leguminous shrubs from the same genus (*Caragana*) to constrain the variations caused by combining different grassland types and shrubs traits in the meta-analysis. Specifically, we investigated how shrubs mediated N dynamics and how shrub-grass interactions varied along a precipitation gradient (147-342 mm) in a temperate steppe of China. At the global scale, nurse plant effects on soil N pool increased with precipitation only for leguminous but not for non-leguminous shrubs. Results from the field study showed that facilitation became relatively weak as water stress increased, indicating a collapse of facilitation. Our results highlight that plant-plant interactions would be altered by precipitation and shrub functional traits (leguminous or not) should be included as mechanisms in conceptual frameworks of plant facilitation.

T3-21-10

How can botanical compostion be managed by grazing in desert steppe

Guodong Han

Inner Mongolia Agricultural University

The desert steppe of Inner Mongolia is traditionally used for year-long sheep grazing. However, overgrazing reduced vegetation production, thereby changing the botanical composition. Grazing is one of important ways to adjust the botanical composition for the development of successful grazing management strategies. A grazing experiment was conducted from 2004 to 2015 in which four stocking rate ranging from 0 sheep equivalent ha^{-1} mo^{-1}, 0.15 sheep equivalent ha^{-1} mo^{-1}, 0.30 sheep equivalent ha^{-1} mo^{-1} to 0.45 sheep equivalent ha^{-1} mo^{-1}. Each stocking rate treatment has 3 replications. Variables measured included dry matter, species number, tiller numbers of dominant species in different treatments. The results showed that the peak dry matter in September has an significant impact on the total dry matter of next June, which means that grass initialed in the grassland will be influenced the dry matter reservation on the previous year. With the reservation dry matter in September increased, total tillers on next May increasing sharply. Plant functional groups (PFGs) changes are good indicators for the evaluation of grassland conditions. Shrubs and perennial grasses were the components most affected by grazing treatments. The higher grazing pressure coincided with the lower the shrub content, and lower species diversity. When we made the management strategies, grazing is a great tool to control the shrub growth in desert steppe. Manage the ration of perennial grasses and shrub fit 2:1 will increase the stability of desert steppe ecosystem.

T3-21-11

The effects of herbivore assemblages on vegetation and soils on steppe grasslands

Deli Wang, *Ling Wang*

Institute of Grassland Science & School of Environment Science, Northeast Normal University, and Key Laboratory of Vegetation Ecology, Ministry of Education

Most previous research has indicated that grazing intensity can exert various impacts on steppe vegetation and soils, but little information is available on how different herbivore assemblages influence potential grassland multifunctionality in grazing contexts, despite the fact that mixed grazing by different species is adopted in many grasslands systems. For this study, we investigated the effects of herbivore assemblages on plant and soil characteristics such as plant diversity and productivity, and soil heterogeneity, nitrogen cycling and carbon flux with a set of grazing manipulation (no grazing, cattle grazing, sheep grazing and mixed grazing) experiments in the Eurasian steppe grasslands. We obtain the following results: (1) Grazer effect on plant diversity strongly depended on herbivore species and combination. Single-species grazing by cattle and mixed grazing of cattle and sheep significantly increased plant diversity, while single-species grazing by sheep significantly reduced plant diversity. Such increasing effects of cattle grazing can be attributed to the fact that cattle had more inhibition on the competitive dominant grasses than sheep. Mixed grazing had a consistently positive effect on plant diversity and it is most likely that the additive effect of mixed grazing enhanced plant diversity through the reduction of competition from grasses. (2) Herbivore assemblages had significantly varying effects on plant biomass. Single-species grazing by cattle decreased plant biomass, while grazing by sheep alone and mixed grazing of cattle and sheep didn't affect plant biomass. Cattle, a large-bodied and less selective herbivore that commonly coped with the low plant nutrient content but requires a higher abundance of energy-rich plants, tended to negatively affect plant biomass accumulations by consuming dominant grasses. (3) Cattle grazing increased the spatial heterogeneity of available N in the soil of *Leymus chinensis*-dominated steppe. However, the effects of sheep grazing and mixed grazing strongly depended on grassland plant diversity, with an increased spatial heterogeneity of available soil N in the high diversity sites. In such grassland systems with high plant diversity, herbivore grazing and plant diversity would jointly improve soil spatial heterogeneity, thus feeding back to maintain higher plant diversity. (4) Herbivore grazing significantly improved soil N availability regardless of herbivore assemblages. Single-species sheep grazing and mixed grazing of cattle and sheep had a greater effect on soil N availability than single-species cattle grazing. It demonstrated that herbivore grazing had a predominantly positive effect on N cycling rate under moderate grazing intensity. (4) Grazing by cattle alone significantly increased net ecosystem CO_2 exchange, but exchange was reduced under sheep grazing. Yet NEE was significantly higher under mixed grazing of sheep than under single-species cattle or sheep grazing. We conclude that, given that herbivore assemblages have varying effects based on different grassland functions, herbivore assemblages should be considered an important affecting factor in future grassland management besides grazing intensity.

T3-21-12

Differential responses of canopy nutrients to experimental drought along a natural aridity

Wentao Luo[1*], *Xiaoan Zuo*[2], *Wang Ma*[1], *Xiyuan Yue*[2], *Cong Xu*[3], *Qiang Yu*[3], *Alan K. Knapp*[4], *Roberto Tognetti*[5], *Feike A. Dijkstra*[6], *Mai-He Li*[1,5], *Zhengwen Wang*[1], *Xingguo Han*[1]

1. Institute of Applied Ecology, Chinese Academy of Sciences, China

2. Northwest Institute of Eco-Environment and Resources, Chinese Academy of Sciences, China

3. Institute of Agricultural Resources and Regional Planning, Chinese Academy of Agricultural Sciences, China

4. Colorado State University, USA

5. Swiss Federal Research Institute, Switzerland

6. The University of Sydney, Australia

Droughts and periods of water limitation are projected to increase in both magnitude and frequency due to climate change. It is, however, poorly understood how canopy nutrient (N and P) concentrations (community weighted by species abundance) will respond to drought and how these responses may vary along a gradient of precipitation and shifting plant community composition. Nutrients in grassland plant canopies can be used to assess and predict the likely responses of fundamental ecosystem processes to environmental changes. At three sites along a natural aridity gradient in the grasslands of northern China, short-term drought manipulation caused irregular effects on canopy nutrient concentrations, and these were driven by combined effects of species turnover and intraspecific variation at each site. In contrast, along the aridity gradient, canopy nutrient concentration was inversely related to precipitation (the arid site had the highest canopy nutrient concentration) and this pattern was driven mostly by species turnover (i.e., an increase in abundance of species with higher nutrient concentrations). Short-and long-term reductions in precipitation directly affected plant nutrient status rather than indirectly via alterations in soil nutrient availability. We conclude that nutrient-rich canopies were more resistant to drought at drier sites, because of the opposing effects of species turnover and intraspecific variability and the higher drought tolerance of nutrient-rich species. Differing responses of canopy nutrient concentration to drought conditions along natural gradients vs. manipulative experiments in the same grassland biome highlights the complexity in predicting how these grasslands will respond to climate change.

T3-22: Atmospheric nitrogen deposition and its impact on plant growth and sustainability of ecosystems

T3-22-01

Nitrogen deposition in China and its environmental impacts

Xuejun Liu

China Agricultural University

Atmospheric nitrogen (N) deposition is not only an important component in the human-accelerated global N cycle but also an indicator of atmospheric N pollution. Excess N deposition has aroused environmental concerns about negative impacts on ecosystem health and services such as loss of biodiversity, forest soil acidification, and increased greenhouse gas emission. Rates of N deposition, especially oxidized N deposition, have leveled off or decreased in Europe and US since the 1980s or early 1990s with the implementation of stricter legislation to limit atmospheric pollution. In contrast, emissions of both NH_3 and NO_x in China have been increasing continuously since the 1980s mainly due to growing agricultural and industrial activities. These increased reactive N emissions to the atmosphere have aroused widespread concern on air pollution in China. However, scientific gaps to understand regional and temporal variability in atmospheric N emissions and deposition still exist. Meanwhile, the ecological and environmental impacts of N pollution and deposition are of great concern in China. In this paper, the authors summarize recent progress on N deposition studies in order to identify: 1) the spatio-temporal variability of N deposition fluxes; 2) the major impacts of N pollution and deposition on air quality and ecosystem health; 3) some future research recommendations and regulatory strategies for mitigation of atmospheric N pollution in China. Recently a number of atmospheric N deposition monitoring networks have been established since the early 2000s. There are increasingly more research on quantification of spatial and temporal variations of atmospheric N deposition and its impact on terrestrial ecosystems, using different atmospheric deposition modelling approaches. The major findings in this paper include three parts: (1) anthropogenic reactive N (e.g., NH_3 and NO_x) emissions contribute largely (up to 40%) to secondary inorganic aerosol formation and haze pollution; (2) dry N deposition is proven to be the same importance to wet N deposition, suggesting the necessity to re-examine the previously identified N critical loads based on wet deposition alone; (3) the enhanced N deposition over China have negatively affected air quality, eutrophication, soil acidification and decline of plant species richness especially in grassland ecosystems. Future research challenges on atmospheric N emission and deposition are discussed as well. China needs to (1) reduce the uncertainties of national N emission inventory; (2) establish national networks for atmospheric N concentration and deposition monitoring; and (3) evaluate ecological and environmental impacts of N pollution and deposition in typical ecosystems. Cross-site N addition experiments along with various forests, grasslands, deserts and aquatic ecosystems are required in order to provide systematic information on the impact of elevated N deposition on both terrestrial and aquatic ecosystems against the background of climate change. In particular, the response of carbon sequestration and/or greenhouse gas balance to elevated N deposition should be taken into account in future studies. Last but not least, N deposition modeling tools should be improved based on localized parameters and further used in future N regulation.

T3-22-02

Recovery from long-term nitrogen deposition

Keith Goulding
Rothamsted Research

The negative effect of increasing atmospheric nitrogen (N) pollution on grassland biodiversity is incontrovertible. Nitrogen deposition from the atmosphere to cropland, grassland and woodland has been measured at Rothamsted Research, UK, for many years. Surprisingly large amounts of N, both wet and dry, oxidised and reduced, were deposited to local ecosystems in the 1990s. However, the recent introduction of cleaner technologies in the UK has led to reductions in the emissions of nitrogen oxides and some reduction in ammonia, with concomitant decreases in N deposition. The effects of these decreases on grassland biodiversity has been investigated in the 160-year old Park Grass Experiment to see the extent to which grassland biodiversity can be expected to 'bounce back', i.e. recover, in response to these improvements in air quality. The hypothesis was that long-term chronic N addition may lead to an alternative low biodiversity state. However, data from the Park Grass Experiment show a positive response of biodiversity to reducing N addition from either atmospheric pollution or fertilizers. The proportion of legumes, species richness and diversity increased across the experiment between 1991 and 2012 as both wet and dry N deposition declined. Plots that have not received inorganic N fertilizer since 1989 recovered much of the diversity that had been lost, especially if lime (calcium carbonate) was applied to overcome acidification. There was no evidence that chronic N addition has resulted in an alternative low biodiversity state on the Park Grass plots, except where there had been extreme acidification. Recovery was probably also facilitated by the twice-yearly mowing and removal of grass, removing much of the N deposited. The fact that natural and semi-natural grassland is not cut and removed might explain why a comparable response of plant communities to reduced N inputs has yet to be observed in the wider landscape.

T3-22-03

Atmospheric reactive nitrogen deposition and effects in the U.S. Rocky Mountain region

Jeffrey Collett
Colorado State University

Increases in reactive nitrogen deposition are a growing concern in the U.S. Rocky Mountain west. Alpine ecosystems in Rocky Mountain National Park (RMNP), Grand Teton National Park (GTNP) and elsewhere evolved under low nitrogen conditions. Increases in inputs of oxidized and reduced nitrogen in recent decades have created new stresses on these ecosystems. Current inputs of reactive nitrogen in this region now substantially exceed the established critical load: the deposition flux at which irreversible changes occur to the ecosystem. For the past decade we have been actively examining the sources, transport and deposition of reactive nitrogen in the Rocky Mountain region. Both RMNP and GTNP are located at high elevation, but can experience periodic, polluted conditions associated with transport of emissions from a variety of regional sources, including urban areas, power plants, agricultural regions, wildfires, and areas of active oil and gas development. While wet deposition of nitrate and ammonium, on which the critical loads for nitrogen deposition are based, are important, our measurements reveal that dry deposition of gaseous ammonia and wet deposition of organic nitrogen are both major contributors to regional reactive nitrogen deposition budgets. Neither form of reactive nitrogen has historically been measured in U.S. air quality or deposition monitoring networks. The presentation will review observed effects of excess reactive nitrogen deposition on regional alpine ecosystems, discuss the establishment of critical loads, summarize findings regarding major sources and deposition pathways, and highlight recent efforts to formulate policies and strategies to reduce reactive N deposition in the future below ecosystem critical loads.

T3-22-04

Modeling atmospheric nitrogen deposition: Sources, processes, and surface ozone responses

***Lin Zhang**, Yuanhong Zhao, Youfan Chen*
Peking University

I will present our recent work on better understanding the sources, processes, and consequences of nitrogen deposition using an atmospheric chemistry model (GEOS-Chem). (1) We will show that satellite observations of atmospheric ammonia concentrations from the Tropospheric Emission Spectrometer (TES aboard EOS Aura) can provide top-down constraints to improve our estimate of ammonia emissions over China. Assimilating TES ammonia observations into the model indicates a strong seasonal variability in the Chinese ammonia emissions with 1.9 Tg N emitted in July compared with 1.2 Tg N in April for the year 2008. This is consistent with surface measurements of ammonium wet deposition flux, as well as an improved bottom-up emission estimate (12.6 Tg N annually) with more practical fertilizer application rates and temperature-dependent livestock ammonia emission factors. (2) We simulate atmospheric inorganic nitrogen deposition over China for 2008-2012, and extensively evaluate model results with available measurements. Total inorganic nitrogen deposition fluxes are simulated to be generally less than 10 kg N ha^{-1} a^{-1} in western China (less than 2 kg N ha^{-1} a^{-1} over Tibet), 15-50 kg N ha^{-1} a^{-1} in eastern China, and 16.4 kg N ha^{-1} a^{-1} averaged over China. Annual total deposition to China is 16.4 Tg N, with 62% (10.2 Tg N) from reduced nitrogen (NH_x), 39% via dry deposition process, and 86% attributed to domestic anthropogenic sources. We estimate that about 15% of the land over China is experiencing critical load exceedances. (3) We will also present an exploratory study to understand how surface ozone pollution would respond to anthropogenic nitrogen deposition using the GEOS-Chem atmospheric chemistry model and the CLM land model. Enhanced nitrogen deposition since industrialization altered the land vegetation coverage and soil mineral nitrogen content, leading to increased biogenic VOC emissions and soil NO_x emissions but also higher ozone uptake via dry deposition. Our results show that anthropogenic nitrogen deposition would overall change the summer mean surface ozone pollution by -2.4 to 2.7 ppbv in the northern hemisphere, with significant regional differences. These values are relatively small but comparable to the previously estimated surface ozone changes driven by changes in land cover and climate.

T3-22-05

Impacts of nitrogen deposition on soil biogeochemical cycles: Perspectives based on enzymology

Dejun Li, *Hao Chen*
Institute of Subtropical Agriculture, Chinese Academy of Sciences

Over the past decades, atmospheric nitrogen (N) deposition has substantially increased worldwide due to increased fossil fuel combustion and widespread use of chemical fertilizer N. Increased atmospheric N deposition has been found to alter processes and functions of terrestrial ecosystems including the biogeochemical cycling of carbon (C), N, phosphorus (P) and sulfur (S). Soil extracellular enzymes control the rate-limiting steps in soil organic matter decomposition and soil nutrient return, and hence play key roles in determining the biogeochemical cycles of the above elements. Accordingly, dynamics of soil extracellular enzymes activities (EEAs) provide useful information of the alteration of soil biogeochemical cycles under global change. In order to get a general view of the responses of soil EEAs relating to C, N, P and S acquisition to N deposition, a meta-analysis of published studies was carried out in the current study. N addition significantly increased cellulase activity by 13.0%, but repressed ligninase activity by 11.2%. N addition had no significant effect on protease activity, but significantly elevated the activities of N-acetylglucosaminidase and urease by 5.5% and 11.6%, respectively. N addition enhanced phosphatase activity by 42.3%, but decreased arylsulfatase activity by 34.4%. The ratios of C:N:P acquisition enzyme activities converged on 1:1:1, and was not affected by N addition. Substrate C quality, estimated using an enzyme-based lignocelluloses index revealed elevated soil C quality under N addition. The stoichiometric relationships between nutrient acquisition enzyme activities also revealed that the terrestrial ecosystems changed from N limited to P limited status under N addition. Our results suggest that dynamics of EEAs provide valuable information of soil biogeochemical cycle responses to N deposition.

T3-22-06

Effects of nitrogen deposition on the carbon and nitrogen cycle of secondary forests in Northeast of China

Yunting Fang[1,2], *Shanlong Li*[1], *Shaonan Huang*[1], *Feifei Zhu*[1], *Dongwei Liu*[1], *Dan Xi*[4], *Weixing Zhu*[1,3]
1. CAS Key Laboratory of Forest Ecology and Management, Institute of Applied Ecology
2. Qingyuan Forest CERN, Chinese Academy of Sciences
3. The State University of New York, Binghamton
4. Fujian Agriculture and Forestry University

With continues increasing in nitrogen (N) deposition for several decades in China, the deposited N has induced large impacts on the plant growth, soil property and elements biogeochemistry cycles on forests ecosystems. Northeastern China was one of the largest areas of plantations and secondary forests. However, we still know little about the exact pattern of N deposition and its integrated environmental effects to the forest ecosystems in this region. Our study was conducted at Qingyuan Forest CERN station, Chinese Academy of Sciences. The station is located in a mountainous region in the eastern Liaoning Province, Northeast China. The objectives of our study were 1) to examine the exact status of N deposition (fluxes of different N forms and patterns of isotope composition) in typical secondary forests, 2) to determine fates of deposited N in two typical secondary forests, using ^{15}N labeling method, and 3) to assess possible effects of increasing N deposition in future on trees growth and litter decomposition. During 2014 to 2015, total dissolved inorganic N (DIN) deposition in precipitation were 18.1 and 20.8 kg N ha^{-1} yr^{-1}, of which NH_4^+ accounted for about 67%. Total DIN leaching via stream water were 6.0 and 9.3 kg N ha^{-1} yr^{-1}, which were higher than the threshold of N saturation (5 kg N ha^{-1} yr^{-1}) reported for European forests. The ^{15}N labeling experiment for 1 year in the two secondary forests shows that 9.7-23.2% of ^{15}N tracer was incorporated into plant tissues. 26.7-49.2% was retained in soil. 35.9-52.6% was not recovered in the ecosystems. Retentions of deposited N at ecosystems scale were also lower than that in many temperate forests in Europe and America. These findings suggested that the forests in the study area may have N saturated. However, the effect of N addition (50 kg N ha^{-1} yr^{-1}) on the basal area of tress was not significant, but the production of litterfall was enhanced significantly in two forests. As TIN concentration in stream water declined obviously in autumn, we concluded that the forests may be seasonally N-limited in the area. We also observe that litter decomposition was inhibited after the N addition in first two years.

T3-23: Tropospheric ozone as part of global change - effects on plant growth and terrestrial ecosystems

T3-23-01

Interaction of ozone exposure of trees with belowground pro-

cesses

Heinz Rennenberg, *Cornelia Herschbach*
Chair of Tree Physiology, Institute of Forest Sciences, University of Freiburg

Ozone exposure of trees does not only interact with aboveground processes, particularly in the leaves, but also indirectly affects belowground processes both, at the tree and the ecosystem level, by disturbing shoot-root interactions. This disturbance is mainly due to ozone-mediated impairment of phloem loading and phloem transport of sucrose. Since sucrose transport is the driver of phloem transport, the shoot-to-root allocation of numerous other intermediates and signalling compounds is affected. Therefore, the consequences of ozone exposure of the shoot for belowground processes can be studied by girdling experiments. The results of such studies show that disturbed anti-oxidant allocation to the roots impairs the homeostasis of reactive oxygen species (ROS) in the roots and, hence, ROS signalling in root growth and development. In addition to this effect at the whole plant level, reduced sucrose transport to the roots causes a decline in rhizo-deposition that in turn affects ecosystem-level processes, such as ectomycorrhizal biodiversity as well as soil microbial abundance in the rhizosphere. As a consequence, e.g., competition for nitrogen is changed in favor of the tree.

T3-23-02

Growth, physiological and biochemical responses of woody species to ground-level ozone in China, with an emphasis on plant functional types comparison

Zhaozhong Feng [1], *Pin Li*[1], *Vicent Calatayud*[2], *Elena Paoletti*[3]
1. Research Center for Eco-Environmental Sciences, Chinese Academy of Sciences
2. Fundación CEAM
3. Institte for Sustainable Plant Protection, National Research Council

The carbon-sink strength of northern hemisphere temperate and boreal forests is decreased by rising ozone concentration [O_3], but lacking of the evidence from China, especially unique subtropical evergreen broadleaved forests. This meta-analysis quantitatively evaluated the effects of elevated [O_3] on growth, biomass, functional leaf traits and plant growth form of woody plants representative of Chinese forests. The database from China consisted of 46 peer-reviewed studies published between 2006 and 2016, taking into account plant functional type, climatic zone, O_3 concentration, experimental duration, and additional treatments such as elevated carbon dioxide concentration, drought and nitrogen addition. The results suggested that 116 ppb [O_3] reduced total biomass of woody plants by 14% compared with control (mean [O_3] of 21 ppb). Significant decreases in root (13%) and shoot biomass (8%) but the root-to-shoot ratios were not significantly affected by O_3. At elevated [O_3], woody plants had significant reductions in height, diameter, leaf mass per area, chlorophyll content, chlorophyll fluorescence parameters, enzymatic antioxidant (ascorbate peroxidase) and soluble sugar which underlie significant damage in photosynthesis. Further, subtropical species were significantly less sensitive than temperate ones while deciduous broadleaf species were significantly more sensitive than evergreen broadleaf and needle-leaf species. Few observations of the interaction of [O_3] with additional treatments limited the understanding of the combined effects of O_3, climate change and/or nitrogen on forests. These findings suggest that carbon-sink strength of Chinese forests has been reduced by rising [O_3]. Given that subtropical broadleaved forests across China are much more resistant to O_3, the suppression strength of the northern hemisphere carbon sink as increases in [O_3] by model results could be likely over-estimated.

T3-23-03

Enhanced tropospheric ozone - hemispheric view in relation to free-air exposure experiments with crop and woody plants in E-Asia

Rainer Matyssek[1], *Zhaozhong Feng*[2]
1. Technische Universität München, Ecophysiology of Plants, Hans-Carl-von-Carlowitz-Platz 2, D-85354 Freising-Weihenstephan, Germany
2. State Key Laboratory of Urban and Regional Ecology, Research Center for Eco-Environmental Sciences, Chinese Academy of Science

An introduction in the rationale of the session "*Tropospheric Ozone as Part of Global Change - Effects on Plant Growth and Terrestrial Ecosystems*" and sequence of presentations is envisaged by this contribution, which will pursue two perspectives: highlighting (i) enhanced anthropogenic ozone (O_3) regimes as a component of global change and air pollutant inhibiting plant growth, and (ii) recent advancements in free-air O_3 exposure experiments as particularly conducted in E-Asia on crop and woody plants. Regarding (i), formation of enhanced tropospheric O_3 will be demonstrated as fostered by high insolation upon precursor release from combustion processes and, under climate warming, in relation to atmospheric CO_2 accumulation. O_3 precursors do not only arise from industrial activities and traffic transportation, but also from land-use changes, often in relation to forest burning, which is a world-wide phenomenon. Hence, O_3 becomes an intrinsic component of global change, the more so, as once being formed, is spread at hemispheric scales and contributes to climate warming. "Hot spots" of O_3 formation under (sub)tropical and Mediterranean climate drive increases in chronic background concentrations worldwide. Interrelationships between O_3 action and such by other drivers of global change (CO_2, warming, drought/flooding, N deposition, BVOCs, N_2O, CH_4) and biotic stressors (competitors, herbivores, pathogens) are addressed, including atmosphere-vegetation feedbacks in O_3 formation versus decomposition. Advanced experimentation and modelling claim enhanced tropospheric O_3 regimes to globally lower carbon sink strength and storage capacities of forest ecosystems (with feedback on climate warming) and reducing yield and quality of agricultural crops (with risks for food security of mankind). However, validation of modeling outcome on carbon and water relations at high spatio-temporal scales is scarce and empirically demanding. On such grounds, perspective (ii) of the presentation will survey three recent free-air O_3 exposure experiments in China and Japan with agricultural crops and forest trees, regarding cause-effect relationships in stress response. 1) In Jiangdu, China, wheat/rice systems were under focus, started in 2007. Effects upon enhancement of the ambient O_3 regime by up to 50% were reported on growth, grain yield and soil microbial communities. Derived O_3 dose-response relationships demonstrated 23% enhancement of ozone has already reduce wheat yield by 20% in four cultivars, reproduced over three growth seasons. Food security was concluded to be at risk in China in the face of further rising O_3 impact. 2) In Sapporo, Japan, monarch birch has been exposed to 60 ppb O_3 during daylight hours since 2011, showing seasonally changing O_3 sensitivity of photosynthesis. The effect was related to inconsistent stomatal O_3 response rather than altered photosynthetic capacity. 3) In Yanqing, China, an O_3 free-air system is currently built for poplar plantations to analyze combined elevated O_3/ fertilization effects on long-term carbon sequestration, water use and N cycling. Clarification is strived for regarding O_3 impact on forest ecosystem functionality and services. Ecosystem-level research approaches are to be strengthened for reconciling O_3 effects on climate, food security, plant health, which altogether are cross-linked through land-use concepts, energy-use and the uncertain role of biodiversity in mitigating O_3 stress.

T3-23-04

Difference in ozone susceptibility of photosynthesis between upper and lower canopy leaves of Siebold's beech

Makoto Watanabe[1], *Kinose Yoshiyuki*[1], *Yasutomo Hoshika*[2], *Takayoshi Koike*[3], *Takeshi Izuta*[1]

1. Institute of Agriculture, Tokyo University of Agriculture and Technology

2. Institute of Sustainable Plant Protection, National Research Council of Italy

3. Silviculture and Forest Ecological Studies, Hokkaido University

Ozone (O_3) in the troposphere is a phytotoxic gaseous air pollutant, and a growing threat for carbon sink strength of forest worldwide. To determine the effect of O_3 on the carbon budget of the whole-canopy, it is necessary to study the distinct responses of individual leaves to O_3 in relation to a gradient of light conditions within the canopy. In this presentation, we will report the results obtained from two O_3 fumigation experiments with Siebold's beech (*Fagus crenata*) and discuss the difference in ozone susceptibility of photosynthesis between upper and lower canopy leaves. We established free-air O_3 fumigation system in 2011. Ozone at 60 nmol mol^{-1} was fumigated to ten-year-old Siebold's beech saplings (as of 2011) during the daytime for the leafy periods. We determined the leaf photosynthetic traits in various canopy positions of Siebold's beech saplings under elevated O_3 during the second growing season. The susceptibilities of photosynthesis and respiration to O_3 were greater in upper canopy leaves than in lower canopy leaves, and the O_3-induced reduction of photosynthetic activity was because of biochemical limitation in the chloroplasts rather than stomatal closure. We then calculated the canopy-level photosynthetic carbon gain (PCG) and respiratory carbon loss (RCL) using a canopy photosynthesis model. The canopy net carbon gain (NCG) was reduced by O_3 by 12.4% during one growing season. The contributions of changes in PCG and RCL under elevated O_3 to NCG were different between late summer and autumn. To clarify the mechanism of difference in O_3 susceptibility of photosynthesis between upper and lower canopy leaves of Siebold's beech, we analyzed the relationships between photosynthetic parameters such as maximum rate of carboxylation (V_{cmax}) and accumulative stomatal O_3 uptake (AF_{st}). Siebold's beech seedlings were grown in greenhouse type O_3 fumigation chamber for two growing seasons. We set three levels of gas treatments (charcoal-filtered air, O_3 at 1.0-time and 1.5-times the ambient concentration) in combination with three levels of soil nutrient treatments (non-fertilized, low- and high-fertilized) as sub-factor. The degree of O_3-induced reduction in V_{cmax} in upper canopy leaves was higher than that in lower canopy leaves in August of second growing season, similarly with above-mentioned free-air O_3 fumigation experiment. The AF_{st} of upper canopy leaves was higher than that of lower canopy leaves. On the other hand, the reduction rate of V_{cmax} per unit AF_{st} in upper canopy leaves was low as compared to that in lower canopy leaves, indicating higher detoxification capacity for O_3 in upper canopy leaves. When we assumed a threshold value of stomatal O_3 uptake rate above which adverse effects of O_3 occurs is proportional to gross photosynthetic rate, reduction rate of V_{cmax} per unit AF_{st} over a threshold became similar between upper and lower canopy leaves. This indicates photosynthate is associated with detoxification capacity for O_3. In conclusion, the susceptibility of photosynthesis to O_3 in upper canopy leaves of Siebold's beech is higher than that in lower canopy leaves, and this is not due to lower detoxification capacity in upper canopy leaves, but to higher stomatal O_3 uptake.

T3-23-05

Ozone impact on crop yields and quality – risk for food security?

Jürgen Bender[1], *Gina Mills*[2]

1. Thünen Institute of Biodiversity, Braunschweig

2. Centre for Ecology and Hydrology, Bangor

Tropospheric ozone is a threat to food production as it has negative impact on the yield and quality of important staple crops. Ozone reduces plant productivity by entering leaves through the stomata, causing oxidative stress and decreasing photosynthesis, plant growth and biomass accumulation and allocation. Even short ozone episodes can cause visible injury symptoms to foliage on leafy vegetables (e.g., lettuce, spinach, salad onions), thereby lowering their market value. However, there is a wide intra- as well as interspecific variation in sensitivity to ozone. Important food crops such as wheat, soybean and rice are highly sensitive to ozone, with potato, maize or barley being less sensitive. Data from ozone exposure experiments have been used to develop flux-based ozone dose-response relationships for crops which formed the basis to estimate regional or global yield losses. For example, in Europe, estimated loss of wheat grain yield from ozone was €3.2 billion in 2000. Even with current legislation to reduce ozone pollution in Europe, economic losses of wheat are predicted to be about €2 billion in 2020. A first ozone flux-based global assessment of economic losses due to ozone effects on wheat yield (9.4% loss) are estimated at $24.3 billion. Highest losses are estimated in areas with medium to high ozone fluxes in combination with medium to high area of wheat cultivation (e.g., Western Europe, SE Asia, China)

T3-23-06

Tree and forest responses to interacting elevated atmospheric CO_2 and tropospheric O_3: A synthesis of experimental evidence

John King[1], *Lingli Liu*[2], *Michael Aspinwall*[3]
1. Department of Forestry and Environmental Resources, North Carolina State University
2. Institute of Botany, Chinese Academy of Sciences
3. Hawkesbury Institute for the Environment, University of Western Sydney

The goods and services provided by forest ecosystems will be needed in greater amounts than ever before in the coming decades, yet are being threatened by climate change and air pollution. We surveyed the peer-reviewed literature on elevated atmospheric CO_2 (eCO_2) and elevated tropospheric O_3 (eO_3) experiments with the goal of determining the general significance of the interaction of these antagonistic air pollutants on forest physiology, biomass production and soil carbon cycling. The review was restricted to eCO_2xeO_3 experiments because the literature on eCO_2 and eO_3 has been extensively reviewed and we specifically wanted to highlight the extent to which this understudied, but important interaction has been investigated. Data were sorted by duration of exposure and vegetation age class (young: 0-5 y, intermediate: 6-20 y, and mature: > 20 y). We were able to locate 58 studies published between 1990 and 2012 reporting primary data on forest responses to eCO_2xeO_3. The total number of observations was greatest from FACE experiments (176), followed by open-top chambers (104), growth chambers (88) and then greenhouses (29), indicating the importance and efficiency of large scale, long-term experiments. Across experiments, average exposure levels were 347 and 603 ppm for ambient and elevated CO_2, respectively; and 35 and 70 ppb, for ambient and elevated O_3, respectively. In young systems, the duration of exposure averaged 3.3 y, with some studies as short as 0.2 y and the longest over 5.0 y. In intermediate age systems, average duration of exposure was 7.7 y, but the single longest running study, The Aspen FACE Project, lasted 12.0 y. We found no eCO_2xeO_3 studies in mature forest ecosystems, representing a corresponding gap in our knowledge. There were commonalities in how eCO_2 and eO_3, alone and in combination, affected physiology, biomass production and litter decomposition, and the net effect of the eCO_2xeO_3 interaction was fairly consistent. In general, eCO_2 enhanced and eO_3 decreased photosynthesis, instantaneous water use efficiency, leaf area index, above- and belowground growth, and tissue concentrations of carbon-based defense compounds and lignin. The interaction treatment (eCO_2xeO_3) was generally intermediate between the control and elevated levels of both factors, although often it was not statistically significant. Contrary to expectation, responses of intermediate age trees were greater than those of young trees. Changes in litter chemistry did not appear to affect decomposition rates or formation of soil organic C under either pollutant, although longer duration experiments are needed to assess the cumulative impact of altered (NPP-driven) soil carbon inputs. Remaining gaps in scientific understanding are identified, key remaining questions proposed, and suggestions made for integrating eCO_2xeO_3 interaction into future experimental designs and integrated research infrastructures.

T3-24: Linking aboveground and belowground interactions under global changes

T3-24-02

Short-term interactive effects of soil warming and precipitation exclusion on fine root production of Chinese fir seedlings in subtropical China

Guangshui Chen
Fujian Normal University

Global warming and shifted precipitation patterns are expected in subtropical China. However, our understanding on how global warming interacted with precipitation changes affecting belowground processes of subtropical forests are lack, which limits our ability to predict the productivity and carbon dynamics of subtropical forests in response to global changes. By establishing a mesocosm experiment in Chenda State-owned Forest Farm in Sanming, Fujian Province, the short-term interactive effects of soil warming and precipitation reduction on fine root production of Chinese fir (*Cunninghamia lanceolata*) seedlings were examined. The design was factorial soil warming (ambient, +5℃) and precipitation exclusion (ambient, -50%), with five replicate plots (2 m × 2 m) for each combination, i.e., control (CK), warming only (W), precipitation exclusion only (P), warming plus precipitation exclusion (WP). Each plot was planted with four 1-year-old Chinese fir seedlings. Fine roots dynamics were monitored by minirhizotron from March 2014 to March 2016. After two years treatment, the WP significantly decreased both seedling height and basal diameter growth, while there was no significant change between other treatments and the CK. Soil warming alone significantly increased fine root production, but there was no significant change in the other treatments. There was significant interaction effect of soil warming and precipitation exclusion on total fine root production. Compared with CK, the W treatment significantly increased fine root production in spring 2014 and winter 2015, the P treatment significantly increased fine root production in autumn 2014, while the WP treatment significantly decreased fine root production in summer 2014. Repeated-measures ANOVA showed there was significant interaction of soil warming, precipitation exclusion and season. Moreover, the interaction of warming and precipitation exclusion was found significant for fine roots at 20-40 cm soil layer. Compared with CK, fine root production increased significantly at 20-40 cm depth in the W and P treatment, but no significant change in the WP treatment. Together, these results showed that soil warming and precipitation exclusion had significant interaction effects on fine root production of Chinese fir seedlings, which varied with season and soil depth.

T3-24-03

Long-term experimental warming have small effect on above and below ground species richness and effective number of species across Alpine Tundra communities, but change correlations among groups and tropic levels

Juha Alatalo[1], *Shengbin Chen*[2], *Karina Clemmensen*[3], *Peter Čuchta*[4], *Annika Jägerbrand*[5], *Jaanis Juhansson*[3], *Björn Lindahl*[3], *Peter Ľuptáčik*[6], *Germán Bonilla Rosso*[3], *Ulf Molau*[7], *Sara Hallin*[3]
1. Qatar University

2. *Nanjing Institute of Environmental Sciences*
3. *Swedish University of Agricultural Sciences*
4. *Academy of Science of the Czech Republic*
5. *Calluna AB*
6. *P. J. Šafárik University in Košice*
7. *University of Gothenburg*

Alpine and arctic tundra are predicted to be highly sensitive to global warming. There is a growing number of studies on the response of species composition and ecosystem function of plant communities. However, soil fauna, fungi, and bacteria, are less studied although they are important in arctic terrestrial ecosystems through their role in symbiotic relationships, decomposition, and nutrient cycling. In 2013 we collected species richness data from above ground vascular plants (deciduous shrubs, evergreen shrubs, forbs, and graminoids), bryophytes, lichens, and belowground soil fauna (collembola and mites), fungi and bacteria from our long-term warming experiments (using open top chambers, OTCs) at Latnjajaure, Northern Sweden. Three contrasting Alpine Tundra plant communities (wet meadow, rich meadow and dry heath) represent a natural nutrient and moisture gradient. The warmed plots had been experiencing two decades of experimental warming with OTCs since 1993 (wet meadow), and 1995 (rich meadow and poor heath). Here we present the effect of the long-term warming and site on 1) above and belowground species richness and effective number of species, and 2) the correlations in species richness, and effective number of species, between vascular plants, bryophytes, lichens, bacteria, fungi, collembola and mites. We found that warming and the interaction of warming and site had no significant effect on species richness or effective number of species for any group. Site had strong effect on species richness and effective number of species of all groups except fungi. The correlations in species richness and effective number of species was found to vary between below and aboveground groups. Notably, fungi had no significant correlation with any species group. Importantly, warming was found to both weaken and strengthen the correlation of species richness and effective number of species between below and aboveground groups. This suggest that while below and aboveground diversity may be resistant to direct effects of climate change, climate change may indirectly affect the linkages between below and aboveground groups, and tropic levels. Furthermore, the effect on species richness correlation may differ among groups.

T3-24-04

Bamboo invasion above and below a treeline ecotone: Seasonal soil moisture dry-down rates and high stress tolerance promote invasion success of *Sasa kurilensis*

Daniel Winkler[1], *Yukihiro Amagai*[2], *Masami Kaneko*[3], *Travis Huxman*[1], *Gaku Kudo*[2]
1. *University of California, Irvine*
2. *Hokkaido University*
3. *Rakuno Gakuen University*

How species invasions impact ecosystem structure and function at important ecotones or boundaries is unknown, but may provide insight into the impacts of climate change and the mechanisms underlying community change. The dwarf bamboo, *Sasa kurilensis*, may be a good system to understand these issues, as the species impacts ecosystem features as it encroaches beyond treeline into alpine systems. We used remote sensing imagery to quantify *S. kurilensis* expansion patterns across its range, measured growth and stress tolerances of *S. kurilensis* above and below treeline, and evaluated components of growth to reveal how shifts in light and water limitations influence the ontogeny of height, branching, and leaf production. We show that *S. kurilensis* more than doubled its abundance across its range, but more than tripled its abundance near and above treeline. Soil dry-down rates were a key driver of invasion above and below treeline, where growth rates decreased with more rapid rates of soil moisture dry-down. We found *S. kurilensis* responds to competition and climate stress by increasing allocation to belowground structures at high elevations. Further, it invests more carbon in fewer—yet taller and heavier—aboveground structures in low-light, low elevation environments. It appears this species' success is driven by considerable morphological and physiological flexibility, coupled with changes in water balance associated with snowmelt that in each habitat results in sites increasingly hospitable to bamboo. Overall, this study links resource allocation strategies and physiological responses to climate change and provides a mechanistic explanation of invasion success.

T3-24-05

Nitrogen competition drives plant-soil microbe relationships along a snow depth gradient in temperate steppe

Weixing Liu[1], *Steven D. Allison*[2], *Yonggang Chi*[1], *Ping Li*[1], ***Lingli Liu***[1,3]
1. *State Key Laboratory of Vegetation and Environmental Change, Institute of Botany, Chinese Academy of Sciences*
2. *Department of Ecology and Evolutionary Biology, University of California, Irvine*
3. *University of Chinese Academy of Sciences*

The effects of rainfall changes on soil microbial communities have been extensively studied, but the effects of snowfall changes on soil microbes and plant-microbe interactions are poorly known. Using snow fences at 8 sites in temperate steppe ecosystems, we investigated how snow depth affects interactions among soil microbes and plants. Both soil microbial biomass (measured as phospholipid fatty acids) and plant biomass increased with increasing snow depth, but the interaction between soil microbes and plants shifted from positive to negative. Soil inorganic nitrogen (N), total N, and microbial biomass N pools were not affected by snow depth, but plant N pools increased with deeper snow. Increased snow depth altered soil microbial community structure and increased the fungal/bacterial ratio. Plant community composition also shifted, with greater grass and lower forb abundance as snow depth increased. Structural equation modelling showed that the direct effects of plant biomass and indirect effects of N (mediated through plants) on soil microbial biomass switched from positive to negative with increasing snow depth, suggesting an increase in competition for N between plants and microbes. Our work implies that changes in snow depth will drive shifts in plant and microbial community composition that are mediated by N availability. Together, our results demonstrate the potential for climate change to alter nutrient competition mechanisms underlying community structure and ecosystem processes.

T3-24-06

Intercropping enhances productivity and sustainability of agroecosystems

Long Li
China Agricultural University

Intercropping advantages are mainly derived from interspecific interactions, including above- and below-ground complementarity, competition and facilitation. There were lots of research on interspecific interactions on above-ground parts, but relatively limited on below-ground interactions between intercropped species. The aim of the study was to examine role of interspecific root-root interactions in overyielding and efficient nutrient utilization of intercropping. Both field and greenhouse experiments were conducted to determine yield and nutrient acquisition advantages by adding sparingly soluble and organic P into soils, interspecific root-root interactions by root barrier, N complementarity by natural or enriched isotope abundance methods. Soil fertility parameters were determined in long-term field experiments. There are significant productivity advantages of intercropping in various legumes/cereals and cereals/cereals intercropping. One-thirds to 100 percentage of intercropping advantages are contributed to root-root interactions. There are two main mechanisms underlying the interspecific root-root interactions, facilitation and complementarity. The interspecific facilitation is that some crop species mobilize unavailable forms of one or more limiting soil nutrients (such as phosphorus, iron and zinc), and improve phosphorus, iron or zinc nutrition for themselves and neighboring non-mobilizing species by releasing acid phosphatases, protons and/or carboxylates, and phytosiderophores into the rhizosphere, respectively. When intercropped with legumes, cereals usually have more root competitive to soil mineral nitrogen, and lead to reduction in soil mineral nitrogen, which facilitates to nodulation and biological N_2 fixation of associated legumes. The root-root interactions maintains soil fertility on a relative fertile soil and enhances soil fertility on a poor soil under continuous overyielding. Our results highlight that interspecific root-root interactions below-ground play an important role not only in overyielding and nutrient acquisition, but also maintaining or enhancing sustainability of agriculture. The mechanisms underlying the root-root interactions include interspecific facilitation on sparingly-soluble soil nutrients and complementary utilization of soil and atmosphere nitrogen in legumes-based intercropping.

T3-25: Forest diversity and degradation in Asia/Pacific

T3-25-01

The genus *Aspidistra* (Asparagaceae) in Vietnam: Diversity, intraspecific variation and distribution patterns

Maxim Nuraliev[1,2], *Nikolay Vislobokov*[1,2], *Leonid Averyanov*[3], *Andrey Kuznetsov*[2,4], *Svetlana Kuznetsova*[2]
1. Moscow State University
2. Joint Russian-Vietnamese Tropical Scientific and Technological Center
3. Komarov Botanical Institute of the Russian Academy of Sciences
4. A.N. Severtsov Institute of Ecology and Evolution of the Russian Academy of Sciences

Aspidistra is one of the most fast-growing genera in the World flora. It currently comprises more than 150 species, about one hundred of which were discovered since 2000. During the last several years, each three weeks a new species of *Aspidistra* was described, mainly from southern China and Vietnam. One can notice the tendency that most of the newly obtained flowering specimens of *Aspidistra* are provisionally regarded to represent undescribed taxa. Indeed, the morphology of flowers in this genus is incredibly diverse. However, quite a lot of its species show only minute differences from their morphologically similar congeners. The comparison becomes especially problematic for the plants being remote geographically from each other, because the species descriptions are commonly based on a single population, and degree of intraspecific variations remains scarcely known for the whole genus. We have performed first analysis of geographical distribution of a number of Vietnamese species of *Aspidistra*, basing on our extensive original collections. Our data show that along with the narrow endemics, there are species with rather large distribution areas, including the recently described *A. bogneri* Tillich, *A. opaca* Tillich, *A. oviflora* Aver. & Tillich, *A. paucitepala* Vislobokov, Nuraliev & D.D.Sokoloff, *A. phanluongii* Vislobokov. Furthermore, *A. connata* Tillich occurs both in northern limestone areas of Vietnam and in the southern mountains. For several species, we have found significant gradual variation of morphological traits, e.g., the contrasting perianth coloration in two varieties of *A. xuansonensis* Vislobokov. On the basis of our data on intraspecific variation, we propose *A. parviflora* N.S.Ly & Tillich as a new synonym of *A. paucitepala*. The other side of the species richness of *Aspidistra* combined with wide distribution range of some of its species consists in a hidden diversity within a taken protected area. As only few plants usually produce flowers simultaneously within a population, all the sterile individuals are commonly assigned to the same species as the blooming material. Meanwhile, species of *Aspidistra* sometimes grow sympatrically. We have registered up to four species inhabiting the same forest. In some cases they form mixed populations, while in others they occur separately, possibly being adapted to different ecological conditions (including examples with elevation-based distribution). Besides, the sympatric species of *Aspidistra* can demonstrate overlapping or strictly opposite flowering phenology. An alternative way of distinguishing species of *Aspidistra* and establishing the species boundaries (and also higher-level taxa) is the DNA barcoding. Our results show that plastid *psb*A-*trn*H and nuclear 5S-NTS regions can be used as such barcodes. Our molecular database will be important for further ecological investigations of *Aspidistra* as it will allow identifying sterile specimens for the full inventory of species number and distribution within a model area. The knowledge of species diversity of *Aspidistra* in the local floras is crucial for forest management and conservation. Efficient species identification and assessing their rarity and ecological niches should greatly improve the interaction between conservation biology and forest industry. This study is supported by the Russian Science Foundation (project 14-14-00250) and Russian Foundation for Basic Research (15-04-00419A).

T3-25-02

Altitudinal gradient analyses of the Balakot Forests of District Swat in the Hindukush range of mountains

***Khan Shujaul Mulk**, Zeb Ahmad*
Quaid-i-Azam University Islamabad

In the Swat District, the Balakot Forests is the part of Hindukush range of mountain that is one of the remote area where vegetation of different areas have not been studied correlated to environmental variables. The recent research were carried out to study the different effects of environmental variables specially altitude on plant species association and variation among plant communities of Balakot Forests. Also this exploration identifies pressures from different environmental factors and recommendation for corrective measures. For valuation of environmental changeability quantitative ecological methods were carried out during summer 2014-15. The technique line transect method were used. Size of the transect was 50 m which were further divided into sub transects 10 m. The stands were planned along 15 altitude on different aspects (South, North, East and West). Total 27 transects were studied in all stands in different station. Phytosociological attribute were observed, i.e. frequency, density and cover for trees, shrubs and herbs in each stands of vegetation communities. A total of 120 plant species were collected in Balakot Forests. After getting initial result, Canonical Correspondence Analysis (CCA) was used to analyze the environmental pitch of vegetation's. The environmental data and species abundance were used in CANOCO software version 4.5. The presence absence data of 120 plant species were elaborated with Cluster and Two Way Cluster Analysis techniques by using PC-ORD version 5 to show different species composition that resulted nine stands. Altitudes, aspect, soil texture show that these are the strongest variables and show significant effect on vegetation and distribution in different communities. It is suggested to protect and use wisely whole of the flora and particularly the rare species.

T3-25-03

An overview of the forest vegetation of Irano-Turanian phytogeographical region

***Volkan Altay**[1], Munir Ozturk[2]*
1. Biology Department, Faculty of Science & Arts, Mustafa Kemal University, Antakya
2. Centre for Environmetal Studies and Botany Department, Ege University, Izmir

The vegetation distribution pattern is of great importance in conservation and biogeographical investigations. The threatened vegetation is one of the most important indicators for decision-making process in the conservation of priority habitats. Irano-Turanian floristic region comprises a vast territory of Southwest and the Middle Asia. It is connected to the Mediterranean, Euro-Siberian (boreal) and Saharo-Sindian regions. The region is chgaracterised by a high rate of plant speciation and endemism, particularly in some large genera like; *Astragalus, Cousinia, Acantholimon, Acanthophyllum* and *Allium*. The heterogeneous topography together with a relatively stable continental climate have lead to its long-term uninterrupted isolation as well as high rates of allopatric speciation. The climate of the region is highly differentiated from adjacent regions, especially in its degree of continentality and seasonality. Vegetation of this area shows an important genetic, biological and ecological diversity. Unique topography, habitat heterogeneity and long vegetation history have resulted in the formation of diverse community types. These include scattered montane forests, widespread *Juniperus* woodlands, wild *Pistacia vera* woodlands, open shrublands and scrub, diverse mountain steppe communities, cliff vegetation, loess and gypsophilous vegetation, semi-desert steppes and halophytic formations, as well as aquatic, ruderal and weed communities. Several forest vegetation types are found in natural and seminatural environments of the area, due to its biogeographical position and diverse topography and habitats. Although a great variation is observed in different sub-phytogeographic regions and altitudes, in general forest vegetation is dominated by the forest tree species like; *Pinus brutia* var. *eldarica, P. nigra* ssp. *pallasiana, P. sylvestris, Quercus aegilops, Q.pubescens, Q. brantii, Q. cerris, Q. infectoria, Q. libani, Q. longipes, Q. macranthera, Juniperus excelsa, J. oxycedrus, Betula pendula, Pistacia atlantica* and *P. khinjuk.* In this study we will try to evaluate the impact of current and potential changes in the near future, the scope of sustainability, and ecological analysis of the forests of this region vis-a-vis the prospects and challenges.

T3-25-04

Protection and economic evaluation of forests: Case study from East Mediterranean, Central and Southwest Asia

*Munir Ozturk[1], Volkan Altay[2], **I. Faridah Hanum**[3]*
1. Centre for Environmetal Studies and Botany Department, Ege University, Izmir
2. Biology Department, Faculty of Science & Arts, Mustafa Kemal University, Antakya
3. Universiti Putra Malaysia, Malaysia

Currently nearly 30 percent of the global land area is forested, more than 1/3 of all forests are considered primary, but around 60,000 km^2 of this is lost or modified annually. In this presentation attempt has been made to evaluate the forest diversity and its conservation priorities in the East Mediterranean, Central /Southwest Asia. The estimates of forest resources in terms of total forest area covered and percentage of land area in the East Mediterranean is 10,854,000 ha (25.7%); Central Asia 39,220,000 ha (51.7%) and Southwest Asia 14,689,000 ha (18%). Important forest types occuring in these regions include predominantly broadleaved forests dominated by non-fruit bearing broadleaf species; Shrublands in arid zones; Riverside forest and Mangroves, and predominantly coniferous forests dominated by spruce, pine or fir species; Juniper forest and Cedar forest. There is an old tradition in the West and Central Asia to cultivate *Populus* as well as other trees like *Morus* and *Platanus* along irrigation channels, for furniture and other purposes. This tradition needs to be kept and maintained, or even enlarged, as part of the economy. These trees give shade in summer and can be used for timber, in the case of mulberries not only for fruit production but even again for silkworm culture and silk production. Overgrazing, industrial activities and housing hand in hand with an expansion in agriculture, particularly into woodlands, coupled with use of different technologies is one of the most devastating causes for deforestation. All these factors lead to a decrease in forest area. A widespread conversion of natural forests into monocultural plantations of commercial value reduces habitat diversity, especially in forest lands where micro-habitat differences are much needed for wildlife. There is a great need to maintain ecosphere for its sustainable use. In spite of our need to manipulate nature we have to understand it in order to be on the proper junction within the ecosphere. A healthy ecosphere needs healthy trees and forests because these have many direct and indirect values to the people in the form of maintaining and improving the soil conditions and the hydrological systems; help in moderating climate by soaking up and storing CO_2; while assuring self renewal processes of the land base and its sustainability. A great stress is being laid on to recognize the factors that threaten our forest ecosystems. We need to take immediate measures to prevent loss of habitats and species diversity, prepare inventory of renewable natural resources on forest lands, develop necessary scientific, technical, legal and administrative measures to improve quality and quantity of forest biodiversity, manage and utilize forest biological resources based on sound ecological and sustainable management principles.

T3-25-05

Forest diversity, ecosystem and landuse changes in Malaysia

A. Latiff

Universiti Kebangsaan Malaysia, 43600 UKM Bangi

Malaysia is endowed with one of the richest forest diversity in the world which is represented by various forest ecosystems and habitats, species and the genomes contained in the plants, animals and microbes. Forest ecosystem diversity is represented by the rich lowland and hill dipterocarp forests, the mangrove forests and the montane forests at Mt. Kinabalu. To-day a total of 1.39 million ha (about 7.6%) of the forest of all types have been set aside for biodiversity protection and conservation. Forest species diversity is represented by about 4,000 species of trees whilst the diversity of animals is represented by more than 6,000 species, excluding that of the insects, the largest and most diverse single group which is unadequately known. As the country is also pursuing industrialisation by the year 2020, conflicts occur between biodiversity conservation efforts and socio-economic development. Measures for sustainable agriculture and forestry are in place but those in biodiversity prospecting, eco-tourism and impacts of urbanisation on biodiversity are yet to be addressed. In the absence of concrete data of biodiversity loss in the country it is difficult to ascertain the lists of endangered or otherwise threatened forest species as the various ecosystems are prone to changes through landuse patterns. Many factors have contributed to forest diversity loss; among them is the rapid socio-economic development of the country that transformed vast forested lands by logging activities, land openings for agriculture and resettlement and subsequently creating new built-up areas such as urban and industrial areas which are relatively poor in biodiversity. These changing landuse patterns are affecting not only biodiversity but the environment per se. Other factors such as over-harvesting and forest plantations have also contributed in small part. These activities had led to significant forest loss, degradation and forest fragmentation. What is needed are frameworks and strategies for biodiversity conservation, some socio-economic strategies for sustainable use and benefit sharing of biodiversity and also some legal measures for sustainable use and protection of forest biodiversity and assessment of landuse patterns.

T3-25-06

Do protected areas effective in preventing land use conversions? A case study in Xishuangbanna, Southern China.

Chaya Sarathchandra[1,2], *Rhett Harrison*[4], *Jianchu Xu*[1,3]

1. Key Laboratory for Plant Diversity and Biogeography of East Asia, Kunming Institute of Botany, Chinese Academy of Sciences, Kunming, Yunnan

2. University of Chinese Academy of Sciences, Beijing

3. World Agroforestry Centre, East and Central Asia, Kunming, Yunnan

4. World Agroforestry Centre, East & Southern Africa Region

Forested areas worldwide are declining in a rapid rate as a result of sustaining increasing needs of growing human population. As many methods and strategies are implemented to prevent this decline it is crucial to evaluate whether these efforts are successful in the conservation agenda. Xishuangbanna is 0.2% of total land of China, harbouring > 20% of China's biodiversity, is now threatened with monoculture rubber plantations expansion and other land use conversions replacing lowland rainforest and shifting cultivation practices. This study was designed to evaluate the effectiveness Protected Areas (PAs) which are specifically designed for species and forest conservation. We measured the success of PAs in preventing expansion of rubber and other non-conservation land uses within their boundaries. We used matching methods to minimize the bias of PAs location and other factors influencing PAs establishment and to compare land use conversion rates within PAs zones and with outside PAs. By 2010, Xishuangbanna had designated an area of 3,455.52 km^2 (~ 18%) as PAs while rubber (22%) showed an expansion rate of 153.4 km^2 y^{-1} and approximately 16% of PAs were deforested by 2010 with an annual rate of 6.8 $km^2 yr^{-1}$ and if this rate continues assuming that the rate is constant, within next 40 years another 10% of PAs land will be deforested but because drivers of deforestation continuously change and accelerate rather than decelerate Xishuangbanna's PAs might lose more forest than we predicted in this analysis. Therefore it should be a timely concern to review or to strengthen current land sparing policies, rules and regulations.

T3-26: Small RNAs and Plant Response to Climate Change and Environmental Stress

T3-26-01

Abiotic stress miRNome in cereals

Hikmet Budak

Montana State University

There are two dilemmas in the world: the first one is a/biotic stress and the second one is malnutrition. Crop species such as wheat and barley, suffer from the effect of many different abiotic stresses such as drought, heat, salinity or nutrient deficiencies and toxicities which cause large yield losses resulting in economic and environmental damage. The destructive effects of a/biotic stresses have increased recently and are predicted to further deteriorate simultaneously with climate change. In order to cope with the changing environmental conditions, generation of more stress-tolerant plant varieties which are also high yielded is vitally important. New generation technologies have discovered an essential tool for improving crop tolerance to the abiotic stresses. miRNAs are small regulators of gene expression that act on many different molecular processes such as development, environmental adaptation and stress tolerance. miRNAs can act at both the transcriptional and post-transcriptional level, although post-transcriptional regulation is the most common in plants where miRNAs can inhibit the translation of their mRNA targets via complementary binding and cleavage respectively. To date, several miRNA families such as miR156, miR159 and miR398 have been detected as responsive to severe environmental conditions by taking charge in the regulation of stress associated molecular mechanisms individually and/or together with their various miRNA partners. Handling of such miRNAs together with their targets may pay the way for improved crop performance under several abiotic stresses. Exploration of further miRNA families together with their functions under stress conditions will improve our knowledge and provide opportunities to enhance plant performance to help us meet with world food demand. We will discuss an outline to set out the workflow to study a/biotic stress miRNome in wheat and barley.

T3-26-02

The sudy of the response and the molecular mechanisms of *Astragalus Membranaceus* Bge. var. *mongolicus* (Bge.) Hsiao to drought stress

***Guilin Chen**, Xin Jia, Chuangshu Sun, Guangyue Li*

College of Life Science, The Good Agriculture Practice Engineering Technology Research Center of Chinese and Mongolian Medicine, Inner Mongolia University

The shortage of the water resource is a common phenomenon in Northwestern China, which seriously affects the development of the local economy and production. *Astragalus membranaceus* Bge. var. *mongolicus* (Bge.) Hsiao (*A. mongolicus*) is primarily is cultivated on a large scale in northwestern China. Understanding the plant responses to drought has important effects on ecological environment recovery and local economic development. Here, we studied the effects of drought stress on the growth, physiological and biochemical responses and secondary metabolites accumulation of two-year-old *A. mongolicus*, and combined transcriptomics (Illumina Hiseq 2000) and metabolomics (^{1}H-NMR) to investigate how the roots of *A.mongolicus* responded to 14 days of progressive drought stress. The results of the present study are as follows: 1) The dried soil reduced the relative water content (RWC) of the leaves and biomass during the 14 days of progressive drought stress. Mild and moderate drought stress enhanced some antioxidative enzyme activity, the content of proline and soluble sugar content, but antioxidative enzyme activity was limited by severe stress. A certain degree of drought stress can enhance accumulation of astragaloside IV, calycosin-7-*O*-*β*-D-glycoside, ononin, calycosin and formononetin. 2) Metabolomics demonstrated that the metabolic processes were significantly affected by drought stress. PCA analysis demonstrated that the sucrose, proline, and malate metabolites contributed greatly to the separation. Strikingly, proline was increased by almost 60-fold under severe stress compared to the control. Some backbone pathways, including glycolysis, tricarboxylic acid (TCA) cycle, glutamate-mediated proline biosynthesis, starch and sucrose metabolism, were significantly affected by drought. 3) Transcriptomics demonstrated that the differential expression of a large fraction of the transcriptome after drought stress. Compared with control, there are 18,292, 15,085, 20,986 differentially expressed genes (DEGs) in mild stress, moderate stress and severe stress. The expression of *P5CS, PDH, MDH*, zinc finger proteins genes and cytochrome P450s genes were different during drought stress, which are more likely to determine tolerance to drought. 4) We obtained the key enzyme genes in the biosynthesis pathway of calycosin-7-*O*-*β*-D-glycoside, ononin, calycosin and formononetin by transcriptional sequencing technique, analyzed the expression of key enzyme genes PAL, C4H, 4CL, PKR and CHI in the drought stress, and combined with the changes in the content of four flavonoids to explore the effects of drought stress on flavonoids accumulation of *A.mongolicus* and the molecular mechanisms of metabolic regulation. Our study presented a comprehensive overview of the changes in the *A. mongolicus* under progressive drought stress and identified important pathways through which this herb adapts to and tolerates drought stress. We found a certain degree of drought stress not affect the biomass and can enhance the accumulation of secondary metabolites.

T3-26-03

MicroRNA-mediated biotechnology for improving plant tolerance to abiotic stress

Baohong Zhang

East Carolina University

MicroRNAs (miRNAs) are an extensive class of endogenous, small RNA molecules that sit at the heart of regulating gene expression in multiple developmental and signalling pathways. Recent studies have shown that abiotic stresses induce aberrant expression of many miRNAs, thus suggesting that miRNAs may be a new target for genetically improving plant tolerance to certain stresses. These studies have also shown that miRNAs respond to environmental stresses in a miRNA-, stress-, tissue-, and genotype-dependent manner. During abiotic stress, miRNAs function by regulating target genes within the miRNA-target gene network and by controlling signalling pathways and root development. Generally speaking, stress-induced miRNAs lead to down-regulation of negative regulators of stress tolerance whereas stress-inhibited miRNAs allow the accumulation and function of positive regulators. Currently, the majority of miRNA-based studies have focused on the identification of miRNAs that are responsive to different stress conditions and analysing their expression profile changes during these treatments. This has predominately been accomplished using deep sequencing technologies and other expression analyses, such as quantitative real-time PCR. In the future, more function and expression studies will be necessary in order to elucidate the common miRNA-mediated regulatory mechanisms that underlie tolerance to different abiotic stresses. The use of artificial miRNAs, as well as overexpression and knockout/down of both miRNAs and their targets, will be the best techniques for determining the specific roles of individual miRNAs in response to environmental stresses.

T3-26-04

Genome-wide reshaping of DNA methylation and its association with gene expression in cadmium-exposed rice (*Oryza*

sativa)

***Zhi Min Yang**, Sheng Jun Feng, Xue Song Liu*
Nanjing Agricultural Univeristy

Cadmium (Cd) is one of the naturally occurring toxic heavy metals in the Earth's crust and negatively affects plant growth and development. Exposure to trace metal cadmium (Cd) affects transcriptional responses. Whether Cd-modified DNA methylation marks are associated with transcription and functional consequences in plants remains unknown. We present the genome-wide single-base-resolution maps of methylated cytosines in Cd-exposed rice, along with global transcriptional change in mRNA. Widespread differences were identified in the composition and patterning of CG and non-CG methylation marks between Cd-exposed and control rice genomes. There are 2,320 non-redundant differentially-methylated regions detected in the genome. RNA-sequencing reveals 2,092 DNA methylation-modified genes differentially expressed under Cd exposure. Rice exposure to Cd led to transcriptional changes in DNA methyltransferases, histone methyltransferases and DNA demethylases. By profiling global DNA methylation and gene transcription, we found more genes hypermethylated than those hypomethylated in CG, CHH and CHG (where H is A, C or T) contexts in the regions of upstream, genebody and downstream under Cd stress. Many of the genes were involved in stress response, metal transport, and transcription factors. Most of the DNA methylation-modified genes were transcriptionally altered under Cd stress. A subset of loss of function mutants defective in DNA methylation/demethylation and histone modification activities were applied to identifying transcript abundance of selected genes. Compared to wide-type, mutation of MET1 and DRM2 resulted in general lower transcript levels of the genes under Cd stress. Transcripts of OsIRO2, OsPR1b and Os09g02214 in drm2 were significantly reduced. A commonly used DNA methylation inhibitor 5-azacytidine was employed to investigate whether DNA demethylation affected physiological consequences. 5-azacytidine provision decreased general DNA methylation levels of selected genes, but promoted growth of rice seedlings and Cd accumulation in rice plant. Fifteen loci generating microRNAs such as OsmiR164a and OsmiR812c were found to be differentially methylated under Cd stress. Thousands of transposable elements that were transcriptionally regulated under Cd stress were also found to be differentially methylated. These results indicate that variation of genome-wide DNA methylation may alter gene expression and biological functions in cadmium-exposed rice plants.

T3-26-05

Small RNA deep sequencing reveals the important role of microRNAs in the halophyte *Halostachys caspica*

Youling Zeng
Xinjiang University

MicroRNAs (miRNAs), an extensive class of small regulatory RNAs, play versatile roles in plant growth and development as well as stress responses. However, the regulatory mechanism is unclear on miRNA-mediated response to abiotic stress in plants. *Halostachys caspica* is a halophytic plant species and a great model for investigating plant response to salinity stress. However, no research has been performed on miRNAs in *H. caspica*. In this study, we employed deep sequencing to identify both conserved and novel miRNAs from salinity-exposed *H. caspica* and its untreated control. Among the 13–19 million sequences generated from both treatments, a total of 170 conserved miRNAs, belonging to 151 miRNA families, were identified; among these miRNAs, 31 were significantly up-regulated and 48 were significantly down-regulated by salinity stress. We also identified 102 novel miRNAs from *H. caspica*; among them, 12 miRNAs were significantly up-regulated and 13 were significantly down-regulated by salinity. qRT-PCR expression analysis validated the deep sequencing results and also demonstrated that miRNAs and their targeted genes were responsive to high salt stress and existed a negative expression correlation between miRNAs and their targets. miRNA-target prediction, GO and KEGG analysis showed that miRNAs were involved in salt stress-related biological pathway, including calcium signalling pathway, MAPK signalling pathway, plant hormone signal transduction and flavonoid biosynthesis, *etc.* This suggests that miRNAs play an important role in plant salt stress tolerance in *H. caspica*. This result could be used to improve salt tolerance in crops and woods.

T3-26-06

Ecophysiological responses of *Camellia japonica* (*Naidong*) to different water and light deposition

***Cuiju Liu**, Xiao Guo, Qingchao Liu, Kuiling Wang, Yingkun Sun, Qinghua Liu*
Qingdao Agricultural University

Camellia japonica (*Naidong*) is the research object in this paper, which is from Qingdao, the city of Shandong province. The paper set two light regimes L1, L2 (65%, 15% of full sunlight), different water supply regimes W1, W2, W3 [75%, 50%, 25% of field capacity (FC)]. The responses of ecophysiological of *Camellia japonica* (*Naidong*) during different water, light and their interactions were studied. The results showed that height (H), basal diameter (BD) and crown area (CA) of *Naidong* were increased with the treatment time under L1 and L2, the morphological index was highter with the L1 treatment as compared to the L2, and the relative growth rate of each index increased rapidly. But with the extending of stress time, the relative growth rate increased slowly. With the decreasing of water regimes, the Photosynthetic Rate (A), Ranspiration Rate (E), Stomatal Conductance (Gs) and Water Use Efficiency (WUE) decreased, Intercellular CO_2 Concentration (Ci) increased gradually, and Vapour Pressure Deficit (VPD) increased firstly and then decreased. The value of Maximal Fluorescence in the dark (F_m) and Maximal Quantum Yield (F_v/F_m) under the L1 conditions were higher than L2 conditions, and the Initial Fluorescence in the dark (F_0) showed the opposite trend. With the decreasing of moisture gradient, the F_0 increased gradually under L1, while showed the opposite trend under L2. Significant differences were found in photosynthetic pigment contents of leaves among various light and water treatments. Under high irradiation, chlorophyll a, chlorophyll b and total chlorophyll were higher than low irradiation, and the Chla/Chlb value showed the opposite trend. The contents of Superoxide Dismutase (SOD), Peroxidase (POD), Catalase (CAT), Malondialdehyde (MDA) and Proline (Pro) were significantly difference at every level of light and water treatment. And the contents of POD, SOD increased firstly and decreased later with the water gradient decreasing. The content of CAT decreased with the decreasing of water content under L2 conditions, and the contents of MDA and Pro increased gradually with the water gradient decreasing. Drought or shade has some restrictive effect to the growth of *Naidong*, certain moisture and light were favorable for plant growing. *Naidong* could tolerate a certain degree of drought resistance and adapt the larger amplitude of light intensity. At the same time, it can change the enzyme activity and osmotic adjustment substances to against damage.

THEME IV: DEVELOPMENT & PHYSIOLOGY

T4-01: Stomatal guard cells: Model systems for signaling and development

T4-01-01

Convergence of signal transduction events in guard cells at reactive oxygen species and calcium channels during stomatal closure by abiotic or biotic stress signals

Agepati S Raghavendra

University of Hyderabad, School of Life Sciences, Hyderabad 500046, India

Rapid regulation of stomatal aperture is essential for terrestrial plants not only to optimize water use/CO_2 uptake, under abiotic stress conditions, but also restricting pathogen entry during biotic stress. Stomatal guard cells evolved efficient ways to sense abiotic and biotic stress signals and respond rapidly to induce stomatal closure. Stomata close when the guard cells lose turgor and become flaccid. When triggered by abiotic/biotic stress signals, osmotically active components, such as potassium ions, anions, malate and sucrose, move out of guard cells causing an efflux of water and decrease in turgor. Among several effectors, the mechanism of ABA (a well-known phytohormone) action on stomata has been studied in detail. ABA induced stomatal closure is mediated by many signalling components like cytoplasmic pH, reactive oxygen species (ROS), nitric oxide (NO), cytosolic free calcium, G-proteins, protein kinases, protein phosphatases, phospholipases, phospholipids and sphingolipids. These signalling components appear to operate during stomatal closure by also methyl jasmonate and elicitors such as chitosan and flagellin. During signal transduction pathway leading to stomatal closure by plant hormones or microbial elicitors, three signalling components stand out as the major converging points. These are ROS, cytosolic calcium and ion channels. Once formed, the ROS and free calcium of guard cells regulate both downstream and upstream events. A major influence of ROS is to increase the levels of NO in guard cells. The rise in ROS and NO would cause an increase in free calcium and modulate ion channels, through a network of events, in such a way that the guard cells lose K^+/Cl^-/anions. The efflux of these ions decreases the turgor of guard cells and then stomatal closure. We propose that ROS, NO and cytosolic calcium act as points of divergence. The interrelationships and interactions of ROS, NO, cytosolic pH, and calcium are quite complex and need further detailed examination. The levels of ROS and NO in guard cells play an important role in not only stomatal closure but also in integrating stimuli from abiotic or biotic stress. The cytosolic pH, G-proteins and phospho-/sphingolipids are also important components during stomatal closure but they may act in parallel. Further work is required to elucidate the perception of signals, such as methyl jasmonate or elicitors and how they activate NADPH oxidase leading to ROS production. Several of the unresolved questions and the ease of using epidermis make the stomatal guard cells an ideal system for studying signal transduction mechanism in plant cells.

T4-01-02

Regulation of aquaporins and plant hydraulics under water and salt stress

Christophe Maurel, *Doan-Trung Luu, Michael Wudick, Alexandre Martinière, Lionel Verdoucq, Olivier Rodrigues, Alexandre Grondin, Zaigham Shahzad, Colette Tournaire-Roux*

Biochemistry and Plant Molecular Physiology, CNRS/INRA, Montpellier

Aquaporins of the Plasma membrane Intrinsic Protein (PIP) subfamily mediate multiple controls of plant tissue water transport. New insights into the mechanisms at work in their regulation in stomata and roots under water, salt and oxidative stress will be presented. Fluorescently tagged *At*PIP2;1 aquaporin expressed in *Arabidopsis* roots was instrumental in studying its subcellular trafficking in response to stimuli (Luu *et al.*, 2012; Hosy *et al.*, 2015; Wudick *et al.*, Mol. Plant, 8: 1103-1114). For instance, salt (150 mM NaCl) triggered AtPIP2;1 internalization in a time frame (15 min) that paralleled the fast inhibition of root hydraulic conductivity (Lp_r). These effects are mediated through a salt-induced accumulation of Reactive Oxygen Species (ROS). The mechanisms and routes of ROS-dependent trafficking of PIPs in root cortical cells and evidence for an enhanced cycling of PIPs between the plasma membrane and endosomes in response to salt stress and ROS will be presented. The dynamic properties of AtPIP2;1 in the plasma membrane in normal, salt and oxidative stress conditions were also addressed by means of single particle tracking and fluorescence correlation spectroscopy. Both NaCl and hydrogen peroxide (H_2O_2) treatment induced an increase in lateral motion and a reduction in the density of a fluorescently tagged AtPIP2;1. In normal conditions, membrane rafts play an important role in the partitioning of AtPIP2;1 and clathrin-mediated endocytosis is predominant. In salt stress conditions, AtPIP2;1 and endocytosis is cooperated by a membrane raft-associated salt-induced pathway and a clathrin-dependent pathway. Based on natural variation of Lp_r in *Arabidopsis* and linkage mapping, we also have identified several genes regulating Lp_r (e.g. Shahzad *et al.*, 2016). One of these encodes a protein kinase that acts as a positive regulator of root hydraulics and, surprisingly, of abscisic acid (ABA) signalling. ABA-induced stomatal movements are also crucial for control of plant water status. A role for AtPIP2;1 in stomatal closure induced by ABA (Grondin *et al.*, 2015) or the Pathogen Associated Molecular Pattern flg22 was recently uncovered. Both ABA and flg22 induced an increase in osmotic water permeability of guard cell protoplasts. HyPer, a genetic probe for intracellular H_2O_2, revealed that ABA and flg22 also induce an accumulation of H_2O_2 in guard cells. These two responses were lacking in *pip2;1* knock-out plants indicating that AtPIP2;1 plays a dual hydraulic and signaling role in guard cells. *In vitro* assays indicate that the

OST1 (SnRK2.6) protein kinase likely mediates AtPIP2;1 phosphorylation during guard cell responses to ABA. Based on expression in *pip2;1* of *At*PIP2;1 phosphorylation mutants, we propose a mechanism whereby ABA and flg22 enhance phosphorylation of AtPIP2;1 at a new site (Ser121), to activate its water and H_2O_2 transport activities and induce guard cell closure

T4-01-03

Predawn xylem water potential and leaf hydraulic conductance influence the stomata behavior of mangrove plants

Huai-Tong Si, Guo-Feng Jiang, ***Kun-Fang Cao***
Guangxi University

Mangroves grow in the circumstances with high salinity and high light, their xylem is constantly under high tension, which could be a risk to induce hydraulic dysfunction. It is important for mangrove plants to control transpirational water loss through rigorously stomatal control. However, it has been hardly understood the stomatal regulation behavior of mangrove plants. Mangrove roots largely exclude salt from sea water. In addition, the leaves of some mangrove species secrete salts. In another study, we found the predawn water potential of mangrove plants are related to salt management mechanisms, with salt secretion species having high water potential than other species with only salt exclusion mechanism. Other researchers have reported the stomatal behavior of terrestrial plants is related to predawn water potential and leaf hydraulic conductance. In this paper, we tested whether these findings also work on the mangrove plants. By examining stomatal conductance (g_s) response to transient change in leaf-to-air vapor pressure deficit (VPD) and photosynthetic photon flex density (PPFD) in intact sun leaves of nine mangrove plants in a mangrove forest, we found linear g_s response to transient increase of PPFD, while three modes of g_s to increasing VPD, i.e. (1) decrease of g_s until to a very low value; (2) initially increase then decrease to a very low value; (3) continuously increase until to a stable value. These different g_s response modes might be related to leaf anatomical structure and water use strategies. The g_s sensitivity to PPFD, as indicated by linear regression slope of the regression between g_s and PPFD, was correlated with predawn water potential. We also found that there were significant linear correlations between the predawn water potential or midday water potential of the leaves and maximum g_s in a clear day, and between the diurnal maximum g_s and maximum leaf hydraulic conductivity across species. Our studies demonstrate the stomatal behavior of mangrove plants are regulated by xylem water potential, which is related to salt management mechanisms, and leaf maximum hydraulic conductance. We are now examining the stomatal conductance, foliar ABA concentration and transcriptome expression during increasing water stress in mangrove plants.

T4-01-04

Diverse stomatal signaling and the signal integration mechanism

Yoshiyuki Murata[1]*, Shintaro Munemasa*[1]*, Izumi C. Mori*[2]
1. Graduate School of Environmental and Life Science, Okayama University
2. Institute of Plant Science and Resources, Okayama University

A pair of guard cells forms stoma in the epidermis of leaves, stems and other organs. The sophisticated cells sense a variety of abiotic and biotic stresses and convert the input signals to turgor pressure regulation through modulation of ion channel activities. The changes in turgor pressure are followed by changes in cell volume and adjustments of stomatal apertures. The stomatal responses allow plants to sustain their growth under the stressful conditions. Guard cells achieve the signal integration to optimize stomatal aperture. In other words, plants encode the input signals to calcium signaling and reactive oxygen species signaling and decode these signaling to stomatal responses. We summarize the details of guard cell signaling regulated by plant hormones. Furthermore, we discuss the low-molecular compounds such as glutathione and reactive carbonyl species involved in the guard cell signaling.

T4-01-05

Stomatal responses to changes in environmental variables

Alistair Hetherington
University of Bristol

Stomata are pores found on the surfaces of leaves that control the uptake of CO_2 and the loss of water vapour. The former is important for photosynthesis and dry matter accumulation while the latter is the driving mechanism for the transpiration stream that brings water and mineral nutrients from the soil to the upper parts of the plant. Restricting water loss also provides the plant with an ability to conserve existing water during times of reduced soil water availability. Plants are able to control loss of water and CO_2 by controlling the aperture of the stomatal pore and by controlling the number of stomata that develop on the surface of leaves. Both stomatal aperture and stomatal development are under the control of environmental signals such as light, atmospheric relative humidity, atmospheric CO_2 and endogenous hormonal signals such as abscisic acid. During the course of this lecture responses to changes in atmospheric CO2 and light will be considered with a focus on signalling and sources of ATP used to support stomatal movements.

T4-01-06

L-Met activates *Arabidopsis* GLR Ca^{2+} channels upstream of ROS production and regulates stomatal movement

June M. Kwak
1. IBS
2. DGIST

Plant glutamate receptor homologs (GLRs) have long been proposed to function as ligand-gated Ca^{2+} channels, but no in planta evidence has been provided. Here we present genetic evidence that *Arabidopsis* GLR3.1 and GLR3.5 form Ca^{2+} channels activated by L-methionine (L-Met) at physiological concentrations and regulate stomatal apertures and plant growth. *glr3.1/3.5* mutations resulted in a lower cytosolic Ca^{2+} level, defective Ca^{2+}-induced stomatal closure, and Ca^{2+}-deficient growth disorder, all of which involved L-Met. Patch-clamp analyses of guard cells showed that GLR3.1/3.5 Ca^{2+} channels are activated specifically by L-Met, with the activation abolished in *glr3.1/3.5*. Moreover, GLR3.1/3.5 Ca^{2+} channels are distinct from previously characterized ROS-activated Ca^{2+} channels and act upstream of ROS, providing Ca^{2+}

transients necessary for the activation of NADPH oxidases. Our data indicate that GLR3.1/3.5 constitute L-Met-activated Ca^{2+} channels responsible for maintaining basal $[Ca^{2+}]_{cyt}$, play a pivotal role in plant growth, and act upstream of ROS thereby regulating stomatal aperture.

T4-02: Recent cell biological insights in plant sexual reproduction

T4-02-01

The roles of MON1/CCZ1-mediated Rab7 activation in vacuolar transport, vacuole biogenesis, and pollen development in plants

Liwen Jiang
The Chinese University of Hong Kong

The plant secretory pathway contains multiple membrane-bound organelles that mediate vacuolar protein trafficking from the endoplasmic reticulum (ER) to vacuole via Golgi apparatus, trans-Golgi network (TGN) and prevacuolar compartment (PVC) or multivesicular body (MVB) (Cui *et al.*, 2016a). We have recently showed that 1) the *Arabidopsis* RABG3f, a member of the plant Rab7 small GTPase family, localizes to PVC/MVB and tonoplast; 2) the proper activation of Rab7 is essential for both PVC-to-vacuole trafficking and vacuole biogenesis; 3) the *Arabidopsis* MON1 (MONENSIN SENSITIVITY1) and CCZ1 (CALCIUM CAFFEINE ZINC SENSITIVITY1) dimeric complex functions as the Rab7 guanine nucleotide exchange factor; and 4) MON1/CCZ1-mediated Rab7 activation is essential for tapetal programmed cell death (PCD) and pollen development. Thus, Rab7 activation by the MON1-CCZ1 complex is critical for vacuolar trafficking, vacuole biogenesis, tapetal PCD and pollen development in plants. Supported by grants from the Research Grants Council of Hong Kong (CUHK465112,466613,14130716, CUHK2/CRF/11G, C4011-14R, C4012-16E, and AoE/M-05/12).

T4-02-02

Membrane lipids and proteins steering pollen tube tip cell polarity machine – exocyst focus.

Martin Potocky[1], Juraj Sekeres[1,2], Lukas Synek[1], Premysl Pejchar[1], Roman Pleskot[1], ***Viktor Zarsky***[1,2]
1. Institute of Experimental Botany ASCR Prague, Czech Republic.
2. Faculty of Science, Charles University, Prague, Czech Republic.

Directionality and speed of growth are crucial parameters of pollen tubes reproduction success resulting in a selection at the level of male gametophytes in Angiosperms. During the pollen grain germination process pollen tube tip growth machinery is assembled aided mechanically by preexisting germination pore or pores formed within the exine during the young pollen grain development within the anthers. But polar growth of transiently stopped or depolarized pollen tubes might be re-established spontaneously from within. Circumstantial evidence implicates small GTPases of RHO family (RHO of plants - ROPs) as important regulators in this process in accordance with the generally accepted model of eukaryotic cell polarization. Small GTPases function as a molecular switches which regulate downstream processes via interacting effectors which orchestrate cytoskeletal and endomembrane dynamics in an intimate cross-talk with specific membrane lipids and their regulators. Exocyst is CATCHR type octameric vesicle tethering complex and was originally described as an effector of exocytosis RAB GTPase in yeast by Peter Novick's lab, but soon not only in yeast, but also in animals RHO/RAC GTPases turned-out to impose regulatory function upon this complex too. While in most organisms exocyst eight different subunits are encoded by single genes, land plants evolved multiple paralogous subunits esp. in the case of EXO70 subunit - from original three basic classes resulting e.g. in 23 *Arabidopsis* EXO70 isoforms. There is an intriguing question - what is the functional significance of this multiplication. Here, using pollen tubes of tobacco and Arabidopsis as a model, we will present evidence, that in single plant cells several EXO70 subunits are not only co-expressed but show specific, non-overlapping localizations and functions. We have also shown that as in other eukryotic models, also plant exocyst subunits are able to directly bind membrane lipids and these exocyst-lipid interactions are vital for vectorial vesicle delivery supporting rapid tip polarized cell expansion in pollen tubes. Acknowledgement – This work was and is supported by the Czech Sci. Foundation/GACR projects 13-19073S (M.P.) and GA15-14886S and GF16-34887L (V.Z) and Czech Min. of Education/ MSMT proj. NPUI LO1417 (V.Z.).

T4-02-03

Male meiotic cytokinesis and pollen ploidy consistency is affected by cold shock and exogenous GA applications through independent signaling

Danny Geelen*, Bing Liu, Nico De Storme*
Ghent University

The genomes of higher plants show multiple incidences of whole genome duplications, which have been suggested to emerge from sexual reproduction involving unreduced gametes. Diploid pollen are for instance formed after a spell of cold temperatures in *Arabidopsis*, underlining the importance of environmental cues affecting ploidy consistency. Recently, we have shown that GA treatment of flowers induces defects in cytokinesis in a subpopulation of male meiocytes, leading to meiotic restitution and the production of diploid pollen grains. Plants depleted of RGA and GAI, two members of the DELLA family that function as suppressors of GA signaling show similar aberrations in meiocyte cytokinesis. As cold- and GA treatment induces similar aberrations in the spatial organization of the phragmoplast-like radial microtubule arrays in dividing meiocytes, we hypothesized that GA mediates the cold response. However, the GA biosynthesis blocker paclobutrazol does not prevent cold-induction of meiotic restitution and GA insensitive mutants are still responsive to cold. Because GAs a transcriptional fusion of *RGA-GFP* is predominantly expressed in the tissue surrounding the meiocytes under GA limiting conditions we hypothesize that meiocyte cytokinesis is controlled by non-cell autonomous GA signaling. This signaling may find its origin from the tapetum as mutants impaired in early tapetum development display cytokinesis defects similar to that observed upon cold and GA treatment.

T4-02-04

Coordination of actin assembly and turnover by ADF and CAP1 regulates the pollen tube growth of *Arabidopsis thaliana*

Yuxiang Jiang[1,2], *Juan Wang*[2], *Yaxian Lan*[1,2], *Ming Chang*[1], *Shanjin Huang*[1,2]

1. Key Laboratory of Plant Molecular Physiology, Institute of Botany, Chinese Academy of Sciences

2. Center for Plant Biology, College of Life Sciences, Tsinghua University

Pollen tube transports the sperm cells from pistil into ovule to finish double fertilization. The rapid growth of pollen tubes (>200 μm/ h in *Arabidopsis*) relies largely on the myosin based vesicle transport mediate by actin filaments. To maintain a dedicate and highly dynamic actin filament structures, various actin binding proteins (ABPs) work in concert in actin assembly, branching, bundling, depolymerizing *etc.* In the hundreds of ABPs, the highly-expressed Actin depolymerizing factor (ADF) and Cyclase associated protein 1 (CAP1) have been reported to play vital roles in actin filament dynamics in various models species. We generated the *adf10* mutant by TALEN technology and carefully examined the changes of actin structures and its effects in the pollen tubes of *adf7*, *adf10* and *cap1* mutants. We found that: 1) Actin filament structures are disordered in pollen tubes of *adf7*, *adf10* and *cap1* mutants. Actin dynamics as well as the velicity of vesicle transport is also downregulated. 2) ADF10 plays a key role in shaping the actin fringe structure to facilitate vesicle transport both to the tip and the base of pollen tube. Disordered actin organization will lead to irregular vesicle movements. 3) Elucidation of the biochemical activity of ADFs and CAP1 *in vitro*. Both ADF7 and ADF10 coworks with CAP1 to promote actin filament turnover, in accordance with *in vivo* results in 1). 4) CAP1 promotes the nucleotide exchange of actin monomers to promote actin assembly, proved by the actin phenotypes in the tip of pollen tube and by CAP1 point mutants, both *in vitro* and *in vivo*. Thus, we provide solid evidence for the working mechanism of how ADFs and CAP1 promote actin turnover, regulate formation of specific actin structures and vesicle transport for the rapid growth of pollen tubes.

T4-02-05

DEEP-mediated modification of tRNAs is essential for endosperm-to-embryo communication in *Arabidopsis*

*Sheng Zhong, Jia Wei, Hongya Gu, **Li-Jia Qu***

The State Key Laboratory of Protein and Plant Gene Research, Peking-Tsinghua Center for Life Sciences, School of Life Sciences, Peking University

Modification of transfer RNAs (tRNAs) is essential for protein synthesis in both prokaryotes and eukaryotes. However, how modified tRNAs affect the development in multi-cellular organisms remains elusive. Here, we report that atRNA modifying enzyme, Defective in Embryo and Endosperm Phenotypes (DEEP), plays an essential role in early embryo and endosperm development in *Arabidopsis*. *DEEP* was expressed both in embryo and endosperm. In the *deep-/-*mutant, embryo and endosperm development were arrested at very early stage. DEEP protein was localized in cytoplasm, mitochondrial and chloroplast and the localization of DEEP in cytoplasm is sufficient for its function in *Arabidopsis* seed development. Endosperm-specific expression of *DEEP* in the *deep-/-* mutant could completely rescue the developmental defects both in embryo and endosperm, while embryo-specific expression of DEEP in the *deep-/-* mutant could only largely rescue the defects in embryo but not those in the endosperm. Our data demonstrate that DEEP-mediated tRNA modification is essential for endosperm-to-embryo communication and thus for embryo/ endosperm development in *Arabidopsis*.

T4-02-06

Kinase partner protein KPP controls pollen tube growth

Weihua Tang, *Haikuan Liu*

Shanghai Institute of Plant Physiology

Pollen receptor kinases from tomato, LePRKs, have been implicated in signaling during pollen germination and tube growth. Kinase partner protein, KPP, is a guanine exchange factor (GEF) for RopGTPases and a downstream signaling component of LePRKs signaling. We found that *KPP* RNAi pollen tubes grew slower both *in vitro* and *in vivo*, and elongated with severe F-actin aggregation close to the germination pores. Overexpression full length of *KPP* promotes pollen tube growth both in tobacco and in tomato and over-expression of truncated *KPP* lacking C-terminal fused with fluorescent proteins triggered branching tips. In addition, we showed that KPP can form dimers during pollen tube growth, that the C-terminal region is necessary for KPP to form homodimers. KPP preferentially activates a pollen-expressed ROP, SlROP4, and the C-terminal region is also necessary for KPP to activate SlROP4. We used affinity column chromatography to identify proteins associated with KPP and thereby identified COPI subunits which co-localized with the ARP2/3 complex. Co-expression assays showed that KPP and SlROP4 recruited coatomer proteins and ARP2/3 to the apical plasma membrane during pollen tube growth, and GST pulldown assays showed that COPI subunits and the ARP2/3 complex associated with KPP and SlROP4. Over-expression of the SHD domain of SlSCAR or dominant negative SlARP2 (G151D) the phenotypes caused by over-expression KPPΔC or KPPΔN. These data suggest that KPP transduces signals from the LePRKs at plasma membrane, activates ROP, recruits COPI and ARP2/3 complex to the apical plasma membrane, and activates ARP2/3 complex to affect actin dynamics during pollen tube growth.

T4-03: Breaking the mold: a role for non-model species in plant development

T4-03-01

Non-model systems for understanding mechanisms of leaf morphogenesis

Hirokazu Tsukaya[1,2], *Naoko Ishikawa*[1], *Ayaka Kinoshita*[1], *Emma Sarath*[1], *Xiaofeng Yin*[1], *Hiroyuki Koga*[1]

1. The University of Tokyo

2. Okazaki Institute for Integrative Bioscience, National Institutes of Natural Sciences

Previously it was nearly impossible to study mechanisms of leaf morphogenesis other than in the model species, *Arabidopsis thaliana.* Even application of our knowledge on the *Arabidopsis* system to non-model species was also challenging. But thanks to the recent development of new technologies including Next-generation sequencing, the gap between the model and non-model is rapidly filled. As typically shown by a historical finding of the contribution of class I *KNOX* genes in compound leaf formation in tomato (Hareven *et al.,* 1996), comparative studies among different systems must be done to reveal the 'real' roles of particular genes on organogenesis. There is awide range of diversity in leaf shape and organogenesis in the world (Tsukaya 2014), and some species with unique morphologies would provide us with important clues to understand mechanisms of leaf organogenesis. Here I will introduce you to some of our present studies on leaf morphogenesis in non-model species. For example, some species of the genus *Callitriche* (Plantaginaceae) are amphibious and exhibit dimorphism in leaf shape according to external environmental cues (mainly water level). We are trying to know what genetic mechanisms are involved in the switching between the dimorphism from RNAseq analyses; how such switching systems were acquired from comparative studies between amphibious and terrestrial species of the above genus. We are also studying indeterminacy of leaves in a one-leaf plant, *Monophyllaea glabra* (Gesneriaceae) by analyzing expression of several key genes for activities of leaf primordia and shoot apical meristem. Comparison between our present data and past reports on another one-leaf plant genus, *Streptocarpus,* suggested that the genetic background for the indeterminacy of leaves differs between these two genera. On the other hand, we are also interested in mechanisms of domatia formation in leaves of an ant plant, *Callicarpa saccata* (Lamiaceae). The domatia are used by symbiotic ants as nests anddomatia formation in this genus is known only in this species. Anatomy indicates that the domatia formation is caused by ectopic activation of cell proliferation at the leaf meristem. We are currently trying to understand the genetic factors involved in it. We are also analyzing flattening mechanisms in unifacial leaves by using *Juncus prismatocarpus* (Juncaceae). Our previous work revealed that the *DL* gene is one of the key factors that make non-polarized terete leaves into a flat shape (Yamaguchi *et al.,* 2010), but our detailed analysis also showed that *DL* action is not cell-or tissue- autonomous. Through our trials, I wish to indicate the fruitful potential of Evo-devo studies on leaf morphogenesis in this era.

T4-03-02

Virus-Induced Gene Silencing, the second best tool for gene function analysis in non-model plants

Annette Becker
Justus-Liebig-University Gießen

Plants are amazingly diverse in their metabolism, vegetative and reproductive morphology which allows them to adapt to ecological niches. Understanding the molecular mechanism orchestrating this biochemical and developmental diversity is one of the major research foci in plant biology. The current model plants examined on the molecular level are, like *Arabidopsis thaliana* (thale cress) or *Oryza sativa* (rice) from highly derived plant lineages, which impedes a direct transfer of knowledge to other species in most cases. With genome sequences available for many species and RNAseq of specific tissues and organs becoming affordable for even smaller labs, the availability of sequencing data is not rate limiting anymore. New model species can be now selected based on e.g. interesting phylogenetic position, important morphological or metabolic traits, and with these new model species at hand, one can study e. g. ecologically relevant traits (such as the origin of zygomorphy) in closely related taxa. However, the limiting step now is the amenability of the new model species to genetic manipulation. Protocols for stable transformation of many species are, if available at all, very labor and time consuming. A feasible alternative is Virus-Induced-Gene Silencing (VIGS), a method that utilizes the plant's reaction towards double stranded RNAs to inhibit viral gene transcription. Many plant viral vector systems have been adopted to serve as VIGS vectors and VIGS-based functional gene analysis in a large number of plant species can be achieved by now. Unfortunately, the VIGS method is transient and lacks standardization. I will provide a brief overview over this technique and its applications and outline the pitfalls. Further, I will present ideas on how this very useful method can me employed in a meaningful way to provide reliable data on gene functions and genetic interactions in new model organisms.

T4-03-03

Exploring floral senescence without a model – The rich rewards of diversity

Hilary Rogers[1], *Tony Stead*[2], *Lara Lombardi*[3], *Sergi Munne Bosch*[4], *Lorenzo Mariotti*[3], *Faezah Mohd Salleh*[1], *Natasha Spadafora*[1], *Anna Price*[1], *Carol Wagstaff*[5], *Laia Arrom*[1,4], *Riccardo Battelli*[1,3], *Piero Picciarelli*[3], *Matthew Casey*[1,2], *Peter Kille*[1], *Angela Marchbank*[1], *Nicholas Kent*[1]
1. Cardiff University
2. Royal Holloway, University of London
3. University of Pisa
4. University of Barcelona
5. University of Reading

Floral senescence does not follow the same pattern in all flowering species. Flowers can be broadly divided between those in which petals abscise while still turgid and those in which petals wilt and may or may not abscise. Furthermore, flower senescence can also be divided into a further two groups. In one group of species ethylene is produced by the flowers as the senescence progresses, and acts as a major regulator of this process. In another group ethylene is not produced, or only at very low levels and does not act as a major senescence regulator. Broadly most species that show rapid abscission without wilting are ethylene sensitive whereas those that wilt are generally ethylene insensitive, although the demarcations are not absolute. Many monocotyledonous flowers including lilies, as well as most Asteraceae are ethylene insensitive, while ethylene sensitive species include orchids, carnations petunia and Brassicaceae including *Arabidopsis.* Given these major differences it has been difficult to find a suitable model to study floral senescence. However new tools such as transcriptomic analyses are offering new approaches to investigate common and divergent regulatory mechanisms. In the obvious model, *Arabidopsis,* senescence proceeds rapidly and within 48h petals are abscised with only moderate wilting. Some progress has been made in identifying genes regulating ethylene-regulated senescence in *Arabidopsis* flowers, and some of these genes may also regulate senescence progression in other ethylene sensitive species. Abscission signals also appear to be shared across different floral senescence groups. A comparison of two lily genotypes that have contrasting senescence programmes identified genes related to abscission previously identified in *Arabidopsis*. However the regulation of senescence in ethylene-insensitive species remains problematic. Recent transcriptomic studies in lily and other species reveal families of transcription factors that may be involved. A further complexity in floral senescence is offered by composite flowers. A transcriptomic analysis of Dahlias is comparing senescence between florets and the whole inflorescence.

T4-03-04

Lace Plant: A novel model to study programmed cell death in plants

Arunika Gunawardena[1], *Adrian Dauphinee*[1], *Gaolathe Rantong*[1,2], *Christian Lacroix*[3]

1. Dalhousie University

2. Botswana International University of Science and Technology

3. University of Prince Edward Island

The lace plant, *Aponogeton madagascariensis* (Aponogetonaceae) and a few genera in the Araceae develop a complex leaf shape with distinct holes, or perforations, formed through the process of programmed cell death (PCD). Although *Monstera deliciosa* and *Monstera obliqua* (Araceae) form holes via PCD during leaf morphogenesis, they are much less tractable than the lace plant for use as a model to study PCD. We demonstrated the outstanding potential of the lace plant as a system in which to study developmentally-regulated PCD. In the lace plant, perforations form at highly predictable locations and specific developmental stages that are accessible for observation and experimental manipulations. PCD begins at the center of each areole in window stage leaves, develops outwards, and stops approximately 4-5 cell layers from the veins. These peripheral cells do not undergo PCD and are therefore referred to as NPCD (non-PCD) cells. The presence of visible changes within an areole provides a convenient gradient of cell death for analysis. The lace plant is an extremely useful system to study developmental PCD for several reasons: the accessibility and predictability of perforations; the existence of an established protocol for propagation of lace plant via sterile cultures; the tractability of the PCD process with live cell imaging *in vivo*; the ability to separate PCD and NPCD cells using laser capture micro-dissection; the ability to grow detached leaves in Petri Plates; the feasibility of performing pharmacological experiments; the possibility to regenerate lace plant via callus; the potential for genetic transformation, and the existence of optimized standard protocols for plant molecular biology research. The challenges in transforming the non-model lace plant species to a potential model species in the study of programmed cell death will be discussed.

T4-03-05

Juncus prismatocarpus (Juncaceae) as an emerging model to study unifacial leaf development

Xiaofeng Yin[1], *Takahiro Yamaguchi*[2], *Hirokazu Tsukaya*[1,3]

1. The University of Tokyo

2. ACEL

3. National Institute for Basic Biology

In recent years we have learned much about leaf development in bifacial leaves, particularly from model plants. However, although evolved repeatedly in the monocot, we know very little about leaf development in unifacial leaves, which lack adaxial side in their leaf blade. One particular reason is that in the past an efficient method to analyze the direction of cell division and thus leaf morphogenesis in non-model plants does not exist. To address this problem, we have first developed a method to efficiently and rapidly quantify the direction of cell division (Yin and Tsukaya, 2016). This method is applicable to all plant species and therefore it is a perfect tool to study leaf morphogenesis in non-model plants. Next we have applied this method to *Juncus prismatocarpus* (Juncaceae), a unifacial plant, to analyze its cell division direction in details. We have found that cell divisions contribute to the thickening growth is most active in the P2 stage. We have also found that cell divisions contribute to the thickening growth is present not only in the putative "adaxial meristem" domain but also in other domains. These data does not support the classical "adaxial meristem" hypothesis (Kaplan, 1975) which asserts that such thickening growth is restricted to the "adaxial meristem". To better understand the underlying mechanism of unifacial leaf mechanism, we have also compared cell division pattern to various key genes expression patterns. Previously we have shown that in *J. prismatocarpus*, *DROOPING LEAF* (*DL*) is responsible for the leaf blade development towards the shoot apical meristem (SAM) in the P2 stage; *PRESSED FLOWERb* (*PRSb*) is responsible for the leaf blade secondary margin development in the P3 stage; and *PRSa* is responsible for the leaf sheath development (Yamaguchi *et al.*, 2010). The comparison of cell division pattern and genes expression patterns indicated that *DL* acts in a non-cell-autonomous manner. In addition, the frequency of cell division in the presumed *PRSa* expression area is higher than that in the presumed *PRSb* area. Our newly developed method is an invaluable tool to study leaf morphogenesis in non-model plants. Using this method and other techniques, we are establishing *J. prismatocarpus* as an emerging model-plant to study unifacial leaf development. We will discuss our latest progress on this topic.

T4-03-06

Enhancing the lace plant model system through genetic transformation

Adrian Dauphinee[1], *Keiko Yoshioka*[2], *Rongda Qu*[3], *Christian Lacroix*[4], *Arunika Gunawardena*[1]

1. Dalhousie University

2. University of Toronto

3. North Carolina State University

4. University of Prince Edward Island

The lace plant (*Aponogeton madagascariensis*) is an aquatic monocot with a unique perforated leaf morphology. The distinct lattice-like pattern of the leaves is created by programmed cell death (PCD), which occurs at a specific point in leaf development known as the window stage. The high degree of predictability of pattern formation, nearly transparent leaves that are ideal for live cell imaging, and the existence of established sterile cultures are the primary features that have led to the emergence of the lace plant as a model for studying developmentally regulated PCD.

Protocol development has been a key focus in the enhancement of the lace plant model system over recent years. This study is the first to report on genetic transformation of the lace plant, also a first within the taxonomic family. Two *Agrobacterium tumefaciens* strains, GV2260 and AGL1 were used in combination with two different plasmids (pJP and pJLU13) containing reporter genes for rapid determination of transformation efficiency. The PJP and pJLU13 plasmids confer constitutive green fluorescent protein (GFP) expression driven by *CaMV 35S* and a rice *ubiquitin* (*rubi3*) promoter, respectively. The plasmids and bacteria were tested on various explants including: inflorescences, leaves, leaf sheaths, embryos, seedlings, calli, and isolated shoot apical meristems (SAMs). Results show that stable transformation has been most successful using callus derived from lace plant corm tissues and isolated SAMs. GFP detection through fluorescence microscopy and PCR analysis indicates that the lace plant is amenable to *A. tumefaciens*-mediated transformation. Current work aims to improve the efficiency of the transformation process and achieve the regeneration of whole plant transformants from SAMs and calli.

T4-04: Comparative genomics perspectives on crassulacean acid metabolism (CAM)

T4-04-01

Origin and mechanism of crassulacean acid metabolism in orchids

Liangsheng Zhang[1], *Zhong-Jian Liu*[2]

1. Fujian Agriculture and Forestry University

2. Shenzhen Key Laboratory for Orchid Conservation and Utilization

Crassulacean acid metabolism (CAM) is a CO_2 fixation pathway that maximizes water-use efficiency (WUE), compared with the C_3/C_4 CO_2 pathway, which permits CAM plants to adapt to arid environments. The CAM pathway provides excellent opportunities to genetically design plants, especially bioenergy crops, with a high WUE and better photosynthetic performance than C_3/C_4 in arid environments. The information available on the origin and evolution of CAM is scant, however. Here, we analyzed transcriptomes from 13 orchid species and two existing orchid genomes, covering CAM and C_3 plants, with an emphasis on comparing 13 gene families involved in the complete carbon fixation pathway. The dosage of the core photosynthesis-related genes plays no substantial role in the evolution of CAM in orchids; however, CAM may have evolved primarily by changes at the transcription level of key carbon fixation pathway genes. We proposed that in both dark and light, CO_2 is primarily fixed and then released through two metabolic pathways via known genes, such as *PPC1*, *PPDK* and *PPCK*. This study reports a comprehensive comparison of carbon fixation pathway genes across different photosynthetic plants, and reveals the importance of the level of expression of key genes in the origin and evolution of CAM.

T4-04-02

Molecular signatures of convergent evolution in crassulacean acid metabolism (CAM) plants

Xiaohan Yang[1,2], *Rongbin Hu*[1], *Hengfu Yin*[1], *Jerry Jenkins*[3], *Shengqiang Shu*[4], *Haibao Tang*[5], *Degao Liu*[1], *Deborah A. Weighill*[1,2], *Won Cheol Yim*[6], *Karolina Heyduk*[7], *Hao-Bo Guo*[8], *Robert C. Moseley*[1,2], *James Hartwell*[9], *Anne M. Borland*[1,10], *Daniel A. Jacobson*[1,2], *Ray Ming*[5,11], *James H. Leebens-Mack*[7], *John Cushman*[6], *Jeremy Schmutz*[3,4], *Gerald A. Tuskan*[1]

1. Biosciences Division, Oak Ridge National Laboratory, Oak Ridge, TN 37831

2. The Bredesen Center for Interdisciplinary Research and Graduate Education, University of Tennessee, Knoxville, TN 37996

3. HudsonAlpha Institute for Biotechnology, 601 Genome Way, Huntsville, AL 35801

4. US Department of Energy Joint Genome Institute, 2800 Mitchell Drive, Walnut Creek, CA 94598

5. Fujian Agriculture and Forestry University, Fuzhou, Fujian

6. Department of Biochemistry and Molecular Biology, University of Nevada, Reno, NV 89557

7. Department of Plant Biology, University of Georgia, Athens, GA 30602

8. Department of Biochemistry & Cellular and Molecular Biology, University of Tennessee, Knoxville, TN 37996

9. Department of Plant Sciences, Institute of Integrative Biology, University of Liverpool, Liverpool, L69 7ZB

10. School of Biology, Newcastle University, Newcastle upon Tyne, NE1 7RU

11. Department of Plant Biology, University of Illinois at Urbana-Champaign, Urbana, IL 61801

Flowering plants of diverse lineages have independently acquired crassulacean acid metabolism (CAM) as photosynthetic adaptations to dry environments. CAM plants have much higher water-use efficiency (WUE) than C_3 or C_4 plants due to an inverted day/night pattern of stomatal closure/opening in CAM species relative to C_3 and C_4 species. Convergence of gene function may play an important role in the evolution of isomorphic phenotypes in phylogenetically independent lineages. We hypothesized that convergent genomic evolution underpins the multiple and independent emergences of CAM from C_3 photosynthesis. Our genome-wide comparative analyses revealed the genomic signatures of convergence shared between eudicot (represented by *Kalanchoë*) and monocot (represented by pineapple and orchid) CAM species. Specifically, approximately 60 genes were found to have undergone convergent evolution in the divergent eudicot and monocot lineages, providing strong evidence that molecular convergence underpins CAM evolution. Engineering of CAM pathway into C_3 crops using synthetic biology approach would have great potential for genetic improvement of WUE and drought tolerance for sustainable production of food and bioenergy in arid, semi-arid and marginal lands. The results obtained from this research provide new insights into the molecular basis of CAM evolution, and will facilitate CAM-into-C_3 engineering to enhance drought tolerance in crops.

T4-04-03

Elucidation of evolutionary transitions between C_3 photosynthesis and CAM in Yuccas, Agaves and related species

Karolina Heyduk, *Jeremy Ray*, *Saravanaraj Ayyampalayam*, *Jim Leebens-Mack*

University of Georgia

The subfamily Agavoideae in the Asparagaceae has three independent origins of Crassulacean acid metabolism (CAM), a photosynthetic modification that allows plants to thrive in water-stressed environments. While multiple origins of CAM have been documented across the flowering plant phylogeny, the mechanism by which C_3 plants transition to CAM is not well understood. We used comparative transcriptomics to describe gene expression changes associated with two independent shifts from C_3 to CAM within the Agavoideae: one within the genus *Yucca* and one on the ancestral lineage leading to *Agave. Yucca* has two main subgenera which diverged roughly 5 Mya from a C_3 ancestor – CAM evolved in one of these lineages encompassing approximately half of extant *Yucca* species. CAM evolved independently in the ancestor of *Agave sensu lato* after divergence from the C_3 *Beschorneria* lineages. However, low-level CAM activity was found in *Polianthes*, which is deeply nested within the *Agave sensu lato* clade, implying a possible reversion from CAM to C_3 photosynthesis. Genes with time-structured expression profiles, as well as canonical CAM genes, were compared within lineages and between C_3 species and independently derived CAM lineages. *Beschorneria* and C_3 *Yucca* exhibit time-structured expression of CAM genes and some of their regulators, perhaps indicating a genomic predisposition in the Agavoideae to evolve CAM. At the same time, analysis of gene interaction networks inferred from expression correlations suggest transitions to CAM in *Agave sensu lato* and *Yucca* involve a common theme of expanded regulatory functions for clock-related transcription factors. These findings highlight the importance of phylogenetically-informed comparative analysis of closely related CAM and C_3 species.

T4-04-04

Engineering crassulacean acid metabolism (CAM) and tissue succulence to improve water-use efficiency of bioenergy feedstocks

John Cushman*, Won Cheol Yim, Jesse Mayer, Sung Don Lim*
University of Nevada

Global warming trends are leading to increased terrestrial soil drying, reduced terrestrial net primary production and carbon sinks, global food security and future biofuel production, and the global expansion of drylands that already cover 42% of the earth's surface. In order to offset these negative effects, an increased reliance upon crassulacean acid metabolism (CAM) crops or the introduction of CAM, a water-wise form of photosynthesis, into C_3 food and bioenergy crops might serve as a useful strategy to improve the water-use efficiency (WUE) of sustainable biomass production systems in the future. CAM features inverse stomatal behavior, in which stomata are open at night for CO_2 uptake when evapotranspiration rates are reduced compared with the daytime and closed during all or part of the day, thereby maximizing WUE. CAM also exploits a temporal separation of nocturnal CO_2 uptake and fixation by phosphoenolpyruvate carboxylase (PEPC), which leads to the formation of C_4 organic acids that are stored in the vacuole. The subsequent decarboxylation of these organic acids during the day releases CO_2 and concentrates it around ribulose-1-5-bisphosphate carboxylase/oxygenase (RUBISCO), suppressing photorespiration, while resulting in carbohydrate production via the C_3 Calvin–Benson cycle. Detailed comparative, functional, and integrative 'omics analyses of several CAM model or crop species have recently defined the basic genetic requirements for CAM. Such information has led to the development of new synthetic biosystem design approaches to modify and transfer photosynthetic modes into different species (http://cambiodesign.org). Facile gene stacking strategies for the assembly of a large number of transcription units (TUs) with appropriate circadian and drought-inducible expression patterns are necessary for the genetic reconstitution of facultative CAM into host C_3 species. We have created a plant-specific position/adaptor/carrier vector system originally designed for engineering mammalian cells that enables the rapid, reliable, and scalable creation of complex gene circuits using the Gibson isothermal assembly process. Design and construction of CAM-specific carboxylation and decarboxylation gene circuits containing 9 and 15 genes has been completed and are in the process of being introduced into Arabidopsis and Poplar. The gene circuits were designed to include mesophyll-specific, drought-inducible, and appropriately timed circadian expression patterns of the transgenes in order to engage the CAM pathway only during water-deficit stress conditions. Lastly, tissue succulence has been successfully engineering in the C_3 photosynthesis model species A. thaliana in order to increased mesophyll cell size for increased malate storage capacity and reduced intercellular air space to limit internal CO_2 diffusion out of the leaf during the day to favor refixation by ribulose 1,5-bisphosphate carboxylase oxygenase (RUBISCO) and thereby increases the capacity to perform CAM. Added benefits of engineered tissue succulence included increased biomass production, increased WUE, and tolerance to water-deficit (drought) and salinity stress. The combination of engineered CAM and tissue succulence is expected to increase the WUE of bioenergy feedstocks and potentially expand their production into more marginal, abandoned, or semi-arid regions. This material is based upon work supported by the U.S. Department of Energy, Office of Science, Genomic Science Program under Award Number DE-SC0008834.

T4-04-05

Evolution and regulation of crassulacean acid metabolism in the genus *Kalanchoë*

Susanna Boxall[1], Louisa Dever[1], Jack Davies[1], Nirja Kadu[1], Richard Heaton[1], Rachel Brenchley[2], Yongxiang Fang[1], Richard Gregory[1], Alistair Darby[1], Neil Hall[3], ***James Hartwell[1]***
1. Dept. of Functional and Comparative Genomics, Institute of Integrative Biology, University of Liverpool
2. School of Environmental and Life Sciences, University of Salford
3. Earlham Institute, Norwich Research Park

Forward engineering of crassulacean acid metabolism (CAM) into non-CAM crop species in order to enhance their water use efficiency requires a comprehensive knowledge of the minimal 'parts-list' for CAM. Here we describe progress with functional genomics research that aims to define and characterize the complete CAM genetic blueprint from model species in the genus *Kalanchoë* (1). A draft assembly of the 246 Mbp *K. fedtschenkoi* genome has recently been further improved through the addition of 100X coverage of PacBio long reads. Quantitative RNA-seq analysis of light/dark time course samples from C_3 and CAM leaves identified candidate CAM-associated genes. Many of these CAM-induced genes also undergo robust oscillations in transcript abundance over the light/dark cycle, consistent with a role in the circadian optimization of the pathway. The RNA-seq data has guided the reconstruction of a comprehensive model of CAM wherein candidate CAM-recruited genes are allocated to each step in the pathway. This in turn has allowed targeted RNAi gene silencing and over-expression approaches to be applied to each candidate CAM gene through the generation of stable transgenic lines of *K. fedtschenkoi* and *K. laxiflora*. Detailed analysis of CAM-associated phenotypes in the transgenic lines is revealing which genes are essential for efficient CAM, and which genes are dispensable (2). Data will be presented for RNAi lines of *K. laxiflora* lacking key CAM genes. Several lines fail to fix atmospheric CO_2 in the dark period, and the phenotypic consequences of this will be described. In addition, there is a wide diversity in the level of the CAM phenotype in the genus *Kalanchoë*, from C_3 with inducible, weak CAM in the most basal/ancestral species in section Kitchingia, through to strong constitutive CAM in the most derived species in section *Kalanchoë*. Section Bryophyllum, which includes *K. fedtschenkoi* and *K. laxiflora*, resides in the centre of the genus and these species possess intermediate levels of CAM. The genus thus represents an excellent system for comparative analysis of CAM gene evolution. Data will be presented on the CAM physiology of a range of species sampled from across the Liverpool *Kalanchoë* diversity set that we have built up for the genus. Furthermore, preliminary insights into *CAM* gene evolution across this diverse genus will be presented based on analysis of a new *Kalanchoë* genome that we have generated for a basal species. *This material is based upon work supported by the U.S. Department of Energy, Office of Science, Genomic Science Program under Award Number DE-SC0008834.*

T4-04-06

Comparative genomics of CAM photosynthesis

Robert VanBuren
Michigan State University

The altered carbon assimilation pathway of crassulacean acid metabolism (CAM) results in up to six times greater water-use efficiency than C_3 photosynthesis, making it an excellent pathway for engineering crop plants with improved drought tolerance. Central to the success of this approach is a deep understanding of the underlying genetic mechanisms controlling CAM photosynthesis. We surveyed detailed temporal and spatial transcriptomic and metabolomic changes in the obligate CAM plant pineapple (*Ananas comosus*) and facultative CAM plant *Sedum album*. Pineapple has a normal number of CAM pathway genes compared to grasses and other monocots, but CAM isoforms are enriched with cis-regulatory elements associated with regulation by circadian clock genes. This suggests CAM in pineapple evolved through the regulatory neofunctionalization of preexisting genes and not through the acquisition of neofunctionalized genes via whole-genome or tandem gene duplication. *Sedum album* shows strong CAM induction under progressive drought stress with strong co-expression of drought and CAM pathways genes. Sedum uses the same orthologs of CAM pathway genes as the model CAM plant Kalenchoe supporting parallel evolution. Together these resources provide targets for functional validation and increase our understanding of this complex trait.

T4-05: Revealing the secrets of angiosperm resurrection plants

T4-05-01

Towards understanding mechanisms associated with desiccation tolerance in Angiosperms

Jill M. Farrant[1], *Maria Cecilia Costa*[1], *H. W. M. Hilhorst*[2]
1. Dept of Molecular and Cell Biology, University of Cape Town, Cape Town, South-Africa
2. Laboratory of Plant Physiology, Wageningen University, Wageningen

The seeds of 90% of Angiosperms are desiccation tolerant (DT), whereas only 0.01% of such species (termed resurrection plants) display vegetative DT. It has been proposed that while vegetative DT first arose with the transition from aquatic to terrestrial life forms, it was lost as plants acquired mechanisms for water deficit (drought) resistance and evolved DT seeds. We propose that resurrection plants subsequently reactivated aspects of mechanisms associated with seed DT in vegetative tissues, through mechanisms not yet completely understood. Using a systems biology approach in which molecular (including all the omics), biochemical, biophysical and physiological studies were utilized, this presentation will provide evidence for similarities and differences in mechanisms associated with DT in seeds and resurrection plants.

T4-05-02

Arabinose-rich polymers as an evolutionary strategy to plasticize resurrection plant cell walls against desiccation: Understanding the woody angiosperm *Myrothamnus flabellifolia*

John Moore[1], *William Willats*[2], *Azeddine Driouich*[3], *Jill Farrant*[4]
1. Institute for Wine Biotechnology, Department of Viticulture and Oenology, Stellenbosch University
2. School of Agriculture, Food and Rural Development, Newcastle University
3. Glyco-MEV Laboratoire, Université de Rouen, Mont-Saint-Aignan
4. Department of Molecular and Cell Biology, University of Cape Town

A variety of desiccation-tolerant resurrection plants from Southern Africa were surveyed using high-throughput cell wall profiling tools (Moore *et al.*, 2009). Hydrated and desiccated leaf (frond) material, were analysed using (Moore *et al.*, 2013). Arabinan-rich pectin and an abundance of arabinogalactan proteins were found associated with the resurrection fern *Mohria caffrorum* and the woody angiosperm *Myrothamnus flabellifolia*. In contrast, *Craterostigma plantagineum* up-regulated wall proteins and osmoprotectants (Moore *et al.*, 2013). The hemicellulose-rich walls of the grass-like *Xerophyta* spp. and the resurrection grass *Eragrostis nindensis* possess highly arabinosylated xylans and arabinogalactan proteins for protection (Moore *et al.*, 2013). Of these species, only the resurrection bush *Myrothamnus flabellifolia* (Moore *et al.*, 2007) is a true woody perennial, displaying the remarkable capacity to be able survive extensive periods (i.e. years) in the desiccated (quiescent) state. Recent research has highlighted a number of anatomical (xylem modification), physiological (photosynthetic adjustment), chemical (antioxidant load), biochemical (cell wall flexibility and membrane stabilizers) and molecular (sugar transport) properties which facilitate survival of this plant under adverse environmental conditions (Moore *et al.*, 2007). This presentation will review the current hypotheses and data that support a general evolutionary mechanism of 'plasticising' the cell walls of resurrection plants (Moore *et al.*, 2007) to desiccation and implicate arabinose-rich polymers (pectin-arabinans, arabinogalactan-proteins and arabino-xylans) (Moore *et al.*, 2013) as the major contributors in ensuring flexibility is maintained and rehydration is facilitated in these remarkable plants.

T4-05-03

Adaptation mechanisms of the resurrection plant *Boea hygrometrica* to dry environment

Xin Deng
Institute of Botany Chinese Academy of Sciences

Vegetative tissues of *Boea hygrometrica*, a member of the Gesneriaceae family, can tolerate severe water loss to desiccated state and fully recover upon rehydration, a capacity termed as desiccation tolerance (DT). Studies have been conducted to identify the physiological, cellular, and molecular mechanisms underlying DT in the last two decades. Both specific responses to water stress, such as cell wall folding and pigment-protein complex stabilizing and common response such as the accumulation of osmoprotectants and antioxidants have been revealed in this plant during dehydration. The desiccation-triggered "dormancy"-like state could benefit the plant to adapt to multiple environmental stresses, such as cold and alkaline. This explains the fact that *B.hygrometrica* is the only species in Gesneriaceae family that distributes widely from the tropics to northern temperate regions in East Asia and grows vigorously in thin, calcareous and alkaline soil on limestone rocks, where dehydration occurs frequently, rapidly, and profoundly. Besides, *B. hygrometrica* is one of the few resurrection angiosperms have been reported to survive rapid dehydration. Such ability is primed by a period of acclimation of slow soil-drying in *B. hygrometrica*. Although phenotypically indistinguishable, principal component analysis revealed that the gene expression profiles in hydrated, acclimated plants resemble those of desiccated plants more closely than those of hydrated, non-acclimated plants. Anti-oxidation, protein quality control, transposons and epigenetic modifications may contribute to the priming and acquisition of DT in *B. hygrometrica*. These studies implied the existence of drought stress-induced transcriptional memory in *B. hygrometrica* that facilitates plant adaptation to the extreme habitat.

T4-05-04

Surviving metabolic arrest: What can we learn from resurrection plants?

Dorothea Bartels
University of Bonn

Resurrection plants can tolerate severe water loss, and mostly adjust their water content with the relative humidity in the environment. The plants grow in ecological niches with seasonal rainfall, they can remain in the desiccated state, comparable to dormancy in dry seeds. When rainfall occurs, the plants take advantage of the conditions, resurrect, grow and produce seeds before other species growing from seeds can do so. The objective of our research is to understand why resurrection plants survive desiccation. Our studies are focused on the desiccation tolerant resurrection plant *Craterostigma plantagineum* and the desiccation tolerant grass *Oropetium thomaeum*. Desiccation tolerance is a complex trait which requires the synthesis of proteins and sugars. To identify essential components of desiccation tolerance close relatives of *C. plantagineum* are being studied on the molecular and cellular level. Some of these species are desiccation tolerant and others are not. This comparative approach demonstrates that expression levels of genes encoding protective molecules are differentially regulated in desiccation tolerant and non-tolerant plants. This differential expression is linked to promoter architecture and the presence of specific regulatory *cis*-elements. The comparative studies are complemented by large scale analyses of gene expression patterns that have been performed to characterize the transcriptomes at different physiological conditions. These studies showed that increased expression of genes encoding protective components, species-specific genes as well as non-protein-coding RNAs are essential components of desiccation tolerance. Desiccation tolerance in resurrection plants and seeds have many metabolic pathways and molecular components in common. However, some components are specific to resurrection plants.

T4-05-05

Identification and characterisation of microRNAs in the resurrection grass *Tripogon loliiformis*

Isaac Njaci[3], ***Brett Williams***[1], *Martin Dickman*[2], *Xiuren Zhang*[2], *Sagadevan Mundree*[1]
1. Queensland University of Technology
2. Texas A&M University
3. Biosciences Eastern and Central Africa

MicroRNAs (miRNAs), a class of endogenous 20-22 nt non-coding RNAs that play critical regulatory roles in plant growth, development and in response to environmental stress have been characterised in many drought sensitive and tolerant plant species. Resurrection plants, a group desiccation tolerant plants can withstand extreme water deficit with the ability to recover on rehydration due to their ability to effectively regulate stress response. It is postulated that post-transcriptional regulatory mechanisms could be involved in desiccation tolerance. *Tripogon loliiformis*, a native Australian resurrection grass that recovers from desiccated state in 72 hrs was used to investigate this phenomenon. High-throughput sequencing of *T. loliiformis* small RNAs libraries under water deficit conditions identified 668 conserved and 119 novel miRNAs. Differential expression showed enhanced miRNA accumulation in the shoots compared to the roots. Interestingly, miRNAs were recorded even in desiccated state suggesting the existence of enhanced regulatory activities as the plant switched towards desiccation with possible repackaging and storage of miRNAs. Predicted targets included transcription factors, kinases, chaperones, carbohydrate metabolism and programmed cell death (PCD) associated genes. The targets were enriched in biological process associated with stress responses, regulation of development, gene expression,

PCD and post-transcriptional gene silencing. Identification and characterisation of *T. loliiformis* miRNAs not only points to their evolutionary conservation but also their indispensable regulatory role during desiccation stress. Understanding of the underlying mechanisms is paramount in devising strategies for the development of genetically improved crop plants with enhanced drought tolerance and water use efficiency.

T4-05-06

Mining the resurrection genome of *Xerophyta viscosa* for a footprint of desiccation tolerance

Maria-Cecília Costa[1], *Mariana Artur*[2], *Eef Jonkheer*[2], *Martijn Derks*[2], *Harm Nijveen*[2], *Wilco Ligterink*[2], *Melvin Oliver*[3], *Jill Farrant*[1], ***Henk Hilhorst***[2]

1. *University of Cape Town*
2. *Wageningen University*
3. *University of Missouri*

Desiccation tolerance (DT) is common in seeds and various other organisms but only a few Angiosperm species possess vegetative DT. These resurrection species may serve as ideal models for the ultimate design of crops with enhanced drought tolerance. Combining genome sequence information with genome-wide gene expression data may provide insight into the 'footprint' of vegetative DT. We produced a high-quality whole-genome sequence for the grass-like resurrection plant *Xerophyta viscosa* and assessed transcriptome changes during its de- and rehydration. We found 126 putative LEA motif-containing proteins divided into eight families, of which 90 were differentially expressed during de- and rehydration of adult plants. The number of LEAs is significantly higher than identified in the genomes of the homoiochlorophyllous resurrection species *Boeia hygrometrica* (67 LEA genes) and *Oropetium thomaeum* (94 LEA genes) as well as in a phylogenetic cross-section of 25 plant species. These differences might relate to the poikilochlorophyllous nature of *X. viscosa* as well as to the habitat in which it occurs. Data revealed induction of transcripts typically associated with DT in seeds, and involvement of orthologues of ABI3 and ABI5, both key regulators of seed maturation. The bZIP transcription factor ABI5 is strongly associated with the acquisition of seed bank longevity in orthodox seeds, as well as with several LEAs. Here we show that ABI5 may be a main regulator of expression of the LEA_4 family, which, together with the LEA-6 family, is significantly expanded in the genome of *X. viscosa*. Thus, this may be an important factor in the extreme longevity of X. *viscosa* in the dry state. Furthermore, we identified two structural orthologs of ABI3, a major regulator of seed maturation and DT along with the majority of the ABI3 regulon expressed in leaves. Again, this exemplifies the strong 'seed character' of vegetative DT in this species. Dehydration resulted in the down regulation of genes located in genomic 'clusters of desiccation associated genes' (CoDAGs), reflecting the cessation of growth that allows for the expression of DT. They may represent genes related to mechanisms that suppress growth and development or energy metabolism in vegetative tissues and may trigger or inhibit the induction of the mechanisms that lead to vegetative DT. The genes in these CoDAGs may also supress the expression of DT in vegetative tissues, allowing the induction of DT only at critical levels of water loss. Their down regulation may thus be part of the reactivated mechanism that evolved to deliver DT in vegetative tissues. We provide strong (molecular) support for the hypothesis that vegetative desiccation tolerance arose by redirection of genetic information from desiccation tolerant seeds.

T4-06: Calcium transport and signaling

T4-06-01

Plant osmosensor OSCA1 calcium channel family

Zhen-Ming Pei[1], *Fang Yuan*[1,2], *Xiaomei Wu*[2], *Yan Xue*[1], *James Siedow*[1]

1. *Duke University*
2. *Hangzhou Normal University*

The increases in cytosolic free Ca^{2+} ($[Ca^{2+}]$) induced by osmotic stress have been speculated to act in the perception of the stress. However, the molecular components remain poorly understood. Using Ca^{2+} imaging-based forward genetic screens we have isolated *Arabidopsis* mutants defective in osmotic stress-induced $[Ca^{2+}]$ increases (osca). The *osca1* mutant displays impaired osmotic Ca^{2+} signaling and attenuated responses to osmotic stress at the cellular and whole plant levels. OSCA1 is a plasma membrane Ca^{2+} channel gated by osmotic stress. OSCA1 belongs to a conserved yet previously uncharacterized gene family in eukaryotes. These genes represent the long-sought components of stimulus responsive $[Ca^{2+}]$ increases, and could provide potential molecular targets for engineering drought resistant crops.

T4-06-02

Oxygen sensing in plant roots linked with ion channels

Feifei Wang

University of Tasmania

Over 17 million km^2 of land is affected by flooding, resulting in substantial yield losses and jeopardising food security across the globe. The crucial role of Ca^{2+} and K^+ in response to abiotic and biotic stimuli has been widely recognized in plants but still poorly understood within specific cell types in different root zones under hypoxia stress. As dependence and reliance of mammalian system on O_2 is much stronger than in their plant counterparts, acute responses to hypoxia must operate within a timeframe of seconds. O_2-regulated ion channels fit this description very well. To further understand the mechanisms by which plants sense low-oxygen stress, we first summarise and identify several known candidates for oxygen sensing in the mammalian literature. We then identify key oxygen sensing domains (PAS; GCS; GAF; PHD) in mammalian systems and use the sequences of those oxygen sensing domains to identify the potential plant counterparts in *Arabidopsis*. Several plasma- and tonoplast based ion channels (such as TPC; AKT; KCO) and oxygen domain-containing proteins with predicted oxygen sensing ability were identified in plants. Whole-plant physiological and tissue-specific Ca^{2+} and K^+ changes were studied using knock-out *Arabidopsis* mutants of Ca^{2+}-ATPase (ACA), Ca^{2+}/proton exchanger (CAX), depolarization-activated guard cell outward-rectifying potassium (GORK) channels and respiratory burst oxidase homologue D (RBOHD) in waterlogging treatment. In the wild-type (WT) plants, the expressions of *ACA8*, *CAX4*, *CAX11*, *GORK* and *RBOHD* were down-regulated by up to 3-fold by hypoxia treatment. Ca^{2+} accumulation in root tissues was much higher in stelar cells in the mature zone of *aca8, aca11, cax4* and *cax11* mutants. In addition, we also show that CAX11 plays a key role in maintaining cytosolic Ca^{2+} homeostasis and/or signalling in root cells under hypoxic conditions. Phenotyping experiments found that *gork1-1* mutant has better K^+ retention ability in both the elongation and mature zone compared with the WT and *rbohD* under hypoxia. Besides, hypoxia induced Ca^{2+} increase in each cell type was much less pronounced in *rbohD* than in the WT. In most tissues except the elongation zone in *rbohD*, the H_2O_2 concentration had decreased after 1 h of hypoxia, but then increased significantly after 24 h of hypoxia in each zone and tissue, further suggesting that RBOHD may shape hypoxia-specific Ca^{2+} signatures via the modulation of apoplastic H_2O_2 production. In summary, Ca^{2+} efflux systems especially CAX11, RBOHD and GORK channel play critical roles in plant adaptive to hypoxia stress by shaping the stress-specific Ca^{2+} and H_2O_2 'signatures' and K^+ homeostasis.

T4-06-03

Nuclear calcium signalling in symbioses and beyond

Myriam Charpentier

John Innes Centre

Nuclear calcium signalling is known to be essential in legumes to promote associations with nitrogen fixing bacteria and phosphate delivering arbuscular mycorrhizal fungi. Legumes are among the world's most important agricultural food crops that are beneficial to billions of farmers and consumers worldwide and provide an essential aspect of natural soil enrichment of organic nitrogen compounds. The mechanisms of plant nuclear calcium signalling were poorly understood. Using a wide range of approaches, we have discovered a number of ion channels located at the nuclear envelope that are responsible for symbiotic nuclear calcium release. Among them, we defined the first plant nuclear-associated calcium channels encoded by cyclic nucleotide gated channels (CNGC15s). The CNCG15s sit at the nuclear envelope in a complex with a potassium permeable channel (DMI1), also required for the generation of the symbiotic nuclear calcium signals. Interestingly, CNGC15s and DMI1 are conserved across all land plants, including non-symbiotic species, strongly suggesting that they have other functions during plant development. The function of this nuclear calcium channel complex in and outside root legume symbioses will be discussed.

T4-06-04

Coordination of the sorting and activity of glutamate receptor-Like channels (GLRs) by cornichon-like proteins

Jose Feijo

University of Maryland

Two decades after the discovery of Glutamate receptor-like channels (GLRs) in the genome of *Arabidopsis*, their molecular properties, localization, physiological relevance and regulation are still under debate. Of relevance, the evolution and apparent explosion of their number and diversity in angiosperms lacks an appropriate rational framework. While this diversification is suggestive of functional specialization, the complexity inherent to redundancy and compensation mechanisms associated with Ca^{2+} signaling have been hindering any generalizing functional model for their physiological functions in plants. Here, we report on a number of approaches to address this enigma, using *Arabidopsis* pollen grains and tubes, and root protoplasts as model cell systems, and a combination of genetic, electro-physiological and imaging methods. Single or multiple mutations in GLRs (up to quadruple mutants) resulted in reduced Ca^{2+} fluxes across the plasma membrane in both pollen and root protoplasts. To our surprise, this was not reflected by the subcellular localization of the channels, as we localized two representative pollen-expressed GLRs at the membrane of the sperm and vegetative vacuole, respectively. With the aim of detecting GLR-interacting proteins mediating the differential targeting of the channels to their target membrane, we identified a family of chaperone-like proteins homologous to the *Drosophila* CORNICHON family. Interestingly, *cornichon* pollen and root protoplasts revealed similarly low Ca^{2+} fluxes as *glr* mutants. Moreover, expression in *cornichon* plants resulted in the mislocalization of GLRs. To understand the specificity of this targeting we localized several integral membrane proteins in *cornichon* single or higher order mutant backgrounds. We observed that most, but not all, had their localization changed in the mutants, while soluble or membrane-attached proteins were unaffected. Remarkably, the heterologous co-expression of GLRs and cornichons lead to increased Ca^{2+} fluxes, suggesting a regulatory effect of the cornichons. We therefore posit that, beyond their implication in the sorting of various protein families, cornichon-like proteins are necessary to regulate the trafficking and activity of GLRs.

T4-06-05

Circadian regulation of calcium signals in *Arabidopsis thaliana*

Timothy Hearn, Maria Marti, ***Webb Alex***

University of Cambridge

The circadian clock is a 24 h time keeper that is present in all multicellular organisms. The structure of these circadian clocks differ between the Kingdoms but broadly they have similar features resulting in persistent rhythms of cellular activity in constant conditions with a period close to 24 h, regulation of a swathe of gene expression, physiology and development and entrainment by light and dark cycles so that these circadian oscillations of activity occur in concert with the light and temperature cycles of the rotating Earth. In *Arabidopsis* the circadian oscillator comprises a translational transcriptional feedback loop but in addition there is circadian regulation of cytosolic calcium signals that peak towards the end of the photoperiod. We have been dissecting the physiological and genetic basis of the regulation of these calcium signals. We have new data suggesting that the circadian regulation of calcium occurs through a coincidence between light and the circadian clock. Additionally we have performed a mutant screen identifying plants with altered responses to nicotinamide, an antagonist of circadian calcium signalling. We have identified genes that are required for the correct regulation of calcium signals. A physiological role for calcium signalling the dynamic adjustment of circadian period is suggested by our phenotypic analysis of nicotinamide over sensitive mutants. Using a new approach for the identification of systems changes based on dynamical modelling of biological networks in different experimental conditions, we have identified the physiological and genetic network that constitutes a calcium signalling based feedback loop in the circadian clock.

T4-06-06

Interaction of calcium and ROS signaling in *Arabidopsis* roots in response to environmental cues

Joerg Kudla
WWU Muenster

Calcium (Ca^{2+}) and reactive oxygen species (ROS) are important second messengers that function in many different biological processes in plants. Spatially and temporally defined changes in their cytoplasmic concentrations represent stimulus-specific signals. However, the modes of interaction and integration of these two important second messengers are largely unknown. Respiratory burst oxidase homologs (RBOHs) are membrane localized plant NADPH oxidases that produce ROS and form a family of 10 proteins in *Arabidopsis*. Cellular calcium signals are detected and transmitted by calcium-binding proteins that function as calcium sensors. In higher plants, calcineurin B-like (CBL) proteins and their CBL-interacting protein kinases (CIPKs) represent important relays in calcium signaling in many different response reactions. In *Arabidopsis*, 10 CBL-type calcium sensor proteins form a dynamic interaction network with 26 CIPKs. Here, I will report in detail how distinct Ca^{2+}-activated CBL/CIPK complexes regulate ROS production of specific RBOH proteins in *Arabidopsis*. In this way I will detail mechanisms that integrate Ca^{2+} and ROS signaling, how activation of ROS signals occur and discuss a model how Ca^{2+} and ROS could function together in plant long distance signaling.

T4-07: Plant responses to flooding

T4-07-01

Ethylene priming: A new mechanism that confers flooding tolerance

Rens Voesenek
Plant Ecophysiology, Utrecht University

The severity and frequency of floods is predicted to increase with changing global weather patterns. These floods adversely affect crop growth and yields. Research aimed at increasing plant resistance to abiotic stresses such as flooding is therefore a high global priority. The volatile plant hormone ethylene accumulates in submerged plant tissues very quickly (within 1 h) and acts as a reliable proxy for submergence and probably therefore this hormone operates upstream of many flood adaptive processes. In *Arabidopsis*, ethylene can 'prepare' (prime) plants for flooding tolerance to later occurring oxygen deficiency (hypoxia). This novel mechanism of flooding tolerance is regulated via increased transcription and stabilisation of group VII Methionine-Cysteine (MC) Ethylene Response Factor (ERF) transcription factors. Upon hypoxia these MC-ERFs regulate the transcription of adaptive core-hypoxia genes. We hypothesize that ethylene-regulated increase of non-symbiotic hemoglobin causes stabilization of MC-ERFs via its capacity to scavenge NO.

T4-07-02

Regulation of root water transport under flooding

Christophe Maurel[1], *Zaigham Shahzad*[1], *Colette Tournaire-Roux*[1], *Alexandre Martinière*[1], *Yann Boursiac*[1], *Olivier Loudet*[2]
1. Biochemistry and Plant Molecular Physiology, CNRS/INRA, Montpellier
2. Institut Jean-Pierre Bourgin, INRA/AgroParisTech, Versailles

Soil water uptake by roots is central for plant growth and survival. Flooding results in root hypoxia, which as many other environmental constraints such as drought, nutrient deprivation or oxidative stress, exerts deep effects on root transport functions and alters root water permeability (root hydraulic conductivity; Lp_r). We previously showed that the early inhibition of Lp_r by hypoxia is mediated through proton-dependent gating of aquaporin water channels (Tournaire-Roux *et al.*, 2003). More recently, quantitative genetics approaches based on the natural variation of Lp_r in the model plant *Arabidopsis* were used to identify Hydraulic Conductivity of Root 1 (HCR1), a Raf-like MAPKKK (Shahzad *et al.*, 2016). This protein kinase accumulates as a function of nutrient and oxygen (O_2) availability and regulates Lp_r accordingly. In brief, HCR1 reduces water entry into the roots when there is a lack of O_2, but only when the soil is also rich in potassium (K^+), a mineral salt essential for plant growth. *HCR1* acts on abundance of RAP2.12, a key transcription factor triggering the core anaerobic response. We propose that *HCR1* delineates a combinatorial signaling pathway integrating two soil signals, K^+ and O_2 availability, for resilience of plants to multiple flooding scenarios including root waterlogging or complete submergence. The adaptive significance of this mechanism is supported by substantial variation of *HCR1* in regulating Lp_r at the *Arabidopsis* species level.

T4-07-03

The mitochondrial retrograde signaling regulates flooding tolerance in rice

Su-May Yu[1], *Kuo-Wei Lee*[1], *Chien-Ru Lin*[1], *Li-Hong Liu*[1], *Tuan-Hua Ho*[2]
1. Academia Sinica/Institute of Molecular Biology
2. Academia Sinica/Institute of Plant and Microbial Biology

Rice is the only cereal crop capable of anaerobic germination and seedling growth, which represents a unique adaptive trait to oxygen (O_2) deficiency under flooding. Our previous study showed that sugar and O_2 deficiency signaling processes converge into a common CIPK15-SnRK1A-dependent regulatory pathway to activate alpha-amylase expression, resulting in hydrolysis of seed starch to sugars to be used for anaerobic energy metabolism in rice seedlings. The underlying mechanism of how rice senses and initiates this anaerobic tolerance response is unclear. Recently, we observed that a coordinated signaling cascade regulates the expression of genes necessary for sugar production, fermentation, and chloroplast biogenesis to sustain anaerobic seedling growth. Studies with mutants of components in the mitochondrial electron transport chain (mtETC) indicates that mtETC monitors O_2 and sugar availability through the two parallel cytochrome c oxidase (COX) and alternative oxidase (AOX) pathways to initiate the retrograde signaling process. Ca^{2+} serves as a second messenger in the signaling pathway, and a Ca^{2+} sensor, CBL, interacts with and translocates CIPK15 from cytoplasm to plasma membrane in a

Ca^{2+}-dependent manner. N-myristoylation & N-acylation of CBL is necessary for the interaction of CBL and CIPK15 in the plasma membrane. Our studies reveal that the mitochondrial retrograde regulatory network and the downstream CBL-CIPK15 interaction regulating alpha-amylase expression confer the flood-adaptive trait in rice seedlings.

T4-07-04

Jasmonate regulates plant responses to post-submergence reoxygenation through transcriptional activation of antioxidant synthesis

Li-Bing Yuan, Yang-Shuo Dai, Li-Juan Xie, ***Shi Xiao***
Sun Yat-sen University

Submergence induces hypoxia in plants; exposure to oxygen following submergence, termed reoxygenation, produces a burst of reactive oxygen species. The mechanisms of hypoxia sensing and signaling in plants have been well studied, but how plants respond to reoxygenation remains unclear. Here, we show that reoxygenation in *Arabidopsis thaliana* involves rapid accumulation of jasmonates (JAs) and increased transcript levels of JA biosynthesis genes. Application of exogenous methyl jasmonate improved tolerance to reoxygenation in wild-type *Arabidopsis*; also, mutants deficient in JA biosynthesis and signaling were very sensitive to reoxygenation. Moreover, overexpression of the transcription factor gene *MYC2* enhanced tolerance to post-hypoxic stress and *myc2* knockout mutants showed increased sensitivity to reoxygenation, indicating that MYC2 functions as a key regulator in the JA-mediated reoxygenation response. MYC2 transcriptionally activates members of the *VITAMIN C DEFECTIVE* (*VTC*) and *GLUTATHIONE SYNTHETASE* (*GSH*) gene families, which encode rate-limiting enzymes in the ascorbate and glutathione synthesis pathways. Overexpression of *VTC1* and *GSH1* in the *myc2-2* mutant suppressed the post-hypoxic hypersensitive phenotype. The JA-inducible accumulation of antioxidants may alleviate oxidative damage caused by reoxygenation, improving plant survival after submergence. Taken together, our findings demonstrate that JA signaling interacts with the antioxidant pathway to regulate reoxygenation responses in *Arabidopsis*.

T4-07-05

The carbon and phosphorus trade market between arbuscular mycorrhizal fungi and rice under different flooding regimes

Xiaozhe Bao, ***Yutao Wang****, Shaoshan Li*
South China Normal University

Arbuscular mycorrhizal fungi (AMF) is arguably the most important and widespread symbiotic microorganisms in nature. They form symbiosis with the majority of terrestrial plants, in which the fungal partner benefits the host plant by facilitating plant access to mineral nutrients (especially phosphorus (P)), in exchange for the plant-assimilated carbon (C). This C-P nutrient exchange between AMF and host plant can greatly influence plant productivity and community structure, and plays a significant role in the global C and P cycling. In recent years, evidences are accumulating that AMF are prevalent in paddy wetlands. AMF are aerobic microorganism, while the soil environment in the paddy field is usually hypoxic. Therefore, the nutrient exchange between them in wetland environment may be different from that in terrestrial ecosystems. Until now, the nutrient exchange in AM symbiosis from wetland environment has not yet been explored. The present study aims to clarify (a) whether the C-P trade, which were commonly present in AM symbiosis in terrestrial ecosystems, also occurs between AMF and its wetland hosts under wetland environments? and (b) the response of the C-P trade to flooding regimes? A pot experiment, using rice (cv. Guinongzhan) and *Rhizophagus irregularis* as the plant and fungal materials, respectively, was conducted. Rice plants were cultivated under three different flooding conditions, and the plant growth parameters and gene expression files relating to P absorption were analyzed. The results showed that under all flooding conditions AMF structure were commonly present in the rice roots with AMF inoculation. Considering the obligate nature of AMF, it suggested that there was C input from rice to AMF under these flooding regimes. The AMF-specific P transporter gene *OsPT11* were detected in rice roots with AMF inoculation, indicating the P delivery from AMF to rice plants. The C-P nutrient exchange was inhibited by increasing flooding intensity, as indicated by decreased AMF colonization rates and down regulation of *OsPT11* expression level. The AMF colonization should also decrease the direct P absorption pathway by rice roots, since the *OsPT2* and *OsPT6* expressions in rice roots were much lower in AMF colonized plants than those without AMF inoculation. The present study provide the first evidence, as far as we know, showing the presence of the C-P trade between AMF and wetland plants in wetland environments. The inhibitary effects of flooding on the nutrient exchange between AMF and rice plants are indicative to the application of AMF in rice cultivation system, but further research were needed to explore, in more details, the mechanisms involved in this inhibitory effects of flooding on the C-P exchange in AM symbiosis at wetlands.

T4-07-06

Elucidating the molecular mechanisms behind adventitious root formation in *Solanum dulcamara*

Xinping Yang[1]*, Eric Visser*[2]*, Peter De Groot*[1]*, Ivo Rieu*[1]*, Mariani Celestina*[1]
1. Institute of Water and Wetland Research, department of Molecular Plant Physiology, Radboud University
2. Institute of Water and Wetland Research, department of Experimental Plant Ecology, Radboud University

Adventitious root (AR) formation is an adaptive trait to partial flooding in many plant species. Phytohormones are extensively involved in this process. Ethylene and ABA were shown to play contrasting roles in AR formation. Ethylene induces AR emergence by activating the activities of cell cycle genes, and promoting programmed death of overlying epidermal cells, while ABA inhibits AR formation in monocot species deepwater rice which mechanisms were unclear yet. In our research, *Solanum dulcamara*, a dicot species, was used. We investigated the interaction between ethylene and ABA in regulating AR formation, and further elucidated the role of ABI5, a bZIP transcriptional factor in ABA-inhibited AR formation. We found that, ethylene treatment, on one hand, highly induced cell wall degrading genes expression which resulted in the loosening of overlying tissues; on the other hand, decreased ABA biosynthesis and thus led to a reduced ABA level in primordia. Locally applying ABA inhibited flooding-induced AR emergence, while removal of ABA signal by fluridone application released primordia from dormancy. These data together confirmed the positive role of ethylene and inhibitory role of ABA in AR emergence. Moreover, One *SdABI5* candidate was figured out by screening in the *Solanum dulcamara* transcriptome database and by blasting with that of *Solanum lycopersicum*. *SdABI5* was highly expressed in primordia and seeds, and was induced by ABA treatment, whereas inhibited by flooding and ethylene treatment. RNA interference (RNAi)-mediated repression of *SdABI5* resulted in an earlier activation of dormant primordia during flooding. To further explore the role of SdABI5, a microarray expression analysis will conducted on *S.dulcamara abi5* mutants.

T4-08: Development and genetics of dioecy and sex chromosomes in plants (two sessions)

T4-08-01

Evolution of flexible sex determination system in polyploid persimmon

Takashi Akagi
1. Graduate School of Agriculture, Kyoto University
2. PRESTO, JST

Although the field of epigenetics has strong roots in plant biology, evidence for its adaptive role is limited. This is surprising because plants are sessile and can be long lived, especially in trees. Epigenetic regulation can add a flexible layer to genetic variation, potentially enabling long term, but reversible cis-regulatory changes of an allele while maintaining DNA sequence. Here, we present a case in which alternative epigenetic states lead to reversible sex determination in hexaploid persimmon *Diospyros kaki*, in an evolution from dioecious diploid *Diospyros* species. Previously, we elucidated the molecular mechanism of sex determination in diploid persimmons and demonstrated the action of a Y-encoded sex determinant pseudo-gene called *OGI*, via the production of small-RNAs against the gene encoding the female-determining factor *MeGI* and resulting in separate male and female individuals (dioecy). On the other hand, in hexaploid Oriental persimmon (*D. kaki*), this regulatory mechanism leads to the formation of both male and female flowers (monoeocy), in genetically male individuals. In these Y-chromosome carrying hexaploid persimmon trees, the *MeGI* and its promoter region exhibits methylation and repression in male flower buds but not in developing female flower buds or any other tissue. Inhibition of DNA methylation by treatment of demethylating agents resulted in the reversion of male developing buds to form semi-feminized flowers, while it did not affect female flowers. In hexaploid persimmon, *OGI* expression is generally undetectable, possibly due to the SINE-like transposon insertion of a transposon in the 5' region of its promoter. This transposon insertion is highly methylated as is reported SINE families in other plant species, and is conserved in all Y-carrying *D. kaki* cultivars, suggesting either natural selection or a strong bottleneck for the silenced *OGI* gene during the establishment of this species. Our findings are consistent with the idea that the evolution of this more flexible system is an adaptation to polyploidization and associated enhanced sex bias. They might also provide an example of an adaptive scenario involving epigenetic plasticity.

T4-08-02

Pattern and process of unisexual flower development

Pamela Diggle
University of Connecticut

Unisexual flowers are all defined by a single attribute: they have only a single set of functional reproductive structures; the androecium or gynoecium. Yet, the developmental patterns and process that underlie the absence of function in the androecium or gynoecium of unisexual flowers are diverse. I will summarize the common processes involved in flower sex determination, the stages of development affected, and examine the distribution of stages and process across the angiosperms.

T4-08-03

Hill-robertson interference causes reduced genetic diversity on a young plant y-chromosome following recent recombination suppression

Josh Hough[1], *Wei Wang*[2], *Spencer Barrett*[2], *Stephen Wright*[2]
1. University of California, Davis
2. University of Toronto

X and Y chromosomes differ in effective population size (*Ne*) rates of recombination, and exposure to natural selection, all of which can affect patterns of genetic diversity. On Y chromosomes with suppressed recombination, selection is expected to eliminate neutral variation and reduce the *Ne* of Y compared to X chromosomes or autosomes. However, non-selective factors including female biased sex ratios and high variance in male reproductive success can also reduce Y-linked *Ne*, making it difficult to infer the causes of low Y-diversity from natural populations. Here, we investigate the factors affecting levels of polymorphism during sex chromosome evolution in *Rumex hastatulus* (Polygonaceae), a dioecious plant with young sex chromosomes. Strikingly, we find that neutral diversity for genes on the Y is on average 2.1% of the value for their homologues on the X, corresponding to a chromosome-wide reduction of 93% compared to the neutral expectation. We demonstrate that the magnitude of this diversity loss is inconsistent with a reduced male *Ne* caused by neutral processes including female-biased sex ratios and high variance in male reproductive success. Instead, using forward simulations and estimates of the distribution of fitness effects of deleterious mutations, we show that diversity loss on the Y can be explained by purifying selection acting over a large number (~ 800 kb) of genetically-linked sites. Our results are in agreement with theory on "interference selection", and provide empirical evidence that this effect can substantially reduce neutral diversity on a non-recombining chromosome. Given the recent origin of *R. hastatulus* sex chromosomes (15MYA), our results imply that Y-chromosome degeneration in the early stages may be largely driven by such interference effects rather than by positive selection for gene silencing followed by neutral genetic drift.

T4-08-04

Towards understanding the gene network controlling sex determination in cucurbits

Abdel Bendahmane
IPS2-INRA, Building 630, rue de Noetzlin, Plateau du Moulon, 91405 Orsay

How the gender of a flower or plant is determined is an important issue in plant developmental biology. Understanding this process also has practical applications in agriculture and plant breeding, as the gender of a flower or plant often limits how it is bred and cultivated. Sex determination is a process that leads to the physical separation of male and female gamete producing structures in separate flowers on the same plant (monoecious species) or on separate individuals (dioecious species). Several species in the *Cucurbitaceae*, including melon, have bisexual floral primordia, but often have flowers limited to a single sex. Sex determination occurs by the selective arrest of either the male stamen or female carpel during development. In melon, sex determination is governed by three genes *andromonoecious* (M), *androecious* gene (A) and *gynoecious* (G) and the interplay of alleles of these three genes results in a range of sexual types. In addition, cucurbit sex expression patterns can be modified by hormones, such as ethylene, and by environmental factors. We have previously shown that the *gynoecious* (G) gene encodes for a zinc finger transcription factor, *CmWIP1*, the *andromonoecious* (M) gene encodes for an ethylene biosynthesis enzyme, *CmACS*-7 and the *androecious* gene (A) encodes *CmACS11*. Now that the identity of the androecy, monoecy and gynoecy sex genes are revealed and genetic pathway controlling sex determination in *Cucurbitaceae* discovered, among the next challenges are to decipher how the sex determination signals are perceived and how the information is translated to cause organ-specific abortion at the flower and the plant level. Our progress in this direction will be discussed.

T4-08-05

New insights about the sex chromosomes in *Coccinia*, a dioecious genus 'between' *Cucumis* and *Citrullus*

Susanne S. Renner
University of Munich (LMU)

The angiosperm species with the relatively largest Y-chromosome is *Coccinia grandis* (Cucurbitaceae), with a male/female difference of 10% of the total genome (male genome size: 0.925 Gb/2C). The Y chromosome arose ca. 3 my ago and is entirely heterochromatic. The chromosomal distribution of repetitive and organellar DNA (detected with probes from high-throughput sequencing) shows that the size of the Y is largely due to an accumulation of certain Ty1/copia and Ty3/gypsy repeats, satellites, and organellar DNA. The centromere of the Y differs from those of all other chromosomes including the X in lacking a highly abundant centromeric repeat (CgCent) and in extended histone marks, raising questions about spindle attachment. PacBio sequencing of *C. grandis* is ongoing. The genus *Coccinia* also includes 24 other dioecious species, and we also tested the repetitive DNA accumulation patterns and performed immunostaining with (peri-) centomere-specific antibodies in three species close to *C. grandis*. Male nuclei of two of the relatives indeed accumulate distinct repeats (in each case on one chromosome pair), and CgCent also labels their centromeres and pericentromeric regions.

T4-08-06

The secrets of sex determination in strawberries

Tia-Lynn Ashman[1], Jacob Tennessen[2], Na Wei[1], ***Aaron Liston[2]***
1. University of Pittsburgh
2. Oregon State University

Sex chromosome evolution is thought to be a universal feature of separate-sexed organisms but our understanding of the process is heavily influenced by animal systems, the majority of which have ancient sex chromosomes. We will present research on species of wild strawberries (*Fragaria*, Rosaceae) demonstrating that plants not only provide valuable insight into the earliest stages of sex chromosome evolution from autosomes but also offer key parallels to dynamics seen at later stages in animal systems. Using comparative approach across multiple species in this young genus we have discovered at least four transitions from hermaphroditism to separate sexes, and multiple genetic avenues to loss of male function. Within the nascent ZW sex determining system in an octoploid and decaploid clade, we revealed novel recombination heterogeneity between the sexes, as well as at least four different genomic locations of a homologous sex determining region. Within this dynamic system, the translocated sex determining region shows surprising fidelity to a single homoeologous set of chromosomes. If this pattern is found in other genera, it could help explain the prevalence of separate sexes in polyploid plants.

T4-08-07

Using common garden and transcriptomics to understand the genomic architecture of sexual dimorphism

Stuart Mcdaniel*, Sarah Carey, Adam Payton*
University of Florida

Sexual dimorphism evolves because selection favors different phenotypes in males and females. However, the trajectory of phenotypic evolution in species with separate sexes is determined by not just nature of selection on the two sexes, but also the correlation between the phenotypic effects of allelic substitutions in males and females. To examine these two phenomena, we studied population differentiation in males and females of the moss *Ceratodon purpureus*, and transcriptomic differences between males and females in early growth rate, a sexually dimorphic trait. We hypothesized that traits in which the difference between males and females evolved among populations may be subject to the strongest selection. In contrast, we found unambiguous evidence for adaptation in development time, a modestly dimorphic trait, but no signature of selection on the most sexually dimorphic traits (leaf size, early development growth). Instead, dimorphism in these traits appears to evolve as a consequence of sexually dimorphic correlated responses to selection on development time. These findings suggest that the genetic architecture of trait variation in males and females is quite different. To begin to understand the genetic basis of sexual dimorphism in early development, we surveyed gene expression and metabolite profile in ten male-female sibling pairs (i.e., two haploid plants derived from the same diploid meiosis) isolated from seven populations spanning the geographic distribution of the species. We identified a core set of ~1000 genes (both autosomal and sex-linked) that were consistently differentially expressed between females and males, of about 15,000 expressed in this tissue, and 25 differentially represented metabolites. These data highlight the importance of pleiotropy and genetic background in governing the phenotypic response to selection. We are currently testing these findings in other tissues to directly link sexually antagonistic selection to its molecular targets.

T4-08-08

Genome-wide association studies reveal different sex determining systems in two poplar species

Xu Zhang[1], *Zeyuan Chen*[2], *Weiwei Wang*[2], *Wenlu Yang*[2], *Jianquan Liu*[1,2], ***Tao Ma***[1,2]

1. State Key Laboratory of Grassland Agro-Ecosystem, College of Life Science, Lanzhou University

2. MOE Key Laboratory of Bio-Resources and Eco-Environment, College of Life Science, Sichuan University

All species of the genus *Populus* are dioecious plants, suggesting a common origin of this trait in their ancestral lineage. Although chromosome XIX was revealed as sex chromosome in all studied poplars, the location of sex determination region differs among species and both male and female heterogametic systems (XX/XY and ZZ/ZW) could be present, indicating that sex determination is highly dynamic in poplar. In the present study, we performed whole genome resequencing of 60 *P. alba* and 60 *P. euphratica* individuals of known sex and used a genome-wide association approach (GWAS) to determine the genomic regions associated with sex in these two species. We identified 114 SNPs located in scaffold 1,514 significantly associated with sex in *P. alba*. Gene syntenic analysis revealed that this sex-linked region were mapped to chromosome XIX in *P. trichocarpa* reference genome. Otherwise, we also identified one unbalanced large duplication in female individuals at this region. The segregation pattern of sex associated SNPs revealed this to be a ZW system in which the female was the heterogametic gender. In contrast, 102 SNPs located in 4 different scaffolds were identified to be associated with sex in *P. euphratica* and we found that the sex of this species was determined by an XY system. Unlike *P. alba*, these regions showed homologous relationships with chromosome I and XIV in *P. trichocarpa*. Results of this study demonstrated a high turnover rate of sex determination mechanisms in poplar and provided valuable information for better understanding the evolutionary processes of sex chromosome in dioecious plants.

T4-08-09

The identification and evolution of sex determination genes in *Asparagus*

Alex Harkess[1], *Kun Huang*[1], *Blake Meyers*[1], *Jim Leebens-Mack*[2]

1. Donald Danforth Plant Science Center

2. Donald Danforth Plant Science Center

3. Donald Danforth Plant Science Center

4. University of Georgia

An elegant model for the conversion of an autosomal pair in a hermaphroditic species to a sex chromosome in a dioecious species was formalized by Deborah and Brian Charlesworth in 1978. Briefly, the conversion from autosome to sex chromosome could require just two loci linked perfectly in non-recombination on a young Y chromosome: one locus must dominantly suppress female (pistil) organogenesis, while another locus must promote the formation of male (anther) organogenesis, but this model has never had strong genic support in any dioecious species. Garden asparagus (*Asparagus officinalis*) is an ideal system to test this question, given that the X and Y sex chromosomes are young and cytologically homomorphic. Leveraging a doubled haploid YY individual and a doubled haploid mapping population, we generated a chromosome-level genome assembly using Illumina, PacBio, and Bionano optical maps and identified a 1Mb region of non-recombination on the Y chromosome that is largely missing from the X. Several independent male-to-hermaphrodite mutants (gamma irradiation, spontaneous SNP) implicate a single gene *SOFF* (Suppressor of Female Function) on the Y chromosome as being responsible for female suppression. While only 12 gene annotations are in this non-recombining region, there exists *Tapetum Development and Function 1*, a gene with an *Arabidopsis* knockout phenotype very similar to Asparagus females. With an additional PacBio and optical map genome assembly for a sibling XX female, we have compared the structure of the X and Y chromosomes and identified ~130kb of X-specific sequence, including several X-specific genes. We are currently performing *in situ* hybridizations to test for the developmental timing and localization of gene expression of our candidate sex determination genes. This is the strongest evidence to date for the Charlesworth's classic "two gene" model.

T4-08-10

Efficient inference of the genetic basis of dioecy in Nepenthes pitcher plants

Mathias Scharmann, *Alex Widmer*

ETH Zürich

Dioecy has evolved multiple times independently in plants, which may facilitate comparative studies of its molecular basis and identification of ecological factors promoting the evolution of separate sexes. However, in the vast majority of plant groups, nothing is known about the genetic basis of dioecy. We present a method to efficiently characterize the genetic sex determination system in plants with homomorphic sex chromosomes without requiring breeding efforts. The method can easily be extended to develop sex-specific markers for application in ecology, conservation or agriculture. We have applied the new method to *Nepenthes* pitcher plants, a genus encompassing approximately 150 dioecious species. We genotyped wild populations of flowering *Nepenthes* from Southeast Asia and the Seychelles using ddRAD-sequencing and developed a novel empirical statistical procedure to robustly distinguish sex-limitation of genomic contigs from stochastic missingness, a major problem inherent to most methods that have to date been used to study the genetic basis of dioecy. We found consistent results in our separate analysis of three *Nepenthes* taxa. Comparison with results from *Silene latifolia*, a model species with heteromorphic sex chromosomes, indicates that the putative sex-specific region in *Nepenthes* is much smaller, as expected for homomorphic sex chromosomes. Characterization of expressed genes located in the sex-specific region reveals that sex chromosomes are shared across the *Nepenthes* radiation and may thus have a single origin.

T4-08-11

Genetic transformation and CRISPR/Cas9-mediated targeted mutagenesis in a dioecious plant model *Silene latifolia*

Vojtech Hudzieczek[1], *Radim Cegan*[1], *Tomas Cermak*[2], *Nela Bacovska*[1], *Daniel Voytas*[2], *Boris Vyskot*[1], *Roman Hobza*[1]
1. Institute of Biophysics, Czech Academy of Sciences, Brno
2. Department of Genetics, Cell Biology & Development and Center for Genome Engineering, University of Minnesota, Minneapolis, Minnesota

Silene latifolia is a dioecious plant model with heteromorphic sex chromosomes. It has been successfully used as a model in studying numerous biological questions including sex chromosome evolution, sex determination, sexual dimorphism, epigenetic processes in flower development, sexually transmitted diseases, invasive species biology *etc.* With the progress of massive parallel sequencing during the past decade, a number of candidate genes for important biological traits increased dramatically. However, a functional evaluation of these genes is difficult because of the lack of methods for gene delivery to *S. latifolia*. We have developed an efficient and reproducible protocol for genetic transformation and *in vitro* regeneration of S. *latifolia*. We demonstrated a successful T-DNA integration in *S. latifolia* genome using GUS assay and standard molecular analyses. Subsequently, we assessed the transgene segregation in T_1 generation and confirmed Mendelian manner of T-DNA transmission that suggests a single T-DNA integration event per T_0 plant. Furthermore, we employed the CRISPR/Cas9 nucleases to disrupt a coding sequence of a sex-linked, MADS-box gene *APETALA3* (*SlAP3*) that is involved in floral organ development. We recovered mutants in both X- and Y-linked alleles of the *SlAP3* gene. To our knowledge, this is the first case of genome editing on plant sex chromosomes.
The work was supported by Czech Science Foundation project No. 17-00567S.

T4-08-12

Different autosomes evolved into sex chromosomes in the sister genera of *Salix* and *Populus*

Jing Hou, Ning Ye, Defang Zhang, Yingnan Chen, Lecheng Fang, ***Tongming Yin***
Nanjing Forestry University

Willows and poplars are dioecious plants that originate from a common ancestor. Sex chromosome in poplar genome was consistently reported to be associated with chromosome XIX. In this study, chromosome XV was identified as sex chromosome in willow, and willow's sex determination was found to occur through a ZW system in which the female was the heterogametic gender. High collinearity was revealed between sex chromosomes and the corresponding autosomes in these two lineages. By contrast, no syntenic segments were found to share between poplar's and willow's sex chromosomes. We proposed that evolution of autosomes into sex chromosomes in *Salicaceae* occurred after the divergence of *Salix* and *Populus*. Syntenic analysis also revealed severe chromosome rearrangements between willow's Z and W chromosomes. Since sex chromosomes in *Salicaceae* are still at their early evolutionary stage, our findings provided valuable information for better understanding the genetics and evolution of sex chromosome in dioecious plants.

T4-09: Foliar vasculature: transport and communication hub

T4-09-01

Observing Real-Time Air Bubble Formation in the Leaves, Petioles and Stems of a Woody Plant

Madelinc R Carins Murphy, *Timothy Brodribb*
School of Biological Sciences, University of Tasmania, Hobart, Tasmania, Australia

The ease with which water flows through different tissues of the leaf to arrive at sites of evaporation near the stomata (defined as leaf hydraulic conductance; Kleaf) is a fundamental determinant of photosynthetic capacity (Brodribb et al. 2007). The magnitude of Kleaf and its interaction with external signals are constrained by the nature of the hydraulic pathway through the leaf xylem and mesophyll tissues (Sack and Holbrook 2006). Xylem embolism during dehydration has been strongly associated with declining Kleaf, and ultimately to tissue death. Understanding the spread of xylem embolism, and its timing relative to critical plant responses such as stomatal closure and tissue damage is fundamentally important. New non-invasive methods allow us for the first time to observe embolism formation in real-time and map where and when they occur in the plant with high resolution. We used the woody tree Magnolia grandiflora to examine embolism formation in the stem, leaf and petiole using a new optical technique developed by our lab group. The timing of embolism formation in these organs was compared with losses in plant hydraulic conductance in the laboratory and field.

Brodribb TJ, Feild TS, Jordan GJ (2007) Leaf maximum photosynthetic rate and venation are linked by hydraulics. Plant Physiology 144, 1890-1898.
Sack L, Holbrook NM (2006) Leaf hydraulics. Annual Review of Plant Biology 57, 361-381.

T4-09-02

TWI1, a transcription factor binding protein, regulates cell-to-cell movement of OSH15 to control bundle sheath and sclerenchyma development in *Oryza japonica* leaves

Xiao Han
Biotechnology Research Institute

Cell pattern formation in plant leaves has attracted much attention from plant biologists and breeders. However, the relevant molecular mechanism remains unclear. Here, we describe the isolation and functional characterization of *TWIST-LEAF1 (TWI1)*, a critical gene involved in development of the vascular bundle sheath and sclerenchyma in the rice leaf. Mutant *twi1* plants have twist leaves due to the compromised development and disordered patterning of bundle sheath and sclerenchyma cells. Expression of *TWI1* can functionally rescue these mutant phenotypes. *TWI1* encodes a transcription factor binding protein which interacts with OSH15, a class I KNOTTED1-like homeobox (KNOX) transcription factor. The cell-to-cell trafficking of OSH15 was shown to be restricted by its interaction with TWI1. During leaf development, widespread distribution of OSH15 impedes cell lignification, an essential step in sclerenchyma formation. Knock-out or knock-down *OSH15* in *twi1* could rescue the twisted leaf phenotype. These studies identified a key factor controlling cell pattern formation in rice leaves.

T4-09-03

Water supply and demand remain coordinated during breakdown of the global scaling relationship between leaf size and major vein density

Julio V. Schneider[1,2], *Jörg Habersetzer*[3], *Renate Rabenstein*[3], *Jens Wesenberg*[4], *Karsten Wesche*[4,5,6], *Georg Zizka*[1,2]

1. Senckenberg Research Institute and Natural History Museum Frankfurt, Department Botany and Molecular evolution

2. Goethe University Frankfurt/Main, Department Diversity and Evolution of Higher Plants

3. Senckenberg Research Institute and Natural History Museum Frankfurt, Department of Paleoanthropology and Messel Research

4. Senckenberg Museum of Natural History Görlitz, Department of Botany

5. German Centre for Integrative Biodiversity Research (iDiv) Halle-Jena-Leipzig

6. International Institute Zittau, Technische Universität Dresden

Vein networks that disobey the global scaling of major vein density with leaf size shed light on functional constraints of vein network formation in dicotyledons. Based on vein traits of 55 species of pantropical Ochnaceae, stomata and climatic niche data, and a dated molecular phylogeny, we unveiled major structural shifts in vein networks through deep time, relationships between leaf size, vein and stomatal traits. Our study shows that the breakdown of the vein-leaf size scaling is associated with vein networks that exhibit densely spaced 2° veins and reduced minor veins, and that this character combination evolved multiple times independently in Ochnaceae. In spite of the major shifts in vein architecture in this venation type, the coordination of vein and stomatal densities remains similar to that in species with normally spaced 2° veins and abundant minor veins.

T4-09-04

Daily dynamics of xylem sap properties and plant hydraulic traits in hybrid aspen

Annika Meitern, *Eele Õunapuu-Pikas*, *Arne Sellin*

Institute of Ecology and Earth Sciences, University of Tartu, Lai 40,51005 Tartu, Estonia

Plants have to cope with very different environmental conditions, which vary circadially enormously in most habitats, shaping daily patterns in physiological processes including long-distance water transport and xylem sap properties. Some of them are poorly documented in natural conditions; especially there is a lack of night-time measurements. The daily dynamics of physico-chemical properties of xylem sap and their covariation with tree hydraulic traits was investigated in hybrid aspen (*Populus tremula* L. × *P. tremuloides* Michx) in field conditions. K^+ concentration ($[K^+]$), electrical conductivity (S_{sap}), osmolality (Osm) and pH of the xylem sap, as well as branch hydraulic traits including whole-branch (K_{wb}), leaf blade (K_{lb}) and petiole hydraulic conductances (K_P) and bare branch hydraulic conductance (K_B), were measured in the field over 24-h cycles. The study aimed also to clarify which environmental drivers influencing the daily dynamics of these parameters. All studied xylem sap properties and hydraulic characteristics including K_{wb}, K_{lb} and K_P presented clear daily dynamics. Air temperature (T_A) and photosynthetic photon flux density (PPFD), but also water vapour pressure deficit (VPD) and relative humidity (RH), had significant impacts on K_{wb} K_{lb}, K_P, $[K^+]$, and s_{sap}. In the same time Osm varied only with light intensity (PPFD), while K_B was influenced by atmospheric evaporative demand expressed as T_A, VPD or RH. Xylem sap pH depended inversely on soil water potential (Ψ_S), and during daylight, also on VPD. Although soil water content was close to saturation during the study period, Ψ_S influenced also $[K^+]$ and S_{sap}. The enhancement of branch hydraulic conductance in the morning was associated with the upregulation of potassium ion concentration in xylem sap. The present study presents evidence of coupling between circadian patterns of xylem sap properties and plant hydraulic conductance providing adequate water supply to foliage under environmental conditions characterised by diurnal variation. A combination of different atmospheric variables generating AED rather than irradiance predominantly controls the circadian patterns of tree's hydraulic conductance and xylem sap ionic content in field conditions. The ionic content of xylem sap also affects plant hydraulic traits, but atmospheric variables mask this effect: the direct impact of environmental conditions on branch hydraulic efficiency outweighs that of xylem sap properties.

T4-09-05

Global analysis of the evolution of leaf venation networks in angiosperms

Christine Scoffoni[1], *Jeremy Beaulieu*[2], *Lawren Sack*[1]

1. University of California, Los Angeles

2. University of Arkansas

Understanding the mechanistic implications of functional trait variation and how this variation evolved through time is key to understanding species adaptation to their environments and how they will respond to global environmental changes. Leaf vein traits are proving especially important to understanding plant performance, environmental adaptation including drought tolerance and even the paleohistory of ecosystems. We quantified leaf morphological and venation characteristics for >5,000 angiosperm species distributed worldwide based on the cleared leaf collections of D. Axelrod and J. Wolfe, to describe the evolution of vein densities and vein architecture types across biomes including their relationship with life form, climate and evolutionary diversification.

T4-09-06

The forgotten side of leaf veins: Phloem hydraulic architecture of Poplar and *Ginkgo*

Monica Carvalho, *Robert Turgeon*, *Tom Owens*, *Karl Niklas*

School of Integrative Plant Science, Cornell University

Over the last decade, research in leaf vasculature has unveiled fundamental form-function properties that govern leaf hydraulics, consequently affecting traits such as drought tolerance and maximum rates for gas exchange among others. Most research deal-

ing with leaf hydraulics has however focused on water transport properties and the xylem, whereas the form-function relationships that affect sugar flow through the phloem continue to be heavily understudied. Part of the discrepancy between the phloem and the xylem research is due to the inherent difficulty in directly measuring flow through a living tissue such as the phloem, as well as the availability of analytical methods required to quantify pressure and sugar translocation *in vivo*. Biophysical inferences based on leaf phloem structure can provide basic principles for understanding phloem flow at the leaf level, including upper boundaries for sugar translocation rates, and xylem–phloem interconnectivity.
We describe and compare the leaf phloem hydraulic architecture of two topologically distinct model species: a species with dichotomous, open leaf venation (*Ginkgo biloba*), and a species with hierarchical, reticulate venation (*Populus x canescens*). Using a combination of electron and light microscopy, we compare the conductive areas of phloem and xylem, as well as sieve and tracheary element size across all vein classes observed in fully mature samples of both leaf types. We have found that the conductive area of phloem scales isometrically with that of xylem in both Ginkgo and poplar. Phloem conductance varies within each of these leaf types, as sieve element or sieve tube diameter and volume are largest in the major veins of poplar and the largest, basalmost veins of Ginkgo. Poplar leaf venation, as in most eudicots, is composed of a system of major and minor veins. The conductive area of minor veins is notably lower than that of the major veins, matching the functional distinction (sugar loading vs. translocation) of minor and major veins. At the whole leaf level, the number and length of minor veins drastically exceed the number of major veins, such that the total conductive area of minor veins to which sugar is loaded drastically exceeds that of major veins and the petiole. Contrarily, the phloem conductive area at the petiole level in Ginkgo leaves exceeds the total phloem conductive area of the collecting veins near the margin of the leaf. In this sense, the total phloem conductive area of Ginkgo tapers from the leaf base towards its margin, despite the characteristic fan-shape of the leaf blade. The application of first principles to our data shows different implications to phloem transport in light of Münch's pressure hypothesis. Using the phloem structural parameters previously presented, we will present a spatially-explicit model for phloem transport at the leaf level to estimate translocation and infer the xylem–phloem interconnectivity required for sugar export. The distinction between major and minor veins in angiosperms may provide faster phloem export rates when only considering structural differences between the studied species.

T4-10: Stem cell and root meristem regulation

T4-10-01

Ribosome biogenesis, cell cycle, and cell size – how an adenylate kinase and its natural genetic variants regulate the root meristem and root growth in *Arabidopsis*

Radka Slovak[1], ***Wolfgang Busch***[1,2]
1. Gregor Mendel Institute of Molecular Plant Biology
2. Salk Institute of Biological Studies

Root growth is dependent on cellular proliferation and expansion, both of which are determined in the apical meristems of roots. It follows that the tremendous variation in root growth between different species, and different strains of one species, originates from the root apical meristem. While many genes involved in stem cell maintenance, cellular proliferation, and differentiation are known, it is not clear whether genetic variation in these genes is responsible for the differences in root meristem activities across and within species. Indeed, many of the important regulators show little genetic variation between *Arabidopsis thaliana* accessions, and are therefore unlikely to be major determinants of natural variation in root growth. We used genome wide association mapping (GWAS) to identify novel regulators of cell proliferation and expansion, where natural genetic variation is responsible for variation in root growth rates between different *Arabidopsis* accessions. One of these regulators is a gene encoding for an adenylate kinase (AK) that regulates cell proliferation and cell size in the meristem. In contrast to many known meristem regulators, loss of this AK does not lead to correlated changes in meristem and mature cell size, but to an opposite effect - meristem size is reduced while mature cell size is increased. At the molecular level, this AK is involved in small ribosomal subunit maturation and ribosome biogenesis. Overall, we have uncovered an unknown regulatory mechanism in the root apical meristem that links ribosome biogenesis, cell cycle, and cell size control. Importantly, genetic variation in a key gene in this process is involved in determining root growth in natural accessions of *Arabidopsis*.

T4-10-02

Glycogen synthase kinases refine the specificity and output of auxin signaling in plant growth and development

Soeun Han[1], *Hyunwoo Cho*[1], *Thomas Greb*[2], ***Ildoo Hwang***[1]
1. Pohang University of Science and Technology
2. Heidelberg University

Peptide-initiated signaling networks play central roles in various aspects of plant growth and development. In particular, ligand TDIF (tracheary element differentiation inhibitor factor) and its receptor TDR (TDIF receptor) modulate lateral root development via a glycogen synthase kinase BIN2 (brassinosteriod insensitive 2) which directly controls the activity of ARF7/19. We also found that BIL1 (BIN2-like 1) acts as a negative regulator of the cambium activity. The loss-of-function *bil1* mutant shows increased thickness of stems and hypocotyl, but not *bin2* loss-of-function mutant, suggesting the regulatory specificity in GSK3-mediated vascular development. TDR directly interacts with BIL1 *in vivo*, and this interaction is sustained after TDIF treatment. In addition, BIL1 increases transcriptional activity of ARF5 which negatively regulates vascular cambial activity. TDIF/TDR suppresses BIL1-mediated ARF5 activation, suggesting that TDIF/TDR could maintain vascular cambial activity by partially inhibiting BIL1-ARF5 signaling module. Taken together, we propose that TDIF/TDR-GSK3 signaling cascade has a critical role to refine ARF activity in diverse plant developmental processes.

T4-10-03

Hemigraphis alternata as a novel model plant for understanding root morphogenesis

Krishna, *Ashalatha S Nair*
Department of Botany, University of Kerala

Developmental processes that underlie plant anatomy and architecture uses root as an excellent model. Its modular structure, lack of cell movement and relative accessibility to microscopic visualization facilitate research in a number of areas of plant biology. It is often difficult to visualize and analyze, the so-called 'hidden half' of plants and requires innovative approaches to understand its developmental processes and responses to external cues. The development of model plants to study these interactions will allow a better knowledge of those mechanisms. Computational plant models or 'virtual plants' are increasingly seen as a useful tool now a days, for comprehending complex relationships between gene function, plant physiology, plant development, and the resulting plant form. However, the mechanisms governing the interaction between various cell layers are yet not well understood, including those genes responsible for lateral rooting. In the present work, Hemigraphis alternata was used as a model to understand the developmental pathways of root regeneration. Hemigraphis alternata, commonly known as red ivy is a creeping, herbaceous, perennial groundcover growing to about 6 inches in height and can spread indefinitely on the ground with its creeping growth habit, rooting at the stem nodes along the way. It is native to India and Indonesia. In vitro culture systems for rooting were standardized using different plant growth regulators and were chosen for further studies. Nodal, internodal regions and leaf petioles were used as explants and inoculated on MS medium supplemented with auxins such as NAA, IAA, IBA and 2,4-D. Rooting observed from all the culture systems showing variations in root morphology, number and branching pattern. Roots developed from culture systems were compared with both acclimatized and ex vitro plant. Development of root from each system was studied at morphological, biochemical and genetic level. The present study also focused on the developmental aspects involved in establishing root anatomy and architecture.

T4-10-04

Auxin mediates Al stress signaling to control root growth through the regulation of local cytokinin biosynthesis in *Arabidopsis*

Zhaojun Ding
College of Life Sciences, Shandong University

Aluminum stress, which inhibits root growth, is one of the widespread environmental cues in acid soils. Our recent studies showed that Al stress induced a local auxin accumulation which is mediated through TAA1 and YUCs in root transition zone (TTZ). The high accumulated auxin in root TZ causes root-growth inhibition. Since it is well known that auxin acts synergistically with cytokinin to control the shoot stem-cell niche, while they act antagonistically to maintain root meristem. However, the role of cytokinin has not yet been addressed. Therefore, we investigated the role of CK in Al-induced inhibition of root growth, and the protential interaction between auxin and CK in this process. Our study showed that CK enhanced root-growth inhibition under stress through mediating Al-induced auxin signaling. Al stress induced a local up-regulation of CK response in the root-apex TZ that is dependent on *IPTs*, which encode adenosine phosphate isopentenyltransferases and regulate CK biosynthesis. *IPTs* are specifically up-regulated in the root-apex TZ in response to Al stress in an auxin-dependent manner, and AUXIN RESPONSE FACTORS, ARF7 and ARF19 play an important role in this process. In summary, different from the current general view in the root meristem, auxin acts with CK in a synergistic way to regulate aluminum-induced root growth inhibition in *Arabidopsis*.

T4-10-05

The SMC5/6 complex couples cell cycles checkpoint and DNA repair to maintain genome stability and stem cell identify

Shunping Yan, *Lili Wang, Hanchen Chen, Chongyang Wang*
Huazhong Agricultural University

Stem cells are hypersensitive to DNA damage. In response to DNA damage, cells activate checkpoints to arrest cell cycle progression and provide sufficient time to repair DNA. Previous studies have identified different sets of checkpoint and repair proteins. How cells coordinate these two processes is not well-understood. The evolutionarily conserved SMC5/6 complex plays multiple essential roles in DNA repair. Here we show that the SMC5/6 complex also regulates G1/S checkpoints by repressing E2F transcription factors. Mutation in SNI1, a subunit of the SMC5/6 complex, results in defective G1/S transition and eventually causes disorganized root meristem and short-root phenotypes, mimicking E2F overexpression. Consistently, mutations in E2Fs suppresses the *sni1* mutant and overexpression of SNI1 suppresses E2F overexpression plants. Remarkably, SNI1 directly binds the activation domain of E2Fs and represses their transcriptional activity. Our study demonstrates that the SMC5/6 complex is the key link to couple checkpoint and DNA repair to maintain genome stability and stem cell identity.

T4-10-06

Temperature-compensated cell production rate and elongation zone length in the root of *Arabidopsis thaliana*

Tobias Baskin[1], *Maura Zimmermann*[1], *Xiaoli Yang*[2], *Guohua Mi*[2]
1. University of Massachusetts
2. China Agricultural University

To understand how root growth responds to temperature, we used kinematic analysis to quantify division and expansion parameters in the root of *Arabidopsis thaliana*. Plants were grown at temperatures from 15 to 30°C, given continuously from germination. Over these temperatures, root length varies more than three-fold in the wild type but by only two-fold in a double mutant for *phytochrome-interacting factor 4 & 5*. For kinematics, the spatial profile of velocity was obtained with new software, Stripflow. We find that 30°C truncates the elongation zone and curtails cell production, responses that probably reflect the elicitation of a common pathway for handling severe stresses. Curiously, rates of cell division at all temperatures are closely correlated with rates of radial expansion. Between 15 to 25°C, root growth rate, maximal elemental elongation rate, and final cell length scale positively with temperature whereas the length of the meristem scales negatively. Non-linear temperature scaling characterizes meristem cell number, time to transit through either meristem or elongation zone, and average cell division rate. Surprisingly, the length of the elongation zone and the total rate of cell production are temperature invariant, constancies that have implications for our understanding of how the underlying cellular processes are integrated.

T4-11: Plant kinase and sugar signaling network

T4-11-01

Transcription factor ANAC060 represses *ABI5* to atteuate the sugar-abscisic acid pathway in *Arabidopsis*

Sheng Teng

Institute of Plant Physiology & Ecology, Shanghai Institute for Biological Sciences, Chinese Academy of Sciences

Arabidopsis seedling establishment is inhibited on media containing high levels (6%) of sugars. A sugar-ABA pathway has been illustrated to render this sugar sensitivity. Previously, the nucleus presented ANAC060 transcription factor was involved in the negative feed back loop in this sugar-abscisic acid pathway. In this study, microarray analysis was undertaken in wild type and *anac060* mutant under sugar (glucose or frucotse) treatment and sorbital control to identify the downstream genes of ANAC060. In wild type, expression levels of 2,122 and 2,509 genes were altered under glucose and fructose treatment respectively. In *anac060* mutant, up to 5,955 and 8,716 genes were regulated by glucose and fructose, respectively. Mapman analysis indicated that the genes involved in light reaction, light respiration, culvin cycle and tetrapyrrol synthesis were supressed by sugar treatment in wild type, and much stronger supression in these genes were found in *anac060* mutant. These results confirmed the attenuater role of *ANAC060* in sugar-ABA pathway on transcriptome level. Further more, ChIP-Seq analysis was undertaken using the line in which the functional ANAC060-GFP fusion protion was over expressed in *anac060* background to identified the targets of ANAC060. A total of 5,631 ANAC060 binding sites were identified. The percentages of the ANAC060 binding sites in premoter, exon, intron TES and intergenic sequence were 66%, 15%, 3%, 6% and 10% respectively. Most of the ANAC060 binding sites in promoter region were in the regions of 3 kb upstream of transcript start sites. The G-Box motif (CACGTG) was identified as ANAC060 binding motif. GO analysis indicated that the ANAC060 binded genes were enriched in chitin response, salt stress, membane target protein, plant hypersensitization response, calcium ion response, abscisic acid (ABA) activated signaling pathway, wouding response, ABA response, cold response, drought response, defense response, endoplasmic reticulum protein folding, programing cell death, jasmonic acid signaling pathway, salicylic acid signaling pathway and insect resistance response. These results indicated that ANC060 was involved in response of biostess and abiostess. Interestingly, *ABA INSENTIVE 5* (*ABI5*) which has been unrevealed to play an important role in ABA signaling and sugar-ABA pathway was identified as a target of ANAC060 by ChIP-Seq. The binding of ANAC060 to *ABI5* promoter was confirmed by ChIP-qPCR and in vitro EMAS assay. The expression of *ABI5* was induced by sugar treatment in wild type. Much stronger glucose upregulation of *ABI5* occurred in the *anac060* mutant. The *anac060 abi5* double mutant showed the sugar insensitive phenotype. These results indicated that ANAC060 atteuates the sugar-ABA pathway partially through repressing of ABI5.

T4-11-02

PRRs act as circadian hubs to regulate TOR pathway in *Arabidopsis*

Bin Li[1], *Yan Wang*[1], *Yuanyuan Zhang*[1], *Jyan-Chyun Jang*[2], ***Lei Wang***[1]

1. Key laboratory of Plant Molecular Physiology, Institute of Botany, Chinese Academy of Sciences, Beijing

2. Department of Horticulture and Crop Science, 4Department of Molecular Genetics, The Ohio State University, Columbus, OH 43210

Circadian clocks are not only the intricate time-keeping mechanisms, also regulating many aspects of plant growth and development, such as cell elongation of hypocotyl and stress response by synchronizing environmental cues. To date, it still largely unknown about regulation of cell proliferation by circadian clock in higher plants. The conserved eukaryotic TOR signaling can promote root cell proliferation and growth by gauging endogenous nutrient and energy status. Here we found Pseudo Response Regulators (PRRs), essential components of circadian core oscillator, redundantly modulates cell division activities in root meristems mediated via TOR signaling. PRRs appear to be positive regulators of TOR signaling, as their null mutants display the reduced sensitivities to reactivation TOR-signaling. We further demonstrated *Arabidopsis Tandem Zinc Finger 1* (*TZF1*), encoding a processing body and stress granule localized RNA-binding protein, is a direct downstream target of PRR proteins in mediating reactivation TOR signaling. PRRs differentially bind to *TZF1* promoter and repress its expression, in accodance with genetic analysis in which TZF1 acts as a negative regulator of TOR signaling. Reciprocally, blocking TOR signaling by either pharmacological approach or genetic knockdown of *TOR* by artificial microRNA significantly lengthens circadian period. Collectively, our studies support a notion that the PRR-TZF1-TOR interacting network plays a crucial role in root meristem activities and root architecture and the reciprocal intersection between circadian clock and TOR signaling is evolutionary conserved in plants.

T4-11-03

Mitogen-activated protein kinases and their downstream WRKY transcription factors in *Arabidopsis* pollen development

Yueping Zheng[1], *Xiangxiong Deng*[1], *Mengmeng Zhang*[1], *Aili Qu*[1], *Juan Xu*[1], ***Shuqun Zhang***[2,1]

1. Key Laboratory of Plant Physiology and Biochemistry, College of Life Sciences, Zhejiang University, Hangzhou 310058, China

2. Division of Biochemistry and Bond Life Sciences Center, University of Missouri, Columbia, MO 65211, USA

Mitogen-activated protein kinase (MAPK, or MPK) cascades are highly conserved signaling modules downstream of receptors/sensors that transduce extracellular stimuli into cellular responses in eukaryotes. *Arabidopsis* MPK3 and MPK6, two closely related MAPKs, are involved in signaling multiple developmental processes, including gametogenesis, embryogenesis, morphogenesis, senescence, abscission, fertilization, and seed formation. We are interested in not only the biological functions of these two MAPKs, but also how the signal specificity is maintained. It is possible that different upstream components such as the receptors and their ligands are involved in spatiotemporal activation of MPK3/MPK6, and/or different downstream targets such as the MAPK substrates are expressed in different cell types, which confers the signaling specificity of MAPKs. MAPK cascades are known to function as molecular switches in animal system. Plant MAPK cascades are likely to be the same and can carry out distinct functions depending on their upstream ligand–receptor interactions and downstream targets, both of which can be spatiotemporal specific during plant growth and development. Previously, we demonstrated that MPK3/MPK6 are critical to *Arabidopsis* oval development. In *mpk6 mpk3/+* plants, cell division of the integuments is hindered during oval development, resulting in a female sterile phenotype as a sporophytic effect. More recently, we found that MPK3/MPK6 also play essential roles in pollen development by regulating WRKY34 and WRKY2, two redundant WRKY transcription factors, in a spatiotemporal-specific manner. WRKY34 can be phosphorylated by MPK3 and MPK6 in both *in vitro* and *in vivo* assays. Phos-tag gel mobility shift assay revealed a spatiotemporal-dependent phosphorylation of WRKY34 by MPK3/MPK6 between bicellular and tricellullar stages during pollen development. *WRKY34* and *WRKY2* are highly expressed during pollen development, and mutation of both genes results in reduced pollen viability, pollen germination, and transmission efficiency. At the meeting, we will report that WRKY2 and WRKY34, and their upstream MPK3/MPK6 cascade, regulate carbon metabolism and storage lipid accumulation during pollen maturation, therefore, influence pollen fitness and plant reproduction.

T4-11-04

Mechanistic studies of light-enhanced translation in de-etiolating *Arabidopsis* seedlings

Guan-Hong Chen[1], *Ming-Jung Liu*[2], *Yan Xiong*[3], *Jen Sheen*[4], *Shu-Hsing Wu*[1]

1. Institute of Plant and Microbial Biology, Academia Sinica, Taiwan

2. Agricultural Biotechnology Research Center, Academia Sinica, Taiwan

3. Chinese Academy of Sciences, Shanghai, China

4. The Department of Genetics, Harvard Medical School, Boston, MA, USA

Photomorphogenesis, or de-etiolation, is a developmental process transforming young plant seedlings to vegetative phase with photosynthetic activities. In addition to transcriptomic adjustments, translation for thousands of genes was enhanced in de-etiolating *Arabidopsis* seedlings. However, the molecular mechanism and signal transduction underlying the translational enhancement remains unclear. Here we show the far-red (FR) light photoreceptor phytochrome A (phyA) and the E3 ubiquitin ligase, CONSTITUTIVE PHOTOMORPHOGENESIS 1 (COP1), positively and negatively regulate translation in photomorphogenic *Arabidopsis*, respectively. The survey of known translational regulators revealed that light could induce the activity of TOR (target of rapamycin) and followed by the phosphorylation of a key ribosomal protein small subunit 6, RPS6. Our results indicated that the phosphorylation of RPS6 depends on light signals but independent of photosynthesis activity or glucose. phyA mediates the phosphorylation of RPS6 under FR, whereas COP1 represses the phosphorylation of RPS6 and TOR activity in darkness. In addition to FR, blue light could also activate TOR-RPS6 pathway, a process requiring functional photoreceptors phyA, cry1 and cry2. Consistent with a delayed cotyledon opening phenotype observed in de-etiolating *tor* mutant, the RPS6 phosphorylation and the *de novo* protein synthesis were markedly reduced in *tor* mutant. The survey of light-regulated and TOR-dependent association of transcripts with polysome fraction revealed a higher representation of ribosome biogenesis genes. We uncovered a new signaling pathway involving phyA, cry1, cry2, COP1, TOR and RPS6 in regulating translation in de-etiolating *Arabidopsis* seedlings.

T4-11-05

Tandem CCCH zinc finger proteins in plant growth and stress response, mRNA decay, and beyond

Ailing Liu[1,2], *J-C Jang*[2]

1. College of Bioscience and Biotechnology, Hunan Agricultural University, Changsha, Hunan, China

2. Department of Horticulture and Crop Science, The Ohio State University, Columbus, Ohio, USA

Tandem CCCH zinc finger (TZF) proteins are evolutionarily conserved regulators of gene expression at the post-transcriptional level. TZFs target AU-rich RNA elements at 3′ un-translated region and recruit catabolic machineries to trigger mRNA degradation. The plant TZF families are over-represented by a class of proteins with a unique TZF domain preceded by an arginine-rich motif (RR-TZF). RR-TZF proteins are mainly involved in hormone-, sugar- and environmental cues-mediated plant growth and stress responses. Numerous reports have suggested that RR-TZF proteins control seed germination, plant biomass, flowering time, and biotic and abiotic stress responses via regulation of gene expression. Despite growing genetic evidence, the underlying molecular mechanisms are elusive. We have found that AtTZF1 binds specific RNA elements and trigger corresponding mRNA degradation. We have also found that RR-TZF proteins are dynamically localized in Processing bodies (P-bodies, PBs) and stress granules (SGs), two cytoplasmic messenger-ribonucleoprotein foci mediating mRNA decay, in response to various internal and external cues. Finally, we demonstrate that RR-TZF proteins preferentially and specifically interact with stress response mediators, particularly protein kinases involved in an array of stress signal transduction.

T4-11-06

Parent-of-origin effect seed mutants regulate endosperm development in maize

Fang Bai[1], *Mary Daliberti*[1], *Miaoyun Xu*[2], *Alyssa Bagadion*[1], *Yubing Li*[1], *Matthew Evans*[3], *W. Brad Barbazuk*[1], *A. Mark Settles*[1]

1. University of Florida

2. Chinese Academy of Agricultural Sciences

3. Carnegie Institution for Science, Stanford

Genomic imprinting in plants is an epigenetic phenomenon by which a subset of genes is expressed in a parent-of-origin-dependent manner. Imprinted gene expression primarily occurs in the endosperm and is thought to influence seed size and embryo development. Although many maize imprinted genes have been identified through transcriptome analysis, imprinted genes with developmental functions in the maize seed have not been identified. We screened rough endosperm mutants from the UniformMu transposon-tagging population for parent-of-origin effects using reciprocal crosses between heterozygous mutant plants and inbred lines. Six maternal rough endosperm (*mre*) and three paternal rough endosperm (*pre*) mutants were identified. When inherited from the female parent, *mre* seeds show a rough, etched, or pitted endosperm surface as well as a reduced seed size and weight. The *pre* mutants show the converse inheritance pattern with pre pollen conferring a seed phenotype after fertilizing wild-type ovules. Characterization of the *mre* and *pre* isolates shows a range of endosperm defects. *mre1* mutants have a general delay in endosperm development with smaller starchy endosperm cells, delayed basal endosperm transfer cell layer (BETL) development, and delayed accumulation of starch granules. The *mre2* mutant shows multiple starchy cell differentiation defects, while *mre3* mutants have an endosperm phenotype consistent with reduced sink strength. In addition to the defects in BETL and starch endosperm development, *pre1* mutant is deficiency in embryo development in consistent with the significant reduction in oil content. Our results showed that imprinting is important for endosperm development as well as the seed size diversity. Molecular mapping experiments have identified four loci: *mre1* is on chromosome 4, *mre2* is on chromosome 6, and both *pre1* and *mre3* are on chromosome 10. Additional mutant alleles obtained from the UniformMu reverse genetics resources appear to identify *mre1*, *mre3*, and *pre1* as imprinted genes in maize. RNAseq analysis from F1 crosses of the *mre1*, *mre2* and *mre3* is in progress and will identify imprinted genes that are differentially expressed in the maternal-effect mutants.

T4-12: Intracellular communication in higher plants (two sessions)

T4-12-01

Choreography of plastidial retrograde signaling network in interorganellar communication

Katayoon Dehesh
UC Riverside

Interorganellar communication is an evolutionary necessity for maintenance of cellular homoeostasis in response to prevailing environment that, in part, is exquisitely controlled via retrograde-signaling pathways. We have identified a novel stress-specific plastidial retrograde signalling metabolite, methylerythritol cyclodiphosphate (MEcPP), previously known solely as an intermediate in the isoprenoid biosynthetic pathway. The additional function of MEcPP as a stress sensor and a coordinator of transcriptional and post transcriptional regulation of key stress-responsive nuclear genes, has unraveled the central role of this metabolite in cellular functions in response to a wide range of environmental and developmental cues. To identify the underlying molecular mechanism of the MEcPP-mediated stress responses, we have performed a multi-omics approach. These studies have led to the identification of a transcriptional hub activated by MEcPP, and have further established a previously unrecognized link between this plastidial retrograde signal and the transcriptional reprogramming of endoplasmic reticulum genes critical for readjustment of protein-folding capacity in stressed cells. Lastly, we have gained an insight into the molecular mechanism by which MEcPP regulates plant growth and development in response to stress. In brief we have advanced our understanding of how MEcPP reprograms a repertoire of intricate networks crucial for coordinating the physiological and metabolic processes required for stress-induced developmental responses and stress adaptation. These studies offer an insight into functional module concept of biological organization and regulation by MEcPP, the plastidial retrograde signaling metabolite and a coordinator of interorganellar communication.

T4-12-02

Roles of GOLDEN2-LIKE 1 transcription factor in the regulation of plastid-to-nucleus retrograde signaling

Takehito Inaba[1], *Mitsuaki Tokumaru*[1], *Tomohiro Kakizaki*[2], *Yasuko Ito-Inaba*[1]
1. University of Miyazaki
2. Institute of Vegetable and Floriculture Science, NARO

Arabidopsis thaliana GOLDEN2-LIKE (AtGLK) transcription factors promote chloroplast development by regulating the expression of photosynthesis-related genes. AtGLK1 is also known to participate in plastid-to-nucleus retrograde signaling. The expression level of *AtGLK1* is tightly correlated with the expression of photosynthesis-related nuclear genes in *Arabidopsis* possessing damaged chloroplasts. Furthermore, the activity of plastid-localized GENOMES UNCOUPLED 1 (GUN1) appears to down-regulate the expression of *AtGLK1* when chloroplasts are dysfunctional. These data suggest that plastid signals repress the expression of photosynthesis-related nuclear genes through the repression of *AtGLK1* expression. In addition to transcriptional regulation of *AtGLK1*, our recent study have demonstrated that ubiquitin-proteasome dependent regulation of AtGLK1 participates in plastid-to-nucleus retrograde signaling. Norflurazon- or lincomycin-treated *gun1-101* mutant expressing normal levels of *AtGLK1* mRNA failed to accumulate AtGLK1 protein. This suggests that plastid signals directly regulate the accumulation of AtGLK1 protein in a GUN1-independent manner. Treatment of the *glk1glk2* mutant expressing functional GFP-AtGLK1 with a proteasome inhibitor, MG-132, induced the accumulation of polyubiquitinated GFP-AtGLK1. Furthermore, the level of endogenous AtGLK1 in plants with damaged chloroplasts was partially restored when those plants were treated with MG-132. Taken together, these data suggest that plastids have evolved multiple mechanisms to regulate the level of AtGLK1 in response to plastid signals.

T4-12-03

GUN1, a jack-of-all-trades in chloroplast protein homeostasis and signaling

Paolo Pesaresi
University of Milano

The *GUN1* (*GENOMES UNCOUPLED 1*) gene has been reported to encode a chloroplast-localized pentatricopeptide-repeat protein, which acts to integrate multiple indicators of plastid developmental stage and altered plastid function, as part of chloroplast-to-nucleus retrograde communication. However, the molecular mechanisms underlying signal integration by GUN1 have remained elusive, up until the recent identification of a set of GUN1-interacting proteins, by co-immunoprecipitation and mass-spectrometric analyses, as well as protein-protein interaction assays. Here, we review the molecular functions of the different GUN1 partners and propose a major role for GUN1 as coordinator of chloroplast translation, protein import and protein degradation. This regulatory role is implemented through proteins that, in most cases, are part of multimeric protein complexes and whose precise functions vary depending on their association states. Within this framework, GUN1 may act as a platform to promote specific functions by bringing the interacting enzymes into close proximity with their substrates, or may inhibit processes by sequestering particular pools of specific interactors. Furthermore, the interactions of GUN1 with enzymes of the tetrapyrrole biosynthesis pathway support the involvement of tetrapyrroles as signaling molecules in retrograde communication.

T4-12-04

Identification of novel auxiliary factors involved in the post-translational regulation of tetrapyrrole biosynthesis in higher plants

Peng Wang, *Andreas Richter, Daniel Wittmann, Boris Hedtke, Bernhard Grimm*

Department of Plant Physiology, Institute of Biology, Humboldt-Universität zu Berlin, 10115 Berlin, Germany

Tetrapyrrole biosynthesis (TBS) is a multiple enzymatic pathway, is exclusively located in plastids, and provides vital cofactors such as chlorophyll (Chl) and heme for a wide range of essential cellular processes. While the biochemical properties of tetrapyrroles and almost all genes encoding enzymes in the TBS pathway have been intensively characterized in higher plants, the post-translational regulation and spatiotemporal organization of the tetrapyrrole metabolism remain to be clarified. To ensure adequate synthesis of tetrapyrrole end-products at any time of developmental and environmental conditions, a wide range of auxiliary factors such as chaperones, scaffold proteins, and proteases have been characterized in tetrapyrrole metabolism. In this study, we aimed to identify a group of novel regulators involved in the post-translational control of TBS pathway in higher plants. To this purpose, the bioinformatics analysis, such as gene coexpression analysis, reverse genetic approach, and multiple biochemical approaches were applied for the identification of the potential candidates. Among them, two types of proteins, chloroplast signal recognition particle 43-kDa protein (cpSRP43) and CAAX-type Endopeptidase-like Protein1 (CPL1), participate in the control of different enzymatic steps of TBS pathway and likely in plastid-to-nucleus retrograde signaling. Firstly, cpSRP43 is well known as an essential component of the post-translational cpSRP pathway, by which the light-harvesting chlorophyll-binding proteins (LHCPs) are transferred from the stroma to thylakoid membranes. Here, we provide evidence supporting the function of cpSRP43 as a chaperone in the regulation of the rate-limiting step of tetrapyrrole metabolism, 5-aminolevolinic acid (ALA) synthesis. Co-fractionation analysis and protein-protein interaction assays suggest that cpSRP43 directly interacts with the first enzyme of ALA synthesis, glutamyl-tRNA reductase (GluTR), and its regulator GluTR-binding protein (GBP). The cpSRP43 deficiency led to reduced GluTR content without altered transcript levels of *HEMA1*, the GluTR encoding gene. Furthermore, cpSRP43, as well as GBP interact with the N-terminal domain of GluTR, which is also the target site at GluTR for the Clp protease. Our data provide a first link between the cpSRP43 chaperone function in the TBS. Secondly, we demonstrated that the previously uncharacterized chloroplastic transmembrane protein CPL1 is required for the activation of the chlorophyll-synthesizing branch of TBS. The *cpl1* null mutants display a significantly reduced ALA synthesis rate, and dramatically reduced accumulation of Chl and its precursors, but not of heme. Moreover, the CPL1 mutation leads to a reduced GluTR and an increased GUN4 (genome uncoupled 4) level. The direct regulation of GluTR and GUN4 was supported by their CPL interaction. In summary, our data suggest that CPL1 serves as a scaffold protein in the thylakoid membrane, and functions in the activation of both ALA synthesis and magnesium chelatase in the Chl branch of TBS.

T4-12-05

Tetrapyrrole-dependent retrograde signalling pathways that function during chloroplast development

Matthew J Terry[1], *Mike T Page*[1], *Sylwia M Kacprzak*[1], *Haruko Okamoto*[1], *Alison G Smith*[2]

1. Biological Sciences, University of Southampton

2. Department of Plant Sciences, University of Cambridge

The development of chloroplasts during de-etiolation is a critical time for a plant seedling as it races towards a photoautotrophic existence in the light. Retrograde signalling between the developing chloroplasts and the nucleus is important during this time to ensure an appropriate supply of chloroplast proteins. We have been investigating two tetrapyrrole-dependent pathways that have been proposed to function during chloroplast biogenesis: a positive, promotive pathway related to heme and mediated by ferrochelatase 1 (FC1) activity and an inhibitory pathway that is dependent on Mg-porphyrin synthesis and mediated by singlet oxygen. To examine the role of a heme-related signal we have constructed a series of transgenic *Arabidopsis* lines in which we alter heme pools through compartment-specific expression of FCs and heme oxygenase (HO1). The results demonstrate that overexpression of *FC1* in the chloroplast results in rescue of nuclear gene expression, but that overexpression in the mitochondrion does not. Analysis of multiple lines with altered FC1 expression showed a strong correlation between changes in a regulatory heme pool and nuclear gene expression supporting a role for FC1-dependent heme synthesis in retrograde signalling. Pharmacological experiments using inhibitors of the tetrapyrroles biosynthesis pathway are also consistent with this interpretation. The positive retrograde signal increases expression of tetrapyrrole genes including *HEMA1* encoding the rate-limiting step for tetrapyrrole synthesis and the Mg-chelatase genes *CHLH* and *GUN4*. What therefore prevents such a signal causing an over accumulation of chlorophyll precursors such as protochlorophyllide? Accumulation of excess protochlorophyllide after transfer from far-red (FR) to white light, after feeding of the tetrapyrrole precursor aminolevulinic acid, or in the *flu* mutant of *Arabidopsis* results in the production of singlet oxygen and rapid and strong inhibition of nuclear photosynthetic genes, and especially tetrapyrrole biosynthesis genes. This inhibitory signal would therefore serve to function as a safety valve to prevent overproduction of photo-toxic tetrapyrroles. We are now identifying mutants in the inhibitory pathway. We have already shown that the singlet oxygen signalling mutants *executer1* and *executer2* show partial rescue of photosynthetic gene expression. In addition, we have undertaken a genetic screen for mutants that show elevated expression of *HEMA1* after a FR pre-treatment. Progress on characterization of these *safe after far-red light* (*saf*) mutants will be reported.

T4-12-06
Intracellular communication of abiotic stress, memory and recovery

Barry Pogson
Australian National University

Abiotic stress such as excess-light and drought cause significant crop losses by reducing photosynthetic efficiency and preventing yield potentials to be realized, conversely, slow recovery from, or constitutive acclimation to, stress can impair growth and yield. Multiple components of excess-light and drought stress induction pathways have been identified. However, we have very little understanding of the reverse processes that are required to signal recovery and for resetting the transcriptome. Systems biology approaches provide valuable tools for assessing these competing interests for energy allocation. Here, several examples of the application of systems biology to understanding stress signaling, recovery and memory will be presented with a focus on gene expression profiling. We are employing large scale biology approaches to investigate the processes and mechanisms enabling both stress tolerance and recovery from stress, including transcriptome, degradome and methylome profiling together with predicting retrograde stress-signal interactomes.

T4-12-07
Mitochondrial biogenesis and signalling

Yan Wang, *Oliver Berkowitz, James Whelan*
La Trobe University

It is widely accepted that in addition to the extensive biochemical metabolic functions of mitochondria, they play an essential role as a cellular signalling hub. Mitochondria act as a sensor to environmental and cellular signals to trigger signal transduction pathways to modify nuclear gene expression. Much of the investigation in this area has focused on stress-induced components of mitochondria, e.g. alternative oxidase (AOX), and the identification of the signalling pathways and components that are involved in induction of AOX under a variety of different conditions. However, in addition to responding to adverse conditions it is hypothesized that mitochondrial signalling is required during normal growth and development. We hypothesize that mitochondrial function and mitochondrial biogenesis must be linked through normal development, to ensure that any change in mitochondrial signalling can be executed. To investigate this we have investigated the changes that occur in mitochondrial function when biogenesis is altered, and vice versa, changes that occur in biogenesis when mitochondrial function is altered. Using knock-out mutants of mitochondrial complex I as an example of altering mitochondrial function, i.e. respiration, and knock-out mutants of the TIM17:23 complex as altering biogenesis, we have uncovered a previously unknown link between protein complex involved in protein import (mitochondrial biogenesis) and respiration (complex I – respiration). Further studies indicate that mitochondrial complex I is a signalling hub in plants and interacts with a variety of signalling protein in the mitochondrial inter-membrane space. Alterations of these proteins not only impacts mitochondrial function, but chloroplast and cellular protein abundance. Results will be presented to show that a variety of proteins interact with complex I and that upon perturbation signalling cascades are initiated to alter total cellular function. The alteration of cellular function occurs at both a transcriptional and post-transcriptional level. Overall these results show that altering mitochondrial function directly impacts the protein complement and function throughout the cell.

T4-12-08
Glycosylphosphatidylinositol-anchored proteins (GPI-APs) as chaperones and co-receptors for FERONIA receptor kinase signaling in *Arabidopsis*

Chao Li[1], *Fang-Ling Yeh*[2], *Alice Cheung*[2], *Qiahong Duan*[2], *Ming-Che Liu*[2], *Hen-Ming Wu*[2]
1. School of Life Sciences, East China Normal University
2. Department of Biochemistry and Molecular Biology, University of Massachusetts

The *Arabidopsis* receptor kinase FERONIA (FER) has been characterized as a critical multifunctional regulator in many aspects of plant growth and development, including hormone signaling, reproduction, and defense. FER perceives the peptide regulator Rapid Alkalinization Factor 1 (RALF1) at the plasma membrane and activate the downstream signaling. LORELEI (LRE), a female reproductive tissue-specific glycosylphosphatidylinositol (GPI)-anchored protein (AP), has been reported to have similar function as FER in mediating pollen tube reception in the ovule. In this study, we show that mutants in vegetative tissue-expressed LRE-like GPI-anchored protein1 (LLG1), *llg1*, resemble *fer* defects in vegetative tissues. Interestingly, FER relies on LLG1 for efficient localization to the cell membrane. Further investigation showed that LLG1/LRE interacts with FER in the endoplasmic reticulum and on the cell surface where LLG1/LRE and FER associate with the RHO GTPase signaling apparatus to mediate multiple downstream processes. Loss of LLG1 function induces cytoplasmic retention of FER, consistent with transport of FER from the endoplasmic reticulum to the plasma membrane in a complex with LLG1. Moreover, we determined that LLG1 binds to the extracellular juxtamembrane region of FER, lacking of which caused retention of FER in the endoplasmic reticulum. Our results support LLG1/LRE acting as a chaperone and co-receptor for FER. These findings elucidate a novel mechanism that underlies FER function and new insights on how GPI-APs might enable the signaling capacity of a cell surface receptor.

T4-12-09

Interaction between brassinosteriod and glucose signaling in regulation of seed germination in *Arabidopsis thaliana*

Chunmei Zhong[1,3,4], Barunava Patra[2], Yi Tang[1], Xukun Li[1], Stitakanta Pattanaik[2], Ling Yuan[2], Xiaojing Wang[1]

1. Guangdong Provincial Key Laboratory of Biotechnology for Plant Development, School of Life Science, South China Normal University, Guangzhou

2. Department of Plant and Soil Sciences, University of Kentucky; Lexington

3. College of Forestry and Landscape Architecture, South China Agriculture University, Guangzhou

4. Key Laboratory of Energy Plants Resource and Utilization, Ministry of Agriculture

Seed germination is a critical biological process that is regulated by multiple phytohormones, as well as metabolic and environmental signals. Brassinosteriods (BRs) and glucose (Glc) are known to regulate developmental process including lateral root growth, shoot gravitropism, and hypocotyl elongation. However, little is known about the involvement of BRs in seed germination. Here, we demonstrate that BR induces seed germination through promoting testa and endosperm rupture. The positive effect of BR is through cell elongation at the hypocotyl-radicle transition region of hypocotyl during endosperm rupture, rather than cell division. Glc appreciably inhibits testa rupture and strongly inhibits endosperm rupture. Moreover, we found that BRs partially counteract the Glc-induced inhibition of cell elongation during endosperm rupture. The role of BR in Glc-induced inhibition of cell elongation is apparently independent of the hexokinase function. We also identified that two key transcription factors (TFs) in BR signaling, BR ENHANCED EXPRESSION2 (BEE2) and HOMOLOG OF BEE2 INTERACTING WITH IBH1 (HBI1), are involved in the regulation of endosperm rupture by direct binding to E-box elements of the *GA-Stimulated Arabidopsis 6* (*AtGASA6*) promoter. Our findings reveal that BRs utilize different signaling cascades to partially reverse the Glc-induced inhibition of seed germination, highlighting the complexity of BR response in regulation of seed germination.

T4-12-10

A novel chloroplast translocase, Sec2, is involved in protein targeting to the chloroplast inner envelope membrane and chloroplast biogenesis

Yubing Li[1], Donna Fernadez[2], Kenneth Cline[1]

1. Horticultural Sciences Department and Plant Molecular and Cellular Biology, University of Florida

2. Department of Botany, University of Wisconsin-Madison

Chloroplasts are essential organelles of plant and green algae. In plant cells, a chloroplast contains about 3,000 proteins and most of them are synthesized in the cytosol and post-translationally imported into chloroplasts. Correctly target these proteins to different sub-compartments within chloroplasts is essential for chloroplast biogenesis and the plant development. After import into the chloroplast, many proteins are further targeted to the thylakoid membrane and the thylakoid lumen by three thylakoid-localized translocases, the Sec1, Tat and SRP/Alb3. Others are targeted to the inner chloroplast envelope membrane through at least two pathways, the "stop-transfer" pathway and the post-import integration pathway. However, the mechanisms of these two pathways are still unknown. Recently, we identified a second Sec translocase system (Sec2) consisting of three protein components: SCY2 (forming the protein-conducting channel), SECE2 (SCY2 binding protein), and SECA2 (ATP-dependent motor protein). SCY2 and SECE2 are integral membrane proteins and localized mainly in the inner envelope membrane where they form a stable complex, while SECA2 is a soluble protein and localized in the stroma. Null mutants of any component in Sec2 in *Arabidopsis* exhibit severe embryo lethal phenotype. To investigate the function of Sec2 system in plants, we used inducible RNAi to knockdown *SCY2* in *Arabidopsis*. Seedlings cultured with inducer were chlorotic with aberrant chloroplasts and undeveloped thylakoids, indicating an essential role for SCY2 in chloroplast biogenesis beyond embryo development. In *SCY2* knockdown seedlings, several thylakoid membrane proteins including SCY1, ALB3 and TATC, were substantially reduced, as well as some inner envelope membrane proteins TIC40, TIC110 and FTSH12, which are through the post-import integration pathway for their targeting. Our data suggested that those down-regulated membrane proteins might be Sec2 substrates. Furthermore, *in vitro* reconstitution of protein integration into chloroplast membranes showed that both SCY1 and ALB3 target directly to the thylakoid membrane and are likely independent of Sec2. FTSH12 was integrated into the envelope membrane in a coupled import/integration reaction that was impaired by the SECA inhibitor sodium azide. The stromal intermediate of TIC40 integrated into the envelope in a reaction that was largely inhibited when antibodies against epitope-tagged SCY2 or SECE2 were applied. These data demonstrate that the Sec2 translocase likely integrates a subset of inner envelope membrane proteins through the post-import integration pathway. More detailed functional studies on Sec2 system are in progress. Our work shed new light on the current understanding of membrane protein trafficking within chloroplasts, which play essential roles in chloroplast biogenesis and plant development.

T4-12-11

ER-localized SNAREs are involved in transport of seed storage proteins

Lixn Li, Xiaonan Zhao, Hailong Zhang, Yun Kong, Xiufen Guo

Northeast Forestry University

Plant seeds accumulate large amount of storage proteins as N resorce for plant inicial growth. Storage proteins are synthesized as large precursors on the ER. The precursors are delivered to the vacuole by vesicle transport and processed to smaller mature forms. In our previous study, we found that ER-localized tethers, MAG2, MIP1, MIP2 and MIP3, form a tethering complex-MAG2 compplex, function on ER-Golgi vesicle transport in *Arabidopsis*. When they are defective, transport of newly synthesized proteins from the ER are affected. To further clarify mechanism underlying MAG2-relating protein transport pathway, we investigated function of predicted ER-localized SNAREs, including Qa-SNARE, Qb-SNARE and Qc-SNAREs. Our results indicated that these ER-localized SNAREs interact with each other to form a SNARE complex. This SNARE complex interacts with MAG2 complex.

When these SNAREs are defective, efficiency of transport of seed storage protein precursors from the ER is decreased. On the other hand, plant growth and development is also affected. Our findings indicate that these ER-Localized SNAREs collaborate with MAG2 complex to regulate ER-Golgi vesicle transport.

T4-12-12

AP-3-dependent vacuolar trafficking of *Arabidopsis* PAT10 involves a subpopulation of COPII and the vacuolar tethering complex HOPS

Qiang-Nan Feng, *Shi-Jian Song, Shi-Xia Yu, Sha Li, Yan Zhang*
Shandong Agricultural University

Plant vacuoles are versatile organelles critical for plant growth and responses to environment. Vacuolar proteins are transported from the endoplasmic reticulum (ER) via multiple routes in plants. Two classic routes bear great similarity to other phyla with major regulators known, such as COPII and Rab5 GTPases. By contrast, vacuolar trafficking mediated by adaptor protein-3 (AP-3) or that independent of Golgi has few recognized cargos and none of the regulators. In search of novel regulators for vacuolar trafficking routes and by using a fluorescence-based forward genetic screen, we demonstrated that the multi-span transmembrane protein, *Arabidopsis* PROTEINS-ACYL TRANSFERASE 10 (PAT10), is an AP-3-mediated vacuolar cargo. We show that the tonoplast targeting of PAT10 is mediated by AP-3 complex but independent of Rab5-mediated post-Golgi trafficking route. We also report that AP-3-mediated vacuolar trafficking involves a subpopulation of COPII and requires the vacuolar tethering complex HOPS. In addition, we have identified two novel mutant alleles of AP-3δ, whose point mutations interfered with the formation of the AP-3 complex as well as its membrane targeting. Results presented here shed new lights on the vacuolar trafficking route mediated by AP-3 in plant cells.

T4-13: Programmed cell death and senescence

T4-13-01

ORE1, an aging regulator feedbacks into the circadian clock and exerts age-dependent doppler effect on the circadian system to sustain rhythmicity in old plants

Hyunmin Kim[1,2], Sukjoon Jung[2,3], Hyo Jung Kim[2], Souk Seo[2], C. Robertson Mcclung[4], ***Sunghyun Hong[2]****, Hong Gil Nam[2,3]*
1. Department of Life Sciences, POSTECH
2. Center for Plant Aging Research, Institute for Basic Science
3. Department of New Biology, DGIST
4. Department of Biological Sciences, Dartmouth College

Aging and the following death are the inevitable fates of most organisms. In plants, aging is associated with age-dependent senescence and death of various organs such as leaf organ. The process of age-dependent senescence of plant leaf organs is critical for plants' fitness and productivity, which is representatively observed in leaf senescence. The physiological effect of aging encompasses the circadian clock system. In animals, aging is associated with disruption of many circadian rhythm, such as change of circadian period, delayed recovery from jet-lag, and reduced amplitude of clock genes. Conversely, the circadian system affects the aging processes. Disruption of circadian rhythms hastens aging process in several animal systems. In plants, the interaction of circadian clock and aging has been rarely explored despite of the importance of the two processes in plant physiology. However, it was recently noticed that plants appear to robustly maintain circadian rhythmicity even upon aging. Here, we show that ORE1, a pro-aging factor is regulated by circadian clock and in turn positively feedbacks into the circadian system, leading to robust circadian rhythmicity in aged plants. We found, in *Arabidopsis*, that most of core circadian components affect aging. Consistently, *ORE1* expression was transcriptionally activated by PRR9, a clock component and posttranscriptionally repressed by clock-controlled miR164. Reciprocally, ORE1 activated transcription of two clock genes, TOC1 and GI, and mediated age-dependent "jet-lag" response and circadian period shortening. ORE1 also stabilizes TOC1 through direct protein-protein interaction. ORE1 controls the age-dependent period shortening, or a temporal Doppler effect through TOC1. We propose that in plants, a cyclic positive feedback loop (PRR9→ORE1→TOC1) between circadian clock components (PRR9, TOC1) and an aging regulator (ORE1) modifies circadian rhythmicity and controls aging.

T4-13-02

Roles of *miR319* and TCP transcription factors in plant growth and senescence

Tomotsugu Koyama
Suntory Foundation for Life Sciences

Leaf development includes the differentiation of leaves from the flank of a shoot meristem and the growth and senescence of leaves. The underlying mechanism of leaf differentiation, growth and senescence consists of many pairs of microRNAs (miRNAs) and their target genes. It has been reported that several miRNAs including *miR164* and *miR319* affect broad ranges of leaf differentiation, growth and senescence, while others affect specific aspects of leaves. Despite the fact that *miR164* and *miR319* respectively target NAC and TCP transcription factor genes, these miRNAs function in common regulatory programs, namely, the specification of the boundary between leaves and the shoot meristem, formation of leaf serrations and onset of leaf senescence. This implies that *miR164*, *miR319* and their target genes comprise a regulatory network for the regulation of leaf differentiation, growth and senescence. The role of the *miR319*-TCP pair in leaf development is obvious, because the downregulation of *TCP* genes in *Arabidopsis thaliana* induce jagged and wavy leaves as well as delayed senescence whereas ectopic expression of the *miR319*-resistant *TCP* genes inhibit the formation of serration in the leaf margin and stimulate precocious senescence. This versatility of the *miR319*-TCP pair in leaf development largely rely on the TCP's ability of regulation of various developmental regulators and senescence-stimulating hormone, jasmonate. Particularly, TCP transcription factors directly regulate *miR164A*, one of three *miR164*-precursor genes, and therefore TCP transcription factors act as master regulators of miR164 and the NAC-related CUP-SHAPED COTYLEDON genes. To obtain a detailed view on the role of the *miR319*-TCP pair, two of three precursor genes for *miR319* were mutated in *A. thaliana*. This *miR319* double mutant inhibited the formation of serration in the leaf margin, which is similar to the plants expressing the *miR319*-resistant *TCP* genes, and thus the mutant gained the functions of *TCP* genes. In this presentation, I will talk about our recent advance in the genetic link between the *miR319*-TCP pair and the regulatory mechanism of leaf differentiation, growth and senescence.

T4-13-03

Hormone crosstalk during rose petal senescence

*Chao Ma, Nan Ma, Changqing Zhang, Bo Hong, **Junping Gao***
Department of Ornamental Horticulture, China Agricultural University

Rose (Rosa hybrid) is one of the most important ornamental plants in the global flower trade. Roses have enriched flower shapes and colors, and attractive fragrance. Trade share of cut rose accounts for 32% of total cut flower trade worldwide. With the globalization of cut flower trade, production of cut rose shows a new trend of development. Long-distance transportation often causes more than 30% value loss of cut rose. Our research aims to understand the causes of quality deterioration and molecular mechanism in terms of influence of flower opening and senescence in cut rose. We firstly established small RNA and RNA transcriptome databases during rose flower senescence, and obtained the ethylene-responsive miRNA and mRNA profiling in rose petals. We then identified an ethylene-responsive *miRNA164*/NAC100 module and found a gene network of cell expansion in response to ethylene in rose petals according to integral analysis of expression profiles of miRNAs and their predicted targets. Intriguingly, our work also found that subfamily I of homeodomain-leucine zipper transcription factor contributes to rose petal senescence. Among them, RhHB1 mediates the antagonistic effect of GAs on ABA and ethylene during rose petal senescence. However, RhHB6 functions as a brake of ethylene promoted senescence through increasing the CTK content. Our results suggest that homeodomain-leucine zipper I transcription factors involve in regulation of petal senescence through hormone crosstalk in rose flower.

T4-13-04

Hormonal regulation of leaf senescence in *Arabidopsis*

Hongwei Guo
South University of Science and Technology (SUSTech), Shenzhen

Leaf senescence or ageing is an intricate and highly orchestrated developmental process that is subject to regulations by numerous environmental and endogenous signals, including plant hormones. Ethylene is known to be an endogenous modulator of senescence, including fruit ripening and flower and leaf senescence. However, the underlying molecular mechanism of ethylene-induced leaf senescence starts to be elucidated until recently. We have reported that EIN3, a key transcription factor in ethylene signaling, is a functional senescence-associated gene (SAG). Loss of EIN3 and its homolog EIL1 function leads to a delay in both age-dependent and ethylene-, JA- or dark-induced leaf senescence. We further define a novel signaling pathway involving EIN2-EIN3-miR164-NAC2 in the regulation of leaf senescence, and provides a mechanistic insight into how ethylene promotes the progression of leaf senescence in *Arabidopsis*. It has long been known that leaf senescence is a progressive natural degenerative process, which can be delayed or accelerated, but not prevented. To understand the progressive feature of leaf senescence at the molecular level, we have screened and collected *Arabidopsis* plants with altered senescence phenotype, from which a positive regulator of leaf senescence, WRKY75, was identified. We found that *WRKY75* transcription was induced by age, ethylene, jasmonic acid (JA), abscisic acid (ABA), salicylic acid (SA), and H_2O_2 treatment. Further studies showed that WRKY75 acted to enhance SA biosynthesis by inducing *SID2* transcription, and enhance tolerance to oxidative stress due to directly repressing *CAT2* transcription. Based on these results, together with the previously observation that SA and H_2O_2 induce each other, WRKY75, SA and H_2O_2 form a tripartite positive feedback loop, which is defined as an aging accelerator that speeds up gradually and irreversibly to accelerate leaf senescence. Further implication of this feedback amplification loop in integrating multiple hormonal signals will be discussed.

T4-13-05

Proteomes and ubiquitylomes analysis reveals the involvement of ubiquitination in protein degradation in petunias

*Jianhang Guo, Juanxu Liu, Qian Wei, **Yixun Yu***
South China Agricultural University

Petal senescence is a complex programmed process. It has been previously demonstrated that treatment with ethylene can extensively alter transcriptome and proteome profiles in plants. However, little is known regarding the impact of ethylene on post-translational modification (PTM) or the association between PTM and the proteome. Protein degradation is one of the hallmarks of senescence, and ubiquitination, a major PTM in eukaryotes, plays important roles in protein degradation. Here, we first obtained reference petunia transcriptome data via RNA sequencing. Next, we quantitatively investigated the petunia proteome, ubiquitylome, and the association between them in petunia corollas following ethylene treatment. In total, 51,799 unigenes, 3,606 proteins, and 2,270 ubiquitination sites were quantified 16 hours after ethylene treatment. Treatment with ethylene resulted in 14,448 down-regulated and 6,303 up-regulated unigenes (absolute log2-fold change > 1 and FDR < 0.001), 284 down-regulated and 233 up-regulated proteins, and 320 up-regulated and 127 down-regulated ubiquitination sites using a 1.5-fold threshold ($P < 0.05$), indicating that global ubiquitination levels increase during ethylene-mediated corolla senescence in petunia. Several putative ubiquitin ligases were up-regulated at the protein and transcription levels. Our results showed that the global proteome and ubiquitylome were negatively correlated, that ubiquitination could be involved in the degradation of proteins during ethylene-mediated corolla senescence in petunias, and that ethylene regulates hormone signaling transduction pathways at both the protein and ubiquitination levels in petunia corollas.

T4-13-06

Molecular mechanisms of DNA damage- induced leaf senescence in *Arabidopsis*

Zhong Hai Li[1]*, Jin Hee Kim*[1]*, Hye Ryun Woo*[1,2]*, Hong Gil Nam*[1,2]
1. Center for Plant Aging Research, Institute for Basic Science (IBS), Daegu, 42988, Republic of Korea.
2. Department of New Biology, DGIST, Daegu, 42988, Republic of

Korea.

Environmental and endogenous sources constantly trigger DNA damage, which has long been suspected to be the cause of aging. However, knowledge about the underlying molecular mechanisms of DNA damage-induced leaf senescence is still lacking. Here, we show that the levels of DNA double strand breaks (DSBs), one of the most cytotoxic forms of DNA damage, increase as leaf aging; by contrast, DNA damage repair efficiency decreases. DSBs triggered by chemicals or inducible expression of endonuclease promote leaf senescence in an age dependent manner. Transcript levels of *ATAXIA TELANGIECTASIA MUTATED* (*ATM*) that signals the existence of DSBs and *SUPPRESSOR OF GAMMA RESPONSE 1* (*SOG1*), a plant-specific NAC transcription factor that plays a central role in DNA damage response (DDR), decline as leaf aging. Loss of functions of ATM and SOG1 accelerate DNA-damaged induced leaf senescence process. In addition, mutations in *ATAXIA TELANGIECTASIA MUTATED AND RAD3-RELATED* (*ATR*), *Ku70* and *Ku80* cause earlier senescence. To make clear how DNA damage affects leaf senescence process, we make a comparison between the transcript profiles affected by DNA damage chemical and senescence-associated genes. Our results suggest that DNA damage induces leaf senescence in part by regulating expressions of senescence-associated genes.

T4-14: Organ Size Control

T4-14-01

Molecular basis of flower-size evolution associated with the transition from outbreeding to selfing

Adrien Sicard[1], *Ushio Fujikura*[1], *Natalia Wozniak*[1], *Christian Kappel*[1], *Stephen Wright*[2], ***Michael Lenhard***[1]
1. University of Potsdam
2. University of Toronto

Understanding the molecular basis of morphological change remains a central challenge in evolutionary-developmental biology. The transition from outbreeding to selfing is often associated with a dramatic reduction in reproductive structures, such as a reduced flower size in plants. We are using the pair of closely related species *Capsella grandiflora* and *C. rubella* to study the molecular basis of evolutionary changes in flower size. The outbreeding ancestor *C. grandiflora* forms large attractive flowers, while the derived selfing species *C. rubella* only shows small, inconspicuous flowers. Using a quantitative-genetic approach, we have identified three genes whose variation between the two species contributes to flower-size reduction in *C. rubella*. One of these encodes the F-box protein STERILE APETALA (SAP), the other a putative transcriptional regulator and the third a cytochrome P450 enzyme involved in brassinosteroid biosynthesis. For the latter, the primary transcript of the *C. rubella* allele is spliced more efficiently than that of *C. grandiflora*, resulting in higher brassinosteroid levels. These restrict organ growth by limiting cell proliferation. More efficient splicing of the *C. rubella* allele results from two *de novo* mutations in the selfing lineage. Splicing variation has recently been suggested as a major contributor to complex traits in humans. Thus, our results highlight the potentially wide-spread importance of differential splicing and higher-than-optimal hormone levels in generating phenotypic variation, and identify the molecular changes underlying flower-size reduction in selfing compared to outbreeding species.

T4-14-02

Leaf size control mechanisms – Systems revealed from studies of compensation

Hirokazu Tsukaya[1,2], *Ali Ferjani*[3], *Kensuke Kawade*[2,4], *Kazune Ezaki*[1], *Gorou Horiguchi*[5]
1. The Univ. Tokyo
2. Okazaki Institute for Integrative Bioscience, National Institutes of Natural Sciences
3. Tokyo Gakugei University
4. National Insititute for Basic Biology, National Institutes of Natural Sciences
5. Rikkyo University

Compensation is a queer phenomenon that is seen in mutants/transgenics having severe defects in cell proliferation in leaf primordia: abnormally enhanced cell expansion triggered by decreased cell proliferation activity. Thanks to this phenomenon, the overall size decrease in leaf area is at least partially recovered than expected from the pure decrease in the number of leaf cells. While few cases can be simply explained by switching from mitosis to endoreduplication, the majority of the compensation-exhibiting mutants do not show enhanced endoreduplication in leaf cells. Moreover, the impact of endoreduplication on cell volume in palisade tissue is much weaker than observed in epidermis. For example, *KRP2* over-expressor in *Arabidopsis thaliana* shows typical Class III compensation and also displays a decrease in endoreduplication. The Class III compensation is unique since increased cell volume occurs even during the mitotic phase. Our genetic studies revealed that the enhanced cell expansion in Class III compensation depends on the activity of vacuolar proton-pump V-ATPase, provided that a mutation in *DET3* cancelled increased cell volume in the *KRP2* over-expressor. Interestingly, the requirement of the V-ATPase activity is not seen in the other classes of compensation, indicating the different nature of the compensation among differed classes. On the other hand, suppressor screening of *fugu5*, that shows Class II compensation characterized by an extended period for post-mitotic cell expansion, suggested that the monofunctional peroxisomal enoyl-CoA hydratase 2 (ECH2) is required for the compensation of *fugu5* mutant. This may be related to biosynthesis of biologically active auxin. Importantly, the peroxisomal ECH2 is not involved in the other classes of compensation. Finally in the *an3* mutant exhibiting typical Class I compensation, that is defined by an enhanced post-mitotic cell expansion rate, we found that compensation signal (s) can be transmitted among cells from analyses of chimeric leaves of *an3*-defective and *AN3*-over-expressing cells. Further detailed analysis demonstrated that the non-cell autonomous induction of the compensated cell enlargement is context-sensitive. Knowledge on how the context affects cell-cell communication in compensation will supply us with clues to identify the factors involved in the Class I compensation.

T4-14-03

Molecular networks governing biomass productivity and

seed yield

Dirk Inze
VIB Center for Plant Systems Biology, UGent

Plant and plant organ growth are regulated by an exceedingly complex interplay of many genes and their interaction with the ever changing environment. The long-term goal of our research is to obtain a holistic understanding of plant organ growth. Numerous genes of which the modified expression enhances plant organ growth have now been identified, and a detailed study of these genes provided novel insights in the molecular machines driving growth. Furthermore, evidence obtained both in the model plant *Arabidopsis* and in maize, demonstrated that the combination of multiple growth enhancing genes can have very profound effects on organ sizes. Field experiments with transgenic maize also show that genes enhancing leaf growth have profound effects on seed yield. Tremendous progress has also been made in understanding how environmental cues, such as mild drought stress, negatively affect plant growth. In unpredictable environments, growth reduction enables plants to redistribute and save resources, ensuring reproduction, even when the stress becomes extreme. However, when the episode of stress does not threaten plant survival, and from the agricultural point of view, growth reduction can be seen as counter-productive, leading to unnecessary yield loss. Limiting growth reduction may thus provide a strategy to boost plant productivity under stress. I will discuss how our insights open up new perspectives for the identification of optimal growth regulatory networks that can be selected by advanced breeding, or for which more robust variants (e.g. reduced susceptibility to drought) can be obtained through genetic engineering. The ability to improve growth of maize and, in analogy other cereals, could have a major impact in providing food security.

T4-14-04
Ubiquitin-mediated control of seed size and weight

*Na Li, Weihuan Jin, Zhibiao Wang, **Yunhai Li***
Institute of Genetics and Developmental Biology, Chinese Academy of Sciences

Seed size is one of the most important yield traits in plants. In angiosperms, a mature seed consists of the embryo, the endosperm and the seed coat, which develop from the zygote, the fertilized central cell and the maternal integument. Thus, the characteristic size of plant seeds is coordinately determined by maternal and zygotic tissues. However, the genetic and molecular mechanisms that set the final size of seeds are largely unknown. We are focusing on understanding how plants determine their final seed size in *Arabidopsis* and rice. We have recently identified several seed size regulators that are involved in ubiquitin-related activities. Here, we will discuss the role of the ubiquitin pathway in seed control.

T4-14-05
New pathways in stem cell regulation and impact on cereal crop yields

David Jackson
Cold Spring Harbor Laboratory

Shoot growth depends upon meristems, pools of stem cells that are maintained in a number of ways, including a negative feedback loop between the *CLAVATA* pathway and *WUSCHEL*. CLAVATA signaling involves a secreted peptide, CLAVATA3 (CLV3), and its perception by cell surface leucine-rich repeat (LRR) receptors, including the CLV1 receptor kinase, and an LRR receptor-like protein, CLV2. We are interested in finding novel players in the CLV-WUS pathway, as well as new regulators that could affect meristem size in parallel pathways. Maize is an excellent model system for these studies, because there are many clavata- type mutants, and their isolation is becoming routine. For example, we isolated the *COMPACT PLANT2* (*CT2*) gene, and found it encodes the predicted a subunit (Gα) of a heterotrimeric GTP binding protein. *ct2* mutants have CLAVATA-like meristem proliferation phenotypes, and genetic, biochemical and functional assays indicate that CT2/Gα signaling transmits a stem cell restrictive signal from a maize CLAVATA LRR receptor, suggesting a new function for Gα signaling in plants. Recent studies have questioned the idea that plant heterotrimeric G proteins interact with canonical GPCRs, and our findings suggest that single pass LRR receptors act as GPCRs in plants, challenging the dogma that GPCRs are exclusively 7TM proteins. We also characterized a new CLV- related receptor, *FASCIATED EAR3 (FEA3),* that appears to function in a different pathway to control meristem size, in parallel to CLV-WUS. We found that FEA3 responds to a CLV3-related ligand that is expressed in differentiating cells in leaf primordia, so it could provide a feedback from these differentiating primordia to the meristem. This pathway is therefore complementary to the known CLV-WUS feedback, and computational modeling suggests the 2 pathways interact. I will discuss the functions of these new pathways, as well as their potential use in improvement of maize yields, for example through enhancing seed number.

T4-14-06
Towards a better understanding of the grain size regulation in crops: cases from the perspective of QTL analyses

Xian-Jun Song
Key Laboratory of Plant Molecular Physiology, Institute of Botany, the Chinese Academy of Sciences, Beijing 100093, China

Enhancing crop production is a matter of cardinal significance in view of the increasing global population. Grain size and weight is a key determinant of crop yield and is determined by multiple naturally occurring quantitative trait loci (QTLs); however, our current knowledge of the underlying molecular mechanisms is still fragmentary, although a dozen such QTLs have been identified. I will describe our work that uncovered two quite different mechanisms, a RING-type E3 ubiquitin ligase and a putative protein phosphatase, underlying major QTLs for, respectively, grain width and grain length (Song *et al.*, 2007; Qi *et al.*, 2012), and a third example of QTL for grain length being due to a chromatin modifier (Song *et al.*, 2015). I will also describe recent work in my lab that shed light on how previously unknown genetic networks and/or signaling pathways operate to fine tune grain size in crops.

T4-15: Plant Regeneration

T4-15-01
Epigenetic control of wound-induced cell reprogramming

Bart Rymen[1], *Soichi Inagaki*[2], *Akira Iwase*[1], *Momoko Ikeuchi*[1], *Tetsuji Kakutani*[2], *Keiko Sugimoto*[1]
1. RIKEN CSRS
2. National institute of genetics

During normal development, a vast rearrangement of the epigenetic landscape is correlated with the progression from undifferentiated cells into differentiated cells. This epigenetic reprogramming serves as a barrier, preventing cells to regress from a differentiated state back into an undifferentiated state. Interestingly, many plant species can overcome this barrier after injury to heal wounds and regenerate new organs. Molecular mechanisms controlling this wound-induced organ regeneration remain largely unknown. We have previously identified *WOUND INDUCED DEDIFFERENTIATION1-4* (*WIND1-4*) as central transcriptional regulators of wound-induced cellular reprogramming in *Arabidopsis*. More recently we demonstrated that WIND1 promotes shoot regeneration by directly up-regulating another transcription factor *ENHANCER OF SHOOT REGENERATION1* (*ESR1*). We have also shown that genes in the WIND-mediated pathway need to be epigenetically repressed by POLYCOMB REPRESSIVE COMPLEX 2 (PRC2) to prevent unscheduled cellular reprogramming in intact plants. Currently, we are investigating the epigenetic and transcriptional rearrangements needed to activate the WIND-mediated pathway.

T4-15-02

Type-B *ARABIDOPSIS* RESPONSE REGULATORs is critical to the specification of shoot stem cell niche by dual regulation of *WUSCHEL*

Xiansheng Zhang
Shandong Agricultural University

Plants are known for their capacity to regenerate the whole body under appropriated cultural conditions. During this process, apical meristems are *de novo* formed from a mass of proliferating cells named callus. Exogenous cytokinin and auxin determine cell fate for the establishment of stem cell niche, which is the vital step of shoot regeneration. However, the mechanisms underlying remains elusive. In the present study, we show that type-B ARABIDOPSIS RESPONSE REGULATORs (ARRs), critical components of cytokinin signaling, activate the transcription of *WUSCHEL* (*WUS*), which encodes a key regulator for maintaining the stem cells. In parallel, type-B ARRs inhibit auxin accumulation by repressing the expression of *YUCCA*s, which encode a key enzyme for auxin biosynthesis, indirectly promoting *WUS* induction. Both pathways are essential for *de novo* regeneration of shoot meristems. In addition, the dual regulation of type-B ARRs on *WUS* transcription is required for the maintenance of shoot apical meristem *in planta*. Thus, our results reveal a long-standing missing link between cytokinin signaling and WUS, and the findings provide critical information for understanding cell fate specification.

T4-15-03

Brachypodium tissue culture as a model system to reveal the functions of the components of the cell wall

Alexander Betekhtin[1], *Magdalena Rojek*[1], *Anna Milewska-Hendel*[2], *Robert Gawecki*[2], *Jagna Karcz*[3], *Ewa Kurczynska*[2], *Robert Hasterok*[1]
1. Department of Plant Anatomy and Cytology, Faculty of Biology and Environmental Protection, University of Silesia in Katowice
2. Department of Cell Biology, Faculty of Biology and Environmental Protection, University of Silesia in Katowice
3. Scanning Electron Microscopy Laboratory, University of Silesia

Brachypodium Beauv. is believed to be one of the oldest genera within the Poaceae. This genus is predominantly distributed in Europe and Asia with disjunctions occurring in Central America and Southern Africa. *Brachypodium distachyon* (Brachypodium) is a model system for functional genomics in grasses. Although there are some studies of *in vitro* Brachypodium cultures including somatic embryogenesis, detailed knowledge of the composition of the components of the main cell wall in the embryogenic callus in this species is lacking. Therefore, we used histological, scanning electron microscopy as well as the immunocytochemical approach against the arabinogalactan proteins (AGP), extensins and hemicelluloses to understand the localisation and possible functions of these cell wall components during the embryogenic mass that appears in Brachypodium callus. We found that the distribution of pectins, AGPs and hemicelluloses can be used as molecular markers of the embryogenic cells. Furthermore, we showed that AGPs and pectins are components of the extracellular matrix. The presented data extends our knowledge about the chemical composition of the embryogenic cells in the Brachypodium callus.

T4-15-04

Auxin and cytokinin interplay controls *de novo* shoot regeneration *in vitro*

Fengning Xiang
Shandong University

Exogenous auxin and cytokinin are required for shoot regeneration under appropriate culture conditions. Some of the gene products participating in *in vitro* shoot formation act within an auxin and/or a cytokinin signal transduction pathway. However, the molecular mechanisms underlying the auxin and cytokinin interplay during *in vitro* shoot formation remain unresolved. Here, we show that the *ARR1-IAA17* and *ARF10-ARR15* regulatory cascades mediate phytohormonal interaction and control shoot regeneration in *Arabidopsis thaliana*. On the one hand, higher numbers of SAMs and shoots are generated in the *arr1* mutant than in the wild type, but lower numbers when *ARR1* was over-expressed. *ARR1* is expressed in nascent shoot apical meristems (SAM). ARR1 promotes expression of *IAA17* by binding to its promoter, thereby lessening the explants' responsiveness to auxin. The loss-of-function of *IAA17* promotes shoot formation, and over-expression of the *VP16-IAA17mImII* construct rescues the shoot regeneration defect shown by explants harboring a p*35S::ARR1* transgene. On the other hand, the shoot regeneration is more prolific in a *miR160*-resistant form of *ARF10* (*mARF10*), but fewer in the *arf10* mutant than that in wild type. ARF10 repressed the expression of *Arabidopsis Response Regulator15* (*ARR15*) via its direct binding to the gene's promoter. The loss-of-function of *ARR15* enhanced shoot regeneration and over-expression of *ARR15* rescues the shoot regeneration defect of *mARF10* plants. Our findings define the ARR1 and ARF10 as key regulators of shoot regeneration and shed light on the molecular mechanisms underlying the auxin and cytokinin interplay during *in vitro* shoot formation.

T4-15-05

Identification and characterization of meiotic recombination suppressors in *Arabidopsis thaliana*

Fan Li*, *Nico De Storme, Danny Geelen*

Department of Plant Production, Faculty of Bioscience Engineering, Ghent University

In plants sexual reproduction, meiotic recombination plays a central role in creating genetic variability, making it an important factor for plant breeding in the background of climate change. However, the molecular mechanisms controlling the frequency and patterning of meiotic recombination events are not well understood. Until now, seven meiotic recombination suppressors have been reported that control the minority type class II crossovers. To identify additional regulators, we initiated a forward genetics screen in *Arabidopsis thaliana*. An extensive EMS mutant screen was performed by using the segregation of fluorescent transgene markers in the male reproduction system as a read-out for male meiotic recombination frequency. Among ~12,000 M2 individual plants obtained from self-fertilization of mutagenized M1 generation, a range of mutants with significant higher (281 lines) and lower (72 lines) recombination frequencies have been isolated, including 22 lines showing more than three-fold increase in male meiotic recombination frequency compared to wild type. Most of the mutations are recessive, except for two, which are dominant. Causative mutations are currently being mapped by whole-genome sequencing. Using highly heterozygous Col-0/Ler hybrids that contain the dominant mutation, we found that meiotic recombination frequency increased up to three-fold in both male and female sporogenesis, and this equally spread across the genome. Our study and research could provide insight into the molecular constraints on meiotic recombination processes, but also could help plant breeders to make fuller use of the genetic variability that is available within crop species.

T4-16: Mechanism of hybrid vigor

T4-16-01

Analysis of transcriptional and epigenetic regulation in hybrid vigor of allopolyploid *Brassica napus* uncovers key roles for small RNAs

Yifei Shen[1], *Shuijin Hua*[2], *Daguang Cai*[3], *Michael Timko*[4], *Qianhao Zhu*[5], *Longjiang Fan*[1]

1. Institute of Crop Science & Institute of Bioinformatics, Zhejiang University

2. Institute of Crop and Utilization of Nuclear Technology, Zhejiang Academy of Agricultural Sciences

3. Institute of Phytopathology, Christian Albrechts University of Kiel

4. Department of Biology, University of Virginia

5. CSIRO Agriculture and Food, Black Mountain Laboratories

Heterosis is a fundamental biological phenomenon characterized by the superior performance of a hybrid compared to its parents. The underlying molecular basis for heterosis, particularly for allopolyploids, remains elusive. In this study, we analyzed the transcriptomes of *Brassica napus* parental lines and their F_1 hybrids at three stages of early flower development. Phenotypically, the F_1 hybrids show remarkable heterosis in silique number and grain yield. Transcriptome analysis revealed that various phytohormone (auxin and salicylic acid) response genes are significantly altered in the F_1 hybrids relative to the parental lines. We also found evidence for previously undescribed decreased expression divergence of the homoeologous gene pairs in the allopolyploid F_1 hybrids and suggest that high-parental expression level dominance plays an important role in heterosis. Small RNA and methylation studies aimed at examining the epigenetic regulation of the gene expression changes in the F_1 hybrids showed that the majority of the small interfering RNA (siRNA) clusters had a higher expression level in the F_1 hybrids than parents, and that there was an increase of the genome-wide DNA methylation in the F_1 hybrid. Transposable elements associated with siRNA clusters had a higher level of methylation and a lower expression level in the F_1 hybrid, implying that the non-additively expressed siRNA clusters resulted in lower activity of the transposable elements through DNA methylation in the hybrid. Our data provide new insights into the role that gene expression pattern changes and epigenetic regulation mechanisms contribute to heterosis during early flower development in allopolyploid *B. napus*.

T4-16-03

Up-regulating the abscisic acid inactivation gene *ZmABA8ox1b* contributes to seed germination heterosis by promoting cell expansion

Yangyang Li[1], *Cheng Wang*[1], *Xinye Liu*[1], *Shuang Fang*[2], *Jinfang Chu*[2], *Qixin Sun*[1], ***Zhongfu Ni***[1]

1. State Key Laboratory for Agrobiotechnology/ Key Laboratory of Crop Heterosis and Utilization (MOE)/ Beijing Key Laboratory of Crop Genetic Improvement, China Agricultural University, Beijing 100193

2. Chinese Academy of Sciences National Center for Plant Gene Research (Beijing), Institute of Genetics and Developmental Biology, Chinese Academy of Sciences,Beijing 100101

Heterosis has been widely used in agriculture, but the underlying molecular principles are still an area for further investigation. Maize F_1 hybrid plants are taller, display increased biomass, grow more rapidly, and have greater yields than their inbred parents. Notably, one of the key aspects of maize heterosis is accelerated seed germination. Here, we reported that the seed germination rate of hybrid B73/Mo17 was significantly faster than that of its parental lines. The T_{50} of parental inbred lines B73 and Mo17 was greatly increased by exogenous ABA treatment compared with the hybrid B73/Mo17, indicating that ABA inactivation plays an important role in seed germination heterosis. Furthermore, we found that the expression level of *ZmABA8ox1b* rapidly increased and displayed an above high parent or high parent expression pattern at the early stage of seed germination. In *Arabidopsis*, the overexpression transgenic lines of *ZmABA8ox1b* showed lower ABA content in dry seeds and faster seed germination rate compared with the wild type. Notably, microscopic observation revealed that cell expansion played a major role in the ABA-mediated maize seed germination heterosis, which could be attributed to the altered expression of cell wall-related genes. Taken together, we propose a simple model for ABA-mediated seed germination heterosis. Briefly, rapid ABA inactivation occurs in hybrids during seed germination due to the up-regulation of *ZmABA8ox1b*. Consequently, the expression levels of genes involved in cell wall loosening and expansion are then altered, which accelerates cell expansion in hybrids and contributes to the observed seed germination heterosis.

T4-16-04

The chromatin remodeler DDM1 regulates hybrid vigor of *Arabidopsis*

Min Wang, *Xiaoyuan Li, Shuang Ma, Linan Xie, Qingzhu Zhang*
Northease Forestry University

Hybrid vigor, or heterosis, refers to the improved performance of hybrid offspring relative to their parents. In *Arabidopsis*, F1 hybrids, derived from various accession combinations, show obvious growth hybrid vigor in early development stage. Although epigenetic modifications, including small RNA and DNA methylation was altered in F1 hybrid progenies after hybridization between parents, and then RNA-directed DNA methylation (RdDM) is proposed to influence heterosis, we identified that RdDM is not involved in regulation of biomass hybrid vigor of Col and C24 accessions using genetic mutants. However, we uncovered that the chromatin remodeler DECREASED DNA METHYLATION 1 (DDM1) promotes early seedling growth heterosis in Col/C24 hybrids by regulating salicylic acid metabolism, suggesting that stress-induced hormesis may underlie specific hybrid vigor traits. We will investigate the epigenetic mechanisms of DDM1 promoting hybrid vigor associated with salicylic acid metabolism.

T4-16-05

Circadian and epigenetic perspectives on heterosis

Z. Jeffrey Chen
The University of Texas at Austin

We investigate mechanisms and evolution for advantages and disadvantages of being polyploids and hybrids and their impact on plant breeding and crop improvement. Most crop plants including wheat, cotton, and canola are polyploids, and many other crops such as maize are ancient polyploids and/or are grown as hybrids. Moreover, polyploidy provides a means to permanently fix the heterozygosity and heterosis in interspecific hybrids. Although polyploidy and heterosis have been widely applied in animal and plant production, the molecular bases remain largely elusive. Using *Arabidopsis*, cotton, and corn as experimental systems, we have deployed integrated approaches of genomics, epigenetics, molecular biology, and systems biology to investigate genetic and epigenetic mechanisms for gene expression changes and phenotypic variation in allopolyploids and hybrids. We found a direct link of epigenetic regulation of circadian clock regulators with the increased levels of chlorophyll biosynthesis and starch metabolism in *Arabidopsis* hybrids and allotetraploids. This link of altered circadian rhythms to growth vigor is provocative but logical because circadian rhythms regulate many metabolic and physiological processes in plants and animals. Moreover, the circadian regulation mediates the adaptation of strains to local environments (stress conditions). Under normal conditions, the basal level of stress response is compromised in the hybrids to promote growth vigor, while under stress conditions, the stress response genes are induced at certain times of the day to combat stress as well as to maintain growth vigor. Recent studies also reveal a role for diurnal repression of exogenous ethylene levels and endogenous ethylene signaling in heterosis. In maize hybrids, the clock-protein binding targets are shifted to morning-expressed genes, promoting early activation of photosynthetic and metabolic activities and leading to biomass heterosis. These mechanistic insights into polyploidy and heterosis will ultimately help us improve the production of hybrid and polyploid crops and plants to meet the growing demands for food, feed, fiber, and biomaterials.

T4-16-06

Molecular mechanisms of heterosis for biomass and bacterial defense in plants

Guangming He, *Hang He, Chengcai An, Xing-Wang Deng*
State Key Laboratory of Protein and Plant Gene Research, The Peking-Tsinghua Center for Life Sciences, School of Advanced Agricultural Sciences and School of Life Sciences, Peking University, China

The rapid development of genomics provides excellent opportunities to explore the molecular mechanism of heterosis in plants. We have made efforts to investigate the genome-wide differences in gene expression and epigenetic modifications between hybrid and their parents, and to identify genetic or epigenetic components potentially contributing to heterosis for biomass and disease resistance in plants. Our studies revealed that insertions/deletions (INDELs), many of which result in the formation/disruption of putative *cis*-regulatory elements is one plausible molecular mechanism underlying heterotic gene action and thus heterosis in rice. Epigenetic components, the regulators of chromatin states and genome activity, were found to correlate with transcriptomic variation in hybrids, and thus also have the potential to impact biomass heterosis in rice, maize and *Arabidopsis*. We identified *Arabidopsis* hybrids showing increased resistance to biotrophic bacterial pathogen, and subsequent transcriptomic and genetic analyses demonstrated that enhanced activation of the salicylic acid biosynthesis pathway play a central role in heterosis for bacterial defense in *Arabidopsis*. Furthermore, combining the identification of candidate genes for heterosis by genome-wide association analysis with the comparative transcriptomic analysis indicate the potential contribution of stimulus-responsive and cell cycle regulation pathways to biomass heterosis in *Arabidopsis*.

T4-17: Hormone biosynthesis, metabolism and transportation

T4-17-01

Cytokinin importer PURINE PERMEASE 14 confines the cytokinin responses

Bruno Mueller
University of Zurich

Multicellular organisms operate via the coordination of specialized cellular identities and functions dictated by signalling systems. However, the mechanisms that define the precise spatiotemporal cell domains and subcellular locales responding to specific signals are not well understood. In plants, cytokinins direct essential cell-to-cell signalling systems to initiate and maintain differential cell functions during gametophyte development, embryogenesis and postembryonic shoot and root development. Guided by the live green fluorescent protein (GFP) imaging of the synthetic reporter, *TCSn::GFP* (Two Component signalling Sensor), we illuminate precise cell populations activating conserved phosphorelay signalling circuitry despite much broader intracellular cytokinin signalling competence and potential. Here, we report a surprising finding that *Arabidopsis* PURINE PERMEASE 14 (PUP14) plays a pivotal role in confining the cytokinin signalling response throughout development by ligand sequestration. The expression patterns of PUP14 are inversely correlated with the cytokinin signalling readout, indicating an antagonistic relation. Indeed, the conditional *PUP14* knock-down by an artificial microRNA (*amiRPUP14*) causes ectopic cytokinin signalling accompanied by aberrant morphogenesis in embryos, roots and the shoot apical meristem. PUP14 protein localises to the plasma membrane and imports bioactive cytokinins as shown by analysing transport in mesophyll protoplasts, seedlings and microsomes. These activities of PUP14 reduce the apoplastic cytokinin pools and extracellular cytokinin perception by cytokinin sensors. The relevance of apoplastic cytokinins in initiating intracellular signalling is corroborated by our findings that signalling output is attenuated by secreted cytokinin-degrading enzyme CYTOKININ OXIDASE 2 (CKX2), but not by intracellular CKX variants. These results uncover the existence of a previously unknown and dedicated transport system that patterns the cytokinin signalling landscape of vascular plants.

T4-17-02

Chemical regulation of plant hormone functions and their cross talk

Tadao Asami

1. raduate School of Agricultural and Life Sciences, The University of Tokyo

2. Department of Biochemistry, King Abdulaziz University

Plant hormones are biosynthesized and perceived by their receptors and then elicit their activity to regulate plant life cycles. Various steps in the life cycles of plants can be modulated by each of hormones, sometimes in a synergistic fashion, suggesting physiological redundancy and/or crosstalk between the different pathways. As small molecules can be used to unravel these plant hormone functions, many chemicals that can dissect plant hormone functions have been developed and accelerate our understanding of plant hormone signaling. In this context, we developed several chemical regulators for plant hormone functions, such as biosynthesis inhibitors, catabolism inhibitors, receptor inhibitors and mimics. Here, we will talk about the development and/or characterization of known or new chemical regulators mainly for GA and SL from the viewpoint of the reduction of crop damage by root parasitic weeds. 1) AC94377 and D671 is a GID1 agonists that bind GA receptors to activate the GA signa. 2) Debranones are SL agonists. 3) KUT15 is an ethylene mimic that partially activates ethylene signal.These chemicals can control damage by root parasitic weed such as *Striga*. AC94377 can reduce SL production in rice. TIS108 inhibits biosynthesis of SLs without affecting branching in rice. KUT15 can induce suicidal germination of *Striga* and can reduce the seedbank of *Striga*.

T4-17-03

NRT1.1 mediated nitrate suppression of asymmetric root growth relies on PIN2- and AUX1-dependent auxin transport

Sen Chai, *En Li, Fu-Rong Ge, Yan Zhang, Sha Li*

State Key Laboratory of Crop Biology, College of Life Sciences, Shandong Agricultural University

Asymmetric root growth (ARG) on tilted plates, in extreme situation root coiling on horizontally-placed plates, is proposed to be a combination of gravitropism, mechanical sensing, and circumnutation, a word designated by Charles Darwin to describe the helical movement that all plant organs make around the growth direction. ARG is a developmentally controlled process in which microtubule-regulating proteins and cell surface proteins such as MLOs participate. Environmental cues, such as nutrient deficiency, also influence ARG.However, it is unclear what nutrients play key roles in regulating ARG that leads to root coiling, what endogenous components are involved in responding to nutrient deficiency for ARG, and how is nutrient deficiency translated into endogenous responses. We report here that nitrate deficiency resulted in a strong ARG in *Arabidopsis*. Nitrate deficiency caused root coiling on horizontal plates, which is inhibited by auxin transport inhibitor, and by mutations at *PIN-FORMED2* (*PIN2*) and *AUXIN RESISTANT 1* (*AUX1*). We further show that suppression of ARG by nitrate is mediated by the nitrate transporter/sensor NRT1.1. Nitrate deficiency and functional loss of *NRT1.1* both resulted in altered membrane dynamics of PIN2 and enhanced asymmetric auxin signaling, leading to root coiling. This study reveals a signaling pathway in root growth by responding to exogenous nitrate and relaying it into intracellular activities through altered auxin transport.

T4-17-04

Long-distance phytohormonal signalling of drought stress

A Fiebig[1], S Donaldson[1], N Mohd Zan[1], J Puertolas[2], F Pérez-Alfocea[2], ***IC Dodd[2]***

1. Lancaster Environment Centre, LEC, University of Lancaster

2. Departamento de Nutrición Vegetal, CEBAS-CSIC, Campus Universitario de Espinardo

Plants respond to drying soil by altering their synthesis and metabolism of phytohormones, thereby changing tissue hormone concentrations and xylem sap composition. While most attention has focused on increased ABA (abscisic acid) concentrations as a putative long-distance signal of drying soil, the advent of multi-analyte phytohormone analyses (Albacete *et al.*, 2009) allows the role of other phytohormones as sensors of soil drying to be investigated. When tomato (*Solanum lycopersicum)* plants were grown in a peat-based substrate and automatically irrigated when soil water content decreased below a threshold value, deficit irrigation increased foliar ABA and ethylene concentrations by *circa* 50% compared to well-watered plants. Although the ethylene precursor ACC (1-aminocyclopropane carboxylic acid) was not detected in root xylem sap that was collected at flow rates matching *in vivo* transpiration, xylem ABA concentration approximately doubled while xylem JA (jasmonic acid) concentrations increased

5-fold. When water was withheld from tomato (*Solanum lycopersicum)* plants grown in a mineral substrate at two different bulk densities (1.4 *versus* 1.74 g cm^{-3}), soil-drying-induced changes in root xylem sap composition were accentuated in plants grown at low bulk density, due to greater soil drying. At the same soil water content, high bulk density increased xylem ABA concentrations. While soil drying increased xylem ABA concentration compared to well-watered plants, xylem JA and cytokinin (trans-zeatin, tZ and isopentenyl adenine, iP) concentrations did not change. When water was withheld from pea (*Pisum sativum)* plants grown in a mineral substrate, there were multiple changes in phytohormone concentration of expanding stem internodal tissue. Soil drying progressively increased stem ABA, ACC, iP and tZ levels, reaching concentrations that were 76-fold, 2.4-fold, 3.4-fold and 2-fold higher compared to well-watered plants respectively, at the time that stem elongation ceased. Moreover, soil drying progressively decreased stem GA1 (gibberellic acid 1) and IAA (indole-3-acetic acid) concentrations by 97% and 80% respectively. Whether these changes in xylem sap and tissue phytohormone concentrations are due to local or long- distance processes will be investigated by self- and reciprocal-grafting of mutants that are deficient in or overexpress selected phytohormone groups.

T4-18: Flavonoid biosynthesis and regulation

T4-18-01

Phenylalanine biosynthetic network: What's left to know?

Natalia Dudareva
Purdue University

In plants, up to 30% of photosynthetically fixed carbon is directed towards formation of aromatic amino acids (AAAs) and their downstream products, making this biosynthetic network one of the most metabolically intensive. Within the AAA network, the highest carbon flux is directed to phenylalanine (Phe), a common precursor of >8000 phenolic compounds, that have profound impacts on plant growth, development, reproduction and defense and constitute approximately 30-45% of plant organic matter. Phe biosynthesis in plants occurs via two alternative routes with either arogenate or phenylpyruvate as a key intermediate. Using a functional genomics approach and petunia flowers, which emit high levels of Phe-derived volatiles as a model system, we have identified genes encoding proteins involved in the arogenate and phenylpyruvate Phe biosynthetic pathways as well as in Phe export from plastids. We have shown that, while plants predominantly synthesize Phe in plastids via the arogenate pathway, the microbial-like phenylpyruvate pathway also contributes to Phe formation. Moreover, flux through the phenylpyruvate route increases when the entry point to the arogenate pathway is limiting. Interestingly, the phenylpyruvate pathway utilizes a cytosolic aminotransferase that strongly favors Tyr as the amino donor. Taken together, our results demonstrate that Phe biosynthesis is not limited to plastids and that there is an interconnection between aromatic amino acid catabolism and biosynthesis *in planta*. This presentation will also discuss the plasticity and complex regulation of Phe and Tyr biosynthetic pathways *in planta*. Another aspect of plant metabolism to be discussed is its compartmentalization and metabolite transport within the cell and how this intracellular transport contributes to the interconnections between and within metabolic networks.

T4-18-02

Flavonoid biosynthesis in pollen: Functional identification of glycosyltransferases responsible for pollen-specific flavonols

Keiko Yonekura-Sakakibara[1], *Ryo Nakabayashi*[1], *Satoko Sugawara*[1], *Eva Knoch*[1], *Kazuki Saito*[1,2]
1. RIKEN Center for Sustainable Resource Science
2. Chiba University

Flavonoids, including flavonols, flavones, anthocyanins and proanthocyanidins, are the most intensely studied plant secondary metabolites and over 9,000 known compounds are widely distributed throughout plant kingdom. Flavonoids play important roles as UV protectants, pigments, attractants of pollinators, phytoallexins, signaling molecules and regulators of fertility and auxin transport. The huge chemical diversity of flavonoids and the intricate distribution patterns in plant tissues and species make it difficult to correlate specific flavonoid structures with their physiological functions. As one of a few exceptions, a relationship between pollen-specific flavonols and pollen fertility is well established in petunia (*Petunia hybrida*) and maize (*Zea mays*) . Pollens of flavonoid-deficient mutants of petunia are unable to germinate, resulting in male sterility. Interestingly, *Arabidopsis* flavonoid-deficient mutants are fertile although petunia and *Arabidopsis* accumulated relatively similar flavonol 3-*O*-diglycosides in pollen. To complete the flavonoid biosynthetic map in pollen of *Arabidopsis* and *petunia*, we identified genes encoding flavonoid glycosyltransferases, *UGT79B6* and *UGT79B31*, responsible for terminal modification of pollen-specific flavonoids in *Arabidopsis* and *petunia*, respectively. *In vitro* characterization of the recombinant proteins and functional complementation of *Arabidopsis* mutants which lacked flavonoid diglycosides, indicated that *UGT79B6* and *UGT79B31* encode flavonoid 3-*O*-glycoside: 2"-*O*-glucosyltransferase. We also discuss the localization of UGT79B6 and UGT79B31 and roles of pollen-specific flavonols.

T4-18-03

Flavone biosynthesis in rice: different biosynthetic routes for *C*-linked and *O*-linked conjugates

Pui Ying Lam[1], *Hongjia Liu*[2], ***Clive Lo***[1]
1. The University of Hong Kong
2. Zhejiang Academy of Agricultural Sciences

Flavones are flavonoids found extensively in land plants with important physiological functions like UV protection, interactions with other organisms, co-pigmentation in flowers, *etc.* They are increasingly popular as dietary constituents or supplements due to their health-beneficial properties. In grasses, flavones are predominantly accumulated as C- or O-linked conjugates in vegetative tissues. The enzymology of flavone biosynthesis in monocot remained largely elusive until recent years. Our recent work established the cytochrome P450 enzymes CYP93G1 and CYP93G2 as key branch point enzymes in rice channeling flavanones to the formation of flavone O-linked conjugates and C-glycosides, respectively. CYP93G1 functions as a flavone synthase II, which generates flavone aglycones for different O-linked modifications.

On the other hand, CYP93G2 is a flavanone 2-hydroxylase, which produces 2-hydroxyflavanones for immediate C-glycosylation, followed by the formation of the flavone nucleus. We further filled the remaining gap in the tricin biosynthesis pathway with CYP75B4. In tricin, the 3', 5'-dimethoxyflavone nucleus is formed before O-linked conjugations. CYP75B4 functions as a unique flavonoid 5'-hydroxylase that converts chrysoeriol to selgin, the immediate precursor of tricin. CYP93G1, CYP93G2, and CYP75B4 homologous sequences are highly conserved in Poaceae, suggesting that they are specifically recruited for the biosynthesis of different flavone-derived metabolites that are prevalent in many grass species today. They also represent good targets for engineering flavones in edible tissues.

T4-18-04

Regulation of flavonoid biosynthesis by UV-B photoreceptor pathway

Ruohe Yin
University of Geneva

Flavonoid biosynthesis is dependent on light. In dark, flavonoid biosynthesis is off. Light induces flavonoid biosynthesis through activation of several transcription factors including HY5, MYB11, MYB12 and MYB111 *etc.* These transcription factors directly bind to promoter regions of genes in flavonoid biosynthesis pathway including *CHS*. Among different light spectra, Ultraviolet-B (UV-B; 280-315nm) is most effective in flavonoid biosynthesis induction. Interestingly, accumulation of flavonoids in plant epidermal tissues can absorb UV-B and protect inner tissues from UV-B damage. Thus, flavonoids can serve as "suncream" for plants.
With the characterization of UV-B photoreceptor UV RESISTANCE LOCUS 8 (UVR8) and its downstream signaling components including CONSTITUTIVE PHOTOMORPHOGENIC 1 (COP1) and HY5, we can now draw a clearer picture from UV-B perception to flavonoid biosynthesis induction. I will present the recent progress from early UV-B signaling biochemical events to downstream flavonoid biosynthesis.

T4-18-05

Arogenate dehydratase isoforms differentially regulate anthocyanin biosynthesis in *Arabidopsis thaliana*

Qingbo Chen[1], ***Cong Man***[1,2], *Danning Li*[1], *Huijuan Tan*[1,2], *Ye Xie*[1], *Jirong Huang*[1]
1. Shanghai Institute of Plant Physiology and Ecology, Chinese Academy of Sciences
2. University of Chinese Academy of Sciences

Anthocyanins, a group of L-phenylalanine (Phe)-derived flavonoids, have been demonstrated to play important roles in plant stress resistance and interactions between plants and insects. Although the anthocyanin biosynthetic pathway and its regulatory mechanisms have been extensively studied, it remains unclear whether the level of Phe supply affects anthocyanin biosynthesis. Here, we investigated the roles of arogenate dehydratases (ADTs), the key enzymes that catalyze the conversion of arogenate into Phe, in sucrose-induced anthocyanin biosynthesis in *Arabidopsis*. Genetic analysis showed that all six ADT isoforms function redundantly in anthocyanin biosynthesis but have differential contributions. ADT2 contributes the most to anthocyanin accumulation, followed by ADT1 and ADT3, and ADT4-ADT6. We found that anthocyanin content is positively correlated with the levels of Phe and sucrose-induced ADT transcripts in seedlings. Consistently, addition of Phe to the medium could dramatically increase anthocyanin content in the wild-type plants and rescue the phenotype of the *adt1 adt3* double mutant regarding the anthocyanin accumulation. Moreover, transgenic plants overexpressing *ADT4*, which appears to be less sensitive to Phe than overexpression of *ADT2*, hyperaccumulate Phe and produce elevated level of anthocyanins. Taken together, our results suggest that the level of Phe is an important regulatory factor for sustaining anthocyanin biosynthesis.

T4-18-06

Key aspects of flavonoid pathway regulation are conserved between angiosperms and liverworts

Kevin Davies[1], *Nick Albert*[1], *William Clayton*[1,2], *Simon Deroles*[1], *Kathy Schwinn*[1], *John Bowman*[3], *Brian Jordan*[2]
1. The New Zealand Institute for Plant & Food Research Limited
2. Lincoln University
3. Monash University

Flavonoids are ubiquitous in extant land plants. The pathway is hypothesized to have evolved for protection against UVB radiation during land colonization around 500 million years ago. In angiosperms, an R2R3MYB-bHLH-WDR (MBW) transcriptional complex controls flavonoid production. Our research on dicot (petunia, antirrhinum) and monocot (onion, cymbidium orchid) species has contributed to the current model for the action of the MBW complex, which proposes that the R2R3MYBs are the key components for environmental and developmental regulatory responses in both leaves and flowers. We have also found that R2R3MYB and R3MYB transcriptional repressors are important for modulating the strength of pathway activation and defining spatial activation domains. However, it is an open question whether the MBW complex for flavonoid regulation is a universal system in land plants, as data are lacking from outside of the angiosperms. As an early diverging lineage of land plants, the liverworts (e.g. *Marchantia* sp.) can help to inform us on systems that have an early evolutionary origin. We are using the model liverwort *Marchantia polymorpha* for our studies. Marchantia produces flavones, a key UV-induced flavonoid of many angiosperms, and a red pigment (ricciodinin) thought to be related to anthocyanins. We used RNAseq analysis of UVB-treated plants and Blast interrogation of transcriptome and genome resources to identify biosynthetic and regulatory gene candidates for the marchantia flavonoid pathway. These were then characterized using transgenic over-expression and CRISPR mutagenesis, combined with tests of the environmental signals that might control flavonoid production in marchantia. The studies revealed that an R2R3MYB gene was a key regulator of flavonoid production. Knockout lines for the MYB had greatly reduced flavonoid levels and lost all red pigmentation, while over-expression conferred production of large amounts of flavonoids, activation of the biosynthetic genes, and constitutive red pigmentation. This suggests that R2R3MYB regulated flavonoid production is a conserved character across land plants and arose early during land colonization.

T4-19: Terpenoid biosynthesis and regulation

T4-19-01

Genome mining for sesterterpenoid discovery in Brassicaceae *plants*

*Qingwen Chen, **Guodong Wang***
State Key Laboratory of Plant Genomics and National Center for Plant Gene Research, Institute of Genetics and Developmental Biology, Chinese Academy of Sciences

Terpenoids represent the largest and most diverse class of plant specialized metabolites, with diverse physiological functions during plant development. In the biosynthesis of terpenoids, short-chain isoprenyl diphosphate synthases (SC-IDS) account for the majority biosynthesis of linear isoprenyl pyrophosphates, which are the direct precursors for terpenoids. *Arabidopsis thaliana* possess a group of SC-IDSs (GGPPS-like proteins, which were supposed to mainly produce geranylgeranyl diphosphate) whose functions are not clearly known. In this study, we found the neo-functionalization of Arabidopsis GGPPS-like gene family: GGPPS6,7, 9 and 10 have GFPPS activities (the predominant product is geranylfarnesyl diphosphate (C25)) and GGPPS8 produces even longer chain isoprenyl diphosphate (>C25). By solving the crystal structures of GGPPS7 (C25), GGPPS8 (>C25) and GGPPS11 (C20), we disclose the product chain-length determination mechanism and describe it as a "three floor" model. Using this model, we discovered the novel GFPPS in various Brassicaceae species. Genome mining further reveal that these Brassicaceae *GFPPS* genes are usually linked to terpene synthase gene and *P450* genes. The *GFPPS-TPS-P450* gene clusters for sesterterpenoids biosynthesis were functional identified and the potential biological functions of these sesterterpenoids in Brassicaceae plants are also discussed.

T4-19-02

Molecular diversity of terpene synthases in the liverwort *Marchantia polymorpha*

*Santosh Kumar[1], Chase Kempinski[1], Xun Zhuang[1], Ayla Norris[2], Sibongile Mafu[3], Jiachen Zi[3], Stephen Bell[1], Eric Nybo[1], Scott Kinison[1], Zuodong Jiang[1], Sheba Goklany[1], Kristin Linscott[1], Xinlu Chen[2], Qidong Jia[2], Shoshana Brown[4], John Bowman[5], Patricia Babbitt[4], Reuben Peters[3], Feng Chen[2], **Joe Chappell**[1]*
1. University of Tennessee, Knoxville, Tennessee
2. University of Kentucky, Lexington, Kentucky
3. Iowa State University, Ames, Iowa
4. University of California, San Francisco, California
5. Monash University, Melbourne

Marchantia polymorpha is a primitive terrestrial land plant, which like most liverworts accumulates a structurally diverse array of terpenes believed to serve in deterring disease and herbivory. Previous studies have suggested that the mevalonate and methylerythritol phosphate pathways, present in evolutionarily advanced plants, are also operative in liverworts. However, the genes and enzymes responsible for the chemical diversity of terpenes have yet to be described. In the current study, we resorted to a HMMER search tool to identify 17 putative terpene synthase genes from *M. polymorpha* transcriptomes. Functional characterization identified 4 diterpene synthase genes phylogenetically related to those found in higher plants and 9 rather unusual monoterpene and sesquiterpene synthase-like genes. The presence of separate mono-functional diterpene synthases for ent-CPP and *ent*-kaurene biosynthesis are similar to orthologs found in vascular plants, pushing back the date of the underlying gene duplication and neo-functionalization of the ancestral diterpene synthase gene family >400 million years ago. In contrast, the mono- and sesqui-terpene synthases represent a new class of enzymes, not related to plant terpene synthases and only distantly so to microbial-type terpene synthases. The absence of a Mg^{2+} binding, aspartate-rich, DDXXD motif place these enzymes in a new, non-canonical family of terpene synthases.

T4-19-03

Triterpenoid saponins in a wild crucifers - biosynthesis, evolution and structure activity relationships in plant-insect interactions

Soren Bak
University of Copenhagen, Department of Plant and Environmental Sciences

It is well known that plants produce specialized metabolites for defense, yet we know little of how they actually evolved. Certain classes of chemical defense compounds have evolved repeatedly in plants, facilitated by recruitment of genes from a few old gene families. This was probably mediated by gene duplications, either locally or by whole-genome duplications, followed by selection on their enzyme substrate and product specificity. The initial assumption that divergent evolution is the major player is being challenged, and it appears that convergent evolution is surprisingly common in specialised metabolite pathways (Pichersky and Lewinsohn 2011). Triterpenoid saponins comprise a highly diverse group of specialised metabolites, derived from 2,3-oxidosqualene, the precursor for sterols in most eukaryotes. They are widespread in different plants families suggesting that their biosynthesis has evolved convergently. We have developed the wild crucifer *Barbarea vulgaris* as a model system to study the biosynthesis of triterpenoid saponins, the genome localization of the genes involved in the pathway, and reconstituted the pathway in tobacco leaves. Structure-activity relationships studies using insects, such as the crucifer pest Diamond back moth, begin to unravel which chemical features of a saponins influences toxicity to specific insects.

T4-19-04

Functional analysis of a promiscuous oxidosqualene cyclases in rice

*Zheyong Xue[1], Zhengwei Tan[1], Ancheng Huang[2], Yuan Zhou[1], Juncong Sun[1], Xiaoning Wang[3], Ramesha B. Thimmappa[2], Michael Stephenson[2], Anne Osbourn[2], **Xiaoquan Qi**[1]*
1. Key Laboratory of Plant Molecular Physiology, Institute of Botany, Chinese Academy of Sciences, Nanxincun 20, Fragrant Hill, Beijing
2. Department of Metabolic Biology, John Innes Centre, Norwich Research Park, Norwich NR4 7UH
3. Department of Natural Product Chemistry, Key Laboratory of Chemical Biology, School of Pharmaceutical Sciences, Shandong University, 44 West Wenhua Road, Jinan

Triterpene synthases, also known as 2,3-oxidosqualene cyclases (OSCs), synthesize diverse triterpene skeletons for the biosynthesis of both essential sterols and also a large number of functionally divergent triterpenoids. These complex tetracyclic and pentacyclic triterpenes are generated by OSCs that can use one of two original carbocations, the prosteryl and dammarenyl cations which adopt chair-boat-chair (C-B-C) and chair-chair-chair (C-C-C) conformations, respectively. Here we report a promiscuous OSC (named OsOS) in rice that produces a pentacyclic triterpene orysatinol having an unprecedented conformation derived from an unknown orysatinyl cation adopting a novel C-C-C conformation different from that of the dammarenyl cation. Phylogenetic and structure-based functional analyses revealed I732, L365, L124 of OsOS are three key amino acids driving the formation of orysatinyl cation, which upon mutagenesis, led to complete conformational switch from orysatinyl C-C-C to prosteryl C-B-C cation which yielded end product parkeol. Our results not only elucidate the detailed process by which OsPS is likely to have functionally evolved from OsOS, but also provide fundamental insights and mechanistic clues about generation of triterpene diversity in nature.

T4-19-05

Applications of plant terpenoids in the synthesis of colloidal silver nanoparticles

Zia-Ur-Rehman Mashwani[1], *Mubarak Ali Khan*[2], *Tariq Khan*[3], *Akhtar Nadhman*[4]

1. PMAS Arid Agriculture University, Rawalpindi
2. Abdul Wali Khan University, Mardan
3. University of Malakand, Chakdara
4. Islamic International University, Islamabad

Green chemistry is the design of chemical products and processes that reduce or eliminate the generation of hazardous substances. Since the last few years, natural products especially plant secondary metabolites have been extensively explored for their potency to synthesize silver nanoparticles (AgNPs). The plant-based AgNPs are safer, energy efficient, eco-friendly, and less toxic than chemically synthesized counterparts. The secondary metabolites, ubiquitously found in plants especially the terpenoid-rich essential oils, have a significant role in AgNPs synthesis. Terpenoids belong to the largest family of natural products and are found in all kinds of organisms. Their involvement in the synthesis of plant-based AgNPs has got much attention in the recent years. The current article is not meant to provide an exhaustive overview of green synthesis of nanoparticles, but to present the pertinent role of plant terpenoids in the biosynthesis of AgNPs, as capping and reducing agents for development of uniform size and shape AgNPs. An emphasis on the important role of FTIR in the identification and elucidation of major functional groups in terpenoids for AgNPs synthesis has also been reviewed in this manuscript. It was found that no such article is available that has discussed the role of plant terpenoids in the green synthesis of AgNPs.

T4-20: Plant response to drought and salt (two sessions: session 1, drought stress; session2, salt stress)

T4-20-01

Regulation of SOS pathway by 14-3-3 proteins

Yan Guo

China Agricultural University

The Salt Overly Sensitive (SOS) pathway regulates intracellular sodium ion (Na^+) homeostasis and salt tolerance in plants. In this pathway, a salt stress-induced transient increase in cytosolic calcium activates two calcium sensors, SOS3 and SOS3-Like Calcium Binding Protein 8, which interact with and activate the SOS2 protein kinase. The SOS3/SCaBP8-SOS2 complex then activates SOS1, a plasma membrane Na^+/H^+ antiporter which transports sodium from the cytosol. So far, the mechanisms that control inhibition of the SOS pathway when plants are grown in the absence of salt stress have remained elusive. We demonstrate that the λ and κ 14-3-3 proteins interact with SOS2 and repress its kinase activity. 14-3-3 λ interacts with the SOS2 junction domain, which has previously been shown to be important for its kinase activity. A phosphorylation site (Ser^{294}) was identified within this domain by mass spectrometry. Mutation of Ser^{294} to Alanine (A) or Aspartate (D) did not affect SOS2 kinase activity in the absence of 14-3-3s. However, the S/A mutation decreased the 14-3-3-dependent inhibition of activity, while the S/D mutation increased this inhibitory effect. $SOS2^{S294A}$ rescued the salt-sensitive phenotype of the *sos2-2* mutant to wild-type levels while $SOS2^{S294D}$ did not. Moreover, we find that anther protein kinase phosphorylates the SOS2 Ser^{294} site in normal condition. Upon salt stress, this kinase activity is repressed by the 14-3-3 proteins, which in turn releases the SOS2 activity for activating SOS1. These results identify 14-3-3 proteins as important regulators of salt tolerance and uncover their role in a mechanism that confers basal repression of the SOS pathway in the absence of salt stress.

T4-20-02

Nuclear-localized SOS3 stabilizes GIGANTEA to ensure flowering under salt stress

Kim Woe-Yeon[1], *Aman Rashid*[1], *Park Hee-Jin*[1], *Irene Villalta*[2], *Francisco Quntero*[2], *Jose M. Pardo*[2], ***Dae-Jin Yun***[3]

1. Division of Applied Life Science (BK21 Program), Graduate School of Gyeongsang National University, Jinju 660–701, South Korea
2. Institute of Plant Biochemistry and Photosynthesis (IBVF), Consejo Superior de Investigaciones Cientificas, Spain
3. Department of Biomedical Science and Engineering, Konkuk University, Seoul 05029, South Korea

The timing of flowering under stress environments is critical for plants to secure reproductive success, and salt stress delays flowering time in *Arabidopsis*. In normal growth conditions, the circadian and photoperiodic flowering time regulator GIGANTEA (GI) makes a complex with and inhibits the salt-tolerance kinase SOS2 (CIPK24). Salt stress causes the degradation of GI protein by the proteasome and the release of SOS2, which is then free to interact with the Ca^{2+}-dependent protein SOS3 (CBL4) to activate the Na^+ efflux protein SOS1 and mount successful adaptation to the saline environment. Degradation of GI delays but does not abrogate flowering in a saline environment. Here we show that SOS3, a regulator of SOS2, also makes a complex with GI in the nucleus to ensure the expression of the flowering activator *CONSTANS* (*CO*) under salt stress. Palmitoylation-dependent nuclear import of SOS3 stabilizes GI in the nucleus to proceed with flowering, while degradation of cytosolic GI releases the protein kinase SOS2 to achieve salt tolerance. Thus, SOS3 is a Ca^{2+}- and palmitoylation-dependent molecular switch required to ensure flowering is a saline environment by selectively stabilizing GI in the nucleus.

T4-20-03

Coordinated regulation of potassium uptake and storage in *Arabidopsis*

Paula Ragel[1], *Belen Rombola*[1], *Zaida Andres*[2], *Javier Perez-Hormaeche*[3], *Ana De Luca*[1], *Imelda Mendoza*[3], *Maria A. Parrado*[1], *Beatriz L. Cubero*[1], *Eduardo O. Leidi*[1], *Francisco J. Quintero*[3], ***Jose M. Pardo***[3]

1. Instituto de Recursos Naturales y Agrobiologia de Sevilla (IRNAS-CSIC)
2. Centre for Organismal Studies, Heidelberg University
3. Instituto de Bioquimica Vegetal y Fotosintesis (IBVF-CSIC)

Potassium (K) is an essential nutrient for all living organisms, but in plants K reaches organismal contents of up to 10% of dry weight. The ability of plants for K uptake and storage is critical for stress tolerance and a key issue in agricultural practices. While cytosolic K is kept at homeostatic concentrations close to 100 mM, surplus K is stored in cell vacuoles in large quantities. An array of K transporters and regulators (protein kinases and phosphatases, interacting proteins, diffusive chemical messengers) have been identified as the basic elements underpinning K fluxes at the plasma membrane and tonoplast, but the regulatory networks involved in controlling K homeostasis in individual cells and whole organisms remain largely unknown. Current evidence indicates that active K uptake (i.e., against the electrochemical K gradient) involves K-H symporters at the plasma membrane (HAK, High-Affinity K transporters) and K/H antiporters (NHX) at the tonoplast. These transporters, together with K-selective channels that fine-tune the membrane electrical potential by K uptake and release, share common regulators that are effectors of K nutrition. We have shown that the Ca^{2+} sensors Calcineurin B-Like CBL1, CBL8, CBL9, and CBL10, together with CBL-Interacting Protein Kinase 23 (CIPK23) activated HAK5 in vivo. CIPK23 is also known to regulate the root specific K-inward channel AKT1. Measurement of K^+ (Rb^+) uptake and growth rate in low-K^+ medium with Arabidopsis mutants bearing mutations *hak5*, *akt1* and *cipk23* in various combinations confirmed the key regulatory role of CIPK23 in K acquisition in planta. Cytosolic K is trafficked further to the vacuolar lumen and the tonoplast-localized K/H exchangers of the NHX family are pivotal to the vacuolar accumulation of K. Hypomorphic and loss-of-function double mutants *nhx1 nhx2* of Arabidopsis had impaired K acquisition, failed to osmoregulate and were unable to create sufficient turgor for cellular expansion. Moreover, they had defective stomatal opening and closure responses. Disruption of K^+ accumulation in guard cells correlated with more acidic vacuoles and the disappearance of the highly dynamic remodeling of vacuolar structure associated with stomatal movements. Notably, vacuolar NHX proteins are also regulated by CIPKs, which are emerging as central regulators of K transport processes occurring at various cells membranes and plant organs.

T4-20-04

High-throughput screening tools for identification of traits contributing to salinity tolerance

Klara Panzarova

1. PSI (Photon Systems Instruments), Drásov, Czech Republic
2 Center for Desert Agriculture, KAUST, Saudi Arabia
3 Institute of Plant Biology, University of Zürich, Switzerland

Non-invasive capturing and interpreting of plant structural and functional phenotypes in controlled or dynamically changing environment is long-standing and necessary requirement for genetic and physiological research by crop breeders, agricultural industry, and academia. To sustain global food security the major challenge global agriculture and plant biology field has been facing is the identification of new high-yielding genotypes of agricultural crops that are adapted to our future climate. Soil salinity is one of the main stress factors that are severely affecting the agriculture land in global scale and results in significant reduction of plant growth and yield. It was shown that plants suffer a rapid growth reduction upon the first exposure of their roots to salt stress, which is occurring prior to the accumulation of ions to toxic concentrations in the shoots. During this early phase, symptoms of growth reduction include slower leaf emergence and a small growth size. The phenotypic traits associated with this type of tolerance can be quantified in the days immediately after imposition of stress using non-destructive image-based phenotyping. To enhance our understanding of the early responses to salinity, we designed an experimental protocol based on using high-throughput and non-invasive imaging technologies developed at Photon Systems Instruments (PSI, Czech Republic). The methodology presented is based on automated integrative analysis of photosynthetic performance, growth analysis and color index analysis at the onset and early phase of salinity stress response of *Arabidopsis thaliana* ecotypes grown in soil. Here we show that stress imposition significantly and rapidly affected photosystem II operating efficiency, subsequently impacted growth dynamics and greening index of Arabidopsis plants at different stages of stress response. Our work provides quantitative insights into early phase of salinity response and provides robust protocol for high-throughput image-based analysis of phenotypic traits associated with this early phase of salinity response. We show that the integrative concept of PlantScreen™ high-throughput phenotyping platform provides a powerful tool for acquisition and selection of morphological, physiological and biochemical parameters, which can be used for identification of various components underlying early plant responses to various environmental conditions.

T4-20-05

Regulation of alternative polyadenylation in *Arabidopsis thaliana* response to salt stress

Lingling Cai[1], *Qingshun Q. Li*[1,2]

1. Xiamen University

2. *Western University of Health Sciences*

Alternative polyadenylation (APA) is an important post-transcriptional regulatory process during eukaryotic gene expression. It has been reported to be involved in abiotic stress such as oxidative stress. However, the roles of alternative polyadenylation in plant response to salt stress remains elusive. In this study, genome-wide studies of polyadenylation sites have been carried out in *Arabidopsis thaliana* seedlings treated with 200mM NaCl using the poly (A) tag sequencing (PAT-seq) approach. We found that a large number of genes were alternative polyadenylated under salt stress and the ratio of genes having more than two poly (A) sites increased. Genomic distributions of poly (A) sites seen only in salt stress condition, unstressed condition and common were quite different, suggesting the genes 3'-end formation pattern being changed under salt stress. We claimed that most of the differential expression (DE) APA genes were also differently expressed genes, which means that APA may cooperated with gene expression regulating in response to salt stress. Besides that, the group of differential expression (DE) APA genes were enriched in salt stress response and plant hormone signal transduction pathway, revealing the effect of APA on salt and hormone response genes. Morever, a dozen of transcription factors and salt response genes were exclusively possessed by APA under salt stress, unveiling a complex loop between APA and transcriptional regulation. Taken together, our study provided new insights into the potential roles of APA in plant response to salt stress.

T4-20-06

Possible roles of RING E3 ubiquitin ligases in drought and proteotoxic stress responses in *Arabidopsis*

Woo Taek Kim
Yonsei University

Ubiquitination is implicated in diverse cellular processes in higher plants. Arabidopsis contains at least 477 RING-motif containing E3 ubiquitin (Ub) ligases. *AtAIRP1* and *AtAIRP2* encode cytosolic C3H2C3- and C3HC4-type RING E3 ligases, respectively, whose expressions were induced by ABA and dehydration stress. Both *AtAIRP1*- and *AtAIRP2*-overexpressing transgenic plants exhibited hypersensitive phenotypes to ABA in terms of stomatal movement. They were highly tolerant to severe drought stress. In contrast, *atairp1* and *atairp2* loss-of-function mutants showed opposite phenotypes in response to ABA and drought stress. Constitutive expression of *AtAIRP1* and *AtAIRP2* in *atairp2* and *atairp1* mutant plants, respectively, reciprocally rescued the loss-of-function ABA-insensitive and drought-hypersensitive phenotypes of the mutant plants. These results suggest that AtAIRP1 plays combinatory roles with AtAIRP2 in ABA-mediated drought stress responses in Arabidopsis. We identified that over-expression of *MPRR1* (Misfolded Protein Responsive RING 1) resulted in the markedly increased tolerance to AZC and arsenate treatments, both of which induced cytosolic proteotoxic stress, in Arabidopsis. Possible role of MPRR1 E3 ligase in response to proteotoxic damage will be discussed.

T4-20-07

Reactive oxygen species in the regulation of stomatal movements

Jaakko Kangasjärvi
University of Helsinki

Water availability is the most limiting factor for agricultural production and insufficient water supply can cause major reductions in crop yields. Guard cells that form stomatal pores regulate plant water balance since plants with more open stomata allow faster water evaporation. Assimilation of CO_2 requires stomatal opening, but also opens the gates for excessive water loss and consequently, several environmental factors, such as light intensity and quality, soil and air water content, and CO_2 regulate stomatal aperture. Thus, guard cells sense and respond to diverse stimuli optimizing photosynthetic carbon dioxide uptake with minimal water loss. Stomatal closure is controlled by a complex network of signaling compounds, including the drought phytohormone abscisic acid (ABA), rise in cytoplasmic calcium (Ca^{2+}) and reactive oxygen species (ROS). ROS are used by the plant as secondary messengers in several abiotic stresses, such as drought, and salt. In responses to these stresses regulation of stomatal aperture requires coordinated activity of ROS-generating enzymes, signaling proteins and downstream executors such as ion pumps, transporters and plasma membrane ion channels that control guard cell turgor pressure. Accumulation of reactive oxygen species (ROS) in the apoplast and chloroplasts is among the earliest hallmarks of stomatal closure. Subsequent increase in cytoplasmic Ca^{2+} concentration governs the activity of multiple kinases that regulate the activity of ROS-producing enzymes and ion channels. In parallel, ROS directly regulate the activity of multiple proteins via oxidative posttranslational modifications to fine-tune guard cell signaling. I will describe recent advances in the role of ROS in stomatal closure as a response to water limitation and discuss the importance of ROS in regulation of signal amplification and specificity in guard cells. Detailed knowledge of how a specific components are employed in constitutive and inducible complexes in response to a specific stimuli, such as drought, at a given time and location will be required for full understanding of the mechanisms underlying stomatal responses and can be used to identify components that regulate plant water use efficiency in the shifting global climate.

T4-20-08

Regulation of clade A type PP2Cs in ABA signaling

Zhizhong Gong
China Agricultural University

The phytohormone abscisic acid (ABA) is quickly accumulated under drought stress, which plays crucial roles in plant development and responses to abiotic and biotic stresses. The accumulated ABA is perceived by PYR1/PYL/RCAR ABA receptors that then interact with and inhibit the clade A protein phosphatases 2Cs (PP2Cs), and release their inhibition on the downstream protein kinases such as CDPKs, SnRKs and GHR1. Thus the core negative regulation PP2Cs play crucial roles in transducing ABA signaling. However, besides ABA receptors, how other factors regulate PP2Cs is not well known. Recently, We found that one of PP2c members, ABI1 (ABA INSENSITIVE1) can be degraded by PUB12/13 E3 ligases only after it interacts with PYR1/PYL/RCAR ABA receptors either in presence of ABA or no ABA. Furthermore, we found that EAR1 (Enhancer of ABA co-Receptor 1), an uncharacterized protein, can interact with N-termini of 6 members of the clade A type PP2Cs and enhance the PP2Cs activities. Removing N-termini of PP2Cs releases their autoinhibition, but EAR1 does not add any enhancement on their PP2C activities. These studies highlights the novel paradigms for regulation of the PP2C activities in *Arabidopsis*.

T4-20-09

The extremophyte *Schrenkiella parvula* as an emerging model to study salt stress adaptations

Maheshi Dassanayake, *Narender Kumar, Kieu-Nga Tran, Pramod Pantha, John Johnson, Guannan Wang, Chathura Wijesinhege, Dong-Ha Oh*

Louisiana State University

Schrenkilla parvula (*Syn Thellungiella parvula)*, a close relative of *Arabidopsis thaliana* and *Brassica* crops, thrives in highly saline soils. It is also drought and cold tolerant as adaptations that mark its lifestyle in the dry and extreme weathers of salt flats in Central Anatolia, Turkey. *S. parvula* compared to *A. thaliana*, presents a unique system to identify naturally selected physiological traits, developmental responses, and genome reorganization behind adaptations to salt stress. Despite the stark differences in adaptations to extreme salt stresses, *S. parvula* and *A. thaliana* show extensive genome-wide macrosynteny. However, explicit genome structural variations, including tandem gene duplications, gene transpositions, and transposable element insertions interrupt the co-linearity observed throughout the genomes and distinguish *S. parvula* from *A. thaliana* and other closely related non-halophytic species. Comparative physiological observations on *S. parvula* support the comparative transcriptomic studies between *S. parvula* and *A. thaliana* that highlight the stress-prepared responses of *S. parvula* even under stress neutral conditions. These studies have led to the identification of key physiological and developmental traits in *S. parvula*, and the gene families involved in ion transport and other metabolic pathways associated with salt stress responses. Genomic structural variants appeared to have led the two species into two distinct lifestyle trajectories by their enrichment in significantly differently expressed homologous genes in root and shoot transcriptomes under neutral and varying salt stress conditions. Furthermore, we observed tissue-specific isoform expression contributing to species-specific salt stress responses. The extremophyte genome of *S. parvula* acts as a repository of genetic changes that have enabled its successful niche adaptation to a hyper-saline environment. Our genomic and transcriptomic dissection offers a framework to identify adjustments of genome architecture and expression that control a set of genes regulating fundamental physiological responses in salt stress adaptations.

T4-20-10

Origin and evolution of the plant succulent syndrome in arid regions of the world

Tania Hernandez-Hernandez

University of Arizona

Of all the major climatic groups, arid climates are the most extensively distributed over the land surface of the earth. Ecosystems in arid environments have low precipitation, highly variable and random, occurring in infrequent discrete events, making water availability the dominant factor controlling biological processes. It is estimated that arid biomes appeared recently, being in most cases no older than the Miocene/Pliocene, and the rapid origin of these environments created the selective pressure under which several plants evolved a diversity of strategies and specialized structures to withstand periods of severe drought. Some drought-adapted lineages went through the most dramatic evolutionary modifications in the plant kingdom by developing extremely modified organs and tissues to store water, i.e. succulence. Some of the succulent lineages are currently the most species rich and structurally diverse plant families on earth. Current studies project increases in the frequency and intensity of drought related to global climate change. Thus, understanding the details of the diversification mechanisms of succulent plants in arid environments could help us develop better predictions about how fast can plants adapt to increasingly harsh conditions, and how they may respond to an even warmer environment in the near future. Moreover, understanding the evolution of the anatomical, physiological and morphological adaptations of these plants, and especially the genetic bases of these adaptations, may provide information that could lead to the development of genetic tools to improve food crops and generate bioenergy for the future. Recent estimates indicate that there are ~12,500 species of succulents, distributed among 690 genera and 83 families. The fact that lineages with the succulent syndrome independently evolved convergent solutions to the problem of water scarcity could indicate that there is a parallel adaptive response to a global trend. It has been proposed that succulent families originated contemporaneously, in response to a global aridification trend and a global scenario of a decline in atmospheric CO_2 concentrations. Here I discuss the most recent knowledge about the morphological and anatomical adaptations of succulents to water scarcity, as well as current knowledge about the genetic and molecular bases of these adaptations. I present estimated dates of origin of major succulent lineages, and discuss the hypothesis of a synchronic origin in response to a global aridification trend. I present data on the diversification patterns of major succulent lineages in different arid regions of the world, and discuss the possible drivers of their diversification. Finally, I discuss the possible outcome of succulent plants under a global warming scenario and the possibility of using information about the origin and evolution of the succulent syndrome to design strategies to understand and improve the future outcome of food crops.

T4-20-11

Revealing the causal relationship between salinity and oxidative stress tolerance in wheat and barley

Haiyang Wang, *Lana Shabala, Meixue Zhou, Sergey Shabala*

University of Tasmania

Drought and salinity are two major abiotic stresses limiting crop production worldwide. Both stresses result in an increased production and accumulation of reactive oxygen species (ROS), causing oxidative stress. The latter disturbs the metabolic balance of the cells and causes damage to major cell structures. ROS also activate several different types of ion channels affecting intracellular ionic homeostasis and thus plant's ability to adapt to adverse environmental conditions. At the same time, more moderate quantities of ROS are essential for a range of signalling pathways mediating a diverse range of processes such as plant growth, development, gravitropism, or hormonal action. Thus, the causal link between ROS production and stress tolerance is not as straightforward as one may expect. In this work, we used non-invasive microelectrode ion flux estimation (the MIFE) technique to establish a causal relationship between salinity and oxidative stress tolerance in two cereal crops: barley (*Hordeum vulgare*) and wheat (*Triticum aestivum*). By screening 80 different genotypes we show that exogenous application of H_2O_2 to 3d hydroponically-grown roots results in massive, dose-dependent K^+ efflux and Ca^{2+} influx in both elongation and mature zones of root epidermis. A strong positive correlation was found between H_2O_2-induced K^+ efflux and overall salinity tolerance in both species in mature root epidermis. The same positive trend was also found in Ca^{2+} influx. Pharmacological experiments suggested that non-selective cation channels (NSCCs) mediated such Ca^{2+} influx and played a major role in controlling K^+ efflux in both species. The above trends were not applicable to cells in the elongation root zone, indicating high tissue specificity of ion flux responses to ROS. Consistent with H_2O_2 data for mature zone, application of a hydroxyl radical (·OH)-generating Cu^{2+}/ascorbate (Cu/A) mixture did not reveal any clear correlation between K^+ and Ca^{2+} flux responses and an overall salinity tolerance. Implication of these findings for plant breeding for salinity stress tolerance are discussed.

T4-20-12

Genome-wide association study of drought tolerance and gene cloning in maize seedlings

Feng Qin
College of Biological Sciences, China Agricultural University

Maize production is frequently threatened by drought stress on a global-scale. Identification of the genetic components underlying the drought tolerance in maize is of great importance. Here, we report a genome-wide association study (GWAS) of maize drought tolerance at the seedling stage that identified 83 genetic variations which were resolved to 42 candidate genes. Firstly, a Miniature Inverted-repeat Transposable Element (MITE) inserted in the promoter of a *NAC* gene (*ZmNAC111*) was found to be significantly associated with the natural variation in maize drought tolerance. We discovered that the 82-bp MITE represses *ZmNAC111* expression via RNA-directed DNA methylation and H3K9 dimethylation. Increasing *ZmNAC111* expression in transgenic maize conferred enhanced drought tolerance by improving water use efficiency and upregulation of drought-responsive genes under water stress. Secondly, the peak signal of GWAS uncovered that the natural variation in *ZmVPP1*, encoding a vacuolar-type H^+-pyrophosphatase, most significantly contributes to maize drought tolerance. Further analysis found that a 366-bp insertion in the promoter, containing three MYB *cis*-elements, confers drought-inducible expression of *ZmVPP1* in drought-tolerant genotypes. Transgenic maize with enhanced *ZmVPP1* expression exhibits improved drought tolerance, probably due to enhanced photosynthetic efficiency and root development. Taken together, this research provides important genetic insights into the natural variation of maize drought tolerance at seedling stage. The identified loci/genes can serve as direct targets for both genetic engineering and selection for the trait improvement of maize.

T4-21: Plant responses to ambient and stress temperature

T4-21-01

Regulation of protein kinases in plant response to cold stress

*Yanglin Ding, Jingyan Liu, **Shuhua Yang***
State Key Laboratory of Plant Physiology and Biochemistry, College of Biological Sciences, China Agricultural University

Cold stress is an environmental factor that affects plant growth, development and distribution. The well-studied cold signaling pathway is the ICE1-CBF transcriptional cascade in *Arabidopsis*; however, the specific protein kinases involved in this pathway remain elusive. We report that SnRK2.6/OST1 (OPEN STOMATA1), a well-known Ser/Thr protein kinase in ABA signaling, acts upstream of *CBF*s to positively regulate freezing tolerance. The *ost1* mutants show freezing hypersensitivity, whereas overexpression of *OST1* enhances freezing tolerance. Cold can activate OST1 kinase activity, and activated OST1 phosphorylates ICE1, which in turn suppresses HOS1-mediated ICE1 degradation under cold conditions, thereby promoting *CBF* expression and plant freezing tolerance. In addition, we isolated another protein kinase CRPK1 (Cold-responsive protein kinase 1) which negatively regulates plant freezing tolerance. The cold activated plasma membrane protein CRPK1 phosphorylated its interacting protein KIP1 (CRPK1 interacting protein 1), which shuttled from the cytosol to the nucleus to promote CBF protein degradation under cold stress, thus fine-tuning CBF signaling during cold response. Our results thus uncover the important roles of different kinases in regulating CBF-dependent cold signaling in *Arabidopsis*.

T4-21-02

Involvement of a subfamily of SAUR genes in growth architecture thermosensitivity in *Arabidopsis*

Jian Hua
Cornell University

We are interested in how diversity in thermosensitivity of growth architecture is evolved. Responses in development and growth architecture to a moderate temperature change differ greatly among accessions of *Arabidopsis thaliana*. Although molecular genetics on the reference accession Col-0 have revealed molecular identities of some regulatory components in temperature responses, it is not known how diversity in thermo-responses is generated. We used natural accessions of *Arabidopsis thaliana* to reveal the major molecular loci conferring the genetic diversity in growth architecture thermosensitivity. Genome Wide Association Study (GWAS) has revealed a major quantitative trait locus (QTL) for variation in thermosensitivity of growth architecture. The major QTL contains a sub-family of auxin response *SAUR* (*Small Auxin Up Response*) genes which likely are central in temperature responses in rosette leaf architecture. The function and diversity of these *SAUR* genes in temperature responses of growth architecture will be discussed.

T4-21-03

HOS15, a WD40 protein, mediated chromatin remodeling in plant freezing tolerance

Dae-Jin Yun

Department of Biomedical Science and Engineering, Konkuk University, Seoul 05029, South Korea

Histone modification in chromatin such as histone acetylation and deacetylation plays key role in gene regulation in eukaryotic cells. Previously, we reported that HOS15, a WD40 repeat protein, is involved in histone modification and cold tolerance in *Arabidopsis* plant (Zhu *et al.*, 2008). However, it remains unknown the molecular function of HOS15, and how HOS15 regulates cold responses in chromatin level. Here, we show that HOS15 functions as a substrate receptor for CULLIN4 (CUL4)-based ubiquitin E3 ligase and regulates C-repeat binding factors (CBFs)-dependent cold responsive genes (*CORs*) expression in chromatin level by degrading histone deacetylase 2C (HD2C) through ubiquitin/proteasome pathway. HOS15 forms complex with CUL4-DDB1 *in vitro* and *in vivo*. A yeast two-hybrid screening identified a HD2C as a substrate of HOS15. Accordingly, HD2C interacts with HOS15 in the nucleus and interacts with histone H3. Cold induced HD2C degradation and increased histone H3 acetylation is dependent on HOS15. Loss of HD2C in plant leads to higher accumulation of *CORs*, such as *RD29a*, *COR15a* and *COR47*, and increased freezing tolerance. Chromatin Immunoprecipitation (ChIP) assays show that a HOS15/HD2C complex is associates with CBFs-dependent *CORs* chromatin in normal condition. However, in response to cold, CBFs binds to *CORs* chromatin in a HOS15–dependent manner and removal of HD2C by HOS15 is essential for CBFs-binding to the *CORs* chromatin and *CORs* expression. Altogether, our data establishes a function for HOS15 as a CUL4-based ubiquitin E3 ligase receptor and uncover a mechanism for cold acclimation and tolerance in plant based on the HOS15/HD2C/CBFs-mediated epigenetic regulation of *CORs*.

T4-21-04

Temperature perception and response in plants

Sureshkumar Balasubramanian

Monash University

Ambient temperature affects several aspects of plant growth and development and the underlying molecular mechanisms are just beginning to be elucidated. Recent studies implicate phytochromes as potential thermosensors in plants. In addition a role for histone modifications have been implicated in thermal response. By using a combination of mutant analysis and natural variation, we uncover the genetic architecture of natural variation in thermal responses of *Arabidopsis*. I will present our recent findings on a novel gene that appears to be essential for themo-morphogenesis in *Arabidopsis*. In addition, I will describe our novel finding that implicates a role for alternative splicing coupled with nonsense-mediated mRNA decay in thermal responses of *Arabidopsis thaliana*.

T4-21-05

Chromatin regulation of heat stress memory

Jörn Lämke, Krzysztof Brzezinka, ***Isabel Bäurle***

University of Potsdam

In nature, plants often encounter chronic or recurring stressful conditions. An increasing number of observations suggest that plants can remember a past exposure to stress in order to be better prepared for a future stress incident. My lab studies heat stress memory in plants as a model case for environmental stress memory. Seedlings acquire thermotolerance through a heat treatment at sublethal temperatures (priming heat stress (HS)) that enables them to survive an otherwise lethal HS. This thermotolerance is actively maintained for several days as indicated by the existence of mutants which are able to establish thermotolerance, but fail to maintain it. We have been exploring the molecular basis of HS memory in the model plant *Arabidopsis thaliana*. We have found that HS induces sustained histone methylation at HS memory-related loci that outlasts the transcriptional activity of these loci and hence marks them as recently transcriptionally active. At certain loci, this is correlated with hyper-induction of gene expression upon a recurring HS, - a signature readout of transcriptional memory. In a forward genetics approach, we have found that regulated nucleosome occupancy is also required for sustained activation of memory gene expression. Sustained low nucleosome occupancy is mediated by the highly conserved FORGETTER1 protein through direct interaction with chromatin remodeling proteins. I will present latest our finding on the role of chromatin structure in stress memory. In summary, the physiologically defined phenomenon of HS memory has a molecular equivalent in the transcriptional memory and changes in chromatin structure at the level of histone modifications and nucleosome occupancy are associated with this. In the long term, our research will provide new approaches to improve stress tolerance in crops.

T4-21-06

Genetic control of thermomorphogenesis in rice

Yongbiao Xue

Institute of Genetics and Developmental Biology, Chinese Academy of Sciences

Global warming is having strong negative impacts on crop productivity. Being sessile, plants have evolved numerous intrinsic strategies to acclimate to ambient temperature increases often referred to as thermomorphogenesis. However, we know little about genetic control of thermomorphogeneis in rice (*Oryza sativa*), a staple crop for more than half of the world population. To dissect its genetic basis, we did a large scale screen for genetic mutations affecting rice thermomorphgenesis and several independent genetic loci were subsequently identified. Genetic and molecular analyses of these loci revealed insights into the molecular control of rice thermomorphogenesis. In this talk, I will present our recent progresses and highlight the challenges and opportunities to breed new heat-tolerant rice cultivars.

T4-22: Plant response to phosphorus and nitrogen (two sessions)

T4-22-01

STOP1-ALMT1 a new signalling pathway inhibiting root growth in low-phosphate

Coline Balzergue[1,2,3], *Thibault Dartevelle*[1,2,3], *Christian Godon*[1,2,3], *Edith Laugier*[1,2,3], *Claudia Meisrimler*[1,2,3], *Jean-Marie Teulon*[1,2,4], *Audrey Creff*[1,2,3], *Marie Bissler*[1,2,3], *Corinne Brouchoud*[1,2,3], *Agnès Hagège*[1], *Jens Müller*[5], *Serge Chiarenza*[1,2,3], *Hélène Javot*[1,2,3], *Noëlle Becuwe-Linka*[1,2,3], *Benjamin Péret*[1,2,3], *Etienne Delannoy*[1,2,3], *Marie-Christine Thibaud*[1,2,3], *Jean Armangaud*[1], *Steffen Abel*[5], *Jean-Luc Pellequer*[1,2,4], *Laurent Nussaume*[1,2,3], ***Thierry Desnos***[1,2,3]

1. CEA
2. CNRS
3. Aix-Marseille Université
4. Université de Grenoble
5. Leibniz Institute of Plant Biochemistry

When grown on a low-phosphate (–Pi) medium, the Arabidopsis primary root growth is inhibited as a consequence of reduced cell elongation and proliferation. We used a classical genetics approach to identify genes involved in these responses. By QTL analysis we identified the *LPR1* gene that inhibits the primary root growth in –Pi; LPR1 is a feroxidase targeted to the apoplast. From a collection of new EMS mutants that display a long root in –Pi, we identified many alleles of two other genes: *STOP1* and *ALMT1*. STOP1 is a transcription factor that directly regulates *ALMT1,* encoding a malate-exuding transporter. The –Pi stress enhances the expression of *ALMT1*, independently of PHR1 and PHL1. This makes the STOP1-ALMT1 module a new pathway of the –Pi signalling. We show that STOP1, ALMT1 and LPR1 rapidly inhibit root cell elongation in the transition zone, whereas LPR1 inhibits of cell proliferation in the meristem. The inhibition of cell elongation and cell proliferation is correlated with deposition of Fe and callose in the cell walls of the elongation zone and the meristem, respectively. By using atomic force microscopy (AFM) indentation, we also demonstrate that the rapid inhibition of cell elongation is correlated with stiffening of the cell wall, in an iron and peroxidase-dependent manner. We will present additional results and a model about the role of these genes in the –Pi-induced primary root growth arrest.

T4-22-02

Phosphate uptake by plant and its regulation

Satomi Kanno[1,3], *Marie-Christine Thibaud*[1], *Pascale David*[1], *Elena Marin*[1], *Jieyu Chen*[2], *Ping Wu*[2], *Tomoko Nakanishi*[3], *Vincent Bayle*[4], ***Laurent Nussaume***[1]

1. CEA-CNRS-Aix Marseille University
2. Zhejiang University
3. The University of Tokyo
4. CNRS

Phosphate (Pi) is a crucial and often limiting nutrient for plant growth. It is also a very insoluble ion heterogeneously distributed in soil triggering numerous morphological and biochemical adaptations. The uptake of this element relies on the presence of nine high affinity transporters (PHT1 family) located in plasma membranes. Multiple and complex steps of transcriptional and post-transcriptional regulations (phosphorylation, degradation) of these proteins were identified illustrating the capacity for plants to tightly control the level of these transporters into the cells. Combining several genetic approaches we reduced 95% of PHT1 family activities offering unique opportunities to investigate the physiological role of these proteins. It highlighted their crucial role for Pi absorption and reveals that at least some (if not all) PHT1 members should exhibit dual affinity properties to improve Plant adaptation to the phosphate concentration present in the environment. Innovative techniques, such as radioisotope live micro-imaging system was used to image the spatial distribution of Pi absorption along the root. It revealed unexpected location of the uptake and provides opportunity to dissect the specific role of different root cell layers.

T4-22-03

The molecular mechanism of ethylene-mediated root hair development induced by phosphate starvation

Li Song[1], *Haopeng Yu*[2], *Jinsong Dong*[1], *Ximing Che*[1], *Yuling Jiao*[2], ***Dong Liu***[1]

1. School of Life Sciences, Tsinghua University
2. Institute of Genetics and Developmental Biology, Chinese Academy of Sciences

Enhanced root hair production, which increases the root surface area for nutrient uptake, is a typical adaptive response of plants to phosphate (Pi) starvation. Although previous studies have shown that ethylene plays an important role in root hair development induced by Pi starvation, the underlying molecular mechanism is not understood. In this work, we characterized an *Arabidopsis* mutant, *hps5*, that displays constitutive ethylene responses and increased sensitivity to Pi starvation due to a mutation in the ethylene receptor ERS1. *hps5* accumulates high levels of EIN3 protein, a key transcription factor involved in the ethylene signaling pathway, under both Pi sufficiency and deficiency. Pi starvation also increases the accumulation of EIN3 protein. Combined molecular, genetic and genomic analyses identified a group of genes that affect root hair development by regulating cell wall modifications. The expression of these genes is induced by Pi starvation and is enhanced in the *EIN3* over-expressing line. In contrast, the induction of these genes by Pi starvation is suppressed in *ein3*and *ein3eil1*mutants. EIN3 protein can directly bind to the promoter of these genes, some of which are also the immediate targets of RSL4, a key transcription factor that regulates root hair development. Based on these results, we propose that under normal growth conditions, the level of ethylene is low in root cells; a group of key transcription factors, including RSL4 and its homologs, trigger the transcription of their target genes to promote root hair development; Pi starvation increases the levels of the protein EIN3, which directly binds to the promoters of the genes targeted by RSL4 and its homologs and further increase their transcription, resulting in the enhanced production of root hairs. This model not only explains how ethylene mediates root hair responses to Pi starvation, but may provide a general mechanism for how ethylene regulates root hair development under both stress and non-stress conditions.

T4-22-04

Identification of root cell-specific regulators of phosphate signalling and their effect on whole plant phosphorus status

Ricarda Jost[1], *Joshua Linn*[1], *Meiyan Ren*[1], *Oliver Berkowitz*[1], *Wona Ding*[2], *Margaretha J. Van Der Merwe*[3], *James Whelan*[1]

1. Department of Animal, Plant and Soil Sciences, Australian Research Council Centre of Excellence in Plant Energy Biology,

School of Life Sciences, La Trobe University, Bundoora, Victoria
2. College of Science & Technology, Ningbo University, Ningbo
3. Australian

Plant roots mediate the acquisition of most nutrients needed for growth and development. The balancing of these nutrient supplies and inherent demands has been shown to be regulated in different tissues across the root, with the root epidermis controlling acquisition and symplastic transport, the endodermis merging symplastic and apoplastic fluxes and the loading of the xylem ultimately determining nutrient supply to the shoot. Recent research indicates that nutrients such as nitrate, potassium or phosphate have different control points along their transport route. Phosphate is a major limiting nutrient for plant growth and we were interested in identifying components of the phosphate signalling network that are expressed in a root-cell specific manner, making them prime candidates for gatekeeping functions of phosphate flux. We identified these candidates by merging transcriptome data on phosphate-starvation induced (PSI) genes with those on root-cell enriched genes. The resulting list of root-cell enriched PSI genes was then expanded by adding predicted interactors and co-expression in a given root cell type. Candidate *PSI-interacting root-cell enriched* (*PRCE*) genes were then selected based on their predicted interaction with phosphate transporters (*WDD1*, *PRCE1*, *PSF2*, *STP10*) or on their potential to alter transport activities. The latter group comprised protein kinase components (*CBL1*, *CIPK2*, *CIPK14* and *S6K2*) as well as other membrane proteins (*MFS1*, *PRCE2* and *PRCE3*). Differential expression across root cell files was confirmed by promoter-GFP lines. Only two candidates, *CBL1* and *CIPK2* were ubiquitously expressed. Analysis of T-DNA insertion lines of the eleven *PRCE* genes revealed altered biomass and organ phosphate accumulation as well as transcriptome profiles in response to limited phosphate supply. The *s6k2* mutant showed a very similar transcriptome profile to the *phr1-1* mutant, a loss-of-function allele of master regulator PHOSPHATE STARVATION RESPONSE 1, a MYB transcription factor controlling the majority of PSI genes. However, in contrast to the *phr1-1* mutant, the *s6k2* mutant showed higher root biomass and shoot phosphate concentration than P-limited wild-type. Root transcriptome profiles of *cbl1* and *wdd1* mutants showed the opposite response to those of *phr1-1*. Together with their sustained biomass accumulation under limited Pi supply, these two genes therefore encode strong negative regulators of the plant's phosphate-starvation response. Consequences of manipulation of these root-cell specific networks for overall plant performance in response to phosphate fertilisation will be discussed.

T4-22-05

Molecular regulation mechanisms of phosphate transporter in rice

Chuanzao Mao, *Jian Yang, Zhili Yang, Jieyu Chen, Zhongchang Wu*
College of life science, Zhejiang University

Phosphorus is an essential macronutrients for plant growth and development. Under current situation of limited phosphate resources, low phosphate fertilizer use efficiency and high demand for green crops, it is important to improve the absorption and utilization efficiency of phosphorus in crops to reduce the amount of chemical fertilizers applied in agricultural production. Phosphate transporters (PTs) are direct executors of phosphorus uptake and transport, which directly involved in plant uptake of phosphate from soils. Phosphate transporters are regulated at transcriptional and posttranscriptional levels which allows plant cells to adapt to changes in cellular phosphorus environment. Rice OsPHF1 is a facilitator protein which facilitates the exit of phosphate transporter (PT) from the endoplasmic reticulum and localized to plasma membrane (PM). Protein kinase CK2 phosphorylates the phosphate transporters under high phosphorus conditions, and the phosphorylated PT does not interact with OsPHF1 thus remains on the endoplasmic reticulum. When plant is moved to low phosphate environment, CK2β subunit would be degraded, it cannot phosphorylated PT anymore. In the same time, the phosphorylated PT would further be dephosphorylated by protein phosphatase. The non-phosphorylated PT exits from the endoplasmic reticulum and localizes to the cytoplasm membrane with the help of OsPHF1, thereby facilitating the transport of phosphate from the extracellular to the intracellular to allow the plant to adapt to a low phosphate environment. The mechanism of the phosphorylation of PT on membrane localization would be discussed. The strategy for improving phosphorus uptake efficiency by using phosphate transporters is put forward.

T4-22-06

Identification of phosphate-stress-induced mRNA-protein complexes involved in phloem-mediated long-distance signaling

Zhaoliang Zhang[1,4], *Jieyu Chen*[1], *Byung-Kook Ham*[1], *Zhangjun Fei*[2], *Leon Kochian*[3], ***William John Lucas***[1]
1. University of California
2. Cornell University
3. University of Saskatchewan
4. Anhui Agricultural University

In plants, phosphate (Pi) is an essential macronutrient, and consequently, it is a main constituent of fertilizers required to sustain agricultural output. Given that rock Pi supplies are finite and non-renewable, achieving the goal of food security will require engineering of crops with enhanced Pi uptake and use efficiency. To this end, our group is exploring the role of the plant vascular system, and the phloem in particular in systemic Pi-stress-associated signaling. Using cucumber as a model system, we have shown that, as part of the plant's early Pi stress response, many hundreds of mRNAs move long distance through the phloem to be targeted to specific sinks, including developing tissues of the root and shoot and young expanding leaves. Proteomics studies conducted on control and Pi-stressed cucumber plants identified 319 phloem sap proteins that are Pi-stress responsive, with 244 being phloem-mobile and, here again these proteins were being delivered to specific sink tissues. In addition, we mapped the response of some 105 phloem RNA-binding proteins (RBPs) that were associated with the imposed Pi-stress condition. These RBPs are being used to characterize the phloem-mobile ribonucleoprotein complexes that mediate delivery of mRNAs to specific tissues, under control and Pi-stress conditions. Transcripts bound by these RNP complexes are also being analyzed to ascertain the sequence/secondary structure motifs, within these mobile mRNAs, that are recognized by each specific RBP. The nature of the surveillance system that appears to operate, within the vascular system, to ensure selective delivery of phloem-mobile transcripts to a specific target tissue is also being investigated. These studies will be discussed in the context of developing breeding strategies to engineer crop plants with the capacity for sustained yields when grown on soils with reduced levels of available Pi.

This work is supported by National Science Foundation grant IOS-1339128.

T4-22-07

Improvement of nitrogen use efficiency in rice: Theory and practice

Chengcai Chu

Institute of Genetics and Developmental Biology, Chinese Academy of Sciences

Nitrogen fertlizer is one of the driving force to increase crop productivity. It is estimated that more than 120 million tons of nitrogen is used as fertilizer annually and will still increase with growing demand for food production. However, overconsumption of nitrogen fertilizer causes severe problems such as soil acidification, eutrophication and other environmental pollution. Thus, reducing fertilizer usage but maintaining crop yield is a pressing challenge and the ideal way to alleviate this problem is to improve nitrogen use efficiency (NUE) of crops. Nitrate is one of the major nitrogen source for higher plants, plants absorb and transport nitrate mainly via nitrate transporters. Our previous work indicated that *indica* rice has much higher nitrate use efficiency over *japonica* rice, and a natural variation at nitrate transporter NRT1.1B/OsNPF6.5 contributes to this nitrate use divergence between *indica* and *japonica*. NRT1.1B-*indica* exhibits higher nitrate transport activity than NRT1.1B-*japonica*, which corresponds well with *indica* having higher nitrate use efficiency. Evolutionary analysis reveals that *NRT1.1B-indica* is a later derived allele and subjected to artificial selection. The large-scale field tests in three different locations across China showed that the NUE and grain yield increase could be achieved by about 30% in *japonica* variety with *NRT1.1B-indica* introgression under 1/2 nitrogen supply conditions. In addition to NRT1.1B, we also found that some other nitrate transporters also play important roles in nitrate uptake and transport in rice. Our results suggested that nitrate transporters might be the potential targets in improved NUE of rice, and in principle, can be also used for NUE improvement of other crops.

T4-22-08

Regulating membrane transporters to increase rice yields and nutrient use efficiency

Guohua Xu

Nanjing Agricultural University

Cellular pH homeostasis is fundamental for life and all cells adapt to maintain this balance. In plants the chemical form of nitrogen supply, nitrate and ammonium, is one of the cellular pH dominators. The activities of plasma membrane transporters of nitrate and ammonium regulate cation/anion balance in cytosol. In addition, NHX-type K^+ (Na^+)/H^+ antiporters play critical roles in establish vesicle and vacuole pH homeostasis. The cycle of K^+ as major accompanying ion for translocation of nitrate and sucrose can act as an important signal for feedback control of nutrient uptake. We report that the rice nitrate transporter *OsNRT2.3* is transcribed into two spliced isoforms with natural variation in their expression ratio. One splice form, OsNRT2.3b is plasma membrane located mainly expressed in the phloem, and which has a regulatory motif on the cytosolic side that acts to switch nitrate transport activity on or off by a pH sensing mechanism. High OsNRT2.3b expression in rice enhances the pH buffering capacity of the plant, increasing N, Fe and P uptake. Interestingly, expression of a vesicle-localized transporter gene *HtNHX2* from *Helianthus tuberosus* could significantly improve rice tolerance to nutrient deficiency by up-regulating some membrane transporters for efficient acquisition of the nutrients. Increased expression of *OsNRT2.3b* or *HtNHX2* could very significantly increase grain yield and nutrient use efficiency. These results indicate that the sensing of cytosolic pH by OsNRT2.3b can function to improve rice NUE and pH balance, providing an explanation for plant adaptation to changes in ammonium-nitrate supply shift between the waterlogged and drained soil environment, while the transporters like HtNHX2 can strengthen vesicle pH and K^+ homeostasis for enhancing K^+-accompanying translocation of nutrients, thus, total uptake and use efficiency.

T4-22-09

A role for bHLHm1 and AMF1 in the AM symbiosis of *Medicago truncatula*

Evgenia Ovchinnikova[1], *David Chiasson*[2], ***Brent Kaiser***[1]

1. The University of Sydney

2. LMU Munich

GmbHLHm1 is a membrane bound basic helix-loop-helix DNA-binding transcription factor. In soybean, it has been localised to the symbiosome membrane, plasma membrane, endoplasmic reticulum and nucleus of nodule cells. *GmbHLHm1* expression is enhanced in nodules relative to roots with the onset of nitrogen fixation and loss of bHLHm1 activity reduces nodule fitness and nodule number. A preliminary study indicated that expression of bHLHm1 increased in roots when Pi and N availability was reduced. In this study, we investigated the activity of bHLHm1 homologs in the model legume *Medicago truncatula* colonised by either rhizobia or the arbuscular mychorrizal (AM) fungal strain, *Rhizophagus irregularis*. We have identified that Medicago homologs of *GmbHLHm1* are expressed in both nodules and AM-colonised roots. Localisation of GFP::MtbHLHm1 indicates membrane localization in arbuscular-containing cortical cells as well as in the nucleus. Loss of *MtbHLHm1* expression through RNAi silencing impairs nodule development and reduces AM colonization and arbuscule development. Plants of AM colonized *bHLHm1* roots displayed a negative growth response. In soybean, bHLHm1 is linked to the activity of the ammonium channel protein GmAMF1. Homologs of *AMF1* in *Medicago* were similarly found localized to arbuscule-containing cells. Loss of *AMF1* via RNAi, disrupted arbuscule development similar to the *bHLHm1* RNAi phenotype. Furthermore, RT-qPCR gene expression analysis indicates that *MtbHLHm1* activity is linked to changes in Pi (PhT) and N (AMT, AMF) transport genes in AM-colonised roots. This research highlights conservation of bHLHm1 activity between soybean and medicago and indicates both bHLHm1 and

AMF1 are intimately involved in the AM fungal symbiosis.

T4-22-10

Nitrogen Limitation Adaptation (NLA) is involved in source-to sink remobilization of nitrate by mediating the degradation of NRT1.7 in *Arabidopsis*

Wenwen Liu, Kai Wang, ***Wen-Xue Li***
Institute of Crop Science, Chinese Academy of Agricultural Sciences

Recent studies on nitrate transporters (NRT) have greatly increased our knowledge of the mechanisms regulating N homeostasis in plants. However, an understanding of how these NRTs are regulated is still lacking. *nla* mutant is hypersensitive to N limitation. In *nla* mutant, ^{15}N-nitrate spotted on old leaves preferentially accumulated in youngest leaves. Analysis of leaf cross sections indicated *NLA* expression was expressed in the sieve element and companion cell system. The results of BiFC, Split-ubiquitin Yeast Two-hybrid and CoIP assays demonstrated that NLA interacts with NRT1.7 in in the plasma membrane. The following findings suggested that NLA directs the ubiquitination of NRT1.7: down regulation of NRT1.7 protein abundance in *35S::NLA/35S::Myc-NRT1.7* double transgenic plants compared to *35S::Myc-NRT1.7* transgenic plants; and up regulation of NRT1.7 protein abundance in *nla* mutant compared to wild-type plants; and direct degradation of truncated NRT1.7 recombinant protein by NLA. Furthermore, analysis of NLA and NRT1.7 protein abundance in *mirna827* knock-out plants showed that N deficiency-guided translational repression of NLA depends on miRNA827. Our findings reveal that plants regulate source-to-sink remobilization of nitrate by the ubiquitin-mediated posttranslational regulatory pathway.

T4-22-11

Root microbiome and plant nutrient uptake

Ting Jiang, Baoyuan Qu, ***Yang Bai***
Institute of Genetics and Developmental Biology

Plants harbor abundant and diverse bacterial microbiota. Culture-independent 16S rRNA gene profiling revealed distinctive bacterial communities associated with plant roots. The availability of comprehensive pure cultures of root- and leaf-derived bacteria is critical to deeply understand functions of root-associated bacteria to a variety of plant physiological processes. We established *Arabidopsis* root-derived microbiota culture collections representing the majority of bacterial species that are reproducibly detected by previous culture-independent community sequencing. Using defined bacterial communities and a gnotobiotic *Arabidopsis* plant system we illustrated that root bacteria facilitate plants to uptake nutrients from natural soil. The comprehensive bacteria culture collections and high quality genomes together with the remarkable reproducibility of the gnotobiotic reconstitution system enable future studies on the cooperation between root bacterial community and plant nutrient uptake ability.

T4-22-12

Constructing a transcriptional regulatory network for nitrogen metabolism in *Arabidopsis*

Peizhu Guan
University of California, San Diego

Humans and animals depend on plants for fitness and survival, which is particularly true in obtaining nitrogen (N) directly from plants. N is the nutrient required in the greatest quantity by plant growth, nevertheless, being the most limiting element for plant productivity due to its poor availability. Nitrate is the predominant form of nitrogen in most agricultural soils. In addition to being an essential nutrient, nitrate is also a key signal for plant growth, development and stress responses. Therefore, deciphering nitrate signaling and regulatory network will provide direct guidance for improving nitrogen use efficiency (NUE) and achieving low-input sustainable agriculture. Much effort has been made to identify nitrate regulatory Transcription factors (TF) and DNA *cis*-elements, but progress has been slowed by the lack of a rapid and large-scale method to study the nitrate regulatory network. In our lab, we have developed a tripartite nitrate-responsive promoter (NRP) containing enhancer fragments from two nitrate-regulated genes, a 109 bp fragment from nitrate reductase (*NIA1*) and a 130 bp fragment from nitrite reductase (*NiR*), and a 35S mini promoter. This construct was used in a forward genetic screen to identify many nitrate regulatory mutants. We also developed a rapid transient expression system in *Nicotiana benthamiana* for analyzing nitrate responses and were able to identify multiple DNA *cis*-regulatory elements in the 109 bp fragment from *NIA1*. In addition, the 109 bp fragment was utilized as bait in a yeast-one hybrid system to screen 2000 clones encoding known and predicted *Arabidopsis* TFs. Among those TFs, we have identified and characterized *teosinte branched1/cycloidea/proliferating cell factor1-20* (*TCP20*), a broad regulator of plant growth and development that plays a key role in nitrate systemic signaling underlying nutrient foraging by roots and discovered TCP20-NLPs (NIN-like proteins) protein interactions and their specific roles in orchestrating the regulation of nitrate assimilation with that of the cell cycle and root meristem growth. The discovery is largely thanks to a DNA *cis*-element- and TFs-based screening system that we have been developing. These regulators (TFs and *cis*-elements) provide important components and information to construct nitrate regulatory network in plants.

T4-23: Germline development and fertilization mechanisms

T4-23-01

FERONIA and related receptor kinases in the regulation of pollen tube growth and pollen-pistil interaction

Alice Y. Cheung
University of Massachusetts, Amherst

FERONIA receptor kinase was first discovered as a critical regulator of female fertility; loss of its function results in a dramatic ovule phenotype of pollen tube pile up inside the female gametophyte and severely suppressed female fertility. It was demonstrated more than a decade ago that FERONIA plays a dual function: one in mediating female gametophyte-induced rupture of a penetrating pollen tube, resulting in sperm release and enabling fertilization; the other in preventing multiple pollen tube entrance to the female gametophyte, ensuring against polyspermy. Loss of FERONIA results in multiple pollen tubes entering an ovule and penetrating the female gametophyte; they continue growth without discharging sperm, thus precluding fertilization. Our earlier work demonstrated that the two functions of FERONIA in regulating pollen tube-ovule interaction are not obligatorily coupled and FERONIA induces a high oxidative environment in the filiform apparatus/micropyle region to induce pollen tube rupture in a Ca^{2+}-dependent process. Here we discuss FERONIA-regulated properties that underlie its role in preventing multiple pollen tube entrance into the female gametophyte. FERONIA is expressed almost ubiquitously except in pollen. We will also discuss the functional role of a FERONIA-related receptor kinase that are specifically expressed in pollen and how it impacts the pollen tube growth process in the pistil.

T4-23-02

How do plant female gametes prevent and allow polyspermy in double fertilization?

Tetsuya Higashiyama

Institute of Transformative Bio-Molecules, Nagoya University

Fertilization in flowering plants is very unique and called as double fertilization. A set of two sperm cells enclosed by an endocytic membrane is conveyed by the pollen tube, which is a tip growing cell precisely attracted to the egg-containing tissue. One of two sperm cells fuses with the egg cell to form an embryo and the other fuses with the central cell on the side of the egg cell to form a nutritive tissue endosperm. Precise one-to-one fusion in the two female gametes is critical for normal seed development. Polyspermy of flowering is likely to be prevented by at least two distinct mechanisms: one is called as blocking polytubey to prevent attraction of supernumerous pollen tubes, and the other is gametic polyspermy block. We have been working on plant reproduction by live-cell imaging analysis (e.g., Science 2001, Nature 2009, 2016, Curr. Biol. 2011, Nat. Commun. 2014). We found double fertilization was intimately related to cessation of additional pollen tube attraction (Curr. Biol. 2012, Dev. Cell, 2013). To our surprise, the fertilized central cell fused with an attractive synergid cell to stop pollen tube guidance (Cell 2015). Secretion of LURE attract peptide by the synergid cell rapidly decreased and the synergid cell nucleus degenerated rapidly due to sudden cell cycle entry. On the other hand, in the step of double fertilization, gametic polyspermy block was also shown by using laser ablation and mutant pollen tubes conveying supernumerous sperm cells. In this symposium, I will talk about both blocks for polytubey and gametic polyspermy.

T4-23-03

Functional analysis of CRPs during reproduction and defense response in *Zea mays*

Liangzi Zhou[1], *Lele Wang*[1], *Georg Haberer*[2], *Klaus F.X. Mayer*[2], *Thomas Dresselhaus*[1]

1. Cell Biology and Plant Biochemistry, University of Regensburg, Regensburg

2. Plant Genome and Systems Biology, Helmoltz Zentrum München, Neuherberg

Double fertilization is a unique sexual reproduction process restricted to angiosperms (flowering plants). Two sperm cells have to be transported to the female gametophyte via a pollen tube and fuse with egg cell and central cell to generate embryo and endosperm, respectively. To accomplish double fertilization, extensive male-female communication and accurate guidance is needed at many steps along the pollen tube journey. Previous studies in different plant species revealed that various highly specific receptor-ligand binding activities are critical to accomplish cell-cell communication between the pollen tube and various tissues of the female flower organ during fertilization. Small secreted peptides, a major class of extracellular signaling ligands in plants, can be classified into two major groups, CRPs (cysteine-rich peptides) and non-CRPs. Functional analysis of these two peptide classes showed that most non-CRPs are involved in signaling pathways of vegetative growth and environment response, while CRPs are more presented in different stages of reproductive growth. Originally CRP subgroups such as defensins were thought to be involved in pathogen defense. RNA-seq data of our group indicates that more than 400 CRPs showed specific expression pattern during fertilization and fungal invasion. To further elucidate the biological functions of candidate CRPs during reproduction and defense responses in maize, RNAi transgenic plants were generated with Agrobacterium-mediated stable transformation and are currently used for phenotypic analyses. For example, we identified three CRP genes encoding rapid alkalinization factors (*RALFs*), which are highly and exclusively expressed in germinated maize pollen tubes. Down-regulation of pollen-specific *RALFs* affects pollen tube growth *in vitro* indicating that *RALFs* could be involved in stabilizing pollen tube growth during fertilization. A CRP subgroup restricted to C4 grasses appears as pollen tube interactor of the egg apparatus-secreted guidance peptide EA1. We will report and discuss these and other functional studies of CRPs in maize to provide a deeper insight into the understanding of the different roles of CRPs during fertilization and stress response, their evolution and their role in speciation (reproductive isolation) in maize and plants in general.

T4-23-04

Generating clonal progeny in maize

Nina Chumak[1], *Mark Williams*[2], *Arco Brunner*[1], *Tim Fox*[2], *Joana Bernardez De Assis*[1], *Wenjing She*[1], *Frederique Pasquer*[1], *Marc Albertsen*[2], ***Ueli Grossniklaus***[1]

1. Department of Plant and Microbial Biology, University of Zurich, Zollikerstrasse 107, CH-8008 Zurich

2. DuPont Pioneer Hi-bred, P.O. Box 1000, Johnston, IA 50131-0184

Apomixis is asexual reproduction through seed. The production of seeds though apomixis, which generates plants that are genetically identical to the mother plant, has considerable agricultural potential to maintain desired complex genotypes, e.g. those of F1 hybrids, over many generations. Gametophytic apomixis deviates from sexual development in three major steps: (1) meiosis is circumvented or aborted, leading to the formation of unreduced, unrecombined embryo sacs (apomeiosis); (2) embryogenesis initiates without fertilization of the unreduced egg cell (parthenogenesis); and (3) developmental adaptations enable the formation of functional endosperm. The aim of our research is to identify mutations that mimic the major components of apomixis, and to combine them to engineer apomictically-reproducing maize plants. In a genetic screen we identified the *n*on-*r*eduction in *f*emale 4 (*nrf4*) mutant, which mimics the first step of apomixis: apomeiosis. Homozygous *nrf4* plants produce up to 95% unreduced embryo sacs. Using SAIFF-by sequencing technology, the mutation was mapped to *GRMZM2G148133* on the long arm of chromosome 7, and the identity of *Nrf4* was confirmed by two additional mutant alleles. To identify whether *nrf4* leads to first or second division restitution (FDR vs SDR), we analyzed maintenance of heterozygosity in the progeny of *nrf4* mutant plants in comparison to mother plants using a SNP array that enabled the analysis of 10-20 SNPs on each chromosome. The effect of the *nrf4* mutation turned out to be more complex than expected and leads to both FDR and SDR. Nonetheless, depending on the genetic background of the mother plant, up to 11% of the unreduced female gametes were genetically identical to the mother. Indeed, pollination of *nrf4* plants by a tetraploid haploid inducer resulted in some clonal individuals. To our knowledge, this is first evidence that production of clonal individuals through seed is possible in maize.

T4-23-05

Chromatin remodeling complex SWR1 coordinates with cytochrome P450 KLU to regulate plant megasporocyte formation

Yuan Qin
Fujian Agriculture and Forestry University

Germline cell specification is essential for sexual reproduction. In the ovules of most flowering plants, only a single hypodermal cell enlarges and differentiates into a Megaspore Mother Cell (MMC), the founder cell of the female germline lineage. The molecular mechanisms restricting the specification of the MMC to a single cell remains elusive. We showed that the transcription factor *AKD28,* exclusively expressed in hypodermal somatic cells surrounding the MMC, is required to repress those cells to differentiate into MMC in *Arabidopsis*. We also showed that the SWR1 chromatin remodeling complex mediates the incorporation of the histone variant H2A.Z into *AKD28* nucleosome and modulates its structure, and that cytochrome P450 *KLU,* expressed in the inner integument primordia, controls H2A.Z deposition at *AKD28* in a non-cell autonomous manner. Our findings show how somatic cells cooperatively use chromatin remodeling to restrict germline cell specification to a single cell in the ovule primordia.

T4-23-06

AP1G mediates vacuolar acidification during synergid-controlled pollen tube reception

Yan Zhang
State Key Lab of Crop Biology, Shandong Agricultural University

Double fertilization in angiosperms requires the delivery of immotile sperm through pollen tubes, which enter embryo sacs to initiate synergid degeneration and to discharge. This fascinating process, called pollen tube reception, involves extensive communications between pollen tubes and synergids, within which few intracellular regulators involved have been revealed. Here we report that vacuolar acidification in synergids, mediated by AP1G possibly through the activity of V-ATPases, is critical for pollen tube reception. Functional loss of *AP1G*, encoding the γ subunit of adaptor protein 1, impaired synergid degeneration and pollen tube discharge, resulting in partial female sterility. Although vacuolar trafficking was compromised in the *ap1g* synergids, genetic and pharmacological interferences suggested that it was not the determinant for defective pollen tube reception. We further showed that vacuolar acidification in the *ap1g* synergids was compromised. Consistently, functional loss of V-ATPases also resulted in the failure of synergid degeneration and pollen tube reception despite that the targeting of V-ATPases was not affected by AP1G. We propose that vacuolar acidification-mediated synergid degeneration might represent a distinct cell death mechanism specifically adopted by the plant phylum.

T4-25: Photosynthetic efficiency and environmental regulation

T4-25-01

Functional analysis of IsiA-containing complexes in *Synechocystis* sp. PCC 6803

Zhan Jiao, ***Qiang Wang***
Key Laboratory of Algal Biology, Institute of Hydrobiology, the Chinese Academy of Sciences

Iron is an essential element that functions as a cofactor in numerous redox reactions and is a key component of the electron transport system of photosynthesis. Iron stress-induced protein A (IsiA) is a major chlorophyll-binding protein in the thylakoid membrane that is significantly induced under iron deficiency conditions. An IsiA-PS I supercomplex, which contains IsiA, trimeric PS I, Slr1128 and HliA/B was detected by sucrose gradient ultracentrifugation in the model cyanobacteria *Synechocystis* sp. PCC 6803, and when trimeric PS I is absent, the other components would form a novel high light-inducible carotenoid-binding protein complex (HLCC). Monitoring the dynamic changes of IsiA-containing complexes in *Synechocystis* sp. PCC 6803 during exposure to long-term iron deficiency discovered two other types of IsiA-PS I supercomplexes, namely IsiA-PS I high fluorescence supercomplex (IHFS) and IsiA-PS I low fluorescence supercomplex (ILFS). Further analysis revealed a role for IsiA as an energy dissipater in the IHFS and as an energy collector in the ILFS. The trimeric structure of PS I mediated by PsaL was found to be indispensable for the formation of IHFS/ILFS. Dynamic changes in IsiA-containing complexes in cyanobacteria during long-term iron deficiency may represent an adaptation to iron limitation stress for flexible light energy distribution, which balances electron transfer between PS I and PS II, thus minimizing photooxidative damage.

T4-25-02

The mechanism and significant of chloroplast gene expression in higher plants

Wei Chi
Institute of Botany, CAS

The chloroplast of higher plants is a unique organelle evolved from a photosynthetically active cyanobacterium by an endosymbiotic event. Strategies and regulation of chloroplast gene expression are distinct from those of eubacterial as well as from those of eukaryotic cytoplasm and reflect a unique chimerical system assembled by multiple origins during endocytobiosis. Proper expression of chloroplast genes relies on coordinated interplay between *cis*-acting elements at the plastid DNA/RNA level and *trans*-acting factors, most of which are nuclear-encoded and specifically bind to their *cis*-acting targets. The significant of chloroplast gene expression has been expanded greatly. The growing evidence has show that it not only regulates the chloroplast biogenesis but also involved in plant development via retrograding signaling. The chloroplast gene expression has been integrated into plant development program. In this presentation, we will summary the advance in chloroplast gene expression and highlight the roles of nuclear-encoded factors e have identified in recent years. Based on our work, we will also discuss how the chloroplast gene expression influence root and leaf development via interaction with auxin signaling.

T4-25-03

Physiological traits mediate bioclimatic segregation of epiphytic CAM bromeliads in Trinidad

Jamie Males, *Howard Griffiths*
University of Cambridge

Trait-mediated divergences in bioclimatic relations could represent an important pathway to the evolution and maintenance of genus-level diversity, but there are relatively few comprehensive case studies of this effect in regional vascular plant florae. To investigate this topic, we mapped the distributions of epiphytic CAM bromeliads of the genus *Aechmea* across the Northern Range of Trinidad, a topographically and climatically heterogeneous continental island in the southern Caribbean. Distinct elevational and latitudinal effects on species presence and abundance were apparent. Bioclimatic and vegetational species distribution modelling was performed in MaxEnt to project the potential ranges of four species across Trinidad, on Tobago and along the Venezuelan Coastal Cordillera. Using the same models, projections of species distributions under an aggressive 2,070 climate scenario revealed particular vulnerability to upslope shifting of bioclimate envelopes in the montane species. To explore the contribution of ecophysiological traits to observed divergences in occupation of environmental niche space, plants of each species were collected and subjected to anatomical and physiological characterisation. Bioclimatic differentiation was associated with contrasting water-use strategies, stomatal sensitivity, temperature relations and drought responses. Measured physiological traits were used to construct a mechanistic species distribution model and probability distribution grid maps were generated for comparison with MaxEnt output. Distributions modelled by correlative and mechanistic approaches showed a high degree of congruence, consistent with a physiological explanation for regional- and landscape-scale patterning in the distributions of *Aechmea* bromeliads. Interspecific variation in quantitative and qualitative traits related to abiotic tolerances is probably an important contributor to the ecological diversity of the Bromeliaceae.

T4-25-04

Effects of close planting on chlorophyll content and photosynthetic function in field grown sorghum

Tao Li[1,2], *Chuang-Dao Jiang*[2], *Wei-Feng Wang*[1]
1. Department of Plant Science, College of Forestry, Shanxi Agricultural University, Shanxi, China;
2. Key Laboratory of Plant Resources, Institute of Botany, Chinese Academy of Sciences, Beijing, China.

To study the effects of mutual shading on photosynthesis in field-grown sorghum. Under absence of root competition, we investigated the effects of planting density on light environment, leaf chlorophyll and photosynthetic function. Increased planting density not only modified the light intensity and spectrum, but also induced a marked increase in chlorophyll content; the chlorophyll increased gradually with increased leaf position in upper canopy at the medium and high density, while it decreased in low density; compared with sorghum grown at high density, the lower canopy leaves at medium density had longer duration of stay-green period and photosynthetic function, and the leaves senescence was quicker at low density. In the absence of mineral deficiency, close planting induce a significant increase in chlorophyll content, and the chlorophyll increase gradually with rise of leaf position in upper canopy; close planting not only changes the duration of stay-green period, but also promotes the optimization of photosynthetic function in field-grown sorghum.

T4-25-05

Regulation of stomatal patterning by CO_2 across *Arabidopsis* tissues occurs during spacing divisions

Miranda Haus[1], *Thomas Jacobs*[2]
1. Michigan State University
2. University of Illinois- Champaign Urbana

Stomatal development is regulated by the atmospheric concentration of carbon dioxide ($p[CO_2]$). Mature leaves can convey developmental cues to immature leaves in response to $p[CO_2]$, but the developmental mechanisms are unknown. We constructed a dual-chamber growth system in which rosette and cauline leaves of *Arabidopsis thaliana* were subjected to differing $p[CO_2]$, to characterize changes in stomatal patterning resulting from signals moving from mature to developing leaves. Young rosette tissue was found to adjust stomatal index (SI) in response to both the current environment and the environment experienced by mature rosette tissue, whereas cauline leaves appear to be insensitive to $p[CO_2]$ treatment. We also found that leaf epicuticular wax across the entire plant is dependent upon only the $p[CO_2]$ experienced by young tissue. The effect of $p[CO_2]$ on stomatal development reappears in cotyledons of the next generation, but only after germination. Moreover, these data suggest that $p[CO_2]$ affects regulation of stomatal development specifically through the development of satellite stomata and not the core pathway. It is likely that cauline leaves and cotyledons deploy mechanisms for controlling stomatal development that are related to, but distinct from, that operating in rosette leaves.

T4-25-06

Transport of photosynthate

Li-Qing Chen[1], *Wolf Frommer*[2]
1. University of Illinois at Urbana-Champaign
2. Carnegie Institution for Science

In order to meet increasing demands for energy and food as the world population grows, there is an urgent need to produce sustainable fuel and increase crop yield beyond existing yield potentials. A better understanding of the mechanisms of fixed carbon allocation from photosynthetic tissues to non-photosynthetic ones is critical to achieve the goal of redirecting sugar flux to storage tissues. Sugar transporters are required to facilitate the carbon allocation. We have recently characterized the SWEET family of sugar transporters, which are involved in different biological efflux processes: *Arabidopsis* AtSWEET11 and 12 export sucrose into the apoplasm from the phloem parenchyma cells prior to phloem loading. *atsweet11;12* double mutant impairs sucrose export out of the leaf. We found more SWEETs are involved in this efflux step before phloem loading. Arabidopsis sucrose transporter AtSWEET11,12 and 15 are required for seed filling. Triple mutant *atsweet11;12;15* exhibits severely delayed embryo development and reduced seed yield. Overexpression expression of *SWEETs* leads to the unexpected results-stunted growth. How to engineer sugar flux will be discussed. The structure-guided functional study of SWEETs offers a valuable foundation for engineering sugar flux to increase crop yields and biofuel.

T4-26: Hormone and plant development

T4-26-01

Structural variations of long-distance translocated cytokinins control the specificity of the action in shoot growth and development

Hitoshi Sakakibara[1], ***Takatoshi Kiba***[2], *Asami Osugi*[1]

1. Nagoya University

2. RIKEN CSRS

Cytokinin (CK) plays an important role in regulation of plant growth and development, and its action is finely controlled by various steps including biosynthesis and metabolism, transport, and signaling. We have demonstrated that IPTs, CYP735As, and LOGs, which are key genes for *de novo* CK biosynthesis, are expressed in various parts during growth and development, and differentially regulate the synthesis of N6- (Δ^2-isopentenyl)adenine (iP) and trans-zeatin (tZ). In addition, ABCG14, a member of ABC transporter family, has been recently identified as a key gene for root-to-shoot translocation of CKs via xylem. Detailed studies on CYP735As mutants show that tZ is important for the normal growth of shoot rather than that of root, suggesting a mechanism that modulates physiological function of CKs by modification of the side-chain structures. This regulation is one of the qualitative controls of CK action involved in shoot growth regulation by root-borne signal. Furthermore, our recent studies suggest importance of translocation of active form CK via xylem (i.e. tZ) for regulation of specific traits in shoot growth and development. These findings suggest that complex action of long-distantly transported cytokinins could be organized by the side chain structure and the dependency of the activation pathway. The dual long-distance cytokinin signaling system would be important for fine-tuning shoot growth manner in response to environmental conditions. We will outline our recent progress in CK study, and discuss the physiological significance of regulation of CK action to optimize growth and development at whole plant level.

T4-26-02

Reprogramming of rice root development under water-deficit stress

Tuan-Hua David Ho[1], *An-Shan Hsiao*[1], *I-Chieh Hsieh*[1], *Su-May Yu*[2]

1. Academia Sinica/Institute of Plant and Microbial Biology

2. Academia Sinica/Institute of Molecular Biology

Rice roots grow longer under mild water-deficit stress, yet their elongation rate is reduced when encountering moderate to severe stress or treated with 2 uM or more ABA, generating shorter and thicker roots. The root biomass is increased in both cases compared to unstressed roots, and consequently the root to shoot ratio is enhanced under stress. Transport of ^{14}C-sucrose from leaves to roots is enhanced under stress conditions, suggesting that the sink strength of roots has become stronger. Storage lipids and starch accumulate to higher levels in root tissues under stress. Transcriptomics analysis reveals that genes for the biosynthesis of fatty acids and starch are preferentially induced in roots by ABA/stress with concomitant increase in the activity of enzymes encoded by these genes. In addition, a novel class of highly proline-rich glycoproteins, RePRPs (Repetitive Proline-Rich Proteins) is also preferentially induced by ABA/stress in rice roots. RePRP proteins contain more than 40% prolines and numerous PEPK (pro-glu-pro-lys) motifs, and are heavily O-glycosylated. Unlike other plant proline-rich proteins, RePRPs are not secreted to the cell walls, but localized on the plasma membrane as well as in the cytosol. Although RePRPs are intrinsically disordered proteins due to the high proline content, they are able to specifically bind to the cell wall polysaccharide arabinogalactan (AG), actin and tubulin. The binding of actin or tubulin is independent of the binding of AG. Overexpression of RePRP leads to shorter and thicker roots in transgenic rice and the reduction of actin filaments, mimicking the effect of ABA/stress. Knocking down the expression of RePRP by RNAi prevents roots from becoming shorter and thicker under ABA/stress treatment. It is suggested that the plasma membrane-localized RePRPs regulate cell wall properties through their interactions with cell wall polysaccharides and cytosolic RePRPs reduce actin filaments, causing the formation of shorter yet wider root cells. The synergism between RePRP expression and storage nutrient biosynthesis alters the architecture and function of rice roots so that they are better suited for coping with severe stress conditions.

T4-26-03

Molecular mechanism underlying rice shoot gravitropism and tiller angle

Ning Zhang, Hong Yu, Hao Yu, Yueyue Cai, Linzhou Huang, Jiayang Li, ***Yonghong Wang***

Institute of Genetics and Developmental Biology, Chinese Academy of Sciences, Beijing

In higher plants, gravitropism is a key factor that affects plant architecture. The auxin homeostasis and distribution are considered to play an important role in shoot gravitropism, however, the mechanism underlying shoot gravitropism is still largely unknown. Rice shoot gravitropism is also regarded as one of important agronomic traits because it can determine tiller angle and thus affect the grain yields. Our previous studies have identified the rice *LAZY1* (*LA1*) gene, which functions as a negative regulator of polar auxin transport (PAT). Loss-of-function of *LA1* greatly enhances PAT and thus alters the endogenous IAA distribution in shoots, leading to the reduced gravitropism and therefore the tiller-spreading phenotype of rice. To elucidate the mechanism of LA1 involved in the regulation of shoot gravitropism and thus tiller angle, we carried out a genetic screening to identify the suppressors of *la1*, and identified multiple suppressors of *LA1* that are components involved in the strigolactones (SLs) biosynthetic or signaling pathway. We showed that SL biosynthetic or signaling mutants could rescue the spreading phenotype of *la1*. SLs are a group of newly identified plant hormones that are essential for shoot branching/rice tillering. Our study revealed that SLs can inhibit auxin biosynthesis and attenuate rice shoot gravitropism mainly through decreasing the local IAA content. We also verified that the SL-mediated shoot gravitropic mechanism is conserved between dicot and monocot plants. Moreover, we used RNA-Seq technology to identify the genes involved in the regulation of shoot gravitropism and tiller angle, trying to reveal the molecular event and dynamic regulatory network of tiller angle at a whole genome level. We found that *Gravitropic Response Gene 4* (*GTR4*) showed asymmetrically expression in the rice shoot base upon gravistimulation, the similar expression pattern as auxin responsive marker gene *OsIAA20*. *GTR4*-RNAi transgenic plants displayed larger tiller angle and attenuated gravitropic response. Further evidence demonstrated that *GTR4* may function downstream of auxin in the regulation of shoot gravitropism and tiller angle.

T4-26-04

Arabidopsis CLE peptide signals contributes to pollen exine development through precisely regulated transcriptomic and proteomic networks

Shuangshuang Wang[1], *Jianan Lu*[1], *Juanying Ye*[1], *Chenjiang You*[1], *Xiufen Song*[2], *Shichao Ren*[2], *Fang Chang* [1], *Chunming Liu*[2,3], *Hong Ma*[1]

1. Fudan University

2. Chinese Academy of Sciences

3. Chinese Academy of Agricultural Science

The CLAVATA3/ESR-RELATED (CLE) peptide signals are required for cell-cell communication in several plant growth and developmental processes. However, little is known regarding the function and working mechanism of the CLEs in the male reproductive process. Here, through a gene expression survey, we identified 7 Arabidopsis *CLE* genes that are specifically/preferentially expressed in the tapetum cells and microsporocytes in the anther; also dominant-negative (DN) mutant plants of each of these genes exhibited significantly reduced male fertility, with abnormal pollen wall formation. Further study with CLE19 as representative and using *cle19*, DN-*CLE19*, and *CLE19-OX* mutant lines as materials, we demonstrated that CLE19 peptide signal affected the expression of more than 1,000 genes at the RNA level and 595 at the protein level, including the important transcription factor gene*AMS*, 280 AMS downstream genes, and other genes involved in pollen coat and pollen exine formation, lipid metabolism, pollen germination and hormone metabolism. Consistently, mutants of the CLE19 downstream genes *GRP20*, *ACOS5* and *MEE48* all exhibited *CLE19-OX*-like pollen exine structure. In conclusion, these findings demonstrate CLE19 and redundant CLE hormone peptides work as negative regulator of AMS and downstream networks in pollen exine formation, and provide insights into studies of both peptide signaling transduction and anther development fields.

T4-26-05

COG1 regulates brassinosteroid biosynthesis via PIF4 and PIF5 in *Arabidopsis thaliana*

Zhuoyun Wei, ***Jia Li***

Ministry of Education Key Laboratory of Cell Activities and Stress Adaptations, School of Life Sciences, Lanzhou University, Lanzhou

Brassinosteroids (BRs) are essential phytohormones regulating various developmental and physiological processes during normal growth and development. Within the last two decades, significant progress has been made in elucidating both BR signal transduction and BR biosynthesis pathways. But how BR homeostasis is regulated by internal and external cues remain poorly understood. *cog1-3D* (*cogwheel 1-3D*) was identified as an activation-tagged genetic modifier of *bri1-5*, an intermediate BR receptor mutant. *COG1* encodes a Dof-type transcription factor previously found to act as a negative regulator of the phytochrome signaling pathway. *cog1-3D* single mutants show an elongated hypocotyl phenotype under light conditions. Loss-of-function by ethylmethanoesulfonate (EMS) mutation or inducible expression of a dominant-negative form of *COG1* results in an opposite phenotype. BR profile assay indicated that BR levels are elevated in *cog1-3D* seedlings. Quantitative RT-PCR analyses showed that several key BR biosynthetic genes are significantly up-regulated in *cog1-3D* compared to those of wild-type. Two basic helix-loop-helix transcription factors, *PIF4* and *PIF5*, were found to be transcriptionally up-regulated by *cog1-3D*. Genetic analysis indicated that PIF4 and PIF5 are required for COG1 to promote BR biosynthesis and hypocotyl elongation. Chromatin immunoprecipitation (ChIP) and electrophoretic mobility shift assays (EMSA) demonstrated that COG1 directly binds to the promoter regions of *PIF4* and *PIF5*; and PIF4 and PIF5 directly binds to the promoter regions of key BR biosynthetic genes, such as *DWF4* and *BR6ox2*, to promote their expression. Previous study revealed that both expression of *COG1* and *PIF4* could be induced by far-red and red light. Our results revealed that COG1 and PIF4 and PIF5 are key regulators mediating BR biosynthesis, supporting that light signaling is critical to BR homeostasis.

T4-26-06

Evolutionary history of a gibberellin receptor, GID1

Miyako Ueguchi-Tanaka

Nagoya University, Bioscience and Biotechnology Center

Gibberellin (GAs) are a large family of tetracyclic diterpenoid phytohormones that induce a wide range of plant growth responses, including stem elongation, seed germination, floral induction and pollen development. Although more than 130 GA derivatives have been identified, only a few GAs, such as GA_4, GA_1 and GA_3, actively function in plant responses. These bioactive GAs essentially have a hydroxyl group at 3-carbon, a carboxyl group at 6-carbon, γ-lactone ring, and no hydroxyl group at 2-carbon. GA receptor can strictly distinguish the structures of bioactive GA from inactive ones. The GA receptor, GID1, has a primary structure of carboxylesterases (CXEs). The crystal structure of the GID1 bound with GA_4 revealed that the GA binding pocket corresponds to the substrate-binding site of CXEs. N-terminal lid of GID1 covers the GA molecule to stabilize at the binding site. The lid also functions for the interaction with DELLA proteins, repressors in GA signaling. Here, we quantified the effect of the amino acids of GID1, which are not shared between *Selaginella moellendorffii* and flowering plant, using physicochemical and phenotypic analyses of transgenic rice plants with various mutant GID1s. We also studied the copy number of *GID1* from comprehensive phylogenetic analysis during plant evolution of GID1 in broad plants species. Based on these observations, we discussed the establishment of GID1 in the process of plant evolution.

T4-27: Light signal and photomorphogenesis

T4-27-01
SPLICING FACTOR FOR PHYTOCHROME SIGNALING promotes photomorphogenesis by regulating pre-mRNA splicing in *Arabidopsis*

Enamul Huq[1], *Ruijiao Xin*[1], *Ling Zhu*[1], *Patrice Salome*[2], *Estefania Mancini*[3], *Carine Marshall*[4], *Frank Harmon*[4], *Marcelo Yanovsky*[3], *Detlef Weigel*[2]
1. University of Texas at Austin
2. Max Planck Institute for Developmental Biology
3. Instituto de Investigaciones Bioquímicas de Buenos Aires–Consejo Nacional de Investigaciones Científicas y Técnicas de Argentina
4. USDA-ARS/Plant Gene Expression Center

Light signal regulates plant growth and development by controlling a plethora of gene expression changes. Post-transcriptional regulation, especially pre-mRNA processing, is a key step in controlling gene expression; however, the molecular mechanisms linking pre-mRNA processing and light signaling are not well understood. Here we report a splicing factor called SPLICING FACTOR FOR PHYTOCHROME SIGNALING (SFPS) that directly interacts with the photoreceptor Phytochrome B (phyB). *sfps* mutants have long hypocotyls, smaller cotyledon angles under monochromatic light and display early flowering. SFPS colocalizes with U2 snRNP associated factors including, U2AF65B, U2A', and U2AF35A in nuclear speckles, suggesting that SFPS might be involved in the 3' splice site determination. SFPS regulates pre-mRNA splicing of a large number of genes enriched in light signaling, photosynthesis and circadian clock under both dark and light conditions. *In vivo* RNA immunoprecipitation (RIP) assay revealed that SFPS associates with *EARLY FLOWERING 3* (*ELF3*) mRNA, a critical gene linking light signaling and circadian clock. Moreover, *PHYTOCHROME INTERACTING FACTORs (PIFs)*, act downstream of SFPS, as knocking out four prominent PIFs masks *sfps* phenotypes. Taken together, these data strongly suggest that SFPS modulates light-regulated developmental processes by controlling pre-mRNA splicing of many light signaling and circadian clock genes.

T4-27-02
Photobody biogenesis in early phytochrome signaling

Meng Chen
Department of Botany and Plant Sciences, Institute for Integrative Genome Biology, University of California Riverside, CA 92521, USA

Photobodies are photosensory subnuclear domains containing the red and far-red photoreceptors, the phytochromes. In *Arabidopsis*, biogenesis of phyB-GFP-containing photobodies is triggered by red light and promoted by increasing light intensity. A central mechanism by which phytochromes control photomorphogenetic responses is to trigger the degradation of a group of antagonistically-acting Phytochrome-Interacting transcription Factors (PIFs). During the dark-to-light transition, phyB-GFP is colocalized with PIF3 before PIF3 degradation, it was therefore proposed that photobodies are sites for PIF3 degradation. However, the mechanism linking photobodies and PIF3 degradation is still poorly understood. We show that PIF3 degradation is tightly correlated with photobody biogenesis. Using a forward genetic screen for mutants defective in photobody biogenesis, we identified HEMERA (HMR), a new phytochrome signaling component required for the biogenesis of large photobodies. HMR is localized to the periphery of photobodies and required for the degradation of PIF1 and PIF3, providing genetic evidence supporting the role of photobodies in PIF degradation. We show that HMR interacts directly with all PIFs. The HMR-PIF interaction is mediated mainly by HMR's N-terminal half and PIFs' conserved active-phyB-binding motif. In addition, HMR possesses an acidic nine-amino-acid transcriptional activation domain (9 aa TAD). Unexpectedly, the 9 aa TAD of HMR is required for both PIF3 degradation and the activation of a subset of growth-related PIF target-genes. These results support a novel phytochrome signaling mechanism, in which HMR mediates downstream photomorphogenetic responses by binding directly to PIFs and coupling the degradation of PIF1 and PIF3 with the transactivation of PIF target genes via HMR's 9 aa TAD. We propose that photobodies are gene hubs for the regulation of a distinct set of light responsive PIF target-genes.

T4-27-03
The reds and the blues of the shade avoidance response

Ullas Pedmale
Cold Spring Harbor Laboratory, 1 Bungtown Road, Cold Spring Harbor, NY 11724, USA.

Plants adjust their body plan in response to ambient environmental conditions. Light is among the most relevant environmental signals because light not only drives photosynthesis but also provides critical information about the local growth environment as well as diurnal and seasonal time. Light is perceived by a complex array of photoreceptors, defined by the color of light they absorb. These include the red/far-red-absorbing phytochromes (phyA-E), the blue/UV-A-absorbing cryptochromes (CRY1-2), phototropins (phot1-2), and ZTL-family, and a UV-B-absorbing receptor. Plants have developed various adaptive responses in interpreting and utilize light directionality, quantity, and quality. One such adaptive response is the ability of sun-loving plants to sense and avoid vegetational shading by nearby plants. In vegetational or canopy shading, plants perceive a decrease in the ratio of red to far-red light (R: FR) and there is also a reduction in blue light and the available photosynthetically active radiation (PAR). CRYs control the phenotypic plasticity of plants in response to changes in blue light. Despite their importance, very little is known about the mechanisms by which CRYs affect growth programs. We present data to show that when plants are deprived of blue light under a shaded canopy, CRY1/2 perceive this change and respond by directly contacting two bHLH transcription factors, PIF4 and PIF5. Activities of these factors are controlled by phytochromes, which are the red:far-red light photoreceptors. Intriguingly, transcriptome analysis shows that genes regulated by PIF4 and 5 in response to low blue light are distinct from those regulated by low red/far-red. Results indicate that CRYs signal by modulating PIF activity, and that different plant photoreceptors can mediate distinct phenotypic responses through shared signaling components.

T4-27-04

Coordinated positive and negative regulatory machinery underlies *Arabidopsis* UV-B signaling network

Hui Ren, ***Xi Huang***
State Key Laboratory of Cellular Stress Biology, School of Life Sciences, Xiamen University

The ultraviolet (UV)-B light is an indispensible component of the sunlight that reaches the earth surface. UV-B irradiation brings about differential physiological responses in plants, dependent on its wavelength, fluence rate and duration. In higher plants, photomorphogenesis is a pivotal developmental process that allows plant seedlings to experience dark-to-light transition during their emergence from soil. Long-wavelength and low-fluence UV-B specifically induces plant photomorphogenesis, which is characterized by the inhibition of hypocotyl elongation, the promotion of anthocyanin accumulation, and eventually the tolerance against UV-B stress. In *Arabidopsis thaliana*, the environmental UV-B light signal is perceived by the photoreceptor UV RESISTANCE LOCUS 8 (UVR8), a plant-specific chromoprotein. UVR8 initiates UV-B specific signaling by interacting with CONSTITUTIVELY PHOTOMORPHOGENIC 1 (COP1) which is a UV-B inducible protein. However, less is known about the molecular mechanism of the negative regulation in UV-B-induced photomorphogenesis. It thus remains poorly understood how the positive and negative regulatory machinery is coordinated in the UV-B signaling network. In this study, we demonstrate that two sets of E3 ubiquitin ligases promote and repress UV-B-induced photomorphogenesis respectively by differentially targeting specific substrates for proteasome-mediated degradation, so as to constitute positive and negative regulatory circuits for balanced signal transduction. Based on the findings, our research aims at understanding the genetic and molecular basis of plant responses to UV-B light, and over a long term, at establishing the linkage between fundamental plant biology and practical outcomes.

T4-27-05

Plant seedling soil emergence: Regulation of the protein machinery

Shangwei Zhong
Peking University

The survival of seed plants in natural environments requires the successful emergence from the soil. A key question in seedling emergence is how plants sense changing soil conditions and adjust their growth accordingly, yet this mechanism is largely unknown. Here, we find that CONSTITUTIVE PHOTOMORPHOGENESIS 1 (COP1), a central repressor of light signaling, is a key component required for seedlings to sense the depth of soil overlay. Mutation of COP1 causes severe defects in penetrating soil, due to decreased level of EIN3, a master transcription factor in ethylene pathway that mediates seedling emergence. We show that COP1 directly targets the F-box proteins EBF1 and EBF2 for ubiquitination and degradation, thus stabilizing EIN3. As seedlings grow towards the surface, the depth of soil overlay decreases, resulting in a gradual increase of light fluences. COP1 channels the light signals while ethylene transduces the information on soil mechanical conditions, which cooperatively control EIN3 protein levels to promote seedling emergence from the soil. The COP1-EBF1/2-EIN3 module reveals a mechanism by which plants sense the depth to surface and uncovers a novel regulatory paradigm of an ubiquitin E3 ligase cascade. Upon reaching the surface, light triggers a dramatic developmental transition termed de-etiolation that requires immediate termination of ethylene responses. We further find that light-activation of photoreceptor phyB results in rapid degradation of EIN3, the master transcription factor in ethylene signaling pathway. As a result, light rapidly and efficiently represses ethylene actions. Specifically, phyB directly interacts with EIN3 in a light-dependent manner and also physically associates with F-box protein EBFs. The light-activated association of phyB, EIN3, and EBF1/EBF2 proteins stimulates robust EIN3 degradation by $SCF^{EBF1/EBF2}$ E3 ligases. We reveal phyB acting as a light-reversible "molecular glue" that directly manipulates the substrate-E3 ligase interactions to control the stability of EIN3. Our findings illustrate a mechanistic model of how plants transduce the light information to immediately turn off ethylene signaling for de-etiolation initiation.

T4-27-06

Blue light- and low temperature-regulated COR27 and COR28 play roles in the *Arabidopsis* circadian clock

Xu Li[1], *Dingbang Ma*[1], *Sheen Lu*[2], *Xinyi Hu*[1], *Rongfeng Huang*[3], *Tongda Xu*[3], *Elain Tobin*[2], ***Hongtao Liu***[1]
1. National Key Laboratory of Plant Molecular Genetics, CAS Center for Excellence in Molecular Plant Sciences, Institute of Plant Physiology and Ecology, Shanghai Institutes for Biological Sciences, Chinese Academy of Sciences. 200032 Shanghai, P. R. China
2. Department of Molecular, Cell, and Developmental Biology, University of California, Los Angeles, CA 90095, USA.
3. Chinese Academy of Sciences Center for Excellence in Molecular Plant Sciences, Shanghai Center for plant stress biology. Shanghai Institutes for Biological Sciences, Chinese Academy of Sciences. 201602 Shanghai, P. R. China

Light and temperature are two key environmental signals that profoundly affect plant growth and development, but underlying molecular mechanisms of how light and temperature signals affect the circadian clock are largely unknown. Here we report that *COR27* and *COR28* are regulated not only by low temperatures but also by light signals. COR27 and COR28 are negative regulators of freezing tolerance but positive regulators of flowering, possibly representing a trade-off between freezing tolerance and flowering. Furthermore, loss-of-function mutations in COR27 and COR28 result in period lengthening of various circadian output rhythms and affect central clock gene expression. Also, the *cor27 cor28* double mutation affects the pace of the circadian clock. Additionally, COR27 and COR28 are direct targets of CCA1, which represses their transcription via chromatin binding. Finally, we report that COR27 and COR28 bind to the chromatin of *TOC1* and *PRR5* to repress their transcription, suggesting that their effects on rhythms are in part due to their regulation of *TOC1* and *PRR5*. These data demonstrate that blue light and low temperature-regulated COR27 and COR28 regulate the circadian clock as well as freezing tolerance and flowering time.

T4-28: Floral organ development (two sessions)

T4-28-01

VRS2 regulates hormone-mediated inflorescence patterning in barley

Helmy Youssef[1,2], *Kai Eggert*[1], *Ravi Koppolu*[1], *Ahmad Alqudah*[1], *Naser Poursarebani*[1], *Arash Fazeli*[3], *Shun Sakuma*[1], *Akemi Tagiri*[4], *Twan Rutten*[1], *Geetha Govind*[1], *Udda Lundqvist*[5], *Andreas Graner*[1], *Takao Komatsuda*[4], *Nese Sreenivasulu*[1,6], *Thorsten Schnurbusch*[1]

1. IPK- gatersleben

2. Faculty of Agriculture, Cairo University, Giza,

3. Department of Agronomy and Plant Breeding, Faculty of Agriculture, Ilam University, Ilam,

4. National Institute of Agrobiological Sciences (NIAS) Tsukuba, 305 8602 Japan

5. Nordic Genetic Resource Center (NordGen), Smedjevägen 3, SE-23053 Alnarp, Sweden

6. International Rice Research Institute (IRRI), Grain Quality and Nutrition Center, DAPO Box 7777, Metro Manila, Philippines

Plant architecture has clear agronomic and economic implications for crops, such as wheat and barley, as it is a critical factor for determining grain yield. Despite this, only limited molecular information is available about how grain-bearing inflorescences, called spikes, are formed and maintain their regular, distichous pattern. Here we elucidate the molecular and hormonal role of Six-rowed spike 2 (Vrs2)— a SHORT INTERNODES (SHI) transcriptional regulator during barley inflorescence and shoot development. We show that Vrs2 is specifically involved in floral organ patterning and phase duration by maintaining hormonal homeostasis and gradients during normal spike developmentand similarly influenced plant stature traits. Furthermore, we establish a link between the SHI-protein family and sucrose metabolism during organ growth and development that may have implications for deeper molecular insights into inflorescence and plant architecture in crops.

T4-28-02

Transcriptome profiling of *Desmos chinesis*: Revealing the molecular basis of dipartite perianth evolution in the early divergent angiosperm family Annonaceae

Amy Wing-Sze Leung[1], *Sangtae Kim*[2], *Boon Leong Lim*[1], *Richard Mark Kingsley Saunders*[1]

1. The University of Hong Kong

2. Sungshin Women's University

During the past 200 million years of speciation and hybridization, angiosperm genome structure has evolved and diversified in size and composition. While many angiosperm taxa experienced genome duplication and different degrees of transposable element proliferation, the homeotic control of floral development has remained conservative. The MIKC-type MADS-domain proteins are transcription factors that dictate organ identity during floral organogenesis. The classic "ABCDE" *MADS*-box gene expression model proposed for the core eudicots is not fully applicable to early-divergent angiosperms, in which a gradual transition is apparent between sepal-like outer tepals and petal-like inner tepals, leading to the proposed "fading-borders" expression model. Flowers in the Annonaceae deviate from other early-divergent angiosperms in possessing a distinctive dipartite perianth. Petals in the Annonaceae ('bracteopetals') have furthermore presumably evolved independently from andropetals in core eudicots. The degree of differentiation of petals from sepals in Annonaceae flowers is generally of functional significance (e.g., enabling the evolution of largely or fully enclosed pollination chambers during anthesis). Little is known of the genome and genetic control of floral development in the Annonaceae. The application of massive parallel sequencing techniques such as RNAseq provides opportunities to profile the transcriptomes of non-model organisms. In addition to a developmental study, we characterized the floral and leaf transcriptome of *Desmos chinensis*, a representative of the Annonaceae that possesses a dipartite perianth and which shares the basic floral Bauplan of the family, in order to understand the genetics of its floral development. Transcriptomes of developing and mature *Desmos chinensis* floral organs and leaves were sequenced using the high-throughput Illumina sequencing platform. After filtering, 16,098 genes and 22,112 transcripts were recovered from *de novo* assembly. We identified possible coding regions and compared them against five public databases. Based on sequence homology against the NCBI non-redundant protein database, transcripts were annotated with 3,991 unique gene ontology (GO) terms and categorized into functional groups. Functional group enrichment analysis was used to identify groups with significant expression level changes during organ development. Transcripts that are commonly differentially expressed among developing and mature floral organs and leaves were identified using the digital gene expression data. We identified 41 putative homeotic MADS-box gene homologs and their respective expression levels in order to determine whether Annonaceae flowers are consistent with the *MADS*-box gene "fading borders" expression model. Other floral developmental regulators of *Desmos chinensis* were also discovered. While many questions remain unresolved at the transcription level, our data emphasises the importance and potential of extending further studies into systems biology, and help understand floral developmental biology as biological networks.

T4-28-03

The *Arabidopsis* CELLULASE 6 and MANNANASE 7 proteins affect cell differentiation in the silique and contribute to silique dehiscence

Hanjun He[1], Mei Bai[1], Yanting Hu[1], Ming Yang[2], Hong Wu[1]
1. South China Agricultural University
2. Oklahoma State University

Cell wall degradation or modification by cellulases, hemicellulases and pectinases is expected to be crucial for many plant developmental processes. Mutations in pectinase genes *ADPG1* and *ADPG2*, for example, cause silique indehiscence in *Arabidopsis*. At present, no mutations of cellulase or hemicellulase genes have been reported to affect silique development and dehiscence in plants. To investigate the roles of cellulases and hemicellulases in fruit development and dehiscence, we first identified *CEL6,* a cellulase gene, and *MAN7,* a hemicellulase gene, as candidate genes encoding cell wall-degrading enzymes in Arabidopsis siliques. Analysis of the expression of these two genes by qPCR in siliques of the wild type and the silique-indehiscent mutants *ind-7* and *alc-3* revealed that their expression was partially dependent on the IND and ALC transcription factors. GUS staining results from transgenic lines containing the promoter-GUS fusion and protein-GUS fusion of these genes confirmed that *CEL6* and *MAN7* were expressed in the valve and the replum, including the dehiscence zone. Such GUS signals were reduced or hardly detectable in the *ind-7* and *alc-3* siliques, consistent with the aforementioned qPCR results. Phenotypic analysis of *cel6* and *man7* loss-of-function mutants indicated that silique development was delayed and dehiscence impaired in these mutants. Secondary cell wall thickening and cell morphology in the valve margin were also affected by the mutations. Most notably, in sections prepared for transmission electron microscopy, cells in the separation layer in nearly mature siliques of the single mutants and the *cel6-1 man7-3* double mutant remained intact whereas they were broken in the control wild-type siliques. Overexpression of *CEL6* and *MAN7* also seemed to slightly increase the dehiscence rates of nearly mature and mature siliques. Phenotypic analysis of the double, triple, and quadruple mutants involving the *adpg1-1, adpg2-1, cel6-1, and man7-3* mutations indicated that the indehiscent phenotypes of the *adpg1-1* single mutant and the *adpg1-1 adpg2-1* double mutant, which were much more severe than those of the *cel6-1* and *man7-3* single mutants and the *cel6-1 man7-3* double mutants, became even more severe by the additions of the *cel6-1* and *man7-3* single mutations or the *cel6-1 man7-3* double mutation. On the other hand, the phenotypes of the *cel6-1* and man7-3 single mutants and the *cel6-1 man7-3* double mutant were epistatic to that of *adpg2-1*. Taken together, our results show that CEL6 and MAN7 act in the cell differentiation process during silique development and contribute to silique dehiscence by promoting cell lysis in the separation layer. Our results also demonstrate that the ability of the silique to dehisce can be manipulated by altering the activities of multiple types of cell wall-degrading enzymes.

T4-28-04

Control of anther development by the EMS1 receptor kinase-linked signaling complex in *Arabidopsis*

Dazhong Zhao[1], Jian Huang[1], Zhiyong Li[1], Yao Wang[1], Nagib Ahsan[2], Gabriel Biener[3], Jay Thelen[2], Valerica Raicu[3], Marisa Otegui[4], Heather Owen[1]
1. Department of Biological Sciences, University of Wisconsin-Milwaukee, Milwaukee, Wisconsin
2. Department of Biochemistry, University of Missouri, Columbia, Missouri
3. Department of Physics, University of Wisconsin-Milwaukee, Milwaukee, Wisconsin
4. Department of Botany, University of Wisconsin-Madison, Madison, Wisconsin

In flowering plants, successful sexual reproduction depends on the specification of distinct reproductive cells that conduct meiosis to form gametes and associated somatic cells that provide nutrition and developmental cues to ensure successful gamete production. Anther, the male reproductive organ in a flower, typically has four lobes (microsporangia). Within each of these lobes in a mature anther, the central reproductive microsporocytes (or pollen mother cells) are surrounded by four concentrically organized somatic cell layers: epidermis, endothecium, middle layer, and tapetum. Microsporocytes give rise to pollen via meiosis, while somatic cells, particularly the tapetum, are required for the normal development and release of pollen. Our previous studies showed that the EMS1 leucine-rich repeat receptor-like kinase (LRR-RLK) plays an essential role in somatic and reproductive cell differentiation during anther development. Recent research in my lab for the first time demonstrates that TPD1 is a novel small secreted cysteine-rich protein ligand for EMS1. The TPD1-EMS1 signaling initially promotes the periclinal division of secondary parietal cells to form a monolayer of tapetal precursor cells and then determines the fate of functional tapetal cells. We also identified the EMS1 co-receptor SERK1/2 and its downstream player CA1. *EMS1* and *SERK1/2* functions in the same genetic pathway. Bimolecular fluorescence complementation (BiFC) and Fluorescence resonance energy transfer (FRET) analyses demonstrate that TPD1 interacts with EMS1 *in vivo*. Trans-phosphorylation between EMS1 and SERK1/2 enhances the EMS1 kinase activity. In addition, we determined phosphorylation sites of EMS1 and their biological significance. Our yeast two-hybrid screening identified the EMS1 downstream player CA1. Besides several lines of evidence supports the biochemical interaction between EMS1 and CA1, our results show that loss-of-function mutants of *CA1* are defective in tapetal cell differentiation, while overexpression of *CA1* results in the formation of additional tapetal cells. A model will be discussed to explain how the TPD1-EMS1/SERK1/2-CA1 signaling pathway controls anther development.

T4-28-05

An ABC transporter, OsABCG26, is required for anther cuticle and pollen exine formation and pollen-pistil interactions in rice

Zhenyi Chang[1], Zhufeng Chen[1], Wei Yan[3], Gang Xie[1], Jiawei Lu[1], Na Wang[1], Qiqing Lu[1], Nan Yao[4], Guangzhe Yang[2], Jixing Xia[2], Xiaoyan Tang[1,3,5]
1. Shenzhen Institute of Molecular Crop Design
2. State Key Laboratory of Conservation and Utilization of Subtropical Agro-bioresources, College of Life Science and Technology, Guangxi University
3. School of Life Sciences, Capital Normal University
4. School of Life Sciences, Sun Yat-Sen University
5. Guangdong Key Lab of Biotechnology for Plant Development, College of Life Science, South China Normal University

Wax, cutin and sporopollenin are essential components for the formation of anther cuticle and pollen exine, respectively. Their lipid precursors are synthesized by secretory tapetal cells and transported to anther and microspore surface for deposition. However, the molecular mechanisms involved in the formation of anther cuticle and pollen exine are poorly understood in rice. Here, we characterized a rice mutant *osabcg26* with male sterility. Molecular cloning and sequence analysis revealed a point mutation in the gene encoding ATP binding cassette transporter G26 (OsABCG26). *OsABCG26* was specifically expressed in anther and pistil. Cytological analysis revealed defects in tapetal cells, lipidic Ubisch bodies, pollen exine, and anther cuticle in *osabcg26* mutant. Expression of some key genes involved in lipid metabolism and transport, such as *UDT1*, *WDA1*, *CYP704B2*, *OsABCG15*, *OsC4* and *OsC6*, was significantly altered in *osabcg26* anther, possibly due to disturbance in the homeostasis of anther lipid metabolism and transport. Interestingly, we found most of the wild-type pollen tubes ceased growth in the upper region of *osabcg26* ovary, and only a few pollen tubes could reach the micropyle for fertilization, leading low seed setting in *osabcg26* cross-pollinated with the wild-type pollen. These results indicated that *OsABCG26* plays an important role in anther cuticle and pollen exine formation and pollen-pistil interactions in rice. Further investigations are needed to decipher the mechanism of pollen tube growth defect in *osabcg26* pistil and to pinpoint the role of OsABCG26 in pollen-pistil interactions.

T4-28-06

Variable cell growth channels reproducible organ development though spatiotemporal averaging

Lilan Hong[1], *Mathilde Dumond*[2], *Satoru Tsugawa*[3,4], *Aleksandra Sapala*[5], *Anne-Lise Routier-Kierzkowska*[5], *Yong Zhou*[1], *Mingyuan Zhu*[1], *Catherine Chen*[1], *Olivier Hamant*[2], *Richard S. Smith*[5], *Tamiki Komatsuzaki*[3], *Chun-Biu Li*[3,4], *Arezki Boudaoud*[2], *Adrienne H. K. Roeder*[1]

1. Cornell University
2. Ecole Normale Supérieure de Lyon
3. Hokkaido University
4. Stockholm University
5. Max Planck Institute for Plant Breeding Research

Organ sizes are strikingly precise, despite the stochastic growth and division of individual cells within them. Because their shapes are highly reproducible and relatively insensitive to environmental conditions, *Arabidopsis* sepals, allow robustness in organogenesis to be assessed within a single plant. To reveal how robust organ sizes emerge from the variable growth of cells, we screened for *Arabidopsis* mutants with variably sized sepals and identified mutations in the mitochondrial i-AAA protease FtsH4. Through live imaging, we observed substantially variable cell growth in wild-type sepals, while in *ftsh4* sepals, the variability of neighboring cell growth is reduced. Through computational modeling and quantitative phenotypic analysis, we found that regular organ size results from spatiotemporal averaging of the cellular variability in wild-type sepals, whereas in *ftsh4* mutants the averaging process is disrupted due to the less-variable cell growth. Reactive oxygen species (ROS) accumulated in *ftsh4* mutants, abnormally increasing both cell wall stiffness and organ size variability. In wild-type sepals, ROS accumulate in the maturing cells and limit organ growth. Enzymatically-lowering ROS produced larger sepals. Our results demonstrate that spatiotemporal averaging of cellular variability is required for robustness in organ size, and there is an inverse relationship between the robustness of plant organ size and ROS. We are currently investigating the mechanism of ROS's regulation of the spatiotemporally variability averaging process, and exploring whether mechanical cues contribute to this process.

T4-28-07

Petal elaboration in *Nigella*: From morphology, micro-morphology to molecular basis

Xu Yao, Rui Zhang, Hong Liao, ***Hongzhi Kong***
State Key Laboratory of Systematic and Evolutionary Botany, Institute of Botany, Chinese Academy of Sciences

As a special type of floral organ, petals show tremendous diversity in shape, structure, color/coloration, and function, and have increasingly been used as a model for studying plant organogenesis and evolution. Petals of the *Nigella* (Ranunculaceae) species are of particular interest because they are highly elaborate, bilabiate structures with extensive modifications. To understand how *Nigella* petals become elaborated and diversified, we first investigated the tempo, mode, and mechanisms of character evolution during petal elaboration within this genus. We found that the degree of petal complexity increased gradually during evolution, involving both modification of existing characters and *de novo* origination of new features. Through transcriptomic analyses, we then identified regulatory modules and genes contributing to petal elaboration in *N. damascena*. We found that *NdLMI1*, a member of HD-Zip I genes, is responsible for the division of the lower lip as well as the initiation of short trichomes. We also found that *NdPSN01* and *NdYAB5-1* play key roles in the development of pseudonectaries, silence of which led to loss of pseudonectaries. Finally, we explored histological evidences and expression patterns of adaxial-abaxial polarity genes during petal development, and propose that the basic structure of *Nigella* petals may have been related to the repatterning of adaxial-abaxial polarity during morphogenesis.

T4-28-08

The origin of floral organ identity quartets

Philip Ruelens[1], *Zhicheng Zhang*[1], *Hilda Van Mourik*[2], *Steven Maere*[3], *Kerstin Kaufmann*[4], ***Koen Geuten***[1]

1. KULeuven
2. Wageningen University
3. Gent University
4. Humboldt University Berlin

The origin of flowers has puzzled plant biologists ever since Darwin referred to their sudden appearance in the fossil record as an abominable mystery. Flowers are considered to be an assembly of protective, attractive and reproductive male and female leaf-like organs. Their origin cannot be understood by a morphological comparison to gymnosperms, their closest relatives, which develop separate male or female cones. Despite these morphological differences, gymnosperms and angiosperms possess a similar genetic toolbox consisting of phylogenetically related MADS-domain proteins. Using ancestral MADS-domain protein reconstruction, we trace the evolution of organ identity quartets along the stem lineage of crown angiosperms. We provide evidence that current floral quartets specifying male organ identity, which consist of four types of subunits, evolved from ancestral complexes of two types of subunits through gene duplication and integration of SEPALLATA proteins just before the origin of flowering plants. Our results suggest that protein interaction changes underlying this compositional shift were the result of a gradual and reversible evolutionary trajectory. Modelling shows that such compositional changes may have facilitated the evolution of the perfect, bisexual flower.

T4-28-09

OsMADS14/15 interact with YABBY6 and specify the perianth organ identity in rice (*Oryza sativa* L)

Feng Wu, Xiaowei Shi, Shaojie Cui, ***Zheng Meng***
Institute of botany, Chinese Academy of Sciences

The well-known ABC model describes the combinatorial interaction of homeotic genes in specifying floral organ identies. While there is evidence that the B and C functions are highly conserved throughout angiosperm evolution, the status of the A function has been hotly debated because in most plants that have been investigated thus far, with *Arabidopsis* being a remarkable exception, one does not find recessive mutants in which the identity of both types of perianth organs is affected. In our study, we present a comprehensive mutational analysis of all four members (*OsMADS14*, *OsMADS15*, *OsMADS18*, and *OsMADS20*) of the *AP1/FUL*-like subfamily of MADS-box genes in rice, and demonstrate that *OsMADS14* and *OsMADS15*, in addition to their function of specifying meristem identity, are also required to specify palea and lodicule identities. Since these two grass-specific organs are very likely homologous to sepals and petals of eudicots, respectively, we conclude that some kind of an floral homeotic A function also exists in grasses. Furthermore, our result of IP-MS showed that OsYABBY6 is an interaction factor of OsMADS14/15. The further experiments, such as Pull-down, BiFC, and Co-IP analyses, confirmed the protein-protein interaction patterns. Our data reveal that, at least to some extent, the mechanism of floret development controlled by multi-gene regulation in rice, and thus expand the understanding of the core theory described by ABC model that are mostly made of MADS-box genes.

T4-28-10

Evolution of the fruit developmental genetic network with particular emphasis on the dehiscence zone bHLH identity genes

Natalia Pabon-Mora[1]*, Clara Ortíz-Ramírez*[1]*, Cecilia Zumajo-Cardona*[2]*, Marco Antonio Giraldo*[3]*, Cristina Ferrandiz*[4]*, Barbara A Ambrose*[2]
1. Universidad de Antioquia, Facultad de Ciencias Exactas y Naturales, Instituto de Biologia
2. The New York Botanical Garden
3. Universidad de Antioquia, Facultad de Ciencias Exactas y Naturales, Instituto de Física
4. Universitat Politecnica de Valencia

Fruit development in *Arabidopsis thaliana* is controlled by seven key transcription factors from different gene families: FUL and SHP1/2 belong to the MADS-box family, IND, ALC and SPT belong to the bHLH family, RPL belongs to the homeodomain (HOX) family and AP2 belongs to the APETALA2/Ethylene Responsive Element Binding Factor gene family, all of which are large transcription factor families. FRUITFULL (FUL) is responsible for proper cell division during valve development; REPLUMLESS (RPL) is involved in controlling replum identity; both FUL and RPL antagonize SHATTERPROOF genes (SHP1/2), which in turn become restricted to the dehiscence zone. SHP1/2 positively regulate INDEHISCENT, responsible for the formation of the lignified layer, as well as ALCATRAZ and SPATULA, which redundantly control the identity of the non-lignified layer. The entire network is negatively regulated by APETALA2. In order to identify the components of the network conserved across angiosperms, we have studied the evolution of each of these gene lineages involved in fruit development. Our data show that: (1) most MADS box gene duplicates are maintained independently in monocots, basal eudicots and core eudicots; (2) the bHLH SPT and ALC paralogs are exclusive to core-eudicots, while paleoSPT/ALC are found in non-core eudicots; (3) IND orthologs are only present in Brassicaceae while HECATE3 proteins are present in all other flowering plants; (4) AP2 homologs are found in all seed plants, with a specific duplication resulting in the paralogous TOE3 genes in Brassicales; and (5) RPL genes have been mostly retained as single copy genes across angiosperms. In order to better understand the evolution of specific genes targeting the formation of dehiscence zones, we studied expression and function of bHLH, ALC and SPT genes in Solanaceae (core eudicot) and Papaveraceae (basal eudicot). In Solanaceae, ALC genes are predominantly retained as single copy, while SPT genes underwent a specific duplication in *Nicotiana*. Both ALC and SPT genes are expressed in stamens, carpels, fruits and leaves of *Nicotiana obtusifolia*, *Capsicum annuum* and *Solanum lycopersicum*. Functional data from *N. obtusifolia* and *C. annum* show that *SPT* and *ALC* play redundant roles in leaf pigmentation, petal fusion and fruit development. Specifically, they seem to negatively control lignification as downregulated plants show smaller fruits with ectopic lignified patches in the pericarp, independently of whether fruits are dry, dehiscent capsules (i.e. *Nicotiana*) or indehiscent berries (i.e. *Capsicum*). Further evidence for the maintained role of these genes in regulating specific tissues in dehiscence zones comes from conserved expression patterns of *paleoSPT/ALC* preduplication genes in *Bocconia frutescens* (Papaveraceae) compared to their core-eudicot counterparts. Altogether our data shows that by comparison to the preduplication genes, SPT and ALC genes have undergone some degree of sub-functionalization in Brassicaceae while maintaining redundancy in Solanaceae. Finally, we propose that co-option of SPT and ALC genes in leaf pigmentation is likely a new role of these genes exclusively in *Solanaceae*.

T4-28-11

The Origin of peloria in Chinese cymbidium (*Cymbidium sinense*)

Shihao Su
Sun Yat-Sen University

Orchid flowers exhibit zygomorphic corollas with highly specified lips. In *Cymbidium sinense*, plenty of peloric varieties with defects in floral zygomorphy have been bred during centuries of domestication since Song dynasty. However, little is known about the molecular basis regulating orchid floral zygomorphy and producing these peloric varieties. Here, we studied the floral morphogenesis of *C. sinense*, transcriptome-wide examined the spatiotemporal expressional patterns of floral homeotic MADS factors, and detected the expression of interested genes among different types of varieties. The floral zygomorphy was established in early developmental process with retarded ventral organogenesis. We found the interactions among AGL6-clade and AP3-clade MADS genes were crucial to the establishment of its floral zygomorphy. *CsAP3-1* was highly expressed in the two inner whorls of floral organs, which coped with a lip-specific factor, *CsAGL6-1*, regulating the differentiation between dorsal lip and ventral petals. The expression of these two genes were absent in varieties with sepal-like petals and lip. The floral sepal whorl was established by another *AGL6*-clade factor, *CsAGL6-2*, whose expression was significantly down-regulated in varieties with defects in sepals and petals. These findings indicate combinatory actions of MADS factors and changes in these factors may have occurred during artificial selection.

T4-28-12

Mechanisms of floral organ specification by MADS domain proteins

Dijun Chen, Wenhao Yan, Jose Muino, Cezary Smaczniak, Suzanne De Bruijn, ***Kerstin Kaufmann***
Humboldt-Universitaet zu Berlin

Flower development is controlled by master-regulatory transcription factors (TFs), which orchestrate gene expression during developmental switches and organ differentiation. However, the molecular mechanisms by which these factors dynamically regulate the expression of their target genes are still poorly understood: how are developmental programmes encoded in plant genomes, and how is this ‘code’ read and executed? We characterized dynamic changes in chromatin structure, gene expression and DNA-binding of floral MADS-domain TFs. Our findings suggest that MADS-domain TFs modulate chromatin accessibility and thereby control target gene expression. MADS-domain TFs control the expression of hundreds of genes in a partly stage- and tissue-specific fashion. In order to study how different MADS-domain TFs achieve their functional specificity, we used an *in vitro* approach to reveal DNA-binding specificities of different heteromeric MADS domain protein complexes that specify distinct organ types in the flower. We are also interested in the evolutionary dynamics of gene-regulatory networks controlling flower development. Using a comparative ChIP-seq approach, we determined the evolutionary variation of DNA binding sites of a developmental key regulatory TF in two closely related species: *Arabidopsis thaliana* and *A. lyrata*. Despite their recent evolutionary divergence of ~10 Mio years, we found a high level of divergence between TF binding site locations. In *A. lyrata*, we found that binding sites can be associated with transposable elements, which likely contributed to the origin of novel binding site locations in this species.

T4-29: Plant branching

T4-29-01

Shoot branching and complex leaf development are controlled by homologous gene modules

Quan Wang, Hernán López, Susanne Rossmann, Bernhard Busch, Wouter Kohlen, Alice Hasson, Gregor Schmitz, ***Klaus Theres***
Max Planck Institute for Plant Breeding Research

Aerial plant architecture is predominantly determined by the pattern of shoot branching and the morphology of leaves, which are governed by underlying developmental processes, axillary meristem formation and complex leaf development, that appear fundamentally unrelated. In tomato, these processes share essential functions in boundary establishment. Boundaries are established in different regions of the plant body and have diverse functions in development. Firstly, they separate different cell groups, for example the differentiating cells of a leaf primordium from the pluripotent cells of the apical meristem. In compound leaved species, like tomato, boundaries separate young leaflets from each other and from their connecting structure, called rachis. Recent experiments have shown that boundary zones between the meristem and leaf primordia show similar properties as boundaries between leaflets. They are characterized by a low rate of cell divisions and specific patterns of gene expression. We identified potato leaf (*C*), a key regulator of leaf dissection, to be the closest paralog of the shoot branching regulator Blind (*Bl*). Expression studies and complementation analyses indicate that these genes have undergone sub- or neo-functionalization due to promoter differentiation. *Bl* is expressed in the axils of leaves whereas *C* transcripts accumulate in the boundary zones of leaflets. The known leaf complexity regulator Goblet (Gob) is crucial for axillary meristem initiation and leaf dissection, acting in parallel to C and Bl. A combination of loss-of-function alleles of *C* and *Gob* results in simple leaves. Secondly, boundary zones initiate new meristems, which requires a low level of the plant hormone auxin. Disruption of polar auxin transport compromises axillary meristem formation. New meristems do not only initiate in the meristem-to-leaf primordium boundary, but also in the distal leaflet boundary (DLB) of tomato leaves. Initiation of these ectopic meristems is dependent on activities of the well-known branching regulators Goblet (Gob) and Lateral suppressor (Ls), which act in hierarchical order. Finally, we have identified a new regulator, Super determinant (Sde), which plays a critical role in axillary meristem formation. *Sde* encodes a protein with similarity to components of the PRC1 complex. A search for interactors indicates that Sde may be involved in an epigenetic mechanism regulating axillary meristem formation.

T4-29-02

Stem cell specification during shoot branching

Yuling Jiao
Institute of Genetics and Developmental Biology, Chinese Academy of Sciences

The homeodomain transcription factor WUSCHEL (WUS) directs stem cell specification in the shoot. Extensive studies have shown how *WUS* homeostasis is maintained and how *WUS* promotes shoot meristems, but the mechanisms underlying the establishment of *WUS* expression remain unknown. Seed plants produce meristematic cells that form axillary meristems in the leaf axil; these meristems may develop into branches, and thus affect plant architecture. Floral meristems are also specialized axillary meristems. Here, we show that cytokinin signaling activates *WUS* expression *de novo* in the leaf axil to promote axillary meristem initiation. Furthermore, type-B ARABIDOPSIS RESPONSE REGULATORs (ARRs), transcriptional activators in the cytokinin signaling pathway, directly bind to the *WUS* promoter to activate its expression. Finally, *WUS* expression requires a permissive epigenetic environment to restrict it to highly defined meristematic tissues. Taken together, these findings establish a mechanistic framework for the postembryonic establishment of the shoot stem cell niche.

T4-29-03

De novo organogenesis in *Arabidopsis*: From wounding to cell fate transition

Lin Xu
Institute of Plant Physiology and Ecology, SIBS, Chinese Academy of Sciences

De novo root organogenesis is a type of plant regeneration for survival after wounding. We established simple methods to study *de novo* root and shoot organogenesis by culturing Arabidopsis explants without added hormones to regenerate adventitious roots and shoots. Wounding is the first event in regeneration. During *de novo* root organogenesis from leaf explants, wound signal spreads rapidly from the wounded site to mesophyll cells and activates auxin biogenesis in mesophyll cells. Auxin is then polar-transported to regeneration-competent cells (i.e. procambium and vascular parenchyma cells) in the vasculature near the wound site. Auxin accumulation in regeneration-competent cells activates *WOX11* and *WOX12*, and this triggers the first step of cell fate transition from regeneration-competent cells to root founder cells. *WOX11* and *WOX12* directly activate *WOX5*, *WOX7* and *LBD16* for the second step of cell fate transition from root founder cells to root primordium. Overall, wound signaling, auxin behavior and cell fate transition are successively involved in *de novo* regeneration of adventitious roots from leaf explants. In addition, *WOX11* and *WOX12* are specifically expressed in founder cells of adventitious roots but not lateral roots, suggesting that the priming and initiation mechanisms for adventitious roots and lateral roots might be different.

T4-29-04

Regulation of rice inflorescence architecture by a cooperative action of *TAWAWA1* and *BOP* genes

Masashi Shindo[1], *Satoshi Naramoto*[1], *Hiroki Tokunaga*[2], *Taiyo Toriba*[1], ***Junko Kyozuka***[1]
1. Tohoku University
2. Riken

The architecture of rice inflorescence is determined by the pattern of shoot branching, which depends on the timing of a change from indeterminate growth to determinate growth of the meristem. Among meristems produced during inflorescence development, the ones produced early acquire an indeterminate branch meristem (BM) identity and iterates shoot branching, while the later produced ones are specified as a determinate spikelet meristem (SM). *TAWAWA1* (*TAW1*) is a negative regulator of the meristem phase change from the BM to the SM. Therefore, higher activity of TAW1 leads to production of higher order branches, which results in the production of an inflorescence with more spikelets. *TAW1* encoded a nuclear protein containing a conserved ALOG domain. TAW1 binds to specific DNA sequences, interacts with an ortholog of *Arabidopsis BLADE-ON-PETIOLE* (*BOP*) and localizes in the nuclear, suggesting it works as a transcriptional regulator. We analyzed target genes of TAW1-BOP by chromatin immunoprecipitation followed by next-generation sequencing. We will discuss a possible molecular function of TAW1-BOP complex in the control of the meristem phase change.

T4-29-05

Branching out into HD-ZIPIII target genes and interacting proteins

Stephan Wenkel
University of Copenhagen

The production of side shoots, commonly referred to as shoot branching, is a major trait determining plant biomass. Plants can initiate new branches by establishing lateral meristems that form in the axils of leafs. Class III homeodomain leucine zipper (HD-ZIPIII) transcription factors are major determinants of shoot patterning and also involved in controlling branching. The activity of HD-ZIPIIIs is strongly regulated by microRNAs and microProteins at both the post-transcriptional and post-translational level.
In order to deeper characterize HD-ZIPIII function, we have used a combination of genomics (ChIP-Seq and RNA-Seq) and proteomics approaches to identify HD-ZIPIII responsive genes as well as interacting proteins. Further genetic analysis revealed several new factors that act downstream of HD-ZIPIIIs in the control of shoot meristem activity, including shoot branching.

T4-29-06

New evidence of complex shoot and leaf meristems in ferns

Rafael Cruz[1], *Gladys Flavia Melo-De-Pinna*[1], *Jefferson Prado*[2], *Barbara Ambrose*[3]
1. University of Sao Paulo
2. Instituto de Botanica de Sao Paulo
3. The New York Botanical Garden

Leaves usually are lateral and determinate organs and these features seem to apply well in most of the cases in seed plants. On the other hand, shoots are characterized by indeterminacy, detected by the expression of Class I *KNOTTED-LIKE HOMEOBOX* (*KNOX*) genes in the shoot apical meristem (SAM). Downregulation of Class I *KNOX* is one of the first signals of the origin of a determinate leaf primordium and plants with defective Class I *KNOX* genes may be unable to maintain the SAM. Ferns are tricky in the sense that most of them produce leaves called fronds that are usually compound with lateral pinnae and a leaf apical meristem (LAM). The LAM and SAM of ferns have a distinctive prominent apical cell, that has been pointed as equivalent to a whole apical meristem by some authors or only the initial cells of a multicellular meristem by other. Organ determinacy seems to

be the key character to understand the evolution and development of fronds. They show intriguing resemblances with indeterminate shoots (e.g. apical meristems presence, lateral organs production, and reduced or absent determinacy) and do not fit on classic morphology concepts of leaves. In order to gather more information concerning fronds and pinnae development, and about LAM and SAM organization, we studied the expression of Class I KNOX in *Mickelia scandens*, a leptosporangiate fern that bears small pinnate fronds in the younger terrestrial form and longer pinnate fronds in the older climbing form. In-situ hybridizations and RTq-PCR experiments were performed in shoot apex and developing leaves and we detected two Class I *KNOX* and one histone genes that were expressed in growth regions of the shoot apex and developing leaves. The expression of these genes shows a significant difference in the smaller developing fronds of the younger form when compared with shoot apices and bigger developing fronds. We also detected a complex meristematic organization, not restricted to the apical cells in the shoot and the leaf apex. Class I *KNOX* genes seems to play a role in the development of pinnae, that are more determinate than shoots and developing leaves; and in the development of the vascular system.
These results challenge the classical view that ferns meristems are simple and very reduced. Instead, they have a complex meristem that allows them to develop one of the most complexes foliar structures in land plants.

T4-30: Molecular and genetic analysis of seed development

T4-30-01

Auxin production following fertilization drives early seed development in *Arabidopsis*

Duarte Figueiredo[1], *Rita Batista*[1], *Pawel Roszak*[2], *Lars Hennig*[1], *Claudia Köhler*[1]
1. Swedish University of Agricultural Sciences and Linnean Center for Plant Biology, Uppsala, Sweden
2. The Sainsbury Laboratory, University of Cambridge, United Kingdom

Seed development in flowering plants is preceded by a double fertilization event, whereby one of the paternal sperm cells fertilizes the female egg cell, giving rise to the embryo, and a second sperm cell fertilizes the central cell, which originates the endosperm. The endosperm is a triploid tissue that supports embryo development and, in species such as *Arabidopsis*, it is consumed as the embryo develops. Surrounding the two fertilization products is the seed coat, which derives from the ovule integuments and is of purely maternal origin. The successful establishment of a seed requires the fine tuning in the development of these three genetically-distinct structures, implicating paternally-derived signals that regulate the developmental processes. Seed development in most plant species is tightly coupled to fertilization, but this requirement can be bypassed in mutants for components of the FERTILIZATION INDEPENDENT SEED Polycomb Repressive Complex 2 (FIS-PRC2), which establish the repressive chromatin mark trimethylation of lysine 27 on histone H3 (H3K27me3). Lack of FIS-PRC2 function results in the formation of autonomous seeds, which develop in the absence of fertilization. We have shown that FIS-PRC2 represses the expression of auxin biosynthesis genes in the female gametophyte, so that the paternal-specific expression of these genes drives both endosperm and seed coat development after fertilization. Our data indicates that auxin production is sufficient to initiate seed development in *Arabidopsis*, but it also has a major role in non-autonomous cell communication between gametophytic and sporophytic tissues. The interplay between auxin and different PRC2 complexes, acting both in the gametophyte and in sporophyte, is therefore a key component for the successful establishment of the distinct seed structures during seed development.

T4-30-02

Are sRNAs involved in the drastic shift of allele-specific expression in hybrid tomato seeds?

Ana Marcela Florez-Rueda[1], *Flurin Fiscalini*[1], *Thomas StÄdler*[2], *Ueli Grossniklaus*[1]
1. University of Zurich
2. ETH ZüRICH

The roles of small RNAs (sRNAs) in plant gametophyte and seed development remain elusive, but recent investigations point to a critical involvement. A transcriptomic study has provided insights into hybrid seed failure in interspecific crosses between wild tomato species. There is a strong postzygotic isolation barrier between *Solanum peruvianum* and *Solanum chilense*, with endosperm-based incompatibilities being responsible for seed failure. We performed laser-capture microdissection coupled with RNAseq of the endosperm and analyzed patterns of gene and allele-specific expression (ASE). We found that ASE is disturbed in hybrid seeds with a genome-wide shift towards increased expression of maternal alleles disrupting normal patterns of ASE. Based on these results, we propose that silencing mechanisms regulating maternal allele transcription are malfunctioning. We investigated whether gene silencing mechanisms involving sRNAs, namely RNA-directed DNA methylation (RdDM), are involved in this process. We hypothesized that there is a lack of sRNA production in hybrid endosperm and thus proper transcriptional regulation/silencing does not take place, leading to the observed increase in maternal ASE, loss of imprinting, and hybrid seed failure. We will present analyses coupling sRNA and methylome sequencing of tomato seeds from intraspecific (viable) and interspecific (failed) crosses. We will present correlates of sRNA expression, methylation, expression and ASE in viable seeds to describe the epigenetic landscape of normally developing tomato endosperm. We will compare and contrast these patterns between viable and inviable seeds and test the hypothesis that sRNAs are down-regulated in hybrid endosperms. We will investigate the proposed link between the malfunctioning of the sRNA machinery and the misexpression of specific candidate genes to test for sRNA involvement in the establishment of reproductive isolation and therefore speciation.

T4-30-03

Temperature modulates seed dispersal in Brassicaceae

Xinran Li, *Vinod Kumar, Lars Ostergaard*
John Innes Centre, Norwich Research Park, Norwich, NR4 7UH, United Kingdom

Environmental signals such as temperature affect plant growth and development. Plants respond to diurnal and seasonal changes in temperature to reprogram developmental pathways critical for reproductive fitness and adaptation. Understanding the molecular mechanisms that define environmental modulation of plant growth and reproduction is important in the context of climate change that threatens crop yield. Here we show that elevated temperature accelerates seed dispersal in the model plant *Arabidopsis* and a number of other members of the Brassicaceae family. We found that *THERMAL RESPONSE FACTOR* 1 (*TRF1*), a valve margin identity gene, is upregulated by temperature-induced chromatin dynamics at the *TRF1* locus. Specifically, we show that temperature-induced changes in *TRF1* expression are associated with temperature-dependent H2A.Z nucleosome dynamics. Our findings connect tissue identity with thermal sensing and provide a molecular framework for understanding how temperature influences seed dispersal. These results will be valuable in the production of crops that are able to withstand the predicted rise in temperature at a global scale.

T4-30-04

Pollen tube contents initiate seed enlargement and enhance autonomous endosperm development without fertilization

Ryushiro Kasahara
ITbM Nagoya University

In animals, numerous sperm cells swim toward the egg and complete fertilization supported by seminal plasma, a fluid medium required for sperm metabolism, function and carrier to female reproductive tract. The SVS2 protein localized only in the seminal plasma protects sperms from uterine attack and is required for sperm capacitation in mice. In angiosperms, pollen tubes carry two sperm cells toward the egg and central cells to complete double fertilization. However, except for being a sperm cell carrier, nothing is known regarding the function of pollen tube content (PTC), a fluid medium, analogous to seminal plasma in animals. To investigate whether the PTC functions male-female interaction in plants, we performed transcriptome analysis for 3 types of ovules (seed primordia), non-pollinated (NP), pollinated and fertilized (PF) and pollinated but non-fertilized (PN; PTC only). Notably, multiple genes associated with cell expansion, cell division or seed coat development were significantly expressed in PN ovules. Based on the transcriptome analysis, we performed phenotypic analysis for PN ovules and the ovules showed cell expansion, cell division and seed coat development without fertilization. Here, we newly identified the PTC functions that are significant for seed development initiation. We termed this phenomenon as Pollen tube-dependent Ovule Enlargement Morphology (POEM) and placed it between pollen tube guidance and double fertilization. As well as seminal plasma in animals, PTC plays crucial roles in male-female interaction in plant reproduction. The other significant finding is that PTC increases endosperm nuclei without fertilization. PTC can dramatically increase endosperm nuclei number without fertilization in the previously identified mutants, capable of autonomous endosperm formation. This finding could be applied in agriculture, particularly enhancing seed formation without fertilization in important crops, which is called apomixis.

T4-30-05

Genetic and molecular regulation of grain chalkiness in rice

Yibo Li
National Key Laboratory of Crop Genetic Improvement, Huazhong Agricultural University

Improving grain quality is one of the most important goals of plant science research. Grain chalkiness in rice, a chalky texture of endosperm, is a highly undesirable trait of the grain for human food that deteriorates milling, cooking, eating and appearance quality, and thus represents a major problem in many rice-producing areas of the world. However, the genetic and molecular basis of this trait is still poorly understood. Here, we performed genome-wide association studies for all grain chalkiness traits with four-year field replications, and detected 29 stable quantitative trait loci (QTL) and 15 major QTLs sensitive to temperature using the difference values of the two responsive chalky traits between the high (33°C/27°C, 2013 at Wuhan) and low temperature (27°C/21°C, 2014 at Wuhan) natural environments. Many NILs for these gain-chalkiness QTLs were constructed, and a major QTL, *Chalk5*, regulates grain chalkiness and also affects head rice yield and many other quality traits. *Chalk5* encodes a vacuolar H^+-translocating pyrophosphatase (V-PPase) with inorganic pyrophosphate (PPi) hydrolysis and H^+-translocation activity. *Chalk5* functions as a positive regulator of grain chalkiness rate. Elevated expression of *Chalk5* increases chalkiness of the endosperm putatively by disturbing the pH homeostasis of the endomembrane trafficking system in developing seeds, which affects the biogenesis of protein bodies and is coupled with a great increase in small vesicle-like structures, thus forming air spaces or spherical granules among endosperm storage substances and resulting in chalky grain. Introgression of the *Chalk5* allele of non-chalky variety into a chalky variety greatly decreases rice grain chalkiness rate by 50.5% and increases head rice yield by 37.9%, equally increasing the grain yield. Our results indicated that two consensus nucleotide polymorphisms among rice varieties in *Chalk5* promoter partly account for the difference in *Chalk5* mRNA levels causing natural variation of grain chalkiness. The tight linkage of *Chalk5* for grain chalkiness and quality, and *GS5* and *qSW5*/*GW5* for grain size, weight and yield, points to the genetic and molecular evidence for the inconsistency between grain quality and yield during modern breeding. Breaking such linkage by recombination may result in genotypes producing wide grain with low chalk to achieve high yield with good quality in breeding programs of *indica* rice.

T4-30-06

Origin of embryo polarity

Thomas Laux
BIOSS Centre for Biological Signaling Studies, Faculty of Biology, University Freiburg

One of the first steps in the development of an embryo from a single celled zygote is the establishment of the main body axis. In model animals the orientation of this axis uses position information provided by surrounding maternal tissue or the sperm entry point. In most flowering plants, the apical-basal body axis is reflected by the asymmetric division of the polarized zygote. However the underlying mechanisms regulating axis formation remained enigmatic. Here we address which factors establish zygote polarity. We demonstrate that both maternal and paternal factors co-operate in the initiation of the embryo axis.

T4-31: Metabolites in biotic and abiotic stresses

T4-31-01

Evolutionary neighbourhood and climate interact to select for heritable defense phenotypes: anthocyanin and flavonoid contents and inducability

Andreas Prinzing, *Maud Deniau, Mickael Pihain*
Université Rennes

Under global change, trees face a double pressure from enemies and harsh climate. Both pressures may be modified by the tree neighbourhood such as increased enemy pressure in a closely related neighborhood. Both, local neighbourhood and regional climates may select for adapted genotypes. Adaptations to such pressures include flavonoids and anthocyanins, suggesting that the adaptation to one pressure may interact with the adaptation to the other, either by a trade-off or by mutual facilitation. We studied oak seedlings in evolutionary proximate and distant neighbourhoods as well as adults of genotypes that have been selected by such respective neighbourhoods. We studied anthocyanin and flavonoid contents as well as morphological defenses and the performance of seedlings and trees in terms of growth and survival. We found evidence for a response to both climate and neighbourhood, and heritability of this response. Tentative results are consistent with inducible chemical defenses and (constitutive structural defenses) being selected by an evolutionary proximate neighbourhood. Sharing neighbourhoods with close relatives seems to require particular levels, compositions and inductions of defense compounds.

T4-31-02

Pseudomonas fluorescens G20-18 derived cytokinins determine the biocontrol of *Pseudomonas syringae* infection in *Arabidopsis* - a novel biocontrol concept

Dominik Grosskinsky, *Thomas Roitsch*
University of Copenhagen

Throughout their life, plants interact with a wide range of pathogenic and beneficial microbes. Beneficial bacteria are described for positive effects such as plant growth promotion, increased tolerance to environmental stress and enhanced disease resistance. Considering future challenges through increasing world population and climate change in combination with restrictions of classic pesticides which urgently demands alternative strategies in plant protection, the biological control of plant diseases by beneficial microbes offers high potential for integrated plant disease management. To facilitate the development of according microbe-based biocontrol strategies, the underlying mechanisms of these effects have to be fully elucidated. Mechanisms contributing to such biocontrol phenomena comprise direct effects on pathogens or virulence factors and strengthening the plant. Recently, the function of the classic growth-stimulating phytohormone cytokinin in modulating plant immunity has been identified in different plant species, complementing the established phytohormone-based defence signaling network. Underlying mechanisms mediating cytokinin-dependent resistance comprise induction of salicylic acid, phytoalexins and reduction of abscisic acid levels in a plant species-specific, cytokinin type-specific, and dose dependent manner. Despite the importance of phytohormones, particularly cytokinins, as essential plant immunity regulators, their microbial production is not yet considered in biocontrol of diseases. We identified the cytokinin production of *Pseudomonas fluorescens* G20-18 as a key determinant of biocontrol of *P. syringae*pv. *tomato* DC3000 infection in *Arabidopsis*. Treatment with this strain strongly suppressed symptom development and spread of the pathogen, thus maintaining tissue integrity, which ultimately sustains biomass yield. While cytokinin-deficient loss-of-function mutants were impaired in controlling the infection, mutants complemented with different cytokinin biosynthetic genes (gain-of-function), exhibiting restored cytokinin production and showed a similar biocontrol effect as the wild-type strain. The efficiency of the biocontrol effect correlated with the cytokinin levels *in planta* resulted from treatments with the different bacterial strains. Analyses of Arabidopsis mutants impaired in defence pathways revealed the necessity of various defence and signalling components. Functional cytokinin perception as well as salicylic acid biosynthesis and signalling were essential to fully establish this biocontrol effect in combination with potential minor involvement of jasmonic acid and ethylene signalling as well as camalexin accumulation. This identification of microbial-derived phytohormones to trigger biocontrol effects offers novel options for the development of successful strategies in plant protection which may be combined, based on cytokinins, with other positive effects such as increased abiotic stress tolerance and plant growth.

T4-31-03

Insect (*Thrips Hawaiiensis* (Morgan)) attacks change the stereochemical configuration of 1-phenylethanol emitted from tea (*Camellia Sinensis*) flowers

Ying Zhou[1], *Lanting Zeng*[1,2], *Yinyin Liao*[1], *Fang Dong*[3], *Ziyin Yang*[1,2]
1. South China Botanical Garden, Chinese Academy of Sciences
2. University of Chinese Academy of Sciences
3. Guangdong Food and Drug Vocational College

Insect attacks change the quantity of volatiles emitted from plants, but little is known about the effects on the volatiles' stereochemical configurations. 1-Phenylethanol (1PE) is a major internal volatile of tea (*Camellia sinensis*) flowers. In the present study, insect-damaged *C. sinensis* flowers in the field emitted a different (*R*)-/ (*S*)-1PE pattern than that emitted from undamaged flowers. Laboratory experiments showed that changes in the (*R*)-/ (*S*)-1PE ratio were caused by *Thrips hawaiiensis* (Morgan), a pest of cut flowers and vegetable crop, attacks. These attacks led to an increase in jasmonic acid (JA) through the activation of allene oxide synthase, which is involved in JA biosynthesis. Exposing flowers to exogenous JA supplements induced an (*R*)-/ (*S*)-1PE emission pattern that was similar to that induced by a *T. hawaiiensis* attack, suggesting that JA was involved in the changes in the stereochemical configuration of 1PE emitted from the flowers after an insect attack. This study provides new insights into the involvement of volatiles in interactions between plants and insects, showing that not only the quantity but also the stereochemical configuration of plant volatiles is affected by insects.

T4-31-04

Molecular and biochemical analysis of high light induced non-uniform anthocyanin pigmentation in fruits of *BrTT8* transgenic tomato System

Yanjie Zhang[1,2], *Yan Li*[1,2], *Jinyong Huang*[1], *Guoping Chen*[2]
1. School of Life Sciences, Zhengzhou University
2. Bioengineering College, Chongqing University

Pigment intensity and pattern are important factors that determinate nutrient and market value of tomatoes. *BrTT8* transgenic tomato plants displayed intense purple anthocyanin pigmentation in leaves, stems, petals and fruits when grown under high-light condition, yet remain acyanic under shade condition. The light-sensitive property of the *BrTT8* transgenic tomato plants makes it a good system that can be used to investigate how light intensity affects anthocyanin biosynthesis and accumulation. The colored phenotypes matched well with anthocyanin production and transcripts abundance of the anthocyanin biosynthetic genes. In *BrTT8* transgenic tomato plants, the constitutive expression of *BrTT8* and *SlTTG1* suggests that high-light induced anthocyanin pigmentation in transgenic tomato plants might be regulated by the temporal and spatial expression of SlAN2. Yeast two-hybrid assay showed that bHLH proteins could interact efficiently with endogenous MYB proteins (SlAN2 and SlMYBL2) and WD40 protein, respectively. Co-expression of *SlAN2* and *BrTT8* induced anthocyanin accumulation significantly in a new set of transgenic plants under normal light condition. These results show that a transcriptional regulation complex consisting of SlAN2, SlTTG1 and exogenous BrTT8 regulates anthocyanin accumulation in *BrTT8* transgenic tomato system under high-light condition. Moreover, a model explaining the non-uniform anthocyanin pigmentation in transgenic tomato fruits was proposed. This work provides more insights into molecular study of accumulation of other secondary metabolites in crops and vegetables with special genotypes under various environmental conditions. Furthermore, the elucidation of the anthocyanin pigmentation pattern in transgenic tomato fruits offers a new angle and cut-in point for the study of non-uniform biological processes in other plant organs.

T4-31-05

Rootstock-specific xylem metabolite signatures in tomato

Alfonso Albacete, Cristina Martinez-Andujar, ***Francisco Perez-Alfocea***
CEBAS-CSIC

Identifying root-derived chemical signals and understanding their specific effects in shoot tissues could open new strategies to improve crop yield stability under abiotic stress conditions (Albacete *et al.*, 2015). Therefore, a commercial tomato cultivar (*Solanum lycopersicum* cv Boludo F1, Monsanto) was either self-grafted (L/L) or grafted onto the interspecific commercial tomato rootstock Maxifort (Monsanto) obtained from a cross between *S. lycopersicum* and *S. habrochaites* (L/H), and onto a recombinant inbred line obtained from a cross between *S. lycopersicum* and *S. pimpinellifolium* from AVRDC (L/P). Grafted plants were cultivated under commercial greenhouse conditions during autumn season in Mazarrón (Spain). Xylem sap, collected by root pressure, was injected into a U-HPLC-MS system (EXACTIVE, ThermoFisher Scientific). The metabolite signature of the '*pimpinellifolium*' L/P graft combination was strongly associated to the first principal component (PC1), which explained 47% of the total chemical variance, while the '*lycopersicum*' L/L combination was linked to PC2, explaining 31% of the variance. The metabolite profile of the '*habrochaites*' L/H plants clustered between the other two combinations. Additionally, we identified 81 putative compounds of both the primary and secondary metabolism. We have shown that the metabolite profile of the xylem sap can be qualitatively and quantitatively modified by changing the root system (rootstock), as a genotypic source of chemical variability. Xylem sap profiling is likely to provide physiological markers to assist root-targeted breeding for resistance to individual or combined biotic and/or abiotic stresses.

T4-32: Stem cell maintenance and cell differentiation in shoot meristem and cambium

T4-32-01

Role of auxin and cytokinin in the secondary vascular tissue regeneration after bark girdling in *Populus*

Jia-Jia Chen[1], *Jing Zhang*[2], *Yrjö Helariutta*[2], ***Xin-Qiang He***[1]
1. School of Life Sciences, Peking University
2. Sainsbury Laboratory, University of Cambridge

The secondary vascular tissues (SVT) can be regenerated after the large-scale bark girdling in trees. However, the mechanisms of SVT regeneration is largely unknown. Here, we explored the roles of auxin and cytokininin in the SVT regeneration process after bark girdling by utilizing an *in vitro* SVT regeneration system in poplar. After the bark girdling, the phloem and cambium were peeled off. With the auxin treatment, the specific callose deposition can be detected at 6-9 days after girdling (DAG) while the expression of phloem specific marker genes is upregulated at the same stage, and the continuous cambium can be observed at 13 DAG confirmed by *pAIL1::UidA* reporter line. With the cytokinin treatment, the phloem can be still regenerated while no cambium regeneration was observed on 13 DAG, suggesting the differential requirement of cytokinin for phloem and cambium regeneration. These results were validated by using transgenetic trees when cytokinin signaling was enhanced. By employing *pDR5::UidA* transgenic line, we were able to visualize the auxin redistribution patterns during the SVT regeneration. New auxin maximum appears along with the formation of the continuous cambium on 13 DAG, indicating that auxin redistribution provides the position information for the cambium recovery. Interestingly, the combined application of auxin and cytokinin failed to display the specific auxin distribution pattern as well as the new cambium. This result implied that cytokinin may block the re-establishment of auxin gradients towards paralyzing the cambium regeneration in this *in vitro* system.

T4-32-02

Feedback from leaves controls shoot apical meristem growth

by modulating auxin transport

Yuling Jiao
Institute of Genetics and Developmental Biology, Chinese Academy of Sciences

Stem cells must balance self-renewal and differentiation; thus, stem cell activities are precisely controlled. In plants, the control circuits that underlie division and differentiation within meristems have been well studied but those that underlie feedback on meristems from lateral organs, such as leaves, remain largely unknown. Here we show that long-distance auxin transport mediates this feedback in a non-cell-autonomous manner. A low-auxin zone is associated with the shoot apical meristem (SAM) organization center, and is required for regulation of SAM size. Using computational model simulations, we show that auxin transport from leaves can inhibit auxin transport from the SAM through an auxin transport switch, and thus maintain SAM auxin homeostasis and SAM size. Genetic and microsurgical analyses confirmed the model's predictions. In addition, the model explains the surprising observation that *yabby* mutants exhibit oscillations of SAM size. Our study shows that plants use a distinct feedback control mechanism for long-distance regulation of stem cell activities.

T4-32-03

Interaction between major regulators WUS and STM controls stem cell specification in shoot apical meristem of *Arabidopsis*

Xiansheng Zhang
Shandong Agricultural University

Plants have evolved a powerful capability to continuously and repetitively generate lateral organs through their meristems. Shoot apical meristem (SAM) gives rise to the aerial part of a plant. Among the regulators identified in SAM of *Arabidopsis*, the homeobox genes *WUSCHEL* (*WUS*) and *SHOOTMERISTEMLESS* (*STM*) encode two major regulators of stem cell specification, yet the molecular mechanisms underlying their function and interaction in stem cell regulation are still under exploration. In this study, we identified STM acting as a direct interacting factor with WUS. STM has synergistic effects with WUS on regulation of their common target *CLAVATA3* (*CLV3*). Moreover, STM synergistically strengthens WUS on *CLV3* transcription, but its function largely depends on WUS. Furthermore, *STM* expression is activated and maintained by WUS, and STM is also critical to maintain *WUS* expression in SAM. Our discoveries thus reveal a new framework for the regulation of stem cell specification during SAM development.

T4-32-04

How do embryos make stem cells?

Thomas Laux
BIOSS Centre for Biological Signalling Studies, Faculty of Biology, University Freiburg

A central process during embryogenesis is the formation of stem cells that provide an organism with new cells during the postembryonic phase. Despite considerable progress in elucidating the regulation of postembryonic stem cell homeostasis, however, the regulatory networks responsible for stem cell formation during embryogenesis remain poorly understood in plants and animals. With regard to their developmental potential and plasticity, stem cells resemble early embryo cells and used to be called embryonic initials by the British botanist C. W. Wardlaw. In this study, we investigate the formation of stem cells during *Arabidopsis* embryo development. Our findings reveal a molecular framework combining transcriptional regulators and hormonal components governing this process. These results link stem cell formation to early embryo patterning and revisit classical experiments by Skoog and Miller on plant regeneration.

T4-32-05

Symplastic communication in vascular meristems

Yka Helariutta
Sainsbury Laboratory Cambridge University

My group is investigating the morphogenesis of phloem, a long distance transporting tissue specified to transport the various products derived from the photosynthesis. Based on identification of the gain-of-function mutations in one of the callose synthase isoforms (responsible for formation of callose, b1,3-glucan polymer that plant uses to regulate symplastic trafficking through plasmodesmata nanochannels) we have recently developed a molecular tool, icals3m, with which we can regulate symplastic molecular trafficking in a time and space specific manner. Using this tool we have identified a family of genes coding for transcription factors that are transcribed in the young sieve element position, and whose protein products move to the adjacent procambial cells to complete procambial/phloem patterning.We are also investigating phloem development (with a focus on sieve element differentiation) in high resolution as a paradigm for how a plant cell progresses from a stem cell to a fully differentiated state. In the *Arabidopsis* root there are some 25 cells from the relatively isodiametric stem cell (touching the quiescent centre of the root) to the elongated cell that loses its nucleus as the final stage of sieve element cell differentiation. Finally, phloem transport is based on the interaction of the sieve element/companion cell unit with their neighbouring cellular domains (in the root culminating at unloading of organic substances to the meristem). We are now analysing other aspects of phloem morphogenesis (and also function) that may involve symplastic communication.

T4-32-06

DWT1 is a key regulator coordinating the panicle growth and culm elongation in rice

Wanqi Liang
Shanghai Jiao Tong University

Rice is one of the major staple food plants around the world. The agronomic performance of rice depends largely on desirable plant architecture. Plant height is one the fundamental component of aerial structure of a rice plant. The elongation of uppermost four or five internodes determines the culm (stem) length, which dose not occur until the transition from vegetative growth to reproductive growth. The coordination of the panicle (inflorescence) growth and the culm elongation ensures the timely emergence of the mature panicle from leaves. However, signals and mechanisms responsible for triggering the internode elongation remain largely unknown. We report that *DWARF TILLER1* (*DWT1*) is involved in initiating the culm elongation in rice. The *dwt1* mutant plant develops dwarf tillers with one to several shorter internodes compared with those of the wild type. Shorter internodes results from reduced cell number and cell length, indicating that DWT1 affects cell division and cell elongation. Map-based cloning revealed that *DWT1* encodes a WUSCHEL-related homeobox (WOX) transcription factor homologous to the Arabidopsis WOX8 and WOX9. The *DWT1* gene is highly expressed in young panicles, but undetectable in the internodes, suggesting that DWT1 expression is spatially or temporally separated from its effect on the internode growth. Transcriptomic analysis revealed altered expression of genes involved in cell division and cell elongation, cytokinin/gibberellin homeostasis and signaling in *dwt1* shorter internodes. Moreover, the non-elongating internodes of *dwt1* are insensitive to exogenous gibberellin (GA) treatment, suggesting that the DWT1 activity in the internode elongation is directly or indirectly associated with GA signaling. Our study reveals a genetic pathway synchronizing the development of the panicle and the culm in rice. Interestingly, the *dwt1* main shoot is much less affected and bears a larger panicle compared with *dwt1* tillers, indicating that DWT1 is also involved in balancing the main shoot and tiller growth in rice. Recent progress on understanding how DWT1 coordinates the panicle growth and the culm elongation will be introduced.

T4-33: Male sterility and EGMS

T4-33-01

Molecular mechanism of *Sa*-related reverse thermo-sensitive male sterility in rice

Letian Chen, *Yongyao Xie, Yaling Zhang, Gousi Li, Yao-Guang Liu*
State Key Laboratory for Conservation and Utilization of Subtropical Agro-Bioresources; College of Life Sciences, South China Agricultural University

Manipulation of male fertility is essential for heterosis utilization in crop production. However, hybrid sterility has long been a major barrier in utilization of heterosis. *Sa* is a major locus for *indica-japonica* . We have revealed that *Sa* consists of two adjacent genes *SaF* and *SaM*. In this study, we identified a reverse thermo-sensitive male sterility (rTMS) rice line G392S, which was male sterile under low temperature condition and male fertile under high temperature condition. The fertility of this rTMS line was controlled by an gene, *ORF13*, adjoining to *Sa*, with an average transit temperature at 31 ℃ . When G392S was treated under high temperature (~35 ℃), the expression level of *SaF* decreased dramatically, suggesting that the *SaF* is involved in the regulation of the rTMS trait. Further, using CRISPR/Cas9 gene editing, we knocked out *SaF*, *SaM* and *ORF13* respectively in G392S, and found that the rTMS trait disappeared in resultant knockout lines. Thus, we proposed a three gene/four-component model for the molecular mechanism of rTMS in rice. The application of this rTMS trait in hybrid rice breeding will be discussed in the presentation.

T4-33-02

RNase Z^{S1} mediates Ub_{L40} mRNA decay controlling thermo-sensitive genic male sterility in rice

Hai Zhou, *Jing Li, Liya Zhu, Yaoguang Liu, Chuxiong Zhuang*
South China Agricultural University

Male sterility is an important trait and plays a key role in hybrid breeding in crops. Thermo-sensitive genic male sterility (TGMS) lines, which are male-sterile at high temperatures but convert to male-fertility at low temperatures, have been widely used in two-line hybrid rice breeding. However, the molecular mechanism underlying TGMS remains unclear. Here, we show that a rice (*Oryza sativa* L.) *thermo-sensitive genic male sterile gene 5* (*tms5*) locus confers the TGMS trait through a loss-of-function mutation of *RNase* Z^{S1}, resulting in failing to mediate the mRNA decay of a temperature-responsive *ubiquitin fusion ribosomal protein L40* (Ub_{L40}). RNase Z^{S1} belongs to a short version of the RNase Z family originally involved in tRNA 3'-end processing in nucleus. RNase Z^{S1} retains the endonuclease activity in tRNA processing *in vitro*, however, it does not play a role *in vivo* due to its localization in cytoplasm. RNase Z^{S1} specifically cleaves three Ub_{L40} genes. Defective RNase Z^{S1} in *tms5* plants causes over-accumulation of Ub_{L40} mRNAs and defective pollen development at high but not low temperatures. Finally, we developed a new method of TGMS line breeding using the CRISPR/Cas9-mediated *TMS5* editing system. Our results not only uncover a novel function of the RNase Z family member and reveal the molecular mechanism of rice TGMS through a new post-transcriptional regulation for ubiquitin homeostasis, but also improve the method of TGMS line breeding. Moreover, this study provides applied potential for hybrid breeding not only in rice but also in other crops.

T4-33-03

Towards controlling fertility for hybrid production in temperate cereals; the target of hybrid wheat - applying lessons learnt from other systems.

José Fernández-Gómez, Behzad Talle, Alison Ferguson, ***Zoe A Wilson***
University of Nottingham

Heterosis or Hybrid vigour can result in increased crop yield (3-15%) and enhanced resilience in many crops, however not all crops are able to capitalize upon these potential gains. This is particularly the case for the temperate cereals such as barley and wheat, which as autogamous crops require self-fertilisation to be controlled via the generation of male sterile plants for crossing and for these lines to then be subsequently maintained. Currently the application of hybrid wheat is extremely limited, with only ~200,000 ha planted in Europe, mostly produced using chemical hybridization agents, and about 35,000 ha in China and India, using photoperiodic sensitivity, or the CMS system from *Triticum timosphevii.* Recently, in 2002, a commercial hybrid barley system was commercialised based on a CMS system, however currently only ~200,000 ha are being grown in Europe, nevertheless this is likely to increase due to the reported increased in yield and disease resistance. However, major advances have occurred in other species, typically rice, of which 20 million ha is hybrid, based upon CMS and environmental sensitivity systems. We have been using knowledge from the model systems of *Arabidopsis* and rice to provide knowledge of pollen development and the control fertility in barley and to subsequently apply this to wheat. We have been investigating the use of key conserved transcription factors that are associated with tapetum function as mechanisms to control fertility. Using specific targets we have been able to generate male sterile plants and then by altering the environmental conditions restore fertility. This is providing valuable information on the process of pollen development and the conservation of gene networks in the anther, but is also providing potential switchable systems for controlling fertility for hybrid development and targeted breeding in temperate cereal crops. The progress towards hybrid development of temperate cereals and understanding of tapetum gene networks will be discussed.

T4-33-04

Comparative analysis of cytoplasmic male sterility developed from various rice.

Tomohiko Kazama, *Keisuke Igarashi, Kinya Toriyama*
Tohoku University

A cooperative relationship between nucleus and organelle is important for a normal development of plant. In rice, a nuclear substitution sometimes causes defects of male organ development, because of an incompatibility between nuclear and mitochondrial genomes. Such a phenomenon is called cytoplasmic male sterility (CMS) and caused by the expression of mitochondrial unique *orf* termed CMS-associated gene. The function of the CMS-associated gene is prevented by the presence of a particular nuclear gene, called *Rf* (*Restorer of fertility*). We have been studying BT-, CW-, and WA-CMS in rice. Recently, we obtained new type CMS rice, RT98- and RT102-CMS developed from wild rice, *Oryza rufipogon*. Those CMS lines possess the same nuclear genome, because they had been developed by backcrossing with the same cultivar, Taichung 65 (T65). Phenotype comparisons of mature pollen among these five CMS lines revealed that the mitochondrial genotype possibly determines the timings of pollen abortion. In the BT-, CW- and WA-CMS line, the mitochondrial *orf79*, *orf307* and *WA352* are known as CMS-associated gene, respectively. To reveal the CMS-associated gene of RT98- and RT102-CMS, we determined the whole sequence of both types of mitochondria. We found *orf113* and *orf352* respectively, as CMS-associated genes. We have cloned *Rf1* for BT-CMS, *Rf17* for CW-CMS and *Rf4* for WA-CMS. These researches revealed that *Rf1* and *Rf4* encode pentatricopeptide repeat (PPR) protein and *Rf17* encodes an unkonwn protein. To identify fertility restorer genes for RT98- and RT102-CMS, we are performing map-based cloning of these genes. Map-based cloning and complementation tests revealed that an *Rf* gene for RT98-CMS, *Rf98*, also encodes PPR protein. We present recent results of our works.

T4-33-05

Hybrid sterility in rice: From genetics to mechanisms

Yidan Ouyang, *Guangwei Li, Jiaming Mi, Qifa Zhang*
National Key Laboratory of Crop Genetic Improvement and National Centre of Plant Gene Research (Wuhan), Huazhong Agricultural University, Wuhan 430070

Understanding the origin and mechanisms of reproductive isolation is of fundamental importance in evolutionary biology. However, little is known about how genomes evolve to form and establish reproductive barriers in the course of evolution especially during domestication, although such reproductive barriers are widely observed and demand understanding for both biological study of speciation and practical application in crop improvement. Studies have shown that the reproductive barrier regulated by S5 features a killer-protector system encoded by three tightly linked genes (*ORFs3–5*). Typical *indica* and *japonica* varieties contain the haplotypes of *ORF3+ORF4-ORF5+* and *ORF3-ORF4+ORF5-*, respectively. The killer is made of functional alleles of *ORF4+* and *ORF5+*, while the functional allele of *ORF3+* confers protection to the gametes, such that female gametes without the protector *ORF3+* are selectively eliminated, causing hybrid sterility and segregation distortion. We find that such trigenic system of S5 originated by gene duplication after the split of the Oryzeae tribe from other grasses, most likely through *Helitron* transposition. A combination of mutational steps generated incompatible *indica* and *japonica* alleles in pre-differentiated rice groups, giving rise to the trigenic reproductive isolation system. Natural selection in *indica* rice and founder effect associated with domestication in *japonica* populations increased the frequencies of incompatible alleles to form a functional reproductive barrier between the *indica* and *japonica* subspecies, suggesting that domestication can lead to the early stages of speciation in crops, and eventually result in genetic differentiation and restructuring of rice genetic composition. This finding provides an unequivocal example of the origin of a complex system by gene duplication and illustrates the evolutionary processes of establishing a reproductive barrier by actions of combinatorial molecular mechanisms and evolutionary forces. The results may have significant implications for understanding genome evolution and for crop genetic improvement.

T4-33-06

Construction of a male sterility system for hybrid rice breeding and seed production using a nuclear male sterility gene

Zhenyi Chang[1], *Zhufeng Chen*[1], *Na Wang*[1], *Gang Xie*[1], *Jiawei Lu*[1], *Wei Yan*[1], *Junli Zhou*[1], ***Xiaoyan Tang***[1,2], *Xing Wang Deng*[1,3]
1. Shenzhen Institute of Molecular Crop Design, Shenzhen 518107, China.
2. South China Normal University
3. Peking University

The breeding and large-scale adoption of hybrid seeds is an important achievement in agriculture. Rice hybrid seed production uses cytoplasmic male sterile lines (CMS) or photoperiod/thermo-sensitive genic male sterile lines (PTGMS) as female parent. CMS lines are propagated via cross-pollination by corresponding maintainer lines, whereas PTGMS lines are propagated via self-pollination under environmental conditions restoring male fertility. Despite huge successes, both systems have their intrinsic drawbacks. Here we constructed a new rice male sterility system using a novel nuclear gene *OsNP1*. *OsNP1* encodes a putative GMC oxidoreductase protein regulating tapetum degeneration and pollen exine formation. It is specifically expressed in the tapetum and miscrospores. The *osnp1* mutant plant displays normal vegetative growth but complete male sterility insensitive to environmental conditions. *OsNP1* was coupled with an a-amylase gene to devitalize transgenic pollen and the *DsRed* gene to mark transgenic seed, and transformed into the *osnp1* mutant. Self-pollination of the transgenic plant carrying a single hemizygous transgene produced non-transgenic male sterile and transgenic fertile seeds in 1:1 ratio that can be sorted out based on the red fluorescence coded by *DsRed*. Cross-pollination of the fertile transgenic plants to the non-transgenic male sterile plants propagated the male sterile seeds of high purity. The male sterile line was crossed with ~1200 individual rice germplasms available. Approximately 85% of the F1s out-performed their parents in per-plant yield, and 10% out-yielded the best local cultivars, indicating the new technology is promising in hybrid rice breeding and production. This new system is fundamentally advantageous over the current CMS and PTGMS systems. Application of this new technology will greatly enhance the effectiveness and efficiency in hybrid rice breeding and production.

T4-34: Hormone perception and signaling

T4-34-01

AtD14 and ShHTL7 are non-canonical receptors for strigolactone

Ruifeng Yao, *Zhenhua Ming, Liming Yan, Suhua Li, Fei Wang, Zihe Rao, Zhiyong Lou, Daoxin Xie*
Tsinghua University

Strigolactones (SLs) are a group of carotenoid-derived small molecules synthesized by plants and act as a very special type of plant hormones. As a class of phytohormones, SLs regulate shoot branching, which is an important agronomic trait to determine crop yield. Intriguingly, SLs are also exuded into the soil as rhizospheric signals for communication with arbuscular mycorrhizal fungi (AMF) and root parasitic weed seeds: SLs facilitate the AMF symbiosis with a wide range of land plants, which improves nutrient and water uptake of the host plants, and stimulate seed germination of the root parasitic weeds such as *Striga* for parasitic growth on host plants, which causes agricultural disaster with the loss of billions of dollars' worth of crops every year. Through analysis of a highly branched *Arabidopsis* mutant *d14-5* (Yao *et al.*, 2016), we show that the AtD14^{G158E} mutant maintains enzyme activity to hydrolyze strigolactone, but fails to efficiently interact with D3/MAX2 and loses the ability to trigger strigolactone signaling *in planta*, indicating that the α/β hydrolase AtD14 has dual functions to hydrolyze and perceive SL. We further demonstrate that strigolactone is hydrolyzed by AtD14 into a D-ring derived intermediate, which serves as the novel active form of SL covalently linked with AtD14 (named as the covalently linked intermediate molecule, CLIM). The crystal structure of the strigolactone-induced AtD14-D3-ASK1 complex reveals that CLIM-bound AtD14 undergoes an open-to-closed state transition to interact with D3/MAX2 for triggering strigolactone signaling. Similarly (Yao *et al.*, 2017), we has further defined *Striga hermonthica* HTL7 (ShHTL7, which is not an ortholog of AtD14) as a non-canonical receptor for SL in root parasitic weeds: In parallel with the SL perception by AtD14, ShHTL7 also hydrolyzes SLs into an active intermediate CLIM and covalently binds CLIM to interact with ShMAX2/MAX2 and other downstream signaling components for stimulating seed germination. These findings uncover a mechanism underlying the allosteric activation of AtD14 by strigolactone hydrolysis into CLIM, and define AtD14 as a non-canonical hormone receptor with dual functions to generate and sense the active form of strigolactone.

T4-34-02

A novel function for host jasmonate signaling in viral transmission

Daoxin Xie, *Dewei Wu, Tiancong Qi, Wan-Xiang Li, Le Kang, Shou-Wei Ding*
1. Tsinghua University
2. University of California, Riverside
3. Institute of Zoology, Chinese Academy of Sciences

Plant hormones are characterized by the property of serving as chemical messengers, which control cellular activities (division, elongation and differentiation) and processes (pattern formation, sex determination, organogenesis and stress responses). Jasmonate (JA) is a class of plant hormone and ubiquitous in plant kingdom. We previously uncovered that the F-box protein COI1 (Science 280:1091) assembles into the SCF complex (Plant Cell 14:1919; Plant Cell 16:5; Plant Cell 25:486) and functions as a JA receptor (Plant Cell 21: 2220) to regulate diverse plant developmental processes and various plant defense responses (Mol. Cell 50:5014; Plant Cell 27:1634; Plant Cell 27:1620; Plant Cell 26:1118; Plant Cell 26: 263; Plant Cell 23:1795; Plant Cell 23:1000; PLoS Genetics 9:e1003653; PLoS Genetics 5:e1000440). Interestingly, we recently found that hormone signaling pathway in plants was targeted by virus to attract insect vectors for efficient viral transmission. We reveal that cucumber mosaic virus employs viral effector protein 2b to target the host JA signaling pathway for inducing odor-dependent aphid attraction. The recent progress will be discussed.

T4-34-03

Ethylene induced combinatorial effect of histone acetylation in ethylene response

Hong Qiao
University of Texas at Austin

My research is focus on how plant hormone affect plant growth and plant abiotic stress response. Ethylene is one of the most important plant hormones. It plays pleotropic roles in many aspects of plant life. A model for hormone signaling has emerged in which the perception of ethylene by the receptors alters the activity of CTR1, which in turn, by an unknown mechanism, functions to relieve repression of EIN2, resulting in activation of EIN3/EIL1-dependent transcription and the activation of an ethylene response. Our recent studies have revealed that EIN2 mediates transmission of ethylene signaling that originates at the endoplasmic reticulum membrane to the nucleus and that the EIN2 C-terminus is cleaved and translocated to the nucleus to initiate the ethylene response. However, the molecular mechanism that nuclear translocation of EIN2, the function of EIN2 C-terminal end in the nucleus and how EIN2 C-terminal end communicate with EIN3 are not understood. We recently found that ethylene specifically elevated acetylation of histone H3K14 and the non-canonical acetylation of H3K23. The up-regulation is positively associated with a set of ethylene regulated transcription activation, and the elevation requires EIN2 and partially EIN3/EIL1. Both EIN2 and EIN3 interact with a SANT domain protein EIN2 Nuclear Associate Protein ENAP1 (ENAP1), of which over expression result in elevation of Histone acetylation. These findings reveal that in the presence of ethylene, EIN2 C-terminus is involved in the regulation of the elevation of acetylation at H3K14 and H3K23. In addition, our study reveals that the plant hormone ethylene induces combinatorial effects of H3K9Ac, K14Ac and K23Ac histone acetylation in gene expression genome widely. Further, for a group of ethylene regulated genes, in the absence of ethylene the levels and the covered breadth of H3K9Ac are the preexist markers for distinguishing up- and down- regulated genes, the change in the levels of H3K14Ac and H3K23Ac are required for the alteration of gene expression in the presence of ethylene.

T4-34-04

Degradation of BES1 through autophagy pathway coordinates brassinosteroid-regulated plant growth and stress responses

Trevor Nolan[1], *Ben Brennan*[1], *Mengran Yang*[2], *Jiani Chen*[1], *Mingcai Zhang*[3], *Zhaohu Li*[3], *Xuelu Wang*[2], *Diane Bassham*[1], *Justin Walley*[1], ***Yanhai Yin***[1]

1. Iowa State University

2. Huazhong Agricultural University

3. China Agricultural University

Plants need to balance their growth and development in response to various stress conditions. BES1 is a transcription factor that functions as a master regulator in the Brassinosteroid (BR) signaling pathway, directing a transcriptional network that controls the expression of thousands of genes involved in plant growth, development and stress responses. Despite extensive characterization of BES1 in driving BR-related gene expression, the pathways and components regulating BES1 protein levels are not well understood. We found that BES1 interacts with the ubiquitin receptor protein BIP5 and is targeted to the autophagy pathway during stress via the interaction of BIP5 with ATG8, an ubiquitin-like protein directing autophagosome formation and cargo recruitment. Additionally, BIP5 is phosphorylated by the GSK3-like kinase BIN2, a negative regulator in the BR pathway. BIN2 phosphorylation of BIP5 flanking its ATG8 interacting motifs (AIMs) promotes the interaction of BIP5 with ATG8, thereby targeting BES1 for degradation. Accordingly, loss-of-function *bip5* mutants accumulate BES1, have altered global gene expression profiles, and have compromised responses to drought and fixed-carbon starvation stresses. Furthermore, we found that E3 ubiquitin ligase BAR1 can ubiquitinate BES1 under stress conditions and is involved in BES1 degradation through autophagy pathway. Our results thus reveal a new mechanism by which plants coordinate growth and stress responses by targeting a central growth regulator to the selective autophagy pathway via an E3 ubiquitin ligase and a phospho-regulated ubiquitin receptor protein.

T4-34-05

Regulation of master growth repressor DELLA by *O*-Glycosylation in *Arabidopsis*

Tai-Ping Sun

Duke University, Department of Biology

Plant development requires constant coordination among complex internal hormonal signaling networks to enhance plant's adaptation to changing environments. The conserved transcription regulators DELLA proteins, originally identified as repressors of phytohormone gibberellin (GA) signaling, play a central role in integrating activities of multiple signaling pathways via direct protein-protein interactions with key transcription factors. We found that DELLA's binding affinity to several interacting proteins is negatively regulated by *O*-linked *N*-acetylglucosamine (GlcNAc) modification by an *O*-GlcNAc transferase (OGT, named SECRET AGENT) in *Arabidopsis*. These results reveal a direct role of OGT in repressing DELLA activity, and indicate that *O*-GlcNAcylation of DELLAs provides a fine-tuning mechanism in coordinating multiple signaling activities during plant development. Interestingly, SPINDLY, a paralog of SEC, plays an opposite role in regulating DELLA function. The biochemical function of SPINDLY will be discussed.

T4-34-06

ABA promotes leaf senescence and bud dormancy to improve plant survival under extreme drought

Yang Zhao

Shanghai Center for Plant Stress Biology, Chinese Academy of Sciences

Drought stress is a condition that in specific climate contexts, results in insufficient water availability and often limits plant productivity through reducing plant growth, development and survival of plants. Plants use senescence of old leaves and dormancy of buds and seeds to survive extreme environmental conditions. The plant hormone abscisic acid (ABA) is key to drought tolerance. ABA is sensed by the PYR/PYL/RCAR proteins. In the presence of ABA, PYLs interact with and inhibit clade-A protein phosphatase type 2Cs (PP2Cs). The inhibition of PP2Cs leads to the activation of the SNF1-related protein kinase 2s (SnRK2s). The activated SnRK2s phosphorylate transcription factors such as ABF/AREBs to regulate expression of ABA-responsive genes, and other substrates related to stomatal movement, production of ROS, cytoskeleton dynamics, RNA binding, miRNA and epigenetic regulation, chloroplast function and many other processes. ABA increases plant survival by inducing quick responses such as stomatal closure, and long-term responses such as extended growth inhibition, osmotic regulation, accumulation of cuticular wax, senescence, abscission, and dormancy. Leaf senescence and abscission of older leaves reduce total plant transpirational water loss and increase the transfer of nutrients to meristems and to some storage tissues. Osmotic regulation favors water consumption in sink tissues, and accumulation of cuticular wax helps to seal the plant surface and limits non-stomatal water loss. ABA is also an important plant hormone controlling plant growth and development. The extreme ABA-insensitive mutants are wilty-dwarf plant with reduced fertility in soil. The senescence and abscission of flower organs are also delayed in these plants. These results reveal ABA is essential for normal plant growth and development under unstressed condition.

T4-35: Signaling by RAC/ROP GTPases and G-proteins

T4-35-01

Regulation of G protein complex in plants

Daisuke Urano

Temasek Life Sciences Laboratory

The heterotrimeric G protein complex, composing of three subunits: Gα, Gβ and Gγ, emerged in a eukaryotic ancestor. Exchange of GDP for GTP on the Gα subunit leads to G-protein complex dissociation to transmit signals to downstream effectors. While animal G-proteins become activated by G protein-coupled receptors, the regulatory mechanisms differ greatly in plants. Unlike in animals where the nucleotide exchange on Gα requires G protein-coupled receptors (GPCRs), G protein spontaneously activates without the aid of GPCRs in plants. The activation of G proteins in plants is controlled by a Regulator of G protein Signaling (RGS) tethered with seven transmembrane helices (7TM-RGS). This form of regulation is present in most vascular plants but excluded in some cereals and bryophytes that lack the 7TM-RGS gene. Additionally, the Gβ and Gγ genes in plants have duplicated and diversified to produce multiple repertoire of G protein complex, conferring signaling complexity to the G protein network. I will talk about (1) the fundamental biochemical property and regulatory mechanism that differ between animal and plant G proteins, (2) cataclysmic evolution of G protein genes and regulatory systems within the plant kingdom, and (3) the current issues in plant G protein science and our current approaches using three model species.

T4-35-02

Heterotrimeric G proteins regulate panicle branching in rice

Xiangdong Fu, *Qian Liu, Jiangqing Zhang, Kun Wu, Shuanshuo Wang*

Institute of Genetics and Developmental Biology, Chinese Academy of Sciences, Beijing 100101, China

The Green Revolution of the second half of the 20th century, which was based on the adoption of semi-dwarf cereals which had an increased harvest index, was responsible for worldwide crop yield increases. However, introduction of semidwarf genes *sd1* and *Rht* into rice and wheat caused the reduction of panicle (or ear) branching. Here we show that a quantitative trait locus *DEP1* is associated with the increases in panicle branching and grain yield in rice. The *DEP1* gene encodes a non-canonical Gβ subunits, which have an N-terminal GGL (G protein gamma-like) domain, a weakly predicted transmembrane helix near in the central region, and a long C-terminal cysteine-rich domain that consists of the vWFC (von Willebrand factor type C) and TNFR (tumor necrosis factor receptor)/NGFR (nerve growth factor receptor) motifs. Loss-of-function mutation at the *DEP1* locus exhibits decreased number of grains per panicle, whereas gain-of-function mutation confers increased number of grains per panicle and grain yield, indicating that *DEP1* is a positive regulator of panicle branching and grain yield. To uncover the role of the DEP1 protein in control of panicle branching, we performed a yeast two-hybrid screening to identify DEP1-interacting proteins. We found that the DEP1 protein physically interacts with both Gβ (RGA1), and Gβ (RGB1), and reduced RGA1 or enhanced RGB1 activity represses panicle branching. Thus, the modulation of G-protein signalling represents new strategies to increase panicle branching and grain yield in crops.

T4-35-03

FERONIA receptor kinase: a multi-tasking cell surface regulator of RAC/ROP GTPase signaling

Hen-Ming Wu

University of Massachusetts, Amherst

Originally identified as a major regulator of female fertility, FERONIA receptor kinase is in fact an almost ubiquitous cell surface regulator with major roles throughout growth, development, and mediating plant-environment interactions. Our earlier work established that FERONIA functions as an upstream regulator of RAC/ROPs, the RHO GTPases and major molecular switch in plants. We demonstrated that FERONIA interacts directly with ROPGEFs, guanine nucleotide exchange factors that activate RAC/ROPs, recruiting NADPH oxidase to produce reactive oxygen species and regulating at least two of its prominent functions, one in supporting polarized root hair growth the other in inducing pollen tube rupture and sperm release. We also demonstrated that FERONIA interacts directly with two related GPI-anchored proteins LORELEI and LORELEI-like GPI-anchored proteins. In seedlings, LLG1 functions as a chaperone to facilitate delivery of FERONIA to its proper functional location in the cell membrane, where they function as a receptor-coreceptor pair, interacting with the RAC/ROP signaling complex to mediate downstream processes. Use of a co-regulator on the cell surface, and use of RAC/ROPs and reactive oxygen species as downstream mediators might underlie how FERONIA manages at least a subset of its multi-faceted biological roles. Here we discuss recent work that provides insights on how FERONIA responds to various signaling cues to mediate diverse downstream processes.

T4-35-04

POLLEN RECEPTOR-LIKE KINASE family and RAC/ROP GTPase signaling in directional tip growth of the pollen tube

Tetsuya Higashiyama

Institute of Transformative Bio-Molecules, Nagoya University

Pollen tube guidance for successful reproduction involves complex cell-to-cell communication (Annu. Rev. Plant Biol. 2015; Plant Physiol. 2017). We have been working on pollen tube guidance in living materials by using two model plant species, *Torenia* and *Arabidopsis*. Defensin-like LUREs are pollen tube attractant peptides (cysteine-rich peptides; CRPs) of these species working in a short distance which are secreted by two synergid cells (Nature 2009; PLoS Biol. 2012). Recent semi in vitro studies lead to discovery of novel intercellular signaling molecules involved in pollen tube guidance. POLLEN RECEPTOR-LIKE KINASE 6 (PRK6) of *Arabidopsis* is a leucine-rich repeat (LRR) receptor-like kinase, which is critical in sensing of AtLURE1 peptide (Nature 2016). PRK6 localizes at the tip surface of the pollen tube and transduce species-preferential LURE signal to the cytoplasm of the pollen tube with other PRKs. As already shown for other

PRKs working in pollen germination and growth, PRK6 interacts with ROPGEFs (guanine nucleotide-exchange factors), suggesting PRK6 is in the ROP1 Rho GTPase-dependent signaling network of the pollen tube. PRK family has been suggested to work as receptors for CRP ligands in pollen germination and growth. Various CRPs derived from pistil tissues might control appropriate pollen tube germination, growth, and guidance spatiotemporally by regulating the ROP1 Rho GTPase-dependent signaling network with PRK family receptors of different functions. Not only PRK6 but other LRR receptor-like kinases, MDIS-MIK heterodimers, have been reported to be involved in pollen tube guidance. I will discuss how multiple ligands and receptors control precise directional pollen tube growth.

T4-35-05

ROP2 signaling to root hair tip growth

*Erfang Kang, Mingzhi Zheng, Lei Zhu, **Ying Fu***

State Key Laboratory of Plant Physiology and Biochemistry, College of Biological Sciences, China Agricultural University,

Root hairs adopt a typical tip growth for cell morphogenesis and proper function. The establishment and maintenance of the polar site are important for root hair tip growth, and are precisely regulated. The Rop GTPase ROP2 is a well known positive regulator of both root hair initiation and elongation in *Arabidopsis thaliana*. We previously reported that *Arabidopsis* MICROTUBULE-ASSOCIATED PROTEIN18 (MAP18) functions in controlling the direction of pollen tube growth and root hair elongation. Interestingly, both loss-of-function of *ROP2* or knock-down of *MAP18* lead to a decrease in root hair length, whereas overexpression of either *MAP18* or *ROP2* causes root hairs with multiple tips or branching hair phenotype. Analyses of root hair phenotypes of mutants and transgenic seedlings demonstrate that MAP18 and ROP2 interact genetically and functionally. MAP18 interacts physically with ROP2 in vitro and in vivo, and preferentially binds to the inactive GDP-bound form of ROP2 protein. MAP18 promotes the ROP2 activity during root hair tip growth. Further investigation reveal that MAP18 competes with RhoGTPase GDP dissociation inhibitor 1 (AtRhoGDI1)/SUPERCENTIPEDE1 (SCN1) for binding to ROP2. The association with MAP18 interferes with the interaction between the inactive form of ROP2 and AtRhoGDI1/SCN1, in turn affecting the localization of active ROP2 in the plasma membrane of root hair tip. Our results reveal a novel regulatory mechanism of ROP2 signaling during root hair growth.

T4-35-06

A control loop in ABA signal transduction pathway

Zixing Li

University of California San Diego

The phytohormone abscisic acid (ABA) is critical to plant development and stress responses. Abiotic stress triggers an ABA signal transduction cascade, which is comprised of the core components PYL/RCAR ABA receptors, PP2C-type protein phosphatases, and protein kinases. Small GTPases of the ROP/RAC family act as negative regulators of ABA signal transduction. However, the mechanisms by which ABA controls the behavior of ROP/RACs have remained unclear. Our results show that an *Arabidopsis* guanine nucleotide exchange factor protein RopGEF1 is rapidly sequestered to intracellular particles in response to ABA. GFP-RopGEF1 is sequestered via the endosome-prevacuolar compartment pathway and is degraded. RopGEF1 directly interacts with several clade A PP2C protein phosphatases. Interestingly, RopGEF1 undergoes constitutive degradation in *pp2c* quadruple *abi1/abi2/hab1/pp2ca* mutant plants, revealing that active PP2C protein phosphatases protect and stabilize RopGEF1 from ABA-mediated degradation. Our findings point to a PP2C-RopGEF-ROP/RAC control loop model that is proposed to aid in shutting off ABA signal transduction. Further new findings will be reported showing a key role for Ca^{2+} and Ca^{2+} binding protein in regulation of the GEF1-ROP-PP2C control loop.

T4-36: Structure, function and regulation of photosynthetic systems

T4-36-01

A pentatricopeptide repeat protein is required for quality controlling of the di-cistronic atpH/F transcripts in chloroplasts

*Lin Zhang[1], Wenjing Li[2], Wen Zhou[2], Congming Lu[2], **Lianwei Peng**[1]*

1. College of Life and Environmental Sciences, Shanghai Normal University, Shanghai

2. Key Laboratory of Photobiology, CAS Center for Excellence in Molecular Plant Sciences, Institute of Botany, Chinese Academy of Sciences, Beijing

Chloroplast gene expression is strictly controlled by a large number of nuclear gene products at various steps. Here, we report that a pentatricopeptide repeat (PPR) protein, BFA2, is essential for accumulation of the *atpH/F* transcripts in Arabidopsis chloroplasts. BFA2 protein belongs to a member of the P subfamily and localized to the chloroplast stroma. Absent of BFA2 leads to a significant reduction in accumulation of the chloroplast ATP synthase, and this phenotype is due to the loss of di-cistronic *atpH/F* transcripts. EMSA analysis demonstrated that the recombinant BFA2 protein binds to the consensus sequence in the *atpH* 3'-UTR and *atpA* 5'-UTR in a sequence-specific manner. We conclude that BFA2 protein binds to the *atpF/H* 3'-UTR and protects it from degradation by 3'-5' exoribonucleases.

T4-36-02

Crystal structure of plant PSI-LHCI supercomplex and its energy transfer mechanism

***Xiaochun Qin**[1,3], Michihiro Suga[2], Tingyun Kuang[1], Jian-Ren Shen[1,2]*

1. Institute of Botany, CAS

2. Okayama University

3. University of Jinan

In higher plants, the photosystem I (PSI) core is surrounded by a large light-harvesting complex I (LHCI), forming a PSI-LHCI supercomplx to capture and transfer light energy efficiently. In order to reveal the mechanism of energy harvesting and transfer within this large pigment-protein complex, it is essential to solve its crystal structure. We have succeeded in improving the crystal quality of the PSI-LHCI supercomplex from pea significantly, and analyzed its structure at 2.8 Å resolution. Our studies showed that the PSI-LHCI supercomplex contains 16 different subunits (including

12 core subunits PsaA-L and 4 LHCI subunits Lhca1-4) and 205 cofactors (143 chlorophylls *a*, 12 chlorophylls *b*, 26 β-carotenes, 5 luteins, 4 violaxanthin, 10 lipids), with a total molecular mass of 600 kDa. Our results identified chlorophyll *a*, chlorophyll *b*, and some carotenoids in the 4 LHCI subunits for the first time, and revealed the differences in the structure of the 4 LHCI subunits, their interactions, and the interactions between them and the PSI core subunits. In particular, the binding sites for violaxanthins in LHCI were identified for the first time, suggesting their possible roles in photoprotection of PSI. Characteristic red chlorophylls were found in the interfaces between LHCI and the PSI core, implying their possible roles in mediating energy transfer. Based on the structure resolved, 4 plausible energy transfer pathways from LHCI to the PSI core complex were deduced. Taken together, this structure provides a solid structural basis for elucidating the energy transfer and photoprotection mechanisms within the PSI-LHCI supercomplex.

T4-36-03

Mechanism of photosynthetic water oxidation based on atomic structure of photosystem II

Jian-Ren Shen
Institute of Botany, CAS/Okayama University

Photosystem II (PSII) catalyzes light-induced water-splitting leading to the evolution of molecular oxygen indispensible for aerobic life on the earth. PSII is a membrane-protein complex consisting of 20 subunits and a number of cofactors, with a total molecular mass of 350 kDa for a monomer. We have solved the crystal structure of dimeric PSII from a thermophilic cyanobacterium *Thermosynechococcus vulcanus* at a resolution of 1.9 Å using synchrotron radiation (SR) X-rays, which revealed a clear picture of Mn_4CaO_5-cluster, the catalytic center for water-splitting. The structure of the Mn_4CaO_5-cluster resembles a distorted chair form, with 3 Mn, 1 Ca and 4 O (oxygen) atoms forming a distorted cubane core, and the 4th Mn connected to the core by 2 oxo-bridges. Some of the inter-atomic distances within the metal cluster, however, were shown to be slightly longer than those obtained by EXAFS and theoretical studies, presumably due to radiation damage caused by the SR X-rays. In order to avoid possible radiation damage and eliminate the uncertainties in the inter-atomic distances, we used femtosecond X-ray pulses from an X-ray free electron laser (XFEL) facility SACLA, Japan, to solve the structure of PSII. In order to obtain a high resolution structure, we used large PSII crystals, and adopted an approach where every point of the crystal was illuminated by 1 XFEL pulse, and every two XFEL pulses were separated at least by 50 μm on the crystal by moving the crystals. This approach required a huge number of large, isomorphous PSII crystals, but allowed us to collect damage-free, high resolution diffraction data, enabling us to solve the PSII structure at 1.95 Å resolution. This structure showed that most of the Mn-Mn and Mn-ligand distances were 0.1-0.3 Å shorter than those observed in the previous SR structure. However, the bond distances of O5, a unique oxo-bridged oxygen, with two of its nearby Mn ions Mn1 and Mn4, are still unusually longer compared with typical Mn-O bond distances, giving rise to the distorted shape and flexible nature of this catalyst. We further used PSII small crystals and serial femtosecond crystallography (SFX) by XFEL to collect diffraction data from some of the reaction intermediates of the catalyst, and gained information on the structural changes of the catalyst during the water-splitting reaction. Based on these information, I will discuss the possible mechanisms for photosynthetic water-splitting. This work is a collaboration of many colleagues, and I thank them for their contributions.

T4-36-04

Regulation of photosynthesis via proton motive force

Toshiharu Shikanai, *Hiroshi Yamamoto, Caiyuan Wang*
Kyoto University

Photosynthetic electron transport through the thylakoid membrane forms proton motive force (pmf) consisting of ΔpH and membrane potential. Both components of pmf contribute to ATP synthesis. ΔpH also down-regulates photosynthetic electron transport by inducing thermal dissipation of excessively absorbed light energy from photosystem II (a qE component of NPQ) and down-regulation of rate of electron transport through the Cytochrome b_6f complex. Cyclic electron transport around PSI generates pmf without accumulating reducing power in the stroma and regulates the size of pmf. Components of pmf are optimized by regulating ion movement across the thylakoid membrane. Thus, the regulation of size and components of pmf fine-tunes photosynthesis by optimizing the tradeoff between ATP synthesis and down-regulation of photosynthesis. We will discuss the molecular mechanism of this regulation by focusing on cyclic electron transport around PSI and the H^+/K^+ antipoter KEA3 localized to the thylakoid membrane.

T4-36-05

What limits the photosynthetic energy efficiency in nature -Lessons from the oceans

Hanzhi Lin[1,2], *Maxim Gorbunov*[1], *Paul Falkowski*[1]
1. Rutgers University
2. University of Maryland Center for Environmental Science

Solar radiation absorbed by both terrestrial and marine plants during photosynthesis can follow three possible paths, i.e. photochemistry, fluorescence and/or thermal dissipation, but constraining photosynthetic energy conversion efficiency in nature is still challenging because in principle, at least any two yield measurements of these three paths must be made simultaneously. We constructed two different, extremely sensitive and precise active fluorometers for phytoplankoton, i.e. photosynthetic microscopic organisms in upper oceans and fresh water on Earth: one measures the quantum yield of photochemistry from changes in variable fluorescence, the other measures fluorescence lifetimes in the picosecond time domain (the lifetime is capable of being converted to actual quantum yield of fluorescence by dividing chlorophyll a's natural lifetime, 15 ns). By deploying the pair of instruments on eight transoceanic cruises over six years crossing almost the global oceans, we obtained over 200,000 measurements of fluorescence yields and lifetimes from surface waters in five ocean basins. Our results revealed that the average quantum yield of photochemistry was ~ 0.35 (i.e. the average photosynthetic efficiecy in the global ocean was ~ 35%) while the average quantum yield of fluorescence was ~ 0.07. Thus, closure on the energy budget suggests that, on average, ~ 58% of the photons absorbed by phytoplankton in the world oceans are dissipated as heat. This extraordinary inefficiency is associated with the paucity of nutrients in the upper ocean, especially dissolved inorganic nitrogen and iron. Our results strongly suggest that, in nature, most of the time, most of the phytoplankton community operates at approximately half of its maximal photosynthetic energy conversion efficiency because nutrients limit the synthesis or function of essential components in the photosynthetic apparatus.

T4-36-06

Evolution of C_4 photosynthesis in succulent lineages: the role of C_3-C_4 intermediate species

Elena Voznesenskaya[1], *Nuria Koteyeva*[1], *Gerald Edwards*[2]
1. *Komarov Botanical Institute of Russian Academy of Sciences*
2. *Washington State University*

In terrestrial plants, including grasses, sedges and dicots, different variants of C_4-Kranz anatomy and biochemistry have evolved independently from C_3 plants at least 66 times. There are 37 families in the eudicot order Caryophyllales (APG III system, 2009), which includes 23 C_4 eudicot lines. C_4 species acquired a C_4 cycle and leaf structure with enlarged chlorophyllous bundle sheath cells surrounded by palisade cells (Kranz anatomy). Stepwise evolution is proposed from C_3 to C_3-C_4 intermediates to C_4 plants. Currently, only ~50 species have been identified as intermediates (based on structure, biochemistry, function) with properties intermediate to C_3 and C_4 (structural, biochemical and functional); 18 intermediate species are in Caryophyllales. Aridization is considered one of the main driving forces in C_4 evolution. Two general paths are proposed for evolution of Kranz anatomy from C_3 ancestors having planar versus succulent leaves. For species with planar leaves, increased vein density is the best supported preconditioning to date, leading to multiple Kranz units surrounding individual veins. There has been less focus on the evolutionary path from C_3 to C_4 in succulent plants. There are many succulent species in Chenopodiaceae and Portulacaceae which develop water-storage tissue in leaves, with 7 structural types where formation of Kranz anatomy is not linked to vein density. We have identified structural and biochemical diversity in Salsoleae which provides a model for the path of evolution from C_3 to C_4 in succulent species. Phylogenetic analysis of the path to C_4 in different lineages is limited by identification of C_3-C_4 intermediates.

T4-37: Regulatory mechanism of flowering time

T4-37-01

Natural Variation of FLOWERING LOCUS T in *Arabidopsis thaliana*

Emmanuel Tergemina[1], *Liangyu Liu*[2], *Johan Zicola*[1], *Korbinian Schneeberger*[1], *Franziska Turck*[1]
1. *Max Planck Institute for Plant Breeding Research*
2. *College of Life Sciences, Capital Normal University*

Flowering time is a complex trait that is crucial for fitness and regulated by seasonal changes and internal cues. In the annual plant *Arabidopsis thaliana,* two life styles have been described; winter accessions flower in response to long exposure to cold temperature (vernalization) and single season accessions flower in a short period of time chiefly in response to long-day (LD). Control of these two lifestyles converges at the level of transcriptional regulation of *FLOWERING LOCUS T (FT),* a florigen encoding gene. Indeed *FT* expression, which is repressed prior vernalization in winter accessions, strongly accelerates flowering mainly in response to LD but also to elevated temperatures in short-day. The photoperiodic induction of *FT* expression requires a 5,7 kb region upstream of the start codon of *FT*. This region contains a proximal core promoter and a distal enhancer with open chromatin structure; both regions are highly conserved within the Brassicaceae. However, the distal enhancer is not required in the regulation of *FT* expression in the siliques, which has been proposed to influence related traits like floral reversion and seed dormancy.
While most of the flowering time quantitative trait loci (QTL) identified in *Arabidopsis* have been associated with variation at coding regions, QTL linked to *FT* are exclusively mapped to its promoter. To test whether *FT* could be a target of selection to modulate flowering time, we dissected Illumina paired-end sequences from the 1001 genomes project for single nucleotide polymorphisms and structural variations at the photoperiod control regions. We identified rare *FT* haplotypes defined by natural variation in *FT* regulatory regions previously characterized in our group. In the rosette leaves, prior the floral transition, allele specific expression (ASE) of *FT* in F1 hybrids revealed non-neutral *FT* haplotypes, which were less responsive to both LD and ambient temperature inductive conditions. Complementation analysis of the *ft-10* mutant with different *FT* promoter accessions constructs revealed *FT* promoter variants that significantly delay flowering time in LD compared to the reference promoter of the Col-0 accession. Interestingly ASE of *FT* in developing siliques of F_1 hybrids remained stable indicating that *FT* haplotypes are specifically affected for the floral transition. Less responsive *FT* haplotypes were enriched in Sweden and coincide with winter accessions. This last observation, led us to investigate the effect of *FT* after extensive vernalization. Surprisingly, extensive vernalization abolished the late flowering effect of the *ft-10* mutant allele compared to wild-type Col-0. These findings redefine the impact of *FT* on flowering time after extensive vernalization. We propose that an accumulation of less responsive *FT* haplotypes could be due to relaxed purifying selection of *FT* in natural population seasonally exposed to long vernalization periods.

T4-37-02

Co-option of shoot maintenance centroradialis and photoperiod regulated FT module on floral reversion and bulbils formation in *Titanotrichum oldhamii* (Gesneriaceae)

Pei-An Wu[1], *Hsin-Yi Huang*[2], *Jen-Yu Chang*[3], *Chen Wei-Yen*[1], *Yi-Wen Ho*[3], *Zhao-Jun Pan*[1], ***Chun-Neng Wang***[1]
1. *Department of Life Science, Insitute of Ecology and Evolutionary Biology, National Taiwan University*
2. *Institute of Plant Biology, National Taiwan University*
3. *Chiayi Agricultural Experiment Branch, Taiwan Agricultural Research Institute*

Floral reversion allows flowers to switch from determinate growth back to indeterminate growth, thus avoiding senescence and promoting ongoing growth, a trait with agronomic value to adjust vegetative and reproductive growth. Naturally occurred *Titano-*

trichum oldhamii (Gesneriaceae) involves reversion of floral primordia to indeterminate inflorescence and subsequently vegetative bulbils in late flowering season when day length becomes short for reproductive assurance. Here we show the expression of inflorescence meristem identity gene and flowering pathway integrator for photoperiod signals, *T. oldhamii* homologues of *CENTRORADIALIS* (*ToCEN*) and *FLOWERING LOCUS T* (*ToFT*), correlated to floral reversion into bulbiliferous shoots, the modified inflorescences arouse from lateral flowers. The expression level of *ToCEN* reached its highest peak during the reversion; and *ToCEN* mRNA localized in inflorescence apices (maintenance of shoot for flowering) and lateral bulbiliferous shoots meristems (floral reversion into shoots) but not in floral buds. Ectopic expression of *ToCEN* in *Arabidopsis* showed enhanced inflorescence architecture with extended growth of lateral shoot (coflorescence) branching and delayed bolting, a phenotype greatly resembling bulbiliferous shoots. Detection of *ToCEN* mRNA in the scions of wild type *Arabidopsis* grafted onto *35S::ToCEN* stocks suggested *ToCEN* acting as transmittable signals. Thus, *ToCEN* not only maintains inflorescence architecture but also a new role on floral reversion to bulbiliferous shoots. Duplication of *ToFT* also seems to play a contributing role to this floral reversion module. While *ToFT2* mainly expressed in leaves and its expression dynamic displayed oscillation in long-day; by contrast, *ToFT1* has high expression level in short-day induced reverting bulbiliferous shoots and gradually declined expression toward leaves, although the phenotypes of transgenic *Arabidopsis* over-expressing *ToFT2* suggest that the *ToFT2* can functionally substitute for *ToFT1* in promoting early flowering. Further works on how *ToFT1* integrate photoperiod signals and induce *ToCEN* for floral reversion and phase transition will help to clarify their interacting roles as a photoperiod regulated module. Also, it would be important to examine whether both *ToFT1* and *ToCEN* are transmittable in *T. oldhamii* when responding to photoperiodic changes. Nonetheless, our results demonstrated that *ToCEN* and *ToFT1* perhaps evolved to constitute a new regulation module in triggering floral reversion into formation of bulbiliferous shoots in *T. oldhamii*.

T4-37-03

Shifting thresholds: the evolution of developmental plasticity to seasonal cues in *Mimulus*

Benjamin Blackman[1], *Nicholas Kooyers*[1,2]
1. Unviersity of California, Berkeley
2. Florida State University

Diverse taxa have adapted to use day length as a cue to promote floral induction because it serves as a reliable indicator of seasonal timing. Indeed, reproduction is obligately dependent on experiencing long days responses in many plants. In such species, floral development is only initiated after calendar dates when photoperiod exceeds a critical minimum day length. Because annual patterns of temperature and precipitation vary regionally, the yearly timing of the growth season varies across latitude and elevation gradients, producing spatially varying selective regimes that maintain variation in flowering time and its plasticity to photoperiod. We have investigated the geographic diversity of critical photoperiod, flowering time, and their genetic basis in annual populations of the common monkeyflower *Mimulus guttatus*. Although critical photoperiod is strongly correlated with elevation and growth season start date, flowering time in inductive conditions is instead correlated with season duration. That these two components of an integrated seasonal phenology can separately evolve not only indicates that different agents of selection drive their evolution, but also predicts that unique genetic architectures govern divergence of each trait. Genetic mapping by multiplex shotgun genotyping in multiple high x low elevation crosses confirms this prediction: critical photoperiod loci do not co-localize with flowering time loci. Surprisingly, the loci contributing to clinal divergence in critical photoperiod are nearly all transect-specific, in contrast to many recent demonstrations of convergence at the genetic level and contradicting a classic prediction about how plasticity evolves. Even when the same genomic region is implicated in critical photoperiod divergence along multiple elevation gradients, patterns of paralog-specific expression divergence at the underlying floral regulatory genes indicate region-specific genetic changes are involved. We have complemented our pursuit of the underlying candidate genes with common garden studies at high and low elevation to understand how allelic variation at two critical photoperiod loci impacts flowering and fitness in native seasonal environments. Notably, our field studies reveal locus-specific patterns of selection. The low elevation allele of one locus is favored in both locations, potentially highlighting this as an allele that will respond to selection as the climate warms and dries through time, while the other locus exhibiting a signal consistent with local adaptation.

T4-37-04

Flowering on time without *CO* and *FLC* in the temperate-llimate legume model *Medicago*

Joanna Putterill, *Geoffrey Thomson, Lulu Zhang, Mauren Jaudal*
School of Biological Sciences, University of Auckland

Flowering time is a major adaptive trait for plants and agriculture and is optimised using environmental cues such as vernalisation (extended exposure to cold) and long day photoperiods. Integration of these signals ensures that plants do not flower and set seed until after winter is past. In eudicots, the flowering-time gene network is best studied in *Arabidopsis*. In *Arabidopsis*, long day floral induction depends on interactions among the circadian clock, photoreceptors and flowering time genes such as *CO*. CO is activated by long days and directly triggers expression of the powerful florigen, the floral activator *FT*. However, in winter-annual *Arabidopsis*, vernalisation is needed for stable silencing of the strong floral repressor *FLC* by Polycomb group proteins to relieve its inhibition of *FT* and *SOC1*, enabling these genes to be upregulated in spring. However, *CO* does not appear to function in the temperate-climate legumes pea and *Medicago* and neither is there a FLC clade. In contrast, the photoreceptor PHYA regulates flowering in *Arabidopsis*, pea and short-day soybean. Work in soybean has also uncovered legume-specific responses downstream of PHYA. For example, in non-inductive long day conditions, soybean PHYA promotes expression of the legume-specific *E1* gene that represses flowering. We are using the powerful genetic and genomic resources that enable reverse genetics in the legume model *Medicago*, to investigate how flowering time is regulated in this reference legume. We will present data indicating that

MtPHYA and *MtE1* both promote *Medicago* flowering. PHYA is a major player that positively regulates *E1* and *FT-like* transcript accumulation and modulates expression of some clock genes. We will also discuss results indicating that *MtVRN2*, a *VRN2-like* Polycomb Group gene, inhibits the transition to flowering via repressing *FTa1* expression, thus having a different function from *Arabidopsis* VRN2 which represses FLC.

T4-37-05

Polycomb-based epigenetic switching

Hongchun Yang, *Scott Berry, Martin Howard, Caroline Dean*
John Innes Centre

We are studying the chromatin dynamics underlying cold-induced epigenetic silencing of *Arabidopsis FLC*. This determines when plants flower and how they respond to different winter conditions. The epigenetic silencing involves a Polycomb-based epigenetic switching mechanism. *FLC* expression is bistable and cold exposure promotes a switch in individual cells form a transcriptional active (marked by H3K36me3) to a Polycomb-silenced state (marked by H3K27me3). The memory of this silenced state is then inherited in cis through many cell divisions until the plant has flowered, after which point the switch is reversed. The talk will describe our latest understanding of this switching mechanism. Our recent work reveals the switching can be separated to mechanistically distinct two phases of Polycomb silencing. Nucleation of specific Polycomb factors holds a metastable memory of cold in the short-term with subsequent transfer to a long-term histone-modification-based memory via general Polycomb factors. This mechanistic understanding is likely to be general relevant to Polycomb targets in all organisms. This mechanistic understanding is likely to be general relevant to Polycomb targets in all organisms.

T4-37-06

Regulation of developmental transitions by microRNAs

Jia-Wei Wang
Shanghai Institute of Plant Physiology and Ecology, CAS

In plants the changes in morphological and physiological traits serve as markers for the developmental transitions. Mutant characterizations and genetic analyses in *Arabidopsis thaliana* delineate an evolutionarily conserved, *microRNA156 (miR156)*-guided timing mechanism that temporally regulates many aspects of biological processes during development. Recent studies now reveal that sugar metabolites, the products of photosynthesis, feed into this developmental timer by regulating *miR156* levels, thereby ensuring that each developmental transition occurs under favorable conditions. We will present our recent progress in understanding how *miR156* and its targets regulate vegetative phase transition, flowering time and regenerative capacity.

T4-38: Fruit physiology and quality regulation

T4-38-01

Tomato *SlVIF* regulates fruit ripening via modulating sucrose metabolism and pigment synthesis

Guozheng Qin, *Zhu Zhu, Weihao Wang, Jianghua Cai*
Institute of Botany, Chinese Academy of Sciences

Fruit ripening is a complex process that involves a series of physiological and biochemical changes to ultimately influence fruit quality traits. Sugar and pigment contribute to flavor and color, and represent the two most important fruit quality traits. Unraveling the interacting mechanisms between sugar metabolism and pigment synthesis is important to understanding the multiple signaling crosstalk of fruit ripening process, but the direct evidence for sugars in regulating pigment biosynthesis and fruit ripening is still lacking. Here, we identify and analyze the expression profiles of all the 36 genes involved in sucrose metabolism during tomato fruit ripening. By using chromatin immunoprecipitation and gel mobility shift assays, we show that *SlVIF* encoding vacuolar invertase inhibitor and *SlVI* encoding vacuolar invertase are directly regulated by the global fruit ripening regulator RIN (RIPENING INHIBITOR). Moreover, we discover that SlVIF physically interacts with SlVI to control sucrose metabolism. Repression of *SlVIF* by RNA interference (RNAi) delay tomato fruit ripening, while overexpression of *SlVIF* accelerate ripening, with concomitant changes in lycopene production and ethylene biosynthesis. An iTRAQ (isobaric tags for relative and absolute quantification)-based quantitative proteomic analysis indicated that a set of proteins involved in fruit ripening changed abundance in the *SlVIF* RNAi fruit, including proteins associated with lycopene generation and ethylene synthesis. These findings provide strong evidence for the direct role of sucrose in promoting pigment biosynthesis and establish meaningful information of *SlVIF* contributed to fruit quality and RIN-mediated ripening regulatory mechanisms, which are of significant agricultural value.

T4-38-02

CmOr is the master switch to trigger beta-carotene accumulation in melon fruits

Li Li[1], *Yaakov Tadmor*[2], *Joseph Burger*[2], *Noam Chayut*[2], *Hui Yuan*[1]
1. Cornell University/USDA-ARS
2. ARO

Beta-carotene is the most potent precursor for vitamin A biosynthesis and determines the nutritional quality of many fruits including melon. *CmOr* is defined as the single master gene that governs melon fruit b-carotene accumulation. Two *CmOr* haplotypes (*CmOr*^{Orange} and *Cmor*^{Nonorange}) distinguish melon fruit flesh color in a broad germplasm collection. Functional analysis of *CmOr* identifies a single 'golden SNP' being responsible for b-carotene accumulation in various orange-flesh melon fruits. This 'golden SNP' alters the evolutionarily highly-conserved arginine^{108} to histidine in CmOR. Recently, an EMS induced melon mutant was identified, which dramatically lowers b-carotene level in melon fruit. This mutant was found to be caused by a *CmOr* nonsense mutation (*Cmor*-lowβ). *Cmor*-lowβ has minimal effect on the expression of carotenogenic genes, but dramatically reduces phytoene synthase (PSY) protein level and enzymatic activity, indicating that an active *CmOr* is required to maintain PSY level for carotenogenesis in melon fruit. While CmOR is a major posttranscriptional regulator of PSY, the high b-carotene accumulation in *CmOr*^{Orange} melon is not due to its enhanced capacity in activating PSY as both PSY protein level and carotenoid metabolic flux are similar in orange and non-orange melon accessions. Instead, *CmOr*^{Orange} was found to stabilize b-carotene and inhibit its turnover, resulting in b-carotene accumulation in chromoplasts. The elucidation of the basis of *CmOr*[Orange] action broadens our fundamental understanding of mechanisms underlying carotenoid accumulation in plants. The discovery of the 'golden SNP' in the melon *CmOr* gene precisely defines the site for genome editing as a new approach toward β-carotene biofortification of crops.

T4-38-03

Glucose sensor MdHXK1 phosphorylates and stabilizes MdbHLH3 to promote anthocyanin biosynthesis in apple

Yu-Jin Hao
College of Horticulture Science and Engineering, Shandong Agricultural University

Glucose is considered as a major regulatory molecule in addition to being essential metabolic nutrients and structural components in higher plants. As is well known, hexokinase1 (HXK1) is a glucose sensor that integrates diverse signals to govern gene expression and plant growth in response to environmental cues. Previously, it is reported that the nuclear HXK1 forms a glucose signaling complex core with the vacuolar H+-ATPase B1 (VHAB1) and the 19S regulatory particle of proteasome subunit (RPT5B), which influences the transcription of target genes. However, it is yet unknown if and how HXK1 directly targets TFs to modulate their function in the nucleus in plants. Glucose induces anthocyanin accumulation in many plant species; however, the molecular mechanism involved in this process remains largely unknown. Here, we found that apple hexokinase MdHXK1, a glucose sensor, was involved in sensing exogenous glucose and regulating anthocyanin biosynthesis. *In vitro* and *in vivo* assays suggested that MdHXK1 interacted directly with and phosphorylated an anthocyanin-associated bHLH transcription factor (TF) MdbHLH3 at its Ser361 site in response to glucose. Furthermore, both the hexokinase_2 domain and signal peptide are crucial for the MdHXK1-mediated phosphorylation of MdbHLH3. Moreover, phosphorylation modification stabilized MdbHLH3 protein and enhanced its transcription of the anthocyanin biosynthesis genes, thereby increasing anthocyanin biosynthesis. Finally, a series of transgenic analyses in apple calli and fruits demonstrated that MdHXK1 controlled glucose-induced anthocyanin accumulation at least partially, if not completely, via regulating MdbHLH3. Overall, our findings provide new insights into the mechanism of the glucose sensor HXK1 modulation of anthocyanin accumulation, which occur by directly regulating the anthocyanin-related bHLH TFs in response to a glucose signal in plants.

T4-38-04

Silencing of *PL*, which encodes a pectate lyase in tomato, confers enhanced fruit firmness, prolonged shelf-life, and reduced susceptibility to Gray Mold

Lu Yang, Wei Huang, Fangjie Xiong, Zhiqiang Xian, Deding Su, Maozhi Ren, ***Zhengguo Li***
School of Life Sciences, Chongqing University

Pectate lyase genes have been documented as excellent candidates for biotechnological improvement of fruit firmness in strawberry and tomato, two typical non-climacteric and climacteric, respectively. However, the application of pectate lyase in regulating fruit postharvest deterioration is still not fully explored. In this report, 22 individual pectate lyase genes were identified in tomato, one *PL* gene (Solyc03g111690), showed dominating expression during fruit maturation and was regulated by ripening related phytohormone. RNA interference (RNAi) of Solyc03g111690 resulted in enhanced fruit firmness and changes of pericarp cells. More importantly, transgenic tomato fruit strengthened the anti-rotting and pathogen-resisting ability. Compared to WT, transgenic tomato had higher content of cellulose and hemicellulose, whereas lower content of water soluble pectin. In line with that, peroxidase (POD), superoxide dismutase (SOD), and catalase (CAT) activity were higher in transgenic fruit, where less malondialdehyde (MDA) level was detected. RNA-Seq results showed numerous differentially expressed genes (DEGs) involving in hormone signaling, cell wall modification, oxidative stress, and pathogen resistance. Collectively, these data demonstrate that pectate lyase play a role in both fruit softening and pathogen resistance, advancing our insights of postharvest fruit preservation in tomato as well as in other fleshy fruits.

T4-38-05

Identification of citric acid and vitamin C associated genes in *Citrus* by genome approach

Qiang Xu*, Lun Wang, Fa He, Yue Huang, Huiwen Yu, Xiuxin Deng*
Huazhong Agricultural University

High-quality genomes of basic citrus species were built up by the third generation sequencing technology, single molecule sequencing data (PacBio long reads) in combination with the second generation sequencing technology short-read (Illumina), i.e. 3+2 strategy. All the genome sequence, annotation files and raw data will be deposited from our website http://citrus.hzau.edu.cn/orange/. We also sequenced wild citrus, semi-domesticated forms and cultivated citrus. Transcriptomes of fruit, leaf, of wild citrus and cultivated citrus were further included to narrow down the genes involved in the citric acid genes. We also find that there are footprints of introgressions of wild citrus in the mandarin population. Recently, we collected several spontaneous mutants of acidless citrus from all over the world and included for genome comparison analysis. A strategy combined genome and bioinformatics to identify the mutated locus was applied and succeeded. A similar approach was also used to identify vitamin C related genes in different citrus types. This approach provides some new insights into the citrus genetic loci affect citric acid metabolism.

T4-38-06

SlBHL1 gene regulates the synthesis of lycopene and chlorophyll in tomato fruit

Daqi Fu*, Lanhuan Meng, Yunbo Luo*
China Agricultural University

During the fruit ripening process, the decomposition of chlorophyll and the accumulation of lycopene were closely related to the quality of fruits. It is of great significance to understand the mechanism of fruit quality formation by the synthesis of chlorophyll and lycopene regulated by transcription factors. We used the virus-induced gene silencing (VIGS) method to screen 30 transcript factors with high expression level during fruit ripening; the results show that *SlBHL1* gene silencing led to the chlorophyll accumulation at mature green stage and lycopene reduction at the red stage in tomato fruit. We also obtained *SlBHL1* RNAi tomato plant, which was consistent with VIGS phenotype. We used RNAi tomato fruits to do RNA-Seq, metabonomics, and Chip-seq analysis, The result shown that the *SlBHL1* gene negative regulate chlorophyll accumulation by directly acting the promoter region of *SlTKN2* genes, but also regulate the ethylene synthesis by directly acting the promoter region of *SlACO* genes in tomato fruit. The functional identification of the *SlBHL1* gene provides a method for regulating the synthesis of lycopene and chlorophyll in fruits.

T4-39: Novel insights into wood research– from evolution to developmental genetics and functional traits

T4-39-01

Evolution of woodiness within flowering plants: Where, why and how?

Frederic Lens
Naturalis Biodiversity Center

Insular floras typically harbour species with a remarkable degree of woodiness. This so-called insular woodiness was described and interpreted correctly by Darwin, but many years later we still do not have a global overview (1) in which flowering plant groups transitions from herbaceousness towards derived woodiness have taken place, (2) why these evolutionary transitions have happened, and (3) what genetic mechanism is causing these convergent shifts. Based on an ongoing review study of derived woodiness, it is evident that this phenomenon is concentrated in the later diverging lineages of flowering plants, and many of these woody taxa are native to islands or island-like regions such as tropical mountain peaks. Surprisingly, this transition has also occurred much more on continents than expected, especially in dry continental regions facing at least a number of consecutive dry months per year, such as savannas or (semi-)deserts. This suggests that increasing wood development and increasing drought stress resistance may go hand in hand, which has not been suggested before. We have obtained experimental evidence in *Arabidopsis* (Brassicaceae) and daisies (Asteraceae), showing that woody species are able to better avoid air bubble formation in their conductive xylem conduits due to drought stress than their closely related herbaceous relatives. However, drought stress is not involved in all derived-woody lineages, and environmental factors driving wood formation are probably taxon specific. As a third item, I want to touch on how herbaceous plant groups develop into woody shrubs and trees during evolutionary time. The massive convergent number of transitions towards derived woodiness suggests that the genetic mechanism should be relatively simple, which is supported by a shrub-like woody mutant of Arabidopsis in which two genes are knocked out, but we do not know what are the key regulatory genes that turn on the wood pathway. Is there only one master gene or are there several regulatory genes within a shared interaction network?

T4-39-02

Forensic timber identification: the authenticity diagnostics of species and geographic provenance using genetic methods

***Lichao Jiao**, Yafang Yin*
Research Institute of Wood Industry, Chinese Academy of Forestry

Deforestation represents a massive threat to global biodiversity with illegal logging and the associated trade in illegally sourced wood products. Timber identification of tree species and their origins is crucial and essential for monitoring and control of illegal activities in timber production and thus enforcement of species trade regulations. The traditional identification relies on diagnostic anatomical features, either macroscopic or microscopic. Examination of wood anatomy generally facilitates identification to genus level, as wood characters tend to be highly conserved within genera. However, it is difficult and even impossible for precise timber discrimination at the species level or tracking the origin of wood. The newly developed genetic methods might overcome these limitations and provide effective information with high resolution. DNA barcoding, a genetic approach based on a short DNA sequence from a standard part of a genome, may be the most appropriate for identifying to the species level or higher. Moreover, population genetics and phylogeography are very powerful tools for geographical traceability. In the paper, we focus on CITES-listed or national protected timber species and provide our recent exploration in relevant fields. Additionally, potential diagnostic applications and novel insights of genetic methods are contributed for forensic timber identification.

T4-39-03

Intervessel pit membranes provide key determinants of embolism resistance in angiosperm xylem

***Shan Li**, Steven Jansen*
Institute of Systematic Botany and Ecology, Ulm University, Germany

Xylem sap in angiosperms is typically transported under negative pressure through vessels that are interconnected by bordered pit pairs. The intervessel pit membrane, which develops from a modified primary wall and middle lamella after cell hydrolysis, is known to provide a major contribution to xylem hydraulic resistance, while limiting at the same time the spread of micro-organisms and embolism. Pit membranes play a crucial role in drought-induced embolism formation via air-seeding, which traditionally assumes that the movement of gas between two vessels is limited by the high surface tension of water and the nanometre-scaled pores of a wetted pit membrane. However, recent findings challenge this view and highlight that air-seeding remains poorly understood because of our limited knowledge of (1) the surface tension of xylem sap, (2) the three-dimensional characteristics of the tortuous pore pathways in wet pit membranes, and (3) the contact angle at the air-water-pit membrane interface. This talk will focus on the ultrastructure of intervessel pit membranes, especially their thickness and porosity. Transmission electron microscopy (TEM) shows a 10-fold variation in pit membrane thickness for fresh samples. Pit membrane thickness is significantly correlated with embolism resistance ($r = 0.78$, $P < 0.01$, n = 37 species), with species from dry environments typically showing thicker pit membranes than those growing in mesic conditions. Assuming that cellulose microfibrils in pit membranes show an equal spatial density, pit membrane thickness is suggested to affect the length of the entire pore pathway, but not the actual pore size. In fact, it is unknown if pit membrane pore sizes show considerable variation across angiosperm species. Perfusion experiments with colloidal gold particles of known size indicate that pit membrane pores are typically 5 to 10 nm wide, which is much smaller than observations based on scanning electron microscopy. This is explained by the finding that dried wood shows shrunken pit membranes with much larger pores than wet, undamaged pit membranes. Depending on the drying method, the thickness of intervessel pit membranes shrinks by 30% to 50% compared to the thickness in fresh samples. Moreover, this shrinkage is irreversible, may occur under natural conditions in the field, and explains air-seeding fatigue. Therefore, TEM observations on pit membrane thickness should be based on fresh samples to avoid preparation artefacts. The shape of pit membrane pores is geometrically highly variable, with many pores showing a slit-like outline, but little cylindrical pores. This means that the air-seeding pressure is typically reduced by ca. 50% compared to all pores being perfectly circular. Also, there is growing evidence that bubble snap-off occurs in a tortuous pore pathway, especially if an air-water meniscus passes a pore volume that has < ½ the diameter of the adjacent pore volume. Future attention should be paid to the role of surfactants that are found in pores of intervessel pit membranes, and their potential role in the coating and stabilisation of nanobubbles after a snap-off event.

T4-39-04

Paving the road to land: gene diversification and arise of water-conducting cells in land plants

Bo Xu[1,2], *Misato Ohtani*[2,3], *Mitsuyasu Hasebe*[4], *Samuel Brockington*[1], *Taku Demura*[2,3]

1. Department of Plant Sciences, University of Cambridge

2. Graduate School of Biological Sciences, Nara Institute of Science and Technology

3. RIKEN Center for Sustainable Resource Science

4. National Institute for Basic Biology

Colonization of land by early land plants coincided with an increase in cellular and morphological diversities. Advances in genome sequence data suggested this increase of diversities was associated with transcription factor duplication and subsequent functional diversification, which played critical roles in innovation of novelties for survival on continental surface. One novelty is related to develop specialized cells, water-conducting cells, for long distance water transport. Recent studies revealed a NAC transcription factors VNS (VND/NST/SND)-based gene regulatory network orchestrating the xylem cell formation in vascular plants. However, it still remains unclear on genetic origin of these cells. Here we investigated in non-vascular plant moss *Physcomitrella patens*, an early land plant lineage, seeking to understand the functional conservation and diversification of VNS family with regard to innovation of water-conducting cells. Eight VNS proteins were identified in *P. patens* and further divided into two groups. By comparing the function of these two groups, it showed that a group of *PpVNS* genes shared the conserved function in water-conducting cell formation with their homologues in vascular plants. *P. patens* mutants with abnormal water-conducting cells tended to a wilting phenotype when exposed to the condition with lower humidity. Our findings indicated a conserved VNS-based genetic network governing the development of water-conducting cells across land plants, and suggested that the diversification of NAC proteins to induce highly specialized cell formation to facilitate water transport was a pivotal event for early plants to adapt to terrestrial environment.

T4-39-05

Tracheid cell wall structure and chemical composition in the earliest woody plants

Christine Strullu-Derrien[1], *Sylvain Bernard*[2], *Alan R.T. Spencer*[3], *Laurent Remusat*[2], *Paul Kenrick*[1], *Delphine Derrien*[4]

1. Department of Earth Sciences, Natural History Museum, London

2. IMPMC, UMR CNRS, Sorbonne Universités, UPMC, IRD & MNHN, Paris

3. Department of Earth and Engineering, Imperial College, London

4. Biogéochimie des Ecosystèmes Forestiers INRA Nancy-Lorraine, Cha

The evolution of tissues composed of cells with robust lignified walls was fundamental to the early diversification of vascular plants, giving rise to forest ecosystems by the Middle Devonian (ca 387 Ma). These developments caused large quantities of recalcitrant organic matter to enter the environment, with consequences for the carbon cycle that are still poorly understood. How was this material decomposed in the environment and what organisms were involved? Here we develop an approach to investigate the wall structure and composition of the tracheids in the earliest fossil wood with the aims of characterizing the nature of decomposition of the cell wall when wood is attacked by fungi. Axes of *Armoricaphyton chateaupannense* (Early Devonian, ca 407 Ma from the Armorican Massif, western France) were imaged using propagation phase contrast X-ray synchrotron microtomography (PPC-SRμCT). The axes of this small plant are preserved as both thin coaly compressions and permineralizations or partial permineralizations in pyrite (FeS2). Tomographic datasets where imported into volume rendering Drishti v2.6 software for visualization, allowing the fine details of the tracheids cell walls (pits, bars, and wall membranes) to be examined in three-dimensions. This was realized through the creation of a three-dimensional reconstruction showing a single (virtually sectioned) tracheid from which digital animations were obtained. High fidelity images of the cell wall and pit structure (p-type) were obtained to a spatial resolution of 0.551 μm per voxel. Further ultrastructural and chemical investigations were performed at the submicrometre scale on focused ion beam (FIB) ultrathin sections extracted from freshly fractured specimens. Scanning and transmission electron microscopies (SEM and TEM) were employed to document the composition of the tracheids. In parallel, synchrotron-based scanning transmission X-ray microscopy (STXM), coupled with X-ray absorption near edge structure (XANES) spectroscopy, was used to determine the carbon speciation of the tracheid to investigate possible lignin signatures. Analyses were realized on both a three-dimensional pyritized specimen and on a specimen preserved in compression allowing comparison between the two samples. Results show that this combination of analytical methods can be highly effective for imaging the cell wall structure and chemical composition of the earleist wood cells. Using these approaches will enable effective comparisons of Devonian Period fossil woods with other fossil and modern woods that have undergone various types of fungal attack (e.g., white rot, brown rot) to shed light on the nature of wood decomposition in early terrestrial ecosystems.

T4-39-06

Diversity of microscopic bark structure revealed and classified

Veronica Angyalossy[1], ***Marcelo R. Pace***[1,2], *Ray F. Evert*[3], *Carmen R. Marcati*[4], *Alexei Oskolski*[5,6], *Teresa Terrazas*[7], *Ekaterina Kotina*[5], *Frederic Lens*[8], *Solange C. Mazzoni-Viveiros*[9], *Guillermo Angeles*[10], *Silvia R. Machado*[11], *Alan Crivellaro*[12], *Karumanchi S. Rao*[13], *Leo Junikka*[14], *Nadezhda Nikolaeva*[15], *Pieter Baas*[8]

1. Departamento de Botanica, Universidade de São Paulo

2. National Museum of Natural History, Smithsonian Institution

3. Department of Botany, University of Wisconsin-Madison

4. Departamento de Ciência Florestal, Universidade Estadual Paulista

5. Department of Botany and Plant Biotechnology, University of Johannesburg

6. Komarov Botanical Institute of the Russian Academy of Science

7. Instituto de Biología, Universidad Nacional Autónoma de México

8. Naturalis Biodiversity Center, University of Leiden

9. Núcleo de Estudos em Anatomia, Instituto de Botanica

10. Red de Ecología Funcional, Instituto de Ecología, A.C.

11. Departamento de Botanica, Universidade Estadual Paulista

12. Dipartimento Territorio e Sistemi AgroForestali, Università degli Studi di Padova

13. B.R.D. School of Biosciences, Sardar Patel University

14. Finnish Museum of Natural History, Helsinki University

15. Karelian Research Centre of the Russian Academy of Science

The importance of bark in plant biology, including structure, physiology, diversity and evolution can hardly be overestimated. Yet its microscopic structure is poorly documented in comparison with wood. The International Association of Wood Anatomists (IAWA) therefore commissioned the production of an IAWA List of microscopic bark features, which sets a new international standard for the definition and recognition of anatomical features of barks in woody angiosperms and gymnosperms. Codes are given for 173 anatomical character states to facilitate databasing the huge diversity of bark structure in trees, shrubs and lianas. It is richly illustrated with 169 colour micrographs. The "IAWA Bark List" is not only indispensable for any future work on comparative bark anatomy and microscopic bark identification, but can also be used as a first inventory of functional traits in this complex tissue. Here we will present some of the enormous diversity present in barks, how the list is organized and can be used both in teaching and in phloem and periderm studies.

THEME V: GENETICS, GENOMICS & BIOINFORMATICS

T5-01: Polyploid genomics

T5-01-01

Genomic analysis of transgressive segregation in autopolyploid sugarcane

Ching Man Wai[1,2], *Jisen Zhang*[1], *Tyler Jones*[3], *Chifumi Nagai*[3], ***Ming Ray***[1,2]

1. FAFU and UIUC-SIB Joint Center for Genomics and Biotechnology, Fujian Agriculture and Forestry University

2. Department of Plant Biology, University of Illinois at Urbana-Champaign

3. Hawaii Agriculture Research Center

Sugarcane is a dual-purpose crop for energy and sugar production. It has the highest biomass production reaching 1.9 billion tonnes in 2014 worldwide. To improve sugarcane biomass accumulation, we developed an interspecific cross between *Saccharum officinarum* 'LA Purple' and *Saccharum robustum* 'MOL5829'. The selected F1 individuals were self-pollinated to generate a transgressive F2 population with a wide range of biomass yields. Leaf and stem internodes of fourteen high biomass and eight low biomass F2 extreme segregants were used for RNA-seq to decipher the molecular mechanism of rapid plant growth and dry weight accumulation. Gene Ontology terms involved in cell wall metabolism and carbohydrate catabolism were enriched among 3,274 differentially expressed genes between high and low biomass groups. Specifically, up-regulation of cellulose metabolism, pectin degradation and lignin biosynthesis genes were observed in the high biomass group, in conjunction with higher transcript levels of callose metabolic genes and the cell wall loosening enzyme expansin. Furthermore, UDP-glucose biosynthesis and sucrose conversion genes were differentially expressed between the two groups. A positive correlation between stem glucose, but not sucrose, levels and dry weight was detected. We thus postulated that the high biomass sugarcane plants rapidly convert sucrose to UDP-glucose, which is the building block of cell wall polymers and callose, in order to maintain the rapid plant growth required for biomass accumulation. The gene interaction of cell wall metabolism, hexose allocation and cell division contributes to biomass yield, expanding our understanding at the molecular level required for energy cane breeding and engineering.

T5-01-02

Genome sequencing supports a multi-vertex model for Brassiceae species

Xiaowu Wang

Institute of Vegetables and Flowers, Chinese Academy of Agricultural Sciences

The economically important *Brassica* genus is a good system for studying the evolution of polyploids. Comparative genomic analysis of *Brassica* genomes has revealed that these have undergone whole genome triplication (WGT), which can be explained by a two-step theory. The diploid progenitor genome of *Brassica* before WGT was deduced to have seven chromosomes that resemble the translocation Proto-Calepineae Karyotype (tPCK). Subgenome dominance phenomenon such as biased gene fractionation and dominant gene expression were observed in *Brassica*. Furthermore, WGT and gene dominant expression were determined to contribute to, or be associated with, morphotype diversification and trait development during domestication and breeding improvement of *Brassica* crops. In addition, the genome of radish (*Raphanus sativus*), which is another important crop of the tribe Brassiceae, was derived from the same WGT event and shows similar subgenome dominance. These findings and molecular dating indicate that radish has a similar position from the evolutionary aspect as that of *Brassica* species. Finally, we extended the *Brassica* "triangle of U" model to an open system as a multi-vertex model to include radish and other Brassiceae species that share the same WGT event. As many Brassiceae species are known to be closely related to *Brassica*, this multi-vertex model describes the relationships or potentials of applying more Brassiceae paleohexaploids in the creation of new allotetraploid oil or vegetable crop species. Finally, further extensive work confirming the relationships of Brassiceae species is warranted.

T5-01-03

Genome evolution in newly formed polyploids: Insights from *Tragopogon*

Douglas Soltis, *Pamela Soltis*
University of Florida

Polyploidy (whole-genome duplication; WGD) is a major force in eukaryotic evolution, and ancient and recent WGDs are well documented in many lineages, including vertebrates, fungi, and, most extensively, angiosperms. Despite repeated WGDs, most eukaryotes are functionally diploid, yet little is known about the earliest stages of polyploidy and the processes that transform a duplicated genome into a diploid one. *Tragopogon* is a textbook model for studying polyploidy with two recently (~80 years old; 40 generations in these biennials) and repeatedly formed natural allotetraploids (*T. mirus*, *T. miscellus*) and their diploid parents (*T. dubius*, *T. pratensis*, *T. porrifolius*). Our cytogenetic, genetic and genomic data suggest that patterns of gene loss, silencing, and chromosomal change in the young polyploids *T. mirus*, *T. miscellus* are repeated across multiple origins of these polyploids in nature, a well as in synthetic lines. These changes are ongoing--we have caught evolution in the act. Furthermore, the ongoing changes observed in these newly formed polyploids appear to be fixed in the older polyploid *T. castellanus*. Thus, patterns of genetic and genomic change may be predictable within just a few generations in young polyploids. With its range of ages, *Tragopogon* affords research opportunities not available in other systems (e.g., cotton, soybean, maize).

T5-01-04

Cytonuclear evolution in allopolyploid plants

Lei Gong, *Bao Liu*
Key Laboratory of Molecular Epigenetics of the Ministry of Education (MOE), Northeast Normal University

Allopolyploid speciation, faces the challenge of stoichiometric disruptions resulted by merging bi-parental nuclear genome yet with only one (always maternal) of the two sets of progenitor cytoplasmic genomes. Rubisco (1,5-bisphosphate carboxylase/oxygenase) is composed of nuclear-encoded small subunits (SSUs) and plastome-encoded large subunits (LSUs), which is an ideal enzyme to explore the evolutionary cytonuclear accommodation process. By studying gene composition and diversity as well as gene expression in natural and synthetic allopolyploid lineages, including *Arabidopsis*, *Arachis*, *Brassica*, *Nicotiana* and *Oryza*, we characterized the cytonuclear co-evolution/coordination in terms of its variation pattern and evolution pace. It is demonstrated that paralogous nuclear-encoded *rbcS* genes within diploids evolved in a "balanced mutation and concerted homogenization" model. Within the hybrids, upon the genome merger, no genomic coordinative responses of allelic *rbcS* gene conversions are made. However, transcriptional coordination could be conditionally achieved via biased expression of maternal *rbcS* nuclear genes; at the initial stage after genome doubling, a pool of allopolyploid individuals harboring variable coordination patterns are generated, in which neither transcriptional nor genomic coordination is established. During afterwards evolution, the functional constraint from the cytoplasmic partner genes and/or other organellar factors could exert selection pressure to the nuclear *rbcS* genes, which could achieve the eventual well-established cytonuclear coordination such that *rbcS* maternal homoeologs are preferentially biased utilized and maternal-like *rbcS* paternal copies carrying the maternal-to-paternal conversions preferentially occur. These results shed lights on the unexplored cytonuclear coordination in plant allopolyploid speciation and evolution, which could be a novel focus for future polyploidy researches.

T5-01-05

Genome organization of autotetraploid potato (*Solanum tuberosum* L.)

Qian Zhou[1,2], *Jue Ruan*[1], *Wu Huang*[1,2], *Guangcun Li*[2], *Canhui Li*[3], *Ruofang Zhang*[4], *Zhonghua Zhang*[2], *Sanwen Huang*[1,2]
1. Agricultural Genomics Institute at Shenzhen, Chinese Academy of Agricultural Sciences
2. 1Institute of Vegetables and Flowers, Chinese Academy of Agricultural Sciences
3. School of Life Sciences, Yunan Normal University
4. Inner Mongolia Potato Engineering and Technology Research Center, Inner Mongolia University

Potato (*Solanum tuberosum* L.), the most important tuber food crop in the world, has been facing many difficulties in its genetic improvement during the past hundred years. Genome sequence of a haploid potato was available, but we know little about the genome organization of cultivated potato, an autotetraploid. Here we reported a chromosome-scale 2.6 gigabase draft genome of a potato cultivar, Cooperation-88. The assembly covered ~87% of the tetraploid genome and was anchored onto 48 pseudochromosomes using a genetic map generated from 1000 S1 progeny from Cooperation-88 . We identified ~1.6 gigabase repeated sequences accounting for 65% of the assembly and annotated 133,733 protein-coding genes. The analyses of the RNA-seq data and Methyl-seq data uncovered the fates of multiple alleles in the autopolyploid genome. The genome landscape of Cooperation-88 provides insights into evolution and domestication of autoplolyploid crop.

T5-01-06

Genetic variation in diploid and polyploid vascular plants

Stacy A Jorgensen, *Michael S Barker*
University of Arizona

Although polyploids often arise from multiple origin events, polyploid speciation is likely a severe bottleneck. We thus expect genetic variation to be lower in polyploids than in diploids. To test this hypothesis, we assembled from the literature an array of expected heterozygosity (He) statistics for vascular plants. This data set includes observations from 54 families and 32 orders representing lycophytes, ferns, gymnosperms, and angiosperms. We made comparisons between diploids and polyploids, between spore- and seed- dispersed plants, and among different methods (AFLP, isozyme, RADseq, RAPD, and microsatellite). We also evaluated the influence of paleopolyploidy on genetic variation in diploids. Contrary to our expectations, we observed a similar level of genetic variation in polyploids and diploids. Spore-dispersed vascular plants have a slight but statistically significant deficit of heterozygosity relative to seed-dispersed plants. Among methods, microsatellite markers consistently estimated higher levels of heterozygosity than the other methods. We observed a slight negative trend between *He* and chromosome number; plants with high chromosome numbers tend to be polyploid ferns with *He* near or below 0.2. Finally, we found higher heterozygosity in diploids with more recent whole genome duplications than in those with more ancient polyploidies; around 5% of the variation in expected heterozygosity can be explained by the age of the most recent paleopolyploidy. These results suggest that whole genome duplication casts a "long shadow" lasting millions of years, with profound, long-lasting effects on the evolution of vascular plant genomes.

T5-02: Gymnosperm genomes and comparative genomics

T5-02-01

DNA methylome of the 20-gigabase Norway spruce

Haifeng Wang
Fujian Agriculture and Forestry University

DNA methylation plays important roles in many biological processes, such as silencing of transposable elements, imprinting, and regulating gene expression. Many studies of DNA methylation have shown its essential roles in angiosperms (flowering plants). However, few studies have examined the roles and patterns of DNA methylation in gymnosperms. Here, we present genome-wide high coverage single-base resolution methylation maps of Norway spruce (Picea abies) from both needles and somatic embryogenesis culture cells via whole genome bisulfite sequencing. On average, DNA methylation levels of CG and CHG of Norway spruce were higher than most other plants studied. CHH methylation was found at a relatively low level; however, at least one copy of most of the RNA-directed DNA methylation pathway genes was found in Norway spruce, and CHH methylation was correlated with levels of siRNAs. In comparison with needles, somatic embryogenesis culture cells that are used for clonally propagating spruce trees showed lower levels of CG and CHG methylation but higher level of CHH methylation, suggesting that like in other species, these culture cells show abnormal methylation patterns.

T5-02-02

Genetic and epigenetic features of gymnosperm genomes. Are the gymnosperms really slow in evolution?

Ales Kovarik[1], *Wencai Wang*[2], *Lu Ma*[2], *Alena Kovarikova*[1], *Ilia Leitch*[3], *Sonia Garcia*[4], *Andrew Leitch*[2]
1. Czech Academy of Science, Institute of Biophysics
2. University of London, Queen Mary College
3. Royal Botanical Garden, Kew, Richmond, UK
4. Institute of Botany, Barcelona

Seed plants today comprise the angiosperms and the gymnosperms. Whilst much attention has been devoted to angiosperm genome structure and evolution, rather little has been expended on gymnosperms. Gymnosperms typically have large genomes exhibiting reduced frequencies of a number of processes (e.g. polyploidy) that have shaped the genomes of other vascular plants. Here we studied genome organization focusing on major chromosome domains (mostly rDNA and telomeres) in gymnosperm lineages including Coniferales, Ginkgoales, Gnetales and Cycadales using combined molecular cytogenetic and genomic (NGS) approaches. While overall genome organization (e.g., retroelement content) is similar to angiosperms wide differences exist in individual genomic components. For example, genus *Cycas* is unique in having extraordinary diversity of rDNA paralogs, high level of DNA methylation and extremely amplified telomeric repeat variants. The unusual linked arrangement of 35S and 5S rDNA has evolved repeatedly in several groups. We also searched for epigenetic genes potentially involved in genome size control discovering that members of a dicer family (*dcl1-4*) underwent multiple losses and gains in evolution of gymnosperms. In conclusion, individual gymnosperm lineages exhibit many unique features not generally found in angiosperms. Thus, gymnosperms genomes may not be as static as previously thought.

T5-02-03

Lack of DNA removal following polyploidization drives genome sizes in the gymnosperm genus *Ephedra*

S.M. Ickert-Bond[1], *A. Sousa*[2], *Y. Min*[2], *I. Loera*[3], *J.S. Metzgar*[1], *J. Pellicer*[4], *O. Hidalgo*[4], *I.J. Leitch*[4]
1. Department of Biology and Wildlife & UA Museum of the North, University of Alaska Fairbanks (UAF), Fairbanks, Alaska, USA
2. Department of Biology, University of Munich (LMU), D-80638 Munich, Germany
3. Biología Evolutiva, Instituto de Ecología A.C., Carretera antigua a Coatepec 351, El Haya, 91070, Xalapa, Veracruz, Mexico

Most gymnosperms are remarkably stable as far as chromosome number, karyotype and genome size are concerned. In this study on *Ephedra*, we investigated the latter three features together with state dependent diversification and species environmental niche breadth in a phylogenetic framework to (1) evaluate the impacts of genome size and polyploidy on speciation rate, (2) identify where shifts in genome size and polyploidy have taken place during diversification both temporally and spatially, and (3) determine whether polyploid taxa have wider ecological niche breadths than their diploid progenitors. To achieve this, we generated four new chromosome counts and measured genome size data for 84% of *Ephedra* species across the phylogenetic tree; used phylogenetic modeling to infer chromosome number, genome size evolution, and ancestral states; tested for state dependent diversification under a BiSSE model, and compared ecological niche breadth using phylogenetic generalized least squares (PGLS) regressions. Our results showed that *Ephedra* is chromosomally highly variable. Chromosome numbers range from $2n = 14$ to $2n = 56$ with rampant polyploidization ($4x$, $6x$, $8x$) affecting 28 taxa (68%). Most of these polyploid taxa (78.6%) were tetraploid. Genome sizes are also very variable with 1C-values ranging nearly 5-fold from 8.09-38.34 pg making Ephedraceae the most variable of all gymnosperm families. *Ephedra* is also notable as it includes the largest genome size so far reported for any gymnosperm (*E. antisyphilitica*, $2n = 8x = 56$). In contrast to angiosperms and most gymnosperms, genome size and chromosome number are strongly correlated in *Ephedra* and C-values can thus be used to predict chromosome numbers and ploidy levels. Consequently, monoploid (1Cx) genome sizes at different ploidy levels are relatively uniform suggesting that polyploidy is not accompanied by significant genome downsizing, as commonly reported in angiosperms. Polyploidy in *Ephedra* is associated with higher speciation rate and is inferred mostly at more recent internal nodes, while the backbone is inferred to be consistently diploid with an ancestral chromosome number of $n = 7$ and ancestral 1C-value of 9.66 pg. PGLS regression results did not find support for polyploid species having wider niche breadths than their diploid counterparts.

T5-02-04

Drifting towards genome gigantisms in seed plants

Andrew Leitch[1*], Lu Ma[1], Wencai Wang[1], Steven Dodsworth[2], Ales Kovařík[3], Petr Novák[4], Jiří Macas[4], Tao Wan[5], Ilia Leitch[2]

1. *School of Biological and Chemical Sciences, Queen Mary University of London, Mile End Road, London, UK*
2. *School of Biological and Chemical Sciences, Queen Mary University of London, Mile End Road, London, UK*
3. *Institute of Biophysics, Academy of Sciences of the Czech Republic, Brno, Czech Republic*
4. *Biology Centre ASCR, Institute of Plant Molecular Biology, Branišovská 31, České Budějovice, Czech Republic*
5. *Key Laboratory of Southern Subtropical Plant Diversity, Fairy Lake Botanical Garden, Shenzhen & Chinese Academy of Science, PR China.*

Most angiosperm genomes are small (modal 1C Genome Size = 0.6pg, mean 1C Genome Size = ~6pg), but some are enormous, indeed the largest known diploid genome in plants is *Fritillaria amabilis* (1C=100.6pg) and largest known genome for a eukaryote is the chromosomal octoploid *Paris Japonica* (1C=152.3 pg). In contrast gymnosperm genomes are typically larger than angiosperms (modal 1C Genome Size = ~10 pg and mean 1C Genome Size = ~ 18.6 pg) with the largest known diploid gymnosperm being *Pinus ayacahuite* (1C = 36 Gbp). Genetics reveal remarkable similarities between large genomes in angiosperms and those of gymnosperms (e.g. diverse heterogeneous repeats) that may be influenced by divergence in the epigenetic machinery. But environmental factors, small population sizes and drift may also play important roles in genome size divergence too. This talk explores genetic, epigenetic, and environmental factors that may contribute towards the drift towards genome giganticism.

T5-02-05

Repetitive elements in the genomes of gymnosperms

Wencai Wang[1], Steven Dodsworth[2], Aleš Kovařík[3], Petr Novák[4], Jiří Macas[4], Tao Wan[5], Ilia J. Leitch[2], Andrew R. Leitch[1]

1. *Queen Mary University of London*
2. *Royal Botanic Gardens, Kew*
3. *Institute of Biophysics, Academy of Sciences of the Czech Republic*
4. *Biology Centre ASCR, Institute of Plant Molecular Biology of the Czech Republic*
5. *Fairy Lake Botanical Garden, Shenzhen & Chinese A*

Extant gymnosperms comprise four distinct lineages, the conifers, cycads, *Ginkgo* and Gnetales. Most species have large genome sizes (1C DNA values mean=18.8pg, mode=10.0pg), in comparison with angiosperms (1C DNA values mean=5.9 pg, mode=0.6pg). These large genome sizes have meant that few studies have addressed genome structure and composition in gymnosperms. Nevertheless, with the advances in next generation sequencing (NGS) technologies, coupled with developments in analytical techniques we can now start to shed light on the genomic composition of the major gymnosperm lineages. Here we use the RepeatExplorer pipeline (www.repeatexploer.com/) to characterize and compare the most abundant repeats comprising the genomes of representative species in each of the major gymnosperm lineages. The analyses show that the repeats are typical of other land plants that have been analyzed comprising abundant satellite sequences and LTR retrotransposons and derivatives, (e.g. *Copia*- and *Gypsy*-like elements). Astonishingly, the repeats in cycads, whose eight genera are estimated to have diverged c. 150 million years ago, are still sufficiently similar that we can group the cycads as a distinct clade from other gymnosperms. However, an analysis of individual gymnosperm repeat families reveals that they can be highly diverse and degraded. Indeed so stable are the repeats in gymnosperms that we can use them to similarly resolve *Welwitschia*/*Ephedra* and Pinaceae/*Ginkgo* lineages. Patterns of slow repeat divergence contrasts with the fast repeat turnover that is typically reported for angiosperms with small genome sizes (1C DNA values <10Gb). Indeed, similar analyses using the phylogenetic signal contained in repeat abundance cannot in our experience group angiosperm taxa above the genus level due to the rapid divergence of the repetitive DNA sequences. Nevertheless, ancient diverse repeats are also observed in the early diverging angiosperm *Amborella*, and fast repeat turnover is therefor likely to be a derived character that evolved after *Amborella* diverged from the rest of the angiosperms. In conifers there is evidence that the large genomes have expanded considerably through the accumulation of repeats, but there is not a proliferation of one or a few repeats that often dominate the genomes of many angiosperms with small genome sizes. Amongst gymnosperms, the Gnetales and in particular *Welwitschia* and *Gnetum* have small genomes, which are likely to have arisen through genome downsizing following their divergence from *Ephedra* (Gnetales), and the other gymnosperm lineages.

T5-02-06

The *Gnetum* genome reveals new insights into seed plant evolution

Tao Wan
Fairy Lake Botanical Garden

The dearth of genomic data from Gnetophytes is preventing resolution of conifer-specific from gymnosperm-specific genomic features. This is inhibiting understanding of evolutionary divergence across major seed plant lineages. Whilst partial assemblies for conifers are available, none exist for Gnetophytes. Here we report the 4.07 Gb fully sequenced and assembled genome of *Gnetum montanum* and find an enigmatic genome, which is distinct from all other seed plants. The contrasting genome dynamics represent a mosaic of characters, including (i) an absence of recent whole genome duplications in contrast to angiosperms and conifers, (ii) a diverse assemblage of retrotransposons (RT), as in conifers but not in most angiosperms, and (iii) a small, dynamic genome, with elevated rates of RT excision and higher rates of gene sequence divergence compared with conifers. Overall, the assembly enables the better resolution of ancestral genomic features within seed plants, the identification of genomic characters that distinguish gymnosperms from angiosperms, and reveals new dynamics in gene family evolution.

T5-03: State-of-the-art phenotyping: deep, high-throughput and field (two sessions)

T5-03-01

CyVerse tools for managing plant data

Ramona Walls
CyVerse, University of Arizona

CyVerse, formerly known as the iPlant Collaborative, is an US National Science Foundation (NSF) funded initiative "to design, deploy, and expand a national cyberinfrastructure for life sciences research, and to train scientists in its use" (http://www.cyverse.org/about). As part of this mission, CyVerse houses over 2 petabytes of data. Much of that data is genetic or genomic, but with the growth of high throughput plant phenotyping, we expect phenotypic and environmental data, including images and geospatial data, to be a rapidly growing part of our catalog. CyVerse includes platforms for managing and analyzing data, including the Discovery Environment (a web GUI), Atmosphere (cloud computing), Bisque (image management and analysis) and multiple APIs (application programming interfaces. The Data Commons (DC) is a new service that strives to build a space where data can live as a searchable, discoverable, and reusable resource. DC development builds on foundational CyVerse infrastructure, while expanding into new areas such as metadata and ontologies, data publication, and federation with external collaborators and repositories. Key components of the DC are the data portal at http://datacommons.cyverse.org/, and newer functions such as metadata templates, permanent identifier requests, data submissions to NCBI, and a Projects Interface (under development). This presentation will introduce the various CyVerse platforms and services available to support plant phenotyping and botanical research more generally.

T5-03-02

LiDAR based platforms for 3D high-throughput crop phenotyping

Qinghua Guo[1], *Fangfang Wu*[1,2], *Shuxin Pang*[1]
1. State Key Laboratory of Vegetation and Environmental Change, Institute of Botany, Chinese Academy of Sciences
2. University of Chinese Academy of Sciences

Molecular designer breeding has been considered as an efficient and effective solution to solve the food crisis caused by the growing population and limited arable land worldwide. However, the inconsistent development of genotype and phenotype delayed the breeding process seriously, which really means that phenotypes cannot be non-destructively measured by high throughput quantification during the whole growth period of crops, especially in the field conditions. Recent progresses in phenomics show that a suite of new technologies, including Light Detection and Ranging (LiDAR) technology, intelligent robot technology, unmanned aerial vehicle (UAV) remote sensing technology, could become the main impetus to accelerate the progress in understanding gene function and environmental responses. Supported by "the Strategic Program of Molecular Module-Based Designer Breeding Systems" and "the Instrument Developing Project of the Chinese Academy of Sciences", we developed a series of phenotyping platforms to characterize agriculturally relevant traits used in greenhouse and field conditions, which were named as Crop 3D. By integrating LiDAR, high-resolution camera, thermal camera and hyperspectral imager, Crop 3D platforms were capable of acquiring multi-sources data through whole growth period automatically and synchronously along single-population-regional crop scales. The platform in greenhouse adopted "sensor-to-plant" working style to avoid interference on plant growth. For single plant in greenhouse, phenotype parameters such as plant height, crown, leaf length and width, leaf angle, *etc.* were extracted. Canopy height, canopy cover, plant area density and other relevant traits were measured to enhance our understanding of relationship between genotype and environment. Comparing the height derived by Crop 3D with the data measured by Electronic Distance Measuring Device, the R^2 was as high as 0.94. Field-based platform precisely measured the expression of physiological traits at canopy level under heterogeneous and fluctuating environment. In large region, UAV platform covered crops phenotype under different growing stages was applied to estimate crop progress, condition and yield. Phenotyping platforms still need to be further optimized in the application process according to the user demands. Transdisciplinary cooperation among breeder, agronomists and engineers will provide a broad stage for the application of phenotyping platforms to play its advantages.

T5-03-03

Remote estimation of leaf area index in wheat using multi-angular imagery acquired from an Unmanned Aerial Vehicle (UAV)

Ning Lu, *Tao Cheng, Jie Zhou, Xia Yao, Yongchao Tian, Yan Zhu, Weixing Cao*
National Engineering and Technology Center for Information Agriculture (NETCIA), Nanjing Agricultural University, Nanjing 210095, China

Leaf area index (LAI) is a critical structural parameter for assessing crop growth status and an important trait in crop phenotyping. Spectral vegetation indices (VIs) represent a common method to estimate LAI from optical remotely sensed data. However, they are known to suffer from saturation when LAI is at moderate to high values (LAI > 5). A possible reason is that the VI-based methods only use spectral data from nadir observations. The flexibility of UAV systems enables the application of multi-angular remote sensing to monitor crop growth, which may mitigate the saturation problem and improve the accuracy of LAI estimation. This study employed a multispectral camera with 5 bands (490 nm, 550 nm, 671 nm, 700 nm, 800 nm) equipped on a multi-rotor UAV to acquire imagery at 7 fixed view angles (0°, ±20°, ±40°, ±60°) in the growing season of wheat at a field station in Rugao (32°16'N, 120°45'E), China. The experiment encompassed 36 field plots representing variations in cultivar, nitrogen fertilization level and sowing density. The UAV was flown at 50m above ground level to acquire multispectral imagery at around a 3 cm resolution, which is high enough to separate soil background from the green vegetation. Concurrent with UAV flights, the field LAI was measured with LI-COR LAI 2200C. A number of VIs selected for LAI estimation from the literature were calculated UAV imagery from different view angles. Empirical models with each VI were established to examine the effect of view angle on LAI estimation. At the end of this study, we draw conclusions on the optimal combination of view angle and VI for estimating LAI and make recommendations for UAV data acquisition strategies for wheat phenotyping applications.

T5-03-04

Combine phenotyping and modeling of canopy photosynthesis to guide breeding for higher crop yield and resource use efficiency

Qingfeng Song[1,2], *Xin-Guang Zhu*[1,2]
1. CAS-MPG Partner Institute for Computational Biology, Shanghai Institutes for Biological Sciences, Chinese Academy of Sciences
2. State Key Laboratory of Hybrid Rice, Shanghai Institutes for Biological Sciences, Chinese Academy of Sciences

Canopy photosynthesis, which is the sum of all leaves' photosynthesis in a field, is a determinator of final yield production as it provides the carbon and energy necessary for plant growth and seeds. However, the development of canopy photosynthesis research is stagnant due to the difficulty to measure it in field. Optimizing canopy photosynthesis is even harder as its susceptibility to many factors, such as plant architecture, micro-climate in the canopy and photosynthetic features of all leaves. To tackle these problems, we developed a new canopy photosynthesis and transpiration system (CAPTS), which can automatically measure gas exchange of plant canopy or soil, for phenotyping canopy photosynthesis and a integrated 3D canopy model, which incorporated a comprehensive three-dimensional (3D) representation of canopy architecture, a ray tracing algorithm and a mechanistic dynamic model of leaf photosynthesis, for optimizing canopy photosynthesis. First, the CAPTS measurement correlated with the calculated canopy photosynthesis using the model, which was detailed parameterized and we showed the previous models using averaged light in each layer of canopy may over estimate canopy photosynthesis up to 25%. Secondly, we theoretically studied the impacts of different plant structures, locations and planting strategies to canopy photosynthesis with validations using CAPTS. Thirdly, we dissected the contributions of environmental and physiological factors to the increasement of canopy photosynthesis under future global change senario, which can hardly be assessed by experiments. Finally, we systematically studied the impact of chlorophyll content in leaves with our model and predicted that proper decrease of leaf chlorophyll content can increase canopy photosynthesis rate and nitrogen use efficiency. This was also confirmed with transgenic experiment. In conclusion, we developed a new equipment for measuring canopy photosynthesis and a new 3D model for studying canopy photosynthesis theoretically. With the two methods combined, we provided a new platform to optimize canopy photosynthesis for higher crop yield and resource use efficiency.

T5-03-05

Exploiting the multidimensional aspect of the radiance measurements for more efficient high-throughput field phenotyping

Fred Baret[1], *Benoit De Solan*[2], *Simon Madec*[1], *Shouyang Liu*[1]
1. INRA - EMMAH-CAPTE
2. Arvalis
3. HIPHEN

Field high-throughput phenotyping is recognized as a major component needed to foster genetic improvement. It is achieved using non-destructive radiance measurements recorded by a number of sensors. Radiance measurements are sampling the total radiance field that extends over at least 6 dimensions, namely the 3 dimensions of the space, time, wavelength and directions. High-throughput field phenotyping measurements should therefore sample optimally the total radiance field of the micro-plots to retain most of the relevant information used to derive the structural and functional traits of interest. This presentation shows examples of the optimal sampling for several traits including the plant density, the green area index, the fraction of light intercepted by the canopy, the canopy height and the leaf chlorophyll content. It involves a number of sensors including RGB cameras, multispectral cameras, spectrophotometers and LiDARs that can be mounted aboard Unmanned Aerial (UAV) working in passive mode (the sun as radiation source) or Ground (UGV) Vehicles allowing to work in active mode with artificial illumination. Emphasis is put on the selection of the optimal spatial resolution, wavebands and directions (view and source) for each of these traits. Further, the temporal sampling is a key element for deriving higher level traits more closely related to the plant functioning. Conclusions are finally drawn on the typical experimental design best suited to extract the most informative traits from high-throughput field phenotyping platforms.

T5-03-06

In-vivo beloground phenotyping using computed tomograpy

*Norman Uhlmann, Joelle Claußen, Norbert Wörlein, **Stefan Gerth***
Fraunhofer Development Center X-ray Technology

During the last years, X-ray computed tomography (CT) has been applied for the non-destructive visualization of belowground structures in pots. Formerly, this technology was only used for medical imaging. Nowadays, CT is used as a standard tool in industrial applications for material analysis. With X-ray CT the 3D volume information of objects can be reconstructed using X-ray projections of the object from different points of view. The scanning geometry we used is the axial 3D-CT, where a conical X-ray beam projects the plant pot on a 2D flat panel detector. The resulting spatial sampling frequency of this setup is about 85 µm using a high-power X-ray source to reduce the measurement time. Due to the non-destructive nature of CT it is possible to track the growth of plant organs such as cassava tubers, potato tubers or root systems in general. Thus, the below-ground development of an individual plant can be observed time dependently in natural soil. To virtually extract the root system out of the soil, new algorithms are used for the automatic segmentation. This allows to search for new traits with an increased throughput compared to manual volume segmentation. We will present the results of time resolved measurements for Cassava storage roots, potato tubers in heat and drought stress and root architectures of two different maize varieties. Within the timeframe of the experiments we periodically analyzed the below ground growth and applied an automated segmentation algorithm. Thus, we are able to characterize the volumetric change of the belowground organs time resolved and track them already at an early stage in a reliable and reproducible manner. For example, this approach allows to calculate the total volume as a function of the depth for each of the different varieties. Due to the time resolved experiment this data can be used to calculate the biomass growth rates, as well. Using this approach, we are able to phenotype for new traits and therefore breed distinct varieties under the aspect of root development.

T5-03-07

Describing the wheat canopy structure dynamics by combining ADEL-Wheat model with green fractions observations

Shouyang Liu[1], *Fred Baret*[2], *Samuel Buis*[1], *Abichou Mariem*[2], *Bruno Andrieu*[2], *Irfan Kamran*[1], *Benoit De Solan*[3]
1. INRA EMMAH, UMR 1114 Domaine Saint-Paul, Site Agroparc, 84914 Avignon Cedex 9
2. INRA-AgroParisTech, UMR 1091 EGC, 78850 Thiverval-Grignon
3. ARVALIS-Institut du végétal, Station expérimentale, 91720 Boigneville

Monitoring the dynamics of canopy architecture during the early stages is highly desirable for breeders to select cultivars of high drought-tolerance, resistance to weed infestation, improved water-use efficiency and potentially high grain yield. These desired traits may be grouped into early vigor trait. A dedicated in silico study was completed to evaluate the feasibility and interest of a phenotyping data assimilation approach. A sequential assimilation scheme was developed to limit the space of canopy realization since the simulations of the 3D canopy architecture as well as the rendering of the phenotyping measurements is computer demanding. The early growth was decomposed into 2 consecutive stages. For each stage, 5 main influential parameters were identified. The phenotyping measurements were consisting into RGB images taken at several dates along the growth cycle and from several view inclination angles. For the first stage, the 5 influential parameters were estimated using a neural network machine learning technique. For the second phase, the 5 additional influential parameters were adjusted using the same neural network based technique while exploiting as prior information the posterior values of the 5 parameters defined during the first phase assimilation process. Results show that the parameters defining the size and rate of change of the green canopy crop elements were estimated with a relatively good accuracy. Conversely, the parameters driving the orientation and inclination of the elements were retrieved with a poor accuracy. However, some important emerging properties were very well retrieved, including the GAI and the number of tillers with more than 3 leaves at the beginning of stem elongation. Further detailed sensitivity analysis highlights the importance of getting enough dates of measurements covering the growth cycle, while dates separated by less than a phyllochron do not provide significant additional information. The sensitivity analysis highlighted also the importance of the inclined observations. The assimilation approach proposed appears very general and could be applied to a number of phenotyping measurements completed along the growth cycle. Nevertheless, because our results were achieved using in silico experiments, further validation is necessary using actual field measurements.

T5-03-08

Plant phenotyping reveals genetic and physiological factors of plant performance

Thomas Altmann[1], *Moses Muraya*[2], *Jianting Chu*[1], *Yusheng Zhao*[1], *Astrid Junker*[1], *Christian Klukas*[3], *Jochen Reif*[1], *David Riewe*[1], *Rhonda Meyer*[1], *Hea-Jung Jeon*[1], *Marc Heuermann*[1], *Judith Schmeichel*[1], *Monique Seyfarth*[1], *Jan Lisec*[4], *Lothar Willmitzer*[4]
1. Leibniz Institute of Plant Genetics and Crop Plant Research (IPK) Gatersleben
2. Chuka University
3. LemnaTec GmbH
4. Max-Planck-Institute of Molecular Plant Physiology

Knowledge of the structural and functional genetic architecture of agricultural traits is a prerequisite for the systematic exploration and utilization of plant genetic resources in plant breeding. To uncover mechanistic links between genetic variation, physiological factors and whole plant performance for development of novel crop improvement strategies, Arabidopsis, maize, and rapeseed are investigated by integrating genotyping, transcript and metabolite profiling, and automated plant phenotyping using dedicated platforms. The three high throughput plant phenotyping facilities at IPK support automated whole plant analyses for small, medium, and large plants including cultivation, transport (plant-to-sensor) and imaging of plants in climate controlled phytotron/glasshouse cabins. They are equipped with camera and illumination systems for visible, fluorescence, and near infrared imaging using top view and side views. Furthermore, 3D laser scanners and LED panels and CCD cameras for functional chlorophyll fluorescence detection and a broad range of environmental sensors are installed. The value of repeated non-invasive/non-destructive monitoring of large plant populations is highlighted by results of the analysis of a collection of 261 maize dent lines using a specifically optimized cultivation and phenotyping setup characterized for biomass accumulation and water consumption, and thus, for water use efficiency. Combined with 50k SNP information, these data have been used to identify QTL of these traits and of the growth dynamics by genome-wide association testing. 12 main effect QTL and 6-pairs of epistatic interactions were detected that displayed different patterns of expression at various individual developmental time points. A subset of them also showed significant effects on relative growth rates in different intervals. Using nonparametric functional mapping and multivariate mapping approaches four additional QTL affecting growth dynamics were detected. Our results demonstrate that plant biomass accumulation is a complex trait governed by many small effect loci most of which act at certain restricted developmental phases. This highlights the need to detect and investigate stage-specific growth control genes operating at different developmental phases. Integrated metabolome analysis and whole plant phenotyping performed in Arabidopsis revealed direct links between a promoter InDel polymorphism of the *FUM2* gene, its mRNA expression, fumarase enzyme activity, and fumarate to malate ratio in leaves. The promoter InDel that coincided with a fumarate QTL found in Arabidopsis Col-0/C24 RILs and ILs was also significantly associated with the fumarate to malate ratio, with malate and fumarate levels, and with dry weight in 174 natural accessions at 15 days after sowing (DAS) and was associated with biomass production in another 251 accessions (at 22 DAS). This supports a role of the cytosolic FUM2, which specifically occurs in *Brassicaceae*, in diurnal carbon storage and point to a growth advantage of accessions carrying the *FUM2* Col-0 allele preferentially occurring in colder climates.

T5-03-09

RhizoChamber-Monitor: A compact, automatic root growth monitoring platform with stretchable shading system

***Jie Wu**, Qian Wu, Yeqing Yuan, Xiaolei Zhang, Mingwei Du, Xiaoli Tian, Zhaohu Li*
China Agricultural University

Roots play a key role in resource acquisition and perception of environmental signals. However, compared with the aboveground, the phenotypic information on root system has been largely neglected due to the inaccessibility of the soil and the complexity of its structure. Here, we provide a novel root growth monitoring platform integrating customized rhizoboxes, rhizobox-loaders (RL), an imaging system, a precise irrigation system and a stretchable shading system as a whole, and a matched software. The rhizobox is assembled as a sandwich leaving two sheets of narrow space between cloth and transparent plastic for root growing and a small groove in the top of rhizobox for mounting irrigation tube. The operational principle of imaging system is similar to the mobile racking system frequently applied in logistics with RL analogous to the storage rack, rhizoboxes to the objects and sensor to the lift. The push-and-pull motion of RL and the three-dimensional and rotational motion of DSLR camera is conducted by one "H" shape mechanical system. The mechanical system crosses the RLs at the bottom of RLs. The cooperation of these two types of motion completes the imaging of both faces of all rhizoboxes. Unlike to the mode - one container with one suit of shading apparatus – that frequently used in HT root phenotyping system, in this platform, neighboring RLs are connected by one long suit of flexible organ shields and thus all rhizoboxes share one shading system. The advantage of using this shading system is that the system does not act as a block between sensor and root during imaging and the change in the size and number of rhizoboxes does not require the modification of the system. The Irrigation system allows the deployment of different water regimes on given groups of rhizoboxes in the same platform simultaneously. Software based on matlab is also developed for processing the time-series images and extracting global and local root traits. However, root crossing and paralleling makes topology analysis an issue. Instead, some specific local traits such as the surface area and number of lateral roots in successive equidistance segment of primary root can be extracted automatically using our software. Root growth of cotton plants under different water treatments was monitored and analyzed to demonstrate the performance of the platform and the software. As cultivation and imaging system are both in the same space and the height space is utilized as traffic aisle, this increases the space utilization ration and allows the deployment of the platform in limited space. The application of the shading system allows the control of root growing temperature independently. The root monitoring platform combined with the capacity for environmental simulation provides a possibility to assess the interaction between genotypes and environment dynamically.

T5-03-10

Plant water stress remote sensing system using UAV with RGB, multiple-spectral and thermal cameras

Wenting Han
1. Institute of Soil and Water Conservation, Northwest A&F University, Yangling, China
2. Institute of Water-saving Agriculture in Arid Areas of China, Northwest A&F University, Yangling, China

Remote sensing technology has been used for determining crop water requirements and irrigation scheduling in recent 20 years. Efficient irrigation scheduling tools and management are based on high spatially and temporally resolution images obtained by remote sensing platforms. Unmanned Aircraft Vehicle (UAV) sensing systems with RGB, multiple-spectral and thermal cameras to obtain high quality images were developed. Two fixed wing and 14-rotors UAVs were developed using open source pilot software. The control system was designed to trigger the camera independent of the pilot system of UAV. The GPS coordinates were obtained and saved in the log file of the camera. The ground resolution can be calculated based on the parameters of camera and the height of UAV. The UAV remote sensing systems were used to obtain the images of crops in 2013,2014 and 2015 with the combination of ground data collection like soil and leaf water contents. Soil water content models were developed using linear regression methods based on the UAV remote sensing images and ground measurements.

T5-03-11

Machine learning methods in plant image phenotyping

***Zheng Xu**, James Schnable, Yufeng Ge, Yumou Qiu, Jennifer Clarke*
The University of Nebraska-Lincoln

The importance of high-throughput and automatic phenotyping of crops has grown in recent years as a way to generate larger and denser datasets to analyze complex plant biological systems. While the automated acquisition of plant image data is becoming more straightforward, extracting interesting features remains a crucial bottleneck. Approaches to increase the accuracy and efficiency of feature extraction, as well as to expand the number and type of phenotypes which can be extracted from complex multimodal image datasets are a critical need for downstream biological analysis. Machine learning methods are promising tools for analyzing large complex data and have great success in applications involving images such as disease diagnosis from CT/PET/MRI, face recognition, pixel segmentation, object labeling and applications not involving images such as voice recognition, language translation and artificial intelligent. This project intends to evaluate and compare the performance of different machine learning tools in plant phenotyping based on matched hyperspectral images and RGB images. The target plant phenotypes include both morphological variables and measures of the plant ionome, including the abundance of nitrogen (N), phosphorus (P) and potassium (K). The raw data come from two experiments conducting in the UNL automated phenotyping greenhouse. The first employed 196 maize (corn) plants, imaged each day from Oct. 10, 2015 to Nov. 9, 2016 and the second includes 504 corn plants imaged every other day from May 16, 2016 to the mid of July 2016. Ground truth manual phenotypic measurements were collected at the end of each experiment. We consider different machine learning algorithms including random forest, supporting vector machine and combinatorial approaches. We implement each algorithm with different settings of kernel functions and data normalization. Our research may assist and serve as a reference for researchers in plant phenotyping.

T5-03-12

Leaf chlorophyll content estimation from multispectral imagery acquired from unmanned aerial vehicle (UAV) over wheat crops

***Tao Cheng**[1], Fred Baret[2], Shouyang Liu[2], Ning Lu[1], Dong Li[1], Jie Zhou[1], Tison Franck[2], Stéphane Jezequel[3], Benoit De Solan[3], Alexis Comar[4], Xia Yao[1], Yongchao Tian[1], Yan Zhu[1], David Han[5]*
1. NETCIA, Nanjing Agricultural University, China
2. INRA-EMMAH, UMR 1114, Avignon, France
3. Arvalis, Avignon, France
4. HIPHEN, Avignon, France
5. PHENOTRAIT, Beijing, China

Estimates of leaf chlorophyll content (LCC) is a highly-desired trait to be characterized in field phenotyping experiments. Current methods are mostly based on tedious destructive sampling or based on leaf-level indirect methods using leaf transmittance or fluorescence measurements achieved by clipping individual leaves. Alternative methods based on canopy reflectance measurements have been proposed. They allow much higher throughput but have been limited in accuracy because of the confounding effect of canopy architecture. This study investigates a new approach allowed by the high spatial resolution images and the inclined views accessible from unmanned aerial vehicle (UAV) observations. A joint experiment between China (Rugao, 32°16'N, 120°45'E) and France (Gréoux, 43°45'N, 5°51'E) was conducted over wheat plots showing a range of nitrogen level fertilization, sowing density, growing conditions and genotypes. Multispectral images were acquired with the AIRPHEN camera aboard a UAV at end of tillering and booting stages of wheat growth. The camera is equipped with 6 bands sampling the chlorophyll absorption features both from nadir and inclined (45° inclination perpendicular to the rows) observations. The UAV was flown at low altitude to acquire high spatial resolution images allowing the separation of the green elements from canopy background. It was flown three times during the day to account for potential influence of the solar position. Ground measurements of LCC, green fractions at 0° and 45° inclination, and green area index (GAI) were concurrently collected. Three types of empirical methods were compared to estimate LCC from the multispectral images acquired over wheat crops at the Rugao and Gréoux sites. The first (respectively second) one is based on the use of vegetation indices (VI) computed with the average reflectance values derived from nadir (respectively inclined) observations. The third one is based on VIs computed with the average reflectance values derived from a fraction of the green pixels observed from nadir. Performances of the three methods are compared with due attention paid to the sensitivity to the green fraction and time of acquisition during the day. Conclusions are finally drawn on the optimal method that would provide the best performances while being practically easy to implement within phenotyping experiments.

T5-04: Evolution of plant reproductive systems: from ecology to genomics

T5-04-01

Identification of causal heterostyly genes in *Primula*

*Cuong Nguyen Huu[1], Christian Kappel[1], Barbara Keller[2], Adrien Sicard[1], Yumiko Takebayashi[3], Holger Breuninger[4], Michael Nowak[5], Isabel Bäurle[1], Axel Himmelbach[6], Michael Burkart[1], Thomas Ebbing-Lohaus[7], Hitoshi Sakakibara[3], Lothar Altschmied[6], Elena Conti[2], **Michael Lenhard**[1]*
1. University of Potsdam
2. University of Zurich
3. RIKEN Center for Sustainable Resource Science
4. University of Oxford
5. University of Oslo
6. Leibniz Institute of Plant Genetics and Crop Plant Research
7. Gartenbau Ebbing-Lohaus

Heterostyly is a wide-spread floral adaptation to promote outbreeding, yet its genetic basis and evolutionary origin remain poorly understood. In *Primula* (primroses), heterostyly is controlled by the *S*-locus supergene that determines the reciprocal arrangement of reproductive organs and incompatibility between the two morphs. However, the identities of the component genes remain unknown. We have recently identified the *Primula CYP734A50* gene, encoding a putative brassinosteroid-degrading enzyme, as the *G* locus that determines the style-length dimorphism.

CYP734A50 is only present on the short-styled S-morph haplotype, it is specifically expressed in S-morph styles, and its loss or inactivation leads to long styles. The gene arose by a duplication specific to the Primulaceae lineage and shows an accelerated rate of molecular evolution. Additional genes with analogous characteristics, such as being present only on the dominant S-morph haplotype, have been identified, representing candidates for other causal heterostyly genes. Their functional characterization is ongoing. Thus, our results provide a mechanistic explanation for the *Primula* style-length dimorphism and begin to shed light on the evolution of the *S*-locus as a prime model for a complex plant supergene.

T5-04-02

Heritability and fitness consequences of architectural effects within inflorescences

Lawrence Harder[1], *Mason Kulbaba*[1,2], *Ilona Clocher*[1], *Hazel Cameron-Inglis*[1], *Colby Regel*[1]
1. *University of Calgary*
2. *University of Minnesota*

Production of multiple flowers allows the reproductive phenotypes of individual angiosperms to include systematic among-flower variation, the details of which could be adaptive in particular environments. Systematic trait variation within inflorescences could arise from resource competition among a plant's flowers, or be a developmentally determined feature of flower position, regardless of resource dynamics. The latter, architectural effect is obvious in individual plants that produce distinct floral morphs (e.g., various types of monoecy, peripheral sterile flowers), but manifests more often as continuous floral variation within inflorescences. For architectural effects to be adaptive, differences in the relations of floral traits to flower position among individuals must both cause consistent variation in reproductive performance and be heritable. We assess systematic positional variation within inflorescences for natural populations *Delphinium glaucum* (Ranunculaceae), which produces vertical racemes with protandrous bisexual flowers. In this species, features of flower size, flowering phenology and female resource allocation (ovule number) and time allocation (female-phase duration) decline from bottom to top flowers. In contrast, anther number is constant among flowers and male-phase duration increases distally. Consequently, upper flowers emphasize relative male function compared to lower flowers. A manipulative field experiment demonstrated that at least the number of new flowers opened per day on inflorescences (anthesis rate) depends on their position, rather than resource competition with lower (earlier) flowers. This trait is an essential control on the number of flowers displayed simultaneously (floral display size), and so contributes to pollinator attraction and among-flower self-pollination. The existence of architecture effects suggests that gradients in floral traits within inflorescences are products of selection and so have functional benefits and are heritable. We examine gradient function by quantifying the association of among-plant variation in overall seed production, siring success and outcrossing rates to gradients of flower size, ovule and anther production and anthesis rate. We specifically characterize phenotypic gradients as spline functions and analyse them as function-valued traits using the developing body of statistical methods known as functional data analysis. We also use SNP-based approaches to evaluate and compare the heritabilities of individual trait means, variances and gradients. These results illustrate that architectural effects within inflorescences affect reproductive performance and are partially genetically determined, rather than simply representing phenotypic plasticity or developmental instability. Such effects demonstrate that the reproductive phenotypes of angiosperms are functional mosaics shaped by adaptation.

T5-04-03

A peculiar marriage arrangement - Evolutionary processes in tristylous populations

Spencer C.H. Brarrent[1], *Christopher Balogh*[1], *Joana Costa*[2]
1. *University of Toronto*
2. *University of Coimbra*

Populations of tristylous plants are reproductively subdivided into three mating types that differ in the reciprocal arrangements of their sex-organs. The mating types are referred to as the long-, mid- and short-styled morphs (hereafter L-, M- and S-morphs). This floral polymorphism is maintained by frequency-dependent selection resulting from disassortative mating. Ronald Fisher first demonstrated that with this mating system a 1:1:1 ratio of morphs (isoplethy) is predicted in large equilibrium populations. Here, we consider the ecological, demographic and reproductive factors that can cause deviations from the Fisherian isoplethic equilibrium, particularly in the context of frequent colonization and biological invasion. We draw examples from the three well known tristylous families (Lythraceae, Oxalidaceae, Pontederiaceae), with a focus on recent comparative investigations of *Lythrum salicaria* (purple loosestrife) in its native European and introduced North American ranges. Founder events, limited sexual recruitment and prolific clonal propagation have played key roles in determining the representation and frequency of style morphs in invasive populations of *Oxalis pes-caprae* (Bermuda buttercup) and *Eichhornia crassipes* (water hyacinth). Until recently, it was widely thought that introduced populations of *O. pes-caprae* were exclusively composed of a sterile pentaploid clone of the S-morph. However, recent studies in the Iberian Peninsula indicate that tetraploid L- and M-morphs capable of sexual reproduction also occur in some populations, and molecular evidence suggests that they have arisen from separate introductions from the native South African range. In the introduced range of *E. crassipes,* the M-morph dominates globally with the L-morph occurring sporadically where sexual reproduction is permitted. Molecular evidence indicates that a striking genetic bottleneck accompanied invasive spread from the New to the Old World. The S-morph is restricted to lowland South America and thus is entirely absent from the introduced range. In contrast to clonal *O. pes-caprae* and *E. crassipes, Lythrum salicaria* reproduces almost exclusively by seed and the majority of populations in both the native and introduced ranges are tristylous. However, recent surveys indicate that invasive populations can often be dimorphic, with ~25% lacking the S-morph. Two large-scale surveys conducted 25 years apart in Ontario, Canada involving >100 sampled populations indicate that despite the rapid increase in abundance of the species, genetic drift and founder events still play a dominant role in governing patterns of morph-frequency variation. In contrast, a survey of 96 populations along a latitudinal gradient in the Iberian Peninsula found that despite increasingly smaller populations towards the more arid southern range margin, 95% of populations were trimorphic. The maintenance of tristyly in European native populations may be aided by the genetic connectivity of populations in the agricultural landscapes where they reside, owing to gene flow by pollen and seed dispersal. Our studies demonstrate that tristyly can be used to provide insights into the relative importance of founder events, genetic drift and frequency-dependent selection in governing patterns of variation at a conspicuous genetic polymorphism of known adaptive significance.

T5-04-04

Integrating micro- and macro-evolutionary perspectives on heterostyly: Case studies from primroses

Elena Conti[1], *Barbara Keller*[1], *Spyros Theodoridis*[1], *Jurriaan De Vos*[2]
1. University of Zurich
2. Royal Botanic Gardens, Kew

The evolution of flowers opened unprecedented opportunities for complex interactions with pollinators, enabling new types of reproductive barriers, generally favoring outcrossing, and promoting diversification. Outcrossing increases genetic diversity, potentially mitigating the detrimental effects of environmental change and favoring long-term resilience to extinction. One of the best-studied floral syndromes enforcing outcrossing is heterostyly. Repeatedly appearing in angiosperm evolution and extensively studied in *Primula* L. (Primulaceae), heterostyly is a genetically-controlled floral polymorphism, whereby individuals in a population produce complementary types of self-incompatible flowers with anthers and stigmas in spatially matching positions. Conversely, homostylous species produce monomorphic, self-compatible flowers. Comprising ca. 90% heterostylous and 10% homostylous species that differ markedly in floral morphology, reproductive strategies, and size of distributional ranges, Primulaceae (ca. 730 species) represent an ideal group to investigate fundamental questions on how specific floral syndromes affect reproductive strategies, evolutionary dynamics, and biogeographic processes. By combining evidence from experimental pollination studies, quantitative morphological analyses of floral traits, different types of modeling approaches, phylogenomics, and biogeographic/demographic analyses, we integrate micro- and macro-evolutionary perspectives on the effects of heterostyly in primroses. It has been observed that several heterostylous groups appear to be especially species-rich, thus we wanted to know whether and how heterostyly might be implicated in explaining this phenomenon. We started to address this question by investigating how heterostyly works at the intra-specific level in the florally dimorphic *Primula* system. We confirmed the general predictions of the disassortative pollination hypothesis, but found that long-styled flowers export more pollen to non-reciprocal than reciprocal stigmas. We then switched to the inter-specific level and asked whether distyly might contribute to reproductive isolation via mechanical barriers. We discovered that distyly supports weak mechanical isolation, but only between high sexual organs, thus documenting, for the first time, the existence of morph-dependent, in addition to species-dependent asymmetries in reproductive barriers. Finally, we moved to the macro-evolutionary level and asked whether the evolution of heterostyly in Primulaceae is associated with an increase of diversification rates. We found that heterostyly accelerates diversification not via spurring speciation, but via decreasing extinction, possibly because it enforces outcrossing, thus favoring lineage survival in varying environmental conditions. Finally, we tested whether repeated switches from heterostyly to homostyly in *Primula* are associated with the so-called selfing syndrome. We discovered that floral traits involved in pollinator attraction and interaction are indeed smaller in homostylous, self-compatible species than in distylous, obligatorily outcrossing species. Most recently, we employed whole-genome analyses to study how the distributional ranges of distylous primrose species respond to climatic oscillations and to generate the first draft genome of any heterostylous species, a tool that has been crucial to progress on the molecular characterization of the heterostyly supergene. Current research exploits the newly available genomic resources to reconstruct a data-rich phylogeny of Primulaceae, the evolutionary steps leading to the assembly of the heterostyly supergene and repeated switches to homostyly, and how hybridization and introgression affect different parts of the genome in heterostylous vs. homostylous primroses.

T5-04-05

Flower number - flower size relations in Montiaceae in the context of reproductive adaptation across environments

Jurriaan De Vos[1,2,3], *Lawrence Harder*[2], *Erika Edwards*[3]
1. Royal Botanic Gardens, Kew
2. Department of Biology, University of Calgary
3. Department of Ecology and Evolutionary Biology, Brown University

Many plant traits do not evolve independently of other traits, rather, suites of traits often evolve in concert. A particularly striking example is the frequently negative relation between the number of flowers a plant produces and the investment in each, leading to flower number-flower size "trade-off"-like relations. Although common across angiosperms, their ecological relevance and environmental drivers remain poorly understood. In this talk, we present results from Montiaceae (Caryophyllales) and compare these to patterns across angiosperms. We employ a baited-phylogenomic marker system to elucidate the phylogeny of Montiaceae and reveal shifts in evolvability of floral traits: in some clades, carpel and stamen numbers have greatly evolved among species and correlate with flower number, in other clades, gynoecium and androecium structure are conserved. We analyse the ecological context of correlated evolution of flower number and size traits and find that evolution of large- (but few-) flowered species is associated with transitions into extremely seasonal environments, such as high-elevation deserts. We compare these results to other taxonomic groups to investigate whether reproductive adaptation at the inflorescence level across environments is a common evolutionary trend.

T5-04-06

Biogeography and ecology of style polymorphisms across spatial and temporal scales in the Mediterranean

Juan Arroyo[1], *Rocio Santos-Gally*[2], *Rocio Perez-Barrales*[3], *Jose Ruiz-Martin*[1], *Daniel Barranco*[1], *Marcial Escudero*[1]

1. University of Seville
2. Institute of Ecology, UNAM
3. School of Biological Sciences, Univ. of Portsmouth

Style polymorphisms are characterized by the presence of two or more discrete morphs within plant populations and have been interpreted as a mechanism to promote disassortative mating. Like most of breeding systems, they have a context-dependent nature and their evolutionary success is variable depending on the shifting ecological conditions in time and space. We will focus here on heterostyly and related polymorphisms, such as stigma height dimorphism, in a specific geographic and ecological setting, the Mediterranean basin, where environmental changes have been the rule due to its transitional stage between temperate and tropical biomes and its strong historical constraints. Particularly, both long- and short-term changes have left a mark in the distribution of these polymorphisms. The geomorphological and climatic scenario, dated back to the late Tertiary, is associated with the evolution of style polymorphism across species in some Mediterranean lineages. More recently, the climatic heterogeneity and oscillation during the Pleistocene has provoked changes in the populations of related style-polymorphic species, expressed in their morph ratios. These changes are associated to deviations from typical heterostyly (perfect reciprocal placement of anthers and stigmas linked to heteromorphic incompatibility and ancillary traits). However, heterostyly in these Mediterranean lineages is rarely broken down to homostyly, which is usually linked to hybridization, polyploidy and selfing in northernmost latitudes. We review the available evidence based on several studies of Mediterranean plant lineages where style polymorphism is present at variable frequency in different conditions. We will focus on both at macro- and microevolutionary levels. *Narcissus* is an old pre-Mediterranean genus that underwent several pulses of diversification in critical periods of the Mediterranean history. This affected differently the geographic and ecological distribution of clades and the representation of style polymorphisms. Species that have experienced considerable range shifts are quite labile in their polymorphism expression at population level. Thus, morph ratios are extremely variable, including the complete loss of polymorphism, due to concomitant changes in pollinator identity and frequency. Boraginaceae is a plant family with a high diversity of breeding systems, which are well exemplified in Mediterranean and European lineages. Studies in *Anchusa*, *Pulmonaria*, *Glandora* and *Lithodora* showed a high diversity of conditions, whose dynamic nature does not fit with traditional models of heterostyly evolution. Despite this fascinating diversity, we are far from understanding its ultimate and proximate causes. *Linum*, flax species in Linaceae, is a sub-cosmopolitan genus with its highest diversity in the Mediterranean. Although it was one of the paradigmatic cases where Darwin put his interest in the topic, recent studies have revealed that this genus includes more variation in the expression of the polymorphism than formerly presumed. As far as we know, the variation affects morphological aspects more than incompatibility systems. This offers an opportunity to study the implications of variations in sex organ deployment affecting reciprocal placement of sex organs and thus the functioning of these style polymorphisms.

T5-05: Plant genome evolution from the very beginning (two sessions)

T5-05-01

How plastids shaped (land) plant genomes and physiology

Jan De Vries[1], *Sven Gould*[2], *Stefan Rensing*[3], *John Archibald*[1]
1. Department of Biochemistry and Molecular Biology, Dalhousie University
2. Molecular Evolution, Heinrich-Heine-University Düsseldorf
3. Plant Cell Biology, Faculty of Biology, University of Marburg

The endosymbiotic origin of plastids is well known as having afforded eukaryotes the ability to harness the energy of the sun. However, the evolutionary impact of plastid evolution extends far beyond photosynthesis. At its origin, thousands of genes were transferred from the cyanobacterial endosymbiont to the nuclear genome of the host, cementing the relationship between host and cyanobiont. While maintaining the plastids' biochemical processes rests upon a tight communication between plastid and nucleus, little is known about this 'cross-talk' beyond land plants. We recently reported that higher branching streptophyte algae share with land plants the lack of a few essential plastid genes that have been transferred to the nucleus. Compared to those of other algae, the plastids of higher streptophytes ought to be more submissive to nuclear control. These organelle-associated changes, we proposed, allowed for an elaborate plastid-nucleus signaling network to evolve. The evolution of such a network aided in dealing with plastid-associated stressors such as high irradiances. Overcoming such stressors might have been a major selective advantage during plant terrestrialization. Ultimately, the increase of nuclear control over plastid function along the trajectory of streptophyte evolution made the plastid of vascular plants plastic; thanks to this plasticity, vascular plant plastids can differentiate into an unparalleled number of different types within a single species. What this suggests is that the embryophytic control over plastid function evolved within the streptophyte algae. Using comparative functional genomics across the diversity of streptophyte algae, we found an increased repertoire of retrograde signaling factors in higher streptophyte (algae). This advanced plastid-nucleus communication has implications beyond the regulation of photosynthesis. Plastids are not only photosynthesis machines; they are at the centre of a vast range of land plant secondary metabolic processes. Many of the canonical phytohormones, such as abscisic or jasmonic acid, are derived from biochemistry occurring within the plastid. The genetic potential to synthesize and sense them is found among streptophyte algae. This prompted us to design a large-scale transcriptomic comparison of both higher and lower streptophyte algae to study how the regulation of streptophyte plastid stress signaling evolved. Our data improves our understanding of their responsiveness to plastid-relevant treatments. Plastids are at the heart of land plant biology. For a comprehensive picture of what defines land plants and what made their success on land possible, it is critical to develop a comprehensive understanding of the lineage-specific adaptations of their plastids and those of their closest relatives.

T5-05-02

The prasinophyte - land plant connection: Opportunities for insights into evolutionary processes within the green lineage

Charles Bachy, ***Alexandra Worden***
Monterey Bay Aquarium Research Institute

Progress in sequencing genomes of photosynthetic microbial eukaryotes in recent years, from model chlorophytes (e.g., *Chlamydomonas reinhardtii*) to ecologically relevant marine phytoplankton (e.g., *Micromonas pusilla* and *Micromonas commoda*), has improved our understanding of algal evolution, biological processes and ecology. Here, we will present highlights from lab and field based studies of marine green algae as well as genomic comparisons to land plants and other algae. The prasinophyte *Micromonas* reveals 'new' suites of genes that now appear to be ancestral among green lineage taxa although lacking in model chlorophytes. In addition to several transcription factors, these include the photosensory protein phytochrome, a master regulator in land plants. Additionally, patchy loss patterns in the peptidoglycan pathway (derived from the cyanobacterial endosymbiont) suggest that plastid division and structure differs among closely related green lineage taxa. Here, we will discuss these and other recent discoveries as well as RNA-seq and proteomic functional studies of prasinophyte algae.

T5-05-03

Insights into land plant evolution garnered from the *Marchantia* genome project

John L. Bowman[1], *Takayuki Kohchi*[2], *Katsuyuki Tom Yamato*[3]
1. Monash Uni
2. Kyoto University
3. Kindai University

The origin of a land flora had profound consequences on the global environment and altered the evolutionary trajectories of other major clades of life. Land plants evolved from an ancestral charophycean alga from which they inherited numerous developmental, biochemical, and cell biological features. Characteristics of extant charophycean algae suggest that at least some of the physiological adaptations to land evolved prior to the origin of land plant. For example, several lineages can occupy terrestrial habitats, tolerate UV radiation and respond to desiccation similarly to land plants. In addition, many gene families found in land plants, but not in the largely marine Chlorophyta algae, have their origins in the charophycean algal grade. However, the origin and diversification of land plants involved dramatic evolutionary changes in life history, physiology, and body plan. The three earliest diverging lineages of extant land plants (liverworts, mosses, hornworts) are likely a paraphyletic assemblage, collectively called the bryophytes, which lack vascular tissues and true roots but collectively acquired all key features associated with land plant evolution (1) the origin of a multicellular diploid sporophyte, (2) a gametophytic apical meristem with an apical cell that produces 3-dimensional tissues, a sporophytic shoot apical meristem (SAM) with a capacity for branching, and (3) cell fate specializations providing morphological and physiological adaptations to life on land. The phylogenetic relationships of the bryophytes remain contentious, with nearly every possible topology proposed based on different molecular datasets. However, fossils assigned to liverworts (Ordovician-Silurian) predate those described as either mosses (Carboniferous) or hornworts (Cretaceous) and evidence from systematics, and phylogenetic studies suggest that the first plants that colonized terrestrial environments possessed attributes of liverworts, and it appears that extant liverworts retain a larger suite of ancestral land plant characters than any other lineage. Due to its ease of growth and genetic manipulation in the laboratory and its extensive historical literature, we have chosen *Marchantia polymorpha* as a representative liverwort and present an analysis of its genome, in comparison to those of charophycean algae and other land plants, including the moss *Physcomitrella patens*.

T5-05-04

Hornworts, a missing link to study fundamental questions of plant biology

Peter Szovenyi
University of Zurich

The monophyletic group of hornworts is believed to represent the immediate sister group of all vascular land plants. However, this traditional view is still debated and cannot be satisfactorily resolved owing to the lack of detailed knowledge on the general biology and genomic features of hornworts. Until now, advancement in this field was primarily hindered by the lack of genomic resources for a hornwort model species. Here we provide an update on the efforts towards a high-quality genome draft of the model hornwort, *Anthoceros agrestis*, and some of its relatives. We show that *A. agrestis* has a remarkably small genome, with few recent paralogs, which makes it approriate for genetic analysis. We also provide an overview of the *A. agrestis* gene space and a preliminary gene expression atlas which shed light on the regulation of morphological and developmental traits that are either shared with other embryophytes or unique to hornworts. Furthermore, we report our first achievements on the genetic transformation of *A. agrestis* using various techniques. Finally, we summarize our achievements and provide a list of issues that need to be resolved in the future.

T5-05-05

Unfurling monilophyte genomics: how polyploidy, transposable elements, and the alternation of independent generations drive fern evolution

Daniel Blaine Marchant[1,2], *Emily B Sessa*[1], *Paul G Wolf*[3], *Pamela S Soltis*[2,4], *Douglas E Soltis*[1,2,4]
1. Department of Biology, University of Florida
2. Florida Museum of Natural History
3. Department of Biology, Utah State University
4. Genetics Institute, University of Florida

Ferns (Monilophyta) are notorious for possessing large genomes (average 1C = 12 pg) and numerous chromosomes (average n = 59). Historically, these features, which characterize all homosporous ferns, were attributed to repeated rounds of ancient polyploidy (whole-genome duplication - WGD). However, this explanation for the origin of large genomes and many chromosomes, long considered emblematic of ferns, is unsubstantiated due to the absence of a sequenced fern genome. The lack of this crucial resource has not only hindered investigations of the evolutionary processes underlying the unusual genome characteristics of ferns, but has also impeded synthesis of genome evolution across land plants. Using the model fern species *Ceratopteris richardii*, or

C-Fern, we obtained the first sequenced and assembled nuclear genome for a homosoporous fern. With this novel resource, we evaluated the possible roles of WGD and the expansion of transposable elements (TEs) in shaping fern genome evolution. In addition, we investigated the ramifications of having two independent life phases (gametophyte and sporophyte generations) at the genome, gene, and transcript level. Not only do these results address the evolutionary processes and genomic changes that have shaped the genomes of ferns, but the assembled data set will also facilitate new insights into evolutionary genomics of land plants as a whole.

T5-05-06

The contributions of hybridization and polyploidization to hybrid establishment in natural populations of homoploid and allotetraploid *Selaginella arizonica* x *eremophila*

Anthony Baniaga[1], *Nils Arrigo*[1,2], *Tara Hall*[1], *Linnea Herbertson*[1], *Michael Barker*[1]
1. Department of Ecology & Evolutionary Biology, University of Arizona
2. Department of Ecology & Evolution, University of Lausanne

At least 13% of plant species are allopolyploids formed via hybridization and genome duplication. Despite the frequency of allopolyploidy, the impacts of hybridization versus genome duplication in shaping allopolyploid genomes remain poorly understood. A major limit to understanding the different impacts of hybridization and polyploidy is the absence of a study system with diploid and polyploid hybrids from the same parental species cross. Here, we describe such diploid and polyploid hybrids from the lycophyte genus *Selaginella* [Selaginellaceae]. In the Sonoran desert of the Southwestern U.S., we analyzed diploid hybrids and allopolyploids formed between crosses of *Selaginella arizonica* and *S. eremophila*. Both types of hybrid populations occupy drier, hotter climatic niches than either of their parents and have different ecophysiological strategies related to water use efficiency and relative growth rate. Using a combination of next-generation sequencing techniques including Illumina, PacBio, and ddRADseq, we dissect the ancestry of these hybrid taxa, the relative contributions of hybridization and polyploidization to the respective genomes, and highlight the evolutionary dynamics of their resurrection phenotypes that may have aided in their establishment and persistence in novel ecological niches.

T5-05-07

Primitive mechanisms to adapt land environments revealed by the genome of a charophyte alga *Klebsormidium flaccidum*

Koichi Hori, *Satoshi Kondo, Kinuka Ohtaka, Yuko Sasaki-Sekimoto, Mie Shimojima, Hiroyuki Ohta*
Tokyo Institute of Technology

We have reported the draft genome sequence of *Klebsormidium flaccidum* (Hori et al 2014). Comparative genome analysis revealed that the genome encodes more than one thousand of land plant type genes which may be associated with its adaptation to harsh land environments. Here, we will introduce their unique intra and extracellular lipid components. By comparing genes involved in the lipid biosynthetic pathways of *Arabidopsis thaliana* with *K. flaccidum* genomes, we identified wax-related genes in this alga. A simple and easy extraction method was developed for the recovery of the surface lipids. Although this alga has wax components, its surface lipids were largely different from those of land plants. We also investigated aliphatic substances in the cell wall fraction of *K. flaccidum*. Many of the fatty acids were determined to be lipophilic monomers in *K. flaccidum*, and a Fourier transform infrared spectroscopic analysis revealed that their possible binding mode was distinct from that of *A. thaliana*. We propose that *K. flaccidum* has a cuticle-like hydrophobic layer composed of lipids and glycoproteins, with a different composition from the cutin polymer typically found in land plant cuticles. We will also introduce our recent study in primitive cellular systems in this microorganism.

T5-05-08

Lycophyte evolution and development: A transcriptomic and genomic approach

Cecilia Zumajo-Cardona, *Barbara Ambrose*
New York Botanical Garden

Genomic research in plants have been focused on angiosperms with 43 sequenced genomes currently available. However, genomic data outside the angiosperms are limited to 3 species: *Selaginella moellendorffii*, *Marchantia polymorpha* and *Physcomitrella patens* where the majority of the studies have been dedicated to *P. patens*. In order to expand the understanding of the genetic mechanisms that underlies the evolution of land plants our research has been concentrated on Lycophytes which occupy a key phylogenetic position within the land plants as sister to all other vascular plants. Features such as vasculature and a sporophyte dominant life cycle arose first in the common ancestor of all vascular plants and by investigating this lineage we will be able to better understand how these innovations arose. Lycophytes evolved several features in parallel to the remainder of the vascular plants, including roots, leaves (microphylls), the tree habit (wood formation), heterospory and endosporic gametophytes. Thus, it is necessary to elucidate the molecular genetic networks underlying the development of these key innovations such as the leaves in Lycophytes because they are integral to understanding the evolution and development of vascular plants-the dominant component of the land flora-and economic botany. In Lycophytes, the genome of *S. moellendorffii* is available and we have sequenced the genome of *S. apoda*. In addition to the whole genome sequence of *S. apoda*, we have transcriptome data from leaves, shoots, roots, cones and rhizophores. Transcriptome analyses will allow us to make comparisons between the two different species of lycophytes but also to investigate the genes specific to root, cone, or rhizophore development. With whole genome analyses in lycophytes we have investigated gene families known to be important for body plan specification such as the MADS-box family of transcription factors. We studied the evolution and expression patterns of MADS-box genes to assess large scale and local duplications and evaluate protein motifs likely related with functional changes in lycophytes. We have found that the MADS-box genes have only 2 independent duplications in lycophytes one for Selaginellales and the other in Lycopodiales. Analyses of the conserved domains show that the lycophytes share the M and K domains with angiosperms and outside those regions we have been able to identify 5 new motifs specific to lycophytes where one is exclusive to Lycopodiales and the other to Selaginellales suggesting that even between lycophytes MADS-box genes are likely involved in different functions. Expression analyses by *in situ* hybridization of the three MADS-box genes in *S. moellendorffii* showed differential expression patterns for each gene, and we are performing similar expression analyses in additional lycophyte species. Genomic and transcriptomic analyses in lycophyte species have allowed us not only to investigate the development of lycophytes but also to provide unparalleled insight into the evolution and development of all vascular plants.

T5-05-09

Exploring algal phylogenomics: Genome sequencing and analysis in the araphid pennate diatom *Psammoneis japonica*

Matthew Parks[1], *Norman Wickett*[1], *Andrew Alverson*[2]

1. Chicago Botanic Garden

2. University of Arkansas

The hyper-diverse and ecologically important diatoms (Bacillariophyta) are unicellular phytoplankton with chimeric genomes derived from two photosynthetic (red and green algae) and one heterotrophic precursor lineages. Complete genome sequences are currently available for three diatom species, representing two of four major morphological types (polar centrics and raphid pennates). Here, we present the genome sequence and analyses for *Psammoneis japonica*, which is a member of a critical third diatom clade, the araphid pennates. The *Psammoneis* assembly comprises approximately 25× PacBio and 100× Illumina coverage and is stringently filtered for contaminant sequences. With a resulting total assembly size of 91.2 Mbp (N50 = 390 kbp) and over 15,000 annotated genes, the *Psammoneis* genome is the largest diatom genome sequenced thus far. Sequencing of this genome enables a more comprehensive, comparative genomic view of diatom ecological (pelagic/benthic), morphological (centric/pennate, araphid/raphid) metabolic (carbon/nitrogen cycles) and reproductive (oogamy/anisogamy/isogamy, dioecy/monoecy) character evolution. General comparisons, including orthologous protein divergence, intron conservation and repetitive element content also provide further insight into rates and patterns of evolution that have shaped genome size and architecture across the diatoms.

T5-05-10

Molecular and evolutionary conservation of ABA signaling for stomatal closure

Zhong-Hua Chen

Zhejiang University

Evolutionary trajectories of land plants have led to structurally complex and functionally active stomata for terrestrial life. The unique morphology, development and molecular regulation of stomata enable their rapid environmental response. Evolution of the molecular mechanism behind stomatal development and membrane transport has clearly drawn on conserved and sophisticated signaling networks common to stomata of all vascular plants and some mosses. ABA-driven stomatal regulation reportedly evolved after the divergence of ferns, during the evolution of seed plants around 360 Mya based on stomata of certain fern species being unresponsive to ABA, but exhibiting passive hydraulic control. However, growing evidence has shown that ABA signaling controlling stomatal closure may have evolved in some species of mosses. Here, we observed that a number of ABA signaling, membrane transporter, and ABA reception protein families diversified over the evolutionary history of land plants. Phylogenetic analysis of the key ABA signaling proteins indicates to an evolutionarily conserved stomatal response to ABA. Moreover, comparative transcriptomic analysis has identified a suite of ABA responsive differentially expressed genes in terrestrial fern species. These genes encode proteins associated with ABA biosynthesis, transport, reception, transcription, signaling, and ion and sugar transport, which fit the general ABA signaling pathway constructed from *Arabidopsis thaliana* and *Hordeum vulgare*. Furthermore, our stomatal assays have shown the first evidence for ABA-induced stomatal closure on epidermal peels of two terrestrial fern species. We thus propose that, despite the advantages of passive hydraulic control for fern stomata, the molecular mechanisms of ABA-induced stomatal closure may not have been lost in all fern species. Understanding this evolutionary trend will inform predictive modeling and functional manipulation of plant productivity and water use and will benefit future efforts towards food security and ecological diversity.

T5-05-11

Conservation and divergence of small RNA pathways and microRNAs in land plants

Chenjiang You[1,2,3], ***Jie Cui***[1,2], *Hui Wang*[4], *Xinping Qi*[5], *Li-Yaung Kuo*[6], *Hong Ma*[5], *Beixin Mo*[1], *Xuemei Chen*[1,3,7]

1. Guangdong Provincial Key Laboratory for Plant Epigenetics, College of Life Sciences and Oceanography, Shenzhen University, Shenzhen, Guangdong 518060

2. Key Laboratory of Optoelectronic Devices and Systems of Ministry of Education and Guangdong Province, College of Optoelectronic Engineering, Shenzhen University, Shenzhen, Guangdong 518060

3. Department of Botany and Plant Sciences, Institute of Integrative Genome Biology, University of California, Riverside, CA 92521

4. Shenzhen Key Laboratory of Southern Subtropical Plant Diversity, Fairylake Botanical Garden, Shenzhen & Chinese Academy of Sciences, Shenzhen, Guangdong 518004

5. Ministry of Education Key Laboratory of Biodiversity Sciences and Ecological Engineering and Collaborative Innovation Center for Genetics and Development, Institute of Plant Biology, School of Life Sciences, Fudan University, Shanghai 200433

6. Institute of Ecology and Evolutionary Biology, National Taiwan University, Taipei, 10617

7. Howard Hughes Medical Institute, University of California, Riverside 92521, CA

As key regulators of gene expression in eukaryotes, small RNAs have been characterized in many seed plants, and pathways for their biogenesis, degradation, and action have been defined in model angiosperms. However, both small RNAs themselves and small RNA pathways are poorly known in other land plants such as lycophytes and ferns, preventing a comprehensive evolutionary perspective on small RNAs in land plants. We selected 25 representatives from major lineages of lycophytes and ferns, most of which lack sequenced genomes, and characterized small RNAs and small RNA pathways in these plants through RNA sequencing. We identified homologs of *DICER-LIKE* (*DCL*), *ARGONAUTE* (*AGO*) and other genes involved in small RNA pathways, predicted over 2,600 conserved microRNA (miRNA) candidates and performed phylogenetic analyses on small RNA pathways as well as miRNAs. We predicted targets of miRNAs and performed degradome sequencing to examine the transcript-cleavage activities mediated by miRNAs. Pathways underlying miRNA biogenesis, degradation, and activity were established early in land plant evolution, but the 24-nt siRNA pathway that guides DNA methylation was incomplete in early-divergent land plants, especially lycophytes. We show that the functional diversification of key gene families such as *DCL* and *AGO* as observed in angiosperms occurred early in land plants followed by parallel expansion of the *AGO* family in ferns and angiosperms. We uncovered a conserved *AGO* subfamily absent in angiosperms. Our phylogenetic analyses of miRNAs in bryophytes, lycophytes, ferns, and angiosperms refined the time of origin for conserved miRNA families in land plants.

T5-05-12

Functional ecology of mycobiome shifts associated with plant senescence - linking environmental and experimental re-synthesis metatranscriptomics

Ko-Hsuan Chen[1], *Hui-Ling Liao*[2], *Hailee B Korotkin*[3], *P Brandon Matheny*[3], *Francois Lutzoni*[1]

1. Duke University

2. University of Florida

3. University of Tennessee

The same fungi are often found in healthy and dead tissues of the same plant species. Because plant tissues at different stages of senescence are usually separated by time and space, tracking functional transitions of their mycobiomes is often confounded with extraneous abiotic factors. The perennial moss *Dicranum scoparium* has a growth form with a continuous senescence gradient, making it an ideal system to study fungal trophic transitions. Our goal is to understand the function of fungi associated with tissues spanning from healthy photosynthetic tissues to advanced states of decomposition. A metatranscriptomic approach was first applied to obtain the metabolic (rRNA) and functional (mRNA) activities of fungi associated with top (photosynthetic), middle (senescent), and bottom (decomposing) layers for the naturally occurring *D. scoparium*. By looking at the top nutrient-related genes expressed in each layer, we found that the highly expressed genes in the top layer are mostly related to carbon exchange, whereas some highly expressed genes in the middle and bottom layers were involved in nitrogen and phosphate transport. In parallel to our metatranscriptomic study of environmental samples of *D. scoparium*, we isolated fungal strains from the three layers of this moss and grow *D. scoparium in vitro* from spores. By comparing and integrating the fungal rRNA sequences generated by metatranscriptome and culture-based approaches, we assigned fungi into three types of association categories: 1) High activity in photosynthetic tissues, 2) High activity in decomposing tissues and 3) Low activity throughout the gametophyte but abundant when using culture-based detection. Based on these association categories, seven fungal strains representing distinct fungal lineages (four Ascomycota, two Basidiomycota, one Mortierellomycotina) were selected to establish fungus-plant re-synthesis pairs in the laboratory with both living and dead gametophytes. For every fungus-plant pair, growth rates were monitored for two months after fungal colonies reached the moss. Metatranscriptomes of these fungus-plant pairs were obtained at the end of this period.

T5-06: Ontologizing botanical knowledge (two sessions)

T5-06-01

Bottom-up phenotype ontology building using OTO and ETC text capture and ontology building tools

Hong Cui[1], *Thomas Rodenhausen*[1], *Yujie Cao*[1], *James Macklin*[2], *Joel Sachs*[2]

1. University of Arizona

2. Agriculture Agri-Food Canada

Vast amount of valuable phenotypic character information has been published in the past, but due to the natural language nature of the character descriptions, they are not directly useful for biological computation. Domain ontologies, as a form of controlled vocabulary, have been recognized as a promising approach to integrate data from different sources, but in the realm of phenotypic characters, the majority of the biodiversity domains never had a standard vocabulary established, leaving authors tremendous freedom of describing phenotypic characters in various ways, for example "[shape] resembling a children's comic idea of a hand-grenade". Such issues have been widely recognized.While recognizing the large variations in terminology usage in published character descriptions, we emphasize that the primary descriptions are the ultimate source that must be covered by phenotype ontologies if we hope to bring these published knowledge to life. Towards this goal, we developed a set of web-based, public-accessible applications: OTO (Ontology Term Organizer) and ETC (Explorer of Taxon Concepts) toolkit, including Text Capture and Ontology Building. ETC Text Capture tools parse morphological descriptions on organisms of various taxon groups and extract phenotypic terms (such as anatomical structures, characters, and character states) in a semi-automated fashion. These terms can be exported to OTO directly from the ETC website. OTO provides a simple drag-and-drop user interface that allows multiple domain experts to categorize various types of terms to reach consensus, which is crucial for ontology building, because for an ontology to be a stable infrastructure for a domain, it must reflects the consensus of the domain. OTO provides a suite of features to support consensus making, including (1) accompanying terms with their source sentences to aid the user in discerning the meaning of a term, (2) visual highlighting of conflicting user decisions, (3) supporting user comments on terms and categorization decisions, (4) referencing definitions in other dictionaries/ontologies (e.g., PATO), (5) recording full user categorization history, and (6) basing the final decisions on input from all users. With categorization decisions being finalized on the terms, the ETC Ontology Building tool can be used by biologists to build a new ontology using the terms, or to add the terms to an existing ontology. A set of design patterns for constructing phenotype ontologies are built into the Ontology Building tool. When users invoke a pattern, a set of assertions will be added to the ontology automatically by software. In this talk we will present the bottom-up phenotype ontology building pipeline and the ontologies built using the pipeline. Through this talk, we hope to generate discussions on ways to improve tools for biologists/biology students to use to create ontologies as a computation infrastructure for systematics biology.

T5-06-02

Merging data across continents and scales with the plant phenology ontology

Ramona Walls[1], *Brian Stuckey*[2], *Kjell Bolmgren*[3], *John Deck*[4], *Ellen Denny*[5], *Robert Guralnick*[2]
1. CyVerse, University of Arizona
2. University of Florida
3. Swedish University of Agricultural Sciences
4. University of California at Berkeley
5. USA National Phenology Network

Plant phenology — the timing of plant life-cycle events, such as flowering or leafing-out — has cascading effects on multiple levels of biological organization, from individuals to ecosystems, and is crucial for understanding the links between climate and biological communities. Plant phenology data are collected and used by many different types of researchers, from plant breeders to ecosystem ecologists. Today, thanks to data digitization and aggregation initiatives, phenology monitoring networks, and the efforts of citizen scientists, more phenologically relevant data is available than ever before. Unfortunately, combining these data in large-scale analyses remains prohibitively difficult, mostly because the individuals and organizations producing the data are using non-standardized terminologies and metrics during data collection and processing. Lack of standardization remains particularly problematic for historical datasets, which are crucial for time-based analyses. Furthermore, phenology data are often recorded as a type of trait, such as whether or not a plant has young leaves or flowers present, making it difficult to associate the data to development stage terms in ontologies like the Plant Ontology (PO). The Plant Phenology Ontology (PPO) is a collaborative effort to help solve these problems by developing the standardized terminology, definitions, and term relationships that are needed for large-scale data integration. The PPO builds on the widely used PO and Phenotype and Trait Ontology (PATO) to promote broad reuse of phenological data. This talk will provide an overview of the PPO, including the history of how it came to be, design patterns used in the ontology, examples of annotating real phenological data, and plans for future development.

T5-06-03

Challenges presented by biodiversity ontologies in the context of building a botanical knowledge portal

James Macklin[1], *Joel Sachs*[1], *Hong Cui*[2], *Heather Cole*[1], *Andru Vallance*[3]
1. Agriculture and Agri-Food Canada
2. University of Arizona
3. Independent

We will describe the role of ontologies in our botanical knowledge portal, and will discuss some of the ontology-related challenges that we have overcome, and some that we continue to struggle with. Our portal uses Semantic Mediawiki (SMW) to store and display structured data extracted from the Flora of North America (FNA), and to integrate this data with an open biodiversity knowledge graph. SMW provides users with a familiar interface for browsing, filtering, and querying semantic data without the need to learn (or even be aware of) semantic web technologies such as RDF (the Resource Description Framework) and SPARQL (the SPARQL Protocol and RDF Query Language). In the case of the portal, this allows users to query taxonomic treatments based on: the characters and character states of taxa, including phenology and habitat; invasiveness and conservation status; occurrences; and any other information contained in the knowledge graph. Broadly, the ontology-related issues we have faced deal with representing, integrating, and reasoning over data about phenotypes, habitats, phenology, and establishment means (native vs. introduced). They fall into a number of categories, including: inter-curator variation in ontology construction; inter-curator variation in the use of ontologies to represent data; idiosyncrasies of applications that consume RDF; and, most broadly, heterogeneous (and sometimes incompatible) approaches to building the semantic web that have been pursued by different communities whose data products we want to integrate. We will describe these in the context of the portal, and will also describe the structure of our biodiversity knowledge graph, demonstrate how human and software agents (such as the portal) can interact with it, and invite collaboration in its continued construction.

T5-06-04

Towards consolidating the plant trait data domain

Jens Kattge[1,2], *Eric Garnier*[3,4], *Ulrike Stahl*[1,2,7], *Marie-Angelique Laporte*[3,4,6], *Ingolf Kuehn*[5], *Lutz Maicher*[8,9], *Stefan Klotz*[5], *Gerhard Bönisch*[1], *Christian Wirth*[1,2,10], *Sandra Lavorel*[11], *Sandra Diaz*[12]
1. Max Planck Institute for Biogeochemistry, Jena
2. German Centre for Integrative Biodiversity Research (iDiv) Halle-Jena-Leipzig
3. Centre d'Ecologie Fonctionnelle et Evolutive (UMR 5175), CNRS – Universite de Montpellier – Universite Paul-Valery Montpellier – EPHE, 34293 Montpellier Cedex 5
4. CEntre for the Synthesis and Analysis of Biodiversity (CESAB-FRB), 13100 Aix-en-Provence
5. Department of Community Ecology, Helmholtz Centre for Environmental Research – UFZ, Halle
6. Bioversity International, via dei Tre Denari, 174/a Maccarese, Rome
7. Julius Kühn-Institut, Bundesforschungsinstitut für Kulturpflanzen, Quedlinburg
8. Computer Science Department, University of Jena, 07743 Jena
9. Competitive Intelligence Group, Fraunhofer MOEZ, Neumarkt 9-19,04109 Leipzig
10. Special Botany and Functional Biodiversity, University of Leipzig, Leipzig
11. Laboratoire d'Ecologie Alpine (UMR 5553), CNRS – Universite Joseph Fourier, 38041 Grenoble
12. IMBIV (CONICET-UNC) & FCEFyN, Universidad Nacional de Cordoba, Cordoba

Plant traits-the morphological, anatomical, physiological, biochemical and phenological characteristics of plants and their organs-determine how primary producers respond to environmental factors, affect other trophic levels, influence ecosystem processes and services and provide a link from species richness to ecosystem functional diversity. Trait data thus represent the raw material for a wide range of research from evolutionary biology, community and functional ecology to biogeography and vegetation modeling. However, so far the value of plant traits is obscured by two roadblocks: ecological research produces a tremendous amount of trait data, but the diversity in topics covered and the ways in which studies are carried out result in large numbers of small, idiosyncratic data sets using heterogeneous terminologies. Such heterogeneity can be attributed, in part, to a lack of standards for acquiring, organizing and describing data. We here present recent progress in consolidating the plant trait data domain by standardized terminology (TOP thesaurus of plant characteristics: top-thesaurus.org) and data integration (TRY database: try-db.org). In combination with other activities these two initiatives may provide a relevant contribution towards ontologizing the plant trait data domain and providing consolidated trait data.

T5-06-05

FLOPO: A tool for integrating morphological traits and phenotypes of vascular plants, and bridging the link to genetics

Robert Hoehndorf[1], *Georgios Gkoutos*[2], *Quentin Groom*[3], *Marco Schmidt*[4], *Claus Weiland*[4]

1. King Abdullah University of Science and Technology

2. Institute of Biological, Environmental and Rural Sciences, Aberystwyth University

3. Botanic Garden Meise

4. Senckenberg Biodiversity and Climate Research Centre (BiK-F)

The systematic analysis of a large number of comparable plant trait data can support investigations into phylogenetics and ecological adaptation, with broad applications in evolutionary biology, agriculture, conservation, and the functioning of ecosystems. Floras, i.e., books collecting the information on all known plant species found within a region, are a potentially rich source of such plant trait data. Floras describe plant traits with a focus on morphology as well as other traits relevant for species identification in addition to other characteristics of plant species, such as ecological affinities, distribution, economic value, health applications, or traditional uses. However, a key limitation in systematically analyzing information in Floras is presented by the lack of a standardized vocabulary for the described traits as well as the difficulties in extracting structured information from free text. A similar problem arises with molecular biology datasets where traits and phenotypes are recorded for a variety of studies including mutant plant model organisms experiment, or plant case-control experiments. To address the data integration problem, the molecular biology community developed a number of ontologies, i.e., controlled, formal vocabularies that capture the basic concepts within a domain and their interrelations. These ontologies are used for data annotation and integration across databases. Combining the information about plant traits and phenotypes resulting from molecular biology studies with the information found in biodiversity texts, such as Floras, would render a large volume of trait data amenable to integration and computational comparative phenomics analyses that can provide insights into the genetics of wildtype plants. Currently such an endeavour is hindered by the lack of a similar ontology for plant traits as well as a method to combine it with the ontologies used in molecular biology. We have developed the Flora Phenotype Ontology (FLOPO), an ontology for describing traits of plant species found in Floras. Similarly to the majority of phenotype ontologies in molecular biology, we used the Plant Ontology (PO) and the Phenotype And Trait Ontology (PATO) to extract entity-quality relationships from digitized taxon descriptions in Floras. We have used a formal ontological approach based on phenotype description patterns and automated reasoning to generate the FLOPO. The FLOPO is primarily intended as a framework based on which plant traits can be integrated computationally across all species and higher taxa of flowering plants. Importantly, it is not intended to replace established vocabularies or ontologies, but rather serve as an overarching framework based on which different application- and domain-specific ontologies, thesauri and vocabularies of phenotypes observed in flowering plants can be integrated. We will outline the current uses of FLOPO in data integration and annotation, and will also showcase our preliminary work on using FLOPO to integrate phenotypes observed in a biodiversity and ecology context with mutant model organism phenotypes. In particular, the combination of genetically well studied crop plants with the ecological and trait diversity of crop wild relatives would be a possible field of application.

T5-06-06

Reasons and effects of the duality of term "woody"

Anna Stepanova

Komarov Botanical Institute RAS

The term "woody" have being used in many botanical disciplines. But special reviews show that this term has a duality of usage (Gatzuk, 1976; Dulin, Kirghoff, 2010; Schweingruber, Büntgen, 2013). 1. Woody plants from ecologo-morphological point of view. This notion has long history, but the main modern criterion of classification is the duration of life of aboveground shoot, positing the plants of the seasonal climates can be determined as woody, semi-woody and herbaceous. 2. From anatomical positions, under woody commonly understand plants with the secondary xylem. At the first look, it correlates with previous proposition (soft herbs without secondary xylem and hard trees and shrubs with secondary xylem). However, detail analyses shows that many herbaceous plants have cambium and secondary xylem. More other the whole ring of the xylem - a common trait of annuals. Under those two topics we will discuss: How the duality of term "woody" inhibits the theoretical generalizations in plant anatomy. How the wood of herbs looks like. What is the best way to use the term "woody".

T5-06-07

IPNI – future vision for providing Plant & Fungal names in a digital world

Abigail Barker[1], *Irina Belyaeva*[1], *Kanchi Gandhi*[2], *Rafaël Govaerts*[1], *Helen Hartley*[1], *Heather Lindon*[1]

1. Royal Botanic Gardens, Kew

2. Harvard University Herbaria & Libraries

For more than a century IPNI has been globally recognized as the primary source for plant name data. The proliferation of plant databases and the advent of electronic publication have radically changed the environment in which IPNI operates. As a response to these technological changes the IPNI partnership (Royal Botanic Gardens Kew, Harvard University Herbaria & Libraries and the Australian National Botanic Gardens), sets out its vision for the future of IPNI ensuring that we continue to be a globally recognized authority on names. IPNI is already widely used as a source of unique identifiers (LSIDs) and names data. In the future we will facilitate automatic updates across platforms for anyone who has linked their data through these identifiers. We will be improving the download and search options over broader ranges of fields with more choices for viewing and sorting results. IPNI will provide a user friendly registration service to speed up the rate at which published and unpublished names are entered. We will also create the possibility within registration to allow for micro-publication of nomenclatural novelties. We aim to improve the completeness of IPNI by integrating data which at the moment reside in separate databases, many of which are at Kew; in particular fungal names and infraspecific plant names. IPNI will then be in a prime position to provide a names backbone, initially for Kew's Plants of the World Online and subsequently for external services. The IPNI partnership want to improve its relationship with users by creating an interactive feedback system and disseminating knowledge on nomenclature by providing online tutorials. We will work more closely with publishers to automate the entry of names from online publications. In formulating this vision of IPNI's future we will be consulting our users as well as the broader community to create a user friendly system fit for the challenges of the digital world.

T5-06-08

The Planteome project: Common reference and application ontologies for plant biology

Pankaj Jaiswal[1], *Laurel Cooper*[1], *Austin Meier*[1], *Justin Elser*[1], *Marie-Angélique Laporte*[2], *Christopher Mungall*[3], *Barry Smith*[4], *Elizabeth Arnaud*[2], *Seth Carbon*[3], *Dennis Stevenson*[5], *Brandon Sinn*[5], *Dario Cavaliere*[5], *Justin Preece*[1]

1. Oregon State University

2. Bioversity International

3. Lawrence Berkeley National Laboratory

4. University at Buffalo

5. New York Botanical Garden

The Planteome project (www.planteome.org) features a suite of reference ontologies for plants, associated with a growing corpus of genomics data in a centralized online plant informatics portal. The species-neutral references, the Plant Ontology, Plant Trait Ontology, and Plant Environment Ontology, (developed by the project), along with those developed by collaborating groups, such as the Gene Ontology and the Phenotype and Trait Ontology, and others, are mapped to species-specific controlled vocabularies for crop plant traits and phenotypes. Traditional plant breeding methods for crop improvement may be combined with next-generation analysis methods and automated scoring of traits and phenotypes to develop improved varieties. Linking these analyses to the growing corpus of genomics data generated by phylogenomics, taxonomy, high-throughput sequencing, transcriptomics, proteomics, phenomics and genome annotation projects requires common, interoperable, reference vocabularies (ontologies) for the description of the data. Data annotations to ontology terms link phenotypes and accessions/germplasm to genomics resources, and provide semantic integration of widely diverse datasets with the goal of plant improvement. Analysis and annotation tools are being developed to facilitate studies of plant traits, phenotypes, diseases, gene function and expression and genetic diversity data across a wide range of plant species. The project database and the online resources provide researchers tools to search and browse and access remotely via APIs for semantic integration in annotation tools and data repositories, providing resources for plant biology, breeding, genomics and genetics.

The project is supported by the National Science Foundation award IOS #1340112

T5-06-09

First steps to build an ontology for flagellate plants using natural language processing tools

Lorena Endara[1], *Gordon Burleigh*[1], *Hong Cui*[2]

1. Department of Biology, University of Florida

2. School of Information Resources and Library Science, University of Arizona

Novel Natural Language Processing (NLP) software has been designed to parse text written in the abbreviated language (i.e., telegraphic syntax) prevalent in taxonomic descriptions. The software greatly facilitates the extraction of phenotypic traits from taxonomic descriptions. Our aim is to create a comprehensive phenomic dataset for the flagellate plants (i.e., bryophytes, lycophytes, ferns, and gymnosperms) using the NLP technology. We are interested in building a formal ontology for flagellate plants, as incorporating a formal ontology to the parsing analyses will increase the number of usable characters in the matrices we build and the precision of the data extracted. We are using the Explorer of Taxon Concepts (ETC) pipeline (http://*etc*.cs.umb.edu/etcsite/start.htm) to parse taxonomic descriptions of gymnosperms and fern taxa. The ETC "Text Capture" tool enables users to load the taxonomic descriptions into the system. The ETC software then transforms the descriptions into XML format, segments the sentences, and identifies terms that describe structures or descriptors. The users must place unrecognized terms into discrete predefined categories, which allows the system to establish "is_a" relationships (e.g., blue is a color, leaf is a structure). The result of this step is downloadable as a .csv file. Once this task is completed, the system parses the text, and generates output files that have detailed-semantically annotated text. The output is then input into the ETC "Matrix Generation" tool to assemble a taxon by character matrix. Using the "Term Review" option of ETC's "Text Capture" tool, we have gathered terms that are not represented in Ontologies. We present the data gathered and discuss how we can integrate our work with other ontology building initiatives.

T5-06-11

An evolutionary continuity principle for evolutionary system of organism divisions

Da-Li Fu[1*], *Hao Fu*[2]

1. *Chinese Academy of Forestry*
2. *Central South University of Forestry and Technology*

Evolutionary research is guided by the scientific basis of organism evolution. To overcome the limitations of being partial and subjective in the tree of life or phylogenetic system, the authors proposed a basic principle of organism evolution, evolutionary continuity principle. Based on this principle, the evolutions can be divided into two basic forms: vertical evolution and horizontal evolution. The vertical evolution is that evolution of the structures and features of organisms from non-existent to entirety, from simple to complex, or from primitive to advanced. The evolutionary course of vertical evolution can be divided into two or three significantly different phases, such as plants: the non-vascular → the vascular, non-seed → Gymnospermophyta D. L. Fu & H. Fu, phyl. nov. → Fructophyta D. L. Fu & H. Fu, phyl. nov., and the animals: Proenteratozoa D. L. Fu & H. Fu, phyl. nov. → Coelenteratozoa → Euenterata, non-chordata and non-gnatha → Chordato-gnathozoa D. L. Fu & H. Fu, phyl. nov. → Vertebrognathozoa D. L. Fu & H. Fu, phyl. nov. The horizontal evolution is that evolution of structures and features from loose to compact, from inefficient to efficient, or from primitive to advanced, which can also cause significant organism changes, such as Nudembryophyta D. L. Fu (Plantae) → Proenteratozoa D. L. Fu & H. Fu, phyl. nov. (Animalia), Bacterophyla D. L. Fu & H. Fu, phyl. nov. → Acytophyla D. L. Fu. Based on the continuity principle, the new evolutionary taxonomical system of divisions was established and the evolutionary diagram was drawn. The new system includes 20 divisions in three kingdoms, and 11 of those divisions are new. These are: I. Regnum Microbia D. L. Fu: 1.Cyanoalgophyla D. L. Fu & H. Fu, phyl. nov., 2. Bacterophyla D. L. Fu & H. Fu, phyl. nov., 3. Acytophyla D. L. Fu, 4. Monoalgophyla D. L. Fu & H. Fu, phyl. nov., 5. Monomycophyla D. L. Fu & H. Fu, phyl. nov., 6. Eualgophyla D. L. Fu & H. Fu, phyl. nov., 7. Fungophyla D. L. Fu & H. Fu, phyl. nov.; II. Regnum Plantae Haeck.: 8. Nudembryophyta D. L. Fu, 9. Bryophyta, 10. Pteridophyta, 11. Gymnospermophyta D. L. Fu & H. Fu, phyl. nov., 12. Fructophyta D. L. Fu & H. Fu, phyl. nov.; III. Regnum Animalia L.: 13. Proenteratozoa D. L. Fu & H. Fu, phyl. nov., 14. Coelenteratozoa, 15. Nematodozoa, 16. Annelidozoa, 17. Arthropodozoa, 18. Echinodermatozoa, 19. Chordato-gnathozoa D. L. Fu & H. Fu, phyl. nov., 20. Vertebrognathozoa. D. L. Fu & H. Fu, phyl. nov. The new evolutionary theory, new evolutionary taxa and new evolutionary system can provide scientific bases for Evolutionomy, a new science of organism evolution.

T5-07: Extremophyte genomics

T5-07-01

Return to the sea: A rare evolutionary event for angiosperms

Jeanine Olsen
University of Groningen, Groningen Institute for Evolutionary Life Sciences (GELIFES)

The genome of *Zostera marina,* the first marine angiosperm to be fully sequenced (*Nature* 18 Feb 16) reveals unique insights into the genomic losses and gains involved in achieving the structural and physiological adaptations required for its fully marine lifestyle, arguably the most extreme habitat shift ever accomplished by flowering plants. Seagrasses returned to the sea some 75-100 Mya. *Zostera* underwent whole genome duplication (WGD) near the Cretaceous-Tertiary boundary thus opening new ecological opportunities in moving from freshwater to fully marine conditions. In returning to the sea, seagrasses found a vast new habitat free of terrestrial competitors, insect pests and drought stress; but they also had to adapt to high salinity, develop flexible leaves able to take up nutrients, and perform photosynthesis in the spectrally reduced light of the submarine environment. Here we present a number of highlights including modified cell walls that are more algal-like than plant-like, loss of all stomata and all of the genes, reduced plant defense genes and volatiles, and modifications for underwater pollination involving exineless pollen. The genome resource will advance a wide range of functional ecological studies from adaptation of marine ecosystems under climate warming, to unraveling the mechanisms of osmoregulation under high salinities that may further inform our understanding of the evolution of salt tolerance.

T5-07-02

Adaptations towards low phosphate soils: Insights from the extremophyte *Eutrema salsugineum* (aka *Thellungiella salsuginea*)

Elizabeth Weretilnyk, *Vera Velasco, Caitlin Simopoulos, Brian Golding, Peter Summers*
Dept. of Biology, McMaster University, Ontario, Canada

Model plants, including *Arabidopsis*, have provided an abundance of information about how plants assimilate and manage phosphate, particularly under phosphate deprivation. However, plants adapted to extreme environmental conditions will likely offer novel insights into mechanisms underlying nutrient assimilation given the challenges to which they have adapted. We applied physiological, lipidomic, and transcriptomic approaches to the study of *Eutrema salsugineum* plants exposed to phosphate-sufficient and phosphate-deficient growing conditions. *Eutrema* is a close relative of *Arabidopsis* but by comparison, *Eutrema* is highly tolerant to extremes in soil salinity, temperature, and water deficits. The Yukon *Eutrema* ecotype is also adapted to highly alkaline soils where phosphate content is typically low and its availability likely further compromised by an unfavorable soil cation composition. We found that Yukon *Eutrema* plants display a remarkable capacity to tolerate low phosphate conditions in that plants showed the same biomass and relative growth rates whether treated with 0 or 2.5 mM Pi. Factors including the capacity to acidify the rhizosphere, exude organic anions, and remodel metabolism, including through altered membrane lipid composition, were examined and found to diverge from mechanisms reported for *Arabidopsis* in showing fewer phosphate starvation-inducible traits. Interestingly, comparisons between the phenotypic features and transcriptomes of phosphate-deprived and sufficient plants suggests that *Eutrema* has a comparatively fixed, and inherently strong constitutive capacity to tolerate low phosphate and that this species likely only invokes "fine-tuning" to its baseline defenses in order to calibrate its response to an actual low phosphate stress. The baseline suite of traits and comparatively modest adjustments under phosphate stress could inform our selection of nutrient efficiency traits for crops as we move towards a challenging future shaped by global climate change. The traits displayed by *Eutrema* have already withstood testing under the challenging conditions of its native habitat and have likely been indispensable for the acclimation and survival of this species on low nutrient soils.

T5-07-03

Life on earth's poorest soils: Applying genomics and transcriptomics to unravel P and N nutrition in *Hakea prostrata* (Proteaceae)

Patrick Finnegan, *Hans Lambers, Albina Ilyasova, Ricarda Jost, Thirumurugen Kuppusamy, M. Asaduzzaman Prodhan, Kosala Ranathunge*
University of Western Australia

The flora of south-western Australia has evolved on the world's most P-impoverished soils. Proteaceae feature prominently, particularly on the most severely P-impoverished soils. We are using genomic and transcriptomic approaches to explore the molecular basis of adaptations in Proteaceae that allow them to thrive in low-nutrient environments, using *Hakea prostrata* as a model species. Proteaceae typically have very high P-acquisition efficiencies. They are able to mine P and other nutrients from soil through cluster roots. These dense clusters of short determinant lateral roots exude high quantities of carboxylates, as well as enzymes, to mobilise nutrients, particularly phosphate, from soil particles in the root zone. The need for the Proteaceae to focus on acquiring phosphate from P-poor soils may have led to a nutrient-acquisition selectivity that is unique among plants. That is, *H. prostrata* appears to be unable to down-regulate its P acquisition as effectively as other plants and easily becomes intoxicated. On the other hand, it seems to very tightly regulate the acquisition of nitrate and sulfate, which is again in stark contrast to most plants. Furthermore, increasing internal P in an artificial hydroponics experiment does not translate into an increased demand for nitrate or sulfate. This lack of response to internal P suggests that evolution has given *H. prostrata* the ability to maintain a low and set requirement for N and S that aligns with the extremely low level of P available in its habitat. Once P, N and S are within the plant, they are used with exquisitely high efficiency. Leaf development in *H. prostrata* displays delayed greening, where young leaves are brownish red or red, instead of green. During delayed greening, P and N investment in ribosomal RNA and protein is spread out over time. Early in leaf development, cytosolic ribosomal RNA predominates, and ribosomes are primarily dedicated to producing proteins for building leaf structure. After full expansion, the leaf greens, and the ribosomal RNA and protein pools are reconfigured to produce an active photosynthetic apparatus. During leaf maturation in Proteaceae, the lipid pool also changes dramatically, decreasing the use of phospholipids and increasing that of sulfolipids and galactolipids. Mature leaves, then, have a remarkably low P and N concentration, while photosynthetic rates are similar to those in plants with much higher leaf nutrient concentrations. This situation translates into very high photosynthetic P-use and N-use efficiencies. Finally, at the end of their organ life, P is highly efficiently and proficiently recycled during senescence. At this time, we have very little knowledge about the molecular mechanisms underlying these remarkable adaptations to a severely nutrient-impoverished environment. We have found that *H. prostrata* has fewer, and perhaps even lacks, genes whose expression responds to variations in nutrient availability in other species. Instead, constitutive orthologues predominate. From these considerations, it becomes apparent that the Proteaceae are ideal models for learning how plants cannot only cope with, but thrive in, nutrient-poor environments.

T5-07-04

Can *Schrenkiella parvula*, an extremophyte relative of *Arabidopsis*, help determine a genomic blueprint in adapting to boron toxicity?

Guannan Wang, *Maheshi Dassanayake*
Louisiana State University

Boron toxicity is a worldwide problem for crop production and significantly limits crop yield especially in Australia, Africa, and Asia. The genetic mechanism behind boron toxicity, as well as genomic signatures for adapting to high borate soils, are largely unknown. *Schrenkiella parvula* (syn. *Thellungiella parvula)* is an extremophyte, naturally adapted to the high borate soils in the shores of Lake Tuz, Turkey. *S. parvula* can survive high borate at levels toxic to most plants including *A. thaliana*. However, the *S. parvula* genome shares remarkable macro-synteny with the closely related genome of *A. thaliana*. Both species also share similar lifecycles. Therefore, *S. parvula* provides an ideal system to study the genomic blueprint for adapting to high borate levels in the soil. To identify the potential key genetic contributors for boron tolerance, we have compared RNA-Seq profiles of borate-treated and control samples of the shoot and root tissues from *S. parvula* and *A. thaliana*. Our goal was to explore the transcriptomic adaptations underlying boron tolerance with the integrated insight from genome comparisons. We observed a significantly reduced level of global expression changes in *S. parvula* to borate treatments compared to the response observed for *A. thaliana*. Clusters of differently expressed genes (DEGs) in *S. parvula* illustrated the stress readiness of the extremophyte transcriptomes compared to *A. thaliana* in a tissue-specific manner. One of the most highly expressed genes at the basal level expression in *S. parvula* was annotated as a putative boron transporter (*SpBOR5*) due to its sequence similarity to the Arabidopsis *AtBOR5*. *AtBOR5* is an uncharacterized gene in the family of putative borate transporters in *A. thaliana*. We found that *SpBOR5* could functionally complement the native yeast BOR1 transporter when expressed in a yeast *bor1* mutant background and SpBOR5 indicated strong boron transporter activity compared to closely related homologs in *A. thaliana* and *S. parvula*. The basal level expression of *SpBOR5* was > 500-fold higher than that of the ortholog, *AtBOR5*, and was even more up-regulated upon borate treatment in *S. parvula*. Interestingly, one of the key genomic structural variations identified from the genome comparisons indicated a 15 kb genomic insertion upstream of the *SpBOR5* gene that is absent in other closely related genomes. This may contribute to the significantly high expression of *SpBOR5*. We have developed loss of function mutants of *SpBOR5* via the CRISPR-Cas9 genome editing system to test whether *SpBOR5* plays a significant role in the borate tolerance exemplified by *S. parvula*. We plan to present the underlying genetic mechanisms of boron tolerance inferred from transcriptionally differently regulated pathways and the functional significance of putative borate transporters using targeted transgenic lines in *S.*

parvula.

T5-07-05

Transcriptome analysis of the *Arabidopsis* Negev desert relative, *Anastatica hierochuntica* (True Rose of Jericho), reveals stress tolerance and developmental determinants that might facilitate an opportunistic extremophyte lifestyle

*Gil Eshel[1], Dong-Ha Oh[2], Ruth Shaked[1], Vered Chalifa-Caspi[1], Shirli Bar-David[1], Aaron Fait[1], Maheshi Dassanayake[2], **Simon Barak**[1]*
1. Ben-Gurion University of the Negev
2. Louisiana State University

The search for novel stress tolerance determinants has led to increasing interest in plants native to extreme environments – so called "extremophytes." One successful strategy has been comparative studies between *Arabidopsis thaliana* and its extremophyte Brassicaceae relatives. Here, we present the Negev desert species, *Anastatica hierochuntica* (True Rose of Jericho), a member of the poorly investigated lineage III Brassicaceae. We show that *A. hierochuntica* has a genome approximately 4.5-fold larger than *Arabidopsis*, divided into 22 diploid chromosomes, and demonstrate that *A. hierochuntica* exhibits tolerance to heat, low nitrogen and salt stresses that are characteristic of its habitat. To investigate molecular mechanisms of *A. hierochuntica* stress tolerance, we generated a high quality *A. hierochuntica* reference transcriptome, and performed a phylogenomic analysis, based on 13,927 orthologous sequences of 17 Brassicaceae plant species including various crop plants, extremophytes and *A. thaliana.* We identified positively selected genes that may have been crucial for adaptation to abiotic stresses, and uncovered evidence suggesting that selection pressures leading to adaptation and speciation focus on common biological processes although the identity of the positively selected genes is specific to each species. We also identified positively selected genes that are specific to *A. hierochuntica*, and may underlie some of the morphological and life history characteristics that allow an opportunistic existence in the harsh desert environment. Lastly, we compared *A. hierochuntica* and *A. thaliana* responses to single and combined, low nitrogen and heat stresses using parallel metabolomic and transcriptomic analyses. Preliminary transcriptome analysis of 16,266 orthologous gene pairs between *A. thaliana* and *A. hierochuntica* revealed species-specific transcriptomes responses, which will be further discussed.

T5-07-06

Local adaptations of neotropical mangrove species discovered through comparative ecophysiology, transcriptomics and genomic scan for selection

***Mariana V. Cruz**[1], Gustavo M. Mori[2], Caroline S. Müller[1], Dong-Ha Oh[3], Maheshi D. Dassanayake[3], Rafael S. Oliveira[1], Anete P. Souza[3]*
1. Campinas State University (UNICAMP)
2. Coastal Campus/São Paulo State University (UNESP)
3. Louisiana State University (LSU)

Mangrove trees inhabit intertidal zones spanning wide ranges of latitude, climate, salinity, nutrients and fresh water availability. The varying environmental conditions to which species are exposed confer different selective pressures, favoring the emergence of locally adapted populations. Recent population genetics studies revealed strong genetic structure in mangroves from the Atlantic coast of South America. However, these investigations used only neutral genetic markers. The variability of functional traits in these populations had never been studied, despite their importance in determining plant survival in a rapidly changing world. Here, we used public data from weather stations of the Brazilian National Meteorological Institute, *in situ* RNA sequencing data (RNA-Seq), and a common-garden experiment for the comparison of environmental pressures and variations in adaptive characters between contrasting natural populations of Neotropical species of the genus *Avicennia.* Additionally, we have combined RNA-Seq and Genotyping by Sequencing data (Next-Seq) for searching for transcribed regions with signatures of selection - i.e. that showed significantly higher or lower genetic differentiation between populations than it would be expected under neutrality. We observed substantial differences in anatomical and physiological traits related to water and carbon balances between individuals from two genetically different populations grown in common-garden. Also, we have found differential expression of transcripts involved in the response to functionally relevant environmental variables contrasting between populations. Genetic markers showing evidence of selection were mapped to putative genes related to the distinct abiotic conditions observed in both sampling sites. Our multi-disciplinary approach revealed an existing overlap between environmental variation and adaptive genetic diversity in the studied species, shedding light on the species' potential to bear novel climate conditions.

T5-08: Chromosome evolution and introgression: past, present and future

T5-08-01

The nature, evolution and large-scale organization of repetitive DNA on the chromosomes of *Chrysanthemum*

***Fengjiao Zhang**[1,2], Nianjun Teng[1], Fadi Chen[1], Trude Schwarzacher[2], Pat Heslop-Harrison[2]*
1. Nanjing Agricultural University
2. University of Leicester

Tandemly repeated DNA and transposable elements make up the majority of most higher plant genomes, and can be identified in clones of sequence assemblies by their structure and similarities to known repetitive DNA motifs. High-throughput sequencing allows survey of all the DNA in a genome, but whole-genome assembly leaves a substantial fraction of highly repeated sequences which cannot be assembled. *Chrysanthemum nankingense* (2n=2x=18; genome size c. 2,500 Mbp) in the Asteraceae is an interesting species with value for breeding and a high-value food and considered as an ancestral species in the genus. We aimed to characterize in detail the major repetitive DNA motifs in *Chrysanthemum* unprocessed raw reads from sequencing using bioinformatics and fluorescent *in situ* hybridization. A combination of graph-based clustering using the program RepeatExplorer and frequency analysis of DNA motifs k-bases long (k-mer analysis where k ranged from 16 to 128 bases) was used to identify repetitive motifs in 12.3 Gb of 250 bp paired-end Illumina reads. Apart from rDNA, a few microsatellites, and telomeric sequences, no highly abundant tandem repeats were identified from the raw reads. There were many transposable element sequences: 40% of the genome had recognizable domains related to TEs. The retroelements showed widespread distribution over chromosomes, while each family had characteristic features such as abundance or exclusion from centromeric or sub-telomeric regions. Another group of very abundant repetitive sequence motifs, including the most abundant families, showed no similarity to known sequence motifs or tandemly repeated elements. Preliminary analysis is suggesting most are related to degenerate and presumably ancient LTRs (Long Terminal Repeats) of retroelements. Better approaches to characterization of repetitive DNA in all species is critical to generate complete genome sequence assemblies. Identifying the presence, nature and genomic organization of major genome fractions, enables inference of evolutionary history, including degeneration and loss, critical to understanding biodiversity and diversification processes in plants.

T5-08-02

Revealing *Brachypodium* chromosome structure and evolution as well as stability after mutagenic treatment using multicolour FISH and BAC-FISH-based comparative chromosome barcoding

Robert Hasterok, *Arita Kus, Jolanta Kwasniewska, Joanna Lusinska, Elzbieta Wolny*

Department of Plant Anatomy and Cytology, Faculty of Biology and Environmental Protection, University of Silesia in Katowice

Modern molecular cytogenetics amalgamates various methodological approaches of cytology, molecular genetics and advanced digital image acquisition, processing and analysis. It focuses on the study of nuclear genomes at the microscopic level. Many plants, including economically and ecologically important cereals and forage grasses, usually have large genomes saturated with repetitive DNA, which hampers detailed molecular cytogenetic analyses. Therefore, the cytomolecular organisation of the vast majority of plant genomes is still rather poorly understood, compared to that of many animals, mammalian genomes in particular. Model organisms possess a set of features, which makes them more amenable to scientific investigation than others. One of relatively recent and rapidly developing model systems contains representatives of the *Brachypodium* genus, particularly *B. distachyon*. This genus comprises 15-20 species with small, and in some cases, sequenced and resequenced genomes with a low repeat content, different basic chromosome numbers, size, morphology and ploidy levels. These species also have interesting phylogeny, short life cycles and simple growth requirements, and are subject to a rapidly and continuously growing repertoire of experimental tools. Here we focus on chromosome rearrangements that have shaped the structure of *Brachypodium* karyotypes, using comparative chromosome barcoding by BAC-FISH (fluorescence *in situ* hybridisation with chromosome-specific bacterial artificial chromosomes as probes). The karyotypes of selected *Brachypodium* species were compared to the model grass *B. distachyon*. Single-locus BAC clones derived from *B. distachyon* genomic libraries were selected from the assemblies of FingerPrinted Contigs that had previously been assigned to the chromosomes of *B. distachyon*. This comparative chromosome barcoding approach can be used to study the organisation of karyotypes and reconstruct mechanisms of the chromosome rearrangements that have shaped the genome structure of extant grasses. Due to its unique cytogenetic properties and well-developed research infrastructure, *B. distachyon* is also a convenient system for analysing 'hot spots' (and 'cold spots') of DNA damage in nuclear genomes, and consequently could be used for environmental monitoring. Chromosome rearrangements are commonly identified using classical cytogenetic techniques. However, combining whole genome sequence information and BAC-FISH-based mapping technology enables comprehensive analysis of mutagenic effects at the chromosomal level and extends our understanding of the mechanisms responsible for chromosomal aberrations. We present the visualisation of mutagen-induced genome changes, including micronuclei formation and alterations of chromosome territories in interphase nuclei using fluorescence *in situ* hybridisation (FISH) with selected chromosome-specific BAC clones, as well as ribosomal DNA and chromosome region-specific (centromeric and telomeric) probes. Financial support from the National Science Centre Poland (grants no. 2012/04/A/NZ3/00572,2014/14/M/NZ2/00519 and 2015/18/M/NZ2/00394) is acknowledged.

T5-08-03

Extraction of the constituent subgenomes of the natural allopolyploid rapeseed (*Brassica napus* L.)

Bin Zhu[1], *Yuqin Tu*[2], *Pan Zeng*[1], ***Xianhong Ge***[1], *Zaiyun Li*[1]

1. Huazhong Agricultural University

2. Jiangxi Academy of Agricultural Sciences

As the dynamic nature of progenitor genomes accompanies the speciation by interspecific hybridization, the extraction of the constituent subgenome (s) from a natural allopolyploid species of long history and then restitution of the progenitor (s) provides the unique opportunity to study the genome evolution and interplay. Herein, the A subgenome from the allotetraploid oilseed rape (*Brassica napus* L., AACC) was extracted through inducing the preferential elimination of C-subgenome chromosomes in intertribal crosses and the progenitor *B. rapa* was restituted (RBR). Then by crossing and backcrossing RBR with *B. napus* donor, the C subgenome was in situ dissected by adding each of its nine chromosomes to the extracted A subgenome and establishing the whole set of monosonic alien addition lines (MAALs, C1~C9). RBR from spring-type *B. napus* genotype "Oro" expressed a phenotype resembling some type of *B. rapa* never observed before, but showed a winter-type flowering habit. This RBR had weaker growth vigor and suffered more seriously from biotic and abiotic stresses compared with Oro. This RBR had weaker growth vigor and suffered more seriously from biotic and abiotic stresses compared with Oro. The phenotypes specific for these MAALs showed the location of the related genes on the particular C-subgenome chromosomes. These MAALs exhibited obviously different frequencies in homeologous pairing and transmission of additional C-subgenome chromosomes, which were associated with the distinct degrees of their relatedness, and even with the possible genetic regulation for meiotic pairing evolved in *B. napus*. Using MAALs, large scaffolds undetermined for sequence assembly of *B. napus* were anchored to specific C-subgenome chromosomes. In different addition lines, varied portions (1.54%-6.52%) of A genome genes were changed significantly in comparison with that of RBR and more genes were down-regulated in all addition lines except in C6. Those genes which are expressed at high and moderate level in RBR were biased to be down-regulated (except in C7) while low expressed genes were substantially up-regulated in MAALs.

T5-08-04

Polyploidy-driven diversification and diploidization: The origin of the Australian and New Zealand crucifers

Terezie Mandakova, *Milan Pouch, Martin A. Lysak*
CEITEC, Masaryk University

We showed previously that three endemic Australian crucifer species (tribe Microlepidieae, Brassicaceae) have undergone a whole-genome duplication (AUS-WGD) followed by diversification and species-specific diploidization, generating some of the lowest chromosome numbers (n = 4 - 7) known for the crucifers (Mandáková *et al.*, 2010, *Plant Cell*). The New Zealand genus *Pachycladon* has undergone either the AUS-WGD or an independent WGD followed by less extensive genome reshuffling towards n = 10 genomes (Mandáková *et al.*, 2010, *BMC Evol Biol*). To further elucidate the origin and fate of Australian/New Zealand mesopolyploid genomes, we analyzed 12 species (9 genera) with variable chromosome numbers (n = 4, 5, 6, and 7) by comparative chromosome painting and gene sequencing. We concluded that all members of the Microlepidieae tribe originated through the AUS-WGD. This allopolyploid WGD most likely spurred the diversification of the group on the Australasia. Cytogenetic signatures and multi-gene (*CHS, PHYA, ndhF)* phylogenies suggest that the ancestral allopolyploid genome had most likely 30 chromosomes (n = 15) and was formed through an inter-tribal hybridization (tribe Crucihimalayeae and Smelowskieae/Descurainieae), dated to c. 10.6 million years ago, followed by a long-distance dispersal from Asia to Australia. The AUS-WGD was followed by independent, multi-speed and massive genome re-diploidizations towards the extant diversity of quasi-diploid genomes (n = 4 - 10) of the Microlepidieae taxa.

T5-08-05

Retrotransposon heterogeneity and their evolution in higher plants

Tony Heitkam[1], *Beatrice Weber*[1], *Katrin Schwichtenberg*[1], *Torsten Wenke*[1], *Kathrin M. Seibt*[1], *Heinz Himmelbauer*[2], *Bernd Weisshaar*[3], *Thomas Schmidt*[1]
1. TU Dresden, Institute of Botany
2. University of Natural Resources and Life Sciences Vienna, Department of Biotechnology
3. CeBiTec & Department of Biology, University of Bielefeld

Repetitive DNA comprises large quantities in all plant genomes studied, with proposed roles in genome structural integrity, as evolutionary motor and transcriptional regulator. Although advances of 'omics' datasets enable unique views on the plant repeat landscape, their assembly and analysis remains challenging. In sugar beet (*Beta vulgaris*), only 567 Mb of its 758 Mb genome are integrated into the reference genome sequence, neglecting a quarter of the genome-retrotransposons and satellite DNAs. Using a combination of next-generation sequencing, comparative genomics and cytogenetics, we identified and characterized the most abundant repeats within the sugar beet genome, generated a FISH karyotype, and derived principles applying to higher plants. In order to survey the sugar beet retrotransposon landscape, we extracted their mobility key enzyme reverse transcriptase (RT). Out of nearly intact 6,000 RTs, 4,476 are typical for long terminal repeat (LTR) retrotransposons of the Ty1-*copia* and Ty3-*gypsy*-type belonging to eight lineages. Prominent clades harboring additional putative envelope ORFs and chromodomains, were investigated in greater detail across the genus *Beta*. We highlight LTR retrotransposon evolution by amplificational bursts, reshuffling, recombination and targeted integration. Additional 1,471 intact long interspersed nuclear element (LINE) RTs form 17 L1 families and one RTE family, differing in sequence, amplification timepoints, heterogeneity and type of structural domains. An increasing species spectrum enables a view into plant LINE evolution: The comparison of 59,390 intact LINE RTs from 23 higher plant genomes indicates a limitation to L1- and RTE-clade-LINEs in higher plants. Whereas RTE LINEs show high homogeneity, even across taxonomical borders, L1 LINEs are highly dissimilar. L1 LINEs are subdivided into seven subclades, reflecting sequence similarity and structural aspects of full-length LINEs. We define these subclades, explain their modular organization and show that each of the plants investigated harbors LINEs of at least two subclades, whereas each subclade occurs in at least five species. In the genus *Beta*, retrotransposons without coding capacity, such as short interspersed nuclear elements (SINEs), are present in high copy numbers (> 34,000) and form 22 families. SINEs often associate with genes, contribute start and stop codons, splice sites, or exon variants, making them strong candidates for gene regulators.

T5-08-06

Phylogenetic models of chromosome number evolution and chromosomal speciation

William Freyman
University of California Berkeley

Chromosomal changes play a fundamental role in evolution. Dysploid gains and losses in chromosome number, as well as polyploidization events, may drive reproductive isolation and lineage diversification. The recent development of probabilistic models of chromosome number evolution in the groundbreaking work by Mayrose *et al.*, (2010) have enabled the inference of ancestral chromosome numbers over molecular phylogenies and generated new interest in studying the role of chromosome changes in evolution. However, these models assume all changes occur anagenetically (along branches), and do not model events that are specifically cladogenetic and may be expected if chromosomal changes result in reproductive isolation. Here I present a new class of models of chromosome number evolution that incorporate both anagenetic and cladogenetic change. These models allow one to determine the mode of chromosome number evolution; is chromosome evolution occurring primarily within lineages, primarily at lineage splitting, or in clade-specific combinations of both? Furthermore, one can estimate the location and timing of chromosomal speciation events over the phylogeny. I test the model's accuracy with simulations and re-examine chromosome evolution in Aristolochia, Carex section Ovales, Helianthus, Mimulus sensu lato, and Primula section Aleuritia.

T5-09: Plant organellar genomics and phylogenomics, driven by the NGS approaches (two sessions)

T5-09-01

Chloroplast phylogenomics of green algae

Linhua Sun[1], *Ling Fang*[1], *David Penny*[2], ***Bojian Zhong***[1]
1. Nanjing Normal University
2. Massey University

The green algal phylum Chlorophyta has six diverse classes, but the phylogenetic relationship of the classes within Chlorophyta remains uncertain. In order to better understand the ancient Chlorophyta evolution, we analyzed a chloroplast genome data set including new chloroplast genomes from class Ulvophyceae (Chlorophyta), and applied a site pattern sorting method to study compositional heterogeneity and the model fit in the green algal chloroplast genomic data. We show that the fastest-evolving sites are significantly correlated with among-site compositional heterogeneity, and these sites have a much poor fit to the evolutionary model. Our phylogenomic analyses suggest that the class Chlorophyceae is a monophyletic group, but the classes Ulvophyceae, Trebouxiophyceae and Prasinophyceae are paraphyletic groups. Our proposed phylogenetic tree of Chlorophyta will offer new insights to investigate ancient green algae evolution, and our analytical framework will provide a useful approach for evaluating and mitigating the potential errors of phylogenomic inferences.

T5-09-02

Development of a reference standard library of chloroplast genome sequences, GenomeTrakrCP

Ning Zhang[1], *Padmini Ramachandran*[1], *Jun Wen*[2], *Andrea Ottesen*[1], *Ruth Timme*[1], *Sara Handy*[1]
1. Center of Food Safety and Applied Nutrition, Food and Drug Administration
2. National Museum of Natural History, Smithsonian Institution

An increasing number of people have turned to herbal dietary supplements for preventing diseases, staying healthy, and as an alternative to medicines after getting sick. In 2015, the total sales of herbal dietary supplements in the United Stated reached $6.92 billion, a 7.5% increase from the year before, and demand for botanicals have increased for 12 consecutive years. Many different chemical techniques have been used to monitor authenticity of supplements and more recently DNA based tools have been included. While DNA barcoding has been found to be a powerful identification tool for many species, questions remain about its utility in plants, especially for processed products and closely related species. Therefore, developing methods targeting smaller diagnostic regions and reference libraries for rapid species identification of plants in foods and dietary supplements would be useful and complementary to chemical methods. In the past, the United States Food and Drug Administration (FDA) has been able to develop species specific assays targeting plant species of interest by utilizing chloroplast genome sequences. Presented here are the details for FDA's whole chloroplast genome sequencing effort and database, known as GenomeTrakrCP. Targeted species include plants found in foods and dietary supplements as well as plants known as toxin producers. Additionally, contaminants or adulterants and closely related species to these targeted species were sequenced. All data will be publically available through a bioproject in GenBank, e.g., PRJNA325670 derived from authenticated specimens and fully annotated. Currently there are 40 complete chloroplast genomes in the database from authenticated specimens. These data can be used by FDA and other government agencies, industry and any other researcher as complete chloroplast genomes or to design species specific assays to target plant species of interest.

T5-09-03

Mitochondrial phylogenomics of vascular plants: Evolutionary relationships and genomic diversity

Jeffrey Mower
University of Nebraska-Lincoln

The mitochondrial genomes of angiosperms are remarkably variable in size, structure and content. Much less is known about the mitogenomic diversity in the other vascular plant lineages, for which few genomes are available. To gain insight into the extent of mitogenomic diversity across vascular plants and the evolutionary changes that have generated such diversity, we have assembled complete or draft mitogenomes for dozens of species of angiosperms, gymnosperms, ferns, and lycophytes. Taxa were chosen to span the breadth of taxonomic diversity in each major lineage, with a particular focus on species from underrepresented groups. We found that each major lineage exhibits a wide range of features, only some of which are shared across lineages. We also performed a comprehensive phylogenomic analysis of vascular plants using mitochondrial genes, and explored the effects of RNA editing on these results.

T5-09-04

Holoparasitism breaks the evolutionary stasis of mitochondrial genome evolution in the broomrape family (Orobanchaceae)

Yanlei Feng[1], *Natalie Cusimano*[2], *Susann Wicke*[1]
1. Institute for Evolution and Biodiversity, University of Muenster
2. Department of Biology, Ludwig Maximilian University of Munich

The parasitic lifestyle in plants has brought about extreme genomic modifications, including dramatic functional and physical reductions to the plastid genome (plastome), accelerated rates of molecular evolution across apparently all genomic compartments, losses of essential mitochondrial genes for respiratory complex I, as well as massive horizontal gene transfers. While the reductive evolution of plastomes is well understood as a consequence of relaxed selective constraints on photosynthesis, we still have little insight into the reorganizations pertaining to mitochondrial genomes (mitogenomes) in parasitic lineages. To infer the evolutionary trajectories and underlying causes of mitogenome reconfiguration in parasites, we completely sequenced and assembled the mitogenomes of 25 species from 17 genera covering all trophic transitions in the broomrape family (Orobanchaceae). We find that mitogenome structure varies greatly in Orobanchaceae, where mitochondrial DNA ranges in size from 316 kb to 2.73 Mb and can be organized into one or more (mini-) circles, several linear fragments with multiple recombination points, or any combination of these. Mitogenome inflation coincides with the transition to a holoparasitic lifestyle, most notably in the broomrape (*Orobanche*) clade. Mitochondrial gene losses in Orobanchaceae are confined to several ribosomal proteins (e.g. *rpl2*, *rps10*, and *rps13*) and the *sdh3* gene, which encodes a cytochrome subunit for mitochondrial complex II. Multiple independent pseudogenizations and differing degrees of degradation of those genes indicate that their losses from the mitogenome likely are the result of ancient functional transfers to the nucleus. Correlating with the degree of plastome reduction, plastid-derived DNA is most abundant in the mitogenomes of photosynthetic parasites. The nature and conservation of plastid inserts suggest that they originated from many independent intracellular transfer events, with such transfers being either more recent or more frequent in hemiparasites. The expansion of mitogenomes is mostly due to an increase of lineage-specific non-coding DNA, whose origin has yet to be determined. However, several lines of evidence suggest that a fraction of these novel DNA segments have been obtained from host plants, implying that recent reports of the sporadic incorporation of foreign genes merely touch the tip of the iceberg. Molecular evolutionary analyses involving trait-rate fusion models show that lifestyle alone has only limited effect on rate elevation in mitochondrial genes of Orobanchaceae and that genomic traits also contribute significantly. Our comparative analysis of completely sequenced mitogenomes of Orobanchaceae species spanning the transition from autotrophy to holoparasitism contributes significantly to our understanding of the genetic reconfigurations in parasitic plants in general, and provides first evidence for a linkage of the structural and molecular evolution between organellar genomes as parasitism unfolds.

T5-09-05

Towards a meta-analysis of plastid genome evolution in heterotrophic flowering plants

Susann Wicke

Institute for Evolution and Biodiversity, University of Muenster

Plastid and mitochondrial genomes (plastomes and mitogenomes, respectively) are rewarding objects for molecular evolutionary studies and useful tools in plant phylogenetics and phylogenomics. Our understanding of the evolutionary trajectories of organellar genomes has benefitted considerably from the broad availability of NGS technologies with which any number of plastomes and mitogenomes can be generated from essentially any group of interest, including the most extreme ones like heterotrophic plants. Available now are the complete plastome sequences from representatives of at least 15 independent transitions to an obligate heterotrophic lifestyle, thus allowing to test (molecular) evolutionary hypotheses in an unprecedented wide manner within and across the various transitions in flowering plants. However, such a "meta-analysis" bears many potential pitfalls. For example, different sequencing technologies or bioinformatic pipelines may leave behind signature errors, taxonomic depths and coverage often differs within and between lineages, and comparing and categorizing trophic specializations across the various heterotrophs could introduce researcher biases. Here, I review our current knowledge of plastid genome evolution in heterotrophic plants and discuss open questions regarding the physical and functional degradation of plastomes under relaxed selective constraints. Using simulations with controlled errors relating to plastome annotations, phylogenetic effects, and trait data allows evaluating the accuracy and robustness of different phylo-statistical approaches. The analyses show that annotation errors are among the most problematic issues, which to correct requires either thorough (manual) re-annotation or solutions that avoid automated parsing of pre-annotated sequence flatfiles. Integrating complete plastome information rather than restricting analyses to (annotated) coding sequences further enables the identification of conserved non-coding sequence elements. Beyond setting the framework for a meta-analysis of plastome degeneration in heterotrophic plants, the results of this study may contribute to better assessing the potential functionality of genes in the absence of both expression data and documented guidelines how to categorize pseudogenes.

T5-09-06

Variable plastid genomic architecture within the *"Adenocalymma-Neojobertia"* clade (Bignonieae, Bignoniaceae) and its phylogenetic implications

Luiz Henrique Fonseca, *Lúcia Lohmann*

Universidade de São Paulo

The chloroplast is one of the most important organelles of green plants, due to its crucial role in photosynthesis. This organelle ranges in size between 25 to 217 kb in flowering plants, and contains a circular DNA with approximately 130 genes. Plastid genomes typically include a quadripartite structure that consists of a small single copy region (SSC), a large single copy region (LSC), and a pair of inverted repeats (IRs). This plastid structure, its gene content and organization is usually highly conserved when compared to plant mitochondria. The use of plastid genomic data in phylogenetic and evolutionary studies became possible with high-throughput sequencing methods, which allowed us to rapidly obtain complete genomes at a reasonable cost. Here, we use high-throughput sequencing to study the "*Adenocalymma-Neojobertia*" clade (Bignonieae, Bignoniaceae), a large lineage (ca. 75 spp.) of Neotropical treelets, shrubs and lianas, that is sister to the "Core Bignonieae" clade. More specifically, we used HiSeq Illumina technology to sequence the plastomes of 46 species. Overall, 12 plastomes were fully sequenced and used in this initial study. For the plastome assembly, we used a pipeline that included Bowtie 2 to select plastome reads, and SPAdes 3.9 to assemble the plastomes using a *de novo* approach. We then used Dogma to annotate contigs, and Mauve 2.4 to compare the genomic structure of all sequenced plastomes with those from closely related taxa (i.e., *Olea europea*, *Crescentia cujete*, and *Tanaecium tetragonolobum*). We further used the complete plastomes to estimate relationships among major clades within the "*Adenocalymma-Neojobertia*" lineage using PhyloBayes 1.7. For the phylogenetic analyses, five different data partitions were used, including the following plastome regions: (1) All genic and intergenic regions; (2) Genic regions exclusively; (3) Intergenic regions exclusively; (4) Intergenic regions excluding poorly aligned portions; and (5) All genic and intergenic regions excluding poorly aligned portions. The analyses of all datasets shows an expansion of the IRs and compression of the SSC region within the "*Adenocalymma-Neojobertia*" clade with the inclusion of the ycf1 and part of the rps15 genes in the IR regions. Major genomic translocations are observed at the LSC and IRs, shuffling gene order among plastomes. However, only small changes in gene content are observed and only two putative pseudogenes (i.e., *acc*D and *ycf*4) are found in the genomes of some species. Phylogenetic analyses of all datasets led to the same topology. Relationships were generally strongly supported, with the highest support values obtained through the analysis of dataset (5), and the lowest support values recovered from the analysis of dataset (3). The "*Adenocalymma-Neojobertia*" clade shows an unusual pattern of plastid structure variation, challenging the current view of conserved plastid genome architecture in terms of gene order, even when closely related species are considered. This variable architecture complicates genomic assemblies using reference genomes and sequence alignments using whole plastomes. This plastic architecture also complicates the alignment of intergenic regions, where rearrangements are most common. Therefore strategies that employ *de novo* assemblies and evaluate sequence alignments are required to prevent assembly and alignment errors.

T5-09-07

The complete chloroplast genome sequence of the CAM epiphyte Spanish moss (*Tillandsia usneoides*, Bromeliaceae) and its comparative analysis

Peter Poczai, *Jaakko Hyvönen*
University of Helsinki

Spanish moss (*Tillandsia usneoides*) is an epiphytic bromeliad widely distributed throughout tropical and warm temperate America. This plant is highly adapted to successfully inhabit its extreme environmental conditions. Striking features of this species include specialized trichomes (scales) covering the surface of its shoots aiding the absorption of water and nutrients directly from the atmosphere and a specific photosynthesis using crassulacean acid metabolism (CAM). Here we report the plastid genome of Spanish moss and present the comparison of genome organization and sequence evolution within Poales. Our study shows that early diverging lineages of Poales did not experience mutation rate acceleration as compared to grasses and plastid genomes of bromeliads show structural features considered to be ancestral in graminids, for example loss of *ycf*, and *accD*. We found that single-nucleotide variation and insertion/deletion frequencies are clearly higher than average in *rps7* revealing that this gene have been positively selected in *Tillandsia usneoides*. The same pattern can be found in the pineapple (*Ananas comosus*), another bromeliad with CAM photosynthesis. The *rps7* gene is encoding the ribosomal protein S7 that is crucial for ribosome assembly and synthesis; processes which have complex regulatory function in the ribosome reprograming pathway in water stress environments. This might suggest that the positively selected *rps7* is subjected to a number of ecological selective pressures and the gene may also be involved in stress acclimation potentially contributing to adaptation of bromeliads in drought conditions.

T5-09-08

Resolving the systematic positions of fence-riding taxa: Manipulating the chloroplast genome data of Paeoniaceae and its alliances

Wenpan Dong[1,2], *Chao Xu*[1], *Ping Wu*[1], *Shiliang Zhou*[1], *De-Yuan Hong*[1]
1. State Key Laboratory of Systematic and Evolutionary Botany, Institute of Botany, Chinese Academy of Sciences
2. Peking-Tsinghua Center for Life Sciences, Academy for Advanced Interdisciplinary Studies, Peking University

Accurately resolving the phylogeny of fence-riding taxa is always a challenge in phylogenetic inference. Such uncertainties could be due to systematic errors (long branch attraction, and compositional heterogeneity) or non-phylogenetic signal artifacts (model violations). Here, we provide an example demonstrating how these factors affect the positioning of Paeoniaceae within Saxifragales based on chloroplast genome data. We newly assembled 14 chloroplast genomes from Saxifragales, and by combining these genomes with those of 63 other angiosperms, three datasets were assembled to test different hypotheses proposed by recent studies. These datasets were subjected to maximum parsimony, maximum likelihood and Bayesian analyses with site-homogeneous/heterogeneous models, different data partitioning strategies, and the inclusion/exclusion of weak phylogenetic signals. These analyses indicated that the phylogenetic position of Paeoniaceae in Saxifragales is conflicting, i.e., site- and substitution model-dependent. Accommodating the site-heterogeneous model after removing the fastest-evolving site yields support for the optimal resolution of Paeoniaceae as a sister to the woody group (A group), which suggests that the robust inference of Paeoniaceae in the core Saxifragales (B or C group) may be due, in part, to evolutionary rate biases among sites. We conclude that although genomic data significantly increase the phylogenetic resolution of fence-riding taxa with high support, phylogenetic results inferred from such data might be analytical or signal dependent. The analytical pipeline outlined here combines phylogenomic inference methods with evaluating the lineage-specific rate of substitution, model section, and assessing systematic error. It would help resolve similar difficult questions in the tree of life.

T5-09-09

Assessing genomic diversity via whole genome resequencing in bean cultivars (*Phaseolus* L. spp.) from Brazil, a putative secondary center of diversity

Ashley N. Egan, *H.C. Lim*, *Theresa L. Miller*
Smithsonian Institution

Pulse crops, such as the lima bean (*Phaseolus lunatus*) and the common bean (*Phaseolus vulgaris*), are critical components of human health and the global economy. With the designation of 2016 by the United Nations as the International Year of Pulses, much research has centered on understanding the evolutionary, nutritional, and agronomic impacts of beans. Both the common bean and lima bean were domesticated by indigenous peoples of the New World and are said to each have two centers of domestication: a Mesoamerican origin and an Andean origin. However, genetic studies have suggested Brazil as another center of diversity. This is supported by anthropological studies of the Canela indigenous culture of northeast Brazil who cultivate an astonishing diversity of beans, with over 60 vernacular-named varieties. To assess the genetic diversity within Brazilian cultivars, we re-sequenced the complete genomes of 28 *P. lunatus* and 18 *P. vulgaris* accessions (including one wild-sourced) as well as the wild species *P. filiformis* collected from Brazil and *P. polystachios* collected from Virginia. All *lunatus* and *vulgaris* accessions were from small garden plots or local markets from areas northern Brazil proximate to Canela lands, with the exception of one proximately collected wild accession of *P. vulgaris*. Single nucleotide polymorphisms were called against the *P. vulgaris* complete nuclear and chloroplast genomes. Genetic diversity of Brazilian accessions was compared against SNPs from each species' two gene pools that represent centers of domestication and population structure as examined for each species. In addition, estimates of divergence times between gene pools and species were calculated. Brazilian germplasm presents a high level of genetic diversity and should be considered for preservation and conservation to help mitigate loss of genetic diversity.

T5-09-10

Phylogenetic relationships within *Lactuca* L. (Asteraceae), based on chloroplast genomics

Zhen Wei[1], *Freek Bakker*[2], *Eric Schranz*[2], *Shixin Zhu*[1]
1. Zhengzhou University
2. Wageningen University

Lactuca L. belongs to the subtribe Lactucinae, tribe Cichorieae of the family Asteraceae. The genus has been the difficult and controversial part in the phylogenetic studies of Asteraceae family for a long time, due to their complex and diverse morphological characters and similar traits to the species from related genera. Domesticated lettuce (*L. sativa* L.) is an important leaf vegetable and consumed by people worldwide, especially in Asia, North and Central America, and Europe. Therefore, the study of *Lactuca* species is of great scientific and economic importance. In recent years, the next generation sequencing (NGS) technologies have been developing very fast, providing an opportunity to study deep phylogenetic relationships within *Lactuca*. This project aims to study the phylogenetic relationships within *Lactuca* by NGS, including *Lactuca* species that have not been sequenced and species from related genera as outgroups. The chloroplast genome sequences will be assembled, annotated and used for constructing phylogenetic trees. Then some important morphological characters, chromosome number and geographical distribution will be plotted on the phylogenetic trees to investigate their patterns on the evolutionary trees. The results of this study will contribute to the circumscription of *Lactuca* and its subgeneric classification and origin. In the meantime, the results will also shed some light on new genetic resources for lettuce breeding.

T5-09-11

The evolution of liverwort mitochondrial genomes

Shanshan Dong[1], *Chaoxian Zhao*[1,2], *Shouzhou Zhang*[1], *Bernard Goffinet*[3], ***Yang Liu***[1,3]
1. Fairylake Botanical Garden, Shenzhen & Chinese Academy of Sciences
2. College of Life Sciences, East China Normal University
3. Department of Ecology and Evolutionary Biology, University of Connecticut

The evolution of the mitochondrial genomes of vascular plants is highly dynamic in terms of structure and composition, resulting in numerous distinct gene orders. By contrast the so-called early land plant lineages show less variation, and in fact the mitogenome of mosses exhibits no change across the moss tree of life. Liverworts may be characterized by similarly stable mitochondrial genomes, considering the synteny among three early diverging liverwort lineages. To better understand the evolution of liverwort mitochondrial genomes, we sampled ~20 liverworts representing members across the liverwort phylogeny. Based on next-generation sequencing (NGS) approach, we have assembled mitochondrial genomes from these liverworts. Preliminary analyses indicate that the structure and genic content of mitochondrial genomes in liverworts are not strictly conserved throughout liverwort evolution, as some genomic rearrangements and gene losses are detected. In this study, we also performed RNA editing analyses based on transcriptome sequence data.

T5-09-12

Plastome evolution in Bignonieae (Bignoniaceae)

Alexandre R. Zuntini[1,2], *Lúcia G. Lohmann*[1]
1. Universidade de São Paulo, São Paulo/SP
2. Universidade de Campinas, Campinas/SP

Bignonieae is the largest tribe of Bignoniaceae, embracing ca. 400 species and 21 genera of Neotropical lianas. A phylogeny that includes 25% of the species known to date is available for the whole tribe. While this phylogeny defined major generic-level clades, the relationship among clades is still moderately supported. In order to improve our knowledge of deeper relationships in the tribe and further understand patterns of plastome evolution, we sequenced whole chloroplast genomes of selected species from most genera in the tribe using genomic pair-ended libraries and an Illumina HiSeq platform. Plastomes were assembled using *de novo* approach in Velvet and the final sequences were annotated using DOGMA and Geneious. Annotated plastomes were aligned using Progressive MAUVE along with plastomes from other Lamiales. Overall, the plastomes of 25 species, representing 16 genera, were completely assembled, with lengths varying between 157 and 181 kbp. In all species, the inverted repeat regions (IRs) extend toward the Small Single-copy region, resulting in the duplication of *ycf1*. Another expansion of the IR, toward the large single-copy region, was observed in most members of the "Multiples of four" clade (M4) examined (i.e., *Amphilophium*, *Anemopaegma*, *Bignonia* and *Pyrostegia*). Fifteen locally collinear blocks (LCB) that ranged from 400 bp to 57 kbp were identified, with variable positions and orientations, leading to two local inversions and four reallocations. One inversion was found in *Bignonia uleana* (1.8 kbp) and another in the most recent common ancestor of the "*Anemopaegma* + *Pyrostegia* clade" (ca. 8 kbp). Given the complementarity of the sequences found flanking the inversions, it seems that both inversions may have resulted from hairpins. Reallocations were found in *Bignonia capreolata* and *Mansoa* (also members of the M4 clade). These reallocations may have resulted from the expansion of the IR in the MRCA of the M4 clade, with two subsequent losses of the duplicated regions. Two other translocations were observed in *B. magnifica* and in *Neojobertia* but the molecular mechanisms the led to these translocations remain unclear. These structural changes in Bignonieae indicate that plastomes may be more variable than previously thought, highlighting the importance of using *de novo* approaches, instead of reference guided approaches, for plastome assembly. An improved knowledge of plastome evolution is crucial for phylogenetic reconstructions, given their impact on rates of molecular evolution in different portions of the plastome.

T5-10: The genomes of orphan crops

T5-10-01

Discovery of genomic and cellular components conditioning pod length provides insights into grain-vegetable cowpea (*V. unguiculata* L. Walp) diversification

Pei Xu[1], *Xinyi Wu*[1], *Maria Munoz-Amatriain*[2], *Baogen Wang*[1], *Xiaohua Wu*[1], *Bao-Lam Huynh*[2], *Close Timothy*[2], *Philip Roberts*[2], *Guojing Li*[1]

1. Zhejiang Academy of Agricultural Sciences

2. University of California-Riverside

Cowpea (*V. unguiculata* L. Walp) is a climate resilient legume crop important for food security. Cultivated cowpea (*V. unguiculata* L) generally comprises the bushy, short-podded grain cowpea dominant in Africa and the climbing, long-podded vegetable cowpea popular in Asia. How selection has contributed to the diversification of the two types of cowpea remains largely unknown. In the current study, a novel genotyping assay for over 50 000 SNPs was employed to delineate genomic regions governing pod length. Major, minor and epistatic QTLs were identified through QTL mapping. Seventy-two SNPs associated with pod length were detected by genome-wide association studies (GWAS). Population stratification analysis revealed subdivision among a cowpea germplasm collection consisting of 299 accessions, which is consistent with pod length groups. Genomic scan for selective signals suggested that domestication of vegetable cowpea was accompanied by selection of multiple traits including pod length, while the further improvement process was featured by selection of pod length primarily. Pod growth kinetics assay demonstrated that more durable cell proliferation rather than cell elongation or enlargement was the main reason for longer pods. Transcriptomic analysis suggested the involvement of sugar, gibberellin and nutritional signalling in regulation of pod length. This study establishes the basis for map-based cloning of pod length genes in cowpea and for marker- assisted selection of this trait in breeding programes.

T5-10-02

Whole genome and transcriptome sequencing unravel the differences between Shum and Gilo groups of the African Eggplant

Damaris Achieng Odeny[1], *Bo Song*[2], *Francesca Stomeo*[3], *Sandra Kamenya*[4], *Elizabeth Balyejusa Kizito*[4]

1. The International Crops Research Institute for the Semi-Arid Tropics, Nairobi

2. BGI-Research, Shenzhen

3. Biosciences eastern and central Africa - ILRI hub, Nairobi

4. Uganda Christian University, Mukono

The African eggplant (*Solanum aethiopicum*) is a phenotypically and morphologically diverse species, divided into four recognized cultivar groups, namely Gilo, Shum, Kumba and Aculeatum. The Gilo and the Shum groups are the most popular and most widely cultivated groups in Africa. Although some previous taxonomic studies have reported the complexity within and across these defined groups of *S. aethiopicum*, there have been no focused genomic studies aimed at enhancing our understanding of the specific groups. We generated whole genome sequences of an inbred landrace, Shum303, belonging to the Shum group using Illumina sequencing technologies. We additionally generated transcriptome sequences from leaves, roots, flowers and fruits of Shum303, Shum307, Gilo524 and Gilo610. The transcriptome sequences were assembled both *de novo*, and aided by the new draft genome of Shum303 as a reference. The draft assembly is 1.03 Gb in size, comprising ~88% of the genome (1.17Gb, estimated using K-mer analysis), with N50 of scaffold and contig of 340 Kb and 21 Kb, respectively. We generated 183,201 and 144,450 unigenes for Shum and Gilo groups respectively. Both Shum and Gilo fruits generated the most unique transcripts in comparison to leaves, flowers and roots. Disease resistance genes were some of the most highly expressed genes in both Shum and Gilo groups suggesting the likely exposure of both genotypes to a disease prior to transcriptome analysis. A comparison of Shum and Gilo with other Solanaceae crops revealed a higher similarity of the African eggplant with tomato than potato. Further comparison between Shum and Gilo revealed that the Shum group was more closely related to tomato than the Gilo group. The Gilo group also showed differential gene expression of a pleiotropic drug resistance gene, known to play an important role in the prevention of serious adverse drug reactions (ADRs). Our work reveals unique group and tissue specific expression patterns in Shum and Gilo and provide excellent opportunities for medical research and resistance breeding.

T5-10-03

AOCC research aided by large scale plant genomics study

Shifeng Cheng[1], *Bo Song*[1], *Prasad Hendre*[2], *Allen Van Deynze*[3], *Alice Muchugi*[2], *Jamnadass Ramni*[2], *Xin Liu*[1], *Xun Xu*[1]

1. BGI-Research

2. World Agroforestry Centre

3. University of California-Davis

Plants play a crucial role in ecosystems, food and nutritional security and resources of medicine. Understanding plant genetics and biodiversity require integrated efforts ranging from distribution of germplasm, germplasm resources collection and storage, genotypic and phenotypic data analysis, transformation and agricultural application. In the past one and half decade, series of innovative technologies especially NGS (Next-Generation Sequencing) technologies have been developed and widely applied in plant research, which has accelerated and reshaped the researches in biodiversity and practices in food and nutrition development. AOCC (African Orphan Crops Consortium) associates a panel of scientists, anthropologists, nutritionists and plant breeders and all aim to make high-nutritional value crops grown by African famers and available to rural and urban consumers. In this consortium, 101 African orphan crops genomes are selected to be sequenced, assembled and annotated. Whole genome sequence, transcriptome sequence, SNP marker arrays/panels can access genetic diversity and also support breeding programs. *De novo* genome sequencing or re-sequencing, as well as transcriptome sequencing of more than 20 crops have been performed to date. The genomics efforts will benefit to accelerate crops domestication, diversify farming landscapes and agribusiness, conserve germplasm, and enhance crops yields for the entire Africa continent with ambition to diversify diets to support healthy nutrition.

T5-10-04

Ecology and community genomics of an important crop wild relative as a prelude to agricultural innovation.

Eric Von Wettberg
Florida International University

Domesticated species are impacted in unintended ways during domestication and breeding. Changes in the nature and intensity of selection impart genetic drift, reduce diversity and increase the frequency of deleterious alleles. Such outcomes constrain our ability to expand the cultivation of crops into environments that differ from those under which domestication occurred. We address this need in chickpea, among the world's most important pulse legumes, by conserving, characterizing and harnessing the diversity of wild crop relatives. We document an extreme genetic bottleneck between modern breeding lines and the wild progenitor species, while also deciphering the genetic history of wild populations. We provide evidence of ancestral adaptations in wild populations for seed coat color crypsis and demonstrate variation between wild and cultivated accessions for agronomic properties including flowering time, seed shattering, seed nutrient density, nitrogen responsiveness and drought adaptation. The assembled germplasm will serve as an essential resource for future chickpea breeding, while our methods may inform collection of other wild crop progenitor species.

T5-10-05

Grapevine rootstock species (*Vitis* spp.): diversity, comparative transcriptomics, and impacts on shoot system phenotypes in grafted plants

Allison Miller[1], *Dan Chitwood*[2], *Laura Klein*[1], *Misha Kwasniewsky*[3], *Laszlo Kovacs*[4], *Mingrui Li*[5], *Jason Londo*[6]
1. Saint Louis University
2. Danforth Plant Science Center
3. University of Missouri
4. Missouri State University
5. Key Laboratory of Plant Resources Conservation and Sustainable Utilization, South China Botanical Garden, Chinese Academy of Sciences
6. United States Depar

Perennial, woody crops including grapevines and many fruit and nut trees are often cultivated using the ancient horticultural technique of grafting. Grafting fuses plant organs together, creating plants where the lower stem and root system is genetically distinct, often a separate species, from the shoot system that produces harvested material. Grafting separates the breeding of shoot from root, permitting the development of domesticated rootstocks selected for superior performance in specific climates and soil types. For many grafted plants, even for well-known economically important crops with extensive genomic resources, species used as rootstocks represent "the lesser known half of the perennial crop equation." For example, grapevines are among the most widely planted berry crops in the world, with total worldwide production exceeding 77M tonnes in 2013. Grafting in grapevine dates to the mid-1800's when the North American aphid *Phylloxera* was inadvertently introduced into Europe where it devastated the roots of cultivated grapevines. Starting with this crisis, native North American grapevine species have been used for their *Phylloxera*-resistant rootstocks. Today the bulk of the global grape industry relies on traditional European grape varieties grafted to North American grapevines and their hybrid derivatives. Mounting concerns about the effects of a changing climate on vineyard locations and crop quality have re-focused attention on the potential contributions of diverse rootstocks for grape crop sustainability. Here we present data from two related studies: 1) morphological, genomic, and transcriptomic diversity of two North American *Vitis* species used as rootstocks; and 2) impacts of hybrid derivatives of these species on shoot system phenotypes in grafted plants. In the first experiment, we established a common garden of *V. riparia* and *V. rupestris* in the Kemper Center for Home Gardening at the Missouri Botanical Garden. Three clonal replicates of three genotypes for each of the two species were grown ungrafted. Leaf and ion concentrations were measured for each individual. We characterized gene expression patterns of each individual at two different time points. These data demonstrate differences in leaf shape, ion concentrations, and gene expression among *V. riparia* and *V. rupestris*. In the second experiment we leveraged an experimental vineyard where a common scion 'Chambourcin' is growing on its own roots and grafted to three different rootstocks. This combination of four individuals is replicated 72 times. Phenotypic data collected in the shoot systems demonstrate significant effects of rootstock on leaf shape, ion concentration, and vine physiology, among other traits. These data underscore the importance and potential of rootstock species and crop wild relatives in general that play an underappreciated role in agriculture and crop improvement.

T5-10-06

Exploring the evolutionary history of *Mangifera* and its domesticated species using ddRAD-seq

Emily Warschefsky, *Eric Von Wettberg*
Florida International University

Among the many domesticated species of the poison ivy family (Anacardiaceae), including pistachio, cashew, pepper tree *(Schinus)* and jocote *(Spondias)*, the mango *(Mangifera indica* L.) provides a novel system in which to study perennial crop domestication. Wild *M. indica* is thought to have originated in the foothills of the Himalayas in what is now northeastern India, Myanmar, Bangladesh, and Nepal. The mango was originally domesticated in India around 4,000 years ago, and today, is one of the world's most important tropical fruits. In addition to *M. indica*, some 26 species of *Mangifera* have edible fruits; these species range from incipiently-domesticated to cultivated landraces, and many exhibit traits of interest to breeders, such as resistance to mango anthracnose, salinity tolerance, and reduced fiber content. *Mangifera* is one of the largest of the Anacard genera, encompassing some 69 species. The native range of *Mangifera* stretches from eastern India to the Solomon Islands, with the highest diversity present in Borneo, Sumatra, and the Malay Peninsula. The most recent taxonomic revision of *Mangifera*, based on morphology, distinguishes between two subgenera, Mangifera and Limus, and six sections, but does not hypothesize species-level relationships. This monograph of *Mangifera* remains the standard for species identification and classification, as no systematic analysis of the genus has been published. Previous work based on morphological evidence proposed the delimitation of *Mangifera* into two subgenera and six sections, but published gene trees have called into question this classification. An understanding of the evolutionary relationships between *Mangifera* species will inform cultivation techniques and could lead to mango cultivars with increased tolerance to both disease and harsh environmental conditions. Here, we present the first multilocus phylogeny of *Mangifera*. Leaf samples were collected from 280 individuals representing 40 taxa grown in living collections in Indonesia, Singapore, Malaysia, and the U.S. Reduced representation genomic data were obtained by double digest restriction site associated DNA (ddRAD) sequencing. Reads were quality filtered, then processed with the software ipyRAD to conduct denovo assembly and call single nucleotide polymorphisms. Phylogenetic reconstruction was performed using Bayesian and Maximum Likelihood frameworks along with SVDquartets. The phylogeny was time-calibrated using a previously published phylogeny for the families Anacardiaceae and Burseraceae. This work challenges some of the previously proposed infrageneric classifications for *Mangifera*, and reveals the evolutionary relationships between the highly domesticated mango and its wild and semi-domesticated relatives.

T5-11: Genomics and domestication of economic crops

T5-11-01

Temporal and spatial variation in a large collection of chickpea (*Cicer arietinum*) landraces.

Eric Von Wettberg
Florida International University

Landraces of crops are important genetic resources, reflecting traditional cultivation practices and potential adaptation to diverse climates and agroecologies. We examine landrace diversity in chickpea, a unique crop with a wild ancestor with an extremely narrow geographic distribution in Southeastern Anatolia but landrace cultivation spanning six continents and conditions from subtropical South Asia and East Africa, to Mediterranean climates worldwide, and to temperate prairie conditions. By assembling a diverse collection of landraces from multiple genebanks spanning nearly century of germplasm collections and the three primary centers of diversification for chickpea, we examine changing patterns of diversity. We find little evidence of declining landrace diversity from the oldest collections in the Vavilov Institute for Research to contemporary collections. We document shifts in diversity in the center of domestication in Turkey to secondary diversification centers in South Asia and East Africa. Comparison to diversity in a recently assembled large collection of the wild relatives allows us to build a demographic model of multiple bottlenecks and diversifications impacting the diversity of chickpea.

T5-11-02

Re-sequenced genomes reveal the genetic diversity of polyploid cotton

Joshua Udall
Brigham Young University

By deep-sequencing many accessions of cotton, we have revealed detailed genetic information and phylogeographic patterns demonstrating the diversity in the cotton genome. We have re-sequenced nearly 600 accessions, including 431 accessions of *G. hirsutum*, 165 accessions of *G. barbadense*, and other cotton species. We detected the variation among these accessions and identified genetic variation between cultivars and wild species, also between the different species. This variation is interpreted against the background of new *de novo* cotton genome sequences. This wealth of sequence information will allows us to identify which specific gene or genome region have been selected during domestication of cotton. Understanding the location of these loci will allow us to further improve the cotton genome for agricultural improvement.

T5-11-03

Domestication origin and breeding history of the tea plant (*Camellia sinensis* (L.) O. Kuntze) in China and India

Muditha Kasun Meegahakumbura[1,2,3], *Moses Cheloti Wambulwa*[1,2], *De-Zhu Li*[1,2], *Lian-Ming Gao*[1]
1. Key Laboratory for Plant Diversity and Biogeography of East Asia, Kunming Institute of Botany, Chinese Academy of Sciences
2. Germplasm Bank of Wild Species in Southwest China, Kunming Institute of Botany, Chinese Academy of Sciences
3. Genetics and Plant Bree

China and India are the largest tea producing countries yet, the breeding history of the tea plant in these countries still unclear. Three distinct lineages of the tea plant, China tea, Chinese Assam tea and Indian Assam tea were defined recently, and nuclear microsatellites (nSSRs) revealed independent domestications in China and India. However, the domestication events need further confirmation and the timing of tea domestication remained unknown. In this study, a combination of 23 nSSRs and three cpDNA were used for genotype of 402 and 101 samples of the tea plant and its wild relative. Multiple analyses of Demographic modeling, NewHybrids and Neighbour joining tree were performed to investigate the domestication origin and breeding history of the tea plant. Three independent domestication centers were confirmed by both nSSR and cpDNA data, and two origins of domestication for Chinese Assam tea were further revealed in Southern and Western Yunnan of China with cpDNA. China tea and Assam tea was firstly diverged 22k year ago during the last glacial maximum (LGM), and subsequently split between Chinese Assam tea and Indian Assam tea in 2770 year ago, corresponding with the early use of tea in Yunnan, China. The three tea types underwent different breeding history and showed a complex breeding patterns. Hybridization may be the most important approach for tea cultivar breeding, and high proportion of the hybrids are F2 and backcross generation. Priority conservation strategy of Chinese Assam tea germplasm, especially the landraces should be implemented, which are very valuable resources for future tea breeding.

T5-11-04

Phylogenomics of the genus *Rosa*: proposal to solve a complex evolutionary scheme

Kevin Debray, *Fabrice Foucher, Valéry Malécot*
IRHS (Agrocampus Ouest, INRA, Univ. Angers)

The rose bush is the most economically important ornamental plant and the model organism for many studies about ornamental traits such as flowering, fragrance or disease resistance. Some of these traits were retrieved from Asian wild *Rosa* which represent the main gene pool, counting for about half of the genus species. The genus *Rosa* has a complex evolutionary history with interspecific hybridization events and polyploidizations. Most of current *Rosa* phylogenies are based on chloroplastic markers despite they are known for being maternally inherited. In addition, such phylogenies show low supports on internal nodes, which may be explained by low mutation rate of these markers.The objective of this project is to identify simultaneously a large number of nuclear and highly variable markers. Recently, as part of an international consortium, co-coordinated by the *Genetic and Diversity in Ornamentals* team from the French Research Institute for Horticulture and Seeds, the first version of the genome sequence of a rose cultivar (*Rosa* 'Old Blush') has been obtained. In addition, other genomic sequences have been recently published or acquired such as *R. damascena*, *R. roxburghii* or *R. palustris*. These sequences provide an unrecorded access to gene variation and organization in *Rosa* species. From comparisons between molecular sequences, we selected the best suitable low copy nuclear genes to reconstruct the phylogeny of the genus as suggested by phylogenies performed with Rosaceae Conserved Orthologous Set of data (RosCOS). About 150 genes have been already identified corresponding to nearly 600Kbp of phylogenetic information. Currently, these genes are tested on an additional representative sample of the four commonly accepted *Rosa* sub-genera. Once this step done, the markers showing the higher sequence variations will be amplified on a broader sample representing about 100 *Rosa* species. We thus expect to obtain more robust phylogenies, with higher supports for internal nodes, making them more accurate to study traits evolution through the genus. Indeed, both current cultivated roses and their wild counterparts show hybrid specimen with mixed characteristics. In this way, a well-resolved phylogeny offers the prospect of understanding how natural and horticultural diversities were generated among the genus. Knowing phylogenetic relationships between roses may also help increasing the cultivated diversity, by introducing new traits through interspecific crosses with wild material.

T5-11-05

Evolution and domestication of the flax (*Linum ustitatissimum*) genome and its further improvement through reverse genetics

Michael Deyholos
University of British Columbia, Okanagan Campus

Flax (*Linum usitatissimum*) has been used for millennia as a source of fiber or oil and is now recognized also as a source many biologically active compounds (e.g. lignans). Flax cultivation decreased dramatically after the middle of the 20th century, however new resources for flax research are supporting renewed interest in this crop, especially in relation to biocomposite materials. These new resources include: a whole genome and transcriptome assemblies from multiple genotypes and wild relatives, high density SNP maps, and mutant populations. The situation in flax therefore presents an interesting opportunity to test whether genomics can effectively accelerate basic and applied research in a previously neglected annual crop. Although flax has great potential to be a dual purpose crop, current varieties are cultivated either for seed (i.e. linseed) or for stem fiber (used traditionally in linen). One of our objectives is to make the stem fiber of linseed available as a feedstock for biodegradable composites and other advanced materials, whereas linseed straw is now seen only as a nuisance. Industrial use of linseed straw is limited by the yield and extractability of the fibers and by their chemical and mechanical properties. We have therefore used the newly available resources for flax to investigate the development of its strong, cellulose rich bast fibers, which have much in common with the specialized cell walls of tension wood produced in trees. To this end, we have characterized the following gene families of flax: beta-galactosidases (BGAL, GH35), cellulose synthases (CesA) and CesA-like proteins, chitinase like proteins (CTL, GH19), pectin methylesterases (PME), and trichome birefringence-like (TBL) proteins. We are using three complementary reverse genetics strategies to identify mutations in genes of interest. Our short term objective is to identify mutants with improved fiber separability (by targeting pectin and metabolism), and improved seed qualities (e.g. taste and meal quality). In the first approach, 10 amplicons, each containing 200 bp of a targeted coding region, are amplified individually from 768 independent M4 individuals, and are combined in 3D-pools for barcoding and subsequent sequencing using an Ion-Torrent 314-chip. In the second approach, a custom olignucleotide library was used to capture exomes of over 30,000 targeted genes. Capture-enriched genomic DNA fragments from individual M4 lines were sequenced using an Illumina HiSeq. Finally, we have also resequenced entire genomes of mutant lines and varieties to discover variants associated with useful traits. Results show that all three approaches are effective in identifying mutations in flax genes of interest, although there are significant differences in cost per-run and cost per-mutation.

T5-11-06

The cotton β-galactosyltransferase 1 (GalT1) that galactosylates arabinogalactan-proteins participates in controlling fiber development

Li-Xia Qin[1], Yun Chen[1], Wei Zeng[2], Yang Li[1], Lu Gao[1], Deng-Di Li[1], Antony Bacic[2], Wen-Liang Xu[1], ***Xue-Bao Li[1]***
1. Hubei Key Laboratory of Genetic Regulation and Integrative Biology, School of Life Sciences, Central China Normal University
2. ARC Centre of Excellence in Plant Cell Walls, School of BioSciences, University of Melbourne, Parkville, VIC

Arabinogalactan-proteins (AGPs) are highly glycosylated proteins which play pivotal roles in diverse developmental processes in plants. Type II AG glycans, mostly O-linked to the hydroxyproline residues of the protein backbone, account for up to 95% w/

w of the AGP, but their functions are still largely unclear. Cotton fibers are extremely elongated single-cell trichomes on the seed epidermis. However, little is known of the molecular basis governing the regulation of fiber cell development. Here, we characterized the role of a CAZy glycosyltransferase (GT) family 31 member GhGalT1 in cotton fiber development. Fiber length of the transgenic cotton overexpressing *GhGalT1* was shorter than that of wild type, whereas in the *GhGalT1*-silenced lines there was a notable increase in fiber length, compared with wild type. The carbohydrate moieties of AGPs were altered in fibers of *GhGalT1* transgenic cotton. The galactose:arabinose ratio of AG glycans was higher in *GhGalT1* overexpression fibers, but lower in *GhGalT1*-silenced lines, compared with that in wild type. Overexpression of *GhGalT1* upregulates transcript levels of a broad range of cell wall-related genes, especially the fasciclin-like AGP (FLA) backbone genes. An enzyme activity assay demonstrated that GhGalT1 is a β-1,3-galactosyltransferase (β-1,3-GalT) involved in biosynthesis of the β-1,3-galactan backbone of the type II AG glycans of AGPs. We also show that GhGalT1 can form homo- and hetero-dimers with other cotton GT31 family members to facilitate AG glycan assembly of AGPs. Thus, our data demonstrate that GhGalT1 influences cotton fiber development via controlling glycosylation of AGPs, especially FLAs.

T5-12: Arabidopsis: genomics and natural variations

T5-12-01

The ecological genetics of seed size and number and adaptive differentiation in *Arabidopsis thaliana*

Tom J. Ellis, Froukje M. Postma, Jon Ågren
Uppsala Univeristy

Local adaptation can be expressed through differences between local and non-local genotypes in several fitness components, and the magnitude of adaptive differentiation will thus depend on which fitness components are considered. Life-history theory predicts that within a given environment, adaptive evolution will be constrained by tradeoffs between allocation to different life-history functions. Reciprocal transplant experiments have demonstrated strong adaptive differentiation between an Italian and a Swedish population of *Arabidopsis thaliana* have shown that differences in fruit production of reproducing plants contribute to the advantage of the local genotype at the Italian, but not at the Swedish site. Here we use a field experiment conducted over five years that involved the parental genotypes and a RIL population derived from a cross between the two populations to examine whether differences in seed production per fruit, and size of individual seeds add to any reproductive advantage of the local genotype arising from differences in number of fruits per reproducing plant, whether any tradeoffs can be detected among these components of reproductive allocation, and to map QTL for the different components of reproductive success at the native sites. In Italy, the local genotype consistently produced more seeds per fruit than did the non-local genotype, whereas in Sweden this was true in only two of five years with no difference recorded in the other three years. Seed production per reproductive plant was 16- to 160-fold higher in the local compared to the non-local genotype in Italy, but only 1- to 9-fold higher in Sweden. In two years in Italy and one year in Sweden, the local genotype produced larger seeds than did the non-local genotype. Seed number per fruit showed weak correlations with seed mass (mean $r = 0.07$), moderate correlations with fruit number (mean $r = 0.18$) and strong correlations with seed production per reproductive plant (mean $r = 0.50$). Alleles at QTL for fruit number favour the local genotype, consistent with observed differences in fecundity between the parental lines. In contrast, alleles influencing seed number per fruit and seed number per reproductive plant show uniform effects at both sites and were just as likely to increase as to decrease seed number. At 10 QTL for seed number per plant, alleles show positive correlations with both seed number per fruit and fruit number among RILs, reflecting the positive relationships between phenotypes. We also identify four QTL where the local allele increases one component of fecundity, but decreases the other, suggesting a trade-off between phenotypes at individual loci.

T5-12-02

Rapid adaptation of *Arabidopsis thaliana* to the Yangtze River basin

Yu-Pan Zou[1,2], Xing-Hui Hou[1,2], Qiong Wu[1], Zi-Wen Li[1], Ting-Shen Han[1,2], Xiao-Min Niu[1,2], Li Yang[1], Jia-Fu Chen[1,2], Yong-Chao Xu[1,2], Jie Zhang[1,2], Fu-Min Zhang[1], Dunyan Tan[3], Zhixi Tian[4], Hongya Gu[5,6], **Ya-Long Guo**[1,2]
1. State Key Laboratory of Systematic and Evolutionary Botany, Institute of Botany, Chinese Academy of Sciences
2. University of Chinese Academy of Sciences
3. Xinjiang Key Laboratory of Grassland Resources and Ecology, College of Grassland and Environment Sciences, Xinjiang Agricultural University
4. State Key Laboratory of Plant Cell and Chromosome Engineering, Institute of Genetics and Developmental Biology, Chinese Academy of Sciences
5. State Key Laboratory for Protein and Plant Gene Research, College of Life Sciences, Peking University
6. The National Plant Gene Research Center (Beijing)

Global climate change has tremendous effect on the human health, food security and biological diversity. To keep pace with a changing environment, organisms need to rapidly adapt. However, how organism could rapidly adapt to the climate change is a crucial biological question to address but largely unknown. An excellent system to study such adaptive processes is the plant *Arabidopsis thaliana*. Its origin is in Europe, but it has spread to the Far East, including the most southeastern edge of its native habitats - the Yangtze River basin, where the climate is very different. We sequenced 91 *A. thaliana* strains from the region flanking the Yangtze River basin, and found that this population is genetically uniform, having been dispersed eastwards recently about 3000 years ago. Here we integrated population genomics and genome wide association study (GWAS) for the both climate parameters and phenotypic trait flowering time together, and found out that genes correlated with responses to stress and flowering time could have contributed to the rapid adaptation to the Yangtze River basin. Especially, two genes, PP2C and ACS8, both involved in either stress response or flowering time, could have been the most important determinant genes during the rapid adaptation to the Yangtze River basin. Overall, our results shed light on how a species can rapidly adapt to locales with a very different climate.

T5-12-03

Exploring *Arabidopsis* natural variation

Maarten Koornneef
Wageningen University

The genetic variation of *Arabidopsis thaliana* in nature provides a useful resource for the functional analysis of genes. The genetic complexity of this type of variation requires Quantitative Trait Loci (QTL) analysis, using various types of mapping populations. In addition to the analysis of gene function, natural variation studies may reveal signatures of selection in nature that may explain local adaptation. To demonstrate the power of natural variation, examples on the analysis of plant architecture (plant length and branching and vein density patterns) will be presented. For plant length, we found ample functional variation for the GA 20 oxidase (*GA5*) gene of which gene mutants have been exploited to generate modern short straw varieties in barley and rice. For branching an example on how to analyse such variation up to the gene level in *Arabidopsis* is the finding of the *AGL6* gene to be involved in this process. The complexity of the genetic interactions among natural variants and its consequences for evolution is shown by genetic incompatibilities that arise in certain combinations of genotypes.

T5-12-04

1001 Genomes Plus (1001G+): From single-reference mapping to the integration of multiple genome assemblies

*Felix Bemm[1], **Fernando A. Rabanal**[1], Magnus Nordborg[2], Detlef Weigel[1]*
1. Max Planck Institute for Developmental Biology
2. Gregor Mendel Institute of Molecular Plant Biology

The 1001 Genomes Consortium, an informal international collaboration, has recently marked an important milestone with the publication of 1135 Illumina short-read resequenced genomes of *Arabidopsis thaliana* from throughout the global range of the species. In that analysis, only an average of 100 Mb (84%) per line were called against the Col-0 TAIR10 reference genome (119 Mb). The missing positions differ greatly between lines, because they are in the vicinity of structural variants (SVs), which are difficult to assess with short read technology. In the next phase, the consortium is building multiple platinum-standard *de novo* genome assemblies, primarily with PacBio long-read technology and BioNano optical mapping, as a further step towards a species-wide pan-genome and to discover a core set of common structural variants in *A. thaliana*. The variants discovered from a comparison of these platinum-standard genome assemblies will then be typed in the 1135 Illumina short-read set to describe the global distribution of SVs. This knowledge will further inform Genome Wide Association Studies (GWAS) in this species, and advance our understanding of how SVs such as gene duplications, transpositions, deletions and transposable elements (TEs) shape the genetics and epigenetics of *A. thaliana*.

T5-12-05

High throughput phenotyping of photosynthesis to identify relevant quantitative trait loci and underlying genes in *Arabidopsis thaliana*

Mark Aarts
Laboratory of Genetics, Wageningen University

Natural genetic variation in plant photosynthesis efficiency is hardly studied even though breeding for photosynthesis would be interesting to maintain increases in crop yields. One of the reasons this is not studied is the notorious difficulty in adequately phenotyping photosynthesis parameters for genetic research. We designed a versatile plant phenotyper, the Phenovator, capable of multispectral imaging of 1440 plants per experiment. Imaging includes chlorophyll fluorescence to determine the light use efficiency of photosystem II electron transport (Φ_{PSII} or Fq'/Fm'), and near infrared reflection to measure projected leaf area. This system has been used to phenotype *Arabidopsis thaliana* recombinant inbred line populations as well as around 350 genetically diverse accessions for genome wide association analysis. Plants were phenotyped at optimal conditions and in response to cold treatment (5 °C) or a switch from 100 to 550 μmol m^{-2} s^{-1} irradiance. The observed genotypic variation was used to identify quantitative trait loci (QTL). Our latest progress in identification and confirmation of QTL for Φ_{PSII}, plant growth and epinastic leaf movement will be presented, including detailed analysis of the allelic variation at the nucleotide level of some of the genes underlying these QTL. Our work has shown that there is sufficient genetic variation for PSII efficiency in Arabidopsis amenable for gene identification, which suggests the same will be the case for crop species. Such would offer interesting opportunities for future crop photosynthesis improvements.

T5-12-06

Populations of *Arabidopsis thaliana* from Tibet, a unique ecotype at high elevation

*Tianshu Sun, Yihao Shi, Chenqi Zhu, Xuefei Wang, **Hongya Gu***
State Key Laboratory for Protein and Plant Gene Research, School of Life Sciences, Peking University

Arabidopsis thaliana is found in different geographic regions with different climatic conditions. Six populations were collected from Tibet, the upper reaches of the Yangtze River, China. They are found in habitats at 3,800-4,400 meters above sea level, and they formed a monophyletic clade based on chloroplast DNA. These populations have distinct morphological and physiological characters such as very short, sometimes multi-branched, low germination rate, very sensitive to freezing temperature, and *etc.* The low tolerance to freezing temperature is partially due to the malfunctioning of a group of C-repeat binding factor (*CBF*) genes which have been found playing an important role in responding to low temperature in plants. The expression level of *CBF1~3* is lower in all 6 populations than that of Col ecotype at 4°C, and a loss-of-function mutation was found in *CBF2* and/or *CBF3* of these populations. An RNA-seq analysis was performed on the plants treated at 4°C for 3 h and 12h. It was found that at 3 h, the 6 populations shared 98 exclusive up-regulated genes which were mainly involved in responses to various phytohormones such as ABA, auxin, and JA, and to cold and salt stress; while Col had 198 exclusive up-regulated genes involving in regulation of gene expression, protein autophosphorylation, ethylene signaling pathway, and systemic acquired resistance. At 12 h, the 6 populations had a totally different set of up-regulated genes from Col. We will discuss about the adaptation significance of these variations, together with the genetic and molecular characterizations of some other interesting traits in this unique ecotype from Tibet.

T5-13: Extremophyte: transcriptomes and beyond

T5-13-01

Transcriptomic analysis reveals molecular mechanisms underlying salt adaptation in *Sonneratia alba*

Jianfang Li, *Renchao Zhou, Suhua Shi*
State Key Laboratory of Biocontrol and Guangdong Provincial Key Laboratory of Plant Resources, Sun Yat-sen University, Guangzhou 510275, China

Salinity in soil is increasing in many areas and presents a widespread and serious environmental risk for most plants. Mangroves are a group of unique plants that well adapt to extreme environments with high salinity in tropical and subtropical intertidal zones. In this study, we performed transcriptomic analysis to investigate the transcriptional regulation underlying salt adaptation in *Sonneratia alba*. Seedlings of *S. alba* were treated by 0 mM, 250 mM and 500 mM NaCl for 7 days, and transcriptome profiles were obtained from leaf and root tissues, respectively. Gene expression pattern analyses suggested an earlier salt-response in roots than in leaves. With the whole genome dataset of *S. alba* as background, we conducted functional analyses for differentially expressed genes (DEGs). Gene Ontology enrichment analysis indicated that GO term "response to stimulus" was overrepresented in up-regulated genes in both leaf and root, suggesting the expression of most known salt-tolerant genes were elevated in response to salt stress. Furthermore, the KEGG pathways of "sulfur metabolism" and "Flavone and flavonol biosynthesis" were significantly enriched in up-regulated genes in leaf tissue. They were both involved in synthesizing osmosis-related metabolites in plants, suggesting the accumulation of osmotic substances might play a major role in salt tolerance in mangroves. Moreover, there were higher percentages of transcription factors (TF) in DEGs than the proportion in the whole genome, and the expression of salt tolerance related TF families were elevated under salt stress, suggesting that TFs could also play a curial role in salt adaptation of mangroves. Our study sheds light on the transcriptional regulation of mangroves to high salinity in the intertidal zones, and provides some useful information for plant genetic improvement.

T5-13-02

Transcript profiles of genes involved in increasing sugar content in leaves and salt exclusion in roots during salt stress in sweet sorghum

Na Sui, *Zhen Yang, Yuanyuan Guo, Shanshan Tian, Baoshan Wang*
Shandong Provincial Key Laboratory of Plant Stress, College of life science, Shandong Normal University

Sweet sorghum (*Sorghum bicolor* (L.) Moench) is a variation of traditional grain sorghum, a C_4 crop, with the characteristics of high photosynthesis and high biomass accumulation. It has been considered as energy and forage crop of high-yield and salt-resistant, which is most suited to grow in the saline-alkali land of Yellow River Delta. However, little is known about sweet sorghum resisting salt stress and increasing its sugar content in response to salt stress. In this study, the transcript profiles of both leaves and roots of two inbred lines of sweet sorghum (salt-tolerant M-81E and salt-sensitive Roma) were analyzed in the presence of 0 mM or 150 mM NaCl. In leaves, the majority of differential expressed genes are involved in photosynthesis, carbon fixation, and starch and sucrose metabolism. Genes important for maintaining photosystem structure and for regulating electron transport were less affected by salt stress in the M-81E line compared to the salt-sensitive Roma line. In addition, expression of genes encoding $NADP^+$-malate enzyme and sucrose synthetase was up-regulated and expression of genes encoding invertase was down-regulated under salt stress in M-81E. In contrast, the expression of these genes showed the opposite trend in Roma under salt stress. In roots, many differentially expressed genes were involved in pathways related to salt exclusion, such as apoplastic barriers, ion transport, aquaporins and ROS scavenging. These results revealed that the salt-tolerant genotype M-81E have a better ability to limit root absorption of Na^+ and to maintain ion homeostasis under salt stress which may be caused by the combination of physical barrier effect of root apoplastic barriers and the transportation of Na^+ out of the xylem into the xylem parenchyma cells by HKT1; 5. Additionally, M-81E leads to increased sugar content under salt stress by protecting important structures of photosystems, by enhancing the accumulation of photosynthetic products, by increasing the production of sucrose synthetase and by inhibiting sucrose decomposition. All these may be regulated by stress-related transcription factors such as bHLH, MYB, ERF, WRKY and NAC.

T5-13-03

Genome, transcriptome and gene function of *Thellungiella salsuginea*

Qi Xie
Institute of Genetics and Developmental Biology, Chinese Academy of Sciences

Thellungiella salsuginea, a close relative of Arabidopsis, represents an extremophile model for abiotic stress tolerance studies. We obtained the draft sequence of the *T. salsuginea* genome, assembled based on ~134-fold coverage to seven chromosomes with a coding capacity of at least 28,457 genes. This genome provides resources and evidence about the nature of defense mechanisms constituting the genetic basis underlying plant abiotic stress tolerance. Comparative genomics and experimental analyses identified genes related to cation transport, abscisic acid signaling, and wax production prominent in *T. salsuginea* as possible contributors to its success in stressful environments. Especially we will present the detail analysis an only copy of a small C2 domain protein exhibit a great stress tolerance ability once it is expressed in vs, both in model and crop plants.

T5-13-04

Genome and transcriptome analysis in the desiccation tolerant grass *Oropetium thomaeum*

Dorothea Bartels
University of Bonn

Limited water availability has resulted in evolution of mechanisms to survive periods without water supply. As a result of this a small group of vascular angiosperm plants termed "resurrection plants" have evolved unique mechanisms to survive desiccation to an air-dried stage. Desiccation tolerance of vascular tissues may have evolved through rewiring seed desiccation pathways. *Oropetium thomaeum*, originally collected in India, is an emerging model for understanding desiccation tolerance. The recently sequenced genome of *Oropetium thomaeum* is the smallest among the grasses. Transcriptome and targeted metabolome analyses demonstrate the dominant presence of seed specific pathways in vegetative tissues. Recently evolved gene duplications are observed for genes encoding protective components. Pathways involved in programmed cell death inhibition, ABA signalling, synthesis of stress protective proteins and seed specific-transcription factors are activated during desiccation. Oleosins and oil bodies are highly abundant in desiccated leaves and may contribute to membrane stability and storage of reserves. Orthologs of seed-specific LEA proteins from other grasses have neo-functionalized in *Oropetium thomaeum* with high expression during desiccation. Accumulation of sucrose, raffinose, and stachyose in leaves during dehydration are similar to those in maturing seeds. The overall transcript analysis suggests that vegetative desiccation tolerance is closely related to seed desiccation. The understanding of desiccation tolerance will provide novel biotechnological targets for improving growth of agricultural plants in environments with restricted water supply.

T5-13-05

Gene expression reaction norms of *Alternanthera philoxeroides* in contrasting hydrological conditions

Ji Yang

Fudan University

Alternanthera philoxeroides is an amphibious invasive weed that can colonize both aquatic and terrestrial habitats. Individuals growing in different habitats exhibit extensive phenotypic variation but little genetic differentiation. Little is known about the molecular basis underlying environment-induced phenotypic changes. Variation in transcript abundance in *A. philoxeroides* was characterized throughout the time-courses of pond and upland treatments using RNA-Sequencing. Seven thousand eight hundred and five genes demonstrated variable expression in response to different treatments, forming 11 transcriptionally coordinated gene groups. Functional enrichment analysis of plastically expressed genes revealed pathway changes in hormone-mediated signaling, osmotic adjustment, cell wall remodeling, and programmed cell death, providing a mechanistic understanding of the biological processes underlying the phenotypic changes in *A. philoxeroides*. Both transcriptional modulation of environmentally sensitive loci and environmentally dependent control of regulatory loci influenced the plastic responses to the environment. Phenotypic responses and gene expression patterns to contrasting hydrological conditions were compared between *A. philoxeroides* and its alien congener *Alternanthera pungens*. The terricolous *A. pungens* displayed limited phenotypic plasticity to different treatments. It was postulated based on gene expression comparison that the interspecific variation in plasticity between *A. philoxeroides* and *A. pungens* was not due to environmentally-mediated changes in hormone levels but to variations in the type and relative abundance of different signal transducers and receptors expressed in the target tissue.

T5-13-06

Dehydration tolerance in a tropical liverwort: Ecological and genomic patterns

Rose A Marks

University of Kentucky

Our work stems from an interest in understanding the strategies by which plants respond to limited and variable water availability. Adaptive mechanisms to cope with water stress are highly relevant in light of current and future climate change. Dehydration tolerance (DhT, also dehydration tolerant) is a relatively rare plant strategy for coping with water shortage that allows tissues to survive substantial cellular drying. This adaptation to water shortage provides a unique opportunity to identify key genes, physiological, and ecological traits that will inform biotechnological advances and management practices aimed at reducing crop and species loss due to drought. Our recent studies indicate that *Marchantia inflexa*, a tropical liverwort, exhibits moderate and variable levels of DhT. Furthermore, *M. inflexa* has the capacity to acclimate to different climatic moisture levels, suggesting that there is a plastic component to DhT in this species. *M. inflexa* is currently undergoing range expansion and colonizing novel habitats, likely facilitated by its ability to acclimate to these novel conditions. However, older populations also exist across a variety of habitat types and these populations could harbor genetic adaptations in DhT. Plants that exhibit adaptive differences in DhT, provide a valuable genetic resource for identifying key genes that contribute to increasing DhT. To test for adaptive differences between *M. inflexa* populations, plants were collected from two habitat types, which differ in their relative water availability. The DhT of these plants was tested immediately after collection to assess acclimation to each habitat. Plants were then cultured in a common garden for four months, and assayed to determine the innate level of DhT in these groups. Plants displayed acclimation to contrasting water regimes. In addition, we recently demonstrated that males of this species are significantly less DhT than females, and that the vegetative meristem survives dehydration more often than the remaining vegetative tissue. Based on these two observations we designed and conducted an RNA sequencing study to characterize changes in gene expression during dehydration and rehydration in these tissues in male and female plants. Consequently, we were able to identify genes with different expression profiles across samples. These genes were implicated as candidate genes that contribute to underlying differences in DhT. By taking a within species approach, we were able to minimize background differences and identify a high-resolution set of candidate DhT genes. Furthermore, the resulting transcriptome provides a valuable resource for other researchers wanting to compare sensitive species or highly desiccation tolerant species to the more intermediate phenotype of a DhT plant.

T5-14: Epigenetics I: histone modifications

T5-14-01

The influence of histone modifications and variants on *Arabidopsis* meiotic recombination

Ian Henderson
University of Cambridge

Meiotic recombination varies extensively in plant genomes, which has important implications for both genome evolution and plant breeding. We are using the model species *Arabidopsis thaliana* to investigate genetic and epigenetic control of meiotic recombination patterns. In this talk I will present our work mapping meiotic DNA double strand breaks, and crossover events, comparing wild type with mutants that have altered histone modifications or histone variants. I will present evidence for both positive (H2A. Z, H3K4me3) and negative (H3K9me2) roles for epigenetic information controlling recombination patterns. Understanding these mechanisms has the potential to modulate meiotic recombination in the context of crop breeding, for example, by releasing heterochromatic recombination which limits improvement of many agriculturally important species.

T5-14-02

Meiotic chromosome remodeling driven by germline-specific Argonaute in rice

Ken-Ichi Nonomura
1. National Institute of Genetics
2. SOKENDAI/Graduate University for Advanced Studies

Phased small RNA (phasiRNA) is a distinct class of small RNAs in plants, and in monocots, abundantly expressed in reproductive organs. We discovered that 24-nt phasiRNA biogenesis was triggered by a transcription factor exclusively expressed in anther tapetum in rice. In addition, a subset of the 24-nt tapetum-of-origin phasiRNAs were bound by germline-specific Argonaute protein, MEL1, suggesting a mobile nature of phasiRNAs from somatic tapetal cells to male germ cells. Downstream of the MEL1 regulatory cascade involves large-scale meiotic chromosome remodeling (LMR), accompanying dynamic alteration of histone H3 lysine-9 dimethylation. The epigenetic relation of LMR to reproductive small RNAs will be discussed.

T5-14-03

Bigdata epigenomics reveals hierarchical regulation of DNA methylation in *Arabidopsis*

Jixian Zhai[1,2], *Yu Zhang*[1], *Qikun Liu*[2,3], *Jake Harris*[2], *Haifeng Wang*[3], *Steven Jacobsen*[2]
1. Southern University of Science and Technology
2. University of California at Los Angeles, Los Angeles; Howard Hughes Medical Institute
3. Fujian Agriculture and Forestry University

Genome-wide characterization by next-generation sequencing has greatly improved our understanding of the "landscape" of epigenetic modifications. Since 2008, whole-genome bisulfite sequencing (WGBS) has become the gold standard for DNA methylation analysis and a tremendous amount of WGBS data has been generated by the research community. However, methods for the systematic comparison of DNA methylation profiles to identify novel regulatory mechanisms have yet to be established. Here we developed a standardized pipeline and re-analyzed over five hundred publicly available *Arabidopsis* WGBS libraries from various mutant backgrounds, tissue types, and stress treatments. In total this collection included more than 3,700 Gb (Giga base-pairs) of sequencing data and a large number of wild-type controls. This enabled us to identify "high-confidence Differentially Methylated Regions" (hcDMRs) with high reliability by comparing each non-wild-type library to each of the 54 controls. We developed two statistical methods, Statistical Measurements on Overlapping of DMRs (S-MOD) followed by Quantitative Measurements on Overlapping of DMRs (Q-MOD), to analyze the hcDMRs, and identified a hierarchical network regulating non-CG methylation, as well as novel connections between methylation pathways. Our bigdata-methylome approach has established a framework for extracting biological insights via large-scale comparison of methylomes, and can also be adopted for other omics datasets. Together, the results demonstrate the effectiveness of data-driven, hypothesis-generating epigenetic research.

T5-14-04

Establishing and interpreting DNA methylation in *Arabidopsis thaliana*

Julie Law
Salk Institute

DNA methylation is a highly conserved epigenetic modification associated with gene silencing in eukaryotic organisms. Although pathways controlling the establishment, maintenance and removal of DNA methylation have been identified, relatively little is understood about how the DNA methylation machinery is targeted to transposons and repeats or how DNA methylation influences gene expression. Using the plant model *Arabidopsis thaliana*, we have recently identified several protein complexes involved in these key aspects of DNA methylation. Using a combination of genetic, genomic and biochemical approaches were are beginning to uncover the mechanisms through which these proteins function, which will provide insight not only into normal biological processes but also into how alterations in chromatin modifications can lead to aberrant gene expression and developmental defects.

T5-14-05

DNA replication-coupled histone deposition and modification maintains polycomb silencing

Danhua Jiang, *Frédéric Berger*
Gregor Mendel Institute

The developmental fate of a cell is specified by its transcriptional program, which is regulated by chromatin-associated mechanisms. Maintenance of repressive transcriptional states by Polycomb group (PcG)-mediated histone H3 lysine 27 trimethylation (H3K27me3) is conserved in animal and plant development. H3K27me3 is deposited by Polycomb repressive complex 2 (PRC2) and recognized by Polycomb repressive complex 1 (PRC1) in a feedback loop that maintains silencing. When cell fate is established in cycling cells, H3K27me3 needs to be properly restored after the dilution caused by DNA replication to maintain transcriptional repression. Recent studies in different animal models have described distinct mechanisms for H3K27me3 inheritance, suggesting a complexity in H3K27me3 transmission. Here we show that histone incorporation and H3K27me3 deposition are coupled at the DNA replication fork during the plant cell cycle. We find that PcG proteins localize at the DNA replication sites and directly interact with the chaperone that deposits replication-dependent histone H3 variant H3.1. Deposited H3.1 at the replication fork serves as a substrate for trimethylation at K27, and safeguards faithful inheritance of H3K27me3 after DNA replication. Moreover, incorporation of H3.1 facilitates H3K27me3 dynamics during developmental transitions. Our results demonstrate coordination between DNA replication and the deposition and modification of histones, shedding light on how epigenetic marks are transmitted in cycling cells.

T5-14-06

Chromatin signatures at distal regulatory regions in plant genomes

William Ricci[1], *Zefu Lu*[1], *Lexiang Ji*[1], *Jaclyn Noshay*[2], *Nathan Springer*[2], *Robert Schmitz*[1], ***Xiaoyu Zhang***[1]
1. University of Georgia
2. University of Minnesota

A handful of long-distance transcriptional regulatory regions have been described in plants (e.g. the enhancers located ~60 kb upstream of the maize *tb1* gene). However, the prevalence, characteristics and functional importance of long-distance regulatory regions in plant genomes remain poorly understood. By profiling and comparing genome-wide patterns of chromatin accessibility, DNA methylation, RNA polymerase activity as well as histone modifications and variants, we identified thousands of putative regulatory regions in the intergenic space in maize with an average size of ~1.2 kb and average distance to the nearest gene of ~30 kb. These regions are devoid of heterochromatic features such as DNA methylation or H3K9me2, but are associated with distinct sets of chromatin signatures that resemble those found at proximal transcription factor binding sites. The identification, characterization, large-scale functional validation and evolutionary analyses of these long-distance regulatory regions will be presented.

T5-15: Biological studies of mosses and algae

T5-15-01

Molecular phylogeny and morphological diversity of inland *Cladophora* (Cladophorales, Ulvophyceae) from China

Huan Zhu, Zhijuan Zhao, ***Guoxiang Liu,*** *Zhengyu Hu*
Key Laboratory of Algal Biology, Institute of Hydrobiology, Chinese Academy of Sciences

Cladophora has a global distribution and provides habitat and food for numerous organisms. However, the taxonomy of *Cladophora* remained confused and poorly defined. In this study, 69 *Cladophora* freshwater and brackish water samples from inland China were collected and their morphological and phylogenetic analysis were performed. The Concatenated molecular phylogenies based on rDNA sequences revealed that these inland samples belong to two groups: freshwater *Cladophora* and non-freshwater *Cladophora*. The phylogenetic results showed that there were 8 clades in the freshwater group. The results of morphological observation showed characteristic similarities among clades and samples in each clade exhibited different degrees of phenotypic plasticity. The prediction model for ITS2 secondary structure showed that freshwater *Cladophora* contains one ring structure with three main helices (Helix I–III), in which Helix III-2 region is most variable. The high phenotypic plasticity and lack of diagnostic characters of freshwater *Cladophora* group made delineating species difficult and challenging. Further sampling from a broader range and combining with more efficient molecular markers would be necessary to clarify the taxonomy of *Cladophora* in the future.

T5-15-02

The application of large datasets to understanding the ecological and evolutionary diversification of diatoms

Andrew Alverson[1], *Matthew Parks*[2], *Teofil Nakov*[1], *Norman Wickett*[2]
1. University of Arkansas
2. Chicago Botanic Garden

Despite their extraordinary diversity and global importance as primary producers, anchors of aquatic food webs, and biogeochemical cyclers of carbon, oxygen, and silica, efforts to understand basic questions about the phylogeny and evolution of diatoms have lagged behind those of many other groups. We present results of two parallel efforts to better understand the phylogeny, diversification, and ecological history of diatoms. The first effort builds upon a growing genomic and transcriptomic dataset for diatoms, leveraging hundreds of molecular markers to resolve parts of the diatom tree that have presented a challenge for analyses based on fewer characters. A complementary effort that compiled all existing sequence data for 11 genes from 1100 diatom taxa was used to to better understand patterns of character evolution, habitat shifts, and species diversification from a much denser sample of diatom diversity. Results from these projects are providing new insights into the historical importance of transitions between the plankton and benthos and marine and freshwaters. They further highlight that shifts in the mode of sexual reproduction and the evolution of active motility set in motion a radiation that produced the majority of present-day diatom diversity.

T5-15-03

The biosynthesis and transcriptional regulation of a novel sunscreen compound in desiccation-tolerant terrestrial cyanobacteria

Jin-Long Shang, Zhong-Chun Zhang, ***Bao-Sheng Qiu***
School of Life Sciences, and Hubei Key Laboratory of Genetic Regulation and Integrative Biology, Central China Normal University

The mycosporine-like amino acids (MAAs) have strong UV absorption and protect cyanobacteria against UV-B damages. However, the molecular mechanisms linking the UV-B signal transduction and MAA biosynthesis remain largely unknown. Here, a gene cluster with five putative members (*mylA-mylE*) for MAA biosynthesis in *Nostoc flagelliforme* was identified and their expression levels were significantly increased under UV-B exposure. A LuxR family protein OrrA was identified as a positive UV-B responsive regulator binding to the promoter region of *mylA-mylE* gene cluster. OrrA functions as a UV-B activator mediating the UV-B induced *mylA-mylE* gene cluster expression and the MAA biosynthesis in *N. flagelliforme*. Heterologous expression of this gene cluster in *Anabaena* PCC 7120 produces the same MAA as that in *N. flagelliforme*. The assigned MAA structure of mycosporine-2- (4-deoxygadusolyl-ornithine) with molecular formula of $C_{34}H_{52}O_{15}N_4$ is confirmed based on proton and carbon NMR of purified MAA from transformant of *Anabaena* PCC 7120. This compound has a molecule weight (M+H) of 757 Da, the largest MAA compound reported to date. The common precursor of MAA (4-deoxygadusol) was synthesized from sedoheptulose-7-phosphate and catalyzed by the products of the first two genes (*mylA-mylB*) in the cluster. The final product of MAA, mycosporine-2- (4-deoxygadusolyl-ornithine), is catalyzed by the three proteins encoded by genes of *mylC-mylE* in the same gene cluster. The new transcriptional mechanisms and novel MAA molecular structure found in the desiccation-tolerant cyanobacterium *N. flagelliforme* suggest the common physiological significance of MAAs and their chemical modifications associated with evolutionary UV protection strategies.

T5-15-04

Insight on the genomics and epigenetics of *Microcystis*

Liang Zhao[1,2], *Lin Li*[1], *Nanqin Gan*[1], *Jean-Francois Humbert*[3], ***Lirong Song***[1]

1. State Key Laboratory of Freshwater Ecology and Biotechnology, Institute of Hydrobiology, Chinese Academy of Sciences, Wuhan, Hubei, PR China

2. University of Chinese Academy of Sciences, Beijing, PR China

3. UPMC INRA, IEES Paris UMR 7618, Paris, France

Microcystis blooms occur in freshwater lakes, estuaries and rives worldwide. They present major challenges in ecosystems, social economy and water resources management, for the accumulation of vast biomass, as well as the potent cyanotoxins they release. The genomic evolutionary strategy of this cyanobacterium is not well-deciphered, which is mainly due to the low syntenic value each isolate' genome shares and the lack of sufficient complete genomic data. In this work, the genome of *Microcystis* model strain PCC 7806 was integrated using Single Molecule, Real-Time (SMRT) sequencing technology. Accompanied with three more newly released high quality *Microcystis* genomes data, comprehensive genomic comparison was conducted and the evolutionary strategy of this genus emerged. We suppose that the *Microcystis* genomes evolution is a process of genome expansion. That is, strains with small genome sizes are closer with ancestral *Microcystis*, expansion process is driven by transposases mediated gene rearrangement accompanied with frequent gene loss and gain through horizontal gene transfer (HGT), and then increased-genomic size strains emerges. We wish that this finding could help in the taxonomy of *Microcystis* and further set global eyes on the cosmopolitan distribution of this genus. Since SMRT can provide us accurate and intact genomic DNA methylation modification profiles (methylomes) as by-product in genome sequencing, DNA methylation modification survey was also carried out in this cyanobacterium. According to our results, *Microcystis* occupies vast and diversified methylomes, as well as restriction-modification (R-M) systems. A total of 200 methyltransferases were found in five strains, while only 20 were shared by all. And NIES-843 strain contains the largest methyltransferases amount when compared with 3,861 prokaryotic collections. We also concluded that the tremendous R-M systems in *Microcystis* mainly caused by 1) intra-genome replication and mutation, 2) insertion and substitution through HGT-mediated mechanisms. At last, the potential regulative role of DNA methylation modification emerged when we conducted RNA-seq investigation in NIES-843 and PCC 7806 under series of nutrient stresses. This study is anticipated to provide epigenetic clues on some critical bloom questions.

T5-15-05

Proteomic analysis of post translational modifications in cyanobacteria

Feng Ge

Key Laboratory of Algal Biology, Institute of Hydrobiology, Chinese Academy of Sciences

Cyanobacteria are a diverse group of Gram-negative bacteria and the only prokaryotes capable of oxygenic photosynthesis. Recently, cyanobacteria have attracted great interest due to their crucial roles in global carbon and nitrogen cycles and their ability to produce clean and renewable biofuels. To survive at various environmental conditions, cyanobacteria have developed a complex signal transduction network to sense environmental signals and implement adaptive changes. The post-translational modifications (PTMs) systems play important regulatory roles in the signaling networks of cyanobacteria. Mass spectrometry (MS)-based proteomic technologies have been applied to identify PTMs in a systematic manner. Although the proteomic studies of PTMs carried out in cyanobacteria were limited, these data have provided clues to elucidate their sophisticated sensing mechanisms that contribute to their evolutionary and ecological success. We will summarize the current status of PTM studies and recent publications regarding PTM proteomics in cyanobacteria, and provide the novel developments and applications for the analysis of PTMs in cyanobacteria. The results of functional studies of different PTMs in cyanobacteria suggested that these PTMs may affect photosynthesis and carbon metabolism in cyanobacteria as well as other photosynthetic organisms.

T5-15-06

Bioaccumulation of selenium and selenoprotein gene cloning in *Haematococcus pluvialis*

Yihong Zheng, *Ming Tao, Ze Li, Zhangli Hu*

Guangdong Technology Research Center for Marine Algal Bioengineering, Shenzhen Key Laboratory of Marine Bioresource & Eco-environmental Sciences, College of Life Science and Oceanography, Shenzhen University

Selenium (Se) is one of the most interesting trace elements because it is both essential and toxic for most species, with a very narrow range between deficiency and toxicity. Algae can uptake and accumulate Se, which can be either essential or toxic to algal growth depending on the dosage and species. In green alga *Haematococcus pluvialis*, an important industrial algal resource for natural astaxanthin, we investigated the effects of Se on this unicellular organism. The results showed that EC_{50} for the algal growth rate was 24 mg/L, and that a low dosage of selenite (3 mg/L) may not hinder *H. pluvialis* cell growth, but selenite at levels higher than 13 mg/L do restrain cell growth. Bioaccumulation experiments showed that *H. pluvialis* accumulated up to 646 g/g total Se and 380 g/g organic Se, dry weight. However, treatment with high concentrations of selenite significantly increased intracellular hydrogen peroxide levels, antioxidant enzyme activity, and the production of astaxanthin. Thus, *H. pluvialis* is an ideal organism for Se bioaccumulation and organic Se transformation. Microalgae assimilated Se can incorporated in the form of SeCys as the backbone of essential selenoproteins. Among the selenoenzymes in *H. pluvialis*, a thioredoxin reductase (TR) has been identified, which is a mammalian-type NADPH TR containing a SeCys residue and type I *cis*-acting Sec insertion sequence (SECIS) element of eukaryotes in the 3′-UTR region. TR as a selenoprotein is involved in many cellular processes and mainly serves as a key member of the thioredoxin system, which is a crucial antioxidant system in defense against oxidative stress. Molecular properties of HpTR1 were analyzed. The full-length cDNA of HpTR1 consists of 2506 base pairs encoding a putative 535-amino acid product. The conformation of HpTR1 has been characterized via sequence analysis and protein structure modelling. A comprehensive phylogenetic tree of TR1 in animals and algae was constructed, which provided an evolution history of selenoprotein from aquatic to terrestrial environment. The expression pattern of TR1 mRNA and TR activity assay indicate that Se is an important regulator of HpTR1 and plays different roles in a dosage dependent manner.

T5-16: New frontiers in genetics and molecular biology

T5-16-01

R-loop is a general chromatin feature in *Arabidopsis* genome

Wei Xu, ***Qianwen Sun***

Center for Plant Biology, Tsinghua-Peking Joint Center for Life Sciences, School of Life Sciences, Tsinghua University

A three-stranded nucleic acid structure is universally formed in the process of transcription, as the nascent RNA molecule will hybridize with the template DNA strand forming a DNA:RNA hybrid and leaving the non-templated DNA single stranded. Such structures are termed R-loops and have been found in various organisms. However, we still have limited understanding of formation and preferences of R-loops in plant genome. By using the newly developed sDRIP-seq, an efficient, unbiased, high throughput and strands-specific method, we have been able to identify thousands of R-loop loci in *Arabidopsis* genome. Comprehensive analysis showed that R-loops are strongly correlated with many chromatin modifications. Our data suggests R-loop is the common feature of *Arabidopsis* genome organization. We will present our results in this meeting about genome wide R-loops characterization, and discuss their biological functions in plants.

T5-16-02

Hybrid seed failure among wild tomato lineages: Perturbed genomic imprinting and expression in the developing endosperm

Morgane Roth[1], *Ana Marcela Florez-Rueda*[2], *Margot Paris*[3], ***Thomas Städler***[1]

1. ETH Zurich

2. University of Zurich

3. University of Fribourg

Endosperm misdevelopment leading to hybrid seed failure (HSF) is a frequent cause of postzygotic reproductive isolation in angiosperms. This phenotype is found in interploidy and homoploid crosses between closely related species. In a variety of studies, histological evidence points to impaired endosperm development at early stages of seed development in crosses yielding inviable mature seeds. Parent-of-origin-dependent expression (that is, genomic imprinting) is thought to be necessary for normal endosperm development. The HSF phenotype might thus be impacted by differences in patterns of genomic imprinting between parental species. The parental conflict theory posits that imprinting is driven by divergent parental interests in outcrossing lineages. Imprinting status may evolve relatively rapidly, thus creating incompatibilities between nascent species. To date, published studies have not yet addressed the potential correlation between imprinting divergence and HSF. The tomato clade (*Solanum* section *Lycopersicon*), encompassing at least 13 closely related species with partial-to-complete HSF phenotypes and diverse mating systems, provides outstanding opportunities to study these issues. For the current study, we employed intra- and interspecific crosses of three diploid, self-incompatible tomato lineages. *Solanum arcanum* var. marañón, *S. chilense*, and *S. peruvianum* were chosen for their recent divergence and partial to near-complete reciprocal HSF. Based on transcriptome sequencing of laser-extracted endosperm, we quantified the extent of shared imprinting among the three reciprocal intraspecific crosses. Candidate imprinted genes have higher mean expression levels compared to non-imprinted genes, implying that specific mechanisms contribute to their apparent upregulation. In the two hybrid crosses with strong HSF, maternal proportions are increased and imprinting is systematically perturbed. Such marked changes were not detected in the hybrid cross resulting in partial HSF (*S. arcanum* var. marañón, *S. chilense*), suggesting that the extent of misexpression covaries with the degree of HSF. Future inclusion of self-compatible tomato species may reveal whether imprinting dynamics is driven by parental conflict. This can be addressed by quantifying the proportions of imprinted genes across a wider range of species, and by studying the molecular evolution of 'conserved' imprinted genes among both self-compatible and self-incompatible lineages.

T5-16-03

Coordinate sister chromatid segregation during pollen development

Roger Schmid, ***Ueli Grossniklaus***

Department of Plant and Microbial Biology, University of Zurich, Zollikerstrasse 107, CH-8008 Zurich

Asymmetric cell divisions play a crucial role during development, for instance to set up distinct cell lineages. The polar localization of internal cellular factors or external signals has often been implied as a mechanism underlying asymmetric cell division.

However, processes affecting sister chromatid segregation during mitosis, generally thought to be random, could also be a decisive factor. For instance, the sister chromatids could carry distinct epigenetic marks or associated proteins, such that their non-random or directed segregation would set up a difference between the resulting daughter cells. Although the directed segregation of marked sister chromatids has been proposed to underlie a variety of biological phenomena, the methods to study such a mechanism has been limited because sister chromatids are genetically identical and mitotic recombination complicates cell biological analyses. We developed a generally applicable method based on labeling sister chromatids, isolating individual nuclei of daughter cells, and single cell genomic sequencing to investigate directed sister chromatid segregation. We applied this method to study whether directed sister chromatid segregation occurs during pollen development, which is characterized by an asymmetric division that gives rise to vegetative and generative cell, the latter producing the two sperm cells and thus contributing to the next generation. During the synthesis phase of the cell cycle both DNA strands (referred to the Watson and Crick strand) of a chromosome serve as a template for the synthesis of the two new DNA strands. Of the resulting sister chromatids, one has an old Watson but a new Crick strand, the other has the old Crick but a new Watson strand. For instance, the immortal strand hypothesis postulates the directed segregation of the replication-error-free template strands into the daughter cell retaining stem cell fate to prevent stem cells from accumulating replication errors. We used strand-specific single cell genomic sequencing of DNA isolated from vegetative and sperm cells, respectively, allowing us to track sister chromatid segregation during pollen development. While the segregation of the old strand to either vegetative or generative cell was random, we observed the old strands of all chromosomes to segregate coordinately into one of the daughter cells. The unexpected finding of coordinate sister chromatid segregation was confirmed using an independent, fluorescence-based method and seems to be evolutionarily conserved in plants and animals.

T5-16-04

Transgene-free genome editing in wheat through transient expression of CRISPR/Cas9 DNA, RNA or RNP

Caixia Gao

Institute of Genetics and Developmental Biology, Chinese Academy of Sciences, Beijing 100101

Crop improvement requires the constant creation and use of new allelic variants. Conventional breeding can be limited in providing the genes and alleles required to meet the agricultural challenges. In the past decade, Genome editing can accelerate plant breeding by allowing the introduction of precise and predictable modifications directly in an elite background. The most promising utilization of both the CRISPR/Cas9 system and TALENs can be used to generate targeted genome modifications including mutations, insertions, replacements and chromosome rearrangements. Typically, TALEN or CRISPR expression cassettes are delivered to plant cells and expressed, which cleaves chromosomal target sites and produces site-specific DNA double-strand breaks (DSBs), leading to genome modifications during the repair process. We developed simple and efficient genome editing approaches in which wheat plants are regenerated from callus cells transiently expressed with CRISPR/Cas9 reagents introduced as DNA, RNA or RNP. The effectiveness of the three methods in yielding specifically targeted, transgene-free mutants in T_0 generation was validated using six different wheat genes. This approach may be widely applicable for producing genome edited crop plants and has a good prospect of being commercialized.

T5-16-05

Using Structure-seq to probe RNA structure-function relationships *in vivo* and genome-wide in *Arabidopsis* and rice

Sarah M. Assmann[1], *Zhao Su*[1], *Laura Ritchey*[2], *Yin Tang*[1], *David Tack*[1], *Yunqing Yu*[1], *Philip C. Bevilacqua*[2]

1. Department of Biology, Pennsylvania State University, University Park, Pennsylvania 16802

2. Department of Chemistry, Pennsylvania State University, University Park, Pennsylvania 16802

Due to the single-stranded nature of RNA, a given RNA molecule can fold into different structures depending on the surrounding physico-chemical environment. We recently developed a method, Structure-seq, which combines chemical probing of RNA structure with high throughput sequencing to allow characterization of RNA secondary structure genome-wide and *in vivo*. Structure-seq relies on covalent modifications by dimethyl sulfate at the Watson-Crick face of unpaired As and Cs which are read out as stops to reverse transcription. This information is then used to restrain structure prediction based on free energy minimization algorithms. Application of Structure-seq to Arabidopsis reveals significant correlations between DMS reactivity and alternative polyadenylation, alternative splicing, and gene function. We are currently applying Structure-seq to both Arabidopsis and rice with the goal of uncovering general principles that govern changes in RNA secondary structure induced by abiotic stresses, particularly temperature and salinity, and their functional consequences.

T5-16-06

Genome-wide fishing expeditions: chromatin features of regulatory regions in *Arabidopsis thaliana* and *Zea mays*

Johan Zicola[1], *Liangyu Liu*[3], *Rurika Oka*[2], *Blaise Weber*[2], *Maike Stam*[2], ***Franziska Turck***[1]

1. Max Planck Institute for Plant Breeding Research

2. Universiteit van Amsterdam

3. Capital Normal University

One fundamental question remains how eukaryotic cells sharing the same genome sequence develop distinct transcriptional responses to external and internal cues. The expression of distinct sets of transcription factors is one part of the answer. Furthermore, the propensity of genes to respond to transcription factors is altered between cell types due to epigenetic variation at regulatory regions, which may be located at considerable distance from their target genes. Recent progress in the characterization of chromatin features allows an approach to pinpoint candidate regulatory regions by making use of conserved chromatin features. I will present examples of distal plant enhancers in *Arabidopsis thaliana* and *Zea mays*. To predict transcriptional enhancers in the crop plant maize, we integrated available genome-wide data on DNA methylation with newly generated maps for chromatin accessibility and Histone 3 Lysine 9 acetylation (H3K9ac) enrichment in young seedling and husk tissue and classified 1,500 intergenic regions as enhancer candidates. Enhancer candidates were linked to putative target genes based on tissue-specific chromatin and expression patterns. Last, I present data illustrating that siRNA induced methylation can be a versatile method to validate enhancer candidates *in vivo*.

T5-17: Genomics and domestication of food crops

T5-17-01

The origin of economically important species of *Cucumis*, *Citrullus*, and *Momordica* – combining Chinese genomics know-how and old European herbaria to find wild relatives

Susanne S. Renner
University of Munich (LMU)

China is World-leading in Cucurbitaceae genomics, having produced the draft genomes of cucumber, *Cucumis sativus* (Huang *et al.*, Nature Genetics 2009; also Qi *et al.*, Nature Genetics, 2013; Shang *et al.*, Science 2014; Zhou *et al.*, Nature Plants 2016) and watermelon, *Citrullus 'lanatus'* (Guo *et al.*, Nature Genetics 2013). Old-fashioned Sanger sequencing over the past 10 years, however, has revealed numerous misclassifications of material of Cucurbitaceae, from simple misidentification of germplasm to serious misplacements of species in unnatural genera, meaning that names do not reflect evolutionary relationships and cannot be trusted to guide sampling for genomic projects. The family Cucurbitaceae is underrepresented in herbaria because wild cucurbits often are climbers in disturbed habitats or desert plants that are hard to find at the right stage. Fruits of the family are difficult to preserve. Only with DNA barcoding has it become possible to figure out what is what (simply put). In the talk I will illustrate how molecular data are revolutionizing our knowledge about the wild relatives of *Cucumis sativus* (in India, Thailand, and tropical China), *Cucumis melo* (in India and Australia), *Citrullus lanatus* (in West or Northeast Africa), *Momordica charantia* (in Africa) and *Luffa* (in India and Australia), using examples of recent and ongoing work in my lab, in collaboration with genomics experts from China.

T5-17-02

Meiotic and epigenetic behaviour of alien chromatin enabling exploitation of novel variation

Trude Schwarzacher[1], *Niaz Ali*[2], *Liu Qing*[4], *Adel Sepsi*[3], *Chetan Patokar*[1], *Pat Heslop-Harrison*[1]
1. University of Leicester, Department of Genetics
2. Hazara University, Masehra
3. Agricultural Institute, Centre for Agricultural Research, Hungarian Academy of Sciences
4. South China Botanical Garden, Chinese Academy of Sciences

Exploitation of the Triticeae gene pool through chromosome engineering and introgression has been the most effective way to introduce novel traits to wheat breeding lines. Transfer and expression of alien chromatin in wheat is critical to access more genetic variability. Many wheat varieties carrying rye or *Thinopyrum* chromatin show increased resistance to biotic and abiotic stresses. Genomic *in situ* hybridization (Schwarzacher *et al.*, 1992) identifies alien chromatin in hybrid and introgression lines and, combined with repetitive DNA probes, we have identified novel translocations, chromatin segments and multiple alien introgressions. The wheat variety 'Mace' shows wheat streak mosaic virus resistance and carries *Th. intermedium* chromatin as a translocation. Sister lines of 'Mace' were found to carry additionally small segments of rye on the short arm of 1B (Ali *et al.* 2016). Other lines investigated carry various intercalary *Th. bessarabicum* translocations involving different recipient wheat chromosomes, wheat-barley addition lines and different rye translocations. Alien chromatin transfer uses wheat x alien hybrids followed by back crossing and loss of most alien chromatin. Loss processes are not controllable but large numbers of lines can be screened by phenotype, markers, and/or chromosomal constitution. It is important to generate recombinant wheat-alien chromosomes to introduce small segments of the alien chromatin so only small numbers of genes carrying desirable traits are transferred. We studied mitotic and meiotic processes to understand stable inheritance of chromosomes to the next cell generation through mitosis; and pairing, recombination and segregation at meiosis. Centromere behaviour is critical for meiotic events and we have investigated epigenetic modification of centromeres using immuno-staining with the centromere specific histone variant CENH3 (Sepsi *et al.*, 2016). We studied DNA methylation and histone modifications to understand the behaviour and epigenetic interactions of alien chromatin in a wheat background. To exploit the diversity of alien species in wheat breeding, we need not only be able to identify alien chromatin segments and specific genes, but also control chromosome stability, meiotic behaviour and gene expression.

T5-17-03

The domestication and modern breeding in rice

Xuehui Huang[1], *Nori Kurata*[2], *Xinghua Wei*[3], *Bin Han*[4]
1. Shanghai Normal University
2. National Institute of Genetics
3. China National Rice Research Institute
4. Institute of Plant Physiology and Ecology, Chinese Academy of Sciences

The domestication of cultivated rice is one of the most important developments in history. However, its origins and domestication processes are have long been debated. We collected a large number of diverse rice accessions from wild rice, landraces and modern cultivars. In the search for signatures of selection, we identified 55 selective sweeps that have occurred during domestication. In-depth analyses of the domestication sweeps revealed that japonica rice was first domesticated from Oryza rufipogon in southern China, and indica rice was subsequently developed from crosses between japonica rice and local wild rice in South East and South Asia. Furthermore, through population genetics approaches we identified 20 selective sweeps during modern breeding in rice, the pattern of which was quite different from that in rice domestication. These works provide an important resource for rice breeding and functional genomics studies.

T5-17-04

Genomic history of vegetable domestication

Sanwen Huang
1. Agricultural genomics Institute At Shenzhen, Chinese Academy of Agricultural Sciences
2. Institute of vegetables and flowers, Chinese Academy of Agricultural Sciences

The majority crop today is domesticated from their wild ancestors. Aid by analyses of 360 tomato genomes, we identified 111.0 Mb sweep regions and re-built the genomic history of tomato. A total of 13 fruit weight QTLs were subjected to domestication and play important parts in a more than 100-fold fruit weight increase in modern tomato than its wild ancestor. As an important resource of vitamins, minerals and other nutritious in daily diet, how breeding have changed metabolome of tomato fruits was largely unknown. By multiple-omics integration of variome, transcriptome and metabolomes, we identified 3,426 mGWAS (metabolites genome-wide association study) signals for 514 metabolites and 42.26% signals were supported by eQTLs (expression quantitative trait locus). We found the genetic basis of how steroidal glycoalkaloids (SGAs) were domesticated and identified a UDP-transferase centered genes clusters functioning during fruit ripe process. Fruit mass is an important domesticated trait, and we take fruit weight *fw11.3* as an example to illustrate how selection for bigger fruits hitchhike neighboring genes that altered multiple metabolites. Color-targeted selection produced pink tomato divergence, and more than 100 metabolites were identified to underlie this phenotype. Gene editing of the transcription factor *Myb12* reveals the comprehensive metabolic and transcriptional networks involving 261 metabolites and 869 genes. To tap the genetic potential and elite traits from wild resource, exotic genomic fragments were introgressed into elite cultivars. We found many metabolic evidences and unexpected changes in fruits through the case of Tm-^{2} resistance genes from *S. peruvianum*, and the changed chemicals might contribute to its pathogen resistance. Wild cucumber plants bear extremely bitter fruits, a trait conferred by the dominant *Bt* gene. Non-bitterness of cultivated cucumber fruit is conferred by *bt*, a domestication gene. We previously mapped the *Bt* locus to a domestication sweep on chromosome 5, and further fine mapping delimited it to a 442-kb region that harbors 67 predicted genes. By RNA-seq, we considered *Csa5G157230* to be a good candidate for *Bt*, given that it is specifically expressed in the fruit of the wild line PI 183967, consistent with the distribution of cucurbitacin C in wild cucumber. In addition, clear positive correlations were observed between the expression levels of *Csa5G157230* and *Bi*, and between fruit bitterness and gene expression, especially in 5 ancient lines, whose fruits are extremely bitter. A local association analysis was next performed using all variants within the 442-kb region to identify further genetic variants associated with the extremely bitter phenotype. We identified two key mutations within the regulatory region of *Csa5G157230*, one structural variant SV-2195 and an SNP-1601. SV-2195 mainly exists in five extremely bitter ancient lines, leading to a dramatic decrease in fruit bitterness. However, we discovered that SNP-1601 is essential for conditional bitterness in cucumber fruits, a phenomenon in some cucumber lines whereby fruits become bitter under stress conditions. When the mutation corresponded to SNP-1601 is 'A', fruits were non-bitter even under low temperature conditions.

T5-17-05

Subgenomic loss, morphorlogical and expression changes during wheat polyploidization events

***Qixin Sun**, Zhongfu Ni, Huiru Peng, Yingyin Yao, Zhaorong Hu, Mingming Xin*
China Agricultural University

Polyploidization has been recognized as a major force in driving and shaping the evolution and speciation of vascular plants. The nascent polyploidy organisms have to reconcile immediate challenges of divergent subgenomes in one nucleus to secure an exclusive intra-genome pairing and orchestrated inter-genomic gene expression in the initial generations to adapt to the environment. However, to what extent polyploids modifying genomic changes and by what molecular means polyploids accommodating environmental constraints are still a challenging question. Hexaploid wheat (BBAADD) is a typical allopolyploid species with three distinct subgenomes and has undergone two separate allopolyploidization events, providing an ideal model for studying polyploid plants. By exploiting three sets of synthetic hexaploid wheats and whole exome capture coupled with RNA-sequencing methods, we observed that the proportion of genomic sequence loss was significantly higher in the inter-genetic region than in other regions during polyploidization events. In addition, subgenome D experienced more sequence loss compared with subgenome A and B. Furthermore, different chromosomes exhibited distinct genomic modification features during wheat hexaploidization scenario, and the sequences with high similarities among wheat subgenomes are preferentially lost. In addition, we also examined gene expression trends of wheat homeologs during polyploidization, and found that up to 20% of genes showed non-additive expression patterns in leaf and root. Based on the high-throughput transcriptome sequencing of hexaploid seedlings under normal and stressed conditions, we presented that *c*. 68.4% of homeologous genes exhibited expression partitioning, which possibly contributes to the wide adaptability and distribution of hexaploid wheat in various environments. 1) The homeologs of *TaEXPA1-A* and *TaEXPA1-D* was expressed while *TaEXPA1-B* was silenced in wheat seedling leaves. Chromatin immunoprecipitation assays revealed that the transcriptional activation of *TaEXPA1-A* and *TaEXPA1-D* was correlated with increased levels of H3K4 trimethylation and H3K9 acetylation, and decreased levels of H3K9 dimethylation in their promoters, respectively. Moreover, a higher level of cytosine methylation was detected in the promoter region of *TaEXPA1-B*, which may contribute to its silencing in leaves; 2) The increased root hair length in allohexaploid wheat originated in the allotetraploid progenitors, and subgenome A biased expression of *TaRSL4* gene by genome interplay shapes root hair length in allopolyploid wheat and consequently lead to enhanced shoot fresh biomass under nutrient-poor conditions; 3) Under salt stress, synthetic hexaploid exhibited higher fitness than both of its parental genotypes, including fresh weight of aerial part, chlorophyll content and K^+/Na^+ ratio. Moreover, the expression of the *TaSOS1* gene, a plasma membrane Na^+/H^+ antiporter, showed an upregulated transcription in the synthetic allohexaploid wheat than its tetraploid and diploid parents. Among of this, *TaSOS1-D* homeolog showed a high expression than A and B homeologs, suggesting that the salt stress induces enhanced expressivity of the D-subgenome homeologs in the synthetic hexaploid wheat.

T5-18: Bioinformatics resources for comparative functional and phylogenetic analysis of plant genomes, pathways and diversity

T5-18-01

Gramene Ensembl Browser: A resource for browsing & comparing plant genomes

Xuehong Wei[1], *Joshua Stein*[1], *Andrew Olson*[1], *Marcela Tello-Ruiz*[1], *Dan Bolser*[2], *Paul Kersey*[2], *Pankaj Jaiswal*[3], *Doreen Ware*[1]
1. Cold Spring Harbor Laboratory
2. European Bioinformatics Institute
3. Department of Botany & Plant Pathology, Oregon State University

The Gramene database (www.gramene.org) is an integrated resource for comparative genomics and pathways in plants. The Ensembl software powered genome browser is the comparative genome portal of Gramene, and is built collaboratively with Ensembl Plants. As of Jan 2017, Gramene Genome Browser hosts 44 complete reference genomes spanning across a wide range of plant kingdom, while maintaining a deep coverage on agronomic importance species, such as grasses. For each genome, Gramene provides a standard set of baseline annotations including protein-coding genes, Interpro domains, GO, cross-reference to other sources, repeats, DNA features such as ESTs. Depends on the data availability, resources to support integration and community interest for some genomes, Gramene also hosts variation data such as SNP and genetic markers (SSR, RFLP), population data with genotype/phenotype, noncoding genes, expression data (RNAseq/microarray), and epigenomic data (methylation). To elucidate evolutionary history of these genomes, Gramene performs genome-wide comparisons at both the protein and DNA level. For protein comparisons, Gramene provides phylogenetic gene trees predicted by EnsemblCompara GeneTrees pipeline. The gene trees provide orthologs/paralogs reference standards for the community. Derived from ortholg prediction, Gramene conducts syntenic regions mapping between a subset of the genomes, using rice, grape and arabidopsis as reference for monocots and dicots. These gene and chromosome comparisons provide a standard reference for cross-species comparisons and characterization conservation of structural and functional features. At the DNA level, Gramene employs Ensembl's Whole Genome Alignment (WGA) pipeline to generate genomic scale alignments between selected genome pairs with the promise to help identify large-scale structure variations and conserved non-coding regulatory features. In addition to serving as a data resource for the community, Gramene also provides many useful tools/functions to help scientists conduct their research. The Gramene browser allows users to add their own data under views such as Location and Karyotype to display it along with other Gramene data; Variation Effect Predictor (VEP) tool allows users to predict the effect of their SNPs in a reference genome; AssembyConverter enables user to transform genomic coordinates across different assembly versions of the same species; BLAST tool serves as an entry point for browsing the region of interest given a pieces of sequence; BioMart tool enables complex queries of sequence, annotation, homology, and variation data, and provides an additional gateway into the genome browsers; Gramene search tool, a recent development from Gramene, provides an unified search against integrated databases of both genomic data and pathway data, results shown interactively in aggregate and gene-context views. Gramene is supported by NSF grant IOS-1127112, and partially from USDA-ARS 1907-21000-030-00D.

T5-18-02

Gramene's Plant Reactome portal: A resource for comparative plant pathway analysis

Sushma Naithani[1], *Justin Preece*[1], *Parul Gupta*[1], *Justin Elser*[1], *Antonio Mundo*[2], *Joel Weiser*[3], *Peter Deustachio*[4], *Lincoln Stein*[3], *Doreen Ware*[5], *Pankaj Jaiswal*[1]
1. Oregon State University
2. European Bioinformatics Institute
3. Ontario Institute of Cancer Research
4. NYU School of Medicine
5. Cold Spring Harbor Laboratory/USDA-ARS

Gramene's pathway portal Plant Reactome (http://plantreactome.gramene.org/) features *Oryza sativa* (rice) as a reference species for the curation of cellular-level pathway networks by employing the Reactome data model that organizes metabolites, proteins, enzymes, small molecules, and various macromolecular interactions into reactions and pathways in the context of their subcellular location within a plant cell. This reference set of curated rice pathways is being used to generate gene-orthology based pathway projections for 66 additional species, including plants and photosynthetic microbes, with sequenced genomes and/or transcriptomes. Currently, the Plant Reactome hosts various types of pathways, including metabolic, transport, genetic/regulatory, developmental and signaling pathways. Users can compare reference rice pathways with the projected pathways from any plant species from the available list. The pathway clustering across the broad phylogenetic spectrum of photosynthetic organisms shows distinct gene-pathway association patterns reflecting evolutionary history and ploidy levels. Plant Reactome also provides analysis tools for the visualization and analysis of omics data, links to external sources that provide detailed information on various pathway entities (e.g., Gramene's species-specific Ensembl Genome browser and gene pages, UniProt, ChEBI, PubChem, PubMed, GO *etc.*). Users can access data programmatically via APIs or download it in various standard formats. The Gramene project is supported by an NSF award (IOS-1127112). The Plant Reactome database is produced with intellectual and infrastructure support provided by the Human Reactome award (NIH: P41 HG003751, ENFIN LSHG-CT-2005-518254, Ontario Research Fund, and European Bioinformatics Institute (EBI) Industry Programme).

T5-18-03

Population genomic analyses of resequenced *Amborella trichopoda* individuals reveal chromosomal regions of selection and genetic diversity associated with local environmental conditions

Richard G. J. Hodel[1,2], *Luke M. Chandler*[3], *Matthew A. Gitzendanner*[1,2], *Pamela S. Soltis*[2,3], *Douglas E. Soltis*[1,2,3]
1. Department of Biology, University of Florida

2. Florida Museum of Natural History
3. Genetics Institute, University of Florida

Amborella trichopoda, an understory shrub native to wet tropical forests in New Caledonia, is the sole living species in the sister lineage to all other extant angiosperms. A draft genome of *Amborella* (*Amborella* Genome Project, 2013) has served as a guide to infer the genome structure of the most recent common ancestor of extant angiosperms. As part of the *Amborella* Genome Project, the genomes of 12 individuals from the native range in New Caledonia were resequenced. Population genomic analyses of the 12 individuals revealed high levels of genetic diversity for such a geographically restricted species. Several population genomic metrics were computed across the scaffolds constructed as part of the original draft genome assembly. Some metrics varied greatly across the genome (e.g., some scaffolds were highly polymorphic and had variation in Tajima's *D* between positions on the scaffold, whereas other scaffolds were virtually invariant and had negative values of *D*), yet the fragmentary nature of the draft genome assembly made it unclear how these population genetic metrics and selection measures varied across chromosomes. The *Amborella* genomic assembly has now been improved; 13 chromosomal assemblies are now available to use as references for mapping reads of the 12 resequenced individuals. We used Bowtie 2 to map the reads from the 12 individuals to each of the chromosomal references and identified variants with SAMtools and FreeBayes. The software package MAPGD was used to estimate various population genomic statistics, such as allele frequencies, heterozygosity and *F*-statistics, across all chromosomes to locate hotspots in the genome that have unusually high or low variation among individuals. The R package PopGenome was used to calculate Tajima's *D* for each position on every chromosome to identify regions of the genome under especially low or high selection. We used locality data from digitized collection records (available from iDigBio; www.idigbio.org) and Bioclim environmental data (available from www.worldclim.org) to build ecological niche models using the software MAXENT to quantify suitable habitat throughout New Caledonia. We then tested for correlation between genomic regions under strong selection or genomic hotspots of variation and modeled environmental conditions in each of the sampling localities. Our analyses will identify regions on each of the 13 chromosomes that have unusually strong or weak selection, and we will pinpoint regions of chromosomes that are hotspots of genetic variation. By modeling environmental conditions, we can test for correlation between local environment and signatures of selection or high genetic variation in the genome. In addition to improving our understanding of selection and genetic variation across each chromosome, our results will serve as a resource for future investigations of the genomic origins of phenotypic variation in *Amborella*.

T5-18-04

Transcriptome single-molecule long-read sequencing of *Amborilla trichopoda*, and a genome-reference sequence independent method for detecting alternative splicing

Xiaoxian Liu[1,2], *Pamela S. Soltis*[2,3], *Douglas E. Soltis*[1,2,3], *W. Brad Barbazuk*[1,3]
1. Department of Biology, University of Florida, Gainesville, FL 32611-8525, USA.
2. Florida Museum of Natural History, University of Florida, Gainesville, FL 32611-7800, USA.
3. Genetics Institute, University of Florida, Gainesville, FL 32610, USA

Alternative splicing (AS) is a major source of mRNA transcript diversity, proteome diversity, and functional innovation, but examining AS in species without well-annotated reference genomes is difficult. Thus, AS studies in non-model plants (e.g., naturally occurring species), which contribute to a majority of diversity, are very rare. We applied Iso-Seq™ to investigate AS in *Amborella trichopoda,* a phylogenetically pivotal species that is sister to all other living angiosperms. Our data show that, compared to RNA-Seq, the Iso-Seq™ platform provides better recovery on large transcripts, new gene locus identification, gene model correction, especially for AS detection. Reference-based AS detection with Iso-Seq™ data identifies AS within a higher fraction of multi-exonic genes than observed for published RNA-Seq analysis (45.8% vs. 37.5%). Using the Iso-Seq-defined transcript collection in *Amborella* as a reference, we also describe a pipeline for detection of AS isoforms from PacBio Iso-Seq™ without using a reference sequence (*de novo*). Results using this pipeline show a 66-76% overall success rate in identifying AS events. This *de novo* AS detection pipeline provides a means to accurately characterize and identify bona fide alternatively spliced transcripts in any non-model plant species that lacks a reference genome sequence. Hence, it has important genetic/genomic implications to the broader plant biology community.

T5-18-05

Biotaphy--Connecting resources to enable large scale biodiversity analyses

Douglas Soltis[1], *Ryan Folk*[1], *James Beach*[2], *Stephen Smith*[3], *Jose Fortes*[1], *Aimee Stewart*[2], *Pamela Soltis*[1], *Alex Thompson*[1], *Jeff Cavner*[2]
1. University of Florida
2. University of Kansas
3. University of Michigan

Extracting biological knowledge from detailed, complex datasets such as those now being compiled requires integrated, powerful tools. Recent developments in phylogenetics coupled with emerging cyberinfrastructure and new data sources provide unparalleled opportunities for mobilizing and integrating massive amounts of biological data, driving the discovery of complex patterns and new hypotheses for further study. We have developed a cyberinfrastructure framework (Biotaphy) linking diverse data (phylogeny, morphology, ecology, fossils, geography, and climate). We are making these linkages across several data-driven community research tools, including Open Tree of Life (and other tree resources), providing a framework to which all other biological data-traits, genes, genomes, and especially specimens are linked, as well as Lifemapper (geospatial data, modeling and analyses), and iDigBio (~1 billion museum specimens that carry locality data and their implicit ecological information). These linkages provide researchers the opportunity to rapidly synthesize rich datasets and use them to perform detailed hypothesis tests addressing diverse evolutionary questions. Biotaphy tools and infrastructure can connect phylogeny with species distribution models, climate projections, genes and traits. As test cases (presented here) we have used: 1) a well-sampled phylogeny of vascular plants of Florida and 2) the hyper-diverse plant order Saxifragales provide excellent test cases for integrative study. Using complete (or nearly so) datasets that cover diverse traits (e.g., morphology, ecology, geography, fossils, and climate) provide motivating biological test cases for tool refinement and their integration. Such integration will be transformative for studies of biodiversity and provides one vision for how a global integration of powerful tools will lead to data-driven discovery in "next generation" biodiversity science.

T5-18-06

Phylogenetic structure (Rc, Rc, fix, Rc, fix (S)) and micro-structural evolution of organellar markers

Michael Krug[1], *Kai Müller*[2], *Philip Testroet*[1], *Sarah Noben*[5], *Yang Liu*[3], *Bernard Goffinet*[3], *Thomas Borsch*[4], *Dietmar Quandt*[1]

1. Nees Institute for Biodiversity of Plants, Bonn

2. Institute for Evolution and Biodiversity, Münster

3. University of Connecticut, Storrs

4. Botanic Garden and Botanical Museum Berlin-Dahlem, FU Berlin

5. University of Zurich

Non-coding plastid DNA sequences can be confidently used to infer evolutionary relationships at far deep level among angiosperms than previously assumed (e.g. Borsch 2003). In comparison with multi-gene approaches, the results even suggested that fast-evolving, non-coding regions are more effective in reconstructing phylogenies at high taxonomic levels. Subsequently this idea was tested by Mueller *et al.*, (2006) who showed that regions evolving close to neutrality are indeed more effective than slowly evolving genes. Later, Barniske *et al.*, (2012) provided more insight into the capability of noncoding regions to resolve phylogenetic backbones (in this case eudicots): spacer and intron regions clearly outperformed coding regions except for *matK* which ranked highest. Two questions remain: 1) does the neutrality assumption hold true for other evolutionary old lineages as well? 2) is this phenomenon also present in mitochondrial data? So far, all studies have been solely based on plastid data. To address these points, the divergence of the major moss lineages was chosen as further test case, as i) mosses represent another evolutionary old clade, and in fact older than angiosperms and ii) various non-coding mitochondrial markers are accessible for all lineages. We used extensions of the phylogenetic structure (R) framework in order to a) correct R for staggered sequence alignments (R_C), b) calculate Rc for a given topology ($R_{C,\,fix}$) and c) characterize the distribution of R on a subset of nodes. ($R_{C,\,fix}$ (S)). Our analyses imply that i) high functional constraints generally lead to low phylogenetic structure that is concentrated in phylogenetically old nodes, ii) evolutionary constraints of organellar non-coding regions of both compartments are similar, iii) plastid markers seem to be more informative than mitochondrial markers in general, iv) an unexpected high number of long SSRs, especially tri-, tetra-, and penta- SSRs are observed, and v) indels are less homoplastic than substitutions.

T5-19: Pan genome studies: Technologies, strategies and beyond

T5-19-01

Advances in genomic big data analysis to boost breeding of more productive crops: Wheat as an example

Gil Ronen

NRGene

Doubling crop plant productivity over the next 30 years is crucial to meet the world's future food requirements. Despite of the vast annual investment in breeding, the year-over-year increase in yield rates is half of that. A shift in breeding from random trial-and-error to a predictive process of defining and selecting favorable DNA combinations, is expected to accelerate yield gain. PanMAGIC is the first ever computational platform that can analyze complex plant genomes and mine DNA data of millions of individual plants in order to find the favorable DNA makeup leading to yield gains. PanMAGIC organizes the broad genomic diversity of any given crop into a pangenome database and uses it to impute the full genomic content of plants from very low-cost genotyping data. Genomic information is then matched with field performance data to discover the set of genes that favorably affect yield. The bread-wheat genome is a hexaploid, 16 Gb genome (five times larger than the human genome) with over 85% repeated sequences, whose assembly and analysis was impossible prior to PanMAGIC. This talk describes the first ever successful construction of the full reference sequence of bread wheat, and its immediate uses to create the wheat pangenome and support active breeding programs of wheat across the globe.

T5-19-02

Histone modifications of diploid and tetraploid cottons reveal global insight into gene activities and genome evolution

Zhiguo Wu, *Gai Huang*, *Xingpeng Wen*, *Yan Yang*, *Yuxian Zhu*

Wuhan University

In the whole processes of eukaryotic cell development and differentiation, functional genes are widely regulated in precise pattern, through various regulatory elements such as promoters, enhancers and suppressors. Histone modifications are highly correlated with the regulatory elements; H3K4me3, H3K27ac and H3K27me3 specifically label promoters, enhancers and suppressors, respectively. Genome wide regulatory elements are unraveled via high through-put Chip-seq experiments and systematic bioinformatics. We carried out 4 histone modification and 2 transcription factor Chip-seq experiments in tetraploid upland cotton Gossypium hirsutum (AADD) wild type and a fiberless mutant, and in its two diploid ancestors Gossypium arboreum (AA) and Gossypium raimondii (DD). We compared the differences of histone modifications in AADD wild type and mutant, found that 678 genes were specifically expressed in wild type ovules while 455 genes were specifically expressed in mutant ovules. We investigated the histone modification signals and transcription factor binding signals in the two diploid cotton (AA and DD) and compared with the tetraploid cotton AADD, found that Chip-seq signals were even stronger in D subgenomes than A subgenomes. Considering that Gossypium raimondii cotton fiber abortion happens before elongation stage, the combination of D genome expressions and A genome expressions enhances upland cotton fiber length and quality. This phenomenon indicated that there were extensive interactions and regulations between D subgenome and A subgenome in the tetraploid.

T5-19-03

De Novo Whole Genome Assembly in Allo-Octoploid Strawberry

Kenta Shirasawa*, *Soichiro Nagano, Hirakawa Hideki, Nagasaki Hideki, Maeda Fumi, Isobe Sachiko*
Kazusa DNA Research Institute

Strawberry (Fragaria × ananassa) is an allo-octoploid (2n=8X=56) and allogomous species. The whole genome sequence of strawberry was published by Hirakawa *et al.*, (2014). Although the data has contributed genetic analysis in strawberry, the assembled sequences consisted of large number of sequences with short N50 length, suggested the necessary of further improvement. Therefore we performed *de novo* whole genome assembly again by using DenovoMAGIC which was developed in NRGene. A Japanese variety, 'Reikou', was used for construction of paired end (PE), mate pair (MP) and 10X genomics libraries. Illumina reads with 250 nt (PE) and 160 nt (MP and 10X) were obtained by HiSeq platforms, and assembled by DenovoMAGIC3.0. The total length of the assembled sequences was 1,406 Mb, consisted of 32,715 sequences and N50 of 3.9 Mb. Meanwhile, a high-density genetic linkage map was constructed with an S1 mapping population of 'Reikou' by mapping SNPs on the IStraw 90K Axiom® Array and previously published SSR markers. The 'Reikou' linkage map was consisted of 11,608 loci spanning 2,827.4 cM of 31 linkage groups. A total of 62 (31 × 2 haplotypes) pseudomolecules were developed based on the linkage map in a total length of 1,125 Mb. By comparison with the linkage map, 40 mis-assembly was found in the scaffolds. The mapping result of the pseudomolecules on the genome of F. vesca, a possible diploid ancestor of strawberry, clearly indicated the different structure of the four homoeologous genomes in strawberry. This is the first report on the pseudomolecules developed in strawberry, which is expected to accelerate genomics, genetics and molecular breeding for strawberry.

T5-20: Non-coding RNAs: the long and the short of it

T5-20-01

Role of long noncoding RNA in RNA-directed DNA methylation

Andrzej Wierzbicki
University of Michigan

RNA-mediated transcriptional gene silencing is a conserved process where small RNAs target transposons and other sequences for repression by establishing chromatin modifications. A central element of this process is long noncoding RNA (lncRNA) which has been proposed to serve as a binding scaffold for small RNAs and associated proteins. In *Arabidopsis thaliana*, this lncRNA is produced by a specialized RNA polymerase known as Pol V. We performed genome-wide identification of Pol V transcripts using RNA immunoprecipitation followed by high throughput sequencing. We show that noncoding transcription by Pol V is controlled by preexisting chromatin modifications located within the transcribed regions. The vast majority of Pol V transcripts is associated with AGO4 but are not substrates for slicing by AGO4. lncRNA-binding proteins remain in close proximity to the Pol V complex and siRNA is likely to base pair with lncRNA exiting Pol V. Finally, Pol V preferentially transcribes into silenced transposable elements. We propose a model explaining several key events surrounding noncoding transcription by Pol V and predicting that determination of heterochromatin boundaries is an important function of noncoding transcription.

T5-20-02

Custodian of RNA induced silencing complexes in plants

Xiuren Zhang
Texas A&M University

RNA silencing is implemented through RNA-induced Silencing Complex (RISC) that is composed of microRNAs (miRNAs) and Argonaute (AGO) proteins. AGOs use miRNAs as guides to slice target mRNAs with highly complementary sequences to produce truncated 5' and 3' RNA fragments. The 5' cleaved RNA fragments are marked with uridylation for degradation to accelerate RISC recycling. However, key enzymes responsible for such degradation and whether the clearance of cleaved targets affects miRNA metabolism remain poorly understood. Here, we identified a novel cofactor of Arabidopsis AGOs, named RICE1 (RISC Implanted Clearing 3'- 5' Exoribonuclease 1). RICE1 and its closest paralog, RICE2, specifically degraded single-strand RNAs but positively regulated miRNA accumulation in vivo. RICE1 exhibited a DnaQ-like exonuclease fold and formed a homohexamer with the active sites located at the interfaces between RICE1 subunits. Whereas dimerization-impaired RICEs were unstable in vivo, ectopic expression of catalytically inactive RICE1 significantly reduced miRNA levels and the retaining of miRNAs in RISC; notably, catalytically inactive RICE1 also increased 5' cleavage RISC fragments with extended uridine tails resulting from RISC activity. Our results suggest that RICEs are engaged in the clearance of the uridylated cleavage products to maintain functional RISC and to secure miRNA sheltering in RISC in vivo. Thus, this study provides new insight into the coordination between decay of cleaved target mRNAs and miRNA stability in eukaryotes.

T5-20-03

DiRNAs reach out for DSB repair

Mingming Liu, Ba Zhaoqing, ***Yijun Qi***
Tsinghua University

Repair of DNA double-strand break (DSB) is critical for the maintenance of genome integrity. We have previously shown that a class of DSB-induced small RNAs (diRNAs) facilitates homologous recombination (HR)-mediated DSB repair in *Arabidopsis thaliana*. How diRNAs mediate DSB repair remains largely unknown. We find that INVOLVED IN DE NOVO 2 (IDN2), a double-stranded RNA (dsRNA) binding protein involved in small RNA-directed DNA methylation, is required for DSB repair in *Arabidopsis*. We find that IDN2 interacts with the heterotrimeric replication protein A (RPA) complex. Depletion of IDN2 or the diRNA-binding ARGONAUTE 2 (AGO2) leads to increased accumulation of RPA at DSB sites and mislocalization of the recombination factor RAD51. These findings support a model in which IDN2 interacts with RPA and facilitates the release of RPA from ssDNA tails and subsequent recruitment of RAD51 at DSB sites to promote DSB repair.

T5-20-04

miRNA-mediated flowering control in temperate grasses

Wu Liang
Zhejiang University

The transition from vegetative growth to flowering stage is crucial for seed propagation in plants. The highly conserved florigen encoded by FLOWERING LOCUS T (*FT*) functions at the core of the flowering pathways. Extensive studies have examined the transcriptional regulation of *FT*; however, other layer of *FT* regulation remains obscure. We identify miR5200, a Pooideae-specific microRNA, in *Brachypodium distachyon*, a model plant of temperate grasses, and find that miRNA is specially expressed in leaves and targets *FT* orthologs for mRNA cleavage. miR5200 is dramatically accumulated in plants grown under short-day (SD) conditions, but is dramatically repressed in plants transferred to long-day (LD) conditions. We also find that the epigenetic chromatin status, specifically the levels of histone methylation marks, at miR5200 precursor loci are changed in response to day-length. Artificial interruption of miR5200 activity by target mimicry in *B. distachyon* alter flowering time in SD but not in LD conditions, indicating a role of miR5200 in photoperiod-mediated flowering time onset. Furthermore, we find that regulation of miR5200 by day-length is highly conserved in temperate grasses. Together, our results show a posttranscriptional regulatory mode of *FT* by miRNA, and provide insights into understanding of the multiple concerted pathways for flowering time control in plants.

T5-20-05

MicroRNA functions in *Arabidopsis* embryos

*Divya Vashisht, Aleksandra Plotnikova, Max Kellner, Magdalena Mosiolek, **Michael Nodine***
Gregor Mendel Institute

MicroRNAs (miRNAs) are a class of small regulatory RNAs that repress key developmental regulators and are essential for plant embryo development. Previously we found that miRNA-deficient embryos exhibit widespread differentiation defects and precociously express maturation-phase genes. This indicates that miRNAs are required for both embryonic pattern formation and the timing of the morphogenesis-to-maturation phase transition. Because plant miRNAs typically repress transcription factors and other key developmental regulators they likely have a large influence on the gene regulatory networks that control plant embryogenesis. Our major goal is to understand how miRNAs shape the gene regulatory networks that govern plant embryogenesis, and I will present our progress towards this aim.

T5-20-06

Epigenetic regulation of the plant micro RNA machinery turnover.

Pablo Manavella
Inst. AgroBiotecnologia del Litoral (IAL)

Micro RNAs (miRNA) are 21 nucleotides molecules essential for post-transcriptional gene silencing. These molecules are generated from a structured RNA precursor and then incorporated into the RNA induced silencing complex (RISC). DICER-LIKE 1, HYPONASTIC LEAVES 1 and SERRATE are the main proteins involved in miRNA biogenesis but several new cofactors have being described recently. After processing of the miRNA precursors, ARGONAUTE 1 (AGO1) binds the mature miRNAs and inhibits targeted-mRNA translation or induces its cleavage. Using a forward genetic approach we have isolated novel mutants deficient in miRNA activity, from which several cofactors of the pathway, such as CPL1, RCF3 and the THO/TREX complex, have been recently described. Here we describe *MIRNA-SILENCING SUPPRESSED 47 (mss47)*, another miRNA deficient mutant identified in our genetic screening. Small RNA blots and RT-qPCR showed that *mss47* mutants present altered level of miRNAs and targets' mRNAs. *MSS47* encodes a protein with a histone-methyltransferase domain. This protein does not directly regulate the transcripts levels of *MIRNA* genes or its targets. Our experiments show that MSS47 epigenetically regulates the abundance of essential factors of the miRNA pathway by modulating their degradation. Such depletion in the miRNA machinery impairs the production and activity of miRNAs and, therefore, the gene silencing and plant development.

T5-21: Epigenetics II: DNA modifications

T5-21-01

Encyclopedia of plant methylomes - from patterns to mechanisms and functions

Robert Schmitz
University Of Georgia

Advances in DNA sequencing technologies have led to the first single-base resolution maps of cytosine DNA methylation. These maps are revealing a diverse set of mechanisms that genomes utilize to control genome stability, transposon silencing, and gene expression. Our previous studies in natural and experimental *Arabidopsis thaliana* populations have uncovered errors in maintenance of DNA methylation that lead to the existence of spontaneous epigenetic alleles. The rates of formation of single methylation polymorphisms are about five orders of magnitude greater than the spontaneous mutation rate, whereas the formation of differentially methylated regions that affect gene expression occurs at a frequency that is similar to the spontaneous DNA mutation rate. To better understand the mechanisms governing maintenance of DNA methylation over longer evolutionary timescales, we sequenced and used publicly available DNA methylomes of 34 flowering plant species and the basal flowering plant Amborella trichopoda. Additionally, we have developed a novel high-throughput analysis method for profiling DNA methylation levels for any species regardless of the availability of a reference genome. We are unraveling the mechanistic underpinnings that evolved to establish and maintain DNA methylation by using this base modification as a molecular phenotype in combination with phylogenetic and molecular analysis of enzymes involved in DNA methylation.

T5-21-02

Critical role of a DNA demethylase in tomato fruit ripening

Zhaobo Lang

Shanghai Center for Plant Stress Biology, Center of Excellence for Molecular Plant Science, CAS

DNA methylation is a conserved epigenetic mark important for genome integrity, development and environmental responses in plants and mammals. Active DNA demethylation in plants is initiated by a family of 5mC DNA glycosylases/lyases (i.e. DNA demethylases). The role of DNA demethylases in transposon regulation, pathogen responses, and gene imprinting and other developmental processes have been studied extensively in the model plant *Arabidopsis thaliana*. Recent studies suggested a role of active DNA demethylation in fruit ripening in tomato. In this study, we generated loss-of-function mutant alleles of a tomato gene, *SlROS1/SlDML1/2*, which is a close homolog of the *Arabidopsis* DNA demethylase gene *ROS1*. In the fruits of the tomato mutants, increased DNA methylation was found in thousands of genes. These genes include not only hundreds of ripening induced genes but also many ripening repressed genes. Our results show that *Sl-ROS1* is critical for tomato fruit ripening, and suggest that active DNA demethylation is required for both the activation of ripening induced genes and inhibition of ripening repressed genes.

T5-21-03

Hybrid and hybrid mimics in *Arabidopsis*

Li Wang[1], *Ian K. Greaves*[2], *Elizabeth S. Dennis*[1,2], *W. James Peacock*[1,2]

1. Faculty of Science, University of Technology, Sydney, NSW 2007, Australia

2. Agriculture and Food, Commonwealth Scientific Industrial Research Organisation, Canberra, ACT 2601, Australia

Hybrids in *Arabidopsis* and crop species are uniform and high yielding. In contrast, the F_2 generation shows a great deal of phenotypic heterogeneity and does not maintain the hybrid yield advantage. The reduced yield and heterogeneity of the F_2 as well as the increased yield of the F_1 are the basis of the hybrid seed industry as farmers have to buy fresh seed each planting season. To address the molecular mechanisms of both the F_1 and F_2 features we carried out recurrent selection using *Arabidopsis* hybrids combination C_{24} and Landsberg *erecta* (*Ler*) in an attempt to "fix "the hybrid vigour phenotype. We generated pure breeding Hybrid Mimic lines with similar properties to the F_1 hybrids and we also selected pure breeding Small Phenotype lines. By comparing the transcriptomes of the selected F6 lines, both Hybrid Mimic and Small Phenotype lines, we determined which changes in gene expression were shared in the mimics and the hybrid, but not shared with Small plant line, these genes presumably are associated with the large rosette phenotype. Basal defence pathways, hormone biosynthesis and response pathways are major pathways altered in the Hybrid Mimic lines and F_1 hybrids. A number of auxin responsive genes promoting leaf growth were upregulated in the F1 hybrids and Hybrid Mimics. The transcription factor PHYTOCHROME-INTERACTING FACTOR (PIF4), may be responsible for these altered activities of auxin-associated genes by targeting the auxin biosynthesis gene *YUCCA8* and regulating auxin signalling through *IAA29*. The Hybrid Mimics are essentially homozygous with chromosome segments from each parent. This finding indicates that the vigour phenotype does not require heterozygosity throughout the genome. 12 conserved chromosomal segments each sourced from one particular parent were present in all of the Hybrid Mimic lines whereas in the F_6 Small Phenotype line the 12 segments were derived from the alternative parent. Loci critical for promoting growth vigor may be contained in each of these conserved segments. In the hybrids segments from the chromosomes of both parents are combined in the one nucleus leading to alterations in gene expression involving trans-interactions. These can be genetic or epigenetic interactions. *Arabidopsis* ecotypes used as parents for hybrids often are more different epigenetically than genetically. In F_1 hybrids Trans Chromosomal Methylation (TCM) and Trans Chromosomal deMethylation (TCdM) occur, resulting in gain or loss of methylation at loci where two parental alleles are differentially methylated. Changed methylation patterns in the F_1 hybrids can be inherited in Hybrid Mimics which though impacting gene expression may contribute to their phenotypes. TCM is dependent upon the RNA directed DNA Methylation (RdDM) pathway and 24 nt siRNAs; in the *RNA polymerase IV* (*pol IV*) mutant TCM is decreased. The *Pol IV-V* pathway is not critical for the hybrid vigor phenotype but the *DECREASED DNA METHYLATION 1 (DDM1)* methylation pathway is required for the full level of hybrid vigor.

T5-21-04

Transcriptomic dynamics after perturbing maize methylome

Qing Li[1,2], *Sarah Anderson*[1], *Peter Hermanson*[1], *Zhaoxue Han*[1], *Nathan Springer*[1]

1. Microbial and Plant Genomics Institute, Department of Plant Biology, University of Minnesota

2. National Key Laboratory of Crop Genetic Improvement, Huazhong Agricultural University

DNA methylation is an important regulator of the expression of both genes and transposons. We previously collected mutant alleles for maize genes that play important roles in controlling DNA methylation. Here RNA sequencing was used to investigate the transcriptomic changes in seedling tissue of plants with altered maize DNA methylation. Expression levels of the majority of genes are not changed in most mutant alleles. However, plants containing two alleles, one loss-of-function allele for chromomethylase (*CMT*) gene and one for *DDM1* gene exhibit strong expression changes in more than 1000 genes compared to their wild-type controls. While no consistent patterns between DNA methylation and gene expression were observed for genes that are down-regulated in the mutant plant, genes with up-regulated expression in the mutants generally have high CHG methylation across gene body and this CHG methylation was greatly lost in the mutants. Consistent with a role of DNA methylation in response to environmental stress, many of the genes with altered expression in the methylome mutants also display expression changes under various stress conditions. In an attempt to study the prevalence of DNA methylation in regulating gene expression in natural populations, we profiled DNA methylation in near 20 diverse maize inbred lines. Interestingly, many of the differentially expressed genes in the methylome mutants have variation in DNA methylation among these inbred lines, many of which also affect natural variation in gene expression. This suggests that many targets of epigenetic regulation in maize are naturally varying, providing a possibility to exploit these epigenetic changes in maize breeding.

T5-21-05

Identification and characterization of DNA methylation readers in *Arabidopsis thaliana*

Jonathan Cahn[1], *Ino Karemaker*[2], *Ozren Bogdanović*[1], *Jahnvi Pflueger*[1], *Owen Duncan*[1], *Harvey Millar*[1], *Michiel Vermeulen*[2], ***Ryan Lister***[1]

1. ARC Centre of Excellence in Plant Energy Biology, The University of Western Australia

2. Radboud Institute for Molecular Life Sciences, Radboud University Nijmegen

Methylation of cytosine bases in genomic DNA (mC) provides an additional layer of information that has the potential to alter transcription and chromatin state. In plants, mC plays an important role in the transcriptional repression of transposable elements, and is established, maintained and removed by a variety of "writer" or "eraser" factors, depending on local DNA sequence context (CG, CHG and CHH, where H = A, C or T). mC can be interpreted by "reader" proteins, some of which are known to be involved in feedback loops with other modifications such as histone methylation. While some plant mC reader proteins have been documented, a comprehensive survey of the proteins that can bind mC in plants is required to identify the range of factors that can interpret this DNA modification. To screen for potential mC readers, we performed a DNA affinity pull-down assay combined with quantitative mass spectrometry, using short methylated DNA probes for each sequence context. We identified 37 nuclear proteins that show a significant enrichment in methylated probes in at least one sequence context, including 12 previously characterized mC readers such as VIM proteins, SUVH2/6/9 and MBD5/6/7. Approximately half of the identified candidates have not previously been reported as potential mC binding proteins. To further explore the function of these mC reader proteins, we are utilizing ChIP-seq to study their DNA binding sites genome-wide, tandem affinity purification coupled with mass spectrometry to identify interacting proteins, and analysis of the transcriptome and DNA methylome in mutant plants to assess their potential role in gene regulation and the epigenome. This work constitutes the first comprehensive assay to identify potential mC readers in plants, and a initial steps toward understanding their molecular functions.

T5-21-06

Epigenetic silencing-mediated alternative polyadenylation (APA) regulates CHG methyaltion by promoting the proper expression of histone demethylase IBM1

Cheng-Guo Duan[1,2], *Xingang Wang*[2], *Mingguang Lei*[2], *Kai Tang*[2], *Daisuke Miki*[1], *Jian-Kang Zhu*[1,2]

1. Shanghai Center of Plant Stress Biology, SIBS, CAS

2. Department of Horticulture and Landscape Architecture, Purdue University

DNA methylation-dependent heterochromatin formation is a conserved mechanism of epigenetic silencing of transposons and other repeat elements in many higher eukaryotes. Genes adjacent to repetitive elements are often also subjected to this epigenetic silencing. Consequently, plants have evolved anti-silencing mechanisms such as active DNA demethylation mediated by the REPRESSOR OF SILENCING 1 (ROS1) family of 5-methylcytosine DNA glycosylases to protect these genes from silencing. Some transposons and other repeat elements have found residence in the introns of genes. It is unclear how these intronic repeat elements-containing genes are regulated. Our recent work identified several new anti-silencing factors via genetic screen and biochemical methods. Genome-wide DNA methylation analysis reveals that these proteins has a similar role to that of the histone H3K9 demethylase INCREASE IN BONSAI METHYLATION 1 (IBM1) in preventing CHG methylation in the bodies of thousands of genes. We found that these protein ensure the proper expression of IBM1 full-length transcript by promoting the distal polyadenylation. Our research also found that the intronic heterochromatin region is required for the usage of alternative polyadenylation (APA). Through mRNA sequencing, we identified many genes containing intronic transposon elements that require these proteins for proper expression. Our study identifies several new regulators of CHG methylation and reveals a unique mechanism by which higher eukaryotes deal with the collateral effect of silencing intronic heterochromatin.

T5-22: Plant genome editing (two sessions)

T5-22-01

A high-efficiency CRISPR/Cas9 system for multiplex genome editing in plants and a sequencing decoding software DSDecode

Yao-Guang Liu

College of Life Sciences, South China Agricultural University

CRISPR/Cas9 genome targeting systems have been applied to a variety of species. We have developed a robust and versatile CRISPR/Cas9 vector system for high efficiency multiplex genome editing in plants. We designed PCR-based procedures to rapidly generate multiple sgRNA expression cassettes, which can be assembled into the binary CRISPR/Cas9 vectors in one round of cloning by Golden Gate ligation or Gibson Assembly, for editing of a number of target sites by single plant transformation. Average 85% or higher rates of mutations were generated in rice. We also developed a web-based tool 'DSDecode' and its updated version 'DSDecodeM' (skl.scau.edu.cn) for decoding sequencing chromatograms with superimposed peaks derived from heterozygous and biallelic mutations of targeted sites. These systems have provided a usable toolbox for studying functions of multiple genes and gene families in plants for basic research and genetic improvement.

T5-22-02

Evidence of high-frequency off-target mutagenesis induced by CRISPR/Cas9 in *Arabidopsis* and improvement of the specificity

Hui-Li Xing, *Qiang Zhang*, *Zhi-Ping Wang*, *Hai-Yan Zhang*, ***Qi-Jun Chen***

State Key Laboratory of Plant Physiology and Biochemistry, Col-

lege of Biological Sciences, China Agricultural University, Beijing 100193, China.

We presented evidence that high frequency off-target effects of CRISPR/Cas9 existed in *Arabidopsis* and thus very possibly in other plants. To improve the specificity, we examined six mutant SpCas9 variants: two eSpCas9 reported in human cells and named for their enhanced specificity, one SpCas9-HF1 reported in human cells and named for its high fidelity, and three not reported combinatorial forms of the two types. We demonstrated that a few variants worked well in plants: showed much higher specificity and maintained on-target activities comparable to wild-type SpCas9. We also demonstrated that on-target activities of the high specificity Cas9 variants were dependent on expression levels of sgRNA/Cas9 complexes. We tested some strategies for improving the expression level of the sgRNA/Cas9 and found some of them useful.

T5-22-03

Double-strand break induced genome engineering in plants

Holger Puchta
Botanical Institute, Karlsruhe Institute of Technology, Karlsruhe GERMANY

Sequence-specific nucleases can be used to induce double-strand breaks (DSBs) in plant genomes. In the past we could show that thus gene targeting (GT) by homologous recombination (HR) can be enhanced and targeted mutagenesis can be achieved by error-prone non-homologous end joining (NHEJ). Moreover, by inducing several DSBs, sequences can be deleted out of the genome or chromosome arms exchanged. In the last years CRISPR/Cas became the major tool for targeted mutagenesis. We were able to demonstrate *Streptococcus pyogenes* (Spy)Cas9 nuclease induced, NHEJ mediated, heritable targeted mutagenesis in *Arabidopsis thaliana* as well as homology dependent *in planta* GT. Off-target effects might be avoided using two sgRNAs and a Cas9 protein that was transformed from a nuclease to a nickase, to induce adjacent single strand breaks (SSBs) in opposite strands. This "paired nickase" strategy has a mutagenic potential at the target site comparable to the nuclease. Interestingly; sequence duplications are a prominent outcome of this approach, hinting to the possibility that the repair of adjacent SSBs is a major cause of sequence duplications during genome evolution of plants. Recently we applied the Cas9 orthologues from *Streptococcus thermophilus* (Sth1Cas9) and *Staphylococcus aureus* (SauCas9) for NHEJ-mediated targeted mutagenesis in *A. thaliana* with efficiencies at least comparable to those of SpyCas9. We were also able to show that the SauCas9 and SpyCas9 proteins only work in the presence of their species-specific single guide (sg) RNAs and show no inter-species interference. Thus, the Cas9 proteins of *S. pyogenes* and *S. aureus* should be appropriate for simultaneously addressing different sequence motifs with different enzyme activities in the same plant cell. The simultaneous use of different Cas9 orthologues will offer the opportunity to control genetic information of plant cells on more complex levels than before and will lay the basis for future synthetic approaches in plant biology.

T5-22-04

A CRISPR-Cpf1 system for efficient genome editing and transcriptional repression in plants

Xu Tang[1], Levi G Lowder[2], Tao Zhang[3], Daniel F. Voytas[4], Yong Zhang[1], ***Yiping Qi[5]***
1. Department of Biotechnology, School of Life Science and Technology, Center for Informational Biology, University of Electronic Science and Technology of China, Chengdu 610054, China
2. Department of Biology, East Carolina University, Greenville, North Carolina 27834, USA
3. Jiangsu Key Laboratory of Crop Genetics and Physiology, Key Laboratory of Plant Functional Genomics of the Ministry of Education, Yangzhou University, Yangzhou 225009, China
4. Department of Genetics, Cell Biology & Development and Center for Genome Engineering, University of Minnesota, Minneapolis, Minnesota 55455, USA
5. Department of Plant Science and Landscape Architecture, University of Maryland, College Park, Maryland 20742, USA

Clustered regularly interspaced short palindromic repeats (CRISPR)-Cpf1 has emerged as an effective genome editing tool in animals. Here we compare the activity of Cpf1 from *Acidaminococcus sp. BV3L6* (As) and *Lachnospiraceae bacterium ND2006* (Lb) in plants, using a dual RNA Polymerase II promoter expression system. LbCpf1 generated biallelic mutations at nearly 100% efficiency at four independent sites in rice T0 transgenic plants. Moreover, we repurposed AsCpf1 and LbCpf1 for efficient transcriptional repression in Arabidopsis, and demonstrated reduction of *miRNA159* transcription >10-fold. Our data suggest promising applications of CRISPR-Cpf1 for editing plant genomes and modulating the plant transcriptome.

T5-22-05

Optimizing gene targeting strategies in plants

Dan Voytas
Dept. of Genetics, Cell Biology & Development and Center for Genome Engineering, University of Minnesota, Minneapolis, MN 55455 USA

The ability to precisely modify plant genomes through homologous recombination (HR) promises to advance both basic and applied plant biology. However, even with the use of sequence-specific nucleases, which stimulate HR by creating targeted DNA double-strand breaks, there are only a handful of studies that report precise editing of endogenous plant genes. Our group has been focusing on two efforts to more effectively modify plant genomes through HR. In one, we are developing new vectors to deliver sequence-specific nucleases and DNA repair templates to plant cells. Specifically, we have been using geminivirus replicons, which function in both monocots and dicots, to amplify nuclease-encoding cassettes and DNA repair templates. In a second effort, we are attempting to achieve HR by either genetically manipulating DNA repair pathways or delivering nucleases and repair templates to cells proficient in HR. Progress on our efforts to optimize gene targeting strategies will be reported.

T5-22-06

Precision mutagenesis in plants via NHEJ or HR pathway

Seiichi Toki
Plant Genome Engineering Research Unit, NARO Institute of

Agrobiological Sciences, National Agriculture and Food Research Organization, 2-1-2 Kannondai, Tsukuba, Ibaraki 305-8602, Japan

There are two types of genome editing technique, non-homologous end-joining (NHEJ) mediated targeted mutagenesis and homologous recombination (HR) mediated gene targeting. In/del and substitution types of mutation can be introduced at the cleavage site of sequence-specific nucleases (SSNs) via imprecise NHEJ. Therefore, targeted mutagenesis is useful if we want to knock-out a targeted gene. The number of papers reporting plant targeted mutagenesis with CRISPR/Cas9 has increased markedly. However, widely used SpCas9 protein from Streptococcus pyogenes uses guanine-rich PAM sequence (NGG). So, it is difficult to mutagenize thymidine-rich sequences. The PAM sequence of Cpf1 from Fransisella novicida is TTN. We have shown that targeted mutagenesis had occurred in transgenic plants expressing FnCpf1. Furthermore, Cpf1 generates DNA ends with 5' overhang, whereas Cas9 creates blunt DNA ends after cleavage. Sticky DNA ends should increase the efficiency of insertion of a desired DNA fragment into the Cpf1-cleaved site using complementary DNA ends via precise NHEJ. In the case of a targeted mutagenesis in seed-propagated crops, once integrated into the genome as a transgene, sequence-specific nucleases (SSNs) genes can be segregated out by self-pollination or back-crossing. However, in the case of crops that are propagated vegetatively, we would like to remove transgene without using genetic segregation. We found that piggyBack transposon can be used as a reversible transgenesis vector in plants to put SSNs IN and OUT without leaving any footprints. Gene targeting enables precise modification of the targeted gene using a specific template DNA which must be supplied exogenously. Usually, a positive selection marker gene is left at the targeted gene after positive/negative-selection-mediated gene targeting. Recently, we succeeded in removing the positive selection marker gene completely using a piggyBack transposition. We can modify any sequence in the rice genome precisely using this system.

T5-22-07

Efficient and heritable genome engineering in barley

Goetz Hensel, *Nagaveni Budhagatapalli, Sindy Schedel, Stefan Hiekel, Twan Rutten, Jochen Kumlehn*
Leibniz Institute of Plant Genetics and Crop Plant Research (IPK)

Genome engineering is a breakthrough technology that facilitates site-directed modifications of genomic DNA in planta. It offers versatile novel possibilities for both the experimental elucidation of gene functions and the improvement of crop performance. Aiming to establish site-directed mutagenesis in *Triticeae* cereals, target gene-specific transcription activator-like effector nucleases (TALENs) were designed and barley mutant plants generated via transfer of a pair of TALEN-coding expression units into haploid cells from which transgenic doubled haploid plants can be produced. Owing to the haploid founder cells and whole genome duplication taking place in the course of the regeneration process, some of the resultant primary mutants proved instantly homozygous, as was indicated by non-segregating progeny. In a second approach, we produced two types of plant lines each carrying only one of the two required TALEN-units. TALEN cleavage activity was then induced in hybrid progeny derived from crossings of complementary pairs of these lines. By this principle, many independent and heritable mutations can be produced even if only a few TALEN-transgenic lines are available, which is a particularly valuable option, especially in plant species that are not readily transformable. Whereas primary mutants are typically chimeric, we demonstrated that chimera dissolution and genetic fixation can be greatly facilitated by generating doubled haploid progeny, i.e. plants which, by definition, inherit no more than one allele each. To allow for optimization of endonuclease construct design and routine prevalidation of target-specific expression units before investing in the laborious creation of stable transgenic plants, a transient expression test was established which indicates cleavage activity via frame shift reconstitution of a reporter gene. More recently, we have also established site-directed mutagenesis using customized RNA-guided Cas9 endonuclease, which derives from the microbial CRISPR/CAS immune system. This genome engineering platform has proved more reliable in generating barley mutant plants as compared with current TALEN technology. We also showed that the proportion of plants carrying genetic modifications at the target site can be as high as 84 percent of the gRNA/Cas9 transgenics produced. Examples will be shown demonstrating targeted mutagenesis using the model cultivar Golden Promise followed by mutant fixation via haploid technology as well as genome engineering of elite breeding material. In addition, homology-dependent genome editing was exemplified at the cellular level using a customized DNA-repair template, which allows to precisely predefine not only the genomic target locus but also the resultant DNA sequence.

T5-22-08

Targeted base editing in plants by Target-AID, the CRISPR/Cas9 deaminase fusion.

Keiji Nishida
Kobe University

DNA deamination is the process by which deoxyribonucleotides turn into other type of bases (ex. deoxycytidine to deoxyuridine). Although most deaminated nucleotides are repaired properly, excess deamination or compromised repair may results in mutagenesis. The activation-induced cytidine deaminase (AID) is responsible for generating genetic variation at the immunoglobulin locus, an essential process for the adaptive immune system in vertebrates. By tethering the DNA deaminase activity to nuclease-deficient CRISPR/Cas9 system, we have developed a genome editing tool that enables targeted point mutagenesis. An AID ortholog PmCDA1 was attached to nuclease-deficient mutant of Cas9 (D10A and H840A). This hybrid system, termed Target-AID, induced cytosine point mutation in 3-5 bases range at the distal site within target sequence. Plant-optimized Target-AID was developed and applied to rice and tomato by agrobacterium mediated transformation to produce base-edited plants to induce loss- and gain-of function mutations.

T5-22-09

Genome-wide target specificities of CRISPR nucleases and deaminases

Jin-Soo Kim

Center for Genome Engineering, Institute for Basic Science & Department of Chemistry, Seoul National University,

Despite broad interest in RNA-guided genome editing, Cas9 and Cpf1 are limited by off-target mutations. We developed nuclease-digested whole genome sequencing (Digenome-seq) to profile genome-wide specificities of these nucleases in an unbiased manner. Digenome-seq captured nuclease cleavage sites at single nucleotide resolution and identified off-target sites at which indels were induced with frequencies below 0.1%. We also showed that these off-target effects could be avoided by using purified Cas9/Cpf1 ribonucleoproteins (RNPs) and modified guide RNAs. Digenome-seq is a robust, sensitive, unbiased, and cost-effective (< USD 1,500) method for profiling genome-wide off-target effects of programmable nucleases.

T5-22-10

Base editing of major cereals using Cas9 nickase-cytidine deaminase fusion

Caixia Gao
State Key Laboratory of Plant Cell and Chromosome Engineering, Institute of Genetics and Developmental Biology, Chinese Academy of Sciences

A plant base editing method (nCas9-PBE), using a Cas9 nickase-cytidine deaminase fusion, has been optimized and validated for genome editing in rice, wheat and maize. Unlike nuclease-dependent genome editing that induce random insertions and deletions, the fusion induces efficient and targeted conversion of cytosine to thymine in both protoplasts and mutant plants, and some of the resulting amino acid mutations have unique and valuable characteristics. The optimized method was tested with five endogenous rice, wheat and maize genes in protoplasts. The frequency of targeted C to T substitution, as estimated by deep amplicon sequencing, varied from 0.57% to 7.07%. Stable rice, wheat and maize mutants carrying desired C to T substitution were obtained, thus validating the value of nickase Cas9-AID mediated base editing for producing mutants useful to the genetic and breeding research of major cereals. This approach will not only technologically advance plant genome engineering, but may also provide better solution for social acceptance of genome-edited crops as they do not require a donor DNA template or chromosomal cleavage.

T5-22-11

Genome editing in agriculture ... an industry perspective on riding the hype-curve for robust outcomes beyond low-hanging fruits

Erik Jongedijk, *Hong Wang*
KWS SAATSE, EINBECK, GERMANY

Recent availability of robust Genome Editing technologies in plants promises to both revolutionize biological research and to facilitate faster and cheaper development of commercial crop varieties with enhanced biotic and abiotic stress characteristics, processing and quality characteristics and yield per se. To date the technology is at the peak of its hype curve ... expectations are sky-high but significant breakthrough developments in a number of adjacent biologic research fields will be required for Genome Editing to achieve its full potential in plant breeding and agriculture. This talk will focus on current opportunities and limitations for Genome Editing technology in modern plant breeding programmes and in particular focus on key developments in plant biology required to overcome current technical, regulatory, public perception and intellectual property hurdles limiting the achievement of its full potential. In addition we will outline KWS' key interests and activities in the field of Genome Editing as well as KWS' open innovation based approach to "research & development" and its consequent unique way of engaging with the scientific community in the academic, the start-up and the corporate research scene.

T5-22-12

Targeted gene activation in plants by a potent dead cas9-based synthetic transcription activator

Zhenxiang Li, Dandan Zhang, Xiangyu Xiong, Binyu Yan, ***Jian-Feng Li***
State Key Laboratory of Biocontrol, Guangdong Provincial Key Laboratory of Plant Resources, School of Life Sciences, Sun Yat-sen University

Targeted activation of gene expression is highly desirable for elucidating and manipulating gene functions in basic and applied plant research. Repurposing the CRISPR/Cas9 technology from a genome editing tool to a targeted transcriptional activation system represents an appealing way to achieve this goal. However, pilot efforts in mammalian cells suggest that the current generation of catalytically dead Cas9 (dCas9)-VP64 transcription activator works with a low efficiency when using a single synthetic guide RNA (sgRNA). Multiple sgRNAs tiling the promoter of target gene need to be used simultaneously to enhance the transcription, leading to increased workload and decreased scalability of the system. Using plant protoplast-based screens, we identified a potent synthetic transcription activator containing dCas9 and a chimeric transcriptional activation module composed of different transcriptional activation domains, each in multiple copies. This improved synthetic transcription activator allows efficient activation of target gene expression using only a single sgRNA in both protoplasts and transgenic plants, which promises multiplex up-regulation of target gene expression in plant cells to rewire regulatory networks.

T5-23: Plant-microbe interactions

T5-23-01

Effectoromics of the citrus Huanglongbing-associated bacteria

Kelley Clark[1], Simon Schwizer[1], Shree Thapa[2], Jessica Franco[2], Nian Wang[3], Gitta Coaker[2], ***Wenbo Ma[1]***
1. University of California Riverside
2. University of California Davis
3. University of Florida

The devastating citrus Huanglongbing (HLB, aka greening disease) has caused billions of dollars of loss in citrus industry worldwide. In the US and China, HLB is dominantly associated with a Gram-negative bacterium *Candidatus* Liberibacter asiaticus (CLas). Transmitted by insect vectors, CLas exclusively colonizes phloem tissues in citrus; importantly, there is no curative treatment after the trees are infected. CLas possesses the Sec secretion system, through which a variety of Sec-delivered effectors (SDEs) could be secreted from the pathogen cells into the citrus phloem. In other bacteria, effectors play an essential role in pathogenicity by suppressing host immunity, facilitating nutrient uptake and causing disease symptoms. We analyzed the genome sequences of ten CLas isolates and determined their SDE repertoire. Each isolate encodes 23-35 SDEs and 13 of them are "core" effectors produced by all the isolates. One of the core effectors, named SDE1, has 40-folds higher expression in citrus than in the insect vector, implicating a possible role in HLB progression. We found that SDE1 directly interacts with several members of papain-like cysteine proteases (PLCPs) in citrus and potentially functions as a broad-spectrum inhibitor. Homologs of PLCPs in Solanaceae plants have been reported to contribute to defense against microbial pathogens. Our proteomic analysis showed that the abundances of specific PLCPs are increased in the phloem of HLB-infected citrus, where the bacteria reside. Furthermore, expression of SDE1 significantly impairs citrus development, supporting a model where SDE1 inhibits PLCP activity to promote HLB. These findings provide key information on HLB pathogenesis and discover promising targets that are subjected to genetic manipulation in order to enhance citrus resistance to this devastating disease.

T5-23-02

RNAi mediated host antiviral defense in rice

Yi Li, *Jianguo Wu, Yongxin Yang, Zhirui Yang, Xiaofeng Cao, Yijun Qi, Yijun Zhou*

1. Peking University

2. Center for Plant Biology, Tsinghua-Peking Center for Life Sciences, College of Life Sciences, Tsinghua University

3. State Key Laboratory of Plant Genomics and National Center for Plant Gene Research, Institute of Genetics and Developmental Biology, Chinese Academy of Sciences

4. Institute of Plant Protection, Jiangsu Academy of Agricultural Sciences

Rice is the most important stable food supply worldwide. RNAi, the evolutionarily conserved mechanism of antiviral defense which is well studied in plants and other eukaryotes, is poorly understood in economically important crops. Rice encodes 19 AGOs, AGO18 is a member of a novel clade of AGOs from rice. We found that AGO18 is highly induced by infections of evolutionary different viruses. Loss-of-function *ago18* mutation abolishes, whereas over-expression of AGO18 increases antiviral activity. Significantly, AGO18 functions antiviral defense through disturbing and sequestering certain miRNAs such as miR168 and miR528 and releasing their targets which are directly involved in antiviral defense. Expression of miR168 resistant AGO1a in *ago18* background rescues rice antiviral activity. Mutated miR528 or expression of miR528 resistant target, *L-ascorbate oxidase* (AO) significantly increases rice antiviral activity. AO localizes in the cell wall and catalyzes apoplastic oxidation of L-ascorbic acid (AsA). We further found that AGO18 sequesters *AO*-targeting miR528 and promotes the role of AO in enhancing antiviral defense response through regulation of basal ROS accumulation. Together with other results, we propose that *miR528* deficiency and AO-mediated plant antiviral defense function in parallel with the AGO18-miR168-AGO1 pathway. While the exact functional mode of ROS in antiviral defense remains to be elucidated, it is well known that ROS are central players in the cellular signaling network in different organisms. To summarize, our study not only reveals a pathway that elegantly regulates the availability of a key defense factor (AO) based on the sensing of virus attack, but also provides strong evidence to support that ROS have an active role in antiviral immunity. We foresee that this unique mechanism could be engineered to improve the antiviral capabilities of crops.

T5-23-03

Interacting calcium pump and BON1 modulate calcium signature, stomatal closure, and plant immunity

Dong-Lei Yang[1,2], Shi Zhenying [1], Yongmei Bao[1,2], ***Jian Hua***[1]

1. Cornell University

2. Nanjing Agricultural University

Calcium signaling is essential for environmental responses including immune responses. Here we provide evidence that the evolutionarily conserved protein BONZAI1 (BON1) functions together with autoinhibited calcium ATPase 10 (ACA10) to regulate calcium signals in *Arabidopsis*. BON1 is a plasma membrane localized protein and the *bon1* loss of function (LOF) mutant has increased expression of immune receptor genes and a consequent autoimmune phenotype. We found that BON1 interacts with the autoinhibitory domain of ACA10 and the *aca10* LOF mutants also have an autoimmune phenotype similarly to the *bon1* mutants. Genetic evidences indicate that BON1 positively regulates the activities of ACA10 and perhaps its homologous protein ACA8 as well. Consistent with this idea, the steady level of calcium concentration is increased in both *aca10* and *bon1* mutants. Most strikingly, cytosolic calcium oscillation imposed by external calcium treatment was altered in *aca10* and *bon1* mutants in guard cells. In addition, calcium and pathogen induced stomatal closure was compromised in the *aca10* and *bon1* mutants. Taken together, this study indicates that ACA10 and BON1 physically interact on plasma membrane and function in the generation of cytosol calcium signatures that are critical for stomatal movement and impact plant immunity.

T5-23-04

Actin remodeling during plant innate immunity requires both capping protein and reactive oxygen species

Jiejie Li

Beijing Normal Univerisy

Plants perceive microbe-associated molecular patterns (MAMPs) and damage-associated molecular patterns (DAMPs) to activate innate immune signaling events, such as bursts of reactive oxygen species (ROS). The actin cytoskeleton remodels during the first five minutes of innate immune signaling in *Arabidopsis* epidermal cells; however, the immune signals that impinge on actin cytoskeleton and its response regulators remain largely unknown. Here, we demonstrate that rapid actin remodeling upon elicitation with diverse MAMPs and DAMPs represents a conserved plant immune response. Actin remodeling requires ROS generated by the defense-associated NADPH oxidase, RBOHD. Moreover, perception of flg22 by its cognate receptor complex triggers actin remodeling through the activation of RBOHD-dependent ROS production. Our genetic studies reveal that the ubiquitous heterodimeric capping protein (CP) transduces ROS signaling to the actin cytoskeleton during innate immunity. Additionally, we uncover a negative feedback loop between actin remodeling and flg22-induced ROS production.

T5-23-05

Argonaute 2 medaited plant resistance to bacteria

Xiaoming Zhang
Institute of Zoology, Chinese Academy of Sciences

Small RNAs (sRNAs) are 20 to 30 nucleotide (nt)-long noncoding RNA molecules that regulate gene expression in eukaryotes through a process generally termed RNA silencing. RNA silencing machinery plays critical roles in plant resistance to virus and other pathogens. To determine the possible function of sRNAs to plant resistance to bacteria, we studied the accumulation of the miRNAs, nat-siRNAs, NBS-LRR siRNAs and other sRNAs at the genome level and found that the accumulations of some sRNAs are specifically regulated by bacteria infection. All sRNAs are loaded into Argonaute (AGO) protein, the key component of RNA silencing pathway, to perform regulation functions. We showed that the accumulation of AGO2 is specifically induced upon bacterial infection. Further analysis uncovered that miR393* loaded into AGO2 can down-regulate the accumulation of MEMB12, a Golgi-localized SNARE gene, and thus increases the secretion of immunity proteins and plant immunity. We also studied the sorting of sRNA to AGOs and found that AGO2 prefers sRNA duplexes without middle and 15th mismatches, while AGO1 can tolerates middle mismatches. Structure remodeling and mutational analyses showed that some conserved motifs within AGO PIWI domain play important roles in this sorting bias.

T5-23-06

Epigenetic regulation of an antagonistic receptor pair confers broad-spectrum and durable disease resistance with yield balance in rice

Yiwen Deng, Keran Zhai, Dongyong Yang, ***Zuhua He***
Shanghai Institute of Plant Physiology & Ecology, CAS

Breeding for broad-spectrum and durable disease resistance with yield balance, i.e. combining high disease resistance and high yield, has been the ideal target of crop breeding. Many resistance (R) genes have been so far identified, however, very few R genes have been adopted for broad-spectrum and durable disease resistance in long-term crop breeding programs due to either resistance breakdown caused by pathogen variation or defense cost with yield penalty. Here we report that the rice *Pigm* locus from an indigenous landrace containing an nucleotide-binding leucine-rich repeat (NLR) receptor cluster confers durable and broad-spectrum resistance to fungal blast (*Magnaporthe oryzae*) without yield penalty. Within the NLR cluster, the NLR PigmR confers broad-spectrum resistance; while another NLR PigmS interacts with and competitively attenuates the homodimerization of PigmR to suppress the PigmR-mediated resistance. The expression of *PigmS* and thus the Pigm-mediated resistance are subjected to tight epigenetic regulation by the RNA-dependent DNA methylation (RdDM) pathway, misregulation of the RdDM pathway altered the PigmR-mediated resistance. Importantly, the yield cost by PigmR-mediated resistance could be counteracted by the PigmS-increased seed production. Therefore, our study reveals a novel mechanism of balancing broad-spectrum resistance and yield through epigenetic regulation of a pair of antagonistic NLRs, providing a tool to develop elite rice with both high disease resistance and yield potential.

T5-24: Genomics and molecular breeding

T5-24-01

A TRIM insertion in the promoter of *Ms2* causes male sterility in wheat

Chuan Xia[1], Lichao Zhang[1], Cheng Zou[1], Yongqiang Gu[2], Jialei Duan[1], Guangyao Zhao[1], Jiajie Wu[1], Liu Yue[1], Xiaohua Fang[3], Lifeng Gao[1], Yuannian Jiao[4], Jiaqiang Sun[1], Yinghong Pan[1], Xu Liu[1], ***Jizeng Jia**[*1], Xiuying Kong[*1]*

1. Key Laboratory of Crop Gene Resources and Germplasm Enhancement, Ministry of Agriculture, The National Key Facility for Crop Gene Resources and Genetic Improvement, Institute of Crop Science, Chinese Academy of Agricultural Sciences, Beijing 100081, China

2. United States Department of Agriculture-Agricultural Research Service, Western Regional Research Center, Albany, CA 94710, USA.

3. Institute of Genetics and Developmental Biology, Chinese Academy of Sciences, Beijing 100101, China

4. Institute of Botany, Chinese Academy of Sciences, Beijing 100093, China

The male-sterile *Ms2* mutant has been known for 40 years and has become extremely important in the commercial production of wheat. However, the gene responsible for this phenotype has remained unknown. Here we report the map-based cloning of the *Ms2* gene. The *Ms2* locus is remarkable in several ways that have implications in basic biology. Beyond having no functional annotation, barely detectable transcription in fertile wild-type wheat plants, and accumulated destructive mutations in *Ms2* orthologs, the *Ms2* allele in the *Ms2* mutant has acquired a terminal-repeat retrotransposon in miniature (TRIM) element in its promoter. This TRIM element is responsible for the anther-specific *Ms2* activation that confers male sterility. The identification of *Ms2* not only unravels the genetic basis of a historically important breeding trait, but also shows an example of how a TRIM element insertion near a gene can contribute to genetic novelty and phenotypic plasticity.

T5-24-02

GW5 acts in brassinosteroid signaling pathway to regulate grain width and weight in rice

Jiafan Liu[1], Jun Chen[1], Xiaoming Zheng[1], Fuqing Wu[1], ***Haiyang Wang**[2], Jianmin Wan[1]*

1. Institute of Crop Science, Chinese Academy of Agricultural Sciences

2. Biotechnology Research Institute, Chinese Academy of Agricul-

tural Sciences

Grain size is a major determinant of grain yield in cereal crops. Previous studies have shown that qSW5/GW5 exerts the greatest effect on rice grain width and weight and is associated with a 1,212-bp deletion, which was selected during rice domestication for increased grain yield. However, the identity of GW5 has remained mysterious. Here we show that a gene encoding a calmodulin binding protein, located ~5 kb downstream of the 1,212-bp deletion, corresponds to qSW5/GW5. GW5 is expressed in various rice organs with highest expression level detected in young panicles. We provide evidence that the 1,212-bp deletion affects grain width through influencing the expression levels of GW5. GW5 protein is localized to the plasma membrane and can physically interact with and repress the kinase activity of rice GSK2, a homolog of Arabidopsis BIN2 kinase, resulting in accumulation of unphosphorylated OsBZR1 and DLT proteins in the nucleus to mediate Brassinosteroid (BR)-responsive gene expression and growth responses (including grain width and weight). Our results suggest that GW5 is a novel positive regulator of BR signaling and a viable target for genetic manipulation to improve grain yield in rice and perhaps in other cereal crops as well.

T5-24-03

The emergence, evolution, and diversification of the miR390-TAS3-ARF pathway in land plants

Rui Xia[1], *Blake Meyers*[2]
1. South China Agricultural University
2. Donald Danforth Plant Science Center

In plants, miR390 directs the production of tasiRNAs from *TRANS-ACTING SIRNA 3* (*TAS3*) transcripts to regulate *AUXIN RESPONSIVE FACTOR* (*ARF*) genes, transcription factors critical for auxin signaling; these tasiRNAs are known as tasiARFs. To understand the evolution of this miR390-*TAS3-ARF* pathway, we characterized homologs of these three genes from thousands of plant species, from bryophytes to angiosperms. We found the lower-stem region of *MIR390* genes, critical for accurate DCL1 (DICER-LIKE 1) processing, is conserved in sequence in seed plants. We propose a model for the transition of functional tasiRNA sequences in *TAS3* genes occurred at the emergence of vascular plants, in which the two miR390 target sites of *TAS3* genes showed distinct pairing patterns in different plant lineages. Based on the cleavability of miR390 target sites and the distance between target site and tasiARF we inferred a potential bidirectional processing mechanism exists for some *TAS3* genes. We also demonstrated a tight mutual selection between tasiARF and its target genes, and characterized unusual aspects and diversity of regulatory components of this pathway. All these data illuminate the evolutionary path of the miR390-*TAS3-ARF* pathway in land plants, and demonstrate the significant variation that occurs in this functionally important and archetypal regulatory circuit.

T5-24-04

New insights into the genetic networks affecting seed lignocellulose content in *Brassica napus*

Liyun Miao[1], *Hongbo Chao*[1], *Li Chen*[1], *Hao Wang*[2], *Maoteng Li*[1]
1. College of Life Science and Technology, Huazhong University of Science and Technology
2. Hybrid Rapeseed Research Center of Shaanxi Province, Shaanxi Rapeseed Branch of National Centre for Oil Crops Genetic Improvement

Oilseed rape or canola (*Brassica napus*) varieties with reduced levels of antinutritive seed phenolic acid and fibre compounds would considerably improve the value of the seed meal after oil extraction. Dissection of the genetic network for lignocellulose biosynthesis is of great importance for improving seed quality. The genetic basis of lignocellulose biosynthesis in B. napus was investigated via quantitative trail locus (QTL) analysis using a doubled haploid (DH) population from a low oil content of oilseed rape line 'KenC-8' crossed with the high oil content of *B. napus* line 'N53-2' with 219 lines. The phenotypic performance and frequency distribution of fiber components in Wuhan and Dali were analyzed. Seed lignin content (LC), cellulose content (CC) and hemicellulose content (HC) were significantly negatively correlated with seed oil content (OC) in DH population, respectively. A total of 85 QTLs were detected for LC, CC and HC. It is very interesting to found that co-localization between QTL for LC, CC, HC and OC was observed on A9 with contribution ranging from 9.87 to 48.50%. A total of 125 homologous genes of *Arabidopsis thaliana* that are involved in lignin, cellulose and hemicellulose biosynthesis were found within the confidence intervals of 75 QTLs. After the candidate gene analysis, a potential regulatory pathway controlling lignocellulose biosynthesis in *B. napus* was constructed including 52 genes. Subsequently, the interaction between the genes and regulatory factors involved in lignocellulose biosynthesis were analyzed. A potential regulatory pathway controlling the lignocellulose was constructed by QTL analysis. 14 candidate genes were identified, including GAPC1, GAPC2, LAC17 and CSLC08. These results enriched our knowledge of QTLs for lignocellulose biosynthesis and provided useful information for understanding the genetics of key seed quality traits in rapeseed.

T5-24-05

Rice Expression Database (RED): An integrated RNA-Seq-derived gene expression database for rice

Lin Xia, Dong Zou, Jian Sang, Xingjian Xu, ***Lili Hao****, Zhang Zhang*
Beijing Institute of Genomics, Chinese Academy of Science

Rice is one of the most important stable food as well as a monocotyledonous model organism for the plant research community. Here we present RED (Rice Expression Database; http://expression.ic4r.org), an integrated database of rice gene expression profiles derived entirely from RNA-Seq data. RED features a comprehensive collection of 284 high-quality RNA-Seq experiments, integrates a large number of gene expression profiles and covers a wide range of rice growth stages as well as various treatments. Based on massive expression profiles, RED provides a list of housekeeping and tissue-specific genes and dynamically constructs co-expression networks for gene (s) of interest. Besides, it provides user-friendly web interfaces for querying, browsing and visualizing expression profiles of concerned genes. Together, as a core resource in BIG Data Center, RED bears great utility for characterizing the function of rice genes and better understanding important biological processes and mechanisms underlying complex agronomic traits in rice.

T5-24-06

Simultaneous identification of multiple causal mutations in rice

Wei Yan[1], *Zhufeng Chen*[2], *Jiawei Lu*[2], *Chunjue Xu*[2], *Gang Xie*[2], *Yiqi Li*[2], *Xing Wang Deng*[2,3], *Hang He*[3], *Xiaoyan Tang*[2,4]

1. College of Life Sciences, Capital Normal University, Beijing, China

2. Shenzhen Institute of Molecular Crop Design, Shenzhen, China

3. College of Life Sciences, Peking University, Beijing, China

4. Guangdong Provincial Key Laboratory of Biotechnology for Plant Dev

Next-generation sequencing technologies (NGST) are being used to discover causal mutations in ethyl methanesulfonate (EMS)-mutagenized plant populations. However, the published protocols often deliver too many candidate sites and sometimes fail to find the mutant gene of interest. Accurate identification of the causal mutation from massive background polymorphisms and sequencing deficiencies remains challenging. Here we describe a NGST-based method, named SIMM, that can simultaneously identify the causal mutations in multiple independent mutants. Multiple rice mutants derived from the same parental line were back-crossed, and for each mutant, the derived F2 individuals of the recessive mutant phenotype were pooled and sequenced. The resulting sequences were aligned to the Nipponbare reference genome, and single nucleotide polymorphisms (SNPs) were subsequently compared among the mutants. Allele index (AI) and Euclidean distance (ED) were incorporated into the analysis to reduce noises caused by background polymorphisms and re-sequencing errors. Corrections of sequence bias against GC- and AT-rich sequences in the candidate region were conducted when necessary. Using this method, we successfully identified seven new mutant alleles from Huanghuazhan (HHZ), an elite *indica* rice cultivar in China. All mutant alleles were validated by phenotype association assay. A pipeline based on Perl scripts for SIMM is publicly available at https://sourceforge.net/projects/simm/.

T5-25: Phenotyping beyond technology - approaching real world problems (two sessions)

T5-25-01

Phenomics, genomics and crop systems model - Three pillars of future crop breeding programs.

Xinguang Zhu

Chinese Academy of Sciecnes

Improving crop yield potential is a major direction of future crop breeding. The rapid advances of genotyping, systems biology and high throughput phenotyping technologies now offer unprecedented opportunities to rapidly advance our capacity to improve crop yield potential. Together with rapidly increasing of genomics information and maturation of crop systems models, modern high throguput phenotyping technologies can help identify the weakest links in an elite cultivar, design the crop ideal-type under a defined environment given certain agronomic practices, and further to identify the molecular markers that can be used in the breeding. Phenomics, genomics and crop systems models are becoming three pillars of future crop breeding programs.

T5-25-02

Crop phenotyping center in HZAU (CPC): Strategies and challenges

Wanneng Yang

1. National Key Laboratory of Crop Genetic Improvement, National Center of Plant Gene Research, Huazhong Agricultural University, Wuhan 430070, China

2. College of Engineering, Huazhong Agricultural University, Wuhan 430070, China

3. Hubei Key Laboratory of Agricultural Bioinformatics, Huazhong Agricultural University, Wuhan 430070, China

The functional analysis of the rice genome has entered into a high-throughput stage, and a project named RICE2020 has been proposed to determine the function of every gene in the rice genome by the year 2020. However, as compared with the robustness of genetic techniques, the evaluation of rice phenotypic traits is still subjective, inefficient, destructive and error-prone. In last decade, the key questions of genomics involved why and how to sequence genomes; now, we are faced with a new challenge: phenomics. To relieve the bottleneck, we established a crop phenotyping center (CPC) in Huazhong Agricultural University, in which we developed a rice automatic phenotyping (RAP), a yield traits scorer (YTS), a high-throughput leaf scorer (HLS), a high-throughput hyperspectral imaging system (HHIS), and a micro-CT. Equipped with multidisciplinary techniques, including photonics, automatics, computers, and mechanics, the integrated facility can extract plenty of morphology-relates traits, biomass-related traits, yield-related traits, and physiological traits. And we also demonstrate that combining high-throughput phenotyping tools and genome-wide association study can unearth more treasures hidden within rice genomics. In addition, the phenotyping had been expanded to maize and rape and we discussed the challenges of rice phenotyping in future.

T5-25-03

Development of next generation phenotyping technologies and application for characterization of genetic resources in pre-breeding scenarios

Fabio Fiorani

Forschungszentrum Juelich GmbH, IBG2 Plant Sciences

Imaging-based phenotyping of crop traits has emerged as an important tool for complementing progress in genomics. A few methods based on imaging time courses of shoot development using visible and non-visible portion of the electromagnetic spectrum are well established and their application demonstrated for quantifying non-destructively leaf area growth as a proxy for biomass accumulation, and reflectance properties linked to pigment composition and to photosynthetic activity. This presentation will first focus on a brief overview of these methods and the current applications for large scale screening of genetic resources in a pre-breeding scenario with examples from different crops (wheat, rapeseed, in particular). Next, the field of plant phenomics however still requires adaptation and of available technologies and automation for plant screening and the development of new sensors. Examples based on our work in this area will focus particularly on sensor development for non-invasive measurements of water content of leaves and developing seeds, including methods that provide volumetric data such as Nuclear Magnetic Resonance (NMR). The case for new approaches and next generation phenotyping technologies will be extended to the study of root architecture and root growth dynamics with the presentation of research phenotyping facilities developed at our IBG2 Plant Sciences Institute that encompass large scale paper pouches and automated rhizoboxes setups, and dedicated Magnetic Resonance Imaging (MRI) and Positron Emission Tomography (PET) for functional studies of root growth and carbon distribution in plants.

T5-25-04

Quantitative information on plants in changing environments - integrated approaches to mechanistic, high-throughput and field phenotyping

Ulrich Schurr, *Roland Pieruschka, Fabio Fiorani, Uwe Rascher*
Forschungszentrum Jülich - IBG-2: Plant Sciences

Plant phenotyping develops rapidly into a bottleneck for progress in basic and applied research. Lack of adequate solutions for quantitative analysis of plant architecture and function as well as their interaction with the dynamic and heterogeneous environment hampers progress in basic sciences as well as in breeding-related research. In recent years significant interdisciplinary approaches have been started to overcome this "phenotyping bottleneck". Techniques were developed to quantify the dynamics and the heterogeneity of plant structure and function as well as of environmental cues. These mostly non-invasive technologies are developed and implemented into biological concepts that allow novel insights in the dynamic characteristics of plants above- as well as belowground. The technologies include high-resolution analysis for mechanistic understanding (like MRI and PET for structure, growth and activity of roots and shoots), the high-throughput approach for analysis of large numbers of genotypes and environmental conditions as well as field approaches, which are the reference to indicate the relevance. The talk will provide an overview on recent developments in technologies as well as conceptual approaches as the basis for a quantitative understanding of plant-environment-dynamics and its application for plant breeding and plant management. The talk will also present recent developments in infrastructure platforms that have been and will be established in Germany, in Europe and globally.

T5-25-05

New phenotyping tools for drought resistance studies in rice

Zilong Guo, Wanneng Yang, Haifu Tu, ***Lizhong Xiong***
Huazhong Agricultural University

Drought severely damages global food production and understanding plant responses to drought benefits drought resistance (DR) breeding. Here, we established a high-throughput phenotyping platform for DR evaluation. More than 50 new image-derived electronic 'traits', which have high heritability and repeatability and can monitor plant dynamic responses to drought, were phenotyped in 507 global rice panel containing 507 accessions at three developmental stages. Using these new traits, we discovered an interesting drought-induced trait association phenomenon. A large number of drought-related loci and candidate genes were identified by genome-wide association study. Integration of the new image-derived 'traits' and traditional DR-related traits reveals a comprehensive genetic network, which led to some significant findings about DR mechanisms. Importantly, we demonstrated that several image traits could be useful for DR evaluation in the field. This study provides new image-derived traits, which are useful in precise and dynamic phenotyping, and demonstrates that an integration of traditional and new image traits is a powerful approach for dissecting complex traits.

T5-25-06

Prototyping of an integrated field phenotyping platform – University of Tokyo field plant phenotyping research laboratory

Seishi Ninomiya*, Wei Guo*
Graduate School of Agriculture and Life Sciences, University of Tokyo

Field Plant Phenotyping Research Laboratory has launched at Institute of Sustainable Agro-ecosystem Services (formerly the University Farm), the University of Tokyo aiming at prototyping of an integrated high-throughput phenotyping platform which will be a part of a high performance breeding pipeline. The laboratory is composed of not only the scientists from several different faculties of the University of Tokyo but also the scientists from other organizations such as NARO, INRA, UQ, IITH, IITB and private sectors accelerating virtual collaboration among them and open science. The campus is located in a very convenient place rather close to the central Tokyo as well as the main campus of U. Tokyo so that the scientist and students not only from U. Tokyo but also visitors even from foreign countries can easily access there. Phenotyping is, of course, to measure crop characteristics. Crop characteristics are controlled and determined by three factors; genotypes, environmental conditions and human managements under fairly complicated interactions among them. To duly evaluate such crop characteristics, we need to understand the complicated relationship among the factors. Namely, we need to develop a model to explain crop performance (phenotype) by genotype, environment and management based on observations; $y = f(g, e, m)$ where y, g, e, m are crop performance (e.g., yield), genotype, environment and managements, respectively. Regarding the data acquisitions for the modeling, field phenotyping is currently a serious bottleneck compared with other factors. Research and development for high throughput plant phenotyping has been becoming popular in Japan. However, the studies still remain in the development of element technologies such as single traits evaluations by image analyses and evaluations of multispectral electromagnetic wave sensors. Thus, we need a study to optimally integrate those technologies to compose a high throughput breeding pipeline. Key research topics in the laboratory are; (1) aerial sensing, (2) ground sensing, (3) image analysis and 3D reconstruction, (4) artificial intelligence and modeling, (5) high throughput training data set collection, (6) plant physiology sensors, (7) environment sensor network, (8) database and data exchange platform, (9) crop performance prediction modeling which can handle G X E X M (8) prototyping of an integrated high-throughput phenotyping platform which will be a part of a high performance breeding pipeline.

T5-25-07

Field phenotyping: Quantifying dynamic plant traits across scales in the field.

***Onno Muller**, M. Pilar Cendrero-Mateo, Mark Mueller-Linow, Beat Keller, Angelina Steier, Benedikt Janssen, Ines Munoz-Fernandez, Lars Zimmermann, Christoph Jedmowski, Luka Olbertz, Anke Schikling, Andreas Burkart, Anna Van Doorn, Roland Pieruschka, Uli Schurr, Uwe Rascher*

Forschungszentrum Juelich

Phenotyping in the field is an essential step in the phenotyping chain from well -defined and controlled conditions in the laboratory and greenhouse to the heterogeneous and fluctuating environment in the field. Field measurements represent a significant reference for the relevance of the laboratory and greenhouse approaches and an important source of information on potential mechanisms and constraints for plant performance to be tested at controlled conditions. Here we present results of a range of methods that are being deployed within the German Plant Phenotyping Network (DPPN, www.dppn.de) focusing on plant architecture, photosynthesis and water relations. Specialized field platforms are established (a) to test innovative phenotyping technologies (b) to provide access to semi-controlled field installation to support breeding approaches for future CO_2-concentrations (breed-FACE) and (c) to study the translation of phenotypic properties from controlled environments to stands in the field. We show that stereo imaging allows for the quantification of canopy structure; active thermography estimates leaf water content and provides information on transpiration conditions, sun induced (SIF) and light induced fluorescence transients (LIFT) techniques allow for remote estimating of photosynthesis at canopy and leaf-to-plant level respectively . SIF is measured by the next European Space Agency satellite Earth Explorer mission as well as with the other methods is incorporated in (semi-)automated systems positioning sensors in the field. This introduces a promising portfolio to measure plant traits for field phenotyping and to enhance our understanding of relevant traits under natural conditions .

T5-25-08

Developing automated field phenotyping platforms to quantify dynamics between key performance traits and environmental factors for bread wheat

***Ji Zhou**[1], Daniel Reynolds[1], Graham Finlayson[3], Simon Griffiths[2]*

1. Earlham Insititute

2. John Innes Centre

3. University of East Anglia

Automated field phenotyping can provide continuous and precise measures of adaptation and performance related traits that are key to today's GxE crop research, breeding pipelines, and agricultural practices. In the talk, I will introduce our multi-layer automated field phenotyping system at John Innes Centre (JIC) and Earlham Institute (EI), including UAV (unmanned aerial vehicles) and fixed-wing light aircraft, 3D scanning crop phenotyping platform (i.e. Phenospex), cost-effective CropQuant workstations and other novel hardware/software solutions that we have developed to facilitate high-resolution and high-frequency crop phenotyping to study performance related traits for bread wheat. In particular, I am going to talk about how to utilise phenotypic information to empower the assessment of genes controlling yield potential and environmental adaptation. I will introduce CropQuant, a cost-effective IoT (Internet of Things) powered platform that has incorporated networked sensors, single-board computers, in-field wireless communication and open high-throughput analysis algorithms to process field experimentation data. Besides performance and adpative trait analysis, we have also established machine-learning based models to explore and predict the dynamics between genotype, phenotype and environment (GxPxE). A proof-of-principle example based on near-isogenic lines (NILs) of wheat such as *Ppd-1* (loss of function), *Ppd-D1a* (photoperiod insensitivity), *Rht-D1b* (semi dwarfing), stay green induced mutants, and *Lr19* (hypersensitive response to the pathogen), and Paragon wild type will be discussed in the talk.

T5-25-09

Using high throughput dynamic phenotyping to uderstand the influence of heat stress on wheat

Jiwan Han

Aberystwyth University

A key driver in plant breeding is to improve crop resilience to unfavourable weather events. Such events are expected to become more frequent. In this project, we examine the response of wheat to combined water and heat stress, two stresses that often occur together. To address how water stress affects subsequent heat stress we subjected plants to one of 2 water regimes at an early stage of development and subsequently to a period of heat stress at booting (meiosis), a stage thought to be particularly susceptible. Water regimes were based on soil water content and heat treatments were 5 days at 25, 30, 35 and 40℃ , with night time temperatures being 5 ℃ lower. Plant growth was monitored by daily imaging and growth stage recording, physiological parameters were taken before, during and after heat stress and mature plants harvested and grain parameters measured using microCT scanning. Image processing segments plant from background and extracts useful feature information. We manually setup the colour range to avoid including background colours into the plant. For each plant, the directly measured information includes the imaging date, bar code, height in pixel, total pixel, total yellow pixel, total pixel of flag leaf area, yellow pixel of flag leaf area, total pixel of second leaf area, yellow pixel of second leaf area, total pixel of remaining leaf area and yellow pixel of remaining area. The derived information include height in mm, area in mm^2, normalized area, overall yellow area percentage, percentage of yellow area of flag leaf area, percentage of yellow area of second leaf area and percentage of yellow area of remaining area. The height and side view area development are direct features to describe growth and senescence. Physiological measurements included measurement of chlorophyll fluorescence, gas exchange and transpiration. CT scanning of harvested mature spikes provided information on grain set and size. Results indicate that there is an interesting interaction between pre-existing water stress and the effects of a heat treatment on both stem elongation and grain yield. Further experiments are being designed to refine the stage at which heat stress has its main effect and will be discussed.

T5-25-10

Using image-based phenotyping to dissect longitudinal and developmental traits in plants

Karen Askew, Roger Boyle, Jason Brook, Anyela Camargo-Rodriguez, Fiona Corke, Despoina Dadarou, Gina Garzon-Martinez, Alan Gay, Jiwan Han, Candida Nibau, Colin Sauze, Richard Webster, Kevin Williams, ***John Doonan***

National Plant Phenomics Centre, IBERS, Aberystwyth University, UK

Plants, as sessile organisms, adapt their growth and development to a wide variety of environmental influences, including abiotic, biotic and competition with neighbours. These stresses may change during their life history and also can change that history. Growth, phenology and stress-induced changes in these processes vary across life history and between individuals and these traits, referred to as longitudinal, can be dissected genetically using natural and induced variation. Until recently, such analyses have been demanding in terms of time and resource. Automated and semi-automated image-based across scales from the sub-cellular to whole organism to the field scale, combined with spacially-dispersed time-resolved and intregrated environmental sensing technologies, allow data acquisition across time, while the development of sophisticated genetically characterised populations provide access to the alleles underlying trait variation. Rapid extraction of information from such data demands innovative computational approaches. Experiments involving both model and crop plants will be used to illustrate progress made towards meeting these challenges. We discuss the potential of novel phenotyping approaches to dissect dynamic and complex traits, ranging from plant shape, response to single and multiple stresses to yield and quality associated characteristics. It is expected that a better understanding of dynamic traits will have an impact in the development of plant breeding strategies to improve crop yield and resilience.

T5-25-11

Phenotyping of the plant architecture in maize: Opportunities and strategies

Weixuan Wang, ***Chunyi Zhang***

Biotechnology Research Institute, Chinese Academy of Agricultural Sciences

Maize is one of the most important cereal crops in the world and the largest grain crop in China. The breeding for ideal plant architecture is always important, especially in near future. Fast, accurate and nondestructive phenotyping has become more and more urgent and necessary. However, the traditional manual measurement which is inefficient, time consuming and error-prone, is mainstream in China, and new strategy and technology in phenotype analysis should be developed soon. Right now, a high throughput phenotyping facility with 456 pots capacity which can automated whole plant phenotyping as well as cultivation for maize has been set up in BRI, CAAS. This facility can extract plenty of morphology-relates traits, biomass accumulation and water content information by visible light imaging and near infrared imaging, followed by high capacity computing at the same time. We have obtained morphology characters of a collection of 222 maize lines during these two years. A lot of QTLs of important traits have been identified with the combination with genome-wide association studies. The phenomics technology will strengthen the understanding of the plant architecture in maize.

T5-25-12

Determination of differential pathogen sensitivity of barley cultivars by multi-reflectance and -fluorescence imaging in combination with deep physiological phenotyping

Dominik Grosskinsky*, Jesper Svensgaard, Svend Christensen, Thomas Roitsch*

University of Copenhagen

Whereas great advances have been made in the cost-efficient and high throughput analyses of genetic information and non-invasive phenotyping, the large scale analyses of the underlying physiological mechanisms is lagging behind. The external phenotype is determined by the sum of the complex interaction of metabolic pathways and intracellular regulatory networks that is reflected in an internal, physiological and biochemical phenotype. These various scales of dynamic physiological responses need to be considered and genotyping and noninvasive phenotyping must be linked to the physiology at the cellular and tissue level. Thus a high-dimensional physiological phenotyping across scales is needed that integrates the precise characterization of the internal phenotype into high-throughput phenotyping of whole-plants and canopies. Thus complex traits can be broken down into individual components of physiological traits. Since the higher resolution of physiological phenotyping by wet chemistry is inherently limited in throughput, high throughput non-invasive phenotyping needs to be validated and verified across scales to be used as proxy for the underlying processes. As a case study we have addressed the infection of four barley cultivars by two fungal pathogens with distinctly different lifestyle, the obligate biotroph *Blumeria graminis* or and the necrotroph *Drechslera teres* by amulti-dimensional, holistic phenotyping approach. The non-invasive phenotyping by multi-spectral and -fluorescence imaging in the novel high throughput phenotyping facility PhenoLab was complemented by metabolic fingerprinting via the determination of activity signatures of key enzymes of carbohydrate and antioxidative metabolism, phytohormone profiles and determination of specific defense marker genes. A differential sensitivity of the tested barley genotypes versus both pathogens could be detected by the sensor based imaging techniques which was further validated and verified by distinct physiological responses and defense reactions. This study proofs that in such an interdisciplinary and multi-dimensional phenomics approach, plant physiology, non-invasive phenotyping and functional genomics will complement each other, ultimately enabling the *in-silico* assessment of responses under defined environments with advanced crop physiology models. This will allow to generate robust physiological predictors also for complex traits to bridge the knowledge gap between genotypes and phenotypes for applications in breeding, precision farming and basic research.

T5-26: Woody plants: Genomics, Transcriptomes and Evolution

T5-26-02

Unravelling the complex regulation of xylogenic organelles

Desre Pinard, *Drew Behrens, Alexander A. Myburg, Eshchar Mizrachi*

Department of Genetics, Forestry and Agricultural Biotechnology Institute (FABI), University of Pretoria

The nuclear genomes of woody plants have evolved over millennia to efficiently sequester atmospheric carbon and deposit it in sink tissues as lignocellulosic biopolymers. In parallel, the plastid and mitochondrial genomes of these plants have had to evolve to function coordinately with the nuclear genome during xylogenesis. Despite being the sites of energy production and important cell wall precursor biosynthesis, the contribution of plastids and mitochondria in wood formation remains almost completely unexplored. We used next-generation sequencing data to analyse the expression and regulation of organellar genes in the main carbon source (mature leaf), sink (developing xylem), and transport (secondary phloem) tissues of *Eucalyptus grandis*, an economically important tree species, in order to better understand how organelles contribute to xylogenesis. We assembled and annotated the plastid and mitochondrial genomes of *E. grandis*, and utilized ribosomal RNA-depleted, polyA-selected, and small RNA sequencing data of leaf, xylem, and phloem, to profile organellar gene expression, with a specific focus on the post-transcriptional regulation of organellar transcripts. We find that organellar gene transcript abundance is regulated at multiple levels between autotrophic and heterotrophic tissues to ensure functional gene products. In plastids specifically, which are primarily geared toward photosynthesis in mature leaf, and towards phenolic precursor biosynthesis in developing xylem, we find that differential gene expression, splicing, and editing contribute to the functional heterogeneity of plastids between tissues. Further, we identify the relationship between nuclear encoded regulators of organellar biology and transcription and the function of organelles during xylogenesis. This study sheds new light on the complex interplay between nuclear and organellar genomes during xylogenesis, opening up new avenues of future research into the regulation and evolution of xylogenesis, as well as the biotechnological improvement of tree crop species.

T5-26-03

Sex in the Salicaceae

Stephen Difazio[1], *Ran Zhou*[1], *Lawrence Smart*[2], *Craig Carlson*[2], *Fred Gouker*[2], *Brian Sanderson*[3], *Sandra Simon*[1], *David Macaya-Sanz*[1], *Ken Keefover-Ring*[4], *Matt Olson*[3], *Gerald Tuskan*[5]

1. *West Virginia University*
2. *Cornell University*
3. *Texas Tech University*
4. *University of Wisconsin*
5. *Oak Ridge National Lab*

The Salicaceae family contains several primarily dioecious genera, most notably *Populus* and *Salix*. These provide excellent opportunities for studying the evolutionary and ecological dimensions of sex dimorphism due to their shared and contrasting characteristics. *Populus* is comprised of about 30 wind-pollinated tree species. In contrast, *Salix* consists of approximately 500 species, most of which are insect-pollinated shrubs. Both genera have excellent genomic and genetic resources and show high collinearity and sequence conservation, thus facilitating comparative analyses of the genomics of sex determination. *Populus* sex determination is conferred primarily by loci on chromosome 19, apparently coupled with other loci on chromosomes 3,6, 9,14, and 18. Some of these may be assembly artifacts, so these auxiliary loci must be validated by improved assembly coupled with dense genetic mapping. In contrast, sex determination in *Salix* is primarily conferred by chromosome 15, coupled with loci on chromosomes 1 and 19. Furthermore, based on genotype configurations in and around the sex determination region (SDR), *Populus* has male heterogamety and *Salix* has female heterogamety. Using long-read sequencing and dense genetic mapping, we have reconstructed portions of the Y and W chromosome regions of each species and analyzed sex-specific sequence expression and sequence motif enrichment. Surprisingly, there is virtually no overlap in gene content of the SDR between the genera. Furthermore, chromosome 19 in *Salix* shows some hallmarks of sex chromosomes, while chromosome 15 in *Populus* does not. This leads to the hypothesis that chromosome 19 is the ancestral SDR, and that *Salix* has undergone a conversion to a ZW system located on chromosome 15. This multi-locus sex determination system may account for the consistently female-biased sex ratios observed in *Salix* species. Functional and population genetic analyses of these regions are ongoing, and should shed light on the evolutionary and ecological factors favoring male and female heterogamety in plants.

T5-26-04

Adaptive landscape and population genomics of silver birch

Jarkko Salojärvi[1], *Olli-Pekka Smolander*[1], *Kaisa Nieminen*[2], *Sitaram Rajaraman*[1], *Victor Albert*[3], *Petri Auvinen*[1], *Yrjö Helariutta*[1], *Jaakko Kangasjärvi*[1]

1. *University of Helsinki*
2. *Natural Resources Institute Finland*
3. *University at Buffalo*

Silver birch (*Betula pendula*) is a pioneer tree species prevalent in boreal forests across Eurasia. In contrast to most other trees, which have generation times from several years to decades, birch can be induced to flower within one year. This, together with a small 440 Mb genome and advanced breeding material, makes birch an attractive target for modern genomics-based biotechnology. We assembled the nuclear genome sequence of a fourth generation self-pollinated individual and anchored it by genetic linkage to chromosomes. Deep-time coding sequence evolution was analyzed by partitioning duplicate genes based on their evolutionary origins. Polyploid duplicates, dating back to the pan-core-eudicot paleohexaploid event, were enriched for transcription factor functions, whereas tandem duplicates were over-represented with environmental responses and secondary metabolism. Population genomic analyses based on 80 resequenced individuals representing most of the geographic species range showed effective population size crashes at well-known points of climatic upheaval during geologic history. Selective sweeps, shallow-time adaptive events, were enriched for genes originating from the ancient polyploidy event, and included genes involved in receptor kinase signaling, cytokinin and light perception, and sexual reproduction. Several candidates showed signatures of recent natural selection correlating with environmental variables. For example, selective sweeps in *PHYC* and *FRS10*, two tightly linked genes encoding light-response proteins, correlated with latitude/longitude and temperature, and in the case of *PHYC*, also with precipitation. Similar associations were also detected for *ARR1*, which acts in growth-promoting cytokinin perception, and for *KAK* and *MED5A*, which are involved in wood development. Overall, swept genes were enriched for key developmental or physiological triggering functions, suggesting that at least in birch, selection has acted by tuning the timing and crosstalk between fundamental plant processes.

T5-26-05

Genomes and evolutions of the endangered and widespread ironwood species

Yongzhi Yang[1], ***Jianquan Liu***[1,2]
1. Lanzhou University
2. Sichuan University

Ironwood species, known for their hard woods, provide an especial resource in our life. But the mechanism of the hard wood formation remains unknown. One of the ironwood species with extremely small population with only five wild individuals also serves as the ideal system to examine genetic consequences of the endangered ironwood species. Here, we report the high-quality genome assemblies of the endangered *Ostrya rehderiana* (366.2 Mb, scaffold N50 = 2.3 Mb) and the widespread *O. chinensis* (386.6 Mb, scaffold N50 = 762.5 Kb). Both assembled genomes cover more than 94%/96% of the estimated genomes (386 Mb/400 Mb) and harbour 27,831/31,152 predicted protein-coding genes respectively. A total of 756 and 1,310 gene families were respectively expanded in *O. rehderiana* and *O. chinensis*. These gene families were mainly associated with the interaction with their local environments, such like signal transduction and ion channel. 606 gene families were expanded in the ancestor lineage of *O. rehderiana* and *O. chinensis*, some of which are related to the lignin metabolic process underlying the ironwood formation. The endangered *O. rehderiana* shows a lower heterozygosity and sequence diversity (π) and higher levels of deleterious mutations and inbreeding than the widespread *O. chinensis*.

T5-26-06

Identification of the molecular processes needed for secondary vascular development in woody plants through transcriptome analysis

Peng Xu[1,2], *Cheng Huang*[1], *Shumin Cao*[1], *Jiayan Sun*[1], ***Laigeng Li***[1]
1. National Key Laboratory of Plant Molecular Genetics, CAS Center for Excellence in Molecular Plant Sciences, Institute of Plant Physiology and Ecology, Chinese Academy of Sciences,
2. Warnell School of Forestry and Natural Resources and Department of Genetics

Usually woody plants are able to grow for many years through perennial cambium activity for development of the secondary vascular system. Specific molecular processes need to be evolved for directing secondary vascular development. Analysis of transcriptome in different vascular systems would help uncover the mechanisms underlying perennial woody plant growth. Genes and processes fundamental to vascular development across different vascular systems were identified. These genes are expressed mainly in vascular tissues, forming a complex of coexpression network with multiple functional connections. Near half of such networks have experimental descriptions, while more fundamental biological processes in controlling vascular system development remains to be elucidated. Transcriptome analyses reveal that approximately one third of the genes that are expressed in secondary vascular development in *Populus* occur alternative splicing. Characterization of the alternative splicing suggests that woody plants may have evolved some alternative splicing events which function particularly in regulating secondary vascular development. In summary, our studies identified a number of molecular networks which are specifically involved in the biological processes needed for perennial woody plant development. Characterization of such processes aids in advancing the understanding of woody plants.

T5-27: GWAS and Genetic Regulatory Networks

T5-27-01

A network-module of *LRR-RLK* genes quantitatively regulates root growth under iron-limited conditions in Arabidopsis

Santosh Satbhai, *Katharina Jandrasits, Samantha Krasnodebski, Claudia Setzer, Santiago Perez, Youssef Belkhadir, Wolfgang Busch*
Gregor Mendel Institute (GMI) and Salk Institute for Biological Studies

Growth responses in plants are orchestrated by complex networks of genes that sense and transduce signals, and regulate specific developmental processes. While many genes have been identified that play a role in growth responses to the environment, very little is known about which genes and gene networks determine the extensive natural variation that can be observed for growth responses to the environment. For instance, the onset of root growth arrest upon iron (Fe) deficiency varies remarkably between different natural accessions of *Arabidopsis thaliana*. We therefore set out to identify networks of genes in which natural genetic variation affects the growth response to Fe deficiency. For this, we used a diverse set of 450 natural accessions of Arabidopsis that showed high variation in their primary root elongation when grown on Fe-deficient medium. We identified more than 20 statistically significant genomic loci that are associated with changes in root growth rate upon Fe deficiency. We used a whole-genome functional gene network model to identify genes in the immediate proximity of these associations that showed evidence for functional interaction. This analysis yielded several network modules of potentially interacting genes. Importantly, one of those modules, a cluster of 3 *Leucine-rich repeat receptor-like protein kinase* (*LRR-RLK*) genes and a protein kinase (*PK*) gene showed a strong signature of epistatic interactions. Each of the single mutant lines of these signaling genes displayed root growth rate reduction on Fe deficient media but not on full media, showing that this gene cluster is involved in growth regulation under Fe limited conditions. Due to their tissue specific expression pattern, we hypothesize that these four genes are coordinating growth responses in different tissues of the root. Consistently, we found that at least two of these LRR-RLKs physically interact at the protein level and we are currently investigating whether the combinatorial interaction of these 4 genes and their alleles are key determinants of the root growth response to Fe limited conditions.

T5-27-02

Global migration and adaptation in *Brachypodium* compilospecies

Justin Borevitz, *Pip Wilson, Jared Streich, Norman Warthmann, Kevin Murray*
Australian National University

To understand the genetic basis for wide adaptability, thousands of accessions were analyzed for genomic variation in ploidy and population structure. The allopolyploid *Brachypodium hybridum* is globally distributed with multiple introductions into Australia and N. America but little subsequent admixture. Diploid *B. stacei* is narrowly distributed in the Mediterranean, subsumed by *B. hybridum*. A core set of diverse diploid *B. distachyon* was selected balancing western and eastern Mediterranean regions of the dominant subspecies for detailed genomic and phenomic characterization. 1.5 M SNPs were identified among 75 accessions phenotyped with replication for early growth and reproductive traits under simulated seasonal and future climates conditions. Major QTL were identified flanking dozens of candidate genes for early vigour, flowering time in photothermal units and other agronomic traits.

T5-27-03

The added value of genome wide association mapping in the elucidation of quantitative trait regulation

Joost Keurentjes, *Johanna Bac-Molenaar, Rik Kooke*
Wageningen University & Research

Over the last decade genome wide association (GWA) mapping has rapidly become a major tool for the identification of natural polymorphisms explaining observed population variation. Especially in *Arabidopsis* the field has rapidly developed and contributed to many aspects of modern biology. Here, we will highlight some of the advancements made and illustrate this with recent case studies. GWA mapping has been instrumental in identifying novel regulators of complex traits. As such, it has contributed to our further understanding of physiological traits, plant metabolism and adaptation and evolution. Now that many studies have appeared it is becoming increasingly clear that natural variation at specific loci can have strong pleiotropic effects on multiple traits. This is especially noticeable in the coregulation of plant metabolism, where often hotspots of trait associations are found. In addition, many examples of epistatic genetic regulation can be seen, further adding to the complexity of quantitative trait regulation. Nonetheless, GWA mapping has allowed the identification of many novel players in the pathway topology of many traits. Some of these show strong signatures of evolution, such as selective sweeps or balancing selection. This is often reflected in the geographical distribution of allelic variation and thus provides indications for ecological adaptation. We will provide examples to demonstrate the contribution that GWA mapping has made to our understanding of the evolution and regulation of quantitative trait variation.

T5-27-04

GWAS in *Arabidopsis thaliana* – Is work on a model organism obsolete?

Arthur Korte
Center for Computational and Theoretical Biology, University Würzburg

Understanding the causal relationship between genotype and phenotype is a major objective both in humans and plants. The main interest is in understanding trait architecture and identifying loci contributing to the respective traits. Genome-wide association mapping (GWAS) is one tool to elucidate these relationships and has been successfully used in many different species. However, most studies concentrate on marginal marker effects and ignore epistatic and gene-environment interactions. These interactions are problematic to account for, but are likely to make major contributions to many phenotypes. Understanding these interactions is crucial to conceive trait architecture and predict phenotypes. Here *Arabidopsis thaliana* serves as an excellent model for developing statistical frameworks to quantify these effects, as a plethora of both genetic and phenotypic data is publicly available. Data about new GWAS models, general considerations about GWAS and data quality will be presented.

T5-27-05

Genome wide association genetics, outlier analysis and RNAseq identify the genetic and genomic basis of drought tolerance in bioenergy *Populus nigra*

Taylor Gail[1], *Smith Hazel*[1], *Allwright Michael*[1], *Scaglione Davide*[2], *Scalabrin Simone*[2], *Paulo Maria-João*[3], *Schabel Sabine*[3], *Wildhagen Henning*[4], *Emiliani Giovanni*[5], *Cohen David*[6,7], *Cyril Buré*[6,7], *Brendel Oliver*[6,7], *Le Thiec Didier*[6,7], *Hummel Irène*[6,7], *Valdes-Fragoso Billy*[1], *Alberti Giorgio*[8], *Zaldei Alessandro*[5], *Keurentjes Joost*[3,9], *Van Eeuwijk Fred*[3], *Morgante Michele*[2,8], *Bogeat-Triboulot Marie-Béatrice*[6,7], *Polle Andrea*[4]
1. University of Southampton
2. IGA Technology Services
3. Wageningen UR Plant Science Crop Group
4. Georg-August-Universität Göttingen
5. CNR-IVALSA
6. INRA NANCY-Lorraine
7. Université de Lorraine
8. University of Udine
9. Wageningen University

More than 500 genotypes from a natural, *P. nigra* mapping population, drawn from across the species' western European range, were cultivated in the field under short rotation coppice (SRC) in Savigliano, Italy. The population was phenotyped for biomass yield and leaf size traits under irrigated and moderate drought treatments. Sequence capture genotyping by sequencing (GBS) targeting the gene space yielded 132,000 single nucleotide polymorphisms (SNPs) representing 29,400 genes. Genome wide association genetics (GWAS) identified candidate genes for biomass, leaf size and drought tolerance while genetic outlier analyses identified additional signatures of adaptive selection to important precipitation and temperature variables. A complementary glasshouse experiment applied a controlled drought to three morphologically and physiologically diverse genotypes from the natural population and these were subject to intensive phenotyping for leaf development and growth traits under control and moderate drought conditions. RNAseq was then employed to elucidate the transcriptomic response of these diverse genotypes to drought stress with over 2700 differentially expressed genes found to be involved in the drought response. When gene-phenotype networks were re-constructed for leaves of distinct developmental stages using the top 100 most differentially expressed genes, five key candidates were identified through gene ontology as being central to the drought response. These microtubule-related genes act as gene hubs in young leaves, playing a central role in the drought response. Subsequently, as leaves mature, these genes are reduced in their connectivity and are situated in the peripheral parts of the networks. Genes of interest overlapping between the GWAS, outlier and RNAseq analyses were identified and these are strong candidates for application in advanced molecular breeding or genome editing for the sustainable intensification of this important lignocellulosic crop.

T5-27-06

Evolutionary game theory reveals the origin of modularity

Wenhao Bo[1], *Menmeng Sang*[1], *Libo Jiang*[1], *Rongling Wu*[1,2]

1. College of Biological Sciences and Technology, Center for Computational Biology, Beijing Forestry University

2. Center for Statistical Genetics, Pennsylvania State University

Developmental modularity, i.e., coherent organization and function of developmentally related organs or traits, has been extensively studied in developmental and evolutionary biology. However, our understanding of how genes modulate developmental modularity has still been very limited. Here we describe and assess a conceptual model to map the genetic architecture of developmental modularity by integrating evolutionary game theory into a genetic mapping design. We view a developmental module as an ecological system in which the constituting components or traits compete for space and resources and cooperate with each other to govern the function and behavior of the module. We embedded mathematical aspects of competition and cooperation within systems mapping, a dynamic framework for genetic mapping, to test whether and how specific genes regulate the coordination of different but interconnected traits toward developmental modularity. The new model provides a powerful tool to study the genetic mechanisms underlying the origin, structure and function of developmental modularity and morphological integration.

T5-28: Genome duplication and variation

T5-28-01

Variations in gene copy numbers in crop wild relatives

Robert Henry, *Peterson Wambugu*, *Agnelo Furtado*

University of Queensland

Gene copy number manipulation may be an under exploited mechanism for crop improvement. Analysis of copy number variation in crop wild relatives may reveal a source of useful variation for many important traits. Advances in genome sequencing and assembly are facilitating a growth in information on gene copy numbers and their variation in plants. Genome sequences for domesticated and wild rice species are now available for most species in the genus making this a good system to explore copy number variation in wild and domesticated plants. Rice grains are mostly starch and this abundant grain component is critical for many of the functional and nutritional attributes of rice. We have examined the genes for starch biosynthesis across the *Oryza* genus to identify any variation in copy number that might have a role in controlling rice starch properties. Most genes of starch metabolism show a high degree of conservation of genome location and copy number in rice and the wild relatives within the *Oryza* genus. A possible second pullulanase gene was detected in the published assembly of the *O. glaberrima* genome. A wild rice from Australia appears to have a second *GBSS* gene. Both the Australian and African taxa have novel starch properties including a high amylose content. The Australian taxa is descendant from a common ancestor of the Asian and African domesticated rice species. Changes in starch genes in the Asian and African wild populations may have been a key step in making the rice easier to cook and resulting in this A genome clade becoming attractive to domesticate.

T5-28-02

Gene expression patterns and RdDM-mediated epigenetic regulations of duplicated genes in *Brassica napus* subgenomes A and C

Chaobo Tong[1], *Xianhong Ge*[2], *Yanqiu Zhou*[1], *Zaiyun Li*[2], *Shengyi Liu*[1]

1. Oil Crops Research Institute of CAAS

2. Huazhong Agricultural University

Plant genome polyploidization and subsequent evolution is a crucial process for trait innovation and new species formation. Building of gene expression patterns of polyploidy duplicated genes is a first step in the process. Previous researches have indicated that these duplicated genes may inherit progenitors' expression patterns or form new patterns. In some polyploids, duplicated gene expression is biased toward one genome (genome dominance) while others are not. In all cases, mechanisms to form these expression patterns are poorly understood. In this study, we used a set of materials including the extracted AA subgenome (AeAe, n = 10) plants and its corresponding tetraploid *Brassica napus* (AnAnCnCn, n = 18) plus the tetraploid progenitors *B. rapa* (ArAr, n = 10) and *B. oleracea* (CoCo, n = 9) genomes to investigate duplicated gene expression patterns and further regulatory mechanisms of the patterns. Through analyses of whole genome re-sequencing and RNA-sequencing data, we found 80% of duplicated genes (in total about 23,000 gene pairs in leaves) inherited progenitors' gene expression patterns and 20% showed altered expression patterns. For the regulatory mechanisms of the altered patterns, we hypothesized that 24-nt siRNAs might be a kind of regulators because A and C have a big difference in transposable elements (TE) producing 24-nt siRNAs which mediate RNA-directed DNA methylation (RdDM) and thus affect nearby gene expression. To test this hypothesis, we performed small RNA sequencing and DNA methylation sequencing. By comparing to Ae, we found that 20% of the genes with expression alteration can be explained by RdDM in which this portion of TEs in the An genome could be targeted and methylated by 24-nt siRNAs from homologous TE copies on the Cn genome. For the inheritable gene expression patterns, we found that TEs which were asymmetrically inserted and then methylated by a small RNA-directed pathway, both already formed in each progenitor genome, have performed fixation on the inheritable patterns. The trends also exist in 15 Mya-triplicated subgenomic expression dominance. Our results revealed expression patterns of *B. napus* polyploidy duplicated genes and provided new insights into the regulatory mechanism of the expression patterns in polyploid evolution.

T5-28-03

Harnessing of untapped genomic variation in *Glycine* subgenus *soja*

Li-Juan Qiu, *Ying-Hui Li*
Institute of Crop Sciences, Chinese Academy of Agricultural Sciences

Soybean has undergone at least two rounds of genome duplication and produced mosaic of duplicated segments. The domesticated soybean (*Glycine max* (L.) Merr.) and its wild progenitor, *G. soja* exhibit substantial morphological difference in growth habits, seed-coat color, seed size and so on. To elucidate the genomic differences, seven representative *G. soja* accessions were *de novo* sequenced and compared with reference genome of *G. max*. We found uneven distribution of genetic diversity across 20 chromosomes. Inter-genomic comparisons identified 1,978 genes with copy number variation, 338 genes with *G. soja*-specific presence-absence variation, 2,094 genes affected by artificial selection during domestication for preferred agricultural traits. Comparison of candidate domestication-related genes with previous QTLs, as well as their homologs, we found some genes may contribute to variation of agronomic traits such as pathogen resistance, seed composition, flowering and maturity time, organ size and final biomass. Accordingly, our study of genome variation has yielded fundamental insights into the genes and biological mechanisms that underlie favorite trait for cultivar development.

T5-28-04

Structural and evolutive analysis of ancestral chromosomes fusion points within the hexaploid wheat genome

Helene Berges[1], *Caroline Pont*[2], *Florent Murat*[2], *Stéphane Cauet*[1], *Nathalie Rodde*[1], *Jérôme Salse*[2], *Arnaud Bellec*[1]
1. INRA-French Plant Genomic Resource Center, INRA - CNRGV, 24 Chemin de Borde Rouge – Auzeville, CS 52627,31326 Castanet Tolosan Cedex, France
2. IINRA GDEC- Genetics, Diversity and Ecophysiology of Cereals (GDEC) Paleogenomics & Evolution Research Group (PaleoEvo) 5 Chemin de Beaulieu 63100 Clermont-Ferrand France

The explosion of genomes sequencing has highlighted whole genome duplications (WGDs) as a major evolutionary driving force shared by multiples branches of the tree of life. However, the genome sizes and chromosome numbers have not increased exponentially over time because WGDs have been balanced by chromosome number reduction (CNR) and gene losses. In plants these two mechanisms have been found through genomes sequences comparisons and ancestral karyotype reconstruction. For monocots models have been proposed explaining CNR by nested chromosome fusions (NCFs). However, molecular mechanisms driving ancestral chromosome fusions that have led to the present day monocots karyotypes are still largely unknown. Based on this ascertainment, we aimed at characterizing the two chromosomes fusion points (FP) of wheat chromosome 1 that was formed by the nested fusion of two ancestral chromosomes. To obtain a high quality sequence of a FP we implemented a strategy based on the reconstruction of the ancestral grass karyotype, the definition of the synteny relationships between wheat and rice (as the modern representative of the ancestral 12 chromosomes grass karyotype) and the screening of wheat chromosome specific BAC libraries. We focused on the two FPs of chromosomal group 1. We investigated this region on chromosomes 1A, 1B, 1D obtaining two FP sequences (one on the short arm and one on the long arm) for each of them. The annotation of genes at these 6 loci allowed us to established orthology relations chromosome 5 and 10 of rice (images of ancestral chromosomes A5/A10) and homeology relations within the 3 subgenomes of wheat. The origin and the order of the genes neighboring the FP were confronted to models of chromosomes fusion of *Triticeae* described in the literature.

T5-28-05

Genome-wide identification, evolutionary analysis of Late Embryogenesis Abundant and Vicinal Oxygen Chelate family genes and their functions in drought stress tolerance and lipid metabolism in *Brassica napus* and *Arabidopsis*

Yu Liang, *Neng Wan*, *Zao Cheng*, *Yufeng Mo*, *Baoling Liu*, *Maoteng Li*
Huazhong University of Science and Technology

Background:
Drought stress is one of the most important abiotic stresses that determines crop yield, geographic distribution of plant species. *Brassica napus* serves as the third largest oil seed crops in the world. Many researches focus on the oil metabolism of *B. napus*. Late embryogenesis abundant (LEA) proteins are a diverse and large group of polypeptides that play important roles in desiccation and freezing tolerance in plants. The vicinal oxygen chelate (VOC) proteins are members of metalloenzyme superfamily that play roles in many biological reactions. Some members of *LEA* and *VOC* gene family have been systematically characterized in other important plants but not *B. napus*. Our study were to identify the members of these two gene families in *B. napus* genome and provided new insights in the evolutionary and functional analysis of *LEA*s and *VOC*s in *B.napus*.
Results:
A total of 108 *LEA* and 38 *VOC* genes were identified based on their conserved domains in *B. napus*. The results revealed that most of *BnVOC* and *BnLEA* genes have few introns, and all of *BnVOC*s contained the typical VOC structure of βαβββ modules. *BnVOC* and *BnLEA* genes are distributed unevenly across 19 chromosomes in *B. napus*, occurring as gene clusters in chromosomes A9, C2, C4, C5 and A6. The synteny of these genes is preserved. Synonymous (Ks) and nonsynonymous (Ka) values show most of

the *BnLEA* and *BnVOC* genes have Ka/Ks ratios less than 1, suggesting their conserve function and structure under evolutionary selective pressure. The expression of most *BnVOCs* and *BnLEAs* was increased in leaves and late stage seeds, further expression pattern results showed that seeds of *B. napus* with high oil content show higher expression levels of *BnVOCs* in drought stress. Different copies of *BnLEA* and *BnVOC* genes and the homologous *AtLEA* and *AtVOC* gene from *Arabidopsis* were cloned and then over-expressed in *Arabidopsis*, the analysis of transgenic seeds showed both the oil content and seed weight were increased 2%-5%, and the fatty acid composition were also affected. In drought stress, *Arabidopsis* T-DNA insert mutant *lea3* and RNA interference plants *atvoc-rnai* showed less drought resistance, exhibiting differences in many physiological parameters, including leaves temperature, chlorophyll content and so on .
Conclusion:
Whole-genome duplication and segmental duplication played a major role in the expansion of the *BnVOC* and *BnLEA* gene family in evolution, *BnVOC* gene family is a consequence of mesopolyploidy in *Brassica* evolution. Expression profiles and functional analysis suggested BnVOCs and BnLEAs might not only response to drought stress but also affect lipid metabolism. This present study gives a comprehensive analysis of the *LEA* and *VOC* gene family, provides new insights into the evolutionary pattern, and links their novel biological function of different aspects in *B. napus*.

T5-28-06

Analysis of Pinaceae gene family evolution suggests accelerated gene turnover in pine trees and lineage-specific amplifications of genetic networks

Claudio Casola, *Weixi Zhu, Tomasz E. Koralewski*
Texas A&M University, College Station, USA

Gene duplications and losses are major determinants of evolutionary change. Comparative genome-wide analyses of gene family evolution are often applied to infer patterns of gene turnover, i.e. gene gains and losses, across a range of species. Such investigations rely on gene annotations, which are often incomplete and partially inaccurate in draft genomes and transcriptomes. This issue is particularly relevant in species with large genomes (>10 Gb), including conifers. In this work we analyze gene family evolution in draft genomes of four conifers (loblolly and sugar pine, Norway spruce and Douglas-fir) using the likelihood framework implemented in the CAFE software, which allows estimating gene turnover rates and patterns in a species phylogeny. After grouping all annotated genes in orthogroups with OrthoFinder, we selected 16,118 gene families (~24,000-28,000 genes per species) occurring in at least two species for downstream analysis. The CAFE analyses indicated that gene turnover is at least twice as high in the two pines (~110 gene gains and/or losses per million years) compared to Norway spruce and Douglas-fir, a result that was robust to changes in the number of pine species used and the model of gene family evolution implemented in CAFE. We also identified 1,192 gene families with significant expansion or contraction changes, and a high proportion of gene duplications vs. gene losses in Douglas-fir. Overall, our results indicate that no less than 1/5 of all gene families experienced gains and/or losses during the evolution of these Pinaceae. However, we observed that up to ~50% of genes are much shorter than the longest gene in the same gene family, likely a result of pseudogenes missannotation. We applied a stringent length cutoff of <75% of the longest gene in the same gene family in order to remove putative pseudogenes from the 1,192 gene families with high gene turnover rates. Despite this conservative length threshold, we were able to identify many genetic pathways in the four Pinaceae with significant expansions in multiple gene families, indicating selective pressure toward retention of gene duplicates and amplifications of entire genetic networks.

T5-29: Genetics and Metablomics Studies of Higher Plants

T5-29-01

Biodiversity and regulation of wheat cuticle waxes in relation to drought

Huihui Bi[1], *Maria Hrmova*[1], *Sergiy Lopato*[1], ***Nikolai Borisjuk***[1,2]
1. University of Adelaide, Australia
2. Huaiyian Normal University

Bread wheat, *Triticum aestivum*, represents about 30% of the world's cereal cultivation area and provides 20% of the calories for the human population. Wheat is cultivated on over 220 million hectares worldwide. Much of this land is often under the influence of limited water supply and heat that significantly impair the crop's yield. Understanding how a plant copes with these environmental challenges has a vital significance for the improvement of crop tolerance and yield. Plant cuticle is a continuous protective sheet that covers all aerial surfaces of the plant body and acts as an interface for the plant's interactions with surrounding environment. The main functions of the cuticle is to conserve water in internal tissues and protect the plant from harmful UV radiation. The goal of the current work is to get insights into cuticle structure, composition, and relatedness to drought tolerance in Australian wheat cultivars. Our analysis of wheat varieties with different levels of drought resistance and glaucousness show that considerable variation exists between genotypes in the composition and structure of the cuticle. In elite Australian wheat cultivars we identify β-diketones to be the main determinant of a glaucous phenotype; this determines the shape of wax crystals on the leaf surface and is an important contributor to efficient tissue water conservation. Our data reveals a complex relationship between the cuticle and drought tolerance, indicating that glaucousness and underlying β-diketones are not a single indicators of drought tolerance. Rather, cuticle thickness, specific cuticle wax composition (in particular the presence of significant contents of β-diketones, primary alcohols and alkanes, in concert with densities of stomata), all influence leaf water loss under the conditions of water deficit. These features of cuticle composition and deposition provide useful selection criteria for further exploration. Further, we demonstrate the drought-inducible expression of four wheat genes encoding *MYB* transcription factor (TF) genes. In wheat cells, the products of three genes (*TaMYB74, TaMYB31*, and *TaMYB24*) operate as the activators of cuticle biosynthetic genes. Functional MYB-responsive elements localized in the promoter region of the *SHN1* gene were specifically recognized by TaMYB74, but not by other MYB TFs. This data, integrated with other observations, allowed us to propose a scheme that links together drought, investigated TFs, downstream cuticle-related biosynthetic genes, and cuticle wax components. The overexpression of wheat homolog of the transcription factor WIN1/SHN1 (TaSHN1), which was previously shown to influence cuticle composition and drought tolerance in several plant species, influences both cutin and cuticule wax composition, plus the performance of selected wheat transgenic lines. Overall, our results extend the knowledge on cuticle biosynthesis regulation in wheat, and could potentially be used for engineering of cereal crops with enhanced tolerance and performance under drought.

T5-29-02

Insights into ROS sensing systems from algae to plants

Ning Shao[1], *Ning Xu*[2], *Qian Zhao*[1], *Guangyou Duan*[3], *Cuimin Liu*[1], *Zheng Zhou*[2], *Ralph Bock*[3], *Jiayang Li*[1]

1. Institute of Genetics and Developmental Biology, Chinese Academy of Sciences

2. Institute of Biophysics, Chinese Academy of Sciences

3. Max-Planck-Institut für Molekulare Pflanzenphysiologie

Reactive oxygen species (ROS), which are continually generated as a consequence of the normal metabolism in aerobic organisms, have recently emerged as important cellular signalling molecules controlling development, initiating cellular stress responses and cell death in a wide range of organisms. An important type of ROS is singlet oxygen ($^1O_2$), which is generated by energy transfer to molecular oxygen. What connects the primary chemical signals generated by $^1O_2$ (oxidized lipids and carotenoids) with the transcriptional responses at the level of nuclear gene expression is currently unknown. Searching for possible ROS sensor proteins with the combination of a ROS-inducible reporter in the unicellular green alga *Chlamydomonas* and reverse genetic analyses in *Arabidopsis*, we identified a small zinc finger protein, METHYLENE BLUE SENSITIVITY1 (MBS1) that is required for induction of $^1O_2$–dependent gene expression and accumulates in distinct granules in the cytosol upon oxidative stress. Using structural techniques in nuclear magnetic resonance (NMR), we firstly demonstrate that the solution structure of the MBS1 reveals the architecture of the single C_2H_2 zinc finger domain, which has emerged as a redox-sensing switch. Moreover, the conformational changes of MBS1 on the zinc binding site undergoes in response to $^1O_2$ exposure, but no significant conformational changes treated with another type of ROS, hydrogen peroxide (H_2O_2) *in vitro*, as shown by NMR and CD spectral. To gain a deeper understanding of the mechanisms involved in ROS sensing and, in particular, of the role of the novel protein MBS in $^1O_2$ sensing and signal transduction in *Arabidopsis thaliana*, point mutants have successfully been generated that eliminate the zinc binding sites in the protein and thus allowed us to directly test the hypothesis that the zinc finger acts as a ROS sensor *in vivo*. The outcome of these experiments was very clear: the zinc finger site of MBS1 is direct and specific $^1O_2$ signaling targets and required for acclimation of *Arabidopsis* to singlet oxygen. These findings provide new insights on a structural for the mechanism underlying the $^1O_2$ sensing of MBS1.

T5-29-03

Genetics and genomics of cotton fiber development

Tianzhen Zhang

Zhejiang University

Gossypium spp. produces two kinds of non-glandular trichomes. The first one is the hair trichome as same as in *Arabidopsis* and another is seed trichomes called fibers which are extremely elongated single cells and the largest source of natural textile grown throughout the world. Cotton fibers can be classified into two types: lint and fuzz. Lint initiates growth between anthesis and two days postanthesis (DPA) and can elongate 2.5-3.5 cm, whereas the fuzz which is approximately 0.5 cm in length initiates growth between 5 and 10 DPA. Based on map-based cloning strategy, for the first time, we isolated industrially important lint fiber production gene and fuzz development gene. We clarified that cotton has developed a unique, highly specialized transcriptional regulatory network responsible for fiber production, which is different from that in Arabidopsis leaf hair trichome development. The characterization of target genes and of the network that regulates fiber production has provided insights into the molecular mechanisms underlying cotton fiber development and underlying the increased lint yield produced by the induction of a greater number of epidermal cells to develop into lint fibers by genetic engineering.

T5-29-04

Duplication and neofunctionalisation of DODA was key to the emergence of betalains

Hester Sheehan[1], *Joseph Walker*[2], *Tao Feng*[1], *Alfonso Timoneda-Monfort*[1], *Samuel Brockington*[1]

1. University of Cambridge

2. University of Michigan

Betalains are plant pigments that produce a palette of pink, purple, yellow and orange colours. They are taxonomically restricted, being found only in the core Caryophyllales, whilst the rest of the flowering plants use anthocyanin compounds to produce a similar palette of colours. Intriguingly, anthocyanins and betalains have never been found together in the same plant. Betalains are derived from tyrosine and the biochemical pathway is very simple, requiring a minimum of two enzymes, DODA and CYP76AD, to produce the core betalain structures – betaxanthins (yellow-orange pigments) and betacyanins (pink-red pigments). The enzyme DODA, 4,5-DOPA-extradiol-dioxygenase, acts on L-dopa to produce betalamic acid, the principle chromophore of betalains. Betalains evolved once only and, just prior to their acquisition, DODA was duplicated. The ancestral DODA lineage (DODAβ) is found throughout the land plants whereas the derived lineage (DODAα) is only found in species with betalain pigmentation. Furthermore, in families that have reverted back to anthocyanin pigmentation, DODAα has been lost or down regulated. This suggests that duplication and subsequent neo-functionalisation of DODA was key to the emergence of betalains. We have isolated genes encoding DODAα and DODAβ from species representing different stages along the phylogeny and around the origin of betalains in order to test their capacity to produce betalain compounds. Only enzymes encoded by *DODAα* genes are able to produce a high level of betalains. Using phylogenies compiled from transcriptome data of 250 Caryophyllales species, we have determined the amino acid residues under positive selection and reconstructed the most likely ancestral DODA sequence preceding the duplication. We are functionally validating the role of these residues in betalain production and determining whether these residues are sufficient, alone or in combination, to convert the ancestral sequence into an enzyme that efficiently catalyses L-dopa. Ultimately, we will be able to determine the critical alterations that allowed an enzyme to specialize on a different substrate and assist in producing a novel class of plant pigments.

T5-29-05

Tracing recent domestication in endemic New Zealand plants.

Lara Shepherd[1], *Peter De Lange*[2], *Mariana Bulgarella*[1], *Robin Atherton*[3], *Pete Lockhart*[3]
1. Museum of New Zealand Te Papa Tongarewa
2. Department of Conservation
3. Massey University

We are examining recent domestication by using genetic analyses of four endemic New Zealand plants that were brought into cultivation by Māori. New Zealand was one of the last places on earth to be discovered therefore cultivated endemic crops offer a unique window into the early stages of the domestication process. Prior to human arrival these species are thought to have been restricted to the northern North Island. Their occurrence in the south of New Zealand and the distant Chatham and Kermadec Island groups, is strongly associated with Māori archaeological sites, and is considered to result from translocations as part of their cultivation. Our research aims to determine the number of times each species was brought into cultivation and the way in which domestication has reduced genetic diversity in the cultivated populations. Another aim is to reveal prehistoric Māori settlement and trade routes. We have used a range of genetic techniques including whole chloroplast genome sequencing, HRM, Sanger sequencing, microsatellite genotyping and ddRAD.

T5-29-06

Application of optical topometry to the plant epidermis

Miranda Haus[1], *Tom Jacobs*[2]
1. Michigan State University
2. University of Illinois-Champaign Urbana

The plant epidermis regulates key physiological functions contributing to photosynthetic rate, plant productivity, and ecosystem stability. Yet, quantitative characterization of this interface between a plant and its aerial environment is laborious and destructive with current techniques, making large-scale characterization of epidermal cell parameters impractical. Here, I present the exploration of optical topometry (OT) for the analysis of plant organ surfaces. OT is a mature, confocal microscopy-based implementation of surface metrology that generates nanometer-scale digital characterizations of any surface. I report epidermal analyses in *Arabidopsis* (*Arabidopsis thaliana*) and other species as well as dried herbarium specimens and fossilized plants. We evaluate the technology's analytical potential for identifying an array of epidermal characters, including cell type distributions, variation in cell morphology and stomatal depth, differentiation of herbarium specimens, and real-time deformations in living tissue following detachment. As applied to plant material, OT is very fast and non-destructive, yielding richly mineable data sets describing living tissues and rendering a variety of their characteristics accessible for statistical, quantitative genetic, and structural analysis.

T5-30: Omics of Plant Stress Resistance

T5-30-01

Gene positioning by light

Meng Chen
Department of Botany and Plant Sciences, Institute for Integrative Genome Biology, University of California Riverside, CA, USA

Plant genomes are extremely sensitive to and can be developmentally reprogrammed by environmental light cues. Increasing evidence in yeast and mammalian models suggests that environmental cues can regulate gene activity by altering the spatial positioning of individual genes within the nucleus. However, if and how the spatial positioning of individual genes contributes to gene regulation in plants is still poorly understood. We have developed a method to label the positioning of individual genes by utilizing rolling-circle amplification of gene-specific circularizable oligonucleotides coupled with fluorescence *in situ* hybridization. Using this approach, we demonstrate that light triggers a rapid repositioning of the light-inducible *CAB1* locus in *Arabidopsis* from the nucleoplasm to the nuclear periphery during its transcriptional activation. *CAB1* repositioning is mediated by the red/far-red photoreceptors, the phytochromes, and is inhibited by the repressors of phytochrome signaling, including COP1, DET1, and PIFs. *CAB1* repositioning appears to be a separate regulatory step occurring prior to its full transcriptional activation. Moreover, the light inducible loci *RBCS*, *PC*, and *GUN5* undergo similar repositioning behavior upon their transcriptional activation. Our results support a novel gene regulatory mechanism in which phytochromes activate light-inducible loci by relocating them to the nuclear periphery. This study provides the initial evidence for the biological importance of gene positioning in plants and shows that plant genes can be spatially reorganized in response to environmental cues.

T5-30-02

Small RNAs contribute to plant resistance to *Phytophthora* pathogens

Yingnan Hou[1], *Yi Zhai*[1], *Jinqiu He*[2], *Duseok Choi*[1], *Jinbiao Ma*[2], ***Wenbo Ma***[1]
1. University of California Riverside
2. Fudan University

A broad range of parasites rely on the functions of effector proteins to subvert host immune response and facilitate disease development. The notorious *Phytophthora* pathogens evolved effector proteins with RNA silencing suppression activity to promote infection in plant hosts. Small RNAs have been shown to regulate immunity, which requires transcriptional reprogramming. The *Phytophthora* effector PSR2 specifically affects the accumulation of phased small interfering RNAs (phasiRNAs), which are secondary siRNAs derived from coding or noncoding loci using microRNAs (miRNAs) as the triggers. *Arabidopsis thaliana* mutants defective in phasiRNA production are compromised in defense against *Phytophthora* infection, indicating that the phasiRNA pathway plays a positive role in plant immunity. Here, we report the identification of specific phasiRNAs that regulate immune responses and discuss the molecular mechanism by which PSR2 suppresses phasiRNA biogenesis and promotes infection.

T5-30-03

H3K9me2-mediated regulation of NB-LRR immune receptor expression control in *Arabidopsis thaliana*

Thomas Eulgem[1], *Tokuji Tsuchiya*[2], *Jianqiang Wang*[1], *Yan Lai*[1]
1. University of California at Riverside
2. Nihon University

COPIA-R7, a transposable element (TE) inserted into the *Arabidopsis thaliana* NB-LRR-type immune receptor gene *RPP7*, has recruited the histone mark H3K9me2 to this locus. H3K9me2 is a suppressive epigenetic signal known to transcriptionally silence TEs in plants. However, at *COPIA-R7* this mark affects the choice between two alternative *RPP7* transcript polyadenylation sites and, thereby, influences the critical balance between *RPP7*-coding and non *RPP7*-coding RNAs. We further demonstrated H3K9me2 levels at *RPP7* to be controlled by EDM2, a novel type of nuclear localized PHD-finger protein. The module of 2 ½ atypical PHD fingers in EDM2 preferentially binds to certain combinations of R2, K4, R8 and K9 marks of histone H3. EDM2-mediated modulation of H3K9me2 levels at *RPP7* in response to defense induction dynamically adjusts the balance between coding and non-coding *RPP7* transcripts and is critical for immunity mediated by this immune receptor. As NB-LRR genes are often organized in clusters containing TEs, recruitment of TE-associated regulatory mechanisms to NB-LRR expression control is likely to be common. Consistent with this idea, we found EDM2-dependent H3K9me2 to control expression of several other TE-associated NB-LRR genes. We further identified additional components contributing to epigenetically-controlled alternative polyadenylation. New results on roles of EDM2 and other chromatin-associated regulators in defense-associated H3K9me2-mediated alternative polyadenylation will be presented at the meeting.

T5-30-04

Harness the power of big data – using gene modules to study plant nutrient uptake and utilization

Shisong Ma
University of Science and Technology of China

The advent of big data in biology offers excellent opportunities while poses daunting challenges for researchers to analyze the massive datasets to derive biological insights. We tackled such challenges by building an *Arabidopsis* gene network based on the graphical Gaussian model, using large-scale transcriptome data. The network contains 18,000+ genes that assemble into 600+ gene modules. These modules function in diverse cell organization, development, nutrients, metabolism, and stress response pathways. As examples, modules involved in the uptake and utilization of phosphate, nitrogen, and iron will be discussed in details, which contain key transcription factors, enzymes, and transporters as well as potential novel regulators for the related nutrient pathways. The network analysis was also extended to major crops and identified gene modules that determine agronomic traits. The modules revealed by our networks provide extensive resources for plant systems biology studies.

T5-30-05

Regulation and manipulation of communication signals in plant response to virus-vector mutulistic interactions

Pingzhi Zhao, ***Jian Ye***
Institute of Microbiology, Chinese Academy of Sciences

Vector-borne pathogens may alter promote the performance of their primary hosts in ways that influence the frequency and nature of interactions between hosts and vectors. The new concept in virology has been called 'Perceptive insect behaviour' and relies on a behavirour change induced by its transmitted viruses. This positive effect of a pathogen on its vector via their shared host plant is termed indirect mutualism. Previous work has reported a phytohormone Jasmonic acid (JA) -mediated terpene biosynthesis is suppressed in begomovirus-infected plants, leading to reduced resistance to whiteflies which transmit these viruses. Recently, we identified the proteins encoded by the monopartite or bipartite begomoviruses as the key genetic factor that suppresses JA-mediated terpene biosynthesis. Our new results suggested that plant innate immunity and other phytohormones also involve in the terpene-based odor resistance to whitefly. Meanwhile, geminiviruses have evolved sophisticated strategies to counter multiple-layer resistances against whitefly. Our results present a novel scenario of how a pathogen regulates host-derived olfactory cues for vector attraction via manipulation of plant host. The results will also help to address similar tripartite interaction systems in plants, animals or humans and will allow developing of innovative control methods through interference with vector transmission.

T5-30-06

Cross-kingdom RNAi and spray-induced gene silencing for plant disease control

Hailing Jin
University of California, Riverside

Small RNAs (sRNAs) are a class of short non-coding RNAs that mediate gene silencing in a sequence-specific manner. We have demonstrated that some sRNAs from eukaryotic pathogens, such as *Botrytis cinerea*, the fungal pathogen that causes grey mold disease on more than 1000 plant species, could be translocated into host plant cells and suppress host immunity genes for successful infection. This finding represented the first example of naturally occurring Cross-Kingdom RNAi in host-pathogen interactions. These sRNAs act as a new class of pathogen effectors that translocate into host cells to suppress host immunity. Recently we have found that transgenic plants expressing hairpin RNAs that targeting *Botrytis Dicer 1* and *Dicer 2* genes could effectively block the generation of fungal sRNA effectors and suppress grey mold disease. This strategy can be adapted to simultaneously control multiple fungal diseases. We show that *Arabidopsis* plants expressing hairpin RNAs that simultaneously target *DCL* genes of *B. cinerea* and *V. dahliae* show enhanced disease resistance to both pathogens. These results also suggest that sRNA trafficking is bi-directional and sRNAs could move from the host cells to the interacting pathogens. Most strikingly, we discovered that *Botrytis* could take up double-stranded RNAs and sRNAs from the environment. Applying sRNAs or dsRNAs that target *Botrytis Dicer* genes on the surface of fruits, vegetables and flowers significantly inhibits grey mold disease. Such pathogen gene-targeting RNAs represent a new generation of environmentally friendly fungicides.

T5-31: Genetically-Engineered Crops: Challenges and Potential

T5-31-01

Genetically-engineered crops: Current status and potential future developments

Peter Davies
Cornell University

Genetically engineered (GE) crops have now been grown for 22 years. They have received overwhelming adoption by farmers in countries where they are permitted. The number of countries where such crops are permitted (mainly the USA, Canada, Brazil and Argentina) is still limited, with further expansion hampered by outdated regulations originating for 30 years ago when the potential effects of genetic engineering were largely unexplored. Despite this there are numerous indications that farmers have taken events into their own hands and promulgated GE crops via "stealth seeds" despite official prohibitions. However the number of traits is still quite limited, being largely restricted to herbicide resistance, especially to glyphosate, and insect pest resistance through Bt. This has been an outstanding success in cotton growing in India with a 90%+ adoption rate, changing India from a cotton importer to a major exporter. Yet insect-resistant aubergine (*Solanum melongena*), which permits a massive reduction in the use of toxic insecticides, remains unutilized except in Bangladesh because of political opposition. Genetic engineering for disease resistance offers a major contribution, with the most notable success being in Hawaii where resistance to papaya ringspot has rescued the Hawaii papaya industry. Yet despite this success and demonstrated efficacy in Thai varieties in Thailand, the Thai government yielded to anti-GMO pressure to ban its adoption. Indeed the opposition to genetically engineered crops is solely political backed up by fear occasioned by the term GMO, coined by opponents, and false science. Consumers have seen little overt advantage to GE food products, as the increase in yields and decrease in pesticide usage is unseen by consumers. However should some food products collapse in availability (e.g., from citrus greening and various banana diseases) we may see more acceptance than at present. Other current and future potential advantages of GE crops include drought tolerance (especially in response to climate change), improved human-nutrient content (e.g., Golden Rice, though Golden Rice remains in limbo for various reasons 17 years after its initial development), improved plant-nutrient uptake, improved photosynthetic efficiency, and improved produce qualities for consumer acceptance (such as non-browning apples and potatoes). It will be up to the plant-science community to engage and inform the public at every opportunity about the productivity, environmental, and nutritional advantages of genetically-engineered crops in order to facilitate more widespread acceptance in future years.

T5-31-02

The development of disease resistance in tropical crops using genetic engineering

Frank Shotkoski
Cornell University

In many of the more developed countries, genetically engineered crops already contribute greatly to agricultural productivity and sustainability. Over the last few years, the largest growth in the adoption of genetically engineered crops has been in developing countries and this trend is expected to continue. The multi-national life sciences companies have been leading the way, but they are focusing primarily on a few crop/trait combinations that have high commercial value and occupy large international markets. The greatest success has been with insect resistant and herbicide tolerant maize, soybean, canola and cotton. Because of the costs and complexity of the issues related to crop biotechnology, many crops and traits of importance to subsistence and resource-poor farmers around the developing world have been overlooked. For example, there are few crops possessing disease resistance traits mitigated by genetic modification. Those that do exist, papaya ringspot resistant papaya and virus resistant squash, are resistant via viral coat mediated resistance. The development of fungal and bacterial disease resistant crops hasn't been met with nearly the success compared to other traits, especially crops important to people in developing countries. More recently, crops genetically modified for fungi and bacteria resistance important to resource poor farmers in developing countries are in the pipeline for commercial release. I will present examples of several of these projects including Banana Xanthomonas wilt resistance in Uganda using the *pflp* and *hrap* genes, potato late blight resistance in Indonesia and Bangladesh using an R gene, *RB*, from a wild-type potato and attempts to generate fungal disease resistant banana using RNAi technology.

T5-31-03

A site-specific gene stacking system for GE crop

Yongqing Li, *Ruyu Li, Chuanghu Wang, Zhicheng Dong, David Ow*
South China Botanical Garden, Chinese Academy of Sciences

A gene stacking system combining the use of the Cre-loxP, Bxb1-att and the CinH-RS2 site-specific systems is being deployed for engineering transgenic crops. In earlier work on the model plant tobacco, we have demonstrated Bxb1 integrase-directed site-specific gene stacking at a genomic target, followed by Cre-directed removal of unneeded DNA. To translate this technology to crop plants, target lines have been generated in both Japonica rice and Indica rice each with a single copy T-DNA insertion that harbors intact recombination sites in the genome. Gene integration into Japonica rice has been demonstrated with reporter genes and we are currently beginning the gene stack with a trait gene that reduces Cd accumulation in grain. Similar approaches in generating target lines are being taken in other important crops such as soybean and cotton. With this gene stacking system, adding new transgenes will not inflate the number of segregating loci when more traits are added; removal of marker genes and other nonessential DNA enables reuse selectable markers for subsequent transformation events as well as help ameliorate biosafety concerns of genetically engineered crops.

T5-31-04

CRISPR/Cas9 mediated base-editing efficiently creates non-GM herbicide resistant plants

Yiyu Chen[1], *Zhiping Wang*[2], *Yong Xu*[3], *Qijun Chen*[2], ***Linjian Jiang***[1], *Hanwen Ni*[1]
1. Department of Plant Pathology, College of Plant Protection, China Agricultural University
2. State Key Laboratory of Plant Physiology and Biochemistry, College of Biological Sciences, China Agricultural University
3. National Engineering Research Center for Veg

Introduction of a point mutation at a desired position remains a great challenge in plant genome engineering. Recently, an improved CRISPR/Cas9 system achieved precise base-editing in mammal cells and T0 rice plants, whether this new system would generate inheritable point mutations in plants needs urgent testing. In this report, we aimed to generate point mutations on ALS gene conferring herbicide resistance in *Arabidopsis* using this base-editing system and evaluate the editing efficiency. We showed that the base-editing efficiency was about 1.7% at T1 generation. Over 75% T2 seedlings of base-edited T1 lines, harboring known and novel herbicide resistant (HR) mutations at the desired position, survived herbicide treatment. Moreover, we identified six non-transgenic T2 HR plants with expected point mutations. These results demonstrated that the base-edited mutations are inheritable at high efficiency. Surprisingly, we discovered that the PAM regions could likely be a new base-editing region for this system, and revealed a second novel HR mutation on ALS gene. Collectively, this new base editing system is a powerful tool to generate known and novel inheritable mutations that bear great values and interests for basic research and crop improvement.

T5-33: Large-scale sequencing of biodiverse plants; beyond model organisms and crop species (two sessions)

T5-33-01

Mining biodiversity for optogenetics; mammalian neuroscience applications of algal proteins

Gane Ka-Shu Wong
1. University of Alberta
2. BGI-Shenzhen

1KP is an international multi-disciplinary consortium that has generated transcriptomics data across the Viridiplantae (green plants). We sequenced exemplars of all major lineages representing a billion years of evolution, including angiosperms, conifers, ferns, mosses, streptophyte and chlorophyte green algae. These efforts have increased plant gene numbers by about a factor of 50 relative to the public databases. One of our many goals was to demonstrate that biodiverse sequencing of species with no previously known utility can generate practical outcomes, at the level of individual molecules and complex pathways. In the former case, we discovered and quantitatively characterized many novel light-sensitive proteins that are used for optogenetics control of genetically defined neurons in conscious mammals. Some of these proteins have been licensed for use in clinical trials. Generalizing the concept, we found that billions of years of evolutionary "tinkering" has created a cornucopia of light-sensitive proteins that potentially allow us to control many other aspects of the cell, not just the electrical signals of the brain, but also the chemical signals that are ubiquitous in molecular cell biology.

T5-33-02

Elucidating the long evolutionary road to plant terrestrialization by comparative genomics

Shifeng Cheng[1], *Shanyun Xu*[1], *Yan Xu*[1], *Sijie Chen*[1], *Linzhou Li*[1], *Jing Qu*[1], *Chang Yue*[1], *Gane Ka-Shu Wong*[2], *Michael Melkonian*[3]
1. BGI-Research, Shenzhen, China
2. University of Alberta, Canada
3. Universität zu Köln, Germany

The origin and diversification of the green lineage (Viridiplantae), in particular the colonization of land, has profoundly changed our planet. However, the long complex stream of genetic innovations and morphological transitions from green algae to land plants is underexplored. I will present the recent progress from a large-scale comparative genomics and phylogenomics analysis incorporating 8 newly-sequenced algal genomes from 2 early diverging prasinophytes and 6 streptophyte algae. Using the new phylogenetic framework from the 1KP project, we determined the core nuclear and organellar genes that define the ancestor of the green lineage. Functional domains and metabolic pathways indicative of terrestrialization were investigated and categorized. In addition, we explored differential expression patterns and alternatively spliced forms in more than 30 transcriptomes from these 8 species. Together with other recent discoveries, it is becoming clear that many of the traits associated with land plants often have a molecular antecedent in the algae.

T5-33-03

Understanding the enigmatic dinoflagellates and their adaptive symbiotic systems with genomics

Senjie Lin
University of Connecticut

Dinoflagellata is a member of the alveolate phylum, a group of protists that also includes Perkinsozoa, Apicomplexa (malaria parasites) and Ciliophora. About 50% of the extant 4,000 species are photoautotrophic while the rest are heterotrophic; some species from both groups are mixotrophic. Dinoflagellates are widely distributed in aquatic ecosystems, important primary producers, a source of essential endosymbionts for coral reefs, and frequently responsible for harmful algal blooms and/or biotoxins in marine ecosystems. They exhibit unusual cytological and genetic characteristics, breaking nearly all of the major rules in biology textbooks, and epitomize the most complex plastid evolution known to science. Although the molecular mechanisms behind these enigmatic characteristics remain elusive, novel insights have begun to emerge from genome and transcriptome analyses. Recent advances will be presented to address what makes these organisms so ecologically successful.

T5-33-04

Deciphering nitrogen-fixing root nodule symbioses Part I – Phylogeny to synthetic biology

Pierre-Marc Delaux
Université de Toulouse

Many of the plant traits that are important for agriculture are quantitative traits that are complex and under the influence of many genetic and environmental factors. This is particularly true for root-nodule symbiosis (RNS), a specific association between a few plant lineages and nitrogen-fixing bacteria. The plant benefits from this association by gaining access to the pool of atmospheric nitrogen, providing an essential nutrient for plant growth. With the rise of synthetic biology, we can envisage engineering desirable traits, such as RNS, in cereal crops for increasing sustainability and productivity in agriculture. However, a comprehensive understanding of the molecular basis of this trait is required before such an undertaking can proceed. I will discuss recent findings supporting the use of comparative phylogenomics to identify genetic innovations behind key plant traits, and how this fits into a global collaborative effort to engineer RNS into cereal crops.

T5-33-05

Algal ancestor of land plants was preadapted for symbiosis: Recent progress and updates

Pierre-Marc Delaux
Université de Toulouse

Physiological studies and fossil records strongly indicate that the ability of the first land plants to form symbiotic associations with beneficial fungi was one of the critical innovations that permitted them to colonize land. In angiosperms, genes required for the perception and transduction of diffusible fungal signals for root colonization and for nutrient exchange have been characterized. However, the origin of these genes and potential correlation with land colonization remain elusive. A comprehensive phylogenetic analysis of 259 transcriptomes generated by the 1KP initiative, as well as 10 green algal and basal land plant genomes, coupled with the characterization of the evolutionary path leading to the appearance of a key regulator (a calcium- and calmodulin-dependent protein kinase) showed that the symbiotic signaling pathway predated the first land plants. In contrast, the downstream genes required for root colonization and their expression patterns appeared subsequent to the colonization of land. Hence we concluded that the most recent common ancestor of extant land plants and green algae was preadapted for symbiotic associations. Subsequent improvement of this precursor stage in early land plants through rounds of gene duplication led to the acquisition of additional pathways and the ability to form a fully functional arbuscular mycorrhizal symbiosis. Here, I will present recent advances made to validate these evolutionary hypotheses.

T5-33-06

A genomic approach to understand the molecular basis of plant-cyanobacterium symbiosis

Fay-Wei Li
Boyce Thompson Institute & Cornell University

Plant symbiosis with nitrogen-fixing cyanobacteria is a unique form of mutualistic association that has independently evolved in diverse yet isolated lineages-from a few species of bryophytes, ferns, and cycads, to a small genus of flowering plants. Compared with other nitrogen-fixing microbes, cyanobacteria are much less dependent on the plant host, and therefore can be an ideal partner for engineering symbiotic nitrogen fixation into crop plants. However, our current understanding of plant-cyanobacterium symbiosis is rudimentary, in large part because of the lack of genomic and genetic resources for their plant hosts. Through a global collaborative effort, we have generated genomic and transcriptomic data for several plant species capable of cyanobacterial symbiosis-*Azolla* (a fern), *Anthoceros* (a hornwort), *Blasia* (a liverwort), and *Gunnera* (a flowering plant). I will highlight our key findings and discuss future directions.

T5-33-07

Discovery and design of hyperstable bioactive cysteine-rich peptides from medicinal plants

James Pingkwan Tam
Nanyang Technological University

Our interest in orally-active peptides has led to a systematic profiling and characterization of the disulfide-rich family of cysteine-rich peptides in medicinal plants and functional foods. Of particular interest are peptides with 2 to 6 kDa molecular weight and 3 to 6 disulfides as framework crossbraces. These highly disulfide-constrained small proteins are super-compact, hyperstable to heat, chemical and enzymatic degradation, i.e. they are an underexploited class of bioactive compounds for the design of orally-active therapeutics. The most well-known are defensins, whose antimicrobial activities contribute to the host defense mechanism. However, profiling of over 1,000 plant species and >100 families show that defensins, which are not constitutively expressed, are not the most common cysteine-rich peptides in medicinal plants. Two themes are emerging. First, each family with a distinctive cysteine motif has a distinctive scaffold with highly evolvable and hypervariable non-cysteine residues. This feature has been exploited for drug design by grafting a metabolically-labile bioactive peptide to a cysteine-rich scaffold, rendering it metabolically-stable and orally-active. Second, molecular diversity is produced from a single gene by two methods, fuzzy bioprocessing from the precursors and post-translational modification, a theme commonly observed in secondary metabolites.

T5-33-08

Deciphering nitrogen-fixing root nodule symbioses Part II - Comparative phylogenomics

Maximilian Griesmann
Ludwig-Maximilians-Universität München

Of all the essential plant nutrients, nitrogen is required in the largest quantity and is frequently the limiting factor in crop productivity. Although plants are surrounded by atmospheric nitrogen, they are unable to use it. Some flowering plant lineages escaped the requirement for fixed nitrogen in soil by forming root nodule symbioses (RNS) with nitrogen-fixing bacteria. This capability is restricted to four closely related orders: Fabales, Fagales, Cucurbitales and Rosales. As a result of this relationship, there is a hypothesis that RNS evolved many times, independently, from a predisposition event that occurred before the divergence of these orders. However, the molecular mechanisms that allowed RNS to evolve remain largely unknown. Through a global collaborative effort, we have sequenced the genomes and transcriptomes of 13 species that together cover the breadth of symbiotic states in the Fabales, Fagales, Cucurbitales and Rosales orders. I will present the key findings from our comparative phylogenomics analysis of this dataset.

T5-33-09

Supporting large-scale genomic studies through open platforms at China National GeneBank

Xin Liu
BGI-Shenzhen & China National GeneBank

China National GeneBank (CNGB) held its opening ceremonies in Shenzhen late in 2016, after five years of construction. The organization was launched in 2011 by the Chinese government as the nation's first genomics repository. With funding from the central and the local government, CNGB was constructed and is operated by BGI. It is focused on establishing open platforms for global scientific research. CNGB has already established a biorepository with the capacity for storing dozens of millions of samples, a database with 20 Pb (petabase = 10^{15} base) of storage capacity, and a living biobank. It houses a sequencing platform with a throughput of 10 Tb per day (terabase = 10^{12} base). Several large-scale genomics studies are already launched, e.g. the avian 10K genomes http://b10k.genomics.cn. Progress towards this and other large-scale genome projects will be discussed. We are actively seeking global collaborations to utilize CNGB's open research platform.

T5-33-10

Was there a Chlamydiae contribution to primary endocytobiosis and the origin of Plantae?

Michael Melkonian
Universität zu Köln

Plantae (Viridiplantae, Rhodoplantae, Glaucoplantae) genomes contain many essential genes that are homologous to genes found in Chlamydiae, a group of intracellular bacteria that function as energy parasites in many eukaryotes, including even humans. However, no extant Plantae are known to harbor intracellular Chlamydiae. Over the last ten years, a hypothesis has been developed that these chlamydial genes are the relics of an ancient endosymbiosis of Chlamydiae that occurred in the ancestor of the Plantae at the time of their acquisition of cyanobacteria, and that these chlamydial genes were essential for the establishment of the primary endosymbiosis and consequently of the origin of plastids in Plantae. This hypothesis has been controversial because the taxon sampling on which the original arguments were based was limited. In the analysis to be presented, we considered 79 chlamydial genes from 362 taxa of bacteria and eukaryotes, many identified from novel transcriptomes and genomes sequenced for this study. These results were subjected to phylogenetic analyses, yielding novel insights into the origin and function of chlamydial genes in Plantae.

T5-33-11

Discovering the molecular basis of C_4 photosynthesis through wide sampling of natural diversity

Steven Kelly
University of Oxford

Much of biology is associated with the evolution of convergent traits. For example, the most productive crop plants (e.g. sugar cane) use C_4 photosynthesis-a more efficient form of photosynthesis than the C_3 form that is seen in 85% of plants (e.g. rice, wheat, soybean). C_4 has evolved independently from C_3 in more than 60 lineages. Although insight into the molecular basis of this trait can be gained by studying model organisms, elucidating the pathway to a level necessary to re-engineer crop plants has been challenging. By comparing the transcriptomes from plants representing 18 independent origins of C_4 photosynthesis, we quantified the extent to which this convergent trait employs identical molecular mechanisms. Nearly all of the established biochemical changes that characterize C_4 species were recovered by an analysis of these data. In addition, we expanded the existing paradigm by four metabolic pathways not previously associated with C_4 photosynthesis. These results show that comparative transcriptomics of non-model species can provide detailed insights into the molecular basis of complex traits.

T5-33-12

Network approaches for plant phylogenomic synteny analysis: Challenges and prospects

M. Eric Schranz
Wageningen University and Research

Network analysis approaches have been widely applied across disciplines. In biology, it is now frequently adopted to organize protein-protein interactions, organize pathways and/or to interpret gene co-expression patterns. In contrast, comparative genomics is still largely reliant on pairwise comparison and linear visualization methods. We present the challenges and prospects for establishing a generalized phylogenomic synteny network approach that is required to interpret the wealth of new and emerging genomics data. This will be illustrated by an analysis of synteny across 101 completed plant genomes. A broad approach like this holds great promise for understanding the evolutionary history of genes and genomes across phylogenetic groups with long divergence times. Additional details are available online at the preprint server http://www.biorxiv.org/content/early/2017/01/17/100990.

THEME VI: PLANTS & SOCIETY

T6-01: Latin American botanical research and the journal impact factor (two sessions)

T6-01-01

Meeting the challenges of impact factors

Tod Stuessy
Ohio State University

Impact factors were designed to aid decisions on which biological and medical journals should be included in *Current Contents*, a popular summary of the major journal literature during the 1960s and 1970s. As time progressed, university administrators realized that these impact factors could be used to evaluate the quality of scientific work published by staff members within their institutions. Because of the difficulty of evaluation of published research in many diverse fields, it is understandable that decision-makers would turn to quantitative and qualitative assessments of productivity. On the one hand, encouraging staff researchers to publish in international and more visible journals is laudable, and this can work to the benefit of the individual as well as the institution. On the other hand, this search for higher impact can sometimes interfere with appreciation and encouragement of research that an individual is best suited to produce. A further problem arises when institutions set impact goals that are also tied directly to salary or other aspects of career development, which can lead to negativism. It is important to realize that the bibliographic monopoly of developed countries is now virtually gone, and if a researcher has Internet connection, then most of the world's literature is at his/her disposal. However, to be able to use the international literature, one must be reasonably fluent in the English language. Although this may seem overly restrictive, it is a natural trend to have one primary scientific language, which initially started with Latin, then French, German, and now English. Accepting this reality, young scientists must become immersed in English to a level where the international journals can be read and understood. In addition to this needed linguistic preparation, the researcher must develop self-confidence to address international topics, instead of only regional ones. Pursuing research topics that are limited geographically in the home country may eclipse the likelihood of addressing questions of broader interest. Sometimes the financial resources are so limited, or so directed to national initiatives, that travel to other countries is difficult. In these cases, search for international funding may be the only avenue for successful global competition. The young researcher must seek out training in laboratories in developed countries, obtain mentoring from international researchers, and select globally significant research topics. Such adjustments in career development will naturally result in publication in higher impact journals, hence relieving pressure from the institution. Achieving this in developing countries is much more challenging than in developed countries, but it can be accomplished with effort.

T6-01-02

Journal Impact Factor and evaluation of researchers in Latin America

Alina Freire-Fierro
IKIAM University and Asociación Latinoamericana de Botánica ALB

Although originally created to evaluate journals, the Journal Impact Factor has nowadays become one of the main factors to evaluate productivity and impact of scientific researchers worldwide. These bibliometrics, though understandable applied to journals published in English, are completely unfair to non-English journals. This situation is worse to Latin America, where the tradition of research/publishing high quality papers, even in Spanish, is relatively new. Additionally, and while for other sciences the impact of journals can be measured in a time span of 2, 3, 20 years, in botanical research many articles/works published more than 200 years ago are still being cited. Adding to these problems, the evaluation of impact factor of journals is made primarily in English language publications, as it has been reported for example, that out of 100 Panamian publications, none had been included in the evaluation of the recently modified science index, the Science Expanded Index (SCI-Expanded). While even in English-speaking countries, many institutions have agreed not to use the Journal Impact Factor as a measurement for a researcher's productivity or impact (e.g. signage of the DORA agreement in the US) or have used other methodologies for this evaluation (e.g. the Research Excellency Framework REF2014 in the UK); in Latin America, many governments/institutions are still using it. The JIF is being used for example to categorize Mexican researchers in the National System of Researchers (SNI), or by Colombian Colciencias, and by the Ecuadorian National Registry of Researchers (RNI). In Perú for example, JIF and H indices count up to 35% of the total score for a researcher's categorization in the National Directorate of Researchers and Innovators (DINA). This unfair evaluation from Latin American governments and institutions towards a researcher's impact really hurt scientific development in the region. Since very good and experienced Latin American researchers want to keep their high academic status, they will continue to publish in English, giving therefore a positive feedback for journals published in English in detriment of journals published in Spanish. Although just as recent as of December 2015, Clarivec Analytics, the current owner of former Thompson Reuters' Institute for Scientific Information (ISI), creator of the IJF, has developed a new index that does not consider the Journal Impact Factor, the Emerging Sources Citation Index. Still though, one of the requirements for journals to be evaluated by this Index is that it has to have the cited literature in English. Taking into consideration the stated above, I consider and encourage to the botanical community and to scientific organizations to sign the DORA agreement, and to follow the methodologies used by REF2014, but modified to reflect Latin America's scientific research reality. Additionally, and as a scientific community, we consider that it is our duty to request the Latin American governments to stop using the JIF as one of the most important elements in their evaluations of researchers in the region.

T6-01-03

The Impact Factor and the Mexican System of Researchers (SNI-CONACyT): Stuk in the "carrot and stick" traditional approach.

***J. Antonio Vázquez-García**, *Miguel A. Muñiz-Castro*
Universidad de Guadalajara

In Mexico, the National System of Researchers (SNI-CONACyT) evaluates the performance of botanists in Area II (Biology and Chemistry) and awards four types of nominations (Candidate, rank I, II, and III), with $ 600 to $ 1200 USD per month (accounting for 25-50% of total revenues). The problems include: 1) In the referees predominate in researchers in the area of chemistry, who ascend more quickly with their publications in journals of greater Impact Factor (IF); 2) the evaluation, supposedly, takes into account many other criteria, however in practice they require you to publish at least one paper per year in JCR journals with rigorous IF (> 0.5). Implications of this "Carrot and stick" approach: 1) Many traditional botanists abandon discipline in pursuit of a more profitable trade, while students prefer to study molecular biology than field botany, since there is almost no support for field research projects. 2) It is more likely that a stressed and sedentary national researcher will first acquire one of the Non-Transmisible Diseases (NCDs) before reaching "the carrot" (a promotion to rank II or III); either that or they first reach their retirement age without receiving their corresponding recognition and encouragement. PROPOSALS: The evaluation of the quality of botanical research can be considerably improved through a better diversity of indicators: 1) INTERNATIONAL PLANT NAME INDEX (IPNI): Consider the number of new species described as author or co-author, or whether the researcher is an authority in a particular plant group, which could be easily generated by IPNI. 2) INDEX HERBARIORUM (IH): Give special value to outstanding or emblematic botanical collectors for each herbarium, this data are available in this data base. 3) GENE & DATA BANKS. Contemplate the number of plant accessions or DNA sequences, recorded in gene banks, databases, botanical or public gardens. 4) Taxonomic keys to identify species in a given group should be specially valued as a tool for curating collections of herbariums, museums and databases and for the formation of new young taxonomists. 5) Geographical and social coverage of research should also be considered. 6) The botanical monographs represent the works of a lifetime of a researcher, must be valued for the quality of their content in a very special way, years of experience in the subject, geographical coverage, scope, degree of complexity, research risks, or rarity of organisms under study. 7) The floristic checklists of a great mountain or of a floristic region with hundreds or thousands of species should also be considered as monographs. 8) Generate algorithms (search engines) to count the number of times that a species or species author has been historically cited in the publications. 9) Consider metrics, additional to the IF: the SNIP, Source Normalized Impact per Paper; the RESEARCH GATE SCORE and total readings that an author has; and the GOOGLE SCHOLAR that provides the h index. 10) Provide greater financial and logistical support to botanical and related journals to improve their capacity for review, editing and dissemination.

T6-01-04

Latin American botanical research and the journal impact factor

Hugo Valdebenito
Universidad San Francsico de Quito

An analysis of the low impact factor of Latin American botanical research is presented as well as future perspectives for Latin American botanical journals are described. It is widely known that the best scientific works carried out in biological sciences, including botany, are published in journals outside the Region and furthermore, published in English. Accordingly, in recent years there has been a growing effort to change to English language the published articles in local journals as means of increasing its level of circulation. However, this process goes to the detriment of the botanical scientific knowledge in the original countries of publication where Spanish or Portuguese language is spoken. In general, there is little information on the impact factor for Latin American journals and, in particular, on journals associated to botany, which are very scarce in the Region. Four Latin American countries have more than one journal registred in the Science Citation Index: Argentina, Brazil, Chile and Mexico, whose average impact factor is less than 1; the remaining countries in the Region do not have botanical journals in this Index. It is clear that the impact factor of Latin American journals is low, however, this factor sometimes takes a second place considering that these journals constitute one of the few if not the only mean of spreading the botanical knowledge in their own countries. Although this last argument is valid, the situation is critical for the botanical research carried out in Latin America and it raises some questions. Should the language of publication be changed to English, considering this language as the main way to spread scientific knowledge, or should be published bilingual journals (Spanish/English or Portuguese/English)? or, alternatively, should be considered more important the dissemination of knowledge in the countries of origin than spreading the knowledge globally? On the other hand, frequently there is a lack of recognition of articles written in Spanish since researchers in other non-Spanish speaking countries do not take the time to read them. Additionally, botanical articles on Latin American topics written in English reach higher degrees of impact and international knowledge. Considering the above, the translation to English language would be a requirement if we really want to improve the impact and dissemination of research published in Latin American journals. However, as stated before, these journals are the main source for in-country botanical education for many students and represent their first contact with research. It is important to note that although the impact factor is an unfair measure for a number of countries, including Latin American ones, it is an universally established scientific measurement that we must consider. Finally, some potential actions to increase the international dissemination of Latin American journals could include their bilingual publication in electronic format as well as the physical document. Another step could be the distribution of Latin American journals by international companies responsible for spreading scientific knowledge, therefore allowing wide availability of articles in full text.

T6-01-05

Citing nomenclatural authors makes a difference: Quantifying effects of alternative citation practices in taxonomy, exemplified by a bibliometric analysis of the literature on cotoneaster

Gerwin Kasperek[1], *Wolf Bernhard Dickoré*[2]
1. Goethe University Frankfurt
2. Botanische Staatssammlung München

Background: An important reason for the crisis in taxonomy lies in the rather modest performance of taxonomic publications when evaluated using current bibliometric tools. Recently, proposals for alternative citation practices in taxonomy with a focus on new approaches on citing nomenclatural authors have been put forward. However, no quantitative data was available on the potential bibliometric effects of such alternative practices. Purpose: The present study was performed in order to quantify bibliometric effects of alternative practices of citation using a representative corpus of taxonomic literature on the genus Cotoneaster (Rosaceae) as a sample. The main question was: What difference would it make if taxonomic publications additionally included all nomenclatural authors in their end references section? Furthermore, we attempted to analyse the basic characteristics of a representative corpus of taxonomic literature and develop qualifiers that might help to reassess various practices of citation, and various practices of bibliometric evaluation. Methods: As it is often difficult to delimit taxonomic literature from literature belonging to other disciplines, pragmatic criteria for doing so were developed. The taxonomic literature making up the sample corpus, pertinent to the genus Cotoneaster, was retrieved as comprehensively as possible. Citation relationships between publications in the corpus were extracted and stored in a database, using a subset of citing publications. Citations were classified, with a particular focus on nomenclatural citations attached to taxon names (Linnaean style references), as opposed to end references. Data analysis was performed using standard software. Results: The taxonomic literature on Cotoneaster comprised some 450 publications. From a subset of 120 citing publications, some 3,600 citation relationships to other publications were recorded. Linnaean style references (ca. 2,400) outnumbered the end references (ca. 900) greatly. The median age of the cited publications was approximately 70 years; almost 90 percent of cited publications were older than 10 years. Linnaean style references tended to be older than end references. Lists of the most-cited publications differed widely depending on whether they were based on Linnaean style references or end references. Conclusions: Using the corpus of the taxonomic literature on Cotoneaster as a sample, bibliometric effects of including all nomenclatural authors in the end references sections of publications could be simulated for the first time. It would make large differences in several bibliometric parameters if Linnaean style references were treated in this manner. Implications of our results for taxonomy as a discipline are discussed with specific regard to current usage of bibliometric parameters, such as impact factors, for research evaluation by academic administrations.

T6-01-06

The discreet charm of publishing an Argentinean botanical journal in the Journal Impact Factor Era

Gabriel Luis Mario Bernardello

Museo Botánico, Facultad de Ciencias Exactas, Físicas y Naturales, Universidad Nacional de Córdoba-CONICET

The *Boletín de la Sociedad Argentina de Botánica* has published 51 volumes since 1945, in all disciplines of Plant Biology and related groups. From the year 2000, it has published regularly at the rate of one volume per year and from 2013 it has published four trimestral issues. In 2010, the journal was included in the Science Citation Index Expanded without requiring use of English.

Why do we publish a scientific journal? 1. To validate and legitimize the ownership of the contributions; the fact that an investigation appears in a prestigious journal, through the filter of a pertinent evaluation, guarantees the intellectual quality of the work. 2. To serve as a way of preserving the heritage of science. 3. To professionally promote the researcher by making his work public. Why do researchers publish their work in journals? 1. To report the results of their investigations. 2. To contribute to the advancement of knowledge. 3. To be properly evaluated, obtain grants and have prestige, in a system where they are evaluated by their publications. 4. To make their work known before other researchers. The publication itself is, then, at the base of the scientific scheme. The JIF has become a chief and widely used tool to rank journals, although there are other systems (SJR, Scopus, Latindex, Scielo, *etc.*). It is not only important to publish, but where to publish. The journal in which a paper is published has become more important to researchers and administrative evaluators than the research itself. Journals, indirectly, endorse the contents of the articles, contents that the administrative evaluators do not care to read. Journals of the Northern Hemisphere have the best rankings whereas Latin American journals, when included, barely reach less than 1. This system indirectly defines what is valid science and it discriminates through low rankings for many science journals. According to Scimago Journal Rank, from ca. 91,000 papers in Plant Science from 2013 to 2015, the USA and the UK published ca. 20,000 each, the Netherlands ca. 14,000, Germany ca. 6,000, and Latin America ca. 4,360 papers (less than 5%; Brazil 2,400, Mexico ca. 600, Chile and Argentina ca. 500 each.) The scientific production of some developing countries is smaller than that of a single USA university department. The publication of a scientific journal is a complex task, more in our countries where they are generally edited by non-profit societies with altruistic goals and meager resources. National universities and research councils fund our research and sustain our jobs, but force us to publish our work in Northern Hemisphere journals, as evaluation committees do not consider papers published in journals from our region. Open access Latin American botanical journals with the best possible quality, both in form and content, should be considered a priority and the work published in them acceptable by administrative evaluators. Our journals are part of our intellectual and cultural heritage and represent us in the community of nations. It may be like swimming against the current, but this is discrimination that must not be left unopposed.

T6-01-07

Measuring the impact of scientific research in Colombia: Towards the globalization of its science

Mauricio Diazgranados

Natural Capital and Plant Health Department, Royal Botanic Gardens, Kew

Research in Colombia has grown exponentially, despite maintaining a low expenditure in Science, Technology and Innovation (ST&I: 0.4% of GDP). Colciencias, the government agency that supports scientific research, keeps a national register of 10,042 researchers working in Colombian institutions, 2,324 of them in natural sciences, mainly in 17 of the 289 universities in the country. Also, it recognizes 4,638 research groups, 855 of which are in natural sciences. In line with the policy of research internationalization, more than a decade ago Colciencias established a registration and evaluation system of both groups and researchers, implemented on the platform ScienTI (with CVLac and GroupLac profiles). Initially, a National System of Indexation and Homologation of Specialized Publications in ST&I (Publindex) had been created. Publindex contains 562 Colombian journals, but only 75 are listed in Thomson Reuters impact factor (JCR) and SCImago Journal Rank (SJR), which means most of the Colombian journals and publications are not even considered for Impact Factor (IF). Moreover, of all those journals, only 14 are in biological sciences, and just a few are in botany (i.e. Caldasia). The WoS reports yearly 8,000 scientific documents, including 6,700 journal articles, from Colombian institutions, being second in South America after Brazil. However, only 600 of them are in the biological sciences, and roughly 50% in botany. After a comprehensive evaluation on the impact of Colombian research, led by the National Observatory of Science and Technology and Colciencias, three conclusions arose: 1) Colombian researchers publish mainly in journals of low impact; 2) Colombian journals have low presence in international indexes; and 3) Colombian science has low impact in the international scientific community. After comparing various metrics, in 2016 Colciencias decided to start classifying publications using the impact factor provided by JCR and SJR, instead of Publindex, with various consequences under hot debate: 1) The system assumes that the main research paradigms in the country are similar to those of interest by the high IF journals, but this is not always the case. 2) Assessment of both researchers and research groups depends on where they publish. Since publishing in high IF journals is expensive, main universities may take advantage against regional universities. 3) Accreditation of graduate programs may be affected by the impacts of the new ranking system. 4) Funding opportunities may remain restricted to the highest ranked researchers and groups, leaving out possibilities for many regional groups and researchers to emerge. 5) Pay scales may depend on the ranking of researchers, based on Colciencias lists, possibly affecting researchers from certain areas or subdisciplines for which impact factors are usually low (e.g. ethnobotany). 6) There might be a loss of social diffusion of scientific knowledge, since most of the high IF journals are in English, and many Colombians (and especially in the regions) do not understand the language. Despite the critiques and problems of this system, the country might see an improvement of the research impact in the scientific international community in the near future.

T6-01-08

Please count taxonomic citations as regular citations: An appeal

Alexey Shipunov[1], *Alexander Doweld*[2]
1. Minot State University
2. National Institute of Carpology (Gaertnerian Institution)

Since their invention by Vasily Nalimov (1969), various citation indexes (CI) became an extremely important tools to measure scientific productivity and success. There is however, one important gap in CI calculations in the field of biological taxonomy: taxonomical citations (which supply scientific names) does not count and therefore the citation level of all taxonomic works is severely underestimated. We appeal to all CI database holders and to all scientific community -- please include taxonomic citations!

T6-01-09

Where do public interest and academic goals converge? Critical shortfalls of narrow endemic plants conservation in Chile

Andres Moreira-Munoz
Pontificia Universidad Catolica de Valparaiso

Chilean flora is composed by around 4,300 species, from which a 45% is endemic. This is one of the highest rates of endemism in the Andean region. Central Chile is globally known as the "Central Chilean biodiversity hotspot". Just a low percentage of the flora has been assessed regarding its conservation status. The taxonomic identity of some groups is still uncertain, and the distributional information is very sparse. These shortfalls are currently known as the Linnean and Wallacean shortfall within the framework of conservation biogeography. These shortfalls are getting more critical nowadays, in the fate of climate change driving distributional changes and high environmental risks, like the unexpected wildfires that dramatically affected almost all central Chile this last summer. An increasing interest within the botanical scientific community towards narrow endemic and threatened species is expected. This is but not the case. A revision of articles published in principal botanical journals, show a minor presence of Chilean plants, and from these, a much more restricted percentage of narrow endemic and threatened ones. Also, botanical journals are not interested in publishing distributional information, which is crucial for the IUCN classification and for taking real actions towards conservation. Some exceptions have been published by *Phytotaxa*. The Chilean highest impact botanical journal, *Gayana Botanica*, is not accepting reports regarding distributional notes anymore. These notes are being still published by Museum low impact journals like the *Boletin del Museo Nacional de HIstoria Natural* or electronic journals like *Chloris chilensis*. The impact of these journals is certainly minor, and these articles are not recognized by the Science Commission Fund (Fondecyt-Conicyt) for projects funding. The disincentive for such publications is evident. We also checked if the Management Plans of the National Park Service (CONAF) do have complete species' lists. The results are dramatic: only half of the protected areas has useful botanical information. And from this half, only another half shows a taxonomic and distributional coherence with the control database. Our results show a high dissociation between the information gaps that need to be filled for taking actions in botanical conservation, and the incentives for research and publication following international "excellence" standards. A profound reflection within the botanical scientific community is needed for building the bridges between academic goals and the real society needs regarding the benefits of botanical conservation and the ecosystem services associated. Not mentioning the important challenges that are related nowadays to climate change and the unexpected impacts like the wildfires that affected central Chiles this last summer and certainly affected important populations of narrow endemic species, that are today, more threated than ever. We emphasize the importance of low impact journals for the continue updating of the distributional knowledge of endangered species, and the need of a long term botanical information program, independent of academic high impact publication needs.

T6-01-10

Neotropical Biodiversity: An open access journal initiative from Ecuador

Nora Oleas[1], *Juan Manuel Guayasamin*[2]
1. Universidad Tecnológica Indoamérica
2. Universidad San Francisco de Quito

Latin America is the most diverse region on Earth. This vast biodiversity is still poorly known at all levels: from it basic description to the interactions at the ecosystem level and its conservation status. A collective effort to facilitate original research and communication is greatly needed in the region. At the same time, publishing in mainstream journals are challenging for Latin American researches. One of the reasons is the regional scope of studies focus in the area. Also the language barrier makes it difficult for researchers to publish in high impact journals. With the aim to expand publication and dissemination of scientific work from Ecuador and Latin America, the Ecuadorian Government and its National Secretariat for Higher Education, Science, Technology and Innovation (SENESCYT in Spanish) launched in 2015 the Neotropical Biodiversity Journal. Neotropical Biodiversity is an open-access and peer-reviewed journal that publishes original research articles on the biodiversity within the Neotropics. The journal publishes research at all scales of biodiversity: population, species, and ecosystem level. It is general in scope and the main criteria for acceptance are originality, scientific rigor and appeal to an international readership. SENESCYT covers all the fees of articles published in the journal, which is distributed through Taylor & Francis, a well-recognized scientific publisher. This eliminates any financial barriers that scientists may encounter when accessing or publishing research and warranties that all papers are widely available to the global scientific community. Even though this endeavor counts with financial support, several challenges are faced in launching a new journal. One of the challenges is making the journal visible to the scientific community. Another issue is convincing scientists to collaborate with the journal as reviewers or authors. Assembling a committed editorial board was difficult, but now the journal counts with 32 associate editors from 11 countries. Also even though Neotropical Biodiversity is free and open access, as a new journal, it is not indexed and has no impact factor. Impact factor measure the yearly average number of citations to papers published in the journal. Impact factor has been criticized as a valid measure of scientific quality; however, it is still used as a measure of a researcher merit in Latin American countries. As a result, researches might not be inclined to contribute their best research in a recently launched journal. The philosophy behind the creation of Neotropical Biodiversity is that scientific information should be available to any interested person and the only criterion to publish research should be quality and no economic factors. The challenges are still current but our expectation is that the journal will become one of the most influential venues to publish research from Latin America.

T6-01-11

How the journal impact factor affects the publication of agriculture and natural sciences research in Argentine journals

Andres Ravelo
CREAN/CONICET

AgriScientia is a periodical publication in the agricultural and natural sciences including Botany in Argentina. It began in 1980 as Revista de Ciencias Agropecuarias and in 1991 changed its name to the current one. Initially its scope was to channel the results of research by the faculty of the academic unit (Facultad de Ciencias Agropecuarias, Universidad Nacional de Córdoba) and later it accepted papers submitted by other nationals and foreign researchers. About two decades ago the National Research Council (CONICET) created a journal advisory body (CAICYT) to set the editorial standards of argentine scientific publications. CAICYT helped AgriScientia with the indexing in Scielo and Scopus among other indexing systems as well as implementing the open journal system. The irony behind this praiseworthy action is that the evaluation committees from CONICET are still using the Journal Impact Factor (JIF) as a measurement for researcher productivity, to allocate project funds, to establish promotions, to award scholarships, *etc.* Therefore, the low JIF of national journals prevent the consideration of the contribution of those papers to enhance the local knowledge. AgriScientia has a low JIF and consequently it is not receiving many contributions from the CONICET scientific community. This has a negative impact on the relevance of published articles and therefore it becomes a vicious circle because the journal is not cited. To attract more contributions, AgriScientia accepts papers written in Spanish, Portuguese and English over a wide range of subjects in agriculture and natural sciences by doing so it became one of the top journals in the country. Unless some changes are introduced in the current evaluation system based on the JIF, most of the scientific journals of Latin America are subject to be ignored or not properly considered. Fortunately, there are some international actions such as the DORA agreement which can reverse the current unfair system for scientific publications in Latin America.

T6-02: Interpretation, interrelationship and representation of plants as symbols in different cultural expressions

T6-02-01

The long journey of South American sacred plants

Gabriel Luis Mario Bernardello
Museo Botánico, Facultad de Ciencias Exactas, Físicas y Naturales, Universidad Nacional de Córdoba-CONICET

South America's ethnic diversity rivals its biological diversity, from the Apurina to the Quichua and from its astonishing tropical forests to its surprising arid environments. Once, this subcontinent was exclusively inhabited by hundreds of diverse indigenous people that made use of its natural resources. Among them, many vascular plants with different psychoactive compounds were prepared by shamans in rituals for divination, prophecy, cure, and ecstasy. The shaman, after consuming them, perceived and expressed the voice of a deity. Thus, these plants were considered to possess divine qualities. The usage, the effects, and the peculiarities of some of these emblematic species used by indigenous people is here discussed, together with the current decontextualized practice of these sacred plants. Ayahuasca has a long history of entheogenic practices among the peoples of Western Amazonia, where it has been revered for millennia. It is basically prepared with *Banisteriopsis caapi* (Griseb.) C. V. Morton (Malpighiaceae) mixed with additional plant species. It has provided humankind with a powerful medicine of healing and transformation, purging of body, mind, and spirit. Ayahuasca healing remains a prominent and accepted shamanic practice in indigenous and mestizo communities throughout tropical South America. Tobacco (*Nicotiana tabacum* L., Solanaceae) was traditionally used by indigenous peoples for religious purposes. It is a spontaneous hybrid native to the Andes of Peru and Ecuador found only in cultivation. It has stimulatory effects of the sympathetic and parasympathetic systems. Shamans consumed it in different ways to see inside the patient and diagnose his illness. Adopted by the Spaniards, tobacco had a rapid global spread and became an industrial product, away from the earlier ritual use. *Echinopsis pachanoi* (Britton and Rose) Friedrich and Rowley (Cactaceae), the San Pedro cactus, is a spineless columnar cactus that generally grows over 1,800 m a.s.l. in Andean warm and temperate regions. It is one of the oldest ceremonially consumed plants with archaeological records from 1,300 years B.C. This intoxicating and hallucinogenic drink is ingested by shamans and healers for divination purposes, diagnosis of diseases, and witchcraft. As Angel's trumpet or floripondio, several tree species of the genus *Brugmansia* Pers. (Solanaceae) are recognized. They are grown as ornamental and medicinal and are used in divination practices, initiation rituals and magic in the Andes and the Amazon for its hallucinogenic and psychoactive properties. There are records of its use in ceramics from 600 B.C. The coca (*Erythroxylum coca* Lam., Erythroxylaceae) is a native shrub to the Andean region. Its leaves have been widely used by native people for thousands of years in their religious cosmology. Chewing coca leaves is common in indigenous communities to reduce pain and as a stimulant against altitude sickness and to overcome fatigue, hunger, and thirst. Its leaves are important in Earth offerings and are also read in a form of divination in shamanic rituals. In the late 19th century, its psychoactive alkaloid (cocaine) was popularized for local anesthesia and was included into different products, such as prescription drugs, medicine, and soda drinks.

T6-02-02

Interpretation, interrelationship and representation of plants as symbols in different cultural expressions

Ana María Planchuelo

Centro de Relevamiento y Evaluación de Recursos Agrícolas y Naturales

All cultures have some myth or tales related with nature and humans. In the context of the social imagination, mankind has taken the plants as symbols and endowed them with an extraordinary and powerful personality. The ancestral memory, the racial memory, as well, as the cultural memory of society, plays a fundamental role in the perception of plants and in the creation of myths about them. Ancestral myths evolve, regress and get new perception through time and come up to our days with the same power they have born. This field of work is very exciting because there are several connections between plants and cultures, such as this example among others. In Greeks and Roman mythology the oaks trees (*Quercus* spp.) are associate with their highest god, Zeus or Jupiter which is the divinity of the sky, the rain, and the thunder. Two different species from different families such the rushes (*Juncus* sp., Juncaceae) and the fennel (*Foeniculum vulgare*, Apiaceae), are the plants that allowed Prometeo to stole the fire from the month Olympus and give it to humans. The bay, (*Laurus nobilis* L, Lauraceae) is the transformation of the dryad Daphne consequently became the symbol of victory and merit. Several genera, species or other taxonomic range were given by botanists in honor to mythological characters such as *Achillea* for Achilles, hero of Troy and *Eragrostis* for Eros, god of love. Also, there are botanical terms related to mythology such as hermaphrodite, which is a flower having both male and female reproductive organs which derived from Hermaphroditus the bisexual child of Aphrodite and Hermes. Also, there are several mythological gods or goddesses associated with nature, trees, gardens and orchards such as Flora goddess of wild plants and Ceres goddess of agriculture. From all of those examples, the myth of Narcissus and Hyacinthus, as well as, the metamorphosis of mythological characters in plants such Philemon and Baucis (transformed on oak and linden trees) are the ones that offers the major impact in human philosophy. Plants names are sometimes in relation with human feelings of joy or sadness such the legend of the sad tree *Nyctanthes arbor-tristis* L. which means "sad tree" which is related with ancient stories in Hindu Mythology. From ancient times till the present plants have relations with different religions cult, such the rose, as attribute of Venus (Aphrodite), symbol of love that was taken by Christianity as a symbol of the Virgin Mary. The Egyptian considered that from the flower of *Nymphaea lotus* L. the God Sun Re or Ra was born. The sacred flower of India (*Nelumbo nucifera* Gaertn.) is the seat of Lakshmi the goddess of luck and good fortune. Also, the lotus flower is the base from where is standing up Guan Yin or Kwuan the goddess of fertility in Taoist myth.

T6-02-03

The myth and uses of three Mexican plants by the Maya and Aztecs

Cruz Elisa Torrecillas Nuñez

Universidad Autónoma de Sinaloa

Mexico has a great variety of native species used for nutritional and medicinal purposes which were also part of the ceremonial rites of Maya and Aztec communities. The traditional uses before the arrival of the Spanish conquerors to the New World were compiled in the Badianus Manuscript or "Little Book of the Medicinal Herbs of the Indians" (1552). In those times the botanical name and chemical composition of the plants were unknown but the knowledge of the Tlazolteolt (Goddess of Medicine) has being related with the uses of the species in current times by several scholars. In this presentation three species were selected to review the uses and relationships with ancestral mythology. One of the species of major uses in current time is maize (*Zea maíz* L.) which was considered in the myth as the plant from which the Maya race originates. According to "*Popol Whu*" (Popol-Vuh the sacred book and the myth of pre-colonial America), the goddess Tepeu and Gucumatz created the first man and first woman from cornmeal dough using yellow and white maize. Kukulcan or Quetzalcoatl the feathered snake is the major figure of Nahua mythology and represents the force of life and, in the form of a black ant, stole maize from the red ants for the benefit of mankind. Xipe-totec, the god of fertility and regeneration, was associated with the crop of maize and several sacrifices were done in his honor to assure a good harvest. The origin of the chocolate started as a fermented beverage used in ceremonial rites. It was prepared by mixing roasted cacao seeds (*Theobroma cacao* L.) with seeds of *Bixa orellana* L., chili pepper (*Capsicum* spp.) and vanilla (*Vainilla planifolia* Andrews). In Nahuatl, the language of Aztecs, the plant was called "kakaw" and chocolate derived from "xocolatl" which meant bitter water. In present time, chocolate has an important international market in the candy and cooking industries. The myth of the pulque or aguamiel which is the alcoholic beverage made from fermented sap of the "manguey cactus" which is not a real cactus, instead is a variety of mezcal (*Agave americana* L.) started when Quetzalcoatl ascended to haven to request the assistance of the goodness Mayahuel to prepare a placebo to bring happiness to humans. In the earth Quetzalcoatl and Mayahuel fell in love and became trees with the branches interconnected. Tzitzimitl, the grandmother of Mayahuel destroyed the tree but, from the debris the manguey cactus was born. This agave plant was used to obtain fibers for clothing and bags and is the raw material for mescal a distilled spirit of Mesoamerica, similar to tequila the traditional alcoholic drink of Mexico.

T6-02-04

Key plants preserve elements of culture: A study over distance and time of fresh crops in Puerto Rican markets in Hartford, CT (U.S.A.) — "A Moveable Feast"

David William Taylor[1], *Gregory J. Anderson*[2]
1. University of Portland
2. University of Connecticut

People retain culinary customs when they migrate. We tested this commitment via study of Puerto Rican fresh produce markets in the continental U.S.A. (Hartford, CT) over time, 18 years, and space, by comparisons of the Hartford markets with the markets in Puerto Rico. A survey of Puerto Rican markets in Hartford in 1993-94 was repeated in 2009-10. A comparative study was made at three major open-air markets in Puerto Rico in 2009. Surveys recorded fresh crops, and consultations with vendors and interviews with Puerto Rican residents in Hartford provided context for the use of the plants in cooking. Altogether, 84 plant crops (64 species; 32 families) were recorded for seven categories. The largest category was 'viandas' (fresh, starchy "root crops" and immature fruits), followed by 'saborizantes' (flavorings). More than 80% of the crops were still present in the second Hartford survey nearly 20 years later. Approximately 90% of the Hartford 1993-94 crops and 75% of the Hartford 2009-10 crops were shared with markets in Puerto Rico. Our results suggest two new concepts. The persistence of these largely tropical foods in a temperate market far removed from insular, and tropical, Puerto Rican markets shows the importance of basic foods as an element of cultural identification. We recognize this stability despite significant time and geographic separation as an example of 'Culinary Cultural Conservation'. Secondly, analysis of these fresh produce markets leads to the conclusion that viandas are the most prominent in diversity, persistence over time and distance, volume, and in terms of consumers' "willingness to pay" (willingness to pay more to forego cheaper local substitute foods). Accordingly, we consider the viandas a good example of a 'Cultural Keystone Food Group', extending the previous concept of 'Cultural Keystone Species' to a group. We posit that the viandas are an example of a culturally elemental food group preference of sufficient strength that migrants bring this preference with them wherever they travel, and that the analysis of this Cultural Keystone Food Group can be used for a type of "rapid assessment" of the degree of culinary conservation in different migrant Puerto Rican communities. Furthermore, our results suggest that the crops that compose a Cultural Keystone Food Group, such as viandas, do not have to be native to a particular area, such as Puerto Rico, in order for them to become iconic of the cultural cuisine, and that several crops once associated with the slave trade to Caribbean islands are now celebrated components of Puerto Rican cuisine.

T6-02-05

The double life of a holy tree: The quishuar in Catholic temples and Andean homesteads

Alexandra Martinez
Universidad Politécnica Salesiana

In many cultures, trees have importance, both, material and spiritual, which represents a potential for their survival. Several ethnographic studies from India, the Philippines, Madagascar and Papua New Guinea, offer accounts of trees being perceived as having a material life and existence that transcends human life: "trees never die; they just morph into a different form". This cultural assertion holds a potential that could prevent the extinction of many arboreal species. The quishuar o kishwar (*Buddleja incana*), from the *Scrophulariaceae* family, is a tree native of the Andes region shared by Bolivia, Colombia, Ecuador, and Peru. It grows at an altitude between 3,000 and 3,500 meters above the sea level. For centuries, Andean people have considered it a sacred tree. At the time of the Spanish conquest, trees had a symbolic and economic importance. During the Inca Empire, trees represented the origin of human life and embodied grandparents and ancestors. The quishuar in particular had multiple uses: as firewood, to manufacture tools, in buildings and to carve out *keros* (sacred vases). Also, this tree was perceived as tree with magical powers because of their capacity to link the underworld with the world of humans. Throughout centuries, quishuar has kept its sacred character. In Ecuador, in two catholic temples built during the 17th century, expose the body of Jesus Christ crucified, sculpted from quishuar, continues being worshiped. This paper analyzes the intertwined condition of the quishuar: its religious symbolism associated with the body of Christ and its magical character as caretaker of the underworld. Based on the analysis of colonial Latin American Indian chronicles and ethnographic work, this paper aims to articulate the current symbolic and religious importance of the quishuar. I suggest that the (*Buddleja incana*), thanks to its usefulness in Andean homesteads continues being perceived as a magical entity and, thanks its homology with the human body, has become part of the Christ crucified during worship services in Catholic temples.

T6-02-06

Cultural variation in the depiction of plants in Indian literature

Madhuri Gokhale[1], *Mahesh Shindikar*[2]
1. Fergusson College, Pune
2. College of Engineering, Pune

'Culture and 'Literature' are closely associated terms as literature is the written portrayal of any culture. In recent times, one of the most important issues in the world in all its diverse forms is the issue of culture. Culture in the traditional sense is considered to be an assemblage of beliefs, practices, traditions, rituals, myths and art of a particular society. Logically, all facets of our life in some or the other way, are confronted with the effects of culture which intern gets manifested in to a local literature at temporal and spatial scales. Since ancient times, man has had an intimate contact with various environmental assets in general and plant resources in particular. The mankind depends on plants not only to fulfill their basic needs and necessities but also for the associated cultural values at personal and social level. In rites and rituals, ceremonies and sacrifices observed across all religions and customs, plants have played a significant role in various forms. Globally it has been observed that while accepting variety of transformations, interestingly literary writers have incorporated the changing phases in plant science in their writings and have reinforced the notion that 'culture' is not a static phenomenon. Depiction of plants and plant parts in Indian literature is also found to be obedient in following this rule. In the present study, a modest attempt to comprehend how culture contributes to the depiction of plants in diverse ways in literature is made. The manner in which plants are depicted in literature bring out the concepts, ideas, thoughts, beliefs, and superstitions of a particular group of society. It is interesting to unravel the threads of good or bad bioclimatic, biogeographic and social conditions as expressed in the respective literature. This study presents examples with reference to plants from literature written in different languages in India, folk literature and also Indian literature in English and argues that references to the plants is the representation of varied cultural and behavioural patterns with customs and traditions in India. It also highlights how value attached to a particular plant differs spatially in different traditions of Indian context as well as the western context. This exercise concludes that it is advisable and would be an enriching experience if the people from literature and plant sciences jointly analyse how various properties and features of plants are reflected in literature. This can provide significant insights about plants to a person from literature and make the reading of literature an enriching and rewarding experience for the person having science background under a variety of cultural settings.

T6-04: Prescriptions for overcoming plant blindness

T6-04-02

Understanding core concepts about photosynthesis - a case study of an experiment-based teaching strategy

Shaoshan Li

School of Life Science, South China Normal University, Guangzhou, China

From a human perspective, photosynthesis is the most important process on earth. During photosynthesis, plants are able to use the solar energy and raw materials-carbon dioxide and water-to build glucose and all organic compounds necessary for survival. All organisms depend on photosynthesis-the main pathway by which carbon and energy enter the web of life. Moreover, photosynthesis releases oxygen (O_2) in the air we breathe, and it is this oxygen that plays such an important role in cellular respiration and the accompanying synthesis of ATP. Thus, without photosynthesis, plants and animals, including humans, would suffocate and starve. Core concepts about photosynthesis are important for students to understand, but they are difficult and there are many misconceptions, not only for middle/high school students, but also for college/university students majoring biology. An experiment-based teaching strategy was used in the present study to improve students' biological literacy about photosynthesis. The results presented are based on a series of efficient and reliable inquiry-based photosynthesis experiments including: (1) Oxygen release from submerged pondweeds (*Elodea*, *Hydrilla*, *Spirogyra*). (2) Microscopic observation of stomata. (3) Paper chromatography of photosynthetic pigments. (4) Investigation of environmental factors affecting photosynthetic rate.

T6-04-03

Planting Science online mentoring program opens students' eyes to plants and plant scientists

Catrina Adams

Botanical Society of America

In this presentation, we share how the Botanical Society of America and 17 other scientific society partners fight plant blindness by bringing plants and plant scientists into secondary school classrooms around the world. Through PlantingScience.org, small teams of students design and execute plant science projects in their classroom with online support from a volunteer scientist mentor. The award-winning program has reached over 24,000 students from 41 U.S. states and five countries. More than 1,000 mentors have registered to participate from 39 countries. As Wandersee and Cleary wrote in 2006, *"The presence of a plant mentor earlier in one's life (someone who helped the mentee observe, plant, grow, and tend living plants) is a key predictor of that person's awareness, appreciation, and understanding of plants throughout the lifespan."*. PlantingScience makes it easy for practicing plant scientists to develop relationships with secondary school teachers and their students with a very small weekly time commitment, regardless of how physically close they are to a school. We will share and discuss examples of interactions that demonstrate how scientist mentors provide students with diverse role models of who scientists are, how they think about problems, how science works, and how they value plants. PlantingScience also provides an opportunity to increase secondary students' exposure to plant biology, and to bridge the gap between secondary and post-secondary education. Most U.S. high school biology teachers have never taken a botany course, and many secondary school teachers feel underprepared to teach plant science to their students. Without support from the plant biology community, biology teachers may frequently choose to use animal examples over plant examples to teach concepts relevant to all life, thus reinforcing plant blindness in their biology students. Through the National Science Foundation-funded Digging Deeper program, PlantingScience brings early career plant scientists together with high school biology teachers for extended, blended professional development. Teachers and scientists work together in-person during a week-long workshop and continue their interactions online as students work through a photosynthesis and cellular respiration investigation theme using the PlantingScience platform. Such partnership programs are mutually beneficial, as many graduate students and early career biologists are underprepared to teach using active learning principles; they are also unfamiliar with secondary school curricula and benefit from relationships with secondary teachers. We share data and examples highlighting benefits of this program for students, teachers, and early career scientist mentors, and lessons learned for facilitating such online partnerships.

T6-04-04

To raise plant appreciation to beta-level through integrative and literary approaches to plants as covered in literary works, cultural context, and economic uses

Paul But
Chinese University of Hong Kong

Getting to know plants is to love plants. The very first attraction to a plant is often induced by its beauty and fragrance, whereas the alpha-level of knowledge comes with the association of the plant with its names. Labels of plants in gardens and color photos and descriptions in pictorial guide books and websites could serve as vehicles to deliver such information, which would enhance amateurs and college graduates, including those with trainings in life sciences, to reach this level. However, to bring them a step further to the beta-level for ingrained admiration and respect of plants, we find it effective to introduce them, in vivid and subtle messages and stories, to the relevant literary works, cultural backgrounds, and economic uses of the plants. Below are a few examples of our approaches presented in a weekly column on a local newspaper. To demonstrate the importance of scientific names, we cite an essay written by the famous Taiwan writer Lung Ying-tai (龍應台) who found the city flower of Hong Kong, Yangzijing (洋紫荊), was actually Yanzijing (艷紫荊) of Taiwan. Some bloggers then mocked at Hong Kong people for this stupid mistake, unearthed by a non-scientist. The truth is both vernacular names refer to the same plant Bauhinia blakeana, which is called by a third vernacular name, Honghua Yangtijia (紅花羊蹄甲), in mainland China. If botanists from the three places meet, it would be very confusing if they do not use scientific names. To introduce lichens to the laymen, we describe the news of the cosmetic disaster of reflective white patches on the face of Peng Liyuan, wife of the current Chinese leader Xi Jinping, during their visit to London in 2015. Such patches are compared with those on the smiling faces of buddha sculptures in Angkor, Cambodia. Further elaboration is then added on lichen botany and their impacts on cultural relicts. Through an announcement of a wedding banquet, we review Chinese greetings, particularly the reference to conjoined images, and elaborate with succinct details on inosculation and the botany of conjoined trees as well as their differences from hugging trees which have no fusion of plant tissues and from stranglers of the genus Ficus. A higher level of knowledge of plants would enhance deeper appreciation of plants as found in relevant literary works, cultural context and wisdom in the economic uses of plants, which would, in turn, enhance higher expectations for conservation of plants and their sustainable uses.

T6-04-05

Overcoming plant blindness with active learning

Bruce Kirchoff
University of North Carolina at Greensboro

Learning to identify plants is extraordinarily difficult. This difficulty leads to what has been called plant blindness "the inability to see or notice the plants in one's own environment." Yet, despite this difficulty, many botanists can quickly and easily make species identifications at a glance, sometimes out of the window of a moving vehicle. They do this without recourse to the use of traditional characters or keys. This talk will present a method for teaching this type of instant plant recognition and overcoming plant blindness. Plant identification is difficult because of variability in the plants' appearance, the difficult of bringing them into the classroom, and because it is often difficult to take students to the field, especially on urban campuses. To solve these problems, we developed and tested a cognitive psychology-based computer program to teach plant identification. The program incorporates presentation of plant images in an active-learning format that was developed to stimulate expert-level visual recognition. *Visual Learning – Plant Identification* (*VL-PI*), and its sister program *VL-Agricultural Plants of the Riverina* (*VL-APR*), is a cross-platform visual learning application that has been experimentally shown to improve plant recognition. In a controlled experimental test against traditional study methods use of the program resulted in up to a 25% improvement in exam scores. The software demonstrates how the use of interleaved examples, spaced repetition, and retrieval practice (SPRINT) can be used to train identification of complex and highly variable objects. The software is open source, and easily customizable. A brief four-minute introduction can be found at https://youtu.be/lfoeldV-_94. A longer introduction, including instructions on how to add your own images is here: https://youtu.be/qMq0n8q8B8Q. The source code and executables can be found here: https://github.com/Jasig/ImageQuiz/releases.

T6-04-06

Biodiversity description simulation: a tool to stimulate interest in plant taxonomy during class

Rafael Medina
Augustana College

Many future biologists complete their college education with just minimal training on plant biology, and the lack of time and resources make impossible a deeper immersion in plant taxonomy practice and the description of biodiversity during an introductory course. If the students are never exposed to this field, it becomes harder for them to gain the appreciation or pursue any interest on a field that would be so needful in the next generation of biologists. The activity proposed here is designed to stimulate the interest on taxonomy with a game in which groups of students simulate to compete to be the first to describe and publish a new plant species. This activity has been tested successfully with classes of about 40 students in a time frame of 75 minutes, which makes it suitable to fit in a single class or lab session. The students are divided in a variable number of groups with three different roles: explorers, editors and curators. Each group receives a different briefing depending on their role: to identify a plant and send a manuscript with its description; to review submitted manuscripts and publish them when they comply with the code of nomenclature; and to accession and database plant specimens in a herbarium. The skills needed to participate in the game can be adjusted to the desired competencies of an introductory botany course. The debriefing discussion may include topics such as the challenges of biodiversity description, the peer-reviewed publication process and the importance of scientific collections.

T6-05: Botanical contributions from the Biodiversity Heritage Library

T6-05-01

The Biodiversity Heritage Library: Empowering discovery through free access to biodiversity knowledge

Martin R. Kalfatovic
1. Biodiversity Heritage Library
2. Smithsonian Libraries

The Biodiversity Heritage Library (BHL) is an international consortium working together to make biodiversity literature openly available to the world as part of a global biodiversity community. Through its extensive network of Members and Affiliates, over 50 million pages of biodiversity literature are now available through the BHL portal. Eight Global BHL Nodes work with institutions in their regions of the world to build capacity for digitization and to promote open access to the biodiversity literature. The BHL was created in 2006 as a direct response to the needs of the taxonomic community for access to early literature. Designed with meeting these needs, the BHL was grant-funded and quickly proved its value to its target users. The original BHL organizational model, based on US and UK partners, provided a template for, first, BHL Europe, and then a series of global nodes, most recently, BHL Africa, BHL Singapore, and BHL Mexico. As the BHL moved from project to a cornerstone of biodiversity infrastructure, sustainability, appropriate expansion, collaboration with national and pan-national digital libraries (Europeana, the Digital Public Library of America) became more important. With significant contributions from many of the major botanical garden libraries and related collections throughout the world, the BHL is an important tool in global botanical research.

T6-05-02

Building botanical literature collections for global use: Highlights from the Biodiversity Heritage Library

Caroyln Sheffield
Biodiversity Heritage Library/Smithsonian Libraries

As an open access digital library, the Biodiversity Heritage Library (BHL) provides extensive and broad coverage of biodiversity literature. Drawn from the collections of libraries from around the world, botanical literature composes both a significant portion of this corpus and a foundational aspect of the consortium. Formed in 2006, four of the original ten founding institutions were the libraries of botanical gardens and herbaria. New partnerships continue to be formed and today there are nine participating institutions making major contributions to the botanical collections including the Chinese Academy of Sciences, Royal Botanic Gardens, Kew, and Singapore Botanic Gardens, among others. From their collections and those of other natural history libraries, botanical literature remains a strong and growing portion of the online corpus. Over the years, special projects have also provided new opportunities to further grow the botanical coverage in BHL. Partners have collaborated across geographic boundaries to build an extensive collection of Seed and Nursery Catalogs. Other projects have focused on scientists' field notes. Thousands of pages of unique, archival materials documenting explorations, discoveries, and botanizing efforts from around the world are now freely available on the BHL website. These, along with the core book and journal collections, enable researchers to investigate and increase understanding of plant life more efficiently than ever before. BHL users have shared some examples of how they're asking big questions in new ways, thanks to the open access literature now available online. Examples include mining BHL metadata to identify agricultural resources and coverage; researching the phytogeography of species; and studying early land plants to inform conservation efforts. This presentation will highlight the breadth of botanical content within the BHL corpus; showcase some of BHL's projects to analyze and build collections of botanical literature within the digital library; and feature some of the real world use cases for BHL's botanical content.

T6-05-03

Real use cases for semantic information in BHL

William Ulate[1], *Riza Batista-Navarro*[2], *Roselyn Gabud*[3], *Nhung Nguyen*[2], *Axel Soto*[2], *Sophia Ananiadou*[2]
1. Center for Biodiversity Informatics, IT Department, Missouri Botanical Garden
2. National Centre for Text Mining, School of Computer Science, University of Manchester
3. Institute of Computer Science, University of the Philippines Los Baños

The Biodiversity Heritage Library (BHL) wants to encourage the reuse of its content by creating an accessible, computational, and servable data store for the purpose of data mining and semantic operations. The Mining Biodiversity project explores the transformation of the BHL into a next-generation social digital library resource to facilitate the study and discussion (via social media integration) of legacy science documents on biodiversity by a worldwide community and to raise awareness of the changes in biodiversity over time in the general public. The resulting digital resource would provide fully interlinked and indexed access to the full content of OCR text library documents, via semantically enhanced and interactive browsing and searching capabilities, allowing users to precisely locate the information of interest to them in an easy and efficient manner. Learning from the experiences of colleagues involved in efforts to enhance the accessibility of biodiversity data through well-defined semantics-rich representations, it was determined that our focus should start first by addressing real questions that BHL users had. However, initial consultations with potential users of this semantically-enhanced information, suggested that the search functionality we had planned would not be fully-seized by taxonomists in their daily use of data. Nevertheless, further consultation with BHL users from other areas allowed our team to identify a valuable set of real use cases for Semantic Information that could be supported. An example search interface implementation will be shown. This talk will demonstrate current real information needs that interoperable semantics for biodiversity data and knowledge discovery can support in BHL. It will also highlight areas challenged by the lack of machine-interpretable semantics that could help addressing and suggest potential solutions. Finally, switching from supporting use cases using text mining for semantic searching over the BHL document collections to support knowledge curation, we will mention two examples of implementation in follow-up projects, one for Philippine biodiversity in general and another one for dipterocarps.

T6-05-04

The development and progress of BHL China

Zheping Xu, *Jinzhong Cui*
Institute of Botany, Chinese Academy of Sciences

We have some development and progress as for the project of BHL China. As BHL has been blocked from China, we have downloaded more than 120,000 volume books from BHL website and put them in the website of BHL China. We also have developed new automatic camera program based on gphoto2 environment (IA scanning machine) to make the scanning machine working again. More support for other biodiversity informatics projects: ABCDNet (Asia Biodiversity Conservation and Database Network), the plant list in Aisa, NSII (National Specimen Information Infrastructure). We also visited some herbaria and libraries in four Asian countries, such as Thailand, Vietnam, Indonesia and Philippines. Some libraries were very interested in BHL project. We hope there are more participation and cooperation from these countries.

T6-05-05

The challenges of a botanical library in the tropics and the need for restoration and digitisation of materials

Nura Abdul Karim
Singapore Botanic Gardens, National Parks Board

The financial stringencies faced by libraries is a major challenge worldwide as the economy slows down and budget cuts becomes a regular exercise in many private and public organisations. The Library of Botany and Horticulture in the Singapore Botanic Gardens (SBG Library) has amassed a large collection of printed scientific materials and an extensive portfolio of serials in areas of tropical botany, taxonomic, horticulture, conservation, ecology, ethnobotany, landscaping and design. Modest budget allocation which has not seen much increase over the years pose challenges to the library to respond to new opportunities as offered by the digital provision of content and to increase acquisition. Another, is how the executive board of its parent organisation views the library as a unit that is "good to have" but need not be extensively staffed and budgeted. The library is expected to sustain existing types and level of services with limited qualified staff to meet the goals of the whole organisation. There is shortage of skilled and trained librarians for employing as such librarians are attracted to and hired by larger libraries that are able to manage and meet the staff's career advancement expectations and training needs as compared to the gardens where the bulk of the budget has always gone into the greening projects and maintenance, horticulture and research staffing, exhibitions and events, equipment purchases, and field trips. The SBG library is also facing constraints in physical space for expansion of its collection and have resorted to using compactors to overcome this issue. With the advent of the internet and availability of e-journals and e-books, the physical use of the library by staff has also seen a dramatic reduction since the late 1990s. Ability of staff to access and obtain information via internet search engines and to use more e-journals and e-books rather than resorting to searches from hardcopy books has meant forgoing the library and the need of librarians except in cases when the access to the needed information is not available freely or not lodged in any electronic form. This has led to an increase in demand by the research staff for the library, with its limited budget, to subscribe to major electronic journals and e-books. However, these e-publishers' policies on pricing and accessibility are another big challenge as this inhibit take-up by this small library. Collections of rare, scholarly printed works and botanical artworks are seeing various stages of deterioration due to poor storage facilities in the past. The environment, poor handling and storage of these collection overtime are unconducive to keeping the literature collections in good state perpetually. The rather exorbitant cost of restoration and digitisation and lack of professional restorers are issues presently faced when the library wishes to engage in preserving increasing number of deteriorating collections. The library will need to review its management and acquisition of information resources and staffing and to take advantage of digital information format and find strategies to keep cost down and still meet demands of the organisation .

T6-05-06

A checklist of the orders and families of medicinal plants in the Philippines

Harold Carag, *Inocencio Jr. Buot*
The Graduate School, University of The Philippines Los Banos

A checklist of the orders and families of Philippine medicinal plants is prepared to provide an up-to-date reference for students, teachers and researchers alike. The accepted names, families and orders in this checklist follow the recent classification of angiosperms largely based on DNA sequence data by the Angiosperm Phylogeny Group (APG) IV. Available and comprehensive references on Philippine medicinal plants were used to come up with the checklist. This was complemented and updated by online resources retrieved from reputable search engines with a focus on ethnobotanical researches conducted on some indigenous peoples in the country. The authors likewise included an alphabetical listing of older and conserved family names. Results indicated a total of 35 orders out of the world total of 45 (77%) and 120 families of the world's 413 (29%) which were classified according to the APG groupings (Magnoliids, Asterids Commelinids, Fabids, Malvids, Lamiids and Campanulids) and further organized into subclasses and classes of the presently used Cronquist system (Magnoliidae, Liliidae, Commelinidae, Rosidae and Asteridae) for convenience and practical use. Moreover, the list recorded more or less 900 medicinal plant species and it also included the common disease each species is utilized for. The checklist does not only prove the rich medicinal flora of our country but as well as the extensive knowledge of our local "herbolarios". Thus, it is imperative for the proper documentation and promotion of these valuable resources through modern platforms to maximize its potential in contributing to our country's health sector.

T6-06: Early land plants: from early adopters to transformative models for citizen science engagement connecting natural history collections to biodiversity research and education

T6-06-01

A three-tier citizen science monitoring scheme for the distribution of plants, bryophytes and lichens

Laurens Sparrius, *Baudewijn Odé*
FLORON, Dutch Botanical Research Foundation

Biodiversity surveys are often carried out to answer specific questions. To overcome the problem of shifting data demands due to changes in endangered species lists and environmental laws and regulations, monitoring all taxa of a certain species group can be an alternative. In this paper the authors describe how a citizen science approach in the Netherlands succeeded in monitoring all wild vascular plants, bryophytes and lichens since 1989. Two different methods are used to monitor common taxa and rare taxa. The third tier assures sufficient spatial coverage. All three methods use data from previous years to calculate when a grid cell or taxon needs to be resurveyed. The resulting data set was used to create live distribution maps. Occupancy modelling was used to calculate national and regional annual species trends. Trends and distribution maps are used as material for National Red Lists and the evaluation of nature management targets. Results over the past ten years showed that the method is robust enough as a long-term monitoring project. The principle of using existing data to assign monitoring tasks is efficient, but it provides also a transparent way of communicating with volunteers participating in citizen science, who will find their work more meaningful.

T6-06-02

Crowd sourced science: A transformative model of engagement connecting natural history collections to biodiversity education and conservation

Matthew Von Konrat[1], *Brian Carstensen*[2], *Laura Trouille*[2], *Dara Arabsheibani*[3], *Audrey Aronowsky*[9], *Laura Briscoe*[4], *Mike Bryson*[5], *Tom Campbell*[3], *Ben Carter*[7], *Steve Cohen*[5], *Charlie Delavoi*[10], *Eve Gaus*[1], *Juan Larraín*[6], *Beth Sanzenbacher*[8], *Caitlin Vaughn*[1], *Zak Zillen*[3], *Taylor Walker*[11]
1. The Field Museum
2. Adler Planetarium
3. Northeastern Illinois University
4. New York Botanical Garden
5. Roosevelt University
6. Pontificia Universidad Católica de Va lparaíso, Campus Curauma
7. San Jose State University
8. Bernard Zell Anshe Emet Day School
9. University of Chicago
10. University of Connecticut
11. Hollins University

Alarmingly, the world's biodiversity is diminishing rapidly and undergoing an extinction crisis. Biological collections of museums and academic institutions, documenting the fossilized and living members of the world's ecosystems and their changes over time, are uniquely poised to inform the stewardship of life on Earth. The National Science Foundation program Advancing Digitization of Biodiversity Collections (ADBC) is enhancing and expanding the national resource of digital data documenting biological and paleontological collections at a rapid pace. We provide a collaborative model founded on these digitization efforts that work towards addressing the critical urgency of the loss of taxonomic expertise and the rapid decline of biodiversity. Scientists and educators are partnering with leaders in online Citizen Science, Zooniverse (see www.zooniverse.org), and have a coordinated network including students and professionals at universities and partnering high schools and middle schools to accelerate the pace of scientific discovery. The project has the specific goal of engaging a broader audience, especially students and citizen scientists, to partner with our efforts in recording critical data sets from digitally rendered images. A web-based tool has been developed (see http://microplants.fieldmuseum.org) demonstrating a model approach to aid in relieving some of the taxonomic impediment, accelerating biodiversity documentation and discovery. To date, over 8,000 participants have generated over 90,000 data points that are being utilized by researchers to help with species delimitation and aiding in uncovering cryptic species. The project focuses on a mega-diverse liverwort genus occurring in Australasia, the Malay Archipelago, and southern South America-representing several global biodiversity 'hotspots'-including the Polynesia-Micronesia hotspot, which is considered the epicentre of the current global extinction crisis by Conservation International. Only scant data exist for liverworts (living descendants of the earliest land plants) compared to many animal and seed plant groups of the region. Data derived directly from the participation of Citizen Scientists, particularly cryptic complexes of species, has a significant impact on our understanding of species distribution patterns and direct implications for conservation of these organisms in the region.

T6-06-03

Bryophytes in Australian ecosystems-opportunities to educate the wider community

Alison Downing[1], *Josephine Milne*[2], *Graham Bell*[3], *D. Christine Cargill*[4], *Cassia Read*[5]
1. Macquarie University, Sydney
2. National Herbarium of Victoria, Melbourne
3. State Herbarium of South Australia, Adelaide
4. Australian National Botanic Gardens, Canberra
5. Friends of the Box-Ironbark Forests, Castlemaine

Australian vegetation is very different from that of the northern hemisphere, featuring plant families and genera, such as Proteaceae, *Eucalyptus* and *Acacia,* which have spectacular, eye catching and colourful flowers; little wonder that minimal attention has been paid to early land plants, in particular bryophytes (mosses, liverworts and hornworts). However, bryophytes play a vital role in ecosystems, from minimising soil erosion in arid regions to humidifying the upper canopies of rainforests. Bryophytes have an intrinsic beauty and once observed, it is easy to generate interest in these exquisite miniatures of the plant world. In Australia, bryologists are few and far between, but in recent years there have been some outstanding initiatives to educate the wider community in the importance of these easily unnoticed plants. Technological opportunities to promote bryophytes are not extensive. The online Atlas of Living Australia has proved to be an immensely useful tool to check distribution of individual species, as well as list species recorded for given locations. The Australian National Botanic Gardens in Canberra has developed web pages designed and written to educate and generate interest, rather than distance the wider community from the scientific aspects of bryology. Other, more traditional methods may be successful in raising the awareness of these plants with the public. A living bryophyte display at the entrance to the Australian National Botanic Gardens' Visitors Centre is ideally placed to capture the interest of visitors. Engagement of the art community has also been an excellent opportunity to showcase bryophytes, including exhibitions resulting from *Artists in Residence* programs or exhibitions of scientific botanical illustrations of bryophytes. The publication of Read and Slattery's *Mosses of Dry Forests in South-eastern Australia,* a community project with the Friends of the Box-Ironbark Forests of northern Victoria, promoted as a '*guide for students and absolute beginners-technically accurate, but free of technical language*' has been a great success. Many Australian herbaria have set up volunteer programs, involving anyone from university students to retirees, to work on bryophyte collections. Inviting local professional and amateur botanists to participate in bryological field trips was a success in a recent bryophyte workshop. Teaching programs prepared for schools and universities developed some years ago are still successfully incorporated in educational programs today. Bryophytes have also featured in tourist magazines in tropical northern Queensland and have captured the interest of locals and also international visitors. Australian institutions may not have the capacity to support high-tech infrastructure to generate citizen science on-line, but the enthusiasm and active participation by Australian bryologists in using a wide range of opportunities to promote bryophytes can go a long way to educate and generate interest in these less-well known, albeit spectacularly beautiful and environmentally essential, group of plants.

T6-06-04

Bryophytes: Easily overlooked but unique plants for public awareness of natural beauty and biodiversity conservation

***Shan Li**, Qin Zuo, Lihui Mao, Li Zhang*
Fairy Lake Botanical Garden, Shenzhen & Chinese Academy of Sciences

Bryophytes, including liverworts and the more familiar mosses, together form the second largest group of land plants after flowering plants. Some bryophyte species are quite conspicuous and common even in urban areas. Unfortunately, they are poorly known by the public generally due to their miniature size and perceived limited influence to human's daily life. On the other hand, there are many unique features of bryophytes that make them appealing to attract the attention of the general public especially the students. The low commercial value of bryophytes provides distinct opportunities for educators and scientists to be free from the commercialization and focus purely on the natural beauty through their exquisite and varied morphological features. For the audience, the pure interest helps to maintain a long-term focus on bryophytes with further study and consideration of the related environmental topics, namely ecological system and biodiversity conservation *etc.* In Fairy Lake Botanical Garden, we have carried out diverse methods to introduce general knowledge of bryophytes to the public in the past decade. These include: publication of popular science books; booklets and a field guide; exhibitions and lectures; and training volunteers as environmental tutors of both regular courses and irregular environmental education activities. We also teach how to look for, observe and identify common local bryophytes, and identify the current threats they face, such as habitat loss. We found that the above multi-pronged approaches have been very successful. People of diverse backgrounds and ages show continued interest towards bryophytes beyond their initial exposure. And 35 volunteers who passed the training exam became the core members in the relevant environmental education projects of the botanical garden. Presently, Fairy Lake is the only botanical garden in China to provide such courses on bryophytes. In the near future, a moss nursery will be established, and this will provide an easier access for the public and students to familiarize themselves with bryophytes. Novel innovations are being considered, for example, to organize art exhibitions and handicraft workshops on bryophytes, to collaborate with schools introducing bryophytes into classroom curricula, and to develop an interactive App about bryophytes and bryological research.

T6-07: Interpretation of plants to the public

T6-07-01

'Searching for Green Treasure in NBG '- an interesting game to learn from plants

Mei Li
Institute of Botany, Jiangsu Province and Chinese Academy of Sciences, China

Botanic gardens are important centres for education, they offer a unique window to the wonders of plant kingdom, and they are uniquely placed to teach the public about the importance of plants in our lifes and in the global ecosystem. Nanjing Botanic Garden (NBG), the first national botanic garden in China, puts emphasis on environmental education since the late 1990s. The environmental education programme of NBG have been carried out in various ways. Scientific guides, printed materials, plant lables, exhibitions, lectures, training courses have been provided; various activities including summer camp and interactive game have been organized. A pant-based game called 'Searching for Green Treasure in NBG', mainly designed for young students is introduced in details in this paper. The game was first held in NBG in 2006, and approximately 3,000 visitors, including adults, took part in the searching activity within 2 months. After that, the game has been redesigned especially for pupils and junior high school students. Involved plants and places (specialized gardens) varied according to weather, students' age and activity duration. With the combination of plant-serching game, team work and competition, the game has become one of the most popular activities ever held in NBG for students, it not only has been interesting, but also has aroused participant's curiosity and encouraged them to observe and discover the uniqueness, beauty and secret of plants, to cultivate a love and respect for them and to create a sense of biodiversity protection.

T6-07-02

Ethnobotanical study of domesticated plants in Ile-Ife, Ilesa and Osogbo areas of Osun state, Nigeria

Abayomi Ezekiel Folorunso
1. Botany Department, Obafemi Awolowo University
2. Botany Department, Obafemi Awolowo University

Ethnobotanical study of domesticated plants in Ile-Ife, Ilesa and Osogbo areas of Osun state was carried out to explore and document the extent of general and medicinal use of the plants; and to also provide ways of conserving various plant species that are useful to man. Questionnaires were structured in such a way that adequate information revealing the familiarity of people with the plants was collected in addition to the usage of their various parts. The returned questionnaires were analysed using IBM SPSS Statistics 21 analytical method. All the domesticated plants are of economic importance and are very useful to people where they are found. *Mangifera indica*, *Newbouldia laevis*, *Spondia mombin*, *Gliricidia sepium*, *Ficus* spp., *Blighia sapida* and *Argyreia speciosa* are used to provide shade apart from their other economic values. The fruits of *Mangifera indica*, *Spondia mombin*, *Annona muricata*, *Terminalia catappa*, *Artocarpus communis*, *Blighia sapida*, *Chrysophyllum albidum* and *Elaies guineensis* are edible. *Polyalthia longifolia*, *Newbouldia laevis* and *Dracaena arborea* are used for landmarks and as boundary. *Gliricidia sepium* was observed to be the most diverse of all the plants in consideration and the most commonly used in various environments. *Calotropis procera* was largely found to be a wild species as it was encountered in various locations where it was not planted, but was not uprooted because of its importance in milk brewing. *Elaeis guineensis* is known to be vast in its importance and its value cannot be over emphasized, since all its parts are useful economically. Ethnobotanical study shows that plants are highly valuable and useful for various purposes such as landmarks, shed, traditional purposes and several medicinal values for treatment of several ailments like fever, cough.

T6-07-03

The Interpretation of volunteer educators: A link between researchers and the public in the botanic garden

***Shan Li**, Weizhe Zhang*
Fairy Lake Botanical Garden, Shenzhen & Chinese Academy of Sciences

Volunteers are essential but overlooked human resource in Chinese botanic gardens for a long time. By the chance of the Environmental Education project, we recruited and trained the first two batches of volunteers, who were ordinary people with limited academic background, as the environmental educators. After a set of training courses and examination, the volunteer educators mainly undertook the informal environmental education in Fairy Lake Botanical Garden. Moreover, these volunteers helped to transfer the scientific knowledge about plants into an easily accessible way to the public and find out the fascinating topics from the perspective of the public. However, there are limited researches about the volunteer team in the botanic garden. To assess the effect of the interpretation of the volunteer educators and improve their performance in the interpretation, we set a series of survey focusing on the volunteer characteristics, motivations, goals, achievements, demands, self-feedbacks and feedbacks from the audience. These surveys enabled us to understand more about our volunteers and gave us suggestions to improve the interpretation given by our volunteers.

T6-07-04

Cultural significance of plants in traditional Chinese gardens

Jinyong Chen
The Museum of Chinese Gardens and Landscape Architecture

Plant is one of the important elements in the traditional Chinese gardens. Plants are integrated with surrounded landscape, not only improve beauty of the scenery, but also give people strong impression by the cultural connotation. From the very beginning, Chinese people live on plants for food, medicine, shelter, and gradually appreciate them. In the Book of Songs published between 1100 B.C. and 600 B.C., about 138 species of plant were mentioned in the 311 songs. For example, 'the young man presents peony flowers to the lady when he made date with her', showing his love. During that time people planted willows surrounding the house as a garden, outside the garden were mulberries and inside the garden was *Pteroceltis tatarinowii*. In Chinese gardens, plants are appreciated not only by the natural beauty, but also the artistry and culture, which deeply rooted in the 3,000 years history. In the Analects of Confucius, Zaiwo stated that the national tree of Xia Dynasty (2205-1782 B.C.) was pine, that of Yin Dynasty (1600-1046 B.C.) was cypress, and Zhou Dynasty (1100 -256 B.C.), chestnut tree. Similarly, Sun Jue in Guweishu (published between Ming and Qing dynasty) stated that pines were planted in the mausoleum of king, cypresses were planted in the duke's mausoleum, chestnut tree were around the official's tomb, *Sophora japonica* around the scholar's, and common people have no tomb but only willows. This defined the hierarchy of different trees, which affected the planting in the traditional gardens. In addition, plants were personalized in Chinese culture. Pine tree, bamboo and Mei flower were recognized as 'Three Friends in the Winter' since they were prosperous in the winter. This beautiful plant landscape was commonly applied in Chinese gardens. Mei, orchids, bamboo and chrysanthemum were praised as 'Four Gentlemen'. The combination of these species was designed with rocks in the classic gardens, forming an elegant cultural landscape. For the royal gardens such as Summer Palace, the Temple of Heaven and Chengde Summer Resort, pine and cypress were extensively planted, not only because the emperor recognized their importance, but also they symbolized eternity due to their evergreen and longevity. In the Summer Palace, different plants have a certain cultural connotation. Cypresses were planted in two rows inside the east gate, standing for the civil and military ministers. Moving forward, near the palace was *Sophora japonica* 'Pendula' representing the emperor by its dragon-like branches. Tree peony, the imperial flowers were planted on the both sides of the palace. *Firmiana simplex* was planted in the empress' yard, meaning phoenix (symbolizing Queen) inside. Magnolia, crabapple and peony were planted in the Summer Palace as well because they symbolized noble and wealthy in Chinese culture. For the private gardens (especially scholar's garden), the design concept usually comes from paintings and poems. For example, Yuanxiang Hall in the Humble

Administrator's Garden originated from an essay Ode to the Lotus Flower, indicating the fragrance of lotus flower in the garden can be smelled from far away. At the same time, plants were bestowed good wishes by their appearance. *Osmanthus fragrans* and *Magnolia denudata* were commonly planted in the courtyard, implying 'gold and jade fill the hall' for they have golden yellow and pure white flowers. Similarly, varieties of yellow flowers and white flowers were planted in the Yuyin Cottage Garden in Guangzhou, which had same meaning. Loquat was frequently planted in the private gardens such as the Humble Administrator's Garden in Suzhou. Its fruits look like gold hanging on the tree. The orange tree has same indication, which is recognized as auspicious in southern China gardens. In the Buddhist temple, plants related to Buddhism are usually planted, such as *Ficus religiosa*, *Saraca dives*, *Shorea robusta*, *Corypha umbraculifea*. Moreover, 'Five Trees and Six Flowers' are commonly planted in Dai temple of Xishuangbanna. In northern China, Ginkgo, horse chestnut, bamboo are commonly planted in the temple garden since they are mentioned in the Buddhist legend. In the Taoist temple garden, plants are somewhat different. Peach trees are commonly planted since peach symbolizes longevity. A famous poet Liu Yuxi in Tang Dynasty described thousands of peach trees in Xuandu Temple. Pine, cypress, lotus and *Camellia japonica etc.* plants are related to the Taoism and are commonly used in the temple gardens. Actually plant culture can be appreciated everywhere in the traditional Chinese gardens. For the pavement, some special pattern such as Meikaiwufu with five petals of Mei flower, means five blessing. The paintings or decoration with plants on the architecture (house, pavilion, corridor *etc.*) are also common. For instance, bottle gourd (Hulu in Chinese) has a special meaning of good fortune. Pomegranate and grape with many seeds indicate successive progeny and prosperous family. All these plant culture will add more interest to the traditional gardens.

T6-07-05

Plants and their symbols: Conservation of cultural knowledge of the Mamanwa tribe of Malimono, Surigao del Norte

Kent Joshua Tangan, ***Muhmin Michael Manting***
Mindana State University-Iligan Institute of Technology

Tribal communities have a closer relationship with the natural environment due to their long-term use of their natural resources. The documentation of their indigenous knowledge regarding the uses of plants is important in this critically degrading environment. This paper presents the ethnobotanical study, more specifically the plants considered as symbolic and those plants used in rituals by a Mamanwa community in Barangay Binocaran, Malimono, Surigao del Norte. An open-ended interview with the use of semi-structured questionaire were done on ten (10) respondents with an age range of 50-78 years to ensure the validity of the information gathered. The respondents provided information on 15 plant species that are used in ritual ceremonies and nine (9) symbolic plant species. This study also identifies the conservation efforts of the tribe towards their cultural knowledge and plant resources.

T6-07-06

The role of botanical garden in science education for young people

Jianjiang Chen
Beijing teaching Botanical Garden

A botanical garden plays an important role in scientific research and public education. It opens to the public and serves as base for species protection. It also makes great function in environmental science and life education for young people. In the past, botanical garden was just a public garden of touring and leisure for primary or secondary school students, and was research site just for the professional plant researchers. However in nowadays, the botanical gardens are trying their best to abstract the public attention to environmental protection through proper ways and means. They want to let the public to understand more about the truth of biodiversity crisis, and aware the importance of the rescue plants protection. Although after several years developing, the environmental education in botanical garden in China is still in the initial stage. There are lots of problems such as single method and form for interpretation, low degree of public participation, the lack of adequate maintenance of educational facilities, lack of a continuous system of environmental education activities, less publications, *etc.* Beijing teaching botanical garden is a special garden, because it was built just for biological and environmental education for the young people aged 6-16. In more than sixty years' development, the garden now makes many efforts in the area of explore experimental activities and gains experience in making full use of botanical garden for student education. The teachers make a theme for the activities all the year after discussion, then design and implement teaching activities. The experiment and activities include but not just "The Climate Change", "Plant Physiology", "Relationship between Soil and Plants". Each lesson lasts about 90 mins. Parts of the lesson held in the garden, and another part in the lab. Those research activities abstract children who living in the city almost all the year to be closer to nature. In Beijing teaching botanical garden they measure the temperature, humidity and other concerns about the causes of climate change and greenhouse effect and coping strategies. With a hand-held magnifier, boys and girls research on plants, insects, soil and water, thinking of social value, ecological value and make their choice. The special experience which cannot be provided by public school will give them chance to thinking about the occurrence of things around life. We have insisted the activities more than 3 years, seven thousand children came and jointed the activities. Multiple use of botanical garden, let the garden to be a platform of environmental and ecological diversity of education for young people will be a good idea and wise choice.

T6-08: Transgene flow in plants: its possible social and environmental impacts

T6-08-01

Transgene flow from herbicide-resistant rice (*Oryza sativa*) and its possible social and environmental impacts

Bao-Rong Lu, *Xiao Yang, Wei Wang, Jia Fang*
Fudan University

Rice is an important world crop providing staple food for nearly one half of the world population. The management of weeds infesting rice fields is challenging. It is even more difficult to control conspecific weedy rice (*O. sativa* f. *spontanea*) that can frequently acquire genes/traits from cultivated and wild rice through gene flow and introgression. Such a process can result in adaptive evolution of weedy rice in diverse rice ecosystems worldwide. The potential commercialization of genetically engineered (GE) herbicide-resistant rice varieties has stimulated great concerns over its possible social and environmental impacts. The commercial cultivation of the non-transgenic rice (Clearfield®) that is resistant to imidazolinone herbicides has already caused more serious weedy rice problems in many rice planting regions of the world, due to crop-weed gene flow, and eventually affected farmers' incomes. Hypothetically, the commercial cultivation of GE herbicide-resistant rice (e.g., glyphosate resistant) will most likely cause the similar problems. It is found that overexpression of endogenous *epsps* (5-enolpyruvoylshikimate-3-phosphate synthase) gene from plants, such as cultivated rice and *Arabidopsis* species, can also increase the resistance of the GE plants to the glyphosate herbicide. Overexpression of endogenous *epsps* gene can provide an alternative method of genetic engineering to produce glyphosate resistant crops. Recent studies demonstrated that overexpression of the endogenous *epsps* gene isolated from cultivated rice has substantially increased the resistance of GE rice plants to the glyphosate herbicides. Unexpectedly, the *epsps* transgene has significantly increased the photosynthesis ratio, tryptophan concentration, and fitness (survival and fecundity) of crop-weed rice hybrid descendants derived from crosses between glyphosate-resistant GE rice and weedy rice accessions, without the application of any glyphosate. Similar results were also obtained in crop-wild rice hybrid descendants. The enhanced survival and fecundity of crop-weed/wild hybrids and their descendants resulted from transgene flow between glyphosate-resistant GE rice and weedy/wild rice may pose great challenges for weedy rice control and environmental biosafety issues, in addition to social consequences related to farmers' incomes, because the problem of transgene flow from GE rice to its coexisting weedy rice or nearby wild rice populations is not avoidable. This might be particularly true for glyphosate-resistant GE rice containing the overexpressed *epsps* transgene that may transfer to wild relatives of rice. This is because the *epsps* transgene may enhance fitness of crop-wild hybrids even without exposure to herbicides. Importantly, EPSP synthase that determines the survival, growth, and development of plants is a key enzyme in the shikimate pathway that is found in all plant and many microorganism species. The glyphosate herbicide can competitively inhibit EPSP synthase, causing weakness and even death of plants at a certain dosage. Therefore, before the application of such a widely used herbicide in GE rice, the social and environmental impacts caused by transgene flow should be thoroughly evaluated.

T6-08-02

Potential social and environmental impacts of plant transgene flow: A 30 year overview

Norman Ellstrand

University of California - Riverside

Concern about unintended transgene flow from engineered plants dates back to the 1980's. At that time, a single concern was voiced, the evolution of more difficult weeds via crop-to-weed hybridization. The initial general view was that relevant gene flow rates were too low to be of much consequence. Relevant data were scarce. Now, over 30 years later, the situation has evolved. Natural rates of gene flow as well as the fitness effects of transgenes in the field have been quantified. Both were sometimes found to occur at significant magnitude to drive introgression. Descriptive and experimental field studies involving deregulated transgenes in field conditions have often yielded surprising results. Novel transgenic traits, from pharmaceutical-production to seed-sterility to gene drive have stimulated the identification of novel potential impacts, both detrimental and beneficial. Furthermore, about two dozen reports of spontaneous transgene flow into unintended populations are informative, illustrating impacts to date. Collectively, the foregoing accumulated data suggest that unintended transgene flow can have idiosyncratic impacts that depend on the recipient plant, the recipient population, the recipient ecosystem, and the values of the human system that interacts with those plants. Like natural gene flow, transgene flow has had and is anticipated to have infrequent significant effects. Whether the rare negative effects are realized will depend on mindful decision-making prior to environmental release. Continued research on case studies, such as those discussed in other presentations in this Symposium, will help guide that decision-making.

T6-08-03

Environmental impact of insect-resistance transgenes flow into rice wild relatives

***Xiao Yang**, Lei Li, Bao-Rong Lu*

Institute of Biodiversity Science, Fudan University

Gene flow from genetically engineered (GE) crops allows novel traits to spread to their sexually compatible wild relatives. Ecological impact caused by transgene flow into wild relatives is largely determined by the fitness effect brought by a transgene. Traits such as resistance to insects may enhance the fitness those weed and wild relatives, but few studies have tested for these effects under natural field conditions. In this study, we established multiple-generation hybrid progeny system between GE rice with insect-resistance transgenes (*Bt*, *CpTI*, and *Bt*/*CpTI*) and its wild relatives including weedy rice (*Oryza sativa* f. *spontanea*) and common wild rice (*O. rufipogon*), to investigate the long-term fitness effects of different transgenes on weedy rice and common wild rice populations in common-garden experiments with different insect environments and planting modes, from 2005 to 2013 in Fuzhou, China. Results from rice crop-weed hybrid descendants (F1-F7) demonstrated that *CpTI* alone did not significantly affect fecundity, although it reduced insect damage. In contrast, under certain conditions with high insect pressure, *Bt*/*CpTI* conferred significantly reduced less insect damage and increased fecundity in transgenic crop-weed hybrid lineages from F2-F7 relative to non-transgenic controls consistently. A positive correlation between insect index and fecundity change was detected, stressing the important role of ambient insect pressures in assessing fitness effects caused by *Bt*/*CpTI* transgenes. However, the increased density of transgenic plants in different cultivation plots could inhibit the insect occurrence and limit the fitness benefit brought by *Bt*/*CpTI* transgene. With the possible commercial cultivation of insect-resistant GE rice in future, significantly reduced ambient insect pressure and limited fitness advantage of insect-resistance transgene by extensive presence of GE rice plants in paddy field, will likely result in minor ecological risk by crop-weed transgene flow. Results from rice crop-wild hybrid progeny (F1-F2) demonstrated no significant differences in insect damages were observed between the wild rice parent and GE hybrid descendants under high insect pressure condition. The wild parent showed significantly greater relative survival-regeneration ratios than its hybrid descendants under different insect pressures. However, more seeds were produced in transgenic hybrid descendants than their non-transgenic counterparts under high-insect pressure. Given that the introduction of *Bt* and *Bt/CpTI* transgenes did not provide greater insect resistance to crop–wild hybrid descendants than their wild parent, we predict that transgene flow from GE insect-resistant rice to wild rice populations may not cause considerable ecological risks.

T6-08-04

Use of case studies of past transgene flow experiences to inform future releases

Carol Mallory-Smith
Oregon State University

Past experiences with gene flow from transgenic crops should be used to direct research, deployment, and regulation of transgenic crops. The frequent gene flow occurrences provide examples of the avenues by which both pollen and seed move in the environment. Studies often report on the occurrence of cross pollination but not on retention of the transgene in subsequent generations nor the consequences of gene flow. Most often the fact that the transgenic species will hybridize is not new information but rather confirmation of a fact that is already known. It is more important to assess the impact that a trait might have in the environment. Gene flow via seed movement which is a more frequent occurrence than via pollen is rarely studied in relationship to gene flow. Seed movement can move a trait globally which can easily impact global markets if a trait is not approved by the importing country. The use of case studies of gene flow from transgenic crops such as alfalfa (*Medicago sativa*), creeping bentgrass (*Agrostis stolonifera*), rice (*Oryza sativa*), and wheat (*Triticum aestivum*) which have different levels of cross pollination, growth habits and markets encompass a variety of issues and impacts which are crucial to consider when releasing other transgenic crops or traits.

T6-08-05

Evolutionary dynamics of weedy rice (*Oryza sativa*) and implications for adaptation through transgene escape

Kenneth Olsen[1], *Lin-Feng Li*[1,2], *Beng Kah Song*[3]
1. Washington University in St. Louis
2. South China Botanical Garden, Chinese Academy of Sciences
3. Monash University Malaysia

Weedy crop relatives pose unique challenges to the commercialization of GE crop varieties. Crop-weed interfertility facilitates transgene flow and can promote introgression of GE traits such as herbicide resistance. In the case of rice, GE crop varieties have existed for well over a decade but have never been commercialized, in part because of concerns over rapid transgene escape into weedy rice populations. Weedy rice is a major agricultural weed in rice fields worldwide and can reduce harvests by >80% if left unchecked. Gauging the risks of transgene flow into weedy rice requires an understanding the weed's origins, population structure, and levels of gene flow with cultivated and wild rice. This presentation describes recent insights into the evolutionary origins and genetic mechanisms of adaptation in US and Asian weedy rice strains. Using a combination of genome-wide SNP analyses, comparative QTL mapping, and genome scans for selection, we find evidence for the following: 1) Weedy rice has evolved multiple times independently through "de-domestication" from different cultivated rice ancestors. 2) Weed evolution has likely occurred both early and late in the history of rice cultivation. 3) Relatively few genetic changes are required for the emergence of weedy traits, and these can evolve independently through more than one genetic mechanism. 4) Despite a high selfing rate, weedy rice can rapidly evolve through adaptive introgression from both cultivated and wild *Oryza* populations. These findings suggest that adaptive GE traits such as herbicide resistance could spread rapidly into weedy rice, and that weed populations could serve as a genetic bridge for transgene movement into wild *Oryza* species. Given the major economic impacts of weedy rice on global rice production, caution is warranted in GE rice commercialization.

T6-08-06

Metabolic redesign of tocopherol biosynthetic pathway: Potential application for human health and abiotic stress alleviation in plants

Deepak Kumar[1], *Mohd. Aslam Yusuf*[2], *Neera Bhalla Sarin*[3]
1. Central University of Jammu, Jammu
2. Integral University, Lucknow
3. Jawaharlal Nehru University, New Delhi

Oxidative stress is one of the major manifestations of unfavourable environmental conditions faced by plants as also of several different diseases, including cancers, in humans. α-tocopherol, the biologically most active form of vitamin E, is a major antioxidant that bulwarks the cells. It constitutes a small fraction of the total tocopherol pool in most oilseed crops. We generated transgenic (TR) *Brassica juncea* plants with ~6 fold higher α-tocopherol levels by overexpressing γ-tocopherol methyl transferase. To better understand the roles of different tocopherol forms in plants we compared the performance of TR plants under conditions of abiotic stresses induced by salinity, heavy metal and drought. This resulted in an increase in total tocopherol levels in both the wild type (WT) and TR plants. Seed germination and leaf disc assay showed that TR *B. juncea* had enhanced tolerance to these stress and that induced by high temperature and methyl viologen. Damage caused by the induced stress was lower in TR plants compared to WT plants as assessed by their higher relative water content, lower MDA and electrolyte leakage. Lesser superoxide and H_2O_2 accumulation was observed in TR seedlings exposed to these stress. The levels of antioxidant enzymes and molecules were higher in TR plants when compared to WT plants under similar stress. Analysis of chlorophyll a fluorescence kinetics showed that there were differential effects of the stress on different sites of the photosynthetic machinery. These effects were found to be alleviated in TR plants. We further checked the efficacy of feeding α-tocopherol enriched seeds in securing the antioxidant defense in mice. We found significant increase in the content of various phase I and phase II enzymes with a corresponding decrease in peroxidative damage. Also, the transgenic seeds were found to have chemopreventive effects against DMBA-induced skin papillomagenesis in mice model. Our results highlight the potential of increased α-tocopherol in transgenic *B. juncea* in health of humans and agricultural crop plants.

T6-09: Turning toxins to treasure using plants

T6-09-01

Carbofuran degradation using iron oxide nanoparticles derived from the bark extract of *Hevea brasiliensis* Muell. Arg.

Abin Sebastian, *Ashwini Nangia*
University of Hyderabad

Hevea brasiliensis is the primary source of natural rubber. Millions of tons of bark are discarded during tapping process for natural rubber, and it is a renewable resource. So usage of rubber tree bark for chemical synthetic process is a sustainable practice. In contrary, pesticide degradation from the agriculture water release and products is a serious health issue need to be resolved. It is well known that iron oxide nanoparticles are used in environmental remediation. Functionalized iron oxide nanoparticles have more reactivity compare with non functionalized particles, and increase efficiency of reactions for which nanoparticles are applied. Carbamate pesticide carbofuran is degraded by iron oxide nanoparticles derived from *Hevea brasiliensis* bark extract in the present study. Synthesis of nanoparticles was carried out by mixing water extract of bark and ferric chloride solution. Surface plasmon analysis of isolated particles confirmed presence of iron oxide in the reaction mixture. SEM-EDS analysis indicated nanoscale size and the presence of iron in the particles. The iron oxide was confirmed as magnetite with help of powder X-Ray diffraction and ESR analysis. The capping agents in nanoparticles were detected using FTIR. Phytochemicals responsible for the formation and capping of nanoparticles were identified with help of a comparative study of bark extract before and after the reaction via GC- MS. The synthesized nanoparticles were utilized for the degradation of carbofuran. It is found that mixing of nanoparticles lead non photo catalytic degradation of carbofuran. Optical absorption studies and nuclear magnetic resonance studies were carried out to describe the reactions involved in the breakdown of carbofuran. In summary, we established methodology for the synthesis of nanoparticles from bark extract of *Hevea brasiliensis.* We also identified phytochemical responsivble for nanoparticle formation, physico-chemical properties of the nanoparticles, and action of the nanoparticles that lead decomposition of carbofuran.

T6-09-02

Morphophysiological, ultrastructural and element uptake in response to cadmium in 2 *Calophyllum brasiliense* Cambess. (*Calophyllaceae* J. Agardh).

Pedro Antonio Mangabeira
Santa Cruz State University

Cadmium (Cd) is a metal known for its genotoxicity and cytotoxicity, much concerned for its potential environmental and human-health impacts. This study evaluates the toxic effect of Cd in *Calophyllum brasiliense* plants. The plants were cultivated for thirty days in full nutrient solution in order to adapt, and after that be dayperiod, for fifteen days in nutrient solution without Cd or with 4, 8, 16 and 32 μmol Cd L^{-1}. After this period under Cd exposure, anatomical analysis of the leaf showed no significant effects of Cd on epidermal thickness in abaxial and adaxial sides, mesophyll, palisade and spongy parenchyma. Contrastingly, changes were noticed in the ultrastructural level in the leaf mesophyll cells as rupture of the membrane of chloroplasts and disorganization of the thylakoid membranes, in starch grains and in mitochondria with rupture of the membrane and invagination of the nuclear membrane. Electrodense materials into cells of the cortex and vascular bundle was also observed. In the cells of the root system, the observed ultrastructural changes were disruption of the cell wall and electrodense material deposition in the cortex cells and vascular region. Cd accumulated in roots with low translocation into shoot. Cd toxicity also affected the photosynthetic activity, inducing stomatal closure and photosynthetic assimilation reduction and the instantaneous carboxylation efficiency, drastically reducing the leaf transpiration. The nutrient content in the stem and root was variable, according to Cd increase in nutrient solution. Based on the experimental evidence, it can be concluded that *C. brasiliense* has potential to bioconcentrate high Cd levels in the root system.

T6-09-03

Cadmium-induced tolerance mechanisms in seedlings of *Avicennia schaueriana* Stapf & Leechman ex Moldenke (Acanthaceae)

Pedro Antonio Mangabeira
Santa Cruz State University

Cadmium (Cd) is a toxic metal with several folds of increase in the environment in recent decades. The response of tropical woody species in contaminated environment is rather scanty. It is therefore proposed to investigate the mechanisms that regulate the uptake and tolerance of Cd in the mangroves. The main objective of this study was to identify possible Cd-tolerance mechanisms of *Avicennia schaueriana* (Acanthaceae) seedlings. Seedlings were collected and placed in pots with nutrient solution for sixty days. After this period, the following concentration of Cd were added to the treatments: control (without Cd), 16,32 and 64 mg L^{-1}Cd in the form of CdCl2.5/2 H_2O, for five days. At the end of the exposure period, no anatomical changes were observed on the leaf blade of *A. schaueriana*. The increasing doses of Cd in the nutrient solution did not significantly alter the thickness of the abaxial and adaxial surfaces, the palisade and spongy parenchyma thickness, and the total mesophyll thickness. Polarized light microscopy revealed an increase of calcium oxalate crystals in the leaf mesophyll in the dose of 64 mg L^{-1} Cd and this is considered one of the tolerance mechanisms of the plant to the metal. At the ultrastructural level, the leaf mesophyll cells showed an increase of starch grains and electron-dense material deposited in the cells of the cortex and vascular bundle at 64 mg L^{-1} Cd. An accumulation of Cd was also observed in the root system, with low translocation to the aerial part. The regression equations showed that Cd affect the levels of macro- and micronutrients in different ways. The present study demonstrates for the first time the adaptive mechanism whereby *A. schaueriana* seedlings tolerates high levels of Cd by eliminating part of the absorbed Cd through the precipitation of calcium in the salt glands and calcium oxalate deposits.

T6-09-04

Plants for prevention of pollution and resource recovery

Narasimha Vara Prasad Majeti

DEPT. of Plant Sciences, University of Hyderbad, Hyderabad, India

Phytodiversity is being used as raw material for environmental decontamination and this field has grown phenomenally in recent years. On the other hand the volume of toxic and contaminated substrates (water, soil and air) are increasing due to anthropogenic and technogenic sources of organics and inorganics. The contemporary world is facing pollution. Sound policies and affordable technologies are required for combating environmental pollution (domestic and industrial). There is increasing demand for food, fodder, fuel, fertilizer and fiber. Cost of phytomass is gradually increasing due to land cost and water scarcity. Therefore, there is a need to address the toxic waste management and curb pollution due to industrial and domestic waste. This presentation will focus recovery of a wide variety of metals from environmental waste which otherwise is extremely toxic. Plant diversity is useful for Prevention of pollution and resource recovery (=phytotechnologies). Plants being photoautotrophs, can meet this demand by providing renewable resources and substitutes for non-regenerable resources. There is a need to bridge the research gap at the interface between pollution mitigation and boosting the biobased economy aiming at United Nations notified 17 sustainable goals. Pollution abatement and the pressing need to create viable value chains from toxic substrates are gaining considerable importance in developing economies. Thus, the efficient recovery and processing of the resources for recycling are significantly gaining importance globally. Awareness and opportunities to obtain valuable resources from process and bleed streams are of paramount importance in recent times. This would attain not only creating revenues but also achieving environmental sustainability. In this presentation various examples of plants and their societal benefits including their role in turning toxins into treasure shall be covered citing examples of plants and their societal benefits. Specific examples taken from a) Mangroves - acts as biofilters of toxic waste and mangrove wood and plants products are invaluable economic products b) Medicinal plants - e.g. Selenium bioconcetration in *Opuntia-ficus indica* c) Air Pollution tolerant plants – Large number of plants with high air pollution tolerance index are appropriate for social, urban forestry e) Ornamentals - Some cut flowers such as marigold (Tagetes spp.) and water lily handle a variety of toxic substances and generate revenue f) Plants for ecocatalysts and green chemistry - Metallophytes serve as invaluable plant resources for Green Chemistry. Recently metal accumulators, viz. *Noccaea caerulescens* and *Anthyllis vulneraria* are used as starting raw materials to prepare novel poly–metallic catalysts g) Phycoremediation and value additions including biodiesel h) Bioeconomy and sustainable environment - promising bioeconomic potential also exists for metallophytes for production of Biofortified nutraceuticals for human health, animal feed/poultry feed.

T6-09-05

The potential of *Etlingera philippinensis* (Ridl.) R. M. Sm., a Philippine endemic wild ginger against *Phytophthora infestans* (Mont.) *de Bary*: prospect as fungicide

***Florfe Acma**, Girlie Guillen*

Central Mindanao University

Etlingera philippinensis of the Zingiberaceae family, is a Philippine endemic wild ginger which grows abundantly in certain areas of the Philippines and with no known use for the local communities. Some species of *Etlingera* were reported to possess antimicrobial and insecticidal activities. Hence, this species was screened against *Phytophthora infestans*, the fungal pathogen of potato and tomato late blight to evaluate its antifungal activity. This study was conducted to isolate *P. infestans* from infected tomato leaves and to test the antifungal efficacy of the ethanolic extracts of *E. philippinensis* leaves and rhizomes against *P. infestans* using poisoned food method. *P. infestans* was isolated from infected tomato leaves and were inoculated into the potato dextrose agar following the tissue culture method. The antifungal assay using poisoned food method showed that both the leaves and rhizome ethanolic extracts of *E. philippinensis* have antifungal activity against *P. infestans*. On the 3rd day up to 9th day of incubation, the leaf extracts with concentrationof 10 mg/mL showed a low level of efficacy compared to the leaf ethanolic extracts 1.0 mg/mL and 0.10 mg/mL. However, on the 12th day of incubation, the ethanolic extracts of the leaf with the concentration 10 mg/mL showed the greatest antifungal activity among the leaf ethanolic extracts. On the other hand, all tested concentrations of the rhizome ethanolic extracts showed greater antifungal activity compared to the activity of the leaf ethanolic extracts. Mycelial inhibition on the 12th day of incubation at 0.10 mg/mL concentration of the rhizome ethanolic extract is comparable with inhibition of the commercial fungicide. It is concluded that both the rhizome and the leaf ethanolic extracts of *E. philippinensis* exhibit antifungal activity against *P. infestans*. Further, it was found out that higher concentration of the leaf extracts exhibited higher antifungal activity while the rhizome extracts exhibited higher antifungal activity at lower concentrations. The study reveals that *E. philippinensis* has a great potential as a fungicide.

T6-09-06

Impact of environmental variables on plant communities of the Thandiani sub forests division of the Western Himalayas of Pakistan

Shujaul Mulk Khan

Department of Plant Sciences, Quaid-i- Azam University Islamabad

In the summers of 2012 and 2013, we surveyed the vegetation of Thandiani in the Western Himalayas of Pakistan. We analyzed how the plant associations differ and develop foreseeable relationships with its respective environmental gradients. We take evidence from links between species and environment and changes in the community structure in fifty analysed stations with 1,500 m^2

plots. Preliminary results showed that the family Pinaceae was the most abundant family with 1892.4 F.I.V, followed by Rosaceae with 14.78.2. Out of 97 plant families, Rosaceae with 20 species was the most dominant family, followed by Asteraceae and Ranunculaceae with 14 and 12 plant species respectively. CANACO and GEO database methods demonstrated strong correlations among species distributions and environmental variables i.e., elevation, topography and edaphic gradients. Our findings show an increas in diversity index and species richness from lower elevation (1,290 m a.s.l.) to higher altitude (2,626 ma.s.l.). It was concluded that soil composition, aspect and altitude were the main factors affecting the vegetational composition of Thandiani sub forests division. The low P value ($P \leq 0.002$) showed that the variation in the vegetation Composition in the study area was highly significant in terms of test statistics. The habitats of the five association types overlapped broadly but still make strong correlation with environmental variables.

T6-10: Capturing biodiversity for food security

T6-10-01

Application of genomics to enhanced utilization of cereal diversity

Robert Henry
University of Queensland

Genomics may greatly enhance utilization of plant genetic resources in support of food security. Analysis of wild and domesticated genetic resources may identify new sources of genetic variation for breeding. Genomics of cereal (especially rice and wheat) genetic resources are key to global food security. Whole genome sequencing and transcriptome sequencing can contribute to discovery of valuable new genes and alleles. Sequencing of the genomes of wild rice (*Oryza*) populations has revealed new ancestral populations that widen the effective gene pool of rice for breeding. These newly identified genetic resources will provide a new source of diversity for use in breeding for disease resistant and tolerant to climate change in rice. Analysis of the transcriptome of developing seeds of diverse wheat germplasm has revealed genetic diversity that explains differences in carbon assimilation, flour yield in milling and bread quality on baking. Together these gene discoveries offer a significant opportunity to accelerate the rate of genetic gain in wheat breeding.

T6-10-02

Genetic dissection of starch biosynthesis diversity for grain quality improvement in rice

***Qiaoquan Liu**, Changquan Zhang, Minghong Gu*
College of Agriculture, Yangzhou University

Rice is one of the most important foods in the oriental world, and more peoples prefer to consume rice with good sensory and health characteristics. Starch is the major component of rice endosperm, and the composition and structure of starch play a dominant role in the quality of the rice grains, as well as their industry application. In the world, especially in China, there are abundant rice germplasms, including lots of local landraces, and the grain quality variation occurred widely among these landraces. Thus, it is very important to mine novel alleles of starch quality-related genes for rice quality improvement. In our studies, we focused on the starch-synthesis related genes (SSRGs), and carried out to mine novel alleles for quality improvement of both *japonica* and *indica* rice cultivars. By using the approaches of re-sequencing and genic molecular markers, the allelic variation of SSRGs among lots of rice germplasms were analyzed and several new alleles were mined. The results showed that there was different combination of these alleles in current cultivars. In order to investigate of the effects of these genes and their alleles on grain quality in rice, a lot of near isogenic lines (NILs) and transgenic lines were developed, and their backgrounds were nearly the same to the receptors. Through determining and comparison of the physiochemical properties of rice grains and the fine structure of starches between the NILs/transgenic lines and their parents, the results showed that there has a certain impact of the allelic variations on grain quality. The resulted and related materials resulted from this study will be very useful not only to rice breeding on quality improvement, but also to understand the genetic diversity of grain quality in rice.

T6-10-03

Large-scale plant genomics studies and CNGB's effort in food nutrition development

***Shifeng Cheng**[1], Howard Shapiro[2], Xin Liu[1], Liu Huan[1], Xun Xu[1]*
1. BGI-Research
2. MARS, INCORPORATED, USA

Plants play a crucial role in ecosystems, development of medicine and particularly as the fundamental resources in nutrition and food productions. Understanding plant genetics and biodiversity is a key to develop new and improved varieties with integrated efforts of germplasm resources collection and storage, genotypic and phenotypic data analysis, transformation and agricultural application. In the past one and half decade, series of innovative technologies (e.g., Next-Generation Sequencing technologies) have been developed and widely applied in sequencing plants at various levels (e.g., genomes and multi-omics data), which has tremendously accelerated and reshaped the researches in biodiversity and practices in food development. With such a remarkable explosion in sequencing technology and analytical tool development, multitudes of fundamental questions in plant genetics, populations and biodiversity were unprecedentedly asked and addressed at genome-wide scale (e.g., the 1KP project: https://sites.google.com/a/ualberta.ca/onekp/). The China National GeneBank (CNGB), with its digitization platform (e.g., BGISEQ-500) and supercomputing platform, is aiming to decode all life forms on this planet especially for plant resources. Furthermore, CNGB is dedicated to developing improved crop varieties, through gene editing and bio-synthesis platforms, and aims to establish a 'super crop' centre focusing in food and nutrition. We call for global collaborations from various research and application areas in food and nutrition with united efforts, and CNGB will provide an important vehicle to the gateway to the new era of global food security.

T6-10-04

Conservation of *Oxalis tuberosa*, "oca", in Ecuador: An iconic Andean tuber

***Hugo Romero-Saltos**, Alex Cabrera*

Yachay Tech (School of Biological Sciences and Engineering)

Oxalis tuberosa, "oca", is an ideal model to understand how the interaction between environment and humans has influenced the origin and maintenance of the diversity of varieties among Andean tuber species. However, in recent decades, cultural and market changes have caused certain varieties of "oca" to disappear in some areas of Ecuador, which can potentially lead to genetic erosion of the species and jeopardize the long-term food sovereignty of the country. During the last two years, Yachay Tech and Yachay Botanic Garden have intensively and extensively collected varieties of "oca" throughout the high Andes of Ecuador. Collections were obtained directly from the farmers themselves (indigenous people and mestizo peasants). Each visit was complemented by a semi-open interview on the agricultural practices related to this tuber. In the near future, we will conduct genetic analyses to measure the genetic diversity and understand the phylogenetic history of the "oca" varieties in Ecuador. Such analyses probably will also include the "oca" collections maintained by Ecuador's Instituto Nacional de Investigaciones Agropecuarias (INIAP), mostly collected a few decades ago. The results of this project can help to design better strategies and management practices for the conservation of Ecuador's native plant genetic resources.

T6-10-05

Diversity of edible aquatic plants from Thailand

Pranee Nangngam

Department of Biology, Faculty of Science, Naresuan University

Aquatic plants are diverse in Thailand, over 400 species occurred with several taxa are edible popularly consumed or sale in the local markets of the region. This is one of key biological asset from the wetland habitats, play important roles of ecosystem services. This study observed edible aquatic plants utilized in Thailand, at least sixty-sixth species including marginal plants thirty-eight species, emerged plant seventeen species, submerged plant two species and floating plant four species. Moreover, five species of ferns are also found. They are consumed in many recipe types including for vegetable, cereal or starch and cooked.

T6-10-06

Utilizing pulses diversity for ensuring food and nutritional security in Asia and Africa

Rajeev K Varshney

International Crops Research Institute for the Semi-Arid Tropics (ICRISAT)

Pulses because of their higher nutrient contents play an important role in providing food and nutrition security in developing countries. Breeding efforts for enhancing crop productivity could not meet the desired targets that are critical to feed the vastly growing global population. This has been partly attributed to low utilization of genetic diversity from genebanks to breeding programs. With an objective to understand the genome architecture and apply genome diversity in breeding, draft genomes of chickpea and pigeonpea were assembled. Subsequently, large scale re-sequencing efforts have been initiated in these crops for identifying new sources of genetic variation and allelic variants of candidate gene (s) associated with beneficial traits. In each of these pulse crops, more than 500 lines including reference set, elite lines, parents of several mapping populations have been re-sequenced. Millions of variants have been identified by aligning the re-sequencing data to their respective draft genomes. Detailed analyses provided comprehensive data on diversity features, gene loss, domestication and selection sweep. Multi-location phenotyping data for high priority traits for breeding along with SNPs was used for genome-wide association studies and several marker trait associations have been established. To accelerate molecular breeding efforts further, "The 3000 Chickpea Genome Sequencing Initiative" has been initaited. In summary, significant advances have been made in the area of assessing and utilizing sequence diversity in breeding programs for improving crop productivity.

T6-11: Exploration of new ornamental plants

T6-11-01

The genetic diversity and pedigree analysis of *Bougainvilleas*

Tao Chen[1], *Yan Zheng*[1], *Yajing Lu*[1], *Xianglan Wang*[1], *Anders Lindstrom*[2], *Cyl Farney Sá*[3], *Suresh Chandra Sharma*[4], *Michael Nee*[5]

1. Fairy Lake Botanical Garden, Shenzhen & Chinese Academy of Sciences, Shenzhen, Guangdong 518004

2. Nong Nooch Tropical Botanical Garden, 34/1 Sukhumvit Highway, Najomtien, Sattahip, Chonburi 20250

3. Instituto de Pesquisas Jardim Botânico do Rio de Janeiro. 22460-030, Rio de Janeiro

4. Botanical Garden, National Botanical Institute, Lucknow 226010

5. New York Botanical Garden, Bronx, New York 10458

Bougainvillea is a genus in the family Nyctaginaceae with 10 to 18 species native to South American countries such as Argentina, Bolivia, Brazil, Ecuador, Paraguay and Peru. Over 300 cultivars have been developed as the result of natural selection and breeding so far. *Bougainvilleas* have become one of the most common ornamental plant groups in the tropical and subtropical regions worldwide for their variegated foliage and especially showy bracts appearing in various shades of many colours except the blue. *Bougainvillea* cultivars are drought-tolerant, warm-requiring, easy to care and long lasting blooms with strong colours, and are widely cultivated for striking soft landscape. They were selected as official flower of several cities and counties in a number of countries and regions as well as the national flower of Zambia. In order to figure out the genetic relationship among the abundant cultivars and their wild relatives, over 60 SSR molecular markers have been developed based on the transcriptomic sequencing of *Bougainvillea arborea* and the validation with more than 500 samples of cultivars, which further provides resolutions for the precise cultivar identification and the germplasm renovation. This study will mainly present the genetic diversity and pedigree analysis of *Bougainvillea* cultivars and the taxonomy and phylogeny of their wild relatives.

T6-11-02

Hoya, the unknown country of beauty

Qi Wei, *Er-Feng Huang*

Guangxi Nanning Ruixi Landscape co., Ltd.

Hoya is a large genus within the milkweeked subfamily Asolepiandiodeae, the dogbane family Apocynaceae. It consists of over 300 species widely distributed from the tropical and subtropical Asia to Oceania. Similar with the well-known epiphytic plants, most of the *Hoya* species also share this special habit. Morphologically, the genus exhibits diverse inflorescence (e.g., umbel, pesudumbel or umbel-like inflorescence), and also is characterized by the highly diversity in color, morphology and the size of single flower, which making it to be an economically and ornamentally important plant group. The Roy Garden, located in Guangxi Province, China, has been endeavoring to collect wild resources of *Hoya* species since 2003, and now over 90% of the *Hoya* species are cultivated there. Based on these obtained materials.The Roy Garden has cultivated *Hoya* plants for more than 10 years, and also summarized the experience of the daily cultivation and conservation about *Hoya* plants, while collecting the wild *Hoya* genus resources. Now, in the Roy Garden, the cultivation and conservation of *Hoya* plants can be controlled with the lighting, temperature, humidity, substrate, fertilizers, moisture and some other conditions. Over 200 species of *Hoya* plants had bloomed in the Roy Garden, every year, from March to October, the Roy Garden looks like an ocean of "umbel" flowers. The Roy Garden has gradually increased for the application of the *Hoya* plants since year 2016, especially in landscape and gardening development. On the basis of exchanging communication and coorperation with public *Hoya* fans in China for several years and in connection with the *Hoya* plants with high potential in landscape and gardening using. The Roy Garden selected about 200 species of *Hoya* plants, and assess them with the plant characteristics, growth patterns, the flower characteristics and fragrance, ease of cultivation and conservation and some other horticultural indicators. Finally, we screened about 40 species of *Hoya* plants from them as the key targets for landscape and gardening using. Now in China, at least 3 botanical gardens have began to cultivated *Hoya* plants as a pilot, the special "Umbel" flower attracted large number of tourists' attention. And the same time, as a new element of family greening, the Roy Garden hope the *Hoya* plants can entered more and more household gradually. From 2014, in order to improve the adaptability of *Hoya* plants cultivate in the subtropical and even temperate environment. The Roy Garden began to develop the new *Hoya* varieties and a new *Hoya* variety named *Hoya henschkeliana* "Roy" independently by using interspecific hybridization. It will be published in 2017. The Roy Garden hopes to make more contribution to the global gardening work, and want to show the world an unknown country of beauty.

T6-11-03

New ornamental magnolias from controlled breeding

Yixia Chai[1], *Donglin Zhang*[2], *Kevin Parris*[3], *Xiaoling Jin*[1]

1. College of Landscape Architecture, Central South University of Forestry and Technology

2. Department of Horticulture, University of Georgia

3. Department of Plant and Environmental Sciences, Clemson University

Controlled breeding has significantly accelerated the speciation and plant evolution. The diversity of *Magnolia* taxa in today's ornamental market and gardens has brought us much more complication in the field of plant sciences, especially delineation of a species. Although "cultivar" has been widely applied for cultivated plants, the bridge between natural species and cultivated plant taxa has not fully constructed. From the oldest hybrid *M.* x *soulangiana* (*M. denudata* x *M. liliiflora*) by Mr. Etienne Soulange-Bodin in 1820, sterile little girl cultivars from *M. liliiflora* and *M. stellata* by Dr. Francis De Vos in 1956, to a unnamed cross between *M. sieboldii* and *M. insignis* by Mr. Kevin Parris in 2014, we had reviewed their morphological characteristics, phonological data, cytological variations, molecular evidences, and adaptation. Ornamental important *Magnolia* taxa have been greatly advanced under controlled hybridization and many outstanding ornamental cultivars from diploid to decaploid had been bred and available on the market. From these various hybrid cultivars, natural species delineation, such as cross-incompatibility (reproductive barriers), should be challenged. Results from our recent intra- and inter-section cross-hybrids of Section *Michelia* provided support to adapt "cultivar" instead of "hybrid species". Further studies should be placed all puzzles from natural species to cultivars, morphological traits to molecular relationships, by botanists, taxonomists, and horticulturists as one big picture.

T6-11-04

Plant hunters enriched the ornamental plant world

Donglin Zhang

Department of Horticulture, University of Georgia, Athens, GA, USA

Plant hunter is a keen person with botanical knowledge and brave mind for achievement. In today's 369,000 flowering plant species known to science, half of them are discovered by plant hunters. In the ornamental plant world, plant hunters almost provided all original species and majority of ornamental plants in gardens, landscapes, parks, greenhouses, and nursery trade nowadays are directly or indirectly from "golden age" for plant exploration and collection in late 19th and early 20th century. From Joseph Banks (1743-1820), who turned plant hunting into a profession and the first promoter of plant expeditions, to Robert Fortune (1812-1880), Ernest Henry "Chinese" Wilson (1876-1930), Frank N. Meyer (1875-1918, losing his life during the collecting trip), Frank Kingdon-Ward (1885-1958), Zhan Wang (1911-2000), Roy Lancaster (1937-), and Daniel J. Hinkley (1954-), the passion for collecting plants had motivated these greatest plant hunters. Their journey of plant collecting trips, especially difficulties and achievement, had been reviewed. Some milestone plants from these plant hunters were examined and their performance in our gardens were summarized. Their plant collection does not only enrich our gardens, but also contain critical genetic resources for our future plant breeding. New plants are new blood for the ornamental industry. We should foster more plant hunters and fully support and manage to fund plant exploration and collecting. There are still 15-30% new flowering plants to be discovered and shared with gardeners. But the most important task for plant hunters is to rescue and conserve endangered and threatened plants under mismanaged and expropriated habitats.

T6-11-05

Morphological characterization of cultivated and wild *Lupi-*

nus (Fabaceae, Faboideae) to be introduced as ornamental and garden plants in Argentina

Ana María Planchuelo
Centro de Relevamiento y Evaluación de Recursos Agrícolas y Naturales (CREAN-IMBIV-CONICET), Facultad de Ciencias Agropecuarias, Universidad Nacional de Córdoba,

Flower plants have been used from antiquity for decoration and grown only for display purposes. Although most species of *Lupinus* qualified as ornamental plants for its colorful papilionoid flowers, the erect racemes and the unique palmately compound leaves, few of almost 600 species are commercialized as garden plants. The most popular of them is the perennial *L. polyphyllus* Lindl., and its hybrids known as "Russell lupins" developed in England by George Russell around 1,911. Other perennial species such as *L. arboreus* Sims (yellow bush lupin), *L. albifrons* Benth. (white-leaf bush lupine); *L. nootkatensis* Donn ex Sims (nootka lupin) and the annuals *L. texensis* Hook. (Texas bluebonnet) and *L. hartwegii* Lindl. (sunrise lupin) are available in the international market, but cultivation are almost restricted to cold areas mainly in Europe and North America. In Argentina, *L. polyphyllus, L. arboreus* and the Russell lupins were introduced in the mountainous region of Patagonia and now are growing wild, mainly in the island of Tierra del Fuego, where the later is the icon flower of Ushuaia city. With the aim of making lupins more popular as garden plants in Argentina, cultivated as well as native species were evaluated by its ornamental and reproductive features as was done for wild species of *Viola* sp. A practical garden guide for lupin cultivation, with the approaches of xerophytic gardening is under preparation. For this presentation, the habitat, habit, plant and inflorescence architecture, flower color and garden uses were analyzed in cultivated lupins (*Lupinus albus* L., *L. angustifolius* L., *L. luteus* L., *L mutabilis* Sweet), in two introduced species (*L. arboreus* Sims., *L. polyphyllus* Lindley) and four wild species natives to Argentina (*L. albescens* Hook & Arnott, *L. gibertianus* CPSm., *L. magnistipulatus* Planchuelo & Dunn and *L. honoratus* CPSm). Recommendation for seed collection of wild species is described and symbols were designed to illustrate plant characteristics, uses and care. As a result of this study, the species analyzed will be presented as ornamental plants in alphabetical order by scientific name, in a booklet with a short description of characters, photographs, ornamental features, habit, uses as garden plant and other possible uses, along with natural origin and distribution. Places and conditions which are needed to be grown, and recommendations for plant care are listed.

T6-11-06

The conservation and utilization of begonias, action by Chinese botanical gardens

***Wenguang Wang**, Shouzhou Zhang*
Fairylake Botanical Garden, Shenzhen & CAS

Begonias are well known ornamentals which are cultivated for thir spectacular blooms and range of leaf shapes and coloration. Presently around 1,800 wild species of Begonia are recognized, making this genus one of the largest among the flowering plants. The species occur in subtropical and tropical regions, with the greatest diversity in America and Asia (>750 species each), while being relatively poor in Africa (160 species) and absent in Australia. China has the largest number of Begonia species in Asian, nearly 200 species. Most Begonias have a single living habitat, and some species distributed only in few sites. Artificial acquisition and environment destruction result in some of them endangered. Living collection in botanical garden is significant and pressing. Fairylake Botanical Garden, Shenzhen & CAS have collected more than 250 wild species and 170 cultivated varieties. In this presentation, we'll introduce the adaptability of cultivation conditions incl. using Begonias in landscape, and review the begonias conservation in China. The morphological and genetic diversity of *Begonia edulis* Lévl. (B. Section. Platycentrum), a wide spreading species in China, will also presented.

T6-12: Challenges of botanic gardens for a sustainable future

T6-12-01

***Crabaple*, the city flower of Baoji**

***Jian Quan**, Ling Guo*
Beijing botanical garden

Crabapple is confirmed as the city flower of Baoji (in Shaanxi province, China) in 2009. "Crabapple" in Chinese is pronounced as "Hai Tang", and "Hai Tang" also stands for many plants in Chinese, such as crabapple (*Malus*), Japanese quince (*Chaenomeles*) or begonia. Since Shaanxi province is the distribution center of the Genus *Malus* in China, then crabapple is the best symbol plant for Baoji. It was called "Xi Fu" (west prefecture) from the north of Wei River to the west of Jing River in Guanzhong area, including Baoji city, in ancientChina. That was why peoples generally recognized "Xi Fu Hai Tang" was determined as the city flower of Baoji. According to Flora of China (English Edition), the scientific name of "Xi Fu Hai Tang", is *Malus* × *micromalus* Makino. But the description in Flora of China (English Edition), was different from the original publication by Makino. It is provide some possibility of advanced discussion. "Xi Fu Hai Tang" is considered as *Malus spectabilis* 'Riversii' in scientific name according to the literature research in this paper. Crabapples with highly ornamental characteristics, looking forward in the future, more cultivars are appearing and could be the symbol of many cities.

T6-12-02

Impacts of climate change on indigenous communities: The role of botanic gardens in biocultural conservation

Christopher Dunn
Cornell University

Global climate change is having a significant, and negative, impact on the biological diversity and, thus, on the integrity of natural systems. What is less well understood, yet just as critical, are the impacts of climate change and of changes in natural systems on indigenous peoples. In other words, as biological diversity is eroding, so too is the cultural and linguistic diversity of the world. In fact, of the approximately 7,000 extant languages still spoken, fully 50% are at risk of extinction, with the vast majority in the tropics and subtropics. This rate of extinction of languages (and thereby of human cultural diversity) is considerably higher than most estimates of extinction risks to plants and animals. In addition, traditional ecological knowledge and livelihoods are being lost. As an example of the latter, many tropical (and other) indigenous cultures rely on phenological or "ecological calendars" to determine appropriate timing of planting, hunting, harvesting, among other necessary activities. Climate change, and consequent impacts on natural systems and resources, is completely disrupting wellbeing of tropical communities. Thus, it is not enough to consider just the effects of environmental change on plant life within the current context of the global conservation initiatives, such as the Convention on Biological Diversity (e.g., Article 8j), the Global Strategy for Plant Conservation (Target 13), and the Aichi Targets (Target 18). Botanic gardens are uniquely positioned to actively engage in understanding the broader impacts of environmental change to biocultural diversity to achieve biological, cultural, and economic resilience. Examples of how botanic gardens in several parts of the world are defining key ways to better understand tropical and cultural conservation will be presented.

T6-12-03

Gardens are inspirational for conservation bloom

Rajesh Rajaselvam
Dalhousie University

Admiring nature's beauty won't be enough to conserve it. Valuing the value of nature should be expressed by allowing nature to run the show. Since the clock of extinction is ticking fast, the message on conservation to educate the younger generation should be faster. Admiration comes through experiencing something good, thus gardens are inspirational for practical education in botany, landscaping and plant and nature conservation. Not only did the school outreach activities on conservation boost, but also the enthusiasm, enrollment, and the success rate increased in undergraduate plant related courses, labs and field activities in Nova Scotia with the active involvement of botanical gardens. My review concluded that garden visits, hands on field activities and commitment of resource personals are found out to be the key in motivation for plant conservation in young generations. Additional garden related activities have been scheduled to confirm this success.

T6-12-04

The introduction, selection and landscape application of native plants in Labagoumen Nature Reserve in Beijing of China

Donghuan Liu[1], *Ling Guo*[1], *Shiwei Zhao*[2]
1. Beijing Botanical Garden
2. Beijing Institute of Landscape Architecture

The introduction and application of native plants from the Labagoumen nature reserve, which is one of the most abundant plant species areas and has the unique geographical environment and the advantage of plant resources in Beijing, were carried out in order to enrich the diversity of native plants and to create a landscape scene with rich geographical characteristics of Beijing. According to the investigation, there are 668 species, belonging to 367 genera and 102 families in Labagoumen nature reserve which accounts for 1.4%, 53.2%, and 38.7% respectively for vascular plants in Beijing. Many of them have high ornamental value, which are good materials for landscaping and greening, including *Caragana pekinensis, Rosa bella, Hydrangea bretschneideri* and so on. 123 taxa of local plants in Labagoumen were introduced, including 15 woody stocks, 58 perennial roots and seeds for 50 plant other species. 50 introduced plants were observed and evaluated basis on growth, ornamental characteristics, heat resistance, drought resistance, light resistance and shading resistance. 30 species of native plants were selected based on comprehensive evaluation, including *Elsholtzia stauntonii, Deutzia grandiflora, Clematis heracleifolia, Clematis tangutica, Dianthus superbus, Hemerocallis lilioasphodelus, Clematis hexapetala, Scutellaria baicalensis, Dracocephalum rupestre, Platycodon grandiflorus, Dictamnus dasycarpus, Paeonia obovata, Scabiosa tschiliensis, Pedicularis resupinala, Geum chiloense, Adenophora hunanensis, Veronicastrum sibiricum, Phlomis umbrosa, Actinidia arguta, Trollius chinensis, Aquilegia yabeana, Lythrum salicaria, Campanula puncatata* and so on. These selected native plants have good ornamental value and strong adaptability to Beijing environment, which can be applicable for Beijing Landscape greening.

T6-12-05

Population ecology of *Pterocarpus angolensis* DC. in the Lowveld National Botanical Gardens, Mpumalanga Province, South Africa

Peter Tshisikhawe
University of Venda

Pterocarpus angolensis DC is a tree species belonging to Fabaceae family. It is one of the tree species that is utilized for various purposes around the world. In South Africa it is mainly used for its medicinal bark and its good quality wood in furniture production. The great demand of the species has resulted in its high level of harvesting wherever it grows. The overexploitation is therefore threatening its population viability outside protected area. The current studies aimed at understanding the status of a population within the protected pristine area of the Lowveld National Botanical Gardens. Data on population parameters was collected from twenty-six line transects of 100 m x 10 m in sizes. Parameters sampled were plant height, crown health estimates, stem circumferences and evidence of debarking. Results showed a healthy population that displayed an inverse J-shaped curve as analysed from some of the parameters sampled. The population in the Lowveld National Botanical should be monitored in order to keep its good health status. Factors that may have negative impact on the population were also identified.

T6-12-06

Ethnobotany and conservation of Okomu Forest Reserve, Edo State, Nigeria

Omokafe Alaba Ugbogu, *Emmanuel Chukwudi Chukwuma*
Forestry Research Institute of Nigeria, Jericho Hill, Ibadan

An ethnobotanical survey of useful plants was conducted in Okomu Forest Reserve, with a view to documenting indigenous knowledge of medicinal plants used by the inhabitants of the communities and enclaves around the reserve. The information were gathered with the use of structurally designed questionnaires. A total of 90 angiosperm species belonging to 45 families were identified to be useful in the management of various ailments within the study area. Euphorbiaceae and the Legumes constituted the highest species occurrence while the largest number of plant families (22) had only one species each, being represented. Further findings showed that the leaves and stem barks are the most useful parts of the plants while the flower and pith were the least useful. In general, the trees were the most useful of all plant habits followed by the herbs and the shrubs. While this work reflects the biodiversity richness of the study area and its environs, it also suggests the enforcement of conservation strategies as a measure to mitigate species loss.

T6-13: Citizen science and conservation: inspirational projects to get the common public into investigating and conserving regional biodiversity

T6-13-01

The Inheritance and Innovation of Chinese Landscape Architecture—On the Past and Present Life of Shenzhen Fairylake Botanical Garden

Fang He
Bejing Forestry University
Shenzhen Yuandao Urban&Landscape Architecture Planning Institute

Over three decades, Shenzhen, having made great achievements in the construction and development of landscape architecture, also created numerous Number One in China in the field of landscape architecture. Led by Professor Sun Xiaoxiang and Academician Meng Zhaozhen from Beijing Forestry University, a group of experts and scholars have practiced the uniquely glamorous Shenzhen Experience with their actions and wisdoms. Since the early 1980s, from site selection to the whole practicing of planning and design as well as the highlighting of the post effect, Fairylake Botanical Garden mainly underwent start-up stage, stable stage developing stage and mature stage, got gradually matured and perfected after years of development and has been enjoying a growing popularity. The start-up, construction and development of Fairylake Botanical Garden, lasting for as long as over 30 years, cannot do without the protracted and unremitting efforts of the landscape architecture practitioners of generations and shows the excellent practice in inheritance and innovation of Chinese landscape architecture.

T6-13-02

The Million Orchid Project: Using citizen science to restore orchids within a large urban area

***Carl Lewis**, Jason Downing*
Fairchild Tropical Botanic Garden

Southeastern Florida (USA) historically supported more than 30 species of native orchids (Orchidaceae). Beginning more than a century ago, the region developed into the Miami Metropolitan Area, which now includes more than 6 million people. As a result of habitat loss and overharvesting, many of the native orchid species are now extinct in the region, and nearly all are reduced to fewer than 100 plants. The Million Orchid Project, established in 2013, is a citizen science program that aims to propagate and restore orchids to the entire Miami Metropolitan Area. The project is a massive experiment, conducted by thousands of citizen scientists, to determine whether densely populated urban areas can support the recovery of rare orchids. Specific research includes (1) experiments on propagation and planting techniques, (2) studies of the genetic diversity of wild and reintroduced orchids, (3) monitoring of survivorship and growth rates in a variety of different urban habitats, and (4) isolation and characterization of fungal endophytes. Volunteers, teachers, and schoolchildren are involved in all parts of the project. More than 100 schools now have orchid propagation supplies and equipment within their classrooms. A specialized mobile micropropagation laboratory, built within the frame of a decommissioned school bus, visits schools throughout the region to give students hands-on experience with sterile orchid culture. To date, eight native orchid species have been propagated from seed, and five have been introduced into urban landscapes. Citizen science data have helped us select optimal growth media, transplanting protocols, and outplanting sites for for each orchid species. As a result, we have seen survivorship improve as the project progresses. Orchids have been succeeding in neighborhoods that were previously thought to be too urbanized. Our results suggest that an engaged community can help restore rare orchids, especially within densely populated urban areas.

T6-13-03

Citizens are supporting the *ex-situ* cultivation of botanic gardens: A community approach in Germany

***Björn Huwe**[1], Alexa Junge[2]*
1. Science Shop Potsdam
2. FreiGarten Potsdam

The ambitions to fight against the global loss of plant biodiversity are high (Secretariat of the Convention on Biological Diversity, 2010). But when it comes to specific provisions, often the societal acceptance is little and the lack of money is high. To protect local biodiversity the cultivation of wild plants in ex-situ collections of botanic gardens is an established tool, jet. For preserving the whole range of local/regional plant biodiversity in accessions, preferably under near natural conditions, the need of space and money becomes a big problem. Particular when it comes to handle more than one population per species, or even several species with many populations. To overcome the mentioned problems of acceptance, space and money the Science Shop of Potsdam (Germany) initiated and mediated between local urban gardening communities, local authorities and the botanic garden of Potsdam an appropriate approach to protect local plant-diversity. Here we will present the experiences of the initiation stage and future steps to evaluate the process.

T6-13-04

Promoting indigenous plants in a pilot project to support more efficient use of water, natural resources and regional biodiversity in the state of Sinaloa, Mexico

***Cruz Elisa Torrecillas Nuñez**[1], Andrés C. Ravelo[2], Antonio Jose Miguel-Rodriguez[3]*
1. Universidad Autónoma de Sinaloa
2. CREAN-CONICET-UNC, Universidad Nacional de Córdoba
3. SAMI Limited

This project aims are to strengthen the interaction between the

University of Sinaloa and the community of the State of Sinaloa, Mexico, through the development of a Green Campus where plants are used to promote diverse methods for the protection of the environment and water resources. In Sinaloa droughts are recurrent and cause depletion of water resources both in surface water and in aquifers. The rational use of water and the sustainable management of natural resources is a social problem that must be addressed in an integrated manner including public policy, infrastructure and reducing consumption by the productive and domestic sectors. In 2013 to 2015 a project was implemented to promote these objectives through a cooperative approach between researchers, students, external consultants and the community as part of the Drought Prevention and Mitigation Program (PMPMS in Spanish) to support the efficient use of water and natural resources and promote biodiversity in the State of Sinaloa. In this context plants can contribute to mitigating droughts as they play a beneficial role by protecting the soil to avoid erosion, filtering contamination and conserving water resources. To demonstrate this fundamental role of plants, a proposal for a "Green University City" was planned for the improvement, adaptation and modernization of university grounds to promote a culture of environmental care and the generation and application of knowledge on relevant issues related to sustainable development. The project includes research into the adaptation of plants to drought conditions, water consumption and their impact on the total water balance, to establish which plants are most suitable for the local climatic conditions. In the university campus the project will include changes to the landscape by making more efficient use of irrigation water and promoting the knowledge of native plants for their conservation. It is proposed that a major part of the green spaces of the campus will be turned into a succulent garden by planting native species of *Opuntia* commonly known as nopales used in many traditional Mexican dishes; *Mammillaria*; *Cephalocereus*; *Cylindropuntia*; *Echinocactus*; *Sclerocactus* and *Selenicereus* between other representatives of the Cactaceae family. Also, species of Agave which is one the plants from which the traditional drink tequila is made, and species of Yucca, will be cultivated in different locations of the campus. The trees to be incorporated include: *Ceiba pentandra* (L.) Gaertn. and Cordia boissieri A.DC. Also patches of native grasses and herbs will be considered as part of the landscape design. Overall it is expected that plants will contribute to conserving at least 30% of the precipitation that at present is lost as surface water flow which in turn causes flooding and limits the recharge of aquifers. Thus it will be demonstrated to the community that plants can have a beneficial use in managing water resources. In turn it is hoped that this example will motivate the community to adopt a more sustainable approach to the selection of plants so that native plants are preferred over introduced species.

T6-13-05

Plant symbols in Turkey and importance of plant symbolizing for nature conservation

Bahar Gurdal[1], *Neriman Ozhatay*[2]

1. *Istanbul University, Faculty of Pharmacy*

2. *Eastern Mediterranean University, Faculty of Pharmacy*

From past to today, plants are used as a symbol for cities. They appear in flags, coins, stamp *etc.* With regard to Turks rose and tulip are abundant during all periods. Tulip usages and senses in Ottoman Empire's and Turkish cultures are very common. Anatolian Seljuki Turks used motifs of rose, rose products in their literary works. Sultan Mehmet II, the conqueror of Istanbul in 1453, once depicted in miniature smelling roses. At the present time Isparta Province is center of rose cultivation in Turkey, and it is called city of rose. The tulip is used as an icon and the festival is organized in Istanbul. In Turkey wild plants are also used as a symbol from the stand point of conservation. During the "Important Plant Areas" project along BTC pipeline in Turkey, 10 plants (*Cousinia sivasica*, *Cyclamen pseudibericum*, *Helleborus vesicarius*, *Hyacinthus orientalis* subsp. *chionophilus*, *Fritillaria alburyana*, *Iris iberica* subsp. *elegantissima*, *Lathyrus karsianus*, *Lilium kesselringianum*, *Orchis anatolica*, *Tchihatchewia isatidea*) were chosen as a symbol of each city. Plants were rare, endemic or economic. Also one of the municipalities in Istanbul chose "*Crocus olivieri* subsp. *istanbulensis*" as their logo. Based on these examples symbolizing flowers/plants make awareness of local people for protection the habits and also plants. Ultimately plants as a symbol are good source to understand culture of nations, history of cities and good instrument for nature conservation.

T6-13-06

Plant species composition and bioculture of urban greenspace in Kunming city, Southwest China

Liangzao Wu, ***Zhaolu Wu***

Institute of Ecology and Geobotany, Yunnan University, Kunming, China

Plant species composition in urban greenspace reflects the natural environment, human management and cultural connotation of the city. But many studies do not discuss the cultural connotation of urban plants. The plant species composition and bioculture were studied based on the field data collected in 22 greenspaces in Kunming City China. A total of 143 families, 448 genera and 716 species of vascular plants was recorded, of which, 649 species were in the campus and 374 species in the parks, indicating the higher species diversity in the campus. In teams of growth form, the herbs, trees, shrubs and vines respectively accorded for 46.51%, 27.79%, 22.49% and 3.21%. The proportion of trees in the parks (33.69%) was higher than that in the campus (28.35%). In terms of species origin, the proportion of alien plants in the total species, campus species and park species were 39.25%, 38.52% and 41.44% respectively. Too many tropical plants introduced caused annually a large number of plant frost damage and management input. The plant composition in the urban greenspace embedded the distinct traditional morality and real aesthetic. Most plants played the roles of afforestation and beautification, reflecting the season changes of colorful flowers and leaves. More local bioculture was showed in the aged greenspaces, compared the dull afforestation and visual sense in the new ones. The key taxa of plant system evolution, state or local protection plants and local endemic species were highlighted in the plant configuration of campus green, promoting the culture service functions for teaching and scientific research.

T6-14: Botanical Gardens and the Exploration of Salt Tolerant Plants

T6-14-01

Finding climate change ornamentals for Galveston Island, Texas, and the strategies needed to deal with a salt challenged environment

David Creech, *Kenneth Farrish, Steve Wagner, Elaine Harris, Elaine Fowler*
Stephen F. Austin State University

In 2008, Galveston Island, Texas, USA, was affected by Hurricane Ike which inundated most of the island with sea water. The storm killed fifty-five thousand large live oaks, *Quercus virginiana*, many of which were over one hundred years old. With funding by the Moody Foundation, a long term platform for salinity research and discovering ornamentals adapted to the island was initiated in early 2016. The full sun one ha plot is located on the west side of Moody Gardens, a nationally known resort on the island, and elevation varies from 1 to 1.5 m above sea level. Soil samples taken across the plot indicate a soil with high SAR at the one meter depth and varying degrees of high salinity depending on proximity to the bay. The research plots are also subject to intense coastal winds for much of the year. This project has three goals. 1) Develop a long term evaluation program for woody ornamental trees and shrubs. A survey of plants in the evaluation program and their performance will be presented. 2) Studying treatments that affect soil amelioration and plant establishment in the project area. In this first study, a randomized block design was utilized with three species, eight treatments, six replications and 2 plants per replication, for a total of 288 plants in the study. The three woody plants tested include *Quercus virginiana*, *Hibiscus hamabo*, and *Taxodium* X 'T406', the latter a hybrid of bald cypress and Montezuma cypress introduced by the Nanjing Botanical Garden. Plant performance and soil analysis change over time will be presented. 3) assess and characterize microbial communities in the salt affected soil on Galveston Island and the opportunities for mycorrhizal colonization of select species to enhance their salt tolerance and performance.

T6-14-02

Analysis of the environmental characteristics and salt-tolerant plant screening involving the issues of coastal area afforestation in Jiangsu Province

Yunlong Yin
Institute of Botany, Jiangsu Province and Chinese Academy of Sciences, Nanjing Botanical Garden Men. Sun Yet-Sen

Jiangsu province is located in the eastern part of China with 954 km of coastline along the Pacific Ocean. With 670,000 hectares of costal land equivalent to 6.5% of the entire area of the province, this land area is almost 25% of the coastal land of China. Silting is enlarging the area at an increasing rate of 1,300 hectares per year. Jiangsu province includes the transition zone of the more temperate north and the south subtropics. Its mild climate and abundant rainfall (900-1000 mm annually), is conducive to the growth of a wide diversity of plants. The utilization of coastal land resources and the restoration of vegetation have drawn increased attention by local governments. However, vegetation restoration and establishing wind-break forests on the bare coastal land are often associated with difficult environmental factors, including soil salinity, nutrient shortage, wind strike and water logging. Plant growth, especially tree growth, is especially vulnerable and high mortalities are common. For many years, the author has engaged in a variety of studies that focus on the introduction and screening of salt-tolerant plant materials. Communication with national and international scholars interested in this common topic is critical to sharing and developing new ideas, finding new plants and discovering the best strategies for restoring vegetation on the salty coastal lands in China and other countries facing similar issues.

T6-14-03

The contribution of economic plant exploration studies of Nanjing Botanical Garden in the past 50 years

Shanan He, *Yin Gu, Hong Yu*
Institute of Botany, Jiangsu Province and Chinese Academy of Sciences, China

The results of researches on introduction, cultivation and breeding of economic plants were briefly reported. Species of *Dioscorea, Castanea, Taxodium distichum, T. mucronatum, Cynodon dactylon, Vaccinium, Rubus (black berries), Lycoris, Stevia rebaudiana, Cistus ladaniferus, Olea europaea, Coreopsis drummondii, Chimonanthus pracox* and *etc.* were included. Research achievement made obvious benefits to national economy and ecological conditions. In the meantime theoretical study on plant introduction and breeding, conservation of biodiversity and environment protection were conducted also.

T6-14-04

The salt tolerance assessment, improvement and utilization of native turfgrass in China

Jianxiu Liu
Institute of Botany, Jiangsu Province and Chinese Academy of Sciences (Nanjing Botanical Garden)

Saline soil is a type of soil widely distributed on the world, accounting for 25% of the total land area. It mainly distributed in the inland arid and semi-arid areas and coastal areas. Along with the development of the industrial construction and ecological construction in saline land, it has brought about a large-scale landscape engineering and ecological restoration project, which required a lot of salinity-tolerant green plant cultivars. In addition, the application of saline reclaimed water and snow melt salt in winter also increased the demand of salinity-tolerant greening plants. Bermudagrass (*Cynodon* spp.) is a kind of salinity-tolerant grass, which has a great potential for application in greening and ecological restoration on saline land. In this research, the regulation mechanism of sodium and potassium ion and the establishment and maintenance technique of salinity-tolerant bermudagrass on heavy saline soil was studied based on the evaluation of salinity tolerance of bermudagrass resources, and then a 'saline soil turf' model was proposed and demonstrated. We preliminary confirmed the genetic variation of salinity tolerance in Chinese bermudagrass resources, and their regulation mechanism of sodium and potassium. Bermudagrass could reduce Na^+ accumulation and maintain K^+ stability in leaves under salinity stress by restricting Na^+ into bundle in roots, selective transport K^+ over Na^+ from root to leaf, and Na^+-selective secretion via leaf salt glands, at the same time maintaining appropriate stomatal conductance to maintain photosynthetic efficiency. We also found salinity stress increased salt gland density, the appropriate increase of K concentration in rhizosphere was beneficial to the competitive absorption of K^+ over Na^+, and the different ion regulation ways were interaction each other. By using cDNA-AFLP differential expression technique, some gene fragments which may be related to ion regulation were obtained. A simplified turf establishment and maintenance in saline soil was proposed based on the model of 'Saline soil-Salinity tolerance turfgrass cultivar-Saline water irrigation' and demonstrated in Jiangsu Province, Shandong Province, and other heavy saline soil including the ecological reconstruction of islands in Chinese South Sea. This research provided a good basis for further mining the salinitytolerant gene, continuingly breeding salinity-tolerant cultivars and demonstration in heavy saline soil in bermudagrass.

T6-14-05

Research on the establishment of coastal shelter forest systems with the function of anti soil erosion and wind breaking

Haibo Hu
Nanjing Forestry University

The suitable trees, allocation patterns, silviculture and management techniques of anti-erosion and anti-fouling forests were studied in the sandy, muddy and rocky coastal areas. Nineteen species of anti-erosion trees had been selected, such as *Fraxinus chinensis, Robinia pseudoacacia, Ascendens mucronatum, Bruguiera gymnorhiza, Kandelia candel* and so on. At the same time, we established the database of south China coastal salt tolerant plants, including two hundreds and forty-six kinds of salt-tolerant plants. We used the silviculture means of lifting isolation layer, cavernous membrane of replacing soil, base plate of replacing soil and the ecological floating bed purifying aquaculture water technology to carry out shelter forest establishment. The soil physical and chemical properties and erosion resistance of coastal shelter forest were researched. In sandy coast, the reunite degree of soil is higher for the stands of *Suliu plant, Fraxinus americana* and *Populus L.*, and the disperse rates were lowest, the resistance to erosion was stronger. The main factor influencing the soil anti-erosion index was the number of fine roots diameter class of trees 0 to 1mm and 1-2 mm of in the rocky coast, and soil anti-erosion index of six forest types were higher than non-forest lands, and the anti-scouring index of 0 to 20 cm soil were higher than that of 20 to 40 cm. On the ditch slopes of the sandy coast, the anti-erosion ability increased 33.2% on average. By means of Land sat TM, World View and Quick bird satellite images and topographic map data, for monitoring the coastal landscape and vegetation dynamics, we develop and apply of the coastal shelter forest vegetation and landform change detection of remote sensing image processing software applications. By the method of layered extraction, site classification and area changes of the muddy coastal mangrove were researched. Using SPOT 5 remote sensing data, we monitored the vegetation and landscape change in sandy and rocky coast of Daishan island. What's more, we accessed a computer software copyright registration "coastal shelter forest vegetation and landform change monitoring and remote sensing image processing software applications (2012SR038689)".

T6-14-06

Interactive effects of sulphur and humic acid on soil properties, and growth, green yield and some chemical constituents of *Pisum sativum* plants grown on newly-reclaimed salt-affected soils

M.M. Rady[1], *A. Osman*[1], *H. Mohamed*[2], *R.M. Nassar*[3], *M. Abd El-Hady*[4], ***Magdi T. Abdelhamid***[4]
1. Fayoum University
2. Ain Shams University
3. Cairo University
4. National Research Centre

About 900,000 ha of Egypt's agricultural lands are suffering from salinity build-up problem. Therefore, two field experiments were conducted on newly-reclaimed moderate saline sandy clay loam soil at the Experimental farm of Faculty of Agriculture, Fayoum University, Fayoum, Egypt during the two successive growing seasons, to examine the positive effects of applied elemental sulphur (S) at the rates of 0, 120, 240, 360 and 480 kg ha^{-1} and humic acid (HA) at the rates of 0.0, 0.5, 1.0, 1.5 and 2.0 g L^{-1} and their combinations on pea (*Pisum sativum* L.) growth, yield of green pods, green seeds and their quality parameters as well as some soil chemical properties after harvesting pea plants. The obtained data showed a significant positive response for the applied treatments of the S and HA as either solely or in combination with a superiority towards the highest levels of combined ones. Also, the obtained results indicate that the greatest values of green seed starch, total soluble sugars and protein contents are more attributed to the positive effect of nutrients. The obtained results reveal that organic humic acid at the rate of 1.5 g L^{-1} and S at rate of 360 kg ha^{-1} should be used to face a great problem of either soil salinity or excessive $CaCO_3$ content. These beneficial effects are more attributed to the achieved soil amelioration in the values of ECe, pH and ESP as well as released nutrients from either native source or applied organic manure, which supported the aforementioned pea traits. So, it could be recommended that organic (humic acid) and elemental sulphur could be used to alleviate the adverse effects of soil salinity.

T6-15: Species richness of the pan-tropical genus Begonia and its conservation challenges at Internet Age

T6-15-01

A better strategy for conservation of Chinese Begonia

Daike Tian, *Yi Tong, Yan Xiao, Naifeng Fu*
Shanghai Chenshan Plant Science Research Center of CAS, Shanghai Chenshan Botanical Garden

Begonia is one of the most diverse plant taxa and the sixth largest angiosperm genera consisting of over 1,800 accepted species names. The species number of this genus sharply increased particularly in the last 20 years, and only in China it increased from 80 to 200 species during this period and is close to that of Brazil with the richest *Begonia* species in the world. Based on recent field survey, the number of *Begonia* species in China is predicted to be between 250 and 300. Most of *Chinese Begonias* are narrowly distributed, and the wild populations are facing shrinking due to unique, isolated and fragmented habitats subject to negative influence of climate change and human's activity. In addition, the exploitation of wild *Begonias* with ornamental and medicinal values has been accelerated by fast developing internet. More and more species should be categorized into rare and endangered class and therefore need immediate protection. Ex situ conservation of *Chinese Begonias* started in 1996, and so far nearly over half of the total species have been introduced by several major botanical gardens belonging to Chinese Academy of Sciences. However, only fewer research institutions, limited funds and human resources have been involved in *Begonia* conservation, and no project has ever been conducted on reintroduction. Therefore, more systematic work and a better conservation strategy remain to be developed. A deep taxonomical study is needed to understand the real situation of *Chinese Begonia* germplasm. A better conservation strategy is proposed for *Chinese Begonias* depending on a solid taxonomical foundation, understanding population diversity, and integrating multiple approaches of protection, such as in situ, quasi situ, near situ and ex situ conservation, seed bank, plant reintroduction, legislation and public education. Although the current protected areas (any types nationally and regionally) provide harbors for the majority of wild *Begonia* species, the species or population targeted conservation zone or micro-reserve should be established for either endangered species or unique populations which are distributed outside the current protected areas. On the other hand, the national and regional level law on plant protection should be extended to prevention of illegal collection and transaction of wild plants particularly the species with unique habitats and small populations.

T6-15-02

Begonia placentas response to the global cooling events via chloroplast genomics

Yu-Min Shui, *Wen-Hong Chen*
Kunming Institute of Botany, Chinese Academy of Sciences

Ice ages are among the most important climatological events in the history of the earth. The multiple global cooling events have been well documented from earth science. During the process, the biotas experience multiple gradual and sudden changes, such as extinction, rapid radiation and evolutionary shift of growth habit. However, it is unknown which definite biotas throughout the world provide clear evidence of the progress of glaciations and what adaptive characters of these biotas are affected by glaciations. Here we show that *Begonia* placentas responses to the global cooling events in polyphyletic, polychromic and polytopic ways via ecotypes. We sampled 110 species in *Begonia* worldwide with diverse placentation and ecotypes and built up the stable phylogenetic tree based on the chloroplast genome. We observe the placental development in ovary of four species from the different linages and confirm the evolutionary direction of placentae from parietal to axile placentation. We also observed systematically the spatial change of placentation respectively in 9 positions in the ovary from 56 species via paraffin section and in three positions from 67 species via cutting of fresh ovary. Here we show that the derived state (axile placentation with single segment in geophytes and therophytes adapted to cold and dry climates) tends to occur at the base of the main clades in the phylogenetic tree. The comparison of development, phylogeny and ecotype shows that *Begonia* has experienced multiple influences from climate freezing, as indicated by the basal groups in the phylogenetic tree with derived characters and cooling-adpative ecotypes during the glacial periods and the main linages with the rapid radiation of species during the interglacial. The evolution of *Begonia* placenta not only responses to global cooling events but also indicated the occurrence time of some tectonic events in global history. It seems that the taxonomic ranks in *Begonia* were gradually formed under the influence of global cooling. The above case seems also happen outside the genus *Begonia* even across the whole biotas, justifying a broader survey of the evolution of entire biotas within the context of global climate change.

T6-15-03

Diversity of the genus *Begonia* L. in Laos

Phetlasy Souladeth[1], *Wen-Hong Chen*[2], *Yu-Min Shui*[2]
1. National University of Laos
2. Kunming Institute of Botany

Indo-Burma region is one of the biodiversity hotspots where more than 7,000 endemic plants live. Laos is located at the centre of this biogeographic region bordering to China, Cambodia, Myanmar, Thailand and Vietnam, thus grasping plant diversity in Laos is crucial to understand how such a huge endemism was developed. However, many endemic species are endangered by recent human activities in Laos. Some projects were urgently launched on the flora of Laos, such as those by Royal Botanical Gardens Edinburgh (RBGE), Muséum National d'Histoire Naturelle (MNHN), Kunming Institute of Botany (KIB) of the Chinese Academy of Sciences (CAS), but much more efforts are required to clarify the whole picture. *Begonia* L. (Begoniaceae) is widely distributed in the tropical and subtropical regions of Africa, America and Asia, and consists of ca. 1,820 species worldwide. Laos is a part of Indo-China, where very few documentation on Begoniaceae has been published. As there is no botanist focuses on *Begonia* of Laos, an obvious gap in Indo-China, while the nearby regions with richness of *Begonia*, such as Yunnan of China (about 100 species), Vietnam (about 80 species) and Thailand (about 70 species). The most account of Lao *Begonia* was published after 2007, including 13 species or 17 taxa. Recently, some new species has been described from this country such as *Begonia afromigrata* J.J.de Wilde, *B. lamxayiana* Souvann. and *B. viscosa* Aver. & H.Q.Nguyen. To put forward the inventory on Lao plants, especially on the genus *Begonia*. We carried out a joint field collection and the examination of the available specimens from Laos in 2016 to explore its potential diversity of *Begonia*. We try to submit the current state of *Begonia* in Laos in Congress via the collaborative study with many botanist in the other countries to fill the gap.

T6-15-04

The diversity and conservation of *Begonia* in continental SE Asia

Ruth Kiew

Forest Research Institute Malaysia

The greatest obstacle to studying any aspect of *Begonia* diversity and conservation in SE Asia is the taxonomic impediment. Among continental SE Asian countries, only Peninsular Malaysia has a monograph for *Begonia*s. From existing specimens in herbarium collections and the rate of published new species and records, it is clear that existing checklists for Myanmar, Thailand and Vietnam include less than half the countries' species. The flora of Laos and Cambodia are woefully incomplete. Habitat diversity in all these countries emphasize the species richness of *Begonia* on karst limestone, particular in northern Vietnam, which is a biodiversity hotspot for *Begonia*. The major problem confronting *Begonia* conservation is the extremely high level of endemism (over 80%) with many species known from a single locality, and secondly the taxonomic impediment. At present, more than half the species fall within the IUCN Data Deficient Category. Major threats throughout the region are deforestation and logging, quarrying of karst limestone, and the construction of hydroelectric dams. Concerted action is required to conserve *Begonias* by (a) revising *Begonia* floras on a national basis, (b) assessing conservation status of species, and (c) because *Begonias* are frequently indicators of biodiverse habitats, identifying and designating areas as Important Plant Areas to form the basis of a network of legally Totally Protected Areas.

T6-15-05

Reconstructing phylogenetic relationship of *Begonia* sect. *Coelocentrum* (Begoniaceae) using targeted enrichment of nuclear genes

Yu-Hsin Tseng[1], *Catherine Kidner*[2], ***Kuo-Fang Chung***[1]
1. Biodiversity Research Center, Academia Sinica
2. Royal Botanic Garden Edinburgh

With estimated more than 1,800 species, *Begonia* L. (Begoniaceae) is one of the largest genera of vascular plants, distributed throughout tropical and subtropical Asia, Africa and America. *Begonia* sect. *Coelocentrum* was first established by Edgar Irmscher in 1939 on the basis of the parietal placentation and geographical distribution. Section *Coelocentrum* comprised ca. 65 species with a variety of leaf shape, texture and variegation and is usually confined to the cave-like microhabitats of karst areas in southern China and northern Vietnam with most species highly localized. A recent molecular phylogenetic study based on DNA sequences of nrITS and chloroplast rpL16 intron indicated that a majority of Sino-Vietnamese limestone *Begonia* (SVLB), including sect. *Coelocentrum* and five species of sect. *Diploclinium*, *Leprosae*, and *Petermannia*, form a strongly supported monophyletic group but deeper nodes are very poorly resolved, which suggests a rapid species radiation across the Sino-Vietnamese limestone karsts in SVLB clade. To construct well-supported molecular phylogeny of sect. *Coelocentrum*, we employed a novel target enrichment method using a set of custom capture probes designed across four *Begonia* transcriptomes, coupled with next generation sequencing (NGS) to gather phylogenetic information from thousands of nuclear loci. We sequenced over 90% of *Begonia* sect. *Coelocentrum* species and used bioinformatic and phylogenetic programs to produce a phylogenetic tree to 1) test the current phylogenetic hypothesis; 2) revise the current taxonomy, and 3) understand the evolutionary history in this group. These data will provide important information for further studies investigating evolutionary mechanisms and plant speciation in limestone flora.

T6-15-06

Vietnam-the history museum of the evolutional platform for *Begonia*-needs to be explored and conserved with priority

Hieu Nguyen
Center for Plant Conservation Vietnam

Begonia L. is comprised of more than 1,800 species, most of which are local endemics with very restricted distribution in tropical and subtropical regions of the world. Inventory of the genus species composition and diversity in local floras, particularly in Indochina remains very far from acceptable completion. When over 175 species are reported from China, with 143 local endemic (82%), very few species known originate from countries of eastern Indochina. Thus, only 36 species of the genus till now have been tentatively reported from Vietnam. Relatively more extensive attempts were undertaken in studies of the genus in Vietnam. Since 1919, French plant taxonomist – François Gagnepain (1866-1952) was the pioneer in exploring *Begonias* in Indochina and a few *Vietnamese Begonias* documented. The first checklist reports 41 native species and 6 cultivated by Vietnamese plant taxonomist, Pham Hoang Ho (1991,1999). Over the last 10 years, 29 species have recently been added by other plant taxonomist as Ruth Kiew, Ching-I Peng, Shui-Yu Min, Nguyen Quang Hieu, Mark Tebbit, Leonid Averyanov, *etc.* Many of them were found in North Vietnam. However, many are isolated in ancient mountain formations, particularly karstic rocky remnant limestone massifs well presented in Vietnam and in Laos undoubtedly provide home to numerous still undescribed as local endemic species. Our explorations during the last few years confirm this supposition and truly outline areas of eastern Indochina as the history museum of the evolutional platform of *Begonia*-needed to be explored and conserved with priority. Our model conservation chooses the right place within local village and engages local community to plant and manage the collections.

Author Index

T

XIX International Botanical Congress (IBC 2017)

Posters and Abstracts

Shenzhen, China
23 - 29 July 2017

Abstract Book II

ISBN: 978-1-5108-6169-5

Printed from e-media with permission by:

Curran Associates, Inc.
57 Morehouse Lane
Red Hook, NY 12571

Some format issues inherent in the e-media version may also appear in this print version.

Printed by Curran Associates, Inc. (2018)

For permission requests, please contact IBC 2017
at the address below.

IBC 2017
Lianhua Zhilu 1004
Futian District Shenzhen
P.R. China

Phone: (+86)755 8307 3632

office@ibc2017.cn

Additional copies of this publication are available from:

Curran Associates, Inc.
57 Morehouse Lane
Red Hook, NY 12571 USA
Phone: 845-758-0400
Fax: 845-758-2633
Email: curran@proceedings.com
Web: www.proceedings.com

ABSTRACT BOOK II

Posters and Abstracts

IBC 2017
XIX International Botanical Congress
Shenzhen China

XIX International Botanical Congress
Shenzhen Convention & Exhibition Center
July 23-29, 2017

Nomenclature Section
Peking University HSBC Business School
July 17-21, 2017

CONTENTS

Posters

T1
P0001
The LIFE project WetFlyAmphibia: When the needs of flora and fauna make a restoration program difficult

Thomas Abeli[1], *Graziano Rossi*[1], *Davide Alberti*[2], *Marcello Miozzo*[3]
1. *University of Pavia*
2. *Parco Nazionale delle Foreste Casentinesi*
3. *DREAM Italia*

The project "Conservation of amphibians and butterflies of open wet areas and their habitats at the Foreste Casentinesi National Park" (WetFlyAmphibia) is a restoration project funded in 2014 in Italy under the EU LIFE scheme. The aim of the project is to increase the reproductive performance of four amphibian species, including the Appenine Yellow-bellied Toad *Bombina pachypus* and the Italian Crested Newt *Triturus carnifex*, and two butterflies (*Euplagia quadripunctaria* and *Eriogaster catax*) in the Northern Apennines (Italy). Here, land abandonment and changed agricultural practices have negatively affected the wetland flora and fauna. The most important action in the project is the restoration of the wetland flora that would enable both amphibia and butterflies to reproduce. In particular, *B. pachypus* and the two butterflies have quite specific requirements in terms of plants that can promote their reproductive performance. After two years of project implementation, results indicate that *B. plachypus* prefers aquatic vegetation with *Chara* sp. and *Myriophyllum* sp. forming little shade, while *T. carnifex* prefers deeper ponds with abundant shade provided by broad-leaved aquatic plants such as *Potamogeton* sp. and *Ranunculus trichophyllus*. The two butterflies largely rely on *Eupatorium cannabinum* as food plant for their caterpillars. Moreover, all the species use large vegetation stands of megaforbs as a connection between wetlands and woodlands. The challenge of the project is to identify areas that match the ecological requirements of both animals and plants to maximize the success of the restoration (reintroduction/reinforcement). Because few natural wetlands are still available with these characteristics an interesting solution is proposed, especially for *B. pachypus*. Old drinking troughs for cattle will be restored and used as unnatural ponds, in which *Chara* sp. and *Myriophyllum* sp. will be introduced from nearby wetlands to facilitate the colonisation by *B. pachypus*. In addition, new ponds will be excavated beside historical ponds and a megaforbs stand planted according to a soil humidity gradient, using Ellenberg indexes to identify the right distance of each plant species from the pond.

T1
P0002
Isolation and identification of fungi associated with some Nigerian bird's nests using internal transcribed spacer primer

Adedotun A. Adekunle[1], *Uchenna C. Kanife*[2], *Abosede M Ebabhi*[3], *Temitope O. Samuel*[1], *Oluwagbenga O. Adeogun*[1]
1. *University of Lagos*
2. *Yaba College of Technology, Lagos*
3. *Distance Learning Institute, University of Lagos*

Studies on identification of fungi associated with Nigerian bird nest was carried out using internal transcribed spacer primers, a method designed for the simple and rapid isolation of DNA from tough to lyse fungi. The fungal samples were rapidly and efficiently lysed with the ultra-high density bashing beads. Sixteen fungal isolates were studied with *Aspergillus niger* occurring most frequently (44%), *Fusarium solani* and *Fusarium verticilloides* occurred at 19%, *Aspergillus flavus* 13%, *Aspergillus tubingiensis, Aspergillus terrus* and *Talaromyces pinophilus* occurred at 24% respectively. One species has no identifiable sequence and another one could not be tacked to any match. DNA fragments from the isolates were amplified by PCR using forward primer (ITS 4TCCTCCGCTTATTGATATGC) and reverse primer (ITS 5 GGAAGTAAAAGTCGTAACG) which demonstrated that the extracted DNA were truly fungi. The phylogenetic tree showed a high level of genetic similarity between these organisms. This study provides a useful data for discovery of unknown fungi and probably species which could be causal organisms of several bird diseases. These results can also lead to understanding the inter-relationship between the bird, nesting material and fungi within the nest. The results also provide information on additional source of thermotolerant fungi.

T1
P0003
Freshwater diatoms diversity of National Parks in Nigeria I: Okomu National Park, South-South, Nigeria.

Taofikat Adesalu
University of Lagos

This study represents the first taxonomic and ecological description of the diatom flora of water and sediments in two freshwater lakes within Okomu National Park Edo State, Nigeria, using light and scanning electron microscopy (SEM). Twenty-nine species of diatoms belonging to fifteen genera, *Achnanthes, Achnanthidium, Chamaepinnularia, Cymbella, Discotella, Eolimna, Eunotia, Frustulia, Gomphonema, Luticola, Nitzschia, Pinnularia, Placoneis Sellaphora* and *Stauroneis* were identified.The taxa were dominated by *Eunotia* spp. Water chemistry of 5.9-6.3 pH and surface water temperature of 28°C probably implies that the diatoms especially *Eunotia* spp. are exploring an ecological niche that is probably favourable for its growth. Okomu National Park appears to be a hotspot for *Eunotia* species diversity, and further study of the oligotrophic waters of this site is certainly recommended.

T1
P0004
Incidence of *Hydnora abysinnica* A. Braun in Nigeria: First report

Otuwose E. Agyeno[1], *Emmanuel I. Aigbokhan*[2], *Adeniyi A. Jayeola*[3], *Emmanuel B. Elisha*[1], *Christiana J. Dawurung*[4], *Yilnya J. Gosomji*[5], *Oluwatobi A. Oso*[3]
1. Department of Plant Science and Technology, Faculty of Natural Sciences, University of Jos, Nigeria
2. Department of Plant Biology and Biotechnology, Faculty of Life Sciences, University of Benin, Nigeria
3. Department of Botany, Faculty of Sciences, University of Ibadan, Nigeria
4. Department of Physiology, Pharmcology & Biochemistry, Faculty of Veterinary Medicine, University of Jos, Nigeria
5. Behind LGEA Primary School, Nekong, Kanke LGA, Plateau State, Nigeria

Hydnora species (Hydnoraceae) are endemic to Africa and until now literature accounts show that their distributions occur only within the Sudano-Zambezian region of Africa. We report here the occurrence of several stands of *Hydnora abyssinica* A. Braun (syn. *Hydnora johannis* Becc.) in Nekong, Wusali ward, a lowland Sudan savanna area in Kanke Local Government Area, Plateau State, Nigeria. It was found in association with the following potential host plant species: *Acacia hockii* De Wild, *Piliostigma thonningii* (Schum.) Milne-Redh., *Acacia seyal* Del., *Boswelia dalzielii* Hutch., *Anogeissus leiocarpus* (DC) Guill. & Perr., *Zizyphus mauritiana* Lam., *Tamarindus indica* L. and *Annona senegalensis* Pers. This represents the first recorded occurrence of any *Hydnora* species in the Guineo-Congolian region which when combined with its well-known Sudano-Zambezian existence, enlarges and establishes its distribution range as an Afrotropical endemic.

T1
P0005

The impact of treated waste water at Khartoum Refinery on plants diversity

Salma Alamin
Environment, Natural Rrsources and Desretification Institute

Khartoum Refinery (KR) is located north of Khartoum State. This work aims to assess the impact of treated water from KR on plant diversity. Twelve field trips were carried out to cover dry and wet seasons. the study area was divided into control site (merely received rain water) and treated one (received rainwater and treated water). In KR, the treated area was subdivided into four units planted with *Eucalyptus camaldulensis* (Ban) based on age of plantation (6,4, 2 years and 2 weeks). Plant specimens were collected, pressed, dried, identified, mounted according to the standard methods and then incorporated in the herbarium. Scientific names were updated from the website of the international plant list version 1.1. Measurement of plant diversity was carried out using the quadrate method (1×1 m) for ground flora. Percentages of the following parameters were calculated: plant frequency, ground cover, bare soil, plant litter, stone and gravel, and plant dominant species. For trees and shrub cover the number of plants was counted in 50 m^2 areas to determine the dominant species. Statistical analysis of data was carried out by statistix 9 program. Sixteen plant species belonging to nine families were recorded in Khartoum refinery. Percentage means of plant ground cover was high in the rainy season compared to the dry season, bare soil was high in both rainy and dry seasons, plant litter was low in the rainy season and high in the dry season; rock was low in both rainy and dry seasons; plant diversity was high in the rainy season and it was low in the dry season. The dominant herbaceous species in the control site was *Aristida adscensionis* while the dominant tree is *Acacia tortilis* ssp. *spirocarpa.* In the treated site, also *Aristida adscensionis* (except for 6 years plantation) and *Acacia tortilis* ssp. *spirocarp* were dominant associated with *A. seyal, Fagonia cretica, Senna alexandrina, Corchours depressu, Panicum turigdum*, and *Chloris virgata.* This study concludes that treated water has indeed effect the biodiversity since many plant species growing and dominating the treated area and not observed in the control area. On the other hand this was prominent from disappearance of *Aristida adscensionis* (Gao) in treated area of six years old *Eucalyptus camaldulensis*. The study recommended periodic monitoring of vegetation.

T1
P0006

Digitization of Mississippi Herbarium specimens aids in understanding plant diversity in the southeastern U.S.A.and improves K-12 education

Lisa Wallace[1], *Mahboubeh Hosseinalizadehnobarinezhad*[1], ***Mac Alford***[2], *Nina Baghai-Riding*[3], *Megan Fedrick*[4], *Lucile Mccook*[5], *Heather Sullivan*[4]
1. Mississippi State University
2. University of Southern Mississippi
3. Delta State University
4. Mississippi Museum of Natural Science
5. University of Mississippi

The southeastern United States is well known for high levels of plant diversity, but availability of natural history collections and associated resources documenting this diversity has been limited for some areas. For example, no flora has been published for the state of Mississippi, and a checklist produced more than 15 years ago was not based on an exhaustive review of herbarium specimens across the state. Missing data, such as these, result in an incomplete inventory of biodiversity for the Southeast and can impact inferences about ecological drivers of biodiversity. A collaboration of six of Mississippi's herbaria has worked to make available the content of Mississippi herbaria through active digitization of more than 200,000 specimen sheets and determination of georeferenced locales. Here, we present an update of these efforts. Specimen data from this project are available through the SERNEC Symbiota project (www.sernecportal.org). We have contributed >207,700 specimen images, skeletal or detailed label information for >310,700 records, and georeferenced locales for >58,500 specimens. The majority of specimens in this dataset were collected in Mississippi or elsewhere in the Southeast. Additionally, some collections contain multiple records from a single locale, which may permit in-depth examination of plant community structure on a local scale. As an example of how this dataset improves on existing resources on plant diversity for the state, we conducted a comparison of species diversity in Cyperaceae (excluding *Carex*) based on specimen records in Mississippi herbaria, Flora of North America, USDA Plants, BONAP, and Flora of the Southern and Mid-Atlantic States. In total, 266 species collected in Mississippi were found in herbarium records, compared to

184–229 species recorded for Mississippi in the other resources. Some of these differences are accounted for by synonyms, but some species were simply not included for Mississippi in these references. Use of the specimens in this manner is aiding in completion of an updated checklist of plants for Mississippi. Through a partnership with the Mississippi Museum of Natural Science, we have also developed a number of outreach projects, including a plant-themed workshop for K–12 teachers and plant ID cards showcasing notable species around Mississippi. We have also been active in using the specimen database in classes we teach at our respective institutions. We welcome use of the specimen data and images in the SERNEC portal and invite experts to contribute to on-going curation of the collections.

T1

P0007

Biodiversity situation in Palestine—West Bank

Banan Alsheikh

National Agricultural Research Center

Palestine is situated in South West Asia in the east of the Mediterranean basin, where the Mediterranean Basin is the third richest hotspot in the world in terms of its plant diversity. Palestine (27,000 square km) extends from the Jordan River and Wadi Arava in the east to the gulf of Aqaba in the south, to the Mediterranean Sea in the west and the Sinai desert in the south west and Lebanon in the north. Palestine is home of 2,600 vascular plant species, while West Bank (5,856 square km) has 1,600 vascular plant species, tremendously high plant diversity for such a small area; it has four phytogeographical regions delineated in this region. These are: Mediterranean, Irano-Turanian, Saharo-Arabian and Sudanian. However, Sudanian elements do not occupy a discrete phytogeographic region in the area but constitute scattered outposts of penetration within Jordan valley. Palestine has a unique dramatically climatic gradient in the Mediterranean Basin. In a distance of 25 km, elevation drops from 914 m above sea level in Jerusalem to -420 m (below sea level) in the Dead Sea shore, i.e. 1,334 meter depth in 25 km distance from Jerusalem to the Dead Sea shore, while mean annual rainfall drops from 540 mm to less than 60 mm. The richness of Palestine in biodiversity is due to its situation as a meeting point between Asia and Africa, wealth of many habitats as a result of climatic transition from Mediterranean in the north to extreme desert in the south, Jordan Rift Valley in the east, different rock types, soil types, long period of human influence. A trading passage between Asia, Africa and Europe, long history of cultivation and grazing which enabled the introduction of many alien species. West Bank has around 500 rare species, some are globally threatened like *Iris lortetii* and *Iris haynei*. Many are on the way of extinction like *Ajuga orientalis, Amgdalus Arabica*. Some extinct in the West Bank like *Avena fatua, Salvia bracteata*. Now we are facing a wave of rapid rate of extinction which is a calm process. What is threatening biodiversity in West Bank? Heavy anthropogenic effects by Palestinians and a long period of continuous implementation of destructive measures by the Israeli occupation like confiscation of land for settlements (establishment of seven settlements in Wadi Qana Nature Reserve), military purposes, bypass roads, over abstraction of ground water, separation wall, urbanization, infrastructure, intensive agriculture, chemical fertilizers, pesticides, heavy grazing, reduced habitat quality, overharvesting of medicinal plants, fire and land reclamation.

T1

P0008

Indigenous edible ferns in southern Philippines: Utilization and conservation

Victor Amoroso[1]*, Rainear Mendez*[1]*, Reggie Dela Cruz*[1]*, Fulgent Coritico*[1]*, Annabelle Villalobos*[2]

1. Central Mindanao University

2. Consultant, Johnson & Johnson (USA) and DOST Balik Scientist

The high incidence of cancer, atherosclerosis, and other degenerative diseases among Filipinos seem to be due to environmental pollutants and to the type of food we eat. Ferns in the Philippines had been used by native people as food, tea and medicine for a long time. We then conducted a study to survey the species of edible ferns in southern Philippines, determine bioactives and proteins in ferns to demonstrate their health and wellness benefits, and conduct *ex situ* conservation. Field and ethnobotanical survey revealed ten species of indigenous edible ferns. These were studied in terms of protein content by the Bradford assay, antioxidant activity by the DPPH assay and relative component proteins by SDS-PAGE. These 10 species of ferns contain quantifiable amounts of antioxidants and proteins. Based on these data, ease in propagation, availability and palatability, six species are chosen as potential alternative food source. These species are *Diplazium esculentum, Marsilea crenata, Stenochlaena palustris, Acrostichum aureum, Angiopteris palmiformis* and *Asplenium nidus*. Laboratory tests showed protein content by the Bradford assay ranging from 0.08 to 4.39 mg/g and antioxidant activity by the DPPH assay as high as 83% that of ascorbic acid or at 143 to 588 ORAC units/g. *Marsilea crenata* gave the highest protein content as well as antioxidant activity. From these 10 edible species, 27 gourmet fern recipes were prepared by the participants during the gourmet cooking contest and evaluated by a sensory panel. We prepared brochures for the propagation of these fern species to educate the public of the health and wellness benefits of ferns in the diet. We held trainings and seminars, which were participated by farmers and local government units. The aim was to share and spread the findings and practices so many Filipinos would benefit from including these species of edible ferns in their food. To conserve these edible ferns, a university fernery was established for mass propagation and as part of the *ex situ* conservation. Moreover, members of the Balinsasayao people's organization were trained on how to identify, utilize and propagate these species of ferns and to establish their own fernery as an income generating project.

T1

P0009

Restoration of Merapi Volcano National Park, Indonesia after the eruption in 2010 using Coconut Fiber Bag and Cocopeat Diamond Method

Sulfiantono Arif

Merapi Volcano National Park, Ministry of Environmental & Forestry

Ecosystem restoration is an important activities to restore ecological services of Merapi Volcano National Park (MVNP) post eruption in 2010. The eruption impacted to the destruction of local people residents, vegetation, water systems and the loss of endemic species. Restoration activities were started through an assesment of the eruption impact to local people, vegetation habitat, wildlife, natural succession and invasive alien species. Some technical approaches such as vegetation analysis, endemic species identification and local people participation. This study aims to determine the efficiency of the technique of planting by seed directly in the area of restoration. The method chosen is growing methods are fast, cheap and produce plants that can grow and live in areas where conditions are extreme. In this study, using methods of planting with Coconut Fiber Bag and Cocopeat Diamond and tested at 8 native species of Merapi Volcano. In the test of Coconut Fiber Bag's method for 600 units per hectare shows Bawangan seed (*Dysoxylum densiflorum*) growing well. While in the test of Cocopeat Diamond's method for 594 units per hectare shows 240 units found Cocopeat Diamond have germinated (a success rate of 57%). The type most widely germinate are Cangkring (*Erythrina fusca*) and Tesek (*Dodonaea viscosa*). These tests also involves the local people to raise awareness of the importance of the region MVNP as a provider of environmental services.

T1

P0010

Air Pollution Tolerance Index (APTI) and dust capturing potential of few selected angiosperms of industrial belt in West Bengal, India for environmental optimization

Swastika Banerjee, *Sumita Mondal, Moumita Das, Abhijit Bandyopadhyay*
The University of Burdwan

Bio monitoring studies are extremely relevant in the field of air pollution science in relation to urban ecosystem restoration. On the onset of this the present study was undertaken to evaluate the air pollution tolerance index and dust capturing potential of 16 angiosperms from industrial area (Burnpur, IISCO) of West Bengal, India. While determining the APTI values, chlorophyll content was found to ranges from 0.1050 mg/g (*Butea monosperma*) to 0.5083 mg/g (*Cassia fistula*), relative water content from 35.68% (*Clerodendrum indicum*) to 93.02% (*Leonotis nepatifolia*); ascorbic acid from 182.432 mg/L (*Pongamia pinnata*) to 1140.541 mg/L (*Swietenia mahagoni*) and leaf pH from 5.00 (*Cassia siamea*) to 7.83 (*Ficus benghalensis*). APTI values ranged from 116.326 (*Cassia siamea*) to 814.855 (*Swietenia mahagoni*), thus indicating all selected species to be highly tolerant to air pollution, hence suitable for greening parks and gardens in polluted areas. Highest dust deposition was found in *Ficus benghalensis* (0.00574 g $cm^{-2}$and lowest in *Eucalyptus globulus* (0.00030 g cm^{-2}) at study site. The present study has suggestions for attenuating and reducing adverse health impacts from constant exposure to air pollutants and also providing regulating ecosystem services thereby modifying the local urban ecology.

T1

P0011

Production, dispersal and predation of diaspores of *Euterpe edulis* Mart. (Arecaceae): Evaluation of natural regeneration in the Atlantic Forest, Brazil

Hector Barros Gomes, *Bruno Takami Fujiwara, Eliana Akie Simabukuro*
University of São Carlos

The regeneration of plants consists on a number of processes that regulates the structure and the distribution of species in vegetation. The study of recruitment of individuals in natural populations is a challenge in landscapes with high environmental heterogeneity, common in tropical forests. In this regard, a study was conducted in a remaining area of the Atlantic Forest in southeastern of Brazil, to evaluate the effect of producing, seed rain and germination of diaspores and the role biotic dispersers on the natural regeneration of *Euterpe edulis* Martius (Arecaceae). The study was conducted in three areas of Atlantic Forest with different densities of seedling stages of *Euterpe edulis* (H- High density; M-medium density L - low density). In a previous survey it was determined that seedling density in the areas was 1.58 (HD), 1.08 (MD) and 0.46 (LD) individual/m^2. The production diaspores presented low synchronism in the three areas (20-60% of trees) and low intensity of fruiting. The area with high density of seedlings (HD), had high production of mature diaspores distributed throughout the year. Low synchrony results in diaspore over a longer period, increasing the possibility of seedling establishment. The amount of propagules in diaspores rain did not vary between locations. However the HD area presented the diaspores with larger diameter and integrity. There was reduced dispersion of the diaspores at long distance, with few records in the distant collectors of the parental plant, characteristic of the species, known to form a seedlings bank next to adult individuals and an aggregate distribution pattern. The removal was higher for the control treatment and permeable exclusion (small mammals access). And the removal of diaspores was higher in MD and smaller in HD, corroborating the information that invertebrates would not be dispersers of *E. edulis*. The seedlings of *E. edulis* had higher mortality at the HD, in amounts of 20% when native and less than 10% when planted. Germination above and under the litter was higher than diaspores buried in all areas. The results indicate that there is variation in the production, availability of diaspore and microscale removal between areas of different seedlings densities, as factors that act synergistically in keeping the conservation of population in natural areas.

T1

P0012

Literature mining on Dipterocarps: Towards better informed regeneration and reforestation in the Philippines

Roselyn Gabud[1,2], *Riza Batista-Navarro*[3], *Vladimir Mariano*[4], *Eduardo Mendoza*[5], *Sandra Yap*[6]
1. Institute of Computer Science, University of the Philippines Los Baños
2. Department of Computer Science, University of the Philippines Diliman
3. School of Computer Science, University of Manchester
4. Centre of Technology, RMIT University
5. Max Planck Institute of Biochemistry
6. Far Eastern University

Access to comprehensive information is critical in Natural Resource Management (NRM). The decision-making process of regulators on conservation policies, for example, is largely shaped by information resources at their disposal. Significant barriers could however present themselves at any point along the NRM information supply pipeline. In this respect, managing the natural regeneration and reforestation of lowland dipterocarp forests in the Philippines is no exception. Dipterocarps are the medium to large forest trees belonging to the tropical plant family Dipterocarpaceae, which are the dominant species and the skeletal backbone of lowland tropical rainforests. About 65 species in six genera have been recorded in the Philippines, and more than 65% are endemic. They are economically and ecologically important: the primary source of timber, the home of Philippine eagle and other wildlife, the stronghold that regulates water releases from the uplands, and a major contributor to ecological and climatic balance. Once the primary base for the logging industry in the Philippines, the dipterocarps have been heavily exploited that the government has begun conservation, protection and reforestation initiatives. The supra-annual mass flowering of dipterocarps occur in irregular intervals of two to ten years, possibly synchronously across Asia. Predicting the likelihood of their regeneration, to subsequently make plans regarding species for reforestation, is thus an extremely complex task. While regeneration patterns could potentially be derived from historical data and inform NRM policy-makers in the implementation of reforestation policies, access to complete and comprehensive information is currently hindered by: (a) failure to maintain data collected over long-term periods, due to interruptions brought about by short-term funding and political cycles; and (b) decentralized data collection resulting in largely fragmented information which come in a variety of formats. In order to fill in the information gaps, we propose to exploit published literature on dipterocarps by means of text mining: the automatic process of extracting structured information from documents written in natural language, e.g., scientific publications, books and agency reports. As a step towards capturing occurrence information from the literature, we have developed a schema for annotating taxon names, geographic locations, dates, habitat descriptions, authorities, and names of herbaria (in the case of collected specimens). Also covered is the annotation of any specimen's reproductive state which will enable the derivation of phenology and masting patterns. We present results based on two domain experts' manual annotation of a document collection which currently contains: (a) documents from the Biodiversity Heritage Library (BHL); (b) agency reports from the Philippines' Department of Environment and Natural Resources (DENR), and (c) article sections from various online journals. The resulting labelled corpus will form the basis of the text mining tools that we will develop as part our next steps, which will ultimately provide NRM policy-makers with a searchable database containing literature-derived information on the distribution and mass flowering of Philippine dipterocarps. In turn, this will help inform their efforts for dipterocarp species regeneration and reforestation.

T1
P0013
Floristic richness in Mulshi watershade catchment area, Northern Western Ghats (NWG), India

Rani Bhagat
PDEA'S Anantrao Pawar College, Pirangut, Tal-Mulshi, Dist-Pune 412115, Maharashtra, India

The Western Ghats (WG) are one of the globally recognized "hot spots" of biodiversity and are a UNESCO World Heritage Site in India. Physiography and climate of the Northern Western Ghats (NWG) support tropical semi evergreen to moist deciduous type of forest vegetation with immense global significance for conservation of biodiversity. Biogeographically the study area comes under Sahyadri ranges and spans varied ecological regions from crestline of Western Ghats, hill ranges and gently sloping part bordering the Deccan Plateau, thus supporting diverse plant communities. Due to high rainfall, processes of soil formation are relatively faster than low rainfall areas on plains, though happening at geological time scale. All these special physical features have formed unique habitats supporting endemic biodiversity. With the backdrop of fragile ecosystems, this biodiversity happens to be extremely essential and valuable. The study area selected is centrally located in NWG at 18°25' N to 18°41' N and 73°20' E to 73°25' E, which covers 240 sq. km from Mulshi tehsil of Pune district, Maharashtra. Extensive field surveys in region reported 1,404 flowering plants species from 774 genera and 154 families from diverse habitats. Number of species belonging to dicotyledones were 1,067 and 337 from monocotyledones, of which 142 species are found to be endemic. Also 40 sacred groves were also reported during the study. In Mulshi region, existing data of plant species is scanty and limited to only representative localities. The state government has declared one wild life sanctuary in Mulshi i.e. Tamhini in year 2013. The important water catchment includes- Mulshi Dam, Temghar and Varasgaon Dam and Pawana Dam in adjoining area. A private planned cities like Lavasa, Sahara and Amby valley lies in the study region. Further, more hill cities are proposed in various parts of Mulshi region threatening rich biodiversity of region. With this background the documentation of floristic diversity was carried out for future management of ecosystem.

T1
P0014
Rapid digitization of herbarium specimen

Jeroen Bloothoofd
Picturae

Picturae has digitized nearly 10 Million herbarium sheets since 2013 and at the moment we have 4 active herbarium digitization projects in the world knowing: Montpellier-France, Washington-US, Meise-Belgium and Geneva-Switzerland. Imaging of herbarium sheets and transcription of herbarium labels has always been a time consuming and costly task with varying results in quality. The objectives was to find a way to make the process of herbarium digitization fast and economical but also safe and reliable. Picturae has proven in the last three years that rapid digitization of herbarium specimen is possible by digitizing almost 10 million individual herbarium sheets that were all validated in quality, all accounted for and all safely returned to the archives. Picturae has developed a conveyor belt system to image the sheets and software and protocols to transcribe the label in close collaboration with Naturalis (Leiden), The Natural History Museum (London) and the Smithsonian National Museum of Natural

History (Washington). With millions of new specimen available in full visual detail and searchable on full label information new discoveries are made in existing collections every day. Less loans are processed and the loans are more accurate. Collections are far better accessible and Big Data analysis gives many new research possibilities. It is very well possible to digitize the vast herbarium collections of the world in our lifetime. And this is just the beginning as we are convinced we can do the same for the other natural history collections that are hidden from the world.

T1

P0015

Phytochemical investigations of *Atraphaxis spinosa* L. (Polygonaceae)

Xinhui Wang, *Robert Borris*

School of Pharmaceutical Science and Technology, Tianjin University, Tianjin, China

The genus *Atraphaxis* L. (Polygonaceae), comprises approximately 25 species of shrubs distributed particularly in North Africa, southwestern Europe and Asia. China has 11 species, including *Atraphaxis spinosa* L. which can be found in Xinjiang, in Northwest China. Previous phytochemical investigations have shown that *Atraphaxis* L. species are a rich source of phenolic and polyphenolic compounds. The major constituents of the *A. spinosa* are flavonoids and flavonoid glycosides. The limited nature of the known phytochemical constituents from this genus makes it an attractive target for further investigation. In the present work, a number of flavonoids, flavonoid glycosides, isoflavones, flavanols, phenylpropanoid amides, fatty acids, monoterpenes and acylphloroglucinol glucosides were isolated from the methanol extract of a sample of the entire plant of *A. spinosa* from the Republic of Georgia. The structures were elucidated by 1D and 2D homonuclear and heteronuclear NMR spectroscopy and by comparison with closely related compounds published previously.

T1

P0016

Is Hengduan Mountains really a biodiversity hotspot?

Hongyu Cai[1], *Nawal Shrestha*[1], *Xiaoting Xu*[2], *Zhiheng Wang*[1]

1. Peking University, Beijing, China

2. Sichuan University, Chengdu, China

Phylogenetic diversity can show the relatedness of species, which can to some extent, reflect the evolutionary history. It is also important in identifying the biodiversity hotspots for conservation priorities. In this study, by showing the patterns of Faith's phylogenetic diversity for 11,163 woody plants in China and correcting the results by a null model approach with different sizes of species pools, we demonstrated that Hengduan Mountains is not a biodiversity hotspot at least on the basis of phylogenetic diversity, which is quite different from previous studies. Despite high species richness, the species in Hengduan Mountains are phylogenetically close to each other. In contrast, Southwest China contains both high species richness and phylogenetic diversity, which highlights the importance of biodiversity conservation in facing the uncertain future.

T1

P0017

Orchid flora in Amapá: Situation and prospect

Patrick Cantuária

Instituto de Pesquisas Científicas e Tecnológicas do Estado do Amapá

The Orchidaceae Family is considered, by various authors, as one of the largest plant families in number of species. The orchids are distributed throughout the planet, with the exception of the polar regions and and desert, with greater diversity in tropical and subtropical regions. The present work aimed to raise the species of family Orchidaceae in the State of Amapá, contributing to the knowledge of the group to the Amazon and, consequently, to Brazil. For the organization of work, were carried out 9 inventories in unknown areas in Amapá, between 2013 and 2016 and consult in the herbaria: AMES, B, BBS, CAY, CEN, CPAP, F, GH, HAMAB, HB, HBRA, HRB, HUEFS, IAC, IAN, INPA, K, L, MG, MICH, MO, NY, P, R, RB, SP, SPF, U, UB, UC, UFP, US. And data from all the geographic coordinates were treated for establishment of the pattern of distribution of Orchidaceae in Amapá. And were still made and synopses of habit and illustrations of important morphological characteristics in the characterization of the species. Were recorded 277 species and 90 genera, being 60 new occurrences for the Amapá, as well as a map showing the distribution pattern of orchids in the State of Amapá in Brazil. We noted three major centers of diversity: Guaiana Plateau, savanna transition-coastal zone and Valley of the Jari. Although many species have increased considerably, there is a need for surveying new areas not yet known.

T1

P0018

Vast amount of *Bletilla striata* (Orchidaceae) seeds cultured in liquid media and seedling establishment in the field quickly

Xiao-Yong Cao, *Huan-Ge Song, Xuan-Ping Hu, Gong-Wei Qin, Yi-Song Zou*

Shaanxi University of Technology, School of Biological Science and Engineering, Shaanxi Key Laboratory of Bio-Resources, Hanzhong 723000, China

Bletilla striata is a type of Chinese traditional herb and its pseudobulbs were used for medicinal purpose. Because its habitat fragmentation and over collection, it is an endangered species under current situation and artificial reproduction is the sustainable way to satisfy the healthcare and medical demands. Currently, there are three different ways for its reproduction: 1. vegetative reproduction (division); cut the pseudobulbs apart and each piece has one or more growing points; disadvantage: low efficiency and accumulation of viral infections over time; 2. sowing seeds in the ground; disadvantage: the result is unstable; necessitates a long time careful management approach to watering and ventilation in protected culture systems before root formation (3 months after sowing) and plantlets are vulnerable to infections; 3. asymbiotic germination; disadvantage: high cost and low efficiency; solidified agar media (expensive ingredient), higher labor cost, pseudobulbs formation will take 3 months, a small number of seedlings cause high density in a glass flask. The seeds sowing in MS liquid media under light developed into a protocorm with its diameter nearly 1.6mm in 60 days, the protocorms were placed on peat directly and the seedlings established in a short term. Based on this method, we have developed a rapid propagation protocol and apply for a patent. Green capsules were rinsed with ethanol 70% (V/V %) for one minute and 1% mercuric chloride solution for 15 minutes, and then washed with sterile distilled water 3 times. The disinfected capsule was cut and released its content (about 10,000 seeds) to 250 ml glass media bottle containing 200 ml MS medium with 3% w/v table sugar. The bottles were kept statically in a temperature-controlled growth room at 25°C with illumination provided by cool white fluorescent tube at a fluency rate of 2000 Lux and a 16-h/8-h light/dark photoperiod. The frequency of seed germination is high, reached more than 95%, and the seeds become swollen in three days, the embryos become green and testa rupture in fifteen days, rhizoid and leaf primordia formed in twenty days; the protocorm grow at average speed of 26 g/d in 60 days, and the weight of protocorm attained to 1.6mg (FW) and its diameter is about 1.6 mm. The protocorms were taken off from bottles and wash it with tap water, packed up with cotton gauze and remove water with spin-dry machine. The protocorms in loose state can be scattered on peat medium (thickness: ~1 cm) evenly and possesses a higher resistance to desiccation. Temperature, humidity, shading and ventilation should be controlled in protected culture systems (cloches or cold frames, polyhouse). The leaf of protocorm grows and elongates immediately and roots appeared within 20 – 40 days with 2 – 3 leaf blades (~1 cm long). After 3 – 7 month planting, the seedlings form small pseudobulbs (diameter: 0.5 - 1 cm) at its base and these pseudobulbs may used for sale or field planting.

T1

P0020

The historical plant specimens collected from Korea by French missionaries, Faurie, Urbain Jean and Taquet, Emile Joseph

Kae Sun Chang[1], *Kyung Choi*[1], *Chin-Sung Chang*[2]

1. Korea National Herbarium, National Arboretum

2. T.B.Lee Herbarium, Department of Forest Sciences and The Arboretum, Seoul National University

Faurie, Urbain Jean (1877-1915) and Taquet, Emile Joseph (1873-1952) were missionaries for Missions Etrangeres de Paris (Foreigh Missions of Paris). Faurie and Taquet collected plant specimens of Korea from 1901 to 1912 and sent them to herbaria mainly in Europe and Japan. They made a great input to the understanding of flora of Korea at the primary stage of its exploration. Those specimens were studed by Leveille, Augustin Abel Hector (1864-1918), who was a French botanist and clergyman, Nakai, Takenoshin (1882-1952), who was taxonomist of Tokyo University, and other Japanese and European Taxonomists. Faurie arrived in Niigata, Japan in 1874. During his stay in Niigata, he has got a request for collecting Japanese plants from French botanist Adrien Franchet (1834-1900) of Museum National D'Histoire Naturelle at Paris (P). Over 40 years, until his death on July 4,1915 in Taiwan, he continued making botanical collections mainly from Japanese archipelago, as well as occasionally his journey East Asia expended southwards to Taiwan, through Korea, and northwards to Sakhalin and Kurile Islands. He visited Korea in 1901,1906, and 1907. The main localities visited were Seoul, Busan, Jinnampo and Wonsan of Hamgyeongnam-do, Mt. Geumgangsan of Kangwon-do, and Jeju-do. In 1907, he visited Jeju-do and collected specimens with Taquet. During his explorations, Faurie collected numerous specimens of Poaceae and Cyperaceae. Based on Faurie's specimens that are deposited at KYO, P, TI and *etc.*, we may suggest that no less than 4,800 plant specimens from Korea. Taquet arrived in Korea in 1898. Based in Mokpo, Korea, he was responsible for missionary work on many islands around the south-western peninsula of the country. He was in Jeju-do from 1902-1915 and collected specimens from 1907-1912. Based on his specimens deposited at E. KYO, TI, A, and K, he collected more than 6,000 specimens of *ca.* 1,600 taxa mainly from Jeju-do. More than 2,500 specimens are deposited in E, because Leveille is an important botanist to examine his specimens. While there are few his Cyperaceae specimens in E, 323 among 378 specimens are Cyperaceae in P.

T1

P0021

Hawaiian seed storage behavior: A case study for tropical native flora

Marian Chau[1], *Timothy Kroessig*[1], *Lauren Weisenberger*[2], *Matthew Keir*[3]

1. Lyon Arboretum - University of Hawaii

2. U.S. Fish and Wildlife Service

3. Laukahi Hawaii Plant Conservation Network

The Lyon Arboretum Seed Conservation Laboratory and the O'ahu Army Natural Resources Program conduct research to determine germination, propagation, and long-term storage protocols for some of Hawaii's rarest native plant species. The seed banks aim to create viable *ex situ* collections to preserve genetic representation of each species and to provide propagules for recovery efforts. Over the last two decades, the seed banks have gathered data from 68 angiosperm families and from taxa representing over 25% of the Hawaiian flora. Seed storage behavior has been classified into two major categories; desiccation-tolerant (which includes species that are orthodox and storable long-term, but also those that are short-lived or sensitive to freezing) and desiccation-sensitive (recalcitrant, not storable by conventional seed banking). Data are available for 303 taxa, and storage classifications have been estimated for another 506 taxa. Approximately 96% of the 809 taxa studied have desiccation-tolerant seeds, representing 55 families. This proportion is higher than the worldwide estimate that 80% of species have desiccation-tolerant seeds. These data support the hypothesis that long-distance dispersal selects against species with desiccation-sensitive seeds. However, another 25% of Hawaiian taxa were either short-lived or freeze-sensitive; therefore a third of taxa investigated cannot be conventionally banked for long-term storage. A high incidence of intermediate seed storage behavior is not known in other regional flora and may be informative for tropical areas. Current and future research efforts of the Hawaii Seed Bank Partnership focus on developing protocols for species in these categories, and expanding our knowledge of other native species' seed storage behavior.

T1
P0022

Information technologies supported citizen science approach for plant diversity survey

Bin Chen[1]*, Jianping Chen*[1]*, Keping Ma*[2]
1. Shanghai Chenshan Botanical Garden, Shanghai, China
2. Institute of Botany, Chinese Academy of Sciences, Beijing, China

Plant species occurrence data are the basis information to map and to monitor the change of plant diversity. Herbarium specimens collected from field surveys are the most important reliable data source of species occurrence. Although there are millions of specimens collection records digitized and shared online, if we look into a certain area, we always encountered with data deficiency of plant distribution information. The high cost and low efficient of traditional specimens collection make it impossible to dramatically increase the data volume in future. With the evolution of information technologies, since from 2006, we have adopted "digital specimens" concept that use digital photography as vouchers of species occurrence. To make this concept practical and efficient in field plant survey, we developed new technologies for field work and data management, including field work designing with Google Earth, field navigating with OruxMaps, digital photography, collecting specimens with digital photography, barcode label and GPS tracklogs recording, online data management with the Chinese Field Herbarium (CFH) platform. On CFH website, users can easily deposit their field survey datasets online, help each other to identify species, reorganized the datasets by classification trees, map the occurrence data on Google Earth and get automatically updated checklists. These technologies are useful for amateurs, researchers and any other people who are interested with plants. Then the CFH users work together and form into an information technologies supported citizen science to survey, identify and study plant diversity. As the result, now there are more than 14,000 users and 8 millions of digital photos on the websites, covering more than 40,000 species, including 20,000 Chinese native plant species in which more than 13,000 species have precise geographic coordinates. According these data, at least 3 new species are described and hundreds of provincial new distributions are observed and thousands of living plant populations are geo-referenced. CFH become one of the largest data centers of original plant occurrence of living plants in the world. In future, we will continue to use more technologies to facilitate plant field surveys, such as artificial intelligence technologies based auto-identification of species from digital photos.

T1
P0024

The effects of environmental stress on sesquiterpenoids accumulation in *Atractylodes lancea* (thunb.) DC. plantlet and mechanism investigation

Fei Chen*, Chuan-Chao Dai*
Nanjing Normal University

Atractylodes lancea (Thunb.) DC. is an endangered traditional Chinese medicinal herb with broad pharmacological activity, reflected by an abundance of sesquiterpenoids. Although the main active components of *A. lancea* have been clarified, there are still several problems existing. On the one hand, the resources of wild *A. lancea* have decreased sharply. On the other hand, the artificial cultivation of *A. lancea* is difficult and the active components of cultured *A. lancea* are different from those of wild ones. Therefore, the regulations of active component have showed a plateau phenomenon. Compared to other species, *A. lancea* originated in the Maoshan region of Jiangsu province where it is the most representative subspecies. In this study, we studied the effects of three kinds of environmental stress of high temperature, drought and microbiological factors on different growth periods of *A. lancea* sterile tissue culture plantlet from Maoshan. The accumulation patterns of tissue-specific sesquiterpenoids in different environmental stress factors were found to be very different. In order to reveal the regulation mechanism of environmental stress in the sesquiterpenoids of *A. lancea*, we first discussed the relationship between stress factors and plant growth from the physiological and biochemical aspects of plants. Then we studied the relationship between stress factors and sesquiterpenoids accumulation from the perspective of differentially expressed genes based on the data of transcriptome analysis (RNA-Seq). Finally, we further validate the key proteins that are susceptible to biotic and abiotic stress regulation at the proteomic level. In this study, we systematically investigated sesquiterpenoid distribution and regulation mechanism in different environmental stress factors of sterile tissue culture plantlets of *A. lancea* via metabolite quantitative, plant physiological and biochemical analysis, and omics technologies. This study provides a central theoretical basis for the manipulation of sesquiterpenoid biosynthesis in *A. lancea*.

T1
P0025

Reinvestigation on processing chemistry of ginseng: Steaming qualitatively and quantitatively changes ginsenosides and carbohydrates by chemical and physical transformation

Shan-Shan Zhou[1,2]*, Jun Xu*[1]*, Song-Lin Li*[2]*, Hu-Biao Chen*[1]
1. Hong Kong Baptist University
2. Hospital of Integrated Traditional Chinese and Western Medicine Affiliated to Nanjing University of Chinese Medicine

The processing chemistry of red ginseng (RG), a ginseng product processed by water-steaming, has been examined, but the relevant studies only focused on ginsenosides in ginseng materials. However, carbohydrates and ginsenosides in ginseng are both important bioactive components. Here we reinvestigated the processing chemistry of RG by simultaneously qualitatively and quantitatively analyzed ginsenosides and carbohydrates (total polysaccharides, oligosaccharides and monosaccharides) in both ginseng materials and steaming liquid. The experimental results revealed that the contents of investigated chemicals in the processed RG were simultaneously and differently altered along with the increment of steaming time. Moreover, the steaming-produced distillate transferred the RG components into the steaming liquid, therefore resulting in the chemical loss of processed ginseng materials. The potential chemical and physical transformation mechanisms involved in the quantitative variation of multiple components were proposed and discussed, i.e. the deglycosylation and dehydration of ginsenosides, the changed water solubility and polymerization degree of polysaccharides, the hydrolysis of polysaccharides and oligosaccharides, the dehydration of monosaccharides, Maillard reaction. The study improved the research on processing chemistry of ginseng, and therefore should be instrumental in elucidation of the differences between RG and white (unprocessed) ginseng. With this illustrative example, we would like to highlight the research on processing chemistry of RG, and it should be helpful for elucidating basis of RG preparation and application.

T1

P0026

Use of automatic species recognition engine of plants in studies on plant diversity and distribution

Mingquan Chen, *Tingting Yao*

Hangzhou Dana Technology Inc.

Xingse App is an automatic plant recognition engine based on deep convolutional neural networks (CNN). Now it's able to identify more than 4,000 kinds of plants, with an accuracy of 88% (top-3 classification). The recognition engine is still being optimized, and it's expected to identify 10,000 kinds of plants with an accuracy up to 90%. By internet crowdsourcing, Xingse mobile App is able to take the public's involvement in plant data collecting, and it has collected about 10 million plant pictures from millions of active users so far. The tremendous amount of pictures for plants with their spatial and temporal distribution information are very helpful for studies on the plant distribution and diversity, as well as endangered and new plant species discovering. In future, Xingse will work closely with non-profit organizations such as Chinese Field Herbarium (CFH) in environmental protection, establishing and completing the plant map of each province, building a comprehensive incentive system, and enlarging the plant database.

T1

P0027

Pine rust fungi of phylogeography and international quarantine

Momei Chen

University of California, College of Natural Resources, Berkeley, CA, 94720.

Pine rust fungi (*Cronartium*) causes economically costly disease throughout the Northern Temperate geography region, resulting in malformation, reduced vigor, and ultimately death of pine plantation and nursery pine plants. *Cronartium* are heteroecious specific plants having coevolved with their obligate hosts. *Cronartium* have the most complicated lifecycles of the fungal taxa in the ecosystem. The geographic pattern is diverse in North America and East Asia. Chen (2006) recently identified 3 major species (available through CABI): *Cronartium ribicola* parasitized 12 white pine species, and 24 *Ribes* or *Pediculris* in the United States, Canada, China, Japan, South Korea, Pakistan, Ireland, and Switzerland; 27 pine species and 28 *Quercus* species susceptible to *Cronartium quercuum* in North America: United States, Canada and East Asia China, Japan, Korea, Russia. *Cronartium coleosporioides* is known only in USA and Canada with seven pines hosts and a number of Scrophulariaceae alternative hosts. Chen (2004) provides new evidence and theories on the famous white pine blister rust flora. It has been suggested that white pine blister rust was introduced to the US from externally source, but the lineage evidence is lacking. Factors in the invasive forest pathogens infestation and establish always resultant with unbalances ecosystem, poor plantation. Design, and lack of integrate pest management. An improved phytogeography will depends on information gathered from molecular evolutionary genetics, natural history, population biology, paleontology and speciation analysis as well as improve the control of tree rust spread and International Quarantine on tree host plants.

T1

P0028

Species delimitation of some *Argyreia* (Convolvulaceae) using phenetic analyses based on anatomical evidence

Natthaphong Chitchak[1], *Paweena Traiperm*[1], *George Staples*[2,3], *Pantamith Rattanakrajang*[1], *Pirada Sumanon*[1]

1. Department of Plant Science, Faculty of Science, Mahidol University

2. Herbarium, Royal Botanic Gardens, Kew

3. Departamento de Biologia, Universidade Federal Rural de Pernambuco

Argyreia Lour. is one of the most taxonomically complex genera of the morning glory family (Convolvulaceae). The number of named species is now 135 and new species are regularly being described. There are several complexes of species that are morphologically similar and difficult to tell apart. Character states intergrade and overlap and the phenomenon of "mosaic evolution" is well documented in the family. Research techniques such as anatomy are required to find new characters to help discriminate cryptic species. Therefore, the aim of the present study is to explore species identification criteria for *Argyreia,* especially new sources for taxonomically informative criteria. Ten populations representing three morphologically similar *Argyreia* species were collected from their natural habitats. Plant materials were investigated using leaf peeling technique and paraffin sectioning method for anatomical characters. Anatomical character states were examined by two phenetic analyses, clustering analysis (CA) and principal component analysis (PCA). Three distinct clusters were clearly separated in both PCA and CA at the internal similarity coefficient of 0.36. Twenty-one effectively distinguishable character states were derived from the high loadings of the first two components such as the width and the length of the epidermal cells, the stomatal density, and the thickness of leaf blade. In conclusion, two separated groups were authenticated as two known species, and the third separated group has high potential to be a new species.

T1

P0029

Aquatic and semi aquatic angiosperms diversity, palynology and their conservation from the wetlands of sub-Himalayan West Bengal, India

Monoranjan Chowdhury[1], *Anurag Chowdhury*[2], *Payel Paul*[1]

1. Taxonomy of angiosperms & Biosystematics Lab. Department of Botany, University of North Bengal, Darjeeling, West Bengal

2. West Bengal Forest Department, Aranya Bhawan, Sector III,

Salt lake cit, Kolkata-700106

Sub-Himalayan areas includes terai, duars and the plains of North Bengal are blessed with various perennial or annual fast flowing streams, rivers, rivulets are coming from the Darjeeling & Sikkim Himalayas are associated with several varied sized palustrine and lacustrine having great biological diversity. Around 436 species representing 239 genera of 72 families of aquatic and semi-aquatic angiosperms are widely growing on those water bodies. These nature wetlands are the store house of various Indian endemic species (*Rotala macrandra* Koehne, *Limnophila pulcherima* (Griffith) Hooker *f.*, *Aponogeton appendiculatus* H. Bruggen, *Aponogeton crispus* Thunberg, *Aponogeton natans* (Linnaeus) Engler & Krause, *Cryptocoryne retrospiralis* Fischer *ex* Wydler, *Crinum defixum* Ker-Gawler), exotic species (*Adenostemma suffruticosum* Gardner, *Potamogeton gramineus* Linnaeus, *Murdania keisak* (Hasskarl) Handel-Mazzetti, *Polygonum hastatogagittatum* Makino, *Ludwigia peruviana* (Linnaeus) H. Hara, *Hygrophila erecta* (N.L. Burman) Hochreutiner, *Soliva anthemifolia* (A. Jussieu) R. Brown and *Carex phacota* Spreng.) and wild relatives (*Oryza rufipogon* Griffith, *Oryza latifolia* Desvaux, *Rosa clinophylla* var. *glabra* (Lindley *ex* Prain) C. Ghora & G. Panigrahi) and a unique place for speciation. These floral rich ecosystems are the life line of wildlife's of various *in-situ* conservatories of this region. In last few years, rapid climate change badly affect on the regularity of rainfall which is crucial for the sustenance of stable wetland ecosystem and during summer dried wetlands creates the severe food scarcity for wildlife's. The Human colonies surrounding various wetlands are continuously taking various benefits including foods, medicines, irrigated water *etc.* Pollen study of 58 species of the aquatic flowering plants (6 rooted with floating leaves, 2 submerged, 2 suspended and 48 emergent species) from different wetlands were investigated and most of the grains are mostly prolate-spheroidal, spheroidal or triangular, circular, rarely prolate or elongated, tectum psilate, granulate, reticulate or echinate. On the basis of aperture types pollen grains of most families are grouped under three distinct types, namely colporate, colpate, sulcate and porate. On the other hand, pollen grains of some species e.g. *Sagittaria guayannensis* & *Eleocharis tetraquetra* are inaperturate. Regarding the evolutionary and phylogenetic point of views the polarity, symmetry, apertural types and exine sculpturing of grains are the most important characters to differentiate between the different pollen types and are extensively used in APG III & IV. Unscientific withdrawal of plants as medicine, food, fodder etc, anthropological (excessive agricultures, urban sewages, monoculture etc) and natural (climate change) threats badly effects the wetlands ecosystem in this area. Scientific conservation measure will be taken design and implement is essential steps to protect these important ecosystems.

T1

P0030

Genome size variation in the grape family of Vitaceae

***Zhao-Fu Chu*[1]**, *Jun Wen*[2], *Ze-Long Nie*[1], *Ying Meng*[1]

1. Jishou University, Jishou, Hunan

2. National Museum of Natural History, Smithsonian Institution, Washington, DC

Genome size variation is of fundamental biological importance and has been a longstanding puzzle in evolutionary biology. The phylogenetic, evolutionary and ecologically significance of nuclear genome size variation in angiosperm have received considerable attention over the years, but little work has focused specifically on Vitaceae. The aim of this study is to analyse the genome size variation within Vitaceae in a phylogenetic framework in order to evaluate the systematic significance of this trait to provide insights into the dynamics of genome size evolution and to assess relationships among DNA content, biogeographic distribution and ecological features within the grape family. In present study, genome size from 61 accessions corresponding to 10 genera and 50 species of Vitaceae and Leeaceae (*Leea guineensis*) is determined using flow cytometry. The DNA 1C-value of 38 species are first reported here, and the largest genome (*Cyphostemma humile*, 1C=3.25pg) is roughly 10.48-fold larger than the smallest (*Vitis vulpina*, 1C=0.31pg). Our results suggested that genome size variation in Vitaceae is largely congruent with the phylogenetic backbone relationships. The three temperate-dominant clades (*Ampelosis-Rhoicissus*, *Parthenocissus-Yua* and *Vitis-Ampelopcissus* clade have relatively small genome sizes (mean=0.59,0.60,0.50pg, std.=0.14,0.13,0.07, respectively). However, the tropical *Cayratia-Cyphostemma-Tetrastigma* group has the largest mean genome size (1.84pg, std.=0.71) with variation region from 0.66pg to 3.25pg. The other tropical clade contains only one genus *Cissus*, with genome size variation ranging from 0.37pg to 1.03pg and a mean size of 0.57pg (std.=0.22). The genome size for the outgroup *Leea guineensis* is 1.13pg. Our data suggested that tropical lineages have larger genome sizes than temperate groups. The highest level of DNA contents found for the *Cayratia-Cyphostemma-Tetrastigma* clade could be correlated with their highly species diversification in tropical areas in the Old World. The result of this study suggests a possible ecological adaption of genome size evolution that was associated with phylogenetic divergence across the biogeographical range of major lineages within Vitaceae.

T1

P0031

Supporting genomics, research and public access through the digitization of national collections

Heather Cole, *Anissa Lybaert*, *James Macklin*

Agriculture and Agri-Food Canada

Agriculture and Agri-Food Canada (AAFC) houses over 19 million specimens of insects, plants, fungi, bacteria and nematodes that are among the most comprehensive in the world and support Canada as a major contributor to biosystematics and nomenclature. The collections are supported by a critical mass of scientists working with them. In 2016, the Treasury Board of Canada awarded 30 million dollars to support research in agricultural genomics, accelerate the digitization of specimens, and improve public access to AAFC collections. As a result, three projects were funded. 1) Data Capture and Imaging: creating a digital inventory of each collection, virtual collections for public access, and enhanced digital resources for specimens, including images, data capture, and georeferencing. 2) Molecular Characterization: producing DNA barcodes and partial or complete genome sequences as appropriate for agriculturally relevant species using protocols developed for high throughput sequencing and data analysis. 3) Data Integration and Sharing: implementation of a centralized collection management system for AAFC collections and a biodiversity portal for sharing specimen and molecular data within AAFC, other government agencies and the public. Efforts are on-going to determine the most effective protocols, technologies and best practices to achieve the goals outlined in each project. Projected outcomes include complete imaging and data capture for the 1.7 million vascular plant specimens in the National Collection of Vascular Plants (DAO).

T1

P0032

Flora NT – an online flora for the Northern Territory of Australia

*Donna Lewis, **Ian Cowie**, David Albrecht*
Northern Territory Herbarium, Department of Environment and Natural Resources

Flora NT - Northern Territory Flora Online is a web based system to access a comprehensive range of information on the NT flora, the NT Herbarium specimen data and the NT vascular plant census. The first production release was November 2014 and is the first of its kind for the NT. Flora NT allows users to search or browse for information on the entire NT flora of approximately 5500 vascular plant taxa and users can also view and extract threatened and introduced species information. Flora NT enables users to download checklists across user defined areas and by family/genus. A factsheet for each plant taxon can be viewed and downloaded. These include floristic descriptions, conservation status, ecological attributes, ethnobotanical information, images of plants and distribution maps generated dynamically from specimen data. Spatial searches of the data can be performed using mapping tools, various spatial layers or geocodes. Identification keys are also available for many families and genera, as well as key regions and habitats such as mangroves, wetlands and a LUCID-based key to the flora of the MacDonnell Ranges bioregion. The content for Flora NT and the 'label' data for over 270,000 vascular plant specimens are maintained in the Herbarium's HOLTZE database and continue to be enhanced by adding new data and images, and by improvements to overall functionality. Flora NT is available at http://eflora.nt.gov.au.

T1

P0034

Novelties of genus *Terminalia* in India

Arvind Dhabe
Dr. Babasaheb Ambedkar Marathwada University, Aurangabad 431004

Genus *Terminalia* is characterized by trees or shrubs, leaves alternate or sub-opposite crowded at the ends of the branches, presence of glands and/or domatia, inflorescence spikes or racemes, petals absent, fruits drupes or samara. It has 150 species distributed in the tropical and subtropical regions of the world and 13 species are reported from India. While working on the genus from India, some novel specimens were collected which are presented in the paper. *Terminalia arjuna* with different kinds of Samara like small thick winged, large triangular winged, wings grow beyond tipare collected from Assam, Maharashtra and Kerala. *Terminalia bellirica* with small spherical drupes with long elliptic leaves collected from Kerala, short and broad winged samara of *T. bialata* from Andaman. About 40 cm long inflorescence of *T. catappa* from Maharashtra. *Terminalia chebul*a with 2 forms - elliptic elongate long stalked, 4.5-7 cm long drupes and ovate ellipticshort stalked, 2.5-4 cm long drupes collected from Gujarat state. Elliptic lanceolate 4-6.5 cm long drupes of *T. citrina* colected from Meghalaya. 3-7 winged samaras of *T. tomentosa* from many places in India. Large oblong drupes with white patches of *T. travancorensis* are collected from Kerala. These are desrving the rank of new species or varieties.

T1

P0035

The Useful Plants and Fungi Portal, an ambitious worldwide collaborative initiative

*Tiziana Ulian, Abigail Barker, **Mauricio Diazgranados**, William Milliken, Robert Turner, Bob Allkin, Nicola Kuhn, Alex Hudson, Mark Nesbitt, Melanie-Jayne Howes, Monique Simmonds*
Royal Botanic Gardens, Kew

Plants and fungi are essential to human wellbeing, providing benefits to society and our economic prosperity, supplying food, medicine, fibres, fuel, building materials and many other products, and supporting important ecosystem services. With a rapidly growing global population, increasing energy, health and food security needs, and a reliance on increasingly scarce natural resources, effective management of plants and fungi becomes a critical component of finding solutions to address the Sustainable Development Goals. Plants and fungi are still significant underexploited natural resources, and research and innovation are fundamental to tackle global challenges such as human and animal health, nutrition, bioenergy, antimicrobial resistance, and industrial biotechnology. However, access to information about these plants and fungi, their properties, their distribution, their identification and their management, remains a major bottle-neck. Although much has been published about useful plants and fungi, these data are fragmented and inconsistent, being stored in any number of local or specialist databases and widely scattered in the scientific and technical literature. Finding all reports of a single plant is further complicated by the use of erroneous or ambiguous names in these published resources and the multiplicity of scientific names associated with many species. For example, medicinal plants on average have more than ten scientific synonyms. Effective research prioritisation and access require information from diverse disciplines being brought together. Facilitating access to all relevant information about useful plants will require partnerships and an innovative joined-up methodology for dynamically sharing and disseminating data. The Useful Plants and Fungi Portal (UPFP) aims to develop a unique information infrastructure capable of supporting a global, authoritative, expert-driven, open access, online resource, providing ready access to information on the economic, potential and traditional uses of plants and fungi. It will be designed to meet the needs of as broad an audience as possible including the public, scientists, policy makers, rural development and conservation agencies. Initially data are to be drawn from existing Kew's databases, including robust taxonomic and nomenclatural frameworks linking all synonyms as well as our plant distribution and use data. We actively seek partners to increase the scope, depth and global coverage of this resource. The UPFP will publish content and provide data access through diverse means (e.g. Kew's Plants of the World Online POWO), including through existing partner portals to address the needs of diverse audiences. The UPFP will respect intellectual property rights, local ownership of knowledge and appropriate benefit sharing as outlined by the Nagoya Protocol and FAO's international Treaty, by using information already available in the public domain and deploying it under CC BY license. The UPFP is one of Kew's 2020 Strategic Outputs aligned with the institutional Science Strategy, which will be dynamically linked to other collections, datasets and portals. Utility and longevity of the portal will depend further upon the full participation of potential users and international partnerships with those holding information in other domains or parts of the world.

T1

P0036

Methodological approach for the implementation of Important Plant Areas (IPAs) in Colombia

Mauricio Diazgranados[1], *Carolina Castellanos*[2]

1. Natural Capital and Plant Health Department, Royal Botanic Gardens, Kew

2. Ciencias Básicas de la Biodiversidad, Instituto de Investigación de Recursos Biológicos Alexander von Humboldt

Plants are essential to life on Earth. Unfortunately nearly two thirds of plant species are under threat, for causes related to human demographic expansion, land use change, deforestation, over-exploitation, invasive species, pollution and climate change. Most countries are committed to have by 2020 at least 17% of their land area under protection, according to the Aichi Biodiversity target 11. However, these areas are often established without the appropriate baseline to maximize the conservation efforts, leaving important species unprotected. To inform conservation plans, a common approach has been the systematic identification of key biodiversity areas (KBA), including Important Plant Areas for conservation (IPAs). The target 5 of the Global Strategy for Plant Conservation aims to have at least 75% of the IPAs in each region protected with effective management in place. However, there is a notable lag in the implementation of IPAs, particularly in some of the most diverse countries. Colombia is arguably the second most diverse country, with 85 general ecosystems and more than 28,000 plant species. Of those, 798 species are in the red list and over 300 are considered keystone species. Currently 13.71% of Colombian continental land is under protection in 640 nature reserves. The government organization responsible for research on biodiversity, the Humboldt Institution, is moving forward to the implementation of IPAs in Colombia. After analyzing various methodologies, a suite of goals, criteria, indicators and methods that fit the interests and needs of the country is now being presented. This methodology, deeply discussed with experts, is based on the general principles of transparence, simplicity, inclusiveness, and coherence. Using ecosystems as units of analysis, it approaches both important species and communities, with no prioritization of criteria. The first implementation is presented here, with the hope of informing future conservation plans for Colombia.

T1

P0037

Studies on phylogeography and industrialization of *Dendrobium officinale* Kimura et Migo

Xiao-Yu Ding

College of Life Sciences, Nanjing Normal University, Nanjin, China

The genus *Dendrobium* could be used as a case study to elucidate the evolutionary history of Orchidaceae in the Sino-Japanese Floristic Region (SJFR) and Southeast Asia region. These evolutionary histories remain largely unknown, including the temporal and spatial distribution of the evolutionary events. The present study used nuclear and plastid DNA to determine the phylogeography of *Dendrobium officinale* and four closely related taxa. Plastid DNA haplotype and nuclear data were shown to be discordant, suggesting reticulate evolution drove the species' diversification. Rapid radiation and genetic drift appeared to drive the evolution of *D. tosaense* and *D. flexicaule*, whereas introgression or hybridization might have been involved in the evolution of *D. scoriarum* and *D. shixingense*. The phylogeographical structure of *D. officinale* revealed that core natural distribution regions might have served as its glacial refuges. In recent years, human disturbances caused its artificial migration and population extinction. The five taxa may have originated from the Nanling Mountains and the Yungui Plateau and then migrated northward or eastward. After the initial iteration expansion, *D. officinale* populations appeared to experience the regional evolutionary patterns in different regions and follow the sequential or rapid decline in gene exchange. In the aspect of industrialization of *D. officinale*, an important research institute named "Jiangsu Provincial Engineering Research Center for Technical Industrialization of *Dendrobiums*" has been established to cooperate successfully with many companies in Jiangsu Provice of China. The germplasm identifying system, the breeding system and the optimized method of tissue culture have been set up by the institute for many years. The organic cultivation methods have been applied for the cultivation of *D. officinale* at mountain regions in southren China, where the adult individuals of *D. officinale* have been cultivated on the surface of stones and the tree trunks. Series of food, health and daily products made from *D. officinale* have being developed for sales. Our researches have played an important role in the conservation and sustainable utilization of *Dendrobium officinale.*

T1

P0038

Can butterflies cope with city life? Butterfly diversity in a young megacity in southern China

Kong-Wah Sing[1], ***Hui Dong***[2], *Wen-Zhi Wang*[3], *John-James Wilson*[1]

1. Institute of Biological Sciences, Faculty of Science, University of Malaya

2. Fairy Lake Botanical Garden, Shenzhen & Chinese Academy of Sciences

3. State Key Laboratory of Genetic Resources and Evolution, Kunming Institute of Zoology, Chinese Academy of Scienc

During 30 years of unprecedented urbanization, plant diversity in Shenzhen, a young megacity in southern China, has increased dramatically. Although strongly associated with plant diversity, butterfly diversity generally declines with urbanization, but this has not been investigated in Shenzhen. Considering the speed of urbanization in Shenzhen and the large number of city parks, we investigated butterfly diversity in Shenzhen parks. We measured butterfly species richness in four microhabitats (groves, hedges, flowerbeds, and unmanaged areas) across 10 parks and examined the relationship with three park variables: park age, park size, and distance from the central business district. Butterflies were identified based on wing morphology and DNA barcoding. We collected 1,933 butterflies belonging to 74 species from six families; 20% of the species were considered rare. Butterfly species richness showed weak negative correlations with park age and distance from the central business district, but the positive correlation with park size was statistically significant ($p = 0.001$). Among microhabitat types, highest species richness was recorded in unmanaged areas. Our findings are consistent with others in suggesting that to promote urban butterfly diversity it is necessary to make parks as large as possible and to set aside areas for limited management. In comparison to neighbouring cities, Shenzhen parks have high butterfly diversity.

T1

P0039

Global Genome Biodiversity Network – Unlocking the hidden treasures of botanical DNA and tissue collections

***Ole Seberg*[1]**, *Katharine Barker*[2], *Jonathan Coddington*[2], *Gabriele Droege*[3]

1. Natural History Museum Denmark

2. National Museum of Natural History, Smithsonian Institution

3. Botanic Garden and Botanical Museum Berlin-Dahlem, Freie Universität Berlin

Genomic research depends upon access to DNA or tissue collected and preserved according to high-quality standards. At present, the collections in most natural history museums do not sufficiently address these standards, making them often hard or impossible to use for whole-genome sequencing or transcriptomics. In response to these challenges, natural history museums, herbaria, botanical gardens and other stakeholders have started to build high-quality biodiversity biobanks. Unfortunately, information about these collections remains fragmented, scattered and largely inaccessible. Without a central registry of relevant institutions, it is difficult and time-consuming to locate the needed samples. The Global Genome Biodiversity Network (GGBN) was created to fill this gap by establishing a central access point for locating samples meeting quality standards for genome-scale applications, while complying with national and international legislations and conventions (e.g. the Nagoya Protocol). The GGBN is rapidly growing and currently has 56 members. Mobilizing currently scattered DNA and tissue samples by providing access to their data through a common platform will boost research, democratize their use, conserve genetic resources, help train the next generation of genome researchers, and make the Convention on Biological Diversity ABS a reality. Molecular research results in an exponentially increasing number of sequences and hundreds and hundreds of publications every year. Knowledge of biodiversity biobank content is urgently needed to enable concerted efforts and strategies in collecting and sampling new material. The GGBN Data Portal together with the GGBN Data Standard provides an infrastructure for making genomic samples discoverable and accessible. Additionally, it will facilitate identification of gaps in the collections, thereby empowering targeted sampling efforts, increasing the breadth and depth of preservation of genetic diversity. The GGBN covers all branches of the Tree of Life, except humans, but here we focus on vascular plants. While current efforts are centred on getting the existing samples of all GGBN members online, a pilot project, GGI-Gardens, has been launched as proof of concept. Over the next 6 years GGI-Gardens aims to add to the GGBN high-quality genetic material from at least one species from each of the approx. 460 vascular plant families and one species from half of the approx. 15,000 vascular plant genera.

T1

P0040

The causes and countermeasures to the degenerate of *Salix matsudana* communities at Xi'an Chan-ba National Ecological District

Xichun Du

Xi'an University

The communities of the *Salix matsudana* at Ba River in Xi'an Chanba National Ecological District were investigated. Comparison and correlation analysis on DBH, lodging, bark cracking and mortality of the *Salix matsudana* in different environments were conducted using the SPSS software. The results showed that: there are no statistically significant differences on DBH between the *Salix matsudana* in different environment, but there are statistically significant differences on lodging, bark cracking and living situation. The correlation analysis showed that there are positive correlation between lodging with skin cracking and mortality. There are also significant negative correlation between cracking and the mortality. So, the lodging and the bark cracking are the direct reasons for the high mortality of *Salix matsudana*. The mortality of *Salix matsudana* had significantly reduced, after the water level of the rubber dam were reduced, the frequency of water storing and draining were reduced, and the speed of water storing and draining were slowed down, which can forming a more stable wetland ecosystems on the river.

T1

P0041

The reaserch of bloom warning of *Bashania fargesii* in Shaanxi Foping National Nature Reserve

Xichun Du

Botanical Society of Shanxi Province

Shaanxi Foping National Nature Reserve is one of the regions with the highest density of giant pandas in China. The *Bashania fargesii*, one of the main food species of Giant Panda. But the death of *Bashania fargesii* will be widespread after flowering, which will have great impact on the growth and breeding of giant pandas. The studies was mainly foucus on the characteristics of early flowering, early warning, and early countermeasures of *Bashania fargesii*, which is of great significance for the protection of the giant pandas. The results showed that, there were no significant differences of the chlorophyll and nitrogen content between bamboo culms of different ages (neonate, annual and perennial) at the same position. There were no significant differences of the chlorophyll and nitrogen contents between the leaves of the same position (the first, secondary and tertiary leaves) on the same branch.But the chlorophyll and nitrogen content of Bashania fargesii were significantly different between the four different flowering status group (vigorous growth group, flowering signs appear group, will flower group, flowering group). The most obvious difference were between vigorous growth groups and the flowering groups, which chlorophyll content was 45.7 and 37.9 respectively, and the mean values of nitrogen content were 3.77 and 3.21 respectively.

T1

P0042

Principles of collections formation and management in the Botanic Garden of the M.V. Lomonosov Moscow State University under present-day conditions

Sergey Efimov, *Galina Degtjareva, Elena Terentieva, Tatiana Varlygina*

M.V. Lomonosov Moscow State University

Plant collections of the botanic gardens have not only scientific significance but also contribute into their conservational and educational roles. University botanic gardens have more opportunities to carry out integrated research projects uniting the efforts of their university researches – not only botanists but of other biological disciplines. The Botanic garden of the M.V.Lomonosov Moscow State University founded in 1706 is one of the oldest in the Russia and possesses successfully developing collections of wild (the arboretum, the alpine section, the taxonomic section, the useful plants section, as well as tropical and subtropical plants) and ornamental plants (roses, peonies, irises, phloxes, lilac, apple-tree, *etc.*). The Botanic garden participates in and coordinates the projects on biodiversity conservation of the wild flora of Russia. Here grow 105 rare and endangered species included in the Red Data Book of Russian Federation, 60 species included in the Red Data Book of Moscow region and 8 species from Appendix I of Bern convention. Since 2014, the Moscow State University realizes the project "National depositary bank of living systems" dedicated to preserve the biodiversity and create new ways of biological material use. The important part of the project is the information system containing comprehensive data about samples from biological collections, including the use of molecular and genomic technologies. The peony plant collection of the Botanic garden can serve as a good model object for work in the given direction. The collection includes 280 cultivars demonstrating all selection trends, as well as 18 species, many of which included in the Red Data Book of Russian Federation. As is known, high intraspecific variability of morphological characters and hybridization events obscure relationships between *Paeonia* species. Collection of samples from different populations and geographic regions, supplemented with information about different morphological characters especially that used as diagnostic, chromosome number, geographical distribution, state of populations, and nucleotide sequences of different DNA regions provides a basis for various fundamental studies. Moreover, the peony plant collection can serve as an example of cooperation between fundamental and applied strategies. The knowledge about species helps to identification of origin of cultivars and its ecological preferences during cultivation. While knowledge about cultivars allows to compare results of natural hybridization events with results of the similar processes passing by creation of cultivars. Conduction of comprehensive inventory of the material deposited in the modern electronic databases can improve the way in which garden plant collection is documented. This helps not only in the everyday work with the collections but also in storage and research of the collected plant.

T1

P0043

The gymnosperm trees of highlands of East Asia and its perspective for introduction at Saint Petersburg (Russia)

Aleksandr Egorov[1], *Aleksandr Afonin*[1], *Vyacheslav Byalt*[3], *Larisa Orlova*[3], *Nikolay Vasiliev*[2]

1. St. Petersburg State University

2. St. Petersburg State Forest Technical University

3. Komarov Botanical Institute Russian Academy of Sciences

The mountains of East Asia are the center of origin of plants, where a high diversity of them exists. Forests and woodlands are located at the upper border of forests in the highlands of East Asia are influenced by the rather harsh climate and can be considered as areas of donor introductions of these plants to the Northern territories of temperate climate. To determine the potential opportunities of the mountains of Eastern Asia, as regions-donors, we have conducted a comparative analysis of the taxonomic diversity of gymnosperm trees, located at the upper border of the forest of these mountains and in the Botanical collections at St. Petersburg. The upper limit of the forest in the mountains of Eastern Asia are located at 3500-4000 m alt. Analysis of the climatic characteristics showed that the average January temperature at 3000-4000 m alt. in the mountains of Eastern Asia is in the range from slightly above 0°C in the southern regions of China to below -10°C in Northern areas. In St. Petersburg January temperature over the past 30 years are -8°C. In Flora of China (2008), the distribution of plants is given, indicating the regions of their growth and height above sea level. Height is not differentiated across regions, so as to identify the particular location of the common species at altitudes in different provinces in the Flora of China (2008) is not possible. Therefore, we carried out the analysis of taxa of gymnosperm trees, at least partially located in the mountains above 3000 m alt. All identified 61 such species. The conditions of the highlands of China at certain altitudes are climate analogues of regions of Northern Eurasia in terms of the growing degree days during the growing season and winter temperatures. However, seasonal changes in day length in the southern highlands is very different from the length of the day in high latitudes, which can lead to out of sync with the rhythm of the development of introduced plants. Therefore, the success of plant introduction in the Northern regions of Eurasia from the highlands ofChinaremains in question. The analysis of the taxonomic diversity of gymnosperm trees showed that the collections of Saint-Petersburg in the open field for last 150 years was tested at least 23 species of 61 (37.7%). Analysis of winter hardiness of these trees allowed us to divide them into the following groups: Winter-hardy and winter-hardy enough (4 species): *Larix sibirica, Picea asperata, P. likiangensis, P. schrenkiana.* Relatively hardy, but in the abnormal winter can significantly damaged (3 species): *Juniperus semiglobosa, Platycladus orientalis, Pinus wallichiana.* Not hardy plants significantly frosted over, sometimes completely on the level of snow or die; also, this group includes taxa that died for an unknown reason (15 species): *Juniperus pseudosabina, J. squamata, Abies delavayi, A. ernestii, A. fargesii, A. forrestii, A. georgei, A. spectabilis, Larix griffithii, L. potaninii, Picea brachytila, P. smithiana, Pinus armandii, P. yunnanensis, Tsuga dumosa.* And 1 species still pass the test. Thus, it can be concluded that of the 24 tested taxa of highlands East Asia at St. Petersburg, winter-hardy and winter-hardy enough are 4 taxa of 16.7%. However, due to climate change can be re-introduced a number of gymnosperm trees.

T1

P0044

Design, testing and evaluation of best practices for *in-situ* conservation of economically important wild plant species in Morocco

Jalal El Oualidi
Institut scientifique, Mohammed V university in RABAT

The present study concerns the wild and semi-wild vegetal species of economic importance. Such a project appears at the appropriate time for Morocco that has several assets from which it can benefit, and which can permit to achieve this project. Indeed, the country has since a few years the results of two very important surveys at its disposal, which are the National Study on Protected Areas and the National Survey on Biodiversity. The main components of the study are: (1) Development of the national strategies and methodologies based on the best *in situ* conservation practices relative to wild vegetal species of economic importance; (2) Setting up of demonstration projects; (3) Setting up of monitoring systems; (4) long term assessment; (5) Dissemination of information. The objectives of the study in Morocco emphasized three points: 1. The presentation of the general data relative to the country, its biological resources, the cultural and socio-economic aspects and the type of legislation adopted concerning plants of economic importance and the protected regions, as well as a brief overview of the programs and projects concerning the *in situ* conservation of the genetic resources of the plants existing inside and outside the protected areas. 2. Identification of target taxa by listing the main wild vegetal or semi-domestic species of economic importance (*s.l.*). 3. The prioritization of target taxa from the potential list by proposing a list of about ten target species and their sites, chosen by taking account the criteria of FAO.

T1

P0045

Weed distribution and abundance index (AI) at New Halfa Agricultural Scheme, eastern Sudan

Mohamed Elkhawad, *Mohamed Khir, Abosofian Osman*
Agricultural Research Corporation

A field survey was conducted during season 2014/2015 to determine the weed flora and weed dominance at New Halfa Agricultural Scheme, eastern Sudan. The scheme divided into three geographical sites, each site consist of two sectors, from which fifty-six fallow fields were selected and ten quadrates (1 m^2) from each field were taken randomly. Individual weed species, mean field density (MFD), frequency (F) and uniformity (U) of each field were determined. Forty-two species belonging to nineteen families (sixteen dicotyledonous and three monocotyledons) were recorded. The most dominant weeds were *Brachiaria eruciformis, Brachiaria reptans, Dinebra retroflexa, Abutilon glacum* and *Alysicarpus rugosus. Brachiaria reptans* recorded highest abundance index (AI) for center and south sites (234.72 and 292.23) and *Alysicarpus rugosus* for north site (215.69). *Alysicarpus rugosus* recorded highest mean field density (MFD) and uniformity (U) values (43.47 and 72.22) for north site, while highest frequency value (100) recorded by *Abutilon glacum, Alysicarpus rugosus, Brachiaria eruciformis, Sesbania sesban, Echinocloa colona, Rhynchosia minima* var. *minima* and *Corchorus* spp. *Brachiaria reptans* recorded highest MFD and uniformity for the center and south sites (76.94,68.89 and 107.79,84.44). Highest frequency values (100) were recorded by *Brachiaria reptans, Corchorus* spp. and *Alysicarpus rugosus* for center site and *Brachiaria reptans, Corchorus* spp., *Abutilon glacum, Xanthium brasilicum, Dinebra retroflexa* and *Cynotis axillaries* for south site. For north site, lowest MFD value recorded by *Ipomoea cordofana* and *Withania somnifera* (0.01), lowest frequency value by *Withania somnifera, Urochloa* sp., *Solanum dubium* and *Phyllanthus niruri* (8.33) and lowest uniformity value by *Withania somnifera* and *Phyllanthus niruri* (0.83). *Ipomoea cordofana* recorded the lowest MFD, frequency and uniformity values (0.01,11.11 and 1.11) respectively, at center site. For south site, lowest MFD and uniformity values recorded by *Withania somnifera, Tribulus terrestris* and *Acalypha indica* (0.01 and 1.11), while lowest frequency values recorded by *Sonchus cornutus, Setaria pallide-fusca, Senna obtusifolia, Withania somnifera, Tribulus terrestris* and *Acalypha indica* (11.11).

T1

P0046

Herbaria in the digital age: Promises and pitfalls of using herbarium records to assess phenological responses to climatic change

Aaron Ellison[1], *Charles Willis*[1], *Charles Davis*[1], *Edith Law*[2], *Alex Williams*[2]
1. Harvard University
2. University of Waterloo

Central events in plant life history, including setting and emergence of buds, leaves, flowers, and fruits, are being affected strongly by climatic change. Contemporary time series of phenological events often are too short to draw strong inferences about relationships between these events and climatic change. Expressions of phenological events, including flowers and fruits, are clearly visible on many herbarium sheets and many recent papers have shown that herbarium data can be used to extend phenological time series. Ongoing digitization efforts at herbaria around the world have expanded greatly the potential database of phenological records, but have raised new challenges for efficient and reliable data collection. We have developed a crowd-sourcing tool, *CrowdCurio*, that can be used to collect phenological data from digitized herbarium specimens. We have tested its utility by having experts and novices (Amazon Turk) collect phenological data (number of flower buds, open flowers, and fruits) from two common members of the New England (USA) flora: *Chelidonium majus* and *Vaccinium angustifolium*. We also created a computer-based mechanism, *Deja Vu*, that distributes at random intervals duplicate tasks to individual workers and assesses their consistency in collecting phenological data. Estimates of flowering and fruiting time derived from digital herbarium data were associated significantly with temperature: both species showed an approximately 2 day per degree C shift in flowering and a 3 day per degree C shift in fruiting. Standard measures of data consensus suggested that data quality differed little between non-experts and experts, but that non-experts collected data more rapidly, efficiently, and cost-effectively than the experts. Our measures of consistency provided by the *Deja Vu* software also provided an estimate of the reliability of data collected by each worker. These measures revealed that differences in estimation of phenophases differed among them and between the species, and that workers experienced fatigue as they evaluated more herbarium sheets, leading to poorer quality data. We conclude that there is great promise in using herbarium data to assess effects of climatic change on plant phenology but that care must be taken to ensure that the data derived from digitized herbarium data is of high quality and leads to reliable and repeatable conclusions.

T1

P0047

Reproductive biology of *Trichocereus bridgesii* (Cactaceae) an endemic cacti from dry valleys (La Paz, Bolivia)

Mariela Escobar-Torrez

Herbario Nacional de Bolivia

In recent years it has been suggested that exist a geographic dichotomy in pollination system for columnar cacti. In extratropical regions the columnar cacti show more open and generalized pollination system, meanwhile, in tropical regions columnar cacti tend to have specialized pollination system. As pollination data for South America are scarce, the aim of this study was to analyze the reproductive biology of *Trichocereus bridgesii* an endemic cactus in Ananta a dry valley located in La Paz (Bolivia). An exclusion experiment was conducted to measure the breeding systems, and the relative contribution of nocturnal and diurnal floral visitors to fruit set and seed set, also, I observe the floral visitors, their frequencies and behaviors. In addition I evaluated if *T. bridgesii* has a specialized pollination system or generalized pollination system. The experiment revealed that *T. bridgesii* exhibits a complementary general pollination system demonstrating that both pollination groups (diurnal and nocturnal) were important for plant reproduction. The importance of floral visitors on the breeding system indicates an obligate xenogamous system, self-incompatible. Seeds produced in the experimental treatments: control, xenogamy and obligate xenogamy were viable, except for the treatment of geitonogamy. During the day native bees (*Ischnomelissa* sp., *Habralictus* sp.) and the introduced *Apis mellifera* were the most frequent visitors and during the night Nitidulidae beetles, moths or bats were not registered. *Trichocereus bridgesii* shows more affinity with columnar cacti from extratropical regions than with those from tropical regions, these results demonstrated that as latitude increases in tropical regions of South America, Cactaceae have a generalist pollination system. Therefore, I recommend to carry out comparative studies from cacti species located within and outside the tropics according to variations in their geographic distribution. It is expected that in very isolated fragments of vegetation, two possible consequences are generated: 1) existence of specialized pollination systems when the main pollinator is abundant and 2) existence of generalized pollination systems when the main pollinator is scarce, in this case comparative studies in disturbed and conserved areas are recommended.

T1

P0048

Distribution pattern of epiphytic mosses from Orthotrichaceae family in China

Lucie Fialová[1], *Lukáš Číhal*[2], *Zuzana Skoupá*[1], *Vítězslav Plášek*[1]

1. University of Ostrava

2. Silesian Museum in Opava

The distribution structure of bryophytes shows considerable variability. One of the main aims of the project is to clarify impact of environmental conditions on distribution pattern of epiphytic bryophytes in China. For evaluation of the distribution patterns of epiphytic bryophytes the data obtained from Chinese herbarium collections and from field survey were used. Collection of these data was carried out during 2014–2016. For analysing, methods of species distribution modelling (SDM), specifically we used the method of Maximum Entropy Modelling (MaxEnt). MaxEnt is now a common SDM tool used by conservation practitioners for predicting the distribution of a species from a set of records and uncorrelated environmental predictors (e.g. Mean Diurnal Range (Mean of Monthly (max tempt – min temp), Min Temperature of Coldest Month, Precipitation of Wettest Month, Precipitation of Driest Month). MaxEnt is a machine-learning method and it calculates a raw probability value for each pixel of the study region. These raw probabilities are scaled to sum to 1, and do not represent probability of occurrence, but rather an index of relative suitability. However, it would be inappropriate to interpret the modelled distributions as actual borders of a species range. Rather, the models identify regions that have similar environmental conditions to those where the species currently maintains its populations. For all species, we considered an initial set of environmental variables representing various candidate predictors that were potentially relevant for the distribution of epiphytic bryophytes. In total, 19 environmental variables (Bioclim) were downloaded from WorldClim (http://www.worldclim.org/), which is a set of global climate layers generated through interpolation of climate data from weather stations. These data were available at the resolution of 30 arc-seconds (approximately 1 km^2 resolution). Because we modelled species that are mainly epiphytic, we also included in our model Percent Tree Coverage (PTC) Global version data, which show the ratio of tree canopy coverage (i.e., density of trees on the ground) (http://www.iscgm.org/, Geospatial Information Authority of Japan, Chiba University and collaborating organizations). These data are based on MODIS data 2008 (Terra & Aqua satellites) and available in 15 arc-second resolutions. All environmental data were acquired in the WGS84 geographical coordinate system (EPSG: 4326). For preparing environmental layers, we used functions from the GDAL library. For the graphics visualization of map output we used geographic information system DIVA-GIS (free computer program). During each next field research, it would be most appropriate to test the models directly in the field, because such research could improve validity and precision of predicted distributions. Until then, models can serve as a guide to future survey expeditions, conservation of the target species, and understanding of their ecology. These methods can minimize substantial financial costs, as well as lot of time consumption, because field survey realized under such a scenario can achieve better and more precise results in shorter time.

T1

P0049

Medicinal plants native in Argentina: Two model species with social inclusion

Renée H. Fortunato[1,2,3], *Hernán G. Bach*[1,4], *Paola A. Faroni*[1,5], *Viviana G. Spotorno*[1], *Nicolás Nagahama*[1], *María M. Manifesto*[1]
1. Instituto Nacional de Tecnología Agropecuaria (INTA)
2. Consejo Nacional de Investigaciones Científicas y Técnicas (CONICET)
3. Facultad de Ciencias Agroalimentarias, Universidad de Morón
4. Facultad de Farmacia y Bioquímica, UBA
5. Facultad de Agronomía, UBA

In Argentina grow more than 10,000 spp., and of which close to 1,500 spp. have records on medicinal use. The information is mainly from ethnobotanical, chemical and biochemical studies, but up to now there are few species in which the chemical compounds have been analyzed and registered towards production of phytopharmaceuticals. During 1993 to 2003 the Institute of Biological Resources, INTA, made screening on search medicinal and agrochemicals active compounds. The proposal was based in agreement with national (Argentina National Constitution, 1994) and international legislation (CBD ratification, 1994). According to this from 2010, we have begun to develop a technological model in two species groups with medicinal background, each of which has importance in the Norwest and South (Patagonia) regions of the country. A) *Bauhinia* (Caesalpinioideae, Leguminosae) "cow's hoof", "cow's paw", "monkey's ladder": Hypoglycemic, antioxidant, analgesic, anti-inflammatory and diuretic. B) North Patagonian species of *Valeriana* (Caprifoliaceae) "valerian": sleep disorders, sedative and anxiolytic effects. To carry out it has been studied morphological characterization, biochemical and molecular (including expression libraries), and after that the germplasm conservation of the populations that showed promising biotypes (Germoplasm Bank INTA). Later on, the cultivation of selected species/populations has been initiated to be available the Biological Resource for domestication and long-term improvement programs. This strategy has made possible to propose designs and technological processes for the quality of life in rural territories and sustainable growth through an integral and rational use of the resource with the preservation of wild populations and their habitat.

T1

P0051

Biomorphological study and *ex situ* conservation of *Aquilegia colchica* Kem.-Nath

Laura Gabedava
National Botanical Garden of Georgia, Department of Plant Conservation

This is the first time we have studied reproductive biology, ontogenesis and growth rhythm, considering the life cycle periods of *A.colchica* - Georgian endemic species, primarily having the status of conservation of the Red List of Caucasus. As a result of the investigations we have estimated that formation of generative sphere and its functioning passes normally. High energetic potential of male sphere. Stamen is four locular. Meiosis passes naturally and 7 bivalents are formed. Tetrad is formed by simultaneous mode. Ready pollen grain is binuclear and pollen fertility is high (90%). Seed bud is anthropous, crassinucelate with two integuments. The plant is entomophilic characterized by protandria. With the purpose to start an *ex situ* collection the quality of seed germination and intergrowth was defined under laboratory conditions. Seeds maintain germination ability for three years. The percentage of germination differs under the controlled regime of temperature and lighting. High percent of germination is attained if seeds are sown as soon as they are collected (92%). Potential and real possibilities of seed formation were defined. *A.colchica* is characterized by high productivity of seed formation (1260 seeds within the sample) and rapid development of its shoots. The anthropogenic stress that leads to a destruction of the habitat must be the reason why it is reduced in nature. *Ex situ* activities were carried out. The collection of seedlings was set up and the Seed Bank was created.

T1

P0052

Systematic study of Acanthochlamydaceae—A new endemic family of China

Baochun Gao
Chengdu Institute of Biology, Chengdu, China

Acanthochlamys bracteata is a Chinese endemic plant that was first discovered, descripted and named by Kao in the West Hengduan Mountains of Sichuan Province. The anatomical characteristics of *A. bracteata* are very unique in monocotyledonous plants, such as the eustele in rhizome, protostele in root, and the leaf-stem compound structure in scape. These characters were found for the first time in monocotyledonous plants. The coexistence of eustele and protostele in vascular bundle system suggests that Acanthochlamydaceae not only retains the ancestral character, but also evolves derived type. It is absent of scattered bundles, which is the most derived bundles type in monocotyledons, but shows a transitional type between protostele and scattered bundles. The family, Acanthochlamydaceae, is a new monotypic family, which contains only one species. Phylogenetic results showed that Acanthochlamydaceae is a sister group of Velloziaceae. The close relationship with the two families indicated that the close floral relationship between Hengduan Mountains and African tropical regions. In this paper, a comprehensive study was performed on the morphological characteristics, the status in the world flora, geographical distribution, phylogeny, origin, and other issues of *Acanthochlamys bracteata*.

T1
P0053
Seed abortion in *M.indicus* (L.) All
Nidhigautam Gautam
Daulat Ram College, University of Delhi

Legumes serve as the best and cheapest source of plant proteins, hence an integral part of our vegetarian diet. These plants play another significant ecological role because of their unique ability of nitrogen fixation maintaining soil fertility. Members of the family Fabaceae therefore, have been a favorite material for embryological, physiological, ecological and molecular studies. Being one of the largest groups of plants it shows a variation in its reproductive biology. The embryological studies on *Mellitus indicus,* a wild fodder legume exhibits many interesting features at light and electron microscopic levels. This fabaceous plant flowers during winter months in Delhi and bears small, yellow papilionaceous flowers in axillary racemes. The small, spindle shaped ovary is covered with tiny glandular trichomes that persist after fertilization and have a defensive function. *M .indicus* produces two superimposed campylotropous, bitegmic ovules on marginal placenta. Embryo -sac development follows the Polygonum- type of development and the embryo development is of Solanad type. Interestingly only one of these ovules matures into a hard seed while the other ovule aborts at an early stage. The hard seed coat derives from outer integument in a sequential manner and the inner integument is consumed during the ontogeny. The pod is single seeded and the seed shows insignificant enlargement. The causes and significance of seed abortion in *M .indicus* will be discussed in the present study.

T1
P0054
Genomic mechanisms of the breakdown of dioecy: A unique case study of the Waterman Mountain elephant tree, *Bursera microphylla*
Xingyue Ge
University of Arizona

Plant sexual and breeding system evolution has long been one of the most intriguing issues to evolutionary biologists. A well-known and commonly accepted breeding system evolutionary pathway is the trend from hermaphroditism (cosexual) to dioecy (separate sexes among individuals). *Bursera microphylla,* it's a semi-succulent tree species native to subtropical dry forest across the Sonoran Desert. Most of the populations of this well-known dioecious species exhibit an unbiased sex ratio of 1:1 throughout its distribution range. Curiously, in an isolated peripheral population in the Waterman Mountains, Ironwood Forest National Monument in Southern Arizona, along the northeastern boundary of the species distribution range, *B. microphylla* has been found to have a strongly female biased gynodioecious (one population that has cosexual and female individuals) sex ratio. Heraphrodites are found in this population with the female individuals. Moreover, by examining the female reproductive success and pollination biology, researchers have shown that the female biased gynodiecious population has lower reproduction success and severer pollen limitation comparing to the traditional dioecious populations with equal sex ratios. All of these empirial ecological patterns indicate a reversal to gynodioecy from dioecy, one breakdown pathway that is theoretically possible yet has never been tested empircally. In this study, we attempted to examine the gemonic signatures of the mechanisms underlie this breakdown of dioecy. We applied novel NGS methods to detect the signals of 1) interspecific hybridization; 2) polyploidization and 3) reverse mutation on Y chromosome. Specifically, we further inquired the three questions: 1) whether this special population result from the hybridization between different species? 2) whether the hermaphrodites result from polyploidization? And 3) whether the hermaphroditic biased genes located on Y chromosomes thus possibly are reverse mutations result in novel sex morph in the Waterman Mountain population? We sequenced and *de novo* assembled 6 flower buds transcriptomes from both sex morphs: hermaphrodite and female individuals in this population. No signal of hybridization or polyploidization was detected from the distribution graph of duplicated gene pairs along evolutionary time. We then calculated and compared the evolutionary rate of sex specific genes found exclusively in each sex morph. Hermaphrodic genes have a significant higher evolutionary rate compare to female genes. Moreover, we conduct the gene annotation of the 93 genes that showed higher evolutionary rate in hermaphrodites compare to TAIR database, 35% of them are associated with biological processes, most of these candidate sex specific genes are involved in reproductive process. Taken together, our results demonstrated the potential genomic mechanism of a breakdown scenario of dioecy: there is very likely to be reverse mutations on the Y chromosomes as the hermaphrodite specific genes are most likely to be located on the Y chromosome. The mutations enable hermaphrodite to re-gain female function as the populations are very pollen limited while the hermaphrodite have a higher reproductive success from selfing. As a unique example of the reverse evolutionary process by breaking down dioecy to gynodioecy, our result has provided the very first empirical evidence to further complete the pathways of plants' breeding system.

T1
P0055
The conservation of critically endangered species Yuanbanshan fir (*Abies yuanbaoshanensis*)—Based on the intergrated capacity building approaches in southern China
Yanjing Geng[1], Dingming Luo[2], Xiaoya Li[1], Tao Ding[3], David Gill[1], Georgina Margin[1]
1. Fauna & Flora International (FFI)
2. Yuanbaoshan National Nature Reserve, Yuanbaoshan, China
3. Guangxi Institute of Botany, Guilin, China

Over 9,600 tree species are threatened with extinction around the world. To safeguard the world's threatened tree species, the Fauna & Flora International (FFI) and Botanic Gardens Conservation International (BGCI) formed the only international programme called the Global Trees Campaign (GTC) in 1999. As part of GTC, FFI's China has been working on the conservation of threatened trees in southern China since 2006. The area is home to more than 200 threatened tree species, many of which are covered by nature reserves. FFI's work mainly focused on empowering the nature reserve staff to protect those threatened tree species to address the major gap that many practitioners lack the skill and experience for the plant conservation. From 2010-2016, FFI supplied six training courses to the technical staff of the nature reserve, In the meantime FFI worked with local experts and stakeholders within four reserves and developed some pilot practices. From 2012 to 2017 *Abies yuanbaoshanensis* is one of the target species. *Abie Yuanbaoshanensis* with a global population of less than 800 individuals, is a Critically Endangered conifer species, endemic to Yuanbaoshan National Nature Reserve, in North Guangxi, China. The main threats are human disturbance and high sapling mortality, young trees were totally absent, the main action constrained by low capacity and poor information. To address the threat and constraint, FFI team gave the training to the staff nature reserve, including involved the staff in the FFI training courses programmer, and improved their capacity through a series of practices together: action plan, monitoring, patrolling, habitat management, sustainable seeds collection, propagation, nursery management, reinforcement, co-community management, awareness raising, eco-tourism relevant to Yuanbaoshan fir. Major results of this work include: (1) the nature reserve upgraded from provincial level to national level, secure more fund and human resource; (2) the main staff and rangers of the nature reserve applied the training to the daily work of nature reserve; (3) some directive impacts including reduced pressure on the core zone in reserve compared to levels before the project started, and reducing ongoing pressure on natural regeneration; increased natural regeneration of both species; increased ownership of these activities (evidenced by work being carried out independent of outside support). Increased involvement of community members in management, evidenced by involvement in patrols, increased dialogue between community members and reserve staff and increased awareness and understanding of the reserve limits and the local and global importance of fir tree populations.

T1

P0056

Taxonomy in big data use cases

Gerald Guala

US Geological Survey

The Integrated Taxonomic Information System (www.itis.gov) is a systematically curated, standardized, machine readable, standard global hierarchy and classification of all organisms. It is global in scope for may groups, including many groups of plants, and comprehensive for vascular plants in North America. In a long term collaborative effort, Species2000 aggregates other data sets with ITIS to produce the Catalogue of Life and it is used inside thousands of other databases in government, industry and academia. Extensive web services allow the direct connection of ITIS to automated workflows and analyses. Examples of this relevant to big data will be illustrated and discussed. These incluse a recent paper (dx.doi.org/10.1371/journal.pone.0162648) showing the substantial deficit incurred when not using ITIS synonyms in literature searches and the role of ITIS in taxonomicly informed searches in BISON (bison.usgs.gov) and other large data repositories and analysis packages.

T1

P0057

Studies on the medicinal plants of gaziantep in Turkey

Salih Gucel[1]*, Mehmet Yilmaz*[2]

1. Near East University

2. Gaziantep Health Directorate

A total of 117 taxa belonging to 42 families were collected from the city of Gaziantep. Out of these 43 taxa were observed to be used for medicinal purposes with 54 applications. Lamiaceae dominated the list with 12 taxa followed by Asteraceae with 5 species and Malvaceae, Liliaceae, Urticaceae with 2 species each.The most commonly used taxa are *Origanum majorona*, *O. onites*, *O. vulgare* ssp. *hirtum*, *Salvia fruticosa*, *Sideritis perfoliata*, *Thymus longicaulis* ssp. *chaubardii* var. *chaubardii*, *Lavandula stoechas* ssp. *stoechas*, *Teucrium polium*, *Urtica dioica*, *Malva sylvestris*, *Portulaca oleracea*, *Urtica pilulifera*, *Tilia rubra* ssp. *caucasica*, *Vitex agnus-castus*, *Vitis vinifera*, *Asparagus acutifol*ius*, *Foeniculum vulgare*, *Juniperus oxycedrus* ssp. *oxycedrus*, and *Hypericum perforatum*. The taxa recorded here have been mostly used for the treatment of cough (18.6%), stomachache (13.4%), kidney ailments (11.6%), cold, analgesic, diüretic and hemorrhoid (9. 3%), injuries, tonic, abdominal pain, laxative and dyspepsia (6.9%).

T1

P0058

Exploration of moss flora along an altitudinal gradient at selected areas in the Central Province, Sri Lanka–preliminary survey

Nadeeka Prasanthi Thenna Gunawardena

National Herbarium, Department of National Botanic Gardens

Sri Lanka's moss flora is very diverse with 60 families, 190 genera, and 561 species of which 11% is endemic to Sri Lanka. As the diversity of moss flora in mountain areas are higher than the other parts of the country, great fraction of this island wide data may represents by the central province moss flora of Sri Lanka. Due to lack of expert knowledge and identification facilities in the country, most of the information regarding this vital group of plants is still at a very fundamental level. The main objectives of the present study are to study the diversity of moss flora at the different micro habitats of the selected location which have higher diversity of moss flora, describe the distribution pattern of terrestrial and epiphytic moss flora along an altitudinal gradient specially in some selected hill top areas in the central province, prepare a check list to the mosses find in Central Province in Sri Lanka and prepare and maintain of herbarium specimens of mosses collected during the study and thereby enrich the national collection of the National Herbarium, Peradeniya, Sri Lanka. The plants were collected from 8 selected localities, which have highest diversity of mosses in Central Province, namely Loolkandura (elevation; 1,200-1,500 m), Dolosbage-Watekelle-Kelabokka (elevation; 1,200 m - 1,350 m), Knuckles-Tangappuwa (elevation; 1,400-1,560 m), Seven Virgin Hills (elevation; 650-815 m), Hantana (elevation; 800-1,200 m), Hakgala Botanic Gardens and adjacent forest reserve (elevation-1,500-1,800 m), Peak Wilderness Area (elevation; 800-1,500 m) and dry phase (Mandaram Nuwara ; elevation-1,200-1,500 m) and wet phase (elevation 2,200-2,250 m) of Pidurutalaga. During the survey twenty field excursions were conducted from April to December 2014. Sampling were done in along non-permanent line transects. All first occurrence of mosses along 10 m x 2 m transects were collected for laboratory verification and field identification. Approximately 2 km in each location along the elevation gradient were sampled. Mosses find on rock surfaces, on soil surfaces and the tree trunks up to 2 m from the ground level, were collected. During this survey, a total of 53 genera and 63 species belonging to 25 families were investigated from 8 different localities of central province of Sri Lanka. Based on the species and generic abundance the most dominant family was Meteoriaceae. The species *Homaliodendron flabellatum* (Sm.) M. Fleisch., *Entodon macropodus* (Hedw.) Müll., *Aerobryopsis longissima* (Dozy & Molk.) M. Fleisch.of and *Circulifolium exiguum* (Bosch & Sande Lac.) S. Olsson, Enroth & D. Quandt were common taxa found in study areas. The highest richness of moss flora has been recorded in Loolkandura forest reserve and the epiphytic forms are found to be the most successful taxa in study areas.

T1

P0059

The structure of mangrove ecosystem in Batangas, Philippines

Mariko Gunchi
Chiba University

This study was conducted in a mangrove forest, Lobo, Batangas, Philippines aiming to evaluate relationships between mangrove vegetation structure and disturbances and to design appropriate rehabilitation strategies. A total of 25 plots of 10 x 10 m was set up over three barangays, Fabrica, Lagadlarin and Olo-Olo in Lobo in three zones: seaward, midward and landward. Dominant mangroves were identified using relative density. The species were further classified using cluster and ordination analyses. Eighteen (18) woody species were encountered and 14 of them were true mangrove species. Three zones were identified and named after the dominant species: ZONE1) Ceriops zone, ZONE2) Avicennia-Ceriops- Scyphiphora zone and ZONE3) Bruguiera- Acacia zone. Canonical correspondence analysis (CCA) illustrated strong relationship between the species density and soil nitrogen and potassium. Other factors influencing species density were soil pH and disturbances. Observed disturbances included encroachments due to agriculture, residential villages, resort construction and bioinvasion of *Acacia farnesiana.* Conservation strategy is necessary to regulate extraction of resources and sustain the livelihood of local people dependent on mangrove forest.

T1

P0060

Chromium resistant cyanobacteria and plant microbe interaction

Abdul Hameed[1], *Shahida Hasnain*[2]
1. Botany Department, Govt. College of Science
2. Women University, Multan

Fourteen chromium resistant cyanobacterial strains were isolated from polluted (*Arthrospira* sp, *Synechocystis* "MK", *S.* "P2A" and non-polluted sites *Lyngbya* sp, *Oscillatoria* "G", *O.*"BJ2", *O.* "BJ", *O.* "A", *O.* "O", *O.*"F.W", *O.* "S.A" *Anabaena* sp, *Gloeocapsosis* sp, *Cyanosarcina* from different of Lahore, Pakistan. All strains showed best growth in BG 11 medium except *Arthrospira*sp. All these strains could resist variable concentrations of hexavalent chromium in liquid medium; however strains with more sheath showed more resistance to different metallic salts and antibiotics. In majority of the strains optimum chromium (VI) reduction was observed at pH 7 and 30 °C. *Cyanosarcina* sp, *Gloeocapsosis* sp, *Synechocystis* "MK (S") and *Synechocystis* "P2A" were very efficient in the reduction of hexavalent chromium into trivalent chromium. Cyanobacterial strains were used to inoculate seeds of *Triticum aestivum*, *Zea mays* and *Vigna radiata*, which were subsequently germinated and grown under chromium stresses (300 g mL^{-1} of Cr Cl3 and K_2CrO_4) for 10 days under controlled light (195 µE m^{-2} s^{-1} for 16 hours) and temperatures (22-25°C). Under chromium stress, different growth and biochemical parameters of all three crops (*Triticum aestivum*, *Zea mays* and *Vigna radiata*) were severely affected. In general cyanobacterial inoculations caused pronounced effect in certain growth, biochemical and other parameters of plant microbe interaction, particularly after more toxin chromium hexavalent treatment. The strains *Arthrospira* sp, *Anabaena* sp, *Gloeocapsosis* sp and *Cyanosarcina* sp strains were more efficient in improving growth parameters of *Triticum aestivum Zea mays* and *Vigna radiata*. These strains were also used for pot experiments under natural conditions for *Triticum aestivum* and *Vigna radiata*. Chromium resistant cyanobacterial strains *Anabaena* sp, *Gloeocapsosis* sp and *Cyanosarcina* sp significantly increased the growth and yield parameters not only in the chromium free but also in the presence of both trivalent as well as hexavalent chromium stresses. Among these isolated cyanobacteria strains, two chromium resistant cyanobacteria strains *Synechocystis* "MK (S)" and *S.* "P2A" were isolated from the chromium contaminated area of tanneries effluents. These were first tested for probable mechanisms of heavy metal detoxification by studying the non-protein thiols, the strain *Synechocystis* "MK (S)" showed very high induction of non-protein thiols after exposure to chromium stress over other strains of same genus. Both strains used for the genetic study of this mechanism and glutaredoxin (accession No. DQ398587 for *Synechocystis* "MK (S)") and glutathione synthetase genes (accession No. DQ381962 and DQ398588 for *Synechocystis* "MK (S)" and *Synechocystis* "P2A") were amplified and sequenced. These two strains *Synechocystis* "MK (S)"and *Synechocystis* "P2A" were, finely selected for 16S rRNA gene sequence analysis and their homologies with reference strains *Synechocystis* sp. PCC 6803 were 98 and 97% respectively. They were identified as *Synechocystis* sp AHZ-HB-MK (accession No. DQ381960) and *Synechocystis* sp AHZ-HB-P2A (accession No. DQ398589). Hence these strains could be easily used in the bioremediation of chromium contaminated waste waters/ soil as well as plant growth promoting tool for various economically important cash crops.

T1

P0061

What makes the red maples and their tropical ally silver? An investigation of the phylogeny and leaf surfaces of red maples, *Acer* section Rubra, and the tropical, Southern Hemi-

sphere maple, *A. laurinum* of sect. Hyptiocarpa.

***A. J. Harris**, Susan Lutz, Jun Wen*
Smithsonian Institution, Department of Botany

Our work unites *Acer laurinum* Hassk. of section Hyptiocarpa with *Acer* section Rubra according to molecular phylogeny and cuticle morphology. *Acer laurinum* is unique among maples for having a distributional range in tropical Asia that includes locations in the southern hemisphere, especially in Indonesia and the Philippines. In contrast, *Acer* sect. Rubra has a temperate distributional range and comprises *A. rubrum* and *A. saccharinum*, which are commonly encountered in eastern North American forests, and *A. pycnanthum*, which has a scattered distribution in the mountains of Japan. Species of *Acer* section Rubra are iconic for their silvery abaxial surfaces, which are the source of the name *A. saccharinum*, or silver maple. In this study, we used chloroplast and nuclear internal transcribed spacer (ITS) to reconstruct phylogenies of *Acer* including *A. laurinum*, all species of sect. Rubra, and representative samples from 19 taxonomic sections previously based on leaf and fruit morphology. Additionally, we examined cuticle morphology of *A. laurinum* and species of sect. Rubra using scanning electron microscopy, because we expected that they would possess similar epicuticular wax configurations, which may partially explain the silvery color of the abaxial leaf surfaces. Our phylogenetic results confirm prior studies based on chloroplast DNA and ITS in showing high support *A. laurinum* as sister to sect. Rubra. Abaxial surfaces of the leaves of all species exhibited roughly similar, dramatic epicuticular wax configurations that were topped by wax splatters. Some late season leaves of *A. pycnanthum* lacked the characteristic silvery color on their abaxial surfaces and the dramatic epicuticular wax configurations. We suggest that *Acer laurinum* can be combined within *Acer* section Rubra based on our results as well as fruit and inflorescence morphology. Section Rubra may have experienced a striking increase in chromosome number associated with its temperate radiation as it has $n = 78,91$, or 108 compared to $n = 26$ in *A. laurinum*. Future work must consider whether *Acer laurinum* represents additional species, especially *A. garrettii* Craib and *A. pinnatinervum* Merrill.

T1
P0062

Metro moss: Integrating ecology and physiology to study the effects of urbanisation on plant diversity

***Alison Haynes**, Kristine French, Sharon Robinson*
University of Wollongong

Urban plant-environment interactions are complex since they involve variability across numerous factors including plant traits, the strength of urban stressors and qualities of local environments. Given the rapid rate of global urbanisation, it is increasingly important to understand these responses, for reasons of conservation, restoration and to improve the quality of urban habitat through urban greening projects. Moss, and the biocrust communities they associate with (liverworts, hornworts, lichen, cyanobacteria, fungi and bacteria) are little studied in urban environments, but may provide important ecosystem services such as preventing soil erosion, enhancing water holding capacity, reducing microdust and cycling nitrogen and carbon. Studies of higher plants suggest up to 75% of native plants of an area become locally extinct when it is urbanised. Numerous urban filters have been proposed to explain this phenomenon: habitat transformation, population fragmentation, novel characteristics of the urban environment such as pollution or disturbance, and human preference. From the limited number of surveys of urban bryophytes, patterns are mixed. Some show a prevalence of species that prefer humid, nutrient rich and disturbed soil; while others find predominant species are those with high tolerance to drought and light. Sexual reproduction tends to decrease, while vegetative reproductive effort tends to increase, especially with increasing pollution. Changes in community assemblage may occur as pollutant tolerant species increase in abundance when their more sensitive competitors cannot thrive; and communities as a whole may represent lower succession stages than in natural environments. In the relatively new discipline of urban ecology, investigation of urban filters has largely been in the form of correlation of environment characteristics with plant traits. An experimental approach is needed to determine the key drivers of urban biodiversity and to identify important thresholds. Key points of such an approach are: 1) Using gradient analyses of cover, diversity and assemblage from natural to urban areas, employing robust simple urbanisation indices such as substrate cover that can be calculated from satellite imagery and used to compare different regions. 2) Since microhabitat is important to moss communities, these analyses should be conducted along a variety of microhabitats such as pavement, grass verge, brick wall *etc.* 3) Manipulative experiments such as reciprocal transplants, common garden studies and colonisation experiments with carefully chosen species. Different stress responses following these planting treatments, combined with altered exposure to contaminants, light, water or heat may explain changes in diversity and abundance along the gradient as well as estimate the extent to which urban habitats promote adaptation and evolution. 4) Investigation of multiple stress responses may explain inconsistencies in current survey findings. Studies of species responses to two or more stresses is rare outside of agriculture - despite this being a common fact of life in the field. As well as aid conservation and restoration, findings from such research could be used to develop novel urban applications that exploit the characteristics of moss such as green roofs, stormwater management and air quality control.

T1
P0063

Influence of climate change on yield and biochemical composition grain of wheat in Tajikistan

A. Abdullaev
Institute of Botany, Plant Physiology and Genetics, Academy Science Republic of Tajikistan

In recent years the attention of many researchers is aimed at the possible impact of global climate change on ecosystems. The reason is that under the influence of global and anthropogenic factors on the environment are substantial changes in climate that may negatively affect the life of the plant. Rising temperatures cause atmospheric drought and long period of time can increase evapotranspiration, resulting in desiccation of the root layer of soil cause drought and increase the salt content on soil. This has negative impact on growth and development of agricultural crops. In this regard, studies on the physiology of resistance of crops to climate change and finding ways to improve sustainability become urgent task of modern crop production. Study of the content of starch, protein and cellulose showed that soil drought and ozone help reduce starch content and high protein content in the grain of both varieties, extreme temperatures contribute to the reduction of the level of starch and protein content. High temperatures, drought, and increased concentration of CO_2 have negative influence on the content of starch and protein content in wheat. Analysis of the cellulose content in the seeds of wheat showed that the highest of its contents was observed in the seeds of wheat, grown under conditions of drought, and it was more by 25%, than in the wheat seeds, grown under optimum water supply. In variant of drought in the seeds of the sort Alex cellulose content increased by approximately 30% than that of the control group of wheat grown under optimal water supply. Increase cellulose content observed in conditions that are unfavorable for plant in stressful conditions. The spectra of the studied proteins gliadin wheat varieties contain up to 21 components, which differ in intensity and electrophoretic mobility in PAGE. Among them are the components with the highest, medium and low intensity. Comparative analysis shown that in the protein component composition Zafar varieties grown in conditions of drought, there are intensive components that are completely absent in the same varieties grown under normal climatic conditions. Comparative analysis of the samples gliadins electrophoregram both varieties showed that, despite the presence of very similar structure, there are variations in the number and intensity of protein components. Component of the protein composition of varieties Zafar varies considerably, depending on the growing conditions, not only in intensity, but also on the presence or absence of individual components. It should be noted that almost all classes, the maximum temperature of large number of protein components is absent or present in trace amounts. The paper discusses the effect of elevated CO_2, heat, water deficit, and air and soil drought on yield and biochemical composition of wheat differing in origin. Also influence of climate change on productivity and qualities of grain wheat growing in different climatic zones is discussed.

T1

P0064

Collection, germplasm introduction and conservation of *Allium* in Xinjiang and its collection garden construction

*Hikmat Hisoriev, Guan Kaiyun Guan, **Jiang Li***

Xinjiang Institute of Ecology and Geography, Chinese Academy of Sciences, Urumqi, China

The very diverse genus *Allim* L. shows a nearly exclusive distribution across the northern hemisphere with a main centre of diversity in Southwest, Central Asia and in China. Since 8,000 years, common onion has been probably used by men kind as edible, spice and vegetable. Currently, rhizomes or bulbs, extracts of those or green parts of several species like *A. cepa, A. sativum, A. stipitatum, A. rosenorum, A. rosenbakhianum, A. oschaninii, A. chinense, A. fistulosum, A. porrum* and *A. tuberosum* are intensively used by the native Asian populations. Also today *Allium* crop species have world-wide importance as spice, medicinal, and food plants. Thus, they own an eminent economical importance and are involved in intensive breeding programs world-wide in order to improve yield, quality, and agriculturally important characters. There are 138 species and 24 varieties (50 endemic, five introduced) of *Allium* species were found in China. Our short investigation on *Allium* collection mission in different regions of Xinjiang Province of China which belong to Tianshan and Altai mountains (around Urumqi, Bozhou, Jinghe, Yining, Zhaosu, Teksi, Xinyuan, Hami, Balikun, Qitai, Fuyan, Altai, Buerjin, Emin, Tacheng and Bole) shows that totally in natural landscapes of Xinjiang Province distributed 50 species of *Allium.* Most of these species (up to 23 species) were found in around Yili, Urumqi and Altai territory. Among them 22 species were found in around Urumqi, belonging to Tianshan Mountains, 23 species – in around Yili (Tianshan territory), and 23 – in Altai Mountains. Between 50 wild species of *Allium* in Xinjiang, several species, such as *A. ramosum, A. atrosanguineum, A. polyrhisum* and *A. przewalskianu, A. polyrhisum, A. senescens* have wide distributed area in China. But next species, as *A. carolinianum, A. hymenorrhizum, A. pallasii, A. grisellum, A. caricoides, A. schoenoprasum* and *A. schoenoprasoides* are presented with a small local populations in several places only in Xinjiang region. By chaotic pasture and anthropogenic pressure distribution area most of them became very small, and at present above mentioned species should be named as rare and endangered plants, and to be included in next Red Data Book of Xinjiang Province or Red Data Book of China. Finally in 2016 more than 70 Accessions of newly collected wild *Alliums* in Xinjiang region and collected also from Tajikistan and Uzbekistan were planted in newly established Yili Botanical Garden for future monitoring, collection their germplasm, and bio-morphological investigations.

T1

P0065

Tajikistan botanical gardens as a valuable fund of the aborigine and introduced plant species

Hikmat Hisoriev

Institute of Botany, Tajik Academy of Sciences, 27 Karamova str., Dushanbe-17, Tajikistan, 734017

There are five botanical gardens in Tajikistan, which situated in different zones: Khudjand Botanical Garden in North Najikistan (8 ha, 350 m above sea level); Kuljab Botanical Garden in South Tajikistan (14 ha, 650 m above sea level); Central Botanical Garden in Dushanbe – the capital of the Republic of Tajikistan (41 ha, 800 m above sea level); Varzob Mountainous Botanical Station in Central Tajikistan (1,000 ha, 1,050-2,600 m above sea level); Pamir Botanical Garden in East Tajikistan (near 100 ha, 2,200 m above sea level). The main scientific research of the gardens are exploration of plant diversity in natural landscapes of Pamir-Alai Range, including endemic, rare, endangered species and ecosystem conservation and investigation on adaptation of introduced from another climatic zones plant species in different botanical gardens. Tajikistan botanical gardens are a permanent research and trial facility designed for multi-faceted research efforts of the National Botanical Institute. It always had its own peculiar highly articulated body of scientific research work, whose core consisted of the study of forest vegetation of Tajikistan and Pamir-Alay as a whole, establishment of floristic (species) composition of the country's dendroflora, the study of morphological and biological features of trees and bushes of Pamir-Alai, development of ways to create on high altitudes highly efficient fruit and forest and other erosion control plantings. A critical reassessment of the native flora has shown that forests of Tajikistan are surprisingly diverse in their species composition. The whole collection of more than 2,000 species of trees and bushes have already collected and cultivated in four parts of Central Botanical Garden: Ancient Mediterranean plants (639 species), East Asia plants (837 species), North American plants (488 species), Middle Europe plants (226 species). Many hitherto undiscovered species have been detected, including *Acer ovczinnikovii, Myrtama elegans, Prunus tadshikistanica, Pyrus tadshikistanica, P. cajon, etc.* But in multiple cases there were no clear-cut boundaries between wild and domesticated plants. This led to a conclusion that the range of fruits cultivated on high altitudes grew on site by means of cultivation of their wild counterparts. The mountains of the Pamir-Alay Range are recognized as one of the global centers of the origins of cultivated, as well as fruit plants. In this regard, the study of the issues and the development of scientific basis for the conservation of food, medicinal and forage plants of the region is one of the major environmental challenges. Up until now, many wild and highly threatened species of food, medicinal and edible plants, are characterized by a narrow ecological amplitude of distribution (as a stenobionts), and are not developed specific actions or recommendations for growing or cultivating in experimental pilots in botanical gardens, and transfer to the place of their natural habitat. In this regard, there is an urgent need to conduct a complete inventory renewal, management and sustainable use.

T1

P0066

Freshwater algal diversity of the South-Tajik Depression in a high-mountainous extreme environment

Mario Boboev

Khatlon Scientific Center, Tajik Academy of Sciences, 12 Shark str., Kulob-25, Tajikistan, 734025

The effects of altitude on freshwater algae distribution is widely discussed in the recent literature but still remains a problem. Our investigation in freshwater algal communities' comparisons and infraspecific levels of comparison gives us a method for a productive approach when there are visible climatic responses in the eco-regional, river basin floras. South-Tajik Depression algae from low- to high mountain habitats of the Vakhsh, Kafirnigan, Kyzylsu, Yakhsu, Tairsu river basins, Muminabad and Selbursk reservoirs have been studied during last 50 years. The aim of the current study was to reveal the response of algal communities in this region to altitude and altitude-related climatic conditions of its habitats, and we tried to implement diverse bioindicational and statistical methods, which represented some new approaches in freshwater algal diversity analysis. The research on algal communities' response to altitude and altitude-related climatic conditions is sparse, and a full, accessible list of algae from the water bodies of the South-Tajik Depression has not been available yet. We compiled an algal species list of 1,190 species and varieties that were revealed by us or listed in references in the South-Tajik Depression water bodies. The altitude gradient of the studied area is about 2,500 m above sea level. An altitude-related algal diversity analysis was done with help of statistics and bioindication. The Willis curve for the distribution of species to genera had a trend line of $R^2 = 0.92$. A multivariate analysis was used for quantitative estimates of the ecological tolerance ranges of critical species and for detecting cardinal factors and trends at local water bodies to global levels of biodiversity evolution. Our analysis shows that species distribution was strongly affected by altitude and altitude-related climatic variables. Statistical methods revealed that temperatures stimulate algal species diversity, while precipitation and altitude suppress it. One of the floristic complexity criteria is the average species richness of the algal genera or the infraspecies to species ratio, which increased from 1.12 to 1.20 and reflects increases in structural complexity with altitude. Therefore, high altitude stresses algal communities and stimulates species polymorphism as a compensatory mechanism for algal species survival. A comparison of the Tajik Depression freshwater algae with other regional algal floras of high mountains shows that altitude and altitude-related climatic variables play a major role in species distribution: temperature stimulated algal diversity, while precipitation and altitude suppressed it. The study of high-mountain algal flora showed increases of structural complexity with altitude. This means that many species per genus in lower-altitude habitats showed evidence of stable development, whereas a high ratio of monotypic genera can be related to climatic, or anthropogenic impacts such as what was seen in the high-altitude communities of the South-Tajik Depression: high altitudes not only stress algal communities but also stimulate species polymorphism as a compensatory mechanism for the survival of algal species.

T1

P0067

Diversity of spore plant in Pamir region (Tajikistan)

***Parvina A. Kurbonova**, Hikmat Hisoriev*

Institute of Botany, Tajik Academy of Sciences, 27 Karamova str., Dushanbe-17, Tajikistan, 734017

The Pamir (which means "roof of the world" in the Tajik language) is unique not only geographically, but its remarkably rich and highly endemic vegetation including many plants of great promise for about 64,000 km^2 (43% of Tajikistan Republic). Tajikistan's Pamir borders on Afghanistan (> 500 km) on the South, China on the East (430 km), Kyrgyzstan on the North. The complex relief of Pamir a result of prolonged history of geological formation – is characterized by Docembrian formations in the Southwest and Paleozoic-Mesozoic formations in the Southeast. Pamir is divided geographically into two parts, western (Badakhshan) and eastern (Pamir). There are more than 10 big and hundreds small lakes and a lot of rivers, springs, thermal and mineralized water bodies in Pamir. In spite of this, the climate of western Pamir belongs to the northern dry subtropical zone of the Mediterranean area and has a maximum annual precipitation of 300 mm, but the climate of eastern Pamir belongs to the Central Asian area, (100 mm), continental high mountainous plateau with cold summers and stern winters. There are four zones of vegetation in western Pamir: ephemeral deserts, wormwood steppes and feather-grass, mountainous xerophytes, and cryptophytes (uplands). There are near 2,000 species of vascular and more than 1,500 species of spore plants (algae, lichens, bryophyta and fungi) in Pamir, including 180 endemic, rare and endared species. Among the endemic plants more than 80 species listed in the Red Data Book of Tajikistan (2015). With the current destruction of habitats occurring at critically rapid pace throughout the western and eastern Pamir, and also the global climate change the degree of potential loss of species makes this region a high priority for study of biodiversity and development of conservation programs. During the 27 years since the first publication of the Red Data Book of Tajikistan (1988), extraordinary economic events in Pamir degraded large areas of the landscape and have adversely affected environmental crises. Many trees and shrubs were cut down for firewood, and many areas were cleared for pastures to raise livestock. As a result, the Pamir region currently faces a critical situation. A few botanical surveys or inventories have been conducted for a last period, and it is feared that many plant species have disappeared or are in danger of extinction. There are 1,100 species with more than 300 varieties of algae were found in water bodies and in soils of Pamir, belonging to 10 Division (*Cyanophyta, Euglenophyta, Cryptophyta, Dinophyta, Chrysophyta, Xanthophyta, Bacillariophyta, Chlorophyta, Charophyta, Glaucophyta, Rhodophyta*), 19 classes 91 families and more than 220 genus. Among the 91 families of whole algae flora, such families as Oscillatiriaceae, Euglenaceae, Naviculaceae, Cymbellaceae, Scenedesmaceae, Desmidiaceae, Closteriaceae and Ulothrichaxeae were dominant. There are more than 15 genus are as a dominants: *Navicula, Synedra, Cymbella, Pinnularia, Euglena, Trachelomonas, Oscillatoria, Phormidium, Ulothrix, Scenedesmus, Cosmarium, Closterium etc.*

T1

P0068

Plant biodiversity conservation in Tajikistan

Hikmat Hisoriev

Plant Biodiversity Conservation in Tajikistan

Tajikistan is a typical mountainous country and covers the whole territory of Pamir-Alai Range with a most complex physical and geo-morphological structure with elevation from 300 to 7,495 m above sea-level and characterized with a wide range of ecological conditions and has a correspondingly rich natural flora (more than 9,500 species of spore and vascular plants) with a wide range of plant ecosystems. There are 2,200 species of Algae, 2,000 - Fungi, 600 species - Lichens, 400 species - Bryophytes and more than 4,500 species of vascular plants found in Tajikistan. Among them more than 800 species are endemic and 267 species (or 2.75%) are included in Red Data Book of the Republic of Tajikistan. The vast majority of Tajikistan plants are edible, medicinal and food species. In the whole pasture and meadows ecosystems occupy more than 24% territory of country; forest ecosystems – 3%; cultivated or agro-ecosystems – 6% (irrigated and non irrigated); nearly 67% are unused natural landscapes, which covered with forests in the past and now damaged by aridization and the anthropogenic pressure, rocky and stone slopes covered with a sparse vegetation, snow-patches and glaciers. Lacks of knowledge about the diversity of life in Pamir-Alai and the effect of scientific activities on this invaluable resource pose the most significant barrier to their management and conservation. The current destruction of habitats occurring at critically rapid race throughout the country, the degree of potential loss of plant species and whole landscapes makes this region a high priority for future monitoring of a whole ecosystems and development of conservation programs. Grazing by animals, as well as cutting of trees, shrubs, uprooting of medicine and food plants have greatly reduced the coverage of the ecosystems and its structure. Notwithstanding the increase of desert and xerophytic ecosystems in the contemporary period and also in near future great significance is put to such ways of realization of species formation as aridization of vegetation in Pamir-Alai Range. During last 50-70 years in the mountainous territory of Zerawshan-Turkestan, Gissar-Darwaz, Karategin-Peter First, in pasture landscapes the population of Traganthoid associations or thorn (bur) plant species, are formatting xrephyllous and hemixerophyllous ecosystems, mainly with thorn and cushion plant species (*Onobrychis echidna, Astragalus* species (sect. Aegacantha), *Acantholimon diapensioides, A. hedinii, Cousinia pannosa, C. franchetii etc.*) have already increase. Furthermore, under the climate change and anthropogenic pressure forest's ecosystem zone in country became narrow: upper forest line came down, but the reverse, lower forest line went up (under anthropogenic pressure). Thus, divers ecological and geographical conditions, ranging from hot deserts and humid subtropical regions to high alpine regions, as a Pamir-Alai Range have favored the establishment of a complex and varied composition and structure of ecosystems, are also greatly influenced by a long history of over-exploitation which has led not only to the almost complete loss of forests but also to widespread degradation of formerly rich woodland and semi-arid ecosystems.

T1

P0069

Climate change, ecological calendars and the Yulong Mountain

Jianchu Xu[1,2], *Pete Hollingsworth*[3], *Antje Ahrends*[3]

1. Kunming Institute of Botany

2. World Agroforesty Centre

3. Royal Botanic Garden Edinburgh

Climate change can have particularly marked effects in mountain systems which may then lead to profound consequences for local communities and livelihoods. Although climate change is a global phenomenon, its impacts occur at all scales and will vary in nature and strength from place to place. Ecological calendars are traditional knowledge systems based on phenological indicators that have served to link biological and cultural activities to seasonal change. In this presentation we will explore the role of ecological calendars coupled with climate change projections as a strategy for integrating traditional indigenous knowledge with climate change science to facilitate dynamic adaptive responses by local communities.

T1

P0070

The spatial distribution of *Primulina* (Gesneriaceae) favorable habitats and the predication of the habitat changing in the future

Xin Hong[1,2], *Fang Wen*[2,3], *Yi-Gang Wei*[2,3]

1. School of Resources and Environmental Engineering, Anhui University

2. Gesneriad Conservation Center of China (GCCC), Guilin Botanical Garden, Guangxi Institute of Botany
3. Guangxi Key Laboratory of Plant Conservation and Restoration Ecology in Karst Terrain,

Thirteen eco-environmental factors, including elevation above sea level, slope, aspect, minimum temperature of coldest month, maximum temperature of warmest month, average temperature of growth, precipitation of growth, Isothermality, mean Temperature of wettest quarter, mean Temperature of direst quarter, precipitation of direst month, precipitation of warmest quarter, precipitation of coldest quarter, were used to build the assessment system of habitat niches for *Primulina* (Gesneriaceae) in South China. Then we calculated the weight of various ecological factors which affected habitat suitability through the application of maximum entropy model. Lastly, the multiply models and geographic information system (GIS) were joined up and used to seek the spatial distribution of suitable habitat range in *Primulina*. At the same time, we also predict and assess the spatial distribution of habitat suitability of *Primulina*, and the optimum regions for optimal growth of the *Primulina* in China in the future.

T1
P0071

Study on distribution and ecological effect of *Usnea* lichens in three parallel rivers based on niche modeling

Lihui Hu, *Deshun Zhang*
Tongji University

Usnea Dill.ex Adans.belongs to the family of *Usnea* sneaceae. Most of the species are widely distributed and the morphological characteristics are highly varied. As the main food source of *Rhinopithecus bieti*, it is of great significance to *Rhinopithecus bieti*, which accounts for more than 60% of the main food sources, and 91% of their feeding time. In addition, *Usnea* lichens is a typical environment sensitive plant, which has obvious reaction to environmental and climatic factors such as air pollution, climate change and altitude. Therefore, by analyzing the geographical distribution pattern of Usnea lichens in the three parallel rivers and predicting its potential suitable area, the influence of environment and climatic factors on its geographical distribution pattern is discussed. Based on its distribution data in China, combined with climate, altitude, vegetation and other environmental data, the use of MAXENT modeling and ArcGIS software to simulate the combination of pine in the three parallel rivers of the potential suitable area. Combined with field investigation and specimen data, the geographical distribution was drawn with the elevation layer and the vegetation cover layer as the base map. Moreover, the predictive data was tested and verified by receiver operating characteristic (ROC), and the influence of the main environmental variables on distribution of *Usnea* lichens was analyzed with Jackknife test. According to the results of the study, the annual average precipitation, temperature seasonal variation, air pollutant index, standard normalized vegetation index, and altitude are the main environmental factors affecting the spatial distribution of *Usnea* lichens. Increase the coverage rate of trees, reduce the habitat fragmentation caused by human disturbance, and ensure that air cleaning is the main measure of *Usnea* lichens protection. According to the above main ecological environment factors, the optimum area, suitable area and unfavorable area of the distribution distribution of the pineapple in the Sanjiang parallel area are divided. With the interference of global climate change and human activities, the potential suitable area is obviously narrowed. Distribution area decreased. According to the above main ecological environment factors, the optimum area, suitable area and unfavorable area of the distribution of *Usnea* lichens in the three parallel rivers area are divided.With the global climate change and human activities of the interference, the potential suitable area was significantly reduced, the optimal distribution area decreased.

T1
P0072

Genetic diversity of the Bahamas *Pinus caribaea* populations in south subtropical zone of China estimated by SSR markers

Yichen Zong[1], *Ping Huang*[1], *Yongqi Zheng*[1], *Furong Lin*[1], *Chengrong Xu*[2], *Shaoying Zou*[2]
1. Research Institute of Forestry, Chinese Academy of Forestry
2. Research Institute of Forestry, Hepu County

Pinus caribaea var. *bahamensis* is native to Bahamas in the Caribbean, and it is a kind of fast-growing conifer in tropical zone. *Pinus caribaea* var. *bahamensis* was introduced to China in 1983, and the provenance trial of this species was conducted 1990s. This species has straight trunk, narrow crown, natural pruning, good resistance, which are advantage of condensed planting. *Pinus caribaea* var. *bahamensis* is a good selection of planted forest in the south subtropical and tropical area in China. Aim to enhance the understanding of *Pinus caribaea* var. *bahamensis* genetic resources which were introduced and conserved in the field. In the study, eight SSR markers was screened and used to analysis the population genetic diversity of *Pinus caribaea* var. *bahamensis*. The result showed that forty-eight alleles were detected in eight loci, and most of loci were Hardy-Weinberg equilibrium. And the fisher test result showed that all loci were genetic linkage disequilibrium and it suggested that genetic linkage between loci did not exist. Moreover, the population genetic parameters indicated that there was not wide variable between populations. However, it was worth noting that the observed heterozygosity was less than expected heterozygosity, this result suggested that heterozygote deficiency was main problem in the exotic populations of *Pinus caribaea* var. *bahamensis*. The results of population genetic differentiation showed that the intrapopulation variation was more dominant to interpopulaiton, the Fst was 0.094 and 0.006 respectively. This indicated that no significant genetic drift happened in the exotic populations. Based on the genetic distance, seven different provenance populations were divided into two groups, and it is helpful to select breeding material. On the basis of existing genetic resource of *Pinus caribaea* var. *bahamensis*, these results were able to guide us to make more crossbreeding work in the future.

T1
P0073

Continuous existence of Chinese pepper suggests limited impact of the cooling trend on climate in southwestern China

Yong-Jiang Huang[1], *Hai Zhu*[1], *Tao Su*[2], *Xue-Ping Ji*[3], *Zhe-Kun Zhou*[1,2]
1. Key Laboratory for Plant Diversity and Biogeography of East

Asia, Kunming Institute of Botany, Chinese Academy of Sciences
2. Key Laboratory of Tropical Forest Ecology, Xishuangbanna Tropical Botanical Garden, Chinese Academy of Sciences
3. Department of Paleoanthropology, Yunnan Institute of Cultural Relics and Archaeology

The late Cenozoic cooling trend has profoundly influenced the plant diversity and distribution worldwide, particularly in middle and high latitude regions, e.g., Europe. Its impact on the vegetation of lower latitude regions, e.g., southwestern China known as a biodiversity hotspot, however, remains debatable. Pantropical plants such as *Zanthoxylum* L. (Chinese pepper) preferably grow in warm areas of southeastern Asia, and thus its biodiversity may reflect a warm climate. In this study, fossil seed remains of *Zanthoxylum* are studied/consulted from five localities in southwestern China that extend from early Pleistocene into early or middle Miocene in age. These fossil records, together with the modern high diversity and wide distribution of *Zanthoxylum* in southwestern China, imply that the genus likely has continuously existed in the region at least since the early to middle Miocene. In contrast to the post-Pleistocene disappearance of *Zanthooxylum* in Europe, in southwestern China the genus has survived through the harsh Quaternary ice house climate to today. This may allow us to further presume that the dramatic Quaternary cooling trend has had limited impact on the climate and vegetation in southwestern China.

T1

P0074

Studies on plant community structure and plant diversity of mountain areas in Shenzhen

Yu-Yuan Huang[1], *Li-Jun Yang*[2], *Xin-Fan Yu*[2], *Kang-Sai Zhao*[2], *Hong Liang*[2], *An-Qiang Dong*[1], *Shi-Zhen Lin*[1], *Yu-Feng Zou*[1], *He-Yin Wang*[1], *Yong-Heng Chen*[1], *Hao Liu*[1]
1. College of Life Sciences, Zhongkai University of Agriculture and Engineering
2. Shenzhen Environmental Monitoring Center

The communities structures and plant diversity in Yangmeikeng of Qiniang Mountain, Chiao of Tianxin Mountain, Lianhua Mountain, Yingrenshi of Yangtai Mountain and Xiaonanshan Mountain, Shenzhen, China were carried out research. The aim is provide theory reference for the plan, management and optimization of ecosystem of local region, and prove theory references to more argument problem on which kind community is biodiversity higher? Is natural forest, artificial forest or more artificial disturbed forest higher? Research according to the several facts chose and divided community: (1) Natural forest; (2) half-natural forest; (3) Artificial disturbed forest or artificial forest. In very community, measured the height of plant, coverage, DBH (tree), density and frequentness *etc.* indices; meanwhile, calculated Simpson index, Shannon–Wiener indices and Odum, Menhinnick richness indices and Pielou evenness index, Sorenson β-diversity, and the richness indices of family and genus. The result showed that the structure and plant diversity of natural forest of Yangmeikeng and Xiaonanshan Mountain obvious better than artificial disturbed forest or artificial forest, the half-natural forest were the middle level of the forests of Lianhua Mountain and a community of Yingrenshi hill area. In the indices of height, DBH and coverage of tree layer, natural forests and half-natural forests were more obvious higher than that of artificial disturbed forest or artificial forest; in some forests of artificial disturbed forest or artificial forest, though their richness indices of family and genus were a slightly higher, but this contribution major was herb layer plants, and some shrub layer plants. Analysed three kinds communities, the tree species more of natural forest, every diversity index value highest, half-natural forest was second, artificial disturbed forest or artificial forest was lowest; in shrub layer, although the some indices of artificial disturbed forest or artificial forest were higher, but the values were not obvious difference to natural forests; in herb layer, some β-diversity indices of artificial disturbed forest or artificial forest were higher than some natural forests, but the integral values of the diversity indices of natural forest and half-natural forest were obvious higher than the former. These results showed that artificial disturbed forest or artificial forest maybe can temporary raise the plant diversity of community, the major contribution is herb and some shrub plants, the tree layer species diversity is very low, although thus, the integral values of diversity indices still are natural forest, half-natural forest, artificial disturbed forest or artificial forest. However, in tree community, because tree biomass is more than 5 times, some more than 10 times than that community was composed by shrub and herb plants, thereby the ecological effects of the former are high many times than the latters. Therefore, author thought that in respect of vegetation protection, restoration and optimization can not only see the value which one layer plant diversity raise, it should be mainly to see whether the diversity raise of high and large tree plant species, and the community structure of tree composition is well in the area where is can grow trees.

T1

P0075

Evaluating the supercritical extraction effects on the chemical compositions of the extracted nutmeg (*Myristica fragrans*) oil using GC/TOF-MS

Ahmad H. Ibrahim[1], *Sawsan S. Abdulhameed Al-Rawi*[1], *Mohd. Omar Abd Kadir*[2]
1. Biology Department, University of Zakho, Duhok, Kurdistan Region, Iraq
2. Biology Department, University of Zakho, Duhok, Kurdistan Region, Iraq

Gas chromatography time of flight mass spectrometry (GC/TOF-MS) is becoming an attractive and alternative full-spectrum technique against others traditional instruments. In this study, the GC/TOF-MS was used to evaluate the effect of supercritical parameters, temperature, and pressure on the chemical composition of the nutmeg extraction yields. A general factorial design of 3x4 was used. The first independent studied variable was extraction pressure (MPa) with 4 levels as follow (20.7, 27.6,34.5 and 41.4 MPa). The second independent variable was a temperature with three levels (40, 50 and 60 °C). The response is the oil yield at each design point was recorded. A total of 12 samples of nutmeg oils were obtained from the extraction and then were analyzed for it chemical components using GC/TOF-MS. A large list of compounds was identified and matched to the NIST library with a hit over 80% on the basis of the specific mass determination of their molecular ions and their major fragments. All the overlaps peaks were separated. The aromatic ethers groups were successfully identified as the main group in the supercritical extracts with the presence of Myristicin as the highest peak. Moreover, all the extracted oils possess similar chemical components but varied in each compound concentration. However, increasing the extraction temperature increased the presence of many compounds in the extracted oil such as elemicin, eugenol, and farnesene. Increasing the extraction pressure increased the presence of pinene, terpinene, thujene, phellandrene. In addition, nutmeg oil was extracted using Soxhlet and was the extracted yield was compared to the supercritical extraction in terms of chemical composition. The result showed the presence of the aromatic ether group in the Soxhlet extraction was lower than the concentration of the same group in the supercritical extractions using GC/TOF-MS. This result reveals that the extraction condition and method has the great effect on the chemical compositions of the extracted oil and on the concentration of each compound. Moreover, the use of GC/TOF-MS confirmed the effect of supercritical extraction parameters such as temperature and pressure represented by the various presences of chemical compounds. This finding indicates and confirms the potential use of the supercritical extraction method in producing richer extract and fractions.

T1
P0076

Bryophyted diaspore bank of peatlands

Nele Ingerpuu, *Kai Vellak, Tiina Samson*
University of Tartu

The area of peatland is declining drastically in the whole world. Therefore conservation and restoration of peatlands has become more actual. The main peat forming plants in temperate regions are bryophytes. Restoration capacity of disturbed peatland depends largely on the available diaspore bank. Our study includes three types of peatland communities in Estonia: rich and poor fens, and raised bogs, three study sites in each community type. The above ground vegetation was registered in four 1 m^2 plots. The thickness of living peat was measured and underneath it diaspore bank was sampled from two depths: upper peat layer (0-7 cm) and deeper peat layer (7-14 cm). From each community type 48 samples were collected. The diaspore bank was grown in controlled conditions in growth chambers for six months. The results showed that the number of species in bryophyte diaspore bank was lowest in raised bog and higher, but quite similar in poor and rich fens. This corresponds well with the species richness of the above ground flora of these communities. The composition of the diaspore bank consisted mainly of the species belonging to the above ground flora of the same community. This indicates that the diaspore dispersal might not be very effective and the quick restoration of the bryophyte cover of disturbed communities relies greatly on the local diaspore bank.

T1
P0077

Evolutionary distinctiveness and conservation priority of the endangered *Najas ancistrocarpa* (Hydrocharitaceae)

Yu Ito[1], *Chigaya Enju*[2], *Jie Li*[1], *Norio Tanaka*[3]
1. Xishuangbanna Tropical Botanical Garden, The Chinese Academy of Sciences
2. Tsukuba University
3. National Science Museum of Nature and Science, Tokyo

Phylogeny is vitally important to evaluate evolutionary distinctiveness and conservation priorities of species. *Najas ancistrocarpa* is among the ca. 30–40 species of the aquatic plant genus *Najas* (Hydrocharitaceae) and is one of the rarest and least known ones in East Asia. We used a recently discovered collection of this nationally endangered species in Japan to assess whether *N. ancistrocarpa* is a distinct species or an abnormal form of other *Najas* species and to evaluate its conservation priority in a phylogenetic context. The sample set of our previous study exploring worldwide phylogeny of the genus with *N. ancistrocapra* in it was used to delimit species boundaries in Najas and obtain conservation priority scores calculated by multiplying evolutionary distinctiveness and global extinction risks converted from Red List categories. Species delimitation analysis objectively assessed that *N. ancistrocarpa* is phylogenetically distinct. Phylogenetic conservation prioritization ranked *N. ancistrocarpa* as the second prioritized *Najas* species. Overall, in terms of biodiversity conservation, *N. ancistrocarpa* represents an important Najas species in Japan and probably in the world.

T1
P0078

Flower color and pigments in *Amherstia nobilis* endemic to Myanmar

Tsukasa Iwashina[1], *Nobuyuki Tanaka*[1], *Mu Mu Aung*[2], *Tomohisa Yukawa*[1]
1. Department of Botany/National Museum of Nature and Science
2. Forest Research Institute/Ministry of Natural Resources and Environmental Conservation

The genus *Amherstia* (Leguminosae) consists of only one species, *A. nobilis* Wallich and endemic to Myanmar. The species is widely cultivated and generally regarded as one of the most beautiful of all flowering trees. However, its wild population in Myanmar is not known. Flower color is orange-red and its pigment has been reported as anthocyanin, pelargonidin (Beale *et al.*, 1941), but detailed data of the flower pigment were hardly known. In this survey, the authors fully analyzed their flower pigments. The flowers were collected in the Tsukuba Botanical Garden, National Museum of Nature and Science, Japan, and University of Yangon, Myanmar. Fresh flowers were extracted with MeOH:HCOOH (97:3). The extracts were applied to prep. PC using solvent systems, BAW (*n*-BuOH/HOAc/H_2O = 4:1:5, upper phase) and 15% HOAc and prep. HPLC and pigments were finally purified by Sephadex LH-20 column chromatography using solvent system, MeOH/H_2O/HCOOH (70:25:5) for anthocyanins or 70% MeOH for flavonols. Each three of anthocyanins and flavonols were isolated from the flowers. Major anthocyanin was cyanidin 3-*O*-glucoside (absorption maxima λmax 325 nm) with minor cyanidin 3-*O*-glucuronide and cyanidin 3-*O*-pentoside, in spite of absorption maxima of intact flower of λmax 506 nm. However, the presence of pelargonidin and carotenoids were not shown in this survey. Three flavonols were identified as isorhamnetin 3,7-di-*O*-glucoside, isohamnetin 3-*O*-glucoside and isorhamnetin

7-O-glucoside by UV, LC-MS, acid hydrolysis and direct TLC and HPLC comparisons with authentic samples. As the results, in spite of flowers of *A. nobilis* expressed orange-red color, anthocyanins in the flowers were cyanidin glycosides (red color). The factor of hypsochromic shift of the flower color may be due to lower pH of the vacuole. We now survey about its factor in progress.

T1
P0079

Circumscription of taxa in *Scutellaria multicaulis* Boiss. complex based on molecular and morphological traits

Keivan Safikhani[1], ***Ziba Jamzad***[2], *Hojatolah Saeidi*[3]
1. Agricultural and Natural Resources Research Center, Agricultural Research, Education and Extension Organization (AREEO)
2. Research Institute of Forest & Rangelands, Agricultural Research, Education and Extension Organization (AREEO)
3. University of Isfahan

Scutellaria L. (Scutellarioideae) is a genus of Lamiaceae and grows both in Old and New Worlds. The Irano-Turanian Region, particularly Central Asia and Afghanistan, is the center of maximum diversity for the genus. However, Eastern Mediterranean and the Andes are the second centers of its speciation. In the Plant List data base there are 465 accepted species names for *Scutellaria*. It is among the medicinal plants list with a wide range of pharmacological actions, such as antitumor, anti-angiogenesis, hepatoprotective, antioxidant, anticonvulsant, antibacterial and antiviral activities. There are 27 species of *Scutellaria* in the flora of Iran of which 9 are endemics. Most species have limited but a few have a wide distribution range in Iran. *Scutellaria multicaulis* Boiss. is one of those with wide range of distribution including south east, centre and west of Iran. Due to the wide range of distribution, rather high variation in morphological features is known for this species. Three subspecies of this taxon were reported in Flora Iranica, including subsp. *loringensis* (Rech. f.) Rech f. endemic to Afghanistan, subsp. *multicauls* endemic to Afghanistan and Iran and subsp. *koelzii* (Rech. f.) Rech f. endemic to Afghanistan and Pakistan. The morphological variation in *Scutellaria multicaulis* was studied with the aid of molecular makers including nr DNA ITS and chloroplast *trn*L-F sequences to be able to define the taxa within the mentioned species complex. Phylogenetic analyses were performed on the aligned data matrices separately using the model suggested by MrModeltest 3.7. The estimations of phylogeny were done with Bayesian inference. Based on morphological data two main groups were recognized, the taxa with yellow flowers and short retrorse hairs in lower and middle parts of stem and simple and glandular patent hairs in inflorescence region (group 1) and those with dark violet flowers with yellow spot in lower lip and short, simple antrorse hairs, intermixed with sessile glands in inflorescence region (group 2). Furthermore, morphological variations could be defined within group 2. The nucleotides sequences of ITS and *trn*L-F markers represent differences between the two groups. Based on morphological and molecular data three new taxa were identified and described.

T1
P0080

Taxonomic study based on database of Iris (Iridaceae) specimens from Middle Asia in LE

Chang Gee Jang[1], *Nu Ree Na*[1], *Jeong Mi Park*[2]
1. Department of Biology Education, Kongju National University
2. Plant Resources Division, National Institute of Biological Resources

The family Iridaceae is a well defined assemblage of approximately 1,800 species belonging to the *Liliidae (*the monocotyledons). The distribution of this family is worldwide, the family has a marked concentration in Africa south of the Sahara, including Madagascar. About genus *Iris*, is composed of ca. 225 to 300 species which is distributed in mainly northern hemisphere. This genus is perennial herb and is taxonomically the member of Iridaceae under Asparagales. The type of rhizome, inflorescense, presence and hairy of outer perianth segments, and shape of fruit and seeds are used as the taxonomical characters in many systematic study. There are about 60 species in the genus Iris L. in the flora of Middle Asia (khassanov & Rrakhimova 2012) and more than a dozen new species were described recently. A. I. Vvedensky (1941,1971) made taxonomical revision of Central Asian Iris, which treated *Juno* was treated as a separated genus. Also *Iris* has been revised in "Conspectus Florae Asiae Mediae" by O. Tscherneva (1971). 36 species of *Iris* are endemic for this area. R. Kamelin (1981) and later on T. Hall and A. Seisums (2011) made several new combinations from *Juno* restoring to *Iris*. Middle Asia can be treated no doubt as one of the largest centers of biodiversity of genus *Iris* L. Herbarium of the Komarov Botanical Institute (LE) is one of the biggest herbarium in the world. LE has specially the largest amount of specimens from Middle Asian area, such as Uzbekistan, Kazakhstan, Tajikistan, Kyrgyzstan, and Turkmenistan. Among the 88 taxa which were recorded in Flora of USSR, the 46 taxa are treated from Central Asia, which means Middle Asia and western part of Kazakhstan territory. Based on results of survey Herbarium of the Komarov Botanical Institute, specimens of 61 taxa were deposited in Middle Asia section. Some did not recorded as Middle Asian taxa in Flora of USSR, others which treated as Middle Asian taxa in Flora of USSR have no specimens in this section. The former included like as *I. sibirica* L., *I. setosa* Pall. ex Link, *I. laevigata* Fisch., *I. pseudacorus* L., *I. ludwigii* Maxim., *I. dichotoma* Pall., *I. pumila* L., *I. potaninii* Maxim., *I. tigridia* Bge in Ldb., *I. germanica* L. *etc.* The latter composed such as *I. sogdiana* Bge Beitr., *I. ewbankiana* Foster, *I. karategina* B.Fedtsch., *I. winkleri* Rgl., *I. maricoides* Rgl., *I. nicolai* Vved., *I. rosenbachiana* Rgl., *I. popovii* Vved., *I. parvula* Vved., *I. tadshikorum* Vved., *I. linifolia* O.Fedtsch., *I. vvedenskyi* Nevski, *I. narynensis* O.Fedtsch., *I. narbutii* O.Fedtsch., *I. kuschakewiczii* B.Fedtsch., *I. subdecolorata* Vved., *I. coerulea* B.Fedtsch., *I. vicaria* Vved., *I. warleyensis* Foster, *I. magnifica* Vved., *I. willmottiana* Foster, *I. orchioides* Carr., *I. maracandica* Vved., *I. tubergeniana* Foster, *I. kopetdagensis* Vved., *I. leptorrhiza* Vved. The other specimens are not treated in Flora of USSR but recorded in PlantList are 30 taxa. Based on taxonomical characters from Flora of USSR, these are redeterminated.

T1
P0081

Maxent modelling of spiked pepper (*Piper aduncum* L.) in Mindanao, Philippines

Japitana Rowena
University of the Philippines Los Banos

Invasive alien species is the biggest threat to biodiversity next to habitat destruction. In Mindanao, *Piper aduncum* is considered as the most invasive alien plant species that affects forest ecosystem and agricultural areas. This study provides insights in identifying suitable areas for *Piper aduncum* in Mindanao using a novel modelling method known as Maxent. Two models were generated: Full Model which is based on the 25 environmental variables and Final Model which is based on the final set of variables maintained after a series of variable reduction method. The relative predictive performance of the two models were evaluated using Receiver Operating characteristic (ROC)-Area under curve (AUC). Result showed that the Final Model performed best with AUC score of 0.825 compared to the Full model (AUC = 0.749). The predicted suitable habitat of *Piper aduncum* was heavily influenced by these top five predictors: soil type, mean temp. of warmest quarter, mean diurnal range, max temp. of warmest month and precipitation of seasonality. We also assess the suitable area coverage in relation to administrative boundaries and protected areas in the study area. Overall, this study will contribute to natural resource managers especially in setting priority areas for current management of the species and predict its potential spread in the future.

T1
P0082

Influences of wetland water factors on the photosynthetic fluorescence characteristics of submerged macrophyte

Minfei Jian*, Sicheng Wang, Houping Yu*
Jiangxi Provincial Key Lab of Protection and Utilization of Subtropical Plant Resources, College of Life Science, Jiangxi Normal University, Nanchang, Jiangxi, China

Submerged macrophyte is an important composition of a lake's wetland ecological system, and plays a crucial role in a lake's productivity and biogeochemical cycle. In this study, the wetland's submerged macrophyte, along with its water environment factors during the spring, summer, autumn, and winter seasons of Poyang Lake in Jiangxi Province were examined. Then we analyzed the characteristics of the temporal and spatial variations, as well as the chlorophyll fluorescence of the submerged macrophyte which grew under natural environmental conditions during the four seasons investigated. The results showed that the species diversity of the submerged macrophytes was higher in the wetland water during the spring and autumn, and the water depth changes were found to range from 25 to 70 cm and from 50 to 170 cm in the spring and autumn seasons, respectively. Also, the water transparency changes ranged from 25 to 65 cm and from 10 to 65 cm, respectively during these two seasons. The water changes ranged from 190 to 250 cm during the summer, with water transparency changes ranging from 45 to 140 cm. The superior species included *Hydrilla verticillata*, *Vallisneria natans*, *Najas minor*, and so on, which were determined to grow rapidly during the summer, with higher photosynthesis activity. The *QY*-max was 0.55 to 0.76, and 0.74 to 0.78, respectively. The photosynthesis activity of the *Vallisneria natans* was found to be higher in the spring, and the *QY*-max was between 0.34 and 0.62. The non-photochemical quenching (*NPQ*_Lss) of the submerged macrophyte, such as *Hydrilla verticillata*, *Vallisneria natans*, and *Najas minor*, which reflected the light protection capability parameters that plants use to avoid photosystem damage in the face of adversity, also showed significant spatial and temporal variations. They were found to reach the highest values during the winter, with mean values of 1.28,0.73, and 1.83, respectively. The wetland's water depth and transparency were determined to be the main factors which defined the diversity distributions and chlorophyll and photosynthetic fluorescence characteristics of the submerged macrophyte.

T1
P0083

Predicting the potential distribution of *Polygala tenuifolia* Willd. under climate change in China

Hongjun Jiang[1,2], ***Jiancheng Zhao***[1]
1. Hebei Normal University, Shijiazhuang, China
2. Hebei Academy of Sciences, Shijiazhuang, China

Global warming has created opportunities and challenges for the survival and development of species. A great many shifts in the distribution and abundances of species occurred. Particularly, those species with high economic values face over-harvesting and environmental disruption. *Polygala tenuifolia* Willd. as an important original plant of Radix Polygalae, was one of them. Thus, determining how climate change may impact multiple ecosystem levels and lead to various species adaptations is necessary for both biodiversity conservation and sustainable biological resource utilization. To seek suitable habitats and evaluate the impact of climate change on *P. tenuifolia,* we employed MaxEnt to predict its changes in the habitat range and altitude under current and future climate scenarios in China. Four representative concentration pathways (RCP2.6, RCP4.5, RCP6.0, and RCP8.5) were modeled for two time periods (2050 and 2070). The model inputs included 732 presence points and nine sets of environmental variables (including ten, land cover, vegetation coverage, topographical variables, and bioclimatic variables) under the current conditions and the four RCPs in 2050 and 2070. The area under the receiver-operating characteristic (ROC) curve (AUC) was used to evaluate model performance. All of the AUCs were greater than 0.80, thereby placing these models in the "very good" category. Using a jackknife analysis, the precipitation in the warmest quarter, annual mean temperature, and altitude were found to be the top three variables that affect the range of *P. tenuifolia.* The total contribution of these three variables was 73.91%. The total contribution of climatic variables was 61.82%. The total contribution of three topographic variables was 19.96%. However, the total contribution of soil variables totaled 12.20%. This showed that climatic variables had a much greater impact than topography, soil, land cover and vegetation coverage on the habitat distribution of *P. tenuifolia.* The MaxEnt predictions showed that the highly suitable habitat of *P. tenuifolia* was mainly located in Heilongjiang, Jilin, Liaoning, Inner Mongolia, Hebei, Henan, Beijing, Shanxi, Shaanxi, Ningxia, Gansu, Qinghai, Sichuan, Hubei, Hunan, Shandong, Anhui, Zhejiang and Jiangxi. These regions were in reasonable agreement with its actual distribution. Our research also found 90% of the highly suitable habitat in the current scenario was below 1,700 m, and less than 10% of medium suitable habitat located above 2,000 m. Our research also revealed that the area of highly suitable habitat for *P. tenuifolia* in all eight future scenarios will decrease, and the medium suitable habitat will increase relative to the current area. In four scenarios (RCP2.6 2050, RCP6.0 2050, RCP8.5 2050 and RCP2.6 2070), the low and highly suitable habitat will decline, and the medium suitable habitat will increase. In the other four scenarios (RCP4.5 2050, RCP4.5 2070, RCP6.0 2070 and RCP8.5 2070), the low and medium suitable habitats will increase, and the highly suitable habitat will decline.

T1

P0084

Plant resources of the transition zone between the Maowusu Desert and the Loess Plateau in China

Lin Jiang
1. Shenmu High School, Shaanxi Province, China.
2. School of Life Science, Yulin University, China

Yulin District in northern Shaanxi Province is located in the transition zone of the Maowusu Desert and the Loess Plateau. It has three different types of landforms: wind-sand region, loess hilly-gully region and river wetland region, retaining diverse vegetation types. Therefore, a systematic investigation of the plant resources across Yulin District is meaningful to understand the drought resistance and drought adaptability in plants. Since the beginning of the 1960s, we adopt the methods of field observation, collection of specimens and literature review to investigate the comprehensive plant resources. We collected ~10,000 wax leaf specimens and ~30,000 plant photos in the ecological environment across this desert-loess transition zone. Then, 101 families, 435 genera, 859 species seed plants were identified. The top four families were constituted of 35.9% of the total number of species, including 95 species from Compositae, 89 species from Leguminosae, 65 species from Rosaceae, 60 species from Poaceae. A scholarly monograph, "Yulin Seed Plants", was published according to the long term investigation, detailed describing the morphological characteristics, distribution, usage and reproduction methods of 801 species across Yulin District, including 14 species with anti-cancer effects, 16 species with strong anti-drought ability, and 22 poisonous species. It indicated lots of different drought resistance mechanisms exist. This study fills the gaps of plant taxonomy in Yulin District and lays a theoretical foundation for the development and utilization of plant resources in the desert-loess transition zone.

T1

P0085

Endangered categories evaluation of *Patis coreana* and *P. obtusa* (Poaceae) endemic to East Asia

Shanwen Jiang[1,2], Zhongshuai Zhang[1], Linglu Li[1], Wenli Chen[1]
1. State Key Laboratory of Systematic and Evolutionary Botany, Institute of Botany, the Chinese Academy of Sciences, Beijing, P.R.China
2. University of Chinese Academy of Sciences, Beijing, P.R.China

The grass family is considered to receive less concern in biodiversity conservation due to the difficulty in identification and distribution information collection. Two east Asia endemic grasses, *Patis coreana* (Honda) Ohwi and *P. obtusa* (Stapf) Romasch., P.M. Peterson & Soreng (Stipeae, Poaceae) are such kind of poor known species, even their taxomomic status were recovered very recently (2011). They distributed in the partly closed habitat in temprate forest, which is quite different from the open grassland habitat in the most of Stipoid grasses. The two species were included in the Least Concern (LC) category in the Red List for Biodiversity in China-higher Plants in 2013. Based on distribution range, current population estimates and habitat status, we evaluated the endangered categories of *Patis coreana* and *P. obtusa* using the IUCN Red List Categories and Criteria (Version 3.1). The data from this study showed that the threatened levels of *P. coreana* and *P. obtusa* were both higher than LC. We also discussed the reason of underestimation and gave some suggestions on their protection.

T1

P0086

Phylogenomics of Orchidaceae based on chloroplast genome

Xiaohua Jin
Institute of Botany, Chinese Academy of Sciences

Orchidaceae are among the largest families of angiosperm. Phylogenetics of Orchidaceae is not well understood. Phylogenomic analysis of relationships of Orchidaceae, based on 78 plastid genes from 85 species in 19 tribes and 5 subfamilies were presented. The backbone of Orchidaceae were well supported, and the interrelationships among tribes and subfamilies are well illustrated. The systematic position of several tribes, such as Codonorchideae, Nervilieae, Tropidieae, Thaieae, is different from previous results. Molecular dating, morphological character evolution and historical biogeography of Orchidaceae were also analyzed and discussed.

T1

P0087

Rapid and simultaneous analysis of five alkaloids in four parts of Coptidis Rhizoma by Near-Infrared Spectroscopy

Jintao Xue
Xinxiang Medical University

Near-Infrared Spectroscopy (NIR) was used to develop a method for rapid and simultaneous determination of 5 active alkaloids (berberine, coptisine, palmatine, epiberberine and jatrorrhizine) in different parts (rhizome, fibrous root, stem and leaf) of Coptidis Rhizoma. A total of 100 samples from 4 main place of origin was collected and studied. With HPLC analysis values as calibration reference, the quantitative analysis of 5 marker components was performed by two different modeling methods, partial least-squares (PLS) regression as linear regression and artificial neural networks (ANN) as non-linear regression. The results indicated that the 2 types of models established are robust, accurate and repeatable for five active alkaloids, and the ANN model was more suitable for berberine, coptisine and palmatine while the PLS model was for epiberberine and jatrorrhizine. The performance of the optimal model was achieved as follows: the correlation coefficient (R) for berberine, coptisine, palmatine, epiberberine and jatrorrhizine was 0.9958,0.9956,0.9959,0.9963 and 0.9923, respectively; the root mean square error of validation (RMSEP) was 0.5093,0.0578,0.0443,0.0563 and 0.0090, respectively. Furthermore, the content of 5 active alkaloids in 4 parts of Coptidis Rhizoma and 4 main origin of place was analyzed and compared by the established NIR model for the comprehensive exploitation and utilization of plant resource of Coptidis Rhizoma. This work demonstrated that NIR spectroscopy may be a promising method as routine screening for off-line fast analysis or on-line quality assessment of traditional Chinese medicine (TCM).

T1

P0088

The biogeography of the flora of arid central Australia: A preliminary study

Peter Jobson

Northern Territory Herbarium, Alice Springs

Contrary to what you would expect, the flora of central Australia is not poor in species number, but rather it consists of over 2300 species. This is due, in part, to the diverse landscape and soil types within this region. Sand dunes, claypans, salt lakes, gibber plains, river flood-outs, gorges-both deep and shallow, hills and mountain ranges all make up the landscape of central Australia. The most speciose part of central Australia is the MacDonnell Ranges region, housing well over 1300 species, with the highest level of endemism as well as providing a refuge to species more often associated with wetter climates. When looking at the floristics of the region, what is immediately obvious are the absences or poor representation of notable and iconic Australian genera. So, if the species one normally thinks of as iconically Australian are missing, but there are still large numbers of taxa present, what is the likely origin of the central Australian plants? The answer is a complex, three-pronged one. The three prongs are: temperate origin, tropical origin, and refugia from a wetter time. This presentation is a preliminary study leading to a more detailed analysis of the biogeography of the flora of arid central Australia.

T1

P0089

Pollen viability of nipa (*Nypa fruticans* Wurmb.) under different storage conditions

Junaldo Mantiquilla, *Merlene Elumba, Jenny Adtoon, Salsabel Ladjahassan, Reynaldo Abad*

University of the Philippines Mindanao

Initial pollen collection of *Nypa fruticans* was conducted in the semi-wild stands of Carmen, Davao del Norte, the Philippines to determine viability levels and storage procedure. Collection of male spikes (staminate rachillae) was best at 5 to 6 AM using paper bags to maintain integrity of pollen samples. Viability levels were effectively determined using iodine potassium iodide (I_2KI) and acetocarmine stain since both clearly identified viable from non-viable pollen. Subjecting pollen at 40°C at different duration (12,24,36 and 48 hours) did not affect general viability levels in spite considering the position of male spikes in the inflorescence (top, middle, bottom region). Percent viability was estimated at 97.80% as determined by acetocarmine method and 98.30% for I_2KI technique. Succeeding tests that included samples from Bago Aplaya, Davao City and Hagonoy, Davao del Sur, the Philippines revealed that oven-dried (40°C) pollen had significantly higher viability than deep frozen (-20°C) pollen using acetocarmine stain. Across storage duration in 75 days, 96.4% viability was recorded for oven-dried samples. The fresh pollen samples from Bago Aplaya produced low viability levels without proper drying. Hagonoy and Carmen samples had relatively higher % viability due to exposure in higher temperature during longer transit. Eventually, viability levels stabilized after 45 days during storage. Hence, nipa pollen requires sufficient drying to achieve higher viability without any signs of decline even after 75 days. *In vitro* germination also showed that pollen broke open by aperture when subjected to different levels of sucrose (2,4, 6,8 and 10%) at different duration (1,2, 3,4 hours) without significant differences among the treatments. In a separate study, 10% sucrose when combined with 0.01% boron and 15% sucrose combined with 0.005% boron achieved significantly higher *in vitro* germination at 60% and 69%, respectively. The latter combination measured the longest knob length at 10.7 μM. The content of the pollen was released during germination and thought to contain sperm nuclei necessary for fertilization instead of developing pollen tube as other plants do.

T1

P0090

The citrus variety collection as a key resource for conserving citrus biodiversity in the USA.

Tracy Kahn

University of California, Riverside

The Citrus Variety Collection was initiated in 1910 as part of the University of California Citrus Experiment Station to establish a broad representation of citrus and related genera from all growing regions of the world. Today, with approximately 4,500 trees of over 1000 accessions of citrus and related genera, the University of California- Riverside Citrus Variety Collection (CVC) is one of the most extensive *ex-situ* collections of citrus diversity in the world. Currently the living collection encompasses approximately two field trees and two potted trees held in screened greenhouses as a backup protected collection cooperatively maintained by UCR and the USDA ARS National Clonal Germplasm Repository for Citrus and Dates. This collection encompasses both commercial citrus cultivars and other genera and species in the Aurantiodeae that is critical for research and extension activities that benefits the California Citrus Industry and is utilized by other researchers throughout the USA. There are accessions in the collection that are not duplicated anywhere else in North America making the CVC a key resource for maintaining citrus biodiversity in the USA.

T1

P0091

EABCN: Research network for plant conservation in Northeast Asian Region

Ho Sang Kang[1], *Miin Bang*[1], *Hyo Cheng Cheng*[1], *Jeong Ho Park*[1], *In Sik Kim*[2], *Hye Jin Kwon*[2], *Cheul Ho Lee*[2]

1. Seoul National University

2. Korea National Arboretum

Upon the background of globally initiated biodiversity conservation goals and strategies, such as the Goal 15 of Sustainable Development Goals and Global Strategy for Plant Conservation 2020, the East Asia Biodiversity Conservation Network (EABCN) was established for the regional cooperation initially in the Northeast Asian Region by six (6) organizations from five (5) countries (China, Republic of Korea, Japan, Mongolia, and Russia) in 2014. The EABCN is institutionalized based on working group (WG) conducting collaborative research and data sharing to establish a mid- and long-term regional plant conservation strategy in East Asia. Currently, four WGs under the EABCN are in operation. The WG on flora of Northeastern Asia (plant checklist) is reviewing plant list collected and re-arranging the scientific name and synonyms in response to the GSPC target of completion of World Flora, with preparation of web-based platform. Currently, it is completed to combine around 100,000 scientific names of plants in Korean peninsula and three provinces of China (Jilin, Liaoning, and Heilongjiang) through BRHAMS software, and it will be completed to combine the scientific names of plants in Mongolia and Russian Far East by the year of 2017. The WG on plant phenological monitoring is considering the common, rare and local endemic plants interacting through the website of East Asia Phenological Network, which is also linked with smart phone application. The WG on joint publication targets endemic plants in Northeast Asia which includes history, botany, ethnic use, cultural relation *etc.* for the public, which currently in preparation for the third volume titled "Important Plants of East Asia: Plants above Forest Limits" (the first volume is titled "Important Plants of East Asia: Plants Tell Stories", and the second volume is titled "Important Plants of East Asia: Endemic Plant Stories"). The objective of WG on vegetation monitoring is to understand the changes of vegetation and species composition in response to climate change. The collaborative researches on invasive, endangered and alpine plants will be developed in 2017.

T1

P0092

New edible fungi from Southeast Asia: Discovery to production

Samantha Chandranath Karunarathna
World Agroforestry Centre, East Asia, Kunming 650201, China

The forests of Southeast Asia have the potential to be a rich source of cultivatable edible fungi. Although significant amounts of research on the taxonomy and phylogeny of edible mushrooms have been carried out, their domestication remains relatively little studied. Today, the most commonly cultivated strains are temperate species, but tropical and subtropical mushrooms are both abundant and highly diverse, with many species having long histories of human consumption. In addition, many new species have recently been introduced to science, including numerous species of high nutritional and medicinal value. The domestication and cultivation of tropical mushrooms therefore provides an enormous opportunity for Southeast Asian countries. Due to the difficulties of cultivating mycorrhizal species, we have concentrated on saprobic species. Most tropical and subtropical mushrooms, if provided with appropriate conditions, grow and produce fruiting bodies more quickly than temperate species. Tropical and subtropical mushrooms can be produced using readily available, cheap waste products such as sawdust, corn cobs, rice straw, sugarcane bagasse, and other forest and agricultural residues, making them an ideal crop for smallholder farmers. We have collected and isolated numerous strains of species of wild mushrooms from Southeast Asian forests, and have published some initial results documenting our progress in domesticating these species. Using a variety of steps including sample collection, isolation, spawn production and fruiting body production in sawdust and compost media, we showed for the first time that it is possible to domesticate *Pleurotus giganteus*; new Thai and French hybrid strain of *Agaricus subrufescens*; *A. flocculosipes*; *A. subtilipes*; *Auricularia thailandica*; *A. cornea* (white) and newly discovered *Ganoderma* species. These advances may create new opportunities for the mushroom growing industry and for smallholder farmers in Southeast Asia in particular.

T1

P0093

The assessment and conservation of rare and threatened species of Boraginaceae Juss. in the Azerbaijan flora

***Vugar Kerimov**, Valida Ali-Zade, Rashad Salimov*
Institute of Botany of Azerbaijan National Academy of Sciences

Azerbaijan is distinguished by the richness and diversity of its vegetation. Majority of vegetation types existed in the world is found in the area of our Republic. The main reason is the diversity of the physico-geographical as well as natural historical conditions that formed under the influence of distance and adjacent floristic region. *Boraginaceae* Juss. has a specific place and comprises 81 species belonging to 26 genera (Flora Azerbaijan 1957). Since 1998, as a result of the scientific research and dozens expeditions to the different botanical and geographical regions of Azerbaijan, the taxon number of this family increased up to 104 species belonging to 32 genera. The research works carried out on *Boraginaceae* Juss. over the years of 2000-2016 allowed us, to gather many valuable materials for justification to assess the current status of many species, particularly endemic, rare and threatened species through the long-term observations. In general each of 16 rare and threatened species of *Boraginaceae* Juss. were assessed separately according to IUCN categories. Herbarium materials of LE, TIB and BAK was analyzed, and literature, maps, the results of monitoring conducted by the various authors in nature were used. The comparative morphological, systematic, botanical, florogenic and other methods were used. Taking into account the species of the family, many of which are represented in the flora are endemics of Azerbaijan, and declining the number in nature, six of them were listed in the 2nd edition of the Red Book Azerbaijan (2013). These species are *Trigonocaryum involucratum* - VU A2c+3c; *Cynoglossum holoceryceum* - VU A2c + 3c; *Solenanthus brachistemon* - EN B1ab (iii)+2ab (iii); *Solenanthus circinnatus* - VU A2c+3c; *Rindera lanata* - EN B1ab (iii)+2ab (iii) and *Omphalodes rupestris* - VU B1b (i, iii). Two new species: *Nonea cyanocalix* Pop. ex V.Karimov and *Nonea bakuensis* Pop. ex V.Karimov were recorded for the science, and two: *Myosotis schistosa* Khokhr., and *Cerinthe alpina* Kit. for the Azerbaijan flora for the first time. In recent years, based on the detailed results, 11 species from the *Boraginaceae* Juss. should be listed to a new edition of "the Red

Book of the Azerbaijan" through assessing them according to the appropriate IUCN categories. These species are as followings: *Heliotropium szovitsii* (Stev.) Bunge; *Heliotropium tzvelevii* T.N.Popova; *Onosma gracilis* Trautv; *Cerinthe alpestris* Kit (*C.glabra* Mill.); *Symphytum peregrinum* Lebed.; *Anchusa officinalis* L.; *Nonea cyanocalix* Pop. ex V.Karimov; *Nonea daghestanica* Kusn.; *Nonea decurrens* (C.A.Mey) G.Don; *Myosotis schistosa* Khokhr.; *Solenanthus biebersteinii* DC.

T1

P0094

Labellum micro-morphology of the *Nervilia plicata* (Orchidaceae) complex in Tropical Asia

***Kanapol Ketjarun*[1], *Stephan W. Gale*[2], *Paweena Traiperm*[1], *Somran Suddee*[3], *Santi Watthana*[4]**

1. Department of Plant Science, Faculty of Science, Mahidol University, Rama 6 Road, Rachathewi, Bangkok

2. Kadoorie Farm & Botanic Garden, Lam Kam Road, Tai Po, New Territories

3. Forest Herbarium, Department of National Parks, Wildlife and Plant Conservation, Chatuchak, Bangkok

4. School of Biology, Institute of Science, Suranaree University of Technology, Nakhon Ratchasima

Nervilia plicata is a morphologically variable terrestrial orchid with a wide geographic distribution ranging from India and South China, through Southeast Asia to New Guinea, Australia and the Southwest Pacific Islands. Several forms have previously been recognised as distinct taxa within this complex, based on differences in the outline, size and colour of floral parts. Most notably, distinct purple and white colour morphs exist. Casual observations in various parts of the range suggest that this species is outcrossing, but the ecological significance of the variable floral morphology in attracting pollinators remains unknown. As a first step towards understanding the links between floral polymorphism, pollination ecology and speciation within the *N. plicata* complex, we undertook a micro-morphological investigation of labellum surface. In particular, we sought to ascertain whether the two distinct colour morphs exhibit any discrete micro-morphological differentiation. The labellum was harvested from flowers collected from 11 populations representing the two colour morphs in Thailand and Hong Kong, fixed in 70% ethanol, critical point dried, and examined under scanning electron microscope (SEM). The results revealed that osmophore cells in the secretary zone of labella from all localities are papillose with cutin sculpturing, but that they differ in cell shape and surface architecture. For example, most were polygonal or round, with a flat surface lacking reticulate ornamentation. However, the osmophore cells of some accessions possessed a papillose surface with a reticulate connective layer. These differences are not correlated with the two morphs and locations. Though these findings are still preliminary, there is therefore some evidence for discrete variation in micro-morphology among different *N. plicata* accessions. Further work, including observations of plant-pollinator interactions and analysis of genetic diversity, is ongoing to help provide taxonomic clarity within the *N. plicata* complex.

T1

P0095

Collection and consumption of wild edible plants: With reference to tribal communities in Kalsubai-Harishchandragad Wildlife Sanctuary

Mahendra Khyade

Department of Botany, S.N. Arts, D.J.M. Commerce and B.N.S. Science College, Sangamner- 422605.Affiliated to Savitribai Phule Pune University, Pune (MS), India.

The present communication deals with the ethnobotanical exploration, identification and future potentialities of the wild edible plant species consumed by the tribal communities in Kalsubai-Harishchandragad Wildlife Sanctuary located in the northern Western Ghats. Data were collected by means of semi-structured open ended interviews with questionnaire in their local language through frequent field visits during 2013 and 2016. Data collected were structured with Use value reports (UV). A total of 52 plant species belonging to 27 families were reported from the study area. Out of 23 families 6 families were belongs to monocotyledon and remaining 21 families were dicotyledons. In dicot, the dominant families were Papilionaceae (6 species), Amaranthaceae, Asclepidaceae (5 species each), Boraginaceae, Cucurbitaceae (3 species each), whereas in monocots, Dioscoreaceae (3 species) Araceae, Liliaceae, Commelinaceae (2 species each) and remaining taxa belongs to one member of each families. Herbs are used more often (23 species), followed by climbers herbs (12 species), Tree (11 species) and shrubs (6species). Wild food plants were classified according to their uses such as salads, cooked vegetables, spices, raw snacks, teas, alcoholic drinks, *etc.*The leaves and fruits were the dominant parts used followed by tuber, flower and root. The most frequent usage was cooked as vegetables followed by raw, snacks and juices. This study provides indigenous knowledge regarding wild edible plants for future generation.

T1

P0096

The state of endangered limestone plants in Peninsular Malaysia

Ruth Kiew

Forest Research Institute Malaysia

The limestone flora of Peninsular Malaysia is relatively well-known botanically. It is a hotspot for biodiversity with 1,300 species that represent 14% of the Peninsula's vascular flora crammed on 540 karst hills that occupy just 0.4% of land area. Not only is it biodiverse but a significant number of species are rare and endangered and they are increasingly threatened by quarrying, burning, cave temples, recreation and tourism. However, no karst hill conserves more than a fraction of the flora. Generally 200-300 species are found on a single hill. In addition, rare species are not concentrated on any particular hill. Of the critically endangered species, over a hundred species are found on a single karst hill and about eighty Endangered species are found on between two to four hills. This represents a challenge for conservation. Most karst hills are not protected because they lie outside State or National Parks. Will the status of the karst limestone as an environmentally sensitive area and individual hills as important plant areas because they harbour rare species carry sufficient weight to protect endangered species from extinction against commercial interests and the need for limestone products, like cement?

T1

P0097

Diversity of economically important seaweeds from Tamil Nadu costs of India

Sahayaraj Kitherian
St. Xavier's College

India has a stretch of about 8,129 km coastline with nine maritime states *viz.*, Gujarat, Maharashtra, Goa, Karnataka, Kerala, Tamil Nadu, Andhra Pradesh, Odhissa, and West Bengal. Tamil Nadu coastline occupies about 13.08% contributing 13 districts namely Thiruvallur, Chennai, Kanchipuram, Villupuram, Cuddalore, Thiruvarur, Nagapattinam, Thanjavur, Pudukottai, Ramanathapuram, Thoothukudi, Tirunelveli and Kanyakumari. Marine algae or seaweeds having four classes namely Chlorophyceae or green algae; Phaeophyceae or brown algae; Rhodophyceae or red algae and Myxophyceae or Blue-green Algae, which are one among the easily as well as readily accessible and available natural resource in considerable quantity world-wide. The first three algae have been utilized as food, fodder, feed additives, green manure besides the sources of enormous compounds like agar, agaroids, algin, proteins, minerals, trace elements (Na, K, Ca, Mg, Chloride Nitrogen, Sulphate, Iron, Copper, Manganese, Boron, Zinc Phosphorus) and Vitamins (A, B12, C, D and E) with values in confectioneries, pharmaceuticals, and various industries. Out of 9,000 algal species reported from all over the world, nearly 12.81% seaweeds (1,153 species) were reported from India. Though Agarophytes and Alginophytes are considered as drifted seaweeds in India; we recorded more than 25 species as drifted seaweeds from Tamil Nadu cost. Porphyra spp. have been utilized for human consumption, whereas species of *Alaria, Laminaria* as animal feed; *Ulva, Porphyra* and *Centroceras* for food, fodder, feed additives. Gracilaria has been utilized for extracting agar, whereas *Hypnea, Spyridia, Sarconema, Acanthophora, Laurencia, Chondria, Gigartina* for algin; *Sargassum, Turbinaria, Cystophyllum, Hormophysa, Dictyota* and Padina used as agaroids sources. Algal proteins mainly obtained from *Ulva fasciata, Ulva rigida, Porphyra vietnamensis, Acanthophora muscoides* and *Centroceras clavulatum. Cymadocea, Diplanthera* and Halophila are utilized as green manure by the farmers. *Ulva fasciata, Codium fragile, Sargassum cinereum, Sargassum ilicifolium, Laminaria digita, Macrocystis pyrifera, Jania adhaerens* have been used by costal people for various medicinal (antioxidant, anticoagulant, antithrombotic, antitumour, immunomodulatory, antipeptic, antiadhesive, hepatoprotective, neuroprotective) uses. Toxic metals copper, nickel, lead, zinc and cadmium from industrial wastewater removing *Sargassum, Laminaria, Ecklonia, Ulva,* and *Enteromorpha* were recorded from Tamil nadu. Seven *Enteromorpha* spp. dwelled in intertidal zones of Tamil nadu coasts utilized for treatment of sewage and some agricultural wastes to reduce nitrogen and phosphorus containing compounds. More than 80 species were recorded from Tuticorin of Thoothukudi district. Rameshwarm of Ramanathapuram district is the hot spot where more than 200 species were observed. Arvee Biotech, Chidambaram; PL Agro Technologies Limited, Chennai; Aachreya Crop Care, ticel bio park, Chennai; Aadrika Organic Farms, Chennai; Anara Trading Company, Chennai; Manidharma Biotech Private Limited, Chennai are exploring available seaweeds for fertilizer manufacturing. BioNature, Salem engaged in seaweed gel production. Seaweed extracts have produced by International Chemicals, Thiruvanmiyur, Chennai; Unitech Chemicals, Ayanambakkam, Chennai; Reddy International, Nellore. Agar Agar (Cellulaids, Chennai), Carrageenan (Vaanan Global Traders, Chennai), Plat growth regulators (APS Biotech, Paranji, Arakkonam), Sea Weed Products, seaweed granules, and dried seaweeds have been produced by various small scale ad medium scale industries at Tamil Nadu. Various industrial activities directly or indirectly altered seaweeds pollutions will also discussed.

T1

P0098

Spatial and seasonal distributional studies on herbaceous plants of lateritic Rocky outcrops of Chaukul (Amboli), Western Ghats, India

Rutuja Kolte, *Malapati Janarthanam*
Goa University

Western Ghats (WG) along with Sri Lanka is a biodiversity hotspot due to high number of endemic species it harbours. Varied types of habitats with their unique environmental conditions have enriched the flora of WG. But one of the neglected habitats among them is lateritic plateaus, which are mostly distributed in northern part of the WG. They are rich in herbaceous species with good proportion of endemics. Most of these endemics are habitat specific and distributed in very small area. In the recent past, sporadic attempts have been made to study the vegetation of these plateaus. However, as most of these endemic plants are ephemeral, intensive studies are needed to understand their composition on each lateritic plateau and their ecology. The present study is an attempt to understand systematic composition, distributional and phenological succession pattern of the herbaceous flora of lateritic Rocky outcrops of Chaukul (Amboli) from Sindhudurg district of Maharashtra. Periodic field visits (10-12 days of interval from June to November 2016 and once in a month in remaining period) were carried out from January 2016 to February 2017. During this period, 135 herbaceous plants belonging to 39 families have been documented from various microhabitats. Out of these, 60 are endemic to WG. It has been observed that as monsoon progresses, vegetation pattern also changes from dominance of tuberous plants to grasses. Family Poaceae top ranks with maximum number of genera and endemic species, which is dominant in post monsoon season. It is followed by family Orchidaceae which is present throughout the year and shows diverse habitat preferences by different members. Species such as *Indigofera dalzelii* and *Lepidagathis clavata* belonging to Fabaceae and Acanthaceae respectively are present throughout the year and show varied degrees of morphological variations. Duration of peak flowering of a given species is proportional to the length of the general flowering period. Overall peak flowering coincides with the receding of monsoon. Species of *Exacum, Senecio* and *Utricularia* occur on wide range of lateritic plateaus but species such as *Adenoon indicum, Ceropegia jainii, Lepidagathis clavata, Merremia rhynchorrhiza* are habitat specific. Few species are locally abundant and they are: *Eriocaulon stellulatum*, *Exacum lawii*, *Habenaria* spp., *Impatiens* spp., *Senecio belgaumensis* and *Utricularia purpurascens* while species like *Ceropegia jainii*, *Drosera burmanni*, *Iphigenia pallida*, *Sonerila scapigera* are found to be very rare. Ephemeral plants present unique adaptive traits such as epiphytic nature, underground perennating organs, carnivory, confining to cryptogamic mat *etc.* Based on rainfall and vegetation dominance four seasons can be recognized on plateaus. They are; pre monsoon period for tuberous plants, monsoon period for ephemeral plants, post monsoon period for grasses and dry period for sturdy plants. However microhabitat plays a very important role in occurrence and phenological succession of herbaceous plants on lateritic plateau. To understand full spectrum of herbaceous plants of lateritic plateaus and their dynamics more studies are required.

T1

P0099

Preliminary investigation of bryophytes in Xizang, China

Jin Kou[1], *You-Fang Wang*[2], *Chao Feng*[3], *Shan-Shan Song*[1], *Li-Li Wang*[1], *Xiao-Tong Song*[1], *Xiao-Ming Shao**[1]

1. Beijing key Laboratory of Biodiversity and Organic Farming, College of Resources and Environmental Sciences, China Agricultural University

2. East China Normal University

3. Inner Mongolia Agricultural University

Xizang is the main part of Qinghai-Xizang Plateau and the representative with significant temperature increasing in China. It creates a special heterogeneous environment for bryophytes by its complex landforms, physiognomy, and climate. Meanwhile, both morphological characters and distribution of bryophytes can be used to reflect the environmental change. The early relevant researches are concentrated from 1970s – 1980s and the investigations were mainly in the Southeast of Xizang. Due to the traffic conditions, field investigations on bryophyte were conducted randomly along the ways. From 2007 to now, our fieldworks were carried out by two types: one was sampling with a size of 25 cm×25 cm or the other size of 50 cm×50 cm, the other one was random sampling. The investigation range is expanded, including Nagqu, the low altitude region of Motuo, and Ali unmanned area. Eight new species were described: *Bryoerythrophyllum pseudomarginatum* J. Kou, X.-M. Shao & C. Feng, *Bryoerythrophyllum zanderi* C. Feng, X.-M. Shao & J. Kou, *Didymodon epapillatus* J. Kou, X.-M. Shao & C. Feng, *Didymodon jimenezii* J. Kou, X.-M. Shao & C. Feng?*Didymodon liae* J. Kou, X.-M. Shao & C. Feng, *Didymodon mesopapillosus* J. Kou, X.-M. Shao & C. Feng, *Encalypta gyangzeana* C. Feng, X.-M. Shao & J. Kou, *Hilpertia tibetica* J. Kou, X.-M. Shao & C. Feng. Three species and one variant were new to China: *Bryoerythrophyllum latinervium* (Holmen) Fedosov & Ignatova, *Tortula brevissima* Schiffn., *Tortula transcaspica* Broth and *Stegonia latifolia* var. *pilifera* (Dicks.) Broth. In addition, *Grimmia indica* (Dixon & P. de la Varde) Goffinet & Greven was firstly discovered in Tibet. Besides, Pottiaceae samples collected from arid and semi-arid regions in 2007,2011 and 2012 were analyzed by GIF and SPSS software and diversity index analysis with climatic variables. The result shows that: 1) Pottiaceae species richness have obvious difference at different altitudes. But their evenness are similar; 2) there are remarkable negative between Pottiaceae diversity and elevation. Instead, Potteaceae diversity has obviously positive correlation with vegetation coverage. There are no significant correlation between it and soil water content; 3) elevation, vegetation type and temperature are the most key environmental factors to affect both Pottiaceae diversity and coverage. Vegetation cover is the second key factor. And there are no significant effect from human disturbance and substrate; 4) To spatial distribution of Pottiaceae and *Didymodon*, human disturbance is the most significant environmental variable. Both climate and vegetation take second place. The influence of topography is the least.

T1

P0100

Taxonomy, ecology and *ex-situ* conservation of potential plant resources

Sanjay Kshirsagar

Shri Shivaji Vidya Prasarak Sanstha's Late Karmveer Dr. P.R.Ghogrey Science College, Dhule, Maharashtra

Present paper communicates potentially important plant resources with reference to their taxonomy, ecology, threats, precise locations, flowering and fruiting period, present status in the field, IUCN status and *ex-situ* conservation. Following plant species are collected from Khandesh regions (Dhule, Nandurbar and Jalgaon districts) and Sahyadri ranges of Western Ghats in Nashik District under UGC (University Grants Commission, New Delhi) PG Diploma in "conservation of medicinal, endemic and endangered plants" and conserved in botanical garden as an aid in education, for research, multiplication and transfer to other botanical gardens. For example, *Curcuma inodora*, *Dioscorea bellophylla*, *Sterculia urens*, *Ceropegia sahyadrica*, *C. bulbosa*, *C. hirsuta*, *Vigna khandalensis*, *Gloriosa superba*, *Rawolfia tetraphylla*, *Pterocapus marsupium*, *Radermachera xylocarpa*, *Entada pusaetha*, *Zingiber devakarianum*, *Michelia champaca*, *Costus picta*, *Costus speciosus*, *Bauhinia vahlii*, *Bauhinia tomentosa*, *Curculigo orchioides*, *Eulophia herbacea*, *Gymnema sylvestre*, *Enicostemma axillare*, *Caralluma adscendens* (Roxb.) R.Br. var. *fimbriata* (Wall.) Grav. & Mayur, *Cissus quadrangularis*, *Acorus calamus*, *Commiphora wightii*, *Amorphophallus bulbifer*, *Tylophora indica*, *Abrus precatorius*, *Notonia grandiflora*, *Terminalia tomentosa*, *Plumbago zeylanica*. *Pinda concanensis*, *Iphigenia magnifica*, *Neuracanthus sphaerostachys etc.*

T1

P0101

The public trial run of PhycoBank - Registration of Nomenclatural Acts for Algae

Wolf-Henning Kusber, *Andreas Kohlbecker*, *Anton Güntsch*, *Walter G. Berendsohn*, *Regine Jahn*

Botanic Garden and Botanical Museum Berlin, Freie Universität Berlin

PhycoBank is to become the registration system for nomenclatural acts (new names, new combinations and types) of algae. Algae are a very diverse group of organisms and are covered by the International Code of Nomenclature for algae, fungi and plants (ICN). PhycoBank will provide a central registration service for the registration of new names and nomenclatural acts submitted either directly by nomenclatural authors or automatically via cooperating publishers. Each nomenclatural act will have globally unique PhycoBank http-based identifiers for new names and typifications, automatically redirecting to corresponding data and metadata in human- and machine-readable formats. Additionally, search functions for taxon names, registration date, category of nomenclatural act, and nomenclatural type specimen will be provided. Feedback mechanisms, annotations, and appropriate treatments of ambiregnal taxa will also be incorporated. Interoperability with the global names infrastructure will be achieved via standardized web services. The public trial run of PhycoBank started in March 2017. Curational data entry of nomenclatural acts is already in place. Data entry for nomenclatoral autors is under construction and an automatic workflow (from publishers) is in the planning phase. Upon registration, PhycoBank will feed evaluated name data into the Global Names Architecture via the Catalogue of Life. PhycoBank will meet all botanical community standards such as IPNI Authors, BPH-S, TL2. In addition, PhycoBank will also cover all specialties needed by the phycological community: treatment of ambiregnal taxa (e.g. bluegreen algae, dinoflagellates, euglenoids), algal group-specific provisions, later starting points, lecto- and epitypes, and figures of algae that cannot be preserved. Phycobank is supported by a DFG Grant (JA 874/8-1) and available at http://phycobank.org.

T1
P0102

Useful plants diversity in Cop Village, Bac Huong Hoa Nature Reserve, Vietnam

Changyoung Lee[1,2], *Trung Thanh Nguyen*[2], *The Bach Tran*[3], *Sang Mi Eum*[1], *Soo-Yong Kim*[1], *Jin-Hyub Paik*[1], *Sang Woo Lee*[1], *Jeong Kwang Park*[1], *Yong-In Kim*[1], *Sangho Choi*[1]
1. Korea Research Institute of Bioscience and Biotechnology
2. VNU University of Science
3. Institute of Ecology & Biological Resources

A study was conducted in Cop Village, Bac Huong Hoa Nature Reserve (BHHNR), East Quang Tri Province to document and collect the useful plants used by local people. We found out 70 plant species distributed in 43 families. Of the total plant species 49% were trees, 24% shrubs, 17% herbs, and 10% vines. Among the collected plants, 57 species were used by local people as medicinal plant, 6 species categorized as food, 4 species categorized as fuelwood, 2 species categorized as dye and 1 species categorized as ornamental. Some of commonly used medicinal plants were *Areca catechu* L., *Litsea cubeba* (Lour.) Pers., *Passiflora foetida* L., *Piper sarmentosum* Roxb., *Schefflera heptaphylla* (L.) Frodin, *Tamarindus indica* L., *Trichosanthes tricuspidata* Lour. 20 different plant parts were used combined together by local people. These medicinal plants were mainly used for the cure of rheumatism, stomachache, dermatitis, cancer, backache, post-natal care, toothache, and wounds.

T1
P0103

Introducing herbarium TUR

Samuli Lehtonen[1], *Seppo Huhtinen*[1], *Sanna Huttunen*[1], *Terttu Lempiäinen*[1], *Mia Lempiäinen-Avci*[1,2], *Nelly Llerena*[1], *Tarja Marsh*[1], *Arto Puolasmaa*[1], *Veli-Pekka Rautiainen*[3], *Mika Toivonen*[1], *Jukka Vauras*[3]
1. Biodiversity Unit, Herbarium, FI-20014 University of Turku
2. Museum of Archaeology, University of Stavanger
3. Biological collections of Åbo Akademi University, Herbarium, FI-20014 University of Turku

The University of Turku Herbarium (TUR) currently holds ca. 1 050 000 specimens, of which 600 000 represent vascular plants, 201 500 fungi, 120 000 bryophytes, 111 000 lichens, 6000 algae, and 15 000 subfossils, respectively. The most important collection is the E.A. Vainio's lichenological collection of over 35 000 specimens and with more than 4 000 types collected from all over the world. The herbarium TUR also incorporates the collections of ?bo Akademi (collection acronym TUR-A) and the two universities work together in maintaining and developing the collections. The collections are currently being databased and type specimens digitized using high-resolution scanning for vascular plants and high-resolution photo stacking for three dimensional specimens such as lichens. As well, the herbarium maintains a botanical library and laboratory facilities for molecular systematic and archaeobotanical studies, both of which are active research topics in the herbarium. The herbarium is internationally networked and actively serves researchers globally. The multidisciplinary University of Turku Amazon Research Team contributes thousands of plant specimens from Amazonia, making the neotropical collections one of the currently most actively studied parts of the collections. The herbarium has an active role in the societal development, for example, by running a citizen science project aiming to gather data on distribution and ecological preferences of Finnish fungi. These data are collected using a customised online database system allowing photo upload and thus, later verification of observations by specialists. Together with museum collections these data aid in the Red List evaluation of threatened Finnish flora, in which herbarium TUR contributes with other Finnish natural history collections. The herbarium also serves the public by providing identification services of plants and fungi, such as the construction damaging rot fungi.

T1
P0104

The molecular identification of bulbous and perennial plants traded at Faraday "hawkers of health" medicinal market, South Africa using DNA barcoding.

Dorcas Lekganyane, *Ronny Kabongo*, *Michelle Van Der Bank*
African Centre for DNA Barcoding, Department of Botany & Plant Biotechnology, University of Johannesburg, PO BOX 524, Auckland Park, 2006, South Africa

The South African medicinal market generates approximately R2.9 billion per annum to the national economy, which provides a fundamental source of income but comes at an expense to the environment. Most plants traded at the market are harvested from wild resources resulting in noticeable levels of species depletion. Adulteration, trade using vernacular names and morphological similarities of plants or the lack thereof, makes identifying samples on a taxonomic level challenging. In this study, DNA barcoding was implemented to rapidly identify the plants, more specifically bulbous plants, traded at the Faraday market. Sixty samples were collected and sequenced for the core barcoding regions (*rbc*La and *mat*K). BLAST results were ambiguous for *rbc*La whereas *mat*K results were more specific, permitting samples to be identified to species level. Overall, 85% of the samples were identified to species level. Two additional approaches, the tree-based and character-based methods were followed to validate *mat*K samples to species level. Results from this study reveal a noticeable increase in the number of species traded with the majority of sought-after species of Least Concern. However, 10% of the species are Declining or Near Threatened in the wild posing a serious conservation issue. Prolonged unsustainable trade of these plants could lead to more Critically Endangered species in the future. An important outcome of the project is a new collaboration with the South African Department of Environmental Affairs to address these problems. Results and the consequences of illegal trade will be discussed.

T1

P0105

Palm studies at Tirimbina Biological Reserve, Costa Rica

Juan Manuel Ley-López
Tirimbina Biological Reserve

Palms are an important component of a tropical forests. However, there has been only limited analysis of basic palm ecology and understorey species have also been neglected. During the last 3 years I have studied the phenology, germination patterns and seedling characteristics of the palm assemblage at Tirimbina Biological Reserve (TBR), located in one of the most diverse areas for palms in the Neotropics. Since October 2013 phenological censuses have been conducted twice a month for 174 individuals belonging to 13 species of understorey palm. Two subspecies of the polymorphic palm *Geonoma cuneata* host in their infrutescence the only gall report for Arecaceae. Phenological censuses regarding the success from every inflorescence in both subspecies was conducted for 138 consecutive weeks in 78 individuals. 5,172 mature fruits where collected and seeds from 220 individuals and 21 species were germinated within 48 hours after collection. Germination was checked weekly under nursery conditions for 57 weeks and the Kaplan-Meier estimate of survivorship was calculated. Finally, I created an identification key with detailed photos and descriptions for the seedlings of all native palms at TBR. While *Asterogyne martiana*, *Chamaedorea tepejilote* and *Prestoea decurrens* are seasonal, all species in the genus *Bactris* and most species in *Geonoma* bear fruit and flowers year round. *Reinhartia gracilis* is a supraannual species with no individuals flowering during the last 4 years. No difference was found with respect to the flowering season in both *Geonoma* subspecies. However, the number of abortions is much higher in *G. cuneata* subsp. *procumbens*. The percentage germination for the palm assemblage (%G) was 65.6%, whereas the mean length of germination and the number of weeks necessary to reach 50% of %G were 16.4 and 14, respectively. Ninety per cent of seeds germinated in 7-25 weeks and by week 40 most species had germinated. Main characteristics for the recognition of seedlings are seed form and size, leaflets arrangement on the rachis, presence of thorns, leaf margins, texture, and measurements of the leaf. The diagnostic characteristics in half of the species are easily recognized. In contrast, the other half require detailed measurements or the development of at least 4 to 8 leaves in order to distinguish them from other species. These studies show that at the community level understorey palms bear fruit and flowers year round and in contrast to the prevalent view that palms have low germination, germination was completed relatively quickly. Most palm seedlings are easily recognized at least at the genus level. Characterizing the phenological patterns, the temporal expression of germination and recognizing seedlings is critical for increasing our understanding of tropical palm ecology, their regeneration and their sustainable management.

T1

P0106

Floristic analysis on the seed plants of Ordos Plateau, a biodiversity hotspot in arid and semi-arid areas of China

***Engui Li**, Yongmei Huang, Huiying Chen*
Beijing Normal University, Beijing, China

Ordos Plateau is one of the biodiversity hotspots in arid and semi-arid areas of China. By integrating the document data and field survey data on Ordos Plateau, 792 species belonging to 346 genera in 85 families were recorded on Ordos Plateau. The area is characterized by herbaceous plants with a total of 607 species (77%), and there are 146 shrubs in this region. The flora of Ordos Plateau is a blend of several geographical elements, reflecting the ancient origin, temperate arid desert characteristic of the flora, and the relevance with broad-leaveas forest region in East Asia. At the generic level, 175 (79.52%) are primarily temperate elements, much higher than the level of China flora (30.95%), While 38 (13.3%) are tropical in nature, much less than the level of China flora (50.60%). The most genus-rich floristic distribution is north temperate (75 genera), contributing 35.49% to the flora. The percentage of Mediterranean, western to central Asia distribution is as much as 12.29% (28 genera), much greater than that of the national average (5.68%). At the species level, including temperate elements (38.64%), East Asia elements (23.86%), desert elements (18.18%), steppe elements (9.09%), and ancient Mediterranean elements (5.18%). Endemic phenomena is not obvious, 9 genera are endemic to China, 43 species are endemic to Alashan-Ordos. We also explored how the climate, geological substrate and the topography varied in this area to illustrate the prominent floristic characters. We discussed the conservation threats from the perspective of anthropogenic impacts, and recommend the endemic relic species be given to more conservation attention in this region. For biodiversity hotspot, we suggest that floristic composition may serve as an approaches to measure the biodiversity, and guide biodiversity conservation of the region.

T1

P0107

How many species of seed plants in Tibetan sacred sites

***Feifei Li**[1,2], Bo Liu[1], Jianqin Li[1,3], Chunlin Long[1,4]*
1. College of Life and Environmental Sciences, Minzu University of China
2. State Key Laboratory of Environmental Criteria and Risk Assessment, Chinese Research Academy of Environmental Sciences
3. Faculty of Forestry, Southwest Forestry University

4. Kunming Institute of Botany, Chinese of Acamedy Sciences

Many researches indicated that religion was an important driving force for biodiversity conservation based on their special protection principle, law enforcement and sanctions regime. The value of sacred natural sites (SNS) was recognized by numerous scholars and international non-governmental organizations (NGOs) in the last decades. Because most SNS had long history and their traditional taboos were similar as some species protection guidelines. Tibetan region was a typical ethnic region that their traditional religion was closely related with natural environment. Tibetan sacred sites (TSS) have been existed for centuries which associated with deities, nature spirits, and spiritual leaders. Under the long-term protecting, TSSs were considered to be ecologically unique and important for conserving various scales of landscape, community, and species. However, the question that "how many species in TSS?" was still remained unanswered. A wider investigation of entire Tibetan region is required to enable a better understanding of the values of TSSs in species protection and nature conservation. We studied the seed plant species diversity of nine TSSs, and the value of TSS for conservation ecology and biotic richness of Tibetan areas. Our study discerned the number of seed plant species, endemic plant genera and endangered plant species in TSSs, as well as to examine whether these areas acted primarily as "plant museums", "plant refuge" or "gene pools". This study also analyzed the urgent conservation value of seed plant species and determined prior protection sacred natural sites, and provided suggestions for protecting these sacred natural sites. Results presented in this report can be summarized as follows: 1. The nine TSSs totally contained 7,630 seed plant species belonged to 200 families and 1,338 genera, in which 910 species are endangered, and 70 genera endemic to China. 2. The area of nine TSSs covered only 3.36% of the Tibetan areas, 0.77% of China's territory. However, they contained 79.83% species, and 80.18% endangered species; 23.00% species, 20.26% endangered species and 28.81% Chinese endemic genera of China. 3. Except the densities of species endemic to Tibetan areas, all of the densities of total species, endangered species and Chinese endemic genera in the nine TSSs were higher than that of the Tibetan areas. 4. Based on the urgent conservation value of seed plant species, the prior protection level of species were determined. Based on principle of complementarity, we proposed the prior TSSs for protection, which were Namcha Barwa Mountains, Baima Mountains and Bita Lake. 5. The human activities of other ethnic groups, ecotourism, changing belief system and management system were the main factors threatened biodiversity protection in TSSs. The key strategies for biodiversity conservation in TSSs were proposed, including action plans of completing surveys of TSSs, establishing protection network, information platform, and forming reasonable management mechanism of TSSs.

T1

P0108

Microsatellite-based genetic diversity patterns in altitudinal gradient populations of *Cypripedium tibeticum*

Jing Li

School of Biological and Environmental Engineering, Xi'an University, Xi'an, China

Cypripedium tibeticum (Orchidaceae) is an unique orchid species with high ornamental and medicinal value. It is endemic to southwest China, including Tibet, Yunnan and Sichuan provinces. Due to habitat changes and human activities, *C. tibeticum* is threatened with extinction and is listed as national protected plant. In this study, we developed eight polymorphic microsatellite markers for *C. tibeticum*, and investigated the genetic diversity and structure of 21 altitudinal gradient populations collected in the distribution range of 3,150 to 3,400 meters above sea level in Huanglong Nature Reserve in Sichuan Province. From calculated values of number of effective alleles (NE), observed number of alleles (NA) and the average of Shannon index, we confirmed that the eight SSR loci we selected are suitable for genetic markers to analyze genetic diversity in the *C. tibeticum* populations. For further analysis, 21 populations we sampled were divided into four groups: Group 1 (G1) to Group 4 (G4) from low to high altitude. The disequilibrium test and Hardy-Weinberg equilibrium test was conducted for each microsatellite locus. The distribution of different alleles in each populations was calculated and the level of genetic diversity in populations of *C. tibeticum* was analyzed. Genetic parameters including *Fst, Gst* and *Nm* (effective gene flow) was calculated between pairwise populations in each altitude group. The results showed that the average *Fst* data display G4 > G3 > G2 > G1, while the average of *Nm* showed G1 > G2 > G3 > G4, which indicated that with increasing altitude the level of genetic differentiation among populations of *C. tibeticum* is increasing, and the level of gene flow is declining. Analysis of molecular variance analysis (AMOVA) was conducted to explore which is the main factors composed of genetic differentiation among *C. Tibeticum* populations. The results of AMOVA showed that proportion of variation component in descending order was "among populations within groups" > "within populations" > "among groups". The results of correlation analysis among pairwise populations between distribution altitude and genetic distance (*Fst*) showed that for small inner scale elevation within groups, there was a certain pattern of correlation. For large-scale distribution of elevation among all the populations tested in this study, there was no obvious (non-significant) correlation between genetic distance and altitude distribution of *C. tibeticum.* This study will help us to understand the genetic diversity patterns in natural environment and make applicable strategy to protect *C. tibeticum* in the future.

T1

P0109

Research and application of landscape water conservation systems in the arid region of Loess Plateau

Ke-Ke Li, *De-Shun Zhang*

Tongji University

As China enter into the 21st century, issues including increased aridity and restoration are brought to the forefront of priorities. As a leading discipline, landscape architecture plays the role of transforming human settlement and sustaining the ecological environment. This research is of significant importance for Loess Plateau to accelerate xeriscape integration and explore native features. This thesis is divided into three parts. Firstly, with the background research of Loess Plateau, and theory literature reviews of ecological restoration, vegetation restoration, site types classification, evaluation system and landscape water conservation theory and practice of local and abroad, theory and analysis foundations are concluded and summarized. Second, the research objects of Loess Plateau arid districts, such as "vegetation", "soil" and "landscape" are defined, together with the relationships between these objects and "water" (e.g., water conservation landscape, vegetation water consumption and soil water content). Thirdly, in order to address the construction methods of landscape water conservation systems, the following principles are studied: "ecology and sustainability", "availability and universality", "quantity and quality", "practicality and operability", "scientificity and accuracy", methods of "vegetation quadrat sample surveys", "soil quadrat sample surveys", "Landscape Function Analysis" and "mathematics model derivation". In order to explore these individual properties, three groups of factors are taken on: "vegetation arrangements", "site types" and "landscape function", respectively. Regressions and clusters are analyzed jointly to study the water conservation system. In the system application portion of this study, the example of Huan County is studied. Local geography, geomorphology, land use, climate, soil and vegetation properties are identified. Based on the framework of part two, conclusions are stated regarding the landscape water conservation system of Huan County; vegetation arrangements affect soil water, site types restrict soil water, and landscape functions predict soil water. Finally, four levels of Huan County vegetation restoration sites are categorized and six restoration formulas are suggested for "recovery" and "reconstruction" in the next phase of restoration. Further research guidelines are provided following the summary of conclusions and innovations.

T1

P0110

Water depth affects reproductive allocation and reproductive allometry in the submerged macrophyte *Vallisneria natans*

Lei Li[1,2], *Zhiping Song*[2]

1. Nanchang University

2. Fudan University

1. The relationship between size and reproductive output is a fundamental aspect of a plant's reproductive strategy: the conversion of growth into fitness. Differences among populations in the reproductive allometry have been attributed to environmental conditions. However, the causes of variation in reproductive allocation are still debated (i.e. due to fixed patterns of development, or plasticity in the developmental trajectory of vegetative and reproductive allocation). 2. Using a mesocosm experiment, we investigated variation in reproductive allocation of a semelparous submerged macrophyte *Vallisneria natans*, one of the dominant species in Poyang Lake (China). To test the hypothesis that allometric trajectories of reproductive allocation are highly plastic in *V. natans*, we grew plants at three levels (shallow, intermediate and deep) of water depth and measured them after 26 weeks of growth. The allometric relationships between reproductive (R) and vegetative (V) biomass among treatments were compared. 3. In deep water treatment, total plant biomass decreased, average height increased and individuals started flowering in smaller size when compared to the two shallower water treatments. Deep water affected the biomass and number of fruits produced per plant leading to less sexual reproduction. Plants in deep water (the more stressful treatment) had relatively high allocation to reproduction, though they were quite small. The log R vs. log V slope in deep water treatment was much lower than those in more favourable treatments, and shallow water plants exhibited a very different log R-log V relationship than intermediate water plants. A possible explanation is that the environmental limits on size in deep water should favour a relatively large reproductive allocation at smaller sizes and a smaller investment in reproduction per additional unit of biomass accumulated. 4. Our results demonstrated that water depth affects the observed pattern of reproductive allometry among experimental *V. natans* populations, and this has important implications for the fitness of macrophytes exposed to significant variations in water depth over time and space due to anthropogenic disturbance (e.g. dam building) and climate change (e.g. changes in local patterns of precipitation). Environmental stress in deep water tends to favor reproduction at relatively small sizes, associated with lower slopes of the log R-log V relationship, supporting that reproductive allometry can be interpreted as an adaptive strategy of plant growth and allocation, rather than as the product of fixed developmental constrains.

T1

P0111

The academic discovery and current conservation status of *Metasequoia* in China

Minghe Li[1], *Shenhou Fan*[2]

1. Central China Agricultural University

2. Lichuan Forestry Institute

For many years, fossil cones and twigs similar to coast redwood were assigned as ancestors or relatives of *Sequoia* and considered to be extinct for over a million years. During World War II, in 1941, a Japanese scientist studied 'redwood' fossils carefully, found that some of those fossils were sufficiently different from *Sequoia,* and thus gave them a new genus name, *Metasequoia.* In July 1943, a Chinese forester collected specimens in Hubei Province from a large old unknown tree. After World War II, in 1946 a Chinese professor concluded that these trees should be assigned to a new taxon that had not been previously recorded. Then, another Chinese scientist examined the literature and found that the specimens closely matched the description of *Metasequoia* fossils. Thus, it was first realized that living members of the fossil genus *Metasequoia* had been found inChina. In 1947, Harvard's Arnold Arboretum sent US$250 to China for additional field work and seed collection. UC Berkeley also sent US$25 to China for seeds. Chinese grantees then sent many specimens and viable seeds to Harvard and Berkeley. American experts re-examined the large collections of 'redwood' fossils from various parts of the world, and concluded that most of them are actually of *Metasequoia.* In February 1948 a two-person American expedition, funded by the Save-the-Redwoods League and the San Francisco Chronicle, traveled to China. They observed the three living trees and the recently-discovered main population in Xiaohe Valley. Their reports confirmed that living members of the fossil genus *Metasequoia* had been found. The news created a stir in the worlds of science. Scientists at Harvard, Berkeley in U.S.and in China quickly sent *Metasequoia* seeds to interested scientists and institutions in at least 76 recorded localities around the world. In May 1948, Chinese scientists published *Metasequoia glyptostroboides* as the scientific name of what the local people had long called shui shan ('water-fir'). It is also now commonly called 'dawn redwood', a name coined and popularized in the western world. The scientific and academic recognition, and then extensive planting of living *Metasequoia glyptostroboidesa* in the wider world had combined the efforts and intelligence of Japanese, Chinese, and American scientists during 1941 - 1948. Chinese governments have since attached great importance to conservation of the pre-1945 *Metasequoia* trees and populations, and added water-fir to the highest-priority list of its tree-species-protection project. Water-fir-protection stations are now established in Lichuan, Shizhu, and Longshan Counties. There are 5,693 old native water-fir trees at present known living in China. The *Metasequoia glyptostroboides* Type Tree in Mo-Tao-chi continues to grow well. Successful plantings of water-fir have been expanded to 26 provinces inChina, and 'dawn-redwood' plantings have now successfully occurred in more than 50 countries. Natural change and human activities have altered the recent native ecological environments of water-fir. Since 1984,84 old water-fir trees have died. This is a serious issue needing to be concerned about.

T1

P0112

The genetic diversity and structure of the reintroduced population of *Cycas debaoensis* and its new offspring generation

Nan Li[1]*, Yiqing Gong*[1]*, Yanzhao Lu*[2]*, Jinjia Gan*[3]*, Yunhua Wang*[1]*, Ting Chen*[1]

1. Fairy Lake Botanical Garden, Shenzhen, Guangdong

2. Mt. Huanglian Nature Reserve, Guangxi

3. Guangxi Institute of Botany

Cycas debaoensis Y.C. Zhong & C.J. Chen published in 1997 is endemic to China, sporadically distributing in a narrow area of Guangxi and Yunnan Provinces. It shapes like bamboo for the bipinnate leaf with dichotomous leaflets, and has significant scientific and horticultural value. Due to the habitat disturbance and stealing digging, the wild populations of the species were seriously destructed Therefore it is listed as the critically endangered species by IUCN (CR B2), and as the first grade protected plant species in China. The *Ex Situ* conservation of *C. debaoensis* has been launched since 2000 at 'China National Cycad Germplasm Conservation Center' at Fairy Lake Botanical Garden, Shenzhen, Guangdong Province, China. For the effective protection of the species with financial support from the China State Forestry Administration, the center reintroduced 500 seedlings of 5-7 years old, selected from the source population with ca. 700 individuals, to the Mt. Huanglian Nature Reserve, Guangxi Province in 2008, where is about eight kilometers away from the type locality of *C. debaoensis*. 440 individuals in the reintroduced population survived according to the survey data in late 2016. The survival rate was 88%. In 2011 eight male cones were found for the first time in the reintroduced population. In 2012 there were 19 individuals with female cones and 67 individuals with male cones in a proportion of 1:3.5, and 1200 seeds were mature in fall of the year. 30% seeds were left in site, while the other 70% seeds were collected for propagation, and 13 offspring were found in the reintroduction area. More female and male cones came into bloom since then. The beetle can implement the pollination during their life history. The natural seed setting rate of *C. debaoensis* was up to 70%, among which 60% can develop into seedlings. The rodents such as the rats and squirrels like to eat the outer seed coat, and carry the seeds into crannies in the limestone. The new generation of offspring in the reintroduction area were sprouted in succession since 2013, with some even 60-80 meters away from the reintroduction area, which indicates that the founder population has achieved self-renewal. By comparing the growth and reproduction of *C. debaoensis* among the *Ex Situ* conservation population at Fairy Lake Botanical Garden, the wild population at type locality in Fuping county and the reintroduced population, the effect of different conservation approaches was evaluated in this study. Meanwhile, 15 SSR molecular markers were used to detect the genetic diversity and structure of the reintroduced population and the new generation of offspring in the reintroduction area. The genetic coverage of the reintroduced population and the genetic background of the new offspring were calculated by comparing their genetic diversity with that of the wild population of *C. debaoensis*. The effective population size in the destination area for reintroduction was also calculated to infer the population dynamics. This study may provide scientific references for the management of reintroduced *C. debaoensis* and the reintroduction of other species.

T1

P0113

Analysis of genetic diversity from different geographical populations in *Bauhinia varigata* with Sequence-Related Amplified Polymorphism (SRAP) markers

Jiuxiang Huang[1,2]*, Miaokun He*[1,2, 3]*, Qun Li*[1,2]*, Xiaoxi Chen*[1,2]*, Zhijiang Feng*[1,2]*, Shaowei Huang*[1,2]*, Tianyi Liu*[1,2]*, Lincoln Fok*[4]

1. College of Forestry and Landscape Architecture, South China Agricultural University, Guangzhou, Guangdong 510642, China

2. Guangdong Key Laboratory for Innovative Development and Utilization of Forest Plant Germplasm, Guangzhou, Guangdong 510642, China

3. Heshan Shaping Experiment Middle School, Jiangmen, Guangdong, 529700, China

4. Department of Science and Environmental Studies, the Hong Kong Institute of Education, HK, China

Bauhinia variegata Linn is a common ornamental plant in South China, The genetic diversity and differentiation among 7 geographic populations of *Bauhinia varigata* were analyzed using sequence-related amplified polymorphism (SRAP) markers in this study. A total of 59 primers were screened and 13 primers were chosen for SRAP polymorphic analysis. The results showed that 112 bands, 66.67% of the total amplified bands were polymorphic. The genetic Nei's distance among different *Bauhinia varigata* populations was 0.0470~0.2548. The general Nei's genetic diversity index was 0.2265. The general Shannon's information index was 0.3405. Our results indicate that the genetic variation was mainly from individuals within groups (54.20%), and the gene flow between populations (Nm) was 0.5917.

T1

P0114

Ethnobotanical survey in multi used zone Antrema in Mada-

gascar

Tian Li[1], *Hasina Sandrah Rahantanirina*[3], *Edmond Roger*[3], *Claire Harpet*[2], *Marcel Hladik*[2], *Annette Hladik*[2], *Charlotte Pioch*[2], *Bernard Riera*[1]
1. UMR 7179. Mecadev, MNHN et CNRS. F 91800 Brunoy, France
2. UMR 7206. Éco-Anthropologie et Ethnobiologie, MNHN, CNRS et Université Paris-7. F 91800 Brunoy., France
3. Université d'Antananarivo, Département Biologie et Ecologie Végétale, Antananarivo, Madagascar

Antrema, as a multi-use protected area, is well-known for its Crowned Sifaka (*Propithecus coronatus*). The local ethnic group respect for this animal created a more in tune with nature and balanced way to utilize the region's resources, which can be a good example for the conservation and the development of traditional culture in local villages. The aim of this research is to clarify how local beliefs can influence people in responsible use of the forest, and how people take advantage of natural resources in their daily life. In this paper, 65 families in 8 villages were investigated in their utilization of plants, and a taxonomy database Xper2 of Antrema was built in order to facilitate the knowledge of local plants' characteristics. A strict hierarchy of divinity was found in Skalavas' religion, and the respect of their "ancestor" Sifaka endows their ability to protect their forest. Natural resources are extensively used in local people's daily life, like the petiole of *Bismarckia nobilis* Hildebrandt & H. Wendl. and *Hyphaene coriacea* Gaertn, used to do wicker-work, to wave baskets and cushions. The axes of *Ravenala madagascariensis*. Sonn., *Rhizophora mucronata* Lam. and *Ceriops tagal* (Perr.) C.B.Rob. and the petiole of *Raphia farinifera* (Gaertn.) Hyl. are frequently used when constructing homes. Thirty-eight medicinal plants were detected and among them, 23 plants are used for women and 9 plants are used for children. Thirteen plants are used to help delivering babies, like *Abrus aureus* R.Vig., *Anisocyclea grandidieri* Baill., 10 plants can help in curing fever, and the others can cure headaches, stomachaches, cholera and diarrhea. Furthermore, the database Xper2 includes all the information about the usage of these plants, including their medical effects and their morphological characteristics. This can be, in turn, used in websites and can be made available in cellphones, even offline, something that can help academics, scientists, locals or anyone interested in getting more information about local flora. This study identified a close relationship between the locals' interests in conserving the forest and the traditional beliefs of local people in which we can root future efforts of conservation of traditional culture in ethnic settlements, helping the preservation and survival of their way of life vis-à-vis the globalization trend.

T1
P0115
Empowering practitioners to save South China's most threatened tree species

Xiaoya Li[1], *David Gill*[2], *Yanjing Geng*[1], *Margin Georgina*[2]
1. Fauna & Flora International China Programme
2. Fauna & Flora International

Trees have evolved over the last 360 million years into a wide range of shapes and sizes, adapted to a huge variety of habitats. Individual species play a myriad of economic, ecological, and cultural roles highly valued by today's society. However, over 9,600 tree species are threatened with extinction around the world. To safeguard the world's threatened tree species, the Fauna & Flora International (FFI) and Botanic Gardens Conservation International (BGCI) formed the only international programme called the Global Trees Campaign (GTC) in 1999. As part of GTC, FFI's China has been working on the conservation of threatened trees in southern China since 2006. The area is home to more than 200 threatened tree species, many of which are covered by nature reserves. FFI's work mainly focused on empowering the nature reserve staff to protect those threatened tree species to address the major gap that many practitioners lack the skill and experience for the plant conservation. From 2010-2016, FFI supplied six training courses to the chief techical staff of the nature reserve, covering status assessment/IUCN red listing, action plan, patrolling, monitoring, reintroduction, habitat management, awareness raising, communities engagement, conservation effectiveness assessment, involved more than 30 priority nature reserves (i.e. reserves with globally important populations of threatened tree species such as *Magnolia sinica, Abies yuanbaoshanensis, Pinus squamata, Firmiana major*). In the meantime FFI worked with local experts and stakeholders within four reserves and developed some pilot practices. Major results of this work include: (1) More than 60 nature reserves personnel supported to develop new conservation skills. (2) Staff from more than 20 nature reserves mentored to implement new conservation actions for 18 threatened tree species. (3) A plant conservation network established with more than 100 experts and practitioners participating, allowing continued exchange of information, experiences and technical support. (4) Fifteen nature reserves were supported to secure co-funding for threatened tree conservation from various sources including Yunnan and Guangxi Forestry Departments. The project is now attracting significant interest from the State Forestry Administration and several Provincial Forestry Departments who are keen to see the extension of this capacity building programme to a larger number of practitioners, sites and species.

T1
P0116
A preview of mangroves' future when the sea level rises and coastal developments intensified

Xinnian Li[1], *Zixiao Guo*[1], *Renchao Zhou*[1], *Cairong Zhong*[2], *Wenqing Wang*[3], *Suhua Shi*[1]
1. Sun Yatsen University
2. Hainan Dongzhai Harbor National Nature Reserve
3. Xiamen University

Mangrove forests, which underpinned intertidal communities in tropics and subtropics, are increasingly threatened by the continuous sea level rising. More and more mangroves' dieback has been reported in recent years, and these events were reflected the vulnerability of mangrove forests to disturbances. The massive dieback of mangroves happened in unexpected ways regionally. Here we report a survey on a mangrove dieback event in Yalong Bay, Hainan, China between 2010 and 2012. Sea level in Yalong Bay rose as much as one meter and lasted for one week several times through a combination of unusual heavy rainfall, and narrowing of the outlet by coastal developments. More than half of mangrove trees died, and the numbers of species of mollusks declined by 77% in this event. A population genetics survey was conducted based on the samples collected before the dieback event. We found genetic diversity of many mangrove species were extreme low, and the diversity level is negatively correlated to the mortality in dieback event. A larger scale of population genetics analysis reveals a general pattern of low population diversity of major mangrove species in Southeast Asia, which implied that mangrove forests have suffered severe population reduction during the historical rapid sea level rising period. These species would be at risk to maintain the evolutionary potential. The combination of sea level rising and coastal developments in specific circumstances could significantly enlarged the threat to mangrove forests. Although the tragedy of the mangrove forest in Yalong Bay happened regionally, the global mangrove forests are under the same threats including sea level rising, low genetic diversity and coastal development intensified. The observation in Yalong Bay may be a rehearsal of the possible future for global mangrove forests. Therefore, urgent mangrove management for the challenge should be required.

T1

P0117

Effects of cultivation management measures on the growth and flower of *Datura arborea*

Yan Li
Xi'an Botanical Garden of Shaanxi Province

The characteristics of *Datura arborea* L. was discussed with respect to its post-introduction. The growing and flowering habits of *Datura arborea* L. under different cultivation conditions was studied. The result can provide theoretical basis on production of *Datura arborea* in North China. With red flower and yellow flower varieties as experiment subject, the effects on the growth and flower of *Datura arborea* were studied under different of fertilization conditions, illumination intensity, medium and PH, temperature and humidity. At the same time, the plant type of *Datura arborea* was observed under different treatment of plant growth regulators. Leaf fresh weight and leaf area increased obviously under fertilizing Jiabao (N:12%, K2O:44%) and compound fertilizer (N:P:K=2:1:1), but the difference between two treatments is not obvious. The latter treatment has more flower buds. There is a good color appearance with 40% sunshade and 17280-25900Lx. Blosom quantity decreased by 8.41%, blossom diameter increased by 18.53%, corolla length increased by 22.02%. The plant performs slow growing at 0-15℃ and robust growing at 15-30℃ or higher than 30℃ . Significant difference between latter treatments is few. The promotion of temperature can enhance growth rate of ground diameter. The plant has a good type and ornamental value at at 15-30℃. Solution PP333 (0.5-1 g/L) can shorten internode length and increases flower buds. The result showed that the cultivation condition was favorable for the growth of *Datura arborea* L. with 40% sunshade, 17280-25900Lx of illumination, RH 60% and PH 6.5-7.5. The main fertilizer is N:P:K at 2:1:1, and were applied 3 to 4 times of Jiabao during growing period. Solution PP333 (0.5-1.0g/L) were applied three times from sprouting period to beginning of jointing stage to guarantee its ornamental value.

T1

P0118

Research progress on germplasm resource and propagational technique of *Rhododendron* L.

Juyang Liao[1,2], *Chunliang Peng*[1,2]
1. Hunan Forest Botanical Garden
2. Hunan Engineering Research Center for Rhododendron L.

The largest genus of Ericaceae, *Rhododendron* L., consists of about 1000 morphologically diverse species. This genus is widely distributed in Asia, Europe, North America andAustralia (2 species). *Rhododendron* L. is the genus modeled on *R. ferrugineum* by Swedish botanist Carl Linnaeus in 1753, and recorded in the *Species Plantarum*. There are 573 species around China except Xinjiang and Ningxia, 409 endemic species are mainly found in the southwest and southern area. There are 58 species found in Hunan, including two new geographical distribution species, *Rhodedendron orbiculare* and *Rhododendron rupivalleculatum*, two new species, *Rhododendron zhangjiajieense* C. L. Peng et L. H. Yan and *Rhododendron tianmenshanense* C. L. Peng et L. H. Yan. *Rhododendron* is widely distributed in southwestern Hunan, the plant diversity of this genus is gradually increased with the increasing altitude. The vertical distribution center is at the altitude of 800-1500 m, the horizontal distribution center is between N 27°-25° and E 108°-114°, among which the subg. *Tsutsusi* has the largest spanning elevation of 50-2000 m.

T1

P0119

Vascular flora research in wetland of Nanshui Lake in Guangdong Province

Weitong Lin, *Mingxuan Zheng, Yonbin Wu*
South China Agricultural University

According to the survey, there were 187 species of vascular plants in Ruyuan Nanshui Lake wetland, belong to 122 genera 67 families; Herbs occupied the largest proportion of all aquatic plants. Gramineae, Cyperaceae, Polypodiaceae, Thelypteridaceae, Urticaceae, Commelinaceae, Polygonaceae and Pteridaceae were dominant families; and Cyperus, Pteris, Polygonum, Selaginella, Hydrocotyle were dominant genus. There were 6 geographical distribution type and1variant in characteristic families, tropical elements also accounted for 83.33%. The geographical elements of all vascular plants were consisted of 13 distribution types and 7 variant, which tropical elements were dominant. Compared with other wetlands at similar latitude, flora of Nanshui Lake showed significant tropical characteristic than higher latitude region; Meanwhile, compared with other region at similar latitude, flora of Nanshui Lake showed more tropical characteristic than inland region. The result of clustering analysis showed that the flora composition of wetland plants from different region depended highly on location and latitude. By analysising the flora of vascular plants in Nanshui Lake wetland, we intended to provide some reliable information in wetland vascular plant resources protection, recovery and sustainable utilization in the future.

T1

P0120

Using species distribution modelling to improve conservation of endangered plants in Southeast Asia

***Bo Liu**, Hongfeng Wang, Zheping Xu, Maofang Luo, Qinwen Lin, Keping Ma**

Institute of Botany, Chinese Academy of Sciences

Southeast Asia is generally defined as the area that includes the Brunei, Cambodia, East Timor, Indonesia, Laos, Malaysia, Myanmar, Philippines, Singapore, Thailand, and Vietnam. It is famous for its complex and diverse topography, and varying climates, provide a partial explanation for the high biodiversity, four global biodiversity hotspots: Sundaland, Wallacea. Philippines and Indo-Burma occur in this region.The natural vegetation and habitats of Southeast Asia are incredibly varied, ranging from the semi-arid scrub of East Nusa Tenggara, Indonesia, to the riverine forests of southern Borneo, (Kalimantan, Indonesia), the rainforests of Peninsular Malaysia, and the arid hills of Thailand. So far, the rich flora of Southeast Asia is under considerable conversion pressure with expanding human destructive exploration activities and economic development. Conservation strategies based on the geographic patterns of endangered plant species richness, including the identification of meaningful floristic regions and priority areas for conservation, could improve the effectiveness of policy and management of flora in this area. In this research, we produce distribution models at 10 km^2 resolution for all endangered plant species integrated by red list data for all Southeast Asia countries, using geo-referenced herbarium collections from this area, corrected for spatial bias using a null model, and detailed environmental variables, also using Bioclim model to predict the potential distribution areas for medicinal plants. As a results, we identified four important aspects of endangered plant species distribution in Southeast Asia: (1) the endangered species diversity hotspots; (2) major floristic rich regions by grid, using a cluster analysis of species presence/absence; (3) priority areas for plants conservation based on the concept of the 'irreplaceability' value of planning units; and (4) the estimated potential distribution for endangered species with the climate data combined with the specimen records under the global warming image. Our maps provide clear priorities for the development of a sustainable and feasible biodiversity conservation strategy for endangered plants in Southeast Asia.

T1

P0121

The outline of taxonomic literatures and herbaria index of Southeastern Asian higher plants

***Bo Liu**, Hongfeng Wang, Zheping Xu, Maofang Luo, Qinwen Lin, Keping Ma**

Institute of Botany, Chinese Academy of Sciences

Plant taxonomy is the essential basis of all relevant disciplines of botany and horticulture. The very beginning for a taxonomical work is to locate the relevant taxonomical literature, and search for herbarium type and non-type specimens for future revision work. While, taxonomic literatures of Southeast Asia countries (Brunei, Cambodia, East Timor, Indonesia, Laos, Malaysia, Myanmar, Philippines, Singapore, Thailand, Vietnam) still remain incomplete and scattered. Many flora books written in local languages don't have an English name, which makes the young students and newcomers a pretty headache to start from volumes and volumes of books and journals from Linnaeus dated back almost 260 years ago.This research aims to serve as the most updated comprehensive guide dealing with the taxonomical literature of higher plants (covering bryophytes, pteridophytes, and seed plants) in Southeastern Asia, and with an thoroughly inventory of herbaria in this area. The first part introduces the basic information of literature, including classification of literature, grades of literature, literature retrieval, specialty of taxonomic literature, and book classification. The second part introduces different kinds of taxonomic books, including index, bibliography, abstracts, dictionaries, Florae, classification systems, collection history, International Code of Botanical Nomenclature, and Latin and types. The third part deals with journals, including local and foreign ones (arranged by different countries, totally 11 countries). The fourth part is appendices which include major herbaria in Southeast Asia.

T1

P0122

The first garden ancient city is in China

***Fuhu Wu**, Haiyan Liu, Chuanrui Pan*

China Shenzhen Lvya Green Gardening Scientific Institute

Archaeological discovery of Sanxingdui site has been worked for 88 years, after the repeated investigation and study, the discovery of her west half city, the total dikes and strange water management facilities, the incomplete ancient city has been restored into the world's first perfect garden ancient city. In the city, we have found the 12 garden types, such as moon palace, Reevesia tree, feng shui garden, indoor garden and tree wall, dating back for 4087 years. It is different from the previous knowledge: the palace and the axis high-end are not in the small Moon Bay, and they are in the large Moon Bay and Qingguan Mountain, although the axis low-end is located between the two treasure pits of Sanxingdui, the axis is about 40 degrees from north to east. Axis is crooked as the city is crooked, the original palace with the neat symmetry and the proposed symmetric city have been split to the half nearly. As the study, the doll weir and the strange water management facilities in Sanxingdui are made for the hydroponic fire, non-defensive wall moat, the tools are fire, water, wood and bamboo. No prehistoric farmland, ancestors have created the "cited cutting", "sun burning" high-tech logging reclamation. The so-called "cited cutting" is to open water ditch, build lake embankment, introduce water, store water, and flood forest in the man-made lake; The so-called "sun burning" is to break up the lake embankment to deepen the drains 4-5 m and widen 40-50 m after flood and cut forest, make the drainage, dry the trees and roots, fire to burn forest, plant with the hard work, the human history has been put forward from the Stone Age into the great farming era. Inadvertently, it also leads to forest fires, melts the glacial and ice cover, raises the sea, floods farmland, settlement and forests, and brings the endless disaster. The land evidence are the strange channel embankment network, lake embankment or massive red burnt soil blocks many prehistoric sites, such as the Sanxingdui culture; the evidences submerged in the sea floor are the massive forests, temples, settlements and pyramids because of the long-term hydroponic fire and melting ice to raise the sea surface. As archaeological discoveries: England Dorgon waters and Norfolk North County with the equivalent of the European area are the forests and settlements thousands of years ago.

T1
P0123

Influence of environmental factors on the active substance production and antioxidant activity in *Potentilla fruticosa* L. and its quality assessment

Wei Liu
Henan University of Science and Technology

Environmental factors may influence types and contents of active substances. This study investigated the influence of environmental factors on the active substance contents and antioxidant activity of Potentilla fruticosa L. from different regions of China. Also, HPLC fingerprint similarity analysis (SA) coupled with hierarchical cluster analysis (HCA) and discriminant analysis (DA) were further introduced for the accurate classification and quality assessment of P. fruticosa. The results showed that altitude was significantly and negatively correlated to the content of tannin ($P < 0.05$). Annual sunshine duration and altitude were significantly and positively correlated to the flavonoids content, rutin content and antioxidant activity ($P < 0.05$). Annual mean temperature was significantly and negatively correlated to the content of total phenolics, while altitude was significantly and positively correlated to the content of total phenolics ($P < 0.05$). Eight samples were unambiguously separated into three groups. Two types of discriminant functions with a 100% discrimination ratio were constructed. All data consistently supported the conclusion that P. fruticosa produced from Kangding, Sichuan Province had high quality among all samples, therefore, Kangding in Sichuan Province with favorable environmentalconditions is recommended as a preferable production location.

T1
P0124

Geographic distribution of species strongly related to the limestone outcrops in Brazil

Pablo Hendrigo Alves De Melo, Tharso Rodrigues Peixoto, ***Julio Antonio Lombardi***
UNESP - State University of São Paulo

Limestone outcrops are places of great phytogeographic importance and are also relevant to research in archeology and paleontology. However, these sites are strongly threatened by mining activity, since limestone is used in the manufacture of fertilizers and in the production of cement and lime. The floristic composition of the limestone outcrops in Brazil is not well known, although occurring in all phytogeographical domains most of the studies are local in scope, are in the *cerrado* biome, and along the watersheds of the Rio São Francisco and Tocantins. Based in a floristic composition survey under way in ten areas at or near the extreme of ranges of the limestone outcrops formations in extra-amazonian Brazil, and distributed in the phytophysiognomies of *caatinga* (semi-arid scrub forests), *cerrado* (Brazilian savannas types; ranging from dense savanna forests to areas of open grass vegetation), and *Atlantic Forest* (humid evergreen to semideciduous rain forests), at the localities of Cafarnaum and Lajedinho (*caatinga*, Bahia State); Formosa (*cerrado*, Goiás State), Funilandia/Prudente de Morais (*cerrado*, Minas Gerais State), Matias Cardoso (*cerrado*, Minas Gerais State), Bela Vista (*cerrado*, Mato Grosso do Sul State), Nobres (*cerrado*, Mato Grosso State), Novo Jardim (*cerrado*, Tocantins State), Rio Branco do Sul (*Atlantic Forest*, Paraná State) and Apiaí/Iporanga (*Atlantic Forest*, São Paulo State), and search in digital databases (gbif.org; jabot.jbrj.gov.br and splink.org.br), we selected ten species strongly related to the limestone outcrops in Brazil to analyzes of geographic distribution: *Begonia exigua* Irmsch. (Begoniaceae), herb, in *cerrado* biome; *Encholirium longiflorum* Leme (Bromeliaceae), herb, in the transition between *cerrado* and *caatinga* biomes, *E. luxor* L.B.Sm. & R.W.Read, herb, in *cerrado* biome; *Cereus jamacaru* subsp. *calcirupicola* (F.Ritter) N.P.Taylor & Zappi (Cactaceae), arborescent shrub, in *cerrado* biome; *Chautemsia calcicola* A.O.Araujo & V.C.Souza (Gesneriaceae), herb, in *cerrado* biome; *Sinningia calcaria* (Dusén ex Malme) Chautems (Gesneriaceae), herb, in *Atlantic Forest*; *Ceiba rubriflora* Carv.-Sobr. & L.P.Queiroz (Malvaceae), tree, in the transition between *cerrado* and *caatinga*; *Ficus bonijesulapensis* R.M.Castro (Moraceae), tree, in the transition between *cerrado* and *caatinga*; *Priogymnanthus saxicolus* Lombardi (Oleaceae), tree, in *cerrado* biome; *Oxalis calcicola* Fiaschi (Oxalidaceae), herb, in *cerrado* biome; *Hemistylus brasiliensis* Wedd (Urticaceae), scrub, in *cerrado* biome. The species occurrences show distribution patterns restricted to small geographical ranges, indicating regionalization of the distribution of these species strongly related to the limestone outcrops in Brazil. (FAPESP #2015/13112-7 to PHAM and #2016/09444-7 to TRP).

T1
P0125

Species richness and genetic diversity in *Dysopsis hirsuta* (Euphorbiaceae) and *Nothomyrcia fernandeziana* (Myrtaceae), endemic species of the Juan Fernández Archipelago, Chile.

Felipe Saéz[2], ***Patricio López-Sepúlveda[1]****, Glenda Fuentes[1], Hellen Montoya[1], Patricio Peñailillo[3], Tod F. Stuessy[4]*
1. Depto. de Botánica, Universidad de Concepción
2. CONAF Juan Fernandez Archipelago
3. Inst. Ciencias Biológicas, Universidad de Talca
4. The Ohio State University

The Juan Fernández Archipelago is located 667 kilometers west of continental Chile and consists of three islands: Santa Clara, Robinson Crusoe, and Alejandro Selkirk, with a geological age estimated between 1-4 my. The Archipelago is a natural laboratory, characterized by a high percentage of specific endemism. Human activity, involving degradation of the forests as well as introduction of exotic animals and alien plants, has caused a significant reduction in population number and size, as well as genetic variation within endemic plant species. The objective of this contribution is to present information on genetic diversity in two endemic species from the Juan Fernández Archipelago, one an herb and another a tree, in conjunction with general species richness values to determine areas of greater interest for conservation. Samples of *Dysopsis hirsuta* (Mull. Arg.) Skottsb. and *Nothomyrcia fernandeziana* (Hook. et Arn.) Kausel in Robinson Crusoe Island were collected from 10 and 18 populations, respectively. After extraction of DNA, AFLPs were obtained for genetic analysis. Scoring was done with the program GeneMarker ver. 1.85, and a binary matrix was generated for each species. The data were analyzed using Genalex and the software Arc 9.3. In parallel, an analysis of richness of thirteen tree species was performed in Diva-Gis using a circular neighborhood method. In *Dysopsis hirsuta* the values of number of effective alleles (Ne = 1.078), percentage of polymorphism (% P = 50.6%), Shannon-Weaver Index (I = 0.237), and expected heterozygosity (He = 0.154) were lower in populations with less species richness (<7 species) and invaded sites. The highest values of genetic diversity (Ne = 1.3, %P = 65, I = 0.272, He = 0.170) were found in sites with high species richness (>9 species) and in fragments of native forest. *Nothomyrcia fernandeziana* does not evidence a relationship between genetic diversity and species richness, and it has high values of genetic diversity in invaded areas (%P invaded = 61.2%, %P native = 54.8%). We discuss the levels of genetic diversity in *D. hirsuta* in relation to forest fragmentation, isolation, low gene flow between populations, and mode of speciation. Homogeneous genetic diversity within *N. fernandeziana* may be associated, among other factors, with the continuous distribution of this species on Robinson Crusoe Island.

T1

P0126

Resource sharing and facilitation of epiphytes in forest canopies

Hua-Zheng Lu[1,2], *Liang Song*[1], *Rob Brooker*[3], *Fei-Hai Yu*[4], *Lawren Sack*[5], *Wen-Yao Liu*[1]

1. Key Laboratory of Tropical Forest Ecology, Xishuangbanna Tropical Botanical Garden, Chinese Academy of Sciences, Menglun, Mengla

2. University of Chinese Academy of Sciences, Beijing

3. The James Hutton Institute, Craigiebuckler, Aberdeen

4. School of Nature Conservation, Beijing Forestry University, Beijing

5. Department of Ecology and Evolutionary Biology, University of California Los Angeles, Los Angeles

Forest canopies house ca.50% of terrestrial biodiversity. One important component of canopy biodiversity is the presence of epiphytes, but how epiphytes adapt to forest canopies remains one of the key questions in canopy ecological research. One notable pattern is that almost all non-vascular epiphytes and many vascular epiphytes are capable of clonal growth. However, how clonal traits such as physiological integration (resource sharing) affect the adaptation of epiphytes to forest canopies is virtually unknown. Epiphytic habitats are usually described as "harsh" because tree crowns are characterized by a limited storage capacity for available nutrients and water, sporadic and dilute nutrient inputs, low physical stability, extreme fluctuations in moisture and temperature, high wind speed, and severe and variable vapor pressure deficits. Significant variation in resource availability can occur at small spatial and temporal scales, and short-term drought can occur even in wet seasons of tropical rainforests. Although it was conceived that competition was the basic ecological process, stress-gradient hypothesis (SGH) implied that the role of competition decreased and facilitation increased with the rise of stresses. However, we know little about how epiphyte species interact with each other and what their effects on community assembly are. The experiments were carried on the clonal integration (with or without resource sharing) of several dominant vascular epiphytes and on its effects on interaction between vascular and non-vascular epiphytes (removed or intact) in a subtropical evergreen broad-leaved forest in Southwest China. Results were as follows, 1) preventing integration (without resource sharing) significantly reduced the ramet survival and biomass of epiphytes, but the magnitude of such integration effects did not vary significantly between species; 2) non-vascular epiphytes (bryophytes and lichens) facilitated the survival and growth of vascular epiphytes; 3) the effects of clonal integration on performance of vascular epiphytes without non-vascular epiphytes were higher than those with non-vascular epiphytes. We concluded that clonal integration (resource sharing) was a general strategy for clonal epiphytes to adapt to forest canopies where resources were limited and heterogeneously distributed in space and time. Meanwhile, our studies provided the first evidence of facilitation between vascular and non-vascular epiphytes and deepened our understanding of habitat adaptation of resource sharing and species interaction of epiphytes in the community assembly in forest canopies. It also shed light on the conservation of canopy biodiversity because of the facilitation cascades and domino effects.

T1

P0127

Antibacterial activity, optimal culture conditions and first successful cultivation of *Ganoderma australe*, a new record to Thailand

Thatsanee Luangharn

Kunming Institute f Botany, Chinese Academy of Sciences, Kunming, China

Ganoderma is a well known genus of medicinal mushrooms that belongs to the order Polyporales. Many members of this genus are used extensively in traditional Asian medicines, and have wide potential for application in bioremediation. Even though 22 species of medicinal *Ganoderma* have been described worldwide, very few *Ganoderma* species have been domesticated. Herein we report a new strain of *Ganoderma australe* collected in Thailand and identified through phylogenetic analysis and an examination of its micro and macro morphological characteristics. We report for the first time in Thailand the antimicrobial activities of *G. australe* and the optimal conditions for mycelia and spawn production and cultivation. Ethyl acetate extracts of *G. australe* mycelia inhibited both Gram positive and negative bacteria including *Micrococcus luteus*, *Bacillus subtilis*, *Staphylococcus aureus* and *Salmonella typhimurium*, while it was not effective against *Escherichia coli* and *Pseudomonas aeruginosa*. The fungus mycelium was able to grow on PDA medium, 25–30°C and pH 7–8, while sorghum and barley were found to be the best grain media for spawn production. Para rubber sawdust with organic and inorganic additives was used as a standard cultivation substrate. The mycelia were fully distributed throughout the sawdust substrate after being incubated at 30°C, 60-75% relative, for 18 days. The first primordia were observed three days after the bags were opened. Three cycles of mushrooms were obtained, with the weight of mushrooms produced decreased each time due to the degradation of the substrate (115.04 ± 48.85 g kg^{-1},72.01 ± 6.00 g kg^{-1} and 36.47 ± 9.04 g kg^{-1}, respectively).

T1
P0128

Wild growing Vacciniaceae in natural populations and prospects of cultivation

***Ekaterina Luginina**, Tatyana Egoshina*
Russian Research Institute of Game Management and Fur Farming

Main wild growing species of *Vacciniaceae* family are cowberry (*Vaccinium vitis-idaea* L.), blueberry (*V. uliginosum* L.), cranberry (*Oxycoccus palustris* Pers.), and bilberry (*V. hirtum* Thunb., *V. ovalifolium* Smith, *V. myrtillus* L.). Biological stock (BS) of raw material of the studied *Vacciniaceae* species in Russia in average fruitful year is 4,733.9 thous. t. The most significant stock of these species is marked for Siberian (BS 1,818.9 thous. t), North-Western (BS 1,117.2 thous. t) and Far Eastern (BS 632.2 thous. t) Federal districts. The largest stocks among separate territories are in Krasnoyarsk Krai (BS 810.5 thous. t), Khanty-Mansiisk Autonomous District (BS 477.3 thous. t), Komi Republic (BS 473.8 thous. t), and Irkutsk region (BS 423.1 thous. t). In the structure of total BS bilberry fruits (40%) and blueberry fruits (27%) prevail. Exploitation rate of *Vacciniaceae* resources in different areas of the country varies from 20 to 98%, and the press on natural populations is increasing. The most significant anthropogenic load is marked for natural populations of berry plants near large cities in the central part of Russia and in boundary regions. In order to decrease exploitation load on natural populations in Russia the introduction, cultivation and selection projects are carried out. The experience of long-term (since 1987) introduction of wild growing berry plants in southern taiga subzone of Russia (Kirov region) allowed revealing the species and cultivars of plants that are highly productive, early ripening, and winter-hardy in conditions of the region with cold climate and short vegetation period. Out of 15 tested cultivars of *Oxycoccus мacrocarpus* Pers. only fruits of Bean Lear cultivar were able to get ripe annually before autumn frosts. Plants of Franklin, Early Black, Wilcox, McFarlin cultivars provided 0.3-1.8 t/ha yield in conditions of Kirov region. Flower productivity reached 40%. In 70% of years underripe fruits were damaged by early autumn frosts. Cultivars of *Oxycoccus palustris* Pers. of Russian selection (Krasa Severa, Dar Kostromy, Alaya Zapovednaya, Khotavetskaya, Severyanka, Sominskaya) gave high yields annually (about 3.6 t/ha), were early ripening (end of August - beginning of September) and had large fruits. Maximum weight of “Dar Kosrtomy” berries reached 3.3 g, “Krasa Severa” - 2.5 g. These cultivars proved to be winter-hardy: frost-killing was not marked during the period of studies. Cultivars of Estonian selection (“Tartu”, “Virusaare”) ripened in the end of September. Their productivity was about 2.8 t/ha. Fruit damage by autumn frosts was registered. These cultivars were the least winter-hardy among all European cranberry cultivars. Experimental plantings of *Oxycoccus palustris* Pers. were consistently fruitful; unfruitful year was registered once during the whole study period. Among 3 tested cultivars of *Vaccinium vitis-idaea* L. (“Coral”, “Rubin”, “Kostromskaya rosovaya”), “Coral” was the least winter-hardy in conditions of experimental ground. The cultivar also had active reflorescence which weakened main spring flowering and fruit setting. There were also trials on *Vaccinium canadense* Kalm. cultivation but in conditions of Kirov region the plants were annually frostbitten at the level of snow cover and never fruit. It was ascertained that only *V. angustifolium* Ait., which is a low shrub in regional conditions and reaches 0.27-0.6 m height on the third year (average 0.43 m), shows suitable winter-hardiness. The level of productivity of 3-year old plants varies from 18.2 to 406.0 g (average 177.5 g, which is about 407 kg/ha). During the latter years, the productivity increases reaching 4-6 t/ha at the age of 5 and more. Average berry weight varies from 0.9 to 5.3 g. It was evident that the least winter-hardy are cowberry “Coral”, American cranberry “Ben Lear”, European cranberry “Tartu” and “Virussaare”.

T1
P0129

Pollen and stamen mimicry: The most speciose mimicry system of the world?

Klaus Lunau
Heinrich-Heine-University Duesseldorf

Many melittophilous flowers display yellow and UV-absorbing floral guides that resemble the most common colour of pollen and anthers. The yellow coloured anthers and pollen and the similarly coloured flower guides are described as protagonists of a pollen and stamen mimicry system. In this study, we investigated the entire angiosperm flora of the Alps, of Crete (Greece), and of Barro Colorado Island (Panama) in regard to visually displayed pollen and floral guides. All angiosperm species were checked by means of colour photos for the presence of pollen- and stamen-imitating structures. For some Flower UV-photos are used to confirm the ultraviolet absorbing properties of pollen and floral guides. More than 70% of the flowering plants of the Alps display yellow pollen, more than 65% of the species display yellow anthers and 28% of the species display pollen- or stamen-imitating structures. The percentage of species displaying pollen- or stamen-imitating structures on Crete is 24%, and on BCI 12%. The most frequent types of pollen and stamen imitations are yellow, UV-absorbing colour patches of petals, stamen-like pistils, and staminodes, and three-dimensional structures such as lower lips and filamental hairs. Dichogamous and diclinous species display pollen- and stamen-imitating structures more often. The visual similarity between the androecium and other floral organs is attributed to mimicry, i.e. deception caused by the flower visitor’s inability to discriminate between model and mimic, sensory exploitation, and signal standardisation among floral morphs, flowering phases and co-flowering species. We critically discuss the deviant pollen and stamen mimicry concepts and evaluate the frequent evolution of pollen-imitating structures in view of the conflicting use of pollen for pollination in flowering plants and pollen eating pollinators such as hoverflies as well as pollen collecting pollinators such as bees.

T1
P0130

Disentangling community assembly mechanisms in distinct layers of plant communities along elevational gradient in subalpine forests

Yahuang Luo[1], *Jie Liu*[1], *Shaolin Tan*[1], *Dezhu Li*[2], *Lianming Gao*[1]
1. Key Laboratory for Plant Diversity and Biogeography of East Asia, Kunming Institute of Botany, Chinese Academy of Sciences, Kunming, China
2. Germplasm Bank of Wild Species in Southwest China, Kunming Institute of Botany, Chinese Academy of Sciences, Kunming, China

Determining how the relative strength of community assembly processes varies along elevational gradients is critical for understanding community responses to environmental change. However, little is known about the different forms of environmental variables that maintain the diversity of distinct vegetation layers. The community assembly processes of three growth forms were inferred by using functional traits and phylogenetic relatedness data in 15 0.1-ha plots across 1,200 m elevational gradients in Yulong Mountains, Yunnan Province, China. Abiotic conditions in terms of topography, soil nutrients, and climatic variables were also quantified. Null model was used to detect the importance of abiotic filtering and biotic interactions along elevational gradient. The effect of spatial and environmental variables on species, functional and phylogenetic diversity within community and across communities were also conducted. Our results revealed the assemblages of overstory tree species tended to be overdispersed at low elevation and clustered at high elevation, and the climatic variables (e.g. minimum temperature and mean annual precipitation) appears to be important factors to determine the species richness and phylogenetic structure. Whereas the understory woody species exhibited opposite trends with elevation, suggesting an important role for environmental filtering (e.g. aspect) in driving phylogenetic convergence at lower elevation. Overdispersion dominated in understory herbaceous assemblages, possibly due to limiting similarity. The functional diversity of leaf area and seed mass were associated with elevation among three vegetation layers. Species in different layers structured by stronger dispersal limitation showed faster species turnover than phylogenetic and functional turnover. The findings suggest the importance of abiotic stress or biotic interactions in shaping the diversity of different group communities at different elevations. Meanwhile, neutral processes may also be a predominant in structuring species composition along elevational gradient. Our study highlight the need for the combination of phylogenetic and trait-based approaches to better understand the species coexistence and community assembly in distinct layers of forests.

T1
P0131
Mapping Asia plants

Keping Ma[1], *Bo Liu*[1], *Hongfeng Wang*[1], *Maofang Luo*[2], *Zheping Xu*[1], *Xuehong Xu*[2], *K Ganeshaiah*[3]
1. Institute of Botany, Chinese Academy of Sciences
2. Biodiversity Committee, Chinese Academy of Sciences
3. University of Agricultural Sciences, GKVK Bangalore

Because of the huge population and rapid growth of economy, biodiversity including plants are being seriously threatened in Asia. Both conservation and sustainable use of biodiversity need species distribution data. Asia is the continent with highest diversity of plants. However, the information of plant taxonomy and distribution in Asia are less available and more scattered, compared with other continents such as Europe and America. So, a better infrastructure for plant distribution in Asia is important. In 2015, Asia Biodiversity Conservation and Databases Network (ABCDNet) started a project entitled Mapping Asia Plants for cataloguing species of plants and collecting distribution data from the following aspects: 1) specimens, 2) observations, 3) species checklists, 4) expert maps, 5) inventories, 6) vegetation plots, 7) literatures and 8) on line datasets. Major achievements so far are as follows: 1) for China, national higher plant checklists including 35,835 species, 9 million records of occurrence; 2) for Southeast Asia, 120 thousands species names, 1.3 million records of occurrence; 3) for North Asia, 6,900 species of higher plants and their distribution in 13 sub-regions. Some of the data collected are shared with the websites of ABCDNet (www.abcdn.org).

T1
P0132
Studies on sesquiterpenes from *Chenopodium album* Linn

Qinge Ma, *Rongrui Wei*
Nanyang Normal University

Chenopodium album Linn is a favorite vegetable in folk, which belongs to the Chenopodiaceae family. In this work, four sesquiterpenes (1-4) were isolated from *C. album* Linn for the first time. They were identified as arecolactone (1), 4α, 8β-dihydroxy-eudesm-7 (11)-en-12,8-olide (2), stereumin P (3), and lyophyllan-etriol A (4) by their extensive spectroscopic data and the reported data in the references.

T1
P0133
Ecological assessment and monitoring of biodiversity in terrestrial and aquatic ecosystems in Didipio Gold Copper Project, Nueva Vizcaya, Philippines

Pastor Jr. Malabrigo[1], *Nelson Pampolina*[1], *Juancho Balatibat*[1], *Crusty Tinio*[1], *Dianne Joy Aguilon*[1], *Keshia Tingson*[1], *Bonifacio Labatos*[2], *Arthur Glenn Umali*[1]
1. University of the Philippines Los Baños, College of Forestry and Natural Resources
2. Oceana Gold Philippines Inc.

The Didipio Gold-Copper Project (DGCP), owned by Oceana Gold Philippines Inc. (OGPI), is situated at the southern part of Palali-Mamparang mountain range, a known key biodiversity area in the region. As mineral reserve co-exists with biodiversity in DGCP site, it is important that mining activities are not affecting biodiversity and the ecological system. The government, hence, requires the conduct of environmental impact assessment and regular monitoring to ensure environmental compliance of the mining company. To comply with it's obligation, OGPI commissioned a group of biologists from the University of the Philippines Los Baños, College of Forestry and Natural Resources (UPLB-CFNR) to conduct the monitoring. In the past three (3) years, UPLB-CFNR diligently conducted the annual ecological assessment and monitoring of biodiversity in the terrestrial and aquatic ecosystems in DGCP. The monitoring conducted was one of the most complete and comprehensive, as it includes all the major groups of organisms (i.e. flora, vertebrate fauna, arthropods, fungi, and lichens). Nine sampling sites representing low, medium, and high impact zones were identified as monitoring stations for terrestrial ecosystems where a combination of transect and quadrat sampling techniques were used. For freshwater ecosystem, macrobenthos, periphyton, and fish assemblages were surveyed in 8 river stations within the mining area. For the three years of monitoring, UPLB-CFNR has accumulated biologically important information that are useful for the company to better conserve and manage the biodiversity in the surrounding area. A total of 484 morpho-species of plants were documented, 30% of which are new province records, 79 species were found to be Philippine endemics and twenty-three (23) species are threatened. More importantly, two undescribed species were discovered. The mountain ecosystem is also home to thirty-seven vertebrate wildlife species comprising of 24 species of birds, seven mammals, three reptiles and three species of frogs. About 51% of the species are Philippine endemics, including 5 threatened species. Beneath the forest ecosystem were terrestrial macrofungi comprising of 37 species, 34 genera and 33 families. Mostly were litter decomposers, some edible and medicinal, while a number were ectomycorrhizal that link roots of dipterocarps and other endangered non-dipterocarps for exchange of carbon and essential nutrients. Across the vertical forest stratum were lichens with 201 records belonging to 42 taxa, 32 genera, and 17 families. In aquatic ecosystem, periphyton assemblage comprised of 22 algal species from divisions of Bacillariophyta (diatoms), Chlorophyta (green algae) and Cyanophyta (Cyanobacteria). The macrobenthos communities list was 285 individuals under 10 taxonomic groups. The top three taxonomic orders were Ephemeroptera (mayflies), Trichoptera (caddisflies), and Hemiptera (water bugs). Trace metals from sampled tissue of freshwater crab were all below the maximum allowable limits set by various international standards. Overall, the three years of monitoring suggests that biodiversity and ecosystem were sustain at a level that allow their normal physiological and ecological processes to function and continue to provide goods and services in Didipio mountains from terrestrial down the river system. Their monitoring and assessment however warrant continued system as guiding parameters by the company and community.

T1

P0134

EDC's ascendancy in the conservation of Philippine Ironwood trees

Xanthostemon Spp. Myrtaceae

Pastor Jr. Malabrigo[1], Ronino Gibe[2], Ernesto Jr. De Guzman[2]

1. Department of Forest Biological Sciences, College of Forestry and Natural Resources, University of the Philippines Los Ba?os
2. Energy Development Corporation

The genus *Xanthostemon* is globally recognized as the densest and strongest among the wood-producing trees. Members of the genus are collectively called Ironwood trees because of their exceptional hardness. The Philippines is blessed to have five unique species of Ironwood trees. All are endemic to the country with very narrow range of distribution. *Xanthostemon* spp. are almost exclusively found in ultramafic forests with soil rich in heavy metal elements. With the long history of mining in mineralized soil of the country, habitat of Ironwood trees became very much devastated putting the fate of these toughest trees into the brink of extinction. Recognizing the urgency to save the remaining populations of the world's hardest trees, Energy Development Corporation (EDC), the Philippines' leading industry in renewable energy production, launched a noble tree conservation program in 2008, called "BINHI" (Filipino term for seedling). Using science-based criteria based primarily on ecological and economic importance, EDC came up with 96 priority threatened trees for conservation including all the Philippine Ironwood species. For the past 8 years, with the partnership with the government agencies and academic institutions, EDC managed to discover and protect several new populations of ironwood species. The company's state of the art vegetative material recovery (VMR) facility has already developed protocols for the mass propagation of species through cuttings with survival rate ranging from 93-97%. Some of the planted seedlings among EDC's 133 partner institutions (mostly high schools and universities) are now source of seeds even at a very young age. Aside from species rescue and mass propagation, BINHI has a very virtuous advocacy and awareness component. EDC sponsors an annual BINHI Youth Conference participated by selected high school students from all over the country. The 3-day conference aims to deepen the level of awareness and appreciation by the young generation to the beauty and importance of the Philippine native trees. In all of the 133 arboreta established, seedlings were properly labeled including the name of the planters. EDC is maintaining a BINHI website wherein photos from regular bi-annual monitoring were continuously uploaded so that the planters can monitor the growth and status of the seedlings they planted even away from the arboretum. Lastly, all botanical information gathered for the 8-years of the project including new distributions, phenology, and propagation technology were recently published into a book "BINHI Tree for the Future" to disseminate the project learning and experiences to a larger public. With the efforts of EDC to mainstream threatened species in urban landscaping and the protection of their remaining natural populations, the Philippines is now assured that the precious ironwood trees are now far from being extinct.

T1

P0135

Azerbaijan endem rare and endangered trees and shrubs

Tofig Mammadov

Institute of Dendrology of Azerbaijan National Academy of Sciences

Azerbaijan is the richest area in biodiversity in the Caucasus region due of a special plant cover distributed by geographical location, relief, soil and climate conditions. The diversity of flora and fauna species in Azerbaijan and classy endemism are associated with environmental conditions in the area due to the historical evolution. Despite of the country richness in flora and fauna in the past 30 years they are close to the destruction as a result of irrational, or various natural and human factors rapidly degradation a variety of plant species. Therefor by impacts of anthropogen factors many of precious trees and shrubs in forests areas *Quercus macranthera, Acer trautvetterii, Betula pendula, B.litvinovii, Paeonia mlokosewitschii, Laurocerasus officinalis, Taxus bacca-*

ta, Pinus eldarica, P. Kochiana, Punica granatum, Cotoneaster saxatilis, Juniperus foetidissima, Pistacia mutica, Rhus coriaria, Rosa spp, *Pyracantha coccinea* are extremely reduced. In addition there are created 9 National Parks, 11 State Nature Reserves and 20 State Nature Preserves in Azerbaijan. In created special protected natural reserves are conserved the majority of the flora and fauna species entered to "the Red Book". There arecarried out research works by ourselves to determine the protection of trees and shrub species in our republic, by accurate preparation of the records, the discovery of these plants belonging to rare plants category. There are researched processes by the distribution of plants, largenes of populations, an ecological resistance, physiological tolerance, reproductive biology, the next stages of developments. There are determined that in 1975 in Azerbaijan Dendroflora were existed 342 species, among them 24 rare and endangered species, in 1984 28 species, in 1989 45 rare and endangered woody trees and shrub species. Currently by the research works are determined that in Azerbaijan Dendroflora are found about 6000 plant species, among them 1600 species are trees and shrub species. They are forming approximately 37,5% of trees and shrubs of Azerbaijan flora. Recently are estimated the protection status of total flora, 300 plants (266 higher and 20 lower), 14 mushrooms are included to Azerbaijan "Red Book". By carried research woks are defined that plant species introduced in Azerbaijan flora are forming 68% species, 87%, genus, 95% families propagated in nature. 25% of trees and shrubs by specifications are relict in natural flora of Azerbaijan. There are determined by carried research works distributed the of rare and endangered trees and shrubs in Azerbaijan, studied results of the diminishing of distribution areas, endangered factors, defined the threat categories. By research works are used IUCN (2001) 3.1 versions categories and criterias, classification of taxas by APG III system. In addition there are comparative analized Azerbaijan Dendroflora and determined rare and endangered trees and shrubs number of species are 38 families, 63 genuses, 188 species. Among them 67 species are Caucasus endemic, 40 species are Azerbaijan endemic, 4 species are relict.

T1

P0136

Medicinal plants among the Meranaos of Lanao, Philippines

Muhmin Michael Manting, *Norfaisah Razuman*
Mindanao State University-Iligan Institute of Technology

Medicinal plants are widely used all over the world to manage diseases and for health maintenance. This has played an important role in human's health for many centuries. Traditional medicinal knowledge, especially using medicinal plants in the developing countries has been in existence and use, and has been a part of therapeutic practices. The traditional use of medicinal plants is widely accepted in the Philippines. Plant species, cultures and practices on using the medicinal plants also vary in every island. Documentation of the indigenous knowledge on plants with ethnobotanical importance must be made. Thus, this study aimed to document and surveys the plants of the Meranaos of Lanao Provinces, Philippines using semi-structured questionnaires through Key Informant and Informal Interviews with a total of 160 respondent ages 60 years old and above. Seventy (70) medicinal plants in Lanao Del Sur and 43 medicinal plants in Lanao Del Norte were documented. Leaves were the most commonly used part in making traditional medicines and decoction method was the most used mode of preparation of medicines. Difference between the two provinces in the uses and the name of some plants varies such as local name of *Plectranthus scutellarioides* L. - "Paer" in Lanao Del Sur and "Mayana" in Lanao Del Norte. In terms of its plant parts and medicinal used, *Zea mays* L. in Lanao Del Sur leaves was used to treat as cure to abscess and cystitis while in Lanao Del Norte the root was used to cure problems in Urinary Tract Infection. The research on ethnobotanical plants can be a great help to a researcher that has ethnoparmaceutical studies and also it helps preserve traditional ethnobotanical knowledge of the Meranao people.

T1

P0137

Trees of Dinagat Islands, Philippines: Basis for biodiversity conservation

May Angelie Paracale, Paul Alfonso Ondoy, Benjam Ray Maligro, Nikkimark Samoya, ***Muhmin Michael Manting***
Mindanao State University-Iligan Institute of Technology

The effect of anthropogenic activities on biodiversity is a global concern. Trees are bio indicators used primarily in the development and implementation of biodiversity conservation programs. Thus, a rapid floristic survey was conducted in Cuarenta Watershed, Municipality of San Jose and Panamaon Watershed, Municipality of Loreto to assess and document the tree species present in the area. The sites were chosen since the area is rich in mineral deposits, thus, mining and overexploitation are the major environmental threats. Data were obtained by means of transect walk on the trail used by the locals in traversing the watershed and photographic documentation. Identification of plant species was based on the given local names, comparison of specimen images with illustrated checklist of the philippineplant.org, the database of Co's Digital Flora of the Philippines, and other published scientific articles. As a result, 88 species were identified and recorded during the assessment period. Thirty percent of the taxa assessed in the study are on list of Threatened Tress Species. In contrast with the assessment of the threatened native flora of the Philippines released by the Department of Environment and Natural Resources, about 22% of the nationally threatened plants are locally threatened. The presence of nationally and globally threatened species calls for the strong development of conservation programs for the long-term protection of the forest ecosystems in the Island of Dinagat.

T1

P0138

Gorilla food plants biomass distribution and variability adjacent to the Volcanoes National Park, Rwanda.

Olivier Jean Leonce Manzi[1,2], *Elias Bizuru*[1]
1. Dian Fossey Gorilla Fund International
2. University of Rwanda/College of Science and Technology

Gorillas ranging patterns are strongly influenced by their food plant distribution. From 2008, an increase of gorillas ranging outside the park in cultivated areas has been observed resulting in a rising conflict between gorilla and local people. This study aimed to determine gorilla food plants and their biomass distribution adjacent to the VNP which will help to determine whether there is a relationship between food availability and the frequency of gorilla visits outside park boundaries. We located about 93 transects perpendicular to the forest edge of 500 m length, each separated by 250 m. A total of 279 plots were established along transects in 250 m intervals. The plot size depended on the dominant plant life form. We identified all plants eaten by gorillas in each plot and calculated the gorilla food plant biomass using standardized methods. Results showed that there are 25 gorilla food plant species near the VNP boundaries. Dominant species were *Eucalyptus* sp., *Y. alpina* and *Rubus* sp. Three gorilla food plant species were found outside but not inside the park (*Eucalyptus* sp., *C. lusitanica,* and *A. glutinosa.)* Among the top 10 consumed plants inside the park, only 3 of them were not found outside the park (*L. alatipes, P.kerstenii,* and *D. erici-rosenii)*. For different life forms, *D. iners, C. nyassanus, Y. alpina*, and *Eucalyptus* sp. showed the highest biomass respectively. Gorilla food plant species diversity and biomass distribution outside the VNP were strongly influenced by both farming intensification and the distribution of Eucalyptus trees. This study shows that gorillas' ranging outside the park is due but not limited to the presence of their food outside the park. The failure to manage effectively the zones of interactions outside the park will result in continuous ranging outside the park and this will result in negative impacts for both gorillas and humans.

T1

P0139

Classification and mapping of the woody vegetation of Gonarezhou National Park, Zimbabwe

Francesco Martini[1,2]*, Robert Cunliffe*[1]*, Alessio Farcomeni*[1]*, Michele De Sanctis*[1]*, Giacomo D'Ammando*[3,4]*, Fabio Attorre*[1]

1. Sapienza University of Rome

2. Guangxi University

3. University of Liverpool

4. University of the Witwatersrand

Within the framework of the Great Limpopo Transfrontier Conservation Area (GLTFCA), the purpose of this study was to produce a classification of the woody vegetation of the Gonarezhou National Park, Zimbabwe, and a map of its potential distribution. Cover-abundance data of woody species were collected in 330 georeferenced relevés across the Park following a plotless approach with a sample area between 0.5-2 ha. These data were used to produce two matrices: the first one using the cover-abundance values as collected in five height layers and the second one based on merging the layers into a single cover value for each species. Automatic classifications were produced for both matrices to determine the optimal number of vegetation types. The two classification approaches both produced 14 types belonging to three macro-groups: mopane, miombo and alluvial woodlands. The results of the two classifications were compared looking at the constant, dominant and diagnostic species of each type. Based on these values and on expert knowledge the classification based on separate layers was considered more effective and retained. A map of the potential distribution of vegetation types for the whole study area was produced using Random Forest at a spatial resolution of 100 m. In the model, the relationship between bioclimatic and topographic variables, known to be correlated to vegetation types, and the classified relevés was used. A comparison with the only known previous assessment and description of the vegetation types of the Park was conducted. Moreover, identified vegetation types were compared with those of other national parks within the GLTFCA, and an evaluation of the main threats and pressures was developed. Vegetation classification and mapping are useful tools for multiple purposes including: surveying and monitoring plant and animal populations, communities and their habitats, and development of management and conservation strategies. Filling the knowledge gap for the Gonarezhou National Park provides a basis for standardised and homogeneous vegetation classification and mapping for the entire Great Limpopo Transfrontier Conservation Area.

T1

P0140

Research of biological peculiarities and *ex situ* conservation of threatened georgian endemic species *Dianthus azkurensis* Sosn

Nina Melia*, Tinatin Barblishvili*

National Botanical Garden of Georgia, Department of Plant Conservation

We present the results of multilateral study of the endangered wild species of Georgia's flora *Dianthus azkurensis* Sosn. (EN) (Caryophyllaceae), included in the Red List of the Endemic Plants of the Caucasus and elaboration of effective measures for its protection. With this aim consecutive phases of ontogenesis- embryogenesis, seed productivity and viability, terms and rate of germination, age stages of seedlings development were studied in this species for the first time. *Dianthus azkurensis* Sosn. grows on rocks, in the middle mountain zone. The species is characterized by protandry. Study of embryonic structures revealed that at the initial stage generative sphere develops normally. The 80% of pollen grains is fertile. Pollen germination was optimal, attaining 15-20%, on artificial medium - 10% saccharose. Number of ovules in the ovary reaches 90, with 2 integuments, arranged in 4 rows on placenta. Embryo sac develops by Polygonum type and is situated on the chalasal end of ovule. Retardation occurs only in the process of fertilization and only small number of ovules finishes development. Number of fertilized ovules though is few, but all develop normally. In the adult seed dicotyledonous embryo occupies the whole seed perimeter. At an early stage endosperm is completely consumed by the embryo. On the next stage the embryo is surrounded by the perisperm. In order to determine seed viability, seed were tested on germination and sprouting capacity on different substrates under laboratory conditions. Spring sowing yields higher germination (91.6%) as compared to autumn sowing (60%), though in both cases germination starts on agar substrate in 3 days after sowing. Seeds were sown in pots to obtain *ex situ* collections of seedlings. While sowing in soil in spring germination starts in 4 days, while in autumn germination takes 2 weeks. On the stage of 5-6 storey of leaves seedlings are too week and auxiliary treatments are required to obtain robust planting material. Seeds of *Dianthus azkurensis*, banked in 2007 in the Caucasus Regional Seed Bank (CRSB) at -20°C were tested on viability. Seed storage behaviour is good and it retains high germination capacity – 70-75%. *Ex situ* conservation activities were carried out. Seed bank of this endangered species - *Dianthus azkurensis* - was created in the CRSB of the National Botanical Garden of Georgia and seed collections duplicated to the Millennium Seed Bank of the Royal Botanic Gardens, Kew, UK.

T1

P0141

The caucasus regional seed bank development within the Millenium Seed Bank Partnership

Tsira Mikatadze-Pantsulaia, *Tinatin Barblishvili, Ana Gogoladze, Tolkha Shetekauri*

National Botanical Garden of Georgia, Department of Plant Conservation

Since 2005 staff of the Department of Plant Conservation of the National Botanical Garden of Georgia (NBGG), working in partnership with colleagues at Institute of Botany Georgia (Ilia State University) have been participating in the global Millennium Seed Bank Partnership, led by the Royal Botanic Gardens, Kew. Priority for the project was creation of the seed bank of endemic, rare, threatened, vulnerable and economically important plants. Seeds of target species are to be necessarily collected in the wild. For each species special data sheet is filled in, which along with geographic and botanical data contains comprehensive information about the sample, its habitat and the process of collecting. At least 500 seeds of each species are deposited at the CRSB and the corresponding labelled herbarium specimen kept at the Herbarium of Georgia (TBI). The duplicates of the seed collection and herbarium voucher are sent to the Millennium Seed Bank and Herbarium of the Royal Botanic Gardens, Kew (K). The Caucasus Regional Seed Banks is designed for the long-term storage of seeds. During the 2005-2016 period, within the scope of MSB-1 and MSB-2, seeds and herbarium samples of more than 1400 plant species and interspecific taxa, belonging to 106 families and 504 genera (more than 33% of Georgia's flora) -265 Endemics of Caucasus and 158 Endemics of Georgia, have been secured in the CRSB. Seed Bank Data are managed in BRAHMS, developed by the Department of Plant Sciences of Oxford University.

T1

P0142

Digitisation and electronic access to19[th] century specimens from China housed in the National Herbarium of Victoria (MEL).

Josephine Milne, *Domenica Rita Macheda*

Royal Botanic Gardens Victoria

The collection in the National Herbarium of Victoria (NHV) at the Royal Botanic Gardens Victoria is estimated at approximately 1.5 million specimens. Approximately 400,000 of these specimens were collected overseas from various geographic areas including Africa, China, Europe and, South America and by many significant nineteenth century botanists and explorers including Odoardo Beccari, Robert Fortune, Joseph D. Hooker, Johann Georg Christian Lehmann, John Kirk (Livingstone Expedition), Franz Wilhelm Sieber, and Karl Zeyher. This significant foreign collection was acquired by Ferdinand von Mueller, appointed as the first government botanist in 1853, through his networking with the foremost herbaria and plant collectors worldwide. A substantial proportion of the foreign collection was also acquired by Mueller through the purchase of the Otto Wilhelm Sonder herbarium in 1883. The Australian collection at the NHV is digitized i.e. databased and accessible on Australia's Virtual Herbarium (AVH) http://avh.chah.org.au/, and type specimens are imaged and accessible on Global Plants on JSTOR https://plants.jstor.org/ . The NHV has now embarked on the Foreign Collection Project which focuses on databasing specimens from its foreign collection and also making the records accessible through AVH. In addition, known type specimens and any new types being discovered are being imaged and delivered to Global Plants on JSTOR. As the Foreign Collection Project progresses, the breadth of specimens collected from China is becoming apparent. The project is revealing specimens collected by a number of nineteenth century botanists and explorers, often described as 'plant-hunters' and material representing taxa from many vascular plant families. Collectors include: Henry Hance (explorer and plant collector) who collected in Guangdong Sheng, Augustine Henry (explorer and plant collector) whose specimens are mainly from Prov. Hupeh [=Hubei] Central China, Edouard Ernst Pere Maire (missionary and botanist) who collected in Yunnan Sheng and Alexander Gustav von Schrenk whose collections are all from Songarei [= Songaria, now Xinjiang Province]. The ongoing digitisation and electronic access to the foreign collection at the NHV will not only provide an invaluable resource for researchers globally, but access to this nineteenth century collection will fulfill, particularly for China, what has recently been described as 'virtual' repatriation of biodiversity records.

T1

P0143

Cosmarium bahianum (Desmidiaceae, Streptophyta) a new desmid species from a phytotelm habitat in the Brazilian Restinga

Carlos Wallace Do Nascimento Moura[1], *Geraldo José Peixoto Ramos*[1], *Carlos Eduardo De Mattos Bicudo*[2]

1. Universidade Estadual de Feira de Santana

2. Instituto de Botanica de São Paulo

Cosmarium Corda ex Ralfs (Desmidiaceae, Streptophyta) includes the largest number of species among the desmids and so far encompasses more than 2,000 registered taxa, 1,100 of which are taxonomically accepted. The genus is characterized by solitary, biradiate cells with a more or less entire outline. Otherwise, *Cosmarium* is such a polymorphic genus that it should be considered to be highly artificial. Despite this, we presented a new desmid species, *Cosmarium bahianum* G.J.P. Ramos, C.E.M.Bicudo et C.W.N.Moura collected from a tank bromeliad (phytotelmata) in an urban area of "Restinga" in Salvador, Bahia State, Northeast Brazil. This species was collected during the rainy season (May 2015) from a *Hohenbergia littoralis* L.B.Smith tank, an endemic bromeliad from the Brazilian "restinga", by using a 50 mL syringe coupled with a polyethylene hose. Cell morphological features and dimensions as well as the generative reproductive stage were observed with a light and scanning electron microscope. *Cosmarium bahianum* is characterized by cells usually slightly longer than broad (20–29.3 x 16.9–23.3 μm), with a moderately deep median constriction. Sinus widely open, about rectangular at the apex. Semicells in frontal view elliptic-rhomboid, in apical view elliptic or subtriangular, in lateral view subcircular. Cell wall coarsely punctate. Chloroplast with 1 pyrenoid. Cells provided with a thick, sharply bounded mucilaginous envelope. Zygospore smooth-walled, globose to elliptical, measuring 16.1–20.7 μm diameter. Regarding morphology, *C. bahianum* is similar to *C. inconspicuum* West & G.S.West, however, this species is distinct from the first one by having smooth cell wall, zygospore irregularly ovoid and smaller cells dimensions (length 13,5–14,5 μm, breadth 9,5–11 μm, isthmus 4,8–5,7 μm, diameter zygospore 14,5–17 μm). The species was extremely abundant in the *H. littoralis* tank, and growth up the abiotic water conditions such as acidic pH (4.8), high temperature (29°C) and low electric conductivity (0.02 mS cm^{-1}), favorable for the development of desmids. *Hohenbergia littoralis* is endemic to Brazil, with the distribution restricted to the north coast of the Bahia State, in the "restingas" of Abaeté and Baixios. This is considered an endangered species due to the eradication of the natural vegetation for the planting of *Cocos nucifera* Linnaeus (coconut trees) near the beach line, together with the urban expansion and development of tourism in the region. If their natural habitat is threatened, *C. bahianum* will also be endangered. Therefore, preservation of *H. littoralis* and other bromeliads from "Restinga" is absolutely essential as potential habitat for certain groups of algae, especially the desmids.

T1

P0144

Field studies on diversity of wetland flora of nanded district in eastern part of Marathwada region in Maharashtra, India

Ramjan Mulani

Department of Botany, DST-FIST, UGC-SAP sponsored School of Life Sciences, Swami Ramanand Teerth Marathwada University, Nanded-431606 (MS), India.

Wetlands are one of the important natural resources which either temporarily or permanently covered by water. This means that a wetland is neither truly aquatic nor terrestrial; it is possible that wetlands can be both at the same time depending on seasonal variability. Thus, wetlands exhibit enormous diversity according to their genesis, geographical location, water regime and chemistry, dominant plants and soil or sediment characteristics. Because of their transitional nature, the boundaries of wetlands are often difficult to define. Wetlands do, however, share a few attributes common to all forms. Wetlands are manly classified as: marine, estuarine, lacustarine, riverine and palustarine. Wetlands support a large variety of plant and animal species adapted to fluctuating water levels, making the wetlands of critical ecological significance. Directly and indirectly wetlands support millions of people in providing services such as food, fiber and raw materials, storm and flood control, clean water supply, natural beauty and educational and recreational benefits (National Wetland Atlas: Maharashtra (2010). According to MoEF report of National Wetland Atlas: Maharashtra (2010), Nanded has Total 1,072 wetlands are mapped including 440 small wetlands (< 2.25 ha) with 30,508 ha area. Rivers/streams occupy 57.21% area. There are 395 Tank/pond mapped with 25.22% area. There are 12 Reservoir/barrage mapped with 4,913 ha area (16.10%). Aquatic vegetation is slightly more during pre monsoon (2,673 ha) than during post monsoon (1,559 ha). Open water spread of the wetlands is significantly higher in post monsoon (26,935 ha) than during pre monsoon (8,311 ha), indicating the rainfall dependence of the wetlands in the state, particularly the dryness of rivers is very prominent. Turbidity of open water is mainly low during post monsoon, while ranged from low to moderate during pre monsoon. Present study deals with the field observations on wetland flora present in various natural and manmade wetlands occur in Nanded district. It is the eastern part of Marathwada where the rainfall is less and it has a forest area towards the border of Andhrapradesh on east and Vidarbha on north. In present work regularly the wetlands were visited and various species were recorded with their onset of various stages of life cycle. Special observations were made on the manmade wetlands with their change of occurrence of species of genera *Aponogeton, Bacopa, Cryptocorine, Ceratophyllum, Hydrilla, Hygrophila, Ipomea, Lemna, Lindenbergia, Lymnophila, Marsilea, Najas, Nelumbium, Nymphea, Nymphoides, Potamogeton, Spirodela, Typha, Utricularia, Vallisneria, Woffia, Zannichelia, etc.* The Government of Maharashtra played an important role in deciding the widening and deepening of both natural and manmade wetlands in 2014. According to this decision, the government has freely allowed removing the silted soil from the wetlands free of cost. This was even continued in 2015-16 also. This ultimately resulted in increase in ground water level as water level in wells. In addition wetland plant flora was also enhanced due to availability of new environment.

T1

P0145

Evaluation of timber and fuel wood consumption and its impact on vegetation cover in northern parts of Pakistan.

Murad Ali Murad

University of Swat

The present study aims to investigate the deforestation and effects of floods on various plants in northern parts of Pakistan. Data was collected from the local people. The data shows that due to drastic and rapid cutting of trees is the major cause of deforestation. The data were recorded in the form of questioner after that, the recorded data were analyzed by SPSS (statistical programmed for social sciences) software. The trees are ruthlessly cut for burning, timber and furniture purposes. The highest ratio of cutting trees is for burning purposes. *Cedrus deodara, Querqus dialata, Betulla utillus, Juglans regia, Picea smithiana, Pinus willichaina.* Land sliding and floods are the second highest reason of deforestation. The deforestation also affects the growth of other plant species as its ground flora. The shade and moisture loving plants (sciophytes) disappear due to deforestation while the halophytes dominate the cleared area. Due to anthropogenic activity and rapid cutting of valuable species of the area affect the climatic condition of the site.

T1

P0146

Herbaceous plant collections in botanical garden, Mongolia

***Enkhtuya Luvsanbaldan**, Ochgerel Nanjidsuren*

Institute of Botany and Experimental Biology, Mongolian Academy of Sciences

Currently 70% of the territory across Mongolia is the subject to desertification and land degradation. Total 11% of species of flora in Mongolia is considered to be extinct and endemic while 26% is on the verge of extinction, yet 37% is considered to be on a high risk of extinction [1]. The Botanical Garden of any country aims to introduce or to transfer wild plants into ex-situ cultivation to conserve the species, to protect their gene pool, and to collect the representatives of various vegetation areas. Flora in Mongolia has a species 149 of endemic plants, 140 of interglacial vegetation, and about 100 of ice age period [2]. Whereas a Nature Reservation and Protection Law [3] officially notes the species 133 of endemic plants and 354 of rare plants. In order to introduce rare species such as perennial herbaceous plants, to evaluate their biological, ecological and adaptive capacity, and to learn the season rhythm from some medicinal and ornamental plants, we designed following tasks: To create an original seed sample collection from plants occurring in different geographic regions and ecosystems across Mongolia. To enrich the seed amounts and numbers from endangered plants of as many as 135 species listed in a Red Book of Mongolia by active plantation. The collections created by sampling of seeds, bulbs, rhizomes, and seedlings. The research and the following collections are started by formerly well used methods [3] Rare, endemic and ornamental herbaceous plants were cultivated in a Botanical Garden from the Institute of Botany and Experimental Biology, distributed in 16 flora-geographical regions of Mongolia. We selected 4,000 samples from 9 flora-geographical regions and have conducted studies on cultivation. The 60% were grown in the areas of original sites were plants were spotted and collected. We conclude that the selected plants from different flora-geographic regions can experience some positive results on introduction process. For 2016, 33048 individuals of 305 species, 141 genera, 59 families (tree, bush, herbaceous *etc.*) were cultivated and reserved in the Botanical Garden, including 70 species of rare and endangered plants and more than 10 sorts imported. We are successfully conducting the trials on introduction and acclimatization of plants from 22 families, 32 genera and 54 species of herbaceous included in the Red Book of Mongolia. As a result of research and introductory works, the endemic and rare species of *Allium altaicum* Pall., A. *sondensatum* Turcz., *A. obliquum* L., *Rhodiola rosea* L., *Adonis mongolica* Simon., *A. sibirica* Patrin, *Paeonia anomala* L., *P. lactiflora* Pall., *Convallaria keiskei* Miq., *Lilium buschianum* lodd., *L.dahuricum* Ker.-Gawl., *Tulipa uniflora* (L.) Besser ex Baker, *Lancea tibetica* Hook.f. et Thoms, *Iris oxypetala* Bunge, *Cypripedium macranthum* Sw., *Incarvillea potaninii* Batalin., *Dictamnus dasycarpus* Turcz. et al were introduced, propagated and taken into protection and moreover favorable conditions and possibilities for creating an experimental garden of rare species plant collection was identified and seed fund of these species was increased. Our conclusion is that the plants selected from different flora-geographical regions are at certain level influenced.

T1
P0147

Altitudinal and temporal varition in mountain gorilla food plants biomass within Volcanoes National Park, Rwanda

Uwitonze Narcisse[1,2], *Bizuru Elias*[2], *Tuyisingize Deogratias*[1]
1. Dian Fossey Gorilla Fund International/Karisoke
2. University of Rwanda

The critically endangered mountain gorilla (*Gorilla beringei beringei*) is the only wild ape subspecies that is growing in number. Extreme long-term conservation efforts have nearly doubled the Virunga population since the 1980s, whereas their habitat remains a small isolated forest island. Improved forest protection is also likely to have resulted in growing buffalo and antelope populations, which share food plants with the gorillas. Higher re-visitation of gorilla groups and other herbivores at feeding sites may lead to a depletion of key gorilla food plants and cause a collapse of the gorilla population. The objective of this study, carried out in 2016, was to assess the altitudinal and temporal variation in biomass of six gorilla food plant species (add list of species here) in Volcanoes National Park. Systematic plant sampling techniques and established phytometric regression equations were used to calculate the biomass in 66 plots that were assessed previously in 1989 and 2010. Small herbs were collected, sun-dried and weighed, while stem length and leaf length were measured from tall herbs to assess biomass. We found that the biomass of five out of the six studied plant species was highest in mid-altitudes (2,900-3,250 m), possibly due to higher carbon sink generated by plant decomposition mainly in this zone. There was no change in *Galium*along the altitudinal range in contrast to the other plant species, which may reflect its current shift from lower to higher altitudes. Tall herbs, with the exception of *Peucedanum linderii*, clearlydecreased over time. Whether this decline in biomass is linked to overharvestingof feeding sites through frequent re-visitation by the growing gorilla population and other large herbivores (buffalo, antelope) who share these food plants needs to be investigated in future studies.

T1
P0148

Testing the Wallace's riverine hypothesis for an autochorous plant species in the Amazon Basin

Alison G. Nazareno[1], *Christopher W. Dick*[2], *Lúcia G. Lohmann*[1]
1. Universidade de São Paulo
2. University of Michigan

The Amazon River and its tributaries have been the focus of several biogeographical studies due to their potential role in population isolation. As biogeographical barriers to gene dispersal and flow, the river and its tributaries may have prompted population differentiation and species diversification through vicariant processes, for a plethora of taxa in the Amazon River Basin. In fact, the main prediction of Wallace's riverine barrier hypothesis is that physical causes of population isolation are responsible for major biodiversity patterns in the Amazon. To test this hypothesis, we sampled a total of twenty-six *Amphirrhox longifolia* (Violaceae) populations in both banks of Rivers Branco and Negro, important waterways in the Brazilian Amazon Basin. We used a high-throughput methodology (i.e., double digest RADseq) to identify SNP markers from anonymous portions of the nuclear genome of 156 individuals. Based on 8,010 unliked SNPs, our data supports the hypothesis that the Negro River – a wide river in the Amazon Basin –

acted as a physical barrier, although a weak one, for *A. longifolia* gene flow. On the other hand, a lack of any significant genetic structure in populations of this plant species in the Branco River – a tributary of the Negro River – indicates that the Branco River did not serve as a barrier to gene flow, as shown by the high rates of migration between opposite river banks. While this study provides the first characterization of genetic differentiation and population structure for *A. longifolia* along two rivers in the Amazon Basin, future research testing Wallace's riverine barrier hypothesis for plant species with distinct biological histories could provide valuable clues for an improved understanding of the origin of the vast biological diversity currently found in the Amazon Basin.

T1

P0149

The Amazonian Herbarium of Ecuador, ECUAMZ

Mercedes Asanza*, David Neill, Julio César Vargas Burgos, Diego Gutierrez Pozo*
Universidad Estatal Amazónica

Ecuador, a country in northwestern South America centered on the equator, is remarkable for its high diversity of plants despite its relatively small size. With a land area of 283,560 km^2 Ecuador covers just 0.2% of the land area of the globe, yet its vascular flora of 18,500 plant species includes about 7% of the world's flora. The Amazon region of Ecuador includes its six eastern provinces which contain the eastern slopes of the Andes mountains as well as the lowlands of the western Amazon basin. In this region, about one-third of the Ecuador' s land area, occur an estimated 10,000 native vascular plant species. The Amazonian Herbarium of Ecuador, with the acronym ECUAMZ, was established in 2012 and is the only herbarium in the Amazon region of the country. The herbarium is part of the research center of the Amazon State University (Universidad Estatal Amazónica); the main campus of the university is in Puyo, the capital of Pastaza province, but the research center, known by the acronym CIPCA, is located 44 km north of the main campus at the junction of the Anzu and Piatua rivers in Napo province. The research center comprises 2858 hectares including 2000 ha of primary forest at the eastern base of the Andes, 550-1200 m in elevation. The Herbarium was created with the intention of strengthening floristic studies and in general the development of the scientific and educational activities of the university and collaborating institutions, and to promote and carry out research and conservation of biological diversity in the Amazon region of Ecuador. The herbarium is currently conducting floristic inventories in several regions of Amazonian Ecuador including the buffer zone of Llanganates National Park on the eastern slopes of the Andes and the Cordillera del Cóndor in southeastern Ecuador. Adjacent to the ECUAMZ herbarium at the CIPCA research center is the Piatua Botanical Garden, covering 7 hectares, with living collections of native Amazonian plants including *ex situ* conservation of endemic and threatened species, as well as selected introduced plants with potential for cultivation in the region. The herbarium collection of ECUAMZ is curated according to international standards and currently contains 12,350 vascular plant specimens and 2850 specimens of non-vascular plants. The herbarium includes the type specimens of a number of recently published species that were described by the ECUAMZ herbarium staff and associates, including *Dacryodes uruts-kunchae* Daly, M.C. Martínez & D.A. Neill (Burseraceae) *Licania condoriensis* Prance (Chrysobalanaceae), *Clethra concordia* D.A. Neill, H. Beltrán & Quizhpe (Clethraceae), *Shuaria ecuadorica* D.A. Nell & J.L. Clark (Gesneriaceae), *Browneopsis puyensis* D.A. Neill & Asanza and *Inga yasuniana* T.D. Pennington (Fabaceae), *Pleurothyrium castaneum* van der Werff (Lauraceae), *Magnolia equatorialis* A. Vázquez, *Magnolia llanganatensis* A. Vázquez & D.A. Neill, and *Magnolia vargasiana* A. Vázquez & D.A. Neill (Magnoliaceae), *Calathea neillii* H. Kennedy (Marantaceae), *Palicourea neillii* C.M. Taylor (Rubiaceae). The symbol of the ECUAMZ Herbarium is the flower of *Meriania pastazana* Wurdack (Melastomataceae).

T1

P0150

Primary result of empowering local communities to engage in conservation and management of Magnolia and conifer trees in Vietnam

Hieu Nguyen[1], ***Hiep Nguyen***[1], *Ngan Tu*[2], *Khang Nguyen*[3]
1. Center for Plant Conservation Vietnam
2. Vietnam National Museum of Nature
3. Institute of Ecology and Biological Resources

Since 2007, with support from Global Trees Campaign (GTC), Flagship Species Fund (FSF), Critical Ecosystem Partnership Fund (CEPF), Center for Plant Conservation Vietnam (CPCVN) and Fauna and Flora International (FFI- Vietnam program) have implemented botanical survey, raised awareness, *in situ* and empowered local communities to engage in conservation and management of *Magnoli*a and conifer threatened species in Vietnam. The aim of our conservation is to focus on taxonomy inventory, assess IUCN, identify and tag mother trees, collect seeds, trial train germination to local people, and bring plants back to their natural habitat. In the last ten years, we have implemented in ten locations in four provinces. We have involved local people in all of the activities as Community Tree-Conservation Teams. Throughout the country, we have established 3 satellite nurseries in North Vietnam (Hanoi, Ha Giang Province), 1 in Highland Plateau (Center Vietnam) and 1 in South Vietnam (Ho Chi Minh City). Up to now, they have produced over 12,000 seedlings of 8 *Magnolia* species, 3,745 seedlings and cutting of 4 conifer species. One of our main focus is *Magnolia* conservation. Fifty seven native magnolia species are recognized as indigenous to Vietnam. Throughout the project, 21 species of *Magnoli*a recorded in Ha Giang Province (Priority Corridor 1 - Sino-Vietnamese Limestone of CEPF) make up 35% in total and 13 species recorded in Lam Dong Province (Greater Mekong Area) make up 23% in total. Over the last five years, thirteen trees have been monitored by the Community Tree-Conservation Team (CTCT). Among those trees, there are two Critical Endangered (*Magnolia grandis, Magnolia sinica*) and three Endangered (*Magnolia coriacea, Magnolia hookeri* var. *longirostrata,* and *Magnolia citrata*). At the time of reporting, 1,005 *Magnolia grandis* seedlings, 134 *Magnolia foveolata*, 45 *Magnolia conifera*, 104 *Magnolia hookeri* var. *longirostrata*, 200 *Magnolia citrata* survive in the commune forest and natural habitat. In terms of the conifer conservation, 34 native conifer species are recognized as indigenous to Vietnam.

Throughout the project, four species identified as flagship species for conservation such as *Xanthocyparis vietnamensis*, *Cupressus tonkinensis*, *Calocedrus rupestris*, *Taiwania cryptomerioides*. At the time of reporting, 2,615 survive in the commune forest and natural habitat.

T1

P0151

Impact of polycyclic aromatic hydrocarbons (PAHs) on physiological and biochemical responses of Coralline algae from Shenzhen, China

Lei Nie[1], *Evelin Peli*[2]

1. Shenzhen Institute of Information Technology

2. Szent István University

Coralline algae (Corallinales and Sporolithales, Corallinophycidae, Rhodophyta) are globally distributed benthic primary producers that secrete calcium carbonate skeletons. In the context of global change, they have received much recent attention due to the potential vulnerability of their calcite skeletons and their many important ecological roles in marine ecosystem establishment and global carbon cycle, especially geniculate corallines. Their important functional roles in ecological systems across latitudes and habitat types, including reef frameworks, carbonate production, larval settlement, fish nurseries, renewable natural resources and accurate paleoenvironmental recorders, thus providing a valuable mechanism for contextualizing global oceanic changes. However, little is known on the effects of human-induced persistent organic pollutants (POPs) on marine calcifying organisms to date. Polycyclic aromatic hydrocarbons (PAHs) are recognized globally as persistent, bio-accumulative and toxic. PAHs contamination can represent a serious threat to humans and ecosystems. Algal polysaccharides, especially extracellular polysaccharides (EPS), have been demonstrated to play an important role as free-radical scavengers in vitro and antioxidants for the prevention of oxidative damage in living organisms. For the purpose of understanding the coralline algae responding to persistent organic pollutants, we carried on an experimental study by exposing *Corallina officinalis* to different levels of typical organic pollutants, anthracene, in order to see the physiological and biochemical responses. The red calcareous alga was collected from the intertidal zone of Dapeng bay, Shenzhen, China (at 114°36'N and 22°25'E). The main results are as follows: Under the conditions of increased anthracene concentration, the relative growth rate (RGR) decreased significantly comparing with that under the control. PAHs treatment also reduced the contents of photosynthetic pigments (Chla, Carotenoid and PE). With the concentration of anthracene in seawater increasing, the permeability of cell membrane and the content of MDA were increased gradually, showed that the capacity of anti-oxidant was also inhibited. The content of extracellular polysaccharides increased at first and then decreased as concentration of anthracene increases, while the content of intracellular polysaccharides continuously going low. The changes of radical scavenging capacity, including superoxide radicals (O_2^-), hydroxyl radicals (-OH) and DPPH radical in extracellular polysaccharides showed a similar first increased-then decreased trend. This study suggests that PAHs pollution has significant negative impact on the physiological activity of Coralline algae. Looking ahead to the further research on coralline algae, particularly in the context of global change and increase of anthropogenic polluted activities, the areas of need including: Molecular studies of algal calcification; Role of multiple stressors on physiology, with an emphasis on integrating studies of pollutants; Mechanics and long-term repercussions of coralline algal bleaching; Importance of coralline algae to reef stabilization under future anthropogenic polluted scenarios.

T1

P0152

Biodiversity conservation status and floristic revised lists of lycophytes and ferns of Myanmar

Thet Yu Nwe

Myanmar Floriculturist Association

Myanmar is a country in the north-western part of mainland South-East Asia. Being topographically highly diverse, Myanmar has species-rich different forest types. Although there is a number of researchers who have done taxonomic and floristic work in other plant groups, surveys on lycophytes and ferns of Myanmar are rather limited and incomplete. Ferns are rather important in Myanmar daily life as native people consume ferns as food and medicine as well as for ornamental and other purposes. The first floristic account of ferns of Myanmar was published by Dickason in 1937. He listed 379 species belonging to 105 genera. Since then several individual reports on ferns from different parts of Myanmar have been made. However, there is no complete assessment of all ferns from a floristic point of view. Therefore, a comprehensive list of native ferns is the basis for a future complete fern flora of Myanmar. In the present work, 622 taxa (51 specise in lycophytes and 572 species in ferns) are reported for new revised lists of Myanmar's lycophytes and ferns. According to the IUCN category, altogether 94 species are considered to be threatened or endangered in Myanmar.

T1

P0153

Biodiversity of host plants of rust fungi (Pucciniales, Basidiomycota) of the island of Hispaniola

Omar Paino Perdomo[1], *Luis Rodriguez De Francisco*[1], ***Ivelisse Montero***[2]

1. Instituto Tecnologico de Santo Domingo

2. Dominican Society of Mycology

The Caribbean region, which is one of the 25 biodiversity hotspots in the world, possesses a unique phytogeographic region within the Neotropics, containing vascular flora plants containing of 58% endemism. Hispaniola is the second largest island in the Caribbean region comprised of several terrains which differ in their origin and geologic history. The landscape has mountainous ranges that are oriented from east to west, and intervening valleys and plains, leading to the highest diversity of climates and ecosystems of the Caribbean. The island has the highest peak of all the Antilles (Duarte Peak, reaching 3.175 m.s.n.m) and its lowest point, Lake Enriquillo, approximately 40 meters below sea level. The combination of these factors have favored a large diversity of plants. The vascular flora plants of Hispaniola have already been well studied and possess about 6,000 species, of which 2,000 are endemic (33% endemism). The flora has five phytogeographic elements: pantropical, Caribbean, endemic, continental and introduced. Currently, a total of 33 families of vascular plants are reported as rust hosts, with 134 genera and 230 species. The plant families with greater diversity of rust are: Asteraceae, Cyperaceae, Euphorbiaceae, Fabaceae, Lamiaceae, Malvaceae, Poaceae, Rubiaceae and Verbenaceae. This represents 3.83% of vascular plants infected by rust fungi. The Poaceae and Fabaceae families are the ones that possess more species of rust host plants. The microflora of the tropical region is still poorly known. Although the Pucciniales are the largest order of plant parasitic fungi among Basidiomycota, including about 9000 species, numerous unknown species supposedly exist, especially in the tropics. Rust specimens have been collected in protected, as well as wild areas of the island. The species of rust fungi collected during fieldwork in Hispaniola are apparently mostly autoecious, i. e. live on a single host-plant species, possible because many tropical plants exhibit green leaves all year. However, this apparent tendency may also be due to a lack of knowledge of the life history of tropical rust fungi. More studies on diversity and ecology, as well as extensive fieldwork regarding taxonomic, morphological, and phylogenetic research are required, necessary in all the chains of Caribbean island, in order to elucidate and understand host plants of rust fungi.

T1

P0154

Onion essential oil induced markedly relaxation on the isolated rat *Corpus cavernosum*

Alev Önder
Faculty of Pharmacy, Ankara University

Introduction: Erectile dysfunction (ED) is a common problem which has become increasingly prevalent in societies. Approximately, 5-20% of the general population in men suffered from moderate or severe ED. It has been considered that many medicinal plants and secondary metabolites alleviate ED. The medieval Persian practitioners recommend some foods to help cure ED like *Allium cepa*. Onion (*Allium cepa* L.) belongs to genus *Allium* from the Amaryllidaceae family and is economically important as a vegetable. It includes sulfur containing active principles decomposed by allinase which are volatile and present in the onion essential oil (OEO) involving various important biological activities. It is also popular in cuisines of the world and has potential medicinal properties such as preventing cancer, reducing symptoms of diabetes, reducing coronary heart diseases, antibacterial activity and anti-inflammatory activity. The present study evaluated the relaxant effect of OEO on rat *Corpus cavernosum* (CC). Methods: Twenty-six adult male Sprague-Dawley rats were used in the experiments. Isolated rat CC strips were placed in organ baths containing Krebs solution, and functional experiments were performed. After preconstruction with phenylephrine (PE, 10^{-5} M), the relaxant response to OEO (25 - 400μL) was investigated in rat CC. The relaxant responses to OEO were repeated in the presence of nitric oxide synthase (NOS) inhibitor L-nitro-arginine methyl ester (L-NAME, 100μM) or soluble guanylate cyclase (sGC), [1H-1,2, 4] oxadiazolo [4,3-a]quinoxalin-1-one (ODQ, 30μM) inhibitor. Results: OEO caused relaxation of isolated CC strips in a concentration-dependent manner (maximum response: 98.0 ± 3.5%). The relaxant responses to OEO were not altered in the presence of L-NAME and ODQ in rat CC. Conclusion: OEO induced relaxation independent from NO/cGMP system in rat CC. We suggest that OEO may have a beneficial effect in ED patients who are non-responders to phosphodiesterase-5 inhibitors. In the future, we would like to investigate major components of onion essential oil and other extracts of the onion because of the health benefits.

T1

P0155

Effects of onion on functional responses of rat *Corpus Cavernosum*

Alev Önder
Faculty of Pharmacy, Ankara University

Erectile dysfunction (ED) is a common healthy problem, affecting more than half of men ages 40 to 70. It has been considered that many medicinal plants and secondary metabolites are used in the management of ED. Onion (*Allium cepa* L.) belongs to genus *Allium* from the Amaryllidaceae family and is economically important as a vegetable. The present study investigated the possible relaxant effect of OEO on rat corpus cavernosum (CC). Therefore, to investigate the relaxant effect, a total of twenty-six male Sprague-Dawley rats were used in this study. Isolated rat CC strips were placed in organ baths containing Krebs solution, and functional experiments were performed. After preconstruction with phenylephrine (PE, 10^{-5} M), the relaxant response to OEO (25 - 400μL) was investigated in rat CC. The relaxation responses to OEO were examined incubation with of nitric oxide synthase (NOS) inhibitor L-nitro-arginine methyl ester (L-NAME, 100μM) or soluble guanylate cyclase [1H-1,2, 4]oxadiazolo[4,3-a]quinoxalin-1-one (ODQ, 30μM) inhibitor. OEO showed relaxant effect on rat CC strips in a dose dependent manner (maximum response: 98.0 ± 3.5%). OEO caused relaxation of isolated CC strips independently of the NO/cGMP pathway. Further investigations are warranted to fully elucidate the mechanistic effects of OEO on ED.

T1

P0156

The genus *Picea* A. Dietr. in Kyrgyzstan

Larisa V. Orlova[1], *Georgy Lazkov*[2]
1. Komarov Botanical Institute RAS
2. Institute for Biology and Soil NAS

The question of the taxonomic status and the origin of *Picea schrenkiana* Fisch. et C.A.Mey. and close related taxa (*P. tianschanica* Rupr., *P. robertii* Vipper and *P. prostrata* Isakov) is one of disputable in taxonomy of *Picea* A. Dietr. Most authors considers *P. schrenkiana* as independent species, close to *P. likiangensis* (Ran *et al.*, 2006), *P. wilsonii* (Lockwood *et al.*, 2013) or *P. smithiana* (Sun *et al.*, 2014; Ran *et al.*, 2015; Li *et al.*, 2015). Our field studies of the *P. schrenkiana* s.l. in different regions of Kyrgyzstan (Terskei range (Barskaun gorge), Kungei range, Susamyr range (Chichkan gorge), Chatkal range (Sary-Chelek Reserve) and comparative morphological studies of herbarium materials in LE, MW, NS, SVER, PE, FRU, K, E, BP, KRAM, *etc.*, allowed to reveal some new diagnostic characters for the determination of *P. schrenkiana* and closely related taxa, and to clarify its geographic distribution. According our investigations, the closest taxa to *P. schrenkiana* are P. *asperata*, *P. crassifolia* and *P. meyeri*. *Picea schrenkiana* differented from *P. obovata* by larger (6-12 cm long and 2,5-3,5 cm wide) cones and seed scales (12-20 mm long, 13-18 mm wide), with a little black rim on the upper margin and slightly wrinkled longitudinally, and also longer (20-35 mm long) needles, bluish-green, with 5-8 stomatal lines abaxially and 3-6 – adaxially, with a little white wax coating only adaxially, and very slighly pubescent annual shoots. *Picea tianschanica* differs by seed scales, more prominent and thickened, appreciably longitudinally wrinkled, dark brown, often with a distinct reddish or purple-brown spots, in the inner part - orange; its terminal buds conical or broadly conical to almost globular, brown; needles on the shoots very crowded (in *P. schrenkiana* seed scales unprominent and more or less stright, almost smooth, brown, monochrome, in the inner part - brown; terminal buds ovoid, light brown; needles uncrowded). This taxon deserves by our view the status of separate species. In the eastern and northern Kyrgyzstan (Terskei and Kungei ranges) occurs *P. schrenkiana* with some characters of *P. tianschanica* (violet color of some seed scales). In Barskaun gorge we found the population of blue spruce intermediate in the characters between *P. schrenkiana* and *P. crassifolia*. The main differences concern the shape of the upper margin of the seed scales, its color, as well as some characters of vegetative organs. We can assume that this blue spruce is possibly result ancient hybridization of *P. crassifolia* with *P. schrenkiana* (which also occurs here, but in smaller amount). Possibility of occurrence of such hybrids in the Eastern Tien Shan pointed out by E.G. Bobrov (1978). Spruce, native to western Kyrgyzstan (Chichkan gorge, Sary-Chelek), is similar to *P. smithiana* by very large cones, wider seed scale, as well as morphology vegetative buds, profills and pulvini of annual shoots, but it possess much more variable characters of needles, as well as noticeably pubescent annual shoots, buds and pulvini, so we assume that they are ancient hybrids of *P. schrenkiana* x *P. smithiana,* and correspond to *P. robertii* Vipper. The geographical distribution of *P. schrenkiana* and close related taxa needs to be further investigation in nature.

T1
P0157

Plants wealth: Power of planet (with specific reference to plant biodiversity, utilization, conservation and biodiversity–climate link)

Vashist N Pandey
DDU Gorakhpur University, Gorakhpur

Plant wealth is super natural power for the biological system on this planet i.e. they govern all biofunctions, distribution and climatic link of different habitat of this world. Biofunctional plants (BPs) are live telecast of our mother planet growing naturally and cultivated in different geographical regions of the world since ancient times. Plants are major component of our biological and Ecological system which deals direct relationship with Biodiversity of plants and Biological system. Plants have specific biofunctional role and growing in different Biospecific Regions (BRs) of the world, commonly known as Biospecific Region Plants (BRPs). In real sense Plants are the super natural wealth and power of our mother earth Planet. Food, Medicine, Health, conservation of Biodiversity, and Biodiversity –Climate link can be achieved by our plant wealth resources. Food security are major concern of our society which fulfilled by some our selected and well-known species of cereals, millets, pulses, vegetables and fruits. There are more 200 plants have been known to use by civilized society throughout world. Out of them approximately 15 plants have been used on large scale. Medicinal plants have been playing pivotal role in providing medical aid and promote positive health in mankind. In recent years there has been a tremendous expansion in the number of the institutions with provision for research and development of various branches of medical systems. The drug plants are collected and prepared in crude indigenous way. The medicinal value of drug plants is due to the presence of some chemical substances in the plant tissues which produce a definite physiological action on the human body. There are about more than 9000 thousand plants are known as medicinal plants but only a few are utilized on large scale by civil society. Health is a comprehensive term used for physical, biological and mental fitness on whole of humans and environmental system, directly or indirectly govern by plants. Considering all benefits derived by plants for human and ecological security, there is an urgent need of Conservation plants in natural state which is done either by *ex-situ* or *in-situ* methods. Fomeag (Fo=Food + Me=Medicine+ Ag=Agricultural/Arboretum/Aquatic Garden) Cafeteria would be a new mile-stone and vision to the nature's health, biodiversity-climate links, utilization and conservation of biodiversity especially for food, medicine and health plants which provides a new impetus in revolutionizing the sustainable and economically feasible development of the country and the whole world.

T1
P0158

Diversity and distribution of ground herbs as indicator of fragmentation effect in tropical forest of Xishuangbanna, Southwest China

Bonifacio Pasion
Community Ecology and Conservation Group, Xishuangbanna Tropical Botanical Garden, Chinese Academy of Sciences, Mengla, China

The tropical forest of Xishuangbanna is one the most biodiverse region in China. However, this forest is being heavily fragmented due to rapid expansion of monoculture rubber plantations. In this study, ground herbs were sampled over the span of three years, which in contrast to trees are short-lived and may show an earlier compositional and diversity responses to the recent fragmentation (~30 years) in Xishuangbanna, Southwest China. Before analysis, herbs were classified into different herb groups (all herbs, monocot, dicot, and ferns), then composition, diversity (Shannon *H*), and species richness were used as response variable in determining fragmentation effect (fragment size, forest cover within 100,500, 1,000 meters around the plot, and plot distance to the forest edge). There were 232 identified herb species consisting of 82 dicots, 76 monocots, and 74 ferns. Non-metric multidimensional scale (NMDS) plot showed strong herb compositional differences between four community types (rainforest, oak forest, limestone forest, mixed forest over limestone). Diversity and species richness were higher in rainforest and lowest in the limestone forest. Fragment size is a poor predictor of herb communities but forest cover within 100,500, 1,000 meters around the plot explains herbs communities better at landscape level while plot distance to forest edge can be a better predictor at community scale. Monocots are most sensitive to fragmentation. Dicots are less sensitive to fragmentation at landscape scale, while ferns are less sensitive at community scale. Lastly, different communities has different responses to fragmentation effect especially for limestone and non-limestone communities, with effect differed among herb groups. This study showed that changes within forest fragments are already substantial after 30 years, calling concerns to protect the larger remaining forest fragments in Xishuangbanna, and initiate management aimed at buffering and, where possible, reconnecting them.

T1
P0159

Solution for dozens of unresolved names regarding the genus *Photinia* (Rosaceae)

***Mitra Lal Pathak**, *Yundong Gao, Xin Fen Gao*
Chengdu Institute of Biology, Chengdu, Post code 610041, Sichuan, China

Nomenclature can be complicated since the authors might have different opinion on species delimitation, as well as the limitation of communications in the past, which caused a lot of redundantnames applied in plant sciences. Accepted name, autonym, synonym, homonym, basionym, unresolved name, conserved name, rejected name, legitimate name, illegitimate name *etc.* are commonly used in taxonomy. Among them, unresolved names are those dubious names which are neither accepted nor synonym. Currently, the one of the widely used online site for nomenclature The Plant List (www.theplantlist.org) includes 1.25 million scientific plant names. Out of which, 33% are accepted names, 44.2% are synonym and about 23% names are "unresolved" since the contributing data sets do not contain sufficient evidence to decide whether they should be accepted names or synonyms. To solve about the seven dozen of unresolved names regarding the genus *Photinia*, the protologue and the type images from all around the world's herbarium were studied. Some herbarium specimens and the live images were also studied. Out of about seven dozens names, 16 names were found used as homonyms. Thirteen names were belongs to other genera. Some of the names were published locally and the description was not in Latin or English, thus made them illegal. Some names (e.g., *Photinia dulongensis* H. Li and *Photinia semiserrata* H.Li) were nomen nudum. The nomenclature and identification confusion was with other genera of family Rosaceae viz. *Eriobotrya, Stranvaesia, Aronia, Heteromeles, Sorbus, Cotoneaster, Malus, Pyrus, Rhaphiodes etc.*The name mentioned in a sheet *Photinia urdanetensis* Elmer was now identified as *Syzygium urdanetensis* of family Myrtaceae. Preliminarily, about fifteen names were now recommended to use as synonyms of the other *Photinia*species in present. Few are under describing as new names (The research is going on). After seriously examination on the protologue and the type specimens, one of the interesting fact was that the twenty different taxon names (*Photinia blumei* Decne., *Ph. dasythyrsa* Miq., *Ph. flavidiflora* W.W. Sm., *Ph. Notoniana* Wall., *Ph. notoniana* W. & A., *Ph. notoniana* W. & A. var. *angustata* Bl. ex Koch in Miq., *Ph. notoniana* W. & A. var. *ceylanica* Hook. *f.*, *Ph. notoniana* W. & A. var. *macrophylla* Hook. *f.*, *Ph. notoniana* W. & A. forma *grandiflora* K. & V., *Ph. integrifolia* var. *notoniana* (W. & A.) J.E.Vidal, *Ph. integrifolia* Lindl. var. *subdenticulata* Miq., *Ph. integrifolia* Lindl.var. *sublanceolata* Miq., *Ph. dulogensis* H. Li., *Ph. eugeniifolia* Lindl., *Ph. lindeyana auct. non* W. & A.: Koch in Miq., *Ph. micrantha* Decne., *Ph. myriantha* Merr., *Ph. sambuciflora* W.W. Sm., *Ph. wardii* C.F.C.Fisch., and *Pyrus integerrima* (Wall.ex D.Don) N.P. Balakr.) described from different locations of the world were conformed as a single species *Photinia integrifolia* Lindl. (under publication). We feel that this kind of taxonomic work is very essential in all genera besides *Photinia*. Otherwise; it will be a great challenge to the taxonomist in the near future.

T1
P0160

Study of native rice cultivars of Wayanad district of Kerala, India cultivated under upland conditions

***Surekha Y. Pawar**, V. V. Radhakrishnan, K.V. Mohanan*
Genetics and Plant Breeding Division, Department of Botany, University of Calicut, Kerala – 673635

Wayanad district, known as the land of paddy fields is in the southern tip of the picturesque Deccan plateau nested among the mountains of the Western Ghats on the Eastern portion of North Kerala, India. The culture of Wayanad is mainly tribal oriented and is very rich in flora and fauna. More than 35% of the area comes under forest cover. Rice is the most widely consumed staple food of the tribal people residing in Wayanad. Several native rice cultivars are phenotypically and genotypically diverse and can be categorized according to ecological and geographical variations. Native rice cultivars cultivated under upland conditions are very important under the present scenario of climate change and increasing rate of water scarcity in the area. It is notable that cultivars grown under upland conditions are water saving or drought resistant with a water requirement of only one third to one quarter of that of wetland rices. It is necessary that upland farming practices are popularized among the tribal farmers of Wayanad. The present paper is an effort to study the performance of some native rice cultivars of the area under upland conditions in terms of their morphological and agronomical characters. The seeds were collected from tribal farmers of Wayanad. The experiment was carried out in the experimental net house of the Genetics and Plant Breeding Division, Department of Botany, University of Calicut, Kerala, India in the first crop season of 2016 under upland conditions. For upland conditions, alternate drying and wetting method was used, but the soil was kept moist. Observations were made on 7 plant characters and 7 yield characters. The paper discusses the potential of some such rice cultivars under upland conditions.

T1
P0161

Species identification among similar *Argyreia* spp. (Convolvulaceae) based on leaf anatomy and phenetic analyses

***Paweena Traiperm**[1], George Staples[2], Janene Chow[1], Possathorn*

Nopun[1], Sasivimon Swangpol[1]
1. Department of Plant Science, Faculty of Science, Mahidol University, Bangkok
2. HRA, Herbarium, Royal Botanic Gardens, Kew, Richmond, Surrey TW9 3AE

The genus *Argyreia* Lour. is one of the taxonomically difficult genera in the family Convolvulaceae. Several species complexes were recognized in which species delimitation was imprecise, especially when examining herbarium materials with scanty flowers. The main goal of this study is to investigate and describe leaf anatomy of some morphological similar species in *Argyreia* using epidermal peeling, leaf and petiole transverse sections, and scanning electron microscopy (SEM). The phenetic analyses including Cluster Analysis (CA) and Principal Component Analysis (PCA) were used to investigate the similarity of these taxa. Differences observed among the species include epidermal cell walls and the trichome types on the leaf epidermis, while the important characters of leaf and petiole transverse sections comprise are the epidermal cell shape of leaf blade, leaf margins and petiole transverse section outline. The phenogram from cluster analysis using UPGMA method represented four groups with an R-value of 0.87. Moreover, the important quantitative and qualitative leaf anatomical traits of the four groups were confirmed by the principal component analysis of the first two components. The results from phenetic analyses also confirmed the differentiation between the similar species. The leaf anatomical features regarded as particularly useful for species differentiation were used in constructing a key to species.

T1

P0162

Modulating role of ROS in re-establishing desiccation tolerance in germinating seeds of *Caragana korshinskii* Kom

Long Peng
National Engineering Laboratory for Tree Breeding, College of Biological Sciences and Biotechnology, Beijing Forestry University, Beijing, China

In close coincidence with visible germination, orthodox seeds lose desiccation tolerance (DT). This trait can be regained under osmotic stress, but the mechanisms are poorly understood. In this study, germinating seeds of *Caragana korshinskii* Kom was investigated focusing on the potential modulating roles of reactive oxygen species (ROS) in DT re-establishment. Seeds with emerged radicles measuring 2 mm long can be rendered tolerant to desiccation by incubation in a polyethylene glycol (PEG) solution (-1.7 MPa). Upon PEG incubation, ROS accumulation was detected in the tip of radicles by nitroblue tetrazolium chloride (NBT) staining and further confirmed by confocal microscopy. The PEG-induced re-establishment of DT was repressed when ROS scavenger was added to the PEG solution. Moreover, ROS act downstream of ABA to modulate PEG-mediated DT re-establishment, and serve as a new inducer to re-establish DT independently. Transcriptomic analysis revealed that DT re-establishment by ROS involved the up-regulation of key genes in phenylpropanoid-flavonoid pathway, and total flavonoids content and key enzyme activity increased after ROS treatment. Furthermore, DT was repressed by phenylalanine ammonia lyase (PAL) inhibitor. Our data suggest ROS play a key role in DT re-establishment by regulating stress-related genes and phenylpropanoid-flavonoid pathway.

T1

P0163

Functional characterization of BnHSFA4a as a heat-shock transcription factor in controlling the re-establishment of desiccation tolerance in seeds

Sirui Lang
Beijing Forestry University, Beijing, China

Desiccation tolerance (DT) is the crucial ability of seeds to resist desiccation. However, the regulatory mechanisms of seed DT are not fully understood. In this study, two heat shock cis-elements (HSE) were identified in the *Brassica napus galactinol synthase* (*BnGolS1*) promoter and shown to bind the heat shock transcription factor A4a (BnHSFA4a). Transcriptional expression of *BnHSFA4a* was induced at the early stage of DT acquisition, prior to increased *BnGolS1* activity and galactinol production. Ectopic overexpression of *BnHSFA4a* (*oxBnHSFA4a*) in *Arabidopsis* enhanced DT, particularly during DT re-establishment. *OxBnHSFA4a* up-regulated the expression of *GolS1*, *GolS2* and *raffinose synthase 2* (*BnRS2*) in *Arabidopsis* and increased the enzymatic activity of *GolS* and RS and the concentration of raffinose family oligosaccharides (RFOs). Additionally, the overexpression lines exhibited increased antioxidant abilities. In contrast, the *Arabidopsis* mutant *athsfa4a* was more sensitive to dehydration, showing decreases in the efficiency of DT re-establishment, RFO contents and oxidation resistance. Complementation analysis indicated that DT was rescued in *athsfa4a/BnHSFA4a* seeds to similar levels compared with those of Col-0. Taken together, these results indicated that *BnHSFA4a* likely functions in the regulation of *GolS* expression and activity and activation of the antioxidative system and other stress response factors to improve DT.

T1

P0164

Knowledge evaluation of characteristics and the state of conservation of three of medicinal plants in Córdoba, Argentina

***Ana María Planchuelo**, María Laura Molinelli*
Centro de Relevamiento y Evaluación de Recursos Agrícolas y Naturales (CREAN-IMBIV-CONICET), Facultad de Ciencias Agropecuarias, Universidad Nacional de Córdoba, CC 509,5000 Córdoba, Argentina

Traditional medicinal plants have been used from ancient times for health care in Córdoba, Argentina, but recently it was observed an increased use of native's species as natural remedies The province of Córdoba has a rich medicinal flora and many of the species are sold in pharmacies and dietary or health shops. The vast majority of the commercialized herbs are not cultivated and are obtained directly from the natural environment where they grow, and this increases the risk of the number of potential endangered species. The objective of the present study was to evaluate the sellers' knowledge of medicinal plants in the city of Córdoba for species being commercialized under the names of "canchalagua" (*Scoparia montevidensis* (Spreng.) R.E.Fr.) "espinillo" (*Acacia caven* (Molina) Molina) and "pulmonaria" (*Croton subpannosus* Müll.

Arg. ex Griseb.) and to establish the sources, and the quality of the marketed product of those species. Surveys were carried out in 75 shops using a questioner to collect information about the uses and the origin of the target species. Samples of these medicinal plants were purchased. The total number of samples analyzed was 50 for "chanchalagua", 23 for "espinillo" and 58 for "pulmonaria". The results showed that the sellers do not have any kind of training in relation to the medicinal plants. The analysis of commercial samples showed two different species (*Scoparia montevidensis* and *Schkuhria pinnata* (Lam.) Kuntze ex Thell.) sold as "canchalagua"; two species (*Acacia caven* and *Cercidium praecox (*Ruiz & Pav. ex Hook.) Harms) named as "espinillo" and three species (*Verbascum thapsus* (Lam.) Kuntze ex Thell.), *Croton subpanosus* and *Chersodoma argentina* Cabrera being sold as "pulmonaria". This work verified the absence of government control in the stages of collection, storage and commercialization. It is also put in evidence that a comprehensive conservation plan should regulate the market and to engage all concerned parties in the implementation of sustainable management guidelines recommended by national and international agencies.

T1

P0165

Ethnobotanical survey of *Okoubaka aubrevillei* (Pelleg et Norman) in Ameere Village of Osun State, Nigeria

Jacob Popoola[1], *Omolara Yakubu*[1], *Conrad Omonhinmin*[1], *Adegoke Adegbite*[2]

1. Covenant University

2. Ondo State University of Science and Technology

Okoubaka aubrevillei (Pellege *et* Norman) (Santalaceae) is a rare tree species traditionally considered sacred with enormous potential for medicine, pharmacology, building, construction, forestry products and tourism. The rarity/near extinction of the species prompted the ethnobotanical survey and search for the species in Osun and Oyo states, Southwest Nigeria, to gather indigenous knowledge use, traditional medicine practices and to enrich the existing traditional knowledge. Ethnobotanical data were collected using the participatory rural appraisals (PRA) method involving the village head, herbalists, traditional worshippers and artisans. Only two stands of the plant were located at Ameere Village of Osun State, while none was found in Oyo State. The two plants stands were considered 'husband and wife' and worshipped as gods, adored during traditional religion ceremonies and to ward off evil spirits. The whole plant parts (leaves, bark, fruits and seeds) are connected with various traditional beliefs and applications including taboos and healing potentials by the people in the studied area. The bark and seeds are the major parts used for traditional medicine. The medicinal potentials of the plant include the treatment of various health conditions such as insanity, convulsion, miscarriage, skin infections, food poisoning, stomach upset, snake bites, leprosy and veneral diseases. The fruit kernel mixed with powdered bark of *Milicia excelsa* in alcohol is employed in the treatment of chronic piles and weak erection. The dried bark of the plant in combination with other plants is also utilized to strengthen children's bones thereby facilitating early walk. The taboos and fears created by conditions and special rituals for assessing and harvesting the plant parts contributed in a large extent to the low level of concerted research efforts to improve the cultivation, management of genetic resources, conservation and utilization of the species.

T1

P0166

The type collection of I.P. Borodin Herbarium of St. Petersburg Forest-Technical University (KFTA), Russian Federation

Alexander F. Potokin[3], *Larisa V. Orlova* [1], *Vyacheslav V. Byalt*[1]

1. St. Petersburg Forest-Technical University

2. Komarov Botanical Institute RAS

3. St. Petersburg State University

Founded in 1865 by S.I. Karelshchikov, the I.P. Borodin Herbarium of St. Petersburg State Forest-Technical University (KFTA) is one of the oldest Herbaria in Russia, and includes about 200,000 specimens. During 2008-2016, we did a critical revision of the collections and as it turned out, the KFTA includes a large number of type specimens. It currently has about 6,000 authentic specimens of different categories. The collection of KFTA consists of two major sections: the General Herbarium and the Russian Herbarium. The General Herbarium (or that of the world flora) counted more than 70,000 specimens belonging to at least 15,000 species. The following exiccatae series were the richest in types: A.H. Curtiss and C.G. Pringle from the United States and Mexico (American Plants, Plantae Mexicanae and Plants of the Pacific Slope), von Eggers from the West Indies (Flora Indiae occidentalis exsiccatae and Flora exsiccatae Indiae occidentalis, ed. A. Toepffer), G.P. Lorentz and Otto Buchtien from South America ('Herbarium Americanum', 'Flora Enteriana'), C. Baenitz ('Herbarium Americanum' and others), G. Zenker from Western Africa ('Flora von Kamerun'), C. Holst from Eastern Africa ('Flora von Usambara'), C. Wilms from South Africa ('Flora Africae australis'), M. Holtze from northern Australia (Port Darwin), E.H. Wilson from China ('Herbarium Veitch Expedition') *etc.* Exsiccatae from Europe and the Mediterranean are especially rich in types. Among these, we shall name C. Magnier's 'Flora selecta exsiccata', C. Baenitz's 'Herbarium Europaeum and Herbarium Dendrologicum', M. Gandoger's 'Flora Gallica exiccata', J. Bornmüller's 'Plantae Anatoliae orientalis', P. Sintenis and J. Bornmüller's 'Iter Turcicum' *etc.* A number of exsiccatae representing certain taxonomic groups also feature many type specimens, for example, M. Gandoger's 'Herbarium Generale. Rosarum Europearum Exsiccatum'—in the genus *Rosa* L.; G. Braun's 'Herbarium Ruborum germanicorum'— in *Rubus*; A. Kneucker's 'Gramineae Exsiccatae', '*Carices* exsiccatae', and 'Cyperaceae' (exclus. *Carices*) et Juncaceae exsiccatae—in the families Poaceae, Cyperaceae, and Juncaceae; *etc.* The Russian Section of Herbarium was started in 1885 and kept separately from the General Herbarium and included collections from various regions of the Former Russian Empire contributed by I.P. Borodin, Yu. F. Bosse, L. von Graff, K.F. Meinshausen, N. Puring, N.I. Bush, K. Golde, S.K. Fedoseev, F.N. Alekseenko, P. Vinogradov-Nikitin, F. Karo, A.N. Krasnov, A. Berg, A.A. Kushakevich, D.I. Litvinov, E. Regel, V.N. Sukachev, A.N. Petunnikov, and many others. Also of particular interest is the section 'Dendrological herbarium' created largely through the efforts of E.L. Wolf (1860-1931) and currently stored separately from the other parts of the Herbarium. The value of the collection of 'Dendrological herbarium' is in the presence of a large number of specimens collected by Wolf himself (historical value), and the presence in its structure of a large number of type specimens (around 500). In 2012–2014 the Borodin Herbarium joined the international 'Global Plants Initiative' for scanning type specimens and secured a grant from the Andrew W. Mellon Foundation intended for scanning and databasing some 3,500 type specimens. As a part of the grant, the Herbarium received a specialized inverted scanner Epson 10000 X producing large-resolution images (up to 200 Mb, with a 600 dpi resolution). During the next years, we scanned more than 5,600 specimens and partly forwarded the obtained images along with metadata to New York and Vienna for their further placement at the 'Global Plants Initiative' (hhtp://www.jstor.org and http://herbarium.univie.ac.at/database/search.php).

T1
P0167

Cytotoxicity and hepatoprotective attributes of arborinine and total alkaloid fraction of *Glycosmis pentaphylla* (retz.) DC. leaves and callus on Hepg2 Cell Line

Pressy P Prakasia, *Ashalatha S Nair*
Department of Botany, University Of Kerala

Glycosmis pentaphylla (Orange Berry), belonging to the Rutaceae family is a shrub or small tree (1.5 – 5 m) widely distributed across India, Malaysia, Southern China and to the Philippine Islands where it occurs in tropical forests at low altitudes. Ethnomedicinally, *Glycosmis pentaphylla* is employed in the treatment of a wide range of ailments, including diarrhoea, cough, rheumatism, anaemia, jaundice *etc.* Leaves of *Glycosmis pentaphylla* is traditionally used for treatment of liver disorders. The current study is aimed at evaluating the cytotoxicity effects and hepatoprotective potential of total alkaloid fraction of *Glycosmis pentaphylla* leaf and callus and arborinine (isolated alkaloid) on human hepatoma HepG2 cell line. The cytotoxicity potential of the total alkaloid fraction and arborinine was assessed on HepG2 cell line and a normal 3T3-L1 mouse adipocyte by cell viability using 3- (4,5-dimethylthiazol-2-yl)-2,5-diphenyltetrazolium bromide assay at the concentration of 12.5,25,50 and 100 μg/ml. Based on the cytotoxicity assay, total alkaloid fractions and the isolated arborinine at the above mentioned doses were assessed for its hepatoprotection potential against CCL_4 induced toxicity in HepG2 cell line by monitoring cell viability. Leaf total alkaloid fraction showed the most potent cytotoxic activity on HepG2 cell line ($p<0.001$) in concentration and time dependent manner with IC_{50} value of 102.40 μg/ml and 5.39 μg/ml after 48 and 72 h incubation respectively than the callus total alkaloid fraction and arborinine. Apoptotic morphological changes in HepG2 cells caused by treatment of total alkaloid fractions and arborinine were observed and showed chromatin condensation, cell shrinkage and apoptotic bodies following AO/EB staining. Toxicity in HepG2 cells was induced by treatment with CCL_4 for 12 h as manifested by a significant ($p<0.001$) decrease in cell viability. Pre-treatment with (prophylactic) with the total alkaloid fractions and arborinine effectively alleviated ($p<0.001$) the toxicity induced by CCL_4 in a concentration -dependent manner. The results evidenced that *Glycosmis pentaphylla,* the rich source of alkaloids has potent hepatoprotective effect on hepatocytes. Moreover, the well established callus cultures can be used as an alternative production system to overcome the limited availability of the medicinally valuable pharmaceuticals.

T1
P0168

Simple and feasible approach to efficient callus establishment in *Glycosmis pentaphylla* (Retz.) DC.

Pressy P Prakasia, *Ashalatha S Nair*
Department of Botany, University of Kerala

Glycosmis pentaphylla (Retz.) DC. (Family: Rutaceae), commonly known as Orange Berry or Gin Berry, is a highly ethanomedicinally important plant widely used by rural and tribal communities in India. The plant is widely distributed across India, Malaysia, Southern China and to the Philippine Islands where it occurs in tropical forests at low altitudes. According to Kerala forest and wildlife department, *Glycosmis pentaphylla* is categorised as endemic plant in India. The plant has been continuously exploited due to its immense medicinal properties such as diarrhoea, cough, rheumatism, anaemia, jaundice *etc* and as well as for its rich source of secondary metabolites, mainly alkaloids. In this study, despite the meticulous efforts involved in callus induction from field grown plants, the success of the establishment of callus cultures was hampered by various obstacles such as bacterial and fungal contaminations, browning of the cultured tissues and the recalcitrant nature to most of the in vitro treatments experimented. In addition to it, the protocols that are dependable in terms of efficiency and repeatability is still rare to find in spite of a volume of work done on phytochemical and pharmacological area. A factorial experiment was conducted with four explants (radicle, hypocotyl, leaf and cotyledon), two light regimes (light and dark) and 4 levels of 2,4-D for callus induction from explants derived from a month-old-seedling grown on basal MS agar-solidified medium. Cotyledon explant gave a surpassing response and recorded the highest in terms of least days to visible callus induction, percentage explants forming callus, callus biomass than the other explants with the treatment of 22.6 μg/ml M 2,4-D augmented MS medium. Furthermore, the quantity of calli produced on a medium under dark condition was higher than that on media under light condition. Averaging the results from all hormone concentrations and light regimes together, highly efficient callus was induced from cotyledon (70.92%) followed by hypocotyl (50.44%). The response from the leaf was lower, 37.39% and the radicle response being the poorest with only 11%. Among the concentrations of 2,4-D taken into account, regardless of other factors, 22.6 μM proved to be the best for callus induction for all the explants. Successful proliferation of callus was achieved when the callus obtained from cotyledon was shifted to proliferation medium consisting of a hormone combination of 2,4-D +NAA and 2,4-D +BAP. Presently optimized callus production protocol holds a promising route for continuous production of plant biomass in laboratory controlled environment, independent of seasonal, climatic and geographical limitations and be considered a bright hope towards conservation and thereby providing a sustainable source for the extraction of therapeutically important chemicals and for the genetic improvement of this pharmaceutically important plant.

T1

P0169

Alkaloid profiling of *Glycosmis pentaphylla* by GCMS and its anticholinestaerase activity

Pressy P Prakasia, *Ashalatha S Nair*
Department of Botany, University of Kerala

Glycosmis pentaphylla (Orange Berry), belonging to the Rutaceae family is a shrub or small (1.5–5 m) tree widely distributed across India, Malaysia, Southern China and to the Philippine Islands where it occurs in tropical forests at low altitudes. Ethnomedicinally, *Glycosmis pentaphylla* is employed in the treatment of a wide range of ailments, including diarrhoea, cough, rheumatism, anaemia, jaundice *etc.* Juice of leaves used against fever, liver complaints, vermifuge *etc.* Paste of leaves used to cater against eczema, skin affections, snake bites, wounds and finally decoction of roots for bilious attacks, intestinal trouble and ascaris. The plant is known for the high diversity of their secondary metabolites which include many groups that represent a rich source of structural diversity and are good candidates as sources of alkaloids that could be useful in pharmacology. The chemical composition of alkaloids from leaves of *Glycosmis pentaphylla* was analyzed by GC/MS revealing the presence of three compounds, belonging to acridone and quinozoline group of alkaloids. Arborinine (acridone group) was isolated using alkaloid acid base protocol and the chemical structure was elucidated by mass-spectrometry and NMR spectroscopic experiments. Dose dependent inhibition of acetylcholinesterase (AChE) was determined by acetylcholinesterase enzyme kit. The results showed a marked inhibition of AChE with an IC50 value of 77.90 μg/ml. The study showed evidence for neuroprotective action of arborinine against stress induced toxicity on SHSY5Y cell lines. Molecular docking revealed the structural insights behind its significant inhibitory activity on AChE. The treatment of neurological disorders and neurodegenerative diseases is related to the levels of acetylcholine (ACh) through the inhibition of acetylcholinesterase (AChE). In this sense, many alkaloids isolated from natural sources such as donepezil, galanthamine, tacrine and rivastigmine have long been recognized as acetylcholinesterase inhibitors. The study indicated promising potential of arborinine as an acetylcholinesterase inhibitor which needs to be further developed and explore its potential lead for Alzheimer's disease.

T1

P0170

Aquatic plants as bioindicator in the wetlands of Mekong Basin and Thailand

Chavalit Vidthayanon
1. Dependent researcher, Bangkok
2. Department of Biology, Faculty of Science, Naresuan University

Wetlands in the Mekong Basin and Thailand are one of the global aquatic plants hotspots. Over 500 species of 32 families have been occurred but less than half were assessed (252 species) and five are reported as threatened taxa. Most wetlands in the Mekong and Thailand region are degraded and disturbed by human activity including habitat alteration, agro-chemical, water extraction, invasive alien species and climate change. Several aquatic plants are disappearing or replaced by tolerance taxa or exotic taxa. The species composition and occurrence in the wetland can indicate the water and habitat integrity. More *In situ* inventory monitoring as well as local market survey are needed approach for assesses wetland habitat indicators in the region.

T1

P0171

Pollination biology of *Hemiboea ovalifolia* (Gesneriaceae), an endangered herb from Guangxi, China

Gaozhong Pu
Guangxi Institue of Botany, Guangxi Zhuang Autonomous Region and Chinese Academy of Sciences

Although information on the reproductive biology of the endangered plant family Gesneriaceae is well known, the pollination mechanisms of these plants in karst regions are poorly understood. To determine the pollination ecology of *Hemiboea ovalifolia*, a rare and highly geographically restricted endemic species in karst regions, flowering phenology, pollinator observations, and pollination manipulations were conducted *in situ*. Findings revealed that the anthesis of *H. ovalifolia* often occurred late, during sunset, or early morning, with a duration of 2–4 days. The most effective pollinators were *Bombus ignitus* and *Anthophora zonata*. The co-existence of spontaneous self-pollination, protandry and herkogamy in *H. ovalifolia* suggested that this species outcrosses with partial self-compatibility, and exhibits late-acting inbreeding depression in seed sets. Controlled pollination indicated that these plants were pollen limited resulting from the seed set. However, despite the co-existence of large numbers of fruit and seed sets, and vegetative propagation in *H. ovalifolia*, a failure in seedling survival and long duration to establishment of first-year seedlings in natural populations suggests that the species does not easily recover from damage.

T1

P0172

Conservation and reintroduction of *Bentinckia nicobarica* in Nicobar Islands- a case study

Pushpa Kumari, *Paramjit Singh, Reshma Lakra*
Botanical Survey of India

Bentinckia, a palm genus is restricted to Indian subcontinent with *B. nicobarica* restricted in distribution to Nicobar Islands. Usually, Island species are quite frequently within their limited area of distribution not having any chance of migration, so that destruction of even a small forest area may render the endemic species high threat over their existence. The Andaman and Nicobar Islands represent 86% of forest coverage including tropical rain forests, tropical moist deciduous forests, lowland forests *etc.* with a total geographical area of 8249 sq.km, while the Nicobar group of Islands cover an area of 1841 sq.km. *B. nicobarica* belonging to the family Arecaceae is one of the hardest, tall palms growing up to ±50 m height with small globose, blackish fruits (when ripen) in big bunches. It specifically grows in moist hills, valleys, plains and slopes of lowland forests of Nicobar Group of Islands mainly in Great Nicobar, Kachal, Nancowry group, Teressa, Chowra and Car Nicobar Islands. In the year 1988, "The IUCN Red List of Threatened Species" assigned this species endangered category. *B. nicobarica* is used in the Islands for housing, furniture, boat making, *etc.* by the tribal of Nicobarese and Shompen. The emerging shoots are eaten by the local people as vegetable and the fallen fruits are eaten by pigs. Due to all these activities over the time the species population has declined very much in these islands. Considering the importance of the species the present study has been taken to investigate its population, multiplication and reintroduction. The method of Ecological Niche Modelling (ENM) map is applied by using the modeling algorithms of DIVA-GIS and MAXENT. Nineteen bioclimatic variables are used as environmental variables for modeling, which includes Annual Mean Temperature, Mean Diurnal Range, Isothermality, Temperature Seasonality, Maximum Temperature of Warmest Month, Minimum Temperature of Coldest Month, Temperature Annual Range, Mean Temperature of Wettest Quarter, Mean Temperature of Driest Quarter, Mean Temperature of Warmest Quarter, Mean Temperature of Coldest Quarter, Annual Precipitation, Precipitation of Wettest Month, Precipitation of Driest Month, Precipitation of Seasonality, Precipitation of Wettest Quarter, Precipitation of Driest Quarter, Precipitation of Warmest Quarter, Precipitation of Coldest Quarter *etc.* Through this modeling, we could predict the potential area of occurrences the species and their reintroduction for the population study and the limiting factors. More than four thousand seedlings of *B. nicobarica* are raised so far in controlled condition in polyhouse of Dhanikhari Experimental Garden Cum-Arboretum, Botanical Survey of India, Andaman and Nicobar Regional Centre and being reintroduced in the identified habitat for improving their conservation status.

T1

P0173

Species diversity of pteridophytes in Ton Nga Chang Wildlife Sanctuary, Southern Thailand

Nutdanai Putthisawong[1], *Chayanin Kraithep*[2], *Sahut Chantanaorrapint*[1]

1. Prince of Songkla University, Songkhla, Thailand.

2. Chulalongkorn University, Bangkok, Thailand.

A survey of pteridophytes at Ton Nga Chang Wildlife Sanctuary was conducted from September 2015 to March 2016. One hundred and twenty five specimens were collected and 97 species were identified, belonging to 50 genera and 19 families. Among these 6 species were lycophytes, and 91 species were monilophytes. Common families are is Polypodiaceae, Pteridaceae, and Thelypteridaceae, with 16,10, and 8 species, respectively. Of these, 56 species are terrestrials, 30 species are epiphytes, and 28 species are lithophytes. Regarding distribution of pteridophytes in the study area, 61 species forest along the stream are the most abundant, followed by 38 species in lower mountain forest, 28 species are in lowland evergreen forest, whilst 14 species are in dry evergreen forest and only 9 species are on rock platform forest. Among them, 37 species can be found in more than one forest types. In addition, an examination of type specimens and several collections of *Bolbitis malaccensis* (C. Chr.) Ching and *B. sinuata* (C. Presl) Hennipman, however, confirm that *B. malaccensis* should be reinstated as a separate species.

T1

P0174

Land-use changes cause rapid loss of seagrass resources: Case in Guangxi Province, China

Guanglong Qiu[1], *Zhinan Su*[2], *Hangqing Fan*[1], *Meiqiu Wei*[2], *Meng Li*[3]

1. Guangxi Key Lab of Mangrove Conservation and Utilization, Guangxi Mangrove Research Center, Guangxi Academy of Sciences

2. Forestry College, Guangxi University

3. School of Geography and Planning, Guangxi Teachers Education University

Seagrasses are the flowering plants which have adapted to exist fully submersed in the sea. They play significant roles in providing nurseries and shelters for marine lives, sequestrating carbon, stabilizing sediments, improving water quality, *etc.* Globally, seagrasses are declining at a accelerating rate mainly due to anthropogenic activities. Here, we present our comprehensive investigation results of seagrass resource in Guangxi, China in 2016. We also accessed the primary causes which induced the losses of Guangxi seagrass resources in the past ~10 years. We used field surveys, combing with Google satellite images interpretations to investigation the area of Guangxi form April to Oct, 2016. Our results show that, total 617.2 hm^2 of seagrasses were found along the coast of Guangxi, of which 20.1 hm^2 were newly found in 2016. Specifically, seagrasses in Beilunhe Esturary, Western Fangchenggang Bay, Eastern Western Fangchenggang Bay, Lianzhou Bay, Inner Yingluo Bay, *etc.*, were all recorded for the first time. In the comprehensive survey of 2016, four seagrass species were found, namely *Halophila ovalis*, *Zostera japonica*, *Halophila beccarii* and *Ruppia brevipedunculata*. Tieshangang and Pearl Bay (Zhenzhugang) were the major seagrass distribution areas in Guangxi, dominated by *Halophila ovalis* and *Zostera japonica*, respectively. Total 354 hm^2 of seagrasses, accounting for 37.6% of seagrasses area in 2008, were lost as a result of coastal land-use practices, like sea reclamations. This leads to -5.5% of annual loss of seagrasses in Guangxi from 2008 to 2016. Besides, the seagrasses *Halodule uninervis*, *Halodule pinifolia* and *Halophila minor* which have been recorded in 2008 in Guangxi, couldn't be found any longer in 2016.

T1

P0175

Ecosystem services of aridlands of District Khushab (Thal Desert), Punjab, Pakistan

Rahmatullah Qureshi

Pir Mehr Ali Shah Arid Agriculture University, Rawalpindi

The aim of this study was to record ethnobotanical information in order to compute ecosystem services of the aridlands of Khushab (Thal Desert), Punjab, Pakistan. For this purpose, field trips were arranged and the study area was surveyed during 2010 to 2013 by using a semi-structured questionnaire to record the uses of plants by the local peoples of this area. A total of 157 plant species belonging to 118 genera and 52 families that served to the natives for fulfilling different needs from them. The use value index (UVI) of species indicated that two species viz., *Acacia nilotica* and *Aza-*

dirachta indica are first in rank which is utilized in fulfilling eight use categories. Since the area is rangeland therefore there is very common of native plants as fodder/forage was very common in the area (25.82%), followed by fuel, medicinal (16.76%), vegetable (7.71%), *etc.* The area sustains high proportion of highly palatable species (81 spp.), followed by medium palatable (44 spp.) and less palatable (19 spp.). With reference to medicinal use, 87 plant species were used in treating 69 different ailments/diseases by the natives. Based on the reults, it can be concluded that the area is rich in plant diversity along with rich heritage of utilization. There is need to promote the customary use of plants by expansion of plant based products obtained from this area. Such efforst not only preseve the tradional lore as well as sustainable development of the area. So efforts are required to conserve/maintain some of the areas depicting diversity of medicinal/valueable plants for their sustainable utilization by the future generation.

T1

P0176

Floristic diversity, threats and conservation isssues: A case of Tharparkar Desert, Pakistan

Rahmatullah Qureshi
Pir Mehr Ali Shah Arid Agriculture University, Rawalpindi, Pakistan

Thar Desert is located in the Sindh Province, Pakistan. This is a hot sandy desert with small hillocks. The aridity is the characteristics of desert with erratic rainfall taking place in clusters. This aridland provides some ecosystem services to human being such as food, feed, shelter and medicine. The plant diversity of this area includes 331 species belonged to 198 genera and 60 families. The vegetation is typified by xerophytic species composed on *Acacia senegal, Aerva javanica, Aristida adscensionis, Calligonum polygonoides, Dipterygium glaucum, Crotalaria burhia, Capparis decidua, Euphorbia caducifolia, Indigofera argentea, Limeum indicum, Leptadenia pyrotechnica, Prosopis cineraria, Salvadora oleoides, Tamarix aphylla, Tecomella undulata* and *Stipagrostis plumosa* mostly found inhabiting on majority of the area. *Abutilon bidentatum, A. pakistanicum, Alysicarpus monilifer, A. tetragonolobus, Caralluma edulis, Cenchrus prieurii, Commiphora wightii, Convolvulus scindicus, Gisekia pharnacioides, Ephedra ciliata, Gynandropsis gynandra, Monsonia heliotropioides, Rhynchosia schimperi* and *Tecomella undulata* are some of the threatened species which are socio-economically used by the inhabitants to meet their various needs. The rising incidence of prolonged drought, desertification, overharvesting of woody species and soil erosion are serious threats to the typical this desert habitat. The pragmatic conservation measures are required to protect such habitat, since it represents threatened/rare and endemic species which are economically and unsustainably used by the local communities of this area.

T1

P0177

The impact of *Lantana camara* and *Chromolaena odorata* on native species diversity in some parts of vhembe district municipality.

Tsumbedzo Leonard Ramalevha, *Maanda Hadzhi Ligavha-Mbelengwa*
UNIVERSITY OF VENDA

Species diversity indirectly increases the productivity of ecosystem and understanding the impact cause by invasive alien plant species at different parts of the municipality is essential in predicting their impact on ecosystem productivity in different localities. Previous studies attributed the impact of invasive alien plant species to their competitive ability, high growth rate, effective reproductive mechanism and other characteristics. Significant progresses have been achieved over the years in accepting and eradicating the alien plant species from new habitats. However, the degree of impact and association of invading plants with the native flora vary from region to region. Rectangular quadrat method will be used for data collection; analysis of variance and multiple regressions will be used to analyze data collected. Impact of alien invasive plant species will be revealed by checking the response of native plant species following the non-removal and removal of alien invasive species in some areas of study.

T1

P0178

Antioxidant, anti-inflammatory and anticancer activities in selected Philippine ferns and lycopods

Reggie Dela Cruz[1], *Aileen Mae Ang*[1], ***Glenda Z. Doblas***[1], *Victor Amoroso*[2], *Heidi Porquis*[1]
1. Central Mindanao University
2. Center for Biodiversity Research and Extension in Mindanao

Ferns and lycopods, collectively called as pteridophytes, are among those neglected plant resources whose medicinal potentials are very high. This study explored the antioxidant, anti-inflammatory and anticancer potentials of selected pteridophyte species. Ethnopharmacological surveys were initially carried out in three ethnic tribes, Talaandig tribe in Mt. Kitanglad, Bukidnon; Higaonon tribe in Mt. Caburakanan, Bukidnon as well as the Mandaya tribe in San Isidro, Davao Oriental. Ten species of pteridophytes were found to be used by these tribes for a total of 13 illnesses which include arthritis and inflammation. A total of fifty-seven methanolic extracts were produced from the fronds and rhizomes of 25 species of ferns and three species of lycopods collected from eight locations in Mindanao. These were studied in terms of their DPPH radical scavenging activity, cytotoxicity against human epidermal keratinocytes, cyclooxygenase inhibition and anticancer activity against lung cancer A549 cell line. The selected species were also identified using morphology and barcoded using the chloroplast *rbc*L gene with an average length of 561 bp. The sequences were compared with the GenBank and BOLD databases to confirm initial identification and phylogenetic analysis was carried out using neighbor joining method. Qualitative phytochemical screening revealed 19 extracts positive to alkaloids, 41 extracts positive to anthraquinones, 46 extracts positive to phenolics, 54 extracts positive to saponins, 22 extracts positive to tannins and 45 extracts positive to terpenoids. Highest total phenolic content was 302.19 ± 28.83 mg gallic acid equivalent/g sample and flavonoid content of 36.74 ± 2.17 µg quercetin equivalent/g sample. The results further showed that crude rhizome extracts from all collected fern samples generally had higher values of total phenolic and total flavonoid contents compared to crude frond extracts.

Antioxidant values of more than 50% radical scavenging activity relative to ascorbic acid were observed in 5 rhizome and 1 frond extracts. All 57 extracts were found to be nontoxic based on MTT cell viability assay using neonatal human epidermal keratinocytes at two doses of 20 and 200 ppm. Assays were conducted to measure inhibition of the two forms of the enzyme cyclooxygenase, *COX*-1 and *COX*-2. Twenty-four crude extracts exhibited more than 50% *COX*-2 inhibition. Moreover, ten extracts gave *COX*-2: *COX*-1 selectivity indices of ≥ 1.2. The highest value was that of extract 20F with 2.64 followed by 21R with 2.34. Moreover, two extracts, 5R and 20F were active against lung A549 cancer cell line. It is interesting that 20F has both anti-inflammatory and anticancer activities. Isolation of the bioactive compounds as well as mechanistic and cell-based assays will be done to provide more conclusive data for the reported bioactivities.

T1

P0179

Efficacy of botanicals derived from indigenous plants to wheat aphid under the field conditions

Hafiz Mahmmod Ur Rehman, *Muhammad Sajjad, Shafiq U Rehman, Javed Akhter, Asif Sajjad, Intazar Ali*
Department of Entomology, University College of Agriculture and Environmental Sciences, the Islamia University of Bahawalpur, Bahawalpur-Pakistan

Aphid damages wheat, *Triticum aestivum* L. by rolling of wheat flag and trapping of emerging heads and awns which results in poor pollination and thus reduction in yield. Present study aimed at to test the efficacy of botanicals derived from some indigenous herbs to wheat aphid under field conditions. The extracts (5% solution) of neem, *Azadirachta indica* (leaves and seed), garlic, *Allium sativum*, eucalyptus, *Eucalyptus globulus* and tobacco, *Nicotiana tabacum* were applied when population of aphid exceeded the economic threshold level (ETL). There was a significant difference in the populations of wheat aphid and its natural enemy (coccinellid beetles) among five treatments after the application of extracts. The population of aphid was significantly lower on neem leaf extract and tobacco. Higher numbers of the aphids were recorded in control where no extract was applied. A significant difference was observed in the population of coccinellid beetle among the different treatments of extracts. Neem (leaf) and tobacco extracts received least while eucalyptus, garlic, neem (seed) and control had significantly higher numbers of beetles. There was a significant difference in numbers of grains per spike. Neem (seed) had comparatively higher numbers than control. No significant difference was observed weight of grain per spike.

T1

P0180

Sleeping beauties awakened by digitization: Herbarium (tu) of the University of Ttartu now digitally *accessible*

Ülle Reier, *Kai Vellak, Meelis Pärtel*
University of Tartu

Most of the old herbaria have undergone many restructurings during their long history.They have survived wars and periods of isolation from the main scientific community. Herbaria without borders is a 21st century phenomenon that has come true thanks to digitization. We intoduce the resurrection of an old herbarium on the example of the herbarium (TU) of the Natural History Museum of the University of Tartu, founded in 1802. Our collections are now available for researchers and citizen scientists at the web page.

T1

P0181

Impcats of REE'S Enrichment on growth and anatomy of eight common mosses from Baiyun Obo

Yeqing Gao, *Caiqin Ding, Dongmei Ren, Xueliang Bai*
College of Life Science, Inner Mongolia University, China

Selecting *Jaffueliobryum wrightii* (Sull.) Ther., *Didymodon constrictus* (Broth.) Saito, *Bryum argenteum* Hedw, *Didymodon tectorum* (C.Muell.)Saito, *Syntrichia sinensis* (C.Müll.) Ochyra, *Didymodon nigrescens* (Mitt.) Saito, *Pterygoneurum kozlovii* Lazarenko and *Tortula atrovirens* (Smith) Lindb from Baiyun Obo mine, the morphological and anatomical characteristics were studied by comparing with the same species from Helan mountain and SangGenDalai. Its main characteristics were taken microscopic pictures and made plates and the specifically important characteristics were measured. Three mosses from different mining area in Baiyun Obo were compared, using statistical analysis methods to explore the variation and variation amplitude of their morphological structure resulted by three different growth environment. the results showed that: (1) the differences between two regions are significant in these index: plant height, stem width, leaf length and leaf width. But non-significant in leaf length-width ratio and leaf cells length-width ratio. (2) three mining area mosses in leaf length and leaf width indicated differences: West mine>Main mine>East mine; and irregular in stem width. (3) the main factor to the mosses growth was the rare earth elements content and influenced mainly through the leaf. A preliminary discussion about the mosses growth influenced by the rare earth elements was made, meanwhile it enriches the content of the moss plant anatomy.

T1

P0182

Red and invisible? Evolution of flower characters in the genus *Fritillaria* L. (Liliaceae) in response to the pollinator shift

Katarzyna Roguz, *Marcin Zych*
University of Warsaw Botanic Garden

Majority of *Fritillaria* species have insect pollinators, however some North American and Asian species are bird-pollinated. The switch from entomophily (insect pollination) to ornitophily (bird pollination) happened at least once in the history of the genus. Such pollinator shifts should be accompanied by a divergence in some floral characters - morphological, phenotypical and/or physiological. New character states should be a response to preferences of the most common and the most effective pollinators. These features may also act as phenotypic filters when flowers become repelling, invisible or discouraging to illegitimate flower visitors. Such characters might be of great significance in case of the ornitophilous *Fritillaria* species where bees can act as the least efficient pollinators. Moreover, flowers characters favouring bird pollination can differ depending on biogeographic context since ornitophilous Fritillaries grow in different continents and their pollinators belong to different taxonomical groups (hummingbirds vs. passerines). The present paper discusses the evolutionary shift from entomo- to ornitophily in *Fritillaria*, illustrating different combination of character states associated with various pollination systems in the genus.

T1

P0183

Plant diversity, management status and research opportunities in the different botanic gardens of Ecuador

Hugo Romero-Saltos

Jardín Botánico Yachay (Yachay Tech & Yachay EP)

Ecuador (in South America) is one of the smallest of the "megadiverse" countries. With a land area of just 0.2% of Earth's terrestrial surface, it is home to approximately 18,400 native vascular plant species—about 7% of global plant diversity—from which 5,400–5,700 species are endemic. In this country, the responsibility for the *ex situ*, at times *quasi in situ*, conservation of the native wild species belongs to the different botanical gardens active in the country, while the responsibility for *ex situ* conservation of agricultural species (crops) is shared with the Instituto Nacional de Investigaciones Agropecuarias (INIAP). At this conference, I will provide general information about the different botanic gardens in Ecuador and will summarize the diversity and management status of their live plant collections. It will become evident that we need to better systematize and use our collections for research and educational purposes.
(Full list of co-authors of this conference is pending for their approval. Given the deadline, I am submitting this abstract with just the first author only).

T1

P0184

Soil substrate constraints the suitable niche of rare plants in Labrador, Canada

Natalia De La Torre-Rodriguez[1], *Sean Bennett*[2], *Shelley Pardy*[3], *Claudia Hanel*[3], ***Julissa Roncal***[1]

1. *Memorial University of Newfoundland*
2. *STANTEC*
3. *Newfoundland and Labrador Department of Environment and Climate Change*

Information regarding rare plant species distribution and their habitat requirements in remote areas of northern Canada is typically lacking. This represents a major challenge to government agencies, environmental consultancies, and indeed all stakeholders when conducting baseline studies under the federal Environmental Impact Assessment (EIA) review process; and to support accurate conservation decision-making on the part of the agencies and project proponents alike. This is particularly true of those rare plant species considered as having both narrow geographical ranges and specialized habitat requirements. According to the Newfoundland and Labrador Department of Environment and Climate Change (NLDECC) there are more than 200 species of flowering plants in Labrador considered "rare", and subsequently of management concern to the Province. The aim of this project was to model the distribution of rare plant species habitats (i.e. realized niche) across Labrador identifying areas where these species could potentially occur based on a set of known environmental attributes. We tested the hypothesis that if the realized niche is sufficiently widespread, then plant rarity is due to other factors like dispersal limitation, or reproductive challenges. This initiative was a collaborative effort between Memorial University of Newfoundland (MUN), in partnership with local government agencies (NL DECC - Wildlife Division) and environmental consultancies. Using locality coordinates acquired from herbarium records, ecological niche models were developed applying the Maximum Entropy (MAXENT) approach to a subset of uncorrelated climatic variables (Worldclim), land cover, altitude and geology (Newfoundland and Labrador Nature Atlas website). Of the 200 plant taxa, five were selected as case studies for niche modelling: *Calla palustris*, *Eleocharis palustris*, *Maianthemum stellatum*, *Potamogetum alpinus* and Viola *renifolia*. The ecological niche models presented here indicate that suitable habitat may exist for rare plant species outside that of their known locations. Overall, the environmental attribute contributing most to the distribution of these five species was bedrock geology (as indicated by current soil substrate), followed closely by temperature. Suitable habitat was widespread for *Calla palustris*, *Potamogeton alpinus* and *Viola renifolia,* suggesting that rarity on the basis of habitat availability is unlikely. Alternatively, suitable habitat for *Eleocharis palustris* and *Maianthemum stellatum* was deemed limited, fragmented, and may pose a challenge to species dispersal and recruitment. Our results contribute to a better understanding of the ecology of rare species, notably their ecological niche, may be an effective tool for new population discovery of rare plant species in Labrador, and has the potential to inform the conservation status assessment and management of these plant taxa in the future. The potential also exists for stakeholders, including our project partners to use these distribution models to guide future fieldwork, make appropriate management recommendations, and prescribe suitable mitigation actions, where feasible.

T1

P0185

Spekboom planting in degraded areas improves soil properties and native species diversity in Eastern Cape, South Africa

Sheunesu Ruwanza

University of Venda

We assessed whether *Portulacaria afra* (spekboom) planting improves both soil physico-chemical properties and native species diversity in the Saltaire Karroid Thicket. We collected topsoil samples from a spekboom planted site and compared them to adjacent natural and degraded sites and quantified soil physico-chemical properties. We further conducted vegetation surveys in plots measuring 10 x 10 m. Results show significant increase in soil P, total C, total N, Ca and soil moisture in the natural and spekboom planted sites compared to the degraded site. Soil compaction and water repellency were significantly high in the degraded site compared to the natural and spekboom planted sites. Species richness and diversity were significantly high in the natural and spekboom planted site compared to the degraded site. We conclude that the planting of spekboom in degraded areas improve some soil properties and native species diversity.

T1

P0186

Vascular plant biodiversity and floristics of the Canadian Arctic

Jeffery Saarela, *Lynn Gillespie, Paul Sokoloff, Roger Bull*
Canadian Museum of Nature

Despite nearly 200 years of exploration, substantial gaps remain in our understanding of the diversity and distribution of the vascular plant flora of the Canadian Arctic, which comprises over one third of the global Arctic ecozone. Detailed information on the diversity and distribution of Arctic plants is urgently needed to understand the potential impacts of climate change on the region's flora. Since 2008 we have been conducting detailed floristic surveys in botanically-understudied regions of the Canadian Arctic. The comprehensive baseline data of our >8000 new collections, all housed in the National Herbarium of Canada and other herbaria in Canada and internationally, adds important knowledge to our understanding of Arctic plant biodiversity. Many of our collections represent first records for specific areas, others represent the second or third collections of poorly-known species at the edge of their ranges in the Canadian Arctic, and many fill in gaps in the known distributions of Arctic species. We will summarize our floristic work to date in the context of current understanding of the Canadian Arctic flora, with a focus on the many particularly noteworthy discoveries we have made in this rapidly changing ecozone.

T1

P0187

The Arctic Flora of Canada and Alaska project

Jeffery Saarela, *Lynn Gillespie, Paul Sokoloff*
Canadian Museum of Nature

The Arctic Flora of Canada and Alaska project aims to produce a new flora treating all vascular plants (ca. 800 species) in the Arctic ecozone in Canada and northern Alaska. Our research team includes botanists from Canada, Norway, and the United States, and is being led by researchers at the Canadian Museum of Nature. We are using a web platform to move the Flora beyond traditional standards, and to produce a treatment that is digital and interactive, taking full advantage of current web and database technologies. The flora will include parallel taxon descriptions, dichotomous keys, detailed nomenclatural data, common names, information on traditional uses, taxonomic comments, and will be richly illustrated with photographs taken in the field and of herbarium specimens showing characters important for accurate identification. Tracking specimens examined will allow future verification of distributional and descriptive data, and will facilitate production of dynamic distribution maps. The Arctic Flora of Canada and Alaska website (http://arcticplants.myspecies.info) will be updated with content on an ongoing basis. The Flora will eventually serve as the key source of information for anybody who requires accurate and up-to-date information on the Arctic flora, including those involved in Arctic terrestrial monitoring, students in the north and elsewhere, and the international botanical community. It will also serve as an up-to-date source of baseline information on plant biodiversity in one of the worlds' most climate-threatened ecosystems.

T1

P0190

The vegetable oils like the Amazon cosmetics basic and development of alternative socioeconomic community

Terezinha De Jesus Soares Santos
Instituto de Pesquisas Científicas e Tecnológica do Estado do Amapá

The Amazon has been a source of inspiration for the latest releases in the cosmetic market due to the richness of its biodiversity, the exoticism of its products and the stimulation factor for sustainable development of local communities. By the year 2010 the main species found on the market were andiroba (*Carapa guianensis*) and the Brazil nut (*Bertholletia excelsa*), from the same year, new species were discovered as excellent raw materials for cosmetics, mostly. But the traditional knowledge holders fail to develop their communities or even their lives from the sustainable use of plant raw materials in abundance in the Amazon rainforest. This work will show some vegetable oils, few popular, but are in favor as a raw material for cosmetic and with the use of machines and low-cost equipment, promoting industrial innovation regarding the quality of raw materials and improvement product quality, can drive, so the input production base for cosmetic in traditional communities in the Amazon.

T1

P0191

Taxonomic inflation caused by philosophical disparities in *species concepts* and peccadilloes in the methodology followed for delimiting new species and recommendation for an International Standard Plant Species Numbering (ISPSN) system

Satheesh George
St. Joseph's College (Autonomous), Devagiri

Organic diversity is divided into different groups of various sizes separated by discontinuity of relationships. These groups are assigned to different categories in a taxonomic hierarchy according to their phylogenetic relationships. The most important category is that of species, the basic unit in taxonomy. But the existence of species in plants is a matter of theory. The judgment of characters may vary between individuals. Usually each taxonomist will have a different opinion on what represents a discontinuous character, what level of variation equals to the species rank and what equals to the ranks below species *etc.* It is the discretion of the taxonomist to fix the individual with discontinuous variation to appropriate category. Hence the criteria for demarcating species is crucial as different methods for diagnosing a species can arrive at different entities. False description of new species/varieties, due to lack of detailed study on intraspecific variations is discussed. Publication of nomenclatural novelties without consultation with authorities and /or publication in ephemeral journals resulted in the increased numbers of species in the checklist of a group. At present there is no proper compiled document on the novelties being published world wide. We propose to introduce the International Standard Plant Species Number (ISPSN) system for ensuring species stability in plant nomenclature. Recommendation for a universal system for delineating new species/sub specific categories is proposed.

T1
P0192
Phylogenetic analysis of resurrection plants on Inselbergs of Madagascar

Juliane Schmiel, *Stefan Porembski*
University of Rostock

Malagasy Inselbergs are geomorphological old isolated rock outcrops whose microclimate differs from the surrounding area. On Inselbergs extreme conditions for plant growth are caused by water deficiency, high light irradiation and temperature variation between day and night and between seasons. A number of plant species have evolved specific adaptations that allow their survival on Inselbergs. These specialists include carnivorous plants, succulents and desiccation-tolerant species. Dominant representatives of desiccation-tolerant plants are mat-forming monocotyledons, which occur in large populations on outcrops. Our model species is *Coleochloa setifera* (Cyperaceae- sedge family), which belongs to these mat-forming specialists. Morphologically there are hardly any differences between populations. Nevertheless, we are assuming that differentiation processes occur between geographically isolated populations of *C. setifera*, since the gene flow is limited beyond the populations. Currently, 19 Inselbergs from different regions of Madagascar have been sampled. From each Inselberg 20 plant leaves samples were taken randomly, dried and stored in silica gel. The aim of the study is the phylogenetic analysis of *Coleochloa setifera* at population level. Due to no detectable morphological differences, genotyping should determine whether a population consists of one single species or varieties. Furthermore, we aim to investigate whether there is any indication of differentiation processes (emergence of new species/ varieties) between Inselbergs of varying geographical distances and isolation levels. We assume that *C. setifera* reproduces clonally within a population. In addition, due to differing microclimatic conditions on each Inselberg, we expect changes in the DNA sequences. Since no genetic sequence information is available, the populations are to be analyzed by genotyping. An economically new method is double digest RAD-Seq (ddRADSeq). DdRADSeq is an inexpensive method for SNP (Single Nucleotide Polymorphism) discovery and genotyping in model and non-model species. Our study has the potential to provide novel taxonomical information on this species. The research findings could serve as powerful tool in terms of conservation issues of the generally threatened Inselbergs and their unique vegetation.

T1
P0193
Collection of naturalized *Helianthus* species from Australia

Gerald Seiler[1], *Gary Kong*[2], *Sue Thompson*[3]
1. U. S. Department of Agriculture, Agricultural Research Service
2. University of Canberra
3. The University of Southern Queensland

Genetic resources are the biological basis of global food security. Collection and preservation of crop wild relatives of important crop species such as sunflower provide the basic foundation to improve and sustain the crop. Acquisition through exploration is the initial step in the germplasm conservation process. Sunflower *(Helianthus annuus L.)* is native to North America and is grown on 20 million ha in 60 different countries. Due to domestication, cultivated sunflower has a very narrow genetic base that has been broadened by the infusion of genes from the crop wild relatives, which have provided a continued source of agronomic traits for crop improvement. The genus *Helianthus* comprises 53 species (14 annual and 39 perennial), all native to North America. However, wild sunflowers have been inadvertently and intentionally introduced into several countries where they have become naturalized, most notably in Australia, Argentina, and southern Europe. The distribution of the wild naturalized sunflower species in Australia occurs primarily within the states of Queensland (QLD), New South Wales (NSW), Victoria (VIC), South Australia (SA), and West Australia (WA). Since these species have been naturalized in Australia, the possibility exists that they may have traits distinct from their North American progenitors due to the different environments and disease and insect pest complexes. Prior to the exploration, there were no populations of wild sunflower from Australia in the USDA-ARS National Plant Germplasm System (NPGS) wild sunflower germplasm collection. This germplasm would greatly increase the available genetic diversity for these species from outside of North America. The objective of this exploration was to collect achenes from as many populations as possible of *H. annuus* and *H. debilis*, and make them available for future research and improvement of cultivated sunflower and *in situ* preservation for the future. The Australian exploration took place from February 22 through March 14,2007. It covered 10,000 km in the states of QLD, NSW, VIC, SA, and WA. Five days were spent exploring WA. In spite of the extreme drought, 14 populations of naturalized *H. annuus* were collected. Six days were spent exploring SA, SW and VIC. These states were especially hard hit with drought, but 21 populations of naturalized wild *H. annuus* were collected. An exploration to Queensland, also suffering from extreme drought, resulted in the collection of two populations of wild naturalized *H. annuus* and two populations of *H. debilis* ssp. *cucumerifolius*. The germplasm collected will be evaluated for oil concentration and fatty acid composition, as well as screened for various diseases and insects as a potential source of genes for the improvement of cultivated sunflower. The origin of the naturalized populations of wild sunflower in Australia is still open for discussion. The species found there are commonly grown in gardens as ornamental flowers and escapes. It is also possible that the seed was inadvertently introduced in imported forages, or possibly mixed in bird seed. Currently, the time of the introduction of the wild species into Australia is not known. Future population studies will help to answer that question.

T1
P0194
Conservation of *Ainsliaea fragrans* in Kochi, Japan

Akihiro Seo[1], *Maeda Ayako*[1], *Yabe Kota*[1], *Hayashida Taishi*[1],

Sugimoto Sayako[2], *Ishikawa Shingo*[2]
1. Makino Botanical Garden
2. Kochi University

Biodiversity in Kochi prefecture has been decreasing predominantly due to human activity and presumably climate change. The protection of plant diversity in the prefecture has been a key priority for The Makino Botanical Garden of Kochi Prefecture (MBK), one of many local botanical gardens in Japan. The garden has initiated several conservation projects, one of which is the conservation of the endangered understory herb *Ainsliaea fragrans* Champ. (Compositae). This plant species is distributed in southern China, and southwestern Japan. In Japan, its easternmost geographical distribution is Kochi Prefecture where it occurs at a single locality and is listed as an endangered species. However, the ecological requirements of *A. fragrans* are poorly known and with the hope of conserving this species, its population dynamics and habitat were observed and analyzed. The habitat of Japanese populations of *Ainsliaea fragrans* is in the understory of forests. Our observations of the habitat at the Kochi study site have shown the vegetation here to be a mosaic of secondary mixed forests dominated by *Castanopsis sieboldii* and deciduous forest dominated by *Vernicia cordata*. The habitat of *A. fragrans* at the study site occurred only under the deciduous forest where seasonal change in light intensity during autumn and winter was higher than in summer. The niche for *A. fragrans* seems to be the relatively bright areas of the forest floor that received higher autumn and winter light. The increase in light during winter seems to be an important factor for the growth of *A. fragrans*. At MBK an *ex-situ* population sourced from the wild population in 1989 was established in the nursery and subsequently provided a source of plants for this study. We initiated two experiments to test our hypothesis; the first started in July 2016 and investigated the effects of different levels of light on plants grown in the nursery. The second involved translocating 100 plants from the *ex-situ* population back to the original wild population at locations under both a deciduous and evergreen canopy in October 2016. Data from the wild population has been collected here since July 2016 and air temperature, humidity, soil moisture and light intensity at the study site were measured and recorded using data loggers. The survival rate of these plants and response to differing light levels *in-situ* was also observed and recorded. Results from the *ex-situ* investigation have shown a higher mortality rate during summer under brighter conditions than under shaded conditions. On the other hand, the mortality during autumn was not clearly different between these conditions. These results did not conflict the hypothesis. The translocated population of *A. fragrans* has not fluctuated under either a deciduous or evergreen canopy, although the observation period was short. Ongoing observations of both the tranlocated population and wild population of *A. fragrans* are required to analyze the correlation between environmental conditions and the survival rate of *A. fragrans*.

T1

P0195

Dendroclimatological investigation of the relict chestnut-leaved oak *(Quercus castaneifolia)* in azerbaijan

Farid Seyfullayev, *Vahid Farzaliyev*
Central Botanical Garden of Azerbaijan National Academy of Sciences

The chestnut-leaved oak (Quercus castaneifolia) is a species of oak in the turkey oak section Quercus sect. Cerris. It is naturally distributed in Azerbaijan and Alborz mountains of Iran, It is a deciduous tree growing up to 35 m tall, with a trunk up to 2.5 m diameter. Tree ring widths were measured in Lintab measuring stage with TsapWin program. Anatomical features were measured under microscope. Investigations were carried out by using dendroclimatological, anatomical methods. Response function results showed that current year May and June precipitation positively, current year June and July temperature negatively influenced annual radial growth. According mean radial growth deviation analysis, during droughts growth was 30-40% less; in wet years were 40-60% more than mean. Calculated Pimm indices showed that even 1-2 years after drought tree didn't reach normal growth level. All these analysis showed that this species is very sensitive to local climate change, depending on the changing precipitation and temperature radial growth strongly changing. As a result of global climate change, there are forming new microclimates which affect radial growth in the natural distribution area of the oak. These new extreme conditions forming frost scars in the wood. In Europe these kinds of scars forming one time in a century. But in our case in samples which were collected from high limit (1450 m) of the distribution oak in Lenkoran region, were observed 7 frost rings during last 60 years. Interesting point that 5 frost rings were formed last 30 year. It means last times local climate changed extremely and that is why oak is not adapted to it. Our investigation showed that effects of the climate change are clear in high limit. If it will continue by this way, during next decades' effect of changes will be visible in the low elevation and will negatively affect growth.

T1

P0196

Biogeography, ethnobotany and conservation status of *Pinus gerardiana* from Pakistan

Amin Ullah Shah, *Sarvat Rahim*, *Arif Muhammad Khan*
University of Sargodha

Pinus gerardiana Wall. ex D. Don (Chilgoza pine) is regarded as a unique, highly specialized and marvelously improbable gymnosperm belonging to family Pinaceae. It has attained the greatest economical value because of its delicious edible nuts (rich in carbohydrates, proteins and fats) throughout the world. Chilgoza pine tree is also unique because of its endemism as it is only restricted to Asia in the world and is native of Afghanistan, India, Pakistan and some parts of China. The distribution of *P. gerardiana* in Pakistan is limited to Zhob, Chilas, Astor, Swat, Chitral, Gilglt, Kurram Valley, Tangir Valley and Kashmir mostly. This species contributes a lot to the economy by providing valuable nuts that find a very briskly market both in the country as well abroad. Reckless cone extraction and almost complete lack of natural regeneration (mostly people pick all the cones from a tree for nuts and the fallen nuts are wiped out by rodents and birds) have led this valuable tree on the verge of extinction which deserves immediate attention for genetic conservation and improvement. Genetic variation is one of the most important components in maintaining population viability and resilience to environmental stress. The basic pre-requisite for genetic improvement and protection of genetic resource is the study of genetic variability. Biogeography and conservation status studies of *P. gerardiana* are urgently needed to properly manage conservation strategies on scientific ground and yet to preserve the species as an industrial plant as well as landscape tree.

T1

P0197

Ethnobotany and floristic diversity in Eastern Ghats of India

Mahammad Khasim Shaik

Acharya Nagarjuna University

The ethnobotanical and floristic diversity studies were carried out in some sacred groves (protected forest patches) of Eastern Ghats of India, from 2011-2015. The ethnomedicinal knowledge was gathered through interviews of tribal physicians. Data collected were processed using Microsoft Excel spreadsheet. The data was analyzed using the fidelity level (FL) of each species determined. Floristic diversity studies of sacred groves were conducted by using random quadrate method (1m × 1m for herbs, 5m × 5m for shrubs and 10m × 10m for trees) for vegetation analysis. Ethnobotanical study revealed that therapeutic applications of 87 plant species belonging to 76 genera and 45 families. These plants are used to cure various ailments such as malaria, jaundice, diabetes, bronchitis, fevers *etc.* The majority of remedies were taken orally in the form of juice extracted from freshly collected plant parts harvested from the wild. Plants that have the highest FL values are *Gymnema sylvestre* (100%) and *Andrographis paniculata* (100%), followed by *Phyllanthus amarus* (91%). The lowest is *Woodfordia fruticosa* (46%). The most used plant parts are leaves. Floristic diversity analysis revealed that a total of 344 species belonging to 101 families were observed in the sacred groves. The total number of species encountered in four sacred groves was 222,139,195, and 186 species in Dummakonda, Korukonda, Peddakonda and Rampa sacred groves respectively. Total density of tree species ranged from 33.2 to 56.1m^2/ha^{-1} and total basal area of all species between 8.61-18.59m^{-2}/ha^{-1}. The Shannon-weiver index of herbs ranged from 2.914 to 3.368, for shrubs 2.617 to 3.154 and for tree species 2.803 to 3.534 respectively. These diversity indices reduced on more disturbed sacred grove (Korukonda) than undisturbed sacred grove (Dummakonda). Over-exploitation, deforestation and Podu-cultivation are main threats for medicinal plant vulnerability. The anthropogenic pressures such as more urbanization, encroachment and modernization of the temple are main cause for loss of sacred grove biodiversity. With the above findings it is concluded that documentation of ethnomedicinal knowledge provides basic data for further studies on pharmacology and conservation practices are immediately needed and recommended for the local community participation for successful conservation.

T1

P0198

A non-destructive DNA sampling technique for herbarium specimens

Lara Shepherd

Museum of New Zealand Te Papa Tongarewa

Herbarium specimens are becoming increasingly popular as a source of DNA for plant research. They provide a ready source of material for rare, extinct and difficult to obtain taxa and can provide historical perspective. However, sampling herbarium specimens for DNA is not without its drawbacks. Herbarium specimens are a finite resource and the removal of part of a specimen for DNA extraction may limit its future use by both geneticists and other researchers. Non-destructive DNA extraction methods have been developed for other museum specimens including teeth, bones and invertebrates. Here I describe a nondestructive method to isolate DNA from the leaves of robust herbarium samples. The sampling for this method is cheap, straightforward and requires no special equipment. Using this method DNA was successfully obtained and PCR amplified from herbarium specimens up to 50 years old.

T1

P0199

Flora and vegetation of the Meskheti and Javakheti (Lesser Caucasus)

Shamil Shetekauri[1], *Davit Chelidze*[2], *Nana Barnaveli*[1], *Tolkha Shetekauri*[1,2]

1. Ivane Javakhishvili Tbilisi State University

2. National Botanical Garden of Georgia

The study area represents provinces of south Georgia and is divided into three floristic provinces: Euxine, Armeno-Iranian and Caucasian (Grossheim, Sosnowski, 1928; Takhtajan, 1986). The flora and vegetation is more or less different from plant species of Mediterranean, Iran-Turkish and northern hemispheric geographical areas. The vascular plants of study area consist of 1,652 species, 561 genera and 131 families. The following vegetation types are represented in Meskheti and Javakheti: riparian, middle and high mountain forests, xerophilic shrublands, halophilous vegetation with *Nitraria schoberi*, vegetation of high mountain meadows and steppes, wetlands, petrophyton and subnival belt vegetation. Phytocoenological composition of wetland vegetation of the Javakheti lakes (Tabatskuri, Paravani, Kartsakhi, Sagamo, Madatafa) and their surroundings is mainly defined by Eurasian, West Asian-Caucasian and European-Mediterranean species including *Carex juncella (=C. wiluica), C. panicea, C. irrigua, C. vesicaria, Calamagrostis canescens, Sagittaria sagittifolia, Menyanthes trifoliata, Utricularia vulgaris etc.* Surroundings of Tetrobi-Chobareti limestone area are distinguished by special botanical-geographic diversity on the Lesser Caucasus. A significant portion of the calcephilous flora is represented by species of Lesser Caucasian-Anatolian geographic type: *Arenaria steveniana, Asphodeline taurica, Scorzonera ketzkhovelii, Cerastium sosnowskyi, Silene dianthoides, etc.* Such diversity of the Tetrobi flora is related to the location of the area on the floristic intersection of the Lesser Caucasus, Armenia and Iran as well as isolation of the Tetrobi limestone massif, which is in turn connected with the history of geodynamic processes on the Javakheti. Subnival zone on the Javakheti vulcanogenic plateau is present only on Mts. Abuli (3,304 m) and Samsari (3,284 m). During the florogenesis of subnival belt of the Lesser Caucasus a decisive role was played by the floristic centre of Anterior Asia and a minor role by the floristic center of the Greater Caucasus and arcto-alpine floristic centres.

T1
P0200
Flora and vegetation of the Meskheti and Javakheti (Lesser Caucasus)
Shamil Shetekauri[1], *Davit Chelidze*[2], *Nana Barnaveli*[1], *Tolkha Shetekauri*[1,2]
1. Ivane Javakhishvili Tbilisi State University
2. National Botanical Garden Of Georgia

The study area represents provinces of south Georgia and is divided into three floristic provinces: Euxine, Armeno-Iranian and Caucasian (Grossheim, Sosnowski, 1928; Takhtajan, 1986). The flora and vegetation is more or less different from plant species of Mediterranean, Iran-Turkish and northern hemispheric geographical areas. The vascular plants of study area consist of 1652 species, 561 genera and 131 families. The following vegetation types are represented in Meskheti and Javakheti: riparian, middle and high mountain forests, xerophilic shrublands, halophilous vegetation with *Nitraria schoberi*, vegetation of high mountain meadows and steppes, wetlands, petrophyton and subnival belt vegetation. Phytocoenological composition of wetland vegetation of the Javakheti lakes (Tabatskuri, Paravani, Kartsakhi, Sagamo, Madatafa) and their surroundings is mainly defined by Eurasian, West Asian-Caucasian and European-Mediterranean species including *Carex juncella (=C. wiluica), C. panicea, C. irrigua, C. vesicaria, Calamagrostis canescens, Sagittaria sagittifolia, Menyanthes trifoliata, Utricularia vulgaris etc.* Surroundings of Tetrobi-Chobareti limestone area are distinguished by special botanical-geographic diversity on the Lesser Caucasus. A significant portion of the calcephilous flora is represented by species of Lesser Caucasian-Anatolian geographic type: *Arenaria steveniana, Asphodeline taurica, Scorzonera ketzkhovelii, Cerastium sosnowskyi, Silene dianthoides, etc.* Such diversity of the Tetrobi flora is related to the location of the area on the floristic intersection of the Lesser Caucasus, Armenia and Iran as well as isolation of the Tetrobi limestone massif, which is in turn connected with the history of geodynamic processes on the Javakheti. Subnival zone on the Javakheti vulcanogenic plateau is present only on Mts. Abuli (3,304 m) and Samsari (3,284 m). During the florogenesis of subnival belt of the Lesser Caucasus a decisive role was played by the floristic centre of Anterior Asia and a minor role by the floristic center of the Greater Caucasus and arcto-alpine floristic centres.

T1
P0201
The Poales Order in the state of Aguascalientes, Mexico
José Carlos Sierra Muñoz[1], *María Elena Siqueiros Delgado*[1], *Rosa Cerros Tlatilpa*[2], *Joaquín Sosa Ramírez*[1]
1. Universidad Autónoma de Aguascalientes
2. Universidad Autónoma del estado de Morelos

The order Poales is one of the groups of plants with greater importance for the human being, currently includes 16 families and presents a cosmopolitan distribution. Mexico is recognized as one of the most biodiverse countries on the planet, however, knowledge of the plant species that inhabit its territory is still limited. In order to contribute to the knowledge of the flora of the state of Aguascalientes, a taxonomic study of the species of the order Poales that are distributed in its territory is carried out. To date, a total of 56 sites have been sampled throughout the State and 1012 samples have been obtained that were properly herborized for taxonomic determination. Also, a total of 2,322 herbarium specimens deposited in the HUAA herbarium. Currently the Poales order in the state of Aguascalientes includes 275 species: 204 of the family Poaceae, 49 of Cyperaceae, 10 of Juncaceae, 9 of Bromeliaceae, 2 of Eriocaulaceae and one of Typhaceae. 26 species have turned out to be new records for the state. It has a preliminary floristic list and a preliminary distribution map that allows to conclude that the order Poales presents cosmopolitan distribution throughout the territory of the State, the families Bromeliaceae, Cyperaceae and Poaceae are those that present the most generalized distribution; Eriocaulaceae, Juncaceae and Typhaceae are characteristic of wetlands and bodies of water throughout the State. Up to now none of the species is in any of the categories of Official Mexican Standard (NOM-059-SEMARNAT-2010) and is not considered vulnerability in their populations.

T1
P0202
Ethno-ecological survey of underutilized plant diversity of Hamirpur District, Himachal Pradesh, India: An edibility assessment
Anand Narain Singh, *Ramchand Bhatti*
Panjab University Chandigarh

The present study was conducted to explore diversity and status of underutilized wild edible plants of Hamirpur district, Himachal Pradesh, India. Total 85 plant species of 66 genera belonging to 39 families under various life forms and origins were recorded; in which, 30 species as trees (25 native and 5 exotic), 11 shrubs (6 native and 5 exotic), 28 herbs (10 native and 18 exotic) and remaining 16 were creepers (12 native and 4 exotic). Edibility index (EI) showed 100% score of edibility by two species, while 80% was shown by 11 species, 40% by 3 species and remaining 69 species showed 10 to 30% only. Across family, maximum utilization efficiency (10 score) was found only in 2 families (Apiaceae and Violaceae), while 5 families (Chenopodiaceae, Commelinaceae, Liliaceae, Oxalidaceae and Portulaceae) have shown 8 and one family (Amaranthaceae) exhibited 5.8 score followed by 31 other families. In this study, it was observed that consumption mode of various plant parts were either fresh or cooked. In conclusion, present study may help in alleviating food scarcity, malnutrition and can be employed further as commercial crop plants for the welfare of human in developing countries.

T1
P0203
Reintroduction of *Bentinckia nicobarica* in Nicobar Islands- A case study
Pushpa Kumari, *Paramjit Singh*, *Reshma Lakra*
Botanical Survey of India

Bentinckia, a palm genus is restricted to Indian subcontinent with *B. nicobarica* reported from Nicobar Islands. Usually, Island species are within their limited area of distribution not having any chance of migration. The Andaman and Nicobar Islands represents

86% of forest coverage including tropical rain forests, tropical moist deciduous forests, lowland forests *etc.* with a total geographical area of 8,249 sq. km while the Nicobar group of Islands covers an area of 1,841 sq. km. *Bentinckia nicobarica*, belonging to the family Arecaceae is one of the tall palms growing up to ±50 m height with small globose, blackish fruits (when ripen) in big bunches. An endemic to the Nicobar Islands, it specifically grows in moist hills, valleys, plains and slopes of lowland forests of Nicobar group of Islands mainly in Great Nicobar, Kachal, Nancowry group, Teressa, Chowra and Car Nicobar Islands. In the year 1988, "The IUCN Red List of Threatened Species" assigned this species endangered category. *Bentinchia nicobarica* is used in the Islands for housing, furniture, boat making *etc.* by the tribal of Nicobarese and Shompen. The emerging shoots are eaten by the local people as vegetable and the fallen fruits are eaten by pigs. Due to all these activities over the time the species population has declined very much in these islands. Considering the importance of the species the present study has been taken to investigate its population, multiplication and reintroduction. The method of ecological niche modelling (ENM) map is applied by using the modeling algorithms of DIVA-GIS and MAXENT. Nineteen bioclimatic variables are used as environmental variables for modeling, which includes Annual Mean Temperature, Mean Diurnal Range, Isothermality, Temperature Seasonality, Maximum Temperature of Warmest Month, Minimum Temperature of Coldest Month, Temperature Annual Range, Mean Temperature of Wettest Quarter, Mean Temperature of Driest Quarter, Mean Temperature of Warmest Quarter, Mean Temperature of Coldest Quarter, Annual Precipitation, Precipitation of Wettest Month, Precipitation of Driest Month, Precipitation of Seasonality, Precipitation of Wettest Quarter, Precipitation of Driest Quarter, Precipitation of Warmest Quarter, Precipitation of Coldest Quarter *etc.* Through this modeling, we could predict the potential area of occurrences the species and their reintroduction for the population study and the limiting factors. More than four thousand seedlings of *B. nicobarica* are raised so far in controlled condition in polyhouse of Dhanikhari Experimental Garden Cum-Arboretum, Botanical Survey of India, Andaman and Nicobar Regional Centre and being reintroduced in the identified habitat for improving their conservation status.

T1

P0204

Moss genera *Leratia, Lewinskya, Nyholmiella* and *Orthotrichum* in China

Zuzana Skoupa[1], *Guo Shuiliang*[2], *Mamtimin Sulayman*[3], *Lucie Fialova*[1], *Vitezslav Plasek*[1]

1. University of ostrava

2. Shanghai Normal University

3. Xinjiang University

In total, the traditionally conceived genus *Orthotrichum* s.l., it currently consists of 45 species in China, including one species of the genus *Leratia*, 15 of the genus *Lewinskya*, one of *Nyholmiella* and 28 of *Orthotrichum* s.str. This study provides new data on the distribution of species based on a critical revision of material from Europaean herbarium, including PC, B, H, W, C and LE, and from four largest Chinese herbaria, namely KUN, PE, IFP (formerly IFSBH) and XJU and field survey of 7 Chinese provinces. One species of the genus *Lewinskya* and two of *Orthotrichum* were recently recorded for the first time in China and moreover 20 species are newly reported by authors from various provinces in China, including Gansu, Heilongjiang, Hubei, Hunan, Inner Mongolia, Jiangxi, Jilin, Qinghai, Shaanxi, Shanxi, Shandong, Sichuan, Xinjiang, Xizang, Yunnan and Zhejiang. Detail distribution and ecological requirements of the species are discussed. Although, species richness of the Orthotrichalean mosses is relatively high in China, it is largely unbalanced in different parts of this country. Species diversity is perceptibly decreasing from the west to the east. Surprisingly, no species within the moss group have been recorded until now in four Chinese provinces (Guangdong, Guangxi, Henan and Ningxia).

T1

P0205

Historical biogeography in Fouquieriaceae (Ericales) with special focus in succulence, and chromosome evolution

Fabiola Soto-Trejo[1], *Susana Magallón*[2], *Ken Oyama*[1]

1. Escuela Nacional de Estudios Superiores Unidad Morelia, UNAM

2. Departamento de Botánica, Instituto de Biología, UNAM

We estimated the biogeographical history of Fouquieriaceae, a small family from arid and semiarid regions of Mexico and southwestern USA, and evaluated hypotheses concerning its biogeographical origin and evolutionary history. We used modern likelihood-based models and methods of statistical model comparison as implemented in BioGeoBEARS for biogeographical inference. The best-fitting biogeographical model showed that the crown node of Fouquieriaceae may have originated in the upper Miocene, around 6.6 Mya in either Baja California desert, the Sonoran desert, or the Tehuacán-Cuicatlán area. This high uncertainty at deeper nodes in the phylogeny could be due to the fact that the split between extant Fouquieriaceae and its sister family Polemoniaceae occurred around 86 Mya, which preclude the inclusion of Fouquieriaceae's relatives that would yield more support in ancestral range estimation. Additionally, the family is a small clade (10 species), some of the species are relatively widespread, and sister species tend to have different geographic ranges, which could affect biogeographic inference. However, we tested if other sources of information such as the succulence and chromosome evolution within Fouquieriaceae could contribute to the identification of the biogeographic origins in this family.

T1

P0206

The plant diversity of the coastal tropical bog, an endangered habitat of South-East Asia Mainland: A case study in the Peninsular Thailand

Kitichate Sridith

Department of Biology, Faculty of Science, Prince of Songkla University

The surveys of plant diversity have been achieved in the tropical bog along the coast in the Peninsular Thailand, which is the part of South -East Asia mainland as a case study of plant diversity in such endangered habitat on the sandbars along the coast of the

Tropical Asia. These coastal sandbars, created by the sea current in the pass, have performed various high ridge dunes as well as lower flat deposition areas between dunes. The ground water permeating in these lower areas between such dunes caused the swampy condition in the place where the water level is higher than the ground, while in the place where the ground water level is about the same as the ground between dunes which comprised of fine silt, it has turned such sandy ground to become permeated areas or "bog". The water of those swamp and permeated areas is weak acidic, due to the peat depositing condition of the developed peat swamp vegetation in the swamp between the dunes. This special acidic condition is allowed only very low nitrate and less microorganism available. This is so called "tropical bog". In which, a special plant community has adapted. The selected groups of plants in such plant community have got Nitrogen from an animal protein in this low Nitrogen condition of soil i.e. the carnivorous plants e.g. genera *Nepenthes* spp., *Utricularia* spp., *Drosera* spp. *etc.* And there is a possibility that this plant community might play some important role in maintaining the balance between nutrient cycling in the coastal dune system which, on the other hand, is not have some impacts to the required plant nutrient level and physical environment conditions among the coastal dunes system only, but to the nearby inland habitats next to the coastal dune system as well as the marine habitats along the coast where the dunes occurred.

T1

P0207

The relation between green tea consumption and CVD (cardiovascular disease) development as action plan for the prevention and control: A systematic review

Fairuza Syarfina, *Aprillia Hardiyani Tanto, Safira Adilla, Fahmi Dimas Abdul Azis, Prima Alifian Hergaputra*
Airlangga University

Background: Tea (*Camellia sinensis*) is one of the most popular beverage in Asia.Tea products can be divided into three categories based on the degree of fermentations, one of them is green tea (unfermented tea). Many researchers have reported antioxidant activity of green tea to prevent chronic diseases. Daily consumption of green tea can reduce the development of atherosclerotic complication related to stroke, myocardial infarction, coronary heart disease, and CVD (cardiovascular diseases). Every year there is increasing number on people die annually from CVD than from any other causes. CVD is the number one cause of death in the world and over three quarters of CVD deaths take place in low and middle income countries. The use of dietary substances in healthy lifestyle takes important role for preventing individuals who have risk factors for developing atherosclerotic complication. There are evidences from clinical studies which prove that green tea can play protective role in the development of CVD. Various components of green tea, especially polyphenol, have antioxidant and antidyslipidemic activity. The utilization of green tea are considered to be potentials in preventing the development of atherosclerotic and CVD. Green tea is also reported with its ability to recover the metabolic abnormalities by various mechanisms. Objectives: The aim of this study was to review the relation between green tea consumption and CVD development. Methods: Published observational studies and meta-analysis (2013-2017) addressing green tea or green tea polyphenol and CVD were identified based on searches of on-line database (Elsevier, Science Direct, PubMed). Results: The number of clinical studies in this field is still limited. Eight research articles met inclusion criteria have been conducted and summarized. The results in clinical studies showed the significant reduction in total and low-density lipoprotein (LDL) cholesterol in subjects who received green tea group compared to the control group. Green tea consumption has found to be inversely associated with the development of CVD. The group received 2-3 cup green tea/ day had less incidence of myocardial infarction, hyperlipidemia risk, and stroke risk compared to the <1 cup/ day group. Taking green tea is effective in reducing atherosclerosis risk associated oxidative stress. Conclusions: Behavioural risk factors such as unhealthy diet affected the development of CVD. The utilization of dietary substances such as green tea have promising potential to prevent the development of CVD complications. In general, green tea consumption is recommended for maintaining human health.

T1

P0208

Biodiversity and ecosystem services in Morocco: Current state, threats and conservation strategies

Mohammed Sghir Taleb
Institut Scientifique, Mohammed 5 University in Rabat

Located at the northwest corner of the African continent between 21° and 36° north latitude and between the 1st and the 17th degree of west longitude, Morocco with a total area of 715,000 km^2 enjoys a privileged position with a coastline of 3,446 km long opening to the Mediterranean and the Atlantic Ocean. Morocco is characterized by an important genetic diversity represented by a rich and varied flora with 5211 species and subspecies and many natural ecosystems. This biodiversity is currently subject to many pressures: overgrazing and deforestation, climate change, including increased drought, urbanization and forest fire. Conscious of the risks that weigh on biodiversity, Morocco set a strategy of management of biodiversity that focus on programs of *in situ* and *ex situ* conservation. This presentation will be focused on the current state of biodiversity, on the different pressures (grazing, timber harvest, harvesting of medicinal and aromatic plants, charcoal making..) and the efforts constructed by Morocco to conserve and sustainably manage biodiversity.

T1

P0209

Plant inventory research in Myanmar: Materials towards the Flora of Myanmar

Nobuyuki Tanaka[1], *Mu Mu Aung*[2], *Shuichiro Tagane*[3], *Akiyo Naiki*[4], *Norio Tanaka*[1], *Hidetoshi Nagamasu*[5], *Thaung Naing Oo*[2], *Tsukasa Iwashina*[1]
1. National Museum of Nature and Science
2. Forest Research Institute, Ministry of Natural Resources and Environmental Conservation
3. Center for Asian Conservation Ecology, Kyushu University
4. Tropical Biosphere Research Center, University of the Ryukyus
5. he Kyoto University Museum

Myanmar (Burma) is located south of the Himalayan region, spans from the mangrove forests and coral reefs of the Andaman Islands in the south, to the snow-capped peaks of Mt. Kakaboradzi (5,881 m) in the north, and Myanmar embraces incredible floral diversity. However, more than half a century has elapsed since botanists previously botanized in the country until 1960's. Due to political situation, there have been nearly 50 years blank of the floristic investigation. Further expeditions were carried out in 2000's, however most of the regions have still not been botanized. Therefore much still remains to be learnt of its flora as well as of its floristic relations with neighboring regions of Asia. National Museum of Nature and Science, Japan (NMNS) and Forest Research Institute (FRI), Ministry of Natural Resources and Environmental Conservation embarked the international joint inventory research project to collect the materials towards the "Flora of Myanmar" under the MoU (Memorandum of Understanding) covering whole regions in Myanmar. Although this memorandum covers not only for the plants, but also fungi, lichens, insects *etc.*, in the whole regions in Myanmar. We are focusing on the hitherto-unexplored areas in Myanmar; especially peninsular part, such as Tanintharyi (Tenasserim), Lenya and Lampi Marin National Park in the Mergui Archipelago. Our inventory research activity is introduced here focusing on the results of vascular plants. In 2016 we performed the Japan-Myanmar joint inventory research in Tanintharyi, Indawgyi Lake Wildlife Sanctuary (Kachin State), Htamanthyi Wildlife Sanctuary (Sagaing Region) and also other regions. Besides dried and pressed herbarium specimens, alcohol preserved specimens, and leaf materials for DNA analyses were collected. The herbarium specimens are deposited in TNS (NMNS) and RAF (FRI), the only one Myanmar herbarium registered in the Index Herbariorum. As the results of the inventory research in 2016, ca. 7,200 vascular plant specimens including duplicates were made. In Annonaceae, Ebenaceae, Euphorbiaceae, Lauraceae, Melastomataceae, Zingiberaceae *etc.*, many taxa were considered to be new or newly recorded from Myanmar. These results will be published as a series paper such as "Contributions towards the Flora of Myanmar". Furthermore the data from our intensive inventory works in the unexplored areas would provide the significant materials for compiling a full flora of Myanmar. In order to further efforts in floristic research, as well as to contribute to the development of human resources in Myanmar in general, it is necessary to provide international assistance in this respect. Part of our project in Myanmar involves cooperation with our partner institutes to help promote the development of human resources.

T1
P0210

An early Oligocene palynoflora from Yunnan, Southwest China

He Tang[1,3], *Shu-Feng Li*[1], *Tao Su*[1], *Zhe-Kun Zhou*[1,2]

1. Key Laboratory of Tropical Forest Ecology, Xishuangbanna Tropical Botanical Garden, Chinese Academy of Sciences, Mengla, Yunnan 666303

2. Key Laboratory for Plant Diversity and Biogeography East Asia, Kunming Institute of Botany, Chinese Academy of Sciences, Kunming 650201

3. University of Chinese Academy of Sciences, Beijing 100049

Following the Eocene-Oligocene transition (~33.9Ma), the Oligocene is an important period marks the establishment of modern vegetation. Until now, fossil-bearing Oligocene sediments with absolute geological dating are relatively rare in China, especially in Southwest China, a biodiversity hotspot worldwide. Recently, a plant fossil-bearing outcrop with several volcanic tuff layers was discovered from Lühe Basin, in central Yunnan Province, Southwest China. The U-Pb dating results show an unequivocal early Oligocene age of this section (33-32 Ma). To better understand the plant diversity of Oligocene flora in this region, we did a combined LM/SEM investigation for the palynomorphs extracted from the outcrop. The early Oligocene palynological assemblage of Lühe was composed of 46 palynomorphs, belonging to 39 genus, covering angiosperms (13%), gymnosperms (73.9%), ferns (8.7%) and other elements (Fungi and Alge, 4.3%). Our result shows that the palynoflora of Lühe was dominated by gymnosperms (41%, such as *Piceapollenites, Pinuspollenites, Tsugapollenites, Ephededripites*), the angiosperm taxa were dominated by (*Alnipollenites, Betulaepollenites, Betulacepollenites, Carpinites, Ulmipollenites, Quercoidites*). Generally, this palynoflora indicate a needle- and deciduous broad-leaved mixed forest with warm and humid subtropical elements in the early Oligocene of central Yunnan. This vegetation type suggest a temperate to subtropical humid climate in this region during the early Oligocene.

T1
P0211

Tackling indonesia's diabetes challenge with Indonesian traditional herbs as dietary therapies to reduce development of cardiovascular complications: Promotion the use of natural product

Aprillia Tanto, *Fairuza Syarfina, Fahmi Dimas Abdul Azis, Safira Adilla*

Airlangga University

Objectives: Indonesia's biggest challenge in the next 10-15 years is non-communicable diseases (NCD) like cardiovascular and diabetes with its complication which are accounted for nearly 50 percent of deaths in the country. As of 2014, Indonesia's diabetes population was the fifth largest in the world, International Diabetes Federation (IDF) predicted more than 14 million Indonesians will suffer diabetes in 2035. Dietary therapy can reduce the risk of diabetes development and its cardiovascular complication. Indonesia as a country that is rich in natural sources has the opportunity to maximize the utilization and benefits of its natural products. Indonesian traditional herbs have been used to treat disease and nowadays they are being scientifically developed for therapies. These natural products are considered to have potentials in delaying the development of diabetes and its complications and to be able to cure the metabolic abnormalities using a variety of mechanism. Methods: Comprehensive literatures were used to review Indonesian traditional herbs used in the management of diabetes. An online database research from Elsevier, Science Direct, and PubMed was conducted to find literature on herbs growth in Indonesia that improves health outcomes in the development of diabetes and its cardiovascular complication. All of the data were analysed with descriptive technique. Results: Indonesian traditional herbs such as *Momordica chatarantia*, *Trigonella foenum-grae-*

cum, *Zingiber officinale*, *Curcuma domestica*, *Allium sativum*, *etc.* have been proven to produce significant improvement of blood glucose, 2 hour post load glucose, A1C, LDL-C, blood pressure level as compared with control intervention which have antidiabetic and cardioprotective activity with different mechanism of action. Conclusions: Indonesian scientifically proven traditional herbs have potentials to be used as dietary therapies to reduce the development of diabetes and its complication. It can be the investment to tackle global challenge for reducing deaths caused by diabetes and its cardiovascular complication.

T1

P0212

Diversity and pharmacology study of selected Zingiberaceae from Peninsular Malaysia

Yannick Tatin

Tropimundo - Erasmus Mundus Master Program

The Zingiberaceae diversity of a restricted area of Kenyir Lake (Terengganu, Malaysia) was characterised using different richness indices. A high diversity was found (α-diversity value H' = 2.04) with sixteen species belonging to five genera. Besides, potential medicinal properties (antimicrobial and cytotoxicity assays) of four *Etlingera* species (*E. megalocheilos, E. metriocheilos, E. punicea, E. pauciflora*) found in the study site were tested using rhizome crude extracts. In vitro antimicrobial assay was performed with the agar disc diffusion method against eight bacteria strains *Vibrio alginolyticus, Vibrio cholera, Eschercia coli, Klebsiella pneumonia, Pseudomonas aeruginosa, Salmonella* sp., *Staphylococcus aureus, Bacillus cereus, Streptococcus uberis* and *Micrococcus* sp. The crude extract of *E. punicea* exhibited the best antimicrobial activity against *S. aureus, S. uberis* and *K. pneumoniae* (inhibition zone of 11.5 mm, 13 mm and 18.75 mm respectively at 100 $mg.ml^{-1}$ of crude extract) whereas *E. pauciflora* was the most efficient against *Microccocus* sp. (inhibition zone of 36mm at 100 $mg.ml^{-1}$ of crude extract). No activity was exhibited against the four other strains of bacteria by any plant. The cytotoxicity against MCF7 cancer cells was tested using the MTT assay method. The crude extract of *E. punicea* exhibited the best cytotoxicity (LC_{50} = 18 $g.ml^{-1}$) followed by *E. metriocheilos* (LC_{50} = 29.2 $g.ml^{-1}$), *E. megalocheilos* (LC_{50} = 33.4 $g.ml^{-1}$) and *E. pauciflora* (LC_{50} > 100 $g.ml^{-1}$).

T1

P0213

International germplasm conservation, database construction and cultivar registration of *Nelumbo*

Daike Tian[1,2], *Fengluan Liu*[1,2], *Dasheng Zhang*[1,2], *Qingqing Liu*[1,2], *Qing Chen*[1,2,3], *Jie Min*[1,2,4]

1. Shanghai Chenshan Plant Science Research Center of Chinese Academy of Sciences, Shanghai Chenshan Botanical Garden, Shanghai, China

2. Shanghai Key Laboratory of Plant Functional Genomics and Resources, Shanghai, China

3. Fujian Agriculture and Forestry University, Fuzhou, China

4. Hunan University of Science and Technology, Xiangtan, China

Lotus (*Nelumbo*) has over 2,500 years of history in cultivation, and it is the national flower of India, as well as one of the traditional Chinese flowers due to its medicinal, edible, ornamental and cultural values. There are only two living species (*N. nucifera and N. lutea*) but nearly 2,000 cultivars in *Nelumbo*. For the purpose of promoting germplasm conservation, scientific research, industry, and public education of lotus, an integrative project is developing at Shanghai Chenshan Botanical Garden of China, China. So far, more than 700 genotypes (species, varieties and cultivars) of lotus has been globally collected and *ex situ* conserved in this garden. The International *Nelumbo* Database or Website is also constructed, from which, the information extensively covered on lotus is available, including literature, database of species and cultivars, scientific research, industry, culture, international lotus cultivar registration, lotus forum, important links, and so on. In addition, the projects are undergoing on lotus cultivar identification, population diversity, mechanisms of double-flower development and petal coloration through the integrated morphological, physiological and molecular methods.

T1

P0214

Privileged-account herbarium plant specimen archival management system

Kun-Chan Tsai[1], *Jenn-Che Wang*[2], ***Chin-Shan Tien***[1]

1. Microtek

2. National Taiwan Normal University

The digitization of herbarium specimen collection consists of two time-consuming processes, one is specimen label recognition and the other is verification of metadata prior to uploading for public access. The latter part is especially vital and usually determined by experienced taxonomists to guarantee the explicit information and image quality (focus, brightness, and contrast *etc.*) of each plant specimen. Given increasing budget vulnerability and limited time availability, the metadata digest rate of herbarium curators and managers can hardly match that of label recognition, further delaying expected digitization progress. The aim of this study is to introduce an image archival management system, which is characterized by (1) assigning privileged-account lists specific to distinct digitization steps, (2) maximizing metadata confirmation efficiency through joint global academic effort, (3) releasing specimen (metadata and image) for public access after professional verification, and (4) intuitive user-interface for a quick-start manipulation. The file structure follows default JSTOR standard, adding significant convenience between global herbaria for metadata sharing and exchanging. By coordinating with the implementation volunteer recruiting-and-training system specialized on specimen digitization and label recognition, herbaria curators and managers can fully devote to backstage management through this system to ensure specimen collection in corrected digitized format.

T1

P0215

Promoting the conservation and creation of wetlands in Mexico through a Low Impact Design Demonstration Project

Cruz Elisa Torrecillas Nuñez[1], *Antonio Jose Miguel-Rodriguez*[2]

1. Universidad Autónoma de Sinaloa

2. SAMI Ltd

Low Impact Design (LID), also known as Green Infrastructure or Sustainable Drainage Urban Systems, is a technology that has been applied in developed countries for more than twenty years but its use is hardly known in Mexico. In Mexico there are 142 Wetlands of International Importance known as Ramsar sites, but there are practically no wetlands in modified environments as they have been destroyed by urban and agricultural land use. Society at large does not appreciate the value of wetlands, hence the need to demonstrate their benefits in cooperation with governmental and non- governmental agencies (NGO) such as CONAGUA, SEMARNAT, CONANP and Fabricas del Agua (a NGO that promotes reforestation in watersheds). This paper presents the results of a project being carried out at the Autonomous University of Sinaloa as part of its Green Campus Strategy with the purpose of demonstrating the feasibility of LID in the arid to semi-arid conditions found in the country. LID is used to control the impacts of deforestation and loss of wetlands and vegetative cover due to changes in land use which affect the balance of the water cycle by altering surface runoff, surface and underground storage and groundwater flows. As a result of the hydrological alternations, there are increases in peak flows and a decrease in the ecological flow, which leads to greater amplitude of flooding and the reduction of the recharge of the aquifers. Another consequence of land development is the generation of silt which smothers plant habitats and the discharge of contaminants which affect ecosystems. Additionally the loss of wetlands due to development leads to the alteration of natural habitats and the removal of wetland-dependent plants. Therefore there is a pressing need to develop and implement strategies to reduce these impacts. LID is a systematic process that begins by protecting ecologically important areas such as wetlands as part of development proposals. LID offers an optimal solution to mitigate the impacts as it addresses the causes by managing the impacts at their source, maintaining wetlands and the natural stream channels, and recovering the capacity of infiltration and storage present in the natural basin. Some of techniques used include: retention and detention ponds, artificial wetlands, rain gardens, bioretention systems, dry wells, permeable pavement, and vegetated swales. Plants play a critical role in LID through evapotranspiration, capturing and treating stormwater; the interaction of soil and plants helps to capture rainfall, reduce flooding and remove contaminants. The inclusion of native vegetation as part of LID ensures that plants are drought resistant and promotes biodiversity. It is expected that this project will have a significant impact as the Autonomous University of Sinaloa is the leading academic institution in Sinaloa State (Mexico), thereby being in a position where it can influence the attitude towards wetlands of the 2,966,321 persons who live in the state.

T1

P0216

Population genotyping of a wild botanical species with the use of SNP TaqMan assays

Sonia Trujillo-Argueta, *Yesenia Pacheco Hernández, Rafael Felipe Del Castillo*

CIIDIR Oaxaca - Instituto Politécnico Nacional

Wild plant species DNA nucleotide sequences are less reported on genebanks, in comparison to economic plant species, so that studies on genetic diversity based on DNA, are more limited. Most nucleotide sequences reported on gene banks of native botanical species, come from phylogenetic relationships studies. We looked for these sequences in order to find SNPs and genotype *Malacomeles denticulata*, a native rosacea that grows from south USA to Central America. We searched for nucleotide sequences on this rosacea on NCBI gene bank. We found 19 sequences and only two transition mutation SNPs, which states for a base change from T to C. We also look for nucleotide sequences on genera closely related to *Malacomeles* and found another transition mutation SNP from *Amelanchier bartramiana*. The three sequences obtained holding an SNP were reported on nucleotide sequences of 5.8S ribosomal gen and internal transcribed spaces 1 and 2. All these sequences were analyzed with Repeat Masker software and Basic Local Alignment Search Tool (BLAST) so that a SNP TaqMan assay could be designed. Genomic DNA was obtained from leaves of *M. denticulate* that grew in seven different localities, in the state of Oaxaca, México, from soil eroded places to pine oak forests. After a PCR-RT, DNA samples could successfully amplified with each designed SNP assay and homozygous and heterozygous individuals, were registered. Since all three SNPs found, were in nucleotide sequences of 5.8S ribosomal gen and internal transcribed spaces 1 and 2, a linkage disequilibrium test was performed. No significant evidence ($P \leq 0.05$) was found that these three SNPs were linked. Genetic diversity analysis performed were percent polymorphic loci (P), mean number of alleles (A), effective number of alleles (Ae), Shannon index (S), genotype frequency, allelic frequency, observed heterozygosity (Ho) and expected heterozygosity (He). Genetic structure was calculated with the coefficient of endogamy (F). Higher values of P were observed on Teotitlán del Valle y Paraje Xía de San Juan Chicomezúchil populations (50% each). Teotitlán del Valle population, was the one that showed the higher mean value of observed heterozygosity (Ho = 0.306), followed by Ixtlán de Juárez (Ho = 0.253), Paraje Xía (Ho = 0.253) and Santo Domingo Yanhuitlán (Ho = 0.066) populations. San Pablo Huitzo, Díaz Ordaz and Santo Domingo Yanhuitlán gulls, hold homozygous populations for the three SNP assays analyzed. Estimates of genetic structure showed that all seven populations studied are genetically different (F_{ST} = 0.337) and, that there is an excess of heterozygotes inside each population (F_{IS}= -0.510). We could successfully genotype seven populations of a native rosaceae, with three SNPs found in the few nucleotide sequences reported on genebank on 5.8S ribosomal RNA gene and internal transcribed spaces 1 and 2.

T1

P0217

Gender variation and sexual system diversity of *Elatostema* (Urticaceae) in Taiwan

Yu-Hsin Tseng[1], *Jer-Ming Hu*[2]

1. Biodiversity Research Centre, Academia Sinica

2. Institute of Ecology and Evolutionary Biology, National Taiwan University

Sexual systems are varied and complex in angiosperms, which has diversified at the level of the individual flower, inflorescence, plant individual, and within or between populations. The diverse sexual expression system in Elatostema provides a special case to study the evolution of sexual system. To investigate the gender expression of Elatostema in Taiwan, we closely scrutinized herbarium specimens for determining the sexual system. Among 15 Elatostema species in Taiwan, sexual systems can be categorized into four types: monoecious, dioecious, both monoecious and dioecious in same species, and female-only/female-biased species. In 52 individuals of five populations monitored over two years, six flower-opening patterns have been recorded in *E. lineolatum* var. *majus*, including pure male, pure female male→stop→male, male→female, male→female→stop→male, and male→stop→female. Additionally, based on the bagging experiment, molecular phylogeny and chromosome observation, three of female-only/female-biased species in Taiwan, *E. cyrtandrifolium*, *E. oblongifolium* and *E. rivulare*, were confirmed to be apomictic plants with polyploidy and evolved independently in different lineages. The results of flow cytometric seed screen and embryogenesis further demonstrated that these species should belong to an infrequent case of obligate autonomous apomixis.

T1

P0218

Patterns of genetic isolation and diversification in limestone karst area: A case study of widespread *Begonia* (Begoniaceae)

Yu-Hsin Tseng[1], *Han-Yau Huang*[1,2], *Ching-I Peng*[2], *Kuo-Fang Chung*[1,2]
1. Biodiversity Research Centre, Academia Sinica
2. School of Forestry and Resource Conservation, National Taiwan University

Begonia sect. *Coelocentrum* is one of the most characteristic plant groups that can serve as a model system of limestone plant speciation. To test the hypothesis of allopatric speciation via isolation-by-distance triggered by extensive karstification, population genetics and phylogeography of a widely distributed limestone *Begonia*, *B. luzhaiensis*, were investigated by using EST-SSR markers and one chloroplast DNA sequence (*trnC-ycf6*), representing 171 individuals in 17 populations. Using next-generation sequencing (NGS) technique, 44,368 contigs were assembled, revealing 5,602 EST-SSR loci and resulting in 16 polymorphic EST-SSR markers for *B. luzhaiensis*. The *F*st analysis and Bayesian clustering approach implemented in Structure revealed a high level of population differentiation and a strong phylogeographical structure. Moreover, Mantel Tests indicated a significant positive correlation between genetic differentiation and geographic distances, concurring to geographic speciation via isolation-by-distance. Haplotype network reconstructed using *trnC-ycf6* sequences showed a highly geographic-correlated phylogeographic pattern, also supporting the hypothesis of population differentiation via geographic isolation. Our data support strongly the importance of extensive karstification in shaping the species diversity on the Sino-Vietnamese limestone karsts. The fragmented karst landscapes and the low dispersal ability of *Begonia* could have been the major driving forces that promoted genetic isolation among populations and resulted in allopatric speciation.

T1

P0219

Biogeography and threatened categories of some Caucasian *Pyrus* L. (Rosaceae) species

Zübeyde Uğurlu Aydın[1], *Ali A. Dönmez*[2]
1. Hacettepe University
2. Hacettepe University

The geographical distribution of pears is concentrated in western Eurasia, especially Caucasia, and eastern Asia. In the study eight Caucasian *Pyrus* taxa, namely *P. caucasica*, *P. georgica*, *P. salicifolia*, *P. oxyprion*, *P. demetrii*, *P. cordata*, *P. elaeagrifolia*, and *P. taochia* were evaluated to show present biogeography and threatened category according to collected specimens from the region, herbarium samples and the relevant literature. Taking into account the collected samples from Turkey, it is clear that *P. georgica* and *P. demetrii* naturally grows further west part than currently known distribution in Caucasia. Turkey plays critical role for this dispersal manner because geographically it is located between Caucasia and Europe and it includes probable mountains as habitat for the genus. Georgia is the most important region of diversity for examined taxa except *P. elaegrifolia* because this species has great morphological variations only in Turkey. However *P. salicifolia* is very similar to *P. elaeagrifolia* morphologically and also they prefer similar habitats, but they are not sympatric. *P. salicifolia* is diversified mostly in Azerbaijan while it was collected from only two close locations in Turkey despite detailed field research. As a result, threatened categories were assigned for *P. georgica*, *P. salicifolia*, *P. demetrii*, and *P. cordata* at first time based on new collections of these taxa. Present distributions were also marked on maps and discussed for IUCN criteria.

T1

P0220

The diversity and commercial potential of the African flora

Ben-Erik Van Wyk
University of Johannesburg

Africa is a continent of exceptional biological and cultural diversity, much of which remains scientifically poorly explored. The continental plant diversity is conservatively estimated at 71,486 species of flowering plants (North Africa: ca. 10,000; sub-Saharan Africa: 50,136; Madagascar: 11,350). The African angiosperm flora therefore represents nearly 20% of the global diversity, recently estimated to be about 391,000 species, of which 369,400 are angiosperms. Valuable and ongoing contributions towards an inventory of the flora have been made since the colonial era. A special highlight was the publication of a flora checklist for sub-Saharan Africa, based on the combined data sets from the Conservatoire et Jardin Botaniques de la Ville de Genève, Switzerland and the South African National Biodiversity Institute. In terms of useful plants, the Plant Resources of Tropical Africa project (PROTA) has provided a valuable collection of detailed monographs on hundreds of tropical African plants. Ongoing ethnobotanical studies are continuously revealing new records of useful plants with commercial potential. Plants have a multitude of direct and indirect uses, and provide a wide range of ecosystem services related to agriculture, horticulture, conservation, education, ecotourism, scientific research, literature, art, designs and symbols. Novelty is an important component of product development and it is likely that many more African plants will reach international markets in one form or another. Perhaps the greatest gift to the world has been coffee, estimated to be second only to petroleum as the most valuable natural product in international trade. African plant species have also contributed to commercial food (119 species out of the global top 900), functional food (10 of the top 56), commercial medicine (83 of the top 900), phytomedicine (eight of the top 43), and pure phytochemical compounds (13 of the top 135 valuable chemical entities extracted from plants). The African flora represents a rich source of inspiration for scientific, cultural and commercial innovations.

T1

P0221

The botanical garden of M.V. Lomonosov Moscow State University and its role in conservation of rare and threatened plants

Tatiana Varlygina*, Sergey Efimov, Galina Degtjareva, Elena Terentieva*

M.V.Lomonosov Moscow State University

The Botanical garden of MSU founded in 1705 by Peter I is one of the oldest in Russia. Since 1977 the Garden studies the questions of natural plant conservation. The researches are conducted in different ways, which are complemented each other and allow to collaborate with other institutes. First of all, there is investigation of native flora of Russia. As a consequence of it, the creation of more than 100 natural plant reserves (important plant areas) was proposed and the composition of Plant Red Lists of the Russia and some its regions was specified. Together with the Ministry of Nature Resources of RF the preparation of the Red Data Book of Russian Federation (2008) was organized and descriptions for some plant species were made. Now it includes 514 rare and threatened species of vascular plants (474 Angiospermae, 14 Gymnospermae, 23 Pteridophyta, 3 Lycopodiophyta), as well as 61 mosses, 35 algae, 42 lichens and 24 fungi species. Among them, 135 species are endemics of Russia and 99 species are near-endemics, growing both in Russia and one of the neighboring countries. Many endemic species belong to the families Fabaceae (16 and 24 relatively) and Asteraceae (10 and 10). 277 (58%) species of Angiosperms from Red Data Book of RF are under protection in zapovedniks. However, the majority of them grow only in one or in just several of them, 14% species grow in national parks and reserves, and 131 species (28%) do not have territorial protection. Cultivation of many rare and endangered species in the Botanical Garden of MSU serves to educational and scientific programs. In our collection there are 115 (22. 4%) species of vascular plants from Red Data Book of Russia and 90 (44%) rare species from Red Data Book of the Moscow region. In addition, investigations of state of population (ecology, age structure, reproductive biology and so on) of rare species in natural conditions are also carried out. Special attention was paid to the Russian orchids. As a result, two books were published. There are Vakhrameeva M.G. *et al.,* Orchids of Russia and Adjacent Countries (within the borders of the former USSR) (2008) and Vakhrameeva M.G., Varlygina T.I., Tatarenko I.V. Orchids of Russia (biology, ecology and protection) (2014). The books include information about species area, ecology, morphology, ontogenesis, seasonal rhythm, and protection. In 2015 together with the Russian Research Institute of Nature Protection methodic recommendations for restoration and translocation of rare plants populations were prepared.

T1

P0222

Polymorthism of chalcone synthase genes as factor of hop breeding

Andrii Venger

Plant Breeding and Genetic Institute

Hop (*Humulus lupulus* L.) is a dioecious plant from Cannabaceae family. One of most important genes in hop's genome are genes *chs*_H1, *chs*2, *chs*3, *chs*4 and *vps* (*chs*5), which encode chalcone synthase 1, chalcone synthase 3, chalcone synthase 3, chalcone synthase 4 and valerophenone synthase (chalcone synthase 5) enzymes, which catalyze biosynthesis of secondary metabolites (some of which impart the bitter taste and aroma to beer), such as bitter acids (alpha-acids humulone and cohumulone, beta-acids lupulone and colupulone) and prenylflavonoids (xanthohumol) accumulated in the lupulin glands. Depending on content of aroma and bitter subtracts hop varieties are derived in aroma and bitter samples. Many hop varieties from the world collection were studied by polymorphism of some of those genes, but there are no varieties studied by polymorphism of all of indicated genes. We have studied polymorphic and non-polymorphic regions of chalcone synthase encoded genes by PCR in Ukrainian hop varieties. There were detected new alleles of polymorphic regions and polymorphism in regions, which were described as non-polymorphic. Data of analysis was processed by Unweighted Pair Group Method with Arithmetic Mean and Maximum Likelihood; Distance Methods. In result there was built dendrogram with two groups one of which contained only aroma varieties whereas the other had only bitter varieties. The use of molecular markers in detection of aroma and bitter type of hop variety is cheaper than usually applied methods and needs lest time than using biochemical methods. Producing of hop genome database is important in studying of hop diversity, evolution and determination of type of varieties, members of population and individual samples of hop.

T1

P0223

Aquatic plant as bioindicator in the wetlands of Mekong Basin and Thailand

Chavalit Vidthayanon

IUCN SSC-Freshwater Fish Specialist Group, Mekong and Southeast Asia Regional Co-Chair

Wetlands in the Mekong Basin and Thailand are one of the global aquatic plants hotspots. Over 500 species of 32 families have been occurred but less than half were assessed (252 species) and five are reported as threatened taxa. Most wetlands in the Mekong and Thailand region are degraded and disturbed by human activity including habitat alteration, agro-chemical, water extraction, invasive alien species and climate change. Several aquatic plants are disappearing or replaced by tolerance taxa or exotic taxa. The species composition and occurrence in the wetland can indicate the water and habitat integrity. More *In situ* inventory monitoring as well as local market survey are needed approach for assesses wetland habitat indicators in the region.

T1

P0224

Vegetation structure and dynamics of tree species along the altitudinal gradient of Ilong Peak, Halcon Range, Mindoro, Philippines

Elaine Loreen Villanueva[1], *Inocencio Jr. Buot*[1,2]
1. University of the Philippines Los Baños
2. University of the Philippines Open University

The vegetation of Mt. Halcon is diverse, unique, and culturally important, but there is a very limited information on its vertical distribution and structure the along the gradient of the Halcon range. This research study analyzed the vegetation structure and dynamics of the tree species distribution along the altitudinal gradient of Ilong Peak, Halcon Range, Mindoro, Philippines. Fourteen plots of 10 m x 10 m were established along the slope from 100-1,400 m elevation. The tree species that were sampled from the plots were subjected to classification and ordination analyses. The cluster analysis identified five zones: Zone I (agroecosystem zone), *Lansium-Nephelium-Mangifera* zone; Zone II (agroecosystem zone), *Artocarpus* zone; Zone III (abandoned swidden zone), *Cratoxylum-Cyathea-Pterocarpus-Saurauia-Litsea* zone; Zone IV (forest zone), *Cyathea-Astronia-Syzygium-Garcinia-Ficus-Psydrax-Diplodiscus-Litsea-Ardisia-Garcinia-Actinodaphne* zone; and Zone V (forest zone), *Agathis*-Suli*-*Ardisia*-Alitugba* zone. The canonical correspondence analysis (CCA) showed that the elevation and the nitrogen level are the environmental variables that have strongly influenced the species distribution in the area, a characteristic that is similar to other tropical mountains.

T1

P0225

New findings and conservation status of the family Magnoliaceae in Vietnam over the past five years

Quang Nam Vu
Vietnam National University of Forestry

The report is a summary and statistics on the family Magnoliaceae's species which have been new findings for the Flora of Vietnam over the past 5 years with some remarkable notes on conservation. Of the total 22 species, belonging to 5 genera added for the Flora of Vietnam, 7 taxa are new species, including: Magnolia *bidoupensis* Q.N. Vu, *M. cattienensis* Q.N. Vu, *M. tiepii* V.T. Tien, N.V. Duy & V.D. Luong, *M. lamdongensis* V.T. Tran, N.V. Duy & N.H. Xia; Manglietia *sapaensis* N.H. Xia & Q.N. Vu, *M. crassifolia* Q.N. Vu, N.H. Xia & Sima; and Michelia *xianianhei* Q.N. Vu. 15 taxa are new records, consisting of Magnolia *championii* Benth.; Manglietia *forrestii* W.W. Smith ex Dandy, *M. lucida* B.L. Chen & S.C. Yang, *M. ventii* Tiep, *M. grandis* Hu & W. C. Cheng, *M. hongheensis* Y. M. Shui & W. H. Chen, *M. longirostrata* (D.X. Li & R. Z. Zhou) Q.N. Vu; Michelia *citrata* (Noot. & Chalermglin) Q.N.Vu & N.H.Xia, *M. flaviflora* Y.W. Law & Y.F. Wu, *M. fulva* Hung T. Chang & B.L. Chen, *M. macclurei* Dandy, *M. mannii* King, *M. coreana* Hung T. Chang & B. L. Chen; Kmeria *septentrionalis* Dandy; and Parakmeria *robusta* (B.L. Chen & Noot.) Q.N. Vu & N.H. Xia. The condensed information on nomenclature, vernacular name, distribution & ecology, remarkable characters is also given for each species. Based on IUCN (2001), 10 species – CR, 10 species – EN and 2 species - VU of the family Magnoliaceae from Vietnam are proposed for conservation.

T1

P0226

Ethnobotanical uses of some plants of family Apocynaceae and Asclepidaceae from north-western region of Ahmedngar district, Maharashtra

Mohan Waman
Dr. D.Y. Patil Arts, Commerce and Science College, Akurdi, Pune. Affiliated to Savitribai Phule Pune University, Pune (MS), India.

Ahmednagar is one of the largest districts in Maharashtra in terms of area and population. The north-west region of this district includes Sangamner and Akole Tehsil, fall in the Sahyadri ranges of Western Ghats which is one of the 34 Global Hotspot. It has diversified topographic and climatic conditions, gives rise to a wide range of flora and fauna along with different ethnic communities. The study was taken up for documentation and analysis of traditional knowledge regarding the use of plants in medicine and food belongs to family apocynaceae and asclepidaceae. Frequent field visits were carried out during 2014–2016 to different villages situated in the hilly regions and the 'key informants' were identified based on their reputation and experience regarding applications of plants for different purposes. The information was collected through semi-structured open ended interviews with questionnaire in their local language. All plants recorded for their use in different ailments and in the food were photographed in the field; voucher specimens were made subsequently and are deposited in the Herbarium, Nature Information Centre, Wildlife Department, Shendi. The information such as botanical name, family, vernacular name, habit and habitat, analysis like parts used, Use value (UV) and Informants Consensus Factor (ICF) are provided. In the present study, a total of 31 plant species belonged to both families used to treat different types of ailments and for food were provided by 47 key informants. Among the 31 plants recorded, 14 species used in medicine, 11 species in food and remaining 06 species used in both food and medicine. The highest number of species belonged to Asclepidaceae (25 species each) followed by Apocynaceae (06 species each). Climbers are used more often (16 species), followed by herbs (09 species), shrubs (3 species) and tree (3 species). The leaves were the dominant parts used followed by tuber, root, flower and fruits. The documentation and data analysis of such traditional knowledge of plants may provide novel compounds from the plants for the treatment of different ailments and new food species for coming generation.

T1

P0227

Community ecology of a 300 mys tropical flora in Inner Mongolia

Jun Wang[1], *Weiming Zhou*[2]
1. Nanjing Institute of Geology and Palaeontology, Chinese Acad-

emy of Sciences, Nanjing, China
2. Nanjing University, Nanjing, China

The peat forming vegetation from the earliest Permian (*ca.* 298 Ma) of Wuda Coalfield in Inner Mongolia buried by ash fall has been called as a Chinese "vegetational Pompeii". Such a T^0 deposits allows palaeoecologists to examine the Late Palaeozoic coal forest ecology in much the same way as neoecologists appraise modern environments. Seven major plant groups consist of the flora, including lycopsids, sphenopsids, noeggerathialeans, ferns and seed ferns, primitive cycadophytes and coniferophytes. Lycopsids are represented by *Sigillaria* cf. *ichthyolepis*, tall pole-like trees. Two sphenopsids were encountered, *Sphenophyllum*, and *Asterophyllites*. Most abundant were the marattialean tree ferns represented by *Pecopteris*. Herbaceous ferns include *Nemejcopteris feminaeformis*, *Cladophlebis*, and *Sphenopteris*. Noeggerathiales, an extinct spore-bearing group of uncertain systematic position, are small trees represented by species of *Tingia* and *Paratingia*. Gymnosperms include possible early representatives of the cycads, *Taeniopteris* and *Pterophyllum*, and tall trees of *Cordaites* that were early coniferophytes, and seed ferns including species of *Sphenopteris* and *Alethopteris*. In addition to the proportional taxonomic composition and individual plant growth habit, landscape heterogeneity (species-area curve), forest structure (ecological assembly, density of individuals) and vegetational recovery are evaluated. This flora growing on peat is also taxonomically distinct from those growing on clastic soils in the same area and during the same time interval. More significant is that a preliminary investigation of ontogeny of the individual taxa of the forest reveals reproductive biological strategy of this tropical swampy vegetation.

T1

P0228

Classification of disturbance regimes of subtropical forest by using forest canopy structure metircs calculated from airborne LiDAR data

Ningning Wang, *Xiangcheng Mi, Haibao Ren, Tianyu Hu, Qinghua Guo, Keping Ma*
State Key Laboratory of Vegetation and Environmental Change, Institute of Botany, Chinese Academy of Sciences

Disturbance is an important factor for shaping forest community pattern and canopy structure. Forest canopy structure provides the physical environment that generates, supports, and maintains forest biodiversity and is considered as important biodiversity indicator. Airborne LiDAR systems (light detection and ranging) could measure three-dimensional forest canopy structure accurately, and have been widely used in wildlife habitat mapping and species distribution modeling. Classification of disturbance regimes using LiDAR-derived variables would be helpful for guiding biodiversity monitoring and conservation assessments of species at regional level. This study conducted an investigation on the effect of different human disturbance regimes on subtropical forest canopy structure. The monitered evergreen, broad- leaved subtropical forest is locate in the Gutianshan Nature Reserve, Kaihua, China (29°15′N, 118°07′E). Nine 1-ha forest plots suffered three disturbance regimes (logged-selective: Logged-clearcut 50 years ago, and logged-selective 20 years ago; Logged-clearcut: Logged-clearcut 50 years ago; undisturbed: Undisturbed more than 100 years) were analyzed. Nine LiDAR-derived variables (Canopy height: 95th percentile height of all returns; SD of height: Standard deviation of canopy height; CV of height: Coefficient Variation of canopy height; Canopy cover: % of total returns above 2 m; Canopy height density-2-5 m: % of total returns between 2 and 5 m; Canopy height density-5-10 m: % of total returns between 5 and 10 m; Canopy height density-10-20 m: % of total returns between 10 and 20 m; LAI and Gap fraction) were calculated and used to classify forest structure. Cluster analysis was applied on the 9 variables grouping the 9 plots into three classes: logged-selective forest charactered as very short and low height variation, high canopy cover and gap fraction; Logged-clearcut charactered as very tall and high height variation, high canopy cover, LAI and low gap fraction forest; Undisturbed forest is tall and with high height variation, low canopy cover and high gap fraction and LAI. This result can be used together with species distribution data for conservation planning.

T1

P0229

Investigation on the effective cmponents of the flowers of *Trollius chinensis* Bunge from the perspective of intestinal bacterial transformation and intestinal absorption

Lina Guo, Shanshan Qiao, Deli Li, Duozhi Shi, ***Rufeng Wang***
School of Life Sciences, Beijing University of Chinese Medicine, Beijing 100102, China

The flowers of *Trollius chinensis* have been used in traditional Chinese medicine for respiratory tract infection due to their antibacterial, antiviral and anti-inflammatory activities. These flowers mainly contain flavonoids, phenolic acids and alkaloids; however, their effective components are debatable because the *in vivo* processes of these compounds are unclear. This study is to investigate the main effective components of these flowers from the perspective of intestinal bacterial transformation and intestinal absorption. The intestinal bacterial transformation model was applied to investigate the biotransformation of aqueous extract of the flowers of *T. chinensis* (AEOF). Both human colon adenocarcinoma cell line (Caco-2) monolayers and everted gut sacs were employed to evaluate the intestinal absorption of the intestinal bacterial transformed AEOF. The results showed 2^2-*O*-b-L-galactopyranosylorientin, orientin, vitexin, quercetin, veratric acid, proglobeflowery acid, and trolline were not transformed by intestinal bacteria while isoquercetin and trollioside were completely transformed by these bacteria to quercetin and proglobeflowery acid. The ratio of flavonoids to phenolic acids to alkaloids changed from 16:10:7 to 9:12:8 before and after absorption. Only glycosides such as isoquercetin and trollioside are transformed into their aglycones by intestinal bacteria, and the dominant position of the flavonioids was replaced by phenolic acids after intestinal absorption. The results of this study improve our understanding of the effective components of the flowers of *T. chinensis.* In addition to flavonoids which used to be considered as the dominant effective ones, phenolic acids and alkaloids are also very important to the efficacy of these flowers.

T1

P0230

Shanghai digital metroflora: Botanical gardens and metroflora

Yuan Wang, *Huiru Li, Jinshuang Ma, Jianping Chen*
Shanghai Chenshan Botanical Garden, Shanghai Chenshan Plant Science Research Center

Shanghai is China's most important economic, financial and trade center. Due to the narrow land, the simple habitat, the huge population, the native plant species is very rare, only recorded more than 800 species. However, Shanghai horticulture industry is very developed, the introduction of plant species up to thousands of species (including varieties), a variety of cultivated plants is very strong demand. Shanghai Chenshan Botanical Garden, as one of the most abundant botanical gardens in China, it also leads the development of horticulture industry and the conservation of rare and endangered plants in Shanghai and East China. In Shanghai, a special place to compile local flora, however it has a very different needs from other provinces in China. We divided the Shanghai area into three hundred areas by 5×5 km^2. The native plants and cultivated plants in each area were investigated in detail, including the introduction of two botanical gardens in Shanghai. For cultivated plants, we rely on the botanical garden's rich data and living plants, most of the detailed identification to the variety, and to solve a lot of long-standing, large-scale use of the names by misuse. According to the survey data, historical specimens and literature, we have established the first provincial digital flora website (http://shflora.ibiodiversity.net) in China, which will show the results of the survey over the years, including the publication of the "Checklist of Shanghai Vascular Plants", " Key of Shanghai Vascular Plants" and "Illustrated Flora of Shanghai" in series of 3 volumes. The site is rich in pictures and information to help the relevant practitioners and ordinary people understand the status of Shanghai plants, rare and endangered plant conservation and plant information, solve the traditional flora by the lack of plant cultivation, stop the confusion names in the Chinese horticultural industry, and promote the health development of horticulture.

T1

P0231

Exploring the evolution of a megadiverse genus on a megadiverse island; herbarium genomics of New Guinean *Begonia*

Hannah Wilson
Royal Botanic Garden Edinburgh/ University of Glasgow

The megadiverse genus *Begonia* L. is one of the world's largest plant genera, comprising over 1,800 species. Globally, *Begonia* species richness mimics angiosperm species richness, so the genus may be used to infer general truths about tropical evolution. With a wide distribution and an estimated 400 endemic species on New Guinea, *Begonia* is an ideal subject for the study of evolution and biogeography in the region. Additionally, *Begonia* displays a preponderance for narrow-range endemism, making it a natural choice for studying species radiation on New Guinea, an island where endemism is remarkably high. The island of New Guinea is one of the last great tropical wildernesses; least explored and poorly understood. New taxa from the region are being published at a rapid rate, yet current knowledge of the flora of New Guinea is still very limited as the region is comparatively under-collected and under-researched. This project aims to understand the evolution of *Begonia* on New Guinea from the arrival of the first ancestor around five million years ago. Initial estimates indicate that the diversification of *Begonia* in New Guinea is possibly the fastest angiosperm radiation on record. To investigate this, firstly we will reconstruct the evolution of these species by using hybrid bait capture on DNA from herbarium specimens, to sequence a representative sample of New Guinea *Begonia* and place them in a time calibrated phylogeny. To further understand the radiation we need to know how many species there are and where they are distributed. Utilising over 1,000 herbarium specimens collected over the past 150 years, complemented by further field exploration, we will document and map the *Begonia* species diversity on New Guinea, then use this data to conduct species distribution modelling. Finally we will use the distribution data and phylogeny to obtain species diversification rates, and see if any rate shifts detected on the phylogeny are correlated with other factors, such as the formation of topographical heterogeneity and promotion of microallopatry by rapid orogenesis in the Pliocene and Pleistocene, or cyclic vicariance linked to frequent fragmentations and reformations of habitat caused by climate and sea-level fluctuations in the Pleistocene. Other areas such as the tropical Andes have long been feted for their massive and recently evolved biodiversity. However, it seems likely that New Guinea more than rivals this diversity and further studies of the region are likely to teach us some general truths about tropical evolution.

T1

P0232

Diversity and human perceptions of insects in tropical East Asian megacities

John James Wilson[1], *Kong Wah Sing*[2], *Hui Dong*[3]
1. China Agricultural University
2. Chinese Academy of Agricultural Sciences
3. Fairy Lake Botanical Garden Shenzhen

The East-Southeast Asia region, currently rich in biodiversity, has experienced rapid urbanisation and improved socioeconomic conditions, but has suffered the greatest loss in species diversity compared to any tropical region over the last 50 years. As urban development replaces native or remnant habitat, resources in the surrounding areas are depleted to support urban economies causing biodiversity loss. However, in urban environments, insects such as bees and butterflies, continue to offer important benefits to humans, especially pollination of plants which contributes to improved well-being and productivity. Studies of urban ecology have suggested city parks can be effective beneficial insect refuges in temperate regions, but their effectiveness as refuges for beneficial insects in East-Southeast Asia regions is unknown. Management schemes and techniques for conserving beneficial insects in urban parks are well- established in temperate countries but are currently lacking for tropical countries. Moreover, it remains to be seen which planting schemes that can be effective in improving the ability of urban parks to sustain populations of beneficial insects in East-Southeast Asia cities.

T1
P0233
Plant ecology and conservation of Desert Ciénegas
Dustin Wolkis[1,2], *Juliet C. Stromberg*[2]
1. National Tropical Botanical Garden, Department of Science & Conservation
2. Arizona State University, School of Life Sciences

Landscape-wide investigations are essential to inform conservation and restoration efforts of rare vegetation types including groundwater-dependent ciénega wetlands of the American Southwest. We asked: How regionally distinct are ciénega floras? How similar are their plant communities? What factors influence species richness and composition? At six study ciénegas, spanning a 1,000-m elevation gradient within the Santa Cruz watershed of Arizona, we sampled vegetation and soils in 30 quadrats, three times. Our results show that ciénegas have a distinct regional floristic signature, with one fourth of species restricted to American Southwest. All were dominated by clonal herbaceous perennials, but floristic overlap was low. Each ciénega supported a distinct suite of rare species. Species richness per site was influenced by salinity (negative) and wetland area (positive); quadrat-level influences included elevation (positive) and moisture (negative). Composition was influenced by water availability, salinity, and canopy cover. *Eryngium sparganophyllum* grew in moderately saline soils (which occurred at sites with high temperatures and negligible flooding), of importance for efforts to restore this imperiled species. Berming has influenced ciénegas, with drought effects apparent only at an unbermed ciénega. The high variability of ciénega wetlands and the rare species they harbor make these vanishing communities high priority for conservation.

T1
P0234
Biologically Agarwood-inducing technique formed Agarwood column in *Aquilaria sinensis*: A novel technique for high-quality and quantity Agarwood producing
Zeqing Wu
Xinxiang Medicial university

Agarwood, a precious resinous trunks or branches derived from wounded trees of *Aquilaria* species. Agarwood had been widely used throughout the world as medicine and fragrances since ancient times. Nowadays artificial agarwood-inducing techniques providing low quality and quantity agarwood had limited the international agarwood supply. Therefore, novel methods on agarwood production are in dire need to provide high-quality and yield agarwood with less time and labor cost. Our study established a novel agarwood-inducing technique in cultivated *Aquilaria sinensis* trees, biologically agarwood-inducing technique. A complete Agarwood column can be taken from the trunk without rotten wood. The whole trunk was divided into 8 parts according to color and length for analysis. Compared with agarwood from natural and Burning-chisel-drilling induced methods, 8 parts were analyzed by HPLC, GC-MS to determine chemical compounds. Other quality characteristics like TLC and alcohol soluble extraction content were also determined. Results showed that three chromones in 8 parts were detected in 6 month after treatment. The compounds from essential oil detected by GC-MS in some parts by Agar-Bit were similar to that of natural agarwood, so were the results of TLC and alcohol soluble extraction content. In conclusion Agar-Bit can significantly produce high-quality and quantity agarwood with much lower cost.

T1
P0235
Alien plants and their diffusion trend in the wetlands of Yunnan Plateau, Southwest China
Zhaoulu Wu, *Xixi Xiang, Yuan Lei*
Institute of Ecology and Geobotany, Yunnan University, Kunming 650500, China

Wetland plants, being important biological resources for ecological stability and social economic development, are facing the simplification and degradation recently. One of the causes is that the rapid decreasing of native species and increasing of alien species. We conducted an investigation of alien plants in 25 wetlands all over Yunnan Plateau, aiming at finding the alien plants and their dispersion. We have three findings: (1) There were 212 alien species belonging to 74 families and 167 genera, of which, 62 were included in the List of Chinese Invasion Plants, and 80 species from America; (2) Almost all of the aquatic alien plants were main artificial introduced to purify polluted water and beautify wetland landscape but some of them escaped and became invasive species, while most of terrestrial alien plants within 100 m around the wetlands were the spread of themselves; (3) The number of alien species decreased as altitude increasing and human disturbance decreasing. Our study indicates that a large number of alien plants (including invasions) appeared at present in the wetlands of the Central Yunnan Plateau, they may spread to the surrounding areas in the nearly future.

T1
P0236
In vitro antioxidant and antidiabetic activities and phytochemical quantification of some apomictic ferns from China
Baskaran Xavier Ravi[1,2], *Zhang Shouzhou*[1], *Liao Wenbo*[2]
1. Fairy Lake Botanical Garden, Shenzhen & Chinese Academy of Sciences, Shenzhen
2. Sun Yat-sen University, Guangzhou

We compared pseudo-viviparous ferns from six genus namely, *Tectaria, Woodwardia, Asplenium, Polystichum, Adiantum, Bolbitis* to compare their *in vitro* antioxidant, antidiabetic activities and phytochemical quantification. We quantified the total phenolics, total flavonoids, tannins and total protein using standard protocols. Several *in vitro* antioxidant methods namely DPPH radical scavenging, metal chelating, phosphomolybdenum, hydrogen peroxide assay, ABTS, superoxide and hydroxyl radical scavenging activities of matured, young leaves, vivipary and rhizome extracts were studied. *In vitro* antidiabetic activity was analysed using α-amylase inhibition assay. Higher amount of total phenolics (320.71 mg GAE/g extract), total flavonoids (49.27 mg RE/g extract), tannins (289.52 mg GAE/g extract), and total proteins (13.41 g BSAE/100 g) were observed in vivipary buds of *W.orientalis.* On the other hand, lower amount of total phenolics (289.76 and 85 mg GAE/

g extract), total flavonoids (28.86 and 2.56 mg RE/g extract), tannins (236.66 and 21.90 mg GAE/g extract) and total proteins (12.12 and 3.97 g BSAE/100 g) were observed in matured and young leaves, respectively. In rhizome of *W.orientalis*, significant amount of total phenolics (278.81 mg GAE/g extract), and total flavonoids (73.95 mg RE/g extract) were exhibited whereas, highest amount of total phenolics (471.90 mg GAE/g extract), total flavonoids (71.92 mg RE/g extract) in rhizome extract of *W.japonica*. Higher amount of total flavonoids (88.75 mg RE/g extract) and tannins (590 mg GAE/g extract) were identified in rhizome of non-viviparous fern, *T.decurrens* while, lower amount of total flavonoids (77.59 mg RE/g extract) and tannins (418.09 mg GAE/g extract) were identified in rhizome of viviparous fern, *T.fauriei*. Higher amount of total flavonoids was observed in vivipary buds (10.11 mg RE/g extract) of *A.bulbiferum* while, 9.67 and 5.70 mg RE/g extract were obtained from matured and young leaves. The vivipary buds of *A.ailaoshanense* showed total phenolics (52.85 mg GAE/g extract), total flavonoids (10.11 mg RE/g extract) and total proteins (12.06 g BSAE/100 g) whereas, 61.90 and 53.80 mg GAE/g extract of total phenolics, 2.16 and 2.38 mg RE/g extract of flavonoids, and 4.13 and 4.42 g BSAE/100 g of total proteins were attained in matured and young leaves, correspondingly. In antioxidant activities, significant metal chelating (17.29 mg EDTAE/g extract) and phosphomolybdenum (193.68 mg AAE/g extract) activities were observed in vivipary buds of *W.orientalis*. Significant superoxide (52.83%) and hydroxyl radical scavenging (46.78%) activities were observed in vivipary buds of *A.bulbiferum*. Higher amount of metal chelating (17.43 mg EDTAE/ g extract), phosphomolybdenum (603.51 mg AAE/ g extract) and hydroxyl radical scavenging (36.84%) activities were observed in vivipary buds of *A.ailaoshanense*. Significant activity of DPPH inhibition (34.86 IC_{50}), metal chelating (17.71 mg EDTAE/g extract), phosphomolybdenum (1221.05 mg AAE/g extract), superoxide (89.57%) and hydroxyl radical scavenging (70.76%) activities were obtained in mature leaf of *Tectaria fauriei* while lower amount of DPPH inhibition (38.16 IC_{50}), metal chelating (17.44 mg EDTAE/g extract), phosphomolybdenum (670.87 mg AAE/g extract), superoxide (82.74%) and hydroxyl radical scavenging (42.69%) activities observed in young leaves, respectively. In conclusion, pharmacological validation using *in vivo* methods, mode of action and their mechanism of the vivipary buds should be studied.

T1

P0237

Medicinal uses of ferns and fern allies

Baskaran Xavier Ravi[1,2], *Zhang Shouzhou*[1], *Liao Wenbo*[2]
1. Fairy Lake Botanical Garden, Shenzhen & Chinese Academy of Sciences, Shenzhen
2. Sun Yat-sen University, Guangzhou

Our present aim was to explore the medicinal uses of fern and fern allies. Pteridophytes (*aka*, ferns and fern allies) are an ancient lineage, and human beings have been utilizing fern taxa for over 2,000 years. Even though more than 12,000 species of pteridophytes distributed all over the world, fewer of them only evaluated for their phytochemicals and its pharmacological studies. We documented several medicinal pteridophytes which belong to following thirty families: Davalliaceae, Equisetaceae, Lygodiaceae, Ophioglossaceae, Polypodiaceae, Psilotaceae, Pteridaceae, Salviniaceae, Selaginellaceae, Tectariaceae, Woodsiaceae, Aspleniaceae, Athyriaceae, Blechnaceae, Cyatheaceae, Dennstaedtiaceae, Dicksoniaceae, Dryopteridaceae, Hymenophyllaceae, Lindsaeaceae, Thelypteridaceae, Marsileaceae, Nephrolepidaceae, Onocleaceae, Marattiaceae, Oleandraceae, Osmundaceae, Cystopteridaceae, Gleicheniaceae, and Huperziaceae. The lycophyte, *Selaginella* sp was exhibited multiple pharmacological activities such as, antioxidant, anti-inflammatory, anti-cancer, antidiabetic, antiviral, antimicrobial and anti-Alzheimer properties. Among all the pteridophytes examined, taxa from Pteridaceae, Polypodiaceae, and Adiantaceae exhibited significant medicinal activities. Many pteridophytes have bioactive principles that could be used in alternative medicine for various human illnesses. Even though several studies reported the medicinal uses of ferns, the possible bioactive compounds were not identified for several pteridophytes. The presence of caffeic acid, chlorogenic acid, rutin and quercetin were reported in many pteridophytes namely, *Pteris multifida*, *P.vittata*, *Pteridium aquilinum*, *Polypodium leucotomos*, *Cyathea phalerata*, *Selaginella tamariscina*, *Onoclea sensibilis*, *Matteuccia struthiopteris*, *Dicranopteris linearis*, *Psilotum nudum*, *Equisetum arvense*, and *Adiantum capillus-veneris*. In addition to biological activities, pharmacokinetics, toxicological profile, multidrug therapies should be developed from early tracheophytes and pteridophytes. Furthermore, their optimal dosage level and treatment strategies still need to be assessed. Moreover, the pteridophyte group of plants should be analysed for their genetic relationship on the basis of secondary metabolites synthesis.

T1

P0238

The conservation of Bromeliaceae germplasm and application in Shanghai Botanical Garden

Meiling Xiu
Shanghai Botanical Garde

We have introduced the Bromeliaceae family plantssince 1998.We have collected more than 1100 species, including horticultural varieties.It is one of the most collection amount in Chinese botanical gardens. We attempted to grow Bromeliaceae plants outdoor and designed the theme garden in flower shows since 2010. We bundle the epiphytes on other plants or rocks in order to show their special characteristics and living habits. We will continue the application for local usage of Bromeliaceae family plants in Shanghai, and select the plants which are suitable for local climate.

T1

P0239

New fern fossils from southern Tibet indicate low elevation of proto-himalaya during the latest Paleocene

Cong-Li Xu[1,4], *Jian Huang*[1,4], *Tao Su*[1], *Shu-Feng Li*[1,3], *Yong-Jiang Huang*[2], *Zhe-Kun Zhou*[1,2]
1. Key Laboratory of Tropical Forest Ecology, Xishuangbanna Tropical Botanical Garden, Chinese Academy of Sciences, Menglun, Mengla 666303, Yunnan, China
2. Key Laboratory for Plant Diversity and Biogeography of East Asia, Kunming Institute of Botany, Chinese Academy of Sciences,

Kunming 650201, China
3. State Key Laboratory of Paleobiology and Stratigraphy, Nanjing Institute of Geology and Paleontology, Chinese Academy of Sciences, Nanjing 210008, China
4. University of the Chinese Academy of Sciences, Beijing 100049, China

Fossil records of Cenozoic ferns are quite rare in Tibet. In this study, a megafossil record of *Cyclosorus nervosus* J. R. Tao is described from the Upper Paleocene sediment of Liuqu Conglomerate, southern Tibet, China, based on well-preserved leaf pinnae with *in situ* sori imprint. The frond fossils have typical characters, e.g., at least 2-pinnate-pinnatifid, catadromous fronds, lanceolate pinnae, ovoid segments with attenuate or obtuse apex and entire margin, relatively thick midveins of segment extending to the apex. The pattern of segment venation is goniopterid, veinlets located in the segments are simple, straight. Some upper lateral veins are curved toward the segment apex, while the lower 6 - 8 pair of lateral veins are connected to each other, forming the 1/2 or less than 1/2 pinnatified. One or two pair of the lowest lateral veinlets of segment and primary vein of the pinna connected forming triangular areolae, occasionally pinnatified connecting vein goes down into triangular areola. Two lines of ovoid sori (~9 sori) are located in the middle of the lateral veins along the midvein of segment from the sori compressions. All these morphological characters confirm the assignment of this species to the genus *Cyclosorus*, which is now distributed mainly in the pan-tropics, with the diversity center in Asia. The discovery of *Cyclosorus nervosus* suggests that *Cyclosorus* occurred in southern Tibet no later than the latest Paleocene. Moreover, according to the floristic composition of the Liuqu flora, it suggests that a warm and humid climate occurred in southern Tibet in the latest Paleocene. Furthermore, proto-Himalaya was at much lower elevation in the latest Paleocene as previously suggested.

T1
P0240

The biological and ecological potential of the genus *Spiraea* (*spiraea* l.) For introduction and cultivation in a garden

Gerelchuluun Yadamsuren
Institute of General and Experimental Biology, Mongolian Academy of Science

Seven species of spirea (*Spiraea* L.) introduced into botanical gardens and evaluate their biological and ecological potential. We determined seedling's developmental periods based on development of the spirea ontogeny stages during the introduction study. Seed germination period – after seed is planted, it continues 7-10 days until sprouting and emergence of seminal roots. Stem emerging period after the germination, stem node emerges with seed cover and extends to the soil surface within 5-10 days. Coleoptile (leaf like structure) emergence period as stem emerges; coleoptile which encloses primary leaves emerges within 6 to 10 days. Bud break and the primary leaf emergence period the first true leaf emerges from the coleoptile tips within 8-12 days. Vegetative organ emergence period seedlings are capable to grow as vegetative organs. Bud growth period after growth of apical buds the plant growth rate decreases to get ready for winter dormancy. Three year old seedlings of spirea met the standard criteria MNS 6255 : 2011 (mean height was 45-50cm with 6-8 branches of twigs, flower shoot emerged, root system well developed, the primary root 25-30cm, and apical bud is grown) and provided required seedlings for the year for city garden development. After planted by seed, *Spiraea aquilegifolia*, *S. media* flowered in the third year, produced seeds in the fourth year, self-reproduced by root, its seeds matured and disseminated in the sixth year and its potential for introduction and cultivation was evaluated by 100 points. Also, *S. alpina*, *S. dahurica*, *S. flexuosa*, *S. salicifolia* were successfully cultivated in the botanical garden and was evaluated between 95-97. Introduction and cultivation of the indigenous species *S. Chamaedrifolia* was evaluated at 90 points. It flowered within three years and matured in 4-5 years after its seeds were planted. The standard of introduction and cultivation of spirea species was issued in 2011 as result of this study. As well as, scientific experimental research based cultivation methodology of spirea species were developed for use of gardening, mining restoration. Invention patent of Mongolia was given for cultivation methodology of elm-leaved spirea S. aquilegifolia Pall. By seeds in 2016 year. There is several advantages cultivation by seeds than vegetative organs including well development of root system, high branching with good shape and flowering evenly but due to minute seeds germination rate is low for cultivation by seeds. It is needed to collect seeds from the nature with same climate condition and geographical region. We collected seeds of *Spiraea aquilegifolia, S. flexuosa, S. hypericifolia S.media, S. alpina S. salicifolia, S. sericea, S. pubescens* from the forest steppe and mountain forest steppe of Mongol-dogar vegetation district of Mongolia. Planting date is crucial for further growth and development of seed and seedlings. According to some researchers, planting late falls or late spring when soils get warm is more productive for spirea. As result of this study, seedling prepared in a greenhouse and transplanting in June was effective way to cultivate spirea. Spirea planted by seeds had well developed root system, highly branched canopy and at the third year seedlings got ready to transplant in a botanical garden.

T1
P0241

Co-administration of Artemisinin and *Ricinodendron heudelotii* leaf extract: Effects on certain antioxidants and Liver Parameters in male wistar rats

Omolara Yakubu[1,2]*, Abiodun Adebayo*[1]*, Jacob Popoola*[1]*, Emeka Okechukwu*[1]*, Oladipupo Adeyemi*[1]*, Emeka Iweala*[1]*, Ying-Jun Zhangyj*[2]
1. Covenant University
2. Kunming Insitiutute of Botany

Startling rate of malaria parasite resistance to Artemisinin and its derivatives has led to possible herb-drug antimalarial combination therapy. This study assessed the effect of coadministration of artemisinin and *Ricinodendron heudelotii* (Euphorbiaceae) leaf extract on certain antioxidant and liver parameters in rats and also phytochemical analysis of different fractions of the extract. Forty male rats were randomly grouped into four and administered: distilled water (Group A), artemisinin only (Group B), artemisinin with extract (Group C) and extract only (Group D). Alanine aminotransferase (ALT), aspartate aminotransferase, alkaline phosphatase (ALP), total protein were determined using standard methods. Reduced glutathione (GSH), catalase and malondialdehyde (MDA) were determined spectrophotometrically while the qualitative and quantitative phytochemical analysis was also carried out. Phytochemical screening revealed the presence of saponin, tannin, flavonoids, alkaloids, terpenoids and steroids. Biochemical analysis revealed total protein increased significantly in the combined and extract only groups (C and D). The activities of ALT and AST were significantly increased ($p \leq 0.05$) in artemisinin only group (B) while their activities were regulated to control level both in the combined and extract only groups (C and D). The antioxidant parameters analyzed in this study reveals that the liver GSH concentration decreased in the artemisinin only group (B) compared to the other treatment groups. Similarly, artemisinin administered singly significantly increase ($p \leq 0.05$) the MDA concentrations in the liver while the extract both in the combined treatment group and when administered alone decrease the MDA and catalase concentration. Histology showed few of the hepatocytes were necrotic in the artemsinin only group while the combined and extract only showed mild to moderate central venous congestion and periportal cellular infiltration. The study indicates that the bioactive constituents in the extract might have a regulatory effect on artemisinin toxicity or collaboratively enhance its activity. Such bioactive constituents can further be isolated and characterized for drug development to tackle *Plasmodium falciparum* resistance.

T1

P0242

Invasive flora of China: Introduction and information

Xiaoling Yan, *Zhanghua Wang, Jing Yan, Huiru Li, Jinshuang Ma*
Shanghai Chenshan Botanical Garden

Invasive species cause serious environmental and economic harm and threaten global biodiversity, so compiling a complete inventory and flora of invasive species is necessary for predicting and hence preventing plants invasion. However, nationwide information on invasive plants in China is still lacking. We set out to investigate how many invasive plant species are currently spreading in China and how their distributed patterns were formed. Based on the literature, field investigations, specimen records and taxonomic research, we compiled a nationwide flora of the invasive plant species of China. Data of taxa, origin, pathway, and environmental impact of invasive plants were analyzed. The flora comprised a total of 515 invasive plant species from 72 families and 285 genera reported from China. The three most dominant families were Compositae, Poaceae, and Leguminosae, accounting for 18%, 14% and 11% of invasive plants, respectively. *Amaranthus, Euphorbia, and Solanum* were the three dominant genera. Over half of all invasive plants were of American origin (56%), followed by those with European (16%) and Asian (15%) origins. Annuals and perennial herbs were prevalent among invasive species. According to their biological and ecological characters, extent of spread within China, and amount of damage caused, each species was put into one of 5 categories: noxious invasion (34 species), serious invasion (69 species), local invasion (85 species), general invasion (80 species), and species requiring further observation (247 species). Most noxious invasive plants were with a perennial life cycle, clonal growth ability and from the American continent. According to our analyses of geographic distributional patterns, Chinese invasive plants were highly concentrated in the southwestern and eastern coastal provinces of China, and were rapidly spreading to inland China. Furthermore, the warm regions, which are the hot spots of local biodiversity, and relatively developed areas of China, require immediate attentions. These categories and distributional patterns of invasive plants could help to assess and control their invasive risk. Our results provide baseline information on Chinese invasive plants; and, at the same time, we put forward preliminary suggestions for further study, public education and the management of the important invasive plants in China.

T1

P0243

Habitat determines fern and lycophyte species composition, diversity and abundance in lowland landscapes in Xishuangbanna, South China

Feng-Chun Yang[1,2,3], *Chaya Sarathchandra*[1,4,5], *Qi Liu*[1,6], *Xue-Bing Wang*[1,7], *Songho Kim*[1,4], *Christos Mammides*[8], *Ferry Slik*[9], *Kyle Tomlinson*[6], *Uromi Manage Goodale*[8]

1. *Program for Field Studies in Tropical Asia (www.pfs-tropasia.org), Xisuangbanna Tropical Botanical Garden, Chinese Academy of Sciences, Menglun, Mengla, Yunnan, 606330, China*
2. *Suranaree University of Technology, Nakhon Ratchiasima*
3. *Honghe University, Mengzi, Yunnan*
4. *Kunming Institute of Botany, Chinese Academy of Sciences*
5. *Rajarata University*
6. *Xishuangbanna Tropical Botanical Garden*
7. *China National GeneBank*
8. *Guangxi University*
9. *Universiti Brunei Darussalam*

Gametophytic and sporophytic physiology and unique aspects of photobiology make ground ferns and lycophytes thrive as ground flora in low-light tropical forest understories. Although these understory plants comprise a major component of tropical forest diversity, they have been relatively little studied. This study examines the ground fern and lycophyte diversity and composition in rainforest (RF), karst forest (KF) and rubber plantation (RU) habitats in Xishuangbanna, Southwest China. We documented species abundance and incidence in four appox. 250 m transects in each habitat using 2.5 m × 1 m plots placed 10 m apart and compared 11 abundance, diversity, rarity and shared species indices for the three habitats. A total of 36 ground fern species and four lycophytes (*Seleginella* species) were observed; 28 species in RF, 20 species in KF and 11 species in RU. *Tectaria subtriphylla* was the most abundant species and was found in all three habitats. The effect of habitat on species accumulation was a significant for 10 indices (all $P < 0.01$), except for number of species represented by only one individual. Rainforest habitats had greater values for all species abundance, diversity, rarity and shared species indices, compared to KF and RU habitats (all $P < 0.01$). The difference between KF and RU habitats were only seen for species abundance, number of species that were only found in two plots and Shannon diversity index (all $P < 0.01$). Permutational Multivariate Analysis of Variance and Analysis of Similarities both showed that species composition measured by abundance and incidence were significantly different among habitats ($F_{2,261} > 22.263$, $P > 0.001$

and $R = 0.2581$, $P > 0.001$, respectively). A common core set of species (*Tectaria subtriphylla*, *Cyclosorus parasiticus*, *Pronephrium gymnopteridifrons*, *Cyclosorus acuminatus*, and *Lygodium flexuosum*) were shared between all three habitats with RU and KF habits containing a subset of the RF species (C-score = 65.79, $P > 0.01$ and C-score = 106.55, $P > 0.05$, respectively). Indicator species analysis with both abundance and incidence data revealed that there were 13,4 and 2 species as indicators of RF, KF and RU habitats, respectively (IV > 0.0481, $P < 0.05$). The top two indicator species for each habitat are: RF: *Tectaria decurrens* and *Archangiopteris henryi*; KF: *Tectaria simaoensis* and *Adiantum soboliferum*; RU: *Microlepia marginata* and *Pronephrium gymnopteridifrons*. Of the 40 species studied here there were four species that were only found in the RU habitat: *Microlepia marginata*, *Pteris semipinnata*, *Pteris fauriei*, and *Schizoloma ensifolium*. Our findings convey that the rainforest habitats maintain a high species diversity of ground ferns and lycophytes while karst forests and rubber habitats also contribute to promoting diversity. Therefore, we conclude that the diversity in habitat types contribute to maintaining a larger group of ground fern and lycophyte species across landscapes and although monocultures, rubber plantations act as refugia for a subset of rainforest fern species while karst species are least likely to be maintained in the rubber understory.

T1

P0244

Biodiversity of the micro-morphologies of the epidermal cells of petals of angiosperms

Shaoyong Yang[1], *Hua Wang*[2], *Hong Hu*[2], *Dixin Wu*[3]

1. Southwest Forestry University

2. Kunming Insitute of Botany, Chinese Academy of Sciences

3. Beijing Forestry University

According to anatomy and microscope photography techniques, the micro-morphologies of the upper and lower epidermals of petals of fresh flowers of different species or cultivars have been researched, a new phenomenon has been discovered: there is a biodiversity among the micro-morphologies of the pigmented cells of the epidermals of petals. The result shows these much-varied pigmented cells appear a lot of different quasi-geometry forms, genus specific and artistic beauty with vivid colors. Such that a brand new research field of a real "Cell Taxonomy" should be established instead of the current "Chromosome Taxonomy" widely used in phytosystematics nowadays. Our preliminary research about 64 species or cultivars is very important to the following three aspects: firstly, plant micro-biodiversity; secondly, a new criteria to taxonomy; and thirdly, a whole new epoch of art history and art creation has been born-these diversified colorful micro-morphologies of the pigmented cells by microscope appear fatastic natural beauties or asethetics, just the so called "Scientific Aesthetics" or "Scientific Beauty" which has been searched long time by Prof. Zhengdao LI, the first Chinese American Nobel Laureate, and many achieved artists such as Shana Chang, the former Principal of Central Academy of Fine Arts, the current Art College of Qinghua University-we successfully build a bridge between science and art. And at last, the cells we studied are quite different from pollens by electroscope, some of them are similar to the cells of the epidermal of leaves, however they are much diversified and prettier than that of the leaves, and such that another necessity to carry on a widespread research to the entomophilous pollination groups and ornithophilous pollenation groups of Angiosperms, of course, a set of terminologies should be worked out to describe these diversified, precious and beautiful colorful cells (We call someone with an "art cell" if he or she is talented on art, no wonder, cell is really artistic. And in Bible Jsus said: Even the King Solomon in his richest time wore is not as magnificent as a lily flower in the wild! Now our result reveals the two secrets!). Besides micro-biodiversity, taxonomy and scientific aesthetics, it is also a new compensation to anatomy.

T1

P0245

On finding potential distributions of threatened plant species using machine learning on specimen data

Yu-Chang Yang[1,2], *Li-Na Zhao*[1,2], *Hui-Yuan Liu*[1], *Hai-Ning Qin*[1]

1. Institute of Botany, Chinese Academy of Sciences

2. University of Chinese Academy of Sciences

A herbarium is not only a treasure house for plant classification specialists, but also an enormous data set which provides valuable information for wild plant conservation - every specimen witnesses the bygone existence of a certain plant species at a certain location, especially those with exact coordinates. Some places recorded on the labels are designated as type localities, where the most typical plant individuals of that species can be discovered. Given longitude and latitude, with some external geographic and climatic data serving as calibration standards, one can estimate the altitude, topography, illumination and precipitation there, which determine temperature, light, water and other environmental factors. As people know, plant species are likely to reoccur at spots environmentally resembling the type localities, which really matters for those thought to be endangered already. To search for these potential distributions, we are now doing a study on a new model based on machine learning, which is designed to train the model itself recursively to better extract features corresponding to environmental indices significant to plant survival, growth and development, from the big data we feed the computers with. We aim to find actual shelters for threatened species which remain unknown to the researchers by using this artificial intelligent model, and acquire new related knowledge necessary to conserve our nature more effectively.

T1

P0246

Red-list assessment for larger fungi based on a database of check-list of fungi in China

Yi-Jian Yao

Institute of Microbiology, Chinese Academy of Sciences, Beijing, China

After the compilation of red-lists of plants and vertebrates in China, the assessment for larger fungi was also carried out from 2016. Although much work has been do for plants and animals, it was the first attempt to assess the threat status of larger fungi in a large scale in China, evolving all the relevant researchers national wide. In order to complete the assessment of about 10 thousands species of lager fungi, including lichens, ascomycetes and basidiomycetes, recorded in China in a limited time, methods had been developed to conduct the task largely based on a database of the Check-List of Fungi in China benefited from the information technology. The database contained the information extracted from publications related to records of non-lichenised fungi from 1970 and of lichenised fungi from 1990 up to 2013 overlaping the literature coverage of pre-1973 for Sylloge Fungorum Sinicorum (Tai, 1979) and pre-1990 for An Enumeration of Lichens in China (Wei, 1991), respectively. Around 2,400 lichen, near 1,000 ascomycetes and around 5,600 basidiomycetes species were determined as larger fungi in China for assessment. All the species were classified as widespread or endemic species with either dispersed or concentrated distribution based on the records in the check-list database. Species of well-recorded as dispersed distribution, no matter of widespread or endemic, were generally considered as Low Concerned. With concentrated distribution, widespread species were assessed as Data Deficient, while endemic species needed further investigation to see if they were threaten or not. However, those widespread species, even if with dispersed distribution in China, needed further attention for assessment if they suffered by human exploitation or by environmental changes. In this way, the large number of larger fungi recorded in China were scanned quickly so that the evaluators were allowed to concentrate on a few hundred species which require close examination. More than 130 Chinese mycologists took parts in the process of assessment. A special website for the project of the Red-list Assessment of Larger Fungi in China have been set up to facilitate the review of initial assessment and two meetings of consultation had been held, each with 60-70 participants from all over the country. Consensus had been reached for almost all the species evaluated. Different opinions for a few dozens of species were recorded in the final report and might become useful in the future assessment. The Red-list of Larger Fungi in China might be officially made public in May 2017.

T1

P0247

Macro-fungi substrate utilization pattern in Greater Mekong Sub-region

Lei Ye

World Agroforestry, CAS Kunming Plant Research

This study aims to raise the following question: (1) Do fungal species have special substrate preferences? (2) What are the differences in substrate utilization patterns in different sites in the GMS region? (3) Can fungal substrate specifically explain fungal diversity pattern? Base on the preciously studies, we hypothesis that: (1) most of the fungal species have their own substrate, less species have diverse substrate; (2) soil, wood, litter and plant-root are the most common substrate for macro-fungi; (3) fungal diversity pattern is correlated to their substrate diversity.

T1

P0248

Assessment and indicator species analysis of Ornamental plant species grown in home gardens of Rustum, District Mardan, Pakistan

Shahab Ali[1], *Shujaul Mulk Khan*[2], ***Zeeshan Ahmad***[2]

1. Department of Botany, Hazara University Mansehra, Pakistan

2. Department of plant science, Quaid-i-Azam University Islamabad, Pakistan

Home gardens are one of the old cultural practices all over the world. It is a system in which variety of foods, vegetables, fruits and medicinal plants are grown in comparatively smaller areas. It is important in terms of biodiversity and thus far their importance is often unrecognized and described as man managed ecosystem with high energy complex structure and multiple functions. Present study was conducted on Ornamentals plant species composition, abundance, distribution pattern and formation of various ornamentals plant communities in the area of Rustum, District Mardan. Special biological (Quadrates) methods were used to assess ornamental plant species distribution in the home gardens of targeted study area. Different stations were recognized at random intervals. Density, frequency, relative density, relative frequency and Importance Values of plant species were measured at each home garden. Preliminary results showed that a total of 43 ornamental plant species belong to 30 families were grown by the people of an area. Presence absence data of 43 Ornamental plant species and 50 home gardens were analyzed using Cluster and Two Way Cluster Analyses via PC-ORD version-5 that resulted four major ornamental plant communities. CANOCO software version 4.5 was used to evaluate the environmental and cultural gradient of ornamental plant through Canonical Correspondence Analyses (CCA). Based on present results it is concluded that further study should be conducted to examine, conserve and manage the ornamental plant species diversity and their contribution towards livelihood in projected area.

T1

P0249

Endangered new genera and species from Wuling Mountains of central China

Dai-Gui Zhang, *Dan Xie*, *Cheng Zhang*, *Ze-Long Nie*

Jishou University, Jishou, Hunan

Wuling Mountains is a mountain range located in Central China, running from Chongqing and east Guizhou to West Hunan and southwestern Hubei. Unique climate and geographical location make this area extremely rich in plant diversity with both old and endemic elements in eastern Asia. Here, alist ofrecent published genera and species from this region is collected to serve as a stimulus to further investigations of plant diversity and protection in this special and rare area. To date, two new genera and 17 new species from 12 family and 14 genera were reported from Wuling Mountains. *Zhengyia* T. Deng, D. G Zhang & H. Sun is new genus of Urtiaceae found recently from Shennongjia. It is monotypic with *Z. shennongensis* Deng, D. G Zhang & H. Sun, an endemic and endangered species only distributed in Wushanhu of Shennongjia. *Changiostyrax* Chen is the other new genus (Styraceae) found from Shimen of Hunan. Its corresponding single species *C. dolichocarpus* (C. J. Qi) C. T. Chen is endemic to China narrowly distributed in Sangzhi and Shimen of Hunan province and is considered as an endangered species according to the IUCN criterion. There are six new species from two genera of Asteraceae: *Sinosenecio baojingensis* Y. Liu & Q. E. Yang, *S. jishouensis* D.

G. Zhang, Y. Liu & Q. E. Yang, *S. hupingshanensis* Y. Liu & Q. E. Yang, *S. albonervius* Y. Liu & Q. E. Yang, *Aster tianmenshanensis* G. J. Zhang & T. G. Gao, and *A. jishouensis* W. P. Li et S. X. Liu. Two new species were reported from Brassicaceae, including *Cardamine hupingshanensis* K. M. Liu, L. B. Chen, H. F. Bai & L. H. Liu and *Eutrema bulbiferum* Y. Xiao & D. K. Tian. Other new species includes *Primula hunanensis* G. Hao, C. M. Hu & X. L. Yu (Primulaceae), *Astragaluswulingensis* Jia X. Li & X. L. Yu (Fabaceae), *Pinellia hunanensis* C. L. Long & X. J. Wu (Araceae), *Epimedium tianmenshanensis* T. Deng, D. G. Zhang & H. Sun (Berberidaceae), *Oxalis wulingensis* T. Deng, D. G. Zhang & Z. L. Nie (Oxalidaceae), *Corybas fanjingshanensis* Y. X. Xiong (Orchidaceae), *Sedum fanjingshanensis* C. D. Yang et X. Y. Wang (Crassulaceae), *Chrysosplenium zhangjiajie* X. L. Yu, Hui Zhou & D. S. Zhou (Saxifragaceae), and *Aspidistra fenghuangensis* K. Y. Lang (Asparagaceae). All the new species are endemic and endangered with a high restricted habitat with a narrow distribution in Wuling Mountains. Moreover, most of them are listed as CR or EN level according to IUCN criterion. It is critical to pay much attention from both scientists and government to protect these endangered species and genera in the Wuling Mountains.

T1

P0250

Molecular evidence for maternal origin of *Sorbus* with pink fruits in North China

Wei Tang, *Xue-Dan Yu, Chuan-Hong Zhang, Xue-Min Ye, Qi-Di Fu*
Research Institute of Forestry, Chinese Academy of Forestry; Key Laboratory of Tree Breeding and Cultivation, State Forestry Administration, Beijing, China

Numerous hybrids in genus *Sorbus* have been documented between different species in England and other countries. Within the genus, there are a number of species, morphologically intermediate between species in overlapped areas which often interpreted as a result of a hybridization event. *Sorbus* individuals with striking pink fruits, which were morphological intermediately between *S. pohuashanenesis* and *S. discolor* in their sympatric area and haven't documented in the flora, were found by members of our research group during their investigation in Mount Tuoliang?Hebei province of North China. The origin of *Sorbus* with pink fruits and its relationship with *S. pohuashanensis* and *S. discolor* are uncertain. Therefore, the altidude distribution pattern of *S. pohuashanenesis*, *S. discolor* and *Sorbus* with pink fruits investigation and molecular biological analysis based on the sequences of four non-coding regions of choloplast DNA were carried out to explore the origin of *Sorbus* with pink fruits found in Beijing Municipality and Hebei Province. The main conclusions are as follows: In addition to Mt. Tuoliang, Sorbus with pink fruits were also found in Mt. Baihua, and Mt. Xiling. *S. discolor* was distributed at an altitude of 1300-2000 m. *S. pohuashanenesis* had an altitude of 1500-2200 m. The minimum altitude and the maximium altitude of *S. pohuashanenesis* were both higher than those of *S. discolor. Sorbus* with pink fruits grow within the overlapping areas were between 1500-2000 m of *S. pohuashanenesis* and S. discolor, with a discontinuous distribution. There were 50 mutation sites in cpDNA sequences including mutation, insertion and deletion of bases. There were 23 cpDNA haplotypes. *S. discolor* had 4 haplotypes (H1, H21, H22 and H23). *S. pohuashanenesis* had 12 haplotypes (H6, H7 and H11-H20). *Sorbus* with pink fruits had 8 haplotypes (H1, H6, H7, H13, H17, H18, H19 and H21), they shared haplotypes H1 with S. discolor, and shared haplotypes H6, H7, H13, H17, H18 and H19 with S. pohuashanenesis. 70% haplotype of individuals with pink fruits shared with S. discolor and 10% with *S. pohuashanensis*, the other is the specific. The cpDNA data provided the first convincing evidence that *S. pohuashanensis* or *S. discolor* may serve as maternal parent of individuals with pink fruits. It is inferred that the *Sorbus* individuals with pink fruits originate from interspecific hybridization between *S. pohuashanensis* and *S. discolor*, and they have multiple origins and might belong to a new species in natural taxa system. Directional hybridization events were responsible for the origin of the individuals with pink fruits. Extensive gene flow was detected between *S. pohuashanensis* and *S. Discolor* in this study, suggesting an important role of reticulate evolution in subgenus. It was concluded that hybridization plays an important role in the evolution of *Sorbus* species.

T1

P0251

Rare and threatened plants of Dawei Mountain National Nature Reserve in Yunnan

Guiliang Zhang[1], ***Jie Cai***[2]
1. Administration Bureau of Yunnan Dawei Mountain National Nature Reserve, Hekou Branch
2. Kunming Institute of Botany, Chinese Academy of Sciences

Dawei Mountain National Nature Reserve, located at Honghe Prefecture in southeastern Yunnan, is a 43,993 hectare of protected area. It extends from Gejiu, Mengzi, Pingbian to Hekou county near the border between China and Vietnam. The reserve's altitude ranges from 100 meters in the valley of Xiaonanxi river to 2,363 meters above sea level at the peak of Dajian mountain. Vegetation of Dawei Mountain is mainly composed of montane rain forest, monsoon evergreen broadleaf forest and montane mossy evergreen broadleaf forest. According to latest statistics, it is estimated that 5,859 species of higher plants belongs to 1,714 genus and 367 families in Dawei mountain, among them there are 538 species, 207 genus, 75 families of bryophyte; 528 species, 151 genus, 55 families of pteridophyte and 4,793 species, 1 356 genus, 237 families of seed plant. Dawei Mountain provides a refuge for many "ancient" plant species, including 9 species of tree ferns (Cyatheaceae), 6 species of Cycads and 35 species of Magnolias, representing as much as 67%, 25% and 31% those of native to China respectively. Monotypic families of seed plant, such as Crypteroniaceae, Bretschneideraceae and Rhoipteleaceae also could be found in this region, which provides invaluable materials for the study of paleogeography and climate changes in the past. Moreover, Dawei Mountain is home to China's rare and endangered plants, there are 122 species endemic to Dawei Mountain and about a quarter of China's national key protected wild plant species (70 species) flourish in this area, many of them are traditional medicinal plants including *Panax stipuleanatus* and *Psammosilene tunicoides*, crop wild relatives such as *Fagopyrum dibotrys*, *Sorghum propinquum* and important timber species like *Garcinia paucinervis*, *Burretiodendron hsienmu*, *Madhuca pasquieri*. In the past two decades, there are 47 new species were discovered and reported from this region. To better understand the plant diversity and its current conservation status, local forestry authority conducted field surveys on rare and threatened plant species in Dawei Mountain. Thirteen rare and threatened species are confirmed new to Dawei Mountain, including 3 of them belong to national key protected species. Another dozen of species collected are potential new to science. Most rare and threatened plants in Dawei Mountain are pictorially documented, which conveyed visual and detailed information to forest rangers and villagers for community-based biodiversity monitoring and conservation, as well as to researchers who are working on the biodiversity of China-Vietnam border area for further investigation.

T1

P0252

Herblabel: An R package for making labels for herbarium specimens

Jinlong Zhang, *Huiling Zhu, Jingang Liu, Gunter Fischer*
Kadoorie Farm and Botanic Garden

Specimens are the basis of plant taxonomy, molecular systematics, ecology and conservation. Labels for herbarium specimens are of great importance as they include field observations and subsequent annotations. With the help of label printing programs, collectors are able to prepare elegant labels for herbarium specimens more efficiently. However, checking the spelling and validity of scientific names, finding out the family for a specific species, as well as manually reformatting the date of collection could be time-consuming and error-prone. Meanwhile, templates adopted by different label printing programs are often not consistent, and data entry in some programs could be troublesome. Here we describe herblabel, an R package to overcome these challenges. herblabel uses an Excel template derived from Darwin Core (http://rs.tdwg.org/dwc/) and CVH5.0 (www.cvh.ac.cn) data exchange standards, thus data entry is straightforward and easy to manipulate. With (1) modern family-genera database, (2) accepted names from The Plant List website V1.1 (http://www.theplantlist.org/), (3) accepted names from Flora of China (http://www.efloras.org/flora_page.aspx?flora_id=2) (4) accepted names from Flora Republicae Popularis Sinicae, and (5) Checklist of Latin words being embedded, the information for creating herbarium labels undergoes thorough evaluation to avoid potential errors. herblabel produces both herbarium labels and the annotation labels in RTF (Rich Text Format) format, which is suitable to be viewed and edited in MS Word. At present, herblabel (version 0.5.6) provides 4 themes for herbarium labels (i.e. PE, KUN, HU and KFBG) and one theme for annotation labels. However users can create their own themes since this package is open source, and is under the GNU GPL-2 license. herblabel is written purely in R, and runs on Windows, MacOS as well as Unix-like systems. Since the template is based on Darwin Core standards, it is not only convenient for generating reports, but also makes data sharing much easier. This package will increase the efficiency and accuracy for data entry, herbarium management, and should play important roles in cataloging plant diversity. herblabel is freely available at https://github.com/helixcn/herblabel.

T1

P0253

Chemical constituents of several *Phyllanthus* spp. and their bioactivities

Ying-Jun Zhang, *Xiao-Hong Zheng, Hong-Tao Zhu, Dong Wang, Chong-Ren Yang*
Kunming Institute of Botany, Chinese Academy of Sciences

The genus *Phyllanthus*, comprising about 600 species, belongs to the family Euphorbiaceae, with most species widely distributed throughout the tropical and subtropical countries of the world. In addition to the edible fruits of some species, most *Phyllanthus* plants have been used medicinally for the treatment of various diseases. Flavonoids, alkaloids, terpenoids, lignans, and tannins were reported from the genus *Phyllanthus*. Some of them showed potent antiinflammatory, antioxidant, antivirus, and antitumor activities. In our continuing efforts to search for the structurally and biologically significant metabolites from genus *Phyllanthus*, more than 200 compounds referring to monoterpenoid, sesquiterpenoid, norsesquiterpenoid, diterpenoids, phenylpropanoids, limonoids, sucrose benzoyl esters, lignan, and simple phenolics were identified from *P. acidus, P. cochinchinensis, P. emblica, P. flexuosus*, and *P. glaucus*. Ninety-five of them are new compounds. Their structures were determined on the basis of detailed spectroscopic analysis and chemical methods. Determination of absolute configurations of these compounds was facilitated by theoretical calculations of electronic circular dichroism (ECD) spectra using time-dependent density functional theory (TDDFT) for the aglycone components, and pre-column derivative/chiral HPLC analysis for the monosaccharides. Most of the isolates were also evaluated for their effects on antivirus against enterovirus 71 (EV71) for hand, foot and mouth disease, influenza A virus strain H3N2, hepatitis B virus (HBV), herpes simplex virus (HSV-1), and Coxsackie virus B3 (CVB3), antifeedant against the larvae of a generalist insect (*Spodoptera exigua*), and cytotoxicity against five human cancer cell lines (breast cancerMCF-7, hepatocellular carcinoma SMMC-7721, human myeloid leukemia HL-60, colon cancer SW480, and lung cancer A-549).

T1

P0254

Development of polymorphic microsatellite loci and study on fine-scale spatial genetic structure of *Pteroceltis tatarinowii*, an endangered plant endemic to China

Yunyan Zhang
Nanjing University

Pteroceltis tatarinowii, the sole representative of the genus Pteroceltis, is an endangered deciduous tree endemic to China. It not only serves as food, medicine and timber, but also is the best source for the manufacture of Xuan paper. Due to the extensive deforestation, the distribution and size of the P. tatarinowii population have been decreasing for years, which makes it a national class III endangered plant in China. With wide distribution in the genome, the microsatellite markers relied on expressed sequence tags (EST) can be linked to genetic traits and applied broadly in population genetics, molecular marker assisted breeding and molecular marker screening of target traits. In this paper, we used

Illumina high-throughput sequencing technique to sequence the P. tatarinowii leaf transcriptome, based on which polymorphic microsatellite loci is developed. The main results were as follows: we obtained a total of 42,477 Unigenes with an average length of 815bp. After aligning with NR, SwissProt, GO, COG and KEGG databases, 23,688 unigenes were assigned. These Among the unigenes, a total of 6,543 EST-SSRs were identified. 130 EST-SSRs were selected for validation as EST-SSR markers by PCR amplification. Of these, 48 EST-SSRs were amplified successfully and 32 EST-SSRs were polymorphic among 47 P. tatarinowii individuals. Additionally, cross-amplifications of EST-SSR were detected in Ulmus gaussenii and Ulmus chenmoui, and the versatility and polymorphism were 25% and 87.5%, respectively. The development of P. tatarinowii microsatellite markers provided the molecular basis for the study of P.tatarinowii and other plants in family Ulmaceae. We also investigate the fine-scale spatial genetic structure of this population. Significant fine-scale SGS was found in the whole population within 28 m, and the Sp was 0.0140. The Sp of class I, II, III, IV and V were 0.0142,0.0158,0.0008,0.0026 and 0.0018, respectively. As a result, P. tatarinowii population was under growing pattern and exhibited high genetic diversity. However, saplings were clumped around adult and aged trees because of limited seed dispersal, resulting in heterozygous deficiency. Furthermore, significant fine-scale SGS was found in class I and II. These results indicated that it was threatened by limited gene flow, increased inbreeding and the risk of inbreeding depression. Therefore, for in situ conservation, artificial promotion of gene flow should be encouraged, while inbreeding depression should be monitored. For *ex situ* conservation, individuals should be sampled at apart to 28 m to reduce genetic similarity.

T1

P0255

Compositions, *in vitro* antioxidant and antibacterial activities of ethanol extracts of *Opisthopappus taihangensis* (Ling) Shih leaves from China

*Jinyong Huang, **Yuqian Zhang**, Junfeng Liu*

Zhengzhou University

Major flavonoids components, as well as the antibacterial and antioxidant activities of ethanolic extracts of leaves (transplanting seedling from tissue culture) from the endemic genus of China-*Opisthopappus taihangensis*, were investigated in this paper. The results were shown as follows: 1. The content of total flavonoids from the 70% ethanol extracts sample was 82.705±3.725 mg Rutin/g dried weight (DW) by sodium nitrite-aluminum nitrate-sodium hydroxide colorimetric method, which was lower (around 33.245 mg Rutin/g DW) than that of wild type detected in the previous study. 2. Three main flavonoids components were analyzed qualitatively and quantitatively by HPLC with the results of chlorogenic acid accounting for 1.18% rutin accounting for 2.07% and quercetin accounting for 0.48%, respectively. It also performed differently compared with each content of the wild type, which were 0.84%, 0.54% and 1.48%, correspondingly. 3. The respective average diameters of inhibition zone of sample extracts to *Escherichia coli* and *Staphlococcus aureus* were 1.245 cm and 1.164 cm, with the corresponding MIC to be 5 mg/ml and 10 mg/ml. It revealed a more significant antibacterial potential to the Gram-negative. 4. The antioxidant potential of *O.taihangensis* was evaluated using three different assays in vitro, they were total antioxidant capacity (T-AOC) kit assay, superoxide anion radical scavenging kit assay and 2,2-Diphenyl-1-picryl hydrazyl (DPPH) radical scavenging assay. The results showed a positive dose-responses relationship between the concentration and bio-activities: the scavenging capabilities were strengthened along with the increasing concentration of ethanolic extracts. It revealed the sample extracts had a remarkable scavenging capacity, especially against DPPH, of which the IC50 calculated by SPSS was 0.101mg/ml. O.*taihangensis* was proved rich in flavonoids components in this study, which is consistent with the previous reports, and it also maintained a favorable antibacterial and antioxidant potential. We made the primary speculation that there may be some close correlation between the components and its bio-activity. Moreover, we analyzed that though the overall content performed to be high, the living environment, cultivating conditions and illumination intensity could be the possible factors contribute to the content differences between the transplanting seedling and the wild type, as the former material was planted in the experimental green house while the latter one was obtained from the cliffs of Taihang Mountain. Hence, *O.taihangensis* can serve as a promising source applied in natural preservatives of food industry and medical care field, *etc.* The study lays a foundation on the further research and applying of this new medicinal plant.

T1

P0256

Mycobiota from karst caves in China

***Zhifeng Zhang**, Lei Cai*

Institute of Microbiology, Chinese Academy of Sciences, Beijing, China

Karst caves, formed from limestone erosion by underground water, have several striking characteristics, such as darkness, constantly low temperature, high humidity and scarcity of organic matters. Studies on cave fungi have a history of more than 50 years, which revealed a high fungal diversity. This is the first time to systematically study fungal resources from caves in China. In the past two years, more than 510 samples were collected from 15 karst caves from five provinces of Southwest China, and 3,763 fungal strains were obtained. They belong to three phyla, 294 genera and 881 species. Based on their characters of phylogeny and morphology, 20 new species were identified and described, and a lot more potential new species awaited further works for naming. This study suggested that karst caves encompass a high fungal diversity, including a number of previously unknown species. Nine of the 15 caves were chosen for metagenomic analysis and four kinds of substrate (air, carbonate, soil, water) were collected. After data processing and annotation, 64.2% OTUs were assigned to the fungal kingdom, while 35.7% OTUs failed to be assigned to any known kingdom. Among the six phyla we observed, Ascomycota is most abundant (53% of the fungal OTUs); followed by Basidiomycota (12%), while about 31% are unculturable fungi. The α-diversity showed that the fungal diversity from water samples is the highest, while that from air is the lowest. Through comparisons of the OTU compositions from different materials, we concluded that fungal diversity from different substrates are generally similar, but large number of habitat-specific species are also identified. The sample cluster analysis and PCA analysis showed that most of the compositions from same sample type clustered very well, and the soil and rock samples outside of cave clustered with samples inside of cave, suggesting that the diversity and distribution of karst cave fungi are highly influenced by the substrate and the fugal flora from the outside environments.

T1

P0257

Research progress on the secondary metabolites of *Vaccinium dunalianum*

Xu-Lu Luo[1], *Na Li*[2], *Ji-Ting Yin*[1], *Dong-Yang Zhu*[1], *Wei Du*[1], *Jian Song*[1], *Xia Zhang*[1], *He Zhang*[1], *Wei-Lin Zeng*[1], *Yong Ding*[1], *Jun-Rong Tang*[1], *Yun Liu*[1], *Chong-Ren Yang*[2], *Ying-Jun Zhang*[2], ***Ping Zhao***[1]

1. Key Laboratory for Forest Resources Conservation and Utilization in the Southwest Mountains of China, Ministry of Education, Southwest Forestry University, Kunming 650224

2. State Key Laboratory of Phytochemistry and Plant Resources in West China, Kunming Institute of Botany, Chinese Academy of Sciences, Kunming 650201

Vaccinium dunalianum Wight (Ericaceae) is an evergreen shrub mainly distributed in the Southwest of China, with leaves used as folk medicine for treating articular rheumatism and the dried leaf buds as herbal tea. Previous phytochemical study led to the identification of 36 compounds including 12 new ones from the leaf buds, leaves and fruits of the plant, referring to acylated arbutin derivatives, flavone glycosides, iridoid glycosides, pyrone glycosides, sucrose benzoyl esters, and simple phenolics. Among them, arbutin, chlorogenic acid and 6'-*O*-caffeoylarbutin (CA) were rich in the leaf buds. CA was also found in our recent study to suppress the melanogenesis, and have hypolipidemic and anti-inflammatory activities in a live zebrafish model. Interestingly, the unusual accumulation of CA, up to 31.76% by yield from the dried leaf buds, suggested that CA and its related biosynthetic key enzymes may possess important biological functions in the titled plant. In order to clarify the high yield mechnism and biological significance of CA in *V. dunalianum*, the potential biosynthetic pathways assocoated with CA was investigated by de novo transcriptome sequencing, and the full-length cDNAs of two glycosyltransferase genes and three acyltransferase genes were cloned by RACE-PCR. The tissue cultures, suspension culture cells, and adventitious root cultures were also established. Meanwhile, investigation on the diversity and secondary metabolites of endophytes in *V. dunalianum*, and the metabolic interactions with host are in progress.

T1

P0258

Duckweeds biodiversity around Hongze Lake and molecular structure of rDNA intergentic spacer

Yuzhen Zhou, *Guiming Chen*, *Borisjuk Nikolai*

Huaiyin Normal University

Duckweed is a group of aquatic plants which can efficiently remediate different types of waste water and is also known as the smallest known flowering plants with accelerated growth rate. The accumulated biomass rich in cellulose, starch and protein can be used for biofuel, bio-fertilizer, animal feed and human food. Systematic characterization of natural biodiversity of duckweed species is essential for the germplasm preservation and various practical applications. We have systematically collected samples of duckweed in Jiangsu province of China, around Hongze Lake and characterized species biodiversity by genotyping using chloroplast (atpF-atpH and psbK-psbI spacers) and nuclear DNA (ribosomal DNA ITS and ETS regions) barcodes. *Spirodela polyrhiza* was revealed as a clear dominant species in all locations of the area with occasional presence *Wolfia globosa* and *Lemna aequinoctialis*. The recent whole genome survey of *Spirodela polyrhiza*, revealed that: 1) duckweed genome has a lowest copy number of rDNA genes coding for ribosomal RNAs, which are essential components for biogenesis of ribosomes, among all the examined plants; 2) the spacer between rDNA genes, which contains the gene promoter, is profoundly undermethylated. Taking into account that ribosome biogenesis is essential for supporting active protein synthesis and organism's growth, we speculated that the transcription of individual rDNA copies in duckweed, an extremely fast growing plant, has to be much more efficient compared to other plants where rRNA supply relies on higher rDNA copies. In order to get insights into rDNA transcription in duckweed, we have cloned and sequenced the rDNA intergenic spacer regions (IGS) from representatives of three duckweed genera: *Spirodela*, *Lemna* and *Wolfia*. The revealed unique features of the duckweed IGS, especially the unusual molecular structure of the rRNA transcription initiation site (TIS), will be discussed in relation to evolution of plant Polymerase I transcription machinery.

T1

P0259

Dynamic changes of photosynthetic properties and chemical compositions of *Eucommia ulmoides* Oliver under two planting models

Ming-Qiang Zhu[1,2], *Wei-Zhou Xu*[1,3], *Jia-Long Wen*[2], *Ya-Hong Zhu*[1], *Yi Li*[1], *Yin-Quan Su*[1], *Qiang Zhang*[1], *Run-Cang Sun*[2]

1. College of Forestry, Northwest A&F University, Yangling, China.

2. Beijing Key Laboratory of Lignocellulosic Chemistry, Beijing Forestry University, Beijing, China.

3. College of Life Science, Yulin University, Yulin, China.

Photosynthesis is a fundamental and complex physiological process for the biosynthesis of secondary metabolites in the green plant species. In this study, the dynamic changes of photosynthetic properties, chemical compositions and cumulative production between the traditional arbor forest model of *Eucommia* planting (AFMEP) and using leaf model of *Eucommia* planting (ULMEP) were systematically investigated. The photosynthetic properties were plotted using a portable photosynthesis system. Chemical composition and structural characterization of the EUO wood were comparatively investigated by a combination of HPAEC, Raman images, Elemental analysis, and 2D-HSQC NMR. Meanwhile, the composition and content of secondary metabolites in EUO leaves from different planting models were determined by the HPLC. Besides, the water content in leaves, barks and wood were calculated by the gravimetric method. Results showed that the light-saturated net photosynthetic rate in ULMEP was significantly higher than that in AFMEP. The content of lignin in AFMEP wood was higher than that of ULMEP wood, while the amounts of β-O-4 linkages were abundant in lignin from ULMEP wood, which was beneficial to utilize the ULMEP wood. By contrast, the content of carbohydrates in ULMEP was slightly higher than that of AFMEP wood. Besides, the contents of various secondary metabolites (chlorogenic acid, geniposidic acid, eucommiol and gutta percha) in the leaves from AFMEP and ULMEP varied with the seasonal change, respectively. Moreover, the cumulative production of dry biomasses in wood, leaves and barks of EUO at the same periods (10 years) were all significantly higher in ULMEP than those in AFMEP, which indicated that the ULMEP was a promising planting model to solve the source of the EUO resources in industrial scenario.

T1

P0260

SpeciesgeocodeR: Automatic cleaning of geographic coordinates for biodiversity analyses

Alexander Zizka[1,2], *Alexandre Antonelli*[1,2,3]
1. Univeristy of Gothenburg, department of Biological and Environmental Sciences, Gothenburg
2. Gothenburg Global Biodiversity Centre, Gothenburg
3. Gothenburg Botanical Garden, Gothenburg

The public availability of large-scale species distribution data has increased dramatically over the last ten years. In particular, the digitalization of collections from museums and herbaria and the aggregation of information in public databases, such as the Global Biodiversity Information Facility (GBIF) have contributed significantly to this development. However, the use of these data for ecological and biogeographic research as well as conservation planning is often hampered by issues regarding data quality. A major issue are erroneous geographical information due to wrong or imprecise or erroneous geographic coordinates. Here we present a new software tools: *speciesgeocodeR* to automatically identify and clean potentially errors in geographic coordinates common to biological collections. Furthermore, we illustrate the use of these tools on GBIF data. *speciesgeocodeR* is available as R-package or shiny app from https://github.com/azizka and www. antonelli-lab.net.

T1

P1409

Ethnobotanical and taxonomical studies of Chamla valley district Buner KPK Pakistan

Anjum Perveen, *Shaukat Ali*
Centre foe Plant Conservation University of LKarachi

Ethnobotany is the direct study of plant interaction of a known area with the local population through its culture. Chamla valley of district Buner is a remote area of Khyber Pakhton Khwa Pakistan. The dependency on scare Hakeems (herbalists) in the valley is reduced by the introduction of allopathic medicines, but still there are many areas where the plants are still used by the locals as medicines. During the study a total of 183 species were collected from the valley. Taxonomic description of all the studies taxa has been also done. Based on ethnobotany medicinal is the largest class represented by 45.3% followed by fodder class 20.5% while the third dominated class is edible and vegetable represented by 14.8% of the total species, the least represented classes were sedative, aromatic and parasitic, the former 2 classes shared 1.41% each while the later one had 0.70% only.

T1

P1419

Study on the variation of spina *Gleditsia* and its grafting products

Tiejie Wang, ***Chang Su****, Xiaoying Guan, Qiaoji Guo, Xiaoxiao Song, Yaoxuan Xie, Jing Li, Shuhong Wang*
Shenzhen Institute for Drug Control

Spina Gleditsia is the thorns of *Gleditsia sinensis*, which is widely used in clinic for anti-tumor purposes. In commercial herbal market, some closely related species such as *Gleditsia japonica* and *Gleditsia microphylla* are commonly used as counterfeits and adulterant of Spina Gleditsiae. Throughout history till today, morphological feature of the thorns is the traditional and reliable method to differentiate the official Spina Gleditsia and its adulterant. But recently, grafting *Gleditsia sinensis* on *Gleditsia microphylla* or *Gleditsia japonica* is gradually popularized, and this planting behavior may cause variation on morphological feature, chemical composition and DNA sequence. In this study we present comprehensive analysis on Spina Gleditsiae by the method of the original planting investigation, morphological identification, microscopic identification, DNA barcoding molecular identification and chromatography identification. The results show that grafting products of *Gleditsia sinensis* do not have significant differences in microscopic features, psbA-trnH DNA barcode, TLC and HPLC chromatography with *Gleditsia sinensis*, but presents morphological variation on thorns and leaves, which is familiar with the feature of the closely related species as the grafting stock. At the same time, *Gleditsia sinensis* as well as its grafting products have significant differences in microscopic features, TLC and HPLC chromatography and psbA-trnH DNA barcode with its closely related species. That means grafting products reserves most of the characteristics of *Gleditsia sinensis.* The results of our work can solve the difficulties in the identification confusion caused by the grafting of Spina Gleditsiae. This study also provides the essential basis for future study to clarify the mechanisms of inheritable variation caused by grafting and the rationality of this cultivated usage in TCM plants.

T1

P1425

Study of rare and endangered plants of Pachmarhi Biosphere Reserve, India

Pramod Patil[1], *Moni Mathur*[2]
1. Institute for Excellence in Higher Education, (IEHE) Bhopal, India
2. Govt. SNGPG College Bhopal, India

India is one of the 17 countries designated as mega biodiversity centers among the 196 countries of the world and geologically has elements of the Palearctic and Indo Malayan Realm. Situated in Central India are the lofty hills and plateaus of Satpura Ranges made up of multilayered sandstone dating from Permian to Cretaceous. Amidst these ranges, at 22° 11' to 22°50'N and 77°47' to 78°52' E meridian is located Pachmarhi, a UNESCO declared Biosphere Reserve over an area of 4928.25 sq. km at an elevation of 1100 meters above mean sea level. Pachmarhi is endowed with moderate tropical summer and up to 200 mm rainfall which foster dense forest. The medicinal plants of Pachmarhi were studied with a view to understand the biodiversity, habitat, availability and their status. Large trees form the upper canopy of the forest with one of the finest timber yielding species of *Tectona grandis* and *Shorea robusta*. The midsized tree, shrubs and climbers together with undercover herbs lodge a treasure trove of medicinal plants. The study revealed more than 600 species of Angiosperms are medicinal; some species which were once common but now rare and threatened are *Aristolochia indica, Berberis aristata, Balanites aegyptiaca, Boswellia serrata, Chlorophytum tuberosum, Gardenia gummifera, Gloriosa superba, Pterocarpus marsupium, Rouvolfia serpentina, Eulophia nuda, Pueraria tuberosa, Mallotus philippensis, Oroxylum indicum, Litsea glutinosa, Terminalia bellerica, T. chebula, Uraria picta*. With the unsustainable harvesting by local inhabitant and over exploitation by profiteers for commercial purposes, their availability has subsequently declined to the extent that many species have become endangered up to critical levels. To salvage this dwindling treasure immediate attention is required for their *in-situ* and *ex-situ* conservation.

T1

P1428

Documentation of medicinal plants in Sabah, Malaysia

Ying Fah Lee, *Doris Jauris Salegin*
Forest Research Centre, Sabah Forestry Department

A project to document the known medicinal plants in Sabah was initiated in 2014. Medicinal plants used by different ethnic groups and various communities are described. Previously uses of medicinal plants have been recorded but not widely publicized. This project aims to collate information on medicinal plants found in Sabah and publish a series of manuals. For each species of medicinal plant, the following information is provided: Group (Pteridophyte, Gymnosperm, Monocotyledon or Dicotyledon), Family, Scientific name, Vernacular name, Habit & occurrence, Taste (Acrid, sweet, sour, bitter or salty), Energy & characteristics (Cold, cool, warm, hot or neutral; toxicity), Function (The role of herb in the healing process), Uses (List of ailments for which the herb is used), Parts used, Folk prescription (Selected traditional uses), Family and vernacular names in Chinese. The objective of thef manuals is to document the known medicinal plants in Sabah to serve as a reference for further research and for information and not to be used to diagnose, treat or cure any disease. Readers are advised to consult qualified health care practitioners for treatment of any ailment. So far two volumes of the series " Medicinal Plants in Sabah" have been published, respectively in 2015 and 2016. Each volume covers 100 species of plants.

T1

P1438

Saving rare wild flora: Lessons from Azerbaijan

Valida Ali-Zade[1], *Gulnara Osmanova*[2], *Naiba Mehdiyeva*[1], *Sevda Ibrahimova*[3], *Rena Abdiyeva*[1], *Rashad Salimov*[1], *Aida Dadashova*[1]
1. Institute of Botany of Azerbaijan National Academy of Sciences
2. Mari State University
3. National Aviation Academy

The flora of the Azerbaijan is estimated to include more than 4,500 species of higher plants of which around 950 are Caucasian and 200 national endemics. Along with its faunal diversity, this botanical wealth makes Azerbaijan as a valuable floristic center within the Caucasus region, which is one of the Biodiversity Hotspots of the world, and of critical conservation concern. No less than 10% of the species are regarded as rare or endangered and are subject to protection. 6% of the flora of Azerbaijan constitutes rare species included in the Red Book of Azerbaijan (II-nd edition, 2013), and listed by IUCN categories as follows: CR - 30, VU - 156, DD - 9, EN- 53, NT – 52. Taking into account continuing negative impact of anthropogenic factors and other drivers on the status of plant populations, in 2015 a new multidisciplinary projectwas initiated at the Institute of Botany of Azerbaijan National Academy of Sciences, to developadynamicintegrateddatabasewith mathematical and computer models for rare species. This ongoing project aims to implement the Targets 1 and 3 of the GSPC that are most closely linked to Aichi Target 19: "by 2020, knowledge, the science base and technologies relating to biodiversity, its values, functioning, status and trends, and the consequences of its loss, are improved, widely shared and transferred, and applied". Database covers indicators such as, taxonomy, life forms and ecological groups of rare species, distribution species according to botanical-geographic regions and altitudinal zonation, rarity status endemic and threatened species, ontogenic status and structure of populations, as well as online-map, photos and their useful properties. By January 2017, database contains data on 180 populations of 41 species.Mathematical model that allows assessing the risk of extinction of species on the basis of population-ontogenetic features is used for the preparation of prognostic scenarios. In addition, since 2013 MSBP, Kew launched the project in Azerbaijan with the Institute of Botany to develop an *ex situ* conservation programme for the saving the wild flora of the Azerbaijan. A key focus of the project was the contribution of seed banking to progress towards GSPC Target 8 and Aichi Target 12, and Azerbaijan's National Biodiversity Strategy in Azerbaijan. Currently, over 193 live seed collections from the 121 Azerbaijan 3E (endemic, endangered and economic) wild species had been seed banked within the framework of the project. The taxonomic diversity of seed collections comprising 46 families and 103 genera. Along with this ongoing project a new RE Garfield Weston Global Tree Seed Bank Project (Azerbaijan) was initiated in 2016 in collaboration with the MSBP, Kew for the most threatened tree species.

T1

P1442

Impacts of domestication on population genetics of a traditional Chinese medicine, *Atractylodes macrocephala* (Asteraceae)

Chuan Chen[1], *Li Zheng*[3], *Chengxin Fu*[2]
1. Hangzhou Botanical Garden
2. Key Laboratory of Conservation Biology for Endangered Wildlife of the Ministry of Education, and College of Life Sciences,

Zhejiang University
3. Key laboratory of Jiaxing Second Hospital

As a medicinal herb, *Atractylodes macrocaphala* experienced centuries of cultivation in China, and *Atractylodes macrocaphala* Rhizoma in the market is generally from cultivated *A. macrocaphala* at present. However, the germplasm resources of all cultivated populations of *A. macrocephala*have degraded over time as a consequence of the domestication processes. This study provided an in-depth analysis by two cpDNA regions and eight microsatellites on genetic and evolutionary consequences of domestication in a broad range of current *A. macrocaphala* germplasm resources. The results revealed that cultivated *A. macrocephala* (except 'Pingzhu' cultivar) exhibited higher genetic diversity (average HE = 0.682) than its wild progenitors (average HE = 0.663) and low level of genetic differentiation occurred between cultivated and wild groups. Both STRUCTURE analysis and UPGMA tree indicated four main genotypic clusters of *A.macrocephala*. We inferred that wild germplasm in Central China (Shanxi and Hunan Province) which is also the main distribution area for natural populations has most likely been involved in the origin of cultivated *A. macrocephala*. Moreover, the wild population from Qimen, Anhui Province and the cultivated 'Pingzhu' variety harbor unique gene pools and rare alleles that could be useful in future breeding efforts. This study represents the first large-scale analysis of population genetics of *A. macrocephala* which has a centuries-long history of human-mediated selection as a medicinal crop and the results will facilitate utilization and conservation of the valuable genetic resources of this species.

T2

P0261

The Genus *Henckelia* Spreng. (Gesneriaceae) in Southern Western Ghats

Janeesha A P, *Santhosh Nampy*
Department of Botany, University of Calicut, Kerala

The genus *Henckelia* was re-established by Weber & Burtt in 1997 to accommodate the species of *Didymocarpus* sect. *Orthoboea* from South India and Sri Lanka, *Didymocarpus* sect. *Didymocarpus* and sect. *Elati* from Malesia and included 180 species. Weber *et al.*, in 2011 re-defined the genus based on molecular data and included species of *Henckelia* sect. *Henckelia*, *Chirita* sect. *Chirita* (excluding the species under *Damrongia* Kerr *ex* Craib), the monotypic genus *Hemiboeopsis* W.T.Wang and exclude the species of *Henckelia* sect. *Loxocarpus*, *Didymanthus*, *Heteroboea* and *Glossadenia*. At present the genus has approximately 60 species world wide, with distribution in North-East and South India, Sri Lanka, the Himalayas and continental South-East Asia excluding Peninsular Thailand and Malaysia. The genus is represented by 30 species inIndia, of which 15 are distributed in southern Western Ghats. Except for *H. humboldtiana* (Gardner) A.Weber & B.L.Burtt (which also occur inSri Lanka) all the others are endemic. The Indian Henckelia's fall under two groups: (1) plants predominantly caulescent, with internodes and leaves in whorls of 2 or 3 andorthocarpic capsules dehiscing along both upper and lower suture. The other group is predominantly acaulescent, with leaves in basal rosette and plagiocarpic capsules opening along the upper suture only. The former occurs in North-East India and Sri Lanka (*Chirita* sect. *Chirita*) while the latter is found in southern Western Ghats (*Henckelia* sect.*Henckelia*). The present paper discusses the taxonomy, ecology, distribution and endemism of *Henckelia* in southern Western Ghats. An artificial key for the identification of species is provided. A phylogenetic tree is also constructed using sequence data from ITS and trn loci.

T2

P0262

Phylogeny and pollination observations of *Amorphophallus* (Araceae) on Borneo

Nor Fathiah Abd Rahim, *Sin Yeng Wong*
Department of Plant Science and Environmental Ecology, Faculty of Resource Science and Technology, Universiti Malaysia Sarawak (UNIMAS)

Amorphophallus Blume ex Decne with *ca* 200 species, belongs to the family Araceae (aroid family). The genus is spread mainly in the tropics from West Africa eastward into Polynesia. On Borneo, there are at least 18 indigenous *Amorphophallus* species, all endemic. This project served to address two main questions. Firstly, to infer the biogeographic history of the entire genus on Borneo whether the dispersal events to Borneo, is more than one event (s). Secondly, to focus on the pollination biology of two unrelated taxa, *Amorphophallus hewittii* Alderw. and *A. eburneus* Bogner. The phylogenetic analyses were carried out by using a plastid region and two nuclear regions: *mat*K, ITS and PhyC. A total of 96 accessions representing 69 species of *Amorphophallus* were included with 36 accessions representing 16 species from Borneo and 60 accessions representing 53 species from other regions. The study showed that the Bornean taxa were placed in three separate clades. *Amorphophallus hewittii sensu lato* is closely related to *A. titanum* (Becc.) Becc. ex Arcangeli. The white-flowered *Amorphophallus* species in Sarawak is related to *A. declinatus* Hett. and *A. dactylifer* Hett. from the Philippines. *Amorphophallus angulatus* Hett. & A. Vogel is not closely related to any of the Bornean taxa. Flowering mechanism and pollination strategy of *A. eburneus* and *A. hewittii* were also presented in this study.

T2

P0263

Phylogenetic analyses of *Shorea* (Dipterocarpaceae) on Borneo

Nur Aina Afiqah Abdul Halid, *Sin Yeng Wong*
Department of Plant Science and Environmental Ecology, Faculty of Resource Science and Technology, Universiti Malaysia Sarawak

Dipterocarpaceae is a key family, both ecologically and commercially, in the forests of Southeast Asia. The distribution is pantropical, confined to Asian tropics between the Seychelles Island and Southeast Asia to New Guinea with the greatest diversity and abundance in western Malesia. Members of the family are predominantly distributed in tropical lowland and hill forests within this region. The family consists of 17 genera with about 680 species in three subfamilies and shows a high rate of endemicity. *Shorea* Roxb. ex C.F.Gaertn. is the largest genus in the family, consisting of 360 species. So far, 138 species are known in

Borneo, of which 91 species are endemic and 130 species occur in Sabah and Sarawak. Previously, the phylogeny of *Shorea* focused on Southeast Asia, but none of these studies were specifically focused on Borneo. Former studies showed that *Shorea* is probably paraphyletic with the genus *Hopea* Roxb., probably belonging within the genus *Shorea*. Therefore, this project is conducted to investigate the phylogeny of the genus *Shorea* on Borneo broadly focusing on the issue of their intrageneric delimitation. There are three objectives in this project: (1) To resolve the phylogeny of *Shorea* species on Borneo with the emphasis on species from Sarawak; (2) To determine the paraphyly of *Shorea* species; (3) To revise *Shorea* species based on the outcome of objective (1). Thirty four accessions are included in this study which comprised of six sections from the tribe Shoreae. Phylogenetic analyses of *Shorea* were carried out in this study by using two gene regions: Internal Transcribed Spacer (nrITS) and Megakaryocyte-associated tyrosine kinase (*mat*K).

T2
P0264
Speciation of the sect. *Camellia* based on pollinator shift - focusing on Japanese *Camellia*

Harue Abe[1], *Hiroki Miura*[2], *Yoshitaka Motonaga*[1], *Yu Matsuki*[5], *Yoshihisa Suyama*[3], *Saneyoshi Ueno*[4]
1. *Niigata University*
2. *Aomori Prefectural Asamushi Aquarium*
3. *Tohoku University*
4. *Forestry and Forest Products Research Institute*
5. *DNAform*

Organisms have the ability to adapt to specific conditions within their environments, resulting in morphological diversification and speciation. It is thought that speciation is more likely to occur as a result of premating isolation rather than premating isolation. Therefore, to elucidate the speciation of plants, it is important to identify the factors that cause premating isolation. In Japan, *Camellia japonica* and *Camellia rusticana* belonging to the sect. *Camellia* are naturally distributed. Despite differences in their habitats and morphologies, they have been classified as variants, subspecies, or species depending on the researchers. Thus, their taxonomic positioning remains unclear, because they are known to hybridize if their habitats are adjacent to each other. *C. rusticana* grows in snowy places, whereas *C. japonica* does not. While *C. japonica* adopts ornithophily, *C. rusticana* adopts entomophily. Both species have adapted to different growing environments and pollinators. Thus, they may have been speciated by premating isolation, which causes morphological mutations in flowers and leaves. Hence, this study aims to examine the degree of speciation of two species by comparing the floral and leaf morphologies and the genetic differentiations. In the results, it was possible to distinguish between these species by the presence or absence of hypodermis in leaf morphology. In flower morphology, PCA analysis clearly distinguished between these species, and both petal and filament colors were also statistically different between these species. The genetic structures based on eight cpSSR loci and 151 SNPs are divided into 3; southern and northern *C. japonica*, and *C. rusticana*. The haplotype diversities are 3.13 in *C. japonica* and 30.8 in *C. rusticana*. These results suggest that morphological and genetic differences between these species are clear. Furthermore we investigated the candidate pollinators quantitatively. They are birds for *C. japonica* and small insects for *C. rusticana*, which means the gene flow between two species might be rare. In conclusion, it is reasonable to think that *C. japonica* and *C. rusticana* are different species and are not closely-related because the difference on morphology is not consecutive and there is no difference in the genetic diversity. It also suggested that premating isolation caused by differences in environments is one of the factors that promote speciation of sect. *Camallia*. It could cause the pollinator shift between birds and insects. The main pollinators of *C. japonica* are birds because it blooms in winter when the availability of fruits and insects is limited in the temperate forests. On the other hand, *C. rusticana* are pollinated by small insects in the early spring season when insect begin to come out. Then the adaptation to different environments might result in the different pollination systems.

T2
P0265
Phylogeny and ancestral state reconstruction of the cool season grasses (Poaceae subfamily Pooideae)

Carmen Acedo[1], *Félix Llamas*[1], *Manuel Pimentel*[2], *Elvira Sahuquillo*[2], *Alicia Alonso*[1], *Ana Molina*[1], *Pilar Catalan*[3]
1. *University of Leon*
2. *University of Coruña*
3. *University of Zaragoza*

The cool season Poacaeae subfamily Pooideae consists of 4234 species belonging to 197 genera and 14 tribes, as recognized current classification, showing great diversity in habit and morphology. In this study, we investigate the phylogeny of main lineages and reconstruct the ancestral character states of ten traits commonly used to build its taxonomy, to better understand their evolution, and compare them with the evolution and diversity of major clades. In grasses, inflorescence architecture can be one of the reasons of the success of some lineages, determining how many flowers and seeds are produced in a single inflorescence. We sequenced two nuclear and five plastid markers of some 170 species representing 120 genera, thirteen of the fourteen tribes, and all the included subtribes. Phylogeny was inferred by maximum parsimony (MP) and bayesian inference (BI). Ancestral states, for traits playing a main role in the taxonomy of the lineages, were reconstructed using MESQUITE by means of parsimony and maximum likelihood (ML) and, full bayesian method in BayesTraits. Differences between topologies obtained by MP and BI of the combined matrices were minimal. The monophyly of the subfamily and the mayor clades were strongly supported. Most of our results agree with previous hypothesis about the phylogeny and, find the same conflictive cases suggesting a more recent diversification of the monophyletic Phaenospermatae after Meliceae and before Stipeae with strong support for all main clades. A perennial habit for the MRCA of subfamily and major clades (the only exception is Ammochloinae and relatives) is not surprising, but a bidirectional origin with reversions is showing. Some traits evolved consistently with adaptive radiation: The branching inflorescence and multi-flowered spikelet are common in genera diversify rapidly. Different ancestral state reconstruction methods do not yield conflicts. As expected, reconstruction found support, for an ancestor of the subfamily and its main lineages, having a branched inflorescence, spikelet with two glumes not enclosing the flowers, and spikelet 1-flowered in the MRCA of the subfamily and early lineages, but multiflowered in the most recent divergent lineages. Our results for the attachment of the central awn, extensively used in to circumscribe lineages, have no certainty about which is the ancestral state suggesting an ambiguous MRCA. The results are clear reconstructing the presence-absence of dorsal awn: MRCA lacks dorsal awns -presence being derived- but results are bidirectional. The reconstruction of geniculate awns is similar. Finally, for dispersion it is suggested a MRCA with diaspore formed by separate florets. In conclusion, our study suggests different relationships among some early Pooideae lineages and corroborates the controversial relationship between traditional subtribes Agrostidinae and Brizinae. It has resolved the morphology of subfamily MRCA and provided information helping us to understand the evolution and success of the main lineages. Branching inflorescence and multi-flowered spikelet reconstruction supports the high diversification of the most specialized annual representatives: A more effective dispersal supports an evolution consistent with the adaptive radiation.

T2

P0267

Tetraploid induction of *Vigna unguiculata*

Olagorite Adetula

National Horticultural Research Institute.P.M.B5432, Idi-Ishin, Ibadan

Cowpea (*Vigna unguiculata* (L.) Walp.) is most important grain legume crop because of its high protein content and economic uses. With increasing economic importance attempts were made to develop tetraploid plants for crosscompatibity breeding between cultivated and wild relatives of Cowpea. Tetraploid plants *Vigna unguiculata* were obtained from both culturing the seed in MS media with different concentration of colchicines (invitro colchicine treatment) and from application of colchicine to the shoot tips. The frequency of tetraploid using 0.05% and 0.1% was 30% and 20% respectively. The survival rate was 90% for 0.05col. and 55% for 0.1col. treatments. Morphological data were collected from both induced tetraploid and diploid plants. Remarkable differences were observed in morphological attributes which are useful traits for breeding work. Effects of colchicine treatment was more pronounced in the plant habit and height ranging between 10-65 cm. Cytogenetic analysis showed the frequency of 80% of the cell of the tetraploid plants with 2n=44. Shoot tips colchicine treatment and invitro colchicne treatment were both effective for producing tetraploid in Cowpea.

T2

P0268

Phenetic analysis of three populations of *Dendropanax arboreus* (L.) Decne. & Planch. (Araliaceae) in the Tuxtlas region, Mexico

***Andrés-Hernández Agustina Rosa**, Tamanis-Coatl Sandra Ivet, Rivas-Arancibia Sombra P., Carrillo-Ruiz Hortensia, Martínez-Moreno David*

Benemérita Universidad Autónoma de Puebla

The species *Dendropanax arboreus* is widely distributed, but there is phylogenetic evidence, based on molecular data, that some studied specimens do not form a monophyletic group. In Mexico, all morphotypes have been labeled with the name *D. arboreus*. Three morphologically different populations have been recognized in different types of vegetation in the Tuxtlas region. This work compares and evaluates the similarities and differences between these populations regarding wood anatomy, and leaf architecture and anatomy of leaf. Anatomical sections of the wood, petioles and laminae were obtained from specimens of *D. arboreus* collected in different locations with different vegetation types: a) high evergreen forest, b) mangrove, c) cloud forest (Populations 1,2 and 3, respectively). The leaves were also diaphanized. Qualitative and quantitative characteristics were analyzed by Principal Component Analysis (PCA), and a dendrogram was constructed. It was found that the specimens from all three locations showed typical characteristics of the Araliaceae family such as diffuse porosity, presence of solitary vessels and in multiples of 2 to 5 with scalariform perforation plates, brochidodromous venation, bifacial leaves and resin canals. The characteristics that allowed us to differentiate the specimens from the three locations were vessel wall thickness, the length of vessel elements and fibers, the type of venation, presence of tracheoblasts and arrangement of the vascular bundles in the petiole. Interestingly, population number 2, collected in a location with mangrove vegetation, has a frequency of vessels of 347 mm^2, which is different than the values reported for other species of the family. PCA results showed that the three populations are separated, and the dendrogram showed that there is a greater similarity between the specimens from mangrove and cloud forest, while the specimens from evergreen forest showed greater dissimilarity. We conclude that there are three different populations based on the anatomy of wood, and suggest that it could be possible to propose two new species.

T2

P0269

Modelling distribution of *Cinchona capuli* L.Andersson: A new report for Peru

***Joaquina Alban**[1], Lizeth Huaman[1], Miguel Durand[1], Eder Chilquillo[2]*

1. Museo de Historia Natural San Marcos

2. Campinas University

The genus *Cinchona* L. is distributed in the neo tropical region, along the slopes of the Andean region from Venezuela to Bolivia. Currently recognized 23 species, of which 17 are reported for Peruvian territory. *Cinchona capuli* L.Andersson is an endemic species of Ecuador, but it was registered in 2006 for Peru (Cusco and Pasco regions). Due this precedent, we developed the present research which main aim is to verify if the species could have a wider distribution.The methodological design was based in examining approximately 700 specimens of herbariums (AAU, AMAZ, CUZ, F, HOXA, HUT, K, LOJA, MO, MOL, NY, QCNE, US and USM). This information was complemented with collection records for Ecuador and expeditions done by the authors between the years 2004-2016. Furthermore, the ecological niche model analysis was conducted in order to identify potentially suitable habitat areas for *C. capuli*. We used the predictive software MAXENT version 3.3.3e and bioclimatic data (WorldClim).The result shows that the specimens of Cusco and Pasco do not correspond to *Cinchona capuli*. However, it was registered in Ayabaca, Piura region. On the other hand, the analyses of the potential niche modelling confirm this; widen the distribution area to northern Peru, based on the mean area under the curve (AUC = 0.987) which indicate an optical performance. It should be mentioned that the predicted model was supporte by 4 of the 19 bioclimatic variables (Isothermality, Precipitation of Coldest Quarter, Temperature Seasonality and Precipitation of Wettest Month).We conclude that *Cinchona capulí* has a distribution up to northern Peru, being not limited to Ecuador as previously believed. Additionally, according to the predictive model, montane forest of San Ignacio, Jaen and Huancabamba could present suitable habitats for the species. To wind up it assumes the presence of 18 species of Cinchona in Peru and reconfirms that Peruvian territory constitutes and the centre of distribution of the genus.

T2

P0270

A diverse Miocene megaflora from the Hattiesburg Formation, Mississippi, U.S.A.

Daniel Mcnair[1,3], *Debra Stults*[2], *James Starnes*[4], ***Mac Alford***[3], *Brian Axsmith*[2]

1. University of California, Davis

2. University of South Alabama

3. University of Southern Mississippi

4. Mississippi Department of Environmental Quality

Our understanding of Miocene plant evolution in the southeastern United States is hampered by the rarity of productive sites with well-preserved fossils. Berry (1916) briefly described a poorly preserved palm (*Sabalites apalachicolensis*) and leaves of *Ulmus floridana* from the Miocene Hattiesburg Formation in southeastern Mississippi. In 1944, a few plant fossils from the Hattiesburg Formation were identified by Roland Brown as a probable *Taxodium* (possibly *T. distichum)*, *Salix*, either *Morus* or *Celtis*, and fragments of some type of monocot. None of these fossils were described in detail or figured. Recently, an extensive megaflora has been recovered from the upper part of the Hattiesburg Formation along the Bowie River in Hattiesburg, Mississippi. The Bowie River locale is typical of the non-marine fine-grained sediments of the Grand Gulf Section. The locale consists of pyritic, gray-green colored, laminated to finely-bedded sediments with alternating clays, silts, and fine-grained sands. This strongly fissile and repetitive sequence is slightly carbonaceous to highly lignitic, with well-preserved fossil leaves and leaf hash along partings and large, imbedded, lignitized logs. The depositional setting is likely an interfluvial over-bank deposit in the lower distributary portion of a coastal river. Ferns are represented by specimens of *Salvinia* with attached sporocarps, a probable *Woodwardia* s.l. based on a pinnatifid frond, and a possible *Osmunda* cf. *regalis* pinna. Conifers are represented by branchlets of *Taxodium* with preserved cuticles. Angiosperms include leaves attributable to the Lauraceae based on venation pattern, oil bodies in the mesophyll, and paracytic stomata. *Platanus* (Platanaceae) is commonly associated with palms and recognized from leaves, fruits, and large logs. Leaflets of *Sambucus* (Adoxaceae) are common. *Cercis* (Fabaceae) is recognized from leaves with palmate venation and pulvini. Leaves of *Quercus* (Fagaceae) sections *Lobatae* and *Quercus* have been recovered. The Juglandaceae include fruits of *Juglans* section *Rhysocaryon* and two species of *Carya* section *Apocarya*. *Morus* (Moraceae) is known from leaves, as are *Populus* and *Salix* (Salicaceae). Of particular interest is an endocarp of *Sargentodoxa* (Lardizabalaceae), being the first record of this Chinese endemic from the coastal plain of the southeastern U.S. and only the third occurrence in North America. Monocots include *Cyperus* (Cyperaceae) based on spikelets with achenes and two types of palm (Arecaceae). One type of palm is costapalmate, the other palmate. Two palmate specimens have robust, recurved spines on the petiole, the first record of this type of fossil palm from North America. The Hattiesburg Formation megaflora is now one of the most extensive in eastern North America, and along with the Brandon Lignite of Vermont, the Alum Bluff flora of Florida, the Brandywine flora of Maryland, and the Gray Fossil site of Tennessee, helps to fill a major temporal and biogeographic gap in our understating of Miocene plant evolution in eastern North America.

T2

P0271

Transatlantic crossing: Revisiting relationships between Neotropics and African-Madagascan ferns and lycophytes

Thais Elias Almeida[1], *Alexandre Salino*[2], *Sabine Hennequin*[3,4], *Jean-Yves Dubuisson*[3,4]

1. Universidade Federal do Oeste do Para

2. Universidade Federal de Minas Gerais

3. nstitut Systématique Evolution Biodiversité (ISYEB), Sorbonne Universités, UPMC Univ Paris 06, MNHN, CNRS, EPHE

4. Centre de Recherche en Paléobodiversité et Paléoenvironnements (CR2P)

Floristic relationships between Neotropics and Africa-Madagascar region (continental Africa, Madagascar and Ocean Indian Islands) are well known in several groups of vascular plants. Compared to seed plants, ferns and lycophytes have a greater capacity of migration (dispersal and establishment) and are less limited by geographical barriers, being more dependent on microhabitats and microclimate for establishment. In this group, a considerable number of disjunct geographic distribution hypotheses have been proposed in the last decades. These usually are represented as species pairs or taxa with morphological affinities occurring in both areas. Given the current available data on phylogenetic relationships for several groups of fern and lycophytes, we revisited some of these hypotheses of phytogeographic relationships. We evaluated species pairs proposed by Moran and Smith (Moran RC, Smith AR. 2001. Phytogeographic Relationships between Neotropical and African-Madagascan Pteridophytes. Brittonia 53,304–351) for which phylogenetic evidence was available. We assessed, in the light of available phylogenetic inference, if these morphological affinities reflect a shared evolutionary history, or simply convergence. In the case of taxa derived from common ancestors we examined the biogeographical history and events that might have led to their current geographic distribution. Among the "species pairs", several represent taxa that are not closely related, as their morphological similarity reflects convergence to similar environments and habitats in both continents. Examples of taxa representing same lineages with disjunct distribution are, indeed, found among the investigated groups. Explanations on how this biogeographic pattern might have emerged include the boreotropics hypothesis, vicariance, and long-distance dispersal. The subjacent biogeographic history of these transatlantic ferns and lycophytes is a complex and composite one, and distinct explanations are required to explain patterns among different families and genera. It involves more commonly migration events through long-distance dispersal, with vicariance playing a subsequent role in radiation following the emergence of local barriers due to geological and climatic changes.

T2

P0272

DNA barcoding and conservation of timber species from Ghana

Phyllis Akua Amamoo, *Gabriel Ameka*

Department of Plant and Environmental Biology, University of Ghana

The Forestry Commission-Ghana estimates that about 50 timber tree species are in commercial trade in Ghana. Of these 31 species were identified and sampled from three biodiversity hotspots and protected areas (Ankasa Resource Reserve, Bia Biosphere Reserve, and Kakum National Park), in the country in the present investigation. The purpose is to provide DNA barcodes for the timber species in the country to enable their accurate identification. Total genomic DNA was extracted from silica-dried leaf samples of each timber species, obtained randomly from the three sampling localities, using the modified CTAB method, and sequenced using standard procedures. To verify the authenticity of the data in the molecular identification of timber species from Ghana, DNA sequences were retrieved from reference database in GenBank for comparison. The results indicate that the combination of the *matK* and *rbcL* gene regions can identify the timber species accurately, thus enabling the tracing of illegal logged timber. Illegal logging is known to be one of the main causes of deforestation and biodiversity loss, and trade with illegal timber and wood products creates market disadvantages for products from sustainable forestry. Sustainable forestry conserves and protects biodiversity and prevents biodiversity loss, by controlling the rates of harvesting timber species. The authors recommend a large scale DNA barcoding of all tree species in Ghana to facilitate and enable the use of DNA barcoding in tracing illegally logged timber and trade with illegal timber and wood products in the country.

T2

P0273

DNA barcoding and conservation of medicinal plants from Ghana

Gabriel Ameka, *Phyllis Akua Amamoo*

Plant and Environmental Biology, University of Ghana

The purpose of this project is to DNA barcode all medicinal plants in Ghana; providing DNA barcode sequences from the *matK* and *rbcL* gene regions. In a preliminary investigation ca. 150 medicinal tree species from protected areas in southern Ghana were sampled. Total genomic DNA was extracted from silica-dried leaf material, and sequenced using standard procedures. The trees are mainly used in the treatment of malaria, general fevers and in post-partum nourishment (nutraceuticals). The DNA barcode sequences will be linked to herbarium voucher specimens, digital images/photos, locality, literature, medicinal uses, will enable the verification and authentication of herbal medicines. The data generated will be available, as a public reference library of DNA barcode sequences of tree species, and associated data in international genomic and DNA barcoding databases, including Barcode of Life data (BOLD) Systems, and EBI/GenBank. This will ensure that substitutes are not used, intentionally or unintentionally, and delivering safer herbal medicines to local and international markets, thus increasing the confidence of the general public in herbal medicines. This will also ensure sustainable harvesting of medicinal trees; and thereby promoting sustainable forestry and thus conserving and protecting biodiversity and biodiversity loss.

T2

P0274

Divergence dates of *Astragalus* L. section *Incani* (Fabaceae) based on nrDNA ITS and plastid *rpl32-trnL* $_{(UAG)}$ sequences

Elham Amini[1], *Shahrokh Kazempour Osaloo*[1], *Aliasghar Maasumi*[2]

1. university

2. Research Institute

Section *Incani* DC. was established with 12 species by De Candolle. Linnaeus introduced the first species of *Incani* as *A. incanus* (exclusion Southern Europe in Spain and France) and Podlech placed this species, as the lectotype of section, for *Astragalus* group with perennial habit, acaulescent, bifurcate hairs and mostly leaves imparipinnate but in a few species tri- or unifoliolate. This section with about 140 species worldwide is one of the richest and the most important sections of *Astragalus*. Section *Incani* distributed throughout Southern Europe, North Africa, Anatolia, Central Asia and the Caucasus. Its main centre of diversity
seems to be in Iran and Turkey with about 120 species. In this study, Bayesian inference (BI) analysis was conducted in BEAST v1.8.3 [Bayesian Evolutionary Analysis Sampling Trees] using the substitution model selected by the AIC for each gene and a Relaxed Clock. Bayesian and Maximum Likelihood phylogenetics analyses applied to nrDNA ITS and *rpl32-trnL* $_{(UAG)}$ sequences for 82 species of section. A. frigidus, A. stocksii and A. aegobromus were selected as outgroup. The present study indicated that sect. Incani forms a monophyletic group. In the divergence time studies that included rates of nucleotide substitutions using *matK* sequences, Wojciechowski provided the estimated *Astragalus* diverged from its sister group, *Oxytropis*, 12–16 Ma, the middle Miocene, with divergence of Neo-*Astragalus* beginning ca. 4.4 Ma. BEAST chronogram is congruent with those resulting from Bayesian analyses. The timing of diversification events is illustrated by a chronogram derived from the all-compatible Bayesian consensus tree with dates for selected nodes constrained to the mean of penalized likelihood estimates obtained from sub-sampling of 1000 trees from the posterior distribution of the Bayesian MCMC tree search. The crown node age of the section *Incani*, is estimated to be 4.1 Mya. The posterior probability distribution of this dating supports a Pliocene origin for sect. *Incani*.

T2
P0275
GCMS Analysis, phytochemical screening, antioxidant and radical scavenging capacities by novel and contemporary methods of different extracts of *Typha domigenesis*

Anjum Perveen[1], *Andleeb Anwar Sardar*[2], *Zaheeruddin Khan*[2]
1. Centre for Plant Conservation, University of Karachi
2. GC University Lahore, Department of Botany

Two innovative, rapid and cost-effective approaches for evaluation of antioxidant potential are developed. N, N- Diphenyl Para Phenylene Diamine (DPPD) decolorization assay and Diamino Naphthalene Decolorization assay are developed for the screening of antioxidant activity. Both of these assays evaluate the free radical scavenging activity of analytes. A chromogenic reaction takes place between DPPD and DAN with potassium persulfate at selective pH produces two distinct radical cations with maximum absorption at 700 nm and 575 nm in its first-order derivative spectrum. A linear inhibition of color production was observed with linearly increasing amounts of antioxidants, with correlation coefficients (R^2) ranging from 0.999 to 0.982. The antioxidant capacity of standard solutions of different hydrophilic and hydrophobic antioxidants was evaluated. Comparison of antioxidant capacity determined with these newly developed DPPD assay and DAN assay with the well-known 2,2'- azinobis-[3-ethylbenzthiazoline-6-sulfonic acid] (ABTS)-persulfate decolorization assay indicated the efficacy and sensitivity of the procedure. The proposed assays are less expensive (costs about US$5 per 100 assays) and require very short time for preparation of radical cation solution in comparison with ABTS assay. These novel assays are also applied for evaluation of antioxidant potential of *Typha domigenesis* and the results obtained are very much in accordance with other contemporary methods. DAN decolorization assay is applied to plasma samples to evaluate the antioxidant activity of plasma or the oxidative stress level of plasma. The results of phytochemical screening and Gas chromatographic analysis reveals that *Typha domigenesis* are enriched with Flavonoids, alkaloids, polyphenols, Terpenoids and other important classes of phytochemicals.

T2
P0276
From Eurasia to the Americas and back: Biogeographic patterns in *Zanthoxylum* (Rutaceae)

Marc Appelhans[1,2], *Niklas Reichelt*[1], *Jun Wen*[2]
1. University of Goettingen
2. Smithsonian Institution

The genus *Zanthoxylum* is the second largest genus within the *Citrus* family (Rutaceae) and its 220+ species are distributed throughout the tropics of the New and the Old World with several species in temperate regions in Asia and eastern North America. Several species are used in traditional medicine and as a spice ('Sichuan pepper') and the genus is usually easily distinguishable by its prickles on trunks and branches ('knobthorns') and sometimes on the leaf rhachis. The genus had been largely neglected by plant sytematists and we are presenting the first molecular phylogeny with a worldwide taxon sampling. Our results based on the markers ETS, ITS, *rps16* and *trnL-trnF* suggest that the Asian genera *Phellodendron* and *Tetradium* as well as the African and Malagasy *Fagaropsis* are the closest relatives of *Zanthoxylum*. *Toddalia*, a monotypic and widespread genus that occurs from Africa to eastern Asia, is not a separate genus but is part of the *Zanthoxylum* clade and is sister to the African and Malagasy species. *Zanthoxylum* was formerly split into the genera *Zanthoxylum* (small flowers with tepals) and *Fagara* (sepals and tepals). Our results show that a distinction between these two taxa is artificial. Four sections used to be recognized within *Fagara*. Out of these, the monotypic section *Mayu* and the American section *Tobinia* are monophyletic. The Australian and Pacific section *Blackburnia* is resolved as monophyletic in the plastid trees, but as paraphyletic in the ETS/ITS trees due to the phylogenetic position of the Hawaiian species. An ILD test revealed that the alignments of the different markers were only congruent after the Hawaiian species had been excluded. A hybridization event prior to the colonization of Hawaii is the most likely explanation for this incongruence. The most species rich section *Macqueria* is largely polyphyletic. Due to the rich fossil record of *Zanthoxylum*, *Tetradium* and *Phellodendron*, this group is well-suited for biogeographic studies. The oldest fossils have been found in the Early Eocene in Europe. The results of our molecular dating and ancestral area reconstructions suggest a Eurasian origin of *Zanthoxylum* in the Early Eocene. From there, one migration to Africa and Madagascar and several migrations to eastern Asia were inferred. One lineage migrated to the Americas and colonized both the northern and southern continent. This lineage also gave rise to a small Asian lineage with tepalous flowers, so that *Zanthoxylum* first crossed the Atlantic Ocean from Eurasia to North America and migrated back to Asia via the Pacific or across the Bering Strait.

T2
P0277
Revised infrageneric classification of *Homalium* (Salicaceae) and taxonomic revisionary studies of Malagasy sections.

Wendy Applequist
Missouri Botanical Garden

The pantropical woody genus *Homalium* (Salicaceae, formerly Flacourtiaceae) includes slightly over 150 currently recognized species. The genus is characterized by two perianth whorls, semi-inferior ovaries, and single large glands at the base of each sepal. It encompasses considerable floral diversity and has previously been divided into two subgenera (characterized by having stamens one per petal or fasciculate) and ten sections. Based on floral morphology, it is possible to conclude with considerable confidence that the subgenera are not monophyletic. Therefore, only sections should be recognized. This author has proposed that all fasciculate-stamened Asian, Malesian and Pacific species formerly included in sect. *Eumyriantheia* should be transferred to sect. *Polyanthera*. The primary center of diversity is Madagascar, where six sections are present, of which five are endemic. The previous taxonomic treatment of *Homalium* in Madagascar, dating to 1973, recognized 36 species. With the much larger volume of material now available, revisionary studies recently completed or now in progress reveal that at least 54 species are present. Underdescription may occur in other regions as well.

T2
P0278

Canalization and modularity of "fit" vs. advertisement traits of triggerplant flowers (Stylidiaceae)

W. Scott Armbruster[1], *Juliet A. Wege*[2]
1. University of Portsmouth
2. Western Australia Department of Parks and Wildlife

In 1960 Raissa Berg suggested that plants with flowers involved in specialized pollination should experience pollinator-mediated canalizing selection, and that such plants should evolve floral modularity, i.e., phenotypic independence of floral traits from variation in plant stature, leaf size, *etc.* She found support for this hypothesis in the correlation patterns within and among floral and vegetative traits. However, the strength of stabilizing and canalizing selection can be expected also to vary among floral traits. For example, one could predict that traits related to floral fit with pollinators will be under stronger canalizing selection than advertisement traits. In addition, modularity is also a possible outcome of genetic-developmental organization rather than selected by pollination function. We might thus expect all floral parts to be similarly integrated into a floral module, with non-floral parts forming a separate module or modules. We used several approaches to test these ideas in 9 species of triggerplant, *Stylidium* (Stylidiaceae), a genus with flowers that transfer pollen via a motile, protandrous columns formed by the fusion of staminate and pistillate tissues. Exploratory factor analyses of 15 traits suggested that traits related to pollination precision and pollinator fit (e.g., column length, corolla-tube length) generally form a module separate from those traits related to advertisement (e.g. petal length and width). Both sets of traits were largely decoupled from variation in a module comprising vegetative traits. We also examined the allometric relationships between size parameters and specific floral traits, with emphasis on contrasting floral traits that can be expected to be under strong (precision and fit) vs. weak canalizing selection (advertisement). We found that, as predicted, traits important in determining floral fit with pollinators (and where anthers and stigmas contact pollinators) varied less and were generally more negatively allometric in relation to variation in non-floral traits than those floral traits involved purely in attraction. Finally we assessed and compared the adaptive accuracies of pollen deposition (column exertion, male phase) and retrieval (column exertion, female phase) to evaluate within- and among-species relationships. Extremely high fundamental pollination accuracy was observed in all 9 species, although realized accuracies are much lower in some.

T2
P0279

The phylogenomic study on mainland Southeast Asian Andropogoneae (Poaceae)

Watchara Arthan[1,2], *Michael R. Mckain*[3], *Paweena Traiperm*[1], *Cassiano A. D. Welker*[4], *Jordan K. Teisher*[3,5], *Elizabeth A. Kellogg*[3]
1. Department of Plant Science, Faculty of Science, Mahidol University, King Rama VI Road, Ratchathewi, Bangkok 10400, Thailand
2. Department of Pharmaceutical Botany, Faculty of Pharmacy, Mahidol University, Si Ayutthaya Road, Ratchathewi, Bangkok 10400, Thailand
3. Donald Danforth Plant Science Center, 975 North Warson Road, St. Louis, Missouri 63132 USA
4. Universidade Federal de Uberlândia, Instituto de Biologia, Rua Ceará s/n, Uberlândia, Minas Gerais 38400-902, Brazil
5. Academy of Natural Sciences of Drexel University, 1900 Benjamin Franklin Parkway, Philadelphia, PA 19103 USA

Andropogoneae evolution has been investigated extensively because of the economic and ecological importance of the tribe and wide native distributions in Africa, America, and Asia. Within these ranges, Southeast Asian Andropogoneae represents extensive and novel diversity from widespread genera, such as *Andropogon* and *Schizachyrium*, to several minor genera such as *Coix*, *Hemisorghum*, *Kerriochloa*, and *Pseudosorghum*, which are characterized by various and unique forms of inflorescences and spikelet structures. However, evolutionary relationships with other Andropogoneae need to be resolved and the taxonomic status of the Southeast Asian taxa needs to be re-evaluated. Full plastomes of Southeast Asian Andropogoneae were analyzed with Maximum Likelihood (ML) and Bayesian Inference. To understand morphological trends, four characters of inflorescences and spikelets were optimized by ML. *Coix*, *Hemisorghum*, *Kerriochloa*, *Pseudosorghum*, and *Andropogon burmanicus* were discovered to diverge outside core Andropogoneae. In core Andropogoneae, *Andropogon* and *Schizachyrium* are polyphyletic genera. *Schizachyrium sanguineum* populations divided into two clades, implying that its populations are being driven by cryptic species differentiation, hybridization, or introgression. However, other species with multiple accessions were monophyletic. In morphological evolution, convergent evolution and reversal are common throughout Andropogoneae, especially in the traits of inflorescence and spikelet. Striking diversity in reproductive parts is evident among minor genera of Southeast Asian grasses. In *Andropogon* and *Schizachyrium*, the traits are so variable that both genera cannot be distinguished from each other using traditional characters.

T2
P0280

Resolving the *Coccineum* complex in *Hedychium* J. Koenig (Zingiberaceae) using molecular and morphometric data

Ajith Ashokan, *Vinita Gowda*
Indian Institute of Science Education and Research- Bhopal

The true species and taxonomic identifications are confusing within *Hedychium* due to the occurrence of intermediate forms (hybrids and polyploids) and clinal variations. Taxa that are not accepted as unique species are considered to be part of a species complex and several complexes have now been identified within the genus *Hedychium*. In Indian subcontinent, *Coccineum* (7 taxa) and *Coronarium* (8 taxa) are the two main species complexes among *Hedychium* that contain species that are undistinguishable based on morphology alone. *Coccineum* complex consists of at least seven species- *Hedychium angustifolium* Roxb., *H. aurantiacum* Roscoe, *H. carneum* G. Lodd., *H. coccineum* Buch.-Ham. ex Sm., *H. longifolium* Roscoe, *H. roscoei* Wall. and *H. squarrosum* Buch.-Ham. ex Wall. All taxa included in this complex are distributed in the Himalayan and/or Indo-Myanmar ecoregions. Of the seven species, six of them are native to the Indian state of Meghalaya (collection location is unknown for *H. roscoei*). Previous treatments of this complex have resulted in lumping all species under one name, *H. coccineum* Buch.-Ham. ex Sm. However, *H. coccineum* has much broader distribution and phenology ranges compared to other members in the complex and we strongly suggest that this group cannot be accommodated under a single name. We believe that many of the species within the *Coccineum* complex show overlaps in their distributional ranges and thus similar ecological conditions might have shaped their morphologies in a more homoplasious manner, and further resulted in the cryptic species status of *H. coccineum* and its allies. During our sampling surveys in the NE Indian states, especially in Meghalaya we observed and collected many of the *Coccineum* and its allies. The phenology data for these species were obtained from the available herbarium specimens, regional floras and last two years field observations. The flowering periods of five *Coccineum* complex species coincide with the monsoon season (June to September) and they overlap at least for a month. Phenology data for *H. roscoei* and *H. squarrosum* are not available since we could not locate any individuals of these species in the field and the protologues were silent about their phenology. Morphological characters were coded in the field and herbaria to understand the variability within and among species. PCA revealed that the morphological characters overlap among different taxa within the *Coccineum* complex. Using field observations we have now identified eight key floral characters (out of 152 original characters) which are believed to vary within and among populations. Many of these characters are not retained in herbarium specimens. This is a huge improvement in our understanding of morphological variations since all of the 8 characters have a potential to be selected by ecological factors such as pollinators, and dispersers and they may contribute significantly towards the plants reproductive success. Earlier confusions generated by the morphological characters had forced the recent studies to focus on the molecular tools for clearly resolving the species boundaries. Interestingly, no representative taxa other than *H. coccineum* were included in any molecular investigations which demands further analysis using molecular tools.

T2

P0281

Studies on morphological variations and evolutionary trends of the genus *Murdannia* Royle (Commelinaceae)

Anna Ancy Antony Attullil[1], *Santhosh Nampy*[2]
1. St. Albert's College, Ernakulam
2. Calicut University

Murdannia Royle (Commelinaceae), with about 52 species worldwide, is represented by 26 species in India. It is the largest genus of Commelinaceae in Asia and has more species in India (50% of the total) than in any other country in the world. The genus is studied in India based on field and herbaria data and the characters are recorded. Entire morphological variations within the genus are considered. Morphometric analysis of the species occurring in India is performed with NTSYS software. Twenty six species occurring in India are selected as Operational Taxonomic Units (OTU's) and a data matrix is prepared on vegetative, floral and seed characters. Sixty one characters are used for the cluster analysis and a dendrogram is prepared. The study shows that the Indian species falls under 6 distinct sub-clades in 3 different homogeneous clades portraying the inter relationships among the species. Variations in morphology including reduction and modification of vegetative and floral parts prove helpful in resolving evolutionary relationships within the genus.

T2

P0282

Phylogenetics of the *Bouteloua gracilis* complex (Poaceae)

Marcela Avendaño-González[1], *Maria Elena Siqueiros Delgado*[1], *Travis Columbus*[2], *Francisco Morales-Dominguez*[1]
1. Universidad Autónoma de Aguascalientes
2. Rancho Santana Botanic Garden

B. gracilis (Kunth) Lag. ex Griffiths is the most common specie in the north american semiarid grassland. The species is distributed natively from southern Canada through the central and western USA to Mexico, and throughout this wide range populations vary appreciably in morphology and chromosome number ($2n$ = 20–84). The basis of morphological differences between populations– genetic and/or environmental– of this species remain unclear and genetic variation among *B. gracilis* populations throughout its distribution is unknown. Nuclear and chloroplast markers sequenced (ITS, *rpL32-trnL* and *rps16-trnK*) resulted invariant in this species populations from throughout the geographic range of the species, indicating that a different approach is needed to understand diversification within *B. gracilis*. Additionally, based on the same data, conflicting topologies, and low clade support, it remains doubtful which of four closely related species is sister to *B. gracilis.* Restriction site-associated next-generation DNA sequencing (RADseq) is going to be employed to reveal relationships among populations of *B. gracilis* and its sisters. Our objective is to resolve relationships in the *B. gracilis* complex, test the monophyly of each currently recognized species and determine if populations within the morphological highly variable *B. gracilis* justify separate species or subspecies recognition as well as examine this species morphology in a phylogenetic context. All potential sister species of *B. gracilis* have been sampled from throughout their distributions in Mexico. Thirty-three populations of *B. gracilis* were sampled, and USA populations will be sampled this summer. Genomic DNA is extracted and RNA purified in order to obtain good DNA quality and quantity. Samples (96) will be sequenced using one lane for RADseq, 87 samples of *B. gracilis*, two samples from each of the four candidate sister species, and one sample of *B. ramosa* (outgroup) will be included. Phylogenetic inference (maximum likelihood and Bayesian) will be carried out after processing the sequences. To evaluate the morphological variation in *B. gracilis* throughout its range, a morphometric analysis will be carried out using specimens from Mexico and USA, including those included in the molecular study. Many vegetative and reproductive traits will be measured and then subjected to various analyses (e.g., ordination and statistical analysis).

T2

P0283

Phylogenomics and spatio-temporal evolution of the Neotropical genus *Amphilophium* (Bignonieae, Bignoniaceae)

Veronica Thode[1], *Isabel Sanmartin*[2], *Lúcia G. Lohmann*[1]

1. Universidade de São Paulo
2. Real Jardín Botánico, RJB-CSIC

The Neotropics represent one of the most biodiverse areas on Earth. Yet, the processes that have shaped such high levels of diversity remain poorly understood. Detailed information on the geographic distribution of taxa combined with time-calibrated phylogenies are key for an improved understanding of the Neotropical history. While previous attempts to reconstruct the spatio-temporal history of Neotropical plants have been based on a few molecular markers, there is growing consensus that phylogenomic approaches based on multiple markers are needed to solve problems of species tree/gene tree incongruence and lack of resolution due to rapid diversification. *Amphilophium* (Bignonieae, Bignoniaceae) includes 47 species with diverse morphology and distribution patterns, representing an excellent model to study the origin and assembly of the Neotropical Biota. In this study, we use 36 newly generated plastid genomes to reconstruct phylogenetic relationships, estimate divergence times and reconstruct temporal and spatial patterns of evolution within the genus. Our dataset consists on alignments with ca. 140 kilobases (Kb), including 78 protein-coding regions that concatenated resulted in a matrix with ca. 62 Kb or 20,600 amino acids when translated. Phylogenetic relationships were estimated using Maximum likelihood (RAxML) and Bayesian inference (MrBayes 3.2.6) as well as site-heterogeneous models in PhyloBayes MPI 1.7. Divergence times were estimated using Bayesian relaxed clock methods implemented in BEAST 1.8.3 and PhyloBayes MPI 1.7, while ancestral ranges and rates of geographic range shifts were inferred using the ML Dispersal-Extinction-Cladogenesis model (DEC) in Lagrange. Our molecular phylogenetic analyses corroborate the monophyly of *Amphilophium* and reconstructs strongly supported clades within the genus that are also well characterized by morphological synapomophies. The spatiotemporal evolution of *Amphilophium* indicates that the genus diversified during the Neogene (23-2.6 My), a period characterized by intense geological activity and climatic changes in the Neotropics. Interestingly, diversification in the Atlantic forest started ca. 21 My, while the Amazonian radiation occurred much more recently, at around 4.6 My. This study shows one of the first phylogenomic studies of a Neotropical plant group. The well-sampled and strongly supported plastid phylogeny of *Amphilophium* provides a solid basis to explore the mechanisms associated with the assembly and evolution of the Neotropical Biota.

T2
P0284

Diversity of pollen types in Apiales – a phylogenetic approach

Jakub Baczyński, *Aleksandra Miłobędzka, Krzysztof Spalik, Łukasz Banasiak*
University of Warsaw Biological and Chemical Research Centre, Department of Molecular Phylogenetics and Evolution

Angiosperm order Apiales comprises almost 5500 species classified in seven families. The largest of them, Apiaceae (umbellifers) and Araliaceae, have long been treated as closely related. The other families - Pittosporaceae, Griseliniaceae, Torricelliaceae, Pennantiaceae, and Myodocarpaceae - have been recognized or reclassified quite recently based on molecular data. Our knowledge on morphological evolution in this group is therefore limited due to the lack of comparative studies conducted within the framework of modern phylogeny. In this study we have analyzed evolution of pollen morphology in Apiales. The results are based upon comprehensive phylogenetic sampling of c. 150 taxa including representatives of all major lineages recognized in the order with particular emphasis on Araliaceae and Apiaceae. For each species included in the study we have defined a set of 17 discrete and 11 continuous traits and subsequently ancestral morphology was reconstructed on a molecular phylogenetic tree. We conclude that most families share a rather uniform (plesiomorphic) morphology characterized by prolate/oblate spheroidal shape, relatively thin wall, long ectocolpus and broad, elongated ora. The advanced morphology resulting from shortening of equatorial axis and gradual reduction of compound aperture is generally apomorphic for umbellifers.

T2
P0285

Genetic variation of Ulleung Island endemic *Saxifraga fortunei* var. *pilosissima* Nakai and its alliance

Sang-Hun Baek[1], *Jin Suk Youn*[1], *Jung Sung Kim*[2], *Jae-Hong Pak*[1,2]
1. Depart of Biology, Kyungpook National University
2. Research Institute for Dokdo and Ulleung-do Island, Kyungpook National University

Saxifraga fortune, a perennial herb species, is widely distributed in Eastern Asia and two varieties have been reported from Korea. One is *Saxifraga fortunei* var. *incisolobata* (Engl. & Irmsch.) Nakai which found in high and humid mountain are of South Korea, the other is *Saxifraga fortunei* var. *pilosissima* Nakai which is morphologically similar to S. *fortunei* var. *incisolobata* (Engl. & Irmsch.) Nakai. The latter one is well-known endemic taxa of Ulleung-island, an eastern volcanic island of Korea. Even though some taxonomist had described the morphological difference of them, it is too difficult to recognize each other due to the overlapped variation between them and it has been still argued the relationship and their taxonomical circumscription. To clear their relationship and understand their evolutionary history, we tried to analyze genetic diversity of them in this study. We collected two population of *Saxifraga fortunei* var. *pilosissima* from Ulleung Island and five populations of *Saxifraga fortunei* var. *incisolobata* from inland of Korea peninsular. We analyzed and compared the sequences of two cpDNA non-coding regions (*trnL-F, psbA-trnH*) and nrITS. Based on sequence variation among them, we found three different haplotypes from each region. Both of Ulleung Island populations of *Saxifraga fortunei* var. *pilosissima* were genetically identical and shared the haplotype against to inland populations of S. *fortunei* var. *incisolobata* which have two different haplotypes. It meant that *S. fortunei var. pilosissima* have genetically diverged from *S. fortunei var. incisolobata* in spite of high morphological similarity. Besides, finding of intermediate type between both taxa suggested the probability of more dynamic relationship among them.

T2
P0287
Do males and females in Pleurocarpous mosses differ in their environmental niches?

Irene Bisang[1], *Johan Ehrlén*[2], *Lars Hedenäs*[1]
1. Swedish Museum of Natural History
2. Stockholm University

Biased adult sex ratios (SR), usually inferred from counts of sexually mature plants, are commonly observed in unisexual bryophyte species. For sex-expressing bryophytes, there is evidence of spatial variation of reproductive traits, including SRs, related to environmental conditions. However, how sexes in rarely fertile or non-fertile species are distributed with regard to environmental factors is largely unknown. We determined genetic sex ratio (GSR), irrespective of sex expression, in natural populations of two pleurocarpous mosses, using a specifically designed female-targeting molecular marker. We measured environmental conditions and assessed whether sexual niche partitioning may contribute to spatial variation in GSR. *Drepanocladus trifarius* and *D. lycopodioides*, two wetland mosses with a documented female bias in overall European GSR, differ from each other in the levels of realized sexual reproduction both at the European scale, and in the study areas (sporophytes absent vs. present). We sampled 214 individual shoots of *D. trifarius* in a completely random design in a population in Central Sweden, and 660 individual shoots of D. *lycopodioides* stratified randomly in 11 local populations in each of three regions in the Swedish Baltic. We confirmed that GSR is female-biased in local and regional populations in both species. In *D. trifarius*, spatial distributions of genetically male and female shoots were not more clustered than expected under random distribution, and environmental factors did not differ between locations with females and males, indicating lack of sexual niche partitioning. In *D. lycopodioides*, GSR varied among local populations and differed among the three study regions. Habitat patch size and Habitat wetness strongly affected GSR and the occurrence of genetically male plants, genetically male plants not occurring in the wettest habitats. We suggest that this is due to higher energy requirements of females due to sporophyte maturation, indicating sexual niche partitioning. We hypothesize that the evolution of sexual niche partitioning between genetic sexes is unlikely in species without or very rare sexual reproduction.

T2
P0288
Phylogenomics of *Ephedra*

Ruben Blokzijl[1], *Niklas Wikström*[2], *Catarina Rydin*[1,2]
1. Stockholm University
2. Bergius Botanic Garden

Relationships among species within *Ephedra* are partly poorly understood, despite previous efforts. *Ephedra* comprises 50-60 species native to arid regions of the Northern Hemisphere and South America. Classification of *Ephedra* based upon morphological characteristics is hard, due to limited character variation and parallel evolution of characters. Besides, results based on molecular data show incongruence with the morphological classifications. The long evolutionary distance to outgroups and the low amount of variation in molecular markers have hampered the inference of *Ephedra*. Previous studies using dense taxon sampling and several molecular markers indicate geographical groupings of species but results are partly weakly supported. We address the phylogeny of *Ephedra* using Next Generation Sequencing data, thus a substantially larger amount of characters, but reduced taxon sampling compared with previous studies.

T2
P0289
Pollen and pollination of *Ephedra* (Gnetales)

Kristina Bolinder, *Catarina Rydin*
Stockholm University

Most gymnosperms are, in contrast to angiosperms, pollinated solely by wind. The pollination biology of the enigmatic gymnosperm group *Ephedra* (Gnetales) is understudied and partly misunderstood. In a series of studies we have therefore investigated pollination biology and reproductive morphology of the group. Through extensive field work in northeastern Greece we found that the sister species to the remaining genus, *E. foeminea*, is insect-pollinated. The pollination system is of a generalist nature but nocturnal moths are among important pollinators. The moths navigate using the moon and interestingly enough the secretion of the pollination drops is correlated with the lunar cycle. Insect pollination is the ancestral state in *Ephedra* and during the long evolutionary history of the genus there has been a shift in pollination mechanism, from insect pollination to wind pollination, which is the prevailing state in the crown-group. In *Ephedra*, the pollen morphology and the composition of the exine is closely related to the pollination biology. Its pollen is ellipsoid and characteristically polyplicate. Insect-pollinated species have smaller pollen grains with a significantly larger number of plica whereas wind-pollinated species have fewer plicae and a branched pseudosulci between each pair of plicae. The ultrastructure of the exine differs between the groups, which has consequences for the settling velocity of the grains and thus for the aerodynamic features of the pollen grains. The density of the granules in the infratectum is higher in *E. foeminea* compared to investigated wind-pollinated species. A higher settling velocity results in a lower flight capability for the insect-pollinated species. Investigated wind-pollinated species have, in contrast, a lower settling velocity and better flight capability. Also, the morphology of the female cones is to some degree related to the pollination biology. In wind-pollinated species the morphology of the female cones and the tilt towards the stem creates and aerodynamic environment that is so specific that is can sort out conspecific pollen grains. In the insect-pollinated *E. foeminea*, however, the morphology of the cone is not aerodynamically adapted for collection of pollen grains.

T2
P0290
Field plant identification at digital time: Potential and limits of automated plant identification based on multimedia data

Pierre Bonnet[1], *Alexis Joly*[2], *Hervé Goëau*[1], *Jean-Christophe Lombardo*[2], *Antoine Affouard*[2], *Samuel Dufour-Kowalski*[3], *Sen Wang*[2], *Remi Knaff*[3], *Nozha Boujemaa*[2], *Jean-François Molino*[4], *Daniel Barthélémy*[1]

1. CIRAD
2. Inria
3. Inra
4. Ird

Pl@ntNet is a free web and mobile platform dedicated to automated, image-based plant identification and to collaborative gathering of plant observations (http://identify.plantnet-project.org/). It relies on crowdsourcing approaches and machine learning techniques for data production, validation and enrichment. First launched in 2013 on 800 French native species, it now covers a large part of European flora (6,200 spp.) and has been extended to other floristic domains, such as the Mascarene Islands, the Guiana Shield and Maghreb. Through its iPhone and Android apps (> 3 million downloads and 10,000-50,000 daily users), Pl@ntNet gathers increasingly large amounts of botanical observations voluntarily contributed by an array of people who are often novice in plant identification. These observations are continually checked and amended (for identifications and image quality) by hundreds of amateur botanists, through Pl@ntNet's collaborative web tools. We present the limits of this approach, as well as the perspectives of improvement, based on both the users feedback, and on analyses of the data already collected. This emphasize on one side, the potential of new technologies for botanical and ecological activities, and on the other side, the capacity of multi-disciplinary projects to address a few topical social needs at large scale.

T2
P0291

Nucleolar dominance in grass allopolyploids – a closer look on interesting epigenetic phenomenon

Natalia Borowska-Zuchowska[1], *Ewa Robaszkiewicz*[1], *Miroslaw Kwasniewski*[2], *Alexander Betekhtin*[1], *Robert Hasterok*[1]
1. University of Silesia in Katowice, Faculty of Biology and Environmental Protection, Department of Plant Anatomy and Cytology
2. University of Silesia in Katowice, Faculty of Biology and Environmental Protection, Department of Genetics

Nucleolar dominance is an epigenetic phenomenon connected with the nuclear-encoded 35S rRNA genes. It is observed in several hybrids and allopoplyploids and consists in preferential suppression of rRNA gene loci inherited from one of the ancestral species. Considerable attention has been paid to discover the molecular mechanisms which stay behind this selective repression of rRNA gene loci. Most experiments using DNA hypomethylating agents, bisulphite-mediated 5-methylcytosine mapping and RNAi knock down of the genes encoding for chromatin modifying proteins proved that the nucleolar dominance has an epigenetic origin (Preuss *et al.*, 2008, *Molecular Cell* 32: 673–684). Despite the fact that more than 80 years have passed since the discovery of nucleolar dominance in *Crepis* hybrids, our understanding of exact mechanisms that determine this process is still fragmentary, especially in the case of the Poaceae family members. In 2008, the presence of this phenomenon has been described in the *Brachypodium* genus, which belongs to the grass family. *B. hybridum* is a natural allotetraploid (2n=30) with putative parental genomes originated from two diploid species, i.e. *B. distachyon* (2n=10) and *B. stacei* (2n=20). Preferential silencing of *B. stacei*-inherited ribosomal RNA genes in this allotetraploid was reported (Idziak and Hasterok 2008, *Genome* 51: 387-391). This presentation outlines the studies on the nucleolar dominance mechanisms in several *B. hybridum* genotypes originated from various geographic locations. The distribution of 35S rDNA loci inherited from ancestral species was determined in both mitotic and meiotic cells. Moreover, we aimed to investigate the epigenetic status of 35 rRNA gene loci in *B. hybridum* and its putative diploid ancestors by the determination of DNA methylation and selected histone modifications (e.g. H4K5ac, H4K16ac, H3K9ac and H3K9me2) immunopatterns. We also show the results of molecular characterisation of intergenic spacers (IGS) between 25S rDNA and 18S rDNA in *B. hybridum* and its progenitors as well as their physical localisation in metaphase chromosomes and interphase nuclei. In all IGS sequences we identified putative transcription initiation sites and spacer promoters followed by subrepeats. Their contribution in selective silencing of 35S rDNA loci is further discussed.

T2
P0292

Pollen morphology of *Dionysia* Fenzl (Primulaceae) in comparison with other Primulaceae genera.

Dmitriy Britski
Komarov Botanical Institute of Russian Academy of Sciences; St.-Petersburg; Russia

Pollen morphology of 9 species, belonging to genus *Dionysia* (*Dionysia aretioides* (Lehm.) Boiss., *D. bryoides* Boiss., *D. haussknechtii* Bornm. & Strauss, *D. hedgei* Wendelbo, *D. hissarica* Lipsky, *D. involucrata* Zapriag., *D. leucotricha* Bornm., *D. michauxii* (Duby) Boiss., *D. microphylla* Wendelbo) was examined by light microscope (LM), scanning electron microscope (SEM) and transmission electron microscope (TEM). The pollen grains of species under study were of medium sizes, spheroidal, polycolpate. Exine patterns were microreticulate, reticulate. Study by TEM showed semitectate exine structure, with short, thick columellae and poorly pronounced foot layer. The results were compared with previous data on pollen morphology of *Primula* and other genera of Primulaceae. Pollen of Primula species can be regarded in frames of 3 morphological types: 1) 3-colporate (*P. farinifolia, P. longiscapa, P. darialica, P. japonica, P.zeylamica, P. forbesii, P.malacoides, P bulleyana*). Endoapertures are small and unclear. Exine pattern is reticulate. 2) Syncolporate (*P. malvacea, P.floribunda, P.verticillata, P. arctica, P. auriculata, P. involucrata, P. turkestanica, P. nivalis, P. pumilio, P. warschenewskiana P.capitellata, P.rosea, P.scotica, P. viali, P.sachalinensis, P.stricta, P. cortusoides, P. eugeniae, P. lactiflora, P. minkvitziae, P. olgae, P. pinnata, P. saxifragifolia, P. algida, P. baernii, P. moorcroftiana, P. tschuktschorum, P. beringiensis, P.laurentiana, P. farinosa, P. borealis, P. cuneifolia, P.elliptica*). Syncolporate type is the most common pollen type in the genus *Primula.* Pollen grains of only one species, *P. malvaceae,* has well pronounced endoapertures, easily visible with light microscope. Small poor pronounced endoapertures are typical for other studied species. In literature pollen grains of these species are usually described as syncolpate but as the more investigations are undertaken sometimes including CLSM method, the more evidence appears for the presence of endoapertures. 3) Polycolpate (*P. amoena, P. egaliksensis, P. elatior, P. finnmarchica, P. heterochroma, P. iljinskii, P. longiflora, P.*

macrocalyx, P. nutans, P. pamirica, P. polyantha, P. pseudoelatior, P. pseudosikkimensis, P. ruprechtii, P. sibirica, P. veris).The last morphological type of *Primula* pollen grain is exactly the same, that was exposed for *Dionysia*. Two others differ from Dionysia pollen significantly. The common feature is weak or absent foot layer. Pollen grains of the rest of Primulaceae are 3-colporate, or, in some cases (Cortusa, Dodecatheon), syncolporate. Pollen of *Lysimachia, Naumburgia, Anagallis, Trientalis, Ardisiandra, Stimpsonia* are medium-sized, 3-colporate, endoapertures equatorially elongated, usually with sharp tapered ends. Exine patterns are perforate, microreticulate and reticulate. Pollen grains of *Androsace, Douglasia, Gregoria* and *Pomatosace* are small to medium size, with small round or slightly elliptical endoapertures and perforate or microreticulate exine. Thus, pollen grains of *Dionysia*, being very similar to pollen grains of some sections of *Primula*, differ significantly from pollen of the majority of Primulaceae species. From the palynological point of view, *Dionysia* and part of *Primula* can be united in one group.

T2

P0293

The early cretaceous coal-forming plants of southern part of East Siberia and Russian Far East

***Eugenia Bugdaeva**, Valentina Markevich*
Federal Scientific Center of the East Asia Terrestrial Biodiversity, FEB RAS

Coal formation in the geological past was linked with the development of plant communities that produced a large biomass and at the same time possessed relatively simple structure. A considerable part of the swamp phytomass became mortmass. The coal seams are particularly the subject of paleobotanical research. Using bulk maceration of coals we obtained the spores and pollen, as well as dispersed cuticle of the plants that gave rise to the Early Cretaceous coals of the Transbaikalia (Khilok, Chita-Ingoda, Bukachacha, and Turga-Kharanor basins), the Amur River Region (Bureya Basin), and the Primorye region (Razdolnaya River and Partizansk basins). The plants of the Khilok Basin that produced coals mainly belong to the extinct ginkgoaleans (*Baierella averianovii*). The coal-forming plants of the Chita-Ingoda and Bukachacha basins are more diverse. They are in the former represented by the bryophytes, lycophytes, cyatheaceous ferns, Umaltolepidiaceae (*Pseudotorellia*), Sphenobaieraceae *(Sphenobaiera starukhiniae*), in the latter– by the cyatheaceous and gleicheniaceous ferns, *Pseudotorellia transbaikalica*, conifers of araucariaceous affinity *Farndalea* sp., *Pityophyllum* sp. having pinaceous affinity. The coals of the Turga-Kharanor Basin consist of remains of the bryophytes, lycophytes, cyatheaceous ferns, *Pseudotorellia kharanorica,* and cheirolepidiaceous plant *Pagiophyllum* sp. The Early Cretaceous coals of the Partizansk Basin, Primorye region, are mostly composed of remains of gleicheniaceous ferns, taxodialean *Elatides asiatica*, subordinate Miroviaceae, rare ginkgoalean *Pseudotorellia* sp., and bennettite *Nilssoniopteris rithidorachis.* The coal-forming plants of the Razdolnaya River Basin mainly belong to the cyatheaceous and gleicheniaceous ferns, to the Miroviaceae (*Mirovia orientalis*, ginkgoalean *Pseudotorellia krassilovii*, to a lesser degree the coals of this basin are composed of other groups of conifers (including scale-leaved *Athrotaxites orientalis*) and bennettites *Nilssoniopteris rithidorachis, N. prynadae,* and *Anomozamites* sp. Thus, the Early Cretaceous mire plant communities of southern part of East Siberia were dominated by bryophytes, lycophytes, cyatheaceous ferns, ginkgoaleans, plants having affinity with Pinaceae and Araucariaceae, whereas those of the Russian Far East – mainly by gleicheniaceous ferns, bennettites, Miroviaceae, and conifers having taxodiaceous affinity. These plants provided a basis for the Early Cretaceous coals.

T2

P0294

Evolution of the process underlying floral zygomorphy development in pentapetalous angiosperm

***Ghadeer Bukhari**, Jingbo Zhang, Wenheng Zhang*
Virginia Commonwealth University

Observations of floral ontogeny indicated that floral organ initiation in pentapetalous flowers most commonly results in a medially positioned abaxial (MAB) petal instead of a medially positioned adaxial (MAD) petal during early development. It was proposed that the conservation of patterns of floral organ initiation might impose developmental constraint that limited the displays of floral zygomorphy that evolved in Asteridae. Here, we has provided the first study of the evolution of developmental processes of floral zygomorphy in pentapetalous angiosperms to investigate whether the patterns of floral organ initiation leads to biases in the evolution of displays of floral zygomorphy. We analyzed patterns of floral organ initiation, displays of floral zygomorphy, and developmental processes that give rise to floral zygomorphy extracted from floral diagrams representing 405 taxa in 330 genera on a phylogeny using ancestral state reconstructions. The results indicated that the MAB petal initiation is the ancestral state of the pattern of floral organ initiation in pentapetalous angiosperms. The remaining taxa, displaying the MAD petal initiation, represent 30 independent origins from the ancestral MAB initiation. There are distinct developmental processes to give rise to floral zygomorphy along diverse lineages of pentapetalous angiosperms, which closely related lineages are likely to share a similar developmental process. We have demonstrated that the development indeed constrains the process that gives rise to floral zygomorphy, while the phylogenetic distance allows the constraint to diverge. Our work provides novel insights on the role that development plays in evolution of floral zygomorphy.

T2

P0295

Phylogenomics of *Atractylodes* (Asteraceae) and domestication of *A. macrocephala* Koidz.

***Minqi Cai**, Pan Li, Li Zheng, Luxian Liu, Chengxin Fu*
The Key Laboratory of Conservation Biology for Endangered Wildlife of the Ministry of Education, College of Life Sciences, Zhejiang University

Atractylodes DC. (Asteraceae) is a small genus endemic to East Asia, containing five to seven species. Species from the *Atractylodes* have been used as the famous Traditional Chinese Medicine "Baizhu" and "Cangzhu". In this study, we used RAD-seq to reconstruct phylogenetic relationships of *Atractylodes*, and to analyze the relationships among different cultivars of *A. macrocephala*. The conclsions of this study are as follows: 1) Phylogenetic and interspecific relationship of *Atractylodes*. Total 115 samples form56 populations used in the study were collected from China, Rassia, Korean peninsula, and Japanese archipelago (including five taxa, one subspecies, one cultivar and one outgroup from Germany). The maximum likelihood method was applied to the dataset including 77,036 unlinked-SNPs loci. The phylogeney of *Atractylodes* was well resolved, showing that this genus contains 6 monophyletic taxa. These were *A. carlinolides, A. macrocephala, A. japonica, A. coreana, A. lancea and A. chinensis*. However, whether the *"luotianensis"* should be treated as the subspecies of *A. lancea* or not, it needs further study. *A. ovata* (Thunb.) DC., widely used in Japan, is proved to refer to the same species with *A. japonica* Koidz. ex Kitam. 2) Evolution history, biogeography and reconstruction of ancestral distribution area of *Atractylodes*. Using six fossil calibrations of Asterales from previous study, BEAST analysis based on four chloroplast fragments revealed that the divergence time among *Atractylodes* was around late Miocene 8.66Ma. Employing this result, we calibrated the lineage of *Atractylodes* using BEAST analysis based on SNPs dataset. The results were also combined with the geographical distribution to carry out the reconstruction of ancestral distribution area. The result revealed that *Atractylodes* most likely originated in sourthern China. Combined with paleogeographic and paleoclimatic events, the ancestor of *Atractylodes* spread along the East Asian mid-latitude arid zone to sourthern China during 13.5Ma ~ 9.7Ma. *A. corlinoides* and *A. macrocephala* were firstly formed in sourthern China, then the rest spread to the Korean peninsula, Japan and northern China. The genus finally formed adaptation to the humid climate, but the niche of *Atractylodes* reflects the preferences of drought. 3) Phylogenetic relationships among different *A. macrocephala* populations. The result showed that the cultivated 'Pingzhu' is the basal clade within *A. macrocephala*, diverged at 3.28Ma, indicating the existence of *A. macrophala* 'Pingzhu'. The wild population "Qizhu" was sister to all other populations of *A. macrocephala* except 'Pingzhu'. The cultivar 'Pingzhu' and wild "Qizhu" haven't been involved in the formation process of widespread cultivated *A. macrocephala*. However, they are important germplasm resources in urgent need of protection. Depend on our sampling result, The cultivated populations of *A. macrocephala* in this study showed the tendency of single-origin.

T2

P0296

Machine learning algorithms improve the power of phytolith analysis: A case study of the tribe Oryzeae (Poaceae)

Zhe Cai, *Song Ge*

Institute of Botany, Chinese Academy of Sciences

Phytoliths, as one of the important sources of microfossils, have been widely used in paleobotany related studies, especially in the grass family (Poaceae) where abundant phytoliths were found. Despite this, several challenges remain when phytoliths are used in various studies, including the accurate description of phytolith morphology and the reliable utilization of phytolith traits in taxon identification or discrimination. In this study, we collected 1063 phytolith samples from 18 taxa representing seven main genera in the tribe Oryzeae (subfamily Ehrhartoideae) and five taxa in the subfamilies Bambusoideae and Pooideae. By focusing on the tribe Oryzeae that has been extensively investigated in terms of taxonomy and phylogeny, we were able to evaluate the discrimination power of phytoliths at lower taxonomic level in grasses. With the help of morphometric analysis and by introducing several machine learning algorithms, we found that 87.7% of the phytoliths could be classified correctly at genus level (support vector machine algorithm, false positive rate 0.02), significantly increasing the resolving power of phytolith evidence in these groups of plants compared with the results by traditional phytolith analysis method. On these bases, we proposed a pipeline of phytolith analyses, including data collection, morphometric analysis and classifier construction for phytolith discrimination. The pipeline should be applied to various studies across different groups of plants. This study provides new insights into the utilization of phytoliths in evolutionary and ecology studies involving grasses and plants in general.

T2

P0297

Does a dynamic genome drive speciation in a mega-diverse genus?

Lucia Campos-Dominguez[1,2], *Catherine Kidner*[1,2]

1. University of Edinburgh

2. Royal Botanic Garden Edinburgh

Speciation events are always driven by genetic changes that cause phenotypic effects, which may be either adaptive, driven by selection, or can be fixed in genetically isolated populations through the action of genetic drift. Such changes can range from small point mutations to the activation of transposons, including large scale chromosome rearrangements such as duplications, deletions and/or inversions. Tropical areas contain the highest levels of biodiversity on Earth, being considered a very species-rich ecosystem. With more than 1900 species, the genus *Begonia* is one of the most species-rich in the tropics. Recent taxonomic studies on *Begonia* provide a robust phylogenetic background for the analysis of evolutionary patterns across the group, and confirm niche specificity cannot be the only factor driving speciation. Furthermore, previous studies have shown wide variation in genome size and repetitive element content across the genus. In this context, we aim to determine the role of genome dynamics and composition in driving the repeated bursts of speciation indicated by the phylogeny. To achieve this, species representing different *Begonia* clades have been selected for genomic structure analysis. Variation in DNA repeats and transposon composition across different radiations has been studied, as well as active transposable machinery using *Begonia* transcriptomes. Such approaches aim to reveal the genome dynamics behind the *Begonia* speciation events, providing new insights on the role of non-coding elements in plant genome evolution.

T2

P0298

Guizhou Province—A holy land of musci flora

Wei Cao, *Yuanxin Xiong, Xianzhen Luo, Hongying Tan*

College of Life Science, Guizhou University, Guiyang, China

Guizhou Province is situated in the Southwest of China. It covers an area of 176 167km2, lying between 24°37′ and 29°13′N and between 130°36′ and 109°35′E. It is located in the eastern slope of the Yunnan-Guizhou Plateau. It belongs to subtropical climate, and in most part of Guizhou, the annual precipitation is ca 1000-1400 mm, and the annual temperature varies 11-19 ℃ . Because of its comfortable climate and complex geomorphic types, it becomes a high biological diversity region. For instance, the total number of bryophytes of Guizhou is just less than Yunnan Province which is one of regions with the highest biological diversity in China. A latest checklist of the Musci for Guizhou Province includes 1058 species, 32 varieties, 6 subspecies in 286 genera and 57 families. In this checklist, 57 species and 5 varieties are endemic to China, including 8 species are endemic in Southeast of China as well as 2 species and 3 varieties are just found in Guizhou. Furthermore, a total of 3 genera, 22 species, 2 varieties and 1 subspecies are newly reported for Guizhou. Moreover, there are some rare species such as Trachyloma indicum var. obtusifolium Y.-X.Xiong & X.-Z.Luo (Pterobryaceae) and Plagiomnium guizhouense Y.-J. Yi & S.He (Minaceae) are new species published past few years and just seen in a very narrow region, Erpodium (Erpodiaceae) is a newly recorded genus in China, and Cytodontopsis leveillei (Thér) P. C. Rao & Enroth (Cryphaeaceae) are recollected from Guizhou and so on. The information in this abstract is based on two monographs "Bryophyte Flora of Guizhou" Vols. I and II and more than 40000 specimens (collected from ca 80 counties of Guizhou, including 9 National Nature Reserves, 6 Provincial Nature Reserves, most of the forest parks and dozens of small nature reserves.) held in GACP, IFSBH and PE herbaria. In fact, the moss flora of Guizhou is still incompletely known by us, and the present list is relatively conservative. Besides, we still have a little knowledge of the liverworts and hornworts of Guizhou, more in-depth research about bryophytes in Guizhou is very necessary.

T2

P0299

Past climate changes facilitated homoploid speciation in three mountain spiny fescues (*Festuca*, Poaceae)

Isabel Marques[1], *David Draper*[2], *Maria Luisa Lopez-Herranz*[1], *Teresa Garnatje*[3], *Jose Gabriel Segarra-Moragues*[4], *Pilar Catalan*[1]

1. Universidad de Zaragoza

2. Universidade de Lisboa

3. Institut Botanic de Barcelona

4. Universitat de Valencia

Apart from the overwhelming cases of allopolyploidization, the impact of speciation through homoploid hybridization is becoming more relevant than previously thought. Much less is known, however, about the impact of climate changes as a driven factor of speciation. To investigate these issues, we selected *Festuca picoeuropeana*, an hypothetical natural hybrid between the diploid species *F. eskia* and *F. gautieri* that occurs in two different mountain ranges (Cantabrian Mountains and Pyrenees) separated by more than 400 km. To unravel the outcomes of this mode of speciation and the impact of climate during speciation we used a multidisciplinary approach combining genome size and chromosome counts, data from an extensive nuclear genotypic analysis, plastid sequences and ecological niche models (ENM). Our results show that the same homoploid hybrid was originated independently in the two mountain ranges, being currently isolated from both parents and producing viable seeds. Parental species had the opportunity to contact as early as 21000 years ago although niche divergence occurs nowadays as result of a climate-driven shift. A high degree of niche divergence was observed between the hybrid and its parents and no recent introgression or backcrossed hybrids were detected, supporting the current presence of reproductive isolation barriers between these species.

T2

P0300

Different colonization routes explain the genetic structure and diversity of the model grass *Brachypodium stacei* in the Western Mediterranean

Valeriia Shiposha[1], *Pilar Catalan*[1], *Marina Olonova*[2], *Isabel Marques*[1]

1. Universidad de Zaragoza

2. Tomsk State University

Mediterranean species of the genus *Brachypodium* are promising energy crops since their selfing nature and annual life cycles are an advantage in breeding programs. Some species, like the false brome *B. distachyon*, have already been sequenced and new genomic initiatives have triggered the *de-novo* genome sequencing of its close relatives such as *B. stacei*, a species that was until recently mistaken for *B. distachyon*. But the success of these initiatives hinges on detailed knowledge about the distribution of genetic variation within and among populations for the effective use of germplasm in a breeding program. Understanding population genetic structure is also an important prerequisite for designing effective experimental populations for genomic wide studies. However, population genetic data are still limited in *B. stacei*. We therefore selected and amplified 10 nuclear microsatellite markers to depict patterns of population structure and genetic variation among 181 individuals from 19 populations of *B. stacei* occurring in its predominant range, the Western Mediterranean area: mainland Iberian Peninsula, continental Balearic Islands and oceanic Canary Islands. Our genetic results support the occurrence of a predominant selfing system with extremely high levels of homozygosity across the analyzed populations. Despite the low level of genetic variation found, two different genetic clusters were retrieved, one clustering all SE Iberian mainland populations and the island of Minorca and another one grouping all S Iberian mainland populations, the Canary Islands and all Majorcan populations except one that clustered with the former group. These results, together with a high sharing of alleles (89%) suggest different colonization routes from the mainland Iberian Peninsula into the islands. A recent colonization scenario could explain the relatively low levels of genetic diversity and low number of alleles found in the Canary Islands populations while older colonization events are hypothesized to explain the high genetic diversity values found in the Majorcan populations. Our study provides widely applicable information about geographical patterns of genetic variation in *B. stacei*. Among others, the genetic pattern and the existence of local alleles will need to be adequately reflected in the germplasm collection of *B. stacei* for efficient genome wide association studies.

T2

P0301

Comparative plastome genomics and phylogenomics of *Brachypodium*: Flowering time signatures, introgression and recombination in recently diverged ecotypes

Ruben Sancho[1], *Carlos Cantalapiedra*[2], *Diana López-Alvarez*[1], *Sean Gordon*[3], *John Vogel4*[3], *Pilar Catalan*[1], *Bruno Contreras-Moreira*[2]

1. Universidad de Zaragoza

2. Estación Experimental de Aula Dei-Consejo Superior de Investigaciones Científicas

3. DOE Joint Genome Institute

Few instraspecific genomic studies have been conducted in plants, and none of them have focused on the intra-specific diversity and evolution of their plastid genomes. We address this issue in *Brachypodium distachyon*, a model system for monocots, and its close relatives *B. stacei* and *B. hybridum*, for which a large genomic data set has been compiled. We analyze inter- and intra-specific plastid comparative genomics and phylogenomic relationships within a family-wide framework. While number of genes of the chloroplast genome of *B. distachyon* is greatly constrained as compared to its nuclear genome, full plastid sequences provided a wealth of genetic variants. Major indel differences were detected between *B. distachyon* and *B. stacei/B. hybridum* plastomes. Within *B. distachyon*, we detected two main lineages, an Extremely Delayed Flowering (EDF+) clade and a Spanish Turkish (S+T+) clade, plus nine chloroplast capture and two ptDNA introgression and micro-recombination events. Early Oligocene (30.9 Mya) and Late Miocene (10.1 Mya) divergence times were inferred for the respective stem and crown nodes of *Brachypodium* and a very recent Mid-Pleistocene (0.9 Mya) time for the *B. distachyon* split. Our comparative genomic study of whole plastome sequences of *B. distachyon* and its close relatives allowed us to detect intra-specific introgressions and other associated evolutionary events (e. g., biparental plastome inheritance, heteroplasmy) that could not be detected with single genes. The observed plastome admixture that goes along with the nuclear genome admixture in the *B. distachyon* Arn1 and Mon3 lines, and the essential swapping of plastomes among the three different *B. distachyon* plastome groups (EDF+, S+, T+), likely resulted from random backcrossing followed by stabilization through selection pressure. The pipeline for the assembly and annotation of plastid genomes is available at https://github.com/eead-csic-compbio/chloroplast_assembly_protocol.

T2

P0302

Biogeographic events are not correlated with diaspore dispersal modes in Boraginaceae

Juliana Chacon[1], *Maximilian Weigend*[1], *Federico Luebert*[1,2]

1. Nees-Institut für Biodiversität der Pflanzen, Universität Bonn

2. Departamento de Silvicultura y Conservación de la Naturaleza, Universidad de Chile

Long-distance dispersal seems to be the main biogeographic event responsible for intriguing distribution patterns in plant groups in which sister taxa are separated by thousands of kilometers of distance across oceans and continents. The biotic and abiotic mechanisms behind such dispersal events are poorly understood and many attempts have been made to explain how plants can manage to disperse and survive these long journeys. The biogeographic history of Boraginaceae, a subcosmopolitan plant family with many disjunct clades, is here addressed and analyzed in the context of the different dispersal modes exhibited by the species. The lack of a clear pattern between the main dispersal events in Boraginaceae and the phylogenetic distribution of the dispersal modes indicates that no single dispersal mechanism can be associated with the events of dispersal in the family. Our study reveals that long-distance dispersal is a very complex process that needs to be analyzed in the context of climatic and environmental changes and the response of plants and their dispersal vectors to these variable conditions.

T2

P0303

Species diversity and reticulate evolution in the *Asplenium normale* (Aspleniaceae) complex

Yanfen Chang

Xishuangbanna tropical boanical garden, Chinese Academy of Sciences

The *Asplenium normale* complex contains the widespread *A. normale* and several geographically restricted species: *A. boreale*, *A. hobdyi*, *A. kiangsuense*, *A. oligophlebium*, *A. pseudonormale*, *A. shimurae and A. pifongiae*. The taxonomy of this group is unclear with some entities treated infraspecifically or as synonyms. Furthermore, the existence of diploids and tetraploids in this species complex is suggestive of reticulate evolution. In order to achieve a natural classification and to investigate the relationships in this complex, phylogenetic analyses of plastid and nuclear sequence data, detailed morphometric analyses and inference of ploidy level were used to assess the distinctiveness of putative taxonomic units and their relationships. Both the evolutionary and morphometric results support the hypothesis that *A. normale* s.l. contains several species. The *Asplenium normale* complex was recovered as a monophyletic group comprising six principal chloroplast lineages. Two of the lineages corresponded to a single taxon: *A. kiangsuense* and *A. oligophlebium. A. normale* s.l. comprised the other four lineages in which except *A. jinpingense* formed a well separated diploid species, the other three lineages comprised both diploid and tetraploid taxa. Species classification recognizing six diploid and seven tetraploid entities in *A. normale* s.l. was outlined and species reticulate relationships in the *A. normale* complex were discussed.

T2

P0304

An account of Hornwort (Anthocerotophyta) in Malay Peninsula

Sahut Chantanaorrapint[1], *Kien-Thai Yong*[2], *Boon-Chuan Ho*[3]

1. Prince of Songkla University

2. University of Malaya

3. Singapore Botanic Gardens

A review of the division Anthocerotophyta in Malay Peninsula is presented based on herbarium specimens and field surveys.

Eleven species are recognized, belonging to 6 genera and 3 families. Of these 5 species are new records for Peninsular Malaysia: *Anthoceros punctatus* L., *Dendroceros javanicus* (Nees) Nees, *D. subplanus* Steph, *Folioceros amboinensis* (Schiffn.) Piippo, and *F. fuciformis* (Mont.) D.C.Bharadwaj. *Anthoceros angustus* Steph. is excluded from this area. The spore morphology through scanning electron micrographs of the species is provided. In addition, notes on the ecology and geographical distribution of the species are presented.

T2

P0305

Cryptic species complexes in Annonaceae: The case of *Marsypopetalum* and *Trivalvaria* (Malmeoideae: Miliuseae)

Tanawat Chaowasku[1], *Anissara Damthongdee*[2], *Daniel C. Thomas*[3]
1. *Chiang Mai University*
2. *Chiang Mai University*
3. *Singapore Botanic Gardens*

Annonaceae are a large pantropical angiosperm family composed of c. 2400 species of trees, shrubs and lianas classified in c. 108 genera. The family is a characteristic and ecologically important element in lowland rain forests. Although subfamilial, tribal and generic delimitations within the family have become more and more settled with the aid of molecular phylogenetic data, revisionary work clarifying species boundaries is still required for the majority of larger genera. In general, species numbers for many genera seem to have been underestimated possibly because only morphology has been taken into account for species delimitation and incomplete material sometimes resulted in merging of two or more distinct and only distantly related species. This can be well exemplified in the sister genera *Marsypopetalum* and *Trivalvaria* belonging to tribe Miliuseae of subfamily Malmeoideae. Using up to seven chloroplast DNA regions (*rbcL*, *matK*, *ndhF*, *ycf1* exons; *trnL* intron; *trnL-trnF*, *psbA-trnH* intergenic spacers; up to c. 7500 bp), phylogenetic analyses indicated that there is considerable cryptic diversity in both *Marsypopetalum* and *Trivalvaria*. Several distantly-related species have been mistakenly identified as *Marsypopetalum pallidum*. Similarly, *Trivalvaria costata* and *T. macrophylla* are polyphyletic. Thorough morphological investigations on these species complexes and putative cryptic species identified a suite of characters distinguishing several species, i.e. inflorescence position, petal morphology, monocarp shape and size, and pericarp thickness for *Marsypopetalum*; petal morphology, sexual system, and monocarp and stipe length for *Trivalvaria*. This study will serve as a basis for modern revisions incorporating comprehensive phylogenetic data in combination with classical morphological examination.

T2

P0306

Novelties in Asian Malmeoideae (Magnoliales, Annonaceae): A new tribe, a second genus of Monocarpieae, and a new genus of Miliuseae

Tanawat Chaowasku[1], *Tri Dung Ngo*[2], *Thai Hung Le*[3]
1. *Chiang Mai University*
2. *Institute of Resources & Environment, Hue University*
3. *Hue University College of Agriculture and Forestry*

Annonaceae are a pantropical flowering plant family prominent in lowland rain forests. The family is composed of c. 2400 species in c. 108 genera. Four subfamilies have been recognized, viz. Anaxagoreoideae, Ambavioideae, Annonoideae, and Malmeoideae. The last two subfamilies constitute the majority of diversity found in the family. Malmeoideae consist of seven tribes, viz. Piptostigmateae, Malmeeae, Maasieae, Fenerivieae, Monocarpieae, Dendrokingstonieae, and Miliuseae. The genus *Phoenicanthus* with two species endemic to Sri Lanka was inexplicably placed in the last tribe because it is a rare genus with only few complete specimens, profoundly hindering inferences on its affinity. Recent expeditions in Central Vietnam resulted in the discoveries of two unknown taxa: on the basis of gross morphology, the first one could belong to the canangoid clade of Ambavioideae or the Monocarpieae-Dendrokingstonieae-Miliuseae clade of Malmeoideae, whereas only a fallen single-seeded monocarp of the second one was available; later the seed germinated, revealing the leaves with a raised midrib on the upper surface. Molecular phylogenetic analyses of up to seven chloroplast markers (*rbcL*, *matK*, *ndhF*, *ycf1* exons; *trnL* intron; *trnL-trnF*, *psbA-trnH* intergenic spacers) revealed that *Phoenicanthus* (only one accession, *P. obliquus*, sequenceable) is recovered as the sister group of the Monocarpieae-Dendrokingstonieae-Miliuseae clade with strong support. Therefore, it undoubtedly deserves recognition as a new tribe in Malmeoideae, characterized principally by its miliusoid and highly reduced number of stamens. The first unknown Vietnamese taxon is recovered as the sister group of the genus *Monocarpia* with strong support. It primarily differs from *Monocarpia* in leaf intramarginal veins (absent or indistinct vs. conspicuous), inflorescence position (axillary vs. terminal), and monocarp appearance (cylindrical with longitudinal ridges vs. subglobose to ellipsoid without ridges). Based on these differences and the relative branch lengths of related lineages, this unknown Vietnamese taxon is better placed in a second genus of the monotypic Monocarpieae. The second unknown Vietnamese taxon belongs to the tribe Miliuseae, but its relationships to other genera are still largely obscure. These important novelties (especially in the case of *Phoenicanthus*) may shed light on the biogeographic scenarios of the large Asian-predominated clade (the new tribe plus Monocarpieae-Dendrokingstonieae-Miliuseae) of Malmeoideae when more sequence data are available.

T2

P0307

Bryophytic diversity in the agricultural landscapes of Pantnagar located in the terai of Kumaun Himalayas, India

Sheetal Rana, ***Preeti Chaturvedi***
G.B.Pant University Of agriculture & Technology

Bryophytes are the second largest group of plant kingdom with about 34,556 species around the world (according to The Plant List, 2013). Although having such high diversity, bryophytes of plains particularly those of agriculture landscapes in the foot hills of Kumaun Himalayas are rarely included in biodiversity surveys. As the major emphasis in the Himalayan Terai is on cultivation of crops, cryptogamic diversity in the region is highly neglected. Moreover, human interference in plains of Terai especially the study area, Pantnagar (longitude, E 79° 28' 33" - 79° 31' 12" and latitudes N 28° 59' 36" - 29° 2' 34", altitude-213-238m) also becomes a limiting factor. Pantnagar is an agriculture dominated area. Natural vegetation of the place has been mostly replaced by huge agricultural farms and cultivated plantations. Still, there are some bryophytes that are surviving, in some microhabitats. Therefore, conservation of bryophytes in these disturbed habitats and identification of those plant species is of utter importance to conserve the regional flora, to maintain the ecological balance and preserve the biodiversity. In the present study, total of 24 species belonging to19 genera of 12 families in 7 orders were recorded from the study area. 19 species in 15 genera of 8 families and 6 orders belonged to mosses whereas liverworts were represented by 5 species belonging to 4 families and 1 order only. Pottiaceae and Bryaceae were the dominant families in the study area with 5 species each followed by Funariaceae, Fissidentaceae and Ricciaceae with 3,2 and 2 species respectively. Dominant genera of the region included *Physcomitrium*, *Fissidens*, *Hydrogonium*, *Gemmabryum* and *Riccia* with 2 species each. Most of the bryophytes were terrestrial, only two were epiphytes.

T2

P0308

A preliminary assessment of DNA barcodes for identifying *Setaria* (Poaceae) species in Taiwan

Chih-Hui Chen, *Hsien-Chun Liao*

Endemic Species Research Institute

There are eight wild species of *Setaria* in Taiwan, including *Setaria viridis* which has been developed as model systems of C4 photosynthesis for a few years. Basically, it is not difficult to distinguish them with examining their morphological characters except an interesting intermediate form between *S. viridis* and *S. verticillata* we found in Taiwan recently. In this study, we assessed the feasibility of chloroplast coding regions *rbc*L and *mat*K, chloroplast intergenic spacers *trn*H-*psb*A and *trn*L-F, and nuclear ITS regions as DNA barcodes for identifying the species. Totally 103 samples of the eight species were analyzed. The average lengths and number of parsimony-informative sites are 669 and 149 for ITS, 876 and 27 for *mat*K, 855 and 30 for *rbc*L, 488 and 10 for *trn*H-*psb*A, 846 and 26 for *trn*L-F. Among them, ITS has the highest percentage of parsimony-informative sites and provides the best resolution for identifying *Setaria* species of Taiwan. Nevertheless, in all phylogenetic trees based on the five DNA barcodes, *S. viridis*, *S. verticillata* and the intermediate form between them were grouped and mixed up in one clade. Regarding the other six species, we could identify them with the phylogenetic trees based on ITS, *mat*K, *rbc*L and *trn*L-F. Due to the poor percentage of parsimony-informative sites, we couldn't use *trn*H-*psb*A as DNA barcode to identify the *Setaria* species. Further studies for elucidating the relationship between *S. viridis* and *S. verticillata* are necessary.

T2

P0309

The taxonomy of *Curcuma* from China

Juan Chen, *Nianhe Xia*

South China Botanical Garden, Chinese Academy of Sciences

Curcuma L. (Zingiberaceae), with at least 120 species, has been used frequently as spices, medicines, dyes and ornamentals. The genus is mainly distributed in south and south-east Asia. The genus has considerable economic importance; however, its phylogeny and taxonomy remain poorly understood. Diagnostic characters used for *Curcuma* species (rhizome color, position of inflorescences, color and shape of bracts and flower parts) are neither unique nor universal to all members of the genus. Meanwile, owing to scanty protologues and a lack of type specimens, many commonly used and cultivated species, are still unclear. Many synonyms, especially those established solely on the study of original discriptions and herbarium specimens, are unreliable. Therefore, the identity of *Curcuma* species form China is rechecked and discussed. Lectotypes are designated for previously untypifications names. Invalid names are discussed. Notes on critical taxa are provided.

T2

P0310

Lineage diversification and evolution of pollinator trapping in *Artabotrys* (Annonaceae)

Junhao Chen[1], *Daniel Caspar Thomas*[2], *Richard Mark Kingsley Saunders*[1]

1. The University of Hong Kong

2. Singapore Botanic Gardens

Evolutionary diversification is influenced by disparate aspects of the biology, ecology and geographical distribution of species. Robust, well-sampled molecular phylogenies have the potential to test which environmental and species attributes likely contributed as evolutionary drivers. Our project investigates lineage diversification and the evolution of pollinator trapping in the species-rich, lianescent genus *Artabotrys* (c. 100 species; Annonaceae), a characteristic element in Old World tropical and subtropical lowland forests. We are reconstructing the phylogeny of the genus based on an extensive taxon sampling (c. 50 species) and a broad range of chloroplast (*matK*, *ndhF*, *psbA-trnH*, *trnL-F*) and nuclear (*AP3*, *LFY*, *ncpGS*, *PhyA*, *RPB2*) DNA markers. Preliminary results indicate that early diversification in the genus resulted in lineages that are markedly asymmetrical in species numbers: the main species-rich *Artabotrys* clade (MAC), characterised by flowers with basally concave petals that form a tightly enclosed pollination chamber relative to the species-poor early-divergent grade (comprising *A. brachypetalus* and *A. thomsonii*). We hypothesise that (i) the floral chamber in the MAC may function as a pollinator trap, with the pollinators attracted by floral scents but subsequently prevented from leaving the flowers by the closure of the chamber, and (ii) that pollinator trapping is synapomorphic for the MAC, and the putative 'xylopioid' traits in *A. brachypetalus* and *A. thomsonii* (separation of inner petals at anthesis and extended duration of the apical connivence of petals during floral development) may be plesiomorphic. To investigate these aspects, pollination ecology studies of exemplar species (*A. brachypetalus* in the early-divergent grade, and *A. hongkongensis* and *A. suaveolens* in the MAC) are being conducted, involving assessments of floral phenology, thermogenesis, scent chemistry, pollinator identification, and pollinator trapping; character mapping will be performed to further elucidate pollination chamber evolution and the morphological basis of pollinator trapping. Our preliminary field studies of *A. hongkongensis* in Hong Kong reveal that the flowers are protogynous, with anthesis over c. 27 hours, and are likely to exhibit pollinator trapping, although the effective pollinator is still to be unequivocally determined. The majority of *Artabotrys* species possess broad, flat, lanceolate petals (e.g., *A. hongkongensis*); narrow, terete, filiform petals have evolved independently in several MAC lineages (e.g., *A. suaveolens*), and may represent an adaptation to a different pollinator guild. Diversification rate analyses will be performed to investigate if pollinator trapping represents a key evolutionary innovation correlated with accelerated diversification in the MAC.

T2

P0311

Genetic differentiation and adaptation of *Sonneratia alba* revealed by population genomics analyses

Qipian Chen, *Jianfang Li, Suhua Shi*

State Key Laboratory of Biocontrol and Guangdong Provincial Key Laboratory of Plant Resources, Sun Yat-sen University

A series of historically demographic events have affected the extant population structure of organisms, which could provide important clues for investigating the population evolutionary history and speciation. Indo-West Pacific (IWP) region is a biodiversity center of mangroves, where the correlation between genetic divergence and complex geography pattern is of biological significance and great interests. In this study, we sequenced the whole genomes of 33 individuals from 7 populations across the IWP region to assess the population structure of *Sonneratia alba*. Our results revealed that, at genome level, *Sonneratia alba* hosts low levels of genetic diversities within each population, as well as strong genetic divergence among populations from different oceanic regions, indicating sea-level fluctuation during the ancient Pleistocene glacial periods. Moreover, population near the Malacca Strait showed an apparent genetic admixture of divergent lineages from the East Indian Ocean and the South China Sea, suggesting the secondary contact at the Indo-Pacific boundary region during interglacial periods. Notably, results of McDonald–Kreitman test revealed excess fixed difference in amino acid change between populations in South China Sea, especially within the highly divergent genes between two Hainan populations. Therefore, those highly divergent genes across the whole genome of *Sonneratia alba* might play a major role in local adaptation. Our study not only provided new insights into genetic variations but also enriched our current understanding of the adaptive evolution of mangroves.

T2

P0312

The comparison and conservation of species diversity of *Begonia* (Begoniaceae) in Dian-Qian-Gui-Tonkin and Borneo regions, two biodiversity hotspots

Wen-Hong Chen, *Radbouchoom Sirilak, Yu-Min Shui*

Key Laboratory for Plant Diversity and Biogeography of East Asia, Kunming Institute of Botany, the Chinese Academy of Sciences, Kunming 650204, P.R. China

Dian-Qian-Gui-Tonkin region which belongs to Holarctic and Palaeotropical Kingdoms is located in south China and north Vietnam, the region is one of the biodiversity centers in the word and dominated by the karst landforms. Borneo region which belongs to the Palaeotropical Kingdom is located in Indonesia, Malaysia and Brunei, and is one of the largest islands and of the biodiversity centers in Asia and worldwide. In this study, we focus on the biodiversity and conservation of the genus *Begonia* L. which occurs in the above two regions. *Begonia* is a mega-diverse genus of Begoniaceae containing more than 1800 species in the world, of which 169 species and 226 species are respectively distributed in Dian-Qian-Gui-Tonkin region and in Borneo region. There are 11 species described in Dian-Qian-Gui-Tonkin region (*vs.* 5 spp. in Borneo region) before 1900,36 species (*vs.* 23 spp.) from 1901 to 1950,9 species (*vs.* 12 spp.) from 1951 to 1980,3 species (*vs.* 8 spp.) from 1981 to 1990,40 species (*vs.* 20 spp.) from 1991 to 2000,47 species (*vs.* 46 spp.) from 2001 to 2010 and 23 species (*vs.* 114 spp.) from 2011 to 2016. With refer to the *Begonia* species in Dian-Qian-Gui-Tonkin region, there are ca. 47 new species described from 1991 to 2000, accordingly because of the fulfilment of "*Flora Reipublicae Popularis Sinicae*" and "Flora of China". In Borneo, from 2011, the discovery of new species of *Begonia* increased from 112 to 226 species, the largely increased species number is mainly due to the extensive exploration of the genus by the local botanists, a unique case of species extension in the tropics. Taxonomically, *Begonia* sect. *Petermannia* (Klotzsch) A. DC. and *B.* sect. *Coelocentrum* Irmscher, the two dominant groups respectively in the tropical forests of Borneo and at the karst habitat of Dian-Qian-Gui-Tonkin region, undergo possible rapid radiation under the different geographical background during the recent global climate optimum. However, the fragility and loss of the karst habitat and tropical forests will considerably threaten their biodiversity so that numerous endemic species will be limited in small distribution and under a small population.

T2

P0313

Molecular systematics of Rosoideae (Rosaceae)

Xun Chen[1,2], *Jinlu Li*[1,2], *Tao Cheng*[2], *Wen Zhang*[2,3], *Yanlei Liu*[2,3], *Shiliang Zhou*[2], *Ling Wang*[1]

1. Northeast Forestry University

2. Institute of Botany, the Chinese Academy of Sciences

3. University of Chinese Academy of Sciences

The four-subfamily subdivision of Rosaceae has been recently overthrown and a new classification of Rosoideae remains to be established on a basis of solid phylogeny. In this study, we sampled all 54 genera presumably belonging to Rosoideae and 12 genera belonging to other subfamilies or families and used twelve chloroplast regions (*mat*K, *rbc*L, *atp*B-*rbc*L, *trn*L, *trn*L-F, *ndh*F, *ycf*1, *trn*C-*ycf*6, *trn*S-G, *trn*S, *psb*A-*trn*H, *rpo*C1, and *trn*S-*ycf*9) to reconstruct their phylogeny. Our results demonstrated (1) exclusion of *Rhodotypos* and *Kerria* from Rosoideae and their inclusion in subfamily Spiraeoideae; and (2) exclusion of *Cowania*, *Chamaebatia*, *Cercocarpus* and *Dryas* from Rosoideae and inclusion in Dryadoideae, a sister subfamily of Rosoideae. There are six strongly supported lineages within Rosoideae which corresponding to six tribes, Potentilleae; Roseae; Sanguisorbeae; Rubeae; Courieae and Filipenduleae. Divergence of the subfamily was dated back to some 69.77 million years ago (Mya; 95% HPD = 61.28-78.33 Mya). And the divergence time within Rosoideae was dated from10.42 to 40.02 million years (Mya; 4.73-59.08 Mya), which formed the six lineages (Potentilleae crown 45.02 (Mya, 95% HPD =38.39-50.12 Mya); Courieae crown 44.94 (Mya, 95% HPD = 28.50-59.08 Mya);Sanguisorbeae crown 40.63 (Mya; HPD = 30.42-49.16 Mya); Rubeae crown 18.37 (Mya; HPD = 9.02-32.3 Mya); Filipenduleae crown 13.47 (Mya; HPD = 5.07-26.32 Mya); Roseae crown 10.42 (Mya; 95% HPD = 4.73-19.74 Mya)). The subfamily is probably of North American origin but thrives in north hemisphere, especially in Asia.

T2

P0314

The complete chloroplast genome sequence of Five *Smilax* species (*Nemexia*): Comparative analysis, highly divergent regions and phylogenetic relationships

***Yao Chen*[1], *Pan Li*[1], *Chengxin Fu*[1], *Kenneth Cameron*[2]**

1. College of Life Sciences, Zhejiang University, Hangzhou, Zhejiang 310012, China

2. Department of Botany, University of Wisconsin, Madison, Wisconsin 53706, USA

Smilax, the only genus in the greenbrier family (Smilacaceae), contains > 200 species of mostly woody climbers around the world. However, a clade of *Smilax* [sect. *Nemexia* (Raf.) A. DC.] are annual understory herbs with a unique eastern Asia (EA)-western North America (WNA)-eastern North America (ENA) disjunct pattern (two, one and eight species in these three areas, respectively). Previous studies revealed a sister relationship between Asian clade (*S. nipponica* + *S. riparia*) and North American clade (WNA *S. jamesii* + ENA clade), but failed to resolve the phylogenetic relationships within the eastern North American herbaceous *Smilax*. In order to reconstruct the relationships among them, here we report five complete chloroplast (cp) genomes of herbaceous *Smilax* (*S. nipponica, S. jamesii, S. herbacea, S. hugeri and S. biltmoreana*). The cp genomes of *Smilax* range from 157,049 bp to 157,656 bp in length, all including a pair of inverted repeats (IRa and IRb) separated by the large single-copy (LSC) and small single-copy (SSC) regions. The genomic comparison showed some differences located in the genomes, and IRs regions were more conservative than LSC/SSC regions. Phylogenomic analyses based on these complete cp genomes strongly support a topology of (*S. nipponica*, (*S. jamesii*, (*S. herbacea*, (*S. hugeri*, *S. biltmoreana*)))). Our findings proved the feasibility of employing cp genomes to infer phylogenetic relationships among closely related species. Besides, eight DNA markers (*psaC–ndhE, ndhC–trnV, ndhG–ndhI, ydfI* intron, *accD–psaI, ccsA–ndhD, petA–psbJ, atpI–rps2*) with number of variable sites greater than 1.0% were identified, and these may be useful for future population genetic and phylogeographic studies.

T2

P0315

Taxonomy of *Plagiochila* in China

Xiafang Cheng

East China Normal University

Plagiochila (Dumort.) Dumort. is one of the largest genera of liverworts and its diversity centers are in Southeast Asia and the Neotropics. It currently contains ca. 700 species in the world, and 80 species occur in China. Molecular studies in recent years have led to new insights into the sectional classification and distribution of *Plagiochila*. Contrary to earlier recognition, many sections are proved to have intercontinental ranges, resulting in the synonyms of some sections. *Plagiochila* are now divided into 29 sections according to recent morphological and phylogenetic analyses, in which 12 sections occur in China. However, there are only 23 Chinese species being included in former molecular studies, which may cause unreasonable sectional classification. Here we conducted phylogenetic analyses of a dataset comprising nrITS and two chloroplast markers (*rps*4, *rbc*L) from 150 species with 57 Chinese species. Combination with the morphological evidence, the results proposed a natural classification of sectional belongings of *Plagiochilla* in China, and suggested several new sections and reinstatements of some synonymized sections in this study.

T2

P0316

Ecological significance of phenology and functional traits of mistletoes in Taiwan, R. O. C.

Shau-Ting Chiu

Division of Biology, National Museum of Natural Science

Most of mistletoes in Taiwan, including 4 genera about 17 species, are dispersed by flowerpeckers and related birds. The phenology of the mistletoes are mutually affected due to the close relationship between the dispersal of mistletoes and the diet of flowerpeckers. For example, *Taxillus tsaii* S.-T. Chiu at low altitude flowers from March to July and bears fruits from June to November. Its major dispersal agent is *Dicaeum concolor* and, in a short season, *D. ignipectus* also participate in dispersal. Certain species at middle elevation, such as *T. rhododendricolus* (Hayata) Danscer., flowers from August to October and bears fruits from October to December. The foraging of their major dispersal agent, *D. ignipectus,* seemed to make for each other with concorde rhythm. In comparison, some seeds contained early emerging seedling when seed dispersal for aggressive invasion to guaranteed effective establishment on host plants. It suggests that the phenoloty, the dispersal and the seed germination in the life cycle of mistletoes are diverse because they have adapted to the different ecosystems with varying temperatures and lengths of the growing seasons. Not only are the phenology and the distribution of the mistletoes in Taiwan highly diversifies, but the hosts species the hosts are also highly varied, ranging from gymnosperms, angiosperms, trees, shrubs and lianas, to even other mistletoes as examples of hyperparasitism. The highest species diversity in Taiwan so far is 4 species per acre and 4 species per host individual. Further investigation may reveal an even higher diversity; especially, functional traits for ecological performance. Establishing morpho-physiophenological traits which impact fitness indirectly via their effects on growth, reproduction and survival, the three components of individual performance can become the foundation of ecological forensics. This study would initiate leaf functional traits about foliar mistletoes along altitudinal gradients for selected habitats. Carrying on the standard protocol of leaf functional traits can input, provide and accumulate long-term ecological forensics data. Mistletoes which grew in distinct micro-environment and included different host functional groups living different habitat, showed varied combination of leaf traits and phenology for the strategy of fitness. It would carry on the important mission of long-term monitoring and also provide renewal conservation information for Botanical Garden's and National Park's management and long-term ecological research in conservation.

T2

P0317

Several molecular markers of chloroplast and nuclear genome reveal the origin and genetic structure of wild *P. yedoensis* (Rosaceae), endemic to Jeju Island, Korea

Myong-Suk Cho[1], *Chan Soo Kim*[2], *Gwanpil Song*[3], *Chul Hyun Jeon*[2], *Seung-Chul Kim*[1]
1. Department of Biological Sciences, Sungkyunkwan University, Suwon, Korea
2. National Institute of Forest Science, Warm-Temperate and Subtropical Forest Research Center, Seogwipo, Jeju-do, Korea
3. Jeju Biological Resources Co., Ltd., Jeju, Jeju-do, Korea

Wild flowering cherry of *Prunus yedoensis* (Rosaceae), naturally occurring on Jeju Island, Korea, has been suggested as of natural hybrid origin from two sympatric *Prunus* lineages on Jeju Island. However, its taxonomic distinctiveness from widely cultivated *P. ×yedoensis* 'Somei-yoshino' remains as highly controversial and enigmatic. This study aims to establish the taxonomic identity of wild *P. yedoensis* by clarifying the phylogenetic relationships among species of *Prunus* as well as to determine genetic structure of wild *P. yedoensis*. We took molecular phylogenetic and nuclear SSR genotyping approaches based on extensive sampling of both wild and cultivated *Prunus* species in Japan, Russia and Korea, including putative parental species. The phylogenetic relationships were elucidated based on nuclear (ITS, G3pdh, POLA1, and several RosCOS putative single copy genes) and chloroplast genome (7 noncoding regions and whole cp genome for two accessions). The cpDNA and nrDNA phylogeny strongly supported the previous study that wild *P. yedoensis* in Jeju Island is of hybrid origin with maternal parent of *P. spachiana* f. *ascencens* and precisely undetermined paternal parent from *P. serrulata* and/or *P. sargentii* complex. The cpDNA haplotype network suggested independent geographical contribution to the origin of wild and cultivated *P. yedoensis* in Jeju Island and Japan, respectively, which was reconfirmed by nrDNA phylogenetic analyses by ITS, POLA1, and RosCOS single copy genes. Several nuclear loci further indicated independent origin of wild and cultivated *P. yedoensis*. Nuclear SSR genotyping based on 83 individuals of wild *P. yedoensis* and 12 of *P. ×yedoensis* 'Somei-yoshino', revealed two genetic groups, corresponding to wild and cultivated *P. yedoensis*. Totally 171 alleles were found at 12 loci, ranging from 6 to 24 alleles per locus, and the population of wild *P. yedoensis* demonstrated much higher numbers of alleles (mean 13.222) comparing to the cultivated *P. ×yedoensis* 'Somei-yoshino' (mean 1.778). Of 83 presumably wild individuals, we found misidentified individuals (2), admixture (9) and unknown genetic contribution (3). We also found significantly higher genetic diversity in wild *P. yedoensis* in Jeju Island than in cultivated one. This study further corroborates our earlier finding that wild *P. yedoensis* in Jeju Island is a distinct taxonomic entity from cultivated *P. ×yedoesnsis* 'Somei-yoshino'. The evidence of gene flow between wild *P. yedoensis* and cultivated *P. ×yedoensis* 'Somei-yoshino' is also detected.

T2

P0318

Phylogenetic relationships of *Pyrrosia* species (Polypodiaceae) inferred from nuclear and chloroplast DNA sequences

Tae Young Choi, *Chong-Wook Park*
Seoul National University, School of Biological Sciences

The genus *Pyrrosia* Mirb. (Polypodiaceae) comprises about 50 species. Most species within the genus are highly variable in morphology, resulting in taxonomic confusion in determining the boundaries and relationships of the species. We analyzed nrDNA *pgi*C region and cpDNA *rbc*L gene, *trn*L– *trn*F IGS, *atp*F –*atp*H IGS sequences from 31 taxa of the genus *Pyrrosia* to elucidate their phylogenetic relationships, and to investigate the parentage of putative hybrid taxa. Maximum parsimony and Bayesian inference analyses of the nrDNA and cpDNA sequence data showed three major clades: 1) a clade containing widespread *P. porosa* and the east Asian species, 2) a clade containing *P. lingua* and species found primarily in southeast Asia, 3) a clade consisting of *P. stigmosa* and *P. costata*. These groupings, however, appear to be incongruent with the traditional infrageneric classification scheme based on blade shape and spore morphology.

T2

P0319

Understanding the diversification of *Eriocaulon* L. in South Asia using molecular phylogenetic approach

Ashwini Mukund Darshetkar[1], *Mandar Nilkanth Datar*[1], *Shubhada Ajit Tamhankar*[1], *Joongku Lee*[2], ***Ritesh Kumar Choudhary***[1]
1. Agharkar Research Institute
2. Chungnam National University

The genus *Eriocaulon* is distributed throughout tropical and subtropical regions of both hemispheres and has more than 525 taxa; about 485 species and more than 40 infra-specific taxa. South Asia harbors almost one fourth of the Eriocaulon diversity, having ca. 100 species. With its rich concentration in Asia, Africa and S America, the genus is known for its high intraspecific variability combined with the limited interspecific differences; producing big challenge to the classical taxonomists to properly identify/characterize/delimit them. Present study is a pioneer attempt to understand the diversification of *Eriocaulon* in South Asia using molecular phylogenetic approach. Further, attempts have been made to understand the pattern of morphological character evolution and phylo-geographic distribution which is an important adjunct to the molecular phylogeny of this genus. Using morphological as well as the nrDNA and cpDNA sequence data, this study has resulted in (i) unravelling the species relationships within this genus, (ii) tracing the geographic distribution pattern with special reference to the endemic and rare species, (iii) elucidating morphological character evolution in the genus (iv) testing the congruence of morphological and molecular data in systematic and phylogenetic assessment, and (v) development of possible DNA barcode of the endemic *Eriocaulon* species for their true identification and delimitation.

T2

P0320

Phylogenetic study of genus *Homalomena* section *Chamaecladon* (Araceae: Homalomeneae)

Kek Shen Chua, *Sin Yeng Wong*
Universiti Malaysia Sarawak

Phylogenetic relationship among species of the genus *Homalomena* section *Chamaecladon* (Araceae) remains unclear as previous study on this genus emphasized on section *Cyrtocladon*. For examining monophyly of section *Chamaecladon*, DNA nuclear gene internal transcribed spacer (ITS), and chloroplast gene *maturase* K (*mat*K) sequences were analyzed. A total of 21 samples including several unknown species, from different localities (Borneo, Peninsular Malaysia, and Sumatera), were assessed in this study. Collection of taxa was conducted in nursery of Semenggoh Botanical Research, Kuching. Throughout the study, thirty nine sequences were newly generated as three unsound ITS sequences were not included. For outgroup species, one species from each sections under genus *Homalomena* (*Homalomena*, *Cyrtocladon* and *Geniculatae*), and one species from genera *Philodendron*, *Adelonema*, *Cercestis* as well as *Culcasia* were selected. Three phylogenetic analyses, namely maximum parsimony (MP), maximum likelihood (ML) and Bayesian analysis, were utilized to construct phylogenetic trees. The resulting phylogeny shown *Homalomena* Schott was monophyletic. However, section *Chamaecladon* was paraphyletic owing to the inclusion of *H. curvata* (sect. *Homalomena*) in the clade. In *mat*K analyses, there were two clades formed in MP analysis of section *Chamaecladon*, Clade I with all ingroup species and Clade II with *H. curvata*, *H punctulata* and other ingroup species. Species in Clade II remained unresolved in ML and Bayesian analysis of *mat*K. In ITS analyses, there were three clades formed in section *Chamaecladon*, namely *Griffithii* clade, *Paucinervia* clade, and *Humilis* clade.

T2

P0321

Polyploidy and biogeography of subtribe Agrimoniinae (Rosaceae)

Kyong-Sook Chung
Jungwon University

Polyploidy is an abundant phenomenon in flowering plants that has provided an important pathway for diversity and evolution. A small taxonomic group, subtribe Agrimoniinae in Rosaceae, exhibits various levels of polyploidy associated with diverse life (habit) forms, morphological characters, and geographic distribution patterns. Inferred from nucleotide markers (*trn*L-F, *trn*G-S, and GBSSI), cytological, morphological, and biogeographic evolution of Agrimoniinae are evaluated. Subtribe Agrimoniinae comprises one widely distributed genus and four monotypic endemic genera: *Agrimonia* L. (about 20 species mainly in Northern Hemisphere), *Aremonia* Neck. ex Nestl. (Europe), *Leucosidea* Eckl. & Zeyh. (Africa), *Hagenia* J. F. Gnek. (Africa), and *Spenceria* Trimen (E. Asia). Two African genera are woody (shrubs or trees) whereas the other three genera are herbaceous. They exhibit polyploidy in various levels from a diploid to octoploids: *Hagenia* (tetraploid, 4x), *Aremonia* and *Leucosidea* (hexaploid, 6x), *Agrimonia* (4x and 8x), and *Spenceria* (diploid, 2x). All the sequence data support that Agrimoniinae is monophyletic, and *Spenceria* is basal to the rest of group. *Spenceria* (2x) is considered to be a relict species, occurring only in high mountains in China and Bhutan. In *Agrimonia*, North American tetraploids (*A. microcarpa* and *A. parviflora*) and an Asian tetraploid (*A. coreana*) form a strongly supported clade, and diverse levels of polyploids (4x, 6x, and 8x) of Asian, American, and European species are clustered. The tetraploid clade shares leaf and fruit morphology and restricted distribution. Unusually in Asia, endemism is found in tetraploids (*A. coreana* and *A. nipponica*) as well as in an octoploid (*A. gorovoii*) species. Fruiting hypanthia with hooked bristly epicalyxes in *Agrimonia* support allopatric speciation in the genus. The result suggests multiple dispersal events in diverse time periods following and/or followed polyploidization.

T2

P0322

Chromosome number variation and species diversity of East Asian *Carex* (Cyperaceae)

Kyong-Sook Chung[1], *Takuji Hoshino*[2], *Tomomi Masaki*[2], *Hyoung-Tak Im Hyoung-Tak Im*[3], *Jong-Cheol Yang Jong-Cheol Yang*[4], *Seong-Jin Ji Seong-Jin Ji*[5], *Hyuk-Jin Kim Hyuk-Jin Kim*[4]
1. Jungwon University
2. Okayama University of Science
3. Chonnam National University
4. Korea National Arboretum
5. National Science Museum

Carex L. (Cyperaceae) is one of the most species-rich genera in flowering plants with more than 2,000 species worldwide, exhibiting a great chromosome number variation ranged from $2n = 12$ to $2n = 168$. The exceptional variation in chromosome numbers has been postulated due to a holocentric chromosomal structure. Previously reported chromosome numbers representing approximately 16% of the species in Cyperaceae manifest an apparent variation patterns for aneuploidy, agmatoploidy (chromosomal fission without gene duplication) and polyploidy; and a series of studies have proposed that chromosome divergence has played an important role in species diversity within the *Carex*. In East Asia vegetation, the genus is an important component occupying various geological and ecological habitats. In addition, recent phylogenetic studies revealed that early *Carex* linages diverged from East and Southeast Asia. To understand species diversity in East Asian *Carex*, chromosome number variations in the regions are evaluated in traditional as well as phylogenetic classifications. Sections with high species such as *Mitratae* (ca. 50 species in East Asia) and *Phacocystis* (ca. 40 species in East Asia) exhibit high chromosome numbers with broad variations ranged from $2n = 32$ to $2n = 90$. However, basal groups of each lineages usually have low chromosome numbers. The early lineage Siderostictae clade (*Siderostictae*, *Surculosae*, and *Hemiscaposae*) demonstrates polyploidy ($2n = 12$ or 24) rather than aneuploidy or agmatoploidy. The clade also shows large chromosomes from 1.5 to 4.0 μm long. The presentation explains diverse chromosome evolution mechanisms (aneuploidy, agmatoploidy, and polyploidy) in *Carex* species diversification, and demonstrate examples of East Asian taxonomic and/or phylogenetic groups.

T2
P0323
Taxonomic revision of the genus *Verbascum* L. (Group D) in Turkey

Burcin Cingay[1], *Ahmet Faik Karaveliogullari*[2], *Birol Baser*[3], *Sezgin Celik*[4]

1. Nezahat Gokyigit Botanic Garden, Department of Science, Istanbul

2. Hakkari University, Yuksekova Vocational High School, Hakkari

3. Bitlis Eren University, Art and Science Faculty, Department of Biology

4. Yıldız Teknik University, Department of Biology Istanbul

In this study, the morphology, ecology, palinology and the seed surfaces of the "Group D" of the genus *Verbascum* L., in Turkey were investigated and new modifications with combinations were carried out for three taxa of the group. Moreover, new version of the descriptions of the genus in the Flora of Turkey has been done. During the field studies, there was no hybrid taxon observed. It is clear that, the number of stamens is not a sufficient character to classify the groups of the genus as given in the Flora of Turkey. The results of the study have showed that the pollen grains belonging to the species in Group D of *Verbascum* L. are tricolpate. The shapes of the pollens are prolate spheroidal and the structure of them is tectate. Furthermore, the ornemantion is reticulate. The narrow colpi of the pollens have recognizable margins. The group shapes of the seeds are varying as ovate and oblong and the seed surfaces are varying as reticulate and alveolate- reticulate, are brown in all the species. It has been seen that, there is a significant difference in terms of the physical characters, that was also calculated as statistically ($p<0.001$). However, there is no significant difference in terms of chemical values of the soils among the soil samples.

T2
P0324
Leaf architecture of selected *Diplazium* species (Athyriaceae)

Jennifer Conda[1], *Inocencio Buot Jr.*[2]

1. Forest Products Research and Development Institute-Department of Science and Technology

2. University of the Philippines Los Banos

The diversity of characters and absence of sori during some season of the year made the identification of *Diplazium* species unresolved. Thus, leaf architecture study of the genus *Diplazium* was conducted to determine the usefulness and effectiveness of leaf characters, particularly, venation pattern in describing and differentiating fern species and to identify leaf architectural characters useful in taxonomic classification of pteridophytes. Herbarium specimen of twenty seven species of *Diplazium* at the Philippine National Herbarium were examined following the work of Hickey (1973), Leaf Architecture Working Group (1999) and book of Palmer (2003). Seven to fifteen pinna/pinnule per species were measure using ruler and protractor. The *Diplazium* species were then described and a dichotomous key was made. Through cluster and principal component analysis *Diplazium* species were divided into two major group namely: free venation group and anastomosing venation group and six clusters. The most unifying features of *Diplazium* included lanceolate pinna or pinnule, acuminate and acute apex, asymmetrical base and pinnate primary vein. Finer venation such as secondary vein angle of divergence and variation in 2° vein angle of divergence, tertiary vein category, 3° vein angle of divergence, variation in 3° vein angle of divergence, 3° vein spacing and lobation were stable and relevant features in differentiating closely related *Diplazium* species. This study proved the usefulness of leaf architecture in describing and differentiating fern taxa specifically *Diplazium* species just like many angiosperms. Leaf architecture, indeed, can be an alternative taxonomic tool of sterile fern species.

T2
P0325
Correcting phylogenetic signal of homoplastic traits: The evolutionary history of perianth fusion in Myrtaceae

Thais Vasconcelos, Gerhard Prenner, Eve Lucas, ***Maria Conejero***

Royal Botanic Gardens, Kew

Shedding a fused perianth in the flowering bud as a calyptra during anthesis is a major evolutionary trend in Myrtaceae. The presence/absence of this character has been of historical systematic interest in at least 15 genera, including the megadiverse *Myrcia*, *Eucalyptus* and *Syzygium*. However, molecular phylogenies, very often do not support the calyptra as a strongly informative character. Evidence suggests that this trait shows low phylogenetic signal and has repeatedly appeared and disappeared during the evolutionary history of the family. In this study, we compare the development and anatomical structure of calyptrate flowers in distinct lineages of Myrtaceae to answer the following questions: 1) Is the calyptra homologous in the family? and 2) What is the variation in the trait's phylogenetic signal before and after the homology test? Twelve species, including representatives from all lineages of calyptrate taxa and their non-calyptrate closest relatives, were sampled and distinct stages of flower development were analysed under SEM and LM. Images were compared between species and are discussed to define the level of homology in the Myrtaceae calyptra. Perianth behaviour during anthesis, a possible source of misperception of this character, was also recorded in the field. Results of the homology test are plotted against a previously published Myrtaceae phylogeny for phylogenetic signal estimation using the *geiger* package in R. Results show that the calyptra is not a homologous structure throughout Myrtaceae lineages. What has been called a calyptra corresponds to at least three anatomically and ontogenetically distinct structures, which are here coined "calycine calyptra", when the calyx presents true gamosepally; "coralline calyptra", when the corolla presents true gamopetally; and "pseudocalyptra", when the free petals overlap during development and attach their surfaces, leading them to fall as a single unit that resembles a fused perianth. The homoplastic pattern of calyptrate structures persists in the phylogeny even after assessment of homology. However, estimates before and after homology tests show that phylogenetic signal increases when homologous structures are clarified, going from λ = 0.634, when presence/absence of calyptra is plotted without any distinction, up to λ = 0.651, λ = 0.782 and λ = 1 when the "pseudocalyptra", "calycine calyptra" and "coralline calyptra" are considered independently. Results presented here demonstrate that careful morphological analysis can improve the evolutionary understanding of a trait even when homoplasy is not completely resolved. Characters previously considered systematic and evolutionary irrelevant can present strong phylogenetic signal, once their structural details are clarified and re-incorporated into an existing phylogenetic framework. Unassessed homoplastic traits might hide phylogenetic signal relevant to understand large scale trait evolution and key-innovation in individual lineages. In Myrtaceae, convergences and parallelisms of a calyptrate structure are perhaps related to the brush blossom display, which reduces selective pressure of pollinator attraction on the perianth and frequently allows it to be disposable at anthesis.

T2

P0326

Consequences of being flexible: The mean and variability of a floral trait have opposing effects on pollination and fitness traits

Can Dai[1], *Xijian Liang*[1], *Jie Ren*[2], *Laura Galloway*[2]

1. Hubei University

2. University of Virginia

Floral traits are essential for assuring successful pollination and reproduction in flowering plants. In particular, style and anther positions are key for pollination accuracy and efficiency. Variation in these traits among individuals has been well studied, but less is known about variation within flowers and plants and its effect on pollination and reproductive success. Style deflexion is responsible for herkogamy and important for pollen deposition in *Passiflora incarnata*. The degree of deflexion may vary among stigmas within flowers as well as among flowers. We measured the variability of style deflexion at both the flower and plant level. The fitness consequences of the mean and variation of style deflexion were then evaluated under natural pollination by determining their relationship to pollen deposition, seed production, and average seed weight using structural equation modeling. In addition, the relationship between style deflexion and self-pollen deposition was estimated in a greenhouse experiment. We found greater variation in style deflexion within flowers and plants than among plants. Variation of style deflexion at the flower and plant level was positively correlated, suggesting variability in style deflexion may be a distinct trait in *P. incarnata*. Lower deflexion and reduced variation in that deflexion increased pollen deposition, which in turn, increased seed number. However, lower styles also increased self-pollen deposition. In contrast, higher deflexion and greater variability of that deflexion increased variation in pollen deposition, which resulted in heavier seeds. Variability of style deflexion and therefore stigma placement, independent from the mean, appears to be a property of individual *P. incarnata*. The mean and variability of style deflexion in *P. incarnata* affected seed number and seed weight in contrasting manners, through the quantity and potentially quality of pollen deposition. This antagonistic selection via different fitness components may maintain diverse style phenotypes.

T2

P0327

The herbarium perspective of biogeographical and ecological aspects of *Syzygium* genus (Myrtaceae) in Papua New Guinea

Kipiro Damas

Papua New Guinea Forest Research Institute

Syzygium is one of the speciose tree genus in the family Myrtaceae having over 500 plus species globally, Frodin, 2004 reported 1041 species in total and was 16th out of 57 most specious genera throughout the world. Craven & Damas, 2015 (unpublished in manuscript) recently reported 1100 - 1400 (-1500) species. Papua New Guinea has more than 4000 tree species and *Syzygium* is the first most speciose tree genus and is known to be widely distributed throughout from the lowlands to the highlands up to 3000 m.a.s.l. The global geographical distribution is sub-saharan Africa, Southern Asia and Indo – Pacific region (Neil Snow and Lynn A. Craven, 2010). The greatest morphological diversity and evolutionary diversity according to Snow and Craven, 2010 occurs in North Eastern Malaysian region. Papua New Guinea's National herbarium in Lae records up to 150 (- 200) known species and almost equal or higher number of species are taxonomically unknown. The enhancement of the taxonomic knowledge depends on more field collections in various forest types. The genus stands out as one of the important tree genus among the Myrtaceae family for its sweet tasting fruits edible to birds, insects, mammals and humans. Several species in the genus *Syzygium* in Papua New Guinea are harvested for timber however there has been confusion in the identifications of species because most species tend to share quiet similar morphological features. The basic or commonly shared characteristics include opposite leaf arrangements, leaf blades having distinct intra marginal venation. The genus is also claimed in an unpublished report (Oavika, 2007) as an indicator of an unproductive forest in terms of timber harvesting. This claim remains to be proven through further research. This is a Masters Research project which is being carried out two components (Laboratory and Field) to establish information on the diversity and composition of the genus in various forest, soil types and other related environmental parameters in PNG. This research intends to find out why this genus is so widely distributed in Papua New Guinea. These information on respective parameters will give us some understanding on why this genus is widely distributed throughout PNG. This paper only discusses all findings of its distribution, composition, diversity *etc.* using specimens stored in the Papua New Guinea National Herbarium in Lae.

T2

P0328

Diversity and habitat analysis of the family Lejeuneaceae (Marchantiophyta) in the State Assam, India

Sudipa Das

Assam University, Silchar

The family Lejeuneaceae is one of the most advanced and complex groups of the order Jungermanniales, Marchantiophyta due to its great extent of morphological variations and scarcity of stable morphological characters. In India, the family is represented by 27 genera with main centers of diversity in Eastern Himalayas and Western Ghats. The state Assam, one of the forest dominated states of north-eastern India, is situated in the rain shadow of Himalayan ranges. Falling under the East Himalayan bryo-geographical territory, the state occupies a major area regarding bryophyte vegetation due to its vast range of topography and tropical-subtropical humid climate. As members of the family Lejeuneaceae prefer to grow in humid tropical-subtropical forests and constitute dominant composition of epiphytic liverworts (both corticolous and foliicolous), the state Assam has become a good abode for Lejeuneaceous members. The present ongoing study is based upon extensive collection of plant materials of the family Lejeuneaceae from different parts of the study area during 2014 – till date. The study reveals the occurrence of 23 species under 10 genera of Lejeuneaceae from Assam till date and it has been found that the family Lejeuneaceae dominates the Marchantiophyta flora of this region by occupying about 35% of the total taxa. Among the 10 genera found, genus *Cololejeunea* (Spruce) Schiffn. is found to be the dominant one with six species followed by genus *Lejeunea* Lib. with four species. Of the total taxa recorded till date, only 9% are found to grow as purely terrestrials, 78% as purely epiphytic (corticolous and/or foliicolous) and 13% do not show any habitat specificity hence grow both as terrestrials as well as epiphytes; among the epiphytic taxa reported, 61% are purely corticolous, 22% are purely foliicolous and 17% are found to grow both as corticolous as well as foliicolous. Both terrestrial and epiphytic taxa are found to grow in extremely moist and shady cool places inside the forests. Some common plants found to act as hosts for corticolous taxa are *Bambusa* sp., *Ficus* sp., *Morus* sp., *Shorea robusta*, etc; plants identified as phorophytes for foliicolous taxa are Ferns, *Gnetum* sp., various Bamboos, *Canna* sp., *Cinnamomum tamala*, *Licuala* sp., *Ophiorrhiza* sp, *Zingiber* sp., *etc.* Among the taxa reported till date, four species viz. *Archilejeunea apiculifolia* Steph., *A. minutilobula* Udar & U.S. Awasthi, *Cololejeunea furcilobulata* (Berrie & E. W. Jones) R. M. Schust. and *C. siangensis* G. Asthana & S.C. Srivast. are endemic to India. Among these, *Archilejeunea apiculifolia* is found only in the state Assam within the Eastern Himalayan bryogeographical region in addition to Kerala and Tamil Nadu of southern India; and *Cololejeunea siangensis* is confined within the Eastern Himalayan territory only in the states of Arunachal Pradesh and Assam. Eight taxa are found rare within the study area as represented by limited collections and two taxa viz. *Caudalejeunea reniloba* (Gottsche) Steph. and *Leptolejeunea epiphylla* (Mitt.) Steph. are found threatened due to habitat destruction.

T2

P0329

Reticulate evolution and parental lineage reconstruction of polyploid moonwort species (*Botrychium*, Ophioglossaceae) using PacBio sequencing technology

Benjamin Dauphin[1], *Jason Grant*[1], *Donald Farrar*[2], *Carl Rothfels*[3]

1. University of Neuchatel

2. Iowa State University

3. University of California at Berkeley

Background — Polyploidy is a major speciation process in vascular plants, and the early divergent moonwort fern genus *Botrychium* well illustrates these recurrent hybridizations at the infrageneric level. However, the near absence of nuclear sequence data has prevented molecular phylogenetic investigations on its biparental patterns of inheritance. Yet, *Botrychium* has unique evolutionary history with an independent lineage from other ferns over 300 mya and the nuclear genomes among the largest known in plants. Henceforth, with the new sequencing possibilities and bioinformatics tools, it becomes accessible to investigate complex polyploid groups and retrace their past history. *Emphasizing main questions and goals* — Our study aims to identify the parental lineages of polyploids, retrace their origins, and characterize the dynamism of polyploidy in *Botrychium* through time and space. *Materials and methods* — We considered a complete taxonomic recognition of the genus *Botrychium* by including 37 taxa from Europe and North America. We designed over one hundred primer sets from three transcriptomes of the sister genera *Botrypus*, *Ophioglossum*, and *Sceptridium*, targeting 10 nuclear regions recently characterized in ferns. We used the PacBio platform to sequence long (up to 1.5 kb) nuclear loci from 48 specimens of 25 diploids, 20 polyploids, and three outgroups. Then, we analyzed raw sequences with the new bioinformatics pipeline PURC to remove PCR-mediated recombinant sequences and generated the final alignments according to the most appropriate parameter settings (of the 10 regimes tested). Finally, we implemented AlloPPNet to infer the multilabeled species tree and reconstruct phylogenetic relationships of the polyploid lineages. *Results* — We amplified and successfully sequenced three nuclear loci (3.5 kb) from a single PacBio run, producing 60151 circular consensus sequences (CCS) of which 33417 were kept after quality control. With the regime *d* retaining a mean coverage of 45.9 CCS per locus, we characterized homologous and homeologous copies for a total of 67,127, and 64 allele variants for *ApPEFP*, *CRY2*, and *Transducine* respectively. In addition, the pipeline PURC detected and removed 50 chimeras from the three final alignments, which were also inspected by eye for comparison and estimation of the accuracy of allele number inference. We found strong support for the monophyly of diploid lineages and we identified most of the parental donors of polyploids, highlighting the multiple origins of several of them. We discuss about the topology incongruence between nuclear and plastid phylogenies and present evidence for an asymmetry in maternal and paternal contributions of polyploids as well as gene silencing that seems to occur in several polyploid individuals. *Conclusions* — The PacBio sequencing technology combined with the use of the pipeline PURC provides new perspectives to study the phylogenetics of *Botrychium* and the major role of polyploidy in their speciation processes. Polyploidization in moonworts is presumably driven by biological mechanisms that foster maternal or/and paternal contributors and further investigations are critically needed to reveal how these parental preferences take place in natural populations.

T2

P0330

New species of *Quambalaria* from *Eucalyptus* seed capsules and eucalypt-feeding insects in South Africa

Sw Morgan, M Schroder, T Paap, S Marincowitz, Mj Wingfield, ***Zw***

De Beer
Forestry and Agricultural Biotechnology Institute (FABI), University of Pretoria

The genus *Quambalaria* at present includes six known species. Some of these species were initially described in asexual Ascomycete genera like *Sporothrix* and *Ramularia*, but DNA sequences confirmed their placement in the *Microstromatales* (*Ustilaginomycetes*). Five of the species are causal agents of leaf, shoot and stem diseases of commercially grown *Eucalyptus* and *Corymbia* spp. in South Africa, Australia, Brazil, China and Southeast Asia. The sixth species, *Quambalaria cyanescens*, is not only an epiphyte of *Myrtaceae* trees, but has regularly been found in bark beetle galleries of hardwoods such as oak, and is also known widely as an opportunistic human pathogen. The aim of this study was to identify *Quambalaria*-like isolates obtained from *Eucalyptus* seed capsules and *Gonipterus* cadavers collected from two areas in Gauteng, using DNA sequences of three gene regions, the ribosomal LSU and ITS, as well as *EF-1α*. Phylogenetic analyses of sequence data revealed that the isolates from seed capsules represented two lineages distinct from all known species in *Quambalaria*. The isolates from *Gonipterus* beetles belonged to *Q. cyanescens*. This represents the first report of the latter species from South Africa. The two new species are in the process of being described. Further research is being conducted to test the pathogenicity of these isolates towards bark beetles and *Eucalyptus* trees and seed capsules, and their possible impact on seed production.

T2
P0331

A new genus and sister species described for the globally important, multi-host root pathogen, *Thielaviopsis basicola*

Wj Nel, Ta Duong, Bd Wingfield, Mj Wingfield, ***Zw De Beer***
Forestry and Agricultural Biotechnology Institute (FABI), University of Pretoria, Pretoria

The ascomycete plant pathogen *Thielaviopsis basicola* (= *Chalara elegans*) is the causal agent of black root rot on various agriculturally important crops, including chicory and groundnuts in South Africa. The fungus was first described in 1850 infecting peas (Fabaceae) and *Nemophila* (Boraginaceae) in the UK, and has since been reported from more than 50 plant families globally. Identifications to date have largely been based on morphology of the asexual chlamydospores and symptoms on the plants. A recent multigene study (de Beer *et al.*, 2014, Stud. Mycol. Vol: 79:187-219) showed that *T. basicola* forms a lineage distinct from well-defined genera such as *Ceratocystis* and *Thielaviopsis* in the *Microascales*. However, the absence of DNA sequences representing the holotype made it impossible to provide a definitive generic placement for the species. In this study, we reconsidered the genus and species level taxonomy of *T. basicola* using sequences for the ribosomal 60S, LSU, ITS, as well the MCM7, Actin, RPBI, and RPBII protein-coding gene regions. The study included 57 isolates from 14 geographical locations and various hosts worldwide. The resulting phylogenetic trees revealed that the *T. basicola* isolates formed a well-supported lineage, distinct from all known genera within the Ceratocystidaceae. Importantly, they also provided clear evidence that represents two cryptic species. An isolate from the Netherlands that closely resembles the holotype morphologically has been selected as epitype for *T. basicola*, and its sequence thus defined one of the two clades as representing that species. Isolates from the second lineage have been described as a novel species with slight morphological differences from *T. basicola*. A new genus in the *Ceratocystidaceae* (*Microascales*) was described to accommodate both species.

T2
P0332

Floral development of *Hymenaea verrucosa*: An ontogenetic approach to unusual floral characteristics (Fabaceae - Detarieae clade)

Fábio Junior Kochanowischi[1,2], Juliana Villela Paulino[3], Simone De Pádua Teixiera[4], Ana Maria Goulart De Azevedo Tozzi[1], ***Vidal De Freitas Mansano[2]***
1. Instituto de Pesquisas Jardim Botânico do Rio de Janeiro
2. Universidade Estadual de Campinas
3. Universidade Federal do Rio de Janeiro
4. Universidade Estadual de São Paulo, Ribeirão Preto

We evaluated the floral development of *Hymenaea verrucosa*, the only African species of the genus. This species has three normal and two vestigial petals. We took scanning electron microscopy images from the initial floral development stage to the final stage and others under light microscopy. The results indicated that sepals have helical initiation, the petals initiate simultaneously, but the two abaxial ones develop rapidly. In the intermediate stage, the abaxial petals stop growing, and the three adaxial ones continue to develop until anthesis. Ten stamens of different sizes are distributed in two whorls, where the antesepalous stamens are initiated before the carpel in a modified unidirectional order and the antepetalous stamens are initiated in a simultaneous order after the carpel. We speculate that the difference between the stamens may be related to changes in floral architecture due to the availability of space provided by the loss of abaxial sepals. The ovary has hairs that overlie a broad base and a stipe with central insertion. Secretory cavities were found in several flower regions, just like mucilaginous idioblasts. Secretory trichomes were found along with non-secretors at the base of the ovary. The calyx is tetramerous, and the three well-developed petals and the position of the stamens characterize a zygomorphic flower.

T2
P0333

Nuclear ribosomal transcribed spacer polymorphism in some *Paeonia* species (Paeoniaceae): Recent divergence or hybridization?

Sergey Efimov, ***Galina Degtjareva****, Elena Terentieva, Tahir Samigullin, Carmen Vallejo-Roman*
M.V. Lomonosov Moscow State University

The genus *Paeonia* L. consists of approximately 35 species of perennial herbs and shrubs disjunctly distributed in the northern temperate region. The genus is taxonomically isolated, having been placed in its own monotypic family (Paeoniaceae), but species boundaries within the genus are difficult to determine due to a great morphological variability. Revealed by both morphological and molecular phylogenetic studies reticulate evolution and different ploidy level further obscure phylogenetic relationships within the genus. The hybrid species were identified by full or partial ITS sequence additivity of the putative parental species or by incongruent positions of the hybrid in multiple gene phylogenies. The focus of the present study is on the nuclear ribosomal transcribed spacer (ITS, ETS) and the plastid *ycf1* gene sequence diversity and species delineation in three groups of herbaceous species, namely, *Paeonia obovata* complex (*P. obovata*, *P. oreogeton*), *Paeonia anomala* complex (*P. anomala*, *P. hybrida*, *P. intermedia*) and *P. lactiflora*. All species were studied from several populations for investigation of infraspecific variation of sequences. The nucleotide sequences were obtained by direct Sanger sequencing of amplicons. The *Paeonia obovata* complex is widely distributed in East Asia and includes both diploid and tetraploid species. The *Paeonia anomala* complex comprises diploid species from Central Asia, Siberia, and adjacent northeastern European regions. *Paeonia lactiflora* is a systematically well isolated diploid species of central and eastern Asia. Our study shows that all species groups are characterized by a large number of nucleotide positions exhibiting additive polymorphic sites. Nucleotide additivity occurs in the ITS and ETS sequences of both diploid and teraploid species and characterizes the group of species as a whole (*Paeonia obovata* complex, *P. intermedia* + *P. hybrida*) or separate species (*P. intermedia*, *P. hybrida*, *P. anomala*, *P. lactiflora*). According to number of polymorphic sites, *Paeonia obovata* complex can be divided into three subgroups, which is not correlated with morphological characters and little correlated only with the geography. The smallest number of polymorphic positions was observed in accessions from the Sakhalin, as well as from the South Sikhote-Alin. Some additive nucleotides support hybrid origin of species postulated earlier, i.e. *P. anomala*. However a lot of additive nucleotides are not found in other species, which could be considered as parent forms. Interestingly, one polymorphic site observed in *P. anomala* did not have a presumably hybrid origin and could be treated as an apomorphy for the species. In many cases, the studied species are not characterized by nucleotide substitutions, but only by additive nucleotides. The groups characterized by additive polymorphic sites were supported by nucleotide substitutions in plastid *ycf1* gene, when variability of the region was sufficient. Different ITS types arise by means of nucleotide substitutions across the loci. The level of ITS variation is high if the rate of concerted evolution is lower than the mutation rate. It is possible, that additive polymorphism observed in nuclear ribosomal transcribed spacers reflects not only hybrid origin of species, but also recent divergence of the group. This work is supported by the Russian Science Foundation (grant #14-50-00029).

T2

P0334

First occurrence of Icacinaceae fossil flower in amber of Le Quesnoy site (Ypresian, Oise, France)

Cédric Del Rio, *Dario De Franceschi*

Museum national d'Histoire naturelle

Since the beginning of the twenty first century, a substantial number of phylogenetic studies, focused on morphological and new molecular data, allowed to clarify the basal relationships within the Lamiid's clade and define the new order of Icacinales Tiegh. This new order includes Oncothecaceae Kobuski ex Airy Shaw and Icacinaceae Miers with 23 genus after its new monophyletic circonscription. This advances precise the evolution and diversification of this group, providing a supported framework for the studies of relevant fossils. Here, we discuss the morphology and affinities of a recently discovered flower, preserved in amber, from Le Quesnoy site (Ypresian, Oise, France). This flower is particularly small (about 2.5 mm diameter), actinomorphic, pentamerous, male or hermaphrodite and hypogynous. The calyx is cupulate, the petals are valvate, fused at base in a short cup and tomentose on inner surface with long, simple flattened and granular hairs. Stamens are alternipetal and erected but lacking anthera (not preserved). The globular ovary is presumed unilocular. Numerous small (20-25 μm) triaperturate and echinulate pollen grains around the flower and on hairs were extracted and have been considered as belonging to this flower. These criteria were used in a phylogenetic analysis with mixed DNA-morphological characters, which leads to place this flower as a member to Icacinaceae family, close to the genus *Hosiea* Hemsl. & E. H. Wilson. However this fossil reveals a combination of features unknown in Icacinaceae's current diversity. A new extinct genus of Icacinaceae was describing, within a clade bringing together 11 genus including *Hosiea*. We also found very similar hair of upper surface of petal on current genus *Nothapodytes* Blume and *Mappia* Jacq. These taxa are basal in core Icacinaceae (excluded *Cassinopsis* Sond.) and this suggests a pleisiomorphic state of this character. Icacinaceae are well known in fossil records during the Paleocene and mostly Eocene. The fossils attributed to this group mainly consist of endocarps, but only a few leaves and wood were also described. However, no fossil flower has ever been described in this clade. Indeed, flower in amber remain rare and only one flower has been described for all the Lamiid's clade from Baltic amber (Gentianaceae Juss.). Exceptional preservation of fossil structures in amber allow a better understanding of evolution combining molecular/morphology phylogeny and description of fragile organs as flowers usually poorly preserved in other kind of fossilization.

T2

P0335

Bipolaris in China

Hui Deng[1], *Yujie Guo*[1], *Yongchun Niu*[1], *Tianyu Zhang*[2]

1. Institute of Agricultural Resources and Regional Planning, Chinese Academy of Agricultural Sciences

2. Shandong Agricultural University

The genera *Bipolaris* Shoemaker is an important plant pathogenic fungal genus, which can cause diseases of plants in Gramineae. Molecular phylogenetic studies of the genus were reported during recent years. According to the reported multi-genes phylogenetic analysis, some species of *Bipolaris* were treated as the new combinations of the genus *Curvularia* and the species of *Bipolaris* were systematically studied. In this study, 255 strains of *Bipolaris* in our lab isolated from the specimens in China and preserved in Agricultural Culture Collection of China (ACCC) were used as materials. All of the strains were sequenced by ribosomal DNA ITS region and phylogenetic analyzed. More than 80 strains with rDNA ITS sequence difference were selected to sequencing the

GPDH and *EF-1α* genes, and the phylogenetic reconstructions of concatenated and individual gene trees were performed. The results show that 251 strains of *Bipolaris* belong to 16 published species, including *B. bicolor*, *B. coffeana*, *B. cookei*, *B. heveae*, *B. heliconiae*, *B. maydis*, *B. oryzae*, *B. peregianensis*, *B. pluriseptata*, *B. secalis*, *B. setariae*, *B. sorokiniana*, *B. victoriae*, *B. yamadae*, *B. zeae* and *B. zeicola*. Among them, the four species *B. coffeana*, *B. heliconiae*, *B. pluriseptata* and *B. secalis* are new records in China. Phylogenetic reconstructions of concatenated and individual gene trees indicate that the four strains do not belong to either *Bipolaris* species. Based on multilocus sequence analysis and morphology, the four strains are assigned to a possible new species of *Bipolaris*. The morphological characters of the representative strain from each species were described in detail, which were grown on the same condition. In this study, most species of *Bipolaris* were isolated from different hosts' species in Gramineae, while *B. sorokiniana* were found on Gramineae, Asteraceae, Cannabaceae, Linaceae, Leguminosae, Magnoliaceae, Malvaceae, Rosaceae, Solanaceae, Ulmaceae 10 families of plants in China. The results will add the knowledge of the species and diversity of the fungi resources *Bipolaris* genus in China, and applied the basis of the further research and utilization of these fungi.

T2

P0336

Glacial refugia as genetic hotspots identified by mapping multi-species haplotype distribution

Tao Deng, *Sergei Volis, Hang Sun*
Kunming Institute of Botany, Chinese Academy of Sciences

Geological and climatic processes that led to creation of refuges could have a similar impact on many species, and refugial regions, compared with areas that have been colonized after the retreat of glaciers, are expected to have distinct patterns of intraspecific genetic variation. Comparative phylogeography, by comparing genetic imprints in multiple independently evolving species, can infer how the geographic ranges of co-distributed species have changed over time, and identify the areas of former refugia where many species survived during Quaternary climate oscillations. We calculated from chloroplast haplotype data for 117 species represented by 2458 populations two genetic diversity parameters, the haplotype genetic diversity and genetic uniqueness, for each species. Then we mapped the cell values on a 1° × 1° latitude/longitude grid map of China. Grid cells within the top 10% of highest values of either genetic diversity or genetic uniqueness were considered genetic hotspots. Although the haplotype diversity pattern matched some previously identified endemic richness hotspots and inferred Quaternary refugia, it suggested existence of many previously unidentified cryptic refugia. Our results suggest important roles of cryptic (or micro-) refugia in evolution of Chinese flora, and that an approach combining and summarizing species-specific haplotype diversity data can be useful for their identification. At the same time we found that, out of the two genetic parameters used, only genetic uniqueness (a proportion of unique haplotypes not found elsewhere) can reliably identify the refugia themselves, while the areas with highest genetic richness (a locally found proportion of haplotypes) are not necessarily the refugia themselves but the areas where genetic lineages arriving from different refugia likely admixed after recolonization.

T2

P0337

Isolation and characterization of *MADS-box* gene from *Magnolia biondii* Pamp (Magnoliaceae)

Tian Deng, *Shouzhou Zhang*
Fairylake Botanical Garden, Shenzhen & CAS

The MADS-box transcription factors are one of the largest families of transcriptional regulators and play diverse regulatory roles in plant growth and development processes, especially in flower development. In this study, the MADS-box DNA-binding domain downloaded from Pfam protein families database was exploited to identify MADS-box genes from the *Magnolia biondii* tanscriptome using HMMER 3.0. The obtained amino acid sequences were analyzed with DNAMAN 6.0, MEGA6.0, DNAStar7.1 softwares. Totally 18 M. biondii MADS-box genes were found in the M.biondii tanscriptome. Phylogenetic analysis revealed that M. biondii MADS-box genes were classified into two Groups: MIKCC and MIKC*. The Group MIKCC was sub-divided into 8 distinct clades, group SEP, AGL6, AGL20, AG, AP1/FUL, SVP, I and AP3. Domain analysis indicated that the MIKCC type MADS-box genes are highly conserved which contained MADS-box region, I region, K-box region and C tail. Using RNA-seq data the expression of genes in four different transcripts (flower buds in pre-meiosis, flower buds in meiosis, completely open flowers and leafs) was analyzed. Most MADS-box genes only expressed in different flowers periods and the expression in AGL20 sub-family, SVP sub-family and MIKC* type MADS-box genes were high in leaf. Different expression always appears in the same family and these data might be used to analyze the conservatism, the diversity and the functional differentiation of MADS-box.

T2

P0338

Abnormal flower phenotype caused by overexpression of the *Magnolia biondii MbAG-1* in *Arabidopsis*

Tian Deng, *Shouzhou Zhang*
Fairylake Botanical Garden, Shenzhen & CAS

Degenerate primers were designed using unigene sequences of *Magnolia biondii* transcriptome data and MbAG-1 gene was cloned by RT-PCR techniques, with flower buds of *Magnolia biondii* as the test material. The full-length sequence of MbAG-1 gene was 991bp with a complete open reading frame (ORF) of 672 bp and encoding a 223 amino acids protein. Molecular phylogeny analysis suggested that, MbAG-1 is an AG homology in Arabidopsis, at the transcriptional activation domain of C-terminal exist two conserved motifs: AG I and AG II. In addition, a sense expression vector of MbAG-1 has been introduced into Arabidopsis via agrobacterium-mediated method then screen positive transformation individuals. Compared with wild type, PBI121-MbAG-1 transgenic individuals appear premature flowering with small flower, short sepals, larger stamen and carpel, and some abnormal phenotype like petal converting into filament-like structure, Incurved leaf with long epidermal hair. Moreover, the abscissions of the outer three floral whorls (sepals, petals, and stamens) were inhibited during development of the siliques, which were bumpy. And the seed set rate are lower than that of normal seeds in the wild type. In summary, overexpression of the *Magnolia biondii* MbAG-1 gene in Arabidopsis could affect floral organ development, so lay the foundation for the floral organ development.

T2

P0339

The status of an overlooked mistletoe genus *Dufrenoya* Chatin, Amphorogynaceae

Mohan Prasad Devkota

Amrit campus, Tribhuvan University

Dufrenoya Chatin (Amphorogynaceae) contains ca. 11 species of aerial parasites found in southeastern Asia. First named as *Viscum*, it has been known from Nepal since 1802 but rarely collected or studied. Two well-defined species, *D. granulata* and *D. platyphylla*, are known from the middle Mountain region in the eastern half of the country. To provide a modern treatment of the genus for Nepal, a study was conducted that examined their distributions (from herbarium records), a review of their complex nomenclatural history, morphological and anatomical descriptions, and a genus-level phylogenetic analysis of Amphorogynaceae using the nuclear ITS rDNA region. The molecular analysis showed that *Dufrenoya* is most closely related to *Phacellaria* (in agreement with Danser 1940) and this clade is distinct from another composed of *Dendromyza* and *Dendrotrophe*. Our results stand in contrast to the taxonomic treatment of Santalaceae for Flora of China where *Dufrenoya* was included within *Dendrotrophe.* These two genera can be distinguished by a number of morphological characters, e.g., *Dufrenoya* has umbellate male inflorescences whereas in *Dendrotrophe* it is racemose. The genera also differ in fruit characters. *Dendrotrophe* has a fleshy mesocarp and a tuberculate endocarp with chambers at the apex and the base as well as transverse chambers. The mesocarp in *Dufrenoya* consists of membraneous strands and its endocarp is smooth with chambers but no transverse chambers. Taken together, both morphology and molecules favor separate generic status for *Dufrenoya* and *Dendrotrophe*. Indeed, attention should be focused upon the generic status of *Dendrotrophe* and *Dendromyza.* We estimate that both species of *Dufrenoya* in Nepal should receive the IUCN classification of vulnerable because of human alteration of natural ecosystems where its host trees occur.

T2

P0340

Geography shapes the phylogeny of frailejones (Espeletiinae Cuatrec., Compositae): A remarkable example of recent rapid radiation in Sky Islands

Mauricio Diazgranados[1], *Janet C. Barber*[2,3]

1. Natural Capital and Plant Health Department, Royal Botanic Gardens, Kew

2. Department of Biology, Saint Louis University

3. Missouri Botanical Garden

The páramo ecosystem, located above the timberline in the tropical Andes, has been the setting for some of the most dramatic recent rapid plant radiations, and it is one of the world's fastest evolving and most diverse high-altitude ecosystems. Today 144+ species of frailejones (subtribe Espeletiinae Cuatrec., Compositae) dominate the páramo. Frailejones have intrigued naturalists and botanists, not just for their appealing beauty and impressive morphological diversity, but also for their remarkable adaptations to the extremely harsh environmental conditions of the páramo. Previous attempts to reconstruct the evolutionary history of this group failed to resolve relationships among genera and species, and there is no agreement regarding the classification of the group. Thus, our goal was to reconstruct the phylogeny of the frailejones and to test the influence of the geography on it as a first step to understand the patterns of radiation of these plants. Field expeditions in 70 páramos of Colombia and Venezuela resulted in 555 collected samples from 110 species. Additional material was obtained from herbarium specimens. Sequence data included nrDNA (ITS and ETS) and cpDNA (*rpl16*), for an aligned total of 2954 bp. Fragment analysis was performed with AFLP data using 28 primer combinations and yielding 1665 fragments. Phylogenies based on sequence data were reconstructed under maximum parsimony, maximum likelihood and Bayesian inference. The AFLP dataset employed minimum evolution analyses. A Monte Carlo permutation test was used to infer the influence of the geography on the phylogeny. Phylogenies reconstructed suggest that most genera are paraphyletic, but the phylogenetic signal may be misled by hybridization and incomplete lineage sorting. A tree with all the available molecular data shows two large clades: one of primarily Venezuelan species that includes a few neighboring Colombian species; and a second clade of only Colombian species. Results from the Monte Carlo permutation test suggests a very strong influence of the geography on the phylogenetic relationships. Venezuelan páramos tend to hold taxa that are more distantly-related to each other than Colombian páramos, where taxa are more closely-related to each other. Our data suggest the presence of two independent radiations: one in Venezuela and the other in Colombia. In addition, the current generic classification will need to be deeply revised. Analyses show a strong geographic structure in the phylogeny, with large clades grouped in hotspots of diversity at a regional scale, and in páramo localities at a local scale. Differences in the degrees of relatedness between sympatric species of Venezuelan and Colombian páramos may be explained because of the younger age of the latter páramos, and the lesser time for speciation of Espeletiinae in them.

T2

P0341

Oligocene fossil winged fruits of the endemic *Dipteronia* (Sapindaceae) from Southwest China and its palaeogeographical and palaeoecological implications

Wenna Ding[1,2], *Jian Huang*[1,2], *Tao Su*[1], *Yaowu Xing*[1], *Zhekun Zhou*[1,3]

1. Key Laboratory of Tropical Forest Ecology, Xishuangbanna Tropical Botanical Garden, Chinese Academy of Sciences

2. University of Chinese Academy of Sciences

3. Key Laboratory of Plant Diversity and Biogeography of East Asia, Kunming Institute of Botany, Chinese Academy of Sciences

The endemic genera, especially the palaeoendemics once beyond the present distribution ranges, play important role in tracking floristic relationships among continents throughout geological time. *Dipteronia* (Sapindaceae) is endemic to China nowadays, and is well known in the Paleogene deposits of western North America. Here, we reported the first fossil record of *Dipteronia* in East Asia, namely *D. brownie* McClain and Manchester, based on diagnostic winged fruit remains from the early Oligocene of Lühe Basin in southwestern China. The fossil species is characterized by samaroid mericarps, with a round wing completely surrounding the seed locule, except for a flat proximal edge representing the attachment scar; primary vein extending from the attachment scar, then deflecting sharply and extending straight to the centrally positioned seed. This new finding significantly expands the historical distribution of *Dipteronia* into southwestern China prior to the early Oligocene. It indicates the close floristic affinity between East Asia and western North America by the Oligocene. Moreover, the southward immigration and extinction of this genus from high latitude are consistent with the global cooling and seasonal intensification of precipitation in the Northern Hemisphere.

T2
P0342
RAD-seq phylogenomics reconstructs a well-resolved phylogeny of a rapid radiation in the genus *Mosla*
Yanqian Ding, *Panan Li, Yingxiong Qiu*
Zhejiang University

The genus *Mosla* (Lamiceae) is a young but morphologically and ecologically diverse group distributed across Eastern Asia. Approximately 12 species occur in China, most of which are sympatrically distributed. They exhibit remarkable variation in floral morphologies and habitats. We hypothesized that habitat specialisation or pollinator specialisation might contribute to the radiation of *Mosla*. To establish the basic background to test this hypothesis, we need to recover a well-resolved phylogeny of the genus. However, despite several efforts, the evolutionary relationships among the species in the genus remain unclear. Here, we use extensive analyses of genome wide restriction-site-associated DNA sequencing (RAD-seq) data to determine the phylogenetic relationships among seven *Mosla* species. Three taxa (*Collinsonia canadensis*, *Keiskea japonica*, *Perilla frutescens*) were selected as outgroups based on previous molecular phylogenetic studies. Two methods of phylogenetic inference, maximum likelihood (ML) and Bayesian inference (BI), were applied to data sets that included as many as 80,648 RAD unlinked SNPs. Both methods yielded similar topologies that included two major clades: Clade I, composed of two species (*M. hangchowensis*, *M. chinensis*) occupying different microhabitats with divergent floral morphologies and overlapping distributions; and Clade II, composed of five species (*M. soochowensis, M. cavaleriei, M. dianthera, M. scabra, M. japonica*) characterized by relatively diversity in floral morphology. Our results show that, for closely related, recently diverged taxa, RAD-seq is likely the most comprehensive approach currently available for phylogenetic inference.

T2
P0344
Floral development of Urticaceae with emphasis on the female diversification
Jia Dong[1], *Louis Ronse De Craene*[1], *Alexandre Monro*[2]
1. Royal Botanic Garden, Edinburgh
2. Royal Botanic Garden, Kew

Urticaceae, as one of the largest wind-pollinated family, has great diversification in the floral morphology, especially the female structures. Molecular phylogeny of Urticaceae has suggested it as a monophyletic lineage in the order of Rosales with four major clades (Urticeae, Elatostemeae, Boehmerieae and Cecropieae). However, the morphological character changes do not always corresponding with the molecular data. Morphological observation and floral development of 11 species from 4 different tribes were investigated to understand the diversity and evolution within a largely understudied and mainly tropical family. Dramatic variations occur among the groups in terms of sexuality, perianth merism and gynoecium morphology. Flowers are rarely bisexual, more often with a high dimorphism between male and female flowers arranged in monoecious or dioecious reproductive systems. We concentrated on the development of female flowers, which show the greatest diversity in the family. Despite the strong reduction of flowers and apparent homogeneity of structures, the details of development reveal an intricate diversity of forms. Particularly, the formation of inflorescence and the differentiation of organs are highly complex. The perianth is generally tetramerous, but has a high tendency for reduction to trimery or dimery. The nature of the tepals is clarified as sepal originated, but the boundaries between tepals and bracts are often unclear. The monomerous nature of the ovary is confirmed, with no evidence for a second carpel and the formation of the ovary, the stigma and the ovule indicates a special, more reduced but also more advanced developmental process within the urticalean families. Different hypotheses for the evolution of unisexuality are discussed based on the developmental evidence and the phylogeny of Urticaceae.

T2
P0345
Systematics and Biogeography of *Carex* section *Laxiflorae*
Jenna Dorey[1,2], *Robert Naczi*[1], *Takuji Hoshino*[3]
1. The New York Botanical Garden
2. CUNY Graduate Center
3. Okayama University of Science

Carex section *Laxiflorae* includes some of the most common sedges in forests of eastern North America, and several narrowly endemic species. These sedges are ecologically important in the forest understory, especially in the prevention of soil erosion. The section is currently comprised of 16 species in North America, but is in great need of taxonomic revision since several species appear to contain previously undocumented diversity, and the sectional boundaries are unclear. The phylogenetic study with most thorough sampling of *Carex* indicates that the *Laxiflorae* are part of a clade that also includes sections *Paniceae* and *Bicolores*, and contains approximately 50-60 taxa from Asia, Europe, and North America. This clade that includes section *Laxiflorae* appears to have two centers of diversity – one in eastern North America, and one in Eastern Asia. We, therefore, hypothesize this clade was once widespread throughout a continuous forest in the Northern Hemisphere and that the present-day patterns of diversity are due to species radiations on both continents. The goals of this research are to test the monophyly and ascertain the sectional limits of *Laxiflorae*, to delimit and circumscribe species within difficult complexes of *Laxiflorae* s.s., and to view the biogeographic patterns that led to the present distribution of sections *Laxiflorae, Paniceae, and Bicolores*. Here, we present results of ongoing field and herbarium work, morphometrics, and molecular systematics. A maximum likelihood reconstruction of 6 gene regions in over 100 individuals of about 65 taxa from North America, Europe, and Asia recovered section *Laxiflorae* as paraphyletic, with a predominantly North American clade separate from a predominantly Eurasian clade. Both molecular and morphological analyses also offer support for undescribed taxa in North America. These important results will lead to a revised classification of section *Laxiflorae,* and description of new taxa.

T2

P0346

Evolution of tropical rainforest species: New insights into the diversification patterns and node dating of the Annonaceae

Eva Drukker[1], *Paul Hoekstra*[2], *Joeri Strijk*[3], *Lars Chatrou*[1]

1. Wageningen University and Research Centre

2. Naturalis Biodiversity Centre Leiden

3. College of Forestry, Guangxi University

Tropical rainforests form the terrestrial biome containing the highest diversity of plant species. Despite extensive ongoing work on floristics of tropical regions and phylogenetics of tropical clades, the evolutionary build-up of this huge diversity is not well understood. One of the approaches to improve our understanding of the origin and maintenance of rainforest hyperdiversity is to unravel the evolutionary history of large clades that are mainly confined to tropical rainforests. The Annonaceae almost exclusively occur in tropical rainforests and, assuming niche conservatism, the temporal and spatial evolution of the Annonaceae can be taken as a proxy for the evolution of the rainforest biome. By far the most species of the Annonaceae belong to two major (sister) clades; the Malmeoideae and the Annonoideae. Due to the different substitution rates of these clades, current models have been unable to produce consistent results with regards to inferring diversification rates and dating the crown nodes of the two clades. It is therefore still unknown whether the huge diversification of the Malmeoideae and Annonoideae clade was the result of a sudden or rather a steady diversification. Insights in these radiations and diversification patterns will help our understanding of the current biogeographical distributions of the Annonaceae and tropical rainforest distributions as a whole. The main goal of this study was to infer a phylogeny, assess diversification rates and date the important nodes of the Annonaceae, based on the most comprehensive dataset and improved modelling. With 7 outgroup species and over 1000 (out of 2400) Annonaceae species, a sampling size of almost 50% of the family was made. All the genera were included, only 5 genera had less than 20% of its species sampled. Of all the other genera more than 20% of the species was represented in the dataset. Newly discovered fossils of the Annonaceae were included for calibration. We applied improved substitution rate models in the programme RevBayes. Due to the flexibility of this programme, many of the issues encountered in previous studies could be accounted for in the RevBayes models.

T2

P0347

Taxonomy and biogeography of the genus *Euonymus* (Celastraceae)

Cheng Du, Yunjuan Zuo, Yuan Wang, ***Jin-Shuang Ma***

Shanghai Chenshan Plant Science Research Center, Chinese Academy of Sciences/Shanghai Chenshan Botanical Garden

Euonymus L. consists of more than 129 species of five traditional sections, and is one of the largest genera of Celastraceae. The genus mainly occur in the temperate regions, with 88.5% species in Asia; however, most genera of the family are distributed in the tropical and subtropical regions of the world. Several taxonomic treatments have been proposed based on various morphological characters, but the taxonomy of *Euonymus* remains controversial due to the limited morphological descrpitions of reproductive organs for several species, the morphological plasticity in vegetative traits, and the lack of a robust phylogeny with extensively sampling to provide insights into the species delimitation. This study aims to infer evolutionary relationships among taxa of *Euonymus*, to test the diversification hypotheses in this rapidly diversified group in eastern Asia, and to provide a new classification system in the phylogeographic context integrated with taxonomic investigation. The phylogenetic relationships of *Euonymus* was reconstructed using the combined dataset of four chloroplast DNA fragments and the ITS regions, with 227 individuals of 87 species sampled. The phylogenetic analysis indicates that assessions of *Glyptopetalum* Thwaites and *Euonymus* clustered together, and seven divergent clades within this genus was identified with strong support. But this genetic grouping is incongruent with the division of five sections based on morphological characteristics of capsule. The fossil-calibrated phylogeny and ancestral area reconstruction analysis suggests that *Euonymus* originated in the temperate eastern Asian zone in Eocene, and diversified in this region during the late Oligocene to the early Miocene. The Antarctic land bridge may have been served as the corridor for the intercontinental dispersal across the continents from eastern Asia to western/eastern North America. The diversification process of *Euonymus* in temperate Asia was triggered by the past climate cooling and habit shifts in Oligocene. A taxonomic revision of the genus *Euonymus* is also presented on the basis of taxonomic literature, field investigation, herbarium specimens, morphological examination and phylogeny reconstruction. *Glyptopetalum* should be incorporated into *Euonymus* as a section to make a monophyletic *Euonymus* sensu lato. The well known five sections based on capsule morphology should be treated as seven sections according to our molecular data. The shape of capsule is no longer a key character for classification and subdivisions of the genus *Euonymus*. The infrageneric divisions of *Euonymus* sensu lato will be recognized, including nineteen new combinations, six new names, one new section and several synonyms.

T2

P0348

A Plastid Phylogeny of the Fern Family Dennstaedtiaceae (Polypodiales)

Yifan Duan[1], *Liang Zhang*[2], *Yongfu Li*[1], *Dekui Chen*[3], *Ngan Thi Lu*[4], *Xianrong Wang*[1], *Libing Zhang*[5]

1. Nanjing Forestry University, Nanjing, Jiangsu 210037
2. Kunming Institute of Botany, Chinese Academy of Sciences, Kunming, Yunnan 650201
3. Chongqing Normal University, Chongqing 401331
4. Chengdu Institute of Biology, Chinese Academy of Sciences, Chengdu, Sichuan
5. Missouri Botanical Garden, P.O. Box 299, St. Louis, Missouri 63166-0299

Phylogenetic approaches with molecular data have resulted in substantial improvement in fern classification by identifying monophyletic groups and rejecting paraphyletic taxa. The family Dennstaedtiaceae is one of the least-known fern families in terms of its circumscription and phylogeny. Currently Dennstaedtiaceae are estimated to contain about 10 genera and 265 species. Previous studies typically had small taxon and character sampling. In the present study, DNA sequences of five plastid loci of 323 accessions representing about 130 species of Dennstaedtiaceae and 20 related genera were used to infer a phylogeny using maximum-likelihood, Bayesian-inference, and maximum-parsimony methods. Our major results included: (i) *Microlepia* is monophyletic and natural hybridization is possibly in this genera; (ii) *Oenotrichia* is resolved within Dennstaedtiaceae; (iii) *Dennstaedtia* is non-monophyletic, with a second clade being more closely related with *Microlepia*; and (iv) *Saccoloma* is polyphyletic, with some falling inside the Dennstaedtiaceae and others outside.

T2
P0349

Further study on the phylogenetic relationships within the Paleotropical clade of the Bignoniaceae

Sutee Duangjai[1], *Rattapoom Ruam-Unyamanee*[1], *Wattanachai Tasen*[1], *Voladol Chamchamroon*[2]
1. Department of Forest Biology, Faculty of Forestry, Kasetsart University, Chatuchak, Bangkok
2. BKF herbarium, Department of National Parks, Wildlife and Plant Conservation, Chatuchak, Bangkok

The Paleotropical clade of the Bignoniaceae contains about 20 genera and approximately 150 species of tree. Out of 20 genera, four of them which are Asian, i.e. *Heterophragma, Pauldopia*, *Pajanelia*, and *Suntisukia*, have never been included in any previous phylogenetic studies. Moreover some genera such as *Dolichandrone*, *Markhamia* and *Stereospermum* have been under-sampled in previous phylogenies. Therefore in this study we aimed to investigate the placement of four Asian genera, *Heterophragma, Pauldopia*, *Pajanelia*, and *Suntisukia*, by using DNA sequence data from *ndhF* and *trnL-trnF* regions. We also tested the monophyly of some genera within the Paleotropical Clade. Four well-supported sub-clades within the Paleotropical clade of the Bignoniaceae were identified. Our results confirm the placement of four genera, *Heterophragma, Pauldopia*, *Pajanelia*, and *Suntisukia*, in the Paleotropical clade. They form a first well-supported sub-clade together with other genera, *Radermachera* and *Tecomella*. The genus *Heterophragma* is close to the genus *Santisukia* and together they are sister to the monotypic genus *Pajanelia*. On other hand the genus *Pauldopia* is close to *Radermachera* and *Tecomella*. The second well-supported sub-clade contains *Dolichandrone*, *Fernandoa*, *Kigelia*, *Markhamia*, *Newbouldia* and *Stereospermum*. Our results also indicate that these genera, *Dolichandrone*, *Markhamia* and *Stereospermum*, are not monophyletic. The third sub-clade contains *Catophractes*, *Rhigozum* and *Spathodea* and the last sub-clade is the member of tribe Coleeae from Madagascar.

T2
P0350

OMG *Boronia* (Rutaceae) is polyphyletic: Coping with large well supported clades that contain few morphologically useful characters.

Marco Duretto, *Margaret Heslewood*
Royal Botanic Gardens and Domain Trust

Boronia (Rutaceae) is a genus of 160 species found in Australia and New Caledonia. *Boronia* is particularly diverse in southern Australia with centres of diversity in both the south-west and south-east. It is found in a wide variety of ecological communities that are usually sclerophyllous in nature, such as heaths and eucalypt woodland or forest, though a few species are found in wetter communities such as swamps and rainforest edges. Species can be herbs, shrubs and sometimes small trees characterised by being sclerophyllous and by having leaves that are opposite decussate or rarely in whorls of three, 4-merous diplostemonous flowers, carpels being fused only along the style and dehiscent fruit. These features are common elsewhere in Rutaceae and a defining morphological apomorphy for the genus is apparently lacking. *Boronia* is classified into eight sections including section *Boronella*, once an endemic New Caledonian genus that was recently reduced to a section of *Boronia*. The subfamily and tribal classification of Rutaceae has not stabilised and the position of *Boronia* is important as it is the type genus of Tribe Boronieae (subfamily Amyridoideae) which has been shown to be polyphyletic. *Boronia* is taxonomically isolated and molecular evidence indicates it is sister to a large clade containing many rainforest genera, including *Melicope* and *Acronychia*, that are largely confined to Australasia, as well as two scleromorphic genera, *Zieria* and *Neobyrnesia*, which are traditionally placed in Tribe Boronieae. The remainder of Tribe Boronieae is sister to this larger clade and the entire clade sister to a clade containing *Flindersia* and relatives. The monophyly of *Boronia* has not been tested even though it has been subjected to a number of generic studies, where it had been assumed to be monophyletic, and been included in several larger studies investigating generic relationships within Rutaceae. These larger studies did not include a representative sample of the diversity found in *Boronia* with each study including species of only one or two sections. Sections *Algidae*, *Alatae*, *Cyanothamnus* and *Imbricatae* were never included in any of these larger studies. Here we present molecular phylogenetic analyses based on two nuclear regions (ETS, ITS) and two plastid regions (*psb*A-*trnH*; the *trn*L-F region, including the *trn*L intron and *trn*L–*trnF* intergenic spacer). Representatives of 42 genera (outgroup includes 8 genera) of the Australasian clade within subfamily Amyridoideae were included. 158 accessions of Boronia were included representing all eight sections plus two of the four species placed *incertae sedis*. Analyses indicate *Boronia* is polyphyletic with section *Cyanothamnus* sister to the *Melicope/Acronychia* clade and this clade sister to a *Zieria/Euodia* clade. *Boronia*, less *Cyanothamnus*, remains in the isolated position sister to the above clades. These clades are strongly supported with molecular data but lacking obvious morphological support; apomorphies are not obvious and/or homoplasious. *Cyanothamnus*, another Australian endemic, will be reinstated. The Australasian clade within subfamily Amyridoideae contains five distinct and strongly supported lineages adapted to sclerophyllous environments as well as three or four distinct rainforest lineages suggesting a complicated history of biome transition in the Australasian flora.

T2
P0351
Some morphological and anatomical features of generative organs of the genus *Hydrocharis* L. (Hydrocharitaceae)

Andrey Efremov[1], *Cezary Toma*[2], *Andrey Filonenko*[3]
1. *Omsk State Pedagogical University*
2. *Kazimierz Wielki University*
3. *Moscow Pedagogical State University*

The species of the genus *Hydrocharis* spreads in tropical and temperate areas of the Old World: *H. chevalieri* (De Wild.) Dandy in Central Africa, *H. dubia* (Blume) Backer in Southeast Asia and the East coast of Australia and *H. morsus-ranae* L. in Europe, Central Asia and North America. This work is an attempt at a critical analysis of literaturary sources and original data on the morphology and anatomy of the generative organs of the genus *Hydrocharis*. The analysis was realized with the native and fixed material, in accordance with standard methods and techniques having used stereoscopic microscope. It is found, that on the whole the species of the genus *Hydrocharis* typically have large (3–5 cm in diam.), 3-membered entomophilic dioecious flowers, although *H. morsus-ranae* and *H. dubia* can occasionally be bisexual. The perianths of both types of flowers are alike. The calyx is segmented, consists of 3 ovate, white, greenish sepals, significantly shorter than the petals. The corolla is dialypetalous, consists of 3 petals, shaped from almost round to broadly ovate. The androecium of the male flower consists of 4–6 alternating circles, a number of fertile stamens varies, elements of androecium are occasionally fused in the base. 6–12 outer elements are fertile, 3–9 inner ones are sterile. There is a wide range of transitions from the typical fertile stamens to the secretory staminodes and the nectaries. The androecium of the female flower is formed by 2 circles, the inner circle consists of concavity nectaries, the outer circle – staminodes. The gynoecium typically consists of 6 carpels immersed in the ¾ or deeper in the receptacle. At the base of the gynoecium 6 isolated locules of the ovary can be clearly distinguished, septs consist of fused tissues of adjacent lateral walls of carpels. All gynoecium zones except the asimplicate one have 40–50 orthotropic ovules, the placentation is laminal diffuse. With that in mind the gynoecium of *Hydrocharis* should be considered as a lower hexamer hemiparacarpe. Male flowers may have a three-blade pistillodium. The fruit wall of *Hydrocharis* is formed by the tissues of overgrown receptacle (extracarpellary part) and the tissues of carpels (carpellary part). During the fruit ripening the covering tissue is destructed and water starts penetrating with the mesocarp aerenchyma, which causes swelling of the mucus and also triggers the break of the wall and dissemination. The fruit further develops in water. In accordance with the morphogenetic classification of fruits they should be best considered as lower hexamer hemiparacarp berries. Seeds are wide-ellipsoid, 1.0-1.3 mm long (not described in *H. chevalieri*), their surface is covered with soft, 1-3-celled, unequally thickened by outgrowths of exotesta. Consequently, the generative organs of *Hydrocharis* have a wide range of morphological adaptations conditioned by entomophily. The primary data on the morphology and anatomy of the gynoecium and fruit of *H. chevalieri* confirm some isolation of this species.

T2
P0352
Flower morphology and development is associated with bee visitation pattern on *Chamaecrista desvauxii* (Leguminosae)

Anselmo Nogueira[1], *Lorena B. Valadão-Menezes*[2], *Poliana Cardoso-Gustavson*[1], *Marília M. Quinalha*[2], *Juliana Villela Paulino*[4], *Elza Guimarães*[2], *Juliana G. Rando*[5], ***Juliana Hanna Leite El Ottra***[3]
1. *Universidade Federal do ABC*
2. *Universidade Estadual Paulista, Campus Botucatu*
3. *Universidade de São Paulo*
4. *Universidade Federal do Rio de Janeiro*
5. *Universidade Federal do Oeste da Bahia*

Chamaecrista (Cassinae, Fabaceae) has floral traits associated with pollination by vibration. Their yellow corollas, poricidal heterantherous anthers, enantiostylous and asymmetric flowers without a nectary are a suite of features considered to have evolved several times within Cassiinae, due to convergent evolution towards buzz pollination mechanism. In such system, anthers are the most valuable tissues concealing pollen exploited by bees as sole floral resource. From the plants perspective, flowers needs to overcome the "pollen dilemma", providing pollen for bee pollinators and also assuring the fertilization success, thought the development of heteranthery and other modifications in flowers, usually associated to the "division of labor" hypothesis among stamens. In this sense, floral morphology and timing of floral organ development have a crucial role in these plants because bees can feed on flowers in distinct floral stages, what could affect pollination. However, detailed studies focusing on differences in the pollen grains in heteromorphic stamens, as well as the relation of floral development to the floral biology has rarely been addressed in Cassinae. Here we investigated the association between floral morphology and growth patterns during flower development with the use of pollen by bees in *Chamaecrista desvauxii* var. *latistipula*. First we evaluated the detailed morphology and patterns of time of development from bud stage to flower opening, considering each inner floral whorl as well as the entire bud. We also evaluated bee visiting behavior, germination of pollen grains from different anther morphs (short and long), and stigma receptivity. We found that flower bud growth is initially fast until reach around 9 mm (elongation phase), gradually slowing down after. Petals and carpel have a delay in their development, showing an exponential growth pattern. Differently, both stamen whorls elongate linearly throughout flower development. Bee species like *Bombus* and *Oxea* had specific behaviors on anthers, and sonicated not only opened flowers but also advanced floral buds. In these flower stages, the stigma was receptive as well as the pollen grains were viable in both short and long anthers. Thus pollination can occur before flower opening, considering that anthers were easily accessed by bees in advanced buds due to the delayed development of petals in this stage. We also found that pollen grains have the same structural and histochemical features regardless their original anther morph, refuting the 'division of labour' hypothesis in this aspect. However, qualitatively the three long anthers had more pollen grains than the short anthers, due to more space available in their thecae. Furthermore, the long anthers always were located facing the inner side of a large modified concave petal, and with similar length to the carpel, suggesting morphological adjustment for pollination. Thus the potential difference in the role of anthers with different sizes could not be associated with the pollen itself, but with relative position of anthers to other floral organs and in the amount of pollen available for transfer. This is one of the first studies in Fabaceae that related the absolute measured time in floral development to the floral biology of a species.

T2
P0353

Floral structure of Neotropical Sapindales: Implications to the systematics and evolution of the order on a global scale

Juliana Hanna Leite El Ottra, *José Rubens Pirani*
Universidade de São Paulo - Instituto de Biociências

Sapindales is a monophyletic order within the malvids clade (core eudicots). The order comprises nine families with ca. 473 genera, mostly distributed in tropical and temperate areas of the world. In the Neotropical region six sapindalean families occur, *i.e.*, Anacardiaceae, Burseraceae, Meliaceae, Rutaceae, Sapindaceae and Simaroubaceae. This region presents ca. 40% of the species of the order, making the neotropics one of its major diversity centres. Sapindalean taxa have well diverse floral morphologies and thus represent an interesting group to address questions on floral evolution. So far, the only previous study focusing on floral evolution in the order was based on a limited sample of taxa (only 13 genera), including few neotropical representatives. As a consequence, to date, there is still an incomplete knowledge on how the floral diversity among the Sapindales has evolved. More recently, a raising amount of detailed studies focusing on the floral structure and ontogeny of several sapindalean families has been carried out (i.e., Nitrariaceae, Kirkiaceae, Anacardiaceae, Burseraceae, Rutaceae and Simaroubaceae). Among the groups with neotropical representatives, only for the Meliaceae there are a few detailed comparative studies on floral structure. Nevertheless, there is a lack of studies describing the inner morphological surface of carpels of Meliaceae, preventing an accurate comparison to the gynoecium structure of other Sapindales where such studies are available, especially in relation to the most closely related families to Meliaceae, the Rutaceae and Simaroubaceae. As new molecular phylogenetic reconstructions of the Sapindales taxa are being made, current hypotheses on the relationships among its members have been changed and improved in several regions of the phylogeny. Still, some relationships among sapindalean families are not yet well supported, as for instance the sister relationship of Meliaceae to Simaroubaceae, and not to Rutaceae. Considering this context, we aim to conduct new structural and evolutionary studies on flowers of Sapindales members, aiming to provide novel characters to support systematic relationships from a structural point of view, and to achieve a better knowledge on the evolutionary patterns within the order. Among our preliminary results we highlight the following ones: structural and ontogenetical studies on the flowers of Meliaceae confirm previous assumptions that their gynoecium is truly syncarpous, with and extensive symplicate and synascidiate regions. This stands out very differently from the most common gynoecium structure reported for the Rutaceae and Simaroubaceae, whose carpels are mostly apocarpous, with an extensive plicate or symplicate zone. Other typical feature in these two latter families is a short remnant of the floral apex at the centre of carpels. We noticed that a similar remnant also occurs in Biebersteiniaceae, which is an early divergent lineage of Sapindales. Thus this feature likely represents a plesiomorphic state, with multiple evolutions within the order. Ancestral character reconstructions are being conducted in order to evaluate the probable evolutionary pattern of selected floral features, such as nectary shape and position, occurrence of tubular structures in the flower, and structural floral sex.

T2
P0354

Sex, drugs and pupusas: Disentangling relationships in Echiteae (Apocynaceae)

J. Francisco Morales[1], *Mary E. Endress*[2], ***Sigrid Liede-Schumann***[1]
1. University of Bayreuth
2. University of Zurich

Apocynaceae are widely used in traditional medicine around the world due to their rich array of bioactive secondary metabolites, including complex indole alkaloids, steroidal alkaloids and cardenolides. In a few genera of apocynoids, a much more rare type of secondary compound is found – pyrrolizidine alkaloids (PAs). Although toxic secondary metabolites in plants have arisen as a means of phytochemical protection for the plant against herbivores, in diverse insect groups, certain lineages have become adapted to the toxic PAs and have evolved a close association with plants containing them. The poisons are ingested by the larvae as they feed on leaves of the host plants, or in a more derived behavior known as pharmacophagy, adult insects imbibe PAs from injured or withered plant parts. PA-adapted insects are able to sequester the toxic compounds and then put them to their own use against predators. Two groups of Lepidoptera – arctiid moths and danaid butterflies – have evolved an even more specialized use of PAs. Males use them as the chemical basis for male-mating pheromones. Experiments have shown that scent-producing glands of males with no PAs in their diet were merely two stunted stalks, whereas those of males that were fed PAs developed four long tubes densely covered with long pheromone-emitting hair-pencils, demonstrating that these species have evolved a dependence on the toxic PAs in their host plants to produce pheromones. At dusk males expand their hair-pencils, emitting pheromones that attract females as well as other males. The result is the formation of a lek, a mating arena in which group sex takes place. In Apocynaceae, the largest number of PA-producing genera is found in Echiteae. In order to determine the distribution of PAs within the tribe, we did a molecular phylogenetic analysis, including representatives of all subtribes and 17 of the 19 recognized genera, onto which we mapped PA-containing species. Echiteae and four of its subtribes

(Echitinae, Parsonsiinae, Peltastinae and Prestoniinae) as currently circumscribed were shown to be non-monophyletic. The fifth subtribe, Pentalinoninae, though monophyletic, was resolved as sister to a clade of Odontadenieae. Two genera, *Fernaldia* and *Peltastes*, were nested within *Echites* and *Macropharynx*, respectively, and were synonymized. Fourteen genera were maintained, divided among five subtribes. Species reported to have associations with PA-adapted insects were found in all subtribes except one, for which no data on secondary compounds was available. Since the characteristic secondary metabolites in the closest relatives of Echiteae, Odontadenieae and Mesechiteae, are cardenolides, we hypothesize that an evolutionary shift took place in the ancestors of Echiteae, in which these were replaced by parsonsine type PAs as the predominant defense compounds. In un-adapted vertebrates, PAs are slow-acting toxins that over time can lead to debilitating chronic medical conditions. In tropical countries, where Apocynaceae often figure prominently in treating a wide number of ailments, herbal teas, dietary supplements and traditional medicines made from plants containing PAs pose a potential health threat to the patients who consume them.

T2

P0355

The 600-year vegetation dynamic record from western Thailand using palaeological approach

Apichaya Englong[1], *Paramita Punwong*[2], *Paweena Traiperm*[1]
1. Department of Plant Science, Faculty of Science, Mahidol University, King Rama VI Road, Bangkok
2. Faculty of Environment and Resource Studies, Mahidol University, Phutthamonthon Sai 4 Road, Nakhon Pathom

The last 600-year vegetation dynamics are reconstructed from pollen and sediment set within a chronological framework from radiocarbon dating of a core (NP3) collected from Nong Pling freshwater spring areas, in Thong Pha Phum, western Thailand. The results show that lowland forest and terrestrial herbaceous were established in this area at least 600 years ago suggesting drier conditions than present until around 340 years before present (BP). After that, wetter condition occurred indicated by a decline in lowland forest and an increase in freshwater spring taxa until around 140 years BP. Oscillations between freshwater spring taxa and lowland forest occurred probably due to climate variability particularly during the last century although these changes may also be due to human activities suggested by the presence of cereal pollen.

T2

P0356

Crocus L. riddle, in the middle of its diversity center: A seek for a resolved phylogeny

Levent Can, *Osman Erol, Orhan Küçüker*
Istanbul University

The genus *Crocus* L. belongs to the family Iridaceae (Asparagales) and is distributed in the Northern Hemisphere, from western China to western Europe and north-western Africa, with a diversity center in Anatolia, Greece and the Balkan Peninsula. The plants belonging to this genus are geophytes with a subterranean corm, which helps the plant to survive the harsh conditions of a year, with a price of having a very short, a considerable cool and rainy time to flower and attract their pollinators. Most of the geophytes overcome this situation by producing flowers with remarkable colours and temptress scents for their pollinators, when not many other plant species are flowering. Consequently, like many geophytes, the crocuses obtained their places in rock gardens, private gardens or botanical gardens. The taxonomy and phylogeny of this genus is quite complex and yet unresolved. One of the reasons for that is the fact that the plant is much reduced and the exiguity of distinctive characters. Prophyll presence, corm tunics, style branching, flower and throat color are the most important characters, but additionally, distribution information and flowering periods are often used to distinguish the taxa. In most cases, the chromosome numbers are also not informative. Another difficulty is the "subspecies problem" in this genus, where many taxa with morphological similarities were ranked and grouped as subspecies. For instance, *Crocus biflorus* Mill. had more than 20 subspecies, which probably does not represent a natural systematics of the taxa. On top of that, the genus *Crocus* has been taxonomically and systematically very active in the last years. Recent studies have published a series of new taxa, and changed the taxonomy and phylogeny of the genus dramatically. Especially in Turkey, nearly as much as half of Turkish taxa were discovered and published in the past sixteen years. For these reasons above, here, we aimed to update the current number of crocuses, and to summarize the developments published recently, along with a brief review on the classification of the genus. We also aim to present the latest taxonomical and molecular systematics results of our target species complexes and to answer following questions: is *Crocus paschei* Kernd. a true species, does the *Crocus olivieri* J.Gay name involve two species as suggested 130 years ago, what is the relatedness of *Crocus flavus* Weston and *Crocus antalyensis* B.Mathew species groups, and what is the variation amongst *Crocus pallasii* Goldb. species group. By doing so, we hope to shed some light on the actual situation of the genus.

T2

P0357

The Turkish endemic *Pseudodelphinium turcicum* (Ranunculaceae): An unusual population of *Delphinium* with peloric flowers that has persisted in the wild for 20 years

Felipe Espinosa Moreno[1,2], *Thierry Deroin*[1], *Kun-Li Xiang*[3,4], *Wei Wang*[3,4], *Myreya Pinedo Castro*[5], *James Byng*[6,7], *Zeki Aytaç*[8], *Sophie Nadot*[2], *Florian Jabbour*[1]
1. Muséum national d'Histoire naturelle
2. Université Paris-Sud
3. Chinese Academy of Sciences
4. University of Chinese Academy of Sciences
5. Pontificia Universidad Javeriana
6. Plant Gateway
7. Naturalis Biodiversity Center
8. Gazi University

The Turkish endemic *Pseudodelphinium turcicum* (Ranunculaceae) was described by Vural *et al.*, in 2012 based on a single population found in the salt lake basin of Tuz Gölü, Konya Province. Although the authors noticed morphological similarities between the specimens and the genera *Delphinium* and *Garidella*, they highlighted unusual features, which led them to describe a new genus name consisting of a single species, *P. turcicum*. We carried out morphological, anatomical, and palynological studies on individuals of this species. We also conducted a molecular phylogenetic analysis to identify the closest relatives of this species. Using this combination of approaches, we showed that the species belongs in *Delphinium* subg. *Delphinium*, distributed (exclusively or not) in Mediterranean and adjacent regions; it is more precisely sister to two species (*D. virgatum* and the Turkish endemic *D. venulosum*). We conclude that it is an unusual population of *Delphinium* presenting peloric flowers that has been maintained in the wild for at least two decades. We provide hypotheses concerning the identity of the floral organs, as well as an explanation for the origin of the floral teratological characters.

T2
P0358

Revision of *Ixora* (Rubiaceae) in Vietnam

Thi Huyen Trang Thieu[1], *Sangho Choi*[1], *The Bach Tran*[2], ***Sangmi Eum***[1]
1. International Biological Material Research Center, Korea Research Institute of Bioscience and Biotechnology,
2. Institute of Ecology and Biological Resources, Vietnam Academy of Sciences and Technology

Ixora is a genus of Rubiaceae widespread in tropical Africa, America, Asia, Madagascar and Pacific islands. Thirty-four species have been reported in Indochina region by F. Ganepain and his colleagues since 1923. However the number of species of *Ixora* and its taxonomic status in Vietnam are still unclear and untouched up to now. In this study, all Vietnamese species of *Ixora* are taxonomically revised on the basis of morphological characters. The results after conducting field trip and examination of specimen in Vietnam enable us to confirm twenty four species and six varieties in Vietnam including a new recorded species, *I. auricularis* Chun & F.C. How ex W.C. Ko, previously has been known only from China and newly founded from northern and center of Vietnam. We also suggest that inflorescence, apex shape of corolla and base shape of leaf can be main key characters to identify for *Ixora* species in Vietnam. Details of distribution, illustrations for main species and an identification key of *Ixora* are provided.

T2
P0359

Phylogeny of *Eremochloa* (Rottboelliinae: Panicoideae: Poaceae): Implications for an endangered species in Australia

Melodina Fabillo, *Tanya Scharaschkin*
Queensland University of Technology

Eremochloa is a genus of grasses, with approximately 12 species, found mainly in the open woodlands of Asia and Australia. *Eremochloa* is characterized by lower glumes bearing apical wings and spiny keels. The three species of *Eremochloa* that occur in Australia are also found in Asia. *Eremochloa bimaculata* and *E. ciliaris* are widely distributed in eastern Australia. *Eremochloa muricata* is known from only two records that are over 2000 km apart; one from Cape Flattery, Queensland (1976) and the other from Byron Bay, New South Wales (1912). *Eremochloa muricata* has been declared an endangered species in Queensland. Identification of the Australian species, using currently accepted keys, is problematic, especially as species descriptions do not consistently agree with the key. The monophyly of *Eremochloa* has not been tested and a genus-level phylogenetic study is lacking. The objective of this study was to conduct a comprehensive morphological and molecular study of the genus, so as to undertake a taxonomic revision of the Australian *Eremochloa* based on phylogenetic results, in conjunction with extensive fieldwork to gain a better understanding of the geographic range of the three Australian species. Preliminary results based on Bayesian inference, maximum likelihood and maximum parsimony analyses of chloroplast sequence data (*atp*B-*rbc*L spacer, partial *mat*K gene, partial *ndh*F gene, *trn*L-F intron and intergenic spacer) show that the genus *Eremochloa* is monophyletic. *Eremochloa bimaculata* is not monophyletic; the Asian accessions are more closely related to other Asian species of *Eremochloa* than to the Australian accessions of *E. bimaculata*. The endangered Australian species, *E. muricata,* is more closely related to Australian *E. bimaculata* than to the Asian *E. muricata.* These findings are consistent with our detailed observations of gross morphological features. The Australian accessions of *E. bimaculata* exhibit extensive morphological variation across the geographic distribution. The 1912 specimens of *E. muricata* has been incorrectly identified whilst the 1976 specimen cannot be differentiated from specimens of *E. bimaculata* collected in the vicinity of Cape Flattery as part of this study. Preliminary results indicate the need for the Australian *E. bimaculata* to be separated from the Asian species. *Eremochola muricata* from Australia should to be treated as a variation of the widespread and variable Australian species *E. bimaculata* and can be removed from the endangered species list. However, if further fieldwork in the Cape Flattery produces specimens that do indeed turn out to be *E. muricata*, these could just as well be first records of a non-native grass, with the potential of being naturalised and invasive, and should be eliminated rather than being listed as an endangered species.

T2
P0360

Divergence times of the tribe Astereae inferred from cpDNA *trn*L/*trn*F and two nrDNAs, ITS and ETS sequences data in Iran

Tayyebeh Farhani
Tarbiat modares university

Astereae with approximately 222 genera and 3100 species is the second largest tribe of Asteraceae distributed worldwide especially in temperate regions with the major centers of diversity within southwestern North America, the Andes, South Africa, Australia, and New Zealand. This tribe comprises 15 genera and 38 species in Iran. The genus *Psychrogeton* with 11 species is the largest genus of tribe in Iran that distributed in higher elevation of Irano-Touranian region. The aim of the present study is to reconstruct the phylogenetic relationships and divergence times in the tribe using the concatenated nrDNA (ITS, ETS) and cpDNA (the *trn*L Intron and *trn*L/*trn*F Intergenic Spacer) data sets, based on Bayesian and BEAST analyses. In this study we used 95 species from 25 genera of Astereae. *Calendula officinalis*, *Chrysanthemum coronarium* and *Chrysanthemum indicum* as outgroups were also included in the analyses. The inferred phylogeny resulted from Bayesian presented a framework to reconstruct the origin of diversification in the tribe. The tree of BEAST analysis is largely congruent with those resulting from Bayesian analysis. Dating analysis of the combined data estimates the age of 58.4 Mya (Paleocene-Eocene boundary) for the stem node. Divergence time estimates indicate that Astereae began to diversify ca. 38.6 Mya (34.7–42.6 95% HPD) in the Middle-late Eocene and extension of the Range of lineage occured mostly during the late-Oligocene to Pleistocene. Our results represent the genus *Erigeron* has undergone rapid diversification and recent speciation in the Late Pliocene and the Early-to-Mid Pleistocene similar to *Erigeron* species in the European Alps, likely during the glaciations in Pliocene and Pleistocene.

T2

P0361

A preliminary infrageneric classification of the species-rich genus *Silene* (Caryophyllaceae): Molecular phylogeny and morphology revisited

Jafari Farzaneh[1], *Gholipour Abbas*[2], *Oxelman Bengt*[3], *Zarre Shahin*[1]
1. University of Tehran
2. Payame Noor University
3. University of Gothenburg

Silene L. with about 700 species is the largest genus of Caryophyllaceae. The delimitation of *Silene* has been controversial. Through merging the genera *Atocion* Adans., *Viscaria* Bernh., *Heliosperma* Rchb., *Pertrocoptis* A.Braun ex Endl., *Eudianthe* Rchb. and *Lychnis* L. in *Silene*, some authors proposed a lumping approach. However, the modern treatments of *Silene* suggest a narrower circumscription of the genus splitting it into distinct genera. The genus is widely spread throughout temperate regions of the North Hemisphere with centers of diversification In the Mediterranean area, SW Asia and Central Asia. The phylogenetic backbone of *Silene* and allied genera has been reconstructed in previous studies. None of the current infrageneric classifications are completely supported by those ones. In the present study, we conducted comprehensive phylogenetic analyses using nrDNA ITS and cpDNA *rps*16 sequences for 262 samples of *Silene* and its allied genera representing of 42 sections recognized by Chowdhuri. Maximum parsimony and Bayesian inference methods were used. STACEY v. 1.2.2 package was implemented in BEAST v. 2.4.4. to estimate species trees and minimal cluster trees. The results support dividing *Silene* to two subgenera, i.e. *S.* subgen. *Behenantha* (Otth) Endl. and *S.* subgen *Silene*, which is in agreement with the previous molecular phylogenetic findings and contradicts Rohrbach's classification. Our study corroborates recognizing following sections in *Silene*: *S.* sect. *Physolychnis* (Benth.) Bocquet, *S.* sect. *Melandrium* (Röhl.) R.K. Rabeler, *S.* sect. *Conoimorpha* Otth. Majority of *Silene* spp. are placed in *S.* subgen. *Silene*. According to the gained phylogenetic results of the present study: (1) *S.* sect. *Lasiostemones* is not monophyletic and majority of its representative are nested within *S.* sect. *Sclerocalycinae* in *S.* subgen. *Silene*. The circumscription of these two sections needs to be clarified; (2) The members of *S.* section *Auriculatae* (Boiss.) Schischkin characterized by perennial habit and auriculate petal claw form a clade with taxa from various sections that might be exauriculate, annual or perennial, gynodioecious or not; (3) Suffruticose plants of *S.* sect. *Spergulifoliae* (Boiss.) Schischkin are closely related to two annual members of *S.* sect. *Lasioclycinae* (Boiss.) Chowdhuri; (4) *S. gallica* L. as the type of the genus forms a core clade with some predominantly annual sections distributed in the Mediterranean area; (5) *S. nizvana* is isolated from other members of *S.* sect. *Auriculatae* sensu Melzheimer; (6) The members of *S.* sect. *Rigidulae* (Boiss.) Schischkin are splitted into several clades, and the circumscription of *S.* sect. *Rigidulae* need to be improved and reduced to exclude.

T2

P0362

Genetic architecture of flower and leaf divergence in a pair of sympatric sister species of *Primulina*

Chen Feng[2], *Chao Feng*[1], *Lihua Yang*[1], *Ming Kang*[1]
1. South China Botanical Garden, CAS
2. University of Chinese Academy of Sciences

Cave-associated plants provide an attractive system for studying phenotypic evolution by the extreme habitats strongly selected a wide range of plant traits through abiotic and biotic properties. The plants in *Primulina* display remarkable interspecific variation in flower and leaf traits and occurs in diverse light regime (from exposed steep cliffs to twilight zones of caves), thus it is an excellent model for studying phenotypic adaptation to extreme cave environments. We conducted QTL analyses of 16 adaptive traits on a pair of *Primulina* sister species which were found sympatric in a cave entrance of Danxia Mountain. We found that interspecific divergence in all traits is due to many QTL with minor to medium effect including highly epistatic interactions among them. QTLST test results showed that flower size was under directional selection, while contrary for flowering time and leaf traits. Our findings are very important for understanding the genetic basis of adaption to cave and cave-like habitats in plants.

T2

P0363

Allopatric divergence, local adaptation, and multiple quaternary refugia in a long-lived tree (*Quercus spinosa*) from subtropical China

Li Feng[1], *Yan-Ping Zhang*[1], *Jia Yang*[1], *Tao Zhou*[1], *Guo-Qing Bai*[1,2], *Jiao Yang*[1], *Zhong-Hu Li*[1], *Ching-I Peng*[3], *Gui-Fang Zhao*[1]
1. Northwest Universtiy
2. Xi'an Botanical Garden
3. Biodiversity Research Center, Academia Sinica

The complex geography and climatic changes occurring in subtropical China during the Tertiary and Quaternary might have provided substantial opportunities for allopatric speciation. To gain further insight into these processes, we reconstruct the evolutionary history of *Quercus spinosa*, a common evergreen tree species mainly distributed in this area. Forty-six populations were genotyped using four chloroplast DNA regions and 12 nuclear microsatellite loci to assess genetic structure and diversity, which was supplemented by divergence time and diversification rate analyses, environmental factor analysis, and ecological niche modeling of the species distributions in the past and at present. The genetic data consistently identified two lineages: the western Eastern Himalaya-Hengduan Mountains lineage and the eastern Central-Eastern China lineage, mostly maintained by populations' environmental adaptation. These lineages diverged through climate/orogeny-induced vicariance during the Neogene and remained separated thereafter. Genetic data strongly supported the multiple refugia (per se, interglacial refugia) or refugia within refugia hypotheses to explain *Q. spinosa* phylogeography in subtropical China. *Quercus spinosa* population structure highlighted the importance of complex geography and climatic changes occurring in subtropical China during the Neogene in providing substantial opportunities for allopatric divergence.

T2

P0364

The high content of tannins in *Rhizophora apiculata* as a mechanism underlying high salinity tolerance

Xiao Feng, *Shaohua Xu, Jianfang Li, Ziwen He, Suhua Shi*
Sun Yat-sen University

As mangroves have colonized and well adapted to the extreme intertidal environment with high salinity, hypoxia and other abiotic stresses, they are attractive models for studying genetic mechanisms underlying environmental adaptations. *Rhizophora apiculata* is a typical mangrove with high contents of tannins which is likely related to salt resistance. Using high throughput RNA-seq data of *R. apiculata* treated with different salt concentrations, we detected tannin biosynthesis pathway significantly enriched of differentially expressed genes. Particularly, *RA_DFR (dihydroflavonol 4-reductase)*, a key enzyme coding gene in the pathway, not only had elevated expression level with salt treatment, but also retained another copy missed in inland relatives. To reveal the up-stream regulatory mechanism, we further identified candidate transcription factors that may regulate these differentially expressed genes in the pathway. The quantification of tannins in the root tissue samples confirmed that the content of tannins increased with salt treatment. Overall, this study identified genes that potentially associated with high tannin content, and shed light on the mechanism of tannin-based high salinity tolerance in *R. apiculata*.

T2

P0365

Comparative analysis of complete chloroplast genomes improves phylogenetic resolution in Smilacaceae

Yu Feng, *Pan Li, Ying-Xiong Qiu, Cheng-Xin Fu*
Key Laboratory of Conservation Biology for Endangered Wildlife of the Ministry of Education, College of Life Sciences, Zhejiang University

Smilacaceae, composed of only one genus, *Smilax*, is a monocot family of lianas, shrubs and herbs widely distributed across the world. The family comprises more than 200 species characterized by climbing habit, reticulate leaf venation, paired petiolar tendrils, unisexual flowers, and superior ovaries. Previous molecular phylogenetic studies have recovered Smilacaceae as monophyletic. However, the deep-level relationships still remain poorly resolved, thus impeding our understanding of the evolution of the family. Plastome sequencing is now an efficient option for increasing phylogenetic resolution in plant phylogenetic and population genetic analyses. Here, the complete chloroplast (cp) genomes of 30 species covering its phylogenetic diversity and almost all geographic distribution of this family were sequenced, to better understand the evolutionary relationships among these plants and the mutation patterns in their cp genomes. The Smilacaceae chloroplast genomes exhibited a typical quadripartite structure and ranged from 157,646 bp to 158,691 bp in length. The cp genomes contained 113 unique genes, including 29 tRNA, 4 rRNA 78 protein coding genes and 3 pseudogenes (*ycf15*, *ycf68* and one copy of *ycf1*). Comparative analyses indicated that the overall cp genome structure among these taxa was quite similar. Sequence divergence in IR regions (Pi = 0.00174) was much lower than that in LSC (Pi = 0.0074) and SSC (Pi = 0.00918) regions. Seven highly divergent DNA regions (*accD*, *rbcL*, *ycf1* for protein coding regions and *trnK-atpA*, *trnT-trnM*, *accD-psaI*, *ndhF-trnL* for non-coding regions), suitable for species identification and phylogeographic analyses, were detected among the 30 cp genomes. Phylogenetic analyses based on whole cp genome data provided robust support for the deep relationships, verifying the ability of whole cp genome to solve deep-level phylogenetic relationships. The findings highlight the potential of the whole cp genomes for improving resolution in phylogeny as well as species identification in phylogenetically and taxonomically difficult plant groups.

T2

P0366

Sorting out the ginger paradise: Phylogenomics of the family Zingiberaceae using Hyb-Seq approach

Tomas Fer[1], *Roswitha Schmickl*[2], *Eliska Zaveska*[1], *Monika Pospisilova*[1], *Jana Leong-Skornickova*[3], *Otakar Sida*[4], *Axel Poulsen*[5], *Mark Newman*[5], *Monica Carlsen*[6], *W. John Kress*[6]
1. Department of Botany, Faculty of Science, Charles University, Prague, Czech Republic
2. Institute of Botany, Czech Academy of Sciences, Pruhonice
3. Research & Conservation Branch, Singapore Botanic Gardens, Singapore
4. Department of Botany, National Museum in Prague
5. Royal Botanic Garden Edinburgh
6. Botany Department, National Museum of Natural History, Smithsonian Institution, Washington, D.C.

Traditional molecular phylogenetic methods utilizing a few nuclear and chloroplast regions are increasingly being replaced by next-generation sequencing (NGS) approaches. Hybridization-based library enrichment (Hyb-Seq) has become a very efficient application allowing targeting several hundreds to thousands of loci with often a small amount of missing data across the whole family or even order. As the method is combined with genome skimming, nearly complete chloroplast sequences and the whole rDNA cistron are also obtained. Such phylogenomic datasets produce well resolved and highly supported evolutionary hypotheses. The Hyb-Seq methodology was applied in the ginger family (Zingiberaceae). 1,180 loci were captured and 694 loci were selected for the analysis of 225 species covering all tribes and almost all genera within the family. The species trees were constructed using concatenation, supertree (MRL) and coalescence-based methods (ASTRAL, ASTRID). Almost all nodes were highly supported except several deep nodes in the phylogeny. Within Zingiberoideae four main lineages were resolved: Globbeae (*Globba*, *Hemiorchis*, *Gagnepainia*), Curcumeae (*Curcuma*, *Camptandra*, *Pyrgophyllum*), *Monolophus*, and Zingibereae (all other genera). Within Alpinioideae there are many supported groups including six *Alpinia* lineages, five *Amomum* lineages, and three *Hornstedtia* lineages. Incongruence among the nuclear loci and the nuclear and plastome dataset pointed out the possible hybrid origin of some genera (i.e., *Newmania*, *Pyrgophyllum*, *Monolophus*, and some *Alpinia* and *Amomum* lineages). In summary, we demonstrated the utility of the Hyb-Seq method for constructing robust and well-resolved phylogenies of the whole family.

T2
P0367
Systematics, diversity and historical biogeography of Neotropical synandrous Mimosoids – the Ingeae tribe

Julia Ferm[1], *Catarina Rydin*[2,3]
1. Uppsala University, Department of Organismal Biology, Systematic Biology
2. The Bergius Foundation, the Royal Academy of Sciences
3. Stockholm University, Department of Ecology, Environment and Plant Sciences

The delimitation of the legume tribe Ingeae, often referred to as synandrous (stamen filaments fused) mimosoids, have long been in a state of flux, with many segregate genera being proposed and/or merged with other genera. This study intends to investigate selected ingoid genera (*Cojoba*, *Punjuba*, *Zygia*/*Marmaroxylon* and *Zapoteca*) with questionable taxonomic status in relation to other genera in the tribe. The project is divided into several parts working at different taxonomical levels, exploring the phylogeny and taxonomy using both morphological and molecular data. With the phylogenetic hypothesis as a basis, the aim is to provide a classification as well as species and subspecies delimitations, that reflect the evolutionary history of the group. The project also aims at exploring the historical biogeography of these species, mainly distributed in the West Indies, Mexico together with Central America, and South America, as well as the evolution of morphological structures and cues.

T2
P0368
Plant taxonomy via geometric morphometrics: Delimiting *Hydrocotyle* L. (Araliaceae) species from South America

Pedro Fiaschi, *Eduardo Nery*
Federal University of Santa Catarina

Geometric morphometrics (GM) is a set of methods that allow an independent quantification of shape and size of biological structures, as well as their statistic evaluation. Although disseminated in other fields of biological inquiry, GM is not applied in plant taxonomy. Nevertheless, GM can reveal morphological gaps along character variation, which could be useful to set boundaries between similar species. The similarity between species is a common issue in *Hydrocotyle*, in which species delimitation is complicated by a large interspecific morphological variation. Among the morphological variable species is *Hydrocotyle quinqueloba* Ruiz & Pav., which currently comprises nine infraspecific taxa, each more or less recognizable by their leaf morphology and indument. However, such treatment may include currently non-recognized species since *H. quinqueloba* displays a disjunct distribution between the Andes and the Brazilian Atlantic forest in South America. Hence, we investigated the morphological variation of *H. quinqueloba* to evaluate whether its current delimitation includes other species or not. We applied landmark-based GM on leaves of 210 specimens of *H. quinqueloba* to assess species' morphospace cohesion, and we compared specimen's morphology to find potential discontinuous and discriminant characters. A principal component analysis (PCA) revealed morphospace gaps among some infraspecific taxa, which could be separated into four groups. A nested analysis of variance (ANOVA) Procrustes indicated a significant effect ($p<.001$) of these groups on the major part (60%) of leaf shape variation. A Hotelling-Lawley T^2 test indicated significant differences ($p<.001$) of means among these four groups along different shape components (PCs). In a similar way, comparative morphology among these groups revealed discontinuous states of leaf margins, number of flowers per inflorescence, petal coloration and fruit size, and discriminant states were found along the indument and pedicel size. Considering the morphospace gaps and discontinuous character states, we proposed the recognition of three infraspecific taxa of *H. quinqueloba* as species. The first species correspond to *H. quinqueloba* var. *asterias*, with small sized leaves, triangular leaf lobes, non-mucronate leaf margins, 12-18 flowers per inflorescence, short pedicels and few large fruits. The second species correspond to *H. quinqueloba* var. *angulata*, with middle sized leaves, unequal and deltate leaf lobes, 20-25 flowers per inflorescence, and small fruits. The third species correspond to *H. quinqueloba* var. *macrophylla*, with large leaves, widely deltate leaf lobes, 45-70 flowers per inflorescence, and red veined petals. Thus, our study shows that GM can be used in taxonomical studies as a way of identifying morphological gaps and groups, which can later be corroborated by traditional comparative morphology.

T2
P0369
Tovomita (Clusiaceae): Shedding light on the systematics of a neglected Neotropical tree genus

Lucas Marinho[1], *Benjamin Gahagen*[2], ***Pedro Fiaschi***[3], *Franciso De Assis Dos Santos*[1], *Cássio Van Den Berg*[1], *André Amorim*[4]
1. Feira de Santana State University
2. Ohio University
3. Federal University of Santa Catarina
4. Santa Cruz State University

Tovomita is the second largest genus of Clusiaceae in the Neotropics; yet, up to three years ago it was taxonomically poorly known when compared, for example, with *Clusia*, *Chrysochlamys*, and other smaller genera. Recent efforts to enhance taxonomical knowledge of Brazilian Atlantic Forest, Central American, and Caribbean species of *Tovomita* have been carried out through collaborative studies and herbarium determinations in Brazil, Europe and U.S.A. Field expeditions to search for poorly known species of *Tovomita* have been intensified in Brazilian Atlantic and Amazon rainforests, to collect samples for phylogenetic studies aiming to test current classification of Clusieae. In the last few years, the number of *Tovomita* species was increased from ca. 45 to 76 (increment of ca. 70%), due to description of 11 new species, re-establishment of eight names under synonymy, as well as proposal of new names for species previously included in *Dystovomita* and *Clusia*. Moreover, five synonyms and 12 lectotypifications were proposed, and the *Tovomita weddelliana* complex was assessed using a multivariate morphometric approach. Regarding morphological data, pollen morphology of 13 species have been studied, and leaf papillae were for the first time registered in the genus. Leaf macro- and micromorphology and a preliminary phylogenetic hypothesis based on nrITS provide support for the late nineteenth century Vesque's infrageneric classification of *Tovomita*. Current studies are underway to investigate the systematics, morphological characters of historical importance, and their value for the generic circumscription in Clusieae, with the purpose of proposing a phylogenetic classification for *Tovomita* and its allies.

T2

P0370

Insights into the diversification and biogeography of *Phlegmariurus* (Lycopodiaceae: Lycopodiopsida)

Ashley Raymond Field[1], *Lucie Bauret*[2], *Germinal Rouhan*[2], *Myriam Gaudeul*[2], *Thaís Elias Almeida*[3], *Danilo Soares Gissi*[4], *Weston Testo*[5]

1. Queensland Herbarium, Department of Science & Innovation and Australian Tropical Herbarium, James Cook University

2. Institut de Systématique, Evolution, Biodiversité, Muséum National d'Histoire Naturelle, Paris

3. Herbário HSTM, Universidade Federal do Oeste do Pará, Santarém, Brazil

4. Department of Botany, University of São Paulo, São Paulo

5. Department of Plant Biology, University of Vermont, Vermont

Phlegmariurus (Lycopodiaceae) is a genus of over 300 species of rainforest epiphytes and montane terrestrials that has a worldwide tropical and subtropical distribution. Extant chloroplast-based phylogenetic investigations of the family have reconstructed an ancient divergence between Neotropical and Palaeotropical clades of *Phlegmariurus* with subsequent long distance dispersal being considered relatively rare. This contrasts with the extant morphology-based systematic classification of *Phlegmariurus* that suggests long distance dispersal is relatively more common. We reconstruct an evolutionary and biogeographic history of *Phlegmariurus* using phylogenetic and phylogenomic data from richly sampled case studies in Oceania, Asia, Afro-Madagascar and the Neotropics. We investigate the roles of dispersal from the Neotropics and Palaeotropics in the diversification of *Phlegmariurus* in Afro-Madagascar and Oceania, how habitat change in the Andes and Brazilian highlands has influenced diversification in Neotropical terrestrials and how epiphytic specialisation has influenced diversification among Palaeotropical epiphytes. In a challenge to the popular concept that lycopod species are relicts from old diversifications, we find that patterns of species diversification and biogeography in *Phlegmariurus* are more likely to be congruent to those found in ferns and angiosperms.

T2

P0371

Inflorescence, flower and fruit morphology of *Hopkinsia anoectocolea* (Anarthriaceae) and multiple origins of one-seeded fruits in the graminid clade of Poales

Constantin Fomichev[1], *Barbara Briggs*[2], *Terry Macfarlane*[3], *Dmitry Sokoloff*[1]

1. Department of Higher Plants, Faculty of Biology, Lomonosov Moscow State University, Moscow, Russia

2. National Herbarium of New South Wales, Sydney, Australia

3. Western Australian Herbarium, Perth, Australia

The majority of wind-pollinated species of flowering plants possess one-seeded fruits and often uniovulate gynoecia. There are several pathways of morphological evolution leading to gynoecia with a single ovule, but such changes have not been investigated in detail in some taxonomic groups. In addition, the same observations are sometimes used to support contrasting morphological interpretations. The graminid clade of Poales is of special interest, as within this group one-seeded fruits evolved in grasses as well as independently in several other lineages. We studied the south-western Australian family Anarthriaceae (included in Restionaceae in APG IV), in which one genus (*Hopkinsia*) has one-seeded indehiscent fruits and two other genera possess three-seeded capsules developing from a tricarpellate syncarpous gynoecium. The female flower of *Hopkinsia anoectocolea* has a trimerous two-whorled perianth (each tepal is vascularized by one bundle), three nonvascularized staminodia and one carpel. Tepal aestivation is unstable. Three vascular bundles depart towards the gynoecium. They are located on the radii of the outer whorl tepals. One of these bundles soon bifurcates tangentially, and each branch fuses with one of two other bundles. As a result, two vascular bundles can be seen at the base of the ovary. These can be interpreted as ventral and dorsal bundles of the carpel. Carpel orientation varies between median (abaxial) and transversal-adaxial; the former case is more common. Identification of carpel orientation relative to the flower-subtending bract required knowledge of inflorescence architecture that was controversially interpreted in earlier studies. Our data show that the basic units of the *Hopkinsia* inflorescence are spikelets, as in many other Poales. The exocarp is represented by cells with tannins. Their outer cell walls are covered by thick cuticle. There are about 13 cell-layers of mesocarp and one of endocarp. No mechanical elements were observed. The seed coat is thin and consists of two well-recognizable cell layers with tannin-like content. We interpreted these layers as exo- and endotegmen. The embryo is disk-like without any signs of differentiation. We interpret the evolutionary origin of the *Hopkinsia* gynoecium as total loss of two carpels of the ancestrally tricarpellate gynoecium (meristic variation). The *Hopkinsia* gynoecium, unlike the grass gynoecium, is therefore monomerous rather than pseudomonomerous. At the same time, we identified the special nature of the innervation of the *H. anoectocolea* carpel in comparison with related groups, which we associate with the new location of the single carpel in the center of the flower, in contrast to carpels of species having trimerous gynoecia. In the tricarpellate *Anarthria* gynoecium, there are three dorsal and three heterocarpellate ventral bundles. Carpel orientation is more labile in *Hopkinsia* than in many other lineages of Poales with one-seeded fruits. Morphological studies of CIF and DDS are supported by the Russian Science Foundation (project 14-14-00250).

T2

P0372

Morphological and cytogenetic studies in genus *Senna* Series *Aphyllae* (Caesalpinoideae, Leguminosae): Taxonomic and evolutionary implications

Federico O. Robbiati[2,3], *Juan D. Urdampilleta*[2,3], ***Renée H. Fortunato***[1,2,4]

1. Instituto Nacional de Tecnología Agropecuaria (INTA)

2. Consejo Nacional de Investigaciones Científicas y Técnicas (CONICET)

3. Universidad Nacional de Córdoba

4. Facultad de Agronomía y Ciencias Agroalimentarias, UM

Series *Aphyllae* Benth. belongs to the genus *Senna* Mill. (Leguminosae) and is comprised of xeromorphic shrubs and subshrubs, with deeply woody roots; the leaves of the adult branch are represented by minute triangular or sublobate scales; the stem is junciform, green, and photosynthetic. In this group the specific and varietal delimitation is still problematic due to the high morphological variation within and among taxa. We assessed the intra and interspecific variability using morphological and cytogenetic analyses (*in situ* hybridization: FISH) to understand the patterns of morphological variation, phylogenetic relationships and to test hypothesis about speciation in the different specific group in series *Aphyllae*. Based on multivariate analyses two specific complexes were recognized: *S. aphylla-S. pachyrrhiza* and *S. crassiramea-S. rigidicaulis*. Our results, suggest that x = 12 is the basic number for series *Aphyllae*. Moreover, this series was revealed as a polyploidy group with 2n = 24,48. The FISH revealed that in the polyploidy species there is not a missing of ribosomal sites indicating a recent origin of these polyploidy. Moreover, our results suggest that some underlying evolutionary mechanisms such as incomplete lineage sorting and/or reticulate evolution (hybridization and introgression) are shaping the relationships in *Aphyllae*. A broader cytogenetic survey of a more representative number of populations, along the geographic distribution of the series *Aphyllae*, should be done to address questions such as if the polyploidyzation is supported by environmental differences.

T2

P0373

A critique of the lineage concept of species and argument for a historical-phenotypic alternative

John Freudenstein[1], *Michael Broe*[1], *Ryan Folk*[2], *Brandon Sinn*[3]

1. Ohio State University

2. University of Florida

3. New York Botanical Garden

A conceptualization of species as lineages that are equivalent to single metapopulations has become common and has formed the basis for some analytical approaches to using molecular data for species delimitation. This approach is equivalent to delimiting species under a particular version of Mayr's "Biological" Species Concept. We argue that such a concept is suboptimal for a number of reasons. First, it only conceptualizes species in "width" (the size of a present group of organisms) and not in "length" (the extent through time of the species). Second, it equates species with another existing unit, the metapopulation, which is redundant. Third, it gives primacy to history with no regard to longstanding ideas about the meaning and relevance of species. We present an alternative to the strict lineage concept that focuses on the notion of biodiversity and its meaning. Species are commonly viewed as the key units of biodiversity; their importance in biodiversity is due to the unique role that each plays. Our approach, which combines elements from previous species definitions, views species as the smallest assemblages of related populations that have a distinct role. Distinct role is equivalent to having a distinct phenotype, since it is primarily phenotypic variability that specifies role. Lineages play a part in this concept, but are not the sole aspect of diversity that is viewed as important.

T2

P0374

Population genetics and phylogeny of *Cryptomeria* Inferred from nSSR and cpDNA data

Qing Ma[1], *Goro Kokubugata*[2], *Chengxin Fu*[1]

1. The Key Laboratory of Conservation Biology for Endangered Wildlife of the Ministry of Education, College of Life Sciences, Zhejiang University, P. R. China

2. Tsukuba Botanical Garden, Department of Botany, National Museum of Nature and Science, Amakubo, Tsukuba, Ibaraki,305-0005, Japan

Cryptomeria D. Don is a genus from the Taxodiaceae family endemic to East Asia. The genus includes two species, *C. japonica* D. Don and *C. fortune* Hooibrenk according to historic records. However, recent studies have considered these two species as a single species, *C. japonica* (Thunberg ex Linnaeus f.) D. Don, with *C. japonica* var. *sinensis* Miquel been regarded as a variety. The phylogenetic relationship and evolutionary history of these two species has largely remained unknown. This study used molecular data from both plastid and nuclear genomes and comprehensive sampling of all known populations of *C. japonica* and *C. fortunei* to investigate the population structure and genetic diversity of *Cryptomeria* populations, infer species differentiation and demographic history of *Cryptomeria* lineages, and provide recommendations for sustainable conservation efforts on these two species. Preliminary results from nSSR data are as follows: 1) nuclear microsatellite genotyping based on eight SSR loci of 18 *C. japonica* populations collected in Japan, 18 *C. fortune* populations collected in China, and two cultivated *C. japonica* populations collected from Jiangxi and Hubei Province in China revealed high genetic diversity (0.717 vs 0.779), allele richness (3.471 vs 3.914), and number of effective alleles (4.154 vs 4.708). AMOVA analyses showed that 88.31% of the variation was within individuals. The global F_{st} value was estimated to be 0.117. 2) STRUCTURE analysis yielded the highest likelihood when all the 38 populations, 479 individuals were clustered into three groups. The analysis of the genetic components carried out at the population level showed that the18 *C. japonica* populations collected in Japan and the two cultivated *C. japonica* populations collected in China were mostly assigned to cluster A. *C. fortune* populations collected from Fujian, Henan, and part of Zhejiang Province in China were mostly assigned to cluster B, while *C. fortune* populations collected from Sichuan, Hubei, Guizhou, Hunan, and other part of Zhejiang Province were generally assigned to cluster C. Most populations of *C. japonica* and *C. fortune* displayed a mixture of both gene pools. 3) IBD analyes indicated evident effect of isolation by distance among all the 38 populations, while no evident effect of isolation by distance was detected in either the *C. fortune* populations from China and *C. japonica* populations from Japan. Mismatch distributions were unimodal for both *C. japonica* and *C. fortunei*. The non-significant SSDH_{Rag} values did not support the spatial and demographic expansions of populations within either China or Japan. 4) The current result supported a single species within the *Cryptomeria* genus, with populations from Tianmu and Wuyi Mountain being the probable early-distributed wild *Cryptomeria* populations in China. These populations display higher genetic divergence with the populations from Japan and may be the earliest diverged clades of *Cryptomeria.*

T2

P0375

A taxonomic revision of Smilacaceae in Asia based on DNA barcoding and morphological traits

Xinjie Jin[1], *Pan Li*[1], *Zhechen Qi*[1,2], *Chengxin Fu*[1]
1. Zhejiang University
2. Zhejiang Sci-Tech University

On the basis of more than twenty years' research on morphology, molecular phylogeny and DNA barcoding, we present here a taxonomic revision of Smilacaceae in Asia, where is the diversification center of the family. Following the standard DNA Barcoding protocol, we sequenced 310 accessions comprising 1-3 individuals of 85 species from East Asia. The performance of five DNA barcodes (*rbc*L, *mat*K, ITS, ITS1 and ITS2) and their combinations were evaluated. The combination of *rbc*L + *mat*K + ITS provids the highest discrimination success (80.86%) in Smilacaceae. Therefore, DNA barcoding using *rbc*L + *mat*K + ITS can be employed as a supplementary means to make taxonomic revision in Asian Smilacaceae. Our revisons are as follows: (1) The genus *Heterosmilax* is reduced to be a section of *Smilax*, which makes *Smilax* the only genus in Smilacaceae; (2) A new taxonomic system for *Smilax* is proposed, which recognizes five subgenera and 22 sections, with about 129 species, 3 subspecies and 5 varieties occuring in Asia; (3) Five new species are discovered: *S. ligneoriparia* C.X.Fu et P.Li, *S. amamiana* Z.S.Sun et P. Li, *S. microdonta* ZS Sun et CX Fu, *S. hirtellicaulis* CY Wu & C Chen ex P.Li, *S. fui* ZC Qi et P Li; (4) A new name, *S. binchuanensis* P Li & CX Fu, is proposed for *H. yunnanensis* Gagnep.; (5) Six names are treated as synonyms: *S. hypoglauca* Benth. as the synonym of *S. corbularia* Kunth, *S. jiankunii* H Li as the synonym of *S. pottingeri* Prian, *S. tsinchengshanensis* FT Wang as the synonym of *S. minutiflora* A DC., *S. munita* SC Chen as the synonym of *S. myrtillus* A DC., *S. nigrescens* FT Wang & Tang ex.PY Li as the synonym of *S. castaneiflora* H Lev. & Vaniot; (6) Two status are changed: *S. luei* Koyama is reduced to be a subspecies of *S. nervo-marginata* Hayata, and *S. umbrosa* JM Xu is raised as an independent species; (7) Four species complexes are revealed: A. *S. chingii* FT Wang & Tang, *S. ferox* Wall., *S. lebrunii* H Lev., *S. ploycolea* Warb. and *S. outanscianensis* Pamp.; B. *S. lanceifolia* Roxb., *S. chapaensis* Gapnep, *S. hemsleyana* Craib. and *S. densibarbata* FT Wang & Tang; C. *S. aberrans* Gagnep. and *S. retroflexa* (FT Wang & Tang) SC Chen. D. *S. trachypoda* JB Norton and *S. vaginata* Decne. Although the phylogenetic relationships within the complexes have not been resolved, statistical analysis of morphological characters can distinguish the species from each other. Thus we tentatively consider them as independent species.

T2

P0376

Allotetraploid cryptic species in *Asplenium normale* in the Japanese Archipelago, detected by chemotaxonomical and multilocus genotype approaches.

Tao Fujiwara[1], *Ayumi Uehara*[2], *Tsukasa Iwashina*[3], *Sadamu Matsumoto*[3], *Yi-Han Chang*[5], *Yi-Shan Chao*[4], *Yasuyuki Watano*[1]
1. Department of Biology, Graduate school of Science, Chiba University
2. Division of Life Science and Engineering, School of Science and Engineering, Tokyo Denki University
3. Department of Botany, National Museum of Nature and Science
4. Department of Biomedical Science and Environmental Biology, Kaohsiung Medical University
5. Hengchun Research Center, Taiwan Forestry Research Institute

Cryptic species are defined as 'morphologically indistinguishable, but reproductively isolated lineages'. Delimitation of cryptic species provides the appropriate understanding of biodiversity and opportunities to elucidate speciation processes. In tetraploid cytotype of a fern, *Asplenium normale*, extensive flavonoid variation has been reported, although the related species have no intraspecific variations in flavonoid composition. We hypothesized that Japanese *A. normale* still harbors multiple cryptic species with different flavonoid compositions, and tested the hypothesis by chemotaxonomical and multi-locus genotyping approaches. We determined the multi-locus genotypes (MLGs) of 230 samples from 37 populations for chloroplast DNA, *rps4-trnS* intergenic spacer, and three nuclear genes, *Leafy, Nia, PgiC*. To delimitating reproductively isolated lineages, population genetic approaches using MLGs were conducted. We also tested the correspondence between the genetically recognized groups and flavonoid compositions. To identify the origins of putative cryptic species, phylogenetic analysis of the DNA markers used in genotyping were conducted. Clustering results based on MLGs and flavonoid compositions showed clear correspondences. We recognized three putative cryptic species in tetraploid *A. normale* in Japan. Phylogenetic analysis revealed that cryptic species 1 and 3 had originated through allopolyploidization between diploid *A. normale* and ancestral diploid progenitor of *A. boreale,* and cryptic species 2 between diploid *A. normale* and *A. oligophlebium.* Our study demonstrated that intraspecific variation of secondary metabolites could be a good indicator of cryptic species in ferns. The two cryptic species with the same progenitor diploids suggests that speciation between allopolyploid lineages of independent origin would be more common than was previously considered.

T2

P0377

Chromosome numbers of the genus *Micranthes* sect. *Rotun-*

difoliatae (Saxifragaceae)

Tomoko Fukuda[1], *Olga Chernyagina*[2], *Igor Bobyr*[3], *Aleksandr Taran*[4], *Hiroshi Ikeda*[5]
1. Dept. of Botany, National Museum of Nature and Science
2. Kamchatka branch of Pacific Institute of Geography
3. State Nature Reserve Kurilskiy
4. Sakhalin Botanical Garden
5. The University Museum, The University of Tokyo

The genus *Micranthes* sect. *Rotundifoliatae* is represented by *M. nelsoniana* (D. Don) Small and its subspecies and varieties, which has large distribution area from N. America to Siberia. The section includes some other species in Asia, as *M. fusca* (Maxim.) S. Akiyama et H. Ohba, from Japan to the Kuril Islands, *M. japonica* (H. Boiss.) S. Akiyama et H. Ohba, endemic to Japan, *M. Ohwii* (Tatew.) T. Fukuda et H. Ikeda (=*S. purpurascens* Kom.), mainly in Kamchatka, and some other species. Among these species, *M. nelsoniana* is reported to have high chromosome numbers up to 2n=84. Preliminary phylogenetic trees using cpDNA and nrDNA show large incongruence, and considered that it is partly due to the presence of hybridization and polyploidization. Therefore, we tried to reveal the chromosome numbers of these species, with special attention to the ploidy level. Though *M. nelsoniana* subsp. *nelsoniana* was reported as having high polyploidy level from Chukotka (W of Bering strait), 2n=ca. 60–84, the plant from south Kamchatka was 2n=28, and those from Esso, the central Kamchatka were 2n=56,60 and 64. There is a tendency that the chromosome number increases from south Kamchatka to the north up to Chukotka. *Micranthes nelsoniana* subsp. *reniformis* occurs in Sakhalin and Hokkaido. The chromosome number of *M. nelsoniana* subsp. *reniformis* in Sakhalin was 2n=26, while 2n=50 from Rishiri Island and 2n=ca. 80 from Mts. Taisetsu, Hokkaido. It seems that polyploidization with aneuploidal variation may take place in Hokkaido. We confirmed the chromosome numbers of *M. japonica* as 2n=28, and *M. ohwii* as 2n=24. The chromosome number of *M. fusca* was 2n=30 in a wide geographical range from Hokkaido to Kyushu, Japan, while 2n=45 and 2n=60 along with 2n=30 in Kunashiri Island, and 2n=40–50 in Etorofu (Iturup) Island. It is possible that polyploidization or hybridization between allied species may occur in the Kuril Islands.

T2
P0378

Downward, bell-shaped flowers are not always bee pollinated: Floral adaptations to nocturnal moths in *Adenophora triphylla* var. *japonica* (Campanulaceae).

Daichi Funamoto[1], *Kazuharu Ohashi*[2]
1. Kobe University
2. University of Tsukuba

Pollination syndromes, i.e., suites of floral traits shared among distantly related angiosperm taxa, have long been hypothesized to reflect convergent adaptations for pollination by specific types of animals. During the last decades, however, many authors have provided evidence for the existence of apparent mismatches between pollinators inferred from floral traits and actual flower visitor spectrum. Here we demonstrate that investigations on relatively concealed floral traits, such as temporal patterns of anthesis and nectar productivity, together with investigations on relative contributions of diurnal and nocturnal pollinators, could greatly improve our understanding of how such apparent mismatches occur as consequences of floral evolution. *Adenophora triphylla* (Thunb.) A.DC. var. *japonica* (Regel) H.Hara (hereafter, *A. triphylla*) has pale blue, tubular-campanulate, pendent flowers without any noticeable sweet fragrance. Flowers with this combination of morphology and color are typically viewed as associated with bee pollination. In contrast to this expectation, recent studies have claimed that flowers of *A. triphylla* are frequented by nocturnal or crepuscular settling moths. To solve this discrepancy between prediction and observation, we conducted 3 years of field survey in two natural populations (Sugadaira and Tsukuba) of *A. triphylla* in central Japan. We investigated (1) temporal traits in flowers (timing of anthesis, timing of change in sexual phase, and diel pattern of nectar production), (2) flower-visitor fauna, and (3) contributions of diurnal and nocturnal visitors to seed set. First, we found that both diurnal insects (mainly hoverflies and skipper butterflies) and nocturnal settling moths (mainly noctuids and pyraloids) frequently visited flowers of *A. triphylla*. The observed visitation rates were similar between day and night in Tsukuba, while diurnal visits were more frequent than nocturnal ones in Sugadaira. Second, in contrast to the above observations, contribution of nocturnal visitors to seed set was much higher than that of diurnal visitors. Third, antheses and sexual phase changes from male to female mostly occurred around sunset, and nectar was produced almost entirely during night. This temporal correspondence between floral traits and moth activity strongly suggests that flowers of *A. triphylla* have specifically adapted to pollination by nocturnal moths rather than diurnal bees. Although it has been often presumed that flowers with downward, bell-shaped corollas are primarily pollinated by bees, this study and other recent studies show that at least some of these flowers are primarily pollinated by nocturnal moths. This growing evidence may suggest the existence of a new, previously overlooked pattern of moth pollination.

T2
P0379

Phylogenetic relationships and biogeographic history of *Pseudognaphalium* Kirp. (Compositae, Gnaphalieae)

Adriel Acosta Maindo, Jessica Requena Ordoñez, ***Merce Galbany Casals***
Autonomous University of Barcelona

The genus *Pseudognaphalium* Kirp. is one of the most species-rich genera of the tribe Gnaphalieae, with c. 90 species. It is constituted by annual, biennial or perennial herbs. Although the vast majority of *Pseudognaphalium* species are distributed in the American continent, the genus has also a few representatives in Africa, Asia, Australasia and Hawaii, and *Pseudognaphalium luteoalbum* is subcosmopolitan. Previous phylogenetic analyses placed *Pseudognaphalium* within the *Helichrysum-Anaphalis-Pseudognaphalium* (HAP) clade, embedded within a paraphyletic *Helichrysum* Mill., but its monophyly was not statistically supported. In addition, incongruence between nrDNA and cpDNA involving *Pseudognaphalium* gave support to the hyphothesised allopolyploid origin of this genus. In the present study, with the use of two nuclear ribosomal (nrDNA) and one plastid DNA markers and an extensive sampling of the genus *Pseudognaphalium*, further evidence relevant to its origin and closest relatives is provided, and additional evidence for incongruence is discussed. A molecular dating of the nrDNA phylogeny and the superposition of distribution areas show that *Pseudognaphalium* originated in South Africa during the Late Miocene to Pliocene, and that subsequent several out-of-Africa migrations caused its striking current disjunct distribution by stepping-stone and long-distance dispersal events to other continents. In addition, the superposition of chromosome numbers on the nrDNA phylogeny suggests at least three independent allopolyploid origins of the genus and provides evidence of the important role of allopolyploidy in the colonization of new territories.

T2

P0380

Macroevolution of the sunflower family (Asteraceae): A morphological explanation for the variation of diversification rates in a large family

Tiangang Gao[1], *Tania Hernandez*[2], *John Wiens*[2]

1. Institute of Botany, Chinese Academy of Sciences

2. Department of Ecology and Evolutionary Biology, University of Arizona

Asteraceae is one of the largest families of flowering plants, with over 1600 genera including 23,000 species. Members of this family are widely distributed in every continent but Antarctica. The subfamily & tribe–level differences in diversity within the family are great. Although some attempts have been made to explain how the family quickly colonized the entire planet and became so incredibly diverse, all of these hypotheses have not been strictly tested and we still don't know clearly how the family quickly became so incredibly diverse. Furthermore, diversification in the Asteraceae should best be explained by focusing on all the subclades (subfamilies or tribes) within the family, rather than treating the family as a unit. Here we studied specific morphological traits of the subclades in Asteraceae that might explain their different success.

T2

P0381

Towards a better understanding to species: A case study on Eastern Asia alpine shrub *Rosa sericea* and *R. omeiensis* (Rosaceae) based on molecular genetics and environment niche modeling

Yun-Dong Gao, *Xin-Fen Gao*

CAS Key Laboratory of Mountain Ecological Restoration and Bioresource Utilization & Ecological Restoration and Biodiversity Conservation Key Laboratory of Sichuan Province, Chengdu Institute of Biolog

The delimitation on closely related taxa is a key to understand the speciation process and biodiversity. Plant species, which showed more ability to hybridize among different taxa within or without a genus, and therefore have much complicate population genetic structure and history. In present we chose a monophyletic alpine shrub of *Rosa sericea* and its sister species *R. omeiensis*, using nuclear single-copy maker and micro-satellites (EST-SSR) as well as plastid sequences based population genetics methods and environment niche modeling (ENM) to test the species boundary within this taxonomic difficult group. The results showed that although all three datasets from molecular only indicated a geographic affinity, these two taxa occupied different ecological niches; however, the niche diversity is not so significant. We speculate that this pattern is conducted by the glacial-interglacial cycles, which promote inter-specific gene communication while sympatric during the last glacial maximum (LGM). The ecological speciation is in process and the notion of species need to revisited since the study in present is just a tip of iceberg. At this stage as our knowledge we should include a time dimension since species is a dynamic identity rather than static.

T2

P0382

Palynolomorphogical and molecular genetic analysis of species and hybrids of the genus *Ribes*

Olga Gavrilova, *Eduard Machs*

Komarov Botanical Institute RAS

In the family Grossulariaceae, which now consist of only one genus *Ribes* have been found pollen with each global and meridional aperture disposition. We have examined pollen morphology and fine structure of 47 species and 10 hybrids with light and scanning electron microscopies, partly with confocal laser scanning and transmissional electron microscopies. Meridional 4-6-colporate grains have been discovered in gooseberries (subgenus *Grossularia*) and one section of currants (subgenus *Ribes*) – *Heritiera*, other currants have 4-porate or 6-12-porate-orate grains with global aperture disposition. We found mainly psilate pollen ornamentation and tectate exine. Species from subgenus *Grossularioides* have 3-4-porate grains with global pore disposition and echinate, atectate exine. Some species have various apertural deviations, four species as well as some hybrids reveal different transitional forms of pollen grains with ectoapertures each colpa and pori. The ultrastructures features demonstrate differences between *Grossularia* and *Ribes* pollen types, between subgenera into the family, sometimes between sections. Genetic distances and specific nucleotide substitutions for parental forms and hybrids of known and unknown origin were evaluated by sequences of spacers and coding region of 5.8S rRNA gene. Comparison of genetic distances shows that all samples are divided into two subclades correlated with the types of pollen grains, but does not allow to solve the problem of the origin of the hybrids. On the other hand, the analysis of nucleotide substitutions indicates a relatively high overall and intragenomic polymorphism, especially in hybrids. Moreover, polymorphic positions of hybrids and presumed parental forms coincide for half of 198 variable positions. Comparison of matching positions and various types of nucleotide substitutions allows with a certain probability to determine parental forms of hybrids of unknown origin, especially in case of a few number of hybridizations. The investigations are supported by Russian Fund of Basic Research N 15-04- 06386a.

T2

P0383

Introduction to the book Detailed Morphology of Chinese

Vascular Plants

Weiliang Ma[1], *Yong Chen*[2], *Hong Zhao*[3], *Hongqing Li*[1], *Kang Ye*[4], ***Binjie Ge***[5]

1. East China Normal University

2. Ningde Normal University

3. Shandong University

4. Shanghai Botanical Garden

5. Shanghai Chenshan Botanical Garden

All come from a creative attempt: the author used the micro photos of live plants' detailed morphology as teaching aids in taxonomic class and got the students' approval, which opened a period of 34 years (1982-2016) of the collection, dissection and shooting. The main content of this book is 7,000 carefully selected images out of 1,102 genera (one species was selected from each genera) taken during the last 34 years' field work which covered the major mountains of 26 provinces and autonomous regions of China, and 16 countries in Eurasia and North America. The most important feature of this book is based on detailed morphology to explore the survival strategies of plants and to understand their adaptability to the environment through micro-perspective. Therefore, it is of great theoretical value for the study of plant taxonomy and plant phylogenetic, and has important application value in botany teaching and plant science education. Through the completion of this book, we are "re-understanding" the plants around us: plants will achieve active pollination by "cleverly" moving the stamens and instantaneously twisting the style; "change" into much smaller shape to survive against the adversity; the evolution of a "pay firstly and then to get the racial life-and-death reward from the others" between the animal and plant symbiotic complex. Behind the discovery of each surprise, are activities hidden in the plant (mostly in the flower) year after year. We just acted as a faithful recorder, willing to share the stories "it is not the strongest of the species that survives, but the most adaptable" with the readers who have the curiosity and awe to the nature.

T2

P0384

Molecular phylogenetics of *Solanum* sect. *Brevantherum* Seithe (Solanaceae), a group of pioneer and potentially invasive neotropical treelets

Leandro Lacerda Giacomin[1], *Gregory A. Wahlert*[3], *Lynn Bohs*[3], *João R. Stehmann*[2]

1. Instituto de Ciências e Tecnologia das Águas e Herbário HSTM, Universidade Federal do Oeste do Pará, Av. Mendonça Furtado 2946, Santarém, PA, 68040-050

2. Department of Biology, University of Utah, 257 South 1400 East, Salt Lake City, UT, 84112

3. Departamento de Botânica, Instituto de Ciências Biológicas, Universidade Federal de Minas Gerais, Av. Antônio Carlos 6627, Belo Horizonte, MG, 31270-901

The systematics of the giant and economically important genus *Solanum* (Solanaceae) has been the focus of several molecular phylogenetic studies in the past two decades, resulting in substantial advances in its understanding. Nowadays, most of the main defined lineages and groups within the genus have comprehensive molecular phylogenies, including the Neotropical *Solanum* section *Brevantherum*. The section, which is part of the larger Brevantherum clade, has long been defined by the plurifoliate, dichasial sympodia of its species, that present erect, terminal, many-branched inflorescences and an indumentum of stellate trichomes. In the last revision (1972), 27 species were assigned to it, a number that has grown since then, with the description of four species. Although easily recognizable and commonly found in disturbed sites and open habitats from Mexico and USA (Florida) to Argentina, the species relationships and character evolution within the section have not yet been investigated using a molecular phylogenetic approach. As three species of the section have become naturalized in the Old World and Australian tropics, it is desirable to understand species relationships in a group with such invasive potential. Considering this, a phylogeny inferred for the section was conducted using plastid *trnT-F* intergenic spacer, nuclear ribosomal ITS and low-copy nuclear *waxy* sequences for 42 accessions of 27 species. Our Bayesian and Maximum likelihood results support the monophyly of section *Brevantherum*. The results also showed that *Solanum bullatum* was not recovered in the Brevantherum clade, supporting earlier taxonomic conclusions that the species should be excluded based on its strongly accrescent calyx, a character not found in any other species of the section. Two well supported clades are recovered within the section and are informally named: i) the Abutiloides clade, containing Ecuadorian-Peruvian species that have lepidote scales (*S. oxapampense* and *S. verecundum*) together with the glandular pubescent *S. abutiloides* from xeric environments in Bolivia and Argentina and ii) the Erianthum clade, a larger clade containing the remaining species with typical porrect-stellate and echinoid trichomes. Within the Erianthum clade, the Brazilian Atlantic Forest species *S. granulosoleprosum*, *S. mauritianum* and *S. stipulaceum* form a strongly supported clade, as do some other Andean and Mexican species groups. Some characters used in the past to separate groups of species such as the few-rayed versus multi-rayed stellate trichomes of stems and peduncles are seemingly homoplasious. Considering that the gene regions used in our analysis contained few parsimony-informative characters (9.5% of 4,106 bp concatenated matrix), high throughput DNA sequencing techniques should be used to better clarify the relationships within sect. *Brevantherum*.

T2

P0385

Hidden robbery: A new approach (δ^2H) discloses organic carbon-gain effectiveness of C_3-hemiparasites from C_3-host plants.

Philipp Giesemann, *Gerhard Gebauer*

University of Bayreuth

Introduction: Previous investigations with stable isotope analyses featuring carbon isotopes (^{13}C) showed for very few 'xylem feeding' C_3-hemiparasites utilizing C_4- or CAM-hosts the usage of two different carbon sources, autotrophy and heterotrophy [1]. This ^{13}C approach, however, failed for the much more frequently occurring C_3-C_3 parasite-host-pairs. To overcome this obstacle, we used hydrogen stable isotope (^{2}H) natural abundances as substitute for ^{13}C [2] within a C_3-Orobanchaceae-sequence (root parasites) graded by morphological features [3] and C_3-Santalaceae (root and stem parasite). Carbon-gain calculations for the parasites help quantifying the influence on host plants and on ecosystem carbon cycles.

Material & Methods: Parasitic Orobanchaceae and Santalaceae (9 species, 67 individuals) and fully autotrophic potential host plants (25 species, 199 individuals) were collected in 2016 in Germany and Austria. Seven parasitic species including their potential host plants were collected once in the vegetation period in five replicates each. Additionally, the Orobanchaceae *Rhinanthus minor* L. and the Santalaceae *Viscum album* L. were sampled several times during the vegetation period to estimate development-dependent differences. A multi-element stable isotope natural abundances approach was realized (EA-IRMS: $\delta^{13}C$, $\delta^{15}N$; TC-IRMS: $\delta^{2}H$, $\delta^{18}O$). All sampled photosynthetic active plants perform C_3-photosynthesis, therefore major differences in $\delta^{13}C$ were not expected. Nutrient-gain efficiencies were calculated with an application of the two-source linear mixing model. Autotrophic reference plants were setting the lower endpoint (0% heterotrophy) and holoparasitic plants ('xylem-feeder': *Lathraea squamaria* L.) the upper endpoint (100% heterotrophy) of the model. The approaches of the intermediate C_3-hemiparasite signatures ($\delta^{2}H$) to one or the other side were assumed to represent the parasitic effectiveness.

Results: Hydrogen isotope abundance in plant tissues is counterbalanced by two effects, transpiration (^{2}H depletion) and gain of host-derived organic matter (^{2}H enrichment). The transpiration effect can be subtracted out of the plant tissue $\delta^{2}H$ values through the plant tissue $\delta^{18}O$ values, which are exclusively driven by the transpiration difference between parasite and host. Based on these calculations a species-specific heterotrophic carbon gain ranging from 13 to 37% was estimated for the investigated hemiparasitic Orobanchaceae species. The sequence in heterotrophic carbon gain mostly met the sequence in root morphological features. Time series revealed development-dependent shifts (young: heterotrophic, adult: autotrophic) for the root parasite *R. minor* and stem parasite *V. album*.

Conclusions: Due to higher transpiration and lower water-use efficiency depletion in $\delta^{2}H$, $\delta^{13}C$ and $\delta^{18}O$ compared to C_3 host plants should be expected for tissues of C_3 hemiparasites. However, stepwise enrichment in ^{2}H stable isotope is caused by the parasites' heterotrophic organic matter gain and can be used to estimate carbon gains from hosts. Time series approaches to evaluate development-dependent isotopic abundances are recommended for further parasitic species.

T2
P0386
Origin of two anagenetically derived endemic species of *Scrophularia takesimensis* (Scrophulariaceae) and *Sorbus ulleungensis* (Rosaceae) on Ulleung Island, Korea

Hee-Young Gil[1], *Masayuki Maki*[2], *Elena Pimenova*[3], *Aleksandr Taran*[4], *Seung-Chul Kim*[1]

1. Department of Biological Sciences, Sungkyunkwan University

2. Division of Ecology and Evolutionary Biology, Graduate School of Life Sciences, Tohoku University

3. Botanical Garden-Institue FEB RAS

4. Sakhalin Branch of Botanical Garden-Institute FEB RAS

Ulleung Island, an oceanic volcanic island in Korea, is home for ca. 500 native species of vascular plants, 8% (40 species) of which are endemic to the island. The Ulleung Island has been recognized as an excellent model system to study the pattern and process of early stages of flowering plant evolutions on oceanic island. Unlike any other oceanic islands in the Pacific and Atlantic Oceans, the predominant mode of speciation on Ulleung Island is considered anagenetic speciation (88%) due to the factors of low vegetation heterogeneity and elevation. In this study, we investigated the origin and evolution of two anagenetically derived endemic species on Ulleung Island to provide an initial framework to further test the genetic consequences of anagenetic species in Ulleung Island. One example includes *Scrophularia takesimensis*, one of critically endangered endemic species in Ulleung Island. We conducted molecular phylogenetic analyses to determine the origin of *S. takesimensis* based on extensive sampling, including two putative progenitor species, from the East Asia. As a nuclear phylogeny, we generated a global scale ITS phylogeny including represented species of *Scrophularia* deposited at GenBank. The ITS phylogeny suggested that *S. takesimensis* is sister to the clade containing eastern North American/Caribbean species rather than to either *S. alata* or *S. grayanoides*. The global scale cpDNA phylogeny demonstrated that the eastern North America/ Caribbean clade is sister to the clade containing three eastern Asian species. In addition, the monophyletic *S. takesimensis* is deeply embedded within paraphyletic *S. alata*, sharing its most recent common ancestor with populations from Japan (2n=94) and Sakhalin (2n=20). Geographically structured two divergent cp haplotype groups within *S. takesimensis* were found, suggesting at least two independent introductions from different source areas. The phylogenetic incongruence between nuclear and chloroplast genomes, polyploidization and aneuploidization of *S. takesimensis* (2n=80) and closely related species, and unusual intercontinental disjunct distribution require further detailed independent nuclear genealogical, cytological, and biogeographic analyses. Another example of anagenetically derived endemic species on Ulleung Island is *Sorbus ulleungensis* (Rosaceae). *S. ulleungensis* has been considered as *S. commixta* but recently described as endemic species in Ulleung Island based on morphological characteristics. As an attempt to investigate the genetic consequences of anagenetic speciation of this taxon, we generated preliminary chloroplast haplotype diversity data of insular derivative *S. ulleungensis* and continental progenitor *S. commixta*. The results suggested that *S. ulleungensis* is most closely related to the populations of *S. commixta* sampled from southern Japan, suggesting possibility of southern geographical origin rather than the Korean peninsula. We also found significant reduction in haplotype diversity of insular derivative *S. ulleungensis* compared to its continental progenitor. Details relationships among haplotypes found in two species and levels of genetic diversity and differentiation within each species will be presented.

T2
P0387
Phylogenetic relationships among *Andrographis* s.l. (Acanthaceae: Andrographinae): Evidences from gross morphology, palynology, seed morphology and anatomy

Gnanasekaran G.[1,2], *Murthy G.V.S.*[2], *Carine M.A.*[3]

1. Madras Christian College (Autonomous)

2. Botanical Survey of India
3. The Natural History Museum

Andrographis Wall. ex Nees (Acanthaceae: Andrographinae) is a tropical Asian genus chiefly confined to southern India (Western and Eastern Ghats) and Sri Lanka. Phylogenetic relationships and generic delimitation in *Andrographis* are investigated using morphological cladistics analysis. The 29 species presently recognized, together with two outgroup species from *Haplanthodes* Kuntze, were coded for 39 morphological characters sourced from gross morphology, palynology, seed morphology and anatomy. Parsimony analysis resolves *A. laxiflora* (Blume) Lindau as sister to all other species of the genus. A clade comprising *A. alata* (Vahl) Nees, *A. chendurunii* E.S.S. Kumar, A.E.S. Khan & S.G. Gopal and *A. elongata* (Vahl) T. Anderson is the next branching lineage. These three species form a monophyletic group that is strongly supported and morphologically well-defined by seven synapomorphies. Within the remaining *Andrographis* clade, six species namely *A. beddomei* C.B. Clarke, *A. echioides* (L.) Nees, *A. glandulosa* (B. Heyne ex Roth) Nees, *A. longipedunculata* (Sreem.) L.H. Cramer, *A. rothii* C.B. Clarke and *A. serpyllifolia* (Vahl) Wight are identified as a distinct well-supported monophyletic subclade, characterized by seven synapomorphies. This clade broadly corresponds to the genus *Indoneesiella* Sreem. The results would support the following generic recircumscription: (i) *Haplanthus* Nees resurrected to accommodate *A. laxiflora*, (ii) *A. alata*, *A. chendurunii* and *A. elongata* to be recognized as a new genus and (iii) all other species to be retained as *Andrographis*. Whilst an '*Indoneesiella'* clade is resolved and is clearly defined, it is not recognized here as a distinct genus since doing so would render *Andrographis* paraphyletic. Molecular data would help to further clarify relationships and resolve poorly supported relationships.

T2
P0388

Ambophily in *Gnetum* (Gnetales) and its implications for the primitive pollination mode of seed plants

Yan-Bing Gong, *Yang Min, Ya-Ru Zhu*
State Key Laboratory of Hybrid Rice, College of Life Sciences, Wuhan University

For the five lineages of extant seed plants, insect pollination is considered to be primitive in angiosperms, cycads and Gnetales; whereas *Ginkgo* and conifers originate from wind polination. Paleontological evidences show that ancient Gnetales, angiosperms and gymnosperms coexist in the early Cretaceous. Some genetic data suggest that Gnetales are sister to all other extant seed plant lineages. Thus, pollination mechanisms of the enigmatic group Gnetales provide an important key to understanding primitive pollination mode of seed plants. Here, we present a two years pollination study on two populations of *Gnetum montanum* (Gnetales), to explore the contributions of insect and wind pollination to this rare and endangered species. Pollinator observations in tropical rain forests of southern China revealed a spatiotemporal varying pollinator assemblage, including diverse insects from Blattodea, Coleoptera, Diptera, Hymenoptera and Lepidoptera. All groups of insects carried pollen grains and thus acted as effective pollinators. Aerodynamic experiment showed the terminal settling velocity to be 3.9 m/s, which was in the range of 1-10 m/s for anemophilous species in forest. Pollen grains of *G. montanum* were not sticky and could be intercepted by pollen traps near the male and female strobili. Pollination treatments applied to *G. montanum* revealed this species to be pollen limited and incapable of apomixis. Seed set in the netted treatment (anemophily) was significantly higher than in the bagged treatment (apomixis) but lower than open pollination (anemophily and entomophily), indicating that both wind and insect pollination contributed to reproduction of *G. montanum*. *Gnetum* (Gnetales) has long been regarded as insect pollinated due to its range-restricted distribution in tropical rain forests, where wind pollination is supposed to be detrimental. The occurrence of ambophily in *Gnetum* and the prevalence of anemophily in *Ehpedra* and *Welwitschia* suggest that wind may also play a role in the pollination of the ancient seed plants.

T2
P0389

Understanding diurnal and nocturnal pollination systems using morphometry, ecological, and molecular tools in the genus *Hedychium* (Zingiberaceae) from the North-East of India.

Vinita Gowda, *Ajith Ashokan, Preeti S*
Indian Institute of Science Education and Research Bhopal

Competition is an important process that structures any community. According to competitive exclusion principle, species competing for the same resources cannot coexist and will evolve traits such as to minimize competition and maximize the use of the limiting resources. Plants can have multiple axes for competition such as flowering phenology, pollinator species, and dispersal species.The genus *Hedychium* is known for its vibrant and fragrant flowers. We studied, two nocturnal (*H. ellipticum*, *H. urophyllum*) and two diurnal species (*H. forrestii* and *H. stenopetalum*) in the North-East of India to understand their floral biology and pollination systems. All the four species are found in the same region (Meghalaya state) and share many similar floral traits (fragrant flowers, night blooming, multi-day, large inflorescence, clonal populations). We proposed that to avoid competition and attract different pollinators, sympatric species will resort to different visual signals (floral morphology) and/or differential rewards (nectar sugar content) to avoid competition. We therefore asked the following questions: (i) Which floral traits are important to distinguish plant species apart? (ii) Does the nectar sugar reward differ among the four species? (iii) How are these species related to each other in a molecular phylogenetic tree. From over 152 morphological characters we found that the color and size of the labellum and lateral staminodes were the most important morphological characters. Nectar sugar varied significantly on the different days for a species, as well as among the four species. Taxonomically, *H. forrestii* showed intermediate traits to *H. stenopetalum* and *H. ellipticum* and behaved as an intermediate or hybrid form of the two. Further, we identified two types of flower types within the *Hedychium* genus based on their ecological characteristics, strict and opportunist nocturnal and diurnal flowers respectively. We propose that these may represent the ecological adaptations that are present to reduce competition among sympatric species and therefore high variability may be expected in these characters on a population scale. Ongoing studies are testing these hypothesis in the wild populations of *Hedychium* in the North-East of India.

T2

P0390

Worldwide systematics and phylogeography of the *Botrychium lunaria* complex (Ophioglossaceae)

Jason Grant, *Benjamin Dauphin, Vinciane Mossion, Alessio Maccagni*
Universite de Neuchatel

The *Botrychium lunaria* complex (Ophioglossaceae) is an assemblage of about a dozen diploid (and a single tetraploid) taxa that are distinguished by both morphology and DNA sequences. These taxa occur largely in mountainous and grassland areas in a continuous band across the Northern Hemisphere from Europe, Asia (Altai, Caucasus, Himalayas, Tien Shan, Turkey, etc), to North America (Alaska, Canada to Mexico), but they also occur on isolated islands and two other continents and including Africa (Atlas Mountains), Australia (Australian Alps, Tasmania), Azores, Falklands, Greenland, Iceland, Japan, New Zealand, South America (Patagonia), and Taiwan. In order to help identify and delineate the taxa in the complex that are often morphologically cryptic, both chloroplast and nuclear markers have been sequenced for hundreds of accessions, which is beginning to give us insight not only on the systematics and taxonomy but also on broader trends on worldwide phylogeography. Morphological characters on the trophophore and sporophore are being studied with the aim to present clear meaningful keys and illustrations so specialists and amateurs can identify taxa. Known published and accepted taxa include *Botrychium crenulatum*, *B. lunaria*, *B. lunaria* var. *melzeri*, *B. nordicum*, *B. tunux*, and *B. yaaxudakeit*. New taxa are known from Europe (the Alps especially) and continental Asia.

T2

P0391

Phylogeny and delimitation of Galipeeae (Rutaceae, Sapindales), based on molecular data: Insights for evolution of zygomorphic flowers and staminodes.

Milton Groppo[1], *Carla P. Bruniera*[3], *Paola Lima Ferreira*[1], *Carolina Ferreira*[1], *José Rubens Pirani*[1], *Jacquelyn Ann Kallunki*[2]
1. Universidade de São Paulo
2. New York Botanical Garden
3. Universidade Federal de São Paulo

Tribe Galipeeae (Rutaceae, Subfamily Rutoideae) is the largest group of Rutaceae in the Neotropical Region, with 33 genera and c. 181 species. The tribe is composed of two subtribes, Galipeeae (28 genera and c.130 species) and Pilocarpinae (5 genera, 51 species), whose typical genus is *Pilocarpus*, a source of pilocarpine, an alkaloid used in the treatment of glaucoma. Subtribe Galipeinae is different from Pilocarpinae by the possession of zygomorphic flowers (v. actinomorphic flowers in Pilocarpinae) and by the constant presence of staminodes (v. all fertile stamens in Pilocarpinae.) A phylogenetic analysis was conducted to test the monophyly of Galipeeae and its two subtribes. A total of 99 terminals, representing 24 genera and 75 species of both subtribes were sampled. Species of *Balfourodendron, Choisya, Helietta, Hortia,* and *Plethadenia,* all from genera traditionally not included in Galipeeae were also used; *Zanthoxylum,* a phylogenetically more distant genus was used as outgroup in all analyzes. Molecular data included four regions, two from the cpDNA (trnL-trnF and rps-16) and two nuclear (ITS1 and ITS2) analyzes. Parsimony and Bayesian analysis were conducted. Results showed all Galipeeae grouped in a strongly supported clade (Bootstrap Percentages BP-100% BP, Posterior Probabilities PP = 1); Subtribes Galipeeae and Pilocarpinae appear as sisters, both clades with sthng support. However, *Balfourodendron* and *Helietta* appear as part of the Pilocarpinae and *Hortia* along with the Galipeinae, As a consequence, the Galipeeae groups and their two subtribes will only become monophyletic if they include *Balfourodendron*, *Helietta* and *Hortia*. Morphological characteristics such as the possession of zygomorphic flowers, presence of staminodes, embryo characteristics and foliar anatomy are also discussed.

T2

P0392

Molecular phylogeny of the large South American genus *Eriosyce* (Notocacteae, Cactaceae): Generic delimitation and profound taxonomic changes

Pablo Guerrero[1,2], *Helmut Walter*[3], *Carol Peña*[1], *Ítalo Tamburrino*[2], *Isabel Larridon*[4]
1. Universidad de Concepcion
2. Institute of Ecology and Biodiversity
3. EX-SIS Conservation Program
4. Royal Botanic Gardens, Kew

Eriosyce is one of the most species-rich genera within Notocacteae (Cactaceae) harboring a variety of stem and flower morphologies, but joined by their fruits with basal abscission. The lack of a well sampled molecular phylogeny contributes to the current taxonomic labiality of the genus, where its circumscription and infrageneric classification has been questioned. Specimens of *Eriosyce* (63 taxa of *Eriosyce* plus 20 outgroups) were analyzed through sequencing three-plastid noncoding introns (*rpl32–trnL, trnL–trnF* and *trnH–psbA*), one plastid gene (*ycf1*), and one nuclear gene (*phyC*). A concatenated matrix was analyzed using maximum likelihood and Bayesian approaches. Phylogenetic analyses retrieved strong support for monophyly of *Eriosyce sensu lato*. Furthermore, seven sections within *Eriosyce sensu lato* were clearly defined based on well-supported branches. Also, several infraspecific taxa appear paraphyletic according to their scattered position in the phylogenetic tree. New combinations are proposed to update species nomenclature according to our inferred molecular phylogenies. Also, since some past taxonomic proposals received low phylogenetic support they should be discarded.

T2

P0393

Determination of molecular phylogenetic properties of some *Ophrys* L. (Orchidaceae) species using the technique of DNA sequencing growing in Natural Flora of European Turkey

Necmettin Guler
Trakya University

The Orchidaceae family which is the most crowded group of the flowering plants (Angiospermae) with 880 genera and 22,000 accepted species as approximate, has a quite problematic systematic so that this family is drawing attention of researchers while applying the classification technics to this group. Thus, this study aims the sequencing of ITS and 18S regions of nrDNA of the genus *Ophrys* taxa belonging to Orchidaceae family, collected from European Turkey. Nowadays the molecular systematic is a method of systematics developing in a fast way and being popular because of the difficulties appeared while a new taxa, founded after the results of research is being included to the previous classifications, the fast and continual development of techniques which are used by molecular biology. Because of the deficiency of anatomic, morphologic, karyologic *etc.* classical methods in the classification of complex plant groups and the error of defining the mixed species. The results of the present study were compared with the molecular characteristics of some species of the genus *Ophrys*, whose DNA sequence analysis are available in literature, and a phylogenetic tree was constructed. The degree of relationships of the subspecies and upper species categories were determined and the distributions of the species within the genera in European Turkey were re-evaluated.

T2

P0394

Morphological and karyological studies of *Lathyrus* taxa belonging to *Pratensis* section (Fabaceae) from Turkey

Fatma Güneş
Trakya University

Morphological and karyological analysis of 5 wild taxa from section *Pratensis* (*Lathyrus*, Fabaceae) growing in Turkey, have been described. These are *L. pratensis* L., *L. laxiflorus* (Desf.) O. Kuntze subsp. *laxiflorus*, *L. laxiflorus* subsp. *angustifolius* (Post ex Dinsm.) Davis, *L. czeczottianus* Bässler and *L. layardii* Ball ex Boiss. Examined taxa is shown the detailed descriptions, general drawning pictures, flower parts and the characterisrics of their fruits. Karyotype analysis was made in detail for the frist time for 2 of these taxa including 3 endemic species. For all species, the chromosome number has been determined as 2n=14. Long arm, short arm and total lengh of the chromosomes were measured, centromeric index and relative lenght were calculated. The photographs, drowing of flower parts and habits and showing metaphase chromosomes of the species, karyotypes and idiograms are included to the study.

T2

P0395

ddRAD sequencing enables unprecedented phylogenetic resolution in Arundinarieae (Poaceae: Bambusoideae)

***Cen Guo**, Zhen-Hua Guo, De-Zhu Li*
Kunming Institute of Botany

Background: Arundinarieae (Poaceae: Bambusoideae) is a large tribe which was resolved as a monophyletic group with 12 and 13 clades in plastid and nuclear phylogeny, respectively. But interrelationships within the tribe were still intricate because of limited taxon sampling and lack of informative markers. And the most scabrous relationships were focus in clades IV, V and VI, which had significant incongruence between the plastid and nuclear topology.

Methods: We have now generated a large reduced-representation genomic data set for 218 individuals in 205 species (200 temperate woody bamboos, plus five tropical woody bamboos as outgroups) using the double digest restriction-site associated DNA sequencing (ddRAD-seq) for our phylogenetic analysis. In order to identify more orthologous RAD tags and eliminate the massive missing data, we finally adopt the reference-guided strategy instead of the *de novo* assembling method for our data assembling.

Key Results: Using maximum likelihood inference, we obtain a most highly resolved and robust Arundinarieae tree with six monophyletic lineages. In this framework, *Ampelocalamus calcareus* is the first branch to diverge from the remainder of Arundinarieae, with the ADH lineage second, followed by *Gaoligongshania megalothyrsa* and the rest of Arundinarieae which can be further divided into the pachymorph lineage, *Indocalamus wilsonii* and the leptomorph lineage. Two new clades established by molecular evidence here, the pachymorph lineage and the leptomorph lineage, can also be well-characterized by their rhizome states using our analyses of character optimizations. Moreover, the deep-level relationships in most clades are well resolved and some systematically ambiguous genera are newly clarified as monophyly, such as *Phyllostachys*, *Chimonobambusa* and *Yushania*. Additionally, further network analyses largely confirmed the currently recognized six clades and conflicting phylogenetic signals are presented at several polyphyletic genera such as *Pleioblastus*, *Pesudosasa*, *Semiarundinaria* and *Indocalamus* in the lepotomorph lineage, which helps us to advance our understanding of some difficult and intricate relationships of Arundinarieae.

Conclusions: To date, using ddRAD data, this study represents the most comprehensive sampling and the largest informative SNPs matrix for Arundinarieae, providing genome-scale support for phylogenetic relationships of Arundinarieae and lays the phylogenetic groundwork for further analysis on the drivers of rapid radiation and reticulate evolutionary patterns across this diverse clade.

T2

P0396

Evolutionary origin of flowers: A genome-wide study of floral development genes in seed plants

***Chunce Guo**[1,2], Claude Depamphilis[2], Hong Ma[1]*
1. State Key Laboratory of Genetic Engineering and Collaborative Innovation Center of Genetics and Development, School of Life Sciences, Fudan University
2. The Huck Institute of the Life Sciences, Penn State University

Darwin famously characterized the origin and rapid diversification of the angiosperms as an abominable mystery. In recent years, new advances in the fields of molecular phylogenetics and evolutionary developmental biology have not only corrected the mistakes in previous theories and models, but also proposed new and testable hypotheses, revealed possible processes through which the flower first originated, and discussed the contributions of the duplication and diversification of floral developmental genes to the origin of flowers. However, previous studies are insufficient for a complete understanding the contributions of genomic changes to those biological innovations. Here, we examined the origin and duplication history of orthogroups containing 424 genes known to play roles in Arabidopsis floral timing and development, and found that upstream genes in the floral developmental regulatory network had higher frequencies of old orthogroups that originated with tracheophytes, and downstream genes were overrepresented by those in new orthogroups originated within derived angiosperms. Furthermore, the similar pattern was found when using 312 core-floral genes, which conserved expressed in flower tissues of water lily, rice, California poppy, and Arabidopsis. These results suggest that although much of the floral development genetic toolkit is ancient, gene recruitment has been an ongoing process in floral evolution, and that different components of the floral regulatory network have recruited genes at different rates through history, with a larger proportion of new orthogroups being represented by genes functioning in organ identity, development and maturation than by genes functioning in the environment sensing and floral timing.

T2
P0397

A mega-phylogeny of the Annonaceae: Taxonomic placement of five enigmatic genera and support for a new tribe, Phoenicantheae

Xing Guo[1], *Chin Cheung Tang*[1,2], *Daniel C. Thomas*[1,3], *Thomas L.P. Couvreur*[4], *Richard M.K. Saunders*[1]

1. School of Biological Sciences, The University of Hong Kong, Hong Kong, China.

2. Current address: School of Science and Technology, The Open University of Hong Kong, Ho Man Tin, Kowloon, Hong Kong, China.

3. Singapore Botanic Gardens, 1 Cluny Road, Singapore 259569, Singapore.

4. Institut de Recherche pour le Développement (IRD), UMR-DIADE, BP 64501, F-34394 Montpellier cedex 5, France.

The Annonaceae, the largest family in the early-divergent order Magnoliales, comprises 107 genera and c. 2,400 species. Previous molecular phylogenetic studies targeting different taxa have generated large quantities of partially overlapping DNA sequence data for many species, although a large-scale phylogeny based on the maximum number of representatives has never been reconstructed. We use a supermatrix of eight chloroplast markers (*rbcL*, *matK*, *ndhF*, *psbA-trnH*, *trnL-F*, *atpB-rbcL*, *trnS-G* and *ycf1*) to reconstruct the most comprehensive tree to date, including 705 species (29%) from 105 genera (98%). This provides novel insights into the relationships of five enigmatic genera (*Bocagea*, *Boutiquea*, *Cardiopetalum*, *Duckeanthus* and *Phoenicanthus*). Fifteen main clades are retrieved in subfamilies Annonoideae and Malmeoideae collectively, 14 of which correspond with currently recognised tribes. *Phoenicanthus* cannot be accommodated in any existing tribe, however: it is retrieved as sister to a clade comprising the tribes Dendrokingstonieae, Monocarpieae and Miliuseae, and we therefore validate a new tribe, Phoenicantheae. Our results provide strong support for many previously recognised groups, but also indicate non-monophyly of several genera (*Desmopsis*, *Friesodielsia*, *Klarobelia*, *Oxandra*, *Piptostigma* and *Stenanona*). The relationships of these non-monophyletic genera—and two other genera (*Froesiodendron* and *Melodorum*) not yet sampled—are discussed, with recommendations for future research.

T2
P0398

Phylogeny of the temperate woody bamboos (Poaceae: Bambusoideas): With an emphasis on the leptomorphy group

Ying Guo

Kunming Institute of Botany, Chinese Academy of Sciences

The temperate woody bamboos are a monophyletic group with diverse morphology in Bambusoideae, recent phylogenetic researches indicate that it is a monophyletic lineage, Leptomorph-rhizome bamboos are a group of the temperate woody bamboos which have two types of rhizomes, Leptomorph rhizome and pachymorph rhizome, and phylogenetic relationships among them have not been clearly resolved till now. In this study, we reconstructed phylogeny of the temperate woody bamboos based on double-digest restriction-site associated DNA (ddRAD) sequencing data of a large sample set with an emphasis on the leptomorph rhizome bamboos. A total of 211 species in 22 genera were sampled, and five species of them were outgroups, i.e. *Fargesia spathacea*, *Yushania niitakayamensis*, *Yushania leigongshanensis*, *Thamnocalamus spathiflorus* and *Chimonocalamus delicates*. We utilized the STACKS software to discover SNP markers for phylogenetic analyses and found 334900 SNPs in the data set for subsequent maximum-likelihood analysis. The preliminary findings show that the leptomorph rhizome bamboos were a monophyletic group, four well supported clades were revealed in this group, i.e. the *Phyllostachys* + *Shibataea* clade, the *Arundinaria* clade, *Indocalamus wilsonii* and the *Chimonobambusa* + *Gelidocalamus* + *Indocalamus* clade. *Indocalamus wilsonii* was first diverged from the rest species .The *Phyllostachys* + *Shibataea* clade, which includes 62 species in 4 genera, and The *Arundinaria* clade, which includes 81 species in 10 genera, were sister groups, and they formed a clade, sister to the *Chimonobambusa* + *Gelidocalamus* + *Indocalamus* clade. At the genus level, many genera (e.g. *Pleioblastus*, *Indosasa*, *Acidocasa*, *Gelidocalamus*, *Indocalamus* and so on) defined according to the morphological features were not monophyletic, only *Arundinaria s.s.*, *Phyllostachys*, *Shibataea*, *Chimonobambusa* and *Ferrocalamus* were revealed as monophyletic. Our results also exhibited close relationships between *Gelidocalamus* and *Indocalamus,* and relationships among most species of the 2 genera were clearly resolved. However, phylogenetic relationships within the *Arundinaria* clade and *Phyllostachys* + *Shibataea* clade were not fully resolved, and further studies are needed to uncover evolutionary histories of them.

T2
P0399

Exploring patterns of genomic divergence among reproductively isolated populations of the Arctic plant *Draba nivalis* (Brassicaceae)

A. Lovisa L. Gustafsson[1], *Michael Nowak*[1], *Siri Birkeland*[1], *Abel Gizaw Seid*[1], *Loren Rieseberg*[2], *Tanja Slotte*[3], *Christian Brochmann*[1]

1. University of Oslo, Natural History Museum

2. University of British Columbia, Department of Botany

3. Stockholm University, Department of Ecology, Environment and Plant Sciences

Understanding the process of speciation requires characterizing genes involved in reproductive isolation. To accomplish this goal, it is important to identify species with populations that are on the verge of reproductive isolation with one another. Previous studies have shown that allopatric populations of the Arctic plant *Draba nivalis* exhibit hybrid incompatibilities suggestive of an early stage in the process of species formation. Hybrid incompatibilities have been identified in multiple crosses between geographically divergent populations throughout the Arctic, and dysfunction in hybrid pollen and seed development has been mapped to 7 loci between Alaskan and Norwegian populations. To explore the patterns of genomic divergence between populations exhibiting partial reproductive isolation in this species, we have produced a *de novo* genome assembly and resequenced populations that exhibit hybrid incompatibilities in experimental crosses. We use these data to examine how sequence divergence between these populations is distributed throughout the genome, and to identify any large-scale chromosomal rearrangements that may be impacting the formation of hybrid gametes. Our production of a genome assembly and generation of comparative population genomic data makes *D. nivalis* an attractive model system for studying the earliest stages in the evolution of reproductive isolation in plants.

T2

P0400

Floral and foliar morphological variation of *Magnolia rzedowskiana* in the Sierra Madre Oriental, México

Marisol Gutiérrez Lozano[1], *Arturo Sánchez González*[1], *J. Antonio Vázquez García*[2]

1. Universidad Autónoma del Estado de Hidalgo, Centro de Investigaciones Biológicas

2. Universidad de Guadalajara, Centro Universitario de Ciencias Biológicas y Agropecuarias

Magnolia rzedowskiana (Magnoliaceae), commonly known as Eloxóchitl ("corn flower") or May flower, is endemic of the mountain cloud forest (BMM) of the Sierra Madre Oriental (SMO), Mexico. Currently, their populations are distributed discontinuously (isolated) and fragmented in six localities in the state of Hidalgo, Queretaro, San Luis Potosi y Veracruz. Locally, the species is known by the beauty of its flowers, which are used in traditional medicine to treat heart problems, nerve or heal the fright, and especially with ornamental purposes. In order to determine the morphological variation of *M. rzedowskiana*, 10 individuals were randomly selected in each locality. Ten leaves, five flowers and two fruits were collected in each individual. Subsequently, 25 morphological characters of flower and leaf, and 7 of dasometric variables were analyzed. Data were analyzed by descriptive statistics, and multivariate methods: cluster analysis, principal components analysis and discriminant analysis. The results indicate that there is a wide morphological variation within and between populations, particularly in the number of stamens and carpels, which is attributed to a high phenotypic plasticity of the species. Because of the change in land use and habitat fragmentation by human activities, in some localities the degree vegetation disturbance is very high, resulting in lower population density (few hundreds of individuals), or in extinction of wild populations, as in Huayacocotla locality (Veracruz) case, where only cultivated individuals exist. It is noteworthy that the description of *M. rzedowskiana* is recent (2016), so little is known about basic ecological and genetic aspects, as to propose its inclusion in a risk category within Mexican Official Standard. To priori, according to the criteria of the International Union for Conservation of Nature (IUCN), this species could be include in the category of "critically endangered".

T2

P0401

Resolving the backbone of *Ptilotus* (Amaranthaceae) using next-generation sequencing

Timothy Hammer[1], *Ian Small*[2], *Kevin Thiele*[1]

1. School of Biological Sciences, Faculty of Science, The University of Western Australia

2. Plant Energy Biology, ARC Centre of Excellence, The University of Western Australia

Family-level phylogenies of Amaranthaceae have identified the 'aervoid' clade, composed of *Ptilotus* (~110 spp.), *Aerva* (11 spp.) and *Nothosaerva brachiata*. A recent phylogeny of *Ptilotus* using ITS and *matK* sequences of 100 species resolved the genus to be monophyletic, but nested within a paraphyletic *Aerva*. The backbone of the *Ptilotus* phylogeny was unresolved, with a polytomy at the base of four large clades. In order to make use of the phylogeny to examine the evolution of novel characters, such as phosphorus hyperaccumulation, a resolved backbone is needed. To resolve the phylogeny, 30 whole-chloroplast genomes using Illumina next-generation sequencing were obtained for species in all major clades identified on the previous phylogeny. The resulting, fully-resolved chloroplast tree was used as a constraint in the reanalysis of the ITS and *matK* phylogeny to obtain a resolved tree of the 'aervoids'.

T2

P0402

Lineage isolation in the face of active gene flow in a typical coastal plant wild radish is reinforced by differentiated vernalization responses

Qingxiang Han[1], *Hiroyuki Higashi*[2], *Yuki Mitsui*[3], *Hiroaki Setoguchi*[2]

1. Linyi University

2. Kyoto University

3. Tokyo University of Agriculture

The respective role and relative importance of natural selection and gene flow in the process of population divergence has been a central theme in the speciation literature. A previous study presented conclusive evidence that wild radish (*Raphanus sativus* L. var. *raphanistroides* Makino) on Japanese islands comprises two genetically isolated lineages: the southern (Ryukyu Islands) and northern (mainland Japan) groups. However, a general understanding of the lineage isolation in the face of frequent seed flow of the coastal plant species is still unclear. We surveyed nucleotide polymorphisms over 14 nuclear loci in 72 individuals across the Japan-Ryukyu Islands Arc to address the demographic history of wild radish using the isolation-with-migration (IM) model. In addition, we investigated the flowering times of the individuals with and without cold exposure to assess the vernalization responses of wild radish. Coalescent simulations suggested that divergence between the southern and northern lineages of wild radish began ca. 17,859 years ago, during the Last Glacial Maximum (LGM; ca. 18,000–22,000 years ago). We further addressed that cold exposure was not required for flowering in the southern group, but could advance the date of flowering, suggesting that vernalization would be facultative. In contrast, the northern group was unable to flower without prior cold exposure, and thus had an obligate requirement for cold treatment. We proposed that temperature profoundly affected the vernalization responses of wild radish, which may repress reproductive success and ultimately drive and maintain intra-specific differentiation between the two groups. Unexpectedly, after population divergence, bidirectional gene flow between southern and northern lineages was found at effective rates, attributing to the combination of high seed longevity and high dispersibility of seeds via sea current. Particularly, the gene flow from the southern to northern groups was significantly higher than that in the opposite direction, indicating effective dispersal of viable seeds via the northward Kuroshio Current. To conclude, divergent natural selection in the southern and northern regions generated trait variation in vernalization requirements as populations adapted to their local environments and in turn probably contributed to lineage isolation in the wild radish, despite the presence of ongoing gene flow. This work increases our understanding of how lineage isolation in flowering coastal plants is maintained by divergent natural selection generated by differing environments in the face of ongoing gene flow. This work also offers a new and fascinating story about wild radish, which is helpful to understand the mechanisms of adaptive evolution, particularly in the case of coastal plants, on a local scale.

T2

P0403

Speciation mechanisms in one of the richest cactus genera: *Mammillaria*

Tania Hernandez-Hernandez[1], *Israel Loera-Carrizales*[2], *Betsabe Ruiz-Guerra*[2], *Andrea Lopez-Martinez*[3]

1. University of Arizona

2. Instituto de Ecologia, A.C.

3. Universidad Nacional Autonoma de Mexico

The genus *Mammillaria* has been considered the richest of the cactus family, including around 145 species. Although the distribution of the genus goes from the south of the United States in the North to a Southern limit in Colombia and Venezuela through the Caribbean Islands, Guatemala, Honduras and México; the latter country concentrates the vast majority of species. *Mammillaria* is practically a Mexican radiation, with species inhabiting all ecosystems present in this megadiverse region. The monophyly of *Mammillaria* has not been confirmed with molecular data, but evidence show it groups together with related genera into the Mammilloid clade. This clade has been reported as having one of the highest diversification rates within the family, a result that is intriguing given that these plants are slow-growing and long lived, with slow population growth rates. In this study, we use the cactus genus *Mammillaria* and related genera included in the Core Mammilloid clade (i.e. *Coryphantha*, *Escobaria*, *Ortegocactus* and *Neolloydia*) to study diversification mechanisms in the Mesoamerican biodiversity region. We focus our study in speciation mechanisms by comparing the ecological niches of sister species. We show evidence supporting the very recent origin of the group, with an allopatric mode as prevalent feature of its diversification. We discuss the possible mechanisms and ecological drivers of *Mammillaria* diversification.

T2

P0404

Insights on the evolution of Mediterranean Island - mainland endemic species: The Iberian-Ibiza endemism *Carduncellus dianius* (Compositae) as example

Sonia Herrando-Moraira[1], *Sergi Massó*[1], *Deborah Zani*[2], *Jordi López-Pujol*[1], *Roser Vilatersana*[1]

1. Botanical Institute of Barcelona (IBB–CSIC–ICUB), Pg. del Migdia s.n., 08038 Barcelona, Spain

2. Millennium Seed Bank, Conservation Science, Royal Botanic Gardens Kew,, Wakehurst Place, RH17 6TN, United Kingdom

Long-term persistence of plant populations through multiple geological and paleoclimatic events has made the Mediterranean Basin a geographical center of species richness and endemism. *Carduncellus dianius* is a good example of a typical Mediterranean narrow endemic species, being restricted to less than 20 populations; about half of these populations are located in the mainland (in the Spanish Province of Alicante, eastern Iberian Peninsula) and half are located in western Ibiza Island (Balearic Islands). We combined molecular techniques with ecological niche modeling (ENM) methods to get insights into the species past evolutionary history, its current levels and distribution of genetic variation, and its potential future distribution. To obtain genetic information, 79 and 143 specimens were analyzed with three cpDNA regions and AFLP markers, respectively. Results from phylogenetic analyses using cpDNA data revealed that this species might have originated in the continental region. The colonization of Ibiza Island could have occurred by a single long-distance dispersal event, with a subsequent back-colonization from the island to the continent. During the Last Glacial Maximum (~20 kya) Ibiza presented higher habitat suitability for the species than the mainland, where harsher environmental conditions could have produced a genetic bottleneck. Reverse results were found in future projections of potential distribution in 2070, that show a considerable loss of suitable habitat in Ibiza, where the effects of climate change could, thus, affect seriously the conservation of island populations.

T2

P0405

A evolutionary framework for *Castanopsis* (D. Don) Spach, a major component of Asian forests

Damien D. Hinsinger[2,1], *Joeri S. Strijk*[1,2]

1. Plant Ecophysiology & Evolution Group, State Key Laboratory for Conservation and Utilization of Subtropical Agro-bioresources, Guangxi University, Nanning

2. Plant Ecophysiology & Evolution Group, College of Forestry, Guangxi University, Nanning

Asia is one of the world's most diversified regions, in terms of environmental variation as well as in biodiversity values. This diversity is mainly due to climatic oscillations during the tertiary and quaternary, as well as to the geological changes that occurred from 30 mya to present. However, a great part of this biodiversity remains unknown, and a phylogenetic framework is still lacking for many taxa in the region. The genus *Castanopsis* (D. Don) Spach (Fagaceae) comprise 110-134 species, which are restricted to tropical and subtropical eastern Asia (from Japan to Indonesia), some of them being very common in subtropical forests. About half of the species are found in China, and 30 are endemic, in a wide range of habitat types, where they are often keystone species. Despite they are sometimes locally abundant, several species are currently threatened and Red Listed. With the recent development of next-generation sequencing methods, it is now possible to study complete genomes to precisely delineate the history and diversity of taxa. However, Whole Genome Sequencing remains unaffordable for many non-model taxa. For these groups, highly repeated regions sequencing, including chloroplasts and nuclear ribosomal cistrons, has been showed to be an efficient method to reconstruct evolutionary history. We obtained complete chloroplasts, nuclear ribosomal cistrons and repeated sequences data from Illumina genome skimming sequencing for a representative portion of the genus. We also generated Genotyping By Sequencing (G.B.S.) data, to get a genome-wide view of the evolution, diversity and taxonomy of the genus. Using both nuclear and chloroplast data, we performed phylogenetic analyses, molecular dating and biogeographical analyses to reconstruct the diversification and phylogeographic history of the genus. The close affinity between the genera *Castanopsis* and *Castanea* was confirmed, as well as the monophyly of the genus. Dating and biogeographic analyses indicated that the Eocene climatic optimum likely drove the divergence between the two genera. We found that climatic and geological changes occurring in Asia during the Cenozoic were a major driving actor in the diversification of the genus, especially the climate changes and eustatic sea level variations. We expect this study will serve as a cornerstone in improving our understanding of the origin and geographic distribution of genetic diversity in tropical trees, thereby aiding ongoing conservation efforts.

T2

P0406

A ddRADseq approach to inform conservation priority decisions for closely related *Grevillea* species.

Gareth Holmes, *Trevor Edwards, Peter Weston, Julie Atkinson, Susan Hoebee*

Ecology, Environment & Evolution, La Trobe University

The Arenaria and Floribunda subgroups of *Grevillea* (Proteaceae) from eastern and south-eastern Australia comprise 16 species (19 taxa) including several that are listed as threatened under federal legislation. No phylogenetic studies have been undertaken to clarify the relationships of lineages within these subgroups. DNA sequence markers traditionally used in plant phylogenetic studies often provide little resolution among closely related Proteaceae. We therefore generated ddRAD data for all taxa and most variants within the subgroups and outgroups, processed the data with the pyRAD pipeline and inferred their phylogeny based on maximum likelihood analysis of organellar and nuclear DNA. Our results suggest that the Floribunda subgroup is not monophyletic with species from another subgroup (Rosmarinifolia) nesting within it. One species, *G. celata*, is nested within *G. chrysophaea* and may require reclassification as a subspecies. Alternatively, *G. chrysophaea* plants from one region could be recognised as a distinct species. There was little phylogenetic structure identified in the most widespread and morphologically variable species of the subgroups (*G. alpina*). Incongruence between phylogenies inferred from maternally inherited cpDNA and mtDNA markers and those of nuclear origin suggests the possibility of hybridisation between species. We conclude that the current informal subgeneric classification of the groups does not represent their evolutionary relationships in all cases and requires revision. The classification of *G. chrysophaea* and *G. celata* (the latter considered vulnerable) also requires revision. Our finding highlights the value of ddRAD in resolving phylogenetic and conservation questions in groups with complex evolutionary histories.

T2

P0407

Taxonomic revision on the genus *Hedychium* J. König (Zingiberaceae) in China

Xiu Hu

Zhongkai university of agriculture and engineering

As a result of extensive studies of the literature, herbarium specimens, living plants, the genus *Hedychium* in China is revised here since eflora of China was published in 2000: (1). Two new species [*H. menghaiense* X. Hu & N. Liu (sp. nov.), *H. longipetalum* X. Hu & N. Liu (sp. nov.)], one new form (*H.* villosum f. *albifiamensum* X. Hu & N. Liu) and two newly recorded species (*H. gardnerianum* Rosc., *H. ellipticum* Ham. ex Sm.) are discovered; (2). For being the same entity and published late, *H. longipetalum* X. Hu & N. Liu (sp. nov.) and *H. efilament* Hand.-Mazz. are treated as the synonym of *H. champasakense* Picheans. & Wongsuwan and *H. wardii* C. E. C. Fisch. respectively; (3). Share the same morphology characters with *H. stenopelatum* Lodd., but differ in polyploidy levels, and published late, *H. puerense* Y. Y. Qian is treated as a variety of *H. stenopelatum* Lodd., i.e. *H. stenopelatum* Lodd. var. *puerense* (Y. Y. Qian) X. Hu; (4). Without obvious differences in morphology, *H. bipartitum* G. Z. Li, *H. tienlinense* D. Fang, H. *panzhuum* Z. Y. Zhu are treated as the synonyms of *H. flavum* Roxb., and *H. menglianense* Y. Y. Qian is treated as the synonym of *H. sinoaureum* Stapf.; (5). Lots of specimens identified as *H. forrestii* Diel. should be *H. puerense* Y. Y. Qian.; (6). *H. villosum* Wall. and *H. tenuiflorum* Wall. ex Bak. (or *H. villosum var. tenuiflorum* Wall. ex Bak.) are mis-identified on each other in most specimens and publications. At last?it is confirmed that there are thirty-two species, two varieties, one form in China, and a key to the species of *Hedychium* in China is given verified by the phylogenetic analysis of SRAP makers.

T2

P0408

Phylogenetic relationships and phylogeography of *Juglans*

Yi-Heng Hu[1], *Peng Zhao*[1], *Ebrahimi Aziz*[2], *Keith Woeste*[2]

1. Northwest University

2. Purdue University

Juglans L. as one of the typical phytogeographic disjunctions in the Northern Hemisphere, the biogeographic disjunction pattern between closely related plant in Eastern Asian and North American has been received considerable attention. This pattern is often explained by the some researches about bore tropical flora. Walnut (*Juglans* L.) is one of the eight living genera belong to the family Juglandaceae, which are including ~21 extant taxa which could be divided into four sections according to fruit morphology, wood anatomy, and foliage architecture. However, a limited number of DNA regions with insufficient phylogenetically informative sites were selected, and low mutation rates of selected regionswere account for the low resolution. Based on multiple markers cannot clearly elucidate the phylogenetic relationships and phylogeography patterns of *Juglans*. We collected the total of 56 individuals to represent all four recognized section of *Juglans* species, and 14 individuals from family Juglandaceae as outgroup. We collected representatives of many genera with ambiguous phylogenetic position. Complete chloroplast genomes and transcriptom used to analyze phylogenetic relationships. Our initial conclusion now is that: 1) Our study has clarified previously unresolved phylogenetic relationships among species of *Juglans* using chloroplast phylogenomic analyses. 2) Whole chloroplast genome data offer promise for generating adequate levels of variation for phylogenetic reconstruction among species, and suggestion *J. cinerea* belong to *Rhysocaryon*. 3) Genus-level analyses incorporating both extant and extinct genera of the Juglandaceae suggested that *Juglans* originated in North America, and migrated to Eurasia during the early Tertiary via the North Atlantic land bridge.

T2

P0410

Out of the Qinghai-Tibet Plateau of *Fritillaria* (Liliaceae) indicated by molecular phylogeny with an emphasis on the East Asian species

Jiao Huang
Sichuan university

The East Asian region is considered the cradle for many cosmopolitan plants and the uplifts of the Qinghai-Tibetan Plateau (QTP) in this region have played a key role in driving diversification and dispersal of numerous northern temperate plants. The hypothesis "out-of-QTP" suggests that the majority of northern temperate plants have originated and dispersed from the QTP and adjacent region. An interesting question is whether the biogeographic history of *Fritillaria*, widely distributed in the northern temperate region, coincides with the hypothesis "out-of-QTP"? In the present study, phylogenetic and biogeographical reconstruction of all eight genera of *Fritillaria* and including most of species endemic to the East Asian region were performed based on ITS, *matK*, *rbcL* and *rpl16* sequences. The results support *Fritillaria* as an"out-of-QTP" genus. The uplift of the QTP in the Miocene facilitated the origin and diversification of *Fritillaria*, orogenies of QTP and Hengduan Moutains and the advent of drying and cooling climates during the Miocene and Pliocene may have fragmented the distribution disjunctly of the genus in the east Asian and Mediterranean-Turanian (MT).The migration from East Asia to North America via the Bering Land Bridge in the late Miocene has been supported. The present study highlights the influence of QTP uplifts on the origin and diversification of northern temperate plants.

T2

P0411

Incipient speciation in a dune sunflower species

Kaichi Huang[1], *Katherine Ostevik*[2], *Rose Andrew*[3], *Loren Rieseberg*[1,4]
1. Botany Department, University of British Columbia
2. Biology Department, Duke University
3. School of Environmental and Rural Science, University of New England
4. Department of Biology, University of Indiana

An ecotype of prairie sunflowers (*Helianthus petiolaris*) has adapted to sand dunes in the Great Sand Dune National Park over the past 5,000 years. The dune ecotype is isolated from its non-dune progenitor due to extrinsic ecological isolation, as well as conspecific pollen precedence. Conspecific pollen precedence is stronger in the contact zone of two ecotypes, which is a hint for reinforcement. Genomic analyses also found some potential "reinforcement" regions on different chromosomes of the species. We are currently using genetic mapping and GWAS approaches to identify the genetic mechanisms underlying the speciation process of this system.

T2

P0412

Multiple origins of *Aquilegia ecalcarata* revealed by the combination of chloroplast and nuclear DNA sequence data

Lei Huang, *Fangdong Geng, Cheng Xue, Juqing Kang, Jianqiang Zhang, Yi Ren*
College of Life Science, Shaanxi Normal University

Floral evolution and its relationship with speciation attract a lot of attention in the past decades. The genus *Aquilegia* has been considered as new model system for plant evolution, ecology and development researches. Several studies have demonstrated that pollinator shift and ecological habitats drove the radiation of some *Aquilegia* species in North America. In these cases, the spur length varied a lot across different closely related species and is under strong selection by pollinators. Of all the approximately 70 species in this genus, *A. ecalcarata* is the only one which is spurless (or very short spur no more than 2mm) with the loss of nectar. *A. ecalcarata*, together with other three Chinese endemic species: *A. yabeana*, *A. oxysepala* var. *kansuensis* and *A. rockii* comprised a monophyly according to the previous phylogeny results based on just one individual for each species. This cannot fully solve the origin of *A. ecalcarata*, given that the uniqueness of *A. ecalcarata* in whole genus, more efforts are required to deeply uncover this question. Here we reported an extensive population study in 42 representative populations of four closely related species using three chloroplast loci and ten single-copy nuclear gene sequence data. The Phylogenetic tree, STRUCTURE and PCoA analysis all indicated the mutiple origins of *A. ecalcarata*, and *A. ecalcarata* I was possibly originated from *A. oxysepala* var. *kansuensis* (with long and curly spur) while *A. ecalcarata* II more likely from *A. rockii* (with long and erect spur). By molecular clock model, the estimated divergence time of each *A. ecalcarata* I and *A. ecalcarata* II originated from ancestors was less than 1 Myr. Genetic differentiation results showed that divergence between *A. ecal-*

carata I and *A. ecalcarata* II was significantly higher (F_{ST} 0.273) than that between *A. ecalcarata* I and *A. oxysepala* var. *kansuensis* (F_{ST}0.090), or between *A. ecalcarata* II and *A. rockii* (F_{ST} 0.092). The differentiation within *A. ecalcarata* I and *A. ecalcarata* II was at intermediate level (F_{ST} 0.439,0.477) and both supported the Isolation by Distance model. The similar habitat and phenotypes implicated the parallel evolution in *A. ecalcarata* I and *A. ecalcarata* II.Pollinator shift and ecological habitat were both the possible driving force for it. Our findings enrich the knowledge about floral especially spur evolution and species divergence.

T2

P0413

Fossil woods of *Castanopsis* (Fagaceae) from the upper Oligocene of Guangxi (South China): The most ancient evidence of this genus in Southeastern Asia

Luliang Huang[1,2], *Jianhua Jin*[1], *Cheng Quan*[2], *Alexei A. Oskolski*[3,4]
1. Sun Yat-sen University, Guangzhou
2. Jilin University, Changchun
3. University of Johannesburg, Johannesburg,
4. Komarov Botanical Institute of the Russian Academy of Sciences, St. Petersburg

Two new species attributed to the genus *Castanopsis* (*C. nanningensis* and *C. guangxiensis*) were described on the basis of well-preserved mummified wood from the upper Oligocene Yongning Formation of Nanning Basin in Guangxi Province, South China. They represent the most ancient reliable fossil record of this genus not only in China, but also in southeastern Asia, the region of greatest diversity of its extant species. This fossil evidence strongly suggest that the species of *Castanopsis* migrated to South China from northeastern Asia in the late Oligocene via the humid margin formed along the ocean coast due to the development of monsoon climate system during this time. The presence of numerous *Castanopsis* woods with faint or absent growth ring boundaries occasionally co-occurred (in *C. nanningensis*) with prominent ring-porous pattern in the Yongning Formation of Nanning Basin confirms that Guangxi (South China) had seasonal (probably monsoonal) tropical climate during the late Oligocene.

T2

P0414

A temporal perspective on Brassicaceae family-wide radiation and diversification

Xiao-Chen Huang, *Dmitry German, Markus Kiefer, Nora Hohmann, Eva Wolf, Marcus A. Koch*
Heidelberg University

Brassicaceae is an important model family for evolutionary and molecular research, comprising 52 tribes, 325 genera, and about 3990 species. It comprises not only the most important model for molecular plant biology (*Arabidopsis thaliana*), but also important crop plants such as rapeseed and cabbage. Evolutionary processes can only be understood if the correlations between environmental dynamics and evolutionary changes are known within a well-resolved temporal context. However, in the past divergence time estimates for Brassicaceae varied among markers as well as calibration points that were used. Consequently, reliable time estimates of all major lineage splits in Brassicaceae are still insufficient and herein tested rigorously using chloroplast phylogenomics. Based on these temporal anchor points from the chloroplast genome backbone, we estimated chronograms on tribal level using ITS datasets comprising more than 2000 species in total. We estimated crown-group ages for these tribes and provided also some evidence for respective stem group ages. Molecular clock estimates revealed that the majority of tribes started diversification between c. 8 to 15 Mya, coinciding with the mid-Miocene climate transition and optimum. This is further correlated with subsequent increasing speciation rates in most species rich tribes indicating radiation throughout the Miocene and Pliocene. Furthermore, it is shown that the Pleistocene also played an enormous role to further increase speciation rates. These results suggest that Brassicaceae evolution and diversification is generally highly correlated with major geological and climatic changes often favored by a transitions towards lower temperatures and drier conditions.

T2

P0415

Asian *Begonia* biogeography tracks mountain building and microplate tectonic movement in Southeast Asia

Mark Hughes[1], *Lakmini Darshika Kumarage*[2], *Adi Mahardika*[1], *Rosario Rubite*[3]
1. Royal Botanic Garden Edinburgh
2. University of Peradeniya
3. University of the Philippines Manila

Begonia is a superb group of plants for tracking the biography of tropical floras. The genus has a high proportion of micro-endemics, is species rich, is found throughout Southeast Asia, and is poorly dispersed. Genetic evidence shows some Begonia populations are very long lived and potentially can remain in-situ for timescales measured in millions of years. Since the first insights into Asian *Begonia* biogeography by Thomas (2011) showed a continental origin for the Malesian species, followed by Miocene east to west dispersal across Wallace's line, three further more focussed regional studies have added detail this picture. Through expanded chloroplast marker phylogenies, increased sampling has shed light on the origin of Sri Lankan, Sunda Shelf and Philippine *Begonia*. We show Sri Lankan endemic *Begonia* to be the result of late Miocene/early Pliocene immigration, followed by in situ Pleistocene speciation. We found no evidence of relict endemics in the Sri Lankan *Begonia* flora. On the Sunda Shelf, Borneo is likely to be an area of origin for the clade containing the bulk of the ca. 500 species in Malesia. Sumatra is also a key area for development of the Malesian flora, with dispersal to Peninsular Malaysia and Borneo in the early Pliocene, following diversification during the Barisan Range mountain uplift in this period. Philippine *Begonia* sect. *Baryandra* (ca. 60 species) are the result of a monophyletic radiation which entered the north of the archipelago during the late Miocene, when the northern islands Luzon, Panay and Palawan were near-contiguous. The *Begonia* floras of these 3 islands show a rapid 3-way phylogenetic split following their separation during the ensuing micro-plate movements in the Philippine region. Mindanao *Begonia* are highly nested in the phylogeny and relatively young compared to species in more northern parts of the archipelago. This corresponds with the very rapid movement of the Mindanao microplate north across the equator to its current position.

T2
P0416
Towards a new alaskan flora: Integrating paper floras, specimens, observations and taxonomic changes

Campbell Webb, Jordan Metzgar, ***Stefanie Ickert-Bond***
Herbarium (ALA), University of Alaska Museum of the North, Fairbanks, Alaska, U.S.A.

Alaska is vast, sparsely populated, and sufficiently under-explored botanically that new species and occurrences are still frequently found. Taxonomic problems abound, partially due to parallel but non-interacting botanical efforts in other Arctic nations. Many of these are highlighted in the Panarctic Flora Checklist. The last comprehensive flora (by Eric Hultén) is almost fifty years old, and there is a pressing need for a new Alaskan Flora to support research. We are beginning to develop a New Alaskan Flora which will integrate i) Hultén's and other paper floras, ii) the 262,000 digitized specimens in the herbarium of the University of Alaska (ALA; http://arctos.database.museum/), iii) taxonomic accounts from online floras (e.g., the Panarctic Flora, the Flora of North America), iv) new, up-to-date taxonomic reviews of some plants groups by collaborating experts, and v) photographs and observations from public biodiversity platforms (such as iNaturalist). Where available, morphological descriptions will be parsed and reconciled to a master phenotype ontology (based on Peter Steven's work at the Angiosperm Phylogeny website), and will be used to generate new descriptions and keys. The core output will be an API (Application Program Interface) that serves the integrated data as XML, JSON and RDF, using Darwin Core biodiversity information standards. This API will be used to dynamically generate attractive taxon pages on a project website which present the viewer with congruent and contrasting information, all carefully linked to data sources. Regular snapshots of the data will also be available in ePUB format, enabling printing of a paper flora, and downloading to portable devices for use in the field. Android and iOS apps may also be developed in the future. At the core of this data integration is a 'taxonomic intelligence engine' that keeps track of evolving and conflicting taxon concepts and taxon names, and manages synonyms in the determinations of specimens. The starting point for name integration is The Plant List from the Royal Botanic Gardens, Kew, but other relevant monographs and revisions will be digitized and integrated into this engine, as will taxonomic data from other online resources. This taxonomic flexibility will permit the user direct access to the full taxon concept history of each species, and crucially, allow the user to see at a glance which species are 'difficult' (and for which any determinations must be viewed with some doubt), and which are 'easy' (monotypic taxa with distinct morphotypes, and little geographic variation). We believe that creating this integrated botanical resource, which will make available the full history of a taxonomic name, present associated images from ALA, and include vetted (research-grade) citizen-science observations, will spur additional botanical research and discovery of new taxa in the Far North.

T2
P0417
Geographic distribution of ploidy level and intraspecific phylogeny of East Asian shrub *Clerodendrum trichotomum* sensu lato

Leiko Mizusawa[1], *Fujii Shinji*[2], *Isagi Yuji*[3]
1. Fukushima University
2. University of Human Environments
3. Kyoto University

Clerodendrum trichotomum is widespread and common shrub of East Asia. This species complex distribute East China, Korean peninsula, Japan archipelago, Taiwan island and Philippine islands. Two varieties of *C. trichotomum* have been known from Southern Japan: *C. trhichotomum* var. *yakusimense* and *C. trichotomum* var. *esculentum*. Moreover, one close relative of *C. trichotomum* have been known from Izu Islands, Japan: *C. izuinsulare*. Here, we call these four taxa "*C. trichotomum* sensu lato". Previous study (Zeng *et. al*., 2011) reported that Chinese var. *trichitomum* are diploid on the basis of karyotypes of materials collected from wide area of China. However, ploidy level of Japanese var. *trichotomum*, var. *yakusimense*, var. *esculetnum* and *C. izuinsulare* are unknown. During the process of development of microsatellite markers on *C. trichotomum* sensu lato, we got results that suggest both diploid and tetraploid lineages exist in *C. trichotomum* sensu lato. Here we report that the results of estimation of ploidy level and phylogenetic relationships in *C. trichotomum* sensu lato based on five microsatellite markers and 1267 pb of chloroplast DNA sequence. Totally, 89 materials were collected from Japan archipelago, China and Korean peninsula. The microsatellite genotypes strongly suggested that var. *esculentum*, Chinese var. *trichotomum* and *C. izuinsulare* are diploid whereas var. *yakusimense* and Japanese and Korean var. *trichotomum* are tetraploid. Neighbor-joining tree based on the microsatellite genotypes showed following three clades: Clade-I [var. *yakusimense* and Chinese, Japanese and Korean var. *trichotomum*], Clade-II [*C. izuinsulare*] and Clade-III [var. *esculentum*]. Chinese var. *trichotomum* (diploid) was completely nested in cluster of tetraploid var. *trichotomum*. On the other hand, Bayesian phylogenetic tree based on chloroplast DNA showed following three clusters: clade-I [Chinese var. *trichotomum*], clade-II [Japanese and Korean var. *trichotomum* and *C. izuinsulare*] and clade-III [var. *esculentum* and var. *yakusimense*]. These results suggest two following possibility for diversification of *C. trichotomum* sensu lato. Firstly, tetraploid lineages have evolved repeatedly in this taxon. Secondly, chloroplast haplotypes have been shared between diploid and tetraploid lineages by past hybridization.

T2
P0418
Systematics of the genus *Kaempferia* L. (Zingiberaceae) in Laos

Oudomphone Insisiengmay[1,3], *Thomas Haevermans*[1], *Mark Newman*[2]
1. Muséum Nationale d'Histoire Naturelle
2. Royal Botanic Garden
3. Ministry of Science and Technology

During the 19th century, the genus was studied by botanists such as P. Horaninow (1862), G.Bentham and J.D. Hooker (1883), and K. Schuman (1900). The oldest specimens of Zingiberaceae form former Indochina (Cambodia, Laos and Vietnam) date from 1742, probably with Loureiro's collections, a Portuguese Jesuit missionary and botanist. He published the first flora of Southeast Asia the "*Flora Cochinchinensis*" in 1790 in which 13 species in tree genera of Zingeberaceae were described including K. galanga. (Skornickova and Newman, 2015). The most recent revision of the Zingiberaceae in Cambodia, Laos and Vietnam has been made by F. Gagnepain for Lecomte's *Flore Générale de l'Indochine*. His account, which was published in 1908, is long now out of date. With the advent of molecular phylogenetics, it became possible to test previous classifications based on macromorphology only. In 2002, Kress and collaborators published a new classification for Zingiberaceae based on molecular evidence. Their results confirmed many of the groupings established by earlier morphologists. Kress et al. did however, make some adjustments for some group ranks and created a few new ones. this classification divides the family into four subfamilies and tribes. My research project is centered on a modern integrative systematic study of genus *Kaempferia*, encompassing new molecular data from complete chloroplasts and selected nuclear genomic regions along with new morphoanatomic characters. This work will also offer the possibility to monograph the genus and evaluate the IUCN statuses for each species, which was never done before.

T2

P0419

Phenetic analysis of floral morphology in the banana family (Musaceae)

Wandee Inta, *Sasivimon C. Swangpol*

Department of Plant Science, Faculty of Science, Mahidol University

It is well-known that the banana family (Musaceae) is the basal clade in the ginger order (Zingiberales). However, its classification into three genera—*Musa*, *Ensete* and *Musella*—and the infrageneric ranking are still ambiguous. Within the *Musa* genus, four formerly separated sections based on seed morphology, molecular data and chromosome number were recently merged and only the *Musa* and *Callimusa* sections have been recognized. Aim of this research is to classify banana taxa in Thailand based on male floral morphology and anatomy using phenetic approach. Fifty-five native and exotic accessions of 24 taxa were investigated. Sixteen quantitative characters were analyzed using Principal Component Analysis (PCA) and Discriminant Analysis (DA), meanwhile, cluster analysis with the Unweighted Pair Group Method with an Arithmetic Mean (UPGMA) was performed on 41 qualitative characters. While the quantitative analysis did not separate *Musa* from *Musella*, the PCA showed that *Ensete* is distinct from the two genera according to width and length of fused tepal lobes, adaxial inner tepal shape and length of style and of ovary. The split was confirmed by the DA based on adaxial inner tepal shape and length of ovary. In contrary, the UPGMA showed that *Musella* was grouped with *Ensete* instead of with *Musa*. Unique characters like having papillate cells on all parts of perianth and capitate stigma placed *Musella* as a distinctive genus. At the sectional level, *Musa* shows variations, e.g. different shapes of papillate cell, adaxial inner tepal and stigma. Nevertheless, only *Musa nanensis* was clearly separated from two other sections by PCA and DA on several characters such as width, length and shape of fused tepals. In conclusion, the quantitative and qualitative characters of flowers supported the arrangement of the three genera within Musaceae and a proposal to add a new section to the *Musa* genus.

T2

P0420

Unique parallel radiations of high-mountainous species of the genus *Sedum* (Crassulaceae) on the continental island of Taiwan

Takuro Ito[1,2], *Chih-Chieh Yu*[3], *Kuo-Fang Chung*[4], *Goro Kokubugata*[2,1]

1. United Graduate School of Agricultural Science, Tokyo University of Agriculture and Technology

2. Department of Botany, National Museum of Nature and Science

3. School of Forestry and Resource Conservation, National Taiwan University

4. Biodiversity Research Cente

Evolutionary radiations, namely, rapid increases in diversity within lineages, provide insight into the dynamics of diversification. Recently, radiations in certain plant groups can be recognized in mountain ranges around the world. Notably, high levels of diversification have been observed in mountains located in tropical-subtropical areas, particularly the Andes and Himalayas. Taiwan, a small continental island (ca. 36,000 km^2) in East Asia located in the tropical-subtropical zone that has exceptionally well-developed mountain systems, including a central range that traverses the island from north to south and features approximately 200 peaks greater than 3,000 m in height. The island shares many characteristics with mountain environments in areas such as the Andes and Himalayas. Previously, several plant radiations have been documented in Taiwan, including *Berberis* L. (Berberidaceae) and *Scutellaria* L. (Lamiaceae). Especially, in case of *Berberis*, the radiation was occurred in high mountain. Considering the geography of Taiwan, it is likely that further radiations have occurred in other plants, particularly in the mountain areas. In the present study, we focused on the plant genus *Sedum* L. (Crassulaceae), which is highly diversified within Taiwan and exhibits extreme ecological and morphological variation among species. Our phylogenetic and ancestral area reconstructions analysis of eastern Asian *Sedum* members based on nrITS and cpDNA sequences showed that Taiwanese *Sedum* comprised two lineages that independently migrated from Japan and Eastern China. Furthermore, the genetic distances among species in these two clades were smaller than those of other East Asian *Sedum* clades; and the Taiwanese members of each clade occupy extremely varied habitats with similar niches shifts in high-mountain regions. These data indicate that species diversification occurred in parallel in the two Taiwanese *Sedum* lineages, and that these parallel radiations could have occurred within a small continental island of Taiwan. Moreover, the estimated time of divergence for Taiwanese *Sedum* indicates that the two radiations might have been correlated to the formation of mountains in Taiwan during the early Pleistocene. In conclusion, we propose that the parallel radiations observed in

Taiwanese *Sedum* could have resulted from, the formation process of the Taiwanese mountains, and high-biological ability of *Sedum* to adapt to newly released ecological niches following multiple migration events to Taiwan.

T2

P0421

Molecular phylogeny of *Eriocaulon* (Eriocaulaceae)

Yu Ito[1], *Norio Tanaka*[2], *Nobuyuki Tanaka*[1], *Anders Barfod*[3], *Seong-Hyun Cho*[4], *Stephan Gale*[5], *Richard Jobson*[6], *Young-Dong Kim*[4], *Isabel Larridon*[7], *Jie Li*[1], *Muthama Muasya*[8], *John Parnell*[9], *Amornrat Prajaksood*[10], *Kotaro Shuto*[11], *Phetlasy Souladeth*[12], *Shuichiro Tagane*[13], *Okihito Yano*[14], *Attila Mesterhagy*[15], *Mark Newman*[16]

1. Xishuangbanna Tropical Botanical Garden, The Chinese Academy of Sciences
2. Tsukuba Botanical Garden, National Museum of Nature and Science
3. Department of Bioscience, Aarhus Universitat
4. Hallym University
5. Kadoorie Farm & Botanic Garden
6. Sydney Botanic Garden
7. Kew Botanic Garden
8. University of Cape Town
9. Trinity College Dublin
10. Khon Kaen University
11. Fukushima University
12. National University of Laos
13. Kyushu University
14. Okayama University of Science
15. University of West Hungary
16. Edinburgh Botanic Garden

Eriocaulon is a genus of about 400 species of monocotyledonous flowering plants in the family Eriocaulaceae. The genus is widely distributed in the world, with the centers of diversity in tropical regions, such as tropical Asia and tropical Africa. A previous molecular phylogeny implied an African origin for *Eriocaulon* as a sister relationship between the genus and an African endemic one was recovered. The species of *Eriocaulon* primarily grow in wetlands while some inhabit shallow rivers and streams with an apparent adaptive morphology of elongated submerged stems. Polyploidy is known from the genus. In this talk, we provide preliminary results of our molecular phylogenetic analysis of the genus aiming to i) assess the biogeographic origin, ii) explore phylogenetic origins of submerged species, and iii) address the evolutionary role of polyploids.

T2

P0422

Updated phylogeny and divergence time estimations of *Centaurea* sects. *Centaurea, Jacea, Lepteranthus* and *Phalolepis* (Asteraceae)

Arnelas Seco Itziar[1], *Perez Collazos Ernesto*[2], *Devesa Alcaraz Juan Antonio*[2], *Lopez Eusebio*[3], *Catalán Pilar*[2]

1. Universidad Tecnica Particular de Loja
2. Universidad de Zaragoza
3. Universidad de Cordoba

Centaurea is one of the most diversified genera of tribe Cardueae (250 taxa) in the daisy family. Sections *Centaurea, Jacea, Lepteranthus* and *Phalolepis* belong to the circum-Mediterranean/Eurosiberian (cME) Cardueae clade. In order to reconstruct an updated phylogeny of this group, we sampled individuals from 30 species (48 taxa), representing nearly 60% of the recognized taxonomic variation of these sections. Haplotypic networks and phylogenetic trees were reconstructed using *trnTL-LF* and ITS sequences. Bayesian dating analysis was performed with the ITS data set. We found two main lineages within cME: (1) a W Mediterranean *C.* sects. *Centaurea* and *Phalolepis* plus *C. diffusa* (CP) clade; and (2) an E Mediterranean Eurosiberian sects. Jacea and *Lepteranthus* (JL) clade. Our analyses provide alternative reconstructions for species of *Jacea* and *Lepteranthus* that differ from those found in recent phylogenetic studies but agree with our new taxonomic findings. The inferred divergence times indicated an early split of the CP/JL clades during the Serravalian (13.6 ± 6.6 Ma), followed by the consecutive splits of *Jacea/Lepteranthus* in the Tortonian (10.85-8.96 Ma), the Late Messinian (7.04-5.07 Ma) divergences of *C. hyssopifolia, C. gr. Janeri, C. exarata–C. corcubionensis–C. stuessyi* and a Pliocene split (3.55 Ma) for *C. linifolia/C. antennata*. Interspecific hybridizations and polyploidizations likely ocurred during Pleistocene (1.48-0.75 Ma) interglacials, when several hybrid lineages apparently originated (e. g. *C.* x *decipiens*). Allopatric speciation might have played an important role in the diversification of the *Centaurea* taxa, since the origins of several endemic species were estimated to have ocurred during the late Pleistocene (2.2-1.31 Ma) in several geographically disjunct Iberian ranges.

T2

P0423

History of dispersal of Monimiaceae: Molecular and morphology data.

Anton Iurmanov

M. V. Lomonosov Moscow State University, Geographical Faculty, Department of Biogeography

The aim of our investigation is reconstruction of the dispersal history of Monimiaceae family based on an integrated morphological and molecular analysis. For the reconstruction of the history of dispersal history of Monimiaceae were analyzed 20 species of 18 genera (from 25 total) from all parts of the family range. We was selected *Gyrocarpus americanus* (Hernandiaceae *s. l.*) as an outgroup. For the molecular analysis the gene 5.8S ribosomal RNA and ribulose-1,5-bisphosphate carboxylase/oxygenase large subunit (*rbcL*) gene. Using data on the nucleic acid sequences from GenBank, in WinClada program were constructed molecular cladogram for genera of Monimiaceae and *Gyrocarpus americanus* using bootstrap analyze. Morphological phylogenetic model, which is based on morphological characters of the original data matrix (27 were studied for morphological and anatomical characters), was built to reflect the most parsimony cladogram. The complex cladograms were obtained as a result of the synthesis of molecular and morphological data. Our conclusions are following. *Xymalos* is basal genus of Monimiaceae, and therefore *Xymalos monospora* – the basalmost genus of the family. The next clade is occupied by Sri Lanka endemic genus *Hortonia*. Intrusion of Monimiaceae in South America through Antarctica occurred not later than the end of the Cretaceous period. At present time the southernmost genus in South America is *Peumus*, it was could be the most resistant to climate change, which allowed him to gain a foothold on the west coast (app. nearly 40 degrees south latitude). Later range of Monimiaceae in South America moved to the north, along with climate change (such genera as *Mollinedia*, *Grazielanthus*, *Hennecartia*, *Macropeplus*). Based on the Tertiary fossils of Monimiaceae in Oman and Germany, we are conclude that for intrusion to Australia Monimiaceae required to make a long way through the coast of East Africa, South Asia (where, in Sri Lanka, remained ancient oligotypic genus *Hortonia*) and after to Malesia (*Kibara*) and Australia (*Hedycarya*, *Palmeria*, *Austromatthaea*, *Steganthera*, *Wilkiea*). New Zealand was intrusion by Monimiaceae from Australia (*Hedycarya*). Madagascar (*Decarydendron*, *Ephippiandra*, *Tambourissa*) and the Mascarene Is (*Tambourissa*) were intrusion by Monimiaceae from New Caledonia (*Kibaropsis*, *Hedycarya*) or Eastern Australia, but it is likely Madagascar could be inhabited by Monimiaceae, which later became extinct.

T2

P0424

Extensive phylogenetical study of the lichen genus *Pyrenodesmia* (Ascomycota, Teloschistaceae) in Eurasia

Ivan Frolov[1,2], *Jan Vondrák*[1,3], *Jiří Košnar*[1]

1. University of South Bohemia

2. Ural Federal University

3. Institute of Botany ASCR

The genus *Pyrenodesmia* was established in XIX century for crustose lichens from the family Teloschistaceae sharing the lack of anthraquinones and the presence of dark brown to black acetone insoluble pigments (Sedifolia-grey). However, in our days it was shown that the genus is closely related to a group of species with anthraquinones in apothecial disk – so called *Caloplaca xerica* group – and probably it is not monophyletic. Pending further study, here we treat *Pyrenodesmia* as a probably polyphyletic group of closely related exclusively epilithic taxa (the group is monophyletic if exclude *Caloplaca xerica* group) always without anthraquinones, but with Sedifolia-grey. The genus is widespread across different geographic regions in the Northern Hemisphere. In Eurasia, it is especially diverse in the Mediterranean and Irano-Turanian regions, inhabiting mainly xerothermic calcareous localities. Currently 24 species and two variations of *Pyrenodesmia* are recognized. During the last several years we collected about 1500 specimens of the genus *Pyrenodesmia* on extensive areas of Europe, Near East and Central Asia. Finally, we made a set of 220 samples which encompasses the whole morphological and geographical variability of the genus in our collection. For this set we sequenced four loci of the nuclear DNA – ITS, β-tubulin, MCM7 and RPB2 – and made a concatenated alignment. Four-locus coverage was complete for 169 individuals, while three loci were obtained for 45 individuals and two loci were obtained for six individuals. We inferred phylogenetic reconstructions by means of both Bayesian and maximum likelihood analyses which were almost identical. We distinguished 28 well-supported clades which have specific recognisable morphology. We call them "species". On our phylogeny, there are also more than 30 single lineages which are combined on the base of morphology neither with each other nor with any clades. We studied type specimens for almost all known *Pyrenodesmia*' s names; compiled list of taxa consists of about 60 names. Currently 26 of these names are accepted. Other names were considered as synonyms to these 26 taxa. We could superpose with currently accepted taxa 15 of our 28 clades. Specimens of *P. oblongula* and *P. ochromela* genetically are not distinguished from *P. pratensis* and *P. diphyodes* respectively. Also, it is not possible to put all clades and lineages of *P. erodens* and *P. variabilis* together to two monophyletic clades, although these samples fit to morphological species concepts of these two taxa. Also, three of more than 30 single lineages could be attributed to known accepted taxa. Thus, we obtained molecular data from 22 of 26 currently accepted taxa and four taxa are lacking. Besides clades that we could superpose with currently accepted taxa we got 13 well-supported clades with specific recognisable morphology, however we could not find any names for them among both currently accepted taxa and synonyms to them. Probably, these clades (as well as about 30 single lineages) represent still undescribed taxa. Eight unknown clades occur in Asia, four – in Europe and one – both in Asia and Europe. The study was supported by RFBR grant 16-04-01488.

T2

P0425

Hybridogeneous species of *Diphasiastrum* in Russia.

Yury Ivanenko

1. Saint Petersburg State University

2. Komarov Botanical Institute

Diphasiastrum Holub (Lycopodiaceae) includes about 20-25 species. The species of *Diphasiastrum* hybridize, and the resulted intermediate forms are considered by taxonomists either as hybrids (Wilce, 1965; Kukkonen, 1967) or as species (subspecies) of a hybrid origin (Pacyna, 1972; Holub, 1975; Stoor *et al.,* 1996; Ivanenko, Tzvelev, 2004). We consider these morphological intermediates to be species of hybrid origin because they are comparatively stable and fertile, at least partially. There are 4 primary or parental species of *Diphasiastrum* in Russia, namely *D. alpinum*, *D. complanatum*, *D. sitchense*, *D. tristachyum*. There is also *D. nikoense* in the Kuril Islands, but the latter taxon is closely related to *D. sitchense* and hardly deserves more than subspecific rank. *D. alpinum* and *D. complanatum* have extensive ranges in Russia, their distribution areas overlap considerably in the northern part of the country, particularly in the Urals and in the mountains of Siberia and the Russian Far East. *D. tristachyum* has much more limited range than the last two species. It is distributed in the southern part of taiga from the northwestern European Russia to the southwestern Siberia. *D. tristachyum* has a disjunctive range in Russia and becomes increasingly rare toward the east. *D. sitchense* is distributed in the Far East of Russia, generally in the Kuril Islands and Kamchatka. There are 3 hybridogeneous species of *Diphasiastrum* in Russia. *D. issleri* (*D. alpinum* × *D. complanatum*) is the most rare species among them. In Russia, *D. issleri* is still known only from one site in the Khibiny Mountains (Jessen, 1991). All the other reports of *D. issleri* findings in the Kola Peninsula (Ivanenko, 1991), in the Altai Mountains (Shmakov, Tikhonov, 2005), or in the Republic of Tyva (Krasnoborov, 2007) are incorrect. These indications really apply either to *D. alpinum*, or to *D. complanatum*. *D. takedae* (*D. alpinum* × *D. sitchense*) is known from the Kuril Islands, Kamchatka and Magadan Oblast. There are more than 20 localities of *D. takedae* in the Russian Far East (Ivanenko, 2006). The most common and widely distributed among the 3 hybridogeneous species is *D. zeilleri* (*D. complanatum* × *D. tristachyum*). The range of *D. zeilleri* generally coincides with the range of *D. tristachyum*, but the eastern limit of *D. zeilleri* distribution in Siberia is located further to the east (in Novosibirsk Oblast and Tomsk Oblast). Furthermore, *D. zeilleri* is much more common than *D. tristachyum*. Comparison of ranges of the parental species and the hybridogeneous species of *Diphasiastrum* demonstrates that the species of hybrid origin are distributed mainly in the territory where the parental species ranges overlap. Hybridogeneous species of *Diphasiastrum* deserve more attention of botanists. There are some unresolved problems concerning the distribution of these taxa in Russia. For example, we do not know why *D. issleri* is very rare there in comparison with a relatively common *D. zeilleri*.

T2

P0426

The effect of mechanical force on the floral development of *Arabidopsis thaliana*

Akitoshi Iwamoto, *Suzuka Kikuchi, Yuna Yoshioka*

Department of Biology, Tokyo Gakugei University

The floral development of *Ceratophyllum demersum* suggested that mechanical force on floral apex may have a key role to determine a phyllotactic pattern in flower (merosity). In particular, the mechanical force on the adaxial side of the flower meristem of *Ceratophyllum demersum* could be an important factor explaining the meristic variation in this species; trimerous, tetramerous and spiral flowers. To demonstrate this theory, we have developed the experimental system to give artificial mechanical force on the abaxial side of young floral apex of *Arabidopsis thaliana* and induce flowers other than tetramerous one. A micromanipulator with silicon device is used to cause the contact pressure on the floral apex. We have obtained some distorted flowers probably due to the mechanical force imposed by the micromanipulator, but we cannot still change the meristic pattern in *Arabidopsis* (i.e. tetramerous). In our poster, the details of our experimental system will be shown and the linkage between the developmental change in the flower of *Arabidopsis* and the mechanical force, will be discussed including the additional results of experiments using our system.

T2

P0427

Carpoanatomical study and its significance in taxonomy of Apiaceae of Northern-Western Ghats

Neeta Jadhav

Department of Botany, B. K. College, Belgaum, Karnataka, India

The family Apiaceae with about 300 genera and 3000 species are distributed throughout world. The members of the family have attracted number of workers for their medicinal, commercial and condimental importance. The Apiaceae are represented by about 68 genera and about 240 species in India. The family is well known for its diversity of fruits. Fruit morphology is of considerable significance in its taxonomy at genus and species level. Therefore attempts have been made to study the morphological and anatomical characters of various species of the family found in Northern-Western Ghats. External and internal anatomy of 17 species belonging to 11 genera has been studied. Observations have been made on size, shape, trichomes and internal anatomical characters of all the species found in Northern-Western Ghats. Significance of fruit characters and their correlation with distribution and habitat preferences of the species have been evaluated. Based on external fruit morphology and anatomical characters a botanical key for identification of various species found in Northern-Western Ghats is provided.

T2

P0428

The Global Flora: A series of global taxonomic treatments

James Byng[1,2], *Maarten Christenhusz*[1,3]

1. Plant Gateway

2. Naturalis Biodiversity Center

3. Royal Botanic Gardens, Kew

Two-hundred and sixty four years since Linnaeus published his *Species Plantarum* our species-level knowledge of many families of vascular plants is still incomplete and is in many cases poorly documented. There are still huge numbers of new species to describe in many groups, whilst others are in need of either minor or major revision as new evidence is generated. Most taxonomic groups are treated in regional floras, sometimes by different authors often decades apart, but few global revisions of taxa are being undertaken which is resulting in laborious paper chases and potentially an error strewn taxonomy. Global treatments of taxonomic groups are urgently needed. Plant Gateway is publishing a series of updatable global taxonomic treatments to document accepted family-level species classifications based on the molecular data available and re-examining the literature and herbarium specimens. Each family will have a description, distribution data and other relevant information, as well as, detailed wood anatomy and pollen descriptions. Each species will have detailed type information listed, a description, ecology and distributional data and all specimens seen in major herbaria. Each treatments will be richly illustrated throughout making them useful to a broad range of users from herbarium taxonomists to horticultralists and conservationists. Notable global treatments to be published first will be the recently named Kewaceae, recently re-circumscribed Metteniusaceae and the species rich genus *Syzygium* (Myrtaceae).

T2

P0429

An educational poster of the Angiosperm Tree of Life

James Byng[1,2], *Greg Kenicer*[3], *Erik Smets*[2], *Maarten Christenhusz*[2,4]

1. Plant Gateway

2. Naturalis Biodiversity Center

3. Royal Botanic Gardens, Edinburgh

4. Royal Botanic Gardens, Kew

Bentinckia, a palm genus is restricted to Indian subcontinent with *B. nicobarica* restricted in distribution to Nicobar Islands. Usually,

Island species are quite frequently within their limited area of distribution not having any chance of migration, so that destruction of even a small forest area may render the endemic species high threat over their existence. The Andaman and Nicobar Islands represent 86% of forest coverage including tropical rain forests, tropical moist deciduous forests, lowland forests *etc.* with a total geographical area of 8249 sq.km, while the Nicobar group of Islands cover an area of 1841 sq.km. *B. nicobarica* belonging to the family *Arecaceae* is one of the hardest, tall palms growing up to ±50 m height with small globose, blackish fruits (when ripen) in big bunches. It specifically grows in moist hills, valleys, plains and slopes of lowland forests of Nicobar Group of Islands mainly in Great Nicobar, Kachal, Nancowry group, Teressa, Chowra and Car Nicobar Islands. In the year 1988, "The IUCN Red List of Threatened Species" assigned this species endangered category. *B. nicobarica* is used in the Islands for housing, furniture, boat making, *etc.* by the tribal of Nicobarese and Shompen. The emerging shoots are eaten by the local people as vegetable and the fallen fruits are eaten by pigs. Due to all these activities over the time the species population has declined very much in these islands. Considering the importance of the species the present study has been taken to investigate its population, multiplication and reintroduction. The method of Ecological Niche Modelling (ENM) map is applied by using the modeling algorithms of DIVA-GIS and MAXENT. Nineteen bioclimatic variables are used as environmental variables for modeling, which includes Annual Mean Temperature, Mean Diurnal Range, Isothermality, Temperature Seasonality, Maximum Temperature of Warmest Month, Minimum Temperature of Coldest Month, Temperature Annual Range, Mean Temperature of Wettest Quarter, Mean Temperature of Driest Quarter, Mean Temperature of Warmest Quarter, Mean Temperature of Coldest Quarter, Annual Precipitation, Precipitation of Wettest Month, Precipitation of Driest Month, Precipitation of Seasonality, Precipitation of Wettest Quarter, Precipitation of Driest Quarter, Precipitation of Warmest Quarter, Precipitation of Coldest Quarter *etc.* Through this modeling, we could predict the potential area of occurrences the species and their reintroduction for the population study and the limiting factors. More than four thousand seedlings of *B. nicobarica* are raised so far in controlled condition in polyhouse of Dhanikhari Experimental Garden Cum-Arboretum, Botanical Survey of India, Andaman and Nicobar Regional Centre and being reintroduced in the identified habitat for improving their conservation status.

T2

P0430

Character evolution, circumscription and taxonomic novelties in the sections *Casparya* and *Semibegoniella* (Begonia, Begoniaceae)

Orlando Adolfo Jara-Muñoz[1], *James E. Richardson*[1,4], *Santiago Madriñán*[2,3]

1. Universidad de los Andes

2. Universidad del Rosario

3. Jardín Botánico de Cartagena Guillermo Pi?eres

4. Royal Botanical Garden Edinburgh

Sections *Casparya* and *Semibegoniella* are some of the richest groups of *Begonia* in the North Andean region. Taken together, both sections are clearly recognizable respect to other sections of *Begonia* because the particular fruit type, but to separate the two is less easy. The gamophyllous perianth and unilamellar placenta has been proposed as diagnostic characters for *Semibegoniella*, however it has been problematic the presence of intermediate species, and the lack of documentation about these characters in several species. Using three chloroplast markers and ITS, we present a phylogeny of the sections *Casparya* and *Semibegoniella* (*Begonia*, Begoniaceae) and representative species of related sections. We demonstrate the paraphyly of section *Semibegoniella* that we now synonymize with *Casparya*. Fruit type is a synapomorphy for *Casparya*, but character evolution analysis reveals evidence of multiple examples of convergence and parallelism in characters frequently used to separate infrageneric groups in *Begonia*. We also include the description of five new species of the section, founded during the course of the study.

T2

P0431

Cytological and reproductive studies of the fern genus *Platycerium* and their hybrids

Tassanai Jaruwattanaphan, *Phannipa Sukmueng*

Department of Horticulture, Faculty of Agriculture, Kasetsart University

Platycerium Desv. is a pantropical fern comprising 18 species. Chromosome numbers of 16 species namely *P. alcicorne, P. andinum, P. bifurcatum, P. coronarium, P. elephantotis, P. ellisii, P. grande, P. hillii, P. holttumii, P. ridleyi, P. stemaria, P. superbum, P. veitchii, P. wallichii, P. wandae and P. willinckii* and 2 hybrid cultivars were counted. The chromosome numbers observed were diploid (2n=2x=74) in all 16 species, this includes 7 species are reported for the first time and for the remaining 9 species matched those previously reported. Chromosome numbers were established for two hybrid cultivars, *P. ×kitshakoodiense* and *P. 'Charles Alford'*, were tetraploid (2n=4x=148) and triploid (2n=3x=111) respectively. The reproductive modes of *P. coronarium* and *P. ridleyi*, which produced only 8 spored sporangia, and their hybrid, *P. ×kitshakoodiense*, were determined in this study. Gametophytes of these three plants produced normal antheridia and archegonia. Young sporophytes were not produced directly on the gametophytic tissue. This result concluded that all these three species are not agamosporous species.

T2

P0432

Phylogeny and biogeography of *Acaena* (Rosaceae): Evidence of long-distance dispersal from the Southern Hemisphere

Javier Jauregui Lazo[1], *Daniel Potter*[2]

1. University of California, Berkeley

2. University of California, Davis

Spatial distribution and history of plants in the Southern Hemisphere have been hot topics in biogeography in recent decades. Ancient Gondwanan connection or subsequent trans-oceanic dispersal events are the primarily drivers of the distribution of extant taxa in Southern Hemisphere. Currently, dispersal events have been proposed as dominant explanations for disjunct distributions because of more accurate phylogenetic reconstructions (San Martin and Ronquist, 2004). *Acaena* Mutis ex L. is an important element in the vegetation of temperate climate regions in Chile, Argentina, New Zealand and Subantarctic Islands (Grondona, 1964; Marticorena, 2006). A few *Acaena* species are also found in South Africa, Central America, California and Hawai'i (Macmillan, 1998). The genus comprises around 45 species that are recognizing by having an unusual reduction of the floral parts (numbers of petals, stamens and carpels) and inflorescence type (raceme or spike) compared to other members of Rosaceae (Potter *et al.*, 2007). However, the most distinctive feature of *Acaena* is the presence of a conspicuous hypanthium, which is commonly armed with hooked prickles (Marticorena, 2006). Due to its exceptional disjunct distribution and extraordinary floral characters, the current study was conducted to show the phylogeny, biogeographic pattern and ancestral character reconstruction of *Acaena*. Phylogenetic analyses of non-coding nuclear (ITS) and chloroplast (trnL-F) gen loci suggest that *Acaena* as currently circumscribed is not monophyletic group. *Margycarpus*, *Tetraglochin*, and *Polylepis* are nested within it. Long-distance dispersal is the main explanation for the Amphitropical and Gondwanan distributions of the genus. Along these lines, it is likely the genus originated in South Africa around 8-5 mya, migrated into South America and subsequently diversified in the temperate regions of Chile and Argentina. From there, the genus migrated to New Zealand and Australia to begin a recent rapid diversification. The species found in California and Hawai'i are more closely related to the species found in Chile and New Zealand, respectively. This highlights an uncommon direction of dispersal from the Southern to the Northern Hemisphere among the temperate taxa. Due to the observed tree topology it is likely that the ancestral state for the inflorescence type was a raceme that evolved into a spike and a head multiple times in the core group of *Acaena*. The apetalous flower seems to be an apomorphic character for *Sanguisorba*, *Cliffortia*, *Polylepis*, and *Acaena* (Sanguisorbinae). An independent evolution of the reduction in number of stamens from five to two stamens is also seen in the genus. Some individuals of *A. magellanica* present genetic variation that may justify their treatment as a separate species. Likewise, *A. pinnatifida var. californica* is not closely related to *A. pinnatifida* as expected and can be treated as different lineage. The evolutionary relationships in *Acaena* combine a history of recent diversification, long-distance dispersal and genetic variation in some particular taxa. Future research should attempt to clarity the infraspecific classification of *A. magellanica*.

T2
P0433
Utility of RosCOS (Rosaceae Conserved Orthologous Set) markers for resolving phylogenetic relationships in recently diverged and closely related species of *Rosa* section *Synstylae* (Rosaceae)

Ji-Hyeon Jeon[1], *Seung-Chul Kim*[1], *Masayuki Maki*[2]
1. Sungkyunkwan University
2. Tohoku University

The genus *Rosa* (Rosaceae) is comprised of approximately 150-200 species and distributed widely in the temperate and subtropical regions of the northern hemisphere. Of four subgenera recognized by Rehder (1940), the subgenus *Rosa* includes most species within the genus and it is further divided into ten sections. Of ten sections, *Synstylae* and *Chinenses* occur mainly in East Asia and include most of the wild ancestors of modern cultivated roses. In particular, section *Synstylae* is one of the most recently diverged section in genus *Rosa*. In previous studies, not only have low support values *Rosa* sect. *Synstylae* species in branches of their phylogenetic tree, but some of those could not be delimited in their phylogeny with conventional molecular phylogenetic markers, such as nuclear ITS or single/low copy nuclear genes like GAPDH. Even in much recently diverged and closely related species in this section, dozens of chloroplast markers showed identical sequences, unable to distinguish those species in their phylogenies. Albeit nuclear markers showed more diverse sequence than chloroplast ones, sequence divergences were not sufficient enough to resolve phylogenetic relationships among the species with good nodal supports. Though recently developed several NGS-based genomic data are extremely valuable to reconstruct phylogenetic relationships of such recently diverged and closely related species, the highly conserved single copy genes could also be quite useful for resolving close species relationships. As Cabrera *et al.*, (2009) developed Rosaceae Conserved Orthologous Set (RosCOS) markers from single copy *Arabidopsis* genes, we tested those RosCOS markers on *Rosa multiflora* and its close relatives in sect. *Synstylae*. Of 33 markers that were selected based on the RosCOS database (Cabrera *et al.*, 2009), we selected a total of eight markers to test their phylogenetic utility in *Rosa* sect. *Synstylae*. The sequence identity values among accessions were ca. 96% and 8% were polymorphic sites, whereas those of ITS region were ca. 99% and 2%, respectively. The concatenated data matrix resolved not only phylogenetic relationships of *R. multiflora* and its relatives but also species delimitation with high support values. In addition, some putative hybrid accessions based on morphological traits were supported based on the RosCOS markers. Our results suggest that RosCOS markers are highly variable and phylogentically informative compared to conventional molecular markers and can provide sufficient resolutions with high support values for complex relationships of recently diverged and highly reticulated lineages of the genus *Rosa*. Detailed phylogenetic relationships among species of sect. *Synstylae* and some cases of reticulate evolution with in the section will be presented.

T2
P0434
Phytogeographic implications of a fossil endocarp of *Diploclisia* (Menispermaceae) from the Miocene of eastern China

Hui Jia[1], *Bainian Sun*[2], *David Ferguson*[3]
1. Xi'an Shiyou University
2. Lanzhou University
3. University of Vienna

Menispermaceae endocarps became common in the Paleogene as the global temperature rose, while *Diploclisia* was present in NE China in the Early Eocene. As the genus is currently known from S China, it could be assumed that the genus continued to be present in China during the Neogene. However, so far there was no evidence to support this supposition. An excavation at Jiahu Village, Zhejiang has revealed the presence of an endocarp affiliated to the Menispermaceae in the late Miocene Shengxian Formation. This endocarp has been investigated using both light microscopy and X-ray microtomography. Comparisons with fruits of living and fossil Menispermaceae, indicate that it is a new species belonging to the genus *Diploclisia*, which is named as *D. praeaffinis*. The present fossil endocarp helps to fill the more than 40 Ma gap in the fossil record by demonstrating that it was indeed present in the region in the late Miocene. As *Diploclisia* was already widely distributed in the Northern Hemisphere by the beginning of the Middle Eocene, it probably originated in the Paleocene (66-56 Ma) at the latest. The past and present distribution of *Diploclisia* shows that the genus has undergone various phases of expansion and contraction afterwards. In the Late Eocene and Oligocene (37-23 Ma), as temperatures fell globally, its distribution became restricted. It was no longer present, or too rare to leave a fossil record, between latitudes 43-58°N. By the Neogene, it seems to have become restricted to East Asia, even on the verge of extinction. *Diploclisia* is currently considered to consist of two species, *D. glaucescens*, which is widespread in S and SE Asia, and *D. affinis*, endemic to S China. Molecular phylogenetics suggest that these species diverged some 17 million years ago. The close resemblance of our fossil to the endocarps of *D. affinis*, would indicate that this species or a similar taxon already existed some 10 million years ago. Based on the greater similarity of the endocarps of *D. affinis* to their Cenozoic counterparts, the location of these fossils and the fan-shaped distribution of extant *Diploclisia*, it would suggest that *D. affinis* is the more plesiomorphic of the two living taxa, which was firstly throughout southern China. Then the more apomorphic species, *D. glaucescens*, probably expanded into S and SE Asia. As it did so, the fruits became increasingly large. This would explain why the largest fruits in the genus are recorded from the Moluccas and New Guinea. This southerly expansion would have been driven by the increasing cold of the Pleistocene glaciations, and facilitated by the low sea-levels (-125±35m) occurring during these pleniglacials.

T2

P0435

Genomic diversity in *Litsea* (Lauraceae) assessed by organelle and nuclear data

Guo-Feng Jiang[1,2], *Damien D. Hinsinger*[1,2], *Joeri S. Strijk*[2,1]

1. Plant Ecophysiology & Evolution Group, Guangxi Key Laboratory for Forest Ecology and Conservation, College of Forestry, Guangxi University, Nanning, Guangxi, PR China;

2. Plant Ecophysiology & Evolution Group, State Key Laboratory for Conservation and Utilization of Subtropical Agro-bioresources, Guangxi University, Nanning, Guangxi, PR China

Lauraceae comprise about 55 genera and more than 3000 economically and ecologically important species, widespread in tropical and subtropical regions. Due to the absence of clear vegetative characters that can aid in successful identification in the family, few data are available to study its evolution or biogeography. *Litsea* species can be distinguished from other Laureae by its umbellate inflorescences, unisexual and trimerous flowers, its fruit with small to rather large cupules, and its leaves usually alternate. *Litsea* species are widely used for industrial oil and charcoal production as weel as for agroforestry (e.g. as Vanilla culture support). *Litsea* represents a middle-size genus in Lauraceae, contains about 200 species, mainly distributed in tropical and subtropical Asia with a few species in Australia and from Central America to subtropical South America; 74 species are found in China (47 are endemics). Despite they are sometimes locally abundant, several species are currently endangered according to the IUCN Red List. The relationships between *Litsea* and related genera remains unclear. Indeed, previous studies resulted in low-resolution phylogenetic hypotheses (mainly due to low-variability loci), hampering our understanding about evolution and biogeography of *Litsea* species. The current emergence of chloroplast-based phylogenomic allows to obtain a highly informative and discriminate marker. In combination with complete nuclear ribosomal cistrons, it provides useful data to assess the genomic diversity, test and reassess the morphological characters used in former classification systems. Here, we used a subsample of *Litsea* species, for which we reconstructed the complete chloroplast genomes, nuclear ribosomal cistrons and repeated sequences data from Illumina genome skimming sequencing. We assessed the diversity found in the different categories of genes and showed that the genomic diversity is unevenly distributed in the plastid genome and among the species, with clusters clearly separating groups of species. We analyzed the plastid repeats and SSRs content and performed a phylogenetic analysis based on the sequences data. The phylogeny retrieved most of the groups corresponding to the diversity clusters. We expect this study will pave the way to other studies – studying both the generic and specific taxonomic level. A better knowledge of the genomic diversity in tropical and subtropical tree species will help us to face the current biodiversity crisis.

T2

P0436

Tracing the dynamics of Neogene Eurasia Tethyan flora: An insight from the biogeography of *Quercus* group Ilex (Fagaceae)

Xiao-Long Jiang, *Min Deng, Ying Li*

Shanghai Chenshan Botanical Garden

Eurasia Tethyan disjunction is commonly occurred in subtropical plants. However, the spatial and temporal diversification pattern of this flora are still not well understood, as the species phylogeny of the keystone families are not resolved and it further complicated by the intense paleogeoclimatic changes in Neogene Eurasia. *Quercus* group Ilex, a monophyletic group with ca. 32 species widespread in Eurasia, is keystone trees in Eurasia subtropical forests with rich fossil records since Oligocene. Therefore, this group offered an ideal opportunities to investigate the dynamics of Tethyan flora. In this study, we employed RAD-seq to reconstruct the phylogeny of *Quercus* group Ilex. The divergent time and ancestor distribution ranges were also analyzed using BEAST and RASP. A total of 38 species (incl. 27 species from Group Ilex and 11 species as outgroup) were analyzed. ML and BI tree yielded identical topology and the species phylogeny in *Quercus* Group Ilex were well resolved. Six lineages with quite different distribution were detected: Himalaya gorgy (I) and Sino-Japan (II); Indo-China (III) were derived accordingly. Heterobalanus (IV) and Engleriana (V) forming a clade is the sister group to the

Europe-West Himalaya (VI) lineage. Divergence time of *Quercus* group Ilex was dated to Oligocene. The ancestor distribution area reconstruction inferred southwest China was the original place of *Quercus* Group Ilex. It is dispersal to East Asia and Indo-China during early Miocene. The dispersal from East Asia-Himalaya, through West Himalaya, then to the Mediterranean occurred in Middle Miocene. The split between Himalaya subalpine lineage to its low-middle elevation relatives occurred at later Miocene. The diversification ratio of *Quercus* Group Ilex was constant as inferred by LTT. Our results demonstrate the Himalaya during Middle Miocene was a corridor rather than an obstacle barrier to the biota between the Eurasia and Central and southern Asia was much humid than the present. The drastic uplifting of Himalaya occurred at later Miocene. Our study provide a critical context to understand the impacts of Neogene paleogeoclimatic changes to shape the subtropical flora in Eurasia.

T2

P0437

Effect of forest fragmentation on epiphyllous liverworts presence in China

Yanbin Jiang
College of Resources and Environment, Huazhong Agricultural University

Forest fragmentation is a continuous process and one of the most important threats to biodiversity and the ecological functions of forest. Epiphyllous liverworts is a special group of bryophytes exclusively inhabiting leaves surface of understorey vascular plants in humid tropical and sub-tropical forest, and very sensitive forest fragmentation. The aim of this study was to investigate of forest types and their fragmentation on epiphyllous liverworts presence in China. MAXENT was employed to model the distribution of epiphyllous liverworts based on species presence data and environmental variables; then applied FRAGSTATS to characterize forest fragmentation based on MODIS Land Cover map. Four forest types related to epiphyllous liverworts presence were extracted in China: mixed forest (MF), evergreen broadleaf forest (EBF), Woody savannas (WS), Cropland/Natural vegetation mosaic (NVM), and Deciduous broadleaf forest (DBF). Epiphyllous liverworts were mostly present in MF, then in EBF. MF covered most land area compared with other forest types, and EBF was the least fragmented forest. Presence of epiphyllous liverworts was positively related to EBF area and all forests area, with EBF explained more than all forest together. At nature reserve level, EBF also has the most explanation ability to epiphyllous liverworts presence, no matter by its total area, or by the largest patch area. The results suggested that forest fragmentation including patch size reduction and isolation are influential on the epiphyllous liverworts. Construction of nature reserves for habitat protection is essential and must take into account the connectivity between forest patches and remaining amount of habitat.

T2

P0438

Complete chloroplast genome of *Lespedeza maritima* (Fabaceae), a Korean endemic shrub

Dong-Pil Jin, *In-Su Choi, Jong-Soo Park, Jong-Won Park, Byoung-Hee Choi*
Inha University

Lespedeza maritima Nakai (Fabaceae; Papilionoideae; Desmodieae) is a Korean endemic shrub, which grows in warm-temperate forests of mountain along southern-eastern coastal line. Although some phylogenetic studies of genus *Lespedeza* were conducted based on sequences of partial chloroplast region, each species was not monophyletic due to frequent hybridization and/or rapid diversification. Analysis using chloroplast genome could be helpful in this case, but there is no information. In the present study, we report complete chloroplast genome (cp genome) sequence of *L. maritima*, and it is first report of *Lespedeza*. The cp genome is 149,022bp in length, including a large single copy (LSC) region of 82,429 bp and a small single copy (SSC) region of 18,939 bp, and a pair of inverted repeats (IRs) of 23,827 bp that separated LSC and SSC. The degree of CG content is 35.0%, and this genome contains 134 unique genes, consisting of 72 protein-coding genes, 18 transcription/translation, 36 tRNAs and 8 rRNAs. Moreover, cp genome characteristics and its phylogenetic significance were also assessed within subtribe Lespedezinae by PCR screening and a genome phylogenetic tree.

T2

P0439

Phylogeny and plastome evolution of the legumes family (Leguminosae)

Jian-Jun Jin[1,2], *Rong Zhang*[1,2], *Yin-Huan Wang*[1,2], *Stephen A. Smith*[3], *Ting-Shuang Yi*[2], *De-Zhu Li*[2]
1. Kunming College of Life Sciences, University of Chinese Academy of Sciences
2. Kunming Institute of Botany, Chinese Academy of Sciences
3. University of Michigan, Ann Arbor

Many parts of the phylogeny of legume family (Leguminosae), the third largest family in angiosperms with great ecologically diversity and economically importance, are still enigmas. The main issues include the relationships among the six subfamilies and the relationships among tribes. Furthermore, the legume plastid genomes (plastomes) show unusual structure variations, mainly the expansion, contraction and loss of IR regions and the large-scale inversions. But the panoramic history and the mechanisms of plastome structure evolution remains unclear. To explore these issues, we assembled 296 complete/near-complete plastomes together with published ones, representing 315 genera of legumes. We reconstructed the phylogeny using multiple dataset combination including coding and noncoding and various methods, including concatenation and coalescence. We conducted divergence time estimation using a penalized likelihood approach and a Bayesian approach. We reconstructed the ancestral plastomes, evaluated the inversions across the evolutionary history of legumes using an adjacency-based maximum likelihood approach. We discussed the correlations between the structure variations and the potential driven roles. The deep relationships of Leguminosae were fully resolved and strongly supported at most nodes, and divergence ages of major clades were estimated. Our results provided a general time-scaled image of the evolutionary history of the legume plastomes, with the mechanisms of the evolution of structure variations addressed. Our results supplied basis for revising the tribal classification of the legume family, and also provided a good sample for investigating the underlying mechanisms of plastomes evolution.

T2

P0440

Utilizing morphological and molecular data to determine if species of *Trema* (Cannabaceae) are forming interspecific hybrids on Hispaniola

Kristina Jordan, *Nancy Garwood, Kurt Neubig, Nicholas Flowers*
Southern Illinois University Carbondale

The Neotropical pioneer tree *Trema* can be found growing in many areas throughout the Neotropics, including Hispaniola. Unfortunately, the taxonomy and phylogeny of species in *Trema* is highly unresolved; the species *Trema micrantha* (L.) Blume encompasses multiple unrelated lineages. Hybridization is likely a contributing factor to this confusion. Multiple distinct clades of *Trema* on Hispaniola grow in close proximity to one another, providing ideal conditions for interspecific hybridization. The purpose of this research is to determine, using a combination of DNA data and leaf morphological characters, if *Trema* is hybridizing on Hispaniola. DNA sequence data was used to evaluate the phylogenetic relationships between species and raw sequence data were used to identify patterns associated with hybridization. Leaf trichomes were imaged using scanning electron microscopy, and overall leaf morphological characters were analyzed in a Principal Components Analysis. We determined that *T. micrantha* and *T. cubensis* (Urb.), as well as *T. micrantha* and *T. lamarckiana* (Roem. & Schult.) Blume are naturally hybridizing on Hispaniola. Analysis of DNA shows that each parent species is a clearly distinguished lineage, and the putative hybrids show phenomena consistent with hybridization. Morphological characters distinguish species and are intermediate in the hybrid offspring. These results are significant as hybridization is a major force in evolution and potentially speciation.

T2

P0441

Phylogenomic investigations of the *Haploclathra* (*Caraipa* + *Kielmeyera*) clade: Taxonomic and biogeographical implications

Rafaela J. Trad[1], *Fernanda N. Cabral*[1], *Benicio Cardoso-Silva*[1], *Zhenxiang Xi*[2], *Volker Bittrich*[4], *Peter Stevens*[3], *Maria C E Amaral*[1], *Charles Davis*[2]
1. University of Campinas
2. Harvard University
3. Missouri Botanical Garden
4. Rua Dr. Mario de Nucci 500,13083–290 Campinas, SP, Brazil

The majority of species in the genera *Kielmeyera* Mart. & Zucc., *Caraipa* Aubl. and *Haploclathra* Benth. (Calophyllaceae) are unique in having alternate leaves and a distinctive geographical distribution. *Kielmeyera* comprises 48 species that grow in Atlantic coastal rainforest and 'cerrado' in Eastern Brazil, while *Haploclathra*, with five species, and *Caraipa*, with 41 species, occur throughout Amazonia and have a few representatives in savannas. Although these genera together form a well-supported clade, relationships within the group have been largely unexplored. An ongoing revision has greatly clarified species delimitations within these genera, including the recognition of 15 new species, thus establishing a solid framework for focused phylogenetic investigation. To resolve their phylogeny we first sequenced four loci (plastid *ndh*F and *rps*3; and nuclear emb and ITS) from ~70% of the species in this clade (~68 species and 200 individuals). Our phylogeny inferred using maximum likelihood and Bayesian Inference confirmed the monophyly of the clade, established the monophyly of each genus, and clarified relationships among the genera. In contrast to previous research, our results identified *Haploclathra* as sister to *Caraipa* + *Kielmeyera*. Relationships within each genus, however, were largely unresolved. To break this impasse, we developed a sequence capture probe set targeting 425 low copy nuclear loci and 36 ultraconserved-like regions. Here we will describe the additional phylogenetic insights gleaned from these new phylogenomic data, and discuss their systematic and biogeographical implications, especially in light of recent monographic efforts. In addition to demonstrating the utility of phylogenomics to resolve rapidly radiating groups, the broad distribution of members of this clade makes it an important proxy for understanding the origin of hotspots of South American biodiversity, including 'cerrado', Atlantic coastal rainforest, and Amazonian forest.

T2

P0442

Global Genome Initiative: Targeting taxonomic DNA barcode gaps in GenBank

Jose Zuniga
Smithsonian Institution

The Global Genome Initiative (GGI) is a Smithsonian Institution program to collect, share, and study genomic samples of non-human species along the major branches of the Tree of Life. Specific goals include preserving genetic material of at least one species of half of the 160-200k estimated genera and sequencing 8k DNA barcodes for taxa for which there is no genetic data available ("dark" taxa). GGI's barcoding strategy is guided by a data-mining approach in which GenBank is periodically queried to detect taxonomic groups that do not have sequences flagged as barcodes, thus allowing GGI to focus all sequencing efforts on lineages that are not represented in this repository. To date, GGI has generated 2k barcode sequences from four loci for more than 500 plant genera that are not currently represented in GenBank. Ongoing work preserving and sampling legacy collections of large plant diversity projects at the Smithsonian will yield at least 3k additional barcodes for more than 900 dark plant genera during 2017.

T2

P0443

Diversity, development and ecological importance of the corolline corona in Apocynaceae

Lize Joubert[1,2], *Linde De Jager*[1], *Beverley Glover*[2]
1. University of the Free State
2. University of Cambridge

The Apocynaceae is characterized by a complex floral morphology with two key innovations, the staminal or gynostegial corona and pollinaria, which are both restricted to and potentially important in the diversification of the milkweed group (Secamonoideae and Asclepiadoideae). However, among the more basal sub-families of the Apocynaceae (Rauvolfioideae, Apocynoideae and Periplocoideae), a corolline corona is wide spread and may occasionally co-occur with the staminal corona in some of the basal milkweed lineages. The five corolline corona lobes (occasionally elaborated to form an annulus) arise from meristematic zones formed after postgenital fusion between five petals in each flower, an evolutionary novelty within the Gentianales. Coronas in the Apocynaceae may have various functions, including attracting, positioning and, in some cases, rewarding potential pollinators with nectar. Recent studies focusing on the pollination biology of the Asclepiadineae (Asclepiadoideae) have shown that, in some members of this tribe, the staminal corona functions as a morphological filter, allowing pollination specialization by limiting access to the reproductive part of the flower based on the overall body size of pollinating insects. In some members of the Periplocoideae the corolline corona may play a similar role in pollination specialization. The Periplocoideae corolline corona is highly labile and is morphologically more diverse than in any of the other four sub-families. We evaluate the evolutionary importance of the corolline corona by comparing various corona morphologies and developmental patterns in a phylogenetic context and testing their correlation with pollen transfer efficiency in representative members of the Periplocoideae.

T2

P0444

Systematics and diversification of southern African genus *Tephrosia* and allies (Millettieae, Fabaceae)

Ronny Kabongo[1], *Anthony Magee*[2], *Michelle Van Der Bank*[1], *James Boatwright*[3]

1. African Centre for DNA Barcoding, University of Johannesburg

2. Compton Herbarium, South African National Biodiversity Institute

3. Department of Biodiversity and Conservation Biology, University of the Western Cape

Comprising of some 350 species, *Tephrosia* Pers. is a large genus of legumes that are mostly African, but also extend to North and Central America, Australia and Asia. Two subgenera are recognized within the genus, i.e. *Tephrosia subg. Barbistyla* Brummit and *Tephrosia subg. Tephrosia.* Relationships within the genus and among its close relatives in the Millettieae (*Apurimacia* Harms, *Chadsia* Bojer, *Mundulea* Benth, *Pyranthus* Du Puy & Labat, *Ptycholobium* Harms and *Requienia* DC., i.e. the *Tephrosia* clade) are not well understood. This study presents results of nuclear internal transcribed spacer (ITS) and plastid DNA sequence data (*matK, rbcLa, ycf1*) generated and analysed for the *Tephrosia*-clade, based on a representative sampling (236 taxa) of these genera across the world, and especially within *Tephrosia.* Preliminary results indicate that *Tephrosia* is polyphyletic, with *Chadsia, Mundulea, Ptycholobium* and *Requienia* embedded within. *Apurimacia* and *Pyranthus* are placed outside of the main *Tephrosia* clade, although the support for this is weak in some analyses. In addition, *Tephrosia subg. Barbistyla* is none monophyletic in most analyses, while *Tephrosia subg. Tephrosia* polyphyletic. These results and their implications for generic and infrageneric classification within the *Tephrosia*-clade will be discussed.

T2

P0445

Polymorphic complex *Trifolium lupinaster*: How many species?

Valentina Kalinkina[1], *Yuliya Mikhaylova*[2], *Dmitriy Kislov*[1]

1. Botanical Garden Institute FEB RAS,

2. Komarov Botanical Institute RAS

The Leguminosa is the largest family of flowering plans, and genus *Trifolium* L. is one of the largest genera in the family. The species of *Trifolium* were found in the old and new world, as well in Africa and South America. The systematic of the genus is a very complicated due to the high degree of polymorphism in some species. Currently the taxonomy of the section *Lupinaster* (Fabr.) Ser. from the *Trifoliastrum* Moench subgenus is still controversial. Main disputes of botanists are focused on systematic position of *T. lupinaster*, because this species is characterized a disjunctive range, wide environmental amplitude and high morphological variability. Different contribution of plant traits (morphometric features of lamina, morphological variability of underground part, biochemical variability), polyploidy and wide distribution area lead to division of polymorphic species *T. lupinaster* to a number of single species. Today, different scientists treated the *T. lupinaster* as an aggregation species (*T. lupinaster* L.) or different variations names (*T. baicalense*, *T. litwinowii*, *T. tundricum*, *T. romanicum*, *T. criswogense*, *T. sprygini*, *T. pacificum, Lupinaster popovii*, *Lupinaster albus*). Statistical analysis showed that the variation of morphometric parameters of *T. lupinaster* L. leaflets linked to habitat conditions of individuals, particularly, influenced by temperature and amount of rainfall during the growing season. Leaf kseromorphism is observed in east-to-west direction, but in Western and Eastern Siberia one can see clear deviations from this trend. Namely, on this territory we can see abundance of isolated *T. lupinaster* forms, variations and species, which is determined mostly by measurable characteristics of a leaflet. However, the results of the statistical analysis we carried out do not support this division of *T. lupinaster*. Despite the fact that some species were previously treated as distinct species and races with different set of chromosomes, there is no morphological or genetic studies exists to consider them as separate species. Molecular phylogenetic analysis showed that the variability of ITS and trnL-trnF sequences in *Lupinaster* section is very low. Sequences of the ITS region of the *T. lupinaster* complex, including highlighted as a separate genus *Lupinaster*, are identical and represented by a single haplotype. Chloroplast haplotype pattern of *T. pacificum* collected from Japan is slightly differed. Statistical analysis of the leaf sheet features of *T. pacificum*, showed there are no significant differences in sizes of a leaflets, but they are significantly different in leaf shape, and they have a quite narrow habitat, specific development of specimens. This may be caused by isolated island state for this population, which, in turn, leads to the accumulation of variability. Against the background of significant morphological plasticity species of *Lupinaster* section, and particularly complex *T. lupinaster*, are surprisingly uniform genetically. After complex morphology and molecular study we consider *T. lupinaster, T.*

pacificum, T. gordejevii, T. eximium as separate species of section *Lupinaster*. Dedicated earlier species *T. baicalense, T. litwinowii, T. criswogense, T. spryginii, L. albus* are not an independent, but represent variants of a life form of *T. lupinaster.*

T2

P0446

Diversity of the genus *Arisaema* Mart. (Araceae) in India

Manudev Kambiyelummal Madhavan[1], *Santhosh Nampy*[2]

1. St. Joseph's College (Autonomous)

2. University of Calicut

The genus *Arisaema*, with about 215 taxa, is the third largest genus of Araceae. It is distributed from Central and East Africa towards South East Asia via Yemen, Oman, Pakistan, Afghanistan, the Himalayan ranges including India, China, Korea, Japan and Siberia eventually to North America with centre of diversity in tropical Asia (Gusman & Gusman, 2006; Govaerts, 2017). A recent taxonomic revision of the genus resulted in enumeration of 50 species and 6 varieties in India under 9 sections, mostly centered in Western Ghats, Eastern Himalayas and Western Himalayas. During the study five new species are described and two taxa *viz. Arisaema ciliatum* H.Li var. *liubaense* Gouda & Gusman and *A. pangii* H.Li, are reported new to India. The protologues of two species are emended based on data acquired during field studies and lectotypes are selected for 19 names conferring to the latest code of nomenclature. A morphometric analysis of species occurring in India is also conducted and a cladogram is drawn using the software NTSYSpc 2.02e based on 69 morphological characters. The cladogram supported the infrageneric classification of the genus proposed by previous workers (Murata *et al.*, 2013) with minor deviations. The present paper discusses the taxonomy and diversity of Indian species of *Arisaema* with updated nomenclature, colour photographs and distribution in India.

T2

P0447

Detection of genetic variation of anthocyanin biosynthesis pathway genes between and within population in *Hepatica nobilis* var. *japonica* involving flower color polymorphism within population

Shinichiro Kameoka[1], *Tomomi Wakabayashi*[1], *Yuki Mitsui*[2], *Keisuke Tanaka*[2], *Tsukasa Iwashina*[3], *Yoshinori Murai*[3], *Hitoshi Sakio*[4], *Harue Abe*[4], *Hiroaki Setoguchi*[1]

1. Kyoto University

2. Tokyo University of Agriculture

3. National museum of nature and science

4. Niigata University

What the factor maintains the polymorphisms both in phenotypic and genetic levels is the fundamental question in evolutionally biology. Some possible mechanisms to maintain the polymorphisms have been suggested: negative frequency dependent selection and so called high polymorphism, the phenomenon in which a specific position significantly deviates from the neutral equilibrium have been recognized as major reason. *Hepatica nobilis* var. *japonica* (Ranunculaceae) involves significant floral color polymorphism within population and can be divided into three types, white, red and blue by the anthocyanin pigment types. SSR analysis showed these flower colors cross randomly and no strong restriction of gene flow exist in natural habitat. Moreover, in the field observation, blue, the minority flower color within the population, might enjoy ecological advantage from the view point of the number of seeds and damage of tepals. In this study, we aimed to identify the candidate gene relating to the flower color polymorphism using transcriptome analysis, and to reveal that of polymorphic sites and nucleotide diversity. We compared the amount of expressed gene among different flower color type and picked up the anthocyanin biosynthesis pathway genes (e.g., *F3'H*, *F3'5'H* and *DFR*). Specifically, *F3'5'H* which required to produce the blue pigment showed higher nucleotide diversity than the other candidate genes. We detect the nucleotide diversity of those genes and test whether the patterns of polymorphisms are compatible with neutral equilibrium, then, conclude about the mechanism of maintenance in genetic polymorphism.

T2

P0448

Pollen morphology of some taxa of *Vicia* L. subgenus *Vicia* (Fabaceae) from European Turkey (Turkish Thrace)

Ciler Kartal, *Necmettin Guler, Ihsan Yalcin*

Trakya University, Faculty of Science, Department of Biology, Balkan Campus, Edirne/TURKEY

In this study, the pollen morphologies of 15 taxa belonging to the subgenus *Vicia* of genus *Vicia* L. (Fabeae, Fabaceae) naturally growing in European Turkey (Turkish Thrace) were studied using Light Microscopy (LM) and Scanning Electron Microscopy (SEM). The taxa studied are *V. galilaea, V. grandiflora* var. *dissecta, V. grandiflora* var. *grandiflora, V. hybrida, V. lathyroides, V. lutea* var. *hirta, V. narbonensis* var. *narbonensis, V. pannonica* var. *pannonica, V. pannonica* var. *purpurascens, V. peregrina, V. sativa* subsp. *incisa* var. *cordata, V. sativa* subsp. *incisa* var. *incisa, V. sativa* subsp. *nigra* var. *nigra, V. sativa* subsp. *nigra* var. *segetalis* and *V. sativa* subsp. *sativa. V. hybrida was found to be the taxa with the longest pollen grains respect to the polar axis* (P = 42.7 µm; E = 22.8 µm) while *V. narbonensis var. narbonensis* has the largest pollen grains in terms of the equatorial axis (P = 41.5 µm; E = 26.9 µm). *V. lathyroides* has the shortest pollen grains (P = 25.4 µm; E = 23.3 µm) with regard to the *polar axis*, although *V. sativa* subsp. *incisa* var. *incisa* has the narrowest pollen grains (P = 33.0 µm; E = 21.1 µm) in terms of the equatorial axis. The pollen grains are prolate-spheroidal in *V. lathyroides* (P/E = 1.09), subprolate in *V. sativa* subsp. *nigra* var. *nigra* (P/E = 1.29) and prolate shape in the other taxa studied. The pollen grains of all taxa examined are trizonocolporate. Colpus length (Clg) ranges from 12.50 µm (in *V. lathyroides*) to 32.20 µm (in *V. peregrina*). Porus length (Plg) ranges from 5.20 µm (in *V. lathyroides)* to 8.60 µm (in *V. narbonensis* var. *narbonensis*). Ornamentation of the mesocolpium is psilate-perforate in *V. lathyroides*, reticulate-perforate in *V. sativa* subsp. *nigra* var. *nigra,* rugulate in *V. narbonensis* var. *narbonensis*, perforate in *V. hybrid* and *V. lutea* var. *hirta* and reticulate in the remaining taxa. The results indicated that the morphological features of the pollens could be useful in distinguishing the examined taxa.

T2

P0449

Taxonomic studies of the family Trichocoleaceae (Marchantiophyta) in Central and South America; *Trichocolea sprucei*, *Trichocolea floccosa*, and *Leiomitra flaccida*

Tomoyuki Katagiri
Hattori Botanical Laboratory

The genus *Leiomitra* Lindb. (Trichocoleaceae, Marchantiophyta) is a small genus including about 10 species and distributed mainly from tropical and subtropical areas of Central and South America and Southeast Asia. *Leiomitra* is distinguished from the other genera of the family (*Trichocolea* Dumort. and *Eotrichocolea* R.M.Schust.) by such characters as succubous leaf insertion, well-differentiated (anisophyllous) leaves and underleaves, and presence of well-developed calyptra and perianth. *Trichocolea sprucei* was described by Stephani in 1909 from Dominica. Schultze-Motel & Menzel (1987) suggested *Trichocolea sprucei* to be a variety of *Leiomitra flaccida*, and Schuster (2000) recognized it in the genus *Trichocolea*. However, the taxonomic position of the species remained tentative, since no reproductive characters were known. In the course of worldwide taxonomic studies on the Trichocoleaceae, it has been possible to study original material of *T. sprucei*, and to find reproductive structures. The presence of perianth lead to the conclusion that the type belongs to the genus *Leiomitra*. *Trichocolea floccosa* Herzog & Hatcher was described in 1957 from Costa Rica. According to Hatcher (1957), *T. floccosa* was distinguished from *T. sprucei* by the leaf disc of 1 cell high with the cells 55 × 18 µm (1–2 cells high with the cells 91 × 29 µm in *T. sprucei*). Schultze-Motel & Menzel (1987) treated *T. floccosa* as a variety of *Leiomitra flaccida*. However, our examination of the holotype shows *T. floccosa* have 1–2 cells high leaf disc and it should be treated as a synonym of *T. sprucei*, although the type specimen lack reproductive characters. *Leiomitra flaccida* have 4–5 cells high leaf disc and can be easily distinguished from *T. sprucei* (1–2 cells high leaf disc). *Leiomitra flaccida* was described by Spruce in 1885 from Ecuador. Spruce did not specify the holotype and the species concept has not been well circumscribed. In the Spruce collection of the herbarium of University of Manchester (MANCH), the specimen MANCH-13001 annotated with "*Trichocolea tomentosa* var. In sylva Canelos. Julis 1857" was selected as the lectotype of *Leiomitra flaccida* because the information on the label and plants completely fit the original description including the information on gynoecia, although this specimen also contains *Trichocolea tomentosa*. Other specimens annotated with "*Trichocolea flaccida*" in his collection have different locality data from the original description: MANCH-10865 with "Tunguragua, ad Pasa"; MANCH-10866 with "Chimborazo"; MANCH-10867 with "Rio Janeiro"; MANCH-10868 with "Campana". The two specimens from Ecuador MANCH-10865 and MANCH-10866 correspond to *T. flaccida*, but the specimen from Brazil, MANCH-10867 turned out to be corresponding to *Trichocolea brevifissa* Steph., in having higher leaf disc being 8–10 cells high. The plants from Campana (Peru), MANCH-10868, was originally described as a forma with the gynoecia restricted to short lateral branches, but was turned out to be a etiolated form of *Leiomitra mastigophoroides* R.M.Schust. (Katagiri 2016).

T2

P0450

Morphological and anatomical characteristics of *Clinopodium graveolens* subsp. *rotundifolium* (Lamiaceae) in Turkey

Ayla Kaya
Anadolu University, Faculty of Pharmacy, Pharmaceutical Botany, 26470 Eskişehir-Turkey

Clinopodium graveolens (M. Bieb.) Kuntze subsp. *rotundifolium* (Pers.) Govaerts, is an erect or ascending annual (in Turkey) herb which grows in 0-2200 m and it is the most widespread in Turkey. In the current study, morphology and anatomical characteristics *C. graveolens* subsp. *rotundifolium*, previously treated as *A. rotundifolius* Pers. in Flora of Turkey, is studied for the first time and detailed descriptions and illustrations of general appearance of plants and their, leaf, bract, flower, calyx, corolla and fruit shapes and anatomical descriptions and illustrations of root and stem are described and illustrated. The plant is 2-34 cm. Leaves are lanceolare-ovate or obovat to orbicular shaped. Corolla is purple-lilac coloured. The root is orbicular in outline in cross-section. The seconder tissues is only observed in root anatomy of species. The stem is almost square in shape and contain non-glandular and glandular hairs in transverse sections. The vascular bundles are well-developed at the corners of the stem.

T2

P0451

A phylogenetic study of the genus *Tanacetum* (Asteraceae–Anthemideae) in Iran

Marzie Kazeminoureini[1], *Ali Sonboli*[2]
1. Tarbiat Modares University
2. Medicinal Plants and Drugs Research Institute, Shahid Beheshti University

With around 160 species, *Tanacetum* L. is the third largest genus of the Compositae–Anthemideae, after the species-rich genera *Artemisia* L. and *Anthemis* L. Its members are found in the circum-Mediterranean region, central and eastern Asia, and parts of northern America. Likely ancestral distribution range of this genus is Eastern Europe, Caucasus and Asia Minor; and dispersal (vs. vicariance) plays a significant role in this genus' distribution to all above mentioned area during mid-Miocene, when likely many of the floras of Iran entered there, from Africa and Anatolia. Despite high morphological diversity of this genus, sequence divergence among *Tanacetum* representatives included in previous studies was found to be very low, leading to unresolved phylogenetic relationships within the genus. In this study, datasets of three molecular markers, (cpDNA *rpL32-trnL*$_{UAG}$, *ndhF-rpL32* spacers and nrDNA ITS) were generated for 67 *Tanacetum* accessions and closely related genera, covering most of the previously recognized sections of this genus in Flora Iranica as well as some accessions from Turkey and Europe. The main aims of our study have been: 1) evaluation the monophyly of *Tanacetum*, 2) phylogenetic reconstruction of this genus to assess its phylogenetic delimitation and to explore the possibility of phylogenetic based infrageneric classification, 3) testing previous taxonomic hypothesis on infrageneric relationships. Furthermore we represent here the first dated ITS sequences-based phylogeny of *Tanacetum* s.l. with special reference to its species and speciation time in Iran, which is housed by at least 36 *Tanacetum* species including numerous enigmatic species complexes. This dating was performed using Bayesian uncorrelated relaxed molecular clock available in BEAST. This analysis revealed that the ancestor of the genus *Tanacetum* was diverged from its sister *Cota-Tripleurospermum* clade and has undergone rapid radiation through the first 1-2 million years after divergence. It seems that diversification and speciation of this genus occurred in Iran mostly through vicariance and followed by subsequent dispersal events to different localities in various directions.

T2

P0452

Molecular phylogeny of some spiny-cushion forming of *Astragalus* species (Fabaceae) based on nrDNA ITS sequences

Zohreh Khalili[1], *Shahrokh Kazempour Osaloo*[1], *Ali Asghar Maassoumi*[2], *Shahin Zarre-Mobarakeh*[3]

1. Department of Plant Biology, Faculty of Biological Sciences, Tarbiat Modares University

2. Research Institute of Forests and Rangelands

3. Department of Plant Biology, Faculty of Biological Sciences, Tehran University

Rhacophorus is one of the major sections of *Astragalus*. The center of speciation and diversity is Iran and Turkey. Delimitation of its boundaries with other spiny sections is too difficult due to extreme variation and high levels of homoplasy in morphological characters. The morphological studies have changed the boundaries in spiny *Astragalous* In addition, displacement occurred among with spiny *Astragalus* sections and intermediate species considered especially among *Adiaspastus and Rhacophorus* which made the boundaries uncertain. It is a very variable section with several Toorly defined species. It is also close to sect. *Prerophorus* and sect. *Macrophyllium*. According to previous studies, sections of *Rhacophorus*, *Macrophyllium*, *Adiaspastus*, *Prerophorus*, *Brachycalyx*, *Hystrix*, *Polystegis*, *Stenonychium*, and *Platonychium* located in Tragacantha subgenera because of producing the valuable tragacanth gum. Members of *Rhacophorus* are characterized by spiny shrubs with paripinnate leaves and spiny rachis; leaflets with white-simple-hairy. Flowers are sessile and forming a dense inflorescence. Bracteoles absent or unlike the calyx lobes. Legume included within the persistent corolla, unilocular, 1 (-2)-seeded. The present study aims to reconstruct phylogenetic relationships in some spiny-cushion forming of *Astragalus* species, using nuclear sequences (ITS). The nrDNA ITS region was amplified using appropriate primers by PCR. Each DNA region was aligned using MUSCLE software. Phylogenetic analyses were performed using maximum parsimony approach as implemented in PAUP* software. Bayesian and Maximum Likelihood phylogenetics analyses applied for 60 species as representatives of sects. *Adiaspastus*, *Rhacophorus*, *Brachycalyx*, *Platonychium*, and *Pterophorus* in Iran and Turkey. *A. stocksii*, *A. boeticus*, and *A. cryptocarpus* were selected as outgroups. The present study represents that studied sections are not monophyletic. Incomplete resolution in sect. *Rhacophorous* indicate that analysis based on chloroplast sequences should be combined to resolve the branches. Relationship among with sects. *Brachycalyx*, *Adiaspastus* and *Platonychium* mostly form the basal branches and are close to members of sect. *Pterophorus*. Our phylogenetic analysis suggest that the current classification of studied sections dose not properly reflect the phylogeny of sect. *Rhacophorus* and allied sections.

T2

P0453

Genetic structuring, dispersal and taxonomy of the high-alpine populations of the *Geranium arabicum/kilimandscharicum* complex in tropical eastern Africa

Tigist Wondimu Kifilemichael

Addis Ababa University

The scattered eastern African high mountains harbor a renowned and highly endemic flora, but the taxonomy and phylogeographic history of many plant groups are still insufficiently known. The high-alpine populations of the *Geranium arabicum/kilimandscharicum* complex present intricate morphological variation and have recently been suggested to comprise two new endemic taxa. Here we aim to contribute to a clarification of the taxonomy of these populations by analyzing genetic (AFLP) variation in rangewide high-alpine samples, and we address whether hybridization has contributed to taxonomic problems. We identified only two genetic groups. One corresponded to *G. kilimandscharicum*, which has been reported as exclusively high-alpine and confined to the eastern Rift mountains in East Africa. The other corresponded to *G. arabicum*, reported from lower altitudes on the same mountains as well as from a wide altitudinal span in Ethiopia and on the western Rift mountains in East Africa. The four populations analyzed of a recently described species from the Bale Mts in Ethiopia were admixed, indicating that they result from recent long-dispersal of *G. kilimandscharicum* from East Africa followed by hybridization with local *G. arabicum* in naturally disturbed habitats. Some admixture between the two genetic groups was also inferred on other mountains, supporting earlier suggestions of introgression based on morphology. We did neither find support for recognition of the recently suggested new subspecies of *G. arabicum* in Ethiopia. Interestingly, the high-alpine *G. kilimandscharicum* lacked clear geographic structuring, suggesting a recent history of colonization of the different mountains or extensive intermountain gene flow.

T2

P0454

Phylogenomic inference and molecular dating of tribe Forsythieae (Oleaceae)

Young-Ho Ha[1], *Changkyun Kim*[1], *Kyung Choi*[2], *Joo-Hwan Kim*[1]

1. Gachon University

2. Korea National Arboretum

The tribe Forsythieae (Oleaceae) is composed of two genera: *Forsythia* and *Abeliophyllum*. The former consists of 13 species distributed in the Far East Asia except *F. europaea* which is located in northern Albania, *Abeliophyllum* is monotypic endemic genus of the Korean peninsula. Most of the species of tribe Forsythieae are endangered in East Asia and their natural habitats have been disappearing. Despite their important roles as medicinal and ornamental resources, there are no consistent systematic studies on this tribe without argument. Here, we report the complete chloroplast (cp) genome sequences of 10 species of *Forsythia* and *Abeliophyl-*

lum using Illumina platform. The chloroplast genome of *Abeliophyllum* showed a highly similar gene content and arrangement in comparison with *Forsythia*, except for 400 bp of the indels. Both phylogenetic analyses (Maximum Likelihood and Bayesian inference methods) based on concatenated 78 protein-coding gene derived identical topology with strong support. *Abeliophyllum* was sister to rest of *Forsythia* species. *F. likiangensis* and *F. giraldiana* were basal lineages followed by *F. europaea* which shared minutely serrate or entire margin. The remaining seven species which distributed in East Asia formed two major clades: one clade of *F. ovata, F. velutina*, and *F. japonica*, which is morphologically supported by broadly ovate leaves, and another clade of *F. suspensa, F. saxatilis, F. viridissima*, and *F. koreana* that have lanceolate leaves. The Bayesian molecular clock dating and biogeographical analyses using four cpDNA sequences (*atpB, matK, ndhF*, and *rbcL*) suggested that Forsythieae and its sister group (*Fontanesia*) estimates at 44.99Mya in the middle Eocene. Forsythieae originated in East Asia with its crown group dated at 12.91Mya in middle Miocene and dispersed to West China and East Europe (northern Albania).

T2
P0455

Phylogeny and historical biogeography of Melanthiaceae (Liliales) : A case of intercontinental dispersal from North America to East Asia

*Sang-Chul Kim, Changkyun Kim, **Joo-Hwan Kim***
Gachon University

Intercontinental floristic disjunction between East Asia and North America in the North Hemisphere has been received much attention during the past three decades, but few studies have focused at the family level. Melanthiaceae, containing 196 species and 17 genera circumscribed in five tribes (Chionographideae, Heloniadeae, Melanthieae, Parideae, and Xerophylleae), is disjunctly distributed in East Asia and North America and it represents on excellent for studying the evolution of biogeographic patterns in this region. Here, we employ phylogenetic, molecular dating, and biogeographic methods to reconstruct the evolutionary history of Melanthiaceae based on comprehensive sampling scheme using four chloroplast DNA regions (*atpB, rbcL, matK*, and *ndhF*). The BEAST analysis produced a well resolved phylogeny of Melanthiaceae indicating the monophyly of the family and the relationships among the five tribes. Melanthieae is found to be sister to the rest tribes of the family and the remaining taxa are divided into two major clades consisting of the Chionographideae + Heloniadeae clade and the Parideae + Xerophylleae clade. The molecular clock dating and the ancestral area analyses suggest that Melanthiaceae most likely originated in North America with its crown group dated at 90.3 Ma (95% HPD 76.0–103.2 Ma) in the late Cretaceous. In the family, five independent migrations into East Asia from North America are inferred to have occurred in the Oligocene and the Miocene-Pliocene via historical paleo-land bridge connections. Cooling trends during the Oligocene resulted in the present East Asia-North America disjunction distribution, while the warm period during the middle Miocene and habitat heterogeneity stimulate the diversification from East Asia.

T2
P0456

Chloroplast DNA assessment for the anagenetic speciation of *Sedum takesimense* (=*Phedimus takesimensis*) (Crassulaceae) in Ulleung Island, Korea

*Hee-Seung Seo, **Seung-Chul Kim***
Sungkyunkwan University

Sedum L., with approximately 470 species, is the largest genus of the family Crassulaceae. The recent molecular phylogenetic study suggests that the genus *Sedum* is highly polyphyletic. Of three endemics among 11 species recognized in Korea, *Sedum takesimense* occurs exclusively in Ulleung-do and Dok-do Island, which is known for unusually high level of anagenetic speciation among oceanic islands. *Sedum kamtschaticum* (=*Phedimus kamtschaticus*), which commonly occurs in northeastern Asia, is considered as continental progenitor species of *S. takesimense* in Ulleung-do and Dok-do Island. The earlier survey of intraspecific cpDNA variation has little value in fully capturing genetic diversity for the progenitor-derivative pairs due to limited sampling in individuals and DNA sequences. In this study, we investigated genetic diversity and structure of *S. takesimense* (derivative) and *S. kamtschaticum* (progenitor) pairs to assess the genetic consequences of anagenetic speciation. We extensively sampled populations of progenitor (*S. kamtschaticum*; 11 pops. from Korean peninsula, 88 individuals) and derivative (*S. takesimense*; 14 pops. 140 individuals from Ulleung-do and 1 pop. 10 individuals from Dok-do) and sequenced five noncoding regions of cpDNA (*atp*I/H, *trn*C/*ycf*6, *trn*L/F, *rps*16/*trn*K, and *ndh*J/*trn*F; ca. 3500 base pairs). We found nearly three times higher haplotype diversity in insular derivative species (21 haplotypes in 93 individuals), which derived from the most common haplotype found in southern and eastern part of Ulleung Island, compared to its continental progenitor species (8 haplotypes in 67 individuals). Only one haplotype found in Dok-do Island was derived from the southwestern population of Ulleung Island. We found no geographical structuring within the island population and very strong genetic differentiation between the island species *S. takesimense* and its continental progenitor *S. kamtschaticum*. These results suggest that the genetic consequences of anagenetically derived species of *S. takesimense* are consistent with other examples of endemic species on Ulleung Island (i.e., *Acer takesimensis* and *Dystaenia takesimana*). The genetic consequences of anagenetic speciation in *Sedum takesimense* based on extensive sampling from northeastern Asia will be further explored using nuclear microsatellite markers. The example of *S. takesimense*, as one representative endemic lineage of Crassulaceae on island, will be of important value in generalizing and synthesizing the genetic consequences of anagenetic speciation in oceanic islands.

T2
P0457

Genetic variation and evolution of *Filibryum* in Yakushima Island

Wonhee Kim[1], *Tomio Yamaguchi*[2], *Hiromi Tsubota*[3]
1. Division of Plant Resources, National Institute of Biological Resources
2. Department of Biological Science, Graduate School of Science,

Hiroshima University
3. Miyajima Natural Botanical Garden, Graduate School of Science, Hiroshima University

Filibryum is a member of Hypnaceae and pleurocarpous moss. This genus consists of three species, which are *F. ogatae* (Broth. & Yasuda) W.Kim & T.Yamag., *F. yakoushimae* (Cardot) W.Kim & T.Yamag. and *F. deguchianum* W.Kim & T.Yamag. *F. yakoushimae* and *F. deguchianum* are Japanese endemic species. On the other hands, *F. ogatae* is widely distributed in Japan and Korea. In the case of *F. deguchianum*, its morphological traits are distinguished from others. However, the circumscription of the interspecific morphological difference between *F. ogatae* and *F. yakoushimae* is unclear. So, this study analyzed the genetic variation between both species in order to solving the problem. Furthermore, this study examined the genetic variation of *F. ogatae* related to the geographical distribution. For analysis of the genetic variation, 30 samples were collected from 10 regions, which are 7 regions from Japan and 3 regions from Korea, and analyzed nrDNA ITS, cpDNA *psaA-rps*4 region and mtDNA *nad*5 sequences. There was *ycf*3 gene between *psa*A and *trn*S, and this gene consists of 3 exons and 2 introns (total 1,930 bp). Especially, interspecific genetic variations between *F. ogatae* and *F. yakoushimae* were verified on *ycf*3 Intron 1 region. Then, samples were categorized as three groups according to distribution regions, which were Japan, Korea and Yakushima population. As the result of analyzing the genetic variations, Yakushima population has distinct genetic differences presented on 4 regions, which are ITS 1 region, *psaA/ycf*3 spacer, *yacf*3 Intron 1 and *rps*4. However, their morphological characteristics were not distinct. Therefore, morphological variations related to geographical distribution were not verified in *F. ogatae*. Therefore, it is possible that Yakuahima population of *F. ogatae* may be a cryptic species derived through anagenetic divergence processes. Such as a cryptic species in Yakushima Island, it was also reported in vascular plants (Kimuara *et al.*, 2014). Recently, it was emphasized that the importance of an anagenetic speciation/evolution in the island (Stuessy *et al.*, 2006). Additionally, *F. yakoushimae* is an endemic species in Yakushima Island, and its genetic variation pattern is similar to the Yakushima population of *F. ogatae*. As the result, it is considered that *F. yakoushimae* was derived from Yakushima population of *F. ogatae* built up genetic variations. Consequently, the present study assumes that both species of *Filibryum* in Yakushima Island are undergoing a cladogenetic divergence process after an anagenetic divergence.

T2
P0458

Taxonomic review of the genus *Catalpa* (Bignoniaceae)

Joseph H. Kirkbride Jr.[1], *Richard T. Olsen*[2]
1. Smithsonian Institution/Department of Botany
2. USDA Agricultural Research Service

Bignoniaceae tribe Catalpeae DC. ex Meisn. is monophyletic and consists of two genera, *Catalpa* Scop. and *Chilopsis* D. Don, and their hybrid ×*Chitalpa* T. S. Elias & Wisura. *Chilopsis* is monotypic and sister to the monophyletic clade *Catalpa*. *Chilopsis linearis* (Cav.) Sweet has been hybridized with *C. bignonioides* Walter, and the hybrids have been named ×*Chitalpa tashkentensis* T. S. Elias & Wisura. The genus *Catalpa* consists of eight species. Four species form the monophyletic clade *Catalpa* section *Macrocatalpa* Griseb., whose members are tropical evergreen and endemic to the Greater Antilles, *C. brevipes* Urb., *C. purpurea* Griseb., *C. longissima* (Jacq.) Dum. Cours., and *C. macrocarpa* (A. Rich.) Ekman, occurring from Jamaica across Cuba and Hispaniola to the Bahama Islands. The other four species form the monophyletic clade *Catalpa* section *Catalpa*, sister to the clade *Catalpa* section *Macrocatalpa*. Those species are temperate deciduous. Two are endemic to the eastern USA, *C. bignonioides* and *C. speciosa* E. Y. Teas, and the other two are endemic to China, *C. bungei* C. A. Mey. and *C. ovata* G. Don. The ancestor of section *Catalpa* dispersed to China, and from there an ancestor of the North American species dispersed to the USA. The Chinese *C. ovata* has been hybridized with the North American *C. bignonioides* and *C. speciosa*, and these hybrids have been named *C.* ×*erubescens* Carrière and *C.* ×*galleana* Dode, respectively.

T2
P0459

Aloe taxonomy and phylogeny: Current status and future prospects

Ronell R Klopper[1], *Olwen M Grace*[2]
1. South African National Biodiversity Institute
2. Royal Botanic Gardens, Kew

Aloes are widespread in Africa, the Arabian Peninsula, Madagascar and some of the smaller Western Indian Ocean islands. These rosulate succulents are very popular worldwide amongst succulent plant collectors and a number of species are revered for their medicinal or other cultural values. Several aloes are considered to be taxa of conservation concern and all aloes (except *Aloe vera*) are thus listed on CITES appendices. Aloes (±630 taxa) and their relatives has been the subject of numerous recent studies to elucidate the phylogenetic relationships among these plants. In this light the family Asphodelaceae has been incorporated into a substantially larger family concept that also includes the Xanthorrhoeaceae s.str. (Australian grass trees) and Hemerocallidaceae (day lilies). Furthermore, molecular and phylogenetic studies have necessitated a reassessment of the classification of the genus *Aloe* s.l. and related genera. This resulted in the establishment of segregate genera for the scrambling aloes (*Aloiampelos*), tree aloes (*Aloidendron*), the fan aloe (*Kumara*), and variegated aloes (*Gonialoe*), as well as the inclusion in *Aloe* of species previously treated in the genus *Chortolirion*. Similar generic adjustments were also made in the genus *Haworthia* where three segregate genera are now recognised, namely *Haworthia*, *Haworthiopsis* and *Tulista*. Phylogenetic analyses of plant utilisation can provide a means to better understand observed patterns in the value of global plant diversity. A phylogenetic framework was used with success to investigate possible links between leaf succulence and medicinal use in aloes. These analyses have also shed light on the origin and diversification of aloes in general. The recent revision of two groups of aloes utilising DNA-based analysis has shed light on the ability of molecular tools to recognise aloe taxa at species level and below. These studies have shown that DNA barcodes might not be able to distinguish between certain species. This can have consequences for conservation-related activities, e.g. identification of traded material. This presentation will provide an overview of the genus

Aloe s.l. with specific emphasis on its current distribution and diversity, as well as its evolutionary history. Recent changes in the family and genus classification of aloes and their relatives will be discussed. The value of molecular tools, especially barcodes, and how informative these techniques are at species level in aloes will be considered. Future research prospects for aloes will also be emphasized.

T2

P0460

A newly discovered oligocene flora from the Maoming Basin, South China: Taxonomy and paleoclimatic implications

Tatiana Kodrul[1,2], *Galina Aleksandrova*[2], *Natalia Maslova*[3], *Robert Spicer*[4], *Jian Yang*[5], *Alexei Herman*[2], *Teresa Spicer*[5], *Yan Wu*[1], *Jianhua Jin*[1]

1. State Key Laboratory of Biocontrol and Guangdong Provincial Key Laboratory of Plant Resources, School of Life Sciences, Sun Yat-sen University, Guangzhou

2. Geological Institute, Russian Academy of Sciences, Moscow

3. Borissiak Paleontological Institute, Russian Academy of Sciences, Moscow

4. School of Environment, Earth and Ecosystem Sciences, The Open University, Milton Keynes

5. State Key Laboratory of Systematic and Evolutionary Botany, Institute of Botany, Chinese Academy of Sciences, Beijing

The fossil Shangcun flora comes from the Maoming sedimentary basin, located in southwestern Guangdong Province, South China. Plant megafossils including leaves, stems, fruits, cones and seeds were recovered from the Shangcun Formation exposed within the Shangcun (21.79778°N; 110.8094°E) and Lishan (21.84417°N; 110.7786°E) quarries. The Shangcun Formation, 300–500 m thick, consists of mudstones and siltstones with minor intercalations of oil shales and coal seams in the lower part (Zhou and Chen, 1988). Plant fossils are distributed in a scattered pattern in greyish brown mudstones and siltstones representing a lacustrine environment. Preliminary taxonomic analysis revealed that the Shangcun flora contains at least 95 taxa belonging to leafy bryophytes (Bryopsida), horsetails (Equisetaceae), ferns (Osmundaceae, Polypodiaceae), conifers (Pinaceae, Cupressaceae, probably Taxaceae) and angiosperms (Platanaceae, Lauraceae, Fagaceae, Malvaceae, Calophyllaceae, Juglandaceae, Fabaceae, Rhamnaceae, Cornaceae, Myricaceae, Menispermaceae, Simaroubaceae and Palmae). Ferns are scarce in the Shangcun flora, but relatively diverse, while mosses and horsetails are very rare. Compared to the older Eocene fossil floras of the Maoming Basin conifers are relatively abundant and diverse in the Shangcun flora, and are composed mainly of pinaceous (*Pinus*, probably *Tsuga)* and cupressaceous (*Calocedrus*, *Sequoia*) taxa. The majority of Shangcun species are angiosperms, dominated by Fagaceae (*Quercus, Castanopsis),* Lauraceae, Myricaceae (*Myrica*), and probably Moraceae. The genera *Palaeocarya* (Juglandaceae), *Paliurus* (Rhamnaceae), and *Ailanthus* (Simaroubaceae) are recognized on the basis of fruit remains. The genus *Burretiodendron* (Malvaceae s. l.) is represented by fruits and associated leaves. Extant species of *Burretiodendron* are mainly distributed in the monsoonal tropical regions of East Asia. *Calophyllum* (Calophyllaceae), recognized on the basis of leaves, is also confined to moist tropical regions of the world. Palynological assemblages from the Shangcun Formation in the Lishan quarry indicate the Eocene-Oligocene transition near the base of the studied section marked by a sudden increase of conifer taxa, loss of megathermal angiosperms and the presence of *Persicarioipollis* sp. pollen and grass pollen. *Persicarioipollis* sp. is characteristic of the early Oligocene in Eastern Asia (Yu, 1983; Shaw, 1998; Yi *et al.*, 2003) and unknown in the Eocene. Higher in the section megathermal angiosperms recover in terms of abundance and diversity, but lack key taxa typical of the Eocene. Palynological assemblages from the upper part of section correspond with the "warm" palynoflora that occurs in the second half of the early Oligocene (Wu *et al.*, 2003). Although many Shangcun plant megafossils were not collected *in situ*, the palynomorph spectra obtained from the same matrix as the leaf fossils are similar to that above the conifer-rich horizon and so the Shangcun megaflora is dated to the second half of the early Oligocene. Based on plant megafossils the Shangcun flora is similar to the Oligocene floras of adjacent regions (e.g. the Oligocene flora from the Ningming Formation, Ningming Basin, Guangxi). As determined by Climate Leaf Analysis Multivariate Program (CLAMP) the early Oligocene Shangcun flora indicates rainfall seasonality approaching that seen in South China today.

T2

P0461

A multilocus phylogenetic reconstruction of *Aconitum* subgenus *Lycoctonum* (Ranunculaceae) in China: Indicating allo- and autopolyploidization in combination with morphological and karyotypic evidence

Hanghui Kong[1], *Yu Hong*[1], *Yu Zhang*[1], *S. Michael Barker*[2]

1. Key Laboratory of Plant Resources Conservation and Sustainable Utilization, South China Botanical Garden, Chinese Academy of Sciences

2. Department of Ecology and Evolutionary Biology, University of Arizona

Polyploidization has long been recognized as one of the most important driving forces of plant evolution. *Aconitum* subgenus *Lycoctonum* (Ranunculaceae) has a wide distribution range and well-known background of polyploidy, thereby providing a potentially valuable model to explore polyploid origin and evolutionary history. In the current study, 29 species including diploid, tetraploid and hexaploid species were sampled in subg. *Lycoctonum*. Using four cpDNA regions (*ndh*F-*trn*L, *psb*A-*trn*H, *psb*D-*trn*T and *trn*T-L) and two nrDNA regions (internal transcribed spacer, ITS and external transcribed spacer, ETS), phylogenetic relationship was reconstructed and showed consistence with previous taxonomical and cytological results. No evidence showed that polyploidization directly led to an increase of diversification rate. Furthermore, in combination with cpDNA haplotype network, S-DIVA analyses as well as morphological and karyotype evidence, polyploidization was interpreted in subg. *Lycoctonum*. Allopolyploidization was inferred for *A. angustius* with intermediate morphologies and heteromorphic chromosomes. *Aconitum finetianum* was suggested to be the maternal progenitor of *A. angustius* due to their highly similar morphologies and overlapping distribution range. Autopolyploidization was inferred for tetraploids in the Hengduan Mountains (HDM) with eight groups of chromosomes of four homologous. Local origin was inferred for tetraploids in the HDM, which diverged approximately 3.00 Ma in the same period with the orogeny of the HDM. The hexaploid *A. apetalum* was inferred to suffer from geographical isolation due to the formation of the Qinghai–Tibetan Plateau (QTP) and the HDM. Hybridization and heterogeneous habitats in the HDM were suggested to play an important role in the polyploidization in subg. *Lycoctonum*.

T2

P0462

Leaf micromorphology of the *Persicaria* sect. *Cephalophilon* (Polygonaceae) and its systematic re-evaluation

***Min-Jung Kong**, Suk-Pyo Hong*
Kyung Hee University

A comparative study of leaf epidermal microstructures in *Persicaria* section *Cephalophilon* was investigated by light microscopy (LM) and scanning electron microscopy (SEM) on 22 taxa in order to evaluate their systematic significance. The leaves are either amphistomatic or hypostomatic in the studied taxa, and the both found with similar number of taxa, and the size range of the guard cells varied among 16.96–41.96 x 13.30–37.30 µm; the small ones was found in abaxial side of *P. strindbergii* (17.34–24.02 x 13.41–17.90 µm) and abaxial side of *P. glacialis* var. *glacialis* (17.98–21.87 x 13.39–18.21 µm), while the largest one occurred to adaxial side of *P. chinensis* var. *chinensis* (32.97–35.84 x 26.14–30.63 µm). Anomocytic and anisocytic stomata complex were usually occurred on the adaxial and abaxial epidermises of same leaf. Paracytic type is a characteristic feature of both *P. palmata* and *P. criopolitana*. The anticlinal wall patterns of subsidiary cells on both sides varied from straight to sinuate. The striation of cuticles was observed in 17 taxa, and it was distributed only in adaxial side (except *P. microcephala* var. *microcephala* – on both sides). Seven different types of trichomes were found: five types of non-glandular trichomes (NGT; stellate trichome with smooth surface, multiseriate trichome with smooth or striate surface, and uniseriate trichome with smooth or papillose surface), and two types of glandular trichomes (GT; peltate gland and long stalked pilate gland). The peltate GT consisted of 4-, 6-, 8- or multi-celled head. The stellate NGT was only found in *P. palmata*. The systematic significance of various stomatal characters and trichome types in *Persicaria* section *Cephalophilon* is discussed.

T2

P0463

Functional differentiation between the inner and outer leaves in a mechanical support by a pseudostem of *Veratrum album* subsp. *Oxysepalum*

***Lingshuang Gu**[1], Yasuo Konno[2]*
1. Henan Polytechnic University
2. Obihiro University of Agriculture and Veterinary Medicine

The pseudostem of *Veratrum album* consists of radical leaves, which have tubular and concentrically aggregated petioles. We examined a hypothesis that the major role of the inner leaves is to raise the leaf blades higher for increased photosynthesis, while that of the outer leaves is to mechanically support the inner leaves at the expense of displaying their leaf blades. The removal of outer leaves resulted in the collapse of the remaining inner leaves, showing outer leaves provide important support for inner leaves. The estimation of mechanical properties for the constituting leaves revealed that inner leaves possess greater mass of leaf blades for their capacity of mechanical support them compared to outer leaves. These results uphold the hypothesis that the primary role of inner leaves is to display leaf blades for photosynthesis, while that for outer leaves is to mechanically support the less stable inner leaves.

T2

P0464

Nuclear and plastid phylogeny of *Lotus* (Leguminosae) and phylogeography of *Lotus corniculatus* complex

***Tatiana Kramina**, Galina Degtjareva, Tahir Samigullin, Carmen Valiejo-Roman, Dmitry Sokoloff*
Lomonosov Moscow State University

Lotus comprises ca. 130 species of Old World herbs, semishrubs and shrubs, including important pasture crops and a model legume, *L. japonicus*. Earlier nrITS-based phylogenies were incongruent with all taxonomic classifications of the genus. In particular, members of the former genus *Dorycnium* were unexpectedly placed near species of *L.* sect. *Lotus*. We present a global phylogenetic study of *Lotus* that uses both plastid (*psb*A-*trn*H IGS and *rps*16 intron) and nuclear (ITS and ETS) markers. The rooting of the *Lotus* phylogeny, its major clades and basic biogeographic patterns are discussed. The placement of the segregate genera *Dorycnium* and *Tetragonolobus* in the synonymy of *Lotus* was confirmed. Analyses of plastid data strongly supported a basal split of *Lotus* into two clades, one comprising species of sect. *Lotus* plus those traditionally placed in *Dorycnium* and the other including the rest of the species. The former clade has a centre of species diversity in Europe and N Asia, and the latter in Macaronesia, Africa and S Asia. Only the "Southern" clade is resolved in analyses of nrITS and nrETS data. Trees inferred from plastid, nrITS and nrETS data shared the occurrence of several smaller clades corresponding to traditionally recognized infrageneric taxa or species groups as well as the occurrence of some well-supported clades that differ from traditional taxonomic concepts. Several instances of incongruence were documented between nuclear and plastid markers and between the two nuclear markers, possibly resulting from reticulate evolution. The extant geographic patterns of *Lotus* are likely biased by at least one round of area fragmentation followed by expansion coupled with extensive speciation associated with the complex history of the Mediterranean biome. *Lotus corniculatus* complex that belongs to the section *Lotus* is the most agriculturally important, but taxonomically problematic group of the genus. The presence of at least two ploidy levels and interspecific hybridization confuses the system of the group. Phylogenetic relationships among *L. corniculatus* complex and other groups of the section *Lotus* vary in reconstructions inferred from different DNA markers. However, all studies clearly demonstrate a highly supported monophyly of the *L. corniculatus* complex. Phylogenetic analysis of the section *Lotus* using *trn*F-L IGS and *trn*L intron of cpDNA revealed taxonomic significance of indels in these DNA regions. A haplotype network of the *L. corniculatus* complex constructed using statistical parsimony method in TCS program allowed to make the following suggestions concerning phylogeography of the group: we revealed a specimen of uncertain morphology with a haplotype related to *Lotus conimbricensis* (the closest relative of the *L. corniculatus* complex), which may shed light on the origin of the complex; a haplotype from one of West European glacial refugia (Systema Central, Spain) found in *L. alpinus* is the closest to a hypothetical ancestral haplotype of the *L. corniculatus* complex; derived haplotypes were spreading eastward from W. Europe and gave two peaks of diversity: high in C. and E. Europe and a smaller one in Asian part of Northern Eurasia. The origin of East African and Japanese lineages of the complex is also discussed.

T2

P0465

Morphobiological features of the nomenclatural type of *Allium nigrum* L., nom. cons.

Lyudmila Krasovskaja, ***Igor Levichev***
Komarov Botanical Institute of the Russian Academy of Sciences

Goal of research. To describe the ontogenetic features of *A. magicum* as an independent taxon not identical to *A. nigrum*.
Methods. Living plants in the nature and under cultivation were used for long-term studies. The main stages of *A. magicum*'s ontogenesis from seedlings to the pre-senile stage are shown (habitually and on sections). The development of seedlings is showcased using transparent containers made of organic glass.
Results. The absence of black ovaries both in annual cycles, and in ontogenesis is demonstrated. Formation of a vegetative bulbil on the tip of a top leaf, a unique feature of *A. magicum* mentioned by Theophrastus and represented by a number of authors before Linnaeus, is steadily combined with many other signs which are absent in related species including *A. nigrum* sensu Linnaei. Demonstrated are features of the replacement shoot and vegetative reproduction, including the addorsed and gorgonoid position of bulbs. The gemmiferous leaf is interpreted to belong to the nature of phylloclade. Explained traits: the ability to embed the base of cotyledon into the soil to 20 cm deep, starting and final localization of a digeotropic growth area, and the seedling structure. Sometimes the cotyledon continues growing, but it unable to push the bulb deeper. Therefore, the bulb rotates by 180 degrees, while the root and cotyledon form a characteristic "S-shaped" curve.
Discussion. Conservation of the name *A. nigrum* as a standard sample of *A. magicum* violates against the basic principle of typification – to serve as a standard of species binomen, and excluded the use of the last name in taxonomy. It is necessary to cancel conservation and select the epitype from among the *A. nigrum* forms which are still cultivated in botanical gardens of European countries.

T2

P0466

Pollen tracking in the food-deceptive, flower colour dimorphic orchid *Dactylorhiza sambucina* showed no predominant switching behaviour of pollinators

Agnes M. Groiß[1], *Andreas Braun*[1], *Josef Greimler*[2], ***Matthias Kropf***[1]
1. University of Natural Resources and Life Sciences, Institute for Integrative Nature Conservation Research
2. University of Vienna, Department of Botany and Biodiversity Research

The non-rewarding (i.e. nectar-free) orchid *Dactylorhiza sambucina* is assumed to cheat its pollinators effectively by showing two flower colour morphs. This hypothesis implies common pollinator switches between yellow- and purple-flowering plants. However, pollen flow within and between morphs has never been tested under field conditions by using a pollen tracking technique. We studied a dimorphic population located in the southern "Wienerwald" (Vienna Woods) in Austria to obtain frequencies of inter- and intra-morph pollinator switches. Pollinia were marked using fluorescent dye powder to observe pollen flow, i.e. the spatial distribution of pollen, pollen export and deposition (i.e. male and female reproductive success) as well as the transport distances from donator to receptor plants. For specifically analysing the switching behaviour of (bumblebee) pollinators, we labelled yellow- and purple-flowering plants morph-specifically using two different dye colours. In our study system, yellow-flowering plants were dominant (70% in 2006; 67% in 2014). In the study year 2014, we labelled 582 pollinia from yellow- and 239 from purple-flowering individuals reflecting the yellow-dominance (71%). In total, 473 marked pollinia were exported by potential pollinators, of which 72.3% originated from yellow flowers. We found that movements of pollinators from yellow to yellow (YY) flowers dominated the four possible pollination events by far (i.e. 93 of all 155 deposits of labelled pollen). Even considering the dominance of yellow-flowering plants in our study population, only 70 respective events (i.e. YY; out of 155) would have been expected, if switching behaviour was by chance. The purple intra-morph movements (i.e. PP) were also higher than expected (i.e. 24 instead of 17 expected), leading to disproportionately high YY- and PP-observation rates (X^2=25.154, df=3, p<0.0001). Therefore, the assumption that pollinators alternate randomly between colour morphs was not supported by our results. The longest pollen transport distance observed was 237m. Moreover, 135 manually conducted cross-pollinations on 47 individuals were evaluated: Seed viability analyses indicated no differences between the four possible crosses. Consequently, the dominance of the yellow morph at our study site could be the primary result of the most frequently observed cross-combination (i.e. YY.).

T2

P0467

Reproductive biology of *Anacamptis* (*Orchis*) *coriophora* (Orchidaceae): A broad pollinator spectrum mediates frequent outcrossing

Susan El-Heliebi, *Monika Kriechbaum*, ***Matthias Kropf***
University of Natural Resources and Life Sciences, Institute for Integrative Nature Conservation Research

In Central Europe, *Anacamptis* (*Orchis*) *coriophora* is a critically endangered species. Over the past decades, populations of this species have almost become extinct in vast areas. One of the few large, still existing populations was studied in the "Donau-Auen" (Danube floodplain) National Park in the Viennese Lobau (Austria). The aim of our study was to draw conclusions from the reproductive system of this nectariferous orchid with respect to its dependence on (different) pollinators. Another aspect discussed in our analysis is the role of cross- and self-pollination in *A. coriophora*, including geitonogamy. Direct pollinator observations showed that *A. coriophora* is attractive for a wide range of pollinating and visiting insects. In addition to (solitary) bees and bumblebees being common (and expected) pollinators of this species, pollinating Coleoptera and Diptera were also frequently identified. Observed plants mainly attracted the following species: *Apis mellifera*, *Lasioglossum xanthopus*, *Bombus terrestris*, and *B. humilis* (Hymenoptera: 52.8% of all observations), but also *Oedemera femorata, Oxythyrea funesta* (Coleoptera: 44.4%) and *Eristalis tenax* (Diptera: 2.8%). Based on pollination experiments (hand pollination and bagging of flowers with fine netting for pollinator exclusion to test possible reproduction mode of agamospermy, spontaneous and manual self- and cross-pollination) and subsequent seed viability tests, self-compatibility and dependence on pollinator action were documented. Furthermore, the species is not capable of asexual reproduction by agamospermy. Due to duration of stay and pollinator behaviour of some of the observed insects (e.g. Coleoptera) and given the short bending times of the species (bending starts already after 18.7 ± 5.8 sec), increased geitonogamy rates must be expected. In summary, allogamy and geitonogamy were identified as the dominant reproductive modes of *A. coriophora* and the optimal reproductive success is a result of cross-pollination by various insects leading to high overall fruit set. Pollinator frequencies, reproductive success, and recent population trend lead to the assumption that the population of *A. coriophora* studied here is currently very vital.

T2

P0468

A pheromone-related switch on mating behaviors in ferns: A case in *Deparia lancea*

Li-Yaung Kuo[1], *Cheng-Wei Chen*[2], *Yu-Ling Hung*[3], *Chia-Wen Ko*[2], *Fay-Wei Li*[1], *Wen-Liang Chiou*[4], *Chun-Neng Wang*[3]

1. Boyce Thompson Institute

2. Division of Silviculture, Taiwan Forestry Research Institute

3. Instiuite of Plant Biology, National Taiwan University

4. Dr. Cecilia Koo Botanic Conservation and Environmental Protection Foundation/ Conservation Center

Ferns are the second largest group among extant vascular plants, and have evolved a complex mating system. Diploid ferns often develop dioecious gametophytes through the regulation of a sex-determination pheromone – antheridiogen. Mature gametophytes secret antheridiogens that induce nearby young gametophytes to differentiate into males (i.e. with only sperm-producing antheridia), and thus promote outcrossing. In contrast, in poylploid ferns, gametophytes tend to be bisexual, having both antheridium and archegonium simultaneously, and under the most extreme situation, a single gametophyte can self infragametophytically, resulting in a complete homozygous sporophyte. However, to date the mechanism for such different modes of gender expression between diploid and polyploid ferns has remained unclear. To shed light on this, we compared the gametangium development in autohexaploidy *Deparia lancea*, and its conspecific diploids. We found that in mix-aged gametophyte populations, the hexaploids are biased being bisexual, while the conspecific diploids displayed a higher degree of dioecy. In the further transplant experiments, we found that for the hexaploids, gametophytes in the later developmental stages, which were matured enough to produce archegonia, were most sensitive to antheridiogen, and, on average, produced more antheridia per individual. By contrast, gametophytes of the diploids were most sensitive to antheridiogen during the earlier stage when the notch meristem has not yet developed. Our results imply that the inbreeding syndrome in polyploidy ferns can be achieved via delaying sensitivity to antheridiogen, and, consequently, their gametophytes can produce both male and female gametangia at the same developmental stage. Our finding is the first to provide direct evidence supporting that a mating system switch in ferns involves a heterochronical regulation on antheridiogen sensitivity.

T2

P0469

Pollen and NPP evidences of the ecosystem's response to environmental changes during the Middle-Late Holocene in the Bale Mountains (East Africa)

Evgeniya Kuzmicheva

A.N. Severtsov Institute of Ecology and Evolution, Russian Academy of Sciences

The study of vegetation and climate dynamics is one of the most topical issues in view of current global changes. Research in this field provide a way to determine trends in natural and anthropogenic environmental changes and use it to predict the ecosystem`s response to global changes in the future. Here we present the results of pollen and non-pollen palynomorphs (NPP) analyses of the zoogenic deposit found in the Bale Mountains (Ethiopia) (06°43′N, 39°44′E; 2700 m a.s.l.). This deposit type is a unique source of palaeodata in default of the common deposit types such as peat and lacustrine sediment. We made an attempt for the NPP analysis, the first known for Ethiopian deposits. The deposit is very interesting in palaeoenvironmental aspect since it is located on the southern slope of the Bale Mountains, characterized by unique forest area (called "Harenna forest"), which is almost gone in East Africa. The age-depth model was built on the basis of the radiocarbon dates; it shows that sedimentation began about 6500 calibrated years (cal yr) BP. Pollen spectra in 6500-4600 cal yr BP are characterized by the predominance of non-arboreal palynotypes (mainly grasses (Poaceae) and sedges (Cyperaceae)), which most probably indicate the absolute dominance of the bamboo. Such vegetation points on wet climate of the African humid period (AHP). At about 4600 cal yr BP vegetation changes indicate the decrease of humidity: Cyperaceae falls off, and Poaceae concentration also drops dramatically in response to the end of the AHP. The accumulation rate of Katcha deposit significantly decreases at about 4600 cal yr BP, which also points on the climate aridization trend. At the same time *Celtis*-type appears in significant concentrations in the spectra. Presumably around 4600 cal yr BP bamboo thickets become sparser in response to the gradual decline in humidity, while *Celtis*, which is considered as a forest species, is rapidly colonizing the areas cleared from forest, capturing all suitable habitats. The climate aridization continues in further period: around 2500 cal yr BP there is extensive development of dry forests type of olive (*Olea*), *Hagenia* and *Podocarpus*. There are no distinct signals of climate change during the last two thousand years. Perhaps the climate signals are masked by other factors. For instance, similar pollen spectra (with numerous and asynchronous changes in pollen concentrations) are recorded for recurrent wild fires. Increase in the concentration of *Gelasinospora* cf. *cratophora* ascospores (HdV-1093) also indicates frequent wild fires. Frequent fires are most likely caused by human activities in the Harenna forest over the past 2000 years. Direct anthropogenic impact was recorded during the last 1000 years with an increase in the concentration of non-pollen palynomorphs such as *Delitschia* spp. (UG-1066), *Podospora*-type (HdV-368), *Cercophora*-type (HdV-1013), and *Sordaria*-type (HdV-55). This indicates an intense development of the cattle breeding. This research was funded by the Joint Ethiopian-Russian Biological Expedition, the Russian Fund for Basic Research (Grant No. 16-34-00292,15-04-04721) and the Russian Programs 'Origin and Evolution of the Biosphere', 'Wildlife' and 'Fundamental Basics of Biological Resources Management'.

T2
P0470
The allopolyploid origin of *Dryopteris dilatata* (Hoffm.) A. Gray and how gene loss can give misleading results

Dennis Larsson[1], *Catarina Rydin*[1,2], *Niklas Wikström*[1,2]
1. Stockholm University, Department of Ecology, Environment and Plant Sciences, SE-106 91 Stockholm, Sweden
2. Bergius Foundation, the Royal Academy of Sciences, Box 50005, SE-104 05 Stockholm, Sweden

The origin of putative allotetraploid *Dryopteris dilatata* (Hoffm.) A. Gray has recently been disputed. The hypothesis that *D. dilatata* is an allotetraploid hybrid between the diploid *D. expansa* (C. Presl) Fraser-Jenk. & Jermy and *D. intermedia* (Muhl. ex Willd.) A. Gray has been challenged by an European team who claims *D. dilatata* is autotetraploid since they only found one expansa-type copy in the nuclear region pgiC. An American team however found two copies in the nuclear regions pgiC and GapCp and claims *D. dilatata* is indeed allotetraploid. In order to settle the dispute a phylogenetic analysis was done on 30 *D. dilatata* specimen (26 herbarium specimen and 4 fresh specimen) using the nuclear regions pgiC and GapCp. By using copy-specific primers designed to mismatch on one of the two copies, both copies in both gene regions were successfully sequenced. The copies found showed either a common origin with *D. expansa* or with *D. intermedia*. The results are consistent with the hypothesis that *D. dilatata* is an allotetraploid hybrid between *D. expansa* and *D. intermedia*. Furthermore, the study show that alleles and entire copy types have been lost in some specimen, most likely through recombination between homeologous chromosomes of the parent genomes. Of the 21 specimen which was successfully sequenced for both loci, 16 had multiple copies in one or both loci and of those, 6 had only one copy-type in one locus while having two copy-types in the other locus. These findings explain the results of the European team. The specimens they collected had either lost theirintermedia-type copies through the before mentioned recombination, or was actually specimens of *D. expansa*, a possibility that can only be conclusively excluded if the ploidy level is known. These findings show that recombination can alter the presence of copies in allopolyploids and that it happens often enough that it can give misleading results and should therefore be taken into account when making phylogenies involving allopolyploids.

T2
P0471
A panmictic Amazonian world?

Alice Ledent[1], *Patrick Mardulyn*[2], *Sylvia Mota*[3], *Rick Overson*[4], *Jairo Patiño*[5], *Marta Pereira*[6], *Norman Wickett*[4,7], *Charles Zartman*[6], *Alain Vanderpoorten*[1]
1. Université de Liège (ULg)
2. Université Libre de Bruxelles (ULB)
3. Naturalis Biodiversity Center (NBC)
4. Chicago Botanic Garden (CBG)
5. Instituto de Productos Naturales y Agrobiología (IPNA-CSIC)
6. Instituto Nacional de Pesquisas da Amazônia (INPA)
7. Northwestern University (NU)

Understanding connectivity over different spatial and temporal scales is fundamental for biodiversity conservation and management. The Amazonian rainforest, one of the most diverse biodiversity hotspots, has experienced dramatic range contractions and expansions due to Pleistocene climate oscillations, and its human-induced fragmentation has accelerated at an unparalleled pace in the course of the Anthropocene. In this context, epiphytes, with their relatively short life-cycles, offer an ideal model to investigate the impact of past and present fragmentation on patterns of genetic structure and diversity. Due to the necessity to switch from one host tree to another, or from one leaf to another, epiphytic bryophytes typically exhibit high dispersal syndromes. In line with such high dispersal capacities, recent metacommunity analyses have arisen the intriguing question that Amazonian epiphytic bryophyte communities are homogeneous across very large spatial scales, ultimately raising the notion that they might behave as a basin-wide panmictic population. Here, we implement fine-scale population genetic analyses to address the following questions: (i) Do Amazonian epiphytes exhibit population structure at regional (< 500 km) scale; (ii) If the hypothesis of a panmictic population is rejected, (iia) at which spatial scale does genetic structuring occur, and (iib) do neutral (isolation-by-distance) or ecological (isolation-by-ecology) processes shape patterns of genetic variation? We sampled exemplars of 15 epiphytic bryophyte species from two ecologically contrasted forest types (lowland rainforest and white-sand forest) in a 50,000 km^2 area in the middle Rio Negro. Genome-wide genetic data were produced using Genotyping By Sequencing. To circumvent severe taxonomic issues in challenging groups, which, like the Calymperaceae, are dominant in the epiphytic flora, we first implemented species delimitation analyses to sort-out specimens taxonomically. We then described the fine-scale genetic structure of each species and performed isolation-by-distance analyses to detect significant spatial genetic structuring. We finally determined whether isolation-by-distance or ecological filtering contribute to the observed patterns of genetic variation. The study will provide key information on the populations dynamics of highly mobile species integral to the iconic Amazonian forest, which may further be employed to refine future conservation policies in the face of accelerating climate change and anthropogenic-mediated deforestation.

T2

P0472

Dynamic evolution of the chloroplast IR borders in Cyperaceae

Bora Lee, *Sangtae Kim*
Sungshin Women's University

Cyperaceae is the third largest family in monocots, comprising ca. 5,000 species in the world. Our previous study, which is the first report of the chloroplast genome sequence in Cyperaceae (*Carex siderosticta*), showed the potential of the IR border regions in the phylogenetic study including DNA barcoding in the family. We explored the chloroplast IR border sequences in 84 Korean Cyperaceae species to test the phylogenetic utility of these regions. We sequenced four IR border regions (B1 for LSC/IRa, B2 for IRa/LSC, B3 for LSC/IRb, and B4 for IRb/SSC) and determined the IR borders by comparison of two complementary IR sequences. Surprisingly, in some cases, there are "transitional zones" in two IRs of which sequences are slightly different. We hypothesized two different scenarios to explain the expansion/reduction process for the IR regions: 1) gradual and random substitutions in both IRs and 2) deletion/insertion of a region of sequence in one IR. We found cases showing both scenarios in our study, indicating that the evolution of the IR borders in Cyperaceae is dynamic and is not explained by only one scenario. Phylogenetic analyses of four IR border regions showed similar clades as previous phylogenetic studies and provided 56.7% species resolution for DNA barcoding. Plotting of the expansion/reduction status of the IR borders on the phylogenetic tree addressed the evolutionary pattern of IR border changes in the process of diversification. Clade-specific expansions and reductions were detected in the tree. For example, expansion of 504 bp in B2 border was found in a clade of *Trichophorum, Eriophorum, Rhynchospora,* and a part of *Scirpus.*

T2

P0473

Reticulate evolution in *Asplenium anogrammoides* complex from Korea

Chang Shook Lee[1], *Kyong-Sook Chung*[2], *Sung Hee Yeau*[3], *Atsushi Ebihara*[4]
1. Department of Science Education, Ewha Womans University, Seoul 03760, Korea
2. Department of Medicinal Plant Science, Jungwon University, Goesan 28023, Korea
3. Department of Science Education, Ewha Womans University, Seoul 03760, Korea
4. Department of Botany, National Museum of Nature and Science, Tsukuba 305-0005, Japan

Asplenium anogrammoides has been hypothesized as the hybrid between *A. sarelii* and *A. tenuicaule* (allotetraploid; Lin & Viane, 2013). A recent cytological study revealed the ploidy level of *A. sarelii* and *A. tenuicaule* as a diploid (2x, n=36). *A. sarelii* (Doldamgosari in Korean) has been recognized in Korea, but it was proposed that some populations of the species should be verified as *A. agnogrammoides* (4x, n=72; counted only from Japanese populations) according to the taxonomic scheme by Lin & Viane (2013). The natural hybrid taxa (autoploidy and/or allopolyploidy) of *Asplenium* occur often in sympatric regions with putative parental species. *A. pekinense* is an autotetraploid of *A. sarelii.* In Korea, great morphological variations in *A. anogrammoides* have been found within populations. Based on their sympatric species and morphological characters, we hypothesize that the hybrid *A. anogrammoides* has undergone additional hybridization events with *A. pekinense*, *A. sarelii*, or *A. tenuicaule*, resulting the hybrid taxa with a series of polyploidy levels from diploids to hexaploids. To evaluation the taxonomic delimitations and phylogenetic relationships of *A. anogrammoides* and relative taxa; and estimate the hybridization origins of them, we present the results of phylogenetic analyses of morphological, chloroplast DNA sequence, and DNA content (flow cytometry) data; and discuss reticulated hybridization processes in the *Asplenium anogrammoides* complex (4x and 6x).

T2

P0474

Reevaluation of two new taxa of *Hypodematium* (Hypodematiaceae) from Korea

Chang Shook Lee[1], *Kanghyup Lee*[2], *Kyong-Sook Chung*[3], *Sung Hee Yeau*[4]
1. Department of Science Education, Ewha Womans University, Seoul 03760, Korea
2. Korea National Arboretum, Pocheon 11186, Korea
3. Department of Medicinal Plant Science, Jungwon University, Goesan 28023, Korea
4. Department of Science Education, Ewha Womans University, Seoul 03760, Korea

About 16 species of the genus *Hypodematium* Kunze are distributed in limestones or rock crevices of subtropical and temperate regions of Asia and Africa with high species diversity of 12 species in China (8 endemic species). In Korea, it has been known only one species, *H. glanduloso-pilosum.* However, we have recently found one new taxon and one unrecorded taxon of the genus *Hypodematium* on rocks at a limestone mountain in Okcheon and Yŏngwŏl, Korea, respectively. For the new taxon, *Hypodematium angustifolium* C.S. Lee & K. Lee is newly given based on its morphology. Its morphology is most similar to *Hypodematium sinense* known from Shantung, north China. In addition, *Hypodematium squamuloso-pilosum* Ching is newly found in Korea, which is widely distributed in only China as an endemic plant. Traditionally, these taxa were often distinguished by several vegetative characters such as pinnatifid lamina, glandular hairs, dense or rare long white hairs or scales on stipes and rachis, and longer and narrower or linear lanceolate scales on stipe bases. As an ongoing attempt to determine the phylogenetic status between two new taxa in Korea and related taxa of *Hypodematium*, we have extensively analyzed two new and relative taxa of *Hypodematium* populations from China, Japan and Korea. The cpDNA phylogenetic analyses and flow cytometry data suggest that *H. angustifolium* and *H. squamuloso-pilosum* are genetically distinct species in a diploid (2x) clade.

T2

P0475

Evidence for narrow species concepts in the subcosmopoli-

tan leafy liverwort genus *Lejeunea*

Gaik Ee Lee[1], *Julia Bechteler*[2], *Matt Renner*[3], *Jochen Heinrichs*[2]
1. University of Malaysia Terengganu, Malaysia
2. Ludwig Maximilian University, German
3. Royal Botanic Gardens and Domain Trust, Sydney

The recognition of all extant plant species is an ongoing challenge despite remarkable progress since Linnaeus. Liverworts are a prime example illustrating difficulties in establishing reliable species circumscriptions, reflected in strongly deviating estimates of the global species diversity. Estimating liverwort species richness may be complicated by their rather conservative morphology and limited range of taxonomically relevant characters. Nevertheless, these characters are required in taxonomic research relying on the identification of diagnostic morphological differences. Molecular investigations have often identified complex genetic structuring within morphologically rather uniform liverworts, suggesting the frequency of morphologically cryptic species is higher than currently appreciated. Molecular phylogenies have also enabled reconsideration of morphological evidence, and have enabled researchers to identify out subtle morphological differences between species that were earlier united under a single binomen. The subcosmopolitan genus *Lejeunea* includes between "more than 100" and 300 species, and is well known for its extensive morphological homoplasy. Molecular data are therefore essential for reconstructing relationships between taxa and deriving robust species hypotheses that reflect actual diversity. In the framework of an ongoing taxonomic revision of *Lejeunea* in Asia, morphology has been critically analyzed and revised within the context of an integrative (molecular-morphological) approach. The investigation has led to a deeper understanding of some of morphologically vague and controversial species complexes and also identified several new species. The newly produced data point to narrow species concepts and suggest that species diversity within *Lejeunea* is likely closer to higher estimates, than to conservative estimates that consider species to be morphologically (and ecologically) variable.

T2

P0476

Systematic study of *Chrysosplenium* L. series *Pilosa* Maxim. (Saxifragaceae)

Yong-In Kim[1], *Seong-Hyun Cho*[1], *Jung-Hoon Lee*[2], *Bo-Yun Kim*[2], *Roman V. Doudkin*[3], *Young-Dong Kim*[2]
1. International Biological Material Research Center, Korea Research Institute of Bioscience and Biotechnology
2. Multidisciplinary Genome Institute and Department of Life Science, Hallym University
3. Botanical Garden-Institute FEB RAS

Chrysosplenium series *Pilosa* Maxim. consists of 11 to 15 taxa, mainly distributed in NE Asia. Many species of the series are known to exhibit severe morphological variations, which led to proposals of many intraspecific taxa and synonyms. To address taxonomic issues pertaining to series *Pilosa*, molecular phylogenetic trees were reconstructed for 11 taxa using nuclear (ITS) and plastid (*psb*A-*trn*H, and *psb*K-I) sequence data. A total of 116 accessions representing the ingroup taxa was collected from China, Japan, Korea, Russia and Taiwan, covering whole distributional range of the series. For some problematic taxa, habit of sterile branches and micro feature of seeds were critically reexamined as well. In general, results from the ITS data was turned out to be useful in the species delimitation of the examined taxa, while the trees based on plastid sequences did not agree with species tree. Lacking the evidence for hybridization among the species, it was postulated that the incongruence between the plastid sequence trees and species delimitation was due to incomplete lineage sorting of the plastome. Several taxa, which have been treated conspecific to other taxon, were found to form distinct clades, respectively. For example, *C. villosum* has long been treated as a synonym of *C. valdepilosum* (=*C. pilosum* var. *valdepilosum*). However, multiple accessions of *C. villosum* collected from its type locality formed a distinct clade, well segregated from a *C. valdepilosum* clade. Likewise, Korean populations of *C. sphaerospermum* (formerly recognized as *C. barbatum*) formed a distinct clade, segregated from Japanese populations of *C. sphaerospermum*. In addition, *C. album* var. *stamineum* was turned out to merit regaining of its species status as it exhibited a sister group relationship with *C. pseudopilosum* rather than with *C. album* var. *album*. In this study, we detected at least four additional novel lineages from Russia and Korea, and some of the lineages were found to have diagnostic morphological characters, especially formerly overlooked feature such as habit of sterile branch. Our data indicated that the morphology of sterile branch at after fruiting as well as seed coat is important for the species delimitation within the series.

T2

P0477

Taxonomic identity of *Astilbe koreana* (Saxifragaceae) based on EST-SSR data

Jung-Hoon Lee[1], *Yong-In Kim*[2], *Seong-Hyun Cho*[2], *Bo-Yun Kim*[1], *Qiang Fan*[3], *Young-Dong Kim*[1]
1. Multidisciplinary Genome Institute and Department of Life Science, Hallym University, Chuncheon
2. International Biological Material Research Center, Korea Research Institute of Bioscience and Biotechnology, Daejeon
3. State Key Laboratory of Biocontrol and Guangdong Provincial Key Laboratory of Plant Resources, Sun Yat-sen University

Astilbe (Saxifragaceae) is a well-known herbaceous genus including various ornamental taxa such as *A. chinensis* and *A. japonica*. *A. koreana* is a Korean endemic species, which can be exploited as an additional source of garden cultivars. The species is commonly distributed in central and southern part of Korean peninsula. Its taxonomic identity, however, has been questioned by various authors due to the paucity of unambiguous morphological and molecular characters distinguishing the species from its closely allied taxa. Currently, *A. koreana* is treated as a synonym of *A. grandis*, *A. chinensis*, and *A. rubra* var. *divaricata* (*A. rubra* sensu Hara), depending on taxonomists. Although recent molecular phylogenetic studies based on ITS data indicate that *A. grandis* is the most distinct species among the four taxa, the relationships and taxonomic identity of the rest species are yet to be clarified. We have developed EST-SSR markers to examine more detailed phylogenetic relationships between *A. koreana* and its three relative taxa. Using the transcriptome data of *A. koreana*, a total of 20 EST-SSR loci, each consists of 2 to 7 alleles, were successfully selected. A total of 199 accessions from 10 populations, representing *A. koreana* (100 accessions - central and southern Korea), *A. chinensis* (20 accessions - China, Jilin; 20 accessions - Russia, Primorsky Krai), *A. rubra* var. *divaricata* (19 accessions - central Korea), and *A. grandis* (20 accessions - China, Hunan; 20 accessions China, Guangdong) were employed to examine the population differentiation and phylogenetic relationships. As a result of genotyping for the 20 loci, *A. koreana* and *A. grandis* turned out to be tetraploid species while *A. chinensis* and *A. rubra* var. *divaricata* were confirmed as diploid taxa. The UPGMA dendrogram based on Nei's genetic distance clustered the 10 populations into two main groups consisted of *A. grandis* populations and populations of the rest three taxa, which is well congruent with previous ITS trees. All accessions of *A. rubra* var. *divaricata* and *A. chinensis* populations shared most of the alleles, being clustered as a single group in Discriminant Analysis of Principal Component (DAPC). The total number of alleles amplified from 20 loci of *A. koreana* was 105 (3 to 9 alleles per locus), among which 52 alleles (1 to 6 alleles per locus) observed from 18 loci were the ones exclusively found within the species. This suggests that *A. koreana* might be a distinct evolutionary unit, and not be originated by direct autopolyploidization of its relative taxa. On the other hand, two populations of *A. grandis* formed two distinct geographical clusters in DAPC, with high degree of genetic differentiation. The clustering result is also in accord with our expanded ITS tree, in which the representative accessions of the geographic clusters positioned in two different clades. These results suggest that utility the of newly developed EST-SSR markers can be extended to address taxonomic issues of neighboring taxa as well as *A. koreana*.

T2

P0478

A phylogenetic study of the genus *Scutellaria* based on multiplex amplicon sequencing with the NGS platform

Yoonkyung Lee, *Sangtae Kim*
Sungshin Women's University

Scutellaria L. is the second largest genus of the Lamiaceae, with approximately 468 recognized species. The current classification system based on the morphological study contains seven sections placed in two subgenera, subgenus *Scutellaria* and subgenus *Apelthanthus*. Twelve morphological species-groups have been suggested in the section *Scutellaria*, the largest section in the genus (ca. 240 species), without any taxonomic rank. Molecular phylogenetic studies to test the monophyly of each subgroup in the classification system have not yet been performed. Advances in Next Generation Sequencing (NGS) and in high-throughput sequencing of multiple amplicons using dual-index sequences have enabled us to perform quick and cost-effective phylogenetic studies. In this study, we selected 224 taxa (198 species) representing each subgroup of *Scutellaria,* which account for approximately 42.3% of the recognized species in *Scutellaria*. Ten fast-evolving DNA regions in the chloroplast genome (*ccsA*, *matK*, *ycf4~cemA*, *ndhG*, and *ndhA*) and in the nuclear genome (ITS, ETS, *LEAFY*, *XDH*, and *APETALA3*) were selected based on the analysis of the three previously reported chloroplast genomes (*Tectona, Ajuga, Premna*) and phylogenetic studies in Lamiaceae. Amplicons from each taxon were labeled with seven bp dual-index sequences. Approximately 2,500 amplicons were combined and sequenced at once using the MiSeq platform. The result showed the first molecular phylogenetic relationship among species of *Scutellaria* and it will be the basis for a new classification system in the genus.

T2

P0479

Different drivers of originations and extinctions in land plants

Samuli Lehtonen[1,2], *Daniele Silvestro*[3,4,5], *Dirk Nikolaus Karger*[2,6], *Christopher Scotese*[7], *Hanna Tuomisto*[2], *Michael Kessler*[6], *Carlos Peña*[2], *Niklas Wahlberg*[2,8], *Alexandre Antonelli*[3,9]
1. Biodiversity Unit, Herbarium, FI-20014 University of Turku
2. Department of Biology, FI-20014 University of Turku
3. Department of Biological and Environmental Sciences, University of Gothenburg, Carl Skottsbergs gata 22B, Gothenburg 413 19
4. Department of Ecology and Evolution, University of Lausanne, 1015 Lausanne
5. Swiss Institute of Bioinformatics, Quartier Sorge, 1015 Lausanne
6. Department of Systematic and Evolutionary Botany, University of Zurich, 8008 Zurich
7. Earth and Planetary Sciences, Northwestern University, Evanston, IL
8. Department of Biology, Lund University, Lund
9. Gothenburg Botanical Garden, Carl Skottsbergs gata 22A, Gothenburg 413 19

Ferns are the second most diverse major plant radiation with over 400 million years of evolutionary history. The fern fossil record is rich and phylogenetic relationships of living ferns are well understood. It has been suggested that the rise of angiosperms during the Cretaceous dramatically affected the fern evolution, either by causing major extinctions as inferred from the fossil record, or resulting in massive adaptive radiation of epiphytic fern lineages. This background makes ferns an ideal group to study macroevolutionary questions, such as the roles of biotic and abiotic factors in diversification through time. In this study, we modelled the changes in fern biodiversity and estimated their origination and extinction rates from the fossil record. We tested whether variation in these rates could be the result of interactions with other plant groups, responses to climate change or other environmental variables, and geological events. We developed and implemented a Bayesian model that jointly estimates 1) the factors that significantly correlate with the diversification dynamics of ferns, 2) whether these factors affect positively or negatively on the origination and extinction rates, and 3) the strengths of each correlation coefficient. Our model indicates diversity-dependent origination rate and environmentally driven extinctions. Angiosperm diversification is insignificant factor in our model, thus contrasting the previous hypotheses of fern diversification drivers. Different taxonomic lineages have succeeded each other as dominant fern groups. We investigated whether lineages that have passed their 'golden age' are less adaptive than currently expanding lineages by modelling the rate of niche evolution over the phylogeny of living species. Despite the dramatic diversification rate shifts observed in the phylogeny, we found no evidence that diversification rate changes correlate with rate of adaptation. We conclude that origination appears to be a largely neutral process unlinked with adaptation, and that environmentally driven extinctions are the main drivers of diversity dynamics.

T2
P0480
Phylogenetic placement of the enigmatic groups within cyperid tree of life

Viktorie Brozova[2], ***Lenka Zaveska Drabkova***[1]
1. Institute of Experimental Botany and Intitute of Botany, the Czech Academy of Sciences
2. University of South Bohemia in Ceske Budejovice

Cyperids encompass around six thousand species from three main families: Cyperaceae, Juncaceae and Thurniaceae (incl. Prionicaceae). It has a primarily pantropical distribution originated in West Gondwana, South America and Africa. Previously, infrageneric classifications have been proposed but none has been based on molecular phylogenetic evidence in this complete view. Moreover, most studies have treated genera from only one geographical region. In this study, chloroplast and nuclear regions were employed from more than thousand species of cyperids covered all infrageneric groups from whole distributional range. *Rbc*L sequence data resolved a part of the supraspecific phylogeny, but many clades remain polytomic (e.g. tribe Cypereae, genus *Luzula*). For this reason, the noncoding regions were included (*trn*L intron, *trn*L-F intergenic spacer and ITS1-5.8S-ITS2). We intended to create hypotheses of relationships among families and within them and to test the classification of the subgenera and sections, but a primary goal was finding stable phylogenetic positions of some taxa or groups with enigmatic relationships (*Juncus capitatus*, genus *Cladium*). Furthermore, we tested the influence of different rooting and ingroup composition on the tree topology. All analyses (MP, ML and Bayesian) revealed several well-supported lineages (genera *Luzula, Cladium, Rhynchospora, Eleocharis* and few more). Additionally, the phylogenetic tree was dated with the Bayesian uncorrelated relaxed molecular clock incorporating cyperid fossil-based calibration. We will discuss the traditionally distinguished taxonomic groups of cyperids in the light of molecular data.

T2
P0481
Evolution of reproductive isolation in outcrossing *Arabidopsis* species

Johanna Leppälä[1], *Martin C. Fischer*[2], *Alex Widmer*[2]
1. Division of Plant Biology, Department of Biosciences, University of Helsinki
2. Institute of Integrative Biology, Department of Environmental Systems Science, ETH Zurich

A major interest in speciation research is to identify reproductive barriers that arise early during divergence, rather than barriers that have evolved after reproductive isolation is complete. Therefore, we have focused on examining post-pollination and intrinsic postzygotic reproductive barriers between outcrossing, perennial *Arabidopsis* species, *A. arenosa*, *A. halleri* and *A. lyrata*, as well as between populations of *A. lyrata*. These three *Arabidopsis* species form genetically distinct entities, but natural hybrid zones and polyploid hybrid species exist. As reproductive isolation is usually formed jointly by several incomplete pre- and postzygotic barriers, we have estimated the strength of these individual barriers, starting from pollination barriers between species all the way to F_1 hybrid fertility. We found that within species, the first barrier that evolves is hybrid male fertility, whereas between species a strong seed developmental barrier exists, in addition to hybrid male fertility reduction. The correlation between genetic distance and the strength of the different postzygotic reproductive barriers is variable. We emphasize that reproductive barriers both within and between outcrossing *Arabidopsis* species are not fixed but instead are highly polymorphic and are thus ideally suited to study the evolution of reproductive isolation in plants.

T2
P0482
Evolutionary modes and trends during origination of angiosperms

Igor Levichev
Komarov Botanical Institute of the Russian Academy of Sciences

Goal of research. To complement aspects of the telome theory of Walter Zimmermann with new results of comparative studying of the shoot structure in diverse groups of *Tracheophyta* and illustrate key aromorphoses of evolution using contemporary models, schemes and paleobotanic reconstructions.
Methods. The author used a set of techniques providing side-by-side comparison of modular units in the structure of sions of diverse taxonomic groups of contemporary and fossil plants with structural organization of sporophyte and gametophyte, pioneers of land of the Silurian period. Rather than ascertaining difference of forms and functions, novelty was provided by searching for similarity and continuity. Comparison was made not from the perspective of telome (an obvious structural ancestor), but its immediate derivative – mesome. More specifically, from the perspective of bipolar mesome and its historically later superposition and juxtaposition unit associations in metameric and colonial soma with individually different structural and modular organization within taxonomical groups. Another methodologic pool of original and new concepts was the system approach to institute of recapitulations within the Haeckel's basic biogenetic law and neotenic divergence. The stopping of organ morphogenesis at an early stage of intrabud development (which corresponds to embryonic condition of any ancestor) and extrabud grandization of such embryonic organ up to the actual sizes of the descendant's soma is the essence of neoteny in plants. The thesis of "Evolution is penetrated by neoteny" belongs to A. Takhtajan and it was not adequately understood by contemporaries. Material for study was mainly provided by live models in the nature and culture.
Results and discussion. On the land, in the support-free air environment, it was not telome but mesome that underwent evolution, and quantitative and qualitative interfaces of mesome based on their ageotropic and geotropic structures. Transformation of a 5-axis mesome with an equal dichotomy up to extreme manifestation of dichopodiality ("monopodium"), at obligatory right or left rotation of axes of each metameric module, reliably ensured vertical stability and 7-dimensional capture of individual scope

of habitat (on the land, in air and in the course of time). The ancestral bipolarity of telom went over to mesome and successively ensured the evolutional freedom of its simpodiya, *etc.* transformations during formation of the root, stele, phyllomes of various nature, geotropic branch of the sympodium, *etc.* At the same time, mesome bipolarity brought about the unprecedentedly ample opportunities for protection of bud and reproductive structures. It was preceded by emergence of the true leaf from a single lateral sporangium. Emergence of dimerism in angiosperms (coupling of ageotropic monopodium and geotropic sympodium in a single metamer) was an outstanding evolutional aromorphosis. In monocotyledonous plants, the superposition organization of innovation shoot from dimeric modules is outwardly not different from mesome structure of pioneers of land, but has different evolution history and specialization. Juxtapositional coupling of two dimer has signaled the emergence of dicotyledonous plants with their most perfect protection all growth localizations, which ensures their present domination on the planet. That said, ovules in both monocotyledonous and dicotyledonous plants always have 2 integuments. This, as well as many other structures (stipules, bicarinate prophyll of sympodium, 2 circles of tepals, intrasheath scales, *etc.*) demonstrate recapitulation of dimer to the embryonic and preceding ancestral fusion of "monopodium" and sympodium. Many of contemporary *Araceae* tend to form, on a single internode, 2 laminas, at a single lateral bud. Perhaps, this paves the way for transition to a new evolutionary level which ?an be notionally named "Gorgonoid angiosperms" in which two dimeric phytomers are consistently formed on a single shoot segment (on a common "podium"). Empirical, inflorescences of the gorgonian type consist of three and from many flowers.

T2

P0483

Molecular evidence for natural hybridization between *Cotoneaster dielsianus* and *C. glaucophyllus*

Mingwan Li[1], *Sufang Chen*[1], *Renchao Zhou*[1], *Qiang Fan*[1], *Feifei Li*[2,3], *Wenbo Liao*[1]

1. Guangdong Key Laboratory of Plant Resources and Key Laboratory of Biodiversity Dynamics and Conservation of Guangdong Higher Education Institutes, Sun Yat-sen University, Guangzhou 510275, China

2. State Key Laboratory of Environmental Criteria and Risk Assessment, Chinese Research Academy of Environmental Sciences, Beijing 100012, PR China

3. College of Life and Environmental Sciences, Minzu University of China, Beijing 100081, PR China

Hybridization between distinct lineages accompanied by polyploidization has variously been viewed as a driving force of evolution and speciation in many plants, particularly in the family of Rosaceae. Interspecific hybrids can form isolated evolutionary units, especially when mechanisms increasing viability and fertility, like polyploidy and apomixis, are involved. The genus *Cotoneaster* is a good example to study the evolutionary process of hybridization associated with both polyploidy and apomixes, which has tetraploid taxa as high as about 70% including most of the apomictic taxa. Despite *Cotoneaster* hybrids are common in cultivation, nonetheless, no sufficient genetic evidence is provided for natural hybridization occurred in this genus, inevitably, relevant mechanism seems far from proven. In this study, we collected all the four *Cotoneaster* species (putative hybrid, *Cotoneaster dielsianus*, *C. glaucophyllus*, and *C. franchetii*,) distributed in Malipo county (Yunnan, China) and five low-copy nuclear genes and six chloroplast DNA fragments were sequenced. Additionally, the ploidy level of each taxon was estimated indirectly by flow cytometry technique, and the results showed that only *C. glaucophyllus* is diploid whereas other three taxa are tetraploidy. *C. dielsianus* and *C. glaucophyllus* were well separated with a total of 36 fixed nucleotide substitutions and one fixed 5-bp indel in the five nuclear gene data sets. All individuals of the putative hybrid showed chromatogram additivity at these 36 fixed sites. In the haplotype analysis, each taxon exhibited a low level of haplotype diversity, and no more than three haplotypes were observed at each of the five genes, and the two haplotypes in putative hybrid at each of five nuclear genes were shared with those of *C. dielsianus* and *C. glaucophyllus*, and no one was unique to putative hybrid. The haplotypes of *C. franchetii* were well separated from those of the three other taxa. Among the six chloroplast regions, *C. dielsianus* and *C. glaucophyllus* were well separated and individuals of hybrid were always identical with *C. glaucophyllus*. Our study provided convincing evidence for natural hybridization between *C. dielsianus* and *C. glaucophyllus*, and all the hybrid individuals were likely F1s. The hybridization was unidirectional and *C. glaucophyllus* was the maternal parent of the hybrid. Combining with the overlapping geographic range and the same flowering periods between biparental individuals and hybrids, we inferred the hybridization followed by polyploidization and produced one F1 individual with reproduction through apomixis. Therefore the genetic isolation might give rise to maintain species integrity between hybrids and the both parents despite hybridization. Disturbance of local habitats was deduced to be the major factor promoting hybridization. Given the small size of hybrid population, conservation strategy has been designed and we will continue to observe the influence of habitat destruction in only hybrid zone we found.

T2

P0484

The life span of leaves in fern species of Taiwan

Pei Hsuan Lee[1,2], *Yao Moan Huang*[2], *Wen Liang Chiou*[2], *Ho Yih Liu*[1]

1. National Sun Yet-sen University

2. Taiwan Forestry Research Institute

Leaves of ferns could be vegetative organs and reproductive ones. Ferns can control the leaf lifespan by getting the return on investment and ensuring plant survival and population renewal. However, leaf lifespan is controlled by internal and external factors, including genes, sex, season and environment. The fern research team in Taiwan calculated leaf lifespan from measuring time intervals between emergence and senescence in the long term monitoring. The leaf lifespan about 30 species of Taiwan native ferns has been accumulated. It show different spans in different species, not exception in close-related species. The ratios of fertile/sterile leaves' lifespan (F/S) in dimorphic species (n=3) were 0.16 – 0.20 and those in monomorphic ones (n=27) were 0.76 – 1.82. The fertile leaves in the dimorphic species were 4 – 5 times shorter in life than the sterile ones, but it's almost the same between the fertile and sterile leaves of the monomorphic species. These situations mean that the fertile leaves of the dimorphy are senescent after spore release in very short time, and the ones of the monomophy are still alive for photosynthesis after spore release. The emerged seasons could affect the leaf lifespans, for example that the spans of sterile leaves of *Osmunda cinnamomea* were longer in the early emergence than in the late one. But the applicable species in this factor appears to be limited. Few studies said that the external environmental factors could be correlated with leaf senescence, so leaf lifespan could be affected by these factors, including in different climate regions. However, it's little evidences for these. In the future, the study of fern leaf lifespan needs to cooperate with phenology and ecophysiology, to understand the reasons of the expression of leaf lifespan.

T2

P0485

Phytologic exchange between India and Pacific Asia prior to the late Oligocene

Qijia Li[1], *Qiongyao Fu*[2], *Jianhua Jin*[2], *Cheng Quan*[1]
1. Jilin University
2. Sun Yat-sen University

The final continental accretion of India to Asia, namely the "hard collision", and the consequent bidirectional biotic exchanges between Gondwanan and low-latitude Oriental regions in the Cenozoic remain an open topic that attracts comprehensive investigations. However, due to the complex nature of tectonic evolution and the fact of lacking suitable indicators at large as conclusive evidence, the accurate time of final terrestrial connections, and the resulted biotic dispersals between India and Asia originally on the opposite sides of the Neotethys Ocean is still highly debated. Here we describe seeds of Annonaceae from the late Oligocene Yongning Formation of Guangxi province, southern China on the basis of ovate to elliptic seed shape, thicker seed testa, lamelliform rumination and the obvious cone-like plug, providing new evidences for botanic exchange between India and Pacific Asia based on fossil records. Furthermore, more than 7 family 8 genus botanic exchange betweenIndiaand Pacific Asia during that period have also been reviewed, nearly negating the possibility of these botanic exchange by random and infrequent dispersals across an ocean barrier. All these biogeographical evidences indicate that the final terrestrial connection which simultaneously triggerred massive terrestrial botanic migrations and invasions from the Gondwana to low-latitude Asia, and vice-versa occurred prior to the late Oligocence.

T2

P0486

Evolution and historical biogeography of Zingiberaceae -- From the perspective of chromonsome evolution

Jian-Li Zhao, ***Qing-Jun Li***
Laboratory of Ecology and Evolutionary Biology, State Key Laboratory for Conservation and Utilization of Bio-Resources in Yunnan, Yunnan University

Zingiberaceae is a pantropical monocot family across Asia, Africa and America, including Alpinioideae (~20 genera), Zingiberoideae (~29 genera), Siphonochiloideae (2 genera) and Tamijioideae (1 genus). Around 95% genera of Zingiberaceae grow in Southeast Asia. Reconstruction of evolution and historical biogeography of Zingiberaceae would be a key to understand the evolution of biodiversity in tropics, especially in paleotropics. Here we reconstructed the phylogeny, historical biogeography and Chromosome number evolution of Zingiberaceae. Alpinioideae has seven types of chromosome counts (90% is 2n=48) and Zingiberoideae has 29 types of chromosome counts (proportion of the most one is less than 16%). Our results indicated Zingiberaceae originated at Africa during the Late Cretaceous. Ancestor of Asian gingers drifted to Asia with India plate from the Upper Cretaceous to early Paleocene. Divergence between Alpinioideae and Zingiberoideae occurred during the Late Paleocene and diversification of Alpinioideae and Zingiberoideae both started during the Middle Eocene respectively. These two events coincided with two similar climate optimums before/after temperature peak during the Early Eocene. Ancestral basic chromosome numbers of Zingiberaceae, Alpinioideae and Zingiberoideae were likely n=12, n=24 and n=11 respectively, indicated ancestor of Alpinioideae underwent a palaeopolyploidization event and ancestor of Zingiberoideae underwent a chromosome loss event when these two subfamilies separated from each other. Ancestor of *Aframomum* (Africa), *Renealmia* (Africa+America) and *Alpinia*-FAX clade (South India+Sri Lanka) came from Eurasian continent from late Eocene to early Oligocene. Divergence between *Renealmia* and *Alpinia*-FAX clade was likely caused by long-distance dispersal event. This research suggested divergence between Alpinioideae and Zingiberoideae may be driven by chromosome shift under past environmental change rather than geological events, diversifications of Alpinioideae and Zingiberoideae were two different evolutionary strategies that one type with polyploidization and diverse chromosome types respectively, *Alpinia*-FAX clade should be listed as a new genus based on its phylogenetic position and evolutionary history.

T2

P0487

The contributions of Chinese botanists to International Botanical Congress

Shuai Liao[1], *Cheng Du*[1], ***Jin-Shuang Ma***[1]
1. Shanghai Chenshan Plant Science Research Center, Chinese Academy of Sciences
2. Shanghai Chenshan Botanical Garden

The International Botanical Congress (IBC) is a global convention that brings together botanists from around the world to discuss scientific research about plants, particular on the *Code*. Since 1900 when first modern IBC held in Paris, France, it has been held nineteen times until now. The first time, Jing-Yue Zhang, a Chinese botanist, attend IBC, is in Ithaca, USA, 1926. Four years later, five Chinese botanists, Chong-Zhen Lin, Huan-Yong Chen [Woon-Young Chun], Jing-Yue Zhang, Ren-Chang Qin [Ren-Chang Ching], and Xing-Jian Si attended the V IBC in Cambridge, UK, 1930. Followed by another five years later, Huan-Yong Chen [Woon-Young Chun] and Ji-Tong Li attended the VI IBC in Amsterdam, Netherlands, 1935. In the next 46 years, Chinese botanists are absent from the VII IBC to the XII IBC for various reason. Since 1970s when China opened their door to the world, the Botanical Society of China sent more than 40 delegates from China to the XIII IBC in Sydney, Australia, 1981. In the past three decades, Chinese botanists paid great enthusiasm to every IBC, from Berlin, Tokyo, Saint Louis, Vienna to Melbourne, with hundreds of delegates. Today, Chinese botanists, with great honor, host the XIX IBC at their home "China: Mother of Gardens". As an important part of the IBC, Nomenclature Section, is organized to discuss and revise (if necessary) the *International Rules/Code of Botanical Nomenclature*, which was renamed as *International Code of Nomenclature for algae, fungi, and plants* (ICN) for better governance of the scientific naming of algae, fungi, and green plants. The formal set of rules for botanical nomenclature is Alphonse de Candolle's *Lois de la Nomenclature Botanique* (Laws of Botanical Nomenclature), which was published in 1867 for the fourth annual IBC in Paris, France, 1867. Since then, 15 official editions of the *Rules/Code* were published for the IBC. As the importance of the *Rules/Code*, several official editions were translated by Chinese botanists for better acquirement and utilization. In order to show the concentration and contribution of Chinese botanists on the *Rules/Code*, here we present the Chinese translation made by Chinese botanists from beginning as below. Details of the Chinese delegates in IBC and full Chinese translation record of the *Rules/Code* can be found in the poster.

T2

P0488

Relicts and refugia: Evidences from some important mountains in the east China mainland, monsoon climate zone

Qiang Fan[1], *Wei Guo*[2], *Wanyi Zhao*[1], *Yunsheng Song*[1], *Jianhua Jin*[1], ***Wenbo Liao***[1], *Hungta Chang*[1]

1. State Key Laboratory of Biocontrol and Guangdong Key Laboratory of Plant Resources, School of Life Sciences, Sun Yat-sen University, Guangzhou, China

2. Department of Horticulture, Zhongkai Agricultural Engineering College, Guangzhou, China

The relict plants can be divided into taxonomic relict species and geographic relict species depending on systematic and geographic characteristics. The relict plants are of great significance in revealing the antiquity of a flora. According to the phyto-geographic theory, the differentiation of relict plants combined with fossil evidence and paleo-geographic features can be used to determine migration routes following the Pleistocene ice ages and to infer the location of refuge. The Chinese seed plant flora contains a great number of ancient gymnosperms and angiosperms, including many basal groups, and many typical relict genera and/or species. First, according to an analysis of 3,200 genera of Chinese seed plant flora, it shows that 162 families and 345 genera and about 943 species was considered as relict genera and/or species, which accounted for 10.78% of the total genera and 3.60% of the total species, including 9 families, 31 genera and 132 species of gymnosperms and 153 families, 314 genera and 811 species of angiosperms. Second, among above 345 relict genera in China, 228 genera belonged to 87 families were discovered owning fossil records, and analyzing by combined with the different geological age and the geologic climate events, those relict genera could be considered that there was 4 important evolution stages, i.e. Cretaceous period, Paleocene-Eocene period, Oligocene-Pliocene period and Pleistocene-Holocene period. Third, based on the analysis of floras, the relict genera and species of several important mountains in China was revealed as following: Luoxiao mountain (LXm) has 142 relict genera and 239 relict species, Wuyishan mountain (WYm) has 117 genera and 169 species, Nanling mountain (NLm) 113 genera and 166 species, Wuling mountain (WLm) 126 genera and 206 species, Mount Shennongjia (SNJ) 100 genera and 149 species, Mount Emeishan (EMS) 99 genera and 141 species, Qinling mountain (QLm) 104 genera and 139 species, Taiwan mountains (TWm) 90 genera and 122 species, Hainan mountains (HIm) 89 genera and 130 species, *etc.* Fourth, as might be expected from the above numbers, the East China mainland, a monsoon climate region also preserves a great of relict genera, and relict plant communities composed by some important relict species or populations such as *Cathaya argyrophylla*, *Abies beshanzuensis* var. *ziyuanensis*, *Amentotaxus argotaenia*, *Cunninghamia lanceolata*, *Fokienia hodginsii*, *Pseudotaxus chienii*, *Taxus wallichiana* var. *mairei*, *Tsuga chinensis*, *Disanthus cercidifolius* subsp. *longipes*, *Exbucklandia tonkinensis* and *Cyclocarya paliurus etc.* So, judging from the present abundant biodiversity and great number of relict plants, some typical mountains of east China mainland, such as Wuling mountain, Nanling mountain, Wuyi mountain, Luoxiao mountain, *etc.* could be considered as the important refuges during the Tertiary and Quaternary ice age.

T2

P0489

DNA barcoding: Genus-level reference library for Korean vascular plants

Ji-Eun Choi, Se-A Ryu, Byoung-Yoon Lee, Min Hwan Suh, ***Chae Eun Lim***

National Institute of Biological Resources

Since 2008, National Institute of Biological Institute (NIBR) have conducted a DNA barcode research of Korean indigenous species for the taxonomical studies of living species and the following purposes: (1) to secure evidence for the intellectual property right or ownership claim of living species; (2) to obtain evidence to determine new or unrecorded species; (3) to confirm food safety, the quality of ingredients, and the authenticity of purity certification; and (4) after judging that it has high applicability as objective and scientific evidence for the ecological, environmental diagnosis and evaluation. Up to date, we produced the DNA barcode information of 7,623 species, which consist 16.2% of the Korean indigenous species. In Particular, we obtained approximately 50% DNA barcode data of Korean vascular plants. Recently, there has been a rapid increase on the cases of misuse, mixed use, forge, and falsification of main biological industrial resources whose species are difficult to be identified. In the present study, we examined the *rbcL* sequences of 1,122 genera for construct the Genus-Level Reference Library for Korean vascular plants. In Korean peninsula, It has been reported that there are 222 Families, 1,155 Genera, and 4,338 taxa of vascular plants (Ministry of Environment, 2011). However, 33 genera, which are distributed only in North Korea, have been excluded in this study. According to the neighbor-joining analysis on 1,122 Korean vascular plants, they are clearly divided into 3 groups; pteridophyta, gymnosperm, and angiosperm by the sequence of *rbcL* region. The analyzed 1,122 taxa had a high resolution of >95% in the genus-level. Based on the results, *rbcL* is very useful to identify Korean vascular plants on the genus-level.

T2

P0490

DNA barcoding of Korean Gentianaceae

Su Min Han[1,2], *Hyosig Won*[2], ***Chae Eun Lim***[1]
1. National Institute of Biological Resources
2. Department of Biological Sciences, Daegu University, Gyungsan, Gyungbuk 38453

Family Gentianaceae is a species-rich plant group, and many species have been used in traditional medicine in Asia and Europe. Gentianaceae is consisted of ~1,700 species in 878 genera, eight subtribes and six tribes, and has worldwide distribution, mostly in the temperate regions and alpine habitats. In Korean peninsula, there are 25 taxa beloning to seven genera, two subtribes (Gentianeae and Swertiinae) of tribe Gentianeae. As a part of building DNA barcode identification system of Korean flowering plants, we have analyzed several DNA barcode regions of Korean Gentianaceae. In this study, we investigated four DNA barcode regions, chloroplast *rbc*L, *mat*K genes, and *trn*L-F IGS, and nuclear ribosomal ITS. The taxon sampling has been expended from herbarium of National Institute of Biological Resources (NIBR). Totally 66 accessions of 21 taxa in seven genera, two subtribes were sampled and sequence variation and characteristics were examined to discrimiate the species. Based on four DNA barcodes, all accessions were clearly distinguished at the subtribe and genus level; (1) subtribe Gentianinae; genus *Gentiana*, and *Tripterospermum* (2) subtribe Swertiinae; genus *Swertia*, *Anagallidium*, and *Halenia*. The chloroplast DNA barcode markers showed high PCR amplification efficiency and sequencing success (86.4~95.4%), while the nrITS the lowest sequencing success (72.2%). The degree of species resolution for each DNA barcode marker ranged from 53.3% (*trn*L-F) to 85.7% (ITS). Among four barcodes, nrITS provided the highest resolution power for species discrimination, although the fragment has the lowest sequencing efficiency. In conclusion, the result from nrITS indicated that each species has unique DNA sequence and the nrITS region can be used as a barcoding marker to distinguish Gentianaceae species in Korea.

T2

P0491

Comparative DNA sequence analyses of five *Hosta* (Asparagaceae) complete chloroplast genomes

Se-A Ryu, Byoung-Yoon Lee, ***Chae Eun Lim***
National Institute of Biological Resources

Hosta (ca. 23-45 species) is restricted to China (two species), Japan (recognized 18 species and seven varieties), Korea (recognized 11 species and two varieties), delimitation has been problematic due to a long history of cultivation, hybridization and selection, from the 18th century onwards. *Hosta* plants are herbaceous perennial, commonly used ornamentally or for medicine, and are thus widely cultivated in parks and botanical gardens throughout the region, know as common name 'Plantain Lily'. Over 100 species, varieties and cultivars of the genus *Hosta* have been discribed globally to date. Genus *Hosta* is taxonomically and phylogenetically regarede as one of the difficult taxa in plant. *Hosta* is related genera, including *Hemerocallis*, *Agave*, and *Yocca*, has been located in the Liliaceae, Agvaceae, or monogenertic Hostaceae (often erroneously called Funkiaceae, it is invalid) but according to molecular phylogenetic studies it has been suggested as Asparagaceae *s.l.*, in the oder Asparagales (APG III). The chloroplast genome sequences are useful genetic markers for DNA barcoding, transplastomic studies and evolutionary studies from the population level, as well as for phylogenetic relationships. Chloroplast genomes are usually 120-170kb in length approximately 100-130 genes are located along the circular genome structure of chloroplasts with a circular quadripartite structure involving two inverted repeat (IR) that divide the rest of the genome into large and small single copy regions. The structure and gene content of chloroplast genomes were relatively conserved in whole land plants with the exception of non-photosynthetic parasitic plants. At present, complete chloroplast genome of *Hosta ventricosa* have been available in NCBI GenBank. Here We report the sequence and characteristics of complete chloroplast genome of five *Hosta* species [*Hosta capitata*, *H. clausa*, *H. minor*, *H. venusta*, *H. yingeri.*, three of them (*H. minor*, *H. venusta*, *H. yingeri*) are as endemic distributed only in Korea] and to reconstruct phylogenetic relationships among the five *Hosta* species using their cp genome sequences. We here sequenced the complete chloroplast genomes of five *Hosta* species using Next-generation sequencing (NGS) data via a combination of *de nove* and reference-guided assembly. The five *Hosta* species chloroplast genomes is a typical double-stranded circular DNA with 156,417-156,756bp in length, exhibit the quadripartite structure common to most land plant chloroplast genomes with large single copy regions and small single copy regions separated by two copies of the inverted repeat. The five *Hosta* species have an identical set of 131 functional genes (85 protein coding genes, 38 tRNA and 8 rRNA genes). Based on comparative analyses of chloroplast genome in five *Hosta* species, *H. venusta* and *H. minor* has a similar to 99.99%, *H. capitata* and *H. yingeri* were confirmed that showed a similarity of 99.6%. Phylogenetic analyses using maximum likehood (ML) methods of a dataset composed of 18 monocotyledon chloroplast genomes.

T2

P0492

Effects of floral color dimorphism on reproduction of *Roscoea cautleoides* (Zingiberaceae)

Tial Cung Ling
Kunming Institute of Botany, Chinese Academy of Sciences

Floral color polymorphism within plant species is considered to be maintained by selective agents such as environmental factors, pollinators, as well as by random genetic drift. Although *Roscoea cautleoides* (Zingiberaceae) shows a sympatric appearance of yellow- and purple-flowered individuals, it is still unknown what are the factors leading to the maintenance of flower color dimorphism in this species. We studied the soil trace-elements, flowering phenology, and morphological characters. We tested the hypotheses that pollinators may discriminate and prefer one of the morphs by monitoring the pollinator's visitation rate and behavior, coupled with foraging time and pollen deposition. We also checked the breeding system and seed production in relation to the morph proportions in five natural sites in purpose to examine whether these proportions affect reproductive success. There were no significant differences in trace-elements, flowering phenology and pollinator visitation rate between the yellow and purple morphs, showing that they are completely sympatric and have a similar flowering season. However, flower density and sugar concentration of the nectar, were higher in the yellow flowers, while the purple flowers had larger ovule numbers and longer corolla tubes and floral lifespan than the yellow ones. The reflectance spectral analysis showed that pollinators could discriminate between the two morphs. The bumblebees, which had shorter tongued-lengths than that of moths, visited the yellow flowers, whereas moths preferred the purple flowers, and therefore their inter-morph visitation rate was significantly low. This result reveals that the selection of pollinator in *Roscoea cautleoides* is based more on flower color and size than on the nectar's concentration. Foraging time and pollen deposition of the bumblebees were significantly higher than that of the moths, showing that bumblebees may be more effective in pollen transfer than moths in this species. Bagged flowers in both color morphs show a low level of spontaneous self-pollination and produced few seeds, in which seed production of the yellow flowers in selfing was significantly less than that of the purple ones. However, reciprocal (inter-morph) hand-pollinated flowers set as many seeds as flowers hand-pollinated with outcross pollen (intra-morph but different flowers) or self-pollen (intra-morph from same flowers = geitonogamy), showing that a high capacity of seed production when the two color morphs are reciprocally pollinated. However, not only bagged and natural control flowers produced significantly fewer seed numbers than that of hand pollinations but also natural seed production in 2016 was significantly fewer than seed production in 2015 as a consequence of lower flower density and visitation rate in 2016. These results suggest that this species mostly depends upon pollinators for pollination. Nevertheless, the overall natural seed numbers of yellow flowers in all morph proportions were significantly higher than that of the purple flowers and showed neither negative nor positive frequency-dependent selection through pollinator preference. In summary, pollinator inter-morphs visitation between the two color morphs may maintain flower color dimorphism in this species.

T2

P0493

Plastids, ploidy and phylogeography of Chinese strawberries (*Fragaria*, Rosaceae)

Aaron Liston[1], *Shannon Straub*[2], *Luxi Chen*[3], *Gao Song*[3], *Yao-Bin Song*[4], *Junmin Li*[3], *Ming Dong*[4], *Tia-Lynn Ashman*[5]

1. Oregon State University

2. Hobart & William Smith Colleges

3. Taizhou University

4. Hangzhou Normal University

5. University of Pittsburgh

China is the center of diversity for the genus *Fragaria* (Rosaceae), with 13 species of wild strawberry. However, species delimitation has not previously been examined with population level sampling and molecular markers. Using published *Fragaria* chloroplast genomes, we determined that the *ndh*A intron is the most variable plastid sequence. This 1200 bp intron was sequenced in 2112 individuals from 148 populations of Chinese *Fragaria*. We identified 96 unique haplotypes. These sequences were aligned, insertions and deletions were scored as binary characters, and a haplotype network was reconstructed. In addition, flow cytometry was used to estimate ploidy in 439 of these individuals. Six putative species, five diploid and one tetraploid, could be diagnosed by one or more haplotypes. We intensively sampled the diploid/tetraploid species pair *F. pentaphylla*/*F. moupinensis* obtaining 1366 chloroplast *ndhA* sequences and 48 haplotypes. We also made 354 ploidy estimates, and found that 51% of the individuals were predicted to be tetraploids. Only four of the haplotypes had mixed diploids and tetraploid estimates, and thus most of the haplotypes were potentially diagnostic for ploidy. Our results indicate that *F. vesca* and *F. orientalis* are absent from southwestern China, in contrast to published taxonomic treatments. We confirmed that *F. viridis* is present in Xinjiang province, and we confirmed the phylogenetic distinctiveness of *F. nilgerrensis* and *F. daltoniana*. *Fragaria nubicola* is monophyletic in the haplotype network, separated from *F. pentaphylla*/*F. moupinensis* by a single mutation. We identified a clade of diploid individuals from Shaanxi province that possess a unique set of haplotypes; we provisionally consider this *F. chinensis*. We apply *F. corymbosa* to a clade of tetraploids that share a unique 23 bp tandem duplication. These haplotypes are primarily distributed in northeastern China with scattered individuals in the southwest.

T2

P0494

A taxonomic revision of four complexes in the genus *Pourthiaea* (Rosaceae)

Binbin Liu, *Deyuan Hong*

Institute of Botany, Chinese Academy of Sciences

The taxonomy of these four complexes (*Pourthiaea pilosicalyx* complex, *P. salicifolia* complex, *P. zhijiangensis* complex, and *P. arguta* complex), distributed in E, S, and SE Asia, is extremely controversial. The lack of proper analysis and evaluation of the characters at population level used previously may have caused these controversies among taxonomists. The present study is the first comprehensive taxonomic revision of these four complexes. We conducted extensive field observations, population sampling, examination of a large number of specimens, and subsequent statistical analysis of the characters. All morphological characters used in previous taxonomic treatments were evaluated one by one, they are length, breadth, margin, base, apex, and shape of leaf blade, length of petiole, number of lateral veins, type of inflorescences, length of pedicel, and density of indumentum on leaf blade and inflorescences. All these characters were found to be extremely variable within a population and continuously varied between populations, and thus they are of no value for delimiting species. However, seven characters, i.e. breadth of leaf blade, lateral veins parallel or not, lower surface of leaf blade glabrescent or not, type of fruits, sepals caducous or not, number of carpels, and ratio of number of fertile seeds to total seeds, were found distinctly different (for qualitative characters) or statistically discontinuous (for quantitative characters); therefore, they are valuable for species delimitation in these four complexes. As a result, we recognized four species and two subspecies, *P. pilosicalyx*, *P. salicifolia*, *P. zhijiangensis*, and *P. arguta*: subsp. *arguta* and subsp. *pustulata*. Forty taxa were reduced as new synonyms, and 22 lectotypes were designated here.

T2

P0495

Phylogeography and intraspecific divergence of *Triosteum himalayanum* (Caprifoliaceae) on the Qinghai-Tibet Plateau and adjacent regions

Hai Rui Liu[1,2], *Qing Bo Gao*[1], *Fa Qi Zhang*[1], *Shi Long Chen*[1]

1. Northweset Institute of Plateau Biology, Chinese Academy of Sciences

2. University of Chinese Academy of Sciences

The extensive uplift of the Qinghai-Tibetan Plateau (QTP) during the Quarternary had great impact on the topography and climate of east Asia following by the evolution of biota. To understand the phylogeographic pattern of *Triosteum himalayanum* (Caprifoliaceae), we sequenced three chloroplast DNA fragments (*rbc*L-*acc*D, *rps15-ycf1*, *trn*H-*psb*A) from 238 individuals representing 20 populations. 19 haplotypes (H1–H19) are identified based on 57 polymorphisms. Most haplotypes were restricted to single population or to neighbouring populations. Analysis of molecular variance for overall populations revealed that most of the genetic variance was found among populations (89.43%), and only 10.57% within populations, indicating the high genetic differentiation across the sampled regions. The difference between G_{ST} (0.781) and N_{ST} (0.929) of overall populations was also significant, showing clear phylogeographic structure. Similar results were also found in two regional groups. We found ecoregions with relatively high genetic diversity or high frequence of private haplotypes, these suggest that this alpine herbaceous plant underwent enhanced allopatric divergence during survived in multiple refugia during the Last Glacial Maximum (LGM) or former glaciations. Our study support the indication that phylogeographic pattern difference between variable species on the QTP could be significant, and searching the evolutionary process details of alpine plants on the QTP could be interesting.

T2

P0496

The ddRAD-based phylogeny of the BDG complex of paleotropical woody bamboo (Poaceae: Bambusoideae)

Jing-Xia Liu[1,2], *De-Zhu Li*[1,2]

1. Kunming Institute of Botany

2. University of Chinese Academy of Sciences

The *Bambusa-Dendrocalamus-Gigantochloa* (BDG) complex mostly consists of *Bambusa, Dendrocalamus, Gigantochloa* and *Melocalamus*. Species of this group occur exclusively in tropical and subtropical areas of Asia, and most of them are of great economic, cultural and ecologic value. The classification of the BDG complex based on morphology and anatomy is long in dispute and still unresolved. In addition, a plethora of molecular studies provide ambiguous and controversial results because of insufficient information sites for such a rapid diversified clade. In order to investigate phylogenetic relationships within the BDG complex, we obtained a large set of SNPs data using the "double-digest" restriction site-associated DNA sequencing (ddRAD-seq) strategy, and performed phylogenetic analyses based on SNPs. Complete chloroplast genomes of some taxa are also sequenced and examined. We collected a total of 93 individuals for ddRAD-seq data collectionincluding 36, 16, 5 and 35 species of *Dendrocalamus*, *Gigantochloa*, *Melocalamus*, and *Bambusa*, respectively. Our sampling covers the major and core groups of the BDG complex. Three species (*Kinobaluchloa* sp., *Neomicrocalamus prainii, Bonia amplexicaulis*) were used as the outgroup based on previous studies. Main conclusions were as follows: 1. Maximum Likelihood (ML) analysis performed on the large data set resulted in a well resolved topology in the BDG complex: 1) Five major clades were well supported, clade I include all the individuals of *Dendrocalamus* and *Gigantochloa*. clade II, clade III and clade IV all consist of species of *Bambusa*. Within clade V, *Melocalamus* was resolved as a monophyletic group with strong bootstrap support, and proved to be sister to all of the rest genera. 2) Within clade I, *Gigantochloa* was a well resolved genera. However, *Dendrocalamus* was not supported to be monophyletic, but clustered with *Gigantochloa* to formed a clade with high statistical support. 3) *Bambusa* was not been well resolved, species of the genera were controversial, and more samples are needed for further researches. 2. The results of Network and Structure showed that there are high genetic identities among the BDG complex, but contain a variety of components especially in *Bambusa*. It demonstrated that the group has undergone a complex evolutionary history. In this context, incomplete lineage sorting and horizontal gene transfer are possible reasons for closely related species and populations to have incomplete reproductive isolation. 3. Additionally, the complete chloroplast genome were used to reconstruct the phylogenetic relationship of BDG complex with 43 representatives, but no robust topology gained. It indicates that the chloroplast genome may be not suitable for the phylogenetic relationships of this complex.

T2

P0497

Drought memory of moss

Li Liu

Kunming Institute of Botaty, Chinese Academy of Sciences

The drought stress memory response, which is induced by pre-exposure to various stresses, could play an important role in stress resistance and recovery in plants. However, whether moss exhibits drought stress memory and what molecular mechanism underlies this process are currently unknown.We discovered 271 novel lncRNAs that appear to function during drought treatment. Interestingly, three lncRNAs, TCONS_00028557, TCONS_00021167and TCONS_00013513, might function in the formation of drought memory. Osa-MIR1428e produced by TCONS_00028567 could activate the ABA-dependent SAPK10 signaling pathway to initiate drought resistance, which indicates that miRNA is likely also an important regulatory factor. Memory genes were compared between moss and other land plants. These results open a new window to understanding the epigenetic regulatory network underlying stress memory mechanisms.

T2
P0498
Revisit the phylogeny of wheat powdery mildew uncovered more relaxed host specialization

*Parivash Shoukouhi, Nicole Beck, Sarah Hambleton, Quinn Eggertson, **Miao Liu***
Agriculture and Agri-Food Canada, Ottawa Research and Development Centre, 960 Carling Ave. Ottawa, ON K1A 0C6, Canada

The species originally described as *Erysiphe graminis*, is an important pathogen that can cause severe damage to various cereal crops. This species was re-classified in a monotypic genus *Blumeria* based on unique morphological characteristics of conidia and chasmothecia, as well as molecular evidence. Host specialization of *B. graminis* was studied by inoculation experiments (Mains 1933, Cherewick 1944, Eshed & Walh 1970), and more recently by analyzing DNA sequences (Inuma *et al.*, 2007). Multigene phylogenetic analyses of the samples from Japan, Iran and Switzerland have revealed nine distinct lineages. In general, samples from individual host genus belonged to a single lineage except for the samples from *Lolium* and *Bromus*. Using historical herbarium specimens housed in the National Mycology Herbarium (DAOM) in Canada, we were able to obtain ITS sequences for isolates on various hosts of mainly North American origin including a few from arctic regions. The phylogeny based on combined sequences generated in this study and those by Inuma et al (2007) largely agreed with previous recognition of nine clades. However, isolates from *Hordeum* spp. were grouped into three different clades, and the host ranges of Dactylis-clade, Poa-clade, and Triticum-clade have expanded by additional 1-4 host genera. In addition, a new lineage was uncovered that include samples on *Deschampsia*, *Elymus* and *Hordeum*.

T2
P0499
Unravelling the evolutionary dynamics of ancient and recent polypoidization events in *Avena* (Poaceae)

Qing Liu[1], *Lei Lin*[1], *Xiangying Zhou*[1,2], *Paul M. Peterson*[3], *Jun Wen*[3]
1. Key Laboratory of Plant Resources Conservation and Sustainable Utilization, South China Botanical Garden, Chinese Academy of Sciences
2. University of Chinese Academy of Sciences
3. Department of Botany, National Museum of Natural History, Smithsonian Institution, Washington, DC 20013-7012

Understanding the diversification of polyploid crops in the Mediterranean biodiversity hotspot is a challenging issue for evolutionary biology. Sequence data of three nuclear genes and three plastid DNA fragments from 109 accessions of *Avena* L. (Poaceae) and those from outgroups was used for maximum likelihood and Bayesian analyses. The evolution of cultivated oat (*Avena sativa* L.) and its close relatives was inferred to have involved ancient allotetraploidy and subsequent recent allohexaploidy events. The crown ages of two infrageneric lineages (*Avena* sect. *Ventricosa* Baum ex Romero-Zarco and *Avena* sect. *Avena*) and the *A. sativa* lineages were estimated to be in the early to middle Miocene and the late Miocene to Pliocene. These periods coincided with the mild seasonal climatic contrasts and the Mediterranean climate establishment in the Mediterranean Basin. Our results suggest that polyploidy, lineage divergence, and complex reticulate evolution has occurred in *Avena*, and this exemplifies the long-term persistence of tetraploids and the multiple origins of hexaploids related to paleoclimatic oscillations during the Miocene-Pliocene interval in the circum-Mediterranean region. This newly-resolved infrageneric phylogenetic framework represents a major step forward in understanding the scenarios for the origin of cultivated oat. Acknowledgements. The research was supported by National Natural Science Foundation of China to Q.L. (31270275), Special Basic Research Foundation of Ministry of Science and Technology of the People's Republic of China to Q.L. (2013FY112100), Chinese Academy of Sciences President's International Fellowship Initiative to Q.L. (2016VBA010), China Scholarship Council Awards to Q.L. (201604910096), Undergraduate Innovation Training Program of Chinese Academy of Sciences (27), and the Endowment Grant Program of the Smithsonian Institution to J.W. The authors thank CN-Saskatchewan, ILRI-Addis Ababa, and USDA-Beltsville Germplasm System for seeds.

T2
P0500
Immediate evolutionary impact of whole genome duplication on RNA alternative splicing, using the recently formed polyploid *Tragopogon* (Asteraceae) as a model

Xiaoxian Liu[1,2], *W. Brad Barbazuk*[1,3], *Pamela S. Soltis*[2,3], *Douglas E. Soltis*[1,2,3]
1. Department of Biology, University of Florida, Gainesville, FL 32611-8525, USA.
2. Florida Museum of Natural History, University of Florida, Gainesville, FL 32611-7800, USA.
3. Genetics Institute, University of Florida, Gainesville, FL 32610, USA.

Plant species that have experienced whole-genome duplication (WGD; polyploidy) during their evolutionary history usually undergo changes in gene number, expression, and regulation. However, it is unclear whether or not such genomic modifications are sufficient to generate the phenotypic novelty that characterizes most polyploids and has allowed many polyploids to occupy new habitats. RNA alternative splicing (AS) is a major regulatory mechanism of gene expression and a major factor promoting functional innovation and proteomic diversity. Yet, no studies have assessed the immediate impact of WGD on AS. The goal of our research is to address the role and extent of AS in recently formed naturally occurring allopolyploids and help elucidate the connections among AS, polyploidy, and diversity. Here, we use allopolyploid species in the flowering plant genus *Tragopogon* (Asteraceae), a model for the study of very recent and recurring allopolyploidy, to study the impact of WGD on AS patterns. Using transcriptomes generated by PacBio Iso-Seq^{TM}, we compare both natural populations of independent origin and synthetic lines of the allopolyploid *T. miscellus* (~80 years old) to its diploid parents (*T. dubius* and *T. pratensis*) to assess the role of AS after polyploidization. The results will provide a better understanding of how AS influences genetic and potentially proteomic diversity in polyploid plants.

T2

P0501

Nitrogen resorption and release by understory mosses in an old-growth fir forest

***Xin Liu**, Zhe Wang, Xiaoming Li, Weikai Bao*
Chengdu Institute of Biology, Chinese Academy of Sciences

Mosses are dominant in understory layer of boreal and subalpine forest, where they strongly affect nitrogen (N) cycling. However, mosses differed substantially from vascular plants in terms of plant structure, which may lead to distinct characteristics in nutrient resorption and release. Here, we measured the N resorption efficiency (NRE), N release rate by leaching and the percentage of N left in litter using stable isotope (^{15}N) techniques. We find that NRE of the studied species, *A. hookeri* and *H. splendens*, was 61.67% and 52.16%, respectively. The mosses release 8.65% and 13.62% of tissue N annually and 23.19% and 33.62% of N from new-growth to senesced by leaching. 15.14% and 14.22% of tissue N was left in the senesced tissue and released by litter decomposition in the following years for respective species. Our results indicate that the forest floor mosses resorb N in senescing tissue at a relatively high efficiency. Given the less N left in senesced tissue and low decomposition rate, N leaching is a more important process that understory moss layer releases N than litter decomposition. Thus, N resorption and leaching release of the forest floor mosses control N flux through the ecosystem.

T2

P0502

DNA barcoding evaluation and implications for phylogenetic relationships in Lauraceae from China

***Zhi-Fang Liu**[1,4], Xiu-Qin Ci[1,4], Lang Li[1], Hsi-Wen Li[2], John G. Conran[3], Jie Li[1]*
1. Xishuangbanna Tropical Botanical Garden, Chinese Academy of Sciences
2. Kunming Institute of Botany, Chinese Academy of Sciences
3. School of Biological Sciences, University of Adelaide
4. University of Chinese Academy of Sciences

Lauraceae are an important component of tropical and subtropical forests and have major ecological and economic significance. Owing to lack of clear-cut morphological differences between genera and species, this family is an ideal case for testing the efficacy of DNA barcoding in the identification and discrimination of species and genera. In this study, we evaluated five widely recommended plant DNA barcode loci *mat*K, *rbc*L, *trn*H–*psb*A, ITS and ITS2 for 409 individuals representing 133 species, 12 genera from China. We tested the ability of DNA barcoding to distinguish species and as an alternative tool for correcting species misidentification. We also used the *rbc*L+*mat*K+*trn*H–*psb*A+ITS loci to investigate the phylogenetic relationships of the species examined. Among the gene regions and their combinations, ITS was the most efficient for identifying species (57.5%) and genera (70%). DNA barcoding also had positive role for correcting species misidentification (10.8%). Furthermore, based on the results of the phylogenetic analyses, Chinese Lauraceae species formed three supported monophyletic clades, with the *Cryptocarya* group strongly supported (PP = 1.00, BS = 100%) and the clade including the *Persea* group, Laureae and *Cinnamomum* also receiving strong support (PP = 1.00, BS = 98%), whereas the *Caryodaphnopsis–Neocinnamomum* received only moderate support (PP = 1.00 and BS = 85%). This study indicates that molecular barcoding can assist in screening difficult to identify families like Lauraceae, detecting errors of species identification, as well as helping to reconstruct phylogenetic relationships. DNA barcoding can thus help with large-scale biodiversity inventories and rare species conservation by improving accuracy, as well as reducing time and costs associated with species identification.

T2

P0503

Floral evolution in Rubiaceae – a morphological overview

***Stefan Löfstrand**[1], Sylvain G. Razafimandimbison[2], Catarina Rydin[1,3]*
1. Stockholm University; Department of Ecology, Environment and Plant Sciences
2. Swedish Museum of Natural History; Department of Botany
3. Bergius Botanic Garden

The coffee family (Rubiaceae, Gentianales) is estimated to have originated in the mid-Cretaceous and is one of the largest flowering plant families with more than 13,000 species. Rubiaceous flowers can be unisexual or bisexual, homostylous or heterostylous, large or small, and all main types of pollination systems are represented. However, there are surprisingly few detailed studies on floral biology within Rubiaceae and few floral synapomorphies have been suggested at or above tribal level. We have reached a point in time where we (mostly) understand the phylogenetic relationships, but not the evolutionary patterns and pathways that lead to the extant structural diversity. More than any other structures in plants (e.g. leaves), flowers provide a direct testimony of evolutionary history, albeit not always phylogenetic relationships, making them highly interesting for the study of distribution patterns of floral characters in extant angiosperm clades. We therefore need to delve deeper into the complex structures of flowers. Are floral traits synapomorphic or ancestral in Rubiaceae? Are they homologies induced by similar biotic and abiotic forces acting on the taxa throughout evolution? In essence, we need to gain more knowledge about what distinguishes rubiaceous flowers beyond general observations. This study aims to, with a broad sampling from all major lineages of Rubiaceae (subfamilies, tribes, genera), investigate general aspects of floral form (e.g. color, calyx lobe number, corolla tube length, and style length), particularly in relation to heterostyly, a trait that appears to have evolved in several lineages in the family. This much-needed review of floral form in Rubiaceae will hopefully result in hypothetical renderings of flowers of the last common ancestor of key clades, such as tribal complexes, subfamilies, and the family as a whole.

T2

P0504

The family Oleaceae in South America

Julio Antonio Lombardi
UNESP - State University of São Paulo

The Oleaceae are trees, shrubs or lianas, monoecious or rarely polygamo-monoecious, trichomes when present generally peltate, leaves are simple or compound, trifoliolate or imparipinnate, unstipulated, petiolate, opposite or rarely alternate (*Jasminum* species), rarely scale like, margins entire, crenate or serrate, pilose domatia often present in the axils of secondary veins on the abaxial face, stomata occasionally in crypts; inflorescences are paniculate, racemiform or fasciculate, or flowers solitary; flowers are actinomorphic, often aromatic, with perianth 2-6-merous, calyx and corolla present and distinct, absent (*Forestiera*), or one whorl missing, either the calyx (*Forestiera*, *Priogymnanthus*) or the corolla (*Forestiera*, *Nestegis*), the calyx is gamosepalous, truncate or lobed, valvar or imbricate, sepals 4 (-15), the corolla is gamopetalous or dialipetalous, imbricate, valvar or contorted, petals 4 (-12), stamens 2 (4), united to corolla or free, all fertile, filaments present or null, anthers dehiscence longitudinal, introrse or latrorse, intrastaminal disc present or absent, gynoecium bicarpellate and bilocular, ovary superior, (1)2 (-4) axillary ovules per locule, rarely numerous (*Forsythia*), style 1, apical, stigma bilobed or capitate, dry; fruits are schizocarpous with circumscissile or indehiscent mericarps (*Menodora*) or loculicidal, or a drupe with a hard or brittle endocarp; seeds are winged or not. The family includes 25 genera and about 600 species, with nearly cosmopolitan distribution. In Neotropics Oleaceae comprises nine genera and over 45 described species, in South America are recognized 27 native species included in five genera: *Chionanthus* L. (18 species, *C. abriaquiensis* Fern. Alonso & Cogollo, *C. colonchensis* Cornejo & Bonifaz, *C. compactus* Sw., *C. crassifolius* (Mart.) P.S.Green, *C. ferrugineus* (Gilg) P.S.Green, *C. filiformis* (Vell.) P.S.Green, *C. fluminensis* (Miers) P.S.Green, *C. greenii* Lombardi, *C. guianensis* (Aubl.) Pers., *C. implicatus* (Rusby) P.S.Green, *C. megistocarpus* Fern.Alonso & Cogollo, *C. micranthus* (Mart.) M.Lozano & Fuertes, *C. parviflorus* Cornejo, Lombardi & W.W.Thomas, *C. pubescens* Kunth, *C. subsessilis* (Eichler) P.S.Green, *C. tenuis* P.S.Green, *C. trichotomus* (Vell.) P.S.Green, and *C. wurdackii* B.Ståhl,), *Forestiera* Poir. (one species, *F. ecuadorensis* Cornejo & Bonifaz), *Menodora* Bonpl. (four species, *M. decemfida* (Gillies ex Hook.) A.Gray, *M. integrifolia* (Cham. & Schltdl.) Steud., *M. pulchella* Markgr., and *M. robusta* (Benth.) A.Gray), *Priogymnanthus* P.S.Green (three species, *P. apertus* (B.Ståhl) P.S.Green, *P. hasslerianus* (Chodat) P.S.Green, and *P. saxicolus* Lombardi), and *Schrebera* Roxb. (one species, *S. americana* (Zahlbr.) Gilg), besides four naturalized species in the genera *Jasminum* L. (one species, *J. fluminense* Vell.), and *Ligustrum* L. (three species, *L. japonicum* Thunb., *L. lucidum* W.T.Aiton, and *L. vulgare* L.). (CNPq - Conselho Nacional de Desenvolvimento Científico e Tecnológico # 307997/2013-8; FAPESP - Fundação de Amparo à Pesquisa do Estado de São Paulo # 2013/06296-9).

T2

P0505

The genus *Zinowiewia* (Celastraceae) and its first record in the Brazilian Flora

Leonardo Biral[1], ***Julio Antonio Lombardi***[2]

1. UTFPR- Universidade Tecnológica Federal do Paraná

2. UNESP - State University of São Paulo

Zinowiewia Turcz. (Celastraceae) is an endemic taxon of the Neotropical region distributed in both South and Central America and Mexico from 250 to 3,150 m elevation. The genus has nine species currently recognized, after some new synonyms were proposed in the recent taxonomic revision for Flora Mesoamericana. The taxon includes trees or shrubs, and can be characterized by glabrous opposite and decussate leaves, with entire margins, inflorescences axillary cymes forked one to several times, flowers bisexual, 5-merous, ovary 2-locular, fruits samaras with one apical wing, and one (rare two) linear-oblong seed. *Zinowiewia* is easily distinguishable from *Plenckia* and *Wimmeria*, the other Neotropical genera with samaras, by its opposite leaves (vs. alternate). It can also be differentiated from *Wimmeria* by flowers 5-merous and samaras with a single terminal wing (vs. flowers 4-merous and samaras with 3-4 longitudinal wings). The genus was created by Turczaninow in 1859 as a new combination for *Wimmeria integerrima*, published by him a year before, but with questions about its generic position. After Turczaninow's delimitation, it took almost 80 years for an additional species to be described in the genus. Of the 17 names published in *Zinowiewia*, 13 were created by Cyrus Lundell between 1938 and 1987, several of them being reduced to *Z. integerrima* and *Z. rubra*, after the treatment on Mesoamerica. The genus was classified in the subfamily Tripterygioideae by Loesener because the winged fruits with one or two seeds without aril, but a phylogeny provided for Tripterygioideae based on morphological and molecular characters showed that *Zinowiewia* is the sister clade of *Microtropis* + *Quetzalia*, sharing with those opposite leaves, ovary two carpelate, and molecular synapomorphies in the ITS region. Currently, Tripterygioideae has been recognized as a highly polyphyletic clade and *Zinowiewia* is nested in Celastroideae. We register here the first record of *Zinowiewia* in the Brazilian flora, a collection of *Z. australis* was located after taxonomic studies were carried out by the authors for Celastraceae in several herbaria. The specimen *G.T. Prance 10015* (F [1772295], INPA [27148], K [001137341], M [s.n.], MG [41717], P [05493140], U [260544]) was collected in the Roraima State, in Serra do Surucutus, NE of Mission station, 1000-1400 m elevation, in a forest at foot of red sandstone cliff, on 17 February 1969. This species is also known by collections from Venezuela, Colombia, Ecuador, Peru and Bolivia, but it was not previously registered for Brazil. With this addition, Celastraceae is now represented in Brazil by 20 genera (eight of them from the Celastroideae subfamily) and 142 species. Several genera of Neotropical Celastroideae, especially those with a small number of species, e.g. *Crossopetalum*, *Gyminda*, *Haydenoxylon*, *Quetzalia*, *Schaefferia*, *Wimmeria*, and *Zinowiewia* have not been the subject of consistent taxonomic studies or these studies have become outdated. Probably new records of occurrence, and even undescribed species, will be registered when more taxonomic studies are carried out for them. (FAPESP - Fundação de Amparo à Pesquisa do Estado de São Paulo # 2011/21879-5 and 2014/22093-3).

T2

P0506

Evolution of floral structure in the transition to pentapetalae (eudicotyledoneae) derived from total evidence analysis of extant and fossil taxa

Andrea Lopez-Martínez[1], *Hervé Sauquet*[2], *Susana Magallón*[3]

1. Posgrado en Ciencias Biológicas, Universidad Nacional Autónoma de México

2. Laboratoire Écologie, Systématique, Évolution, Université Paris-Sud
3. Instituto de Biología, Universidad Nacional Autónoma de México

The Eudicotyledoneae (eudicots) is a solidly recognized clade that contains approximately 75% of all angiosperm species, and includes a wide diversity of floral morphology. Phylogenetic results have consistently identified a grade of independent lineages at the onset of eudicot diversification, corresponding to orders Ranunculales, Proteales, Trochodendrales, Buxales, and Gunnerales, which are successive sister taxa to a species-rich, Pentapetalae. Each of the lineages of the early-diverging eudicot grade includes a small to moderate number of species, and together, they encompass substantial floral structural diversity, including, for example, spiral and whorled phyllotaxy, variable and fixed merosity; and different degrees of fusion among floral organs. In contrast, most of the flowers of Pentapetalae are structurally stable, displaying whorled phyllotaxy, fixed merosity with usually 5 (or 4) organs per whorl in the perianth and androecium, and a two-whorled perianth differentiated into calyx and corolla. In this study, we investigate the evolution of floral structure in the transition between the early-diverging eudicot grade and the initial diversification of Pentapetalae, with a special focus on the establishment of the fixed meristic pattern and bipartite perianth of Pentapetalae. To conduct this study, we combined nucleotide sequences of three plastid markers (*atp*B, *rbc*L, *mat*K) of extant species, and 79 floral morphological characters for 54 extant and 16 three-dimensionally preserved fossil flowers. We estimated phylogenetic relationships among extant and fossil taxa with maximum likelihood and Bayesian phylogenetic inference. The temporal context of floral evolution was then investigated by two different means of fossil calibration. First, we conducted a conventional node-calibrated molecular dating analysis under bayesian relaxed clock models, using only on extant taxa as terminals and incorporating the identified phylogenetic placement of fossils to calibrate internal phylogenetic nodes. We conducted a total evidence 'tip-dating' analysis in which fossils were included as terminals with fixed age, together with extant species. Based on the resulting timetrees, we investigated the evolution of floral structure during the transition to Pentapetalae by estimating the ancestral condition and character state shifts of perianth differentiation and floral merosity. Our results confirm previously suggested affinities of most fossils, but in a few instances, a new phylogenetic placement was obtained. Stochastic mapping for floral characters indicates numerous structural changes in the early eudicot grade, and suggest that merosity was variable in the most recent common ancestor of Pentapetalae.

T2
P0507
Highly efficient pollination system in *Aeschynanthus acuminatus* (Gesneriaceae) reveals the potential role of generalized passerines in floral evolution of East Asia

Jing-Yi Lu, *Kai-Hsiu Chen, Chun-Neng Wang*
National Taiwan University

Background and aims: The interaction between angiosperms and their pollinators is regarded as a major driving force leading to the evolution of diverse floral characters. Pollinator assemblages, thus, exert selective pressures on pollination syndromes and affect the species composition among different geographic regions. A well-known example is the extremely low diversity of nectar-specialized avian in East Asia northern of the Himalayans, which gave rise to the long belief of lacking plant-bird relationships with only few detailed investigated cases. *Aeschynanthus*, a Gesneriaceae genus with more than 160 species in subtropical SE Asia, has red and tubular flowers associated with sunbird pollination syndrome. Inhabiting in the northern edge of *Aeschynanthus* distribution where the putatively specific pollinator sunbird is absent, *A. acuminatus* has changed to reddish green flower and widely opened corolla tube in Taiwan. We thus investigated the pollination biology of *A. acuminatus* to increase our understandings on extremely rare observation of ornithophily in Asia, north of Himalayas region. Methods: Pollination efficiency was quantified as natural fruit set and conspecific pollen transfer rate on stigmas. Floral morphology of co-flowering species sharing the same pollinators as *A. acuminatus* were compared to evaluate the effect of interspecific pollen placement. The reflectance spectra of flowers were measured and nectar properties were recorded with nectar sugar composition further obtained by high performance anion-exchange chromatography (HPAEC). *Results:* Two species of passerines, *Yuhina brunneiceps* and *Alcippe morrisonia*, were observed effectively visiting *A. acuminatus*. The high natural fruit set (60%) and conspecific pollen transfer rate (95%) indicate high reproductive success which appears to be the result of accurate pollen placement on the pollinators. In addition, the existence of a major peak in long-wavelength (red) spectrum of floral reflectance and the copious (58.7 μl) and diluted (6.8%) nectar with abundant hexose sugars are consistent to generalized-bird pollination syndromes. *Conclusion:* The complement of generalized-passerines as pollinators in *A. acuminatus* with high pollination efficiency appears to play an important role in successful colonization to Taiwan as we discovered its pollination syndromes are closely matching the generalized-bird pollination syndromes. *Discussion:* In comparison with other members of *Aeschynanthus*, the distinct pollination syndromes of *A. acuminatus* relate to the different avian pollinator assemblage in East Asia; Ecologically, co-flowering species of this generalized-passerine pollination system with different floral morphology could co-exist with high efficiency by pollen placement, reaching higher species diversity in the community. Our research reveals the successful adaptation of generalized-passerine pollination system, which provides a direction of future studies on pollination in East Asia, the potential role of generalized-bird pollination interactions in driving the evolution of species and diverse floral traits.

T2
P0508
Pollen spectrum for the recognition of desert vegetation types in NW China, Eastern Arid Central Asia

Kai-Qing Lu[1], *Gan Xie*[1], *Ming Li*[1], *Jin-Feng Li*[1], *Anjali Trivedi*[2], *David K. Ferguson*[3], *Yi-Feng Yao*[1], *Yu-Fei Wang*[1]
1. Institute of Botany, the Chinese Academy of Science
2. Birbal Sahni Institute of Palaeosciences
3. Institute of Palaeontology, University of Vienna

The vast temperate desert in Arid Central Asia, as an ancient Silk

Road, bred splendid human civilization, provided an important channel for economic and cultural exchanges between East and West. The coupling relationship between human existence and the dynamic change of desert ecosystem in this region is a hotspot concerned by governments and scientific communities. However, lacking of a pollen spectrum about the dominant and indicative taxa within temperate desert vegetation and a corresponding relation between pollen assemblages and specific desert vegetation types prevents us from further understanding the formation and maintain of this desert ecosystem. Here we depict and establish a pollen spectrum of dominated plants in desert vegetation, covering 14 families, 36 genera, 56 dominant and indicative species in eastern Arid Central Asia, and provide a key for identifying the dominant and indicative taxa, and recognize four main temperate desert vegetation types, i.e., semi-dwarf arbor desert, shrubby desert, semi-shrub and dwarf semi-shrub desert and cushion dwarf semi-shrub desert based on the characteristics of pollen assemblages. The present study gives an important reference on distinguishing pollen grains from surface and sub-surface samples in this region, and has a great potential applying for tracing the past change of desert vegetation succession and pattern formation.

T2

P0511

Insights into genome structure of two pentaploid dogroses (*Rosa* sect. *Caninae*) harbouring a unique asymmetrical meiotic system

Jana Lunerová[1], *Veit Herklotz*[2], *Christiane M. Ritz*[2], *Aleš Kovařík*[1]

1. Department of Molecular Epigenetics, Institute of Biophysics of the CAS v.v.i., Czech republic

2. Department of Botany, Senckenberg Museum of Natural History, Germany

Members of genus *Rosa* section *Caninae* (also known dogroses) are multihybrids harbouring variable ploidy levels ranging from four to six. All dogroses evolved an unusual matroclinal meiotic system characterised by asymmetrical distribution of genetic material between the female and male gametes: In pentaploid dogroses, (2n=5x=35), the pollen transmits one basic genome (x=7) derived from the seven segregating bivalents, whereas the egg transmits four basic genomes (4x=28) one set derived from the segregation of seven bivalents and three sets of univalent-forming chromosomes. In this study, we explored the genomes of two pentaploid dogrose species, *Rosa canina* and *R. inodora,* which belong to different subsections and are abundant shrubs in central Europe. We sequenced their genomes by Illumina technology and analysed the genomic and chromosomal composition using bioinformatic, molecular and cytogenetic approaches. The clustering analysis using RepeatExplorer revealed that the repeatom (sum of satellite repeats, RNA and DNA transposons, simple repeats and rDNA) is similar between the species with little variation in repeat copy number. We cloned a major satellite called CANR4 which constitutes 1.5 and 2.7% of *R. canina* and *R. inodora* genomes, respectively. Fluorescent *in situ* hybridisation (FISH) revealed its mostly pericentromeric location on several chromosomes in both species. FISH in meiotic pollen mother cells showed that the bivalent-forming chromosomes are relatively deficient in CANR4 in *R. canina*, at least. Additionally, we reconstituted chloroplast genomes of *R. canina* and *R. inodora* revealing specific SNPs and haplotypes. We hypothesize that *R. canina* and *R. inodora* formed from independent hybridisation events involving common closely related progenitors. Thus, unusual asymmetrical meiosis arose at least twice in evolution of Caninae dogroses.

T2

P0512

The evolution of genetic diversity hotspot: Adaptation and ongoing speciation in a mangrove tree, *Avicennia Marina*

Haomin Lyu, *Zhengzhen Wang, Xinnian Li, Ziwen He, Lishi Huang, Suhua Shi*

Sun Yat-sen University

Geographical changes have been considered one of the most important factors to account for both demographic history and adaptation. Documenting the admixture events and positive selection can clarify the role of geographic features in speciation. Three varietas of *Avicennia marina*, inhabiting across the Indo-West-Pacific (IWP) region, provide a good example to insight how coastal and marine organisms in the biodiversity hotspots evolve and adapt during environmental changes. Here we present the global patterns of genetic diversity by the resequencing of 13 populations from all the three varietas. The results show that the intermediate varietas (*A. marina* var. *eucalyptifolia*) bears much more genetic variations than the others. The integrative inference of admixture history demonstrates that this pattern is almost independent from the admixture among varietas. The high-frequency transformations between opens and closes of Torres Strait during sea level fluctuations may cause the cycles of migrations and isolations in the interval of 1.806 ~ 5.96 million years ago, suggestive of the accumulation of genetic variations. The demographic model and phylogenetic discordance of loci under selection indicate that these accumulated variations could contribute to the rapid adaptation and high divergence among three varietas and different populations. We also infer the genetic basis of adaptive evolution and differentiation in *A. marina* var. *australasica*, such as the genes regulating the photoperiod and the flowering time, while the differentiation of *A. marina* var. *marina* might be attribute to more complicated geography and isolations. Although the standing genetic variations facilitate the adaptation to changing environments, the accumulation of genetic variations almost halts, and effective population size persistently reduces during the repeated glaciation in Pleistocene. However, following the acceleration of global warming and sea level rising, the mangroves, even other marine and coastal organisms in IWP region, would face unprecedented hardship.

T2

P0513

Reduction of transposable element load when hosts invade stressful habitats: A common preemptive strategy in mangroves

Haomin Lyu[1], *Zhengzhen Wang*[1], *Ziwen He*[1], *Zhou Renchao*[1], *Lishi Huang*[1], *Jianhua Ouyang*[1], *Zixiao Guo*[1], *Cairong Zhong*[2], *David Boufford*[3], *Chung-I Wu*[1,4], *Suhua Shi*[1]

1. Sun Yat-sen University

2. Hainan Dongzhai Harbor National Nature Reserve

3. Harvard University Herbaria
4. University of Chicago

Transposable elements (TEs) are generally known to be parasitic and deleterious to the host. Evolution of TE load as hosts invade stressful habitats should shed light on genomic consequences of adaptation to new environments. Woody plants living at the interface between land and sea, collectively referred to as mangroves, are excellent examples of invasion of inhospitable habitats. TEs are also the most common components of the genomes of woody plants. We took advantage of mangrove genomes recently generated by our group to examine TE evolution in plants. The birth vs. death rate of LTR-RTs can be estimated because the death of retrotransposons would leave a trace in the form of solo-LTRs or truncated LTR-RTs. We found that (i) all lineages of mangroves in our study independently and massively reduced their TE load in comparison to their non-mangrove relatives; (ii) genome size reduction has happened in all six true mangrove lineages; (iii) the TE load reduction in all examined true mangroves can largely be attributed to the paucity of young TEs; (iv) the rarity of young TEs is a consequence of the reduction in new TE birth, rather than a death rate increase. Finally, we conclude that many mangroves employ a common strategy of TE load reduction and repressing proliferation when adapting to a new environment.

T2
P0514
Peristome patters and spore morphology of *Bryum* Hedw. (Bryaceae) from the Altun Mountain National Nature Reserve in Xinjiang, China

Sulayman Mamtimin[1], *Sabiram Esrayi*[1], *Atigul Molla*[1], *Yong-Ying Liu*[2]
1. College of Life Science and Technology, Xinjiang University, Urumqi 830046, China
2. Department of Biology, Jiaozuo Teachers College, Jiaozuo 454001, Henan, China

The Altun Mountain National Nature Reserve is located in the Kumukule Basin of East Kunlun Mountain, south of Xinjiang Uyghur Autonomous Region, China. The area is the first built, having the largest range for protected plateau desert ecosystem. It was regarded as one of the rare gene banks of plateau species by International Union for Conservation of Nature (IUCN) and World Wildlife Fund (WWF). Unique geographic position and natural conditions of the Altun Mountains cause bryophyte of the area can bear bitter cold on highland. Bryophyte is one of the important roles in the vegetation of the area. Moss flora of the Altun Mountains has an important position in China, attract the attention of many bryologists in the world. More than twenty species of genus *Bryum* Hedw. have been recorded in the area during our studying of bryophyte diversity, it is about 2/3 of Xinjiang. The paper aims to figure out the taxonomic value of the peristomial structures and spore morphology in *Bryum*, to discuss the relationship between the natural environmental condition and the diversity of *Bryum* at the Altun Mountain National Nature Reserve. Morphological characters of peristome and spores from 18 species of the genus *Bryum* in the area were observed by the scanning electronic microscope (SEM). The results showed that: (1) The characters, such as, the exostomial teeth shape, pattern of exostomial dorsal suface, width of perforation on the endostomial segments, the length of the endostomial basal membrane, development of the cilia and exine ornamentation of the spore, are different among the taxa in *Bryum*, and they are stable for each species in the genus. So their presence often is considered highly diagnostic and taxonomically in infrageneric classification combining with other sporophyte characteristics and/or gametophyte characteristics. (2) When most highly developed or perfect peristome, it consists of longer exostome teeth, a high endostomial basal membrane, broad segments and appendiculate cilia. Otherwise, modified or reduced peristome has shorter exostome teeth, lower endostomial basal membrane, narrower segments and shorter and fewer cilia or absent. (3) It seems that the rich diversity of *Bryum* in the Altun Mountain National Nature Reserve is related with the peristomial structures and spore morphology being adaptive for spores dispersing in dry environment.

T2
P0515
Morphometrics of the giant sedges of *Carex* section *Rhynchocystis* (Cyperaceae): Overlooked vs. overvalued characters and unnoticed species

Mónica Míguez[1], *Berit Gehrke*[2], ***Santiago Martín-Bravo***[1], *Pedro Jiménez-Mejías*[3]
1. Universidad Pablo de Olavide
2. Johannes Gutenberg-Universität
3. New York Botanical Garden

Carex section *Rhyncocystis* (Cyperaceae) is a group of five species of giant sedges disjunctly distributed in the Western Palearctic (*C. microcarpa* and *C. pendula*) and sub-Saharan Africa (*C. bequaertii, C. mossii, C. penduliformis*). Recent phylogenetic studies revealed that *C. bequaertii* and *C. mossii* are not mutually monophyletic, and that *C. pendula*, despite monophyletic, actually embraces two strongly differentiated lineages. In order to objectively assess the morphological limits and taxonomic circumscription of the species and lineages, we performed a morphometric study of macromorphological characters based on 109 herbarium vouchers, as well as a nomenclatural survey. Micromorphology of achenes was also studied. Our results suggest the existence of five species, of which only *C. microcarpa* and *C. penduliformis* were supported in their traditional circumscriptions. The two *Carex pendula* lineages were found to correspond to two morphologically distinct, mostly allopatric taxa: one from Western Europe (*C. pendula* s.s.), and another from Central and Eastern Europe, to which the long neglected 18th century name *C. agastachys* should apply. On the contrary, *C. bequaertii* and *C. mossii* were found to be weakly morphologically differentiated and thus better treated within a single species with two subspecies (*C. bequaertii* ssp. *bequaertii* and ssp. *mossii*). Our analyses revealed clear-cut diagnostic characters neglected by the traditional taxonomy of the group, whereas we found that other characters have been overvalued and are not diagnostic. This study illustrates that even within a relatively well-studied area as Europe, plant taxonomy may benefit from objective and careful revaluations.

T2

P0516

New fossil flora from Colombia, a window to the Eocene climate and biogeography in the Neotropics

Camila Martinez[1], *William Crepet*[1], *Carlos Jaramillo*[2]
1. Cornell University
2. Smithsonian Tropical Research Institute

An important step in the evolution of our modern glacial climate state occurred during the middle Eocene to early Oligocene, when the global temperature cooled, and the first Antarctica ice sheet appeared. This dramatic climate change caused a significant global turnover in marine and terrestrial biotas, however, more evidence from tropical regions is still needed to document these changes at low latitudes. A new macroflora from the Esmeraldas Formation of Colombia provide plant fossil evidence to understand some of the biotic changes that could have occurred in tropical regions. The Esmeraldas Flora has been collected from two localities and has more than six hundred specimens that include compressions and impressions of leaves mostly, but also seeds, fruits and seldom flowers. Detailed stratigraphic data shows that the Flora was deposited in floodplains of braided and meandering rivers. The age of the Esmeraldas Formation ranges from middle to late Eocene. Here, we present results of paleoclimatic analyses based on leaf fossil material from the Esmeraldas Formation. The obtained data are suggesting drier conditions during deposition compared to those from the present. Leaf cuticle analyses that include anatomical features and isotopic values? ^{13}C are also being performed to provide estimates of atmospheric CO_2 concentration for the Esmeraldas Formation. Finally, we propose taxonomic affinities of some fossil leaves, flowers, fruits and seeds, that suggest the presence of families like Fabaceae, Malvaceae, Menispermaceae, Myrtaceae, Passifloraceae, Rhamnaceae and Solanaceae. Our preliminary results are showing that the Esmeraldas Flora exhibit a great diversity, a predominance of microphyll leaf sizes and a great preservation of cuticles, that in general are suggesting that the Esmeraldas Flora could have represented a dry tropical ecosystem.

T2

P0517

Polyphyly of Japanese *Carex* section *Mitratae* and its chromosome evolution based on nuclear and chloroplast DNA sequences

Tomomi Masaki[1], *Takuji Hoshino*[1], *Chun-Kuei Liao*[2], *Marcia Waterway*[3]
1. Okayama University of Science
2. Da-Yeh University
3. McGill University

Genus *Carex* L. has 2,000 species in the world and the largest number in Cyperaceae (Reznicek, 1990). About 493 taxa and 26 genera were reported in the Japanese Cyperaceae, and Japanese *Carex* comprises 311 taxa and 53 sections (Hoshino and Masaki, 2011). Section *Mitratae* is the largest section in Japanese *Carex* and adjacent regions of Japan (Oh, 2006, Dai *et al.*, 2010), containing 80 taxa, 53 taxa is endemic to Japan (Hoshino and Masaki, 2011). This section has separate staminate and pistillate spikes, the width of leaves 0.2-15 mm, and style base of achenes annulated. Akiyama (1955) separated this section into six subsections including adjacent regions of Japan, subsects. *Cryptostachydeae* (Franch.) Ohwi, *Lageniformes* Ohwi, *Nervatae* Akiyama, *Leucochlorae* Akiyama, *Pisiformes* Akiyama, and *Foliosissimae* Akiyama. The chromosome numbers of the species of this section were reported 2n=30-94 (Tanaka, 1948; Okuno, 1940; Hoshino, 1981; Ohkawa et al., 1998; 2000; Yano *et al.*, 2007a; 2007b; 2008; 2010). In this study, we focused on the phylogenetic relationships among the species of section *Mitratae*. We analyzed ITS, ETS-1f, *trn*L intron, and *trn*L-*trn*F intergenic spacer regions for 143 species in 46 sections of Core *Carex* clade (subgenera *Carex* and *Vigneastra*). Our results showed that the Japanese species of section *Mitratae* do not form a monophyletic group. Three major clades were found in section *Mitratae*: 1) *C. breviscapa* (subsect. *Lageniformes*) and *C. pudica* (subsect. *Nervatae*), 2) subsects. *Lageniformes*, *Nervatae* and *Leucochlorae*, and 3) subsects. *Foliosissimae* and *Pisiformes*, sensu Akiyama (1955). Subsects. *Lageniformes*, *Nervatae*, and *Leucochlorae* had characteristic insertions in ETS-1f region. Subsects. *Pisiformes* and *Foliosissimae* clade were divided into two groups, and showed the same tendency in the chromosome numbers.

T2

P0518

Reinterpretation of *Cynara baetica* based on a principle of integrative species delimitation

Sergi Massó[1,2], *Jordi López-Pujol*[1], *Roser Vilatersana*[1]
1. Botanic Institute of Barcelona (IBB-CSIC- ICUB)
2. BioC-GReB, Laboratori de Botànica, Facultat de Farmàcia, Universitat de Barcelona

Cynara is a Mediterranean genus from the family Compositae that currently includes nine species. One of them is *Cynara baetica*, a diploid (2*n* = 34) perennial herb distributed in both sides of Strait of Gibraltar (the gateway between Atlantic Ocean and Mediterranean Sea). Historically, it was described as two different species: (1) *Cynara alba*, that grows in open areas in deep and nitrogenous soils on calcareous bedrocks at altitudes between 500 and 1700 m in southern Spain, and (2) *Cynara hystrix*, that occurs in roadside ditches and open areas under the same soil conditions at somewhat higher altitudes (from 900 to 2100 m) in northern Morocco. In 1992, a morphologic study carried out by A. Wiklund merged these two species in one (*Cynara baetica*) although separated into two subspecies: *C. baetica* subsp. *baetica*, for the Spanish populations; and *C. baetica* subsp. *marocanna,* for the Moroccan ones. The aims of the present study are to: (i) analyse the genetic diversity within and between *Cynara baetica* populations, including samples from the two subspecies (subsp. *baetica* and subsp. *maroccana*); (ii) infer the phylogeographic pattern of *C. baetica* and the role of Strait of Gibraltar as a barrier; (iii) test whether there are morphologic, genetic and/or ecological differences between both subspecies; and (iv) provide conservation guidelines. To achieve these goals, and following the most recent approximations on species delimitation, phylogenetic and population genetic analyses (based on both nuclear—three AFLP primer combinations—and plastid—two cpDNA regions—DNA markers) were combined with ecological niche modelling (ENM) and morphologic studies.

Results obtained showed a clear genetic, morphological and ecological differentiation between the two subspecies. According to these results the current taxonomic treatment is modified and the conservation guidelines are proposed according this taxonomical change.

T2
P0519
Morphological studies of trichomes in the family Malvaceae

Rubeena Mattummal
University of Calicut

Plants may have hairs on their axial surfaces, superficially similar to the hairs of human body. They are called trichomes. Trichomes affect the plant in number of ways. There are mainly two types of trichomes: Glandular and non-glandular. Glandular trichomes contain or secrete a mixture of chemicals that have been found to have an enormous array and can be used as pesticides, pharmaceuticals, *etc.* Some trichomes may give resistance to plant growth. Trichomes are small protrusions of epidermal origin on the surfaces of leaves and other organs many plants. They range in size, shape, number of cells and morphology, as well as composition, from long cotton fibers, which are seed trichomes, to tiny bumps onthe surface of a leaf. Trichomes are well known to humans, even those who do not normally think about plant morphology and biochemistry. This is because they give many plants their distinctive look and smell: anyone familiar with the smell released when you handle the leaf of a tomato plant, crush fresh or dried herbs such as basil, mint or thyme, or who notices the stickiness or fuzziness of plant leaves or stems has experienced the impact of trichomes on our everyday perception of plants. They exhibit tremendous diversity, be it in their shape or the compounds they secrete. This diversity is expressed between species but also within species or even individual plants. The industrial uses of some trichome secretions and their potential as a defense barrier, for example against arthropod pests, has spurred research into the biosynthesis pathways that lead to these specialized metabolites. Because complete biosynthesis pathways take place in the secretory cells, the establishment of trichome-specific expressed sequence tag libraries has greatly accelerated their elucidation.The family Malvaceae is notable due to their higher economic importance and also due to the presence of mucilage. To know the variant distribution of trichomes on them 10 different species were selected and the study was conducted. The plant for the present study are collected from different localities, in and around the campus of farook college. The morphology of trichomes on various parts of the selected members of the family, malvaceae are observed. Identification and morphological study of the different types of trichomes are done using light microscope and taxonomic keys. Cleared photographs of the various types of tricomes on different parts of the plants are taken. The plants selected for the work are: 1. *Abelmoschus esculentus* (L.) Moench; 2. *Hibiscus lanarifolius* Willd.; 3. *Hibiscus mutabilis* L.; 4. *Hibiscus rosasinensis var. rosasinensis* L.; 5. *Hibiscus rosasinensis var. schizopetalus* Dyer, Gard; 6. *Malvaviscus penduliflorus* DC.; 7. *Sida acuta* Burn.; 8. *Sida alnifolia* L.; 9. *Sida cordata* (Burn.f.) Bross.; 10. *Urena lobata* L.

T2
P0520
Biogeography and chromosome number evolution of tribe Polygonateae (Asparagaceae: Nolinoideae)

Ying Meng[1], *Jia-Jian Wang*[1], *Yong-Ping Yang*[2], *Hang Sun*[2], *Jun Wen*[3], *Ze-Long Nie*[1]
1. Jishou university
2. Kunming Institute of Botany, Chinese Academy of Sciences
3. National Museum of Natural History, Smithsonian Institution

Polygonateae is a tribe of monocots (Asparagaceae: Nolinoideae) characterized by a sympodial rhizome, an elongated aerial stem and fleshy berries. The tribe includes four genus and ca. 100 species distributed worldwide in the Northern Hemisphere, but mainly from subtropic to temperate areas. The monophyly of the tribe has been supported by most molecular studies and usually two groups are recognized in Polygonatae: one including *Polygonatum* Mill., *Disporopsis* Hance and *Heteropolygonatum* M.N. Tamura & Ogisu and the other comprising *Maianthemum* Wigg. s.l. (including *Smilacina* Desf.). The tribe is one of the few lineages in Asparagaceae with classic disjunctions in Asia, Europe and North America and a wide variation of chromosome numbers and therefore is an ideal group to investigate the timing and origin of the disjunctions associated with chromosome evolution in the Northern Hemisphere. Here we first provide a phylogenetic framework of the evolutionary relationships among species of Polygonateae by analyzing DNA sequences from five loci: ITS, *rbc*L, *trn*K, *psb*A–*trn*H and *trn*C–*pet*N. The resulting phylogenetic hypothesis is then used to infer divergence times and historical biogeography of the tribe, as well as the chromosome number evolution. Temporal evolution in Polygonateae was estimated using a Bayesian divergence time analysis (BEAST) with fossil calibrations reported from literatures. Biogeographical patterns are explored using a dispersal – extinction – cladogenesis (DEC) model in Lagrange & RASP based on a maximum likelihood method. The evolution of chromosome numbers is investigated using likelihood method for tracking the pattern of chromosome number change along a phylogeny using ChromEvol. The origin of the two lineages within Polygonateae is inferred to have a tropical affinity from the late Oligocene to early Miocene during a time frame coinciding with paleoclimatic events in the Northern Hemisphere. Different biogeographic patterns are found between the two clades in the tribe. The early divergence of the *Maianthemum* clade is estimated in the middle Miocene with possible origin in the tropical America, while the *Polygonatum – Heteropolygonatum – Disporopsis* clade is suggested in the early to middle Miocene with possible ancestral area in the tropical to subtropical Asia. Our biogeographic results also reveal that the modern distribution of the tribe was shaped by multiple recent migrations/exchanges among areas of eastern Asia, Europe (including central Asia), and North America from the late Miocene to the Pliocene. Analyses of chromosome number evolution suggested that $x = 18$ as the ancestral base number for Polygonateae. The chromosome number evolution of the tribe is dominantly on the diploidy level with both trends of descending and ascending chromosome numbers and it seems that dysploidy has played an important role in the evolution of Polygonateae, especially in *Polygonatum*. Polyploids are found in all genera in Polygonateae, except for *Disporopsis*, but most polyploidization events are associated with infraspecific diploidy. Our results suggest that polyploids seems to be less and recent in the tribe and there is no obvious correlation between the frequency of polyploidy and their biogeographic distributions.

T2

P0521

Cetrarioid lichens (Parmeliaceae) from India

Gaurav Kumar Mishra, *Dalip Kumar Upreti*

CSIR-National Botanical Research Institute, Rana Pratap Marg, Lucknow-22001

Among the different plant groups in India lichens exhibit their rich diversity as both macrolichens (foliose and fruticose growth form) and microlichens (crustose and squamulose growth form) are found growing luxuriantly in different habitats. Based on the type of growth forms, habitat, location and type of fruiting bodies and secondary metabolites present in thallus, lichens can be categorized in different communities/groups. Among the macrolichens together with Parmelioid, Cyanolichens and Cetrarioid are the most common group of lichens in India. Cetrarioid lichens are characterized by an erect foliose or subfruticose growth form of the thallus being loosely attached to the substrate, presence of marginal apothecia and pycnidia. A total of 149 species and 25 genera of the group are reported from the world, out of which 46 species belonging to 11 genera are known from India. *Cetreliopsis hypotrachyna* (Müll. Arg.) Randlane & Saag, *Nephromopsis isidioidea* (Räsänen) Randlane & Saag, *Parmotrema subthomsonii* (D.D. Awasthi) A. Crespo, Divakar & Elix and *Nephromopsis sikkimensis* (Divakar & Upreti) Randlane & Saag are endemic to India. *Nephromopsis* with 12 species is the most dominant genera of cetrarioid lichens followed by *Cetrelia* and *Cetraria* with 10 and 7 species respectively in India. *Cetrelia cetrarioides* (Delise ex Duby) W.L. Culb. & C.F. Culb., *Parmotrema thomsonii* (Stirt.) A. Crespo, Divakar & Elix, *Cetrelia braunsiana* (Müll. Arg.) W.L. Culb. & C.F. Culb., *Nephromopsis pallescens* (Schaer.) Park and *N. stracheyi* (C. Bab.) Müll. Arg., are the most widely distributed species of this group. Among the Indian states, the state of Sikkim exhibit maximum diversity of cetrarioid lichens represented by 38 species followed by Uttarakhand with 29 species. A more intensive survey in the Indian Himalayan region will definitely add few more species to the present list as some of the difficult terrain of the area are still not explored systematically for collection of lichens.

T2

P0522

Phylogenetic reconsideration of family Phrymaceae and its relatives (Lamiales)

Hye-Kyoung Moon, *Suk-Pyo Hong*

Kyung Hee University

The family Phrymaceae had been traditionally known as a monotypic family. Recent phylogenetic studies, however, have extended its delimitation with inclusion of 12 genera (*ca.* 200 species) transferred from former Scrophulariaceae and showed the close link with Paulowniaceae and Mazaceae. Nevertheless, phylogenetic relationships between those genera in redefined Phrymaceae and with other families in Lamiales remain to be resolved. In this study, we conducted phylogenetic analyses using the sequence data of multiple genes (nuclear ITS, plastid *mat*K, *ndh*F, *rbc*L, *rps*16 and *trn*L-F) collected extensively across Lamiales to resolve their relationships. Our preliminary results from multiple genes consistently indicated that Phrymaceae is a monophyletic group in Lamiales with Paulowniaceae to be a close sister group. However, the number of major clades within the family and phylogenetic position of Mazaceae in Lamiales may differ according to the markers and sample size applied. Overall, our results support the delimitation of the current Phrymaceae, albeit Paulowniaceae required to be reevaluated. We will further discuss the phylogenetic consistency among the phylogenetic trees generated differently by respective genetic markers and their combination (NRF-2014R1A1A2058195).

T2

P0523

The phylogenetic potential of pollen and orbicules in Lamiaceae

Hye-Kyoung Moon

Kyung Hee University

Study on pollen diversity has improved our understanding on the evolution of flowering plants since pollen variation connote phylogenetic history. However, the phylogenetic potential of palynological data should be reevaluated because the traditional classification of angiosperms has been significantly changed with molecular phylogeny. Here, I revisit the phylogenetic potential of palynological information in Lamiaceae, delimitation of which has been greatly expanded as with the inclusion of many genera from Verbenaceae. As a result, I will present pollen morphological data and the presence of orbicules among representative taxa of newly proposed Lamiaceae and will discuss their phylogenetic significance (NRF-2016R1D1A1B04934156).

T2

P0524

A new phylogenetic hypothesis brings light to morphological diversity in *Chresta* (Asteraceae, Vernonieae)

Carolina Siniscalchi[1,3], *Benoit Loeuille*[2], *José Rubens Pirani*[1], *Jennifer Mandel*[3]

1. Universidade de São Paulo

2. Universidade Federal de Pernambuco

3. University of Memphis

Chresta encompasses 18 species of Compositae, mostly distributed in the Brazilian *Cerrado* and *Caatinga*. Plants of this genus present great morphological diversity, both in vegetative and reproductive structures, despite a relative small number of species. Species of *Chresta* are herbs to shrubs, usually with well-developed underground structures, occurring in a number of environments: rock outcrops, savannahs, quartzitic soils, riverbanks, wet soils and in disturbed areas. The heads are always fused in syncephalia, which can be spherical, with strongly fused capitula, or hemispherical, with the heads loosely grouped. In three species the secondary heads present indeterminate growth of their axis, a highly unusual feature within the Vernonieae. Two of the *Cerrado* species present red flowers and are hummingbird-pollinated, another rare character in the tribe. The *Caatinga* species have deeply lobed leaves, silver-green in most of the species. Palynological analysis has shown that there are three different pollen types in the genus, relative to *Vernonia* types A, C and F. This morphological diversity lead to the controverted taxonomic history of the genus, which either has been considered part of *Eremanthus*, due to the secondary heads, or split up into five different genera, namely *Argyrovernonia, Chresta, Glaziovianthus, Pithecoseris* and *Pycnocephalum*. A previous phylogenetic analysis, based on 3 molecular markers and a morphological dataset including 10 *Chresta* species, recovered the monophyly of *Chresta,* but did not present enough support to understand relationships within the genus, neither brought new insights on how morphological diversity evolved. In this work we used target capture combined with high-throughput sequencing, with a previously developed set of 1,000 nuclear markers, to recover the relationships within *Chresta,* aiming to achieve a better understanding of its morphological diversity. The obtained matrix contains ca. 600,000 characters for each terminal. Parsimony analysis was run in TNT, generating one tree, with maximum bootstrap support in 1,000 replicates. Maximum likelihood analysis used RAxML, with 1,000 bootstrap replicates. Both analyses recovered the same topology for internal relations in *Chresta,* with maximum bootstrap support at all nodes, except for one. *Chresta martii* is sister to all other species, which are grouped in two clades, representing *Caatinga* and *Cerrado* species, both of which are composed by two clades each. The ancestral syncephalium type appears to be hemispherical, with loosely joined capitula, with the spherical, tightly joined capitula type arising twice, once in each clade. The two red-flowered species are separated in two different clades, indicating that this character evolved twice in the group history, as well as the pollen type A that is exclusive from these two species. The clades obtained loosely follow the generic divisions previously proposed by other authors, with the important distinction that *Argyrovernonia* and *Glaziovianthus* would not be monophyletic.

T2

P0525

Taxonomic study of *Pedicularis* ser. *Verticillatae* (Orobanchaceae) in Taiwan and Japan

Kaori Murayama[1], *Richard H. Ree*[2], *Kuo-Fang Chung*[3], *Chih-Chieh Yu*[4], *Noriyuki Fujii*[1]

1. Kumamoto University

2. Chicago Field Museum

3. Academic Sinica

4. Taiwan University

Pedicularis ser. *Verticillatae* in Taiwan (Taiwanese populations) has often been classified as *P. verticillata* L. However in Flora of China, the Taiwanese populations are classified as *P. transmorrisonensis* Hayata., otherwise they are classified as *P. refracta* var. *transmorrisonensis* (Hayata) Hurus. by Hurusawa (1948). To taxonomically reexamine the Taiwanese populations, we performed molecular phylogenetic analysis using the internal transcribed spacer (ITS) regions of nuclear ribosomal DNA and the trnK-matK region of chloroplast DNA. We also compared nine characteristics to determine morphological differences among the Taiwanese populations, *P. verticillata* and *P. refracta*. In both ITS and cpDNA trees, the Taiwanese populations were not included in the clade of so-called *P. verticillata*, and instead formed a clade at high bootstrap probabilities (95-100%) with *P. refracta* (Maxim.) Maxim., which is an endemic species of Kyushu, Japan. Furthermore, in the morphological analysis, significant differences were observed between the Taiwanese populations and *P. verticillata* in the following characteristics: angle of the apex of the galea (upper lip of the corolla), corolla length (excluding the calyx tube), width between rachis and sinus of the pinnae and width between rachis and longest pinna *etc.* On the other hand, there were two significant differences were observed between the Taiwanese populations and *P. refracta.* They were the corolla length (the Taiwanese populations = approximately 9 mm, *P. refracta* = approximately 13 mm) and the plant length (the Taiwanese populations = approximately 18.3 cm, *P. refracta* = approximately 31.4 cm). These results suggest that the Taiwanese populations are not *P. verticillata* but are very closely related to *P. refracta.* However there were also significant morphological differences between the Taiwanese populations and *P. refracta*, and their habitat were also different. The Taiwanese populations occur in the alpine zone of high mountains (alt. approximately 3000-3500m) but *P. refracta* occur at grassland habitat of lower mountains (alt. approximately 400-800 m). Therefore, we concluded that *P.* ser. *Verticillatae* in Taiwan (the Taiwanese populations) should be treated as *P. transmorrisonensis* Hayata, as described by Bunzo Hayata in 1915 (Icones Plantarum Formosanarumosa). *P. transmorrisonensis* is an endemic species in Taiwan closely related to *P. refracta.*

T2

P0526

Phylogeny of the *Bouteloua repens* complex (Poaceae: Chloridoideae): A preliminary study based on nuclear ribosomal internal transcribed spacer and chloroplast *trn*T-L-F sequences.

Geraldine Murillo-Pérez[1], *María-Elena Siqueiros-Delgado*[1], *Gilberto Ocampo-Acosta*[1], *J. Travis Columbus*[2]

1. Universidad Autónoma de Aguascalientes

2. Rancho Santa Ana Botanic Garden

The *Bouteloua repens* complex (Poaceae: Chloridoideae) has been treated as a group of five species distributed from south USA to Brazil. Due to considerable morphological variation, putative hybridization, polyploidy, and apomixis, circumscription and relationships among taxa have been uncertain. To infer the phylogeny of this complex, two non-coding regions –the chloroplast *trn*T-L, *trn*L-F intergenic spacers, and *trn*L intron, and the nuclear ribosomal internal transcribed spacer (ITS1+5.8S+ITS2), were sequenced and analyzed by Bayesian inference method. The ITS phylogeny was used to carry out a character state reconstruction analysis to test the utility of morphological characters used for species circumscription, and to reconstruct the historical biogeography. Nuclear and plastid data revealed similar phylogenetic patterns. Both support the monophyly of the *Bouteloua repens* complex, but not the species monophyly, except for *B. americana* and *B. repens*, which form a strongly supported clades in the ITS, *trn*T-L-F and combined results. ITS and combined phylogenies show the *B. alamosana* samples from north and south in different clades. However, the *trn*T-L-F hypothesis resolves it as monophyletic. The character-state reconstruction, shows that all characters used as diagnostic are homoplastic, except the shape of the upper floret, which could be the synapomorphy of the *B. americana* clade, since is the only species showing an upper floret reduced to a column. According to the results of historical biogeography, the *B. repens* complex originated in central Mexico, and it was later dispersed until it acquire its current distribution.

T2

P0527

Classification and phylogeny of the genus *Dianella* (Asphodelaceae)

Karen Muscat

The University Of Melbourne

The genus *Dianella* (the flax lilies) is a group of petaloid monocots (c.60 taxa) that mostly occur in Australia (more than half of the taxa), throughout South-East Asia, the Pacific (east to Hawaii) to New Zealand and across to the Indian Ocean to Madagascar. Current infrageneric and species-level classification is based on morphology without detailed molecular phylogenetic analyses. The first detailed global phylogeny will be presented with closely related genera using chloroplast and nuclear markers. Sampling includes the majority of Australian *Dianella* and extra Australian species from New Zealand, New Caledonia, New Guinea, Norfolk Island, Lord Howe Island, Hawaiian islands, Japan, Taiwan, Brunei, Malaysia, Bangladesh, Madagascar and Mauritius. Results show a good correlation between morphological patterns and molecular relationships. Biogeographic patterns across the global range and within Australia are identified, indicating an early divergence between Eastern and Western Australian lineages. The study also confirmed numerous new species for Australian and extra Australian *Dianella*.

T2

P0528

Multiple-biomineralization in trichomes of Boraginales

Adeel Mustafa, *Hans-Jürgen Ensikat, Maximilian Weigend*

Nees-Institut für Biodiversität der Pflanzen, Universität Bonn

Biomineral deposits are widely prevalent in various types of trichomes of higher plants. The occurrence of mineralized trichomes on plant surfaces is considered to be crucial for plant-environment interactions, particularly to deter herbivores and mitigate irradiation and water stress. Our recent study has demonstrated the presence of calcium phosphate in trichomes of Loasaceae. The present study investigates biomineralization in the trichomes of a range of different, closely allied plant groups in the order Boraginales (Boraginaceae; Codonaceae; Hydrophyllaceae; Namaceae). Scanning Electron Microscope (SEM) and Energy Dispersive X-ray (EDX) analyses were employed for that purpose. The trichome tips were found to consist of either silica or calcium phosphate in high concentrations, providing sharpness, hardness, stiffness and toughness. The basal part of the trichome cells can contain relatively high concentrations of silica or calcium phosphate, whereas the median part, the shaft, is mainly mineralized with calcium carbonate; calcium phosphate and silica occur are here only found as a minor components. Calcium phosphate tips were found in several plant families of Boraginales, but not in Boraginaceae themselves. The process and complex patterns of biomineralization in the trichomes of Boraginales provide a new way of looking at trichomes in the light of systematic relationships and also regarding their functionality.

T2

P0529

Phylogeny and taxonomy of *Leptopetalum biflorum* (Rubiaceae)

Akiyo Naiki[1], *Kensei Akai*[2], *Hidetoshi Kato*[3], *Yoshiteru Komaki*[4], *Keiko Mizunashi*[4], *Tetsuo Ohi-Toma*[4]

1. Tropical Biosphere Research Center, University of the Ryukyus

2. Okinawa Churashima Research Center

3. Graduate School of Science and Engineering, Tokyo Metropolitan University

4. Botanical Gardens, Graduate School of Science, The University of Tokyo

The genus *Leptopetalum* (Rubiaceae), recently re-circumscribed and divided from *Hedyotis* s.l., comprises three herb and four shrub species. The herb species (*L. biflorum*, *L. pteritrum*, and *L. strigulosum*) are widely distributed in Asia, Southeast Asia, and Pacific islands, whereas the shrub species (*L. foetidum*, *L. grayi*, *L. mexicanum*, and *L. pachyphyllum*) are limited in some oceanic islands of Japan (the Bonin Islands), Marianas, and the West Pacific regions. One of the herb species, *L. biflorum*, has large variation in plant shape and size, and habitat. We recognized at least two morphological types of the plants. One [type A] has thick coriaceous leaves and is well-branched and procumbent on coastal limestone rocks. The other [type B] has chartaceous leaves and is few-branched and erect in inland open habitat. The present molecular phylogenetic analyses using nucleotide sequences of nuclear ribosomal ITS and ETS regions, and chloroplast *rps16* intron and *petB-petD* region indicated that the shrub species formed one clade and were nested into herb species lineages, and that plants of *L. biflorum* are divided into two polyphyletic clades, which correspond to the type A and B. These two types can be treated as different species. Characteristics of type B plants correspond to those of the type specimen of *L. biflorum*. Type A plants should be treated as another species based on *Hedyotis albido-punctata*, *H. paniculata*, *H. racemosa*, or *H. taiwanense*, or as a new species.

T2

P0530

The family Gesneriaceae in India- An overview

Santhosh Nampy, *Janeesha A P*

Department of Botany, University of Calicut, Kerala

Gesneriaceae Rich. & Juss. in DC., with ca 3400 species, has roughly an equal distribution in the Neotropics and Paleotropics. The plants are mostly perennial herbs with conspicuous showy flowers. The family is easily distinguished by opposite deccusate (sometimes anisophyllous or alternate) leaves, pair-flowered cymes, pentamerous bisexual flowers with zygomorphic corolla, usually four or two stamens, presence of an annular or a one-sided disc and 1-locular ovary with parietal placentation. The Indian

Gesneriaceae belongs to the subfamily Didymocarpoideae and its two tribes, Epithemateae and Trichosporeae. The family is represented by 22 genera and 109 species in India. Since 2007, and essentially based on molecular data, many genera have been redefined, several new genera established and a considerable number of genera sunk into synonymy and a number of additional taxa has been described or recorded. The present paper discusses the diversity, distribution and endemism of the family in India.

T2

P0531

Origin and evolutionary history of freshwater Rhodophyta: Further insights based on phylogenomic evidence

Fangru Nan, *Jia Feng, Junping Lv, Qi Liu, Kunpeng Fang, Chaoyan Gong, Shulian Xie*

Shanxi University

Freshwater representatives of Rhodophyta were sampled and the complete chloroplast and mitochondrial genomes were determined. Characteristics of the chloroplast and mitochondrial genomes were analyzed and phylogenetic relationship of marine and freshwater Rhodophyta were reconstructed based on the organelle genomes. The freshwater member *Compsopogon caeruleus* was determined for the largest chloroplast genome among multicellular Rhodophyta up to now. Expansion and subsequent reduction of both the genome size and GC content were observed in the Rhodophyta except for the freshwater *Compsopogon caeruleus*. It was inferred that the freshwater members of Rhodophyta occurred through diverse origins based on evidence of genome size, GC-content, phylogenomic analysis and divergence time estimation. The freshwater species *Compsopogon caeruleus* and *Hildenbrandia rivularis* originated and evolved independently at the inland water, whereas the *Bangia atropurpurea*, *Batrachospermum arcuatum* and *Thorea hispida* are derived from the marine relatives. The typical freshwater representatives Thoreales and Batrachospermales are probably derived from the marine relative *Palmaria palmata* at approximately 415-484 MYA. When comparing the organelle genome content, it was discovered that *trx* and *grx* genes in chloroplast genome probably involved in the polysaccharides synthesis of Rhodophyta. Gene group substitution rates analysis showed that the ribosomal protein (*rps*), RNA polymerase (*rpo*) and ATPase (*atp*) genes in chloroplast, ATPase (*atp*) and succinate dehydrogenase (*sdh*) genes in the mitochondria were substantially elevated. The primary inference on the origin and evolutionary history of freshwater Rhodophyta proposed in this study needs to be testified with more organelle genome sequences and wider global sampling.

T2

P0532

Family Commelinaceae in India: New taxa, infraspecific taxa and new additions

Mayur Nandikar[1], *Rajaram Gurav*[2]

1. Naoroji Godrej Centre for Plant Research (NGCPR), Shirwal, Satara, Maharashtra, INDIA 412801

2. Department of Botany, Shivaji University, Kolhapur, Maharashtra, INDIA

Since last decade authors are working on family Commelinaceae in India. In the journey we screened thousands of live populations, made consultation of herbaria in search of different taxa in the family. It is a noble experience, resulted in documentation of about 15 genera and about 95 species. While revisiting family, numbers of taxa show so much variation that it has not been possible to divide them satisfactorily into new taxa or infraspecific taxa. However, some are consistent in their characters and have been recognized as a distinct species or variety. Authors described three new taxa in the genus *Murdannia* viz. *M. brownii*, *M. saddlepeakensis*, *M. ugemugei* and one variety *M. spirata* var. *flavanthera*. New reports have been made for Sri Lankan taxa *M. striatipetala* and *M. spirata* var. *parviflora*. Likewise, a new species has been added in the genus *Commelina*, i.e. *Commelina badamica*. Besides, four new taxa are being proposed and communicated. The present paper highlighted all these additions and a note on their distinctness.

T2

P0533

Divergence times of *Onosma* L. (Boraginaceae) inferred from nrDNA ITS and two cpDNAs, *trn*H-*psb*A and *rpl32*-*trn*L $_{(UAG)}$ sequences data

Fatemeh Nasrollahi

Tarbiat Modares University

Onosma L. a taxonomically difficult group consisting of some 150 species is the largest genus of the tribe Lithospermeae. The Asia and Mediterranean regions particularly Iran and Turkey are the main centers of diversity of the genus. They usually grow in open, sunny habitats, and are well adapted to dry and rocky soils and steppes. The name onosma is derived from a Latin word "osma" originated from a Greek "osmê" (smell) and Stearn reported that *onosma* is treated as feminine gender. They have scorpioid inflorescence and tubular corolla. Hair characters has major taxonomic importance in *Onosma*. Corolla colour, shape and size are used as taxonomic characters in this genus. Divergence times were estimated from Bayesian Inference (BI) analysis using BEAST ver. 1.7 [Bayesian Evolutionary Analysis Sampling Trees]. We applied the substitution model selected by the AIC for each gene and a Relaxed Clock. Bayesian and Maximum Likelihood phylogenetics analysis applied to nrDNA ITS and *rpl32-trn*L $_{(UAG)}$ and *trn*H-*psbA* sequences for 42 species (40 ingroups and *Echium italicum*, *E. amoenum* were selected as outgroup). Our result showed that *Onosma* forms a monophyletic group. Since there are no reliable fossils of *Onosma*, the clock was calibrated using the estimate of mean age 30 Mya for a node, encompassing from *Echium* through *Onosma*. BEAST chronogram is congruent with those resulting from Bayesian analyses. The timing of diversification events is demonstrated by a chronogram derived from the all-compatible Bayesian consensus tree with dates for selected nodes constrained to the mean of penalized likelihood estimates obtained from sub-sampling of 1,000 trees from the posterior distribution of the Bayesian MCMC tree search. Our results illustrated that mean nodal Bayesian divergence time estimates indicated that stem–node (the root) is estimated to be middle Oligocene, ~30 Mya. The overall estimated mean ages and mean rates of evolution were obtained.

T2
P0534

Disentangling complex evolutionary relationships in the sun orchids (*Thelymitra*, Orchidaceae)

Lars Nauheimer[1], *Mark Clements*[2], *Rowan Schley*[3], *Katharina Schulte*[1,4]
1. Australian Tropical Herbarium
2. Centre for Australian National Biodiversity Research (CANBR)
3. Imperial College London
4. CSIRO National Collections and Marine Infrasructure

Thelymitra is a genus of terrestrial orchids in the tribe Diurideae, which comprises over 118 species occurring in Australasia, with the majority of species in Australia. Despite considerable taxonomic work over the past two decades, evolutionary relationships and species delimitation within *Thelymitra* are still poorly understood. This is mainly due to the high degree of morphological variation found within the genus and complex evolutionary relationships due to past hybridization events. For a first assessment of the infrageneric relationships and the degree of genetic diversity within *Thelymitra*, we conducted a molecular phylogenetic study based on one nuclear and three plastid loci including the majority of the genus' species. This traditional sequencing approach resolved major clades, which in part corresponded well with prominent morphological features, e.g., of the column. Nevertheless, relationships between these major clades or between closely related species remained largely unclear due to lack of phylogenetic signal. Furthermore, incongruences between the plastid and nuclear phylogenies were detected that indicate hybridization events in recent times as well as in the early evolution of the genus. Thus, our four loci study provided the first molecular phylogenetic insights into relationships within *Thelymitra,* albeit with limited resolution. We have now embarked on applying high-throughput sequencing on an Illumina platform combined with target enrichment using target capture probes. Nuclear genes for probe development were selected based on transcriptome sequences across the tribe Diurideae, including Thelymitrinae. This wide sampling was chosen to ensure future combinability of data sets across the tribe. The plastid genes for the development of target capture probes were selected based on whole plastome sequences of Thelymitrinae generated via shotgun high-throughput sequencing. This approach will yield high numbers of nuclear and plastid regions and allow the reconstruction of phylogenetic relationships at finer scales, including the assessment of species delimitation in different species complexes. The increased phylogenetic resolution will then allow us to reconstruct ancestral-character states, to carry out divergence time estimations and ancestral range reconstructions, and thus will increase our understanding of the spatial and temporal evolution of this iconic group of Australasian orchids.

T2
P0535

Phylogenomics of *Dichanthelium* (Paniceae: Poaceae): Biogeography and the problem of high sympatry

Kurt Neubig[1], *Lucas Majure*[2], *Robert Naczi*[3], *J. Richard Abbott*[3], *Walter Judd*[4], *W. Mark Whitten*[4]
1. Southern Illinois University Carbondale, Plant Biology, Carbondale, IL, 62901
2. Desert Botanical Garden, Department of Research, Conservation, and Collections, Phoenix, AZ, 85008
3. The New York Botanical Garden, Bronx, NY, 10458
4. University of Florida, Florida

Dichanthelium is distributed throughout the Americas and Hawaii, with its highest species diversity in the eastern United States. Species delimitation within the genus is contentious, with most floristic treatments differing drastically in the number and circumscription of species. Low levels of sequence divergence with conventional molecular markers result in poor resolution at deep phylogenetic levels. Therefore, we used a "genome skimming" approach to reconstruct phylogenetic relationships in *Dichanthelium* for over 80 different samples. *Dichanthelium* is monophyletic and although many deep phylogenetic relationships are highly resolved with these data, many species in eastern North America show very little DNA sequence divergence. Hawaiian *Dichanthelium* species form a clade and one species there is of hybrid origin. However, despite high sympatry, clear hybrids are very rare in *Dichanthelium* of North America. Many complex nomenclatural problems are ameliorated with these phylogenetic data, but new problems of polyphyly have arisen, such as in *D. commutatum*. Using ultrametric analyses, the timing of major biogeographic patterns will be discussed.

T2
P0536

Convergent recruitment of new pollinators triggered by independent hybridization events in *Narcissus*

Isabel Marques[1,2], *Andreas Jürgens*[3,4], *Javier Fuertes-Aguilar*[5], ***Gonzalo Nieto Feliner***[5]
1. High Polytechnic School of Huesca, University of Zaragoza
2. University of British Columbia
3. University of KwaZulu-Natal
4. Technische Universität Darmstadt
5. Real Jardín Botánico, CSIC. Madrid

Hybridization is recognized as an important pathway to generate new diversity in plants. But hybrids need to achieve isolation to prevent gene flow and avoid assimilation into the parental gene pool. The recruitment of novel pollinators by hybrids has been hypothesized to be one way in which such reproductive isolation can be achieved. We tested whether pollinators contributed to isolation between two natural *Narcissus* hybrids and their progenitors using pollination experiments, observations, plus morphological and floral-volatile measurements. These hybrids share the same maternal but different paternal progenitors. We found that only the hybrids were visited by ants whereas the progenitors whereas the progenitors were pollinated by butterflies and bees, respectively. The two hybrids exceeded their progenitors in floral-tube aperture size and nectar production. The emission of floral volatiles by hybrid plants was not only equal to or higher than the progenitor species, but also contained some new compounds not produced by the progenitors. The recruitment of ants as novel pollinators in the hybrids involved the combination of increased nectar secretion and the production of novel floral scent compounds. A breakdown of chemical defense against ants may also be involved. This study provides support for the hypothesis that the recruitment of novel pollinators can contribute to reproductive isolation between hybrids and their progenitors.

T2

P0537

When ribosomal loci go mad in *Anacyclus* (Asteraceae, Anthemideae): High levels of 45S rDNA site-number diversity in a diploid plant genus

Marcela Rosato[1], *Inés Alvarez*[2], *Gonzalo Nieto Feliner*[2], ***Josep Antoni Rosselló***[1,3]

1. Jardín Botánico, ICBIBE-Unidad Asociada CSIC, Universidad de Valencia

2. Real Jardín Botánico, CSIC, Madrid

3. Marimurtra Botanical Garden, Blanes

Indications that ribosomal DNA (rDNA) plays a more important role in cell machinery than classically acknowledged have raised interest in understanding all aspects of this family of repetitive DNA. One of these aspects is the structure of rDNA within the genome. The number of rDNA sites and their distribution has been assessed in a significant sample of angiosperms revealing clear regularities but also intra-specific variation, particularly in polyploid and cultivated plants. We aim to test if the structure of the 45S rDNA has been affected by intense contemporary hybridisation in a mostly weedy, Mediterranean genus. The number and location of rDNA 45S sites was assessed in 196 individuals from 47 populations, belonging to all nine species in the genus *Anacyclus*. We used fluorescence in situ hybridization and karyotype analysis. The number of 45S rDNA sites ranged from 4 to 11. This along with the chromosomal location (terminal or interstitial) and other features allowed the recognition of 14 different karyological ribosomal phenotypes that were arranged in a network to facilitate the inference of patterns of cytogenetic evolution. The variation (and inferred dynamism) of rDNA sites found in *Anacyclus* is outstanding: (1) The range of variation for rDNA sites spans a large part of the total range found per diploid complement across all angiosperms. At the Asteraceae level, *Anacyclus* shows (2) the largest range of 45S rDNA site-number variation and the highest number of rDNA sites per diploid chromosome complement. The distribution of karyological phenotypes across species suggests that the main factor in the dynamism found in *Anacyclus* is genomic shock effects resulting from contemporary hybridization.

T2

P0538

Phylogeography of *Orostachys spinosa* (L.) C.A. Mey (Crassulaceae) based on chloroplast DNA markers

Arthur Nikulin, *Andrey Gontcharov*

Federal Scientific Center of the East Asia Terrestrial Biodiversity FEB RAS

We used the sequence data of *trn*H-*psb*A, *trn*Q-*rps*16 and *rpl*32-*trn*L cpDNA intergenic spacers to investigate genetic differentiation within and among populations of *O. spinosa*, crassulacean species characterized by a wide distribution range, spanning from the Siberia's Pacific coasts to the Southern Urals, to analyze the genealogical relationships between haplotypes, and to reconstruct the possible ways of the species range extension. This study provides the first molecular data of the subsection *Appendiculatae* for a better understanding of the evolutionary history of the Far-Eastern crassulacean floras. A total of 621 sequences of three cpDNA loci were obtained from 207 plants representing 25 populations. The length of alignment was 2,345 bp, (ca. 1,4–1,6% of the plastid genome). A number of structural autapomorphies and synapomorphies (indels and inversions) were observed in the data set. Nucleotide substitutions and indel variations revealed 86 haplotypes. Most of these (59; 68%) were found in a single specimen (private haplotypes) and no haplotype sharing between populations was detected. The highest diversity was observed in populations from Altai Mts. yielding 32 haplotipes: almost each individual here carried a private haplotype. Somewhat increased haplotype diversity was also noted in populations from the south-western shore of Lake Baikal and the Tatar Strait shore. The median-joining network revealed two groups of haplotypes, Eastern and Western, separated by 11 mutation steps, of these 2 mutations were "synapomorphic". The Eastern group comprised 27 haplotypes (92 accessions) from populations of the south-eastern part of *O. spinosa* range. The Western group included a comparable number of accessions (115) but was characterized by a significantly higher number of haplotypes (59). Analysis of molecular variance showed that 76.36% of the genetic variation was due to differences between populations. For the among groups analysis, only 21.6% of the total genetic variance resided within populations of one group. The majority of molecular variance was attributable to differences among populations within groups (62.12% of the total), and those differences were significant again. Based on cpDNA, the genetic structure of *O. spinosa* was characterized by relatively low genetic variation within and high genetic differentiation between populations suggesting a restricted gene flow between the populations. The high genetic diversity and differentiation observed in our study is likely a result of isolation of *O. spinosa* populations. Despite quite extensive distribution range the species is characterized by an "island" distribution pattern because its populations are restricted to discontinuously scattered rock crevices, dry slopes and similar mostly xeric habitats. The high levels of genetic diversity parameters of the Altai populations suggest that they have a long evolutionary history. We hypothesize that the South Siberian Mountains is a likely center of *O. spinosa* origin from which the species was spreading in several directions: westwards to the Southern Urals and to the east by several lines: to the lake Baikal area; north-east Asia (Yakutia, Magadan, *etc.*), and the south-east (China and Primorsky Territory). This research was supported by the Russian Foundation for Basic Research grants (16-34-00176 and 15-29-0250515).

T2

P0539

Assessment of phylogenetic relationships between *Sedum* L. and related genera (Sempervivoideae, Crassulaceae) by ITS rDNA sequence comparisons

Vyacheslav Nikulin, *Andrey Gontcharov*

Federal Scientific Center of the East Asia Terrestrial Biodiversity FEB RAS

Phylogenetic relationships in the most species-rich and taxonomically complex crasulacean genus *Sedum* and related genera were analyzed using 223 ITS rDNA sequences. To identify positional homology between divergent sequences, secondary structure models were generated for each sequence by mfold web server (http://mfold.rna.albany.edu) and used to guide the alignment. Our putative model of the ITS1 and ITS2 secondary structures generated with the thermodynamic-based folding conformed models predicted for other plant groups. Our molecular phylogenetic data corroborated results of previous studies based on other markers and limited datasets and supported the monophyly of the tribes Aeonieae, Semperviveae, and Sedeae. In the tree species were grouped into four clades/clusters: Acre, Leucosedum, Sempervivum and Aeonium. Sedeae accommodated majority of *Sedum* species either in the robust Acre clade or paraphyletic Leucosedum cluster. Besides *Sedum*, these lineages comprised a number of genera that were mostly embedded among *Sedum* taxa. Our study established only a few significant subclades in Leucosedum, and neither a morphological nor a lucid biogeographical pattern could be inferred from the phylogenetic pattern. Nevertheless, the extended taxon sampling confirmed the affinity of the North American genus *Dudleya* to Leucosedum and revealed its close relationship with *Sedum* subgen. *Gormania* members from the same continent. This affinity was supported not only by bootstrap and posterior probability thresholds, but also by a molecular synapomorphy in ITS rDNA sequence secondary structure. In the most species- (> 500 spp.) and phenotypes-rich *Acre* clade ITS rDNA data provided support for a sister relationship between Eurasian and American taxa forming respective lineages. Placement of *S. nudum* and *S. bressemoretii* from Madeira in the American subclade was the only exception suggesting either long-distance dispersal of these species from America to Macaronesia or Macaronesian ancestry of the American subclade in *Acre*. The Eurasian subclade of *Acre* was the only large lineage in the tree that accommodated no other genera except *Sedum* and predominantly distributed in Asia species. We revealed two subclades (A1 and A2) comprising taxa co-occurring in East China, Taiwan, Korea, and Japan, but despite largely overlapping distribution, there was no indication of their common ancestry. The presence of species occurring in the Himalayas and adjacent regions in the A2 subclade traces its origin to these territories, while the A1 subclade currently includes only East-Asian taxa. Complex relationships between *Sedum* spp. and other genera in the American subclade of Acre, within Leucosedum and Aeonium clades highlight the inadequate genus concept in the tribes Aeonieae and Sedeae. In the former, this problem relates only to a few distinct species occupying a deviant phylogenetic position that strongly suggests their alternative generic assignment. In the otherwise monophyletic Sedeae, *Sedum* represents a grade of clades/lineages with at least 10 genera embedded into it and often also being para- or polyphyletic. This pattern of relationship calls for re-evaluation of the genus concept in the family Crassulaceae and the tribe Sedeae in particular. This research was partially supported by the Russian Foundation for Basic Research grants (16-34-00176 and 15-29-0250515).

T2

P0540

Ecotypic adaptation to serpentine soil and acidic wetland in *Aster rugulosus*

Takako Nishino[1], *Shota Sakaguchi*[2], *Naoko Ishikawa*[3], *Keitaro Fukushima*[4], *Yuji Isagi*[2], *Motomi Ito*[3]

1. Osaka Prefecture University
2. Kyoto University
3. The University of Tokyo
4. Tokyo Metropolitan University

Ecotypic adaptation to edaphic environment has promoted the study of speciation and many reports revealed that rapid physiological response was followed by gradual ecological differentiation. Serpentine soils contain toxic metals and poor basic nutrition for plants, so that serpentine areas restrict novices from entering. *Aster rugulosus* is a perennial species endemic to Japan, which has insect-pollinated flowers and disperses achenes with pappus. It is thought to be an endangered or a vulnerable herb in several districts due to abrupt decrease in habitats. *A. rugulosus* populations are growing in two contrastive edaphic conditions: non-serpentine, acidic wetland with clay and fine sand in lowland, and low mountainous regions with dry and fragile serpentine soils. All populations have some rhizomes in autumn in the swamp zone, whereas only a few populations localized in narrow serpentine zone near the Shizuoka-Aichi border, the central part of Japan, do not develop their rhizomes in the crumbly serpentine soils. Here, we studied phenotypic differences between locally adapted serpentine and wetland populations of *A. rugulosus*. Some life-history traits, especially related to reproduction were measured in natural populations and germination properties were tested under controlled artificial conditions. Furthermore plants from serpentine and non-serpentine sites were grown in reciprocal soil transplant experiments within a common nursery. The present study also examined the relationship of genetic diversity between serpentine and non-serpentine 20 populations using of allozyme and microsatellite polymorphism. The minimum size for flowering of serpentine populations was smaller than that of non-serpentine ones. The entities inhabiting serpentine sites bore achenes with the larger body and shorter pappus so that seed dispersal occurs near their mother plants surrounded by vacant germination sites. This strategy of life history in serpentine populations could compensate for rhizome loss and adapt to physical disturbance of the crumbly serpentine soils. Seed germination was indicative of variation in serpentine tolerance. On serpentine soils, the germination rate, the subsequent growth and the seedling survival in non-serpentine populations was much inferior to which of the serpentine populations. 50% individuals of non-serpentine were survived after transplanting to serpentine soils. It would be suggested that non-serpentine *A. rugulosus* with potential tolerance to serpentine succeeded in entering the serpentine area. Genetic diversity peculiar to serpentine populations has not occurred yet so that lapses of time from establishment of the serpentine ones would be similarly situated to the local non-serpentine ones. So each of the serpentine populations would be originated from the neighbor non-serpentine ones by seed dispersal and as the generations go by, the survivors with serpentine tolerance succeed in ecological adaptation, especially life history traits.

T2

P0541

Allelopathy as a biochemical marker: allelopathic effect of the Egyptian *Marrubium* L. Species extracts on seed germination and seedling growth of *Raphanus sativus* L.

Salama El-Darier, Ream Marzouk, Sania Kamal, ***Iman Nour***
Botany and Microbiology Department, Faculty of Science, Alexandria University, Alexandria, Egypt

A numerical study were carried out to investigate the allelopathic potential of two *Marrubium* species (donor species); *M. alysson* L. and *M. vulgare* L. distributed in 11 locations along the western Mediterranean coastal region of Egypt. Aqueous extracts with different concentrations were tested on *Raphanus sativus* L. (recipient species). The results obtained were subjected to numerical analysis to discriminate among the eleven locations. The use of the germination characters such as plumule (PL) and radicle (RL) lengths, plumule (PGR) and radicle (RGR) growth rates and radicle/ plumule (RL/PL) ratio demonstrated a reliable dendrogram. *M. vulgare* samples were separated in a major group, whereas, the other major group containing *M. alysson* samples were further distinguished into four groups. The performed cluster analysis finally supported the obtained dendrogram from SDS-PAGE protein pattern and RAPD-PCR. Subsequently, the allelochemical compounds belonging to a class of secondary metabolites may be served as chemotaxonomic markers at both inter- and intraspecific levels which indicated the presence of particular biosynthetic pathways that conserved among them.

T2
P0542

Evaluation of repetitive-DNA-sequence-based PCR fingerprinting for discriminating of two *Nostoc* strains

Bahareh Nowruzi
Department of Biology, Science and Research Branch, Islamic Azad University, Tehran, Iran

The identification and taxonomy of cyanobacteria species is still problematic and confusing, often being based on current morphological and physiological studies, which generates confusing classification systems based on plastic characters that vary with the environmental and cultural conditions. The present research aimed to investigate through a polyphasic approach, the differences in morphological and genotypic features of two cyanobacteria strains isolated from paddy fields of Iran, belonging to the family Nostocaceae (subsection IV. I). Challenges arose when the two *Nostoc* strains could not be discriminated by cloned 16S rRNA and ITS genes sequencing. As a result, the PCR-based techniques based especially on fingerprinting of repetitive DNA fragments such as Highly iterated palindromes (HIP1) has been used for assessing better resolution amongst closely related two *Nostoc* species. Interestingly, the DNA fingerprints generated using four HIP markers viz. HIP-AT, HIP-CA, HIP-GC, and HIP-TG showed 100% polymorphism in two *Nostoc* strains and each marker produced unique and strain-specific banding pattern. This study highlights the efficiency of DNA fingerprinting technique for assessing proximity of small number of strains, but probably for larger number of strains belonging to very closely related orders, the inclusion of DNA fingerprints in conjunction with other molecular tools would be necessary for better resolution of genetic diversity, genetic proximity, and taxonomic affiliation.

T2
P0543

Taxonomic studies of the genus *Indigofera* Linn., in Nigeria

Onyebuchi Ephraim Nwankwo
Ebonyi State University, Abakaliki, Nigeria

Taxonomic studies were carried out on the macro-morphology, foliar anatomy and palynology of ten species of the genus *Indigofera* Linn., in Nigeria. The ten species have odd-pinnate compound leaves except *I. nummulariifolia* and *I. subulata* with simple and trifoliate leaves respectively. The leaflets' surfaces are pubescent in *I. hirsuta* and glabrous in other species studied. The leaflets' size ranged from 0.7×0.3 cm^2 in *I. geminata* to 3.2×1.7 cm^2 in *I. macrophylla*. The epidermal cell shapes are mostly irregular on both surfaces of all the species studied except in *I. capitata*, *I. nummulariifolia* and *I. oblongifolia* in which they are polygonal. Sizes of the pollen grains ranged from 22.9×20.5 µm^2 in *I. subulata* to 35.2×28.5 µm^2 in *I. nummulariifolia* while the mean of exine thickness ranged from 0.9 µm in *I. congolensis* to 4.4 µm in *I. capitata*. The grains are tricolpate and tricolporate, subprolate and spheroidal except in *I. spicata* in which it is prolate. Marked variation exists in the number of leaflets per petiole and pollen grain aperture in *I. macrophylla* collected from two different locations. The leaf type and arrangement, number of leaflets per petiole, flower colour, stomatal type and epidermal cell shapes are significant in the taxonomy of the genus *Indigofera*. A distributional map showing the diversity of the species in Nigeria is provided.

T2
P0544

Patterns of hybridization in sympatric populations in *Goodyera* (Orchidaceae)

Sang-Wook Han, Seon Won Seo, ***Sang-Hun Oh***
Daejeon University

Goodyera is a relatively small genus in Orchidaceae with about 40 species distributed worldwide and characterized by creeping rhizomes, rosette leaves with often marked lines, and single anthers with two sectile pollinia. In Korea, six species (*G. schlechtendaliana*, *G. biflora*, *G. repens*, *G. rosulacea*, *G. maximowicziana*, and *G. velutina*) and one hybrid, *G.* x *tamnaensis*, are distributed. All of these species, except for *G. rosulacea*, occur on Jeju Island, and some of which grow together at one site, forming sympatric populations. As part of comprehensive systematic study of *Goodyera*, we located sympatric populations of *G. maximowicziana* and *G. velutina* on Jeju Island. To understand patterns of morphological variation between the species, we examined diagnostic characters for *G. velutina* and *G. maximowicziana* such as stem color, leaf venation, color, and margin, and flower color in 53 individuals of six species. We found that at least three intermediate forms occurred in sympatric populations of *G. velutina* and *G. maximowicziana* on Jeju Island, suggesting hybridization may have played a role in generating morphological diversity within populations. Phylogenetic analysis using both nuclear and chloroplast DNA sequences suggests that the hybridization events associated with *G. velutina*, and *G. maximowicziana*.

T2

P0545

Phylogenetic and biogeographic perspectives on pollination systems in the Apocynaceae: An update

Jeff Ollerton[1], *Sigrid Liede-Schumann*[2], *Ulrich Meve*[2], *Mary Endress*[3], *Adam Shuttleworth*[4]
1. University of Northampton
2. University of Bayreuth
3. University of Zurich
4. University of KwaZulu-Natal

The Apocynaceae is a large, diverse family of more than 5,000 species in at least 360 genera, distributed from temperate to tropical environments. The family has representatives in every major biome except arctic tundra, and is particularly species rich in the dry and wet tropics, though there are some subtropical centres of diversity such as the upland grasslands of eastern South Africa. The family received early attention because of the highly derived pollination mechanisms of some subfamilies, particularly the Asclepiadoideae and these asclepiads remain the best studied members of the family. In many ways the asclepiads parallel the orchids in the complexity of their floral structure and the presentation of pollen as packaged pollinia. Unlike the Orchidaceae, however, life forms in the Apocynaceae cover almost the whole spectrum of plant types, including vines, robust (high-climbing) lianas, herbs, tubers, epiphytes, stem and leaf succulents, shrubs, and trees. The pollination ecology of the family is equally diverse and some groups have been relatively well studied, in depth if not in breadth, e.g. North American *Asclepias*. Since the last review of pollination systems within the subfamily Asclepiadoideae (Ollerton & Liede 1997 - *Biological Journal of the Linnean Society* 62: 593-610) our knowledge of other asclepiad genera and other Apocynaceae subfamilies has greatly increased, with important papers published on genera such as *Angadenia*, *Apocynum*, *Ceropegia*, *Mandevilla*, *Oxypetalum*, *Pachycarpus*, *Periploca*, *Tylophora* and *Vincetoxicum*. However, there are still major groups of Apocynaceae that are only beginning to be studied, and there are whole clades for which we have no published data on their pollination ecology. In this presentation we review what is currently known about pollination systems in the family, using recent advances in our understanding of the major phylogenetic relationships as a framework for interpreting evolutionary patterns and answering questions about ecological, biogeographic, and evolutionary patterns and processes within the Apocynaceae.

T2

P0546

Monophyly of the Philippine endemic myrmecophytic *Neonauclea* Merr. (Rubiaceae) species including the description of two novel species

Jorge Anton Ordas
University of Santo Tomas

The Rubiaceous genus *Neonauclea* comprises 70 species of diverse trees and shrubs predominantly distributed across Southeast Asia. The genus is the largest within tribe Naucleeae and is well regarded for its extensive radiation of myrmecophytic species, with 19 recorded species and four are found in the Philippine archipelago. With no Philippine representative species investigated in previous molecular studies, this presents a large dearth on phylogenetic data on the genus itself. In this study, we performed a combined Bayesian and Maximum Likelihood analysis of myrmecophytic samples collected from different localities in the Philippines using ITS, *rbc*L, and *trn*T-F datasets to test the monophyly of the myrmecophytic *Neonauclea* species from the Philippines. In contrast to the independent acquisition of myrmecophytism of Bornean myrmecophytic *Neonauclea*, our results suggest a monophyletic ancestry for these Philippine myrmecophytes, exclusively distributed in the Visayas and Mindanao island groups of the archipelago, demonstrated by a strongly supported clade (BS=100; PP=1.00) in the phylogenetic analysis. In addition, two new endemic species are described and illustrated in this study, namely *Neonauclea connicalycina* and *N. viridiflora*, both are closely allied with the endemic *N. formicaria*. A taxonomic key and a map for these species' distribution is also included in the study.

T2

P0547

A new delimitation of the species *Chionoloma bombayense* (Müll. Hal.) P. Sollman (Pottiaceae, Bryophyte) based on molecular and morphological data

Marta Alonso, Juan Antonio Jiménez, ***Jose David Orgaz****, María Jesús Cano*
Universidad de Murcia

Chionoloma bombayense (Müll. Hal.) P. Sollman is a moss belonging to the family Pottiaceae. According to Sollman (2000, 2005), it is extremely variable and, for many years, it was considered the only species of the genus *Pseudosymblepharis* Broth., and later *Chionoloma* Dixon (Sollman 2,001). All the species of these two genera in Asia and Africa were regarded as conspecific with *C. bombayense*. It was justified by the extensive plasticity shown by the morphological characters, implying that the different recognized morphospecies were connected by intermediate specimens. In the last years, we have been focus on the revision of *Chionoloma*. The phylogenetic study carried out included many species considered as synonyms of *C. bombayense*, such as *C. angustatum* (Mitt.) M. Menzel, *C. crassicostatus* (D.H. Norris & T.J. Kop.) M. Alonso, M.J. Cano & J.A. Jiménez, *C. dubium* (Thér.) M. Alonso, M.J. Cano & J.A. Jiménez, *C. duriuscula* (Mitt.) M. Menzel and *C. induratum* Dixon (Alonso *et al.*, 2016). However, these species were not grouped together in the phylogenetic tree but, they appeared in different clades, suggesting independent lineages. Although the current results are not sufficient to estimate the number of species in this genus, they do strongly suggest that more than one species should be recognized. Moreover, collections of *C. bombayense* s. str. share some morphological synapomorphies that allow their recognition as a separate species. They are characterized by unbordered and short leaves, the percurrent leaf costa with two stereid bands, with the dorsal band larger than the ventral band, and the basal laminal cells forming a differentiated area like an inverted "U", and not extending up along the margins to form a V-shaped transition to the upper cells. Therefore, based on molecular and morphological data, the concept of *C. bombayense* is here redefined. A complete description of the species is present, its distribution is updated, and the principal distinctive characters that separate it from the nearest species of *Chionoloma* are analyzed and illustrated.

T2

P0548

The genus *Sciuro-hypnum* (Hampe) Hampe (Brachytheciaceae, Bryophyta) in Japan

Jose David Orgaz[1], *Tomio Yamaguchi*[2], *Juan Guerra*[1]
1. Universidad de Murcia
2. Hiroshima University

Sciuro-hypnum (Hampe) Hampe is a widespread moss genus distributed mainly in temperate and boreal regions. Even thought our knowledge about *Sciuro-hypnum* has increased since Ignatov and Huttunen (2002) published the circumscription and nomenclature of this genus within the Brachytheciaceae, the taxonomy of this genus remains poorly understood, especially in East Asia. In Japan, the circumscription of this genus has not been properly studied since Takaki (1955) obviously included all the current *Sciuro-hypnum* species under *Brachythecium* Schimp. Based on the revision of more than 2000 herbarium specimens and different fieldtrips taken out during 2014 and 2015 in order to collect fresh samples, we can conclude that 13 species of *Sciuro-hypnum* are present in Japan. Main taxonomical characters to distinguish the different species are discussed. For each species, we also provided microphotographs as well as distribution maps in order to a better circumscription of the species.

T2

P0549

The taxonomy and geography of Russian Pinaceae

Larisa V. Orlova
Komarov Botanical Institute RAS

The family *Pinaceae* - is the largest amongst gymnosperms, it includes about 230 species in 10 genera, of them in Russia 27-29 wild species (Orlova, 2012). During 2009-2016 complex comparative systematic, morphological and molecular investigations in LE, MW, SVER, NSK, VLA, BP, C, E, K, KRAM, PE, PR, W, Z *etc.*) and in nature – Far East (Kamtschatka, South Primorie), Eastern and Western Siberia, Ural, Murmansk region, Karelia *etc.*) were done. Our investigation allowed to reveal new important features of reproductive (the morphology and sizes of bract scales, the ratio of its length and seed scales) and vegetative organs: terminal buds and its scales – the shape, sizes, color, presence of white waxy coating and pubescence, the ratio of length of profills (thickened basal scales at the base) and bud; annual and older shoots – its thickness, color and pubescence, the sizes and morphology of pulvini; needles – sizes, color, shape of needles and its apex; arrangement, color and the number of stomatal lines and white wax coating on its sides. For *Larix* also – the shape of epidermal cells, the structure of cuticle and wax touch (Orlova *et al.*, 2014) and for *Pinus* –sizes and morphology of sheaths of brachyblasts, its scales, and also the profills at its base (Orlova, 2001; Orlova, 2012). Our research has allowed to clarify the taxonomic status and the geographical distribution of some critical taxa, such as *Pinus funebris* Kom., *P. friesiana* Wichura, *P. sosnowskyi* Nakai, *Abies gracilis* Kom., *A. semenovii* B. Fedsch., *L. olgensis* A. Henry, *L. cajanderi* Mayr, *L. dahurica* Laws., *L. komarovii* Kolesn., *L. kamtschatica* (Rupr.) Carr., *L. archangelica* Laws., *Picea fennica* (Regel) Kom. *etc.* (Orlova, 2006; Orlova, Egorov, 2011; Orlova, 2012; Orlova *et al.*, 2016 *etc.*). Determined that in the Southern Urals two morpologically different groups of *Picea obovata*, confined, apparently, to a certain altitude zones, that confirmed the results of molecular genetic studies by Z.H. Shigapov (2005) and his conclusion about the existence of several genetically distinct populations in this territory. In the Southern Primorye (Russian Far East) have established the availability of a wide strip of contacts of *L. olgensis* and close *L. komarovii*, within which there are hybrid individuals (*L. olgensis* x *L. komarovii*). Microsatellite analysis of 78 samples of *L. olgensis* and *L. komarovii* compared with *L. czekanowskii* samples and *L. cajanderi* of Eastern Siberia, has allowed to establish population-genetic differentiation of *L. olgensis* and *L. komarovii*, as well as to identify possible hybrid forms. In natural populations of Western Siberia – the Yamalo-Nenets Autonomous Okrug and Republic of Komi (Vorkuta) – near Salekhard, Nadym, Pangody *etc.*, we discovered new hybrid larch *L. archangelica* Laws. x *L. dahurica* Laws., with combine features of *L. archangelica* Laws. and *L. dahurica* Laws. Such hybrids are common here, while a typical *L. archangelica* Laws. found to be rarer. Further investigations confirmed the distribution of this hybrid larch in Western Siberia to the south up to the boundary with the Khanty-Mansiysk Autonomous Okrug (village Uvat, Tyumen region) (Lebedeva, Orlova, 2014; Orlova *et al.*, 2014). Now we are working to establish the exact boundaries of the distribution of different *Pinaceae* taxa to be uploaded into an electronic database and electronic maps of distribution to be created.

T2

P0550

Phylogeography, ecological paleodistribution and morphological divergence in *Pinus patula*

Alfredo Ortiz Martínez[1], *David A. Moo Llanes*[2], *Juan P. Jaramillo Correa*[3], *Lev O. Jardon Barbolla*[4], *Xitlali A. Aguirre Dugua*[5], *Eduardo Ruiz Sanchez*[6], *David Gernandt S.*[5]
1. Posgrado en Ciencias Biológicas, Universidad Nacional Autónoma de México
2. Facultad de Medicina, Universidad Nacional Autónoma de México
3. Instituto de Ecología, Universidad Nacional Autónoma de México
4. Centro de Investigaciones Interdisciplinarias en Ciencias y Humanidades, Universidad Nacional Autónoma de México
5. Instituto de Biología, Universidad Nacional Autónoma de México
6. Departamento de Botánica, Universidad de Guadalajara

In Mexico, little is known over diversification processes in Pinus due to their large genome with highly repetitive elements, large effective population sizes, weak reproductive barriers and the recent origin of the species. Mexico harbors 49 species and is considered a center of diversification of the genus, especially due to complex geological and climatic history. The aim of this study was to evaluate the morphological divergence, geographical delimitation and phylogeographic history of *P. patula*. Our results indicate that this taxón is indeed formed by two lineages, which are traditionally recognized as varieties based on morphological characters is supported by multivariate analyses (analysis of principal components, discriminant analysis and analysis of grouping). Most morphological variation was found within *P. patula* var. *patula*, which is distributed in the Sierra Madre Oriental, the most ancient mountains range from Mexico (540 million years). Ecologically, the paleodistribution shaped from bioclimatic variables indicates that these two lineages have incurred in expansions and contractions since the Pleistocene to present, being most noteworthy their expansions from the mid-Holocene to the present. The genomic information shows a more complex history than was expected, a number 123 of Single Nucleotide Polymorphisms correlated with morphology and ecological differentiation, but without showing a clear genetic or phylogeographic structure. The cold conditions during glacial periods favored the processes of historical demography and the configuration of the architecture of the genetic variation of the populations. The study of morphology, environmental differentiation, genetic structure and inference of past climatic scenarios, are useful for understanding, how historical processes have molded the structure of the genetic variation of *P. patula* in Mexico.

T2

P0551

Phylogenetic relationships of *Gomphrena* and related genera (Amaranthaceae)

Teresa Ortuño[1], *Thomas Borsch*[2]

1. Museo Nacional de Historia Natural (MNHN) - Herbario Nacional de Bolivia (LPB),

2. Botanischer Garten und Botanisches Museum Berlin

The phylogenetic relationships of the subfamily Gomphrenoideae using *rpl16* and *trn*LF; Reveal that Gomphrena is a polyphyletic group with species that belong to different ancestral lineages. One of these lineages binds Gomphrena globosa (*Gomphrena* type str.) To the genus *Blutaparon*, *Lithophila*, and *Gossypianthus*. To elucidate the generic limits of Gomphrena s. Str., And its close relationship with these allied genera, we present a phylogenetic reconstruction with a broad sampling of taxa and characters. The matrix includes combined sequences of the chloroplastdial regions *mat*K-*trn*K and + *trn*L-F and rpl16, core sequences (nrITS), and morphological characters of 29 species. In addition to the analysis of a matrix close to 80 samples based on matK-trnK and nrITS regions to represent the high diversity of *Gomphrena* in the Neotropics, with emphasis in Bolivia, where it is the most diverse genus of the Amaranthaceae family. These sequences were analyzed with parsimonic (MP), maximum likelihood (ML) and Bayesian inference (BI) methods. We also analyzed the evolution of 20 macro and micro morphological characters and the C3 and C4 photosynthesis pathway of the species included in the molecular analysis. The results confirm the systematic position of the species of the genus *Blutaparon*, *Lithophila* and *Gossypianthus* within *Gomphrena* str. The relationship between the coastal species of *Blutaparon* and *Litophila* with the species of Gomphrena of Australia. Gomphrena species with C3 photosynthesis of the cerrado (G. *mollis*, *G. rupestris*) have a position as a sister group to the rest of the species of the *Gomphrenas* lineage. Tr. Which have a C4 metabolic pathway including Blutaparon, Litophila and Gossypianthus. Phylogenetic reconstruction shows different lineages in the tropical South Andes within Gomphrena s.tr. Where some clades include new species. We review the nomenclature to elaborate a new circumscription of the genus Gomphrena based on a monophyletic concept.

T2

P0552

Disentangling putative inflorescence and floral homoplasies between New World and Old World Loranthaceae.

Favio González[1], *Natalia Pabon-Mora*[2]

1. Universidad Nacional de Colombia, Facultad de Ciencias, Instituto de Ciencias Naturales, Bogotá

2. Universidad de Antioquia, Facultad de Ciencias Exactas y Naturales, Instituto de Biologia

With c 77 genera and 950 hemiparasitic species, the Loranthaceae is the most successful and diversified family of the order Santalales. Flowers in the family are exceedingly variable in terms of ground plan, shape, size and pollination syndromes. Recent phylogenetic analyses based on molecular data have shown high degrees of putative morphological homoplasies especially among Old World and New World clades. The aim of this study is to explore in detail the developmental and morphological traits of several New World (NW) taxa that appear at first glance as homoplasic with Old World (OW) lineages. These traits are: the reduction of dichasia to single flowers conforming racemes (e.g. NW *Peristethium* and *Tristerix* versus OW *Dendrophthoe*); the acquisition of unisexual flowers (e.g. NW *Passovia* and *Peristethium* versus OW *Loranthus*); the reduction of floral size associated with shifts in pollinators (e.g. NW insect pollinated *Passovia* and bird pollinated *Psittacanthus* versus OW insect pollinated *Loranthus* and bird pollinated *Lepeostegeres*); the fenestrate corolla anthesis (e.g. NW *Tristerix* versus OW *Actinanthella*); the ligulate petals (e.g. NW *Psittacanthus* and *Ligaria* versus OW *Plicosepalus*); the polysporangiate anthers (e.g. NW *Aetanthus* versus OW *Amyema*); and the cleistogamy (e.g. NW *Aetanthus* versus OW *Helixanthera*). Our results indicate that, although superficially similar, most of these putatively homoplasious features exhibit differences in terms of developmental pathways or morphoanatomical characters.

T2

P0554

Comparative genomics and transcriptomics of representative holoparasitic species of Apodanthaceae in the Americas

Natalia Pabon-Mora[1], *Juan Fernando Alzate*[2], *Laura Marcela Arias Agudelo*[2], *Sebastian Guzman*[3], ***Favio González***[3]

1. Universidad de Antioquia, Facultad de Ciencias Exactas y Naturales, Instituto de Biologia

2. Universidad de Antioquia, Facultad de Medicina

3. Universidad Nacional de Colombia, Facultad de Ciencias Exactas, Instituto de Ciencias Naturales, Bogotá

With two genera and thirteen species, the Apodanthaceae are the only parasitic lineage within the order Cucurbitales. Species of both genera (*Apodanthes* and *Pilostyles*), exhibit an achlorophyllous, holoparasitic lifestyle, with a micellium-like endophyte, and gregarious, sessile flowers that emerge from the parasitized stems.

Whereas the monotypic *Apodanthes* is restricted to the Americas and parasitizes preferentially Salicaceae, *Pilostyles* is distributed worldwide and grows parasitizing several legumes. With the aim to compare genome evolution across Apodanthaceae, we have generated and assembled the first plastome of *P. boyacensis,* a species from the Andes of Colombia. Phylogenetic analyses indicate that *P. boyacensis* is more closely related to *P. thurberi* from Southern United States and Mexico, than to the widely distributed *P. blanchetti,* from Panama to Argentina, Brazil and Chile. Such relationship appears to be driven by the *Dalea* host specificity rather than to the geographic proximity between species, as *P. boyacensis* is the only South American species that parasitized Dalea. MiSeq Illumina DNA sequencing from *P. boyacensis* flowers resulted in 8,515,547 paired end reads. Data processing was followed by extensive merging using FLASH and assembly by Newbler v2.9 where a subset of 1,000,000 filtered representative reads resulted in two contigs. Further reassembly by Newbler 30x resulted in a single contig that was then compared with the available plastomes of the African *P. aethiopica* (11 Kb) and the Australian *P. hamiltonii* (15 kb). De novo contig re-assembly by selecting two different reference points indicates that *P. boyacensis* has a circular plastome of 12 Kb. Despite maintaining some synteny with the reported plastomes of *P. aethiopica* and *P. hamiltonii*, mutation rates outside coding sequences are extremely high finding virtually no match in the three species. Comparative annotation shows that the *P. boyacensis* plastome maintains seven functional genes, namely, *rrn16*, *rrn23*, *accD*, *rps3*, *rps12*, *rpl2* and a putative oxidoreductase (*PbOX*). RNA-seq data generated in parallel from flowers and fruits of *P. boyacensis* show transcriptional activity for all genes and indicate that *PbOX* has five functional alternative transcripts. By comparison with the other two species of *Pilostyles*, the plastome of *P. boyacensis* lacks an ORF for *rps4*, while maintaining functional loci for *rpl2* and the Oxidoreductase. Our results point to independent gene loss and differential selection during the past 23-33 Myr in the three species inhabiting different continents. Our data further shows that while the plastome is extremely reduced, the mitogenome and the nuclear genomes have not experienced the same reduction. Our preliminary data on the mitogenome size yields a 2.0 Mb estimate similar to the *Cucumis sativus* mitogenome estimated in 1.7Mb. Although nuclear genome estimates are still under investigation, we generated a preliminary search for the key regulatory network of nuclear encoded transcription factors involved in flower development. Our results show an intact network of ABCDE *MADS*-box genes with two interesting trends: (1) there are fewer active copies of A-class and E-class genes than typically found in other core eudicots and (2) there are alternative spliced transcripts for the B, C and D-class genes.

T2

P0556

Phylogeny of the *Bouteloua barbata* complex based on molecular data

Pacheco-Jimenez Marisol[1], *Maria Elena Siqueiros-Delgado*[1], *Rosa Cerros-Tlatilpa*[2], *Amanda Fisher*[3], *Ernesto Flores-Ancira*[1]

1. Universidad Autonoma de Aguascalientes

2. Universidad Autonoma del Estado de Morelos

3. California State University, Long Beach

Bouteloua barbata complex is a group integrated by *B. barbata* var. *barbata*, *B. barbata* var. *rothrockii*, *B. barbata* var. *sonorae*, *B. parryi* var. *parryi*, *B. parryi* var. *gentryi*, *B. elata*, *B. ramosa* and *B. breviseta*. Even though the phylogenetic relationships of the genus *Bouteloua* has been studied due to its high forage value and wide distribution, the evolutionary relationships between the species of the *Bouteloua barbata* complex had been poorly explored, and the monophyly of their taxa and the evolution of their morphological characters are still unknown. The objectives of this study were (1) to estimate the monophyly of the species complex, (2) to define the diagnostic morphological characters that allow the species circumscription, and (3) to define the morphological apomorphic and plesiomorphic characters of the group. Specimens were obtained through field and herbarium collections, covering their whole distribution range from the United States, Mexico and Argentina. Nuclear and chloroplast DNA sequences were used to infer the complex phylogeny. Maximum parsimony (MP) and maximum likelihood (ML) analysis, of individual and combined data, were performed to estimate phylogeny, and the bootstrap was used to get statistical support. To estimate the evolution of the morphological characters used in species circumscription, samples with the highest variation were selected, analyzed and reconstructed using the MP hypothesis in Mesquite software.

T2

P0557

Diversity, evolution and biosystematics of Legumes of U.P. India

Vaibhav S Pandey

Ddu Gorakhpur University

Plants came on this plant about 450-500 million years ago. Legumes came on this planet about 60 million years ago and developed symbiotic association about 58 million years ago. Non-nodulation remains common in Caesalpinioideae, with smaller numbers in Mimosoideae and Papilionoideae These plants revolutionized the mother earth planet for fixing the nitrogen and established natural nitrogen fixing industry on this planet. These plants are natural asset for our biological system and govern all biofunctions, distribution, and biodiversity climate link of the world. Legumes belongs to family Fabaceae or Leguminosae. These plants have specific biofunctional quality and growing in different Biospecific regions (BRs) and commonly known as Biospecific regions plants (BRPs). Legumes are live telecast of mother earth planet growing naturally as well as cultivated in different parts of the world since ancient times. India and Indian sub-continent have green and lush vegetation with natural biological industry, with their great biological supporting system to vegetation. Legumes are a significant source of protein, dietary fiber, carbohydrates and dietary minerals. In India, Uttar Pradesh is the largest state, comprises many legumes with their biological industry. Considering these facts diversity of legumes was explored for their documentation and distribution in Uttar Pradesh. The legumes of Caesalpinioideae, Mimosoideae and Papilionoideae were explored. Maximum number of plants were recorded in Papilionoideae followed Mimosoideae and Caesalpinioideae. The diversity, evolution and biosystematics of these Legumes were recorded and discussed.

T2
P0558
Bryophytes gemmae biodiversity and related environmental adaptability studies in different vegetation communities of dolomite subtropical

Jia Peng Pang[1], *Zhi Hui Wang*[1], *Zhao Hui Zhang*[2]
1. School of Life Sciences, Guizhou Normal University
2. Key Laboratory for Information System of Mountainous Area and Protection of Ecological Environment of Guizhou Province, Guizhou Normal University

From October 2015 to April 2016 biodiversity of bryophytes gemmae and study on the adaptability to the environment by sampling method and field investigation method on Guiyang Luchongguan Forest Park dolomite subtropical in different vegetation communities. The results show that 100 samples were collected from sampling sites, and 7 families, 16 genera and 30 species were identified, Among them had 6 families, 15 genera and 29 species of Musci, Hepaticae only had *Marchantia polymorpha* L., Barbula and Brachymenium is the dominant families of bryophytes is the number of species in bryophytes indicated arbor and shrubs community>lianas, boscage and climbing shrubs community>boscage and low shrubs community>bryophytes community on bare dolomite. Species diversity and evenness index of arbor and shrubs community community is the highest, the lowest is exposed bryophytes community on bare dolomite, nevertheless the dominance index was contrary. Among them have 2 families 7 genera and 9 species had gemmae, accounting for 30% of the total, which belongs to the Pottiaceae and Bryaceae, The differences of gemmae from size and number, morphological structure, insertion position, frequency of occurrence, special construction. *Barbula ditrichoides* Broth. And *Trichostomum barbuloides* (Broth.) Chen were common bryophytes with gemmae. Common with all vegetation communities, arbor and shrubs community, but the highest frequencies of bryophytes with gemmae were boscage and low shrubs community. Diversity analysis indicated that the bryophytes communities were significant to vegetation communities, coverage, and vegetation communities and coverage, the variance was significant; For bryophytes with gemmae, the groups were extremely significant among humidity. This research can be used as a basis that the environmental adaptive reproductive strategy of bryophytes.

T2
P0559
Morphological variation in Australian and Papua New Guinean *Polyosma* Blume (Escalloniaceae)

Oliver K. Paul
PNG Forest Research Institute

The morphological variation within and between species of *Polyosma* Blume from Australia and Papua New Guinea (PNG) was investigated. The taxonomic usefulness of morphological characters was determined. A total of 161 Operational Taxonomic Units (OTU) of *Polyosma*, comprising 35 known species, three undescribed taxa and nine un-identified herbarium specimens were assessed by the use of 30 characters. Of these characters, 20 were qualitative (non-overlapping) and 10 were quantitative (overlapping). The variability within these morphological characters proved to be useful for circumscribing the currently recognised species. In the analysis, two main groups were obtained based on whether the characters were hairy or glabrous, which were further down had resolved into thirteen groups. Twenty four species were recognised from Papua New Guinea. Three un-published names were discovered. Of these, two species grouped well with published names where as one was significantly different to the published named species it was seen grouped with. Australia has six species, with a possible new species from Mt. Spurgeon, Queensland. More materials are required to analysis the Mt. Spurgeon species. The characters and character-states used were consistent within the entire taxon. In future, few more characters will be included and a molecular phylogeny to compliment this morphology study of the taxon in the two study areas.

T2
P0560
Ancient geographical barriers drive differentiation among *Sonneratia caseolaris* populations and recent divergence from *S. lanceolata*

Yuchen Yang[1], *Norman C. Duke*[2], ***Fangfang Peng***[1]
1. State Key Laboratory of Biocontrol and Guangdong Provincial Key Laboratory of Plant Resources, Sun Yat-Sen University
2. Trop WATER, James Cook University, Townsville, QLD

Glacial vicariance is thought to influence population dynamics and speciation of many marine organisms. Mangroves, a plant group inhabiting intertidal zones, were also profoundly influenced by Pleistocene glaciations. In this study, we investigated phylogeographic patterns of a widespread mangrove species *Sonneratia caseolaris* and a narrowly distributed, closely related species *S. lanceolata* to infer their divergence histories and related it to historical geological events. We sequenced two chloroplast fragments and five nuclear genes for one population of *S. lanceolata* and 12 populations of *S. caseolaris* across the Indo-West Pacific (IWP) region to evaluate genetic differentiation and divergence time among them. Phylogenetic analysis based on sequences of nuclear ribosomal internal transcribed spacer and a nuclear gene *rpl9* for all *Sonneratia* species indicate that *S. lanceolata* individuals are nested within *S. caseolaris*. We found strong genetic structure among geographic regions (South China Sea, the Indian Ocean, and eastern Australia) inhabited by *S. caseolaris*. We estimated that divergence between the Indo-Malesia and Australasia populations occurred 4.035 million years ago (MYA), prior to the onset of Pleistocene. BARRIERS analysis suggested that complex geographic features in the IWP region had largely shaped the phylogeographic patterns of *S. caseolaris*. Furthermore, haplotype analyses provided convincing evidence for secondary contact of the South China Sea and the Indian Ocean lineages at the Indo-Pacific boundary. Demographic history inference under isolation and migration (IM) model detected substantial gene flow from the Sri Lanka populations to the populations in the Java Island. Moreover, multi-locus sequence analysis indicated that *S. lanceolata* was most closely related to the Indian Ocean populations of *S. caseolaris* and the divergence time between them was 2.057 MYA, coinciding with the onset of the Pleistocene glaciation. Our results suggest that geographic isolation driven by the Pleistocene ice age resulted in the recent origin of *S. lanceolata*.

T2

P0561

Natural hybridization and adaptive introgression in New-World red mangroves (*Rhizophora*)

Zixiao Guo[1], ***Fangfang Peng***[1], *Norman C. Duke*[2]

1. State Key Laboratory of Biocontrol and Guangdong Provincial Key Laboratory of Plant Resources, Sun Yat-Sen University

2. Trop WATER, James Cook University, Townsville, QLD

Hybridization in plants can result in adaptive introgression between species and thus contribute to rapid adaptive evolution. It is generally considered that the New World red mangrove genus (*Rhizophora*) consists of two species (*R. mangle* and *R. racemosa)* and a natural hybrid between them (*R.* × *harrisonii*). Another taxon, *R. samonensis*, is regarded as a synonym of *R. mangle.* To test if there is adaptive introgression between the two species and clarify the taxonomic status of *R. samonensis*, here we collected 119 samples of *Rhizophora* of these taxa across the Atlantic-East Pacific (AEP) region by sequencing seven nuclear genes (SBE2, 23056, C49, 22274, 22454, 23777 and 23714) and one chloroplast DNA spacer (*trn*L-*trn*F). Our results showed that *R. mangle* and *R. racemosa* diverged by 1.11%-5.00% at these nuclear genes. The individuals assigned as the hybrid *R.* × *harrisonii* are heterozygous at each of these nuclear genes, with one allele matching the haplotype of *R. mangle* and the other *R. racemosa*, confirming their hybrid status. Chloroplast *trn*L-*trn*F sequencing indicated unidirectional hybridization, with *R. mangle* being the maternal parent. Moreover, *R. samonensis* collected in Peru was found to be close to but slightly divergent from *R. mangle*. (Atlantic, except for Hawaii). Although *R. racemosa* and *R.* × *harrisonii* show morphological variation (two forms were found for each taxon), little genetic differentiation was detected between the two forms of each taxon. Strikingly, about half of the *R. mangle* samples collected in Nigeria are homozygous with the same haplotype of *R. racemosa* in the gene 23777, while there was no signal of introgression at all other six nuclear genes, which suggests adaptive introgression from *R. racemosa* to *R. mangle* at this gene. Our study clarifies the taxonomic problems in the New World *Rhizophora* and provides convincing evidence for adaptive introgression in mangroves for the first time.

T2

P0562

The phylogeny of *Youniga* based on plastid and nuclear loci

Yulan Peng

Chengdu Institute of Biology

The genus *Youngia* Cass. is estimated to contain about 35 species distributed in eastern Asian. Previous molecular studies were based on showed that *Youngia* is polyphyletic. In the present study, DNA sequences of three plastid and one nuclear locus from 520 accessions representing approximately 30 species of *Youngia* worldwide were used to infer a phylogeny using maximum likelihood, Bayesian inference and maximum parsimony methods. The study identifies strongly supported major clades and well resolves relationships among them. Major results include: (i) two deep-level clades are discovered representing the deep splits of *Youngia*; and (ii)four major clades representing four major evolutionary lineages are identified.

T2

P0563

New records on orchid pollen vectors and visitors from the tropical lowland forests of Mexico and Panama

Oscar Alejandro Pérez-Escobar[1], *David Roubik*[2], *Günter Gerlach*[3]

1. Royal Botanic Gardens, Kew

2. Smithsonian Tropical Research Institute

3. Botanischer Garten München

Orchids are one of the most megadiverse, prominent and endangered components of the Neotropical flora. The Neotropical Orchidaceae have evolved an astounding array of interactions with a diverse set of organisms, upon which they entirely rely essential biological services. This is particularly true for pollination, which is predominantly carried out by insects. Despite the importance of pollination success on orchid population survival, very little is still known about the pollinator identity of Neotropical orchids. In this study, we observed and sampled *in-situ* visitors to flowers and pollen vectors of seven orchid species (*Cycnoches guttulatum, C. dianae*, *Mormodes* sp., *Gongora truncata*, *G. unicolor*, *Sarcoglottis sceptrodes*, and *Trichopilia* sp.) distributed in the lowland tropical forests of Southern Mexico and Panama. These taxa exhibit adaptations for euglossine pollination (e.g. production of volatile compounds, complicate mechanisms for pollen package emplacement). We collected additional data on orchid pollen vectors by using chemical baits known to attract a wide range of male euglossine bees. Morphological comparisons, DNA sequencing, and BLAST searches allowed a reliable identification of pollinia recovered from bees captured on flowers and chemical baits. Visitors of four orchid taxa were recorded for the first time, and pollen vectors of four orchid species were further reported. Five bee species were identified as orchid pollen vectors, namely *E. cyanura, E. obtusa, E. villosa,* and *Eulaema polychroma*. Strikingly, *E. cyanura* appears to be a shared pollen vector of *C. guttultum* and *C. dianae*, two morphologically distinctive taxa that share the same distribution range (i.e. lowland wet forests of Panama). Our study contributes to the knowledge of Neotropical orchid-insect interactions and opens interesting questions regarding the role of fragrance profile specificity as an isolative barrier in male euglossine-bee pollinated orchids.

T2

P0564

Systematics studies on the Hymenaea clade (Leguminosae)

Rafael Pinto[1], *Vidal Mansano*[2], *Giulia Serbin*[1], *Boris Domenech*[3], *Makeli Lusa*[4], *Juliana Mayer*[1], *Benjamin Torke*[5], *Anne Bruneau*[3], *Eliana Forni-Martins*[1], *Ana Tozzi*[1]

1. Universidade Estadual de Campinas (UNICAMP)

2. Jardim Botânico do Rio de Janeiro (JBRJ)

3. Université de Montréal

4. Universidade Federal de Santa Catarina (UFSC)

5. The New York Botanical Garden

The Hymenaea clade is included in the resin-producing clade of subfamily Detarioideae (Leguminosae). It is composed of three genera distributed mainly in tropical regions of Africa and South America: *Guibourtia* (15 species, two in the Neotropics and 13 in Africa), *Hymenaea* (16 species in the Neotropics and one in Africa) and *Peltogyne* (21 Neotropical species). These genera share an arborescent habit (few species of *Hymenaea* are predominately shrubs – *H. eriogyne*, *H. rubriflora* and *H. travassii*) and bifoliolate leaves but differ morphologically in the presence of a corolla (absent in *Guibourtia*), stamen length (heteromorphic in *Peltogyne*) and fruit dehiscence (indehiscent in *Hymenaea*). The goal of this study was to test the monophyly of these three genera and to provide an updated taxonomic treatment, supported by phylogenetic, morphological, cytogenetic and anatomic investigations. Preliminary molecular phylogenetic analysis is based on two informative markers - ITS (nuclear) and *trn*K (plastid) - suggest that both *Hymenaea* and *Peltogyne* are monophyletic, although some species, such as *H. courbaril* and *P. paniculata* appear to comprise multiple lineages with separate origins. *Guibourtia* is resolved as paraphyletic and is composed of at least two distinct lineages, one containing only African species, the other both African and Neotropical species. In contrast to the widely held view that Neotropical *Guibourtia* comprise a single species, our taxonomic studies support the recognition of two species, corresponding to Cuban (*G. hymenaefolia*) and South American populations (*G. chodatiana*). We also found that species diversity in *Hymenaea* was underestimated, and we an in the process of describing three new species several new combinations in the genus. Anatomical studies corroborate previous findings of glandular cavities in leaflets of most species of *Hymenaea*, and for the first time document the presence of these cavities in *Guibourtia* and most species of *Peltogyne*. However, such cavities were not found in *H. velutina* and *P. catingae*. Morpho-anatomical characterization of the midrib and internerval region of leaflets and histochemical analyses of the epidermis proved useful in delimiting species in the three genera. The marginal position of the cavities in the leaflet lamina was found to be useful for differentiating the three genera of this clade from other representatives of subfamily Detarioideae. Cytogenetic analyses have shown that most taxa of the Hymenaea clade have predominantly $2n=24$, the most common condition in subfamily Detarioideae. Exceptions are *G. coleosperma* ($2n=48$) and *G. chodatiana* ($2n=12$). Both *Hymenaea* and *Peltogyne* have two pairs of CMA^+ sites, while the species of *Guibourtia* analyzed have just one pair. FISH analyses are still being conducted, and results for species of *Guibourtia* and *Hymenaea* show one pair of 18S and one pair of 5S rDNA sites while *P. pauciflora* presented two pairs of 5S rDNA sites.

T2
P0565
The biodiversity hotspot as evolutionary hot-bed: Spectacular radiation of *Erica* in the Cape Floristic Region

Michael Pirie[1,2], *E.G.H. Oliver*[2], *Ana Mugrabi De Kuppler*[3], *Berit Gehrke*[1], *Nicholas Le Maitre*[2], *Martha Kandziora*[1], *Dirk Bellstedt*[2]
1. Johannes Gutenberg-Universität, Mainz
2. University of Stellenbosch
3. Rheinische Friedrich-Wilhelms-Universität Bonn

The disproportionate species richness of the world's biodiversity hotspots could be explained by low extinction (the evolutionary "museum") and/or high speciation (the "hot-bed") models. We test these models using the largest of the species rich plant groups that characterise the botanically diverse Cape Floristic Region (CFR): the genus *Erica* L. We generate a novel phylogenetic hypothesis informed by nuclear and plastid DNA sequences of c. 60% of the c. 800 *Erica* species (of which 690 are endemic to the CFR), and use this to estimate clade ages (using RELTIME; BEAST), net diversification rates (GEIGER), and shifts in rates of diversification in different areas (BAMM; MuSSE). The diversity of *Erica* species in the CFR is the result of a single radiation within the last c. 15 million years. Compared to ancestral lineages in the Palearctic, the rate of speciation accelerated across Africa and Madagascar, with a further burst of speciation within the CFR that also exceeds the net diversification rates of other Cape clades. *Erica* exemplifies the "hotbed" model of assemblage through recent speciation, implying that with the advent of the modern Cape a multitude of new niches opened and were successively occupied through local species diversification.

T2
P0566
Targeted NGS for species level phylogenomics: "Made to measure" or "one size fits all"?

Malvina Kadlec[1], *Dirk Bellstedt*[2], *Nicholas Le Maitre*[2], ***Michael Pirie***[1,2]
1. Johannes Gutenberg-Universität, Mainz
2. University of Stellenbosch

Targeted high-throughput sequencing using hybrid-enrichment offers a promising source of data for inferring multiple, meaningfully resolved, independent gene trees suitable to address challenging phylogenetic problems such as in species complexes and rapid radiations. The targets in question can either be adopted directly from more or less universal tools (e.g. "Anchored Hybrid Enrichment"), or custom made for particular clades at considerably greater effort. We applied custom made scripts to select sets of homologous sequence markers from transcriptome and WGS data for use in the flowering plant genus *Erica* (Ericaceae). We tested the resulting targets by obtaining an empirical dataset, and compared them to those that would be selected both using different available tools (Hyb-Seq; MarkerMiner), and when optimising for broader clades of more distantly related taxa (Ericales; eudicots). Approaches comparing more divergent genomes (including MarkerMiner) delivered fewer, shorter potential targets than those targeted for *Erica*. The latter may nevertheless be effective across the wider family Ericaceae. We would conclude that rather than searching for "one size fits all" universal markers, we should be improving and making more accessible the tools necessary for developing our own "made to measure" ones.

T2
P0567
Female gametophyte and ovule ultrastructure in apomictic and sexual members of Asteraceae (genera: *Taraxacum*, *Hieracium* and *Pilosella*)

Bartosz Jan Płachno[1], *Piotr Świątek*[2]
1. Jagiellonian University in Kraków

2. *University of Silesia*

In apomictic plants there is an evolutionary tendency to reduce "unnecessary costs" associated with the sexual reproduction of their amphimictic ancestors (e.g. lack of pollen grains production), which may result in greater reproductive success. Thus, it can be assumed that a cost reduction may also take place in the development of an unreduced female gametophyte and also in ovule tissues which are responsible for the attraction and growth of the pollen tubes. The aim of our studies were to compare: both the egg apparatus and the micropyle structure and also general ovule anatomy and ultrastructure in amphimictic and apomictic species from Asteraceae family (genera: *Taraxacum*, *Hieracium* and *Pilosella*). We found that ultrastructural characters of egg apparatus were similar in both amphimictic and apomictic species. Structural reduction of ovular transmitting tissue was also not observed in apomicts. The only difference between amphimictic and apomictic species (*Taraxacum*), which we found, is that both synergids persisted for a long time without any degeneration. We propose that the persistent synergids in apomicts may play a role in the transport of nutrients to the developing embryo. It seems that there are some evolutionary benefits of retaining the functionality of all of the necessary structures such as synergids and transmitting tissue that are connected with pollination and fertilization in apomicts. Another explanation is that studied taxa are so (too evolutionary young and the reduction of "unnecessary costs " has not occurred yet. Acknowledgements: This study was funded by National Science Centre, Poland. Contract grant number: DEC-2013/09/B/NZ8/03308.

T2

P0568

Phylogeny and revision of the Neotropical *Adiantum raddianum* group (Pteridaceae)

Regina Yoshie Hirai, ***Jefferson Prado***
Instituto de Botanica

With more than 200 species, the maidenhair fern genus *Adiantum* is among the top ten most diverse fern genera. *Adiantum* is pantropical in distribution and, due to the presence of a unique synapomorphy (sporangia borne on indusia rather than laminae), perhaps the most easily recognized fern genus. Many of its members, including numerous cultivars derived from *A. raddianum*, are grown as ornamentals. Because of its size, a comprehensive taxonomic study of *Adiantum* is difficult and the genus is perhaps better approached through a series of narrower studies. Here, we focus specifically on *A. raddianum* and putative allies. We find a newly defined *A. raddianum* group to be strongly supported as monophyletic and segregated from other maidenhair ferns on the basis of genetic as well as morphological characteristics. Bayesian inference and maximum likelihood analyses of plastid *atp*A, *chl*L, *chl*N, *rbc*L, and *rpo*A sequences support the *A. raddianum* clade as sister to *A. poiretii* and its allies. We identify round-reniform indusia to be a characteristic of the *A. raddianum* group (vs. lunate in the *A. poiretii* group). Additionally, we find species in the *A. poiretii* group to differ in having a unique deletion of 66 nucleotides in our *chl*N gene alignment. For the Neotropical *A. raddianum* group 16 species were recognized. Some of them are widely distributed and one was recently described (*A. alan-smithii*).

T2

P0569

Phylogeny of *Osbeckia* (Melastomataceae) in India

Prashob Pulpra Prashob Pulpra, *Sibichen M. Thomas Sibichen M. Thomas*
Plant Systematics Division, Department of Botany, St. Joseph's College (Autonomous), Devagiri, Calicut 673008, Kerala

The genus *Osbeckia*, endemic to the North Eastern Himalayas and Western Ghats, comprises twenty five species in India. The present work is an attempt to unveil the phylogenetic relationship among the species of the genus and to establish the circumscription of the genus. Morphological and molecular database have been used separately as well as combined for phylogenetic analysis. The nuclear DNA (internal transcribed spacer, ITS) and-morphological characters were considered for the analysis. The study revealed the character evolution of the clade and also supported its monophyletic origin.

T2

P0570

Trends in peristome architecture in the Dicranales (Dicranidae)

Amanda Guerreiro[1], ***Michelle J. Price***[2]
1. *University of Geneva*
2. *Conservatory and Botanical Garden of the City of Geneva*

The Dicranidae, a well-supported monophyletic group of mosses, are thought to have undergone rapid diversification after the evolution of the haplolepidous peristome (a single ring of articulate teeth surrounding the capsule mouth that is derived from the inner peristome layer during the developmental sequence). Based on early observations of moss capsules at various stages of development, and specifically of mature capsules at the capsule mouth, by Georg Boyung Scato Lantzius-Beninga (1815-1871), we recently established protocols for observing and comparing taxa with the aim of testing his theory that peristome architecture, seen in longitudinal section, may hold information of taxonomic-systematic interest in the peristomate mosses. By comparing peristome architecture of the capsule mouth – peristome teeth junction region (the peristome attachment region) in 82 samples representing 19 species from 13 genera in 5 families of the Dicranales (Dicranaceae, Ditrichaceae, Fissidentaceae, Leucobryaceae, Rhabdoweisiaceae), and *Grimmia pulvinata* (Hedw.) Sm. (Grimmiaceae, Grimmiales) used as an comparative outgroup, some specific areas or traits were identified that are potentially relevant in taxon discrimination (the form and thickness of the exothecium, form and thickness of the connective tissue, connection between the peristome base and teeth, cell wall thickening patterns), but also in the tooth insertion position, features of the capsule wall and mouth shape, as revealed in longitudinal sections. Clear and often quantifiable differences in the composition and the architecture of the peristome attachment region were observed between the 2 different orders as well as between each of the different families studied, and in some cases also between genera and/or species. Our findings also supported several recent observations on the phylogenetic placement of species included in the analysis (for example, *Paraleucobryum longifolium* shares more peristome architectural features in common with the Dicranaceae seen in the study than the other members of the Leucobryaceae included). Our preliminary investigations form a solid foundation for further exploratory research on peristome architecture within the Dicranales, with the aim of identifying, describing and defining taxonomic unit specific traits using traditional methodologies (hand sectioning and SEM) combined with the application of new and innovative methods for observing whole structures and their three-dimensional composition and form.

T2

P0571

Techniques for exploring peristome architecture in haplolepidous mosses: Reviving old innovations

Amanda Guerreiro[1], ***Michelle J. Price***[2]

1. University of Geneva

2. Conservatory and Botanical Garden of the City of Geneva

The peristome of the arthrodontous mosses is an evolutionarily conserved structure composed of a concentric ring or rings of articulate teeth that surround the mouth of the capsule (sporophyte generation). The peristome plays a key role in spore release and dispersion, and is unique to this group of plants. Peristomes have long fascinated bryologists and have been used, at various different taxonomic levels, in the classification of this group of plants. In the 18th Century, the German botanist, Georg Boyung Scato Lantzius-Beninga (1815-1871), investigated the developmental anatomy of moss sporophytes, including of the peristome. Using comparisons of longitudinal sections of moss capsules from mature, but not dehisced, capsules from a selection of peristomate species he also affirmed that species are distinguishable by their unique peristome architecture. His observations of the capsule mouth – peristome teeth junction region of the capsule revealed that taxonomic information can be found within this region based on variations in the developmental patterns of the specific structures, layers or cells that constitute the peristome teeth themselves (cell wall deposition, orientation of the deposition, trabeculae development, *etc.*). His early findings have obvious implications for moss systematics but, likely due to a lack of explanations on the techniques in the original text combined with the complexities involved with sectioning small objects and an incomplete vocabulary associated with the original observations, his theory has not been widely tested within the mosses. Here we give details on simple protocols for specimen treatment, handling and sectioning as well as for specimen observation, illustration and description. Our technique of image preparation is also outlined with the aim of providing baseline information for researchers interested in exploring peristome architecture using traditional techniques that may subsequently reveal some innovative findings in specific groups of mosses.

T2

P0572

Juncus quartinianus A.Rich. (Juncaceae, sect. *Ozophyllum*), a Neglected Species from Ethiopia, Eritrea and Somalia

Anna Faltyn[1], *Anna Jakubska-Busse*[2], *Paweł Jarzembowski*[2], ***Jarosław Proćków***[1]

1. Institute of Biology, Wrocław University of Environmental and Life Sciences

2. Institute of Environmental Biology, University of Wrocław

Juncus quartinianus was described by A. Richard in 1851 from Ethiopia. Type specimens were collected by R. Quartin-Dillon and A. Petit in Chiré (Ethiopia) in July 1844. Richard distinguished this species on the basis of the following features: sessile, lateral single head, lanceolate, acute, subcastaneous tepals, and the inner ones slightly longer than the outer, six stamens one-third as long as the tepals and pyramidal capsules longer than the perianth. Later authors have treated this taxon as a synonym of *J. fontanesii* J.Gay ex Laharpe or *J. oxycarpus* E.Mey. ex Kunth. According to the latest available monograph of the Juncaceae family, specimens of *J. oxycarpus* from Ethiopia and Eritrea, that were initially classified as *J. fontanesii*, have longer capsules and are similar to *J. fontanesii* subsp. *pyramidatus* (Laharpe) Snogerup. Indeed, in the course of studying the herbarium material of *J. oxycarpus* and *J. fontanesii* subsp. *pyramidatus*, we found some unusual specimens from Ethiopia, Eritrea and Somalia. These were previously identified mainly as *J. oxycarpus*, but differ in their capsules that are longer than the perianth and their quite different seed coat surface. Some specimens were initially classified as *J. fontanesii*; however, we found that they do not belong to the latter species as their anthers are shorter than the filaments or rarely, the anthers and filaments are equally sized (anthers of *J. fontanesii* are distinctly longer than filaments). Our detailed morphological analyses of flowers, fruit and seeds show that these specimens represent a distinct species from both these taxa. After careful analysis of the literature and herbarium material, it was found that this species was named *Juncus quartinianus* A.Rich., 1851 and thus we proposed to restore this name. In addition, revision of the herbarium material revealed that Richard's description of this species is inaccurate and it has been corrected. "The project was supported by the Leading National Research Centre (KNOW) programme of the Wrocław Center of Biotechnology for years 2014-2018"

T2

P0573

Epitypification of the J. E. Smith's Name *Epipactis purpurata* (Orchidaceae, Neottieae)

Anna Jakubska-Busse[1], *Elżbieta Żołubak*[1], ***Jarosław Proćków***[2]

1. Institute of Environmental Biology, University of Wrocław

2. Institute of Biology, Wrocław University of Environmental and Life Sciences

The identity of the holotype of *Epipactis purpurata* Sm. is shown to be demonstrably ambiguous. The specimen is not representative for identification because all its flowers are at the budding stage (pre-anthesis) – thus it is not possible to identify the specimen (holotype) to species based on the recently analyzed features (i.e. gynostemium and floral segments morphology, especially details of the lip). The taxon has been misidentified many times due to its resemblance to some forms of morphologically variable *E. helleborine* (L.) Crantz, which we discovered based on the analysis of herbarium materials from different herbaria (BR, BRNU, FR, G, KTU, M, MA, S, STU, WRSL, ZT, Z). Many plants of *E. helleborine* can have similar combinations of features like the holotype of *E. purpurata* (i.e., robust stem even in the inflorescence and a small leaves). Considering this ambiguity, an epitype for the name is proposed (M 257866) in order to enable its precise taxonomic interpretation. The proposed epitype is composed of two ramets (two shoots) at the full flowering stage. The flowers are wide open, which allows to perform biometric measurements or to look at the details of sepal and petal morphology as well as the lip. *Epipactis purpurata* is the important James Edward Smith's name because the orchid is widely distributed in western and central Europe. "The project was supported by the Leading National Research Centre (KNOW) programme of the Wrocław Center of Biotechnology for years 2014-2018".

T2

P0574

New characters useful in determination of *Epipactis purpurata* Sm. s.str. (Orchidaceae, Neottieae) based on the analysis of herbarium material and field research data

Anna Jakubska-Busse[1], *Elżbieta Żołubak*[1], *Paweł Jarzembowski*[1], ***Jarosław Proćków***[2]

1. Institute of Environmental Biology, University of Wrocław

2. Institute of Biology, Wrocław University of Environmental and Life Sciences

The genus *Epipactis* is a difficult and controversial taxon due to its wide but poorly recognized range of phenotypic variability (particularly ontogenetic), which impedes proper identification of the species. The available determination keys do not allow to identify the plants in the pre-flowering and fruiting stages, but only in open flowers. New reports on describing new species of the *Epipactis* genus, e.g. from Iberian Peninsula or Crimea made us undertake studies on revision of the range of variability of *Epipactis purpurata* Sm. Our detailed analysis of the herbarium collections on loan (BR, C, FR, G, K, KTU, M, S, STU, Z, ZT) and field investigations (Poland, Czech Republic and Slovakia) showed that they still contain numerous erroneously identified specimens. An important character in identification of *E. purpurata* is the colour of the shoot, dark-green to magenta in nature, sometimes purplish, rarely green. It must be, however, remembered that the drying method may influence the retention of colour. The aim of our project was to demonstrate full variability of *E. purpurata* Sm. *sensu stricto* and compare it with previous studies. We established new ranges of morphological character variability for the taxon, including diagnostically significant features. The variability refers to flower features, i.e. the shape and pigmentation of the lip, sepals and petals, also the shape and size of the gynostemium as well as the size, shape and colouration of leaves, which, in turn, makes it difficult to determine the diagnostic features for the species. The detailed morphological and anatomical studies of leaf characters, including the shape and size of papillae, confirmed that this feature should be carefully considered in taxa identification as we present different types of papillae in examined specimens. "The project was supported by the Leading National Research Centre (KNOW) programme of the Wrocław Center of Biotechnology for years 2014-2018"

T2

P0575

An overview of the genus *Radula* subgenus *Amentuloradula* Devos *et al.* (Radulaceae, Marchantiophyta)

Chatchaba Promma[1], *Matt A. M. Renner*[2], *Rui-Liang Zhu*[1]

1. East China Normal University

2. Royal Botanic Gardens and Domain Trust

The genus *Radula* Dumort., is an isolated genus belonging to family Radulaceae, with about 250 currently accepted species and mainly distributed in tropical and subtropical regions. Subg. *Amentuloradula* Devos *et al.,* is accepted in this genus together with other six subgenera. Most of the species in this subgenus are well characterized by having the microphyllous branches and/or amentulose branches, single lobular hyaline papilla located on the apex of the leaf-lobule, ovoid capsule, spores comparatively large with echinate with brachiate tip ornamented. The preliminary study of subg. *Amentuloradula* is presented and twenty-six species are enumerated, based on available literature, field surveys and specimens examination from various herbaria. This subgenus is mainly distributed in Australasia, most of the species are greatly restricted distribution in Australia, New Caledonia, and New Zealand. Japan and South China seem to be its northernmost limit. The majority of the species are epiphytes on tree trunks, branches and twigs occurring from near sea level to over 4,000 m in elevation.

T2

P0576

Taxonomic revision and phylogeny of the genus *Leucas* s.str (Lamiaceae: Lamioideae)

Sunojkumar Purayidath Kandy

Department of Botany, University of Calicut

The genus *Leucas* R.Br. is considered in the strict sense here. Species included in the genus are characterised by the dorsal side of calyx exceeding or equalling the ventral side, anterior corolla lip longer than posterior lip and absence of glands on ovaries. Species previously included in the genus from Africa and Arabia are omitted as they are paraphyletic (Scheen and Albert, 2009). *Leucas s.str.*, thus, form an Asian genus, distributed in India, south east Asia up to Australia. Taxonomic revision based on fresh specimen and herbarium data identified 41 species and 9 infraspecific variants. Almost 63% of the species are found as endemic to India with a high diversity and endemism in Westrn Ghats area. Main characters at species level are the nature of calyx mouth, teeth, shape of lower lip of corolla and nature of bracteoles. All species are described with regard to distribution, flowering season and habitat. Comparative morphology identified 3 infra generic sections with *Leucas* s.str. High level of morphological plasticity and small flowers led different views on the infrageneric relationship within the genus. Phylogenetic analysis of the morphospecies using combined sequence data from four plastid regions (*trn*L–*trn*F, *trn*S–*trn*G, *mat*K and *rps16* intron) were conducted. Maximum likelihood and Bayesian approaches were used to assess phylogeny of multi accessions of 37 species and several infraspecific variants. African *Leucas* and related genera were used as outgroups. These approaches proved highly informative in delimiting relationships among species of *Leucas*. Phylogenetic studies confirmed monophyly of *Leucas s.str*. Three well supported clades, corresponds well with the three infrageneric sections recognised by classical taxonomists (Bentham, 1830) are noticed. Clade 1 consists of perennials with ovate leaves and axillary loose inflorescence. Clade 2 consists of perennials with oblong-lanceolate leaves, terminal compact inflorescence and the dorsal side of calyx equalling the ventral side. Clade 3 consists of annual herbs with linear-lanceolate leaves and oblique calyx mouth. A fourth clade consists of two herbaceous species with axillary inflorescence are also found. Clade one corresponds to the primitive section Leucas Benth.; clade 2 to section Astrodon Benth. and clade 3 corresponds to the advanced sect. Plagiostoma Benth. Clade 4 is a new section recognised in this genus. Infrageneric relationships are indicated in the trees provided.

T2

P0577

Comparative transcriptomics of strawberries (*Fragaria* spp.) provides insights into evolutionary patterns

Qin Qiao[1], *Jiajun Lei*[2], *Ticao Zhang*[3]

1. School of Agriculture, Yunnan University, Kunming

2. College of Horticulture, Shenyang Agricultural University, Shenyang

3. Key Laboratory for Plant Diversity and Biogeography of East Asia, Kunming Institute of Botany, Chinese Academy of Sciences

Multiple closely related species with genomic sequences provide an ideal system for studies on comparative and evolutionary genomics, as well as the mechanism of speciation. The whole genome sequences of six strawberry species (*Fragaria*) have been released, which provide one of the richest genomic resources of any plant genus. In this study, we first generated seven transcriptome sequences of *Fragaria* species de novo, with a total of 48,557–82,537 unigenes per species. Combined with 13 other species genomes in Rosales, we reconstructed a phylogenetic tree at the genomic level. The phylogenic tree shows that *Fragaria* closed grouped with Rubus and the *Fragaria* clade is divided into three subclades. East Asian species appeared in every subclade, suggesting that the genus originated in this area at ~7.99 Mya. Four species found in mountains of Southwest China originated at ~3.98 Mya, suggesting that rapid speciation occurred to adapt to changing environments following the uplift of the Qinghai–Tibet Plateau. Moreover, we identified 510 very significantly positively selected genes in the cultivated species *F.* × *ananassa* genome. This set of genes was enriched in functions related to specific agronomic traits, such as carbon metabolism and plant hormone signal transduction processes, which are directly related to fruit quality and flavor. These findings illustrate comprehensive evolutionary patterns in *Fragaria* and the genetic basis of fruit domestication of cultivated strawberry at the genomic/transcriptomic level.

T2

P0578

Supplementary descriptions of flowers for four species of *Bambusa* (Poaceae: Bambusoideae)

Qiaomei Qin[1,2], *Jingbo Ni*[3], *Rushun Lin*[3], *Nianhe Xia*[1]

1. South China Botanical Garden, the Chinese Academy of Sciences

2. University of the Chinese Academy of Sciences

3. South China Botanical Garden

The morphology of reproductive organs is very important to the modern plant taxonomy. The flowering habit of bamboo is very special. The flowering of bamboo is an intriguing phenomenon because it is a unique and very rare occurrence. Most bamboos flower once every 60-130 years. And as a result, there are rather a percentage of documented bamboo species which were validly published based on the vegetative characters. The genus *Bambusa* is a big genus of Bambusoideae (Poaceae). More than 30 species documented in *Flora of China* are still without descriptions of flowers in this genus. Here, supplementary descriptions of flowers for *Bambusa subtruncata* L. C. Chia & H. L. Fung, *B. contracta* L. C. Chia & H. L. Fung, *B. corniculata* L. C. Chia & H. L. Fung and *B. cornigera* McClure are given, according to the fresh specimens collected from the bamboo garden in South China Botanical Garden. Photographs and illustrations of flowers for these four species are presented. *B. subtruncata*: one spikelet containing 9-10 florets, the intermediate 4-7 florets perfect; prophylls about 3 mm long, 2-keeled; rachilla segments flat, inflated and ciliolate at apex; empty glume 1-3, oblong; lemmas oblong-lanceolate, 11-14 mm long, glabrous; paleas lanceolate, 10-12 mm long, 2 keeled, sparsely ciliolate towards tips of keels; lodicules 3; stamens 6, anthers yellow to purple, 5-6 mm long; ovary obconiform, style 3, stigmas 3 (4), plumose. Caryopsis obconiform when young. *B. contracta*: one spikelet containing 4-5 florets, the intermediate 2-3 florets perfect; rachilla segments flat and glabrous, inflated at apex; empty glume1-3, oblong; lemmas oblong, 5-9 mm long, glabrous; paleas lanceolate, 6-8 mm long, 2 keeled, glabrous; lodicules 3; stamens 6, anthers yellow to purple, 3.5-5 mm long; ovary obelliptic, style 3, stigmas 3, plumose. Caryopsis obelliptic when young. *B. corniculata*: one spikelet containing 6-7 florets, the intermediate 2-5 florets perfect; prophylls about 1mm long, 2-keeled; rachilla segments flat and glabrous, exceedingly inflated at apex; empty glume 2-3, ovate-oblong; lemmas broadly elliptic, 8-12 mm long, glabrous; paleas lanceolate, 8-12 mm long, 2-keeled, glabrous; lodicules 3; stamens 6, anthers yellow to purple; ovary obelliptic, style 3, stigmas 3, plumose. Caryopsis obelliptic when young. *B. cornigera*: one spikelet containing 5-12 florets, the intermediate 5-7 florets perfect; rachilla segments flat and glabrous, exceedingly inflated and ciliolate at apex; empty glume 1-2, oblong; lemmas broadly elliptic, 10-12 mm long, glabrous; paleas lanceolate, 8-11mm long, 2-keeled, ciliolate towards tips of keels; lodicules 3; stamens 6, anthers yellow, 4-8 mm long; ovary very broadly ovoid, style single, stigmas 3, plumose. Caryopsis very broadly ovoid when young.

T2

P0579

Sub-localization of paralogous genes duplicated from alternatively-spliced, dual-targeting single-copy genes

Yichun Qiu[1], *Yii Van Tay*[1], *Yuan Ruan*[2], *Keith Adams*[1]

1. University of British Columbia

2. California Institute of Technology

As a major contributor to the angiosperm genomes, gene duplication can lead to the divergence of gene expression patterns and functions. Among all possible fates of duplicated genes, subfunctionalization refers to that duplicates take on different parts of the pre-duplicated function, and this partitioning event could be reflected by subcellular localization of the protein products.

Sub-localization happens after gene duplication when the original gene was localized to two subcellular locations and the duplicates are each localized to one of those locations. Alternative splicing is one mechanism by which the products of a single gene can be localized to two subcellular compartment, as the peptides from alternative spliced transcripts may vary in signals. After duplication of an alternatively spliced, dual targeting gene, one potential type of subfunctionalization is partitioning of the alternatively spliced forms between the duplicates, such that the product of each duplicate is localized to one of the original locations. Here we report the ascorbate peroxidase (APX) genes in land plants, which experienced more than ten lineage-specific sub-localization events.

T2

P0580

Plant pollinator networks and evolution of floral traits in Ipomoea

Martin Santiago-Hernandez[1], *Silvana Marten-Rodriguez*[1], *Martha Lopezaraiza*[2], *Francisco Parraguirre*[1], *Ken Oyama*[1], ***Mauricio Quesada***[1]

1. Universidad Nacional Autonoma de Mexico

2. Universidad Autonoma de Guerrero

The interactions between plants and pollinators have been important drivers of Angiosperm diversification. Network analysis has provided a comprehensive tool to understand the role of plant-pollinator interactions at the community level; however, most studies have analyzed this biotic interaction in terms of floral visitation and at specific times of the year. To understand how community-level interactions and phenology may influence floral evolution, it is important to assess how pollinators influence plant fitness on particular plant lineages throught the year. This is the first study that evaluates plant-pollinator networks, phenology and plant reproductive success on a group of coexisting congeners that exhibit different pollination syndromes including ornitophily, melitophily, sphingophily and chiropterophily. To evaluate the role of floral visitors on plant fitness, we conducted different network analyses of an *Ipomoea* community in the Pacific coast of Mexico, including: (1) all flower visitors, (2) visitors that contact the reproductive organs, (3) visitors that deposit pollen on stigmas, and (4) visitors that effect fruit and seed production. We conducted a series of pollination exclusion experiments to test the effectiveness of pollinators and also determined the anthesis and flowering phenology of the Ipomoea species studied. Results show the following: (1) floral visitors of most Ipomoea species included the pollinator functional group that corresponded to the floral syndrome as well as other visitors; (2) effective pollinators, in terms of pollen deposition on stigmas and fruit and seed production, corresponded to the floral syndromes and were the most frequent visitors in all cases; (3) secondary pollinators were also present in most syndromes; (4) coexisting Ipomoeas species sharing the same floral syndrome have staggered flowering phenology or different patterns of anthesis. In conclusion, the fitness network of plant-pollinators of Ipomoea is determined by the floral syndrome, providing strong support for Stebbins Principle that most frequent and effective floral visitors are the primary drivers of the evolution of floral traits. Furthermore, flowering phenology determines pollination network dynamics suggesting avoidance for competition in plants with the same pollination syndrome.

T2

P0581

KNOX genes and differences in shoot development between monocots and eudicots: A hypothesis

Geeta R

University of Delhi

Shoot developmental featurs such as indeterminate growth are conserved across vascular plants. However, other features may differ among vascular plant lineages. Differences between monocots and eudicots include leaf initiation and node formation, early leaf development, internode identity and elongation, stem widening and vascularization. Early emphasis in 'Evo-Devo' was to identify genetic mechanisms underlying conserved traits; the role of *KNOTTED*-like homeobox (*KNOX*) genes in regulating indeterminate growth of shoot apices across angiosperms is such an instance. It has been only relatively recently that attention was turned to investigation of differences. Here, I review known morphological and anatomical differences in shoot development between grasses and eudicots; interpret known variation in expression patterns and functions of the Class 1 group (*KNOX1* genes) and suggest that differences in their expression may be implicated in differences in shoot development (specifically, primary thickening and vascularization) between grasses and eudicots. A comparative phylogenetic analysis of this variation helps to identify taxa that may provide suitable systems to test the hypotheses posed here.

T2

P0582

Genetic consequences of geological and ecological factors in the Qinghai-Tibet Plateau: Three closely related *Primula* species as a case study based on RAD sequencing

Guangpeng Ren

University of Lausanne

The mechanisms of the interplay between geological and ecological factors in driving speciation, population divergence and shaping current distributions remains poorly understood, especially for species-rich radiations inhabiting the world's biodiversity hotspots. The toolbox of population genomics holds great promise for addressing these key issues, especially if genomic data are analysed within a spatially and ecologically explicit context. We used three closely related *Primula* species (*Primula nutans, P. tibetica* and *P. fasciculata*) that mainly occur in the Himalayas, the Hengduan Mountains and Northeast Qinghai-Tibet Plateau (QTP), respectively, to investigate the genetic consequences of past geological and climatic events in the QTP. Genomic data were obtained for 770 samples of the three species using restriction site-associated DNA (RAD) sequencing. We used a combination approach of population structure analyses, approximate Bayesian computation (ABC) modeling, Bayesian generalized linear mixed modeling (GLMM) and species distribution modeling (SDM) to examine the effects of past geological and climatic events on the three species. The ABC results indicate that the three species originated in the Himalayas and diverged from each other following the uplifts of the Hengduan Mountains and the Northern QTP during the Pliocene. After a long period of divergence, all the three species have retreated to different glacial refugia and expanded their distributions during the Quaternary glacial cycles as shown in the SDMs. For *P. nutans*, GLMM reveals that geographic distance is the major factor in driving population divergence, while for the other two species, geographical distances, climatic variables, pairwise altitude differences among populations and/or edaphic variables are involved in driving genetic variation of different genomic fractions (i.e. negative, neutral and positive genomic regions). Higher population differentiation and more distinct lineages were identified in *P. tibetica* and *P. fasciculata* than in *P. nutans*. The divergence times of these distinct lineages estimated by ABC modeling coincide well with the past glacial and interglacial periods, indicating that past climatic events were the major factors in shaping their large-scale spatial genetic structure. Furthermore, the response to climatic changes of populations of a species depends on its specific ecological preferences, which may show different ways of response to climatic changes compared with the ancestral populations as shown in *P. tibetica*. These results illustrate how geographical and ecological factors affected the evolutionary history of alpine plants. Our findings highlight the significance of combining the use of population genomics with environmental data when evaluating the effects of past geological and climatic events.

T2

P0583

Historical biogeography of endemic seed plant genera in the caribbean

Maria Esther Nieto-Blazquez[1], *Alexandre Antonelli*[2], ***Julissa Roncal***[1]

1. Memorial University of Newfoundland

2. University of Göteborg

The aim of this study was to better understand the historical assembly and evolution of endemic seed plant genera in the Caribbean, by first determining divergence times of endemic genera to test whether the GAARlandia landbridge played a role in the archipelago colonization, and second by testing South America as the main colonization source as expected by the position of landmasses and recent evidence of an asymmetrical biotic interchange. To accomplish this we gathered DNA sequences from four loci (18S rDNA, *atp*B, *mat*K and *rbc*L) for 610 species, including 41 seed plant genera endemic to the Caribbean (out of 180 in total). We reconstructed a dated molecular phylogeny using Bayesian inference and ten calibrations. To estimate the range of the ancestors of endemic genera we performed a model selection between a null and a more complex biogeographical model that included timeframes based on geological information and dispersal probabilities among regions. Divergence times for endemic genera ranged from Early Eocene (53.1 Ma) to Late Pliocene (3.4 Ma). Only the origin of four endemic genera occurred within the GAARlandia timeframe (35 to 33 Mya), whose ancestors were already present in the Antilles or derived from the Old World. For the entire data set, Central America appeared as the main source of colonization for endemic seed plant genera. Only two genera had a South American origin. In contrast with recent patterns shown for vertebrates and other organisms, GAARlandia did not act as a main colonization route for plants between South America and the Antilles. Divergence of endemic genera occurred mostly during the Miocene, i.e. after the proposed existence of GAARlandia. Contrary to expectations, Central America was the main colonization source, outplaying South America. A species-level synthesis on Caribbean plant dispersal will be required to reveal finer-scale patterns and mechanisms.

T2

P0584

Biogeographic patterns of *Haplopappus* (Asteraceae, Astereae), a genus of shrubs of the Arid Diagonal of Southern South America.

Marcelo Rosas, *Pablo Guerrero*

Universidad de Concepcion

The biogeographic patterns and historical processes of endemic arid plants in South America are scarcely known. In this study we used a genus of endemic shrubs, *Haplopappus* (Asteraceae, Astereae), which includes 68 species from the Arid Diagonal of the Southern Cone. Our goals were: i) to describe the distribution patterns, ii) to test ecological hypotheses of distribution patterns, iii) identify areas of high endemism, iv) to explore temperature and precipitation as explanatory variables of species richness. We used a list of 1058 specimens with taxonomic and geographic validation, from 21° S to 50° S and from 8 to 4103 m.a.s.l., the data were grouped in bands of 1° latitude and in bands of 200 m of altitude. Latitudinal and altitudinal richness gradients were evaluated according to the adjustment to nonlinear models. The latitudinal structure patterns were assessed using a similarity profile routine using a resemblance matrix, and a cluster analysis. The Mid Domain effect, the Rapoport's rule, and the source-sink dynamic were also tested. A Bayesian analysis of Endemicity (BAE) was used to evaluate hierarchical relationships of endemism between different latitudinal bands. The climatic drivers of species richness were evaluated using spatial regressions. Latitudinal and altitudinal pattern fit Gaussian distribution. The BAE indicated two areas of endemism: (1) the northern area between 21° S - 31° S, and (2) the southern area between 32° S- 50° S. The Mid Domain showed an important effect, except in the central part of the humpback curve and in Patagonia. The nesting analysis showed a significantly nested matrix, which can relate to a source-sink dynamic. The explanatory variables were temperature seasonality and mean temperature of wettest quarter. Our results support the existence of non-random pattern of richness gradients in *Haplopappus*, an integrative hypotesis derived from our results suggest that the genus diversification may be molded by climatic variables related to mediterrean biome and the expansion of some species to colder steppe-like habitats.

T2

P0585

Generic delimitation and biogeographic history of the complex *Carthamus - Carduncellus* (Cardueae, Compositae).

Roser Vilatersana[1], *Juan Antonio Calleja*[2], *Núria Garcia-Jacas*[1], *Alfonso Susanna*[1]

1. Institut Botànic de Barcelona (IBB – CSIC – ICUB)

2. Unitat de Botànica, Dept. Biologia Animal, Biologia Vegetal i Ecologia, Facultat de Biociències, Universidad Autónoma de Barcelona

The complex *Carthamus-Carduncellus* (Cardueae, Compositae) is formed by aproximally 60 taxa widely distributed throughout the Mediterranenan basin and western Asia. Generic delineation in the complex has been always controversial. The three main taxonomical proposals are (1) the complex is formed by a single genus *Carthamus*, as suggested by Greuter (2003), (2) two genera, *Carduncellus* and *Carthamus*, as proposed by López González (2012); and (3) four genera, *Carduncellus*, *Carthamus*, *Femeniasia* and *Phonus*, by Vilatersana *et al.*, (2000). Our major aim is to provide a robust and comprehensive systematic proposal for the complex *Carthamus-Carduncellus*. To do so, we face the specific goals: (1) to build nuclear and plastid phylogenies encompassing most of the richness of all the putative genera; (2) to propose and test the generic delimitation of the complex and, (3) to infer the biogeographic history of the complex by estimating divergence times and ancestral distribution areas and lineages. To address these issues, we constructed a phylogeny with a nearly-complete taxon sampling based on two nuclear regions (ITS and 3'ETS) and four plastid regions (5'*trn*T-*trn*L r*pl32*-*trn*L, *trn*H-*psb*A, and 3'*ndh*F). For the generic delimitation, we used both cpDNA and nrDNA datasets, we discussed the incongruence between both datasets, and we tested the three hypothesis on the systematic of the complex *Carthamus-Carduncellus* by using the "taxa delimitations" tool implemented in Beast. We used the nrDNA dataset because of its greater resolution for dating and biogeographic analysis performed with BEAST and BioGeoBears. The discussion is complemented with morphological and kariological data.

T2

P0586

Lineage divergence of a temperate forest species in eastern Asia, *Smilax sieboldii* (Smilacaceae): Vicariance vs. adaptive divergence

Yalu Ru[1,2], *Ruijing Cheng*[1,3], *Hanyang Lin*[1,2], *Yunpeng Zhao*[1,2], *Pan Li*[1,2], *Chengxin Fu*[1,2]

1. The Key Laboratory of Conservation Biology for Endangered Wildlife of the Ministry of Education, College of Life Sciences, Zhejiang University

2. Laboratory of Systematic & Evolutionary Botany and Biodiversity, Institute of Ecology and Conservation Center for Gene Resources of Endangered Wildlife, Zhejiang University

3. College of Agriculture & Biotechnology, Zhejiang University

Elucidating how various evolutionary factors interact to affect speciation is an essential theme in evolutionary biology. Insights of incipient speciation can be gained from studying lineage divergence. Eastern Asia harbors world richest temperate flora. The mechanisms generating such a rich species and genetic diversity have been fascinating evolutionary biologists for decades while considerable gaps remain in particular whether adaptive divergence also plays a role and its importance relative to vicariance. *Smilax sieboldii* is a typical woody climbing plant inhabiting temperate forests in eastern Asia with a disjunct distribution in mainland China and Taiwan, Japanese Archipelago and Korean Peninsula, providing us an ideal study system to test the relative role of vicariance and adaptive divergence. We employed three data sets including plastid genomes of 12 haplotypes previously suggested by two cpDNA fragments, EST-SSRs for 400 individuals from 38 populations, and RADseq tags for 45 representative individuals (one accession per population). Multiple data sets will be analyzed to confirm geographic lineages, to reconstruct their phylogenetic relationships and demographic histories, e.g., divergence times, dynamics of population sizes and gene flow. We will test competing hypotheses explaining the lineage divergence in relation to historical climate oscillation, sea level fluctuation, emergence and submergence of East China Sea Landbridge, and range shifts. Finally, canonical correlation and redundancy analysis will be utilized to compare the relative contribution of environment and geography to genetic divergence of the focal species. Our findings will imply the evolutionary process of *S. sieboldii* and indicate the relative role of geographical isolation and differential selection which shaped the diversity and phylogeographic structure of temperate forests plants in eastern Asian.

T2

P0587

A taxonomic revision of genus *Ceratostylis* (Orchidaceae) in Thailand

Wirata Rujichaipimon[1,2,3], *Santi Watthana*[4], *Paweena Traiperm*[2]

1. Master of Science Program in Plant Science, Faculty of Graduate Studies, Mahidol University, Bangkok

2. Department of Plant Science, Faculty of Science, Mahidol University, Bangkok

3. Department of Pharmaceutical Botany, Faculty of Pharmacy, Mahidol University, Bangkok

4. School of Biology, Institute of Science, Suranaree University of Technology, Nakhon Ratchasima

Ceratostylis is a genus in the orchid family (Orchidaceae) distributes from tropical and subtropical Asia to western pacific with approximately 153 species. The aim of this study was to revise the species of this genus found in Thailand. We collected and gathered the data from previous literatures, as well as studied the voucher specimens held at the main herbaria in Thailand and Europe including BCU, BK, BKF, BM, C, K, L, QBG and PSU. In addition, the type specimens were examined and compared from the herbaria where visited, as well as consulted the digitized specimens that are available online. In Thailand, six *Ceratostylis* spp. have been recorded i.e., *C. ampullacea* Kraenzl., *C. pleurothallis* (C.S.P. Parish & Rchb.f.) Seidenf., *C. radiata* J.J. Sm., *C. siamensis* Rolfe ex Downie, *C. subulata* Blume and *C. thailandica* Seidenf. Among these six species, *C. siamensis* and *C. thailandica* are endemic to Thailand. All species are dense tuft plant and exhibit only epiphytic form, except *C. subulata* which having both epiphytic and lithophytic habits. A key to species is provided together with descriptions, illustrations, photos, ecological data and notes.

T2

P0588

Phytogeography, endemism and conservation of the genus *Sonerila* Roxb. (Melastomataceae) in Western Ghats

Resmi S, *Santhosh Nampy*

Department of Botany, University of Calicut, Kerala

Sonerila Roxb. (Melastomataceae - tribe Sonerileae) with about 175 species is distributed from Sri Lanka and southern India to the Indo-Pacific region. The genus is easily distinguished in having uniparous (scorpioid) cymes and trimerous flowers. The members are caulescent/acaulescent herbs of shady habitat, often with basal rosettes of large, turgescent leaves, sometimes with tubers and mostly seen in moist rocky crevices. The genus is represented by 51 species and two varieties in India. Western Ghats has the highest species diversity with about 41 species and 2 varieties (81%). Most of them are endemic. Considerable work was conducted in the Sonerilas of neighbouring countries by various authors. However, no similar attempt is so far initiated in this group in India. In spite of greater diversity and endemism, the genus has been largely neglected taxonomically in India. The present paper discusses the diversity, distribution, endemism, conservation and status of the genus in Western Ghats.

T2

P0589

Multi-gene phylogenetic analysis reveals the multiple origin and evolution of mangrove physiological traits through exaptation

Sunil Kumar Sahu[1], *Kathiresan Kandasamy*[2]

1. Sun Yat-sen University

2. Centre of Advanced study in Marine Biology, Annamalai University

Mangroves are taxonomically diverse group of salt-tolerant, mainly arboreal, flowering plants that grow in tropical and sub-tropical regions and have adapted themselves to thrive in such obdurate surroundings. While evolution is often understood exclusively in terms of adaptation, innovation often begins when a feature adapted for one function is co-opted for a different purpose and the co-opted features are called exaptations. Thus, one of the fundamental issues is what features of mangroves have evolved through exaptation. We attempt to address these questions through molecular phylogenetic approach using chloroplast and nuclear markers. First, we determined if these mangroves specific traits have evolved multiple times in the phylogeny. Once the multiple origins were established, we then looked at related non-mangrove species for characters that could have been co-opted by mangrove species. We also assessed the efficacy of these molecular sequences in distinguishing mangroves at the species level. In addition by using the AFLP, chloroplast and nuclear markers we resolved the taxonomical controversy between *Rhizophora* hybrids *R.* × *annamalayana* and *R.* × *lamarckii*. This study revealed the multiple origin of mangroves and shed light on the ancestral characters that might have led certain lineages of plants to adapt to estuarine conditions and also traces the evolutionary history of mangroves and hitherto unexplained theory that mangroves traits (aerial roots and viviparous propagules) evolved as a result of exaptation rather than adaptation to saline habitats.

T2

P0590

DNA barcoding of *Hippophae*

Saloni Malik[1], *Soom N. Raina*[2], *Shashi B. Babbar*[1]

1. Department of Botany University of Delhi

2. Amity Institute of Biotechnology Amity University

Species of *Hippophae,* reported to inhabit high altitudes of Himalayas across Jammu and Kashmir (J&K), Himachal Pradesh (HP), Uttarakhand (UK) and Sikkim (SK) states in India, are *H. rhamnoides* subsp. *turkestanica, H. tibetana* and *H. salicifolia.* It is difficult to identify these species correctly if available in vegetative state. The present investigation was undertaken to see if DNA barcodes distinguish these species. A total of 56 accessions of *Hippophae* collected with tentative identifications as *H. salicifolia*, *H. rhamnoides* ssp. *turkestanica*, and *H. tibetana* from the states listed and 24 accessions of *H. rhamnoides* subsp. *carpatica*, *H. rhamnoides* subsp. *mongolica*, *H. rhamnoides* subsp. *rhamnoides* procured from Ukraine, Russia, Germany and Finland were analyzed. The loci used for DNA barcoding were ITS, *mat*K, *rbc*L and *rpoC1* and the number of their sequences obtained were 72,75,80 and 79, respectively. In the neighbor joining (NJ) trees of all the tested loci, accessions of *H. rhamnoides* subsp. *turkestanica* from UK, co-segragated with *H. salicifolia* from HP, UK and SK. Likewise, accessions of *H. salicifolia* from Kinnaur (HP) grouped with *H. rhamnoides* subspecies *turkestanica* and four of *H. tibetana* from HP co-segregated with *H. salicifolia* collected from HP, UK and SK. These observations raised doubts about the identification of some of accessions of *H. salicifolia* (HP), *H. rhamnoides* subspecies *turkestanica* (UK) and *H. tibetana* (HP). Therefore, sequences of all the loci were BLAST searched on NCBI and BOLD (only *mat*K and *rbc*L). BLAST1 analysis revealed that the accessions collected as *H. rhamnoides* subspecies *turkestanica* from UK were, in fact, *H. salicifolia*. Likewise, the accessions collected as *H. salicifolia* from HP turned out to be *H. rhamnoides* and the correct identity of four accessions collected from HP as *H. tibetana* was *H. salicifolia*. To reveal the inter-relationships among species, NJ trees were constructed again. ITS NJ tree had two main clades, one comprising only *H. tibetana*, while the other included *H. salicifolia* and all sub-species of *H. rhamnoides.* The latter clade futher segregated in two groups, one of which was represented by *H. rhamnoides* ssp. *turkestanica* and the other comprised rest of the sub-species *of H. rhamnoides.* NJ tree based on *matK* too revealed two clades. However, in this tree, *H. tibetana* clustered with *H. salicifolia* in one clade, but in two distinct sub-clades. The other clade too had two sub-clades, one of which had all accessions of *H. rhamnoides* subsp. *turkestanica* along with four of eight accessions of *H. rhamnoides* subsp. *rhamnoides* from Germany. The other sub-clade included rest of the sub-species of *H. rhamnoides*. In the *rbc*L NJ tree, there was no clear-cut segregation of species as well as sub-species of *H. rhamnoides*. The three species could be easily distinguished on the basis *rpoC1* sequences. However, no distinction could be made among the sub-species of *H. rhamnoides.* The study highlights the power of DNA barcoding for identification of *Hippophae* species and also points towards the distinctive nature of *H. rhamnoides* subsp. *turkestanica*, which could possibly be raised to the species level.

T2

P0591

The Zingiberaceae in the Flora of Peninsular Malaysia

Yen Yen Sam

Forest Research Institute Malaysia

Peninsular Malaysia has an exceptionally rich ginger flora with a total of 21 genera and close to 190 species recorded so far. The warm and humid climate plus the wide range of habitats in the evergreen tropical rainforest contributed to the high diversity and endemism observed in the family. Though located in the Malesian region, its ginger flora is very distinct from the neighbouring floristic regions where approximately 60% of the species are endemic to the peninsula. Peninsular Malaysia has a long history in the documentation and taxonomic study of the ginger species. However, there are still many new discoveries in recent times, e.g. three new genera and 32 new species are described in the last 15 years. The Flora of Peninsular Malaysia project currently undertaken by the Forest Research Institute Malaysia has amassed a great amount of herbarium collections for a comprehensive revision of the family. Based on the experiences of the current and past Flora projects, the collaboration and contribution from both local and foreign experts are very crucial in revising such big family which is estimated to exceed 200 species when the treatment is completed.

T2
P0592

Diversification correlates of Neotropical plants biodiversity

Andrea Sanchez Meseguer, *Frederic Delsuc, Fabien Condamine*
CNRS

Due to its unique geological history and tropical ecology, the South American biota bears many of the known biodiversity hotspots in the world, including the richest (the Tropical Andes, where many rapid radiations have occurred). The establishment of complex riverine systems, the uplift of the Andes, and Cenozoic climatic changes have all influenced Neotropical plant diversity, but the historical causes and regional determinants remain debated. The ensuing question '*What gave rise to the Neotropical staggering biodiversity?*' is one of the most debated topics in evolutionary ecology. Despite an increasing number of evolutionary studies on Neotropical groups, the relative contribution of different factor in Neotropical diversification is still poorly understood. Most studies in the field focus to detect changes in diversification trajectories that have occurred through time and/or across clades. If rate variation is detected, the potential impact of different factors on the observed pattern is generally discussed based on temporal congruence between diversification changes and the potential factor. The recent development of environment-explicit phylogenetic birth-death models now allows explicitly testing the relative contribution of various factors on Neotropical diversification. Here, we present a meta-analysis of diversification correlates for more than 45 Neotropical plant clades, including angiosperm (e.g. orchids, palms, grasses) and gymnosperm lineages (cycads), that span different taxonomic levels (families, tribes, genera, sections), ages of origin (from the Eocene to the present), as well as Neotropical regions (e.g. Andes, Amazonia, Central America). The global plant dataset is further classified according to the main distribution of lineages (Andean-centered vs. Amazonian-centered), and altitude (lowland, montane or highland). According to this division, we evaluate the relative contribution of Andean orogeny and Cenozoic climate change on diversification trajectories of Neotropical plants. We compare the fit of a series of diversification models that incorporate mountain building and/or climate change, against other constant and time variable null models. The overall pattern from our analyses is that, in more than 50% of the phylogenies, temperature-dependent models fit the data better than other time- or altitude-dependent models. Most lowland and Amazonian clades show positive correlation with temperature (i.e. speciation rates decrease over time), while mountain-highland clades present a mixed pattern: some clades are positively correlated while others are negatively correlated with temperature (i.e. speciation increases with climate cooling). Mountain building appears to have directly influenced diversification of lowland Neotropical taxa form the Andes and the Amazonian regions, generally towards increasing diversification. Unexpectedly, the effect of mountain uplift over other montane-highland clades is less pronounced, which are mostly influenced by changing temperatures. For less than 20% of the phylogenies, time-dependent models fit the data better than the other environmental-dependent models, which might suggest that other factors not considered in this study are at play. These results suggest that Cenozoic climate cooling had differential effects on Neotropical diversity, promoting diversification for some clades but hampering diversification for others, while the Andean orogeny has always spurred Neotropical species richness.

T2
P0593

New insights on the floral evolution in the Gomphrenoideae (Amaranthaceae), based on molecular phylogenetics and floral morphological and developmental data

Ivonne Sánchez-Del Pino[1], *Alexander Vrijdaghs*[2]
1. Centro de Investigación Científica de Yucatán, A.C. (CICY)
2. Ecology, Evolution and Biodiversity Conservation KU Leuven

Androecium and gynoecium are key characters in the systematics of Gomphrenoideae (Amaranthaceae). SEM and LM based floral ontogenetic and morphological studies provide new insights in the evolution of the flower in Gomphrenoideae. A sampling of 37 species was studied for morphology of gynoecium, androecium and pollen characters whereas ontogenetic development was observed for 13 species. Character optimization was conducted in a parsimony tree using Mesquite software. Seventeen characters were optimized, of which eight were related to androecium, three to stigmas, and six to pollen. In this study we aimed to perform homology assessments for the androecial tube and its appendages of Gomphrenoideae; to test Eliasson´s hypothesis for the evolution of the androecium appendages; to re-evaluate pollen and gynoecial characters in a phylogenetic context.Two types of androecial appendages, not homologous, and hypothesized to be de novo organs were characterized. Based on these findings, we can suggest that none of Eliasson´s hypothesis can be supported. Four pollen characters support major clades within Gomphrenoideae. A perygynous hypanthium appears twice in Amaranthaceae sensu lato and it is here for the first time well documented. At early development stages, a central ovule primordium slightly fixed at one side to the annular ovary wall, and dome-shaped flower receptacles are hypothesized to be synapomorphies for respectively Amaranthaceae sensu lato and Gomphrenoideae.

T2
P0594
Genetic diversity analysis of *Curcuma* by Random Amplified Polymorphic DNA (RAPD)

Pattamon Sangin, *Pranee Nangngam, Audsharaporn Jai-Eun*
Naresuan University

Genetic diversity analysis of *Curcuma* and intraspecific level of *Curcuma longa* was studied using RAPD technique. One out of ten primers (MI1) was found to give polymorphic DNA patterns. A total of 49 bands were scored with the size of 0.30-3.0 Kb. Genetic similarity was estimated by the Jaccard coefficient from NTSYSpc 2.20e version which ranged between 0.612-1.000. A dendrogram was also constructed by using Unweighted Pair Group Method with Arithmetic Mean (UPGMA) which divided *Curcuma* into two major clusters. First group comprised all *C. longa* L. 11 accessions, which *C. longa* L. (Cu19) and *C. longa* L. (Cu20) were grouped together with the high similarity index (1.000). The second clade was 8 samples of *Curcuma*; *C. manga* Valeton & Zijp (Cu10), *C. amada* Roxb. (Cu11, Cu12 and Cu15), *C. rubescens* Roxb. (Cu17), *C. aromatica* Salisb. (Cu14), and *Curcuma sp.* (Cu3 and Cu16).

T2
P0595
Phylogenetic relationship of *Jatropha curcas* using *atp*B – *rbc*L intergenic spacer

Pattamon Sangin, *Sudarat Kasemcholathan*
Naresuan University

The plants in genus *Jatroph*a belong to the family Euphorbiaceae, which are distributed in tropical and subtropical region around the world. There are only 5 species of *Jatropha* in Thailand; *J. curcas*, *J. gossypiifolia*, *J. multifida*, *J. podagrica* and *J. integerrima*. However, *J. curcas* is the only one which is attracted as an alternative to biodiesel. It is widely distributed in many areas in Thailand, which is difficult to identify by morphological properties because their characteristics are similar. Therefore, in this research, DNA sequences data from *atp*B–*rbc*L intergenic spacer were used to elucidate the evolutionary relationship in 15 accessions of *J. curcas*. The result showed that the length of of *atp*B-*rbc*L was 857 to 873 bp. The phylogenetic tree was reconstructed by maximum likelihood method revealed that *J. curcas* accessions were divided into 2 groups as *J. curcas* (11) was closely related to *J. curcas* (8). Furthermore, *J. multifida*, *J. podagrica* and *J. gossypiifolia* revealed a close phylogenetic relationship.

T2
P0596
Genetic diversity analysis of *Jatropha* by Random Amplified Polymorphic DNA (RAPD)

Pattamon Sangin, *Sagaoduen Moungyu*
Department of Biology, Faculty of Science, Naresuan University

Genetic diversity analysis of *Jatropha* and intraspecific level of *J. curcas* was studied using RAPD technique. Six out of one hundred and seven primers (2075 2391 2393 2777 OPA4 and OPA18) were found to give polymorphic DNA patterns. A total of 144 bands were scored with the size of 0.25-3.0 Kb. Genetic similarity was estimated by the Jaccard coefficient from NTSYSpc 2.20e version which ranged between 0.46-1.000. A dendrogram was also constructed by using Unweighted Pair Group Method with Arithmetic Mean (UPGMA) which divided *Jatropha* into two major clusters. First group comprised all *J. curcas* 18 accessions which Hongsa, Lumpang and Jomthong were grouped together with the high similarity index (1.000) as well as Mukdahan and India also clustered together. The second clade was 5 samples of *Jatropha*; *J. gossypifolia*, *J. podagrica*, *J. multifida*, *J. integerrima* (1) and *J. integerrima* (2) which *J. integerrima* (1) and *J. integerrima* (2) were clustered together with the genetic similarity value of 0.792 according to morphological character.

T2
P0597
Pollen morphology of some entomophilous trees from Terai – Dooars region of west Bengal, India

Baishakhi Sarkar[1], *Subir Bera*[2], *Abhaya Prasad Das*[1]
1. University of North Bengal
2. University of Calcutta

The sub-Himalayan West Bengal, also known as the Terai-Dooars region is a part of the Himalaya Biodiversity hotspot and is well-known for its floristic and faunal diversity. Documentation and identification of floristic elements involves the collection and preservation of plant parts of which pollens provide very comfortable option using little space but exposing numerous but quite useful characters those can be used for identification as well as for phylogenetic analysis. The flowering plants have evolved different modes of pollination mechanisms to facilitate the fertilization process. Although the morphology of the grains is not always directly affected by the mode of pollination, but it reflects the environmental and ecological conditions. Present study deals with pollen morphology of 100 randomly selected entomophilous tree species comprising 77 genera under 39 families from Terai and Dooars regions of West Bengal, India and portrays diversification of pollen characters. Although the grains were mainly in monad condition, tetrads and polyads were also found especially in Mimosoideae. The extremely diversified exine patterns like granulate, bacculate, rugulate, verrucate, scrabrate, various types of reticulation and also psilate tends to be characteristic to not only at the family level but also can be species specific. Other morphological characters showed distinct variations. The shape varied from oblate (*Moringa olieferae, Erythrina* sp.) to perprolate (*Tecoma stans, Combretum* sp.). The size also varied from very small grains of *Syzygium operculatum* (8.78 µm) to very big grains of *Bauhinia acuminata* (117 µm). Although the dominant aperturate condition was tricolporate type, other apertures like tricolpate, triporate, polycolpate, polyporate, polycolporate can also be noticed. Using these morphological characters the phylogenetic relationship of the taxa were examined. This study may help in establishing the insect plant relationships and tends to be important as it lays the basic foundation for future investigation of the late quaternary vegetation and climatic condition of the area along with other palynological aspects like pollination ecology, pollen biology, melissopalynology and aeropalynology.

T2

P0598

Evolutionary and systematic studies in south American endemic genus *Tristagma* (Amaryllidaceae)

Agostina Sassone
Instituto de Botánica Darwinion

Subfamily Allioideae is one of three within family Amaryllidaceae (Chase *et al.*, 2009), and of great economic importance. The closest relatives of *Allium* L. in South America are represented by the endemic genera of the tribes Gilliesieae and Leucocoryneae. The tribe Leucocoryneae includes 6 genera exclusively distributed in South America, except of *Nothoscordum* Kunth, which is widespread in South America with one species occurring in North America. Most of the genera have a controversial taxonomy, species and genus delimitation has been considered problematic. *Tristagma* was described by Poeppig (1833: 369) with *Tristagma nivale* Poepp. as type species. Circumscription of *Tristagma* species has been a source of confusion for various reasons; e.g., generic concepts, nomenclatural inconsistencies, and doubtful species. The aim of this research is to present a multidisciplinary study of *Tristagma*, in order to define the number of species, contributing to a better knowledge of the subfamily Allioideae in South America. Morphological, phylogenetic, a complete bibliographic, taxonomic revision, and geographical distribution analyses were conducted. Phenetic analyses were performed including 200 specimens of *Tristagma* on the base of 60 vegetative and reproductive morphological characters. Through multivariate analyses 3 complexes of species and diagnostic morphological characters are identified.Fifty-seven specimens representing species of *Tristagma* and allied genera were examined. Three molecular markers (ITS, *mat*k and *ndh*F) were sequenced and analyzed with Parsimony and Bayesian inference analyses, to elucidate relationships within *Tristagma* and related groups. Exploring molecular marker ITS, *Tristagma* resulted monophyletic, when analyzing cpDNA phylogenetic hypothesis the position of *Tristagma* was not defined. The combined analysis supports *Tristagma as* paraphyletic being necessary the inclusion of *Ipheion* to be considered as a monophyletic genus. Also, the optimization of DNA content shows genome size variation among Leucocoryneae genera and within *Tristagma*. As a result of the bibliographic compilation, the identification of different species with fresh material and the study of available herbaria specimens, it is concluded that *Tristagma* is composed by 12 species. The genus *Tristagma* can be characterized by uni or plurifloral inflorescences, flowers with fused tepals forming a tube, and staminal filaments disposed in two series. Also, preliminary biogeographic studies were conducted in order to define the environmental requirements of the different species of *Tristagma*. Available pecimens of all species, covering the entire range of distribution of the genus, where georeferenced. Bioclimatic variables and altitude from the WorldClim dataset were associated with each locality of the studied specimens. By the correlation of climatic and geographical information it is concluded that precipitation and the mean temperature of the winter are the most influencing variables of the actual distribution of *Tristagma* species.

T2

P0599

Taxonomical identity of *Encalypta* species with gemmae and genetic diversity in Japan

Takumi Sato, *Tomio Yamaguchi*
Hiroshima University

The genus *Encalypta* Hedw. (Encalyptaceae, Bryophyta) includes 36 species worldwide, and is mainly characterized by the markedly thickened transverse walls of the basal laminal cells and the campanulate calyptrae completely covering the capsules. The genus is predominantly recorded in montane and arctic regions throughout the world, and the species are approximately restricted to alpine regions in Japan. Seven species have been reported from Japan: *E. alpina* Sm., *E. asperifolia* Mitt., *E. ciliata* Hedw., *E. microstoma* Bals.-Criv. & De Not., *E. rhaptocarpa* Schwaegr., *E. streptocarpa* Hedw. and *E. trachymitria* Ripart. However, it has been discussed about taxonomical identity of Japanese plants of *Encalypta* with gemmae. The Japanese plants bearing gemmae was firstly identified as *Merceya serratinervis* Takaki by Takai in 1951, but later Takaki (1958) recognized it to be *Encalypta stretocarpa*. Horton (1983) concluded that Japanese plants previously identified as *E. streptocarpa* were misidentifications of *E. procera*, and she recognized *E. streptocarpa* to be restricted to Europe in her revision of *Encalypta* from North America. Iwatsuki (2008) re-examined all specimens collected from Japan and identified as either *E. streptocarpa* or *E. procera* focusing on gametophyte morphology and concluded that only *E. streptocarpa* is present in Japan. In the present study, molecular analyses were performed for the first time to reveal the taxonomic identity of Japanese specimens previously identified as *E. streptocarpa* in Japan. Phylogenetic analyses based on *trn*L-F and *ITS* support the monophyly of each Japanese *Encalypta* species and revealed that all Japanese samples bearing gemmae were nested within *E. streptocarpa* clade. We also provided the genetic diversity of *E. streptocarpa* from Japan based on haplotype network analyses.

T2

P0600

The preliminary study of the genus *Alysicarpus* (Desmodieae, Fabaceae) in Thailand

Jiratthi Satthaphorn, *Charan Leeratiwong*
Department of Biology, Faculty of Science, Prince of Songkla University

The preliminary study of the genus *Alysicarpus* (Desmodieae, Fabaceae) in Thailand was studied between May 2016 and January 2017 based on field surveys and herbarium specimens especially from Thailand. A total of three species of Thai *Alysicarpus* were found namely *A. bupleurifolius* (L.) DC., *A. rugosus* (Willd.) DC. and *A. vaginalis* (L.) DC. The genus is significantly characterized by leaflet shapes, stipules, calyx lobes, corolla colors and legume indumentums. Morphological descriptions, distribution, ecological aspects and taxonomic treatments of the genus are provided.

T2

P0601

Phylogeny of the grass subtribe Rottboelliinae (Poaceae):

Why such a rotten mess?

Tanya Scharaschkin, *Melodina Fabillo*
Queensland University of Technology

Rottboelliinae is one of six subtribes in the grass tribe Andropogoneae. The number of genera recognized varies with taxonomic treatments, ranging from 10 to 18. *Coelorachis, Hackelochloa, Heteropholis, Ratzeburgia* and *Rottboellia* were synonymized with *Mnesithea* on the basis of morphological characters. Recent molecular phylogenetic studies partially support this decision, but results do not concur. Some authors recognise *Heteropholis, Ratzeburgia* and *Rottboellia* as separate from *Mnesithea* whilst others suggest that these three genera should remain synonymised with *Mnesithea* but *Hackelochloa* be recognised as a distinct genus. The genera *Elionus* and *Ophiuros* are included in the subtribe by some authors but not others. Taxonomic instability in the subtribe appears to be due to overlapping variability in gross morphological characters used to delimit genera and the lack of comprehensive taxonomic sampling in previous phylogenetic studies. In Australia, Rottboelliinae is represented by ~20 species in six genera that occur mainly in mesic eucalypt woodlands and savannahs. Seven of these species are endemic to Australia and two are rare. *Hemarthria uncinata* var. *uncinata* and *H. uncinata* var. *spathacea* show great morphological diversity, raising the question as to whether *H. uncinata* includes cryptic species or is a single variable species. Morphological characteristics of species of *Eremochloa* in Australia do not consistently agree with taxonomic descriptions and identification using published keys is problematic. The primary objective of this study was to test previous hypotheses of relationships and taxonomic decisions regarding the number of genera included in the subtribe Rottboelliinae. A secondary objective was to determine the phylogenetic position of the Australian taxa belonging to this subtribe. In order to resolve taxonomic issues, the following strategies were used: comprehensive and extensive taxon sampling, multi-gene sequencing, inclusion of morphological characters and careful examination of vouchers. Preliminary phylogenetic results based on separate and combined analyses of chloroplast sequence data (*atp*B-*rbc*L spacer, partial *mat*K gene, partial *ndh*F gene, *trn*L-F intron and intergenic spacer) using Bayesian inference, maximum likelihood and parsimony provide some interesting results. The genera *Eremochloa, Elionuros, Glyphochloa* and *Thaumastochloa* appear to be monophyletic. The genus *Mnesithea* is polyphyletic and previous recommendations of synonymy are not fully supported. The genus *Hemarthria* is not monophyletic. One possible cause for conflicting molecular phylogenetic results leading to contradictory recommendations for taxonomic decisions appears to be incorrect identification of species. The importance of careful examination of vouchers and re-examination of vouchers for Genbank sequence data cannot be overstated. This is the first comprehensive phylogenetic study of Rottboelliinae at a global and Australian level.

T2
P0602

Comparative diversification analyses of Hydrangeaceae and Loasaceae: The role of continental dispersal in generating species diversity

John Schenk[1], *Sarah Jacobs*[2], *Larry Hufford*[3]

1. Georgia Southern University
2. University of Idaho
3. Washington State University

Hydrangeaceae and Loasaceae are sister clades that likely originated in arid regions of Mesoamerica but subsequently experienced different biogeographic and speciation histories. Hydrangeaceae became particularly diverse in Asia, with additional centers of diversity in western and eastern North America. Loasaceae, in contrast, has two centers of diversity, one in western North America and a second in the South American Andes Mountains. We tested the hypothesis that these clades share a common origin in Mesoamerica, and then compared how they diversified along both biogeographic and phylogenetic axes. Because these lineages experienced different biogeographic transitions that expanded their ranges into novel continental areas, we were able to examine and compare the consequences of colonization to ecological opportunity. Under an ecological opportunity model, we predicted that clades will experience an increase in speciation rates following the first colonization of new continental areas. We constructed a time-calibrated, phylogenetic hypothesis consisting of approximately 64% of the species diversity, and then applied the phylogeny to diversification and historical biogeographic analyses. The Hydrangeaceae-Loasaceae split occurred at least 95.34–72.94 mybp. Speciation rates were approximately constant from that time until about 18 mybp, after which they sharply increased. Historical biogeographic analyses supported a Mesoamerican origin of Loasaceae and Hydrangeaceae, followed by numerous biogeographic transitions in both clades. Lineages of Hydrangeaceae diversified early into Asia from western North America, followed by colonizations into eastern North America. A significant diversification rate increase was identified in the *Philadelphus* clade, associated with its initial colonization of Eurasia. A second increased diversification rate shift, not associated with a geographic change, was identified in the so-called Hydrangea-2 clade. Loasaceae experienced greater early lineage diversification in western North America than Hydrangeaceae, followed by a biogeographic transition into South America at the origin of Loasoideae. The latter range expansion into South America was not associated with a significant increased diversification rate; however, increased diversification was inferred in the *Cajophora* clade of Loasoideae, a change likely tied to ecological opportunities created with the uplift of the Andes. Two increased shifts were associated within the genus *Mentzelia*: one in *Mentzelia* section *Trachyphytum* and a second in *Mentzelia* section *Bartonia*. Neither of the two *Mentzelia* shifts were associated with continental-scale biogeographic transitions. We conclude that geographic range expansions have been important in generating species diversity in both families, but climate change, including aridification of the American West during the Miocene and glaciations during the Pleistocene, may have been equally important in accelerating cladogenesis in *Mentzelia*.

T2
P0603

Nomenclatural novelties for *Mimosa* subseries *Obstrigosae* from South America

Fernanda Schmidt Silveira, Silvia T.S. Miotto, ***João R. V. Iganci***

Universidade Federal do Rio Grande do Sul (UFRGS)

Brazil is a center of diversity and endemism of *Mimosa* genus, represented by ca. 360 species, particularly on South and Central regions. Nevertheless, only from the last 20 years, we have observed an increasing in taxonomic revisions, Floras and specialists training, especially for Brazil Central. *Mimosa* section is the second in number of species, having circa 180 spp., distributed in three series and 37 subseries. Taxonomy of *Obstrigosae* subseries is considered by specialists of South America really difficult and uncertain. It is commoly find exsicates incorrectly identified. In this context, typification and name validation plays an important role, once these assign the correct application of the name for a species. Regarding this, our goals were: (1) lectotypification of *M. adpressa* Hook. & Arn and *M. obstrigosa* Burkart to the correct use of these names and (2) revalidation of *M. subinermis* Benth. as a distinct species from *M. rupestris* Benth. to allow the evaluation of its conservation status and improvement of the knowledge of *Mimosa* diversity. As lectotypus of *M. adpressa* was chosen the exsiccate from Argentina of the Province of Entre Rios, collected by Tweedie and deposited at Kew herbarium. For *M. obstrigosa*, the chosen exsiccate to be the lectotypus was from Argentina of the Santa Ana city, collected by Rodriguez and deposited at SI herbarium. *Mimosa subinermis* species can be readily distinguished from *M.rupestris* by its habit, presence of aculei, branches sparsely covered with strigose trichomes, paleaceous calyx covering fast half of the corolla length, fruits covered with strigose adpressed trichomes.

T2

P0604

Resolving phylogenetic relationships of pantropical Ochnaceae s.l. using targeted sequence capture

Julio V. Schneider, *Thomas Berberich, Georg Zizka*
Senckenberg Research Institute

Ochnaceae s.l. (including Medusagynaceae and Quiinaceae) is a pantropical family of about 550 species and 32 genera largely distributed in rainforests and savannas. To date, intergeneric relationships could be only partially resolved due to incomplete taxon and molecular sampling as characteristic for many tropical plant families which rely on DNA extraction from poorly preserved herbarium specimens. To overcome limitations of traditional Sanger sequencing, we have chosen a gene capture approach, targeting 660 kbp of the nuclear genome. Here, we present the pipeline from bait development to sequence assembly as well as new insights into the phylogenetic relationships of Ochnaceae.

T2

P0605

Names matter: *Fagonia* sp. and breast cancer

Melanie Schori[1], *Allan Showalter*[2]
1. US Department of Agriculture, Agricultural Research Service
2. Ohio University, Department of Environmental & Plant Biology

Plants are an importance source of many medicines, and researchers around the world continue to investigate traditionally used plants for their therapeutic effects. Collaboration between systematists, taxonomists, and medical researchers is important to ensure that tested material is correctly identified before it is used and to highlight closely related species that might have similar medicinal properties. *Fagonia* is a genus of plants in the caltrop family (Zygophyllaceae) and includes 34 species, many of which were lumped into the species *F. cretica* L. in the 1860s. The name *F. cretica* continues to be widely used in primary, secondary, and tertiary literature, even though Beier revised the genus in 2010. In 2012, an article on cytotoxic properties of an aqueous extract of *Fagonia* from Pakistan on breast cancer cells reported the species as *cretica*, which does not grow in the country. DNA sequencing has since been used to distinguish *Fagonia* species from Pakistan, and the cultivation requirements of the widespread *F. indica* are now being studied. Several other species with limited distribution may also be effective against breast cancer, but their traditional use has not been well documented and they still need to be evaluated in the laboratory. In the future, *Fagonia* may prove to be an important source for breast cancer treatment, and correct identification of the plant material will be essential for ensuring efficacy and conservation of genetic resources.

T2

P0606

Atlas Florae Europaeae – the greatest ever effort for mapping floristic diversity in Europe runs across borders and projects

Alexander Sennikov[1,2], *Arto Kurtto*[1], *Raino Lampinen*[1], *Pertti Uotila*[1]
1. University of Helsinki
2. Komarov Botanical Institute

There is a pressing need to accelerate biodiversity research and provide for comprehensive availability of its results because of the rapid degradation of the environment and the high risk of human-induced extinctions in all parts of the world. The development of unified global databases is a major step forward in biodiversity studies and conservation efforts. Whereas taxonomic databases (and electronic database-derived monographs) are becoming commonplace in biodiversity research, databases of distributional data are much less common and are largely limited to countries with lower levels of plant diversity and higher levels of resources. In addition, distributional data derived from taxonomic monographs are fragmented and thus poorly available to the scientific community. *Atlas Florae Europaeae* (AFE) is a long-term programme of mapping the diversity of vascular plants in Europe, aiming at covering the whole flora of Europe with grid maps of species and subspecies distribution. With 16 printed volumes already published, AFE currently covers about 25% of the European flora. To date, the quality of its coverage far exceeds that of the data available from other sources, e.g. GBIF. Established as a supplement to *Flora Europaea* (FE), AFE has surpassed its original purpose of providing the technical background of distributional data to taxonomic treatments in FE. At present, AFE collaborates with Euro+Med PlantBase (EMPB) by contributing some treatments and checklists and by providing map visualisation, hosting and maintenance. To respond to conservation needs, AFE collaborates with IUCN on assessment of European plant species. Based on the AFE database and with input from the AFE staff, the main floristic elements of Europe were defined and the main correlations between plant, mammal, climate and environment data were uncovered. We have identified two main problems with the existing data and its acquisition and handling. Re-mapping of many taxa is required because of historical data deficiency in East Europe and some of the earliest treatments being outdated. Secondly, traditional data-collection is too slow and no longer lives up to the current demand for scientific development. As a solution, AFE is being converted from a static data collection into a live, online database, which allows flexible data additions and corrections and expands taxonomic coverage without limits. The taxonomic checklist of AFE will be largely based on EMPB, with on-demand addition of further preliminarily accepted taxa which makes the checklist more flexible and open to accommodate original research. A revised mapping programme is being developed, operating with distributional information at the level of hierarchical territories, grid squares and precise records, with emphasis on the grid level. Data-collection and processing are to be no longer limited by recognised taxa or events. The next step forward is a direct link enabling further integration of EMPB and AFE, with bi-directional data flow. As well as collaboration and integration with other large-scale projects, AFE aims to embrace an unlimited number of smaller-scale side mapping and taxonomic projects utilising the AFE mapping workflow, offering the possibility of establishing new regional mapping programmes. This will open a pathway for the acceleration of more comprehensive data coverage in Europe.

T2

P0607

The diverse leaf division patterns of yarrows are under effects of differential expression patterns of the *NAM* gene

Sha Sha, *Duo Chen, Yanping Guo*
College of Life Sciences, Beijing Normal University

Understanding the genetic mechanisms of phenotype divergence among related species can inspire hypotheses about the process and relevant evolutionary forces of speciation. The leaves of the yarrow species (*Achillea*, Asteraceae) show a great deal of diversity. Pinnate leaves, divided once to three times, are common in this genus, while some species have serrate leaves. Changes to phenotypes among organisms should be mainly attributed to historical signals, or a more appropriate term would be evolutionary processes. In model plants, serrated leaves develop with a program centered by the transcription factor NAM (CUC1/2). Whether such regulatory "tools" also work in yarrows? What makes leaves of the congeneric species different? We isolated orthologs of the *NAM* gene from *Achillea acuminata* and *A. asiatica* characterized by serrated and pinnate leaves, respectively. We verified that the *Achillea NAM* of is functionally equivalent to *AtCUC2* in the development of leaf serrates using VIGS experiments and a complementation assay applied to *Arabidopsis cuc2* mutants. We checked the patterns of expression of this gene by the whole mount *in situ* mRNA hybridization and two quantitative gene expression assays, the droplet digital PCR (ddPCR) and quantitative Real-Time PCR (qPCR). *In situ* hybridization showed *NAM* expression signals at leaf margins of both species, and furthermore, in *A. asiatica* on the margins of the elongated leaf lobes corresponding to the positions of the secondary dissection. Expression signals disappeared when the blades were fully expanded. Both ddPCR and qPCR revealed a continuous decline of *AacNAM* expression from the early to the late developmental stages of a single leaf of *A. acuminata*, whereas, a relatively long maintenance and fluctuation of *AasNAM* expression in a leaf of *A. asiatica*. The leaf transformation between the yarrow species is correlated to the altered expression patterns of the *NAM* gene, especially its differential expression dynamics.

T2

P0608

Physiological and biochemical study of *Racomitrium canescens* under extreme high temperature stress

Wei Sha, Man Wang, Meijuan Zhang, ***Tianyi Ma***
Collage of Life Science, Agriculture and Forestry, Qiqihar University, Qiqihar, Heilongjiang

The Bryophytes are well regarded as pioneer terrestrial plants, distributed everywhere all over the world, including many kinds of extreme environmental conditions. So, bryophytes were considered possessing variable stress tolerance. *Racomitrium canescens* is belonging to Grimmiaceae, Grimmiales, Bryatae, Bryophyta, mainly growing on rock surfaces at lower altitudes, thin soil on rock surfaces, sand land, or stone walls and base of trees, distributed over China, North Korean, Japan, Vietnam, southeast of Siberia and Australia. In this study, *R. canescens* was discovered being able to be survived in extreme high temperature stress of 60 ℃ , 80℃ and 100℃ for at least 24 hours in wet condition. Moreover, *R. canescens* can grow normally more than 10 days after being transferred back to regular condition. Physiological and biochemical indexes were determined in different high temperatures and treatment time points. Changes of proline contents, soluble sugar contents, soluble protein contents and malondialdehyde contents of *R. canescens* were found not regularly depending on treatment temperature and time. However, chlorophyll contents, net photosynthesis rates and transpiration rates were in the same tendency under 60 ℃ and 80 ℃ stress, but performed differently under 100℃ treatment, supposing different responding mechanisms will play roles when *R. canescens* suffers to different high temperature stresses. This study provides theoretical proofs for researches of plants in response to extreme high temperature stress.

T2

P0609

A systematic study of the southern African endemic genus *Sisyranthus* (Apocynaceae: Asclepiadoideae–Ceropegieae)

Khanyisile Shabangu[1], *Stoffel Petrus Bester*[2], *Michelle Van Der Bank*[3]
1. African center for DNA barcoding, University of Johannesburg
2. National Herbarium (PRE), South African National Biodiversity Institute
3. African center for DNA barcoding, University of Johannesburg

Sisyranthus E.Mey. species are placed in the tribe Ceropegieae, subtribe Anisotominae (Apocynaceae–Asclepiadoideae). Plants have their fertile parts hidden in the tube of the flowers, and are cryptic in both their habit and small flower size making it difficult to find them in the wild. The genus was first described by Meyer in 1837 and last revised in Flora Capensis (1908) with only one new species described since then. Currently the genus comprises

13 recognised species found in the grasslands of southern Africa, with one species restricted to Zimbabwe. Many of the species are range-restricted and poorly known. The existing key is difficult to use thus leading to confusing identifications. In existing DNA phylogenies, the subtribe Anisotominae has been under sampled and a broader sampling of the southern African taxa is required in order to resolve relationships within and between *Sisyranthus* and its close allies. A revision will lead to a better understanding of the group and its affinities. The aims of this study was: (1) to produce a molecular phylogeny for *Sisyranthus* and its close relatives using sequence data from two nuclear markers (ITS and ETS) and five plastid regions (*mat*K, *ndh*F, *rbcLa, trn*L-F, and *ycf1*); (2) to map morphological characters onto the resulting phylogeny; (3) to revise the genus; and to (4) produce a key to properly identify the species. Preliminary results will be discussed.

T2

P0610

Biosystematic study of *Tamarix* L. species in Semnan Province, Iran

Tahmineh Shagholi[1], *Maryam Keshavarzi*[2], *Masoud Sheidai*[3]
1. Alzahra University
2. Alzahra University
3. Shahid Beheshti University

Tamarix L. with approximately 54 Species is the largest genus of the Tamaricaceae (Baum, 1978). This genus consists of shrubs and semi-shrubs which grow in dry and temperate coastal areas. Iran as one of the major centers of the speciation of this genus has 25 species which are distributed in different Regions of the country (Mobayen, 1996). In this research *Tamarix* species of Semanan province has been studied. Totally, 16 populations of 8 species were gathered through 2015 to 2016. Studied species were: *T. arceuthoides*, *T. mascatensis*, *T. ramosissima*, *T. chinensis* (previously were only reported from China, Mongolia and Japan (Gaskin & Shaal, 2003)), *T. szowitsiana*, *T. meyeri*, *T. androssowii* and *T. aucheriana*. In order to clarify the identifications of the studied species ITS phylogenetic marker and blasting the DNA sequences with NCBI accessions were used, in addition to morphological characters. Using phylogenetic analysis, Maximum Likelihood, UPGMA, Neighbor Joining and HGT tree of species relationships and their gene flow were analyzed. Based on results of present study, morphological and molecular were not in concordant and agreement. Morphological and molecular variation of 6 populations of two species as *T. szowitsiana* and *T. androssowii* were studied using ISSR marker. The result of each primer after PCR and differentiation on agarose based on presence or absence of band, were numbered on 0 and 1. UPGMA cluster analysis and MDS shows the absence of a clear limit between accessions of these two species and their high gene flow. Species relationship has been discussed.

T2

P0611

Studies on seed morphometry in some Epidendroideae (Orchidaceae)

Mahammad Khasim Shaik
Acharya Nagarjuna University

SEM (Scanning Electron Microscope) studies on seed morphometry of nine orchids, such as *Malaxis densiflora* (A.Rich.) O.Kuntze, *Oberonia arnottiana* Wight., *O.ensiformis* Lindl., *Calanthe triplicata* Lindl., *Acampe praemorsa* Blatt.&Mc. C., *A.rigida* Lindl., *Luisia zeylanica* Lindl., *Vanda testaceae* (Ldl.) Reich. F., *V.tessellata* Hk. F. have been carried out. The quantitative data related to the length and width of the seed and embryo, seed and embryo volume, percentage of air space, and numbers of testa cells were presented here. The seed truncation character can be used to differentiate between species in the genera such as *Calanthe, Malaxis, Oberonia, Vanda, Acampe* and *Lussia*. This data indicate that the seeds of *Calanthe* species are more truncated than that of the other studied taxa. Seeds with higher ratio of seed volume/embryo volume (more than 2.2) especially in *C. triplicata, V. tessallata, V. testaceae* are expected to be more buoyant than those with a lower ratio of seed volume/embryo volume. These are widely spread out species in Western Ghats of south India. The buoyancy of seeds could be attributed to the dispersal of seeds to vast areas as well as widely distribution of the species.

T2

P0612

Taxonomic study of the genus *Populus* L. from Pan-Himalaya

Ce Shang, *Limei Zhang, Zhixiang Zhang*
Laboratory of Systematic Evolution and Biogeography of Woody Plants, College of Nature Conservation, Beijing Forestry University

Plants of *Populus* are distributed and cultivated worldwide, but some taxonomic problems remain unsolved within this genus. The considerable morphological variation both within and among populations have posed challenging obstacles in species delimitation and classification. In the Himalayas and adjacent regions, 57 species and varieties have been recorded, mostly from China. Based on herbarium specimens and field observation among populations, we have performed morphological character analysis for species circumscription. Number of capsule valves was found to be a key character to differentiate taxa. Triploidy *P. davidiana* Dode, which leaves are quite similar to those of *P. rotundifolia* Griff., was found using flow cytometry, suggesting the independence of both the species. Phylogenetic relationship between sections was discussed on the basis of chloroplast genomes. In this work, 22 species, 3 hybrid and 1 variety are recognized in four sections; Sect. *Tacamahaca* Spach is reduced to be a synonymy of Sect. *Aigeiros* Duby; 3 new combination and 27 new synonymy are proposed.

T2

P0613

Evolutionary developmental genetics of capitulum architecture variation in Asteraceae

Chuze Shen[1], *Jie Chen*[1], *Yanping Guo*[2], *Guangyuan Rao*[1]
1. School of Life Sciences, Peking University
2. College of Life Sciences, Beijing Normal University

The rich morphs of angiosperm flowers and inflorescences are intimately linked to their efficiency of reproduction and modes of speciation. In the large daisy family, Asteraceae, many florets assemble into a head-like inflorescence called capitulum that functionally acts as a flower. The formation of such a pseudanthium has helped the composite plants gain an increased fitness and thus an evolutionary radiation. There are a variety of types of capitulum differing in the morph, number and arrangement of ray florets. *CYC2* members are involved in the formation of the rays. We have identified a new *CYC2* member (*CYC2g*) in the more derivative lineages of the daisy family and confirmed a set of nested gene duplications across the Asterales. Thereinto, the *CYC2c* and *CYC2g* appear to be ray-specific and express alternatively in different lineages/species. To test whether these two genes have similar phenotypic effects on ray morphogenesis, we have developed *Chrysanthemum lavandulifolium* as a model system. Overexpression of *ClCYC2g* can significantly increase the number of rays in a flower head as *HaCYC2c* does in the sunflower mutants. Whilst ectopic expression of *ClCYC2d* severely inhibits the petal growth of ray flowers and make the corolla tube split at the adaxial side. Altogether, our data clarified the evolutionary trajectory of the *CYC2* genes across Asterales and put forward an evolutionary developmental mode of ray evolution and variation in Asteraceae.

T2

P0614

Stipe anatomy of Thelypteridaceae, new insight for the classification

Kuanyu Shen[1], *Zhaorong He*[2], *Yuehong Yan*[3], *Yeachen Liu*[1,4]

1. Department of Biological Resources, National Chiayi University, Chiayi

2. Institute of Plant Science, School of Life Science, Yunnan University, Kunming

3. Shanghai Chenshan Plant Science Research Center, Chinese Academy of Sciences, Shanghai Botanical Garden, Shanghai

4. Author for corresponding

Thelypteridaceae is one of the largest families in Eupolypods II. This family is certainly confirmed as monophyly, however, several genera of Thelypteridaceae are paraphyly or polyphyly. In the past, these genera were recognized as taxonomic units by pinna shape, venation, position and morphology of trichomes. These chosen characteristics might be affected by convergence evolution or improper weighting, cause several taxonomic units to be unnatural. Otherwise, some separated taxa should be recognized as a clump in the view of molecular phylogenetic study. Anatomical structures reveal important new insights of classification and evolution of ferns. The amounts, shapes of vascular bundles and xylem bundles morphology could be applied to the characters for specific taxa, thus the similarities and differences between the taxa are the clues for the evolution. Unfortunately, very few monographs discuss about these anatomical characters in the classification of Thelypteridaceae. In the present study, 50 more species represented 17 genera have been studied. The fresh materials had been fixed in the 70% alcohol and slicing by microtomes with paraffin and free-hand methods. The measurement of cell size, the xylem and phloem cells arrangement in the vascular bundles, existence of fiber and circumendodermal band (CB) between the genera, leaf size and habitat types are also considered as water usage strategy of each species. The results show CB is absent in *Macrothelypteris pro parte*., *Phegopteris*, *Pseudphegopteris pro parte*., *Metathelypteris*, *Parathelypteris pro parte*., *Christella pro parte*. and *Pronephrium*. The thicken types of CB cell walls, the diameter of xylem cells, channels embedded in the vascular bundles are assumed relating to habitat and leaf size.

T2

P0615

Stipe anatomy of Thelypteridaceae, new insight for the classification

Kuanyu Shen[1], *Zhaorong He*[2], *Yuehong Yan*[3], *Yeachen Liu*[1,4]

1 Department of Biological Resources, National Chiayi University, Chiayi

2. Institute of Plant Science, School of Life Science, Yunnan University, Kunming

3. Shanghai Chenshan Plant Science Research Center, Chinese Academy of Sciences, Shanghai Botanical Garden, Shanghai

4. Author for corresponding

Thelypteridaceae is one of the largest families in Eupolypods II. This family is certainly confirmed as monophyly, however, several genera of Thelypteridaceae are paraphyly or polyphyly. In the past, these genera were recognized as taxonomic units by pinna shape, venation, position and morphology of trichomes. These chosen characteristics might be affected by convergence evolution or improper weighting, cause several taxonomic units to be unnatural. Otherwise, some separated taxa should be recognized as a clump in the view of molecular phylogenetic study. Anatomical structures reveal important new insights of classification and evolution of ferns. The amounts, shapes of vascular bundles and xylem bundles morphology could be applied to the characters for specific taxa, thus the similarities and differences between the taxa are the clues for the evolution. Unfortunately, very few monographs discuss about these anatomical characters in the classification of Thelypteridaceae. In the present study, 50 more species represented 17 genera have been studied. The fresh materials had been fixed in the 70% alcohol and slicing by microtomes with parafilm and free-hand methods. The measurement of cell size, the xylem and phloem cells arrangement in the vascular bundle, existence of fiber and circumendodermal band (CB) between the genera, leaf size and habitat types are also considered as water usage strategy of each species. The results show CB is absent in *Macrothelypteris pro parte*., *Phegopteris*, *Pseudphegopteris pro parte*., *Metathelypteris*, *Parathelypteris pro parte*., *Christella pro parte*. and *Pronephrium*. The thicken types of CB cell wall, the diameter of xylem cell, channel embedded in the vascular bundle are assumed relating to habitat and leaf size.

T2

P0616

Diversity, vertical distribution patterns and drivers of epiphytic bryophytes in tropical and subtropical forests, SW China

Shen Ting[1], *Corlett Richard*[1], *Guo Xinlei*[2], *Song Liang*[1]

1. Xishuangbanna tropical botanical garden

2. Northwest Institute of Eco-Environment and Resources

Epiphytic bryophytes play significant roles in forest ecosystems, including water retention, nutrient cycling, interaction with epiphytic vascular plants, and providing habitat for a variety of small animals. Their diversity and distribution patterns in forest canopies are little known because of access problems. However, with the innovation of canopy techniques—from telescopes, ladders, single rope techniques, canopy walkways, hot-air balloons, and inflatable rafts, to canopy cranes— forest canopy studies have broken through the previous barrier of easy access to the different positions in the canopy. This study surveys the diversity patterns and species distributions of epiphytic bryophyte communities along a vertical gradient in the tropical rain forest in Xishuangbanna and the subtropical montane moist forest in Ailao Shan, both in Yunnan, China, using canopy cranes. We will investigate how the differences between the two sites in elevation, climate, and forest type influence their bryophyte communities, and how microclimate factors (relative humidity, PAR, mean temperature) and host characteristics (tree diameter, bark pH, bark roughness, water holding capacity of bark) influence the vertical distribution of epiphytic bryophytes at each site. The results of this study will help us to understand how bryophyte diversity is maintained in three-dimensional space and to assess the conservation significance of these patterns. This project is the first to explore canopy epiphytes in China, so it will provide fundamental data for the Chinese forest biodiversity monitoring network as well as contributing to international cooperation in the forest canopy.

T2

P0618

Phylogeny and systematics of *Prunus* L. subgenus *Cerasus* (Mill.) A. Gray (Rosaceae)

Shuo Shi[1,2], *Yu-Xin Li*[2], *Jin-Lu Li*[1], *Jia-Hui Sun*[1], *Jing Yu*[1], *Shi-Liang Zhou*[1], *De-Yuan Hong*[1]

1. Institute of Botany, the Chinese Academy of Sciences

2. Hebei Normal University

Subdivision of *Prunus s. l.* (Rosaceae) has been a conundrum for about 300 years. The purpose of this study was to solve this problem with a special emphasis on taxonomy of true cherries by integrating both morphological and molecular evidence. The first step was taken to reconstruct the phylogeny of *Prunus s. l.*, and the subsequent subdivisions of *Prunus s. l.* were based on the phylogeny. By doing so the circumscription of true cherries was followed and phylogeny of true cherries was built. Taxonomic treatments of true cherries were provided. In total, fifteen DNA fragments, including twelve chloroplast fragments (*atp*B-*rbc*L, *mat*K, *ndh*F, *psb*A-*trn*H, *rbc*L, *rp*L16, *rpo*C1, *rps*16, *trn*S-G, *trn*L, *trn*L-F and *ycf1*) and three nuclear genes (ITS, *s6pdh* and *SbeI*) were used to reconstruct the phylogeny of seventy-seven species of *Prunus s. l.* representing all 10 entities of the genus in broad sense. The molecular marker *ycf1* was used for the first time for the genus, and *SbeI* was developed for the genus. The resulting phylogeny indicated that *Prunus s. l.* is a monophyletic group that has diverged into three clades. Considering close relationships among the species belonging to the three clades, we treat *Prunus s. l.* as one genus, *Prunus.* The genus was subdivided into three subgenera, subgenus *Padus* (Mill.) Petem., subgenus *Cerasus* (Mill.) A. Gray and subgenus *Prunus,* which correspond to the three clades. The subgenus *Prunus* were further subdivided into seven sections based on the phylogeny, namely, section *Amygdalus*, section *Armeniaca*, section *Emplectocladus*, section *Microcerasus*, section *Persicae*, section *Prunocerasus* and section *Prunus*. Phylogeny of thirty-five species of the subgenus *Cerasus* were reconstructed using three chloroplast fragments (*atp*H-*atp*I, *pet*A-*psb*J and *pet*N-*psb*M) and the nuclear *SbeI-a*. The topologies created by the maximum parsimony, maximum likelihood and Bayesian inference were nearly identical. Exclusion of section *Microcerasus* from the subgenus and inclusion of *P. mahaleb* L., *P. maackii* Rupr. within the subgenus were confirmed. After literature verification, specimen examination and living plants observation, taxonomic treatments of subgenus *Cerasus* of China were made. *Prunus tianshanica* (Pojarkov) S. Shi., comb. nov. and *Prunus* L. subgenus *Prunus* section *Persicae* (T. T. Yü & L. T. Lu) S. L. Zhou were proposed.

T2

P0619

A revision of the phylogeny of the Thermopsideae: Data from *cp*DNA and ITS markers

Wei Shi[1], *Peiliang Liu*[2], *Borong Pan*[1], *Zhihao Su*[1]

1. Xinjiang Institute of Ecology and Geography, CAS

2. Northwest Agriculture and Forestry University

In this paper we resolved conflict in the phylogeny of the clade Thermopsideae, using internal transcribed spacer (ITS) and four chloroplast intergenic spacer (*mart*K, *trn*H–*psb*A, *trn*L–*trn*F and *rpc*L) marker data. We genotyped 110 individuals, including six distinct haplotypes of individuals in the genus *Ammopiptanthus,* as well as 85 individuals from other key genera in the Thermopsideae. Traditionally, the Thermopsideae were not considered monophyletic group, and the wider Thermopsideae were divided into two subclades. However, we report that molecular data indicates that species in the genus *Sophora* are nested within *Ammopiptanthus* and thus form an isolated clade. Furthermore, we found that the genera *Piptanthus* and *Baptisia* are monophyletic, while the genus *Thermopsis* is not. *Piptanthus*, *Anagyris*, *Baptisia,* and *Thermopsis* are clustered together into an uncontested clade, during which North American *Thermopsis* spp. form a side branch within which *Baptisia* is nested. *Thermopsis* show a geographical division between Old World and New World populations. Our results can either strength to just the four 'core genera' in Thermopsideae with *Ammopiptanthus* excluded, or confirm the sister group relationship between *Sophoreae* and *Ammopiptanthus*.

T2

P0620

Observation of flower development and chromosome analysis of *Osmanthus fragrans*

Yaya Shi, *Xiaojuan Wang*

Shanghai Sceince and Technology Museum

Osmanthus fragrans is an evergreen small tree which originated from Asia, and 75% of the varieties of *O. fragrans* are distributed in China. *O. fragrans* has unique flower development and the rare breeding system of androdioecy (coexistence of males and hermaphroties). It has been reported that in hermaphrodites flowers, two carpel primordia can heal up gradually and form a normal pistil, however in male flowers, two carpel primordia can not heal up and ultimately form bractlike degenerated pistil, through paraffin section assay. In our researches, the male and hermaphrodites flower buds of *O. fragrans* were gathered separately from the end of July until the end of October, the whole process of male and hermaphrodites flower buds' development were analyzed by scanning electron microscopy assay, the different development process of carpel primordia in hermaphrodites and male flowers were observed. The chromosome of *O. fragrans* was also analyzed to reveal the mechanism of sex determination mechanism. Because the androdioecious phenomenon is very rare in nature (about 3%), It has been predicted to be a intergradations from bisexual to unisexual flower. Our research on the development and differentiation of *O. fragrans* sexual organs can be used not only to understand the mechanism of flower differentiation of this species, but also the evolution of flower sexual differentiation.

T2

P0621

Molecular phylogenetics and character evolution in Vochysiaceae

Gustavo Shimizu[1], *Deise Gonçalves*[2], *Amy Litt*[3], *Kikyo Yamamoto*[1], *André Simões*[1]

1. University of Campinas (Unicamp)
2. The University of Texas at Austin
3. University of California, Riverside

Vochysiaceae is a small family within the order Myrtales, comprising eight genera and about 240 species. This study presents the largest molecular dataset to date for this group, using both plastid (*ndh*F gene) and nuclear (ITS1) DNA sequence data. One hundred and sixty new sequences and eight others retrieved from GenBank were analyzed using maximum parsimony and Bayesian inference. The combined analyses with seven genera and 64 species of Vochysiaceae aimed to test the monophyly of its tribes and genera, as well as start to unravel infrageneric relationships, and to find morphological synapomorphies for major clades. Three main clades were obtained: Erismeae (*Erisma* and *Erismadelphus*), QRC (*Qualea*, *Ruizterania* and *Callisthene*) and VS (*Vochysia* and *Salvertia*), but there is no strong resolution in their relationships. *Erisma* is sister to *Erismadelphus*, *Callisthene* is uncertain, *Qualea* and *Ruizterania* lineages form a polytomy, and *Salvertia* is sister to *Vochysia*. *Ruizterania* is considered a synonym under *Qualea*. Several of the morphology-based infrageneric categories were corroborated as monophyletic by the molecular data, and some intriguing taxa, like *V. discolor* Warm. and *V. pseudopumila* Rizzini & Heringer, had their placement clarified. Among the morphological traits mapped, four have states represented as potential synapomorphies for the family: absence of translucid glands in the leaves, presence of spur or bursiform prominence in the fourth sepal, erect stamen in the flower bud and only one fertile stamen. Other nine character states are also potential synapomorphies for major clades. New in-depth questions inside some groups can be driven based on the present results, but efforts to better resolve the backbone of the family will continue as well.

T2

P0622

Phylogeny of Plantagineae (Plantaginaceae, Lamiales)

Alexey Shipunov[1], *José Luis Fernández Alonso*[2], *Gustavo Hassemer*[3]

1. Minot State University
2. Real Jardín Botánico CSIC
3. University of Copenhagen

Plantagineae (Plantaginaceae, Lamiales) consist of well-known, worldwide distributed plantains (Plantago), small aquatic Littorella and Andean páramo shrubs Aragoa. This assemblage is unusual in many ways. First, it is a clear example of evolutionary reduction. Second, Plantago exhibits amazingly high diversification rates. Third, the worldwide distribution of plantains raises numerous questions related to vicariance and dispersal. In addition, there is a striking contrast between morphological disparity of these three genera. Frequent hybridization in Aragoa is also an important aspect of this genus evolution. Here we present the most complete draft phylogeny of the group and discuss its evolutionary, morphological and biogeographical implications.

T2

P0623

Udaria – A new liverwort genus of Lophocoleaceae from Eastern Himalaya, India

Devendra Kumar Singh Devendra Kumar Singh, ***Shuvadeep Majumdar Shuvadeep Majumdar***, *Devendra Singh Devendra Singh*

Botanical Survey of India

The family Lophocoleaceae currently includes 21 genera in the world, 10 of which are monotypic, with the taxonomic circumscription of the genera yet not fully resolved. In India the family is represented by 30 species and infraspecific taxa, belonging to genera *Chiloscyphus* Corda (06 species), *Heteroscyphus* Schiffn. (14 species) and *Lophocolea* (Dumort.) Dumort. (09 species and 01 variety), mostly distributed in the Indian Himalayan region. Nine species, or 30 per cent of the hitherto recorded taxa, are endemic to the country. The taxonomy of the family in India is well understood. The Eastern Himalaya, which is the most species rich bryogeographical region in India, exhibits the maximum diversity of the family with 26 taxa known from this territory, followed by Western Himalaya and the Western Ghats with 16 taxa each, Central India (06 species), Eastern Ghats & the Deccan Plateau (03 species) and the Andaman & Nicobar Islands (02 species). While, some taxa are restricted to their respective bryogeographical region in Indian bryoflora, a number of them are common to one or more than one of the above territories. During our on-going studies on the liverworts of the Eastern Himalaya, we came across some interesting plants from Arunachal Pradesh and Sikkim which, in the presence of undifferentiated stem; 2-lobed, sub-transverse to succubous leaves with thin-walled cells; 2-lobed underleaves with 1–2-dentate margin, occasionally with additional tooth along the inner margin of the lobe; rhizoids fascicled at underleaf base; gynoecia on the main shoot; large female bracteoles more or less equal to the female bracts, and long exserted, oblong trigonous perianth with dorsal third keel and 3-lobed mouth, are referable to the family Lophocoleaceae. The plants with 1–2-celled or linear multicellular gemmae, female bracts and bracteoles larger than leaves and underleaves respectively and long exserted, symmetrically trigonous perianth with 3-lobed mouth show closer affinity with Lophocoleoid elements of the family. However, in the presence of longitudinal, 1–12 cells high lamelliform strips on stem surface; striolate – rugulose leaf cuticle; gynoecia with 1–2 sub-floral innovations and gemmiparous female bracts, these plants are remarkably distinct from hitherto all the known genera of the family, and hence described here as new genus *Udaria* D.K.Singh, S.Majumdar & D.Singh, *gen. nov.* The new genus, at present, is monotypic and represented by *Udaria lamellicaulis* D.K.Singh, S.Majumdar & D.Singh, described here as new species. The paper provides detailed description and illustration of the new taxon and discusses features underlining its taxonomic delimitation.

T2

P0624

A novel phylogenetic classification for *Schinus* L. (Anacardiaceae) and insigths into biogeographic patterns in southern South America

Cíntia Silva-Luz[1], *José Rubens Pirani*[1], *John D. Mitchell*[2], *Douglas Daly*[2], *Susan K. Pell*[3], *Gregory M. Plunkett*[2]
1. São Paulo University
2. The New York Botanical Garden
3. United States Botanic Garden

Schinus belongs to the Anacardiaceae and comprises about 33 species mostly restricted to southern South America. Species are divided into two subgenera defined using a combination of characters that is not always consistent, and species delimitations, especially in *Schinus* subg. *Duvaua* remain controversial. In South America, *Schinus* represents the richest genus and its species can be found in a great diversity of habitat as Patagonia, Chaco, *Monte*, Seasonally Dry Tropical Forest, *Yungas, Prepuna*, Chilean Sclerophyll Forest and Atlantic Forest. Phylogenetic studies in Anacardiaceae are still scarce, especially for South America. A phylogenetic study of *Schinus,* combined with analysis of Anacardiaceae fossil records might provide a deeper understanding of the biogeographic history of the genus. Moreover, these biogeographic hypotheses can increase our knowledge of South America biogeography, since *Schinus* is a remarkable element from vegetation formations of this region. In order to obtain a more thorough understanding of infrageneric relationships of the genus, we sampled 47 *Schinus* taxa, and 13 genera of Anacardiaceae as outgroup. We performed phylogenetic analyses from DNA sequences using plastid (*trn*L-*trn*F and *rps16*) and nuclear (*ETS*, *ITS*, AT2G44760 and AT2G20790) regions. Bayesian and ML analyses supported the monophyly of *Schinus* (BPP: 1.00, BS: 100%). According to phylogenetic analyses, characters traditionally used in the genus circumscriptions probably evolved multiple times or they are plesiomorfic. For example, *S. terebinthifolia, S. weinmanniifolia, S. molle* and *S. lentiscifolia*, compound leaved species, were the first lineages to branch out. However these taxa were placed in *S.* subg. *Schinus* they appeared in our phylogenetic reconstruction as a basal grade. Therefore, this subgenus might not be considered as monophyletic, and compound leaves are probably a plesiomorfic character. *Schinus lentiscifolia* is sister to a clade comprising all simple leaved species. Our results suggested that sections of *Schinus* subg. *Duvaua* need new circumscription and diagnostic characters, since spine-ending branches character probably evolved multiple times, and therefore it is not suitable to define the sections proposed by Barkley. Phylogenetic reconstruction showed interspecific relationships in *Schinus* that had never been demonstrated before. Besides, it also suggested possible cryptic species within some clades, in which was not evidenced by morphological studies. In addition, our analysis showed five well supported clades within simple leaved species: 1 - *Yungas Clade,* 2 - *Sclerophyll Forest - Patagonian Alliance,* 3 - *Patagonian - Prepuna Alliance,* 4 - *Arid Areas Clade,* and 5 - *Atlantic Forest Clade.* These biogeographical patterns are similar to other studies for southern South America biota. For instance, relationship between species from "clade 2" provides strong evidence for past floristic connections because it may have resulted from dispersal events between Central Chile subregion and the northern part of the Subantarctic subregion. Previous study suggested a possible association between climatic niche evolution and diversification because sister species or sister clades in Anacardiaceae share the same geographic region but inhabit different climatic niches. Our analyses also highlight that some sister clades in *Schinus* phylogeny presented similar pattern of climatic niche evolution, and this may be associated with relatively great diversity in *Schinus*.

T2

P0626

Polyploidy as a factor in the evolution of the *Bouteloua curtipendula* complex (Poaceae: Chloridoideae)

María Elena Siqueiros Delgado[1], *Amanda Fisher E.*[2], *J. Travis Columbus*[3]
1. Universidad Autónoma de Aguascalientes
2. Department of Biological Sciences
3. Rancho Santa Ana Botanic Garden and Claremont Graduate University

In this paper we assess the impact of polyploidy and hybridization in the *Bouteloua curtipendula* species complex (BCC). The BCC is a monophyletic group of perennial grasses in the Chloridoideae subfamily. We tested for evolutionary signatures of hybridization and polyploidy in the BCC by obtaining 77 chromosome counts from anther mother cells ($2n = 20$ to >100) and comparing the phylogenetic pattern of diploids and polyploids in nuclear and chloroplast trees. We sequenced ITS and *trn*T-L-F regions for 96 and 70 individuals, respectively, resulting in 150 nuclear ribosomal ITS sequences, including 54 cloned sequences. We found no evidence for recombination between ITS sequences with a PHI test. Maximum parsimony and Bayesian analyses were used to estimate the ITS phylogeny. Diploid samples were found in all clades, while most of the polyploid samples were concentrated in a single clade. Cloned diploids contained 1-3 copies of ITS with >99% sequence similarity. The tetraploids *B. purpurea* and some *B. curtipendula* samples had a low amount of variation among ITS copies, while the cloned polyploids possessed several highly divergent ITS copies. Pollen size correlates with ploidy-level in the BCC, but is not a clear indicator of genome size. We found evidence that the BCC lineage has a complicated evolutionary history that has included autopolyploidy and allopolyploidy.

T2

P0627

Ultra structural characters of calcium oxalate (CaOx) crystals and Idioblasts of some selected angiospermic taxa of the

family Araceae from India: A systematic approach.

Imam Saadi Sk Md Abu
Department of Botany and Forestry, Vidyasagar University, West Bengal

Crystallization is a common phenomenon in many plant groups. In Angiosperms, only the family Araceae, CaOx crystals are very abundantly present in their different plant parts (leaf, petiole, stem/storage organ, roots) within parenchyma/aerenchyma cells to form crystals idioblast. These crystals idioblast contain the composition of calcium and oxalic acid to form different crystal diversity like- prismatics, crystal sands, druses and raphides. Present study divided raphide on the basis of their termination like- SWC, SWP1, SWP2, ThTT, and TnTT and other different micro-morphological characters (shapes of barbs, length of bridge and grooves, idioblast morphology and their arrangement within cell). These characters may apply in taximetrics to solve the taxonomic problems in the different levels viz. families, genus, and species.

T2
P0628

Different ways to obtain similar results. The development of the corolla and epipetaly in temperate and tropical herbaceous *Rubioideae* (*Rubiaceae*)

Alexander Vrijdaghs[1,2,3], *Petra De Block*[2], ***Erik Smets***[1,3]
1. Naturalis Biodiversity Center
2. Botanic Garden Meise
3. KU Leuven

It is commonly assumed that flowers in *Rubiaceae*, one of the largest angiosperm families after *Asteraceae*, *Fabaceae* and *Orchidaceae*, have a calyx with basally fused lobes and a tubular corolla with epipetalous stamens. In several, but not all, tropical *Rubiaceae* species studied in literature, floral developmental investigations revealed that the corolla is early sympetalous, developing a stamen corolla tube from an annular primordium. Other species showed late sympetaly with a rudimentary or absent stamen corolla tube. Consequently, the stamens are inserted at the very base of the corolla tube or in the hypanthium. Payer observed in herbaceous European *Rubioideae* that the corolla originates from individual petal primordia. Later in the development, the petals and the basal part of the filaments fuse postgenitally, forming together a floral cup with only the distal parts of the filaments remaining free. In this study, we want to test the hypothesis that rubioid tubular corollas are developmentally compound floral cups resulting from three processes; 1) the development of a stamen-corolla tube from a common annular primordium, 2) the development of a corolla tube sensu stricto from an annular intercalary meristem, and 3) postgenital fusion of petals or petal lobes. Following Goethes "*Prinzip der variablen Proportionen*" we hypothesize that these processes are differently expressed in the species studied. We investigated the development of the perianth and androecium in temperate (European) and tropical herbaceous species within Rubioideae. Our observations on temperate *Asperula*, *Galium* and *Rubia* species confirm those of Payer. In these cases the floral cup results from postgenital fusion of initially free petals and filaments. In the temperate *Phuopsis* and tropical *Richardia* species studied, a stamen-corolla tube is formed, resulting in a corolla tube with epipetalous stamens. In *Phuopsis*, the stamens are inserted halfway between the base of the corolla tube and the corolla lobes, due to the development of a corolla tube sensu stricto, which is absent or rudimentary in *Richardia*. The developmental patterns in these species concur with our hypothesis. In contrast, the developmental patterns in *Asperula*, *Galium* and *Rubia* differ due to the absence of a stamen-corolla tube and corolla tube sensu stricto. Moreover, the tubular corolla-like structure results from postgenital fusion between filaments and petals. Consequently, the tube is a compound structure composed of the fused parts of filaments and petals and the observed epipetaly is actually a pseudo-epipetaly.

T2
P0629

Biodiversity, distributions, and adaptations of epiphytic bryophytes in the context of global changes in Yunnan, SW China

Liang Song[1], *Xian-Meng Shi*[1], *Xi Chen*[1], *Wen-Yao Liu*[1], *Wen-Zhang Ma*[2]
1. Xishuangbanna Tropical Botanical Garden, Chinese Academy of Sciences
2. Kunming Institute of Botany, Chinese Academy of Sciences

Yunnan Province, Southwest China, an high-elevation and low-latitude inland province, covers a territory with complex terrain and diversified climate zones and functions as refugia for diverse and unique nature resources. Unfortunately, species composition and distribution patterns of epiphytic bryophytes at multiple scales and their links to environmental changes remain poorly studied. Three investigations had been carried out to address biodiversity and distributions of epiphytic bryophytes in Yunnan, while ecophysiological measurements and manipulation experiments had been conducted to assess adaptations of epiphytic bryophytes and their fates in a changing world. The three protected forests in Yunnan harbored 226 epiphytic bryophyte species belonging to 101 genera, 46 families. The subtropical montane cloud forest located in the Ailao Mountain National Nature Reserve of Southwest China, which is generally characterized by persistent and frequent cloud cover at the canopy level, is especially rich in epiphytic bryophytes. In total, 176 epiphytic bryophyte species have been recorded and the total biomass of the epiphytic bryophytes is 6.7 tons per hectare. Bole epiphytic bryophytes adapt to the shady understorey microhabitats through a series of correlations and trade-offs between functional traits. However, even slight changes in climate resulted in remarkably reduced rates of growth and detrimental effects on the health of the transplanted epiphytic bryophyte species over 2 years. The low water retention capacity and high dependence of net photosynthesis to water content of epiphytic bryophytes implied high risks of imbalance of carbon budget for them under the increasing severity of drought. Epiphytic bryophytes could take up N from the bark and wet deposition in both organic and inorganic N forms. The process of carbon metabolism and the chemical stability of epiphytic bryophytes are strongly influenced by N addition levels, and presenting strong correlations with the bryophytes loss. In addition, a general decline in community heterogeneity and biomass production of bryophytes induced by high N loads can be expected, which will probably alter the C and N cycles of subtropical montane forest ecosystems and arising their risk of the function and structure stability under intensive climate changes.

T2

P0630

Host–hemiparasite–arbuscular mycorrhizal fungi interactions vary with different host plants

Xiao-Lin Sui, *Ai-Rong Li, Kai-Yun Guan, Yan Chen*
Kunming Institute of Botany, Chinese Academy of Sciences

Although arbuscular mycorrhizal fungi (AMF) are ubiquitous and well known for improving plant growth and regulating plant-plant interactions, their roles in regulating interactions between root hemiparasitic plants and their host plants remain less known. Leguminous and gramineous plants are both considered good hosts for root hemiparasitic plants, showing robust growth promoting effects once attached. However, the two functional groups tend to have different responses to AMF inoculation, with leguminous plants often showing positive responses while gramineous plants showing little or even negative responses. We therefore hypothesized that AMF inoculation might have different effects on the host-hemiparasite interactions between a leguminous host and a gramineous one. We conducted greenhouse pot cultivation experiments to test effects of *Glomus mosseae* on the growth and photosynthesis of gramineous (*Elymus nutans*) and leguminous (*Trifolium repens*) parasitized by a root hemiparasite *Pedicularis kansuensis*. The results showed that shoot and root dry weights (DWs) of both hosts were significantly reduced when parasitized by *P. kansuensis*. The ETR (electron transport rate) of *E. nutans* was significantly reduced, whereas that of *T. repens* was significantly improved when parasitized by *P. kansuensis*. For hosts parasitized by *P. kansuensis*, inoculation with *G. mosseae* significantly increased shoot and root DWs, ETR and chlorophyll content, decreased N:P ratio of *T. repens*. In contrast, inoculation with *G. mosseae* only increased the root DW, decrease ETR and chlorophyll content of parasitized gramineous plants. These results show that the effect of AMF on regulating the relationship between hosts and hemiparasite varies in different host species.

T2

P0631

A phylogeny of *Mastigolejeunea* (Marchantiophyta: Lejeuneaceae) -- Evidence for reduction of *Mastigolejeunea* to subgeneric rank in *Thysananthus* and transferring *M. florea* to *Spruceanthus*

Phiangphak Sukkharak[1], *Robbert Gradstein*[2]
1. Department of Biology, Faculty of Science, Burapha University, Thailand
2. Dept. Systématique et Evolution, Muséum National d'Histoire Naturelle, France

The pantropical *Mastigolejeunea* is morphologically close to *Thysananthus* and the separation of these two genera has long been controversial. The relationship between *Mastigolejeunea* and *Thysananthus* is assessed based on Bayesian inference and maximum likelihood analyses of plastid *psb*A-*trn*H, *trn*G and *trn*L-F, and nuclear ITS. *Mastigolejeunea* was resolved as monophyletic and sister to *Thysananthus* with moderate support, but *M. calcarata* was nested in *Thysananthus* subsect. *Anguiformes* and *M. florea* was resolved in *Spruceanthus*. The molecular evidence strengthens the intermediate position of the subsect. *Anguiformes*, a group largely restricted to New Zealand and New Caledonia, between *Thysananthus* and *Mastigolejeunea*. The results underline the morphological overlap of *Mastigolejeunea* and *Thysananthus* and indicate that the two groups do not merit recognition as separate genera. *Mastigolejeunea* is thus reduced to subgeneric rank under *Thysananthus*. The latter genus becomes the second largest genus of Lejeuneaceae subfam. Ptychanthoideae with 30 species, and is one of the largest genera of liverworts that has been monographed worldwide.

T2

P0632

Species delimitation in the Norway-Siberian spruce complex (*Picea*: Pinaceae) using genome-wide data

Alexis Sullivan, *Xiao-Ru Wang*
Umeå Plant Science Centre, Department of Ecology and Environmental Science, Umeå University

The combined ranges of Siberian (*Picea obovata* Ledeb.) and Norway spruce (*P. abies* (L.) Karst.) cover a vast expanse of Eurasia from the Atlantic to the Pacific coasts. Although relatively well-studied among plants, the species boundary between Norway and Siberian spruce is still unclear. Recent studies have shown Fennoscandian *P. abies* populations have chloroplast and mitochondrial genomes more closely related to Siberian spruce than to continental European populations of *P. abies*. Because the organelle genomes are differentially inherited from the seed and pollen parent in *Picea*, this result could indicate the deep population structure observed in Norway spruce is not due to glacial subdivision but rather represents two separate species. However, organelle genomes may also be polyphyletic due to organelle capture or incomplete lineage sorting and nuclear microsatellite data gives little indication of a polyphyletic Norway spruce. To more conclusively address species limits within the Norway-Siberian species complex, we are using exome capture data from 40,000 probes to infer coalescent phylogenies and implement Bayes Factor species delimitation tests. We genotyped 35 Norway-Siberian spruce individuals evenly distributed across the entire range of the complex. In addition, we genotyped 15 *Picea* species representing the phylogenetic breadth of the genus to test the hypothesis that Norway and Siberian spruce are sister species. After stringent filtering to minimize sequencing errors and paralogous loci, we obtained ~50,000 high-quality SNPs. Our first results from coalescent species trees do not support the commonly-assumed sister relationship between *P. abies* and *P. obovata*. Second, we find that both mitochondrial and chloroplast genomes are discordant with nuclear loci: most Fennoscandian Norway spruce populations are more closely related to their southerly conspecifics than to Siberian spruce. Third, we detected a lineage in extreme northern Fennoscandia most closely related to Siberian spruce. Ongoing analyses are testing if this extreme northern lineage represents a distinct species or if genomic data have confirmed the presence of Siberian spruce ~2,000 km further east than its described distribution.

T2

P0633

A new case on bio-exchange between Eurasia and India subcontinent

Bin Sun

Institute of Botany, Chinese Academy of Sciences

The bio exchange between Gondwana and Laurasia happened after the collision and docking between the Indian subcontinent and Eurasia. The uplift of Tibet had profoundly affected this bio exchange process. Today, Indian Tertiary floras increasingly show that tropical elements in the subcontinent originated in the southern hemisphere, whereas the origin of those temperate elements in northeast India is still unclear. Based on fossil records, a former biogeographic hypothesis suggested that the temperate elements in Northeast India came from southwest China, and migrated westward across southern Tibet in the middle Miocene before the Himalayas arose, then entered India through Kashmir in the late Pliocene to Pleistocene before spreading eastwards along the southern slope of the Himalayas. Here we provide a new potential migration route of temperate plant, *Berberis*, from eastern Asia to India subcontinent based on a new fossil link. The find, differing from the former hypothesis, suggests that the northern Tibet, but not the southern Tibet, had play the role as the pathway when it was not so high (ca. 2,000 m a. s. l.) as to block any migration of temperate elements between two continents.

T2

P0634

Comparative genomic in situ hybridization between sweetpotato (*Ipomoea batatas* (L.) Lam) and its wild related species

Jianying Sun, *Zongyun Li*

School of Life Science, Jiangsu NormalUniversity

Sweetpotato (*Ipomoea batatas* (L.) Lam) is the world's food security bottom line crop, and it is also an important feed and industrial raw material crops, while plays an important role in China's energy security. Sweetpotato is highly heterozygous in heredity, and there is a wide range of cross incompatibility in interspecies, which restricts its breeding. As a result, the genetic diversity of cultivars is reduced, which seriously restricts the genetic improvement of sweetpotato varieties. The wild relatives of sweetpotato are one of the important sources of new genes for disease resistance, quality and stress resistance in cultivar breeding. They have very rich genetic background and carry a large number of excellent trait genes. The introduction of these excellent traits can be carried out to solve the narrow genetic basis of sweetpotato. Therefore, understanding the genome structure and genetic diversity of hexaploid sweetpotato and its related wild species is important for us. In this study, comparative genomic *in situ* hybridization (cGISH) technique was used to find the relationships between *I. batatas* and its related species. The hybridization signals reflect the degree of homology of the repeats. Suggested by previous researches, 13 species were considered. The wild species genome were made into probes to study the homology between *I. batatas* cultivar 'Xushu 18' and wild species with FISH. According to the signals, the 13 wild species were divided into 4 groups. The genome of the first group of species have strong signals on all mitosis chromosomes of cultivated sweetpotato with only several weak signals. In some species, there genome have no signals on 8-15 chromosomes, while on the others the signals are strong. *I. umbraticola* is grouped into the third, and the hybridization signals looks wildly exists but weak. The last group contains only one species named *I. grandfolia*, it shows the furthest relationship with cultivar with only a few signals. In all FISH between wild species and cultivar, the signals were concentrated in the middle or the end of the chromosomes. It indicated that the *I. batatas* has different degrees of repeat sequences with these related species. Suggested by the results presented here, *I. tabascana* is very close to the sweetpotato in addition to the previously thought that *I. trifida* and *I. triloba* had the most relatively relationship with the sweetpotato.

T2

P0635

Using fluidigm amplification and NGS sequencing data to explore biodiversity across an ancient floristic disjunction

Miao Sun[1,2], *W. Mark Whitten*[1], *Matthew A. Gitzendanner*[1,2,3], *Douglas E. Soltis*[1,2,3], ***Pamela S. Solits***[1,3]

1. Florida Museum of Natural History, University of Florida

2. Department of Biology, University of Florida

3. Genetics Institute, University of Florida

The multidimensional biodiversity of the forests in eastern Asia (EA) and eastern North America (ENA) has fascinated scholars and researchers worldwide for decades. Advances in high-throughput sequencing (HTS) have offered an excellent opportunity to re-evaluate these two anciently connected forest communities with a shared evolutionary and ecological history, exploring the drivers of biodiversity across geographic space and through evolutionary time. The Fluidigm Access Array System uses sets of PCR primers, simultaneously amplifying 48 target regions across 48 samples (or 192 loci x 24 samples in a new configuration) using microfluidic amplification reactions, incorporating sample-specific barcodes and HTS adapters. Based on this system, here we present an efficient workflow from DNA sample quantification, primer design, PCR validation, and primer-pooling to sequence data assembly and bioinformatics pipeline. We designed 143 universal microfluidic PCR primer pairs to amplify the most variable regions of the plastome, as well as commonly sequenced mitochondrial (*atp1*, *rps3*, *nad5* and *mat*R) and nuclear rDNA loci, generating a large phylogenetically informative multi-locus dataset across all disjunct genera between EA and ENA sites. This promising approach permits us to compare the disjunct forests of EA and ENA through reconstruction of species-level community phylogenies at multiple sampling locations in both regions with maximal genomic-scale data and hundreds of samples per site.

T2

P0636

Exploring the phylogeny and diversification of rosids through a supermatrix approach with limited sampling

Miao Sun[1,2], *Charlotte Germain-Aubrey*[2], *Stephen A. Smith*[3], *Pamela S. Soltis*[2,4], *Zhiduan Chen*[5], ***Douglas E. Soltis***[1,2,4]

1. Department of Biology, University of Florida

2. Florida Museum of Natural History, University of Florida

3. Ecology and Evolutionary Biology, University of Michigan
4. Genetics Institute, University of Florida
5. Institute of Botany, the Chinese Academy of Sciences

Molecular phylogenetics relies on adequate and representative sampling of taxa and genes, although what constitutes adequate sampling may not always be clear. Biased or incomplete sampling may result in the estimation of a strongly supported but incorrect topology. Perhaps no major clade of angiosperms plays a more significant role in many modern terrestrial ecosystems than rosids, an enormous clade of ~90,000 species, representing approximately 25% of all angiosperms and exhibiting tremendous overall diversity in morphology, habit, reproductive strategy, and life history. However, our knowledge of this clade is remarkably limited along any metric. With our bioinformatics pipeline, we have found that less than 34% of all named rosid species have DNA sequences usable for phylogeny reconstruction, and these sampled data are highly biased toward specific subclades. Here we use the rosid clade as an exemplar to explore and evaluate how the analyses of phylogeny and diversification are influenced by limited gene and taxon sampling. We employed two supermatrices: 1) a 9,300-species, four-gene (*atp*B, *rbc*L, *mat*K, and *mat*R) matrix; and 2) a 5-gene 20,974-taxon matrix assembled from five widely used genes (*atp*B, *rbc*L, *mat*K, *mat*R, and ITS) with maximum coverage of all rosid species deposited in GenBank. Our results generally agree with previous estimates of rosid phylogeny, but with a few significant differences. Comparisons of rosid diversification analyses based on the two supermatrices show that rosids have undergone two diversification rate shifts, confirming that other lineages (ants, amphibians, fungi, ferns) radiated in response to the rise of rosids. However, the downstream analyses are relatively limited when the molecuar data for phylogenetic analyses are limited, and when other biological data for species and all other traits are limited, especially for such a big clade.

T2
P0637

R. M. Schuster's theory on the origin and biogeography of liverworts tested

Yu Sun
Botany Unit, Finnish Museum of Natural History, University of Helsinki

Inferring the age and origin of liverworts as a whole and their major lineages is greatly hampered by the scarcity of the fossil records and the poor state of preservation of most of the early fossils. The long-lasting and most influential theory on the origin and biogeography of liverworts was that of R.M. Schuster in the early 1980s, who assumed that liverworts originated in the Gondwana and the patterns of biogeography across most taxonomic groups were resulted from tectonic events and stepwise dispersals. These interpretations have been greatly challenged by the recent studies based on molecular data, and currently both fossil data and reconstructed phylogenies suggest that liverworts are the earliest diverged lineage of land plants that possibly dates back to the Ordovician. In the present study, a broad sampling of a multigene data was compiled to test the origin and biogeographical patterns of liverworts. This includes 300 samplings from three chloroplast genes (*rbc*L, *rps*4 and *trn*L-F), including 287 species of liverworts as ingroup and 13 species mosses as outgroup. A time-calibrated, molecular-based phylogenetic analysis and different approaches for ancestral range inference were conducted. Nine biogeographic areas were defined according to palaeogeographic history: Africa (AF), Australasia (AU), southern South America (SSA), Neotropics (NE), Europe (EU), North America (NA), East Asia (EA), India (IN) and Southeast Asia (SEA). The results show that liverworts likely originated from Australasia in the early Ordovician. Totally, 60 dispersal events were detected throughout the tree. Most of the dispersal events (43 dispersals) occurred after 100 MA, only 4 dispersal events occurred before 200 MA and 13 dispersal events occurred between 200 MA to 100 MA. During the Paleozoic Era, liverworts dispersed from AU to EA twice and formed the Haplomitriopsida and Marchantiopsida, and dispersed from AU to NE which is the ancestor of Frullaniaceae, Jubulaceae and Lejeuneaceae. Subsequently, liverworts dispersed to EU and NA from both EA and AU before Cretaceous, and distributed rapidly all over the world during Cretaceous. It supports the Gondwanan origin of the group, but instead of the previously assumed continental drift and stepwise dispersal, long-distance dispersal has played an important role in resulting the distribution patterns of liverworts. The frequent dispersal events since Cretaceous may well have been conditioned by the evolution and diversification of angiosperms.

T2
P0638

Molecular phylogenetics of two new species related to southwestern China buckwheat based on nuclear and plastid DNA sequences

***Zhanmin Sun**[1], Yu Tang[2], Yanmin Wu[1], Yixiong Tang[1], Jirong Shao[3], Meiliang Zhou[1]*
1. Biotechnology Research Institute, Chinese Academy of Agricultural Sciences
2. Department of Food Science, Sichuan Tourism Universit
3. School of Life Sciences, Sichuan Agricultural University

During the 2004-2010, our research group investigated the wild buckwheat distributed throughout the Sichuan Province, in the southwest of China. During, last 10 years, *Fagopyrum pugense*, *F. crispatifolium*, *F. qiangcai*, *F.wenchuanense*, *F. luojishanense* and *F. hailuogouense* are discovered as new species in fagopyrum by our group. Here, phylogenetic analysis of two recently identified *Fagopyrum* wild species, *F.luojishanense* and *F.hailuogouense*, related to southwestern China buckwheat was examined using the sequence of internal transcribed spacers (ITSs) and the *maturase K* (*matK*) gene. 14 *Fagopyrum* species including the 2 new speicies were chosed to build the evolutionary tree, The neighbor-joining (NJ) and maximum-likelihood (ML) trees were almostly congruent with each other, and in all analyses, *Fagopyrum* species except *F. hailuogouense* and *F. luojishanese*were divided into the cymosum group and urophyllum group, agree with the previous study. Interestingly, *F. luojishanese* belonged to cymosum group in *matK* trees while belonged to urophyllum group in ITS trees, which suggested that it was a transitional type of two groups. As for *F. hailuogouense*, it had a separate cluster in both ITSs and *matK* trees, showing a far genetic distance with others, this recent data may provide a new way to inference the classification of *Fa-*

gopyrum. Our results provide a framework for investigating the patterns of morphological evolution and geographical distribution of the *Fagopyrum* genus and a better understanding of evolutionary mechanisms and genetic relationships of 2 new species related to southwestern China *Fagopyrum* sepecies.

T2
P0639
A preliminary study of the genus *Calymperes* Sw. (Calymperaceae, Bryophyta) in Thailand

Orawanya Suwanmala, *Sahut Chantanaorrapint*
Prince of Songkla university

A review of genus *Calymperes* in Thailand is presented based on field surveys, herbarium specimens, and data from previous publications. Twenty taxa have been recorded. Most species are corticolous epiphytes growing on tree trunks, but are also found on decaying wood, soil and rocks, including limestone and sandstone. They are found in various forest types, from dry to moist deciduous or evergreen forests, i.e. coastal, mangrove, spring, lowland evergreen and montane forests, from sea level to over 1500 m alt. The southern part of the country exhibits great diversity of species with 19 taxa, followed by 13 species in the southern-eastern region. The most common species of *Calymperes* in Thailand is *C. afzelii* Sw., distributed in 5 floristic regions including northern, eastern, southern-western, southern-eastern and peninsula.

T2
P0640
Actinomorphic flower: A rare autapomorphic case in Zingiberales

Sasivimon Chomchalow Swangpol, *Wandee Inta*
Department of Plant Science, Faculty of Science, Mahidol University

The Zingiberales order is described as having bilaterally symmetric (zygomorphic) flowers with five stamens or less. In the banana family (Musaceae), which is considered the most primitive taxon in the order, the perianth typically appears as two whorls of three outer tepals and two inner ones united into a compound tepal and the third inner tepal free with five free stamens. However, a new species, *Musa nanensis*, recently found in northern Thailand near the Thai-Laos border, possesses a vastly unique floral structure of two united whorls of tepals composed of three outer and three inner tepals. The similar shapes, though different sizes, of these tepals indicates that the merging of both whorls is not only a simple fusion of a compound tepal and a distinct free tepal, but an exceptional floral structural change. Six fertile stamens fused at base also do not indicate an infrequent structural aberration. Although molecular data and chromosome number do not support the separation of this taxon from *Musa*, its distinctive floral morphology adds new description to the genus and family. Modification from zygomorphic to actinomorphic flowers, and from five to six stamens, indicates a unique reversal event in the Zingiberales approaching basal monocotyledons and may have been influenced by pollinator shifts. In order to understand this uncommon autapomorphy, further parallel investigations on pollination biology, floral development and regulation of floral homeotic genes of *M. nanensis* are being further studied.

T2
P0641
Research of flower color diversity in Taiwan, seen through the eyes of bee pollinators

King-Chun Tai[1], *En-Cheng Yang*[2], *Chun-Neng Wang*[1,3]
1. Department of Life Science, National Taiwan University, Taiwan
2. Department and Graduate Institute of Entomology, National Taiwan University, Taiwan
3. Institute of Ecology and Evolutionary Biology, National Taiwan University, Taiwan

There are more than 4,000 angiosperm species native to Taiwan, at least 60% of them are speculated to be insect pollinated. Color signal plays a major role in attraction and decision making of insect pollinators. But experimental data about the flower color diversity of Taiwan plants in color vision of honeybee, the most important pollinator worldwide, remains unknown. We conducted large scale measurements on flower reflectance spectrum and summarized them into bee color hexagon followed by best discrimination fit analysis, in order to answer following two questions: (1) How colorful are flowers native to Taiwan, in the eyes of honeybee? (2) Would color diversity of subtropical Taiwan flora emerge specific patterns, compared to the results of well-studied region, such as Australia and Israel? Mapping of 824 spectrum data collected from 584 Taiwanese native flowers, the result showed that loci of the flower colors form two obvious clusters in uv region and blue-green region, where 57% of the loci makes up a distinct 'White Arm' pointing to 60°. It seems plausible that the dominance of bee pollinators in Taiwan and their innate preferences led to such a tight clustering. And the best discrimination fit analysis showed that the frequency of slope midpoints in long wavelength is also much lower in Taiwan than that of Australia and Israel, which revealed that few pollinators in Taiwan can detect flower color with rapid spectral changes in long wavelength region. In conclusion, we suggested that specific pattern of flower color diversity in Taiwan comparing to Australia and Israel. This may result from the adaption of flower color of plants to different pollinator composition, especially the fact that almost none avian pollinators were documented in Taiwan island. This is also the first color diversity research in East Asia with bee color space analyzed. Our successful approach shall be beneficial to many studies follow on.

T2
P0642
Evolutionary history of Series *Sakawanum* (genus *Asarum*: Aristolochiaceae) showing interspecific clinal variation of calyx lobe lengths

Daiki Takahashi[1], *Tsutomu Teramine*[2], *Shota Sakaguchi*[1], *Hiroaki Setoguchi*[1]
1. Kyoto university
2. Kochi Gakuen College

Clines, the gradual variation in measurable traits along a geographic axis, play major roles in organism evolution. Series

Sakawanum (genus *Asarum*; Aristolochiaceae) shows a typical interspecific cline in floral morphology. This series comprises three species (*A. minamitanianum*, *A. sakawanum*, and *A. costatum*) and all species are parapatrically distributed in southern Japan. Within the series, the lengths of the calyx lobes increases gradually among the species inhabiting regions from east to west in southern Japan. *A. minamitaniaum*, which distributed in western-most extent of the range has the longest calyx lobe lengths (50-150 mm), and *A. costatum*, which distributed in eastern-most extent has the shortest lengths (8-20 mm). *A. sakawanum*, which distributed between *A. minamitaniaum* and *A. costatum* has the intermediate length (20-40 mm) between the other species. As clinal variation can be formed by the neutral processes (e.g. isolation by distance, range expansion and secondary contact), to infer the evolutionary history of this series would be very useful for considering the formation of clinal variation in this series. In this study, we used 17 EST-SSR markers with approximately Bayesian computation (ABC) and ecological niche modeling to infer the evolutionary history of series *Sakawanum*. Our ABC analysis showed the secondary contact with gene flow model was more probable model than other tested models (blanching models and radiation models) under the all examined tolerance rates. Following the model, *A. sakawanum* (morphologically intermediate species) would be originated by the results of secondary contact of *A. minamitanianum* and *A. costatum* after the Last Glacial Maximum period. We considered that this secondary contact and subsequent gene flow would likely have formed the intermediate morphology of *A. sakawanum* and interspecific clinal variation of this series.

T2

P0643

Monograph of the *Daphniphyllum* (Daphniphyllaceae)

Mo Shih Tang[1], *Yuen-Po Yang*[2], *Chiou-Rong Sheue*[3]

1. Chung Hwa University of Medical Technology, Tainan, Taiwan

2. National Sun Yat-sen University, Kaohsiung, Taiwan

3. National Chung Hsing University, Taichung, Taiwan

Daphniphyllum is a small-sized monogenus of the Daphniphyllaceae distribution from Himalayas toJapanand from Central China toNew Guinea, including south of India and Sri Lanka. The plants of *Daphniphyllum* have limited morphological features, especially leaf and fruit that strictly constraints the identification of the genus. Based on the examination of more than 4500 specimens and the comparative study of leaf anatomy and pistillate flower, this genus comprises 28 species, including 4 varieties. *D. lanyuense stat. nov.* was described and mapped.

T2

P0644

Effective application of next-generation sequencing (NGS) approaches in biogeography and population genetics: Case study in *Abies alba*

Anass Terrab[1], *Francisco Balao*[1], *Juan Luis García Castaño*[1], *María Teresa Lorenzo*[1], *Jose Manuel Sánchez Robles*[1], *Paun Ovidiu*[2]

1. University of Seville

2. University of Vienna

Abies (Pinaceae) is a widely distributed monophyletic genus of Northern Hemisphere conifers (firs) comprising around 50 species, being the silver fir (*Abies alba*) the widespread species in Europe. Many molecular phylogenetic researches were conducted for *Abies alba*. All of these studies used only data from few markers. Here we present the molecular phylogeography of the species, using Next-generation sequencing (RADseq) from populations belonging to 13 countries. The application of NGS by RADseq method in this study has provided great information for the relationship and evolution of *Abies alba* in Europe. The evolutionary history of the European silver fir is discussed in the context of the paleogeological and postglacial history of the region.

T2

P0645

Floral morphology of Sabiaceae: Contribution to understanding floral diversity and evolution in Proteales

Pakkapol Thaowetsuwan[1], *Dietmar Quandt*[2], *Herve Sauquet*[3], *Louis Ronse De Craene*[1]

1. Royal Botanic Garden Edinburgh

2. University of Bonn

3. Universite Paris-Sud

Recent floral developmental and anatomical studies have provided detailed information about all three genera of Sabiaceae, i.e., *Meliosma*, *Ophiocaryon* and *Sabia*. The family presents several unusual floral traits among angiosperms, e.g., flowers with sepals opposite petals and stamens, and unusual disporangiate monothecal anthers. Within the family, *Meliosma* and *Ophiocaryon* were found to be more closely related with the presence of two fertile stamens and three staminodes contrary to the five fertile stamens in *Sabia* and this was supported by recent molecular phylogenetic studies. Earlier classifications mostly assigned Sabiaceae to Sapindales/Rutales, and a few suggested an affinity with Ranunculales. Molecular phylogenetic studies have confirmed a position of Sabiaceae within the basal eudicots grade close to Ranunculales and Proteales. Sabiaceae show several unique characters that are unusual among the grade, e.g., pentamery, a bipartite perianth, disporangiate monothecal stamens, tricolporate pollen grains, a receptacular nectary and a syncarpous gynoecium with two carpels. However, there are conflicts in determining the exact position of Sabiaceae on the phylogeny with some studies suggesting a secondary branching from Ranunculales or a close relationship to Proteales, a thesis mostly accepted nowadays (e.g. APG IV). This conflict encouraged our team to further investigate morphology and development of this family, including the study of the third lesser known genus *Ophiocaryon*. Earlier studies in *Meliosma* and *Sabia* found that Sabiaceae share many common characters with trimerous Ranunculales, e.g., Menispermaceae and Lardizabalaceae, and three hypotheses have been presented to interpret the origin of pentamery in Sabiaceae either from spiral, trimerous or dimerous precursors. However, comparison of flowers of all four families of Proteales, viz., Sabiaceae, Nelumbonaceae, Platanaceae, and Proteaceae, is difficult because of their historic divergence and highly different pollination systems. The results from a comparative study of the three genera of Sabiaceae with other families in Ranunculales and Proteales has revealed several morphological characters that are potential synapomorphies of all Proteales, e.g., a (nectar) disk, a tanniferous cell layer in the ovary wall and orthotropous ovules. Molecular phylogenies using fossil calibration suggested that Proteales is a very old clade that appeared more than 100 million year ago; therefore, the great diversity of floral morphologies may be due to the early separation of all four families. More extensive floral morphological investigations in Proteaceae and the distinctive *Platanus kerrii* should be done to gain more information about floral evolution and diversity within Proteales.

T2

P0646

A preliminary study on the phylogeny of the Chinese Vaccinieae (Ericaceae)

Yi Hua Tong, *Nian He Xia*

South China Botanical Garden, Chinese Academy of Sciences

Vaccinieae is the largest tribe of Vaccinioideae, which consists of about 1,250 species of 34 genera distributed all over the world but concentrated in the tropical regions of Asia and America. There are few phylogenetic studies on the generic relationships of blueberry tribe, so the boundary of many genera is very unclear. *Vaccinium* is one of the most complex genera of this tribe, with 400 species widely distributed in the world, and close to *Agapetes*, *Gaylussacia*, *Orthaea*, *Notopora*, *Costera*, *etc.* According to the recent preliminary phylogenetic analyses of Vaccinieae using nuclear and chloroplast DNA sequence data, *Vaccinium* was found to be polyphyletic. Many sections recognized within *Vaccinium* are considered more distinct than some genera. Chinese Vaccineae represents by only two genera, *Vaccinium* and *Agapetes*, with 91 species and 54 species respectively. At present, some evidence showed that *Agapetes* is close to some Asian sections of *Vaccinium*. Some researchers even merged *Agapetes* into *Vaccinium*, which created a more complicated *Vaccinium*. However, due to lack of samples from East Asia, which sections of *Vaccinium* are closest to *Agapetes* is unknown and whether it is appropriate to treat these two genera as one also lack of convincing evidence. Thus, to figure out the relationships of *Agapetes* and *Vaccinium*, we conducted a preliminary phylogenetic analyses with nuclear ribosomal ITS, and plastid *mat*K and *rbc*L as markers, using a sampling of 62 *Vaccinium* species (covering 20 sections) and 28 species of *Agapetes* species. The results showed that both *Agapetes* and *Vaccinium* are not monophyletic. The studied *Agapetes* species contains three groups: (1) The monotypic section *A.* sect. *Pseudagapetes* Airy Shaw, i. e. *A. scortechinii* (King et Gamble) Sleumer, a species endemic to Malay Peninsula. The phylogenetic analyses revealed that it is distantly related to other *Agapetes* species and should be excluded from the genus. However, which genera it should be positioned in has not been resolved yet. (2) *Agapetes bracteata*. Evidence from molecular systematics also showed that rather than the species of the same series, *A. bracteata* is closely related to those of *V.* sect. *Conchophyllum*. Thus, it seems to be the best treatment to transfer it to *Vaccinium* for now. (3) The core *Agapetes* group. Including most species of the genus, the core *Agapetes* group in phylogenetic relationship is near with four sections of *Vaccinium*, i. e. sect. *Epigynium*, sect. *Aëthopus*, sect. *Calcicolus* and sect. *Monanthos*. However, there will be still much work to be done to clarify the relationship between the core *Agapetes* group and its affinities. In addition, our molecular systematic study indicates that many currently circumscribed sections of *Vaccinium* are not monophyletic, such as sect. *Epigynium*, sect. *Aëthopus*, sect. *Calcicolus*, sect. *Conchophyllum*, sect. *Bracteata*, sect. *Eococcus*, *etc.*

T2

P0647

Discovery and description of two new species using DNA barcoding

Hironori Toyama, *Tetsukazu Yahara*

Department of Biology, Faculty of Science, Kyushu University

Tropical forests contain over 90% of the world's tree diversity, where we are still finding many new species by recent floral surveys. However, traditional procedures of new species description require us a well-trained skill of taxonomy based on morphological observations. Additionally, the number of taxonomists well-trained for tropical plants are declining and thus we need to develop a more efficient way to discover and describe new species. DNA barcoding can provide such an efficient way of taxonomic discovery. Here, we show two examples for discovery and description of new species using DNA barcoding. *Eustigma* is a small genus of Hamamelidaceae, distinguished from the other genera by small auriculate petals and greatly enlarged stigma. Previously, only three species have been known from China, Laos, Taiwan and Vietnam: *E. balansae* Oliv., *E. lenticellatum* C.Y.Wu and *E. oblongifolium* Gardner & Champ. During our DNA barcoding studies on sterile collections of trees from Hon Ba Nature Reserve in South Vietnam, we found a new species, *Eustigma honbaense* H.Toyama, Tagane & V.S.Dang. The *matK* sequences between *E. honbaense* and *E. oblongifolium* differed in 6 bases of the 781 total. The *rbcL* and *matK* sequences between *E. honbaense* and *E. balansae* differed in 1 base of the 517 total and 5 bases of the 761 total, respectively. Another example is a new species, *Euphorbia bokorensis* H.Toyama & Tagane (Euphorbiaceae), we found in Bokor National Park, Cambodia. It has very unique morphological features including subsucculent stems, gland-like stipules and conspicuous cyathophylls and we could not confidently identify its genus by morphology. Phylogenetic analyses using *mat*K, *ndh*F and ITS regions revealed that it is a member of *Euphorbia* subgenus *Euphorbia* exhibiting the greatest diversity in both species richness and growth forms among the four subgenera of *Euphorbia*. Previously, 21 sections were recognized in this subgenus, but *E. bokorensis* was not included in any of them. Thus, we described a new section *Bokorenses* H.Toyama & Tagane. Bayesian phylogeny showed that it is sister to the Malagasy clade, including sections *Denisophorbia*, *Deuterocalli* and *Goniostema*, and also to the northeastern African clade, including sect. *Rubellae*. While standard cpDNA barcodes (*rbc*L and *mat*K) are useful for species discrimination within relatively old clades like *Eustigma*, barcodes of the nuclear Internal Transcribed Spacer (ITS) are useful for species discrimination among closely related species like *Euphorbia*. Although relatively fewer efforts of DNA barcoding have been made for plants of Southeast Asia, our study showed that it provides an efficient way of discovery and description of new species. Further efforts of DNA barcoding for Southeast Asian plants will enable us to efficiently document undescribed diversity of species in tropical forests that is under threats of rapid loss.

T2

P0648

The role of long distance dispersal syndromes in the diversification and biogeography of the hyperdiverse Cynoglossoideae (Boraginaceae)

Ana Otero[1], *Pedro Jiménez-Mejías*[2], ***Virginia Valcárcel***[3], *Pablo Vargas*[4]

1. Real Jardín Botánico de Madrid, Consejo Superior de Investigaciones Científicas

2. National Museum of Natural History, Smithsonian Institution

3. Universidad Autónoma de MAdrid

4. Real Jardín Botánico de Madrid, Consejo Superior de Investigaciones Científicas

Long distance dispersal syndromes (LDD), especially endozoochory, have been hypothesized to facilitate plant colonization boosting diversification. However, few studies have analyzed the influence of epizoochorous fruits in diversification rates and biogeography. We evaluate the effect of epizoochory in the diversification patterns of cosmopolitan Boraginaceae (Cynoglossoideae). Phylogenetic reconstructions of 57 genera (227 spp.) and estimates of divergence times are used to infer evolutionary pathways. We employed trait-dependent diversification analysis (HiSSE) and a biogeographic reconstruction approach (Lagrange). The geographical range of Cynoglossoideae lineages is not necessarily related to fruit structures. Differentiation of Cynoglossoideae lineages in the Paleocene reveals an origin in eastern Paleartic and that epizoochorous traits (glochids) have been long-term maintained through different groups from an ancestral condition. This supports the idea that complex traits are not necessarily derived and do not necessarily imply the endpoint of fruit differentiation. Two events of diversification rate increases occurred in two major epizoochorous lineages at approximately the same place (eastern Palearctic) and same time (16-36 myr). However, there are four epizoochorous lineages with an unexpected lower diversification rate, which is revealing a hidden effect. These results suggest that the diversification of Cynoglossoideae has been driven by many different variables rather than the effect of one key innovation.

T2

P0649

Independent placements of temperate genera across the phylogeny of the tropical Asian Palmate group (Araliaceae) since its early evolution

Virginia Valcárcel[1], *Rong Li*[2], *Jun Wen*[3]

1. Department of Biology, Universidad Autónoma de Madrid

2. Kunming Institute of Botany, Chinese Academy of Sciences

3. National Museum of Natural History, Smithsonian Institution

The Asian Palmate group is one of the main clades of the Araliaceae. It gathers 50% of the diversity of the family with 21 generic-lineages and nearly 1,500 species. The diversity center of the group has been located in SE Asia, with 19 generic lineages occurring in Asia (13 in SE Asia) plus three in the neotropics. The Asian Palmate clade is a heterogeneous group in terms of morphology, diversification patterns and niche affinity. On one hand it includes both moderately to highly diversified generic lineages (e.g. *Dendropanax*, c. 85 spp.; Neotropical *Schefflera*, c. 200 spp.; Asian *Schefflera*, c. 350 spp.) and scarcely diversified to monotypic genera (e.g. *Chengiospanax,* 2 spp.; *Gamblea,* 4 spp.; *Kalopanax,* 1 sp.; *Sinopanax,* 1 sp.; *Tetrapanax*, 1 sp.). On the other hand, while most of its generic lineages are tropical or subtropical, seven of them display temperate affinity. Interestingly, five of the seven temperate genera are scarcely diversified (*Chengiopanax,* 2 spp.; *Gamblea,* 4 spp.; *Kalopanax,* 1 sp.; *Oplopanax,* 3 spp.; *Tetrapanax*, 1 sp.). In this study we analyse the complete plastome of 20 generic lineages of the Asian Palmate group to investigate the role of niche shifts in the evolution of the group. Our results indicate that the diversification of the Asian Palmate group took place in Asia during the late Cretaceous and the Paleocene, with the emergence of all generic lineages before the Oligocene. The temperate taxa appear scattered across the Asian Palmate plastome tree, right from the root of the tree, as inferred from the basal placement of *Oplopanax*. Long branches and early divergence from tropical closest relatives also characterize the evolution of the temperate genera. Also, the temperate lineages display less diversification than their tropical sister-groups; except for the *Hedera-Merrilliopanax* pair where the temperate *Hedera* is more diversified (12 spp.) than the tropical *Merrilliopanax* (4 spp.). These results may challenge the idea that temperate taxa derived from tropical counterparts in the Asian Palmate group and raises the question of whether the temperate affinity constitutes an evolutionary dead-end in the Asian Palmate group or a diversification driving force.

T2

P0650

Gregor Mendel's constant hybrids

Peter Van Dijk

Keygene N.V.

Gregor Mendel's (1822-1884) crossing experiments in *Hieracium* (hawkweed) are not well-known compared to his experiments in *Pisum*, although they continued for seven years. Geneticists generally assume that they were intended to verify his *Pisum* findings. Because most *Hieracium* species are apomictic, this would be doomed to fail. Therefore the *Hieracium* studies are considered as a frustrating failure. In contrast, historians of science generally assume that Mendel's main interest was not the inheritance of traits, but the formation of new species by hybridization. In this view, the *Hieracium* experiments were more important than the *Pisum* experiments. However, after careful analysis of the surviving letters of the correspondence between Mendel and Carl Nägeli, it becomes obvious that Mendel was looking for 'constant hybrids', in other words apomicts, in *Hieracium*. Mendel was the first to show the inheritance of apomixis in crosses between sexual and apomictic *Hieracium*. The misunderstanding that *Hieracium* was chosen to confirm the *Pisum* results is probably caused by a missing page in one of Mendel's letters to Nägeli. We could not find support for the interpretation of the historians of science. On the contrary, Mendel's later pollen competition experiments suggest that Mendel was above all a geneticist.

T2
P0651
Population genetics and phylogeography of the tropical African tree *Staudtia kamerunensis* using newly developed microsatellite markers.
Samuel Vanden Abeele[1], *Steven Janssens*[1], *Franck Monthe*[2], *Armel Donkpegan*[2], *Olivier Hardy*[2]
1. Botanic Garden Meise
2. Université Libre de Bruxelles

Despite the high levels of tree biodiversity in African rainforests, current logging is mainly focused on a selected number of tree species, which causes these species to be under high pressure. To facilitate the conservation and regeneration of these valuable timber species, knowledge on the genetic population structure is extremely valuable. In the current study, the genetic structure of *Staudtia kamerunensis* populations from Central Africa was analyzed with a set of newly developed microsatellite markers. Both variants of the species were included to verify if there is a genetical component underlying the morphological differences. The software QDD v2.1 was used to detect microsatellite loci of interest and to select suitable primers from a microsatellite enriched genomic library of *Staudtia kamerunensis*. Selected primers were then tested with different PCR conditions to ensure DNA amplification and readability after which polymorphic loci were retained and combined in multiplexed reactions using Multiplex Manager 1.2. Multiplexed reactions were carried out on populations from Cameroon, Gabon, Republic of the Congo and Democratic Republic of the Congo. Allele lengths of the different loci were analyzed with Peak Scanner Software 2 and Geneious 9.1.6. Heterozygosity levels, allele frequencies and the number of alleles were then used to assess genetic diversity within and between populations and clustering analyses were carried out with Structure 2.3.4 and the *adegenet* package for the R software. Finally, these results were discussed in a historic climatological and biogeographical context.

T2
P0652
Phylogenomics of the phaseoloid and millettioid legumes using a target enrichment approach
Mohammad Vatanparast[1], *Adrian Powell*[2], *Sue Sherman-Broyles*[2], *Jeff J. Doyle*[2], *Ashley N. Egan*[1]
1. Smithsonian Institution
2. Cornell University

The legume family (Leguminosae) is the third largest family of flowering plants and is the second most important family in economic value. The phaseoloid (ca. 2,070 species in 112 genera) and millettioid (ca. 1,100 species in 56 genera) legumes form a single large clade of the papilionoid subfamily and include various important crops such as soybean, common bean, cowpea, pigeon pea and winged bean. The majority of taxa in this clade correspond to the Phaseoleae sens. lat. and Millettioid sens. stric. subclades; however, in spite of multiple molecular phylogenetic studies, relationships within this clade remain unresolved or with low statistical support, particularly along the backbone. To understand the evolutionary history of these legumes, we used transcriptomes of 24 taxa from this group as well as six outgroup legume genera to develop thousands of hybridization probes. Our probes include representatives of all subtribes and major, delineated clades determined in previous studies within the millettioid clade. We obtained sequences for hundreds of genes to build a robust phylogenetic tree encompassing phaseoloid and millettioid legumes. Our results reveal robust phylogenetic relationships of multiple clades and subclades within the millettioid clade and provide well-supported phylogenetic evidence useful for the classification within this important legume clade.

T2
P0653
Assembling Brazilian subtropical highland vegetation over space and time
João R. V. Iganci[1], *Josimar Kulkamp*[1], *Cássia Plá*[1], *Mariane Dornelles*[1], *Tiina Sarkinen*[3], *Gustavo Heiden*[2]
1. Universidade Federal do Rio Grande do Sul
2. Embrapa Clima Temperado
3. Royal Botanic Garden Edinburgh

A high number of endemic species is being reported for the Brazilian subtropical highland grassland and mixed forest biomes. Both formations occur in highland areas mainly in Southern Brazil, under similar climate and environmental conditions. Bayesian inference on biogeography using multiple lineages have being proving to be a good approach for assembling biomes evolution over space and time. Here we apply this method to assemble both subtropical highland grassland and mixed forest evolution. The subtropical highland grasslands are a species rich formation found over basalt rocky outcrops and peatlands, while the mixed forest is an Ombrophilous Forest dominated by the gymnosperm *Araucaria angustifolia*. Both formations behave as continental vegetation islands, and encompasses tropical and temperate lineages, sharing species with other sourrounding biomes. Dominant and endemic species from each formation were mapped and their distribution as comunities were modeled to identify the most suitable areas for both grasslands and mixed forests. Specific lineages represented by species rich taxa in each formation were chosen to compose the Bayesian analyses. Each biome represents a complex history of niche evolution represented by lineages reaching new habitats or local especiation events.

T2
P0654
Polyploids on islands: A NGS study of Macaronesian endemic *Lavatera acerifolia* (Malvaceae)
Irene Villa-Machío, *Gonzalo Nieto Feliner*, *Javier Fuertes Aguilar*
Real Jardín Botánico

By increasing dramatically the sampling of genomes, Next Generation Sequencing (NGS) techniques opened a wide avenue towards reconstructing species evolutionary histories thus representing a crucial source of information for phylogeography. However, polyploidy --a very common process among vascular plants— is a major challenge for the NGS techniques due to difficulties in genotyping individuals resulting from the existence of multiple paralogous loci, which is even more complex in plants without a reference genome. In addition, population-base studies require a highly reliable set of homologous fragments and SNPs. Plant evolution in oceanic islands provides a suitable framework where genetic groups can be directly related to discrete geographic areas. *Lavatera acerifolia* (2n=44) is one of the two endemic entomophilous species of *Malvaceae* family occurring in the Canary Islands, sister to *Lavatera maritima*, also polyploid (2n=44) and distributed along the West Mediterranean basin, including SW Morocco. Our study applies GBS to reconstruct the origin and evolution of *L. acerifolia* in the Canary Islands. To overcome the complexity derived from the polyploid condition of this species we have done sensitivity analyses comparing different settings in critical steps of the filtering process. Specifically, de-novo assembly and mock-reference workflows to SNP discovery have been performed and compared. Our analyses group populations into 2-4 genetic groups with a clear geographic structure that is consistent with an east-west colonization complex pattern. Methodologically, we confirm the critical influence of missing data, paralog filtering and multiple heterozygous loci, when analyzing data under phylogenetic approaches.

T2

P0655

Long-term evolutionary effect of climate change explored by combining micro and macroevolutionary levels

Tamara Villaverde[1], *Lisa Pokorny*[2], *Ricarda Riina*[1], *Sanna Olsson*[1], *Isabel Sanmartin*[1]

1. Real Jardín Botánico, CSIC

2. Royal Botanic Gardens, Kew

To understand the long-term evolutionary effect of climate change, an approach that combines the micro and macroevolutionary levels, from within populations to across species, is needed. Bridging this gap is difficult because it requires the integration of many disciplines and analytical approaches, but it has become recently feasible through the advent of Next Generation Sequencing techniques (e.g. target enrichment with genome skimming, i.e. HybSeq) by scaling up the molecular data available. We focused on Africa, a continent that has received relatively little attention in the global biodiversity research effort. We chose two non-tropical lineages exhibiting a common disjunct distribution known as Rand Flora to explore if high extinction rates were associated to historical and ongoing drought. This enigmatic floristic disjunct pattern occurs in unrelated plant lineages with similar large-range disjunct distributions across the continental margins of Africa and adjacent islands (Macaronesia-northwest Africa, Horn of Africa-Southern Arabia and Southern Africa). Although Rand Flora lineages belong to different angiosperm families and exhibit different life-history traits and climatic preferences (from sub-humid to semi-xeric affinities), they share the tropical lowlands of Central Africa and the Saharo-Arabic deserts areas in the north as effective climatic barriers to the dispersal and persistence of populations. To explore the effect of extinction at micro (populations) and macro (species) levels, we reconstructed the phylogenetic relationships in *Euphorbia* section *Balsamis* and in the genus *Camptoloma* using nuclear and chloroplast molecular data. We inferred speciation and extinction rates as well as possible declines in population size. Our preliminary results in *Euphorbia* show phylogenetic resolution for the section and within species and a biogeographic structure that might indicate dispersal from the African continent to Canary Islands, which is related to climatic aridification waves in Africa.

T2

P0656

Taxonomy, nomenclature and endemism in genus *Ipomoea* (Convolvulaceae) from India

Vinod Shimpale

Department of Botany, The New College, Kolhapur, 416 012 (MS) India. Affiliated to Shivaji University, Kolhapur

India is one of the 12 mega diversity regions of the world. As India shows wide and richest floristic composition there are many large and diverse families of Angiosperms of which the family Convolvulaceae is one of them. There are about 55 genera and 1,600 species of the family Convolvulacae throughout the world, while in India the family is represented by about 20 genera and 180 species. The genus *Ipomoea* is one of the larger genera of family Convolvulaceae. It comprises near about 650 species distributed throughout the world mainly in tropical and warm temperate region. As per the previous Indian literature in India the genus *Ipomoea* is represented by 60 species. Among the Indian Convolvulaceae the genus *Ipomoea* is the dominant with high number of species followed by *Argyreia*, *Convolvulus etc.* During revisionary studies on the genus *Ipomoea* it is observed that according to ICN few species needs typifications, few are not collected after type locality and few are not hitherto reported for Indian flora. During present investigation about 40 species of genus *Ipomoea* are collected from different phytogeographical regions of the India. All the aspect regarding endemism, nomenclature, distribution and taxonomy are discussed in present work.

T2

P0657

Be (e)ing with blossoms in the Outer Himalaya: Using pantraps as floral functional units to explore patterns in bee-plant interactions across land uses in Doon Valley, Uttarakhand, India.

Preeti Virkar[1,2], *V. P. Uniyal*[1], *Qamar Qureshi*[1]

1. Wildlife Institute of India

2. Navdanya Trust

Currently, human use of land and its intensification is regarded one of the most significant threats to species, including wild pollinators. Alteration of habitat (not just loss) can also modify network structure emphasizing a landscape approach for pollinator conservation. Besides, pollinators, chiefly bees respond to commonly found floral colours, which is associated with the rewards they offer. However, traditional plant-bee interactions are assessed with active methods that are either phytocentric (observing flowers for pollinator visitations) or zoocentric (investigating pollen loads from bees), which requires skilled surveyors. Constraints of funds and time, furthermore, result in low and incomplete sampling, indicating an inability of active methods to record maximum interactions. Declining pollinator populations worldwide, insist future investigations should approach the mutualistic interactions with attention to the "species functional groups". For the first time, we use the passive method of pantrap captured bee samples to build interaction matrices with floral functional units based on (pantrap) colours. Considering the robustness of plant-bee interactions in different habitats should vary depending on the availability of diverse floral functional groups, we tested in agroecosystems and forests of Doon Valley, (1) whether real mutualistic (bee-plant communities) interactions varied compared with three randomized networks (using null models), and (2) how bee interactions with floral functional groups affect network attributes in different selected habitats. Network attributes viz. connectance, interaction evenness, linkage density and nestedness were selected to assess their patterns and ecological implications across agroecosystems and forests habitats in Doon Valley. Real plant-bee interactions varied from null models. Highest bee interactions were seen in the white floral functional groups (n = 236), and the least were in blue floral functional groups (n = 75). Bee families showed differential floral colour preferences in forests and agroecosystems. Plant-bee communities were more robust in forests than agroecosystmes. Network attributes such as weighted linkage density and weighted nestedness confirmed benefits of quantitative data over only presence-absence records for the conservation of mutualistic interaction. A detailed investigation into the types of forests revealed riverine habitats were robust to invasive species than Sal forests, indicating forest managers to take appropriate steps in vegetation management. Network interactions in agroecosystems indicated to agriculturists adopt a polyculture approach to cultivation to maintain robust mutualistic interactions for better productivity in crops. The study also highlighted that pantraps could be reliably used by numerous unskilled surveyors, in different areas simultaneously for long-term investigation and monitoring of plant-bee interactions.

T2

P0658

Pollination of two sympatric species of *Balanophora*: *B. fungosa* and *B. harlandii*.

Nikolay Vislobokov

M.V. Lomonosov Moscow State University, Biological faculty, Department of Higher Plants.

Balanophora is a genus of holoparasitic plants, which inhabit mainly in tropical regions of Asia. The genus divided to two subgenera (subgen. *Balanophora* and subgen. *Balania*) by merism of male flowers, while its female flowers are simplified to filiform structures. Species of the both subgenera inhabit the same territory (sympatric). In addition to morphological, ecological differences between the sympatric species are expectable, e.g. different system of pollination. Pollination biology of some species from subgen. *Balanophora* was investigated earlier. It was showed that *B. abbreviata* is pollinated by bees. Pyralid moths were recognized as pollinators of *B. kuroiwai*. Data on pollination of species from subgen. *Balania* are poor: ants and cockroaches were noted as visitors of *B. tobiracola*. The present research is dedicated to flowering biology of two species of *Balanophora* from different subgenera. *Balanophora fungosa* (subgen. *Balanophora*) and *B. harlandii* (subgen. *Balania*) were investigated in the same season and habitat in northern Vietnam. Inflorescences of the both species open acropetally. Within an inflorescence of *B. harlandii*, 8–16 flowers can be anthetic simultaneously; each flower remains anthetic for 1 day. Within an inflorescence of *B. fungosa*, 10–20 flowers can be anthetic simultaneously; each flower remains anthetic for 2–5 days. Inflorescences of both species were visited in daytime by flies (Calliphoridae, Drosophilidae, Syrphidae), wasps (*Vespa sp.*), ants (Formicidae, Mirmicidae), harvestmen (Phalangidae) and others. Drosophilid flies were the most frequent visitors of *B. fungosa*, while *B. harlandii* was visited predominantly by the wasps. Pollen grains of *Balanophora* were found at the bodies of the drosophilid flies and the wasps. On average, 102 pollen grains of *Balanophora* were found on the body of each wasp, while only 2 pollen grains were found on the body of each fly. Drosophilid flies visit male inflorescences of *B. fungosa* on average 433 times per day and visit female inflorescences 144 times per day. Wasps visit male inflorescences of *B. harlandii* on average 62 times per day and visit female inflorescences 18 times per day. It can be concluded, that drosophilid flies are the main pollinators of *B. fungosa* and wasps are the main pollinators of *B. harlandii*. However drosophilid flies can take minor part in pollination of *B. harlandii* as well as wasps can pollinate *B. fungosa*. The flies carry far less pollen than the wasps, but higher frequency of visits offsets this difference. As a result, sufficient effectiveness of pollination is ensured for the both species. Mechanism of plant-pollinator interaction is somewhat similar for *B. fungosa* and *B. harlandii*: sweet smell attracts insects and they feed on nectar as a reward. Probably, some differences in chemical composition of volatiles led to situation, when flies prefer one species but wasps prefer another one. Chemical change of volatiles and followed change of pollinators range can served as factors of divergence of this species during sympatric evolution. This study was funded by the Russian Foundation for Basic Research (project 16-34-00330).

T2

P0659

Dynamic evolutionary history of the East Asian Tertiary relict *Platycarya* (Juglandaceae) based on species distribution models and genetic data

***Qiuchi Wan**[1,2], *Zhuo Zheng*[1], *Marta Benito-Garzón*[2], *Kangyou Huang*[1], *Erwan Guichoux*[2], *Rémy Petit*[2]

1. Sun Yat-sen University

2. BIOGECO, INRA, Univ. Bordeaux

Subtropical East Asia harbours a large plant diversity and was an important glacial refugia for plant species throughout Quaternary glacial-interglacial cycles. However, the impact of past climate change on the dynamics of tree species in this region remains poorly understood. Here we use observations of *Platycarya*, a widespread subtropical Asian tree genus currently treated as monotypic (although two species have been formerly described), to explore the consequences of past climate change on species' evolutionary and population demographic history in subtropical China. We relied on species distribution models (SDMs) and genetic data (two cpDNA intergenic spacers, nuclear internal transcribed spacer regions and random genomic single nucleotide polymorphisms). By compiling distribution data and performing SDMs, we found that the two putative species, *P. strobilacea* and *P. longipes*, are sympatric over part of southern China and have similar climatic envelopes. Their past distributions as inferred by mapping their climatic envelopes at the Last Glacial Maximum did not differ much from their current distributions. Using nuclear genetic markers, we found interspecific genetic divergence matching well with previously recognized taxa and geographically structured intraspecific variation. Using cpDNA markers, we also found geographically structured intraspecific variation. Nevertheless, we also detected several cases of interspecific and intraspecific genetic admixture in southwest China. Integrating SDMs with phylogeographic studies suggests that *Platycarya* populations in subtropical China did not undergo significant range contraction during glacial episodes, allowing allopatric divergence, but the limited amount of range expansion resulted in secondary contacts and subsequent genetic exchanges across porous barriers. These findings help understand the origin of the large plant diversity found in subtropical China through allopatric speciation as well as stabilization of new hybrid populations. This study is supported by National Natural Science Foundation of China (No 41230101) and by the 7th Framework Program of the EC under the Trees4Future project (No 284181).

T2

P0660

Revision of series Gravesiana (*Adiantum* L.) based on morphological characteristics, spores and phylogenetic analyses

Aihua Wang, ***Faguo Wang***
South China Botanical Garden, Chinese Academy of Sciences

Since the adoption of some ambiguous and quantitative characters in Flora Republicae Popularis Sinicae 3 (1), species identifications of the series Gravesiana have been in disarray, requiring clarification. Two hundred and fifty-nine individuals from 47 different populations were collected for the estimation of morphological characters and phylogenetic analyses. Spores of 26 populations were observed through scanning electron microscope. Our results were different from those of previous research: (1) six identifiable species, rather than five species observed previously, were confirmed in the series Gravesiana, they are *A. gravesii*, *A. juxtapositum*, *A. mariesii*, *A. dentatum*, *A. longzhouensis* and *A. obovatum*, of which the latter three are newly recognized species. (2) Thirteen characters were measured and estimated through the program Mesquite v. 2.71. The character whether the pinna stalks were 1/3-1/2 times longer than the pinna was used to distinguish *A. gravesii* and *A. lianxianense* previously and was found to be unreliable here, whereas such characters as the height of the plant (H), pinna aligned forms (FP), number of pinna (NP), pinna margin (M), number of veins flabellate at base (NV), sori number and shape per pinna (NSS), pinna texture (T), and powder-covered or not on the abaxial surface of the pinna (P) are estimated to be stable and reliable characters useful for identification. Descriptions of new species and their retrieve keys are also listed. (3) Surface ornamentations and spore sizes are helpful for us to distinguish species in series Gravesiana.

T2

P0661

Phylogeography and integrated taxonomy of *Diabelia* (Caprifoliaceae), a Sino-Japanese disjunct and Asian endemic genus

Hua-Feng Wang[1], *Sven Landrein*[2], *Shota Sakaguchi*[3], *Masayuki Maki*[4], *Zhi-Xin Zhu*[1], *Kun-Kun Zhao*[1]
1. College of Tropical Agriculture and Forest, Hainan University, Haikou 570228, China
2. Xishuangbanna Tropical Botanical Garden, Chinese Academy of Sciences, Menglun 666303, China
3. Graduate School of Human and Environmental Studies, Yoshida-nihonmatsu-cho, Sakyo-ku, Kyoto 606-8501, Japan
4. Botanical Gardens, Tohoku University, Kawauchi 12-2, Aoba, Sendai 980-0862, Japan

Diabelia (Caprifoliaceae) is an Asian endemic genus with a Sino-Japanese disjunct distribution pattern. The genus has potential to be developed and used in the horticulture industry with its diverse flower forms and colors. Furthermore, some taxa within this genus are listed as endangered in China and Japan; studying its genetic diversity would help conservation efforts. However, due to possible inter-specific hybridization and introgression, inter-specific differentiation and/or recent convergent evolution, the species delimitation within the genus is disputed; moreover, the phylogeography within the genus and its speciation processes are not clear. In this study, we collected 37 populations (335 individuals) in the field in Zhejiang (China) and Japan. We studied its morphological characters and sequenced ITS, *trn*L-*rpl32*, *trn*H-*psb*A and Expressed Sequence Tags-Simple Sequence Repeats (EST-SSR) for all individuals. We developed 13 pairs of Tags-Simple Sequence Repeats (EST-SSR) based on transcriptomes. After analyzing the population diversity and population genetic structure, we found 40 haplotypes based on *trn*L-*rpl32 a*nd *trn*H-*psb*A markers. Japan Ehime (H1), Shiga (H2) and Shimane (H16) population expanded whereas Tokushima population (H13) shrank. Chinese *D. spathulata* was sister to the other haplotypes, Chinese range might represent vicariance, but this was not conclusive as we did not have samples from Kyushu, the closest island to E China. The phylogeny work showed that Chinese haplotypes were embedded within the other Japanese ones, and it was more likely dispersed from Japan to E China based on cpDNA data. Based on molecular dating, we speculated that Diabelia originated in Wenzhou, Zhejiang, China, and then it spread to the west of Hongshu (e.g. Shiga) of Japan ca. 20 Ma ago (e.g. *D. spathulata*). However, cpDNA was prone to introgression, and just represented one gene genealogy. In the future we would use microsatellite markers to have more information on the allele diversity and distribution. Comparing different genetic markers would be much more informative in future.

T2

P0662

Comparison of floral morphology and pollinator behavioral pattern among different sexes in *Eurya Japonica* Thunb.

Hui Wang*, Peiming Zheng, Qiang Li, Xiao Liu, Ning Wang, Fenghua Bai, Weihua Guo, Renqing Wang*
Shandong University

As reproductive organ, plant flower can attract the pollinators during pollination process. Generally, morphology of entomophilous flower is different among different sexes, which may attract different pollinators, thus will further affect the reproductive characteristics of different sexual individuals. *Eurya japonica* Thunb. (Pentaphylacaceae) is a subdioecious shrub, and both the female and male reproductive success of hermaphrodite individuals are outperformed by females and males, respectively. This study was conducted to compare the floral morphology and pollinator behavioral pattern among different sexes. The characters of floral morphology, and the species and the number of visiting insects in certain time were observed. The results showed that the flower of different sexes of *E. japonica* have many obvious differences in respect of the floral morphology and flower visiting insects. The flowers colour of different sexes is similar by visual inspection. The flowers amount of male individuals was the highest, while that of hermaphrodites was the least. The male flower is bell-shaped with non-reflexed petals; the female flower is cup-shaped with reflexed petals; while the hermaphrodite flower presents some transitional traits between the male and female flowers. The stigmas number of female flowers and the stamens number of male flowers are more than those of hermaphrodite flowers, respectively. In addition, no difference was found in the flowers nectar volume and sugar content between female and hermaphrodite flowers, whereas male flower has the least nectar volume. The number of visiting insects in male individuals was the highest in certain time. The visiting insects of female flower were mostly Diptera, while those of male and hermaphrodite flowers were Thrips. The floral morphology and pollinator behavior adapt to each other during the evolution. It is also thought that the differences of floral morphology and pollinator behavioral pattern influence the reproductive success of *E. joponica.*

T2

P0663

A systematic revision of pantropical genus of *Caudalejeunea* (Lejeuneaceae, Marchantiophyta)

Wang Jian

East China Normal University

The genus of *Caudalejeunea* (Steph.) Schiffn., with the capacity of the species to reproduce asexually by means of discoid gemmae, is the only frequently epiphyllous genus and abundant in humid tropical regions in Ptychanthoideae. A world-wide revision of the genus leads to the recognition of eleven species. Morphological characters of *Caudalejeunea* are include the presence of gemmiparous branches, trigones cordate and more conspicuous cell-wall thickenings and the homogeneous oil bodies. *Gradsteinianthus* R.L.Zhu & Jian Wang bis gen. nov. is newly described based on molecular and morphological evidence. Based on the distinct important morphological characters, *C. latigynolobula* Jian Wang bis sp. nov. is described new to science. *Caudalejeunea grolleana* Gradst. is transffered to *Acrolejeunea* and a new combination *A. grolleana* (Gradst.) Jian Wang bis is proposed. *Caudalejeunea recurvistipula* (Gottsche) Schiffn. is resurrected as a distinct species and *C. africana* is reduced to a synonym of the species. *Caudalejeunea mauritiana* Horik. is reduced as a synonym of *Dicranolejeunea axillaris* (Nees & Mont.) Schiffn. and *C. acutifolia* Gerola is excluded from the genus. A key to the species of the genus *Caudalejeunea* is provided and each species is fully described and illustrated, with complete synonymy and detailed data on habitat and distribution. *Caudalejeunea* is pantropical in distribution; the centre of diversity of the genus is tropical Africa with highest endemics, and some species exhibit intercontinental disjunctions between tropical regions of Africa, America, Asia and Oceania. To understand the reason of bryophytes with intercontinental disjunctions and, we analyse the patterns of beta diversity of tropical bryophytes to determine if they possess high dispersal capacities based on the distributions of tropical bryophyte species.

T2

P0664

Five species newly recorded herbaceous plants of Metachlamydeae in Northeast China

Meng Wang, *Shu-Mei Zhang*

Dalian Natural History Museum

Northeast of China, located in the East of Eurasia, is vast in territory, and has varieties of plants species, a total of 164 families, 928 genera and 3,103 species of vascular bundle plants. It is an important region for the origin, evolution and development of flora. Based on the results of field investigation and related data, five species newly recorded herbaceous plants of metachlamydeae are reported (*Cynanchum hancockianum* (Maxim.) Al. Iljinski, *Lamium amplexicaule* Linn, *Mentha sachalinensis* (Briq.) Kudo, *Solanum photeinocarpum* N. et O, *Pedicularis artselaeri* Maxim). As newly recorded species plants in Northeast China, *Cynanchum hancockianum* (Maxim.) Al. Iljinski is distributed in Zhuanghe, Dalian. *Lamium amplexicaule* Linn is distributed in Dalian. *Mentha sachalinensis* (Briq.) Kudo is distributed in Liaoning. *Solanum photeinocarpum* N. et O is distributed in Jinzhou, Dalian. *Pedicularis artselaeri* Maxim is distributed in Lingyuan, Chaoyang. The finding of many newly recorded plants distributed in Liaoning confirms that the plants in Liaoning province have obvious features of transitional zone between north China flora and northeast China flora and provides new evidences for the relationship between them. The voucher specimens were preserved in the herbarium of Dalian natural history museum. These newly recorded plants have a great significant meaning to study the plant species diversity and provide new materials for investigation of their geographical distributions in China.

T2

P0665

Newly recorded herbaceous plants of Archichlamydeae in Northeast China

Meng Wang, *Shu-Mei Zhang*

Dalian Natural History Museum

Northeast of China, located in the East of Eurasia, is vast in territory, and has varieties of plants species, a total of 164 families, 928 genera and 3,103 species of vascular bundle plants. It is an important region for the origin, evolution and development of flora. Based on the results of field investigation and related data, one family (Aizoaceae), three genera (*Mollugo* L., *Achyranthes* L., *Dicranostigma* Hook. f. et Thoms.) and four species newly recorded herbaceous plants of Archichlamydea are reported (*Mollugo stricta* L., *Achyranthes bidentata* Blume, *Dicranostigma leptopodum* (Maxim.) Fedde, *Erodium cicutarium*). As a newly recorded family in Northeast China, aizoaceae is distributed in Tieling. As newly recorded genera in Northeast China, *Mollugo* L. is distributed in Tieling. *Achyranthes* L. is distributed in Dalian and Fengcheng. *Dicranostigma* Hook. f. et Thoms. is distributed in Jinzhou, Dalian. As newly recorded species plants in Northeast China, *Mollugo stricta* L. is distributed in Tieling. *Achyranthes bidentata* Blume is distributed in Dalian and Fengcheng. *Dicranostigma leptopodum* (Maxim.) Fedde is distributed in Jinzhou, Dalian. *Erodium cicutarium* is distributed in Dalian. The finding of many newly recorded plants distributed in Liaoning confirms that the plants in Liaoning province have obvious features of transitional zone between north China flora and northeast China flora and provides new evidences for the relationship between them. The voucher specimens were preserved in the herbarium of Dalian natural history museum. These newly recorded plants have a great significant meaning to study the plant species diversity and provide new materials for investigation of their geographical distributions in China.

T2

P0666

Phylogenomics of *Betula* (Betulaeae) species based on RAD-seq

Nian Wang[1,2], *Richard Buggs*[1,3]

1. School of Biological and Chemical Sciences, Queen Mary University of London

2. School of Forestry, Shandong Agricultural University

3. Royal Botanic Gardens Kew

Discordance among molecular phylogenetic trees has been frequently observed and attributed to various factors, such as different evolutionary histories, deep coalescence, hybridisation and introgression or horizontal gene transfer. New approaches have been used to construct accurate phylogenetic trees that take account into the relationships between many separate gene trees and the underling history of species divergence. *Betula* provides an excellent model to study species relationships. In this study, we developed restriction site associated (RAD) markers for most diploid *Betula* species and used a concatenation method and a method based on summary statistics to infer relationships among diploid species of the genus. After filtering and trimming, the total number of reads for each species ranged from 1,065,196 in *B. hainanensis* to 2,560,486 in *B. populifolia* whereas the number of mapped reads was from 904,705 in *Alnus inokumae* to 2,439,447 in *B. populifolia* with the percentage of mapped reads varied from 66.0% in *A. inokumae* to 96.8% in *B. pendula* ssp. *mandshurica*. With a reference genome for mapping, we obtained up to 16,213 reads with a minimal length of 300 bp in *A. inokumae* whereas this figure was 32,977 in *B. maximowicziana*. 4,890 loci with a minimal length of 300bp were used in the analysis. The results show that *B. corylifolia* and *B. michauxii* should be incorporated into subgenus *Aspera* whereas *B. maximowicziana* should be within subgenus *Betula*. Based on our phylogenomic approach, we proposed a new classification of *Betula* of three subgenera and seven sections: subgenera *Acuminata* [section *Acuminatae*], *Aspera* [sections *Asperae* and *Lentae*], and *Betula* [sections *Betula*, *Costatae*, *Dahuricae* and *Maximowiczianae*]). The study shows that RAD-seq is suitable for phylogenomic analysis especially with an available reference genome (albeit fragmented). This dataset will also allow future investigation of the parental origins of polyploid species.

T2

P0667

Phylogeography of *Scrophularia incisa* and its close relatives (Scrophulariaece): Revealing the evolutionary history of desert plants in northwestern China

Ruihong Wang[1], *Pan Li*[2], *Zhaoping Yang*[2], *Chengxin Fu*[2]

1. Zhejiang Sci-tech University

2. Zhejiang University

Pleistocene climate oscillations appear to have had an important effect in shaping intra- or inter-specific genetic variation and evolutionary history in many arid-adapted species. To understand the process involved and reveal evolutionary relationships, phylogeographic analysis were examined from *S. incisa*, and two other species also from *Scrophularia* section *Caninae*, including *S. kiriloviana*, *S. dentata*. We collected 522 individuals from 32 populations of the three closet relatives of *Scrophularia incisa* species complex and sequenced two maternally inherited chloroplast (cp) DNA fragments and also the bi-paternally inherited nuclear ribosomal ITS and microsatellite loci. We recovered four major clades in phylogenetic trees reconstructed from cpDNA and internal transcribed spacer (ITS) sequence variation. Most sampled individuals that are distributed in Inner Mongolia, QTP, Tibet and Xinjiang, respectively, comprised monophyletic clades nested within those found in Central Asia. *Scrophularia incisa* in Inner Mongolia and Qinghai-Gansu were paraphyletic, while the placement of *Scrophularia incisa* from Inner Mongolia was in the Xinjiang gene pool revealed in the SSR structure, differing with the ITS and cpDNA phylogenetic trees. Our phylogeographic analysis supported the importance of Pleistocene climate change, in this case the increase of aridification and enlargement of deserts, as a cause of divergence in *Scrophularia incisa* complex as a result of allopatric speciation, hybridization and introgression. These findings highlight the complexity of closest relatives evolutions and the importance of the Pleistocene climate change in shaping the genetic diversity.

T2

P0668

Accounting the events on the evolutionary pathway from early land plants to angiosperms

Xin Wang

State Key Laboratory of Palaeobiology and Stratigraphy, Nanjing Institute of Geology and Palaeontology

Ovules are megasporangia that are retained on their mother plants until seed maturation and recruit additional protective parts in the evolution of seed plants. Establishing the correlation between ovules in angiosperms with those in gymnosperms and spores in other vascular plant groups is the foundation for integrated systematic analysis of land plants. Gymnosperms have invented various ways to protect their seeds as seen in the cones of Cycadales and Coniferales. Protection of megasporangia or ovules is initially provided by integuments. This is followed by various laminar structures derived from formerly branches that form the cupular walls in various taxa. This trend is further enhanced to a new level by angiosperms that protect their seeds even before the seeds are born (in ovule stage). Different from their gymnospermous peers, angiosperms have developed ovule-enclosing as a competitive strategy in their survival struggle, ensuing the protection by integuments and cupular walls in various seed plants. Besides ovule-enclosure (angio-ovuly), angiosperms have developed angiocarpy in various taxa independently. Associated with or after these stepwise enhanced protections, stepwise ensured nutritional supply, outcross-enhancing reproductive characteristics, improved fruit and seed dispersing features give angiosperms additional edges in their evolution. Ensurance of plenty of nutrition for the offsprings is a major advantage for plants in their evolution starting from *Rhynia* on. The first step in this direction is the occurrence of heterospory in various taxa, which allocates more resources to the female spores. Following heterospory, megaspores become retained in their mother plants, giving rise to ovules that can absorb extra nutrition from the mother plants for their healthy development. In addition, these changes and advantages are coupled with outcrossing enhancing self-incompatibility. Extended nutritional bond with sporophyte after meiosis is further extended to a new level of vivipary, namely, germination of seeds on the sporophytes. Different extent of megaspore protection and enclosure (angio-spermy) and the spatial relationships between the ovules and their protecting laminar structures define and distinguish various fossil or extant plant groups, including ferns, gymnosperms and angiosperms. As the nutritional and protecting relationships evolve, different fructifications emerge in various plant groups throughout times, and, correspondingly, various gene dispersal and communication mechanisms evolve. The coupling evolution between animals and plant fructifications plays a crucial role in the radiation of angiosperms that accelerates the evolution in both groups.

T2

P0669

Genome-wide evidence for sympatric introgression and speciation with gene flow of two *Rhizophora* mangrove species

Xinfeng Wang, *Zixiao Guo, Suhua Shi*

State Key Laboratory of Biocontrol and Guangdong Provincial Key Laboratory of Plant Resources, Sun Yat-sen University

The two species in the mangroves genus *Rhizophora*, namely *R. mucronata* and *R. stylosa* are very closely-related, estimating as diverging at approximately 0.66 million years ago. Owing to their overlapping distribution in Indo-west Pacific region and observations of inter-species hybridization and gene flow, their independent species status were doubted. We firstly characterized the divergence and gene exchanges between the two species on the whole genome scale by resequencing 26 individuals of *R. mucronata* and 17 individuals of *R. stylosa*. Comprehensive analysis of the 5.4 million single nucleotide polymorphisms (SNPs) obtained by mapping to the reference genome of *R. apiculata* supported their independent species status, while observed dramatically reduced divergence when they are sympatric. The introgression found in sympatric populations in Daintree River, Australia and Tg.piai, Malaysia were proved to exist from ancient to recent time, and further found to be bidirectional in Daintree River and unidirectional from *R. mucronata* to *R. stylosa* in Tg.piai. These findings are useful to study the role of gene flow played in speciation or species maintenance after speciation.

T2

P0670

Taxonomic revision on subsect. *Oyama*

Yaling Wang[1], *Shouzhou Zhang*[2]

1. Xi'an Botancial Garden of Shaanxi Province

2. Laboratory of Southern Subtropical Plant Diversity, Fairylake Botanical Garden, Shenzhen & Chinese Academy of Sciences

Magnolia section *Rhytidospermum* subsection *Oyama* is famous ornamental group for its pure white pendant flowers with showy red or purple androecium. Although they have most wild distribution at latitude (25°32′-41°55′N), longitude (85°20′-119°42′E) and altitude (330-3400m), and quiet similar morphological characters within and among populations, the systematic position and internal classification of this group has still been controversial. In this study, 116 individuals from 12 populations representing 4 taxa in subsect.*Oyama* were studied for the taxonomic revision of sect. *Oyama*. The morphology variation and geographical distribution, as well as molecule data analysis on the combine data of three universal DNA regions (matK, rbcL, trnH-psbA) were conducted to detect the phylogenetic relationships. The results revealed that subsect. *Oyama* include three species and two varieties: *Magnolia globosa* Hook.f. & Thoms., *M. sieboldii* K. Koch, *M. sieboldii* ssp. *japonica* K. Ueda, *M. wilsonii* (Finet. & Gagnep.) Rehder, *M.wilsonii* var. *sinensis* (Rehder & Wilson) Y.L. Wang & S.Z. Zhang.

T2

P0671

Systematic studies on the *Umbelopsis*

Ya-Ning Wang, *Xiao-Yong Liu, Ru-Yong Zheng*

Institute of Microbiology, Chinese Academy of Sciences

The fungi of *Umbelopsis* (Umbelopsidaceae, Mucorales, Mucoromycotina) are saprobes and widely distributed in soils and plant residues. In recent years, some species of this genus have been found to be endophytes. *Umbelopsis* species are important oleaginous fungi due to their high oil contents. Compared with other taxa of the Mucorales, *Umbelopsis* is characterized by forming: 1) velvety and often colored colonies, 2) a low and dense layer of aerial hyphae; 3) sporangiophores which are often arising erectly from the vesicles; and 4) small columellae or not forming at all. *Umbelopsis* is a basal monophyletic group within the Mucorales clade in phylogenetic analyses. A total of 128 *Umbelopsis* strains comprising 121 strains from China and 27 ones from other countries have been studied. Based on the results of morphological characteristics, maximum growth tempearature and multilocus phylogenies (partial 18S and 28S, internal transcribed spacer regions of nuc rDNA, and two protein-coding genes, *act1* and *MCM7*), fourteen species were recognized in the genus *Umbelopsis*. In this study, *U. changbaiensis* was described as a new species; *U. angularis*, *U. dimorpha* and *U. nana* were reported as new record species from China; *U. longicollis* was recombinated from *Mortierella longicollis*; *U. nana*, *U. roseonana* and *U. versiformis* were treated as heterotypic synonyms; *U. ramanniana* and *U. isabellina* were revealed to be heterogenetic and may be two species complexes. In addition, lectotypes and epitypes for *U. longicollis*, *U. nana* and *U. vinacea* were also designated.

T2

P0672

Variable architecture of legume plastomes beyond Papilionoids

Yin-Huan Wang[1,2], *Xiao-Jian Qu*[1,2], *Si-Yun Chen*[1], *De-Zhu Li*[1], *Ting-Shuang Yi*[1]
1. Kunming Institute of Botany, Chinese Academy of Sciences
2. University of Chinese Academy of Sciences

It is universally known that legume plastomes experience significant structural and size variation. However, most studies have focused on subfamily Papilionoideae and limited studies on the mimosoid clade. In order to explore structural variation of plastomes beyond Papilionoids, we sequenced plastomes of 20 species from representing genera. Integrated three previously published plastomes with 11 new ones, we firstly addressed the architecture of mimosoid plastomes. The mimosoid plastome presented typical quadripartite structure and contained 111 unique genes. Their IRs (inverted repeats) experienced multiple expansion/contractions. A ca. 13 kb IR expansion into SSC (small single copy) was detected in plastomes of a clade formed by the tribe Ingeae and *Acacia* sensu stricto (s.s.) and to be the synapomorphy of this clade. A ca. 1.7 kb IR expansion into and a ca. 1.9 kb contraction out of LSC (large single copy) were found in *Pithecellobium flexicaule* and *Acacia dealbata*, respectively. Linear regression analysis showed decreased synonymous substitution rates of genes relocating from SSC into IR. A loss of both introns of *clpP* occurred in *A. dealbata* and *Faidherbia albida*, and a duplicated *clpP* copy was detected in *A. dealbata*. Furthermore, a 421 bp inversion that containing *rps18* was found in *A. dealbata*. Size of mimosoid plastomes was found significantly affected by IR-SC boundary shift and also associated with repeat content. Plastomic coding and noncoding regions with variable sequence divergence may supply valuable markers for molecular evolutionary and phylogenetic studies at different taxonomic levels. We have also detected a variety of structural variations in plastomes of seven species from subfamily Cercidoideae except *Adenolobus garipensis* and *Cercis glabra* representing two basal branches of this subfamily. These variations included IR expansion and contraction, gene or intron loss and sequence inversion. Our results would be crucial for further studies on the variation and evolution of legume plastome architecture.

T2

P0673

Discrepancy between nuclear and chloroplast DNA variations reveals alpine valleys to be the arena for new species under iterative climate changes

Zimeng Wang
Peking University

Rhodiola (Crassulaceae) has experienced rapid diversification on the Qinghai-Tibet Plateau (QTP) since 21 million years ago. For a better understanding of this evolutionary process, this study took a species pair of *R. stapfii* and *R. prainii* as a miniature, 25 populations were sampled from the whole distribution area of these two species, mainly in the southern part of QTP. 13 nuclear SSR loci and five unique chloroplast sequence markers generated from high-throughput sequencing reads were analyzed as well as traditional markers such as internal transcribed spacer (ITS). SSR analyses show that *R. stapfii* and *R. prainii* have different population structures. Phylogeny based on ITS sequences conflicts severely with that of chloroplast sequences. The cpDNA phylogenetic tree braches are dense but those based on ITS are more dispersed. Tree topology conflicts happens as well, indicating during certain periods some populations were distributed closely so that they had the strong gene flow by pollens, not seeds. Also, niche modeling shows that these two species were sympatric during glacial periods and became parapatric or even allopatric during interglacial periods. Thus, a evolutionary scenario can be expected that during warm periods populations are "pumped out" from their glacial refuges to adjacent regions, while during glacial periods they are squeezed into valleys of Yarlung Zangob River, where their possible incomplete reproductive isolation were challenged by pollen flows. Special landscape along Yarlung Zangob River as well as iterative climate changes in Quaternary period provides an arena for new speciation of the genus *Rhodiola*.

T2

P0674

Systematics and phylogeny of Apiaceae subfamily Apioideae major lineages related to *Angelica*

Jun Wen
Sichuan University

Apiaceae is a large and taxonomically complex family (include more than 3,700 species spread across about 455 genera) of angiosperms that include many economically and medicinally important plant families? The typical representative medicinally genus is *Angelica* L. In the previous study we know that *Angelica* is not a monophyletic group, and it is scattered in four large lineages. However, in the previous study they didn't have got a definite generic boundaries of *Angelica*. Our research will be based on richer sampling, more suitable DNA barcode (including ITS, ETS, some chloroplast genes and low copy nuclear conserved ortholog set (COS) genes or even single copy genes) as well as a clearer morphological analysis, including leaf epidermis micro-morphological fruit anatomy and morphology, pollen morphology and so on) of these four lineages with Apioideae subfamily as the framework to discuss the systematics and phylogeny of *Angelica*. Some species should be reduced to synonyms, and we should define the composition of *Angelica* sensu stricto. In addition, we also expect to solve the problem of some groups of genus *Angelica* at the transcriptome level.

T2

P0675

Why don't all flowering plant male gametophytes have extremely fast pollen tube growth rates?

Joseph Williams
University of Tennessee

Pollen tube growth rate (PTGR) is an important aspect of male gametophyte performance because of its central role in the fertilization process. Theory suggests that under intense competition, PTGRs should evolve to be ever faster, especially if PTGR accurately reflects gametophyte quality in the sporophyte generation. A comparative analysis of PTGRs in 457 seed plants revealed a strong phylogenetic signal, even when slow-growing gymnosperms were removed. Species with increasingly faster PTGRs were increasingly nested in many independent groups of angiosperms. Species with the fastest PTGRs (> ~5,000 μm/h) were found in at least 16 independent groups that display a number of convergent features thought to select for rapid PTGR. Similarly, species with exceptionally slow PTGRs (< ~50 μm/h) were restricted to lineages which lack features that promote high levels of pollen competition. Given the basic life history benefits of rapid growth rate (greater competitive ability, shorter lifespan), it is surprising that we still lack a basic understanding of costs and tradeoffs associated with evolving rapid PTGR. Pollen tube growth rate is ultimately determined by how effectively the work of tube construction is translated to elongation (growth and growth rate). Do pollen tubes grow with similar costs and do they grow equally efficiently? I compared the scaling of wall production rate (WPR) to PTGR in three distantly-related water lilies that flower concurrently in the same ponds: *Nymphaea odorata*, *Nuphar advena* and*Brasenia schreberi*. Flowers were pollinated and carpels were collected at two time points. PTGR was calculated as the average increase in tube length over a growth period of about 45 minutes. Tube circumferences (C) and wall thicknesses (W) were then measured on glycol methacrylate-embedded material collected at the end of that time period. Wall production rate (WPR) was calculated as: mean ($C \times W$) × mean PTGR for each pollination. The ratio of WPR to PTGR is a measure of growth efficiency - efficient growth occurs when wall production rate is low relative to PTGR. Within species, pollen tubes maintained a constant WPR to PTGR ratio over a range temperatures and growth rates. Slow growing tubes were equally efficient as fast-growing tubes. However, fast-growing tubes had significantly thinner walls, but the effect was offset by a nearly opposite pattern for tube diameter. Species had significantly different growth efficiencies, however. Whereas *N. odorata* and *N. advena* had similar PTGRs, for any given PTGR, they had significantly different WPRs. In conclusion, PTGR is a fundamental aspect of plant life histories that can evolve for many reasons. PTGR efficiencies evolve by changes in the volume of wall material used for growth and in how that material is partitioned between lateral and length dimensions. Rapid growth can be achieved by increasing energetic costs and/or by increasing growth efficiency. The economics of pollen tube growth are determined by tube design, which is consequent on tradeoffs between efficient growth and other pollen tube functions.

T2

P0676

Relationships of the tribe Metrosidereae and realignment of the tribe Kanieae (Myrtaceae) based on sequence and morphological data

Peter Wilson[1], *Margaret Heslewood*[1], *Myall Tarran*[2]

1. National Herbarium of New South Wales

2. The University of Adelaide

Myrtaceae is a large family of tropical and temperate trees and shrubs currently divided into 2 subfamilies and 17 tribes. The tribe Metrosidereae, which includes the widespread Pacific genus *Metrosideros*, has a strongly southern distribution and is presently the only tribe in subfamily Myrtoideae that is absent from mainland Australia. The tribe Kanieae, as presently constituted includes genera restricted to Australia and Malesia. The tribal classification was based on a phylogenetic analysis of the chloroplast *trn*K intron that showed good support for most tribes but little for any intertribal relationships. Sequence of chloroplast *psb*A–*trn*H and nuclear ribosomal ITS and ETS regions was added for representative species in the tribes Kanieae and Metrosidereae. This included sequence of 3 accessions of the genus *Kania*, type of the tribe Kanieae, that had been represented by one very poor sequence in the previous study. Analyses of this extended dataset now show a strong sister relationship between *Kania* and the tribe Metrosidereae but not with the other genera presently assigned to the Kanieae. This relationship between *Kania* and the tribe Metrosidereae is strongly correlated with a morphological feature, hypanthium vascularisation, that has come to light through recent research on fossil fruits of *Metrosideros*. A consequence of this finding is that a new tribe will be required to accommodate the majority of other genera presently assigned to the Kanieae. Since *Kania* only occurs in Papua New Guinea and the Philippines, there are now 2 sister clades that do not occur naturally on the Australian mainland.

T2

P0677

Are we splitting hairs? Resolving infrageneric relationships in the Australian *Plectranthus* (Lamiaceae)

Trevor Wilson[1], *Paul Forster*[2], *Barry Conn*[3], *Maurizio Rosetto*[1], *Murray Henwood*[3]

1. National Herbarium of New South Wales

2. Queensland Herbarium

3. The University of Sydney

Plectranthus (Lamiaceae) is well-known for its use in medicine, agriculture and horticulture. From its highest centre of diversity in Africa, the distribution of the genus extends throughout the old world tropics. Over the past few decades the described number of Australian *Plectranthus* has more than doubled to 48, making eastern Australia another area of high diversity for the genus. Identification of Australian *Plectranthus* can be challenging due to a troublesome combination of delicate macro-morphological characters (e.g. corolla shape) and cryptic micro-morphological characters (e.g. trichomes). Furthermore, phenotypic plasticity and the sympatric distribution of similar-looking species are also common. This combination has assisted with masking the high diversity of Australian *Plectranthus*, and it has continuously facilitated misidentifications. As of yet no quantifiable approach has been implemented to understand the relationships of Australian *Plectranthus*. For the many descriptions of Australian *Plectranthus* currently in the pipeline, this lack of knowledge hampers the provision of useful diagnosable characters. A careful integration of molecular and morphological data is therefore necessary to test the usefulness of current taxonomic characters and institute an understanding of relationships. We describe the preliminary phylogenetic results for Australian *Plectranthus* based on our analysis of chloroplast DNA acquired from high throughput sequencing methods. These results are also combined with data of extra-Australian *Plectranthus* to improve our perspective on relationships and biogeography of the entire genus.

T2

P0678

eFlora of New Zealand

Aaron Wilton, *Ilse Breitwieser*

Landcare Research

The eFlora of New Zealand initiative aims to deliver a resource on the wild New Zealand vascular and cryptogrammic flora that is up-to-date and engages a wide range of users both inside and outside the taxonomic community. Development of the eFlora has required implementation of a process that combines important components of traditional floras with an information system that integrates data from a range of sources. For example, this approach combines traditional editorial and review processes with a team approach for managing nomenclatural and taxonomic information. The eFlora currently integrates data from specimen, taxonomic and nomenclatural, descriptive and image systems to deliver a website (http://nzflora.info) and PDFs of treatments. The eFlora system provides a means for authoring new taxonomic treatments – including taxonomic and nomenclatural novelties. Since 2010 over 50 new family level treatments have been published using this method. This new taxonomic content is combined with taxonomic treatments from earlier New Zealand Floras and more recent taxonomic treatments published in journals to publish the web site within which information is provided as different profiles - templates that combine different subsets of information to address differing audience needs. We will present our experiences developing, and future plans for, the eFlora of New Zealand.

T2

P0679

Inferring the evolutionary history of *Penstemon* (Plantaginaceae) from phylogenetic analysis of high-throughput sequence data

Andi Wolfe

The Ohio State University, Department of Evolution, Ecology and Organismal Diversity

Penstemon is the largest genus of plants endemic to North America. Its 280+ species are distributed from coast-to-coast and from Alaska to southern Mexico and the highlands of northern Guatemala, excluding most of the Canadian Shield region. Most species occur west of the Rocky Mountains, with the largest diversity of species in the Intermountain Region. The current taxonomy of *Penstemon* circumscribes six subgenera, divided into 12 sections and 23 subsections. Most species are xerophytic, and more than half occur in montane to subalpine habitats. Edaphic endemics are found in all sections of the genus, with many species restricted to calcareous, volcanic, shale, clay, and deep sand substrates. *Penstemon* exhibits considerable morphological diversity. Species may be woody, suffrutescent, or herbaceous, and exhibit extreme variation in vegetative characters, inflorescence architecture, vestiture, and floral morphology. We have used targeted amplicon sequencing for 48 loci to reconstruct the phylogeny of >90% of the species in this large and very diverse genus. Our goals for this phylogenetic study were to: 1) build a robust phylogeny for *Penstemon* that will be used to examine rates of diversification in the context of testing adaptive radiation theory; 2) examine taxonomic relationships in the context of an evolutionary framework; 3) elucidate trends in the evolution of morphological traits; 4) determine if there is a phylogenetic pattern for edaphic specialization; and 5) test biogeographic hypotheses proposed in previous studies of the genus.

T2

P0680

Genomic consequences of gene flow between two *Senecio* species on Mount Etna, Sicily

Edgar L.Y. Wong, *Bruno Nevado, Simon J. Hiscock, Dmitry A. Filatov*

Department of Plant Sciences, University of Oxford

S. aethnensis and *S. chrysanthemifolius* are two sister species which range above 2000 m and below 1000 m respectively on Mount Etna, Sicily. It was estimated that the species diverged around 100-150 thousand years ago and the divergence might have been driven by the rise of Mount Etna, which provided novel niches for the morphological and genetic divergence. Ongoing gene flow was found between the two species as they maintain a hybrid zone between their respective ranges. However, some hybrid breakdown was observed to manifest in F_2 hybrids, which suggested that these species have already evolved some degrees of postzygotic isolation. Previous analyses of gene expression have identified a set of differentially expressed genes. Some of the differentially expressed genes are involved in primary amine oxidase activity and magnesium ion binding – functions that are likely important for adaptation to volcanic soils. These genes have been evolving under diversifying selection. This study reports DNA polymorphism and divergence analyses in the differentially expressed genes in order to test whether they have been subject to diversifying selection in the recent past. For this purpose, we sequenced differentially expressed genes from multiple populations at different elevations on Mount Etna. Genomic clines were analysed for these genes to investigate the selective pressures keeping the two species distinct while maintaining a hybrid zone. Overall, this project aims to reveal the evolutionary forces during on-going ecological speciation driven by adaptation to contrasting conditions at high and low altitudes.

T2

P0681

Evolution of WRKY transcription factors in Poaceae family

Kangcheng Wu

Fujian Agriculture and Forestry University

WRKY gene family is one of the largest transcription factor families in higher plants, which plays important roles in regulating plant development and resistance. Poaceae is a family covering various economic-valued plants. In this report, a total of 17 Poaceae species, spanning 6 subfamilies, from basal to crown Poaceae, were studied to systemically research the evolution of *WRKY* gene family. Totally, 5 WRKY groups were classified according to their phylogeny, Group I, IIa, IIb, IIc, IId, IIe, III. It was found that the large proportion of Poaceae WRKY was mainly produced by several recent-independent polyploidization events and tandem duplication. In addition to species that experienced recent polyploidization events, other Poaceae also showed various WRKY numbers, such as sugarcane, millet, which were contributed by large tandem duplications of Group III. The analysis of selection pressure indicated that Group III of WRKY evolved faster than other groups. Some genes showed fast evolutionary rate in Group I, IIc and IIe. Positive selections and functional divergence were found in these fast-evolved genes. It was found that there was a branch with fast evolution in the Group I, and specific functions were unknown. The divisions of Poaceae plants were divided into BOP clade, Chloridoideae clade, and PACMAD clade (except the subfamily of Chloridoideae), which were different from previous classifications.

T2

P0682

Leaves and seed cones of *Dacrycarpus* (Podocarpaceae) from the Miocene of South China

Xinkai Wu[1], *Xiaoyan Liu*[1], *Tatiana Kodrul*[1,2], *Cheng Quan*[3], *Jianhua Jin*[1]

1. State Key Laboratory of Biocontrol and Guangdong Key Laboratory of Plant Resources, School of Life Sciences, Sun Yat-sen University

2. Geological Institute, Russian Academy of Sciences

3. Research Center of Paleontology and Stratigraphy, Jilin University

Dacrycarpus (Podocarpaceae) has an abundant macrofossil record in southeast Australia, New Zealand and Patagonia, but no record was reported previously in the Northern Hemisphere. In this study, we described a new species of *Dacrycarpus* recovered from the Miocene Erzitang Formation of Guiping Basin, Guangxi, South China. The specimens contain various forms foliage and three-dimensional preserved female cones, with well-preserved cuticles. The new materials have dimorphic foliage including bifacially flattened "adult" leaves on long shoots and bilaterally flattened "juvenile" leaves on short shoots. All foliage have a single midvein and incurved apex, as well as resin ducts. Bilateral leaves are distichous, falcate, straight to slightly apically curved, decurrent, stongly keeled with a single prominently raised vein. Maximum length per shoot is 3.4-7.4 mm. Maximum width per shoot is 0.6-1.1 mm. The ratio of leaf length and width is 4:1-7:1. Short shoot is about 12-26 mm long × 5-10 mm wide. Leaves are amphistomatic, with two longitudinal bands of stomata occurring from leaf base to apex on each surface. Each band contains 2-5 rows of stomata. Stomatal axes parallel to the midvein. Florin rings are well developed, sunken below leaf surface. Stomatal complex is paratetracytic. Polar subsidiary cells usually shared between adjacent stomata of row. Epidermal cell walls are straight, smooth and thin. Bifacial leaves are small, 2-3.5 mm in length, 0.4-0.8 mm in width, awl-shaped, spirally arranged, loosely imbricate, appressed, strongly keeled and amphistomatic. Seed cones are terminal, obovate, 4.2-5.9 mm long, 3.6-5.2 mm wide, deployed singly on a warty receptacle. Based on the architectural features of leaves and seed cones, the present fossils are assigned to the genus of *Dacrycarpus*, showing the most resemblance to *D. imbricatus* (Blume) de Laubenfels, the only extant species in China. The discovery demonstrates that the genus has spread into South China at least by the Miocene, and the migration may have occurred with the collision between the Indian and Eurasian plates. The new species of *Dacrycarpus* builds the first reliable fossil link for this genus between two hemispheres.

T2

P0683

Understanding functional diversity and mechanisms of plant community assembly using non-target metabolite profiling: A test of HPLC-MS method.

Wenbin Zhou, *Sirius Li*, *Qiu-Yun (Jenny) Xiang*

Department of Plant and Microbial Biology, North Carolina State University, Raleigh, NC 27695-7612, USA.

Secondary metabolite diversity represents a special dimension of biodiversity important to ecological functions and community assembly. However, there have been few studies evaluating secondary metabolite diversity and the relationship between chemical diversity and phylogenetic diversity within and among plant communities. Recent studies have demonstrated the value of analyzing the distribution and variation of functional traits to gain insights into primary ecological processes (niche differentiation or environmental filtering) shaping community assembly. We attempt to characterize the pattern of secondary metabolite variation in the eastern Asian and eastern North American disjunct forests to provide new insights into the biodiversity and community assembly. As a first step, we tested the utility of the HPLC-MS non-target profiling method. A set of 30 leaf samples with various habits from three plant communities including species from distantly related clades of seed plants as well as species from the same family, same genus, or same species for a pilot study. The samples were from the Ordway-Swisher Biological Station, Talladega National Forest, and Mountain Lake Biological Station. Our goal is to see if sufficient data would be obtained from dry leaf extracts and useful for comparative analyses. We obtained 6174 chemical compounds for the 30 species from the profiling using Aglient MassHunter Profinder B.06.00 software. Of which, 27% were rare (present in < 5% of the samples). Data sets shared by 30 species, >25 species, >20 species, >10 species, and >1 species, were constructed and transformed by cube root, normalized by sum, and scaled by mean centering for further analyses. Partial least square (PLS) test revealed significant differences among sites, suggesting the chemical composition is correlated with sites (MANOVA, $F_{(2,27)} = 5.7223$, $P = 2.762*10^{-06} < 0.05$). Fifteen chemical markers were identified to explain most of the patterns distinguishing among sites. Hierarchical cluster analysis (HCA) with ward's method in R for samples within each site showed samples from the same genus or the same species were clustered together, but not samples from the same family or order. Comparison of phylogenetic diversity measured by *rbc*L-*mat*K trees and chemical diversity will be conducted. Patterns of chemical clustering of species and phylogeny will also be compared within and among sites to gain insights on relationship of chemical diversity to phylogenetic diversity and on processes of community assembly. The preliminary study demonstrated HPLC-MA is useful for the community level analysis of chemical diversity.

T2

P0684

On the generic circumscription of the complex thalloid liverwort family Corsiniaceae (Marchantiophyta) based on molecular and morphological evidence

You-Liang Xiang, *Lei Shu, Rui-Liang Zhu*

School of Life Sciences, East China Normal University

The most recent classification divided Corsiniaceae into three genera, *Corsinia* Raddi, *Cronisia* Berk. and *Exormotheca* Mitt. A combined molecular phylogenetic analysis of three chloroplast genes (*rbc*L, *atp*B and *psb*A) and one nuclear gene (the ribosomal LSU) was carried out to assess possible alternative treatments of the family, using *Targionia hypophylla* and *Monosolenium tenerum* as outgroups. All the *Exormotheca*, *Corsinia* and *Cronisia* species included in our analyses formed a well-supported clade and all *Exormotheca* species were nested within the other two genera. In combination with morphological data, here we propose to subsume all species in *Cronisia* and *Exormotheca* into the earlier published genus *Corsinia*. *Exormotheca* and *Cronisia* are treated as new synonyms. Eleven new combinations are proposed: *Corsinia bischlerae* (Furuki et Higuchi) R.L.Zhu, Y.L.Xiang & L.Shu comb. nov., *Corsinia brevipedunculata* (Kashyap)R.L.Zhu, Y.L.Xiang & L.Shu, comb. nov., *Corsinia bulbigena* (Bornefeld, O.H.Volk et R.Wolf) R.L.Zhu, Y.L.Xiang & L.Shu, comb. nov., *Corsinia ceylonensis* (Meijer) R.L.Zhu, Y.L.Xiang & L.Shu, comb. nov., *Corsinia fimbriata* (Nees) R.L.Zhu, Y.L.Xiang & L.Shu comb. nov., *Corsinia gollanii (*Steph.) R.L.Zhu, Y.L.Xiang & L.Shu, comb. nov., *Corsinia holstii* (Steph.) R.L.Zhu, Y.L.Xiang & L.Shu, comb. nov., *Corsinia pustulosa* (Mitt.) R.L.Zhu, Y.L.Xiang & L.Shu, comb. nov., *Corsinia tuberifera* (Kashyap) R.L.Zhu, Y.L.Xiang & L.Shu, comb. nov., *Corsinia weddellii* (Mont.) R.L.Zhu, Y.L.Xiang & L.Shu comb. nov., and *Corsinia welwitschii* (Steph.) R.L.Zhu, Y.L.Xiang & L.Shu, comb. nov.

T2

P0685

Stable carbon isotope variation in recent and fossil *Liquidambar* leaves as palaeoenvironmental indicator

Liang Xiao[1,2], *Xiangchuan Li*[1,2], *Junfeng Guo*[1], *Nan Sun*[1], *Xiaoyong Yao*[1]

1. School of Earth Science and Resources, Chang'an University

2. State Key Laboratory of Palaeobiology and Stratigraphy, Nanjing Institute of Geology and Palaeontology, Chinese Academy of Sciences

Quantitative reconstruction of ancient environments and climates has been a major goal for studies in geobiology. Fossil leaves are increasingly used as indicators of palaeoenvironmental parameters such as temperature, atmospherical CO_2 concentration, precipitation. It is demonstrated that the carbon isotope composition of leaf remains can also reveal ecophysiological information on plant's response to past environmental change. In this study, stable isotope compositions of recent *L. formosana* leaves collected in 6 different sites as well as fossil *L. miosinica* leaves sampled from late Miocene Shengxian Formation, Zhejiang Province, China, were obtained. According to variations of stable isotopic discriminations on six recent *L. formosana* leaves with the climates of their living areas, the correlativity between $\Delta^{13}C$ and environmental factors was established. Among all correlativity, three relativities are notable between $\Delta^{13}C$ and annual extremely warmest temperature (EWT), mean annual precipitation (MAP), annual solar radiation hours (SRH). The relativity values are 0.60, 0.75, 0.80, respectively. Thereby, the three regression equations are used to infer palaeoenvironments. The $\Delta^{13}C$ value is measured and then changed as 20.82. When the $\Delta^{13}C$ value is substituted into three regression equations, the EWT, MAP, SRH in the late Miocene, Zhejiang, are counted as 39.2 °C, 1,329 mm, 1,938 hrs, respectively. The ancient SRH is obtained for the first time. Here, the ancient MAP is within the range reconstructed from other methods, such as the coexistence approach (CA), climate analysis of endemic species (CE), which suggests that the palaeoenvironmental results based on $\Delta^{13}C$ of fossil leaves are reliable. In addition, comparing the results to recent environments, it is found that the late Miocene palaeoenvironments of Zhejiang, China are similar to that of the same place. It indicates that the late Miocene climate is warm and moist, which is ascribed into subtropical climate.

T2

P0686

Genetic evidences of Karst-habitat adaptation in genus *Urophysa* (Ranunculaceae): A comprehensive transcriptome analysis

Deng-Feng Xie

Sichuan university

The Karst environment characterized by low soil water content, periodic water deficiency, and poor nutrient availability provides an ideal natural laboratory for studies on adaptive evolution. However, few studies have been conducted on how plants adapt to the Karst habitat. *Urophysa* is a Chinese endemic genus with high degree of Karst-specific distribution, and is hoped to be a new model system to study speciation and ecological adaptation in the special habitat. In this study, we performed transcriptome sequencing of flowers and leaves of *U. henryi* (HF and HL) as well as those of *U. rockii* (RF and RL). A total of 249,191 unigenes were obtained and 54% of which are homologous with sequences in public protein databases (Nr, Swiss-Prot, Pfam-A, GO, and KEGG). The differential gene expression analysis indentified 233 highly expressed genes among *U. rockii* and *U. henryi* transcriptomes, including 97 genes between flowers (HF vs. RF) and 136 genes between leaves (HL vs. RL). After genes annotation, theses differentially expressed genes were enriched in functions involved in photosynthesis, stress response, biological process regulation and metabolism. Our findings indicate that species of *Urophysa* evolved complex strategies for adapting to extreme environments on Karst and provide novel insights into genetic mechanisms of Karst adaptation in plants.

T2

P0687

Floral ontogeny of *Illicium lanceolatum* (Illiciaceae), with emphasis on carpel, ovule and floral apex development

Xin Zhang

College of Forestry, Northwest A&F University

Illiciaceae is a member of the ANITA grade, which has been identified as the most basal among extant angiosperms based on molecular investigations. The family includes a single genus with apocarpous gynoecium, demonstrating various ancestral morphological characteristics especially in terms of carpel, ovule and floral apex. Various aspects of Illiciaceae morphology have been investigated before, however, with the new APG system of flower arising, the details of the carpel, ovule and floral apex of Illiciaceae need to research for understanding of the flower evolution. Here the development of carpels and ovules of *Illicium lanceolatum* were studied using LM and SEM. Based on comparisons with other taxa, the evolutionary implication of a persistent floral apex is an apical meristem residuum of a shoot which is the relic of flowering plants ancestor. The carpel is a leaf like structure that will encircle the ovule and the ovule primordium originates from floral axis in Illiciaceae. The micropyle is formed by both integuments in contrast to previous results and shows the diversity within Illiciaceae can also be used for taxonomic purposes.

T2
P0688
Reconsideration of the carpel in angiosperms
Zhang Xin
Norwest A&F University

Angiosperms distinguish themselves from gymnosperms by their ovules that are enclosed before pollination. However, how the ovules get enclosed in angiosperms remains a mystery, especially for Magnoliaceae. The only key to this mystery is finding a series of carpels transitional from fully closed with enclosed ovules to open with naked ovules.We use routine paraffin section technology, LM, SEM to document the morphology and anatomy of carpel variation in *Michelia figo* (Magnoliaceae). A series of carpel variations within a single flower of *Michelia figo* (Magnoliaceae) are documented, in which the ovules are exposed in atypical carpels. These atypical and typical carpels for the first time demonstrate clearly how the naked ovule get enclosed. Each atypical carpel, with naked ovules subtending foliar part and branches bearing ovules, suggesting that a typical carpel is actually an end-product of the fusion between the ovuliferous branches and subtending foliar parts. The only difference among these carpels is the extent of fusion . This is generalization is in full agreement with the molecular genetic studies on angiosperm flowers. Therefore, the formerly huge gap between angiosperms and gymnosperms appears diminishing.

T2
P0689
ITS non-concerted evolution and rampant hybridization in the legume genus *Lespedeza* (Fabaceae)
***Bo Xu*[1], *Xin-Fen Gao*[1], *Li-Bing Zhang*[2]**
1. Chengdu Institute of Biology, Chinese Academy of Sciences
2. Missouri Botanical Garden

The internal transcribed spacer (ITS) as one part of nuclear ribosomal DNA is one of the most extensively sequenced molecular markers in plant systematics. The ITS repeats generally exhibit a high level of within-individual homogeneity, though relatively small-scale polymorphism of ITS copies within individuals has often been reported in literature. In this study, we identified large-scale polymorphism of ITS copies within individuals in the legume genus *Lespedeza* (Fabaceae), which suggests incomplete concerted evolution of ITS in *Lespedeza*. Divergent paralogs of ITS sequences, including putative pseudogenes, recombinants, and multiple functional ITS copies were detected in the same individual in our study. Thirty-seven ITS pseudogenes could be easily detected according to nucleotide changes in conserved 5.8S motives, the significantly lower GC contents in at least one of three regions, and the lost ability of 5.8S rDNA sequence to fold into a conserved secondary structure. Two recombinations of ITS are discovered in our putative hybrid species *L. jiangxiensis* and are resolved as basal to the parental lineages, indicating that they may result in the loss of clade resolution in the phylogenetic tree and may obscure the real phylogenetic relationships. Interestingly, the distribution patterns of the putative functional clones were very different between the traditionally recognized two subgenera. Within *L.* subg. *Macrolespedeza*, the ITS clones from most species were mixed together, suggesting low rate of concerted evolution. In contrast, the ITS clones from each species of *L.* subg. *Lespedeza* each formed a monophyletic group or unresolved, implying a higher rate of concerted evolution. This great difference between the resolution patterns of the ITS clones in the two subgenera could be attributed to their different extents/frequencies of hybridization. This inference is further corroborated by the analysis of the single-copy nuclear gene PGK which indicates that rampant hybridization has played a critical role in the evolution of *Lespedeza*, especially in *L.* subg. *Macrolespedeza*. Homogenization of parental polymorphism has been detected in our putative hybrid samples of *L. daurica*/*L. potaninii* and *L. virginica*. Different directions of homogenization are found between *L. daurica*/ *L. potaninii* and *L. virginica*. These findings of our study have significant implications in using ITS marker for reconstructing phylogeny and studying hybridization and polyploidy evolution; if not well recognized, ITS data could lead to incorrect tree inference and erroneous estimates of evolutionary relationships.

T2
P0690
The symbiotic relationship between *Cypripedium tibeticum* and its specific mycorrhiza Tulasnellaceae along altitudinal gradient
Lingling Xu
Xi'an University

Cypripedium tibeticum, which is an endemic and widespread perennial herb in China, belongs to the family Orchidaceae. *C. tibeticum* forms specific mycorrhizae with Tulasnellaceae. Due to habitat changes and human activities, *C. tibeticum* is threatened with extinction and is listed as national protected plant. In this study, roots of *C. tibeticum* along altitudinal gradient between 3270-3400 m were collected, and two method were used to obtain internal transcribed spacer (ITS) sequences of orchid mycorrhizal fungi. First method, isolated mycorrhizal fungi from the roots of *C. tibeticum*, extracted mycorrhizal DNA fragments and amplified ITS sequences with primer pair ITS5/ITS4. Second method, *C. tibeticum* roots' DNA were extracted and amplified potative orchid mycorrhizal ITS sequences using primer pair ITS1OF/ITS4OF. This primer pair has the advantage that it does not exclude *Tulasnella* species and thus should give an accurate view of orchid associations within the Basidiomycota, representing the vast majority of the mycorrhizae found on orchid species. A threshold of 97% similarity between ITS sequences was applied to circumscribe operational taxonomic units (OTUs) among the mycorrhizal taxa. To identify the different OTUs, representative sequences for each OTU were queried against GenBank by using BLAST. ITS phylogeny of the *C. tibeticum* mycobionts associated with the Tulasnellaceae family were constructed. Correlation analysis were used to reveal the effect of altitudinal gradient on the diversity of orchid mycorrhizal Tulasnellaceae. The result showed that 433 strains were isolated from roots of *C. tibeticum*, of which 107 (106 Tulasnellaceae and one Ceratobasidiaceae) strains were assigned to putatively orchid mycorrhizal OTUs. The 743 high-quality orchid mycorrhizal ITS sequences were obtained straight from roots' DNA of *C. tibeticum*, including 742 Tulasnellacea and one Sebacinaceae. Fifteen different OTUs were distinguished. Only 2 OTUs were shared between orchid mycorrhizal strain-isolating and straight amplifying ITS sequences method. Three OTUs of Tulasnellaceae were observed to show significant patterns of occurrence in relation to altitude. This research will elucidate the symbiotic relationship between *C. tibeticum* and the Tulasnellaceae at altitudinal gradient, and provide theory evidence for resource protection, propagation and sustainable utilization of *C. tibeticum*.

T2
P0691

Genome-wide convergence during evolution of mangroves from woody plants

Shaohua Xu[1], *Ziwen He*[1], *Zixiao Guo*[1], *Zhang Zhang*[1], *Gerald Wyckoff*[2], *Anthony Greenberg*[3], *Chung-I Wu*[1,4], *Suhua Shi*[1]
1. Sun Yat-sen University
2. University of Missouri-Kansas City
3. Bayesic Research
4. University of Chicago

When living organisms independently invade a new environment, the evolution of similar phenotypic traits is often observed. An interesting but contentious issue is whether the underlying molecular biology also converges in the new habitat. Independent invasions of tropical intertidal zones by woody plants, collectively referred to as mangrove trees, represent some dramatic examples. The high salinity, hypoxia and other stressors in the new habitat might have affected both genomic features and protein structures. Here, we developed a new method for detecting Convergence at Conservative Sites (CCS) and applied it to the genomic sequences of mangroves. In simulations, the CCS method drastically reduces random convergence at rapidly evolving sites as well as falsely-inferred convergence caused by the mis-inferences of the ancestral character. In mangrove genomes, we estimated ~400 genes that have experienced convergence over the background level of convergence in the non-mangrove relatives. The convergent genes are enriched in pathways related to stress response and embryo development, which could be important for mangroves' adaptation to the new habitat.

T2
P0692

Genetic variation and population structure of 'Zangli' pear landraces in Tibet revealed by SSR markers

Lei Xue, *Jun Wu*
Nanjing Agricultural University

Pear (*Pyrus* L.) is one of the most important fruits in temperate regions. Even in the mountainous border region of Tibet, Sichuan, and Yunnan provinces (China), a variety of highly adapted pear landraces were discovered in Tibetan Autonomous Prefecture and were named 'Zangli pears'. Here, we analyzed genetic diversity of 29 genotypes of Zangli distributed in Tibetan region, together with 30 genotypes of sand pear (*P. pyrifolia*) in border regions, and 8 genotypes of *P. ussuriensis* as out groups, using simple sequence repeats (SSR) markers. Twenty-eight SSR markers displayed high polymorphism with a total of 202 alleles. Cluster analysis revealed a close genetic relationship between Zangli and sand pears, while population structure analysis revealed that gene flow between the western and eastern side of the Jin-sha River was restricted by geographic isolation. Diversity statistics for five geographic groups implied a migration event from Sichuan and Yunnan provinces to the border region of western Sichuan, eastern Tibet, and northwestern Yunnan provinces. Six specific genetic alleles in Zangli pear landraces might be related to their adaption to severe environmental conditions.

T2
P0693

Genetic structure and population demography of widespread sea-dispersal legume *Vigna marina* in the Pacific

Takashi Yamamoto[1,2], *Yoshiaki Tsuda*[3], *Koji Takayama*[4], *Reiko Nagashima*[5], *Yoichi Tateishi*[2], *Tadashi Kajita*[2]
1. Kagoshima University
2. University of the Ryukyus
3. Tsukuba University
4. Museum of Natural and Environmental History, Shizuoka
5. Chiba University

Sea dispersal plants like mangrove plants produce specifically adapted diaspores (mainly seeds and fruits) to sea dispersal and can disperse them over the vast distance by ocean currents. As sea dispersal is often effective for long distance dispersal, many of the sea dispersal plants have extremely wide distribution range in tropics and subtropics worldwide. Distinct spatial genetic structures are common in many mangrove species across apparent geographic barriers (for example, *Bruguiera gymnorhiza* across Malay Peninsula and *Rhizophora mangle* across Central Panama Isthmus), and they are explained in the context of vicariance caused by land barriers that prevented migration of seeds by sea dispersal. However, some recent works on mangroves using genetic markers also reported clear genetic structure within oceanic regions where no apparent geographic barriers exist, which may imply the presence of "Cryptic Barriers". Although these genetic structures may promise to provide new insight into historical biogeography of mangroves, detail studies have not been well performed yet. In this study, we employed a widespread sea dispersal legume *Vigna marina* (Burm.) Merr. as a model species to test the

"Cryptic Barrier" hypothesis in the Pacific. This species distributes in IWP region and has wider distribution range than all other IWP mangroves. We conducted wide-scale phylogeographic analysis to understand detail patterns of population genetic structure using both cpDNA and nuclear microsatellite (SSR) markers for more than 300 individuals collected from 16 countries. Moreover, we employed Approximate Bayesian Computation (ABC) based approaches to infer past demographic history that formed the contemporary genetic structure in the focal region. *CpDNA* sequences showed high levels of genetic diversity and clear population genetic structure within the distribution range. We recognized clear genetic structure divided into three groups; North Pacific (NP), South Pacific (SP) and Indian Ocean (IO), strongly differentiated each other (F'CT=0.954 to 1.000). Although the genetic structure was less distinct for nuclear microsatellite, regional tendency of genetic diversity was similar to the one for cpDNA. These results supported the presence of the "Cryptic Barrier" within the Pacific Ocean. Additionally, ABC analysis estimated the divergence time between NP and SP at last glacial maximum (LGM), which indicates a strong influence of past climatic events on the population genetic structure of sea dispersal species including mangroves. These finding may provide new insight into biogeography and conservation of coastal plants including mangroves.

T2

P0694

Allopolyploid origin in *Rubus* (Rosaceae): Evidence from nuclear Granule-Bound Starch Synthase (*GBSSI*) sequences

Yan Wang[1,2], *Tao Chen*[1,2], *Jing Zhang*[1,2], *Qing Chen*[1], *Bo Sun*[1], *Haoru Tang*[1,2], *Xiaorong Wang**[1,2]

1. College of Horticulture, Sichuan Agricultural University

2. Institute of Pomology and Olericulture, Sichuan Agricultural University

Rubus L. is a large and taxonomically complex genus exhibiting apomixis, polyploidy and frequent hybridization. Most of Chinese *Rubus* are mainly concentrated in two major sections, *Idaeobatus* and *Malachobatus*. Our previous phylogenetics suggested that polyploid sect. *Malachobatus* probably originate from common ancestors, via polyploidization of hybrids between *R. pentagonus* ($2n = 2x = 14$) of sect. *Idaeobatus* and sect. *Cylactis* species. To confirm this hypothesis, we investigate the copy number and subparalog structure of *GBSS*I-1 gene within *Rubus* at different ploidy levels. For diploids, we found *GBSS*I-1α existed in species of subsections *Thyrsidaei*, *Idaeanthi*, *Pileati*, *Stimulantes*, *Pungentes*, and *Wushanenses*, while *GBSS*I-1β existed in sects. *Rosaefolii*, *Leucanthi*, and *Corchorifolii*, respectively. Both *GBSS*I-1α and *GBSS*I-1β were identified in diploid *R. pentagonus* and *R. peltatus*, as well as in majority of *Malachobatus* polyploids. *GBSS*I-1α has a longer intron 4 than that of *GBSS*I-1β. We further constructed the phylogenetic tree and phylogenetic network based on multiple *GBSS*I-1 copies. Based on these results, we confirmed that these sect. *Malachobatus* polyploids are allopolyploidy origin. They had experienced adaptive radiation, resulting species explosion in a short time. In addition, it was evidenced that *R. pentagonus* plays an important role in the speciation from diploids to polyploids in *Rubus*.

T2

P0695

Structural and developmental features of secretory canals in secondary xylem of *Sindora glabra* trunk

Jinchang Yang

Research Institute of Tropical Forestry, Chinese Academy of Forestry

Popularly known as diesel tree in China, *Sindora glabra*, a tropical species of Leguminosae, subfamily Caesalpinioideae, is characterized by the exudation of yellowish oleoresin or amber oil when wounded in the trunk, having a good prospect on producing essential oil, medicine and cosmetics. Despite the multipurpose use of this species, the knowledge on the oleoresin producing mechanism remains inadequate and the information on anatomical feature is very limited. The aim of this study was to describe structural and developmental features of secretory spaces present in the trunk of S. *glabra*, based on the usual techniques of wood anatomy. Three naturally distributed plants with the diameter of 28.6-32.7 cm were selected to study the origin, development and structural features of secretory canals in secondary xylem of *S.glabra* trunk, and variation of density, diameter, ratio of thin-walled and thick-walled epithelial cells in radial direction were analyzed. Only axial secretary canals were observed in the secondary xylem in the trunk, distributed along the marginal parenchyma bands which delimited growth layers and arranged in a tangential direction. They originated in the cambial zone from the initial parenchyma cells derived from fusiform initials or ray initials. The secretory canals were formed by schizolysigenesis and the process of development could be divided into five stages, namely initial, formative, expanding, mature and senescent stage. At the initial stage, 4-8 initials were arranged in a radial direction and had dense cytoplasm and clear visible nucleus. At the formative stage, the canal space appeared after the middle lamella of central cell walls degraded and the initials separated. At the expanding stage, the secretory lumen can be enlarged by the tangential extension and the anticlinal and periclinal division of the living epithelial cells. At the mature stage, secretory canals were constituted by a locally biseriate epithelium, lumen and sheath cells, and had 7-19 epithelial cells, measured 44.55 to 167.07 μm in tangential diameter; the epithelium composed of two-thirds of thin-walled cells and one-third of thick-walled cells. At the senescent stage, partial epithelial cells degraded in their walls, some cells filled with gum in their cavities, or expanded to form the tylosoids by blocking secretory lumen. Within the secondary xylem, the density of secretory canals increased from the close cambium towards the near pith in general, but the tangential diameter of secretory canals and the ratio of thin-walled cells showed the opposite trend. The average diameter and density of secretory canals were 58.83 μm and 6.73 per mm respectively, and the proportion of thin-walled epithelial cells at all developmental stages was above 50%. The initiation of secondary secretory canals and the fusion between two adjacent canals in the trunk resulted from the separation of the initial cells and the adjacent radial cells. The formation and development of secretory canals in secondary xylem gave an important hint to probe the mechanism of oleoresin secretion in the trunk of *S. glabra*, and the high percentage of epithelial cells with thin and unlignified walls might be of value in terms of sustainable exudation of the oleoresin.

T2

P0696

Origin and evolution of *Rubus takesimensis* (Rosaceae) in Ulleung Island, Korea

Jiyoung Yang[1], *Woong Lee*[1], *Jae-Hong Pak*[1], *Seung-Chul Kim*[2]
1. Research Institute for Ulleung-do & Dok-do, Kyungpook National University, 80 Daehak-ro, Bukgu, Daegu, 41566, Korea
2. Department of Biological Sciences, Sungkyunkwan University, 2066 Seobu-ro, Suwon, Gyeonggi-do 16419, Korea

On Ulleung Island, Korea, about 88% or even all endemic species (nearly 40) are considered to be anagenetically derived. However, we are still far from fully understanding the genetic consequences and anagenetic speciation in diverse endemic lineages of vascular plants on Ulleung Island. The family Rosaceae includes the most number of endemic species (i.e., *Cotoneaster wilsonii*, *Potentilla dickinsii* var. *glabrata*, *Prunus takesimensis*, *Rubus takesimensis*, and *Spiraea insularis*), yet the genetic consequences of them rarely exist, hampering generalization of anagenetic speciation in endemic species rich lineage. In this study, we have chosen one of endemic species *Rubus takesimensis* and compared pattern of genetic diversity and population structure with its continental progenitor species *R. crataegifolius*. We extensively sampled the progenitor-derivative pairs, *R. crataegifolius* (212 accessions from 42 populations in Korea, Japan, Russia and China) and *R. takesimensis* (113 accessions from 15 populations) and sequenced five noncoding regions of chloroplast DNA. We found that island endemic *R. takesimensis* was not formed monophyletic taxa and was deeply embedded within the continental progenitor *R. crataegifolius*, making it paraphyletic. In addition, several geographically diverse continental populations were responsible for the origin of *R. takesimensis*; majority of *R. takesimensis* accessions were sister to the clade containing accessions of *R. crataegifolius* primarily from the Korean peninsula, while rare accessions from Dodong, Jeodong and Seok-po shared their common ancestor with the ones from southern part of Korean peninsula, Jeju Island and Japan. There was also an evidence for few accessions from Chusan population originated seperately from the Korean peninsula. Of 129 haplotypes, we found 81 and 48 haplotypes in *R. crataegifolius* and *R. takesimensis*. There were no haplotype sharing between two species and most of haplotype were in low frequency. Higher haplotype diversity found in Korean peninsula populations than Japanese populations. Also, Chusan region that have been observed as bird habitat and Dodong and Jeodong region are located in near a port for passenger ships showed high genetic diversity. The AMOVA results showed very little genetic differentiation between the progenitor-derivative pairs and substantial genetic differentiation among populations in insular endemic *R. takesimensis*. Also, reduction of haplotype diversity found in *R. takesimensis*. Therefore, genetic consequences of *R. takesimensis* showed unusual case of anagenetic speciation that was expected patterns.

T2

P0697

A new genus of *Paralysimachia* F. Du, S. Yang & J. Wang gen. nov. and it's new species of *Paralysimachia xueshanensis*, *Primulaceae* in Yunnan Province, China

Fan Du, *Shaoyong Yang, Juan Wang*
Southwest Forestry University

A new genus *Paralysimachia* of *Primulaceae* has been identified in Yunnan Province, China. This is the first time to identify a new genus without corolla, most similar to genus *Lysimachia.* The new species of *Paralysimachia xueshanensis* are distributing at the alpine meadow steppe at the alt. of 3,390~3,420 m, on the top of the Big Snow Mountain of Bangdong county in Lincang city, within a habitat of 200-300 square meters area, it's anatomy and taxonomy has been described and illustrated. The differences between *Paralysimachia* and *Lysimachia have been compared and studied.* The rich plant species in Yunnan Province has produced many new species, and the isolated geographical ecology is the important factor to originate new species and genus, such as the new genus *Paralysimachia*, the new species *Paralysimachia xueshanensis* is distributing in a habitat of 200-300 square meters area.

T2

P0698

Genetic diversity and phylogeography of *Gymnocarpos przewalskii* (Caryophyllaceae) based on EST-SSR makers

Zhao-Ping Yang[1,2], *Ma Qing*[1], *Pan Li*[1], *Cheng-Xin Fu*[1]
1. Key Laboratory of Conservation Biology for Endangered Wildlife of the Ministry of Education, College of Life Sciences, Zhejiang University, Hangzhou
2. College of Life Science, Tarim University, Alaer

The genus *Gymnocarpos* (Caryophyllaceae) contains only ten species, ranging across arid and semi-arid regions of Africa and Asia, with the Southern Arabian Peninsula as the estimated ancestral area. *G. przewalskii* Maxim. is an endangered shrub species which is endemic to Northwest China. Due to lack of efficient molecular markers, the genetic diversity and historical demography of this endemic species is poorly understood. In this study, we performed high-throughput transcriptome sequencing of *G. Przewalskii* leaves using the Illumina HiSeqTM 2000 sequencing platform, and generated large transcript sequences for functional characterization and development of gene-associated SSR markers. A total of 92,039 unigenes (mean length 1057 bp) were assembled from about 113.5million reads, and 40,971 (44.51%) of them showed similarity to public databases. From the 92,039 unigenes, 11,752 expressed sequence tag–simple sequence repeats (EST-SSRs) were identified. Mono-nucleotide SSRs were the most abundant motif (46.3%), followed by tri- (29.7%) and di-nucleotide (19.1%) repeats. From the 11,752 EST-SSRs, 3611 primer pairs were designed for marker mining. Among them, 65 pairs were selected randomly for further validation, and twelve of which were identified as polymorphic SSR loci. Then, these twelve EST-SSR markers were applied to study the phylogeography of *G. przewalskii.* We sampled in total 353 individuals of *G. Przewalskii* from 24 populations, which covers the whole range of its distribution in China. The sampling sites were scattrd across the Tarim Basin (eight), the Hami Basin (seven), the Hexi Corridor (seven), Zhongwei (one) and the Urad Rear Banner (one). Nine to fifteen individuals were sampled for each population. Across the twelve

EST-SSR markers surveyed, the range was 2.3–8.4 for mean number of alleles (NA), 2.814–4.203 for mean allelic richness (RS), 0.430–0.612 for mean expected heterozygosity (HE) and 0.428–0.616 for genetic diversity (GD) of 24 populations. These diversity estimates were highest in Yumen of Ganshu and lowest in Luntai of Xinjiang. A Bayesian clustering analysis suggested that the populations of *G. przewalskii* were divided into two genetic clusters, the western and eastern lineages. The former contains all the populations from Tarim basin, and the latter contains the other populations. This is largely congruent with the previous findings revealed by cpDNA sequences. 78.38% of the variations found within populations, whereas, 8.83% are found among populations within linaeages, and only 12.29% are found between lineages. It is likely due to short divergence time of the lineage and long lifespan of shrub species. As a result, the EST-SSR markers developed in our research will be an informative resource for future studies on ecological, evolutionary, and conservation gentics of *G. Przewalskii,* as well as other relatd taxa. It will also help us to improve our understanding of the microevolutionary processes of this endangered species.

T2

P0699

Phylogeny and chromosomal evolution of the *Schoenus apogon* complex (Cyperaceae): Evidence for multiple polyploidization

Okihito Yano[1], *Yu Ito*[2], *Hiroshi Ikeda*[3], *Norio Tanaka*[4], *Jeremy Bruhl*[5], *Karen Wilson*[6]

1. Okayama University of Science
2. Xishuangbanna Tropical Botanical Garden
3. The University of Tokyo
4. National Museum of Nature and Science
5. University of New England
6. National Herbarium of New South Wales

A previous phylogenetic study of the genus *Schoenus* (Cyperaceae) discerned a species complex centred on *S. apogon*, comprising five species distributed in Australia, Japan, and New Zealand. Chromosome variations, 2n=8 from New Zealand and 2n=18 from Western Japan, has been observed in the complex, but its correlation with phylogeny has yet to be explored. Here we apply a molecular phylogenetic approach to explore whether and how polyploidy characterizes evolutionary lineages. Our taxon sampling covers three of the five species (*S. apogon*, *S. caespitans*, and *S. yarrabensis*) in the complex from the three biogeographic regions. Our chromosome observations using the sample set confirms and complements both aneuploidy and polyploidy (2n=8 from Australia, Japan (East), New Zealand; 2n=10 from Australia; 2n=16-18 from Japan (West); 2n=20 from Australia). Our analyses of chloroplast (*rbc*L and *trn*L-F) and nuclear (ITS) DNA data sets produced mostly congruent results, in which the 2n=10 cytotype was consistently placed sister to all the rest with strong support. The 2n=16-18 cytotypes from Western Japan and an unknown cytotype from Australia formed a monophyletic clade with moderate support. The morphology of the 2n=16-18 lineage matches that of *S. yarrabensis*. The 2n=20 cytotype from Australia is distantly related to the 2n=16-18 cytotypes in Western Japan, indicating multiple origins of polyploidization, for which different lineages of 2n=8 cytotypes have been involved.

T2

P0700

The making of elaborate petals through evolutionary innovation and developmental repatterning

Xu Yao[1,2], *Wengen Zhang*[1,2], *Xiaoshan Duan*[1,2], *Rui Zhang*[1], *Hongyan Shan*[1], *Hongzhi Kong*[1]

1. State Key Laboratory of Systematic and Evolutionary Botany, CAS Center for Excellence in Molecular Plant Sciences, Institute of Botany, Chinese Academy of Sciences
2. University of Chinese Academy of Sciences

As a special type of floral organ, petals show tremendous diversity in shape, structure, color/coloration, and function, and have increasingly been used as a model for studying plant organogenesis and evolution. Petals of *Nigella* (Ranunculaceae) species are of particular interest because they are highly elaborate, bilabiate structures with extensive modifications. To understand how *Nigella* petals become elaborated and diversified, we investigated the morphology, micro-morphology and development of petals in seven representative species (i.e., *N. integrifolia*, *N. nigellastrum*, *N. orientalis*, *N. damascena*, *N. sativa*, *N. hispanica*, and *N. arvensis*). We found that the degree of petal complexity increased gradually during evolution, involving both modification of existing characters (e.g., the elongation of stalk and the alteration of the upper and lower lips in shape) and *de novo* origination of new features (e.g., pseudonectaries, short trichomes, and conical cells). By investigating and comparing the processes of petal morphogenesis in these species, we also demonstrated the duration of important developmental events that led to the diversification of mature petals in gross morphology and micro-morphology. We found that developmental repatterning has played key roles in generating petal complexities during evolution.

T2

P0701

Discrepancy between chloroplast and multiple nuclear sequences signify multiple refugia in *Lindera aggregata* (Lauraceae) in Subtropical China

Jun-Wei Ye

Natural History Research Center of Shanghai Natural History Museum, Shanghai Science & Technology Museum, China

Subtropical evergreen broad-leaved forest (EBLF) is one of the most important vegetation types in East Asia. Palaeo-biome reconstruction (PBR) and predominant phylogeography studies were controversial on range shift history of the EBLF during Last Glacial Maximum (LGM). Recently, few studies on relatively lower altitude distributed plants (ca. < 1,500 m) using a combination of chloroplast DNA (cpDNA) and ecological niche modeling (ENM) show consistent pattern with the PBR. However, maternal inherited marker only is partial in exploring the evolutionary history. Thus, in this study, *Lindera aggregata*, a widespread low altitude distributed (200 - 1,000 m) dominant plant was selected, cpDNA is combined with multiple nuclear sequences to investigate whether it conform to contraction-expansion pattern or not. Nine haplotypes were identified in 131 individuals from 18 populations using four cpDNA fragments. The most recent common ancestor (MRCA) of all haplotypes was around the Pliocene-Pleistocene boundary estimated by second calibrations with four calibration points. No significant phylogeography structure ($N_{ST} = 0.860$, $G_{ST} = 0.833$, $P > 0.05$), widespread haplotype H3 at northern 24°N and star-like network suggest possible postglacial expansion. While, the signal were not detected by mismatch analysis and neutrality test. Private haplotype in Wuyi Mt. may be a result of *in situ* glacial persistence. Meanwhile, three distinct clusters were derived through Bayesian clustering using haplotype of 14 nuclear sequences genotyped by self-developed primers. South cluster located mainly at southern 24°N, two clusters at northern 24°N were roughly located at east (central cluster) and west (north cluster) side of Luoxiao Mt. *BEAST show two northern clusters have a common ancestor, this ancestral population and south cluster have a common ancestor. Central and north clusters, all three clusters were respectively diverged at 0.55 Ma and 0.76 Ma estimated by isolation by migration (IMa). Low migrations among the three clusters indicate long-term isolation. The Extended Bayesian Skyline Plot (EBSP) suggest both central and north cluster experience gradually expansion since 0.1 Ma. Thus, Despite possible northward postglacial expansion indicated by cpDNA, *L. aggregata* populations have multiple refugia and even experienced localized expansion at LGM (ca. 0.02 Ma). ENM of different models also show different patterns. Potential habitat predicted by Model for Interdisciplinary Research on Climate (MIROC) was almost the same as the present, while Community Climate System Model, version 4 (CCSM4) predict contraction to western Nanling Mt. and Wuling Mt. in central China in LGM. Consequently, low altitude distributed plants may also conform to multiple refugia pattern. More informative nuclear genome must be included in future study on controversy of range shift history of the EBLF.

T2

P0702

Gaolejeunea, a new genus of subtribe Echinolejeuneinae (Lejeuneaceae, Marchantiophyta) from China

Wen Ye[1], *Rui-Liang Zhu*[2]

1. South China Botanical Garden

2. East China Normal University

Over the past few decades, *Cheilolejeunea gaoi* has always been a peculiar member of Chinses Lejeuneaceae. A narrowly endemic species known only from the type locality in Shangsi Co., Guangxi, it has previously been treated as a species of *Cheilolejeunea*. However, the presence of Allorgella-type denticulations that formed by two adjacent cells on leaf and underleaf margins, the small, ovate leaf lobules without distinct teeth, the ental hyaline papilla, and the finely segmented oil bodies, all suggest somewhere else. Maximum likelihood, maximum parsimony, and Bayesian analyses of a three marker dataset (nrITS, cpDNA *rbc*L and *trn*L-*trn*F) derived from 85 accessions of Lejeuneeae and 15 outgroup species revealed *C. gaoi* belongs to Echinolejeuneinae. Based on the results of phylogenetic analyses, with combination of morphological data, we suggest here that *C. gaoi* should be established as a new genus, *Gaolejeunea*.

T2

P0703

Diversification of alpine bamboos since the late Miocene

Xiaying Ye[1,2], *Guoqian Yang*[1,2], *Pengfei Ma*[1], *Dezhu Li*[1]

1. Germplasm Bank of Wild Species, Kunming Institute of Botany, Chinese Academy of Sciences

2. University of Chinese Academy of Sciences

Fargesia and *Yushania* (alpine bamboos), two genera of Arundinarieae which mainly distributed at high elevation of the Hengduan Mt, are the major dominant species of the cloud forest and have ecological importance. However, due to the extremely low level of sequence divergence in plastid genome and low-copy nuclear regions, as well as less sampling, the relationship between species of this group have been proven difficult to elucidate. Here, we adopted ddRAD-seq method, which is a powerful tool for non-model organisms to compare interspecific relationship, to illustrate the phylogenetic relationship for alpine bamboos with widely sampling. We also clarified the cause of complex phylogeny of alpine bamboos, dated the divergence time of major clades for Arundinarieae, particularly the alpine lineages, and explored the correlation between geological events and the diversification of alpine bamboos. Both concatenation (maximum-likelihood) and coalescent (SVDquartets) methods reconstructed a robust topology for alpine bamboos, treated *Yushania* as a monophyletic lineage and *Fargesia* as polyphyletic groups. In this phylogeny, we total received eleven clades and six were new established in our study. The new clades were named after the lineages of plastid topology, including V-Y clade comprising of the *Yushania* genus, four parallel clades V-F1, V-F2, V-F3 and V-F4 which separated from *Fargesia*, and alpine *Bashania* clade (V-S). For *Yushania*, interspecific relationship was basically solved, species with numerous branches per node and high culm grouped at the basal, which is identical to the section division of traditional classification. *Y. crispata*, *Y. violascens*, *Y. puberula* formed a strong support clade and sister to the remainder of *Yushania*. The phylogenetic relationship within and between four clades of *Fargesia* all exhibited substantial bootstrap support, and the distributions for four parallel clades have a trend from low elevation to high elevation in the Hengduan Mt, but no correlation to the traditional classification. According to network and genetic clustering analyses, gene exchange was happened frequent at the initial divergence of alpine bamboos and its relatives, especially between V-F3 clade and the basal group of *Yushania* which had minor difference when altered reconstruction methods. Meantime, alpine bamboos is a recent origin group, its stem age was dated to the late Miocene, c. 7.18 Ma, and experienced a rapid divergence. The crown age of four clades of *Fargesia* – V-F1, V-F2, V-F3 and V-F4 – was 4.01 Ma, 2.8 Ma, 2.63 Ma and 2.73 Ma, respectively. *Yushania* separated from *Fargesia* at 3.31 Ma and diverged into the species-rich subclades in the Pliocene (3.13-0.89 Ma). Owning to recent origin, insufficient information accumulation and gene exchange may be responsible for the intricate phylogenetic relationship of alpine bamboos. Additionally, diversification analyses demonstrated two speciation rate shifts happened along the branch, one shift occurred about 7-8 Ma when the alpine bamboos just origin initially, another one happened at 2-4 Ma which was coincident with the time of alpine bamboos spreading into higher mountains. All the two shift events are related to the putative uplift phase of the Hengduan Mt and indicate mountain uplift trigger the diversification of the alpine bamboos.

T2

P0704

An owerview of the genus *Nonea* in Turkey

Yeter Yeşil

Istanbul University

The genus *Nonea* Medik. (Boraginaceae tribe Boragineae Bercht. & J. Presl) includes about 36 taxa. The distribution of the genus ranges from central Asia to the Iberian Peninsula and Morocco. Its main centre of diversity stretches from the Pontic Caucasian mountain system to the Anatolian and Irano-Turanian highlands, while fewer species occur in Europe and in the Mediterranean, especially along the North African coast (Selvi & Bigazzi, 2002; Selvi *et al.*, 2006). Turkey is the richest country, with approximately 23 species representing more than 60% of the whole genus Baytop 1978; Selvi & Bigazzi, 1999; 2001; 2002; Bigazzi *et al.*, 2004; Bilgili *et al.*, 2012; K.rüklü, 2012. In this study morpholgy of Turkish *Nonea* species were examined using light and scanning electron microscopy. The most important characteristics of *Nonea* species such as nutlet shapes and surfaces, and corolla colour and shapes were investigated. Nutlet shapes are reniform and transversly reniform, obliquely erect or transversely ovoid; corolla colour white, yellow, pink, orange-red to violet or dark purple or variegated; corolla shapes are actinomorphic, zigomorphic, infundibular or irregularly infundibular. An identification key is provided according to morphology, especially nutlet and flower characteristics to better evaluate the similarities and relationships between taxa.

T2

P0705

An investigation on seed micromorphology of some *Aethionema* R.Br. species (Brassicaceae) grown in Turkey by Scanning Elektron Microscope

Gulden Yilmaz[1], *Mehmet Tekin*[2]

1. Trakya University, Faculty of Science, Department of Biology

2. Cumhuriyet University, Faculty of Pharmacy, Depaertment of Pharmaceutical Botany

Aethionema R.Br. (Brassicaceae) is a taxonomically complex genus and a few macromorphological characters are available for species delimitation. The Turkish flora comprises of about 41 *Aethionema* R.Br. species, of which 20 species are endemic to Turkey. The aims of the present study are to illustrate the utility of derivated seed morphology features in the identification of most of the species considered, and to relate such characters to the systematics of the genus. So, testa micromorphology of *Aethionema membranaceum* (Desv.) DC., *Aethionema schistosum* Boiss. & Kotschy and *Aethionema oppositifolium* (Pers.) Hedge grown around Middle Anatolia in Turkey were given for the first time with this study. Seeds were examined under Zeiss ESEM and SEM micrographs were taken. The results were compared with those from previous studies of different species of the same genus. Micromorphological characters may provide additional taxonomic information in the genus. The importance of ultrastructure of seed surface, as a reliable approach for solving taxonomic problems has been well recognised.

T2

P0706

Plastid genomes revealed three lineages of *Ginkgo biloba* diverged in the Mid- to Late-Pleistocene

Wen-Bin Zhou[1], *Ping-Ping Yin*[1], *Chun-Xiang Xie*[1], ***Yun-Peng Zhao***[1], *Marcus Koch*[2], *Cheng-Xin Fu*[1]

1. The Key Laboratory of Conservation Biology for Endangered Wildlife of the Ministry of Education, College of Life Sciences, Zhejiang University

2. Department of Biodiversity and Plant Systematics, Centre for Organismal Studies, University of Heidelberg

Two refugia of *Ginkgo biloba* in China were confirmed by our previous phylogeographic studies using AFLP, nuclear microsatellites and three cpDNA fragments. However, the phylogenetic relationship between the common haplotype occurring across its range and the rest ones, as well as the timing of their divergence, was not resolved due to insufficient informative loci. In the present research, we addressed these two questions using plastid genome data of ten representative cpDNA haplotypes of *G. biloba*. The phylogeny recovered three strongly-supported clades, the widespread haplotype clade (H1 clade), east-specific haplotype clade (E clade) and west-specific haplotype clade (W clade), of which the latter two are sisters. Timing inference in a context of gymnosperm phylogeny with ten fossil calibrations suggested the H1 clade initially diverged at 0.603 mya (95% HPD: 0.043~0.809 mya), followed by the divergence between E and W clades at 0.286 mya (95% HPD: 0.016~0.752 mya). Our results provide a well-estimated temporal hypothesis to be tested using co-occurring species inhabiting warm temperate forests in eastern Asia, and emphasizing the importance of both application of plastid genome data and fully-resolved phylogeny of all recovered genetic lineages to timing estimation.

T2

P0707

Topography and climate explain bias of plant diversity in the genera disjunctly distributed in eastern Asia and eastern North America

Xue Yin

Sun Yat-sen University

Disjunct distributions of the same plant clades (e.g. genera) and biogeographic bias in different continents are critical to understand the ecological relationship between plants and environment such as topography and climate. A long-recognized biogeographic mystery is the disjunct plant genera between eastern Asia (EAS) and eastern North America (ENA), first observed by Carl Linnaeus and has puzzled generations of botanists. Although it has been argued that differences in physiography and climate between the two regions are the primary causes of bias in species diversity in these disjunct genera, this hypothesis has not been tested. We compiled data of topography, contemporary and historical climate to quantitatively assess contributions of these factors to the Asian bias of EAS-ENA disjunct plant diversity. We showed that contemporary climate explained 25% of the spatial variation in diversity of the EAS-ENA disjunct genera, 4% more than the variation explained by topography (21%). In contrast, on the species level, topography explained 21% of the variation in species richness while climate only explained 8%. Historical climate (Last Glacial Maximum climate) was highly correlated with current climate and showed no significant effects on diversity at both genus and species levels. We further found that topography accounted for about 29% of the variation in species richness in topographically more heterogeneous EAS but only 5% of that in ENA. We concluded the diversity patterns of EAS-ENA disjunct genera, to a large extent, could be a product of topographic heterogeneity and contemporary climate, although the importance of each factor depends on taxonomic level and region.

T2
P0708
The contribution of Neogene plateau uplift and Quaternary climate fluctuations to endemic plant diversity of the Tibetan Plateau

Haibin Yu
Sun Yat-sen University

Aim To test the effects of Neogene plateau uplift and Quaternary climate oscillations on the formation of extant species and phylogenetic diversity. Location Tibetan Plateau (TP). Methods We collated data on endemic species distributions based on county-level mapping from several monographs and online databases. We calculated both species richness and phylogenetic diversity for all species, herbs, shrubs and trees from these data for a 0.5 × 0.5-degree grid covering the TP. We posed eight biogeographic hypotheses associated with plateau uplift or climate fluctuations and quantified. We used partial regression analysis and mixed conditional autoregression (CAR) model to assess the relative contribution of each hypothesis to species diversity. Results Endemic species diversity is concentrated in the southeast part of the plateau and declined from the southeast to northwest. Hypotheses based on plateau uplift independently explained three-times more variance in diversity than Quaternary climate changes. The full CAR model containing all hypotheses explained 35 – 75% of the total variation in diversity, while each predictor explained at least 5% on average, and the montane museum hypothesis explained the most variation averaged across diversity measures and life forms (17%). Main conclusions Extant endemic seed plants diversity is concentrated in the Hengduan and East Himalayan mountains. Both the Neogene TP uplift and Quaternary climate fluctuations had large impacts on current patterns of species diversity, but the influence of plateau uplift seemed to be more important than climate changes. Our study provides insights for explaining the diversity patterns of alpine species and has implications for the conservation of montane species under the context of future climate changes.

T2
P0709
Evidence for the involvement of *TEOSINTE BRANCHED1*-Like Genes in the asymmetric floral development of *Canna indica* L.

Qianxia Yu[1,2], *Xueyi Tian*[1,2], *Jingping Liao*[1]
1. South China Botanical Garden
2. University of Chinese Academy of Sciences

Asymmetric flower which lacks any symmetry plane is rare across angiosperm. *Canna indica* (Cannaceae, Zingiberales) has conspicuously asymmetric flowers. In its flower, all the whorls contribute to the floral irregularity, especially the irregular androecial petaloidy which makes the flower more elaborate. In the androecium, a single half of the stamen primordium develops to the fertile anther, while the rest abort or develop to petaloid structures. The *CYCLOIDEA/TEOSINTE BRANCHED1* (*CYC/TB1*)-Like genes have long been identified as the main factor to control floral symmetry in numerous lineages. They have functions both in aborting stamen and promoting petal growth. In order to test their possible roles in regulating floral symmetry in this species, we analysed the spatio-temporal expression patterns of them. We identified three homologs of *TB1*-Like (*TBL*) genes and named them *CiTBL-1a*, *CiTBL-1b* and *CiTBL-2* according to phylogenetic analysis. RNA *in situ* hybridization analysis showes that *CiTBL-1a* and *CiTBL-1b* have identical expression patterns. In the young inflorescence, both *CiTBL-1*s are expressed in the bracts, the bottom of the floral primordia and the cincinnus primordia. While in the developing flower, their expressions are restricted to the outer two wholes--sepals and petals. In contrast, *CiTBL-2* transcripts are mainly found in the floral primordia and the inflorescence primordia. During floral development, the expression become stronger in the androecium and gynoecium. The expression regions of *CiTBL-2* are exactly complementary to those of *CiTBL-1*s. The expression data indicates that compared with *CiTBL-1*s, *CiTBL-2* possibly plays a more important role in the floral development of *C. indica*. In addition, we have observed that the flower is bilateral symmetric before the differentiation of the petal-stamen common primordia, and the axis of symmetry just passes through the second floral primordia of the cincinnus. This phenomenon indicates that the inflorescence architecture may have impact on the floral symmetry of *C. indica*. However, the expression patterns of *CiTBL*s almost remain unchanged during the process of symmetry shift from bilateral symmetry to asymmetry. Further explorations of their function are needed to explain the irregular floral patterning and origin of androecial petaloidy.

T2
P0710
Diversity and ecology of bryophytes in the headwaters region of Urumqi River, Xinjiang, China

Jia-Xing Guo, Wei Sheng, ***Yu-Huan Wu***
Hangzhou Normal University

The headwaters of the Urumqi River is located in the central Tianshan Mountains of Xinjiang, and the elevations of the glacier terminuses are 3,650-3,700 m. Based on the field investigation and specimen collection, 148 species in 70 genera and 32 families of bryophytes are reported. Among them, *Bryum stirtonii* Schimp. and *Jungermannia japonica* Amak. are new to China.*Platydictya subtilis* (Hedw.) Crum, *Anomobryum yasudae* Broth., *Didymodon acutus* (Brid.) Saito and *Dumortiera hirsuta* (Sw.) Reinw. are new to Xinjiang province. One liverwort and 25 arcocarpous mosses are found in the foreland of the reatreating Tianshan No.1 Glacier since since the end of the Little Ice Age. The results supported the hypothesis that arcocarpous mosses are more adapted to harsh environments than pleurocarpous mosses. The way of sexual reproduction is employed by most mosses instead of asexual reproduction. The breed system of mosses in glacier foreland since the end of the Little Ice Age are mainly dioecious (69.23%), much more than those of monoecious (26.92%) and polyoicious (3.85%) indicating the way of unisexual reproductive may be more advantage in extreme climate. Canonical Correspondence Analysis (CCA) was also applied to study the relationships between the bryophytes and their environmental factors. The results showed that the influence of herb coverage, altitude, soil humidity, and tree canopy are the most important factors in distribution pattern of bryophytes.

T2
P0711

Relationships of *Atraphaxis*, *Persepolium* and *Bactria* (*Polygoneae, Polygonaceae*) based on molecular and morphological data, and plastid phylogeny of *Atraphaxis*

Olga V. Yurtseva[1], *Evgeny V. Mavrodiev*[2]
1. M.V. Lomonosov Moscow State University
2. Florida Museum of Natural History

An updated molecular phylogeny of Polygoneae was obtained with Maximum Likelihood (ML) and Bayesian analyses (BI) applied for the DNA sequence data (ITS1&2 regions of nrDNA and combined *trnL* intron $^{(UAA)}$ + *trn*L-F IGS and *rpl*32–*trn*L $^{(UAG)}$ IGS regions of plastid genome). *Knorringia* is an isolated lineage sister to the rest Polygoneae. The clade including *Fallopia, Muehlenbeckia*, and *Reynoutria* is sister to the clade including *Atraphaxis, Duma, Polygonum*, and *Polygonella*. *Duma* is grouped with *Polygonum* and *Polygonella* in plastid topology, and is a separate lineage in ITS-based topology. The clade of *Polygonum* and *Polygonella* is sister to *Atraphaxis* s.l. in which a new genus *Persepolium* (=*Polygonum sect. Avicularia subsect. Spinescentia*) was established forming the clade immediately sister to *Atraphaxis* s.s. Based on molecular and morphological data, a new genus *Bactria* was described as a sister to the clade (*Atraphaxis* s.s. plus *Persepolium*). *Bactria ovczinnikovii* from Pamir and *B. lazkovii* from Tien Shan are grouped in plastid topology, but demonstrate paraphyly in ITS-based topology. The intra-individual ITS1-2 polymorphism and strongly different secondary structure of ITS1 detected in *B. lazkovii* are arguing for the hybrid origin from an unknown *Bactria* species and an unknown donor (donors) of different ITS variants, one of which demonstrates partial pseudogenization. Striking intra-individual polymorphism of sporoderm surface detected in *B. lazkovii* agrees with this suggestion. The standard maximum parsimony (MP) analysis and three-taxon statement analysis (3TA) were applied to the comprehensive morphological dataset of 22 accessions of *Bactria, Persepolium,* and *Atraphaxis*. s.s. In total, 27 characters, including the life history, habitus, morphology of the shoots, leaf blades, ochreas, perianth, achene, the ultrasculpture of perianth and achene surfaces, as well as the pollen morphology were analyzed. All the characters were conceptualized with a certain number of states and treated as unordered. The obtained patterns of the relationships are congruent to the ITS-based topologies resulted from the molecular treatments of the same taxonomical complex. MP reconstructions of perianth and pollen morphology were performed for Polygoneae using topologies that resulted from ML analyses of molecular data sets. MP reconstructions of selected morphological traits were performed for the clade including *Bactria*, *Persepolium*, and *Atraphaxis* s.s., using topologies that resulted from MP analysis of morphological data set. The campanulate perianth of 5 equal-sized segments and microreticulato-perforate sporoderm surface detected in *Bactria* and *Persepolium* are symplesiomorphies. The perianth with accresent inner segments and long filiform tube is an apomorphy appeared in *Atraphaxis* s.s. from the campanulate perianth. Striato-perforate sporoderm surface shared by *Atraphaxis* and *Persepolium dumosum* is treated as a homoplasy. Preliminary molecular phylogeny of *Atraphaxis* was built with ML and BI applied for 65 tips combined plastid matrix based on 3-loci (cpDNA *trn*L $^{(UAA)}$ intron, *trn*L–*trn*F IGS, and *rpl*32–*trn*L $^{(UAG)}$ IGS regions). The section *Atraphaxis* with dimeric flowers forms a distal subclade, the members of sections *Tragopyrum, Physopyrum* with specialized pentamerous perianth and the species with campanulate perianth of 5 equal-sized segments form no subclades. The ITS-based topology is incongruent to combined plastid phylogeny due to probable allopolyploid origin of many species.

T2
P0712

Evolution of Perianth morphology and sporoderm ornamentation in and *Atraphaxis*, *Persepolium* and *Bactria* (*Polygoneae, Polygonaceae*)

Olga V. Yurtseva
M.V. Lomonosov Moscow State University

Atraphaxis, Bactria, and *Persepolium* are close related in phylogenetic reconstructions based on molecular and morphological data. *Atraphaxis* is sister to *Persepolium*, their common clade is sister to *Bactria,* what let to trace evolution of their morphological features. *Bactria* combines two rare endemic shrubs from Pamir and Tian-Shan. They have terminal frondulose thyrses of 5-10 spaced cymes of 2-3 flowers. The campanulate perianth has five equal segments covered adaxially and abaxially with elongate-rectangular sinuate epidermal cells with random cuticular striation. Eight filaments are deeply inserted in receptacle lined with nectar-secreating tissue. These characteristics rather indicate entomophily. *Persepolium* combines five dwarf caespitose undershrubs or shrubs endemic for SW Iran. They have terminal frondose or frondulose thyrses of 10-20 well spaced cymes of 2-3 flowers, urceolate or campanulate perianth divided in 5 equal coriaceous segments densely velutino puberulent. The outer perianth surface is covered with elongate-rectangular epidermal cells forming short trichomes, alternating with stomata. Nectar-secreating tissue lines the receptacle and the bases of filaments inserted deeply in receptacle. These features are usual for self-pollinators. *Atraphaxis* numbered ca 35 shrubs or undershrubs is distributed from SE Europe to South Siberia and China. Atraphaxis has compact bracteose thyrses of 6-20 congested cymes of flowers and diverse perianth which is 1) campanulate with 5 equal segments and short funnel-form tube, 2) specialized with long filiform tube, two or three accrescent inner segments and two outer segment; 3) specialized with short filiform tube, three inner segments spherically surrounding the ovary and two reflected ones. Due to the campanulate perianth with five equal segments and short tube is shared by some *Atraphaxis* species, *Bactria* and *Persepolium,* it can be treated as initial to the other types of perianths in *Atraphaxis*. Dimeric flowers can be treated as the last novelty appeared in *A.* section *Atraphaxis* which formed distal subclade in phylogenetic reconstructions. Showy petaloid perianth of *Atraphaxis* has white or brightly colored segments covered outside and inside with rectangular-elongated cells sinuate in outline, with random cuticular striation. The expanded part of tube is covered with polygonal cells with random cuticular striation. The filiform basal part joint to long pedicel serves for better exhibition of the perianth in compact thyrse. Nectar-secreting tissue lines short receptacle and the inner surface of the filaments, which are inserted at the edge of receptacle to ease exposure of pollen. This morphological complex was formed in Atraphaxis as an adaptation to entomophyly. Pollen grains in all the taxa are prolate to subprolate, spheroidal to oblong-spheroidal, tricolporate, elliptical in equatorial view, trilobed-circular in polar view. *Bactria* has microreticulato-foveolate or foveolate-perforate sporoderm surface. *Persepolium* has foveolate, foveolato-perforate to microreticulato-foveolate sporoderm surface, excepting *P. dumosum* which has striato-perforate sporoderm surface. *Atraphaxis* has striato-perforate sporoderm surface originated from reticulate-perforate surface present in *A. toktogulica* by arranging of lumina in rows, rising of lateral edges of lumina, and emerging of grooves with rows of perforations. Striato-perforate sporoderm surface possibly reduces the weight of pollen and facilitates transfer by pollinators.

T2

P0713

Plastid phylogenomics and Ranunculaceae diversification after K-P boundary

Wei Zhai[1], *Xiaoshan Duan*[2], *Rui Zhang*[2], *Chunce Guo*[2], *Lin Li*[2], *Hongzhi Kong*[2], *Yi Ren*[1]

1. College of life sciences, Shaanxi normal University

2. Institute of Botany, CAS

Ranunculaceae is a basal eudicotyledonous family with about 53 genera and 2,500 species. The family has been the focus of systematic studies for centuries for its unstable position in angiosperm phylogeny and its tremendous diversity in vegetative and reproductive structures. Recently, the family is increasingly being used as a model to answer the important evolutionary questions that cannot be easily addressed by using any other existing models species. Yet, because of the uncertainties in the phylogenetic relationships among genera, it has been difficult to infer the evolutionary histories of many characters. Recent molecular phylogenetic studies have shown that the Ranunculaceae, as a monophyletic group, included Glaucidium, Hydrastis and the core Ranunculaceae. *Glaucidium* was the basal most in Ranunculaceae, followed by *Hydrastis* as sister to the core Ranunculaceae. Within the core Ranunculaceae, the Coptideae was the first diverged clade, and the remaining genera were divided into 11 well supported lineages/tribes, namely Isopyreae, Adonideae, Delphinieae, Nigelleae, Helleboreae, Cimicifugeae, Caltheae, Asteropyreae, Callianthemeae, Anemoneae and Ranunculeae. However, the relationships among the 11 clades were unresolved. To figure out the phylogenetic relationship of Ranunculaceae, 81 coding genes (77 protein-coding and 4 RNA-coding sequences) from 37 taxa of this family were used to reconstruct the phylogenetic trees. In the ML (maximum likelihood) tree, all of the relationships were strongly supported, including several internal nodes which are previously weakly supported in different topologies of the family. Glaucidium was resolved as the basal most lineage of the family, followed by Hydrastis and Coptis (Coptideae), confirming several other studies. The remaining genera form 11 strongly supported lineages, each of which corresponds to the tribes recognized by previous studies. Unlike previous studies, the relationships among the 11 lineages, however, are different. Specifically, tribe Adonideae appears to be sister to tribe Isopyreae, and they together are strongly supported as sister to a group containing the remaining 9 tribes. Of the remaining 9 tribes, Ranunculeae is the closest sister of Anemoneae, followed by Helleboreae, Callianthemeae and Cimicifugeae, whereas Asteropyreae, Caltheae and Nigelleae are successive sister of Delphineae. The molecular dating results indicated that the origin of Ranunculaceae is 105.2 Mya (HPD=96.2-116.1); the divergence times for the Glaucidium, Hydrastis, and Coptideae are estimated to be at 89.3 Mya (HPD=75.8-103.4), 83.5 Mya (HPD=69.8-96.3) and 72.6 Mya (HPD=59.8-85.1), respectively. The remaining 11 lineages of core Ranunculaceae began to diversify around the K-P boundary (66.3 Mya; HPD=54.1-78.2), and each ancestor of these 11 lineages have been originated within 15 million years (65-50 Mya). These implied that the extant 11 lineages may split out rapidly within a short period after the K-P boundary. The LTT (lineage-through-time) plot analysis also highlighted a rapid radiation after the K-P boundary, at which the diversification rate has increased substantially.

T2

P0714

Implications of speciation as demonstrated by *Magnolia* hybrids

Kevin Parris[1], ***Donglin Zhang***[2]

1. Department of Plant and Environmental Sciences, Clemson University

2. Department of Horticulture, University of Georgia

Controlled crosses *Magnolia sieboldii* K. Koch 'Colossus' (2n=2x=38) × *M. grandiflora* L. 'Kay Parris' (2n = 6x = 114), *M. foveolata* (Merr. Ex Dandy) Figlar 'Shibamichi' (2n=2x=38) × *M. laevifolia* (Law & Y.F. Wu) Noot. 'Gail's Favorite' (2n=2x=38), and *M. sieboldii* 'Colossus' × *M. insignis* Wall. 'Anita Figlar' (2n=2x=38) were performed in 2013, 2010, and 2014, respectively. Genetic variations and assessed inheritance of parents and F1 sibling populations were analyzed using ISSR markers. A total of 35 siblings (9, 11, and 15, respectively), 6 parents, and 5 comparator males were included in the molecular work. 79.3% of bands were polymorphic in the first cross. UPGMA tree revealed that all siblings were closer to their hexaploid pollen parent (*M. grandiflora*) and more distant from their diploid parent (*M. sieboldii*), supporting foliage morphology observations. In the second cross, phenotypic differences in foliage and flower morphology among the siblings were apparent by the second year. Branch lengths in the UPGMA tree revealed greater separation among the siblings and less separation from their Section Michelia parents. In the third cross (intersectional) that was analyzed, phenotypic differences were not yet apparent, but UPGMA tree indicated branch lengths with less separation among the siblings and greater separation from parent species than the previous cross (intrasectional). The implication of speciation from these hybrids was discussed in details with both morphological traits and molecular data. Further studies should include selection of the best individuals for future controlled crosses and determination of their reproductive fecundity.

T2

P0715

Pollinator shift drives floral divergence and reproductive isolation in two close related species of *Cimicifuga* (Ranunculaceae)

Jianqiang Zhang, *Yi Ren*
Shaanxi Normal University

Traditionally the great diversification of animal-pollinated angiosperms has been proposed to be promoted by divergence in pollinators. However, the underlying evolutionary and ecological processes are still poorly understood, in part because most studies have focused on floral diversification and pollination relationships in a phylogenetic context or pollinator-mediated phenotypic selection within populations. Here we report a intriguing case of pollinator shift driving floral divergence and speciation in two close related species *Cimicifuga japonica* and *C. purpurea*. Both species belong to plant family Ranunculaceae, and the former species possess white and radial symmetry with caducous sepals and nectariferous, while the flowers of *C. purpurea* form a monosymmetric feature, the sepals are purple, chartaceous, erect and lately deciduous, the petals (nectary leaf) are yellow, erect and hidden inside of the sepals to hide the nectar in the flower, the stamens have purple filaments and yellow anthers. Through field observations, flower morphological measure and experimental analysis, we found an obvious pollinator shift between the two species. For *C. japonica*, pollen foraging flies are main visitors, among which some large sized flies can touch the anthers and stigma when crawling and collecting pollens on the flowers. Among the visitors of *C. purpurea*, *Vespula rufa rufa* (Linnaeus), the red wasp, has obvious higher visiting frequency and pollen carrying efficiency, thus is the main pollinator for the species. Interestingly, morphological analysis showed that the flower morphology are exactly fitted for *Vespula rufa rufa* pollinator: the purple sepals and yellow anthers of *C. purpurea* color could be easily recognized by the red wasp comparing to *C. japonica*'s white ones. In addition, the purple-yellow color of the inflorescence of *C. purpurea* is similar to the yellow and red-brown stripes of red wasp that might avoid the red wasp to be found and attacked by other predators when sucking the nectar. Besides, the red wasp has larger size and powerful, chewing mouthpart with developed and hard palate can open the wrapped sepals, and its hard palate length with close petal can get the nectar reward. The natural distribution area of *C. japonica* and *C. purpurea* are largely allopatry, which provides alternative drivers for the floral divergence of the two species. Using cpDNA *trn*L-F and *trn*S-G spacers, we detected very low gene flow between the two species. Molecular dating analysis suggested that the two species diverged at ca. 0.3 Mya (Million years ago). Such a shallow divergence should be characterized by extensive gene flow and incomplete lineage sorting, yet not seen in the case of *C. japonica* and *C. purpurea,* suggesting strong selective pressure acting on the divergence of the two species. We thus represent this intriguing case of pollinator shift-driven speciation, and further studies on process and underlying mechanism of this process would greatly strengthen our understanding of pollinator-driven speciation in angiosperms.

T2

P0716

Floral symmetry genes and the development of zygomorphy in Solanaceae

Jingbo Zhang, *Ghadeer Bukhari, Wenheng Zhang*
Department of Biology, Virginia Commonwealth University

In core eudicots, the origins of floral zygomorphy from ancestral actinomorphy in several lineages are always associated with the repeated deployment of *CYCLOIDEA2*-like (*CYC2*-like) genes, which belong to the *CYC/TB1* clade of the TCP gene family. We have sought to characterize the genetic basis of the independent origin of floral zygomorphy in Solanaceae. To this end, we cloned the *CYC2* genes from Solanaceae and Convolvulaceae, and reconstructed the gene phylogeny for these sequences. The differential expressions of the *CYC2* paralogs were examined by qRT-PCR and RNA *in situ* hybridization in plants with three types of floral symmetry: actinomorphic flowers (in *Solanum lycopersicum* L.), zygomorphy in androecium and calyx (in *Nicotiana obtusifolia* Martens & Galeotti), and strong zygomorphy in androecium, corolla, and calyx (in two species of *Schizanthus* Ruiz & Pav.). Differential gene expression was further tested in *Schizanthus* by using RNA-seq. We identified two major sequence types of *CYC2* genes, *CYC2-I* and *CYC2-II,* which resulted from a gene duplication in the common ancestor of Solanaceae. Furthermore, a genus-specific gene duplication identified in *Schizanthus*, which represents the most basal clade of the family, produced two *CYC2-I* paralogs, *scCYC2-Ia* and *scCYC2-Ib*. The expression of *CYC2* genes was detected at both early and late stages during floral development in the tested species. In particular, in *So. lycopersicum, CYC2* genes were broadly expressed early on across the floral meristem. During floral organ initiation, the expression of *soCYC2-I* was concentrated in stamens and ovules, while the expression of *soCYC2-II* was concentrated in sepals. A similar pattern of gene expression was observed in *N. obtusifolia* except for the expression of *niCYC2-II,* which exhibits an asymmetric pattern correlated with its zygomorphic calyx. In *Sc. grahamii* Gill. ex Hooker, the expression of *scCYC2* was detected mainly in zygomorphic androecium and calyx. Specifically, *scCYC2-Ia* was found strongly expressed in the aborted dorsal and ventral stamens, while *scCYC2-Ib* was expressed complementarily in the functional stamens. *scCYC2-II* showed a dorsoventral asymmetrical expression pattern in the zygomorphic calyx. Furthermore, the RNA-seq results obtained with *Sc. pinnatus* Ruiz & Pav. indicated no expression of *CYC2* genes in corolla. Therefore, our results suggest that *CYC2* genes are likely involved in the development of floral zygomorphy in androecium and calyx, but genes other than *CYC2* are involved in the development of zygomorphic corolla in Solanaceae.

T2
P0717

A taxonomic revision of *Radula* in china based on molecular and morphological evidence

Li-Na Zhang[1], *Rui-Liang Zhu*[2]
1. Institute of Tropical Agriculture and Forestry, Hainan University, 58 Renmin Road, Haikou 570228, China
2. School of Life Sciences, East China Normal University, 500 Dongchuan Road, Shanghai 200241, China

Radula Dumort., a monotypic genus in Radulaceae with more than 250 currently accepted species, is mainly distributed in subtropical and tropical regions. Over 40 species placed in three subgenera have previously been recognized in China. Although the genus is well delimitated and easily recognized, the infrageneric taxonomy is considered to be very contentious. In the present study, a taxonomic revision of *Radula* in China is presented based on both molecular phylogenetic and morphological-anatomical analyses. The materials used in this research were fresh collections and herbarium specimens collected from all over the country. The morphological investigation focused primarily on three re-evaluated characters (viz., stem cross-section, oil body and papilla). Four chloroplast DNA sequence (*trn*G, *trn*L-F, *rps4* and *atp*B-*rbc*L) data generated from 33 species in China were analyzed applying different phylogenetic approaches including the maximum parsimony, maximum likelihood and Bayesian inference. Our revision confirmed the occurrence of 38 species in six subgenera in China: *Amentuloradula* (2 spp.), *Cladoradula* (4), *Metaradula* (9), *Odontoradula* (5), *Radula* (17), *Volutoradula* (1). Two new species, *R. hainanensis* L.N.Zhang & R.L.Zhu and *R. subacuminata* L.N.Zhang & R.L.Zhu, both belonging to subg. *Metaradula*, were described. Five species, *R. fauriana*, *R. mittenii*, *R. stenocalyx*, *R. tabularis* and *R. ventricosa* are newly reported for China. One species, *R. inouei*, is hitherto known only from China. The highest diversity of species was found in Hainan island with 24 species, and the second in Taiwan with 21 species. Two sections, sect. *Epiphyllae* Castle ex Grolle and sect. *Chinenses* Yamada were proposed to be resurrected. Based on the morphological analyses, we found that oil body and papilla can be divided in three types respectively, and they also play important roles in taxonomy of *Radula*, especially in the infrageneric classification.

T2
P0718

A study of *Cryptocarya* (Lauraceae) in southeast Asia based on morphology and phylogeny

Meng Zhang, *Yahara Tetsukazu, Mase Keiko, Tagane Shuichiro*
Center for Asian Conservation Ecology, Kyushu University

The genus *Cryptocarya* within the Lauraceae family is widely distributed throughout the tropic forests including the Southeast Asian region. The total number of *Cryptocarya* is estimated between 200 and 250. However *Cryptocarya* has not been fully studied and still remains difficult to identify. Thus we are carrying on a series field survey of plant biodiversity in Southeast Asia, including Cambodia, Indonesia, Laos, Malaysia, Myanmar, Thailand, and Vietnam. We used both plots survey and general specimen collecting methods at different elevations of the tropical mountains. In addition, we applied the both morphology observation and phylogenetic analysis based on DNA barcode ITS to identify the *Cryptocarya*. As a result, within 83 collecting *Cryptocarya* samples, 20 *Cryptocarya* species were identified and 9 species are remaining for identify including 3 new candidate species, which are *Cryptocarya bokorensis* sp. nov., *Cryptocarya kaengkrachanensis* sp. nov., and *Cryptocarya honbaensis* sp. nov.

T2
P0719

Biogeographic history of *Saxifraga* sect. *Irregulares* (Saxifragaceae) disjunct between Eurasia and western North America

Meng-Hua Zhang, *Xiao-Shuang Zhang, Dai-Gui Zhang, Ze-Long Nie*
Jishou University

Saxifraga is the largest genus in Saxifragaceae and a characteristic component of temperate to polar climatic zones and of alpine vegetation belts in mountain ranges of the Northern Hemisphere, especially diversified in high mountains of SW China. Sect. *Irregulares* is a special group in the genus usually characterized by zygomorphic flowers, except for *Saxifraga mertensiana* from North America with actinomorphic flowers. This section includes 15-20 species, most of which are found in eastern Asian and only one species disjunctly distributed in North America. Phylogenetic studies suggested that it is the first early-diverged group in the genus. In order to understand its biogeographic history under a robust phylogeny, we employed both nuclear gene (ITS) and chloroplast DNA regions (*psb*A-*trn*H, *trn*L-F) to reconstruct the species' relationships within sect. *Irregulares*. Biogeographical inferences were based on a Bayesian uncorrelated log-normal relaxed molecular clock together with fossil calibrations and geographical distribution data to address relationships, divergence times, and historical biogeography. Our results indicate that sect. *Irregulares* is monophyletic and composed of three main lineages: (1) Ser. *Stonoliferae*, a group with long-creeping rhizome; (2) *Saxifraga mertensiana* from western North America characterized by actinomorphic instead of the zygomorphic flowers; and (3) the remainder species without stolons were grouped together (ser. *Rufescentes*). Sect. *Irregulares* was estimated to have split with *Saxifraga* at 20.17 Ma (95% HPD: 12.26–28.78 Ma) in the early Miocene, and the split between eastern Asia and North American lineagesat 16.24 Ma (95% HPD: 9.68–23.43 Ma) in the middle Miocene. Our biogeographical and analyses suggested that both Bering land bridge and North Atlantic land bridge are likely to explain the expanding of sect. *Irregulares* from Eurasia towards western North America. In addition, the analyses supported that *Saxifraga kegangii* D. G. Zhang, Z. L. Nie & M. H. Zhang, a new species of *Saxifraga* sect. *Irregulares* from Hunan province of central China. This new species can be easily distinguished from the other related taxa by its leaf shape and disc semiannular around the side of the ovary.

T2
P0720

Anther and ovule development in *Camellia japonica* (Naidong) in relation to winter dormancy: Climatic evolution consider-

ations

Qiang Zhang, *Qing Hao, Xiao Guo, Qinghua Liu, Kuiling Wang*
Landscape and Forestry College, Qingdao Agricultural University

The *Camellia japonica* (Naidong) population in Qingdao, Shandong province is the only one in temperate areas in China. While this population has been present in this area since the tertiary, its has evolved independently after that. To date, there are no reports on its embryology, and no research on the connection between embryology and climate. Here, we studied anther and ovule development of *C. japonica* (Naidong) using light and scanning electron microscopy, in order to determine the characteristics of early flower development, microsporogenesis, microgametogenesis, megasporogenesis and macrogametogenesis in this plant. We observed basic anther wall formation, glandular tapetum and successive microsporocyte cytokinesis. Moreover, microspores tetrads were mostly tetrahedral and occasionally symmetrical, we did not observe pseudopollen grains as previously reported for Camellia. The ovary was superior and presented axial placentation and anatropous, bitegmic ovule, allium type of embryo-sac, integumentary tapetum and hypostate which may provide winter cold temperature resistance. The early flower development sequence placed *C. japonica* (Naidong) in a most primitive branch of the phylogenetic tree compared to other species. The embryo characteristics and the development of *C. japonica* (Naidong) are significant for phylogeny research of Theoideae, as the taxonomy of this family is currently problematic at all levels. Moreover, this information is also useful flowering induction as a possible horticultural application. Flower buds went in dormancy during winter as gametophyte state, a characteristic evolved in *C. japonica* (Naidong) in response to the climates changes.

T2

P0721

Diversification of Rosaceae since the Late Cretaceous based on plastid phylogenomics

Shu-Dong Zhang[1], *Jian-Jun Jin*[1], *Si-Yun Chen*[1], *Mark W. Chase*[2], *Douglas E. Soltis*[3], *Hong-Tao Li*[1], *Jun-Bo Yang*[1], *De-Zhu Li*[1], *Ting-Shuang Yi*[1]
1. Kunming Institute of Botany, CAS
2. Science Directorate, Royal Botanic Gardens, Kew
3. Florida Museum of Natural History, University of Florida

Phylogenetic relationships in Rosaceae have long been problematic because of frequent hybridization, apomixis and presumed rapid radiation, and their historical diversification hasn't been clarified. With 87 genera representing all subfamilies and tribes of Rosaceae and six of the other eight families of Rosales (outgroups), we analyzed 130 newly sequenced plastomes together with 12 from GenBank in an attempt to reconstruct deep relationships and reveal temporal diversification of this family. Our results highlight the importance of improving sequence alignment and the use of appropriate substitution models in plastid phylogenomics. Three subfamilies and 16 tribes (as previously delimited) were strongly supported as monophyletic, and their relationships were fully resolved and strongly supported at most nodes. Rosaceae were estimated to have originated during the Late Cretaceous with evidence for rapid diversification events during several geological periods. The major lineages rapidly diversified in warm and wet habits during the Late Cretaceous, and the rapid diversification of genera from the early Oligocene onwards occurred in colder and dryer environments. Plastid phylogenomics offers new and important insights into deep phylogenetic relationships and the diversification history of Rosaceae. The robust phylogenetic backbone and time estimates we provide establish a framework for future comparative studies on rosaceous evolution.

T2

P0722

DNA barcoding of *Oryza* (Poaceae)

Wen Zhang[1,2], *Jia Liu*[1,4], *Chao Xu*[1,2], *Haifei Zhou*[1], *Yanlei Liu*[1,2], *Xun Chen*[1,3], *Shiliang Zhou*[1]
1. Institute of Botany, the Chinese Academy of Sciences
2. University of Chinese Academy of Sciences
3. Northeast Forestry University
4. Sichuan Agricultural University

Rice (*Oryza sativa*) is one of the most important crops in the world, which supports half of world's population. There are about 24 species in rice genus. The genetic germplasms of rice genus have been collected, conserved in seed banks and utilized in varying degrees. It is not easy to identify seeds of different species and there are a certain proportion of misidentifications of seed materials housed in seed banks. In this study, we used 85 samples representing all species in the genus and evaluated 35 the most variable chloroplast DNA regions to identify rice species using the BLAST method, Best Close Match method, and tree-base method. The combinations of the 35 barcodes provided the highest species discrimination success in BCM method (70.73%) and maximum likelihood tree (62.96%). DNA barcoding provide effective discrimination in most genome types (C, CD, E, F, G, HJ, HK) except species of A, B and BC genome type, which can help differ whole species in five series of Oryza (*Ser. Australianses*, *Ser. Brachyanathae*, *Ser. Meyerianae*, *Ser. Ridleyanae*, *Ser. Schlechterianae*)

T2

P0723

How many ferns and lycopods and where are the hotspots in China?

Xian-Chun Zhang
Institute of Botany, Chinese Academy of Sciences

The flora of China is one of the richest in the world, as to the ferns and lycopods the number of species is ranked the first by country. The number of species was 2,742 (including species, infraspecific taxa and hybrids) in the Chinese national flora (Flora Reipublicae Popularis Sinicae), but revised to be 2,270 in the English version Flora of China. Meanwhile the number of recognised families and genera also changed from 63 and 220 to 40 and 178 respectively. The exact number of species of ferns and lycopods is still uncertain. New discoveries and taxonomic revsions do make changes of the species numbers. Especially in some difficult groups, such as *Angiopters*, *Dryopteris*, *Polystichum*, *etc.*, the number of species might be reduced by extensive taxonomic revisions and molecular DNA barcode analysis. Hotspots are biodiversity rich geographical areas where the endemic species and endangered species number are high. Facing the problem of global changes and biodiversity losses in this century, we should pay special attention to the conservation of these hotspots and its biodiversity. Besides the two main biodiversity hotspots in China, the Southwest China Mountains and the East Himalayas, there are smaller "hotspots" across China in many provinces and districts. Here, I analysis the endemic species of ferns and lycopds in each region of China in order to find the hotspots for conservation.

T2
P0724

Phylogenomic synteny network analysis reveals an ancient MADS-box transcription factor tandem duplication and lineage-specific transpositions

Tao Zhao[1], *Rens Holmer*[1], *Suzanne De Bruijn*[1], *Harrold A. Van Den Burg*[2], *Eric Schranz*[1]
1. Wageningen University
2. University of Amsterdam

Conserved genomic context (or synteny) provides critical information for comparative evolutionary analysis, such as the inference of ancient polyploidy events, recurrent genomic rearrangements across species and gene ancestry. With the increase of sequenced and assembled plant genomes, we have an opportunity to use synteny to analyze the dynamics of gene family conservation, expansion and transposition across broad phylogenetic groups. Here we present an integrated approach to organize gene synteny networks using *k*-clique percolation. As an example, we analyzed gene synteny networks for the MADS-box transcription factor gene family based on fifty-one completed plant genomes. Several novel evolutionary patterns can be inferred and visualized from our analyses. For example, we identified several lineage-specific gene clusters derived from transposition events including floral morphology (*AP3* and *PI*) and flowering-time (*FLC*) in the Brassicales and root-developmental genes (*AGL12* and *AGL17*) in Poales. We identified two highly connected gene clusters containing many key phenotypic regulatory Type II MADS-box gene clades (*SEP1, SEP1, SQUA, TM8, SEP3, FLC, AGL6 and TM3*). Synteny results support that these genes thus evolved from an ancient tandem gene duplication likely predating the radiation of seed plants, which then expanded by polyploidy events and sub-functionalization. Our phylogenomic synteny network approach can be applied to any group of species to gain new insights into the evolution and dynamics of any set of genes.

T2
P0725

Morphological study on the asexual reproduction by deciduous leaves of *Takakia lepidozioides*

Tianxiong Zheng
Graduate School of Sciences, Hiroshima University

The class *Takakiopsida* consisting only two species, *Takakia lepidozioides* and *T. ceratophylla*, represent the earliest lineages within the mosses and has been attracting researchers due to enigmatic morphologies such as, (1) both leafy and leafless rhizomal shoots, (2) cylindrical shape of leaves, (3) the archegonia and the antheridia without protective structure, (4) the absence of rhizoids, (5) Symbiotic fungal hyphae in mucilage hairs on the rhizomal shoots. In Japan, *T. lepidozioides* is known only as female plants and the deciduous leaf seems to serve as a means of asexual reproduction. We observed that the development of leaves at the growing shoot, plant regeneration from detached leaves and the process of reinfection of symbiotic fungi. To detect proliferative region in leaves, we detect DNA synthesis by using 5-ethynyl-deoxyuridine (EdU) detection system. In young leaf primordia on growing shoots, the single apical cell with two cutting faces was observed. Visualization of DNA synthesis by EdU assay indicated that whole cells of young primordia including the apical cell had mitotic activity and the apical cell play an important role for morphogenesis of cylindrical shape of leaf. As the leaves grow, apical cell lost mitotic activity and intercalary region in leaf remain proliferative. In mature leaves, all of cells have no mitotic activity and the apical cell often died. Mature leaves easily detached at base from the shoot by slight mechanical shocks. In our culture stocks, many of regenerated shoots repeatedly arose around the plants. Morphological observation and EdU assay showed that the beak shaped mucilage hairs initially arose on the detached leaf. Subsequently, the apical cell producing strap-like or cylindrical meristematic region was directly regenerated without differentiation of filamentous protonema like other mosses. The mucilage hairs on the detached leaf were hardly distinguishable to those of rhizomal shoots. The cylindrical leaves with a single apical cell of *Takakia* seem to properties resembling individual shoot itself, although it show determinate growth unless detaching from main shoot. Co-culturing of sterilized detached leaf and symbiotic fungi isolated from mucilage hairs showed that the symbiotic fungi immediately enter the inside of the newly formed mucilage hairs through the mucilage pore and established the labyrinth of hyphae in mucilage hairs between the cell wall and cytoplasm. The appearance of mucilage hairs in detached leaves seems to promote the re-establishment of symbiotic relationship with fungi and contribute to survive and grow of regenerated shoots.

T2
P0726

Evolution of WRKY transcription factors in Brassicaceae

Qinqin Zhou, *Fei Chen, Yue Hu, Liangsheng Zhang*
Fujian Agriculture and Forestry University

Brassicaceae, containing many edible plants, medicinal plants and the mode plant Arabidopsis, is a highly diverse plant family with vast adaptations. The recently released and various Brassicaceae genomes provide the great resources for understanding the molecular basis for their adaptations. WRKY gene family is one of the largest transcription factor families in higher plants with important roles in regulating plant development and resistance. In this report, a total of 22 Brassicaceae genomes, spanning 10 tribes, from basal to crown Brassicaceae, were studied to systemically analyze the evolution of *WRKY* genes in Brassicaceae. In total, 5 WRKY groups were classified (Group I, IIa, IIb, IIc, IId, IIe, III). By comparing the number of *WRKY*, we found that the large expansion of Brassicaceae *WRKY* was mainly contributed by recent-independent polyploidization events. Except the species with recent polyploidization events, all other core Brassicaceae showed conserved *WRKY* numbers, which did not contribute to the diversity and adaptability of Brassicaceae. By analyzing the selection pressure, we found that the Group III *WRKYs* tended to evolve faster than other groups. However, some genes in Group I, IIc and IIe also showed fast evolutionary rate. Positive selections and functional divergence were found in these fastly evolved genes. By comparing the salt-treated expression of *WRKY*, more Arabidopsis than Thellungiella genes were up-regulated. However, more down-regualed genes were found in Thellungiella compared with Arabidopsis. Both orthologs and paralogs had significant different evolutionary rate and the tissue expression of these genes also showed significant variations, which could explain the diversity and adaptability of Brassicaceae. These results uncovered the characteristics of *WRKY* evolution in Brassicaceae, which were believed to provide not only clues about the diversity and adaptability of Brassicaceae but also a good foundation for studying the diversity and adaptability of other highly adaptive plant lineages such as Asteraceae, Leguminosae, Poaceae and Orchidaceae.

T2

P0727

Occurrence of earliest gigantopterid from the basal permian of the North China Block and its bearing on evolution

Weiming Zhou[1], *Jun Wang*[2]
1. Nanjing University
2. Nanjing Institute of Geology and Palaeontology, Chinese Academy of Sciences

Gigantopterids are a morphological group consisting of a number of enigmatic fossil plants with angiosperm–like leaves and reticulate venation that are of uncertain systematic affinity. Gigantopterid plants were abundant and a characteristic floral element in the Late Palaeozoic Cathaysian floras. However, in China, their oldest occurrence was much later than that in North America (Artinskian) and Indonesia (Asselian to Sakmarian). Here we document the gigantopterid *Gigantonoclea* cf. *mira* from basal Permian (Asselian) strata in the North China Block that represents the oldest unequivocal evidence for the gigantopterids. The foliage is characterized by complex venation composed of four orders, small isodiametric meshes and intercalary subsidiary veins arising directly from the rachis, being a distinctive morphological type of leaf venation. In addition, a non–marginal feeding trace is detected on the lamina, extending the earliest record of plant–animal interaction between gigantopterid plants and arthropods to the earliest Permian, and supporting the selective feeding preference of *Gigantonoclea* leaves within Cathaysian gigantopterids. Our new observations show that the venation characters of *Gigantonoclea mira* and *Gigantonoclea* cf. *mira* more closely resemble *Gigantopteris dictyophylloides* than other members of gigantopterids including the evolutionary *Callipteridium* sequence to which it was previously assigned.

T2

P0728

Characteristics of vascular-plant diversity of Mt. Kenya, East Africa

Yadong Zhou[1,2,3], *Qingfeng Wang*[1,3]
1. Wuhan Botanical Garden, Chinese Academy of Sciences
2. University of Chinese Academy of Sciences
3. Sino-Africa Joint Research Center, Chinese Academy of Sciences

Mt. Kenya is one of the highest mountains in the tropical regions of the world. The uniqueness of plant compositions and patterns of species diversity at Mt. Kenya has attracted botanists from all over the world to examine and evaluate.In the past 3 years, we had collected and identified about 1,500 plant specimens of Mt. Kenya. Combining our data and other collections of more than 7,000 specimens from herbaria around the world, we concluded that, there were about 1,550 species in total (including exotic plants) at Mt. Kenya, belonging to 726 genera, 157 families. Specifically, 9 lycophytes (belonging to 5 genera, 2 families), 131 ferns (belonging to 55 genera, 20 families), 6 gymnosperms (belonging to 5 genera, 3 families) and 1,404 angiosperms (belonging to 661 genera, 132 families) were classified. We have also identified 3 new species and 75 new records at Mt. Kenya. With regard to life forms, 53.24% of total plant species were herbs, 10.40% woody herbs, 9.25% small trees, 8.34% lycophytes and ferns, 6.42% shrubs, 5.07% trees, 3.40% lianas and 3.04% climbers. More than half of plant species (60.49%) were endemic in Africa, whereas 25.13% cosmopolitan, 9.29% endemic in tropical East Africa, 3.63% endemic in Kenya, and only 1.36% endemic in Mt. Kenya. With the increase of altitude, the plant diversity showed a quadratic pattern with the peak of diversity from 2,000 to 2,600 m a.s.l., which were largely composed of forests (Fa & Fb). However, communities such as the bamboo zone (B, including bamboo-*Hagenia* forest, *Hagenia-Hypericum* forest), heath zone (E), Afromontane vegetaion (A) and Afromontane desert (Ad) have relatively low diversity values.

T2

P0729

Phylogenomics based on RAD-seq and whole chloroplast sequencing recovers ancient hybridization and asymmetric introgression between *Cercidiphyllum japonicum* and *C. magnificum*

Shanshan Zhu[1], *Yonghua Zhang*[2], *Yingxiong Qiu*[1]
1. Zhejiang University
2. Sun Yat-sen University

Chloroplast capture - the apparent introgression of a chloroplast (plastid) genome from one plant species into another, is commonly suggested as an explanation for topological incongruity. However, the cases of incongruence between phylogenetic trees are usually presented based on few nuclear loci and organellar fragments. Chloroplast capture occurred in *Cercidiphyllum* species is of a typical example. With the development of next generation sequencing (NGS), it provides new opportunities to get high quantity and quality data with low cost. Here, we obtained genomic data from 21 samples of *Cercidiphyllum* species to infer the phylogenetic relationships using restriction site-associated DNA sequencing (RAD-seq). Meanwhile, using a combination of reference-based mapping and *de novo* assembly of whole genome Illumina data, we assembled the complete chloroplast genomes of *C. japonicum* (5 individuals respectively from North China, South China, South Japan, Central Japan and North Japan) and one individual of *C. magnificum*. Phylogenetic analyses based on RAD-seq data clearly supported the distinctiveness of two species of *Cercidiphyllum*. By constrast, phylogenetic analyses of the complete chloroplast genome revealed that the *C. japonicum* individuals from certral and north Japan were clustered with *C. magnificum*. Comparative chloroplast genome analyses showed few genetic and structural variation between the *C. magnificum* and *C. japonicum* (from central Japan and north Japan), and the differences were indeed smaller than that with other individuals of *C. japonicum* (from North China, South China, South Japan). This is the first time to confirm the two species of *Ceicidiphyllum* at genome-wide. It also provides the first documentation of chloroplast capture between species of *Cercidiphyllum* based on whole chloroplast genome sequencing.

T2
P0730
Phylogenetic relationships in Chinese *Anaphalis* (Asteraceae), based on chloroplast and nuclear DNA sequences
*Xuemin Xu, Zhen Wei, Junyu Gu, **Shixin Zhu***
Zhengzhou University

Anaphalis is the largest genus in Asian Gnaphalieae tribe (Asteraceae), containing about 110 species. There are about 54 *Anaphalis* species intensively distributing in Hengduan Mountains in China. The Chinese *Anaphalis* is a taxonomically problematic group, especially in the terms of species delimitation. The phylogenetic relationships of *Anaphalis* and its closely related genera in Gnaphalieae, such as *Helichrysum*, *Gnaphalium*and *Pseudognaphalium*, are not clear as well. In this study, three cpDNA sequences, including *psb*A-*trn*H, *trn*L-*trn*F and *rpl32*-*trn*L intergenic spacer, and one nrDNA sequence, ITS, were sequenced for 73 *Anaphalis* taxa representing 25 species. The aim of this study is to provide the phylogenetic relationships within the Chinese *Anaphalis* species, examine the monophyly of the genus and discuss the relationship between *Anaphalis* and closely related genera. The results indicated that two clades within *Anaphalis* are well recognized. One clade involved most *Anaphalis* species, and the other one only included 4 *Anaphalis* species (*A. nepalensis*, *A. margaritacea*, *A. triplinervis* and *A. viridis*). However, all Chinese *Anapalis* species we used are nested with *Helichrysum* in nrDNA phylogenetic tree, while in cpDNA phylogenetic tree *Anaphalis* are nested with *Helichrysum* and *Pseudognaphalium*. Based on nrDNA and cpDNA phylogenetic trees, the current *Anaphalis* is not monophyly and need to be revised. Several important morphological characters were then plotted on the phylogenetic trees and discussed.

T2
P0731
Systematics of East-Mediterranean *Raphanus* L. (Brassicaceae)
Jotham Ziffer-Berger
Herbarium of the Hebrew University of Jerusalem, the National Natural History Collection

An evaluation of the two sections of *Raphanus* (Brassicaceae) *sensu* O.E. Schulz, as given in Engler's *Pflanzenreich,* on the basis of multi-species morphological and ITS data, supported that Raphanus is a polyphyletic group embedded in the Oleracea lineage of the tribe Brassiceae. Section Raphanis, which includes *Raphanus raphanistrum*, *R. pugioniformis* and *R. sativus*, is a strongly supported monophyletic lineage. Section Hesperidopsis is embedded in a different lineage together with *Brassica deflexa* and *B. aucheri* . We propose to reinstate the genus *Quidproquo* in place of *Raphanus* sect. Hesperidopsis, to reflect the polyphyletic origins of the genus *Raphanus* sensu O.E. Schulz.

T2
P0732
Research progress on taxonomy of *Xanthium* L.
***Liping Yin*[1]**, *Ning Zhang*[1], *Manyu Zou*[2], *Shuiliang Guo*[2]
1. Shanghai Entry-Exit Inspection and Quarantine Bureau
2. Shanghai Normal University

The species of this genus *Xanthium* L.is distributed globally, however are most frequently found in the tropical and sub-tropical regions. Taxonomy of *Xanthium* L.is confused that there have a variety of classification methods. For example, Flora of China and Flora of North America considers that there are two or three species in the world, while Plant List records sixteen accepted names of the species of *Xanthium*. Löve and Dansureau revised the genus *Xanthium* and reduced the number of species to only two. Weaver and Lechowich though there are 20 species belonging to this genus. Caius reported that the genus *Xanthium* includes 25 species. According to lots of diverse literature of *Xanthium*, there is a controversy that whether this genus should split into more species groups, or unit the smaller entities into larger or very large groups. USDA classifies the *Xanthium* L. into two species and three varieties in line with Cronquist's classification method. And Fan reported 17 species of the genus. Predecessors mainly draw on several morphological traits of *Xanthium*, such as the leaf shape, involucre shape, involucre volume, spines length and the density of spines, to classify species. However, there are few type specimens to prove its correctness. Morphological traits are so variable that the taxonomic study of the members of this genus is sometimes difficult. The cytological study showed that the karyotype analysis would provide strong support for authentic taxonomic diversification of some uncommon species of *Xanthium*. It was concluded that chemistry analysis including ultraviolet spectrometry and infrared spectrometry could help classify the species. Molecular markers analysis with the results that the genetic material among several species has a certain difference. As a consequence, morphology, cytology, chemistry and molecular markers study could only differentiate among two or several species. Research progress above help in making a clear understanding on the classification of the genus.

T2
P0733
Similar morphology but different origins: Hybrid status of two more semi-creeping taxa of *Melastoma*
Peishan Zou, *Shuqiong Wang, Renchao Zhou*
State Key Laboratory of Biocontrol and Guangdong Provincial Key Laboratory of Plant Resources, School of Life Sciences, Sun Yat-sen University, 510275, Guangzhou, Guangdong, China

Inferring the origins of hybrid taxa based on morphology alone is difficult because morphologically similar hybrids can arise from hybridization between different populations of the same parental species or be produced by hybridization of different parental species. In this study, we investigated the origins of two semi-creeping taxa in *Melastoma*, which are similar to a natural hybrid *M. intermedium*, by sequencing one chloroplast intergenic spacer, nuclear ribosomal internal transcribed spacre (*nr*ITS) and two low-copy nuclear genes (*tpi* and *cam*) in these taxa and their putative parental species. Our sequence analysis results provide compelling evidence for hybrid status of the two semi-creeping taxa. The two hybrids originated from hybridization between *M. dodecandrum* and *M. malabathricum*, and between *M. dodecandrum* and *M. normale*, respectively. Therefore, they have different origins from *M. intermedium* and morphological similarity for the three hybrids might be due to origins from hybridization between the same creeping species *M. dodecandrum* and an erect species in each of the three cases. We also observed low rate of introgression from *M. normale* to *M. dodecandrum*, which may contribute to rapid adaptive evolution of *M. dodecandrum* to different environments. Rare occurrence of these two hybrids may be due to relatively small range overlap between parental species in one case and flowering period differentiation between parental species in the other.

T2
P0734

Orchidantha yunnanensis (Lowiaceae), a new species from China, and notes on the identity of *Orchidantha laotica*

Pu Zou
South China Botanical Garden, Chinese Academy of Sciences

Orchidantha yunnanensis, a new species from Malipo county in Yunnan, China, is described and illustrated here. The identity of *Orchidantha laotica*, originally described from herbarium material, is discussed here, and a colour plate of this species based on recent collections from Laos is included.

T2
P0735

Unraveling the evolutionary dynamics of ancient and recent polypoidization events in *Avena* (Poaceae)

Qing Liu[1], *Lei Lin*[1,2], ***Xiangying Zhou***[1,2], *Paul M. Peterson*[3], *Jun Wen*[3]
1. Key Laboratory of Plant Resources Conservation and Sustainable Utilization, South China Botanical Garden, Chinese Academy of Sciences, Guangzhou 510650, China
2. University of Chinese Academy of Sciences, Beijing 100049, China
3. Department of Botany, National Museum of Natural History, Smithsonian Institution, Washington, DC 20013-7012, USA

Understanding the diversification of polyploid crops in the circum-Mediterranean region is a challenging issue in evolutionary biology. Sequence data of three nuclear genes and three plastid DNA fragments from 109 accessions of *Avena* L. (Poaceae) and the outgroups were used for maximum likelihood and Bayesian analyses. The evolution of cultivated oat (*Avena sativa* L.) and its close relatives was inferred to have involved ancient allotetraploidy and subsequent recent allohexaploidy events. The crown ages of two infrageneric lineages (*Avena* sect. *Ventricosa* Baum ex Romero-Zarco and *Avena* sect. *Avena*) were estimated to be in the early to middle Miocene, and the *A. sativa* lineages were dated to the late Miocene to Pliocene. These periods coincided with the mild seasonal climatic contrasts and the Mediterranean climate established in the Mediterranean Basin. Our results suggest that polyploidy, lineage divergence, and complex reticulate evolution have occurred in *Avena*, exemplifying the long-term persistence of tetraploids and the multiple origins of hexaploids related to paleoclimatic oscillations during the Miocene-Pliocene interval in the circum-Mediterranean region. This newly-resolved infrageneric phylogenetic framework represents a major step forward in understanding the origin of the cultivated oat.

T2
P0736

Deep phylogenetic incongruences between plastid and nuclear genomes in a clade of tropical trees

Sophie A. Zwartsenberg[1], *Jenifer C. Lopes*[1], *Rita D. Brandão*[2], *Lars W. Chatrou*[1]
1. Wageningen University and Research
2. Maastricht University

Evolutionary plant biologists have for a long time only used a limited number of markers to infer the phylogeny of plant families or genera. With sequencing now becoming increasingly cheap, the usage of whole genome data is becoming more commonplace. The data used for inferring phylogenies has often been derived from the chloroplast genome as it was long assumed that the evolution of the chloroplast was a good indicator of the evolution of the entire plant species. This study gives an example of why using only a single plant genome for phylogenetics might obscure important evolutionary patterns. The evolutionary history of the plant family Annonaceae has, over the years, been quite well established aside from a few nodes in the phylogeny which are difficult to resolve. One of these unresolved nodes involves the genus *Oxandra* from the tribe Malmeeae. In previous phylogenetic analyses, monophyly of the genus was not demonstrated whilst (floral) morphologically it is a homogeneous genus. This study aimed at resolving the phylogeny of *Oxandra* using genomic scale chloroplast data as well as nuclear ribosomal DNA. We generated high-throughput sequence data for 33 species of the Malmeeae clade, including 20 species of *Oxandra*. From the sequence data we obtained a data set of 61,000 bp of chloroplast sequences, and 5,800 bp of nuclear rDNA sequences.Based on chloroplast DNA, the genus *Oxandra* is polyphyletic. However, the genus was inferred to be monophyletic using nuclear ribosomal DNA. These incongruences between the different phylogenies affect nodes deep in the phylogeny, placing all *Oxandra* species in one clade or placing the species in two different clades with genera such as *Klarobelia* and *Ephedranthus* separating them. The incongruent nodes all receive high posterior probabilities (> 0.98). The mechanisms causing this strongly incongruent pattern is yet unknown and the species phylogeny of *Oxandra* remains unresolved as it can be both a monophyletic or polyphyletic genus depending on what genome you are looking at. The results of this study show that it might be wise for phylogenetic studies to sample multiple genomes that are available within the plant as these might shed a different light on the organismal evolution.

T2
P0737
Is *Angelica sylvestris* (Apiaceae), a supergeneralist plant, locally adapted to its pollinators?

Marcin Zych[1], *Katarzyna Roguz*[1], *Małgorzata Stpiczyńska*[1], *Massimo Nepi*[2], *Robert R. Junker*[3]
1. Botanic Garden, Faculty of Biology, University of Warsaw
2. Department of Life Sciences, University of Siena
3. Department of Ecology and Evolution, University of Salzburg

In terms of the pollination system umbellifers (plants of the Carrot family, Apiaceae) are regarded generalists, since their (usually dichogamous) flowers are visited by a wide range of insects representing several taxonomic orders. However, recent analyses of insect effectiveness revealed that the plants may be effectively pollinated by a narrow assemblage of flower visitors. In our study we focused on *Angelica sylvestris* L., a common European species pollinated by a taxonomically diverse insect assemblage. In three populations, located along approx. 500 km long transect, over three years we analysed floral visitors, assessed insect effectiveness, studied nectar and scent composition and performed transplantation experiment. We showed that our study populations differ in nectar and scent profiles and despite similar taxonomic composition of insect visitor assemblages, they are effectively pollinated by different pollinator guilds. Although this suggests local adaptations to the most effective pollinators, our transplantation experiment showed that reproductive success is not related to the origin of experiment plants and insect do not exhibit preferences towards local genotypes. Our paper discusses possible evolutionary pathways leading to the observed diversity of local variants of the *A. sylvestris* pollination system. The study was financially supported by the National Science Centre, Poland grant 2011/01/B/NZ8/03146 and by Faculty of Biology, University of Warsaw DSM grant 112630.

T2
P1403
The taxonomic treatment of the *Dryopteris varia* complex in Japan

Kiyotaka Hori[1], *Atsushi Ebihara*[3], *Noriaki Murakami*[2]
1. Makino Botanical Garden
2. Makino Herbarium, Tokyo Metropolitan University
3. National Museum of Nature and Science

The *Dryopteris varia* complex (subg. *Erythrovariae*, sect. *Variae* Fraser-Jenk.) is an apogamous fern complex. Most species of the complex are triploid apogamous, show morphological and genetic variation, and have taxonomic problems especially in their species classification. Recent molecular researches suggested that the complex experienced reticulate evolution through hybridization between diploid sexual and triploid apogamous species, and that is the reason why each apogamous species are so variable. In this study, we summarized the taxonomic treatment of 11 species within the *D. varia* complex on the basis of their genome constitutions revealed by the molecular analyses.

T2
P1404
XII Latin American Botanical Congress, Quito, Ecuador October, 2018

Alina Freire-Fierro
IKIAM Universidad Regional Amazónica, and Asociación Latinoamericana de Botánica ALB. Vía Tena-Atacapi, Ecuador

Latin American Botanical meetings have been organized since 1938, and in 2018 we will be celebrating 80 years of these congresses. Next year will be also the 20th anniversary of the publication of the first Angiosperm classification based on molecular data published by the Angiosperm Phylogeny Group. In order to reflect these two important events, we have chosen as a theme for the congress the reality and virtual development in Latin American botany, and the logo for the congress represents an Asteraceae species endemic to the Andes, Chuquiraga jussieui J.F.Gmel. The congress, to be held in Quito, Ecuador, during October 21-28,2018, is being organized by the Latin American Botanical Society (ALB), the Ecuadorian Botanical Society (ABE), and IKIAM Universidad Regional Amazónica. For this event, we expect to host 1500-2000 participants, and the guest speakers are: Tod Stuessy (USA), Jorge Crisci (Argentina), Danna Leaman (Canada), Fausto Sarmiento (Ecuador/USA), Lucia Lohmann and Mónica Carlsen (Brasil/Venezuela), Ernesto Medina (Venezuela), Mark Chase (UK), Katya Romoleroux (Ecuador), Neill (Ecuador-USA), and Pamela Soltis and Douglas Soltis (USA). The ALB was founded in 1986 and nowadays includes more than 300 members. The board is formed by Ecuadorian and Latin American botanists and has, as one the main objectives, the organization of Latin American botanical congresses and the ABE was recently formed to allow legal transactions incurred during the congress in Ecuador. IKIAM is a young emblematic research university created by the Ecuadorian government. It is located in the eastern Andean slopes towards the Amazon, and it is the responsible for managing the +93.000 Ha Colonso-Chalupas Biological Reserve. In addition to ALB, ABE and IKIAM, for the XII CLB, there are many national and international collaborating institutions, among them the IAPT, BSA, SENESCYT, *etc.*

T2
P1410
Phylogenetic analysis of the genus *Convolvulus* L. (Convolvulaceae) using chloroplast DNA sequences from *trn*L-*trn*F region

Perveen Anjum, *Kousar Shazia*
University of Karachi

The family Convolvulaceae consists of species rich genera from which *Convolvulus* is next to *Ipomoea*. *Convolvulus* is a large and diverse genus, It is a cosmopolitan genus but the Mediterranean and East Asia are the major points of diversity of *Convolvulus*.
Chloroplast non-coding regions are now frequently use in Phylogenetic studies. A consensus tree based on chloroplast DNA sequences from the *trn*L-*trn*F region was examined by evaluating phylogenetic relationships in 12 species of *Convolvulus*. Result indicated that Phylogenies based on *trn*L-*trn*F sequences incongruent with traditionalmorphological classification in which, the

genus *Convolvulus* is divided into different sections based on spine and growth forms. The present study revealed that all species are delimited into three distinct lineages. In the first lineage, *C. kotscyanus*, and *C. arvensis* showed sister association while *C. leiocalycinus*, *C. acanthocladus* and *C. psedocantabricus* were grouped together. Within the second lineage, *C. virgatus* and, *C. aitchisonii* were closely related while *C. spinosus*, *C. scindicus* and *C. prostratus* occupied position on second branch. The third branch made the largest cluster of *C. lineatus*, *C. rhyniospermus*, *C. cephalopodus*, *C. fruticosus* and *C. glomeratus*, *C. pyrrhotrichus* located distantly in the third lineage.

T2

P1414

Jekyll existing in two major allelic variants is a lineage-specific gene for the closely related Triticeae and Bromeae tribes

Volodymyr Radchuk[1], *Rajiv Sharma*[1,2], *Elena Potokina*[1,3], *Ruslana Radchuk*[1], *Eberhard Munz*[1], *Martin Mascher*[1], *Benjamin Kilian*[1,4], *Ljudmilla Borisjuk*[1]

1. Leibniz-Institute of Plant Genetics and Crop Plant Research (IPK)

2. Division of Plant Sciences, School of Life Sciences, James Hutton Institute, University of Dundee

3. Vavilov Institute of Plant Industry (VIR)

4. The Crop Trust

Due to economic value and intriguing genetic and agricultural histories of Triticeae species, search for lineage-specific, domestication and agronomically-relevant genes for the tribe are of high importance. Earlier we have discovered the *Jekyll* gene which is expressed in nucellar projection of the developing barley (*Hordeum vulgare*) grains and in tapetum of developing anthers, whereupon *Jekyll* protein is localized on the pollen surface. The *Jekyll* plays pivotal role in terminal cell differentiation of nucellar grain tissues and influences nutrient delivery to the endosperm during grain feeling. In the tapetum, *Jekyll* is important for pollen maturation and anther dehiscence. Repression of *Jekyll* resulted in a significant loss of fertility and thus proves the role of this gene in sexual reproduction. We found also two additional sequences, from which *Jek3* was established as an allelic variant of *Jekyll* (further *Jek1*) despite of low identity at nucleotide and amino acid levels. The *Jek1* and *Jek3* proteins share identical signal peptides, conserved cysteine positions in the variable middle regions and several direct repeats at C terminals. The *Jek1/Jek3* sequences are located in the same locus on chromosome 3H and inherited by monogenic Mendelian fashion. *Jek3* has similar expression patterns to *Jek1* and can also complement the *Jek1* function in *Jek1*-defficient plants. Analyzing characteristic genetic background in sequenced genome of barley, we suppose that huge allelic dimorphism betweenJek1 and *Jek3* is originated by retrotransposition. Analysis of *Jek1/Jek3* molecular diversity in 494 accessions of barley core collection showed that both allelic variants were almost equally distributed among cultivated and wild barleys. All cultivated barleys with *Jek1* allele belong to single haplotype indicating genetic bottleneck during domestication. Cultivated barleys with *Jek3* allele distributed among three haplotypes.

Phylogenetically related groups of species contain lineage-specific genes, which exhibit no sequences similarity to any genes outside of lineage. We found *Jekyll*-like sequences in all analyzed species of closely related Bromeae andTriticeae tribes (constituting Triticodae lineage) but not in other Poaceae lineages. Comparative grain analysis using Nuclei Magnetic Resonance Imaging demonstrates that *Jekyll* appearance is linked with the formation of nucellar cavity and modified aleurone cells in Triticodae. We postulate that emergence of *Jekyll* during molecular evolution of Poaceaewas a driver for separation of the Triticodae lineage and herewith the *Jekyll* is a lineage-specific gene. The acquisition of new *Jekyll* gene in course of Triticodae separation also leads to rapid innovation in grain phenotype, namely in a formation of a maternal/filial conduit for assimilate transfer towards endosperm consisting of the nucellar projection and modified aleurone cells. Aleurone cells faced this cavity are delayed in differentiation and retain their assimilate transfer function over whole grain development. In contrast to typical aleurone layer (like Brachypodium), modified aleurone layer in Bromus and species of Triticeae does not acquire abilities for accumulation of lipids and storage proteins. Therefore, *Jekyll* has become an essential genetic component controlling endosperm development and storage accumulation.

T2

P1420

Genetic identification of *Amomum villosum* from different planting areas based on internal transcribed spacer of ribosomal DNA

Chang Su[1], *Xiaoying Guan*[1], *Yaoxuan Xie*[1], *Siyue Liu*[2], *Xiaoxiao Song*[1], *Shuhong Wang*[1]

1. Shenzhen Institute for Drug Control

2. Saint Edward's School

Yang Chun Sha is the dried ripe-fruit from *Amomum villosum*. It is one of the four most famous herbs in Southern China, and has been used on clinic for more than 1,300 years. According to the ancient herbal records, *Amomum villosum* was discovered from the city of YangChun in Guangdong province, however about 40 years ago, *Amomum villosum* were introduced and artificial planted to Guangxi and Yunnan province to increase the yield, and are extended to surrounding Southeast Asian countries (Laos and Myanmar) recent years. But traditionally *Amomum villosum* from Guangdong is considered as the best in medicinal quality and has much higher market values than the herbal products from other planting areas. In this study we investigate the genetic diversity of *Amomum villosum* from different planting areas by analyzing the nucleotide sequences of internal transcribed spacer (ITS) of ribosomal genes. Phylogenetic trees were constructed using several clustering methods, including unweighted pair-group method with arithmetic average (UPGMA), neighbor-joining (NJ) and maximum evolution (ME). The present work also provides the original planting investigation on two planting area in China and morphological identification on its herbal products from other two Southeast Asian countries (Laos and Myanmar). The results of the phylogenetic trees constructed using UPGMA, NJ and ME method shows the same grouping pattern, all 21 testing samples from different countries form separate clades respectively, but samples from different areas in China are closely related and nested in the tree. Herbal products of *Amomum villosum* from China and Laos show no difference in morphological features, but products from Myanmar present significant differences in the part of peel and the arrangement of the seed mass. Therefore, the ITS region could be adopted as a molecular marker for differentiating the herbal products of *Amomum villosum* from different planting areas.

T3
P0738

Reproductive adaptations of *Lycium ruthenicum* and *L. qingshuiheense* in sandstorm environment, at Kasghar, China

Aysajan Abdusalam
Key Laboratory of Ecology and Biological Resources in Yar kand Oasis, College of Life and Geography Sciences, Kashgar University

Temporal variation in flowering phenology and pollination in sandstorm environment may completely switch mating pattern and show some adaptation in flowering plants. We investigated this prediction by exploring the reproductive adaptations of conducted in a natural population of homostyly plant *Lycium ruthenicum* and approach herkogamy plant *L. qingshuiheense* in sandstorm environment, at Kashgar China. We investigated the variations of floral traits, pollinator activity, self-incompatibility and mating patterns in both species along a sandstorm and sunshine spring environment on the south Xinjiang. There were significant differences in flowering phenology, floral traits, nectar volume, self-incompatibility and mating patterns in both sexual morph speceis. Flowering phenology of *L. qingshuiheense* was shorter than *L. ruthenicum*, and flowering time of *L. qingshuiheense* was finished before summer. Sandstorm harsh environment can be negatively effect female reproductive success of self-incompatible approach herkogamy plant *L. qingshuiheense*. Pollen limitation increased in *L.qingshuiheense*, but harsh environment have no effect on self-compatible homostyly *L. ruthenicum* female reproduction and pollen limitation. Our results suggested that pollinator activity and floral traits in *L. qingshuiheense* plants is relatively more important to seed set and pollen limitation than that of *L. ruthenicum*, and self-compatible homostyly of *L. ruthenicum* is highly adaptable than self-incompatibility of approach herkogamy of *L. qingshuiheense* in long term sandstorm climate.

T3
P0739

Plant-pollination interactions in generalized food-deceptive genus *Dactylorhiza*

Ada Wróblewska[1], *Izabela Tałałaj*[1], *Andrzej Bajguz*[1], *Lech Szczepaniak*[2], *Beata Ostrowiecka*[1], *Emilia Brzosko*[1], *Edyta Jermakowicz*[1], *Paweł Mirski*[1], *Agata Kostro-Ambroziak*[1]
1. University of Białystok, Faculty of Biology and Chemistry, Institute of Biology, Poland
2. University of Białystok, Faculty of Biology and Chemistry, Institute of Chemistry, Poland

In the animal and plant kingdoms, deception is the most effective and widespread strategy for survival. Although, there are differences between animals and plants in deception strategy, the fitness was always important in estimation with being a deceiver. *Dactylorhiza* genus belongs to generalized food-deceptive orchids. This genus complex is a good study object of food-deception strategy because evolution by recurrent hybridization and polyploidization particularly have resulted in a wide spectrum of taxa with different flower signals, and in consequences of specific-pollinator associations as well reproductive fitness. Here, we analyzed and compared the interaction between pollinating insects and floral display (different floral morphs and floral scent) of four *Dactylorhiza* taxa including two coloured morphs *D. incarnata* ssp. *incarnate* and *D. incarnata* ssp. *ochroleuca* as well as *D. fuchsii* (diploid plants) and *D. majalis* – allotetraploid taxa, which have evolved repeatedly by hybridization between two broadly defined parental lineages: *D. incarnata* and the *D. fuchsii*. We questioned what are differences between these four *Dactylorhiza* taxa taking into account their floral display (different floral morphs and especially floral scent) and the pollinating species/their behavior with respect to food-deceptive strategy.

T3
P0740

Calcium improves the leaf physiology of salt treated *Limonium stocksii*: A floriculture crop

Muhammad Zaheer Ahmed, *Muhammad Yousuf Adnan, Tabassum Hussain, Salman Gulzar, Muhammad Ajmal Khan*
Institute of Sustainable Halophyte Utilization, University of Karachi

Calcium acts as a signaling molecule to improve resistance during unfavorable environment. *Limonium stockii* seedlings were grown in 0, 300 and 600 mM NaCl and in 0 and 600 mM NaCl with additional Ca^{++} in the nutrient solution for 15 days. The effects of these treatments were studied on plant growth, leaf water relations, MDA content, Na^+ secretion rate, ion flux and photosynthesis. Plant biomass reduced by 50% and MDA content increased 50% when treated with 600 mM NaCl for 15 days. Decline in leaf WC, RWC and increased in sap osmolality began from 3rd day after 600 mM NaCl treatment. Leaf Na^+ and Na^+/K^+ ration were higher 7th days after salinity treatment while Na^+ secretion rate increased with the increase in NaCl concentration and duration of treatment. Decrease in the rates of photosynthesis was coupled with lower stomatal conductance and intercellular CO_2, however, WUE improved under salinity. Efficiency of PSII, ETR, Pq and NPQ were reduced but photosystem was intact with higher Y (NO) under salinity. The Ca^{++} application improved 25% plant FW at high salinity (600 mM NaCl) and lowers the membrane damage. In addition, Ca^{++} enhanced leaf WC and reduces tissue OP. The Na^+ accumulation in leaf reduced while the secretion increased in the presence of additional Ca^{++}. However, CO_2 assimilation and light harvesting complex remained unaffected by Ca^{++} application. The result obtained from our experiment indicate that the decrease WC, down regulation of photosynthesis and respiration, higher leaf Na^+ and energy intensive defense prevent *L. stocksii* development at 600 mM NaCl. Application of Ca^{++} enhanced salinity resistance by improving water balance, salt-secretion and membrane integrity.

T3
P0741

Phytoremediation ability of *Calotropis procera* in reducing air

pollution in Jeddah city-Kingdom of Saudi Arabia

Alkhazzan Mulook Mohammed
King Abdul-Aziz University

This study was conducted in Northeastern part of Jeddah city. *Calotropis procera* plant was used as bioindicator and phytoremediator of contamination with heavy metals arising from the cement factory located in this area. Three study sites were chosen, the first site (control) where *C. procera* plants growing 4.88 km far away from the polluted area (the cement factory) and the other sites at the cement factory has also been divided into two sites the first one include *C. procera* plants growing around the factory (location 2), the second site include *C. procera* plants growing 150 meters far away from the factory (location 1). Samples of plants and soils were collected during summer and winter seasons (2013-2014) for laboratory works. Amount of some heavy elements (Aluminum, Chromium, Boron, Barium, Copper, Manganese, Iron, Lead and Zinc), were estimated, in plants and soil. Results revealed that, *C. procera* has the ability to have reasonable amount of heavy elements at normal ranges. Worth mentioning is that soil heavy elements were less than in plants, indicating the phytoremediation ability of this plant in cleaning the soil, meaning that, this plant absorbs the pollutants from the atmosphere. It is noticeable that heavy elements accumulated more during winter season in plants and during summer season in soil. Therefore, *C. procera* can be used as protective belts near polluted areas in Jeddah.

T3
P0742
Importance and conservation status of forest tree species in Turkey

Volkan Altay[1], *Munir Ozturk*[2]
1. Biology Department, Faculty of Science & Arts, Mustafa Kemal University, Antakya-Turkey
2. Centre for Environmetal Studies and Botany Department, Ege University, Izmir, Turkey

Nearly 90 percent of the forests in Turkey are natural in origin and contain over 450 species of trees and shrubs. These are spread over an area of around 21.2 million ha. A major part of these forests are located in the Euro-Siberian and the Mediterranean phytogeographical regions. The dominant broad-leaved species are *Quercus* spp. (22.7%), *Fagus orientalis* 82 (3.3%), *Alnus* spp. (0.2%), and *Castanea sativa* (0.1%). Others include mixed broad leaved forests (18.5%), maquis (8.4%), coniferous species (60%), *Pinus* spp. (30%), *Juniperus* spp. (4.6%), *Abies* spp. (0.9%), *Picea orientalis* (0.7%), *Cedrus libani* (0.5%) and mixed coniferous forests (5.5%). The broad leaved trees contribute to 32 percent and coniferous species 68 percent of the timber resources. *Quercus* species occupy 6.4 million ha, *Pinus brutia* 5.4 million ha, *Pinus nigra* 4.2 million ha, *Fagus* 1.7 million ha, *Pinus sylvestris* 1.2 million ha, *Abies* 0.6 million ha, *Juniperus* 0.5 million ha, *Cedrus* 0.4 million ha and *Picea* 0.3 million ha of forest area. The biodiversity resource utilization by local people from these forest species has been elucidated in terms of tree use such as medicinal, food, handcrafts, firewood, dye, fodder and other uses. A total of 130 naturally distributed tree species are utilised economically. The most common economic uses are: medicinal (99 species), followed by food (87 species), handcrafts (65 species), firewood (60 species), dye (38 species), fodder (30 species) and other uses (56 species). Terrestrial biodiversity has decreased mainly because of the loss and/or deterioration of forest ecosystems. There is a great need for developing a system to utilize forest resources while conserving biodiversity. This study aims to elucidate the socio-economic background causing forest decrease, its effects on biodiversity, and ecological services that might be lost. We also evaluate the forest use options both from ecological and socio-economical aspects to develop a sustainable utilization system.

T3
P0743
Clusia - one genus, many pollination mechanisms

José Elvino Nascimento Jr.[1], *Ana Cláudia A. Silva-Santos*[1], *Volker Bittrich*[2], *Maria C. E. Amaral*[1,3]
1. Programa de Pós graduação em Biologia Vegetal, Institute of Biology, University of Campinas - UNICAMP, Campinas, SP
2. Rua Dr. Mario de Nucci 500, Campinas, SP
3. Departamento de Biologia Vegetal, Institute of Biology, University of Campinas - UNICAMP, Campinas, SP

One of the remarkable features in *Clusia* (Clusiaceae) is the great variation in floral morphology, especially in staminate flowers. There are flowers with four and others with hundreds of stamens, an enormous variety of androecial morphology, with stamens free or fused in various ways, and normal anthers or athecate multilocellate ones. *Clusia* has some unusual features in relation to reproduction, such as dioecy, present in almost all species, and the secretion of resin by stamens and staminodes, a feature widely dispersed throughout the phylogeny of *Clusia*, but lacking in several species. The pollination biology of most *Clusia* species studied involves floral resin, and showed some very unusual mechanisms. Species that secrete floral resin are pollinated by bees that use the resin for nest construction. Only three species not offering floral resin were already studied: *Clusia pusilla* and *Clusia criuva*, pollinated by bees and beetles, respectively, in deceit systems; and *Clusia sellowiana*, pollinated by cockroaches in a system for which the possibility of attraction by floral scent similar to sexual pheromones of cockroaches was suggested. In order to verify if cockroaches and beetles could carry out the pollination through other mechanisms, we studied the pollination of *Clusia burle-marxii*, *Clusia obdeltifolia*, and *Clusia sellowiana*. The first two are pollinated by small beetles of families Chrysomelidae and Curculionidae. Unlike *Clusia criuva*, the flowers are not pollinated by deceit. Beetles feed mainly on pollen of staminate flowers, also using these as place of aggregation and copulation. For oviposition, female beetles fly to pistillate flowers and lay eggs at the ovary base. During this process, pollen grains are deposited on stigmas. After hatching, the larvae penetrate the ovary, where they remain for some time feeding on the developing seeds. Before the pupal stage, larvae perforate the wall of the immature fruit and drop to the soil, where they conclude their development. Although the larvae are able to pierce the outer fruit wall, they do not pierce septa dividing locules, thus feeding on young seeds of only one locule, leaving the other locules intact. Pollination of *Clusia sellowiana* is performed by cockroaches, practically the only flower visitors. Cockroaches are attracted to the flowers by acetoin, a volatile compound released also by decaying leaves, which are the main food of cockroaches in natural environments. It was previously suggested that cockroaches would be lured to the flowers through sexual attraction, since acetoin is part of the pheromone composition of some cockroach species. However, almost half of the visitors are nymphs, which are not sensitive to sexual signaling. In other words, cockroaches are attracted to food, not sex. In addition, acetoin possibly repels pollen thieves. Our studies show even more unusual pollination mechanisms in *Clusia* and that the same groups of pollinators may be involved in different mechanisms of pollination.

T3

P0744

Prediction of the *Acacia decurrens* Willd. invasion using habitat suitability index (HSI) modelling in The Mount Merapi National Park

Susilo Ari Wibowo
Gunung Merapi National Park

Acacia decurrens . is a foreign species in The Mount Merapi National Park (TNGM). In its development after the eruption of Mount Merapi in 2010, the plant is actually growing rapidly in the open areas caused by the eruption. In this area, *Acacia decurrens* . be a threat to the recovery of mountain flora. This study aims to identify factors that affect the distribution and how wide of the *Acacia decurrens* . invasion. The approach in this study using HSI spatial modeling with binary logistic regression analysis. This analysis predicted the presence or absence of characteristics of the response variable based on the prediction of predictor variables. Then the model equation is transformed into a spatial form with ArcGIS 10.1 software. This model was validated in order to obtain a model that is totally appropriate. Based on this model, habitat suitability classes can be classified based on environmental factors that influence and wide of the *Acacia decurrens* . invasion in TNGM region can be known. Results of this study concluded that *Acacia decurrens* . tend to be invasive and is influenced by: the distance of the settlement as a function of the community activities that tend to hinder the process of natural succession in the invasion area, the distance from the seedbank as a function of source of seed for the *Acacia decurrens* . invasion, and the Normalized Difference Vegetation Index (NDVI) as a function of vegetation density of the area of the *Acacia decurrens* . invasion in TNGM region. While the invasion has evolved from the original planting (years 1968-1993) covering an area of 1,080.66 ha into 1,676.25 ha and for 47 years, has invaded outside the planting plots areas, covering an area of 783.79 ha or 72.53% of the total planted area.

T3

P0745

The role of inbreeding depression in selecting for the link between selfing and dispersability: The case of *Hypochaeris salzmanniana*

Montserrat Arista
University of Seville

The evolutionary link between breeding system and dispersability has been proposed due to both traits affect genetic variability, gene flow and colonizing capacity. In plants, the few empirical data up to date show contrasting results, sometime associating selfing and non-dispersal and others selfing and dispersal. Consistently, theoretical models that take into account range position also result in opposite associations making necessary to test that link in model species. We test the association between selfing and dispersability in the heterocarpic *Hypochaeris salzmanniana*. We used both an experimental approach and surveys during two years in five natural populations. These populations differ in breeding system and occurr in a gradient of pollinator availability. Traits affecting dispersability, and the heritability of dispersability were also studied. Dispersability had a high narrow-sense heritability, as h2 for mid-parent regression was 0.56. Non-beaked fruits were thinner and slighter and showed higher dropping velocity and lower dispersal distance than beaked fruits. Dispersability and selfing ability were linked at all the scales studied. Experimentally, self-incompatible plants showed lower dispersability than self-compatible plants, and dispersability in self-compatible plants after outcrossing was lower than after selfing. This link was also supported in nature because populations with high proportion of self-compatible plants showed higher dispersability than self-incompatible populations. The link between selfing and dispersal seems to be a consequence of inbreeding depression that causes both, lower production of the costly non-beaked fruits and more lighter dispersible fruits. The higher dispersal of selfing fruits would be selected as a way of avoiding inbreeding depression and sib-competition. The high dispersability of selfers improves colonizing capacity in *H. salzmanniana*, as self-compatibility confers reproductive assurance in pollen limiting conditions.

T3

P0746

Global survey of the dryland forest

Fabio Attorre[1], *Danilo Mollicone*[2], *Chiara Patriarca*[2], *Mario Messina*[1]
1. Sapienza University of Rome
2. Food and Agriculture Organization of the United Nations

The extent of forest in the world's drylands is a major source of uncertainty in estimates of global forest area, and hence in estimates of global carbon stocks and fluxes too. A convenient proxy estimate for forest area in drylands has long been the 734 million hectares (Mha) of tropical open canopy forest in 1980; this accounted for about a fifth of global forest area, though not all of it was in the drylands, not all drylands are in the tropics and some dry forest has a closed canopy. Few estimates have been made subsequently, since drylands forest is rarely surveyed, even though its overuse contributes to climate change and desertification. Recent global maps based on medium resolution (30 metre) satellite images provided new estimates, but such images are too coarse to accurately assess open canopy forest. Here, we fill this gap by reporting the first global estimate of forest area in drylands based on visual interpretation of very high resolution ($\leq$ 1 metre) satellite images, supported by time series of vegetation indices from high temporal resolution satellite imagery. Using Collect Earth software, the Google cloud computing system and Microsoft Bing Maps, we found that 1,327 Mha of land in the drylands had $\geq$ 10% tree canopy cover in 2010, of which 1,079 Mha constituted forest as defined by the UN Food and Agriculture Organization. Surprisingly, the majority of all dryland forest was closed canopy, while only one third was open canopy. These findings raise the estimate of tree canopy cover in one recent Landsat map by 427 Mha (47%), equivalent to forest area in Amazonia; suggest that recent estimates of world forest area might be underestimated by at least 9%; and could help resolve current uncertainty around terrestrial carbon sink estimates.

T3
P0747

Effect of salt stress on Rubisco enzyme and chlorophyll content of some cowpea [*Vigna unguiculata* (L) Walp] cultivars grown under irrigation in Hadejia-Nigeria

Mohammed Isa Auyo
Federal University, Dutse-Nigeria

The current climatic changes experienced globally is undoubtedly increasing salt content of our soils threatening plants growth and making them to thrive poorly and making agricultural practices challenging, costly and difficult. It is in this regard that, a research work was conducted in the field under irrigation in the Phase II Irrigation Scheme of the Hadejia-Jamaare River Basin Development Authority in Auyo and Ayama communities in Jigawa State, Nigeria, irrigating plants with surface water and ground water, to determine the effect of varying concentrations of NaCl, $CaCl_2$ and $CaCO_3$ with ECe values of 2.0 dSm^{-1}, 3.0 d Sm^{-1} 4.0 d Sm^{-1} and 5.0 dSm^{-1} in addition to 0.0 dSm^{-1} which served as control. These were evaluated on Rubisco Enzyme Concentration and Leaf Chlorophyll content of IT97K-452-2, IT04K-332-1, IT98K-503-1, and *Dan Wuri* varieties of Cowpea (*Vigna unguiculata)* Linn, (Walp). Result from the experiment revealed that both Rubisco Enzyme Concentration and Leaf Chlorophyll were inhibited by the salts, particularly at higher concentrations of 4.0 dSm^{-1} and 5.0 dSm^{-1}. $CaCO_3$ and $CaCl_2$ proved to be the salts that greatly suppressed these parameters. Local varieties of cowpea were observed to thrive more poorly in saline soils than improved cultivars. Rubisco Enzyme Concentration and Leaf Chlorophyll content were significantly ($P \leq n0.05$) inhibited as the concentration of the salt increased from 3.0dSm^{-1} to 4.0dSm^{-1}, with 5.0dSm^{-1} showing the greatest inhibition. Variety, IT97K-452-2 appeared most tolerant and *Dan wuri* were most susceptible. Concentration of 4.0 dSm^{-1} and 5.0 dSm^{-1} were most inhibitory, killing some of the plants involved. The salt concentration tolerated by the varieties was generally in the range of 0.0 to 2.0 dSm^{-1}. The research reveals that cowpeas varied in their response to salinity and salt stress inhibited important photosynthetic characters thus inhibited photosynthesis and therefore crop yield, as such it is wiser cowpea is better grown in non saline soil and other varieties to be used in large scale production ought to be screened for salinity tolerance prior to their cultivation.

T3
P0748

Community of arbuscular mycorrhizal fungi in different ecosystem on the east of Qinghai-Tibetan Plateau

*Ali Bahadur, **Zhongcai Jin**, Yongjun Liu, Huyuan Feng*
School of Life Sciences, Lanzhou university

The community of arbuscular mycorrhizal fungi (AMF) is designated for the first time in different ecosystems on the east of Qinghai-Tibetan Plateau. Arbuscular mycorrhizal fungi have mutualistic linked with the roots of over 80% of all the plant species, as they regulate nutrient and carbon cycles, and influence ecosystem multifunctionality and soil structure. We explored the diversity and structure of AMF communities in forest, shrubs, alpine meadow, and farmland. Soil and roots from different ecosystems were sampled for the analysis of soil characters and AM fungal parameter. In farmland significantly lowest glomalin-related soil protein (GRSP) and AM fungal hyphal length density was found, while highest was indicated in shrub and forest. As predicted, soil moisture content was dominant factor revealing the difference of AM fungal spore density and hyphae length in each ecosystem. Soil phosphorus, carbon and pH were significantly correlated with AMF colonization, representing that these abiotic variables could be responsible for community composition differences among ecosystem. Water-stable soil aggregate, hyphae length and spore density were positively associated with glomalin, while the negative correlation between spore density. On the other hand, soil pH showed the opposite effect on the variables. AM fungal spore density and hyphae length increased with soil pH, while GRSP decreased. Our data reveal a sharp differentiation of AMF communities between forested areas, shrubs, alpine meadow and periodically disturbed farmland. Our results highlighted and emphasized the divergence in AM fungal diversity with different ecosystems.

T3
P0749

Identification of an environmental-dependent mutant

Xue Bai[1,2]*, Shengyang Wu*[1]*, Chunming Liu*[1]
1. Key Laboratory of Plant Molecular Physiology, Institute of Botany, Chinese Academy of Sciences
2. University of Chinese Academy of Sciences

Unlike animals, plants cannot move to avoid environmental damage. Environmental stress can be divided into biotic and abiotic stress. It is estimated that abiotic stress, like drought, heat, cold and nutrient deficiency, is responsible for 50% decrease in crop yield worldwide. Deciphering the mechanism that how plant sense and respond to environmental stress would greatly benefit for crop yield improvement. We used rice as a model crop plant, and planted an EMS-mutagenized rice population in Hainan and Beijing to screen mutants which exhibit different phenotype in both places. Fortunately, we identified a mutant that shows similar phenotype with wild type when planted in Beijing, however, shows severely dwarf and sterile phenotype in Hainan. It would be interesting to characterize what kind of environmental signal contributes to the phenotype difference.

T3
P0750

Spatial structure of rhizobial communities in sandy beach environments

Masaru Bamba[1]*, Seishiro Aoki*[2]*, Koji Takayama*[3]*, Motomi Ito*[2]*, Masaki Miya*[4]*, Tadashi Kajita*[5]
1. Graduated School of Science, Chiba University
2. Department of General Systems Studies, University of Tokyo

3. Museum of Natural and Environmental History, Shizuoka
4. Natural History Museum and Institute, Chiba
5. Iriomote Station, Tropical Biosphere Research Center, University of the Ryukyus

The symbiotic relationship between leguminous plants and root nodule bacteria is a well-known example of mutualism. Host plants provide photosynthetic products and safe habitats to their bacterial symbionts, and the symbionts return products of nitrogen fixation to the host plants. With the proliferation of bacteria in nodules, legumes can get enough nitrogen products and can grow on the soil with the low level of nitrogen. Symbiotic relationships with rhizobia significantly enhance the fitness of legumes in this way; However, legumes do not develop symbiotic relationships with every rhizobium due to host-symbiont specificity. When legumes are dispersed into a novel habitat where no suitable symbionts exist, colonization may not be successful because of lack of the symbiotic relationships. The presence of rhizobial bacteria in soils are influenced by local environments in a small scale, although the spatial distribution of rhizobia community has not been well understood. In this study, to understand the spatial response of rhizobial communities in sandy beach environments in which gradient change of environmental factors is expected, we conducted metagenome analysis using soil samples collected along line transects from tideline to inland. Using soil samples collected at two locations in Iriomote Island, Japan, environmental DNA were studied by amplifying *glnII* (glutamine synthetase II protein) gene using the primers specifically designed to target *Rhizobium*, *Ensifer*, *Mesorhizobium*, and *Bradyrhizobium*. *GlnII* gene amplicons were sequenced by an Illumina Miseq platform and metagenomic data were processed and analyzed to recognize OTUs. To extract solely rhizobial OTUs from the metagenome data, BLAST searches were performed using the center sequence of OTU cluster as a query. Using the 2,609 rhizobial OTUs detected from the metagenome data, diversity analyses suggested that alpha diversities of the rhizobial communities are correlated with the distance from tidelines. Also, beta diversity analysis among rhizobial communities showed that there are significant differences between seaward and landward communities. These results suggest that the rhizobial communities in sandy beach environments are spatially influenced by environment gradient in a small scale.

T3
P0751

Inventorization of lianas in north-eastern India with notes on their climbing modes and reproductive attributes

Debjyoti Bhattacharyya, *Lalawmkima Darlong, Malemnganbee Longjam Chanu*
Assam University, Silchar

Lianas, the woody vines, are one of the most curious growth forms with diverse morphology and twinning patterns. They are abundantly distributed in the tropics of the globe and are one of the key components of tropical forests. North-eastern region of India, which falls under Indo-Burma biodiversity hotspot, shows great diversity in liana flora both in terms of species composition and richness. A study on inventorization of lianas in northeastern India (particularly Assam, Manipur and Tripura) based upon morpho-taxonomic investigation is going on since 2013. Their twining patterns and reproductive attributes have also been taken into consideration. The study reveals the occurrence of 95 species of lianas under 58 genera and 32 families from the region. Fabaceae is found to be the largest family with 11 genera and 23 species. Some other dominant families found are Apocynaceae (eight species), Vitaceae (six species) and Combretaceae (five species), *etc.* Genus *Combretum* is the largest with five species. Gymnospermous liana is represented only by the genus *Gnetum* (Gnetaceae) with two species viz. *G. montanum* and *G. oblongum*. The taxa enumerated show four distinct types of climbing modes, viz. stem twinner, tendril climber, hook climber and thorn climber among which stem twining is the chief climbing pattern (85%). Study of reproductive attributes reveals the presence of 27% taxa with inconspicuous (< 1 cm) flowers and 73% with conspicuous (> 1 cm) flowers. Two flowering peaks have been observed with a major peak in March-April and a minor peak in October while fruit ripening and dispersal of seeds take place mainly before the monsoon. The fact of major peak of flowering during March–April can be correlated with entomophilous nature of pollination in most of the species which is driven by enhanced insect activity in warm and dry months. For seed dispersal, four major types of diaspores are recorded viz. Barochore, Sarcochore, Pogonochore and Pterochore among which Barochory is the most recurrent attribute (45%). It has been observed that the species richness and abundance of lianas in the region is diminishing gradually, mostly due to anthropogenic activities. Habitat destruction by eradication of supporting host tree species, practice of jhum cultivation, rubber and areca nut plantation, and tea and pineapple cultivation, extension of villages and agricultural fields due to human population explosion are some of the alarming threats for lianas growing in the region. Many IUCN red-listed taxa viz. *Gnetum montanum*, *G. oblongum*, *Derris cuneifolia* and *Spatholobus parviflorus* occurring in the region are facing serious threats of extermination in near future. Genus *Tetrastigma* (Vitaceae) is proposed to be considered as threatened as three species of this genus viz. *T. lanceolarium*, *T. obovatum* and *T. serrulatum* could not be traced in the field in recent times. *Beaumontia grandiflora* (Apocynaceae), *Parabarium micranthum* (Apocynaceae) *Pueraria bella* (Fabaceae) and *Uncaria sessilifructus* (Rubiaceae) are also rare in the region. So, conservation of lianas in the region is a matter of concern and a proper policy needs to be framed immediately.

T3
P0752

Finding the best match: Comparing preferences of *Nicotiana attenuata* and its pollinators

Julia Bing, *Danny Kessler, Ian Thomas Baldwin*
Max Planck Institute for Chemical Ecology

The self-compatible wild tobacco plant *Nicotiana attenuata* grows in genetically diverse populations and is visited by a variety of pollinators, such as hummingbirds, hawkmoths and bees. These floral visitors show differences in their behavior and flower handling. They also vary in their response to floral traits, such as scent or nectar, which the plant uses to allure potential floral visitors. Pollinators choose based on their adaptations to different floral traits. But not only pollinators make a choice, tobacco does as well. If pollen of different paternal genotypes has been transferred to the stigma of *N. attenuata*, the plant is able to select its mates from this diverse pool of pollen. For this, two questions have to be considered: 1) what floral traits are important for the different pollinators? 2) which pollinators match the plant's preferences? To answer this, we conduct natural and semi-natural experiments with transgenic lines and native accessions that show differences in floral traits. Since ethylene plays an important role in enabling the mate selection process we used transformed plants with silenced ethylene production (ACO) in comparison to empty vector control plants (EV). Pairs of EV and ACO plants were planted in a population of different native accessions that vary in floral traits important for pollinator attraction. Seeds produced after pollinator visitation will be used for genotyping. The comparison of the seed set from both transformed lines will shed light on which pollen was brought by the pollinators (ACO plant) and what pollen was selected by the plant to set seeds (EV plant). That allows us to uncouple the events of pre-pollination sexual selection (pollinator choice) and post-pollination sexual selection (mate choice of plant) to get an idea if there are pollinators that transfer exactly the pollen genotypes to the flowers that also the plant would choose based on its mate selection pattern. This will give us further insight into the ecological coadaptation between different genotypes and there best pollinators.

T3
P0754

Lichen as a biomonitor of nitrogen and sulfur depositions in petrochemical, agricultural and forested areas in Thailand

Chaiwat Boonpeng[1], *Wetchasart Polyiam*[1], *Chutima Sriviboon*[2], *Tunsinee Jhampasri*[2], *Kansri Boonpragob*[1]
1. Lichen Research Unit, Department of Biology, Faculty of Science, Ramkhamhaeng University
2. Department of Chemistry, Faculty of Science, Ramkhamhaeng University

Nitrogen oxides (NO_x) and sulfur oxides (SO_x) are mainly originated by fossil fuel combustion. These compounds are toxic to human health and ecosystems, thus amounts of them are necessary to measure in order to evaluate human and environmental impacts. These gases can be monitored by air monitoring equipment, but only a few stations could be achieved due to financial constraint especially in developing countries such as Thailand. Alternatively, lichens are widely used as biomonitors of atmospheric deposition, moreover; this technique is cost effective. Therefore, the main goal of this research was to use lichen as a biomonitor of atmospheric deposition of nitrogen (N) and sulfur (S) in petrochemical, agricultural and forested areas in Thailand. Intact thalli of the epiphytic foliose lichen *Parmotrema tinctorum* (Despr. ex Nyl.) Hale were collected from a relatively unpolluted area, and then transplanted at one forested site in the Khao Yai National Park, at two agricultural sites near Wang Chan District, Rayong Province, and at eight petrochemical industrial sites in Map Ta Phut, Rayong Province. The transplantation was lasted for 8 month (October 2013 to June 2014) and the lichen samples were periodically collected for analyzing amounts of N (as NO_3^-) and S (as SO_4^{2-}) using Ion Chromatography (IC). The result showed that mean concentrations of both N and S in the lichens in the petrochemical industrial area (304 mg/kg for N and 664 mg/kg for S) were evidently higher than those from the agricultural (193 mg/kg for N and 437 mg/kg for S) and the forested areas (159 mg/kg for N and 381 mg/kg for S). N content showed strongly positive correlation with S content ($r = 0.82$, $p < 0.01$), and both compounds trend to decrease with increasing distance from the main industrial area. This study reinforced the utilization of lichen as a biomonitor of atmospheric N and S depositions, which is appropriate for developing countries where air quality monitoring instruments are insufficient.

T3
P0755

Comparison of herbivory intensity and impacts on populations of *Trichocentrum undulatum* in southern Florida and neighboring Cuba

Haydee Borrero[1,2], *Hong Liu*[1,2,3]
1. Florida International University
2. International Center for Tropical Botany
3. Fairchild Tropical Botanic Garden

With a rapidly changing climate it is important to study the role of plant-animal interactions and species' persistence within their distributional margins. The mule-ear orchid *Trichocentrum undulatum* is endemic to the Caribbean region. Southern Florida is the species distribution northern limit, with only one surviving population, in the Everglades National Park (ENP). A rare (possibly endemic) herbivorous fly, *Melanagromyza miamensis*, which is able to halt the development of the orchid's inflorescence, has incurred an alarmingly high rate of attacks in recent years. Flower development was halted by the herbivore in 94% of the new flower stalks with only 3% producing any fruit in the 2013 flowering season at a site in ENP and no fruit being produced in either 2014 or 2015. As a consequence, the population has largely not been able to reproduce for several years. The future of this species may be at risk due to its inability, or significantly reduced capacity, to produce either flowers or fruits. In this study, we wish to compare the nature and intensity of interactions between the herbivorous insect and the rare orchid in both southern Florida (the species' marginal distribution range) and Cuba (the core range). On a recent exploratory trip to the field in Cuba, populations were visited to determine the presence of this particular herbivore. Indeed, all populations showed similar signs of inflorescence stalk herbivory by the fly. We are not aware of any report prior to ours on the inflorescence herbivory of the *T. undulatum* in Cuba. Further evaluating and comparing the impacts of the insect on populations throughout the island and those in southern Florida would shed light on this complex plant-animal interaction.

T3
P0756

Species composition, stand strusture and environmental interpretation in the forest community dominated by *Pinus kwangtungensis*, a Chinese endemic

Linying Cai, *Xinyuan Zhang, Ding Ma, Lu Zhang*
South China Agricultural University

Pinus kwangtungensis is the pines with five needles located in the montane areas. Global warming and long-term disturbance threaten its existence, and *Pinus kwangtungensis* is listed as vulnerable in the Species Red List in China. To analyse whether and how major environmental factors influence the species composition and stand structure in the subtropical forest community dominated by *Pinus kwangtungensis*, a total of seventy-two 100 m^2 plots were laid out in the southern slope of the Nanling Mountains, South China (1,100–1,900 m a.s.l.). We found that 227 species in 138 genera and 76 families of vascular plants were recorded in the plots. Both family and genus present a highly significant relationship with the species richness ($P < 0.001$). Contrary to previous studies, abundant saplings of *Pinus kwangtungensis* were found in the forest community and the analysis based on diameter class of the endangered species showed a normal pyramidal shape, indicating that the species was not a regressive species at the population and community level. Canonical Correspondence Analysis and Cluster Analysis were used to analyze and portray the relations of *Pinus kwangtungensis* to other species in the forest community and the relations of species patterns to their habitat. Spatial distribution of *Pinus kwangtungensis* in the forest community showed a significant pattern along an altitudinal gradient ($P < 0.001$). The endangerment status of *Pinus kwangtungensis* can be ascribed to large-scale events such as habitat fragmentation, natural and human disturbance, which has important implications for species conservation at the landscape or regional level.

T3
P0757

The shape is more important than we ever thought: Plant to plant interactions in a high mountain community

Marcelino De La Cruz, ***Julia Chacón-Labella***, *David S. Pescador, Adrian Escudero*
Universidad Rey Juan Carlos

Working with fully mapped communities provides the ecologist the opportunity to estimate precisely who interacts with who, how frequently and which of those interactions are important for the assembly of the community. Most analyses of fully mapped communities have employed point pattern analysis techniques, as these extract the maximum information available in the dataset. However, when dealing with high mountain communities, an important part of the spatial structure of the community is lost by approximating plant individuals as points when most chamaephyte and hemicryptophyte individuals have non-symetric, irregular sizes that are best represented by planar shapes. Based on this, we have developed a new method to test for the existence of interactions between individuals. Shapes of the individuals are mapped in the field and intersections (i.e., interactions) between them are quantified and tested against the expectations of a null model of spatial configuration of shapes (e.g., random rotation). We applied our new method to study community assembly in a two-phase pasture-like alpine Mediterranean community in central Spain analysing pair-wise interactions between the 11 most abundant species and compared the results to those obtained by a traditional point pattern analysis (ppa). We found that our new method increased the number of significant interactions from 21% (ppa) to 55% and changed the sign of around half the interactions detected with ppa. Most significant interactions were negative. Therefore, our method provides a new efficient tool to study community assembly in alpine communities.

T3
P0758

What is driving species' independence in a species-rich shrubland? Evidences for a stochastic geometry of biodiversity

Julia Chacón-Labella, *Marcelino De La Cruz, Adrian Escudero*
Universidad Rey Juan Carlos

Most ecological theories seeking to explain coexistence in megadiverse communities share a set of three rules describing a stochastic geometry of biodiversity: (1) individuals show intraspecific clustering; (2) species abundances vary following a log-normal distribution; (3) the spatial arrangement between species is independent. The first two rules have received strong empirical support, however the third one remains highly unexplored. To fill this gap, we evaluated the independent species arrangement rule in a species-rich shrubland, and its potential drivers: the levels of species richness and intraspecific clustering experienced by a given species at different scales and the relative abundance of such species in the community. Here we show that interspecific associations were rare and that independence was positively related to species richness and intraspecific clustering, and negatively related to relative species abundances. Our results concur with the independent species arrangement rule and provide empirical support to a stochastic geometry of biodiversity. In a context of species-rich plant communities, the likelihood of two species to encounter become really small. However, what our study reveals as a novelty in this context, is that both, intraspecific clustering (as the result of dispersal limitations) and the relative species abundances are playing a fundamental role determining the probability of two species to encounter and interact, especially at very fine spatial scales.

T3
P0759

Individual species affect plant traits structure in their surroundings: Evidence of functional mechanisms of assembly

Julia Chacón-Labella, *Marcelino De La Cruz, David S. Pescador, Adrian Escudero*
Universidad Rey Juan Carlos

Evaluating community assembly through the use of functional traits is a promising tool for testing predictions arising from Niche and Coexistence theories. Although interactions among neighbouring species and their inter-specific differences are known drivers of coexistence with a strong spatial signal, assessing the role of individual species on the functional structure of the community at different spatial scales remains a challenge. Here, we ask whether individual species exert a measurable effect on the spatial organization of different functional traits in local assemblages. We first propose and compute two functions that describe different aspects of functional trait organization around individual species at multiple scales: individual weighted mean area relationship and individual functional diversity area relationship. Secondly, we develop a conceptual model on the relationship and simultaneous variation of these two metrics, providing five alternative scenarios in response to the ability of some target species to modify its neighbour environment and the possible assembly mechanisms involved. Our results show that some species influence the spatial structure of specific functional traits, but their effects were always restricted to the finest spatial scales. In the basis of our conceptual model, the observed patterns point to two main mechanisms driving the functional structure of the community at the fine scale, "biotic" filtering meditated by individual species and resource partitioning driven by indirect facilitation rather than by competitive mechanisms.

T3

P0760

What is driving species' independence in a species-rich shrubland? Evidences for a stochastic geometry of biodiversity

Julia Chacón-Labella, *Marcelino De La Cruz, Adrian Escudero*

Universidad Rey Juan Carlos

Most ecological theories seeking to explain coexistence in megadiverse communities share a set of three rules describing a stochastic geometry of biodiversity: (1) individuals show intraspecific clustering; (2) species abundances vary following a log-normal distribution; (3) the spatial arrangement between species is independent. The first two rules have received strong empirical support, however the third one remains highly unexplored. To fill this gap, we evaluated the independent species arrangement rule in a species-rich shrubland, and its potential drivers: the levels of species richness and intraspecific clustering experienced by a given species at different scales and the relative abundance of such species in the community. Here we show that interspecific associations were rare and that independence was positively related to species richness and intraspecific clustering, and negatively related to relative species abundances. Our results concur with the independent species arrangement rule and provide empirical support to a stochastic geometry of biodiversity. In a context of species-rich plant communities, the likelihood of two species to encounter become really small. However, what our study reveals as a novelty in this context, is that both, intraspecific clustering (as the result of dispersal limitations) and the relative species abundances are playing a fundamental role determining the probability of two species to encounter and interact, especially at very fine spatial scales.

T3

P0761

Strong effects of slope aspects on arbuscular mycorrhizal fungal communities: Evidence from differences in composition on Qilianshan Mountains, Northwest China

Yuxing Chai, *Huyuan Feng*

School of Life Sciences, Lanzhou University

Arbuscular mycorrhizal (AM) fungi are root-associated microorganisms which play an important function in the terrestrial ecosystem. The diversity of AM fungi can be affected by series of biotic and abiotic factors. As an important topographical factor, slope can change the biodiversity and community structure via changing the illumination, temperature and soil physical and chemical properties. But for underground microbes, especially for arbuscular mycorrhizal (AM) fungi, its distribution characteristics along slope is rarely reported. The Tibetan Plateau is considered to be particularly vulnerable to climate changes. Meanwhile there are many natural many similar ridges and valleys in which the vegetation types are significant different. To reveal the distribution of AM fungi along the different slopes aspect, an investigation about the diversity of AM fungi was conducted on the Qilianshan Mountains. Significant difference of vegetation types existed in the different slopes, and the North-facing slope contained the higher species richness. The vegetation in South-facing slope mainly consisted of some grasses and compositae, such as *Stipa capillata*, *Heteropappus altaicus,* while the vegetation in North-facing slope was dominated by weeds, such as *Polygonum viviparum*, *Anaphalis lactea.* Inconsistent to the plant distribution pattern, the North-facing slope showed lower root length colonization (% RLC) and AM fungal extraradical hyphal length density. A total of 32 AM fungal phylotypes were described in our study using 18S rRNA clone library, which belonged to 7 genus: *Glomus*, *Rhigophagus*, *Funneliformis*, *Septoglomus*, *Claroideoglomus*, *Diversispora*, *Gigaspora* . There were 22 phylotypes found in North-facing slope and 23 phylotypes found in South-facing slope. Furthermore, significant difference of the AM fungal community composition also existed between different slope, and soil pH, moisture, organic C, total N and available N were significantly correlated with the AM fungal community. Besides, the variance decomposition results show that 29.82% of the variation of community structure of AM fungi was explained, and revealed that plant properties (12.42%) explained the largest variance in AM fungal composition. And the similarity of plant properties and composition, which is usually determined by climatic and soil conditions, was positively correlated with the similarity of the AM fungal composition. Our data provide important insights that host plants and mutualistic fungi may respond to slope aspect change idiosyncratically, potentially altering carbon and nutrient cycles in relation to the plant–fungus associations.

T3

P0762

Natural colonization of plant species in sites deranged by opencast mining in Raniganj colliery, Burdwan District, West Bengal

Priyanka Chatterjee, *Ambarish Mukherjee*

The University of Burdwan

The present work reveals the floristic assemblage and the physico-chemical characters of the over-burdens (OB) resulting from dumping of residues (spoil) from the opencast mining of coal at Raniganj Colliery of Burdwan district, West Bengal. Values of pH of the spoils collected from six sites ranged from 6.6 (Nidha) to 8.63 (Bhuianpara). The organic carbon content of the spoils ranged from > 0.5% (site Nathman colliary; SiteIIA &IIB) to 0.5% (Site Nidha, Bhuianpara and Mahabir chanak). Negligible amount of nitrate–nitrogen could be found in the spoil of Sahebkothi which was very low in case of Nathman Colliary (Site IIA). Interestingly the nitrate-nitrogen content was found to be very high in Nathman colliery (Site IIB) and high in case of others (Nidha, Bhuianpara and Mahabir chanak). Spoils were with low contents of ammoniacal nitrogen in all cases in general and that of Sahebkothi in particular. The phosphate content of spoils of Sahabkothi was negligible and those of Nidha and Bhuimpara were low. Medium levels of phosphate were found in the spoil samples collected from Mahabir chanak and the second site of Nathman colliary (SiteIIB).

The first site of Nathman colliary (Site IIA) was exceptional in having high content of phosphate in their spoil samples. From the study sites as many as 37 species of angiosperms belonging to 15 families were detected. Interestingly all the individual plants were highly reduced in size under extreme xeric conditions. So far the dominant families are concerned Poaceae with six species and followed by Acanthaceae with four species. The most prevalent species is *Croton bonplandianum* which is present in 4 out of 5 sites studied (80%) which is followed by *Tephrosia purpurea* with prevalence value of 60%.

T3

P0763

Overexpression of a *Chorispora bungeana CBF* gene in *Arabidopsis thaliana* enhances freezing tolerance and affects growth

Jian Chen, *Lixiang Ma, Zhiguang Zhao, Huyuan Feng*
School of Life sciences, Lanzhou University

Freezing stress is a major environmental factor that limits plants growth and development. CBFs play important roles in plant freezing tolerance. In this work, we analyzed the functions of CbCBFx1 from a subnival alpine plant, Chorispora bungeana. The results showed that the CbCBFx1 was localized to the nucleus and expressed in roots, leaves. Further analysis showed that the expression of CbCBFx1 was influenced by the circadian clock and cold stress. CbCBFx1 overexpression in Arabidopsis resulted in freezing tolerance increasing, cold-regulated (COR) genes expression, vegetative growth inhibition and flowering delay. Furthermore, our results showed that overexpression of CbCBFx1 in *Arabidopsis* leads to multiple biochemical changes associated with cold acclimation. These results suggest that CbCBFx1 might play a role in cold acclimation and freezing tolerance of *C. bungeana* in subnival alpine environment.

T3

P0764

Rising variability as a leading indicator of a stochastically driven abrupt transition in a dryland ecosystem

Ning Chen[1], *Ciriyam Jayaprakash*[2], *Kailiang Yu*[3], *Vishwesha Guttal*[4]
1. Lanzhou University
2. The Ohio State University
3. University of Virginia
4. Indian Institute of Science

Complex systems can undergo abrupt state transitions near critical points. Theory and controlled experimental studies suggest that the approach to critical points can be anticipated by critical slowing down (CSD), i.e., a characteristic slowdown in the dynamics. Validity of this indicator in field ecosystems, where stochasticity is important in driving transitions, remains unclear. We analyzed long-term data of a dryland ecosystem and showed that it underwent an abrupt transition from a nearly bare to a moderate grass-cover state. Prior to the transition, the system showed no (or weak) signatures of CSD, but exhibited expected increasing trends in variability of the grass-cover, quantified by variance and skewness. We argue that these surprising results are consistent with the theoretical expectation that abrupt transitions occurring away from critical points due to increasing driver stochasticity; Indeed, a driver of vegetation - annual rainfall - showed rising variance prior to the transition. Our study suggests that rising variability can potentially serve as a leading indicator of stochastically driven transitions in real world ecosystems.

T3

P0765

How do ontogeny, temperature and precipitation impact on biomass allocation for desert ephemeral plants and shrubs in Northwestern China?

Renfei Chen
Lanzhou University

Biomass allocation plays a key role in precisely predicting global carbon sink. In this paper, we observed the biomass allocation pattern for vegetation in Northwestern China (including ephemeral plants and shrubs in deserts) as well as global forests respectively. Furthermore, we explored the variation of Root Shoot Ratio (R/S) along with total biomass and climate. In addition to this, we also studied the consistency between optimal theory and allometric approach. Lastly, we discussed the effect of ontogeny and climate on the scaling exponent between above- and belowground biomass for different plant types. Our results show that (1) the scaling relationships between above- and belowground are allometric for ephemeral plants and shrubs in desert of Northwestern China. (2) For both ephemeral plants and shrubs, there is no consistency observed between two biomass partitioning theories. (3) Moreover, unimodal curve occurs which indicates that scaling exponent varies in logistic rule along with plant ontogeny. According to climate factors analysis, plant taxon plays an important role in biomass allocation.

T3

P0766

Studies on *Platycarya* (Juglandaceae) highlight the boundary between floristic sub-kingdoms in the East Asia flora and adaptive genetic variation

Shichao Chen[1], *Douglas E. Soltis*[2], *Pamelas Soltis*[2], *Li Zhang*[1]
1. Tongji University
2. University of Florida, Florida Museum of Natural History

The flora of Eastern Asia is distinctively rich compared with other floristic regions in the northern hemisphere. However, knowledge of its evolution is limited. The genus *Platycarya* (Juglandaceae) is endemic to Eastern Asia. It was once widespread and provides an excellent opportunity to investigate the role of Pleistocene climatic change on the distribution of Eastern Asian biodiversity. We employed AFLP and cpDNA and nuclear sequence data, and sampled plants over the present distribution, to reconstruct phylogenetic relationships among haplotype and their geographical distribution patterns - we sought to highlight its origin, speciation, migratory route and demographic history. The geographic pattern of cpDNA suggests a long-term fragmentation and multiple refugia survival during the Quaternary climatic oscillations; no recent expansions were suggested. Previous study revealed two divergent lineages in a line with the boundary between the Sino-Japanese Floristic Region and Sino-Himalayan Region. In Platycarya, we discuss the possible formation of two Eastern Asiatic Floristic Sub-Kingdoms. The new RAD-seq technique produced high-resolution genomic data sets and available environmental databases promise to increase our knowledge of adaptive genomic variation of Platycaryaon the landscape.

T3

P0767

Ammonia emissions from agriculture over China

Youfan Chen[1], *Lin Zhang*[1], *Yuanhong Zhao*[1], *Binxiang Huang*[2]

1. Peking University

2. China Agriculture University

This study presents an improved bottom-up estimate of ammonia emissions from fertilizer use in China at a horizontal resolution of 0.5° × 0.5° and monthly variability. We collect the data of crop planting area, fertilizer application time and rate for 18 main crops to estimate the fertilizer application amount in China. Ammonia emission factors from fertilizer use are estimated by soil properties such as soil pH, cation exchange capacity (CEC), and agricultural activity information such as crop type, fertilizer type, application mode, *etc.* We further consider ambient temperature and wind speed to account for the meteorological influences on ammonia emission factors. The derived ammonia emission from fertilizer use in China is about 4.75 Tg NH_3 for the year 2008. Henan province and Jiangsu province are the two largest emitting areas (621 Gg NH_3 and 469 Gg NH_3). The ammonia emissions from fertilizer use have distinct seasonal variability; the largest emission is in June with nearly 816 Gg NH_3 emitted, and the minimum is in January (90 Gg NH_3). We also estimate the ammonia emission from livestock is 5.31 Tg using method adjusted from Huang *et al.*, (2012). Combining with other ammonia source (e.g. human waste and transport) estimates from the REAS v2.1 emission inventory, we obtain a total ammonia emission inventory in Asia for the year 2008. The total ammonia emissions in Asia are 30.5 Tg NH_3 in 2008. China emits 12.6 Tg NH_3, accounting for 41.3% to the total Asia ammonia emissions. The most important ammonia sources are fertilizer use and livestock waste, which contribute respectively 46.9% and 29.1% in Asia, and 37.8% and 42.1% in China. With this improved Asian ammonia emission inventory, we use the GEOS-Chem chemical transport model to simulate the ammonium wet deposition fluxes over Asia, and compare with measurements from EANET. The model simulated ammonia wet deposition fluxes show a good agreement with measurements (r = 0.62-0.91). What's more, the temporal and spatial distribution of our ammonia estimate is consistent with ammonia column from satellite (IASI).

T3

P0768

Mechanism of plant invasion in relation to performance, chemical defense and herbivore resistance: A case study of *Senecio vulgaris* L. (Asteracea) in China

Dandan Cheng[1,2], *Jiwen Ge*[1], *Viet -Thang Nguyen*[1], *Noel Ndihokubwayo*[1]

1. School of Environmental Studies, China University of Geosciences (Wuhan)

2. State Key Laboratory of Biogeology and Environmental Geology, China University of Geosciences (Wuhan)

The research of biological invasion is valuable in relation to the theoretical and practical aspects. The reason for success of invasion is one of the hotspots of the researches about biological invasion. As an extension of the Evolutionary of Increase Competition Ability (EICA) Hypothesis, the Shift Defense Hypothesis (SDH) argues that the success of plant invasion can be explained by the evolution of chemical defense in the invasive populations. The chemical defense of plants can be divided into two kinds: quantitative defense and qualitative defense. In invasive ranges alien plants are released from specialist herbivores, increase the qualitative defense and reduce the quantitative defense. Thus, they allocated more resource to growth and reproduction, which leads to the more fitness and successful invasion of the plants. *Senceio vulgaris* L. (Asteraceae) is an invasive species wildly distributed in China. Few previous studies have been conducted to investigate the invasive history, mechanism and risk management of this species in China. We conducted experiments with *S. vulgaris* seeds collected from the native and invasive populations, in order to figure out whether the invasive populations of *S. vulgaris* in China have evolved the shift defense and improved the fitness. Comparison of plants grown in green houses and common gardens in native and invasive ranges showed that the invasive *S. vulgaris* plants outperformed than their conspecific plants in relation to the growth and reproduction traits. Moreover, the relative dry matter content, considering as quantitative defense, was no different in these two kinds of plants. We determined pyrrolizidine alkaloids (PAs) from shoot and root of *S. vulgaris* plants originated from native and invasive populations using Liquid Chromatography–Tandem Mass Spectrometry (LC-MS/MS). There was a slight trend indicating lower PA diversity and lower total PA concentration in invasive *S. vulgaris* populations than native populations. A non-choice feeding bioassay was conducted in a laboratory using generalist herbivore (White jade land snail, *Achatina fulica*) fed with leaves from native and invasive *S. vulgaris* plants. The result indicated that native and invasive *S. vulgaris* plants are resistant to *A. fulica*, and there were no significant differences between native and invasive *S. vulgaris* plants in relation to the resistance against *A. fulica*. We confirmed the outperformance of *S. vulgaris* plants originated from invasive populations compared to those originated from native populations. But we did not regard that this was attribution to the chemical defense shift, because the PAs, the main qualitative defense compounds, in the invasive plants were not more than those in the native plants. Another important result is that the invasive *S. vulgaris* were not more resistant again a general snail than the native ones. As a summary, we regarded that the invasive *S. vulgairs* might retain increased completive ability, as expected by EICA. However, not as predicted by SDH, this is not necessarily caused by enemy release or defense shift.

T3

P0769

Habitat environment and population characteristics of *Megaleranthis saniculifolia* Ohwi

Seung Ho Choi, *Jin Su Seo, Kyeong Bae Nam, Bon Yeol Koo, Hye Jin Kwon, Yong Chan Cho, In Sik Kim*

Korea National Arboretum

Megaleranthis saniculifolia is a perennial herb belong to family Ranunculaceae. It is a monotypic species and only found in Korea (rare and endangered species; EN class). Although interest on conservation and restoration of biodiversity is increased, researches for understanding species and habitat characteristics are relatively insufficient. Through examining population characteristics, we can understand rationally the environment, vegetation change and bio-interaction processes of a species in a particular area. Thus, this study was conducted to evaluate habitat environment and population characteristics of *M. saniculifolia* in natural habitat such as Mt. Dosolsan and Mt. Jumbongsan in Gangwon-do and Mt. Sobaeksan in Chungcheongbuk-do. The distribution pattern of *M. saniculifolia* was mostly formed a group of discontinuous populations in the shaded sites along the valley. The average irradiance in study areas was 40 ± 15.8% (Mt. Dosolsan), 29 ± 7.4% (Mt. Sobaeksan) and 25 ± 14.1% (Mt. Jumbongsan). The variation in specific lear area (SLA) of mature individuals was significantly different among populations; Mt. Sobaeksan (n = 12, 350 ± 63.1 $cm^{-2}\,g^{-1}$), Mt. Dosolsan (n = 10, 290.7 ± 57.0 $cm^{-2}\,g^{-1}$), and Mt. Jumbongsan (n = 7, 277.7 ± 35.2 $cm^{-2}\,g^{-1}$). Relatively high SLA in Mt. Sobaeksan is indicative of low leaf nutrient efficiency and water stress tolerance of individuals in the area. Leaf area and SLA were negatively correlated each other. SLA was positively correlated with altitude in all three areas. The average biomass was not significantly different among populations; Mt. Sobaeksan (0.6 ± 0.5 g), Mt. Jumbongsan (0.5 ± 0.8 g). Mt. Dosolsan (0.4 ± 0.1 g). The biomass ratio (above/underground) was 3.0 ± 1.9 in Mt. Jumbongsan, 2.4 ± 0.9 in Mt. Dosolsan and 0.5 ± 0.0 in Mt. Sobaeksan ($p > 0.05$). 11 populations of *M. saniculifolia* showed a stable population structure of the inverted J type, which number of individuals gradually decreasing with increasing the diameter of the stem. Overall result indicated that small individuals of the species are constantly settling down.

T3

P0770

Responses of seed size variation on the germination of a coastal tree species: *Calophyllum inophyllum*

Jin Chang Yang[1], ***Fung Yun Chong***[2], *Ying Fah Lee*[2], *Roland Kueh*[3]
1. Research Institute of Tropical Forestry, Chinese Academy Of Forestry
2. Sabah Forestry Department
3. Universiti Putra Malaysia

As part of the China-Malaysia bilateral research collaboration, a study on seed characteristics of the coastal tree species, *Calophyllum inophyllum* was conducted in Sabah, Malaysia. This study was aimed to trace the response of seed characteristic toward seed germination. *C. inophyllum* is a coastal tree that helps in stabilizing the soil erosion, and sometime being used in medicinal and biodiesel oil purposes. The *C. inophyllum* seeds were collected from 6 trees for each study site (N = 24) along the coastal areas representing Eastern (1 plot) and Western parts of Sabah (3 plots), Malaysia. The seed size variations, namely seed length, seed diameter and seed weight were measured and further germination test was conducted at the nursery of the Research Institute of Tropical Forestry, Chinese Academy of Forestry. Our findings showed that the seed germination was significantly related to seed weight, while there were no relationship to the seed length and width. With respect to germination, optimum seed weight, 3.5-4.0 g gave seedlings higher germination compared to lighter seeds or heavier seeds. These seed traits variation can help in improving to the knowledge of their ecology, especially in regards to conservation and establishment of excellent germplasm resources.

T3

P0771

Germination characteristics of two desert species and its ecological significance in the low temperature condition

Guangming Chu
Shihezi university

Germination behaviour played an important role for seedlings regeneration and the sustainable development of population in arid and semiarid areas. The objectives of the study were to explore the low temperature germination characteristics of Haloxylon ammodendron and Anabasis aphylla and its ecological significance in Junggar basin of China. Germination was studied in the illumination incubators with six temperature gradients of 20/15 ℃, 15/10 ℃, 10/5℃, 5/2 ℃, 2 ℃ and 2/-3 ℃ for 16 days. Final germination percentages of two species among temperatures were not obviously different and all were greater than 90%. Meanwhile, germination percentages of two species surpassed 50% in the eighth day under low temperature condition. In conclusion, final germination percentages of *H. ammodendron* and *A. aphylla* between low temperature and high temperature were not obviously different and only reduced germination speed. Germination behavior of two desert species in the low temperature habitat may take advantage of favorable edaphic conditions in the snowmelt period, and promote seedling growth and lignification before experiencing desiccation of early spring. Moreover, *H. ammodendron* and *A. aphylla* had different germination strategies in the low temperature habitat.

T3

P0772

Fruit production adjustment in a dynamic coastal environment: Insights from *Corema album*, a dioecious shrub

Ana Martins[1], ***Joana Costa***[1], *Helena Freitas*[1], *Sofia Costa*[2,3]
1. Centre for Functional Ecology, University of Coimbra
2. Mountain Research Center, Polytechnic Institute of Bragança
3. Centre of Molecular and Environmental Biology, University of Minho

Flowering plants usually produce large floral displays, but only a small fraction of flowers develop into mature fruits. Available empirical evidence suggests that resource availability is fundamental in regulating fruit maturation and consequently, female reproductive success. We used *Corema album*, a dioecious, wind-pollinated, coastal shrub, to investigated Lloyd's serial adjustment hypothesis (that is, plants adjust the maternal investment in reproduction at three sequential stages: flower development, ovary production, and fruit and seed maturation, depending on resource availability). Because male and female functions are separated in different individuals, *C. album* allows a comprehensive analysis of patterns of flower and fruit production from the maternal viewpoint. To test the serial adjustment hypothesis, we examined flower and fruit production in two habitat types: coastal semi-fixed dunes, and woodland further inland, from three selected sites inhabited by *C. album* along the Portuguese West coast in a north to south cline: Tocha (T), São Pedro de Moel (S) and Vila Nova de Santo André (A). In sites T and S, five males and five females were randomly marked for sampling in each of three plots established in each habitat type. In site A, sampling took place only in the semi-fixed dune. In addition, flower and fruit related variables were measured in ten inflorescences/infructescences per marked individual. Overall, the site had a stronger impact on flower and fruit production variables than habitat type. Nevertheless, fruit weight and diameter, and seed set showed no variation between habitat types or among sites. We detected a strong positive correlation between fruit weight and diameter ($r = 0.95$), and fruits originating from larger infructescences were also larger ($r = 0.34$) and heavier ($r = 0.33$). Fruit set was also larger for larger infructescences ($r = 0.66$). Female plants displaying more flowers per inflorescence also produced more ($r = 0.45$) and larger fruits ($r = 0.27$), but had a lower fruit set ($r = -0.29$) and the seeds were lighter ($r = -0.46$). *Corema album* inhabits in a dynamic environment, and several factors conditioning resource availability (e.g., solar irradiance, water availability, salinity) differ considerably between flower development and fruit maturation times. Differences in flower production among sites suggest intra-specific genetic variation and/or environmental heterogeneity. Collectively, our results provide evidence supporting Lloyd's serial adjustment hypothesis in *C. album*. Females produced a large number of flowers per inflorescence, thus maximizing the probability of successful mating. Subsequent maturation of a reduced number of fruits supports sequential maternal adjustment according to limiting factors, such as resource availability. This conclusion is further supported by the fact that plants that yielded more fruits per infructescence were also able to produce larger and heavier fruits. In conclusion, the adjustment of fruit development under dynamic environmental conditions is likely to play an important role in the overall reproductive success of *C. album*.

T3
P0773

Antioxidant capacity in sedges on the Qinghai-Tibetan Plateau is adequate for yaks

Guangxin Cui[1,2], *Fuping Tian*[1], *Yuan Lu*[1], *Yu Hu*[1]
1. Lanzhou Institute of Husbandry and Pharmaceutical Sciences of Chinese Academy of Agriculture Science
2. International Centre for Tibetan Plateau Ecosystem Management, Lanzhou University

Traditionally, yaks (*Poephagus grunniens*) raised on the Qinghai-Tibetan Plateau, an area characterized by strong UV radiation, hypoxia and severe cold, are often subjected to poor nutrition and used as pack animals. The extremely harsh environment on the Qinghai-Tibetan Plateau is associated with high oxidative stress to the animals and with an increase in altitude there is an increase in oxidative stress. Keeping a proper balance between oxidants and antioxidants is crucial for biological systems to maintain a healthy status. As antioxidant production in animal cells is limited, consumed feed is the basis for the elimination of ROS. Under traditional husbandry, yaks graze only natural pasture and receive little or no feed supplement. In fact, yaks lose close to 25% of their bodyweight during the cold season because of insufficient feed intake. Study shows that protein, fat, total phenols, tannins contents and trolox-equivalent antioxidant capacity (TEAC) within plant species tended to increase with increasing elevation. Sedges are the main forage consumed by yaks, accounting to 64.1% of dietary dry matter intake in summer, although three (*Kobresia humilis, Kobresia capillifolia* and *Kobresia royleana*) of four sedges available to yaks on the Qinghai-Tibetan Plateau were classified as containing a medium level of TEAC and one (*Kobresia tibetica*) as low. Yaks are well adapted to the harsh conditions both anatomically and physiologically. Thus it is conceivable that the antioxidant capacity in the high altitude sedges was adequate for yaks .

T3
P0774

Combined effects of water and nitrogen addition on nitrogen recovery in *Haloxylon ammodendron* based ecosystems in the Gurbantunggut Desert

Xiaoqing Cui[1], *Ping Yue*[2], *Xuejun Liu*[1]
1. China Agricultural University
2. Xinjiang Institute of Ecology and Geography, Chinese Academy of Sciences
3. China Agricultural University

Desert ecosystems are prone to respond to global climate change and reactive nitrogen deposition in biological and biogeochemical respect. The effect of increase precipitation and nitrogen deposition in deserts largely depend on its fate in the plant- soil ecosystems. However, there had limit researches about the fates of nitrogen deposition in desert ecosystems, and quantifying plant and soil retention response to water and nitrogen increase. Here we reported the results using ^{15}N labeling experiment from the *Haloxylon ammodendron* based Gurbantunggut Desert in northwest China. We chose $^{15}NH_4NO_3$ as ^{15}N tracer and added water and nitrogen twelve times every year evenly distributed in three month of growing seasons in the year 2015 and 2016. The herbs exhibited a lager sink of nitrogen retention than *Haloxylon ammodendron* forests, and the nitrogen retention had different response to water and nitrogen addition in 2015 and 2016 relative to precipitation. The different parts of Haloxylon ammodendron forests had distinct recovery with the branches largest retention. Soil as the majority sink of nitrogen retention, most of nitrogen accumulated in the topsoil layer and the topsoil and subsoil had different response to water and nitrogen addition in different years. The whole ecosystem recovery ranged from 40% to 60%. The whole ecosystem retention significantly increased with precipitation increase and adding nitrogen deposition decreased the whole ecosystem recovery inverse. All this indicated that in future climate change the desert ecosystem recovery will increase.

T3
P0775

Within-inflorescence resource reallocation patterns following differential pollination in *Sagittaria trifolia*

Can Dai, *Lu Miao*

Hubei University

Resource allocation to reproduction through time and space have long attracted ecologists and yielded many ecological principles. Despite its popularity and importance, resource reallocation, the possibility of resource movement among flowers during fruit production, has been rarely examined and largely taken for granted. In this study, we aimed to empirically explore resource redistribution patterns among flowers within inflorescences in response to differential pollination intensity. Using a common herb *Sagittaria trifolia*, we conducted supplemental and controlled pollination for single, partial, or all the flowers in simple and complex inflorescences. The fruit set, seed production, and average seed weight between flowers of different manipulations or different treatment levels were compared within each inflorescence type, in order to reveal the possibility of resource reallocation and the limiting factors for reproduction on a fine scale. Pollen supplementation on a single flower significantly promoted fruit set; however, the same manipulation on an inflorescence did not result in any increase in reproduction. Comparing the fruit set following supplemental pollination at flower and inflorescence levels, single flowers had a higher percentage fruit set than the inflorescences. In complex inflorescences, there was no difference between supplemental and controlled pollination for flowers on main branches, neither for lateral branches. The results pointed out that single flower manipulation would exaggerate pollen limitation and lead to biased understanding on individual's reproductive status. There was evidence of resource reallocation among flowers in simple inflorescence following differential pollination. However, resource was unlikely to move between main and lateral branches in complex inflorescences, suggesting a flowering branch as an integrated physiological unit (IPU) in *S. trifolia*.

T3

P0776

16S rDNA fingerprinting of rhizosplane soil bacterial communities associated with *Elephantopus scaber* L., a reputed medicinal plant

Moumita Das, *Abhijit Bandyopadhyay, Ambarish Mukherjee*
The University of Burdwan

A growing body of research on plant–microbe interactions in soil has been contributing to the development of new microbial perspectives of community-ecology. Soil-dwelling microorganisms are diverse and their interactions with plants vary with respect to specificity, environmental heterogeneity, and fitness impact. Plant-associated microorganisms fulfill important functions for plant growth and health. Direct plant growth promotion by microbes is based on improved nutrient acquisition and hormonal stimulation. Diverse mechanisms are involved in the suppression of plant pathogens, which is often indirectly connected with plant growth. The species is rich in different types of secondary metabolites many of which are bioactive, thus reflecting efficacy in adaptive and defensive strategies. The present study focuses on association of bacteria in rhizoplane with *Elephantopus scaber* L., a plant traditionally popular in treatment of cancer, diabetes, rheumatism, fever, scabies and many other ailments. The plant usually prefers sandy-loam to lateritic soil with alkaline pH ranging from 7.97 to 8.47, concentrations of organic carbon ranging from 0.64 to 1.89% ; phosphorus ranging from 0.92 to 6.1 mg/L; nitrogen from 31 to 50 mg/L and potassium from 105 to 245 mg/L. Molecular identification of the bacteria was done by 16S-rDNA sequencing which enabled to record three bacterial entities in data point 1 i.e. Golapbag campus of Burdwan University and 4 in data point 2 i.e. Shibpur forest beat of Durgapur forest range. Capability of producing extracellular enzymes like catalase and citrate were checked. For better visualization of the isolates fluorencence microscopy and scanning electron microscopy were done. It was *Psychrobacillus psychrodurans* which showed psychrophilic nature. Sensitivity to antibiotics viz. Ampicilin, Streptomycin and Chloramphenicol was cheeked against all the isolates. Interestingly only one isolate, *Psychrobacillus psychrodurans*, was found to be resistant to all the three type of antibiotics used. Gram properties of the isolates were also visualized. The need for intensive exploration of beneficial plant–microbe interactions was realized since such a study offers promising environment friendly strategies for conventional and organic cultivation of plants worldwide.

T3

P0777

Fungal symbionts of the conifer-infesting bark beetle *Ips typographus* in China

Rl Chang[1], *Ta Duong*[1], *Sj Taerum*[1], *Mj Wingfield*[1], *Xd Zhou*[2], ***Zw De Beer***[1]
1. Forestry and Agricultural Biotechnology Institute (FABI), University of Pretoria
2. FuturaGene Biotechnology

Ips typographus (Coleoptera, Scolytinae) infests several species of conifer trees, including *Pinus* (pine), *Larix* (larch), and *Abies* (fir), but it is primarily known for the considerable damage it can cause to *Picea* (spruce) forests. This happens especially when populations of the beetle build up where trees were damaged or stressed by external factors like wind or drought. Although some studies have shown that populations of *I. typographus* in Europe, Japan and China are genetically distinct, the populations are biologically similar, including a strong association with ophiostomatoid fungi, some of which might assist the beetle in overcoming host responses of the tree. To date, only two *Leptographium* spp. have been reported from *I. typographus* in China, while more than 40 and 10 ophiostomatoid species are known associates of the beetle in Europe and Japan respectively. In this study, we collected and identified the ophiostomatoid fungal associates of *I. typographus* in northeastern China, and determined whether the fungal assemblages reflect the different geographical populations of the beetle. Field surveys in Jilin and Heilongjiang provinces yielded a total of 1,046 isolates from 145 beetles and 178 galleries. Isolates were grouped based on morphology and representatives of the groups were identified using DNA sequences of the ribosomal LSU, ITS, β-tubulin and elongation factor 1-α gene regions. A total of 24 species of ophiostomatoid fungi were identified, including 12 previously described species and 12 novel species. The dominant species were *Ophiostoma bicolor*, *Leptographium taigense* and *Grosmannia piceiperda* representing 40%, 28% and 18% of the isolates respectively. Making comparisons between the species from China, Europe and Japan is complicated by the fact that some of the European and all the Japanese identifications were based on morphology only and DNA sequences are lacking for these fungi. However, assuming that previous identifications are correct, only five species (*Endoconidiophora polonica, G. cucullata, G. penicillata, O. ainoae* and *O. bicolor*) are shared between Europe, Japan and China which matched with the population structure of the beetle. Only one fungal species is shared between China and Japan (*O. japonicum*), five species (*G. olivacea, Graphium fimbriisporum, L. taigense, O. brunneolus* and *O. floccosum*) are shared between Europe and China, and two species (*Ceratocystiopsis minuta,* and *O. piceae*) are shared between Europe and Japan. The results from our work showed that the symbionts of the bark beetle could only reflect the population structures to some degree, and the use of fungal symbiont assemblages to infer population structures and invasion history of its vectors need to be carefully interpreted. In addition, our results highlight the risks of moving infested logs or timber of the tree species involved between different countries and geographical regions, as new fungal species that might pose a risk to tree health can inadvertently be introduced together with the beetles. As has been the case with other conifer-infesting bark beetles, climate change will have an impact on the spread of *Ips typographus*, but to fully comprehend this impact, more research on this beetle and its fungal symbionts is needed.

T3

P0778

Re-evaluating the genus *Grosmannia* (Ophiostomatales, Ascomycota), with special emphasis on conifer-infecting species from China

M Yin[2], *Mj Wingfield*[1], *Xd Zhou*[1], ***Zw De Beer***[1]

1. Forestry and Agricultural Biotechnology Institute (FABI), University of Pretoria, Pretoria

2. College of Forestry and Landscape Architecture, South China Agricultural University, Guangzhou, Guangdong Province

The fungal genus *Grosmannia* (Ophiostomatales, Ascomycota) includes numerous ecologically and economically important species that stain and degrade wood. The genus is typified by *G. penicillata*, and includes an additional 19 species, most of which are associates of conifer-infesting bark beetles. Many of these species were previously treated in the genus *Leptographium,* but the two genera were recently separated based on a comprehensive multigene phylogenetic study of the Ophiostomatales. During a survey of ophiostomatoid fungi associated with bark beetles and sapstained wood in the spruce forests of western China, we obtained 45 isolates from three bark beetle species (*Ips shangrila, I. nitidus* and *Polygraphus poligraphus*) infesting *Picea purpurea* and *P. crassifolia*. The aim of this study was to identify these isolates based on multigene (ITS, βT, TEF1-α, and Calmodulin) phylogenetic analyses and morphological comparisons with other species in the genus. At the same time it provided the opportunity to revise the taxonomy and produce a robust phylogeny including all species in the genus for which cultures were available. The results revealed four novel *Grosmannia* species from spruce in China that are in the process of being described. Some isolates from pine in the USA that had previously been identified as *L. abietinum*, also represented a distinct taxon that is being described. An interesting outcome of the study was that conifer isolates were separated into three well-supported lineages that respectively included six European and Asian species from spruce, one Norwegian species from spruce, and five North American species from spruce, fir and larch. A fourth lineage included three species from hardwoods in East Asia and Africa. From our results it is clear that the diversity of ophiostomatoid fungi occurring in Chinese forests need to be explored further, as some of these fungi and their beetle vectors might pose threats to forests when introduced into new environments as a result of human activity or movement due to a changing climate.

T3

P0779

Pollen limitation and reproductive effort: The role of the quantity and quality factors on fitness consequences and reproductive costs

Rafael F. Del Castillo*, Sonia Trujillo-Argueta*

Instituto Politécnico Nacional

Understanding the factors limiting reproduction and their consequences is fundamental for answering essential questions in biology such the evolution of sex, breeding systems, and life history. Pollination often limits the female reproductive function. However, pollen limitation has often been estimated only in terms of seed numbers; thus, its fitness consequences remain elusive. Life history theory posits that organisms can succeed in a given environment through the combination of resources allocated to survival, growth, and reproduction that optimizes fitness. By competing for the same resources, fitness components are constrained by trade-offs. Consequently, resource allocation affects not only the processes to which plant resources are allocated but also the processes to which resources are not allocated. This requires assessing not only the increases in seed quantity and quality gained by supplemental pollination but also their costs. Such costs instead can be diverted to other fitness components. If pollination levels are suboptimal, provisional commitment to seed production can be revoked. Decreases in fruit or seed set may increase fitness at least in two ways: by improving offspring quality or by saving resources that can be diverted to other vital functions. We test these ideas in a sexually polymorphic population of *Opuntia robusta* (Cactaceae) two of which sex morphs have contrasting sensitivities to pollen limitation. Females are entirely dependent upon external pollination; while hermaphrodites, by virtue of their autonomous selfing, depend partially on external pollination. We estimate the quality and quantity of pollen limitation by applying to randomly selected females and hermaphrodites in the field two kinds of pollen treatment to different flowers: pollen from males' plant and pollen from hermaphrodites' and estimated their consequences, not only in terms of fitness but also in terms of reproductive costs, by comparing supplemental pollination with natural pollination. We found that the quality and quantity component of pollen limitation affect significantly both plant fitness and reproductive effort. Fitness estimated in terms of fruit set, seed set, seed germination and seedling survival increases with supplemental pollination. Further, male's pollen tends to boost fitness more than hermaphrodite's pollen particularly in females. We conclude that in females, fitness can be enhanced above the levels expected for plants with lower dependence on external pollination, such as hermaphrodites. Full pollination can make that female seed quantity and quality exceed substantially those of hermaphrodites, but also increased significantly reproductive cost. Under conditions of pollen limitation, however, females may invest less in reproduction than hermaphrodites. Such savings may contribute to explain why females tend to exceed hermaphrodites in clonality under conditions of pollen limitation. Our results suggest that understanding the consequences of pollen limitation requires considering not only their reproductive effects of their quantity and quality components on plant reproduction but also their costs. We hypothesize that individuals more susceptible to pollen limitation might be more plastic in their responses before environmental uncertainty.

T3

P0780

Zonation patterns of pteridophytes along the northeastern slope of a secondary forest in Mt. Makiling, Philippines

Marjorie Delos Angeles, *Inocencio Buot Jr.*
University of the Philippines Los Banos

Pteridophyte zonation patterns as well species composition along the northeastern slope of Mt. Makiling during the wet season were determined. The plot technique method was employed along the altitudinal gradient of Mt. Makiling resulting to 9 sampling sites during the months of December 2010 and January 2011. A total of 31 fern species were identified belonging to 19 genera and 14 families during the wet season. There were three zones identified during the wet season using a dendogram by average clustering: (i) Zone 1: *Bolbitis* – Ferns sp.1 (150-250 m.a.s.l.); (ii) Zone 2: *Lygodium* – *Christella* - *Sphaerostephanos* (350-650 m.a.s.l.); (iii) Zone 3: *Sphaerostephanos* – *Selaginella* (750-1050 m.a.s.l.). Moisture content, pH, and substrate contributed to a higher diversity of ferns during wet season. There is an increasing trend in diversity along the altitudinal gradient.

T3

P0781

Soil organic carbon storage and pattern in Northwest China

Longwei Dong, *Jianming Deng*
Lanzhou university

The desert ecosystem accounts for the 20% of territorial area in China, at the same time soil organic carbon plays a very important role in the terrestrial ecosystem . In this paper, we choose the desert ecosystem located in the northwest China arid areas as the research object. From the investigation in the field, we find that in the depth of one meter the soil organic carbon storage is 7.02 Pg, the average soil organic carbon density is 5.64 Kg C m^{-2}. Among them, the average soil organic carbon density in the desert steppe (6.22 Kg C m^{-2}) is remarkable higher ($P < 0.01$) than that in the desert shrub (3.00 Kg C m^{-2}). The main factors which influence the average soil organic carbon density from zero to twenty centimeters are different for different vegetational types. NDVI and the content of clay respectively explain the variation of soil organic carbon density by 39% and 24% in the desert steppe ecosystems, meanwhile, the content of sand explains that by 31% in the desert shrub ecosystems .

T3

P0782

Potential effects of climate change on geographic distribution of the Tertiary relict tree species *Davidia involucrata* in China

Cindy Q Tang[1], ***Yi-Fei Dong[1]****, Sonia Herrando[2], Tetsuya Matsui[3], Haruka Ohashi[3], Long-Yuan He[4], Katsuhiro Nakao[5], Nobuyuki Tanaka[6], Mizuki Tomita[7], Xiao-Shuang Li[8], Hai-Zhong Yan[1], Ming-Chun Peng[1], Jun Hu[9], Ruo-Han Yang[1], Wang-Jun Li[1], Kai Yan[10], Xiuli Hou[11], Zhi-Ying Zhang[1], Jordi López-Pujol[2]*
1. Institute of Ecology and Geobotany, Yunnan University, Kunming 650091, China
2. Botanic Institute of Barcelona (IBB-CSIC-ICUB), Passeig del Migdia s/n, Barcelona 08038, Spain
3. Center for International Partnerships and Research on Climate Change, Forestry and Forest Products Research Institute, Matsunosato 1, Tsukuba-shi, Ibaraki-ken, 305-8687, Japan
4. Kunming Institute of Forestry Exploration and Design, the State Forestry Administration of China, Kunming 650216, China
5. Kansai Research Center, Forestry and Forest Products Research Institute, Momoyama, Kyoto, 612-0855 Japan
6. Tokyo University of Agriculture, 1-1-1 Sakuragaoka, Setagaya-ku, Tokyo 156-8502, Japan
7. Tokyo University of Information Sciences, 4-1 Onaridai Wakaba-ku, Chiba 265-8501, Japan
8. Yunnan Academy of Forestry, 650204 Kunming, China
9. Key Laboratory of Mountain Ecological Restoration and Bioresource Utilization and Ecological Restoration Biodiversity Conservation, Key Laboratory of Sichuan Province, Chengdu Institute of Biology, Chinese Academy of Sciences, Chengdu 610041, China
10. Center for Mountain Ecosystem Studies, Kunming Institute of Botany, Chinese Academy of Sciences, 650201 Kunming, China
11. Department of Life Science and Technology, Kunming University, Kunming 650214, China

The conservation and management of Tertiary relict plants will eventually depend on knowledge of their geographic distribution and the environmental changes that threaten them. This study, using species distribution modeling (involving a new approach that allows for uncertainty), predicts the distribution of climatically suitable areas prevailing during the mid-Holocene, the Last Glacial Maximum (LGM), and at present, and estimates the potential formation of new habitats in 2070 of the endangered and rare Tertiary relict tree *Davidia involucrata* Baill. The results regarding the mid-Holocene and the LGM demonstrate that south-central and southwestern China have been long-term stable refugia, and that the current distribution is limited to the prehistoric refugia. Given future distribution under six possible climate scenarios, only some parts of the current range of *D. involucrata* in the mid-high mountains of south-central and southwestern China would be maintained, while some shift west into higher mountains would occur. Our results show that the predicted suitable area offering high probability (0.5–1) accounts for an average of only 29.2% among the models predicted for the future (2070), making *D. involucrata* highly vulnerable. We assess and propose priority protected areas in light of climate change. The information provided will also be relevant in planning conservation of other paleoendemic species having ecological traits and distribution ranges comparable to those of *D. involucrata*.

T3

P0783

Convergence of carbon and nitrogen isotopes in terrestrial bryophytes across global nitrogen deposition and climate gradients

Yuping Dong, *Xueyan Liu, Wei Song, Xinchao Sun, Congqiang Liu*

Institute of Surface-Earth System Science, Tianjin University

Bryophytes, a widespread and functionally important vegetation in almost all terrestrial ecosystems, have been recognized to be one of the most sensitive biomes to reactive nitrogen (N) deposition. It has been reported that increased N deposition can influence photosynthesis, N utilization, species composition and productivity in bryophytes. However, the knowledge on C and N biogeochemistry in terrestrial bryophytes has not been well-established, which prevented a mechanical understanding of N-deposition effects on terrestrial bryophytes. Natural C and N isotopes are unique tools to provide *in situ* evidences on the responses of C and N dynamics in terrestrial bryophytes to N deposition. This work complied worldwide data of C and N concentrations, and stable isotope compositions (i.e., $\delta^{13}C$ and $\delta^{15}N$ values) of terrestrial bryophytes, as well relevant data of environmental and climatic factors. We examined correlations among bryophyte C and N concentrations, $\delta^{13}C$ and $\delta^{15}N$ signatures, mean annual temperature (MAT), mean annual precipitation (MAP), and N deposition levels. No significant correlation was found between bryophyte $\delta^{13}C$ and MAT or MAP. N deposition enhanced bryophyte N uptake, which explained more variations of bryophyte C concentrations and $\delta^{13}C$ values than species differences. The negative relationship between $\delta^{15}N$ values and N concentrations in bryophytes indicates that bryophytes assimilated more ^{15}N-depleted N sources that dominated in high N deposition. Moreover, we proposed a level- and N form-dependent mechanisms for N-deposition effects on bryophyte C and N fixation, which should be considered in future species- and ecosystem-specific studies. In summary, this study furthered the interpretation of C and N geochemical records in terrestrial bryophytes, which is useful for evaluating influences of anthropogenic N deposition on C and N cycles in bryophytes and moss-dominated ecosystems.

T3

P0784

Community nitrogen indicator: An new tool to indicate ecosystem nitrogen availability

Enzai Du

Beijing Normal University

Nitrogen (N) deposition has been increasing globally and has arisen concerns of its impacts on terrestrial ecosystems. Ecological indicators play an important role in ecosystem monitoring, assessment and management in the context of an anthropogenic transformation of the global N cycle. By integrating species composition and leaf N stoichiometry, a new community N indicator was defined and tested in the understory plots of an N enrichment (as NH_4NO_3) experiment in an old-growth boreal forest in Northeast China. Three-year N additions showed no significant effect on the understory species richness, but an obvious shift in species composition occurred. The response of leaf N content to N additions was generally positive but varied by species. Overall, the community N indicator increased significantly with higher N addition level and soil available N content, being in the shape of a non-linear saturation response curve. The results suggest that the community N indicator could be an effective tool to indicate changes in ecosystem N availability. Critical values of the community N indicator for specific vegetation type could potentially provide useful information for nature conservation managers and policy makers.

T3

P0785

Facilitation or competition? The effects of the shrub species *Tamarix chinensis* on herbaceous communities are dependent on the successional stage in a coastal wetland of North China

Ning Du[1,2], *Pan Wu*[1,2], *Franziska Eller*[3], *Dayou Zhou*[1,2], *Jian Liu*[4], *Wenhao Gan*[1,2], *Ruirui Yang*[4], *Min Dai*[1,2], *Yandong Chen*[1,2], *Renqing Wang*[1,2,4], *Weihua Guo*[1,2]

1. Institute of Ecology and Biodiversity, Shandong University

2. Shandong Provincial Engineering and Technology Research Center for Vegetation Ecology, Shandong University

3. Department of Bioscience, Aarhus University

4. Institute of Environmental Research, Shandong

The responses of herbaceous species to shrubs are crucial for community succession, but are not fully understood. Three sites experiencing degraded, early positive and late positive successional stages in Laizhou Bay coastal wetland of North China were selected to investigate herbaceous responses to the dominant shrub *Tamarix chinensis* at both community and species level. Plots both inside and outside the shrub canopy were surveyed. The effects of *T. chinensis* on plant diversity and growth of herbaceous community were negative in the degraded succession but facilitative in the positive succession. Solar irradiance was the only significant environmental factor for species' proximity to the shrub in the degraded succession, while both aboveground (solar irradiance) and belowground environmental factors (soil conductivity, water content, nitrogen concentration) played an important role in the positive succession. The performances of dominant herbaceous species in interaction with *T. chinensis* were both species-specific and traits-specific. Solar irradiance and soil conductivity, both significantly affected by the shrub, were the key factors for plant growth and leaf nutrient concentration at the species level. Thus, *T. chinensis* played an important role in the succession processes of the coastal wetland, but mainly in the positive succession stage and less in the degraded succession.

T3

P0786

Investigation on siderophore production by the arbuscular mycorrhizal fungi, *Glomus* species

Sunanda Dutta, *Medhatithi Sen, Sikha Dutta, Jayanta Chakraborty*

The University of Burdwan

Siderophores, the low molecular weight (<10 KD) iron chelating compounds are produced by Arbuscular mycorrhizal fungi (AM) (Boer-de *et al.,* 2003; Aliasgharzad *et al.,* 2009). These compounds are ferric specific ligands and are produced by microorganisms as agents for iron scavengeing in order to combat low iron stress. Siderophores solubilize the insoluble iron present in the soil and transfer it into the microbial cell. The ferric specific free ligands of siderophores produce intricated complex compound with ferric iron (Fe^{3+}) to form ferrated siderophores which is then transported to microbial cell *via* specific transport receptors located in the plasma membrane. Dhanya and Potty (2007) reported that inside the cell, siderophores turn into free form, from Fe^{3+} to repeat and the cycle. Depending upon the functional group iron siderophores are classified into three types *viz.,* hydroxamat type, catecholate type and carboxylate type (Jalal and Dick, 1990; Dutta *et al.,* 2006). Siderophore production by microbial agents has long been identified as an important antagonistic characteristic of several biocontrol agents including arbuscular nycorrhizal fungi which are efficient antagonists against many soil borne phytopathogens (Neilands, 1984; Leong, 1986; Bossier *et al.,* 1988; Barash, 1990; Boer-de *et al.,* 2003). In consequence, the ability of siderophores production by many VAM fungi, may certainly be one of the important reasons for reduction in infection and disease severity of many economically important plants. Present study emphasizes on the production of siderophores by the VAM fungi, *Glomus mossae* and *G. intraradices.* It also transact with the exploration of the threshold value of the ambient concentration of ferric Fe (Fe^{3+}) at which siderophore production by the test fungi is maximum. The effect of pH on siderophore production by *G. mossae* and *G. intraradices* was also investigated.

T3

P0787

Pre-mycorrhizal seedlings: An alternative for the cultivation of *Jatropha curcas* L. in salinized soils

Dilliani Felipe Barros Oliveira, Jose Vieira Silva, ***Lauricio Endres***
Universidade Federal de Alagoas

The cultivation of *Jatropha curcas* to produce biodiesel is a possibility to use saline areas; however biomass production is limited in these soils. Arbuscular mycorrhizal fungi (AMF) are a promising alternative for bioremediation in salinized soils. However, salinity also affects the AMF and, in this case, not always the symbiosis is established. Therefore, this study aimed to verify whether seedlings previously inoculated with AMF in non-salinized soil, subsequently transplanted to salinized soil, are more tolerant to salt stress. The experiment was carried out in a greenhouse, using a randomized blocks design in a 4 x 2 factorial scheme. The first factor consisted of soil salinized with electrical conductivity of 2, 5, 8 and 10 dS/m, and the second one was the presence or absence of AMF, with eight repetitions. Salt had a deleterious effect on the AMF non-inoculated plants, which had a reduction in the growth and photosynthetic activity, and an increase in stress indicators such as lipid peroxidation, proline accumulation, and high antioxidant enzyme activity. On the other hand, pre-mycorrhizal plants were less affected by the presence of salt in the soil. The AMF pre-inoculated seedlings had no reduction in photosynthesis and growth in saline soil. No increase in lipid peroxidation and antioxidant enzyme activity was observed in the AMF pre-inoculated seedlings, even under a high salt stress. The results suggest the existence of a mechanism that mitigates salt stress effects in pre-mycorrhizal seedlings. Therefore, previously mycorrhizal *J. curcas* seedlings may be an alternative to the cultivation of this species in salinized soils.

T3

P0788

Vegetation changes and perturbation events by anthropic activity based on core peat bogs in Escalerani (Bolivia)

Katerine Escobar-Torrez[1]*, Teresa Ortuño*[2]*, Marie-Pierre Ledru*[2]
1. Herbario Naciona de Bolivia
2. Museo Nacional de Historia Natural
3. Institut de recherche pour le développement

We realized a palynological study from 1 m depth sediment core. The study area is in the Paramo yungueño, nearest a *Polylepis* forest remnant, in the eastside of the Cordillera Real, Escalerani. Results show us relationship Poaceae/Asteraceae pollen before 7,505 cal. yr BP, we have had a wetter period than the actual, followed by a gradual change. The study of Sandoval (2011) find that *Polylepis* pollen do not allow long distance than 100m, then we can conclude that the presence of *Polylepis* pollen around 5,600 cal. yr BP, could give us insights about a major extension of the *Polylepis* forest near the peat bog. Additionally, there is another charcoal peak 3,655 cal. yr BP that is coincident with a Poaceae, probably Festuca-type, these results could show us the beginning of the camelid harvest in a near place. Furthermore we find charcoal peaks from 450 cal. yr BP, with an increase in their frequency until the present; These results are also supported with a reduction of pollen taxa concentration, and are coincident with the beginning of the Columbian colonization. At the top of the core we see high values of *Plantago* pollen, actually is the major taxa represented in the peat bog. Actually we are doing studies of $^{13}C/^{15}N$ and XRF to have a better understand of Climate Change in this particular area, and its responses to major climate drivers in the continent.

T3

P0789

Assessment of contribution of allelospoly and allelopathy to interference of *Medicago sativa* with *Raphanus sativus*

Salama El-Darier[1]*,* ***Esraa Fawzy***[2]*, Eman El-Kenany*[3]
1. Botany and Microbiology Department, Faculty of Science, Alexandria University, Alexandria, Egypt
2. Botany and Microbiology Department, Faculty of Science, Alexandria University, Alexandria, Egypt
3. Botany and Microbiology Department, Faculty of Science, Alexa

A simple mathematical model was applied to quantify the relative contribution of allelopathy to interference against *Raphanus sativus* by *Medicago sativa.* The interference by *M. sativa* was assessed and the relative contributions of allelopathy and allelospoly to overall interference were modeled. Competitive interference by *M. sativa* was eliminated and there were high R^2 values (0.937, 0.939) in dry weight of shoot and root of *R. sativus* respectively.

The contributions of competition and allelopathy to the interference by *M. sativa* were averaged as 13.5% and 86.5% at 6R, 23.74% and 75.37% at 12R and 3.9% and 96% at 18R in shoot dry weight, while 85% and 14.9% at 6R, 98.59% and 1.4% at 12R and 44.27% and 55.72% at 18R in root dry weight respectively. The agglomerative cluster analysis of the four samples (12R, 12R + E, 12R + 12M and 12R + 12M + E) showed two major clusters. This study suggested the complex effect of both components of interference (allelopathy and allelospoly) on the composition of vegetative proteins, and confirms the modeling results

T3

P0790

Bryophytes in field culture as excellent model organisms for ecological experiments

Belén Estébanez[1], *Nagore G. Medina*[2], *Belén Albertos*[3], *Mónica Tamayo*[1], *María Leo*[4], *Cristina Branquinho*[5], *Sara Lobo-Dias*[5], *Victoria González-Cascón*[6], *M. Cruz Díaz Antúnez-Barradas*[7], *Pedro Aragón*[6], *Joaquín Hortal*[6]

1. Universidad Autónoma de Madrid
2. University of South Bohemia
3. Universitat de València, Spain
4. Real Jardín Botánico, RJB-CSIC
5. Faculty of Sciences, University of Lisbon
6. Museo Nacional de Ciencias Naturales, MNCN-CSIC
7. Universidad de Sevilla

Research on bryophyte ecology has grown steadily during the last years. Indeed, their dimensions and relatively easy culture in field and lab conditions can make them particularly useful to evaluate diverse ecological hypotheses. Also, they can be considered as sentinels of ecosystem health. Bryophytes are small, allowing multiple experimental replicates in limited spaces. Many species are also quick and easy to grow, and gardening techniques are developing rapidly, so results from common garden and/or reciprocal transplant experiments can be obtained in short periods of time. Besides, they can be easily subjected to simple growth measurements (for instance, total biomass production), or effective proxies of it coming from spectral reflectance characteristics (including NDVI). We present an example of how to combine field and growth chamber experiments to determine the optimal environmental conditions, growth periods and rate of growth for some common moss species. We set up experiments in three localities along an altitudinal gradient in Central Spain, plus additional ones near the coast at Portugal and Eastern Spain. We evaluated the effects of two factors (sun vs. shadow conditions, and entire shoots vs. ground moss material) on six species, with ten replicates per experiment, totalling at least 280 culture plots per locality (including control plots). In Eastern Spain, two additional factors (substrate and propagule density) are evaluated on three species with ten replicates per factor (total 381 plots including controls). Such experimental setup needed a total space of less than 10 m^2 per locality (mostly for the shaded area). Preliminary results will be used to develop a similar experimental setup in growth chambers at several key temperatures, representing the whole range of variation in the conditions between field localities. Besides the difficulties of adjusting bryophyte culture protocols developed in temperate climates to Mediterranean environments, preliminary results are indeed promising. We detected differences in growth in less than one month, as well as some previously unexpected results, such as some species thriving outside their natural area, or the importance of shading even for heliophilous species. We remark the advantages of these organisms for large-scale culture experiments, and their potential usefulness for tackling many research questions, such as community-level experiments on coexistence, dispersal, recruitment, effects of community structure on ecosystem functioning or *ex situ* conservation.

T3

P0791

The effect of soil copper on moss patches of *Ptychostomum capillare*, a resistant moss species

Nuria J. Elvira[1], *Nagore G. Medina*[2], *María Leo*[3], *Victoria Cala*[1], ***Belén Estébanez***[1]

1. Universidad Autónoma de Madrid
2. University of South Bohemia
3. Real Jardín Botánico, RJB-CSIC

The mining industry produces diverse and profound disturbances in the environment, and is one of the main sources of heavy metal pollution in the soil, with long-lasting toxic effects. Among these heavy metals, copper considerably affects the growth, development and interspecific relationships of plants. Bryophytes have shown contrasted sensitivity to copper pollution, and could play an interesting role in biomonitoring or in bioremediation. This work aims at studying the effect of copper mining on *Ptychostomum capillare (= Bryum capillare)*, a moss species known to be resistant to several pollutants. In order to assess the effect of this metal on reproductive and developmental parameters, and on other coexisting moss, lichen and algal species, we established nine plots (1 x 1 m) along a copper pollution gradient in the mine tailings of an obsolete mine in Lozoyuela (Madrid, Central Spain). In each plot, we recorded the abundance and size of all *P. capillare* patches, its growth parameters, the observed damage in vegetative and reproductive structures of this moss, the presence of other vegetal organisms in the plots, and the incidence of epiphytic algal communities. We observed a significant effect of soil copper concentration on several factors, especially shoot density, size of the rhizoidal gemmae, tissue yellowing percentage, and presence and covering of coexisting organisms. We also noticed some additional damages in *P. capillare*, such as the persistence of protonematal stages and elongated caulidia in the plots with the highest copper concentrations. The positive relationship between copper concentration and shoot density might be due to the successfull competitive interactions of this resistant moss. Finally, we remark the interaction of other environmental factors, such as humidity and organic matter in the soil, with the effect of copper.

T3

P0792

Impact of climate variability on dispersal and distribution of airborne pollen and fungal spores in Nsukka (South-east Nigeria); implication on public health

Dimphna Ezikanyi, *Catherine Nnamani*

Ebonyi State University, Abakaliki, Nigeria

Airborne pollen and fungal spores are major triggers of allergies and their abundance and seasonality depend on plant response to climatic variable and change. Studies on seasonal prevalence of airborne pollen and fungal spores in Nsukka (South-east Nigeria) and their relationship to climatic variables were carried out from June 2015-May 2016. The aim of the study was to determine the seasonal variations of airborne pollen and fungal spores and ascertain the impact of weather variables on their spatial distribution. Airborne samples were collected using a Tauber-like pollen sampler modified to a height of 5 ft above the ground, aerosamples were collected monthly and acetolysed. Dominant pollen recorded were those of Poaceae, *Elaeis guineensis* Jacq., *Aspilia africana* (Pers.) C. D. Adams, *Olax subscorpioides* Oliv., *Alchornea cordiforlia* (Schum. & Thonn.) Mull.-Arg, *Pentaclethra macrophylla* Benth. *and Casuarina equisetifolia* Forst & Forst. Total pollen count correlated positively with temperature and wind, though not significant and negatively with rainfall, relative humidity and wind. Fungal Spores dispersed by *Fusarium, Nigrospora, Puccinia, Spadicoides, Alternaria and Helminthosporium* were more abundant. A positive correlation occurred between total spores count and monthly rainfall. Rainy season (May–August) had lower records of pollen whereas the principal pollination period was during the dry season (September–November). The allergenic and pathogenic fungal spores were more abundant in rainy season (June–August) and late rainy season (September–November). The study revealed a strong influence of climatic variables on abundance and spatial distribution of pollen and fungal spores.

T3

P0793

Genome-wide association mapping of growth response to flooding stress in sunflower (*Helianthus annuus* L.)

Lexuan Gao, *Joon-Seon Lee, Loren Rieseberg*
University of British Columbia

Sunflower, as a globally important oil crop, whose growth and production are limited by flooding stress. However, the understanding of how growth is affected by flooding stress is lacking. In this study, a large scale phenotyping experiment was performed to evaluate the variation in growth responses to flooding stress in a diverse collection of cultivated sunflower accessions. Flooding responses in multiple traits, such as shoot biomass reduction, chlorosis and hypocotyl hypertrophy were collected as an estimate for flooding resistance. The genetic architecture underlying the growth responses upon exposure to flooding stress was investigated by performing a genome-wide association (GWA) mapping on the different growth related phenotypes. No locus with major effect was found to be associated with differential growth responses to flooding stress, while many promising peaks of association with small phenotypic effects were identified. Many candidate genes underlying the associations were suggested. Further functional analyses of allelic variation in candidate genes will provide insights into the genetic basis of variation in flooding resistance in sunflower.

T3

P0794

Chinese management status of exotic invasive plants and quarantine

Zhenshuan Zhao, *Jun Wang, Lie Qian, Guiming Zhang*
Shenzhen Entry-exit Inspection and Quarantine Bureau

With trade globalization, exotic plant invasion has become a worldwide problem, not only bringing serious harm to the ecosystem and environment, but also threatening human survival and development. Compared with invasive animals and other species, invasive plants have more species, ecological adaptability and stronger ability to reproduce and spread. China has so far confirmed 291 species (including varieties) of invasive plants, with common means of spreading through improper human introduction, being carried via various forms of transportation, international trade of goods and natural transmission. According to statistics, the inspection and quarantine department yearly intercept over one thousand kinds of exotic weeds, hosted mainly in products such as imported grain, wool, and crop seeds, and prevent their introduction and spread through their destruction and regular supervision of domestic storage places and processing plants, effectively alleviating the pressure of propagule invasion. At present, there are many problems in the research and prevention of alien invasive plants, such as bull management, imperfect laws and regulations, management lag, lack of in-depth study, lack of early warning, low public awareness and so on. In this regard, the entry-exit inspection and quarantine departments cooperate with related departments, strengthen the research and prevention of invasive plants, through strengthening the evaluation of imported seeds, perfecting the risk early warning system, exotic weed monitoring, strengthening the on-site inspection and quarantine identification technology research, and improving the processing and control. Comprehensive measures such as these have made staggered results in the prevention of exotic plant invasion. Further measures to prevent exotic plant invasion requires perfecting the system of the rule of law, enhancing the promotion of national security, and to launch the power of the whole society to participate.

T3

P0795

The weed seeds quarantine and analysis of imported agriculture product in Shenzhen port

Dongmei Chen, *Caiyu Xiang, Xingyuan Ling, Ruifang Gao, Ying Wang, Guiming Zhang, Lin Kang*
Shenzhen entry-exit inspection and quarantine bureau

Each year Shenzhen imported from abroad a large number of agricultural products. According to the statistics during 2014-2016, a total of 8,228 batches of food and oil products, which were 16.4158 million tons and $6.66 billion US dollars, were imported. The mainly involving food types are soybean, wheat, corn, canola, sorghum, oats, barley, rice, *etc.* and the mainly producing areas are the United States, Argentina, Brazil, Canada, Australia, Thailand, Vietnam and other countries. According to the national quarantine law on the entry and exit animal and plant, pest quarantine on above large number of imported agricultural products were carried out, and the weed quarantine results were statisticed and analyzed. Totally, 34 kinds of quarantine weed seeds that were prohibited from importing, such as *Sorghum halepense* (L.) Pers., *Ambrosia artemisiifolia* L., *Lactuca serriola* L, *Euphorbia dentata* Michx. and *Crotalaria spectabilis* Roth *etc.*, 49 kinds of poisonous weed seeds such as *Datura stramonium* L., *Phytolacca americana* L., *Crotalaria mucronata* Desv., *Agrostemma githago* L., *Euphorbia heterophylla* L., and 396 kinds of other harmful weed seeds that were first intercepted or new listed weed seeds such as *Leucospora multifida* (Michx.) Nutt., *Kochia californica* S. Watson, *Pilea microphylla* (L.) Liebm., were intercepted. These quarantine weed seeds, poisonous and harmful weed seeds and other ones that have no risk assessment, and were not distributed or not widely distributed in our country are extremely dangerous. Once invasion and colonization, they will seriously affect the city's ecological environment, production of agriculture, forestry and animal husbandry, and people's life safety. According to the data in the nearly three years, in 2014, 14,197 times of 334 species weeds, including 993 times of 24 species quarantine weeds and 32 species poisonous weeds, were intercepted; in 2015, 28,942 times of 367 species weeds, including 2,275 times of 31 species quarantine weeds and 43 species poisonous seeds, were seized; in 2016, 24,433 times of s 422 species weeds, including 1,504 times of 32 kinds of quarantine weeds and 36 kinds of toxic seeds, were captured. Data also showed that among all weed seeds intercepted, according to the species of agricultural products, 175 kinds of weeds seeds were intercepted from imported rapeseed, 262 from wheat, 280 from soybean, 78 from corn, 75 from sorghum, 130 from barley, 50 from oats and 77 from rice; according to the importing country, 390 species of weeds were intercepted from the products imported from the United States, 209 from Canada, 115 from Argentina, 148 from Brazil, 157 from Australia, 71 from Thailand and 45 from Vietnam, indicating the situation of pest invasion was very serious. To stop the distribution and spread of foreign poisonous and harmful weeds in our country, measurements like strengthening the quarantine treatment technology research, standardizing operating procedures, strengthening the management of source, enforcing stringent examination and approval of the imported agricultural products, stating detailed inspection and quarantine requirements, carrying out further pest risk analysis of the exporters' agricultural products, and formulating corresponding management measures, should be carried out. Except for to handle the outbreak in time and do a good supervision on the epidemic situations, regular outbreak investigation and control work within the respective jurisdictions should be carry out, so that the spread of the epidemics were prevent and the safety of the city's ecological environment and production of agriculture, forestry and animal husbandry are protected.

T3

P0796

Late holocene hydroclimatic variability in the Darjeeling Himalaya, India and corresponding vegetation response

Ruby Ghosh[1], *Oindrila Biswas*[2], *Dipak Kumar Paruya*[2], *Shailesh Agrawal*[1], *Anupam Sharma*[1], *C. M. Nautiyal*[1], *Subir Bera*[2]
1. Birbal Sahni Institute of Palaeosciences
2. University of Calcutta

Hydroclimatic variability in the Darjeeling Himalaya over the past ~2400 years has been explored using a range of palaeoclimatological tools combining pollen, phytoliths, non-pollen palynomorphs (NPPs), stable carbon isotope ($\delta^{13}C$), TOC and grain size data from a lacustrine deposit. It is evident that this climatic variability also induced changes in the regional vegetation succession. The region was wet harbouring a dense broad-leaved evergreen forest between ca. BC 364 and AD 131 while, a comparatively drier condition prevailing between ca. AD 131 and 624 might be the reason behind the thinning in the forest cover. A recovery wet phase was observed after ca. AD 624. Signature of an intensified monsoon was noticed during ca. AD 1,367 that persisted till ca. AD 1,800. For the two late Holocene transient rapid climatic phenomena, i.e. Medieval Warm Period (MWP) and Little Ice Age (LIA), which are broadly regarded as wet and dry phases respectively present data further shows departures from general trends. Considerable variations exist globally for these warm (moist) and cool (dry) periods with respect to their timing, duration, and hydroclimatic dynamics. Our results identified a humid climatic phase at the beginning of the last millennium, a pre-MWP less humid phase, while MWP was similar to present day with a minor change in rainfall pattern and a wet LIA in the Darjeeling Himalaya. Reviewing the records from other proxy data, plausible reasons of these departures are addressed which indicate centennial scale variations in frequencies of "active dominated" and "break-dominated" periods are the key forcing mechanisms behind the differential behavior of the ISM (Indian summer monsoon) over the Himalayas and the peninsular India.

T3

P0797

Host plant alters soil bacterial community through arbuscular mycorrhizal fungi in subtropical forest soils during litter decomposition

Heng Gui[1,2], *Peter Mortimer*[1], *Jianchu Xu*[1], *Kevin Hyde*[2]
1. Kunming Institute of Botany, Chinese Academy of Sciences
2. Mae Fah Luang University

Interrestrial ecosystems, most of plants cooccur with arbuscular mycorrhizal fungi (AMF) and AMF play a crucial role on linking host plant and belowground soil bacterial community. However, the interaction between these AMF and soil bacterial community and their effects on litter decomposition remains largely unexplored. In order to help overcome this knowledge gap we investigated the effect of arbuscular mycorrhizal fungi (AMF) on soil bacterial communities found in a subtropical rainforest, in southwestern China. Our experimental set up included a dual microcosm unit and two treatments: with (AM) and without (NM) AMF. Illumina sequencing was used to assess any changes in the soil bacterial communities. We found that most of the obtained operational taxonomic units (OTUs) from both treatments belonged to the phylum of Proteobacteria, Acidobacteria and Actinobacteria. The community composition of bacteria at phylum and class levels was slightly influenced by both time and AMF. However, time and AMF significantly influenced the structure and dynamics of the bacterial communities at the genus and OTU levels. Opposite to the bacterial community composition, we found overall soil bacterial OTU richness and diversity are relatively stable and were not significantly influenced by either time or AMF inoculation. OTU richness at phylum and class levels also showed consistent results with overall bacterial OTU richness. Our study provides new insight into the influence of AMF on soil bacterial communities at a fine scale resolution.

T3
P0798

Estimating vegetation biomass using very high resolution satellite image in a karst watershed of Guizhou province, southwestern China

Yinming Guo
State Key Laboratory of Environmental Geochemistry, Institute of Geochemistry, Chinese Academy of Sciences

Vegetation biomass in karst terrain has been rarely measured because the steep and harsh mountainous limestone landscape limited our ability to sample woody plants. Satellite observation, especially in high spatial resolution, is therefore an important surrogate to estimate biomass of karst forest and shrubland. In this study we build an Artificial Neural Network (ANN) model using very high resolution (VHR) satellite imagery and field biomass measurements to estimate the aboveground biomass (AGB) in the Houzhai River Watershed, a typical plateau karst basin in central Guizhou province, southwestern China. A Back Propagation ANN (BP ANN) model was developed. Seven vegetation indexes including the Atmospherically Resistant Vegetation Index (ARVI), Enhanced Vegetation Index (EVI), Normalized Difference Vegetation Index (NDVI), Difference Vegetation Index (DVI), Green Normalized Difference Vegetation Index (GNDVI), Ratio Vegetation Index (RVI), Soil-Adjusted Vegetation Index (SAVI), two spectral bands of Pléiades-1A, and one geomorphological parameter were selected as inputs while the AGB as output. Vegetation biomass at 78 plots estimated by allometric functions was used as training data (58 plots, 75%) and validation data (20 plots, 25%). Data-model comparison showed that the ANN model performed well with the absolute RMSE of 8.22 t/ha, which is 6.51% of the average AGB. In addition, seven land cover types, with a total accuracy of 95.33% and the Kappa coefficient of 0.93, were classified based on the Pléiades-1A satellite image. On the base of the newly developed ANN model, a AGB map in the Houzhai River watershed was produced. The estimated AGB of the main forest type in the watershed, the secondary evergreen and deciduous mixed forest, was 147.85 t/ha in average. The large distributed shrubland, tussock and farmland had averaged AGB of 41.24,9.08 and 12.44 t/ha, respectively. This finding indicated that the spatial distribution pattern of AGB estimated by the new ANN model in the karst basin is consistent with the field investigation. This model can be further used to estimate vegetation biomass in the regional scale of karst lanscapes distributed in the surface of the Yun-Gui Plateau.

T3
P0799

Transcriptomes provide new insights into community assembly in a subtropical forest

Baocai Han[1], *María Natalia Umaña*[2], *Xiangcheng Mi*[1], *Xiaojuan Liu*[1], *Lei Chen*[1], *Yunquan Wang*[1], *Yu Liang*[1], *Wei Wei*[1], *Keping Ma*[1]
1. Institute of Botany, The Chinese Academy of Sciences
2. Department of Biology, University of Maryland

Differences in seedling survival of tree species have a lasting imprint on community structure. Therefore, identifying the underlying mechanisms of these differences is a critical research objective regarding how organismal functions interact with the local environment that influences survival rates. In tree communities, differences in light use strategies are frequently invoked to explain differences in seedling demographic performance. For example, shade tolerant species grow slowly and have higher survival rates, whereas light demanding species grow quickly and have lower survival rates. Thus, functional traits related to photosynthesis/light should be strong predictors of seedling species survival. Here we provide the community-wide inventory of transcriptomes in a subtropical tree community. This information is utilized to determine the degree to which species share homologous genes related to gene ontologies for light harvesting and use. These species similarities are used in neighborhood models of seedling survival and we ask whether survival increases or decreases in the presence of transcriptomically similar or dissimilar species. We also compared the effects of phylogeny, functional traits and transcriptomes related to light and examined the effects of conspecific neighbor density, heterospecific neighbor desity on the seedling survival of 85 seedling species in a subtropical of China. Our results show that the gene ontologies related to the light quantity (GO:0010196) and light quality (GO:0010201) significantly affect the seedling survival. However, there is not significant effect for phylogeny. For functional traits, only the length to width ratio (LWR) has significant effect on seedling survival both for all species and for shade tolerant species. We found that the coefficient of the correlation between LWR and GO:0010201 is high (mantel $r = 0.185$, $p = 0.021$). Our results indicate that light is a limit factor for seedling survival in the subtropical forest and the new approach could detect the light use strategies not only in the light quality but also in the light quantity. However, the phylogeny could not detect the light use strategies. For functional traits, only LWR could detect the light use strategies. The high correlation between the LWR and GO:0010201 reconfirms the reliability of the transcriptomic approach. Synthesis: The present work takes a novel approach by sequencing the transcriptomes of naturally co-occurring tree species in a subtropical forest in China. The study demonstrates that exploring the functional genomic similarity of non-model species in natural community would greatly increase the breadth and depth in quantifications of functional similarity among coexisting species.

T3
P0800

Plant community patterns in the hot-dry valleys of Southwestern China

Cindy Q. Tang, ***Peng-Bin Han***, *Diao-Shun Huang*, *Li-Qin Shen*, *Yi-Fei Dong*
Institute of Ecology and Geobotany, Yunnan University, Kunming 650091, China

The hot-dry valleys are steep gorges created by major rivers in the subtropical area of SW China, particularly the deeply incised valleys along the Nu River (Nujiang, upper reaches of the Salween, below 1,200 m asl), the Lancang (Lancangjiang, upper reach of the Mekong, below 1,000 m), the Jinsha (Jinshajiang, the upper section of the Yangtze, below 800–1,200 m), the Yuan (Yuanjiang, the upper stretch of the Red River, below 1,000–1,400 m), and the Nanpan (Nanpanjiang, below 1,000 m) in Yunnan, also in the down-stream stretch of the Jinsha river, as well as its tributaries including the Yalong (Yalongjiang), Anning (Anninghe), Dadu (Daduhe), Min (Minjiang), Baishui (Baishuihe) rivers of western Sichuan. Some hot-dry valleys are also found in the Southeast (Wushan, Fengjie, Xuyong, Gulan and Pingshan in Chongqing Municipality) along the Yangtze and Chishui rivers of the Sichuan Basin. In Guizhou, hot-dry valleys are confined to the southwestern part of the province, as in the Nanpanjiang and Beipanjiang valleys. A foehn effect caused by the rain shadow on the leeward slopes of the mountains accounts for their hot-dry climate. Many genera of plant communities in the hot-dry valleys have affinities with the xeric flora of tropical Asia and tropical Africa, including Woodfordia, Calotropis, and Adinandra, which are disjunctly distributed in areas of tropical Asia (including India, Malaysia and southern Yunnan) and tropical Africa. Ecologically the major vegetation in the hot-dry valleys is comparable to that of savannas of India and Africa. However, these valleys are densely inhabited by humans because their high temperatures favor good crops, as well as grasses for cattle grazing. Human activities have greatly modified the vegetation and changed these into what we now call semi-savannas (semi-natural savannas) or secondary savannas. These semi-savannas typically include the sparse woody grassland and the thorny succulent shrubland. Besides the semi-savanna, some scrublands are also found. In this study, plant communities are differentiated and vegetation-environment relationships are analyzed. For each community, we identify indicator species. Ecological preferences of each community are examined.

T3

P0801

Impacts of long-term nitrogen addition on biodiversity and nutrient ecological stoichiometry in a temperate steppe

Tianxiang Hao[1], *Ling Song*[2], *Xuejun Liu*[1]

1. China Agricultural University

2. Institute of Mountain Hazards and Environment, CAS

Atmospheric nitrogen (N) deposition has been verified to decrease biodiversity and alter nutrient cycle in terrestrial ecosystems. Nitrogen deposition, as an important N input to natural ecosystems, has significantly increased across China over last three decades. As aggravated phenomenon was concerned, N deposition was supposed to increase or decline via non-controlled or controlled anthropogenic emissions of reactive N, just as China (increase), European Union (decrease) and the United States (stabilization) experienced. However, the response of species richness and nutrient ecological stoichiometry to long-term N enrichment in Chinese grassland is still lacking. To answer this crucial question, we did an 11-year grassland N addition experiment with N rates varying from 0 (N0), 30 (N30), 60 (N60), 120 (N120), 240 (N240), and 480 (N480) kg N ha^{-1} yr^{-1} in Inner Mongolia, China. In this study, we ceased N addition in N480 treatment since the fifth year (2009) of the experiment to observe the potential restoration from an extremely N-enriched condition. The species richness of grasses (dominant species) did not suffer from all N addition treatments, while the species richness of forbs (rare species) showed significant reduction in species richness at N30 and N60 (part years) or N120 and higher N treatments (almost all years).This indicates that forbs are more sensitive to N enrichment compared grasses in this grassland region. In N480, significant decrease in plant species richness since the first year of the experiment but a step by step restoration of biodiversity was observed when we ceased N addition in this treatment since the fifth year (2009), suggesting the ecological restoration from an extremely N-enriched condition. Meanwhile, 9 elements (carbon / N / phosphorus / potassium / calcium / magnesium / iron / sodium / aluminum) in shoots after 9 years (2013) fertilization presented three different N-induced variations. As increasing N addition, the concentrations of C and N significantly increased and saturated at N addition rates of 30 and 60 kg N ha^{-1} yr^{-1}, respectively; for Ca, Na, Fe and Al, the concentrations almost decreased in stages; the concentrations of P, Mg and K were always stable. It is noticed that, after 5 years cessation of N addition, N480 was still in line with the above rules of elements' concentration. Furthermore, plant species richness increased partly after ceasing N addition in N480 for last three years, but it seems unlikely to recover to the original status for dominant species and plant nutrients (except N) in N0. Our findings suggest that the critical threshold for N-induced species loss in this temperate steppe is between 30 and 120 kg N ha^{-1} yr^{-1}. Even the N addition rate as low as 30 kg N ha^{-1} yr^{-1} can significantly alter shoot nutrient concentrations. As reduced N inputs, the grassland would recover much of the plant species that had been lost, but the changes in plant nutrients would be hardly back to the original level. This work highlights the ecosystem recovery that exist in assessing the impacts of N deposition on biodiversity and nutrient ecological stoichiometry.

T3

P0802

Population genetic analysis of invasive Kudzu (*Pueraria montana* var. lobata) throughout Asia and the United States with emphasis on varietal delimitations

Matthew S. Haynsen[1,2], *Mohammad Vatanparast*[2], *Lu-Xian Liu*[3,4], *Cheng-Xin Fu*[4], *Keith A. Crandall*[1,5], *Ashley N. Egan*[2]

1. The George Washington University, Department of Biological Sciences

2. Smithsonian Institution National Museum of Natural History, Department of Botany

3. Henan University, College of Life Sciences

4. Zhejiang University, College of Life Sciences

5. The George Washington University, Computational Biology Institute

Pueraria montana var. lobata (Kudzu), one of the most invasive plant species in the United States, is native to southeastern Asia. The twining vine was initially introduced into the United States from Japan during the Centennial Exposition in 1876 with subsequent mass introductions throughout the southeastern United States through the 1930s to 1950s. The native geographic origins of these introductions remain unknown, which poses a concern due to the wide native range of *P. montana*, a species containing three recognized varieties that share overlapping ranges and delimiting morphological characteristics. It is possible that the invasiveness of Kudzu found in the United States can be ascribed to not only admixture from geographically and genetically distinct populations of *P. montana* var. lobata, but hybridization among two or more varieties of *P. montana*. Only through the use of high resolution molecular markers on populations of *P. montana* across both its native and introduced ranges will it be possible to identify the geographic and varietal sources of genetic diversity within introduced populations of invasive Kudzu. In this study we used Genotyping-by-Sequencing, a next generation sequencing methodology, to assess the genetic variation of over 700 individuals of *P. montana* from China, Japan, Thailand, and the United States. Through the collection of hundreds of SNPs per individual we assessed both the among and within nation genetic diversity, native and introduced range structuring of genotypes, and models of introduction pathways. We also combined morphological measurements with genotypic assignments to determine the native range varietal boundaries and to assess areas of varietal hybridization within the native and introduced ranges.

T3

P0803

Genetic analysis of inherited reduced susceptibility of *Fraxinus excelsior* L. seedlings in Austria to ash dieback

Berthold Heinze[1], *Alexandra Wohlmuth*[1], *Franz Essl*[2]

1. Austrian Federal Research Centre for Forests, Department of Forest Genetics

2. University Vienna, Division of Conservation, Vegetation and Landscape Ecology

Hymenoscyphus fraxineus causes massive dieback of common ash (*Fraxinus excelsior*). Previous studies have revealed differences in susceptibility among individuals, suggesting a genetic basis for reduced susceptibility to the pathogen. The aim of the study was to identify any correlation between damage intensity of mature trees and their offspring in natural ash stands. Crown and shoot damage of naturally infected trees and saplings were assessed in two isolated stands in Austria, and parentage analysis was carried out with molecular markers. No significant correlation could be verified using Spearman's rank correlation analysis, suggesting a lack of evidence for genetically inherited resistance at the sites. We thus support the results of other studies, i.e. that resistant individuals occur only at low frequency within European ash populations. While most of the previous studies were conducted in progeny trails or seed orchards and suggested a fairly strong genetic component, results from our investigation support a more complex mechanism of susceptibility differences under natural, heterogeneous conditions. Further analyses are needed to obtain a better understanding of gene-environment interactions and individual infection pressure of ash dieback in natural environments. Identification and propagating of non-susceptible ash trees is an important challenge to halt large-scale dieback of *Fraxinus excelsior.*

T3

P0804

Removing the sun and shade confounding factor from fossil leaf palaeoenvironment interpretations.

Kathryn Hill, *Robert Hill, Alice Jones, Fiona Mcqueen*

The University of Adelaide

Leaf anatomy has been used as a palaeoclimate predictor for decades. Anatomical features such as stomatal and epidermal cell measurements change in response to climate and environment changes to modify maximum potential physiology such as gas exchange and hydraulic conductivity. We can optimise the use of anatomical features to more accurately predict past climate changes by aiming to remove the confounding factor of light availability. The aim of the project is to determine how leaf anatomy changes both between sun and shade, and along an environmental cline. Thus we can determine what leaf anatomy measurements constitute a sun or shade (or part-shade) leaf and thus determine what light environment a leaf grew in. Having removed light availability as a factor driving leaf anatomy change we can determine palaeoclimate by comparing (for example) living shade leaf anatomy from varying environment types with fossil shade leaves to determine to what paleoclimate the fossil shade leaves were exposed.

T3

P0805

The warmest habitat where plants linger on? - Climate change vulnerability assessment of vascular plants in Taiwan

Huan-Yu Lin[1,2], *Wei-Chun Chao*[3], *Chang-Fu Hsieh*[1], *Tze-Ying Chen*[4], ***Jer-Ming Hu***[1]

1. Institute of Ecology and Evolutionary Biology, National Taiwan University

2. Taiwan Forestry Research Institute

3. Department of Forestry and Nature Resources, National Chiayi University

4. Department of Forestry and Nature Resources, National Ilan University

Taiwan is a continental island on the Tropic of Cancer with an area of 35,980 km^2. More than 73% of the land is hills and mountains, with the highest central ridge close to 4,000 m, while 60% of the land is covered by forest. Because of the diverse topography and warm-humid climate, this island supports over 4,200 vascular species, of which 1,052 (22.9%) are endemic to Taiwan. The annual mean temperature in Taiwan showed an increase by 1.4°C from 1911 to 2009, which is twice higher than the global warming rate of 0.07°C per decade. It is thus urgent to investigate habitats that are vulnerable to global warming and make conservation strategies in advance. In 2015, we complied a geo-referenced database with 1.4 million records of the flora of Taiwan for analyzing areas that are sensitive to climate change. Projected temperature tolerance range were estimated for each species based on their geo-reference information. Based on the upper distribution limit of each species' optimal temperature range, we found 19 areas showed more upper-limited individuals aggregated, and could be highly sensitive to global warming. We also found that 7 of 19 sensitive areas are isolated mountain tops represented by fragmented cool-temperate vegetation. Some of these areas suffered climatic change or extreme natural disasters (e.g. weakening of northeast monsoon, landslides, typhoons and droughts) during the past 30 years, and induced an evident recession of scarce populations, such as *Fagus hayatae, Lithocarpus formosanus* and *Rhododendron simsii.* For the sake of making adequate conservation strategies, we established an evaluation scheme that integrates vegetation characteristics and environmental factors to assess conservation priority among sensitive habitats. Lower cost and feasible conservation strategies, including in-situ conservation, building corridors and assisting migration, are suggested for populations that are not endangered critically. Furthermore, immediate ex-situ conservation is essential for rare species suffering critical recession.

T3
P0806
Sexual system of *Daphne arisanensis* Hayata (Thymelaeaceae)

*Wei-Jie Huang, **Jer-Ming Hu***
Institute of Ecology and Evolutionary Biology, National Taiwan University

Angiosperms exhibit great diversity of sexual systems, combining different degrees of male and female function at flower, inflorescence, and individual level. As the majority of extant species are hermaphrodite, the evolution of unisexuality has drawn attention since Darwin's age. Here we documented the rarely known sex expression patterns for the flowers of *Daphne arisanensis* Hayata (Thymelaeaceae), an endemic shrub in Taiwan. Both morphologically bisexual and male-sterile flowers were found on different individuals among *D. arisanensis* populations, hence the species might be gynodioecious. However, we demonstrated clearly that it is not the case after our thorough investigation of the spatial and temporal expression of sexuality. The gynoecia of morphologically bisexual flowers displayed receptivity to pollens from either itself or other plants, but they hardly set fruits among all pollination treatments. We also found some male-sterile flowers had the ability to set fruit even without fertilization, i.e. apomictic. In addition, there is no individual altered the flowering type during the three-year observation. These findings suggest that the sexual system of *D. arisanensis* is a cryptic and leaky dioecy, and the evolution toward dioecism probably underwent gynodioecy pathway.

T3
P0807
The genome of *Pugionium* and climatic adaptation

***Quanjun Hu**, Tao Ma, Kun Wang, Huan Hu, Xinyi Guo, Yazhen Ma, Jianquan Liu*
Sichuan University

Understanding the mechanisms of speciation is a fundamental problem in evolutionary biology. To identify genome-wide patterns underlying ecological speciation, we assembled a high-quality reference genome of *Pugionium dolabratum*, an endangered psammophyte. We conducted genome-wide population surveys of natural populations to investigate the evolutionary processes and the genetic basis underlying speciation. Our findings highlight the prevalent role of geographic isolation and climate change in speciation. We identified several genes related to stress tolerance, leaf morphology and root growth. Our work highlights the value of genome-wide population surveys of natural populations in non-model species to understand the genetic basis for evolutionary change.

T3
P0808
The relationships between above- and belowground biodiversity and ecosystem multifunctionality, and its drivers in the drylands of China

*Jianming Deng, Qiajun Du, **Weigang Hu***
Lanzhou University

Biodiversity is the foundation for the maintenance of ecosystems, which are valued for their ability to provide multiple functions and services simultaneously (ecosystem multifunctionality; EMF). However, we have limited knowledge of the links between biodiversity and multifunctionality as well as drivers of these relationships in large-scale natural ecosystems. This gap in our understanding is particularly true for drylands, areas covering 41% of Earth's land surface and supporting over 38% of the global human population, which are among the most sensitive ecosystems to climate change. Here, we present a regional-scale field study conducted along natural aridity gradients across 157 dryland sites from northwestern China. We surveyed plant community by measuring 30 m × 30 m plots, characterized bacterial, archaeal and fungal communities in the soil surface (top 20 cm) by using Illumina Miseq profiling of ribosomal genes and internal transcribed spacer (ITS) markers, gathered a range of key abiotic factors (i.e. climate, elevation, and soil texture), and assessed 14 ecosystem function indicators (aboveground biomass, plant carbon, plant nitrogen, plant phosphorus, soil organic carbon, soil nitrogen, soil phosphorus, nitrate, ammonium, soil available phosphorus, β-glucosidase, urease, phosphatase, and soil microbial biomass) related to soil biogeochemical cycles, soil fertility, biological productivity, and the build-up of nutrient pools. In this study, we aim to elucidate the relationships between above- and belowground biodiversity and multifunctionality, and to illustrate the interactions among abiotic factors and biotic attributes (i.e. above- and belowground biodiversity) in driving multifunctionality. We also evaluate how increases in aridity, as defined by the aridity index, affect biotic attributes and multifunctionality, and further emphasize the role of biological feedbacks on the responses of ecosystem multifunctionality to increasing aridity. This knowledge is essential to predict responses of dryland ecosystems to ongoing global change, and guide conservation and sustainable management efforts in drylands of China.

T3
P0810
Studies on tracheary element of several native cycad species in Australia and two American plants

***Yu-Yuan Huang**[1], Harvey Ottley[2,3], Yolande Yep[4,5], Sharon Wilson[2], David Griffiths[2], Nikeeta O'Sullivan[4,5], Jane Hempel[2]*
1. College of Life Sciences, Zhongkai University of Agriculture and Engineering, Guangzhou, Guangdong 510225, China
2. George Brown Darwin Botanic Gardens, Darwin, NT 0820, Australia
3. Palm and Cycad Society of the Northern Territory, Darwin, N T 0820, Australia
4. School of Environment, Charles Darwin University, Darwin, NT 0909, Australia
5. Research Institute for Environment & Livelihoods, Charles Darwin University, Darwin, NT 0909, Australia

Cycas angulata, Cycas amstrongii and *Cycas conferta* which are native species of Australia were collected in Darwin city of tropical region, *Zamia erosa,* and angiosperm *Guaiacum officinale* which are introduced collected in the Darwin Botanic Gardens, and were carried out electron microscope observed research to tracheary element and parenchyma tissue cells of leaflet and rachis. The results showed that there were more vessel elements in their xylems. The length of vessel element of Cycadaceae with Zamiaceae were not obvious difference, the length of vessel element of angiosperm *Guiacum officionale* was shorter, showed its more evolutional characteristic. These vessel were annular vessel, spiral vessels, scalariform vessels, reticular vessels, pitted vessels and scalariform-pitted vessels. In the transverse section, the vessel elements of Cycadaceae and Zamiaceae were circular, polygonous, more similar to vessel characteristics of petiole and stem of some angiosperms; however, the number of circular vessel of *Guiacum officionale* were more, about occupied 35%, others were polygonous. The diameter of vessel were that annular vessel, spiral vessel were least, scalariform vessel was medium, reticular vessel and pitted vessel were biggest, the characteristic was same as angiosperm. In *Cycas conferta etc.* plants often seen several large perforations in the end walls, some species were several big scalariform or pitted perforations formed multiple perforation plates, in many vessel element lateral walls, there were some perforations formed a horizontal transport avenue of aqueous solution. The scale of diameter of *Cycas* vessel element were similar to that of *Zamia*, and was bigger than that of *Guiacum officionale*, although *Cycas* plants are more primitive than *Zamia* plant, and *Guiacum officionale* is more evolutionary, but the scale of vessel diameter of *Cycas* plants is similar to *Zamia*, even larger than *Guiacum officionale*, these showed that cycads have more evolutionary and more developed characteristics in aqueous solution transport system. In the respect of observed parenchyma tissue, any parenchyma tissue cells were without perforation, all cell walls were smoothed and like a membrane. Aimed at only one or two people pointed the query about Jeffery's method, some scholars have made comparative researches, their research used as follow methods: (1) The fresh materials were cut sections 1–2 mm thick by hand (the query person provided and thought right method), were examined with SEM. (2)The materials were treated by Jeffrey's Fluid and observed with SEM and so on. The results showed that the pit membrane, the remnants in the perforation and the structural characteristics of perforations were no different. These results justified that Jeffrey's method is reliable and reasonable. Our research result showed that only vessel element can formed perforation, and in other all parenchyma tissue cells can not formed perforation are also proved that the perforations of vessel element are natural inherited character. The research is important significant and theory reference to understand cycads evolutionary aqueous solution transport system, and understand the ecological adaptation mechanism which why present cycads of 3 families in world major distributed in tropical or subtropical regions, even many drought and barren areas.

T3
P0811

The comparative studies on plant diversity and their ability of absorbing and reducing to PM2.5 of Xiaonanshan with Dananshan Mountain areas

Yun-He Lu[1], *Ruo-Yu Wei*[2], *Hai-Yang Wen*[2], *Zhen Ye*[2], *Xin-Fan Yu*[3], *Fan Wang*[2], ***Yu-Yuan Huang***[2], *Kang-Sai Zhao*[3], *Hong Liang*[3], *Long He*[3], *Zhi-Ming Yang*[2], *Ya-He Li*[1]

1. Environmental Monitoring Station of Nanshan District, Shenzhen, China
2. College of Life Sciences, Zhongkai University of Agriculture and Technology, Guangzhou, China
3. Shenzhen Environmental Monitoring Center, Shenzhen, China

The plant community structure and plant diversity of Dananshan and Xiaonanshan Mountain areas were carried out measured studies, meanwhile measured PM2.5 concentration indices of outside and inside of forests of 17 communities, comparative analysed the ability of absorbing and reducing of forests with different structures, and explored the argument problem on which kind community is biodiversity higher? Is natural forest, artificial forest or is more artificial disturbed forest higher? The results showed that: there were 195 species in 12 communities of natural forests in Xiaonanshan, in these species, ferns, gymnosperms, and angiosperms were 14, 2 and 179, respectively. Because Dananshan Mountain areas almost all areas were the artificial forests, and several document species in tree layer, so chose 6 communities made research. There are total 137 species, and ferns, gymnosperms, and angiosperms were 16, 2, 119, respectively. The number of species of Xiaonanshan was obvious higher than that of Dananshan. The values of Simpson indices, Shannon–Wiener indices and Odum, Menhinnick richness indices and Pielou evenness indices of Xiaonanshan were also higher than that of Dananshan artificial forests. In respect of community structure, the forests of several species of *Eucalyptus* and *Acacia auriculaeformis* in Dananshan, the tree layer had only 5-6 species, and the individual number of document species occupied most proportion, other species were mainly shrubs and herbs. However, in the communities of Xiaonanshan, the tree layer usually was richness with 15-22 species. And the height of shrub layer was higher and more species, the herb species were also more than that of Dananshan. In the community distribution, the plant diversity of lower altitude places were lower than middle altitude places in Xiaonanshan, maybe the reason was more artificial disturbed, and more pollutants emission. In higher altitude of near 265 m, the plant diversity were also below, probably where were approach top of the mountain, wind was stronger. The situation of Dananshan was similar to Xiaonanshan, but the difference was relative little because of artificial forest. In aspect of absorbing and reducing air pollutant of PM2.5, in the 17 plant communities of Dananshan and Xiaonanshan, the concentration of forest inside were lower 6-7 ppm than that of outside of forests, but the effects of Xiaonanshan communities were stronger than Dananshan. In the southern slope of Xiaonanshan, it is toward the see, in the foot of hill, there was a big heat and power plant, it 24 hours per day continuous emitted pollutant of PM2.5. This caused the concentration of PM2.5 of forest outside usually was 65-69 ppm, but 8 communities of Xiaonanshan could still reduced 6-8 ppm concentration of PM2.5. Only several low altitude communities were a little difference of PM2.5 concentration where were near the heat and power plant, because the high concentration and continuously emitted pollutants. So that these plants the absorbing speed equation or slower than that of pollutants entry forest inside, showed that complex structure and high plant diversity of community have better effects to absorbed and reduced PM 2.5 pollutants.

T3

P0812

Conservation genetics/genomics of *Oxera* species (Lamiaceae) diversified in New Caledonia – The speciation process and implications for appropriate conservation

Yuji Isagi[1], *Yoshihisa Suyama*[2], *Kentaro Uchiyama*[3], *Gildas Gâteblé*[4], *Ayu Narita*[1], *Kohei Kamasaka*[1], *Ryohei Nishijima*[1]
1. Kyoto University
2. Tohoku University
3. Forestry and Forest Products Research Institute
4. Institut Agronomique Neo-Caledonien

New Caledonia is a global biodiversity hotspot, where the flora is characterized by high levels of endemism and diversity. The flora includes ca. 3,400 vascular plants, and 77% of them are considered endemics, which makes New Caledonia as the third highest ranked island for endemism, after Hawaii (82%) and New Zealand (82%). Further, the flora contains very unique taxa such as the most basal angiosperm (*Amborella trichopoda*) and parasitic gymnosperm (*Parasitaxus ustus*). Unfortunately, however, many of these endemic plants are threatened with extinction by various factors as mining, fires, logging, urbanization, and introductions of exotic species. Being a small area size (18,575 km^2), New Caledonia is inhabited by many genera of diversity which are composed of over 20 species. Typical examples of these are *Phyllanthus* (Phyllanthaceae, 113 species), *Psychotria* (Rubiaceae, 85 species), *Syzygium* (Myrtaceae, 70 species), *Pycnandra* (endemic genus, Sapotaceae, 57 species), *Eugenia* (Myrtaceae, 54 species), and *Tapeinosperma* (Primulaceae, 53 species). These taxa are suitable for ecological genetic/genomic research of adaptive speciation, and *Oxera* (Lamiaceae) comprising ca. 32 species in New Caledonia (including several undescribed species) is one of the potential genera for the research. The *Oxera* species in New Caledonia are all endemic, showing rich diversity in size (ten centimeters to several meters), plant form (trees, shrubs, forbs, vines), and flower (color, shape, position). Occurrence frequency of each species is also diverse throughout the country, ranging from micro-endemic to omnipresent. To understand the diversification processes in the genus *Oxera*, we selected several species with different traits to perform extensive genetic/genomic analysis by RAD-seq, RNA-seq, and MIG-seq. Consequently, a significant genetic differentiation was found between subpopulations of a species with wide distribution range (*O. baladica*) along with a speciation process of its sister taxon (undescribed species). We searched for loci with the dN/dS ratio of more than zero among the species with distinct traits, and have identified several candidate genes involved in the speciation. As for species with limited distribution ranges such as *O. rugosa* and *O. pancheri*, we inferred optimum ESU (Evolutionarily Significant Unit) and MU (Managing Unit) by analyzing their demography and genetic differentiation among remnant populations.

T3

P0813

Real-time analysis of the variation of volatile components in *Polygonum hydropiper* by GP-MSE/GC-MSafter the induction of *Plutella xylostella*

Shicui Jiang, *Ziyu Guo, Donghao Li*
Key Laboratory of Natural Resources of Changbai Mountain & Functional Molecules (Yanbian University), Ministry of Education

The volatile components of plants are essential chemical signals of the transmission of information between plants and insects, which also act as thedirect expression of repellent response from plants. Among the volatile compounds in plants, the composition and concentration of terpenes exert significant influence on the repellent effect. The volatile compounds of *Polygonum hydropiper* were analyzed by using the GP-MSE/GC-MS technique. A total of 50 compounds were qualitatively identified, among which were 9 species of terpene compounds. Due to the different amount of volatile compounds being released at different times, this study aimed to analyse the variation of constituents and concentration of volatile compoundsin *Polygonum hydropiper* under normal conditions from 0 to 48 hours after the induction of *Plutella xylostella*. By examining the gene expression profile based on time differences, the candidate genes of terpene weredetermined and were subsequently verified by employing the real-time PCR quantification approach. A novelgene expression pattern of the variation of constituentsin *Polygonum hydropiper* was developed. This study will provide the basic data to research further on the repellent effect of *Polygonum hydropiper* and for the future development of biopesticides.

T3

P0814

Paleoenvironmental evolution since 45ka BP based on sedimentary pollen records of 121712 core in the Northeastern Indian Ocean

Weiming Jiang[1], *Chuanxiu Luo*[2]
1. School of Geographical Sciences of Guangzhou University
2. CAS Key Laboratory of Marginal Sea Geology, South China Sea Institute of Oceanology, Chinese Academy of Sciences

Based on sporo-pollen assemblage characteristics and AMS 14C dating results of 121712 core in the Indian Ocean, the authors determined four pollen assemblages from A to D from bottom to top by using Tilia software. The following results can be found in the analysis of the 4 zones: 1. The pollen and spore composition of the surface sediments were complex. We identified 50 different pollen and spore taxa. The spores included *Microlepia, Cyathea, Pteris,* Polypodiaceae, *Cibotiumbarometz, Hicriopteris, etc.* Herbaceous pollen were from Chenopodiaceae, *Artemisia,* Compositae, Ericaceae, Poaceae, Polygonaceae, Cyperaceae, Caryophyllaceae, *etc.* The tree pollen included *Pinus,* Palmae, *Picea,* Tiliaceae, *Alnus,* Betula, Juglans, Quercus, Myrtaceae, *etc.* 2. Pollen diagram in core 121712: A zone, about 45~42 ka BP: fern (45.32%), tree pollen (32.61%), herb (only 22.08%), pollen concentration (632 grains/g). B zone, about 42~36 ka BP: fern (49.37), herb (26.14%), tree pollen (only2 4.48%), pollen concentration (964 grains/g). C zone, 36~26 ka BP: fern (53.75%), herb (28.10%), tree pollen (only 18.15%), pollen concentration was very low (73 grains/g). D zone, 26~5 ka BP: fern (50%), herb (31.58%), tree pollen (only 18.42%), with a lowest pollen concentration (22 grains/g). 3. Comparing the percentage of spores: In the A zone and B zone, the *Castanopsis* and Evergreen *Quercus* have a high pollen percentage, and there is a large amount of pollen of the Chenopodiaceae and Poaceae. On the contrary, In the C zone and D zone, the

Castanopsis and *Evergreen* Quercus have a low pollen percentage, and the pollen percentage of *Alnus* and *Betula* is gradually increasing. At the same time, the number of pollen of the Chenopodiaceae and Poaceae in this zone was gradually decreasing. About 18ka BP, the pollen number of Chenopodiaceae, Caryophyllaceae, Euphorbiaceae, Alnus and Pinus reached the maximum. In summary, we can find the conclusions as follow:
1). About 45~36 ka BP, there is a large amount of pollen of *Castanopsis* and Evergreen *Quercus*. Combined with high pollen concentration, it can reflect warm and humid climate of the resource areas. 2). About 36~5 ka BP, The percentage of pollen of *Castanopsis* and Evergreen *Quercus* was gradually decreasing, but the percentages of pollen of *Betula* and *Alnus* (belong to the temperate zone) increased, which reflects the climate at that time was turning cold. Herbaceous pollen percentage (such as Chenopodiaceae, Poaceae) increased, it is reflecting the climate gradually becoming cold and dry, the vegetation becoming scarce and drought. About 18 ka BP, the pollen number of Chenopodiaceae, Caryophyllaceae, Euphorbiaceae, *Alnus* and *Pinus* reached the maximum, which reflects the climate was coldest and dryest. And this moment coincides with the Last Glacial Maximum.

T3

P0815

Relationships between aboveground biomass and plant cover at two spatial scales and their determinants in northern Tibetan grasslands

Yanbin Jiang
College of Resources and Environment, Huazhong Agricultural University

Questions: (1) How does aboveground biomass (AGB) of alpine grassland relate to plant cover at different spatial scales? (2) Does response of AGB–cover relationship to biotic and abiotic factors depend on scales? Location: Northern Tibet, China. Methods: Community survey (species, cover, height, and abundance) was conducted within 1 m × 1 m plots in 70 sites. Linear regression, Pearson correlation analyses, and hierarchical partitioning were employed to identify contributions of community features (total cover of community, cover of cushion species, height, and species richness) and environmental factors (mean annual precipitation, mean annual temperature, soil bulk density, and carbon and nitrogen contents) to AGB. Results: AGB exponentially increased along a precipitation gradient and was positively and linearly correlated with community cover for all grasslands without alpine meadow. AGB was also linearly correlated to species cover at both regional and landscape scales. Community attributes and environmental factors to AGB significantly depended on spatial scales. Total community coverage was decisive factor for most alpine grassland AGB except for alpine meadow. Cushion plants and species richness significantly affected AGB–cover relationship at both spatial scales. Mean annual precipitation (MAP), soil bulk density and soil nitrogen content influenced the regional AGB variation. Conclusion: Present study indicated generally positive and linear relationships between AGB and cover are at both regional and landscape scales. Spatial scale may affect ranges of cover and modify contribution of cover to AGB. AGB–cover relationships are influenced by many other variables. Species composition, especially presence and cover of cushion plants, and species richness were important community variables that controlled landscape AGB pattern. By contrast, at regional scale, AGB was significantly affected by MAP, soil bulk density, and soil nitrogen content. Therefore, in deriving AGB patterns at different spatial scales, community composition and related environmental factors should be considered to obtain acceptable accuracy.

T3

P0816

Biogeographical evolution of *Toxicodendron* (Anacardiaceae) mixed with pantropic to temperate disjunctions in the Northern Hemisphere

Ying Jiang[1], *Ying Meng*[1], *Jun Wen*[2], *Ze-Long Nie*[1]
1. Jishou University
2. National Museum of Natural History, MRC 166, Smithsonian Institution

Biogeographic diversification of flowering plants in the Northern Hemisphere is extremely complicate due to the dynamic exchanges between tropic to temperate regions. It is necessary to includes close relatives from the pantropic to fully elucidate the evolution of the temperate disjunctions in the Northern Hemisphere. *Toxicodendron* contains approximately 24 species disjunct distributed between eastern Asia and North America, extending to both southeastern Asia and the neotropics, which is an excellent model for investigating the complexity of biogeographical evolution in the Northern Hemisphere. Phylogeny of the genus have been studied, but with limits of tropical samplings. The aim of this study is to reconstruct the phylogenetic relationships in the genus using two nuclear ribosomal DNA (ITS, ETS) and two chloroplast (*ndhF*, *trnL-F*) sequences to examine the evolutionary history of biogeography in *Toxicodendron* in the phylogenetic context of its closely related tropical taxa. Twenty-two species of *Toxicodendron* are sampled, covering the whole taxonomic and biogeographic distributions, especially including samples from tropical SE Asia. Divergence times of disjunct lineages are estimated with relaxed Bayesian dating with fossil constraints. Phylogenetic analysis strongly support *Toxicodendron* as a monophyletic group. *Toxicodendron borneense* (sect. *Simplicifolia*) and *T. nodosum* (section *Toxicodendron*) from SE Asia form a clade sister to *T. trichocarpum* from central China. *Toxicodendron lamprocarpum* from New Guinea is sister to sect. *Griffithii*. Two temperate disjunct lineages were well supported with disjunction in the late Miocene, one from section *Toxicodendron* and the other between the eastern North American *T. vernix* and the eastern Asian *T. vernicifluum.* The two temperate disjunctions likely favored the migration route by way of the Bering land bridge. The pantropical disjunction between section *Griffithii* (warm temperate to tropical Asia) and *T. striatum* (from the neotropic) was also supported and their divergence time was estimated to be 24.73 (10.19–30.58) Ma. This pantropical disjunct between the Old and New World could be exchanged by North Atlantic land bridge corresponding to global warm in the late Oligocene. However, it may have also resulted in range expansion of tropical forest belt via long distance dispersal. Two disjunct lineages are found between tropical Pacific and the mainland Asia (*T. lamprocarpum* - section *Griffithii, T. borneense* and *T. nodosum* - *T. trichocarpum*), consistent with the southward retreat of many tropical taxa into the equator due to global temperatures declined markedly after Eocene.

T3

P0817

Leaf functional traits reveal that fern's living strategies have diversified in response to angiosperm-dominated forests since Upper Cretaceous

Dongmei Jin[1], *Harald Schneider*[2,3], *Xile Zhou*[1], *Guohua Zhao*[1,4], *Ying Wang*[1,5], *Yuehong Yan*[1]

1. Shanghai Chenshan Plant Science Research Center, Chinese Academy of Sciences/ Shanghai Chenshan Botanical Garden

2. College of Ecology and Evolution, Sun Yat-Sen University

3. Department of Life Sciences, Natural History Museum

4. Key Laboratory of Plant Biology, Harbin Normal University

5. College of Life and Environmental Sciences, Shanghai Normal University

Ferns (monilophytes), the sister group of seed plants, first appear 360 million years (My) ago in the Late Devonian Period. With estimated 11,929 extant species globally, they are the second most diverse group of vascular plants. About 80% of extant fern species, mostly Polypods, have derived in recent 100 My after angiosperm-dominated forests became prominent in terrestrial ecosystems. These findings suggest that ferns have diversified in response to the new environments shaped by angiosperms. However, few evidence from the change of living strategies in ferns have been reported. Leaf functional traits, which shaped by both evolutionary and environmental drivers, can reflect plant living strategies. Analyzing leaf functional traits of abundant fern species against a phylogenetic background may reveal the changes of living strategy in ferns and may detect major changes of the environment during evolution. Thus, we may find evidence about whether and how angiosperms have driven the evolution of ferns since Upper Cretaceous. We sampled sporophytes of 240 fern species covering main linages in fern phylogeny during growing seasons from five natural reserves in China. Five leaf functional traits which are closely related to photosynthesis were measured, namely, leaf mass per area (LMA), chlorophyll content (SPAD), nitrogen content (Narea), nitrogen concentration (Nmass), and leaf area (Area). Leaf traits of ferns were compared to documented traits of seed plants; differentiation of leaf traits among five major groups in fern phylogeny were analyzed using multiple comparisons; and ancestral state of fern leaf traits since 350 My ago were estimated using Phylocom. Our results showed that, in general ferns have lower LMA and Narea but higher Nmass than seed plants, suggesting a preferred shade tolerant strategy in ferns. Within ferns, Early monilophytes, Early polypods, and their ancient nodes were moderate in the leaf traits suggesting a moderate shade tolerance strategy in general, except that Hymenphyllaceae species form an extremely shade tolerant clade. Derived ferns of Eupolypds II were low in SPAD, LMA, and high in Nmass and Area, suggesting an increased shade tolerance. The mainly epiphytic Polypodiaceae clade and part of Dryopteridaceae clade showed increases in SPAD, Narea and LMA, and reduces in Nmass and Area, suggesting an adaption to light-exposed and dryer environment. The expanded variation of fern leaf traits coincided with the divergence of derived ferns since Upper Cretaceous. Our results support that fern's living strategies have diversified in response to the rise of angiosperm-dominated forests, most species belong to Eupolypds II have adapted to the shady, humid forest floor, while most Polypodiaceae species and part of Dryopteridaceae species have adapted to the xeric, light-exposed forest canopy.

T3

P0818

Environment effects on trait distributions in forests of a fragmented landscape

Yi Jin[1], *Jingjing Yu*[1], *Shilu Zheng*[1], *Zuhua Wang*[1], *Guang Hu*[2], *Jinfeng Yuan*[1], *Jinliang Liu*[1], *Mingjian Yu*[1]

1. Zhejiang University

2. Zhejiang Sci-Tech University

Understanding environment effects on trait distributions of forest communities is meaningful to inferring the underlying assembly processes. Here we investigate how the trait distributions of local assemblages of the secondary Masson pine (*Pinus massoniana*) forests are affected by environment in a fragmented landscape (i.e. the Thousand Island Lake) of Eastern China. We established forest dynamics plots (FDPs) in the Masson pine forests on 29 land-bridge islands that largely differ in size, shape and isolation. We conducted the first survey of all woody stems with diameter at breast height (1.3m; DBH) $\geq$ 1 cm in these FDPs during 2008–2009, and recorded a total of > 180,000 stems belonging to 77 species. We measured ten functional traits (e.g. leaf chloroplast concentration, leaf stomata density, specific leaf area, *etc.*) for these 77 species, and built species trait dendrograms for the traits together and each trait separately. We then calculated trait distributions (i.e. community mean trait value, $-1 \times SES.D_{pw}$ and $-1 \times SES.D_{nn}$) of local scale forest assemblages. By adopting a linear mixed effects modeling framework, we predicted trait distributions using environmental factors (e.g. island area, distance to mainland, distance to forest edge, elevation, slope, soil phosphorous availability, *etc.*). We found the environmental factors largely determined trait distributions in most cases. The results suggest that environment, possibly through regulating habitat filtering, to control the assembly of Masson pine forest communities in the study fragmented landscape.

T3

P0819

Abiotic factors responses of invasive *Solidago canadensis* populations in China: Plasticity versus local adaptation

Li Junmin

Taizhou University

Plasticity and local adaptation are both important to help plants adapt to environmental heterogeneity. *Solidago canadensis*, which is invasive in China, has spread widely. We sampled 15 populations of *S. canadensis* in China and conducted common greenhouse experiments to test the plasticity of *S. canadensis* populations in response to abiotic factors treatment, including temperature, water, shade and salt. Among the four traits with significant variation in plasticity among populations in response to temperature, plasticity of leaf length-to-width ratio was significantly positively correlated with latitude and temperature seasonality of the populations. In addition, root/shoot ratio and water-use efficiency showed significant variation in plasticity among populations in response to water availability, and plasticities of these two traits were significantly negatively correlated with longitude and positively correlated with precipitation seasonality. The observed geographic clines in plasticity suggest that phenotypic plasticity of *S. canadensis* may have evolved rapidly in regions with different climatic conditions. Compared to control, shadetreatment significantly reduced growth and content of defense metabolites in *S. canadensis*. Three-way analysis of variance revealed geographical origin to significantly affect the basal diameter of *S. canadensis,* while genotype significantly affected plant height, intercelluar CO_2 concentration (C_i), transpiration rate (T_r), and proline content. Most of the phenotypic variations can be found among individuals within populations. Phenotypic selection analysis revealed that fitness was significantly positively related to plant height, basal diameter, C_i, total flavonoid content, as well as the plasticity of plant height, leaf length, leaf width, g_s, C_i, total flavonoid content, and malondialdehyde content under the control condition. Salt treatment significantly decreased the growth of *S. canadensis*, including the rates of increase in the number of leaves and plant height; the root, shoot, and total biomass; and the root/shoot ratio. Furthermore, salt stress significantly reduced the net photosynthetic rate, g_s, T_r and relative chlorophyll content but significantly increased peroxidase activity and the proline content of *S. canadensis*. Two-way ANOVA showed that salt treatment had a significant effect on the physiological traits of *S. canadensis*. Most of the variation in plasticity existed within populations, not among populations. *S. canadensis* populations from soil with moderate/severe salt levels grew similarly to *S. canadensis* populations from soils with low salt levels. No significant correlation between salt tolerance indices and soil salinity levels was observed. The plasticity of proline content, C_i and the chlorophyll content had significant correlations with the salt tolerance index. Rather than local adaption, these results suggest that individual plasticity played a more prominent role in the shade and salt response of the invasive *S. canadensis.* In summary, both plasticity and local adaptation to abiotic factors may contribute to the spread of this invasive species in novel invaded regions.

T3

P0820

Interactive effects of disturbances on grassland soil and vegetation

Katlego Mashiane[1], *Natalie Haussmann*[1], *Peter C. Le Roux*[2]
1. Department of Geography, Geoinformatics and Meteorology, University of Pretoria, South Africa
2. Department of Plant and Soil Sciences, University of Pretoria, South Africa

Disturbances by alien plant species pose a significant threat to indigenous plant biodiversity and host ecosystems. Similarly, large herbivores can act as disturbance agents, in that their actions can trample plants, and modify resource availability and physical soil properties. The purpose of this study was to gain a greater insight into the effects of these two potential disturbance agents and their interactive effects on soil and vegetation characteristics in a conserved grassland (Rietvlei Nature Reserve, Pretoria, South Africa). Soil and vegetation data were collected from 22 sites, and comprised four treatments: 1) trampled, and invaded by the alien plant species *Campuloclinium macrocephalum*, 2) trampled and uninvaded, 3) untrampled and invaded and 4) untrampled and uninvaded. The two disturbance agents did not interactively affect soil organic matter and soil organic carbon. However, trampled sites had significantly higher soil compaction values than untrampled sites, with this difference being larger at invaded sites. Species richness and diversity were largely unaffected by trampling and invasion, while vegetation cover was lower under trampling and/or *C. macrocephalum* invasion. Of the biotic characteristics analyzed, only species composition showed a significant (albeit weak) response to the interaction between trampling and invasion, with results suggesting that trampling reduces the difference in composition between invaded and uninvaded sites. Thus, while the effects of trampling and alien plant invasion were interactive for some variables, for most soil and vegetation characteristics the impacts of these two disturbances were independent of each other.

T3

P0821

Does the flowering period differ between trees and herbs? A test based on phylogenetic relationship

Ai Kawakubo, *Tetsukazu Yahara*
Kyushu University

There have been many comparative studies about the relationship between growth form and flowering phenology of plants. However, no consensus has been obtained in previous studies. This lack of consensus may be partly due to the lack of statistical tests based on phylogenetic relationship in previous studies. In this study, we tested difference of the flowering period between tree and herbaceous species using independent contrasts calculated from a phylogenetic tree. We observed flowering phenology of 144 plant species growing in the Biodiversity Reserve of Kyushu University from April to July in 2015 and from March to July in 2016. We recorded the presence or absence of flowering for each species along an observation route every three to seven days. The flowering period of *Salix triandra* was the shortest among the 56 tree species, being only 9 days. In contrast, the flowering period of *Rubus hirsutus*, 105 days, was the longest among the tree species. The flowering period of *Cardiocrinum cordatum* was the shortest among the 88 herbaceous species, being 8 days. In contrast, the flowering periods of *Trifolium repens*, 138 days, were the longest among the herbaceous species. To test difference of the flowering period between trees and herbs, we constructed a phylogenetic tree for 144 species based on DNA sequences (*rbcL* and *matK*) and reconstructed ancestral states of growth form and flowering period in each internode of the phylogenetic tree. Then, each speciation event was classified into three categories; 'tree-tree', 'tree-herb', and 'herb-herb', and the average of the contrast of flowering period (difference of flowering period evolved in each speciation event) was compared among three categories. Speciation events where ancestral states are ambiguous are excluded from the analysis. As a result, the average of tree-tree contrast was significantly smaller than that of herb-herb contrast, and there was no significant difference between tree-tree and tree-herb, or tree-herb and herb-herb contrast. The result suggested that flowering period changes tend to be larger in herb-herb speciation events than in tree-tree speciation events, resulted in relatively longer flowering periods in herbaceous species than in trees. This difference of flowering periods between trees and herbs may be associated with the difference of pollinator species. Understory-flowering herbs are usually pollinated by endothermic species such as bumblebees, and a longer flowering period may be advantageous because of their high flower constancy. In contrast, canopy-flowering trees are mostly pollinated by ectothermic insects such as syrphid flies and small bees, and mass flowering with shorter flowering periods may be advantageous to attract those opportunistic pollinators.

T3
P0822
MPRR1 is a quick response E3 ligase that immediately eliminates emergent misfolded proteins bypassing chaperone-mediated triage decisions

Jong Hum Kim[1], *Seok Keun Cho*[1], *Oh Tae Rin*[1], *Ryu Moon Young*[1], *Yang Seong Wook*[1,2], *Kim Woo Taek*[1]

1. Department of Systems Biology, College of Life Science and Biotechnology, Yonsei University

2. Section of Plant Biochemistry, Department of Plant and Environmental Sciences, Faculty of Sciences, University of Copenhagen

Proteotoxic stress induced by misfolded proteins is one of the most important causes of cell disorder throughout whole eukaryotic organisms. UPS (Ubiquitin-Proteasome System) is well-known defense mechanism via degradation of misfolded or disordered protein in eukaryotic cells. Besides animal systems, plant also use UPS for maintaining and regulating crucial cellular processes such as signal transduction, cell death, and nutrient regulation. Here, we show that MPRR1 (Misfolded Protein Responsive RING protein 1) is a regulator of 26S proteasome activity under proteotoxic stress. We investigated transcript expression profiled under proteotoxic stress and generated transgenic plants, which over-expressed arabidopsis E3 ligases responsive to proteotoxic stress. Among the transgenic plants, MPRR1 over-expressor showed enhanced growth and reduced Ub-positive aggregates compared to wild-type plant. Knock-down transgenic plant of MPRR1 and Stable over-expressor of MPRR1^{C153S} RING domain mutant protein showed impaired growth under proteotoxic stress. Because we found that proteasome inhibitor MG-132 decreases tolerance of over-expression transgenic plant of MPRR1 under proteotoxic stress, we hypothesized that enhanced fitness of the transgenic plant is originated from enhanced proteasome activity. The activity of 26S proteasome was up-regulated compared to wild-type plant and the amount of 26S proteasome was maintained when it of wild-type was decreased. In addition, artificial substrate clearing activity was enhanced in the over-expressing plant compared to wild-type plant. Because MPRR1 interacted with 19S subunits physically, regulation of 26S proteasome activity might be direct. We also found that MPRR1 protein is stabilized under proteotoxic stress post-translationally and the regulation of the protein stability was dependent to its RING E3 ligase activity. Collectively, a cytosolic RING E3 ligase, MPRR1 improves plant fitness and an aberrant protein clearance under proteotoxic stress condition through stabilization of 26S proteasome activity by physical interaction and ubiquitination.

T3
P0823
A geographic cline in pollen color may be driven by abiotic selection

Matthew Koski, *Laura Galloway*

Department of Biology, University of Virginia

Floral pigmentation functions in both attracting pollinators and ameliorating abiotic stress experienced by pollen and ovules. The study of floral color diversity is often restricted to petal pigmentation though many floral reproductive parts can display marked variability in color. Here, we explore the degree of pollen and petal pigmentation diversity across the range of the North American herbaceous plant, *Campanula americana*, and address whether geographic patterns may be explained by abiotic factors. Pollen color ranges from white to deep purple, and there is a strong geographic cline whereby eastern populations have light pollen while western populations have darker pollen. After accounting for latitudinal variation in temperature, UV and precipitation, both temperature and UV are higher in the western portion of the range. Thus, both temperature and UV are associated with pollen color diversity. Experiments show that dark purple pollen tends to germinate better in higher temperature and UV conditions, but that light and dark pollen perform similarly under lower temperature and UV conditions. Dark pollen may therefore be adaptive under high heat and UV conditions, but the fitness benefits of light pollen remain unclear. Our results contribute some of the first data on geographic clines in pollen color and the adaptive nature of pollen pigmentation with respect to abiotic stress. Further studies of neutral genetic diversity will elucidate whether selection drives the observed geographic cline in pollen color.

T3
P0824
A phylogeographical transect study of Central European steppe grassland species from the Hungarian Pannonicum towards the westernmost limit

Kristina Plenk[1], *Katharina Bardy*[1], *Maria Höhn*[2], *Mike Thiv*[3], ***Matthias Kropf***[1]

1. University of Natural Resources and Life Sciences, Institute for Integrative Nature Conservation Research

2. Szent István University, Department of Botany and Botanical Garden of Soroksár

3. Stuttgart State Museum of Natural History

In our study, genetic and performance data were used to analyse genetic structuring and diversity, performance parameters, and biogeographical history of different “steppe grassland plants” in an east-west transect representing their distribution range periphery. This transect explicitly covers the absolute distributional limit of most investigated taxa, which typically show a more widespread continuous occurrence in the eastern region (Pannonian Basin, Hungary), compared to populations at the margin of this continuous Pannonian occurrence (Eastern Austria) and the (north) westernmost exclaves in western Germany (especially in Rhineland-Palatinate). Referring to theories like the “Abundant

Centre" hypothesis and the "Central-Marginal-Model" we assume more isolated and individual-poor populations at the (north) westernmost absolute distribution limit and therefore, reduced fitness, increased among population differentiation as well as decreased within population diversity. All study species presented here, namely *Adonis vernalis* (Ranunculaceae), *Carex supina* (Cyperaceae), *Inula germanica* (Asteraceae), *Linum flavum* (Linaceae), *Oxytropis pilosa* (Fabaceae), and *Poa badensis* (Poaceae), are rare and endangered in Central Europe. To analyse patterns of genetic diversity and structuring, Amplified Fragment Length Polymorphisms (AFLPs) and chloroplast (cp) DNA sequence data were generated usually for four representative populations per study region and species. Plant fitness was investigated based on the same populations using different field performance parameters like fruit set and mass, as well as germination rates under standardized conditions. Results from the fitness analyses indicate marked differences among study species and regions, but do not clearly support the idea of declining performance towards the absolute distribution range limit. Depending on the species surveyed, other factors like life history traits (notably, clonal growth) and/or variation in population size seem to have greater impact on plant performance. Genetic data (AFLPs & cpDNA), currently available for *C. supina*, *L. flavum*, *O. pilosa*, *P. badensis*, and *A. vernalis* (only AFLPs) partially show expected patterns: particularly, AFLP data group samples predominantly according to their geographical origin (i.e. representing the three study regions), interpreted as vicariance. However, with respect to the extent of population differentiation and genetic diversity patterns, again no study species follows all of the hypotheses as assumed, even though several species fulfil single expectations. In summary, we have therefore to note that phylogeographical patterns observed in Central European steppe grassland plants are mostly species-specific.

T3

P0825

Plant-microbial interactions: Depth-dependent effects of roots on microbial activities in arable soil

Amit Kumar, *Muhammad Shahbaz, Yakov Kuzyakov, Johanna Pausch*

Georg-August University of Göttingen

In rooted soils, microbially mediated soil organic matter (SOM) decomposition is of special interest as roots are characterized by inputs of high quality of substrates for microbial uptake. A direct expression of soil microorganisms to available nutrients and metabolic requirements are extracellular enzyme activities (EEA). Production of extracellular enzymes by microorganisms is an important mechanism to obtain energy and nutrients from soil via SOM decomposition; however, various biotic and abiotic factors regulate microbial activities. Furthermore, energy sources (root-derived organics) decrease with soil depth and consequently microbial activities are hampered. In addition, application of inorganic N-fertilization also affects microbial activities and consequently affects SOM decomposition. Therefore, we hypothesized that planting increases EEA via microbial activation and N-fertilization reduces EEA because microorganisms are less dependent on nutrient acquisition through SOM decomposition. A field experiment was established on an arable soil in Gottingen, Germany. Maize was grown in plots (5 x 5 m^2) with- or without N-fertilization and the plots were separated with 2 m wide buffer strip to avoid neighboring effects. Control plots (without maize) with- and without N-fertilization were also established. Plants biomass (root and shoot biomass) and microbial biomass C and N, and potential activity of C-, N-, and P-acquiring enzymes were measured with fluorogenically-labeled substrates at 4 soil depths and two sampling times (72 and 130 days after maize planting). We found that roots enhanced microbial biomass C and N and potential activities of C-, N-, and P-acquiring enzymes, especially in the rooting zone at 5-15 and 15-25 cm soil depth. Moreover, N-fertilization had minor effects on microbial biomass C, and C-, and P-acquiring enzyme activities; however, it increased shoot biomass and decreased microbial biomass N and N-acquiring enzyme activities. Higher microbial activity in rhizosphere provides evidence of root-mediated microbial activation resulting in increased SOM decomposition and consequently release of nutrients into the soil. It is noteworthy finding that in short-term, the root-mediated microbial activation and not N-fertilization will accelerate nutrient cycling in arable soil.

T3

P0826

Testing the scaling effects and mechanisms of N-induced biodiversity loss: Evidence from a decade-long grassland experiment

Zhichun Lan[1,2], *G. Darrel Jennerette*[3], *Shuxia Zhan*[2], *Wenhuai Li*[2], *Shuxia Zheng*[2], *Yongfei Bai*[2]

1. Institute of Watershed Ecology, Nanchang University

2. State Key Laboratory of Vegetation and Environmental Change, Institute of Botany, the Chinese Academy of Sciences

3. Department of Botany and Plant Sciences, University of California, Riverside

Although extensive studies demonstrate that nitrogen (N) enrichment frequently reduces plant diversity within small quadrats (0.5–4 m^2), only a few studies have evaluated N effects on biodiversity across different spatial scales. We conducted the first experimental test of the scale dependence of N effects on species richness from a 10-year N treatment (1.75-28 g N m^{-2} yr^{-1}) in a typical steppe. We used species area relationship (SAR) to analyze the scale dependence of species loss with power model $S = cA^z$ (S is species number, A is area, c is intercept, and z is slope). We found that: (1) absolute species loss decreased at sampling area > 8 m^2; (2) proportional species loss (compared to control) decreased and critical threshold (N_{crit}) for biodiversity losses increased with sampling areas; (3) these scale dependences were quantified as increasing slope (z-value) of SAR with N addition. Through SAR decomposition, we found that this overall positive effect was in response to positive effects of changes to the species abundance distribution over negative effects of overall species richness losses. As N enrichment typically occurs at scales much larger than individual plots, understanding how N enrichment affects the scaling patterns of biodiversity is necessary for biodiversity conservation and ecosystem management in response to anthropogenic N deposition.

T3
P0827
The potato (*Solanum tuberosum*) rhizosphere and novel methods for controlling potato cyst nematodes (*Globodera pallida* and *G. rostochiensis*)

Eoin Lettice, *Rachel Hearne, Peter Jones*
University College Cork, Ireland

Potato cyst nematodes (PCN; *Globodera pallida* and *G. rostochiensis*) are known to be significant pests of potato crops in agriculture and are a threat to global food security. The pests are involved in a tritrophic interaction with potato (*Solanum tuberosum*) and the soil microbiome. Central to this interaction is the release of allelochemicals ('hatching factors') from the plant roots which trigger a break in dormancy in the plant pathogenic nematodes. A series of experiments were conducted in order to elucidate the role played by each element, including the host plant, in this tritrophic interaction. Soil samples from potato ridge and bulk areas of a commercial potato field were collected and the impact of the potato rhizosphere effect was assessed using community-level physiological profiling. Soil bacterial activity and diversity were shown to be greater in potato ridge soil compared to bulk soil, indicating a direct effect of potato root leachates on the soil microbiome. The use of soil bacteria, isolated from the rhizosphere of a non-host plant (sugar beet; *Beta vulgaris* ssp. *vulgaris*) as a biological control agent for PCN was examined. Two of three isolates assessed were found to be competent colonisers of the sugar beet rhizosphere and were able to significantly increase the hatch of *G. pallida* in the absence of the host plant. The use of such a trap-crop system to induce the hatch of PCN in the absence of the host plant is proposed as a potential control option for the pest. In addition to further elucidating the tritrophic interaction between PCN, potato and soil microorganisms, the interaction between the sibling species of PCN was also examined. PCN-susceptible potato cultivars were inoculated with single- and mixed- species populations of PCN. This experiment demonstrated that the maturity class of the potato cultivar has a significant influence on PCN multiplication with *G. rostochiensis*, for example, performing better on late maincrop cultivars Kerr's Pink and Golden Wonder. Overall, *G. pallida* was found to outcompete *G. rostochiensis* in mixed-species populations. This work has demonstrated the increasing problems posed by PCN and *G. pallida* in particular. It also confirms that the host plant is crucial to our understanding of the pest lifecycle. By understanding the potato rhizosphere we can better understand and control this important plant pest.

T3
P0828
Environmental heterogeneity and mechanism of stoichiometry properties of vegetative organs in dominant shrub communities across the Loess Plateau

Danfeng Li
Institute of Botany, Chinese Academy of Sciences

Exploring variation patterns of C, N, P contents and their ratios of plant vegetative organs along environmental gradient can provide insight on the resource allocation strategy of plants. In order to search the environmental heterogeneity and mechanism of stoichiometry properties of plant vegetative organs (including root, stem and leaf), we investigated the influences of different soil nutrients conditions on the element allocation among plant vegetative organs in the dominant shrub communities along a hydro-thermic gradient across Loess Plateau. The species compositions and structures of 123 small quadrats (5 m × 5 m) of dominant shrub communities in 41 sites were investigated by standard community sampling method along the declined hydro-thermic conditions from southern Gansu Province to northern Ningxia Hui Autonomous Region and western Gansu. The data about height, coverage and number of the dominant shrub were recorded. At least 369 biological samples of plant vegetative organs (root, stem and leave) of dominant shrubs were collected and their fresh and dry mass were measured, respectively. About 123 soil samples were taken by earth boring auger and ring knife. C, N and P contents of all biological and soil samples were measured after lapping and sieving. Important findings: 1) Organic matter content (mean concentration of leaf C ((424.11 ± 42.21) $mg \cdot g^{-1}$)) in Gansu and Ningxia Hui Autonomous Region was relatively low, and nitrogen (mean concentration of leaf N ((23.00 ± 8.09) $mg \cdot g^{-1}$)) was higher while phosphorous (mean concentration of leaf P ((1.18 ± 0.50) $mg \cdot g^{-1}$)) was relatively deficient. 2) Along the increasing growing season temperature and declining annual precipitation from southern Gansu to northern Ningxia Hui Autonomous Region and western Gansu, soil nutrients content decreased, and the storage of C, N and P in root, stem and leaf decreased. The shift trends of C:N agreed with the N storage's shift trends in root and stem while N:P in all organs were contrary to the P storage's shift trends. Meanwhile, the allocation patterns of C, N and P among vegetative organs among the three regions dominated by shrub communities were much different. 3) In the region with high soil nutrients, the nutrient storage in vegetable organs were similar among different species in the same dominant shrub regions, but in the regions with low soil nutrients, the nutrient storage in vegetable organs displayed high variations. The variations of nutrient storage of vegetable organs for the same shrub communities from different regions were not significant.

T3
P0829
Water depth affects reproductive allocation and allometry in the submerged macrophyte *Vallisneria natans*

Lei Li[1,2], *Jiakuan Chen*[2], *Zhiping Song*[2]
1. Center for Watershed Ecology, Institute of Life Science, Nanchang University
2. Institute of Biodiversity Science, Fudan University

The relationship between size and reproductive output is a fundamental aspect of a plant's reproductive strategy: the conversion of growth into fitness. Differences among populations in the reproductive allometry have been attributed to environmental conditions. However, the causes of variation in reproductive allocation are still debated (i.e. due to fixed patterns of development, or plasticity in the developmental trajectory of vegetative and reproductive allocation). Using a mesocosm experiment, we investigated variation in reproductive allocation of a semelparous submerged macrophyte, *Vallisneria natans*, growing at three levels (shallow, intermediate and deep) of water depth, to test the hypothesis that allometric trajectories of reproductive allocation are highly plastic in *V. natans*. Plants were measured after 26 weeks of growth. The allometric relationships between reproductive (R) and vegetative (V) biomass among treatments were compared. In deep water treatment, total plant biomass decreased, average height increased and individuals started flowering in smaller size when compared to the two shallower water treatments. Deep water affected the biomass and number of fruits produced per plant leading to less sexual reproduction. Plants in deep water (the more stressful treatment) had relatively high allocation to reproduction, though they were quite small. The log R vs. log V slope in deep water treatment was much lower than those in more favourable treatments, and shallow water plants exhibited a very different log R-log V relationship than intermediate water plants. A possible explanation is that the environmental limits on size in deep water should favour a relatively large reproductive allocation at smaller sizes and a smaller investment in reproduction per additional unit of biomass accumulated. Our results demonstrated that water depth affects the observed pattern of reproductive allometry among experimental *V. natans* populations, and this has important implications for the fitness of macrophytes under environmental changes due to anthropogenic disturbance (e.g. dam constructions) and climate change (changes in precipitation pattern). Environmental stress in deep water tends to favor reproduction at relatively small sizes, associated with lower slopes of the log R-log V relationship, supporting that reproductive allometry can be interpreted as an adaptive strategy of plant growth and allocation, rather than as the product of fixed developmental constrains.

T3

P0830

The effectiveness of *Wedelia trilobata* of man-made slopes in Hong Kong

Lu Li[1], *Sai Leung Ng*[2]

1. Baoan Urban Management Bureau, Shenzhen

2. Department of Geography and Resource Management, The Chinese University of Hong Kong

With a dense population, Hong Kong often suffers from geological hazards of landslide in the rainy season. Slopes are widely concerned by the public and government. Safety and aesthetics are two major considerations when man-made slopes constructed. The objective of this study is the assessment on *Wedelia trilobata* performance of man-made slopes with the hard surface. In this study, three surveys were conducted in 2007, 2008 and 2009 on man-made slopes with the hard surface applied with fifteen products of slope bioengineering techniques. Substrates of different products were sampled and analyzed. Results of this study indicated that successful greenery on man-made slopes with the hard surface can be achieved by monotonous use of fast-growing landscaping taxa, such as *Wedelia trilobata*, which is an exotic species with a wide range of environmental tolerance. This species inhibited the development of other species although it provided lush vegetation coverage. The main benefit of using fast growing landscaping taxa stems was from their ability of forming large colonies over the slope in a very short time period. However, the drawback of landscaping taxa was the out-competition or even elimination of other plant species. Therefore, selection of landscaping plants for greening would have significant effects on the vegetation development. Once the slope was dominated by fast growing landscaping taxa, processes and patterns of natural succession can be highly disturbed and altered. In the study, biodiversity could not be improved by increasing vegetation cover. Simple species composition of our man-made slopes calls for room for improvement in this area. Therefore, slope engineers or landscape managers, who used to emphasize the establishment of green cover, can be reminded that the consideration of urban biodiversity should be incorporated into the management of urban slopes. Slope engineers or landscape managers are encouraged to put more efforts into stimulating and accelerating natural succession.

T3

P0831

Consequences of drought for recovery from defoliation in *Robinia pseudoacacia* and *Sophora japonica*

Ming-Yan Li[1,2], *Wei-Hua Guo*[1,2], *Xiao Guo*[3], *Ning Du*[1,2]

1. Institute of Ecology and Biodiversity, College of Life Sciences, Shandong University

2. Shandong Provincial Engineering and Technology Research Center for Vegetation Ecology, Shandong University

3. College of Landscape Architecture and Forestry, Qingdao Agricultural University

Climate change may result in the increase of both pest insect damage and the frequency of drought. Therefore, there is a need to better understand the combined effects of defoliation and drought on tree growth, and physiological processes to predict the effect on forest ecosystem productivity in response to climate change. Artificial defoliation was often used to simulate the pest insect damage. In this study, we conducted a manipulative experiment in a greenhouse, where 1-year-old seedlings of both *R. pseudoacacia* and *S. japonica* were exposed to three soil moisture conditions (75%, 55%, 25% of saturated soil moisture content) and three defoliation regimes, i.e. control, removed upper canopy leaves (approximately 48% of total leaf area) and all leaves (100% defoliation). We found that after 14 days of treatment, photosynthetic capacity of *R. pseudoacacia* was higher than that of *S. japonica* regardless of water treatments. Photosynthetic capacity of the two species was significantly increased under defoliation treatments, which indicated that plants have already started to compensate growth by increasing photosynthetic capacity, especially in *S. japonica*. After a period of growth, photosynthetic capacity and chlorophyll content of the two species under defoliated treatment had no significant difference between all treatments. However, the parameter of growth, biomass accumulation and non-structural carbohydrate were lower under 100% defoliation treatment regardless of water condition, which means seedlings of both species are difficult to recover under any water regime. In addition, most of measured parameters were inhibited by drought. Meanwhile, our results showed additive effects of defoliation and water stress on most of parameters, which suggests drought inhibited the recovery capacity after defoliation. We also observed that biomass accumulation of *S. japonica* had no significant difference among defoliation treatments, while biomass accumulation of *R. pseudoacacia* under 100% defoliation treatments was significantly lower than the control treatments. Meanwhile, gas exchange of *S.*

japonica was higher than that of *R. pseudoacacia* after 14 days of treatment. According to our results, we concluded that recovery capability of *S. japonica* is stronger than that of *R. pseudoacacia*. By studying recovery capability, *S. japonica* may have a better ability to adapt to future global change than *R. pseudoacacia*. Thus, *S. japonica* suits for vegetation restoration and reconstruction in this warm temperate zone.

T3

P0832

Responses to successional water deficit and recovery in four warm-temperate tree species

***Qiang Li**, Ning Wang, Shuna Liu*
Shandong University

To detect the responses to successional drought stress and recovery, studies of physiological attributes and growth of the seedlings of four warm-temperate trees (*Robinia pseudoacacia* L., *Amorpha fruticosa* L., *Vitex negundo* L. var. *heterophylla* (Franch.) Rehder, *Quercus acutissima* Carruth.) were conducted in a greenhouse. In the study, 3-month-old seedlings were subjected to a progressive drought and a following rewatering treatment at two different drought stages. Stomatal conductance (g_s) was measured daily to show plant water status, and seedlings of well-irrigated (control), mild drought stress (MD, with g_s 50% declined), severe drought stress (SD, with g_s declined to the lowest point), rewatering after MD (RMD) and SD (RSD) were sampled. Soil water content, stem water potential, xylem hydraulic conductivity, leaf photosynthetic rate and osmoprotectant contents were measured. The treatment significantly influenced many aspects of seedlings performance, but water-stressed seedlings reposessed most physiological functioning after the recovery period. *Q. acutissima* and *V. negundo* var. *heterophylla* seedlings exhibited a greater degree of isohydry with drought, since they controlled stomata timely from the onset of dehydration. However, *R. pseudoacacia* and *A. fruticosa* behaved near-anisohydric, because they kept stomata open longer during drought. Our results support mounting evidence that xylem hydraulic conductivity interacts with leaf gas exchange and stem water potential, which might be the final recovery in the hydraulics-photosynthetic attributes continuums. We also demonstrate trade-offs and coordination among attributes influencing plant water use, photosynthesis and compounds regulation.

T3

P0833

The role of arbuscular mycorrhizal fungi in plant nitrogen uptake of temperate grassland

***Yawen Li**[1], Yongfei Bai[1], Ping Wang[2], Baoming Ji[2]*
1. Institute of Botany, the Chinese Academy of Sciences
2. Beijing Forestry University

Arbuscular Mycorrhizal Fungi (AMF) are ubiquitous microbes in the ecosystem, which can form symbiont with 80% terrestrial plant species. The significant function of AMF is to help plants get water and mineral elements from soil. Traditional researches are more focused on phosphorus uptake, and we know less about nitrogen uptake. Especially in grassland ecosystem, Nitrogen (N) is a key factor for its productivity and stability. Thus, we use a split-compartment cultivation system and ^{15}N-labeling technique to test the uptake of different forms of N (NH_4^+, NO_3^-, GLY) in different N levels by AMF. *Leymus chinensis*, *Cleistogenes squarrosa*, and *Stipa grandis*, which are constructive species of temperate typical grassland, are inoculated with indigenous field soil. The results show that ^{15}N uptake capability vary from plant species to N levels. *Cleistogenes squarrosa* are more rely on AMF, especially with high N content in the soil. No significant effects are observed on dry weight of *Leymus chinensis* and *Stipa grandis* with AMF inoculation. And the ^{15}N uptake capability enhances with the increasing N content in soil among all these species. However, the ^{15}N uptake capability of AMF is not influenced by N forms. The efficiency of AMF to transfer nitrogen is from 0.2%~27%.

T3

P0835

Early positive effects of tree species richness on soil carbon sequestration in a large-scale forest biodiversity experiment

Yin Li
State Key Laboratory of Vegetation and Environmental Change, Institute of Botany, the Chinese Academy of Sciences

Tree species richness has been reported to have positive effects on above-ground biomass and productivity, but little is known about the effects on soil organic carbon accumulation. To close this gap, we made use of a large biodiversity ecosystem functioning experiment in subtropical China and tested for tree species richness and environmental effects on soil organic carbon accumulation. In 2008 and 2015, soil samples were taken by depth increments to the depth of 30 cm from 57 plots ranging in tree richness from one to eight species. Redundancy analysis, linear mixed-effect models and structural equation models were used to test for correlation of plant-related and topographic variables with soil organic carbon accumulation, to assess the relative importance of these predictors and to compare direct and indirect effects of plant-related and topographic variables on the soil organic carbon accumulation. Soil organic carbon stocks decreased by 253.7 g C m^{-2} in the top 0-5 cm, but increased by 309.3 g C m^{-2} in the 10-30 cm depth, resulting in an increase of 6.2 g C m^{-2} from 2008 to 2015 over the whole depth profile. Soil organic carbon accumulation significantly increased with tree species richness, slope and leaf litter carbon content. While tree species richness and slope had direct effects on the change of soil organic carbon stocks, there were only weak or absent indirect effects of the other variables tested. Our results indicate that tree species richness has direct positive effects on soil organic carbon stocks, which however, are only weakly mediated via leaf litter biomass, fine root biomass and leaf litter carbon content. Future studies should make more effects on the impact of leaf litter turnover rates, leaf litter quality and soil microbial community structure on soil organic carbon accumulation.

T3

P0836

Canopy trees of dominant species shape fine-scale spatial heterogeneity of soil nutrients in Xishuangbanan tropical rainforest

***Zhipeng Li**[1,2], Shangwen Xia[1], Luxiang Lin[1], Douglas Schaefer[1], Xiaodong Yang*[1]*

1. Key Laboratory of Tropical Forest Ecology, Xishuangbanna Tropical Botanical Garden, Chinese Academy of Sciences, Mengla 666303, China
2. Department of Agricultural Soil Science, Georg-August University of Göttingen, Busgenweg 2, Göttingen 37077, Germany

The highly spatial heterogeneous of soil properties plays an important role in species coexistence and biodiversity maintenance via leading niche partition in tropical rainforest. Although plants effects (PEs) have been shown to be one of the most important mechanisms influencing soil heterogeneity, the relative importance and characteristic scales of them on fine scale soil heterogeneity (in 1 ha area) still remain unclear. In this study, we combined mapping trees information and high resolution soil nutrients data (scale is 5 meters), using multivariate statistical analysis and wavelet analysis tool to analysis relationship between trees biomass, topography and soil nutrients at multiple scales. Among five selected tree species, only PARACH had strong and positive correlation with all measured macronutrients. The magnitude of its effects on soil properties were P > NO_3-N > Total N > K which may reflect the degree of demand in tree growth. Wavelet analysis showed that topography and plants significant driven soil heterogeneity at multiple scales. Topography influenced soil pH at whole scale, inorganic N at small scale and P at large scale. Effects of PARACH influenced soil NO_3-N at whole scale and P at large scale. The effects of PARACH were independent of topography. Our result indicates that canopy trees of dominant species play a crucial role in soil heterogeneity generation in tropical rainforest, and more efforts needed to be devoted to understanding the relation between traits of dominant tree species and ecological consequences which they probably cause.

T3

P0837

Hunan Chang-Zhu-Tan City Cluster National Research Station of Forest Ecosystem

Juyang Liao[1,2], *Youjun He*[1,2]
1. Hunan Forest Botanical Garden
2. Hunan Chang-Zhu-Tan City Cluster National Research Station of Forest Ecosystem

To be approved by the panel of CFERN, "Hunan Chang-Zhu-Tan City Cluster National Research Station of Forest Ecosystem" is the first urban agglomeration forest ecosystem observation research station in China. The main station is located in Hunan Forest Botanical Garden (N 28°06′40″, E 113°01′30″), the slave station is located in Zhaoshan Scenic Spot. The station is located in downstream of Xiangjiang river. It has a humid subtropical monsoon climate. The main soil types: red soil, yellow soil and yellow brown soil, the main vegetations: *Pinus massoniana* forest, Chinese fir, bamboo and *Camellia* forest. It has the typical characteristics of subtropical evergreen broad-leaved forest. The station aims to analyze and study the interaction between forest ecosystem and human activities, economic and social development, then provide scientific and accurate data for ecological environment improvement and forest ecosystem self-recovery of Chang-Zhu-Tan City Cluster, and offer the scientific decision-making basis for ecological environment construction, economic sustainable development; To provide data collection, processing, sharing services for the national ecological environmental science research and the majority of users.

T3

P0838

The investigation of potential interactions between chosen grass plants in savanna ecosystem in the Nylsvley Nature Reserve, Limpopo Province, South Africa

Maanda Hadzhi Ligavha-Mbelengwa
University of Venda

Potential interactions between grass plants was investigated in the Nylsvley Nature Reserve. Nearest neighbour analysis technique was used to study the interactions between grass individuals and it was possible to prove that within and between species interactions, particularly competition was occurring between grasses investigated. Two phases were employed in the removal experiments where the first phase entailed pre-removal bombing wherein the xylem water potentials of all the individual species still neighboured were determined. A month later post-removal pressure bombings were conducted wherein three nearest-neighbour grass individuals of the same species were removed to isolate the fourth similar individual species. On the other hand three similar species different from the one isolated were also removed to isolate the individual species different from those three similar individual species. Removal of plants from both monospecific and mixed stands resulted in less negative soil water potentials as was observed by an improvement in xylem water potentials on cleared stands. Removal experiments also led to changes in plant growth, production and reproduction. Plant individuals in cleared stands showed out-of-season leaf flowering tillers production. It was concluded that there was an increase in soil moisture content and prolongation in the period of soil moisture availability for non-neighboured plant individuals. Root of some plant species of each of the two species on both cleared and uncleared stands were excavated. Some root extensions into new soil volumes were observed; and such might be proof enough that water and space are some of the resources necessary for the wellbeing of plants.

T3

P0839

Ephemeral root modules widely exist in 12 perennial herb, shrub and tree species

Bo Liu[1,2], *Fanjiang Zeng*[2], *Arndt Stefan K.*[3]
1. Linyi University
2. Xinjiang Institute of Ecology and Geography, Chinese Academy of Sciences
3. University of Melbourne

The terminal branch orders of plant roots have been proposed as a short-lived 'ephemeral' modules specialized for resource absorption. The occurrence of ephemeral root modules has so far only been reported for a temperate tree species and it is unclear if the concept widely exists in other woody (shrub, tree) and herb species. Fine roots of 12 angiosperm perennial herb, shrub and tree species were monitored using branch-order classification, sequential sampling and improved rhizotron in a warm temperate China Taklimakan desert, and their median life spans, anatomical traits, nitrogen concentrations and respirations were also compared through two years' experiments. Two different classes of root modules existed in all monitored plant functional types. The first two (perennial herbs, shrubs) or three (trees) root orders were of ephemeral nature and had a primary structure, consisted mainly of cortex cells, had no secondary growth, high nitrogen concentrations and respiration rates and very short life spans of one to four months. The last two branch orders (third and fourth orders in herbs, shrubs; fourth and fifth orders in trees) in all plant functional types were of a perennial nature, with roots of greater diameters, no cortex, distinct secondary growth and cork cells, and these roots had lower nitrogen concentrations and respiration rates but much longer life spans. Ephemeral and perennial branch orders probably form two different functional modules: a primary ephemeral module that has mainly absorptive functions and a secondary perennial module that has mainly transport, storage and structural functions. The distinct short-lived distal cohorts of roots seem to be widely distributed and maybe even a general feature amongst many plant functional types. Ephemeral root modules method will simplify root differentiating and sampling, and enriched method of root system classification compared to coarse–fine root dichotomies and branch orders.

T3

P0840

Warming-determined spatial and temporal patterns of forest dieback in Inner Asia

Hongyan Liu
Peking University

Temperate and boreal forests border Inner Asia, the world's largest dryland, and form a long and diversified forest-steppe ecotone. Forest dynamics in this circum-gobi ecotone strongly contribute to regional climate feedback, carbon pool, as well as environmental conditions in the northern hemisphere. Geographical heterogeneity was stressed in studies on vegetation response to climate change at site scale albeit water deficit was documented to be commonly constraining tree growth. Here we report a MAP-associated forest growth pattern with dieback of all forest types near the xeric treeline, reduction in forest-growth within the forest-steppe ecotone and increase in forest-growth in continuously distributed forests. Tree-growth as indicated by ring-width and basal area index (BAI) is significantly associated with Palmer drought severity index, proxy of soil moisture, in this region. We also identified a threshold of drought duration of 15 dry months during the growing season of the prior three year for forest dieback, regardless of contribution from winter water storage. The border between water surplus and water deficit for tree-growth has shifted and area of water deficit has been enlarged, which is clearly caused by raised temperature, during the last 50 years. To monitor forest dynamics in this region, transect across the ecotone is proposed.

T3

P0841

Landscape tree species selection and climatic adaptability evaluation using distribution mapping in Shanghai, China

***Ming Liu**, Deshun Zhang*
Tongji University

Landscape trees play a fundamental role in urban green spaces such as the living environment improvement, landscape beautification, microclimate regulation, disaster prevention and other benefits and services. However, urban landscape trees are demanded stronger resilience and resistance in the severe environmental stresses higher than their natural habitats. Furthermore, ongoing climate change will bring new challenges in urban landscape trees selection and landscape design. Shanghai is located in east China, the west coast of the Pacific. And Shanghai City belongs to China's high-density urbanized metropolis, affected by climate change in addition to the urban climate impacts, where exacerbated by the difficulty of planning urban landscape trees. In the recent 30 years, the temperature increasing, precipitation rising, sunshine diminishing and wind fluctuations hazards are new four major characteristics of Shanghai climate change: (1) Urban heat island (UHI) effect is more significant, and the trend of UHI has spread to the eastern region. (2) The urban rain island (URI) effect has enhanced with the average annual precipitation rising. (3) The air pollution and haze are more and more serious due to the increased days of cloudiness and rainy. (4) The southeast wind is still the driven wind possibly lead to hotter summer and warmer winter. However, there are virtually little native plants and endemic vegetation as in the alluvial plain of the Yangtze River Delta in history. The vast majority of plants are imported from domestic provinces or around the world. At present, the main plants include trees, shrubs, vines and bamboos, having 738 species belong to 215 genera and 78 families. After climatic adaptability assessment by using climate envelopes models, the result shows that 85.12% of the tree species can well adapt to the future climate change in Shanghai. Yet, about 72 tree species may not be in a good state under climate change. These species are mostly limit to some climatic factors for the normal metabolic function, some species are probably due to extreme cold temperature in the winter, some trees required a more appropriate habitat for the growth in the cold would not grow well in warmer and humid climates in Shanghai. Finally, these selected landscape tree species shape the landscape of green land in Shanghai, building a rich diversity of ecological communities, maintain the stability and sustainability of urban ecosystem.

T3

P0842

Random species loss underestimates dilution effects of host diversity on foliar fungal diseases under fertilization

***Xiang Liu**, Shurong Zhou*
Ministry of Education Key Laboratory for Biodiversity Science and Ecological Engineering, School of Life Sciences, Fudan University

With increasing attention being paid to the consequences of global biodiversity losses, several recent studies have demonstrated that realistic species losses can have larger impacts than random species losses on community productivity and resilience. However, little is known about the effects of the order in which species are lost on biodiversity-disease relationships. Using a novel removal experiment manipulating plot-level plant diversity, and multi-year nitrogen addition and artificial warming experiments, in natural assemblages of alpine meadow vegetation on the Qinghai-Tibetan

Plateau, we explicitly compared the effect sizes of random versus realistic species losses under fertilization and/or warming on plant foliar fungal diseases. We found that realistic species losses under fertilization had greater effects than random losses on fungal diseases, and that species identity outweighed nested structure itself in driving the diversity-disease relationship. The plant species most prone to foliar fungal diseases were also the least vulnerable to extinction under fertilization, demonstrating the importance of protecting low competence species to impede the spread of infectious disease. In contrast, there was no difference between random and realistic species loss scenarios on the diversity-disease relationship under experimental warming (or the combination of warming and fertilization), an important reminder that the functional consequences of species losses may vary under different drivers.

T3
P0843

A comparative research on ecological adaptability between alien and native woody plants in warm temperate region

Xiao Liu, *Qiang Li, Ning Wang, Ning Du*
Ecology and Biodiversity Institute

Under the background of global change, warm temperate region is becoming drier and drier. In order to compare the ecological adaptability between alien and native woody plants in warm temperate region, via researching water utilization strategy in different time and season, a field experiment was designed in Fanggan Village, where there is an ecological experimental base of Shandong University. In this experiment, *Quercus acutissima* Carruth., *Vitex negundo* L. var. *heterophylla* (Franch.) Rehd., *Robinia pseudoacacia* L., *Rhus Typhina*, were chosen as the research objects. This research shows that when the climate becomes drier, different kind of plants has different strategy in ecological adaptability. As arbors, *Quercus acutissima* Carruth. grows slower than *Robinia pseudoacacia* L., but the former is abler to resist the drier envoriment, so in *Robinia pseudoacacia* L.- *Quercus acutissima* Carruth. mingled forest, with the help of the global change, *Quercus acutissima* Carruth. may occupy more advantage, and at the end, it may be the only dominant species in the forest. In terms of shrubs, the ability of drought resisting in V*itex negundo* L. var. *heterophylla* (Franch.) Rehd. and *Rhus Typhina* shares on a fifty-fifty basis. But compared with V*itex negundo* L. var. *heterophylla* (Franch.) Rehd., the ability of metabolism in *Rhus Typhina* is higher when the weather is a little dry. On the contrary, when it is a little moist, *Vitex negundo* L. var. *heterophylla* (Franch.) Rehd. is better. In the future, V*itex negundo* L. var. *heterophylla* (Franch.) Rehd. - *Rhus Typhina* mixed shrub may be faced with niche separation, *Rhus Typhina* may move to the dry place while the other may stay in the wet ground.

T3
P0844

Size fractions of air dust and associated metals removed by plant leaves in water soluble and liposoluble states at roadside of Beijing urban area

Yanju Liu, *Zheng Yang, Minghao Zhu*
Beijing Milu Ecological Research Center

Phytoremediation is a cost-effective and environmentally friendly method to improve air quality. Previous study observed that plant leaves of *Ulmus pumila* (S2), *Prunus cerasifera* f. *atropurpurea* (S3), *Jasminum nudiflorum* (S5), *Euonymus japonicas* (S6), *Populus tomentosa* (S11), *Salix babylonica* (S12), *Parthenocissus tricuspidata* (S13), *Ailanthus altissima* (S20), *Robinia pseudoacacia* (S21), *Koelreuteria paniculata* (S24), *Weigela florida* cv. Red Prince (S25), *Magnolia denudate* (S26), *Ligustrum vicaryi* (S29), *Malus micromalu* (S30), *Syringa oblate* (S32) could efficiently remove air dust. However, it is unknown what size fractions of dust were accumulated. In this study, the above 15 plant species leaves were collected at the roadside of Beijing urban area and their size fractions of water soluble and liposoluble dust (WLD = WD + LD) and associated metals Cd, Cr, Cu, Fe, Mn, Pb, Zn were analysed. In total, plant leaves could remove 321-2,757 (mean 1,159) mg m^{-2} WLD with the highest values for S6 and S30. Higher proportions of leaf dust were observed as water soluble, and the highest WD value occurred in S6, with a range of 193-1,971 (mean 817) mg m^{-2}. Varied size fractions of WD were found from different plant leaves, showing higher ratios of WD >100 μm in S2, S26, S11, S12, S24, S20 (57.8%, 55.6%, 47.6%, 47.2%, 42.9%, 32.9%); WD at 10-100 μm in S30, S13, S25, S29, S6, S32 (67.0%, 53.6%, 49.1%, 46.1%, 38.4%, 36.1%); and WD at 2-10 μm in S3, S5, S21 (41.3%, 39.1%, 39.1%) respectively. Low ratios of WD at 0.2-2 μm were observed in all species leaves. Liposoluble dust (LD) ranged from 51 to 965 (mean 342) mg m^{-2} with the highest value also for S6. Generally, LD occurred mostly in the size fraction over 10 μm. Values of dust associated metals and their size distributions were also different for individual plant species and metals. WLD Cd was in the range of 321-2,757 (mean 1,159) μg m^{-2} with higher values and larger size fractions occurred in S6, S30 and S2. Similar WLD ranges were found for Cr as 438-4,348 (mean 1,725.6) μg m^{-2} and Pb as 104-4,974 (mean 890.0) μg m^{-2}, but the highest WLD values were observed for Cr in S12, S6, S29 and for Pb in S24. The highest WLD Cu occurred in S29 with a range of 103-2,717 (mean 795.7) μg m^{-2}. Both highest WLD values of Fe and Mn were observed in S6 with larger particle size fractions, apart from the differed value ranges (Fe: 7,300-117,086, mean 41,587.1 and Mn: 49-2,030, mean 727.7) μg m^{-2}. WLD Zn ranged as 294-3,411 (mean 1,340.7) μg m^{-2}, with higher values for S5, S6 and S29, whilst higher fractions of 10-100 μm WD Zn for all three species and 3-10 μm LD Zn for S6 and S29 were observed.

T3
P0845

Evaluation of airborne dust and heavy metal pollution at seven sites in Beijing by analyzing five common tree species leaves

Zheng Yang, *Yanju Liu, Minghao Zhu*
Beijing Milu Ecological Research Centre

Fast urbanization has been damaging the environment and bringing heavy pollution to the atmosphere, soil and water. Phytoremediation has been used to reduce pollution, and to evaluate overall environmental quality. Chinese capital city Beijing, located at the core area of Beijing-Tianjin-Hebei, is suffering from increasing traffic pressure, air pollution and water safety risks. To meet current ecological need of city construction, further trials are required to explore the roles of plants in urban pollution control and evaluation. This study aims to assess airborne dust and heavy metal pollution status by analyzing plant species leaves. Five common plant species leaves were sampled from seven sites (traffic roads and parks), including *Ailanthus altissima*, *Sophora japonica*, *Populus beijingensis*, *Ulmus pumila* and *Prunus cerasifera* f. *atropurpurea* in September 30, 2015 (no rain for 18 days prior to collection). Three sample duplicates were collected for each species. Dust accumulation per leaf area was measured for all species at all sites. Both unwashed and washed leaves were analyzed for metal concentrations of Cd, Ni, Cr, Pb, Cu, Zn, Mn by Atomic absorption spectrometry. The average concentration of each metal in leaves was calculated for every species to identify the best monitoring plant species for certain metals. Metal Accumulation Index (MAI) was used to evaluate multi-metal pollution status at each site. In result, *A. altissima* accumulated the highest amount of dust (39.4 g/m^2), and *S. japonica* the lowest (5.22 g/m^2). Average leaf dust accumulation of all species for individual site follows the order from high to low as (g/m^2): Wangjing Park (27.9) > Chaoyang Park (26.5) > Yuting Bridge (25.2) > Yuegezhuang Bridge (22.6) > Niantan reservoir (21.8) > Milu Park (21.2) > Jingcheng Highway (19.3). Also, dust accumulation per leaf area was significantly related to the mean PM_{10} value in 30 days just before sampling date. Different species leaves have different element accumulation preference. *A. altissima* accumulated the highest Cd (unwashed/3.71 mg/g and washed/3.67 mg/g) and Cu (unwashed/28.8 mg/g and washed/27 mg/g), whilst *P. beijingensis* accumulated higher amounts of Pb (unwashed/10.9 mg/g and washed/6.39 mg/g), Zn (unwashed/75.3 mg/g and washed/72.1 mg/g) and Mn (unwashed/72.2mg/g and washed/70.6mg/g). The highest accumulations were observed for Ni in *U. pumila* (unwashed/7.31 mg/g and washed/6.03 mg/g) and for Cr in *S. japonica* (unwashed/9.06 mg/g and washed/7.86 mg/g). Higher geometric mean MAI of unwashed leaves of all plant species occurred at sites Jingchen Highway (4.89) and Yuting Bridge (4.32), reflecting heavy metal pollution in atmospheric dust, which is mainly affected by traffic activities, whereas higher geometric means of MAI in washed leaves, observed at sites Niantan Reservoir (7.22), Chaoyang Park (4.19) and Yuting Bridge (4.10), reflected heavy metal pollution status on a general basis due to probably water sources nearby.

T3

P0846

From greenhouse to the wildlife: Invasion patterns of *Kalanchoe × houghtonii* using ecological niche modeling

*Pilar Cachón[1,2], Sonia Herrando-Moraira[1], Neus Nualart[1], **Sergi Massó**[1,3], Daniel Vitales[1], Jordi López-Pujol[1]*

1. Botanic Institute of Barcelona (IBB-CSIC- ICUB)

2. Facultat de Biologia, Universitat de Barcelona

3. BioC-GReB, Laboratori de Botànica, Facultat de Farmàcia, Universitat de Barcelona

Alien species are currently considered one of the main direct threats to global biodiversity in our planet. Some alien plant species can become invasive and damage ecosystems, leading to negative effects on the local and global economy and society. Here, we have studied *Kalanchoe × houghtonii* D. B. Ward (Crassulaceae), a hybrid species artificially created by the horticulturist A. D. Houghton with gardening purposes in the 1930s. It was a result of the crossing between *K. daigremontiana* Raym.-Hamet & H. Perrier and *K. delagoensis* Eckl. & Zeyh., two endemic species from Madagascar. Soon, the hybrid taxon showed large colonizing capacity, escaping outside the cultivation spots and rapidly expanding its distribution area. Now, *K. × houghtonii* is currently found in all continents except Antarctica. Therefore, considering the well-known background of the species, as well as its strong invading abilities, this plant represents an attractive model to investigate the expansion of alien species. The aims of the present study are evaluating the potential worldwide habitat of *K. × houghtonii* at different time frames, from present to future, employing various scenarios of climate change. To reach these goals, we first carried out documentary research, involving the finding of localities where the species is present, through online databases (such as GBIF), citizen science web portals, as well as many published academic papers. With the obtained occurrences, and using the maximum entropy algorithm implemented in MaxEnt, we modeled the potential habitat of the species to the present, which was later projected to the future under different scenarios. Results derived from this study would allow us to better understand the invading behavior of species with high potential invasion such as *K. × houghtonii*, and, at the same time, inferring possible range contractions or expansions of the species across its whole distribution area in the context of climate change.

T3

P0847

Past, present, and future of the invasion of *Senecio pterophorus*

*Cristina Rubió, Sonia Herrando-Moraira, Neus Nualart, Roser Vilatersana, **Jordi López-Pujol***

Botanic Institute of Barcelona (IBB-CSIC- ICUB)

Senecio pterophorus DC. is a perennial shrub of the family Compositae native to eastern South Africa (Eastern Cape and KwaZulu-Natal provinces) that inhabits forest margins, grasslands, and fynbos. One hundred years ago, *S. pterophorus* expanded into Western Cape in South Africa and was introduced cross-continentally in Australia as well. A more recent expansion occurred 25–30 years ago in Europe, concretely in Liguria (NW Italy) and Catalonia (NE Spain). In the introduced ranges, it mainly appears in disturbed areas such as railroads or roadsides, although it can also occur in natural areas. During the late 19th and the early 20th centuries, *S. pterophorus* was recorded in Belgium and the United Kingdom, respectively, although it is now considered extinct in both countries. We used ecological niche models [geographic (G) space] and niche comparisons analyses [both in geographic and environmental (E) spaces] aimed to: (1) determine whether the species has changed its niche during the invasion process, (2) ascertain why the species has not succeeded neither in Belgium nor in United Kingdom, and (3) assess possible expansions and/or contractions of the species under different climate change scenarios by comparing habitat suitability between present and future models. In the G-space we used the maximum entropy algorithm (implemented in MaxEnt) to create niche distribution models, which are projected and evaluated in the native area vs. the invaded ones, and vice versa. Possible differences in niches of native and invaded areas were also tested by comparing background divergence with the observed niche divergence in the Principal Components Analysis (PCA)-reduced axes. The results of this study would contribute to understand in which conditions *S. pterophorus* is behaving as an invasive plant.

T3

P0848

Montjuïc Mountain (Barcelona, Spain): A "natural" laboratory for plant invasions?

Paula Farelo[1], *Josep Maria Montserrat*[2], *Samuel Pyke*[2], ***Jordi López-Pujol***[3], *Neus Ibáñez*[3]

1. Unitat de Botànica, Facultat de Biociències, Universitat Autònoma de Barcelona

2. Jardí Botànic de Barcelona, Consorci del Museu de Ciències Naturals de Barcelona

3. Botanic Institute of Barcelona (IBB-CSIC-ICUB)

Invasive alien species are currently considered the second most important direct driver of global biodiversity loss. Urbanization is a major factor promoting biological invasions worldwide, because (i) cities contain heterogeneous, highly disturbed habitats that provide ecological niches for alien species (where native species cannot be able to adapt to the new conditions), and (ii) cities often act as gateways to alien invasions through airports, harbors, highways, and train stations, whereas gardens, urban parks and other tracts of suitable habitat (streetscapes, vacant lots, front and backyards, courtyards, balconies, terraces, rooftops, fences, hedges, walls, and so on) can become propagule sources for spreading to natural areas. We have chosen one of the main European cities, Barcelona, as a case study to illustrate how such a city can become a major pathway for the introduction and establishment of plant invasions. Specifically, we have selected Montjuïc, an urban mountain (the top is at 183 m) of ca. 3.6 km^2 (the core area) that is located in the sea front, just behind the commercial port (albeit there is an expressway and a railway between the mountain and the docks). Whereas the northern and western outskirts are occupied by residential areas, most of the mountain is occupied by public gardens and urban parks (including the Barcelona Botanical Garden), cultural (museums, art galleries, and music and theater halls) and sport facilities (the mountain was home of a large part of the 1992 Olympic venues). There are also many historical assets, such as a military fortress (dating from 18th century), Roman and Iberian quarries, and even remains from the Paleolithic period, which illustrate the deep transformation of the mountain. Together with the fact that the mountain is visited yearly by nearly 20 million people, all these factors make Montjuïc a candidate to be a "natural" laboratory for plant invasions. We present herein the first results of a project aimed to obtain a comprehensive catalogue of the mountain's alien flora. The study area was divided into Operational Geographic Units (OGUs), which were UTM (Universal Transverse Mercator) squares of 0.25 km^2, and extensive fieldwork was carried out (combined with database/literature search) to identify all alien plant species thriving in the mountain, both as casual and naturalized (that is, fully established). Areas with evident management, such as botanic gardens, have been avoided for obvious reasons. Vouchers for some species have been deposited in the BC Herbarium, with special focus on floristic novelties and the potentially most harmful species. At the completion of the catalogue, the following information will be available for all the identified alien species: (i) distribution within Montjuïc Mountain (by OGUs); (ii) habit (annual, biennial or perennial); (iii) life form (herb, shrub, or tree); (iv) habitat type in Montjuïc; (v) native range; and (vi) whether they are listed as alien in any of the reference catalogues at regional or state levels.

T3

P0849

Characterisation of the sporophytic self-incompatibility (*S*) locus in *Senecio squalidus* (Asteraceae)

Ying-Qi Lou[1], *Batstone Tom*[2], *Navado Bruno*[1], *Quillet Marie-Christine*[3], *Hendriks Theo*[3], *Filatov Dmitry*[1], *Hiscock Simon J*[1]

1. Department of Plant Sciences, University of Oxford

2. School of Biological Sciences, University of Bristol

3. Unité Evolution, Ecologie, Paléontologie, UMR 8198, CNR

Around 80% of all plants in the wild take outbreeding as the normal mode of reproduction, which allows genetic diversity to protect against pressures from the environment. In flowering plants, outcrossing is frequently imposed by self-incompatibility. Self-incompatibility (SI) is defined as the inability of a fertile hermaphrodite seed plant to produce zygotes after self-pollination. It is a pollen discrimination system, usually controlled by a single *S*-locus with multiple *S*-haplotypes (*S*-alleles) which carry genetically tightly linked male (pollen-expressed) and female (pistil-expressed) determinant genes. SI is classified as either gametophytic SI (SSI) or sporophytic SI (SSI). Research on the molecular mechanisms of SI systems in the Brassicaceae, Papaveraceae and Solanaceae have shown that a collection of divergent molecular mechanisms control SI in different families. Thus to better understand the molecular diversity of SI and the evolution of SI in angiosperms, it is vital to characterize more SI systems in diverse families. The Asteraceae is the second largest angiosperm family containing 1,620 genera, approximately 23,600 species, many of which are economically important such as sunflowers, lettuce, chicory and Artemisia. One UK species, *Senecio Squalidus* (Oxford Ragwort) has undergone a unique invasive history to become a successful colonizer of the British Isles since its introduction from Sicily, Italyin the late 17th century. It was confirmed to have strong sporophytic SI with a low *S* allele number. It also exhibits common features of SSI systems such as dominance interactions and pseudo-self-compatibility. *S. squalidus* is an ideal model to study the SI system as a member of the Asteraceae. Here we conduct bioinformatics analysis of the genome sequence of *S. squalidus* to screen for specific male and female *S*-determinants linked to the *S*-locus on the basis of recently mapped *S*-locus region. And with the help of preliminary comparative transcriptome data analysis of *S. squalidus* and other SI plants, we characterize the tissue-specificity of expression of candidate genes and test-for polymorphism in the alleles of putative *S* genes. Overall, the purpose of the project is to elucidate the molecular genetic basis underlying the SI mating system of the Asteraceae.

T3
P0850
Effects of silicate application on photosynthesis and transpiration in winter wheat under elevated UV-B radiation
Yunsheng Lou
Nanjing University of Information Science and Technology

A field experiment was conducted to investigate the effects of silicate application onphotosynthesis and transpiration in winter wheat under elevated UV-B radiation. The experiment had two UV-B radiation levels, i.e. ambient UV-B (A, ambient) and elevated UV-B radiation (E, elevated by 20%); and two silicate application levels, i.e. a control (Si0, 0 kg SiO_2/hm^2) and added silicate (Si1, 200 kg SiO_2/hm^2). The results indicated that, compared to ambient UV-B radiation, elevated UV-B radiation decreased the net photosynthesis rate (Pn), stomatal conductivity (Gs), intercellular CO_2 concentration (Ci) and transpiration rate (Tr) at different stages by 9.21%-18.94%, 7.56%-23.66%, 1.24%-12.81% and 13.88%-27.85%, respectively, increased stomatallimitation (Ls) and water use efficiency (WUE) by 5.07%-19.67% and 2.35%-27.86%, respectively, in the no silicate treatment (Si0). Under elevated UV-B radiation, silicate application significantly increased Pn, Ls and WUE at different stages by 5.23%-12.23%, 4.61%-45.32% and 18.15%-61.33%, respectively, decreased Gs, Ci and Tr by 6.35%-10.94%, 5.41%-11.01% and 6.75%-30.51%, respectively. In conclusion, elevated UV-B radiation significantly decreased the net photosynthesis rate (Pn) and transpiration rate (Tr), increased water use efficiency (WUE) in winter wheat leaves, however, silicate application could alleviate the depressive effects of elevated UV-B radiation on net photosynthesis rate and further decrease transpiration rate and water use efficiency.

T3
P0851
Investigate how plants response to local climate change at an organic tea farm in Nantou, Taiwan
Pei-Luen Lu, *Yi-Shuo Liang*
Department of Life Science, National Taitung University

According to mounting pressure on the available water resources is occurring as the global temperature increases, there is a risk of a truth in the intensity and duration of extreme weather, causing flooding and droughts in the future. Transpiration of vegetation is the most difficult estimate in the global water cycle and is an important process in water cycle of ecosystem for forests and agriculture. Sap flow is one of the proper indicators to understand transpiration in plants. Tea tree is one of the best candidates to investigate this question, including their response to the being frequently extreme weather in Taiwan. Tea is one of the most important cultural and economic drinks in the world. Taiwan has native tea tree species and commercially grows a lot of tea trees in various mountains and hills for domesticated consuming and export. Tea trees' growing habitat and environment is similar to cloud forests. Thus, tea trees play an important but unclear role in transpiration. For improving the sustainable water usage and understanding the response to local climate change, we conducted this research at an organic tea farm in central Taiwan, with a heat balance method. The evolution of the daily and severe weather sap flow was observed and recorded in the dynamics for twelve months based on the plant *Camellia sinensis* L. varieties Jin Xuan (TTES#12). The results were calculated by Penman-Monteith equation. Ecosystem of organic tea farm for local climate change and recommendation made for further work on using the technique and tea agriculture were addressed.

T3
P0852
Distribution and transportation of modern pollen in the northeastern Indian Ocean and their paleoenvionmental application
Chuanxiu Luo
South China Sea Institute of Oceanology, Chinese Academy of Sciences

In order to analyze the terrestrial area, thirty-three seabed surface sediment samples from the northeastern Indian Ocean were selected and thirty-one airborne pollen samples were collected from the study area for pollen analysis. Then pollen analysis in two cores from north east Indian Ocean were conducted and compared with fossil pollen results from South China Sea. Firstly it was found that spores comprised the highest percentage in the surface sediment samples in the northeastern Indian Ocean, while the overall percentage of herb pollen was higher than that of tree pollen. Utilizing the 90 °E Indian Ridge as a dividing line, more *Microlepia* spores and herbaceous pollen were present on the west or northwest of the line, possibly originating in Sri Lanka and India. More Polypodiaceae spores and tree pollen (*Pinus* excluded) were present on the east or southeast of the boundary; therefore, it is inferred that Polypodiaceae pollen and tree pollen may originate from Sumatra Island. Based on PCA analysis, it was found that two main factors affect pollen percentage: ocean currents from southeast to northwest and the distance from landmasses around the study area. In the northeastern Indian Ocean, airborne pollen types and concentrations showed a close relationship with wind direction and distance from the sampling point to the continent. Combined with pollen analysis from surface sediment pollen samples, inferences that Fagaceae (including *Castanopsis)* pollen and Chenopodiaceae pollen in surface sediment were transported mainly via ocean currents instead of early summer or spring wind. Based on fossil pollen analysis in two cores from north east Indian Ocean, we found that warm pool can be taken as an climate humidity index both in South China Sea and Indian Ocean based on the fossil pollen of the cores. *Pinus* pollen can be taken as an index of sea level change, its percentage increase in the high sea level period only in the cores near continent shelf, its percentage decrease in the cores in the north east Indian Ocean because they are almost in the deep sea and the pollen transporting distance did not change.

T3
P0853
Autonomous selfing provides potential reproductive assurance in *Conandron ramondioides* (Gesneriaceae)
Hong-Wen Ma[1], *Kai-Hsiu Chen*[2], *Kuan-Ting Hish*[1], *Chun-Neng Wang*[2]
1. Institute of Ecology and Evolution and Evolutionary Biology,

National Taiwan University
2. Department of Life Science, National Taiwan University

Conandron ramondioides is a perennial herb survived in patches of high humidity habitat in dense shade forests where pollinators are scare. Observations of high proportion (ca. 78%) of fruit set in the field and caged flowers imply that autonomous selfing is possible for *C. ramondioides*. To understand the mechanism of autonomous selfing in *C. ramondioides*, we conducted pollination experiments to answer the following two questions: (1) Whether autonomous selfing is its major reproductive strategy? (2) When during anthesis does autonomous selfing occur? (3) If so, does selfing evolve due to less prone to inbreeding depression? (4) Whether selfing provides reproductive assurance for *C. ramondioides* survival? In pollination experiments, the seed set resulted from emasculation (allow insect pollination) is 523.5 and direct bagging (prevent insect but selfing) is 541.9, which suggests that *C. ramondioides* had well-developed capacity for autonomous selfing. The timing of autonomous selfing can continuously occur during anthesis because seed sets continuously increased from bagged flowers emasculated at different time intervals, which indicates the mode of autonomous selfing is competing selfing. The calculated inbreed depression is low ($\delta = 0.29$) based on proportional of seed germination rate differences between outcrossing and selfing. The proportion of autonomous selfing seed set among intact flower seed set is an indication of reproductive assurance (RA). We found autonomous selfing in *C. ramondioides* provided middle level of reproductive assurance (RA = 0.22). According to our pollination experiment, we suggest that *C. ramondioides* evolved an efficient strategy of mixed breeding system. The well-developed capacity for competing autonomous selfing, low inbreeding depression and high effect of insect-pollination allowed *C. ramondioides* to survive in isolate and patchily natural habitats. This is the first study to reveal autonomous selfing mechanism and its implication of mating system in Taiwan.

T3
P0854

The physiological function of chloride and its possible molecular basis in the xerophyte *Pugionium cornutum* in response to salt and osmotic stresses

*Yannong Cui, Xiaoting Li, Chenghang Yang, Huan Guo, Suomin Wang, **Qing Ma***
Lanzhou University

Pugionium cornutum, as one kind of succulent xerophyte species that is widely distributed throughout the desert regions of Northwestern China, can well cope with salt stress, moreover, has an excellent adaptability to arid environments. Our previous investigations showed that *P. cornutum* could accumulate high Cl^- in shoots, which contributes to enhancing the osmotic adjustment capacity under salt treatments. To further reveal the physiological function of Cl^- in *P. cornutum* under salt and osmotic stresses, 4-week-old seedlings were treated with 50 mM NaCl together with the inhibitors of Cl^- channels (20 μM DIDS and Nif); seedlings were also subjected to sorbitol-induced osmotic stress (-0.3 MPa) in the presence or absence of additional 20 mM NaCl. The result indicated that under 50 mM NaCl treatment, the addition of 20 μM DIDS and Nif significantly decreased the net Cl^- uptake rate, Cl^- concentration in shoots and the contribution of Cl^- to leaf omostic potential in *P. cornutum*, and as a result, inhibited plant growth. Compared with control, the net Cl^- uptake rate, Cl^- concentration in shoots and its contribution to leaf omostic potential were significantly increased under -0.3 MPa osmotic stress. The addition of 20 mM NaCl significantly increased the shoot water content and net photosynthesis rate under -0.3 MPa osmotic stress, and the further addition of 20 μM DIDS significantly reduced Cl^- concentration in shoots and the contribution of Cl^- to leaf osmotic potential, and consequently, counteracted the mitigative effect of 20 mM NaCl on the osmotic stress-induced growth inhibition of *P. cornutum*. These findings suggested that *P. cornutum* could accumulate high concentrations of Cl^- in shoots and use the Cl^- as an osmoregulator to cope with salt and osmotic stress. We further investigated the possible molecular basis of Cl^- transport by performing transcriptome sequencing after *P. cornutum* were treated by 50 mM NaCl for 6 and 24 h. The results showed that in roots, chloride channel genes *PcCLC-a* and *PcCLC-g* were significantly induced by 9.4- and 1.1-fold at 6 h, respectively, and the transcripts of *PcCLC-g* and *PcCLC-f* were significantly increased by 9.3- and 3.1-fold at 24 h, respectively; while in shoots, 50 mM NaCl triggered a 8-, 7.7- and 2.4-fold increase in the transcripts of *PcCLC-f*, *PcCLC-b* and *PcCLC-g* at 6 h, respectively, and increased the expression levels of *PcCLC-f*, *PcCLC-g* and a cation-chloride co-transporter gene (*PcCCC1*) by 1.7-, 1.2- and 7.2-fold at 24 h, respectively. These genes may probably play crucial roles in Cl^- transport in *P. cornutum.*

T3
P0855

Responses of forest understorey plants and soil microbial community to multiple global environmental changes - The key role of light availability

***Shiyu Ma**, Pieter De Frenne, Kris Verheyen*
Ghent University

Multiple global environmental changes are affecting ecosystems around the world. In forests, stratified canopy structures result in shading, which can buffer the effects of both nitrogen deposition and global warming on understorey communities. However, the effects of multiple environmental changes, and the role of light, on co-occurring ecosystem compartments above- and below-ground are barely studied together. To better understand the co-occurring response and the link of plant and soil microbial communities to multiple global environmental changes, we manipulated light with fluorescent tubes, nitrogen fertilization and temperature (with experimental warming) in a full factorial design in an ancient temperate forest from 2011 to 2015. Our results show that light has a strong effect on plant community diversity but nitrogen addition and warming had no such effect. Illuminated plots experienced a substantial decrease of soil microbial biomass. However, the microbial structure (defined as the ratio of fungi to bacteria) remained unchanged based on phospholipid fatty acids (PLFA) analyses. High throughput sequencing illustrated that the bacterial community diversity responded significantly to the treatments of light and warming but did not respond to nitrogen addition, although a large amount of unique operate taxonomy units (OTUs) were observed. The effect of light on the bacterial community diversity depended on nitrogen addition and warming. Co-inertia analyses demonstrated that plant and microbial communities were co-structured and light explained most of the deviance in the co-structures. Our results provide robust evidence of light as a driver for changing plant and microbial communities' composition, and advocate for more multi-factor experiments that scrutinize the direct and indirect effect of light in changing the plant – soil continuum in the context of global change.

T3

P0856

Long-term nitrogen deposition negatively affects physiological traits of light-preferring species in the understory layer of a tropical reforested forest

Qinggong Mao[1], *Xiankai Lu*[1], *Cong Wang*[2], *Kaijun Zhou*[2], *Jiangming Mo*[1]

1. Chinese Academy of Science

2. University of Chinese Academy of Science

Reforested forests are becoming the dominant forest cover type in the tropics where N deposition has been elevated. However, less is known about long-term effects of N deposition on diverse understory plants in tropical forests. To test the following hypothesis: long-term N additions negatively affect the understory plant by decreasing the light penetration from canopy in reforested forests. Physiological responses (photosynthetic capacity, physiological stress and foliar N storage) of seven understory species to a decade gradient N additions were examined in a typical reforested forest in south China. Experimental N additions were conducted at the following levels from July 2003 to July 2013: no addition (Control); 50 kg N ha^{-1} yr^{-1} (N 50); 100 kg N ha^{-1} yr^{-1} (N 100). Results showed that six of seven understory species did not physiologically respond to N addition, whereas the most light-preferring species (*Alchornea trewioides*) responded significantly to N addition by showing decreased maximum photosynthesis (Pmax). Furthermore, the decreased Pmax of *Alchornea trewioides* was significantly correlated to the increased canopy closure with N addition ($P = 0.05$). Our findings suggest that the understory species are less limited by N availability in tropical reforested forests under high ambient N deposition, and long-term N deposition could suppress understory light-preferring species by decreasing the light penetration from canopy.

T3

P0857

The role of sandstone refugia in the distribution and diversification of andromonoecious and functionally dioecious *Solanum* in the Australian Monsoon Tropics

Christopher T. Martine[1], *Jason T. Cantley*[1], *Emma S. Frawley*[1], *Daniel Hayes*[1], *L. Mae Lacey*[1], *Peter Jobson*[2], *Ingrid E. Jordon-Thaden*[3]

1. Bucknell University

2. Northern Territory Herbarium, Alice Springs

3. University of Calfornia, Berkeley

The Australian Monsoon Tropics (AMT) is one of the least disturbed tropical biomes on the planet. Surprisingly little is known about the evolution of its flora or how past geologic and climatic events may have shaped it, particularly during the dynamic climate fluctuations of the Quaternary that ushered in numerous periods of hyper-aridity. We present the initial results of a project to investigate the biogeographic origins and patterns of historical distribution for two lineages of narrowly endemic AMT *Solanum* clades, one functionally dioecious and one andromonoecious, across both spatial and temporal scales. These lineages occur, in part, along the escarpment edges of three major sandstones, which support a sporadically distributed microclimate maintained by shaded gorges and some orographic precipitation. We hypothesize that: a) the same natural breaks occur within the geographic ranges of these lineages (and potentially other unrelated angiosperm lineages), and that b) these breaks reflect barriers formed during unfavorable glacial periods of hyper-aridity when escarpment microclimates acted as habitat refugia. To uncover current AMT biogeographic patterns of angiosperms, we collected populations of taxa and constructed molecular phylogenies for each lineage to understand phylogenetic relationships and time of lineage divergence obtained from molecular clock techniques. Descriptive statistics and demographic modeling allow for the inference of historical bottlenecks, effective breeding sizes through time, reticulate evolution, and periods of range expansion. Phylogenetic and demographic data are then correlated with niche models of each taxon's present day distribution via robust geospatial data, which in turn are combined with Global Circulation Models to illuminate historical areas of suitable habitat during the Last Glacial Maximum (LGM) when the AMT was last postulated to have undergone a period of hyper-aridity. Upon this biogeographic background we then overlay reproductive biology studies examining rates of *ex situ* hybridization within and between clades, comparative measures of genetic diversity between obligately outcrossing species (dioecious) and potential selfers (andromonoecious), and efforts to identify and describe novel taxa.

T3

P0858

What is the extend of calotropis procera along road verges in the mutale local municipality within the Vhembe Biosphere Reserve, Limpopo, South Africa

Sipho Mbambala, *Peter Tshisikhawe*

University of Venda

Biological invasions are considered the second major cause of the world biodiversity loss and, even more worrying, is the irreversibility of the impacts caused by invasion, which worsens within the time, as the exotic species disseminate. Aggressive invasion of road verges by alien invasive plant species such as calotropis procera may have negative impacts on road verges biodiversity. Road verges have long been seen as being of potential biological value and are believed to make a major contribution to the conservation of indigenous flora in various parts of the world. Road verges are considered to be important in the conservation and management of landscapes used for agricultural purposes. Sites will be selected, transects of 10 m by 10 m will be constructed running parallel to the road, will be examining the plant cover, species richness, frequency and abundance on road constructions compared with places without road construction impact. For the fenced sites, transects will be located at 1 m from the road edge, 2 m from the fence defining the road verge on the road side of the fence. Signs of disturbance including evidence of erosion, grading and vehicle tracks will be recorded.

T3

P0859

Effect of the light habitat on the structure, diversity and dynamics of the tropical dry forests of the Magdalena river high valley, Colombia, South America

Omar Melo[1], *Nathaly Rodriguez*[2]

1. Tolima University

2. Universidad Austral de Chile

Tropical dry forests are complex and fragile ecosystems with high anthropogenic intervention and restricted reproductive cycles. They are rich in unique structural, physiological and phenological diversity. This research was carried out in the upper Magdalena river high valley, in four fragments of tropical dry forest with different successional states. The flat coordinates of the zone are: 4° 59'53.48" N and 74° 55'38.87" E. The study area has an average annual temperature of 27°C, an average annual precipitation of 1,750 mm and a relative humidity of 71%. The height above the sea level oscillates between the 475 to 580 m of altitude. In each fragment four permanent plots of 1.0 hectare were established. And the richness, relative abundance and species rarity were evaluated, as well as the forest dynamics that included mortality, recruitment and diametric growth for a period of 10 years. To evaluate the light habitat that indicates the amount of energy used by the forest cover to perform functional processes of assimilation and productivity, the direct measurement of leaf area index (LAI) was used with a LI-2200TC leaf analyzer. The different levels of light absorbed by the canopy of the evaluated forests were expressed in relative values of the photosynthetically active radiation (PAR), which was measured as the non-absorbed radiation on the forest floor surface expressed as the density of photosynthetic photon flux ($\mu mol.m^{-2}.s^{-1}$), between 400 and 700 nm of the light spectrum, for which the LI-191SA - Line Quantum Sensor and LI-190SA - Quantum Sensor sensors were used and a collector Of LI 1400 data (LI-COR Inc., Lincoln, NE, USA). Likewise, the coefficient of extinction of the light (K) was derived, by means of the application of the Beer & Lambert Law. The results showed that there is a direct relationship between the availability of the light resource and the dynamics of the tropical dry forest, the higher growth rates of the tropical dry forest are associated to habitats with high availability of PAR that is absorbed by a canopy with high LAI, a high growth that generates an increase in the mortality rate. The liberated spaces are occupied by new individuals who are waiting for the opportunity for growth. For more advanced successional state such as the mature forest, the canopy that characterizes the forest structure has a lower LAI, which allows greater availability of resources in the understory allowing a greater diversity of habitats that are occupied by several species types, hence the greater heterogeneity and structural complexity this generates a smaller growth and greater stability between the mortality and the recruitment, which owes their dynamics to factors endogenous to the biotic community.

T3

P0860

Impact of nitrogen fertilizer on foliar endophytic fungal communities in an alpine meadow ecosystem

Yiming Meng, *Qi Zhang, Yongjun Liu, Huyuan Feng*

Ministry of Education Key Laboratory of Cell Activities and Stress Adaptations, School of Life Sciences, Lanzhou University

Endophytic fungi are present in all plants in natural ecosystems, yet our understanding of their community dynamics and processes is very limited. Here we examined the foliar endophytic fungal communities associated with three grass species along a four-year nitrogen fertilizer gradient in an alpine meadow ecosystem. Ascomycota and Basidiomycota were the dominant phyla, and the proportion of Ascomycota and Basidiomycota fungi significantly shifted with high fertilizer inputs, indicating that nitrogen changed the role of endophytic fungi in plant. Besides, water-soluble carbohydrates, and carbon/nitrogen ratio of leaves also affected fugal community assemblage. Our research first revealed endophytic fungal community structure and offered clues of their nitrogen acquisition strategies across this cold terrain, indicating that variations in diversity and community composition of endophytic fungi measured here may feedback and influence nitrogen cycling and dynamics in this alpine meadow ecosystem.

T3

P0861

Study on ecological effects of indoor vertical greening plants

Hui-Zhi Mo, *Xu-Rong Luo, Jing Zhang, Liang Zhao, Jian-Xin Xu*

Shenzhen Techand Ecology & Environment co., Ltd., Shenzhen 518026, China

To investigate the ability of vertical greening plants on improving indoor environment (air quality and micro-climate), the efficiency of carbon sequestration, oxygen release, cooling, humidification and dust-retention of 6 indoor vertical greening plant species have been quantitative measured and comprehensive evaluated. The results suggested that all the vertical greening plants showed significantly improvements on indoor environment quality, with various effects by different plants. *E. aureaum* 'Virens' was the best plant species of carbon sequestration and oxygen release per unit of plant green area, which absorbed 5.48 g CO_2 and released 3.98 g O_2 every daytime. While *P. mandaianun* 'Royal Queen', released 581.34 g water vapor and absorbed 1415.85 KJ heat every daytime per unit of green area, as well as reducing temperature of 0.12 ℃ , was the best plant with cooling and humidification capabilities. *A. andraeanum* has the best capability of daily dust-retention per unit of green area, which retained 0.01 g dust daily and 3.64 g annually. The ranking of comprehensive ecological effect of the 6 vertical greening plants per unit of green area was *A.* 'Sliver Queen' > *E. aureaum* 'Virens' > *P. mandaianun* 'Royal Queen' > *A. andraeanum* > *P. tetraphylla* > *S. podophyllum* 'While Butterfly'. This study implies building vertical greening system should consider the integrated ecological effect instead of purely landscape effect. Enhancing indoor light intensity and planting density are two effective approaches to improve the ecological effects of indoor vertical greening.

T3

P0862

Isolation, screening, and characterization of bacteria associated with rhizosphere soil of *Holoptelea integrifolia* Roxb. : A tropical medicinal plant

Sumita Mondal[1], *Moumita Das*[2], *Swastika Banerjee*[2], *Abhijit Bandyopadhyay*[2]

1. Vivekananda Mahavidyalaya

2. The University of Burdwan

The interaction of plants with microbes is of great importance for growth and productivity of plants in agriculture and natural ecosystem. Rhizosphere means the soil region nearest to plant root system. Exploiting microbial synergism is one of the popular methods of soil dynamics and associated changes in nutrient environment of rhizosphere. Microbial activity of rhizospheric region affects pH, oxygen, carbondioxide concentration as well as nitrogen to the benefit of the plant. The present study focuses on association of bacteria in rhizosphere with *Holoptelea integrifolia.* It is a road side tree species belongs to the family Ulmaceae. From ancient time, various part of this tree had been reported to use because of its ethno-medicinal properties such as anticancer, antidiabetic, antiemetic, antioxidant, hepatoprotective, wound healing and many others aliments. The present study reveals the presence of four bacterial entities in the rhizospheric soil at Golapbag Campus of Burdwan University. Various enzyme production capability like Lipase, citrate, amylase of 4 bacterial strain has been checked. Thermal death points of 4 isolates are also determined. Sensitivity to antibiotics like Ampicillin, Streptomycin, chloramphenicol are also checked against all isolated bacteria. Gram properties of the isolates are also visualized. Fluorescence microscopy for each bacterium is done for better visualization. The plant usually prefers sandy-loam to laterite soil. The data obtained for pH, organic carbon, nitrogen, phosphorus and potassium are: 8.46, 1.405 (mg/l), 49 (mg/l), 1.2 (mg/l), 150 (mg/l) respectively. Molecular identification of the bacteria was done by 16s-rDNA sequencing. Plants microbe interaction provides a rational basis for developing sustainable strategies for disease resistance in next generation crops.

T3

P0863

The European alpine seed conservation and research network

Andrea Mondoni[1], *Elinor Breman*[2], *Christian Berg*[3], *Jacqueline Detraz-Meroz*[4], *Brigitta Erschbamer*[5], *Noemie Fort*[6], *Catherine Lambelet-Haueter*[4], *Vera Margreiter*[5], *Florian Mombrial*[4], *Thomas Abeli*[1], *Konrad Pagitz*[5], *Francesco Porro*[1], *Graziano Rossi*[1], *Patrick Schwager*[3], *Margherita Cesarini*[1], *Jonas Muller*[2]

1. University of Pavia, Department of Earth and Environmental Science (Italy)

2. Millennium Seed Bank, Conservation Science, Royal Botanic Gardens, Kew (United Kingdom)

3. Karl-Franzens-Universitat Graz, Botanischer Garten (Austria)

4. Conservatoire et Jardin Botaniques de la Ville de Geneve (Switzerland)

5. University of Innsbruck, Institute of Botany (Austria)

6. Conservatoire Botanique National Alpin, Gap (France)

Alps are one of the 24 centres of plant diversity in Europe, with almost 4,500 taxa of vascular plants, representing 148 families. Yet the ecosystems supported by these mountains are under threat, mostly due to changes of land use and climate warming. Consequently, many plant species require urgent conservation action. To this end, a new project has been developed by RBG Kew (UK), with funding from the David and Claudia Harding Foundation, to conserve European alpine flora and to raise awareness of its increasing vulnerability. The Alpine Seed Conservation and Research Network (http://www.alpineseedconservation.eu/) currently brings together six plant science institutions from five countries housed at leading universities and botanic gardens. The aim of the project is to set up a network of seed conservation and research in the European Alps and improve the conservation status of endangered plant species and communities in their habitats in the European Alps. Over three field seasons seeds of 500 vascular plant species will be collected from subalpine, alpine and nival altitudinal belts and adjacent plant communities, giving priority to endemic and threatened species. The project got underway in 2016 with a highly successful collecting season. In addition to the conservation actions, one MSc and three PhD research projects will investigate the evolutionary patterns and processes of alpine species resulting from global warming, identify niche adaptation and gather new data on the conservation status of endemic species in the target areas. Additionally, public outreach, involving stakeholders and schools across the alpine region, is being developed.

T3

P0864

Response of soil microbial communities to the conversion of natural forests to rubber plantations

Jutamart Monkai[1,2,3], *Stefanie D. Goldberg*[1,3], *Peter E. Mortimer*[1,3], *Kevin D. Hyde*[4], *Jianchu Xu*[1,2,3], *Rhett D. Harrison*[5,6]

1. Centre for Mountain Ecosystem Studies, Kunming Institute of Botany, Chinese Academy of Science, Kunming, Yunnan, China

2. University of Chinese Academy of Sciences, Beijing, China

3. World Agroforestry Centre, East and Central Asia, Kunming, China

4. Institute of Excellence in Fungal Research, Mae Fah Luang University, Chiang Rai, Thailand

5. Key Laboratory for Plant Diversity and Biogeography of East Asia, Kunming Institute of Botany, Chinese Academy of Sciences, Kunming, Yunnan, China

6. World Agroforestry Centre, East & Southern Africa Region, Woodlands, Lusaka, Zambia

The conversion of tropical forests to monoculture rubber plantations throughout Southeast Asia threatens to have widespread negative impacts on ecosystem services. Despite this, little is known about the effect of this land use conversion on soil microbial communities. This study addresses the questions: (1) How do soil microbial communities respond to conversion from natural forests to rubber plantations, and how do responses change as rubber plantations mature? (2) Assuming that changes do occur, what are the main environmental factors which drive changes in soil microbial communities after conversion of forests to rubber plantations? Using phospholipid fatty acid analysis of soil samples taken from natural forests, young and old rubber plantations, we found that total microbial biomass significantly decreased following conversion from forests to rubber plantations. This was true for young and old rubber plantations tested in both dry and wet seasons. We also found that the composition of microbial communities differed among natural forests, young rubber plantations and old rubber plantations. These differences were influenced by soil pH, total phosphorus, soil texture, soil moisture and soil temperature. Our results also show that seasonal variation is an important driver of changes in soil microbial community composition in forests and plantations. The loss of soil microbial biomass and the changes in soil microbial community composition in rubber plantations observed in this study highlights the potential negative consequences for soil health of converting forests to rubber plantations.

T3
P0865

Local reproductive neighborhood, flowering synchrony, and female reproductive success of the dioecious neotropical tree *Virola surinamensis* (Rol. ex Rottb.) Warb. (Myristicaceae) in a human-modified landscape

Juan Isaac Moreira Hernandez[1,2,3], *Pablo Riba Hernandez*[1,2], *Jorge Lobo Segura*[2]
1. Proyecto Carey, Peninsula de Osa
2. Universidad de Costa Rica
3. University of Missouri-St. Louis

Understanding plant reproductive patterns in disturbed environments is crucial to evaluate population level responses to disturbance and to determine current and future reproductive potential to persist in human-modified landscapes. For dioecious species, in which sex expression occurs in separate individuals, seed production by females could decrease if post-disturbance conditions result in diminished local availability of male plants as sources of pollen in space and time. We evaluated local reproductive neighborhood structure (to 100 m), flowering synchrony between sexes at the local and population level, and seed production of 18 *Virola surinamensis* (Myristicaceae) female trees during two years in a human-modified landscape in southern Costa Rica. Focal female neighborhoods included an average of one male tree and two other female trees. Local sex ratio in neighborhoods was skewed towards females, showing that female *V. surinamensis* trees occur in clumps at the disturbed sites studied with few or no males in close proximity. Average distance to nearest male was 87.10 ± 86.89 m (range: 8-302 m), while mean diameter at breast height of closest male neighbor was 70.9 ± 20.5 cm, and it did not differ from that of the focal female (76.1 ± 16.0 cm). In both years, the population of male trees started flowering earlier and a higher proportion maintained > 50% coverage by inflorescences with open flowers for a more extended period compared to the female population. Males' flowering peak always occurred between late December and early January. Females exhibited a shorter, more synchronous flowering period on the first year with a flowering peak that occurred about two weeks later than that of males, while on the second year they flowered for a longer time but at a lower intensity, and their flowering peak coincided with the males. Total fruit production by the 18 focal females was influenced by the total number of flowers produced per year, and by flowering synchrony with local males during the second year only. The ratio of matured versus initiated fruits was also correlated with female tree crown size suggesting a role of resource availability on female reproduction. Taking together, our results show that female investment in flower production is generally sufficient to secure high fruit production in *V. surinamensis* at our study landscape, but also that local reproductive neighborhood effects on female reproduction in disturbed landscapes can manifest via reduced flowering synchrony between sexes in some years.

T3
P0866

Reproductive biology and pollinator sharing in two unrelated (from different families), but morphologically similar flowers; *Thunbergia atriplicifolia* and *Sebaea grandis*

S.M. Msweli, *S.D. Johnson*
School of Life Sciences, University of KwaZulu-Natal

Co-existing, morphological similar flowers that flower simultaneously, may interact with the same pool of pollinators, therefore sharing pollinating species and individuals. Pollinator sharing may be detrimental (i.e. competition), beneficial (i.e. facilitation or mimicry) or have no effect on the interacting species. This study was conducted to determine the reproductive biology of *Sebaea grandis (S. grandis)* and *Thunbergia atriplicifolia (T. atriplicifolia)* and document pollinator sharing between the two species. Investigated aspects of reproductive biology are: flowering phenology, nectar rewards, floral colour, scent analyses and breeding system. The study was conducted at Vernon Crookes Nature Reserve. Weekly survey of flowers indicated that both plants flower simultaneously but *S. grandis* is at low density. They provide similar amount of nectar but *S. grandis's* has a high sucrose concentration. Both flowers have similar spectral reflectance and the colour hexagon showed that they are indistinguishable to a bee pollinator. Colour similarity in the two species was found to be coincidental and not adaptive. The scent profile of the two species is similar, rendering them undistinguishable by scent. Hand-pollinations established that *T. atriplicifolia* is self-compatible and self-autogamous but self-pollen results to reduced seed set. From literature, *S. grandis* has a heteromorphic self-incompatibility system (Wolfe 2009). Both species are pollinated by solitary bees from the Apideae and Hactilidae family. Pollinator observations and stigma analysis show that the two species share pollinators but those pollinators are generalist in the community. The striking similarity between floral signals and evidence for pollinator sharing suggests a potential positive interaction between the two species. More research is required to understand the consequences (if there are any) of pollinator sharing between the two species.

T3
P0867

A strong phylogenetic signal in Mongolian seed weight data

Amartuvshin Narantsetseg, *Karl Hülber*
1. Institute of General and Experimental Biology
2. Department of Botany and Biodiversity Research, University of Vienna, Austria

The evolution of seeds is one of the most important and successful innovations in the history of plants. Seeds enable vascular plants to migrate, to survive long periods of hostile environmental conditions and to establish even under unfavourable environmental conditions. The high functionality is reflected by a huge variation in seed weight spanning 9 to 10 orders of magnitude representing differences in phylogenetic history, reproduction strategy and habitat conditions. However, our knowledge on the drivers of the evolution of the observed variation is still limited. Several studies found correlations between mean seed weight or seed weight variation and climatic factors like temperature, precipitation and aridity, but even the direction of climate effects remain a matter of debate. Based on a huge collection of seed weight data (241 species from 118 sample sites collected from 1978 to 2007) covering the major vegetation zones of Mongolia, we aimed at quantifying the relative effects of climatic conditions and the phylogenetic history. We evaluated the correspondences between seed weight and growth form or dispersal type and detected a strong phylogenetic signal in seed weight data, but only weak effects of climatic parameters. Furthermore, we found seed weight to be related to growth form, trees seeds are heavier, and dispersal type, seed weight of anemochorous plants Iower than that of hydrochorous or zoochorous species. Our results suggest seed weight to be a highly phylogenetically conserved functional trait.

T3

P0868

Invasive plant *Cassia atala* root traits and its competition with neighbor species

Ming Ni, *Suqin Fang, Yu Liu*

SYSU-Alberta Joint Lab for Biodiversity Conservation, Sun Yat-sen University

Biological invasion causes huge impacts on ecosystem and human society. Knowing what kind of species are more invasive can help us establish effective prevention program. But traditional trait comparative studies usually have different conclusions. To solve this inconsistence, we should consider the invasion process and traits' roles in invasion, especially the competition with native species. Limited by technique, seldom research studied invaders' root traits and root competition. In this study, three plant species including invasive herb *Cassia atala*, naturalized plant *Corchorus capsularis* and native plant *Desmodium reticulatum* were selected and grown together in the transparent plant growth system to *in situ* compare their competition ability dynamically. Our results showed that *C. atala'* s roots grew much faster than *C. capsularis* and *D. reticulatum* in all three weeks. *C. atala* had significantly larger total root length, root volume, root surface area than *C. capsular* and *D. reticulatum* while there were no significant differences between the latter two species on these three traits. *C. atala* demonstrated an apparent root competition advantage compared with two other species. Competing with *C. capsularis* or *D. reticulatum* didn't cause *C. atala* significant biomass decrease. While *C. atala* had huge impacts on *C. capsularis* and *D. reticulatum*. Especially competing with two *C. atala* individuals made *C. capsularis* and *D. reticulatum* about 77% and 81% total biomass decrease respectively. Rapid root growth and large root system may help invaders outcompete natives. We should extend this research by combining with other techniques, like minirhizotrons and X-ray computed tomography, to investigate invasive plants' root in future. Our finding also indicates that we should also consider root traits of potential invasive plants to prevent biological invasion.

T3

P0869

Low flower-size variation in bilaterally symmetrical flowers

Aoi Nikkeshi[1], *Daiki Kurimoto*[2], *Atushi Ushimaru*[2]

1. Faculty of Life and Environmental Science, University of Tsukuba, Japan

2. Graduate school of Human Development, Kobe University, Japan

The evolutionary shift from radial to bilateral symmetry in flowers is generally associated with the evolution of low flower-size variation. This supports the hypothesis that the lower size variation in bilateral flowers could be attributed to low pollinator diversity. To test the three hypotheses, we examined the relative importance of pollinator diversity, composition, and bilateral symmetry itself as the selective forces on low flower-size variation. We examined pollinator diversity and composition and flower-size variation for 36 species in a semi-natural ecosystem with high bee richness and frequent lepidopteran visitation. Bilateral flowers were more frequently visited by larger bees compared to radial flowers, but functional group diversity of pollinators did not differ between symmetry types. Although bilateral flowers had significantly lower flower-size variation than radial flowers, flower-size variation did not vary with pollinator diversity and composition but was instead related to bilateral symmetry. Our results suggest that the lower size variation in bilateral flowers might have evolved under selection favoring the control of pollinator behavior on flowers to enhance the accurate placement of pollen on the body of the pollinator, independent of pollinator type. Because of the limited research on this issue, future work should be conducted in various types of plant-pollinator communities worldwide to further clarify the issue.

T3

P0870

Effect of crude oil on coastal plants

Yanisa Olaranont[1,2,3], *Paweena Traiperm*[2], *Alyssa Stewart*[2]

1. Master of Science Program in Plant Science, Faculty of Graduate Studies, Mahidol University, Bangkok

2. Department of Plant Science, Faculty of Science, Mahidol University, Bangkok

3. Department of Pharmaceutical Botany, Faculty of Pharmacy, Mahidol University,

Nowadays, industries are an important part of human society and can help drive a country's economy. However, they can be harmful to the environment and living organisms. Oil pollution is one hazardous effect of industry development, and oil spills occur in many countries around the world, including Thailand. However, few studies have examined the detrimental effects of oil on plants. Therefore, the objectives of this study were (1) to determine the impacts of crude oil on the anatomy and growth of two coastal plant species, and (2) to investigate their potential as an indicator of oil pollution. Two common coastal plants, *Ipomoea pes-caprae* (L.) R. Br. (Convolvulaceae) and *Ischaemum muticum* L. (Poaceae), were treated with four concentrations of crude oil (1%, 2%, 3% and 4% volume/weight). Physiological parameters such as greenness content and chlorophyll fluorescence were measured for four months. Leaves of the two species were also collected in order to study micromorphology using epidermal peeling and transverse sectioning. Moreover, plant tissue was stained with Oil red O to detect crude oil. The effect of crude oil on plant physiology and micromorphology over time will be presented. This research addresses an understudied topic and is important to understanding how plants respond to oil pollution.

T3

P0871

Recent genetic differentiation of *Phaedranassa dubia* and *P. brevifolia* in Ecuador

Nora Oleas[1], *Javier Francisco-Ortega*[2], *Alan Meerow*[3]
1. Universidad Tecnológica Indoamérica
2. Florida International University
3. United States Department Of Agriculture

Phaedranassa is a genus of Amaryllidaceae restricted to Costa Rica, Colombia and Ecuador. Seven out of the eight species described in Ecuador are endangered or vulnerable to extinction. *Phaedranassa dubia* is the only species found in Ecuador and Colombia. Both *P. dubia* and *P. brevifolia* are located at northern Ecuador. We genotyped 11 microsatellite loci in 12 populations for both species. We used a Bayesian approach to identify genetic groups. We estimate population history with an Approximate Bayesian computation analysis (ABC). Also we estimate species distribution models based on bioclimatic variables and species geographic coordinates. Bayesian structure clustering analysis suggests that *Phaedranassa* populations at northern Ecuador correspond to different genetic groups. The assignment of individuals of these two species together shows a higher degree of admixture between the geographically closer populations of the two. ABC estimates suggest a recent differentiation between species. Vegetatively, the most evident difference between this species and *P. dubia* is the length of the leaves. *Phaedranassa brevifolia* is the species with the smallest leaves (ca. 12 cm long) while *P. dubia* leaves length are ca. 20 cm long. *Phaedranassa dubia* has larger flowers with a slightly inflated perianth base and darker red color (Fig 1). Moreover, the Species Distribution Models (SDM) for these two species do not overlap, indicative of niche differentiation. The results show that these two species seem to be an example of peripatric speciation and may be a result of a recent divergence. The topographic complexity of the Andes, which creates local microhabitats that might explain the patterns of *Phaedranassa* species differentiation.

T3

P0872

Determination of heavy metal and mineral elements in flowers of some medicinal plants widely used in Turkey

Ibrahim Ilker Ozyigit[1], *Cengiz Colak*[1], *Sedat Karadeniz*[1], *Goksel Demir*[2], *Ibrahim Ertugrul Yalcin*[3]
1. Department of Biology, Faculty of Science and Arts, Marmara University, Istanbul, Turkey
2. Department of City and Regional Planning, Faculty of Architecture, Kirklareli University, Kirklareli, Turkey
3. Department of Molecular Biology and Genetics, Faculty of Engineering and Naturel Sciences, Bahcesehir University, Istanbul, Turkey

In this study, flowers of some widely used medicinal plants were obtained from the herbalists of three different districts (Kadıköy, Bakırköy and Güngören) of Istanbul-Turkey, and their mineral element and heavy metal amounts were analyzed by using ICP-OES apparatus. Among these plants; strawflower, chamomile, common marshmallow, pomegranate, jasmine, high mallow, elderberry, basswood, pot marigold, common hops, sage, yarrow, klamath weed, European centaury and lavender were analyzed. The quantities of copper (Cu), boron (B), zinc (Zn), iron (Fe), cadmium (Cd), mercury (Pb), magnesium (Mg), manganese (Mn), nickel (Ni), potassium (K) and sodium (Na) were determined. Copper in the flower samples were measured as the highest rate in common marshmallow with 9.833 mg/kg and the lowest rate in pomegranate with 0.500 mg/kg. The rates of other plants are as follows; Boron, the highest rate in sage with 16.339 mg/kg, the lowest rate in Klamath weed with 1.697 mg/kg zinc, the highest rate in sage with 35.279 mg/kg, the lowest rate in lavender with 1.504 mg/kg, iron, the highest rate in sage with 276.056 mg/kg, the lowest rate in pomegranate with 31.216 mg/kg, calcium the highest rate in basswood with 6,265.794 mg/kg, the lowest rate in jasmine with 1,858.219 mg/kg, magnesium, the highest rate in elderberry with 1,653.869 mg/kg, the lowest rate in strawflower with 636.962 mg/kg, manganese, the highest rate in lavender with 34.744 mg/kg, the lowest rate in jasmine with 3.709 mg/kg, nickel, the highest rate in lavender herb with 2.245 mg/kg, the lowest rate in Klamath weed with 0.216 mg/kg, potassium, the highest rate in chamomile with 8,167.861 mg/kg, the lowest rate in elderberry with 2,328.003 mg/kg, sodium, the highest rate in pot marigold with 1,648.676 mg/kg, the lowest rate in basswood with 25.743 mg/kg is measured. When we compare the maximum and minimum rates of elements, heavy metal densities of analyzed samples were not at toxic level. As a result of this study we reach the fact that plants which are sold as medicinal purpose in selected districts have no risk for human health.

T3

P0873

Effect of nano-bentonite in compound fertilizer on rice growth

Fuyun Pei, *Tianye Wang, He Ren, Qin Peng, Xiaoyan Feng, Luming Tan*
Shenzhen Techand Ecology & Enviroment CO., Ltd.

Silicon is one of the beneficial element for rice growth, and natural bentonite is a convenient resource of silicon. The research investigates the effect of bentonite with different particle size in conpound fertilizer an rice growth. The compound fertilizer was prepared by tower granulation process. Bentonite was appended according to different weight ratios as 1%, 3% and 5%. Root application of different compound fertilizer was evaluated for their effects on promoting growth and the quality of rice via pot experiment. The dry weight of plant was increased with addition of bentonite with different partilce size. A comparison of several treatments showed that addition of both bentonite and nano bentonite increased the dry biomass, fresh weight, and root system. With the 3 g/100 g addition of bentonite and nano bentonite in the compound fertilizer, dry biomass increased by 9.49% and 34.40%, respectively. Compared to the control, the total nitrogen, phosphorus and potassium contents were increased by 6.13% and 21.47%, respectively, and the root dry weight increased by 16.10% and 61.25%, respectively. The statistical results revealed that nano bentonite possess high activity and bentonite improved the slow-release of fertilizer because of its lamellar structure.

T3

P0875

Evaluating biodiversity losses and exotic species invasion in the Sierra de Cordoba ecosystems

Ana María Planchuelo

Centro de Relevamiento y Evaluación de Recursos Agrícolas y Naturales

The province of Cordoba (Argentina) has a continental location with a landscape which includes flat land, hill and mountain regions with a land cover varying from grassland, xerophytic forest, wetlands, high mountainous ecosystems and agricultural areas. In the last decades many natural areas have been degraded by deforestation, urbanization, invasion of exotic plants, overgrazing and rural fires. These biotic and abiotic processes caused changes in the landscape, accelerated biodiversity losses and created dispersion barriers for many valuable species. The study area is located in the Paravachasca valley (31.70° to -32.62° S and 64.40° to -65.10° W), in the eastern side of the Sierras Chicas. Field observations were carried in five geo-referenced sampling sites of 10.5 ha. Vegetation cover was classified as natural forest, shrub land and prairies. Three land uses were identified: natural undisturbed areas, cultivated land, and disturbed areas by rural fires and/or anthropogenic activity. The analysis of vegetation cover changes were carried out using the program Land Change Modeler and selected LANDSAT images (2009-2016), after the area was affected in 2009 by extensive rural fires, which affected significantly the native vegetation. Type and abundance of species were recorded and they were classified using a priority conservation index. Results indicates that rural fires, as well as, agricultural activities caused, in short time, a massive ecological impact; while overgrazing caused in a slow process changes in the flora composition and urban development created disjunctive patches of natural vegetation. All the disturbances, changed the flora biodiversity and many native species were replaced by weeds and patches of invasive trees such as: *Ligustrum lucidum* W.T. Aiton (Oleaceae) an evergreen tree native to the Southern China known as privet or locally "ligustro". Although it has some ethnomedical uses it is difficult to eradicate; *Melia azedarach* L. (Meliaceae) native to Indomalaya and Australasia, known by several common names including Chinaberry tree, Persian or Indian lilac, and locally as "paraiso". Although it is used as timber and other economic uses it is an invasive species.; *Gleditsia triacanthos* L. (Fabaceae) native to central North America known by the names of honey or thorny locust and locally as "acacia negra" (black acacia). Although it is cultivated as ornamental tree and produce quality wood it is well known that has become a significant invasive in many regions of the world; *Pyracantha* spp. (Rosaceae) natives to the Southwestern Europe and Southeastern Asia are evergreen thorny shrubs known with the common names of pyracantha, firethorn and crataegus in Argentina. Although it is a valuable ornamental plant for gardens and green fences, its rapid dispersal makes this species very invasive along water streams. Several approaches are considered to eradicate or to find no negative environmental impacts of the invasive species.

T3

P0876

Juncus fontanesii J. Gay ex Laharpe (Juncaceae) as a new host plant for *Livia junci* (Schrank, 1789) (Hemiptera: Psyllidae)

Paweł Jarzembowski[1], *Anna Faltyn*[2], ***Jarosław Proćków***[2]

1. Institute of Environmental Biology, University of Wrocław

2. Institute of Biology, Wrocław University of Environmental and Life Sciences

In the herbarium material from Portugal we found the specimens of *Juncus fontanesii* with characteristically transformed shoots, so-called "witches' brooms". Inside their structures we observed a few larvae of *Livia junci* (Hemiptera: Psyllidae). Similar changes were found in the specimens collected in Iraq, however due to the feeding of insects, the plants had been deformed to the extent, that we were unable to certainly confirm if they belong to *J. fontanesii*, especially that the taxon is highly variable morphologically. In southern Europe, it grows mainly in Mediterranean humid grasslands of tall grasses and rushes, and alongside streams and in temporarily inundated sites, and it also occurs in northern Africa and western Asia. "Witches' brooms" is a disease or deformity in a plant, where the natural structure of the plant is changed, i.e., a dense mass of shoots grows, usually from a single point, with the resulting structure resembling a broom. The direct causes of the enormous changes of the plant have not been known so far, although they have been related to the presence of parasite Livia junci. To date, the Hemiptera insect has been already found on different rushes mainly belonging to the Ozophyllum section (and also others). To date, Livia junci has not been mentioned from *J. fontanesii*. Therefore, this is the first report of the feeding of this parasite on this host plant species. Additionally, this taxon formed galls in response to the feeding, similar to other representatives of the genus *Juncus*. *Livia junci* is the only representative of the genus *Livia* that feeds on Juncus species in the Western Palearctic area. Thus the mass occurrence of this insect can cause a decrease in the yields of different rush species. Namely, in such situations the plants cannot flourish, so it is impossible to produce fruit and seeds to be spread. Taking into account the character of these changes as well as the manner of feeding of the insect (pricking tissues near vascular bundles), we claim that *L. junci* is only a vector, while the factor responsible for producing galls is phytoplasma.

T3

P0877

First report of jointleaf rush (*Juncus articulatus*) as a new host plant for *'Candidatus* Phytoplasma asteris' associated with "witches'-brooms"

Paweł Jarzembowski[1], *Hanna Berniak*[2], *Anna Faltyn*[3], *Anna Jakubska-Busse*[1], ***Jarosław Proćków***[3]
1. Institute of Environmental Biology, University of Wrocław
2. Research Institute of Horticulture, Skierniewice
3. Institute of Biology, Wrocław University of Environmental and Life Sciences

Juncus articulatus L. (Juncaceae) is a species of rush occurring in Eurasia, Canada and the United States of America. In 2013 symptoms of "witches'-brooms", similar to those associated with phytoplasma infection in other plants, were observed on jointleaf rush plants in Lower Silesia, with some pests feeding on them. "Witches' brooms" is a disease or deformity in a plant, where the natural structure of the plant is changed, i.e., a dense mass of shoots grows, usually from a single point, with the resulting structure resembling a broom. The direct causes of the enormous changes of the inflorescence have not been known to date, although they were related to the presence of the insect. *Livia junci* (Liviinae, Hemiptera), is a small plant-feeding sap-sucking insect that affects monocotyledonous plants. To confirm the presence of phytoplasma in fifteen examined plants, total DNA was extracted from 100 mg of inflorescence and leaf samples collected on July 2013, in Bogatynia, Poland, from six symptomatic and six asymptomatic plants. Additionally, three leaf samples from asymptomatic rush plants, collected from a location where the disease was not observed (Wrocław, Poland), as well as water blank samples were included as negative controls. Moreover, thirty-two insects were collected from symptomatic plants. DNA from *L. junci* specimens (the imago and the last larva stage) was extracted. Extracted nucleic acids were used as templates for PCR employing: a) phytoplasma universal rRNA primer pairs P1/P7 followed by R16F2n/R16R2 (1), b) primers rp1-rp2 followed by rp3-rp4 allowing amplification of fragments of ribosomal protein rpl22 and rps3 genes (2), and c) primers AYsecYF1/AYsecYR1 (3) for amplification of the secY gene. The phytoplasma was detected in insects and plant samples. No amplification products were detected in negative control samples from Wrocław or in water blanks. The fact that we detected the pathogen in some asymptomatic plants indicated that a low titre may have been prior to the development of disease symptoms. Amplicons representing three genetic loci were sequenced. To avoid sequencing errors, all DNA samples were sequenced twice in both directions. The obtained sequences were nearly identical, and representative sequences isolated from two plants and one insect, were deposited in GenBank. BLAST analysis of the sequenced 16S rDNA fragments revealed that tested strains shared more than 99% sequence identity to the sequences of phytoplasmas from the aster yellows group. The same analysis performed for ribosomal proteins and secY genes confirmed the highest identity (99%) of analyzed sequences with those of '*Candidatus* Phytoplasma asteris'. Thus the mass occurrence of this insect can cause a decrease in the yields of different rush species which are usually common components of the meadow communities of various parts of Europe. Namely, in such situations the plants cannot flourish, so it is impossible to produce fruit and seeds to be spread.

T3
P0878

Current and future potential distribution of *Selinum alatum* – Plausibly another invasive Apiaceae in Europe

Kamil Konowalik[1], *Małgorzata Proćków*[2], ***Jarosław Proćków***[1]
1. Institute of Biology, Wrocław University of Environmental and Life Sciences
2. Museum of Natural History, University of Wrocław

Selinum alatum (M. Bieb.) Poir. is native to the Caucasus and the Armenian Highlands. Recently a few established populations of this plant have been found in the north-eastern part of the Carpathians. Currently the species is spreading particularly in Poland where it outcompets native plants in several areas. In this study a potential climatic niche of the plant is described with the goal to also highlight future spreading of the species. Our results demonstrate that the extent of the favourable habitat of this plant is broader than currently known. This suggests that *S. alatum* may become a potential new component in some semi-natural or disturbed ecosystems associated with mountainous areas, especially in Central and Southern Europe. Future (2070) models mostly rendered similar suitability maps, but showed slight differences over particular areas and a contraction of suitable habitats, mainly in the northern part of the non-native range.

T3
P0879

The late holocene mangrove dynamics in the upper Gulf of Thailand

Paramita Punwong[1], *Katherine Selby*[2], *Rob Marchant*[2]
1. Faculty of Environment and Resource Studies, Mahidol University
2. Environment Department, University of York

The mangrove dynamics in Samut Songkhram along the upper Gulf of Thailand was reconstructed using pollen analysis from a 460-cm sediment core. The results show that initial colonization of the area was made by true mangroves dominated by *Rhizophora* sp. and back mangroves characterised by *Suaeda* sp. This indicates that this area has been influenced by sea water since at least AD 1,200 until AD 1,250. After that, *Rhizophora* sp. gradually declined and were replaced by grass, indicating lower in sea level allowing mangroves to retreat seaward until AD 1,800. An increase in cereal pollen is also present suggesting possible human activities occurring around this area from AD 1,250 to the present day. Mangroves dominated by *Avicennia* sp. re-colonised at the top of the core suggesting a shift of mangroves landward probably due to a sea level rise until the present day.

T3
P0880

Bryophyte preferences in bryophagous insects – What matters?

Petr Pyszko, *Katerina Kuravova*, *Zuzana Skoupa*, *Michaela Drgova*, *Petr Kocarek*, *Vitezslav Plasek*, *Pavel Drozd*
University of Ostrava

Which features predetermine moss to be consumed by insects?

Despite bryophytes represent favorable habitat for great diversity of arthropods, their dietary utilization is marginal, seemingly due to toxicity and poor nutrition quality. We focused on three largest consistent bryophagous taxa with chewing mouthparts: Tetrigidae (Orthoptera), Scopariinae (Lepidoptera) and Byrrhidae (Coleoptera). All groups have been neglected by ecologist for a long time, and almost nothing have been known about their host specificity and preferences. Unlike vascular plants, bryophytes are determinable according to phylloid fragments also in gut content. Insect specimens were collected by individual sampling or sweeping in Central Europe, Borneo and China and some samples were reared in growing chambers. The gut content was disected or gained from frass, and analyzed to identify and quantify mosses. Host preferences were also tested by measuring survival and fitness on different host species. Whereas mosses comprise only about 15% of diet in Tetrigidae, Byrrhidae and some Scopariinae are strictly bryophagous, despite mosses are defecated seemingly undigested. Majority of bryophagous taxa are surprisingly highly polyphagous, consuming plenty of mosses with preferences for particular species. Shared traits of preferred bryophytes are discussed, but physical properties as cell wall thickness or structure of moss cushions seem to be essential.

T3

P0881

What determines Theaceae diversity in China: Ecological or evolutionary processes?

Mide Rao[1,2], *Manuel Steinbauer*[3], *Xiangcheng Mi*[2], *Xiaoguo Xiang*[2], *Keping Ma*[2], *Jingtun Zhang*[1], *Jens-Christian Svenning*[3]

1. Beijing Normal University

2. IBCAS

3. Aarhus University

The geographic variation of species richness is one of the most obvious patterns on earth. Ecological and evolutionary factors have been thought to be major drivers. However, the relative important of ecological and evolutionary factors remains to be determined, and the role of two hypotheses (time-for-speciation and evolutionary rate) that integrate evolutionary and ecological factors are still far from universal. In this study, we used the data both on species richness and phylogeny of tea family, and environmental variables (the mean temperature of the coldest month, mean annual precipitation and soil pH) to reveal the processes underling species richness patterns of Theaceae *s.s.* across China. We found: (1) Both ecological and evolutionary factors have effects on species richness patterns of Theaceae *s.s.* (2) The former has effects at both family and tribe level. However, evolutionary effect takes place at family level. Species richness and phylogenetic relatedness show the responses to minimum temperature and pH which is based on time for diversification hypothesis, while precipitation reflects the prediction of evolutionary rate hypothesis. (3) At tribe level, annual precipitation and soil pH are related to species richness but not to phylogenetic relatedness. On the contrary, minimum temperature has effect on phylogenetic relatedness rather than species richness. These results suggest that the species richness patterns of Theaceae *s.s.* appear to be predominantly the result of both ecological and evolutionary drivers. Moreover, different environmental factors relate to different mechanisms in species richness pattern, and different taxon levels also reflect different underling processes.

T3

P0882

Worldwide hemisphere-dependent lean in Cook pines

Jason Johns[1], *Jenn Yost*[1], *Dean Nicolle*[3], *Boris Igic*[2], ***Matt Ritter***[1]

1. Department of Biology, California Polytechnic State University, San Luis Obispo, USA

2. Department of Biological Sciences, University of Illinois at Chicago, Chicago, USA

3. Currency Creek Arboretum, Currency Creek, Australia

A tree's form is determined by a complex interaction of genetics and environmental stimuli. Under most conditions, trees grow vertically in response to the opposing influences of light and gravity. In challenging environments, where competition for light or mechanical stress is intense, trees may grow in a non-vertical fashion. Although these growth responses have been studied for over 100 years, there is still much to learn about the mechanisms driving them. Here we describe a novel leaning growth habit in *Araucaria columnaris* (Cook pine), a conifer endemic to New Caledonia and cultivated throughout the world. We document for the first time a remarkable pattern of directional hemisphere-dependent leaning. Specifically, in a large sample of individuals from around the world, we show that Cook pines display a consistent lean, commonly several meters away from the base of the trunk. This incline away from the vertical axis is highly non-random—trees in the northern hemisphere lean south, and those in the southern hemisphere lean north. Additionally, the magnitude of the lean is more pronounced at higher latitudes. The Cook pine's unique growth pattern may be the first-documented species-wide disruption of the balanced interactions between gravitropism, phototropism, and other environmental factors that influence tree architecture.

T3

P0883

Growth, mortality and wood density of liana communities in fragmented forests

Mareike Roeder

Xishuangbanna Tropical Botanical Garden

High abundance of liana stems is an indicator for forest disturbance, as they often occur on forest edges, secondary forest succession or gaps. Forest fragmentation was found to increase liana abundance due to edge effect. However not all liana species benefit from disturbance and it is rarely studied how diversity and traits of liana communities change with disturbance. For example can climber types of lianas change with forest succession, as tendril climbers tend to depend on trees of small diameter for climbing, whereas stem twiners were associated with larger trunks for support. Large lianas are indicators for old growth forest, due to their slow diameter increment. For a more accurate picture of forest communities, it is crucial to go beyond stem counts for lianas. We include growth, mortality, wood density, liana size, and climber type to examine how liana communities change with disturbance and in different habitats. We wanted to know, if fragmented (disturbed) forests tend to have liana communities of more pioneer like growth forms, e.g. more lianas species with lower wood density, increased growth, high mortality rates, more tendril climbers, small but many stems; or if different habitat types rule the liana communities. We conducted a repeated survey (2014 & 2016) of lianas (> 0.5 cm diameter) in 47 forest fragments around a township in SW China and gathered wood density data for 67% of species and 90% of all stems. The fragments were distributed in several forest types that are characterized by different bedrock (limestone yes or no), elevation and topography. We tested how growth, mortality and average wood density, number of small stems (< 2 cm) and climber type changes with fragmentation and habitat. Predictors for the models were fragmentation (edge distance OR forest area in a 500 m circle around the plot), canopy cover and habitat (elevation, topography and limestone). In general, the natural habitat explained most variation in liana communities: valleys had high growth and mortality rates of lianas and high abundance of small liana stems. Forest on limestone hosted more tendril climbers and less small stemmed lianas. Fragmentation explained less variation: growth and mortality increased with fragmentation, average wood density decreased. Generally the explanatory power of the models for growth, mortality and wood density was low (Pseudo-$R^2 < 0.35$). Nonetheless, the trends found in our study of lianas confirm patterns found in other fragmentation studies on different growth forms such as trees, e.g. a higher turnover with increased fragmentation.

T3

P0884

Efficacies of Northern Shakti fertilizer on the yield and quality of mango

Durlave Roy

Northern Agro Services Ltd

A field trial was conducted to evaluate the efficacies of Northern Shakti fertilizer on yield and quality of mango in acidic soil of Bangladesh. Three treatments -T0 untreated control, T2 = Recommended dose (Cowdung 26-30 kg + Urea 1000 g + TSP 500 g + MOP 350 g + Gypsum 350 g + Zinc Sulphate 15 g), T1= Northern Shakti fertilizer 1.5 kg + Urea 200 g + Northern organic fertilizer 3.5 kg). Results revealed that the treatment T1 significantly increased the mango per plant. It was 45.8% higher yield over control. It is concluded that Northern fertilizer with recommended urea is superior sustainable production of mangoes.

T3

P0885

Lineage divergence of a temperate forest species in Eastern Asia, *Smilax sieboldii* (Smilacaceae): Vicariance vs. adaptive divergence

***Yalu Ru*[1,2], *Ruijing Cheng*[1,3], *Hanyang Lin*[1,2], *Yunpeng Zhao*[1,2], *Pan Li*[1,2], *Chengxin Fu*[1,2]**

1. The Key Laboratory of Conservation Biology for Endangered Wildlife of the Ministry of Education, College of Life Sciences, Zhejiang University

2. Laboratory of Systematic & Evolutionary Botany and Biodiversity, Institute of Ecology and Conservation Center for Gene Resources of Endangered Wildlife, Zhejiang University

3. College of Agriculture & Biotechnology, Zhejiang University

Elucidating how various evolutionary factors interact to affect speciation is an ssential theme in evolutionary biology. Insights of incipient speciation can be gained from studying lineage divergence. Eastern Asia harbors world richest temperate flora. The mechanisms generating such rich species and genetic diversity have been fascinating evolutionary biologists for decades while considerable gaps remain in particular whether adaptive divergence also plays a role and its importance relative to vicariance. *Smilax sieboldii* is a typical woody climbing plant inhabiting temperate forests in eastern Asia with a disjunct distribution in mainland China and Taiwan, Japanese Archipelago and Korean Peninsula, providing us an ideal study system to test the relative role of vicariance and adaptive divergence. We employed three data sets including plastid genomes of 12 haplotypes previously suggested by two cpDNA fragments, EST-SSRs for 400 individuals from 38 populations, and RADseq tags for 45 representative individuals (one accession per population). Multiple data sets will be analyzed to confirm geographic lineages, to reconstruct their phylogenetic relationships and demographic histories, e.g., divergence times, dynamics of population sizes and gene flow. We will test competing hypotheses explaining the lineage divergence in relation to historical climate oscillation, sea level fluctuation, emergence and submergence of East China Sea Landbridge, and range shifts. Finally, canonical correlation and redundancy analysis will be utilized to compare the relative contribution of environment and geography to genetic divergence of the focal species. Our findings will imply the evolutionary process of *S. sieboldii* and indicate the relative role of geographical isolation and differential selection which shaped the diversity and phylogeographic structure of temperate forests plants in eastern Asian.

T3

P0886

Floristics and distribution of spiny bamboo (*Bambusa spinosa* Roxb., Hort. Beng. [1814]; Poaceae: Bambusoidea) stands in Cebu Island, Philippines

Val Salares

University of San Carlos

The woody bamboos of the subfamily Bambusoideae play a critical role in the ecology of their forest habitats and have long been of economic importance to humans. This study followed an observational/descriptive research design using naturalistic observational methods of field measurements and utilized random sampling method along the boundaries of fragmented forest remnants of Cebu Island. Quadrat sampling method was employed with an area of 20 m by 20 m and a total of thirty (30) quadrats were randomly surveyed per area with a total of four municipalities. A survey of different areas in Cebu Island was done from August 2014 to March 2015 to observe the occurrence of natural bamboo stands in the island and the municipalities of Balamban, Tuburan, Barili, and Moalboal with only *Bambusa spinosa* Roxb., Hort. Beng. [1814] stands included. In all four areas, *B. spinosa* Roxb., Hort. Beng. [1814] is one of those with the highest Importance Value which means that is contributes much to the productivity and biomass of the area. For the similarity index, Tuburan is the most dissimilar to all the sampling areas. Species richness (D_{mg}) and diversity (H') were relatively low. Low values were also obtained for Simpson's Diversity Index (D) but since this focuses on dominant species, diversity increases as the value nears zero. Floral association values show tendencies towards indigenous species although the values are low. The Floristic Quality Index show values between 3 to 5, which shows a trend towards disturbed environments and association with flora that are either invasive or higher tolerance to disturbance. All the natural bamboo stands exhibited heights ranging from 17 to 20 m and DBH ranging from 5 to 8 cm. The number of culms was very varied from 10 to almost 60. Basal area were usually more than 10% of the 20 × 20 m quadrat. Biomass ranged from 150 to 250 kilograms where Moalboal had the highest values. MODIS products showed that bamboos stands have low LAI as the bulk of their bodies are the culms but with variable FPAR, GPP and Net Photosynthesis. Many bamboos prefer soil with good water holding ability (high SMI) and water permeability for optimum rhizome growth but cannot tolerate water logging (low NDWI and MNDWI). High NRI values denote presence of nitrogen in high levels. Multiple correlations were analyzed and of all the physiognomic characters, biomass had the most correlations with the other factors especially with species richness and diversity. Principal Component Analysis was utilized as a dimension reduction procedure. The community, physiognomic and physico-chemical parameters were reduced to six principal components namely (1) reflectance-based (soil physico-chemical) parameters, (2) productivity, (3) temperature regime, (4) species richness and diversity, (5) floristic quality, and (6) water availability. These parameters contribute most to the distribution of natural bamboo stands.

T3

P0887

Phytochemical analysis of a novel medicinal plant *Clerodendrum indicum* L. under in-vitro condition

Avishek Sarkar, *Anindita Mulkerjee, Abhijit Bandyopadhyay, Sikha Dutta*

The University of Burdwan

Plants have great potential use especially as traditional medicine and pharmacological drugs. A bulk of the world population still depends on traditional medicines. Medicinal plants have provided the modern medicines with numerous plant derived therapeutic bioactive compounds. Natural products play pivotal role in the discovery of new drug resources for the cure and prevention of diseases. In India large number of plant species had been screened for their pharmacological properties but still a vast range of endangered species remain unexplored. *Clerodendrum indicum* L. – a little explored ethnomedicinal plant belonging to the family verbenaceae in India and China contain a number of therapeutic chemicals. The alkaloids, steroids and phenolics present in this plant were reported to possess significant antimicrobial property (Venkata *et. al.*, 2010; Ravindranath *et. al.*, 2003; Rahaman *et. al.*, 2000). In present communication an attempt has been made to assay the phytochemicals present in the calli of *Clerodendrum indicum* L. produced under in vitro condition. Phytochemical assay of the calli was done both under controlled condition as well as under pathogenic stress exerted by *Curvularia lunata* (Wakker) Bodiejn. To identify the novel phytochemicals GC-MS and NMR studies were performed. A number of compounds were identified from both the control and treatmen sets and the treatment sets showed some more novel compounds in contrast to the control set, like, benzene1, 4dichloro compound which has been recommended as halogenated benzene and used as insecticide and pesticide. 3- (1-Ethoxy-ethoxy)—4,4, 4trifluro butyric acid which is suggested to be flurobutyrate benzene used to reduce triglyceride and cholesterol level in blood. Ethaneperoxic acid, 1-cyano-1-[2-(2-phenyl-1,3-dioxolan-2-yl) ethyl] pentyl ester which was reported to be a potent antimicrobial agent (Janakiramann *et.al.*, 2012). The $_1$H NMR analysis of the methanolic extract of *C. indicum* exhibited presence of aromatic group of compounds which might be responsible for the bioactive properties of the plant extract. A phenolic compound Cleroindicin (A, B, C, D, E, F) was identified by NMR studies.

T3

P0888

Effect of strong phenotypic selection on *Pinus sylvestris* genomic variation

Outi Savolainen[1], *Jaakko Tyrmi*[1], *Rami-Petteri Apuli*[1], *Sonja Kujala*[1], *Tanja Pyhäjärvi*[1], *Katri Kärkkäinen*[2]

1. University of Oulu, Dept Ecology and Genetics

2. Natural Resource Center Finland

Scots pine has a very large range spanning diverse environmental conditions. Climatic variation subjects conifers to strong selection for climatic adaptation in different life stages. Extensive provenance trials and other experiments demonstrate adaptation to local climatic conditions, as seen in patterns of phenotypic variation correlated with climatic conditions. These data also allow quantification of selection. We examine patterns of sequence variation in exome capture data along two north-south transects, to search for patterns consistent with local adaptation. On the other hand, after self-fertilization, there is high mortality at especially early life stages due to the high numbers of lethal equivalents. The effects of selection due deleterious recessives on genomic patterns of variation have also been predicted. Genetic data from exome capture in Scots pine can be contrasted against some of these predictions.

T3

P0889

Evaluation of heavy metal accumulation capacity and mineral nutrient status of aquatic plant *Vallisneria nana* in artificially contaminated media

Ugur Sen[1], *Ibrahim Ilker Ozyigit*[1], *Mehmet Emin Uras*[1], *Ibrahim Ertugrul Yalcin*[2], *Elif Yilmaz*[1]

1. Marmara University/Faculty of Science and Arts, Department of Biology, Istanbul, Turkey

2. Bahcesehir University, Faculty of Engineering and Natural Sciences, Department of Molecular Biology and Genetics, Istanbul, Turkey

Heavy metal pollution recently increased because of anthropogenic effects due to rising usage of heavy metals in different area of industry. The major heavy metal sources are mining, metal industry, fuel combustion, and production of some goods like paints, battery. This type of pollution is a serious problem due to toxic effects of heavy metals to every members of every ecosystems.

Heavy metals are very persistent and remains in ecosystem for a long time. Aquatic organisms are also suffered from heavy metal accumulation caused by disposal of industrial waste water into their ecosystem. This study was investigated the heavy metal accumulation and mineral nutrient element conditions of *Vallisneria nana* plats that were treated with several concentrations (100, 200 and 400 μM) of Cd, Cu, Ni and Pb elements as artificially contaminated solutions including 0.2% Hoagland solution). Some heavy metals and mineral nutrient elements (Ca, Cd, Cu, Fe, K, Mg, Mn, Na, Ni, Pb and Zn) were determined by using ICP-OES. According to results, levels of Ca, Cd, Cu, Ni and Pb increased in plant root and stem/leaf while mineral elements decreased such as Fe, K, Mg, Mn, Na and Zn. Measured lowest and highest (dw-mg/kg) element values, respectively include 11,860.091 ± 230.062 (stem/leaf; 100 μM)-16,031.537 ± 329.276 (root; 400 μM) for Ca; 9.519 ± 0.120 (stem/leaf; 100 μM)-42.669 ± 0.676 (root; 400 μM) for Cd; 83.292 ± 1.618 (stem/stem; 100 μM)-151.786 ± 2.536 (root; 400 μM) for Cu; 425.668 ± 8.268 (stem/leaf; 400 μM)-736.138 ± 14.299 (root; 100 μM) for Fe; 15,080.103 ± 407.911 (stem/leaf; 400 μM)-21,732.795 ± 420.366 (root; 100 μM) for K; 2,060.485 ± 39.990 (stem/leaf; 100 μM)-3,189.129 ± 61.888 (root;100 μM) for Mg; 125.712 ± 2.440 (stem/leaf; 400 μM)-268.455 ± 5.215 (root; 100 μM) for Mn; 669.329 ± 13.001 (stem/leaf; 400 μM)-928.905 ± 18.043 (root; 100 μM) for Na; 27.225 ± 0.529 (stem/leaf; 100 μM)-60.013 ± 1.166 (root; 400 μM) for Ni; 55.135 ± 1.071 (stem/leaf; 100 μM)-98.141 ± 1.906 (root; 400 μM) for Pb; 181.387 ± 3.523 (stem/leaf; 400 μM)-263.734 ± 5.123 (stem/leaf; 100 μM) for Zn. *V. nana* could accumulate Cd as 646.9 times than control in stem/leaf and 710.5 times than control in root. Therefore it can be suggested that *V. nana* can be called as a good accumulator for Cd.

T3

P0890

Assessing and predicting biodiversity response to impacts of climate change: A case study Western Soutpansberg Mountain Range; Vhembe Biosphere Reserve Limpopo, South Africa

David Sengani[1], *Natasha Potgieter*[1], *Peter Tshisikhawe*[1], *Benis Egoh*[2]

1. University of Venda; School of Mathematical and Natural Sciences; Department of Botany

2. Council for scientific and Industrial Research (CSIR) Biodiversity and Ecosystem Services Group

Assessing and predicting biodiversity responses to climate change and its impacts on human well-being are high priority targets for this study. The climate is changing, yet to which degree and in what pattern remains uncertain in many areas (Rasche, 2014). Temperature is a key driver of organism function and ecosystem processes (Forsman *et al.*, 2016). Previous studies projected global warming as an effect to world's biodiversity. Accurately predicting climate change effects at the ecosystem level is complex and anxious with difficulty (Clark *et al.*, 2017). However, such knowledge is critical to understanding future biodiversity and the impact on ecosystem services. Understanding the controls of geographic distribution of species is a central issue in ecology (Overgaard *et al.*, 2010). Little is known regarding the driving factors of the species distribution including its limiting factors. Notably, the world's biodiversity is foreseen to be critically threatened not only by land-use changes and contingent habitat loss but also by climate change (Overgaard *et al.*, 2010). Why Vhembe Biosphere Reserve (VBR)? Vhembe Biosphere Reserve is one of the largest reserves in South Africa rich in biological diversity, and is recognized as a centre of endemism with at least 57 endemic plant species. VBR is subjected to a wide range of environmental stress including changes in climate. Climate change is widely recognized as a significant threat to biodiversity and ecosystem services. The main objective of this study is to provide answers on how biodiversity response to temperature shifts due to climate change effects. To identify the factors that lie behind the sensitivity or resilience of a range of species. Understanding species' responses to environmental change underpins our abilities to make predictions on future biodiversity under any range of scenarios (Clark *et al.*, 2017). These assessments and predictions are essential inputs to defining and optimizing adaptation and mitigation strategies for climate change. The findings from this study will thus be useful for policy and decision making processes relevant to biodiversity conservation in a changing climate and sustaining ecosystem services to support for human well-being.

T3

P0891

Potassium retention in leaf mesophyll as an element of salinity tissue tolerance in halophytes

William Percey[1], *Lana Shabala*[1], *Qi Wu*[1,2], *Nana Su*[1,2], *Michael Breadmore*[1], *Roseanne Guijt*[1], *Jayakumar Bose*[1,3], ***Sergey Shabala***[1]

1. University of Tasmania

2. Nanjing Agricultural University

3. University of Adelaide

Soil salinity remains to be a major threat to the global food security, and the progress in crop breeding for salinity stress tolerance may be achieved only by pyramiding key traits mediating plant adaptive responses to high amounts of dissolved salts in the rhizosphere. This task may be facilitated by studying natural variation in salinity tolerance among plant species and, specifically, exploring mechanisms of salinity tolerance in halophytes. The aim of this work was to establish the causal link between mesophyll ion transport activity and plant salt tolerance in a range of evolutionary contrasting halophyte and glycophyte species. Plants were grown under saline conditions in glasshouse, followed by assessing their growth and photosynthetic performance. In a parallel set of experiments, net K^+ and H^+ transport across leaf mesophyll and their modulation by light were studied in control and salt-treated mesophyll segments using vibrating non-invasive ion selective microelectrode (the MIFE) technique. The reported results show that mesophyll cells in glycophyte species loses 2 to 6 fold more K^+ compared with their halophyte counterparts. This decline was reflected in a reduced maximum photochemical efficiency of photosystem II, chlorophyll content and growth observed in glasshouse experiments. In addition to reduced K^+ efflux, the more tolerant species also exhibited reduced H^+ efflux, which is interpreted as an energy-saving strategy allowing more resources to be redirected towards plant growth. It is concluded that the ability of mesophyll to retain K^+ without a need to activate plasma membrane H^+-ATPase is an essential component of salinity tolerance in halophytes and halophytic crop plants.

T3

P0892

Increased precipitation and nitrogen alter shrub architecture in a desert shrubland: Implications for primary production

Weiwei She, *Yuqing Zhang, Shugao Qin, Bin Wu, Yuxuan Bai*
Yanchi Research Station, School of Soil and Water Conservation, Beijing Forestry University

Shrublands are one of the major types of ecosystems in the desert regions of northern China, which is expected to be substantially more sensitive to global environmental changes, such as widespread nitrogen enrichment and precipitation changes, than other ecosystem types. However, the interactive effects of nitrogen and precipitation on them remain poorly understood. We conducted a fully factorial field experiment simulating three levels of precipitation (ambient, + 20%, + 40%) and with two levels of nitrogen deposition (ambient, 60 kg N ha^{-1} yr^{-1}) in a desert shrubland in the Mu Us Desert of northern China. We used plant architectural traits (plant cover, volume, twig size and number) as proxies to predict aboveground net primary productivity (ANPP) of the dominant shrub (*Artemisia ordosica* Krasch), and assessed the responses of plant productivity and architectural traits to water and nitrogen addition. We found significant differences in twig size and number of *A. ordosica* under water and nitrogen treatments but not in shrub cover/volume, which suggest that twig size and number of the shrub species were more sensitive to environmental changes. The productivity of the overall community was sensitive to increased precipitation and nitrogen, and shrubs played a more important role than herbaceous plants in driving productivity in this ecosystem. Precipitation- and nitrogen-induced increases in vegetation production were positively associated with increases in twig size and number of the dominant shrub. Water addition enhanced the twig length of *A. ordosica*, while nitrogen addition resulted in increased twig density (the number of twigs per square meter). Water and nitrogen interacted to affect twig length, but not twig number and shrub ANPP. The trade-off, defined as negative covariance between twig size and number, was likely the mechanism underlying the responses of twig length and shrub ANPP to water and nitrogen interactions. Our results highlight the sensitivity of twig size and number as indicators to estimate shrub production and the mechanism underpinning desert shrub ANPP response to global environmental changes.

T3

P0893

Ammonia deposition in the neighbourhood of an intensive cattle feedlot in Victoria, Australia

Jianlin Shen[1,2], *Deli Chen*[2], *Mei Bai*[2], *Jianlei Sun*[2], *Yong Li*[1]
1. Institute of Subtropical Agriculture, Chinese Academy of Sciences
2. Faculty of Veterinary and Agricultural Sciences, The University of Melbourne

We conducted the first study in Australia to measure ammonia (NH_3) deposition within 1 km from a commercial beef feedlot in Victoria. NH_3 concentrations and deposition fluxes decreased exponentially with distance away from the feedlot. The mean NH_3 concentrations decreased from 419 μg N m^{-3} at 50 m to 36 μg N m^{-3} at 1 km, while the mean NH_3 dry deposition fluxes decreased from 2.38 μg N m^{-2} s^{-1} at 50 m to 0.20 μg N m^{-2} s^{-1} at 1 km downwind from the feedlot. These results extrapolate to NH_3 deposition of 53.9 tonne N yr^{-1} in the area within 1 km from the feedlot, accounting for 8% of the annual NH_3-N emissions from the feedlot. This high NH_3 deposition rate nearby the cattle feedlot had caused the increase of soil inorganic nitrogen content, especially for NO_3^- (from 33 mg N kg^{-1} at 1000 m from the feedlot to 124 mg N kg^{-1} at 50 m from the feedlot). Higher N content (4.0% to 5.7%) in the above-ground part of grassland species and high cover rate of single species (e.g., a cover rate of 31% to 42% at 50 to 200 m from the feedlot for *Cymbonotus lawsonianus*) were found in the grassland transect to the southeast of the feedlot. Our results suggest that NH_3 deposition is significant nitrogen (N) nutrient input for surrounding croplands and natural ecosystems.

T3

P0894

Ecophysiological responses of urban plants to increasing O_3 concentrations

Sheng Xu
Institute of Applied Ecology, CAS

By using open-top chambers (OTCs), the effects of elevated O_3 concentration on the growth and photosynthesis of urban trees were studied. The results showed that elevated O_3 concentration significantly inhibited the growth of *Catalpa ovata*. Compared to the ambient air control (OA), relative water content, dry weight, fresh weight, specific leaf weight, dry weight increment and fresh weight increment were significantly decreased in leaves of *C. ovata* after 20 days exposure to O_3 (160 nmol·mol^{-1}). The net photosynthetic rate (P_n) of *C. ovata* gradually decreased with the extension of time. The decrease in photosynthesis of *C. ovata* was related to non-stomatal factors. The change trends of intercellular CO_2 concentration (C_i) and transpiration rate (T_r) and stomatal conductance (G_s) were consistent, with decline first and then increase. The maximal photochemistry efficiency of PS II (F_v/F_m) and quantum yield of PS II electron transport ($\Phi_{PS\,II}$) decreased significantly over the time ($P < 0.05$), the decline of *C. ovata* photosynthesis correlated markedly with the excess of light energy, and photoinhibition phenomenon under O_3 stress occurred. Superoxide dismutase (SOD) and ascorbic acid peroxidase (APX) activities showed no significant change during the whole O_3 stress, which implied that it was quite difficult for *C. ovata* to tolerate excessive O_3 concentration, and two antioxidant enzymes did not make positive physiological response to the acute stress of O_3.

T3

P0895

Effects of grazing on CO_2, CH_4 and N_2O fluxes in three temperate steppe ecosystems

Huiqiu Shi[1,2], *Longyu Hou*[1], *Liuyi Yang*[1,2], *Dongxiu Wu*[1], *Lihua Zhang*[1], *Linghao Li*[1]
1. Institute of Botany, CAS
2. University of the Chinese Academy of Sciences

Terrestrial ecosystems play a critical role in regulating the emission and uptake of the most important greenhouse gases (GHGs) of CO_2, CH_4 and N_2O. However, the effects of grazing on these GHG fluxes in different steppe types remain unclear. Here we compared the effects of grazing on seasonal CO_2, CH_4, and N_2O fluxes in meadow (MS), typical (TS), and desert (DS) temperate steppe ecosystems in Northern China. CO_2 emission rates increased from 311.4 ± 73.2 to 349.6 ± 55.4 mg m^{-2} h^{-1} in MS, but decreased in TS (from 341.3 ± 93.0 to 239.5 ± 81.9 mg m^{-2} h^{-1}) and DS (from 212.1 ± 53.7 to 163.0 ± 83.4 mg m^{-2} h^{-1}) in response to summer grazing. N_2O emission rates increased in MS from 4.7 ± 2.2 to 8.1 ± 3.4 μg m^{-2} h^{-1}, but no significantly changed in TS (9.2 ± 4.2 VS 8.4 ± 2.4 μg m^{-2} h^{-1}) and DS (6.3 ± 1.5 VS 5.7 ± 1.6 μg m^{-2} h^{-1}) by summer grazing. CH_4 uptake rates increased in MS from 33.0 ± 11.7 to 47.1 ± 10.4 μg m^{-2} h^{-1} and decreased from 64.4 ± 7.6 to 56.2 ± 5.9 μg m^{-2} h^{-1} in TS in response to summer grazing. In MS and DS, N_2O emissions were positively related to seasonal CO_2 emissions and negatively related to CH_4 uptakes. No significant relationships were found between GHG fluxes in TS. Summer grazing did not affect the relationship between CO_2 and N_2O emissions in MS, but reduced the relationship by enhancing the effect of aboveground biomass (AGB) on N_2O emission in DS. The significant negative relationship between CH_4 uptake and N_2O emission in MS and DS could be attributed to the significant relationship between soil temperature (ST) and AGB in MS and to the significant effects of soil moisture (SM) on both CH_4 uptake and N_2O emission in DS. The decrease in the magnitude of the correlation coefficients between CH_4 uptake and N_2O emission by summer grazing was due to the negative relationship between ST and AGB simultaneously in MS and DS. Our results suggest that effects of summer grazing on GHG fluxes were varied in different steppes and the relationship among GHGs was steppe-dependant and summer grazing also changed the relationship by affecting GHG fluxes induced by varied soil and environmental factors.

T3
P0896
Proposed model for understanding evolution of mangroves along the central West Coast of India based on ecological and palaeoecological studies.

Mahesh Shindikar
College of Engineering, Pune

Mangrove represents one of the sensitive ecosystems to analyze climatic, edaphic and biotic changes along the continental margins of tropics and subtropics. West Coast of India is continuously subjected to naturally occurring geological and climatic changes in its past and anthropogenic pressures in recent times. Mangroves are found to be very sensitive to such challenges in all time scales. An attempt has been made to use past and present mangrove floristics in addressing various environmental changes occurred along the Maharashtra coast at the spacio-temporal scale. Palynological assemblages available at Neogene, Late Pleistocene–Holocene and recent past are correlated with contemporary geological episodes, climatic alterations and anthropogenic interventions. Based on available fossil records, presence of sheltered habitats, extensive mangrove cover, diverse floristic elements and conducive wet-humid climate has been inferred during Neogene time from the Sindhudurg Formation (or Ratnagiri beds). The geological activities like tectonic upliftment and subsidence along the coast that has resulted into subsequent geomorphological changes *viz.* change in the river courses (freshwater diversions) or loss of sheltered bays. As such these geomorphic alterations have caused loss of habitat of specific species like *Nypa* and *Eugeissona* from the coast. Contemporary climatic changes in the hinterland have resulted in loss of wet evergreen species like *Cullenia*, *Ctenolophon* from the catchment areas. In addition to this, the transgression-regression of sea level has significantly affected the extent of mangrove habitats along the coast. In the second phase, the geochronological information from available cores along the coast is mostly restricted to the Late Pleistocene–Holocene period after the Neogene. In the present investigation, recovery of mangrove dominated palynological assemblage from the lignite material that is evident for tectonic upliftment of near shore deposits at higher altitudes along the coast. Quaternary deposits along the coast, though very limited in their occurrence, provide excellent spatial pattern of distribution of species which can be correlated to present day distribution also. Palynological data, chronology and environmental magnetism studies on the sediments obtained from the subsurface levels of mangrove deposits are found to be significant and useful while considering their palaeoclimatic implications during the Late Quaternary. Further, the anthropogenic threats provide recent picture of forces on these ecosystems. Thus, on this background of past, the knowledge about present status of mangroves from the central west coast of India becomes useful as a proxy for tracing their antiquity and probable changes in the future.

T3
P0897
Cold tolerance in two species of mangrove, *Bruguiera gymnorhiza* and *Rhizophora apiculata*

Aidan William Short, *Alison K.S. Wee*
Plant Ecophysiology & Evolution Group, College of Forestry, Guangxi University

As the climate warms the distribution of species is shifting, the general movement being towards higher latitudes and elevations. Cold stress tolerance is vital to the prediction of a species' future range limits, because cold stress limits the poleward distribution of a species (MacArthur 1972). Mangroves are a useful system for the study of clinal trait variation because they are found throughout the world in tropical and subtropical regions, and have recently displayed northward range expansion across the world. In this study, we examined the cold tolerance of, and measured changes in gene expression in response to cold stress in two mangrove species, *Bruguiera gymnorhiza* and *Rhizophora apiculata*. *Bruguiera gymnorhiza* and *Rhizophora apiculata* were chosen as focus of this study because they frequently co-occur and are widely distributed throughout the eastern Indo-West Pacific area, yet display differing latitudinal limits. *Bruguiera gymnorhiza*'s latitudinal limit being farther north (Kagoshima prefecture, Japan) than that of *Rhizophora apiculata* (Hainan, China). This study experimentally examined whether *Bruguiera gymnorhiza* and *Rhizophora apiculata* varied predictably in freeze tolerance, whether physiological traits could explain this variation, and whether changes in genetic expression level would occur in response to cold stress. Three hypothesis were tested (1) freeze tolerance will vary with latitudinal limit, (2) variation in mortality, biomass and photo-inactivation will be linked to freeze tolerance, and (3) cold stress response is the result of changes in the expression of genes. The prediction is that the species with the more northern limit range will have a higher freeze tolerance than that of the more southern species, that the higher freeze tolerance of the more northern species will be the result of differences in the functional traits between the species, and that exposure to cold stress will induce changes in the expression of genes related to cold tolerance. Through the contrast of the physiological and molecular response of each species to cold stress, a better understanding of the mechanisms of cold tolerance can be gained, providing a basis for the prediction of plant movement in response to climate change.

T3

P0898

Functional attributes of planted woody species on coal mine spoil for ecological restoration in a dry tropical environment, India: An implication for C management in a changing environment

Anand Narain Singh
Panjab University Chandigarh

Due to mining operation significant areas of land are degraded and existing ecosystems replaced by undesirable waste materials in the form of dumps or mine spoil with low concentration of C, N and P nutrients. Therefore, the primary aim was to rehabilitate the soil component by planting certain desirable plant species as well as counteracting emission of CO_2 to the atmosphere. Thus, present study was conducted on young plantations of four native species (*Albizia lebbeck, Albizia procera, Tectona grandis* and *Dendrocalamus strictus*) raised on mine spoil. We examined both component of the ecosystem (plant and soil). From soil part, total SOC (soil organic carbon) and their accumulation rates in deferent soil depth, bulk density, WHC, and microbial biomass C were analyzed. Whereas from plants, total plant biomass production and standing state of C nutrient stock in different plant components were estimated at ages 3, 4, 5 and 6 years respectively. The soil bulk density, SOC and accumulation rates at 0-10 and 10-20 cm soil depth were significantly varied among plantations, therefore, nutrient stock C (kg ha^{-1}) was also varied in the corresponding way but values were substantially low in the plantation of *T. grandis* followed by *A. lebbeck, D. strictus and A. procera.* Total tree layer C stock (t ha^{-1}) significantly increased in all plantations as they aged from 3-yr to 6-yr. Plantations of *A. lebbeck* and *D. strictus* supported much more biomass than that of *A. procera* and *T. grandis* which indicates different species have different sequestering pattern. Biomass allocation in different components (leaf, stem and root) was widely different in both species. However, larger partitioning of leaf biomass contributed high amount of C input to the soil surface indicated high quality of C deposition in the soil layer that could be formed as organic layer of the soil in long term. Therefore, rapid biomass accumulation efficiency and faster growth rates in the plant species evidently indicates rapid deposition of C. Furthermore, it might be more beneficial step if we consider as in the way of ecological economics of managing C as well as restoring degraded ecosystems in top priority approach in the current scenario.

T3

P0899

Investigation of rare endemic *Verbascum calycosum* Hausskn. ex Murb. in terms of soil-plant relationship

*Muhip Hilooğlu[1], **Emel Sözen**[1], Ersin Yücel[1], Ali Kandemir[2]*
1. Anadolu University
2. Erzincan University

Verbascum calycosum Hausskn. ex Murb. (Scrophulariaceae) is a rare and endemic plant species that shows narrow distribution around Erzincan, Turkey. *V. calycosum* was first collected from Kemaliye/Erzincan district in 1890 by Sintenis. Later, this species was recorded as extinct (EX) in Turkish Plants Red Data Book. However, it was rediscovered from its original locality by different investigators. At present, only two populations of different size (5,437 and 2,512 individuals) exist in Kemaliye. In this study, physical and chemical characteristics of soils that *V. calycosum* naturally distributed were determined. The soil types of this species were serpentine and loamy, between 2.5Y-10YR in color, very dry, saltless, slightly alkaline and low in organic matter and lime content. In addition, correlation between micro and macro element (N, K, Mg, P, Ca, Fe, Mn, Zn, Cu) contents of plant and soil samples were studied. Obtained results will provide valuable data for ex-situ conservation studies for rare and narrow endemic species of *V. calycosum.*

T3

P0900

Bee pollination increases yield and quality of cash crops in Burkina Faso, West Africa

*Katharina Stein[2], Drissa Coulibaly[3], Souleymane Konate[3], Dethardt Goetze[1], **Stefan Porembski**[1], Karl Eduard Linsenmair[2]*
1. University of Rostock
2. University of Wuerzburg
3. University Nangui Abrogoua Abidjan

Insect pollination constitutes an ecosystem service of global importance, providing significant economic benefits to human society alongside vital ecological processes in terrestrial ecosystems. A growing human population especially in developing countries places rising demand for food and income security under rapidly changing environments. 75% of agricultural crop species rely, to some degree, on animal pollination. Bees are the most important pollinators worldwide; a complete loss of their pollination service could reduce yields by ca. 40% for fruits. The study aimed to investigate the contribution of pollination by bees for yield and quality in cotton and sesame. Field research was carried out in 2015 in south-west Burkina Faso. Pollination experiments were conducted to determine the rate of self-compatibility and the pollinator dependence of the crops. On 11 fields for each crop pollinator exclusion and outcrossing experiments were conducted on 50 flowers per field. The efficiency of various bee pollinator species for fruit set and quality was investigated. Germination experiments with seeds resulting from self or outcross pollination were conducted to test for seed quality in terms of inbreeding depression. Honey bees and one wild bee species were the most effective pollinators.

Exclusion of pollinators caused a yield gap of 37% in cotton and 59% in sesame. Pollination by bees increased the economically most important fiber weight of cotton by 62%, and seed weight of sesame was tripled compared to pollinator exclusion. The germination rate of seeds resulting from self-pollination decreased significantly in both species, which is a clear sign of inbreeding depression and economical important, since the seeds are used for the next sowing season. The gratis pollination service by bees was thus beneficial, contributing to cotton and sesame production by enhancing the quantity and quality of these major cash crops in Burkina Faso.

T3

P0901

Evolution of labellar secretory structures and pollinator food-rewards in Old World *Bulbophyllum Thouars*

Małgorzata Stpiczyńska[1], *Kevin L. Davies*[2], *Marcin Zych*[1], *Bartosz Płachno*[3]

1. University of Warsaw, Faculty of Biology, Botanic Garden

2. School of Earth and Ocean Sciences, Cardiff University

3. University of Warsaw, Faculty of Biology, Botanic Garden

4. Department of Plant Cytology and Embryology, Jagiellonian University in Kraków

Floral food-rewards are relatively uncommon amongst orchids, the most frequent being nectar. Some species, however, provide insect pollinators with other food-rewards, such as pseudopollen, oil, volatiles and resin-like secretions. *Bulbophyllum Thouars* is widely distributed, and is the largest genus of orchids, estimated to contain some 2,200 species. Its great size and highly specialized ecological relationships make this genus a rich source of new information concerning plant adaptations. It includes some of the smallest plants in the orchid family, but has also some of the most intricate flowers, often displaying very diverse and complex pollination strategies. Remarkably, given the enormity of the genus, very little is known about its floral, and in particular, its labellar anatomy and associated secretory activity. Furthermore, detailed micromorphological studies of *Bulbophyllum* flowers are scarce. Our studies have shown that the labellum of *Bulbophyllum* has the capacity to produce a greatly diverse range of food-rewards and associated compounds, such as proteins, lipids, mucilage and nectar. For example, unlike the labella of most species of *Bulbophyllum* investigated to date, which secrete lipid-rich material, those of representatives of section *Racemosae* produce a protein-rich, mucilaginous secretion. This, like the secretion of many other species of *Bulbophyllum*, is produced in specialized, palisade-like epidermal cells which line the labellar groove or sulcus. Although lipids are absent from this secretion, lipid droplets were observed in both secretory cells and sub-secretory parenchyma, and our more recent work has shown that not only do these secretory cells produce food-rewards, but also fragrance. Here we extend our investigations to embrace a number of other florally diverse species, and this represents the first detailed comparative account of labellar secretory tissues and their associated secretions for such a large range of Asiatic *Bulbophyllum* species assigned to a number of sections. Histochemical analyses used in conjunction with micromorphological and ultrastructural investigations enable us to compare the structure of secretory tissues, the composition of the secretory product and details of the secretory process. Given its size, diversity and very wide distribution, it is hoped that our investigations will shed light on the evolution of food-reward production in *Bulbophyllum* in response to pollination pressure and pollinator selection.

T3

P0902

Nectaries on inflorescences and leaves of Malpighiaceae endemic genera from Brazilian dry and wet environments: How different are they?

Isabel Guesdon[1,2], *Ítalo Antônio Cotta Coutinho*[3], ***Renata Maria Strozi Alves Meira***[1]

1. Universidade Federal de Viçosa, UFV, Departamento de Biologia Vegetal, Viçosa, Minas Gerais, Brazil.

2. Universidade Federal do Amazonas, UFAM, Instituto de Ciências Exatas e Tecnologia, Itacoatiara, Amazonas, Brazil.

3. Universidade Federal do Ceará, UFC, Departamento de Biologia, Campus do Pici., Fortaleza, CE, Brazil

Nectaries may be classified as floral (FN) and extrafloral (EFN), according to their position, or as nupticial and extranuptial, based on the involvement of nectar with pollination. However, FN have been used as a synonymous with nuptial nectaries while EFN with extranuptial nectaries. Such approach is not appropriate and might be misleading, especially for nectaries that clearly occur on floral parts (e.g., axis of the inflorescence, pedicel, bracts and petals) but are not involved with pollination. In Malpighiaceae, the mcvaughioid clade is a Neotropical lineage composed of three genera *Burdachia*, *Glandonia* and *Mcvaughia* with a peculiar biogeography as Burdachia and Glandonia are endemic trees to the Amazon while *Mcvaughia* is represented by endemic shrubs to "Caatinga" (Brazilian savanna) and "Restinga" (sandy coastal plains). Species of the mcvaughioid clade present both leaf and bracteolar nectaries what makes such a clade a model for comparative studies regarding the secretory and ecological roles of nectaries. This study aims to characterize the anatomy, the nature of secretion and record insect visitation on leaf and bracteolar nectaries in the mcvaughioide species. Samples of leaves and inflorescences at different stages of maturity were collected in field and submitted to standard anatomical techniques for light and scanning electron microscopy. Histochemical tests were performed to detect polysaccharides, proteins, lipids and phenolic compounds. The glucose concentration of the secretion was measured in the field with aid of glucose reagent strips. Leaf nectaries are immersed, sessile or short-stalked while bracteolar are sessile or subsessile. The morphoanatomy and nature of the secretion are similar in both leaf and bracteolar nectaries. Nectaries are vascularized, present a secretory epidermis with thick cuticle and a nectary parenchyma and show positive reaction to polysaccharides, proteins and phenolic compounds, including tannin. Droplets of lipids in the secretory epidermis and nectary parenchyma were also detected. Bracteolar nectary development is precocious, when flowers are not completed formed, and the secretory activity is persistent throughout fructification. The concentration of sugar was higher in bracteolar nectar than in the leaf ones. Ants of eight genera were identified being *Dolichoderus* the most common. Nectar produced by nectaries on the inflorescences was generally consumed by the ants that foraged each one of the nectaries on the inflorescence axis. The differences found among the nectaries could not be related to the biogeography of species. (CAPES, CNPq, FAPEMIG, FAPEAM)

T3

P0903

The spatial heterogeneity of light effect on micro-environment and specific leaf area within the gap, subtropical forest, China

Supaporn Buajan[1], *Jinfu Liu*[1], *Zhongsheng He*[1], *Xueping Feng*[1], *Abrar Muhammad*[2]

1. College of Forestry, Fujian Agriculture and Forestry University

2. Fujian Provincial Key Laboratory of Insect Ecology, Department of Plant Protection, Fujian Agriculture and Forestry University

Each gap has its own environment, which effect on species growth. Light is the most changes when gap creates. Hence, this study explored whether light has effect on micro-environment and determined the effect of light on specific leaf area (SLA) of species regeneration layer within nine sampling gaps and non-gap. In growing season, environment factors such as light intensity (LI), air temperature (AT), air relative humidity (RH), soil temperature (ST), and soil water content (SWC) were collected. Leaf samples were collected to measure the SLA, nitrogen (N) and phosphorous (P) content. We found that LI had a positive significant correlation with the ST and AT but negative significant correlation with the RH, which these correlation were inversed in non-gap area. The significant different of LI between gap size were found, which the highest average LI occurred in the large gap size meanwhile, the lowest average LI occurred in the small gap size. The spatial heterogeneity of LI differ among the gap depends on the topography, slope direction, and location of trees at each direction. In medium and large gap size, LI had a negative significant correlation with the SLA ($r = -0.248$, $p = 0.04$ and $r = -0.264$, $p = 0.008$). At all directions, the SLA increases with a decreasing LI. Our results concluded that LI had effect on micro-environment in different gap size and SLA in each direction within the gap when compared with non-gap. In addition, SLA had a positive significance with N and P content.

T3

P0904

Allelopathic effects of *Trianthema portulacastrum* L. on seed germination and early growth of Jute (*Corchorus olitorius* L.) seedlings

Jhuma Sutradhar[1], *Abhaya Prasad Das*[2], *Adani Lokho*[1]

1. Department of Botany, Visva-Bharati, Santiniketan, West Bengal

2. Taxonomy & Environmental Biology Laboratory, Department of Botany, University of North Bengal, Siliguri, West Bengal

The Bardhaman district of West Bengal, India, is generally referred as granary of the state. Large part of the state is under the cultivation of wide range of crops. These include cereals, pulses, potato, sugarcane, fibre and oil-crops, *etc.* A detail work on survey of major crop field weeds of Bardhaman district was carried out during the years 2013–2015 and a taxonomically rich weed flora has been recognized. It was predicted that some of these weeds might have allelopathic effects on one or more crop plants under cultivation. Earlier, the allelopathic effects of *Trianthema portulacastrum* L., a major weed of the area, was confirmed on a oil-crop, *Sesamum indicum* L. Jute (*Corchorus olitorius* L.) is another important cash crop in Bardhaman and its nearby districts. *Trianthema portulacastrum* is a major weed also with jute all over the district. Now, to understand the allelopathic effects of *this weed* on jute, if any, the seed germination and the early growth of root and shoot of its seedlings were studied under the aqueous extract and leachate of the whole plant in different concentration gradients (1:2.5,1:5,1:10 and 1:20 w/v). The jute seeds were allowed to germinate separately in petridishes with different concentration of test solution. Distilled water was used to maintain a control set. Germination percentage, root and shoot length of each set was compared with control. In high concentration of *Trianthema portulacastrum* aqueous extract and leachate revealed a significant inhibitory effect on Jute seed germination. In 1:2.5 concentration (i.e highest concentration of extract and leachate) showed the lowest value for germination percentage, mean root, shoot and seedling length. Allelopathic effects on germination and thus root and shoot growth increases with increasing concentration of test solutions. Inhibitory effects of extract is significantly severe than the leachate.

T3

P0905

Detection of distribution changes of evergreen broad-leaved trees in an old-growth forest in a vegetational transition zone over 44 years for assessment of climate change impacts

Masatsugu Yasuda[1], *Etsuko Nakazono*[2], ***Nobuyuki Tanaka***[3], *Hiromu Daimaru*[4]

1. Asia Air Survey

2. The University of Tokyo

3. Tokyo University of Agriculture

4. Forestry and Forest Products Research Institute

The globally averaged combined land and surface temperature showed a warming of 0.85℃, over the period 1880 to 2012 (IPCC, 2013). More rapid temperature rise is predicted in the next 100 years due to greenhouse gas emission. There have been a number of studies on predicting potential habitat shifts of plants/vegetation under future climate change scenarios. However, studies on detecting real distribution shifts driven by past climate change are still a few (cf. Yasuda *et al.*, 2007, Lenoir *et al.*, 2008, Shimazaki *et al.*, 2012). Especially, distribution shifts of tall forests are not so reported, probably because technical and ecological problems. The distribution shift of tall forests, which are multi-story communities, are technically more difficult to detect than that of low vegetation such as swamp and alpine vegetation. Tall forests are usually changing under vegetation succession, which obscures the magnitude of climate change impact. To detect distribution changes of evergreen broad-leaved trees (EBTs) in an old-growth forest in the transitional zone between cool-temperate deciduous broad-leaved forest zone and warm-temperate evergreen broad-leaved forest zone, we used orthophoto images converted from aerial photos taken between 1961 and 2005. The study area we selected is an old-growth forest preserved by a religious reason on the southern slope of Mt. Tsukuba in Japan. The crowns of EBTs in plot A (125 ha) increased from 1,820 pieces to 2,424 pieces (1.3 times). The total crown area in plot B (20 ha) located within plot A increased from 28,336 m^2 to 46,480 m^2 (1.6 times). The upper limit elevation, below which ratio of total crown area of EBTs is more than 40%, moved upward from 627 m to 677 m a.s.l. over 44 years. According data of two meteorological stations on the mountain, three critical boundaries, that is, the lower limit elevation of the deciduous broad-leaved forest zone estimated by 85 of Kira's warmth index, the upper limit elevation of the evergreen broad-leaved forest zone estimated by -10 of Kira's coldness index, and the upper limit elevation of Quercus acuta, a single dominant EBT species, estimated by -4.7 of minimum temperature of the coldest month, move upward by 98 m, 147 m and 146 m over 100 years. These past changes occurring in this old-growth forest, which had been not under anthropogenic pressure and was proximate to being a climax forest, could be caused by past climate change much more than by vegetation succession. Thus, we conclude that the distribution shift and increase of EBTs on Mt. Tsukuba are explained by the past climate warming.

T3

P0906

Report of fungal endophyte in anthers of *Stigmaphyllon* subgenus *Ryssopterys* (Malpighiaceae)

David William Taylor[1], *Gregory J. Anderson*[2]
1. University of Portland
2. University of Connecticut

In summer 2016 we carried out Scanning Electron Microscopy analysis of pollen grains in *Stigmaphyllon* subgenus *Ryssopterys* (Malpighiaceae), a group of woody vines endemic to island chains in Southeast Asia and Oceania, and found the presence of basidiomycete hyphae as well as basidia and spores. The hyphae were found wrapped around pollen grains. Furthermore, fungal spores were found on the surface of pollen grains. Given that this pollen was collected from anthers in closed flower buds, we propose that this fungal species is an endophyte that is producing spores in association with the pollen grains of its host, presumably, as in some bird and bee pollinated plants, in order that the spores be dispersed along with the grains of pollen. Due to the showiness of the flower and the pollen grains apparently adhering in groups to fungal hyphae, it is likely that the pollen and spores are dispersed by an animal pollinator, whose identification is a part of ongoing studies.

T3

P0907

Cryptic dioecy in *Stigmaphyllon* subgenus *Ryssopterys* (Malpighiaceae)

David William Taylor[1], *Gregory J. Anderson*[2]
1. University of Portland
2. University of Connecticut

Stigmaphyllon subgenus *Ryssopterys* (Malpighiaceae) is a group of 22 species of woody vines endemic to coastal and other low-elevation sites in parts of island chains in Southeast Asia and Oceania. These species have been characterized as being androdioecious, but studies have not been carried out to test if these species do indeed manifest this rare sexual system, or if the apparently hermaphroditic flowers are functionally female. If confirmed, the latter would recast these species as functionally dioecious. In the summer of 2016 we carried out Scanning Electron Microscopy analysis of pollen grains of 11 of these species and found results that correspond to the presence of cryptic dioecy, with pollen grains in morphologically hermaphroditic flowers bearing inapeturate, and thus sexually nonfunctional, pollen grains. This is the first known observation of dioecy in *Stigmaphyllon. Stigmaphyllon* subgenus *Ryssopterys* is nested within *Stigmaphyllon* subgenus *Stigmaphyllon*. The latter is endemic to South America and has functionally hermaphroditic flowers. Thus, our discoveries suggest that the dioecious condition in this genus evolved after dispersal to island chains in Southeast Asia and Oceania. The pollen-bearing sexually-nonfunctional stamens in subgenus *Ryssopterys* female flowers might serve as attractants for pollinators, as landing platforms for pollinators (as is the case for functionally dioecious species of *Solanum*) or fodder for pollinators, whose identification is a part of ongoing studies.

T3

P0908

Foliar extrafloral nectaries of *Melia azedarach* (Meliaceae)

Patricia Tilney[1], *Magda Nel*[2], *Abraham Van Wyk*[2]
1. University of Johannesburg
2. University of Pretoria

Melia azedarach is a medium-sized to large deciduous tree with bipinnately compound leaves, native to Asia and Australia but has become naturalized in several other parts of the world including South Africa where it is invasive. Despite its wide distribution, there appears to be very little known about the structure or development of the extrafloral nectaries (EFNs). Several minute, circular, green EFNs were found to be present on both sides of the leaflets. These have not previously been reported. In transverse sections of actively secreting EFNs, the glandular tissue is seen to be composed of cells which stain dark blue in the outermost portion and green-blue in the innermost with Schiff's reagent and toluidine blue. Stomata are absent. The cuticle appears to be absent in the middle of the EFN. However, material staining the same as the cuticle is concentrated at the periphery where it extends amongst cells of the dark blue layer. Druses of calcium oxalate are present in the vicinity of the EFNs in small numbers. The innermost portion of the EFN is very close to, if not in direct contact with, the vascular tissue. An analysis of the exudate confirmed the presence of glucose, fructose and sucrose, sugars characteristic of nectar. EFNs, occurring in pairs, in recesses at the petiole base have previously been reported but not studied anatomically or developmentally. We found each petiolar 'EFN' to consist of a group of several individual EFNs. They are structurally similar to the laminar EFNs but differ in usually protruding somewhat beyond the surface of the recess and being composed of more anticlinally elongated cells. Their initial development appears to differ in that cells of both the epidermis and subepidermal tissue divide in various directions in the formation of the laminar EFNs whereas, in the formation of the petiolar EFNs, the subepidermal cells divide mainly periclinally. Both laminar and petiolar EFNs attract ants of different species. Members of the Meliaceae (including *M. azedarach*) are well known sources of limonoids, compounds with insect antifeedant properties. However, it is generally accepted that plants with EFNs, by providing food for ants, get protection from herbivores. It is therefore noteworthy that despite already having a chemical defense against herbivores, a further considerable investment is made by the plant to produce and sustain EFNs. It is suggested that EFN's provide a first line of defense, resulting in the removal or deterring of potential herbivores even before any damage is inflicted on the plant.

T3

P0909

Fungi involved in pelotons colonization in *Vanilla* species from different cultivation systems

Ma. Del Carmen González-Chávez[2], *Andrea Porras-Alfaro*[1], ***Terry J. Torres-Cruz***[1]

1. Western Illinois University

2. Colegio de Postgraduados, Campus Montecillo

Vanilla is an orchid native to Mexico but is widely cultivated in tropical regions; for example, India, China and especially in Madagascar. Despite of the economic importance and pantropical distribution of *Vanilla*, little is known about its endophytic and mycorrhizal fungi. This study involved the evaluation of fungal colonization of *Vanilla* species (*V. planifolia*, *V. pompona*, *V. insignis*, *V. planifolia* var. 'rayada'), the identification of fungi associated to pelotons in terrestrial roots of *Vanilla* plants under flowering stage through culture dependent (fungal isolation from pelotons), and culture independent (sequencing directly from pelotons) techniques. Plants were established in crop systems located in Totonacapan (Puebla and Veracruz states). DNA was extracted directly from 40 pelotons for molecular identification based on ITS rDNA sequencing. Fungi identified directly from pelotons were mainly *Fusarium*, *Pyrenophora seminiperda* and *Scleroderma aerolatum*. The later fungus has been reported as a typical ectomycorrhizal fungus associated with trees and shrubs. All *Vanilla* species were colonized by pelotons without any apparent influence by crop system. Pelotons in different degradation stages were confirmed by electron microscopy. This research enhances the knowledge related to fungal colonization in *Vanilla* species established under different management systems in Mexico.

T3

P0910

Surface pollen deposition in Keibul Lamjao National Park and Loktak Lake of Manipur, Northeast India and their bearing on paleoclimate assessment

Swati Tripathi[1], *Yengkhom Raghumani Singh*[2], *Kamlesh Kumar*[1]

1. Birbal Sahni Institute of Palaeosciences, Lucknow, India.

2. Department of Earth Sciences, Manipur University, Imphal, India.

Manipur is one of the most sensitive area to climate change of the Indo-Burma Range, owing to its unique topographic features and ecosystem. Soil pollen analysis is an important component of paleoecological research, while pollen preservation and the relationship between pollen and vegetation can influence the correct interpretation of fossil pollen spectra. In this paper, 32 surface pollen samples from more or less open-land (Loktak Lake, Ramsar site) and forested (Keibul Lamjao National Park) areas, have been analyzed to determine the relationships between pollen, vegetation and pollen preservation in Bishnupur district of Manipur. The Keibul Lamjao National Park is the only floating national park in the world and famous as a solitary habitat of endangered brow-antlered deer, *Cervus eldi*. The palynodata reflect the tropical moist deciduous and semi-evergreen forest comprising *Syzygium*, *Schleichera*, *Mimusops*, Dipterocarpaceae, *Terminalia*, *Dauabanga*, *Ilex*, *Sterculia* and *Lagerstroemia* under a warm and humid climate in response to rainfall variations in the region. Presence of Euphorbiaceae, *Strobilanthes*, *Adhatoda*, Moraceae and *Syzygium* inferred high monsoonal activity in and around the region. We observe a strong relationship between local herbaceous vegetation and pollen data. However, arboreal pollen tends to weaken this 1:1 relationship, due to the differential dispersal and deposition pattern. Physical parameters of the soil (pH, temperature, salinity, conductivity and TDS) were used to understand the preservation of modern pollen taxa, and to identify pollen indicators of tree cover and anthropogenic influence. Pollen frequency decrease sharply when the soil pH values are over 7.1. Frequent pollen clumping is highly significant and suggests their native genesis and entomophilous nature. Thus, an attempt has been initiated to precisely observe the behavioural pattern of modern pollen deposition which could in turn help in assessing the expanse of pastoral practices and the depth of deterioration of pristine forests as well as in the reconstruction of past climate and vegetation shifts in central valley region of Manipur. A correlation of regional palyno-data with other parts of the Indo-Burma Range is very needful for better interpretation.

T3

P0911

Late Holocene vegetation change, monsoonal fluctuations and anthropogenic influence in Barak Valley of Assam, Northeast India based on pollen data synthesis

Swati Tripathi

Quaternary Palynology Department, Birbal Sahni Institute of Palaeosciences

Paleoclimate records traversing the past millennium unveil manifestations of two distinct climate anomalies – the Medieval Warm Period (MWP), also known as the Medieval Climate Anomaly (MCA) in the context of tropical and sub-tropical regions, followed by the Little Ice Age (LIA). The occurrence of these warm and cool periods differs from region to region, in terms of timing, duration and magnitude of the temperature anomalies. However, attempts to trace the signature of these anomalies in Southeast Asia are still sparse, despite the fact that this region contains numerous lakes and wetlands that may hold potential sedimentary archives. Thus, the pollen analyses of soil samples from a 90-cm deep sedimentary core from Chatla floodplain of Southern Assam has provided insight into the changing vegetation and climatic scenario in the region during the Late Holocene. The Palyno-investigation has revealed that between 1,470 and 1,200 cal yrs BP, the occurrence of major riparian tree taxa like *Barringtonia*, *Syzygium*, *Duabanga* and Sapotaceae along with marshy taxa like

Cyperaceae and *Polygonum* indicates humid climate attributable to strong southwest monsoon. An evidence of agricultural practices evinced in the presence of cereal and other cultural pollen taxa. Later, more humid climate took over during 1,200 and 602 cal yrs BP, as evinced by an exceptional increase in major arboreal group (*Syzygium-Schleichera-Duabanga-Barringtonia-Terminalia*-Moraceae), marshy and aquatics (*Impatiens, Polygonum* and *Lemna*) and annual herbs (Euphorbiaceae and Convolvulaceae) suggesting the consolidation of vegetation around the lake under warm and increased humid climate, which is well corroborated to global MWP that is known between AD 740 and 1,150 worldwide. Between 602 to 387 cal yrs BP, inclination in cereal and other heath land taxa is observed indicating relatively little humid climate, reduction in southwest monsoon precipitation is invoked to explain this which is further supported by a steep fall in major arboreal group, also suggesting an anthropization. From 387 cal yrs BP onwards, decline in arboreal continued with an exponential rise in cereal and shrubby pollen displaying the ongoing landscape changes due to human settlement under warm and relatively dry climate. Thus, in the tropical regions where monsoon plays an important role in determining the regional climate, these temperature anomalies (MWP and LIA) primarily impact monsoonal dynamics and hence are expressed as hydroclimatic variability.

T3

P0912

Habitats and growth characteristics of cluster-root-forming *Helicia cochinchinensis* (Proteaceae) in Japan

Hiromi Tsubota[1], *Tomonori Moroishi*[2], *Tadashi Okamura*[3], *Shinji Uchida*[4], *Jun Wasaki*[3]

1. Miyajima Natural Botanical Garden, Graduate School of Science, Hiroshima University

2. Department of Biological Science, Graduate School of Science, Hiroshima University

3. Graduate School of Biosphere Science, Hiroshima University

4. Technical Center, Hiroshima University

The genus *Helicia* (Proteaceae) consists of ca. 100 species of trees and shrubs with its predominant distribution in the Southern Hemisphere. Most species of the genus naturally grow in rainforests in tropical Asia and Australia. Only a single species of the genus *Helicia*, *H. cochinchinensis* Lour., is present in Japan. This represents the northernmost distribution of the genus with the species occurring from southwest Japan to the Indochina Peninsula. *H. cochinchinensis* is a rare species in Japan and listed in some prefectural Red Lists with the status of Near Threatened (NT) or Vulnerable (VU). Recently, we found cluster roots formed by *H. cochinchinensis* at one of the UNESCO World Heritage sites, Miyajima Island (alias Itsukushima), Hiroshima, Japan (Yamauchi *et al.*, 2015). It is still unknown what environmental determinants define the distribution of this cluster-root-forming species. We present here habitats and growth conditions of *H. cochinchinensis* to investigate the growth conditions, morphological properties of roots, and phosphate mobilizing abilities of this species growing in Miyajima Island. Phytosociological data with or without *H. cochinchinensis* trees were obtained from a forest reserve at Miyajima Island and its surrounding protected area. The data were used for the analyses were obtained from over 500 phytosociological stands (Suzuki *et al.*, 1975). Total phosphorus (P) and Olsen P content of soils for *Helicia* habitats were also measured; plant P and N (nitrogen) of leaves were also measured after chemical digestion. Cluster roots and their rhizosphere soils were collected to investigate root morphology and exudates. Ordination and clustering analyses of the phytosociological data showed that the habitat of *H. cochinchinensis* is in the company of some evergreen trees and shrubs such as *Symplocos* spp. Available P in soils with *H. cochinchinensis* ranged from 0.46–3.7 mg-P/kg-soil (Olsen-P). We found cluster roots from P deficit *H. cochinchinensis* in natural and hydroponic conditions. It was shown that acid phosphatase activity was elevated in the rhizosphere soil. Activity staining revealed a higher activity of acid phosphatase in root tips of each rootlet. A strong decrease of pH in the rhizosphere of matured cluster roots was also shown, suggesting that the cluster root enhanced organic acid exudation like other Proteaceae plants. P concentration of leaves of *H. cochinchinensis* was low; those of mature and senesced leaves ranged from 0.34–0.69 mg-P/gDW and 0.15–0.29 mg-P/gDW, respectively. We conclude that the P mobilization by cluster roots supported P uptake from unavailable P in the soil at Miyajima Island. P remobilization from senesced to mature leaves may sustain the growth of *H. cochinchinensis* under P deficient conditions. The ecological and physiological significance will be discussed by the comparison between *H. cochinchinensis* at Miyajima and other Proteaceae plants at Western Australia and Chile.

T3

P0913

Seed germination traits in alpine grasslands play a part in species' assembly

Maria Tudela-Isanta[1], *Malaka Wijayasinghe*[1], *Graziano Rossi*[1], *Simone Orseningo*[3], *Eduardo Fernández-Pascual*[2], *Hugh W. Pritchard*[2], *Andrea Mondoni*[1]

1. University of Pavia

2. KEW

3. University of Milan

Alpine environments are characterized by rich and diverse habitats, resulting in many plant species along with their many adaptation mechanisms. Seed germination has shown to play a crucial role in the regeneration and maintenance of such plant communities, though it has been difficult to identify a typical behavior for "alpine germination". There is not any bioregion in the world that we find the steep environmental gradients that are typical in the alpine habitats. These different gradients create a mosaic of microenvironments covered by different plant communities. For this reason, alpine habitats provide an ideal context to assess the role of germination niche in plant distribution. To this end, seeds of 53 species from alpine grasslands belonging siliceous (#6230 Directive 92/43) and calcareous (#6170 Directive 92/43) habitats were exposed to different temperature treatments in a controlled laboratory conditions (i.e., 15/5°C and 25/15°C, 12-h daily photoperiod after 0, 3 and 5 months of cold wet stratification, and with gibberellic acid). Then, phylogenetic least squares correlations (PGLS) were fitted to assess relationships between germination traits and habitat, chorology, microenvironment and ecological traits. Also, the germination strategies in each habitat were identified by clustering with k-means. As a result, we identified that germination traits are correlated with the species' ecological traits, chorology and microenvironments. In particular, high dormancy scores were identified in the seeds produced by species inhabiting in heaths, growing in moist and low or medium nutrient soils contents and having an Artic-Alpine distribution. Meanwhile, the species with high germinability scores belong mostly to the generalist and inhabit in medium nutrient soils. The habitat analyzed revealed higher germination in the siliceous habitat and three common germination strategies and one habitat-related, though species from calcareous germinate more during the cold stratification. These results highlight the importance of considering germination traits when studying alpine grasslands, because alpine species show different regeneration strategies depending on habitat provenances, species' ecological traits and chorology. The patterns found here may have important implication in developing good practices for restoration and conservation activities in alpine habitat and also for a better understanding of the response of alpine plants in the regeneration from seeds due to global warming.

T3

P0914

Quantitative and qualitative foliar flavonoid survey of *Suaeda maritime*, which sink in the sea on high tide

Ayumi Uehara[1,3], *Hiroaki Setoguchi*[2], *Tsukasa Iwashina*[3]

1. Division of Life Science and Engineering, School of Science and Engineering, Tokyo Denki University

2. Graduate School of Human and Environmental Studies, Kyoto University

3. Department of Botany, National Museum of Nature and Science

Suaeda maritima (L.) Dumort (Amaranthaceae) is an annual halophytic herb, which is widely distributed in the Northern Hemisphere. This plant grows on coastal sandy land in which partial or whole plant body stay sunk in the seawater on high tide. This plant continually exposed to UV and/or salinity stresses. Flavonoids are important secondary metabolites and those in the leaves act as the protectants to various environment stresses, e.g., UV, poor nutrition, lower temperature and/or seawater. In this study, we investigated the flavonoids accumulation in the leaves of *S. maritime* plant that did not sink in the seawater on high tide. We collected the leaf samples of *S. maritima*, which sink area (sink) or not (non-sink) on high tide, in June, August and November. The MeOH extracts were surveyed by UV-Vis spectrophotometer for total flavonoid content, and were also qualitatively surveyed each flavonoid by HPLC. Their flavonoids were isolated by various chromatographies (CC, TLC, PC and HPLC) and identified by various spectra such as UV and LC-MS *etc.* Moreover, their leaf segments were stained by 2-aminoethyldiphenylborinate, for the observation of flavonoid accumulation with fluorescence microscope. The environmental factors as such EC value and accumulated solar radiation were also surveyed. The flavonoids were significantly accumulated at samples of "non-sink" area. Three major flavonoids were identified as flavonol glycosides, quercetin 3-*O*-rutinoside, 3-*O*-glucoside and isorhamnetin 3-*O*-glucoside, and three flavone glycosides, luteolin 7-*O*-glucuronide, apigenin 7-*O*-glucuronide and chrysoeriol 7-*O*-glucuronide. Their flavonoids were reported from the leaves of *S. maritima* for the first time. They were formerly reported as effective antioxidants in sea buckthorn (*Hippophae* sp.), buckwheat (*Fagopyrum esculentum*) and so on. Flavonoids were qualitatively the same between "non-sink" and "sink" areas. Observation of flavonoids with fluorescence microscope showed the existence of intracellular flavonoids in adaxial side. Soil of "sink" area showed the higher EC values, which mean salt concentration. On the other hand, "non-sink" area showed the higher solar radiation. Thus, we presumed that foliar flavonoids of *S. maritima* were accumulated as UV protectants.

T3

P0915

Effects of some heavy metal mixture on mineral nutrient status of an aquatic plant *Tonina* sp. Belem

Mehmet Emin Uras[1], *Ibrahim Ilker Ozyigit*[1], *Ugur Sen*[1], *Recep Vatansever*[1], *Gamze Gerez*[1], *Ibrahim Ertugrul Yalcin*[2]

1. Marmara University, Faculty of Science and Arts, Department of Biology, Istanbul, Turkey

2. Bahcesehir University, Faculty of Engineering and Natural Sciences, Department of Molecular Biology and Genetics, Istanbul, Turkey

Heavy metal pollution puts heavy pressure on water and land ecosystems. In water ecosystems, this pressure may negatively affect organism, e.g., reducing the number of organisms especially autotrophs. To determine pollutant types and sources are initial steps in terms of remediation and rehabilitation of polluted areas. In this sense, heavy metal accumulation and mineral element status of *Tonina* sp. Belem have been investigated. Aquatic environment of *Tonina* sp. Belem was artificially contaminated with 100, 200 and 400 μM concentrations of Cd, Cr, Cu and Zn for each and also included a non-treated control group. All plants were supplied by 0.2% Hoagland solution as nutritional media. Then, Ca, Cd, Cr, Cu, F, K, Mg, Na and Zn levels were analyzed by using ICP-OES. It was revealed that concentrations of heavy metals (Cd, Cr, Cu and Zn) were increased in plant leaves and stems while mineral elements (Ca, Fe, K, Mg and Na) were decreased. The lowest and highest levels (mg/kg) for Ca were as 3,021.963 ± 70.11 in stem (400 μM - treated samples) and 9,422.902 ± 199.375 in leaf (in control groups), for Cd as 0.386 ± 0.021 in stem (control) and 52.283 ± 1.338 in leaf (400 μM), for Cr as 1.629 ± 0.032 in stem (control) and 90.007 ± 3.005 in leaf (400 μM), for Cu as 35.082 ± 0.866 in stem (control) and 128.307 ± 4.561 in leaf (400 μM), for Fe as 421.235 ± 6.605 in stem (400 μM) and 1,340.082 ± 22.110 in leaf (control), for K as 3,898.766 ± 45.529 in stem (400 μM) and 1,1286.552 ± 129.367 in leaf (control), for Mg as 1,025.369 ± 26.611 in stem (400 μM) and 5,114.951 ± 135.554 in leaf (control), for Na as 355.835 ± 8.881 in stem (400 μM) and 1,352.811 ± 37.625 in leaf (control), for Zn as 56.924 ± 1.035 in stem (control) and 176.934 ± 4.118 (leaf/400 μM). Treatment with heavy metals has notably affected the uptake and accumulation of other mineral elements. Accumulation of macro elements like Ca, Fe, K and Mg were reduced with accumulation of Cd, Cr, Cu and Zn. In addition, based on control groups, Cd demonstrated the higher accumulation levels in 400 μM-treated leaf (127 times higher) and stem (115 times higher) tissues. In light of obtained results, *Tonina* sp. Belem stood as a good accumulator for Cd.

T3

P0916

Salt tolerance mechanism in *Urochondra setulosa*

Elena Voznesenskaya[1], *Nouria Koteyeva*[1], *Anna Berim*[2], *David Gang*[2], *Gerald Edwards*[2]
1. Komarov Botanical Institute
2. Washington State University

Urochondra is a genus of plants in the grass family with the only one known species *Urochondra setulosa*, native to northeastern Africa and southwestern Asia. This species grows in coastal sand dunes, salt marshes and estuaries and has high tolerance to salt. In this study mechanisms of tolerance to salinity in *Urochondra* were analyzed in comparison to *Spartina anglica,* a serious invasive species endemic to natural saltmarsh ecosystems of Britain. The leaf anatomy and ultrastructure of salt glands, pattern of salt excretion, features of gas exchange, accumulation of key photosynthetic enzymes, leaf water content, Na^+ and K^+ accumulation, and osmolality along with levels of some osmolytes and betaine aldehyde dehydrogenase expression, were compared in the two species grown without salt, with 200 mM NaCl, and with 200 mM KCl. Measurements of gas exchange showed photosynthesis in both species had even some stimulation under NaCl treatment while some suppression of photosynthesis was shown under KCl treatment. Both species have distinct ridges on the adaxial leaf side and C_4 type of photosynthesis with Kranz anatomy; abaxial side is flat in *S. anglica* while it is undulated with small ridges in *U. setulosa*. Both species have salt glands on both leaf sides; they are mostly located on the sides of leaf ridges but in *Urochondra* large salt glands are also located on the adaxial leaf side between the ridges above the band of colorless cells. In both species glands are bicellular with basal and cap cells but they have some differences in their positioning and structure. Thus, salt glands are sunken under the leaf epidermis in *S. anglica* and only cap cells are located on the epidermal level and hardly could be seen from the surface. In *Urochondra* all salt glands are located on one level with epidermal cells and represent their modification. Some differences also could be seen in the internal structure of gland cells: while in both species basal cells have intensive development of endoplasmic reticulum and numerous mitochondria in the distal part, the distinctive feature of basal gland cells in *Spartina* is the presence of the longitudinal cell wall labyrinths absent in *Urochondra*. Sodium and potassium ion contents increase about twice during corresponding treatments. Both species increase osmolality in response to compatible solutes accumulation. Under salt treatments they accumulate glycine-betaine (with increasing of betaine aldehyde dehydrogenase expression) and proline while the proline content is significantly higher in *Spartina*. In summary, *U. setulosa* and *S. anglica* show high tolerance to salt in function of photosynthesis, maintenance of water content, and in excretion of salt from leaves; but, they have distinct differences in the positioning and structure of salt glands.

T3

P0917

Climate change causes secondary contact of willow species in the Alps – radseq data reveal insights in recent hybridization and introgression patterns on glacier forefields

Susanne Gramlich, *Natascha D. Wagner*, *Elvira Hoerandl*
Department of Systematics, Biodiversity and Evolution of Plants; Albrecht-Haller-Institute for Plant Sciences; Georg-August-University of Goettingen

Little attention has been paid to the evolutionary consequences of colonization dynamics and succession processes triggered by climate change. The retreat of glaciers due to climate change leaves behind devastated glacier forefields as a novel territory for colonization by pioneer species like willows. Another effect of global warming is that lowland species start to move upwards and colonize higher elevations resulting in overlapping ecological niches of lowland and alpine species. Given the lack of postzygotic crossing barriers due to previous ecological isolation by altitude, this often results in the formation of hybrids. Transgressive segregation in the F2 or later hybrid generations may lead to the formation of novel genotypes that are able to establish in extreme habitats outside the ecological optimum of their parents. Additionally, if hybrids and the parental species co-occur in so-called secondary contact zones, backcrossing can facilitate introgression. This study investigates these processes in the model genus *Salix* by analyzing hybrids that have recently formed on glacier forefields of the European Alps. In the secondary contact zone of the lowland colonizer *S. purpurea* and the subalpine species *S. helvetica* hybrids as well as the parental species were collected. RADseq data were used to analyze the composition of the hybrid population and to study segregation patterns in hybrid offspring raised from seeds that had been collected from naturally pollinated catkins. Our results revealed that most hybrid individuals are F1 hybrids, probably due to the recent emergence of the hybrid population. The offspring of those hybrids mainly consisted of backcrosses with both parental species. Thus, the most likely evolutionary trajectory is introgression of genes between the hybridizing species probably facilitating adaptation to a changing environment.

T3

P0918

Reconstruction about the exhibition greenhouse in Shanghai Botanical Garden

Ninghai Wan
Shanghai Botanical Garden

With the floor area of nearly 5,000 square meters, Greenhouse of Shanghai Botanical Garden opened to the public in 2001. For long time in the environment of high temperature and high humidity, the existing facilities, such as conditioning, ventilation and automatic control system, often failed and could not meet the original requirements. This reconstructioncontains as followed: maintenance of housing structure; air conditioning, ventilation, automatic control system; rockery reconstruction; fire control; the scenery tower reconstruction; water supply and drainage *etc.*

T3

P0919

Genetic validation of Janzen-Connell recruitment patterns of dominant species (*Castanopsis eyrei*) in a subtropical forest of China

Fang Wang[1,2], *Yu Liang*[1], *Walter Durka*[3], *Jianhua Xue*[1], *Wubing*

Xu[1,2], *Xiangcheng Mi*[1], *Keping Ma*[1]
1. State Key Laboratory of Vegetation and Environmental Change, Institute of Botany, Chinese Academy of Sciences
2. University of Chinese Academy of Sciences
3. Department Community Ecology, Helmholtz Centre for Environmental Research - UFZ

Recruitment is an important process affecting community composition and dynamics. The Janzen-Connell hypothesis, as a model describing the process of recruitment, has been considered as an important mechanism of species co-existence and diversity maintenance, which hypothesizes that the offspring dispersed near the mother trees have lower survival rate due to the existence of host-specific enemies and thus make space for other species. Most studies testing Janzen-Connell hypothesis in natural ecosystems, however, based on homo-specific data rather than verified mother-offspring relationship and focused only on seedlings or saplings separately. In this study, twelve microsatellite markers were used to assign parentage of offspring to evaluate the Janzen-Connell effect on seedling recruitment of *Castanopsis eyrei* in a natural subtropical forest. Our results showed that, (1) Janzen-Connell effect was strong in the early life stages of *C. eyrie*; (2) offspring dispersal distance kernels shifted across life stages, implying a decreased Janzen-Connell effect from seedlings to adults; (3) the spatial genetic autocorrelation was highest for seedlings and decreased for later offspring life stages, which indicated that offspring with close genetic similarity tended to be excluded during early life stages.

T3
P0920
Vascular plants in tourist area of Lushan Mountain: Status, threats and conservation

Hui Wang
Huazhong Agricultural University

Lushan Mountain was declared World Geopark and hotspot of biodiversity. As a great tourist resort for natural landscape and historic culture, Lushan may be disturbed by human activities especially tourism and plant introductions. In order to illustrate the conservation status of rare species and influence of human disturbance on plant diversity, we carried out an extension survey on vascular plants along main tourism routes in the Lushan Mountain. A total of 572 species were recorded, belonging to 118 families. Dominant species and constructive species belong to Lauraceae (13 species), Fagaceae (11 species), Camelliaceae (11 species) and Magnoliaceae (9 species). Ancient tree species, such as *Liriodendron chinense*, *Ginkgo biloba* and *Metasequoia glyptostroboides*, were well preserved. At least 158 species were endemic to China and 3 of them were endemic to Lushan. 7 species were endangered and 15 species were near threatened. 76 introduced species had established in surveyed area (represented by species of Compositae, Iridaceae and Scrophulariaceae), and 26 of them were invasive species. For the surveyed area, most of the forest community are well conserved, but more attention should be paid on the possible influences of plant introduction and plantation on native plants.

T3
P0921
Patterns of urban development and socio-economic factors drive plant diversity- A case study in Shanghai, China

Meng Wang[1], *Guanhai Xiang*[1], *Haiyu Li*[1], *Junxiang Li*[1,2]
1. East China Normal University
2. Shanghai Key Laboratory of Urbanization and Eco-restoration

Urbanization is one of the major causes for plant diversity loss. Cities have been studied as hotspots for local species conservation. The patterns of urban sprawl is drastic and different depending on urban development and urban structure. In this paper, we try to test the effects of urban development and urban structure on plant diversity. 251 sample sites generated using stratified random sampling method covered the entire city of Shanghai which is composed of 10 districts. Changxing Island (belongs to Baoshan) and Chongming Island are all excluded in our study. Plant diversity index including α diversity (Species richness, Shannon-wiener diversity and Heip evenness) and β diversity (species accumulation rate) were calculated. Each administration districts were used as a city unit. We choose the distance to each district center from samples points and the socio-economic factors of each districts as the proxy to measure the urban development and urban structure. Linear Regression analyses, stepwise regression analyses, Sample-based rarefaction curve, ANOVA (Followed by a Tukey's test) analyses were conducted between plant diversity and variables factors. Our findings are: 676 Vascular plant species were recorded in 251 plots and the percentage of exotic species richness is 60.65%. The plant diversity indices are different in different administration districts of Shanghai. Of all the samples in different district, all plant richness (62.85 ± 12.66) and exotic plant richness (31.46 ± 8.62) both are the highest in Minhang districts. The richness of all plant (46.47 ± 12.36) and perennial herbaceous (11 ± 3.73) both are the lowest in Qingpu district. Woody plant richness (18.87 ± 7.33) and perennial herbaceous richness (16.9 ± 4.72) both are the highest in Urban center district. On the contrary, the annual herbaceous richness (15.03 ± 7.01) is significantly lowest in urban center district ($P < 0.05$). Native plant richness didn't show any significant differences in all districts, but it is lowest in urban center (26.97 ± 7.89). Shannon-wiener diversity (22.18 ± 8.05; 11.61 ± 5.72) and Heip evenness (0.37 ± 0.08; 0.19 ± 0.11) both are highest in Jinshan district and lowest in Jiading district. β diversity of the whole shanghai is lowest (β diversity = 0.2917) compare to each districts and showed a severe homogenization phenomenon. Followed by Pudong district (β diversity = 0.3877), the area of which is the biggest. The fitting between the distance to district center and α diversity indices exhibited significant linear, log and quadratic polynomial relationship depends on different plant groups. It showed that the pattern of urban development is a force driving plant composition and distribution in different districts. Moreover, of all the socio-economic factors, we also demonstrated that the total areas and built-up area of each district, GDP, Total Investment in fixed assets, Investment in real estate both significantly influence the β diversity respectively ($P < 0.05$) and GDP is the most explanatory variable. As a case study, this is the first time to explore the effects of urban development, urban sprawl thus the socio-economic factors on urban plant composition and homogenization process in regional scale. We suggest that the decision makers and urban designers should recognize the integral role of cities in driving the composition and distribution of plant diversity in future urban planning and investment

T3
P0922

Environmental divergence and difference of stomatal characteristics among mainly *Picea* species in China

Minghao Wang, *Changming Zhao*
Lanzhou University

Ecological speciation played a prominent role in speciation and reflected as closely related species developed both environmental vicariance and morphological differences. But previous analysis of ecological speciation seldom took phenotypic data into account. With an example from Chinese mainly *Picea* species, we collected their geolocation and built phylogenetic tree, used SEEVA to test the correlations between the ecological divergences and the phylogenetic splits. And then, we discussed the differentiation of stomatal characteristics by leaf anatomy and the relationship between stomatal features and environment divergences for *Picea* species in China. The result indicated that temperature variables (maximum temperature of warmest month and temperature annual range) splited at basal nodes, precipitation variables (annual precipitation and precipitation of driest quarter) splited at terminal nodes following the uplift of Qinghai-Tibet Plateau. Linear stomatal densities (LSD), number of stomatal rows (N), pore depth (L) and mean maximum stomatal theoretical area (S_{max}) of species displayed significant difference on both surfaces in common garden. LSD was positively correlated with N on both two surfaces. Furthermore, relationship between stomatal characteristics and annual precipitation of *P. crassifolia, P. meyeri, P. asperata* and *P. koraiensis* reflected select pressure of stomata that species diverged to drier environment with higher stomatal density and smaller stomata. These results might imply that the climate influnce of QTP uplift could explain the ecological divergence of species during this geological time. Stomatal characteristics might be selected by environment variables at least in *Picea*.

T3
P0923

Accumulation of heavy metals in current-year old needles of *Pinus tabuliformis* Carrière. and *Pinus bungeana* Zucc. in different atmospheric environment

Runxi Wang, *Xin Wang, Hongrui Wu, Lihong Zu, Fuchen Shi*
College of Life Sciences, Nankai University

Heavy metals in different atmospheric environment had adverse impact on green-plants growing in varying degrees. In this study, the concentrations of Cu, Mn, Zn and Cd in current-year old needles of *Pinus tabuliformis* Carrière. and *Pinus bungeana* Zucc. that are widely planted in northern China, were determined. The resistance indexes: malonic aldehyde (MDA), soluble protein, soluble sugar, and free proline levels were also determined. The Pearson coefficients between the resistance indexes and the concentrations of heavy metals were analyzed to compare the abilities of two plants to accumulate heavy metals and their resistance characteristics. The topsoil and unwashed and washed needles were from urban, suburban and rural sites in Tianjin where the atmospherc conditions are significantly different according to the environmental monitoring results. The results indicated that there is a significant trend among the concentrations of heavy metals in current-year old needles and topsoil as follows: urban areas > suburban areas > rural areas ($P < 0.05$). In urban areas, the Mn concentration in *P. tabuliformis* and *P. bungeana* were as high as 122.72 $mg \cdot kg^{-1}$ and 128.91 $mg \cdot kg^{-1}$ which were almost as 20 times as in rural areas. Compared with *P. tabuliformis*, the accumulation of Mn and Zn were higher in *P. bungeana*, but as to Cu and Cd no significant differences were found between the two species. Pearson correlation coefficients among the absorption of Cu, Mn, Zn and Cd in needles are at quite a high level ($P < 0.01$). The concentrations of heavy metals in the unwashed needles were higher than that in the washed needles in the urban and suburban sites, which indicated that the surface of the needles could adsorb many heavy metals. The MDA, soluble protein, soluble sugar, and free proline contents increased as the concentrations of heavy metal rose along the rural-urban gradient, and were highly correlated with the concentrations of plant heavy metal in both *P. tabuliformis* and *P. bungeana*. In conclusion, high concentrations of Cu, Mn, Zn and Cd in aerial parts were mostly a consequence of atmospheric deposition, though other factors could affect this accumulation in plants. *P. tabuliformis* and *P. bungeana* could absorb and accumulate heavy metals, mainly through increased physiological resistance to stress.

T3
P0924

NaCl improved the nutritional status in *Zygophyllum xanthoxylum* responding to drought compared with *Arabidopsis thaliana*

Wenying Wang, *Weiwei Chai, Changyu Zhao, Qing Ma, Suomin Wang*
State Key Laboratory of Grassland Agro-ecosystems, College of Pastoral Agriculture Science and Technology, Lanzhou University

Zygophyllum xanthoxylum is a salt-accumulating xerophytic species with excellent adaptability to adverse arid environments. Our previous studies showed that *Z. xanthoxylum* absorbed, from low-salinity soil, a great quantity of Na^+ that transported to leaves and efficiently compartmentalized into vacuoles as an osmoregulatory substance under arid field conditions. Further investigations revealed that *Z. xanthoxylum* responded to salt with increased growth and moreover, became more tolerant to drought in the presence of moderate salinity, with improved photosynthesis and water status, is a totally different strategy from the glycophytes. To further reveal the special mechanisms underlying how Na^+ accumulation stimulate the growth and improves the drought resistance of *Z. xanthoxylum*, in this work, the effects of NaCl on the nutritional status in *Z. xanthoxylum* subjected to osmotic stress were investigated compared to that in the typical glycophyte *Arabidopsis thaliana*. Results showed that compared with control, either 50 mM NaCl or osmotic stress (-0.5 MPa) significantly increased the Na^+ concentration in leaves and roots in *Z. xanthoxylum*; while 5 mM NaCl had no significant effect on the Na^+ concentration in *A. thaliana*, and osmotic stress significantly decreased the Na^+ concentration in shoots in *A. thaliana*. Under osmotic stress, the addition of 50 mM NaCl significantly increased Na^+ concentration in leaves and roots by 6.2 and 3.7 fold in *Z. xanthoxylum*, resulting in improved growth with enhanced biomass and tissue water content; nevertheless, such beneficial effect was not observed in *A. thaliana* treated with 5 mM NaCl under osmotic stress. Compared to the control, root and leaf K^+ concentrations remained unchanged in *Z. xanthoxylum* exposed to osmotic stress with or without additional 50 mM NaCl; in contrast, osmotic stress disturbed cellular K^+ homeostasis in *A. thaliana*, with a significant decrease in shoots K^+ concentrations in the presence of 5 mM NaCl. Moreover, 50 mM NaCl promoted the absorption and accumulation of N, P, Fe, Si, Ca^{2+} and Mg^{2+} in *Z. xanthoxylum*, whereas these nutrients in *A. thaliana*, treated with 5 mM NaCl, almost unchanged. Furthermore, compared with -0.5 MPa osmotic stress, osmotic stress plus 50 mM NaCl significantly increase the concentrations of N, P and Si in whole plant, the Ca^{2+}, Mg^{2+} in leaves and Fe in roots of *Z. xanthoxylum*. However, no differences on the concentrations of these nutrients were observed in *A. thaliana* treated with osmotic stress plus 5 mM NaCl, compared with that under osmotic stress. These results further highlighted the positive roles of NaCl in drought resistance of *Z. xanthoxylum* probably due to the beneficial effect on the nutritional status, including maintaining stable intracellular K^+ and enhancing the concentrations of N, P, Fe, Si, Ca^{2+} and Mg^{2+} in plants.

T3

P0925

Aerosol increases both sunlit and shaded leaf photosynthesis rate but with different mechanisms

Zhenhua Wang[1,2], *Bin Wang*[1,2], *Xin Wang*[1,2], *Lingli Liu*[1,2]

1. Institute of Botany, Chinese Academy of Sciences

2. University of Chinese Academy of Sciences

Aerosol pollution is one of the most serious environmental challenges facing by many regions around the world. Previous studies found aerosols could increase carbon sequestration in many ecosystems, but the underlying mechanism was still in debate. We conducted a field experiment to assess the response of sunlit and shaded leaf carbon assimilation to aerosol-loadings during the growing season in 2014 and 2015 in Beijing. Our study found that averaged over daytime hours, aerosols decreased the sunlit leaf temperature by 2.71 ± 0.22°C, increased the photosynthetically active radiation (PAR) by 113.3 ± 18.6% within the canopy but reduced the global radiation outside the canopy by 10.7 ± 1.2%. Meanwhile, both sunlit and shaded leaf net photosynthesis rate (Pn) increased during most of daytime, with the mean daily Pn of sunlit leaf and shaded leaf enhanced by 2.08 ± 0.23 $\mu mol\ CO_2\ m^{-2}\ s^{-1}$ and 0.87 ± 0.15 $\mu mol\ CO_2\ m^{-2}\ s^{-1}$, respectively. And the light use efficiency of both sunlit and shaded leaf increased significantly. Our study further demonstrated that, for sunlit leaf, the increment in mean daily Pn was mainly attributed to the mitigations of the high temperature inhibition, whereas for shaded leaf, the high Pn was due to the alleviation of the light limitation. Overall, our study suggests that the aerosol's cooling effect via enhancing plant C assimilation should be taken into account seriously as well as its radiation effect in the future climate change research.

T3

P0926

Asymmetrical mating pattern and reproductive success of heterodichogamouos tree *Machilus thunbergii.*

Shuntaro Watanabe[1], *Yuko Kaneko*[2], *Koh-Ichi Takakura*[3], *Naohiko Noma*[3], *Nishida Takayoshi*[3]

1. Kyoto University

2. Toyo University

3. The University of Shiga prefecture

A substantial proportion of plant species arrange the female and male organs at different locations or timings. Such species sometimes exhibit two or more floral sexual morphs that differ in reproductive traits, which can promote pollination between morphs. Heterodichogamy is defined as the presence of two flower morphs that exhibit male and female functions at different times among individuals within a population, and is regarded as a temporal analogy to heterostyly. As is the case with heterostyly, heterodichogamy is regarded as an adaptation to promote outbreeding through enhanced inter-morph pollination, together with 1:1 ESS morph ratio. However, in highly fragmented populations, the morph ratio may be more likely to be biased by chance alone. In such situations, we address the following hypotheses: (1) the intra-morph mating is more frequent within majority-morph trees than within minority-morph trees, (2) the proportion of selfing is greater in majority-morph trees than in minority-morph trees, and (3) the average seed-set rate is greater for the minority morph trees based on pollen availability of the opposite morph. After confirming heterodichogamy by observing flowering behavior and phenology, we tested these hypotheses in a highly fragmented population of *Machilus thunbergii*, a broad-leaved evergreen laurel tree, using paternity analysis and direct observation. Our observations confirmed that *M. thunbergii* is heterodichogamous, consisting of two types of protogynous and bisexual flowers; a morning female (MF)–afternoon male morph and a morning male (MM)–afternoon female morph at the individual level. Sexual expression of the two morphs was highly synchronized and reciprocal. The estimated flower number was 2.02 times biased to MM in the study site. Paternity analysis revealed that heterodichogamy in *M. thunbergii* promoted outbreeding via inter-morph mating and restricted selfing. The probability of occurrence of intra-morph mating was 40% of that of inter-morph mating. We also confirmed following results (1) the intra-morph mating rate was significantly higher in mother trees of the majority-morph than in those of the minority-morph, (2) the selfing rate observed in the seed stage was not significantly different between the two sexual morphs, but the percentages of immature seeds was higher in mother trees of the majority-morph than those in the minority-morph, (3) the minority morph trees showed significantly higher seed-set rate than did the majority morph trees. These results indicate that (1) heterodichogamy in *M. thunbergii* efficiently promotes outcrossing and restricts selfing, (2) owing to the early-stage inbreeding depression which acts on the gamete to seed life stages, the majority morph mothers, suffering a shortage of the opposite morph pollen, could compensate partly the reduced reproductive success by intra-morph mating rather than by selfing, (3) negative-frequency dependent selection may be involved for maintenance of the two morphs.

T3

P0927

Identification of an ABC transporter involved in cuticular wax accumulation for xerophyte *Zygophyllum xanthoxylum* responding to abiotic stresses

Li Wei*, Jing Zhang, Pei Wang, Ye Tian, Qing Ma, Jin-Lin Zhang, Ai-Ke Bao, Suo-Min Wang*

State Key Laboratory of Grassland Agro-ecosystems, College of Pastoral Agriculture Science and Technology, Lanzhou University

Cuticular wax is a key component of the plant cuticle, which covers the outer surface of all the primary aerial organs of plants, and serves as a waterproof barrier that protects plants from abiotic stresses, such as salt, drought and heat. However, most researches have focused on glycophytes with less developed cuticular wax. *Zygophyllum xanthoxylum*, a succulent xerophyte growing in desert areas of northwest China and Inner Mongolia, possesses a thick cuticle in their leaves. Therefore, this xerophyte species is an ideal material to study the function of cuticular wax in stress adaptation. Several members of ATP-binding cassette (ABC) transporters have been proved to be involved in the export of cuticular lipid from the plasma membrane into the apoplast in glycophytes. In this study, we cloned and characterized *ZxABCG11* from *Z. xanthoxylum*. The results showed that the full length of *ZxABCG11* was 2,416 bp, including ORF 2,100 bp, which encoded 700 amino acids. The amino acid sequence contains ABC transporters characteristic motif, the nucleic acid binding domain, transmembrane domain composed of six hydrophobic alpha helixes and shared high similarity (73%-85%) with the identified ABCG11 of the other plant species. The quantitative real-time PCR analysis revealed that *ZxABCG11* was preferentially expressed in young leaves. And salt treatment (50 mM NaCl), osmotic (-0.5 MPa) and heat (35 °C) stresses significantly induced the expression of *ZxABCG11* in a short time period (0.5-6 h after treatments). The transcript abundance of *ZxABCG11* raised to its maximum amount under the dual stresses of osmotic and heat, which was 8 times compared to the control, indicating that ZxABCG11 possibly plays an important role in response to different stress. The SEM analysis showed wax crystal density increased by salt treatments (50 and 150 mM NaCl), osmotic and heat stresses in the middle layer leaves. Taken together, our results indicate that the expression levels of *ABCG11* are strongly associated with cuticular wax accumulation in *Z. xanthoxylum* under abiotic stresses.

T3

P0928

The U.S. National Vegetation Classification for dry coniferous forests and woodlands of the southern Appalachian Mountains: A reassessment

Thomas Wentworth[1]*, Raafiah Haroon*[2]*, Kimberly Israel*[3]*, Michael Lee*[4]*, Brooklynn Newberry*[1]*, Robert Peet*[4]*, Michael Schafale*[5]*, Alan Weakley*[4]

1. North Carolina State University

2. Perll Diagnostics, Inc.

3. North Carolina Division of Parks and Recreation

4. University of North Carolina at Chapel Hill

5. NC Department of Natural and Cultural Resources

Xeric to subxeric coniferous forests and woodlands of the southern Appalachian Mountains with a substantial component of yellow pines (*Pinus echinata, P. pungens, P. rigida*, and *P. virginiana*) have declined in area during recent decades because of fire suppression, drought, and outbreaks of southern pine beetle. The net result of these processes has been a shift in many yellow pine communities to dominance by drought-tolerant broadleaved trees and shrubs. The possibility of a change in the global conservation status of yellow pine communities from vulnerable to imperiled prompted us to reconsider their existing classification, given the important supporting role of classification in conservation efforts. Even prior to widespread decline of yellow pine communities, classification efforts were hampered by their dynamic nature and their tendency to intergrade with a variety of other communities. We have been exploring numerical classification of a dataset consisting of over 1,000 permanent plot records extracted from the Carolina Vegetation Survey database, selected using criteria of location in the southern Appalachian region and having at least 10% combined cover of the four yellow pines. After homogenizing species' taxonomic treatments and identifying outlier stands, we have used a combination of a priori assignments and quantitative methods (hierarchical cluster analysis and fuzzy clustering) to identify subsets of plot records corresponding best to the principal associations currently in use for these communities in the US National Vegetation Classification (USNVC). Our current focus is on the utility of a core group of 11 USNVC associations currently recognized and the potential need to modify the existing classification structure. Ongoing analyses include numerical classification, ordination, determination of diagnostic species, and the identification of environmental constraints and dynamic trends affecting the distribution of southern Appalachian yellow pine communities.

T3

P0929

To be on the safe site – ungroomed spots on the bee's body and their importance for pollination

Petra Wester*, Laura Koch, Klaus Lunau*

Institute of Sensory Ecology, Heinrich-Heine-University

Bees, by far the most important group of pollinators, visit flowers to collect nectar and pollen in large quantities. Pollen, on the one hand, serves as food provision in bees, and on the other hand, pollen is essential for the plants' reproduction, serving as a transport container for male genetic information. Pollen is expensive, and, different to nectar, emptied pollen resources cannot be refilled. Collected pollen that has been stored in the bees' transport organs is lost for the flowers' pollination. It has been hypothesized that specific body areas, that bees cannot groom, serve as 'safe sites' for pollen transfer between flowers. For the first time, we experimentally demonstrate the position, area and amount of pollen grains on safe sites at the examples of the Western Honeybee *Apis mellifera* and the Buff-tailed Bumblebee *Bombus terrestris*. For that purpose, we combined artificial contamination of the bees' body with pine or sunflower pollen and observations of the subsequent bees' grooming. We found localised safe sites on the forehead, the dorsal thorax and waist, and the dorsal and ventral abdomen. These areas were not or less groomed by the bees' legs. A particularly large amount of pollen was found on the waist, followed by the dorsal areas of thorax and abdomen. At the example of *Salvia pratensis*, *S. officinalis* and *Borago officinalis*, we experimentally demonstrated using fluorescent dye as a pollen substitute that the flowers' pollen-sacs and stigma contact identical distinct safe sites. This provides evidence that pollen deposition on safe sites is more likely to be transferred to intraspecific floral stigmas by pollen-collecting bees. Future research will demonstrate the significance of safe sites under field conditions.

T3

P0931

Structural equation modelling reveals winter sensitivity influencing species richness and community composition of mangrove forests across China

Yingtong Wu, *Zijian Huang, Shixiao Yu*

Sun Yat-sen University

Mangrove forests have been considered as "coast guard" and essential carbon sink. However, ecological conservation of mangrove ecosystems is challenged by global climate change. Thus, it is urgent to understand the effects of climatic conditions on both species richness and community composition in mangrove forests. In this research, we investigated the distribution pattern of true mangrove species across coastal China and explored the effects of abiotic environmental factors on species richness and community composition. Our study aims to address the following questions: (1) What is the spatial pattern of mangrove species richness and composition in China; (2) what and how environmental factors dictates its local species richness and community composition in mangrove ecosystems. To answer these questions, we collected occurrence records of true mangrove species in China from peer-reviewed articles. Species composition was indicated using non-metric multidimensional scaling (NMDS) and structural equation modelling was used to study the interactive effects of geological location, seasonal air temperature, precipitation, sea water temperature and salinity on species richness and community composition. Our results revealed the decrease of species richness with increasing latitudes and decreasing mean annual temperature. Structural equation modelling identified direct negative effect of latitude on species composition, as well as its indirect effects through winter air temperature, winter precipitation and winter sea water temperature. The model also indicated that both the richness and composition of mangrove forests were sensitive to winter environmental conditions, rather than summer ones. Our findings implies the potential shift of species richness and species composition in mangrove forests within the context of climate change. On the other hand, properly-managed reforestation of mangrove forests should be supported by species distribution modelling and future climate predictions.

T3

P0932

Phenotypic selection on flowering phenology and pollination efficiency trait among *Primula* populations

Yun Wu[1], *Qing-Jun Li*[2]

1. Xishuangbanna Tropical Botanical Garden, Chinese Academy of Sciences

2. Laboratory of Ecology and Evolutionary Biology, Yunnan University

Phenotypic selection on floral traits has largely been attributed to in plant-pollinator interactions. However, the strength of this link has rarely been ascertained with actual pollinators. We conducted pollinator observations and estimated selection through female fitness on flowering phenology and floral traits among two *Primula secundiflora* populations. We quantified pollinator-mediated selection by subtracting estimates of selection gradients of plants receiving supplemental hand pollination from plants receiving open pollination. There was directional selection for earlier flowering start at population where the dominant pollinators were syrphid flies, and flowering phenology was subjected to stabilize quadratic selection, too. However, later flowering start date was significant selected at population where the dominant pollinators were legitimate bumblebees and 'nectar robbers', and flowering phenology was subjected to disruptive quadratic selection. Wider tube entrance diameter was selected at both populations, but the strength of selection was stronger at population where the dominant pollinators were syrphid flies. Pollinator-mediated selection explained most of the among-population variations in the strength of directional selection on flowering phenology and tube entrance diameter. Our results suggested the important influence of pollinator-mediated selection on floral evolution. Variations in pollinator assemblages not only resulted in variation in the direction of selection, but also the strength of selection on floral traits.

T3

P0933

Biomass allocation ratios in temperate lianas: A test of the Darwin's hypothesis

Tomasz P. Wyka

Adam Mickiewicz University in Poznań, Faculty of Biology, ul. Umultowska 89,61-614 Poznań, Poland

Climbers are a diverse group of plants that are incapable of autonomuos vertical growth and therefore must rely on external support. For this reason they have been nicked "structural parasites". According to Darwin, the evolution of this growth form from self-supporting plants has been driven by the benefits of increasing biomass investment in leaves. Such shift in allocation was made possible by reduced investment of biomass in the construction of stems leading to lower cost of support. Although available studies indeed show higher leaf to stem biomass ratios in climbing vs. self-supporting plants, they often fail to account for the effect of plant size on biomass allocation patterns. Moreover, there is a shortage of studies that would account for biomass of the entire plant, including the root system. To test the Darwin's hypothesis on a whole-plant basis, we cultivated 5 species of temperate lianas and 5 species of shrubs belonging to the same families in large 120 L containers. Entire plants were harvested over three growth seasons and separated into principal organ fractions. Our analysis, strenghten by literature data, confirmed that climbers often achieve larger leaf mass fraction in comparison with self-supporting species. Combined with lower leaf-mass-per area (LMA; g m^{-2}) index of individual leaves in climbers, this leads to a greater ratio of plant leaf area to whole-plant biomass (leaf-area-ratio, LAR; m^{-2} g). Such biomass allocation pattern, together with the high foliar nitrogen content is conducive to high biomass productivity and fast growth, and facilitates the dominance of climbers in sufficiently productive habitats.

T3
P0934
Sequence polymorphisms at the pseudophosphatase *RDO5* underlie natural variation in dormancy caused by the *DOG18* QTL in *Arabidopsis*

Yong Xiang
Shenzhen agricultural Genomic Research Institute, CAAS

Seed dormancy controls the timing of germination, which regulates the adaptation of plants to their environment and influences agricultural production. The time of germination is under strong natural selection and shows variation within species due to local adaptation. The identification of genes underlying dormancy quantitative trait loci (QTL) is a major scientific challenge, which is relevant for agricultural and ecological goals. In this study, we describe the identification of the *DELAY OF GERMINATION 18* (*DOG18*) QTL, which was identified as a factor in natural variation for seed dormancy in *Arabidopsis thaliana*. *DOG18* encodes a member of the PP2C gene family, which we previously identified as the *REDUCED DORMANCY 5* (*RDO5*) gene. RDO5 alters the seed phosphoproteome but does not have detectable phosphatase activity, suggesting that it functions as a pseudophosphatase. *DOG18/RDO5* shows a relatively high frequency of loss-of-function alleles in natural accessions restricted to Western Europe. The loss of dormancy in these loss-of-function alleles can be compensated by genetic factors like *DOG1* and *DOG6*, and by environmental factors such as low temperature.

T3
P0935
Reconstructing the vegetation succession and climatic change in Beijing since Holocene

Gan Xie
State Key Laboratory of Systematic and Evolutionary Botany Institute of Botany, Chinese Academy of Sciences

Beijing, the ancient and modern capital of China, is a hot spot for studying the oriental human evolution and development of civilization. Undoubtedly, Holocene climatic fluctuations and vegetation successions played a key role in the survival and advancement of humans lived there. However, lack of quantitative climatic data inhibits us to understand the degree of progress on climate change and its ecological impacts in Beijing during Holocene. Here we have reviewed and reconstructed the vegetation succession and climate change, and illustrated ecology landscapes in plain and montane parts of Beijing during the Holocene by applying published pollen data. New climatic data portray that the mean annual temperature in Beijing decreased from the beginning of Holocene, then increased after reached its nadir at 8,000-5,500 cal yr B. P.; while mean annual precipitation increased from the earliest Holocene, and decreased after reached its peak at 9,500-8,000 cal yr B. P.. Quantitatively comparison with other units in China, MAP in Beijing, East Asian Monsoon Region, was 400-600 mm comparatively lower than Haligu and Xingyun Lake, Yunnan Provence, South Asian Monson Region, and was 200-400 mm higher than Tianchi Lake, Gansu Province, Gonghai Lake, Shanxi Provence and Daihai Lake, Inner Mongolia which falls in the Semi-arid Region in China. The increment of mean maximum monthly precipitation (about 25% higher in the current than in the late Holocene) indicated, the rainfalls in Beijing is strengthening in summer and suggested the Beijing government to prevent geological disasters in the montane region of Beijing in summer. These findings not only promote the comparing and understanding of the differences in climate changes between Beijing and other units in China and even in global since Holocene, but have potential for guiding the Beijing government to deal with the climatic and environmental changes in the future.

T3
P0936
Large- and fine-scale population structure of ectimycorrhizal fungal *Floccularia luteo-virens* in Qinghai-Tibet Plateau

Rui Xing
Northwest Institute of Plateau Biology, Chinese Academy of Sciences

Ectomycorrhizal fungi (ECM fungi) plays a key role in the terrestrial ecosystem. Currently, the population genetic structure study of ECM focuses on the number of genotypes and their spatial distribution. The underlying mechanisms that lead to the generation and maintenance of these genotypes remain unclear. The ECM fungal life cycle is generally composed of two processes: sexual reproduction and vegetative mycelial extension. The balance between those two processes is one of the most important factor that influence the distribution pattern of ECM. The most effective way to dissect the secret of ECM reproduction is to examine the genets. In this proposed grant, we developed new SSR markers for *Floccularia luteo-virens*, a wildly distributed ECM fungi on the meadow of Qinghai-Tibet Plateau to reveals the distribution pattern among large- and fine-scale. For large-scale study 404 individuals of *F. luteo-virens* from 23 geographically separate populations were analyzed using eight pairs of SSR markers. The average number of alleles among nine populations was 28.65 with a range from 20 to 38. The mean observed (H_O) and expected (H_E) heterozygosity ranged from 0.446 to 0.501 and from 0.499 to 0.554, respectively. An analysis of molecular variance (AMOVA) showed that most genetic variation occurred within populations (83.32%). The mean value of F_{ST} (0.176) suggested low genetic differentiation among populations. The 23 studied populations were divided into two main clusters based on the NJ dendrogram, which was also confirmed by STRUCTURE analysis. For fine-scale research more 200 individuals were collected in two years from three plots. The relative between above- and blow-ground genets, the reproductive characteristic, genets dynamics and gene flow of this species were invested. This research will shed light on the ecological function, reproductive strategy and conservation. Furthermore, the knowledge gained from this study can be useful for understanding the function and diversity of the ECMs under the grassland ecosystem.

T3
P0937
Lead tolerance mechanism in roots of *Typha orientalis* presl:

subcellular distribution, oxidative effects and nutrients

Xiaoying Xu, *Yikun Xu, Yingen Jin*
Yangzhou University

In order to investigate the phytoremediation potential of *Typha orientalis* to remove Pb^{2+} from contaminated water, the laboratory experiments were designed to evaluate the responses induced by Pb^{2+} (0.25-2 mM) in *T. orientalis* roots over a period of 10 days. The accumulation of Pb in subcellular fractions decreased in the order of cell wall > organelle > soluble fraction. Most of Pb in cells was located in cell wall and membrane system. The contents of K and Ca declined in *T. orientalis* roots under Pb^{2+} stress, but there was an opposite effect for some mineral elements (Mg, Cu, Zn and Fe). Meanwhile, the level of H_2O_2 increased in a concentration dependent manner, which induced oxidative stress. However, significant reduction in the levels of $O_2^{\cdot-}$ and MDA were observed for all Pb^{2+} concentrations. These results indicated that Pb^{2+} was toxic to *T. orientalis* in consideration of inducing oxidative stress and breaking the balance of mineral elements. But *T. orientalis* was also able to resist Pb^{2+} caused damage through the immobilization mechanism, which involved in the cell wall and membrane system.

T3

P0938

Ecological stoichiometry and habitat heterogeneity of different land use types in a subtropical karst trough valley of Guizhou Province, southwestern China

Xin Xu, *Yue Xu, Jian Ni*
State Key Laboratory of Environmental Geochemistry, Insitute of Geochemistry, Chinese Academy of Sciences

A 30 m × 200 m plot was established on both sides of the karst trough valley along an altitudinal gradient in Langxi small watershed, Yinjiang county, Guizhou province. Five land use types were chosen from the top to the bottom of the valley: the evergreen and deciduous broadleaved mixed forest, *Pinus massoniana* plantation, throne shrubland and tussock, orchard garden and rice crop field. Key characteristics of plant community, soil property, topography and micro-habitat were chosen to study the relationship between the spatial variability of soil nutrients and plant ecological stoichiometry by using the Kriging interpolation, principal component analysis (PCA), cluster analysis and canonical correspondence analysis (CCA). Results showed that the contents and spatial distribution of nutrients were different in different land use patterns and different types of micro-topography. The C, N and P contents of both plant and soil had significant differences among different vegetation types. With the degradation of vegetation, the pool and spatial distribution of soil nutrients changed correspondingly. High correlation existed between vegetation types, soil physicochemical properties and micro-topography, indicating that spatial distributions and variability of soil nutrients were mainly affected by topographic factors and habitat heterogeneities. In addition, the spatial distribution of soil nutrients has an important impact on plant ecological stoichiometry and tree species distribution.

T3

P0939

Early Pleistocene floral connection between Japan and Taiwan across the Ryukyu Islands—Plant fossils from the Katanoyama site in Tanegashima Island

Atsushi Yabe[1], *Toshiyuki Fujiki*[2], *Eunkyoung Jeong*[3]
1. Department of Geology and Paleontology, National Museum of Nature and Science
2. Faculty of Science, Okayama University of Science
3. Division of Biological Sciences, Chonbuk National University

The Ryukyu Islands, comprising approximately 190 islands, are a chain of islands stretching from southwest Kyushu to Taiwan along the Okinawa Trough to the north-west and the Ryukyu Trench to the south-east. These islands are known to be connected to Eurasian continents, Taiwan, as well as Japanese mainland before the rifting of the Okinawa Trough during the Pliocene–Pleistocene and have played important roles in the dispersal of land animals and plants. Based on molecular phylogeographic studies, extant faunas and floras of these islands are assumed to be created by a mixture of several different origins with different dispersal routes during different times. However, fossil records of these islands are still limited and do not contribute much to the discussion on the origin of their faunal and floral compositions. We investigated the Early Pleistocene plant fossil assemblage from the Katanoyama Member of the Masuda Formation in Tanegashima Island, which is located at the northern end of the archipelago, to understand the floristic connection between Japan and Taiwan across the Ryukyu Islands. The age of the plant-bearing deposits are assumed to be 1.3 Ma, a time period slightly after the opening of the Okinawa Trough. The Katanoyama Member has been well known to yield fishes (fresh water and marine), land mammals, reptiles, crabs, shells, as well as land and freshwater plants, which were transported to the coastal shallow marine environment. Reinvestigation of previously and newly collected materials revealed various plant mega- and microfossils (pollen and spores). A preliminary study of these materials revealed the presence of three different components in this assemblage: 1) those common to the modern day Tanegashima; 2) those common to mountain vegetation in Kyushu and Yakushima Islands; and 3) those common to mountain vegetation in Taiwan. The first component includes evergreen species belonging to the family Fagaceae (*Quercus* (*Cyclobalanopsis*) spp.) and Lauraceae (*Machilus thunbergii*), which together with deciduous oaks (*Quercus phillyraeoides*) and two-needle pine (*Pinus* cf. *thunbergii*), comprise the modern day coastal vegetation. The second component contains several conifer species such as *Abies firma*, *Tsuga sieboldii*, *Picea* sp. (Pinaceae), *Cryptomeria japonica*, *Thuja* sp. (Cupressaceae), and *Torreya nucifera* (Taxaceae), which are not present in this island nowadays. Most of these species or genera occur in cooler temperate forests in the mainland of Japan and their southern distribution limit is located at higher altitudes (over 1,000 m) of Yakushima Island. The third component includes *Fagus* cf. *hayatae*, *Liquidambar* cf. *formosana*, *Taiwania* sp. and *Cunninghamia konishii*. Some of these are known as relict species from the Neogene period in Japanese islands. The co-occurrence of the first and second components implies a similar or slightly cooler climate conditions than today; moreover, the presence of a mountain area with high altitude behind the basin, which is completely absent today (current highest point, ca. 280 m a.s.l.), is suggested. The third component can be derived either from the south or the north. More detailed systematic studies are needed to understand their origin.

T3

P0940

Current re-vegetation patterns and restoration issues of degraded geological phosphorus-rich mountain: A synthetic analysis of Central Yunnan, SW China

Kai Yan, *Jianchu Xu*

Kunming Institute of Botany, Chinese Academy of Sciences

China has the largest inland geological phosphorus-rich (GPR) mountain, where vegetation restorationis significant for China's environmental safety.We reviewed the published literature and collected raw data to analyze the re-vegetation patterns and plant communities status in central Yunnan, where have the most studies around China. The aim of the study is to suggest future improvements in restoration in GPR mountain. The results showed that spontaneous recovery is the majorrestoration type. N-fixation species, e.g., *Coriaria nepalensis*, *Alnus nepalensis*, play a vital role in the succession. Monoculture tree plantation was the primary mode of afforestation practices in the past. In the recent years, several methods of restoration have experimented including agroforestry system. For practical restoration, the spontaneous recovery could have the best results. However, restoration results are affected by high erosion risk in the early stage, and the origin of propagates and environmental variation. In contrast, human-made communities were better in the early stage, but higher cost and vulnerable to degradation and erosion were constraints. Moreover, N-fixation species e.g., *Alnus nepalensis*, *Acacia mearnsii*, used for plantations reported less success. Thus there are fundamental restoration issues as follows. (1) Understanding merit and demerit of current natural and human engineered restoration, and their respective advantages. (2) The feedback mechanism between phosphorus-rich soil and species selected for restoration, especially N-fixating species. (3) The introduction of market incentives to vegetation restoration, e.g. specific agroforestry model, and improving the industry value chain.

T3

P0941

The paleovegetation and paleonvironment reconstruction during the quaternary of the Khorat Plateau based on the palynological records

Feng-Chun Yang[1,2,3], *Paul Grote*[1,3]

1. Suranaree University of Technology

2. Honghe University

3. Northeastern Research Institute of Petrified Wood and Mineral Resources Nakhon Ratchasima Rajabhat University

The Khorat Plateau is located at the center of mainland Southeast Asia. It is a relatively high altitude tableland. The local flora is closely related to the Indo-Himalayan-Chinese region in the north and the Malesian region in the south. This plateau is an important bridge connecting the biotas of the two large regions. However, it is still unclear how the Khorat Plateau has been affected by global changes including climate change. An exposed sand pit along the ancient fluvial terrace at the village Ban Som, Tha Chang Subdistrict, Chaloem Phra Kiat District, Nakhon Ratchasima Province, Thailand was selected for palynological research. A 14 m depth exposed section was analysed based on the stratigraphy and sedimentology. According to the results, three different stratigraphic groups were classified from the bottom to the upper layers, shallow lake and flood basin, developed river channel, and dry land with and an abandoned river terrace. Abundant pollen grains were discovered from the sediments, from which approximately 100 fossil palynomorphs were recognized and classified. Most of the pollen grains were concentrated in the channel river stage. The main paleovegetation varied from forest to the savannah. Mangrove components such as *Acrostichum, Barringtonia, Ceratopteris, Nypa, Oncosperma*, and Rhizophoraceae were recognized in a continuous deposit in layers 11, 10B, and 10A. These three layers contained pollen indicating deciduous and mixed forest in the ages from 52,296 ± 6,800 yr BP to 28,150 ± 7,860 yr BP. A period with a relatively high sea level was suggested. Temperate and subtropical savannahs were prevalent in cooler periods 16A (172,739 ± 22,400 yr BP), 13B (88,661 ± 10,600 yr BP), and 9B (27,332 ± 3,000 yr BP). These three periods corresponded well with cold events occurring elsewhere. The cold winter monsoon had influenced the Khorat Plateau even though it was far from the ice sheets presented during glacial periods. The pollen grains of confiners were not regarded as local deposits, but were probably transported from the north by wind.

T3

P0942

The special floral traits and pollination adaptation of psammophytes ephemeral plant *Hypecoum erectum* L.

Shanlin Yang, *Xiang Shi, Jiashu Liu, Fanxiang Meng*

Shihezi University

In order to donate pollen to recipient stigmas efficiently, animal-pollinated plants have evolved a variety of floral characteristics. To adapt to the pollination is the fundamental power to floral traits. The evolution of floral characteristics and pollination adaptation is always one of the core problems in the field of evolutionary ecology. The study was carried out in the natural population of *H.erectum* found on the edge of the Gurbantunggut Desert. The floral traits and pollination adaptation were investigated both in the field and laboratory. The results show that: According to the external petal opening situation, the typical single flowering process for a flower can be divided into four stage, we observed the four stage of single flower including external petal color changes, the relative position of the median lobe and anthers, the pollen quantity changes in median lobe, the relative position of the stigma and anthers, the relative position of the median lobe and stigma, stigma dehiscence, and the situation of anther dehiscence. *H.erectum* has elaborate floral characteristics. Two outer petals white or yellow with obovate and 3 low crack, and the two inner petals were composed of lateral lobe and middle lobe. The middle lobe is specialized as a yellow cucullate structure. The middle lobe will reflexively wrap the anther as soon as the anther dehiscence, obtaining pollen and growing upward. The cucullate structure is the key functional organ in attracting pollinators secondary pollen presentation. "Intrafloral stamen differentiations" is a kind of significant differentiation phenomenon that between stamens and stamens. There are intrafloral stamen differentiations exist in *H.erectum*. The pollen viability of the two lateral stamen is more higher than the two middle stamen in all the flowering stage. And the seeds setting of cut the middle anther and bagged is significant higher than the operation on the lateral anther. The number of the pollen in the middle stamen is higher than lateral pollen. *Colletes vestitus* is the only effecity pollinators of *H.erectum*, the visiting frequency showed double-climax: 12: 00-13: 00 and 19: 00-20: 00. To analyze the breeding system type and the mating system under the desert habitat through artificial pollination experiment to explore the adaptation and the evolution significance. The duration of stigma receptivity about 48 h, the stigma always have receptivity along the flower phase. The breeding system type of *H.erectum* is mixed mating system. A part of floral shows self-compatible and another shows outcross. The pollen of geitonogamy is contribute to their reproductive success. Under the selection pressure for a long time, the *H.erectum* tend to mixed mating system, could be a response to adapt the adverse the environment conditions in desert.

T3

P0943

Meet the balance of forest conservation and land use productivity - Case study from nature reserve in China

Xueqing Yang[1,2,3], *Hongxi Liu*[1], *Sergey Blagodatsky*[1], *Carsten Marohn*[1], *Reza Golbon*[1], *Jianchu Xu*[2,3], *Georg Cadisch*[1]

1. University of Hohenheim, Germany

2. Kunming Institute of Botany, Chinese Academy of Sciences

3. World Agroforestry Centre (ICRAF)

In Mainland of South East Asia, land use change from natural forests to intensified economic plantations decreased biodiversity, led to habitat loss and increased carbon emission. In Southwestern China, the dramatic expansion of rubber monoculture plantations has been blamed for high rates of deforestation and forest degradation, and even for threatening food security if converted from swidden land. Against global warming background, some studies analyzed the possible future distribution of rubber plantations, while information about gradual rubber trees' physiological response to climate change is still unclear. In order to better understand the dynamic process of biomass partitioning and latex production in rubber system, and various response of lowland rubber (located in suitable area) and highland rubber (located in marginal area) to future climate change, we combined empirical and process-based approaches by applying the Land Use Change Impact Assessment (LUCIA) model. Parameterization and validation of plant growth parameters in the model was completed using field dataset from destructively sampled thirty rubber trees, and non-destructively measured 882 rubber trees sampled from 27 plots in langdscape of Naban Watershed National Nature Reserve (NRWNNR), Xishuangbanna. The monthly measured leaf area index (LAI) and litterfall also used to calibrate model. In addition, for nature reserve carbon dynamic evaluation from 1989 to 2012, the RaCSA method (Rapid Carbon Stock Appraisal) was applied based on tree, plot, land use and landscape level. According to model simulation, lowland rubber plantation produced more total biomass (60 Mg ha^{-1}) and latex yield (18.3 Mg ha^{-1}) within 35-year rotation length if compared with highland rubber plantation. Lowland rubber plantations' total biomass and cumulative latex yield decreased by 3.6% and 5.7%, respectively, if current CO_2 emission scenario is substituted by high CO_2 emission scenario (RCP 8.5, 2050). Highland rubber plantations, in opposite, had an increase of 23% and 27% accordingly due to climate change. Based on instinct nature of tree development, modeling results could guide future cultivation area of various rubber trees' genotypes, and can be integrated with social-economic information to identify tradeoffs between optimization of tree growth and maximization of economic benefit under climate change. Time series analysis of land-use and land-cover maps (1989, 2007, 2012) demonstrated that during 23 years, the whole landscape of the nature reserve (26,574 ha) gained 0.644 Tg C. The effective functional zoning and reforestation activities serve as practical strategies to cope with wide expansion of rubber plantations, most importantly, to balance forest conservation and land use productivity in landscape perspective.

T3

P0944

Size-mediated interaction between cushion species and other plant species at high elevations of the Hengduan Mountains, SW China

Yang Yang, *Jian-Guo Chen, Hang Sun*

Kunming Institute of Botany, Chinese Academy of Sciences

Plant size or stage is an indicator of a wide range of ecological and evolutionary processes including responses to neighbours. Plants experience distinct differences in tolerance to extrinsic factors at different stages and therefore has evolved different physiological status. This then manifests itself as variations in morphological traits of plant, which in turn mediates plant-plant interactions by altering microhabitat conditions for other species. Within alpine habitat, species that have cushion morphology often acts as a 'nurse' and provide facilitation to other non-cushion species, size of which is easy to be measured. However, little is known about the mechanism link of size-dependent morphological and physiological variability of cushion species and the variation in balance between competition and facilitation so far. In this study, we assessed a series of functional traits in different size (i.e. 10 cm, 15 cm, 20 cm and 30 cm in diameter) of *Arenaria polytrichoides*, a common cushion plant occurs at high elevations of Himalayan Hengduan Mountains, SW China. Furthermore, we analysed how these size-dependent functional traits related to soil properties beneath cushions and the number of associated species and individuals compared with open areas. Large cushions showed good physiological; they were compact, had higher organic matter content and soil nutrient contents (N and P) compared with open areas and showed facilitation effects on other non-cushion species. By contrast, physiological data indicated stressful abiotic conditions for small size cushions of *A. polytrichoides* in the study site. These cushions showed less improved soil conditions and had reduced facilitation as compared with those larger ones. Our data, thus, offer a clear demonstration of a size-dependent mediates interactions occur among cushion species and other plant species, because the distinctive variations in morphology and vigour of this nurse plant at different size. This variability may affect the small-scale environmental created by cushion plant and finally have a relevant effect on its facilitative effects on other species.

T3

P0945

Mangrove aerial plants and an update of the IUCN critically endangered and data-deficient mangroves

Jean Wan Hong Yong[1], *Wei San Wong*[2], *Jin Eong Ong*[3], *Chiou-Rong Sheue*[4]

1. School of Plant Biology, University of Western Australia, Department of Environment and Agriculture, Curtin University, Perth, Australia

2. Nanyang Environment and Water Research Institute, Nanyang Technological University, Singapore

3. School of Biological Sciences, Universiti Sains Malaysia, Penang, Malaysia

4. Department of Life Sciences & Center for Global Change Biology, National Chung Hsing University, Taichung, Taiwan

Little attention has been given to the aerial plants (epiphytes, mistletoes and climbers) in Old and New World mangrove forests globally. Epiphytes, mistletoes and climbers generally have a wide range of hosts, including both mangrove and non-mangrove host trees. In the mangroves, by using other host plants to perch high above the ground, mangrove epiphytes avoid the salt water sprays and attain favourable light conditions for growth in the dense canopy. Ferns, orchids and rubiaceous ant-plants constitute the majority of mangrove epiphytic plant life forms. Unlike the parasitic plants, epiphytes do not acquire water and mineral nutrients from the hosts. Hence, only the mistletoes appear to be more selective in colonizing and parasitize certain mangrove trees and shrubs. Conversely, epiphytes, and also climbers, are less 'picky' with their phorophyte host and several species of epiphytes and climbers can colonize a single host tree. Unlike epiphytes and mistletoes, climbers are plants that twine and curl round other plants, mainly trees while remaining rooted to the ground. Will the demise of mangrove aerial plants – both from the main and back mangroves, affect the pollination biology and dispersal of certain mangrove taxa? What will be the fate of aerial plants in the mangrove forests in view of increasing anthropogenic disturbances and climate change? An on-going effort is now in progress to compile the extant and extinct mangrove-linked aerial plants from past herbarium records, literature reviews and contemporary field surveys around Asia. Some salient observations of our current research is discussed using new data collected from the old (e.g., Myanmar, Malaysia, Indonesia, Singapore) and New World (Costa Rica, Panama) mangrove forests. of special concern are the two mangrove species that are listed as Critically Endangered (CR), the highest probability of extinction measured by the IUCN Red List. The rare *Sonneratia griffithii* (Perepat daun lebar) is distributed in parts of India and southeast Asia, where the loss of many mangrove areas have occurred within its patchy range over the past 60 years, with significant losses in Myanmar, Thailand and Malaysia, primarily due to agriculture, aquaculture and coastal development. New updates about *S. griffithii* and *Bruguiera hainesii* (Berus mata buaya) in Asia and especially from Myanmar and Malaysia will be discussed. Several rare Data Deficient (DD) *Brownlowia argentata* (Tiliaceae) trees was recently discovered at Sungai Kerian (along the Kedah-Penang state borders, Malaysia) and these new observations (field and herbarium records) suggested that this rare species has an unusual and specific preference for a certain mangrove habitat type: periodic and alternate freshwater and seawater inundation, and with certain periods of "drying out". Sungai Kerian is a river with suitable geomorphic features that provided these set of hydrological conditions that supported the continual occurrence of *B. argentata*. Perhaps, these specific environmental and especially hydrological requirements may explain in part the rarity of *Brownlowia argentata* throughout its biogeographical range.

T3

P0946

Breaking the rules of allometry allows host tracking in a plant insect community

Hui Yu[2], *Finn Kjellberg*[1]

1. South China Botanical Garden, the Chinese Academy of sciences

2. CEFE, CNRS - Université de Montpellier

While the communities constituted by phytophageous insects and their parasites and parasitoids may represent half of all terrestrial animal species, understanding how they have diversified remains a major challenge. Phylogeographic studies show numerous instances when parasites have failed to track their hosts when they colonised new habitats and also cases of delayed host tracking, but provide no mechanistic explanation of the observed patterns. Phytophages may escape from their parasites following variation in host plant morphology, by colonising new parts of plants or by producing differently shaped galls. In this study we investigate how parasitoids of a *Ficus*-pollinating wasp have responded to reduced accessibility of the larvae they feed on in part of the figs distribution. We show that an increase in ovipositor length from 1.7 to 2.8 times body size has allowed a single population of a species of parasitoid to invade the geographic area in which pollinators previously escaped parasitism. The shift in ovipositor length results from the evolution within a population of new allometric rules. Such strong modification of allometry has never been reported within any species but is frequent among species. It is a first step that could lead to ecological speciation following rapid morphological evolution. The results also evidence a period of latency before one species of parasitoid managed to track its host while another species has failed to evolve longer ovipositors. Our results support the idea that small incremental mutations are sometimes not sufficient to bridge adaptive valleys. Populations may have to wait for the chance appearance of rare large effect mutations to diversify by colonising a new dimension of their potential habitat (ref mimicry circles). We show that, in multitrophic interactions, phenotypic variation at one trophic level may often directly affect other trophic levels by generating new ecological opportunities, thus triggering diversification.

T3

P0947

Transcriptome analysis of mangrove plants *Bruguiera gymnorrhiza* under drought stress

***Tianhui Yu**, Huaitong Si, Kunfang Cao, Alison K.S. Wee*
Guangxi university

The mangrove species, *Bruguiera gymnorrhiza*, is widely distributed throughout the Indo-West Pacific and is the dominant species of mangrove in China. Mangroves are a taxonomically diverse group of plants capable of thriving within tropical and subtropical intertidal zone, an area characterized by regular inundation by salt water and high temperatures due to sun exposure. To cope with the frequent physiological drought stress that occurs as a result of regular exposure to fluctuating salinity and high temperatures, mangrove plants have evolved unique physiological and morphological adaptations to survive. Abscisic acid (ABA) is a major response factor in the regulatory response to drought stress, through the initiation of leaf stomatal closing and other processes related to water conservation and retention. In this study, we performed transcriptomic analysis on *B. gymnorhiza* under drought stress, in conjunction with measurements of leaf ABA content. We subjected six one-year-old *B. gymnorhiza* seedlings to physiological drought stress for 14 days. We measured photosynthetic rate throughout the drought treatment and collected leaf samples for ABA content measurements and high-throughput RNA sequencing (RNA-seq). We performed RNA-Seq on Ilumina Hi-Seq platform, followed by transcriptome assembly, differential expression analysis and functional annotation, to identify the differentially expressed genes during drought stress. We identified 33,254 unique sequences annotated to 66 GO terms that were differentially expressed, including 466 up-regulated genes and 971 down regulated genes. The up-regulated genes were mostly functionally related to cytoplasm, response to pressure and abiotic stimulation. Our results showed that drought stress enhanced the expression of antioxidant genes and stress response in *B. gymnorhiza*. Following this, we tested candidate genes related to the ABA response pathway for expression validation. By understanding the drought stress regulation in mangroves, this study sheds light on the physiological adaptation of terrestrial plants to a coastal setting.

T3

P0948

Multi-stage marginal impact of meteorological factors on rice productivity in China

***Wen Yu**[1], Shiwei Xu[1], Jiajia Liu[1], Jiayu Zhuang[1], Yu Wang[1], Yongen Zhang[1], Muhammad Luqman[2], Liwei Gao[1]*
1. Agricultural Information Institute, Chinese Academy of Agricultural Sciences, Beijing
2. University College of Agriculture, University of Sargodha

Climate changing have an important impact on crop yield in the world. Many researches have mentioned the possible impact of climate change on crop production using simulation models. In this research we use China provincial panel dataset for different season rice during 1979-2014, to estimate the meteorological impact on Chinese rice yield. We find that higher temperature in north is better for the yield, and in the filling stage, but it is not good for the planting stage; the sunshine duration and precipitation also have significant impact to the rice yield in different rice growth stage. In the yield-climate function, we also consider that material input including fertilizer, labor, so as to have an accurate estimation of meteorological impact on rice yields.

T3

P0949

A meta-analysis of experimental warming effects on woody plant growth and photosynthesis in forests

Yongge Yuan
Shanghai Science and Technology Museum

Increasing field manipulation experiments have been conducted in forest to better understand the responses of woody plant growth and photosynthesis to climatic warming. However, it is still less known whether there is a general pattern in relation to how and to what extent warming impacts woody plants in forest ecosystems. Here, a meta-analysis was conducted with 522 observations from 52 studies in forests to investigate the direction and magnitude of warming effect on the traits of woody plant growth and photosynthesis. When the temperature increase was between 0.33 and 10°C, warming significantly increased specific leaf area (SLA) by 5.94%, plant height by 7.75%, plant biomass by 21.88%, foliar calcium (Ca) and manganese (Mn) concentrations by 20.67% and 39.62% and net photosynthetic rate (Pn) by 9.92%. Enhanced growth and Pn may have a relationship with the changing SLA, efficiency of PSII, photosynthetic pigment concentrations and foliar nutrient. Our results showed that warming was beneficial to woody plants, and the positive response magnitude to warming on forests was higher than the global average (12.3% for growth, 4.5% for Pn). Our findings will be useful to understand the underlying mechanisms of forests responding to global warming.

T3

P0950

Landscape plant selection for green infrastructure (GI) response to climate change

Deshun Zhang
Tongji University

Green infrastructure is established by utilizing natural condition and natural law with the least changes to the nature. Urban GI which refers to an area covered by greenery in the city is a real flow of biological system. According to the dimension, the urban GI can be classified into three categories: place scale, built-up area scale and city area scale. The benefits of the urban GI are conducted through ecosystem service function including consensus establishments, exploitation of social public service of multiple values and assessment strategies building. In order to build urban GI with high ecology elasticity, the investigation and determination of stress-resistant physiological indexes of flood resistance, wind resistance and salt resistance of landscape plants are performed in Shanghai urban green space system as case study. Optimization matrix of future tree species selection was constructed based on plant resistance and scenarios of climatic change. Additionally, master planning strategies of thematic gardens for plant response climate change are introduced in several new Chinese botanical gardens.

T3

P0951

Different foliar uptake of unsaturated water vapor under humidification conditions of desert plant *Tamarix ramosissima*

in different months

Huiwen Zhang[1], *Honglang Xiao*[2]
1. State Key Laboratory Breeding Base of Desertification and Aeolian Sand Disaster Combating, Gansu Desert Control Research Institute, Lanzhou
2. Key Laboratory of Ecohydrology of Inland River Basin, Cold and Arid Regions Environmental and Engineering Research Institute, Chinese Academy of Sciences, Lanzhou

Plant can absorb moisture from the atmosphere water vapor through the leaves. This process is called "reverse transpiration". It is difficult to form the dew and fog in extremely arid desert area because that wind is strong near the ground. But the atmospheric water vapor content increases a higher degree, or even close to saturation near the ground when temperature drops in the evening and morning. The unsaturated atmospheric water vapor is a special and precious source for desert plant when it is difficult to use precipitation and occult precipitation. Previous researches about foliar uptake of water have been reported for numerous species and almost focused on using saturated atmospheric water vapor (liquid water), but the researches about desert plants absorbing and utilizing unsaturated atmospheric water vapor (vapor water) have rarely not been reported, especially about differences in water vapor uptake by leaves in different months. This research taked *Tamarix ramosissima*, the key species of desert ecological system, as the research object and carried out humidification control field experiment with unsaturated ultrapure water vapor in different months. Stable isotope mixing models were used to quantify source contribution to a mixture of stem water and thus quantify the water vapor absorbed by leaves. We found that the amount of water vapor absorbed by leaves was different in different months. In the humidifying experiment from June to August, absorption of the ultrapure water in stem water accounted for the largest proportion of more than 50%, followed by soil water between 100 cm to 200 cm. Humidification of ultrapure water vapor was the main source of stem water, and might be involved in important physiological processes such as photosynthesis. In the humidifying experiment in September, absorption of the ultrapure water in stem water accounted for only around 20%, soil water between 50 cm to 80 cm was the main water source of stem water, followed by soil water between 100 cm to 200 cm. Humidification of ultrapure water vapor only eased the water shortage degree of plant, and could not be used as the main source of stem water. This research certified the phenomenon that the plant can absorb unsaturated water vapor through leaves. And humidifying experiments of three years at different months showed that changes of the climate conditions, water shortage degree, physiological structure and photosynthetic capacity of plant, soil profile water content in different months would have great impact on plant water use pattems. This research was supported by the National Natural Science Foundation of China (No. Y411381001) and the Major Research plan of National Natural Science Foundation of China (No. 91125025).

T3

P0952

Hydrologically contrasting environments induce genetic but not phenotypic differentiation in *Solanum dulcamara*

Qian Zhang[1,2], *Janny Peters*[2], *Eric Visser*[2], *Hans De Kroon*[2], *Heidrun Huber*[2]
1. Taizhou University
2. Radboud University Nijmegen

In contrasting habitats, locally adapted populations are expected to evolve through directional selection. Hydrological gradients provide a scenario where strong selection forces have led to species segregation in communities along the hydrological niche axes due to specific adaptations to their ambient environment. However, much less is known about the processes enabling species with wide ecological amplitudes, such as *Solanum dulcamara* L., to occupy the contrasting niches at the far ends of the gradient. Here, we investigate the question whether local adaptation is indeed the driving mechanism enabling this species to inhabit contrasting wet and dry habitats. Using molecular analyses and performing a common garden experiment, we examined whether genetic and phenotypic differentiation had occurred between populations originating from flooding- and drought-prone habitat pairs from nine different locations. We also tested whether plants were better adapted to their home site stress. We found a significant genetic variation but hardly any phenotypic differentiation between populations from contrasting habitats for plants grown under optimal conditions. Surprisingly, comparison of the genetic differentiation in neutral markers with that in phenotypic traits using the QST-FST (estimated by ΦPT) approach revealed that the QST values were consistently and significantly smaller than the ΦPT values. This is in contrast to the hypothesis that directional selection is an important component enabling *S. dulcamara* to survive under contrasting habitat conditions. In addition, no home site effects were detected for responses to flooding and drought treatments. All plants were characterized by high levels of plasticity for traits associated with flooding and drought tolerance, such as rapid adventitious root formation upon flooding and increased root to shoot ratio in response to drought, irrespective of habitat of origin. Moreover, no trade-offs in such flooding and drought responses were found. Our results show that *S. dulcamara* has not evolved locally adapted populations in response to flooding and drought stress, despite genetic differentiation and despite the presumably strong selection gradient. The generally high levels of adaptive plasticity in traits increasing flooding and drought tolerance may be the main mechanism allowing *S. dulcamara* to occupy hydrologically contrasting habitats.

T3

P0953

Response of physiological ecology characteristics to simulated climate change factors

Wei Zhang, *Li Zhuang*
Shihezi University

In recent years, with the greenhouse effect intensifying, global atmospheric temperature rising and nitrogen deposition and precipitation pattern changing, the relationship between global climate change and terrestrial ecosystems has become the important content and hot spot of global change research at home and abroad. As an important part of terrestrial ecosystem, plant photosynthesis is one of the most important biochemical reaction on earth and is the basis of terrestrial ecosystems of material circulation and energy flow, which is vulnerable to development state of plants and changing exogenous environmental influences. Many recent studies demonstrated that atmospheric temperature rise both have promoted effect and inhibitory effect on photosynthetic physiological characteristics of plant, even no effect; Atmospheric nitrogen deposition has led to soil acidification, which will be helpful to the plants growth in a short-term, however, which will result in the decrease of plant species richness and ecosystem function in a long-term. In order to maintain their physiological ecological balance and adapt to global climate change, plants also have produced a series of feedbacks: they are mainly reflected in the instruction role of plant physiological traits (such as changes in stomatal regulation, photosynthetic structure adjustment and photosynthetic pathway and changes in plant photosynthesis, respiration rate, osmotic substance content or protective enzyme activity, *etc.*) and phenological characteristics to global climate changes. However, most researches on plant photosynthetic physiological characteristics under climate change are still limited in single or double factors coupling effect of raising temperature, add nitrogen and increasing precipitation. The purpose of this study is to discuss the photosynthetic physiological mechanism and adaptive strategy under the multi-factor coupling conditions of simulated raising temperature, add nitrogen and increasing precipitation, in order to provide more scientific and reasonable theoretical basis for further study on plants growth and distribution under the future global climate.

T3
P0954

Impacts of alien black locust on species diversity: A comparison of native and alien plant communities in Shandong Province, China.

Wenxin Zhang*, Qiang Li, Xiao Liu, Ning Du, Renqing Wang*
Institute of Ecology and Biodiversity, Shandong University

The invasion of alien species is the main biological threats to biodiversity all over the world. Black locust (*Robinia pseudoacacia* L.) is a widespread alien species commonly thought to influence natural ecosystems. However, the effects of black locust on the species diversity of plant communities are still controversial. The main aim of this study was to compare the species richness, species composition, phylogenetic diversity and phylogenetic structure between black locust communities and native communities dominated by *Quercus acutissima* to study the possible impacts of black locust on native ecosystems. 17 black locust and 17 paired native communities were selected and all plant species were surveyed in the 20 m × 30 m plots in mountainous area of Shandong Province, China. In this study, there was no significant difference in the species richness between black locust communities and native communities. However, we found that alien plants in black locust communities are more diverse than in native communities. Black locust may increases soil nitrogen stores enhancing the establishment of alien plants. Black locust communities have relatively low level of phylogenetic diversity. We also found that plant assemblages tended to become more phylogenetically clustered in black locust communities. This result indicates that environmental filtering is the main driver of plant community assembly in black locust stands. Black locust could influence community structure, soil properties and nutrient cycling, which makes the habitats become more homogenous. Plant species surviving in black locust communities are from few lineages that have evolved the ability to tolerance these habitats.

T3
P0955

Ammonia emissions are substantially underestimated in China

Xiuming Zhang[1]*, Yiyun Wu*[1]*, Baojing Gu*[1]*, Xuejun Liu*[2]*, Stefan Reis*[3]*, Jiaxin Jin*[4]*, Ulli Dragosits*[5]*, Martin Van Damme*[6]*, Lieven Clarisse*[6]*, Simon Whitburn*[6]*, Pierre François Coheur*[6]
1. Zhejiang University
2. China Agricultural University
3. NERC Centre for Ecology & Hydrology
4. Nanjing University
5. University of Exeter Medical School
6. Université libre de Bruxelles (ULB)

China is a global hotspot of atmospheric ammonia (NH_3) emissions and, as a consequence, very high nitrogen (N) deposition levels are documented in particular in areas of intensive agricultural production. However, previous estimates of total NH_3 emissions in China were much lower than inference from observed deposition values would suggest, highlighting the need for further investigation. We have reevaluated NH_3 emissions based on a mass balance approach, validated by N deposition monitoring and satellite observations, for China for the period of 2000 to 2015. Our results show that total NH_3 emissions in China increased from 12.1 Tg N yr^{-1} in 2000 to 15.5 Tg N yr^{-1} in 2015 at an annual rate of 1.9%, which is approximately 40% higher than existing studies suggested. This difference is due to more emission sources now having been included and NH_3 emission rates from mineral fertilizer application and livestock having been underestimated previously. Our estimated NH_3 emission levels are consistent with the measured deposition of NH_x (including NH_4^+ and NH_3) on land (11-14 Tg N yr^{-1}) and the substantial increases in NH_3 concentrations observed by satellite measurements over China. These findings provide a new perspective and substantially improve our understanding on NH_3 emissions in China, implying that future air pollution control strategies have to consider the challenges and potentials of reducing agricultural NH_3 emission to improve air quality in China overall.

T3
P0956

Genome-wide transcriptome analysis revealed photosynthetic reaction responses to light variations in *Solidago canadensis*

Yang Zhang*, Yongge Yuan, Xiaojuan Wang, Liang Jin*
Shanghai Natural History Museum, Branch of Shanghai Science & Technology Museum

Solidago canadensis, the invasive plant in Central and East China seriously, was little known about its transcriptome of photosynthesis. In this research, we compared the transcriptome data of native species (*Solidago decurrens*) and invasive species (*S.canadensis*) with different illumination (shading 0%, 40%, 60% and 80%), to revealing the cause of the phenomenon of invasion from the molecule perspective. With the increase of shading intensity, the photosynthetic indexes (Fv/Fm, photosynthetic rate, α value and biomass) appeared a decreasing trend, both in the native species and invasive species. However, the light adaptation of invasive species was stronger than native species, except the one under the shading 80%. Coincidentally, the RNASeq results were similar to the photosynthetic data. In the process of photosystem I and within the same species, the genes which closely related to electron transfer (*PsaA*, *PsaD*, *PsaL*, *et al.*,) was negative regulation. In the photosystem II, LHCII gene family also had obvious decline. In addition, the expression level of partial LHCII gene family of invasive species was higher than local species' based on RT-PCR. We thought that the response to light might be an obvious strength of invasive species.

T3

P0957

Curbing desertification development referenced to the determinants for plant distribution pattern

Zhi-Feng Xiong[1], *Ming-Hui Han*[1], *Lin Li*[1], ***Yi Zhang***[1], *Shuang-Xing Guo*[2]

1. College of paleontology, Shenyang Normal University;Key Laboratory for Evolution of Past Life in Northeast Asia, MLRC

2. Department of Palaeobotany and Palynology, Nanjing Institute of Geology and Palaeontology, Chinese Academy of Sciences

Traditionally, the determinants for plant distribution pattern are regarded as climate factors, especially temperature and humidity. But recently, based on the comparison between the distribution patterns of *Metasequoia* and *Populus*, we found that the distribution patterns of different plants have different determinants. *Metasequoia*, originated from the Late Cretaceous warm temperate zone in the Northern Hemisphere, was widely distributed in the Northern Hemisphere during the Early Cenozoic. But along with the global cooling and catastrophic climate changes during the Middle and Late Cenozoic, *Metasequoia* was forced to migrate southward. Due to migration routes in North America and middle Eurasia curbed by arid zones, only the route in East Asia was suitable for the *Metasequoia* migration. In Pleistocene, *Metasequoia* only survived in South China and Japan under the control of warm temperate climate. Extant monotypic *Metasequoia*, is only distributed in the area bordered by Chongqing, Hubei and Hunan. On the contrary, *Populus*, originated from the Eocene warm temperate zone in the Northern Hemisphere, experienced the similar changes during the Middle and Late Cenozoic, but today it is widely distributed in the warm and cool temperate zones, even colonized in cold and tropical zones, including 92 species. In this species, *Populus euphratica* is adapted to survive in the subtropical and temperate deserts. *Metasequoia* and *Populus*, both of them were originated from the similar warm temperate zones of the Northern Hemisphere, and experienced the same catastrophic climate changes, but their today's diversities and sizes of the distribution area are completely different, indicating their abilities in heritable variation very different. Plants, such as *Populus*, with very higher abilities in heritable variation to adapt to the environmental changes, are able to spread to the different climate zones; vice versa, such as *Metasequoia*, will be gradually eliminated. At the present time, it is an important strategic task for China to curb the desertification development in northern China, especially in Northwest and North China. Based on the above understanding, we think that although we have no way to change the fact that the ongoing desertification takes place in Northwest China resulted from the uplift of the Qinghai Tibet Plateau, it is possible for us to curb desertification development if we can take below measures: applying modern biological engineering to select drought-resistant, salt-tolerant, and fast-growing species, such as the *Populus* plants; introducing the drip irrigation technology to expand the area of shelter forests; establishing three-dimensional ecological farms; and taking efficient agricultural and forestry management.

T3

P0958

The ecological divergence and projected the future suitable habitats for homoploid hybrid species *Picea purpurea*

Changming Zhao

Lanzhou University

Picea purpurea is the endemic and dominant tree species of alpine dark coniferous forest in the northeastern of Qinghai Tibet Plateau (QTP) in China. It is demonstrated that *P. purpurea* derived from the natural hybridization of *P. wilsonii* and *P. likiangensis* without the change of chromosomal number. The habitat of *P. purpurea* is characterized by higher latitude and altitude than *P. likiangensis* and *P. wilsonii*, respectively. In this study, we assessed the ecological divergence between *P. purpurea* and its progenitors via a series of statistical methods based on the current distribution information and ecological characters of three spruce species firstly. Then estimated the distribution patterns for three spruce species under different climate change scenarios using species distribution models (SDMs) for two future decades: the 2050s and 2080s. Additionally, we also compared the ecophysiological differences between *P. purpurea* and its progenitors in terms of leaf water relations and leaf photosynthetic apparatus frost tolerance via a common garden experiment. Our results indicated that *P. purpurea* occupied different niche from its parental species. Moreover, the low winter minimum temperature and high soil moisture were the key factors that contribute to this niche divergence. And the future suitable area of *P. purpurea* would increase by an average of 10% under all three scenarios in the 2050s, and increase by an average of 28% under medium stabilizing and very heavy emission scenarios (RCP 4.5 and RCP 8.5) in the 2080s. In contrast, the future suitable habitats of *P. wilsonii* were severe threated, whereas the *P. likiangensis* showed the slight shrinkage in 2050s (~6%) and had no difference between current stage and 2080s and exhibited the tendency of migrate from southeast to east of QTP. For the ecophysiological analysis, significant difference was detected in osmotic potential at full turgor (π_o), water potential at turgor loss point (Ψ_{TLP}) and the temperature causing a 50% reduction in initial *Fv*/*Fm* (FT_{50}) between the hybrid species and the progenitors. These results indicate that the ecological separation in *P. purpurea* from its progenitors was promoted by enhanced capacity of cell withstand desiccation and photosynthetic apparatus frost tolerance, thus facilitated its successful colonization and establishment in high altitude and latitude habitats.

T3

P0959

The relationship between seed reserve compound contents and germination of six grassland species

Ming Zhao[1], *Hongxiang Zhang*[2]

1. College of Life Science, Northeast Normal University

2. Northeast Institute of Geography and Agroecology, Chinese Academy of Sciences

Reserve compound and content of seeds is one of the inner factors that affect the germination characteristics. Previous studies on seed germination focused mainly on the effects of external environmental factors or the internal molecular mechanism. The aims of this study are to test the relationship between germination characteristic and seed reserve compound content, as well as the reserve dynamics during germination. We measured the contents of starch, protein and fat in dry seeds of six herbaceous species (*Kochia scoparia*, *Chloris virgata*, *Astragalus adsurgens*, *Lespedeza hedysaroides*, *Leonurus Artemisia*, *Dracocephalum moldavica*) with different seed size (thousand seed weight 0.454 - 2.069 g) in Songnen grassland. We also tested the contents of water, starch, fat, soluble protein and soluble sugar during different stages of germination, including imbibition, 1% germination, 50% germination, total germination, early seedling stage (with visible cotyledons). Results showed that seeds of *C. virgate* are high starch content type, seeds of *K. scoparia*, *L. hedysaroides,* and *A. adsurgens* are high protein types and seeds of *L. Artemisia, D. moldavica* are high fat content types. Except *D. moldavica*, the three reserves took more than 80 percent of the seed weight of other five species. Germination percentages of all the six species were higher than 80%. During germination periods, starch was first decomposed and utilized most for the high starch type *C. virgata* seeds. For the high protein type *K. scoparia*, *A. adsurgens* and *L. hedysaroides*, soluble protein was first decomposed and utilized most. Fat was first decomposed and soluble protein started to be used higher than fat from the mid-germination period for the high fat type *D. moldavica*. Soluble protein and starch were utilized during the whole germination periods for *L. Artemisia*. The soluble sugar increased during the germination periods for all species except *A. adsurgens*. No significant correlation was found between the seed reserve compounds content and germination characteristics. This study was funded by the National Natural Science Foundation of China (41571055).

T3

P0960

Changes in germination responses of buried seeds in nature of the cold desert shrub *Zygophyllum xanthoxylon*

Xiaoying Zhao[1], *Jiao Wang*[1], *Carol C. Baskin*[2]

1. School of Life Science, Xinjiang Normal University

2. Department of Biology, University of Kentucky

Germination responses and dormancy of soil seed in nature can reveal dynamics of soil seed bank and predict trend of population regeneration, and thus information is needed on this aspect of the seed biology of a species before it is selected for use in habitat restoration projects. *Zygophyllum xanthoxylon* (Bunge) Maxim is a super-xerophilous shrub that is a potential pioneer species for use in revegetating highly disturbed areas of the cold deserts of northwest China. We study seasonal changes of germination and dormancy of buried seeds in nature of the species for two years in its cold desert habitat. Fresh fruits were put on surface and buried 2-5 cm in soil in August 2015 in Dabancheng of Xinjiang, China and exhumed and tested germination every 6 weeks in first year, every 2 months in second year over a range of temperature. Most seeds of *Z. xanthoxylon* exhibited an conditional dormancy in autumn in nature. Fruits exhumed in December (buried for 18 weeks) had a high germination percentage and become non-dormancy during winter in nature. A little fruit (about 20%) undergo high temperature of summer in June and July re-enter into secondary conditional dormancy which they can't germinate at summer temperature in nature. Fruit exhumed at 23 December 2016 (buried for 70 weeks) undergo second cold winter germinated over 90%, and there are still more than 93% of the seed had vitality both on surface and in soil. Fruit on surface had higher germination percentage than in soil. These results maybe result from seasonal temperation changes, rainfall and special pericarp.

T3

P0961

Dormancy breaking of buried seeds in nature of the cold desert shrub *Calligonum junceum*

Juma Aypaxia[1], *Xiaoying Zhao*[1], *Ying Hai*[2]

1. College of Life Science, Xinjiang Normal University

2. College of Geography Science, Xinjiang Normal University

Changes of dormancy statuaton of buried seed contributes to a better understanding how the timing of dormancy breaking is controlled in nature, and thus information is needed on this aspect of the seed biology of a species before it is selected for use in habitat restoration projects. *Calligonum junceum* (Fisch. et Mey.) is a super-xerophilous shrub that is a potential pioneer species for use in revegetating highly disturbed areas of the cold deserts of northwest China. We study seasonal changes of germination and dormancy of buried seeds in nature of the species for about two years in its cold desert habitat. Fresh fruits put on surface and buried 2-5 cm in soil in August 2015 in Dabancheng of Xinjiang, China and exhumed and tested germination every 6 weeks in first year, every 2 months in second year over a range of temperature. Over 95% freshly seeds of *C. junceum* exhibited an primary dormancy in autumn in nature. Fruits exhumed in March (buried for 29 weeks) had a high germination percentage and become non-dormancy during winter and Spring in nature. A Most of fruit (about 60%) undergo high temperature of summer in June and July re-enter in to secondary dormancy which they can't germinate at summer temperature in nature. Fruit exhumed at 23 December 2016 (buried for 70 weeks) undergo second cold winter germinated over 80%, breaking the secondary dormancy; thus, the species exhibited an dormancy/non-dormancy cycle and there are still more than 93% of the seed had vitality both on surface and in soil. When the snow in the Field Fruit on surface had higher germination percentage than in soil. When it rains in the field Fruit in soil had higher germination percentage than on surface. These results maybe result from seasonal temperation changes, rainfall and special pericarp.

T3
P0962
Atmospheric nitrogen deposition to China: A model analysis on nitrogen budget and critical load exceedance

Yuanhong Zhao[1], *Lin Zhang*[1], *Youfan Chen*[1], *Xuejun Liu*[2], *Wen Xu*[3], *Yuepeng Pan*[4], *Lei Duan*[5]
1. Laboratory for Climate and Ocean-Atmosphere Sciences, Department of Atmospheric and Oceanic Sciences, School of Physics, Peking University
2. College of Resources and Environmental Sciences, China Agricultural University
3. State Key Laboratory of Urban and Regional Ecology, Research Center for Eco-Environmental Sciences, Chinese Academy of Sciences
4. State Key Laboratory of Atmospheric Boundary Layer Physics and Atmospheric Chemistry (LAPC), Institute of Atmospheric Physics, Chinese Academy of Sciences
5. State Key Joint Laboratory of Environmental Simulation and Pollution Control, School of Environment, Tsinghua University

We present a national-scale model analysis on the sources and processes of inorganic nitrogen deposition over China using the GEOS-Chem model at 1/2°×1/3° horizontal resolution. Model results for 2008-2012 are evaluated with an ensemble of surface measurements of wet deposition flux and gaseous ammonia (NH_3) concentration, and satellite measurements of tropospheric NO_2 columns. Annual total inorganic nitrogen deposition fluxes are simulated to be generally less than 10 kg N ha^{-1} a^{-1} in western China (less than 2 kg N ha^{-1} a^{-1} over Tibet), 15-50 kg N ha^{-1} a^{-1} in eastern China, and 16.4 kg N ha^{-1} a^{-1} averaged over China. Annual total deposition to China is 16.4 Tg N, with 10.2 Tg N (62%) from reduced nitrogen (NH_x) and 6.2 Tg N from oxidized nitrogen (NO_y). Domestic anthropogenic sources contribute 86% of the total deposition; foreign anthropogenic sources 7% and natural sources 7%. Annually 23% of domestically emitted NH_3 and 36% for NO_x are exported outside the terrestrial land of China. We find that atmospheric nitrogen deposition is about half of the nitrogen input from fertilizer application (29.6 Tg N a^{-1}), and is much higher than that from natural biological fixation (7.3 Tg N a^{-1}) over China. A comparison of nitrogen deposition with critical load estimates for eutrophication indicates that about 15% of the land over China experiences critical load exceedances, demonstrating the necessity of nitrogen emission controls to avoid potential negative ecological effects.

T3
P0963
Functional traits of dominant tree species in a karst forest of southwestern China

Qiaolian Zhong, *Xin Xu*, *Jian Ni*
State Key Laboratory of Environmental Geochemistry, Institute of Geochemistry, Chinese Academy of Sciences

Plant functional traits are attributes of plant adaptation to the environment. In order to study ecological strategies of dominant tree species in adapting the harsh Karst habitats, a comprehensive survey of functional traits was conducted in a 30 m × 100 m plot of the Tianlongshan Mountains in Puding county, Guizhou province, southwestern China in the summer 2016. Choosing five dominant tree species (*Carpinus pubescens, Machilus cavaleriei, Itea yunnanensis, Platycarya strobilacea, Lithocarpus confinis*), three shrubs (*Zanthoxylum ovalifolium, Stachyurus obovatus, Rhamnus heterophylla)* and two vines (*Rosa cymosa, Dalbergia hancei*) in this evergreen and deciduous broad leaved mixed forest, all above ground and below ground traits from leaf, stem, branch and root, as well as habitat heterogeneity of soil and rock were investigated and measured. Relationships between above ground and below ground plant traits, and between traits and environmental factors were analyzed. The preliminary results showed that: 1) the specific root length (SRL) between trees, shrubs and vines have no significant differences, but significant differences in the specific leaf area (SLA) and root tissue density were found; 2) there are significant positive relationships between tree SRL and twig dry matter content, leaf dry matter content and fine root tissue density ($p < 0.05$). SRL is negatively related with SLA and leaf thickness (LT) ($p > 0.05$); 3) in contrast, SRL of shrubs and vines increases with increasing LT ($p < 0.01$), but decreases with increased SLA ($p < 0.01$), along with very weak relationships to other traits. These findings revealed that plants adopt different survival strategies to adapt to the karst environment.

T3
P0964
Leaf morphology shift with climate change of 7 dicots in China

Dongting Zou, *Zhiheng Wang*, *Yaoqi Li*
Peking University

Leaf is a climate sensitive organ of plants. Leaf size and shape have been found to vary along geographical and climatic gradients and this phenomenon has been widely acknowledged. However, intraspecific variation of this pattern is largely lacking. Similarly, we do not know if leaf size and shape carry any signal of climate change. In the present study we investigated how leaf size and shape varies with geographical gradients and how they respond to climate change. We measured 6,063 specimens from Chinese Virtual Herbarium (CVH) of 7 dicots in China collected between 1910 and 2008. The measurements included leaf length and width, leaf size and leaf length – width ratio. We used loess regressions, linear models, ANOVA and redundancy analysis (RDA) models to investigate the geographical patterns of leaf size and shape and the relationship between leaf size and shape and climatic variables isolated from occurrence localities of the specimens. The results suggested that 1) for different species, leaf size and shape show similar patterns along geographical gradients, with leaf size and leaf length – width ratio usually increasing as latitude, longitude or elevation increases; 2) Spatially leaf size and leaf length – width ratio increases with temperature, precipitation and annual evaportranspiration for most species, and the relationship is consistent among different climatic variables; 3) Leaf size and shape have similar time trends for different species, and the trends of leaf length, leaf width and leaf area are similar; 4) the response of leaf size and shape to teperature and annual evaportranspiration change is various among species. Nevertheless, as precipitation increases, shrubs tend to grow larger leaves while leaves of trees and herbs do not change much; 5) For different species the same leaf morphological trait may be controlled by different climatic variables; 6) for the same species, different leaf morphological trait may mainly be affected by varied climatic variables. We conclude that specimens collected at a larger temporal scale may carry useful signal of climate change and could be a good resource for future climate change studies.

T3

P1407

Powdery mildew and rust of *Salix* and *Populus* in Inland Alaska

Momei Chen, *Katherine Wu*
University of California, Berkeley

In these verbal accounts, I learned about the social history of Alaska before and after it was sold to the U.S. but its natural history as same as the evolution of other Taiga forest. The study of fungi has increased exponentially in the past 100 years, but Alaska forest pathology and fungi are still being ignored and neglected by many fields of study. Fifteen of which *Salix* in Alaska inland, among them seven the above is resistant to Powdery mildew and rust of *Salix*. It's as a candidate choice for willow species plantation.

T3

P1408

Identification of fungi species in forest ecosystem of internal Alaska

Momei Chen, *Ivy So Yeon Kim*
University of California, Berkeley

This study was primarily focused on identification with the traditional method which uses the fungus morphological data to identify each specimen. Current identification act upon the nomenclature and with techniques in applied molecular reference level. Total 310 specimens are still on process for preservation internal Alaska fungi specimens.

T3

P1412

Species richness and composition along microtopography and snow cover gradient in Hardangervidda National Park, Southern Norway.

Prakash Bhattarai, *Kuber Prasad Bhatta, John-Arvid Grytnes, Ole Reidar Vetaas*
University of Bergen

The species richness and composition of vascular plants, bryophytes and lichens in alpine area are determined by local microtopography, snow cover gradient and strong winds. The aims of this study are to show the species richness pattern and species composition with respect to micro topography and snow cover gradient. We surveyed lichen and vascular plant species in 151 plots of size 4 m^2 within 23 transects. The transects ran from the top of the small hills (ridge) and ends on snowbeds. Weighted average snow indicator value (WASI) and relative distance (RD) from ridge to snowbeds of each plot were calculated. DCA analysis was performed to show the species composition with respect to microtopography. One way anova was performed to show the species richness with respect to micro topography and GLM was performed to show the pattern of species richness with WASI and RD. All together 69 species were recorded, of which 52 species were vascular plants and 17 species were lichens. Plant species associated with ridge (e.g., *Alectoria ochroleuca*, *Alectoria nigricans*, *Sphaerophorus globosus* and *Vaccinium vitis-idaea*) were ordinated in the negative part of the DCA axis I whereas snow bed species (e.g., *Anthoxanthum odoratum*, *Bistorta vivipara*, *Festuca spicata*, *Pyrola alpine* and *Rumex acetosa*) were ordinated on the positive side of the same axis. The species richness of total plant species, vascular plant species and herbaceous plant species followed the same pattern, highest in snow bed and least in ridge whereas, dwarf shrub followed the different trend in species richness in comparison to other life forms. Species richness of dwarf shrub remain almost constant with micro topographical change, while lichens species richness was least in snowbeds than ridge and mid. Regressionanalysis (GLM) revealed a linear relationship in species richness for total plant species and all other life forms except for dwarf shrub species when regressed against WASI and RD, while dwarf shrub showed unimodal pattern. We conclude that the microtopography is an important environmental gradient that shaped the pattern and composition of plant species in space.

T3

P1421

Did grassland extensively migrate southward into the present-day forested eastern China during the Last Ice Age? - A new pollen record

Junwu Shu
Key Laboratory of Economic Stratigraphy and Palaeogeography, LPS, Nanjing Institute of Geology and Palaeontology, Chinese Academy of Sciences

Very little is known about the vegetation pattern covered the widespread low plain in east China during the Last Glacial Maximum (LGM) mainly to due to the loess-related sediment archive unfavorable for pollen preservation in this area. Fortunately, a LGM alluvial sediment sequence (23-16 ka) was occasionally newly recovered. Pollen results show first that herbaceous taxa are surprisingly abundant dominated by *Artemisia* and Poaceae (about 70% in average percentage) which are characteristic taxa in grassland in northern China. In addition, grassland taxa such as *Aster*-type, *Echinops*-type, *Taraxcum*-type, Apiaceae and *Thalictrum* are also common in low abundance. Arboreal taxa are poorly present such as *Quercus*, *Pinus,* and *Alnus.* The pollen flora suggests grassland or grassland with sparse trees distributed in the lowland under a dry and cold continental climate during LGM in sharp contrast to the present-day warm-humid broadleaved forest in the study area. The extensive migration of grassland (now in northern China) more than 1,000 km southward into SW China is probably attributed to the strengthening winter monsoon during the glacial period. Furthermore, the spread of loess into the south China support the grassland migration. The study will put insight into climate modeling and loess south-limit during LGM.

T3

P1431

A new compound identified in smoke promotes seed germi-

nation of fire-chaser plants

Dechang Cao, *Matthias Schöttner, Lan T Baldwin*
Max-Planck Institute for Chemical Ecology, Jena, Germany

Smoke-stimulated seed germination is an adaptive trait for plants to colonize post-fire habitats. Regardless that karrikins have been reported as active compounds responsible for smoke-stimulated seed germination, some smoke-sensitive seeds do not respond to karrikins. Thus, the mechanisms underlying smoke-stimulated seed germination remains illusive. Here, we used a bioassay-guided fractionation on HPLC and UPLC to separate liquid smoke which can promote seed germination of a fire chaser, *Nicotiana attenuata*. An active fraction was obtained and a new germination cue identified in the active fraction. The new germination cue differs from karrikins in chemical characteristics and involves differential mechanisms in promoting seed germination. The new findings may open a new window to our understanding of plant adaptation to smoke in fire-prone ecosystems.

T3
P1432

Floral nectary morphology and proteomic analysis of nectar of *Liriodendron tulipifera* Linn. plant polyphenols and their dynamics during leaf development under nutrient limitation: A potential nutrient conservation strategy for mangrove plant

Haichao Zhou[1,3*], *Xue Song*[1,3], *Pingping Wei*[1,3], *Hualin Xu*[2], *Qijie Zan*[1,2]
1. College of Life Sciences and Oceanography, Shenzhen University, Shenzhen, China
2. Guangdong Neilingding Futian National Nature Reserve, Shenzhen, China
3. Futian-CityU Mangrove Research and Development Centre, City University of Hong Kong, Shenzhen, China

Mangroves are a group of shrub and tree species that live along shores, rivers, and estuaries in the tropical and subtropical areas. The diversity of plant polyphenols in mangrove plants was reported as an evolutionary advantage adapting to harsh environments. However, there is very limited research on the chemistry and dynimics plant polyphenols in mangroves. Nutrient availability, especially nitrogen (N) and phosphorus (P), is reported as an important factor responsible for mangrove growth. N:P ratios have been applied to identify thresholds of nutrient limitation in wetlands. Thresholds of leaf N:P ratios were found to be16 for P-limitation. Although a increasing knowledge of nutrient limitation for mangrove plants and their adaptation on low nutrient availability, there are many different species of mangrove plants and still litter information about the importance of N:P ratio and variety of leaf polyphenols in nutrient conservation. In this study, we fully investigated the related nutrients (N, P and N:P ratio) and plant polyphenols of mangroves in Dongzhai Harbor national nature reserve. It is the first systematic research on the dynamics of plant polyphenols and nutrients during leaf senescence of 22 mangrove species. Among these species, 12 (mainly Rhizophoraceae), 7 (mainly Sonneratiaceae) and 3 (*Avicennia marina*, *Xylocarpus granatum*, *Excoecaria agallocha*) are condensed tannin (CT), hydrolysable tannin (HT) and low molecular weight polyphenol (LMWP) as the main type of plant polyphenol, respectively. Most of species remained high level of plant polyphenols during leaf senescence, with an increase in the CT species, but a decrease in the HT and LMWP species. Leaf N and P concentrations decreased significantly during leaf senescence. Based on the theory of N:P ratio, most of the mangrove forests are more or less N-limited with N:P < 16 in China. High levels of CTs in mangrove leaf were not only significant in reduction of leaf litter decomposition rate, but also played an important role in N immobilization, which provided a vital nutrient conservation strategy for mangrove plants to survive in mangrove swamps with low nutrient availability. The results from this research will enhance our current understanding of potential nutrient conservation strategies used by mangrove plants and improve management of mangrove wetland ecosystem.

T3
P1434

The emergence of the hyper-invasive vine, *Mikania micrantha* (Asteraceae), via admixture and founder events inferred from population transcriptomics

Ming Yang[1], *Ziwen He*[1], *Yelin Huang*[1], *Lu Lu*[1], *Yubin Yan*[1], *Lan Hong*[2], *Hao Shen*[3], *Ying Liu*[1], *Qiang Guo*[4], *Lu Jiang*[4], *Yanwu Zhang*[4], *Anthony J. Greenberg*[5], *Renchao Zhou*[1], *Xuejun Ge*[3], *Chung-I Wu*[1,6], *Suhua Shi*[1]
1. Sun Yat-Sen University, State Key Laboratory of Biocontrol and Guangdong Provincial Key Laboratory of Plant Resources
2. College of Horticulture and Landscape Architecture, Zhongkai University of Agriculture and Engineering
3. South China Botanical Garden, Chinese Academy of Sciences
4. Shenzhen Wildlife Protection Administration
5. Bayesic Research
6. Department of Ecology and Evolution

Biological invasions that involved well-documented rapid adaptations to new environments provide unequalled opportunities for testing evolutionary hypotheses. *Mikania micrantha* Kunth (Asteraceae), a perennial herbaceous vine native to tropical Central and South America, successfully invaded tropical Asia in the early 20th century. It is regarded as one of the most aggressive weeds in the world. In an effort to elucidate the molecular and evolutionary processes underlying this invasion, we extensively sampled this weed throughout its invaded range in Southeast and South Asia, and surveyed its genetic structure using variants detected from population transcriptomics. Clustering results suggest that more than one source population contribute to this invasion, and the computer simulations using genome-wide genetic variation support a scenario of admixture and founder events during invasion. The genes differentially expressed between native and invasive populations were found to be involved in oxidative and high light intensity stress responses, pointing to a possible ecological mechanism of adaptation. Our results provide a foundation for further detailed mechanistic and population studies of this ecologically and economically important invasion. This line of research promises to provide new mitigation strategies for invasive species as well as insights into mechanisms of adaptation.

T3
P1437

Role of wood traits and phylogeny in wood decomposition: A

common garden experiment in a botanical garden

Gbadamassi G.O. Dossa[1,2,3,5], *Ekananda Paudel*[1,2,3], *Kunfang Cao*[2,4], *Douglas Schaefer*[2], *Jianchu Xu*[1,3,5], *Rhett D. Harrison*[3,6]
1. Kunming Institute of Botany, Chinese Academy of Sciences, China
2. Key Laboratory of Tropical Forest Ecology, Xishuangbanna Tropical Botanical Garden, Chinese Academy of Sciences, China
3. Centre for Mountain Ecosystem Studies, Kunming Institute of Botany, Chinese Academy of Sciences, China
4. Ecophysiology and Evolution Group, State Key Laboratory for Conservation and Utilization of Subtropical Agro-Bioresources, and College of Forestry, Guangxi University, China
5. World Agroforestry Centre, East & Central Asia Regional Office, Kunming, CHina
6. World Agroforestry Centre, East & Southern Africa Region, Lusaka, Zambia

Disentangling specific factors effects on wood decomposition remains a challenge given the complexity and the interaction among factors affecting wood decomposition. However, one would better draw meaningful conclusions if one designs an experiment that controls for certain factors/variables but leaves others to vary. One way of realizing such event is to have litter bed experiment where the decomposing substrates are exposed to the same environmental factors but substrates would vary in their respective innate functional traits. Even so, there is still the during life environmental "imprint". Therefore, to push this constraint one step higher, one could collect substrates from species that had experienced similar environment. Doing so, this would greatly reduce the "during life environmental 'imprint'" among species. Botanical gardens provide an ideal institutional environment to conduct manipulative experiments on diverse living collections using a common garden approach. At Xishuangbanna Tropical Botanical Garden we conducted a three years long experiment on the role of phylogeny in the decomposition of woody debris (WD). WD is an important terrestrial carbon pool contributing an estimated 10-20% globally, and can over 50% of the carbon is some forests. Using the garden's plant collections we sampled fresh wood logs (5 cm diameter) from 25 species selected across the phylogeny of higher plants. In each major clade, we selected species with both high and low wood density to maximize the range of wood traits represented at this scale. These were incubated in a patch of secondary rain forest within the garden, monitored for mass loss periodically over three years. In addition to phylogeny we collected data on wood density, bark and wood chemistry, and analyzed the fungal communities development and dynamics in the decomposing wood using both traditional techniques and next generation sequencing. We are currently analyzing the preliminary results. Through the phylogenic approach, we hope to produce a predictive model that incorporates microbial diversity effect on wood decomposition and can be used to calibrate WD decomposition globally based on knowledge of forest tree species composition.

T4

P0965

Coordination assessment of glutathione with other antioxidants in halophyte *Halimione portulacoides* L. exposed to nanoscale-copper

Anjum Naser, *Duarte Armando, Pereira Eduarda*
CESAM-Centre for Environmental and Marine Studies, University of Aveiro

As a major component of ascorbate (AsA)-glutahione (GSH) pathway, key soluble redox-compounds such as AsA and GSH significantly contribute to cellular redox homeostasis by metabolizing various reactive oxygen species (ROS), major cause of oxidative stress. This study evaluated the glutathione coordination with other antioxidants in halophyte *Halimione portulacoides L.* exposed to nanoscale-copper oxide (nano-CuO) (<50 nm) concentrations (0,20,40,80,160 and 320 mg L^{-1}) under hydroponic set-up. *H. portulacoides* exhibited its differential sensitivity to tissue-Cu-burdens in terms of cellular redox system violators (such as $O_2^{\cdot-}$ and H_2O_2) and its managers (such as ascorbate, AsA; glutathione, GSH; AsA/DHA; GSSG/GSSG; GSH reductase; GSH peroxidase; GSH-sulfo-transferase; ascorbate peroxidase, APX; catalase, CAT). A close relationship among Cu ion release, tissue-Cu-burdens and cellular redox homeostasis was noted. This presentation will discuss and interpret the previous results to get insights into the nano-CuO-accrued phytotoxicity, and adaptive responses of *H. portulacoides*, and potential use of this plant in nano-CuO remediation in estuarine system.

T4

P0966

The periderm structure in the leaves of *Agathis* species (Araucariaceae)

Alisa Arbicheva, *Anatoly Pautov*
Saint Petersburg University

Periderm is a dermal tissue of secondary origin which replaces the epidermis and takes over the protective function. Periderm regularly develops along surfaces that are exposed after abscission of leaves and branches. Mechanical damage causes the formation of wound periderm. This tissue develops regularly in the scales of wintering buds of some species. Photosynthetic leaves do not normally produce periderm. Nonetheless, a number of species do form periderm in the leaves. We studied periderm of the leaves of *Agathis brownii* (Lem.) L. H. Bailey and *Agathis vitiensis* Benth. & Hook.f. ex Drake using methods of light and transmission electron microscopy, experimental wounding and histochemical staining. Additionally we studied the periderm initiation in the *Agathis* stems. *Agathis* leaves are long-lived and remain alive for more than 25 years. The species studied have leaves of a similar structure. Periderm occurs regularly and can be initiated in the epidermis, mesophyll, phloem parenchyma of the leaf veins and in the ground tissue of the petiole. Periderm structure is either typical, consisting of phellogen, multilayered phellem and phelloderm, or disordered. Periderm arrangement in the leaves is highly diverse. It can be initiated in the epidermis and cover a small area of the leaf. If numerous cells undergo intense divisions, a lenticel-like structure may form. Phellogen may form in a paradermal plane when initiated in the mesophyll. There are cases of simultaneous periderm formation both on the abaxial and adaxial surfaces of the lamina, when only the central cell layers remain alive. We have found rare cases of circular phellogen initiation in the leaves of *A. vitiensis*. At the junction of the petiole and the lamina the oldest leaves may form a large number of transverse planes of periderm initiation. It leads to the splitting of the base of lamina, some of the areas undergoing necrosis. *Agathis* leaves are able to form true wound periderm, which has been shown experimentally. We believe that at least some of the factors inducing periderm initiation are identical both in leaves and stems. Phellogen initiation in the thickening stem was found exactly in the regions of cuticle breaks. These breaks point to the presence of a significant mechanical tension or epidermis deformation which exceed cuticle hardness. Such deformations and mechanical tensions arise in petioles in the part near the lamina, which is usually curled to maintain the lamina in a horizontal position. Mechanical deformations in petioles are indirectly evidenced by changes in epidermal cell shape. These data support the point of view that mechanical forces play an essential role in morphogenesis processes. Periderm initiation deep in the mesophyll leads to the isolation of cell groups or dissection of leaf parts. Cells of isolated parts are filled with tannins. The cells of the leaf and petiole are the storage of crystals, the amount of crystals substantially increasing with age. Leaf tissues isolated due to the suberinization process may function as a storage site for the substances not involved in the plant metabolism or excluded from it.

T4

P0967

MOD1 modulate cell and organ growth by regulating endore duplication in *Arabidopsis thaliana*

Usman Aslam, *Yunhai Li*
Institute of Genetics and Developmental Biology, University of Chinese Academy of Sciences

How plants modulate growth and definite size of their organs is a fundamental question in plant biology. In agronomic context, organ size provides ultimate forecast of yield and total biomass which are key determinants of food and biofuel crop production. Typically, organ growth is controlled by coordinated maneuver of cell proliferation, expansion and differentiation into particular tissues. Many studies have highlighted the role of endoreduplication in development and physiology, however, the underlying mechanisms linking endopolyploidy with cell growth and final organ size are still nature's secrets. Previously, UBIQUITIN-SPECIFIC PROTEASE 14 (UBP14) encoded by *DA3* ("*DA*" means big in Chinese) was described to function with anaphase-promoting complex/cyclosome (APC/C) ubiquitin ligase through interacting its inhibitor ULTRAVIOLET-B INSENSITIVE 4 (UVI4) to modulate endoreduplication and organ growth in *Arabidopsis thaliana*. The *Arabidopsis da3-1* mutant possesses larger cotyledons, leaves and flowers with higher ploidy levels. Using Next Generation Mapping (NGM) approach, I genetically identify several modifier genes of *da3-1* from EMS mutagenized *Arabidopsis* populations that influence the ploidy and growth phenotypes of *da3-1*. I selected one of them for functional analysis and designated as *mod1-1*. The cellular and genetic studies suggested that *mod1-1* strongly represses the cellular ploidy and synergistically regulate the organ growth phenotypes of *da3-1*. The DNA and protein sequence analysis revealed that *MOD1* is a key players of splicing complex in higher eukaryotes, however, their role in endoreduplication or any organ growth related mechanisms has never been described. The expression pattern analysis of *MOD1* revealed that they are highly expressed in transcriptionally active and differentiating cells. Bimolecular fluorescence complementation (BiFC), co-immunoprecipitation and pull down experiments demonstrate that MOD1 physically interacts with UBP14 and CDKB1;1 *in vivo* and *in vitro*. Further studies proves that UBP14 influences the stability of MOD1. We already knew that *cdkb1;1-1* enhanced the ploidy phenotype of *ubp14-1* and *uvi4-1*, however, in contrast our results showed that *mod1-1* does not suppress the cellular ploidy of *uvi4-1* and *cdkb1;1-1* suggesting *MOD1* is a downstream components of UBP14 in endocycle pathway. Overall, these findings define a genetic and molecular mechanism of MOD1 function in endoreduplication and plant organ growth regulation.

T4

P0968

The mechanism and molecular basis underlying salt secretion in *Reaumuria soongorica*

Tianhui Bai, *Shuyi Zhao, Suomin Wang, Aike Bao*
Lanzhou University

Reaumuria soongorica, a salt-secreting super xerophyte belonging to Tamaricaceae which is a perennial fodder semi shrub at desert steppe in the northwest of China. It lives in the semiarid regions that endows the plant with salt-resistant, drought-resisitant and sand trapping inhere character. *R. soongorica* is also significant for the ecology protection. Our previous studies showed that secreting salt via salt glands is an important strategy for *R. soongorica* adapting to high salinity environments. However, the mechanism underlying salt-secretion of *R. soongorica* is still unclear. In this study, we analyzed the real-time Na^+ secretion dynamics via salt gland of *R. soongorica* using Non-invasive microtest technology (NMT).The results showed that under either salt (100,300 mM NaCl) or osmotic stress (-1.0 MPa), the net Na^+ efflux via salt glands showed significantly higher than control, and interestingly, the net H^+ influx across the salt gland was correspondingly induced by salt and osmotic stress, suggesting that there is active Na^+/H^+ exchange in the process of salt secretion on the plasma membrane of the secretory cells (SeC) in salt gland of *R. soongorica*. To further investigate how Na^+ was accumulated into collecting cells (CoC) and then secreted from SeC, we cloned *RsSOS1* and *RsNHX1* genes, which encode plasma membrane and tonoplast Na^+/H^+ antiporters of *R. soongorica*, respectively. The expression pattern analysis indicated that both *RsSOS1* and *RsNHX1* genes predominantly expressed in shoots, and were significantly induced by salt (50,200 mM NaCl) or osmotic (-0.5, -1.0 MPa) stress treatments. These results suggest that *RsSOS1* and *RsNHX1* may be involved in salt-secretion of *R. soongorica*. We thus proposed a model in that Na^+ may be firstly sequestered into vacuole of CoC mediated by *RsNHX1*, then transported into SeC and finally excluded by *RsSOS1*.

T4

P0969

Gemination and initial establishment of *Euterpe edulis* Martius (Arecaceae)

Hector Barros Gomes, *Bruno Takami Fujiwara, Eliana Akie Simabukuro*
University of São Carlos

The germination of *Euterpe edulis* Mart. (knowing as palm jussara) it's a key aspect for population maintenance process of the species. In this respect, understanding the conditions required to maintain viability of seeds and growth of seedlings is critical for a specie like *E. edulis* that presents recalcitrant seed and seedlings with slow and uneven growth. The aim of this study was to verify the seed germination of *E. edulis* at different stages of maturation and the development of their seedlings. The species presented four maturation stages: immature (green pericap); in maturation (dark green); mature (purple pericarp) and post-maturation (black pericarp). The dry biomass had increased during maturation (0.82 to 0.91 grams), and the lowest relative humidity was observed in mature diaspores, with 50.2%. Diaspores with purple pericarp (mature) showed higher germination percentage (84%) and germination rate index (0.96) when compared to other maturity stages. When stored for a week (5°C), the germination percentage (G.P.) for purple pericarp increased to 98%. The best diaspore germination was observed at temperatures of 20-30°C and 20°C with 90% of G.P., and better results for seeds that have been stored for a short period (5°C). Seedlings originating with green and purple pericarp showed better growth (stalk with 15 cm and root with 22 cm). We can claim that the environmental conditions of the parent plant can predict the needs of progeny for *E. edulis* and purple diaspores had higher survival at the end of the experiment indicating the best maturation stage.

T4

P0970

Analysis of the protein domains of the major carpel developmental regulator CRABS CLAW

Thomas Gross, *Kai Pfannebecker, Annette Becker*
Justus-Liebig-University Gießen

One of the most crucial steps in the life cycle of angiosperms is the development of carpels. They are the most complex plant organs, harbor the seeds, and later develop into fruits and are thus an important ecological and economic trait. CRABS CLAW (CRC), a Yabby protein and putative transcription factor, is one of the major carpel developmental regulators in *A. thaliana.* It is involved in various processes during floral development, such as carpel fusion, floral meristem termination, and nectary formation. While its genetic interactions with other carpel development regulators are well described, its biochemical properties and molecular way of action remain unclear. Here, we reveal the molecular base of CRC function with e.g. protein interaction and localization assays to assign functions to the three domains of CRC (a zinc finger like domain, a serine-proline rich domain, and the protein family specific Yabby domain), in respect to subcellular localization, nuclear transport, protein-protein interactions, and DNA binding. We can show that CRC regulates carpel fusion and nectary formation by activating transcription and regulates floral meristem termination by repressing transcription of target genes. In addition, we present our recent progress in the elucidation of the regulation of CRC during carpel development.

T4

P0971

Investigating the role of regulatory B" subunits of protein phosphatase 2A in responses to salinity stress

Mallesham Bulle[1,2,3], *Amr Ramzy Abass Kataya*[2], *Kirti Pulugurtha Bharadwaja*[1], *Sadanandam Abbagani*[3], *Lillo Cathrine*[2]
1. University of Hyderabad
2. University of Stavanger
3. Kakatiya University

Protein phosphatase 2A (PP2A) is a serine/threonine-specific protein phosphatase that plays an important roles in regulation of most cellular processes of eukaryotes. Canonical PP2As are heterotrimeric complexes consisting of a catalytic (C1-5), scaffolding (A1-3), and regulatory (B) subunits. The regulatory B subunits play an important role in conferring substrate specificity and determining subcellular localization to the holoenzyme complex and are classified into B55, B', and B" non-related families. In *Arabidopsis*, 17 regulatory subunits are present that can lead to 255 possible combinations. At present, little is known about the functions and substrates of plant PP2As. In this study, we identified a new function for PP2A in salt stress tolerance. Interestingly, *AtPP2A B"α* is targeting to mitochondria and we fused all the remaining B" subunits with a reporter protein, and their localizations are under investigation. Expression pattern analyses revealed that *AtPP2A B"α* is strongly expressed in roots, and responds to NaCl stress at the transcription level. From the preliminary data, transgenic Arabidopsis plants overexpressing *AtPP2A B"α* developed more lateral roots, whereas, the seedlings of these knockout mutants show sensitive to high salt concentrations. The results indicating that PP2A plays an important role in salt stress tolerance. However, to fully understand the role of PP2A and its regulatory subunits in salt stress, it is of uttermost importance to identify proteins that are substrates of PP2A complexes containing the various B-subunits.

T4

P0972

Methodology for analysis of multiple phytohormones

Bao-Dong Cai
Wuhan Greensword Creation Technology Co., Ltd.

Phytohormones are special small molecules that play important role in plant growth and development at trace levels. Quantification of multiple interacting phytohormones will be great helpful for research about cross-talk that plant hormones regulate the responses of plants against both biotic and abiotic stresses by means of synergistic or antagonistic interaction. However, it is still a challenge for simultaneous analysis of multiple phytohormones in one small sample because of the structural and chemical diversity of phytohormones. It is difficult to simultaneously enrich them.
We have been committed to improve the methodology for analysis of multiple phytohormones. Based on UPLC-MS/MS, more than 60 plant hormones could been analyzed from plant tissues with improved smaple pretreatment procedure.

T4

P0973

Regulation of plant metabolism by the target of rapamycin (TOR) kinase pathway

Carolina Monte Bello[1], *Marina Cmm Soldi*[1], *Viviane Da Silva*[1],

Valeria Cota[1], *Ana Beatriz Verdugo*[1], *Luis Guilherme De Abreu*[1], *Juliana Aricetti*[1], *Lucia Wolf*[1], ***Camila Caldana***[1,2]
1. Brazilian Bioethanol Science and Technology Laboratory, Brazilian Center for Research in Energy and Materials
2. Max Planck Partner Group at Brazilian Bioethanol Science and Technology Laboratory

Plant growth is controlled by metabolic network regulation, development and carbon partitioning between sink and source organs. There are now considerable genetic and biochemical evidences that the target of rapamycin (TOR) kinase is one of the key pathways in sensing and transducing sugars to control growth. We used genetic and pharmacological approaches to further investigate the kinetics of TOR inhibition in plants under different photoperiods to identify the mode of action of this pathway in controlling C metabolism. By combining metabolic profile, biochemical assays and gene expression analyses, we showed that down-regulation of TORC leads to accumulation of starch in the light, but this was not linked neither to allosteric regulation nor for the redox regulation of ADPG pyrophosphorylase (APGase). Surprisingly, the levels of maltose, the main starch breakdown product, were lower in plants exposed to short-term inhibition of TORC when compared to the control, suggesting an impaired starch breakdown at dawn. In addition, massive changes in organic acids from the TCA cycle clearly show that TOR influences not only C metabolism related to storage and use of sugars but also affect the energetic status of the plants. Furthermore, massive changes in the levels of transcripts and proteins involved in starch metabolism and sugar transport revealed a role of this pathway in keeping sucrose homeostasis to sustain plant growth.

T4
P0974

Effect of shoot-to-root abscisic acid (ABA) transport on xylem sap ABA concentration, water status and stomatal conductance of plants in drying soil.

Pedro Castro Valdecantos, *Ian C. Dodd*, *Jaime Puertolas*
Lancaster University

Apoplastic ABA concentration is critical to determine stomatal aperture. Experimental evidence suggests that root sourced ABA is not necessary to increase apoplastic ABA levels and trigger stomatal closure under drought conditions. Recent findings indicate substantial transport of ABA from the shoots to the roots, with root synthesized ABA of secondary importance compared to shoot ABA biosynthesis in response to water stress. Our study aimed to understand the impact of shoot-to-root ABA transport on ABA concentration in the xylem sap extracted from different parts of the plant, and ultimately its effect on stomatal conductance. Soybean (*Glycine max* cv. Siverka) were grown in special pots designed to fit in a pressure chamber and subjected to a factorial experiment with irrigation (WW: well-watered; DR: droughted) and girdling (NG: intact plants; G: Girdled plants) as factors Girdling was achieved surgically in 30 days old plants (after sowing) by excising the phloem tissue with a sharp razor blade. Four days after girdling and withholding water, stomatal conductance, root, shoot and leaf water potential were measured and ABA was determined in xylem sap extracted from the leaf, shoot and root. Soil drying decreased water potential in all plant tissues. Root and shoot water potential were not affected by girdling, but leaf water potential in DR was slightly higher (by -0.1 MPa) for girdled plants. For non-girdled plants, xylem sap ABA concentrations remained stable in WW plants and approximately doubled in DR plants. Girdling diminished the increase in leaf and shoot xylem ABA concentration, and completely prevented it in root xylem sap and tissue. On the contrary, girdling increased ABA concentration of WW plants but only in leaf xylem sap. Stomatal conductance was best related to leaf xylem sap ABA concentration, and approximately halved for all the treatments except for WW-NG The results support previous evidence indicating the predominant role of shoot-sourced ABA explaining root ABA accumulation in response to drought and suggest that shoot to root ABA transport might be a mechanism to avoid apoplastic ABA accumulation in the shoots in order to maintain stomatal opening under optimal soil water content.

T4
P0975

Diversity and association of phenotypic and metabolomic traits in the close model grasses *Brachypodium distachyon*, *B. stacei* and *B. hybridum*

Diana Lopez-Alvarez[1], *Hassan Zubai*[2], *Manfred Beckmann*[2], *John Draper*[2], ***Pilar Catalan***[1]
1. Universidad de Zaragoza
2. Aberystwyth University

Morphological traits in combination with metabolite fingerprinting were used to investigate inter and intra species diversity within the model annual grasses *B. distachyon*, *B. stacei* and *B. hybridum*. Phenotypic variation of 15 morphological characters and 2,219 nominal mass (*m/z*) signals generated using Flow Infusion Electrospray ionisation - Mass Spectrometry (FIE-MS) were evaluated in individuals from 174 circumMediterranean wild populations and 6 inbred lines, and 12 Iberian lines of the three species, respectively. Basic statistics and multivariate Principal Component Analysis (PCA) and Discriminant Analysis (DA) were used to differentiate inter- and intraspecific variability of the two types of variables, and their association was assayed with *rcorr*. Basic statistics and ANOVA detected eight phenotypic characters [(stomata) leaf guard cell length, pollen grain length, (plant) height, second leaf width, inflorescence length, number of spikelets per inflorescence, lemma length, awn length] and 434 tentatively annotated metabolite signals that significantly discriminated the three species. Three phenotypic traits (pollen grain length, spikelet length, number of flowers per inflorescence) might be genetically fixed. The three species showed different metabolomic profiles. DA significantly discriminated the three taxa with both, morphometric and metabolome traits and the intraspecific phenotypic diversity within *B. distachyon* and *B. stacei*. The populations of *B. hybridum* were considerably less differentiated. Highly explanatory metabolite signals together with morphological characters revealed concordant patterns of differentiation of the three taxa. Intraspecific phenotypic diversity was observed between northern and southern Iberian populations of *B. distachyon* and between E Mediterranean – SW Asian and W Mediterranean populations of *B. stacei*. Significant association was found for pollen grain length and lemma length and 10 and 6 metabolomic signals, respectively.

These results would guide the selection of new germplasm lines of the three model grasses in on-going GWAS experiments.

T4
P0976
Leaf life-span dynamics of woody species in tropical dry forests of India

Ravi Kant Chaturvedi[1], *As Raghubanshi*[2]
1. Xishuangbanna Tropical Botanical Garden
2. Banaras Hindu University

Phenology is sometimes, concurrently or at different times of the year, controlled by multiple factors, therefore, it is important to understand the factors influencing the foliar phenology in the tree species of tropical dry forests (TDFs). In the present study, we analyse the diversity of leaf life-span in the woody species of Vindhyan highlands and group the plant species according to their leaf flushing period, leaf life span and leaf fall period. We also study the factor influencing the foliar phenology of the woody species. The study was conducted on five sites (Hathinala, Gaighat, Harnakachar and Ranitali, listed in order of decreasing SMC) within the tropical dry deciduous forest in northern India. Fifteen individuals each, of the woody species were randomly selected from the three 4 ha plots at each site and were marked. Plant species which were not present at all the five sites were selected from only those sites where they were present. On each marked individual, one twig (currently growing shoots of last-order branches) on each of four major branches (one in each direction was marked with metal tags). On these twigs monthly count of leaf number was made from January 2005 to December 2006. Soil moisture content was measured at 10 locations, randomly in each plot, at each site, as percentage by volume every month, at a depth of 10 cm at 1-month intervals for 2 years. Site wise variations of soil moisture content in each month was analysed by multivariate ANOVA. Across the 12 months in a year, July and August showed higher soil moisture content. Lower soil moisture content was observed in November and December. Across the 12 months, site wise variations in soil moisture content were not significant only in July, August and September. Leaf life-spans of the woody species were also highly variable. About 67% species had peak of their leaf flush intiation during summer period (pre-rain leaf flushing) and rest species had their peak leaf initiation during rainy season (post-rain leaf flushing). The peak period of leaf flushing initiation at all the sites was May when 52% of the species initiated their leaf formation and the peak period of leaf flushing completion was August when 38% of the plant species completed their leaf formation. The peak period of leaf fall initiation in maximum species (54%) at all sites was November and the peak period of leaf fall completion at all the sites was February when 63% of the species shed their leaves. The peak period of leaf flush initiation as well as leaf fall initiation in most of the species at dry sites was one month before as compared to that of moist sites.

T4
P0977
The structural types of secondary xylem of perennial herbs

Ksenia Chebotareva[1], *Anna Stepanova* [1], *Shagdar Tsooj*[2]
1. Komarov Botanical Institute RAS
2. Institute of general and experimental biology MAS

Anatomical studies of wood or secondary xylem of main root of perennial herbs show that distinct annual rings are characteristic for many species. Their boundaries can be marked by the different ways: row of flattened elements, marginal parenchyma, variances in vessel diameter *etc.* But the regularities of diversity of these characters still remain unclear. More other, some peculiarities of wood structure of herbs result in terminological difficulties. We have studied in detail wood anatomy of main root on 35 herbaceous plants, relating to 14 families. The material were collected in Mongolia in 2009-2010 and analyzed with methods and terminology commonly using in wood studies. Three structural types of secondary xylem were proposed. 1. Fibrous wood, similar to secondary xylem of majority of woody plants, consists from vessels, fibres and axial and ray parenchyma. 2. Parenchymatous wood, formed mainly by vessels and parenchyma cells. 3. Zonate wood, there combination of fibrous and parenchymatous zones have been observed. This typification seems to be useful for descriptions and analyses of wood of herbs. Also some suggestions about factors, correlating with those types are proposed.

T4
P0978
Agronomic nitrogen-use efficiency of rice can be increased by driving *OsNAR2.1* or *OsNRT2.1* expression with the *OsNAR2.1* promoter

Jingguang Chen[1,2], *Xiaorong Fan*[1,2], *Guohua Xu*[1,2]
1. State Key Laboratory of Crop Genetics and Germplasm Enhancement, Nanjing Agricultural University
2. Key Laboratory of Plant Nutrition and Fertilization in Low-Middle Reaches of the Yangtze River, Ministry of Agriculture, Nanjing Agricultural University

The importance of the nitrate (NO_3^-) transporter for yield and nitrogen-use efficiency (NUE) in rice has been demonstrated previously using map-based cloning. First, we used the *OsNAR2.1* promoter to enhance *OsNAR2.1* gene expression in transgenic plants. Transgenic lines expressing *pOsNAR2.1:OsNAR2.1* constructs exhibited increased *OsNAR2.1* expression, and transcription of *OsNRT2.1*, *OsNRT2.2*, and *OsNRT2.3a* in transgenic plant roots was increased, respectively, by 117,121, and 129% in 0.2 mM NO_3^-; by 105,118, and 110%, in 2.5 mM NO_3^-; and by 76,68 and 73% in 1.25 mM NH_4NO_3, compared with WT. And ^{15}N influx in roots increased 21–32% in response to 0.2 mM and 2.5 mM $^{15}NO_3^-$ and 1.25 mM $^{15}NH_4{}^{15}NO_3$. Grain yield, agricultural NUE (ANUE), and dry matter transfer (DMT) of *pOsNAR2.1:OsNAR2.1* transgenic plants grown in the field increased 21,22, and 21%, respectively, relative to WT plants. So next we enhanced the expression of the *OsNRT2.1* gene, which encodes a high-affinity NO_3^- transporter, using a ubiquitin (*Ubi*) promoter and *OsNAR2.1* promoter to drive *OsNRT2.1* expression in transgenic rice plants. Transgenic lines expressing *pUbi:OsNRT2.1* or *pOsNAR2.1:OsNRT2.1* constructs exhibited increased total biomass including yields of approximately 21% and 38% compared with wild-type (WT) plants. The ANUE of the *pUbi:OsNRT2.1* lines decreased to 83% of that of WT plants, while the ANUE of the *pOsNAR2.1:OsNRT2.1* lines increased to 128% of that of WT plants. The DMT into grain decreased by 68% in the *pUbi:OsNRT2.1* lines and increased by 46%

in the *pOsNAR2.1:OsNRT2.1* lines relative to the WT. The expression of *OsNRT2.1* in shoot and grain showed that *Ubi* enhanced *OsNRT2.1* expression by 7.5-fold averagely and *OsNAR2.1* promoters increased by about 80% higher than the WT. Interestingly we found that the *OsNAR2.1* was expressed higher in all the organs of *pUbi:OsNRT2.1* lines, however for *pOsNAR2.1:OsNRT2.1* lines *OsNAR2.1* expression was only increased in leaf sheaths and inter nodes. The ratio of *OsNRT2.1* to *OsNAR2.1* expression was altered in the transgenic lines with ratios of approximately 11.3:1 and 4.7:1 in the *pUbi:OsNRT2.1* and *pOsNAR2.1:OsNRT2.1* lines, compared with a ratio of 7.2:1 in WT plants. We also compared the phenotypes of *pOsNAR2.1:OsNAR2.1* and *pOsNAR2.1:OsNRT2.1* transgenic lines in the field and found that post-anthesis N uptake differed significantly between them, and in comparison with the WT. PANU increased approximately 39% and 85%, in the *pOsNAR2.1:OsNAR2.1* and *pOsNAR2.1:OsNRT2.1* transgenic lines, respectively, possibly because *OsNRT2.1* expression was less in the *pOsNAR2.1:OsNAR2.1* lines than in the *pOsNAR2.1:OsNRT2.1* lines during the late growth stage. We show that increased expression of *OsNAR2.1* or *OsNRT2.1*, especially in combination with a relative lower expression ratio with its partner gene *OsNAR2.1*, can improve yield and NUE in rice.

T4

P0979

MYB89 transcription factor represses seed oil accumulation in *Arabidopsis*

***Mingxun Chen**, Dong Li, Changyu Jin*
Northwest A&F University

In many higher plants, seed oil accumulation is precisely controlled by intricate multi-level regulatory networks, among which transcriptional regulation mainly influences oil biosynthesis. In Arabidopsis, the master positive transcription factors for seed oil accumulation are *WRINKLED1* (*WRI1*), *LEAFY COTYLEDON1-LIKE* (*L1L*), and *LEC2*. We found that an R2R3-MYB transcription factor, MYB89, was predominantly expressed in developing seeds during maturation. Oil and major fatty acid biosynthesis in seeds was significantly promoted by *myb89-1* mutation and *MYB89* knockdown; MYB89 was thus an important repressor during seed oil accumulation. RNA sequencing revealed remarkable upregulation of numerous genes involved in seed oil accumulation in *myb89* seeds at 12 days after pollination. Posttranslational activation of a MYB89- glucocorticoid receptor fusion protein and chromatin immunoprecipitation assays demonstrated that MYB89 inhibited seed oil accumulation by directly repressing *WRI1* and five key genes and by indirectly suppressing *L1L* and 11 key genes involved in oil biosynthesis during seed maturation. These results help us to understand the novel function of MYB89 and provide new insights into the regulatory network of transcriptional factors controlling seed oil accumulation in *Arabidopsis*.

T4

P0980

Arabidopsis remorins modify root gravitropism against plasmodesmata-mediated virus infection

***Xu Chen**[1], Meiyu Ke[1], Yanbiao Sun[1], Dingquan Huang[1], Deyan Wang[1], Zichen Chen[1], Liwen Jiang[2]*
1. Haixia Institute of Science and Technology, Fujian Agriculture and Forestry University, Fuzhou, China
2. School of Life Sciences, Centre for Cell and Developmental Biology and State Key Laboratory of Agrobiotechnology, The Chinese University of Hong Kong

Plasmodesmata (PD) are nano-sized membrane-lined channels that connect neighboring cells for cell-to-cell communication in plants. PD are also essential passages for intercellular spread of plant viruses which move through leaf-root-leaf route, while the role of root in regulation of PD-mediated virus transportation is still unclear. So far, only a few PD proteins were characterized and the mechanisms in control of PD permeability are largely unknown. In our study, we used *Arabidopsis* root as study system, aiming to demonstrate the role of root-specific expressed Remorins (REMs) which are PD-located proteins involved in PD-mediated virus spread, and further influencing root gravitropic response. Our results showed that the four members of *AtREM1* subfamily are individually expressed in different cell layers of root. Single mutants and multiple mutants of *rems* are hypersensitive to Cytomegalo virus (CMV), and overexpressed *REMs* significantly block PD permeability. Sterol biosynthesis of overexpressed *REM* lines is overall reduced, suggesting that *REMs* maintain PD open/closure possibly through the regulation of sterol enrichment. *REM* overexpression impairs endomembrane trafficking of auxin transporter PIN2, and thus disrupting auxin gradient during gravitropic response. Based on the data presented here, we hypothesize that REM1 proteins are important regulators in control of PD permeability in the process of virus infection, meanwhile function essentially in auxin-dependent root gravitropism.

T4

P0981

SKU5 and SKS1 function redundantly to regulate cellulose synthesis in *Arabidopsis*

***Lihong Li**, Dandan Zheng, Chenjin Wen, Yuting Qiu, Dingquan Huang, Zengyu Liu, Tongda Xu, Xu Chen*
Haixia Institute of Science and Technology, Fujian Agriculture and Forestry University, Fuzhou, China

Cell wall is a fundamental plant constituent, composed of the polysaccharides cellulose, hemicellulose, pectin and lignin. Cellulose synthases (CESAs) complexes localize at plasma membrane, being the critical components for cellulose biosynthesis, while regulation of this process is poorly understood. In our study, we found a glycosylphosphatidylinositol (GPI)-anchored glyco protein SKU5 which is highly co-expressed with all CESAs at transcription level. Further study showed that *sku5 sks1* double mutant (SKS1 is a homolog of SKU5) decreases crystallized cellulose level and exhibits typical cell wall-related defects. CESA3 speeds and trajectories were severely altered in *sku5 sks1* double mutant, indicating SKU5 and SKS1 are important regulators of CESAs involved in cellulose synthesis. As GPI-anchored proteins, we hypothesize that SKU5 and SKS1 might affect phospholipid-mediated membrane stability which is required for CESAs embedding. Hence, we have uncovered a mechanism for how SKU5 and SKS1 modulate CESA-dependent cellulose synthesis, which helps us to better understand the signals upstream of CESA regulation.

T4
P0982

Researches on some structural genes in flavonoid metabolic pathway of *Dryopteris erythrosora*

Xuefei Chen[1,2], *Jianguo Cao*[1], *Xiling Dai*[1], *Quanxi Wang*[1], *Youfang Wang*[2]
1. Shanghai Normal University
2. East China Normal University

Flavonoids are a kind of important secondary metabolites in plants. In recent years, the synthesis of flavonoids and their molecular mechanisms have been extensively studied in flowering plants. However, the molecular mechanism of flavonoid biosynthetic pathway in ferns are less characterized. In this study, the unigenes involved in flavonoid metabolic pathway were identified from the transcriptome database of *Dryopteris erythrosora*. These unigenes include chalcone synthase (28 unigenes), chalcone isomerase (6 unigenes), flavanone 3-hydroxylase (8 unigenes), flavonol synthase (64 unigenes) and flavanone 4-reductase (18 unigenes). We screened candidate genes for functional investigation by cluster analysis and redundancy deletion from the unigenes respectively. The expression of these genes has been detected by qPCR with different light intensity treatment. Changes in expression levels are inconsistent from different genes. Among them, *DeCHS1* and *DeCHI1* have high levels in buds. The expression levels of *DeF3H1* and *DeF3H2* were low and decreased under the bright light. *DeFLSs* were highly distributed in buds; *DeFLS4* and *DeFLS6* were mainly distributed in buds and affected by light, may involved in plant development. Only *DeDFR2* showed very high expression of *DeDFRs*. In addition, by using the prokaryotic expression and enzyme activity analysis, we found all of three DeCHI proteins can not only catalyze naringenin chalcone to naringenin, but also catalyze isoliquiritigenin to liquiritigenin. This function was only reported in CHI-type II genes of legume plants and some bryophytes. It was not fully consistent with the evolutionary relationship, and need to be further verified. The results of enzyme activity tests of DeDFR1, DeDFR2 and DeDFR4 in vitro show that different products can be obtained under different DeDFR. Ferns are extremely important groups in the process of plant evolution. The current research results also show that the ferns have their particularity in the process of flavonoid biosynthesis. This study provides scientific data for elucidating the flavonoid metabolic pathway of ferns and flavonoid gene evolution.

T4
P0983

Effect of controlled deterioration on seed vigor of 4 oat (*Avena satuva L.*) cultivars

Hang Cheng, *Lingling Chen, Ye Zhang, Mingya Wang, Peisheng Mao*
China Agricultural University

Seed deterioration occurs during storage. This would lead to the decline of seed quality and ultimately the loss of seed vitality, which caused huge economic loss in the agricultural production. In order to explore the deterioration tolerance capacity of different oat (*Avena satuva L.*) cultivars of Qinghai Province, 4 main cultivars were studied in this research. All the oat seeds with 16% moisture content were deteriorated at 45°C for 0,1, 2,3, 4,5 d. Then the germination percentage, germination potential, germination index and the average germination time were determined. The results showed that the germination of all these cultivars were inhibited after seed deterioration. The germination percentage, germination potential, germination index were declined and the average germination time was increased. The germination percentage of Baiyan 7 and Linna shapely declined with deterioration and lost the vitality after deteriorated 4 d. However, at the same time, the germination percentage of Qinghaitianyanmai and Jiayan 2 were 38% and 3%, respectively. After 2 d deterioration, the germination potential and germination index were the maximum in Qinghaitianyanman and minimum in Baiyan 7. The average germination time of Qinghaitianyanmai was the lowest among all these cultivars during seed deterioration. As a conclusion, Qinghaitianyanmai had better deterioration tolerance than Baiyan 7, Linna and Jiayan 2 and could be prior to use in agricultural production.

T4
P0984

Effect of waterlogging on the growth and photosynthetic characteristics of *Cynodon dactylon* in Three Gorges Reservoir area, China

Ruimei Cheng
Institute of Forest Ecology, Environment and Protection, Chinese Academy of Forestry, Beijing

To explore the effects of water levels in the Three Gorges Reservoir in the growth, physiological and chlorophyll fluorescence traits of *Cynodon dactylon,* four different water treatments were applied to the plants to study the responses of photosynthetic physiology to waterlogged regimes, including normal growth group, root submergence group, semi-plant submergence group and whole plant submergence group. Results indicated that: (1) The basal diameter, stem height and non-photochemical quenching tended to increase in different waterlogging treatments for 45 d, illustrating that *C. dactylon* had adaptive response to different waterlogged environments. (2) The responses of the leaf length, leaf width and aboveground biomass in different waterlogged environment were different. Compared with the normal growth group, the semi-plant submergence group had an increase in the leaf length, leaf width and aboveground biomass, whereas the root submergence group and the whole plant submergence group had a decrease in those parameters. (3) The specific leaf area, net photosynthetic rate, stomatal conductance, transpiration rate, maximal photochemical efficiency, electron transfer rate and photochemical quenching of all waterlogged groups were significantly lower than thc normal growth group. The above-mentioned parameters in the root submergence group and the semi-plant submergence group were able to maintain at a relatively high level, while that the whole plant submergence group declined to extremely lower. This study suggested that *C. dactylon* can adapt to waterlogged environment well (especially in partial waterlogged environment), and thus it can be applied to the re-vegetation and landscape restoration of Hydro-Fluctuation Belt in Three Gorges reservoir area in future, especially in the low altitude region (145-160 m) which suffers from the long-term water level fluctuation. Meanwhile, those tolerate-submergence characteristics of *C. dactylon* also provide a basis for screening riparian plants of Three Gorges Reservoir, China.

T4

P0985

Elucidation of mechanism of formation of amino acids in dark-treated tea (*Camellia sinensis*) leaves

Sihua Cheng[1], *Yiyong Chen*[1], *Xiumin Fu*[2], *Ziyin Yang*[2]

1. South China Botanical Garden, University of Chinese Academy of Sciences

2. South China Botanical Garden, Chinese Academy of Sciences

Effect of dark on crop growth has been widely studied, while less attention has been paid to effect of dark on quality-related metabolites in crops. Shade management (dark treatment) on tea (*Camellia sinensis*) plants is a common approach to improve free amino acids in raw materials of tea leaves. However, the reason for accumulation of free amino acids in dark-treated tea leaves is still unknown. In the present study, dark treatment significantly increased content of free amino acids and reduced content of soluble proteins in tea leaves. Quantitative proteomics analysis showed that most enzymes involved in biosyntheses of amino acids were downregulated by dark treatment. Chloroplast numbers reduced in dark-treated leaves and the content of soluble proteins reduced in the chloroplasts isolated from dark-treated leaves compared to control. These suggest that proteolysis of chloroplast proteins contributed to amino acid accumulation in dark-treated leaves. Two chloroplasts proteases, ATP-dependent Clp protease proteolytic subunit 3 and protease Do-like 2, were upregulated in dark-treated leaves. This study firstly elucidated the mechanism of accumulation of amino acids in dark-treated tea leaves.

T4

P0986

In vivo calcium dynamics in plant cells: A holistic view

Alex Costa

University of Milan, Department of Biosciences

In plants, increases in cytosolic Ca^{2+} concentration ($[Ca^{2+}]_{cyt}$), occuring in response to both biotic and abiotic stimuli, work as a key components of different signal transduction pathways. Depending on the stimulus, Ca^{2+} rises can display the form of a single transient or repetitive Ca^{2+} oscillations and are commonly designated as "Ca^{2+} signatures". Generation and shaping of Ca^{2+} signatures depends on fine-tuning of Ca^{2+} influxes and effluxes occurring at both the plasma membrane (PM) and membranes of the different subcellular compartments. The opening of PM Ca^{2+}-permeable influx channels (e.g GLRs, CNGCs, OSCAs....) in response to different stimuli will release Ca^{2+} into the cytosol and cause the generation of a Ca^{2+} spike, while activity of active Ca^{2+} efflux transporters (e.g H^+-Ca^{2+} antiporters, Ca^{2+}-ATPases...) will return the $[Ca^{2+}]_{cyt}$ to resting concentrations. Not only cytosol but also organelles and other subcellular compartments (e.g. chloroplasts, mitochondria, endoplasmic reticulum...) experience Ca^{2+} transients, hence putatively participating in the cellular Ca^{2+} homeostasis and potentially in the Ca^{2+} signature shaping process. In order to precisely define how cytosolic and organellar Ca^{2+} dynamics are generated and shaped in different plant cell types, we have followed two complementary strategies: i) the development of new molecular tools for the *in vivo* study of the Ca^{2+} transport through plasma and organelle membrane/s; ii) the analyses of Ca^{2+} transients in cells of complex tissues and organs, in natural context and in close-to physiological conditions. To fulfil the above requirements, the following evidence will be presented: a) *in planta* use of the genetically encoded FRET (Förster Resonance Energy Transfer)-based Ca^{2+} Cameleon sensors. Attention will be given to the developed Cameleon probes for the analyses of Ca^{2+} dynamics in several subcellular compartments resulting in the identification of organellar Ca^{2+} transport mechanisms; b) development of a new microscopy solution, based on Light Sheet Fluorescence Microscopy (LSFM), for FRET-based Ca^{2+} imaging in root; c) use of the presented technologies for the study of mutants impaired in the generation and regulation of cytosolic Ca^{2+} dynamics involved in development processes. Our approach has allowed to identify and study components of PM, mitochondria and ER localized Ca^{2+} transporters which are effective in regulating Ca^{2+} dynamics in the different compartments in response to external stimuli or developmental processe.

T4

P0987

Suppression of *OsIBR*, encoding an E3 ubiquitin ligase, results in the enhanced Brassinosteroid responses in rice (*Oryza sativa* L.)

Hye Jo Min, ***Li Hua Cui***, *Woo Taek Kim*

Yonsei University

Ubiquitination is a critical post-translational modification of diverse cellular proteins. In sequential steps of ubiquitination, E3 ubiquitin ligase provides substrate specificity by transfer ubiquitin (Ub) to target proteins. In higher plants, E3 ubiquitin ligases are involved in various cellular processes, including hormonal and environmental stress responses. In this study, we elucidated possible role of a novel rice E3 Ub ligase IBR in brassinosteroid (BR) signaling. We identified a T-DNA inserted rice mutant, *osibr (increased brassinosteroid response)*. The *osibr* mutant plant showed increased BR responses in terms of leaf morphology, vegetative growth, and BR-related gene expressions. *osibr* displayed an increased lamina joint bending phenotype at both seedling and vegetative stages. When *osibr* was grown in the presence of BR or propiconazole (Pcz), an inhibitor of BR synthesis, it exhibited BR-sensitive and Pcz-insensitive phenotypes, respectively. In addition, the expression levels of BR biosynthesis genes were lower in the mutant plants than those of wild-type rice plants, suggesting an increased level of BR signaling in the mutant plant. Bacterially expressed OsIBR contained an E3 ubiquitin ligase activity and displayed dual-localization patterns to nucleus and cytoplasm. OsIBR directly interacts with OsBZR1, a positive regulator of BR signal transduction pathway, as evidenced by the results of yeast two-hybrid and in vitro pull-down assays. A higher amount of OsBZR1 was detected in the protoplasts prepared from *osibr* as compared to the wild-type protoplasts. Consistent with these results, OsBZR1 was poly-ubiquitinated by OsIBR and degraded by the 26S proteasome system. Collectively, our results suggested that OsIBR participates in the BR response as a negative factor by down-regulating OsBZR1 in rice plant.

T4
P0988
Optimization of regeneration and transformation in *Sesamum indicum* L. cultivar JK-1 for studying tissue specific promoters
***Anirban Jyoti Debnath**, Debabata Basu, Samir Ranjan Sikdar*
Bose Institute

For the crop improvement of *Sesamum indicum*, heterosis breeding can be an approach in which a male sterile/restorer line of sesame is necessary. Male sterility can be achieved by silencing of anther specific genes. For this approach, study of tapetum specific promoter is a perquisite phenomenon. Since sesame is an extremely recalcitrant plant, regeneration protocol of this plant was standardised at first for this study. Various combinations of 6-benzylaminopurin, thidiazuron, α-napthalinic acetic acid, indole acetic acid, indole butyric acid and gibberellic acid were used to standardise each step of organogenesis, shoot elongation and root induction. Among the different types of tried explants, best regeneration frequency (94.82±1.34%) was obtained from 4 days old cotyledon. To find an anther specific gene, sesame anther specific cDNA-subtracted library was created and screened thoroughly. One EST having homology with β-1,3 glucanase was isolated. Upstream element of the gene was identified by promoter walking, cloned in pPZPY112 vector and was fused with GUS reporter gene. TA29 is a known anther specific gene of *Nicotiana tabacum.* Promoter of TA29 served as control in this study. The promoter region of TA29 was amplified from *Nicotiana* genomic DNA and cloned in pCAMBIA2301 and was fused with GUS reporter gene. *Agrobacterium tumefaciens* A4404 was transformed by both the constructs. Both constructs were used for *Agrobacterium*-mediated transformation of sesame, taking *Nicotiana* as control. Transgenic lines of sesame and *Nicotiana* were raised and confirmed by PCR and southern hybridisation. Anther-specific GUS expression of the both promoters were visualised by sectioning of anther.

T4
P0989
How does fast or slow drought affect the drought-induced woody seedling mortality?
***Honglang Duan**[1], Yue Xu[2], Houbao Fan[1], Juxiu Liu[2]*
1. Jiangxi Key Laboratory for Restoration of Degraded Ecosystems & Watershed Ecohydrology, Nanchang Institute of Technology
2. Key Laboratory of Vegetation Restoration and Management of Degraded Ecosystems, South China Botanical Garden, Chinese Academy of Sciences

The worldwide forest mortality has been attributed to severe droughts and increasing temperatures. Despite substantial progress in unravelling physiological mechanisms underlying drought-induced tree mortality, details about carbon and water relations towards tree mortality remain still unresolved. In particular, it is currently poorly understood how the intensity and duration of drought would impact the relative contributions of hydraulic failure and carbon depletion to mortality of subtropical woody species. In the context of global climate change characterizing with changing precipitation patterns, bridging this knowledge gap will help to improve the predictions in forest responses to future climates in the subtropical regions of China. In this study, one year old potted seedlings of three native woody species in south China, i.e. *Schima superb* (a tree species), *Castanopsis chinensis* (a tree species) *and Syzygium rehderianum* (a shrub species), were well watered and fertilized until the commencement of soil drought treatments (i.e. fast drought and slow drought). Water was completely withheld in the fast drought treatment, while 80-90% of the whole plant water loss was rewatered to seedlings in the slow drought treatment. We measured leaf gas exchange, growth, plant hydraulics, non-structural carbohydrates (NSC) as well as rate of mortality regularly until the death of seedlings, enabling to distinguish the responses of carbon and water relations to fast or slow drought. We are also able to compare drought strategies among the three species. Results showed that leaf gas exchange, water potential and xylem hydraulic conductance of the three species exhibited more profoundly declines in the fast drought treatment than slow drought treatment, indicating that fast drought had greater impacts on physiological responses than slow drought. *Syzygium rehderianum* had greater drought sensitivities in leaf gas exchange, leaf water potential and xylem embolism, associated with less negative P50 (i.e. the xylem potential at which 50% of xylem hydraulic conductance was lost) and turgor loss point, while *Schima superb* had higher drought tolerance. Organ NSC will also be reported. In combination with NSC data, we are able to determine the relative contribution of hydraulic failure and carbon depletion to seedling mortality and generate a better understanding of effects of drought intensity and duration on carbon and water relations. Furthermore, implications of shifts in forest composition and functioning in subtropical China will be obtained.

T4
P0990
Genes sufficient for synthesizing peptidoglycan are retained in gymnosperm genomes, and MurE from *Larix gmelinii* can rescue the albino phenotype of *Arabidopsis MurE* mutations
*Yilan E[1], Ningning Li[1], Kudo Hiromi[2], **Xiaofei Lin**[1], Zhang Zhe[3], Takano Hiroyoshi[2]*
1. College of Life Sciences, Inner Mongolia University
2. Graduate School of Science and Technology, Kumamoto University
3. College of Biological Science, China Agriculture University

The endosymbiotic theory states that plastids are derived from a single cyanobacterial ancestor that possessed a cell wall. Peptidoglycan (PG), the main component of the bacteria cell wall, gradually degraded during plastid evolution. PG-synthesizing *Mur* genes have been found to be retained in the genomes of basal streptophyte plants, although many of them have been lost from the genomes of angiosperms. The enzyme encoded by bacterial *MurE* genes catalyzes the formation of the uridine diphosphate-N-acetylmuramic acid (UDP-MurNAc) tripeptide in bacterial PG biosynthesis. Knockout of the *MurE* gene in the moss *Physcomitrella patens* resulted in defects of chloroplast division, whereas T-DNA-tagged mutants of *Arabidopsis thaliana* for *MurE* revealed inhibition of chloroplast development but not of plastid division, suggesting that AtMurE is functionally divergent from the bacterial and moss MurE proteins. Here, we could identify 10 homologs of bacterial *Mur* genes, including *MurE*, in the recently sequenced genomes of *Picea abies* and *Pinus taeda*, suggesting the retention of the plastid PG system in gymnosperms. To investigate the function of gymnosperm MurE, we isolated an ortholog of *MurE* from the larch, *Larix gmelinii* (*LgMurE*) and confirmed its presence as a single copy per genome, as well as its abundant expression in the leaves of larch seedlings. Analysis with a fusion protein combining GFP and LgMurE suggested that it localizes in chloroplasts. Cross-species complementation assay with MurE mutants of *A. thaliana* and *P. patens* showed that the expression of *LgMurE* cDNA completely rescued the albefaction defects in *A. thaliana* but did not rescue the macrochloroplast phenotype in *P. patens*. The evolution of plastid PG and the mechanism behind the functional divergence of *MurE* genes were discussed in the context of information about plant genomes at different evolutionary stages.

T4

P0991

Genetic mechanisms underlying the evolutionary switches between pigments in *Caryophyllales*

Tao Feng[1,2], *Samuel Brockington*[1]

1. University of Cambridge

2. Wuhan Botanical Garden, CAS

Three groups of pigments, carotenoids, flavonoids including anthocyanins, and betalains are responsible for the colouration of flower, fruit, seed and vegetarian organ. Whereas anthocyanins are the most common pigments and broadly distributed across flowering plant, betalains replaced anthocyanins in the order of the Caryophyllales. However, along with the diversification of Caryophyllales, betalains have been lost and the pigment type reversed back to anthocyanin in at least three lineages. Intriguingly, these two kinds of pigments are mutually exclusive and have never been found in one single species. In betalain pigmented species, it is known that the structural enzymes within anthocyanin pathway are evolutionarily maintained but downregulated at the regulatory level. However, the fate of the betalain synthesis pathway, when the pigment type reverses to anthocyanic pigmentation is not known. Here we examine evolutionary fate of the betalain synthesis pathway across multiple reversions to the anthocyanic condition, and show that invariably the betalain pathway, is in various stages of degeneration and pseudogenisation, suggesting that mutual exclusion of the two pigment types is in part the result of wholesale loss of the betalain pathway.

T4

P0992

Genetic mechanisms underlying the evolutionary switches between pigments in caryophyllales

Tao Feng[1,2], *Samuel Brockington*[1]

1. University of Cambridge

2. Wuhan Botanical Garden, CAS

Three groups of pigments, carotenoids, flavonoids including anthocyanins, and betalains are responsible for the colouration of flower, fruit, seed and vegetarian organ. Whereas anthocyanins are the most common pigments and broadly distributed across the flowering plant, betalains replaced anthocyanins in the order of the Caryophyllales. However, along with the diversification of Caryophyllales, betalains have been lost and the pigment type reversed back to anthocyanin in at least three lineages. Intriguingly, these two kinds of pigments are mutually exclusive and have never been found in one single species. In betalain pigmented species, it is known that the structural enzymes within anthocyanin pathway are evolutionarily maintained but downregulated at the regulatory level. However, the fate of the betalain synthesis pathway, when the pigment type reverses to anthocyanic pigmentation is not known. Here we examine the evolutionary fate of the betalain synthesis pathway across multiple reversions to the anthocyanic condition, and show that invariably the betalain pathway, is in various stages of degeneration and pseudogenisation, suggesting that mutual exclusion of the two pigment types is in part the result of the wholesale loss of the betalain pathway.

T4

P0993

Secondary woodiness in African representatives of Tordylieae (Apioideae, Apiaceae): Paedomorphic traits are confined to rosette shoots

Kamil Frankiewicz[1], *Alexei Oskolski*[2,3], *Anthony Magee*[4], *Łukasz Banasiak*[1], *Krzysztof Spalik*[1]

1. Department of Molecular Phylogenetics & Evolution, Institute of Botany, Faculty of Biology, University of Warsaw, Biological & Chemical Research Centre, ul. Żwirki i Wigury 101, 02-089 Warszawa

2. Department of Botany and Plant Biotechnology, University of Johannesburg, PO Box 524 Auckland Park 2006, Johannesburg, South Africa

3. Botanical Museum, Komarov Botanical Institute, Prof. Popov 2, 197376 St.Petersburg, Russia

4. Compton Herbarium, Biosystematics and Collections, South African National Biodiversity Institute (SANBI), Compton Herbarium, Kirstenbosch Research Centre, Kirstenbosch National Botanical Garden, P/Bag

Tordylieae is mostly herbaceous clade of Apiaceae subfamily Apioideae. However, several species of its African lineage have stems with woody base while the genus *Notobubon* is shrubby. Character mapping on molecular tree suggests that this genus is secondarily woody, i.e. wood habit was derived from herbaceous ancestor. According to Carlquist's theory of paedomorphosis such plants should demonstrate juvenile wood traits prolonged into secondary xylem. To verify if the theory is applicable to woody Tordylieae, we have examined secondary xylem in shrubby *Notobubon capense* and herbaceous representatives of closely related genera: perennial *Afrosciadium magalismontanum*, annual *Capnophyllum macrocarpum* and perennial *Dasispermum suffruticosum*.

We found that the wood of *N. capense* as well as the outer regions of secondary xylem in *D. suffruticosum* and *C. macrocarpum* have the suite of characters that is common for other woody Apioideae (e.g. *Bupleurum*, *Heteromorpha* etc) without any paedomorphic traits. However, inner regions in two latter species are paedomorphic, as expressed by scalariform pitting on lateral vessel wall. Both species are similar in forming lateral plagiotropic polycarpic shoots branched off from the rosette shoot with compressed internodes topped with terminal umbel. *Afrosciadium magalismonta-*

num also forms rosette, but topped with orthotropic monocarpic shoot. Wood in this species resembles inner regions of wood in *D. suffruticosum* and *C. macrocarpum* and is completely paedomorphic. Our observations allow us to speculate that development of paedomorphic wood in these taxa is confined to rosette shoots with short internodes. Apparently, the onset of formation of lateral polycarpic shoots as in *Notobubon* can trigger the deposition of advanced wood lacking paedomorphic traits. The shrubby habit in *Notobubon* could be result of longer persistence of lateral polycarpic shoots in its perennial herbaceous ancestor.

T4

P0994

Study on interspecific and intraspecific allelopathy of *Pinus thunbergii, Pinus densiflora* and *Pinus tabuliformis*

Yue Lin Hou, *Yuan Gao*

Linyi Scientific Exploration Laboratory

The photosynthetic pigments (chlorophyll a, chlorophyll b, carotene and anthocyanin) and organic acids (citric acid, oxalic acid and malic acid) of the seedlings of *Pinus thunbergii, Pinus densiflora* and *Pinus tabuliformis* were studied by the water leaching solution of *Pinus thunbergii, Pinus densiflora* and *Pinus tabuliformis*. The results show: 1) The water leaching solution of *Pinus thunbergii* had no significant effect on *Pinus thunbergii, Pinus densiflora* and *Pinus tabuliformis*. 2) The water leaching solution of Pinus densiflora had significant effect ($P < 0.05$) on *Pinus thunbergii* and *Pinus densiflora*, and had no significant effect on *Pinus tabuliformis*. 3) The water leaching solution of *Pinus tabuliformis* had significant effect ($P < 0.05$) on *Pinus densiflora* and *Pinus tabuliformis*, and had no significant effect on *Pinus thunbergii*. 4) In this study, the promoting allelopathy of *Pinus densiflora* and the inhibitory allelopathy of *Pinus tabuliformis* were very significant ($P < 0.05$), but the allelopathy of *Pinus thunbergii* was not significant.

T4

P0995

Study on kin recognition and physiological strategies of adjacent *Pinus thunbergii*

Yuan Gao, *Jiahui Guan*

Linyi Scientific Exploration Laboratory

The kin recognition of plants is a plant that adjusts its own growth ecological strategy and promotes the survival and reproduction of the relatives by identifying the relationship between the surrounding individuals and their own. At present, there is still much controversy about whether the genetic identification of the plants is real. This study used Chinese typical gymnosperms and afforestation-*Pinus thunbergii*. High performance liquid chromatography (HPLC) was used to detect the content of chlorophyll (chlorophyll a and chlorophyll b) in the primary production capacity and phenolic acid (ferulic acid, vanillic acid and syringic acid) in the secondary metabolites, to try to reveal the existence of kin recognition of gymnosperms. The results of the study showed that pine needles of *Pinus thunbergii* of chlorophyll a, chlorophyll b, ferulic acid, vanillic acid, syringic acid and ferulic acid showed is a group of far-genetic relatedness was significantly higher than a group of near-genetic relatedness. This study reveals that under natural conditions, *Pinus thunbergii* can identify neighbor strains phylogenetic distance, and correspondingly adjust physiological strategy, such as primary production and secondary metabolite distribution differences. This is the first study to discover and validate kin recognition of *Pinus thunbergii* is a real existence, as the world's first reported in gymnosperms, global the second case reports of woody plants.

T4

P0996

Dynamic localization of ROP2 GTPase in root hairs requires the trans-Golgi network/early endosome-mediated vesicle trafficking in parallel to GDI1

Fu-Rong Ge[1], *Sen Chai*[1], *Qiang-Nan Feng*[1], *En Li*[1], *Jian-Wei Pan*[2], *Sha Li*[1], *Yan Zhang*[1]

1. State Key Laboratory of Crop Biology, College of Life Sciences, Shandong Agricultural University

2. College of Chemistry and Life Sciences, Zhejiang Normal University, Zhejiang

Plant tip growth is regulated by spatiotemporally restricted localization of activated ROPs. Extensive studies suggested evolutionarily recurrent mechanisms on the dynamic localization of activated ROPs by guanine nucleotide exchange factors (RopGEFs) and GTPase activating proteins (RopGAPs). However, whether guanine nucleotide dissociation inhibitors (GDIs) and vesicular trafficking play roles in the dynamic localization of ROPs, like their counterparts in other phyla, was unclear. Here, we examined the roles of GDI1 and vesicle trafficking on the dynamic targeting of ROP2 during *Arabidopsis* root hair growth. We show that the dynamic targeting of ROP2 relies on brefeldin A (BFA)-sensitive, ADP-ribosylation factor1 (ARF1)-mediated post-Golgi trafficking. Specifically, ROP2 recycling depends on Adaptin Protein 1 (AP1)-regulated protein sorting at the trans-Golgi network/early endosomes (TGN/EE), but not on RabA4b-mediated post-Golgi secretion or RAB5-mediated vacuolar trafficking. Functional loss of GDI1 significantly reduced ROP2 recycling at the apical plasma membrane while genetic interference of vesicle trafficking significantly contributed to ROP2 recycling only at the absence of GDI1. Results presented here demonstrate a crucial role of GDI1 and TGN/EE-mediated protein sorting in ROP2 targeting during root hair growth and reveal an evolutionarily conserved regulatory mechanism underlying the establishment and maintenance of a single and unique polarization site during tip growth.

T4

P0997

Functional analysis of kation transport proteins in salt-tolerance of the halophyte *Nitraria sibirica*

Xin Geng, Li Wang, Chaoran Li, Yukun Ma, ***Xiaofei Lin***

College of Life Sciences, Inner Mongolia University

The mechanism of that saline plants adapt to salt-stress environment, involves in complex factors, such as cell structure, physiological processes and metabolism regulation. Among these complicated processes, the kation transporters play important roles in plant salt-tolerance. Na^+/H^+ antiporters are integral proteins located in vacuolar membrane and cytomembrane. The cytomembrane Na^+/H^+ antiporter proteins (SOS1) can transport cytoplasmic Na^+ to the exocellular, and the vacuolar Na^+/H^+ antiporter proteins (NHX1) can deliver Na^+ into the vacuoles against the electrochemical gradient. The compartmentation of Na^+ provides an effective mechanism to avert the deleterious effects of Na^+ on cytosol, and maintain an osmotic potential by Na^+ in vacuoles. On the other hand, high-affinity potassium transport protein (*HKT1*) located on the plasma membrane, uninstalls excessive Na^+ from the xylemby to the phloem by the way of symplasmic, then transports back to the roots, to prevent the excessive accumulation of Na^+ in aboveground tissues. *Nitraria* L. is a genus of flowering plants in Zygophyllaceae family, which grows primarily in deserts and semideserts, and exhibits strong salt-resistance. Therefore, to elucidate the molecular mechanism of salt-tolerance and to exploit salt-tolerant genes in *Nitraria* is very necessary. In the study, we isolated orthologs of *NHX1*, *SOS1* and *HKT1* genes from *Nitraria sibirica Pall* using homology-based cloning methods, and analyzed the molecular features and expression characteristics of the *NsHX1*, *NsSOS1* and *NsHKT1*.These results showed that the expression of *NsHX1*, *NsSOS1* and *NsHKT1* were induced by abiotic factors, such as drought, cold and salt. On the other hand, the function and application value of these genes in improving salt tolerance of plants were investigated by transforming *Arabidoposis thiliana* and *Populus tomentosa*. The results indicated that over-expression of *NsHX1*, *NsSOS1* and *NsHKT1* enhanced salt-tolerance of transgenic plants, suggesting that these genes play important roles in adapting to salt-stress environment of *N. sibirica*, and will likely be useful for improving salt-tolerance of forage grass agricultural crops and forest trees.

T4

P0998

Pectin modification and promotion of root elongation by Aluminum in *Camellia sinensis* L. seedlings

Masoumeh Safari

Tarbiat Modares University (TMU)

Aluminum (Al) toxicity is a major factor limiting crop productivity in acidic soils. The cell wall is the first point of contact when plant roots are exposed to Al, and serves as a major pool for the metal. In Al-sensitive plants, Al bound to the cell wall negatively affects wall structure and function by increasing its rigidity and reducing cell expansion and mechanical extensibility, thus inhibits root elongation. The major Al binding site in cell walls are pectic polysaccharides. Demethylated, negatively charged carboxylic groups of pectin, bring about from the activity of pectin methylesterase (PME), have a high affinity for Al. Different from most of terrestrial plants, tea (*camellia sinensis* L.) growth is stimulated by Al and especially its root elongation is promoted. In this study the effects of Al on root elongation, the mechanical extensibility of the cell wall, and the modification of pectin fractions in the roots of tea were examined.Ten day old tea seedlings were treated with 400 µM Al in an aerated modified ¼ strength Hoagland solution (without phosphate, pH 4.5), for 8 days. Mechanical change of root apex of seedlings was monitored by freeze–thawing method. monoclonal antibodies (JIM5 and JIM7), which are specific for cell wall pectin differing in the degree of methylation, were used for immunofluorescence localization of cell wall pectins. Uronic acid content and the activity of PME were measured. Exposure to Al significantly promoted root cells extensibility and increased cell elongation of tea seedlings. Using immunofluorescence localization of pectin with specific antibodies bright fluorescence of antibody JIM7 was observed in sections obtained from Al-treated root apices demostrating high-methyl-ester pectins, compared to the control root sections where the fluorescence intensity of JIM5 (stains low-methyl-ester pectins) was high. Moreover, in comparison with the control, Al treatment resulted in a significant decrease of PME activity. Analyses of pectin fractions showed elevation of the ratio of total sugars to uronic acids in Al-exposed roots up to twice of the control ones. The result suggest that structural modifications of pectin in elongation zone of tea roots prevent insertion and accumulation of Al and its inhibitory role. However the mechanism by which elongation is triggered is yet to be clarified.

T4

P0999

Cell culture of *Acanthophyllum glandulosum* L. as an alternative source for saponin

Mitra Jamshidi

1. Research and Development Section, Sehat Industrial and Commercial CO., Tehran, Iran,

2. Department of Plant Biology, Faculty of Biological Science, Tarbiat Modares University (TMU), Tehran, Iran

Saponins are a large family of secondary metabolites which are structurally-related compounds of steroid or triterpenoid aglycone (sapogenin) linked by glycosidic linkage to one or more oligosaccharide moieties. Because of their surfactant and foaming properties, saponins are used in production of shampoos, liquid detergents, toothpastes. Due to critical function of foam in removal of oils and fats from surfaces, foaming characteristic of detergents has an important role in their acceptability. Saponins are also used as emulsifier and long-lasting foaming agents in beverages. *Acanthophyllum glandulosum* L. is one of the endemic species of the genus Acanthophyllum (Caryophyllaceae) in Iran, which because of the prevalence of saponin in its roots has been traditionally used as a detergent. Despite the wide use, the plant has never been cultivated, neither it's various properties e.g., seasonal variations of growth are well studied. Biotechnological approaches have been recently provided researchers with the opportunity to find alternatives for production of secondary metabolites, other than the intact and sometimes endangered plants. It is widely accepted that plant tissue cultures have potential to produce bioactive compounds, in the amounts comparable with the intact plant, with the great advantage of sustainability of the cell cultures. Moreover, there is no need for harsh methods for extraction. In the present study a rapid growthing cell line of *Acanthophyllum glandulosum* L. was established from the seeds in a modified LS medium, supplemented with 3 mg L^{-1} of NAA, 3 mg L^{-1} of IAA, and 1 mg L^{-1} 2,4-D. The latter was omitted as soon as callus was emerged. Saponins were extracted from callus as well as intact plants root with diethyl ether and n-butanol subsequently. Steroid saponins were assessed spectrophotometrically using anisaldehyde-ethyl acetate and triterpenoid saponins with vanillin-sulfuric methods. The contents of saponin in calli was evaluated as 3.25 mg g^{-1} DW, which was noticeable vs. of the intact roots, 0.4 mg g^{-1} DW. The results clearly introduce the callus of *Acanthophyllum glandulosum* L. as an appropriate natural source for saponins. It is also noteworthy that saponins can be easily and efficiently extracted from this source.

T4

P1000

Crucial role of fructan in the maintenance of membranes of wheat seedlings during severe drought stress

Farnoosh Nemati
Tarbiat Modares University (TMU)

The plasma membrane is thought to be the major target for drought stress damage due to changes in composition and structures of water and phospholipids. Removal of water from the membrane disrupts the normal bilayer structure and results in the membrane becoming exceptionally porous when desiccated. Fructan has the capacity to stabilize membranes and prevent leakage during water cessation. In the present study the relationship between fructan content and membrane damages was evaluated in 4-day old seedlings of a drought-tolerant (Sirvan) and a drought-sensitive (Marvdasht) wheat cultivar exposed to 7 days water cessation and subsequent re-watering. In comparison with sensitive cultivar, the tolerant one accumulated more fructan (3.56 ± 0.3 µg/g FW) meanwhile decreased the level of membrane leakage and lipid peroxidation. Freeze–thawing experiment showed that drought stress remarkably increased freeze-disrupt coefficient of sensitive cultivar (4.5 fold higher than of tolerant one) and severely damaged the epidermal and outer cortex cells of their leaves. The results suggest fructan as a flexible soluble carbohydrate with a direct protective effect on membrane integrity which plays a crucial role in the tolerance of wheat seedling against drought stress conditions.

T4

P1001

Generation of a Humidity-sensitive Male Sterility (HSMS) system for hybrid wheat breeding

*Changfeng Guo, Song Bo, Yingchun Zhang, **Xiaoquan Qi***
Institute of Botany, Chinese Academy of Sciences, Nanxincun 20, Fragrant Hill, Beijing

Heterosis has been widely used in crops such as rice and maize for production of higher grain yield. Wheat (*Triticum aestivum*) is one of the most important cereals in the world. But heterosis in wheat has not been extensively explored due to lack of easy-to-use conditional male-sterile germplasm. The commonly used photoperiod-sensitive male sterility (P SMS) suffers from unpredictable changes of temperatures, causing failure in hybrid production. Our previous study on rice showed that a gene (*OsOSC12*) encoding 2,3-oxidoapualene cyclase for biosynthesis of a triterpene poaceatapetol involves in pollen coat formation. Lost-function mutants of *OsOSC12* are male sterile at normal relative humidity (RH) and fertile when RH is higher than 80%. This humidity-sensitive male sterility (HSMS) has the advantage of using two-line hybrid breeding system for hybrid production. In this study we identified three conserved homologs of *OsOSC12* in the hexaploid wheat (*Triticum aestivum*) genome. Each wheat genome has one conserved gene, and the three annotated genes from the A, B and D genomes were named as *TaOSC12A*, *TaOSC12B*, and *TaOSC12D*, respectively. By using TILLING (Targeting Induced Local Lesions IN Genomes) technology and Sanger sequencing, 9 mutations in *TaOSC12A*, 5 mutations in *TaOSC12B*, and 14 mutations in *TaOSC12D* were identified. GC-MS analysis of *Nicotiana benthamiana* heterologous expressing the three wheat genes revealed that *TaOSC12B* and *TaOSC12D* produced poaceatapetol, while *TaOSC12A* failure to product this triterpene. We expect that crosses between loss-of-function mutants of *TaOSC12B* and *TaOSC12D* will generate HSMS plants for hybrid seed production in wheat.

T4

P1002

NaCl alleviates the impact of drought stress in the halophyte *Atriplex canescens*

***Huan Guo**, Ya-Qing Pan, Yan-Nong Cui, Suo-Min Wang, Ai-Ke Bao*
State Key Laboratory of Grassland Agro-ecosystems, College of Pastoral Agriculture Science and Technology, Lanzhou University

Drought and salinity are the major abiotic factors threatening to agricultural productivity and ecological environment all over the world. Halophytes have evolved multifarious adaptation strategies to deal with saline surroundings. One of those is capable of secreting great quantities of salt in epidermal bladder cells, which sequester excess of Na^+ deviate from the photosynthetic tissue and allow plants survive and grow well in a low water potential environment. *Atriplex canescens* is a C_4 perennial fodder shrub with excellent tolerance to salinity and drought. Our previous study showed that moderate salinity (100 mM NaCl) has a positive effect on the growth of *A. canescens* seedlings, and especially, Na^+ exhibits high contribution to the leaf osmotic potential (Ψs) by increasing accumulation in tissues and salt bladders under salinity treatments, implying that Na^+ accumulation might also facilitate the osmotic adjustment of *A. canescens* seedlings in response to drought conditions. To test this hypothesis, in the present study, five-week-old *A. canescens* seedlings were subjected to water deficit (30% of field water capacity) in the presence or absence of additional NaCl (100 mM) with pot experiments. The results showed that the biomass, leaf net photosynthetic rate and relative water content of *A. canescens* seedlings decreased significantly under drought stress. However, the addition of 100 mM NaCl enhanced photosynthesis capacity and increased Na^+ concentration in all tissues and Na^+ sequestration in salt bladders. The contribution of Na^+ to Ψs significantly increased by 72% in the presence of additional NaCl compared to drought stress alone, while the contribution of K^+ to Ψs maintained a stability. In addition, the concentrations of three compatible solutes (free proline, betaine and soluble sugar) in leaves exhibited a significant increase in the presence of 100 mM NaCl than that of the drought stress. As a result, *A. canescens* is able to accumulate a high concentration of Na^+ and osmocompatible solutes and use them directly for osmotic adjustment, which was coupled with an improvement in leaf hydration and photosynthetic activity. These findings suggest that, 100 mM NaCl resulted in optimal growth of *A. canescans* seedlings and alleviated the deleterious impacts on the growth of plants under arid environments.

T4

P1003

Molecular characterization of floral left-right asymmetry

Wuxiu Guo

Sun Yat-sen University

Left-Right (LR) asymmetry widely manifests in nature. In animals, previous works reported LR asymmetry is regulated by factors including Myo31DF in *Drosophila* and Nodal in snails. However, it is poorly studied in plants. In legume subtribe Phaseolinae, species are diversified with flowers in symmetrical, sinistral (left-handed), and dextral (right-handed) patterns. Here, we selected mung bean (*Vigna radiata*) as a model, which bears sinistral flowers, to unveil the molecular basis regulating floral LR asymmetrical development. We identified and characterized mutated mung bean *sinistral flower* (*sif*) processing flowers inverted into dextral patterns by large-scale γ-ray mutagenesis together with map-based cloning. *SIF* encodes a microtube-associated protein and is specifically enriched in developing flowers. Interestingly, we found Phaseolinae floral patterns exhibiting geological-associated distributions, and closely related with unique *SIF* homologues statuses. Our study for the first time unveils a conserved factor SIF recruited in flower LR asymmetrical development, suggesting its function possibly gaining or losing in different areas during the evolutionary process of subtribe Phaseolinae.

T4

P1004

Identification of nitrogen use efficiency genes in barley: Searching for QTLs controlling complex physiological traits

Mei Han[1,2], *Julia Wong*[1,3], *Tao Su*[1,2], *Perrin Beatty*[1], *Allen Good*[1]

1. Department of Biological Sciences, University of Alberta

2. Co-Innovation Center for Sustainable Forestry in Southern China, College of Biology and the Environment, Nanjing Forestry University

3. School of Biotechnology and Biomolecular Sciences

Over the past half century, the use of nitrogen (N) fertilizers has markedly increased crop yields, but with considerable negative effects on the environment and human health. Consequently, there has been a strong push to reduce the amount of N fertilizer used by maximizing the nitrogen use efficiency (NUE) of crops. One approach would be to use classical genetics to improve the NUE of a crop plant. This involves both conventional breeding and quantitative trait loci (QTL) mapping in combination with marker-assisted selection (MAS) to track key regions of the chromosome that segregate for NUE. To achieve this goal, one of initial steps is to characterize the NUE-associated genes, then use the profiles of specific genes to combine plant physiology and genetics to improve plant performance. In this study, on the basis of genetic homology and expression analysis, barley candidate genes from a variety of families that exhibited potential roles in enhancing NUE were identified and mapped. We then performed an analysis of QTLs associated with NUE in field trials and further analyzed their map-location data to narrow the search for these candidate genes. These results provide a novel insight on the identification of NUE genes and for the future prospects, will lead to a more thorough understanding of physiological significances of the diverse gene families that may be associated with NUE in barley.

T4

P1005

A potential innovative technology for *in-situ* wood identification based on image processing and artificial intelligence

Tuo He, *Lichao Jiao, Min Yu, Xiaomei Jiang, Yafang Yin*

Wood Collection and Department of Wood Anatomy and Utilization, Research Institute of Wood Industry, Chinese Academy of Forestry

Illegal logging and timber trade is becoming serious problem domestically and internationally, threatening individual wood species, and even entire ecosystem. The need for accurate and rapid recognition of wood to combat illegal logging and illegal timber trade around the world is outpacing the ability to train personnel to perform this task. The traditional identification methods based on the wood anatomical characteristics rely on the experience and proficiency of inspectors, however, can only identify wood to the level of genus or category in general, despite of the non-anatomical methods (e.g. genetic, spectroscopic and chemical analysis). The recent rapid developing of computer technology and artificial intelligence suggests that it is high time to move towards automatic wood inspection systems and methods for *in-situ* wood identification. In this work, an automatic wood identification system based on image acquisition, image processing, feature extraction and artificial neural networks was proposed. The images based on wood macroscopic features were acquired from processed wood samples by an image acquisition system. The image processing including image binarization, medium filter, image enhancement and histogram equalization was then carried out. The wood anatomical structures, such as vessels, rays, axial parenchyma and fibers, were extracted using the discrete wavelet transform (DWT) and grey level co-occurrence matrix (GLCM) algorithms. Furthermore, a multi-layered network based on the support vector machine (SVM) was adopted for training to realize the classification of wood species. The result indicates that the application of this technique could provide an *in-situ* inspection for wood identification.

T4

P1006

Molecular cloning and analysis of a novel gene encoding a KUP/HAK/KT potassium uptake transporter in *Cryptomeria japonica*

Yoshihiro Hosoo, *Hirokazu Nishiwaki*

Graduate School of Agriculture, Shinshu University

Potassium ion (K^+) is an essential nutrient for plants and plays crucial roles in various aspects of plant physiology, such as enzyme activation, osmotic regulation, and homeostasis maintenance. Many of the roles of K^+ are closely linked to K^+ transport across biological membranes, such as the plasma membrane and the tonoplast. Such K^+ transport is mediated by membrane proteins, which are classified as transporters and channels. The molecular mechanism of K^+ membrane transport has been studied in detail in herbaceous plants such as *Arabidopsis thaliana* and rice, whereas information on this mechanism in trees, especially in conifers, is limited. We have recently isolated and characterized several genes encoding K^+ transporters from *Cryptomeria japonica*, which is one of the most commercially important conifers in Japan. In this study, we isolated a novel gene encoding a K^+ transporter (CjKUP3) from *C. japonica*, and investigated its function and expression pattern. Samples of current-year needles, inner bark, differentiating xylem, male and female strobili, and roots were collected from *C. japonica* saplings and mature trees, and total RNA was extracted from these samples. Based on the sequences of *C. japonica* expressed sequence tags homologous to known K^+ transporters, primers were designed to amplify *C. japonica* cDNA containing the predicted full-length open reading frame (ORF) of a K^+ transporter gene. The target cDNA (*CjKUP3*) was amplified by RT-PCR and its sequence was analyzed. Subsequently, complementation tests using the *Escherichia coli* LB2003 strain, which is deficient in K^+ uptake and cannot grow on low-K^+ media, were conducted to examine the K^+ uptake function of the protein encoded by *CjKUP3*. A pPAB404 vector carrying the *CjKUP3* cDNA was transformed into the LB2003 cells and the transformants were tested for their ability to grow on solid media containing different K^+ concentrations. Additionally, expression levels of *CjKUP3* in various parts of *C. japonica* trees were quantified by real-time RT-PCR. The *CjKUP3* expression levels were quantified relative to actin gene expression levels as an endogenous control. The *CjKUP3* cDNA was predicted to contain an ORF of 2,358 base pairs encoding a protein of 786 amino acids. The translated sequence of *CjKUP3* was homologous to KUP/HAK/KT-type K^+ transporters derived from other plant species. The predicted membrane topology of CjKUP3 was similar to that of other plant KUP/HAK/KT transporters. The LB2003 cells expressing CjKUP3 were capable of growing in medium supplemented with a low level of K^+. This indicates that CjKUP3 is able to compensate for the defect in K^+ uptake in this mutant, and therefore that it plays a functional role in K^+ uptake. *CjKUP3* expression was detected in all samples examined in this study. *CjKUP3* expression levels in developing male strobili and female strobili were higher than those in other parts of *C. japonica*. These results indicate that CjKUP3 is involved in K^+ membrane transport in the cells of various parts of *C. japonica* trees, especially in the reproductive organs.

T4

P1007

Dynamic change of HG methylesterification patterns in the procambium-cambium continuum

Jundi Liu[2], ***Jie Hou*** [1], *Xinqiang He*[1]
1. Gansu Agricultural University.
2. College of Life Sciences, Peking University.

Pectin is a major component of primary cell walls of all land plants and encompasses a range of galacturonic acid-rich polysaccharides. Homogalacturonan is one of the main components of pectin in cell walls in poplar. Homogalacturonan (HG) is known to be synthesized and secreted in a highly methylesterified form. The procambium and cambium is considered to represent sequential developmental stages of the same vascular meristem. The former has been further subdivided into procambium, initiating layer, and metacambium based on anatomical observations. In this work, dynamic change of HG methylesterification patterns in the procambium-cambium continuum was studied in *Populus tomentosa* by immunolocalization with monoclonal antibodies (JIM5 and JIM7) against HG with different methylesterification patterns. JIM5 optimally binds to at least four contiguous unesterified residues between or adjacent to a methyl-ester group while JIM7 binds to a relatively highly methyl-esterified epitope. As a result, JIM5 signal was most abundant in the differentiated phloem, but is faint in the cell walls of procambium, initiating layer, metacambium and cambium. While the JIM7 signal equally distributed in all the cell walls. Furthermore, we characterized the expression patterns of pectin methyl esterase (PME) genes which removed methyl ester groups from HG in this process and found that different PMEs had different expression patterns in the procambium-cambium continuum. In conclusion, pectin showed different methylesterification patterns in the procambium-cambium continuum, and further experiments need to be performed to characterize the different roles of PMEs in this process.

T4

P1009

Variations in volatile oil yields of *Magnolia biondii* Pamp flower buds at different growth stages

Mingli Hu[1], *Mei Bai*[1], *Wei Ye*[2], *Hong Wu*[1]
1. State Key Laboratory for Conservation and Utilization of Subtropical Agro-bioresources, South China Agricultural University
2. Xi'an Botanical Garden of Shanxi Province

The dried flower bud of *Magnolia biondii* Pamp, commonly known as Xin-yi in China, is offcially listed in the Chinese Pharmacopoeia. To investigate the variations of the contents of volatile oil in *M. biondii* flower, biological characteristics observation, modified version of the tissue clearing method, semi-thin sections and volatile oil extraction were employed to detect the volatile yields of *M. biondii* flowers at various growth stages. The results indicated that the volatile oil yields in *M. biondii* flower at different growth stages had significant differences, which were closely related to the dry weight of flower, the oil cell densities and oil accumulation degree. The yields of volatile oil first increased and later decreased with the growth of flower buds. Flower buds of *M. biondii* were formed in June of 2015, the flower buds not only had the lowest dry weight (0.13 g) and the lowest density of oil cells (100.83 n/mm^2), but also had the maximum percentage of oil cells at the oil less stage, resulting in the lowest oil content (0.8%). In February of 2016, the flower buds had the highest dry weight (0.53 g), the maximum percentage of oil cells at the oil saturation stage and the highest density of oil cells (217.92 n/mm^2), which coincided with the highest oil yield (3.25%). In the full flowering stage in March of 2016, the flower buds had a lower dry weight and higher percentage of oil cells at the oil degrading stage and the lowest oil cell density (145.42 n/mm^2), resulting in the decrease of oil yield (1.45%). These results provided reference information for the economic value evaluation and rational utilization of the *M. biondii* flower.

T4

P1010

Spatial and temporal regulation of petal growth by transcriptional networks

Hongyang Yu[1], *Jing Li*[1], *Wei Wang*[1], *Yanzhi Wang*[1], *Yongxia Zhang*[1], *Vivian Irish*[2], ***Tengbo Huang***[1,2]
1. Shenzhen University
2. Yale University

Plant organ growth entails the regulation of cell division, cell expansion, cell type differentiation, and patterning of the organ as a whole. Recent studies have uncovered a number of genes involved in controlling these critical processes during plant organ development, however, our understanding of the gene network regulating the temporal and spatial changes in organ growth is still limited. The *Arabidopsis* petal is an excellent model for dissecting the temporal and spatial regulation of plant organogenesis. In this study, we focus on investigating key transcription factors that conduct temporal and spatial-specific roles during petal growth. We characterized the interactions of the downstream genetic networks of these transcription factors by utilizing genomic, genetic, developmental biology approaches. We found that they coordinately regulate important genes involved in plant organ development, particularly those associated with plant hormonal pathways. We hope these results will help elucidate how these key genes function together in the temporal and spatial regulation of petal growth, and in the long term, further understand the general mechanisms that establish the patterns of plant organ development.

T4

P1011

Optimization of the genetic transformation system for *Sinningia speciosa*

Yu-Ling Hung[1], *Chwan-Yang Hong*[2], *Chun-Neng Wang*[3]
1. Institute of Life Science, National Taiwan University
2. Department of Agricultural Chemistry, National Taiwan University
3. Institute of Ecology and Evolution Biology, National Taiwan University

Sinningia speciosa is a popular houseplant because of its big flower with a remarkable diversity in colors, patterns and shapes. *S. speciosa* has a small genome size (300 Mb), short life cycle, self-compatible, easily propagated in tissue culture. Therefore, it is emerging as a model plant for flower development studies. However, a reliable genetic transformation system is not available in *S. speciosa*. To this end, the *Agrobacterium*-mediated transformation and particle bombardment transformation were tested in this study. The overall transient transformation rate reached to 78.3% when three-weeks old seedlings were pre-cultured for 3 days on medium supplied with 1 ppm BA, followed by 5 days of co-culture with *Agrobacterium* strain EHA105. We further tested whether particle bombardment transformation can achieve better transformation rate. It was found that the transient GUS transformation rates under helium pressure 900 psi reach to 58.1% and 21.6% at distance 6 and 9 cm, respectively. Comparing the transformation efficiency of *Agrobacterium* and particle bombardment (78.3% vs. 58.1%), *Agrobacterium*-mediated transformation is chosen for further optimization. Under above pre-culture and co-culture conditions, the shoot regeneration rate is 17.2%, and the transformation rate is up to 2.1% after evaluating by Southern blotting. However, these seedlings take 8-12 months to regenerate. It is known callus grows rapidly with high regeneration rate, we also studied callus induction condition. Embryogenic callus was successfully induced with 0.1 ppm 2,4-D and 2 ppm BA plus 25 or 50 mM sorbitol in the medium. Callus transformation efficiency will be tested further. This study provided easy to follow and high efficiency *Agrobacterium*-mediated transformation protocol for studying gene regulation and functions in *S. speciosa*.

T4

P1012

Photosynthesis, phototropins and myosin XI-i regulate mitochondrial behavior in *Arabidopsis thaliana* leaf palisade cells

Md. Sayeedul Islam[1], *Van Toan Nguyen*[1], *Shingo Takagi*[1], *Motoki Tominaga*[2], *Yusuke Kato*[3], *Wataru Sakamoto*[3]
1. Osaka University
2. Waseda University
3. Okayama University

Light-induced mitochondrial redistribution in *Arabidopsis thaliana* mesophyll cells occurs along with that of chloroplasts. Genetic approaches identified phototropin as the photoreceptor involved in blue light dependent chloroplasts photo-relocation movement. There are two phototropins in *A. thaliana*, termed *phototropin 1* (*phot1*) and (*phot2*). Our recent study proposed that the intimate association of mitochondria with chloroplasts is precisely regulated through photosynthesis- and phot2- dependent signals. To dissect how non-uniform distribution patterns of mitochondria are established, we analyzed movement of individual mitochondria on the outer periclinal cytoplasm under weak blue light by time-lapse fluorescence microscopy. In the dark-adapted wild-type (WT) leaves, mitochondria moved smoothly one time, suddenly paused, and resumed the movement. We examined two parameters; 1) the velocity of movement of each mitochondrion, and 2) the probability of stoppage of each mitochondrion. Results revealed that phot1 and phot2 function non-redundantly in the early acceleration of the velocity of mitochondrial movement, while phot2 predominantly controlled the late immobilization of mitochondria. We were further interested in factors involved in the regulation of mitochondrial behavior. Higher plants have two classes of myosin, VIII and XI. Myosin XI members have been suggested to play important roles in intracellular trafficking of Golgi stacks, peroxisomes, and mitochondria. In *myosin XI-i* disruptants, the early acceleration of mitochondrial movement was not observed after exposure to weak blue light, while mitochondrial movement in the dark appeared to be normal. The late immobilization of mitochondria also took place normally. On the other hand, using the photosynthetic inhibitors (DCMU and DBMIB) and the *yellow variegated 2* (*var2*) mutant showing leaf variegation, a requirement of intact photosynthesis for the immobilization of mitochondria was confirmed. To reveal the effects of abnormal mitochondria-chloroplasts co-localization on photosynthesis, chlorophyll fluorescence measurements were carried out. Photosystem II was more susceptible to high-intensity white light in *phot2* and *phot1phot2* mutants, and *myosin XI-i* disruptants than in WT plants. These results indicate that phot2 and myosin XI-i contribute to the optimization of photosynthesis through regulating the intracellular positioning and interaction of mitochondria and chloroplasts.

T4

P1013

Post-storage peel colour development strategies for cultivar samar bahist chaunsa mango under reefer and controlled atmospheric (CA) condition

Hafiz Umer Javed[1], *Aman Ullah Malik*[2], *Muhammad Amin*[2], *Muhammad Fiaz*[2]

1. China Agriculture University, Beijing

2. University of Agriculture, Faisalabad

Mango fruit colour is one of the main criteria of consumer's choice. Post-storage peel colour development is a serious issue in our commercial mango cv. S.B. Chaunsa. This study was aimed at evaluating the effect of pre-cooling and ripening temperature on quality and peel colour development under reefer and controlled atmosphere storage conditions. The fruit were harvested from a commercial orchard at physiological mature stage. After harvesting, fruit were physically de-sapped and subjected to prochloraz (0.5ml/L) dip followed by air drying, packaging and transport to PRTC in a reefer van at 18°C. At PRTC, the fruit subjected to pre-cooling for 8 hrs. at 12°C or 17°C, subsequently stored at 17°C under reefer and CA condition (4% CO_2 & 6% O_2) for different storage periods (10 & 20 days under reefer) and (10,20,30 and 35 days under CA). At removal, fruit were artificially ripened at two different temperatures (24°C or 35°C). The fruit response was significantly different under different treatment regimes. The experiment was laid down under CRD along with factorial design. Under reefer storage conditions (without CA), the fruit removed after 10 days of storage had 4-5 days of shelf life and in CA storage condition, the fruit removed after 10 & 20 days had 5-6 days of shelf life (including ethylene treatment period) at both ripening temperatures; whereas, the fruit removed from reefer (20 days) and CA (30 & 35 days) were found ripe (eating soft) at the time of removal (with 0 days shelf life), thereby not requiring any ethylene treatment. The peel colour development increased significantly at ripening temperature 35°C than 24°C and after removal to final ripe stage while pre-cooling temperature had non-significant effect. Significant difference was found among pre-cooling and ripening temperature regarding to physiological weight loss and respiration rate of reefer and CA stored fruit. Chilling injury, side rot and stem end rot were increased as the storage duration extends in CA stored conditions. Under reefer, the physico-chemical and organoleptic quality of fruit were found meritorious than the CA condition. This paper provides detail about treatment impact on physico-chemical, physiological and organoleptic quality attributes.

T4

P1014

Identification and functional analysis of microRNAs involved in the rice flower development

Hyojin Heo, ***Dong-Hoon Jeong***

Hallym Univeristy

MicroRNAs (miRNAs) are non-coding small RNAs that regulate target gene expression at the posttranscriptional level. Plant miRNAs have been identified that contribute to a variety of biological process, including flower development. Functional studies examining *Arabidopsis* miRNAs have identified several miRNAs that play crucial roles in flower development by repressing key regulatory genes in floral organ identity. Although these miRNAs are well conserved in rice, their role in rice flower development had not been well characterized. To better understand the role of rice miRNAs in flower development, small RNA libraries were constructed from rice floral organs including glume, palea/lemma, lodicule, anther, and pistil. We also constructed PARE (Parallel Analysis of RNA Ends) degradome libraries from anther and pistil. Both small RNA and PARE libraries were deeply sequenced by Illumina technology and analyzed to identify differentially regulated miRNAs and to investigate their target cleavage functions. From the small RNA libraries, we have identified the floral-organ preferentially expressed miRNAs. In addition, analysis of PARE data revealed the cleavages of target RNAs in the particular floral organ where these miRNAs are expressed. Of floral-organ preferentially expressed miRNAs, *miR5179*, which targets an *OsMADS16* that is involved in anther identity, was pistil-preferentially expressed. To determine the biological function of *miR5179*, transgenic rice plants over-expressing *miR5179* were generated. The transgenic plants exhibited the multiple-pistil phenotypes, which is due to down-regulation of *OsMADS16*. Additional studies are underway to investigate the association of floral-organ-preferential miRNAs with the expression of their target genes.

T4

P1015

Genetic analysis of pseudopeloric mutation in *Habenaria radiata*

Akira Kanno, *Mai Mitoma, Yumi Kajino, Risa Hayashi*

Graduate School of Life Sciences, Tohoku University

Orchids are commercially important plants with flowers that are unique and very specialized in shape and color. The flowers consist mostly of sepals, lateral petals, lip (labellum) and column. *Habenaria radiata* grows in wet grasslands in both warm and cold temperate regions of Honshu, Shikoku and Kyushu in Japan. This species has greenish sepals in the first whorl and white lateral petals and lip in the second whorl. Among *H. radiata* cultivars, one pseudopeloric cultivar 'Hishou' is known. In the 'Hishou' cultivar, the greenish median sepal is homeotically altered to a white-petaloid structure and the two lateral sepals are modified to white lip-like organs. In order to investigate the inheritance of 'Hishou' flower character, we performed intraspecific crosses using wild-type and 'Hishou' cultivars of H. radiata. Since 'Hishou' is female sterile, crosses were made using three lines of wild-type (W1, W2 and W3) plants as the female parent and one line of the *H. radiata* 'Hishou' (H) as the male parent. We obtained 230 progenies by these crosses. Among these F_1 progeny, 204 plants had flowers and 108 had 'Hishou' type flowers (F_1[H]) and 96 had wild-type flowers (F_1[WT]). We obtained 268 progenies by self pollination of F_1[WT] plants. Among 268 F_2 plants, 100 plants had flowers and all of these flowers were wild type. F_1[H] was female sterile as 'Hishou' cultivar, crosses were made using wild-type as the female plants and F_1[H] as the male plants. We obtained 208 BC_1 progenies from this cross. Among 208 progenies, 138 plants had flowers and 70 had 'Hishou' type flowers and 68 had wild-type flowers. These results suggest that the characteristics of pseudopeloric mutation found in the 'Hishou' cultivar were inherited dominantly and that the underlying mutation is a gain-of-function mutation. Since half of the intraspecific hybrids (F_1) and half of the back cross progenies (BC_1) had wild-type flowers, the locus of pseudopeloric mutation is likely be heterozygous in the 'Hishou' parent.

T4

P1016

Temporal responses of two wheat cultivars under phosphate starvation

Mohammad Rezaul Karim, *Ping Lan*
State Key Laboratory of Soil and Sustainable Agriculture, Institute of Soil Science, Chinese Academy of Sciences

Phosphorus (P) is an essential macronutrient and often acts as a limiting factor for plant growth and development due to its low solubility and fixation in soils. Wheat is one of the most important cereal crops worldwide and the production is seriously affected by P status. The present study was executed to analyze the comprehensive temporal traits of wheat plants to P-starvation. Two wheat cultivars, Atlas 66 and Scout 66, were grown hydroponically under P–sufficient (250 μM) and –deficient (0 μM) conditions. Morpho-physiological and molecular characterization were examined at early (6 h to 1 d), mid (1 d to 7 d) and long (7 d to 21 d) time periods. Root length and biomass of both cultivars were increased steadily while shoot biomass was decreased at long time periods under P-deficient conditions. Notably, the Scout 66 exhibited higher root biomass and root volume, and more root tip numbers under P-starvation as compared to the Atlas 66. Inorganic phosphate (Pi) concentrations were tended to reduce from early time periods in both roots and shoots while the total P contents of roots and shoots were reduced from 3 d and 7 d time periods, respectively. Overall, the Scout 66 had higher concentrations of Pi and total P over the Atlas 66. In addition, phosphate deficiencies significantly affected the contents of iron (Fe) and manganese (Mn) in roots and decreased Zinc (Zn) content in both roots and shoots. Likewise, the Scout 66 possessed significantly higher amount of Fe, Mn, and Zn content in its roots and Mn content in shoots in both P-levels. The marker genes for P stress, *TaIPS1*, *TaSQD2* and *TaSPX3*, were significantly induced in both roots and shoots from early time periods till 21 d of P-depletion. But the magnitude of induction of these Pi-responsive genes varied between two cultivars. Taken together, the wheat cultivar Scout 66 might be more P-efficient cultivar and maintained higher level of Pi homeostasis and trace metal elements by the increased root biomass, root volume and more root tip number. These results enlarged our knowledges regarding the time course related responses of wheat cultivars under phosphate starvation.

T4

P1017

Flower morphology and development in *Polyscias* (Araliaceae: Apiales): Meristic variations, multiple gains of monomerous gynoecium and the nature of epicalyx

Polina P. Karpunina[1], *Maxim S. Nuraliev*[1], *Alexei A. Oskolski*[2,3], *Dmitry D. Sokoloff*[1]
1. Biological Faculty, M.V. Lomonosov Moscow State University
2. Botanical Museum, V.L. Komarov Botanical Institute
3. Department of Botany and Plant Biotechnology, University of Johannesburg

Most eudicots and all monocots have cyclic flowers. While some groups of angiosperms are characterized by isomerous flowers (with equal organ numbers in all whorls), anisomery is also a common pattern. In many cases, anisomery is expressed in gynoecium merism that differs from the merism of other floral whorls. However, counting carpel number in syncarpous gynoecia is sometimes problematic due to the congenital nature of intercarpellary fusion. A convenient group for studying evolution of flower merism is the genus *Polyscias* (Araliaceae) that comprises about 160 species distributed throughout the Indian and Pacific Ocean basins from Africa and Madagascar to Australia and Hawaii (Lowry, Plunkett, 2010; Plunkett, Lowry, 2010). Members of the genus vary considerably in floral merism, sometimes even within a given species, which is generally unusual for core asterids. Although pentamerous flowers with syncarpous bicarpellate gynoecium are the most common type in the genus, gynoecium merism shows a considerable variation. Both polymerization of gynoecium (e.g., 10 carpels in *P. duplicata* from Madagascar) and its oligomerization up to the unilocular condition occurred in several lineages of *Polyscias*. Shifts to unilocular gynoecium with single fertile ovule took place in four groups of this genus, i.e. in a biphyletic group of species from Madagascar previously segregated as *Cuphocarpus* and in two lineages within the Australo-Malesian clade (Karpunina *et al.*, 2016). In angiosperms, reduced gynoecia with single fertile ovule can appear via several evolutionary pathways, including monomery and pseudomonomery. In an evolutionary context, pseudomonomery can be understood as a result of gradual sterilization of all carpels but one in syncarpous gynoecium (or rarely in gynoecium with postgenitally united carpels). In contrast to pseudomonomery, monomery is viewed as a result of abrupt change in carpel number with no intermediate stages. Our examination of morphology, anatomy and development of unilocular gynoecia in different clades of *Polyscias* convincingly shows that these gynoecia are truly monomerous. Both unicarpellate and pluricapellate species of *Polyscias* have unstable orientation of flower organs relative to flower-subtending bract. In two examined Australo-Malesian species with monomerous gynoecia and pentamerous outer whorls, the median sepal is either adaxial or abaxial. In the flower of the Malagasy *P. compacta* with monomerous gynoecium, two sepals of tetramerous calyx can be in transverse position, or all sepals are in oblique positions. In all three species, orientation of ventral side of the carpel is not fixed (adaxial, abaxial, transversal or oblique). Another Malagasy species, *P. aubrevillei*, has 4-5-merous outer whorls and 2-3 carpels which also show unstable orientation. A feature of some Malagasy species is the presence of a distinctive rim at the flower base. This rim is commonly described as "epicalyx", but its nature remains obscure. The epicalyx of *P. compacta* and *P. schatzii* ined. is initiated at late stages of flower development by extension of the peduncle, and it is not innervated by vascular bundles. Some Malesian *Polyscias* species have a similar, but less pronounced extension at the base of the flower, which is covered with trichomes.

The study is supported by RFBR (project 15-04-05836).

T4
P1019
Prediction physiological changes in *trigonella* (*in vivo* and *in vitro*) in response to physical mutagenesi (gamma rays)

Mahmoud Khater
National Research Centre (NRC)

This investigation was carried out during two successive seasons 2013/2014,2014/2015 at the Greenhouse and tissue culture laboratory, Botany Department, National Research Centre, Giza, Egypt. Molecular studies were carried out at Genetics Department, Faculty of Agriculture, Zagazig University, Egypt. to study the physiological and molecular changes in fenugreek (*Trigonella foenumgraecum* L.) *In vivo* and *In vitro* as a response to gamma rays with doses (0,100,150,200,250 and 300 Gy.). Results revealed that callus induction frequency increased in response to the increase of 2,4-D concentrations and varied from one gamma-ray dose to another. The best interaction between gammaray and 2,4-D concentrations was between 2.0 mg/l 2,4- D and all doses of γ-rays, especially 200 Gy, which recorded the highest percentage of callus (above 90%). On the other hand, in RAPD analysis there were 44 bands (total bands) divided to polymorphic bands (25), monomorphic bands (2) and unique bands (17) with an average percentage 88.00% polymorphism and numerous molecular sizes that ranging from 265.122 to 1911.225 bp, Moreover, results of ISSR analysis illustrated that there were 22 bands (total bands) divided to polymorphic bands (3), monomorphic bands (11) and unique bands (8) with an average percentage 76.00% polymorphism and different molecular weights that ranging from 260.582 to 1333.384 bp.

T4
P1020
Developmental sources of variation in floral organ numbers: Floral phyllotaxis and organ fate determination

Miho Kitazawa, *Koichi Fujimoto*
Osaka University

Floral organ number is one of the fundamental features that characterises floral morphology. It is largely conserved in the major clade of angiosperms: Core eudicots show pentamerous or tetramerous flowers with multiple of five or four floral organs, whereas monocots show trimerous flowers. On the other hand, the number has been changed multiple times during evolution, as indicated by the fact that the closely related species show flowers with different numbers of floral organs. The variation can be a key to address the question of what developmental mechanism underlies such co-existence of robustness and flexibility. Although the floral organ number is a hallmark of the species, it can distribute stochastically even within an individual plant. Especially in the family Ranunculaceae, which branched near the monocot-eudicot bifurcation, the flowers show remarkable variations in floral organ numbers.

Here we suggest two candidate sources for the variation in floral organ numbers, through field observation of Ranunculaceae flowers and theoretical models. One is the stochasticity in the boundary of floral organ fate, and the other is the changes in floral organ arrangement, i.e., floral phyllotaxis. The former is suggested by the probability distribution of the floral organ numbers within a population. The probability distributions of floral organ numbers of Ranunculaceae are usually asymmetric and steep, indicating that they cannot be explained by the normal distribution. We compared several statistical models, and found that the most plausible model to explain the distribution is the one assuming stochasticity of the expression boundary of organ fate determinants. In addition to such stochastic homeotic transformation of floral organs, we can find variation of floral organ arrangements that affect floral organ numbers in many species. To measure this, we observed perianth organ arrangements in the genus *Anemone* (Ranunculaceae), which contains species with various merosities, i.e., the basic number of floral organs. We observed three species, *A. flaccida*, *A. scabiosa*, and *A. nikoensis* that are normally pentamerous but stochastically show higher perianth organ numbers. Quincuncial arrangement stably appeared in the flowers with five perianth organs, whereas the arrangements were more variable in the flowers with more than five perianth organs. Especially, we found positional arrangement of the excessive sixth perianth organ indicating the possibility of a transition from pentamerous to trimerous arrangement. The phyllotaxis model for initiation of the sixth organ showed that the arrangements found in the *Anemone* occur at a divergence angle <144°, indicating the spiral nature of floral phyllotaxis rather than a perfect penta-radial symmetry of 144°. The model further showed that selective occurrence of trimerous arrangement is mainly regulated by the organ growth rate, which may play a central role in the transitions between pentamerous and trimerous flowers.

T4
P1021
The tissue and cell specific expression characteristics of the key enzyme regulated chlorogenic acid biosynthesis in *Lonicera japonica* flower organ at different growth stages

Dexin Kong[1], *Yanqun Li*[1,3], *Hanjun He*[1], *Mei Bai*[1], *Haiyang Wang*[1], *Hong Wu*[1,2,3]
1. State Key Laboratory for Conservation and Utilization of Subtropical Agro-Bioresources, South China Agricultural University, Guangzhou 510642, China
2. Guangdong Technology Research Center for Traditional Chinese Veterinary Medicine and Natural Medicine, South China Agricultural University, Guangzhou 510642, China
3. Guangdong Key Laboratory for Innovative Development and Utilization of Forest Plant

Hydroxycinnamoyl-CoA quinate transferase (HQT) is one of the key enzymes in the biosynthesis of chlorogenic acid in *Lonicera japonica*. Although some researches have investigated *HQT* gene expression and regulation at the transcriptional level, how the tissue-and cell-specific patterns of protein synthesis and gene expression regulated chlorogenic acid (CGA) change during the different growth stages of *L. japonica* and how they correlate with the biosynthesis and distribution of CGA remain largely unknown. The expression of HQT and its tissue and subcellular localization at different growth stages of *L. japonica* ((young alabastrum (S1), green alabastrum (S2), slightlywhite alabastrum (S3), whole white alabastrum (S4), silvery flower (S5), and golden flower (S6)) were investigated by using the techniques of western blotting, histochemical localization, in situ hybridization and immune-electron microscopy in this study. Whilst, we systematically examined the dynamic accumulation of chlorogenic acid at six growth stages using an HPLC-DAD-base method. The results indicated that the variation of HQT protein accumulation level and the dynamic accumulation of CGA exhibited a positive correlation with the temporal accumulation patterns of *HQT* gene transcript at different growth stages in *L. japonica*, respectively. Interesting, the results of in situ hybridization and histochemical localization revealed that transcript expression of *HQT* in tissue specificity or cellar specificity significantly related to the distribution of CGA in *L. japonica* corolla tube cells at different growth stages. For example, in S1-S3, the *HQT* was presented in the inner and outer epidermal cells, and parenchyma cells and vascular bundle, the CGA was also found to occur in these tissues. *HQT* and CGA were only distributed in vascular bundle and inner epidermal cells at S5-S6. The results of the rabbit anti-HQT Immunogold (lgG) and the laccase colloidal gold demonstrated that anti-HQT-lgG and laccase-lgG were mainly located in chloroplasts, plastids and cytoplasm during S1-S2. In S3-S4, while the weakening of structure and function of chloroplast, anti-HQT-lgG particles were reduced and mainly distributed in degrading cytoplasm, the laccase-lgG particles were transferred from plastids or cytoplasm to vacuole, accompanied by the process of vacuolization. After the buds fully opened, the large central vacuole almost occupied the whole cell, only few anti-HQT-lgG particles were observed in cytoplasm. The vacuole and cell wall were not found with the anti-HQT-lgG particles throughout the all growth stages, laccase-lgG particles were mainly located in vacuole in the S5, S6. Therefore, we suggested that chlorogenic acid was mainly synthesized in chloroplasts, plastids and cytoplasm. In middle and later stages of *L. japonica*, the chlorogenic acid was mainly synthesized in cytoplasm, and stored in the vacuole. This research will give further insight for studying the biosynthesis and regulation of Cholorgenic acid in *L. japonica*.

T4
P1022
Genome-wide analysis of *FWL* gene family in *Brassica* plants

Chen Kuang[1], *Jun Li*[1], *Hong Fang Liu*[1], *Zhe Dong*[1], *Wei Hua**[1], *Jing Chu Luo**[2]

1. Oilcrops research institute, chinese academy of agricultural sciences

2. College of Life Sciences, Center for Bioinformatics, National Laboratory of Protein Engineering and Plant Genetic Engineering, Peking University

Plant and organ size can vary in natural populations. The individual differences in plant and organ size result from changes in the expression level of the specific genes. After the successful pollination, fruit size depends mainly on cell number and cell size, which determines the final organ size and plant growth and, ultimately, crop yield. A lot of genes associated with fruit size traits have been isolated by molecular biology technology. In tomato, the *FW2.2* gene is hypothesized to control 30% of the variance in fruit weight by negatively regulating cell division. In maize, the *CNR* genes are identified, which affects the plant size by restricting the cell division duration to reduce cell number rather than cell size. To expand investigation of how related genes may impact other crop plant or organ size, we identified the Brassica plants gene family of putative *fw2.2* orthologs, naming them *Brassica Organ Size Regulator (BOSR)* genes. This family represents an ancient eukaryotic family of Cys-rich proteins containing the PLAC8 or DUF614 conserved motif. To date, no genome-wide overview of the *Brassica Organ Size Regulator (BOSR)* gene family has been performed. Therefore, as the first step and as a helpful strategy to expound the structure and function of *BOSR*, a genome-wide study for this gene family is necessary. In this study, we used some bioinformatical equipment, such as BLASTp, tBLASTn, and Hmmer to identify *BOSR* sequences in Brassica plant databases. Using TMHMM trans-membrane domain prediction tool, we identified trans-membrane structure of *BOSR*. The subcellular localization of *BOSR* was forecast by targetp and their gene sequence were analyzed by MEME and WEBLOG. We investigated the evolutionary relationship of the *BOSR* gene family by phylogenetic analysis and the 3D structure of *BOSR* was forecast by phyre2. To verify gene expression changes, we further used quantitative real-time reverse transcription-polymerase chain reaction. Additionally, *BOSR* may have functions similar to tomato *FW2.2* or maize *CNR*. The genome-wide identification, bioinformatic analysis and expression analysis of *BOSR* genes can facilitate research of this gene family and polyploidy evolution. As previously proposed in maize, this study may contribute to the potential application to Brassica breeding and crop improvement.

T4
P1023
Cytokinins in the light of plant evolution. Are there any evolutionary links within the plant kingdom?

Lenka Zaveska Drabkova[1,2], *Vaclav Motyka*[1], *Eva Zizkova*[1], *Petre Dobrev*[1]

1. Institute of Experimental Botany, the Czech Academy of Sciences

2. Institute of Botany, the Czech Academy of Sciences

Evolutionary physiology represents an explicit fusion of two approaches, evolution and physiology. In our research, we address one of important questions regarding physiological evolution: What phylogenetic patterns are reflected in the metabolism of plant hormones cytokinins (CKs) during evolution of land plants? Bioactive CKs represent essential molecules regulating many aspects of plant growth and development but a wide spectrum of non-active and/or weakly active CK derivatives exists and proportionally differs among plant species. However, not all of CK forms have been examined yet in the vast majority of plant species, tissues or developmental stages, therefore it is difficult to state that CK regulatory pathways are similar throughout the whole plant kingdom. Available data demonstrate ubiquitous occurrence of numerous CK metabolites including *O*- and *N*-glucosides as well as *cis*-zeatin-type CKs in the plant kingdom. We analyzed the occurrence of endogenous CK derivatives in cyanobacteria, algae, mosses, gymnosperms and angiosperms in order to perform phylogenetic analyses with the aim to test two main hypotheses, that (1) *O*-glucosylation has originated as a regulatory step from preexisting elements of the primary metabolism already in evolutionarily older plants while *N*-glucosylation has been acquired independently in different plant families during further evolution

and whether (2) *cis*- vs *trans*-derivates of zeatin are more prevalent in the evolutionarily older plants, and also more significant here with respect to CK biological pathways. Moreover, (3) we elucidated evolutionary processes of *N*- and *O*-glucosyltransferase pathways based on phylogenetic and molecular analyses of UGT and *cis*ZOG1 genes in selected groups of plants. In our comprehensive screening throughout the plant kingdom, CK-*N*7- and CK-*N*9-glucosides were found ubiquitous in vascular plants. In contrast, their only rather low levels or a total absence was shown in non-vascular plants such as algae and mosses. Surprisingly, fungi representatives showed similar spectra of CK derivates with a lack of CK-*N*-glucosides and a prevalence of *cis*-zeatin types as lower plants. It is possible that the absent or sparse *N*-glucosyltransferase pathway is substituted here by other down-regulating mechanisms, e.g. by enhanced formation of *cis*-zeatin derivatives, as considerably higher *cis*-zeatin/*trans*-zeatin ratios were revealed for most of non-vascular than vascular plants.
Supported by the Czech Science Foundation (16-14649S)

T4
P1024

The function of phosphorylation sites of BvM14-STPK in response to salt stress

Haiying Li[1,2], *Bing Yu*[1,2], *Chunquan Ma*[1,2], *Duanmu Huizi*[1,2], *Li Hongli*[1,2], *Mengdi Yu*[1,2], *Xiang Liu*[1,2]
1. Key Laboratory of Molecular Biology of Heilongjiang Province, College of Life Sciences, Heilongjiang University, Harbin, P. R. China, 150080
2. Engineering Research Center of Agricultural Microbiology Technology, Ministry of Education, Heilongjiang University, Harbin, P. R. China, 150080

In our lab, Guo *et al.*, obtained the sugar beet monosomic addition line M14, which contains the *Beta vulgaris* L. genome with the addition of No.9 chromosome of *B. corolliflora* Zoss. Sugar beet M14 line is a unique germplasm that contains genetic materials from *Beta vulgaris* L. and *Beta corolliflora* Zoss, and it exhibits tolerance to salt stress. Comparative proteomics analyses in the leaves and roots of M14 line had been performed under 0, 200 and 400 mM NaCl treatment. 40 differentially expressed proteins in leaves and 36 differentially expressed proteins in roots were identified, in which one serine-threonine protein kinase designated as BvM14-STPK is involved in signal transduction. Under salt stress, the expression of BvM14-STPK protein in roots was increased 3.24 times. So, BvM14-STPK protein might respond to the salt stress in sugar beet M14 line. Here the function of phosphorylation sites of BvM14-STPK (one serine/threonine protein kinase in sugar beet M14) was introduced. Based on the transcriptome database of M14 under salt stress, the full-length cDNA named BvM14-STPK was obtained by PCR. Sequence analysis showed that BvM14-STPK had a ser/thr binding domain, which belonged to the family of Protein Kinase C Family. The Real-Time PCR result showed that *BvM14-STPK* gene was responsive to salt stress. BvM14-STPK fusion protein was expressed in prokaryotic expression system and identified by Western blot. After gel digestion, desalting, enrichment, the peptides of BvM14-STPK were identified by mass spectrometry. The phosphorylation sites of BvM14-STPK protein were analyzed. In order to reveal the function of phosphorylation sites in BvM14-STPK, 8 key serine phosphorylation sites were divided into five segments to research. By site-directed mutagenesis technology, each of five Ser phosphorylation sites were mutated into aspartic acid in BvM14-STPK, which meaned that each of five Ser phosphorylation sites became sustained phosphorylated. Simultaneously, each of five Ser phosphorylation sites were mutated into phenylalanine in BvM14-STPK, which meaned that each of five Ser phosphorylation sites became non-phosphorylated. Each cDNA full length of ten *BvM14-STPK* genes by site-directed mutagenesis technology was obtained. Ten plant expression vectors containing the above ten mutant genes were constructed and transformed into the tobacco separately. The function of phosphorylation sites of BvM14-STPK were further clarified in transgenic tobacco. The transgenic seedlings of T1 generation were transplanted on the MS medium containing different concentrations of NaCl (0,150mM). After 20 days salt treatment, the result showed that the fresh weight and root length of all transgenic plants were inhibited after salt treatment. In transgenic plants containing non-phosphorylated site, the effect of 230 Ser site on inhibition of transgenic plants was obviously higher than that of other sites. The possible reason is that 230 Ser site is in the active domain of this kinase BvM14-STPK. The expression of mutant *BvM14-STPK* gene of transgenic plants under salt stress was analyzed by the real-time PCR technique. The results showed that the expression level of *BvM14-STPK* gene of transgenic plants containing sustained phosphorylated site was significantly higher than that of wild-type tobacco and the expression level of *BvM14-STPK* gene of transgenic plants containing non-phosphorylated site was obviously decreased than that of wild-type tobacco. In transgenic plants containing non-phosphorylated site, the 230 Ser site and 216 Ser, 217 Ser site had a greater inhibition effect on the plant. Analysis of the function of phosphorylation sites of *BvM14-STPK* under salt stress has provided insight into specific response mechanisms underlying salt stress tolerance. The knowledge gained can be potentially applied to improving crop salt tolerance.

T4
P1025

The *robin1* mutation increases the BRI1 abundance to enhance BR signaling in *Arabidopsis*

Shasha Ren, *Xiaolei Liu, Jianming Li*
Shanghai Center for Plant Stress Biology, Chinese Academy of Sciences

Brassinosteroids (BRs) are a unique class of plant polyhydroxylsteroids that play important roles throughout the plant life cycle. BR-deficient mutants or BR-insensitive mutants display a characteristic set of phenotypes that include dark green leaves, dwarfed stature, delayed flowering and senescence. To fully understand the BR signaling mechanism, we performed a large-scale forward genetic screening for extragenic suppressors for the *Arabidopsis* mutant *bin2-sf,* carrying a dominant single amino-acid change (Ser^{308}-Phe) in an *Arabidopsis* GSK3-like kinase BRASSINOSTEROID-INSENSITIVE2 (BIN2). Our genetic and biochemical studies have shown that *revertant of bin2-sf 1 (robin1)* partially recovers BR responsiveness. We have recently cloned *ROBIN1* gene and found that the ROBIN1 protein is mainly localized in the nucleus. The *robin1* mutations also partially rescue the growth phenotypes of *bri1-9* (a weak BR receptor mutant) and heterozygous *bin2-1*/+ (another gain-of-function *bin2* mutant). By contrast, overexpression of *ROBIN1* could enhance the dwarfism of another BR receptor mutant *bri1-5*. Our biochemical studies revealed that the *robin1* mutation increases the abundance of BRASSINOSTEROID INSENNSITIVE 1 (BRI1, the cell-surface localized BR receptor). Additional studies aiming to link the robin1-increased BRI1 abundance with enhanced BR signaling will be reported at the conference.

T4

P1026

BBP1 and its homolog negatively regulate BIN2 to stimulate brassinosteroid signaling in *Arabidopsis thaliana*

Xiuwen Rao, *Xuan Du, Xiaolei Liu, Jiamu Du, Jianming Li*
Shanghai Center for Plant Stress Biology, Chinese Academy of Sciences

Brassinosteroid (BR) is one of the most important plant hormones that regulate many aspects of plant development and growth. Extensive studies in the past 20 years have revealed a protein phosphorylation-mediated signaling cascade to transmit the extracellular BR signal into the plant nucleus, which involves a key regulatory step of inactivating an *Arabidopsis* GSK3-like kinase BIN2. However, the biochemical mechanism for BIN2 regulation remains controversial. Using a biochemical approach, we have discovered a previously uncharacterized *Arabidopsis* protein BBP1 for BIN2-binding protein 1. We found that BBP1 interacts with BIN2 at the plasma membrane to prevent the BIN2's phosphorylation activity towards its substrates BES1 and BZR1. BBP1 overexpression inhibits the phosphorylation of BES1/BZR1 and leads to enhanced BR signaling in a dosage-dependent manner. More importantly, simultaneous elimination of BBP1 and its homolog reduces BR signaling and enhances the dwarf phenotypes of weak BR receptor mutants. Genetic and/or chemical manipulation of the plasma membrane and nuclear localization of BIN2 and/or BBP1 could enhance and reduce BR signaling, respectively. Based on these findings, we propose that BBP1 and its closest homolog positively regulate BR signaling by controlling the subcellular of BIN2 to inhibit its phosphorylation activity.

T4

P1027

Assessment of drought tolerance in urban trees and planting site selection: An application of multiple correspondence analysis and association rules

Ke-Ke LI, *De-Shun ZHANG*
Tongji University

Urban trees will be exposed to severe drought stress due to climate change and therefore require a high tolerance facing the coming years and decades. It is known that good quality of sites increases probability of maintaining healthy trees and survival whereas low quality sites diminish tree's health and survival. In fact soil characteristics (such as soil water content, soil mechanical composition, drainage and aeration), site characteristics and location (for example site size, air pollution, streets, sidewalks, parks, utility lines) and environment (like wind speed, temperature, population stress) seem to be factors linked to tree survival. Not all trees adapt to any site and some are more adequate for a given site than others. This study shows the results of extensive research and evaluation of reactions, parameters and characteristics of urban trees linked with drought tolerance. Our research target is to use successive steps of multiple correspondence analysis and handling missing attribute values for some data about the drought of urban tree This resulted in 42 possible parameters including such as leaf morphology, stomatal densities, wilting symptoms, wood anatomy and physiology. The estimations values are validated with a clustering approach. By detailed counterchecking, seven parameters with the highest significance for urban trees could be selected, followed by a ranking according to the significance. This ranking also allows the evaluation of "new", i.e. uncommon tree species and native tree species considering their drought tolerance. In the second part, we present an exploration of association rules determine planting sites considering urban tree's characteristics. In first step item sets and rules are generated using the unsupervised algorithm Apriori. They are rapidly characterized in terms of tree planting sites. In a second step planting sites are fixed as target values to establish rules (a supervised version of the a priori algorithm). An original approach is also presented and validated for the prediction of the planting site of the species.

T4

P1028

ER-localized SNAREs are involved in transport of seed storage proteins

Lixin Li, *Xiaonan Zhao, Hailong Zhang, Yun Kong, Xiufen Guo*
Northeast Forestry University

Plant seeds accumulate large amount of storage proteins as N resorce for plant inicial growth. Storage proteins are synthesized as large precursors on the ER. The precursors are delivered to the vacuole by vesicle transport and processed to smaller mature forms.In our previous study, we found that ER-localized tethers, MAG2, MIP1, MIP2 and MIP3, form a tethering complex-MAG2 compplex, function on ER-Golgi vesicle transport in Arabidopsis. When they are defective, transport of newly synthesized proteins out from the ER are affected. To further clarify mechanism underlying MAG2-relating protein transport pathway, we investigated function of predicted ER-localized SNAREs, including Qa-SNARE, Qb-SNARE and Qc-SNAREs. Our results indicated that these ER-localized SNAREs interact with each other to form a SNARE complex. This SNARE complex interacts with MAG2 complex.When these SNAREs are defective, efficiencyoftransport of seed storage protein precursors from the ER is decreased.On the other hand, plant growth and development is also affected.Our findings indicate that these ER-Localized SNAREs collaborate with MAG2 complex to regulate ER-Golgi vesicle transport.

T4

P1029

Differential regulation of protochlorophyllide oxidoreductase abundances by VIRESCENT 5A (OsV_{5A}) and VIRESCENT 5B (OsV_{5B}) in rice seedlings

Qingzhu Li

Shanghai Academy of Agricultural Sciences

A low-temperature virescent mutant (*osv*$_{5a}$) was isolated from a ^{60}Co-irradiated rice (*Oryza sativa*) population. At early seedling stage, the mutant exhibits chlorosis phenotype with reduced pigment content at a low temperature (22°C), but it produces green leaves at normal growth temperature (28°C). Chlorophyll accumulation is gradually restored in the mutant at 22°C as it is developing into five-leaf stage. Map-based cloning revealed that OsV$_{5A}$ is a J-like protein with four transmembrane domains. A close homolog, OsV$_{5B}$, was also identified in the rice genome. Both OsV$_{5A}$ and OsV$_{5B}$ are localized in the chloroplast envelope and thylakoid membranes. We demonstrated that they function as chaperone proteins of protochlorophyllide oxidoreductase (POR), which catalyzes a light-dependent reaction in the chlorophyll biosynthesis pathway. OsV$_{5A}$ and OsV$_{5B}$interact with two rice PORs (OsPORA and OsPORB) inside chloroplasts and they stabilize OsPORB in vitro under oxidative stress. Differential protein abundances of OsV$_{5A}$ and OsV$_{5B}$ in rice seedlings at different leaf developmental stages were also revealed. OsV$_{5A}$ apparently developed a specialized role for regulating POR abundances at the leaf pre-emergence stage, while the same function is performed by OsV$_{5B}$ during leaf emergence and expansion. Deficiency of OsV$_{5A}$ and OsV$_{5B}$ occurred in pre-emerged and emerging leaves in *osv*$_{5a}$ seedlings at 22°C, leading to reduced POR accumulation and chlorophyll content. Duplication of V5, which is not common among dicots, may have allowed the diversification of their differential roles in regulating chlorophyll biosynthesis in rice and other grass species.

T4

P1030

Identification and functional analysis of downstream targets of the MKK7-MPK6 cascade involved in shoot branching in *Arabidopsis*

***Shujia Li*[1]**, *Weiyan Jia*[1,2], *Baohua Li*[1,3], *Xiaowei Wu*[1], *Yan Liang*[1], *Mei Ma*[1], *Jiayang Li*[1], *Yonghong Wang*[1]

1. Institute of Genetics and Developmental Biology, Chinese Academy of Sciences, Beijing

2. University of Wisconsin, School of Medicine and Public Health, Madison

3. University of California, Davis, California

Mitogen-activated protein kinase (MAPK) cascades play an important role in transducing environmental and developmental signals into adaptive and programmed responses. In higher plants, MAPK cascades are involved in a broad spectrum of signals, which include biotic and abiotic stresses, and hormone-mediated development. The *Arabidopsis* genome encodes a large number of MAPK signaling components, with more than 60 members of MAPKKK, 10 members of MKK and 20 members of MAPK. The least number of MKKs suggests that MKKs serve as the convergence and divergence points in the MAPK signal transduction. The complexity of MAPK cascades raise challenges not only to identify the MAPK module *in planta* but also to define the specific role of an individual module. Our previous study have characterized an *Arabidopsis bushy* and *dwarf1* (*bud1*) mutant, in which the MKK7 was constitutively activated, resulting in multiple phenotypic alterations. It is indicated that MKK7-activated cascades involve in many developmental and hormone signaling pathways through phosphorylating different substrates. Recently, we demonstrated that MPK6 and MPK3 are two major downstream targets of MKK7, and most strikingly, polar auxin transport are transmitted by MKK7-MPK6 signalling pathway specially. To elucidate molecular mechanisms of polar auxin transport regulated by MKK7-MPK6 cascade, we further identified the MKK7-MPK6 downstream specific substrates and studied the function of them. So far, our data have provided evidence to explain how the MKK7-MPK6 signaling pathway regulates polar auxin transport through the specific substrate to determine shoot branching in *Arabidopsis*. Shoot branches develop from axillary meristems initiated in the boundary zone. We therefore wondered whether the MKK7-MPK6 cascade regulate shoot branching through affecting axillary meristem initiation. In-depth genetic and biochemistry verification are underway.

T4

P1031

Overexpression of maize phosphoenolpyruvate carboxylase gene enhanced the catabolism of endogenous saccharide under drought conditions

Xia Li [1,2], *Xiaolong Liu*[1], *Yafei He*[1], *Jinfei Zhang*[1], *Ting Yan*[1], *Chen Zhang*[1,2]

1. Jiangsu Academy of Agricultural Sciences

2. Nanjing Agricultural University

We compared the drought tolerance of wild-type (WT) and transgenic rice plants (PC) over-expressing the maize C_4 *PEPC* gene, which encodes phosphoenolpyruvate carboxylase (PEPC, EC 4.1.1.31) gene, and evaluated the roles of saccharide and sugar-related enzymes in the drought response. Pot-grown seedlings were subjected to drought conditions outdoors, and the yield components were compared between PC and untransformed wild-type (WT) plants. The stable yield from PC plants was associated with higher net photosynthetic rate under the drought treatment. The physiological indices of WT and PC seedlings under a simulated drought treatment (25% (w/v) polyethylene glycol-6000 for 3 h; PEG 6000 treatment) were analyzed in detail for the early response of drought. The relative water content was higher in PC than in WT, and PEPC activity and the *C_4-PEPC* transcript level in PC were elevated under the simulated drought conditions. The endogenous saccharide responses also differed between PC and WT under simulated drought stress. The higher sugar decomposition rate in PC than in WT under drought stress was related to the increased activities of sucrose phosphate synthase, sucrose synthase, acid invertase, and neutral invertase, increased transcript levels of *VIN1*, *CIN1*, *NIN1*, *SUT2*, *SUT4*, and *SUT5*, and increased activities of superoxide dismutase and peroxidase in the leaves. The greater antioxidant defense capacity of PC and its relationship with saccharide metabolism was one of the reasons for the improved drought tolerance of PC. In conclusion, PEPC effectively alleviated oxidative damage and enhanced the drought tolerance of rice plants, which were more related to the the increase of the saccharide decomposition. These findings show that components of C_4 photosynthesis can be used to increase the yield of rice under drought conditions.

T4
P1032
Seed germination in *Artemisia frigida* Wild. under the drought stress
Yang Li
Xi'an Botanical Garden of Shaanxi Province

Artemisia frigida is a dominant species in the typical steppe of northern China, and highly adapted to the local edaphic conditions. As water is the main limiting factor in arid and semi-arid regions, we hypothesis *A. frigida* could endure drought stress. However, little is known about the effects of drought stress on seed germination characteristics. In our study, PEG-6000 was applied to soak the seeds of *A. frigida* to simulate drought stress. We applied 4 level of PEG-6000: CK, 5%, 10%, 15%. The results showed that after soaking with PEG-6000, there were no significant change in seed germination rate, germination index and germination time among different treatments. Our results showed that *A. frigida* could endure drought stress according to seed germination characteristics. It is one of reasons that *A. frigida* could dominant in arid and semi-arid regions.

T4
P1033
Identification and characterization of MYB-bHLH-WD40 regulatory complexes controlling anthocyanin biosynthesis in flowers of a monocot plant, *Freesia hybrida*
Yueqing Li, Xiaotong Shan, Ruifang Gao, Song Yang, Li Wang, ***Xiang Gao****
Key Laboratory of Molecular Epigenetics of MOE and Institute of Genetics & Cytology, Northeast Normal University, Changchun, China

The emergency of anthocyanin biosynthetic pathway has played vital roles in plant development. In the past several decades, an anthocyanin-promoting ternary protein complex (designated as MBW complex), composed of R2R3-MYB, basic helix-loop-helix (bHLH) and WD40 repeats (WDRs) has been well characterized in plants. However, divergent regulatory mechanisms were found in dicot and monocot plants represented by *Arabidopsis thaliana* and *Zea mays*, respectively. Presently, a whelming proportion of MBWs have been well elucidated in core dicot plants, whereas they were studied randomly in monocot plants, which are of crucial importance to decipher their functional diversities among angiosperms during evolution. *Freesia hybrida*, a monocotyledonous genus of herbaceous perennial flowering plants in the family Iridaceae, is grown universally as a cut flower plant with a wide range of flower colors including white, yellow, red, blue, purple and bicolors. In our previous studies, five kinds of anthocyanin aglycons, i. e. delphinidin, petunidin, malvinidin, peonidin, and cyanidin, were detected in *F. hybrida* "Red River®". A strategy combining transcriptomic data, expression profiles, gain-of-function experiments and transient protoplast transfection assays was undertaken to identify *F. hybrida* anthocyanin regulators. Among the candidate genes isolated, five were similar to *AtPAP1*, *AtTT8* and *AtTTG1* (*FhPAP1L1*/*FhPAP1L2*, *FhTT8L*/*FhGL3L* and FhTTG1L, respectively) and two encoded putative negative MYB regulators (FhMYB4L and FhCPCL). *FhPAP1L1*/*FhPAP1L2*, *FhTT8L*/*FhGL3L* and *FhTTG1L* were found to positively regulate anthocyanin biosynthesis when heterogeneously expressed in *Arabidopsis* or tobacco respectively. In addition, they interacted and activated the anthocyanin biosynthetic gene promoters when co-expressed in *Arabidopsis* leaf protoplasts. Further experiments revealed sets of the MBW complexes involved in anthocyanin biosynthesis were organized in a hierarchical and partially redundant manner in *F. hybrida*. Moreover, R2R3-MYB FhMYB4L and R3-MYB FhCPCL were deemed as negative regulators as both of them could inhibit anthocyanin biosynthesis. Genetic and molecular evidences showed that FhMYB4L functioned as part of the MBW complex by interacting with bHLH proteins and repressed transcription through its C-terminal EAR motif, which changed the positive MBW complex into negative one. Comparatively, FhCPCL functioned as negative regulators by competing bHLH regulators with anthocyanin-promoting MYBs, which titrated the formation of anthocyanin-promoting MBW complex. Actually, the precise regulatory cascade and feedback regulation in the control of anthocyanin biosynthesis in *Freesia* flowers were much similar like that in core eudicot plants, *Arabidopsis* and petunia. However, more and more evidences exhibited there supported the finds that divergent regulatory manners were evolved after the division of monocots and dicots. For example, FhPAP1L1 or FhPAP1L2 was sufficient to activate *EBGs* (early biosynthetic genes) as well as *AtTT8*, and further interacted with AtTT8 to regulate the expression of *LBGs* (late biosynthetic genes), whereas in *Arabidopsis*, the *EBGs* were activated by co-activator independent and functionally redundant R2R3-MYB regulatory genes (*MYB11*, *MYB12*, *MYB111*). Moreover, AtPAP1 could not activate *AtTT8* or *LBGs* independently. In conclusion, it can be deduced that the regulatory mechanisms between monocots and dicots were divergent in many aspects, and the finds will benefit for the understanding of the genetic basis of the regulatory mechanism accounting for flower pigmentation in monocot plant.

T4
P1034
The effect of flooding on clonal growth and regeneration of *Populus pruinosa* Schrenk riparian forest
Yaqiong Zheng, Zhengshuai Zhao, ***Zhijun Li***
College of life science, Tarim University

This study explored the effect of flooding on the clonal growth and regeneration of *Populus pruinosa* Shrenk riparian forest along the upper reaches of the Tarim River in Xinjiang, China. In 2014 and 2015, from April to October in both years, we surveyed *P. pruinosa* Shrenk clonal growth as well as soil physical and chemical factors before and after flooding. The results showed that: (1) From April to June, before flooding, there was a gradual increase in adventitious buds, unearthed/earthed clonal ramets, the number of ramets in each cluster, the ramet height and the growth of their basal diameter; while from August to October, after flooding, the numbers of buds and ramets decreased and the growth rate gradually declined. (2) From April to June, before flooding, there was a gradual increase in soil moisture, field capacity, and soil porosity, while soil bulk density gradually decreased. During the same period, soil organic matter, available nitrogen, and available phosphorus, soil available potassium, total salt content, and pH value all increased gradually. From August to October, after flooding, these chemical factors (soil organic matter, available nitrogen, soil available phosphorus, available potassium, total salt content, and pH value) continued to increase gradually. (3) Flooding had an effect on soil moisture which in turn led to significant changes in field capacity, soil bulk density, total salt content, and pH value. Correlation analysis also showed that the variation of four physical factors and six chemical factors exerted positive or negative effects on adventitious buds, unearthed/earthed clonal ramets, number of ramets in each cluster, height of ramets, and growth rate of ramet basal diameter. Thus, we can conclude that flooding not only nourished clonal growth during the flooding period, but also sustained clonal growth from April to June of the following year by replenishing soil moisture through the freezing and thawing of soil water in winter. Variations in soil physical and chemical factors due to flooding also influenced *P. pruinosa* Schrenk clonal growth. Therefore, flooding played an important role in the clonal reproduction and regeneration of *P. pruinosa* Schrenk riparian forest.

T4
P1035

The relationship between the spatial and temporal characteristics of flower occurrence and the heterophylly of *Populus euphratica*

Jiahao Li, Peipei Jiao, ***Zhijun Li***
College of life science, Tarim University

We choose the *Populus euphratica* with different diameters on the same site as the research object, focus on the occurrence of flowers and heterophyll, and their vertical distribution range as diameters classes increased and the relationship among them. The results showed: (1) linear leaves occupied the entire crown before diameter class 2, lanceolate leaf, ovate leaves and broad ovate leaves began to appear from the canopy top in order, lanceolate leaves and ovate leaves occurred from diameter class 2, broad ovate leaves began to emerge from diameter class 4. There is a dynamic change of their frequency and the proportion of vertical distribution region in crown height as the diameter increased, the proportion of linear and lanceolate leaves in crown was significantly negative correlated with diameter classes ($P < 0.01$); while the proportion of ovate leaves and broad ovate leaves in the crown were significantly ($P < 0.05$), and very significantly ($P < 0.01$) positive correlated with diameter classes. (2) As same as the appearance of broad ovate leaves, flowers occurred from the top of crown and diameters class 4; the maximum distribution region and the proportion of flower concentrated region in crown height showed a very significant correlation with diameter classes ($P < 0.01$). (3)The flower maximum distribution region and flower concentrated region overlapped with the distribution region of broad ovate leaves, the vertical spatial distribution range of flowers and broad ovate leaves widened as the diameters classes increasing from diameter class 4-18, and processed simultaneously, and the occurrence time and spatial distribution range of flowers also overlapped with that of broad ovate leaves. Correlation analysis showed, the flower maximum distribution region in vertical space, and the proportion of concentrated distribution region in crown height of flowers were very significant positive correlated with the proportion of concentrated distribution region in crown height of broad ovate leaves, the results illustrated that the appearance of broad ovate leaves is a morphological characteristic and sign for *Populus euphratica* to transfer from vegetative to reproductive development.

T4
P1036

Functional characterizations of β-glucosidases involved in aroma compound formation in tea (*Camellia sinensis*)

Yinyin Liao[1]*, Ying Zhou*[1]*, Lanting Zeng*[1,2]*, Jiadong Gui*[1,2]*, Fang Dong*[3]*, Ziyin Yang*[1,2]
1. South China Botanical Garden, Chinese Academy of Sciences
2. University of Chinese Academy of Sciences
3. Guangdong Food and Drug Vocational College

Tea (*Camellia sinensis*) aroma is an important factor affecting tea quality. Many tea aroma compounds are present as glycosidically conjugated forms in tea leaves for better storage, and the glycosidically conjugated aromas can be hydrolyzed by beta-glucosidase (b-Glu) and beta-primeverosidase to release free tea aromas. b-Primeverosidase has been identified and functionally characterized, while b-Glu has not been identified in tea leaves. In the present study, we established a yeast expression system to recombine CsGH1BG1, CsGH3BG1, and CsGH5BG1, which belonged to GH1, GH3, and GH5 families in plants, respectively. These three recombinant Csb-Glus hydrolyzed the b-glucosidically conjugated aromas to form free aromas, suggesting that Csb-Glus were not specific for the hydrolysis of b-glucosidically conjugated aromas *in vitro*. Furthermore, subcellular localization of the Csb-Glus indicated that CsGH1BG1 and CsGH3BG1 were located in the cytosol and vacuole, respectively, while CsGH5BG1 was located in the cell wall. As b-glucosidically conjugated aromas occurred in the vacuole, CsGH1BG1and CsGH3BG1 might be responsible for the hydrolysis of b-glucosidically conjugated aromas in tea leaves during the tea manufacturing process. This study provides the first evidence of Csb-Glus in tea leaves, and will advance understanding of tea aroma formation.

T4
P1037

Physiological indice study of the resurrection plant *Boea Hygrometrica* using weighted gene co-expression network analysis at different hydrating stages.

Tao Xu, ***Chih-Ta Lin***
Institute of Botany, The Chinese Academy of Science

Resurrection plants are defined as the plants that rapidly restore their biological processes after severe water deficit. In this course, the physiological indices and gene expression of the plants are varied to adapt the desiccation. To study the differentiation in gene expressions of plants in dehydrated and rehydrated status systematically, co-expression networks of *Boea Hygrometrica* have been constructed using weighted gene co-expression network analysis (WGCNA) method. Through modular analysis and the correlation between modules and variation in physiological indices, the constructed networks were found high quality and aggregation. Seven networks were found specific and significant from GO enrichment and KEGG annotation in both biological functions and cellular components, as well as their high physiological correlation compare to others. Modular analysis showed the highest correlation of the hub genes from individual network is with the ubiquitin-related genes, and the next is the bZIP-like genes. The ubiquitination levels in *Boea Hygrometrica* at different hydrating stages was further investigated to verify its involvement and importance in desiccation tolerance.

T4

P1038

Endocytic mechanisms of membrane proteins in *Arabidopsis*—a single molecule analysis approach

Jinxing Lin

1. Beijing Forestry University

2. Institute of Botany, Chinese Academy of Sciences

The plasma membrane (PM) is highly dynamic and PM proteins play important roles in regulating various transport processes. Measuring the mobility and interactions of proteins is key to understanding cellular signaling mechanisms. However, bulk approaches have been unable to unambiguously monitor the behaviors of the individual PM molecules without disrupting their membrane environment. Besides, transient molecular interactions cannot be identified by conventional fluorescence imaging-approaches. In the past few years, we have developed variable-angle total internal reflection fluorescence microscopy (VA-TIRFM) to image PM proteins in intact plant cells, applied single-particle techniques to protein tracking and subunit counting, provided new information on the spatiotemporal dynamics of specific molecules and their interactions. Endocytosis is an essential way for entry of membrane proteins, lipids, and extracellular molecules into the cell. Clathrin-mediated endocytosis plays an essential role in many cellular and developmentalprocesses. Based on single-molecule and genetic approaches, we demonstrated that *Arabidopsis* AP2σ is closely associated and physically interacts with the clathrin light chain (CLC), and AP2σ-mCherry fluorescence appears and disappears before CLC-EGFP fluorescence. In addition, the density and turnover rate of the CLC-EGFP spots are significantly reducedin the *ap2σ* mutant. These findings led us to conclude that AP2 is involved in the CCV initiation, assembly and maturation stages, which is required for clathrin-mediated endocytosis. In recent years, the structure, composition, and possible functions of plant plasma membrane raft-like domains have been described. Similar to animal cells, plant cells have additional endocytic pathways besides clathrin-mediated endocytosis. In our investigations, we found that *Arabidopsis* flotillin1 (AtFlot1) was associated with the membrane microdomains, and participated in clathrin-independent endocytosis. By using VA-TIRFM, we demonstrated that the dynamic behavior of GFP-Flot1 puncta was different from that of CLC-mOrange puncta. Analysis of amiRNA AtFlot1 transgenic *Arabidopsis* plant lines established that a reduction in AtFlot1 transcript levels gave rise to a reduction in shoot and root meristem size and a retardation in seedling growth. Colocalization analysis and immunogold labeling revealed that AtFlot1 was not colocalized with CLC. These findings suggested that AtFlot1 is involved in a clathrin-independent endocytic pathway and functions in seedling development. Further studies based on single-molecule approaches and other methods revealed that clathrin and AtFlot1-associated membrane microdomains cooperatively regulate membrane protein dynamics and endocytosis, such as PIP2;1 (plasma membraneintrinsic protein), AMT1;3 (ammonium transporter), RbohD (respiratory burst oxidase homolog D), and BRI1 (brassinosteroid receptor) and the membrane microdomain played a role in altering its activity of these proteins by positively or negatively affecting their clustering and signal transduction.

T4

P1039

A regulatory cascade for auxin regulated root growth under aluminium stress

***Guangchao Liu**, Shan Gao, Zhaojun Ding*

Shandong University

Auxin acts synergistically with cytokinin to control the shoot stem-cell niche, while both hormones act antagonistically to maintain the root meristem. In aluminum (Al) stress-induced root growth inhibition, auxin plays an important role. However, the role of cytokinin in this process is not well understood. In this study, we show that cytokinin enhances root growth inhibition under stress by mediating Al-induced auxin signaling. Al stress triggers a local cytokinin response in the root-apex transition zone (TZ) that depends on IPTs, which encode adenosine phosphate isopentenyltransferases and regulate cytokinin biosynthesis. IPTs are up-regulated specifically in the root-apex TZ in response to Al stress and promote local cytokinin biosynthesis and inhibition of root growth. The process of root growth inhibition is also controlled by ethylene signaling which acts upstream of auxin. In summary, different from the situation in the root meristem, auxin acts with cytokinin in a synergistic way to mediate aluminum-induced root growth inhibition in *Arabidopsis*.

T4

P1040

A study of drought tolerance of ectomycorrhizal *Acacia confusa* seedlings

***Fengming Song**[1], Jianhua Liu[1], Jiangli Lei[2]*

1. Shenzhen Techand Ecology & Environment Company Limited

2. Fairylake Botanical Garden, Shenzhen & Chinese Academy of Sciences

Soil improvement and drought tolerance improvement in plants are the keys to successful recovery and reconstruction of slope vegetation. In this study, seedlings of *Acacia confusa*, a tree commonly used to afforest slopes, were inoculated with different quantities of the ectomycorrhizal fungus *Pisolithus tinctorius* for the purpose of soil improvement and drought tolerance improvement. Half a year later, the seedlings' mass growth and mycorrhizal colonization rates were measured. After that, the ectomycorrhizal seedlings and the control seedlings (not inoculated with Pisolithus tinctorius) were both exposed to natural drought stress. Leaf photosynthesis and chlorophyll fluorescence in the two groups were measured before and after the treatment in order to observe changes in the two indexes. The results show that, compared to the control seedlings, the ectomycorrhizal seedlings had significantly higher mass growth, photosynthesis and chlorophyll fluorescence under drought stress and experienced much slower decreases in the three indexes. These indicate that the inoculation with *Pisolithus tinctorius* can increase photosynthetic efficiency and water use efficiency of *Acacia confusa* and thereby improve its drought tolerance. Moreover, both large and small quantities of *Pisolithus tinctorius* were found to be able to boost growth of *Acacia confusa* seedlings, but large quantities of the fungus can result in higher mycorrhizal colonization rates and are thus more effective in promoting growth and enhancing drought tolerance.

T4

P1041

Study on the effect of three arbuscular mycorrhizal fungi on the growth and drought resistance of *Zenia Insignis*

Jianhua Liu[1], *Fengming Song*[1], *Zhengjun Shi*[2]
1. Shenzhen Techand Ecology & Environment Company Limited
2. Fairylake Botanical Garden, Shenzhen & Chinese Academy of Sciences

In order to improve the growth of plants in the greening of soil spray sowing, the survival rate of the plants in the initial stage of planting as well as the drought-resistance in the state of low maintenance in the later stage, three arbuscular mycorrhizal fungi, namely *Glomus mosseae, Glomus etunicatum* and *Glomus intraradensis*, are selected to study the effect on the growth and drought resistance of *Zenia Insignis* through mycorrhizal inoculation and natural drought stress. The results show that the rate of infection from *Glomus mosseae* is the highest (96.7%). The inoculation of three species of arbuscular mycorrhizal fungi could all significantly increase the plant height, ground diameter and biomass accumulation. Through natural drought experiment, it is concluded that the relative water content of mycorrhizal seedlings is significantly higher than that of the control group, while the relative conductivity and MDA content of the mycorrhizal seedlings are significantly lower than that of the control group, and the soluble sugar content is significantly higher than that of the control group, indicating that there is relatively little dehydrated behaviors of leaves of mycorrhizal seedlings, and the degree of damage on the cytomembrane is less, and more osmotic adjustment substances is produced at the same time. In the process of drought resistance, the photosynthesis and chlorophyll fluorescence of the control and the mycorrhizal fungi were the same in terms of the index change, and the inoculation of arbuscular mycorrhizae could increase the photosynthetic capacity and water use efficiency, decrease the damage degree of ΦPSII reaction center, enhancing the efficiency of solar energy utilization and the drought-resistant ability of seedlings, among which the Glomus mossea has the best performance in drought resisting.

T4

P1042

Formation and emission of linalool in tea (*Camellia sinensis*) infested by tea green leafhopper (*Empoasca* (*Matsumurasca*) *onukii* Matsuda)

Xiaoyu Liu[1,2], *Xin Mei*[1], *Ying Zhou*[1], *Xiumin Fu*[1], *Lanting Zeng Zeng*[1,2], *Fang Dong*[3], *Ziyin Yang*[1,2]
1. South China Botanical Garden, Chinese Academy of Sciences
2. University of Chinese Academy of Sciences
3. Guangdong Food and Drug Vocational College

Herbivore-induced plant volatiles play important roles in defense against stresses, and also can be used as quality components of agricultural plants. As an example, the famous oolong tea (Oriental Beauty), which is manufactured by the tea (*Camellia sinensis*) leaves infected by tea green leafhoppers with piercing-sucking mouthpart, has characteristic volatile monoterpenes, which were derived from linalool. However, effect of tea green leafhopper attack on formation and emission of linalool is unknown. In the present study, tea green leafhoppers inducing production of characteristic volatiles was identified as *Empoasca* (*Matsumurasca*) *onukii* Matsuda. *E.* (*M.*) *onukii* attack did not significantly affect internal linalool content, but induced higher emission of linalool from tea leaves, which was due to accumulation of high levels of *linalool synthases* (*CsLIS1 and CsLIS2*) mRNA. *CsLIS1 and CsLIS2* recombinant proteins exhibited ability of transformation from geranyl diphosphate to (*S*)- linalool with high ratio. An automatic machine was employed to simulate continuous mechanical damages from *E.* (*M.*) onukii attack on tea leaves. In contrast to single mechanical damage, the continuous mechanical damages had similar patterns of changes in emitted linalool, precursors of linalool, expressions of genes involved in formation of linalool with those of *E.* (*M.*) *onukii* attack. These suggested that continuous mechanical damages from *E.* (*M.*) *onukii* attack was one key factor affecting formation and emission of linalool from tea leaves. The information can be essential for the future utilization of stress-responses of plant secondary metabolism for improvement of quality components of agricultural products.

T4

P1043

Molecular genetic mechanism of hybrid male sterility by the *Sc* locus in rice

Yao-Guang Liu
College of Life Sciences, South China Agricultural University

Hybrid sterility is a major form of postzygotic reproductive isolation between the species and subspecies in *Oryza* genus. A number of gene loci are involved in hybrid sterility in rice. So far, several hybrid sterility genes in five loci (or locus pairs) have been cloned, including *S5* (Chen *et al.*, 2008, Yang *et al.*, 2012), *Sa* (Long *et al.*, 2008), *S27/S28* (Yamagata *et al.*, 2010), *DPL1/DPL2* (Mizuta *et al.*, 2010), and *HSA1* (Kubo *et al.*, 2016). Our recent results reveal that a pollen-essential gene in the *Sc* locus in rice has evolved into a speciation gene between *indica* and *japonica* by allele divergence via genomic structural variation. The allelic interaction of this locus in *indica-japonica* hybrids results in epigenetic variation in the *japonica* allele, leading to its silencing thus pollen abortion. These researches demonstrate that diverse molecular mechanisms have been produced for reproductive barriers in rice speciation.

T4

P1044

Galactinol Synthase (*AnGolS1*) of *A.nanus* can improve the seed germination ability in tomato under low-temperature

Yudong Liu

Shenyang Agricultural University

Galactinol synthase (GolS) is important in the physiology of plant stress resistance, photosynthate translocation and seed physiology. A galactinol synthase (*AnGolS1*) which was isolated from *Ammopiptanthus nanus* (*A.nanus*) seedlings was transformed into tomato. In previous studies, we found that *AnGolS1* could affect *SlGolS* genes expression in the *AnGolS1* overexpression tomato whether or not low-temperature treatment. In addition, *AnGolS1* can also improve the germination ability and rate of overexpression tomato seeds under room temperature and low-temperature. It explains that *AnGolS1* maybe anticipate in the growth and development of plant and helps plant to resist cold damage. Thus, it is important to define its metabolic functions that will help us to further understand the molecular mechanism of cold tolerance of tomato.

T4

P1045

Phenotypes of 57 terrestrial tree species under waterlogging

Ling Ma
South China Agricultural University

Both hydroponic experiments and waterside ecological restoration need abundant information about plant species that adapted to waterlogging environment during the planning stage. In this study, we evaluated the growth conditions and phenotypes of 57 plant species from 22 families and 33 genus with several different provenances grown hydroponically. We found that most plant species could survive in hydroculture for more than 10 weeks with survival rate >60%, and showed new shoot growth in hydroculture. While there were very interesting phenotypes expressed by some species, such as all leaves withered rapidly then new leaves grew out gradually for *Anthocephalus chinensis*; leaves kept alive but stopped growth for *Phoebe sheareri*, and *Acacia falcata*; all leaves fell down rapidly and had kept naked branches for at least 2 weeks for *Cassia barkeriana*, *Pithecellobium clypearia* and *Mytilaria laosensis*, *etc.* In addition, the 57 plant species also showed different fresh biomass growth rate. The phenotypes and growth condition information of 57 plant species were addressed in this study, it would be very useful for planning hydroponic experiments or waterside ecological restoration, although the waterlogging condition in waterside would be quite different from hydroculture.

T4

P1046

Role of calcium and nitric oxide in enhancement of therapeutic potential and oxidative stress management in *Trigonella foenum-graecum* L.

***Palash Mandal**, Saran Kumar Gupta*
University of North Bengal

Trigonella foenum-graecum being rich source of various bioactive phytochemicals has been used for the treatment of wide spectrum of diseases and disorders since ancient times. Considering such diverse implication of fenugreek, the present study was designed aiming the enhancement in the phytochemicals especially associated with the antioxidant and antidiabetic activity as these two therapeutic properties are the major point of concern of the present research work. The elicitors of two important signalling molecules (nitric oxide and calcium) were implemented in the present research work, to investigate their role in the elicitation of therapeutic potential of fenugreek sprouts. The technique applied in the study was priming; this method has been reported to be one of the important strategies in enhancement of crop productivity. Accordingly, the seeds were pre-treated with various elicitors of nitric oxide and calcium namely, sodium nitroprusside (SNP) as an exogenous source of nitric oxide; c-PTIO (CP), a nitric oxide scavenger; calcium chloride (CC) as an exogenous source of calcium ion; EGTA (EG) a calcium chelator; and lanthanum chloride (LC): a calcium channel blocker. As a result, the present study revealed significant increase in free radical scavenging activity and *in vitro* anti-diabetic activity along with associated bioactive phytochemicals by elicited sprouting in fenugreek with exogenous sources of nitric oxide and calcium ion. Additionally, a significant correlated relationship between the therapeutic potential and the phytochemical contents was obtained, indicating these phytochemicals as efficient antioxidant and antidiabetic agents. Further, the negative effects of the antagonists of calcium and nitric oxide indicate that the flux of these signalling molecules within the cellular system is essential during germination phases to regulate the mechanisms responsible for the biosynthesis of these bioactive compounds. The presence of trigonelline as a major hypoglycemic component was further confirmed by IR and NMR analysis. After successful elicitation of bioactive phytochemicals and their related therapeutic properties, the exogenous sources of nitric oxide and calcium were applied for studying their effect on the alteration of various biochemical parameters along with their antioxidant defence system during germination stages. The pre-treated seeds were exposed to saline condition and their various morphological, physiological and biochemical attributes were determined. Our results suggest that priming of fenugreek seeds with exogenous source of nitric oxide and calcium enhanced the morphological and biochemical attributes along with the antioxidant defense system under saline condition, which was further substantiated by the occurrence of adverse effects of salinity on the seeds which were unprimed and also those primed with the antagonists of these signalling molecules. Therefore, the enhancement in the enzymatic as well as non-enzymatic components might be due to the involvement of calcium and nitric oxide leading to tolerance towards salinity accompanied with better growth and development. Additionally, an uninterrupted influx of these signal molecules is very much essential for the better nutraceutical quality and growth of fenugreek seedlings, which can be hypothesized on the basis of the tremendous deterioration of the therapeutic potential as well as oxidative stress management status of the fenugreek seedlings revealed in the present study.

T4

P1047

Development of stay-green chickpea technology to benefit small-holder farmers in the semi-arid tropics

***Edward Marques**[1], Jana Kholova[2], Ramachandra Varma Penmetsa[3], Eric Von-Wettberg[1]*
1. Florida International Institution
2. International Crops Research Institute for the Semi-Arid Tropics

3. University of California, Davis

Chickpea is the second most grown pulse legume worldwide and is a primary source of plant protein for 15% of the world's population living in semi-arid tropical (SAT) regions. In these regions, fluctuations in chickpea production and consumption threatens the nutritional and economic status of the subsistent farming communities. In many of these SAT regions, chickpea crops are typically raised during the low-income post-rainy season in variable rain-fed conditions which largely affects its production. Thus, finding and introducing chickpea adapted to these conditions is a practical strategy to enhance chickpea production value chains of the semi-arid tropics. One viable option to overcome this post-rain season and limited irrigation environment is "cosmetic" stay-green chickpea technology. This stay-green chickpea type carries a deleterious mutation of the *StGR1* gene that inhibits enzymes of the chlorophyll catabolism pathway, which leads to extended chlorophyll retention in seeds and leaves. Consequently, elevated levels of chlorophyll may extend the lifespan of light harvesting antennae (and increases production of reactive oxygen species (ROS) in plant tissues, which in turn, requires additional maintenance in the form of higher carotenoid levels, the ROS scavengers. Higher levels of carotenoids involved in photo-oxidative protection are very likely to enhance the nutritional value of the crop since these are also precursors of vitamin A. Similar findings have been documented in other green-seeded legumes such as the garden pea (*Pisum sativum*), common bean (*Phaseolus vulgaris*), and Alfalfa (*Medicago sativa*) which have lead to development of commercial products. Since the stay-green technology has not been investigated in detail, the goal of this study is to investigate its effect on key agronomic parameters and response to growth in SAT (drought-prone) regions. Our previous study of *StGR1* germplasm confirmed that these genotypes, indeed, maintain approximately a 33% higher carotenoids content in the seeds compared to other varieties. Additionally, it was found that the *StGR1* gene product might be specific to chlorophyll b since these *StGR1* lines maintained chlorophyll b at significantly higher levels. High levels of chlorophyll b appear to affect leaf thickness and efficiency of water use. A drawback in this study involved the use of a very diverse gene-bank germplasm which didn't allow us to rigorously compare agronomically important parameters. Therefore, we aim to repeat the study with *StGR1* (from donor parent ICC16340) introgressed on the background of 2 cultivars (KAK2, JGK1) fixed for phenology and harvest index. We are currently assessing the *StGR1* mutation effect on basic agronomy traits, plant water use-related traits, responsiveness to soil and atmospheric drought, chlorophyll contents, and carotenoid concentrations using the imaging phenotyping platform, LeasyScan, and analytical scales for gravimetric transpiration measurements. This study is currently ongoing at the International Crops Research Institute for the Semi-Arid Tropics (ICRISAT). Ultimately, the data gathered will allow us to determine the potential of stay-green technology to enhance crop nutrition and the main agronomic parameters important for small-hold farmers in the semi-arid tropics.

T4

P1048

Allocation of resources in sweet potato plants (*Ipomoea batatas* (L.) Lam.) In the community of San Félix Hidalgo, Atlixco, Puebla, México

David Martínez-Moreno[1], *Francisco Basurto-Peña*[2], *Agustina Rosa Andrés-Hernández*[1], *Jenaro Reyes-Matamoros*[1], ***Sombra Patricia Rivas-Arancibia***[1]

1. Benemérita Universidad Autónoma de Puebla

2. Universidad Nacional Autónoma de México

At present there is a growing interest in traditional agricultural production systems, and the value of indigenous technology, developed over centuries and based on empirical practices and accumulated experiences has been recognized for decades. This is the case of sweet potato (*Ipomoea batatas*), which has been studied from the following points of view: origin and history, ethnobotanical, taxonomic, taxonomical characterization through morphological and isoenzimatic descriptors, industrial, processes post-harvest, chemical characterization, industry and genetic engineering. Therefore, the present research aimed to record the allocation of resources to the different organs of three sweet potato varieties (White, Purple and Yellow) through 10 destructive crops in the community of San Félix Hidalgo, Atlixco, Puebla. The differences of the dry weights in white and purple sweet potato inflorescences were in general very similar and the energy of these structures perhaps affected the allocation of energy to the anthropogenic organ that is sweet potato, a situation that was not present in the yellow sweet potato variety, since the dry weight percentage of the different structures shows this. Finally, there are no differences in the last harvest, this was due to the producer's strategy of not harvesting the crop in harvest nine, because the price of sweet potato was very low at that time, this shows that many times this type of decisions is not correct since leaving the crop for a longer time and not harvesting it at the right time can reduce its dry weight, as occurred in this work, and as a consequence the producer lost money, this decision many times causes the crop gets lost, since the risk that the sweet potato was decomposing or attacked by animals was enormous. In conclusion, the yellow sweet potato variety does not elaborate reproductive structures and the percentage of biomass allocation to sweet potato is high, compared to its vegetative structures. These three varieties do not develop seeds, which indicates that asexual reproduction probably accumulates unfavorable mutations or that pollinators specific to varieties may not be found. Keeping the sweet potato in the field for more time than programmed for its life cycle, results in a reduction of its dry weight and it gets decomposed or attacked by insects, and this situation would generate losses.

T4

P1049

Centrin in *Marchantia polymorpha*: characterization, localization, and implications in the evolution of microtubule system in land plants

Kaori Nomura[1], *Kimitsune Ishizaki*[2], ***Masaki Shimamura***[1]

1. Department of Biological Science, Graduate School of Science, Hiroshima University

2. Department of Biology, Graduate School of Science, Kobe University

The intracellular localization of centrin, a centrosomal-associated protein, was investigated by immunofluorescence microscopy in a liverwort, *Marchantia polymorpha.* In vegetative tissue, mitotic spindles were initiated from the spherical "polar organizers" (POs) lacking the centrioles. Although gamma-tubulin, another typical centrosomal protein, clearly localized on the POs, centrin did not localized there. The typical centrosomes including centrioles arise *de novo* just before the production of spermatid mother cells. Centrin was clearly detected at the spherical MTOCs corresponding to the centrosomes in spermatid mother cells. Through the spermiogenesis, centrin was detected at basal bodies and multilayered structure (MLS) of sperm cells. Unexpectedly, centrin also localized on the developing cell plate in phragmoplasts. Phylogenetic analysis and genetic analyses using gene targeted mutagenesis indicated that the three homologous genes of centrin in *M. polymorpha* have specific function respectively. Although angiosperms have no centrosomes through the life cycle, they also have centrin homologs. Functional diversification of centrin through the evolution of land plants seems to be associated with evolutional changes of microtubule organizing system concerned with the formation of mitotic and cytokinetic apparatus.

T4

P1050

A transcriptomic approach to explore fruit formation and maturation in *Arabidopsis thaliana*

Chiara Mizzotti[1], *Lisa Rotasperti*[1], *Bianca Galliani*[1], *Moretto Marco*[2], *Kristof Engelen*[2], ***Simona Masiero***[1]
1. Università degli studi di Milano
2. Fondazione Edmund Mach

Given the fundamental nature of both the dietary and biological significance of fruits, molecular dissection of fruit maturation has, certainly, considerable interest. The yield and quality factors associated with fruits are of key importance to agricultural production, and future improvements of fruit characteristics will rely on the comprehension of the mechanisms controlling fruit development and maturation. Here we report our efforts to shed light into the molecular networks controlling fruit development in the model plant *Arabidopsis thaliana*. In order to identify genes whose products control *Arabidopsis* fruit development and maturation, a transcriptome analysis by RNA-deep-sequencing has been performed, comparing wild-type (WT) siliques, devoid of seeds, at 3, 6, 9 and 12 DPA, thus covering all the phases of silique development and maturation. The bioinformatics and statistical analysis of the data led to the identification of about one thousand genes differentially expressed between early and late stages of silique development. Among all the different clusters, we are currently exploring the role of the NAC transcription factors.

T4

P1051

Overexpression of UDP-glucose pyrophosphorylase from *Larix gmelinii* enhances vegetative growth in transgenic *Arabidopsis thaliana*

Lili Mei[1], *Ningning Li*[2], *Xiaofei Lin*[2], ***Wenbo Zhang***[1]
1. College of Forestry, Inner Mongolia Agricultural University
2. Inner Mongolia University College of Life Sciences

UDP-glucose pyrophosphorylase (UGPase) is an important regulatory enzyme in carbohydrate metabolism, that catalyzes the reversible production of glucose 1-phosphate (Glc-1-P) and the conversion of uridine triphosphate to uridine diphosphate glucose (UDPG) and pyrophosphate. UDPG, the substrate/product of UGPase, is the key precursor for synthesizing sucrose, mannose, cellulose, cutin, glucoprotein, glycolipid, and carbohydrate. Therefore, the expression of *UGPase* may affect the timber qualities of the forest trees by regulating the biosynthesis of plant cellulose. Dahurian larch, *Larix gmelinii* (Rupr.), one of three major timber conifers in northeast China, has high wood density, strong mechanical properties, and a clear wood texture. In addition, larch fiber cells are long and the nonfiber cell content is low, so *L. gmelinii* has important economic value in the construction, decoration, paper production, and chemical fiber industries. Therefore, investigating the role of *LgUGPase* in cellulose biosynthesis is important to improve growth and timber qualities by controlling *LgUGPase* gene expression. In this study, a larch UGPase (*LgUGPase*) gene was isolated from *L. gmelinii* using reverse transcription-polymerase chain reaction (RT-PCR). The 1443-bp open reading frame encodes a protein of 480 amino acids with a predicted molecular weight of 53.7 kDa and shows striking sequence similarity to UGPase proteins from *Pinus taeda* and *Picea sitchensis*. Semiquantitative reverse transcription-polymerase chain reaction showed that the *LgUGPase* gene was expressed primarily in the larch stem in addition to in root and leaf. Southern blot analysis indicated that *LgUGPase* is encoded by two genes in the *L. gmelinii* genome. Overexpression of *LgUGPase* enhanced vegetative growth in transgenic *Arabidopsis* and increased the contents of soluble sugars and cellulose, and thickened parenchyma cell walls. These results revealed that *L. gmelinii* UGPase participates in sucrose/polysaccharide metabolism and cell wall biosynthesis, suggesting that *LgUGPase* may be a good candidate gene for improvement of fiber cell development in plants.

T4

P1052

Relationship between light habitat and functionality of the seedling bank of *Calophyllum mariae* Planch. & Triana, in tropical rainforests of the pacific coast of Colombia, South America.

Omar Melo, *Edgar Andrey Sosa*, *Leidy Carolina Mendez*
Universidad Del Tolima

The functional traits are morphological, physiological and phenological characteristics that directly or indirectly influence the fitness of the plant against the environment affecting its growth, reproduction and survival and generating phenotypic differences between species that play an important role in the maintenance of diversity within communities. The research was carried out in secondary forests located in the Bajo Calima region of the, Biogeographical Chocó of Colombia. Coordinates 3°57'2.11" N and. 76° 59'9.39" W. The average annual rainfall is 7467 mm. The average annual temperature is 26 °C. This region is one of the most humid places in the world and with high values of biodiversity. Twelve *Calophyllum mariae* seedlings banks were placed, which were classified into three light environments (100%, 60% and 40%). We using the photosynthetic active radiation (PAR). Populations were monitored for three years by conducting monthly assessments. In each bank, the following variables were recorded: leaf area index (LAI), light extinction coefficient (K), relative growth rate (RGR), root/shoot ratio (R/S), specific leaf area (SLA). The values of biomass were generated from the seedling harvest and determination of their dry weights. Likewise, a foliar analysis was generated at the end of the experiment and the contents of nitrogen, total proteins and total phenols were evaluated. The hypothesis of compromise between adaptation and competition was raised. When *C. mariae* seedlings bank are subjected to contrasting light environments, their stability in the differential distribution of biomass is altered, modifying both physiological and morphological components, in such a way that the seedlings prioritize the organs that facilitate the best adaptation and/or competition against them. By subjecting the seedlings to environments with light constraints, the relative growth rate accelerates and reaches its highest values, for which it must increase the SLA, to maximize the availability in the capture of photons. This is why biomass is prioritized towards aerial components. This strategy allows the natural regeneration of the species to be established in the undergrowth, under low radiation light environments, where their seedlings compete strongly for space and light mainly, to reach the canopy in the shortest possible time. In contrast, when *C. mariae* seedling banks are established under high illumination conditions, the seedling life strategy prioritizes the resources towards the growth components that guarantee their survival and that is why the underground biomass reaches the highest values in these. It also modifies its SLA, allowing it to reduce its growth rates and adapt better to an environment different from its natural regeneration. These results show the high plasticity of the juvenile populations of this species, which allows their adaptation to different types of growth environments.

T4

P1053

L-drying preservation of cultured plant cells

Tomoyo Mikuni[1], ***Hiroto Kinoshita***[1], *Hideyuki Yamazaki*[2], *Akira Kuriyama*[1]

1. Graduate School of Science and Engineering, Tokyo Denki University

2. National Institute of Technology and Evaluation

Vacuum-drying is a popular method for preservation of microorganisms, because long-term viability excellent in most cases and the storage distribution requirements are simple. For example, strains of microorganisms enclosed in glass tubes can be in cold storage and transported easily. Liquid-drying is a useful method of vacuum-drying for the preservation of microorganisms that are particularly sensitive to the freezing stage of the normal lypholization process. Liquid-drying has several advantages over freeze-drying and effectively used for preserving fungi also.
Cultured plant cells become gene resources together with seeds, spores, pollens *etc.* Cultured cells and tissues are commonly cryopreserved for the long-term storage. Since Nag and Street (1973), Sugawara and Sakai (1974) successfully preserved cultured carrot cells and sycamore cells in liquid nitrogen respectively, cultured cells of various plant species have been preserved. However, there is no report of successful preservation of cultured plant cells by the vacuum-drying method. In this study, we tried to preserve actively growing cultured cells by liquid-drying (L-drying) method.
A suspension culture of Pogonatum inflexum was established and maintained in CI medium supplemented with 10 μM BA, 5 μM 2,4-D, and 3% sucrose. Cells were subcultured onto fresh medium for a period of 7 days and used for experimental materials. Actively growing cells were exposed to L-drying after washing with and resuspending in liquid culture medium with or without torehalose. L-drying itself decreased survival rapidly with little further effect from temperature of storage, although survival was markedly improved by addition of torehalose into suspension medium. Cells suspended in the torehalose containing medium survived L-drying at rates comparable to controls (>86%). Dropping the torehalose suspended cells on a piece of aluminum foil before L-drying was important for their subsequent survival through these procedures because the cells dropped on filter paper did not survived L-drying. Cells prepared by L-drying were successfully regrown after 4 weeks storage at 5°C.
L-drying method, widely used for preservation of microorganisms, appears to be a technique which could be used for the preservation of cultured plant cell.

T4

P1054

Extremely low nucleotide diversity in the X-linked region of papaya

Robert Vanburen[1,2], *Ching Man Wai*[1,2], ***Ray Ming***[1,2]

1. FAFU and UIUC-SIB Joint Center for Genomics and Biotechnology, Fujian Agriculture and Forestry University

2. Department of Plant Biology, University of Illinois at Urbana-Champaign

The papaya Y-linked region showed clear population structure, resulting in the detection of the ancestral male population that domesticated hermaphrodite papayas were selected from. The same populations were used to study nucleotide diversity and population structure in X-linked region. Diversity is very low for all genes in the X-linked region in wild dioecious population, with nucleotide diversity π_{syn} = 0.00017, tenfold lower than the autosomal π_{syn} = 0.0017 and twelvefold lower than the Y-linked region π_{syn} = 0.0021. Analysis of the X-linked sequences shows an undivided population, suggesting a geographically wide diversity-reducing event, whereas two sub populations were observed in the autosomes separating gynodioecy and dioecy and three sub populations in the Y-linked region separating three male populations. The extremely low diversity in papaya X-linked region was probably caused by a recent, strong selective sweep before domestication, involving either the spread of a recessive mutation in an X-linked gene that is beneficial to males, or a partially dominant mutation that benefitted females or both sexes. Nucleotide diversity in the domesticated X samples is about half that in the wild Xs, probably due to the bottleneck when hermaphrodites were selected during domestication. The extreme low nucleotide diversity in papaya X-linked region is much greater than observed in humans, great apes, and the neo-X chromosome of *Drosophila miranda*, which show the expected pattern of Y-linked genes < X-linked genes < autosomal genes, whereas papaya shows an unprecedented pattern of X-linked genes < autosomal genes < Y-linked genes.

T4
P1055

Pseudoperolic mutation is caused by the insertion of B-class gene in *Habenaria radiata*

Mitoma Mai, *Kajino Yumi, Kanno Akira*
Graduate School of Life Sciences, Tohoku University.

Habenaria radiata (Orchidaceae) has two whorls of perianth, consisted with three greenish sepals, two white petals and one lip (labellum). In contrast, pseudopeloric cultivar of *H. radiata*, whose name is 'Hishou', forms a white petaloid organ instead of dorsal sepal, and the two lateral sepals replaced to lip-like structures. According to the orchid code and P-code model, *DEFICIENS-like* and *AGL6-like genes* have important role for morphological difference of tepals in orchid. Therefore, we isolated and characterized these genes from H. radiata to clarify the molecular mechanism of pseudopeloric mutation in *H. radiata*. We isolated and characterized four *DEFICIENS*-like (*HrDEF-C1, HrDEF-C2, HrDEF-C3* and *HrDEF-C4*) and two *AGL6*-like genes (*HrAGL6-1* and *HrAGL6-2*) from this plant. Comparative expression analysis revealed that many of the genes showed similar expression patterns in wild type and 'Hishou' cultivar except *HrDEF-C3*. *HrDEF-C3* gene was expressed in petal and lip in wild type, but *HrDEF-C3* transcript was observed in the sepal, petal, lip, leaf, root and bulb in 'Hishou'. The genomic sequence of *HrDEF-C3* gene was compared in wild type and 'Hishou', we found retrotransposon-like insertion in the promoter region of one of the *HrDEF-C3* gene in 'Hishou'. We also investigated the genetic inheritance of pseudopeloric mutation by crossing 'Hishou' cultivar with wild-type plants, and found that the characteristic of pseudopeloric mutation was inherited dominantly and this character is linked to the insertion found at the promoter region of *HrDEF-C3* gene. These results demonstrated that pseudoperolic phenotype is caused by the mutation of *HrDEF-C3* gene and the expanded expression of *HrDEF-C3* gene due to the insertion of retrotransposon-like structure in the promoter region. The expression of *HrAGL6-2* gene, whose product has interaction with *DEF-C3* and related to lip development, is limited to dorsal sepals and lip. This differential dorsoventral expression of *HrAGL6-2* gene is correlated with the lateral-specific change of sepal to lip-like structure.

T4
P1056

Transcriptome analysis of genes involved in sex determination of male and female in *Arisaema serratum*

Yuko Miyazaki, *Miyu Adachi, Tetsuya Matsumoto, Miwa Okada*
Okayama University

The genus *Arisaema* is a dioecious perennial plant belonging to the family Araceae, and its sexual expression varies with a year even if it is the same individual. There is a correlation with individual size in the sex change of *Arisaema*, and it tends to become a female when the size is large, and a male when it is small. Since the individual size can be thought of by replacing it with the amount of resources in the individual, it is possible that some resource may be involved in sex determination. However, the trends seen between the individual size and sex is only to detect the correlation, it does not refer to the mechanism and the mechanism of determining sex has not been clarified yet. In this study, we conducted comparative analysis of transcriptome data between male and female flowers of *Arisaema serratum* in order to clarify the genes that cause a difference in the sexual phenotype of *Arisaema*. Male and female inflorescences of *Arisaema serratum* growing on the forest floor were sampled from 3 individuals and used for analysis by the next generation sequencer, Hiseq 2500. As a result of comparison with the microarray results of each floral organ of *Arabidopsis thaliana*, male flowers of *A. serratum* contained a statistically significant number of genes highly specifically expressed in *A. thaliana* male organs (stamens, pollen). Similarly, in the female flowers of *A. serratum*, a statistically significant number of genes highly expressed in *A. thaliana* female organ (stamen) were included. These results suggest the validity of the results of this transcriptome analysis. The genes with high expression in male flowers of *A. serratum* contained a statistically large number of genes whose expression was high under the limitation of nitrogen and sucrose. On the other hand, the genes with low expression in male flowers of *A. serratum* contained a statistically large number of genes whose expression was low under the limitation of nitrogen and sucrose. As a result of this transcriptome analysis, male flowers of *A. serratum* may be deficient in the amount of nitrogen and sucrose, and it seems that resources such as nitrogen and sucrose are involved in the formation of male flowers and female flowers.

T4
P1057

Ecological signals in the wood anatomy of the hemiparasite *Krameria lappacea* (Krameriaceae) in the high elevation cold desert of Peru

Samantha Moody[1], *Carole T. Gee*[1], *Maximilian Weigend*[2]
1. Steinmann Institute of Geology, Mineraology, and Paleontology, University of Bonn
2. Nees Institute for Biodiversity of Plants, University of Bonn

Although *Krameria lappacea*, commonly known as rhatany, is a medicinal shrub of global economic importance, little is known of its ecological wood anatomy. The antiseptic properties of the species' roots have led to overharvesting and pertinent questions of future sustainability, while basic questions about its role as a keystone species in the cold deserts of Peru remain unknown. Previous studies have shown that *K. lappacea* is hemiparasitic on multiple hosts of various species. Thus, the issue is if the ecological signals inherent in the wood of *K. lappacea* reflects its dependence on host plants or are instead tied to abiotic factors. Here we show that the growth dynamics in the wood of *K. lappacea* are clearly not affected by temperature and precipitation, but by hemiparasitism. Measurement of cell dimensions and densities in the thin sectioned wood of eight individual shrubs were taken, and calculations of standard wood indices were done. The wood of *K. lappacea* has short vessel elements (mean of 117 μm) and narrow vessels (mean of 20.25 μm), like most woody xeric plants. It also has a low vulnerability index (VI=0.25) and mesomorphic ratio (MR=15.3) based upon the vessels, which show the ability to withstand water stress. In addition, *K. lappacea* lacks the helical thickenings and grouped vessels that are found in other clearly xerophytic woods. Curiously, it has a non-complacent mean sensitivity (MS=0.39), which exhibits a degree of influence on growth from outside influences. Comparisons of growth ring widths with mean annual temperature and precipitation for 25 years (1985-2010) showed no significant correlation, meaning the steady rates of growth observed in the growth rings of *K. lappacea* are not due directly to climate. Another species of hemiparasite, *Santalum album* (Santalaceae) of dry deciduous forests in the paleotropics, has many xeromorphic wood characteristics, such as helical thickenings, narrow vessels, and shorter vessel element lengths. Apart from the helical thickenings, the wood anatomy of *S. album* is like that of *K. lappacea*, i.e., they both possess tracheids and have indistinct growth rings, which means ecological signals do occur in the wood anatomy. Because *K. lappacea* wood exhibits seasonally affected growth and presence of indistinct growth rings, it can be inferred that its growth is influenced by their parasitization of neighboring species for water and nutrients. The ecological signal of hemiparasitism is recognizable through the characteristics that support xeromorphy e.g., numerous vessels per millimeter squared, narrow vessels, shorter vessel element lengths, and the presence of vasicentric or vascular tracheids, and by growth that shows variance from year to year that is not influenced directly by climate.

T4

P1058

How do plants age and die? – Some new insights

Hong Gil Nam[1,2], *Hyojung Kim*[1], *Junhyun Lim*[1], *Timilsina Rupak*[2], *Ji-Hwan Park*[3], *Sunghyun Hong*[1], *Sukjoon Jung*[2], *Jin Hee Kim*[1], *Hye Ryun Woo*[2], *Pyung Ok Lim*[2], *Daehee Hwang*[1,2]

1. Center for Plant Aging Research, Institute for Basic Science, Daegu

2. Department of New Biology, DGIST, Daegu

3. Department of Chemical Engineering, POSTECH, Pohang, Gyeongbuk

Aging followed by death is an inevitable fate in most organisms including human. Our group has been studying the molecular mechanisms of aging and death over two decades to solve the question 'How does living organisms know when and how to die?' We have been revealing that aging and death in plants are highly regulated at the chromatin, transcriptional, post-transcriptional, and epigenetic levels. In this presentation, we will report some new insights on the regulatory mechanisms of aging and death in plants with a comparative view on aging in animals.

Aging involves transitions of complex regulatory networks across lifespan. Yet, limited is understanding of causal relationship between age-associated networks and aging processes. Here, we built age-evolving causal regulatory networks of *Arabidopsis* NAC transcription factors across lifespan using mutants of 49 age-associated NACs. These age-evolving networks revealed a regulatory inversion at pre-senescent stage from activating to repressive regulatory modes, which was governed by three hub NACs that negatively regulated key pro-aging pathways, ROS and salicylic acid pathways. Knockout and overexpression of the three hub NACs (the NAC troika) delayed and advanced the regulatory inversion, respectively, rendering earlier and delayed aging. Thus, the network transition by the NAC troika at pre-senescent stage determines initiation and rate of the following aging process in *Arabidopsis*. Meristems are small population of pluripotent stem cells located at shoot and root tips of plants. They have remarkable ability to maintain themselves for a long period and serve as the constant source of cells for newly developing tissue and organs, allowing the long lifespan often observed in many tree species. We are striving to reveal the mechanisms behind the long lifespan stemmed from the long-time maintenance of meristem cells, employing the root apical meristem (RAM) as a model. *Arabidopsis* RAM consists of mitotically inactive quiescent cells (QC) and surrounding active stem-cells. While active stem-cells are susceptible to various stresses, QC cells are highly stress resistant. We have found that *ORE15* has important role in maintenance of RAM. We will report the underlying mechanisms. Light is one of the most important environmental factors which affect plants growth and development. The critical roles of red light through phytochrome-mediated light signaling has been reported recently. We examined the role of far-red light on leaf senescence and found that leaf senescence is modulated by red (R): far-red (FR) ratio (R/FR). We also found a strong correlation between FR-responsive leaf senescence of various ecotypes and latitudinal cline, suggesting that FR signal is used for latitudinal adaptation of *Arabidopsis* ecotypes. 'Climatome' analysis using climate information around ecotypes' habitat revealed that temperature in winter and spring is strongly associated with FR-responsive leaf senescence. We will report a possible mechanism on how *Arabidopsis* adapted to latitude-dependent environmental changes. We are also comparing the regulatory mechanisms of aging between plants and animals. We will report the differential roles of tRNA fragments in regulating aging in *Arabidopsis*, *C. elegans* (worms), and mouse.

T4

P1059

Elicitation of *Andrographis paniculata* (Burm. f.) Nees suspension culture for andrographolide and its derivatives production

Khoa Van Nguyen[1,2], *Benyakan Pongkitwitoon*[1], *Thanika Pathomwichaiwat*[1], *Sompop Prathanturarug*[1]

1. Department of Pharmaceutical Botany, Faculty of Pharmacy, Mahidol University, 447 Sri-Ayuthaya Road, Rajathevi, Bangkok

2. Department of Plant Science, Faculty of Science, Mahidol University, Rama VI Road, Rajathevi, Bangkok

Since *in vitro* plant cell culture is a platform for secondary metabolites production which can be enhanced by various elicitors. This biotechnology would offer sustainable and economically feasible alternative to propagated plants. In this present investigation, we aimed to establish callus and cell suspension cultures of *Andrographis paniculata* and elicit the cultures for andrographolide and its derivatives (e.g. neoandrographolide, 14-deoxy 11,14-didehydroandrographolide) production and analyze their secondary metabolites by HPLC. The first part of study was to induce callus and cell suspension culture from *A. paniculata* explants. Leaf, stem and root parts of *in vitro Andrographis paniculata* seedlings were used for callus induction. They were cultured on MS media supplemented with 6-benzylaminopurine (BA, 0.5 mg/L) and α-napthaleneacetic acid (NAA, 0.0 – 4.0 mg/L). The optimum combinations for callus induction from three different explant types were 0.5 mg/L BA and 4.0 mg/L NAA. Browning was observed, but could be reduced when NAA concentration increased. Leaf-derived callus was green, hard, and compact; callus derived from stem was greenish yellow, hard, and compact; whereas root-derived callus was soft and friable texture with creamy white color. Therefore, root-derived callus was the most suitable for cell suspension culture initiation. Suspension culture was established using 500 mg fresh weight of root-derived calli. They were grown in liquid MS media supplemented with 0.5 mg/L BA and 4.0 mg/L NAA. The suspension culture reached stationary phase at day 25th - 30th of culture. They were friable with creamy white color. No browning was observed during culture time. Abiotic elicitors were used to elicit *A. paniculata* cell suspension culture for andrographolide and its derivatives production. HPLC chemical analysis of *A. paniculata* cell suspension cultures will be discussed.

T4

P1060

Comparative evaluation of the proximate and phytochemical composition of preserved tomatoes

Aliyu Ahmad Nuhu[1], *Hauwa Tasiu*[1], *Faiza Inuwa Hussain*[1], *Fatima Batul Mukhtar*[2]

1. Northwest University, Kano

2. Federal University, Dutse

Tomatoes are consumed widely throughout the world. The edible part of the fruit is known as the power house of nutrition. Many forms of preserved tomato today are available in the market. They range from dried, paste and some other forms. These are to ensure nonstop supply of the fruit throughout the year and to prevent spoilage. A comparative evaluation of the Proximate, mineral and vitamin analysis conducted on five samples of tomato (Fresh, Mutilated, Paste, Sun dried and Oven dried) show that, the fresh tomato has high percentage composition of moisture and least percentage of ash, protein and carbohydrates than the other tomato samples. The Paste has the high percentage composition of pH, carbohydrate and protein, compared to samples of tomato with significant difference ($p < 0.05$). Phytochemical analysis was conducted and it indicate that lycopene (36.90 ± 0.00), saponin (3.26 ± 0.01), alkaloid (1.28 ± 0.00) and glycoside (4.12 ± 0.02) were significantly higher in sun dried tomato, while the tannin was found to be significantly higher in mutilate tomato. Beta-carotene content of fresh tomato is higher while that of vitamin C is higher in gino tomato paste. Five species of fungi were isolated and identified thus *Aspergillus niger, Rhizopus* spp., *Penicillium* spp., *Fusarium* spp. and *Mucor* spp. Mutilate tomato has the highest percentage occurrence with 36.59% and theTomato paste with the lowest percentage occurrence with 9.76%. This result revealed that Sun dried, Oven dried and paste tomato contain more nutrient than Fresh tomato.

T4

P1061

Discovery of small molecules display ethylene activity through compound library screening

Keimei Oh

Akita Prefectural University

Ethylene is a gaseous plant hormone that plays key roles in plant growth and development. By now, great efforts have been made to develop chemicals with ethylene activity and one of the most successful example is ETHEPHON®, a prodrug of ethylene which degrades and subsequently releases ethylene in plant tissues. Because of the gaseous nature of ethylene restricted the extensive usefulness of ethylene, we conducted a triple response assay based chemical library screening for searching such chemicals which are non-gaseous at normal atmosphere but with ethylene activity. Among 9600 compounds, we found N-[(1,3, 5-trimethyl-1H-pyrazol-4-yl) methyl]-N-methyl-2-naphthalenesulfonamide (EH-1) displayed promising biological activity on inducing triple response in Arabidopsis seedlings. To determine the structure-activity relationships of this synthetic series, chemical synthesis of EH-1 and its analogues was carried out. Data obtained from the biological studies of synthesized compounds indicating that 3,4-dichloro-N-methyl-N- (1,3, 5-trimethyl-1H-pyrazol-4-ylmethyl) benzenesulfonamide (compound 8) exhibits the most potent biological activity on induce triple response morphological characteristics in *Arabidopsis* seedlings.

T4

P1062

Wood structure as manifestation of climatic niche: A case study

Alexei Oskolski[1,2], *Pavel Alexeev*[2], *Anna Kostikova*[3]

1. University of Johannesburg

2. Komarov Botanical Institute

3. InsideDNA Ltd.

Wood structure of a particular species is rather uniform within its members: only few traits show considerable infra-specific variation influenced by local environmental fluctuations. Accordingly, adaptive or functional significance of many wood traits is linked to parameters of ecological niche of a species rather than to individual responses to local environments. Quantitative parameters of specific niches remain, however, beyond the scope of ecological wood anatomy: most studies are focused on correlations of wood traits with local ecological parameters. Open data sources on biodiversity and climates foster new opportunities for addressing this question. Georeferenced species occurrence records combined together with the corresponding climatic records allow to estimate climatic tolerances of their ecological niches. These data allow to assess correlations between wood structure and specific niche dimensions to reveal ecological patterns of the interspecific wood diversity. We used climatic niche modelling and analysis of wood structure to assess the relationships between interspecific variation of wood traits and climatic parameters within 30 species belonging to the Neotropical clade of *Schefflera* (Araliaceae). The species of this group occur in both rainforests and savanna-like habitats in more seasonal climates, such as the cerrados and campos rupestres. The coordinates of the species occurrences were obtained from the GBIF database. For each occurrence record, we extracted the values of 19 bioclimatic variables from the WorldClim climate layers. We used the wood anatomical data published by Kotina *et al.*, (2013). Relationships between the values of bioclimatic variables and quantitative wood features for each species were analysed by correlation analysis, also by PCA and CCA. Besides, we investigated the correlations between bioclimatic data for collecting localities of the wood samples and their anatomical traits. We found that the lengths of vessel elements, length of libriform fibers, number of bars on scalariform perforation plates and ray height within the species of Neotropical *Schefflera* show significant positive correlations with the precipitation parameters of climatic niches as well as negative correlations with the indicators of annual inequality of temperature and precipitations. Infraspecific variations in vessel diameter depend positively only on precipitation, whereas vessel frequency correlates negatively with temperature and its seasonality. Intervessel pits size, ray width and vulnerability index show positive correlation with seasonality and/or annual range of temperature. Fibre wall thickness correlates negatively with parameters of precipitation and annual inequality of temperature. Mesomorphy index depends negatively of diurnal range of temperature and precipitation seasonality. Similar, but weaker trends in variation of length of vessel elements and libriform fibers, bar number on scalariform perforation plates and intervessel pit size have been revealed in relation to the bioclimatic parameters for collecting localities of the wood samples. Proposed approach allows to distinguish between the effects of temperature, precipitations, and their temporal inequality on the wood structure occurred in the course of diversification of the Neotropical *Schefflera* species. Our results suggest that this methodology can be used for clarification and explanation of ecological patterns in diversity and evolution of secondary xylem. The study was supported by the Russian Foundation of Basic Research (grant 16-04-00725).

T4

P1063

Molecular mechanism of Hd1 functional switch in rice photoperiod flowering

Xinhao Ouyang
Xiamen University

The molecular mechanism by which Hd1 reverses its activity in response to day length is largely unknown. Here, we demonstrate that the repression of flowering in LD mediated by Hd1 is dependent on *DAYS TO HEADING 8* (*DTH8*). Loss of function of *DTH8* results in the activation of *Hd3a* by *Hd1*, activate leading to early flowering. We also show that Hd1 directly interacts with DTH8 and the DTH8-Hd1 module is necessary for the inhibitory activity of Hd1 in LD. Further, we reveal that DTH8 associted with *Hd3a* promoter to modulate trimethylated H3K27 levels. our results showed the H3K27me3 levels are increased by the presence of the DTH8-Hd1 module in *Hd3a*. However, Hd1 attenuated the H3K27me3 levels in *Hd3a*, when DTH8 was absent. Therefore our data demonstrate that Hd1 represses flowering in LD by constituting the DTH8-Hd1 module which enhances the H3K27me3 levels within the *Hd3a* and in turn negatively regulates *Hd3a* expression. In contrast, Hd1 promotes flowering in LD through reduction of the H3K27me3 levels within the *Hd3a* and in turn positively regulates *Hd3a* expression, when DTH8 is absent in LD.

T4

P1064

Cambial variants in Malpighiaceae: Diversity, evolution and their impact on diversification rates in the family

Marcelo Pace[1,2], *Pedro Acevedo*[1], *Veronica Angyalossy*[2]
1. Department of Botany, National Museum of Natural History, Smithsonian Institution
2. Departamento de Botanica, Instituto de Biociências, Universidade de São Paulo

Although Malpighiaceae contains numerous genera and species with cambial variants, the nature and diversity of these structures are still not well understood. Their ontogeny, common developmental stages, and their significance on the evolution and diversification of the family are among the many questions that remain to be answered. Working toward this end, we have sampled 99 species in 42 genera of Malpighiaceae where we are applying standard anatomical procedures to classify these structures according to their ontogenetic pathways and to map their distribution upon a well-supported phylogeny for the family using a maximum likelihood approach. Our preliminary results support the recognition of 8 different types of secondary growth as follows: 1) regular, 2) interxylary phloem, 3) interxylary cambia, 4) phloem wedges furrowing the xylem, 5) fissured stem without inner xylem partition, 6) fissured stem with inner xylem partition, 7) fissured stem without complete inner xylem partition, and 8) asymmetrical stems. Interxylary phloem is exclusive of *Dicella* and involves the periodical formation of phloem by the cambium toward the inside and outside, ultimately generating phloem islands embedded in secondary xylem; interxylary cambia result from the differentiation of new cambia deriving from non-lignified xylematic axial parenchyma, a novel, undescribed type of cambial variant; phloem wedges are very common and appear in several genera, including all species with fissured stems, and are derived from cambial portions that have differential production of phloem-and xylem. Fissured stems, which can develop in 3 different ways, are formed by the combination of phloem wedges and non-lignified xylematic axial parenchyma in bands that merge, either by the formation of new cambia and/or by intense proliferation of axial parenchyma. In *Alicia* and *Callaeum* it involved the complete partition of the inner xylem due to the proliferation of disruptive parenchyma from the pith, xylem, phloem, pericycle and cortex. In other genera, a partition may be partial or inexistent. From ancestors with regular secondary growth, these seven cambial variants evolved independently at least 8 times within Malpighiaceae and are conserved within genera or sometimes between sister genera. This study also shows how different ontogenies may result in almost identical mature anatomies and their importance in characterizing clades previously solely circumscribed by molecular sequence data. Our study shows wood anatomical characters to be conservative and essential in recognizing ancient phyletic relationships.

T4

P1065

Environmental control of *de novo* root regeneration in *Arabidopsis*

Jing Pan, Jie Yu, ***Lin Xu***
Institute of Plant Physiology and Ecology, Shanghai Institutes for Biological Sciences, CAS

Plants have the ability to regenerate a new plant from many organs after detachment or wounding via *de novo* organogenesis. Adventitious roots formation can be induced by appropriate types and concentrations of plant hormones in the medium, but endogenous hormones contributes more to this progression in natural conditions. By culturing *Arabidopsis thaliana* leaf explants on medium without additive hormones, we imitate natural conditions for adventitious root formation, and find that the regeneration ability of the leaf explants depends on the developmental stage of the leaf. Our results showed that different nutrients can have obvious influences on adventitious root formation, and so does light condition. Our method could provide a simple procedure to investigate the regeneration of adventitious roots, and these results suggest that leaf developmental stage, together with sugar and plant hormones, should have complex and composite effects on adventitious root formation.

T4

P1066

Mirroring establishment: molecular mechanisms governing zygomorphic flower development in *Sinningia*

Zhao-Jun Pan, *Hao-Chun Hsu, Jen-Chih Chen, Chun-Neng Wang*
National Taiwan University

Multiple evolutionary transitions of symmetry in zygomorphic flowers contribute to the flowering plant diversification. This syndrome shift is hypothesized to associate with the co-evolutionary relationships between bilateral symmetry flowers and theirs pollinators. Most Gesneriaceae species possess zygomorphic flowers with dorsal-ventral asymmetric petals and stamens. In this study, we analyzed and compared the global gene expression profiles among dorsiventral petals of zygomorphic flowers and dorsal petals of actinomorphic flowers in *Sinningia* by RNA-seq. A total of 399 genes showed as differentially expressed genes (DEGs) from the dorsiventral petals of zygomorphic flower transcriptomes. Based on extensive data analysis, 151 genes were assigned to hormone or hormone-related genes and 57 transcripts were predicted as transcription factors (TFs). Of which, the *CYCLOIDEAE* (*CYC*)-clade class II *TCP* gene, *SsCYC*, was identified as a dorsal petal-specific gene, while a class I *TCP* gene showed ventral petal expression. In addition, other predicted TF genes belong to families of *B3*, *MYB*-related, *TIFY*, *TCP*, *ARF* and *ERF*. The dorsiventrally asymmetric transcripts of these newly identified candidates and the well-known floral symmetry genes were validated their spatial-temporal co-expression patterns. Sequence analysis revealed one or more predicted TCP binding sites within the promoter regions of candidate genes. This provided the first insight into the genetic regulation governing *Sinningia* flower symmetry and implied *SsCYC* and other DEGs may control distinct developmental pathways in dorsiventral petals.

T4

P1067

Regulation of ethylene-responsive SlWRKYs in the color change during tomato fruit ripening

Ling Wang[2], *Yanan Tian*[1], *Xuelian Zhang*[1], *Zhaoqi Zhang*[2], ***Xuequn Pang***[1]
1. College of Life Science, South China Agricultural University
2. college of Horticulture, South China Agricultural University

WRKY transcription factors are specifically and widely present *in planta*. They have been found playing important roles in plant responses to biotic and abiotic stresses as well as developmental regulation. Tomato is a typical climacteric fruit, where rapid respiration and ethylene production, chlorophyll degradation and lycopene accumulation occur during fruit ripening process. To study the role of *SlWRKY* family transcription factors in regulation of fruit ripening, tomato fruit ripening was manipulated by treatment with ethylene and 1-MCP, an ethylene receptor inhibitor. Based on the similarities to 24 ethylene responsive *WRKY* genes from *Arabidopsis*, rice, *Gossypium hirsutum* and *Brassica napus*, 23 *SlWRKY* genes were selected from the whole tomato *SlWRKY* gene family that includes 81 members. The expression profiling of these *SlWRKYs* was analyzed in the ripening system along with 8 fruit ripening related genes, including 5 color change related genes. Eight *SlWRKYs* genes displayed over-lapping up-regulation expression patterns as the ripening related genes and were detected to localize in nucleus, and showed potential transcriptional activation in yeasts. Yeast One-Hybrid and *in vivo* regulation assays in tobacco leaves showed that SlWRKY16, 17, 22, 25, 31, 33, 53, 54 were found to bind and activate the promoters of the color change related genes, *SlPPH* (*PHEOPHYTINASE*), *SlPAO* (*PHEOPHORBIDE A OXYGENASE*), *SlPSY1* (*PHYTOENE SYNTHASE1*) *and SlPDS* (*PHYTOENE DESATURASE*). Furthermore, yeast two-hybrid assay indicated that obvious interaction may occur between WRKY16, 17, 33, 53, 54 and with other ripening regulators including, SlRIN, SlERF2b and SlERF7. Our study provides for the first time direct evidence that the WRKY transcription factors may function in fruit ripening, particularly in color change, and may connect to the intricate regulatory network with other ripening key regulators.

T4

P1068

Microsporogenesis and pollen formation in *Zingiber officinale* Roscoe

Shaik Sha Valli Khan Patan
Yogi Vemana University

The present study provides the first description on microsporogenesis and pollen formation in *Zingiber. Officinale* Roscoe (family Zingiberaceae). This species is one of the most important vegetable condiment with proven antioxidant activity. Ginger widely cultivated in the tropics and sub-tropics, in India particularly in North-east. This species produce very rarely seeds and it is very difficult to breed new genotypes of this species through sexual hybridization. Classical cytological squash technique, and staining protocols with acetocarmin, FDA, DAPI were employed for the study of microspore development as well as assessment of embryogenic window and microspore viability. The study revealed that the general pattern of microsporogenesis and pollen formation in this species starting from PMCs to binucleate mature pollen grain. The microspore development was divided into four basic stages viz. (1) pre-meiotic, (2) the tetrad stage, (3) free microspores and (4) the mature pollen grains. This information enriches the database of cytology, pollen viability estimations and can be useful for breeding program hybridization, also contributes with basic information that will hopefully allow the development of doubled haploid protocol for production of androgenic haploids to accelerate breeding and genetic improvement of ginger. Further research is continuing on evaluating several media and other cultural factors influencing microspore embryogenesis and plant regeneration in this valuable species.

T4

P1069

Role of the outer stomatal ledges in the mechanics of guard cell movements

Anatoly Pautov[1], *Svetlana Bauer*[1], *Olga Ivanova*[1], *Elena Krylova*[1], *Yulia Sapach*[1], *Galina Gussarova*[2,1]

1. Saint Petersburg University

2. University of Oslo

Methods of light, scanning and transmission electron microscopy were used to study stomata of the leaf epidermis in evergreen *Carissa spectabilis* (Sond.) Pichon (Apocynaceae), *Acokanthera oblongifolia* (Hochst.) Codd (Apocynaceae), *A. oppositifolia* (Lam.) Codd (Apocynaceae), *Exbucklandia populnea* (R.Br. ex Griff.) R.W. Brown (Hamamelidaceae), *Trochodendron aralioides* Sieb. et Zucc. (Trochodendraceae). The stomata are located on subsidiary cells, have large outer ledges, and lack inner ledges. To elucidate the role of the ledges, we applied dynamic modelling using the finite-element method. In modelling, we took into account the shape of the guard cells, uneven thickness of their walls, positioning and size of the outer ledges. Obtained geometric data were used as parameters for the construction of the finite-element models. The turgor pressure was simulated by creating the load distributed on the inner surface of the guard cells. To estimate the effect of the outer ledges of the guard cells on stomatal mechanics, the following models were constructed: Model 1. The guard cells with the walls of uneven thickness without the outer ledges; Model 2. The guard cells with the walls of uneven thickness and the outer ledges. For stomata having no ledges, the modelling has demonstrated that the turgid guard cells, located on subsidiary cells, are bended above the leaf surface. The opening stomatal pore also moves above the leaf surface. The outer ledges prevent wide opening of the stomatal pore and its lifting above leaf epidermis. The more the stiffness of the ledges, the less the movements of the pore. The results of the modelling are supported by the observed deformations in the guard cells of the real stomata. The modern concepts of the stomatal mechanics do not consider the presence of the ledges on the guard cells (Aylor *et al.*, 1973; Cooke *et al.*, 1976, 2008; Sharpe *et al.*, 1987; Sitte *et al.*, 2008). In contrast, our modelling results showed decisive influence of the outer ledges on the guard cell movements. The discovered mechanics takes place in stomatal complexes with guard cells lying on subsidiary cells. We found such stomatal mechanics in the families, which according to APG III system (Bremer *et al.*, 2009), belong to several taxonomic groups occupying different positions in the flowering plant system. This means that discovered by us stomatal mechanics either has an ancient origin or has arisen in evolution repeatedly and independently in several taxa. All studied species show similar leaf anatomy. It has much in common with the leaf anatomy of species connected in their origin with subhumid Tertiary laurophyllous forests.

T4

P1070

RTFL peptide family: How it regulates the morphogenesis in land plants

Pin Guo[1], *Tomoyuki Furuya*[1], *Takayuki Kohchi*[2], *Tomoko Abe*[3], *Hirokazu Tsukaya*[1,4]

1. Department of Biological Sciences, Graduate School of Science, The University of Tokyo

2. Graduate School of Biostudies, Kyoto University

3. RIKEN Nishina Center

4. Bio-Next Project, Okazaki Institute for Integrative Bioscience, National Institutes of Natural Scie

The land plant-specific RTFL family encodes small peptides functioning in the regulation of positional cues and polarized cell proliferation of multiple developmental processes in *Arabidopsis thaliana*. Up to now, our understanding of above biological function of *RTFL* family is mainly based on the pleiotropic phenotypes of the lateral organs among transgenic overexpressors in *A. thaliana*. The loss-of-function lines, however, provide little functional information of *RTFL* family, indicating a high level of its genetic redundancy in *A. thaliana*. In this study, a double *RTFL* overexpressor in *A. thaliana*, which was constructed with two *RTFL* members from *A. thaliana* and *Oryza sativa*, was mutated by heavy-ion beam irradiation. Up to now, two suppressors were obtained after screening of 18,680 M_2 populations. Meanwhile, we also used *Marchantia polymorpha* to study the function of *RTFL* family, regarding for the non-redundancy of *RTFL* (*MpRTFL*) in its genome and its important evolutionary position in land plants. We checked the transcriptional patterns of *MpRTFL* in the haploid generation of *M. polymorpha*. In the young gemmae and thallus, *MpRTFL* expression was found in apical notch areas and the epidermis. After the far red-light induction of reproductive organs, *MpRTFL* expression was found in archegoniophores and antheridiophores. The overexpression lines of *MpRTFL* driven by the promoter of *ELONGATION FACTOR-1* alpha showed discernible phenotypes compared with *MpRTFL*-knockout lines which were made by CRISPR-Cas9 in *M. polymorpha*.

T4

P1071

Histochemical screening of terpenoid in secretory structures of *Pogostemon* (Lamiaceae)

Kanjana Pramali[1,2,3], *Bhanubong Bongcheewin* [2], *Paweena Traiperm*[1]

1. Department of Plant Science, Faculty of Science, Mahidol University

2. Department of Pharmaceutical Botany Faculty of Pharmacy, Mahidol University

3. Master of Science Program in Plant Science

Plants in the genus *Pogostemon* Desf. are the important source of commercial patchouli oil and perfume industry, especially *P. cablin* (Blanco.) Benth. The patchouli oil is mainly composed of sesquiterpenes, one type of terpenoids. Leaves are the main material for essential oil extraction since they possess various secretory structures involving in the accumulation of essential oil such as glandular trichome, internal gland and oil body. Therefore, leaves of twelve *Pogostemon* spp. i.e. *P. auricularius* (L.) Hassk., *P. cablin* (Blanco.) Benth., *P. glaber* Benth., *P. globulosus* Phuong ex S. Suddee & A.J. Paton, *P. helferi* (Hook.f.) Press, *P. hispidus* (Benth.) Prain, *P. menthoides* W.W. Smith, *P. myosuroides* W.W. Smith, *P. nudus* Bongcheewin & Pramali, *P. parviflorus* Benth., *P. trinervis* Chemsri. ex Press, and *Pogostemon* sp. were investigated by histochemical methods using NADI reaction in order to preliminarily screen for the presence of terpenoids. From this study, capitate glandular trichomes of the most of *Pogostemon* species, except *P. helferi* and *P. myosuroides*, showed positive results to NADI reaction, indicating that capitate glandular trichomes are the main site for storing the patchouli oil. Moreover, peltate glandular trichomes of five species such as *P. auricularius*, *P. helferi*, *P. menthoides*, *P. myosuroides*, and *P. trinervis* are also positive to NADI reaction revealing the presence of terpenoids in these structures. Furthermore, internal glands of *P. trinervis* and oil bodies of three species i.e. *P. auricularius*, *P. menthoides* and *P. trinervis*, present positive results with NADI reaction. Since, terpenoids have been known as the secondary metabolite in protecting plant from herbivores, the accumulation sites of terpenoids in *Pogostemon* from this study are allocated in both external and internal parts of leaves. Therefore, terpenoids inside those structures may be the first step of mechanisms in preventing plants from herbivores through irritated scent of terpenoids. In addition, it may also an indirect method of protection for the plants by affecting the alimentary system of herbivores.

T4

P1072

Physiological and biochemical response of wheat (*Triticum aestivum* L.) genotypes to water stress environments

Abdul Qayyum[1], *Yamin Bibi*[2], *Ahmad Sher*[3], *Ayaz Mehmood*[1]
1. UNIVERSITY OF HARIPUR
2. Pir Mehr Ali Shah, Arid Agriculture University Rawalpindi
3. Bahauddin Zakariya University, Bahadur Campus Layyah

Water stress is a major threat to production of agricultural crops including wheat world-wide in arid and semi-arid regions. Lab and greenhouse experiments were conducted to evaluate wheat genotypes (GA-2002 and Chakwal-50) for water stress tolerance. Under laboratory condition modified MS medium solution supplemented with polyethylene glycol (PEG-6000) to induce osmotic stress of various concentrations (-2, -4, -6 and -8 bars) was used and data was recorded on germination indices, biochemical parameters and enzymes assay. The results revealed significant ($p<0.05$) differences among both genotypes for various traits under osmotic stress levels. Speed of germination, shoot, root and coleoptile length, root-shoot fresh and dry weight decreased with the increase in osmotic stress. Root-shoot dry weight susceptibility index, membrane stability index and relative water content decreased with the increase in osmotic stress. Free proline, total soluble sugar, total soluble protein, superoxide dismutase, catalase, peroxidase, malondialdehyde activity increased with the increase in osmotic stress. In greenhouse experiment various levels (80,60,40 and 20% field capacity) of water stress were created by gravimetric method and data was recorded for gas exchange measurements and yield attributes. The results revealed significant ($p<0.05$) differences among both wheat genotypes for various traits under different levels of field capacity. Photosynthesis, transpiration, stomatal conductance, leaf osmotic and water potential, number of grains spike^{-1},1, 000 grain weight and grain yield plant^{-1} decreased significantly with the increase in water stress however epicuticular wax content increased. Overall Genotype Chakwal-50 performed better as compared to GA-2002 under water stress conditions.

T4

P1073

Dual-source nuclear monomers of UV-B light receptor direct photomorphogenesis in *Arabidopsis*

Chongzhen Qian[1], *Weiwei Mao*[1], *Yan Liu*[1], *Hui Ren*[1], *On Sun Lau*[2], *Xinhao Ouyang*[1], *Xi Huang*[1]
1. State Key Laboratory of Cellular Stress Biology, School of Life Sciences, Xiamen University
2. Department of Biological Sciences, National University of Singapore

Ultraviolet (UV)-B light of long wavelength and low fluence directs photomorphogenic development and UV-B acclimation in *Arabidopsis*. UV RESISTANCE LOCUS 8 (UVR8), a plant-specific chromoprotein, is responsible for UV-B light perception. Upon the exposure to UV-B light, homodimeric UVR8 performs rapid monomerization and associates with CONSTITUTIVELY PHOTOMORPHOGENIC 1 (COP1) to mediate light signaling cascades in the photomorphogenic developmental process. UVR8 localizes in both the cytoplasm and the nucleus, and UV-B illumination stimulates its nuclear accumulation. However, limited information is available with regard to light regulation of UVR8 in terms of its subcellular localization and physiological significance. Here we explore the functional relevance among the subcellular distribution of UVR8, its photobiological activity and UV-B signal transduction. We provide a biochemical presentation of UVR8 localization in combination with cell biology analysis, and dissect physiological roles of subcellular UVR8 proteins using a conditional localization system. We reveal that in the UV-B-induced photomorphogenesis, light signal transduction is essentially mediated by nuclear UVR8 monomers whose accumulation is UV-B-light enhanced and dual-source derived. Two types of UVR8-interacting proteins, COP1 and REPRESSOR OF UV-B PHOTOMORPHOGENESIS 1 (RUP1)/RUP2, play positive and negative roles respectively in the establishment and maintenance of nuclear UVR8 pool to ultimately control photomorphogenic UV-B signaling output.

T4

P1074

Epigenetic variation of *ICE1* mediates CBF cold-response transcription pathway conferring cold-tolerant phenotype polymorphism of a plant

Sheng Qiang, *Hongjie Xie*
Nanjing Agricultural University

Cold stress is a major abiotic factor encountered by plants during their life cycle. Plants often demonstrate phenotypic polymorphisms for cold tolerance, but the innate mechanisms are poorly understood. DNA methylation is considered to be mainly involved in regulation of gene transcription and expression and determination of phenotypic plasticity. However, empirical evidences still lack to establish causal relationship between phenotypic polymorphisms and gene methylation variations. Cold/freezing damage indexes showed that there were all distinct cold/freezing-tolerance differentiations among geographical populations of *Arabidopsis thaliana*, Crofton weed (*Ageratina adenophora*), and weedy rice (*Oryza sativa*). It was found that only the cold response regulator *ICE1* in mainstream of CBF cold-response pathway had significant site variation of methylated cytosines, significantly negative relating to cold tolerance levels. Methylation variations were found in the promoter region, or coding region or both of ICE1 gene, which indicates that regulation of gene expression through DNA methylation is complex. The significant negative relation between the transcription levels of the primary CBF pathway members, except for *CBF2*, and the methylation levels among distinctly geographical populations demonstrated that the *ICE1* demethylation-upregulated transcription level of the CBF pathway caused this adaptive cold/freezing-tolerance evolution consequence. Those discoveries advance our understanding of the roles of DNA methylation in regulation gene expression and it may also suggest that the plasticity of any stress response in plants may be attributed to variations in methylation of a key gene in related pathways.

T4

P1075

In vitro regeneration from organogenic callus and assessment of clonal fidelity in *Hemigraphis alternata* (Burm.F.) T. Anderson

Krishna R Nair, *Ashalatha S Nair*

University of Kerala, Thiruvananthapuram, Kerala, India

Hemigraphis alternata commonly known, as red flame ivy plant, belongs to the family Acanthaceae. It is a prostrate herb chiefly grown as an ornamental plant. In folk medicine the plant is used for its wound healing activity. Traditionally, the leaves are consumed to mend gall stones, excessive menstruation and as a contraceptive also it has immense power to pacify vitiated pitta, fresh wound, cuts, ulcers, inflammations and skin complaints. An efficient callus induction and plant regeneration system has been standardized for this potent ethnomedicinal plant. Two explants i.e. nodal and leaf segments were used for callus induction. Murashige and Skoog (MS) medium supplemented with 5.0 μM 2,4- dichlorophenoxy acetic acid (2,4-D) gave the optimum frequency (91%) of callus induction from leaf explant. Nodal segments inoculated on 2,4-D obtained basal compact callus. The results showed that the highest response in terms of percent callus regenerating (86%) and number of shoots (23) per culture was recorded on MS medium supplemented with 5.0 μM N6-benzylaminopurine (BA) and 2.5 μM α naphthalene acetic acid (NAA). The regenerated shoots were transferred to lower BAP concentration resulted in further elongation of shoots and rooting. On BAP supplemented medium, 100% of the shoots produced roots with a mean number of 5.6 roots per shoot. The rooted plantlets obtained were hardened successfully in the green house condition with 100% survival rate. Inter-simple sequence repeats (ISSR) were used to establish the clonal fidelity of regenerated plantlets and the banding profiles from callus derived plants were monomorphic and similar to those of mother plant, thus ascertaining the true-to-type nature of these plants.

T4

P1076

Programmed cell death in maternal seed tissues controls grain size in barley

Volodymyr Radchuk[1], *Van Tran Thi Thuy*[1], *Ruslana Radchuk*[1], *Mercedes Diaz-Mendoza*[2], *Joerg Fuchs*[1], *Goetz Hensel*[1], *Manuel Martinez*[2], *Ljudmilla Borisjuk*[1]

1. Leibniz-Institute of Plant Genetics and Crop Plant Research (IPK)

2. Centro de Biotecnologia y Genomica de Plantas, Universidad Politecnica de Madrid, Instituto Nacional de Investigacion y Tecnologia Agraria y Alimentaria

Angiosperm embryo and endosperm are limited in space because they grow inside maternal seed tissues. Besides, maternal seed coat provides protection against environment and delivers nutrients to the developing endosperm and embryo. The elimination of cell layers of maternal seed coat by programmed cell death (PCD) could provide space and nutrition to the filial organs. Using barley seed as a model, we elucidate the role of pericarp-expressed vacuolar processing enzyme 4 (VPE4) in cereals by using an RNAi approach and targeting the enzymatic properties of recombinant protein. The recombinant VPE4 protein exhibited strong legumain and caspase-1 properties *in vitro*. A comparative characterization of transgenic versus wild type seeds includes transcriptional and metabolic profiling, flow cytometry, histology and nuclear magnetic imaging of grains. Pericarp disintegration was delayed in the *VPE4*-repressed grains leading to an increased number of vital cells in maternal tissues. Although the *VPE4* gene and enzymatic activity was decreased in early developing pericarp, the size of endosperm and embryo and accumulation of starch and lipids were reduced in the mature VPE4-repressed grains indicating that PCD in pericarp during early seed development affects later grain filling and grain size. The coordinated decrease of several caspase-like activities following *VPE4* repression indicates that the co-action of caspase-like protease activities is required to execute and/or regulate PCD processes in plants analogous to those that occur in animals. The persistence of pericarp in the *VPE4*-affected grains constrains endosperm and embryo growth and leads to transcriptional reprogramming, perturbations in signalling and adjustments in metabolic architecture. Cell number in endosperm of the *VPE4*-repressed grains was not changed. However endosperm cells were enclosed in smaller volume of tissue. Transgenic endosperm accumulated also less starch. We concluded that predominant role of pericarp PCD is to provide space for endosperm cell expansion. The delayed cell elimination in *VPE4*-repressed pericarp relieves a physical restraint for the growing endosperm, thereby prompting the early cellularization of smaller cells with lower levels of endoploidization and resulting in reduced starch accumulation later in development. Thus, endosperm cell expansion and the final grain filling and seed size are determined by PCD in maternal pericarp. We conclude that VPE4 repression impairs seed filling and size by executing PCD in pericarp and argue for the role of PCD in maternal control of seed size in cereals.

T4
P1077
Preliminary study on the relationships between inflorescence architecture and optimal fruit and seed production in *Leucaena leucocephala* (Lam.) de Wit.
Jose I Ramirez Domenech
Interamerican University of PR - Ponce Campus

Implications of inflorescences architecture among mimosoid legumes are significant to predict the optimal fruits and seeds production in populations of plants among different eco-environments. Our research is focused on the relationship between inflorescences, number of fruits and seeds in a population of *Leucaena leucocephala* located in a hypersaline lagoon in Ponce, P.R. Specimens were collected from 5 line transects of 50 meters along the lagoon area. Inflorescences were examined at during early, mid-, and late developmental stages. A total of 65 fruits per plant and 24 seeds per fruit was counted from 20 inflorescences during late developmental stages during one season from August to November. Developmental studies show a pattern of division of labor conducive to optimize seeds and fruit production. Studies on populations of *L. leucocephala* are needed to develop management strategies on the invasive threats of this species in local ecosystems.

T4
P1078
Leaf folding in *Bauhinia*
Rita C.K. Chung, ***Lawrence Ramsden***
University of Hong Kong SPACE, Kadoorie Centre

Leaf folding is a characteristic of several species of Bauhinia trees and lianas. The closing of leaves occurs along the central axis with the two halves of the heart shaped leaves folding upwards in contrast to the nyctinasty or sleep movement in many other plants where the leaves commonly fold downwards.This folding response appears to show variation in response to environmental conditions, it can occur at night with low light or also at high illumination in exposed sites at midday. Here we examine the hypotheses that this movement is related to 1) water stress responses, 2) reduction of photostress, or 3) a mechanism to avoid herbivory.
Bauhinia trees (*B. purpurea*) and a liana (*B. glauca*) were tested for leaf folding under various illumination conditions to examine rate of folding movements in response to stimulus. To establish the value of the folded state for the leaf, open and closed leaves were compared for transpiration and photosynthesis rates. In field trials the effect of permanently opening leaves or permannently closing leaves was examined for effect on leaf survival. The degree of herbivory was also compared between the opened and closed leaves. Our results indicate that high light stress rather than water stress is likely to drive leaf folding at midday as there was no appreciable difference in transpiration rates between open and closed leaves. However this cannot easily explain night time closing and it appears that herbivory avoidance is the main potential benefit by reducing the exposed leaf surface area. In this case it is possible that the response only uses a light dependent mechanism and may not need to respond to water stress signals, which will be the subject of further investigation.

T4
P1079
Reproductive morphology and development of Thai endemic *Remirema* (Convolvulaceae)
Pantamith Rattanakrajang[1,2,3], *Paweena Traiperm*[2], *George Staples*[4]
1. Master of Science Program in Plant Science, Faculty of Graduate Studies, Mahidol University, Bangkok
2. Department of Plant Science, Faculty of Science, Mahidol University, Bangkok
3. Department of Pharmaceutical Botany, Faculty of Pharmacy, Mahidol University, Bangkok
4. Herbarium, Royal Botanic Gardens, Kew, Richmond

Remirema Kerr is a monospecific genus in the morning glory family (Convolvulaceae). A single species endemic to Thailand, *R. bracteata* Kerr, is a herbaceous climber characterized by epipetalous scales at the filament insertion on the corolla, a deeply 4-lobed ovary, a gynobasic style insertion, and a unique fruit with 4 free cells, each containing one seed. These diagnostic traits have been used to distinguish *Remirema* from all other genera in the family. In order to understand more about some generic characters, flower-to-fruit development from various stages of the monospecific genus was investigated. Plant samples were collected from the natural habitat in Kanchanaburi province, Thailand. Paraffin method was applied for material sectioning and investigation using light microscopy. The results presented *R. bracteata* morphological features, its inflorescence consists of 2-5 flowers and it is covered with large bracts. Each flower is enclosed with 5-aposepalous with peltate glandular trichomes on the abaxial epidermis of calyx. Corolla is tubular-campanulate with five pointed lobes. The androecium comprises 5-epipetalous stamens. A gynoecium shows apocarpus pistil comprising 4 ovules in 4-lobed ovary which correspond to the previous literatures. Fruits are encircled in accrescent calyx and displayed at most 4 free nutlets. The early developmental stage shows 2 incompletely fused styles and 2 incompletely fused ovaries. In the later stages, the 2 styles are completely fused, forming one simple long style with a biglobose stigma, and 2 deeply bilobed ovaries. In the final fruit stage, the 2 bilobed ovaries appear as 4 distinct locules, each with a single nutlet. These developmental characters are revealed for the first time and might be additional evidence to support the recognition of *Remirema* as a distinct genus.

T4
P1080
Sexual dimorphism and the evolution of sex-biased and sex-limited genes in the dioecious tree *Populus balsamifera*
Brian Sanderson[1], *Li Wang*[2], *Peter Tiffin*[3], *Zhiqiang Wu*[2], *Matthew Olson*[1]
1. Texas Tech University
2. Iowa State University
3. University of Minnesota

Animals have been shown to exhibit sexual dimorphism in gene expression in as many as 3/4 of their genes, and male-biased genes generally exhibit elevated rates of molecular evolution. Although ~5-6% angiosperm species are dioecious, little is known sexual dimorphism in gene expression or whether patterns of evolution differ for male- and female-biased genes. *Populus* exhibits clear sexual dimorphism in floral color, but few dimorphisms in non-floral traits. We characterized transcriptomes of two tissues (leaf and flower) of five male and five female individuals of *Populus balsamifera* from Fairbanks, Alaska. Five times as many genes exhibited sex limited expression in male floral tissues (476) than female floral tissues (76), and no genes exhibited sex limited expression in leaves. 11,068 genes were differentially expressed between male and female flowers, and only one gene exhibited significant differential expression between male and female leaves. Functional analyses of these genes revealed 15 enriched gene pathways. Notably, male and female flowers differed in energy-related pathways. Transcripts in photosynthesis and tetrapyrrole synthesis pathways were over-expressed in female flowers, whereas mitochondrial electron transport and TCA/org transformation pathways were over-expressed in male flowers. Surprisingly, the distribution of dN/dS ratios for sex-biased and sex-limited genes did not differ from that of randomly selected genes. Taken together, these results highlight that the expectation that sex-biased genes exhibit elevated rates of molecular evolution may depend on the mating system or biology of the species of interest. Sexual dimorphism in *P. balsamifera* primarily manifests in differences in short- and long-term energy pathways. These genes may have a limited opportunity for sexual selection, given their critical functional roles.

T4
P1081
Biosynthesis and function of betalain pigments in the New Zealand native iceplant, *Disphyma australe*

Kathy Schwinn[1], *Gagandeep Jain*[2], *Tanja Karl*[2], *Ella Grierson*[1], *Yanfei Zhou*[1], *Simon Deroles*[1], *Ken Ryan*[2], *Kevin Davies*[1], *Kevin Gould*[2]
1. The New Zealand Institute for Plant & Food Research Limited
2. Victoria University of Wellington

Intriguingly, betalains replace the near ubiquitous anthocyanin pigments within the core Caryophyllales. Betalains (red or yellow) are formed in a pathway unrelated to that of the anthocyanins, having a tyrosine-derived central chromophore. Betalain taxa include species that survive in harsh environments such as sand dunes and salt marshes. The New Zealand native iceplant *Disphyma australe* has coastal sympatric red- and green-leaved morphs. The red morph accumulates red betalains (betacyanins) in response to salt stress, and it was observed to be more abundant than the green morph in areas with higher salinity. The green morph is unable to synthesise betacyanins unless it is fed a biosynthetic precursor. We have found that the betacyanins confer salt tolerance, and we are studying the mechanism behind this. Also, we are studying the biosynthesis of betalains and its regulation in this plant, and determining why the green morph does not produce betalains in vegetative tissue.

T4
P1082
Photosynthesis efficiency under phosphorus and nitrogen limitation in wheat

Sarah Seco[1], *Maïté S. Guignard*[2], *Joseph J. Wolfensohn*[1], *Mark Trimmer*[1], *Ilia J. Leitch*[2], *Alexander V. Ruban*[1], *Paula J. Rudall*[2], *Andrew R. Leitch*[1]
1. Queen Mary University of London
2. Royal Botanic Gardens, Kew

Nitrogen (N) and/or phosphorus (P) availability in the environment limits productivity across many ecosystems and agricultural settings. Because N and P frequently restrict plant growth, N fertilizers are used in vast quantities to sustain crop resources. Photosynthesis is demanding in N-usage because of its requirement for the protein Rubisco, which incorporates CO_2 into sugars in the dark reaction of photosynthesis. There is also a significant demand for N in building light harvesting pigment-proteins, which represent another major N-demand of photosynthesis. Growth also requires N, not only in the production of proteins, but also to build nucleic acids, these being amongst the most N and P demanding biomolecules of the cell. Given these competing demands, under nutrient limitation there is a potential trade-off between photosynthesis, growth and reproduction. To investigate this hypothesis, we have studied the impact of N and P availability in dwarf wheat (*Triticum aestivum* L. cv. USU-Apogee). The plants were grown in 16 different fertiliser conditions, comprising four concentrations of phosphate and four concentrations of N. By measuring chlorophyll fluorescence parameters, gas exchange and growth parameters, we investigated the impact of these nutrients on the efficiency of photosynthesis, growth and reproduction. We found that under nutrient limitation, wheat plants have a decreased efficiency of photosynthesis, under N limitation Vcmax is affected, and under P limitation Jmax, NPQ, qP and Fo are affected and, consequently there is a decrease in biomass production and seed production. The interaction between N and P is also essential for cell proliferation, growth and reproduction and we explore why the reasons for this.

T4
P1083
Specification of vascular patterning in maize

Olga Sedelnikova, *Jane Langdale*
University of Oxford

Maize is a vital crop for food production and utilises the highly efficient C_4 photosynthetic pathway. The patterning of cell types in maize leaves (Kranz anatomy) is instrumental in allowing this biochemical pathway to function. One characteristic feature of Kranz anatomy is high vascular density, therefore vein initiation is a crucial step during Kranz development. Little is known about the genetic regulation of monocot vascular development, and whilst the ontogeny of early maize foliar development has been well characterised, the genetic factors that lead to Kranz anatomy are currently unknown. *In situ* hybridisation was used to investigate the expression of candidate regulatory transcription factors during the earliest stages of maize foliar development, leading to the identification of genes expressed specifically at the beginning of procambium development. The expression dynamics of these transcription factors suggest regulatory roles during vein development and allow their use as markers of procambium initiation in maize. Identification of Kranz developmental regulators would strongly contribute towards harnessing C_4 photosynthesis for crop improvement of major C_3 crops such as rice, with a predicted increase in photosynthetic efficiency of 50%, improved nitrogen use efficiency and doubled water use efficiency.

T4
P1084
Exogenously applied 24-epibrassinolide (EBL) ameliorate detrimental effects of salinity by reducing K^+ efflux via depolarization-activated K^+ channels

Nazila Azhar[1], *Nana Su*[1,2], *Lana Shabala*[1], ***Sergey Shabala***[1]
1. University of Tasmania
2. Nanjing Agricultural University

This study has investigated mechanisms conferring beneficial effects of exogenous application of 24-epibrassinolides (EBL) on plant growth and performance under saline conditions. Barley seedlings treated with 0.25 mg/L EBL showed significant improvements in root hair length, shoot length, shoot fresh weight and relative water content when grown in the presence of 150 mM NaCl in the growth media. In addition, EBL treatment significantly decreased the Na^+ content in both shoots (by ~50%) and roots. Electrophysiological experiments revealed that pretreatment with EBL for 1 and 24 h both suppressed or completely prevented the NaCl-induced K^+ leak in the elongation zone of barley roots, but did not affect root sensitivity to oxidative stress. Further experiments using Arabidopsis loss of function *gork1-1* (lacking functional depolarization-activated outward-rectifying K^+ channels in the root epidermal cells) and *akt1* (lacking inward-rectifying K^+ uptake channel) mutants showed that NaCl-induced K^+ loss in the elongation zone of roots was reduced by EBL pretreatment over 200 fold in wild type Col-0 and *akt1*, but only 10 fold in *gork1-1* mutant. At the same time, EBL treatment shifted H^+ flux towards net efflux. Taken together, this data indicates that exogenous application of EBL effectively improves plant salinity tolerance by prevention of K^+ loss via regulating depolarization-activated K^+ channel.

T4
P1085
Plant ionic relation and whole-plant physiological responses to waterlogging, salinity and their combination in barley

Zhinous Falakboland[1], *Meixue Zhou*[1], *Fanrong Zeng*[2], *Ali Kiani-Pouya*[1], *Lana Shabala*[1], ***Sergey Shabala***[1]
1. University of Tasmania
2. Zhejiang University

Waterlogging and salinity stresses significantly affect crop growth and global food production and are often interrelated, as waterlogging can lead to land salinization by transporting the salts to the surface. While the physiological and molecular mechanisms of plant responses to each of these environmental constraints have been studied in detail, much fewer studies dealt with mechanisms underlying plant tolerance to their combined stress. This gap in knowledge is jeopardizing the success of breeding programs and has to be bridged. In this work, we have studied physiological and agronomical responses of 12 barley varieties contrasting in salinity stress tolerance to waterlogging (WL), salinity (NaCl), and combined (Wl/NaCl) stresses. Stress damage symptoms were much stronger in plants under combined WL/NaCl stress compared to separate stresses. The shoot biomass, chlorophyll content, maximum photochemical efficiency of PSII and shoot K^+ content were significantly reduced under WL/NaCl conditions, while shoot Na^+ content increased. Plants exposed to salinity stress showed less damage indexes compared to WL. Chlorophyll fluorescence Fv/Fm value showed the highest correlation with the plant damage index under WL/NaCl conditions ($R = -0.751$) compared to other measured physiological traits and was nominated as a good parameter to rank the tolerance of varieties. The average fresh weight was reduced by 73 ± 2, 52 ± 1 and 23 ± 2 percentage of control upon NaCl, WL and combined WL/NaCl treatments, respectively. Generally, the adverse effect of WL/NaCl stress was much stronger in salt-sensitive varieties compared to more tolerant varieties. Na^+ content under control conditions was 97 ± 10 μmol/g, and increased to 1519 ± 123,179 ± 11 and 2733 ± 248 μmol/g under NaCl, WL and combined WL/NaCl stresses, respectively. K^+ content was 1378 ± 66,1260 ± 74,1270 ± 79 and 411 ± 92 μmol/g under control, NaCl, WL and combined WL/NaCl stresses, respectively. Overall, the reported results indicated that K^+ reduction in the plants under combined WL/NaCl stress but not stress-induced Na^+ accumulation in the shoot was most critical feature determining the overall plant performance under combined stress conditions.

T4
P1086
Structural and biochemical analyses of pollen and stigma in sunflower

Rashmi Shakya[1], *Satish Chander Bhatla*[2]
1. Miranda House, Department of Botany, University of Delhi, Delhi
2. Department of Botany, University of Delhi, Delhi

Detailed structural and biochemical analyses of pollen and stigma in sunflower (*Helianthus annuus* L.) have been carried out in order to understand the process of initial events of pollen-stigma interaction. The exine pattern is of 'Helianthoid type' which allows easy communication between the pollen surface and the caveae, thus facilitating the exchange of water and physiologically active substances. Sunflower stigma bears extracellular secretions in the basal region of the papillae. The plasmodemic connections between papillae and the transmitting tissue indicate symplastic transport of various metabolites to the papillae and stigma surface. In the tryphine fraction of sunflower pollen grains, unsaturated and saturated free fatty acids make upto 55 and 30% of the total fatty acid content, respectively. The low lipid content in the stigma (2.7% of fresh weight) could be attributed to its semidry nature and absence of apparent stigmatic secretions. Six isoforms of acyl ester hydrolase activity in pollen grains and eight isoforms in the stigma have been detected. Lipase activity has been localized on the pollen coat which might be involved in the degradation of cuticle present on the surface of stigma. The nature of lipases reported in stigma, is yet to be elucidated and it is considerably less on the stigmatic papillae as compared to that in tryphine. These observations indicate the possible interaction of lipase localized on the tryphine zone, with triacylglycerides (TAGs) present on the stigmatic papillae, during pollen-stigma interaction.

T4
P1087
Overexpression of receptor-like kinase *ERECTA* improves thermotolerance in rice and tomato

Hui Shen[1,2], *Xiangbin Zhong*[1], *Zuhua He*[1]
1. Institute of Plant Physiology and Ecology, SIBS, CAS
2. Shanghai Chenshan Plant Science Research Center, CAS, Chenshan Botanical Garden

The detrimental effects of global warming on crop productivity threaten to reduce the world's food supply. Although plant responses to changes in temperature have been studied, genetic modification of crops to improve thermotolerance has little success to date. Here we demonstrate that overexpression of the *Arabidopsis thaliana* receptor-like kinase ERECTA (ER) in *Arabidopsis*, rice and tomato confers thermotolerance independent of water loss and that *Arabidopsis er* mutants are hypersensitive to heat. A loss-of-function mutation of a rice *ER* homolog and reduced expression of a tomato *ER* allele decreased thermotolerance of both species. Transgenic tomato and rice lines overexpressing *Arabidopsis ER* showed improved heat tolerance in the greenhouse and in field tests at multiple locations in China during several seasons. Moreover, *ER*-overexpressing transgenic *Arabidopsis*, tomato and rice plants had increased biomass. Our findings could contribute to engineering or breeding thermotolerant crops with no growth penalty.

T4
P1088
Molecular cloning and functional analyses of two genes encoding dihydroflavonol 4-reductase (DFR) from orange-flower gentian (*Gentiana lutea* L. var. *aurantiaca*)

Yanmin Sheng[1], *Xiuzhen Ni*[1], *Xin Jin*[2], *Lianxuan Shi*[3], *Gemma Farre*[2], *Teresa Capell*[2], *Paul Christou*[2,4], *Changfu Zhu*[1,2]
1. Changchun Normal University
2. University of Lleida-Agrotecnio Center
3. Northeast Normal University
4. Catalan Institute for Research and Advanced Studies

Flower color is an important characteristic that determines the commercial value of ornamental plants. Orange- to red-colored flowers were difficult to produce by conventional breeding techniques in some floricultural plants, i.e. petunia, cymbidium, and irises. This is due to the deficiency in the formation of pelargonidin-derived anthocyanins, which confers orange to red colors in flowers. One reason for absence of pelargonidin is due to dihydroflavonoid 4-reduactase (DFR) substrate specificity. DFRs of petunia, cymbidium and irises do not catalyze dihydrokaempferol (DHK) to leucopelargonidin, the direct precursor of pelargonidin (Johnson *et al.*, 1999, 2001; Sasaki and Nakayama, 2015). Thus, cloning of *DFR* gene encoding a DHK-specific DFR is crucial to generate pelargonidin via genetic engineering. *Gentiana lutea* L. typically produces yellow corollas, but at the southwestern tip of its distribution range (Iberian Peninsula), the species blossoms with orange flowers. *G. lutea* with orange or yellow flowers are classified as *G. lutea* L. var. *aurantiaca* (hereafter abbreviated to *aurantiaca*) and *G. lutea* L. var. *lutea* (hereafter abbreviated to *lutea*), respectively. We found only *aurantiaca* petals accumulated exclusive pelargonidin glycosides, whereas no any anthocyanins were detected in *lutea* petals (Berman *et al.*, 2016). Pelargonidin glycoside synthesis in *aurantiaca* petals appears to reflect the higher steady state levels of pelargonidin synthesis gene [*CHS* (chalcone synthase), *F3H* (flavonone 3-hydroxylase), *DFR*, *ANS* (anthocyanidin synthase) and *3GT* (anthocyanidin-3-*O*-glucosyltransferase)] transcripts as compared to the *lutea* petals (Berman J *et al.*, 2016). The both *aurantiaca* and *lutea* petals expressed similar levels of *F3'H* (flavonoid 3'-hydroxlase) and *F3'5'H* (flavonoid 3', 5'-hydroxlase) mRNA, encoding enzymes involved in cyanidin and delphinidin biosynthesis, respectively. CHS, CHI (chalcone isomerase) and F3H must be functional in both varieties due to the accumulation of similar levels of flavonols in petals. Comparisons of deduced amino acid sequences encoded by the isolated anthocyanin cDNAs indicated that ANS1 and ANS2 are identical in both *aurantiaca* and *lutea* varieties whereas minor amino acid differences of the deduced DFRs between both varieties were observed. DFR1 copies between *aurantiaca* and *lutea* varieties differ by one amino acid, while DFR2 copies differ by three amino acids. Thus, possible changes in the substrate specificity of DFR enzymes may represent additional mechanisms for producing pelargonidin glycosides in petals of *aurantiaca*. We are investigating whether the *aurantiaca* DFR enzymes are able to preferentially reduce DHK, as well as identifying amino acid (s) in *aurantiaca* DFRs that potentially determines the substrate specificity through the stable transgenic tobacco (*Nicotiana tabacum*) overexpressing of *aurantiaca* DFRs or amino acid mutagenesis-based DFRs. This work was supported by the National Natural Science Foundation of China (31270344) and MICINN, Spain (BIO2014-54441-P; BIO2014-54426-P).

T4
P1089
Secretory structures in *Casearia sylvestris* Sw. (Salicaceae): Diversity, mechanisms of secretion and exudate complexity

Valéria Ferreira Fernandes[1,2], *Marcela Thadeo*[3], *Vadnéa Casagrande Dalvi*[4], ***Renata Maria Strozi Alves Meira***[1]
1. Universidade Federal de Viçosa, Departamento de Biologia Vegetal, Programa de Pós-Graduação em Botânica, Viçosa, Minas Gerais
2. Universidade Estadual de Santa Cruz, Departamento de Ciências Biológicas, Programa de Pós-Graduacão em Botânica, Departamento de Ciências Biológicas, Ilhéus, Bahia
3. Universidade Estadual de Maringá, Departamento de Biologia, Maringá, Paraná
4. Instituto Federal de Educação, Ciência e Tecnologia Goiano, Rio Verde, Goiás

Casearia sylvestris Sw. is a shrub or tree, popularly known as "guaçatonga", which is widely used in traditional medicine for treatment of burns, wounds, herpes, snakebite and minor skin injuries. It has been suggested that such medicinal properties occur due to terpenoid and phenolic compounds present in the leaves. Idioblasts, cavities, ducts and colleters have been described in leaves of C. sylvestris, although the mechanisms and compartments involved in the secretory processes of these structures have not been clarified. This study aims to describe the chemical nature of the secretions produced by all secretory structures of C. sylvestris leaves and to identify the compartments and mechanisms of synthesis, storage and release of secretion. Shoot samples were collected and analyzed using standard techniques for both light and electron microscopy (transmission and scanning). Histochemical tests were performed in order to identify the main classes of exudate compounds. The subcellular compartments involved in the secretory process of the idioblasts, cavities, ducts and colleters were identified with all organelles being consistent with the chemical nature of secretion. Tannin secreted by the idioblasts is stored within the vacuole by fusion of vesicles filled with osmiophilic deposits that displaces the cytoplasm and nucleus to the periphery of the cell. As the tannin secretion increases, the vacuole of idioblasts may take on a large volume and the chloroplasts present well-developed membrane system and starch grains. In the colleters, plastids with osmiophilic compounds are related to the production of oil while dictiossomes are associated with syntheses of polysaccharide secretion. The smooth endoplasmic reticulum, periplastidial reticulum and plastids present in the secretory epithelium of the cavities and ducts produce the oleoresin, while dictiossomes are involved with hydrophilic secretion. Ultrastructure variation among the organelles of the four types of secretory structures studied, as well as the chemical complexity of exudates produced by them, revealed particularities and enabled the description of models for the secretory mechanism of each structure. The mechanisms involved in the release of secretion through the cell wall were presented and they brought new insights about our knowledge of the secretion biology. (FAPEMIG, CNPq, CAPES)

T4

P1090

Tissue culture and rapid propagation technique of superior tree from *Paulownia fortunei*

Jiang Su, *Jin-Xiang He, Kang-Hua Xian, Chuan-Ming Fu, Hui-Jin Huang, Ning-Zhen Huang*

Guangxi Institute of Botany, Chinese academy of Sciences

Through tissue culture test, using excellent stem segments of *Paulownia fortunei* as explants, The research has expored that the optimal sterilization time of 0.1% $HgCl_2$ was 7 minutes; the suitable primary induction medium was MS + 6-BA 2.0 mg L^{-1} + IBA 0.2 mg L^{-1} + sucrose 30 g L^{-1} + agar 3.5 g L^{-1} (pH 5.8), 30 days later, bud induction rate was 70%; the suitable method for multiplication culture was MS + 6-BA 4.0 mg L^{-1} + IBA 0.4 mg L^{-1} + sucrose 30 g L^{-1} + agar 3.5 g L^{-1} (pH 5.8) and MS + 6-BA 0.4 mg L^{-1} + IBA 0.04 mg L^{-1} + sucrose 30 g L^{-1} + agar 3.5 g L^{-1} (pH 5.8) alternately, with this method multiple shoots were health and vitrification was lower than 5%, multiplication factor was higher than 6.0/25d ; the optimal medium for rooting was 1/2 MS + NAA 0.2 mg L^{-1} + sucrose 20 g L^{-1} + carrageenan 3.4 g L^{-1} (pH 5.8), 14 days later, the rooting seedlings were obtained, rooting number were 5~10, root length was 3~5 cm, the rooting rate was 98% and whose root was white, with less and short hair, easy to clean. Through hardending off, the rooting seedlings were transplanted in the greenhouse, 50 days later whose height was 1.0m averagely and ground diameter was 1.0~2.0 cm.

T4

P1091

Suppression of extracellular invertase inhibitor gene expression improves seed weight in soybean (*Glycine max*)

Xiaofei Tang[1], ***Tao Su***[2,3], *Mei Han*[3], *Hongbin Wei*[2], *Steffen Greiner*[2], *Thomas Rausch*[2], *Lijun Liu*[1]

1. Soybean Research Institute, Academy of Agricultural Sciences

2. Plant Molecular Physiology, Centre for Organismal Studies (COS), Heidelberg University

3. Co-Innovation Center for Sustainable Forestry in Southern China, Nanjing Forestry University

Cell wall invertase (CWI) and vacuolar invertase (VI) play multiple functions in plant growth. As well as depending on transcriptional and post-transcriptional regulation, there is growing evidence that CWI and VI are also subject to post-translational control by small inhibitory proteins. Despite the significance of this, genes encoding inhibitors, their molecular and biochemical properties, and their potential roles in regulating seed production have not been well documented in soybean (*Glycine max*). In this study, two invertase inhibitor isoforms, GmCIF1 and GmC/VIF2, were characterized to possess inhibitory activities in vitro via heterologous expression. Transcript analyses showed that they were predominantly expressed in developing seeds and in response to ABA. In accordance with this, surveys of primary targets showed subcellular localizations to the apoplast in tobacco epidermis after expressing YFPfusion constructs. Investigations using RNAi transgenic plants demonstrated marked elevations of CWI activities and improvements in seed weight in conjunction with higher accumulations of hexoses, starch, and protein in mature seeds. Further co-expression analyses of *GmCIF1* with several putative *CWI* genes corroborated the notion that GmCIF1 modulation of CWI that affects seed weight is mainly contingent on post-translational mechanisms. Overall, the results suggest that post-translational elevation of *CWI* by silencing of *GmCIF1* expression orchestrates the process of seed maturation through fine-tuning sucrose metabolism and sink strength.

T4

P1092

Immunogold scanning electron microscopy can reveal the cell wall polysaccharide architecture of xylem elements

Qiang Sun[1], *Mimo He*[1], *Yuliang Sun*[2], *Kevin Juzenas*[1]

1. University of Wisconsin-Stevens Point

2. University of Wisconsin-Madison

Many of the functions that cell wall performs are closely related to its polysaccharide composition and architecture. Currently, immunofluorescence microscopy and immunogold transmission electron microscopy (TEM) are the two main techniques used to detect polysaccharides in cell wall. Despite their important role in localizing cell wall polysaccharides, both have some major limitations including detection of polysaccharides at either low resolution or in highly restricted region. In this study we have developed a novel technique to study cell wall polysaccharide architecture. This technique combines immunocytochemistry with scanning electron microscopy (SEM). It not only makes possible the visualization of polysaccharides across a large wall surface and up to a high resolution but also generates tremendous advantages over the other two methods in studying cell wall polysaccharide architecture. When tested with multiple cell wall monoclonal antibodies (mAbs), the technique was demonstrated to be effective and reliable in localizing multiple groups of hemicellulosic and pectic polysaccharides in the cell walls of five xylem structures: vessel element, fiber, axial parenchyma cell, ray parenchyma cell and tylose. In addition, this technique was demonstrated to be effective in revealing details of the three-dimensional polysaccharide distribution in a cell wall as well as in analyzing polysaccharide compositional and structural modifications of cell walls during cell differentiation and pathogen-host plant interactions *etc.* Therefore, this technique is expected to be valuable in better understanding the polysaccharide composition, architecture and functions of cell wall.

T4

P1093

Metabolic regulation during drought acclimation contribute to the acquisition of desiccation tolerance in resurrection plant *Boea hygrometrica*

Run-Ze Sun, *Xiao-Fei Zhang, Xin Deng*
Institute of Botany, the Chinese Academy of Sciences

Drought is a major environmental stress seriously compromises plant development, distribution and productivity. Most higher plants are unable to survive desiccation. Interestingly, however, a unique group of vascular angiosperm plants, referred to as 'resurrection plants' which can tolerate an extreme water deficiency of their vegetative tissues and resume normal physiological and metabolic activities after rehydration. The homoiochlorophyllous resurrection plant *Boea hygrometrica* is able to survive slow drying and possess inducible rapid desiccation after drought acclimation, thus can serve as a model system for understanding the mechanisms of dehydration tolerance acquisition in plants. Previous transcriptome analyses have revealed that transcripts involved in multiple metabolic pathways are favored during dehydration acclimation, yet, the metabolic changes underlying this adaptive process still remain practically unknown. Here, a gas chromatography-mass spectrometry (GC-MS) based global untargeted metabolite profiling analysis was conducted on leaves of *B. hygrometrica* plants that were hydrated (fresh, F), fast-dehydrated (FD), soil-dehydrated (SD), rehydrated after SD (acclimated, A) and fast-dehydrated after acclimation (AD). A total of 138 metabolites, including sugars, organic acids, amino acids, nitrogen compounds, hydrocarbons and other carbohydrates, were characterized from plant materials under different physiological conditions. The metabolite profiles between acclimated (SD, A and AD) and non-acclimated (F, FD) plant samples were well distinguished based on the principal-component and hierarchical cluster analysis, indicating that the drought acclimation response required for dynamic changes in metabolic adjustment. By using partial least-squares discriminant analysis, 26 marker metabolites were identified showing significant differences in abundance between the non-acclimated and acclimated plants. Notably, the content of some metabolites, such as maltose, L-Tryptophan (L-Tyr) and Vitamin E (VE) in acclimated plants was significantly higher than those in non-acclimated plants, suggesting that the accumulation of these compounds might lead to increased drought tolerance. Pathway analysis of transcriptome data revealed that the up-regulated expression of genes encoding starch branching enzyme (SBE) and isoamylase (ISA) required for maltose biosynthesis, 3-deoxy-7-phosphoheptulonate synthase (DHS), 3-phosphoshikimate 1-carboxyvinyltransferase (aroA) and 3-dehydroquinate synthase (aroB) required for L-Try biosynthesis during drought acclimation, as well as homogentisate phytyltransferase 1 (HPT1) and tocopherol cyclase (VTE1) in the VE biosynthetic pathway in both acclimation and rehydration plants was strongly associated with the metabolite differences among samples. Taken together, our observations indicate that the accumulation of important metabolites correlates with the transcriptional activity of biosynthetic related enzyme genes during drought acclimation might contribute to the desiccation tolerance acquisition of *B. hygrometrica*.

T4

P1094

FtSAD2 and FtJAZ1 regulate the activity of the FtMYB11 transcription repressor in of the phenylpropanoid pathway in *Fagopyrum tataricum*

Meiliang Zhou[1], *Zhanmin Sun*[1], *Yanmin Wu*[1], *Yixiong Tang*[1], *Maria D. Logachevab*[2], *Ivan Kreft*[3]
1. Chinese Academy of Agricultural Sciences
2. Department of Evolutionary Biochemistry, AN Belozersky Institute of Physico-Chemical Biology, MV Lomonosov Moscow State University
3. Nutrition Institute, Tržaška 40, SI-1000 Ljubljana

Little is known about the molecular mechanism of the R2R3-MYB transcriptional repressors involved in plant phenylpropanoid metabolism. Here, we describe one R2R3-type MYB repressor, FtMYB11 from *Fagopyrum tataricum*. It contains the SID-like motif GGDFNFDL and it is regulated by both the importin protein 'Sensitive to ABA and Drought 2' (SAD2) and a JAZ protein, a repressor from the Jasmonates (JAs) signaling cascade. Yeast two hybrid (Y2H) and Bimolecular Fluorescence Complementation (BiFC) assays demonstrated that FtMYB11 interacts with SAD2 and the JAs-responsive repressor FtJAZ1. Protoplast transactivation assays demonstrated that FtMYB11 with FtSAD2 or FtJAZ1 act synergistically and directly represses its target genes via the MYB-core element AATAGTT. Changing the Asp122 residue to Asn in the SID-like motif results in cytoplasmic localization of FtMYB11 because of loss of interaction with SAD2, while changing the Asp126 residue to Asnresults in the loss of interaction with FtJAZ1. Overexpression of FtMYB11 or FtMYB11D126N in *F. tataricum* hairy roots resulted in reduced accumulation of rutin, while overexpression of FtMYB11D122N hairy roots did not lead to such a change.The resultsindicate that FtMYB11 acts as a regulator via interacting with FtSAD2 or FtJAZ1 to repress the phenylpropanoid biosynthesis, and this repression depends on two conserved Asp residues of its SID-like motif.

T4

P1095

Characterization of autophagy pathway and its response to environmental stresses in *Cucumis sativus*

Pawaree Thanapipatpong, Supachai Vuttipongchaikij, ***Anongpat***

Suttangkakul
Kasetsart University

Autophagy is an important catabolic process used for recycling nutrients and eliminating unwanted substances within cells. Since the discovery of the genes responsible for this process in yeast, orthologs of autophagy (*ATG*) genes have been identified and characterized in various organisms, allowing better understanding of the mechanisms and functions of this pathway. In plants, this process is critical for responses to nutrient starvation, as well as biotic and abiotic stresses. *Cucumis sativus* (cucumber) is an important crop and also used as a model system for the Cucurbit family. Rapid environmental changes could have adverse effects on cucumber growth, and genetic improvements to increase tolerance to environmental stresses is important for the sustainability of cucumber production. However, potential effects of autophagy in responding to environmental stresses is not known in cucumber. We, therefore, identified and characterized key *ATG* genes in *C. sativus* as well as assessing the autophagic level in response to biotic and abiotic stresses including pathogen infection, oxidative stress, and nutrient starvation. The homologs of *ATG* genes in the conjugation pathway were identified based on protein sequence similarity to the yeast and *Arabidopsis* ATG proteins. The majority of *C. sativus ATG* genes were identified as gene families similar to what previously found in other plants. In addition, transcript variances were also identified. Quantitative RT-PCR showed that the expression of *ATG* genes and their transcripts variances were changed in response to downy mildew infection, hydrogen peroxide treatment and nitrogen starvation. However, the immunoblot analysis of ATG8, a marker for autophagy levels, did not reflect the changes in transcripts abundance. In summary, we identified autophagy genes in *C. sativus* and found that autophagy process is regulated at various levels, including transcription, mRNA splicing, protein expression and protein degradation, in response to environmental stresses. Understanding how autophagy is regulated in *C. sativus* may help with generating cucumber variety with increase tolerances to environmental stresses.

T4
P1096

Developmental features of microspore tetrad and tapetum in male sterile mangosteen (*Garcinia mangostana* L.) compared with male fertile seashore mangosteen (*G. hombroniana* Pierre.)

Pornsawan Sutthinon[1], *Lacey Samuels*[2], *Upatham Meesawat*[1]
1. Prince of Songkla University
2. University of British Columbia

Pollen abortion in mangosteen (*Garcinia mangostana* L.) has been considered to be related with tapetum activity during microgametogenesis. Developmental events at cellular and sub-cellular level of tapetal cells at microspore tetrad stage were therefore needed to investigate by comparing with fertile anther of seashore mangosteen (*G. hombroniana* Pierre.). Ultrastructural features of microspore tetrad and tapetum were observed by transmission electron microscope (TEM), callose wall deposition around microspore tetrad were also obtained by aniline blue staining. The results revealed that tetrad cells of both species were contained remarkable amounts of stacked ER and mitochondria. In *G. mangostana* tapetal cells, mitochondrial breakdown was suggested by the presence of their unusual internal structures. This mitochondrial aberrance presumably indicated an abnormality in tapetum cell, which may initiate pollen abortion leading to an early degeneration of microspore tetrads in male sterility of mangosteen (*G. mangostana*).

T4
P1097

Disruption of the *Arabidopsis* defense regulator genes *SAG101*, *EDS1*, and *PAD4* confers enhanced freezing tolerance

Weijuan Tan
Sun Yat-sen University

In *Arabidopsis*, three lipase-like regulators, *SAG101*, *EDS1*, and *PAD4*, act downstream of resistance protein-associated defense signaling. Although the roles of *SAG101*, *EDS1*, and *PAD4* in biotic stress have been extensively studied, little is known about their functions in plant responses to abiotic stresses. Here, we show that *SAG101*, *EDS1*, and *PAD4* are involved in the regulation of freezing tolerance in *Arabidopsis*. With or without cold acclimation, the *sag101*, *eds1*, and *pad4* single mutants, as well as their double mutants, exhibited similarly enhanced tolerance to freezing temperatures. Upon cold exposure, the *sag101*, *eds1*, and *pad4* mutants showed increased transcript levels of C-REPEAT/DRE BINDING FACTORs and their regulons compared with the wild type. Moreover, freezing-induced cell death and accumulation of hydrogen peroxide were ameliorated in *sag101*, *eds1*, and *pad4* mutants. The *sag101*, *eds1*, and *pad4* mutants had much lower salicylic acid (SA) and diacylglycerol (DAG) contents than the wild type, and exogenous application of SA and DAG compromised the freezing tolerance of the mutants. Furthermore, SA suppressed the cold-induced expression of DGATs and DGKs in the wild-type leaves. These findings indicate that *SAG101*, *EDS1*, and *PAD4* are involved in the freezing response in *Arabidopsis*, at least in part, by modulating the homeostasis of SA and DAG.

T4
P1098

Aquaporins are involved in the amelioration of hypoxic aspen (*Populus tremuloides*) root hydraulics and *in planta* oxygen by exogenous ethylene

Xiangfeng Tan, *Janusz J. Zwiazek*
University of Alberta

Hypoxia caused by flooding is a great challenge for terrestrial plants. Ethylene plays a profound role in plant responses to flooding, but we still know very little about the effects of ethylene on aquaporin-mediated water transport in hypoxic plants. In the present study, the physiology and gene expression of aquaporins was investigated in aspen (*Populus tremuloides*) seedlings exposed to hypoxia and treated with exogenous ethylene. Roots of hydroponically-grown aspen plants were subjected to aeration, aeration with exogenous ethylene, hypoxia, and hypoxia with exogenous ethylene treatments. Physiological measurements included gas exchange, photosynthetic light responses, and leaf chlorophyll fluorescence measured with an infrred gas analyzer. Root hydraulic conductance was measured with a high-pressure flow meter, and root oxygen concentrations were determined with an oxygen microsensor. Root porosity was estimated by a pycnometer method and mRNA expression profiling was measured by qRT-PCR. Hypoxic seedlings showed lower gas exchange, light-saturated net photosynthetic rate (P_n), effective quantum yield, root hydraulic conductance and root oxygen concentrations. Exogenous ethylene also significantly enhanced P_n and transpiration rates. qPCR results showed that hypoxia had significant effect on *PtPIP1;2*, *PIP2;1*, *PIP2;2* and *PIP2;5* transcript levels and significant effects of exogenous ethylene on *PtPIP2;4* expression was detected by the two-way ANOVA. Root oxygen concentrations also increased with ethylene treatment compared to non-ethylene treated hypoxic plants, and this may partly be due to increased P_n. Our results suggest that exogenous ethylene may enhance root water transport in hypoxic plants through the effect on aquaporin expression. The enhanced root water transport by ethylene is likely responsible for the increase in P_n which, in turn, may have a positive effect on plant aeration.

T4

P1099

RopGEF1 mediates polar auxin transport in early *Arabidopsis* development

Yuting Liu[1,2], *Qingkun Dong*[1,2], *Daniel Kita*[3], *Hen-Ming Wu*[3], ***Lizhen Tao***[1,2]

1. State Key Laboratory for Conservation and Utilization of Subtropical Agro-bioresources,South China Agricultural University,Guangzhou 510642, China

2. Guangdong Provincial Key Laboratory of Protein Function and Regulation in Agricultural Organisms,College of Life Sciences, South China Agricultural University, Guangzhou 510642, China

3. Department of Biochemistry and Molecular Biology, University of Massachusetts, Amherst, Massachusetts 01003, USA

Polar auxin transport, facilitated by the combined activities of auxin influx and efflux carriers to maintain asymmetric auxin distribution, is essential for plant growth and development. Here, we show that *Arabidopsis RopGEF1*, a guanine nucleotide exchange factor and activator of ROPs (Rho GTPases of plants), controls polar distribution of auxin influx carrier AUX1 and differential accumulation of efflux carriers PIN7 and PIN2, and is important for embryo and early seedling development when RopGEF1 is prevalently expressed. Knock-down or knock-out of *RopGEF1* induces embryo defects, cotyledon vein breaks and delayed root gravity responses. Altered expression from the auxin response reporter *DR5rev:GFP* in the root pole of *RopGEF1*-deficient embryos and loss of asymmetric distribution of *DR5rev:GFP* at their gravistimulated root tips suggest that auxin distribution is affected in *ropgef1* mutants. Importantly, the polarity of AUX1 is altered in *ropgef1* embryos and roots, shifting from the normal apical membrane location to a basal location in embryo central vascular and root protophloem cells. *RopGEF1* mutations also reduced PIN7 accumulation at embryos and PIN2 distribution in gravistimulated roots. In establishing a role for RopGEF1 in AUX1 polarity and PIN differential accumulation, our results implicate that RopGEF1 is critical for cell polarity and polar auxin transport could impact auxin-mediated plant growth and development.

T4

P1100

Isolation and characterization of PSI-LHCI super-complex and their sub-complexes from a red alga *Cyanidioschyzon merolae*

Lirong Tian[1,2], *Zheyi Liu*[3], *Fangjun Wang*[3], *Liangliang Shen*[1,2], *Jinghua Chen*[1,2], *Lijing Chang*[1,2], *Songhao Zhao*[1,2], *Guangye Han*[1], *Wenda Wang*[1], *Tingyun Kuang*[1], *Xiaochun Qin*[1,4], *Jian-Ren Shen*[1,5]

1. Institute of Botany, Chinese Academy of Sciences

2. University of Chinese Academy of Sciences

3. Dalian Institute of Chemical Physics, Chinese Academy of Sciences

4. University of Jinan

5. Okayama University

Photosystem I (PSI)-light-harvesting complex I (LHCI) super-complex and its sub-complexes PSI core and LHCI, were purified from a unicellular red alga *Cyanidioschyzon merolae* and characterized. PSI-LHCI of *C. merolae* existed as a monomer with a molecular mass of 580 kDa. Mass spectrometry analysis identified 11 core subunits (PsaA, B, C, D, E, F, I, J, K, L, O) in the core complex and three LHCI subunits, CMQ142C, CMN234C and CMN235C in LHCI, indicating that at least three Lhcr subunits associate with the red algal PSI core. PsaG was not found in the red algae PSI-LHCI, and we suggest that the position corresponding to Lhca1 in higher plant PSI-LHCI is empty in the red algal PSI-LHCI. The PSI-LHCI complex was separated into two bands on native PAGE, suggesting that two different complexes may be present with slightly different protein compositions probably with respective to the numbers of Lhcr subunits. Based on the results obtained, a structural model was proposed for the red algal PSI-LHCI. Furthermore, pigment analysis revealed that the *C. merolae* PSI-LHCI contained a large amount of zeaxanthin, which is mainly associated with the LHCI complex whereas little zeaxanthin was found in the PSI core. This indicates a unique feature of the carotenoid composition of the Lhcr proteins and may suggest an important role of Zea in the light-harvesting and photo-protection of the red algal PSI-LHCI complex.

T4

P1101

Distinct *cis*-regulatory regions participate in modulating the transcriptional response of *FLOWERING LOCUS T* to photoperiod

Johan Zicola[1], *Tingting Ning*[1,2], *Emmanuel Tergemina*[1], *Tingting Yang*[1], *Liangyu Liu*[1,3], ***Franziska Turck***[1]

1. Max Planck Institute for Plant Breeding Research

2. Zhejiang University

3. Capital Normal University

FLOWERING LOCUS T (FT) encodes the mobile flowering hormone florigen that regulates the floral transition in many plant species by integrating environmental seasonal signals and internal cues at the transcriptional level. We have previously shown that two interdependent regulatory regions are necessary and sufficient to convey photoperiod responsiveness to *FT* expression in

Arabidopsis thaliana leaves. We now show that other *cis*-regulatory modules, located in introns and a newly discovered *FT* downstream enhancer, also participate in the regulation of *FT*. Furthermore, we explore the effect of natural variation located at regulatory regions on *FT* expression.

T4

P1102

Expression of cyanobacterial bicarbonate transporters in *Arabidopsis* chloroplasts and their effects on the growth and photosynthetic performance

Susumu Uehara[1], *Yasuko Ito-Inaba*[3], *Izumi C Mori*[2], *Takehito Inaba*[1]
1. Fac. Agr., Univ. Miyazaki
2. IPSR, Okayama Univ.
3. TT Org., Univ. Miyazaki

It has been thought that photosynthesis in C_3 plants are inefficient due to photorespiration. In contrast, cyanobacteria have evolved CO_2-concentrating mechanism, including bicarbonate transporters, and are capable of decreasing photorespiration. Hence, installation of cyanobacterial bicarbonate transporters to the inner envelope membrane (IEM) of chloroplasts in C_3 plants has been thought to improve photosynthesis. However, the method to deliver cyanobacterial bicarbonate transporters to the chloroplast IEM remains to be established. In this study, we provide evidence that the cyanobacterial bicarbonate transporters, BicA and SbtA, can be specifically installed into the chloroplast IEM using the chloroplast IEM targeting signal in conjunction with the transit peptide. We fused the transit peptide and the mature portion of Cor413im1 to either BicA or SbtA isolated from *Synechocystis* sp. PCC6803. We confirmed that four chimeric bicarbonate transporters, designated as BicAI, BicAII, SbtAII, and SbtAIII, were expressed in *Arabidopsis*. Furthermore, these chimeric transporters were specifically targeted to the chloroplast IEM. They were also resistant to alkaline extraction but can be solubilized by Triton X-100, indicating that they are integral membrane proteins in the chloroplast IEM. Taken together, our results indicate that the addition of IEM targeting signals, as well as the transit peptide to bicarbonate transporters, allows us to efficiently target nuclear-encoded chimeric bicarbonate transporters to the chloroplast IEM. Moreover, we also report the effects of the expressed bicarbonate transporters on the growth and photosynthetic performance of *Arabidopsis*.

T4

P1103

Arabidopsis PROTEIN S-ACYL TRANSFERASE4 mediates root hair growth

Zhi-Yuan Wan, *Sen Chai, Fu-Rong Ge, Qiang-Nan Feng, Yan Zhang, Sha Li*
Shandong Agricultural University

Polar growth of root hairs is critical for plant survival and necessitates fine tuned ROP signaling. Multiple ROP regulators participate in root hair growth. However, protein Sacyl transferases (PATs), mediating the Sacylation and membrane partitioning of ROPs, are yet to be found. By using a reverse genetic approach, combining fluorescence probes, pharmacological drugs, sitedirected mutagenesis, and genetic analysis with related roothair mutants, we have identified and characterized an Arabidopsis PAT, which may be responsible for ROP2 Sacylation in root hairs. Specifically, functional loss of PAT4 resulted in reduced root hair elongation, which was rescued by a wildtype but not an enzymeinactive PAT4. Membrane-associated ROP2 was significantly reduced in *pat4*, similar to that of Sacylation deficient ROP2 in wild type. We further showed that PAT4 and SCN1, a ROP regulator, additively mediate ROP2 stability and targeting. Results presented indicate that PAT4-mediated Sacylation mediates the membrane association of ROP2 at the root hair apex and provide novel insights into the dynamic ROP signaling during plant tip growth.

T4

P1104

Maize microtubule-associated protein ZmGLR regulates leaves development by Brassinosteroids

Chenchen Wang, *Hua Zhang, Qian Zhao*
China Agricultural University

Brassinosteroids (BRs) is a major kind of growth-promoting plant hormone, participating in diverse plant growth and development process such as promotion of stem elongation and cell division, increase of tolerance to various stresses. Recent researches have illuminated BR signaling pathway and revealed major mechanisms of BR crosstalk with other signaling pathways. However, the exact executors of BR signaling is unclear. *ZmGLR* (*Zea mays L.* Glutamic acid and Lysine Rich) was isolated from maize which encodes a glutamic acid and lysine rich protein including a DREPP conserved domain. Co-sedimentation assay revealed that the ZmGLR recombinant protein can bind to microtubules *in vivo*. Moreover, the fluorescent co-location and negative staining assay indicated that ZmGLR bundled microtubules. Therefore, we concluded that ZmGLR is a microtubule associated protein from maize. The subcellular localization showed that ZmGLR was associated with the plasma membrane, and 25 amino acid of N-terminal sequence have an important effect on the location. Interestingly, the GFP fluorescence occurred in cytoplasm in condition of high Ca^{2+} concentration. Therefore, Ca^{2+} may have an impact on the subcellular localization of ZmGLR. To investigate the function of ZmGLR, we obtained *ZmGLR* overexpression and RNAi plants. Both *ZmGLR* overexpression and RNAi transgenic maize showed a severe dwarf phenotype and developed severely malformed leaves. *ZmGLR* overexpression maize showed totally curly leaves which even became a cylindet shape, and RNAi maize appeared tortuous and wrinkled blades. Moreover, blade epidermal cells in mutants became short and abnormal, whereas the wild-type cells were a uniform rectangular shape. Stomas in mutants were deformed, and the amount of stomas in overexpression plants have obviously decreased compare with wild-type, while the RNAi plants have increased stomas. Real-time PCR results showed *ZmGLR* expression enhanced by eBL treatment and depressed by IAA treatment. Then, we found the ZmBES1, a key transcription factor in BR signal pathway, can bind the promoter of ZmGLR and increase its activity. These results demonstrate that ZmGLR participates in both BR and auxin pathways as a final executor. BRs promote the correct arrangement of microtubule to ensure cell expansion and elongation by regulate the expression of microtubule-associated protein ZmGLR. However, how auxin regulates ZmGLR need to be investigated.

T4

P1105

Functional diversity of ribosomal proteins in plant mitochondria

Chuande Wang, *Martine Quadrado, Céline Dargel-Graffin, Hakim Mireau*

Institut Jean-Pierre Bourgin, INRA

Expression of mitochondrial genes implies a tight cooperation between degenerated prokaryotic-like features and nuclear-encoded eukaryotic-derived factors that appeared throughout the evolution. This dual origin has resulted in complex mRNA expression processes that are still poorly understood at the molecular level. This is particularly true for mitochondrial translation, which has remained largely unexplored notably in higher plants. Yet, we know that the mitochondrial translation machinery differs profoundly from its bacterial counterpart in many essential aspects. To better understand the functioning of the plant mitochondrial machinery, we prepared and analysed mitochondrial ribosome footprints from Arabidopsis inflorescences. This allowed us to determine the extent of the *Arabidopsis* mitochondrial translatome and reveal interesting new features about the dynamic and the functioning of the mitochondrial translation apparatus in plants. In complement to this approach, we classified and analysed nuclear genes encoding potential mitochondria-targeted ribosomal proteins in *Arabidopsis thaliana*. This exhaustive survey led to us to observe that most ribosomal and ribosomal-like proteins are encoded in small gene families comprising up to 10 members. Further analyses suggested that several of such families might not correspond to multigene families *sensus stricto*. To explore the activity and the functional diversification of ribosomal proteins in plants, we selected and analysed in details several of the identified families. The sub-cellular distribution of selected proteins was determined by GFP fusion and Arabidopsis mutants affected in the corresponding genes were characterized at the molecular level. Very interestingly, our results revealed several cases of molecular exaptation and the acquisition of new functions in mitochondrial mRNA expression for certain ribosomal-like proteins. I will present our very last results on these approaches and discuss the evolution of RNA expression systems in higher plant mitochondria.

T4

P1106

Phytochrome-interacting factors directly suppress *MIR156* expression to enhance shade avoidance response in *Arabidopsis*

Yurong Xie, Yang Liu, Xiaojing Ma, Baobao Wang, Hai Wang, ***Haiyang Wang***

Biotechnology Research Institute, Chinese Academy of Agricultural Sciences

As sessile organisms, plants have evolved a repertoire of strategies collectively termed the shade avoidance syndrome (SAS) to avoid canopy shade and compete with their neighbors for light. SAS, albeit beneficial for the survival of plants in the natural environment, is detrimental for crops to adapt to their high-density stands in modern agricultural practices. However, the signaling mechanism governing the adaptive changes of adult plant architecture to shade is not well understood, even in the model species *Arabidopsis thaliana*. Here we show that in *Arabidopsis,* several PHYTOCHROME-INTERACTING FACTORS (PIFs) overexpressors (*35S::PIF1-OE*, *35S::PIF3-OE*, *35S::PIF4-OE* and *35S::PIF5-OE*) all display constitutive SAS under normal high R:FR ratio conditions, and are less sensitive to the simulated shade, compared to the wild type plants. In contrast, *MIR156* overexpressors exhibit reduced SAS under normal high R:FR ratio conditions, and are more sensitive to the simulated shade, compared with the wild type. In addition, the simulated shade rapidly induces PIF protein accumulation, accompanied by reduced expression of multiple *MIR156* genes and a concomitant elevated expression of *SQUAMOSA-PROMOTER BINDING PROTEIN-LIKE* (*SPL*) family of transcription factors. Moreover, *in vivo* and *in vitro* assays indicate that these PIF proteins directly bind to the G-box motifs in the promoters of several *MIR156* genes and suppress their expression. Our results provide a direct link between the phytochrome-PIFs and miR156-SPL regulatory modules in mediating SAS during plant vegetative growth and development. Implications of our findings in modulating crop plant architecture for high density planting are discussed.

T4

P1107

High-throughput assays for oxidative stress tolerance in cereals

Haiyang Wang, *Sergey Shabala, Lana Shabala, Meixue Zhou*

University of Tasmania

Soil salinity is a global issue and a major factor limiting crop production worldwide. One side effect of salinity stress is an overproduction and accumulation of reactive oxygen species (ROS), causing oxidative stress and leading to severe cellular damage to plants. While the major focus of the salinity-oriented breeding programs in the last decades was on traits conferring Na^+ exclusion, breeding for oxidative stress tolerance has been largely overlooked. The importance lies in its occurrence in almost all the environmental stresses, not only in salinity stress. This calls for the study on exploring the relationship between oxidative stress and abiotic stress and a need for finding a reliable and less time-consuming screening method for fine mapping of related quantitative trait loci (QTL) in cereals. In this work, we developed and compared several high-throughput assays for oxidative stress screening in a representative cereal crop: barley (*Hordeum vulgare*). In root growth assay experiment, we showed that the biggest difference of root elongation rate between salinity tolerant and salinity sensitive cultivars happened when plants were grown in the presence of 1 mM H_2O_2 in the growth media. Under these conditions, a significant negative correlation was found between root elongation rate and the overall salinity tolerance, whereas no difference has been found between cultivars in viability staining experiments on both root elongation zone and mature zone. By applying the MIFE technique for non-invasive ion flux measurements, we also found a significant positive correlation between ions (Ca^{2+} and K^+) fluxes and the overall salinity tolerance under 10 mM H_2O_2 treatment and also a clear difference in the matter of correction red fluorescence intensity (dead cells) in viability staining experiments among cultivars. Reliability of these high-throughput assays for plant breeding for oxidative stress tolerance are discussed.

T4

P1108

UDP-glucose: anthocyanidin 3-*O*-glucoside-2"-*O*-glucosyltransferase is involved in further glucosylation of anthocyanins in purple sweet potato

Hongxia Wang, *Weijuan Fan, Peng Zhang*

Institute of plant Physiology and Ecology, SIBS, CAS

Further glycosylation contributes to the diversity and stability of anthocyanins in plants. In purple sweet potato Ayamurasaki, anthocyanin 3-O-sophoroside and its derivatives constitute the major anthocyanin compounds. Here we report a sweet potato UDP-glucose: anthocyanidin 3-O-glucoside-2''-O-glucosyltransferase (Ib3GGT) that catalyzes the conversion of anthocyanidin 3-O-glucoside into anthocyanidin 3-O-sophoroside using UDP-glucose predominantly as the glycosyl donor. Homology-based protein modeling and site-directed mutagenesis of Ib3GGT suggested that Thr-138 is a key amino acid residue for UDP-glucose recognition and plays an important role in sugar donor selectivity. *Ib3GGT* overexpression in *Arabidopsis* produced the new component cyanidin 3-O-sophoroside without altering the existing anthocyanins, and complemented the purple phenotype in *ugt79b1* mutants. *Ib3GGT* expression was associated with anthocyanin accumulation in different organs during plant development, mainly in immature leaves and storage roots of Ayamurasaki plants; in addition, the novel enzyme was regulated by the IbMYB1 transcription factor. Localization assay of Ib3GGT showed that further glucosylation occurred in both the endoplasmic reticulum and Golgi apparatus. In sweet potato, *Ib3GGT* overexpression resulted in increased anthocyanin accumulation, while its silencing reduced anthocyanin accumulation and affected anthocyanin release from anthocyanic vacuolar inclusions into vacuoles. Overall, Ib3GGT is the key glucosyltransferase responsible for further glycosylation in a UDP-glucose dependent manner, and causes anthocyanin accumulation in purple sweet potato.

T4

P1109

Duplicated *TT2-like* MYB genes from *Freesia hybrida* play divergent roles in the biosynthesis of proanthocyanidins

Xiaotong Shan, Yueqing Li, Song Yang, Ruifang Gao, Li Wang,* ***Xiang Gao****

Key Laboratory of Molecular Epigenetics of MOE and Institute of Genetics & Cytology, Northeast Normal University

Flavonoids are one of the largest classes of plant secondary metabolites involved in plant development and defense, which mainly include flavonols, anthocyanins, and proanthocyanidins. In our previous studies, a considerable amount of the three kinds of flavonoids were detected in the flower of "Red River®", a representative cultivar of a perennial monocotyledonous plant, *Freesia hybrida*, and they showed diverse temporal-spatial accumulation patterns during the development of the flowers and in 5 different flower tissues. Previous studies have focused on the regulation of anthocyanin biosynthesis. However, the transcription regulation mechanisms for other two branches still remain to be elucidated. In this study, four R2R3-type MYB transcription factors phylogenetically clustered with *AtTT2* were isolated from the flowers of "Red River®", tentatively designated as *FhTT2L1*, *FhTT2L2*, *FhTT2L3* and *FhTT2L4*. Two groups of expression patterns were observed for the four *TT2-like* MYB regulatory genes aforementioned. *FhTT2L1*, *FhTT2L2*, and *FhTT2L4*, which were clustered together and separated with *FhTT2L3*, showed an identical expression profiles significantly correlated the *FhLAR*, the gene involved in the biosynthesis of catechin-derived proanthocyanidins. In contrast, the expression of *FhTT2L3* was found to be synchronous to *FhANR*, which is responsible for epicatechin-derived proanthocynidins biosynthesis. Therefore, it is reasonable to deduce that the duplicated *TT2-like* genes in *Freesia* might function divergently to control the biosynthesis of the two categories proanthocyanidins. However, it is not always clear whether the functions of *FhTT2L1*, *FhTT2L2*, and *FhTT2L4* are partially or completely redundant because of their differential expression levels. In accordance with TT2-like proteins from model plant as well as other extensively studied plant species, FhTT2Ls could interact with bHLH proteins participating in the MBW complex (mainly comprising of AtTT2, AtTT8 and AtTTG1 in *Arabidopsis*) and they were capable to activate the expression of late anthocyanin biosynthetic genes (LBGs) as well as proanthocyanidin related *AtBAN* gene when co-transfected with *AtTT8* and *AtTTG1* in *Arabidopsis* leaf protoplasts. Unexpectedly, FhTT2Ls could also activate the expression of *AtCHS* gene when transiently expressed alone, whereas AtTT2 could not, indicating a plausible divergent transactivation capacity of TT2-like regulators between the *Freesia* and *Arabidopsis*. Furthermore, the four *FhTT2L* genes were successfully ectopically expressed in tobacco plants, hyper-accumulation of proanthocyanidins was observed in individual transgenic lines, which was consistent with the up-regulation of the expressions of *NtDFR*, *NtANS*, *NtLAR*, *NtANR* genes. In conclusion, it can be speculated that the duplicated *TT2-like* MYB genes might play divergent roles in the biosynthesis of proanthocyanidins in *Freesia hybrida*, and their transcription regulation manners might also evolve differentially compared with *Arabidopsis*. After the confirmation of target genes and existence of MBW complexes in *Freesia*, the conclusion will be further strengthened. To our knowledge, this is the first reports of the *TT2-like* genes isolated from flowers of *Freesia* and even in Liliales plants, the finds will not only provide new insights into the transcriptional regulation of proanthocyanidin biosynthesis in monocot plants, they will also pave the way to dissect the crosstalk networks controlling multiple flavonoids biosynthesis which are simultaneously accumulated in the same plant organs with diverse patterns.

T4

P1110

The SMC5/6 complex couples cell cycles checkpoint and DNA repair to maintain genome stability and stem cell identify

Lili Wang, *Shunping Yan*

College of Life Science and Technology, Huazhong Agricultural University

Stem cells are hypersensitive to DNA damage. In response to DNA damage, cells activate checkpoints to arrest cell cycle progression and provide sufficient time to repair DNA. Previous studies have identified different sets of checkpoint and repair proteins. How cells coordinate these two processes is not well-understood. The evolutionarily conserved SMC5/6 complex is believed to play multiple essential roles in DNA repair. Here we show that the SMC5/6 complex also regulates G1/S checkpoints by repressing E2F transcription factors. Mutation in *SNI1*, a subunit of the SMC5/6 complex, results in defective G1/S transition and eventually causes disorganized root meristem and short root phenotypes, mimicking *E2F* overexpression. Consistently, mutations in *E2F*s suppresses the *sni1* mutant and overexpression of *SNI1* suppresses *E2F* overexpression plants. Remarkably, SNI1 directly binds the activation domain of E2Fs and represses their transcriptional activity. Our study demonstrates that the SMC5/6 complex is the key link to couple checkpoint and DNA repair to maintain genome stability and stem cell identity.

T4

P1111

CYP86A1, a key enzyme in suberin biosynthesis, is pivotal for salt tolerance in *Arabidopsis*

Pei Wang, *Li Gao, Haili Yang, Qing Ma, Yannong Cui, Suomin Wang*

State Key Laboratory of Grassland Agro-ecosystems, College of Pastoral Agriculture Science and Technology, Lanzhou University

Soil salinity is one of the major environmental stresses restricting agricultural production, higher plant developed several protective strategies to adapt to salt stress, including forming apoplastic barriers to limit Na^+ radial transport in roots. Suberin, a macromolecular heteropolymer combined by lipid and phenolic compounds deposited on the cell wall of root endodermis and periderm, is an important component of apoplastic barriers in roots, prevents the free apoplastic diffusion of ions and water, however, it is still undefined how suberin functions in responding to salt stress. Cytochrome P450 ω-hydroxylase encoded by *CYP86A1* (*HORST*) catalyzes mid-chain fatty acid to create ω-hydroxy acid and α, ω-diacid, which are the most major monomers of suberin. In this study, the functional mechanisms of suberin in plant adaptation to salt stress were investigated by using *Arabidopsis cyp86a1* mutant. Results showed that moderate NaCl induced suberization and the expression of *CYP86A1* gene in *Arabidopsis* roots. *cyp86a1* mutant displayed a salt-hypersensitive phenotype in the adult stage, including growth inhibition and disordered leaf membrane system under salt stress compared with wild-type (WT) plant. The level of suberization declined in *cyp86a1* mutants through FY088 tracing. GC-MS analysis also showed suberin monomer contents decreased sharply in *cyp86a1* mutants than WT under normal condition, and NaCl increased suberin content in WT while did not change in *cyp86a1*. Loss function of CYP86A1 resulted in more Na^+ accumulation in shoots via trans-cellular pathway transport, accompanied by shoots K^+ content decreased. In addition, mutation of *CYP86A1* led to an increase in root hydraulic conductivity and photosynthetic capacity of rosette leaves under normal condition, while showed an opposite trend under 30 mM NaCl. *cyp86a1* had a lower water use efficiency than WT with or without NaCl. Furthermore, transcript profiling showed that genes involved in cell wall metabolism, phytohormones, nutrient uptake, lipid metabolism and cellular signaling network were inhibited in *cyp86a1*, compared with WT. Taken together, our results indicate that suberin affects Na^+ accumulation, root water permeability and leaf photosynthetic capacity under salt stress by altering its own monomer content, meanwhile, functions in growth and metabolic regulation, thereby changes the salt tolerance of *Arabidopsis*.

T4

P1112

C_4 photosynthesis in C_3 rice: A theoretical analysis of biochemical and anatomical factors

Shuyue Wang[1], *Danny Tholen*[2], *Xinguang Zhu*[1]

1. CAS-MPG Partner Institute for Computational Biology, Shanghai Institutes for Biological Sciences, Chinese Academy of Sciences

2. Institute of Botany, Department of Integrative Biology, University of Natural Resources and Applied Life Sciences, BOKU Vienna

Engineering C_4 photosynthesis into rice has been considered a promising strategy to increase photosynthesis and yield. A question that remains to be answered is whether expressing a C_4 metabolic cycle into a C_3 leaf structure and without removing the C_3 background metabolism improves photosynthetic efficiency. To explore this question, we developed a 3D reaction diffusion model of bundle-sheath and connected mesophyll cells in a C_3 rice leaf. Our results show that integrating a C_4 metabolic pathway into rice leaves with a C_3 metabolism and mesophyll structure may lead to an improved photosynthesis under current ambient CO_2 concentration. We analysed a number of physiological factors that influence the CO_2 uptake rate, which include the chloroplast surface area exposed to intercellular air space, bundle-sheath cell wall thickness, bundle-sheath chloroplast envelope permeability, Rubisco concentration and the energy partitioning between C_3 and C_4 cycles. Among these, partitioning of energy between C_3 and C_4 photosynthesis and the partitioning of Rubisco between mesophyll and bundle-sheath cells are decisive factors controlling photosynthetic efficiency in an engineered C_3-C_4 leaf. The implications of the results for the sequence of C_4 evolution are also discussed.

T4

P1113

Effects of short-term osmotic stress on leaf hydraulic conductivity and plasma membrane intrinsic proteins genes mRNA accumulation in maize seedlings

Wei-Feng Wang[1,2], *Liang-Min Wang*[1], *Sui-Qi Zhang*[2]

1. College of Forestry, Shanxi Agricultural University

2. State Key Laboratory of Soil Erosion and Dryland Farming on the Loess Plateau, Northwest A&F University

Plants maintain water balance by varying hydraulic properties, and plasma membrane intrinsic proteins (PIPs) may be involved in this process. Leaf xylem and root hydraulic conductivity and the mRNA contents of four highly expressed ZmPIP genes (*ZmPIP1;1*, *ZmPIP1;2*, *ZmPIP2;2* and *ZmPIP2;5*) in maize (*Zea mays*) seedlings were investigated. Under well watered conditions, leaf hydraulic conductivity (K_{leaf}) varied diurnally and was correlated with whole-plant hydraulic conductivity. Similar diurnal rhythms of leaf transpiration rate (E), K_{leaf} and root hydraulic conductivity (K_{root}) in well-watered plants can be important for maintaining whole-plant water balance. After 2 h of osmotic stress treatment induced by 10% polyethylene glycol 6000, the K_{root} of stressed plants decreased but K_{leaf} increased, compared with well-watered plants. The mRNA contents of four *ZmPIP*s were significantly up-regulated in the leaves of stressed plants, especially for *ZmPIP1;2*. Meanwhile, *ZmPIP2;5* was significantly down-regulated in the roots of stressed plants. After 4 h of osmotic stress treatment, the E and leaf xylem water potentials of stressed plants unexpectedly increased. The increase in K_{leaf} and a partial recovery of K_{root} may have contributed to this process. The mRNA content of *ZmPIP1;2* but not of the other three genes was up-regulated in roots at this time. In summary, the mRNA contents of these 4 *ZmPIP*s associated with K_{leaf} and K_{root} change in maize seedlings during short-term osmotic stress, especially for *ZmPIP1;2* and *ZmPIP2;5*, which may help to further reveal the hydraulic resistance adjustment role of *ZmPIP*s.

T4

P1114

Small RNA and transcriptome deep sequencing provides insight into pistil mutations in alfalfa possibly derived from an infection of RNA virus

Xiaojuan Wang[1], *Haichen Zhou*[2]

1. Shanghai Science & Technology Museum/Natural History Research Center

2. Lanzhou University/School of Pastoral Agriculture Science and Technology

Alfalfa (*Medicago sativa* L.) is considered as one of the most studied agronomical plants. Its multi-pistil mutant MP-1 exhibited two to three pistils compared with its wild-type plant, which is highly valuable in the study of floral development in alfalfa. Our previous study revealed that MP-1 was partial-female-sterile, significantly decreasing the production yield. In this study, we conducted high-throughput sequencing of the miRNAs and transcriptomes of the alfalfa mutant florets (MP-1) and its wild-type florets (WT) to gain insights into pistil development in alfalfa. A total of 151,359 putative unigenes with mean length of 631 bp were generated from WT and MP-1 transcriptomes. Comparative profiling showed that 2306 differentially expressed genes were identified, including 1100 up-regulated unigenes and 1206 down-regulated unigenesin MP-1 vs WT. The up-regulated genes in MP-1 belong to biological processes such as RNA-directed DNA polymerase activity-dependent DNA replication, DNA metabolic process, DNA polymerase activity, DNA integration and DNA replication by GO enrichment analysis. In addition, the expression of genes encoding capsid protein and RNA dependent RNA polymerase were significantly abundant, which were found highly similar to a RNA virus (white clover cryptic virus). Besides, a total of 148 known miRNAs, belonging to 72 families, were identified from two small RNA libraries of MP-1 and WT. Comparative profiling revealed that 110 miRNAs exhibited differential expression between MP-1 and WT, 43 of which were novel. The correlation analysis of small RNA and transcriptome revealed that 18 highly differentially expressed miRNAs and their 28 target genes were closely associated withpistil development. KEGG pathway and GO annotation indicated that high-ranking miRNA target genes were involved in the metabolic and signal transduction pathways. This implied that the feature of the mutant might origin from the over-duplication and assembly of the related genes in the plants after the infection of the two-chain RNA virus. Our work provides a comprehensive genetic resource which can be used to better understand alfalfa pistil development and to identify candidate genes for female sterility in alfalfa.

T4

P1115

Chicory R2R3-MYB transcription factors CiMYB3 and CiMYB5 regulate fructan 1-exohydrolase expression in response to abiotic stress and hormonal cues

Hongbin Wei[1], *Hongbo Zhao*[2], *Tao Su*[3], *Thomas Rausch*[1]

1. Heidelberg University

2. South China Agricultural University

3. Nanjing Forestry University

In the biennial *Cichorium intybus*, inulin-type fructans accumulate in the taproot during the first year. Upon cold or drought exposure, fructans are degraded by fructan exohydrolases, affecting inulin yield and degree of polymerization. While stress-induced expression of *1-FEH* genes has been thoroughly explored, the transcriptional network mediating these responses has remained unknown. In this study, several R2R3-MYB transcriptional regulators were analyzed for their possible involvement in *1-FEH* regulation via transient transactivation of *1-FEH* target promoters and *in vivo* co-expression with target genes under different stress and hormone treatments. CiMYB3 and CiMYB5 selectively enhanced promoter activities of *1-FEH1*,*1-FEH2a* and *1-FEH2b* genes, without affecting promoter activities of fructosyltransferase genes. Both factors recognized the MYB-core motif (C/TNGTTA/G) abundantly present in *1-FEH* promoters. *In vivo*, *CiMYB5* displayed co-expression with its target genes in response to different abiotic stress and phytohormone treatments, whereas correlations with *CiMYB3* expression were less consistent. Oligofructan levels indicated that the metabolic response depended on the balance of the relative expression levels of fructan exohydrolases and fructosyltransferases. In summary, CiMYB5 and CiMYB3 act as positive regulators of fructan degradation in chicory.

T4

P1116

In vitro spore germination, polyembryony induction and plant regeneration of *Pteris tripartita* Sw., a critically endangered fern from south India

Baskaran Xavier Ravi

Fairy Lake Botanical Garden, Shenzhen & Chinese Academy of Sciences

An efficient protocol was developed for *in vitro* spore germination, polyembryony development and micropropagation of *Pteris tripartita* Sw. Spores were germinated after 15 days in MS medium fortified with pH, sucrose and hormones. Significant spore germination (84%) was exhibited in 70 g/L of sucrose and 79.33% in pH

5.7 of culture medium. In both pH 5.7 and 6.7, maximum size of gametophyte lengths (484.39 and 507.72 μm) and widths (846.58 and 1270.98 μm) were displayed. At 4 mg/L of GA_3, highest rate of spore germination (95%) and rhizoid formation were observed. Highest length of rhizhoid (787.69 μm) with maximum gametophyte length (307.74 μm) and width (834.36 μm) were noticed in 1 mg/L of GA_3. In our present study, Di to Octa numbers of polyembryony or juvenile sporophytes was induced in the midrib of gametophytes after 5 months culture period. The numbers of juvenile sporophytes per gametophyte, its lengths, and the total number of polyembryony or juvenile sporophytes per culture treatment were counted, and their lengths were measured after 8 months of culture period. Among five concentrations of BAP (1–5 mg/L), Di, Tri, Tetra, Hexa, and Octa numbers of polyembryony or juvenile sporophytes were observed from a single gametophyte at 3 mg/L of BAP. Of them, significant lengths (1.00,0.70,1.16,1.00,1.13 cm) were observed in Di, Tri, Tetra, Hexa, Octa polyembryony or juvenile sporophytes per gametophyte, respectively. Octa numbers of juvenile sporophytes were induced per gametophyte with 1.13 cm of length and were observed only at 3 mg/L of BAP. Three months old gametophytes were sub-cultured in MS medium supplemented with different concentrations of BAP, KIN and GA_3. Among five concentrations of BAP, highest numbers of sporophytes were induced with 4.16 cm of height and 1.51 cm of root length at 4 mg/L. In both 3 and 5 mg/L of BAP; 55 and 54 numbers of sporophytes were developed with 3.84 cm and 3.57 cm of shoot lengths along with 1.45 cm and 1.35 cm of root lengths, respectively. The MS medium supplemented with Kinetin (4 mg/L) induced sporophytes (56.03 mean numbers) after five months and reached about 2.61 cm height along with 1.39 cm length of root. Among various concentration of GA_3, highest mean numbers of juvenile sporophytes (59.66) were induced with 2.64 cm mean lengths of sporophytes along with about 2.23 cm of root length at 4 mg/L of GA_3 in MS culture medium. Finally, plantlets were transferred to small cups containing a mixture of sterilized soil, sand and vermiculite (1:1:1) in growth chambers with controlled temperature and light. Plantlets were sprayed regularly with water and covered with a porous polyethylene bag to maintain high humidity (90%–95%). The potted plantlets were kept under *in vitro* conditions and acclimatized with 90% of a survival rate. Our present protocol could be use useful for the rehabilitation of this critically endangered fern and also reduce the pressure on natural population.

T4
P1117

Functional analysis of *SecF* and *SecY* homologs in rice and maize

*Meiling Xia, Yuan Zhou, Bo Song, Yingchun Zhang, **Xiaoquan Qi***
Institute of Botany, Chinese Academy of Sciences, Beijing

A vast majority of proteins are secreted to cytoplasmic membrane via the Sec system in bacteria. The Sec system consists of the SecYEG, SecA and SecDF complex that forms an evolutionarily conserved membrane channel for protein trafficking. However, the function of Sec system in plant is largely unknown. In our previously study two genes, *HvSecF* and *HvSecY*, which control a quantitative trait locus *Rphq4* for resistance to leaf rust were cloned. The phylogenetic analysis suggested that *HvSecF* and *HvSecY* are co-evolved and only exist in Poaceae family. There are 3 pairs of homologs, *OsSecF1/OsSecY1*, *OsSecF2/OsSecY2*, and *OsSecF3/OsSecY3* in rice, and 2 pairs, *ZmSecF1/ZmSecY1*, *ZmSecF2/ZmSecY2* in maize. We are creating mutants for each gene in rice and maize by use of CRISPR/Cas9 system. So far we obtained multiple mutant lines for *OsSecF1* (2 lines), *OsSecY1* (2 lines), *OsSecF2* (4 lines), *OsSecY2* (2 lines), *OsSecF3* (2 lines) and *OsSecY3* (3 lines). The increase of plant height for ca. 20% was observed in the mutant lines of *OsSecF2*, *OsSecY2*, *OsSecF3* and *OsSecY3* comparing to the transformed plants with Cas9 empty vector. These results suggest that this Sec system may play an important role in plant development.

T4
P1118

Tissue culture and rapid propagation of *Chirita wentsaii*

*Kanghua Xian, Chuanming Fu, Fengluan Tang, Yunping Shi, Jinxiang He, **Ningzhen Huang***
Guangxi Institute of Botany, Guangxi Zhuangzu Autonomous Region and Academia Sinica

Tissue culture and rapid propagation of Chirita wentsaii were studied by using the fleshy leaves as explants. The results showed that the initial medium for the callus induction and seedling regeneration was MS + 6-BA 0.1 mg L^{-1} + NAA 0.1 mg L^{-1}. The subculture media for bud proliferation and sound seedling cultivation were MS + 6-BA 0.5 mg L^{-1}+ NAA 0.2 mg L^{-1}+ banana 10% and MS + 6-BA 0.5 mg L^{-1} + NAA 0.5 mg L^{-1} + banana 10%, and their proliferation coefficients were 6.7 and 4.8 every 60 d respectively. The optimal rooting medium was MS, and the rooting rate was 100%. The rooting plantlets were transplanted in greenhouse and the highest survival rate was about 94%.

T4
P1119

Functional anaysis of *Torenia* ALOG gene family reveals underlying regulation of petal fusion

Wei Xiao
State Key Laboratory of Biocontrol and Guangdong Key Laboratory of PlantResources, School of Life Sciences, Sun Yat-sen University, Guangzhou, China

The ALOG (*ArabidopsisLSH1* and *OryzaG1*) gene family, widely existingin angiosperm, is recruited in the regulation of flower architecture. Flower architecturecan be specialized by adopting different strategies including fusion of petals. Previous studies demonstrate divergent developmental mechanismshave been involved in the establishment of petal fusion. However, whether ALOG gene family participates in this procedure is still largely unknown. Here, we isolated seven ALOG family members from *Toreniafournieri*, and characterized the evolutionary process, subcellular localization and expressional patterns of these factors. We further explored the function of a petaltube specific gene *TfALOG1* by transgenic. We found the angiosperm ALOG genes divided into two clades, an ancestral ALOG clade and a core eudicots clade. Most of *Torenia* ALOG members were localized in nuclear region except *TfALOG7*. *TfALOG1* and *TfALOG3* were highly expressed in the boundary regions of petal tubes indicating their possible functions in petal development. Ectopic expression of *TfALOG1* exhibits abnormal flowers manifested in elongated of petal tube region together with enlarged epidermal cells, suggesting potential role of *TfALOG1* in the regulation of petal fusion. Our resultssupport the idea that ALOG family proteins widely function inflower architectureformation such as petal fusion.

T4

P1120

Probing the effects of mutation of PsbV-Y137 on photosystem II

Yanan Xiao[1], *Guangye Han*[1], *Jianren Shen*[1,2]
1. Institute of Botany, the Chinese Academy of Sciences
2. Okayama University

Photosystem II (PSII) performs light-induced water-oxidation, leading to the splitting of water into electrons, protons and oxygen. The structure of cyanobacterial PSII has been studied extensively, which showed that water-oxidation is catalyzed by a Mn_4CaO_5-cluster bound to the protein matrix of PSII. In order to perform the water-splitting efficiently, it is important to pump the protons generated at the reaction site to the outside of the protein complex. The latest high-resolution structure of PSII has revealed the presence of several well-defined H-bond networks leading from the Mn_4CaO_5 cluster to the lumenal bulk solution, which may be important for the transport of protons out of its reaction site. PsbV-Tyr137 is located in the exit of one such H-bond network starting from the D1-Try161-His190 residue pair close to the Mn_4CaO_5-cluster, and located in the interfaces between the D1, CP43, and PsbV subunits. In order to probe the function of PsbV-Y137, four mutants, PsbV-Y137A, PsbV-Y137F, PsbV-Y137G, and PsbV-Y137W, were generated with *Thermosynechococcus vulcanus* (*T. vulcanus*). Photoautotrophic growth measured at a light intensity of 40 μmol of photos $m^{-2} s^{-1}$ indicated that these mutants showed slightly slower growth rates than that of the wide type (WT) strain; however, their oxygen-evolving activity was apparently lower than that of the WT strain, and reached to about 43%, 64%, 40% and 68% of the oxygen-evolving activity of the WT strain, respectively. In addition, the thylakoid membranes isolated from PsbV-Y137A, PsbV-Y137G mutants lost the oxygen-evolving activity almost completely. On the other hand, PSII-complexes purified from the PsbV-Y137F and PsbV-Y137W mutants contained two PSII fractions, one of which lacks the three extrinsic proteins PsbO, PsbV and PsbU and showed almost no oxygen-evolving activity, whereas the other one contained the three extrinsic proteins and showed a lower oxygen-evolving activity than that of the WT-PSII. These results suggested that the PsbV-Y137 residue is required for the stable binding of the extrinsic proteins to PSII and also has a role in maintaining the oxygen-evolving activity.

T4

P1121

Unsaturation of very-long-chain ceramides protects plant from hypoxia-induced damages by modulating ethylene signaling in *Arabidopsis*

Li Juan Xie, Qin Fang Chen, ***Shi Xiao***
Sun Yat-sen University

Lipid remodeling is crucial for hypoxic tolerance in animals, whilst little is known about the hypoxia-induced lipid dynamics in plants. Here we performed a mass spectrometry-based analysis to survey the lipid profiles of *Arabidopsis* rosettes under various hypoxic conditions. We observed that hypoxia caused a significant increase in total amounts of phosphatidylserine, phosphatidic acid and oxidized lipids, but a decrease in phosphatidylcholine (PC) and phosphatidylethanolamine (PE). Particularly, significant gains in the polyunsaturated species of PC, PE and phosphatidylinositol, and losses in their saturated and mono-unsaturated species were evident during hypoxia. Moreover, hypoxia led to a remarkable elevation of ceramides and hydroxyceramides. Disruption of ceramide synthases LOH1, LOH2 and LOH3 enhanced plant sensitivity to dark submergence, but displayed more resistance to submergence under light than wild type. Consistently, levels of unsaturated very-long-chain (VLC) ceramide species (22:1,24:1 and 26:1) predominantly declined in the *loh1*, *loh2* and *loh3* mutants under dark submergence. In contrast, significant reduction of VLC ceramides in the *loh1-1 loh3-1* knockdown double mutant and lacking of VLC unsaturated ceramides in the *ads2* mutants impaired plant tolerance to both dark and light submergences. Evidence that C24:1-ceramide interacted with recombinant CTR1 protein and inhibited its kinase activity *in vitro*, enhanced ER-to-nucleus translocation of EIN2-GFP and stabilization of EIN3-GFP *in vivo*, suggests a role of ceramides in modulating CTR1-mediated ethylene signaling. The dark submergence-sensitive phenotypes of loh mutants were rescued by a ctr1-1 mutation. Thus, our findings demonstrate that unsaturation of VLC ceramides is a protective strategy for hypoxic tolerance in *Arabidopsis*.

T4

P1122

Ice active proteins in New Zealand tussocks, *Chionochloa* species

Haoyu Xiong[1], *David Wharton*[2], *Janice Lord*[3], *Craig Marshall*[1]
1. Department of Biochemistry, The University of Otago
2. School of Zoology, The University of Otago
3. Department of Botany, The University of Otago

Unlike fishes and insects that secrete cryoprotectants to depress the freezing point of body fluids, overwintering plants produce ice active agents with much lower thermal hysteresis but higher ability to inhibit ice recrystallization. Plants also minimize the damage from freezing by triggering ice formation at high sub-zero temperatures with ice nucleation agents. *Chionochloa* are tussocks mainly endemic to New Zealand and dominate the alpine grasslands. Previous studies indicated significant ice nucleation activity (INA) (~-5°C), in two alpine *Chionochloa* species, *C. macra* and *C. rigida*, but provided no more information about other ice activities in this genus. We investigated 17 of 34 *Chionochloa* species and found seasonal variations of ice activities in most collections with winter collections showed the highest activities. We show here that intrinsic proteins are essential in maintaining all three types of ice activity, which showed different responses to treatments including heat, pH, high salt, proteinase K, lysosome, reducing and oxidizing agents. Ice affinity purification indicated the presence of antifreeze proteins in *C. macra*, which was confirmed by mass spectrometry and transcriptome analysis from overwintering *C. macra*. Amino acid alignment analysis showed highly conserved ice-binding domains in the recrystallization inhibition proteins (RIPs) among *C. macra* and grass species including *Lolium perenne* and *Deschampsia antarctica*. Phylogenetic analysis of cold stress related genes, IAPs, C-repeat binding factors and fructosyltransferase, in *C. macra* and other grass species indicated these genes from *C. macra* evolved differently to the core *Pooideae* family and genes from *Brachypodium distachyon*. Protein expression in *E. coli* confirmed ice activities of codon optimized ice active genes from *C. macra*.

T4

P1123

Comparative transcriptome analysis of axillary buds in response to the shoot branching regulators gibberellin and cytokinin in the perennial woody plant *Jatropha curcas*

Jun Ni[1,2], *Mei-Li Zhao*[1,3], *Mao-Sheng Chen*[1], *Bang-Zhen Pan* [1], *Yan-Bin Tao* [1], ***Zengfu***[1] ***Xu***

1. Key Laboratory of Tropical Plant Resources and Sustainable Use, Xishuangbanna Tropical Botanical Garden, Chinese Academy of Sciences

2. Institute of Technical Biology & Agriculture Engineering, Hefei Institutes of Physical Science, Chinese Academy of Sciences

3. College of Life Sciences, University of Chinese Academy of Sciences, Beijing

Cytokinin (CK) is the primary hormone that positively regulates axillary bud outgrowth. However, in many woody plants, such as *Jatropha curcas*, gibberellin (GA) also promotes shoot branching. The molecular mechanisms underlying GA and CK interaction in the regulation of bud outgrowth in *Jatropha* remain unclear. To determine how axillary buds respond to phytohormones GA and CK, we performed a comparative transcriptome analysis of the axillary buds of *Jatropha* seedlings treated with GA_3 or benzyladenine (BA, a synthetic CK). Multiple GA_3- or BA-responsive genes were identified, and 255 genes were shown to be co-regulated, with similar expression profiles, in response to GA_3 or BA. Seven *NAM/ATAF1/CUC2* (*NAC*) family members were down-regulated after treatment with both GA_3 and BA, whereas these genes were up-regulated after treatment with the shoot branching inhibitor strigolactone. Furthermore, the bud dormancy-related genes *Dormancy-Associated Protein1* (*DRM1*), *DRM1-LIKE* and *BELL1* were negatively regulated by GA_3 or BA. Additionally, the expressions of the cell cycle genes *Cell Division Control 6* (*CDC6*), *Cell Division Control 45* (*CDC45*) and *Growth Regulation Factor 5* (*GRF5*) were up-regulated after treatment with both GA_3 and BA, suggesting that these hormones may promote bud outgrowth, in part, via regulation of the cell cycle machinery. Overall, the comprehensive transcriptome data set provides novel insight into the responses of the axillary buds to GA and CK.

T4

P1124

The maize chromatin remodeler ZmCHB101 plays an essential role in osmotic stress response via both transcriptional regulation and alternative splicing

Xiaoming Yu, Xinchao Meng, Yutong Liu, Bao Liu, ***Zheng-Yi Xu***

Key Laboratory of Molecular Epigenetics of the Ministry of Education (MOE), Northeast Normal University

Drought and osmotic stresses are recurring and severely constraint crop production. Accumulated evidence in *Arabidopsis thaliana* suggests that components of the SWI/SNF chromatin remodeling complexes play essential roles in abiotic stress responses. Here we show that *ZmCHB101*, a homologue of *A. thaliana SWI3D* in maize, plays an essential role in osmotic and dehydration stress responses. Transcriptome profiling revealed that ZmCHB101 modulates transcriptional expression landscape of osmotic stress responsive genes. Co-immunoprecipitation assay showed that ZmCHB101 associates with RNA polymerase II (RNAPII) *in vivo*, which might be a prerequisite for the proper occupancy of RNAPII on the promoter and proximal regions of transcription start sites of stress response genes. In addition, ZmCHB101 plays a role in the occurrence of alternative splicing of stress response genes under normal and osmotic stress conditions, probably mediated by alternation of chromatin states and changes in transcriptional elongation rates. However, the regulatory roles of ZmCHB101 in gene expression and alternative splicing are largely uncoupled, pointing to diverse molecular functions of ZmCHB101 in transcriptional and posttranscriptional regulations of gene expression. Taken together, our findings have unraveled novel regulatory roles of ZmCHB101 in affecting gene expression and alternative splicing by remodeling chromatin states and directly associating with RNAPII.

T4

P1125

Phytoplasma infection altered photosynthetic and sugar metabolism in Chinese jujube (*Ziziphus jujuba* Mill.)

Chaoling Xue[1], *Zhiguo Liu*[2], *Jiaodi Bu* [1], *Zihui Jiang* [1], *Mengjun Liu*[2], *Li Dai* [2], *Jin Zhao**[1]

1. College of Life Science, Hebei Agricultural University

2. Research Center of Chinese Jujube, Hebei Agricultural University

Phytoplasma is a class of cell-wall-free prokaryotic organisms that are parasitized in plant phloem tissues. Its genome is lack of the genes required to complete their own metabolic energy, but with well-developed sugar transporters to absorb nutrients from the host. Jujube witches' broom disease (JWB) caused by phytoplasma is a destructive disease of Chinese jujube. Its typical symptoms include yellow leaves, phyllody and witches' broom or condensed branch with tiny leaves. The primary symptom of jujube leaves caused by phytoplasma infection is yellowing, meaning that its photosynthesis is significantly affected. To further reveal the pathogenic mechanism, the influence of JWB phytoplasma on photosynthetic and sugar metabolism of jujube trees were systematically studied at the transcriptional, physiological and structure levels. Transcriptome data revealed that the expression of the genes related to starch, sucrose and glucose synthesis (the *1,5-diphosphate ribulose carboxylase* (*RuDPCase*), *phosphoenolpyruvate enzyme* (*PEPC*), *sucrose phosphate synthase* (*SPS*), *sucrose synthase* (*SuSy*), *invertase* (*INV*), *glucose-1-phosphate adenylyltransferase* (*GPAT*), *starch synthase* (*SS*) and *granule-bound starch synthase II-1* (*GSBBII-1*)) in diseased leaves were significantly increased, suggesting that the sugar metabolism in jujube was enhanced by phytoplasma infection. Accorrdingly, the higher contents of starch, sucrose and glucose in diseased leaves were examined by starch kit and HPLC, respectively, which were consistent with the transcriptome data. Furtherly, the result of transmission electron microscope (TEM) showed that the number of starch granules in mesophyll cell of diseased leaves increased, the grana and stroma lamellar of chloroplast were destroyed structurally and formed the original lamellae. The content of chlorophyll, PSII activity and the expression of photosynthesis-related genes in diseased leaves were also significantly lower than that in healthy ones. In addition, the result of TEM found that the P protein agglutinated and callose deposited in the sieve wall of diseased young stem, implying that the transport of nutrients was inhibited in diseased trees. On the whole, phytoplasma infection disturbed the operation and formation of photosynthate of jujube trees, caused the disorder of sugar metabolism and particularly stimulated the abnormal accumulation of starch in chloroplast. The increasing of sugar content is benefitical to phytoplasma proliferation temporarily, however the excessive damage in structure and metabolism of host jujube lead to the exhaustion of jujube.

T4

P1126

Tudor-SN, a component of stress granules, regulates the growth under salt stress by modulating GA20ox3 mRNA levels in *Arabidopsis*

Chunxia Yan

College of Life Sciences, Inner Mongolia University

The Tudor-SN protein (TSN) is universally expressed and highly conserved in eukaryotes. In *Arabidopsis*, TSN is reportedly involved in stress adaptation, but the mechanism involved in this adaptation is not understood. Here, we provide evidence that TSN regulates the mRNA levels of *GA20ox3*, a key enzyme for gibberellin (GA) biosynthesis. The levels of *GA20ox3* transcripts decreased in TSN1/TSN2 RNA interference (RNAi) transgenic lines and increased in TSN1 over-expression (OE) transgenic lines. The TSN1 OE lines displayed phenotypes that may be attributed to the overproduction of GA. No obvious defects were observed in the RNAi transgenic lines under normal conditions, but, under salt stress conditions, these lines displayed slower growth than wild-type (WT) plants. Furthermore, two mutants of *GA20ox3*, *ga20ox3-1* and *ga20ox3-2*, also showed slower growth under stress than WT plants. Moreover, a higher accumulation of *GA20ox3* transcripts was observed under salt stress. The results of a Western blot analysis indicated that higher levels of TSN1 accumulated after salt treatment than under normal conditions. Subcellular localization studies showed that TSN1 was uniformly distributed in the cytoplasm under normal conditions but accumulated in small granules and co-localized with RBP47, a marker protein for stress granules (SGs), in response to salt stress. The results of RNA immunoprecipitation experiments indicated that TSN1 bound *GA20ox3* mRNA *in vivo*. On the basis of these findings, we conclude that TSN is a novel component of plant SGs that regulates growth under salt stress by modulating levels of *GA20ox3* mRNA.

T4

P1127

Type B ARRs play an integrative role in the transcriptional regulation of *TAA1* in response to developmental and environmental cues

Zhenwei Yan[1], *Xin Liu*[1], *Karin Ljung*[2], *Shuning Li*[1], *Wanying Zhao*[1], *Meiling Wang*[1], *Yi Tao*[1]

1. School of Life Sciences, Xiamen University, Xiamen Plant Genetics Key Laboratory

2. Swedish University of Agricultural Sciences, SE-90183, Umeå

Auxin and cytokinin are key hormones required for plant development. During embryogenesis and organ formation, establishing proper gradient is critical for auxin function, which is achieved through coordinated regulation of both auxin metabolism and transport. Expression of auxin biosynthetic genes is often tissue specific and is regulated by environmental signals. However, the underlying regulatory mechanisms remain elusive. Here, we investigated the transcriptional regulation of a key auxin biosynthetic gene, *TAA1* (L-Tryptophan aminotransferase of *Arabidopsis* 1). Two cis-elements, required for tissue specific expression of *TAA1*, were identified in the promoter and the second intron of *TAA1*. C-terminal of ARR1, a type B ARRs (*Arabidopsis* response regulators), directly binds to both cis-elements and activates the expression of *TAA1*. Mutation in ARR1,10 and 12 not only affects cytokinin induced expression of *TAA1* but also reduces auxin levels, suggesting multiple type B ARRs participate in the transcriptional regulation of *TAA1*. We further demonstrated that the ARRs also mediate the transcriptional regulation of *TAA1* by SCARECROW and light, and DELLA proteins interact with ARR1 and promote its transcriptional activity. We propose that the ARR proteins act as key transcriptional regulators of *TAA1* by integrating hormonal, environmental and developmental signals.

T4

P1128

Pollen tube guidance in fowering plants: The interplay between male and female gametophytes

Weicai Yang, *Hongju Li, Tong Wang, Mengxia Zhang*

Institute of Genetics and Developmental Biology, University of Chinese Academy of Sciences, Chinese Academy of Sciences

Novel reproductive structures and mechanisms have evolved to adapt to terrestrial land environment in plants during evolution. In flowering plants, such evolutionary development is manifested by the flower, multicellular gametophyte, double fertilization, loss of sperm motility, and siphonogamy in which the immotile sperm was delivered to the egg by a pollen tube produced by the male gametophyte (pollen), a process named pollen tube guidance (PTG). Previous studies suggested that PTG requires the intimate interactions between the pollen tube and maternal tissue of the pistil and the female gametophyte respectively. Although signaling molecules, such as LURE1 peptide produced by the synergid cell of the female gameptophyte (embryo sac) to attract and guide the pollen tube growth, were identified (for review see Higashiyama and Takeuchi, *Annu Rev Plant Biol*, 2015), how the pollen tube recognizes and responds to the guiding molecule is not yet clear. Through genetic screen, we isolated a number of *Arabidopsis* mutants that disrupt PTG processes. *CCG*, a central cell-specifically expressed gene, is required for the female gametophyte to attract the pollen tube. *CCG* encodes a nuclear protein that regulates the expression of a number genes important for PTG via CBP1 which interacts with RNA polymerase II, the Mediator complex and AGL transcription factors in the central cells and also *LURE1* expression in the synergid cells indirectly (Chen *et al.*, Plant Cell, 2007; Li *et al.*, Plant Cell, 2015). *POD1*, a pollen tube-expressed gene, is required for the male gametophyte to respond to the female signals. *POD1* encodes a ER protein that interact specifically with CRT3 which is implicated to control the foldin gof Leucine-Rich Repeat Receptor-Like Kinases (LRR-RLKs). These data suggest that POD1 might play a role in the protein folding of putative receptor proteins. Recently, we identified the male MDIS/MDIK receptor complex that recognizes the female attracting signals (Wang *et al.*, Nature, 2016). At the same time, another LRR-RLK PRK6 has also been reported to be LURE1 receptor in *Arabidopsis* (Takeuchi and Higashiyama, *Nature*, 2016). These findings provide novel insight to mechanims controlling PTG and will be discussed.

T4

P1129

Competition between branch components of *Haematococcus lacustris* for control of algal carotenoid biosynthesis

Lao Yong Min[1], *Jin Hui*[2], *Zhou Jin*[1], *Zhang Huai Jin*[2], *Cai Zhong Hua*[1]

1. The Division of Ocean Science and Technology, Graduate School at Shenzhen, Tsinghua University

2. School of Life Sciences, Tsinghua University

Cyclization of acyclic lycopene by cyclases marks an important regulatroy point in carotenoid biosynthesis. Though some algal lycopen epsilon cyclases (LCYEs) have been predicted computationally, very few have been functionally identified. Little is known about the regulatory mechanisms of algal LCYEs. Recent comparative genomic analysis suggested that Haematococcus lacustris contained only the beta type cyclase (HlLCYB). However, in this study, carotenoid profiling found trance alpha-carotene in the salt-treated cells, indicating the *in vivo* activity of HlLCYE, a missing component for alpha-branch carotenoids. Thus, genes coding for HlLCYB and HlLCYE were isolated and functionally complemented in *Escherichia coli*. Expression pattern studies, comparative promoter analysis, and site-directed mutagenesis assays showed that both cyclases were differentially regulated by the regulatory *cis*-acting elements in promoters and the key regulatory residues in proteins to correlate with primary and secondary carotenoid biosynthesis under environmental stresses. Substrate specificity assays revealed an exclusive cyclization order of HlLCYE to HlLCYB for the biosynthesis of heterocyclic carotenoids. Identification of the branch components in algal carotenoid biosynthesis revealed a mechanism for control of metabolic fulx into alpha- and beta-branch by the competition and cooperation between HlLCYE and HlLCYB, and supplied a promising route for molecular breeding of cyclic carotenoid biosynthesis.

T4

P1130

Selection and application of DNA barcodes for wood identification of *Dalbergia* species: A case study on vouched samples of wood xylarium

Min Yu, *Lichao Jiao, Juan Guo, Tuo He, Xiaomei Jiang, Yafang Yin*

Chinese Academy of Forestry

The increasing in illegal logging and timber trade of CITES-listed tropical species necessitates the development of unambiguous identification methods at species level. DNA barcoding has been promoted as a promising tool for species identification. However, the main barrier and obstacle of extensive application of DNA barcoding is the lack of a comprehensive and reliable DNA reference library. Vouched wood xylarium samples are potentially an enormous and reliable resource of materials. In this study, the wood specimen of nine species were selected from *Dalbergia* genus stored in the Wood Collection of the Chinese Academy of Forestry and DNA was then extracted from them for further PCR amplification. Eight potential DNA barcode sequences and their different combinations were used to test discrimination ability by three methods, i.e., the character based, tree-based and TAXONDNA method. It was shown that the discrimination ability of DNA barcodes combination was higher than single DNA marker at the species level. The present results demonstrate the applicability of the DNA barcoding to species identification from wood xylarium samples.

T4

P1131

Effects of exogenous hydrogen peroxide on flower bud differentiation in tobacco

Yuanyuan Li, Yingying Huang, Shixin Zhu, Jinyong Huang, Zhen Wei, Jian Du, ***Caipeng Yue***

School of Life Sciences, Zhengzhou University

A preliminary study on reactive oxygen species (ROS) metabolism and flower bud differentiation in tobacco was completed through pot experiment, by spraying exogenous hydrogen peroxide at different time points. The results showed that: (1) Spraying H_2O_2 significantly changed the ROS metabolism and launched an endogenous protective enzyme system in tobacco within 24 hours. H_2O_2 content rapidly increased in tobacco after spraying H_2O_2. Sequentially, the O_2^- content and MDA content rose as well. The activity of POD enzyme was quickly activated in tobacco. Later the activity of CAT was also activated. The CAT activity was slightly weaker than POD enzyme at beginning stage, then it increased much greater than POD activity. The response of SOD enzyme to H_2O_2 was later than the POD and CAT. In mid and late stage, O_2^- content remained at a high level, but H_2O_2 content significantly decreased. (2) In long term response to H_2O_2 (measured at every five days till 30 days), exogenous H_2O_2 made ROS metabolism into an unbalanced state in tobacco, relatively higher content of O_2^- and lower content of H_2O_2. This unbalanced state influenced the transition of the growth and development of tobacco apical bud. The anatomical observation of tobacco apical bud and the number of leaves during flowering stage showed that the floral development was earlier than control when spraying 5 d-25 d of

H_2O_2. The leaf number at flower budding stage was related to the changes of H_2O_2 content in tobacco. Therefore, we assumed that the changes of ROS metabolic balance affect the flowering transition of tobacco and promote flower bud differentiation, to a certain extent. This article only investigated the relationship between reactive oxygen metabolism and tobacco floral development from a physiological aspect. The molecular mechanisms and signaling pathways, which reactive oxygen species regulate flowering, will be further explored.

T4
P1132
The *Arabidopsis MUR4* contributes to cell wall strength and salt tolerance

Omar Zayed[1], *Chunzhao Zhao*[1,2], *Jian-Kang Zhu*[1,2]
1. Purdue University
2. Shanghai Center for Plant Stress Biology

Salinity is one of the widely-studied plant abiotic stresses that substantially limits crop yield and production. Salt affects plant growth and development by different manners including reduction of plant cell wall flexibility. Maintaining cell wall elasticity and integrity is an important criterion for enhancing salt stress tolerance. Presence of wall arabinose-containing polymers had been suggested to be the key structural component responsible of the unique abiotic stress tolerance characteristic of plants. Screening for *Arabidopsis* mutants, defective in growing in sodium chloride medium, was applied to identify genetic loci involved in salt tolerance. One of these mutants, *mur4* showed hypersensitivity to salt, with around 50% reduction in cell wall arabinose content. *MUR4* encodes a UDP-xylose-4-epimerase, which catalyze the conversion of UDP-xylose to UDP-arabinose, the main pathway of arabinose biosynthesis in plant. It was shown that arabinose-containing polymers confer different contribution to plant salt tolerance. Arabinan and extensin are considered to be the major cell wall structures required for salt tolerance. In addition to the reduction of arabinose and arabinose-containing structures in the cell wall, reactive oxygen species (ROS) and hydrogen peroxide H_2O_2 were shown to be over-accumulated in *mur4* plants after salt treatment compared to wild type. Our data suggest that *MUR4* play a key role in salt tolerance by decreasing ROS in plants grown under salinity conditions. The increment of H_2O_2 in mur4 after salt treatment, was coincide with downregulation of cell wall peroxidases genes along with Xyloglucan endotransglucosylase/hydrolase (XTH) and Expansins (EXP) genes, that may cause reduction in root expansion. PLAT (Polycystin, Lipoxygenase, Alpha-toxin and Triacylglycerol lipase) are plant-stress proteins that are found to promote salt tolerance. Nevertheless, little information is known regarding their specific functions and regulation in plant species. It is suggested that *PLAT* regulate salt tolerance by affecting the catalytic activity and substrate specificity of some membrane proteins. However, the mechanism and the associated proteins are still unknown. Our data showed that both *MUR4* and *PLAT1*, which are induced by salt stress, positively regulate salt tolerance. Using Co-immunoprecipitation, Split-LUC and yeast two hybrid, we confirmed that *MUR4* directly interacts with *PLAT1*. Altogether, our data suggest that arabinose content is an important factor in salt tolerance and *PLAT1* is involved in the regulation of *MUR4* activity.

T4
P1133
Formation of a characteristic volatile constituent in tea (*Camellia sinensis*) leaves exposed to multiple stresses

Lanting Zeng[1,2], *Ying Zhou*[1], *Xiumin Fu*[1], *Xin Mei*[1], *Ziying Yang*[1]
1. South China Botanical Garden, Chinese Academy of Sciences
2. University of Chinese Academy of Sciences

Indole not only serves as an important signal substance and precursor of indole-3-acetic acid (IAA), but also is a characteristic volatile constituent in tea (*Camellia sinensis*), especially oolong tea, a famous semi-fermented tea. Our previous study indicated that indole was mostly accumulated at the turn over stage of oolong tea manufacturing process. However, formation of indole in tea leaves remains unknown. In this study, one tryptophan synthase α-subunit (TSA) and three tryptophan synthase β-subunits (TSBs) from tea leaves were isolated, cloned, sequenced, and functionally characterized. Combination of *CsTSA* and *CsTSB*2 recombinant protein produced in *Escherichia coli* exhibited the ability of transformation from indole-3-glycerol phosphate to indole. *CsTSB*2 was highly expressed during the turn over process of oolong tea. Continuous mechanical damage, simulating the turn over process, significantly enhanced the expression level of *CsTSB2* and amount of indole. These suggested that accumulation of indole in oolong tea was due to the activation of *CsTSB*2 by continuous wounding stress from the turn over process. Black teas (fully fermented tea) contain much less indole, although wounding stress is also involved in the manufacturing process. Stable isotope labeling indicated that tea leaf cell disruption from the rolling process of black tea did not lead to the conversion of indole, but terminated the synthesis of indole. Our study provided evidences concerning formation of indole in tea leaves for the first time.

T4
P1134
Physiological responses of the inoculation seedlings of *Dalbergia odorifera* to drought stress

Hong-Xing Zhan, *Xue-Gang Zhou*, *Xue-Ying Zhuang*, *Yong-Bin Wu*
South China Agricultural University

The impacts of drought stress on the physiological characteristics of the mycorrhizal seedlings of *Dalbergia odorifera* were studied with inoculation growth experiment. The results showed that variations of the leaf relative conductivities and MDA contents of the seedlings with inoculated mycorrhizal fungi were less than the non-inoculated seedlings (CK), but their proline contents and soluble protein contents were significantly higher than CK under drought stress. The variations in chlorophyll contents among the inoculation treatment and CK were not significant. The inoculation with the endophytic fungi could enhance the seedling drought adaptability of *Dalbergia odorifera* than the non-inoculated seedlings (CK). The drought tolerances of the seedlings inoculated with *Glomus margarita* was higher than those inoculated with *Glomus etunicatum*.

T4

P1135

OsRACK1A, a circadian clock-regulated WD40 protein, is a negative regulator of salt-stress adaptation in rice

Dongping Zhang[1], *Yuzhu Wang*[1], *Jiansheng Liang*[1,2]

1. Yangzhou University

2. Southern University of Science and Technology

The receptor for activated C kinase 1 (RACK1) is a WD40 type protein that is involved in multiple signaling pathways and is strongly conserved from prokaryotes to eukaryotes. Here, we showed that rice *RACK1A* (*OsRACK1A*) is regulated by circadian clocks and plays an important role in the salt stress response. *OsRACK1A* was found to follow a rhythmic expression profile under circadian conditions at both the transcription and translation levels, although the expression was arrhythmic under salt stress. Analysis of plant survival rates, fresh weight, proline content, malondialdehyde, and chlorophyll showed that suppression of *OsRACK1A* enhanced tolerance to salt stress. The ion concentration in both roots and leaves revealed that *OsRACK1A*-suppressed transgenic rice could maintain low Na^+ concentrations and high K^+. Furthermore, *OsRACK1A*-suppressed transgenic rice accumulated significantly more abscisic acid (ABA) and more transcripts of ABA- and stress-inducible genes compared with the wild-type plants. Transcriptome analysis revealed that many stress-response genes were up-regulated in the *OsRACK1A*-suppressed transgenic rice line, including AP2/ERFs transcription factors. Real-time PCR analysis confirmed that many stress-related genes were also up-regulated in the *OsRACK1A*-suppressed transgenic rice line. These results suggest that *OsRACK1A* is involved in the regulation of circadian rhythm, ABA, and salt stress responses.

T4

P1136

A rice heterotrimeric G protein γ subunit positively regulates development of endosperm cells and starch accumulation of grains

Dongping Zhang[1], *Yuzhu Wang*[1], *Bing Lv*[1], *Jiansheng Liang*[1,2]

1. Yangzhou University

2. Southern University of Science and Technology

Heterotrimeric G protein-mediated signal transduction is one of the most important and highly conserved signaling pathways in eukaryotes, which involves in the regulation of many important biological processes. As an atypical Gγ protein, rice qPE9-1 controls grain size and yield in rice, but its molecular mechanism remains elusive. Here, we confirm the interaction between qPE9-1 and rice G protein β subunit RGB1, and find that qPE9-1 is mainly localized to the plasma membrane. Wuyunjing 8 (WYJ8), which contain qPE9-1 allele (qpe9-1), have smaller grain size, less endosperm cells and lower starch content than its qPE9-1-covering transgenic line WYJ8 (qPE9-1). In WYJ8 (qPE9-1), the content of zeatin riboside (ZR), a kind of cytokine, is more than those in WYJ8 at the early stage of grain development. Furthermore, the expression levels of genes encoding grain filling-related enzymes are significantly higher in WYJ8 (qPE9-1) than those of WYJ8 at the late stage of grain filling. Contents of two phytohormone, abscisic acid (ABA) and indole-3-acetic acid (IAA), which were reported to involve in grain filling regulation, are also significantly higher in the WYJ8 (qPE9-1) grains than those in WYJ8 at the late stage of grain filling. In addition, both exogenous ABA and IAA treatments enhance the starch accumulation and expression of genes encoding grain filling-related enzymes in the grains of WYJ8 (qPE9-1), whereas only IAA has the same effect in WYJ8. Based on these results, we conclude that, as a G protein γ subunit, qPE9-1 promotes endosperm cells proliferation through cytokine (ZR) at the early stage of grain development, and also positively regulates starch accumulate largely through ABA and IAA, which enhance the expression of genes encoding several starch biosynthesis at the late stage of grain filling.

T4

P1137

Two tonoplast MATE proteins function as turgor-regulating chloride channels in *Arabidopsis*

Haiwen Zhang[1,2,3], *Legong Li*[2], *Sheng Luan*[4]

1. Beijing Academy of Agriculture and Forestry Sciences

2. College of Life Sciences, Capital Normal University

3. School of Life Sciences, Nanjing University, Nanjing

4. Department of Plant and Microbial Biology, University of California, Berkeley

Turgor pressure is the driving force for cell growth in plants and the large central vacuole provides the major space for turgor regulation. The vacuolar membrane (tonoplast) contains a large number of transporters that mediate fluxes of solutes and water thereby adjusting cell turgor in response to developmental and environmental signals. However, the molecular identity and function of many tonoplast transporters remain unknown. Here, we showed that two tonoplast Detoxification efflux carriers (DTX)/ Multidrug and Toxic Compound Extrusion (MATE) transporters, DTX33 and DTX35, function as chloride channels essential for turgor regulation in *Arabidopsis*. Ectopic expression of each transporter in *Nicotiana benthamiana* mesophyll cells elicited a large voltage-dependent inward chloride current across the tonoplast, demonstrating that DTX33 and DTX35 each constitute a functional channel. Both channels are highly expressed in *Arabidopsis* tissues including root hairs and guard cells that experience rapid turgor changes during root hair elongation and stomatal movements. Disruption of these two genes, either in single or double mutants, resulted in shorter root hairs and smaller stomatal aperture with double mutants showing more severe defects, suggesting that these two channels function additively to facilitate anion influx into the vacuole during cell expansion. In addition, *dtx35* single mutant showed lower fertility as a result of defect in pollen tube growth. Indeed, patch-clamp recording of isolated vacuoles indicated that the inward chloride channel activity across the tonoplast was impaired in the double mutant. As MATE proteins are widely known transporters of organic compounds, finding MATE members as chloride channels expands the functional definition of this large family of transporters and opens new doors to understanding their physiological roles in plants.

T4

P1138

Expansion and functional divergence of flowering locus T

genes in monocots

Liangsheng Zhang, *Xing Liu*
Center for Genomics and Biotechnology, Haixia Institute of Science and Technology, Fujian Agriculture and Forestry University

Flowering locus T (*FT*) gene is crucial in regulating flowering time. How it evolved and contributes to flowering diversity remain elusive. In this study, through phylogenetic analysis of *FT* genes in all angiosperm clades, we show that few *FT* genes exist in eudicots and more *FT* genes are encoded in monocots, especially in grasses. Furthermore, monocot *FT* genes can be grouped into three clades: M1, M2 and M3. M1 clades are conseverd and under purifying slection, contains mostly florigen functioning genes, showing the M1 clade is the ancestor of FT genes in monocots .For M2 and M3, many gene duplication events have contributed significantly to the FT copy number increase in monocots. Compare with M1 clades, we identified 6 and 8 positive selection sites in M2 and M3 clades, respectively, suggesting the functional divergence of these genes. Gene expression analysis reveals diverse patterns for different *FT* genes across tissues, further supporting functional divergence of these genes. Our study provides a global picture of *FT* gene evolution in angiosperm, particularly in monocots.

T4
P1139

Influence of drought stress and semi shade environment on health of 10 ground cover seedling species during roof greening

Peng-Qian Zhang, *Jia-De Bai, Yan-Ju Liu, Zheng Yang, Tian Liu*
The Department of Ecology Research, Beijing Milu Ecological Research Center

Roof greening, the "Fifth surface greening of city", can mitigate the urban heat island effect, improve air quality and enrich urban biodiversity. Beijing, a typical northern super-city in China, has to develop the roof green industry to improve the environment. Due to the climate limitations in Beijing, the plants applied to the roof green are suffered with restricted rainfall in the winter, spring and some times in the summer. According to the "Beijing local standards-roof greening specification guidelines", there are 10 recommended ground cover seedling species applied in the roof green, including *Paeonia lactiflora*, *Hemerocallis dumortieri*, *Physostegia virginiana*, *Iris lacteal*, *Hylotelephium erythrostictum*, *Sedum lineare*, *Iris germanica*, *Cosmos bipinnata*, *Hosta plantaginea* and *Dianthus barbatus*. This research is trying to select the plant species with strong drought resistance through drought stress test. This research sets 3 different drought stress degrees by controlling the soil relative water content (RWC), including moderately drought stress (40% ± 2% < RWC < 45% ± 2%), strong drought stress (RWC < 30% ± 2%) and control group (RWC > 75% ± 2%). When the seedlings survived the stress test after 5 days, the damaging rate of permeability (DRP), total chlorophylls concentrations (Chl), Superoxide Dismutase (SOD), Peroxidase (POD) and Ascorbate Peroxidase (APX) of seedlings will be measured. Finally, a subordinate function method was applied to assess these species' drought resistance. The result shows the species of *Cosmos bipinnata* and *Physostegia virginiana* was dead after having suffered with moderately drought stress and strong drought stress, respectively. Although other species survived, the individual variation was huge especially for physiological and biochemical index. The species *Hemerocallis dumortieri*, *Iris lactea* and *Hosta plantaginea's* DRP had little change when they lived in the normal water condition and suffered with drought stress. The semi-shaded condition reflects the change of Chl obviously. 6 species' Chl increased when they suffered with moderately drought stress compared with the control group. This may indicate these plant seedlings have strong adaptation to poor light. The change of SOD activity is sensitive to drought stress. The most of the species (except *Paeonia lactiflora* and *Sedum lineare*) showed a lower SOD activity during moderately drought stress compared with the sufficient soil water condition and strong drought stress condition. The changes of plants' POD activity and APX activity are very similar: when drought stress enhanced, the activity of protect enzyme reduced. According to the subordinate function method, the order of plants' resistance to the moderately drought is as follow: *Iris germanica* > *Hemerocallis dumortieri* > *Hosta plantaginea* > *Iris lactea* > *Sedum lineare* > *Dianthus barbatus* > *Hylotelephium erythrostictum* > *Physostegia virginiana* > *Paeonia lactiflora*. And the order of plants' resistance to the strong drought is as follow: *Hosta plantaginea* > *Sedum lineare* > *Iris germanica* > *Hylotelephium erythrostictum* > *Paeonia lactiflora* > *Dianthus barbatus* > *Hemerocallis dumortieri* > Iris lacteal > *Ph*ysostegia virginiana = *Cosmos bipinnata*.

T4
P1140

Genome-wide characterization of the MADS-Box gene family and their contribution to organ development in Carnation (*Dianthus caryophyllus* L.)

Xiaoni Zhang, *Dandan Li, Xiaopeng Fu*
Key Lab of Horticultural Plant Biology (MOE), College of Horticulture & Forestry Sciences, Huazhong Agricultural University

The MADS-box gene family is key regulator in flower development, especially in flowering time regulation and floral organogenesis formation. The systematic identification and characterization of MADS-box gene family has been reported in many plant species such as *Arabidopsis thaliana*, *Oryza sativa*, Orchids, and *etc.*, while it has not been reported in Carnation (*Dianthus caryophyllus* L.). In this study, 39 MADS-box genes were identified from the carnation genome and transcriptome data by the phylogenetic analysis between carnation and *Arabidopsis thaliana*, which were categorized into four subgroups (30 MIKC^c, 2 MIKC*, 2 Mα, 5 Mγ). Meanwhile, the MADS-box domain, gene structure and conserved motif compositions of the carnation MADS genes were analyzed. The MADS-box domain is highly conserved in TypeII class, while duplicated in Type I . Additionally, the expression of DcaMADS genes were systematically investigated in stems, leaves, flower buds and individual flower organs. Some crucial DcaMADS genes correlated with its putative function were validated. The subcellular localization of the 16 DcaMADS proteins in each subclass was investigated. This report is the first to describe genome-wide characterization of the MADS-box gene family in Carnation, and these results could provide valuable information to explore the potential functions of these subfamily.

T4

P1141

Biochemical elucidation of the tailing processes for isoflavonoid biosynthesis in kuduzu plant

***Yansheng Zhang**, Jia Li, Xin Wang, Changfu Li*
Wuhan Botanical Garden, Chinese Academy of Sciences

As a traditional Chinese medicine, kuduzu root is used for the treatment of cardiovascular diseases and diabetes for a long history in China and other Asia countries as well. The active compounds attributable to those pharmacological activities of *P. lobata* are believed to be isoflavonoids in its root. Previous phytochemical studies revealed that most isoflavonoids in kuduzu roots are derived from daidzein, which is glucosylated at C-7, 8, and 4' positions while the O-methylation takes place at its C-4' and C-3' positions. Despite of important benefits to human health, the knowledge regarding how these daidzein derivatives are formed in kuduzu roots is lacking at present. By the combinatory usage of molecular biological and biochemical techniques, we report here the biochemical mechanism of the tailing processes for the biosynthesis of these unique isoflavonoids in kuduzu roots. This report helps people better understand the biosynthetic pathways to isoflavonoid production in kuduzu plant, and opens a window for a large-scale production of these unique isoflavonoid compounds in microorganisms using synthetic biology techniques in the future.

T4

P1142

Mean germination time of Sudangrass seed lots (*Sorghum sudanense*) predicts seed vigor and germination performance in soil

***Ye Zhang**, Huifang Yan, Hang Cheng, Peisheng Mao*
China Agricultural University

With the development of animal husbandry, the quality of Sudangrass (*Sorghum sudanense*) seed is becoming increasingly important. But research shows that lab routine germination rate test can not accurate represent germination performance in soil . It is essential to develop a new method to assess the vigor of Sudangrass seed. Previous work on maize and other crops has showed that measurements of mean germination time (MGT; 2mm radicle) is a possible alternative assessment of vigor. Five lots of Sudangrass seeds and two methods (germination in rolled paper test and emergence test simulating field condition) were used in this experiment. The objective of this study was to estimate that weather mean just germination time (MJGT) and mean germination time (MGT) in rolled-paper test could be used to predict the seed vigor of Sudangrass by measuring the related indicators of germination. In the rolled paper germination test of five lots, the result showed that MJGT and MGT were significant negatively correlated with the rate of seed germination, the average length of seedlings and emergence rate at both 18°C and 20-30°C, but were positively related to mean emergence time.It is concluded that the MGT can be used as a reliable method to predict the seed vigor of Sudangrass seed. Meanwhile, the method is time saving and environmental protection.

T4

P1143

Middle Triassic Linjia flora from eastern Liaoning, China and its paleoclimatic implication

***Yi Zhang**[1], Shao-Lin Zheng[1], Yong-Dong Wang[2]*
1. College of Paleontology, Shenyang Noraml University; Key Laboratory for Evolution of Past Life in Northeast Asia, MLRC
2. Nanjing Institute of Geology and Palaeontology, Chinese Academy of Sciences

The Middle Triassic flora of eastern Liaoning is represented by fossil plants collected from the Linjia Formation of the Linjiawaizi in Benxi of eastern Liaoning, China. The Linjia flora is very spectacular because its components are complicated, representing the only one well-known fossil flora of the Middle Triassic age in NE China. Both typical Paleozoic elements, such as *Lobatannularia, Taeniopteris, Pterophyllum, Psygmophyllum, Pecopteris lativenosa, P. orientalis, P. candolleana* and *Cordaites* as well as typical Mesozoic ones, such as *Pleuromeia, Neocalamites, Danaeopsis, Symopteris, Scytophyllum, Thinnfeldia, Sinozamites, Yabeiella, Sphenobaiera* and *Podozamites* are documented in this flora. Among them, *Lobatannularia, Pleuromeia, Scytophyllum* and *Sinozamites* are typical elements from the Northern Hemisphere. The flora also includes some interesting taxa, for example, *Benxiopteris*, similar to the typical element of the Southern Hemisphere, *Dicroidium* in its basic leaf structure, but different in reproductive organs; *Yabeiella*, as a typical element from the Southern Hemisphere; *Glossopteris*-like plants very like *Glossopteris* in its leaf morphology but their reproductive organs unclear. The flora also includes components which are difficult to be assigned, such as *Symopteris*, which is very similar to the Late Paleozoic element, *Fascipteris*. In this flora, two specimens with their counterparts preserved, probably represent the *Gigantopteris*. In one of these two specimens, scars possibly representing the areas of ovule or seed attachment are also preserved. Cathaysian elements, e.g., *Lobatannularia, Taeniopteris, Gigantopteris* and *Pecopteris lativenosa* mixed with typical Mesozoic components in this flora indicate that the distribution area of the Linjia flora probably represents the known final extinction area of the relict Cathaysian elements in the world. Because the distribution area of the flora was located in the northeastern North China Plate in Middle Triassic, which was controlled by the special warm and humid coast climate, it is reasonable for the relict Cathaysian elements to survive in that time.

T4

P1144

Growth and antioxidant enzyme activities of *Hemarthria altissima* and *Cynodon dactylon* in the water-level-fluctuating zone responses to water depth gradient

***Zhiyong Zhang**[1,2], Chengyan Wan[1]*
1. Institute of Hydroecology, Ministry of Water Resources and Chinese Academy of Sciences
2. College of Resource and Environment, Huazhong Agricultural University

Water-level- fluctuating zone (WLFZ) is most common and important in regulated rivers. The Three Gorges Project reached the normal water level of 175 m on October 26, 2010, which resulted in the formation of the WLFZ (up to 30m in elevation) with exposed in summer and submerged in winter. The original vegetation, especially trees and shrubs, could not survive because they could not tolerate the winter submergence and summer drought, while herb was the most abundant life form. Vegetation in the WLFZ zone was dominated by perennial herbs such as *Hemarthria altissima* and *Cynodon dactylon*. Therefore, it is important to examine the ecophysiological adaption mechanisms the dominant species in submergence duration for vegetation restoration in this reservoir. To investigate clone growth and antioxidant enzyme activities of dominant plants in the WLF of the Three Gorges Reservoir, height, cover, sprout number, biomass, content of total protein and Malondiadehyde (MDA), enzyme activity of superoxide dismutase (SOD) and peroxidase (POD) of root system of *H. altissima* and *C. dactylon* were detected at four water depths (0,2, 5 and 15 m) in situ experiment from December in 2014 to May in 2015. Both plants were cultivated in plastic pots hanged on a floating platform. The correlation coefficients of among total protein and antioxidant enzyme activities of root system were also analyzed. The results showed that height, cover, sprout number, biomass, total protein, enzyme activity of SOD, content of MDA of root system declined compared with non-flooded (0 m). However, the activity of POD of root system increased. Height, cover, sprout number, the content of total protein and the activity of SOD decreased with increasing water depths. However, the activity of POD increased significantly. The results of regression analysis showed that significant positive correlation was observed between the content of MDA and the activity of SOD. Submergence may play a role in correlation coefficients of between total protein and antioxidant enzyme activities. The coefficients among SOD and POD, MDA and SOD of root system were lowest at the water depth of 15 m. The coefficients of among enzymes of *C. dactylon* were higher than those of *H. altissima*.

T4

P1145

Genome-wide identification, characterization of the MADS-Box gene family in Chinese jujube and their involvement in flower development

Liman Zhang[1], ***Jin Zhao***[1], *Weilin Gao*[1], *Mengjun Liu*[2], *Jiurui Wang*[3]

1. College of Life Science, Hebei Agricultural University

2. Research Center of Chinese Jujube, Hebei Agricultural University

3. College of Forestry, Hebei Agricultural University

MADS-box genes encode transcription factors that are involved in plant development control (particularly in floral organogenesis) and signal transduction pathways, though a comprehensive analysis of MADS-box family proteins in Chinese jujube (*Ziziphus jujuba* Mill.) is still missing. Here, we report a genome-wide analysis of the MADS-box gene family in Chinese jujube. Based on phylogenetic analyses, 52 jujube MADS-box genes were classified into 25 MIKC^{C}-type, 3 MIKC*-type, 16 Mα, 5 Mβ and 3 Mγ genes. 37 genes were randomly distributed across all 12 putative chromosomes. We found that the type II genes are more complex than the type I genes and that tandem duplications have occurred in three groups of MADS-box genes. Meanwhile, some gene pairs in the same clade displayed similar or distinct expression profiles, suggesting possible functional redundancy or divergence. MIKC^{C}-type genes exhibited typical temporal and spatial expression patterns in the four whorls of floral tissues. The expressions of B, C/D and E-type genes were significantly suppressed in phyllody as compared to flower, providing valuable evidence for their involvement in flower development. This study is the first comprehensive analysis of the MADS-box family in jujube, and provides valuable information for elucidating molecular regulation mechanism of jujube flower development.

T4

P1146

Effect of shading manipulations and growth regulators on shell differentiation in walnut (*Juglans regia*)

Shugang Zhao[1], *Hongxia Wang*[2], *Jing Wen*[2], *Shuang Wang*[2], *Zhihua Zhang*[2]

1. College of Life Sciences, Hebei Agricultural University, Baoding, Hebei

2. Mountain Area Research Institute, Hebei Agricultural University, Baoding, Hebei

Walnut (*Juglans regia* L.) shell, is often overlooked, plays an important role in nut development, transportation, washing and storage. In particular, kernel quality has close relations to shell seal grade and thickness. Walnut shell structure is influenced by cultivars, illumination intensity and harvest period, effective mechanism is not well understood. In this study, the morphologic characteristics and composition of nutshell were studied with applications of different plant growth regulators and two levels of shading. Result show that mature shell is composed of three parts, sclereid layer, sclerenchyma cell layer and fiber-like cell layer, from outside to inside respectively. 30% and 70% shading treatment caused shell thickness to thin and number of sclereid layer to reduce, content of lignin and polyphenol to decrease, number of fiber-like cell layer to increase. In addition, 70% shading treatment significantly caused shell seal grade to reduce. 50mg/L GA_3 treatment dramatically caused shell thickness and number of sclereid layer to increase. 100mg/L paclobutrazol treatment significantly caused shell thickness and number of sclereid layer to thin, content of lignin and polyphenol and seal grade to reduce. From these data light and plant regulators are important affected the nutshell structure.

T4

P1147

The impact of inoculation with mycorrhizal fungi *in vitro* on growth and resistant enzymes of *Rhododendron moulmainense*

Yujie Zhou, *Jiuxiang Huang*

College of Forestry and landscape Architecture, South China Agricultural University

We conducted a study of *Rhododendron moulmainense* inoculation with mycorrhizal fungi to select the better strains, and to provide the reference for improving breeding efficiency, also to reveal the mechanism of improving plant resistance by mycorrhizal fungi. 7 kinds ericoid mycorrhizal fungis, separated from wild *Rhododendron moulmainense*, were used to inoculate *Rhododendron moul-*

mainense tissue culture seedling. Growth situation was observed regularly, and height growth rate, biomass growth rate, mycorrhizal infection rate, disease index, PAL and LOX activity were determined. The result showed that: (1) Plants of control group were wizened after 2 weeks, but plants of mycorrhizal inoculation treatment were viable. (2) Height growth rates and biomass growth rates of NO.7 treatment were the highest, 71.81% and 83.62% respectively, NO.17 and NO.14 treatment fllowed. Mycorrhizal infection rates of NO.16 treatment were the highest, NO.1 treatment were the lowest. Disease indexes of NO.7 were the lowest, NO.5 and NO.3 were the highest. PAL and LOX activity of each mycorrhizal inoculation treatment all higher than control group. PAL and LOX activity of NO.16 treatment were the highest, 4.82 U/mg and 0.88 U/mg respectively, PAL activity of NO.1, NO.5, NO.17 were the lowest, 1.90~2.20 U/mg, LOX activity of NO.1, NO.3 were the lowest, 0.29~0.38 U/mg. The correlation analysis showed that, colonization intensity were positively correlated with PAL and LOX activity significantly, but disease index were negatively correlated with PAL and LOX activity. Each of the 7 kinds ericoid mycorrhizal fungi could form mycorrhizal symbionts with *Rhododendron moulmainense*, but different fungis had different impact on growth of host plants, the result showed that, NO.7 (*Bionectria ochroleuca*) was the optimal fungi, NO.14 (*Aspergillus sydowii*) and NO.17 (*Paecilomyces javanicus*) were the second choices. Higher mycorrhizal infection rate help to improve PAL and LOX activity. It may have an impact on resistant and growth of plants, but resistant and growth of plants were more impacted by species of fungi.

T4

P1148

Regulation of β-carotene biosynthesis and accumulation in maize endosperm

Changfu Zhu[1,2], *Chao Bai*[2], *Xin Jin*[2], *Judit Berman*[2], *Gemma Farre*[2], *Teresa Capell*[2], *Paul Christou*[2,3]

1. Changchun Normal University

2. University of Lleida-Agrotecnio Center

3. Catalan Institute for Research and Advanced Studies

Over the last two decades, many carotenogenic genes have been cloned and used to generate plants accumulating higher levels of carotenoids. However, comparatively little is known about the regulation of endogenous carotenogenic gene expression in higher plants, and this limits our ability to predict how engineered plants will perform in terms of carotenoid content and composition. Maize (*Zea mays* L.) is one of the most widely grown crops in the world, and is an important source of human food, animal feed and bioenergy. Maize β-carotene hydroxylase 2 (BCH2 also termed crtRB1, or HYD3) governs critical steps in the carotenoid biosynthetic pathway in endosperm. *ZmBCH2* alleles associated with reduced transcript expression correlate with higher β-carotene accumulation. Here we report that *ZmBCH2* is preferentially transcribed in maize endosperm and is regulated developmentally, further confirming the important role of ZmBCH2 in carotenoid biosynthesis in the endosperm. Moreover silencing of maize *BCH2* resulted in increased β-carotene content in the endosperm in different maize genetic backgrounds. In order to understand the regulatory mechanisms controlling *ZmBCH2* expression, the full-length *ZmBCH2* promoter and a series of its 5′-deletions were fused to the β-glucuronidase (*GUS*) reporter gene and transferred into rice (*Oryza sativa*) to test the promoter activities. The results showed that the 2,036 bp upstream region from the translational start site drove GUS expression preferentially in the endosperm, but not in leaves of transgenic rice plants which is consistent with the *ZmBCH2* expression pattern in maize. We also demonstrated that the 5′-untranslated region containing the first intron in *ZmBCH2* plays a critical role in terms of conferring high levels of expression in endosperm. We found that the 774 bp 5′-flanking region upstream from the translation start site contained all the necessary *cis*-regulatory elements responsible for higher GUS expression in transgenic rice endosperm. This basal promoter region of 774 bp 5′-flanking sequence of *ZmBCH2* contained a single prolamin box and one AACA motif which are specifically recognized and bound by endosperm-specific prolamin-box binding factor (PBF) and GAMYB transcription factors, respectively. The expression patterns of isolated *ZmPBF* and *ZmGAMYB* genes were similar to that of *ZmBCH2* which is preferentially transcribed in maize endosperm and is regulated developmentally. We are currently in the process of determining whether ZmPBF and ZmGAMYB act individually or synergistically to regulate seed-specific *ZmBCH2* expression.

T4

P1149

Physiological and molecular mechanisms mediating xylem Na^+ loading in barley in the context of salinity stress tolerance

Min Zhu, *Meixue Zhou, Lana Shabala, Sergey Shabala*

School of Land and Food, University of Tasmania

Control of xylem sodium loading is crucial to salinity stress tolerance in plants. In this study we screened a large number of barley genotypes for their ability to exclude Na^+ for its loading into the xylem and delivery to the shoot. Using four genotypes contrasting in salinity stress tolerance, we then studied kinetics of xylem Na^+ and K^+ loading at multiple time points and linked it with overall plant performance and sequestration of Na^+ and K^+ in leaf tissues. Our results showed that both leaf and xylem K^+/Na^+ ratios had positively correlated with overall plant salt tolerance after prolonged (5 weeks) exposures to salinity stress. Interestingly, it was Na^+ but not K^+ content that determined this correlation. In shorter terms, accumulation of Na^+ in the xylem sap of salt-tolerant genotypes TX9425 and CM72 was higher than in salt-sensitive genotypes Naso Nijo and Gairdner. It reached its peak 5 days after salt application and then declined. Salt-sensitive varieties were less efficient in controlling xylem Na^+ loading and showed a gradual increase in the xylem Na^+ content over the time. The MIFE (microelectrode ion flux measurement) technique was then applied to study some aspects of salt stress signalling and Na^+ loading into the xylem, through measuring net fluxes of Ca^{2+}, K^+ and Na^+ from xylem parenchyma tissue of control- and salt-grown plants in response to two known second messengers H_2O_2, and ABA. Our results indicates that NADPH oxidase-mediated apoplastic H_2O_2 production acts upstream of xylem Na^+ loading and is causally related to ROS-inducible Ca^{2+} uptake systems in the xylem parenchyma tissue. It was also found that ABA regulate (directly or indirectly) the process of Na^+ retrieval from the xylem. Transcript levels of *HvHKT1;5_like* and *HvSOS1_like* genes in root stele was observed to decrease after salt stress while there was an increase in *HvSKOR_like* gene, indicating that these ion transporters are involved in primary Na^+/K^+ movement into/out of xylem.

T4

P1150

Probing the efects of mutation of *PsbO-D158* in photosystem II

Qingjun Zhu[1], *Yanyan Yang*[1], *Guangye Han*[1], *Jian-Ren Shen*[1,2]
1. Institute of Botany, the Chinese Academy of Sciences
2. Okayama University

Photosystem II (PSII) is a large membrane-bound protein complex containing 20 subunits, which performs light-induced charge separation and a series of electron transfer reactions that lead to water oxidation and plastoquinone reduction. Mn_4CaO_5 cluster is the catalytic heart where water oxidation takes place. The protons generated by water oxidation need be transferred to the lumenal side of thylakoid membrane, in order to form trans-membrane proton gradient for the synthesis of ATP. It is generally accepted that protons are transported by hydrogen bond networks formed by water molecules and amino acid residues. Based on the high-resolution structure of PSII, *PsbO-D158* participates in one of the hydrogen bond networks starting from the Mn_4CaO_5 cluster. In order to study the role of this site in proton channel, three mutants were generated where *PsbO-D158* was changed to K, N or E, respectively. The photoautotrophic growth of these three mutants was found to be roughly the same, but slower than that of the wild type. The oxygen evolving activity of three mutant cells was approximately 35% of the oxygen evolving activity of wild type. However, the isolated thylakoid membrane of three mutants showed no oxygen evolving activity, which was found to be due to the loss of peripheral proteins during the isolation process. These results suggest that *psbO-D158* may play an important role in maintaining the binding of *PsbO* with the PSII complex.

T4

P1151

Evaluating antioxidant activity and some secondary methabolits in wild almond (*Amygdalus scoparia L,*)

Lale Ziaebrahimi, *Somayeh Fariabi*
Islamic Azad University, Kerman Branch

The purpose of this study was to evaluate the antioxidant activity, total phenol and flavonoid content in methanol extracts of fruits and flowers of three Almond subspecies. The content of total phenol and flavonoid pigments was measured with colorimetric method. Antioxidant capacity of plant extracts were evaluated by DPPH method. In the present study the highest antioxidant activity measured in Archen. IC50 values in the methanol extract of three sub-species of bitter almond, sweet almond and Archn was 126,115 and 85 μg/ml respectively that in comparison with BHT, indicating a relatively high antioxidant capacity. The highest level of Phenol and flavonoid was in Archen, then in Sweet Almond and Bitter almonds Respectively. The results indicate that green exocarp of the three subspecies have higher antioxidant effect than the flowers.

T4

P1152

Advancing high-throughput visualization of leaf venation networks using new 2D and 3D X-ray imaging approaches

Julio V. Schneider[1,2], *Renate Rabenstein*[3], *Jens Wesenberg*[4], *Karsten Wesche*[4,5,6], ***Georg Zizka***[1,2], *Jörg Habersetzer*[3]
1. Senckenberg Research Institute and Natural History Museum Frankfurt, Department Botany and Molecular evolution
2. Goethe University Frankfurt/Main, Department Diversity and Evolution of Higher Plants
3. Senckenberg Research Institute and Natural History Museum Frankfurt, Department of Paleoanthropology and Messel Research
4. Senckenberg Museum of Natural History Görlitz, Department of Botany
5. German Centre for Integrative Biodiversity Research (iDiv) Halle-Jena-Leipzig
6. International Institute Zittau, Technische Universität Dresden

Leaf venation traits are important for many research fields such as systematics and evolutionary biology, plant physiology, climate change, and paleoecology. In spite of an increasing demand of vein trait data, studies are often data-delimited because traditional destructive methods for imaging vein networks are slow. Recently, non-destructive X-ray technology proved useful as an alternative. However, the utility of rapid 2D X-ray was partially limited because image resolution was often insufficient to visualize the smallest veins. Here, we advance X-ray technology to increase image resolution and thereby the utility of high-throughput visualization of leaf venation together with new standards for very high resolution imaging using microfocus X-ray computed tomography. For 2D microradiography, we developed a new method, which allowed us to achieve an image resolution of 7 μm, i.e. almost a 4-fold increase compared to the industrial standard. In the majority of the samples, there was a significant improvement in estimates of minor vein density even without applying a contrasting agent. In X-ray computed tomography, our approach outcompeted traditional imaging of chemically cleared venation in terms of vein resolution. In conclusion, our study demonstrates that the here established 2D X-ray imaging approach is a significant advancement for high-throughput visualization of leaf venation networks. Additionally, with our methodological advancement, microfocus X-ray tomography becomes a valuable alternative to traditional methods of very-high resolution imaging of leaf venation.

T4

P1402

Interaction of microfilaments with microtubules mediated by AtFH14

Jiaojiao Wang, Xiuhua Xue, Xiangfei Wang, Bailing Hu, ***Haiyun Ren***
Key Laboratory of Cell Proliferation and Regulation Biology of Ministry of Education, College of Life Science, Beijing Normal University

Arabidopsis formin 14 (AtFH14) decorates preprophase bands, spindles, and phragmoplasts, and may function as a linking protein between microtubules and microfilaments. Here, we compare the binding activity of AtFH14FH1FH2 to actin filaments (AFs) and microtubules (MTs), and quantitative analyses reveal that the binding affinity of AtFH14FH1FH2 to microtubules is 20-folds higher than to actin filaments. However, AtFH14FH1FH2 cannot interact with monomeric tubulin and the microtubule-binding AtFH14 FH1FH2 can bind to the side of actin filaments and inhibits the nucleation, suggesting that the microtubule-binding AtFH14FH-1FH2 cannot interact with the end of actin filament. Furthermore, we employ TIRFM and single-molecule imaging technology to find that AtFH14 FH1FH2 can recruit AFs to microtubules, and the AFs can slide on microtubules physically randomly. The quantitative assays indicate that the elongation and sliding rates of AFs bind to microtubules are decreased by AtFH14FH1FH2. AtFH-14FH1FH2-RFP and microfilament /microtubules fluorescence marker are transiently co-expressed in onion epidermal cells. The results indicate that AtFH14 FH1FH2-RFP co-localize with cortical microtubules marked by MAP4-GFP, that is identify with the in vitro result of higher ability to bind with microtubules. The binding of FH1FH2-RFP to microtubules were revealed to be dynamic by fluorescence recovery after photobleaching experiment. Time-lapse confocal microscopy showed that FH1FH2-RFP could display a dynamics similar to the microtubule dynamic instability. Pharmacological assays on the double fluorescence labeled cells with latrunculin B indicate that reorganization of microfilaments is easier to bind with microtubules marked by FH1FH2-RFP than MAP4-GFP. Our results suggest that AtFH14 function as a cross-linker between microfilaments and microtubules in regulating dynamics of actin and microtubule cytoskeletons by recruiting actin filament to microtubules via binding the side of actin filament in plant cells.

T4

P1405

Studies of the mechanisms that AtFH14 bundles microfilaments and microtubules

Zhanhong Ren, Yunqiu He, Shanwei Li, Xiuhua Xue, ***Haiyun Ren***

Key Laboratory of Cell Proliferation and Regulation Biology of Ministry of Education, College of Life Science, Beijing Normal University, Beijing, China

Microtubles and microfilaments are two major cytoskeletal networks found in plant cells. Both of them play important roles in many aspects of fundamental processes of plant cell growth and development, such as cell division, cell expansion, intracellular organization and motility. AtFH14, a type II formin from *Arabidopsis thaliana*, was found to bundle and crosslink microtubule and microfilament arrays, these processes of which are vital to plant cell growth and morphogenes.However, the mechanism by which AtFH14 bundle microtubule and microfilaments in plant is still not well understood. In this study, we found that AtFH14FH2 contains three highly conserved basic amino acids which are considered to interact with the negative charge surface of microfilaments and microtubules in mDia1. Compared with AtFH14, the bundling ability of the three sites mutagenesis protein (Lys845Ala, Arg850Ala and Lys855Ala in AtFH14) is weaken, which indicates that these three amino acids have an effect on the bundling ability of AtFH14, but may not act as critical roles in actin filament bundling. Then we purified AtFH14FH2 △ N which was acquired via protease digestion of α-chymotrypsin. Although the post subdomain of AtFH14FH2 △ N contains Lys845, Arg850 and Lys855 which may constitute a binding interface with other proteins such as F-actin, it lacks the amino terminal. And the result of high and low speed co-precipitation experiments indicate that AtFH-14FH2 △ N neither bind nor bundle microfilament and the amino terminal of AtFH14 FH2 plays an important role in actin filament bundling. In summary, we propose that the amino terminal (1-93) of FH2 play a critical roles in the binding and bundling of AtFH14 to microfilaments and microtubules.

T4

P1406

Characterization of a putative microtubule-associated protein of *Arabidopsis thaliana*

Pingzhou Du, Lu Deng, Lu Liu, Xiuhua Xue, ***Haiyun Ren***

Key Laboratory of Cell Proliferation and Regulation Biology of Ministry of Education, College of Life Science, Beijing Normal University

Highly dynamic microtubules, regulated by many microtubule-associated proteins (MAPs), play acritical role in many important physiological processes, such as cell morphogenesis, cell expansion and cell division. We derived a possible microtubule-associated protein 96 (MAP96) from *Arabidopsis thaliana*. Our preliminary results showed that AtMAP96-GFP localized in a transverse filamentous network manner at the pericellular, distributed radially around the nucleus, or simultaneously had the former two forms of distribution in stable transgenic BY-2 cells in interphase. During mitosis, the localization of AtMAP96 changed in consistent with specific microtubule or microfilament arrays, such as the preprophase band, the spindle apparatus, phragmoplasts. To verify whether AtMAP96 is co-localized with microtubule or actin filaments, pharmacological experiments were applied. After treatment with a microtubule depolymerizing drug, oryzalin, we found that AtMAP96-GFP filamentous network became disorder as thedepolymerization of intracellular microtubules. While AtMAP96-GFP filamentous network was stillcomparative integrity treated with the latrunculin B, a microfilament depolymerizing drug. In addition, immunofluorescence results showed that AtMAP96 and microtubules were co-localized in different stages during the cell cycle progression. These data suggest that AtMAP96 co-localizemainly with microtubules in living plant cells and it may regulate the dynamics of microtubule cytoskeleton, thereby participating in cell growth and division.

T4

P1417

Microscopic study of chloroplast development in the albino mutant *bzs20* in *Arabidopsis*

Wing Shing Wong, *Na Li, Zheze Zhao, King Shing Wong, Junxian He*

School of Life Sciences, State Key Laboratory of Agrobiotechnology, The Chinese University of Hong Kong, Sha Tin, New Territories, Hong Kong SAR, China

Chloroplasts are essential organelles which dedicate to undergo photosynthesis to provide energy for plant survival and development. In the early stage of seed development, chloroplasts already exist in embryo. During embryo maturation, photosynthetic active chloroplasts dedifferentiate to etioplasts which are non-photosynthetic and colorless because of loss of pigments (chlorophylls and carotenoids). While upon seed germination, light exposure converts the etioplasts into chloroplasts which contain activated chlorophylls required for photosynthesis. Normal biogenesis and development of chloroplasts ensure plants to obtain optimal amount of energy form photosynthesis. Previous studies in our laboratory have discovered a novel albino mutant *bzs20* in *Arabidopsis thaliana*. *bzs20* is a recessive T-DNA insertion mutant which is seedling lethal and shows albino cotyledons and pale green true leaves. Thus, *bzs20* mutation might have caused some defects in chloroplast development or synthesis of the green pigment chlorophyll. However, the detailed mechanisms causing these defects remain unknown. Therefore, the aim of my thesis project is to perform further detailed phenotypic studies to the *bzs20* mutant, particularly to the chloroplast ultrastructure during seedling development. To this end, different approaches and methodologies have been applied to illustrate the chloroplast development defects in the *bzs20* mutant plant. Ultrastructural analysis with TEM reveals that the mutant is destructive in thylakoid network and lacking starch grains in chloroplast. Phenotypes of *bzs20* embryos during embryogenesis have been observed with light microscopy. However, no significant morphological difference was observed between *bzs20* and wild-type embryos, indicating that the albino phenotype is not formed during embryogenesis. Therefore, we will test whether the albino phenotype is formed during seedlings process or during or after seed maturation processes. These structural studies of chloroplast development in the novel albino mutant *bzs20* will help understanding the function of *BZS20* gene in regulating the normal development of chloroplasts in plants. This work was supported by the GRF grants (14121915,14148916) and an AoE grant (AoE/M-05/12) from the Research Grant Council of Hong Kong, the Shenzhen Science & Technology Research & Development Funding - Peacock Scheme, and the Direct Grants from The Chinese University of Hong Kong.

T4

P1422

An *Arabidopsis* mutant deficiency in sphingolipid desaturase is used as a tool to investigate function of plasmodesmata

Ning-Jing Liu[1], *Pei-Shuang Chen*[1], ***Yong-Mei Qin***[1], *Yu-Xian Zhu*[1,2]

1. The State Key Laboratory of Protein and Plant Gene Research, College of Life Sciences, Peking University

2. Institute for Advanced Studies, College of Life Sciences, Wuhan University

Sphingolipdis along with sterols are essential constituents of plasma membrane located raft-like domains, detergent-resistant membranes (DRMs) suggested by the presence of glycosylphospatidylinositol (GPI)–anchored proteins and receptor-like kinases. DRMs contribute to the sorting, trafficking and targeting of specific plasma membrane proteins, which are involved a range of cellular processes. Plasmodesmata (PD) mediate intercellular transport of nutrients, RNA, metabolites, plant viruses, and plays vital roles in plant developmental events and cooperative responses to environmental stimuli. A link between DRM and PD has been suggested based on the identification of the first DRM marker protein, Remorin. Proteomic approaches identified Remorin and GPI-anchored proteins on PD, suggesting the presence of plasmodesmata DRMs. Here, we showed that *Arabidopsis* double mutant *sld1/sld2* lacking sphingolipid long chain-base delta 8 desaturase (SLD), exhibiting leaf yellowing and spontaneous cell death. A fast increase of free phytosphinganine (t18:0) in the leaves was detected after the plants were challenged by fungal wilt pathogen *Verticillium dahlia* or bacterium *Peudomonas syringae* pv *tomato*. Exogenous application of t18:0 promoted the wild-type plants to enhance their resistance against *V. dahliae* or *Pst* DC3000. In consistent with this observation, the *sld1/sld2* plants accumulated in t18:0 displayed enhanced resistance to these pathogens. Further, we extracted and purified PD-enriched membrane fractions. Lipid profiles indicated that t18:0-based glycosylinositolphosphoceramides were enriched in the *sld1/sld2* compared with the wild-type. The transcripts of the genes encoding callose synthase significantly accumulated, which correlated with the increased callose synthase activity in the *sld1/sld2*. Results of Aniline blue staining and dye-loading indicated that the *sld1/sld2* plants increased callose deposition and lowered PD trafficking activity. Our data suggests that phytosphinganine acts as a key component of complex sphingolipids for regulation of PD structure and function during immune-responses to pathogen infections in *Arabidopsis* plants.

T4

P1423

Ultrastructure of plastids in secretory cells in relation to secretion composition in *Cleome* (*Cleomaceae*)

Alexandra Ivanova[1,2], ***Nuria Koteyeva***[1], *Elena Voznesenskaya*[1]

1. Komarov Botanical Ibstitute

2. St. Petersburg State University

Plastids and endoplasmic reticulum in secretory cells are the main sites of secondary metabolite secretion biosynthesis. The structure of plastids is very multifarious in secretory cells; it was concluded to be related to secretion composition. Ultrastructure, secretion composition and development of glandular trichomes was studied in 15 *Cleome* species to expand our knowledge of plastid functioning and ontogeny. Glandular trichomes of four main types were distinguished in *Cleome*. In the most studied species glandular trichomes produce more or less hydrophilic secretion that consists of terpenoids, carbohydrates, phenylpropanoids. Secretion accumulates in subcuticular space in C. *viscosa*, C. *paradoxa* and in *C. droserifolia* large trichomes. In other species and trichome types secretion the secretion trickles down the trichome to the plant surface. It is well known that plastids in secretory cells producing monoterpenes are leucoplasts with undeveloped thylakoid system. In contrast, developed grana are contained in plastids of secretory cells in *Cleome* species except *C. viscosa* that produce hydrophobic oleoresin. Analogous to *Artemisia annua*, lower cell layer in the glandular head of *Cleome* trichomes contain chloroplasts, upper layers contain leucoplast-like plastids. Endoplasmic reticulum (ER) is abundant, smooth ER prevails usually except trichomes located over veins on the leaves of *C. monophylla*, *C. gynandra* and *C. viridiflora* where granular ER is developed. Plastids in *Cleome* secretory cells have multiple and/or long contacts with smooth ER elements that is characteristic to resin producing cells. Opaque granules within reticular cisternae are thought to be terpenoids on their way via ER to periplasmic space and than to subcuticular space. The sections of plastids in secreting cells are multiple, they are ameboid in the most species. Sections in the same cell are so different in *C. droserifolia* that they seem to be plastids of different types, chloroplasts and leucoplasts. But serial sections and reconstruction of plastids provide evidences that plastids in secretory cells are large branched structures; in *C. droserifolia* they contain different compartments. Number of plastids is significantly less than in progenitor tissues and moreover in the secretory cells at presecretory stage. Plastid degradation by autophagy during maturation of secretory cells was shown in glandular trichomes of *C. africana* and *C. hassleriana*. In conclusion, glandular trichomes in *Cleome* species as well as and plastids in their secretory cells have the common characters with terpenoid secreting ones and special features (like chloroplast in secretory cells) that are interesting for the future research. The study of plastid morphology in closely related species may be useful to reveal the correlations between specific characters of ultrastructure and the synthesis of certain groups of metabolites in secretion. The research was partly supported by core facility "Molecular and cell technologies" of St. Petersburg State University.

T4

P1426

A putative MYB-like transcription factor involved in secondary cell wall formation in *Arabidopsis*

Chunming Wang, *Meihui Yu, Junxian He*

School of Life Sciences and State Key Laboratory of Agribiotechnology, The Chinese University of Hong Kong. Shatin, New Territorries, Hong Kong SAR

Secondary cell walls (SCWs) are important for plants to support their growth and adapt to changing environments. SCWs are composed mainly of three biopolymers, namely lignin, cellulose and hemicellulose. Studies have suggested that the biosynthesis of these biopolymers is regulated by a complex transcriptional network. However, the details of this transcriptional network have not been fully uncovered. From an activation tagging based genetic mutant screen in *Arabidopsis*, we identified a new mutant, *rif1* (*Respressor of Interfascicular Fibers 1*), which has severe defects in secondary wall formation in stems, making the plant cannot stand erectly. Microscopy study indicated that the SCW deposition in interfascicular fibers of *rif1* is completely missing, and SCW components assay showed that the levels of the three biopolymers are all significantly decreased in the *rif1* mutant. *RIF1* encodes a MYB like transcription factor in *Arabidopsis* which has not been characterized before. RIF1 is localized in nucleus and has putative transcriptional regulatory activity. From a protoplast transcription transient assay, we demonstrate that RIF1 acts as a transcriptional repressor and negatively regulate several important SCW biosynthesis genes. Our results suggest that RIF1 may define a novel negative transcription regulator in SCW formation in plants. This work was supported by the GRF grants (14121915 and 14148916) and the AoE grants (AoE/M-05/12, AoE/M-403/16) from the Research Grant Council of Hong Kong, the Shenzhen Science & Technology Research & Development Funding - Peacock Scheme, and the Direct Grants from The Chinese University of Hong Kong.

T4

P1433

Calli of Tartary buckwheat as a unique *in vitro* plant system to study the stability of nuclear genomes

Alexander Betekhtin[1], *Magdalena Rojek*[1], *Joanna Jaskowiak*[1], *Anna Milewska-Handel*[4], *Jolanta Kwasniewska*[1], *Yulia Kostyukova*[2], *Ewa Kurczynska*[4], *Natalya Rumyantseva*[2,3], *Robert Hasterok*[1]

1. Department of Plant Anatomy and Cytology, Faculty of Biology and Environmental Protection, University of Silesia in Katowice

2. Kazan Institute of Biochemistry and Biophysics of Kazan Science Centre of the Russian Academy of Sciences, Laboratory of Physiology and Genetics of Plant Cell Cultures

3. Kazan Federal University, Institute of Fundamental Medicine and Biology, Department of Botany and Plant Physiology

4. Department of Cell Biology, Faculty of Biology and Environmental Protection, University of Silesia in Katowice

The data that was obtained during the cultivation of tissue cultures on different plants demonstrated an increase in genetic variability. The morphogenic calli of *Fagopyrum tataricum* (Tartary buckwheat), which together with non-morphogenic clones make a unique system for studying the mechanisms that are responsible for this genetic instability/stability, are an interesting exclusion from this rule. The morphogenic calli of Tartary buckwheat are able to preserve genetic stability and high morphogenic potential through ten years of cultivation. This feature appears to be unique in the plant kingdom. Tartary buckwheat is a wild species in Europe and is treated as a weed. However, it is widely used as a food crop in Asia. Tartary buckwheat is a source of rutin and other bioflavonoids and these compounds are powerful antioxidants that can be used to prevent cardiovascular diseases and cancer. In this work, we used histological sections and transmission electron microscopy to identify and describe the micro- and ultrastructure of the nuclei in morphogenic and non-morphogenic callus lines. We demonstrated that the embryogenic callus cells had prominent round nuclei in contrast to the non-morphogenic callus lines, in which we found multilobed nuclei. Flow cytometry analysis revealed significant differences in the relative DNA content between the analysed calli. The results of flow cytometry showed that the non-morphogenic calli are of an aneuploid nature. The TUNEL test enabled us to visualise the nuclei that had DNA fragmentation in both the morphogenic and non-morphogenic lines. We revealed significantly higher frequencies of positively labelled nuclei in the non-morphogenic callus lines than in the morphogenic ones. These results clearly indicate greater genome stability in the morphogenic callus lines.

T4

P1443

Effect of ethyl methanesulphonate on broad bean (*Vicia faba* L.)

Moumita Basu

University of Gour Banga

Broad bean or Fava bean is widely used in the Mediterranean region as source of protein in both human and animal nutrition. Broad bean (*Vicia faba* L.), referred to as broad bean, horse bean or field bean – is a major food and feed legume because of the high nutritional value of its seeds. But modern agronomic demand is of high yielding cultivars of the crop. There are certain limitations of producing high yielding cultivars through traditional hybridization method. Mutation breeding is such a technology which can fulfill the mission by overcoming all sorts of agronomic hazards. Now-a-days the mutation breeding is no doubt neglected to some extent avoiding laborious field experimentations and monumental skills. Not only this, it is a chance factor to achieve a desirable promising line. In present study, attempt has been made to ascertain the magnitude of induced genetic variability followed by mutagenesis with EMS (Ethyl methanesulphonate) and also magnitude of associations between yield and its component characters in M1 generation. The broad bean varieties differed in response to the mutagenic treatments regarding quantitative traits, viz; plant height at 50% flowering phase(cm),days to 50% flowering phase, total number of branches per plant, length of leaf(cm), total number of pods per plant, length of pod(cm), total number of seeds per pod, plant height at maturity(cm),100 seeds weight(g), yield per plant(g). Analysis of variance (ANOVA) showed highly significant differences among the genotypes for almost all the characters studied, there by suggesting the presences of considerable amount of variability among the six genotypes of broad bean under study. It is evident that there were very little differences in GCV and PCV for length of leaf(cm), length of pod(cm), total number of seeds per pod, plant height at maturity(cm), 100 seeds weight(g),- yield per plant(g) among the character studies. This indicates that these characters changed a little due to effect of the environmental factors, while other characters showed higher degree of environmental influences. Higher GCV values in major characters like total number of pods per plant, yield per plant(g) proved the existence of justifiable genetic variation among the genotypes under study. High heritability combined with high genetic advanced over mean was observed for 100 seeds weight (g), yield per plant(g), length of leaf(cm) showing additive gene effect. Therefore, it is logical that selection based on the phenotypic performance on those traits would be effective. The inter relationship among the nine characters were studied in the population of six genotype (to control and four M1 lines) through 'correlation co-efficient analysis'.The results revealed that the positive correlations of one-character pair e.g, 100 seeds weight(g) vs. plant height at maturity(cm), indicates that increasing one character would result in increase in other character. Thus, the results of co-heritability estimates and evaluations of genotypic and phenotypic correlations speak the same language as for as important characteristics

T5
P1153

Transcriptome and chloroplast genome sequencing decipher the origin and diversification of allotetraploid *Brassica napus*

***Hong An*[1]**, *Xinshuai Qi*[2], *Michael Barker*[2], *J. Chris Pires*[1]
1. University of Missouri
2. University of Arizona

Brassica napus is a worldwide multiuse crop, especially as a very important edible oil resource. It is an allopolyploid species naturally hybridized from *Brassica rapa* (AA=20) and *Brassica oleracea* (CC=19) about 7,500 years ago. Now, it mainly has three groups: *B. napus subsp. oleifera*, *B. napus subsp. rapifera* and *B. napus subsp. pabularia.* To study its origin and diversification, we collected 187 *B. napus*, 137 *B. rapa*, 43 *B. oleracea* and 22 other C genome related accessions, and then generated high density genome-wide nuclear and chloroplast SNPs data based on these collections from high-throughput RNA-seq and GSS data. According to our nuclear genome data, we confirmed it originated from Europe and the *B. napus subsp. oleifera* group was diversified into spring type, European winter type and East Asian winter type. Besides that, we also found that A sub-genome and C sub-genome had different nuclear phylogenies and nucleotide diversity, which suggests A and C sub-genome may have experienced different selection. We will use whole chloroplast genomic SNPs to construct chloroplast phylogeny among all our collections to explore the multi- or single- origin hypothesis and the maternal species of *B. napus*. The selective sweep regions and founder haplotypes will also be studied to get a better understand of their diversification. These results have implications for understanding the evolutionary patterns and processes after their initial domestication and will facilitate further studies of Brassica genomics and *Brassica napus* crop improvement.

T5
P1154

Isolation and screening of viable mutants for yield contributing traits in groundnut (*Arachis hypogaea* L.)

Jeevakoti Apparao Boddu
Jijamata College of Science and Arts, Bhende.

In the present investigation, an attempt was made to induce genetic variability in yield contributing traits, viz., plant height, number of primary branches per plant, number of pods per plant, number of seeds per pod, days to maturity, days to flowering, hundred seed weight and seed yield per plant, in groundnut (*Arachis hypogaea* L.), employing gamma radiation (γ). Germplasms of a locally adopted cultivar of groundnut, JL-24, were irradiated with different doses (10, 15, 20, 25 and 30 KR) of gamma rays and were sown in the field to raise M_1 progeny. Subsequently M_2 and M_3 progenies were raised using M_1 and M_2 seeds. The progeny plants in M_2 generation were screened for presence of mutations in yield contributing traits. They showed presence of 6 different types of plant type mutations viz., tall mutants, pod mutants, early maturing mutant, high yielding mutant and seed coat colour mutants, as compared to control. All these 6 different types of mutations were recovered even in the M_3 generation, as viable mutants. Remarkable achievement of the present investigation is isolation of early maturing and high yielding mutants. The early maturing mutants were at least 8-10 days early in flowering and in attaining maturity. The high yielding mutants showed significant increase in almost all yield contributing traits as compared to control. Some of these viable mutants seemed to be very promising and can be used in breeding programs aimed at genetic improvement of groundnut.

T5

P1155

Comparative transcriptome analysis and phenotypic monitoring of *Trifolium pratense* (Fabaceae) under land use scenarios

Denise Herbert, *Annette Becker*
Justus Liebig University Giessen

Red clover (*Trifolium pratense*) is an important forage crop worldwide and of great interest to agriculture, offering a high nutritional value and soil improving characteristics, which can improve intake in livestock farming and reduce the need of fertilizer. Unfortunately, its persistence to environmental perturbation is low such that red clover shows drastic phenotypic plasticity, affecting the whole plant morphology, architecture and growth behavior in response to grazing and mowing. We address the question on the exact nature of these morphological changes in response to mowing/grazing as well as which genetic pathways lead to the induced phenotypic plasticity. A combination of phenotypic monitoring, including the measurement of a large scale of morphological important traits to describe the temporal changes in response to grazing/mowing andtranscriptome analysis from plants grown on different field sites and treatments. This will provide a deeper insight gene expression changes underlying the observed morphological plasticity. This allows the identification of candidate genes for further functional analysis including transgenic approaches to ultimately understand the complex molecular mechanism of the red clover's phenotypic plasticity in response to mowing/grazing. We will provide results of the digital gene expression analysis between combinations of different transcriptomes and their preliminary functional analysis. Genes acting as chromatin remodelers and others involved in the gibberellin pathway may be responsible for the phenotypic changes during regrowth.

T5

P1156

snRNA 3' end processing by a CPSF73-containing complex essential for development in *Arabidopsis*

Liu Yufeng, ***Bin Yu***
Center for Plant Science Innovation and School of Biological Sciences, University of Nebraska, Lincoln, NE 68588–0660, USA

Uridine-rich small nuclear RNAs (snRNAs) are the basal components of the spliceosome and play essential roles in splicing. The biogenesis of the majority of snRNAs involves 3′ end endonucleolytic cleavage of the nascent transcript from the elongating DNA-dependent RNA ploymerase II. However, the protein factors responsible for this process remain elusive in plants. Here, we show that DEFECTIVE in snRNA PROCESSING 1 (DSP1) is an essential protein for snRNA 3′ end maturation in *Arabidopsis*. A hypomorphic *dsp1-1* mutation causes pleiotropic developmental defects, impairs the 3′ end processing of snRNAs, increases the levels of snRNA primary transcripts (pre-snRNAs), and alters the occupancy of Pol II at snRNA loci. In addition, DSP1 binds snRNA loci and interacts with Pol-II in a DNA/RNA-dependent manner. We further show that DSP1 forms a conserved complex, which contains at least four additional proteins, to catalyze snRNA 3′ end maturation in *Arabidopsis*. The catalytic component of this complex is likely the cleavage and polyadenylation specificity factor 73 kDa-I (CSPF73-I), which is the nuclease cleaving the pre-mRNA 3′ end. However, the DSP1 complex does not affect pre-mRNA 3′ end cleavage, suggesting that plants may use different CPSF73-I-containing complexes to process snRNAs and pre-mRNAs. This study identifies a complex responsible for the snRNA 3′ end maturation in plants and uncovers a previously unknown function of CPSF73 in snRNA maturation.

T5

P1157

Measurement of stomata and pollen as an indirect indicator of polyploidy in five species of the genus *Polylepis* (Rosaceae)

Joselin Caiza, *Dominique Vargas, Camila Olmedo, Mario Arboleda, Anahí Boada, Osmar Acurio, María Claudia Segovia-Salcedo*
Fuerzas Armadas-ESPE University

Polylepis forest (Family: Rosaceae), are originated in the Andes and considered one of the most threatened forest ecosystems. In Ecuador, eight of the 30 species have been described: one endemic *P. lanuginosa*, one introduced *P. racemosa* and six natives (*P. sericea, P. incana, P. weberbaueri, P. pauta, P. microphylla*, and *P. reticulata*). Throughout their evolutionary history they have shown several episodes of polyploidy affecting their process of speciation. Cell structures such as pollen and occlusive cells are highly influenced by the number of genes in each species, representing morphometric parameters to estimate the degree of ploidy. In this study, the relationship of pollen grain size, length of occlusive cells was analyzed with genome size and chromosome number. Additionally, the pollen viability of species of *P. pauta, P. sericea, P. racemosa, P. incana*, and *P. microphylla* was analyzed using staining techniques. Pollen grains and stomata were photographed at a magnification of 400×, their length was measured in micrometres, using ImageJ program 1.49v. The results show that there is a positive correlation $r = 0.56$ and $r = 0.79$ for stomata, $r = 0.71$ and $r = 0.48$ for pollen, therefore it serves as an indirect analysis method for polyploidy. These data are important for reforestation, restoration and conservation projects of *Polylepis* in Ecuador, allowing prior identification of potential hybrid individuals and cytotypes along the Andes quickly and inexpensively. Additionally it opens the possibility of the use of this data in studies of climate change. It is intended to extend this project to the rest of species of this genus.

T5

P1158

Cloning and functional characterization of epidermis-specific promoter *MtML1* from *Medicago truncatula*

Weiwei Chai, *Ye Tian, Mengci Chen, Li Wei, Hongju Yin, Suomin Wang*
State Key Laboratory of Grassland Agro-ecosystems, College of Pastoral Agriculture Science and Technology, Lanzhou University

Constitutive and tissue-specific promoters driving transgene expression specifically in vascular bundles, roots and floral organs have been widely used for the genetic improvement of alfalfa.

However, ectopic expression of specific functional genes such as the cuticle-related gene need epidermis-specific promoters with few studies focused on. The *MERISTEM LAYER 1* (*AtML1*) promoter from *Arabidopsis thaliana* is believed to drive gene expression specifically in the epidermis, both in the shoot apical meristem (SAM) and leaves. Nevertheless, the epidermis-specific promoter from the model legume *Medicago truncatula* has not been reported. In this study, we firstly cloned a 5′ flanking sequence from the upstream -2150 bp to the start ATG codon of *MtML1* gene. The sequence analysis showed this region contained promoter feature elements such as TATA box and CAAT box, as well as elements involved in light response. Compared with the promoter of *Arabidopsis AtCER6* expressed specifically in the epidermis, *MtML1* promoter had a plurality of 5′ UTR Py-rich stretch motif conferring high transcription levels identical with *AtML1* promoter. The GUS histochemical analysis showed *MtML1* promoter could drive *GUS* gene expressed transiently in *Nicotiana tabacum* leaves as the same as *AtML1* and *AtCER6* promoters. Furthermore, tissue-specific expression analysis of *MtML1* promoter in *A. thaliana* revealed the spatiotemporal expression patterns involving in high expression levels in leaves and siliques rather than in roots. In addition, the leaf cross-section observations indicated that *MtML1* were exclusively expressed in leaf epidermis. The results confirmed the epidermis-specific promoter *MtML1* was the foundation for driving the cuticle-related gene to express in epidermis, which could dress the alfalfa in waxy coat withstanding the drought, UV, heat, frost and pathogens.

T5

P1159

Identification of differentially expressed proteins on large and small rubber particles in rubber tree (*Hevea brasiliensis*)

Chada Chantrachamnong[1,2], *Panida Kongsawadworakul*[1,2], *Onrapak Reamtong*[3], *Jarunya Narangajavana*[4], *Unchera Viboonjun*[1]

1. Department of Plant Science, Faculty of Science, Mahidol University

2. Department of Pharmaceutical Botany, Faculty of Pharmacy, Mahidol University

3. Department of Molecular Tropical Medicine and Genetics, Faculty of Tropical Medicine, Mahidol University

4. Department of Biotechnology, Faculty of Science, Mahidol University

Natural rubber is a biopolymer which could be economically produced from rubber tree (*Hevea brasiliensis*). The rubber biosynthesis occurs on the surface of rubber particles. In vitro assays revealed that small rubber particles (SRPs) contained higher levels of rubber biosynthesis activity than the large rubber particles (LRPs). However, its underlying molecular mechanism is still elusive. In this study, LRPs and SRPs protein expression patterns were investigated by 2-dimensional gel electrophoresis (2-DE). The differentially expressed protein spots were analyzed by ESI-LC-MS/MS coupled to protein database search. The results demonstrated that rubber elongation factor (REF, 14.7 kDa) and small rubber particle protein (SRPP, 22.3 kDa) were almost equally localized on both LRPs and SRPs whereas another isoforms of REF (19.6 kDa) and SRPP (25.6 kDa) were higher expressed on LRPs. Additionally, hevamine, a component involving in a latex coagulation was higher accumulated on LRPs leading to the possibility that LRPs may play important role in the latex coagulation. More information on proteomic data should help to unravel molecular mechanisms of rubber biosynthesis and might serve as a novel biomarker for rubber tree breeding program.

T5

P1160

Genetic dissecting of seed oil and protein content and identification of network associated with oil content in *Brassica napus*

Hongbo Chao[1], *Hao Wang*[2]

1. Huazhong University of Science and Technology

2. Shaanxi Rapeseed Branch of National Centre for Oil Crops Genetic Improvement

High-density linkage maps could improve the precision of QTL localization. A high-density linkage map which contained 3106 SNP-bins and 101 non-SNPs covering 3072.7 cM of *Brassica napus* genome was constructed in KN DH population. A total of 164 and 68 identified QTLs were detected for seed oil content in 12 trials and seed protein content in 11 trials and could highest explain 22.24% and 27.48% of the phenotypic variation respectively. After meta-analysis, 67 and 38 consensus QTLs for seed oil and protein content were identified with an average confidence interval of 5.26 and 4.38 cM respectively. Thirty-eight associated genomic regions from BSA overlapped with and/or narrowed the SOC-QTLs, which further confirmed QTL mapping based on the high-density linkage map. Potential candidates related to lipid metabolism and seed storage protein underlying SOC and SPC respectively were identified and analyzed, while six among of them were checked and showed expression differences between the two parents during different embryonic development periods. Meanwhile, a whole genome search was implemented to acquire distribution of orthologs related to lipid metabolism, and a total of 3058 orthologs of *A. thaliana* lipid-related genes were identified in the whole genome. Minor acyl-related genes enrichment/prevalence in the CI of SOC-QTLs was found as compared to whole genome. In addition, a magnificent primary carbohydrate pathway based on potential candidates underlying SOC and SPC-QTLs, and interaction networks based on potential candidates underlying SOC-QTLs, were constructed in order to dissect complex mechanism from the aspects of metabolism and gene regulation, respectively. Accurate QTL mapping and potential candidates identified based on high-density linkage map and BSA provide new insights to the complex genetic mechanism of oil and protein accumulation in seeds of rapeseed.

T5

P1161

Authentication of the medicinal plants in Smilaceae by DNA barcoding technique

Chaozhi Chen, *Yuanjing Duan, Yue Yu, Zhentao Wang, Shuyuan Li*

Guangdong Pharmaceutical University

Smilaceae is one of the hot plant populations with medicinal value and food nutritional value, which are widely used in the world, especially in China. However, the morphological characters of this family are similar, even geographical distance of this family may lead to a small range of genetic varia. So we chose four sequences (ITS2, psbA-trnH, matK, rbcL) to establish the DNA barcoding system of Smilaceae. The related data of those sequences were analyzed using the software Codon Code Aligner V 3.7.2 and MEGA 6.0. The Kimura-2-Parameter distances were calculated and the phylogenetic tree was constructed using the Neighbor-Joining method, NJ tree shows that they can be distinguished from molecular level obviously. We have identified potential barcodes of the medicinal plants in Smilaceae by comparing DNA barcoding sequences. For better comparison, the Wilcoxon signed rank test was employed to test the difference of those samples by choosing psbA-trnH, matK and rbcL sequences. In conclusion, DNA barcoding technique not only can be used for authentication of the medicinal plants in Smilaceae, but also has established a model platform for the effective authentication of crude drug in Smilaceae furtherly.

T5

P1162

Evolution of MIKC^c-type MADS-box genes in gymnosperms and angiosperms implicates the origin of flowers

***Fei Chen**, Xingtan Zhang, Xing Liu, Liangsheng Zhang*
Fujian Agriculture and Forestry University

MIKC^c-type MADS-box genes encode transcription factors that control floral organ morphogenesis and flowering time in flowering plants. Here, in order to determine when the subfamilies of MIKC^c originated, we sampled and analyzed the genomes and large-scale transcriptomes representing all the Orders of gymnosperms and basal angiosperms. Through phylogenetic inference, the subfamily MIKC^c-type MADS-box genes were subdivided into 14 monophyletic clades based on sequence and functional similarity. Among them, the gymnosperm orthologs of *AGL6-SEP-AP1*, *GMADS*, *SOC1*, *AGL32-AP3/PI*, *SVP*, *AGL15*, *ANR1*, *AG* were identified. We identified and characterized the origin of a novel subfamily *GMADS* within gymnosperms, and lost orthologs in monocots and Brassicaceae. ABCE model prototype genes were relatively conserved in terms of gene number in gymnosperms, and expanded in angiosperms, whereas *SVP*, *SOC1*, and *GMADS* had dramatic expansions in gymnosperms, but conserved in angiosperms. In summary, our results provide the evolutionary history of all MIKC^c gene members in gymnosperms and angiosperms. Although the near complete set of MIKC^c genes evolved in gymnosperms, the duplication of MIKC^c genes in the ancestor of angiosperms triggered the first flower.

T5

P1163

How much positive selection exists in a certain genome —— Joint analysis of McDonald–Kreitman test and site-based LRTs in *Arabidopsis*

***Qipian Chen**[1], Ziwen He[1], Hao Yang[1], Chung-I Wu[1,2], Suhua Shi[1]*
1. State Key Laboratory of Biocontrol and Guangdong Provincial Key Laboratory of Plant Resources, Sun Yat-sen University
2. Department of Ecology and Evolution, University of Chicago, Chicago, Illinois, USA

Estimating the extent of positive selection is a fundamental problem in the study of molecular evolution. The fraction of adaptive substitutions (α) could be estimated based on the McDonald–Kreitman (MK) test that contrast polymorphism and divergence between selectively and neutrally evolving sites (McDonald and Kreitman, 1999). While site-based likelihood-ratio tests (LRTs) in PAML (Phylogenetic Analysis by Maximum Likelihood) software uses codon substitution models to identify amino acid sites under diversifying selection (Yang *et al.*, 2000). In this study, we choose 1,135 *Arabidopsis thaliana* genomes along with other four cruciferous plants' genomes to infer positive selection signature in a total of 15,064 orthologs. We find that a small proportion of adaptive nonsynonymous fixations in *A. thaliana* vs *A. lyrata* (α=0.05) based on MK test, whereas average advantageous amino acids sites is 4.66 per gene based on site-based LRTs. Weak positive correlation between these two methods are found based on a few evidences. First, on a gene by gene basis, positively selected genes that are detected by MK tests and site-based LRTs have a very poor overlap (341 vs 830 genes that are significant, with only 38 shared) across the genome. Second, in a broader scale, a similar tendency was found between these two methods when genes are ranked and pooled together based on one method and send to the other for a test. For their lack of consistency, one explanation is that the two methods focus on different aspects of the different evolutionary processes. Hence, we did a series of simulations that allow positive selection happened in different time-scales with different selection coefficient distributions. The result can be partly explained by it. Since no method are available to detect the full selection pressures, the positive selections might be substantially underestimated.

T5

P1164

CIN-clade TCP transcription factor TCP4 modulates leaf morphogenesis through associating with histone deacetylase HDA6 in *Arabidopsis*

***Kai Cheng**[1], Song-Guang Yang[1], Ke-Qiang Wu[2]*
1. Key Laboratory of Plant Resources Conservation and Sustainable Utilization, South China Botanical Garden, Chinese Academy of Sciences
2. Institute of Plant Biology, National Taiwan University

Leaf morphogenesis is a strictly programmed proceed and is regulated by various transcription factors, such as CIN-clade TCPs, notably TCP4, which connects cell proliferation to leaf development. However, the TCP4-mediated transcription regulatory networks remain largely unknown. In this study, we found that TCP4 genetically interplayed with a histone deacetylase HDA6. Compared with *tcp4* and the *hda6* mutant line *axe1-5*, *tcp4/axe1-5* double mutant plants displayed more severe curved and serrated leaves. Real-time quantitative PCR revealed that the expression of leaf development and cell cycle-related genes were changed in all these mutant plants. Global transcriptome analysis indicated that TCP4 and HDA6 coordinately regulated the expression of genes that were involved in cell proliferation and expansion, leaf morphogenesis, and phytohormone signals. Furthermore, we demonstrated that TCP4 interacted with HDA6 in vitro and in vivo, and the transcriptional activity of TCP4 was affected in the presence of HDA6. Taken together, these results suggested that TCP4 may modulates leaf morphogenesis by collaborating with HDA6.

T5
P1165
Comparative analysis of three *Zabelia* species (Linnaeaceae) by complete chloroplast genome

Kyeong-Sik Cheon, *Kyung-Ah Kim, Ki-Oug Yoo*
Kangwon National University

The complete chloroplast genome sequences of three *Zabelia* speices (Linnaeaceae), *Z. coreana*, *Z. insularis*, and *Z. tyaihyoni*, were determined in this study. The total genome length of three *Zabelia* species ranged from 157,961 (*Z. tyaihyoni*) to 158,416 (*Z. insularis*). Each genome contains 128 known genes, including 78 protein coding genes, 37 tRNA genes, and 8 rRNA genes. Among these genes, 16 genes contained one or two introns. In comparison of genome structure, gene order between two species (*Z. coreana* and *Z. insularis*) was confirmed to be identical. However, one inversion of large gene block (*trnT-UGU-rbcL*) in the LSC region was identified cp genome between *Z. tyaihyoni* and the aforementioned two species. The total number of SSRs were 49 in two plastomes (*Z. conreana* and *Z. insularis*), and 51 in *Z. tyaihyoni* cp genome. Also, tandem repeats were 49 in all three species. A comparison of coding genes and non-coding regions among the three species revealed divergent site, the divergent value was 0.01 or more in 9 regions. The complete chloroplast genome sequence information regarding the three *Zabelia* will be helpful for elucidating phylogenetic relationships of *Zabelia* and Linnaeaceae.
This study was supported by grants from Scientific research (KNA1-2-13,14-2) of Korea National Arboretum.

T5
P1166
Identification and analysis of circular RNAs in plants

Qinjie Chu, *Xingcheng Zhang, Li Chen, Chuyu Ye, Longjiang Fan*
Institute of Crop Science and Institute of Bioinformatics, Zhejiang University

Circular RNAs (circRNAs) are covalently closed loops derived from back-splicing of precursor mRNAs. Recently, a large number of circRNAs have been detected among diverse organisms. For the first time in plants, we have identified 12,037 and 6,012 circRNAs in *Oryza sativa* and *Arabidopsis thaliana*, respectively, with 56% (10/18) of the sampled rice exonic circRNAs validated experimentally. We demonstrated that circRNAs are widespread in plants and revealed the common and distinct features of circRNAs between plants and animals. Additionally, a software for prediction of plant circRNAs, termed PcircRNA_finder, was developed, which is more sensitive and precise in detecting plant circRNAs than other frequently used programs, such as find_circ and CIRCexplorer, by analyzing simulated and real rRNA-depleted/RNase R-treated RNA-Seq data from *A. thaliana* and *O. sativa*. Another software, termed circseq-cup, was also developed, which can assemble the full-length sequences of circRNAs using the back-splicing RNA-Seq reads and their corresponding paired-end reads. We for the first time identified full-length sequences of nearly 3,000 circRNAs in *O. sativa* by applying "circseq-cup" to an rRNA-depleted/RNase R-treated RNA-Seq dataset. We further showed that alternative circularization of circRNA is a common feature in *O. sativa* and, surprisingly, found that the junction sites of a large number of circRNAs are flanked by diverse non-GT/AG splicing signals in *O. sativa* while most human exonic circRNAs are flanked by canonical GT/AG splicing signals. Moreover, we have created a database (PlantcircBase, http://ibi.zju.edu.cn/plantcircbase/) of plant circRNAs. We have collected publicly available circRNAs identified in recent years by bioinformatics prediction and/or experimental validation, as well as circRNAs newly identified by our own lab. Totaily, PlantcircBase includes 40,314,35,884,496,854 and 47 unique circRNAs in *O. sativa*, *A. thaliana*, *Zea mays*, *Solanum lycopersicum* and *Hordeum vulgare*, respectively. Apart from the detailed and comprehensive information for each circRNA entry, PlantcircBase also provides basic tools including browsing, searching and downloading as well as advanced tools, including visualizing structures of circRNAs and predicting circRNAs.

T5
P1167
Effects of colchicine exposure time on ploidy level of *Globba williamsiana* 'White Dragon' *in vitro*

Rodjanacorn Chuengpanya[1,2], *Ngarmnij Chuenboonngarm*[2], *Thaya Jenjittikul*[2], *Kanchit Thammasiri*[2], *Puangpaka Soontornchainakseang*[2], *Atchara Muangkroot*[2]
1. Ph.D. Program in Botany, Faculty of Graduate Studies, Mahidol University
2. Department of Plant Science, Faculty of Science, Mahidol University

Globba, a member genus of the Zingiberaceae, has gorgeous inflorescence for using as ornamental plant. The demand of *Globba* from floral industry is increasing every year. However, new appearances such as big flower size, long stem, and long vase life of ornamental plant are always need and so do *Globba*. Searching of new *Globba* sp. couple with breeding program for dealing the floral markets are necessary. *Globba williamsiana* 'White Dragon', which has big white bract with long beautiful yellow inflorescence and is sold as cut-flower in some floral market of Thailand, was selected as studied plant for *in vitro* plant mutation breeding program. To increase ornamental quality, tetrapliod induction through colchicine treatment is generally done by plant breeder. In this study, the effects of colchicine on ploidy level of *in vitro G. williamsiana* were investigated. Young leafy-shoots, 1.5 cm high, were treated with 0.2% (w/v) colchicine at 0,4, 8, and 12 days. After transferring the treated explants onto MS gelrite medium for 8 weeks, the highest survival percentage (50%) was obtained from the 4-day treatment. The survival percentage decreased when colchicine exposure time increased. The increase of colchicine exposure time reduced the formation of leafy-shoot and root, amount of green leafy-shoot, and leaf maturation of the colchicine-treated M_1V_1 plants. At M_1V_3 generation, although all of treated plants grew as the same rate as control (0-day treatment), some of them showed wrinkled leaf morphology. This abnormal morphology was found in every colchicine exposure time. After investigation the ploidy level of M_1V_3 plants by flow cytometry, four ploidy levels, which were diploids, tetraploids, and 2 types of mixoploids

(2x+4x and 2x+8x), were found. Percentage of each ploidy level was 43.11, 33.16, 18.88, and 4.85, respectively. The highest tetraploid percentage was obtained from the 4-day treatment. All abnormal leaf morphological plants in M_1V_3 plants were 2x+8x. Size of stomata and density of epidermal cell and stomata were also observed from M_1V_4 plants. The results showed that the tetraploid plants had stomata size significantly larger than the diploid plants. Lower density of epidermal cell and stomata of the tetraploid plants were significantly observed when compared with the diploid plants. In conclusion, colchicine exposure time affected on ploidy level, morphological characters, and anatomical characters of *G. williamsiana* 'White Dragon'.

T5

P1168

Marama, an orphan crop with potential for resource poor farmers, genome sequencing and analysis

Christopher Cullis

Case Western Reserve University

Marama bean (*Tylosema esculentum* (Burchell) A. Schreiber: Caesalpinioideae) is a perennial non-nodulating legume of southern Africa (Namibia, Botswana and South Africa), growing in poor, arid soils surrounding the western, southern and eastern Kalahari and is being developed as a possible crop for resource poor farmers in arid regions of Southern Africa. Marama produces large (2.5 g per seed without shell), protein- and oil-rich edible seeds, and extremely large (up to 200 kg) tubers, containing protein and carbohydrate. This species has been identified as an orphan crop. Whole genome sequencing using both Illumina and PacBio platforms have yielded a nuclear genome assembly, a complete chloroplast molecule and a mitochondrial genome. In addition to the whole genome sequence of one individual, lower coverage Illumina sequencing for 39 individuals across the range of Namibia and South Africa has been obtained. The genome assembly has been interrogated for the possible loss of nodulation genes as the reason for this species being non-nodulating. Also, since marama has a novel protease content, the genes for proteases and their inhibitors have also been identified. The chloroplast genome has been assembled from next generation sequencing using both Illumina and Pac-Bio data. The genome is of typical organization with a large single copy and a small single copy region separated by a pair of inverted repeats and covers 161,537 base pairs. The inverted repeat is expanded compared with other legumes and the small single copy region reduced. However, the marama chloroplast genome contains a unique inversion not present in any other legumes, even in the closest relatives for which the complete chloroplast genome is available, and two complete copies of the ycf1 gene. The sequencing of multiple individuals has identified variation in the chloroplast genomes of marama. The genomes from individuals collected in South Africa differ from those collected from Namibia. There is a more limited variation among the Namibian samples, than that between the South African and Namibian samples, but that which does occur appears appears to be geographically restricted. However, the sequence data form these individuals indicated that marama is heteroplasmic with respect to the chloroplast. The equivalent variation in the mitochondrial genome is also being characterized to determine if this is also heteroplasmic.

T5

P1169

Living at the edge: Diversity of Chinese Dipterocarps

Tijana Cvetkovic[1], *Joeri Strijk*[1,2], *Damien Hinsinger*[1]

1. Plant Ecophysiology & Evolution Group, Guangxi Key Laboratory of Forest Ecology and Conservation, College of Forestry, Guangxi University, Nanning, Guangxi 530005, PR China

2. Plant Ecophysiology & Evolution Group, State Key Laboratory for Conservation and Utilization of Subtropical Agro-bioresources, Guangxi University, Nanning, Guangxi 530005, PR China

Tropical forests in the South-East Asia are one of the most important repositories of terrestrial biological diversity, providing both goods and services. Therefore, a better understanding and assessment of the biodiversity is essential for sustainable development in this region. From 2005-2015, total plywood production averaged more than 73 million m^3 in China, Indonesia and Malaysia. Among them, Dipterocarpaceae (with more than 550 species), as the economically most important native tree family for wood production, generated over 900 million US$ revenue for China alone. They form the dominant structural component of Asian tropical wet lowland forests, including both wide ranging and endemics species. Living dipterocarps *sensu lato* are spread over the tropical belt of three continents of Asia, Africa and South America. In China, Dipterocarpaceae are represented by five genera and 12 species; southern China also represents the northern limit for this solely tropical family. In spite of high diversity and ecological dominance of Dipterocarpaceae, very little is known about the genomic diversity and phylogenetic relationships within the family, and the diversity of the species that can be found in China. Here, we assembled a dataset comprising the 12 Dipterocarps species occuring in China, in addition to local suspected synonyms. Using N.G.S. and a genome skimming approach, we assembled the chloroplasts and the nuclear ribosomal cistrons for each species. In addition, we generated a genome-wide dataset of the repeats relative occurrences for these species and performed several diversity analyses to delineate the genomic features of these "edge" species. Our results indicate the genomic diversity is not evenly distributed among both the species and the chloroplasts sequence, different genes and genes families showing strikingly different substitution patterns. We will also discuss the phylogenetic diversity of the chinese species of Dipterocarps. Our research provides the scientific community with important resources for further studies on intraspecific, interspecific and intrageneric levels of genomic diversity in Dipterocarpaceae, as well as biodiversity, origin and evolution of Dipterocarpaceae and evolution of Asian tropical forests.

T5

P1170

Charophyte algae as model systems

Charles Delwiche

University of Maryland

Genetic model systems are organisms that have been studied in great detail, and for which there are molecular genetic techniques and tools available. In their best application, model systems provide insights that can be extended to other organisms. Phylogenetic distance is one of several factors that influence such transferability; closely related organisms are more likely to share properties. Unfortunately, despite recent progress with gymnosperm and bryophyte model systems, the phylogenetic diversity of plant model systems remains relatively poor. Outside of the land plants, *Chlamydomonas* and *Volvox* stand out as enormously important model systems, but they diverged from land plants well over a billion years ago, and are substantially different in some important properties. In contrast, the charophyte green algae (basal streptophytes) are known to be much more closely related to land plants, and as such are potentially valuable for understanding fundamental plant properties and the origin of a terrestrial flora. Among the charophytes are several organisms that seem promising as model systems, but substantial development is still needed. Challenges include availability of high-quality cultures, techniques to consistently complete the life cycle in culture, methods for haploid genetics, and development of a diversity of molecular genetic tools.

T5

P1171

Proteomic analysis of differentially expressed proteins of *Nicotiana benthamiana* triggered by INF1 elicitin from *Phytophthora infestans*

Juan Du[1], *Xiao Guo*[1], *Lin Chen*[2], *Conghua Xie*[1], *Jun Liu*[1]
1. Huazhong Agricultural University
2. South China Agricultural University

Pathogen-associated molecular pattern (PAMP)-triggered immunity (PTI) is considered to be durable, given that PAMPs are conserved in entire classes of microbes. Elicitins are structurally conserved extracellular proteins in oomycete species and are well characterized as having features of PAMPs. INF1 is an elicitin protein secreted by the late blight pathogen *Phytophthora infestans.* A cell surface receptor-like protein that mediates INF1 response was recently cloned in potato. In addition, some other genes are reportedly involved in INF1-triggered immune responses; however, the molecular mechanisms of INF1-triggered immunity remain poorly understood. Here, we used isobaric tags for relative and absolute quantification-based quantitative proteomics to analyze proteins involved in INF1-triggered cell death responses in *Nicotiana benthamiana.* Our approach identified 2964 proteins, 32 of which were significantly altered in abundance after INF1 induction. Two of eight selected upregulated proteins, namely, ATP-dependent transporter and 60S ribosomal protein L15 were shown to be essential in INF1-triggered cell death responses by virus-induced gene silencing analysis. This study represents the first proteomic analysis of INF1-triggered cell death responses in plants and provides the basis for further work to elucidate molecular mechanisms into oomycete PTI in host plants.

T5

P1172

Genetic engineering of CymMV-resistant *Phalaenopsis*

Shuguo Fan
School of Chemistry and Life Sciences, Chuxiong Normal University

The orchid has been one of the most economic floricultural crops in China. Among them, *Phalaenopsis spp.* takes first place in commercial value in the orchid industry. However, viral infections result in the great reduction in both quality and quantity of flowers, hence becoming a limiting factor in the orchid industry. *Cymbididum mosaic virus* (CymMV) is one of the most prevalent plant viruses in orchids. The aim of this research is to generate CymMV-resistant *Phalaenopsis amabilis* overexpressing either the sense or anti-sense strand of the CymMV coat protein (CP) genes using particle bombardment. Hopefully, the transgenic orchids will be resistant against viral infection. Calli have been induced from *P. amabilis.* Plant regeneration system has been established from callus. The gene comprising the capsid proteins of CymMV has been constructed, optimum conditions for particle bombardment have been investigated. Resistant calli and transgenic plantlets have been screened on selection medium. The selection medium was optimized for shoot induction with L-methionine sulfoximine (MSO) added as selection agent. Selection was done in both solid and liquid medium. It was found that liquid media and experiment showed that selection in liquid medium was faster and more efficient. PCR amplification of the inserted CP genes were used for genotyping, and PCR Southern assay confirm the insertion of anti-sense strand of CymMV CP genes into several lines of *P. amabilis* transformants. Northern blot and Western blot have also been done for the putative transformed plants, and viral inoculation and detection will be performed on the transgenic *Phalaenopsis* when they grow to be big. Our results showed that the transgenic *P. amabilis* has been successfully created, and hopefully, the transgenic orchids will be resistant against viral infection.
Funding: This work was supported by grants from the program for Innovative Research Team (in Science and Technology) in University of Yunnan Province (IRTSTYN), Yunnan Province Universities Key Laboratory of Applied Biology, Applied and Basic Research Foundation of Yunnan Provincial Science and Technology Commission (2013FB061), and Key Discipline Construction of Yunnan Province and Chuxiong Normal University (05YJJSXK03).

T5

P1173

Genetic engineering of ORSV-resistant *Phalaenopsis*

Shuguo Fan
School of Chemistry and Life Sciences, Chuxiong Normal University

Odontoglossum ringspot virus (ORSV) is one of the most prevalent plant viruses in orchids. The aim of this research is to generate ORSV-resistant *Phalaenopsis equestris* var. *alba* overexpressing either the sense or anti-sense strand of the ORSV coat protein (CP) genes using particle bombardment. In this research, calli have been induced from different explants of *P. equestris* var. *alba.* Plant regeneration system has been established from callus. The gene comprising the capsid proteins of ORSV has been constructed, optimum conditions for particle bombardment have been investigated. Resistant calli and transgenic plantlets have been screened on selection medium. The selection medium was optimized for shoot induction with L-methionine sulfoximine (MSO) added as selection agent. Selection was done in both solid and liquid medium. It was found that liquid media and experiment showed that selection in liquid medium was faster and more efficient. PCR amplification of the inserted *CP* genes were used for genotyping and PCR Southern assay confirm the insertion of anti-sense strand of ORSV *CP* genes into several lines of *P. equestris* var. *alba* transformants. Northern blot and Western blot have also been done for the putative transformed plants, and viral inoculation and detection will performed on the transgenic *Phalaenopsis* when they grow to be big. Our results showed that the transgenic *P. equestris* var. *alba* has been successfully created, and hopefully, the transgenic orchids will be resistant against viral infection.

Funding: This work was supported by grants from the program for Innovative Research Team (in Science and Technology) in University of Yunnan Province (IRTSTYN), Yunnan Province Universities Key Laboratory of Applied Biology, Applied and Basic Research Foundation of Yunnan Provincial Science and Technology Commission (2013FB061), and Key Discipline Construction of Yunnan Province and Chuxiong Normal University (05YJJSXK03).

T5

P1174

Influential factors on *Phalaenopsis* transformation by biolistic bombardment

Shuguo Fan

School of Chemistry and Life Sciences, Chuxiong Normal University

Phalaenopsis equestris var. *alba* is a plant of the orchid genus *Phalaenopsis*. *Odontoglossum ringspot virus* (ORSV) is one of the two most prevalent viruses infecting orchids by mechanical transmission and no suitable control measurements are available thus far. Studies on various plants and viruses show that overexpression of a virus CP in plants makes the plants confers resistance against this virus and close-related viruses. In this study, *ORSV CP* gene was introduced into *P. equestris* var. *alba*'s genome, the factors affecting bombardment transformation of this *Phalaenopsis* species was investigated. *P. equestris* var. *alba.* was used in this study. Protocorm-like bodies (PLBs) were induced and proliferated in KC medium supplemented with 15% (v/v) coconut milk, 2% (w/v) sucrose and 3 mg/L BA. PLBs were subcultured into fresh medium every 2 weeks. The optimal concentrations of MSO for selection were determined by plating 150 wild-type *P. equestris* var. *alba.* PLBs onto the solid media without phytohormones, and with various concentrations of MSO (0.1,0.2,0.3,0.4,0.5,0.6,0.7,0.8,0.9 ,1.0 μM) added. To generate the vector *pORSV-CP* used for genetic transformation in this study, the coding region of *ORSV-CP* was cloned into a *pGreen 0229-35S* binary vector under the control of two CaMV 35S promoters. Phalaenopsis transformation was done by particle bombardment and putative transformants were selected with MSO. PCR and PCR southern have been done for the putative transformants. Gene transformation was performed by biolistic bombardment with a 3-day recovery period on MSO-free medium and two selection stages on media with increasing amounts of selection agent, using concentrations of 0.5 and 2 μM MSO. Helium gas pressure of 1350 psi and a distance of 9 cm from the stopping screen to orchid PLBs on plates were optimal parameters for biolistic bombardment of *P. equestris* var. *alba*, as transformation under these parameters gave the highest number of putative transformants after 3 months of MSO selection. Independent transgenic lines were obtained and the presence of the transgene was confirmed by PCR and PCR Southern blot analysis, respectively. A reliable protocol for *Phalaenopsis equestris* var. *alba* transformation by particle bombardment have been created.

Funding: This work was supported by grants from the program for Innovative Research Team (in Science and Technology) in University of Yunnan Province (IRTSTYN), Yunnan Province Universities Key Laboratory of Applied Biology, Applied and Basic Research Foundation of Yunnan Provincial Science and Technology Commission (2013FB061), and Key Discipline Construction of Yunnan Province and Chuxiong Normal University (05YJJSXK03).

T5

P1175

Fine-mapping of quantitative trait loci (QTLs) by use of an ultra-large F_2 rice population

Laibao Feng[1], *Aimin Ma*[1], *Bo Song*[1], *Sibin Yu*[2], ***Xiaoquan Qi***[1]

1. Institute of Botany, Chinese Academy of Sciences, Beijing

2. Huazhong Agricultural University, Wuhan

Understanding the genetic basis of yield-related traits is important for rice breeding. Although many yield-related QTLs are detected, only a few have been cloned or fine-mapped. Construction of large traditional mapping populations, such as recombinant inbred lines and chromosome segment substitution lines, is costly and time-consuming. Here we genotyped ca. 1,500 F_2 progeny derived from two *indica* rice cultivars, Minghui 63 and Zhenshan 97 using genotyping-by-sequencing method. A genetic map containing 3,230 bin-markers was constructed, and 7 QTLs for plant height, 5 QTLs for heading date, 4 QTLs for grain length, 5 QTLs for grain width were detected respectively. Of them 16 QTLs have been previously reported or cloned. Three QTLs corresponding to *Hd1*, *GS3* and *GW5* were mapped in small intervals ~200 kb, ~350 kb and ~300 kb, respectively. Our study provides an efficient approach for rapid mapping QTLs with high resolution in a cost-effective manner.

T5

P1176

Genome-wide comparison of *Picochlorum* (Chlorophyta, Trebouxiophyte) species

Fatima Foflonker[1], *Hwan Su Yoon*[2], *Debashish Bhattacharya*[1]

1. Rutgers University

2. Sungkyunkwan University

Members of the *Picochlorum* genus are considered candidates for biofuel, nutraceutical, and wastewater applications due to their robustness, high biomass productivity, protein and lipid content. The genome of *Picochlorum sp.* strain SENEW3 was previously studied and found to have a small streamlined (13.5 Mb) genome characteristic of extremophiles. It also has an interesting genome organization with 'gene neighborhoods' of colocalized genes that were coexpressed under stress. Physiological, photosynthetic, and genomic differences were compared for five *Picochlorum* species. Difference in genome organization, gene family expansions, bacterial horizontal gene transfer, and evolution of salinity tolerance will be examined. Of these five species, closely related and highly salt tolerant species, *Picochlorum sp.* SENEW3 and *P. oklahomensis* were shown to have differences in genome size and the genomes will be compared for evidence of positive selection on genes to identify key functions that may be involved in adaption to each respective habitat.

T5
P1177
Prevalence, mechanisms and importance of duplicate gene divergence in exon–intron structure

Xuehao Fu[1,2], *Hongyan Shan*[1], *Peipei Wang*[1,2], *Guixia Xu*[1], *Hongzhi Kong*[1]
1. State Key Laboratory of Systematic and Evolutionary Botany, CAS Center for Excellence in Molecular Plant Sciences, Institute of Botany, Chinese Academy of Sciences, Beijing 100093, China
2. University of Chinese Academy of Sciences, Beijing 100039, China

Gene duplication plays key roles in organism and genome evolution. Understanding how duplicate genes evolve and diverge through time is critical for elucidating the mechanisms underlying the origins of new characters and new organisms. Previous studies have shown that, at least in plants, some duplicate genes have diverged in the exon-intron organization, suggestive of structural divergence. However, because the species and gene pairs sampled were very limited, it is still unclear whether this phenomenon is widespread and, if yes, how prevalent it is. In this study, by conducting a genome-wide study on closely related duplicate genes from four representative species of plants (*Arabidopsis thaliana*), animals (*Drosophila melanogaster*), fungi (*Saccharomyces cerevisiae*), and protists (*Paramecium tetraurelia*), we found that structural divergence occurred prevalently in every examined species but with different proportions, ranging from 70.9% in *P. tetraurelia* to 91.9% in *A. thaliana*. Three mechanisms, including exon/intron gain/loss, exonization/pseudoexonization, and intra-exonic insertion/deletion, are detected to be responsible for structural divergence. Similar to non-synonymous substitutions, the probability of duplicate genes to diverge in structure increases with evolutionary time. We used Pearson correlation coefficient (PCC) and Euclidean distance (ED) to evaluate the difference of expression patterns between a pair of genes, and observed that PCC is lower and ED is higher in diverged than in undiverged duplicate genes, suggesting that structural divergence may be coupled with expression divergence. Using d_N/d_S values as a measurement of functional constraint, we found that duplicate genes with structural changes have higher d_N/d_S values, indicative of weaker functional constraint on these genes. This is in concordance with their lower expression levels. Further function enrichment analysis revealed that these genes are involved in unconserved biological processes, including biological regulation, signal transduction, and response to stimuli. Our findings show that the modes of structural divergence of duplicate genes are generally consistent in different eukaryotic species, implying that structural divergence is an important contributor to the evolution of duplicate genes.

T5
P1178
Gray Herbarium Card Index: Then & now (historical perspective)

Kanchi Gandhi
Harvard University Herbaria

The Harvard University Herbaria made the nomenclatural contents of the *Gray Herbarium Card Index* (GCI) available online via a Gopher server in 1991. One website hailed this accomplishment as: "This grandfather of online checklists is a searchable database of names applied to the New World plant species." The origins of GCI can be traced to Josephine Clark (1892), the librarian at the U.S.D.A., who published a compilation of newly described and renamed vascular plants under the title, "Systematic and Alphabetic Index of New Species of North American Phanerogams and Pteridophytes, in 1891." Thus the progenitor of the GCI predates the Index Kewensis (IK; 1893). Clarke went on to collaborate with Mary Day, librarian at the Gray Herbarium, to extend coverage to the entire New World and include non-vascular plants. The names were printed on cards and distributed in sets. In late 1903 Clarke handed over the responsibility of the index to the Gray Herbarium. The initial scheme was to include all post-1885 names at the ranks of genus and below pertaining to the New World plants. Cards sets were usually issued every three months, and sets of cards with new names appeared within six months of publication. The indexing of non-vascular plant names was stopped in 1927 because of certain difficulties. In 1945 it was decided to include names of all vascular plant infraspecies published from 1753-1885 for the New World. The GCI served as a source of nomenclature with links to homotypic names. In contrast, from 1893 through 1965 the IK was published in book format and served as a global taxonomic source of seed plant species. Infraspecies names were added from 1970. After the first two volumes published in 1893 and 1895, supplements were issued every five years, and a total of 21 supplements were issued from1902 through 2002. The production of GCI transitioned from cards to microfiche in 1986 to save on costs. Using both the cards and microfiche together to have full access of the GCI data often posed problems. As access to computers and programming became more widespread, it was decided to put the entire content into a database system. The complete index was digitized in 1991, and the Harvard University Herbaria website was established in 1993 to make the GCI content available. The new scheme allowed to record type data. Similar to the GCI's effort, the IK's content was issued as a CD-ROM in two editions (1: coverage 1753-1992 ; 2: coverage 1753-July 1996). In mid-1990s, GCI and the IK agreed to collaborate to avoid duplication and make the indexing process more efficient. The *Australian Plant Name Index* was also invited to join this venture. All the three indices merged their data together to constitute as single database and launched it online in 2000: the International Plant Names Index (IPNI).

T5
P1179
Main regulatory pathways, key genes, and microRNAs involved in flower formation and development of moso bamboo (*Phyllostachys edulis*)

Wei Ge, Ying Zhang, Zhanchao Cheng, Dan Hou, Xueping Li, ***Jian Gao***
International Centre for Bamboo and Rattan

Moso bamboo is characterized by infrequent sexual reproduction and erratic flowering habit; however the molecular biology of flower formation and development is not well-studied in this species. We studied the molecular regulation mechanisms of moso bamboo development and flowering by selecting three key regulatory pathways: plant-pathogen interaction, plant hormone signal transduction, and protein processing in endoplasmic reticulum at different stages of flowering in moso bamboo. We selected *PheDofl*, *PheMADS14,* and six microRNAs involved in the three pathways through KEGG pathway and cluster analysis. Subcellular localization, transcriptional activation, western blotting, *in situ* hybridization and qRT-PCR were used to further investigate the expression patterns and regulatory roles of pivotal genes at different flower development stages. Differential expression patterns showed that *PheDofl*, *PheMADS14* and six miRNAs may play vital regulatory roles in flower development and floral transition in moso bamboo. Our research paves way for further studies on metabolic regulatory networks and provides insight into the molecular regulation mechanisms of moso bamboo flowering and senescence.

T5

P1180

The origin and diversity of tarap (*Artocarpus odoratissimus*, Moraceae), an underutilized tree crop from Borneo: Insights from HybSeq and SSRs

Elliot M. Gardner[1,2], *Joan T. Pereira*[3], *Aida Shafreena Ahmad Puad*[4], *Nyree J.C. Zerega*[1,2]
1. Northwestern University
2. Chicago Botanic Garden
3. Forest Research Centre, Sabah Forestry Department
4. Universiti Malaysia Sarawak

Target enrichment sequencing (HybSeq) is applicable to a wide range of questions and sample types. Here, we present insights from HybSeq-based phylogenomic analyses and nuclear SSRs into a question at the intersection of phylogenetics and population genetics: the origin and diversity of tarap (*Artocarpus odoratissimus*, Moraceae), an underutilized tree crop from Borneo. We combine a HybSeq data set of ca. 450 nuclear genes based on fresh and herbarium samples from across the species range with SSR analysis of more densely-sampled populations in Sabah and Sarawak. We discuss strategies for library construction and hybridization for degraded samples up to 100 years old. Phylogenomic analyses and population structure analyses shed light on the relatioship of cultivated tarap to morphologically diverse wild forms. Results elucidate the genetic diversity, intra-species relationships, and domestication history of this locally prized but underutilized crop.

T5

P1181

Identification and characterization of microRNAs at different flowering developmental stages in moso bamboo (*Phyllostachys edulis*) by high-throughput sequencing

Jian Gao[1], ***Wei Ge***[1], *Ying Zhang*[1], *Zhanchao Cheng*[1], *Long Li*[1], *Dan Hou*[1], *Chenglin Hou*[2]
1. International Centre for Bamboo and Rattan
2. Capital Normal University

Researching moso bamboo (*Phyllostachys edulis*) flowering has been difficult because of its unknown flowering interval and the rarity of florescent samples. To identify microRNAs (miRNAs) and study their expression patterns during the flower developmental process of moso bamboo, small RNAs from non-flowering leaves and four flower developmental periods were sequenced using Illumina technology. In total, 409 known miRNAs and 492 differentially expressed novel miRNAs were identified in moso bamboo. Of the known miRNAs that were differentially expressed between non-flowering and flowering samples, 64 were predicted to have a total of 308 targets. Among the miRNAs, seven known and five novel miRNAs were selected, as were four of their target genes, and their expression profiles were validated using qRT-PCR. The results indicated that the miRNA expression levels were negatively correlated with those of their targets. The research comprehensively revealed that the differentially expressed miRNAs and their targets participated in diverse biological pathways and played significant regulatory roles in moso bamboo flowering. The data provide a significant resource for understanding the molecular mechanisms in moso bamboo flowering and senescence, and serve as the primary foundation for further studies on metabolic regulatory networks that involve miRNAs.

T5

P1182

Production of red flower rapeseed by ectopic expression of *OvPAP2* from *Orychophragmus violaceus*

Wenqin Fu[1], *Daozong Chen*[1], *Qi Pan*[1], *Fengfeng Li*[1], *Zhigang Zhao*[2], ***Xianhong Ge***[1], *Zaiyun Li*[1]
1. Huazhong Agricultural University
2. Qinghai University

Oilseed rape is both an important oleaginous crop and agriculture sightseeing crop whereas has relatively scanty flower color (yellow flowers). As natural flavonoids, anthocyanins are responsible for the attractive red, purple, and blue colors of various tissues in higher plants. One *Brassica napus-Orychophragmus violaceus* disomic addition line (M4) obtained previously exhibits red petals. Transcriptome analysis of M4, *B. napus* (H3), *O. violaceus* with purple (OvP) and white petals (OvW) revealed that most of structural genes for anthocyanin synthesis were up-regulated in both M4 and OvP. Reads assembling and sequence alignment indicated that one of unigenes identified as *PAP2* in M4 was from the transcript of *O. violaceus* (*OvPAP2*). Overexpression of *OvPAP2* via CaMV35S promoter in *A. thaliana* leads to different levels of red/ purple pigments accumulation in most of organs, including petals. However, overexpressed *B. napus* exhibited restricted accumulation of anthocyanins primary in stamens. Interesting, when driven by the petal-specific prompter XY355, transgenetic plants of *B. napus* showed both red anthers and petals. RT-PCR analysis indicated that the relative expression values of *OvPAP2*, which influenced by different promoters and genetic background, are closely related to the expression of *DFR* and *ANS* as well as the red color formation in different organs. Metabonomics analysis found abundance and specific anthocyanins accumulated in red petals of M4 and XYP3 transgenic plants. Present study gives a successful case for flower color modification of the agriculture sightseeing rapeseed and provided materials for further expounding the regulation of anthocyanin biosynthesis in petals of *Brassica*.

T5

P1183

Profiling of plant lncRNA encoded polypeptide and their functions under stresses

Sisi Geng, *Zhi Lu*

MOE Key Laboratory of Bioinformatics, Center for Plant Biology, School of Life Sciences, Tsinghua University

Recent studies have found that plant long non-coding RNAs (lncRNAs) play many roles in important biological processes such as development, flowering, biotic and abiotic stress responses. With the development of high-throughput sequencing technology, more and more studies have accumulated large amounts of unannotated plant lncRNA. However, except for several studies that unfolded the detailed molecular mechanisms of several lncRNAs, functions of most lncRNAs remain elusive. Studies on lncRNAs using ribosome profiling and sequencing (Ribo-Seq) technique in both animals and plants indicate that lncRNAs have similar translational efficiency when compared to protein-coding mRNAs. But only a few lncRNA encoded polypeptides have been detected so far either due to limitations in detection methods or because they are degraded quickly. This work integrates Ribo-Seq and proteomics based on liquid chromography coupled mass spectrometry (LC-MS) optimized for short peptide detection to study the coding potential, as well as the change of lncRNA coding potential under stress conditions to demonstrate the function of lncRNAs encoded polypeptides in plants under stresses.

T5

P1184

Time-resolved metabolomics and proteomics reveals guard cell high and low CO_2 response mechanisms

Sisi Geng[1,2], *Biswapriya Biswavas Misra*[2], *Bing Yu*[3], *Ning Zhu*[4], *Evaldo De Armas*[5], *David Huhman*[6], *Hans Alborn*[7], *Lloyd Sumner*[6], *Craig Dufresne*[5], *Sixue Chen*[1,2,4]

1. Plant Molecular and Cellular Biology Program, University of Florida, Gainesville, FL 32610

2. Department of Biology, Genetics Institute, University of Florida, Gainesville, FL 32610

3. Key Laboratory of Molecular Biology of Heilongjiang Province, College of Life Sciences, Heilongjiang University, Harbin 150080, China; Engineering Research Center of Agricultural Microbiology Technology, Ministry of Education, Heilongjiang University, Harbin 150080

4. Proteomics and Mass Spectrometry, Interdisciplinary Center for Biotechnology Research, University of Florida, Gainesville, FL 32610

5. Thermo Fisher Scientific, 1400 Northpoint Parkway, West Palm Beach, FL 33407

6. Samuel Roberts Noble Foundation, Ardmore, OK 73401

7. Chemistry Research Unit, Center of Medical, Agricultural, and Veterinary Entomology, Agricultural Research Service, United States Department of Agriculture, Gainesville, FL, 32610

Foliar stomatal movement is critical for regulating plant water status and gas exchange. Elevated carbon dioxide (CO_2) is known to induce stomatal closure, and vice versa, low CO_2 induces stomatal opening. However, current knowledge on CO_2 signal transduction in stomatal guard cells is limited. Here we report metabolomic responses of *Brassica napus* guard cells to different CO_2 concentrations using a hyphenated metabolomics approach. During the one-hour CO_2 treatment, the metabolome showed a generally increasing trend under elevated CO_2, and a decreasing trend under low CO_2. This may be explained by the availability of carbon source. Interestingly, jasmonic acid (JA) biosynthesis was significantly increased under elevated CO_2 treatment. Yet under low CO_2, a branching pathway of JA biosynthesis leading to green volatiles and traumatic acid production was induced. Together with results obtained from JA biosynthesis and signaling mutants as well as CO_2 signaling mutants, we discovered that jasmonic acids partially mediates high CO_2 induced stomatal closure, but not low CO_2 induced stomatal opening. Under low CO_2, we observed other phytohormones including cytokinins, auxins, brassinosteroids, as well as melatonin increased at early time points, indicating possible roles of these hormones in mediating stomatal opening under low CO_2. Sucrose, malate, mannitol, citric acid, as well as some other organic acids were found to be significantly altered under both high and low CO_2 treatments, and their trends correlate well with the osmotic changes during stomatal opening or closure. In addition, proteomic level changes are currently under investigation to reveal how enzyme amount or modification changes may relate to metabolite changes in guard cell responses to different CO_2 levels.

T5

P1185

Heterogeneous plant phenotyping big-data aggregation for crop breeding with diverse data-acquiring platforms and technologies

Liang Gong

Shanghai Jiao Tong University

It plays an important role for high-throughput phenotyping in cutting-edge crop breeding field, and this automation generates heterogeneous measuring data for subsequent meta-analyses, modeling, and ground-truth dataset building. However, data-processing problems arise due to the lack of an effective data modeling technique and tool to manage and scale them. In order to quantitatively use the phenotyping data for crop breeding, a data aggregation methodology is proposed to tackle the heterogeneous data obtained by means of digital text, image, or point cloud. First, plant organ-specific phenotyping parameter index sets for crop breeding are optimally determined, and corresponding phenotyping instrumentation are introduced. Second, an entity-relationship data aggregation model is built to organize and present the phenotyping big data; Third, a paradigm of creating a phenotyping database is proposed to facilitate crop breeding. Finally, a formal paradigm for constructing a crop breeding phenotyping database is established, which highlights the plant morphometric data retrieval and data mining. This data aggregation scheme provides an effective tool and exemplary template for dealing with big plant phenotyping data acquired by different devices and equipment under user-defined resolution. The case study for creating a GEOP phenotyping database is step-by-step investigated to show the feasibility and effectiveness of plant phenotyping big-data aggregation.

T5

P1186

Tissue- and stage-specific processes in plant fiber development: View from the level of transcriptome

Tatyana Gorshkova, *Tatyana Chernova, Natalia Mikshina, Oleg Gorshkov*

Kazan Institute of Biochemistry and Biophysics, Kazan Scientific Center, Russian Academy of Sciences

Plant fibers are the most widely spread cell types in mechanical tissues of vascular plants. For the plant itself, fibers are important as a source of mechanical support, in defense from herbivores, and in some cases as elements with contractile properties, resembling those of muscles. Being the structural components of timber and the final product of fiber crops, plant fibers constitute one of the most important renewable resources. The major distinguishing features of fibers are the extreme cell length that can reach many centimeters and thickened cell wall. Correspondingly, the especially important stages of fiber development are cell elongation and cell wall thickening. Fiber elongation is mainly performed by intrusive growth - the type of plant cell enlargement characterized by the excessive rate of a cell elongation as compared to the surrounding tissues. During intrusive elongation, a cell splits the middle lamellae of the cells on the way and squeeze in between the surrounding tissues. Intrusive growth is very important both for the development of plant organism architecture and for the yield and quality of plant fibers, because this very process provides the extreme length to the individual fibers and also forms the cell surface for the subsequent thickening of the cell wall. However, the data on the intrusive growth are extremely scarce, since intrusive elongation occurs within the depth of plant tissues, making it extremely difficult to isolate fibers at this stage of development in quantities sufficient for wet biochemical analysis. We have isolated the intrusively growing fibers from flax stem by laser microdissection and performed the RNA-Seq analysis to obtain the transcriptome landscape. Similar type of analysis was applied for fibers isolated at advanced stage of specialization – deposition of G-layer. There are two types of plant fibers differing in design and properties. One, developing seconadary cell wall, provides static reinforcement, helping to stabilize the organism and its parts in space. The other develops contractile properties that may permit to move the plant parts and as such can be considered as plant "muscles". The latter deposits fiber-specific G-layer (tertiary cell wall). As distinct from the secondary cell wall that consists of several layers of helicoidally oriented cellulose microfibrils interlaced with xylan and lignin, G-layer contains mainly of axially oriented cellulose microfibrils and specific version of rhamnogalacturonan I. To characterize the general picture of specialized plant cell during its development in planta and to reveal the genes specifically expressed at key stages of fiber development we compared the whole transcriptome portraits of fibers during intrusive elongation and G-layer deposition and of the surrounding tissues in several stem parts. Besides, the obtained results were compared to the published data on transcriptome landscapes during wood development in tree species. Transcriptome analysis permitted to elucidate the important players involved in the key stages of fiber development, like transcription factors, ion transporters, glycosyltransferases, cell wall proteins, and provided a solid basis for the in-depth analysis of fibre-specific processes. The work was supported by the grant from Russian Scientific Foundation (project #16-14-10256).

T5

P1187

Numerous foreign mobile elements shape the mitochondrial genome of the fern *Gleichenia dicarpa*

Felix Grewe[1], *Andan Zhu*[2], *Prisca Viehoever*[3], *Bernd Weisshaar*[3], *Volker Knoop*[4]

1. The Field Museum, Chicago

2. University of Lincoln-Nebraska

3. University of Bielefeld

4. University of Bonn

Largely due to their size and complexity, only two complete fern mitochondrial genome sequences are available to date, both from early diverging eusporangiate species. To gain more insight into the evolution of the mitogenome of ferns, we sequenced the mitochondrial genome of the leptosporangiate fern *Gleichenia dicarpa* by a hybrid fosmid sequencing and high-throughput genome skimming approach. The assembled fern mitochondrial genome is very repetitive and characterized by a high frequency of integrated foreign mobile elements, such as promiscuous DNA from the plastid and nuclear genome, DNA and RNA polymerases derived from linear mitochondrial plasmids and mitoviruses, a recent intron transposition from the *rpl2* gene into the *rps1* gene, and many highly conserved fragments of retroelements. In addition, Fosmid cloning identified identical fragments of the mitochondrial genome integrated into the nuclear genome of *G. dicarpa*. Some of these nuclear mitochondrial regions were disrupted by retroelements indicating a recent massive expansion of retroelements in the nuclear genome. Therefore, the sequencing of *G. dicarpa* demonstrates how mobile elements can strongly shape mitochondrial genome patterns in addition to their effect on the nuclear genome.

T5

P1188

Cytogenetic relationships among *Citrullus* species inferred from rDNA distribution patterns

Yonghua Han

Jiangsu Normal University

Background: Comparative mapping of 5S and 45S rDNA by fluorescent *in situ* hybridization (FISH) technique is an excellent tool to determine cytogenetic relationships among closely related species. Results: In this study, the number and position of 5S and 45S rDNA loci in all *Citrullus* species and subspecies were determined. The cultivated watermelon (*C. lanatus* subsp. *vulgaris*), *C. lanatus* subsp. *mucosospermus, C. colocynthis* and *C. naudinianus* (or *Acanthosicyos naudinianus*) had two 45S rDNA loci and one 5S rDNA locus which was located syntenic to one of the 45S rDNA loci. *C. ecirrhosus* and *C. lanatus* subsp. *lanatus* had one 45S rDNA locus and two 5S rDNA loci, each located on a different chromosome. *C. rehmii* had one 5S and one 45S rDNA locus positioned on different chromosomes. The distribution of 5S and 45S rDNA in several species belonging to other genera in Benincaseae tribe was also investigated. The distribution pattern of rDNAs showed a great difference among these species. Conclusions:The present study confirmed evolutionary closeness among cultivated watermelon (*C. lanatus* subsp. *vulgaris*), *C. lanatus* subsp. *mucosospermus* and *C. colocynthis*. Our result also supported that *C. lanatus* subsp. *lanatus* was not a wild form of the cultivated watermelon instead was a separate crop species. In addition, present cytogenetic analysis suggested that *A. naudinianus* was more closely related to *Cucumis* than to *Citrullus* or *Acanthosicyos*, but with a unique position and may be a link bridge between the *Citrullus* and the *Cucumis*.

T5
P1189
Dissecting grass genome organisation at the cytomolecular level using the model genus *Brachypodium*

Robert Hasterok[1], *Alexander Betekhtin*[1], *Natalia Borowska-Zuchowska*[1], *Agnieszka Braszewska-Zalewska*[1], *Karolina Chwialkowska*[2], *Marta Hosiawa-Baranska*[1], *Dominika Idziak-Helmcke*[1], *Arita Kus*[1], *Jolanta Kwasniewska*[1], *Miroslaw Kwasniewski*[2], *Joanna Lusinska*[1], *Ewa Robaszkiewicz*[1], *Magdalena Rojek*[1], *Rakesh Sinha*[1], *Aleksandra Skalska*[1], *Marta Sowa*[1], *Elzbieta Wolny*[1], *Karolina Zubrzycka*[1]

1. Department of Plant Anatomy and Cytology, Faculty of Biology and Environmental Protection, University of Silesia in Katowice
2. Department of Genetics, Faculty of Biology and Environmental Protection, University of Silesia in Katowice

In contrast to animals, the organisation of plant genomes at the cytomolecular level is still relatively poorly studied and understood. However, the *Brachypodium* genus in general and *B. distachyon* in particular represent exceptionally good model systems for such study. This is due not only to their highly desirable 'model' biological features, such as small nuclear genome, low chromosome number and complex phylogenetic relations, but also to the rapidly and continuously growing repertoire of experimental tools, such as large collections of accessions, whole genome sequence information, large insert (bacterial artificial chromosome; BAC) libraries of genomic DNA, *etc.* Advanced cytomolecular techniques, such as fluorescence *in situ* hybridisation (FISH) with evermore sophisticated probes, empowered by cutting-edge microscope and digital image acquisition and processing systems, offer unprecedented insight into chromatin organisation at various phases of the cell cycle. A good example is chromosome painting which uses pools of chromosome-specific BAC clones, and enables the tracking of individual chromosomes not only during cell division but also during interphase. Here we outline our current projects and their future prospects, using *Brachypodium* species for research on various aspects of grass genome organisation using cytomolecular approaches and focuses on following topics: (i) karyotype structure and evolution, (ii) distribution of chromosome territories and domains within the interphase nucleus, (iii) dynamics of epigenetic modifications of chromatin during embryo development and cell differentiation, (iv) true nature of selective rRNA gene inactivation in some *Brachypodium* allopolyploids, (v) intraspecific epigenetic variation in natural populations of *B. distachyon*, (vi) instability of a small grass genome subjected to mutagenic treatments, and (vii) meiosis and chromosome pairing control in *B. hybridum*. Financial support from the National Science Centre Poland (grants no. 2012/04/A/NZ3/00572,2014/14/M/NZ2/00519 and 2015/18/M/NZ2/00394) is acknowledged.

T5
P1190
Sequencing genomes of *Blasia* and *Coleochaete* to discover keys to embryophyte success

Xiaolan He[1], *Charles Delwiche*[2], *Mari Miranto*[1], *Elina Peltomaa*[3], *Jaakko Hyvönen*[4]

1. Finnish Museum of Natural History, University of Helsinki
2. Department of Biology, University of Maryland
3. Department of Environmental Sciences, University of Helsinki
4. Finnish Museum of Natural History and Department of Biosciences, University of Helsinki

Colonization of terrestrial habitats by plants is one of the most important events in the history of life on Earth. Extant embryophytes (land plants) share a common ancestor with a grade of freshwater charophyte green algae (basal streptophytes). A range of features of charophyte green algae have been shown to be orthologous to those found in embryophytes, and bryophyte-grade plants share a common ancestor with polysporangiophytes. However, many key questions regarding the origin of embryophytes, transition from a single-celled zygote to a multicellular embryo, and development of alternating life cycles, have remained ambiguous. Challenges surrounding these questions are for example the paucity of the fossil record, the depth of phylogenetic divergence of streptophytes, as well as the limited genome-scale data of bryophytes and charophytes. We aim to sequence complete genomes of a simple thalloid liverwort *Blasia pusilla* and a charophyte *Coleochaete orbicularis* and establish them as model species for further, more detailed studies. These two species occupy a critical position in the streptophyte evolution. We aim to conduct a comparative study on the evolution of streptophytes, particularly on the molecular mechanisms underpinning the origin and success of the embryophytes. We will present our preliminary data on the axenic cultures, techniques to complete the life cycle in laboratory conditions, and the NGS techniques we are planning to use in whole genome sequencing.

T5
P1191
SINEs in Solanaceae – SINE family characteristics, gene association and utilization as molecular markers

Kathrin M. Seibt[1], *Torsten Wenke*[1], *Katja Muders*[2], *Bernd Truberg*[2], ***Tony Heitkam***[1], *Thomas Schmidt*[1]

1. TU Dresden, Institute of Botany
2. Norika GmbH, Groß Lüsewitz

Short interspersed nuclear elements (SINEs) are non-autonomous retrotransposons widespread in the plant kingdom. SINEs are short in length, non-coding and contain promoter motifs derived from cellular RNAs, primarily tRNA. Due to their lack of coding domains and high sequence heterogeneity, SINEs are difficult to identify and annotate. We identified ten Solanaceae SINE families and subfamilies using our SINE-Finder tool which enables the targeted identification based on the few structural characteristics of SINEs. Using five Solanaceae reference genome sequences including potato, tomato, and pepper, we annotated 82,900 SINEs. The copy numbers per genome of the SINE families range from tens to over eight thousand. An analysis of the sequence divergence in comparison to reference sequences revealed different age structure of the SINE family populations. The presence of highly similar copies indicates a recent activity of some SINE families in potato. Fluorescence in situ hybridization and mapping along the assembled chromosomes showed that SINEs are dispersed across all chromosomes with a preference for distal euchromatic regions similar to the distribution of the annotated genes in the genome assemblies. Our analysis revealed a strong association of some SINE families with genes. 10% of the annotated Solanaceae genes contain up to 16 SINE copies, mainly in introns and untranslated regions. SINEs can contribute splice sites, start and stop codons as well as exons to genes. Moreover, 2% of the SINEs occur in tandem-like organization and SINEs are involved in the transduction of genomic sequences across the genome. Based on the analyzed Solanaceae SINEs, a PCR-based marker system called inter-SINE amplified polymorphism (ISAP) for genotyping potato varieties was developed for applications in breeding and genebank management. Our results indicate the particular suitability of high copy number families with recent activity as the derived markers are highly polymorphic enabling us to discriminate a collection of 364 European varieties and breeding lines with only two primer combinations.

T5

P1192

Golden SusPtrit - a well transformable accession to study barley nonhost resistance

Goetz Hensel[1], *Freddy Yeo*[2], *Rients Niks*[2], *Jochen Kumlehn*[1]

1. Leibniz Institute of Plant Genetics and Crop Plant Research (IPK)

2. Wageningen University and Research Centre

Plant-microbe interactions could be roughly divided into host- and nonhost resistance mediated by different mechanisms of which most are not yet well understood. Candidate genes are at best examined in a genotype which is either resistant or susceptible for the type of resistance under study. Consequently, the genotype 'SusPtrit' would be the line of choice in studies on nonhost and partial resistance in barley-rust fungus interactions. Seedlings of this experimental line have an extraordinary high susceptibility to rust fungi to which barley is normally a nonhost. Unfortunately, this line is not amenable to *Agrobacterium*-mediated genetic transformation. To combine susceptibility to non-adapted rust fungi of 'SusPtrit' with high transformability of the model cultivar 'Golden Promise', doubled haploids (DHs) were created from hybrids between these two accessions. Four DH lines showing a level of non-adapted rust infection similar or higher than that of 'SusPtrit' were selected and tested for transformation efficiency using a construct that harbours the *gfp* reporter gene under control of the maize *UBIQUITIN-1* promoter. DH line SG062N, designated 'GOLDEN SUSPTRIT', proved to be almost as transformable as 'Golden Promise'. Also, copy numbers, expression and inheritance of the transgene were well comparable with results obtained in other transformable barley accessions. 'GOLDEN SUSPTRIT' is deemed highly valuable for the functional validation of genes putatively involved in plant-fungus interactions.

T5

P1193

Production of an anti-human immunodeficiency virus antibody in the barley endosperm

Goetz Hensel[1], *Doreen M. Floss*[2], *Twan Rutten*[1], *Udo Conrad*[1], *Jochen Kumlehn*[1]

1. Leibniz Institute of Plant Genetics and Crop Plant Research (IPK)

2. Heinrich-Heine-University

Molecular Farming is a useful option for the production of valuable recombinant proteins. As a major prerequisite, highly efficient and cost effective expression systems are required. The cereal grain is an attractive vehicle for producing recombinant proteins, because its evolution as a protein storage organ assures an environment in which proteins accumulate to high abundance and remain stable over long periods without any need for refrigeration or aseptic conditions. The anti-Human Immunodeficiency Virus antibody 2G12 is a well-studied model which had been used in many different expression systems including plants. Here, the 2G12 antibody was used to demonstrate that the endosperm of barley grains is a particularly valuable Molecular Farming platform.

T5

P1194

The first DNA genes and their ubiquitous modern derivatives: Reverse Transcriptase enabling catalytic and self-assembling RNA to become stable and regulated

Js (Pat) Heslop-Harrison, *Trude Schwarzacher*

University of Leicester

The first catalytic and self-assembling molecules were almost certainly based on RNA. In the transition to DNA, some 3.5 billion years ago, reverse transcriptase functions must have been involved. We see the likely signature of this primitive enzyme as a motif including three aspartic acid residues -D-n-D-D- (where n is a run of other amino acids and D is an aspartic acid residue) in retrotransposons, retroviruses and telomerase in modern plants. Now, retrotransposons (Class I transposable elements) and their reverse transcriptase genes, often degenerate, are the most abundant component of many genomes. Using high-throughput sequence raw-read and assembled sequence data from various crops and wild plants, with novel informatic analyses to extract motifs, we will discuss the nature of reverse-transcriptase genes, including those found in pararetroviruses, LTR- and non-LTR retrotransposons and the telomerase enzymes in plants. We will also show data from examination of their evolutionary modes including evidence for horizontal transfer at various phylogenetic levels, and their distribution, amplification, loss and homogenization in chromosomes and related genomes or plant families.

T5
P1195
Comparative genomics and phylogenetics of the moonseed genus (*Menispermum*, Menispermaceae)

Faiza Hina, *Yao Chen, Pan Li*
Laboratory of Systematic & Evolutionary Botany and Biodiversity, College of Life Sciences, Zhejiang University, Hangzhou 310058, P. R. China

Menispermum L. (moonseed) is a small genus of deciduous climbing woody vines in the moonseed family (Menispermaceae). There are three species in this genus, namely *M. canadensis* L., *M. dauricum* DC., and *M. mexicanum* Rose. They exhibit a unique eastern Asia-eastern North America-Mexico disjunct pattern (one species each in these three areas, respectively). *M. mexicanum* is thought to be the same species of *M. canadensis* by some authors, whereas some other botanists think it is a synonym of *M. dauricum* instead. The rhizomes of *M. dauricum* ("Bei-Dou-Gen") have been used as Traditional Chinese Medicine (TCM) for a long time, to treat sore throat, colitis, dysentery and rheumatic arthralgia, while some Native American tribes used *M. canadensis* medicinally as dermatological, gastrointestinal, gynecological, and venereal aids, and as remedies for various other complaints. Although *Menispermum* species have medicinal and horticultural uses, studies related to species identification and molecular phylogenetic analysis of this genus have not been reported. Here, we report the complete nucleotide sequences of two *Menispermum* chloroplast (cp) genomes (*M. canadensis* and *M. dauricum*) using Illumina paired-end sequencing technology. We aim to: (1) characterize and compare the cp genomes of two *Menispermum* species in order to gain insights into their evolutionary patterns; (2) screen and identify the most rapidly evolving cpDNA regions and polymorphic SSR markers of the *Menispermum* genome for species identification and future phylogeographic studies of the genus; (3) resolve the phylogenetic relationships within *Menispermum* using these novel markers.

T5
P1196
Genetic polymorphism for *Dreb 1* loci in bread wheat varieties with different drought resistance

Anastasiia Honcharova, *Olga Kolesnyk*
Plant Breeding and Genetics Institute – National Center of Seed and Cultivar Investigation

Drought is a major problem throughout the world, which seriously affects crop yield and quality. *Dreb* genes are transcription factors. Their active expression can lead to the regulation of the entire series of genes controlled by them. There is evidence that *Dreb* genes are involved in the formation of resistance to a number of stresses, including drought. The aim of our work was to detect the polymorphism in *Dreb 1* genes for wheat varieties of home and foreign breeding and associate this polymorphism with drought resistance of wheat varieties. To determine the genetic polymorphism for *Dreb 1* genes there was performed DNA isolation from home bred wheat varieties that are highly drought-resistant: Odes'ka napivkarlykova, Al'batros odes'kyy, Bezosta 1, Stepnyak, Odes'ka 16, Zastava, Antonivka, Dal'nyts'ka, Odes'ka 132, Hoduval'nytsya, Vatazhok, Chayka, Viktoriya; from home bred varieties of wheat that are low drought-resistant: Hostianum 237, Odes'ka 3, Karlyk 1. As material with potentially low drought resistance there were used wheat varieties and lines of foreign breeding: Dromos, Florus, Faqus, Ortequs, Platin, 114013, WW4734, Armada, Alhambra, Ascott, Artagnen, Anapurna, T153, T154, T158, LCS News, LCS 3114. The study was performed with use of PCR analysis according to recommendations of Wei *et al.,* Husyeynova observed that 717 bp amplification product was detected in the PCR analysis of DNA drought-resistant variety Barakatli-95, and indicates the presence of the *Dreb 1* gene in B genome. But 717 bp amplification fragment was not found for other drought-resistant varieties studied by Husyeynova. We have detected 717 bp amplification product of *Dreb 1B* locus for the following varieties and lines: Odes'ka napivkarlykova, Al'batros odes'kyy, Bezosta 1, Stepnyak, Odes'ka 16, Zastava, Antonivka, Dal'nyts'ka, Odes'ka 132, Hoduval'nytsya, Vatazhok, Chayka, Viktoriya, Hostianum 237, Odes'ka 3, Karlyk 1, Dromos, Florus, Faqus, Ortequs, Platin, Ascott, Artagnen, Anapurna, T154, LCS 3114. In varieties and lines Alhambra, Armada, T153, T158 and LCS News there was found amplification fragment of 715 bp for *Dreb 1B* locus. Thus, the 717 bp amplification product is present both in the drought-resistant varieties and in the varieties which are low drought-resistant. For *Dreb 1D* locus there was detected amplification fragment of 1193 bp for the following varieties and lines: Odes'ka napivkarlykova, Al'batros odes'kyy, Bezosta 1, Stepnyak, Odes'ka 16, Zastava, Antonivka, Dal'nyts'ka, Odes'ka 132, Hoduval'nytsya, Vatazhok, Chayka, Viktoriya, Hostianum 237, Odes'ka 3, Karlyk 1, Dromos, Florus, Faqus, Ortequs, Platin, Ascott, Artagnen, Anapurna, T154, LCS 3114. Additionally amplification fragment of 1191 bp was found in varieties and lines: Alhambra, Armada, T158 and LCS News. According to our research amplification fragment of 1193 bp was detected for both drought-resistant varieties and for those varieties which are not drought-resistant.

T5
P1197
Comparative transcriptome analysis of *Klebsormidium nitens* NIES-2285 between solid and liquid culture.

Koichi Hori[1], *Hiroyuki Ohta*[1,2]
1. Tokyo Institute of Technology, School of Life Science and Technology
2. Tokyo Institute of Technology, The Earth-Life Science Institute

The colonization of land by plants was a key event in the evolution of life, making the modern terrestrial environment habitable by supplying various nutrients and sufficient atmospheric oxygen. During the colonization of land, the transition species of aquatic algae must have acquired a range of adaptive mechanisms to cope with the harsh features of terrestrial environments, such as drought, high-intensity light, and UV radiation. It is generally accepted that the ancestor (s) of current terrestrial plants was closely related to present-day charophytes. The charophyte algae *Klebsormidium* is an early diverging lineage of charophytes, and usually consist of multicellular and non-branching filaments without differentiated or specialized cells. *Klebsormidium* species therefore have primitive body plans, and most species that have adapted to land also can survive in fresh water. We provided 17,060 revised gene models (Version 1.1) including UTR sequences of *Klebsormidium nitens* NIES-2285 (formerly identified as *K. flaccidum*) based on new Illumina RNA sequence data. In this presentation, we report RNA-seq data of *K. nitens* NIES-2285 in solid and liquid culture. We will discuss the factors involved in adaptation to land environments.

T5

P1198

Geophytic Organisms — Ontogeny & Phylogeny (GOOP)

Cody Coyotee Howard[1,2], *Nico Cellinese*[1,2]
1. Florida Museum of Natural History
2. University of Florida

Geophytes are a fascinating group of plants that can spend a large part of their lives underground, hidden from dangers such fire or unfavorable weather. This ability to withstand unconducive growth conditions by retreating underground has allowed species with this capability to inhabit landscapes across the globe ranging from deserts to alpine regions. Their belowground structures come in a wide array of morphological forms and include types such as bulbs, corms, tubers and rhizomes, and taxa exhibiting these structures can be found throughout the plant tree of life. While researchers have focused on select geophytic groups of interest from across the globe, the combined efforts of interested scientists would greatly increase our overall knowledge of these diverse, curious taxa. We hope to bring together the geophytophiles of the world with the Geophytic Organisms-Ontogeny & Phylogeny (GOOP) working group. The main goal of presenting at IBC is to interact with interested researchers and gauge the sciencitific community's interest in collaborative research on geophytes in order to broadly investigate their complex evolution and development.

T5

P1199

Linked and co-expressed QTLs in syntenic genomic regions for resistance to *Sclerotinia* stem rot and flowering time in *Brassica napus*

Fengqi Zhang[1,2], ***Junyan Huang***[1,3], *Mingxing Cheng*[1], *Xiaohui Cheng*[1], *Yueying Liu*[1], *Chaobo Tong*[1,3], *Jianguo Chen*[3], *Shengyi Liu*[1,3]
1. Oil Crops Research Institute of CAAS, Key Laboratory of Biology and Genetic Improvement of Oil Crops, Ministry of Agriculture
2. Cereal Crops Research Institute, Henan Academy of Agricultural Sciences
3. Hubei Collaborative Innovation Center for Green Transform

Sclerotinia sclerotiorum stem rot is a major disease in *Brassica napus* oilseed rape worldwide, particularly in China where annual yield loss is more than 10%. A close association of disease incidence with flowering time is widely observed in oilseed rape fields or germplasm. Up to now there are no reports on cloning of resistant QTLs and its definition of the association. To reveal the genetic relationship between the two traits, we investigated QTLs of both traits in a recombinant inbred line population derived from a cross between 888-5 (a susceptible and early flowering line) and M083 (a highly resistant and later flowering line). The population was genotyped by employing a *B. napus* 60K SNP array which produced a high density genetic map comprising 9,278 SNPs and spanning 4,071 cM of the genome. QTL mapping was conducted with phenotypic data of stem lesion size of artificial inoculation and both the disease index and flowering time which were evaluated in eight environments. A total of 31 QTLs for resistance were detected. Of them, the major QTL *qSSE2-1* in A2 explained 10.50%-27.30% of the phenotypic variation across the eight environments, but with no interactions with environments. For flowering time, 21 putative QTLs were found. Of them, the stable major QTLs *qFTE2-1* and *qFT2-6*, contributing to 16.40%-34.50% of the phenotypic variation in six environments, co-located with stable major QTLs *qSSE2-1* and *qSS2-9* in a narrow A2 region, but they are closely linked, rather than pleiotropy of a single gene. A syntenic analysis approach was used to search co-expressed QTLs in syntenic regions between the subgenomes A and C. Of syntenic regions found, *qSS2-10* in the A2 region is syntenic to *qSS12-1* in a C2 region and within the two QTL regions, two pairs of orthologous genes (in total, five genes and of them, two are tandem duplicates) were identified as preferential candidates which were annotated as hormone biosynthesis and transcription factor, respectively. Our results on the linked QTLs and the co-expressed syntenic QTLs will promote their fine mapping and help improve efficiency of breeding for *Sclerotinia* resistance and early maturation in oilseed rape.

T5

P1200

The population genetics of *SODA1* indicated a possible divergent selection in two closely related wild rice—*Oryza rufipogon* and *O. nivara*

Lei Huang[1], *Yi Ren*[1], *Song Ge*[2]
1. College of Life Science, Shaanxi Normal University
2. State Key Laboratory of Systematic and Evolutionary Botany, Institute of Botany, Chinese Academy of Sciences

Ecological speciation accentuates the role of divergent selection in the process of two species adapting to different habitats. Identifying the candidate genes which may be the target of divergent selection or play a key role in ecological speciation is of great importance in evolutionary researches. *Oryza rufipogon* and *O. nivara* are two incipient species at the early stage of speciation with distinct differences in morphology, life history traits and habitat preference. Previous studies showed that *O. rufipogon* and *O. nivara* were good model of ecological speciation. In this study, we identified a candidate gene-*SODA1* by outlier approach based on whole genome re-sequencing data. The F_{ST} was very high in *SODA1* and we found 15 highly differentiated SNPs in this gene between *O. rufipogon* and *O. nivara*. In this study, we sampled six pairs of sympatric natural populations to verify the genetic differentiation of *SODA1* and carried out neutrality tests to detect possible selection signals in gene full length, 3 kb upstream and 3 kb downstream. In the 120 individuals sequenced, we identified a total of 36 highly differentiated SNPs in all six species pairs, and 3 SNPs were in coding region while 2 SNPs were nonsynonymous mutations. Interestingly, the two nonsynonymous mutations were found in all six species pairs, in contrast with other highly differentiated SNPs which were detected only in part of samples. Secondly, the average F_{ST} was relatively higher in *SODA1* gene region than those in upstream or downstream. The neutrality tests showed that *SODA1* gene might undergo positive selection in gene region. However, the selection signals were not detected in either upstream or downstream region. This indicated that the two nonsynonymous mutations would be more likely to the target of positive selection, and divergent selection might drive the genetic differentiation between *O. rufipogon* and *O. nivara* on *SODA1*. Previous gene function researches proved that *SODA1* was induced by drought and salinity stress. Since *O. nivara* adapted to seasonally dry environment from deep water habitat of *O. rufipogon*, the divergent selection of *SODA1* provided us new insights into the ecological speciation of *O. rufipogon* and *O. nivara*. Further work on the gene function may help us uncover the puzzle of how the divergence of *SODA1* is related to the speciation process. Furthermore, the different pattern of genetic differentiation on *SODA1* in 6 species pairs implied multiple origins of *O. nivara*, which was in accordance with previous studies.

T5

P1201

Using CALUX bioassay to determine the dioxins in the effluent and the sludge of the pulp and paper enterprises in Guangxi Province, China

Ning Huang
Guangxi Environmental Monitoring Center

Chemical activated luciferase gene expression (CALUX) could be utilized for quantification of dioxins, and possesses advantages over high resolution gas chromatography/high resolution mass spectrometer (HRGC/HRMS) for short testing period, low testing cost, large quantities of samples analyzed in parallel and the total toxic potency of samples determining. It is applied specially for large-scale high throughput screening of large quantities of samples. In order to collect data for the "China-dioxins reduction from the pulp and paper industry project" sponsored by the global environmental facility (GEF), both samples of the effluent and the sludge collected from processing departments of several pulp/paper production enterprises in Guangxi Province were analyzed with CALUX bioassay. Water samples were collected at 40 L, and net weights of sludge samples 20 g. The analyzed results demonstrated that (1) The value of toxic equivalent quantities of dioxins in the effluent were within the range of 3.0 ~ 167 pg-TEQ/L, while the value of toxic equivalent quantities of dioxins in the sludge were within the range of 3.3 ~ 5.8 pg-TEQ/g. (2) Most dioxins content in processing effluent discharged by the pulp/paper production enterprises was higher than the discharged control limit value 30 pg-TEQ/L referenced as the discharge standard of water pollutants for pulp and paper industry (GB3544-2008), while the levels of dioxins in sludge were relative lower. (3) The dioxins content in the factory applied for chlorine bleaching process was higher than that for elemental chlorine-free bleaching process. This work was first conducted in effluent and sludge for the pulp/paper industries in Guangxi Province, and the testing results can provide valuable data for the treatment of dioxin pollution to the government of Guangxi.

T5

P1202

Polyploids are not the genetic and ecological additivity of their progenitors

Stella Huynh[1], *François Felber*[1,2], *Olivier Broennimann*[3], *Antoine Guisan*[3], *Christian Parisod*[1]
1. Laboratory of Evolutionary Genetic, University of Neuchâtel
2. Musées et jardins botaniques cantonaux vaudois
3. Laboratory of Spatial Ecology, University of Lausanne

Polyploid organisms contain the chromosome sets of their progenitors and are therefore expectedly characterized by the additivity of their genetic backgrounds and ecological properties. Previous studies have long rejected this assumption, showing genome restructuring and ecological niche shifts following polyploidy. However, evolutionary processes shaping parental subgenomes in polyploids and their influence on the polyploid success remain unclear. We investigated here four diploids *Aegilops* species that differentially combined in four derivative tetraploids with 361 representative accessions that were genotyped at 35 single-copy nuclear loci. Amplicon sequencing with overlapping paired-end reads provided individual haplotypes which dosages were inferred based on their proportions. All four polyploids showed genome composition that is coherent with their selfing reproductive system, with homologs usually homozygous while heterozygosity is mostly fixed among homeologs. Some loci however presented alleles from only one of the parental subgenomes, highlighting asymmetrical genome reorganization with only part of the genetic diversity present in the diploid progenitors being retained in polyploids. To address whether this asymmetry is reflected in species ecological requirements, the niche of polyploid species was further compared to the addition of their diploid progenitors. Niche modeling based on 5821 georeferenced occurrences and 20 bioclimatic factors supported parallel ecological shifts with niche expansion towards moister and colder environments for three polyploids but substantial niche retraction for the last one. Niche comparisons between polyploids and their respective diploid ancestors show a symmetrical overlap with their progenitors for two polyploids while two others present asymmetrical overlap toward their common progenitor species. The reported non-additivity of *Aegilops* polyploids is thus characterized by asymmetrical changes of genetic and ecological traits, providing new insights on processes driving long-term evolution following polyploidy.

T5

P1203

What have we learnt about plant drought tolerance mechanisms from high throughput phenotyping platforms?

Puertolas Jaime, *Ryan Annette C, Rothwell Shane A, Davies William J, Dodd Ian C*
Lancaster Environment Centre. Lancaster University

Selecting drought tolerant genotypes is one of the main research targets to increase crop yield under limited water availability, but plant responses to water deficit are complex and the traits controlling those responses can have different impacts depending on the drought scenario. Technologically advanced high-throughput screening platforms equipped with automated irrigation systems can measure a wide range of morpho-physiological variables in a large number of genotypes, to link phenotype to gene. At Lancaster, we have installed a "water use efficiency" phenotyping platform within our greenhouses, comprising 200 balances that continuously monitor crop water use. Maize plants (comprising 100 different genotypes) were grown with well-watered conditions under naturally varying atmospheric vapour pressure deficit (VPD), and transpirational responses to VPD and whole plant water use efficiency (WUE, defined as the accumulation of above ground biomass per unit of water used) were measured. While some genotypes showed a linear response of transpiration to VPD, others showed a non-linear response with a pronounced change point. Variation in the transpiration response to VPD was linked to genotypic ranking for WUE. When plants were grown under water-limited conditions, the response of xylem ABA concentration (a measure of soil water availability) was not uniquely related to transpiration rate or stomatal conductance according to the irrigation technique. At the same whole pot soil water content, maintaining sub-optimal soil water by frequent irrigation resulted in larger soil water content gradients, lower root and xylem ABA concentrations, along with higher transpiration rates or stomatal conductance, compared to plants from which water was withheld. Since physiological responses to soil drying depend on the irrigation method, genotypic variation in "drought tolerance" derived from phenotyping platforms must be carefully interpreted.

T5

P1204

Phenotyping from the lab to the field and back

Marcus Jansen
LemnaTec GmbH

Plant phenotyping activities frequently originated from laboratory test, but since plants strongly respond to the environment, field based measurements gained high importance in recent times. At the same time, demand for gaining physiological data together with the morphometric properties came into play. Physiological measurements frequently are more complex than measuring plant dimensions, so that the laboratory environment gained importance for driving the development of new tools for physiological phenotyping. Physiological processes, as part of the development or in response to the environment or other organisms, cause changes in the plants that can be detected by means of non-invasive sensing. LemnaTec approaches such challenges by improved image and data processing for measuring disease and stress responses and by establishing tools for measuring early stages of plant development. Involving wavelengths beyond the visible spectrum as well as spectrally resolved images these developments will allow new insights in plant physiology and help to address new types of phenotypic data that complement classical laboratory analyses.

T5

P1205

Research on gene clone and its expression, function in *E. coli* of *PselF5A* from peony

Changhua Jiang[1,2], *Yan Gao*[1,2], *Kan Ye*[1,2], *Yao Song*[1,2], *Guangqi Zhao*[1,2], *Shucheng Feng*[1,2]
1. Shanghai Botanical Garden
2. Shanghai Engineering Research Center of Sustainable Plant Innovation

Eukaryotic translation initiation factor 5A (eIF5A) is the only cellular protein known to contain the unusual amino acid hypusine. It is a highly conserved protein found in all eukaryotic organisms. Although originally identified as a translation initiation factor, recent studies suggest that eIF5A is mainly involved in translation elongation, mRNA turnover and decay, cell proliferation, and programmed cell death. *RcelF5A* cloned from *Rosa chinensis* expression is up-regulated in *Rosa chinensis* under high temperature, and oxidative and osmotic stress conditions. In this research, *PselF5A*, a gene induced by heat stress, had been cloned from '*Fengdanbai*', one heat-tolerance variety of *Paeonia suffruticosa* under 38 ℃ /3h high temperature and its ORF has 480bp, with high homology comparison of nucleotide and amino acid sequence in *VvelF5A* from grape and *MaelF5A* from *Mongolian ammopiptanthu*. Recombinant *PselF5A* had detected in recombine *E. coli* BL21 by SDS-PAGE and western blot and its expression had made recombinant *E. coli* increase the abiotic stresses, such as high temperature (50℃), low temperature (4℃), heavy metal (300mM LiCl and 350mM $CdCl_2$), salt (550mM NaCl), high Ph (15mM Na_2CO_3) and oxidation (400um H_2O_2). So, this research result had indicated *PselF5A* can response to various abiotic stresses.

T5

P1206

Mysterious beauty: Evolution of R2R3-MYB transcription factors in plants

Chen-Kun Jiang, *Jing-Dan Han, Guang-Yuan Rao*
School of Life Sciences, Peking University, Beijing 100871, China

R2R3-MYB proteins are one of the largest groups of paneukaryotic transcription factors (TFs) in plants. Although many studies have made the classification of plant R2R3-MYB proteins using phylogenetic approaches, there is no consensus on the classification system due to lack of reliable and robust criterion in the identification of homologous sequences. More importantly, little attention was paid into deep phylogenetic relationships among R2R3-MYB TFs, which means that the origin and evolution of plant R2R3-MYB proteins remain unclear. In this study, we establish a new criterion to identify R2R3-MYB proteins, and use genome sequences from 38 species of algae and land plants to explore the relationships between these sequences. Phylogenetic analyses indicate that land plant R2R3-MYB proteins constitute nine subfamilies, and eight of them present in all of the major lineages, including bryophytes, lycophytes, gymnosperms and angiosperms. No member of R2R3-MYB subfamilies is identified in glaucophytes, rhodophytes and chlorophytes, whereas a stepwise increase in the number is found in streptophytes before terrestrialization of embryophytes. The subsequent increase of R2R3-MYB TFs in seed plants mainly result from the diversification of subfamily VIII, the largest subfamily. Interestingly, our study shows that DNA binding profiles are highly variable within subfamily VIII, while DNA binding profiles of other subfamilies are relatively conserved. We conclude that the R2R3-MYB TFs specific to land plants emerged in early embryophytes, and their diversification is attributed into intrinsic features of subfamily VIII MYB proteins as well as increasing complexity of biological processes.

T5

P1207

ODE reconstruction of high-dimensional genetic networks through game theory with application to dissecting tree salt tolerance

Libo Jiang[1], *Rongling Wu*[1,2]
1. Beijing Forestry University
2. The Pennsylvania State University

Ordinary differentiation equations (ODE) have proven to be powerful for reconstructing precise and informative gene regulatory networks (GRNs) from dynamic gene expression data. However, joint modeling and analysis of all genes, essential for the systematical characterization of genetic interactions, are challenging due to high dimensionality and a complex pattern of genetic regulation including activation, repression and antitermination. Here, we address these challenges by unifying variable selection and game theory through ODE. Each gene within a GRN is co-expressed with its partner genes in a way like a game of multiple players, each of which tends to choose an optimal strategy to maximize its "fitness" across the whole network. Based on this unifying theory, we designed and conducted a real experiment to infer salt tolerance-related GRNs for Euphrates poplar, a hero tree that can grow in the saline desert. The pattern and magnitude of interactions between several hub genes within these GRNs were found to determine the capacity of Euphrates poplar to resist to saline stress.

T5

P1208

Transcriptome analysis reveals the morphological difference of *Opisthopappus* Shih.

*Juan He, Wei Chen, **Yiling Wang***
College of Life Science, Shanxi Normal University

Opisthopappus (Asteracceae) were consisted of *Opisthopappus longilobus* shih and *Opisthopappus taihangensis* (Ling) Shih. Deepen our understanding of phylogenetic relationships ultimately assist morphological difference of the *O. longilobus* and *O. taihangensis* through molecular methods. However, the molecular methods are lack of genomic resource; it only provided limited information on full-length genes to compare the *O. longilobus* and *O. taihangensis* on the phylogenomics analysis. In our study, samples of *O. longilobus* and *O. taihangensis* were collected and sequenced, from which total RNA were ready for RNA-Seq. The transcriptome library was constructed, and first strand cDNA and second strand cDNA were synthesized. As result of sequence, average 35 million clean 150 bp-paired-end reads were obtained from three *O. Longilobus* and *O. taihangensis* tissue libraries, generated a total of 12.9 giga-bases (Gb) of sequences. A average 306,839,788 clean reads was obtained from 319,668,420 raw reads with percent of 99.99% To annotate the Trinity-assembled unigenes, the 33,974 unigenes were subjected to Blastx searches against six public databases. Based on the protein annotation results from the NR database homology search, 7,533 unigenes (Biological process: 5,231 unigenes; Cellular component: 4,689 unigenes; Molecular function: 6198 unigenes) could be assigned to 50 GO groups. The number of reads mapped back to the orthologous region of each gene in proper pairs were 45,161,366 (71.54%) and 47,663,788 (69.42%) in the *O. taihangensis* and *O. longilobus*, respectively. To validate gene expression, 52 DEGs related to somatic embryogenesis were selected for qRT-PCR analysis. Totala 3410 differentially expressed transcripts were detected, with 1925 up-regulated and 1485 down-regulated genes. The conclusion showed that the information obtained in this study will undoubtedly provide genomic resource and insights into the molecular mechanisms of leaf forms diversity of *Opisthopappus* Shih.

T5

P1209

A specialised information service for biodiversity research, involving large-scale data mobilisation by mining German biodiversity literature

Gerwin Kasperek[1], *Marco Schmidt*[2]
1. University Library Johann Christian Senckenberg
2. Senckenberg Gesellschaft für Naturforschung

Biodiversity research heavily relies on recent and older literature, and the data contained therein. Despite great effort, large parts of the literature and the data it holds are still not available in appropriate formats needed for efficient compilation and analysis. As a part of the current funding strategy of the German Research Council (Deutsche Forschungsgemeinschaft, DFG), and resulting from an extensive dialogue with the scientific community in Germany, a "Specialised Information Service" (Fachinformationsdienst, FID) for Biodiversity Research will be established with the objective of making further segments of literature about biodiversity available in up-to-date formats. This project, starting 2017, is conducted by the University Library Johann Christian Senckenberg (Frankfurt/Main, Germany) together with the Senckenberg Gesellschaft für Naturforschung and the Text Technology Lab of the Goethe University (Frankfurt/Main). The new Specialised Information Service for Biodiversity Research (FID Biodiversitätsforschung) comprises four core elements: An essential focus is on the mobilisation of biodiversity data from the literature via a text mining approach which encompasses advanced text technologies and a large body of 20th century literature (Module 1). The aims of this module are twofold: to mobilise data on selected keystone groups (vascular plants, butterflies, birds) in the Central European literature, and, to develop reusable text mining tools which are adapted to the domain-specific contents and formats of the biodiversity literature. The chosen approach involves integration of botanical, zoological and ecological ontologies (such as PO, FLOPO, ENVO) as well as ontology development, maintenance and enhancements. The digitisation of German literature (Module 2) provides a significant part of the text corpus which is a necessary prerequisite for Module 1. The online availability of 20th century literature regarding the selected keystone groups will be substantially improved. Furthermore, Module 2 provides a basis for the journal platform (Module 3). A platform für Open Access journals (Module 3) will be established as a long-term service for biological academic societies and not-for-profit editors. This service also comprises the transfer of print-only titles into digital formats. The acquisition of specialised print literature (Module 4), combined with efficient services for document delivery, ensures nation-wide availability for researchers in Germany, with delivery also available to other countries. Built on the foundation of the comprehensive holdings of biological literature supported by the DFG for decades, this service encompasses the entire bandwidth of biodiversity literature.

T5

P1210

Identification of novel QTLs for tea quality traits in black tea (*Camellia sinensis*)

Kipyegon Robert Koech
University of Pretoria

The tea plant is indigenous to southern and eastern Asia and has been cultivated in China since ancient times. Tea, a product made up from the leaf and buds of the plant *Camellia sinensis*, is one of the most widely consumed beverages in the world, and has become an important agricultural product. Tea being a popular beverage, has attracted attention of both the consumers and the scientific community because it contains a number of bioactive compounds, such as the flavonoids, caffeine, L-theanine and γ-aminobutyric acid (GABA). It has been demonstrated to have a wide variety of pharmacological properties including antioxidant, anti-inflammatory, anticancer, antidiabetic, neuroprotective effects. These health benefits associated with tea intake have led to high consumption of tea products in regards to tea extracts in pharmaceutical, food and beverage and cosmetic industries. In addition, to these health benefits, tea quality is a complex phenomenon and depends mainly on the biochemical components that determine colour, aroma and taste, attached to the tea infusions. Tea quality is a complex and subjective trait, and breeding for this trait is time consuming and depends on knowledge of the genetics of its components. In a study conducted in Kenya, we have developed an integrated genetic map of tea (*Camellia sinensis*) using DArTseq platform on a segregating 251 F_1 population derived from a reciprocal cross between two highly heterozygous commercial tea cultivars (TRFK 303/577 and Gw Ejulu). High-density linkage maps of both parents were constructed based from the F_1 population that was taken from cross-pollinated population using integrated approaches (regression algorithm). The map consisted of 14 linkage groups that spanned 1,162.80 cM with mean interval of 1.2 cM between markers. A total of 15 phenotypic traits were assessed in the two tea plant populations. Both interval and multiple QTL mapping revealed a total of 33 putative QTLs in the 14 LGs associated with tea quality at a significance genome-wide threshold of 5%. In total, six caffeine QTLs, 23 catechins QTLs, three theaflavins QTLs and one QTL for liquor brightness were detected. Of these 33,13 major QTLs were identified for five traits in three main regions on LG01, LG03, LG11, LG12 and LG13. QTLs associated with individual catechins, caffeine and individual theaflavins were mostly located clustered in LG01 and LG03 but in different regions of the map. The population variability explained by each QTL was predominantly at moderate-to-high levels and ranged from 5.5% to 56.6%, with an average of 10.7%. The identification of QTLs that are tightly linked to these markers at the seedling stage and using UPLC technique coupled with genetic markers would greatly accelerate development of elite tea cultivars. These will also contribute to development of marker-assisted selection (MAS) for tea quality improvement.

T5

P1211

Molecular genetic polymorphism of bread wheat varieties and associations of microsatellite loci alleles with agricultural valuable traits

Olga Kolesnyk, *Andrii Venger*
Plant Breeding and Genetics Institute – National Center of Seed and Cultivars Investigations (PBGI)

The executed researches are devoted to estimation of allelic diversity of microsatellite loci, differentiation and identification of modern bread wheat varieties (*Triticum aestivum* L.) originated in Plant Breeding and Genetic Institute and registered in State Register of plant varieties suitable for dissemination in Ukraine during different years. There were analyzed the intravarietal heterogeneity and distribution of allele frequencies of microsatellite loci that have selective and adaptive value for the South region of Ukraine. Based on the results of SSR analysis of studied varieties, molecular genetic registration certificates of all varieties were created. To ensure the unambiguous differentiation and identification of modern bread wheat varieties with a common origin from a single breeding center there were created the minimally required working sets of microsatellite markers that have close differentiating abilities and consist of six-eight microsatellite loci. Determining of specific genomic regions which control important agronomical traits is considered to be one of the main wheat breeder's goals. The detection of significant associations between microsatellite markers and such traits as heading time, plant height, awn size, colour and size of ears, which can be used for applying in plant molecular breeding to improve the efficiency of marker assisted selection (MAS) for finding important quantitative trait loci (QTLs). 34 marker trait associations (MTAs) for heading time, 37 for plant height, 32 for awn size, 18 for ear colour and 8 for size of ears were proved to be stable and significant in all growing seasons. The SSR markers *Xgwm186-5A* and *Xgwm3-3D* were significantly associated with all analyzed traits and showed stability in all four years for plant height. Analysis of microsatellite markers associated with grain colour was performed in order to identify the important regions involving in this trait. As a result, 36 MTAs were found to be significant in one-three growing seasons, of which 19 MTAs were significantly associated with the lighter grain colour (CI_g) while 17 MTAs showed association with the darker Cl_g. Among studied 17 microsatellite markers we have found alleles of *Xgwm357-1A*, *Xgwm18-1B*, *Xtagl-gap-1B*, *Xgwm155-3A*, *Xgwm389-3B*, *Xgwm3-3D*, *Xgwm186-5A*, *Xgwm190-5D*, *Xgwm325-6D*, *Xgwm577-7B*, *Xgwm44-7D* and *Xwmc405-7D* associated with the value of CI_g. During the analysis of microsatellite markers associated with 1000 kernel weight (TKW) 26 MTAs were found to be significant in one – three growing seasons, of which 12 MTAs were significantly associated with the larger value of TKW while 14 MTAs showed association with the smaller value of TKW. Among studied 17 microsatellite markers we have found alleles of *Xgwm18-1B*, *Xgwm095-2A*, *Xgwm389-3B*, *Xgwm3-3D*, *Xgwm165-4A*, *Xgwm186-5A*, *Xgwm408-5B*, *Xgwm325-6D* and *Xwmc405-7D* associated with TKW. Microsatellite markers which showed significant associations and stability in different seasons can be useful and suitable for improving the efficiency of MAS in Ukrainian wheat breeding programs.

T5

P1212

Cytogenetic and epigenetic analysis of rDNA Loci in invasive *Spartina anglica* allopolyploid and the homoploid *S. × townsendii* hybrid

Alena Kuderová[1], *Ilia Leitch*[2], *Dalibor Húska*[1], *Julie Ferreira De Carvalho*[3], *Andrew R. Leitch*[4], *Malika Ainouche*[4], *Aleš Kovařík*[1]

1. Institute of Biophysics, Academy of Sciences of the Czech Republic, Královopolská 135, 612 65 Brno, Czech Republic

2. Jodrell Laboratory, Royal Botanic Gardens, Kew, Richmond, Surrey TW9 3AB UK

3. 3UMR 6553 ECobio Université de Rennes 1, Bât. 14A Campus de Beaulieu, France

4. Queen Mary University of London, School of Biological and Chemical Sciences,, Mile End Road, London E1 4NS, UK

Interspecific hybridisation between the hexaploid *Spartina alterniflora* (♀, 2n=6x=62, genomic composition AA) from North America and the native British hexaploid *S. maritima* (♂, 2n=6x=60, genomic composition MM) produced the largely infertile homoploid hybrid *S. × townsendii* (2n=6x=62, genomic composition AM) first collected in 1870 at Hythe, Southampton, UK. The derived fertile allopolyploid *Spartina anglica* C.E. Hubbard dodecaploid (2n=12x=120-124), genomic composition AAMM) is thought to have formed by spontaneous duplication of chromosome sets soon after the hybridisation event. Invasive *S. anglica* has rapidly expanded in distribution and occurs on several continents. In evolutionary ecology, *S. anglica* has become recognized as a model system of recent allopolyploid speciation. Cytogenetically it represents a difficult object due to a large number of chromosomes and limited availability of tissues containing mitotic cells. Samples were also collected from 11 French (mostly Brittany) and 3 British (Southampton water) populations of *Spartina.* The localities were repeatedly visited during 2011-2016. The ploidy levels were determined by flow cytometry. Real-time RT-PCR using locus-specific primers allowed us to quantify homeologous 35S rRNA transcripts. rDNA loci were visualized on chromosomes by fluorescent *in situ* hybridization (FISH) in four individuals collected from the British populations using the 18S and 5S rDNA probes. The *S. × townsendii* hexaploid hybrid inherited two sites of each 35S and 5S rDNA. The derived *S. anglica* dodecaploid (two individuals) inherited four sites of each 35S and 5S rDNA. However, the third *S. anglica* individual from the same locality (Ealing Marchwood, UK) harboured only two 35S sites, and Southern blot hybridisation analysis revealed that the M-genome loci were lost. At the expression level, large survey of more than 60 individuals collected from English and French populations revealed: (i) The A-genome genes were always expressed and frequently dominant. (ii) The M-genome genes were silenced in most populations except of two populations from Southern Brittany where codominance was observed. (iii) *S. × townsendii* and *S. × neyrautii* (independent hybrid from the south of France) exhibited strong silencing of M-genome units indicating that interspecific hybridisation is sufficient to trigger epigenetic inactivation of rDNA homeologs. Despite large number of chromosomes and high ploidy levels (up to 12x) *Spartina* allopolyploids tend to keep diploid numbers of rDNA loci. This is explained by rapid locus loss following polyploidisation events. Though most *S. anglica* individuals we analysed here retain locus and gene additivity we identified one individual that lost the M-genome locus suggesting that the process of rDNA "diploidisation" may have already started in these recently formed allopolyploids. We hypothesize that locus inactivity may eventually lead to its elimination. Thus, there might long-term genetic consequences of immediate and directional epigenetic changes triggered by interspecific hybridisation.

T5

P1213

Multicolour FISH-based analysis of mutagen-induced micronuclei in *Brachypodium distachyon*

Arita Kus, *Jolanta Kwasniewska, Robert Hasterok*

Department of Plant Anatomy and Cytology, Faculty of Biology and Environmental Protection, University of Silesia in Katowice

Higher plant bioassays are well established systems for screening and monitoring environmental genotoxicity. *Brachypodium distachyon* is an internationally accepted model grass, which possesses numerous, highly desirable 'model' biological features, such as one of the smallest (~350 Mb) nuclear genomes described to date in Poaceae with low (x = 5) chromosome number. This grass is also characterised by a small stature, self-fertility, short life cycle and undemanding growth requirements. Due to well-studied organisation of the *B. distachyon* nuclear genome both at the molecular and cytogenetic level, we used this species as a reference organism, which would be a convenient system in mutagenesis to analyse "hot spots" of DNA damage in a small genome grass. The advances in molecular biology enabled the progress in methods of the detection and estimation of genotoxicity of different agents. Here, we quantitatively analyse the maleic hydrazide (MH)-induced micronuclei in order to examine the involvement of specific chromosomes or chromosome fragments in their formation. Fluorescence *in situ* hybridisation (FISH) with four different and used simultaneously DNA probes (5S rDNA, 25S rDNA, *Arabidopsis*-type (TTTAGGG)n telomeric sequence and *Brachypodium* originated centromeric BAC (Bacterial Artificial Chromosome) clone CB33J12 was applied in the quantitative and qualitative analysis of the micronuclei. The results allowed to distinguish ten different types of micronuclei, which were characterised by the presence or absence of specific FISH signal (s). We demonstrated that the most of the MH-induced micronuclei originated from the distal regions of chromosomes. Furthermore, for the detailed analysis of mutagen-induced genome changes we used FISH with selected single-copy chromosome-specific BAC clones. Our findings demonstrate a promising potential of *B. distachyon* to be a valuable model to analyse the effects of various genotoxic agents on the plant nuclear genome stability. This work was supported by the National Science Centre, Poland (grant no. 2012/04/A/NZ3/00572).

T5

P1214

Transcriptome analysis of *Michelia maudiae* petals: Putative

genes associated with flower colour formation

Xiaoan Lang[1,2,3], Lingfei Li[3], Leilei Yang[3], Jianhua Qi[1], Tiecheng Cui[2], Shouzhou Zhang[3]
1. Southwest Forestry University
2. Zhaoqing University
3. Fairylake Botanical Garden, Shenzhen & Chinese Academy of Sciences.

Michelia maudiae Dunn is one of the most highly valuable medicinal and ornamental plants with white colour flowers. However, in the wild there also exist another mutant *M. maudiae* Dunn var. *rubicunda* Yi et J. C. Fan, whose flowers can be naturally pink, thus it provides an opportunity to unravel the complex regulation networks of colour formation in Magnoliaceae. White petals cDNA library of *M. maudiae* and pink petals cDNA library of *M. maudiae* var. *rubicunda* in two individual developmental stages were used for transcriptome sequencing using Illumina HiSeq 2500 platform. Among all 109,729 uni-transcripts isolated. A total of 60,538 unigenes were annotated by comparison with non-redundant National Center for Biotechnology Information (NCBI) protein (Nr), non-redundant NCBI nucleotide (Nt), Gene Ontology (GO), and Kyoto Encyclopedia of Genes and Genomes (KEGG) databases. Nr annotation results revealed that the squences of *Michelia* showed more similarity to Nymphaeaceae than to other families, with 37.64% match to *Nelumbo nucifera*. Total 41,231 differential experssion genes (DEGs) were identified by comaprison between different groups. GO term enrichment analysis of these DEGs showed that ATP binding, intergral component of membrane, and metabolic process were dominant; KEGG pathway enrichment analysis revealed they were mainly involved in spliceosome, starch and sucrose metabolism, biosynthesis of amino acids. Futhermore, total 258 unigenes were identified as putative homologues of colour-related genes in other species, of which 142 showed differential expression between white and pink petals, account for 55.0%. Interestingly, *4CL*, *F3H*, *DFR*, *UFGT* showed low expression in white while high expression in pink petals which could be the direct reason led to transtion of petals from white to pink, although the detail mechanism still need to clarify. This study will provides a new theoretical basis and novel gene resources for tree breeding.

T5
P1215
Cis-regulatory variations of a single copy gene associate with root growth adaptation to high iron levels in *Arabidopsis thaliana*

Baohai Li, Wolfgang Busch
Gregor Mendel Institute, Austrian Academy of Sciences, Vienna Biocenter, 1030 Vienna, Austria

The geographic distribution of heavy metals in soils is a critical factor that determines the diversity and population size of plant species in a local region. Heavy-metal tolerance in plants is therefore important for local adaptation. While, Iron (Fe) is the most important transition metal in all living organisms, it is highly toxic to cells when present in excess. In past decades, several important molecular components involved in Fe uptake, transport and homeostasis have been identified, however, which genes and mechanisms are responsible for the natural genetic variation of plant sensitivity to Fe toxicity are largely unknown. Here, we conducted genome-wide association studies (GWAS) in 319 *Arabidopsis thaliana* accessions using root growth responses to a high level of Fe. We identified one single copy gene, encoding for an alcohol dehydrogenase that is highly conserved among different species to be the major locus responsible for the variation of root growth responses to high Fe in the *Arabidopsis* population used for the GWAS. Compared with the reference (Col-0) allele, the variant is associated with increased root sensitivity to high Fe and a low expression level. Interestingly, the lowly expressed variant occurs predominantly in accessions derived from areas with a higher soil pH; a growth environment with decreased Fe bioavailability. In agreement with the hypothesis that activity of this gene is important for growth under high Fe, the root growth of a knockout mutant line is negatively correlated with the Fe level supplied in growth medium, and an overexpression line displays enhanced resistance to high Fe. As the variation of its expression level seems key for its role in natural variation, we aimed to map the expression quantitative trait loci that underlie this expression variation, by making use of a GWAS with its transcript levels in 665 *Arabidopsis* accessions. Our analysis revealed that natural variation of our candidate gene's expression is mainly due to cis-regulatory variation in multiple independent SNPs in its promoter region. In summary, we have identified cis-regulatory elements that most likely confer intraspecies expression variation of an otherwise highly conserved alcohol dehydrogenase that according to its expression level modulates root growth in response to high Fe levels. This suggests that the modulation of enzyme levels counteracts the detrimental effects of high Fe may be an important adaptation to high Fe environments in *Arabidopsis*.

T5
P1216
Development of incompletely fused carpels in maize ovary revealed by miRNA, target gene and phytohormone analysis

Hongping Li[1], Ting Peng[1], Yufeng Wu[2], Qun Wang[1], Jianfeng Chang[1], Moubiao Zhang[1], Guiliang Tang[1,3], Chaohai Li[1]
1. Agronomy College, Collaborative Innovation Center of Henan Grain Crops, Henan Agricultural University, Zhengzhou 450002
2. Bioinformatics Center, National Key Laboratory of Crop Genetics and Germplasm Enhancement, Nanjing Agricultural University, Nanjing 210095
3. Department of Biological Sciences, Michigan Technological University, Houghton, MI 49931

Although the molecular basis of carpel fusion in maize ovary development remains largely unknown, increasing evidence suggests a critical role of microRNAs (miRNAs). Here, a combination of miRNA sequencing, degradome and physiological analysis was used to characterize carpel fusion development in incompletely fused carpels (IFC) ovary and completely fused carpels (CFC) ovary in maize. A total of 162 known miRNAs distributed across 33 families were identified, of which 20 were differentially expressed. In addition, 53 miRNA candidates were identified, and 10 of them were found to be differentially expressed in IFC and CFC ovaries. In degradome analysis, a total of 113 and 11 target genes were predicted for known and novel miRNAs, respectively. Thereinto, 24 (60%) target genes of the differentially expressed known miRNAs were found to code transcription factors, including auxin response factors (ARF), TB1-CYC-PCFs (TCP), APETALA2 (AP2), growth regulating factor (GRF), MYB, NAC, NF-YA and so on, which have been proved to have roles on regulating carpel fusion development. Correlation analysis of differentially expressed known miRNAs and their targets to phytohormone signals revealed that either of the miRNA or of its target had a significant correlation to at least one phytohormone signal, the main regulator of carpel fusion development. These results suggested that incomplete carpel fusion is partly the result of differential expression of certain miRNAs and their targets. Taking together, the findings improve our knowledge about the effect of miRNA regulation on target expression, providing a useful resource for further analysis of the interactions between miRNAs, target genes and phytohormones during carpel fusion development in maize.

T5
P1217

Comparative study on the complementary distribution of simple sequence repeats and intron polymorphism in plant

Yi Wang[2], ***Huayang Li***[1], *Long Yang*[1]
1. Agricultural Big-Data Research Center, College of Plant Protection, Shandong Agricultural University
2. Beijing Key Laboratory of Grape Science and Enology and Key Laboratory of Plant Resource, Institute of Botany, Chinese Academy of Sciences

Molecular marker is an important tool in the study of comparative genomics. Due to the rapid development of biotechnology, it is more urgent need some new and efficient markers that with high polymorphism, high cover rate or functional. In this study, we large-scale developed two kinds of markers, simple sequence repeats (SSRs) and intron polymorphism (IP) in 259 plants' genome. Using a pipeline program by Perl script, we had developed a total of 1,715,915 SSR, 1,232,445 IP and 2,326,806 PIP. Among all the markers, 1,063,896 SSR and all IP were based on the model plants. The overlap rate of SSR and IP markers in all species was in a very lower level. The moss had the highest overlap rate (9.41%), and most of others were less than 5%, even 1 or 2 percent minimum. That's mean that SSR and IP is complementary distribution in plant. PSID (Plant SSR and IP Database) is a platform that provides SSR and IP markers and other related information of most sequenced plants to the biologist. It is freely available on the web (http://biodb.sdau.edu.cn/psid/home.html). The result would cause a revolution that the combined utility of SSR and IP markers in the area of molecular plants.

T5
P1218

Physiological and epigenetic analyses of *Brassica napus* seed germination in response to salt stress

Jian Li, *Yujie Fang, Youping Wang*
Yangzhou University

Salinity stress significantly affects plant growth and development because of osmotic stress, ion toxicity, and nutrient imbalance. Therefore, salinity stress becomes a serious threat to rapeseed production in agriculture. Epigenetic regulations, including DNA methylation and histone modification, play a major role in tuning gene expression into plant response to environmental stimuli. Although many progresses have been reported in plant response to salt stress, the epigenetic changes in *Brassica napus* under salt stress are far from being understood. A series of physiological parameters, including water content, proline content, malondialdehyde content, electrolyte leakage, and antioxidant enzyme activities, under different concentrations (0, 25, 50, and 100 mM) of NaCl treatment in "Yangyou 9" was determined at the germination stage. Immunofluorescent staining and high-performance liquid chromatography-assisted quantification were conducted to analyze the level and distribution patterns of DNA and histone methylation under salt stress. Results of morphological and physiological analyses under salt stress indicated that 25 mM NaCl treatment promoted the growth of "Yangyou 9" seedlings, whereas 50 and 100 mM NaCl treatments inhibited the growth of "Yangyou 9" seedlings. Epigenetic investigations showed that 25 mM NaCl mediated the enrichment of H3K4me3, as well as decreases in H3K9me2 and 5-methylcytosine (5-mC), whereas 50 and 100 mM NaCl induced increases in H3K9me2 and 5-mC and a decrease in H3K4me3. Overall, this study offers new insights into the epigenetic changes in salt stress response in rapeseed, and this information would be propitious to engineering crops for salt tolerance.

T5
P1219

A new variety breeding and study of purple V-marking white clover

Jinbo Li
ShenZhen Techand Ecology & Ennironment CO., LTD

White clover (*Trifolium repens* L.), perennial and creeping, is a kind of good leguminous forage and ideal flowering and foliage cool-season turfgrass, belonging to family Leguminosae and genus *Trifolim*. One hundred and eighteen accessions of white clover samples around the world were collected. a mutant strain with distinguishing purple V-marking on leaves was obtained from 118 accessions of white clover samples around the world collected. Through years of breeding, the new variety of purple V-marking white clover, which was the first variety used of turf and forage, approved by the Inner Mongolia Committee of Grass Variety examination in 2015, filled the gap of lawn with white clover breeding in China. Relationship analysis showed that it originated from "QZH2004-3" and genetic analysis showed that purple is controlled for single gene. Then, the purple candidate gene *TrMYB1* was determined through the RNA-seq transcriptome sequencing and pH differential method. The main results we got are as follows. 1. One hundred and eighteen accessions of white clover samples around the world were collected to build a germplasm nursery, based on this nursery, we evaluated the white clover germplasm resources and bred new variety. After selfing and hybridization using mutant strain, a new landscaping variety of white clover integrated ornamental, covering and turf value with a distinguishing purple V-marking on leaves was obtained. This variety approved by the Inner Mongolia Committee of Grass Variety examination in 2015, widely used in the production. 2. The results of genetic diversity of ISSR and ITS using cluster analysis revealed that white clovers showed a relatively regional distribution law: the samples from Asia, Oceania and America have a relatively close relationship, while the European and Western Asia ones in relationship were near. purple v-marking white clover built a separate branch with QZH2004-3 from Guizhou, China in cluster, suggesting that they have close relationship, and the former is probably a mutant of the latter. 3. RNA-sequencing was proceed in both QZH2004-3 and purple V-marking white clover. In the new variety many differential expression gene about purple V-marking were identified. To provide candidate of V-marking related gene, the gene involved in KEGG pathway and those differential expression gene were mutual authenticated, and 6 candidate genes were further studied. 4. pH-differential method were used in the qualitative and quantitative analysis of anthocyanin in purple V-marking white clover. The results were remarkalbe, the specific metabolites of white clover with purple V-marking is anthocyanin, and the content reached 5.43mg/100g. Conserved sequence of MYB were used to design primers and the transcription factor *TrMYB1* gene regulated the anthpcyanin biosynthesis was isolated and homology-based cloned, the genetic and corresponding protein structure were also analysed.

T5

P1220

Drought-induced gene expression changes in tissues of loblolly pine clones

Jingjia Li[1], *Jason West*[1], *Carol Loopstra*[1], *Tom Byram*[1], *Jean-Christophe Domec*[2], *Jeff Puryear*[1], *Matthew Smith*[1], *Claudio Casola*[1]

1. Texas A&M University

2. Bordeaux Sciences Agro

Loblolly pine plantations represent a major economic and ecological resource in the Southeastern United States, a region bound to sustain a dryer and warmer climate in the near future. In this study, we combine physiological and transcriptome measurements to 1) identify variation in sensitivity of loblolly pine clones to low water availability; 2) identify genes involved in drought tolerance in needle and root tissues. This study takes advantage of a greenhouse experiment with seedlings from several loblolly pine clones grown under control and low-water treatments. Several traits related to water use were measured in these seedlings to determine physiological responses to water stress. One of two clones used for transcriptomic analysis exhibited drought-tolerance physiological responses in preliminary experiments, although the same two clones showed no significant difference in subsequent analysis of seedlings response to low water availability. To investigate on the genetic basis of drought tolerance, RNA was extracted from needles and roots of seedlings from these two clones under control and low water treatments. Analyses of the RNA-Seq data showed 71,600 expressed genes, of which ~92% were transcribed in both tissues, although root samples exhibited higher transcription levels and more expressed genes than needles. Using the DESeq2 we identified 560 up-regulated genes and 2,173 down-regulated genes in tissues from low-water treatment seedlings compared to control plants. Gene Ontology analyses revealed a functional enrichment for genes involved in response to stimulus in both groups of genes, and a specific abundance of up-regulated transcripts involved in response to abscisic acid and potassium transmembrane transport. Additionally, we identified 7,817 genes with differential expression between the two clones. Ongoing analyses will point to the genetic networks involved in drought tolerance across the examined tissues and clones of loblolly pine. These results will be integrated with the physiological measurements to provide novel insights into the genetic architecture of traits implicated in drought response in loblolly pine.

T5

P1221

RNA polymerase III dependent *AtR8* long non-coding RNA relate with *WRKY53* and *WRKY70* on plant defense in *Arabidopsis*

Shuang Li[1], *Saraswati Nayar*[2], *Sanjay Kapoor*[2], *Juan Wu*[3], *Yasushi Yukawa*[1]

1. Grad. Sch. Nat. Sci., Nagoya City Univ.,

2. Dept. Plant Mol. Biol., Univ. Delhi South Camp.,

3. Alkali Soil Nat. Env. Sci. Cent., Northeast Forestry Univ.,

An *AtR8* RNA transcribed by RNA polymerase III (Pol III) was discovered as a highly abundant long non-coding RNA (lncRNA) in *Arabidopsis*. The AtR8 RNA is a 260 nt lncRNA observed in root tips and localized in cytosol. And in *atr8* mutant, early flowering-stalk elongation was observed. In the previous study, this RNA was responsive to hypoxic stress and salicylic acid (SA) treatment. SA is one of plant hormone that play important roles in responses to environmental conditions. Microarray assay with *atr8* mutant showed some WRKY transcription factor (FT) genes were up regulated. WRKY TFs are one of the largest families among transcriptional factors in plants. When WT, *atr8*, *wrky53* and *wrky70* mutant were growing in the medium with or without SA (20 μM), different root elongations were observed. And then each RNA accumulation levels were assayed by real time RT-PCR. *WRKY53*, *WRKY70* mRNA and *AtR8* RNA expression levels were also quite immediately elevated by plant pathogen *Pseudomonas syringae* infection. To sum up, *AtR8* RNA expression could relate with *WRKY53* and *WRKY70* expressions through plant defense.

T5

P1222

Transcriptome analysis provides insights into wood formation during larch tree aging

Wan-Feng Li[1], *Wen-Hua Yang*[1], *Shou-Gong Zhang*[1], *Su-Ying Han*[2], *Li-Wang Qi*[1]

1. State Key Laboratory of Tree Genetics and Breeding, Research Institute of Forestry, Chinese Academy of Forestry

2. State Key Laboratory of Tree Genetics and Breeding, Research Institute of Forest Ecology, Environment and Protection, Chinese Academy of Forest

Tree age affects wood formation and yield. However, the underlying mechanisms are poorly understood, particularly at the molecular level. In this study, we investigated the transcriptomic changes of the uppermost main stems of *Larix kaempferi* in an entire rotation period using the RNA-Seq method. In total, ~151 million reads were obtained from the stems of 1-, 2-, 5-, 10-, 25-, and 50-year-old *L. kaempferi* trees. Combining these with the published Illumina sequencing reads, 299,637 assembled transcripts were generated, of which 161,232 were annotated. 12,927 transcripts were identified as differentially-expressed genes (DEGs) by time-series expression profiling; function enrichment analysis of these DEGs showed that 459 gene ontology terms in the biological process category were enriched. These terms were associated with the processes of wood formation, such as cell differentiation, growth and death, and its hormonal regulation. Based on the expression patterns of *L. kaempferi* homologues of genes associated with ethylene, calcium, and cell-wall expansion and synthesis, the regulatory network of tracheid growth was outlined. Altogether, the comparative transcriptomic analysis reported here demonstrated the molecular aspects of aging effects on *L. kaempferi* wood formation. The identification of genes involved in the regulatory network of tracheid growth provides a means of investigating the regulation of wood formation in gymnosperm trees and also offers potential targets for genetic manipulation to improve the properties of xylem fibers.

T5
P1224
Gene duplicability of core genes is highly consistent across all angiosperms

Zhen Li[1,2], *Jonas Defoort*[1,2], *Setareh Tasdighian*[1,2], *Steven Maere*[1,2], *Yves Van De Peer*[1,2,3], *Riet De Smet*[1,2]
1. VIB
2. Ghent University
3. University of Pretoria

Gene duplication is an important mechanism for adding to genomic novelty. Hence, which genes undergo duplication and are preserved following duplication is an important question. It has been observed that gene duplicability, or the ability of genes to be retained following duplication, is a nonrandom process, with certain genes being more amenable to survive duplication events than others. Primarily, gene essentiality and the type of duplication (small-scale versus large-scale) have been shown in different species to influence the (long-term) survival of novel genes. However, an overarching view of "gene duplicability" is lacking, mainly due to the fact that previous studies usually focused on individual species and did not account for the influence of genomic context and the time of duplication. Here, we present a large-scale study in which we investigated duplicate retention for 9,178 gene families shared between 37 flowering plant species, referred to as angiosperm core gene families. For most gene families, we observe a strikingly consistent pattern of gene duplicability across species, with gene families being either primarily single-copy or multicopy in all species. An intermediate class contains gene families that are often retained in duplicate for periods extending to tens of millions of years after whole-genome duplication, but ultimately appear to be largely restored to singleton status, suggesting that these genes may be dosage balance sensitive. The distinction between single-copy and multicopy gene families is reflected in their functional annotation, with single-copy genes being mainly involved in the maintenance of genome stability and organelle function and multicopy genes in signaling, transport, and metabolism. The intermediate class was overrepresented in regulatory genes, further suggesting that these represent putative dosage-balance-sensitive genes.

T5
P1225
A new factor containing WD-40 domain positively regulated the primary miRNAs processing in *Arabidopsis thaliana*

Chao Liang[1], *Qiang Cai*[1], *Beixin Mo*[1], *Xuemei Chen*[1,2]
1. Shenzhen University
2. University of California, Riverside

MicroRNAs (miRNAs) are a class of 20-24 nucleotide (nt) non-coding RNAs, which negatively regulate gene expression through cleaving their target RNA or repressing translation. We have chosen a genetic screening for *Arabidopsis* miRNA biogenesis pathway mutants, utilizing vein beaching phenotype caused by phloem specific expression of an artificial miRNA (amiR-SUL) targeting *CH42* gene as a visible selection marker, and isolated a suppressive mutant named *ab57*. Both exogenous amiR-SUL and the endogenous miRNAs levels are reduced in *ab57* seedlings, along with the decrease of their primary transcripts. Chromatin Immunoprecipitation (ChIP) result shows that the occupancy of RNA polymerase II (Pol II) at *MIR* promoters is reduced in *ab57*. Here we report the *ab57* mutant which is defective in a WD-40 domain protein. The WD-40 proteins are evolutionarily conserved in plants. We reveal that this WD-40 domain protein interact with DICER LIKE 1 (DCL1) which is efficient for miRNAs generation from pri-miRNAs. Besides, we further reveal the interaction between the WD-40 domain protein and CDC5, which regulates the *MIR* transcription and the pri-miRNAs processing. Based on the results, we propose this WD-40 domain protein links transcription of *MIR* genes by Pol II to processing of pri-miRNAs by DCL1.

T5
P1226
Phylogenomics of Lauraceae: Complete chloroplast genome sequence of *Neolitsea sericea* (Blume) Koidz. and clarifying the phylogenetic position of *Neolitsea*, an important Asian tree genus

Xijian Liang[1], *Damien D. Hinsinger*[1], *Joeri S. Strijk*[1,2]
1. Plant Ecophysiology & Evolution Group, Guangxi Key Laboratory of Forest Ecology and Conservation (under state evaluation status), College of Forestry, Nanning, Guangxi,
2. State Key Laboratory for Conservation and Utilization of Subtropical Agro-bioresources

Genomic, evolutionary and biogeographical studies of complex, diverse and widely distributed plant groups are hampered by a number of technical and practical difficulties. This is especially true for major plant families in the tropics, which are often characterised by amorphic and widespread tropical genera supported by weak taxonomical and nomenclatural frameworks. This has major consequences for our efforts to understand, study and conserve tropical biodiversity. Unlike most other Magnoliids, Lauraceae are highly species rich with over 3000 species worldwide, making it one of the dominant plant families encountered in predominantly warm temperate and tropical regions. This makes them an important structural and evolutionary component of wet tropical forests. Many species are locally and commercially used for their wood, fruits and high concentration of essential oils which are valued for spices and perfumes. The loss of untouched habitat in tropical regions, coupled with overexploitation of commercially attractive species has put many species in danger of extinction, as evidenced by the growing number of Lauraceae on the IUCN Red List. The Red List currently holds 295 species, with the majority (72%) in the top three categories and the most important threat reported being logging and wood harvesting. Regrettably, despite their abundance and widespread occurrence, Lauraceae remain poorly known and difficult to identify, primarily due to the absence of clear vegetative characters that can aid in successful identification. This absence of diagnostic characters perturbs also into a poor delimitation of genera, prohibiting the construction of congruent classifications and the continued existence of large paraphyletic and polyphyletic genera in Lauraceae to this day. The genus *Neolitsea* consists of nearly 100 species distributed widely from Indo-Malaysia to East Asia and Australia. They occur mostly as small shrubs and trees, forming an important component of low- and mid-elevation forests. China is a centre of diversity for the genus with nearly 50 species distributed across the southern provinces. Little is known about the internal relationships of the genus or its exact position in the family, as higher generic relationships in Lauraceae remain unclear. In our study we aim to resolve internal relationships by employing whole chloroplast sequencing, enriched with additional sequencing of portions of the nuclear genome. Here we present the first complete plastome for the genus *Neolitsea*. The chloroplast of *N. sericea* showed the structure and composition usually found in angiosperm (LSC, IRa, SSC, IRb ; GC = 37.8%), and contained 129 genes (29 tRNA, 8 rRNA, 92 coding genes). Using complete chloroplasts for 15 related genera produced by our ongoing project, we clarified the position of the genus *Neolitsea* in Lauraceae and confirmed its inclusion in the Laureae group. We also provided a first estimate of its genomic divergence from the rest of family. As demonstrated by our study, the use of complete chloroplasts opens exciting opportunities to address evolutionary questions in Lauraceae. Finally, better knowledge of this ancient family, successfully facing major environmental changes since its origin, will help us to face the ongoing global climatic changes.

T5
P1227
Genomic imprinting and DNA methylation in dicotyledonous seeds with persistent endosperms

Wei Xu, ***Aizhong Liu***
Kunming Institute of Botany, Chinese Academy of Sciences

Genomic imprinting causes parent-of-origin specific differential expression of maternally and paternally inherited alleles. Triploid endosperm tissue is the main place where gene imprinting occurs. In this study, we inspected genomic imprinting in castor bean endosperm where gene imprinting persists throughout seed development using high throughput RNA sequencing. Total, we identified 209 genes with potential parent-of-origin specific expression, including 200 maternally expressed genes (MEGs) and nine paternally expressed genes (PEGs) in reciprocal endosperms, and seven MEGs and three PEGs in the embryos. To dissect the potential molecular mechanism underlain gene imprinting arise in endosperms, we comprehensively analyzed the distribution of genomic DNA methylation between embryo and endosperm tissues, we uncovered 1) a significant hypodemethylation in the endosperm and an enrichment of DMRs around the earlier identified imprinted genes; 2) the CHH methylation levels in endosperm and embryo were, unexpectedly, substantial higher than known plants; 3) the endosperm hypodemethylation was caused by the global reduction in the CG and CHG methylation level relative to its embryo, but the CHH methylation occurring in endosperm did not exhibit a significant reduction; 4) combining with the expression profiles of DNA methyaltion-related genes in endosperm and embryo, we proposed a novel genomic DNA methylation pattern. Taken together, we discussed that emergence of gene imprinting, biological interests and its potential driver power of DNA methylation. This study might add new insights into understanding the genomic imprinting and DNA methylation occurring in dicot seeds.

T5
P1228
SNP genotyping by exome-seq localized 9.5K genes for development of a high density genetic map in limber pine (*Pinus flexilis* James)

Jun-Jun Liu[1], *Anna W. Schoettle*[2], *Richard A. Sniezko*[2], *Fupan Yao*[1], *Benjamin Rancourt*[1], *Hao Chen*[1], *Arezoo Zamany*[1]
1. Canadian Forest Service
2. USDA-FS

Limber pine is one of keystone conifer species of high elevation forest ecosystems in North America. Many of its populations are seriously impacted by the invasive, exotic fungal pathogen *Cronartium ribicola*, which causes white pine blister rust (WPBR) on five-needle pines of the subgenus Strobus. Genetic maps with localization of functional genes at the genome level provide essential knowledge for understanding genetic resistance and local adaptation to changing climates in all living organisms. Because limber pine possesses a huge genome (~30.5Gbp), exome-seq was used to explore genetic variations such as single nucleotide polymorphisms (SNPs) and genotype two seed families with 91 and 99 individuals. About 14 K unique gene sequences from a limber pine reference transcriptome de novo assembled in a previous study (Liu *et al.*, BMC Genomics, 2016) were targeted by probe design to enrich the exome. Analysis of exome-seq data found that a total of 87,374 SNPs from 12,389 genes (~85% of total genes selected for exome enrichment) segregated at least in one mapping seed family with an expected segregation ratio. A consensus limber pine genetic map was constructed with localization of 9,510 expressed genes (including >600 NBS-LRR and >200 RLK genes) on 12 linkage groups (LGs), corresponding to the number of chromosomes in *Pinus* species. The total genetic distance was calculated at 2,043 cM with ~4.6 genes per cM. Of all genes co-segregated with phenotypic traits, we identified functional candidates that putatively control host resistance against WPBR. Syntenic relationships among *Pinus* species were further determined, revealing both conservative and divergent regions in genome structure among *Pinus* species. Using genome sequences of two other *Pinus* species released recently, we investigated how genetically co-localized genes are distributed across the genome sequence scaffolds. This high density genetic map provides genomic resource and practical tools for application in breeding and genetic conservation programs, and will facilitate final characterization of functional genes contributing to adaptive traits and help sequencing and assembly of the full-length genomes in limber pine and related *Pinus* species.

T5

P1229

Genomics study of five-needle pine major gene resistance to white pine blister rust

Jun-Jun Liu[1], *Anna Schoettle*[2], *Richard Sniezko*[3]
1. Canadian Forest Service
2. USDA Forest Service, Rocky Mountain Research Station
3. USDA Forest Service, Dorena Genetic Resource Center

Molecular breeding incorporates efficient tools to increase rust resistance in five-needle pines. Susceptibility of native five-needle pines to white pine blister rust (WPBR), caused by the non-native invasive fungus *Cronartium ribicola* (J.C.Fisch.), has significantly reduced wild populations of these native conifers in North America. Four five-needle pine species have been found with major gene resistance against specific avirulent pathotypes. In this study we screened genic SNP markers by comparative transcriptome and genetic association analyses. Saturated SNP linkage maps were constructed for the *R* (*Cr2* and *Cr4*) loci in western white pine (*Pinus monticola*, Douglas ex D. Don) and limber pine (*Pinus flexilis* E. James). Comparative SNP mapping among *Pinus* species anchored *P. monticola Cr2* locus and *P. flexilis Cr4* locus on the Pinus consensus linkage group1 (LG1) and LG8, respectively. GO annotation identified a set of NBS-LRR and other resistance-related genes as *R* candidates in five-needle pines. Practical molecular tools for marker-assisted selection (MAS) were developed by association genetics of non-synonymous SNP loci of the NBS-LRR genes, which will shorten the breeding cycle of resistance screening and aid in the restoration and management of WPBR-disturbed and threatened forest ecosystems.

T5

P1230

Profiling methyl jasmonate-responsive transcriptome for understanding induced systemic resistance in whitebark pine

Jun-Jun Liu[1], *Williams Holly*[1], *Xiao Rui Li*[1], *Anna W. Schoettle*[2], *Richard A. Sniezko*[2], *Michael Murray*[3], *Arezoo Zamany*[1], *Hao Chen*[1]
1. Canadian Forest Service
2. USDA-FS
3. Ministry of Forests, Lands and Natural Resource Operations, BC Government

Whitebark pine is an endangered conifer species due to its high susceptibility to mountain pine beetle and white pine blister rust fungus. Resistance mechanisms, such as induced systemic resistance (ISR), are not well characterized in conifers, hindering forest conservation and resistance breeding programmes. Exogenous jasmonates are well known to trigger ISR by manipulating the plant immunity system through regulation of global gene expression to the production of chemical defense compounds. In the present study whitebark pine transcriptomes were profiled post methyl jasmonate (MJ) treatment using RNA-seq. MJ-responsive transcriptome was *de novo* assembled and sequent transcriptome profiling identified a set of differentially expressed genes (DGEs) in response to MJ application, revealing 1,421 up- and 999 down-regulated transcripts with at least two-fold-changes (FDR corrected $p < 0.05$), respectively. GO analysis revealed that these MJ-responsive DEGs have putative functions in plant defense signalling, transcription regulation, biosyntheses of secondary metabolites, and other biological processes. Comparison with MJ signalling in model plants detected whitebark pine lineage-specific defense-related genes. It was further found that MJ-triggered transcriptome re-programming highly overlapped with the defenses responses against either herbivore or fungal pathogen as reported in close Pinus species, suggesting that MJ has potential application to improve whitebark pine resistance/tolerance to MPB and WPBR. Our study provides new insights into molecular mechanism and metabolic pathways involved in whitebark pine ISR. The candidate genes for improving resistance may benefit genetic dissection of genomic variation contributing to host immunity and breeding selection of elite genotypes with better adaptive fitness to environmental stresses in this endangered tree species.

T5

P1231

Study the function of *AtFbox* on flowering time in *Arabidopsis*

Wei Zhu, *Zhixue Dong*, *Junyan Huang*, *Shengyi Liu*
Oil Crops Research Institute, Chinese Academy of Agricultural Sciences

Plants react to flower formation controlled by complicated molecular pathway, and have many alternative splicing variants (ASVs) in these genes controlling flowering time. However, why there are so many ASVs and whether there is function differentiation involved in traits such as flowering time are unclear. Here we characterize a alternatively splicing variants gene controls flowering time and its protein interacts with AtFbox2, named *AtRDR* producing two splicing variants of *AtRDR1* and *AtRDR2* which promote flowering. Through yeast library screening and yeast two hybridization, we identified one candidate protein AtFbox whose function is unknown and whose gene *AtFbox* that produced 4 ASVs was chosen to study its function and further check whether specific interactions occur between ASVs of the *AtRDR* and *AtFbox* . The BIFC (bimolecular fluorescence complementation) identified that AtRDR1 and AtRDR2 interacted with AtFbox2 respectively in cell membrance and cell nuclei, but did not detected for interaction with AtFbox1. AtFbox2 has been identified to regulate photoperiod-responsive growth by integrating the clock with light perception pathways through modulating daily phyB-signaling. *AtFbox* belongs to the F-box family, and produces three extra transcripts *AtFbox2*, *AtFbox3* and *AtFbox4* as well as a normal full length transcript *AtFbox1*. *AtFbox2* only contains the first two exons and lacks the sequence encoding the F-box functional domain required for protein ubiquitination in different functional pathways. Genetic analysis showed that negative regulation of *AtFbox* on flowering time, and over-expressed *AtFbox1* and *AtFbox2* in *Arabidopsis* wild type exhibited later flowering and shorter hypocotyl compared to WT. The T-DNA insertion mutant *fbox* was earlier flowering and longer hypocotyl, which is consistent to the SAMs of *fbox* differentiated earlier than WT. The *fbox* mutant will be partially recovered WT phenotype by over-expression of *AtFbox1* or *AtFbox2* in the genetic complementation test.

T5

P1232

Linked and co-expressed QTLs in syntenic genomic regions for resistance to *Sclerotinia* stem rot and flowering time in *Brassica napus*

Fengqi Zhang[1,2], ***Junyan Huang***[1,3], *Mingxing Cheng*[1], *Xiaohui Cheng*[1], *Yueying Liu*[1], *Chaobo Tong*[1,3], *Jianguo Cheng*[3], *Shengyi Liu*[1,3]

1. Oil Crops Research Institute of CAAS, Key Laboratory of Biology and Genetic Improvement of Oil Crops, Ministry of Agriculture, Wuhan, China.

2. Cereal Crops Research Institute, Henan Academy of Agricultural Sciences, Zhengzhou, China.

3. Hubei Collaborative Innovation Center for Green Transformation of Bio-Resources, Hubei University, Wuhan, China

Sclerotinia sclerotiorum stem rot is a major disease in *Brassica napus* oilseed rape worldwide, particularly in China where annual yield loss is more than 10%. A close association of disease incidencewith flowering timeis widely observedin oilseed rape fields or germplasm. Up to now there are no reports on cloning of resistant QTLs and its definition of the association. To reveal the genetic relationship between the two traits, we investigated QTLs of both traits in are combinant inbred line population derived from a cross between 888-5 (a susceptible and early flowering line) and M083 (a highly resistant and later flowering line). The population was genotyped by employing a *B. napus* 60K SNP array which produced ahigh density genetic map comprising 9,278 SNPs and spanning 4,071cM of the genome. QTL mapping was conducted with phenotypic data of stem lesion size of artificial inoculation and both the disease index and flowering time which were evaluated in eight environments. A total of 31 QTLs for resistance were detected. Of them, the major QTL *qSSE2-1* in A2 explained 10.50%-27.30% of the phenotypic variation across the eight environments, but with no interactions with environments. For flowering time, 21 putative QTLs were found. Of them, the stable major QTLs *qFTE2-1* and *qFT2-6*, contributing to 16.40%-34.50% of the phenotypic variation in six environments, co-located with stable major QTLs *qSSE2-1* and *qSS2-9* in a narrow A2 region, but they are closely linked, rather than pleiotropy of a single gene. A syntenic analysis approach was used to search co-expressed QTLs in syntenic regions between the subgenomes A and C. Of syntenic regions found, *qSS2-10* in the A2 region is syntenic to *qSS12-1* in a C2 region and within the two QTL regions, two pairs of orthologous genes (in total, five genes and of them, two are tandem duplicates) were identified as preferential candidates which were annotated as hormone biosynthesis and transcription factor, respectively. Our results on the linked QTLs and the co-expressed syntenic QTLs will promote their fine mapping and help improve efficiency of breeding for *Sclerotinia* resistance and early maturation in oilseed rape.

T5

P1233

Dissecting QTLs for plant height in a set of dwarf and semi-dwarf *Brassica napus* mutants

Chuanji Zhao, *Junyan Huang, Chaobo Tong, Khorshed Alam Muhammad, Yueying Liu, Xiaohui Cheng, Shengyi Liu*

Key Laboratory of Biology and Genetic Improvement of Oil Crops, the Ministry of Agriculture, Oil Crops Research Institute of CAAS, Wuhan 430062, China.

Oilseed rape (*Brassica napus*) in the Changjiang valley is a tall crop and mechanized management such as chemical spray in the crop late growth stage and mechanical harvestare difficult, and furthermore its harvest index is rather low. An ideal plant architecture should be developed to meet mechanization requirement and to increase harvest index. To this end, understanding controlling genes of plant height is one of crucial elements. In this project, five lines of four dwarf and one semi-dwarf (*dwf* and *s-dwf*) mutants and one taller mutant, all derived from a bulk of EMS mutagenized seeds of *B. napus* cv. zhongshuang9 (ZS9), has been investigating. In the mature stage, in comparison with ZS9, the height of the *dwf* and *s-dwf* was respectively reduced by about 58% and 30% evaluated in multiple years in fields. For genetic analysis of the two mutants, we constructed two F_2 populations from the ZS9×*dwf* and ZS9×*s-dwf*, which showed an expected Mendelian inheritance ratio of 7:9 and 15:1 (tall-dwarf plants, $\chi^2 < \chi^2_{0.05} = 3:84$; $P < 0.05$), indicating that the phenotype of the two mutants was respectively controlled by two gene loci. Allelic test showed the loci are not allelic between the two mutants. Genome wide SNP profiling byre-sequencingof BSA pools from theF_2 populations identified two significant genomic regions, one on A04 (3.3 Mb) and another on A04 (0.1 Mb) controlling dwarf. Similarly, two significant genomic regions, one on C03 (7.6 Mb) and another on C03 (1.5 Mb) were identified for semi-dwarf. Further studies are in progress.

T5

P1234

Arabidopsis BREVIPEDICELLUS interacts with the SWI2/SNF2 chromatin remodeling ATPase BRAHMA to regulate *KNAT2* and *KNAT6* expression in control of inflorescence architecture

Minglei Zhao, ***Xuncheng Liu***

South China Botanical Garden, Chinese Academy of Sciences

BREVIPEDICELLUS (BP or KNAT1), a class-I KNOTTED1-like homeobox (KNOX) transcription factor in *Arabidopsis thaliana,* contributes to shaping the normal inflorescence architecture through negatively regulating other two class-I *KNOX* genes, *KNAT2* and *KNAT6*. However, the molecular mechanism of BP-mediated transcription regulation remains unclear. In this study, we showed that BP directly interacts with the SWI2/SNF2 chromatin remodeling ATPase BRAHMA (BRM) both *in vitro* and *in vivo*. Loss-of-function *BRM* mutants displayed inflorescence architecture defects, with clustered inflorescences, horizontally orientated pedicels, and short pedicels and internodes, a phenotype similar to the *bp* mutants. Furthermore, the transcript levels of *KNAT2* and *KNAT6* were elevated in *brm-3*, *bp-9* and *brm-3 bp-9* double mutants. Increased histone H3 lysine 4 tri-methylation (H3K4me3) levels were detected in *brm-3*, *bp-9* and *brm-3 bp-9* double mutants. Moreover, BRM and BP co-target to *KNAT2* and *KNAT6* genes, and BP is required for the binding of BRM to *KNAT2* and *KNAT6*. Taken together, our results indicate that BP interacts with the chromatin remodeling factor BRM to regulate the expression of *KNAT2* and *KNAT6* in control of inflorescence architecture.

T5

P1235

Mitochondrial genome of mycoheterotrophic *Hypopitys monotropa*: unusual structure and extensive RNA editing.

Maria Logacheva[1], *Mikhail Schelkunov*[2], *Viktoria Shtratnikova*[1], *Aleksey Penin*[1]

1. Lomonosov Moscow State University

2. Institute for Information Transmission Problems

The capability to photosynthesis is the distinctive feature of plants, which is of highest importance for biosphere. However, there are plants, including several thousands of flowering plant species that obtain organic substances from sources other than photosynthesis. These plants get the organics either from associated fungi (myco-heterotrophy), or by parasitising on other plants. While plastid genomes of non-photosynthetic plants are well characterized, little is known about mitochondrial genomes. We report the structure of mitochondrial genome in *Hypopitys monotropa*, a mycoheterotrophic member of Ericaceae, and the expression of mitochondrial genes. In contrast to highly reduced plastid genome, the mitochondrial genome of *H. monotropa* is larger than that of its photosynthetic relative *Vaccinium macrocarpon*, its complete size is ~ 810 Kb. We found an unusual long repeat-rich structure of chondriome that suggests the existence of linear fragments. Despite this unique feature, gene content of *H. monotropa* chondriome is typical for flowering plants. Notably, we did not found any acceleration of substitution rates in mitochondrial genes, in contrast to previous observations on non-photosynthetic plants. Transcriptome sequencing revealed trans-splicing of several genes and extensive RNA editing.

T5

P1236

De novo assembly and characterization of the root transcriptome of two traditional chinese medicine species, *Scrophularia ningpoensis* and *S. burgeriana* (Scrophulariaceae)

Jocelyn Losh, *Chengxin Fu, Chuan Chen*

Zhejiang University

Scrophularia ningpoensis (Ningpo figwort) and *Scrophularia buergeriana* (Buerger's figwort) occur throughout southern and northern China, respectively. While *Scrophularia ningpoensis*' roots are used extensively in Chinese medicine to treat arthritis and inflammation, *S. burgeriana*'s roots are less-commonly used, even though modern studies have shown that the species contains the same bioactive compounds. The diversity between the two species is not yet fully understood due to the lack of genome and SSR data. Here we report the results of *Scrophularia ningpoensis* (XS-L) and *S. buergeriana*'s (XS-3Y) *de novo* root transcriptome sequencing, characterization, and a comparative transcriptome analysis between the two species. Our goals were to create a foundation of genome sequence data and to identify genes potentially expressed differently between the root transcriptomes of these two economically important species. In this study, the sequencing of cDNAs resulted in a total of 46,918,622 (*XS-L*) and 48,007,868 (*XS-3Y*) clean reads. These clean reads were assembled into 58,893 and 56,235 unigenes, respectively. These unigenes were annotated through comparison with known protein databases NR, Swiss-Prot, KEGG and COG. For XS-L, 50,161 (85%) were annotated, and for XS-3Y, 49,988 (89%) were annotated. Based on NR annotation, 32,093 unigenes from XS-L and 32,396 from XS-3Y were assigned gene ontology (GO) with 14,034 shared functional terms. We performed transcriptome sequencing of *S. ningpoensis* and *S. buergeriana* and generated 6724 orthologous genes. To find the differently-expressed genes (DEGs), we employed the RPKM method to compare the transcriptomes. The differently expressed genes between *S. ningpoensis* and *S. buergeriana* result in up-regulation or down-regulation of various biological processes such as in metabolism or response to stimuli, as well as in differing cellular components and molecular functions. However, the greatest percentage of DEGs (14.72%) relate to metabolic pathways the result of differential expression of genes can be varying levels of secondary compounds such as cinnamic acid, harpagoside, angroside C, and acteoside, important factors in *Scrophularia*'s medicinal effects. Additionally, we developed EST-SSR molecular markers from single nucleotide polymorphisms (SNPs) in the unigenes. The unigenes were searched for SSRs using the MISA program and primers were designed using the Oligo 7 software. Around 50 primers were successfully designed with polymorphic amplicons. To validate the markers, 12 *S. ningpoensis* individuals (10 wild, 2 cultivated), 2 *S. buergeriana* individuals (1 wild, 1 cultivated) and one individual each from the closely-related *S. kakudensis* and *S. yoshimurae* were tested for amplification. The markers may be used in the future for phylogenomic studies of the *Scrophularia* genera. This study provides valuable resources for the field of Traditional Chinese Medicine as well as for future phylogenomic endeavors.

T5

P1237

Genome-wide survey reveals transcriptional differences underlying the contrasting trichome phenotypes of two sister desert poplars

Jianchao Ma, *Xiaodong He, Xiaotao Bai, Zhimin Niu, Bingbing Duan, Ningning Chen, Xuemin Shao, Dongshi Wan*

State Key Laboratory of Grassland Agro-Ecosystem, School of Life Sciences, Lanzhou University, Lanzhou 730000, China

Trichomes, which are widely used as an important diagnostic characteristic in plant species delimitation, play important roles in plant defense and adaptation to adverse environments. In this study, we used two sister poplar species, *Populus pruinosa* and *Populus euphratica*, which have, respectively, dense and sparse trichomes to examine the genetic differences associated with these contrasting phenotypes. The results showed that 42 and 45 genes could be identified as candidate genes related to trichomes in *P. pruinosa* and *P. euphratica* respectively; most of these genes possessed high degrees of diversification in their coding sequences but they were similar in intron/exon structure in the two species. We also found that most of the candidate trichome genes were expressed at higher levels in *P. pruinosa*, which has dense trichomes, than in *P. euphratica*, where there are few trichomes. Based on analyses of transcriptional profiles, a total of 195 genes, including many transcription factors, were found to show distinct differences in expression. The results of gene function annotation suggested that the genes identified as having contrasting levels of expression level are mainly associated with trichome elongation, ATPase activity and hormone transduction. Changes in the expression of these and other related genes with high sequence diversification may have contributed to the contrast in the pattern of trichome phenotypes between the two species.

T5

P1238

Constitutive over expression of *HSP24.4* gene for the development of salt tolerance in cultivated tomato

Upender Mahesh[1], *Mamidala Praveen*[2], *Rao Pjm*[3], *Hari Yadla*[3], *Kirti P.B.*[1], *Ramaswamy Nanna*[2], *Aragao Francisco Jl*[4]
1. *University of Hyderabad*
2. *Kakatiya University*
3. *Prof. Jayashankar State Agricultural University*
4. *National Centre for Genetic Resources and Biotechnology*

Tomato (*Solanum lycopersicum* L.) is sensitive to high salt conditions at almost all the stages of its growth and development. Considering the important role of heat shock proteins in imparting tolerance against high salt to the cells and tissues, for the first time, we isolated small *HSP24.4 (MasHSP24.4)* cDNA from wild banana (*Musa accuminata*) and introduced into the cultivated tomato cv PKM1 via *Agrobacterium tumefaciens* Stable integration and expression of the transgene in the tomato genome was demonstrated by southern, northern and western blot analyses. The genetic analysis of the transgenic T_2 tomato lines showed the Mendelian ratio in transgene segregation. The *MasHSP24.4* showed expression in root, shoot and stem tissues at 200mM and 300mM NaCl treatments and conferred tolerance to high salt stress by showing increased seed germination, healthy vegetative growth and normal fruit and seed setting. The transgenic tomato plants showed significantly the best growth performance in the recovery phase following the stress could be due to over expression of *HSP24.4* gene. Thus, the development of transgenic tomato plants for high salt tolerance could provide advantage to use the semi-arid tropics and also drought-prone areas in a better way.

T5

P1240

Computational approach to study the role of active site amino acid in thermostability of alpha amylases

Arooma Maryam, *Abdul Rauf Siddiqi*
COMSATS Institute of Information Technology (CIIT), Islamabad-45550

Pakistan experiences a huge production cost in detergent, textile processing and leather industry due to import of alpha amylases. This study provides insight into the structural determinants of thermostability in alpha amylase family to aid the process of protein engineering. We have undertaken a comparative sequence and structural study on seven homologous mesosphilic and thermophilic microbial species to decipher the thermostability features in alpha amylase family. Sequence Analysis and Molecular docking studies revealed presence of charged residues in primary structure and strong network of ionic and stacking interactions within the active site and disulphide bridges involving cysteine residues (Cys153, Cys154, Cys388, Cys432) as main factors of thermostabilityof *Pyrococcusfuriosus* alpha amylase as compared to its prokaryotic and eukaryotic homologues. One of the promising factors contributing to the thermostability in *Pyrococcusfuriosus* that was deficient inits homologs is presence of numerous charged residuesin the active site of thermo tolerant alpha amylases. These charged residues are making strong hydrogen bonds and vander wall interactions with the substrates. Among non-covalent interactions aromatic interactions (arene- arene&arenecations) interactions seems a determinant factor for the overall stability in this enzyme family. Analysis of protein protein interactions in hyperthermostable *Pyrococcusfuriosus* had shown the involvement of residues from neighbouring domain to stabilize its catalytic domain. These structural features might be helpful for in house-production of thermo stable enzymes for industrial application.

T5

P1241

A defence pathway linking plasma membrane to chloroplasts is co-opted by a virus to suppress salicylic acid signaling

Laura Medina-Puche, *Tabata Rosas-Diaz, Liping Wang, Tamara Jimenez-Gongora, Dan Zhang, Xue Ding, Chanhong Kim, Yuanyuan Zhang, Shanshan Lv, Renyi Liu, Rosa Lozano-Duran*
Shanghai Center for Plant Stress Biology, Chinese Academy of Sciences, Shanghai

Plant pathogens need to suppress plant defence responses in order to establish a successful infection. RNA silencing is considered the main plant anti-viral defence; in recent years, however, it has become increasingly clear that plants have additional strategies against viral invasion, which include production of defensive hormones such as salicylic acid (SA) and jasmonic acid. C_4 is a small protein encoded by *Tomato yellow leaf curl virus* (TYLCV) which is essential for infectivity, but of which the molecular function remains obscure. Our results indicate that C_4 is localized in two different subcellular compartments, namely plasma membrane and chloroplasts. This double localization correlates with the presence in the C_4 protein of two targeting signals, an N-myristoylation motif required for plasma membrane localization, and a chloroplast transit peptide. Interestingly, we have found that these two targeting signals are present in a number of pathogen effectors, as well as in a subset of plant proteins, many of which play a role in defence. These data suggest that a pathway may exist in plants linking plasma membrane and chloroplasts to regulate defence, and that this putative pathway is hijacked by plant pathogens, presumably to suppress these responses. Strikingly, transcriptome analysis of transgenic *Arabidopsis* plants expressing C_4 shows a clear repression of SA biosynthesis and responses. Treatments with pathogen-associated molecular patterns (PAMPs) to activate defence trigger a re-localization of C_4 from plasma membrane to chloroplasts, and production of SA in response to this treatment is halved in plants expressing C_4 or a non-myristoylable version of C_4, which localizes to chloroplasts exclusively. Given that the biosynthesis of this defensive hormone occurs in the chloroplast, our current hypothesis is that activation of plant defence leads to the re-localization of C_4 from plasma membrane to chloroplasts, where it inhibits SA biosynthesis to promote the viral infection.

T5

P1242

Genetically encoded FRET-based biosensor for *in vivo* monitoring of glycine betaine in living cells

Mohammad Ahmad[1], *Tariq Omar Siddiqi*[1], *Altaf Ahmad*[2]

1. Jamia Hamdard

2. Aligarh Muslim University

Glycine betaine (GB) accumulates in the cell at exceedingly high level under the conditions of high salinity and plays a crucial role in the maintenance of osmolarity of the cell without affecting the physiological processes. Since, analysis of stress-induced physiological conditions in living cells requires real-time monitoring of cellular GB level. A genetically encoded FRET-based biosensor has been developed in this study that allows the real-time monitoring of GB levels inside living cells. GB binding protein (ProX) was sandwitched between the cyan fluorescent protein (CFP) and yellow fluorescent protein (YFP), which are known as Förster resonance energy transfer (FRET) pair. This trimeric sensor protein was named as Glycine Betaine Optical Sensor (GBOS). Conformational change in ProX, which was used as sensory domain, reported the change in the level of this compatible solute under *in vitro* and *in vivo* conditions. Binding of the GB to the sensory domain fetches close to both the fluorescent moieties that result in the form of increased FRET ratio. So, any change in the concentration of GB is correlated with change in FRET ratio. This biosensor also reported the GB cellular dynamics in real-time in *Escherichia coli* cells after the addition of its precursor, choline. GBOS was also expressed in yeast and mammalian cells and live cell imaging was performed to monitor the intracellular GB level, quantitatively. Therefore, the GBOS represents a unique FRET-based nanosensor which allows the non-invasive ratiometric analysis of the GB in living cells.

T5

P1243

Comparative organellar genomics of the European Calypogeiaceae

Kamil Myszczyński[1], *Monika Ślipiko*[1], *Katarzyna Buczkowska*[2], *Alina Bączkiewicz*[2], *Monika Szczecińska*[1], *Jakub Sawicki*[1]

1. University of Warmia and Mazury in Olsztyn

2. Adam Mickiewicz University

Calypogeiaceae is liverwort family distributed worldwide, primarily in subtropical and tropical climates. Nowadays only a few chloroplast and mitochondrial genomes of liverworts have been completely sequenced. Further neither plastome nor mitogenome sequence of Calypogeiaceae have been revealed. In this study we present phylogenomic analyses of sequenced chloroplast and mitochondrial genomes of all known European Calypogeiaceae species. The plastome and mitogenome of Calypogeiaceae display the typical structure of most liverworts, however few exceptions were found including introns loss. SNP and indels were also identified among analyzed species. Comparative genomics analyses of the species were conducted with main focus on possible reasons of differences in reproductive morphology of Calypogeiaceae. Identification of differences between monoecious: *Calypogeia integristipula* and *C. neogea* and dioecious: *C. arguta* and *C. suecica* revealed that presence of one or both sex organs is not due to chloroplast or mitchondrial genomes diversity.

T5

P1245

Molecular and chemical evidences for the poisoning of a duck by Anatoxin-a, Nodularin and Cryptophycin at the coast of the ShoorMast Lake (Mazandaran province)

Bahareh Nowruzi

Department of Biology, Science and Research Branch, Islamic Azad University, Tehran, Iran

The toxic cyanobacteria in aquatic ecosystems are a big concern especially in warm summer months. Animals are incapable of avoiding toxic blooms near the shore and might get poisoned. There have been many cases, when ducks, sheep, cows and dogs have died for poisoning caused by cyanobacteria. In this study the aim was to find a cause of death for a duck found near the coastal area of the ShoorMast Lake. Cyanotoxin poisoning was suspected as the cause of fatalities, as a result, the water samples were collected and examined with phenotypic and genotypic analyses in order to find if the duck is poisoned by the toxic cyanobacteria. Different molecular and chemical methods such as phylogenetic analysis based on 16S rRNA gene sequence, polymerase chain reaction (PCR), disc diffusion bioassays and liquid chromatography-mass spectrometry (LC-MS) were used. Based on the results, three different cyanobacteria strains were found in duck´s stomach and three diverse groups of potentially bioactive compounds (Anatoxin-a, Nodularin and Cryptophycin) might account for the death of this animal. The cyanobacteria strains were identified and registered under DDBJ as *Anabaena* sp. BN KY303913, *Nodularia* sp. BN KY303914 and *Nostoc* sp. BN KY303912. This case is the first documented incident of toxicity from fresh water cyanobacteria related intoxication in Iran.

T5

P1246

Primitive auxin response without TIR1 and Aux/IAA in the charophyte alga *Klebsormidium nitens*

Kinuka Ohtaka[1], *Koichi Hori*[2], *Yuri Kanno*[3], *Mitsunori Seo*[3], *Hiroyuki Ohta*[1,2,4,5]

1. Graduate School of Bioscience and Biotechnology, Tokyo Institute of Technology

2. School of Life Science and Technology, Tokyo Institute of Technology

3. Center for Sustainable Resource Science, RIKEN

4. Japan Science and Technology Agency (JST), Core Research for Evolutional Science and Technology (CREST) program

5. Earth-Life Science Institute, Tokyo Institute of Technology

The phytohormone auxin regulates many aspects of growth and development in land plants; however, the origin and evolution of auxin signaling and response mechanisms remain largely unclear. Auxin research in more primitive plants is therefore essential to address these issues. The charophyte alga *Klebsormidium nitens* is an intermediate organism between green algae and land plants, and its draft genome revealed some putative genes involved in auxin response (Hori *et al.*, 2014).We detected auxin (indole-3-acetic acid, IAA) in *K. nitens* by hormonome analysis. In this study, we aimed to clarify primitive auxin response in *K. nitens*. Exogenous IAA treatment inhibited growth of *K. nitens*. Therefore, we analyzed cell division and cell elongation with exogenous IAA treatment in detail. The results suggested that IAA affected on not only cell division but also cell elongation. Furthermore, we tested for auxin mediated-regulation of gene expression by analyzing *K. nitens* transcriptome. Here, we discuss primitive auxin response in *K. nitens* based on these results.

T5

P1247

In Silico Comparative Analyses of *Phytochelatin Synthase* (*PCS*) Genes in Selected Plants: Employing Bioinformatics Approaches

Ibrahim Ilker Ozyigit[1], *Ertugrul Filiz*[2], *Recep Vatansever*[1]
1. Marmara University, Faculty of Science and Arts,
2. Duzce University, Department of Crop and Animal Production, Cilimli Vocational School

Heavy metal-based environmental pollution in soil and water damages to living organisms related to its toxicity and accumulation. Toxic heavy metals displace endogenous metal cofactors from their cellular binding sites, react with thiol groups of proteins and induce reactive oxygen species formation. There are several defense strategies in plants to mitigate the negative effects of heavy metals and re-establish the metal balance. Like algae and fungi, higher plants also overcome heavy metal stress by increasing the synthesis of sulfur containing defense compounds called cysteine, glutathione (GSH) and phytochelatins (PCs). Phytochelatins promote the vacuolysosomal sequestration of heavy metals like As, Ag, Cd, Cu, Hg, Pb and Zn, and also contribute to their detoxification. Although phytochelatins are activated by several heavy metals, Cd is one of the most effective activator for them.
In this study, 14 PCs homologs were identified from the genomes of selected 10 plant species by using bioinformatics tools. PCSs were seen as polypeptides of 452-545 residues and characterized both phytochelatin and phytochelatin_C domains. C-terminal site of proteins was less preserved while N-terminal site was highly conserved. Additionally, earlier reported full-conserved Cys-residues in heavy metal binding were identified. In association with metal binding, other preserved cysteines, with minor substitutions Cys (C) or Ser (S) or Tyr (Y) or Trp (W) were also identified in PCS sequences. Cys56, His162 and Asp180, characterized catalytic triad residues from *Arabidopsis thaliana* were conserved in respective sites of PCSs. As a result of phylogenetic analysis, a clear monocot/dicot separation was revealed and further corroborated by exon-intron organizations of *PCS* genes. *Cis*-regulatory motif, miRNA, *co*-expression network and GO term analyses showed the complex and dynamic organization of *PCSs*. In addition to metal detoxification, data showed their roles in different metabolic pathways such as signaling, defense and stress. Furthermore, protein structure variations were also implicated to confer the functional divergence in PCS proteins.

T5

P1248

Transcriptomic and expression analysis indicate involvement of NO in *Camellia sinensis* pollen tube growth at low temperature

Junting Pan[1], *Weidong Wang*[2], *Zaifa Shu*[1], *Xiaoli Ye*[1], *Yuhua Wang*[1]
1. Nanjing Agricultural University
2. Northwest A&F University

Nitric oxide (NO) functions as a critical signaling molecule in the low-temperature stress responses in plants, including polarized pollen tube growth in *Camellia sinensis*. We investigate alterations to gene expression in *C. sinensis* pollen tubes exposed to low-temperature stress and NO using RNA-Seq technology, in order to find the potential candidate genes related to the regulation of pollen tube elongation by NO under low-temperature stress. We produced three libraries from pollen tubes under control (CK), low-temperature stress (LT) and NO stress (NO), including a dataset containing 45,432 unigenes from the *C. sinensis* pollen tube transcriptomes using paired-end and *de novo* sequencing with the Illumina HiSeq™2000 platform. Among these unigenes, 36,097 were annotated with descriptions from the NR, NT, Swiss-Prot, KEGG, COG and GO databases. Our dataset includes comprehensive analysis in sequence and DEGs profiling data that provide a dynamic perspective on transcriptomic variations caused by NO and low-temperature in *C. sinensis* pollen tube elongation. Many DEGs were simultaneously found in both the CK-VS-LT and CK-VS-NO libraries in different functional pathways, such as plant hormone signaling (ABA, BRs, ET and auxin), TFs (MYB, ZFP, CAMTA and MADS-box), vesicle polarized trafficking and cell wall biosynthesis (PI4K and COBRA-like), the ubiquitination machinery of the ubiquitin system (E3s, 26S proteasome, U-box and BTB/POZ) and caffeine and theanine biosynthesis (caffeine synthase and glutamate receptor). Combining our findings and previous reports, we propose a crosstalk network which may indicate the probable mechanisms underlying NO mediating *C. sinensis* pollen tube growth during the response to low-temperature stress. Plant hormone signaling pathways (ABA, BRs, ET and auxin) respond to pollen germination and tube elongation at low temperature, and then TFs (MYB, ZFP, CAMTA and MADS-box) modulated relevant proteins expression. Moreover, the UPS (E3s, 26S proteasome, U-box and BTB/POZ) probably involves in affecting pollen germination and tube growth in response to low-temperature stress by modulating the activity, localization or stability of different proteins. Additionally, Ca^{2+} gradient, vesicle polarized trafficking as well as cell wall biosynthesis might participate in pollen tube elongation under low-temperature stress through the NO signaling pathway by CAMTA, PI4K and COBRA-like. Taken together, these findings offer a more profound understanding of the candidate genes employed by *C. sinensis* pollen tubes to integrate low temperature signals and finely tune their tolerance responses, present a fundamental transcriptomic resource to further explore the molecular mechanisms of reproductive tissue tolerance to low-temperature.

T5
P1249
The complete chloroplast genome sequence of Korean endemic species *Corydalis filistipes* Nakai (Papaveraceae: Fumarioideae)

Kyu Tae Park, *Seonjoo Park*
Department of Life Sciences, Yeungnam University

Corydalis is a large genus of approximately 470 species in Papaveraceae .and mainly native to temperate Northern Hemisphere. Most species of genus *Corydalis* are diverse in China and the Himalayas. One of this genus, *Corydalis filistipes*, is endemic species in Korea. *C. filistipes* is restricted only to Uleng island. And the genetic and molecular data is deficient. Therefore we conducted analysis through NGS and determined the complete chloroplast genome sequence of *C. filistipes*. The chloroplast genome of *C. filistipes* is a circular molecule of 155,782bp containing two single copy regions and two inverted repeat regions, a large single copy region (LSC, 86,884bp) and a small single copy region (SSC, 18,348bp), separated by inverted repeat regions (IRs, 25,275bp each). In addition, the chloroplast genome of *C. filistipes* has 40.6% GC content and contains 129 genes, of which 83 are protein-coding, 38 are transfer RNA, and four are ribosomal RNA. A phylogenetic analysis based on 71 protein coding genes indicate that *Corydalis* is related to Papaveroideae species.
Acknowledgments: This study was supported by grants from Scientific research (KNA1-2-13.14-2) of Korea National Arboretum.

T5
P1250
DNA-binding domains of dehydration responsive element binding proteins of bread wheat (*Triticum aestivum* L.)

Svitlana Proskurina[1], ***Olga Kolesnyk***[2], *Andrii Venger*[1], *Anastasiia Honcharova*[2]
1. International Humanitarian University
2. Plant Breeding and Genetic Institute

Drought is a common adverse environmental condition that significantly influences plant growth and productivity worldwide. In the last years there has begun the intensive study of the functions of genes encoding transcription factors (TFs), which can be directly or indirectly involved in the regulation of processes of plants resistance in response to drought. Dehydration responsive element binding (DREBs) are important plant transcription factors that regulate the expression of many stress-inducible genes and causes a strong influence in improving the abiotic stress tolerance of plants by interacting with a DRE/CRT cis-element present in the promoter region of various abiotic stress-responsive genes. According to structural characteristics, DREBs proteins are subdivided into six subgroups (A-1-A-6), which may perform different functions. The precise study of domains of different DREB subgroup proteins will give the opportunity to compare their functions in order to understand the mechanism of response to stressful conditions. However at the moment the function of DREBs proteins of bread wheat are studied insufficiently. The purpose of this research was to compare the domains of 6 subgroups of DREBs proteins in order to reveal similarities and differences in their functional features as these proteins are associated with the mechanisms of response to stressful conditions. Protein sequences of DREBs proteins (accession number of DREB1 is AAL01124.1; DREB2 – ADM89078.1; DREB3 – AAX13277.1; DREB4 – AAX13283.1; DREB5 – AAX13286.1; DREB6 – AAX13289.1) were taken from National Centre of Biotechnology Information and were aligned according to database of protein sequences using online program "BLAST" by Smith-Waterman algorithm. Protein domains and biochemical properties of DREBs were figured out by Domain Enhanced Lookup Time Accelerated BLAST online program. Consequently according to delta-BLAST online program each of six subgroups of DREBs proteins consists of three domains: 1) AP2 (accession number cd00018)-DNA-binding domain found in transcription regulators in plants such as APETALA2 and EREBP (ethylene responsive element binding protein). In EREBPs the domain specifically binds to the 11bp GCC box of the ethylene response element (ERE), a promotor element essential for ethylene responsiveness. EREBPs and the C-repeat binding factor CBF1, which is involved in stress response, contain a single copy of the AP2 domain. APETALA2-like proteins, which play a role in plant development contain two copies of the AP2 domain. 2) AP2 (accession number smart00380)-DNA-binding domain in plant proteins such as APETALA2 and EREBPs. 3) AP2 domain (accession number pfam00847). This 60 amino acid residue domain can bind to DNA and is found in transcription factor proteins. The mentioned domains of different subgroups of DREB proteins differ by the number of amino acid residues as well as they are located in different parts of the DREB protein which means that although they have partially similar structure these proteins may have the functional peculiarities. The presence of different subgroups of DREB genes in the genotypes of bread wheat varieties of South Ukraine makes it possible to lead to different ways of plant response to drought conditions.

T5
P1252
Using a chloroplast genome reference library for the identification of cultivated peanut

Caroline Puente-Lelievre, *Sara Handy*, *Ning Zhang*, *Anne Eischeid*
US Food and Drug Administration

The advancement of high-throughput sequencing technologies has expedited the progress of plant genomics, particularly chloroplast genomics. To date, the National Center for Biotechnology Information (NCBI) organelle genome database harbors more than 300 chloroplast genomes from crops and trees. These genome data have been used mainly for studies in phylogenetics, breeding, domestication, and conservation; but could also be used to develop specific DNA markers to identify allergenic plants such as peanuts. Domesticated peanut (*Arachis hypogea*, Fabaceae) is one of the most consumed legumes in the world, and is widely cultivated in the tropics and subtropics (annual production of ~46 million tons). While peanuts have a key role in human nutrition, they are one of the most common allergy-causing foods in the United States. Food allergy reactions can be fatal and are a leading cause of anaphylactic events. Consumers with food allergies must rely on correct food labeling to identify the presence of allergens. To help protect consumers in the U.S., the Food Allergen Labeling and Consumer Protection Act of 2004 requires the presence of any of the major food allergens (including peanut) be declared on the label. It is critical for the enforcement of this regulation to have reliable analytical methods that accurately identify and detect these plant allergens. DNA markers have become increasingly common and effective tools to achieve this goal. The *Ara h 2* and *Ara h 3* (allergenic protein) genes and the internal transcribed spacer (ITS) have traditionally been used for peanut identification, but their level of sensitivity has been limited so far. The number of copies of nuclear markers in the cell is inherently lower than the number of plastid copies, which significantly restricts their use for reliable peanut identification in processed samples that often contain small DNA concentrations. We propose an alternative application of chloroplast genomes as a source of unique DNA markers to identify domesticated varieties of peanut. We are building a chloroplast genome reference library of economic crops to support the development of highly sensitive DNA-based assays to detect the presence of peanuts in a broad range of commercial products. We also compare and discuss the effectiveness of these newly developed chloroplast markers against nuclear markers previously published.

T5
P1253

Genetic programming based models as an alternative to the traditional statistical methodology in plant research

Meenu R.M., *Ashalatha S. Nair*
Department of Botany, University of Kerala, Kariavattom

In this paper we compared the efficacy of observation based modelling approach using genetic algorithm with the usual statistical analysis as an alternative methodology in plant research. The preliminary experimental data pertaining to *in vitro* rooting in *Wrightia tinctoria* was taken for this study; with an aim to understand the effect of activated charcoal (AC) and naphthalene acetic acid (NAA) on successful rooting. The two variables were also optimized for maximum result. Observation based modelling as well as traditional approach could identify the presence of charcoal in the culture medium along with NAA as a critical factor in rooting and survival of the plantlets under the experimental conditions employed. The results suggested that NAA has a significant positive effect on rooting provided AC is present in the medium. In the present study AC had a positive and stimulatory effect in rooting of shoots by reducing basal callus. Thus absence of callus initiation enhanced rooting in auxin containing medium. However, with traditional statistics, estimating the combination/s of both variables that producing the best results can not be achieved. Symbolic regression analysis using the software employed here, optimized the treatments studied and was successful in identifying the complex non linear interaction among the variables with minimalistic preliminary data. It provides both optimum model function as well as the optimum variable values in the model. Such an approach is advantageous for establishing in vitro culture protocols as these models will have significant potential for saving time and expenditure in plant tissue culture experiments at commercial level laboratories.

T5
P1254

Establishment of a consensus map for fatty acids and oil content QTLs, identification of candidate genes for seed oil alteration in *Brassica napus*.

Nadia Raboanatahiry
Huazhong University of Science and Technology

Introduction: Deciphering the molecular architecture of a species is a good way to understand its evolutionary history, but also to tailor its profile for breeding elite cultivars with desirable traits. Combining diverse populations in one consensus map and utilizing it as a basis for multiple analyses assure a stronger accuracy to understand the genetic system that works on a given phenotype. In this study, we unraveled the genetic architecture of fatty acid (FA) and triacylglycerol (TAG) biosyntheses in rapeseed *Brassica napus*. *B. napus* (AACC, 4n = 38) is a well-known source of vegetable oil and is the preferred oil crop for biodiesel production in Europe, it was produced from hybridization of *Brassica rapa* (AA, 2n = 20) and *Brassica oleracea* (CC, 2n = 18), and both of them are closely related to the model plant *Arabidopsis thaliana*. Our objectives were to establish a consensus map for FA and oil content QTLs in *B. napus*, and identify potential candidate genes for seed oil alteration.
Methods: The genes involved in FA biosynthesis and TAG formation in *Brassica* were identified based on homology with *A. thaliana* genes. A comparative study of these genes were performed to highlight a brief evolution synopsis among this tribe. Then, a consensus map representing oil content QTLs (OC-QTLs) and fatty acids QTLs (FA-QTLs) was built based on the QTLs identified in nine populations: *DY, KN, M201xM202, PT, RNSL, SG, SO, TN* and *Z5*. The physical positions of the markers were determined by using of E-PCR. Finally, potential candidate genes underlined by these QTLs were identified, and their genetic network interaction and pathway were analyzed.
Results: A total of 438 genes involved in the FA and TAG biosyntheses were identified in *B. napus*, they were homologous to 224 genes of *B. rapa* and 173 genes of *B. oleracea*. Synteny analysis of these genes was made. Compared to parents *B. rapa* and *B. oleracea*, *B. napus* genome showed mix of loss and insertion. While settling the QTLs for oil content and 10 fatty acids traits identified from *DY, KN, M201xM202, PT, RNSL, SG, SO, TN* and *Z5* on the physical map of *Darmor-bzh*, the map was covered by 217 molecular markers and 335 FA-QTLs and OC-QTLs could be highlighted ; 101 QTLs were co-localized, and 78 among them were OC-QTLs. Chromosome C6 hosted the highest amount of co-localized QTLs (27) and C3 contained a total of 22 co-localized QTLs with all trait studied unless for C18:3. Co-localized QTLs were found in populations with similar environments. Besides, 218 candidate genes were identified in 162 QTL intervals, 77 among them were underlined by 61 co-localized QTLs, and they were majorly potential candidate genes for oil content. FAE genes represented potential genes that might affect all traits unless C18:3. Analysis of network interaction and pathway of these candidate genes indicated ten genes that strongly interacted with the other genes and might affect them. Traits might be affected by these genes in multilevel of oil biosynthesis.
Conclusion: Our findings provide new information on genetic

characteristics of *B. napus*. New track showing genes that might alter seed oil are now accessible, and might be suitable for breeding improvement purpose.

T5

P1255

Heterologous expression of *Lolium perenne* antifreeze protein confers chilling tolerance in transgenic tomato

Sathishkumar Ramalingam, ***Balamurugan Srinivasan***
Bharathiar University

Being sessile in nature, plants constantly encounter several abiotic and biotic stresses that adversely affect plant growth, development and survival. Among them, abiotic stresses are serious, as they are uncontrollable and unpredictable. The low temperature stress imposed on plants has huge impact on agriculture in particularly in the tropical or sub-tropical plants, which will potentially limit plant survival leading to reduced yield. Antifreeze proteins (AFP) are produced by certain plants, animals, fungi and bacteria that enable them to survive during chilling conditions. Perennial rye grass, *Lolium perenne* was reported to possess AFP, which protect them from cold environment. Here, we isolated *AFP* gene from *L. perenne* and expressed in tomato, the second most consumed chilling sensitive vegetable crop to confer chilling tolerance. The T_1 transgenic tomato lines were subjected to molecular, biochemical and physiological studies. Stable integration and transcription of *LpAFP* in the host genome was confirmed by Southern blot hybridization and RT-PCR, respectively. Physiological analyses under chilling conditions showed that the chilling stress induced physiological damage was observed in wild type lines, while the transgenics remained healthy phenotypically. Total sugar content was increased gradually in both WT and transgenics throughout the stress period. Interestingly, transgenics exhibited remarkable alterations in terms of relative water content and electrolyte leakage index than that of WT during the chilling treatment. Relative water content was increased significantly by 3-fold and the electrolyte leakage was reduced by 2.6-fold in transgenics than WT. Overall, these findings demonstrate that *LpAFP* gene confers chilling tolerance in transgenic tomato plants by upholding the membrane integrity. Hence, we propose *LpAFP* could be a potential candidate to extrapolate the chilling tolerance on other chilling sensitive food crops. The outcome will significantly help the high altitude farming of economically important food crops in long term.

T5

P1256

Patterns of cyto-nuclear disequilibrium and the influence of heteroplasmy in wild carrot, *Daucus carota* (Apiaceae)

Adam Ramsey[1], *Dviti Mody*[2], *David Mccauley*[2], *Massimo Iorizzo*[3], *Philipp Simon*[4], *Jennifer Mandel*[1]
1. University of Memphis
2. Vanderbilt University
3. Plants for Human Health Institute
4. USDA-Agricultural Research Service

Cyto-nuclear linkage disequilibrium (LD) is the non-random association of alleles in the nuclear and cytoplasmic organellar genomes (cpDNA or mtDNA). Patterns of cyto-nuclear LD can indicate admixture of divergent populations or of two species, positive or negative cyto-nuclear selection, inheritance of organelles in a non-traditional fashion, or the occurrence of more than one mitochondrial or plastid genome within an individual (heteroplasmy). Cyto-nuclear interactions have been shown to play a role in cytoplasmic male sterility (CMS) which occurs in gynodioecious plant species. These species are also implicated in instances of paternal inheritance of organelles. We used the gynodioecious species, Daucus carota ssp. carota, or wild carrot, to investigate cyto-nuclear LD. We genotyped a total of 255 individuals from two North American regions, 129 from the Olympic Peninsula, Washington State, and 126 from Nantucket Island, Massachusetts. We used 15 nuclear microsatellites, two mitochondrial SNPs and one plastid insertion-deletion (indel) to calculate nuclear-nuclear LD, cyto-nuclear LD and LD between organelles both among regions and within regions. We were further able to identify over 100 individuals heteroplasmic for one of the mitochondrial SNPs and over 30 individuals heteroplasmic for the plastid indel. Thus we were able to calculate the same sets of cyto-nuclear LD values for individuals that are heteroplasmic and homoplasmic (non-heteroplasmic) for the mitochondrial SNP. Differences were found between several of the LD values between the heteroplasmic and homoplasmic groups.

T5

P1257

Solving the puzzle: Using DNA barcoding and chemical analysis for authenticating *Pelargonium* species in commercial herbal products

Ryan Rattray[1], *Michelle Van Der Bank*[1], *Alvaro Viljoen*[2]
1. University of Johannesburg
2. Tshwane University of Technology

Several *Pelargonium* species indigenous to South Africa are highly valued for their horticultural value and medicinal properties. For hundreds of years various ethnic groups have used root extracts of *P. sidoides* DC as a remedy to treat coughs, upper respiratory tract irritations and gastrointestinal conditions. An ethanolic extract is used in the proprietary herbal tincture known as Umkaloabo® that is currently successfully marketed in Germany with sales that have escalated over 700%. The use of commercial herbal products (CHP) have surged globally in recent years and as a result the phytomedicine industry is under immense pressure to develop rapid, accurate and economical methods for quality control, especially for the positive identification of raw materials. The taxonomic delimitation of the two taxonomic allies *P. sidoides* and *P. reniforme* Curt. has been debated in literature. Most consumer products explicitly refer to the botanical active as *P. sidoides* posing quality control concerns as *P. sidoides* and *P. reniforme* have not been proven to be pharmacologically equipotent. Here, we used DNA sequence information generated through standardized DNA barcoding techniques for the authentication of *Pelargonium* sidoides products. Furthermore, a DNA barcode reference library for the section Reniformia was added to the Barcode of Life Database and several herbal medicines tested, showed not to contain DNA material of *Pelargonium* species, indicating adulteration of the said products. Results as well as recommendations from the study will be presented.

T5
P1258
Genomic diversity in *Vaccinium vitis-idaea* and its association with environmental conditions and biochemical variables

Zobayer Alam, Lourdes Peña-Castillo, ***Julissa Roncal***
Memorial University of Newfoundland

Partridgeberry (*Vaccinium vitis-idaea* L.) is rich in nutrients and biochemical properties beneficial for human health, and is one of the least studied crops of the *Ericaceae* family. The cultivation of this fruit crop is popular in Europe, in early development in the United States, but in Canada harvesting for local consumption and exportation is from the wild. Genetic markers are needed to facilitate the selection of berries with desirable agronomic and nutritious characteristics to start a cultivation program. In this study, we estimated the total phenolic content (TPC) and antioxidant capacity (AC) of 56 partridgeberry fruit samples collected from different regions in Newfoundland and Labrador, the province with the largest amount of export in Canada. We applied next generation sequencing technology, specifically the genotyping-by-sequencing method to discover single nucleotide polymorphisms (SNPs) across the 56 samples. We identified 1,586 high-quality putative SNPs using the UNEAK pipeline available in TASSEL. To search for an association between adaptation to environmental conditions and population-level genetic diversity, we obtained the correlation of each of the identified SNPs with eight environmental variables. We also searched for an association between the identified SNPs and TPC and AC. We found 260 SNPs likely to be associated with at least one of the environmental and biochemical variables. The functional annotation of the 260 SNP-containing sequences, based on BLAST similarity searches, identified 128 genes with functional roles likely to be related to phenolic compound biosynthesis, transcription regulation, transport, and drought. The SNP markers discovered in this study might provide a useful resource for future molecular genetic studies, such as high-density quantitative trait locus analysis, and association mapping. The identified candidate genes containing the SNPs need further studies on their adaptation ability in partridgeberry.

T5
P1259
Genetic dissection of agronomic performance, yield and quality of durum wheat through genome-wide nested association mapping

Yuefeng Ruan[1]*, Bianyun Yu*[2]*, Wentao Zhang*[2]*, Prabhath Lokuruge*[2]*, Peng Gao*[2]*, Christine Sidebottom*[2]*, Janet Condie*[2]*, Ron Knox*[1]*, Richard Cuthbert*[1]*, Curtis Pozniak*[3]*, Andrew Sharpe*[2]*, Pierre Fobert*[2]
1. Swift Current Research and Development Centre, Agriculture and Agri-Food Canada
2. Aquatic and Crop Resource Development, National Research Council Canada
3. Crop Development Center, College of Agriculture and Bioresources, University of Saskatchewan

Durum wheat is an important crop in Canada. The rate of genetic gainin yield and quality traits in Canadian durum cultivars have been achieved greatly in the past 50 years. However, Climate change and an increasing world population pose challenges to the yield with current rate of genetics gain.Recently, advances in genomic assisted breeding provide promising technologies to meet these challenges and to speed up the genetic gain on Canadian durum wheat. The efficiency of such an approach relies on the precision of QTL mapping and robust molecular markers linked to the trait. However, most important crop traits are complexly inherited and controlled by multiple genes with minor effects. Current mapping strategies - linkage mapping or association mapping alone has less statistical power to detect these minor alleles and less precision to estimate their position and effects. Nested association mapping (NAM), combining the advantages of both linkage analysis and association mapping has recently been proved as a promising approach to address these challenges with current mapping strategies on complex traits. To explore the usefulness of this NAM based approach in improving the genetics gain on Canadian durum wheat, a durum NAM panel comprised of 14 sub-populations with a common reference parental line "Strongfield" crossed with 14 other durum cultivars was assembled. Each sub-population has 50 lines, resulting in a total of 700 inbred lines. The NAM population, parents and checks were grown and evaluated foragronomic traits, yield and yield components, protein content and foliar disease leaf spot in two growing seasons at two locations. The Illumina infimum wheat 90K SNP chip was used to genotype the NAM population. Combining the genotypic data with field data, genome wide association study is being carried out to detect QTL for each trait with NAM based mapping approach. The efficiency, precision, effectiveness and usefulness of NAM approach to capture the multiple alleles with minor effects underling these complex traits in Canadian durum wheat will be discussed.

T5
P1260
Rhopalocnemis phalloides possesses one of the most reduced and mutated plastid genomes known

Mikhail Schelkunov[1,2]*, Maxim Nuraliev*[2,3]*, Maria Logacheva*[1,2]
1. Institute for Information Transmission Problems
2. Lomonosov Moscow State University
3. Joint Russian–Vietnamese Tropical Scientific and Technological Center

Holo-heterotrophic plants are plants, completely incapable of photosynthesis - instead they feed by parasitizing on fungi or other plants. Several hundreds of holo-heterotrophic plants are known, representing several dozen cases of independent transition from autotrophic to holo-heterotrophic lifestyle. Their plastid genomes ("plastomes") are usually highly reduced, mainly due to loss of genes with functions related to photosynthesis. To date, only one higher plant is supposed to have lost its plastid genome completely - *Rafflesia lagascae*. This work is dedicated to an analysis of a plastome of *Rhopalocnemis phalloides* (family Balanophoraceae) - a plant which we initially suspected to have also lost a plastome. Instead, deep sequencing has shown that, in fact, it possesses a plastome, but its sequencing coverage was depressed due to an extremely high AT-content of 87% - the highest among all sequenced plant plastomes. The plastome has a length of 18 622 base pairs and contains a gene set typical for holo-heterotrophic plants, with most of retained genes encoding ribosome components. Although all of the retained genes are likely functional, they have accumulated an enormous amount of mutations, with proteins encoded by some genes having sequence identity of about 25% to the closest homologs in GenBank. Due to the high AT-content, sequences of several genes converged with sequences of genes from Apicomplexa, which also have AT-rich plastomes. To sum up, *Rhopalocnemis phalloides* represents one of the most extreme cases of plastid genome reduction.

T5

P1261

Identification and characterization of N-alpha acetylated proteins in *Ostreococcus tauri*

Rashmi Shakya[1], *Thierry Lebihan*[2]

1. Miranda House, University of Delhi

2. The Institute of Quantitative Biology, Biochemistry and Biotechnology, SynthSys-Systems and Synthetic Biology, School of Biological Sciences, University of Edinburgh

Protein Nα-acetylation is one of the most common modifications found in living organisms and plays a significant role for the fate of proteins in terms of stability and activity. In spite of being so abundant these modifications remain poorly described and badly characterized, especially in plants. Recent advances in N-terminal positional proteomics have made it possible to unravel the nature of N-terminal modifications. By employing N-terminal enrichment and no enrichment-based strategies we have identified 228 non-redundant Nα-acetylated peptides in the smallest known eukaryote, *Ostreococcus tauri,* and showed that the effect of penultimate amino acid on the efficiency of methionine cleavage is conserved. Proteome-wide comparison has revealed that bias for serine residue at ultimate and aspartic residue at penultimate position seems to be a characteristic of Nα- acetylome in *O. tauri.* Analysis of experimentally identified Nα-acetylome has led to the prediction of four methionine amino peptidases (MAPs) and seven N-acetyltransferases (NATs) orthologs which probably play a crucial role in N-terminal processing mechanisms. In agreement with higher eukaryotes, NatA was found to be the most abundant NAT followed by NatB and NatC. We report redundancy in terms of NATs orthologs predicted in *O. tauri* protein sequence database. Based on proteogenomics analysis of 48 downstream Nα-acetylated (dNAA) peptides which had methionine residue before them, it has been suggested that probably 12 proteins (corresponding to these 12 dNAA peptides) have not been annotated correctly. Finally, we report a possibility of 103 novel proteins in *O. tauri* (which are found in *O. lucimarinus*) as indicated by proteomics data.

T5

P1262

Identification of potential genes that contributed to the variation in the taxoid contents between two *Taxus* species (T*axus media* and *Taxus mairei*)

Chunna Yu[1], *Hong Guo*[1], *Yangyang Li*[1], *Erxu Pi*[1], *Lei Zhang*[2], ***Chenjia Shen***[1]

1. Hangzhou Normal University

2. Washington State University

Taxol is currently a valuable anticancer drug; however, the accumulated mixture of taxoids can vary greatly among *Taxus* species. Genomic information for the genus *Taxus* is currently unavailable. Thus, transcriptome analysis is a powerful approach to explore the different regulatory mechanisms underlying taxoid biosynthesis pathways in *Taxus* species. First, we quantified the variation in the taxoid contents between *Taxus media* and *Taxus mairei* using an high-performance liquid chromatography-tandem mass spectrometry method. Then, the transcriptome profiles of *T. media* and *T. mairei* were analyzed to investigate the altered transcripts. From six complementary cDNA libraries, ~350.1 million short reads were generated, and 88,194 unigenes were found. A total of 20,704 significantly differentially expressed genes (DEGs), including 9,865 unigenes highly expressed in *T. media* and 10,839 unigenes highly expressed in *T. mairei*, were identified. In total, 120 jasmonic acid-related DEGs were annotated, suggesting variation in jasmonic acid's metabolism and signaling pathway between *T. media* and *T. mairei*. Furthermore, a large number of genes related to the precursor supply, taxane skeleton formation and hydroxylation, and C13-side chain assembly were also identified. The differential expressions of the candidate genes involved in taxoid biosynthetic pathways may explain the variation in the taxoid contents between *T. media* and *T. mairei*.

T5

P1263

Comparative proteomic analyses of two *Taxus* species (*Taxus × media* and T*axus mairei*) to reveal the variations in the metabolisms that associated with Taxol and other metabolites

Juan Hao[1], *Hong Guo*[1], *Lei Zhang*[2], ***Chenjia Shen***[1]

1. Hangzhou Normal University

2. Washington State University

Taxus species are well-known for their taxol, which exhibits antitumor activities and is used for the treatments of various cancers. Although most *Taxus* species are widely spread in many areas, few studies have characterized the variations in the metabolisms of different *Taxus* species. Using an integrated approach involving TMT labeling and LC-MS/MS, we investigated the proteomes of *T. media* and *T. mairei*, and 4,078 proteins were quantified. The screening and classification of differential expressed proteins revealed a large number of metabolism associated proteins. In detail, four enzymes involved in flavonoid biosynthesis pathway predominantly expressed in *T. mairei*. Four enzymes associated with supplying precursors for Taxol biosynthesis and three cytochrome P450 taxoid oxygenases were preferentially expressed in *T. media* rather than in *T. mairei*. Furthermore, variations between the taxoid contents of *T. media* and *T. mairei* were determined by HPLC-MS/MS analysis. Variations between the flavonoid contents of *T. media* and *T. mairei* were determined by HPLC analysis. A number of differential expressed enzymes may provide an explanation for the variation in the metabolisms of different *Taxus* species.

T5

P1264

CHD1 (CHR5) and CHD3 (PKL) act antagonistically to regulate seed maturation program in *Arabidopsis*

Yuan Shen

Xinxiang Medical Univeristy

Chromatin modification and remodeling are the basis for epigenetic regulation of gene expression. *LEAFY COTYLEDON1 (LEC1)*, *LEAFY COTYLEDON 2 (LEC2)*, *ABSCISIC ACID-INSENSITIVE 3 (ABI3)*, and *FUSCA3 (FUS3)* are key regulators of seed maturation and are expressed during embryo development but epigenetically repressed after seed germination. The CHD3 protein PICKLE (PKL) is involved in the epigenetic silencing of the genes. However, the chromatin mechanism that establishes the active state of the genes during embryogenesis and seed development is not clear.Our work showed that the *Arabidopsis* CHD1-related gene, *CHR5*, displayed a similar expression pattern as *ABI3* and *FUS3* during embryo development. The mutation of the gene led to decreased expression of *LEC1*, *ABI3* and *FUS3* in developing seeds and reduced accumulation of seed storage proteins. In addition, we found that CHR5 was directly associated with the promoters of *ABI3* and *FUS3* and *chr5* mutations led to increased nucleosome occupancy near the transcriptional start site. To study the functional interaction between *CHR5* and *PKL*, we generated *chr5/pkl* double mutants. Analysis of double mutants showed that CHR5and PKL have an opposite function to regulate embryo gene promoter activity and seed storage protein accumulation. CHR5 regulates embryo identity genes expression by modulating nucleosome occupancy, whereas PKL represses the expression of the geneslikely by promoting H3K27me3.The antagonistic function of *CHR5* and *PKL* on embryo gene expression by acting on different aspects of chromatin modifications suggests important but complex roles of chromatinin the regulation of seed maturation gene expression.

T5

P1265

Full-length transcriptome sequences and gene co-expression network analysis applied to the two medical components of naringin and neoeriocitrin biosyntheses in *Drynaria roosii*

*Jingyi Li[1,2], Meiyu Sun[1], Dong Li[1], Fengjie Huang[1], Di Wang[1], Hui Li[1], Quan Xing[1], Yongzhen Pang[1], **Lei Shi**[1]*

1. Key Laboratory of Plant Resources and Beijing Botanical Garden, Institute of Botany, Chinese Academy of Sciences

2. University of Chinese Academy of Sciences

Drynaria roosii (Nakaike), a large epiphytic fern of the family Polypodiaceae, is a source of traditional Chinese medicine commonly known as 'Gu Sui Bu', wherein its rhizome is particularly valued. The corresponding effective flavonoids components include naringin and neoeriocitrin, the biosyntheses of which are the subjects of considerable interest. In our study, we found that the contents of naringin and neoeriocitrin varied in different tissues and years tested by HPLC. However, the molecular mechanism of these two medical components biosyntheses in *D. roosii* with different tissues and years has remained elusive. Moreover, the investigations of the *D. roosii* transcriptome have never been reported before. Here, we applied a combination of short-read next-generation sequencing (NGS) and single-molecule real-time (SMRT) sequencing to various tissues (leaf, vein, young rhizome, old rhizome) and rhizomes from different years (2,3, 5,6, 8,10 years old) to generate a more complete/full-length *D. roosii* transcriptome. From the SMRT data, 4,364,566 raw reads, 272,121 full-length non-chimeric reads were obtained; moreover, 59,294 CDS sequences were predicted. As with the iterative clustering for error correction (ICE) algorithm, 67,147 high-quality full-length transcripts were assembled with the accuracy of more than 99%. 59,100 transcripts (88.02%) were annotated for their function, thereinto, 52,439 transcripts, 41,844 transcripts, 43,863 transcripts, 37,028 transcripts, 24,876 transcripts, 29,012 transcripts were allocated to Nr, Nt, Swiss-Prot, KEGG, COG, and GO databases, respectively. With using of the NGS data, we detected the unigenes related to phenylpropanoid biosynthesis and flavonoid biosynthesis exhibited marked differential expression in various tissues and rhizomes from different years of *D. roosii*. By analyzing dynamic changes of differentially expressed genes, many potentially interesting genes (*PAL, CA4H, 4CL, CHS, CHI, F3'H, UGT*) were identified and suggested that they were responsible for biosynthesis pathways of naringin and neoeriocitrin in *D. roosii* indeed. Furthermore, weighted gene co-expression network analysis (WGCNA) has been widely used as a method to cluster gene expression and metabolite data and identify hub genes in several non-model plant species, and genes with correlated expression patterns can be best identified by using coexpression network analysis of transcriptome data. Thus in our study, we analyzed the temporal-spatial coordination of gene expression in various tissues and rhizomes from different years of *D. roosii* and constructed the gene network using WGCNA to get further insight into the biosynthesis pathways of naringin and neoeriocitrin in *D. roosii*. This work was supported by the National Natural Science Foundation of China (no. 81573519) and the Key Projects of the Chinese Academy of Sciences (no. KZCC-EW-103-3).

T5

P1266

The origin, diversification and adaptation of a major mangrove clade (Rhizophoreae) revealed by whole genome sequencing

*Shaohua Xu[1], Ziwen He[1], Zhang Zhang[1], Zixiao Guo[1], Wuxia Guo[1], Haomin Lyu[1], Jianfang Li[1], Ming Yang[1], Zhenglin Du[2], Yelin Huang[1], Renchao Zhou[1], Cairong Zhong[3], David Boufford[4], Manuel Lerdau[5], Chung-I Wu[1,2,6], Norman Duke[7], **Suhua Shi**[1]*

1. Sun Yat-sen University

2. Beijing Institute of Genomics, Chinese Academy of Sciences

3. Hainan Dongzhai Harbor National Nature Reserve

4. Harvard University Herbaria

5. University of Virginia

6. University of Chicago

7. James Cook University

Mangroves colonize the most marginal habitats for woody plants - at the interface between land and sea. Their origin, diversification and adaptation are of scientific significance, particularly with the impending sea level rises. In this study, a combination of single-molecule long reads and the conventional short reads are generated from *Rhizophora apiculata* for *de novo* near-chromosome level assembly. Genomic sequences of six other species from this major mangrove clade have also been generated. Using these data, we find that the ancestor of Rhizophoreae experienced whole genome duplication, which was soon followed by the intertidal colonization and species diversification 50-70 Myrs ago. In the saline environments, mangrove genomes have evolved unusual amino acid (AA) compositions and experienced rapid protein evolution associated with unique AA substitution patterns. Some of these proteins are found to be associated with traits that are specialized for the intertidal habitats (e.g., vivipary and red bark). Despite the extensive adaptations, mangroves are among the least genetically-diverse plants, likely the result of rapid habit turnovers in the recent geological past, caused by repeated sea level changes. Thus, mangrove genomes can inform about their past evolutionary successes but their low diversity may also portend a possibly difficult future.

T5

P1267

The evolution of plant microRNAs—Insights from a basal eudicot sacred lotus

Tao Shi

Wuhan Botanical Garden, Chinese Academy of Sciences

MicroRNAs (miRNAs) are important noncoding small RNAs that regulate mRNAs in eukaryotes. However, under which circumstances different miRNAs/miRNA families exhibit different evolutionary trajectories in plants remains unclear. In this study, we sequenced the small RNAs and degradome from a basal eudicot, sacred lotus (*Nelumbo nucifera* or lotus), to identify miRNAs and their targets. Combining with public miRNAs, we predicted 57 pre-eudicot miRNA families from different evolutionary stages. We found that miRNA families featuring older age, higher copy and target number tend to show lower propensity for miRNA family loss (PGL) and stronger signature of purifying selection during divergence of temperate and tropical lotus. Further analyses of lotus genome revealed that there is an association between loss of miRNA families in descendent plants and in duplicated genomes. Gene dosage balance is crucial in maintaining those preferentially retained *MIRNA* duplicates by imposing stronger purifying selection. However, these factors and selection influencing miRNA family evolution are not applicable to the putative *MIRNA-likes*. Additionally, the *MIRNAs* participating in lotus pollen-pistil interaction, a conserved process in angiosperms, also have a strong signature of purifying selection. Functionally, sequence divergence in *MIRNAs* escalates expression divergence of their target genes between temperate and tropical lotus during rhizome and leaf growth. Overall, our study unravels several important factors and selection that determine the miRNA family distribution in plants and duplicated genomes, and provides evidence for functional impact of *MIRNAs* sequence evolution.

T5

P1268

Metabolomic analysis of alfalfa (*Medicago sativa* L.) root with symbiotic rhizobia under alkali stress

Tingting Song, *Hua Cai*

Northeast Agricultural University

With respect to salinity stress, alkaline salt causes more severe damage to plants than neutral salt, due to the additional effect of pH stress. However, the mechanism by which plants respond to alkali stress is not fully understood. Here, we studied the effects of root nodule symbiosis on alkali stress in alfalfa using comparative metabolomics. *Rhizobium*-nodulized (RI group) and non-nodulized (NI group) alfalfa were treated with 200 mmol/L $NaHCO_3$ and root samples were subjected to malondialdehydyde (MDA), proline, glutathione (GSH), superoxide dismutase (SOD), and peroxidase (POD) content analysis, as well as metabolite profiling with gas chromatography combined with time-of-flight mass spectrometry (GC/TOF-MS). Phenotypically, the RI alfalfa exhibited a greater resistance to alkali stress than NI plants. Physiological analysis and metabolic profiling revealed that RI plants accumulated more antioxidants (SOD, POD, GSH), osmolytes (sugar, glycols, proline), organic acids (succinic acid, fumaric acid, and alpha-ketoglutaric acid), and metabolites that are involved in nitrogen fixation. Our pairwise metabolomics comparisons revealed that RI alfalfa exhibited a distinct metabolic profile in conferring alkali resistance relative to NI alfalfa. Our study provides new information about the relationship between non-nodulized, rhizobium-nodulized alfalfa and alkali resistance, giving valuable insight into the mechanisms mediating alkali resistance in legume crops.

T5

P1269

Spiciness in the age of discordance: Genome-scale patterns of incongruence in the phylogeny of peppers and relatives (*Capsicum/Lycianthes*, Solanaceae)

Daniel Spalink[1], *Kevin Stoffel*[2], *Allen Van Deynze*[2], *Lynn Bohs*[1]

1. University of Utah

2. University of California-Davis

Discordance among gene trees can have important consequences for our understanding of the evolution of key traits in organisms of ecological or economic importance. For example, the cultivation of peppers (*Capsicum spp.*, Solanaceae) is a global, multi-billion-dollar industry, yet our understanding of the origin, diversification, and significance of their pungency has been confounded by genes with either poor resolution or discordant evolutionary histories. The genus *Capsicum* includes species with extremely pungent, variable, or non-pungent fruits. The closest relatives of the peppers belong to the genus *Lycianthes*, which has non-pungent fruits. Our previous work, based on a concatenation of two chloroplast and two nuclear gene regions, suggests that *Lycianthes* is paraphyletic in relation to *Capsicum.* This backbone phylogeny, however, is only weakly supported. Thus, with our current dataset we are limited in our ability to investigate when or where pungency originated, how it has changed through time and space, or its ecological significance. In this study, we sequenced transcriptomes of four species of *Capsicum* and *Lycianthes* that span the major clades from our four-gene phylogeny. Our primary objectives were to identify genome-scale patterns of discordance among gene trees, to determine the underlying causes of this discordance, and to estimate a representative species tree. We constructed hundreds of gene trees and compared the resulting topologies. We then mapped the genes onto the twelve chromosomes of a complete *Capsicum annuum* genome to visualize the genomic structure of gene history and determine if the physical proximity of two genes on a chromosome can predict their level of discordance. We found 24 distinct topologies among the eight species and 890 nuclear genes included in our analysis, with each topology supported by at least six genes. Nearly 300 genes supported a topology that placed *Lycianthes biflora* sister to a clade of *Lycianthes* and a monophyletic *Capsicum*. Over 100 nuclear genes placed *Lycianthes biflora* as sister to *Capsicum* but not the remaining *Lycianthes*, while the plastome and 82 nuclear genes supported reciprocally monophyletic genera. All three possible topologies among the three pungent peppers included in this study were supported to varying degrees, the most common of which was recovered by 539 genes. We found no pattern between the physical position of genes on the *Capsicum annuum* genome and their evolutionary history, as the 24 topologies were supported by genes scattered within and among each of the pepper's 12 chromosomes. Within each chromosome, we found no significant relationship between the physical distance between genes and their level of discordance. To further explore these issues, we have used our newly sequenced transcriptomes to design baits to target over 2401 genes, which we are currently using to expand our dataset to include all members of *Capsicum* and *Lycianthes*.

T5
P1270
Identification of $1R_S.1B_L^{PavonMA1}$ wheat-rye chromosome translocation in F1, F2 hybrids of bread wheat using cytological and molecular genetic methods

Liudmyla Sudarchuk, *Ivan Motsny, Oleksandr Rybalka*
Plant Breeding and Genetics Institute – National Center of Seed and Cultivars Investigations (PBGI)

One of the common methods of enriching the gene pool of wheat is the introgression of genetic material from related species. The best practical results were obtained using the short arm of chromosome rye 1R *Secale cereale L.* in wheat breeding. Segment of chromosome derived from Petkus variety contains genes that are closely linked to the short arm of chromosome rye $1R_S.1B_L$ and that improve the adaptability of bread wheat (*Pm8*, *Yr9*, *Lr26* and *Sr31*). Importing of 1RS chromosome to wheat genome contributes to wheat yields, although reduces the quality of flour. Wheat line with modified wheat-rye translocation ($1R_S.1B_L$*PavonMA1*) has wheat cluster instead of *Sec*-1 rye cluster, and thus in the short arm of translocation there had been moved positive wheat quality genes, in particular allele *Gld1B2* (*Gli-B1d*). The main goal of this scientific work was the investigation of chromosome behavior in meiosis of F_1 hybrids obtained by crossing of line $1R_S.1B_L$*PavonMA1* which has a modified wheat-rye translocation with bread wheat variety Kuyalnik and identification of wheat-rye modified translocation in F_2 hybrids. A significant decrease in the level of homologous chromosome pairing causing aneuploidy with the frequency 5.3% has been revealed in the wheat line $Pavon^{MA1}$, which has a modified wheat-rye translocation $1R_S.1B_L$*PavonMA1*. The "hybrid desinapsis" which is premature complete or partial disjunction of bivalents was observed in F_1 hybrids (Kuyalnik×Pavon $^{MA1)}$. The hybrids were not cytologically constant; their meiotic index ranged from 77.8% to 95.5%. $Pavon^{MA1}$ line does not contain any reciprocal intergenomic translocations with respect to Kuyalnik variety. Any case of forming 21^{II}_{Closed} was not observed among 571 pollen mother cells (PMCs) studied, that indicated the presence of pairing between the short arms of the translocation and intact 1B chromosomes. The F_2 hybrids had been investigated for presence of specific PCR-amplification products. Molecular markers were selected for this study: rye microsatellites *Xrems*1303-$1R_S$, secalin-specific STS-marker *ω-secalin* and wheat microsatellites *Xgwm18*-$1B_S$, *Xgwm550*-$1B_S$. In the result of PCR-analysis for F_2 hybrids there was detected amplicon of 290 bp size by *Xrems*1303 locus. The mentioned fragment was observed in line $Pavon^{MA1}$ and Petkus variety. 400 bp fragment of amplification by *ω-secalin* locus was detected only in Petkus variety which contains in its genome locus *Sec*-1. F_2 hybrids with modified wheat-rye translocation $1R_S.1B_L^{PavonMA1}$ originated from Petkus rye were analyzed with the molecular genetic markers *Xrems*1303 and *ω-secalin*. The conducted investigation will be useful for the creation of new wheat varieties which hold in their genome resistance genes to different fungal deseases.

T5
P1271
HpQTL: A geometric morphometric platform to compute the genetic architecture of heterophylly

Lidan Sun[1], *Qixiang Zhang*[1], *Tangren Cheng*[1], *Rongling Wu*[2], *Jing Wang*[2], *Mengmeng Sang*[2], *Libo Jiang*[2]
1. Beijing Key Laboratory of Ornamental Plants Germplasm Innovation & Molecular Breeding, National Engineering Research Center for Floriculture, Beijing Forestry University
2. Center for Computational Biology, College of Biological Sciences and Technology, Beijing

Heterophylly, i.e., morphological changes in leaves along the axis of an individual plant, is regarded as a strategy used by plants to cope with environmental change. However, little is known of the extent to which heterophylly is controlled by genes and how each underlying gene exerts its effect on heterophyllous variation. We described a geometric morphometric model that can quantify heterophylly in plants and further constructed an R-based computing platform by integrating this model into a genetic mapping and association setting. The platform, named *Hp*QTL, allows specific quantitative trait loci (QTLs) mediating heterophyllous variation to be mapped throughout the genome. The statistical properties of *Hp*QTL were examined and validated via computer simulation. Its biological relevance was demonstrated by results from a real data analysis of heterophylly in a wood plant, mei (*Prunus mume*). *Hp*QTL provides a powerful tool to analyze heterophylly and its underlying genetic architecture in a quantitative manner. It also contributes a new approach for genome-wide association studies aimed to dissect the programmed regulation of plant development and evolution.

T5
P1272
Large-scale digitization of herbarium specimens using an automated, high-throughput conveyor system

Patrick Sweeney[1], *Charles Davis*[2], *Paul Morris*[3], *Binil Starly*[4]
1. Division of Botany, Peabody Museum of Natural History, Yale University, New Haven, CT, USA
2. Department of Organismic and Evolutionary Biology and Harvard University Herbaria, Harvard University, Cambridge, MA, USA
3. Museum of Comparative Zoology, Harvard University, Cam-

bridge, MA, USA
4. Edward P. Fitts Department of Industrial and Systems Engineering, North Carolina State University, Raleigh, NC, USA

The billions of specimens housed in natural science collections provide a tremendous source of under-utilized data that are useful for scientific research, conservation, commerce, and education. Digitization and mobilization of specimen data and images promises to increase their utilization. While digitization of natural science collection specimens has been occurring for decades, the vast majority of specimens remain un-digitized. If the digitization task is to be completed in the foreseeable future, innovative, high-throughput approaches are needed. To create a data set for the study of global change in New England, we designed and implemented an industrial-scale, conveyor-based digitization workflow for herbarium specimen sheets. The workflow is a variation of an object-to-image-to-data workflow that prioritizes imaging and the capture of collections-level data. The workflow utilizes a novel conveyor system developed specifically for the task of imaging flattened herbarium specimens. Using our workflow, we digitized almost 350,000 specimens over a 131-week period. Our project has demonstrated that it is possible to capture both an image of a specimen and a core database record in 35 seconds per herbarium sheet plus additional overhead for container-level data capture, barcode application, and movement of specimens between the collections space and the conveyor system. This rate was in line with the pre-project expectations for our approach. Our throughput rates are comparable with other similar, high-throughput approaches focused on digitizing herbarium sheets and is as much as three times faster than rates achieved with more conventional non-automated approaches used during the project. We report on challenges encountered during development and use of our system and discuss ways in which our workflow could be improved. The conveyor apparatus software, database schema, configuration files, hardware list, and conveyor schematics are available for download on GitHub.

T5

P1273

Characterization of lignin monomers biosynthesis genes in rice (*Oryza sativa* L.) by using CRISPR/Cas9 system

Muhammad Hassan Askri Syed, *Lingqiang Wand*
Biomass and Bioenergy Center, College of Plant Sciences and Technology, Huazhong Agricultural University

Lignin is second most abundance biopolymers among all vascular plant and plays substantial role in stress response, mechanical support and water transport in plants. The expression, evolutionary and functional study of genes related with lignin biosynthesis provides an excellent model to study the coordinative evolution of a biochemical pathway in plants and helps to develop better strategies for bioenergy feedstock improvement. In this regard, CRISPR/Cas9 technology which is a welcoming advancement in genome editing, owing to its efficiency, precision, ease of adoption and versatility. This study proposes that the agricultural fitness of the rice plant can be improved by modifying the H, S and G lignin monomers ratio. In addition, it is aimed to characterize seven of the twenty seven lignin biosynthesis genes identified in rice by our laboratory, evaluating individual as well as collective events of reduced lignin, checking productivity of transformed plants and to know secondary functions of selected genes. Furthermore, this study will significantly play role in improving knowledge on composition of secondary cell wall specifically in rice and developing improved approaches for biofuel production.

T5

P1274

Prediction of protein–protein interactions by evidence combining methods

Ji-Wei Chang[1,2], *Yan-Qing Zhou*[1,2], ***Muhammad Tahir Ul Qamar***[1,2], *Ling-Ling Chen*[1,2], *Yu-Duan Ding*[1,2]
1. National Key Laboratory of Crop Genetic Improvement, Huazhong Agricultural University, Wuhan 430070, Hubei
2. College of Informatics, Huazhong Agricultural University, Wuhan 430070, Hubei

Most cellular functions involve proteins' features based on their physical interactions with other partner proteins. Sketching a map of protein–protein interactions (PPIs) is therefore an important inception step towards understanding the basics of cell functions. Several experimental techniques operating *in vivo* or *in vitro* have made significant contributions to screening a large number of protein interaction partners, especially high-throughput experimental methods. However, computational approaches for PPI predication supported by rapid accumulation of data generated from experimental techniques, 3D structure definitions, and genome sequencing have boosted the map sketching of PPIs. In this review, we shed light on *in silico* PPI prediction methods that integrate evidence from multiple sources, including evolutionary relationship, function annotation, sequence/structure features, network topology and text mining. These methods are developed for integration of multi-dimensional evidence, for designing the strategies to predict novel interactions, and for making the results consistent with the increase of prediction coverage and accuracy.

T5

P1275

Genome sequencing and preliminary assembly of tung tree (*Vernicia fordii*)

Xiaofeng Tan
Central South University of Forestry and Technology

Tung tree (*Vernicia fordii*) is an economically important non-edible woody oil tree originated in China. It is mainly distributed in subtropical areas of China and used to produce tung oil for centuries. Argentina, Africa, Paraguay and the United States have introduced and cultivated the tree. Tung tree seeds contain 50-60% oil with about 80 mole α-eleostearic acid (9cis, 11trans, 13trans octadecatrienoic acid). Because of the three special conjugated double bonds in eleostearic acid, tung oil is easily oxidized and forms a unique polymer after being exposed to air, to form a tough, glossy, waterproof, insulative and anticorrosive surface on the object it is coating. Tung oil is widely used in paints, varnishes, coatings and finishes due to these excellent properties. Tung oil is also a high-quality raw material for producing biodiesel and composite functional material. Besides, tung tree is an important ornamental with bright-color flowers before leaves. The objective of this study was to sequence the tung tree genome. We obtained the low heterozygosity tung tree plant from inbred lines through SSR marker selection. DNA libraries with different insertion fragments (180 bp, 500 bp, 800 bp, 3 Kb, 10 Kb, 15 Kb, 20 Kb) were constructed from the tung tree genome. These fragments were sequenced at both ends using paired-end sequencing method by Illumina Hiseq 2000. PacBio library was constructed and sequenced by the third generation sequencing. The *Aleurites cordata* Thunb., male and female plant of *Vernicia Montana*, close relatives of *Vernicia fordii*, were used for resequencing of the genomes. We obtained the whole genome sequencing data followed by assembly and bioinformatics analysis. The total amount of genome sequencing was 326.60 Gb with a coverage of 249.46X. Preliminary analysis showed that the tung tree genome was 1.3 Gb, not heterozygous but contained highly duplicated sequences. We obtained the preliminary assembly result with 60.6 Kb of Contig N50 and 803.8 Kb of Scaffolds. Total length of the assembled sequence was 1.12 Gb, approximately 85% of the predicted genome size. Transcriptome data were used to assess the genome data. The results showed that the coverage of unigenes was 94.97% and that the matching rate of reads achieved 88.3-95.6%. These data indicate that the assemble result of the tung tree genome is of high integrity. Based on the HI-C technology, a total of 1107.82Mb length sequences were located on 22 chromosomes, accounted for 98.92% of the whole genome sequences. This study provides fundamental information for evaluating GC distribution and sequencing depth distribution, annotating genes, and analyzing evolution and biological functions in subsequent studies.

T5

P1276

Population transcriptomics uncovers expression variation pattern of differentially expressed genes between natural population and transplanted population of *Miscanthus lutarioriparius* (Poaceae)

Chengcheng Tao[1,3], *Qin Xu*[1], *Zhihong Song*[1,3], *Caiyun Zhu*[2,3], *Wei Liu*[2], *Tao Sang*[1,2]

1. Key Laboratory of Plant Resources and Beijing Botanical Garden, Institute of Botany, Chinese Academy of Sciences

2. State Key Laboratory of Systematic and Evolutionary Botany, Institute of Botany, Chinese Academy of Sciences

3. University of Chinese Academy of Sciences

Miscanthus lutarioriparius, a perennial species distributed in the middle and lower reaches of Yangtze River of China is a candidate of promising energy crops due to huge biomass and high cellulose content. Previous studies of this plant suggested the expression variation of differentially expressed genes (DEG) played an important role in acclimation among the individuals under stress environment. However, it is unclear whether there was a random change of expression diversity under different environments for each gene. To answer the question, we identified 1886,2417 and 685 DEG by comparing the population transcriptome of leaves between samples from natural habitats (NP) and from transplanted fields which located in Jiangxia Hubei Province (JH), Qingyang Gansu Province (QG) and Dongying Shandong Province (DS), respectively. E_d was used to measure the level of gene expression variation and the mean E_d of DEG tends to be lower than that of background genes under all environments. In contrast to expression variation, the average genetic diversity of DEG represented by π was not significantly different from that of background genes in DS and was very close to the value of background genes in JH and QG. Results of GO enrichment analysis suggested the functional classifications of DEG were found to be similar in each comparative analysis between natural habitats and transplanted field while the subset of DEG among them were in the minority (103 DEG). These results implied that the DEG between natural population and transplanted population might be conservative on expression level among individuals under the same environment. At the same time, they were sensitive to changes of environment no matter how it change to make contribution for plant acclimation. Further studies on these genes and their expression patterns among populations would provide some advice for introduction and domestication of *M. lutarioriparius* or even other energy crops.

T5

P1277

Genetic basis of nodulation in *Elaeagnus umbellata*

Alexander Taylor, *Yin-Long Qiu*

University of Michigan

Nodulation, a mutualistic symbiosis in which plants exchange nutrients with nitrogen-fixing bacteria in specialized nodule organs on the plant root, plays a critical role in global nutrient cycling. Nodulation appears in several lineages of the nitrogen-fixing clade (NFC) of rosid angiosperms (Fabales, Fagales, Cucurbitales, and Rosales), but does not occur in any other land plants. The evolutionary origin of nodulation in these lineages involved the recruitment of pre-existing pathway – common to all land plants – that mediate the arbuscular mycorrhizal symbiosis. However, specific patterns of gene duplication and recruitment vary between nodulating lineages. These independent, lineage-specific evolutionary histories can provide insights into the genetic requirements for the evolution of nodulation, as well as the evolution of morphological and developmental differences between nodulation in these different lineages. Here we report on an RNA-Seq study of nodulation in *Elaeagnus umbellata,* a species in the Elaeagnaceae (Rosales) that forms coralloid nodules with *Frankia* actinobacteria. We examine the evolutionary history of genes involved in nodulation in *E. umbellata* and other nodulating lineages, using both public and our transcriptomic data to assess the patterns of gene evolution, duplication and recruitment during the origin of nodulation genes in different lineages.

T5

P1278

Rhizosphere and endophytic bacteria diversity of *Senecio vulgaris* (Asteracea)

Zhongsai Tian[1], *Lin Xu*[1], *Liang Feng*[1,2], *Dandan Cheng*[1,2]

1. School of Environmental Studies, China University of Geoscience (Wuhan), Wuhan, China

2. State Key Laboratory of Biogeology and Environmental Geology, China University of Geosciences (Wuhan), Wuhan, China

Endophytic bacteria are widely distributed in plants and involve in co-evolution with their host plants. Currently studies on invasive mechanism of alien plants have made great progress, but they mainly focus on the relationship between plants and their natural enemies or competitors, and few previous studies reported the interaction between invasive plant and its endophytic microorganism (especially endophytic bacteria). Research work on endophytic and rhizosphere bacteria of invasive plants helps to understand the relationship between endophyte and plant invasion. *Senecio vulgaris* (Asteraceae), originated in Europe and a widely-distributed weed in temperate regions, has been listed as one of the invasive plants in China. We investigated bacteria communities in the rhizosphere, leaf and root endosphere in 4 *S. vulgairs* populations in Shennongjia, Hubei, China. By selecting specific primers, we successfully amplified the sequence of V5-V7 zone of 16S rDNA of bacteria, and avoided the pollution of plant chloroplasts. After homogenization, the effective sequence of each sample was 12,511, averagely. On this basis, we analyzed the α-diversity and β-diversity of bacteria in rhizosphere and endophytic bacteria, and compared the diversity of bacteria in different plant parts and different populations. As for the α-diversity, the rhizosphere bacteria communities were at the highest level, followed by root endophytic bacteria, and leaf endophytic bacteria diversity in the lowest. The dominant phyla of the three kinds of bacterial communities were also different: Proteobacteria (64.00%), Bacteroidetes (14.42%) and Actinobacteria (9.34%) were dominated in rhizosphere; Proteobacteria (84.01%), Bacteroidetes (12.41%) and Actinobacteria (1.91%) were in roots ; Proteobacteria (70.30%), Firmicutes (13.48%) and Actinobacteria (10.20%) in leaves. The bacterial community in the rhizosphere of different *S. vulgaris* populations shared similar composition, but components and abundance of bacteria in leaves and roots differed between populations.This study reveal that plant compartment affect microbiome composition and suggested that plant genotype may have an important influence on shaping the leaf and root microbiomes of an invasive plant.

T5

P1279

Quadruplex DNA in long terminal repeats of maize LTR retrotransposons affects an expression of a reporter gene *in vivo*

Viktor Tokan[1], *Matej Lexa*[2], *Boris Vyskot*[1], *Michaela Vorlickova*[3], *Eduard Kejnovsky*[1]

1. Department of Plant Developmental Genetics IBP CAS

2. Faculty of Informatics, Masaryk University Brno

3. Department of CD Spectroscopy of Nucleic Acids IBP CAS

Guanine rich sequences with pattern $G_3N_{1\text{-}8}G_3N_{1\text{-}8}G_3N_{1\text{-}8}G_3$ can form secondary DNA/RNA structures called G-quadruplexes (G4s). Many cellular processes are influenced by G4s: transcription, translation, splicing, replication, recombination and many others. Recently potential quadruplex-forming sequences (PQS) were found close to promoters within long terminal repeats of plant LTR retrotransposons, the major component of plant genomes. In order to test a potential effect of G4s on retrotransposon expression, we cloned long terminal repeats of maize LTR retrotransposon into vector upstream of lacZ reporter gene. Then, we measured lacZ transcription and translation in transformed yeasts. We found that G4s had an inhibitory effect on translation *in vivo*. Reporter gene expression increased in "mutants" where guanines were substituted in G tracks and so G4s formation was disrupted. In parallel, we confirmed the presence of quadruplexes *in vitro* by circular dichroism and by primer extension. Our results indicate that quadruplex DNA can play a regulatory role in LTR retrotransposon life-cycle and thus affect genome dynamics.

T5

P1280

Genetic characterization of endemic and rare plant species in Kazakhstan

Yerlan Turuspekov, *Shyryn Almerekova, Alibek Zatybekov, Saule Abugalieva*

Institute of Plant Biology and Biotechnology

Kazakhstan is the ninth-largest country in the world by territory, and it is home to more than 6,000 plant species. In 2015, a new project was launched with the main goal being to study genetic variation of endemic, rare, and economically important plant species in National State Reserves and National Nature Parks of Kazakhstan. The main goal of the project is DNA barcoding of collected endemic, rare and important plant species by using different types of DNA markers, including universal markers of nuclear and chloroplast genomes. The project combines the efforts of botanists and geneticists from National State Reserves and National Nature Parks, National Universities, Botanical Gardens, and two Biotechnology Research Institutes. Currently 664 populations representing 401 species were collected within this project. DNA samples were isolated and purified for each collected plant and two representatives of each species were genotyped using ITS and *mat*K markers. DNA sequences of studied species were used for phylogenetic analysis and assessment of molecular taxonomy. The information related to the place of the collection, images of herbarium, DNA isolation status, DNA sequences of the ITS and *mat*K markers, and results of phylogenetic analyses were documented and organized in single database of the project.

T5

P1281

The evolution and molecular genetics of apomixis in dandelions (*Taraxacum officinale*)

Peter Van Dijk, *Diana Rigola, Charles Underwood, Rik Op Den Camp, Marcel Prins*

Keygene N.V.

Outcrossing, selfing and apomixis are the three major plant breeding systems. Apomixis (clonal reproduction by seeds) is the least common (~0.1 percent of the angiosperm species). Apomixis is relatively unknown, although Mendel already knew about "constant hybrids". Apomixis fixes the full maternal genome, maintaining genome-wide heterozygosity and epistasis. Although apomixis does not occur in in major crops, apomictic crops would have many benefits for breeding, seed production and agriculture. Knowledge of the molecular genetic basis of natural apomixis genes may help to develop apomictic crops. In this presentation I will first discuss some commonalities in the distribution of apomixis across the flowering plants. After that I will report on the molecular genetics of apomixis in the common dandelion. The fact that apomixis is much less common than selfing may be because it depends on two or more gain-of-function mutations, whereas a single loss-of-function mutation is sufficient for the shift from outcrossing to selfing. Nevertheless, apomixis is phylogenetically widespread suggesting that it evolved many times during the evolution of flowering plants. Apomixis is found typically at the twigs of phylogenetic trees. This suggests that apomicts are of relatively recent origin, possibly because they become extinct soon. Apomixis is strongly correlated with polyploidy, but this could be a secondary association. In contrast to strong selfers, apomicts do not show clear modification of traits that promote outcrossing. This may be due to either evolutionary young age or to lack of adaptive potential due to lack of genetic variation, or both. We studied the molecular genetic control of apomixis in common dandelions. In this species sexuals are diploid and apomicts triploid. By crosses (using apomicts as pollen donors) we could show that apomixis is controlled by two unlinked dominant loci, one controlling apomeiosis, and one controlling parthenogenesis. Apomeiosis, the avoidance of meiotic reduction and recombination during female reproduction, is caused by first division restitution (diplospory). Using two successive forward genetic mutation screens we succeeded in identifying the diplospory gene within the diplospory locus. We show that the dominant diplospory allele is perfectly associated with apomixis in a diverse panel of apomictic and sexual plants.

T5

P1282

Genetic diversity of tomatillo (*Physalis philadelphica* Lam.) in Central Western Mexico assessed with SSR markers

Ofelia Vargas-Ponce
Universidad de Guadalajara

Genetic diversity and structure of thirteen populations of *Physalis philadelphica* were analysed with seven microsatellite loci. It is a species of cultural and economic importance in Mexico because of its edible fruits. This taxon can grow under wild, weedy or cultivated conditions. Genetic parameters were determined on 298 individuals from thirteen populations representing a management gradient, including three wild, three weedy and seven cultivated populations. Diversity estimators reveal moderate to high genetic diversity in the total gene pool (P = 100%, n = 14.6, Ae = 7,457, Ho = 0.44, HT = 0.84). This pattern is also observed at the gradient of management. Weir and Cooker estimator reveal a moderate genetic structure value ($\theta = 0.268$) to the total pool a low between management categories ($\theta = 0.079$). AMOVA show a similar pattern ($\Phi st = 0.144$, $\Phi ct = 0.007$). The Bayesian analysis showed that the most likely number of genetic groups is K= 12. From the results we can argue that life history traits of the species, particularly life form, the self-incompatible breeding system, wide distribution, pollination by bees, seed dispersion by birds, mammals, and humans favour genetic diversity and gene flow. The moderate values of differentiation for the species, suggest that anthropogenic management has had slight consequences on the genetic structure of *P. philadelphica*. However, actions as different selection criteria and preferences by cultivate particular morphological phenotypes, some of them restricted to certain geographic areas, contributes to the differentiation of some particular local variants as show the analysis of Bayesian assignment analysis. It is necessary to extend the analysis, including the seven races recognized for this species, in order to prove this hypothesis of differentiation between races of *P. philadelphica.*

T5

P1283

Cytogenetic characterization of *Amaranthus caudatus* L.

Prajitha V*, Thoppil J E*
University of Calicut

The present study is aimed for the cytogenetic characterization of *Amaranthus caudatus* an economically important species, cultivated mainly for grain production. Karyomorphological studies in *Amaranthus* are scarce, probably due to higher number of small sized chromosomes. Karyomorphological studies were conducted using mitotic squash preparation of young healthy root tips by acetocarmine and DAPI staining. Karyological parameters and karyotypic formula were established using various software programs and tabulated the karyomorphometric and asymmetry indices *viz*., Disparity index, Variation coefficient, Total forma percentage, Karyotype asymmetry index, Syi index, Rec index, Interchromosomal and Intrachromosomal asymmetry index and Degree of asymmetry of karyotypes. The mitotic chromosome number observed for *A. caudatus* was 2n = 32 with a gametic number n = 16. In *A. caudatus* the chromosome length during somatic metaphase ranged from 0.8698 μm to 1.7722 μm with a total length of 39.1412 μm. Various karyomorphometry and asymmetry indices analyzed revealed the extend of intraspecific variation and evolutionary status of the species studied.

T5

P1284

Molecular biotechnology of hop variety type, sex and agrobacterial infection detection

Andrii Venger*, Olga Kolesnyk*
Plant Breeding and Genetic Institute

This work is devoted to usage of molecular genetics and bioinformatics methods for developing the molecular biotechnology of hop variety type, sex and agrobacterial infection detection. Nucleotide sequences of hop genes *chs*_H1, *chs*2, *chs*3, *chs*4, *vps* encoding chalcone synthase were researched for polymorphism. There were obtained indels and single nucleotide polymorphisms in different regions of these genes. The molecular-genetic polymorphism of genes *chs*_H1, *chs*2, *chs*3, *chs*4, *vps* in different regions of 20 hop varieties of Ukrainian breeding was analyzed by polymerase chain reaction (PCR). Polymorphism of fragment length in different parts of genes encoding chalcone synthase was described. Polymorphism analysis of genes encoding chalcone synthase of Ukrainian hop varieties was conducted. The dependence between gene polymorphism and the level of bitter substances in hop cones was shown. Reliability and validity was calculated by Spearman coefficient. The possibility of usage of molecular markers that show the polymorphisms of genes encoding chalcone synthase for determination of hop variety type (aroma of bitter) was shown. Polymorphism of 20 hop varieties of Ukrainian breeding was investigated by microsatellite locus HlAGA7; correlation between

HlAGA7 alleles and hop sex was absent. Y-chromosome specific STS-marker for hop sex determination was confirmed. Nucleotide sequences of agrobacterial Ti-plasmids were researched by bioinformatic and molecular-genetic methods. The possibility of applying of molecular markers of *vir*D2 and *ipt* genes was shown for all types of Ti-plasmid determination. The system of detection of *A. tumefaciens*, which causes crown gall, was optimized using molecular markers of *vir*D2 and *ipt* genes. Genetics formulas of Ukrainian hop varieties were supplemented according to polymorphism of genes *chs*_H1, *chs*2, *chs*3, *chs*4, *vps*. On the basis of bioinformatic and molecular genetic analysis results the molecular biotechnology of hop variety type, sex and agrobacterial infection detection was obtained.

T5

P1285

Taming microalgae for transformation and transgene expression

Anongpat Suttangkakul, Anchalee Sirikhachornkit, Piyada Juntawong, ***Supachai Vuttipongchaikij***
Department of Genetics, Faculty of Science, Kasetsart University

Microalgal transformation and transgene expression face many difficulties such as low transformation efficiency, transgene cleavage, low transgene stability, and transcription suppression. This limits the improvement and utilization of microalgae especially for the production of bioproducts. Recent development in the model microalga *Chlamydomonas reinhardtii* has shown a number of improvements for transgene expression through the uses of strategic vector systems and mutagenesis of the host cells. However, this is currently limited to *Chlamydomonas*, whereas other microalgae with potential industrial uses still suffer from this limitation. Here, we tested the transformation efficiency on a number of oleaginous microalgae isolated from natural environments to identify a potential host species and then performed strategic improvements for transgene expression. A strain of *Scenedesmus* species was selected as a host. We tested the expression using various constructs including a reporter-coding gene with or without introns from *Chlamydomonas.* Constructs expressed as transcript fusion with zeocin selectable maker and 2A peptides were also tested. These constructs yielded different levels of improvement for the protein expression. We performed UV mutagenesis and identified three mutants for high transgene expression. Furthermore, to reduce the use of antibiotic for a large scale application, we developed a strategy for selecting transformants using *PSY* gene and light selection. White mutants with *PSY* mutations were generated by UV irradiation and identified by target sequencing of the *PSY* gene. The mutants were tested for complementation and transgene expression. Complemented strains (green colonies) can be selected on agar media using light exposure. The expression of the transgene was confirmed at protein levels. This study provides a proof of concepts for applying knowledges based on *Chlamydomonas* to improve wild microalgae.

T5

P1286

Detecting loci contributing to local adaptation through flowering time variation in *Lotus japonicus* in Japan

Tomomi Wakabayashi[1], *Sachiko Tanaka*[2], *Stig Andersen*[3], *Shusei Sato*[4], *Masayoshi Kawaguchi*[2], *Hiroaki Setoguchi*[1]
1. Graduate school of Human and Environmental Studies, Kyoto University
2. Division of Symbiotic Systems, National Institute for Bacis Biology
3. Department of Molecular Biology and Genetics, Aarhus University
4. Graduate school of Life Sceinces, Tohoku University

The mechanism of local adaptation through flowering time is one of the important factors to understand the local adaptation in plant species. Improving the recent sequencing technology enables to detect causable candidates for various kinds of phenotypes with genome-wide association study (GWAS). *L. japonicus* is a legume model plant and distributed in whole range of Japan. The archipelago lies in the wide range of latitude and harbors various environments including high mountain ranges and coastline. Based on previous studies, this plant exhibits wide variation of flowering time depends on habitat's latitude. In this study, GWA analyses were conducted with flowering time phenotype and environmental factors of each population's collection sites. Considering the overlaps of the detected genes, several candidate loci for local adaptation through flowering time were found.

T5

P1287

Transcriptomic analysis of the regulation of seed germination under salt stress in two sister species (*Populus euphratica* and *P. pruinosa*)

Zhang Caihua, ***Dongshi Wan***
State Key Laboratory of Grassland Agro-Ecosystem, School of Life Sciences, Lanzhou University

Soil salinity is thought to be the most significant abiotic stresses limiting plant growth, development, productivity in semiarid and arid areas. *Populus euphratica* and *P. pruinosa* (Salicaceae) are naturally distributed in desert regions and extraordinarily well adapted to the local salt stress. Seed germination is highly related to seedling survival rate and subsequent vegetative growth, and further directly affects poplar quality. To address the molecular regulatory mechanisms of the seeds germination process underlying salt stress tolerance, a time course (0h, 4h, 12h, 24h, 48h and 72 h) transcriptomic data set was produced and analyzed by the protocol of HISAT, StringTie and Ballgown, coupled with related physiological analysis of two species. We identified two sets of differential expression gene, including 12831 DEGs across time courses samples in the two species, and 8071 DEGs between the two species across time courses samples (the species is host variables and the time across is covariates). The transition of major biochemical processes along the germination time axis of the seed in the poplar is produced partly by more and more highly coordinated transcript dynamics. We suggested that the flavonoid and brassinosteroid are the two most important regulatory pathways for the phenotypic and salinity tolerance differences between *P. euphratica* and *P. pruinosa* during seed germination process under salt stress.

T5
P1289
Epigenetic alteration under nitrogen-deplete and nitrogen-replete in photosynthetic microorganisms

Jiangxin Wang[1,2], *Peng Xiao*[1], *Lang Hu*[1], *Qin Shao*[1], *Huan Qin*[1]
1. Guangdong Technology Research Center for Marine Algal Bioengineering, Guangdong Key Laboratory of Plant Epigenetic, College of Life Sciences and Oceanography, Shenzhen University
2. Nanshan District Key Lab for Biopolymers and Safety Evaluation, College of Materials Science and Engineering, Shenzhen University

Nutrients are important to microbes in the environment, while deplete and replete of nutrients are also common in nature. Epigenetics, the study of cellular and physiological traits that are heritable by daughter cells and not caused by changes in the DNA sequences, is well known as connected with many important biological processes, such as development and transgenerational heredity. Cyanobacteria and microalgae play very crucial role in the earth, as the primary producers in the ecosystem. There are lots of studies explored physiological and biochemical changes, and altered gene expressions at different levels such as RNAs and proteins against environmental changes. However, no knowledge about epigenetic information in microalgae, especially under any environmental stress, is reported so far. After treated with nitrogen starvation for 72 h and transferred back to normal complete culture media for 2 continuous generations, DNA samples from the model cyanobacteria and green microalgae, *Synochecystis* PCC 6803 and *Chlamydomonas reinhartii* CC849, were bisulfite converted for next generation sequencing. Interestingly, compared with those of control (CT), DNA methylation sites and patterns in nitrogen starvation (NS) and nitrogen repletion (NR) were found much more similar in both cyanobacterium and eukaryotic green alga. In *Synochecystis*, the methylation levels of CG, CHG and CHH were significantly increased in NS and remained high in NR. Among them, very high CmG levels (~80%) were observed in both NS and NR samples, while CmHG and CmHH levels were relatively low at range of 25-45%. In *Chlamydomonas*, a total of 16,6874 individual mC sites were hypermethylated in NS cells, compared to CT, and among with 40,920 mC were remained in NR cells. Different from *Synochecystis*, only methylation levels of CG were dramatically increased by nitrogen depletion and repletion in *Chlamydomoans*. GO and KEGG enrichment analysis of differentially methylated regions (DMR) indicated that those genes were mainly involved in photosynthesis. This study revealed the global DNA methylation pattern of microalgae between normal and nutrient stress conditions. The result of this study might contribute to a better understanding of epigenetic regulation in microalgae adaptation to the ever changing environment.

T5
P1290
Induction of *Xa10*-like genes in rice cultivar Nipponbare confers disease resistance to rice bacterial blight

Jun Wang[1,2], *Dongsheng Tian*[1], *Keyu Gu*[1], *Xiaobei Yang*[1], *Lanlan Wang*[1], *Xuan Zeng*[1], *Zhongchao Yin*[1,2]
1. Temasek Life Sciences Laboratory, 1 Research Link, National University of Singapore, Singapore 117604
2. Department of Biological Sciences, 14 Science Drive, National University of Singapore, Singapore 117543

Bacterial blight of rice, caused by *Xanthomonas oryzae* pv. *oryzae*, is one of the most destructive bacterial diseases throughout the major rice growing regions in the world. The rice disease resistance (*R*) genes *Xa10* confers race-specific disease resistance to *X. oryzae* pv. *oryzae* strains that deliver the corresponding transcription activator-like (TAL) effectors AvrXa10. Upon bacterial infection, AvrXa10 binds specifically to the effector binding element (EBE) in the promoter of the *R* gene and activates its expression. *Xa10* encodes an executor R protein that triggers hypersensitive response and activates disease resistance. Rice cultivar Nipponbare carries two *Xa10*-like genes in its genome, of which one is the susceptible allele of the *Xa23* gene, a *Xa10*-like TAL effector-dependent executor *R* gene isolated recently from rice cultivar CBB23. However, the function of the two *Xa10*-like genes in disease resistance to *X. oryzae* pv. *oryzae* strains has not been investigated. Here we designated the two *Xa10*-like genes as *Xa10-Ni* and *Xa23-Ni* and characterized their function for disease resistance to rice bacterial blight. Both *Xa10-Ni* and *Xa23-Ni* provided disease resistance to *X. oryzae* pv. *oryzae* strains that deliver the matching artificially designed TAL effectors (dTALEs). Transgenic rice plants containing *Xa10-Ni* and *Xa23-Ni* under the *Xa10* promoter provided specific disease resistance to *X. oryzae* pv. *oryzae* strains that deliver AvrXa10. *Xa10-Ni* and *Xa23-Ni* knockout mutants abolished dTALE-dependent disease resistance to *X. oryzae* pv. *oryzae*. Heterologous expression of *Xa10-Ni* and *Xa23-Ni* in *Nicotiana benthamiana* triggered cell death. The 19-amino acid residues at the N-terminal regions of XA10 or XA10-Ni are dispensable for their function in inducing cell death in *N. benthamiana* and the C-terminal regions of XA10, XA10-Ni and XA23-Ni are interchangeable among each other without affecting their function. Like XA10, both XA10-Ni and XA23-Ni locate to the endoplasmic reticulum (ER) membrane, show self-interaction and induce ER Ca^{2+} depletion in leaf cells of *N. benthamiana*. The results indicate that *Xa10-Ni* and *Xa23-Ni* in Nipponbare encode functional executor R proteins, which induce cell death in both monocotyledonous and dicotyledonous plants and have potential of being engineered to provide broad-spectrum disease resistance to plant pathogenic *Xanthomonas* species.

T5
P1291
Identification of endogenous small peptides involved in rice immunity using transcriptomics and proteomics analyses

Pingyu Wang, *Ken-Ichi Kosami, Shaolun Yao, Renyi Liu, Yoji Kawano*
Chinese Academy of Sciences

Plant small peptides are important signaling components in plant growth and development. Some plant peptides also regulate plant immunity and function as endogenous elicitor peptides. However, the number of known endogenous peptide families involved in plant immunity is limited. Here, in order to identify peptide involved in rice immunity, we combine transcriptome and proteome analyses of rice plants treated with rice blast fungus *M. oryzae* and rice suspension cells treated with fugal cell wall component chitin. Our combination approach identified 2454 upregulated genes in

RNA-seq and 3327 proteins by MS. Next, among those genes and proteins, we screened small secreted proteins (SSPs) based on two criteria: whose protein length is less than 250 amino acid (aa) and which contain an N-terminal secretion signal peptide sequence. As a result, we isolated 236 SSPs, including several known SSPs related to defense response, such as Bowman-Birk inhibitors. A total of 52 unannoted SSPs were found and considered as candidates of new peptide family precursors. Among them, we found one protein, which we named immune response peptide-1 (IRP-1), putatively containing a 66 aa proteins with a 29 aa N-terminal signal peptide sequence and a conserved C-terminal sequence. Its expression was induced by not only chitin but also bacterial peptidoglycan. Furthermore, overexpressing of *IRP-1* activated the expression of defense gene *PAL-1*. The 52 unannotated SSPs will be useful to further elucidate rice immunity.

T5

P1292

Codon usage bias and cluster analysis of plant *CPR* genes

Li Cheng, Yikui Li, Xiaodan Li, Binbin Qian, Sheng Xu, Bing Xia, ***Ren Wang***

Institute of Botany, Jiangsu Province and Chinese Academy of Sciences

Cytochrome P450 reductase (CPR) plays an important role in plant growth, development, biosynthesis of natural compounds and protection against external stimulants. Based on thirty-six *CPR* genes in twelve plants from GenBank, codon usage bias was analyzed by CodonW software, and cluster analysis based on gene codon and coding sequence was performed with SPSS and MEGA5.0 respectively. The effective number of codons (ENC) of *CPR* genes in monocotyledon and dicotyledon ranged from 52.92 to 58.17 and 46.85 to 53.58 respectively, indicating that codon usage bias of plant *CPR* genes was weak. The ENC-plot revealed that all *CPR* genes diverged from the ENC expectation curve, indicating that codon usage bias of plant *CPR* genes was mainly affected by natural selection pressure. The synonymous codon bias of *CPR* genes was different between monocotyledon and dicotyledon, and also was of difference even from the same plant. Based on the relative synonymous codon usage (RSCU), the *CPR* genes in monocotyledon and in dicotyledon were clustered together respectively, which was closer to the traditional classification of twelve plants, reflected the evolution relationship of species, and could be used as a supplementary of phylogenetic analysis. While in the phylogenetic tree based on the coding sequence, *CPR* genes from the same species were clustered into two groups, which could be used for gene classification and function prediction. In this study, codon usage bias and cluster analysis of plant *CPR* genes were investigated to facilitate further research in gene classification and evolution, gene function prediction and protein expression, and plant breeding.

T5

P1293

Comparison of DNA methylation in the developing seeds of yellow- and black-seeded *Brassica napus* through MSAP analysis

Xiaqin Wang, *Jinjin Jiang, Youping Wang*

Yangzhou University

DNA methylation is a crucial modification process implicated in the epigenetic regulation of gene expression, which may be responsible for agronomic trait variation. However, the epigenetic regulation of the yellow- and black-seeded characters of *Brassica napus* has not been reported. In this study, methylation-sensitive amplification polymorphism (MSAP) analysis was performed to investigate the target regions of seed coat variation in *B. napus*; in the analysis, the yellow-seeded rapeseed line derived from the somatic hybrids of *B. napus–Sinapis alba* and the black-seeded rapeseed line were used as materials. Extensive methylation changes were observed between yellow- and black-seeded *B. napus*. In particular, 10% of demethylation and 5% of hypermethylation were detected in yellow rapeseeds compared with black seeds. Nonetheless, this variation was barely identified among different development stages. The relative expression value and 20 polymorphic fragments in MSAP profiles were also analyzed. The gene expression of the demethylated fragments in yellow rapeseeds was upregulated. For instance, bHLH, a transcription factor regulating flavonoid biosynthesis, was upregulated three to five weeks after flowering of yellow seeds. Therefore, epigenetic changes among rapeseed lines with different seed colors help elucidate the formation of yellow seed character.

T5

P1294

A gold standard to deconvolute complicated structures of plant genomes with recursive polyploidizations

Xiyin Wang

North China University of Science and Technology

Plants often have complex genomes, due to recursive polyploidizations and genome repatterning. This makes it difficult to deconvolute their genome structures, and barrier the understanding their formation and the exploration of gene functional evolution. It would be a great pity if failing to decipher a newly sequenced genome structure when enormous amount of money and time invested. However, such failures occurred quite often in last several years. Here, we propose our approach to perform the genome structural analysis, adopted by quite several plant genome sequencing efforts, which we suggest be taken as a gold-standard to analyze a new genome sequence.

T5

P1295

Genome-wide association analysis of nine agronomic traits in alfalfa (*Medicago sativa* L.) using genotyping-by-sequencing

Zan Wang[1]*, Lin Ma*[1]*, Haiming Zhao*[2]*, Congjun Jia*[1]

1. Institute of Animal Sciences, Chinese Academy of Agricultural Sciences

2. Institute of Dry-Land Farming, Hebei Academy of Agriculture and Forestry Sciences

Alfalfa (*Medicago sativa* L.) is the most important and widely planted forage legumes for hay and silage in the world. Understanding the genetic basis of agronomic traits in alfalfa will provide valuable information for improving alfalfa. In the present study, a broad collection consisted of 322 genotypes from 75 tetraploid alfalfa accessions were characterized phenotypically for nine agronomic traits under three environments. Phenotypic data analysis revealed that nine traits were significantly influenced by genotypes with different traits having various heritabilities. Using genotyping-by-sequencing (GBS), we identified 115,654 SNPs in the collection, and of those, 44757 SNPs were uniquely physically mapped on *M. truncatula* genome. Population structure and linkage disequilibrium analysis showed this collection could be divided into two sub-populations consistent with their origins and the average LD decay for the entire genome was estimated as 51.4 kb at $r^2 < 0.1$. The genome-wide association (GWAS) study by LD identified 42 putative significant marker-trait associations related to the agronomic traits except biomass yield (BY). Candidate genes assigned with the SNPs were also identified and characterized in this investigation. This study demonstrates that integrating GBS and GWAS could be used as a powerful approach for dissecting agronomic traits in alfalfa and providing candidate genes related to important agronomic traits in alfalfa breeding.

T5

P1296

Transcriptomic analysis uncovers the molecular mechanisms of the interactions between *B. napus* and *S. sclerotiorum*

Meiling Xu, Li Lin, ***Jian Wu***
Yangzhou University

Sclerotinia stem rot caused by *Sclerotinia sclerotiorum* is one of the most devastating diseases in many important crops including *Brassica napus* worldwide. Quantitative resistance is the only source for genetic improvement of *Sclerotinia*-resistance in *B. napus*, but the molecular basis for such a resistance is largely unknown. Here, we performed an in-depth transcriptomic analysis to understand the differential defense responses to *S. sclerotiorum* in a resistance line (J964, R-line) and a susceptible line (J902, S-line) of *B. napus* at 0,24,48 and 96 hpi. Both the numbers of and fold changes in differentially expressed genes in the R-line were larger than those in the S-line at 48 hpi. To better understand these genes with different expression patterns or fold changes between the R- and S-lines, we introduced the concept of relative differentially expressed genes (RDEGs). In total, 5,910 up-regulated RDEGs and 3,091 down-regulated RDEGs in the R-line were identified. After monitoring the RDEGs, we found that a set of genes involved in immune responses were induced more intensely in the R-line, including pathogen recognition, MAPK signaling cascades, WRKY transcription regulation, jasmonic acid/ethylene signaling pathways, defense-related protein (PR2, PR3, PR4, PR5-like, PDF1.2b, chitinase, lectin, β-1,3-glucanase and PGIP) production and indolic glucosinolate biosynthesis. Our data also reveal that the degree of differential expression tends to become larger between the R- and S-lines along with the defensive signal transduction chains. The change of the chitinase, β-1,3 glucanase, PGIP activity and the glucosinolates content in the R- and S-lines after *S. sclerotiorum* inoculation were measured. We found that the activities of chitinase, β-1,3 glucanase, PGIP were indeed higher in the R-line than that in the S-line after inoculation, and indole glucosinolate content in the R-line were dramatically increased after inoculation. These results were consistent with the ones obtained by transcriptomic analysis. Thus, the differences between R- and S-line were associated with the intensity of defense responses. Interestingly, we found that the difference of defense responses between R- and S-lines at the early time of infection (before 24 hpi) may be associated with their physical barriers. The stem of R-line developed a more compact vascular and higher content of lignin in the pith, which contributes to limiting the invasion and expansion of *S. sclerotiorum* at the early time of infection. Moreover, we found that the endo-polygalacturonase gene (*SS1G_10167*), cellobiohydrolase gene (*SS1G_09020*) and oxaloacetate acetylhydrolase gene (*SS1G_08218*) and eight candidate effectors might play important roles in *S. sclerotiorum* pathogenicity by monitoring the transcriptomic changes in *S. sclerotiorum*. Thus, above results provide novel information for understanding the complex molecular mechanism of the *B. napus-S. sclerotiorum* interactions.

T5

P1297

Pseudomonas syringae pv. *tomato* (*Pst*) DC3000 type three secretion system (TTSS) effector AvrPto1 targets and interrupts the SOBIR1-involved resistance signalling pathway

Jinbin Wu
Wageningen University

Leucine-rich repeat receptor-like proteins (LRR-RLPs) and LRR-receptor-like kinases (LRR-RLKs) are cell surface receptors that are essential for detecting invading pathogens and subsequent activation of plant defence responses. In contrast to RLKs, RLPs lack a cytoplasmic kinase domain to trigger downstream signalling leading to host resistance. The LRR-RLK SOBIR1 constitutively interacts with Cf-4 and is required for Cf-4-mediated resistance to *Cladosporium fulvum*. Accumulating evidence shows that SOBIR1 is broadly required for RLP-involved resistance to fungal, oomycete and bacterial pathogens. *Pseudomonas syringae* pv. *tomato (Pst)* DC3000 injects approximate 30 effectors into host cells via the type three secretion system (TTSS) and causes bacterial speck disease on host plants. Here we show that *Sl*SOBIR1 and its paralogs are all targeted by AvrPto1. And the kinase activity of *Sl*/*At*SOBIR1 is not required for the interaction with AvrPto1. Moreover, the auto HR triggered by over-expression of *AtSOBIR1* is suppressed by AvrPto1. Furthermore, Cf-4-mediated HR by recognition of Avr4 is suppressed by AvrPto1, but the effect of AvrPto1 on Cf-4/SOBIR1/BAK1 complex formation is remain unknown.

T5

P1298

Transgene by pollen-tube pathway of *Dendrobium officinale*

Kanghua Xian*, Chuanming Fu, Jinxiang He, Qingfang Gong, Jiang Su, Ningzhen Huang*
Guangxi Institute of Botany, Guangxi Zhuang Autonomous Region and Chinese Academy of Sciences

In this paper, the transgenic technology of *Dendrobium officinale* was studied by using pollen-tube pathway with plasmid and agrobacterium vectors as target genes delivery. The main results were summarized as follows: 1. The minimum lethal concentration of kanamycin to seed germination and protocorm growth were 90 $mg \cdot L^{-1}$ and 150 $mg \cdot L^{-1}$ respectively. Further study showed that 100 $mg \cdot L^{-1}$ and 150 $mg \cdot L^{-1}$ of kanamycin were well for selecting transgenic seeds and protocorms during selective culture *in vitro* respectively. 2. Plasmid and agrobacterium containing GFP or GUS gene was transferred into *D. officinale* via pollen-tube pathway. The technology and method were summarized as follows: Collected target plasmid (pSuper1300 and pBI121) and agrobacterium which containing GFP or GUS gene; resuspended the plasmid pSuper1300 and pBI121 to concentration of 100 $ng \cdot \mu L^{-1}$ with thrice distilled water, while resuspended the agrobacteria which carrying plasmid pSuper1300 and pBI121 to $OD_{600} = 0.7\sim0.8$ with the solution of 2% sucrose + 1/2MS + 0.1% silwet-77 + 0.1% AS or 5% sucrose + 0.1% silwet-77 + 0.1 $mmol \cdot L^{-1}$ AS, respectively; during 0.5~2.5h after artificial pollination, used stigma dripping method to transfer into plasmids and agrobacteria which containing target gene; collect the mature seeds; selective culture and PCR certificated that the resistant materials of almost all treatments were integrated GFP and GUS gene. Moreover, compared with agrobacterium as target gene delivery, the plasmid gene delivery could harvest more fruits.

T5

P1299

Genetic diversity and population structure of Golden wood (*Phoebe zhennan*) inferred from RAD-seq-derived SNP markers: effects of sequence clustering parameters

Jianhua Xiao[1,2], *Xin Ding*[1,2], *Jie Li*[1]

1. Xishuangbanna Tropical Botanical Garden, Chinese Academy of Sciences

2. University of Chinese Academy of Sciences

Restriction-site-associated DNA sequencing (RAD-seq) and related methods are revolutionizing the field of population genomics in non-model organisms as they allow generating an unprecedented number of single nucleotide polymorphisms (SNPs) even when no genomic information is available. Yet, RAD-seq data analyses rely on assumptions on nature and number of nucleotide variants present in a single locus, the choice of which may lead to an under- or overestimated number of SNPs and/or to incorrectly called genotypes. Using the Golden wood (*Phoebe zhennan*), as case study, here we explore the difference of genetic diversity and sensitivity of population structure inferences to three crucial aspects in RAD-seq data analysis: the minimum depth allowed to merge reads into a stacks, the maximum number of mismatches allowed to merge reads into a locus and the relatedness of the individuals used for genotype calling and SNP selection. Our study estimates the genetic diversity and resolves population structure of the *Phoebe zhennan*, but, most importantly, provides insights into the effects of alternative RAD-seq data analysis strategies on population structure inferences that are directly applicable to other non-model species.

T5

P1300

Identifying differentially expressed genes connected to seed shattering in *Elymus sibiricus*, an important forage grass in Qinghai-Tibet Plateau

Junchao Zhang, *Wengang Xie, Xuhong Zhao, Zongyu Zhang, Yanrong Wang*

State Key Laboratory of Grassland Agro-ecosystems, College of Pastoral Agriculture Science and Technology, Lanzhou University

Elymus sibiricus is an important forage grass in semi-arid regions, but it is difficult to grow for commercial seed production due to high seed shattering. To better understand the underlying mechanism and explore the putative genes related to seed shattering, we conducted a combination of morphological, histological, physiochemical and transcriptome analysis on two *E. sibiricus* genotypes (XH09 and ZhN03) that have contrasting seed shattering. The results show that seed shattering is generally caused by a degradation of the abscission layer. Early degradation of abscission layers was associated with the increased seed shattering in high seed shattering genotype XH09. Two cell wall degrading enzymes, cellulase (CE) and polygalacturonase (PG), had different activity in the abscission zone, indicating their roles in differentiation of abscission layer. cDNA libraries from abscission zone tissue of XH09 and ZhN03 at 7 days, 21 days and 28 days after heading were constructed and sequenced. A total of 86,634 unigenes were annotated and 7,110 differentially expressed transcripts (DETs) were predicted from "XH09-7 vs ZhN03-7", "XH09-21 vs ZhN03-21" and "XH09-28 vs ZhN03-28", corresponding to 2,058 up-regulated and 5,052 down-regulated unigenes. The expression profiles of 10 candidate transcripts involved in cell wall-degrading enzymes, lignin biosynthesis and phytohormone activity were validated using quantitative real-time PCR (qRT-PCR), 8 of which were up-regulated in low seed shattering genotype ZhN03, suggesting these genes may be associated with reduction of seed shattering. The expression data generated in this study provides an important resource for future molecular biological research in *E. sibiricus*.

T5

P1301

Transcriptome analysis of *Chionochloa macra* reveals biological features

Haoyu Xiong[1], *David Wharton*[2], *Janice Lord*[3], *Craig Marshall*[1]

1. Department of Biochemistry, University of Otago

2. School of Zoology, University of Otago

3. Department of Botany, University of Otago

The genus *Chionochloa* (tussocks), is mainly endemic and dominates the alpine grassland of New Zealand. *Chionochloa macra*, the slim snow tussock, is one of the most common high mountain tussocks, and is mainly found in the South Island. Previous studies of *Chionochloa* have focused on their ecological features and seasonal variations in chemical composition. Recent research has found species specific features including ice activities and water collection mechanisms in *C. macra* but there have been few studies analyzing the molecular genetics of these characters. Moreover, there are only a few studies on the phylogeny of the *Chionochloa* group based on rDNA and/or mitochondrial markers and the lack of genetic information has hampered further research. In this study, the first transcriptome of *C. macra* was achieved by sequencing the total RNA from four *C. macra* plants with an Illumina Hiseq 2000 platform. A total of 351 million high quality reads

(Q20) were generated and assembled into 901,869 contigs by Trinity assembler. This was finally filtered to 297,695 contigs with a N50 value of 1009. The filtered contigs were searched against NCBI non-redundant database and Swiss-Prot database (E-value<e^{-5}). </eGene Ontology annotations were given to 109,850 contigs by Blast2GO and 42,570 contigs were mapped to KEGG pathways. Essential proteins and metabolic pathways related to C. macra's biological features including high polysaccharide content, triterpenoid expression and water conservation were investigated. Genes coding ice-binding proteins were mapped with mass spectrometry data from ice affinity purification. Alternative splicing events, SSRs markers, SNPs, miRNAs were reported and gene expression profiles were provided for further genetic research in C. macra and other species in this genus.

T5

P1302

Comparative transcriptome analysis of the female and male inflorescences of monoecious plant pecan (*Carya illinoinensis*)

*Gang Wang, **Jiping Xuan**, Jiyu Zhang, Xiaodong Jia, Zhongren Guo*
Institute of Botany, Jiangsu Province and Chinese Academy of Sciences

Pecan (*Carya illinoinensis*) is a popular long-lived diploid tree in silviculture and horticulture. As pecan is a wind pollinated monoecious tree exhibiting heterodichogamy, the location and timing of functioning female and male organs are different, often leading to fruit-setting problem and thus yield decreases. The limited availability of pecan genomic information has hindered research on the mechanisms underlying its flower development. In this study, we obtained the first de novo assembly of pecan transcriptome and performed a comparative analysis of pecan female and male inflorescences using RNA-seq technology. A final unigene dataset containing 53,894 unigenes in the female pecan inflorescence was obtained, with an N50 length of 1,411 bp. To obtain insight into the functions of pecan unigenes, the unigenes were then annotated with Nr, Nt, Swiss-Prot, TrEMBL, KEGG and KOG databases. Using the bioinformatics approach, we identified 11,813 SSRs in unigenes and developed primers for 7,725 of them. A total of 26,321 differentially expressed genes (DEGs) were identified between pecan female and male inflorescences. A large number of them were linked to plant hormone regulation, especially gibberellin. In addition, out of 20 MADS-box transcription factors, one was more highly expressed in the female inflorescence whereas eight were expressed at a lower level in the male inflorescence, illustrating that the MADS transcription factors are involved in the flower development. In summary, this first unigene assembly of pecan could contribute to enhancing understanding of the gene specialization in flowers of difference sexes, and also be particularly useful for pecan germplasm management and breeding programs.

T5

P1303

A ddRAD-based linkage map and comparative genomics of *Dendrocalamus latiflorus* Munro

***Guoqian Yang**, Dezhu Li, Zhenhua Guo*
Kunming Institute of Botany, Chinese Academy of Sciences

Construction of linkage maps is a first step in exploring the genetic basis for adaptive phenotypic divergence in closely related species by quantitative trait locus (QTL) analysis. Linkage maps are also useful for comparative genomics in non-model organisms. Advances in genomics technologies make it more feasible than ever to study the genetics of adaptation in natural populations. Restriction site-associated DNA sequencing (RAD-seq) in next-generation sequencers facilitates the development of genetic markers and genotyping. We aimed to construct a linkage map of the Dendrocalamus latiflorus Munro of the genus Dendrocalamus (Gramineae) for comparative genomics with rice, Brachypodium distachyon and maize (all belong to the grass family) and for the future QTL analysis of the genetic architecture underlying adaptive phenotypic evolution of bamboo. We constructed the first high-density genetic linkage map of D.latiflorus Munro using a 190 selfing (S1) progenies of a wild species in southwest China. Based on 3,627 RAD-tag markers, a linkage map spanning 3,114.98 cM with 36 linkage groups and an average marker interval of 0.87cM was constructed. We also identified regions of segregating distortion involving transmission ratio distortion (TRD). Synteny and collinearity were extensively conserved between bamboo and rice and other grass species. The dense SNP-based linkage map presented here serves as a basis for future QTL analysis. It will be helpful for transferring genomic information from a monocotyledonous model species, rice, to screen candidate genes underlying ecologically and economically important traits of the bamboo.

T5

P1304

Difference of the annual roots of *Malus hupehensis* in transcriptome level

*Chunran Zhou, Hui Cao, Weiwei Zhang, **Hongqiang Yang***
College of Horticulture Science and Engineering, Shangdong Agricultural University

The roots of fruit tree contain perennial roots and annual roots grouped by growth years. Annual roots were divided into absorbing roots, extensive roots and conductive root by function. These differences are the adaption of root to the environment in specific genetic background, which is not only influenced by the environment but also controlled by gene expression. The experiment performed the transcriptome analysis of the annual roots of *Malus hupehensis* Rehd. using Illumina HisSeq 2500 RNA-seq. The 52,217,996 clean reads were obtained after filtering, which contained 10.54G Base in total. 54,396 unigenes were generated after assembling original reads. The unigenes were further blatsed using the BLAST and Blast2GO software in Nr, Swiss-Prot, GO, COG, and KEGG database for homology comparison. 31,530 unigenes were found to be homologous with the genes in these databases and accounted for 57.96% of the total unigenes. There are significant differences among the expression level of a large amount of genes of the absorbing roots, extensive roots and conductive roots. It shows that the expression level of genes producing ammonium transporters protein, subtilisin-like protease, aquaporin protein and casparian strip membrane protein is higher in absorbing roots than that in the extensive roots or conducting roots. The expression level of xyloglucan endotransglucosylase/hydrolase gene is higher in extensive roots than that in conducting roots. In addition, the expression levels of many extension genes are higher in extensive roots than those in absorbing roots or conducting roots. The expression level of genes producing heat shock protein, sugar transport protein and pectate lyase is higher in conducting roots than that in the extensive roots or absorbing roots. The difference of these genes expressions coincides with the functions and growth characteristics of these three types of roots.This work was supported financially by the National Natural Science Foundation of China (No. 31372016).

T5

P1305

RNase H1 cooperates with DNA gyrases to restrict R-loops and maintain genome integrity in *Arabidopsis* chloroplast

Zhuo Yang, *Lingling Cheng, Quancan Hou, Wei Xu, Yantao Hong, Shuai Li, Qianwen Sun*

Center for Plant Biology and Tsinghua-Peking Joint Center for Life Sciences, School of Life Sciences, Tsinghua University, Beijing 100084, China

Maintaining the integrity of organellar genomes is essential for eukaryotic cells. The chloroplast genome in plants is threatened by DNA stresses such as oxidative or photochemical damage from the altered surroundings, but the mechanisms whereby chloroplast genome integrity is maintained are not entirely understood. Here we found that plastid R-loop levels are related to chloroplast genome instability, and this is regulated by an R-loop removing enzyme AtRNH1C. AtRNH1C is predominantly localized in chloroplasts and necessary for chloroplast R-loop homeostasis and development. Interactome analysis revealed that AtRNH1C associates with multiple chloroplast-localized DNA- and RNA-metabolism related proteins, including the core DNA gyrases complex. The interaction between AtRNH1C and AtGyrases was critical for R-loop homeostasis in chloroplast and essential to release the transcription-replication conflicts in the highly transcribed and replication originated cp-rDNA regions, and thus to reduce the DNA damage. Together, our results reveal the plastid R-loop accumulation is the primary cause of cpDNA instability, and provide insight into the maintenance of genome integrity in chloroplasts, in which the evolutionarily conserved RNase H1 and DNA gyrases proteins are involved.

T5

P1306

The alternative splicing differentiation of the homeologous genes between the A and C subgenomes in *Brassica napus* L.

Shengli Yao, *Ming Hu, Dong Caihua, Huang Junyan, Tong Chaobo, Liu Shengyi*

Oil Crops Research Institute, Chinese Academy of Agricultural Sciences, Key Laboratory of Biology and Genetic Improvement of Oil Crops, the Ministry of Agriculture

Polyploidization is known to be prominent and significant force inducing diversification and speciation in plants. Many duplicated genes evolved and diverged accompanying following polyploidization, and their evolution is an important foundation of the polyploid formation. *Brassica napus* is ideal model for studying the formation of allopolyploid. Alternative splicing (AS) is an important mechanism for posttranscriptional modulating gene expression and proteins diversity in higher eukaryotes. Second-generation sequencing technologies have been extensively used to analyzetranscriptomes. However, a major limitation of short-read data is that it is difficult to accurately predict full-length splice isoforms. Here we sequenced the *B. napus* transcriptome using Pacific Biosciences single-molecule real-time (SMRT) long-read isoform sequencingtechnology. We start with all duplicated genes originated from *B. napus*, which have been constructed from chromosome syntenic analysis between two subgenomes in *B. napus*. Through Illumina RNA-Seq sequencing technology, we identified total 109,579 alternative splicing events that occurred at least from 2 biological replicates in 6 tissues in *B. napus* (cultivar "ZS11"). Almost 60% of alternative splicing events occurred in homeologous genes only from one of homeologous genes which belong to the gene in A or C subgenome. Total 26,687 expressed multi-exon genes were highly enriched in GO categories related to binding, regulation, location, transporting, metabolism, transcription, cell component and so on. It shows that intron retention is the most abundant AS type in higher eukaryotes as reported, the proportions of four main AS types in different tissues are similar and most AS events are tissue-specific. The AS gene highly enriched in metabolism, regulation and related pathways, which suggested that alternative splicing plays important roles in posttranscriptional regulation. Through SMRT sequencing we identified 217,963 AS events from 5 pooled the same tissues as Illumina sequencing. Only nearly 30% SMRT isoforms wered recovered with the assembled transcripts from Illumina sequencing. The work suggests that SMRT sequencing is highly powerful in AS discovery and provides a rich data resource for later functional studies of variant isoforms.

T5

P1307

Plastid genome comparative analysis between *Platycladus orientalis* and *Thuja koraiensis*

Dong-Keun Yi[1,2], *Hyuk-Jin Kim*[2], *Keum Seon Jeong*[2], *Min Jung Joo*[2], *Eun Mi Sun*[2], *Kyung Choi*[2]

1. International Biological Material Research Center, Korea Research Institute of Bioscience and Biotechnology

2. Division of Forest Biodiverstity and Herbarium, Korea National Arboretum

Cupressaceae is most recent speciated family among gymnosperm. The family includes 27~30 genera with about 130~140 species in total. Sequenced plastomes of Cupressaceae show various plastome structure and some genus have very different structure when compared with its related genera. *Platycladus orientalis* was often included in genus *Thuja* as to be called "oriental thuja". But there are many morphological differences between two genera, *P. orientalis* was identified to different genus. Here we have sequenced two complete chloroplast genomes of *P. orientalis* and *T. koraiensis* (Cupressaceae). The genomes are 127,113 bp and 130,381 bp in length, respectively. Comparative chloroplast DNA sequence analyses with two species reveal that gene and intron contents are similar to the Cupressaceae plastome except a few minor modifications. The genome of *T. koraiensis* includes 118 genes (78 peptide-encoding genes, 32 tRNA genes, four rRNA genes, 4 open reading frames) and the genome of *P. orientalis* hs one more tRNA gene (trnP-GGG). The major portion (63.1%) of the *T. koraiensis* plastome consists of gene-coding regions (57.7% protein coding and 5.4% RNA regions), and the major portion (64.7%) of the *P. orientalis* plastome consists of gene-coding regions (59.1% protein coding and 5.6% RNA regions). The overall A-T content of *T. koraiensis* and *P. koraiensis* plastomes are 65.7% and 65.3%, respectively. The A-T contents in the non-coding (70.7% and 70.3%) are higher than in the coding (63.1% and 64.7%) regions. Compared between structures of *T. koraiensis* and *P. orientalis*, plastome structure of *P. orientalis* are similar with structure of genera *Callitropsis*, *Cupressus*, *Hesperocyparis* and *Juniperus* which speciated in comparatively recent date, but *T. koraiensis* have distinctive plastome structure. Structure comparison between *T. koraiensis* and *P. orientalis* shows one inversion and one rearrangement modification. Genome structure comparison among Cupressaceae will provide information of relationship between structure modification and evolution, and scenario of structural modification.

T5

P1308

Application of CRISPR-Cas9 and CRISPR-Cpf1 in rice towards engineering altered stomata

Xiaojia Yin[1], *Akshaya K. Biswal*[1,3], *Jacqueline Dionora*[1], *Kristel M. Perdigon*[1], *Christian P. Balahadia*[1], *Shamik Mazumdar*[1], *Caspar Chater*[2,4], *Hsiang-Chun Lin*[1], *Robert A. Coe*[1], *Tobias Kretzschmar*[1], *Julie E. Gray*[2], *W. Paul Quick*[1,5], *Anindya Bandyopadhyay*[1]

1. International Rice Research Institute

2. Department of Molecular Biology and Biotechnology, University of Sheffield

3. Department of Biology, University of North Carolina at Chapel Hill

4. Departamento de Biologiá Molecular de Plantas, Instituto de Biotecnologiá, Universidad Nacional Autónoma de Mexico

5. Department of Animal and Plant Sciences, University of Sheffield

The development of programmable, sequence-specific nucleases such as Zinc-finger nucleases (ZFNs), transcription activator like effector nucleases (TALENs), clustered regulatory interspaced short palindromic repeat (CRISPR)-Cas9 and CRISPR-Cpf1 systems has provided scientists with easy-to-use tools for genome manipulation. The advantage of these tools is that they target a precise location within the genome facilitating the production of transgene-free homozygous heritable gene edited plants. The most widely used tool, the CRISPR-Cas9 system, recognizes and cuts specific sites that have a protospacer adjacent motif (PAM) of NGG, suitable for editing GC rich regions. Whereas the CRISPR-Cpf1 system recognizes and cut sites that have a PAM of TTTV, which is more suitable for editing AT rich regions. In this study, we have used both CRISPR-Cas9 and CRISPR-Cpf1 to knock out a rice (*Oryza sativa*) ortholog of the *Arabidopsis* Epidermal Patterning Factor like-9 (EPFL9) gene, a positive regulator of stomatal development. Germ-line mutants with edited OsEPFL9 gene were generated by CRISPR-Cas9 system and advanced to T_2 generation, when Cas9-free homozygous mutants were obtained. There were no detectable off-target effects in these plants when we sequenced the potential off-target sites. These plants had a more than eightfold reduction in stomatal density on the abaxial leaf surface. Knockout of the same gene using the CRISPR-Lb-Cpf1 (Lachnospiracae bacterium Cpf1) system led to heritable genome edits and similar phenotypic changes in T_1 generation plants. This study demonstrates the application of both CRISPR-Cas9 and CRISPR-Cpf1 to precisely target genomic locations and develop transgene-free homozygous heritable gene edited plants. It also confirms that the loss of function analysis of the candidate genes emerging from different systems biology based approaches can be performed with these tools. These systems add value to gene function studies.

T5

P1309

FY of *Arabidopsis* comprehensively regulates alternative polyadneylation

Zhibo Yu[1], *Juncheng Lin*[1], *Qingshun Quinn Li*[1,2]

1. College of the Environment and Ecology, Xiamen University

2. Graduate College of Biomedical Sciences, Western University of Health Sciences, Pomona

Alternative polyadenylation (APA) plays a crucial role in gene expression regulation across eukaryotes. This process requires a complex set of protein factors. In mammals, the 3'-end processing factor WDR33, a subunit of the Cleavage and Polyadenylation Specificity Factors (CPSF), directly binds to AAUAAA, the most conserved polyadenylation signal in eukaryotic. Arabidopsis FY is the homologue of WDR33. Previous studies have demonstrated that FY was involved in poly (A) site choice of flowering related genes. However, it remains to be clarified whether FY comprehensively involves in poly (A) site choices of other genes in Arabidopsis. Here, using Poly (A) Tag Sequencing (PAT-Seq) approach, we investigated transcriptome-wide poly (A) site profile in a mutant (fy-1) lacking a PPLPP domain of FY. The results showed that poly (A) site choice in a large number of Arabidopsis genes were significantly changed between wild type and fy-1 mutant with adjusted $p<0.05$, suggesting that FY functioned in the process of APA in Arabidopsis thaliana. Functional enrichments analysis of differentially expressed poly (A) site clusters indicates that these genes are significantly enriched in several biological processes, including response to osmotic stress, response to reactive oxygen species, defense response, cell death, among others. To further elucidate if FY mutation affects these biological processes, we assessed the sensitivity of fy-1 mutant to osmotic and oxidative stresses. Comparing to wild type, fy-1 mutant had higher rate of seed germination under salt or drought stress. The results suggested that fy-1 mutant had more tolerance to osmotic stress. Furthermore, the root length of fy-1 mutant was shorter than wild type under methyl viologen conditions, indicating that fy-1 mutant was more sensitivity to oxidative stress. Remarkably, we found that intergenic PATs significantly increased in fy-1 mutant compared to the wild type. To explain this, the expression level of CPL3, a phosphotase for the RNA polymerase II C-terminal domain leading to transcription termination, was determined by qRT-PCR. We found that the expression level of CPL3 significantly decreased in fy-1 mutant. Thus, we speculated that low expression level of

CPL3 is unable to dephosphorylate Ser2 leading to increase intergenic PATs in fy-1 mutant. Taken together, these results suggest that FY regulates poly (A) site selection of *Arabidopsis* genes and is involved in transcription termination.

T5

P1310

The pineapple genome and the evolution of CAM photosynthesis

Ray Ming[1,2, 3], *Robert Vanburen*[1,2, 3,4], *Ching Man Wai*[1,2, 3], *Haibao Tang*[1,2, 5], *Robert E Paull**[6], *Qingyi Yu**[7]

1. Fujian Agriculture and Forestry University and University of Illinois at Urbana-Champaign–School of Integrative Biology Joint Center for Genomics and Biotechnology, Fujian Agriculture and Forestry University, Fuzhou, China.

2. Fujian-Taiwan Joint Center for Ecological Control of Crop Pests, Fujian Agriculture and Forestry University, Fuzhou, China.

3. Department of Plant Biology, University of Illinois at Urbana-Champaign, Urbana, Illinois, USA.

4. Donald Danforth Plant Science Center, St. Louis, Missouri, USA.

5. iPlant Collaborative/University of Arizona, Tucson, Arizona, USA.

6. Department of Tropical Plant and Soil Sciences, University of Hawaii, Honolulu, Hawaii, USA.

7. Texas A&M AgriLife Research, Department of Plant Pathology and Microbiology, Texas A&M University System, Dallas, Texas, USA.

Pineapple (*Ananas comosus* (L.) Merr.) is the most economically valuable crop possessing crassulacean acid metabolism (CAM), a photosynthetic carbon assimilation pathway with high water-use efficiency, and the second most important tropical fruit. We sequenced the genomes of pineapple varieties F153 and MD2 and a wild pineapple relative, Ananas bracteatus accession CB5. The pineapple genome has one fewer ancient whole-genome duplication event than sequenced grass genomes and a conserved karyotype with seven chromosomes from before the ρ duplication event. The pineapple lineage has transitioned from C3 photosynthesis to CAM, with CAM-related genes exhibiting a diel expression pattern in photosynthetic tissues. CAM pathway genes were enriched with cis-regulatory elements associated with the regulation of circadian clock genes, providing the first cis-regulatory link between CAM and circadian clock regulation. Pineapple CAM photosynthesis evolved by the reconfiguration of pathways in C3 plants, through the regulatory neofunctionalization of preexisting genes and not through the acquisition of neofunctionalized genes via whole-genome or tandem gene duplication.

T5

P1311

Genetic and expressional variations of the genes in gibberellin metabolic pathway during japonica rice domestication

Chunyan Jing[1,2], *Sha Zhang*[1,2], ***Fumin Zhang***[1,2]

1. State Key Laboratory of Systematic and Evolutionary Botany, Institute of Botany, Chinese Academy of Sciences

2. University of Chinese Academy of Sciences

Molecular evolution of metabolic pathways is very important to understanding the molecular mechanism of phenotypic differentiation between species, which is also a key to the fundamental problem of biological evolution on phenotypic diversification and adaptation. Japonica rice (*Oryza sativa subsp. japonica*) was domesticated from its wild ancestral species *O. rufipogon* 10000 years ago. During the domestication, evolution pattern and underlying mechanisms of gibberellin metabolic pathway under artificial selection remain unclear in spite of importance of the pathway to plant growth and development. To reveal the pattern and underlying mechanisms, we conducted a comparison on population genetic and expressional variations of all the 21 genes of gibberellin metabolic pathway between japonica rice and *O. rufipogon* occurring in China. The sequences of all the 21 genes in gibberellin metabolic pathway were retrieved from next generation sequencing (NGS) data of japonica rice and *O. rufipogon* samples. Using the sequences, we investigated population genetic variation of all the 21 genes within and between japonica rice and *O. rufipogon*. Our results showed that diversities of five genes of japonica rice had significantly decreased as compared to their wild ancestor, indicating that the genes experienced strong artificial selection during domestication process of japonica rice. Moreover, these genes were mainly located in the branch points of downstream in the gibberellin metabolic pathway. Our analysis of population genetic differentiation displayed that genes with high F_{ST} values were mainly in the branch points of downstream in the pathway, too. The neutral test indicated that the genes that significantly deviated from the standard neutral model were also mainly located in the branch points of downstream in the pathway. On the other hand, we conducted qPCR experiments to compare the expressional variation of all the 21 genes at two developmental stages of seedling and elongation between japonica rice and its wild ancestral species *O. rufipogon*. The experimental results showed that differentially expressed genes (DEGs) identified at both stages were mainly located in downstream of the pathway, including those in the branch points. Based on the results, we conclude that the genes in the branch point of downstream in gibberellin metabolic pathway are main targets of artificial selection during japonica rice domestication process. Because the variation of genes in branch point of downstream in pathway affect mainly the distribution of metabolic flux, we infer that a gene with a higher effect on the distribution of metabolic flux usually is the target of selection in metabolic pathway.

T5

P1312

Genetic diversity and domestication footprint of Chinese cherry (*Cerasus pseudocerasus* (Lindl.) G.Don) as revealed by nuclear microsatellite markers

Jing Zhang[1,2], *Tao Chen*[1,2], *Yan Wang*[1,2], *Qing Chen*[1], *Bo Sun*[1], *Hao-Ru Tang*[1,2], *Xiao-Rong Wang**[1,2]

1. College of Horticulture, Sichuan Agricultural University

2. Institute of Pomology and Olericulture, Sichuan Agricultural University

Chinese cherry (*Cerasus pseudocerasus* (Lindl.) G. Don) is a commercially and culturally important fruit crop in China, but the domestication history and genetic relationship of the cherry landraces still remains unclear. To address these questions, we employed nineteen nuclear microsatellite markers to genotype 532 Chinese cherry landrace and wild accessions from 57 populations of 12 provinces in China. All of our samples covered the whole natural distributions in Sothwest China (SWC), Longmenshan Fault Zones (LFZ), Qinling Mountains (QLM) and North and East China (NEC). 16 *Cerasus* relative species (118 wild relative accessions) from 17 natural populations were took as out group in our study. We identified the population structure in Chinese cherry with Bayesian clustering method in STRUCTURE. Our results showed three genetic clusters in Chinese cherry. Wild accessions were assigned into one genetic cluster (WC). Landrace accessions were divided into two different genetic clusters (LC1 and LC2). LC1 almost consisted of the landraces from LFZ and SWC, and LC2 mainly included those from QLM and NEC. We calculated seven genetic diversity indices within populations and within genetic clusters, and conducted the Kruskal-Wallis test for allelic richness (AR) and gene diversity (Hs) among all Chinese cherry populations and between pairwise genetic clusters. Significant loss of genetic diversity were detected in most landrace populations and both landrace clusters during domestication. With Garza-Williamson index and two-phased model of mutation (TPM), we detected significant fingerprint of recent genetic bottleneck in LC1 (LFZ and SWC) but not in LC2 (QL and NEC). Based on historical evidences and our genetic results, we estimated the most probable demographic model for Chinese cherry with the approximate Bayesian computation (ABC) under pruned data (without recent admixed and misclassified accessions identified by STRUCTURE) and full data (all Chinese cherry accessions). Our results under two different datasets showed strong evidence for the independent domestication for two landraces clusters. The parameters estimation in ABC model indicated that the LC2 (QL and NEC) was diverged from the wild ancestors about 3900 years ago, which was earlier than LC1 (LFZ and SWC). Combined with microsatellites and phenotypic data, we found that long-term independent domestication process and different artificial selection have led to two significantly different domestication footprints on the two landrace clusters. To analyze the genetic relationship and recent demographic process among landrace populations and between landrace and wild populations, we estimated the admixture coefficients (*mγ*), correlated coefficients (*rxy*) and genetic draft (*F*) between pairwise populations. Close genetic relationship was detected between landrace accessions and populations from the neighboring populations and areas. Frequent gene flow from wild accessions was detected in the landraces from QLM and SWC. Our study contributes to the knowledge about the domestication history and genetic relationship for Chinese cherry landraces. As a long-lived perennial fruit crop with high inbreeding, the domestication patterns of Chinese cherry were different from those of some out-crossing perennial fruit crops.

T5
P1313
Evolution and diversity of E (z) homolog genes in green plant

***Zihua Chen**, Liangsheng Zhang*
Center for Genomics and Biotechnology, Fujian Agriculture and Forestry University

Polycomb group proteins (PcG) methylate lysine 27 of histone H3 and play important roles in the maintenance of silent states of genes. These core PcG proteins, such as Enhancer of Zeste (E (z)), are conserved in animals and plants. By a comparative phylogenetic analysis, we present a potential framework of E (z) gene evolution in green plant lineage. We found the E (z) genes are duplicated in the ancestor of seed plants resulting in two clades: SWN and CLF. Then, E (z) genes are highly conserved in angiosperms. However, Brassicaceae contains two clades by whole genome duplication, which forms MEA and SWN. Similarly, Fabaceae also contains two clades, named MEA-like and SWN-like. Interestingly, codon-based branch analysis among taxa suggests that within the MEA-like, but not SWN-like, continues to evolve under positive selection in Fabaceae. We analyze the expression of MEA-like genes and found that most tissues did not express in soybean, but there is little expression in the heart-shaped embryo, suggesting that it might be specifically expressed in embryonic development. From the sequences analysis and expression, we propose that MEA-like lead to neofunctionalization in soybean or Fabaceae .

T5
P1314
Origin and diversity of *WRKY* gene family in green plants

***Hu Yue**, Zhang Liangsheng*
Center for Genomics and Biotechnology; Key Laboratory of Genetics, Breeding and Multiple Utilization of Corps, Ministry of Education; Fujian Agriculture and Forestry University

WRKY gene family is a kind of important transcription factor, which plays important roles in regulating plant development and resistance. WRKY gene can be divided into Group I, Group II and Group III. In order to clarify the evolution of the WRKY gene in plants, we used a large number of genome and transcriptome data, including a variety of species from Chlorophyta, Charophyta, and land plants. According to the phylogenetic tree topology, we re-divided the WRKY family into Group I, Group IIa + IIb, Group IIc, Group IId + IIe and Group III. 12 Chlorophyta species contained WRKY genes, which distributed in Group I and Group III. 16 species of Charophyta have WRKY genes and are distributed in all groups except the family Group IId + IIe. We used three moss genome data, namely *Sphagnum fallax*, *Takakia lepidozioides* and *Physcomitrella patens*, as well as the genome of kind of liverwort species, *Marchantia polymorpha*. There were 36 and 34 WRKY genes in the *T. lepidozioides* and *P. patens*, while only 14 and 11 were found in *M. polymorpha* and *S. fallax* respectively, each of them had a similar distribution pattern of WRKYs. From the topological structure of the Group III phylogenetic tree, four species form a monophyletic group, while the seed plants were divided into another one, and the *M. polymorpha* was further distant from the other three. *P. patens* only had Group IId members, while the other three are distributed in both Group IId and Group IIe. In Group IIa + IIb, their WRKY genes were gathered together with angiosperms instead of gymnosperms. Besides, from the figure Group IIc could be divided into two clades and all the moss were concentrated in one branch. The number of Gymnosperms WRKY is the least in Group III, with no more than 3 in all species. In the GroupIIa + IIb, the *Pinus taeda* and the *Picea abies* are monophyletic, indicating that the WRKY gene of GroupIIa + IIb could be specifically expand in the Pinales.

T5

P1315

The genetic architecture of shoot-root covariation during seedling emergence of a dessert tree, *Populus euphratica*

Miaomiao Zhang[1], *Wenhao Bo*[1], *Fang Xu*[1], *Huan Li*[1], *Meixia Ye*[1], *Libo Jiang*[1], *Chaozhong Shi*[1], *Yaru Fu*[1], *Guomiao Zhao*[1], *Yuejiao Huang*[1], *Kirk Gosik*[2], *Dan Liang*[1], *Rongling Wu*[1,2]

1. Center for Computational Biology, College of Biological Sciences and Technology, Beijing Forestry University

2. Center for Statistical Genetics, The Pennsylvania State University

The coordination of shoots and roots is critical for plants to adapt to changing environments by fine-tuning energy production in leaves and the availability of water and nutrients from roots. To understand the genetic architecture of how these two organs covary during developmental ontogeny, we conducted a mapping experiment using Euphrates poplar (*Populus euphratica*), a so-called hero tree able to grow in the desert. We geminated intraspecific F_1 seeds of Euphrates poplar individually in a tube to obtain a total of 370 seedlings, whose shoot and taproot lengths were measured repeatedly during the early stage of growth. By fitting a growth equation, we estimated asymptotic growth, relative growth rate, the timing of inflection point and duration of linear growth for both shoot and taproot growth. Treating these heterochronic parameters as phenotypes, a univariate mapping model detected 19 heterochronic quantitative trait loci (*h*QTLs), of which 15 mediate the forms of shoot growth and four mediate taproot growth. A bivariate mapping model identified 11 pleiotropic *h*QTLs that determine the covariation of shoot and taproot growth. Most QTLs detected reside within the region of candidate genes with various functions, thus confirming their roles in the biochemical processes underlying plant growth.

T5

P1316

Genome expansion and the origin of major plant innovations

Ticao Zhang[1], *Qin Qiao*[2], *Jinling Huang*[3]

1. Key Laboratory for Plant Diversity and Biogeography of East Asia, Kunming Institute of Botany, Chinese Academy of Sciences

2. School of Agriculture, Yunnan University

3. Department of Biology, East Carolina University

Major organismal lineages are defined by distinctive structural and physiological innovations. How these innovations evolved over time as a result of increase in gene repertoire remains unclear. In this study, we adopt a new phylogenomic framework to understand the origin and evolution of structural and physiological innovations of major plant lineages based on core genome expansions, each of which includes a synapomorphic genome with shared derived genes for the focal lineage. Our analyses identified two major expansion events coincident with the origin of streptophytes and angiosperms respectively. The synapomorphic genome of streptophytes evolved in a much slower tempo, but is enriched in genes related to adaptation to terrestrial environments. We show that the phenylpropanoid metabolism evolved in charophyte green algae, rather than in land plants. The establishment of the complete flowering network was determined by the origin of several key genes in angiosperms, although the majority of flowering genes evolved at earlier stages. These findings are critical for understanding the interplay of innovations at different levels (e.g., gene, genome, physiological, structural) and the origin of major plant lineages.

T5

P1317

Precision automation of tissue and cellular phenotyping classification and quantification based on microscopic images

Ying Zhang, *Jianjun Du*, *Xinyu Guo*, *Chunjiang Zhao*

Beijing Key Lab of Digital Plant, Beijing Research Center for Information Technology in Agriculture, Beijing Academy of Agriculture and Forestry Sciences, Beijing, Germany

Plant phenotyping is defined as the application of methodologies and protocols to measure the specific traits, ranging from the cellular level, organ level, to the whole plant or canopy level, related to plant structure and function (Ghanem *et al.*, 2014). Whole plant phenotyping systems are highlighted combination with advances in machinery manufacturing and automation in image acquisition and analysis that enable notable throughput increase. There have been many progresses on the whole plant phenotyping research, which for example acquire a large number of traits including plant growth, physiological traits, biotic or abiotic stress markers. Organ level phenotyping researches are focus on high-dimensional phenotypic data at elevated spatial and temporal resolution. Such as various imaging technologies have been applied to monitor the root phenotyping non-disruptively and throughout a time period. More importantly, many commercial whole plant or organ phenotyping solutions are available for indoors and outdoors use and which have dramatically changed plant breeding programs and functional genomics in the last decade (Rahaman *et al.*, 2015; Hall *et al.*, 2016). Compared with rapid, precise, and accurate phenotypic traits data from organ level to the whole plant level high-throughput phenotyping (HTP) solutions, the potential of automated phenotyping with cellular resolution is largely unexploited. Operating at a lower throughput, phenotyping at cellular level requires more complex produces, such as extensive sample preparation and advanced microscopic acquisitions. Recent advances in microscope automation provided new opportunities for high-throughput cell level phenotyping, i.e. Micro-CT, LSCM, and SEM imaging. However, Quantitative analysis of those anatomical characteristics based on micro-images is a very tedious work, which mostly relies on manually examination. Obviously, manual examination is labor intensive and underlies human subjectivity and inconsistency that may severely limit the throughput and precision of micro-phenotypic traits analysis, hence high-throughput and accurate analysis methods are strongly needed. A particular focus of our work is the investigation of image process pipeline for different microscopic images, from slide scanning images obtained by classical paraffin embedding and microtome techniques to CT scanning images, which (1) automatically segment maize stems into tissue regions or individual cells, (2) assigns each segment to a vascular tissue or cell type, (3) quantifies phenotypic traits, and also (4) extracts a wealth of morphometric data. Short processing times permit large data sets, which will promote the application of micro-phenotyping in plant breeding programs and functional genomics.

T5

P1320

Identification and analysis of candidate genes for the varietal characters of *Phyllostachys edulis* 'pachyloen' by whole-genome resequencing and RNA-seq

Yun Zhang[1], *Ying Zhang*[1,2], *Hong Ran*[1,3], *Yun Feng*[1], *Tao Hu*[1], *Qi-Rong Guo*[1]

1. International Center for Bamboo and Rattan

2. Guangdong Academy of Forestry

3. Chongqing Academy of Forestry

Phyllostachys edulis (moso bamboo) originated from China and is one of the best economic bamboo species with the largest distribution area, most accumulation on volumes, highest production, and best research and utilization degree. There are more than 20 varieties had been released, but *Phyllostachys edulis* 'Pachyloen', one of the cultivar of moso bamboo, is the only accession germplasm with the bamboo wall thickened obviously. It owns desirable traits for its wood and shoots, and has the characteristics of stable heredity. Exploring the formation mechanism of its good traits by whole-genome re-sequencing and RNA sequencing based on the reference genome of moso bamboo, especially for its thick wood, will provide important theoretical basis for bamboo breeding. A total of 260.5 million reads produced approximately 24.5 Gb of sequence for *Phyllostachys edulis* 'Pachyloen' with 12X genome coverage, and mapping of 245 million (97.43%) reads to the reference genome by the whole-genome re-sequencing. The variation including SNP, InDel and CNV have great impact on the change of genomes and the biological characters. In addition, we identified 6631 differentiating genes which were performed the analysis of KEGG pathway, and these genes fall into some metabolic pathway such as the metabolism of starch, carbohydrate and hormone, biosynthesis of carbohydrate, protide, and lipid in the ribosome and endoplasmic reticulum. According to the developmental stages of bamboo from bud to young culm a total of eight times, we gathered the eight shoots of *Phyllostachys edulis* 'Pachyloen' and moso bamboo which were selected as the control group, respectively, and they were proceed the RNA sequencing. For *Phyllostachys edulis* 'Pachyloen' and moso, there were 2.4-4.3 Gb and 1.7-5.6 Gb of sequence respectively. The unique genes and differential expression genes which were gained from the experimental and control group and maybe correlated with the specific qualities could be screened. The unique genes of *Phyllostachys edulis* 'pachyloen' were ranged from 666 to 1839, and the differential expression genes ranged from 1719 to 11033 which were screened based on the value of RPKM (Reads Per Kilobases per Million reads) against the gene co-expression. The genes we selected were processed the GO functional and KEGG pathway analysis. Considering the results of whole-genone re-sequencing and RNA sequencing, the candidate genes for the varietal characters of *Phyllostachys edulis* 'pachyloen' can be identified and analyzed, and the conclusion maybe the scientific basis for the further research.

T5

P1321

Development and characterization of polymorphic EST-SSR of *Pteroceltis tatarinowii*, an endangered plant endemic to China

Yunyan Zhang

Nanjing University

With wide distribution in the genome, the microsatellite markers relied on expressed sequence tags (EST) can be linked to genetic traits and applied broadly in population genetics, molecular marker assisted breeding and molecular marker screening of target traits. However, the lack of the *Pteroceltis tatarinowii* genomic and transcriptional dataset limits the application of molecular markers in genetic studies of its populations. In this paper, we used Illumina high-throughput sequencing technique to sequence the *P. tatarinowii* leaf transcriptome, based on which polymorphic microsatellite loci are developed. A total of 42,477 unigenes with an average length of 815 bp were obtained. After aligning with NR, Swiss-Prot, GO, COG and KEGG databases, 23,688 unigenes were assigned. These data provided the foundation for gene expression and function analysis of *P. tatarinowii*. Among the unigenes, a total of 6,543 EST-SSRs were identified and 130 EST-SSRs were selected for validation as EST-SSR markers by PCR amplification. Of these, 48 EST-SSRs were amplified successfully and 32 EST-SSRs were polymorphic among 47 *P. tatarinowii* individuals. Additionally, cross-amplifications of EST-SSR were detected in *Ulmus gaussenii* and *U. chenmoui*, and the versatility and polymorphism were 25% and 87.5%, respectively. The development of the *P. tatarinowii* microsatellite markers provided the molecular basis for the study of the *P. tatarinowii* and other plants in family Ulmaceae.

T5

P1322

Improved NGS library preparation methods for high efficiency and low cost genome *de novo* assembly and resequencing

Sheng Zhao[1], *Xiaojuan Deng*[2], *Ying Lu*[2], *Ke Liu*[1], ***Yuxiao Chang***[1]

1. Agricultural Genomics Institute at Shenzhen, Chinese Academy of Agricultural Sciences

2. CAAS Biological Breeding Technology Co., Ltd (CBBT)

The DNA sequencing cost has decreased rapidly in the past years, however, the cost for library preparation part did not have big change. With the launch of Illumina Hiseq X and Novaseq with even more cheap sequencing price, there is an urgent need to decrease the cost for library preparation step, especially for large sample population projects. Here we described two improved next generation sequence (NGS) library preparation methods for either genome *de novo* assembly or resequencing with high efficiency and low cost. The methods introduced here are especially suitable for large sample population project with small budget. The first method we developed is the N-Size-Fragment (NSF) MP library preparation technique, which could generate one library contained up to 10 large DNA fragments ranged from ~2 to 12 kb simultaneously. We tested the performance of the method in rice genome de novo assembly and prepared a library contained 10 fragments with very sharp size distribution. A total of 80Gb Paired-end 250 bp sequence data generated from this library achieved the de novo assembly scaffold N50 more than 0.8Mb. The total reagent cost for NSF-MP library preparation and Hiseq 2500 PE250 bp sequence was only ¥30,000 (~ $ 4,300) for a rice genome. The second method we improved is a high-throughput shotgun sequence library preparation workflow, in this workflow, we omitted some unessential steps and mix the barcoded library before size selection and quantification with special calculation, by this modification, the total time/hand-on time were decreased ~50% with no compromise of the sequence data quality and data distribution among the mixed individual library. By using this method, it need only ¥500 (~$70) for one rice genome resequencing with 10-fold coverage.

T5

P1323

Genome-wide identification of QTL for seed yield and yield-related traits and construction of a high-density consensus map for QTL comparison in *Brassica napus*

Weiguo Zhao

1. Shaanxi Rapeseed Branch of National Centre for Oil Crops Genetic Improvement

2. Huazhong University of Science and Technology

Seed yield (SY) is the most important trait in rapeseed, is determined by multiple seed yield-related traits (SYRTs) and is also easily subject to environmental influence. Many quantitative trait loci (QTLs) for SY and SYRTs have been reported in *Brassica napus*; however, no studies have focused on seven agronomic traits simultaneously affecting SY. Genome-wide QTL analysis for SY and seven SYRTs in eight environments was conducted in a doubled haploid population containing 348 lines. Totally, 18 and 208 QTLs for SY and SYRTs were observed, respectively, and then these QTLs were integrated into 144 consensus QTLs using a meta-analysis. Three major QTLs for SY were observed, including *cqSY-C6-2* and *cqSY-C6-3* that were expressed stably in winter cultivation area for three years and *cqSY-A2-2* only expressed in spring rapeseed area. Trait-by-trait meta-analysis revealed that the 144 consensus QTLs were integrated into 72 pleiotropic unique QTLs. Among them, all the unique QTLs affected SY, except for *uq.A6-1*, including *uq.A2-3*, *uq.C1-2*, *uq.C1-3*, *uq.C6-1*, *uq.C6-5* and *uq.C6-6* could also affect more than two SYRTs. According to the constructed high-density consensus map and QTL comparison from literatures, 36 QTLs from five populations were co-localized with QTLs identified in this study. In addition, 13 orthologous genes were observed, including five each gene for SY and thousand seed weight, and one gene each for biomass yield, branch height and plant height. The genomic information of these QTLs will be valuable in hybrid cultivar breeding and in analyzing QTL expression in different environments.

T5

P1324

Regulation of the RNA-directed DNA methylation pathway

***Ming Zhou**, Ana Marie S Palanca, Clara Bourbousse, Julie A. Law*

The Salk Institute for Biological Studies

Cytosine DNA methylation is a heritable chromatin mark that plays crucial roles in gene regulation and transposon silencing in eukaryotic organisms. In plants, DNA methylation is established via the RNA-directed DNA methylation (RdDM) pathway, wherein both 24nt small interfering RNAs (siRNAs) and long non-coding RNAs are required to guide methylation at cognate genomic loci. Production of these RNAs is dependent on two plant specific RNA polymerases, Pol IV and Pol V. Pol IV transcripts are converted into double-stranded RNAs by RNA-DEPENDENT RNA POLYMERASE 2 (RDR2), sliced into 24nt siRNAs by DICER-LIKE 3 (DCL3), and then loaded into ARGONAUTE (AGO) effectors. Pol V transcripts serve as templates for the recruitment of siRNA-loaded AGO complexes as well as other downstream RdDM factors, including the *de novo* methyltransferase DOMAINS REARRANGED METHYLTRANSFERASE 2 (DRM2), to facilitate the establishment of DNA methylation and gene silencing. Here we report the roles of several uncharacterized factors in mediating DNA methylation in a locus-specific manner, revealing an additional layer of complexity to the RdDM pathway.

T5

P1325

Identification and characterization of stress responsive lncRNAs in plants

***Chao Di**, Jiaopei Yuan, Yumin Zhu, Zhi Lu*

MOE Key Laboratory of Bioinformatics, Center for Plant Biology, School of Life Sciences, Tsinghua University

Recently, in addition to poly (A)+ long non-coding RNAs (lncRNAs), many lncRNAs without poly (A) tails, have been characterized in mammals. However, the non-polyA lncRNAs and their conserved motifs, especially those associated with environmental stresses, have not been fully investigated in plant genomes. We performed poly (A)- RNA-seq for seedlings of Arabidopsis thaliana under four stress conditions, and predicted lncRNA transcripts. We classified the lncRNAs into three confidence levels according to their expression patterns, epigenetic signatures, and RNA secondary structures. Then, we further classified the lncRNAs to poly (A)+ and poly (A)- transcripts. Compared with poly (A)+ lncRNAs and coding genes, we found that poly (A)- lncRNAs tend to have shorter transcripts and lower expression levels, and they show significant expression specificity in response to stresses. In addition, their differential expression is significantly enriched in drought condition and depleted in heat condition. We have also tested lncRNA expression levels in rice and have seen similar patterns. Overall, we identified 245 poly (A)+ and 58 poly (A)- lncRNAs that are differentially expressed under various stress stimuli in Arabidopsis. The differential expression was validated by qRT-PCR, and the signaling pathways involved were supported by specific binding of transcription factors (TFs), phytochrome-interacting factor 4 (PIF4) and PIF5. Moreover, we found many conserved sequences and structural motifs of lncRNAs from different functional groups (e.g. a UUC motif responding to salt and an AU-rich stem-loop responding to cold), indicated that the conserved elements might be responsible for the stress-responsive functions of lncRNAs.

T5

P1326

Chromosome numbers of the flora of Germany - a new online database

*Thomas Gregor[1], Juraj Paule[1], Marco Schmidt[1,2], Eva-Maria Gerstner[2], Stefan Dressler[1,3], Karsten Wesche[4], **Georg Zizka**[1,3]*

1. Senckenberg Research Institute and Natural History Museum Frankfurt, Department Botany and Molecular evolution

2. Senckenberg Biodiversität und Klima Forschungszentrum (BiK-F), Frankfurt am Main

3. Goethe-Universität, Frankfurt am Main

4. Senckenberg Museum für Na

Chromosomal speciation processes gain increasing attention in plant systematics and evolution, and new approaches revealed a high diversity in chromosome numbers even within recognized taxa. Reliable counts linked to known accessions are thus needed yet often hardly available. We present a new online database of chromosome counts and flow cytometric ploidy estimates of the flora of Germany with a detailed documentation of the examined material, and its sampling location. The chromosome database builds upon a relational database and includes standardized taxon identification, study date, georeferenced locality and additional collection as well publication details from which the karyological information was extracted. In order to reach the best compatibility with other botanical publications of the study region, taxonomic concepts and nomenclature follow the "Rothmaler" (Jäger *et al.*, 2011), the widely accepted field flora of vascular plants in Germany. Our continuously updated database currently comprises over 11 000 datasets and 1678 taxa and is available at http://chromosomes.senckenberg.de.

T5

P1411

Transcriptome analysis of the hybrid origin of *bauhinia blakeana*

***Wm Tang**[1], Sl Kwok[1], Rl Davidson[2,5], Hs Kwan[3], Sc Edmunds[2,5], Skw Tsui[1,4]*

1. School of Biomedical Sciences, The Chinese University of Hong Kong

2. Beijing Genomics Institute, Hong Kong

3. School of Life Sciences, The Chinese University of Hong Kong

4. Hong Kong Bioinformatics Centre, The Chinese University of Hong Kong

5. GigaScience

Bauhinia blakeana Dunn, commonly known as Hong Kong Orchid Tree, was adopted as the city flower of Hong Kong in 1965 and became the flora emblem of HKSAR in 1997, appearing in local coins and banknotes. It was first discovered in Pokfulam in the 1880s and is now planted all over the city. It has long been suspected that *Bauhinia blakeana* was the hybrid from two related species, *B. purpurea* and *B. variegata* as it is a sterile plant without seeds and unable to propagate under natural condition. Also, very limited genetic information of this plant was available before the era of molecular biology. Lau *et al.*, (2005) determine the paternity of B. blakeana by using analysis of variation in inter-simple sequence repeat (ISSR) markers to study the genetic polymorphism and morphology based methods, while Mak *et al.*, (2008) conducted the identification study for Bauhinia species by sequencing the internal transcribed spacer 1 (ITS1), chloroplast rbcL and atpB-rcbL intergenic spacer among them. They provided the evidence to support the hypothesis while sequencing data revealed that genetic distance between B. blakeana and B. variegate was closer than B. blakeana and B. purpurea, while B. purpurea is the maternal lineage of B. blakeana. However, the genomes of the Bauhinia family were still not available. Next-generation sequencing (NGS) technology including RNA sequencing (RNA-seq), provided the gateway for scientists to have better understanding of gene expression, biochemical pathways of metabolites synthesis, diversity characterization and evolution study without relevant prior sequencing data. In this study, whole-transcriptome sequencing of leaves from B. blakeana, *B. purpurea* and *B. variegata* plus flowers from *B. blakeana* was performed using next generation sequencing. Approximately 5.4 Gbp high quality (HQ) reads were obtained from each tissue and unigenes of *B. blakeana* (94,755), *B. purpurea* (111,976) and *B. variegata* (81,757) were obtained after de novo assembly. We also pooled the reads of flowers and leaves from *B. blakeana*, resulting in 123,668 unigenes. We investigated the number of very similar transcripts between *B. blakeana* and each of its parents, *B. blakeana* and *B. purpurea* yield 11,082 clusters, while *B. blakeana* and *B. variegata* yielded 8,390 clusters, suggesting that *B. blakeana* is closer to *B. purpurea*, in contrast with Lau's findings. These initial results support that transcriptomics data mining is an effective way to discover similarities and differences of genetic make-up among species with the same genus. We will further investigate how gene expression affects seed germination, sterility, defense against herbivores and pathogens and even medicinal value.

T5

P1413

Genetic analysis fails to distinguish the "new" diploid African Baobab

***Alice Muchugi**, Hendre Prasad, Samuel Muthemba, Robert Kariba, Simon Kangethe*

World Agroforestry Centre

Until recently, there were eight species documented in the monophyletic genus *Adansonia*; six species endemic to Madagascar, one in Australia and one in Africa. However, a second diploid species *A. kilima* was recently described and its presence confirmed in mainland Africa. Unlike the other previously recorded Adansonia species which are diploid, the African baobab, *A. digitata* is tetraploid. The existence of diploid *A. kilima* was confirmed on basis of floral morphology, pollen and chromosome numbers. According to the species description *A. kilima* and *A. digitata*co exist in the same ecology. However, *A. kilima* was said to be restricted to moderate elevations (650-1500 m), while *A. digitata* is widespread throughout Africa in elevations below 800 m. Interestingly, genetic analysis using SSR and ITS markers of samples from a 200 m to 1300 m transect failed to resolve the two species distinction. The proposal that the tetraploidy evolved rather relatively recently contrasts *A. digitata* wide distribution across mainland Africa.

T5

P1415

Comparative analysis of alternative polyadenylation between *japonica* and *indica* of rice during developmental gene expression regulation.

Qian Zhou[1], *Haihui Fu*[1], *Dewei Yang*[2], *Xinfu Ye*[2], *Xiaohui Wu*[1], ***Qingshun Q. Li***[1]

1. College of the Environment & Ecology, Xiamen University, Xiamen 361102, China

2. Rice Research Institute, Fujian Agriculture Academy, Fuzhou, China

3. Department of Automation, Xiamen University, Xiamen 361005, China

4. Graduate College of Biomedical Sciences, Western University of Health Sciences, Pomona, CA 91766, USA

Rice is one of the most important food staple in the world, which also is the model plant of monocots. Two subspecies of rice, *Oryza sativa* L. ssp *indica* and *japonica*, which are widely planted throughout Asia, exhibiting apparent phenotypic variations in development and environmental responses. Alternative polyadenylation (APA) can regulate gene expression by producing multiple transcripts of different length from a single gene, and it has been found play an important role in rice growth and development. However, APA-mediated transcriptional regulations underlying the physiological and biochemical differences between the two subspecies have never been reported. In this study, we employed a poly (A) tag sequencing approach (PAT-seq) to compare the genome-wide landscapes of APA and poly (A) signals (PAS) in 14 different tissues and developmental stages of the cultivate line 93-11 of O. sativa ssp. indica and Nipponbare of O. sativa ssp. japonica. A total of 40,697 Poly (A) Clusters (PACs) in the two subspecies were resided in 25,068 genes, 88.26% of which had the same PACs numbers both in 93-11 and Nipponbare. About 1151 and 5845 unique PACs were found to be specific in 93-11 and Nipponbare, respectively. Five 93-11-specific PACs in eight genes encoding NBS-LRR disease resistance protein were located in CDS region, which may give rise to non-function transcripts, whereas the Nipponbare-specific PACs in 16 NBS-LRR resistance protein encoding genes were located in 3' UTR. It indicates that the different resistance to bacterial infection between the two subspecies may be regulated by APA. We further examined APA switching events, and found that more switching genes of Nipponbare mainly enriched in the plant growth biosynthetic process such as photosynthesis and carbohydrate metabolic process tend to use distal PACs in booting stage, whereas the switching gens during the mature and germinating stages preferred to use distal poly (A) sites in 93-11. Most of differentially expressed (DE) - PACs between the two subspecies were identified in the booting stage including the leaf 60-days, stem 60-days, root 60-days and anther, which suggested that the significant differences of gene expression occurred. Co-expression analysis of DE-PACs suggested that darked, lightcyan and yellow modules related to reproductive development and metabolic process had strong correlations with mature pollen of rice, which showed specific single nucleotide profile around PAS. Our results offer a new insight into the differentiations in transcriptional dynamics and functional significance of APA during growth and development processes of the two rice subspecies and provide potentially useful information for future breeding work.

T5

P1418

Arabidopsis JMJ24 reinforces transcriptional gene silencing by stabilizing RDR2

Shulin Deng, *Nam-Hai Chua*

Temasek Life Science Laboratory, National University of Singapore

RNA-directed DNA methylation (RdDM) is the major pathway that initiates *de novo* DNA methylation in *Arabidopsis* and sets up a self-reinforcing silencing loop between DNA methylation and H3K9me. However, a key issue is the requirement of a basal level transcript from the target loci to initiate the RNA based silencing. How the heterochromatic silenced loci are transcribed remains largely unknown. Here, we show that JMJ24, a JmjC domain-containing protein counteracts H3K9me to promote basal level transcription of endogenous silenced loci in *Arabidopsis*. JMJ24 functionally resembles the fission yeast JmjC protein Epe1. The transcript promoted by JMJ24 is, at least in part, processed to small RNA to initiate the RdDM. Genome-wide transcriptome profiling indicates that transcript levels of TEs are more likely regulated by JMJ24, compared with protein-coding genes. Our data suggest that JMJ24 plays a conserved role in promoting basal level transcription of endogenous silenced loci to reinforce the silencing. We also provide evidence of JMJ24 stabilizes RNA-dependent RNA polymerase 2 (RDR2).

T5

P1429

Potential horizontal gene transfers in *Nymphaea colorata* genome

Chaoxian Zhao[1,2]*, *Ruiliang Zhu*[2], *Liangsheng Zhang*[3], *Yang Liu*[1]

1. Shenzhen Fairy Lake Botanical Garden, Shenzhen, China, East China Normal University, Shanghai, China

2. East China Normal University, Shanghai, China

3. Center for Genomics and Biotechnology, Fujian Agriculture and Forestry University, Fuzhou, China

Horizontal gene transfer (HGT), also known as lateral gene transfer, is one of the paths for sharing genetic information among different organisms. It is well known that HGT is an important driving force for the evolution of prokaryotes, such as bacteria, but HGT is rare in eukaryotes. Aquatic plant lives in fresh water, is easily exposes to bacteria, and hence has more chance to obtain HGT genes. Here, based on the newly sequenced genome (unpublished data) of *Nymphaea colorata*, a floating aquatic plant and one of the earliest angiosperms, we simply test the hypothesis that aquatic plants carry micro-organisms originated horizontal transferring genes. We used a local python script that newly developed in the HGT-inferring pipeline (https://github.com/Chao-Z/Horizontal_gene_transfer_project), to calculate HGT AI, HGT Index as well as HGT Identity value of the top BLASTP hit of each *Nymphaea* protein sequence among different microbe and plant lineages. Preliminary analysis showed that 1,194 *N. colorata* genes do not appear in both algae and protist lineages, and are candidate HGT genes from microbial donors. The other 572 genes appearing in algae and/or protist are also potential HGT genes, which are all belonging to class C genes with three evaluating values equal to or higher than 30. In addition, the difference of determination of three values suggests that combining HGT AI, HGT Index, as well as HGT Identity can get a more comprehensive set of potential transferring genes.

T5
P1430
The mitochondrial genome of *Nymphaea colorata* is highly repetitive with active repeat-mediated homologous recombinations, and largest organellar introns

Shanshan Dong[1,2]*, *Chaoxian Zhao*[3], *Shouzhou Zhang*[1], *Hong Wu*[2], *Liangsheng Zhang*[4], *Yang Liu*[1]
1. Fairylake Botanical Garden, Shenzhen & Chinese Academy of Sciences, Shenzhen, China
2. College of Life Sciences, South China Agricultural University, Guangzhou, China
3. Department of Biology, School of Life Sciences, East China Normal University, Shanghai, China
4. Center for Genomics and Biotechnology, Fujian Agriculture and Forestry University, Fuzhou, China

Flowering plant (angiosperm) mitochondrial genomes are dynamic in genome structure. Mitogenome of the earliest angiosperm *Amborella* is remarkable for its carrying rampant foreign DNAs; in contrast to *Liriodendron*, the other only known early angiosperm mitogenome, described as fossilized. To better understand the characteristics of early angiosperm mitochondrial genomes, we assembled the complete mtDNA for *Nymphaea colorata* using PacBio long reads. The assembly received an average depth of 585×, and yields a circular chromosome of 617,195 bp. The genome encodes 41 protein coding genes, 20 tRNA and 3 rRNA genes. With 48.89% of the total length represented by repeated sequences, *Nymphaea* has one of the most repeat-rich mitogenome among angiosperms. Repeated sequences contributed substantially to its intron size, making *Nymphaea* possess one of the longest mitochondrial introns among angiosperms. It has been reported that repeats can result in mitogenome intragenomic recombination. Benefit from the long PacBio reads, we evaluated homologous recombinations among repeated sequences with at least 85% similarity within *Nymphaea*; altogether 1,187 repeats ranging from 50 to 3,293bp were examined. Recombined reads were detected from 913 repeats with an average recombination frequency of 5.97%. Our result revealed that smaller repeats tend to induce more recombinations, reflecting the importance of small repeats in evolution of angiosperm mitogenome structures. A total of 755 RNA editing sites were identified in *Nymphaeamt*DNA, and 84.24% of which are shared across *Amborella*, *Liriodendron* and *Nymphaea*; Extensive gene order changes were detected in the three mitogenomes, but 12 ancient collinear gene clusters were identified among them. *Although* many foreign DNA insertions were found from *Amborella* mitogenome, none was observed fromNymphaea.

T5
P1435
Plant grafting: How genetic exchange promotes vascular reconnection

Jing Wang[1], *Libo Jiang*[1], *Rongling Wu*[1,2]
1. Beijing Forestry University
2. The Pennsylvania State University

Grafting has been widely used to improve horticultural traits. It has also served increasingly as a tool to investigate the long-distance transport of molecules that is an essential part for key biological processes. Many studies have revealed the molecular mechanisms of graft-induced phenotypic variation in anatomy, morphology and production. Here, we review the phenomena and their underlying mechanisms by which macromolecules, including RNA, protein, and even DNA, are transported between scions and rootstocks via vascular tissues. We further propose a conceptual framework that characterizes and quantifies the driving mechanisms of scion-rootstock interactions toward vascular reconnection and regeneration.

T5
P1436
Synthetic microProteins as a tool to regulate flowering in *Oryza sativa*

Tenai Eguen, *Stephan Wenkel*
University of Copenhagen CPSC

Changes in the photoperiod of plants serve as an important environmental que for the transition from vegetative to reproductive state. Flowering in short-day plant, *Oryza sativa* is controlled by Hd1, a homolog of *Arabidopsis* flowering regulator CONSTANS. Under non-inductive long day conditions, Hd1 represses the transcription of *Ehd1* and florigens *Hd3a* and *RFT1*, thereby delaying flowering. In order to control flowering in *O. sativa* in long day conditions, synthetic microProteins were expressed in rice as an approach to regulate flowering time. MicroProteins are small proteins containing protein-protein interaction domains that negatively regulate larger, related proteins by engaging them in novel complexes which often alters their biological function. The synthetic microProtein rice plants show a conversion from short day rice plant to long day plant without compromising grain yield. This research will reveal a comprehensive approach involving the use of microProteins as biotechnological tools to control important traits in crop plants.

T5
P1439
Genetic epistasis interaction between *BrFLC1* and *BrFLC2* in regulating flowering time variation of *Brassica rapa*

Jian Wu
Institute of Vegetables and Flowers, Chinese Academy of Agricultural Sciences

Brassica rapa is an mesopolyploidy sepcies. FLOWERING LOCUS C (FLC), a repressor of flowering, encodes a MADS-domain transcription factor in *Arabidopsis*. FLC has four orthologs in *B. rapa* as a consequence of the whole genome triplication. The four copies of the FLC gene increase the potential variation in flowering time in *B. rapa*. While all the four copies have been reported to be functional in flowering time control, little is known on their interaction manners in flowering regulation. To analyze the interactions among the multiple BrFLCs, we developed a set of near-isogenic lines (NILs) of *B. rapa* carrying nine allele combinations of BrFLC1 and BrFLC2 by introgressing the functional allele (late allele) or non-functional allele (early allele) of BrFLC1 and BrFLC2 into the genetic background of recurrent parental line L58 (ssp. parachinensis). The flowering phenotype were mostly determined by the genotype of BrFLC2 and almost independ of genotype of BrFLC1. The NILs flowered significantly earlier as long as loss-of-functional allele of BrFLC2 exists, consistantly the NILs harbouring functional allele of BrFLC2 flowered latest. The phenotypic effects of BrFLC1 were visible only when genotype of BrFLC2 fixed. The results indicated that BrFLC2 has an epistasis effect on BrFLC1 in flowering time regulation of *B. rapa*. Transcriptional analysis showed among the four copies of FLC, BrFLC3 expressed on the highest level. BrFLC1 expressed higher than BrFLC2, and the expression of mutiple FLCs did not show dosage compansation when BrFLC1 or BrFLC2 or both of them loss-of-function.

T5

P1440

The non-brittle rachis of domesticated barley independently evolved by loss-of-function mutations in *BTR1* and *BTR2*

Goetz Hensel[1], *Mohammad Pourkheirandish*[2], *Nils Stein*[1], *Takao Komatsuda*[2], *Jochen Kumlehn*[1]

1. Leibniz Institute of Plant Genetics and Crop Plant Research (IPK)

2. National Institute of Agrobiological Sciences (NIAS)

Grain crops were essential for humans to transition from hunter-gathering to agrarian societies. Barley (*Hordeum vulgare* L.) was a founder crop in this process, and the most important step during its domestication was the occurrence and selection of plants that retained mature grains on their inflorescence and thus allowed effective harvesting. Classicial genetic studies have established that the non-brittle rachis of domesticated barleys is associated with a genotype of either btr1BTR2 (btr1-type) or BTR1btr2 (btr2-type), suggesting that loss-of-function mutations in either of these two closely linked genomic loci were sufficient to prevent the formation of constriction grooves at the rachis nodes. These disarticulation zones themselves are a major prerequisite for the free dispersal of mature grains in barley's wild ancestor *Hordeum spontaneum*. In this study, the functional verification of candidate genes of the two BTR loci was achieved through transgenic complementation of btr1-type barley using BTR1-candidate ORF1 driven by its native promoter and, likewise, by complementation of a btr2-type accession by BTR2-candidate ORF3 driven by the maize UBIQUITIN1 promoter. Both approaches independently resulted in the brittle rachis character, which co-segregated with the respective transgene in progeny of the primary transgenic plants. The functionally confirmed *BTR1* and *BTR2* genes share no significant similarity with one another at either the nucleic acid or the amino acid sequence level, which supports the hypothesis that they act complementarily. We further conclude that the non-brittle rachis type independently evolved more than once during barley domestication.

T5

P1441

The first illumina-based *de novo* transcriptome sequencing and analysis of sumac (*Toxicodendron vernicifluum* (Stokes) F. A. Barkley)

Guoqing Bai[1,2], *Yun Jia*[2], *Sifeng Li*[1]

1. Institute of botany of Shaanxi Province

2. College of Life Sciences, Northwest University

Toxicodendron vernicifluum (Stokes) F. A. Barkley is widely used as an economic plant due to its abundance of raw lacquer, however the limited transcriptomic and genomic data hinder research on the regulatory mechanisms underlying its raw lacquer formation. In this study, RNA sequencing technology was employed to generate the transcriptomic profiling of *T. vernicifluum*. A total number of 148,093,436 clean reads were obtained and used for *de novo* assembly by Trinity program, which produced 87,856 unigenes with an average length of 701.61 bp and an N50 length of 1206 bp. Through similar comparison with known protein databases, 46,522 (52.95%) unigenes were annotated in Nr, Swiss-Prot, Kyoto Encyclopedia of Genes and Genomes (KEGG), Cluster of Orthologous Groups (COG) and Gene Ontology (GO). Accordingly, potential simple sequence repeats (SSRs) and transcription factors (TFs) were identified in *T. vernicifluum*. Furthermore, we identified five candidate genes associated with the lacquer composition in three varieties. This study is the first report on transcriptome information in the *Toxicodendron* species and provides rich gene transcript resources for further studies on understanding the molecular basis of raw lacquer formation, and serving the genetic improvement and resource utilization in *Toxicodendron* plants.

T5

P1444

Field-based Phenomics platform: image-based methods for analyzing morphological traits and agronomic performance under different Nitrogen treatments for sweet potato

Yu Zhang

The University of Tokyo

Spatial variability in a crop field creates a need for precision agriculture. Early detection and management of problems associated with crop yield indicators can help increase yield and subsequent profit. Remote sensing can be used to assess spatial variability in crop yield. While highly predictable spatial variability may be amenable to multiple approaches of site-specific N management, strong temporal variability is much more difficult to manage. Remote sensing techniques that use the crop to indicate its N status show considerable promise for improving N management. Infrared and near-infrared (NIR) imaging systems are advantageous in their ability to visualize plant grown in field, but are currently limited by their small analysis volume and often low resolution and precision, as well as their cost, accessibility, and low-throughput. With further advancements, IR/NIR technologies have the greatest scale-up potential for the eventual non-destructive imaging and phenotyping of field-grown plant root systems. Although these current root growth systems and imaging technologies are still unable to accurately visualize and quantify complex, mature plant root systems grown under field conditions, they have contributed greatly to increase the precision and efficiency of 2D and 3D spatial. Management of fertilizer and watering in agriculture is very difficult. A part of N is lost in the atmosphere or leached into groundwater, lakes, and rivers, which causes severe environmental pollution. Despite that major crops can utilize only 30 – 40% of the applied N. In addition, plant response to each N source show stage specific difference and very sensitive. To solve this case of problem, development of remote sensing technology is useful to evaluate agronomic performance. In this study, we set up three different N application fields to make different yield performance with diverse genotypes of sweet potato as one simulation of different agronomic performance.

T5

P1445

3D Modeling and Estimating of Leaf and Stem Shape of Nursery *Paprika* Plants by a Novel Multi-camera Photography System

Yu Zhang

The University of Tokyo

In plant growth and phenotyping, accurate measurements of plant structure parameters are very crucial. Over the past three decades, 2D imaging has been applied to a variety of aspects, ranging from structural analysis, stomata movements and photosynthetic diagnosis, herbicide control, disease detection and yield estimation. However, as 2D imaging is not robust against the occlusion of plant organs, such as overlapping leaves and stems and changing shapes during the measurement, most of these applications are either for large-scale analysis or for simple plantvariables' measurements at the early growth stage. Recently, 3D imaging technologies for the robust measurement of plant growth and phenotype parameters have emerged and been applied for solving this problem owing to the advance of sensing technology. Among the most noticeable applications of 3D imaging technology in agriculture are the 3D model construction for plant and the properties estimation for plant growth. In this research, a novel solutionwas proposed to improve surface representation and construct 3D plant models through automated feature extraction and key point matching across images based on the Structure from Motion (SfM) algorithm. Our method could solve some problems or limitations of other methods, such as high cost, self-occlusion, low accuracy, information missing, noises, and so on. This study aimed at estimating and comparing plant parameters with different cameras and lenses for 3D imaging. We have, therefore, developed a high-efficiency multi-camera photography (MCP) systemcombining Multi View Stereovision (MVS) with Structure from Motion (SfM) algorithm. In this study, we measuredsix variables of nursery paprika plants and investigated the accuracy of 3D models reconstructed from photos taken by four lens types at four different positions. The results demonstrated that error between the estimated and measured values was small and the root mean square error (RMSE) for leaf width/length and stem height/diameter was 1.65mm ($R2=0.98$) and 0.57mm ($R2=0.99$), respectively. The accuracies of 3D model reconstruction of leaf and stem by 28 mm lens at the first and third camera positions were the highest and the percentages of the number of leaf and stem whose 3D model shape surface area reconstructed over 95% to real shape were about 47% and 100%, respectively. The results confirmed the feasibility of the proposed new methodology for the reconstruction of fine-scale plant model and accurate estimation of the plant parameters. It also displayed that our system is an ideal systemfor capturing high-resolution 3D images of nursery plants with high-efficiency.

T6

P1327

Ethnobotanical studies and floristic diversity assessment through the Geographical Information System (GIS) in Khonjerab National Park of Pakistan.

Qamar Abbas

Karakoram International University, gilgit, Pakistan

The Gilgit-Baltistan of Pakistan is famous in the world on the basis of its floristic biodiversity, massive glaciers out-side the polar region, glorious rivers, streams, forest, range lands, fabulous valleys, and high mountainous ranges. The area is consider as a home of world highest mountain ranges, i-e Junction of Karakoram, western Himalayas, and Hindukush. The Khonjerab National Park is geographically lies in the Karakoram ranges, extreme northeastern part of the Pakistan, at Pak-China border, north end of the Hunza and Nager district covers an area of 6150 km^2. The diversified climatic and topographic conditions, a large number of medicinal, aromatic and commercially important plant species are existing in this region. The frequent and detail field visits conducted in this area during May to September 2014-15 to explore the natural treasures of the area. The total 78 medicinally and economically important plant species were reported, belonging to 59 genera and 45 families. The adequate information about the biodiversity distribution evaluated through the geographical information system (GIS 10.2), and traditional uses of these plant species collected from the buffer zone communities of the national Park through questionnaires and interviews. This study elaborated the effective assessment of floral status and future threats for the biodiversity of the region. The major threats and challenges for biodiversity conservation and the Park management are overgrazing, road and buildings constructions, anthropogenic activities of the traders, and tourists.

T6

P1328

Quinoa (*Chenopodium quinoa* Willd.), A potential new crop for Egypt

Magdi T. Abdelhamid

National Research Centre

Quinoa (*Chenopodium quinoa* Willd.) is one of the main food crops of the Andean mountains, where it has been a staple crop for thousands of years. It is characterized by an outstanding protein quality and a high content of a range of vitamins and minerals, and for that reason it has been selected by FAO as one of the crops destined to offer food security in this century. Quinoa is grown in a wide range of environments in the South American region, however Quinoa requires temperate temperatures especially during flowering and for most genotypes short day-length to flower and produce seeds. Therefore Quinoa fits to the Egyptian winter climate. The genetic variability of quinoa is huge, which makes it possible to select, adapt and breed cultivars for growth under Egyptian conditions. Other very important properties of Quinoa for its successful development in Egypt are its good growth in poor soils, drought resistance, high water use efficiency and salt tolerance. The high nutritional quality and multiple uses in food products make Quinoa ideal for utilization by the food industry as well as for local consumption and even consumption at the farmers own house, thus providing a huge market. Furthermore there exist high price markets in Europe because of shortage of the high quality seeds that can be produced in the dry Egyptian climate. Developing Quinoa in Egypt require further dissemination of information among farmers and consumers, proper marketing and efficient post-harvest technologies, as well as continued development of better Quinoa varieties. Quinoa could be an important new crop for Egyptian agriculture, providing highly nutritive and versatile food products for the population, and a new raw material for the industry. In particular, it could be cultivated in many of the marginal environments afflicted by drought or salinity stress, which currently suffer from very low productivity.

T6

P1329

Ornamentals and energy crops for environmental and economic sustainability in Nigeria.

A.A.Warra Aliyu Ahmad Warra

Kebbi State University of Science and Technology, Aliero

Nigeria is faced with environmental erosion such as deforestation, water, soil and air pollution dumping of toxic waste and industrial pollution. Further, Nigeria is also witnessing periodic instances of floods, drying up of rivers and lakes drop in agricultural yield, loss of bio-diversity. All these problems have contributed to rapid negative effects to climate, sanitation and health. In this paper ornamental plants and energy crops of Nigeria, their role in environmental remediation and resource recovery are presented. These plants are used as biofences to protect crop fields from grazing animals and to prevent erosion. Ornamentals plants haves gained recognition in the developing of mega/smart cities of Nigeria. They are used for decorative purposes in landscaping; they beautify the home and places such as streets in general public places and institutions. They contribute in the environmental and socio-economic development. Cultivation of energy crops have found wide application for economic sustainability and rehabilitation of environmentally constrained lands. In some ancient and modern communities, respondents claimed that ornamental plants arrest dust, supplies oxygen and help as wind break. These plants produce value added non-edible products and are traditionally used as cosmetics for e.g. manufacturing of soap. Researchers suggest that the detoxification or complete removal of the toxicity is essential before their use in industry or medicinal applications. Products from *Jatropha curcas, Ricinus communis*, *Thevetia peruviana*, *Terminalia mantaly* and *Polyalthia longifolia* and their environmental and economic applications presented in the paper. This would widen the market scope in agri-biobusiness for value chain and value additions.

T6

P1330

Biosynthesis, characterization and evaluation of some medicinal activities of *Cinnamomum iners* bark extract mediated biocompatible silver nanoparticles

Geo Vigila Antony Varuvel[1], *Sahayaraj Kitherian*[1], *Jansi M*[2]

1. St. Xavier's College (Autonomous), Palayamkottai, Tirunelveli, Tamil Nadu

2. ST Hindu College, Kottar, Nagercoil, Kanayakumari district, Tamil Nadu

Cinnamomum (Lauraceace) is an economically important tree which includes 250 species distributed in Asia and Australia. Cinnamon is used as flavouring agent in teas, bakery products and an important ingredient in several Indian cuisines. There has been a great interest in recent years on biosynthesis of nanoparticles using plant extracts for their bioactivity due to less toxicity. In our present investigation, we have synthesized the silver nanoparticles (Ag-NPs) using *Cinnamomum iners* bark extract, *in vitro* antioxidant and antimicrobial activities have been evaluated. To biosynthesize Ag-NPs, the reaction mixture contained aqueous bark extract of *C.iners* and an aqueous solution of $AgNO_3$, which was initially colourless, turned into reddish-brown, indicating the biosynthesis of Ag-NPs. Their characteristic surface plasmon absorption was observed at 429 nm. Biosynthesized Ag-NPs was characterized using UV-visible spectroscopy, Fourier transform infrared spectroscopy, X-ray diffraction (XRD), zeta potential, scanning electron microscopy, and high-resolution transmission electron microscopy (HRTEM). The *in vitro* antioxidant assays such as 2,2-diphenyl-1-picrylhydrazyl (DPPH), 2,2'-azino-bis (3-ethylbenzothiazoline-6-sulphonic acid) (ABTS), Ferric Reducing Antioxidant Power (FRAP) and superoxide scavenging activities were performed using standard procedures. The XRD pattern depicts the prominent peaks at 35.77°, 39.17°, 52.71°, and 54.82° which corresponding to (100), (102), (104) and (006) reflections, respectively. Various shapes spherical, rod, hexagonal of Ag-NPs were identified using HRTEM. Among various solvents, ethanol bark extract exhibited higher total phenolics (545.94±11.54 mg GAE/g extract), flavonoids (723.33±6.40 mg RE/g extract) and tannins (44.11±8.40 mg GAE/g extract). Significant ABTS free radical scavenging activity was observed in ethanol bark extract (1092.81±2.94 µM TEAC/g extract) of *C. iners* whereas, 1015.19±3.57,739.80±5.40 and 472.50±23.88 µM TEAC/g extract) of lower activity were displayed in acetone, Ag-NPs and water extracts, respectively. Higher ferric reducing activity was exhibited in ethanol bark extract (5840.05 mM Fe (II) E/g extract) while 5642.60, 2896.67 and 1731.07 mM Fe (II) E/g extract) were displayed in acetone, water and Ag-NPs, correspondingly. Higher superoxide (54.87%) and DPPH (3.25 IC_{50}) radical scavenging activities were observed in biosynthesized Ag-NPs. In disc diffusion method, biosynthesized Ag-NPs (10mg/ml) exhibited higher antibacterial activity against *Escherichia coli* (1.33±0.03cm), *Staphylococcus aureus* (1.06±0.03cm), *Serratia marcescens* (0.83±0.03cm), *Klebsilla pneumoniae* (1.06±0.03cm), *Pseudomonas aeruginosa* (1.53±0.03cm) and antifungal activity against *Candida albicans* (1.46±0.06cm), and *Aspergillus fumigatus* (1.23±0.03cm), respectively. The ethanol bark extract of *C.iners* displayed only least activity against *E. coli* (0.86±0.03cm),

S. aureus (1.00±0.00cm), *K. pneumoniae* (0.83±0.03cm). Due to significant bioactivity of Ag-NPs, this may be effectively used as a natural plant-based drug and their mechanism of action should be investigated in further studies.

T6

P1331

Taxonomic revision of the *Ficus* section Malvanthera (Moraceae) in New Guinea.

Billy Bau Bieso
New Guinea Binatang Research Center

There are over 800 *Ficus* species occurring as hemi-epiphytes, terrestrial trees, shrubs and creepers in tropical rainforests and in sub-tropical region. The genus is classified based on their morphological characters into six subgenera; *Ficus, Synoecia, Sycidium, Sycomorus, Pharmacosycea* and *Urostigma*. Past studies in New Guinea still showed inconsistencies in the section Malvanthera (*Urostigma*) and this study is important to define the species concepts within this taxa. The main objectives are to revise taxonomic concepts within *Ficus* section Malvanthera in New Guinea, and to re-describe all species with synonyms using original type images and circumscribe potential new species, based on morphological and genetic characters. Fieldwork had included making botanical collections and involved studying its ecological features and gathering of proper botanical specimens with at least ten duplicates which had already been distributed to other overseas herbaria. Laboratory compilation of 605 digital images of 202 herbarium specimens from BRC, LAE, BRI, CANB, SING, K, and BM had their key distinguishing characters measured digitally using the *ImageJ* software. Preliminary results included 22 fertile botanical collections from the PNG LNG Vegetation Monitoring project and Mt. Wilhelm altitudinal *Ficus* plots near Bundi Station, in the Madang Province of Papua New Guinea. In addition, 605 digital images had 195 duplicates with 108 of their duplicate images have already been databased into the Malvanthera Access Database. A quarter of these images have already been measured digitally using the *ImageJ* software and tested using the Statistical software package *Canoco v.5*. It is highly likely from the field collection samples that one set of botanical collection from the Bundi area along the Mt. Wilhelm transect showed significant morphological variations to the others that have already been described.

T6

P1332

A review of the phytochemistry and ethnopharmacology of *Tridax procumbens*

Samantha Beck, *Olga Kopp*
Utah Valley University

Originating in Central and South America, *Tridax procumbens*, a member of the Asteraceae family has great ethnobotanical potential. It is a perennial plant with a capitulum inflorescence that results in abundant production of achenes, making it a noxious weed if not controlled. The purpose of this paper is to review the traditional uses of *T. procumbens*, along with the documented biological activity in order to highlight the importance of continuing research on this species. An in-depth review of the literature of this species was done using different sources, including ScienceDirect, Web of Science, Nature, PubMed and other databases. *T. procumbens* produces secondary metabolites that have been reported to have a variety of medicinal uses such as: anti-anemic, anti-inflammatory, anesthetic, pain relief, antimicrobial, antioxidant, anticancer, and immunomodulating properties. Different methods of extraction were compiled to show the wide range of various *T. procumbens* extracts in a variety of biochemical research. This paper shows that more exploration is needed for the potential of *T. procumbens*' secondary metabolites as medicine or preventive treatment, making it a promising ethnobotanical resource.

T6

P1333

The application research of ferns on slope restoration

Jingru Cai, *Jianxin Xu, Wenzhu Liu, Yanhui Shen, Tanghuang Qian*
Shenzhen Techand Ecology & Environment Co., Ltd.

The Herb plants are always used as pioneer species on slope restoration, and which always combined with seeds of shrubs and trees in hydroseeding for reforestation quickly. But Herbs would decay when upper levels'crown density increase as time goes by. While ferns can grow well in the lower layer as shade-tolerance plants group, which is good choice to fill up the undergrowth vacancy. Researchers of Shenzhen Techand Ecology&Environment Co., Ltd have been working for vegetation diversity and stability of slope many years. In recent years nine kinds of good slope growing ferns were collected from south-China and west-China areas such as *Nephrolepis auriculata, Nephrolepis biserrata, Cyclosorus parasiticus, Blechnum orientale, Pteris fauriei, Pteris cretica* var. *nervosa, Woodwardia unigemmata, Pteris vittata, Macrothelypteris torresiana*. Incomplete culture technique and artificial seed technology were used for propagation of ferns, as well ferns transplanting and vegetation carpet were researched for practicability and application way on slope restoration. Results were as follow: (1) The general breeding system of ferns was established. Aseptic prothalli of ferns were obtained by spore induction, and of which proliferation rate was 4-5; The optimal combination of artificial seed coating for inhibiting bacteria and promoting germination was screened out by orthogonal experiment, and artificial seeds could germinate and grow normally. (2) Ferns transplants had a high survival rate on slope, and dued to slow growth, transplants had little effect on growth of shrubs and herbs at the beginning of spray-seeding. From the density point of view, it is appropriate to plant *Cyclosorus parasiticus, Pteris Fauriei* and *Macrothelypteris torresiana* with 40cm*40cm density, while *Nephrolepis biserrata* reproduces by runner tips and crown diameter of *Blechnum orientale* seedlings is more than 100 cm, which can be configured by point planting. (3) Ferns' roots could pass through coconutfibre and grow well, so that ferns vegetation carpet was workable. Aseptic prothalli of ferns could grow after sowing 18 days on coconutfibre, and roots of which could grow into it with juvenile sporophyte growed after 48 days; while using 2-months seedlings as planting materials, ferns'roots could pass through coconutfibre after 48 days. The suitable thickness of coconutfibre is 0.5 cm considering uniformity and weight. After being spread on slope, ferns' vegetation carpet should keep enough water within 30 days for ferns'roots growing into soil.

T6

P1334

Biological characteristics and propogative technique of *Euonymus Grandiflorus*

Yan Chen[1,2], *Yi Sun*[1,2]

1. Beijing Botanical Garden

2. Beijing Floriculture Engineering Technology Research Centre

Euonymus grandiflorus is small arbor of *Euonymus* L. in Celastraceae, which is distributed mainly in northwest and southwest area of China. *Euonymus grandiflorus* with better resistance, bright-colored and longtime autumn leaves is optimum color-leafed plant while seldom applied in urban landscape. Biological characteristics of *Euonymus grandiflorus* had been observed. The result showed that phenological periods including sprouted stage and leaf expansion stage of *Euonymus grandiflorus* in different years had the difference, while blooming stage and leaf color change were similar, early blooming stage was in the middle of June, late blooming stage was in the middle of July, the whole flower period lasted nearly one month; leaf color change started on 10th October, defoliation stage was close to late November, leaf color period from early October to late November reached more than 50 days, during that period the leaf cholorophyll content decreased while anthocyanin content rised sharply, which caused the bright amaranthine leaf. The ornamental value of *Euonymus grandiflorus* was much better than other color-leafed plants. The different methods of propogation, including sowing seeds from different provenances, cutting, grafting and tissue culture at different time were tested. The optimum way of rapid propogation on *Euonymus grandiflorus* in Beijing area was grafting before branch burgeoning in spring, stock should be 2 or 3 years old seedling of *Euonymus bungei* and scion should be one year old healthy branches of *Euonymus grandiflorus*. Cultivation technique and application of *Euonymus grandiflorus* in Beijing area were studied and discussed. The conclusion indicated that *Euonymus grandiflorus* was suitable to grow in the place with full sunlight, well-drained and loosen soil. It was better to apply *Euonymus grandiflorus* in block or group planting methods, matched with tall color-leafed arbors, evergreen trees and some low color-leafed shrubs in order to increase landscape level, enrich landscape color and extend the ornamental period.

T6

P1335

Studies of leaf functional traits and foliar vasculature for montane and subalpine woody plants of Yushan National Park in Taiwan, R. O. C.

Shau-Ting Chiu[1], *Wei Cheng*[2]

1. Botanical Garden and Division of Biology, National Museum of Natural Science, Taichung, Taiwan, R.O.C.

2. Department of Biology, National Chung Hsing University

"Functional traits" capture fundamental tradeoffs that determine species' ecological roles. Establishing morpho-physiophenological traits which impact fitness indirectly via their effects on growth, reproduction and survival, the three components of individual performance can become the foundation of ecological forensics. This study initiated leaf functional traits about Ericaceae in Nanhsi area along altitudinal gradients from Nanhsi Forest Dynamics Plot (Nanhsi FDP) to the subalpine Mount. Yushan. Carrying on the standard protocol of leaf functional traits can input, provide and accumulate long-term ecological forensics data and renewal conservation information for National Park's management and long-term ecological research. Sampling area from 1,800 m to 3,900 m above sea level in Nanhsi area, 1,210 leaf samples were collected from 242 shoots for 178 individuals of 9 species and 5 genera in Ericaceae. Laminar density of the sampling rhododendrons was usually low toward the high altitude. Leaf dry matter content (LDMA) of most rhododendrons was between 0.3 and 0.5. Only those LDMA values of *Gaultheria itoana* Hayata and *Vaccinium randaiense* Hayata increased along their enlarging leaf areas. Leaf thickness inclined along the LDMA but declined in specific leaf area (SLA), except *Lyonia ovalifolia* (Wall.) Drude. The smallest closed areoles in leaf vasculature occurred in *Rhododendron morii* Hayata and the largest ones occurred in *R. rubropilosum* Hayata. Short terminal veins with obvious bundle sheath of *R. morii* displayed probably the C4 photosynthetic pathway. On the contrary, the bundle sheath of *R. rubropilosum* was not dominant. Results are the first time to show the leaf functional traits applied to the same rhododendron species in different habitats, sun- or shade-leaves of the same individuals and the adaptation to special montane ecosystem. Nanhsi FDP with the largest plot area at the middle elevation of Taiwan had diversed of plant functional groups, including *Castanopsis carlesii* - *Machilus japonica* evergreen and *Alnus formosana* deciduous broad-leaved forests. Among leaf traits of four functional groups of leaf habits and growth forms, the LCC of tree growth form was significantly higher than that of shrub. Evergreen trees had the highest LDMC and LD. However, deciduous shrubs had the lowest. Deciduous trees had higher LNC, SLA and lower C:N than evergreen trees. Temperature, humidity, illuminance, and canopy openness were significantly different between deciduous forest and evergreen broad-leaved forest. LNC, C:N and LDMC significantly correlated with canopy openness. Although the elevation was the most identical environmental factors for species distribution and leaf functional traits by canonical correspondence analysis and redundancy analysis, the elevation actually reflected microclimate trend of the temperature gradient. Microclimate effect was indeed the key environmental factors. Due to leaf lifespan, leaf size, vein density and other leaf traits of different species had different invested proportion between the main and the fine veins. Leaf functional traits reflected varied plant survival strategy for adaptation. Deciduous and evergreen broad-leaved forests with different species composition display that the main effect on leaf functional traits was at the species level. The varied plant community implied the habitats of fitness that had close interaction with leaf functional traits. It would fulfill the important mission of long-term monitoring and also shows the efforts and contribution of Botanical Garden and National Park in conservation.

T6

P1336

Searching for the *Podaxis* fungus on the trails of early explorers in Southern Africa with the help of rural communities

Zw De Beer[1], *B Conlon*[2], *Hh De Fine Licht*[3], *Sme Vreeburg*[4], *Dk Aanen*[4], *M Poulsen*[5]

1. Forestry and Agricultural Biotechnology Institute (FABI), University of Pretoria, South Africa
2. Molecular Ecology, Institute of Biology/Zoology, Martin-Luther-University Halle-Wittenberg, Halle an der Saale, Germany
3. Section for Organismal Biology, Department of Plant and Environmental Sciences, University of Copenhagen, Thorvaldsensvej 40,1871 Frederiksberg C
4. Laboratory of Genetics, Plant Sciences Group, Wageningen University, Droevendaalsesteeg 1,6708 PB, Wageningen, Netherlands
5. Centre for Social Evolution, Section for Ecology and Evolution, Department of Biology, University of Copenhagen

Podaxis pistillaris (Agaricales, Basidiomycota) is the name often given to the conspicuous, torpedo-shaped fungal fruiting bodies that form during the rainy season on termite mounds across southern Africa. The species was described initially in 1,871 in the genus *Lycoperdon* by Linnaeus based on a specimen from India. In 1881, Linnaeus described a second species, *L. carcinomale*, from a specimen sent to him by Carl Thunberg who collected it during his travels in South Africa. In 1812, William Burchell made a painting of the fungus during his exploration of Southern Africa, of which a black-and-white picture was printed in Doidge's 1954 encyclopaedic lists of all South African fungi. By 1,933,33 *Podaxis* species have been described from Africa, Asia, Australia and the USA, but Morse, an American mycologist, lumped all as synonyms of *P. pistillaris*, and although another 12 species, including two from South Africa, were subsequently described, most authors conveniently referred to all these fungi as *P. pistillaris.* Despite its frequent association with termite mounds, virtually nothing is understood of the biology of this fungus. In a quest to better understand its association with termites and to resolve the taxonomy of the fungus, we studied a collection of Southern African specimens from the National and Schweickerdt herbaria in Pretoria, and we had access to several specimens from the USA, Mexico, India, and Africa, including the type specimens of the two species described by Linnaeus. In addition, we collected fresh specimens from several locations. We visited the sites where Thunberg collected his specimen in the Western Cape Province, and where Burchell made his painting in the Northern Cape Province, but we could not find fresh specimens at these sites that could assist in resolving the taxonomy of these species, as these areas experienced a severe drought during the preceding year. At least we could confirm that the original collections were probably from *Trinervitermes* (*Termitidae, Nasutitermitinae*) termites. However, we distributed flyers to local farming communities in both the Western and Northern Cape with photographs of the fungus and requests to contact us should these be observed. Good summer rains fell soon after our visit and we received fresh specimens within six weeks after distributing the flyers. One of these specimens came from a farm close by the camping site where Burchell made his painting. DNA was extracted and the ribosomal LSU and ITS regions successfully sequenced from all the fresh and almost all herbarium specimens, including some that was older than 100 years. Phylogenetic analyses showed that the southern African specimens forms at least five well-supported lineages that probably represent distinct species. Four of these lineages seem to be associated with termite mounds, while the fifth seems to be free-living on open ground. Our study not only provided the first ever phylogeny for the genus, but emphasized the value of herbarium specimens for taxonomic and ecological studies. The one fresh specimen collected by a farmer from the Burchell site will serve as type for a new species resembling Burchell's painting.

T6
P1337

Cape citizen science: Public engagement for plant disease research in a biodiversity hotspot

*Jm Hulbert, **Zw De Beer**, Mj Wingfield*
Forestry and Agricultural Biotechnology Institute (FABI), University of Pretoria, Pretoria

The Cape Floristic Region of South Africa is an important biodiversity hotspot. It is important to conserve plant communities in this region for many reasons including their exceptional endemism. Many species in these communities are also threatened with extinction. Cape Citizen Science is an initiative to involve the public in research to assess the threat of plant-killing *Phytophthora* species. The initiative has pioneered many methods of public engagement in plant disease research. These include a targeted sampling campaign and various environmental education activities that could be implemented in other communities threatened with plant disease. Public contributions have broadened the collection of samples included in the research and reports of dying plants have generated new areas of interest. Cape Citizen Science has demonstrated that citizens can be invaluable for plant disease research and that the act of research can serve as a platform for public engagement and environmental education.

T6
P1338

Using big data to unleash the drug discovery potential of ethnobotany

John De La Parra
Northeastern University

A large database, covering thousands of years of human-plant interactions and containing culturally and geographically diverse ethnobotanical information on the selection and use of medicinal plants for neglected protozoan diseases has been created and mined for provable drug discovery leads. Candidate species were prioritized and subsequent phytochemical characterization and bioassays were completed. This study's large-scale accounting for indigenous plant knowledge indicates that medicinal plants are regularly chosen because of specific environmental conditions. Additionally, plants subjected to attack by insect herbivores are often observed to have higher levels of more diverse specialized metabolites, many with unique bioactivities. By querying the database, an initial test candidate species was chosen. *Pentalinon luteum*, a traditional Seminole medicinal plant chosen with careful attention to environmental conditions, regularly undergoes heavy seasonal herbivory by the larvae of the Oleander moth, Syntomeida epilais. These larval predators consume and store unique cardiac glycosides that are produced by their plant host. While these compounds are highly toxic to their predators, as the ethnobotanical database in this study shows, plants with high cardiac glycoside content have had a long history of medicinal value to humans.

In addition, a very closely related raditional medicinal species (*Pentalinon andreuxii*) mined from the database, was shown to be effective against the neglected protozoan disease Leishmaniasis. To control for this variability of indigenous selection and herbivory-induced stress responses, in vitro tissue cultures have been created to investigate the effects of the addition of stress-signaling hormones as correlated to indigenous plant selection. By scanning a large dataset of available ethnobotanical information, a robust study of the relatively unexplored aspects of how humans choose medicinal plants is presented.

T6

P1339

Allelopathic effect of the weed *Pennisetum polystachion* (L.) Schult. against the commonly cultivated pulses

Archana E R, *John E. Thoppil*
University of Calicut

Allelopathy is the direct or indirect effect of a plant on other plants of its vicinity through the production of phytochemicals that are released into the environment. *Pennisetum polystachion* (L.) Schult., commonly known as mission grass is a vigorous annual or perennial weed in tropical region, that are capable of spreading along roadsides into both agricultural and natural habitats by producing large numbers of seeds. The present study was designed to reveal the phytoconstituents in the plant extract and to investigate its allelopathic effect. 500 g/500 mL stock was prepared. Various concentrations of 5%, 25%, 50%, 75% and 100% were taken and distilled water was kept as the control. The seeds of three commonly cultivated pulses were used for the study *viz. Phaseolus, Cicer* and *Vigna*. The effect of the plant extract on the germination of the pulses were studied by noting various parameters like number of seeds germinated, length of the root, fresh weight and dry weight of the seedlings. The germination and growth rate of the cereal seeds were found to be directly proportional to the concentration of the weed extract. The germination was found to be 83%, 57% and 90% in *Phaseolus, Cicer* and *Vigna* respectively, in the lowest concentration while it was found to be 53%, 23% and 23% in the pure extract. In contrary, 96%, 60% and 93% of germination was seen in the distilled water control. Root length was measured on the day 2 which was found to be in the ranges of 7.3 cm-2.9 cm in *Phaseolus*, 3.54 cm-1.56 cm in *Cicer* and 11.7 cm-7.78 cm in *Vigna*, according to the increasing concentration. Average root length in the control seeds were obtained as 7.78 cm, 4.43 cm and 11.84 cm, respectively. Similar gradation was obtained in the fresh weight and dry weight of the seedlings. The fresh weights showed a range of 2.48 g-1.47 g, 6.75 g-5.8 g and 3.87 g-3.47 g; dry weight showed the range of 0.413 g-0.328 g, 2.99 g-2.64 g and 0.72 g-0.64 g in *Phaseolus, Cicer* and *Vigna*, respectively. The control showed fresh weights of 2.64 g, 6.95 g and 4.43 g while dry weights of the same were found to be 0.453 g, 3.09 g and 0.74 g. The allelopathic effect of the aqueous extract of the weed *Pennisetum polystachion* is clearly evident from the above results. The plant also shows considerable presence of the secondary metabolites like phenols, flavanoids, terpenoids, alkaloids and other phytochemicals. This might be the reason for the allelopathic effect of the plant. Thus, it is important to eliminate or control the vigorous spreading of the weed in to the agricultural lands.

T6

P1340

Journal Impact Factor and evaluation of botanical researchers in Latin America

Alina Freire-Fierro
IKIAM Universidad Regional Amazónica/Asociación Latinoamericana de Botánica-ALB

Although originally created to evaluate journals, the Journal Impact Factor (JIF) has nowadays become one of the main factors to evaluate productivity and impact of scientific researchers worldwide. These bibliometrics, though understandable applied to journals published in English, are completely unfair to non-English journals. This situation is worse in Latin America, where the tradition of research/publishing high quality papers, even in Spanish, is relatively new. Additionally, and while for other sciences the impact of journals can be measured in a time span of 2, 3, 20 years, in botanical research many articles/works published more than 200 years ago are still being cited. Adding to these problems, the evaluation of impact factor of journals is made primarily in English language publications, as it has been reported for example, that out of 100 Panamian publications, none had been included in the evaluation of the recently modified science index, the Science Expanded Index (SCI-Expanded). While even in English-speaking countries, many institutions have agreed not to use the Journal Impact Factor as a measurement for a researcher's productivity or impact (e.g., signage of the DORA agreement in the US) or have used other methodologies for this evaluation (e.g., the Research Excellency Framework REF2014 in the UK); in Latin America, many governments/institutions are still using it. As examples, JIF is being used for evaluating researchers by Mexican National System of Researchers (SNI), Colombian Colciencias, and Ecuadorian National Registry of Researchers (RNI). And in Perú for example, JIF and H indices count up to 35% of the total score for a researcher's categorization in the National Directorate of Researchers and Innovators (DINA). This unfair evaluation from Latin American governments and institutions towards researchers' productivity, really hurt scientific development in the region. Since very good and experienced Latin American researchers want to keep their high academic status, they will continue to publish in English, giving therefore a positive feedback for journals published in English in detriment of journals published in Spanish. As recent as of December 2015, Clarivec Analytics, the current owner of former Thompson Reuters' Institute for Scientific Information (ISI), creator of the IJF, has developed a new index that does not consider the Journal Impact Factor, the Emerging Sources Citation Index, in order to solve this situation. Still though, one of the requirements for journals to be evaluated by this new Index is that the articles in these journals have to have English cited references. Taking into consideration the stated above, I consider and encourage to the botanical community and to scientific organizations to sign the DORA agreement, and to follow the methodologies used by REF2014, but modified to reflect Latin America's scientific research reality. Additionally, and as a scientific community, we consider that it is our duty to request the Latin American governments to stop using the JIF as one of the most important elements in their evaluations of researchers in the region.

T6
P1341
Practical botanical activities: The use of images for teacher aid

Bruno Takami Fujiwara, *Hector Barros Gomes, Carolina Patricia Costa Soares Dos Santos, Eliana Akie Simabukuro*
Universidade Federal de São Carlos

The use of practical activities as teaching resource is an experience of scientific investigation and discoveries, relevant to approximate the students of it study object and to mean abstract concepts. The teaching of botany presents these abstract conceptualizations, mainly in the area of vegetal physiology due to the absence of the plants in the students' daily life. In order to facilitate the process of development of practical activities and to bring students closer to botany, this work developed practical activities botany protocols with the additional of having comments and images of the experiments to assist the teacher in the development of the class, facilitating the preparation and execution of the activity. The experiments developed morphological, anatomical and physiological analysis of the plants at different stages of development, with different species, to demonstrate the existing biodiversity. All the practices were previously elaborated to obtain the images and to allow suggesting more than one option for the development of the activity, in order to ensure greater success in the development of the teacher's class. The experiments resulted in eight protocols with the following themes: "Fruit evaluation", "Seed germination", "Photomorphogenesis", "Asexual reproduction", "Geotropism", "Effect of luminosity", "Effect of substrate" and " Effect of density ", which respectively try to analyze the variety of fruits as to the shape, color, layers, edible parts, quantity and arrangement of the seeds; Germination of different types of seeds, with varied germination rates; Germination and development of seedlings under conditions of incidence and absence of light; Reproduction of several species by cutting; Development of plants on the influence of gravity; Effect caused in the development of seedlings submitted to different light incidences; Effect caused on the development of seedlings in different types of substrates; And effect caused on the development of seedlings grouped under different density conditions of individuals. The protocols cover botanical contents (morphology, anatomy and physiology) and ecology that pass through all the years of elementary school 2 and high school, and can be worked more than once in different years, with different approaches and deepening, without the need following a pre-established sequence. In this way, these activities allow teachers and students to work on some broader themes using plants as examples, rather than keeping the picture just in animal diversity. Thus, the teacher has several options for differentiated activities to insert in his planning, while continuing to work on content suggested by the National Curricular Parameters of Science and Biology, established by the Ministry of Education (Brazil) and the State of São Paulo Education Department.

T6
P1342
Tracking the long-term dynamics of mountain biodiversity with a network of citizens and rangers. Winners, losers, and steady organisms

Maria Garcia, *Iker Pardo, Pablo Tejero, Jose Luis Silva, Daniel Gomez*
Spanish National Research Council (CSIC)

Mountains shelter the highest biodiversity in temperate ecosystems. While many species are common or shared by different mountain ranges, others are endemic, rare, at the limit of distribution, or trapped on top of high summits. But we do not have many evidences on the real and precise effect of climate change and land use change at the species level. Such knowledge constitutes a challenge that can only be undertaken with a large network of people. Four years ago, with the support of a European LIFE project, we launched a program called "adopt a plant" to monitor the density and abundance of populations of many kinds of different plant species in the NE of Spain, an area covering from the semi-desert till alpine summits. The key of the program is the personal training of volunteers and rangers in the field, after agreeing on a simple but scientifically robust design. The network is now made of 150 people from 21-76 years old, and is fast expanding. Since they get the commit of visiting the adopted plant for the next 10 years, they are providing us with the most accurate information on the demographic performance of more than 250 populations of 150 plant species, representing all kinds of life histories (short- to very long-lived herbs and shrubs), habitats (forests, wetlands, grasslands, rocky areas..) and altitudes. This information is key to determine population trends, and forecast extinction risks in the short and mid-term, as well as to test the effect of temporal fluctuations and natural perturbations. Small devices recording the precise temperature at the populations will help us to understand sensibility to climatic fluctuations, as well as information on how the particular habitat has changed over the last 60 years at each monitored population. As far as we know, this is a unique project disentangling the effect of climate and land use change on different kind of species, and demonstrating the value of coordinating researchers, administrative staff, and citizen science.

T6
P1343
The novel nortriterpenoid glycosides from *Salicornia bigelovii* and their effects on STAT3 signaling pathway

Fuqin Guan, *Qizhi Wang, Yu Shan, Yu Chen, Min Yin, Fei Liu, Ming Wang, Youyi Zhao, Xu Feng*
Institute of Botany, Jiangsu Province and Chinese Academy of Sciences

Salicornia L. (Chenopodiaceae) is a genus of annual apparently leafless halophytic herbs that have articulated, succulent stems. *Salicornia bigelovii* was evaluated as an oil seed crop and seasoned vegetable with direct seawater irrigation. Although the economic value of this plant has already been discovered and developed, its medicinal value has been ignored for a long time in China. Investigation of characteristic constituents of *Salicornia bigelovii* led to isolation of thirty-five compounds, including seven 30-nortriterpenoid glycosides and nine oleanane-type triterpenoid glycosides, with four of them new. Then twenty-one triterpenoid glycosides, isolated from *Salicornia bigelovii* and *Salicornia europaea*, were evaluated for their cytotoxicity and their effect on STAT phosphorylation. Eight compounds showed cytotoxicity against breast carcinoma (MCF-7, MDA-MB-231,

MDA-MB-468), liver carcinoma (HepG2) and lung carcinoma (A549), with IC_{50} values of 10-70 μM. A new 30-nortriterpenoid glycosides, named Bigelovii A, inhibited interleukin-6 inducible STAT3 and STAT5 activation in dose-dependent and time-dependent fashions. Treatment with the protein tyrosine phosphatase (PTP) inhibitor pervanadate reversed the Bigelovii A-induced down-regulation of STAT3, suggesting the critical role of both SHP-1 and SHP-2 in its possible mechanism of action. Bigelovii A also downregulated the expression of STAT3-regulated gene products that mediate apoptosis, cell proliferation, and invasion. These results suggested that the novel nortriterpenoid glycoside from *Salicornia bigelovii* may impart health benefits when consumed, and should be regarded as potential chemopreventative agents for cancer.

T6

P1344

Micropropagation as an alternative to the growth of *Dionaea muscipula*.

José Antonio Guzmán-Nava
Ecovoraz S. de R.L de C.V

In this study we develop a protocol was developed on for the micropropagation of *Dionaea muscipula*. Dionea is a monotypic genus of carnivorous plant in the family Droseraceae, *Dionaea muscipula*. The only species of *Dionaea muscipula*, of whose common name is Dionea flytrap or Venus flytrap, that catches live prey. Due to its popularity, the demand for these plants is increasing, however, growing large quantities of these plants is difficult. They are reputed to be difficult to grow. Thus, the cultivation of plant tissues by means of the micropropagation technique could be seen to be the way in which a larger number of individuals could be obtained in a shorter time and without having seeds. Ecovoraz has given itself to the task of growing these plants by means of of being able to make a medium of culture, for the invitro growth of *Dionaea muscipula*, obtaining very favorable results. Plants from the market of Madre Selva Xochimilco city of Mexico at Mexico City were used. They are given a method of disinfection disinfected under greenhouse conditions with quaternary ANIBAC salts. Subsequently, the Murashige and Sook medium was were used as the base for the concentration of micronutrients and macronutrients. Stock solutions were then used and were used at a concentration of 25%, 50%, 75% and 100%. The culture medium was supplemented with Indol 3 butyric acid 1.24 and 6-bencial aminopuria 1.5 mg. Our results indicate that forAs a result in which the for a salts concentration of 25%, we obtained on average concentration of salts at 25% averaged 18 shoots, the concentration of for 50% was obtained on average 32 shoots, while at nutrient concentration of 75% an average of 46 outbreaks were developed, and in at the 100% concentration an average of 26 shoots were obtained. This implies that, in order to obtain the greater. For which protocol to establish a greater number of outbreaks, we require is the concentration of 75% concentration of MS salts, adding plus 1.5 mg of 6-benzylaminopurine and 1.24 mg of indole-3-butyric acid.

T6

P1345

Leaf microscopic features of some local vegetables found in Nong Khai province, Thailand

***Piriya Hassa**, Paweena Traiperm*
Department of Plant Science, Faculty of Science, Mahidol University

Local vegetables are plant grown for the consumption in local areas such as field, orchard or nearby house, which locals can easily harvest and reduce costs of production. Local people, especially the Northeastern Thai people, routinely consume various local vegetables for nutrition and health maintenance. In Nong Khai, a northeastern province of Thailand, people have received a number of food cultures from different races, including Thai, Laos and Vietnamese. Thus, Nong Khai province is suitable to study the local vegetables consumption. In forensic science, plant remains normally appear as evidence in a crime scene. For example, incompletely digested food in a stomach of dead bodies can be promising evidence to approximate time after the food had been eaten before death and to predict the locations where the food was prepared. For some cases, leaf fragments can facilitate the process of case conclusions by comparing the information with a testimony by a witness in that the specific origin of victims can be identified. However, the information about leaf anatomy of Thai local vegetables have been scarcely published because of the locally specific consumption of plants. Therefore, leaf epidermal peeling and leaf transverse sectioning methods were applied to examine leaf microscopic features of 17 species of local vegetable found in Nong Khai in order to find specific characters of each species for identification criteria. Stomatal pattern from leaf epidermal peeling on the adaxial surface consisted of anomocytic, tetracyclic, paracytic, and anisocytic types. While the abaxial surface exhibited anomocytic, actinocytic, diacytic, and cyclocytic patterns. Anticlinal cell walls vary from straight, straight to curve, and wavy. Three types of indumentum: unicellular trichomes, multicellular trichomes, and multicellular gland were found. The leaf transverse sectioning shows cuticle thickness, mesophyll layer, indumentum and vascular bundle characters. The important characters of each species were extracted and used to construct the key to species of 17 plants.

T6

P1346

Potato wild relatives (*Solanum* sect. *Petota*, Solanaceae) from Brazil: Filling the gaps of knowledge in herbaria and gene banks

***Gustavo Heiden**[1], Luís Henrique Dal Molin[2], Laísa Barcelos[2], Guilherme Klasen[2], Daiane Vahl[2], Caroline Castro[1]*
1. Embrapa Clima Temperado
2. Universidade Federal de Pelotas

The potato (*Solanum tuberosum*, Solanaceae) is one of the most important crops and one of the most complete foods, providing high quality protein, vitamins and minerals, as well as being a source of energy supplied by carbohydrates. Plant breeding has contributed positively to the Brazilian potato production, with the launching of varieties adapted to the different producing regions of the country. However, the process of selection of more productive genotypes present as a setback the narrowing of the genetic basis of the cultivated species. In this context, it is essential that we dedicate sufficient efforts to the collection and conservation of germplasm of wild potato relatives with potential for use in crop breeding. To fill the gaps of knowledge and collections in herbaria and gene banks of genetic resources of wild potato relatives (*Solanum* sect. *Petota*, Solanaceae) for crop breeding, we are proceeding the taxonomic revision of wild species of potato from Brazil; mapping the geographical distribution of herbaria samples and gene banks; identifying the collection gaps in ex situ collections of germplasm; realizing fieldworks to fill the sampling gaps of germplasm and evaluating the morphological and agronomical characteristics of the accessions maintained in the Embrapa Clima Temperado gene bank of potato and its wild reatives. The main results are the taxonomical revision of herbaria collections and gene bank accessions, the morphological characterization of Embrapa wild potatoes genebank accessions using potato descriptors; the mapping of geographical distribution of wild potatoes herbaria and genebank samples and the identification of priority areas to collect germplasm, providing new material for breeding programs of the cultivated potato. A total of 655 occurrence data were collected for the native potato species in Brazil. The taxonomic revision of wild potato samples demonstrated that there are more than two taxa native to the country, a number higher than currently recognized. Through the morphological characterization of genebank accessions, we verified a great phenotypic variation in vegetative and reproductive traits, including tuber characteristics of agronomic value. Despite the occurrence of phenotypic diversity, it is necessary to increase variability through new collections, besides the application of other characterization methods to improve the use of conserved genetic resources. The mapping analysis revealed that there are areas located in four Brazilian states (RS, SC, PR, and MG) that are not yet sampled through the collection and introduction of new accessions in the gene bank, especially in the states of Santa Catarina and Paraná. Our results highlighted new priority areas for the collection of wild potato germplasm in order to increase the genetic variability in the gene bank of Embrapa Clima Temperado with the introduction of new accessios from Brazilian areas previously not sampled. The Embrapa genebank of potato and its wild relatives currently has 76 accessions of wild relatives from collections in the Southern and Southeastern regions of Brazil. This germplasm can be used as a source of resistance genes to biotic and abiotic stresses which are not found in the potato gene pool (CNPq).

T6
P1347
Conservation and public education about *Fissidens macaoensis*

Pou Ieng Hong[1], *Li Zhang*[2], *Kun Fong Leong*[1], *So Mui Cecilia Cheung*[1]

1. Civic and Municipal Affairs Bureau (Instituto para os Assuntos Cívicos e Municipais), Macao SAR, China

2. Fairy Lake Botanical Garden, Shenzhen & Chinese Academy of Science, Shenzhen, China

Macao is located at the south-western tip of the Pearl River Delta of China and has a total area of 30.5 square kilometers. Macao is a prosperous city which supports over 640,000 residents and attracts more than 30 millions of tourists every year. It is one of the cities that have highest population density around the world. Before 2006, we knew almost nothing about the biodiversity of bryophytes in Macao. In order to present the whole picture of Macao's plant diversity and also arouse the general public's concern on these small but valuable natural resources, the Civic and Municipal Affairs Bureau (Instituto para os Assuntos Cívicos e Municipais, IACM) of Macao SAR has collaborated with Fairy Lake Botanical Garden on bryophytes since October 2006. This collaboration has brought various fruitful results, including scientific research and science educations about bryophytes. In 2011, *Fissidens macaoensis* bas been published as a new species for science. As the first species of bryophyte and the second species of plant named with "Macao", *F. macaoensis* is significant to Macao. Since *F. macaoensis* is very miniature and its populations are restricted in small and narrow area, we are now employing multiple measures for saving this species, including public education, setting up strict protection measure and propagation in lab via spore culture.

T6
P1348
Diversity and utilization of indigenous vegetables and edible plants in Bang Krachao, Samut Prakarn Province, Thailand

Tassanai Jaruwattanaphan, *Pongson Wongsong*

Department of Horticulture, Faculty of Agriculture, Kasetsart University

Fifteen villages in the Ehlanzeni District within three local municipalities were visited for the study. Thirty eight participants were identified on the basis of referrals, which included a traditional health practitioner (THP), herbalist, elderly people, middle age and youths. Ethnobotanical data was collected by means of interviews using semi-structured questionnaires. Data analyzed revealed that infants within the communities were treated with traditional medicine for ailments such as colic, nail biting, sunken and bulging fontanelles. *Hypoxis hemerocallidea* Fisch., C.A.Mey. & Ave-Lall, *Lippia javanica* (Burm.f.) Spreng. *Vachellia karroo* (Hayne) Banfi & Gallaso and *Annona senegalensis* Pers. subsp. *senegalensis* were amongst plants species which were frequently reported to be used medicinally. Decoction were mostly prepared from roots (79%) followed by bulbs (8%). In terms of administration of the medicines most of the plants were taken orally. The use of medicinal plant species in maternal healthcare has been perceived to be effective and reliable in preventing illness amongst communities within this district municipality. Validation for scientific efficacy will be needed as the search of new drug development continues.

T6
P1349
Bioactivity and chemical profiling of rhizome methanolic extract of *Amomum masticatorium* Thwaites (Zingiberaceae)

Sinitha K, *Dr John E. Thoppil Thoppil*

University of Calicut

Rhizome of Zingiberaceae plants have been used in traditional medicine since ancient times. They also form ingredients of many Ayurvedic and Unani preparations. *Amomum,* an important genus of Zingiberaceae is used in traditional medicine. *Amomum masticatorium* rhizome is used to chew with betel in Srilanka. The anti-inflammatory and antiproliferative activities of the methanol extract of *Amomum masticatorium* rhizome were observed in various experimental models. *In vivo* anti-inflammatory activity was evaluated in carrageenan and formalin induced paw edema test in mice and compared with that of standard drug diclofenac. In both the models the extract showed significant reduction in edema. The extract at 200mg/kg b. wt showed significant reduction in carrageenan induced paw edema at 3rd hour and chronic edema on 6th day. The *in vitro* anti-inflammatory activity of the extract was investigated by various methods *viz.* albumin denaturation, membrane stabilization assay and proteinase inhibitory activity. Results showed that, the extract exhibited significant anti-inflammatory activity by inhibiting the heat induced albumin denaturation with IC50 value 34.15 ± 0.404 µg/ml, effectively stabilized RBC membrane with IC50 value of 51.47 ± 1.56 µg/mL and significantly inhibited proteinase activity with IC50 value of 52.8± 0.61 µg/mL. The anti proliferative activity was evaluated by MTT assay. The extract showed significant activity with IC 50 value 85.01 µg/mL. The extract was also subjected for GC-MS analysis to reveal their phytoconstituents like 4-Isobutylphenol (28.58%), 2,3-Dihydro-1-benzofuran (11.68%), Octadecyl palmitate (6.81%), γ-Sitosterol (5.73%) *etc.* From the results, it is concluded that, methanol extract of *Amomum masticatorium* rhizome exhibited significant anti-inflammatory and anti-proliferative activity, which may be due to the presence of active phytochemicals in it and can be used as a source for potential drug.

T6

P1350

Utilization of ethnobotanical knowledge of *Aphanamixis polystachya* (Wall) Parker in production of beneficial phytochemicals

***Rakhesh K V**, Ashalatha S Nair*
Department of Botany, University of Kerala

Aphanamixis polystachya (Wall) Parker belongs to the family Meliaceae is a medicinally important tree. Traditional healers used various plant parts for treatment for liver and spleen disorders, tumours, rheumatism, ulcer, dyspepsia, intestinal worms, skin diseases, leprosy, diabetes, eye diseases, jaundice, haemorrhoids, burning sensation, arthritis and leucorrhoea. Proper utilization of the ethnic knowledge of this plant will provide the key for finding cure for various diseases in our community. Very few biological works of medicinal interest have so far been carried out on this plant. The petroleum ether extract of the leaves yielded a neutral compound designated aphanamixol. The bark extract yielded a high quantity of aphanamixinin. From the ethanolic extract of powdered stem bark, a new saponin, poriferasterol-3-rhamnoside has been isolated and characterized. The Isolation and characterization of roots ethanolic extract yields limonoid and flavonoids along with aphanamixinin. A lignin polystachyol, two lignan glycosides, lyoniside and nudiposide, and a sterol have been isolated from a MeOH extract of the dried bark of *Aphanamixis polystachya.* Its principle secondary metabolite components, two terpenoids amooranin, aphanamixin and an alkaloid rohithukine isolated from bark were well studied for treatment of cancer, ulcers, *etc.* in modern medicine. Development of these promising drugs demands alternative method for production of these secondary metabolites. TLC analysis on bark and leaves revealed the presence of these secondary metabolite fractions. Alternative *in vitro* methods were also developed to produce these compounds. The invitro methods will help in conservation of wild germplasm and boost modern medicine production for various ailments.

T6

P1351

Investigation on sacred plants, in North Maharashtra: Diversity, ethno economics and conservation

***Swapnil Khare**, Shubhangi Pawar*
Pratap College Amalner

North Maharashtra is comprised of three districts Viz. Jalgaon, Dhule and Nandurbar. It is situated at the northern border of Maharashtra and lies between 22° 38' to 22° 3' Northern latitude and 73° 47' to 76° 28' East latitude and located in the upper Tapi basin and Deccan traps covering all most the whole area. The range of western ghat (Sahydri Mountains) extends in the western part of Khandesh where as Satpura mountain extends in the north eastern direction. The forest in general dry deciduous type. Tribal community such as Bhil, Pawara, Tadavi, Kokani, Mavachi, Gavit inhibit the hilly forest. The ancient Indians worshiped nature. They considered the plant, animals, lakes, reveres are sacred. The sacred groves are the pockets of climax vegetation preserved on religious ground. Such several patches are present in remote tribal areas of North Maharashtra which are still unexplored. The presence of sacred groves is due to certain taboos, strong beliefs with mystic folk lore. Due to civilization in remote area, culture of tribal's is changing to urban pattern. Much beliefs folk lore are being ignored and forgotten in near future. The urban folk have started exploiting these sacred plants, groves and tribal culture. The region of North Maharashtra constitutes a vast diversity in sacred plants and groves which are playing the role of reservoirs of biological diversity. It is richer in number of plant species. It is often contains many endangered plant species. A large population living in and around the forest areas is extremely poor. They almost entirely dependent on the forest product and renewable resources from these forest. The raw material such as leaves, branches, flowers and fruits are utilized for various byproducts. Some fruits of wild trees are sold in market for various purpose. Though the raw material is comparatively less. Some of them play important role in economy of local people. The sacred plants and flora of sacred groove not only used for religious purpose but Bhagat, Vaidus, some Doctors used those plants for health care. The practice of the nature conservation is very ancient tradition in India. The protection provided to many plants such as People, Umber, Wad trees is well known. A patch of forest receiving the protection because of their association some deity. With the weakening of religious beliefs plants wealth is now in danger of losing this traditional protection and under pressure due to deforestation.

T6
P1352

The potential of useful plants as flagship species to counteract plant blindness

Peter Pany[1,2], *Christine Heidinger*[2], ***Michael Kiehn***[1]
1. Core Facility Botanical Garden, University of Vienna, Vienna
2. Austrian Educational Competence Centre for Biology (AEC-Cbio), Vienna

One option to counteract plant blindness (the phenomenon that people overlook plants in everyday life) is to use plants for teaching and learning which are interesting for students. A questionnaire was developed to explore students' interest in useful plants, subdivided in five subgroups: medicinal plants, stimulant herbal drugs, spice plants, edible plants and ornamental plants. The analysis of data for 1,299 students aged between 10 and 19 years showed (for all ages and both genders) that medicinal plants and stimulant herbal drugs trigger high interest, whereas spice plants, edible plants and ornamental plants raise less interest. As mean values cannot be used to draw conclusions on an individual level, a frequency analysis of individual interests was performed. Based on the identified interest profiles, recommendations are made which plants to use from different subgroups of useful plants in order to best concur with students' interest.

T6
P1353

International Association for Plant Biotechnology: Meeting global challenges

Eoin Lettice[1,2], *Barbara Doyle Prestwich*[1,2]
1. International Association for Plant Biotechnology
2. University College Cork, Ireland

The International Association for Plant Biotechnology (IAPB) was founded in 1963 and has as its objectives to promote plant biotechnology, in all its forms, including cell and tissue culture applications. From the early work of Gottlieb Haberlandt (1854-1945), Gautheret and Nobecourt, through to the work of Norman Borlaug and others, plant biotechnology has become an essential tool for plant scientists to expand the agronomic and commercial usefulness of plants. The IAPB hosts a quadrennial conference, the 14th of which will take place in Dublin, Ireland from 19th-24th August, 2018. This will be the first time since 1994 that the conference has taken place in Europe. Topics and themes that will be covered in the IAPB 2018 programme include food security; new plant breeding techniques; *in vitro* culture and morphogenesis; plant nutriomics; biopharmaceuticals and bioproducts; roots and tubers; crops and the bioeconomy; science communication and outreach. Given that many of the global challenges of our time are connected to or can be alleviated by plants (e.g., food security, climate change, energy production, *etc.*), it has never been more important to advance the science of plant biotechnology and our understanding of the botanical world. Agriculture currently occupies approximately 38% of the Earth's terrestrial surface and yields of commonly grown crops have been steadily increasing in recent decades. Despite this, various studies have predicted that global agriculture will need to increase production significantly to keep pace with growing demand and population increases. This will place added pressure on existing agroecosystems and our natural environments. Plant biotechnology is one element of an integrated responses to these global agricultural challenges and the IAPB are committed to supporting the advancement of plant biotechnology and plant science in general. This is achieved by our regular activities, our quadrennial conferences and our peer-reviewed journal *In Vitro Cellular & Developmental Biology-Plant.*

T6
P1354

Research progress of wind resistance and evaluation system for landscape trees

Ke-Ke Li, *Ling-Lu Li*
Tongji University

Trees may suffer different damages in areas that storm happens often, the one that happens in forest may cause a series of economic losses, the one that happens in city then brings risks to the life of humans, causing traffic jams, and also economic losses, which has a bad impact on the whole city's normal operation and the eco-system. There is a good chance that climate change may make the extreme weathers happen more often, under this situation we should find the proper factors to deep analyze and evaluate trees' stability in wind. As we know, trees' stability in wind has always been studied in Forestry, Tree cultivation, disaster management and some other fields, multi-disciplinary knowledge is needed for this, like the morphology, mechanics, the physics and the aerodynamics. The thesis focuses on the characteristics of the single tree, and sum up the methods, strategies, conclusions and evaluation system about tree stability, in order to fully understand the internal and external factors that matter most. Thus some conclusions can be used in the future tree selection and management in cities which suffer frequent storms.

T6
P1355

(±)-Ganoapplanin, a pair of polycyclic meroterpenoid enantiomers from *Ganoderma applanatum*

Lei Li[1], *Huan Li*[2], *Xing-Rong Peng*[1], *Bo Hou*[1], *Mu-Yuan Yu*[1], *Jin-Run Dong*[1], *Xiao-Nian Li*[1], *Lin Zhou*[1], *Jian Yang*[2], *Ming-Hua Qiu*[1]
1. Kunming Institute of Botany, Chinese Academy of Sciences
2. Kunming Institute of Zoology, Chinese Academy of Sciences

(±)-Ganoapplanin (1), a pair of novel meroterpenoid enantiomers featuring an unprecedented dioxaspirocyclic skeleton constructed from a 6/6/6/6 tetracyclic system and an unusual tricyclo [4.3.3.0$^{3',7'}$] dodecane motif, was isolated from *Ganoderma applanatum*. Its structure and absolute configurations were determined by spectroscopic analyses, X-ray crystallography and ECD (electronic circular dichroism calculations). A plausible biogenetic pathway, involving a key Gomberg-Bachmann reaction, was also proposed for (±)-1. Biological studies showed that (±)-1 and its enantiomers exhibited different inhibitory activities on T-type voltage-gated calcium channels.

T6
P1356

The suggestions on construction of slow life recreation area

in Maluanshan based on biodiversity conservation

Weiying Gu[1], ***Lu Li***[2]
1. Pingshan New District Urban Planning and Land Affairs Center of Shenzhen
2. Baoan Urban Management Bureau, Shenzhen

Maluanshan is a beautiful hilly area undisturbed by human beings in Shenzhen which is belonged to Pingshan District. This area is one of the best preserved areas in ecological environment. There are plenty of species and good ecological base, where vascular plants belonging 161 families, 561 genus and 1,048 species were founded. This study is based on the perspective of urban ecological balance and biodiversity conservation. A series of suggestions were produced to the construction of slow life recreation area including projects of ecological conservation, environmental education and ecotourism.

T6
P1357
Women's contribution to plant species discovery: a different use for historic botanical nomenclature data

Heather L. Lindon[1], *Lauren M. Gardiner*[1], *Abigail Brady*[2], ***Maria S. Vorontsova***[1]
1. Royal Botanic Gardens, Kew
2. 79B Mill Street, Oxford

How has women's contribution to science developed over multiple generations? We present the first quantitative analysis of the role played by women in publishing botanical species names, and the first complete analysis of women's contribution to any field of science with a timeframe of more than 260 years. The inclusion of dates with nomenclatural information allows us to get a view of how naming plant species has changed over time. The International Plant Names Index and The Plant List were used to analyze the contribution of female authors to the publication of land plant species names. Authors of land plant species were automatically assigned as male or female using Wikipedia articles. This was followed by manual research for authors marked as female with over 100 names or publishing before 1900. Female authors make up 12.05% of the total number of authors, and they published 2.59% of names. Female contribution has accounted for more than 1% of new species names per decade since 1900, and stood at 11.99% in the first half of this decade. The difference in productivity between male and female authors has declined over time, and female authors are now 80% as productive as their male counterparts. We determined that Elizabeth Blackwell was the first woman to publish a plant name, Amomum verum Blackw., in the 1750's. The most prolific female plant author was Harriet Margaret Louisa Bolus and one of the top 10 most prolific women, Dr. Charlotte Taylor, is still working today. There are only 28 women authors of plant names before 1900 accounting for about 176 species names. In spite of botany's traditional image as a feminine pursuit, women's contribution was not significantly reflected in species authorship until the twentieth century, around the same time as in other branches of science. This study illustrates how such large and nearly complete datasets of people, dates and plant names can be used to answer questions beyond their original intent and scope. The major limitations of the data as it stands now is the difficulty in assigning gender to some of the authors, in particular those of authors with missing or Chinese first names. This means that our analysis is somewhat restricted geographically. However, despite these difficulties we can see a clear trend of the closing of the gap in productivity between men and women when it comes to describing new species today. Studies such as these illustrate the great strides women have made in science over the past 260 years and hopefully will inspire the current generation of taxonomists to close the gender gap.

T6
P1358
Assessment, innovation and breeding of Chinese main warm season turfgrass conducted in NBG-CAS

Jianxiu Liu
Institute of Botany, Jiangsu Province and Chinese Academy of Sciences (Nanjing Botanical Garden)

With the aim to nationalize China's turfgrass industry, a lot of work have been conducted on turfgrass germplasm resource collection, assessment, genetic improvement and extension since 1994 in China. They include the following specific topic areas: (1) turfgrass germplasm collection and assessment; (2) studies on stress tolerance of turfgrass including cold, salinity, drought, low pH and flooding; (3) genetic analysis and molecular marker development of important agronomic traits for turfgrass; (4) isolation and functional characterization of genes related to stress tolerance; (5) turfgrass breeding to improve the stress tolerance and decrease the maintenance cost; (6) key technical development of turfgrass establishment and maintenance. On the basis of these works, 1,300 accessions of warm season turfgrasses, belonging to 20 species of 10 genera, were collected. More than 200 research papers and two books were published mainly on the assessment, genetic improvement and utilization of turfgrass germplasm in China. Five cultivars of bermudagrass and zoysiagrass were registered with the Department of Agriculture, China, one cultivar was registered with OECD, and one bermudagrass cultivar was licensed to Barenbrug USA to be marketed globally. Five patents were granted and a first prize of science and technology award at the provincial-ministry level was presented to us. Several research platform were established including National Herbage Cultivar Evaluation Station (Nanjing), Jiangsu Engineering Research Center for Turfgrass Improvement and Propagation, Sport Turf Research and Development Center of Jiangsu Province, and Chinese Test Base of Royal Barenrbrug Group. New cultivars have been used successfully in sports field, ecological restruction in coastal region and erosion control in China.

T6
P1359
The research about restoration plants' niche in damaged slope and arrangment model explortion of Southern China

Wenzhu Liu
hen Techand Ecology & Environment Co., Ltd.

In order to construct reasonable plant community and ensure the stability of the vegetation restoration and reconstruction in South China, ecology investigation and data measure of typical slope restoration projects in Shenzhen were carried out. Data were collected for the important value, niche breadth and niche overlap value analysis. The results showed that most of the plants belong to 53 families. Most of them were Euphorbiaceae, Asteraceae, Papilionaceae, Leguminosae. *Leucaena leucocephala*, *Tephrosia candida*, *Mimosa sepiariia* occupied a large proportion of important value in sample plots. The niche breadth value had almost the same sequence with niche overlap value, *Leucaena leucocephala*, *Heteropanax fragrans*, *Tephrosia candida* ranked higher. *Cyclosorus parasiticus*, *Pteris semipinnata* and other ferns had a relatively higher niche breadth in herb layer. Analysis of niche overlap value indicated that competition among *Leucaena leucocephala*, *Heteropanax fragrans* and *Syzygium cumini* was severe, while competition among *Trema cannabina*, *Rhaphiolepis Schefflera octophylla* and other plants was low. When conduct slope restoration, plants with wider niche breadth value and stronger vigor can be used as pioneer species, such as *Leucaena leucocephala*, and *Acacia mangium*. In addition, plants with lower niche overlap value between dominant species should be selected reasonably, for instance, *Rhaphiolepis indica*, *Trema cannabina* and *vines*. In conclusion, both plants' adaptability of the slope environment and competition and succession of species should be taken into consideration to establish a stable plant community of slope. According to the results of investigation, researches on arrangement model and seed quantity of slope restoration plants were carried out by simulating an artificial slope. Tentative combinations of species and quantity were obtained after half a year of observing and data analysis. This study attempted to estimate quality of a slope restoration project from the viewpoint of ecology and explored a new method to establish plant communities.

T6

P1360

New ornamental aroids in China

Chunlin Long[1,2], *Bo Long*[3], *Hang Shu*[1], *Yuanyuan Ji*[1], *Yutong Han*[1], *Runfu Li*[1], *Bo Liu*[1], *Hairu Guo*[2,4], *Yu Ji*[2,4], *Li Wang*[4], *Yujing Liu*[1,5]

1. Minzu University of China

2. Kunming Institute of Botany, Chinese Academy of Sciences

3. Yunnan University

4. College of Horticulture and Landscape, Yunnan Agricultural University

5. College of Agricultural Equipment and Engineering, Jiangsu University

Aroids, or plants in Araceae, have become very popular and important potted indoor ornamentals in China. They are also cultivated for cutting flowers and gardens or landscapes. Many exotic ornamental aroids are introduced in recent decades, such as *Aglaonema*, *Anthurium*, *Caladium*, *Dieffenbachia*, *Monstera*, *Philodendron*, *Spathiphyllum*, *Syngonium*, *Zamioculcas*, and *Zantedeschia*. However, there are rich biodiversity of Araceae in China, of which most groups are potential ornamentals. Among its 26 genera and 181 species native to China, 72 are endemic to the country. The biodiversity at ecosystem, species and genetic levels of China's aroids has been reported in the present paper. In China's tropical and subtropical regions, aroids demonstrate great diversity including epiphytic, epilithic, aquatic, and terrestrial habitats. Their propagation strategies are also various, by producing seeds, stolons, tubercles, bulbils and other vegetative organs. China is the diversity center of *Arisaema*. Almost half species (ca. 80 species) of *Arisaema* grow in China, particularly in China's eastern Himalayan region. Native aroid species used as ornamentals in China are from *Alocasia*, *Leucocasia*, *Rhaphidophora*, *Epipremnum*, and *Colocasia*. The most common ornamentals include *Alocasia odora*, *Alocasia macrorhizos*, *Leucocasia gigantea*, and *Rhaphidophora decursiva*. Many other aroids native to China are very potential for future horticulture and gardens. In addition to ornamental values, about 50% of Araceae species in China are medicinal plants. Some species of *Amorphophallus* and *Colocasia* are used for food or industrial purposes. Wild populations of some aroids have been threatened by over harvesting. Conservation strategies had been proposed in this paper. In particular, most species in *Arisaema* should be protected.

T6

P1361

Clonal integration mediates the heavy metal tolerance of ferns

Jing-Qi Sun[1,2], *Hua-Zheng Lu*[1], *Hang-Yu Li*[1,3], *Liang Song*[1], *He-De Gong*[2], *Jiao-Lin Zhang*[1]

1. Key Laboratory of Tropical Forest Ecology, Xishuangbanna Tropical Botanical Garden, Chinese Academy of Sciences, Menglun, Mengla

2. Faculty of Ecotourism, Southwest Forestry University, Kunming

3. Puer University, Puer, Yunnan

As an environmental friendly and cost-effective technique, phytoremediation (i.e. an ecological restoration technique using plant hyperaccumulator to remove pollutants from environments, e.g., soil, water and atmosphere), becomes more and more popular. Although there are many researches focused on the phytoremediation of toxic heavy metals, the researches about clonal plants, especially clonal ferns were so limited. We carried out two experiments on heavy metal tolerance of two common clonal ferns, *Bolbitis heteroclita* which is a terrestrial fern in tropical seasonal rainforests and *Nephrolepis auriculata* which is an epiphytic fern on the border trees along avenues. One experiment was to survey the tolerance of Pb and Zn of single ramets, while the other was to test the effects of clonal integration on the Pb and Zn tolerance of pairs of ramets. The first one showed that the toxic damages were increased with the concentration of heavy metal and increased from one heavy metal to two metals on both ferns. The second one showed that physiological integration minimized the damages of heavy metals and mediated the tolerance of Pb and Zn of ramets. Our findings clarified that the stresses of single heavy metal on ferns can hardly represent those of multiple heavy metals in the simulated greenhouse studies. Furthermore, we first verified that clonal integration had played an important role in the heavy metal tolerance of ferns. This may imply that the ecological adaptation and strategy of ferns relates to the clonal growth in the long history.

T6

P1362

Gungzhou 366 Popular Green Plants Manual

Ling Lu, *Zhiwu Chen*, *Daping Liang*, *Jianzhong Huang*, *Jianji Li*

Humanity Ecology Professional Committe of Guangdong Provin-

cial Ecology Society

With nickname as "Flower City", Guangzhou is located in the warm and livable Tropic of Cancer belt that is scattered with diverse living things. Growing ornamental plants have become the tradition for people since ancient times. All kinds of plants with various shapes and colors are seen around the houses, displaying a harmonious scene of human living with plant, and also greatly improving the living quality of people. Ornamental plants not only beautify people's living environment, but also cultivate people's love to the nature. In recent years, with advances in technology and economy, as well as further popularization of ecological health culture, people not only care about the plant outlook and cultural connotation, but also care about the plant impact on people's living environment and health. The selection of ornamental plants and growing ways are quietly changing. In order to effectively use the plants around to purify the environment around us, cultivate an ecological civilization featuring human harmoniously living with plants, and build a healthier and beautiful home and community, the project team is working hard to create a way that human and plants can harmoniously live together. After widely analyzing thousands of indoor plants in Guangzhou City, we select the plants according to certain proportion and diverse models, e.g. ornamental vegetables and fruits, ornamental Chinese medicines, and we also introduced new models to finally create the Manual of Popular Green Plan in Guangzhou, providing brief overview of 366 plants that are suitable for indoor and garden growing. Including: Herbs, Shrubs, Vines and Herbaceous Plants and Arbor Plant *etc.* The plants we selected are characterized by ornamental value, health to human, eliminating pollution, medicinal and edible value. Our purpose is to guide the people to grow these plants through organic growing way and fully utilize the plan resources, and we also give warning in the manual of the bad plants that will negatively affect people's health. Manual includes the following main parts: Plant and Human Health, Concept of Green Plant, Indicators of Green Plant, Chinese Plant Culture. (This project was supported by the Guangzhou science and Technology Innovation Committee in 2015-2016)

T6

P1363

Regional "Plant eco-calendar" theoretical framework

Ling Lu

Guangdong Provincial Ecological Society

This thesis intends to elaborate the essential of modern ecology according to the basic principles of modern ecology, the origin and universal development law of the calendar, as well as the examples of the lunar calendar and Gregorian calendar. It points out: calendar is a basic law that human handles grasp their own basic laws of nature and the ecological relationships: a kind of "ecological balance rule" effective and simple theoretical tools. It can not only act as the basic coordinate for human to understand the characteristics of the area and the natural ecological activities at different times, but also can act as guideline for human to figure out correct ecological measures according to the unity of the Earth's ecosystem and the cycle response/law of changes. Plant Eco-calendar is a calendar system and its applications that are established based on the unity of celestial body periodic motion and the periodic change law of the Earth's Plant ecosystems, and combine the Earth and the characteristics of regional ecosystems in order to realize the harmonious and healthy development between human and natural environment system. It includes elements such as basic theory, basic calendar law, system and application, *etc.* The establishment of Plant eco-calendar is based on the astronomy, geography and plant ecology, and supported by Chinese lunar calendar, Gregorian calendar, world calendar and farmer calendar, *etc.* The application of plant eco-calendar should not only respect the traditional almanac form, but also innovate calendar form. In the basis of calendar system, set the science popularization content of ecological culture, such as astronomy, geography, phenology, custom, health, planting and breeding, this is an effective way to promote social harmony and cultural science, people's health and industrial development. This thesis also discusses the theoretical framework including technical route, natural foundation, style and format of the regional "Plant eco-calendar" design, showing broad prospects for future Plant eco-calendar applications.

T6

P1364

Origin of plant life unity of meteorite strike energy model and bioenergetic Modem

Ling Lu

cological society of guangdong province

"Meteorite strike is a common and extensive geological process in the solar system", the Earth is no exception. Based on the causes of meteorite impact on the earth's crust, the cognition to celestial spiral motion and particle structure in the galaxy, as well as the comparison between meteorite strike energy model and various biomorph on the earth, matching between meteorite strike structure on ground rock stratum and plant and biota, it is inferred that the morphological characteristics of the life on the Earth, including fungi, plants and animals is highly consistent with the development of meteorite strike energy model. This study suggests that the said fact provides the key clue proving the mechanism of the origin of life, including plants, on the Earth. In the course of celestial bodies hitting the earth's surface, the superior potential energy converted into high-energy physics process such as kinetic energy, thermal energy and nuclear energy, causing the fission and fusion of the particles, the basic material objects of falling celestial bodies, when the structure of these particles is swelling, and the energy information exchanges, the space-time order of energy movement during the meteorite strike process can be "memorized". As the germ of life on the Earth and under certain conditions, the information they have memorized can express in sequence, such as "gene models." Plant energy model corresponds to the phase before the main clue of meteorite strike energy model ruptured, while the animal energy model extended to a later phase. The life embryo created during the meteorite strike process continued to develop and evolve under the specific ecological environment on the Earth.

T6

P1365

False colour photography of flowers in bee view

***Klaus Lunau**, Christian Verhoeven*

Heinrich-Heine-University Duesseldorf

Bees see colours different as compared to humans. Like Humans bees have three types of photoreceptors, but the sensitivity maxima of the bee' photoreceptors are in the ultraviolet blue and green range of wavelengths. Thus, bees are sensitive to ultraviolet, but insensitive to red light. To illustrate the bees' view of flowers, traditionally a colour photo and a UV-photo from the same position are presented next to each other. Some observers are able to imagine how the UV-patters overlays the colour pattern, but most observers fail to imagine a subtraction of the red pattern at the same time. One problem is that the flowers must not display red colour, since also white, yellow and purple colours reflect in the red range of wavelengths. False colour photos solve this problem: from three separate photos taken in the ultraviolet, blue and green range of wavelength, matching the sensitivity range of the bees' photoreceptors, a false colour photo can be digitally reassembled. In doing so ultraviolet is displayed as blue, blue is displayed as green, green is displayed as red, and red is discarded. Thus the colour contrast is displayed in a bee-subjective correct manner. False colour photos of flowers in bee view are able to demonstrate the colour of tiny and protuberant structures of which spectral reflectance measurements cannot be taken due to the minimal size of the measuring spot required. We will explain a cheap and simple technique to produce false colour photos of flowers in bee view.

T6

P1366

A review of *Vhavenḓa* ethnobotany

Khathutshelo Magwede

University of Venda

Vhavenḓa are indigenes of the *Vhembe* District within Limpopo province in South Africa. They have a rich social culture that is intrinsically embedded to the natural environment where their livelihood depend on sustainable utilization of plant derived resources. *Vhavenḓa* ethnobotany had been documented in several studies that had reported on different use categories ranging from medicinal, vegetables, fruits, beverages, firewood, building material, craftwork and several minor use categories. Some plant species had been recorded to be utilized in more than one use category. To access information on *Vhavenḓa* ethnobotany, several studies were consulted followed by an ethnobotanical rapid rural appraisal survey within Vhembe District Municipality. Data collected was analysed to determine use category trends of plant species. A total of 512 plant species from 325 genera made out of 110 families were recorded. The list include 34 climbers, 20 grasses, 4 sedges, 149 herbs, 130 shrubs, 1 parasite, 1 epiphyte and 173 trees. Out of a total of 110 plant families, those contributing the largest number of useful plant species were singled out as most ten contributing families in descending order: *Fabaceae* (61 spp.), *Malvaceae* (30 spp.), *Asteraceae* (27 spp.), *Poaceae* (20 spp.), *Euphorbiaceae* (19 spp.), *Curcubitaceae* (17 spp.), *Rubiaceae* (17), *Solanaceae* (17 spp), *Apocynaceae* (16 spp.), and *Anacardiaceae* (14 spp.).The overwhelming majority of plants were reported to be used as medicine for humans and animals healthcare, followed by vegetables, then building material, edible fruits, firewood, craftwork, magic and beverage. Ethnobotanical rapid rural appraisal survey revealed an informative data on plant trend uses from earlier times to present. Data collected from interviews show less plant uses from earlier time to present. Ethnobotanical Knowledge was highest in senior citizens above 60 years of age, followed by adults between the age of 35 and 59 years and lastly the youth below 35 years of age. This empirical finding warrants for immediate recording of *Vhavenḓa* plant uses before it disappears forever. Some enigmatic *Vhavenḓa* plants such as *Muṱavhatsindi (Brackenridgea zanguebarica Oliv.) and Muangaila (Millettia stuhlmannii Taub.)* are facing local extinction in the Limpopo Province because the traditional system of myths and taboos is no longer effective as conservation measure to ensure sustainable utilization.

T6

P1367

Medicinal plants used in treatment of maternal healthcare related problems by the Mapulana of Ehlanzeni District, Mpumalanga Province, South Africa

***Shalom Mashile**, Peter Tshisikhawe, Ndivhaleni Masevhe*

University of Venda

Fifteen villages in the Ehlanzeni District within three local municipalities were visited for the study. Thirty eight participants were identified on the basis of referrals, which included a traditional health practitioner (THP), herbalist, elderly people, middle age and youths. Ethnobotanical data was collected by means of interviews using semi-structured questionnaires. Data analyzed revealed that infants within the communities were treated with traditional medicine for ailments such as colic, nail biting, sunken and bulging fontanelles. *Hypoxis hemerocallidea* Fisch., C.A.Mey. & Ave-Lall, *Lippia javanica* (Burm.f.) Spreng. *Vachellia karroo* (Hayne) Banfi & Gallaso and *Annona senegalensis* Pers. subsp. *senegalensis* were amongst plants species which were frequently reported to be used medicinally. Decoction were mostly prepared from roots (79%) followed by bulbs (8%). In terms of administration of the medicines most of the plants were taken orally. The use of medicinal plant species in maternal healthcare has been perceived to be effective and reliable in preventing illness amongst communities within this district municipality. Validation for scientific efficacy will be needed as the search of new drug development continues.

T6

P1368

Sustainable national network

Gloria Matiella

HuertosMX

In an era where sharing information through the Internet is a must and a sustainable life is becoming a global goal, is it possible to integrate technology, design and horticulture? This is the main question that pushed a thesis work from an idea into an emerging national social program. *Huertos MX* was born to create a national state of mind on the importance of a sustainable and cooperative community where people could be able to communicate and share experiences, tips or even to make a business out of their garden. Urban horticulture has emerged as a new mindset in Mexico. As we live in one of the most polluted cities in the world, it is imperative to search for the best quality products that we can afford. It is known that there are plenty of institutions that promote a sustainable way of life. They offer a variety of courses and workshops where people can get involved with the agriculture community or reestablish their previous knowledge about the subject. Maybe you knew about these courses because you saw their announcement on a flyer, your neighbor told you about them, or you may have seen it on the web. But what if there was a way for you to be able to know most of the members of the urban agriculture community in your city? From the smallest to the biggest corporative enterprise that makes gigantic installations on the rooftop of the skyscraper? *Huertos MX* is a project that searches all current urban, personal, scholar and institutional gardens in Mexico. To make this possible, technology was the way to go. We made a mobile application that helps our users locate people and institutions that share a common goal: sustainability. By doing this, we are creating a national movement to build a network where people can communicate and share anything they want. Imagine a map where the gardening community could locate their peers or costumers, a map of schools that could help parents who are aware of the importance of garden-based learning in their children's lives make an informed decision. These are just some of the benefits that a national map of urban gardens may provide to society.

T6

P1369

Perception gaps that may explain the status of taro as an "orphan crop", despite global distribution and value as a starchy root crop and green vegetable

Peter Matthews

National Museum of Ethnology

The present distribution of taro (*Colocasia esculenta*), as a cultivated food plant, extends from southern to northern Africa, western Asia to eastern Asia, throughout Southeast Asia and the Pacific Islands, and through the Americas, from the USA to Brazil. Despite its vast geographical range, high nutritional value, and considerable trade as fresh and processed crop, there has been relatively little interest in the crop and its wild relatives among research funding agenices, and little effective or large-scale assessment of production, trade and usage. There are many historical, cultural, and practical reasons why taro continues to be under-supported as a subject of academic and practical research. Given the proven ability of this crop to grow under diverse climatic regimes, from the equatorial tropics to northern and southern temperate zones, it may be useful to consider perception gaps that contribute to disregard of the crop. Here, a range of perception gaps are suggested, including an apparent gross under-estimation of global production. The FAO is the main organisation responsible for publishing global statistics on food crops, but depends on national-level reports of taro that are generally unreliable, incomplete, or entirely lacking. Production for geographic areas not represented in FAO data can be extrapolated from published estimates for ecologically and culturally comparable areas. This method of first approximation can be used to encourage more accurate data collection, and greater awareness of the present and potential contributions of taro to local, regional, and world food security.

T6

P1370

An ethnobotanical survey of wild edible vegetables with additional uses: a study of the Vhembe region in the Limpopo Province, South Africa.

Mokgaetji Georginah Mokganya, *Milingoni Peter Tshisikhawe*

University of Venda

Studies conducted globally evident that rural communities are faced with food insecurity and are chronically malnourished. The use of the wild edible vegetables can be of pivotal use to mitigate the stress of food insecurity. On the other hand, wild edible vegetables can serve multiple purposes such as 1) medicinally 2) and as condiments in special traditional cuisines. A total of 160 participants were interviewed in 16 different villages of the four local municipalities found in the Vhembe district municipality. The interviews aimed at gathering data on ethnobotanical aspects such as local names of wild edibles vegetables, parts used, life forms, mode of consumption, taste and mode of storage. The taxonomic diversity of wild edible vegetables with additional uses is 24 species belonging to 20 genera and 15 families. Thirty six percent (36%) of the wild edible vegetables listed during the survey were herbs. Some of these vegetables were reported to play paramount roles in the traditional therapeutic practices. As for the taste of the vegetables, it was explained by many participants that the ones with bitter taste are capable of lowering the blood pressure. All in all the study indicated that wild edible vegetables can not only serve as vegetables that supplement the maize-based diets, but can also serve as food-medicine, spices to be used in preparations of special Venda traditional cuisines like the one called *bovhola*. The dried fruits of some vegetables can be used as kitchen utensils.

T6

P1371

Synergistic effect of Amphotericin B and essential oils against fungal strains known to cause mucormycosis

Karaleen Anderson, Mariel Hatch, ***Olga Kopp***

Utah Valley University

Mucormycosis is a serious but rare fungal infection that occurs in immuno-compromised individuals, infecting the sinuses, brain, lungs and other areas. It is caused by fungi of the order Mucorales, including Mucor, Absidia, Rhizopus and Cunninghamella, among others. The common treatment of mucormycosis includes the use of Amphotericin B, but this treatment can cause adverse side effects, including, chills, fever, headaches, loss of appetite, muscle pain, nausea, and weight loss, as well as fatal syndromes such as hepato and nephrotoxicity. For this reason, it is essential to study new compounds with antifungal activity. In this research, we used a combination of Amphotericin B and essential oils. This will provide an alternative treatment that could allow clinicians to lower the dose of Amphotericin B, thus reducing the side effects while still maintaining and utilizing the effective antifungal properties of both the oil and the drug. Essential oils and their antifungal properties have been thoroughly studied, however, little research has been done on the synergetic effect of essential oils and Amphotericin B on the three most common species isolated in mucormycosis infections, namely *Rhizopus oryzae, Abisidia corymbifera* and *Rhizomucor Pusillus*. In this study, we used fungi biofilms to evaluate the synergistic effects between selected essential oils and the antifungal compound Amphotericin B against mucormycosis-causing fungi, in an effort to find alternatives for the high dose of amphotericin B that is currently used to treat this disease.

T6

P1372

A comparison between ethnobotanical information and preliminary biological activity test result from Indonesian medicinal plants

Jin Hyub Paik[1], *Jung Gwang Park*[1], *Sei-Ryang Oh*[1], *Paryanto Imam*[2], *Juniarti Fifit*[2]

1. Korea Research Institute of Bioscience and Biotechnology (KRIBB)

2. The Agency for the Assessment and Application (BPPT)

The joint research project for Indonesian Medicinal Plants between KRIBB and BPPT has been conducted since 2008 until 2016. We selected plant species which have ethnobotanical information and compared with the preliminary biological activity test result to use for the basic information of development of natural drug or nutraceuticals. The joint project produced not only high quality of herbarium & plant extracts (more than 2,500 samples), publications, and database but also provided highly potential samples to be successful in development for the benefit of both countries.

T6

P1373

The account of *Begonia* flora in Thailand and rediscovery after hundred years of four endemic limestone taxa in Thailand and Myanmar

Thamarat Phutthai[1], *Mark Hughes*[2]

1. Faculty of Environment and Resource studies, Mahidol University (Salaya campus), Nakhon Pathom, Thailand

2. Royal Botanic Garden Edinburgh, Scotland

Thailand is one of high biodiversity in Southeast Asia. One of the most beautiful plant family in Thailand is *Begoniaceae* and can be found all types of vegetation from the mean sea levels to the highest summit of Thailand. Karst vegetation is found many endemic and rare species of flora and fauna in SE Asia. Karst limestone in Thailand has an archipelago-like distribution and a rich and highly endemic flora. Many areas of this habitat are under-explored botanically, and are providing a constant source of new species. During the studied family *Begoniaceae* for the Flora of Thailand project since 2007. Approximately 60 species of *Begonia* spp. are found in Thailand. Most of them are tuberous species and highly level of endemism in one or two localities. Endemic species of *Begonia* spp. are mostly restricted to karst vegetation in Thailand. *Begonia* section *Diplocinium* (Lindl.) A.DC. and *Parvibegonia* A.DC. are commonly found with many members in the mention sections. A new section record for Thailand; *Begonia* section *Apterobegonia* Warb. has been reported for the first time with a new species that could be found in the Northern Thailand. Moreover, the extensive studied of literatures, specimens in various herbaria and extensive field surveys so *Begonia procrifolia* Wall., *B. tricuspidata* C.B. Clarke, *B. surculigera* Kurz and *B. adenopoda* Lem. are new recorded species in Thailand. The original type localities of four new record species are found only in Myanmar with more than 100 years ago of the original publications so the recent study of *Begoniaceae* for the account of the flora of Thailand has been reported the novelties distribution and rediscovery of narrow endemism species after century of four new species record in Tenasserim Range, Thailand. At present, the flora account of *Begoniaceae* are going up over 70% since 2007 and six new species have been described in 2016 from Thailand and Myanmar within three publications.

T6

P1374

The *Channapatana* toys and the Indian ivory wood tree

Meenu R. M., *Ashalatha S. Nair*

Department of Botany, University of Kerala, Kariavattom

*Channapatna*toys are traditional craft protected as a geographical indication under the World Trade Organization, administered by the Government of Karnataka. Traditionally, the craft involved lacquering the wood of the *Wrightia tinctoria*-the Indian ivory wood tree. Artisans use ivory wood because of its softness and clean white colour, and the fact that the bright colours of lacquer show easily on its surface. The socio-cultural success of *Channapatna*toys is intriguing. However this unprecedented success has taken its toll on the species. The population of ivory tree is declining since it is put to a number of economic use. The tree is harvested from the wild as a medicine and source of a dye and wood. Leaves are extracted as stock for phyto-remadies. The leaves, flowers, fruits and roots are sources of indigo. Hence the tree is identified and probed as a prospective silviculture species and its micropropagation potential is highly sought after. Earlier attempts at *in vitro* culture of the species identified basal callus mass formation, lack of morphogenetic competence in callus and high cost of production. This study proposes a rapid micropropagation strategy for commercial production of this deciduous tree. Under the regime used in the study, the most critical factor for rooting was activated charcoal (0.05%) along with 6 μM Naphthalene acetic acid. Simultaneous shooting and rooting reduced the time and cost of the protocol. The plantlets hardened with longer photoperiods had better survival percentages when compared with those acclimatised over shorter photoperiods. Hardened plantlets obtained within 45 days, and established in soil with 86.8% survival. This *in vitro* method can be used for rapid mass clonal propagation *W. tinctoria* for conservation strategies, commercial production irrespective of season and for the production of phytomedicines. This method can also be used in gene transformation studies of this obligate autogamous tree species.

T6

P1375

Begonia diversity in China and relationships with neighboring countries

Sirilak Radbouchoom[1,2], *Wen-Hong Chen*[1,3], *Yu-Min Shui*[1,3]

1. Key Laboratory for Plant Diversity and Biogeography of East Asia, Kunming Institute of Botany, the Chinese Academy of Sci-

ences, Kunming, P.R. China
2. Karst Conservation Initiative of Yunnan, Kunming, Yunnan, China
3. University of Chinese Academy

Mega-diverse genus *Begonia* L. covers over 1,800 species in the world and seems to be discovered constantly. Asian *Begonia* includes over 650 species, among which, about 200 *Begonia* species or more than 10% of the genus member happen to China, which are endemic of 60% Chinese *Begonia*. Chinese *Begonia* is mostly distributed to South China. The provinces of species richness about *Begonia* diversity in China are Yunnan (102 species) and Guangxi (71 species). In China, *Begonia* sect. *Platycentrum* and *B.* sect. *Coelocentrum* is the two dominant sections in two provinces (Yunnan and Guangxi). Furthermore, South China borders with a few countries, such as India, Myanmar, Laos, and so on. So many species should be shared among these countries. As the two biodiversity hotspots, Himalaya and Indo-China possess high richness about plant species diversity as *Begonia*. The recent data indicated that the amount of *Begonia* species distributing in both area as China and another country as following 23 species with Vietnam, 14 species with India, 12 species with Myanmar, 7 species with both Thailand and Nepal, 6 species with Bhutan, 4 species with Laos, 2 species with Japan, and 1 species with Bangladesh, Indonesia, Malaysia and Philippine respectively. Especially Chinese *Begonia* species share with Vietnam are maximum species, where 22 species in Yunnan, Guangxi and Vietnam, only one species in Hainan and Vietnam. However, as a part of the Indo-China biodiversity hotspot, Laos only shares 4 species with China, which are dispersed species, viz. *B. acetosella, B. handelii, B. longifolia, B. palmata*. It is predicted that more new discovery would occur in Laos with the further herbaria and field work. And as a part of Himalaya biodiversity hotspot, Tibet (Xizang) of China, India and Nepal maybe need more research and surveys to discover more undescribed species. The geographical and taxonomic study on *Begonia* would be useful for supporting conservation of this genus and the evaluation of biodiversity priority.

T6
P1376
Orchids and their cultural and historical significance

Andrea Ravelo, *J. Chris Pires*
University of Missouri - Columbia

Orchids have a widespread distribution, so it is not surprising that they have had a myriad of opportunities to impact human culture and sometimes even affect historical events. Their influence ranges from medicinal and culinary uses to broad horticultural appeal to literary inspiration. Some of the earliest mentions of orchids originate in East Asia where references to a plant called *che lan* by Confucius (551- 479 BCE) seem to refer to *Cymbidium* orchids. The name of the family (Orchidaceae) derives from the genus *Orchis,* which was first used by the Greek philosopher. Theophrastus (370-285 BCE), who described the effect the orchid tubers had on determining the gender of offspring when consumed by a man or a woman. This explanation was apparently inspired by the shape of the tubers, which bear resemblance to human testicles. This visual similarity also influenced the medicinal applications of orchids in other parts of the world. Although orchids are not known for their agricultural importance, one genus, *Vanilla*, stands out for its culinary significance. The vanilla bean, which is used as a flavoring in a wide variety of dishes and cuisines, was first known to be used by the Aztecs and Mayans as part of the drink Chocolatl. The desire to cultivate and produce vanilla outside of the Americas was a driving force in the discovery and application of hand pollination methods, which later impacted the horticultural industry by allowing the creation of thousands of new hybrids. These hybrids helped augment the orchid collections that were starting to form all around the world as travelers and enthusiasts brought back specimens, some which were quite rare and of conservation concern. The resulting (often-illicit) trade activities have been the subject of books (e.g., Orchid Fever) and even Hollywood movies (e.g., Adaptation) as they delve into the sometimes-fanatical dedication of orchid collectors and enthusiasts. Even as current day scientific research starts to clarify some of the questions regarding the processes and interactions that impact orchid biology, the fascination with the family still holds strong and will most likely continue to enrich human culture for many more years.

T6
P1377
Plant based quorum sensing inhibitor molecules for the development of antipatogenic drug for resistant pathogens

Sushil Kumar Shahi
Bioresourse Tech Laboratory, Botany Department, Guru Ghasidas Vishwavidyalaya

Antimicrobial and antiquorum sensing (AQS) activities of ethanol, methanol and benzene extracts of different parts of twelve plants were screened against *Pseudomonas aeruginosa* POA1 and *Chromobacterium violaceum* 12472. Depending on the plant part extract used and the test microorganism, variable activities were recorded by well diffusion method. During antimicrobial screening of plants extracts against *Chromobacterium violaceum* 12472 the maximum inhibition showed 21.5±1.5 mm by methanol extract of *Psoralea corylifolia* seed, and against *Pseudomonas aeruginosa* PAO1 the maximum inhibition was showed 5±1.5 mm by methanol extract of *Punica grantum* pericarp. The Antiquorum sensing activity of methanol extracts of *Terminalia bellerica* fruit against *Chromobacterium violaceum* 12472 was recorded 7.75±0.5 mm. The three plant extracts were tested to explore their effect on quorum sensing-related swarming motility of *Pseudomonas aeruginosa* POA1. Maximum reduction on swarming motility of plant extract was recorded the benzene extract of *Mangifera indica* leaf showed 15± 1.5 mm zone. However, the value of minimum inhibitory concentrations (MIC's) in case of *Pseudomonas aeruginosa* PAO1 ethanol and methanol extract of *Terminalia bellerica* fruit, benzene and methanol extract of *Mangifera indica* leaf and ethanol and benzene extract of *Punica grantum* pericarp was found 12.5×10^{-3}. The MIC of ethanol extract of *Mangifera indica* leaf was found 3.75×10^{-5} and in case of *Chromobacterium violaceum* 12472 the MIC value of both ethanol extract and methanol extract of *Terminalia bellerica* fruit, and both ethanol and benzene extract of *Mangifera indica* was found, $6.25\times10^{-}4$. Methanol extract of *Mangifera indica* leaf was 3.75×10^{-5}, and the MIC of ethanol extract of *Punica grantum* pericarp was found 6.25×10^{-4} and benzene extract of *Punica grantum* pericarp was found 1.25×10^{-6} and methanol extract of *Punica grantum* pericarp was found 3.75×10^{-5}. In conclusion, results indicate the potential of these plant extracts in treating microbial infections through cell growth inhibition or quorum sensing antagonism, validating their medicinal use. Thus the drug having quorum sensing molecules may be effective for resistant pathogens.

T6

P1378

Varietal development and Horto-taxonomy of Bougainvilleas in India

Sc Sharma[1], *Tao Chen*[2]

1. National Botanical Research Institute, Lucknow

2. Fairy Lake Botanical Garden, Shenzhen

Floriculture is a very lucrative business amounting to 50 billion US$ per year with 10% annual increase. In the floriculture trade, there is always demand for the newer, floriferous and dwarf varieties of ornamental crops. All the cultivated plants have arisen from their wild relatives. *Bougainvillea* is one of the important ornamental climbing shrubs. It was first collected from Rio de Janeiro, Brazil. Bougainvilleas achieved importance in 1920 with the introduction of the cultivar 'Mrs. Butt' in Agri-Horticultural Society, Kolkata, (India) from the Royal Botanic Gardens, Kew. Out of ten *Bougainvillea* species, only *B. glabra*, *B. spectabilis*, *B. peruviana* and a natural hybrid *B.* x *buttiana*, are of horticultural importance. Today, a large number of *Bougainvillea* cultivars (*ca* 500 cvs.) are available from three to multibracted bracts in various color and variegated foliage as a result of the natural hybridization, selection, bud sports, plant and mutation breeding. The Botanic Garden of CSIR-National Botanical Research Institute (NBRI), Lucknow, India is maintaining rich germplasm collection of *Bougainvillea* (*ca* 220*species*/cvs), which serves as the Reference Centre for the R&D work on Bougainvilleas. In the last 50 years, NBRI has developed 25 novel *Bougainvillea* cultivars. 'Shubhra', 'Shweta', 'Hawaiian Beauty', 'Parthasarthy', 'Palekar', 'Zinna Barat', were evolved by spontaneous mutation, 'Arjuna', 'Aruna', 'Los Banos Variegate' were developed by gamma irradiation while 'Chitra' and 'Tetra Mrs. Mc.Clean' were evolved by chemical mutagens. Other institutes which have contributed in the varietal development of Bougainvilleas are Indian Agricultural Research Institute, New Delhi, Indian Institute of Horticulture Research, Bengaluru, Bhabha Atomic Research Centre, Mumbai and KSG & Sons, Bengaluru. 30% of the *Bougainvillea* cultivars have been evolved in India and the country has been recognized as the International Registration Authority for Bougainvilleas. There is lot of confusion in the identification of Bougainvilleas. For boosting the sale of Bougainvilleas, nurserymen give fancy names. In order to check the corruption in the floriculture industry, Horto-taxonomical parameters are employed for the authentic identification and classification of the *Bougainvillea* cultivars.

T6

P1379

Empirical analysis on economic sustainable of rice rain-fed area in rural Cambodia

Darith Siek[1,2], *Shiwei Xu*[1], *Wen Yu*[1], *Abdul-Gafar Ahmed*[1]

1. Agricultural Information Institute, Chinese Academy of Agricultural Sciences, Beijing, China

2. Regional Polytechnic Institute Techo Sen Battambang, Ministry of Labor and Vocational Training, Battambang, Cambodia

This paper investigates the economic sustainability of Cambodian rural rice farmers. Rice in Cambodia is a stable crop just as in many other South Asian countries. The rice plants not only provide food to people but also a source of feed for livestock. Based on 2015 survey about of 6 districts of the rural household in Battambang of Cambodia, a total 204 rice farmers were interviewed and a multiple regression models was applied using the method of Ordinary Least Squares (OLS) to determine the factors affecting the Crop-livestock farmers' disposable income. The study aggregated the livestock of 3 species including cattle, pig and chicken in cattle unit standard known as Livestock Unit (LU). The result of the empirical analysis shows that many household demography factors as well as the socioeconomic factors influences the disposable income. The findings in the study provide an insight to the government and extension offices on effective development of economic sustainability of rice farming and livestock production system in the regions.

T6

P1380

Phoenix dactylifera ecosystem in Morocco: ecology, socio economic role and constraints to its development

Mohammed Sghir Taleb

Institut Scientifique, Mohammed 5 University in Rabat

The date palm (*Phoenix dactylifera* L.) represents the essential element of the oasis ecosystem for Saharan and pre-Saharan regions of Morocco. It plays an important role, not only due to its economic importance, but also its ecological adaptation to, firstly, to ensure necessary protection for crops against underlying warm and dry sales, and secondly to contribute to the fight against desertification. This is one of the oldest cultivated plant species best adapted to difficult climatic conditions of the Saharan and pre-Saharan regions, because of its ecological requirements and economically most suitable for investing in oasis agriculture. Furthermore, the date palm produces fruits rich in nutrients, provides a multitude of secondary products and generates needed revenue for the survival of oasis populations. In Morocco, the development and modernization of the date palm sector face, both upstream and downstream of the industry, several major constraints. In addition to climate constraints (prolonged drought), in its environment (lack of water resources), to the incessant invasion of disease Bayoud, Moroccan palm ecosystem suffers from a low level of technical and traditional practices prevail and traditional, from the choice of variety and site preparation up to harvesting and recycling of products. In this presentation the author will focus on the role of the date palm ecosystem in the local and national development and on any problems that hinder the development of the date palm cultivation in Morocco.

T6

P1381

HSCCC separation of the two iridoid glycosides and three phenolic compounds from *Veronica ciliata* and their *in vitro*

antioxidant and anti-hepatocarcinoma activities

Qiuxia Lu[1,2], *Yiran Sun*[1,2], *Yueyue Shu*[1,2], *Shancai Tan*[1,2], *Li Yin*[1,2], *Yiran Guo*[1,2], ***Lin Tang***[1,2]

1. Key Laboratory of Bio-Resources and Eco-Environment of Ministry of Education, College of Life Sciences, Sichuan University, Chengdu, Sichuan, China

2. National and Local Joint Engineering Laboratory for Energy Plant Bio-Oil Production and Application

Five main compounds, including two iridoid glycosides (catalposide, verproside) and three phenolic compounds (luteolin, 4-hydroxy benzoic acid, 3,4-dihydroxy benzoic acid), were separated and prepared from the crude extract of *Veronica ciliata* by high-speed countercurrent chromatography.n-Hexane/n-butanol/water (1.5:5:5, v/v/v) was used for the separation of catalposide and verproside.n-Hexane/n-butanol/water (3:2:5, v/v/v) was used for the separation of luteolin, 4-hydroxy benzoic acid and 3,4-dihydroxy benzoic acid. The head-to-tail elution mode was used with a flow rate of 5.0 mL/min and a rotary speed of 800 rpm. Finally, a total of 1.28 mg luteolin, 6 mg 4-hydroxy benzoic acid, 2 mg 3,4-dihydroxy benzoic acid, 2 mg verproside and 10 mg catalposide with purities of 98%, 99.1%, 99.5%, 99.8% and 99%, respectively, were obtained from 200 mg of crude extract. In addition, their structure was identified using MS, 1H-NMR and 13C-NMR. To the best of our knowledge, this is the first report of the separation and purification of iridoid glycosides and phenolic compounds from *V. ciliata* by high-speed countercurrent chromatography (HSCCC). Among these compounds, luteolin, 4-hydroxy benzoic acid and 3,4-dihydroxy benzoic acid were separated from *V. ciliata* Fisch. for the first time. The results of the antioxidant activity show that protocatechuic acid and luteolin have strong antioxidant activity compared to 2,6-di-tert-butyl-4-methylphenol (BHT) and vitamin C (Vc). Five compounds also exhibited strong anti-hepatocarcinoma activities.

T6

P1382

Iridoid glycosides fraction isolated from *Veronica ciliata* Fisch. protects against acetaminophen-induced liver injury in mice

Shancai Tan[1,2], *Qiuxia Lu*[1,2], *Yueyue Shu*[1,2], *Yiran Sun*[1,2], *Fang Chen*[1,2], ***Lin Tang***[1,2]

1. Key Laboratory of Bio-Resources and Eco-Environment of Ministry of Education, College of Life Sciences, Sichuan University, Chengdu, Sichuan

2. National and Local Joint Engineering Laboratory for Energy Plant Bio-Oil Production and Application, Chengdu

Veronica ciliata Fisch. has traditionally been used in Tibetan medicine for the treatment of hepatitis, cholecystitis, rheumatism, and urticaria. We analyzed the chemical composition of the iridoid glycosides fraction (IGF) isolated from V. ciliata and evaluated the antioxidant and hepatoprotective properties. Te IGF was separated by high-speed countercurrent chromatography (HSCCC) and the main compounds were identifed by ultra-performance liquid chromatography coupled to a photodiode array. We determined the in vitro antioxidant ability of the IGF through radical scavenging assays and assessed the in vivo hepatoprotective potential in an acetaminophen- (APAP-) induced acute liver injury murine model. Te IGF was separated by HSCCC and three major iridoid glycosides (verproside, catalposide, and amphicoside) were identifed as potent antioxidants and hepatoprotective compounds. Treatment with the IGF signifcantly suppressed the APAP-induced elevation in serum alanine aminotransferase, aspartate aminotransferase, and tumor necrosis factor-alpha (TNF-; improved serum total antioxidant capacity; decreased malondialdehyde formation; elevated superoxide dismutase and glutathione activity; and decreased expression of proinflammatory factors (TNF- nuclear factor kappa B) in the liver. Finally, we examined the histopathology of resected livers for evidence of hepatoprotection. The protection conferred by the IGF may be related to the reinforcement of antioxidant defense systems.

T6

P1383

Bioactivity-guided isolation of antioxidant and anti-hepatocarcinoma constituents from *Veronica ciliata*

Li Yin[1,2], *Qiuxia Lu*[1,2], *Shancai Tan*[1,2], *Lisheng Ding*[3], *Yiran Guo*[1,2], *Fang Chen*[1,2], ***Lin Tang***[1,2]

1. College of Life Sciences, Sichuan University, Key Laboratory of Bio-resources and Eco-environment, Ministry of Education, No.24 South Sect. 1, Yihuan Road, Chengdu,

2. National and Local Joint Engineering Laboratory for Energy Plant Bio-oil Production and Application, Chengdu, Sichuan

3. Key Laboratory of Mountain Ecological Restoration and Bioresource Utilization, Chengdu Institute of Biology, Chinese Academy of Sciences, Chengdu

Veronica ciliata Fisch., widely distributed in western China, has been traditionally used in Tibetan Medicine as a treatment for hepatitis, cholecystitis, rheumatism, and urticaria. However, *V. ciliata* Fisch. has not been subjected to detailed chemical constitution analysis and the bioactive studies were restricted to its crude extracts. It is necessary to investigate the active chemical components of these extracts and identify their biological effects. Four iridoid glycosides, (veronicoside, cataposide, amphicoside, and verminoside) were isolated from the ethyl acetate fraction. Among these compounds, veronicoside and verminoside were isolated for the first time from this plant. These compounds exhibited strong antioxidant activity and inhibitory activity on HepG2 cell proliferation. The antioxidant activity of verminoside was equal to Vc. Cataposide, amphicoside and verminoside had stronger antihepatocarcinoma activity than 5-fluorouracil. Four iridoid glycosides, (veronicoside, cataposide, amphicoside and verminoside) were isolated from the extract of *V. ciliata* Fisch. using bioassay-guided screening. Among these compounds, veronicoside and verminoside were isolated for the first time from this plant. The above results indicated that these compounds were the active chemical components responsible for the antioxidant and anti-hepatocarcinoma properties of *V. ciliata* Fisch. The underlying mechanism of their bioactivity is worthy of further investigation.

T6

P1384

Antioxidant and hepatoprotective activity of *Veronica ciliata* Fisch. extracts against carbon tetrachloride-induced liver

injury in mice

Li Yin[1,2], *Lei Wei*[1,2], *Rao Fu*[1,2], *Lisheng Ding*[3], *Yiran Guo*[1,2], ***Lin Tang***[1,2], *Fang Chen*[1,2]
1. College of Life Sciences, Sichuan University, Chengdu
2. Sichuan Key Laboratory of Resource Biology and Biopharmaceutical Engineering, Key Laboratory of Bio-resources and Eco-environment, Ministry of Education, Chengdu
3. Key Laboratory of Mountai

Veronica ciliata Fisch. has been traditionally used in Traditional Chinese Medicine prescriptions due to its curative effects for hepatitis, cholecystitis, rheumatism, and urticaria. The present study was focused on investigating the role of ethyl acetate and aqueous extracts of *Veronica ciliata* Fisch. Furthermore, in vitro antioxidant activity (scavenging of DPPH, ABTS, superoxide, and nitrite radicals; reducing power; β-carotene bleaching) and the hepatoprotective effect of the ethyl acetate extract by means of CCl4-induced oxidative stress in mice were investigated. The ethyl acetate extract of *Veronica ciliata* Fisch. displayed more noteworthy in vitro antioxidant activities than the aqueous extract. Moreover, it significantly prevented the increase in serum T-AOC, ALT, AST and ALP level in acute liver damage induced by CCl4, decreased the extent of MDA formation in liver and elevated the activities of SOD and GSH in liver. This activity was found to be comparable to that of bifendate. Histopathological observation of the liver was also performed to further support the evidence from the biochemical analysis. The results indicated that strong antioxidant activities and a significant protective effect against acute hepatotoxicity induced by CCl4 of *Veronica ciliata* Fisch. were concentrated in the ethyl acetate extract. The results suggested that this activity may be due to free radical-scavenging and antioxidant properties.

T6
P1385
Development of chrysanthemum without pollen contamination

Xinhua Zhong, Lichun Hua, Jiangsong Fei, Lulu Huang, Weimin Fang, Fadi Chen, ***Nianjun Teng***
College of Horticulture, Nanjing Agricultural University, Nanjing, China

Pollen contamination of many flowers not only significantly reduce their ornamental value and quickly shorten their vase life, but also cause severe allergic reactions to people, in particular those with pollen allergies. Thus, it is urgent to develop flowers without pollen contamination. Chrysanthemum is among the ten most popular traditional flowers in China and one of the four most popular cut flowers in the world, but the excessive quantity of pollen dispersal during its flowering stage can lead to serious pollen contamination. In the past several years, we have focused on the development of chrysanthemum without pollen contamination. It was found that two types of chrysanthemum cultivars do not cause pollen contamination during flowering stage, the first one is there is no pollen in anthers due to pollen abortion, and the second one is there is a failure in anther dehiscence. Pollen abortion mainly occurs at tetrad phase stage when tapetal cells have been in degradation in advance, and several key protein and genes related to programmed cell death (PCD) were found to possibly result in degradation of tapetal cell. In addition, abnormal hypertrophy of endothecium cell serves as the cellular basis of anther indehiscence; and anther dehydration, the enhancement of anther cell toughness due to a high level of ions may be the determining physiological factors of anther indehiscence. Furthermore, more than 10 new chrysanthemum cultivars without pollen contamination have been developed through crossbreeding programs.

T6
P1386
Utilization of *Wolffia arrhiza* extracts for improvement of dry weight and number of nodules in lentil crop

Amit Tiwari
D.B.S. College Kanpur

The dynamic features of aquatic habitats in South-East Asia in general and India in particular have supported luxuriant growth of aquatic weeds. Studies have been made for their utilization in multiple ways in developing countries. *Wolffia arrhiza* is known as noxious weed and rated as one of the ten major weeds growing in South-East Asia, commonly infesting natural impounded waters in India. Studies on effect of *Wolffia arrhiza* extract on dry weight and number of nodule in per plant of lentil crop (*Lens culinaris* D.P.L.-15). Pre-soaking seed treatment for 12 and 24 hrs. were given with 1, 2 and 5 percent ether and water extracts of *Wolffia arrhiza*. Influence of dry weight per plant and consequently number of nodules were increased in treatments of 12 and 24 hrs. with 5 percent ether and water extracts. Results are statistically significance at 5 percent error probability. Present finding are of both academic and economical significance of agriculture of lentil by growers of commercial crop.

T6
P1387
The application research of the *epiphytes* on vertical gardening

Jia Wang
Shenzhen Techand Ecology and Environment Co., Ltd.

In order to break through the application of the *epiphytes*, the key technical problems, such as the plant arrangement, supporting structure and irrigation systems *etc.*, were solved to develop products for office, indoor and outdoor environments. Focusing on the *epiphytes*, as one of the important native plant resources and local greening plant resources with great development potential in Shenzhen, we first created the complete systems compatible with plant characteristics to ensure the landscape effects and quality of vertical gardening. We consequently broke the routine in vertical greening development to improve the artistry and plant diversity of vertical greening.

T6
P1388
Analysis on light response curve characteristics of ten common vertical greening plants

Jia Wang
Shenzhen Techand Ecology and Environment Co., Ltd.

In order to research photosynthetic capacity of common vertical greening plants, taking ten vertical greening plants as materials, different photosynthetic parameters were determined by using Li-6400 portable photosynthesis system, and the light response curves were drawed. The results showed that *Schefflera heptaphylla* was the strongest adaptability to bright light and the highest light saturation point, which was 281.55 mol·m^{-2}·s^{-1}. The maximum photosynthesis rate of *Schefflera heptaphylla* was the highest, 13.31 mol·m^{-2}·s^{-1}, which showed that *Schefflera heptaphylla* was the strongest photosynthetic capacity and the strongest adaptability to the light, followed by *Nephrolepis cordifolia, Peperomia obtusifolia, Chlorophytum comosum, Spathiphyllum kochii, Nephrolepis cordifolia* and *Anthurium andraeanum*had lower light compensation points, 7.47 mol·m^{-2}·s^{-1},7.76 mol·m^{-2}·s^{-1},8.2 mol·m^{-2}·s^{-1},9.90 mol·m^{-2}·s^{-1} respectively, which indicated they were stronger adaptability to low light. The lowest transpiration rate of *Anthurium andraeanum* indicated the best water saving and drought resistance capacity, followed by *Nephrolepis cordifolia* and *Dieffenbachia seguine*. The apparent quantum yield of *Nephrolepis cordifolia* was the highest, 0.1059, followed by *Neoregelia carolinae*.

T6
P1389
Chemical constituents from *Evodia rutaecarpa* and their activities and chemotaxonomic significance

Qizhi Wang, *Fei Liu, Yu Chen, Yu Shan, Shu Xu, Bi Wang, Xu Feng*
Institute of Botany, Jiangsu Province and Chinese Academy of Sciences/Nanjing Botanical Garden Mem. Sun Yat-Sen., Nanjing, PR China.

Evodia rutaecarpa is an original plant of several crude drugs. The *Evodia* fruits (Wu Zhu Yu) have been used in traditional Chinese prescriptions for the treatment of headache, abdominal pain, migraine, chill limbs, postpartum hemorrhage, diarrhea, hyperbaropathy, dysmenorrhea, nausea, nociception, inflammatory and cancer. In order to discover novel bioactive molecules, we investigated the chemical constituents of *Evodia rutaecarpa*, which led to isolation of five novel indole alkaloids (1-5), together with fifteen known indole quinoline alkaloids (6-20), seven quinolone alkaloids (21-27), ten limonins (28-37), nine flavonoids (38-46), seven triterpenoids (47-53), and thirteen other compounds (54-66). The structures of novel compounds were identified on the basis of MS and NMR spectroscopic analysis, and compound 2 and 5 were confirmed by X-ray crystallographic analysis. The known compound structures were established based on chemical and spectral methods. Nine compounds were isolated for the first time from *Evodia rutaecarpa*. Compounds 2, 4 and 6 showed good cytotoxic activity. The compound 4 showed obvious effect on easing pain, reducing blood viscosity and improving immunity in vivo. Compound 1 and 10 have significant antifungal activity of inhibition enzyme activity on the Lanosterol 14α Demethylase (CYP51). Based on the previous studies, this chemical composition of Subfam. Rutoideae is in accordance with the chemical profile of other species of Rutaceae, and complies with the rule that indole quinoline alkaloids are the characteristic constituents of *Zanthoxylum* and *Evodia* genus plants, which provide reliable evidence for their chemotaxonomy.

T6
P1390
Using nature journal to achieve EE outcomes

Ximin Wang
Xishuangbanna Tropical Botanical Garden

Nature journal is a popular activity to inspire kids to appreciate nature. During the journaling, the participants can observe any object, which could be an insect, flower or leaf, very carefully and reflect on their thoughts and feelings. Many natural education organizations in China are using nature journal in their programs. However, how nature journal affects the participants? No serious evaluation has been done yet. In 2016, Xishuangbanna Tropical Botanical Garden organized a nature journal activity in the whole Xishuangbanna. More than 500 students from 14 local schools involved it. After analyzing the objects in the paintings, it shows that the most participants could observe local environment, especially plants to draw. During a questionnaire for 96 students, including participants and non-participants, it found that the participants' attitude and willingness have been improved. But in knowledge, knowledge and skills, no significant differences were found between two groups. Interestingly, after interviewing 48 participants, they said that they did improve their knowledge from nature journal, awareness and attitude, as well. The results showed that nature journal was helpful increasing participants' interests about nature. How to implement it is important.

T6
P1391
Functions of service system of natural history museum on plant research and protection

Xin Wang, *Shumei Zhang*
Dalian Natural History Museum

The service function of natural history museum relies on its carrying capacity in the natural science field. As an important research field of natural science, botany conforms to an important component of natural history museum service function system. The paper takes Dalian Natural History Museum as an example to explore the basic role on plant research and protection in the museum series of service function system, such as collection, research and education. From the museum's collection function, the paper use nearly 40,000 pieces of plant collections from plant fossils to existing plant specimens to establish the digital museum system, which provide the basic data and information. A large number of plant collections can provide scientific basis for the classification and identification of plants. Especially precious conservation such as key protection and collections of endangered plant species can provide valuable reference for plant protection. From the perspective of the research function of the museum, the botanical researchers at Dalian Natural History Museum will carry out botanical classification research as a long-term and continuous work, and introduce the new concept of biodiversity into the traditional fields of plant basic science research. The study of plant diversity in this region can be used to understand the distribution of ecological resources of plant resources by means of plant resource investigation, to grasp the current status of plant resources, and to provide effective science for sustainable utilization of plant resources, plant diversity protection and ecological environment protection. From the perspective of the educational function of the museum, the design and implementation of the botanical-based educational curriculum and activity system, the diversified exhibitions based on plants, will not only expand and popularize botanical knowledge content effectively, but also the botany and natural science related disciplines and even the field of art integration, promote research in new areas of botany exploration.

T6

P1392

Non-random transmission of parental alleles into artificially grouped crop-wild and crop-weed hybrid lineages in rice (*Oryza*)

Zhe Wang, *Lei Wang, Zhi Wang, Xingxing Cai, Bao-Rong Lu*

Ministry of Education Key Laboratory for Biodiversity Science and Ecological Engineering, School of Life Sciences, Fudan University

The worldwide cultivation of genetically engineered (GE) crops has stimulated great concerns over the undesired environmental impacts caused by transgene flow to crop wild relatives. Therefore, it is important to assess the environmental impacts before the commercial release of GE crops. To analyze the fitness effect of a transgene is a part of the risk assessment for transgene flow. Normally, the F_2 (selfing) or BC_1 (backcrossing with wild/weed) hybrid lineages with or without the transgene (s) derived from artificial crosses between a GE crop and its wild/weedy relatives are used to assess the fitness effect in a common-garden experiment. This is based on the assumption that the artificially formed transgene-present and -absent lineages should have the same genetic background. Therefore, the detected differences between the GE and non-GE lineages are merely resulted by the transgene. However, whether a selective-advantageous transgene influences the random transmission of alleles into transgene-present and -absent lineages, resulting the same genetic background of the lineages, has never been properly demonstrated. To address this question, we produced F_2 and F_3 crop-wild/weedy hybrid lineages in rice with or without the herbicide-resistance transgene (*epsps*, 5-enolpyruvoylshikimate-3-phosphate synthase) to examine the frequencies of alleles transmitted from both parents (GE rice and wild/weedy rice) at multiple microsatellite (SSR) loci in these hybrid lineages. Our results showed significantly increased frequencies of crop alleles in transgene-present hybrid lineages, suggesting that grouping of hybrid lineages using a transgene as a marker can result in the non-random transmission of parental alleles into their hybrid descendants. To confirm such a sampling effect on the non-random transmission of parental alleles, we further examined the effect of artificial grouping on allelic transmission into hybrid lineages, using three neutral SSR loci as markers. We obtained similar results in which selecting neutral markers to form hybrid groups (with or without the markers) also caused the preferential transmission of the parental alleles into different hybrid lineages. Together, these results indicated that using a marker, regardless of a transgene or neutral marker, to form hybrid lineages can cause non-random transmission of parent alleles, which leads to the differential genetic background of hybrid lineages/groups. The non-random transmission of parent alleles into the transgene-present or -absent hybrid lineages may mislead the prediction of the fitness effect of a transgene for the risk assessment of transgene flow from a GE crop to its wild relatives.

T6

P1393

The brief history of Camellia hybridization and its new era on creating outstanding cultivars

Zhonglang Wang, *Yunguang Shen*

Kunming Institute of Botany, Chinese Academy of Sciences, Kunming Yunnan

The Camellia genus includes over 260 species with great important values, juvenile leaves and buds of *Camellia sinensis* are used as tea, seed oil *C. oleifera* and other species in this genus is used as edible oil or hair and skin care, and many Camellias are amongst the world's best flowering plants. So far, camellias species have originated over 25,000 cultivars. In recent decades in China, many outstanding camellia cultivars were produced, like many summer blooming cultivars derived from *Camellia azalea*, which make camellias could have flowers year round. Flowers range from simple singles, semi-double, anemone and peony form to fully formal double bloom. Many are large and showy enough to be competition-worthy. And some are very fragrant. Colors are also very abundant than ever. Many new cultivars are more resistant to cold temperatures than those cultivars used to be. In the field of Tea, many new cultivars were derived, like 'Zijuan' which produces higher contain of anthocyandin, and many new cultivation were used from the species in Sect. Thea were used like *C. taliensis* and *C. crassicolum*. In seed oil field, many good cultivars were created and selected with very good characteristics, like high oil content, high quality. These progress will greatly promote the development of camellias' industory.

T6

P1394

Investigation of the origin of rice agriculture in lower Yangze river area based on the phytolith record

Yan Wu

Institute of Vertebrate Paleontology and Paleoanthropology

Using rice cultivation as a method to adapt to environment is one of the most important processes in the human civilization history. There is a continuous debate about whether rice remains from archaeological sites in early Neolithic is collected wild rice or early cultivated rice. One of the key research topics is to find out solid scientific evidence of human intervention in the rice growth and plantation. It is obvious that the change of human behavior will cause the change of associated flora to some extent. Here, we will focus on the ancient rice in the lower Yangze river area, and use phytolith distribution pattern as a new research method, which is then combined with the results from AMS^{14}C identification. Then we can investigate whether change of phytolith distribution has occurred due to the manual intervention, which can provide a novel method to research on the origin of rice agriculture.

T6

P1395

The compilation of *Plants of China* — an illustrated toolbook for 20 thousand species

***Min Li**, Jing Xuan*
Institute of Botany, CAS

The Plant Photo Bank of China (PPBC) is a full-time agency for plant picture management in Herbarium (PE) of Institute of Botany, CAS. Since founded in 2008, PPBC has collected about 3 million photographs of over 25,000 species from 4,146 genera in 538 families. Based on the photograph data, we've started to compile a set of books named *Plants of China*, which is divided into 65 volumes. The set of books is expected to cover all genera and families of native, invasive or naturalized and introduced plant species in China and the species involved are based on the latest classification system. Besides species photographs, their information including the common name, scientific name, distribution map, morphological characters is described in the book. The compilation work is conducted through online editing system which can realize online image selection, text editing and digital typesetting, *etc.* All books are expected to be completed by 2018.

T6

P1396

"iFlowers" — an image-based immediate plant identification system

***Min Li**[1], Jing Xuan[1], Zhi Liang[2]*
1. Institute of Botany, CAS
2. Robusoft (Beijing) Co., Ltd.

"iFlowers" is an APP for image-based plant identification developed by Institute of Botany, CAS and Robu soft Company. It is built by taking advantage of deep learning techniques and based on about 3 million classified plant photographs of over 25 thousand plant species from the database "Plant Photo Bank of China" (PPBC). Through comprehensive use of deep convolution network, image semantic analysis, fine-grained identification and other fields of technology to achieve the image recognition training and detection, the program has realized the immediate identification of plant species. The program's image recognition training has been completed for China's common wild or cultivated plant species, and the real-time shooting recognition has been achieved for more than 11,000 species that covering most of the common species. Not like other APPs, the images used for identification in "iFlowers" are not limited to specific parts of a plant, and the program have realized the identification of all types of vascular plants, such as ferns, gymnosperms and angiosperms. This program can provide service through APP on android and iOS, WeChat public platform (cn-flora) and the website (http://stu.iplant.cn).

T6

P1397

Effects of drought stress and calcium levels on some traits of seed corn (Ksc704)

Ali Zaree
Damghan branch Islamic Azad University, Damghan, IRAN

Drought stress reduces the relative water content of plant tissues, limits plant growth, and causes some physiological and metabolic changes in plants. This research compared the effects of calcium on seed yield and on seed oil, protein, and calcium contents of seed corn grown under conditions of underirrigation. A split-plot experiment was conducted using the randomized complete block design with three replications. The experimental treatments in the main plots included four irrigation levels (conventional irrigation, irrigation cessation at the 7-9 leaf stage for 12 days, irrigation cessation at the 7-9 leaf stage for 20 days, irrigation cessation at the 7-9 leaf stage until corn leaves curl). Three levels of calcium spray (the control sprayed with pure water, 4% calcium spray, and 8% calcium spray) were the treatments in the minor plots. Results showed that the maximum seed oil and protein contents were achieved under the mutual effects of irrigation and calcium spray. The highest and lowest seed oil contents were observed in the treatments of conventional irrigation and 8% calcium spray (7.1%) and irrigation cessation at the 7-9 leaf stage until the leaves curled and also in the control treatment (4.81%). However, the treatment of irrigation cessation at the 7-9 leaf stage until the leaves curled and 8% calcium spray resulted in the highest seed protein content (9.43%), and the treatment of conventional irrigation and the control treatment had the lowest seed protein content (8.83%). Results indicated seed oil and protein contents had an inverse relationship: increases in one led to decreases in the other. Considering these results, we can conclude that calcium application under conditions of underirrigation leads to stability of cell walls and increases plant tolerance to drought stress.

T6

P1398

Protection and utilization of endangered species, *Acer pentaphyllum*

***Shang Gao**[1,2], Dan Li[1], Yaqiong Cao[1], Jun Zeng[1], Shaokui Liu[1], Fei Li[1], Jiaojun Xu[1], Yanbo Hu[1]*
1. Sichuan Engineering Laboratory of Breeding & Cultivating for Colorful Plants
2. Sichuan Colorful Forestry Co., Ltd

Acer pentaphyllum, as a deciduous tree of maple species is endemic in southwest China. It is regarded as one of the world most ornamental valued tree of maple species for its unique leaflets and colorful leaves. In western horticultural field, it is as famousas the Chinese noted dove tree (*Davidia involucrata*). *Acer pentaphyllumis* an endangered species which needs retrievable protection now, and it is also very important to research the origin and successions of maple species. Meanwhile, it's a rare ornamental tree which can make the urban environment more colorful and comfortable. With the protection and utilization, the protection awareness of *Acer pentaphyllum* will be improved further. The propagation of *Acer pentaphyllumis* mainly by seeds, but the population is tiny, which produces certain influence to the breeding efficiency. In this study, a rapid tissue culture system of *Acer pentaphyllumhas* been established. In direct organogenesis way, axillary buds were firstly induced from the stem segments with 0.1 mg/L 6-BA and 0.05 mg/L TDZ, then the axillary buds were proliferated using 0.1 mg/L 6-BA and 0.5 mg/L CPPU, and the regenerated plants were obtained from buds by 0.5 mg/L NAA finally. The study provides a way for saving the endangered species, furthermore, the research finds an approach for landscape application of *Acer pentaphyllum*.

T6

P1399

The role of botanical garden in the orchids protection

Ruyao Zhang, *Shucheng Feng*
Shanghai Botanical Garden

The orchids are international endangered plants, which are all recorded in the CITES list. Not only China but the world is also focusing on orchid protection. We reviewed the application of botanical gardens to the collection, domestication and promotion of orchids, and analyzed the problems which exised. The botanical garden can play an important role in the protection of orchid plants.

T6

P1400

A quick method to access high quality plant images in the wild and indoor

Xin Zhong
Shanghai Chenshan Botanical Garden

We have established a set of shooting methods that can be used to quickly obtain high-resolution images of plants in the field and indoor. In the study of plants, we need to use pictures to show the morphological characteristics of plants. In the past, too shallow depth of field has been a problem: the image outside the focal plane in any case will have a certain degree of fuzzy. Shrinking the aperture and using the flashlight can alleviate the problem to a certain extent, but it can not be solved fundamentally. So, in the digital age, all this has a whole new solution: depth of field stacking, blending all the sharp areas together and produce a completely sharp image. But in the past, this technology can only be used for indoor plant shooting with tripod. We try a number of software and established a set of method that a technician can work in the field, and then get the panoramic picture of plants. All the steps based on modern computers and digital cameras, which take focus stacking to a whole new level, automatically stacking tens or hundreds of frames to make astounding images from microscopy that once could never capture in a single shutter click. These images thus obtained can be used for large format output and printing, fully reflecting the microstructure of the plant, and even replace the dissecting microscope with camera.

T6

P1401

Selection of medium for cultivation of *Idesia polycarpa* Maxim. var. *vestita* Diels

Jingbo Zhou[1], *Zirui Zhao*[2], *Hong Chen*[1], *Yuan Jiang*[3]
1. Anhui Vocational & Technical College of Forestry
2. Shanghai Bohao Biotechnology Corporation
3. Anhui Agricultural University

Idesia polycarpa Maxim. var. *vestita* Diels is a kind of important energy and ornamental plant that has the very high utilization value. In order to improve the seed germination rate and cultivate the high quality seedlings, there is need to find an optimal cultivation medium for seedling growth of *Idesia polycarpa* in a plastic cave tray (plug seedling). In this study, four types of media including M1 (soil), M2 (peat : perlite = 2:1), M3 (peat), and M4 (peat : perlite : vermiculite = 2:1:1) were tested in industrialized plug seedling. The seedling growth of *Idesia polycarpa* including plant height, crown diameter, leaf counts, ground diameter, the relative chlorophyll contents of leaves, *etc.* was measured. The results indicated that medium M2 was the best one for the increases in seedling height, crown diameter, leaf counts, ground diameter and the relative chlorophyll contents of leaves. Therefore, medium M2 is recommended in the industrialized nursery process for the seedling growth of *Idesia polycarpa.*

T6

P1416

Archived lichen and fungi (1852 – 2010) as biomonitors of atmospheric lead contamination in Australia

Liqin Wu
1. Department of Environmental Sciences, Macquarie University, Sydney
2. Department of Environmental Science and Engineering, Guangzhou University, Guangzhou

Archived lichens (*Cladonia* and *Usnea*) and fungi (*Trametes*) obtained from various herbaria were used as reliable biomonitors of atmospheric lead contamination throughout the Greater Sydney area (New South Wales, Australia) and central and southern Victoria. Lead (Pb) concentrations and their isotopic compositions were determined for the lichens and fungi that had been collected and archived over a 150-year period. The findings in this study show that early metal mining activities have been responsible for elevated Pb levels observed in lichen and fungi archives dating back to the 1880s, particularly in central and southern Victoria. Elevated median Pb values were also measured in archival samples from the leaded petrol era (1932–2001) for both the Greater Sydney and central and southern Victoria regions. Lower median Pb levels were found in samples collected post 2002, following the cessation of leaded petrol consumption in 2002. Lead isotope data of lichens and fungi indicated that the most likely sources of atmospheric lead contamination were natural background lead, leaded gasoline combustion and industrial lead emissions, inter alia other secondary sources. Despite the removal of Pb from petrol, the contemporary lichen and fungi samples investigated in this study demonstrate that emissions from lead sources are persistent in the environment.

T6

P1424

Studies on chemical constituents in the roots of *Suaeda glauca*

Yu Shan, *Qizhi Wang, Ming Wang, Fuqin Guan, Fei Liu, Yu Chen, Min Yin, Shu Xu, Xu Feng*
Institute of Botany, Jiangsu Province and Chinese Academy of Sciences, Nanjing Botanical Garden Mem. Sun Yat-Sen

Suaeda glauca (Bge.) Bge. belongs to the family Chenopodiaceae, which is one of the wild resources in saline soil. It is also a dominant species of halophyte in developing the seawater-irrigation agriculture. As a high-quality vegetable and plant resource with the value of health care, there is a broad prospect for its development, which is lack of system research. The chemical constituents and its activities in the roots of *S. glauca* were systematically studied in this paper. In our study, 21 compounds were isolated from the ethanol extract of *S. glauca* by a variety of modern chromatography and spectroscopic methods. Their structures were elucidated by spectroscopic methods, including 1D-, 2D-nuclear magnetic resonance (NMR) and high-resolution time of flight electrospray ionization mass spectrometry (HR-TOF-ESI-MS), including Fourteen flavones: 6,7, 2′-trimethoxy- 5-hydroxyisoflavanone (SG-2), Rutin (SG-8), 5,8-dimethoxy-7,2′-dihydroxy isoflavanone (SG-9), 5,7, 8,2′-tetrahydroxyisoflavanone (SG-10), 5,8, 2′-trimethoxy-7-hydroxyisoflavanone (SG-11), 5,6-dimethoxy-5,2′-dihydroxyisoflavanone (SG-12), 5,6-dimethoxy-7,2′-dihydroxyisoflavanone (SG-13), Quercetin 3-O-β-D-glucoside (SG-14), 7,8, 2′-trimethoxy-5-dihydroxyisoflavanone (SG-15), 2′-hydroxy-6,7-methylenedioxy-S-methoxyisoflavone (SG-16), Kaempferol (SG-17), 5,6, 7,2′-tetrahydroxyisoflavanone (SG-18), 8,2′-dimethoxy-5,7-dihydroxyisoflavanone (SG-19), Quercetin (SG-23); Five sterols: Ergosterol (SG-1), β-daucosterol (SG-3), β-sitosterol (SG-7-a), Stigmasterol (SG-7-b), Spinasterol (SG-7-c); One triterpenses: β-Amyrin (SG-4); One lignan: (-)-syringaresinol (SG-21). Among which, SG-11 and SG-13 are two new compounds, SG-2, SG-10, SG-15, SG-18 are four new natrual products, SG-9, SG-12, SG-16, SG-19 are isolated from the genus *Suaeda* for the first time. The pharmacological activities (cytotoxic effects, antioxidant effects) of isoflavones were evaluated in vitro. SG-2 and SG-12 show antiproliferation activity on human breast cancer cell lines (MDA-MB-231), with an IC50 value of 27.28 μM and 38.31 μM, respectively. The study provides essential chemical information for further development of *S. glauca*.

T6

P1427

Evaluation and optimization strategies of green spaces in suburban forest parks of Shanghai, China

Ting Zhang
Shanghai Botanical Garden

Nowadays, Chinese society is facing the challenge of urbanization. With such a fast-paced urbanization in China, the suburban forest park, as the live public infrastructure, is increasingly seen as the new direction in the development of forest park in China. During last few decades, lots of suburban forest parks have been built to meet increasing population recreation requirements in Shanghai, China. In this study we investigate landscape characteristics of 52 sample spots distributed in 7 typical suburban forest parks of Shanghai. We evaluate the quality of the spots varying in 5 types of vegetation structure using the methods of Delphi and AHP. The evaluation system involves twelve indicators (species number, local identify, community structure, health status, ornamental characteristics, individual forms, spatial pattern, seasonal change, the pro-green border, environmental quality, activity quantity and recreation facility), which are belonged to three factors (ecological naturalness, scene beauty and recreational support). The results show the following conclusions. First, tree-shrub-herb structure accounts for 68% of all samples, while tree-herb and shrub-herb layer accounts for 32%. Branch height of horizontal canopy vegetation and highly enclosed vegetation is about 3.2 m, which are higher than those of open grass, semi-open grass with moderate vegetation and vertically enclosed space (2.3 m). Second, semi-open grass with moderate vegetation receives a higher score (7.45) than that of the other four types, followed by open grass (6.89), vertically enclosed vegetation (6.78), horizontal canopy vegetation (6.44) and highly enclosed vegetation (5.66). Third, the horizontal canopy vegetation and highly enclosed vegetation, accounting for nearly 2/3 of all the samples, receive high score in ecological naturalness factor but low in recreational support. Accessible boundary and clean environment in the interior forest landscape should be created to improve their recreation suitability. Meanwhile, open grass and vertically enclosed vegetation are estimated as scenic beauty but get lower score in ecological naturalness. More variation in species and more naturally and densely vegetated area should be able to increase the ecological stability and wide attractiveness.

Abstracts

T1

A003

The ecological mechanism of bryophytes diversity restoration on Volcanic Lava landforms

Xueliang Bai, *Yu Wang, Le Zhang*

College of Life Science, Inner Mongolia University, Hohhot

The classification and determination of 3247 bryophytes specimens, which were collected from Wudalianchi Volcanos, present 47 families, 128 genera, 286 species and varieties. It is notable in dominant family phenomenon, and the xeric moss Pottiaceae is the largest family. Liverworts is not one of 10 dominant families. Four species are new to China. The variety of species is the highest in the new-period volcanos. Along with the increasing age of the volcanos, the number of bryophytes species, genera and families shows a decreasing trend. Bryophytes are the preponderant vegetative cover in the new-period volcanos, and the microhabitat plays an important role in distribution pattern of bryophyte. The plant modular theory and method are used to determinate the number of archegonium, capsula, reproductive branch and vegetative branch in 15 dominant species. Most species of the ramet ratio in young moss is bigger than the ratio in old moss, which has a growth age structure. The amount of vegetative branches is more than reproductive branches, most species are mainly vegetative reproduction. We found that *Rhytidium rugosum* and *Racomitrium canescens* have a high speed in spreading, but slow in thickening, which leads to great area species. *Ptychomitrium sinense* and *Grimmia pilifera* grow in an opposite way, which leads to small hill. The study on age structure of population will contribute to the prediction of population development, succession trend, restore potential of species, and restoration of plant diversity in volcanic landforms. The study is significant for the invasion, expansion and maintaining mechanism of bryophytes on the volcanic bare land. It also can provide scientific information for the protection of diversity of botanical species and scientific management in volcano natural reserve. In addition, the study could assist long-term monitoring of volcanic activities.

T1

A004

Prelinnean specimens kept in Herbarium of St. Petersburg State University (LECB), St. Petersburg, Russian Federation

Valentina Bubyreva[1], *Vyacheslav Byalt*[2]

1. Saint Petersburg State University

2. Komarov Botanical Institute RAS

Herbarium of the Botany Department of the Saint Petersburg State University (LECB) is one of the largest university herbarium of Russia. Its funds contain at least 800,000 herbarium sheets of higher plants. Now Herbarium collection is unique in the structure of the University, both in historical and in scientific terms by geographical coverage and systematic collection of funds, the existing type specimens (more than 3,500 types in the present, as we found). The type collection includes quite a significant amount of rare and original authentic specimens collected by various botanists such as E. Boissier (1810–1885), V. Borbas (1844–1905), H. Cuming (1791–1865), A.P. Decandolle (1778–1841), J. Forster (1754–1794), W. Hooker (1785–1865), G. Schimper (1804–1878), V. Tineo (1791–1856), W. Besser (1784–1842), A. Bunge (1803–1890), R. Hohenacker (1798–1872), G. Karelin (1801–1872), I. Kirilov (1822–1842), St. Krasheninnikov (1711–1755), C. Ledebour (1785–1851), R. Maack (1825–1886), K. Meinshausen (1819–1899), C. Meyer (1795–1855), A. Nordmann (1803–1866), P. Pallas (1741–1811), Ch. Steven (1781–1863), N. Turzaninov (1793–1863), A. Shrenk (1816–1876) *etc.* This Herbarium is probably one of the oldest as Russia storage facilities. The oldest herbarium specimens kept in LECB, belong to the middle of XVII century, or even the end of XVI century. Prelinnean specimens found by us in G. Bongard (1786–1839) and E. Lindemann (1825–1901) collections. Thus, according to Lindemann, in his collection during his lifetime were kept for more than 60 specimens collected by prelinnean J. Breyn (1637–1697) (36), C. Bauhin (1) (1560–1624), A. Clegero (1) (he collected *Juniperus japonica* from Japan in 1688), J. Gmelin (1709–1755) (20). Many specimens found in Bongard's herbarium, but how many relics it contains is difficult to say, since Bongard did not kept accounting herbarium revenues. Currently LECB, revealed more than 60 prelinnean specimens. The most of them were found in herbaria Bongard – specimens belonged to J. Commelin (1629–1692), C. Commelin (1668–1731), Breyn, P. Hermann (1646–1695), G. Steller (1709–1746), Gmelin *etc.*, as well as in the herbarium Lindemann to Bauhin, Clegero, Breyn, Gmelin, *etc.* Currently, the oldest specimen found in LECB, is a specimen of C. Bauhin (1560–1624) – *Jucca gloriosa* L., without an exact date on the label, but, undoubtedly, belonged to Bauhin. Some of the old specimens have only historical interest, other are syntypes for species of C. Linnaeus (1707–1778), P. Pallas (1741–1811), J. Lamarck (1744–1829) and others, which is why value these specimens is particularly large. Examples of these types: *Melanthium sibiricum* L., *Anthericum calyculatum* L., *Convolvolus ammannii* Desr., *Convolvulus lineatus* L., *Yucca gloriosa* L., *Andromeda stelleriana* Pall., *Claytonia arctica* Adam, *Claytonia sibirica* L., *Alyssum hyperboreum* L., *Phlox sibirica* L. Thus, LECB is the third in Russia after Moscow State University Herbarium (MW) and Komarov Botanical Institute RAS (LE), which revealed prelinnean specimens, and some of which refers to syntypes of Linnean taxa.

T1

A005

Geographical distribution and conservation of the rare and endangered plant *Madhuca pasquieri* (Dubard) Lam.

Linying Cai, *Xinyuan Zhang, Ding Ma, Lu Zhang*

South China Agricultural University, Guangzhou, China

As an important component of biodiversity, rare and endangered plants are one of the core contents of conservation biology. *Madhuca pasquieri* (Dubard) Lam. as a rare species currently, its endangered level is Vulnerable (VU) in IUCN Red List. In China listed as national key protected wild plants (II) and wild plants of extremely small populations, it is a kind of rare oil tree species and precious timber species with high medicinal value. With global climate change and habitat fragmentation, it is rather important to study the current situation of *Madhuca pasquieri* (Dubard) Lam. and formulate relevant conservation strategies. In the paper, based on the data of conservations and managements of natural distribution and specimen records of herbariums, the database of the natural geographical distribution was established. At the same time, the population and community characteristics of the main distribution sites were compared, and the current research and protection at domestic and overseas were systematically analyzed. The results showed that: 1) The natural distribution of wild populations of *Madhuca pasquieri* was mainly concentrated in the northern Vietnam, the southern Guangxi, the southwestern Guangdong and the southeastern Yunnan. The vertical distribution range of *Madhuca pasquieri* is wide. In Yunnan, the wild populations grow well above 1,000 m above sea level. In Guangxi and Guangdong, *Madhuca pasquieri* are mainly distributed in the low altitude of 100-600 m mountains or hills; 2) Wild individuals of *Madhuca pasquieri* are scarce, mainly scattered in the nature reserve with less human disturbance. The climate is warm and humid, and the zonal soil are krasnozem and laterite soil. The species composition of the community of *Madhuca pasquieri* forest is richness. *Madhuca pasquieri* is the dominant species in the main forest layer or important companion species in acid rain forest in Guangxi, but the saplings and seedlings of *Madhuca pasquieri* are rare; 3) The artificial cultivation, growth characteristics and chemical composition of *Madhuca pasquieri* has been research. However, these research are still in primary stage. We need to combine the field investigation and long-term experiment, and then use the habitat distribution model to predict the suitable field for the reintroduce of *Madhuca pasquieri* based on *in situ* conservation and *ex situ* conservation. Combining with molecular biological techniques, breeding and cultivation techniques of *Madhuca pasquieri*, expanding its artificial population, we can actively develop its medicinal value, the oil value and timber value, popularize its application in landscaping. In addition, understanding of the origin and evolution of *Madhuca pasquieri* at the molecular level and exploring the endangered mechanism can provide theoretical basis for the effective protection of *Madhuca pasquieri*.

T1

A006

Physiological response of marine diatom *Phaeodactylum tricornutum* to iron concentration in laboratory condition

Zhuoping Cai
Ecological Society of Guangdong Province

Iron supply may influence the biomass, growth rate, composition and primary productivity of phytoplankton in the ocean. The phenomenon of high nutrient low chlorophyll (HNLC) has implied the restriction effect of iron. Most studies have been concentrated in the effect of iron form on the growth, proliferation and structure composition of phytoplankton; yet additional studies on the response of phytoplankton to the iron concentration are urgently required. In the present study, optical density of algal culture, specific growth rate, cell density, cell biomass, chlorophyll a and protein of *Phaeodactylum tricornutum* were characterized in a laboratory simulative experiment by employing a range of varying iron concentrations including 3.15,6.30,9.45,18.90 and 34.65 $mg \cdot L^{-1}$, so as to investigate the growth characteristics and some important biochemical compositions of marine microalgae in response to iron concentrations. Our results indicated that algal growth differed greatly under the different iron concentrations. The highest optical density at 450 nm (OD450) was 0.877 under the iron concentration of 6.30 $mg \cdot L^{-1}$ in the end of trial; it was decreased with continually higher iron concentrations. Specific growth rate and cell density increased obviously with the increasing iron concentrations within the range of iron concentration from 3.15 to 9.45 $mg \cdot L^{-1}$, whereas much higher iron concentrations resulted in a decrease of specific growth rate and cell density. In addition, the cell biomass also appeared to be concentration-dependent, which was increased gradually over the entire range of iron concentrations. Iron concentration also caused significant changes in the chlorophyll a content and protein content of marine microalga *Phaeodactylum tricornutum*. For the chlorophyll a content, it rose under iron concentration ranging from 3.15 to 18.90 $mg \cdot L^{-1}$, but declined under continually higher iron concentrations. For the protein content, it was increased by 9.45 $mg \cdot L^{-1}$ iron concentration but decreased by much higher iron concentrations, with the lowest protein content of 0.104 $mg \cdot L^{-1}$ under the iron concentration of 34.65 $mg \cdot L^{-1}$. The results suggest that iron concentration is playing a role in the phytoplankton proliferation and metabolism; while there is an optimal concentration range for the growth and physiology of marine microalga *Phaeodactylum tricornutum*. These findings can provide a theoretical base for the control of red tides along the coastal waters and protection of marine biological resources.

T1

A007

Uncaria rhynchophylla (Rubiaceae): A medicinal dietary plant used by Gannan Hakkas in Southern Jiangxi (China)

***Limin Cao**, Hua Xu, Xuexiong Wang*
Gannan Normal University

Wild medicinal dietary plants provide food as well as economic income for the indigenous people and are of great importance in treating human diseases. Gannan Hakkas area, Jiangxi Province, southeastern China, is known for its cultural diversity and biodiversity. The Gannan Hakkas (*kejiaren,* belonging to the Han branch, has a thousand-year history), have accumulated traditional knowledge of the utilization of wild plants, especially the medicinal dietary plants, such as *Uncaria rhynchophylla* (Miq.) Miq. ex Havil., *etc.* As a woody lianas of the family Rubiaceae, *Uncaria rhynchophylla* is mainly found in Jiangxi, Hunan and other regions of southernChina. *U. rhynchophylla* is a medicinal dietary plant used by the indigenous people to treat a wide range of body disorders. The hook-like stems of *U. rhynchophylla* are called the world's oldest medicine, mainly used as oral medicine, while its other medical uses remain unknown. In addition to medical applications, its dried stems, roots and root barks are traditionally used by the Gannan Hakkas as food and substitutes of tea. So far, its chemical composition has been documented, but the nutritional elements are unclear. In this paper, we collected the ethnobotanical data of *Uncaria rhynchophylla* from May 2013 to October 2015. The nutritional components (trace elements: Cu, Zn, Fe, Mn, Se and amino acidsthe Gannan Hakkas have provided ethnobotanical information for the modern world. The nutritional component information may indicate that the plant has enormous potential for development as a medicinal dietary plant from Gannan area. Both safety and economic development of the plant should be conducted to allow them to be used to improve health and prevent diseases in the further studies. *Uncaria rhynchophylla* is endangered with economic development, such as deforestation for agricultural purposes, flood, over-harvesting and firewood collection, *etc.* At the same time, the related traditional knowledge is in danger of being lost. Thus, conservation and sustainable utilization of this plant in this area are of the utmost importance.

T1

A009

Genetic diversity and relationship analysis of *Phalaenopsis* germplasm resources

He-Ming Chen[1,2], *Fu-Bing Lv*[1,2], *Zuo Li*[1,2], *Wen-Fang Xiao*[1,2]

1. Environmental Horticulture Research Institute of Guangdong Academy of Agricultural Sciences

2. Guangdong Provincial Key Laboratory of Ornamental Plant Germplasm Innovation and Utilization

Phalaenopsis is the plant of orchids (Orchidaceae) which was the earliest discovered in 1750, and it has been found more than 70 native species which most distribute in the hot and humid regions of Asia, but *Phalaenopsis* of artificial breeding is widely planted in different parts of the world which is one of the world popular tropical orchids and is deeply loved by the masses, known as the orchid queen.At present, the study of *Phalaenopsis* mainly focused on cross breeding, tissue culture and rapid propagation, cultivation and management, disease and insect control, *etc.* But the reports were very few about the genetic diversity and relationship of germplasm resources. Therefore, the genetic distance and R cluster and Q cluster were analyzed through cluster method for 213 germplasm resources of*Phalaenopsis* in order to provide the reference for breeding, classification, application, and genetic relationship. From the frequency analysis, the genetic distance was mainly focused on 0.0 - 0.6 and 0.6 - 2.4 which accounted for 54.93% and 32.86% of germplasm resources respectively, that is, 87.79% focused on 0.0 - 2.4 which showed that the genetic distance was relatively narrow and genetic diversity was relatively low.So hybridization or hybrid of native species with germplasm will make germplasm resources rich. R type cluster analysis showed that 4 traits of leaf (leaf shape, leaf shape, leaf attitude, leaf anthocyanin coloration, leaf texture) and 11 traits of flower (peduncle color, inflorescence type, flower length in front view or flower diameter, flower petals thickness, number of flowers, flower fragrance, flower color, lateral sepal shape, petals shape, lip pubescence and type) played an important role in classification. Q type cluster analysis obtained that 213 resources could be divided into two types in binding line L1= 4.667, namely hybrid and native species, the two types were subdivided into 20 small classes according to the various varieties of similar traits which basicly found out relationship of *Phalaenopsis* germplasm resources.

T1

A010

Synchronous characterization secondary metabolites and carbohydrates in *Bai-Hu-Tang*

Wei-Fang Zhong, Wing-Sum Tong, Shan-Shan Zhou, ***Jun Xu****, Hu-Biao Chen*

Hong Kong Baptist University

Bai-Hu-Tang (BHT) consists of four traditional Chinese medicines (TCM), i.e. Gypsum Fibrosum (GF, Shigao), Anemarrhenae Rhizoma (AR, Zhimu), Glycyrrhizae Radix et Rhizoma Praeparata cum Melle (GR, Zhigancao) and non-glutinous rice (NR, Jingmi). In order to characterize the elusive chemical composition of BHT, a combination of analytical approaches including ultra-performance liquid chromatography coupled with quadrupole time-of-flight mass spectrometry (UPLC-QTOF-MS/MS) and ultra-performance liquid chromatography couple with photodiode array detector (UPLC-PDA) was developed and validated to simultaneously qualitative and quantitative determine secondary metabolites and carbohydrates in BHT. The selected ion monitoring mode of QTOF-MS was used to quantify 10 major secondary metabolites, totally, 42 compounds in BHT were tentatively or definitely identified based on the quasi-molecular ions, molecular ions and fragment ions characteristic to each analyte. Meanwhile, pretreatment-UPLC-PDA was selected to characterize sugar composition and content analysis of carbohydrates in BHT. The results demonstrated that the developed methods were instrumental in quality control of BHT. Therefore, the study strengthens the research on the chemical characterization of secondary metabolites and carbohydrates of BHT, which provided a comprehensive analytical strategy for the further chemical and biological evaluation of TCM formulas.

T1

A011

Characteristics of *Platycladus orientalis* natural forests in Yin Mountains

Long Chen[1], *Liqing Zhao*[2]

1. Natural Energy Research Institute

2. Inner Mongolia University, Hohhot, China

Platycladus orientalis natural forests in Yin Mountains are the northwest boundary of Chinese *P. orientalis* natural forests and also one of the most important forest communities, with well solid slope, soil conservation, water conservation and the important role of improve habitats. We selected *P. orientalis* natural forests as the research object, through quadrat investigation, the *P. orientalis* natural forests divided into 8 associations. Studied the characteristic of the flora of *P. orientalis* natural forests, structure characteristic of each association, and the diameter structure of *P. orientalis*. And analysised the ecological series between associations space. The results showed that: (1) There are 96 vascular plants belonging to 30 families, 70 genera in Yin Mountains' *P. orientalis* natural forests; depending geographic component of species, they are contain 8 types and 17 subtypes, the most is the East Pan-North Pole distribution, followed by the distribution of East Asia; (2) The diameter structure of *P. orientalis* follows a typical left partial normal distribution which is a typical growth population, average plant height increased with diameter overall linear growth model; (3) From the analysis of community structure, well-developed side of the *P. orientalis* natural forests showed significant tree layer, shrub layer and herb layer, partly because of human disturb, shrub coverage is generally low, especially disappeared; (4) Disturbed by human extremely, these *P. orientalis* natural forests are in the early stages of development, the average diameter of *P. orientalis* is smaller, and they are also more bifurcation. So we suggest that regarding these *P. orientalis* natural forests as ecological forest and ornamental plantation.

T1

A012

The flowering biological characteristics of caltrop (*Tribulus terrestris* L.)

Panfeng Dai, *Bo Yang*
College of Agricultural/ College of Tree Peony, Henan University of Science and Technology

In order to research the biological characteristics and pollination of caltrop (*Tribulus terrestris* L.), the details of its dynamic flowering, pollen viability and stigma receptivity, pollen count and the number of ovule ratio (P/O value), insects visit flowers behavior and frequency were tested, and *T. terrestris* have been processed in different ways of mating. The results showed that: (1) Population blossom lasts for about 4 months, single flower open lasts one day. (2) Pollen activity continuous downward from high level in the beginning and only be held for about two hours. (3) pollen - ovule ratio of 173, and the outcrossing index (OCI) 2 combining with the results of different mating process showed that the presence of caltrop facultative autogamy self-pollination phenomenon. (4) The most effective pollinators of *T. terrestris* were bees of Apidae and butterfly of Pieridae. Research shows that the breeding system of *T. terrestris* belongs to facultative autogamy. Pollinating animals have an important role to significantly improve their seed rate.

T1

A013

Flowering characteristics and pollination biology of *Dicranostigma leptopodum*

Panfeng Dai, *Zhen Meng, Qiqi Luo*
College of Agricultural/ College of Tree Peony, Henan University of Science and Technology

Wild *Dicranostigma leptopodum* in Luoyang was introduced and blossom characteristics was observed through early spring period. Pollen viability, stigma receptivity, pollen and ovule ratio (P/O) and outcrossing index (OCI) were examined, and artificial pollination test was developed. The main results are as follows: the anthesis of *D. leptopodum* was about 5 days, the population flowering continued for about 56 days. Pollen viability achieved the maximum about 91% at 1-2 h after four petals opening. And no pollen viability was detected after 3 days. Stigma achieved the maximum at 48 h and sustained acceptance about 8 days after opening. The pollen ovule rate (P/O) was 5193.8, and the outcrossing index (OCI) was 5. According to the artificial pollination test, the fruit set rate and seed set rate of *D. leptopodum* were 18.54% and 6.43%, respectively. The results mean that the breeding system was obligate outcrossing and need pollinators as pollinating agents. Insects about 8 species were observed to visit the flower of *D. leptopodum*. The most effective pollinators among these floral vistors were bees (*Apis dorsata*) and (*Sphaerophoria* sp.).

T1

A014

Ethnoveterinary investigation of medicinal plants in Tibet autonomous region

Songjia Dai[1], *Dexin Xu*[1], *Long Chunlin*[1,2]
1. Minzu University of China, Beijing, China
2. Kunming Institute of Botany, Chinese Academy of Sciences, Kunming, China

Tibetan ethnoveterinary knowledge, as an important part of Tibetan traditional culture, is declining under the influence of the western medicine. Through the combination of field work and literature research, the common diseases of Tibetan livestock, the Tibetan people's cognition of these diseases and the treatment with ethnoveterinary medicinal plants are gathered and analyzed through the approaches of anthropology, ethnobotany, and linguistics. On this basis, 38 plant species which have been used in the treatment of common livestock diseases in Tibet are analyzed by means of phytochemical and pharmacological information reported previously. The results showed that 30 species of them can be scientifically explained based on the existing literatures, while the remained 8 species need to be further studied in the laboratory. Although the cognition and practice of Tibetan ethnoveterinary medicinal plants have distinct regional features, they are demonstrated scientific implies to some extent.

T1

A015

Collections of the Main Botanical Garden of Russian Academy of Sciences as a base of plant biodiversity conservation

Alexander Demidov, *Mikhail Shustov, Svetlana Potapova*
Main Botanical Garden named after N.V. Tsitsin, RAS

Collections of the Main Botanical Garden is a national and world heritage. Currently the living collections comprise 18,259 taxa: 10,729 species, varieties, forms and 7,530 sorts of plants. They are placed in exposition and collection sites, "The Natural Flora" (1,759 taxa), "Arboretum" (1,564 taxa), "Ornamental Plants" (5,690 taxa), "Cultivated Plants" (2,478 taxa), "Japanese Garden" (100 taxa), "Stock Greenhouse" (6,768 taxa). Cheboksary Branch collections include 2,810 names of plants. The main scientific direction is the development of theoretical bases and the general issue of introduction and acclimatization of plants: analysis of flora and attraction of plant resources in Russia and from abroad; initial assessment of new plants; variation of plants in the nature and under the influence of the transfer from nature to the culture; development of methods of cultivation. The next area - the scientific basis of conservation of plant genetic resource of natural flora *ex situ*: identification in nature and study of rare and endangered plant species; formation of collections of species to be protected; development of methods of cultivation and reintroduction of endangered plant species back to natural plant communities; storage of seeds and plant meristems. One of the key areas – development of scientific bases of ornamental gardening and landscaping: the introduction and study of ornamental plants of natural flora; attracting and testing of diversity of ornamental plants; variety assessment, selection and implementation of valuable varieties in landscaping. Development of theoretical bases and methods of protection of introduced plants from pests and diseases – scientific problem, which is of great practical importance. It is especially relevant for institutions working with introduction of large number of various plants that differ in their biological characteristics and resistance to pests and diseases. An important area of research is to develop the theoretical bases of remote hybridization to create new species, forms and varieties of plants: the scientific basis of hybridization with wild crop; search for new ways of management the processes of speciation and intermutation at remote hybridization. Establishing a National Center for Genetic Resources of wild and cultivated plants will help to fulfill some actual tasks: - organization of the Central Bank of seed and meristems of wild plant flora of Russia; - development of effective methods for clonal micropropagation and conservation of the gene pool of rare and economically valuable plant species in vitro; - creation of online databases "Biodiversity of plants of Russia", "Rare species of Red Book of the Russian Federation, stored in the botanical gardens of Russia? - conducting experimental work on the introduction of less-known and less common fruit, berry, ornamental, new medicinal and aromatic plants; - organization on the basis of inter-institutional ecosystem monitoring center under the stress of anthropogenic pressures; - development of environmental protection methods from diseases and pests. So, collections of The Main Botanical Garden to named after N.V. Tsitsin are able to make a major contribution to strengthening global environmental and food security.

T1

A016

Checklist of local spermatophyte above the slope and artificial slope in urban areas of Lanzhou

Pin Ding, *Jian-Rong Wu, Jin-Xia Li*

Gansu Provincial Academy of Sciences Institute of Biology

Objective: Through investigating, sampling, identifying the slopes and artificial slopes in urban areas of Lanzhou, to provide reference for the selection of the appropriate local plants in artificial slope, and for the data support to study the changes of the plants at slopes and artificial slopes in Lanzhou; Mothed: Begin with June 2013, October 2015 to end, to research, collect and identify the plants at slopes and artificial slopes which gradient are more than 20 degree, and record the plants species; Results: Recorded vascular plants 26 families, 65 genus, 86 species; Conclusion: The numbers of plants species at slopes and artificial slopes were less than in the whole Lanzhou. There were the largest number of species in Gramineae at the slopes and artificial slopes in urban of Lanzhou, Chenopodiaceae and Compositae were the next. For slope and artificial slope greening, it can consider Gramineae or the mode of Gramineae combination of Compositae and Chenopodiaceae.

T1

A017

Novel *Cynomorium songaricum* polysaccharides with enhancement of T cell immune response

Haiyan Fan, *Guilin Chen*

Inner Mongolia University

Cynomorium songaricum, known as "suo yang" in China, is a wild perennial plant mainly parasitizing on the roots of genera *Nitraria* which widely distributes across the deserts of Northwestern China. Fresh *C. songaricum* has been used as edible food resources because of its medicinal and nutritious value. Dried fleshy stems of *C. songaricum* is generally used for reinforcing immune system, treating lumbar weakness and enhancing sexual ability in traditional Chinese medicine. Modern research revealed that polysaccharides from *C. songaricum* displayed a range of biological effects, including anti-fatigue, anti-ulcer, anti-oxidation-related disease, and hypoglycemic activity. So far, there is few reports upon *C. songaricum* polysaccharides (CSP) boosting immunological function because of difficulty to get them. In the current study, we have developed novel CSP that could enhance T cell immune response. The main factor affecting the yield of polysaccharides is the extraction and separation process used. Enzyme-assisted extraction (EAE) attracted much attention due to its high-efficiency, environmental compatibility and simplified manipulation overcoming the shortcomings of the conventional hot water extraction (HWE). Protease commonly has specific cleavage sites to cleave peptide bonds between hydrophobic and amino acids, and has no effect on fats and carbohydrates. EAE was applied for CSP to optimize extraction yield. The study comprised of three distinct stages, a single factor experiment design, a screening procedure using Plackett-Burman design, and an optimization response surface methodology using Box-Behnken design. The optimal extraction conditions were as follows: pH 1.5, liquid-solid ratio 108:1, and temperature 40°C. The optimal extraction conditions resulted in the highest CSP yield, i.e., 23.63±0.21%, which represented an increase of 233% compared to the yield obtained using HWE method. Preliminary characterization indicated that the total sugar, protein, reducing sugar?and uronic acid of the product were 90.63±0.77%, 1.07±0.19%, 1.46±0.029%, and 14.42±0.87% (w/w), respectively. Compared to CSP-HWE, CSP-EAE provided higher contents of polysaccharide as well as lower contents of protein. The CSPs were characterized by Fourier transform infrared (FT-IR) spectroscopy, scanning electron microscopy (SEM), atomic force microscopy (AFM) and gas chromatography (GC). The results showed that CSP-EAE differentiated from CSP-HWE in functional groups and surface morphology leading to its distinct properties. Further research on the precise chemical structures and the unique biological functions of CSP-EAE are currently under investigations in our laboratory. Furthermore, CSP-EAE only showed moderate dose-dependent antioxidant activity *in vitro*. Encouragingly, a combination of CSP-EAE and tumor-derived autophagosome-based therapeutic vaccine (DRibbles) can induce functional 4T1 tumor-reactive $CD8^+$ T cells enhancement. It showed that CSP-EAE could mediate potent autitumor efficacy.

These findings will lay a foundation for future pharmacological and biochemical studies of *Cynomorium songaricum* polysaccharides.

T1

A018

Definitions of botanic garden and conservation of terminological diversity

Andrei Filimonov

Peter the Great St. Petersburg Polytechnic University

Analysis of laws, regulations, reference books and scientific publications reveals that the concept "Botanic garden" has constantly changed throughout more than two centuries (Filimonov, in print). Despite a number of discussions about the concepts definitions, this problem is still very actual (Wyse Jackson, 2009). From our point of view, the main reason for the observed terminological diversity is that most definitions of Botanic Garden were inspired by a local sociohistorical background, rather than by basic features of these institutions. To eliminate the discrepancy mentioned above, we propose the following definition: Botanic garden is a piece of ground, whose boundaries are determined in accordance with local legislation, which is used for the purpose of long-term cultivation and monitoring of properly documented and labelled specimens of wild and cultivated flora (mostly non-native), representing a wide range of taxonomic groups and intended for scientific research and (or) the education.

T1

A019

Research on nutritional components of root of *Apios fortune* Maxim

Shunhua Fu

Zhejiang A&F University

In China, *Apios fortunei* Maxim is rich and is used as food and medicine. But most of it is still in wild and lack of basic research. Here, the general nutrient components, 8 mineral elements and 17 amino acids were studied by the experiment material from 4 sites of different provinces. The results show: The contents of crude protein, crude fiber, polysaccharide and soluble protein, sugar and starch content in it are determined. The results show that the contents of soluble sugar, crude fiber and polysaccharide in different areas are very different, while the difference of starch and protein content is relatively small. The content of soluble sugar, polysaccharide, starch and protein is decreasing with the increase of growth age, and starch is increasing with the increase of year in Pan'an, Zhejiang Province and Shan xian, Shandong Province. The contents of 8 mineral elements in it are determined. The results show that the contents of P, Ca, Mg and K in root of it are rich. P, Mg and Cu are the characteristic elements of it in Shanxian, Shandong Province, and Ca and Cu are the characteristic elements of it in Lanping, Yunnan Province. The element contents of P and K in it of Pan'an, Zhejiang Province, and Shanxian, Shandong Province are significantly decreased with the increase of growth age, and the contents of Ca and Mg are contrary.The microelement content of Fe and Mn increase with the growth age, and the content of Cu and Zn is oppositely trend. In it 15 kinds of amino acids is rich, is lack of Cystine, Methionine. The total amount of amino acids in different sites is ranked as follows: Shanxian, Shandong Province>Pan'an, Zhejiang Province> Lanping, Yunnan Province >Nantong, Jiangsu Province. The total amount of non-essential amino acids is: Shanxian, Shandong Province> Lanping, Yunnan Province >Pan'an, Zhejiang Province>Jiangsu Province; The total amount of essential amino acids in the order: Pan'an, Zhejiang Province>Shanxian, Shandong Province>Nantong, Jiangsu Province> Lanping, Yunnan Province. Among all kinds of amino acids, the Aspartic acid (Asp) is rich mostly and the Lysine is the first limiting amino acid. The ratio of essential amino acids to total amino acids (E/T)and the ratio of essential amino acids to nonessential amino acids (E/N) of 1 year average content of amino acids in Shanxian county in Shandong Province, Pan'an in Zhejiang Province are 32%, 48%, more than 2 year' 8.70%, 12.89%, respectively, which near the FAO/WHO standard (E/T is 40%and E/N is about 60%). According to the result of comprehensive evaluation by principal component analysis, the order of amino acid nutritional value is Shanxian, Shandong Province, Pan'an, Zhejiang Province, Nantong, Jiangsu Province and Lanping, Yunnan Province declinely. In summary, it is rich in nutrient content, beneficial amino acids and abundant mineral elements and it bring the broad development prospects for this resource plant.

T1

A020

The structure and dynamics of natural population of the endangered plant *Tetracentron sinense* Oliv.

***Xiaohong Gan*[1], *Huaichun Li* [1], *Shan Li*[1]**

1. College of Life Science, China West Normal University

2. Key laboratory of southwest China wildlife resources conservation (Ministry of Education)

Tetracentron sinense Oliv. appears to have survived largely unchanged since the early Tertiary, like *Ginkgo* and *Mestasequoia*, and belongs to the family of Tetracentraceae. This species, the East Asian endemic species, a threatened deciduous broad-leaved canopy tree, with a distribution restricted to central and southern China, mainly lived in moist temperate deciduous, or mixed evergreen and deciduous forests or forest edge, above 1100-3500 m sea level. Due to the high demand for medicine, timber, ornamental and scientific research, the destruction of *T. sinense* resources recently reached a very serious degree, resulting in the continuous decrease of remaining wild population and individuals. As a consequence, *T. sinense* was listed in CITES, and listed as a national second-grade protected plant in China. Up to now, the natural population of this species is still at risk of extinction. Previous studies had focused on its systematic status, chemical composition, limiting mechanism of natural regeneration of natural population, but little is known about the survival status and dynamic of natural population of *T. sinense* so far. In order to explore the survival status of natural population and predict the population dynamic in the future, and discuss the protection countermeasures, the age structure and quantitative dynamics of the endangered plant *Tetracentron sinense* Oliv. natural population in Meigu Dafengding Nature Reserve in southwest China were studied through establishment of the time-specific life stable and survival analysis. The results are as follows. (1) Due to the strong screening effect of the environment, there was lack of seedlings, and only a few saplings and arboret, and vast majority of mid-arbors and arbors in natural population, indicting the decline tye of age structure with inverted pyramid character. (2) The life table of *T. sinense* population showed that the life expectancy of *T. sinense* is higher in first and 7th age class, and the survival curve of the population belonged to Deevey type II. The mortality rate and vanish rate curve of *T. sinense* population all showed two peaks in the 6th and 12th age class, so did the mortality density and hazard rate curve. The population survival rate of *T. sinense* decreased monotonically, whereas the cumulative mortality rate increased monotonically. (3) Survival analysis showed that population quantity of *T. sinense* population decreased steadily at earlier stage, and changed smoothly at middle stage, but declined in older stage, resulting in the sharp decrease of population quantity, which may be one of important factors limiting the natural regeneration of *T. sinense*.

T1

A021

Prioritizing fodder species based on traditional knowledge: A case study of mithun (*Bos frontalis*) in Dulongjiang area, Yunnan Province, Southwest China

Yanfei Geng[1,2,3], *Guoxiong Hu*[4], *Sailesh Ranjitkar*[2,5], *Yuhua Wang*[2], *Dengpan Bu*[5,6,7], *Shengji Pei*[2], *Xiaokun Ou*[1], *Yang Lu*[1,2,3], *Xuelan Ma*[2,3], *Jianchu Xu*[2,5]

1. Yunnan University

2. Kunming Institute of Botany

3. Universtiy of Chinese Academy of Sciences

4. Guizhou University

5. World Agroforestry Centre East and Central Asia

6. Institute of Animal Science, Chinese Academy of Agricultural Sciences

7. Hunan Co-Innovation Center of Animal Production Safety, CICAPS

Livestock rearing is one of the oldest and most important types of smallholder farming worldwide. The sustainability of livestock production depends on the efficient utilization of locally available resources. Some traditional methods of raising livestock may offer valuable lessons in this regard. This study documented and evaluated local knowledge of wild forage plants in the Dulongjiang area in Southwest China in the context of rearing mithun (*Bos frontalis*) in order to provide a sound evidence base for tree fodder selection and the establishment of integrated tree-crop-livestock systems. The snowball technique was used to identify key informants with specific knowledge about the topic. Free listing and semi-structured interviews were conducted with 58 households. Participatory investigation and transit walks were used to investigate potential fodder species. Ethnobotanical information was collected, documented and organized. Overall, 142 wild forage plants from 58 families and 117 genera were identified. Species of the Poaceae, Rosaceae and Urticaceae families were most abundant, with 16,14 and 11 species respectively identified as fodder plants. Our results indicated that tree/shrub forage plays a major role in the diet of mithun, unlike that of other ruminants. Mithun prefers to browse and move around the forest in search of food, particularly rough and even barbed leaves. Tree species like *Debregeasia orientalis*, *Saurauia polyneura* and *Rubus* species were identified as being important fodder sources. Farmers in this area have traditionally relied on common property resources such as community-managed forests and grasslands to feed their livestock. Farmers have strong incentive to raise mithuns rather than other livestock species due to Dulong people's cultural preferences. The wide variety of plants cited by the informants demonstrates the importance of traditional knowledge in gathering information about forage resources. This diversity also offers the prospect of identifying promising species which could be used as fodder plants. Identifying such species and tree fodder species in particular could help smallholder farmers to integrate trees, livestock and crops as part of a sustainable farming system.

T1

A022

Regeneration system of *Ammopiptanthus mongolicus* nodes

Qi Guo[1,2], *Peiya Wang*[1,2], *Jun Zhang*[1,2], *Tao Yang*[1,2], *Hui Yang*[1,2]

1. Institute of Biology, Gansu Academy of Sciences

2. Key Laboratory of Microbial Resources Exploition and Application, Gansu Province

Ammopiptanthus mongolicus (Masxim.) Cheng f. belongs to the Leguminosae plants, is a national third protection plant, is unique to the desert region of northwest China super xerophytic evergreen broad-leaved shrub, is also a kind of special constructive plants in Alaskan desert, plays a important role in maintaining the local ecological balance. Xinjiang province, Inner Mongolia, Gansu provinc are its main distribution areas. Narrow distribution range and human activities are two of the main causes driving this species to a risk of extinction. At present, many researchers tried to use the tissue culture technology to accelerate the breeding speed to get a number of tube seedlings in a short term. It is a way to protect its germplasm resources and expand its population. Young stem, stem segments and root as explants of *Ammopiptanthus mongolicus* were used in the callus inducition research. But callus differentiation was difficult accompanied by severe browning. Long-term subculture can cause the death of callus to making it impossible to be conserved in vitro. A protocol has been development for the establishment of regeneration system *in vitro* by using *Ammopiptanthus mongolicus* nodes. The number of shoots was highest when the 5-day-old explant cultured on B_5 medium supplemented with 1~2.0 mg•L^{-1}BA. Roots were efficiently established on B_5 medium supplemented with IAA 1.0~2.0 mg•L^{-1} and NAA 0.5 mg•L^{-1}.

T1

A023

Study on the population genetic structure of *Disanthus* (Hamamelidaceae)

Xueying Guo

Guangzhou University

Disanthus (Hamamelidaceae) is an East Asia endemic genus with the distribution pattern of East China and Japan disjunction. *Disanthus cercidifolius* subsp. *longipes* mainly distributes scatteredly in Chinese provinces of Guangdong, Hunan, Jiangxi, and

Zhejiang. *Disanthus* cercidifolius mainly distributes in Central Honshu and Southwest Honshu of Japan. In this study, the population genetic structure of 16 natural populations of *Disanthus* sp. based on the sequences variations of three chloroplast DNA fragments (*psbA-trnH, rps16F-R2, psaA-ycf3*) and nrDNA ITS were investigated. AMOVA analysis base on cpDNA and ITS sequence indicated that obvious genetic differentiation between the Chinese and Japanese subspecies. The result base on cpDNA sequence revealed that genetic variation occurred mainly between groups (53.08%), while 24.11% variation occurred among populations within groups and 22.80% variation occurred among individuals (within populations). The result base on ITS sequence revealed that genetic variation occurred mainly between groups (90.03%), while 5.01% variation occurred among populations within groups and 4.96% variation occurred among individuals (within populations). The total genetic diversity of *Disanthus* sp. is higher (Ht = 0.725), while the average genetic diversity within populations is lower (Hs = 0.148). As Nst is significantly larger than Gst (P<0.01), it demonstrats highly significant phylogeographic structure across the distribution range of *Disanthus* sp.

T1

A024

A floristic geography study of vascular plants in Guanshan mountain

Yi He, *Quanru Liu*
Beijing Normal University

Guanshan Mountain is located in the Loess plateau, China. As this area was governed by three different provinces separately, historically, there has been fewer integrated floristic investigations. Therefore, a detailed floristic research of this region is of great significance to the improvement of floristic data in Northern China, it also helps to solve the controversy on small-scale flora district division and protect plant diversity. In order to figure out the floristic composition and geographical component in this area, we investigated 14 comprehensive line transects, collected 1285 specimens, combining with 19 forest quadrats of 600 m^2 and 6 shrub quadrats of 100 m^2. Based on literatures and field works, we built a database of vascular plants of 12 typical mountain areas in Northern China for comparative study. The main results are summarized as follows: 1. Guanshan Mountain is a comparatively species-rich area. There are 1166 species (including infraspecies taxon) of vascular plants, belonging to 118 families and 480 genera. There are 34 fern species belonging to 22 genera in 11 families, 8 gymnosperm plant species belonging to 6 genera in 4 families and 1124 species of angiosperm plants belonging to 452 genera in 103 families; Families are significantly differentiated, the first 11 largest families contain 586 species; Genera are also differentiated: Big genera are highly developed, while middle and small genera occupy an important position; There are 20 mono-species genera and 46 minor-species genera, comparatively rich; The endemism of this area is pretty low: Only 2 species are endemic, combining with 8 genera, 325 species and no family endemic to China. 2. The geographical components are relatively complicated and diverse. Pteridophyte genera are divided into 6 areal types, mainly worldwide. The number of tropical types is more than that of temperate types; Seed plant families are divided into 11 areal types and 8 subtypes, mainly worldwide. Tropical types are more than temperate types; Seed plants genera are divided into 15 areal types and 7 subtypes, the proportion of temperate genera is as much as 70.79% (except the worldwide types), especially the north temperate type. 3. Comparing with other 11 nearby mountains, Guanshan Mountain is most closely related to Liupan Mountain, rather than Ziwuling Mountain and Huanglong Mountain. At species level, the gini coefficient and species composition analysis show that Guanshan Mountain is the most similar to Liupan Mountain. Cluster analysis and ordination analysis support this as well; At genus level, floristic spectrum shows that it has a much closer relationship with Taibai Mountain (northern slope) and Zhongtiao Mountian, but not quite similar to the Ziwuling Mountain and Huanglong Mountain.

T1

A025

Study on construction of core collection and genetic diversity of mulberry germplasm in northern china

Dechang Hu[1], *Ping Zhang*[1], *Jincheng Zuo*[1], *Zhongli Wang*[1], *Zhaohong Wang*[2], *Chuanjie Chen*[2], *Yangfang Zhang*[3]
1. Ludong University
2. Sericulture Research Institute of Shandong Province
3. Fujian Academy of Agricultural Sciences

Mulberry (*Morus* L.), an economically important plant used for sericulture, is widely distributed in China with complex genetic backgrounds. In this study, more than 40 phenotypic traits in 560 mulberry samples mainly in northern China from 7 species and 2 varieties were surveyed and complemented. Statistical analysis showed that the mean value of each trait was 2.187 with a standard deviation of 0.922, an average variance of 1.003 and an average coefficient of variation of 47.391, indicating that the mulberry germplasm variation was abundant. The highest genetic diversity index was 1.303 for *M.alba*, followed by 1.077 in *M multicaulis,* 0.869 for *M. atropurpurea*, 0.845 for *M. mizuho*, 0.768 for *M. bombycis* and 0.593 for *M. australi*. The principal component values of the first, second and third phenotypic traits in mulberry were 3.032,1.920,1.609, respectively, which accounted for 10.829%, 6.856% and 5.746% of the total variance. The cumulative contribution of principal components from the first to the eleventh was up to 59.103% of the total variance. Comparison of CV, RPR, VPF, I, Dmax and Dmin of core collections by different strategies could be used as an indicator of the candidate core collection. The result showed that a mulberry core collection had been developed using the following sampling strategy: 5% as suitable total sampling ratio, logarithm proportion as sampling proportion within group and least distance stepwise clustering with group. The DNA of more than 117 the primary core collection of mulberry were collected and used to investigate the genetic diversity by SSR and SRAP Molecular markers. The results showed that many significant genotypic variations in the tested materials. According to the statistical data, the genetic similarity coefficients were calculated and the dendrogram had been constructed by using UPGMA of NYSYS software to take cluster analysis and PCA analysis. The genotypes of the species *M.alba*, *M. atropurpurea* and *M. multicaulis* were in a tangle in the dendrogram, which suggested the level of intraspecific variation was rich among different mulberry species. Both *M. alba* and *M. multicaulis* had close genetic relationship because their similarity coefficients were above 0.82. The current complex genetic background was mainly caused by frequent interspecific hybridization and genetic mutation. Based on the genetic information of molecular markers, 34 genotypes were extracted as the core collections of mulberry, which showed that this sample was a good representation of the population in northern China. With the establishment of core collection for the genus *Morus*, the preservation of mulberry germplasm, evaluation and innovative utilization of the genus, as well as the related researches will be readily accessible.

T1

A026

The botany, traditional uses, phytochemistry and pharmacology of *Abrus* species (Leguminosae) in China

Baokang Huang[1], *Xiao Xiao*[1], *Zhongyuan Xu*[2]

1. School of Pharmacy, Second Military Medical University

2. Nanfang Hospital, Southern Medical University

The genus *Abrus* (Leguminosae) is widely distributed in the tropical and subtropical areas of the world with about 18 species. There are four species of this genus in China, including *Abrus canescens* Baker (*Abrus cantoniensis* Hance), *Abrus madagascariensis* R.Vig. (*Abrus mollis* Hance), *Abrus precatorius* L. and *Abrus pulchellus* Thwaites, which mainly grow in Guangdong, Guangxi, Taiwan, Yunnan, Hainan, and some *Abrus* species in China are important medicinal plants with nutraceutical potential and pharmaceutical functions. The published phytochemical studies of *Abrus* species have shown the presence of more than 243 secondary metabolites belonging to flavonoids, alkaloids, triterpenoid saponins, organic acids, amino acids and aliphatic compounds. A widely range of pharmacological activities of plant extracts and fractions focus on antioxidant, anti-inflammatory, anti-parasitic, antimicrobial, anti-trypanosomal and hepatoprotective function. The active ingredients of flavonoids, alkaloids and triterpenoids are useful marker for the quality evaluation of the herb by HPLC. The toxin abrin possesses both bioactivities and toxicity. The genus *Abrus* possesses chemical diversity and biological activities, and many traditional uses of the Chinese *Abrus* species have been validated by modern pharmacology researches. *Abrus cantoniensis* Hance and *Abrus mollis* Hance, as folk hepatic protective herbs, are promising plants which have attracted much attention owing to their multiple functions. The whole plant of *Abrus cantoniensis* Hance is a valuable traditional Chinese medicine with multiple pharmacological effects. It was firstly recorded in Herbalism Record of Lingnan with Chinese name Jigucao to treat jaundices effectively. Nowadays, it has been widely used both in folk and in clinical to treat jaundice, hypochondriac pain, epi-gastric pain, acute and chronic hepatitis, mastitis and other diseases, especially hepatobiliary disease.

T1

A027

Orthodox behavior of *Camellia sinensis* L. seeds from Jiujiang, China.

Ganesh Jaganathan, *Danping Song, Guorong Wu, Wei Liu, Baolin Liu*

University of Shanghai for Science and Technology

Tea (*Camellia sinensis L.*) is the largest and economically most important plant in the Theaceae family. Tea seeds are mostly classified as recalcitrant, i.e. desiccation-sensitive because drying to lower moisture content in the range of 20-30% moisture content on fresh weight basis (MC) results in mortality. However, tea plants growing in different provinces of China are reported to be survive desiccation, therefore they are also classified as orthodox. The present study was undertaken to verify this suggestion. Fruits of tea from Jiujiang, China were collected at the point of natural dispersal in mid-December, and dried over freshly regenerated silica gel (1:5 ratio) in air-tight containers. The moisture content of the seeds at the time of collection was 35.6% on a fresh weight basis. These seeds survived desiccation to 7.3% with 50% viability, indicating the seeds were not recalcitrant. Dried seeds stored for one-month at various low temperatures lost viability. Seeds stored in liquid nitrogen for one week and three months had 23.3 and 13.3% germination respectively, suggesting that some seeds are able to survive cryopreservation. However, fresh seeds stored at 4℃ for 5 months showed 70% germination, but no seeds germinated after 10 months. The tea seeds from Jiujiang, China are certainly not recalcitrant, and are suggested to be orthodox. Nevertheless, further studies are required to strongly conclude the success of cryopreservation.

T1

A028

Ethnobotanical study on Clusiaceae in China

Yuanyuan Ji[1], *Xinbo Zhang*[1], *Bo Liu*[1], *Ping Li*[1], *Hang Shu*[1], *Chunlin Long*[1,2]

1. College of Life and Environmental Sciences, Minzu University of China

2. Kunming Institute of Botany, Chinese Academy of Sciences

There are about 40 genera and 1000 species in Clusiaceae around the world, in which 8 genera and 87 species occur in China, mainly *Garcinia* and *Hypericum*, distributed in Southern China including Yunnan, Guizhou, Guangxi, Guangdong, Hainan, Fujian, Taiwan and other provinces. The plants of Clusiaceae can provide edible fruits, drinks, herbal medicines, dyes, oil-bearing seeds and hard wood, widely used by many ethnic groups in China. However, less systematic ethnobotanical study on Clusiaceae in China was conducted. The traditional knowledge of utilization, and diverse culture about Clusiaceae should be investigated. Six fields investigation were ethnobotanically carried out in Yunnan, Guangxi, Guizhou and Hainan in 2012-2016. Informants were from different ethnic groups including Yi, Dai, Hani, Bai, Jinuo, Zhuang, Bulang, Yao, Buyi, Dong, She, Miao and Shui. Literature research, structured interviews, semi-structured interviews, direct observation, key informant interview, focal group discussions and participatory rural appraisal were used to collect the data and information. The main results and conclusions are as below: 1. During the field investigations on Clusiaceae, 17 species including *Garcina hanburyi, G. bracteata, G. mangostana, G. oblongiflia, G. paucinervis, G. xanthochymus, G. xipshuanbannaensis, Hypericum beanii, H. bellum, H. japonicum, H. monogynum, H. patulum, H. sampsonii, H. henryi, H. stellatum, H. kouytchense*, and *Mesua ferrea* had been surveyed. For each species, the distribution, traditional utilization, important values including food, medicine, wood, and ornamentals were recorded. Most Clusiaceae plants in China are wild with limited distribution. The conservation of germplasm resources and cultivation are required. 2. A number of Clusiaceae species have widely been used as folk medicine for a long history in many ethnic groups, especially the genera *Garcinia* and *Hypericum*. The *Garcinia* species are famous traditional Chinese medicine for the treatment of skin ulcer, hemorrhoids, tumor, dense tinea, bruises, bleeding and burns. The *Hypericum* herbs have the effect of heat-clearing, detoxifying, activating blood circulation to dissipate blood stasis, and treat dysentery, hepatitis, jaundice, irregular menstruation, and traumatic injury. Based on photochemical and pharmacological studies, the chemical constituents from Clusiaceae mainly including benzophenones, xanthones, flavones, naphthodianthrone, acylphloroglucinol, coumarin, and volatile oil, have been proven to be with strong biological activities, which support the traditional use of Clusiaceae as herbal medicine in ethnic societies. 3. Biological and cultural diversity have positive correlation. Because of wide utilization and cultural significance, the germplasm resources of Clusiaceae can be saved and developed better if local participation would be adopted. Some local people have learnt rich parataxonomic knowledge of Clusiaceae and recognized ethnotaxa. Such knowledge should be studied further. Based on ethnobotanical investigations and literature studies on Clusiaceae in China, the traditional knowledge had tentatively been recorded. Findings may be helpful for promoting the application and conservation of traditional knowledge of Clusiaceae. Modern ethnobotanical researches on Clusiaceae are recommend for better protecting and developing Clusiaceae plants in China.

T1

A029

The ecological survey and characteristics analysis of *Avicennia marina* and *Excoecaria agallocha* mangrove community in shenzhen dapeng peninsula

Xiaodong Jin

Guangzhou University, Guangdong Province

Mangrove is an important vegetation type occurred in the tropical and subtropical intertidal line, which plays an important role in maintaining ecological balance of coast, and it is of high ecological significance. *Avicenna marina* is a common dominant population in mangroves. In this paper, the basic mangrove community investigation in Baguang of Shenzhen Dapeng Peninsula was conducted, and its ecological characteristic of *Avicenna marina* community was analyzed. Results showed that the species diversity is very low *in Avicennia marina* community, only found a small amount of *Aegiceras corniculatum*, and *Kandelia candel*. In the investigated 1000 m^2 quadrat, the quantity of *Avicennia marina* is more? *Avicennia marina* is absolute dominant in the community, the important value is 291.582. In the age diameter class of community structure: the number of strong trees is 102, the number big trees is 128, the number of deadwood or dead trees are 54; the number of seedlings or saplings is only 5; While other species (*Aegiceras corniculatum* and *Kandelia candel*) have a relatively small number, the number of the young trees, strong tree and the tree are 20,11 and 16 respectively, which spreading to the sea muds of the forest fringe.We can inferred that the *Avicennia marina* community is in recession. The result would provide a guideline for the protection of *Avicennia marina* community and the construction of *Heritiera littoralis* nature reserve. *Excoecaria agallocha* is a common mangrove associates. In this paper, the *E. agallocha* mangrove in Shenzhen Dapeng peninsula was investigated, and the characteristics of communities was analyzed. *E. agallocha* mangrove in Dapeng peninsula was divided into two vegetation types: *E. agallocha - Kandelia candel* community, *E. agallocha - Aegiceras corniculatum* community. *E. agallocha* was the dominant species; the sequence of community from the water to the embankment was *K. candel* ? *E. agallocha-A. corniculatum* ? *E. agallocha*; The species diversity of community was relatively low, the Shannon diversity index range between 0.214-0.591? the Simpson index range between 0.380-0.500; Community buck level structure distribution was showed recession periods, the community age was most in big trees, old trees and strong tree. *A. corniculatum* seedlings was spreading further deep into the sea. The results were showed that *E. agallocha* mangrove in Dapeng Peninsula was one of the rare typical communities in our country, the environment has obvious been seriously polluted by fishponds, reclamation, tourism development and the living garbage pollution threat, and we must start ecological zoning plan to *Excoecaria agallocha* mangrove ecosystem in Dapeng Peninsula as soon as possible.

T1

A030

Rapid and simultaneous analysis of five alkaloids in four parts of coptidis rhizoma by near-infrared spectroscopy

Jintao Xue

Xinxiang Medical University

Near-Infrared Spectroscopy (NIR) was used to develop a method for rapid and simultaneous determination of 5 active alkaloids (berberine, coptisine, palmatine, epiberberine and jatrorrhizine) in different parts (rhizome, fibrous root, stem and leaf) of Coptidis Rhizoma. A total of 100 samples from 4 main place of origin was collected and studied. With HPLC analysis values as calibration reference, the quantitative analysis of 5 marker components was performed by two different modeling methods, partial least-squares (PLS) regression as linear regression and artificial neural networks (ANN) as non-linear regression. The results indicated that the 2 types of models established are robust, accurate and repeatable for five active alkaloids, and the ANN model was more suitable for berberine, coptisine and palmatine while the PLS model was for epiberberine and jatrorrhizine. The performance of the optimal model was achieved as follows: the correlation coefficient (R) for berberine, coptisine, palmatine, epiberberine and jatrorrhizine was 0.9958,0.9956,0.9959,0.9963 and 0.9923, respectively; the root mean square error of validation (RMSEP) was 0.5093,0.0578,0.0443,0.0563 and 0.0090, respectively. Furthermore, the content of 5 active alkaloids in 4 parts of Coptidis Rhizoma and 4 main origin of place was analyzed and compared by the established NIR model for the comprehensive exploitation and utilization of plant resource of Coptidis Rhizoma. This work demonstrated that NIR spectroscopy may be a promising method as routine screening for off-line fast analysis or on-line quality assessment of traditional Chinese medicine (TCM).

T1

A031

Alternative approaches to cryopreserve Aster altaicus Kitam, an endangered species

Haeng-Hoon Kim
Sunchon National University

Aster altaicus var. *uchiyamae* Kitam is an endemic and endangered species in urgent need of a comprehensive conservation strategy. The objective of this study was to develop complementary cryopreservation protocols, i.e., cryopreservation of dried seeds and droplet-vitrification of in vitro shoot tips. Seeds were cryopreserved after been desiccated to 4.6% moisture content (MC; 19 h over activated silica gel), and seedling emergence was recorded after 2 and 4 weeks. Shoot tips of in vitro plants were cryopreserved using a droplet-vitrification method following optimization of preculture, osmoprotection, and vitrification solution (VS) treatment steps. The standard protocol included preculture with 10% sucrose and 17.5% sucrose for 55 h and 17 h, respectively, followed by osmoprotection with C4-35% (17.5% glycerol + 17.5% sucrose) for 30 min, and cryoprotection with either B5-80% (40% glycerol + 40% sucrose) for 60 min or A3-90% (37.5% glycerol + 15% dimethylsulfoxide + 15% ethylene glycol + 22.5% sucrose) for 20 min. The first count germination rate of cryopreserved seeds dried for 2-19 h, at seed MC of 11.3-4.6%, was comparable to that of the fresh seeds (52.4%). But the final count emergence of cryopreserved seeds dried for 6-19 h, at MC of 7.2-4.6%, was lower than that of fresh seeds. Hence, the hydration window for cryopreserving *A. altaicus* seeds was 8.0-11.3%. In droplet-vitrification methods, preculture and osmoprotection were not critical for the regeneration of cryopreserved *A. altaicus* shoot tips. B5-80%, 80% dilution of PVS3, was comparable to PVS3 in liquid nitrogen survival and regeneration, which indicated that *A. altaicus* shoot tips were sensitive to the osmotic stress induced by highly concentrated VS. A3-90%, a modified PVS2 produced significantly higher regeneration of cryopreserved shoot tips than did the original PVS2 (54.5% vs. 38.8%, respectively). Several ex situ complimentary conservation approaches were successfully developed, including cryopreservation of desiccated seeds and a droplet-vitrification method for in vitro shoot tips.

T1

A032

Traditional plant-derived pesticides to control rice pests in Dong communities communities, Guizhou, China

Qiyi Lei[1,2], *Jiangju Zhou*[2], *Xingyuan Li*[2], ***Chunlin Long***[1,3]
1. College of Life and Environmental Sciences, Minzu University of China
2. College of Life and Environmental Sciences, Kaili University, Guizhou
3. Kunming Institute of Botany, Chinese Academy of Sciences, Kunming

Chemical pesticides have resulted in serious environmental pollution, while botanical pesticides have received widespread attention because of their advantages. More than 1000 plant species have been reported with insecticidal effects. However, only 20 species have been used. In 2015-2016, we conducted field works in three Dong villages (Kengdong, Huanggang and Kendong) in Liping County, southeast Guizhou, China. Traditional knowledge of plant pesticides against rice pests had been collected using semi-structured interview, key informant interview, participatory observation and field test. Results showed that the Dong people have used 15 plant species for controlling pests of vegetables and rice. They are mainly *Derris trifoliata, Cinnamomum camphora, Brucea javanica, Nicotiana tabacum* and others. The local people prepared traditional botanical pesticides according to the proportion of different plant materials to control different pests of vegetables and rice. These plant pesticides showed strong activities against rice planthopper (*Nilaparvata lugens*, *Sogatella furcifera* and *Laodelphax striatellus*), *Chilo suppressalis* and *Tryporyza incertulas*, in which control rate in rice planthopper reached 97.82% while for *Chilo suppressalis* and *Tryporyza incertulas* it reached 95.64%. The utilization of traditional knowledge of local plant-derived pesticides in ethnic communities has shown great development potential. Unfortunately, such local plant-derived pesticides and associated traditional knowledge are becoming endangered. It is of great significance and urgencies to carry out further investigation of and studies on these plant-derived pesticides for sustainable development of agriculture and the safe production of food.

T1

A033

The comparison of the soil physical and chemical properties among different cinnamomum camphora forest types

Hanbing Leng, *Shucheng Feng*
Shanghai Botanical Garden

Urban soil is the foundation of plant growth, affecting the growth status and landscape of plants directly. However the ecological function of soil is disappearing due to increasing urbanization after the long-term interruption of human activity. The natural physical and chemical property of soil has been damaged arising from soil pollution, variation of soil characteristics and *et al.*, Artificial forest with the function of restoration and reconstruction to urban ecological system can strengthen the integration of vegetation, soil and other organisms. Therefore 3 different *Cinnamomum camphora* forest types (pure forest on bare soil, pure forest only with grass, mixed forest planting with grass and shrubs), common species communities in Shanghai, were analyzed, based on field sampling and laboratory test. We explored the effect of different forest types on the change pattern of soil physical and chemical properties. The aim is to provide scientific guidance for urban landscaping application. The result showed that: 1) Significant differences were found in bulk density, non-capillary porosity, maximum water capacity, capillary water capacity, field capacity and soil water storage capacity under different forest types. Mixed forest planting with grass and shrubs presented better aeration and water-holding capacity in soil. 2) There were significant differences in organic carbon contents, total nitrogen contents, hydrolysis nitrogen contents and available phosphorus contents among different forest types. Mixed forest planting with grass and shrubs contained highest carbon and nitrogen contents in soil while more available phosphorus contents was distributed in pure forest only with grass. 3) The observation in this study showed a significant negative correlation between bulk density and non-capillary porosity, maximum water capacity, capillary water capacity, field capacity, soil water storage capacity, soil organic carbon contents, total nitrogen contents, total phosphorus contents and hydrolysable nitrogen contents. Soil organic carbon had a positive significant correlation with total porosity, maximum water capacity, capillary water capacity, field capacity, total nitrogen contents and hydrolysable nitrogen contents. These findings indicated that the compound plantation structure (arbor-shrub-herb type) had the plant root more uniformly distributed in the vertical and horizontal level and increased the input of understory vegetation litter fall to the most extent. And the lower bulk density and higher organic carbon contents caused by this kind of structure played an important role in improving soil physical and chemical properties as well as promoting the soil nutrient cycling.

T1

A034

Effects of progressive drought stress on the accumulation of flavonoids in the growth of radix astragali

*Guangyue Li, Yuehua Liu, **Guilin Chen***
College of Life Science, The Good Agriculture Practice Engineering Technology Research Center of Chinese and Mongolian Medicine, Inner Mongolia University

The plants defense environment changes by producing secondary metabolites and physiological and biochemical reactions. Flavonoids play an important role in plant response to abiotic stress and have tissue specificity. Radix Astragali is a traditional Chinese herb, it has successfully evolved in arid and semi-arid environments. The plant growth usually be suppressed in humid environment. It is mainly distributed in the north and northwest of China. Now, many farmers have begun to cultivate them because of shortage of wild resource. Therefore, it is important to understand the accumulation of flavonoids in different organs of Radix Astragali under drought stress, and to understand the main synthetic sites of isoflavones glucoside, which is of great significance to improve the comprehensive utilization of Radix Astragali. The dynamic of physiological state and content of calycosin-7-o-β-D-glucoside and calycosin in root, stem and levaes were studied, use the seedlings of *Astragalus membranaceus* var. *mongholicus* (Bge.) Hsiao (*A. mongholicus*) and Astragalus membranaceus Bunge (A. membranaceus), under the continuous drought stress, by pot culture simulated drought conditions, at greenhouse in Inner Mongolia University. calycosin-7-o-β-D-glucoside and calycosin are distributed and obvious difference in the three different parts, which the content of calycosin-7-o-β-D-glucoside was leaf > stem > root. The content of calycosin-7-o-β-D-glucoside (0.4mg/g DW) in the leaves of *A.membranaceus* was higher than that of *A. mongholicus* (0.25mg/g DW). However, calycosin are only distributed in stems and leaves of seeding in two different species, which the content of calycosin (0.22mg/g DW) in the leaves of *A. membranaceus* was higher than that of *A. mongholicus* (0.13mg/g DW). What's more, the contents of calycosin-7-o-β-D-glucoside and calycosin were increased during 8-12 d in leaves, but the calycosin-7-o-β-D-glucoside was no significant change in roots of two different species under drought stress. Those results indicated that the responses to drought stress were different in *A. membranaceus* and *A. mongholicus*, and the distribution and content of calycosin-7-o-β-D-glucoside and calycosin were different in different organs of two different species. What's more, moderate drought stress could promote the accumulation of calycosin-7-o-β-D-glucoside. In addition, some studies have shown that the key enzymes and glycosyltransferase gene expression of calycosin-7-o-β-D-glucoside in the leaves of Astragalus membranaceus were significantly higher than those in other organs. Moreover, the differences in the content of different organs in the three-year-old *A. mongholicus* were found. Thus, the leaves may be the main synthetic site of the calycosin-7-o-β-D-glucoside.

T1

A035

Cloning and analysis of Phenylalanine Ammonialyase Gene (*CiPAL*) in *Clematis intricata* Bunge

*Ting Liu, Hongjiao Li, Xingjiao Sun, Tebuqin Borjigin, **Xiao Li***
Inner Mongolia Medical University

Clematis intricata Bunge was one of the traditional Mongolian medicinal plants with the drug name "Temuri-aoriyanggu"and it commonly used in compound preparations which have the function against lump, tumor, dyspepsia, edema, wound, hemorrhoids, *etc. Clematis intricata* contained several flavonoids with the pharmacological action of liver protection, anti-inflamatory, antioxidant, anti-tumor, *etc.* Phenylalanine ammonia-lyase, a key enzyme of plant secondary metabolism, catalyzed the first reaction in the biosynthesis from L-phenylalanine to a wide variety of natural products (flavonoids included) based on the phenylpropane skeleton. Phenylpropanoids metabolic products could further form flavonoids via the metabolic pathway. In this study, phenylalanine ammonialyase gene, named asCiPAL, was isolated from *Clematis intricata* Bunge.The full-length of CiPALis 2386 bp obtained by RT-PCR and RACE, which includes 78 bp of 5'untranslated region, 190 bp of 3' untranslated region and 2118 bp complete open reading frame (ORF) encoding 705 amino acids. The protein formula was C3382H5395N937O1035S26, the molecular weight and theroreticalisoelectric point (pI) of the deduced CiPAL protein were 76.6kDa and 6.10, respectively .The amino acid sequence analysis showed that there are three domains of PAL-HAL, PLN02457 and pheamlyase, belonging to the superfamily of cleavage enzymes. And an enzyme active site sequence (GTITASGDLVPLSYIA) at position 187-202 of CiPAL which belongs to a typical PAL family member. It was highly homologous to other PAL proteins from different species, up to 85%. Phylogenetic tree analysis indicated that CiPAL was closely related to PAL of Caprifoliaceae and Berberidaceae. Real-time PCR analysis showed that CiPAL expressed in root, rhizome, stem, leaf, bud, flower and seed tissues at different levels, and the expression was highest in seed that is more 10 times than that in leaf. And we will make further plant secondary metabolism research of *Clematis intricata* Bunge based on the result in this study about the characteristics of CiPAL.

T1

A036

Assessment on forest fragmentation of National Forest Na-

ture Reserves in China

Yao Li[1], *Xiangming Xiao*[1,2]
1. Fudan University
2. University of Oklahoma

Forests produce half of net primary production in the world, and play an important role in the global carbon cycle, water and heat fluxes, biodiversity, and soil and water conservation. The information of forest area, spatial distribution and fragmentation of forest is needed to assess the biodiversity conservation and ecosystem services. In this study, using the forest fragmentation model and 50-m forest cover generated by analyses of MODIS and PALSAR images, a forest fragmentation map of China in the year of 2010 was built at six different fragmentation levels including interior forest, perforated forest, edge forest, patch forest, transitional forest and undetermined forest. We explored the forest fragmentation status in 114 national forest nature reserves of China. Our results show that in national forest nature reserves of China, the patch forest accounts for the largest proportion (16.20%), and the perforated forest area accounts for the second largest proportion (15.82%), while interior forest is the smallest (3.11%) (except the undetermined forest). In most of protected area, the areas proportion of six different fragmentation levels forest accounts for 30% (in entire protected area, core area, buffer area and experimental area). Overall, our results clearly show that the forest fragmentation occurs extensively and varies substantially in national forest nature reserves of China, and therefore provide basic information to reduce forest fragmentation and avoid further losses of forests in the future.

T1

A037

Preliminary study on vascular plant specimen collected by emile licent in North China (1914-1936)

Li Yong
Tianjin Natural History Museum

Emile Licent (1876-1952), a french Jesuits missionary, doctor of zoology in French Academy of Sciences and a biologist, created the first comprehensive natural science museum in north China--The Hoangho Paiho Museum (former Tianjin Museum of Natural History), was known for its abundant collections. In 1914, Emile Licent came to China and since then, worked in China for 25 years. Travelling almost every region of north China, he collected a large amount of specimens of geology, paleontology and flora and fauna, among which the paleontological fossils in Cenozoic era won him international fame. In 1938, he returned to France and settled down in Paris. Most Valur specimens collected by E.Licent in Northern China are kept in Tianjin Natural History Museum. Others were collected in Paris, London and America. There are about 40,000 specimen collected in TIE (Herbarium code of Tianjin natural history museum), which belong to 1201 generas and 210 families. Some of them were identified by Handel-Mazzetti, parts of them were collected and research by France and Russia' botanist.

T1

A038

Gap Analysis with spatial conservation prioritization for Conservation initiatives in Southeast Asia region

Bo Liu*, Hongfeng Wang, Zheping Xu, Maofang Luo, Qinwen Lin, Keping Ma**
Institute of Botany, Chinese Academy of Sciences

The Southeast Asia region contains four global biodiversity hotspots: Sundaland, Wallacea. Philippines and Indo-Burma; however, these hotspots and the biodiversity they support are under the constant threats of human destructive exploration activities and economic development. Conservation strategies based on the geographic patterns for conservation gaps could improve the effectiveness of policy and management of nature reserves conservation initiatives for Southeast Asia at the broad scale. This paper aims at establishing and optimizing an ecological network for conservation in this region based on large-scale gap analysis. Gap analysis has been used to identify priorities for species and habitat types. We also replicated the analysis by using Species Distribution Models (SDMs) as biodiversity features to evaluate the robustness and utility of our environment-based analysis. Based on current available red list data and nature reserve list in this region, results showed, the 100 most threatened animal and plant species in this region are under comprehensive protection. However, there are still some protection gaps, and these identified gap areas will provide a scientific basis for further planning and development of nature reserves.

T1

A039

Cryopreservation of shoot-tips of *Populus euphratica* domant buds by encapsulation-vitrification

Yanping Liu
Tarim University College of life Science

Shoot-tips of *Populus euphratica* domant buds were successfully cryopreserved by using encapsulation-vitrification. Domant shoot-tips were encapsulated 3% calcium alginate beads. The main factors influencing the viability such as sucrose concentration, preculture time, different vitrification solution and the time beads in vitrification solution were studied. The results show that the high viability can be obtained when alginate beads were precultured for 24h in the 0.8mol/L sucrose. The beads were dehydrated with vitrification solution [40%Gly + 45% (0.4mol/L) Suc + 10%PEG (4000)+5%DMSO]for 30 min at 25℃ .Under optimum conditions, the viability of domant shoot-tips can reach 77% at the most.After cryopreservation, the dorman t shoot-tips could regrow directly and the genetic stability was maintained.

T1

A040

Efficient *in vitro* propagation method on shoot multiplication of *Paphiopedilum parishii* and *P. charlesworthii*

Bo Long[1], *Sven H. Wagner*[2], *Wenjun Su*[1], *Jichen Wei*[1], *Dexi Kong*[1]
1. School of Life Sciences, Yunnan University, Kunming, China
2. Sails For Science Foundation

Paphiopedilum plants needs four to five years to generate a new plant in the wild. The native plant supply cannot satisfy the horticultural market demands. In the China's Biodiversity Red List Species—Higher Plant, *P. parishii* (Rchb. f.) Stein is treated as Critically Endangered, *P. charlesworthii* (Rolfe) Pfitzer is treated as Endangered. *Paphiopedilun parishii* is also evaluated as Critically Endangered in China's Red List Species, only have one location, is seriously threaten species. Urgent protection and more in depth investigation for its propagation is now needed. In order to conserve these species and avoid further destruction on remaining plants at natural habitats, propagate *in vitro* is rather needed. The effects of various plant growth regulators (PGRs) on the frequency of shoot organogenesis in *P. parishii* and *P. charlesworthii* were investigated. The combination of 6-BA 1 mg/L with NAA 0.5 mg/L is optimal for the shoot multiplication on *P. charlesworthii*. An average shoot length of 156%. The longest shoot was achieved at an average of 1.60 cm through a combination of 6-BA 8 mg/L with NAA 0.5 mg/L with just 56% shoot multiplication. *P. parishi* incubated only in the presence of 2 mg/L ZT + 0.5 mg/L NAA, showed a 33% increase in the frequency of shoot organogenesis, the new shoot reached up to 0.4 cm after 10 months of incubation. Our methods may prove to be of value for both, horticultural and conservation purposes.

This research was funded by a grant from the National Natural Science Foundation of China (31100249).

T1

A041

Selection of native tree species for subtropical forest restoration in Southwest China

Yang Lu

Kunming Institute of Botany, Chinese Academy of Sciences

The use of native species in forest restoration has been increasingly recognized as an effective means of restoring ecosystem functions and biodiversity to degraded areas across the world. However, successful selection of species adapted to local conditions requires specific knowledge which is often lacking, especially in developing countries. In order to scale up forest restoration, experimental data on the responses of native species to propagation and restoration treatments across a range of local conditions are required. In this study, the restoration potential of 34 native tree species was evaluated based on nursery research and field planting experiments at a highly degraded site in a subtropical area of southwest China. We examined species performance in terms of germination rates as well as survival rates and growth two years after planting. Of the 34 species examined, 25 had a germination percentage greater than 50%. Survivorship ranged from 0 to 97% across species and was greater than 50% for 20 species. Mean monthly growth increments varied between species. Pioneer species performed well, and 14 mid- and late-successional species performed reasonably well to very well in this study. However, the remaining 16 mid- and late-successional species performed poorly. These results indicate that carefully selected mid- and late-successional species can be effectively incorporated into mixed species plantings. This data can be used to inform restoration planning, helping to identify suitable species and so enhance the biodiversity and resilience of restored forests.

T1

A042

Diversity of Tibetan incense plants in China

Binsheng Luo[1], *Feifei Li*[1,2], *Chunlin Long*[1,3]

1. College of Life and Environmental Sciences, Minzu University of China

2. Chinese Research Academy of Environmental Science

3. Kunming Institute of Botany, Chinese Academy of Sciences

Burning incense plants for smoke to worship and communicate with gods is a traditional religious custom in Tibetan area in China. In order to comprehensively and deeply understand the relative culture from the ethnobotanical angle, several times of field investigations during 2012 to 2013 were carried out in some Tibetan areas including Muli county, Kangding county, Danba county in Sichuan province and Lhasa, Mainling County, Nyingchi County and Gongbo'gyamda County in Tibet. Semi-structured interviews and random interviews were conducted to 53 people including local Tibetans, monks, incense merchants. In total, 97 species of incense plants, from 34 families and 20 genera, and their relative information were collected. In addition, 14 species were endangered, 27 species were medicinal plants and 13 species were endemic species of Tibetan. Otherwise, the importance, protection and sustainable utilization of incenses plants had been discussed, the difference of incense plants among Tibetan nationality and other ethnic groups were analyzed in the end as well. This research not only helped to record the endangered traditional Tibetan Knowledge, but also laid a foundation for future development of incense plants.

T1

A043

Dormancy and germination of *Firmiana danxiaensis*

Xiaoying Luo[1], *Qianmei Zhang*[2], *Hai Ren*[2], *Guohua Ma*[2]

1. Shaoguan University

2. South China Botanical Garden, CAS

Firmiana danxiaensis H. H. Hsue & H. S. Kiu (family Sterculiaceae) was listed as a Second Class Key Protected Wild Plant of China in 1999 and in the Conservation Program for Plants with Extremely Small Populations in China in 2012. Its distribution is restricted to Danxia landform, including Danxiashan National Nature Reserve (DX) and Cangshizai County Nature Reserve (NX) at Shaoguan City, South China. *Firmiana danxiaensis* grows on red cliffs which formed by sandy conglomerates and shallow soil layers. In order to conserve this species, we studied its germination methods. We collected ripe seeds of *F. danxiaensis* from DX and NX on August, 2016. The seeds were globose and yellowish-brown, ca. 6.23 mm (NX) to 6.89 mm (DX) in diameter, 92.28 g (NX) to 141.79 g (DX) in 1,000-grain weight, 4.92% (DX) to 8.44% (NX) in the percentage of empty seeds, 11.01% (NX) to 11.36% (DX) in water content, and 58.44% (NX) to 82.89% (DX) in seed viability. Germination test was carried out with total 12 treatments, including four temperatures (25,25/35,15/25 and 10/20℃ (night /day)), and three photoperiods (24 h light, 24 h dark, 12 h light /12 h dark). The highest average germination rate was 1.48% at 25/35°C, which was significantly lower than the seed viability (58.44%-82.89%). According to this result, we deduced that the seed was dormant. The results also indicated that illumination had no effect on seed germination and 25/35 ℃ treatment was the suitable germination condition with the shortest germination time lag and the fastest growth rate. The water absorption rate (WAR) of CK (intact seed) was about 20% when it achieved constant weight in 48 h. The WAR of ZK (puncture the seed hole) reached 120% (DX)-140% (NX) in 96 h and saturation has not yet reached. The WAR of QK (shuck the seed coat) was about 130% (DX)-170% (NX) in 48 h. These results suggested that the seed coat was a physical obstacle on water absorption, and resulted in seed dormancy. To break dormancy, six treatments were conducted, including 1) soak the seeds into 98% H_2SO_4 for 1 h; 2) hot and cold water alternation (90℃ /5 min transfer to 0℃ /5 min); 3) 4℃ storage for 7 days; 4) soak the seeds into 500 mg/L GA_3 and 500 mg/L 6-BA for 24 h at 50℃ , respectively; 5) soak into water at 50℃ ; 6) CK. The 98% H_2SO_4 treatment turned out to be an effective way to break the dormancy of the seeds. Those germination methods are helpful for the conservation of this species.

T1

A044

Flexibility of resource allocation in a hermaphrodite-gynomonoecious herb through deployment of female and male resources in perfect flowers

Jannathan Mamut[1], *Ying-Ze Xiong*[2], *Dun-Yan Tan*[1], *Shuang-Quan Huang*[2]

1. Xinjiang Key Laboratory of Grassland Resources and Ecology and Ministry of Education Key Laboratory for Western Arid Region Grassland Resources and Ecology, College of Grassland and Environment Scienc

2. Institute of Evolution and Ecology, School of Life Sci

Premise of the study: It has been hypothesized that two flower types permit flexible allocation of resources to female and male functions, yet empirical evidence for the sex allocation hypothesis remains scarce in gynomonoecious species. To characterize resource allocation to pistillate and perfect flowers and allocation of perfect flowers between gynomonoecious and hermaphrodite individuals, we examined the flexibility and whether female-biased allocation increases with plant size in the hermaphrodite-gynomonoecious herb Eremurus anisopterus. Methods: Frequency of gynomonoecious individuals, flower production and plant size were investigated in different populations. Floral allocation was compared among the three flower types of E. anisopterus. Key Results: Frequency of gynomonoecious plants varied from 2% to 17% in nine populations. Only larger plants produced female flowers at the bottom of racemes. Both female and perfect flower production tended to increase proportionately with plant size in gynomonoecious individuals. Female flowers did not produce less biomass than perfect flowers from hermaphroditic or gynomonoecious plants. However, both female and perfect flowers from gynomonoecious individuals had lighter stamen but larger pistil mass than perfect flowers from hermaphrodites. Conclusions: Although the prediction of an increase in female flower number with plant size was not observed in E. anisopterus, the flexibility of sex allocation in gynomonoecious species was confirmed in that gynomonoecious individuals had a female-biased floral allocation compared to hermaphroditic individuals. Such comparisons of gynomonoecious to hermaphroditic individuals permit us to unveil a sexual adjustment strategy, flexibility of sexual investments within plants.

T1

A045

Population of rare species *Alcea hyrcana* (Grossh.) Grossh. in Azerbaijan

Naiba Mehdiyeva [1], *Gulnara Osmanova*[2], *Rena Abdiyeva*[3]

1. Institute of Botany of Azerbaijan National Academy of Sciences

2. Mari State University, Russian Federation

3. Institute of Botany of Azerbaijan National Academy of Sciences

Alcea hyrcana (Grossh.) Grossh. - Hyrcanian hollyhock (Malvaceae family), rare species of Azerbaijan is classified as ?endangered? (EN B2ab (ii, iii, iv) according to the IUCN Red List. The height of perennial herb is 35-85 (100) cm. Leaves are petiolate, floccose, rounded, heart-shaped, 5-7 lobed with irregular dentated blunt lamina. The nimbus is pink and petals on the top are daedalous, up to 400 mm long. Seeds are brown, covered with rare short hairs. Flowering in June-August, fruiting in July-August. Reproduced by seeds. Xerophyte, heliophyte grows in lowland and mountainous parts of the Lankaran area, on seaside sands in psammophyte-littoral groupings and in the foothill forests. The plant has decorative and medicinal value. Researches of coenopopulation (CP) *A. hyrcana* were carried out in the territory of Hirkan National Park in 2015-2016. Characteristic type of vegetation of the Hirkan National Park is forest with edificators of *Quercuseta, Parrotieta, Populuseta, and Pterocarieta*. Individuals of *Alcea hyrcana* were studied in the composition of association *Parrotia persica+Zelkova carpiniefolia+Punica granatum*. Components of association are *Albizzia julibrissin* Durazz., *Ficus hyrcana* Grossh., *Filago arvense* L., *Pimpinella peregrina* L., *Sedum lencoranicum* Grossh., *Allium lenkoranicum* Miscz. ex Grossh., *Celtis caucasica* Willd., *Pteris cretica* L., *Pteridium tauricum* V. Krecz., *Plantago media* L. *etc.* According to the classification ?delta-omega?, CP of this species in this area is young (Δ - 0,078, ω - 0,144) and presented by individuals of pregenerative (97%) and generative periods (3%) with domination of individuals of the pregenerative period (juvenile (j) - 39 individuals, immature (im) - 23, virginile (v) – 2). High values of the recovery index (Ir) 32.0 and substitution (Is) 30.0 evidences of active self-maintenance of this CP. Ontogenetic spectrum of CP is left-sided. Evaluation of ontogenetic state *A. hyrcana* indicates that CP of plant develops normally and as it grows in a protected area it is not subjected to the negative anthropogenic influence. However, due to the affection of the surrounding vegetation with fungal diseases there is high probability of destruction of vegetative and generative organs. *A. hyrcana* could adversely affect the favorable development of this type CP in the future.

T1

A046

Comparative biogeography of Australian plants

Daniel Murphy[1], *Darren Crayn*[2]

1. Royal Botanic Gardens Victoria

2. Australian Tropical Herbarium

The Australian flora has been the subject of many historical biogeographic analyses undertaken using a comparative approach. In this talk we review and summarise the main results of these analyses, illustrate the broad ecological and historical patterns that are emerging, and discuss the future directions and value of a comparative approach, using plant groups in Australia. Our review highlights the existence of five ecologically-based biomes; a Northern Australia monsoon tropical savanna biome; a centrally located and massive arid biome; mesic biomes of peripheral southern parts of the continent; and an aseasonal wet biome along the eastern coastal parts of Australia. Furthermore we point to some consistent patterns and results for Australian plants including: (1) temperate southeast-southwest disjunct pattern; (2) the arid zone has a complex history; (3) the McPherson-Macleay zone (central eastern coast) zone is contentious; (4) Tasmania is sister to the rest of Australia; (5) northern and southern Australia show an early break in connectivity; and (6) immigration into northern Australia is strong and mostly recent. However, we also noted that within Australia there are multi-layered and complex historical distributional patterns of endemism of plants, including some human-mediated, pre-human and ancient (relictual) distributions apparent. Australian plant historical biogeography now sits at a crucial point in history. Two major recent developments are driving a step change in the discipline. First, the availability of enormous geospatial datasets based on digitised collections data (e.g. Australia's Virtual Herbarium and the Atlas of Living Australia) have catalysed methodological developments that utilise these data for defining areas (phytogeographical regionalisation) and discovering patterns of phylogenetic endemism and diversity. Secondly, new genomic approaches have increased both the range of herbarium specimens that are amenable to DNA analysis and the volume of characters for analysis which is chipping away at one of the largest impediments to historical biogeography—the availability of phylogenies for many key Australian plant lineages.

T1

A047

Quantitative assessment of priority for rare and endangered plants in Yerengu Nature Reserve of Hubei

Qianqian Peng

Hubei University

Based on sample-plot survey and sample-line survey, rare and endangered plants resources in Yerengu Nature Reserve of Hubei Province was investigated, and the priority conservation of rare and endangered plants was quantitative analyzed by 5 synthetical indicators like threatened coefficient, genetic coefficient, species value coefficient, comprehensive evaluation value, protect status coefficient and breeding difficulty coefficient, and basing on 15 individual indicators. The results showed that: There were 31 species of rare and endangered plants belonging to 30 genra of 26 families; 5 species was ranked as the first grade for conservation priority, 14 species as the second grade, 8 species as the third grade, and 4 species as the fourth grade; Most rare and endangered plants distributed at the altitude of 1,000-1,600 m, and the distribution is concentrated.This research result is more realistic and effective, which provides a scientific basis for the management of the protected area to make the protection plan of the rare and endangered plants in the area.

T1

A048

Study on spatial distribution of *Lithocarpus* in China

Lichuan Qiu

Taiyuan Normal University

Fagaceae has many species which are important in subtropical evergreen broad- leaved forest. The character and range of the forest in Chinaare more prominent and larger than those of other place in the world. Fagaceae has 7 genus. *Lithocarpus* is one of them. The number of its species is more than that of other genus inChina. Spatial distribution characters of *Lithocarpus* will be helpful for protecting and restoring Fagaceae and vegetation. The paper provides basis of plant spatial distribution with GIS. Its spatial data is gotten from Flora Republicae Popularis Sinicae, Scientific Database of China Plant Species and so on. The data is made and analysed in ArcGIS 9.2, including map and attributes inChina. The map is composed of counties. The attributes are species of *Lithocarpus*. There is 136 species in *Lithocarpus*. *Lithocarpus* is mainly distributed in the south ofChina. It is from the most northern Wenxian to the south of the Sanya and from the most eastern Ningbo to the west of the Nielamu. It spans latitude from 33°20' to 18°9'N and a large longitude from 122°8' to 85°27'E. *Lithocarpus* accounts for 24% of China's counties, 51% of Fagaceae's counties. The number of species in Pingbian is more than that in other counties. The south of Yunnan is the centre of species diversity. Ranges of species are different in *Lithocarpus*. A few species are widely distributed, such as *L. hancei*, *L. harlandii*, *L. henryi*, *L. litseifolius*. Their distribution is close to the range of *Lithocarpus*. These species are more adaptable to environment than others. Fagaceae and vegetation will be restored by these species. Most species are distributed in a limited range. Some species distribute mainly in Yunnan. They are *L. amygdalifolius*, *L. fenestratus*, *L. truncates*, *L. dealbatus*, *L. grandifolius*, *et al.,* Some species are only in Yunnan, such as *L. fohaiensis*, *L. mairei*, *L. microspermus*, *L. xylocarpus*. Some species are limited in a place. Most of these species are in the south of Yunnan. They are *L. cryptocarpus*, *L. tabularis*, *L. leucodermis*, *L. tephrocarpus*. Yunnan is the most important place of *Lithocarpus* inChina. It should be protected. Some species are limited in island, such as *L. elaeagnifolius*, *L. obovatilimbus* in Hainan and *L. kawakamii*, *L. lepidocarpus*, *L. nantoensis* inTaiwan. We should protect these species which are in limited range. Spatial distribution of *Lithocarpus* is studied in quantification with GIS. Its range may be showed by longitude and latitude. The distributed counties and diversity centre of *Lithocarpus* may be showed and gotten by spatial distribution map.

T1

A049

Current status of *ex situ* conservation of medicinal plants in China

Ling Que, *Guang Yang*

(State Key Laboratory of Dao-di Herbs, National Resource Center for Chinese Materia Medica, China Academy of Chinese Medical Sciences

Ex situ conservation of plants is an emergency protection method to prevent endangered species from extinction. In order to balance the relationship between resource conservation and exploitation, it is necessary to grasp the status quo of *ex situ* medicinal plants in China. Based on Chinese Union of Botanical Gardens and Chinese Union of Medicinal Plant Gardens, this paper studied the *ex situ* conservation medicinal plant types. The results showed that there are more than 5,700 species of medicinal plants in China, and the *ex situ* conservation and breeding mechanism of medicinal plants are still immature. There are a large number of medicinal plants subject to natural environmental restrictions are not *ex situ* conservation, but many common medicinal plants are over-protected. This paper is of great reference value for improving the protection system of medicinal plants and improving the efficiency of *ex situ* conservation.

T1

A050

Diversity of the *Cynomorium songaricum* rhizosphere microbial community under natural condition

Mengyun Ren

Hainan University; Chinese Research Academy of Environmental Sciences

Many studies have shown that rhizosphere soil microbial communities are crucial to plant growth, health, and stress resistance and numerous studies have implicated desertification as a cause for loss of biodiversity. In this study, 16S ribosomal RNA amplicon sequencing was used to determine the taxonomical structure of the microbial communities of the typical holoparasite psammophyte *Cynomorium songaricum*. The α-diversity and β-diversity were then calculated using QIIME. The gallic acid, protocatechuic acid, catechin, and phloridzin contents in *C. songaricum* from 18 fields were determined using high-performance liquid chromatography. The relationship between the chemical components and the microbial genetic composition is critical for understanding how microbes on the *C. songaricum* community affect its medicinal function. *Proteobacteria*, *Actinobacteria*, *Euryarchaeota*, *Bacteroidetes*, *Firmicutes*, *Gemmatimonadetes*, *Nanohaloarchaeota*, *Acidobacteria*, *Thaumarchaeota*, and *Saccharibacteria* were the most abundant phyla in the rhizosphere of *C. songaricum*. The gallic acid, protocatechuic acid, catechin, and phloridzin contents showed regional differences. However, principal coordinates analysis indicated strong interspecific patterns of microbial rhizosphere communities. Variance partitioning canonical correspondence analysis indicated that environmental factors and chemical composition exerted strong influences on the structure of the rhizosphere microbial community and explained 38.18 and 24.68% of community variances between samples, respectively. Our results showed strong interspecific clustering of the microbial rhizosphere communities of *C. songaricum*. Annual precipitation explained most of the variation in the composition of microbial rhizosphere communities of *C. songaricum*, followed by altitude and pH, annual mean temperature, gallic acid, catechin, protocatechuic acid, and phloridzin. Desertification was shown to have complex effects on microbial communities in the northwest of China.

T1

A051

Hotspots within hotspots: Identifying plant diversity and endemism centers in tropical Asia

Ke Tan, ***Ming-Xun Ren***

Hainan University

Southeast Asia is the home of diverse tropical plants and is the main part of Indo-burma biodiversity hotspot. This region is mostly consisted with islands and peninsulas, which makes the habitats and species distributions largely fragmentedly. The information about geogaphical locations of species diversity, especially the endemic plants, is crucial to to detect biodiversity hotspots for conservatio priority. Collection data from two species-rich tropical families, i.e. Dipterocarpaceae (560 spp.) and Malpighiaceae (80 spp.) were used to explore patterns of biodiversity at 1 × 1 latitude/longitude grid resolution, with the main goal of identifying the occurrence of centers of species richness and endemism. For each grid cell, species richness (total count of species), weighted richness (down-weighting each species by the inverse of its range) and endemism proportion were calculated. Grid cells within the top 10% of highest values of weighted richness were considered centers of endemism. Our results identified six centers of species diversity and endemims, two are islands, i.e. Sri Lanka and Borneo, two are in Thailand, respectively in north (upper reaches of Chao Phraya River) and south (Siam Bay), the other two are in Southwest China, roughly corresponding to river valleys of Hongshuihe River (upper reaches of the Pear River) and Nujiang River (upper reaches of Salween River). The river and the monsoon climate can facilitate northward spreads of tropical plants, which facilitated the process of speciation near river valleys in Thailand and China. This research can provide an overview of approaches to identifying specific biodiversity hotspots of high conservation priorities.

T1

A052

Carotenoid profiling in the peel and pulp of selected banana (*Musa* spp.) varieties

Ou Sheng

Institute of Fruit Tree Research, Guangdong Academy of Agricultural Sciences

Bananas (*Musa* spp.) are not only one of the most popular fruits but also important staple food in some tropic countries. Carotenoid in pulp of banana fruits is an important source of vitamin A. While little is reported on carotenoid in peel, which is traditionally recognized as byproduct of food industry and may be a potential source of antioxidant compounds. Here we determined and reported carotenoid content and composition in both peel and pulp of 36 different banana cultivars. A total of 3 and 8 carotenoid components were identified in pulp and peel, respectively. The results show that carotenoid ranged from 0.18 to 36.82μg/g FW in pulp and 0.9 to 16.2 μg/g FW in peel. Vitamin A values as retinol activity equivalents (RAE) in the pulp ranged from 0.003 to 1.85 μg/g. This report represents the first comprehensive assessment of carotenoid profiling in both fruit peel and pulp of 36 white to orange types of bananas and demonstrates banana peel product is also a valuable source with great nutritional properties.

T1
A053
Proposal to popularize *Acorus* as indoor ornamentals

Hang Shu[1], *Yuanyuan Ji*[1], *Shuang Zhang*[1], ***Chunlin Long***[1,2]
1. College of Life and Environmental Sciences, Minzu University of China, Beijing, China
2. Kunming Institute of Botany, Chinese Academy of Sciences, Kunming, China

Plants in *Acorus* are not only important traditional herbal medicines, but also beautiful ornamentals. In China, the history of *Acorus* used as ornamental can be tracked back to 2000 years ago. Initially, *Acorus* was a courtyard ornamental. Until Tang and Song dynasties, it was grown as an indoor ornamental such as potted plant. There are 3 species in the genus *Acorus*, in which *Acorus tatarinowii* is widely used for ornamental purpose. *Acorus* plants are perennial and evergreen, with aromatic leaves and rhizomes. Because of their unique biological characteristics, they are regarded as symbolic and holy plants which have been endowed with the exorcism roles in traditional Chinese culture. They are also praised as elegant ornamental plants symbolized the moral integrity of indifferent to fame and wealth and resist severe coldness. With strong viability and resistance to stress, *Acorus* can be transplanted at any time and grown easily in adequate humid and shady environments. The reserves of them are extremely rich. *Acorus* plants have the functions of purifying environment by absorbing the formaldehyde, benzene and other indoor decorating pollution. Therefore it is very suitable to grow *Acorus* indoors. According to their horticultural characters and cultural values, we strongly recommend *Acorus tatarinowii* and *A. gramineus* being popularized as indoor ornamentals, while *A. calamus* can be cultivated as a courtyard ornamental plant.

T1
A054
Park "Presidio" (San-Francisko, USA) - unique botanical object

Mikhail Shustov[1], *Vladimir Shatko*[1], *David Gunn*[2], *Betty Withrow*[2], *Gideon Dollarhide*[3], *Anna Erst*[4]
1. Main Botanical Garden named after N.V. Tsitsin, RAS
2. Missouri Botanical Garden
3. Botanical Garden, The University of California, Berkeley
4. Central Siberian Botanical Garden, SB RAS

In the framework of the Russian-American project "Conservation of rare and endangered species of plants and a comparative study of the North American and Eurasian Flora" in August 2016 in San Francisco (USA), we visited the park, "Presidio", located on the northern tip of the San Francisco Peninsula and included in Golden Gate National recreating Area since 1994. The park was founded in 1776 on the territory of the military base and the US Army hospital. Plants were brought from all subtropical regions of the world available to the US Army. For 240 years it was formed unique botanical object of world importance. Because the jurisdiction of the US Army, its territory was not studied by botanists properly. It is found out from scientific literature and the Internet information available. The list of plants is far fromcomplete. According to our preliminary assessment there are more than 200 taxa. The age of some exotic species (such as *Cupressus macrocarpa* Hartw., *Quercus ilex* L., *Eucalyptus globosus* Labill., *Cryptomeria japonica* (L.) K. Koch is 200 - 250 years, and many of them are almost never found in the territory of California. Sothe park is not only of historical, but the scientific importance. The most common species are: *Cedrus deodara* (Roxb.) G. Don, *C. atlantica* (Endl.) Manetti ex Carriere, *Eucalyptus globosus* Labill., *E. oblique* L'Her., *E. robusta* Sm., *Cupressus macrocapa* Hartw., *Pinus radiata* D. Don, *P. contorta* Douglas, *P. canariensis* C. Sm., *Araucaria araucana* (Molina) K. Koch, *Phoenix canariensis* Chabaud, *Jubaea chilensis* (Molina) Baill., *Ficus macrophylla* Desf. ex Pers., *Hydrangea macrophylla* (Thunb.) Ser, species and varieties *Pittosporum, Cistus.* Also there some exotics, rarely met in landscaping of California: species, forms and varieties *of Quercus, Banksia, Hebe, Magnolia, Ceanothis, Ilex, Cotoneaster, Ligustrum, Arbutus, Iochroma, Abelia, Hydrangea, Agave, Pyrus, Malus, Prunus, Lagerstroemia, Washingtonia, Melaleuca nesophila* (F.Muell.) Kuntze, *Cyathea cooperi* (W. J. Hook. ex Muell.) Domin. and *etc.* So according to assortiment of plants and their age the park "Presidio" is unique botanical object.The park deserves the scientific investigations and conservation efforts.

T1
A055
The lichens in Red Books of the Ulyanovsk and Samara regions of Russia

Mikhail Shustov
Main Botanical Garden named after N.V. Tsitsin, RAS

The Red Book of the Ulyanovsk region includes 32 species of lichen. Some of them are located at the southern borders on the plain: *Phaeophyscia constipata* (Norrl & Nyl.) Moberg, *Cladonia acuminata* (Ach.) Norrl, *C. decorticata* (Flörke.) Spreng., *C. caespiticia* (Pers.) Flörke, *C. turgida* Hoffm., *Lecanora cenisia* Ach., *Melanelia panniformis* (Nyl.) Essl., *M. sorediata* (Ach.) Goward et Ahti, *Immersaria cupreoatra* (Nyl.) Calat & Rambold, *Umbilicaria deusta* (L.) Baumg., *Bryoria capillaris* (Ach.) Brodo et D. Hawksw., *B. furcellata* (Fr.) Brodo et D. Hawksw., *B. fuscescens* (Gyeln.) Brodo et D. Hawksw., *Hypogymnia tubulosa* (Schaer.) Hav., *Platismatia glauca* (L.) W.L. Culb. et C.F. Culb. Some lichens are found at the northern borders: *Cladonia subrangiformis* Sandst, *Lecanora bolcana* (Pollin.) Poelt, *Neofuscelia ryssolea* (Ach.) Essl, *Xanthoparmelia camschadalis* (Ach.) Hale, *Rinodina terrestris* Tomin, *Fulgensia fulgens* (Sw.) Elenkin, *Ramalina capitata* (Ach.) Nyl. in Cromb, some species – at the western border: *Aspicilia transbaicalica* Oxner, *Lasallia pensylvanica* (Hoffm.) Llano, *Lecanora crustacea* (Savicz) Zahlbr. Some lichens characterized by widely dispersed distributions have isolated habitats in the territory of the Ulyanovsk region: *Flavopunctelia soredica* (Nyl.) Hale, *Cladonia portentosa* (Dufour) Coem, *Bryoria subcana* (Nyl ex Stizenb.) Brodo et D. Hawksw, *Ramalina polymorpha* (Lilj.) Ach., *Dimelaena oreina* (Ach.) Norman, *Rinodina milvina* (Wahlenb.) Th. Fr., *Lasallia rossica* Dombr. The Red Book of Samara region consists of 7 species of lichens and 27 species are recommended. The main part of them on the territory of the Samara region have isolated areas of their habitat on the southern borders on the plain: *Phaeophyscia constipata*, *Rinodina turfacea* (Wahlenb.) Körb, *Leptogium tenuissimum* (Dicks.) Körb, *Physconia muscigena* (Ach.) Poelt, *Romjularia lurida* (Ach.) Timdal, *Rinodina oxydata* (A. Massal.) A. Massal. Some species: *Glypholecia scabra* (Pers.) Müll. Arg., *Rinodina lecanorina* (A. Massal.) A. Massal., *R. terrestris* are located on the northern borders of their habitats. Of particular interest has the lichen *Lobaria pulmonaria* (L.) Hoffm. In 30-ies of XX century, it was found in the Zhiguli occasionally, but now it is not found. Some lichens in Samara region have isolated areas of their habitat on the southern borders of the spread on the plain: *Cladonia arbuscula* (Wallr.) Flot, *C. rangiferina* (L.) Weber ex F.H. Wigg., *Cetraria islandica* (L.) Ach. At the same time *Psora decipiens* (Hedw.) Hoffm. is located on the northern border. Such lichens as *Rusavskia elegans* (Link) S. Kondr. & Kärnefelt and *Dermatocarpon miniatum* (L.) W. Mann have single habitat in the Samara region. All these lichens in Ulyanovsk and Samara regions are rare in need to be protected. Of particular importance to the protection is the fact that most of them are climatic relicts of floras of different periods. Thus, in the Red Book of the Ulyanovsk and Samara regions includes 39 species and 27 species of lichen are recommended. The conservation of them is the most important task of the state.

T1

A056

The lichens of the Privolzhskaya upland

Mikhail Shustov

Main Botanical Garden named after N.V. Tsitsin, RAS

The Privolzhskaya upland is a part of the Eastern - European (Russian) plain whish looks like a plato spreaded in the meridian direction along the Volga river with an area of more than 165 thousand square kilometers. Flora and vegetation of the Privolzhskaya upland attracted attention of the greatest Russian botanists of the XIX - XX centuries in relations to the fact that the region territory was covered only partly by the Quaternary glaciers. The possibility of preserving the Tertian flora and vegetation or their elements on the Privolzhskaya upland is of a primary importance for understanding historical processes of forming flora and vegetation of the Russian plain. Lichens flora of the Privolzhskaya upland represents a complicated heterogeneous complex of different geographical lichen groups. There were distinguished 9 geographical elements in the investigated flora. Lichens flora of the Privolzhskaya uplands can be characterized as nemoral - boreal with a great extent of arid, hypoarctomontanous, montanous and arctoalpine as well as suboceanic and alpine species. There were distinguished 12 arealogic groups in the regional flora of the lichens. The areological analysis exposes a number of species, which deserve a special study. The results of the geographical analysis of the Privolzhskaya upland lichens flora, to a certain extent, do not correspond to modern conditions of the regional nature. It is indicative of a long and complicated history of the investigated flora formation process that took place from the end of Paleogen and was not interrupted by the Quaternary glaciers. The Privolzhskaya upland lichens flora include species, the specific features of their spreading all around the world and growing on the Privolzhskaya upland at present time, as well as paleogeographical data allow to specify those lichens as climatic relicts of the next geological periods: the Early - Middle Miocene, the Late - Miocene, the Early - Middle Pliocene, the Late Pliocene, Eopleistocene, the Early - Middle Pleistocene, the Late Pleistocene, Holocene. The Privolzhskaya upland lichens flora has been continuously developing since the Late Paleogene. The specific features of the lichens spreading all around the world and growing on the Privolzhskaya upland at present time, as well as paleogeographical data allow to specify the following main stages of regional lichensflora formation: the Early - Middle Miocene, the Late - Miocene, the Early - Middle Pliocene, the Late Pliocene, Eopleistocene, the Early - Middle Pleistocene, the Late Pleistocene, Holocene. There are reasons to suppose that further evolution of lichens flora of the Privolzhskaya upland will occur in the way of developing desert-steppe-xerophytic elements and decreasing forest mesophytic elements on the basis of its general degradation, impoverishing and simplification. The unique lichens flora of the Privolzhskaya upland is subject to the negative effect as a result of economic activity. The modern condition of the regional lichens flora represents a serious concern. It is vitally important to work out the Federal programme for preserving natural objects located on the territories, including the Privolzhskaya upland, that served as refuges during the Quaternary glaciers and conserve relicts of different periods.

T1

A057

Map the distribution of *Phragmites australis* based on Generlized Additive Models

Chuangye Song[1], *Chong Huang*[2]

1. Institute of Botany, Chinese Academy of Sciences

2. Institute of Geographical Sciences and Nature Resources, Chinese Academy of Sciences

This study aims to present a vegetation mapping approach based on generalized additive model and then map *Phragmites australis* in the Yellow River Delta. Reflectance values of Spot and Landsat images were respectively employed as predictive variables to calculate the occurrence probability of *Phragmites australis*. The predictive variables were selected into the GAMs based on F-test. The fitness of the GAMs was assessed by model deviance and area under curve of receiver operating characteristic curve (ROC). Maximized sensitivity approach was used to select the threshold to transform the projected occurrence probability to presence/absence map. The results indicated that near infrared band and visible red band were the most important variables in the fitted GAM of Spot 5. And for Landsat 5, visible blue, thermal infrared and near infrared bands were the most crucial predictive variables in the fitted GAM. The fitted GAMs gave relatively high sensitivity compared with previous researches (the maximized sensitivity are 0.74 and 0.82 for GAM based on Spot 5 and Landsat TM respectively). We hope this research could provide some references for regional vegetation mapping.

T1

A058

Identification of *Senecio vulgaris* L. (Asteracea) by DNA Barcodes

Xindong Hou[1,2], *Dayong Han*[1], *Zhongsai Tian*[1], ***Dandan Cheng***[1,2]

1. School of Environmental Studies, China University of Geosci-

ence (Wuhan), Wuhan, China
2. State Key Laboratory of Biogeology and Environmental Geology, China University of Geosciences (Wuhan), Wuhan, China

A lack of species identification tools has hindered progress in understanding and managing biological invasions. Compared to traditional methods, DNA barcodes have proved to be efficient for plants species discrimination and identification using short and standardized genomic regions. Five DNA regions, namely, rbcL, matK, ITS, ITS2, and psbA-trnH, have been recommended as primary DNA barcodes for plants. *Senecio vulgaris* (Asteraceae), originated in Europe and a widely-distributed weed in temperate regions, has been listed as one of the invasive plants in China. Plants of *S. vulgaris* are morphologically similar to native plant species such as *Senecio dubitabilis*. Hence, it is important to discriminate *S. vulgaris* from other *Senecio* species by DNA barcoding. Here we used three plants from each of 6 invasive (European) and 6 invasive (Chinese) *S. vulgaris* populations and isolated ITS, trnH - psbA and rbcL sequences from the 36 *S. vulgaris* plants. Then we downloaded related sequence of 5 composite species from the NCBI database. Using MEGA we built NJ phylogenetic tree with the three kind of sequences, respectively. The result showed that ITS correctly identified plant samples at genus level and roughly at the specie level. In addition, *trnH - psbA* displayed a variable ability to discriminate related species within different genera. However, *rbcL* was unable to distinguish plant from different genera. Generally, neither of the three genes could identified *S. vulgaris* plant samples from different populations. This study provides an empirical paradigm in identification of invasive plants species and barcoding analysis in composite species.

T1

A059

Pollinators and their foraging behaviors on a medicinal parasitic plant, *Cynomorium songaricum*

*Dehui Wang, Le Feng, **Guilin Chen***
College of Life Science, The Good Agriculture Practice Engineering Technology Research Center of Chinese and Mongolian Medicine, Inner Mongolia University

Cynomorium songaricum, belonging to genus *Cynomorium*, family Cynomoriaceae, is a commonly used medicine in traditional Chinese medicine and Mongolian medicine. It is aholoparasitic plant that mainly parasitizes the root of *Nitraria. C. songaricum* is mainly distributed in dry, rocky, or sandy soils of northern hemisphere and plays an important ecological role in windbreak and sand fixation. It is distributed in China, Afghanistan, Mongolia and Iran. In recent years, wild *C. songaricum* resources are threatened by extinction due to its slow growth rate and overexploitation. In addition, people pay more attention to a higher economic value of *Cistanche deserticola* Y.C.Ma which has a same living environment with *C. songaricum*. So its survival becomes more difficult. It is important to understand the pollination biology for its reproduction. *C. songaricum* is a cross flower pollination of insects that has been mentioned, but who pollinators are involved in pollination and how they pollinate is still poorly understood. To reveal the pollination of *C. songaricum,* we carried out two periods of 3 days investigation of pollinators during flowering in May 2016 in Zuo County, Alashan City, Inner Mongolia Municipality, China (N39°36'E105°05'). We used photo, video and visual measurement to observe *C. songaricum* populations visitors. We recorded the species, quantity and visiting frequency of all pollinators from 07:00 hours until 18:00 hours in sunny days, and made specimens of pollinators to identify. In order to identify which volatile components of the flowering *C. songaricum* has an chemical temptation effect on the pollinators, we used petroleum ether to extract the base, stem and inflorescences of flowering *C. songaricum.*, respectively. The extract was divided into three portions, and the chemical constituents of each portion were determined by GC-MS. The result shows that Main pollinators of *C. songaricum* included Coleopteran (Tenebrionidae), Diptera (Calliphoridae, Sarcophagidae, Muscidae, Tachinidae) and Hymenoptera (Formicidae), total of 9 species of insects, sometimes we can see a few *Phrynocephalus versicolor* on it. The change trend of the number of pollinators exhibits as follows: an average of 6.65; numbers/individual/h at 7:00-9:00, an average of 3.18 numbers/individual/h at 9:00-11:00, an average of 3.4 numbers/individual/h at 11:00-12:00, an average of 6 numbers/individual/h at 12:00-14:00, an average of 5.26 numbers/individual/h at 14:00-16:00, an average of 4.18 numbers/individual/h at 16:00-18:00. The duration of pollinator exhibits as follows: Coleopteran's duration is 9s-2min, Diptera's duration is 4s-7min, Hymenoptera's duration is 5s-4min. Petroleum ether extract of *C. songaricum* mainly contains alkanes, esters, ethers, aldehydes and ketones, total of 56 substances. The following studies, investigation of *C. songaricum* pollinators, confirmation of pollination insects, tendency of pollination insects on petroleum ether extract, and other research, will be carried out by us in this year.

T1

A060

Influence of Aitrogen addition on the primary production in Inner Mongolia degraded grassland

Jing Wang
Institute of Geographic Sciences and Natural Resources Research, Chinese Academy of Sciences

Irrational utilization and global climate change have caused degradation of grassland ecosystems in northern China with low soil fertility, decreased vegetation coverage and productivity. Nitrogen addition has been suggested an effective way to enhance restoration of those degraded grassland. In this study, we selected a typical steppe with three different degrading levels, including lightly, moderately and heavily degraded communities, in East Ujimqin, Inner Mongolia. Our objectives of this study were to examine how nitrogen (N) addition can enhance restoration of those degraded grassland. Treatments with four levels of N addition (0,5.0,10.0 and 20.0 g $N \cdot m^{-2} \cdot a^{-1}$) were conducted to each of the three degraded communities from 2014 to 2015. Nitrogen was applied as urea in June of both years. Aboveground biomass was collected at the species level in 1 m × 1 m plot in August each year, all species biomass was summed as net primary production, and biomass of plant functional groups was calculated by perennial rhizome grasses, perennial bunchgrasses, perennial forbs, shrubs and semi-shrubs, annuals and biennials. Our results showed that the high (20.0 g $N \cdot m^{-2} \cdot a^{-1}$) and medium level N addition (10.0 g $N \cdot m^{-2} \cdot a^{-1}$) significantly increased the aboveground biomass of the slightly degraded community by 53.1% and 51.6% compared with no N addition. N addition had no significant effects on the moderately and heavily degraded communities. N addition with high and medium levels increased aboveground biomass of perennial rhizome grasses by 45.1% and 47.7%, but decreased that of perennial forbs by 37.4% and 42.1% at the slightly degraded community. Our results indicated that N addition could increase the growth of perennial rhizome grasses, and the growth of perennial forbs was suppressed consequently. Our results suggest that even the application of N fertilizers can only be helpful to restoration of those slightly degraded grasslands. Besides, N addition had no significant effects on species richness in different degraded communities indicating the fact that the study may not last long enough. For the purpose of increasing aboveground biomass of degraded grassland, we should not only consider the type and quantity of fertilization, but also the attribute of the degraded communities. In addition, the response of degraded community in biomass may strongly be impacted by degrading level of studied grassland.

T1

A061

The response of vegetation to the Younger Dryas event in the Hengduan Mountains region

Xia Wang[1,3]*, Yifeng Yao*[1]*, Alexandra H. Wortley*[2]*, Yufei Wang*[1]*, Stephen Blackmore*[2]*, Chengsen Li*[1]

1. State Key Laboratory of Systematic and Evolutionary Botany, Institute of Botany, Chinese Academy of Sciences, Beijing, P.R. China

2. Royal Botanic Garden Edinburgh, 20a Inverleith Row, Edinburgh, United Kingdom

3. University of Chinese Academy of Science

The Hengduan Mountains region, located at the southeastern margin of the Tibetan Plateau, is one of the 34 biodiversity hotspots in the world. It is a hot topic whether the climate fluctuation caused by Quaternary glaciation and inter-glaciation has affected the distribution pattern of plants and vegetation in this region. Here we present a high-resolution pollen and loss on ignition record from Haligu wetland on the Jade Dragon Snow Mountain, one of the highest peaks in the Hengduan Mountains region, to reveal the dynamic vegetation change during late glaciation. The result shows that the vegetation was dominated by needle-leaved forest during 12.9 and 12.4 cal. ka BP with high relative abundance of *Abies* and *Picea*, indicating a cold and dry climate. Between 12.4 and 11.9 cal. ka BP, it was needle-leaved forest (mainly *Pinus*, *Abies*), with slightly decreasing of *Abies* and *Picea* and distinctly increasing of *Tsuga*, implying a cold and humid condition. From 11.9 to 11.2 cal. ka BP, the vegetation was still dominated by needle-leaved forest, while the relative abundance of *Picea* increased and *Tsuga* decreased. The climate became cold and dry again. The period of 12.9 to 11.2 cal. ka BP corresponds to the YD event. The quantitative climate analysis shows that the climate fluctuated frequently during this period, with several striking cooling events and a maximum decreasing amplitude of 7.2℃. From 11.2 to 10.0 cal. ka BP, the vegetation was replaced by *Pinus* forest. Coniferous trees (*Abies*, *Picea* and *Tsuga*) decreased sharply, along with large increasing of broad-leaved trees (*Betula*, *Corylus*, *Carpinus*, *Carya* and *Quercus*), suggesting a warm and dry climate. Between 10.0 and 9.1 cal. ka BP, the vegetation was dominated by *Pinus* forest. *Abies* pollen continued to decrease, accompanied by increasing of *Tsuga*, indicating a warmer and wetter climate compared to the preceding stage.

T1

A062

Bacterial diversity and community structure along different depth in the rhizosphere soil of *Ferula*

Xiuling Wang

College of life science, Shihezi University

Ferula is an important national herbal medicine, the vast majority of Ferula are grow in the wild. There is little artificial large-scale cultivation of history. The diversity of rhizosphere microorganisms had an important influence on the growth and development of *Ferula*. Rhizosphere microorganisms generally several times or dozens of times of the none rhizosphere microorganisms, and rhizosphere microorganisms promote the growth of plants. Different types of plants generally have specific root microbial communities. In order to explore the specificity of different kinds of *Ferula* rhizosphere microorganisms, we analyzed the 4 kinds of *Ferula* species (*Ferula lehmannii*, *Ferula ferulaeoides*, *Ferula gracilis*, *Ferula syreitschikowii*) and different depth rhizosphere soil microbial diversity that distributed in northern of the Xinjiang Uygur Autonomous Region, China. In this paper, 4 kinds of the *Ferula* are all belong to the class of ephemeral plants. Stratified sampling (3-5cm, 20cm, 40cm) was carried out in the rhizosphere soil of different kinds of *Ferula*, and the rhizosphere microbial diversity of different kinds of Ferula roots was studied. The experiment was carried out by high throughput sequencing of the V4 region of the rhizosphere soil microbial 16SrRNA to obtain the information of the phylum, the class, the order, the family, the genus and species of the rhizosphere soil. We have compared the rhizosphere soil microbial diversity index with the soil simple depth and species of *Ferula*, and find that the richest six Phylum are *Proteobacteria, Actinobacteria, Gemmatimonadetes, Bacteroidetes, Acidobacteria, Cyanobacteria* in ferulic rhizosphere soil bacterial community, no matter relative or absolute abundance. The distribution of *Gammaproteobacteria* is stratified Evidently and the least distribution is the first layer.

T1

A063

A simple method for lichenized fungi preservation

Youzhi Wang

Institute of Microbiology Chinese Academy of Sciences

Lichens are symbiotic association formed by fungi and algae or cyanobacteria, and traditionally treated as lower plants. The lichenized fungi mostly belong to ascomycota and a few to basidiomycota. Lichens are difficult to culture artificially and its slow growth is a major obstacle for further study of lichens on other sides such as experimental biology. First of all, the focus was spotted on the isolation and culture and preservation of lichenized fungal strains. A simple method for preservation was introduced here. The lichenized fungus was inoculated in liquid malt extract media (Malt extract 3%, Mycological peptone 0.5%, Distilled water) for shaking cultivation for at least one month till mycelial pellets were formed. The mycelial pellets were then transferred to a tube with 8 cm high and 2 cm in diam. and with 5 ml of 10% glycerin in it, after then kept it at 4 °C. Because of the lichenized fungi are usually not sporulated, their thalli are used for preservation. Lichenized fungi usually grow slowly. With a long time of flask shaking culture, dense mycelial pellets were formed. A fungus can be kept alive for at least 3 years in a protectant solution of 10% glycerin in a laboratory refrigerator. Non sporulated and fast growing fungi can grow in 10% glycerin at 4°C, therefore are not suitable to preserve by this method. They are usually preserved in liquid nitrogen. So, this method is suitable for the non sporulated and slow growing fungi.

T1

A064

Protozoa communities of bryophytes and their biomonitoring for heavy metall pollution in the Carlin Gold Mines in Guizhou

Zhihui Wang[1], *Qimei Wu*[1,2], *Zhaohui Zhang*[1,2]
1. Guizhou Normal University
2. Guizhou Key Laboratory for Mountainous Environmental Information and Ecological Protection

The protozoan communities of bryophytes in mines are unique and special, but they have been little studied from the Carlin Gold Mines in China. Supported by the National Natural Science Foundation of China (NSFC. No. 31360035; No. 31360043), we carried out some ecological surveys at Carlin Gold Mines of karts area in Guizhou. The results showed that 56 species were identified in bryophytes and 25 species in top soils, and the community composition had great differences. The similarity of protozoan community in bryophytes and topsoils in NGM (Nipu Gold Mine) and ZGM (Zima Gold Mine)was less than the similarity in NGM (Nipu Gold Min) and WGGM (Wangjiawan Gold Mine), and the similarity was higher in NGM and WGGM. It indicated that different district and types of gold mine affected their community structure. Colpoda simulans is a very common species on both on bryophytes and topsoils. By using a community diversity index and redundancy analysis (RDA), the protozoa communities living on mosses and their relationships with heavy metal pollution were investigated from slags, orebodies, abandoned ores and nearby naked limestones in the Carlin gold mineralized area in the southwestern Guizhou. The protozoa species had an abundance range from three individuals/g to 482 individuals/g fresh weight of bryophytes. The slag sites were most severely contaminated by heavy metals. The structure of the protozoa community of bryophytes was dramatically affected by heavy metal pollution. Some species were found to be sensitive or resistant to heavy metal pollution. Therefore, the protozoa communities of bryophytes could be used to biomonitor heavy metal pollution in gold mineralized ecosystems.

T1

A065

High genetic diversity and structural differentiation in fragmented populations of wild barley (*Hordeum brevisubulatum*)

Wanli Guo
Zhejiang Sci-Tech University

Hordeum brevisubulatum (Trin.) Link., tolerated diverse abiotic stresses, is a wild barley species widely distributed in Songnen Plain of China. However, many populations of the wild barley were fragmented and/or dispeared during the last few decades mainly due to anthropogenic effects. The decrease in fragmented populations may affect the genetic diversity and structure of the populations, and in turn their survival potential. Five fragmented and four unfragmented populations (FP and UFP) in Songnen Plain were analyzed using two types of DNA marker namely, amplified fragment length polymorphism (AFLP) and sequence-specific amplified polymorphism (SSAP). The genetic diversities (0.162_{AFLP} and 0.239_{SSAP}) of five FPs [SB1, SB2, DBS, QZJ and QG] were higher than those (0.126_{AFLP} and 0.20_{SSAP}) of four UFPs [SS1, SS2, FY1 and FY2], although the later one (the UFPs) had large population size. Moreover, the 5 FPs also showed higher population genetic differentiation (G_{ST}, 0.197_{AFLP} and 0.192_{SSAP}) comparing with 4 UFPs (0.086_{AFLP} and 0.108_{SSAP}). The habitat fragmentation could improve the genetic variability, and lead to heterogeneous impact on different genome regions in *H. brevisubulatum*.

T1

A066

Dynamic comparative analysis on antioxidant activity and phenolic compounds of *Chrysanthemum Indicum* and *Opisthopappus shih*

Dongwei Wei[1], *Wuyong Sun*[2], *Jing Sun*[1], *Yaping Zhou*[1]
1. Henan Agricultural University
2. Henan Entry-Exit Inspection and Quarantine Bureau

Chrysanthemum indicum (*C. indicum*) is widely used as a health food and as a popular herb in China for many centuries. *Opisthopappus Shih* (*O. shih*) often takes the place of *C. indicum* in functional tea or medicine prescriptions in place of origin. In this paper, dynamic comparative analysis on phenolics and antioxidant activity of aqueous ethanol extracts of *C. indicum* and *O. shih* were investigated. Antioxidant properties of plant extracts were investigated using DPPH and ABTS radical scavenging capacity assays. Total phenolics and total flavonoids were measured using Folin-Ciocalteau and aluminum chloride colorimetric methods respectively. An HPLC method was used to simultaneously quantify five phenolic compounds including chlorogenic acid, luteolin, rutin, quercetin, and apigenin. Results indicated that Trolox equivalent antioxidant activity (TEAC) values had extremely large variation during plant growth. The most abundant phenolics and potent antioxidant activity appears at vegetative growth stage. Antioxidant activities and phenolics of *O. shih* were higher than that of corresponding parts of *C. indicum* at the same collection time. Health-promoting properties of *O. shih* may be much better than its related genera *C. indicum*. The whole plants of *O. shih*, especially its leaves and flowers are good candidates to obtain nutraceuticals and functional food ingredients.

T1

A067

Observations on female gametophyte and embryo of the critically endangered species *Isoetes orientalis*

Hongyu Wei, Jinghong Liu, ***Baodong Liu****

Harbin Normal University, Key Laboratory of Plant biology, College of Heilongjiang Province, Harbin

Isoetes orientalis was special ancient vascular plants of China, which was a new species isolated from *I. sinensis*. *I. orientalis* was critically endangered and had great significance in vascular plant evolution. However, its reproduction and endangered mechanism was unclear. With the advantages of implementation of artificial breeding and acquisition of adequate materials, the development of female gametophyte and embryo using thin sectioning were observed. The main results were as follows: Megaspore germinated after 3-5d, female gametophyte was circular with internal structure of uniform and formed a number of archegonias. There were neck cells, two neck canal cells and one egg in the archegonia. The abdominal wall cells were transformed from vegetative cells. Most of the female gametophytes only developed into one embryo, occasionally the coexistence of two embryos. With the first primordium rapid growth, primitive cells of ligula began to cleft horizontally. The ligula of the first primordium not only protected itself but also provided nutrition for the second and third ones. The early primary axis in embryonic development rotated to 30°-40°. It were speculated that the block of gametophyte development, the instability of genital differentiation, egg or reproductive helper cell abortion and embryo differentiation stopped and so on might be the causes of abortion. The double also reported and discussed the protective function of early leaf tongue on embryonic development and its phylogenetic significance.

T1

A068

The population genetic structure and conservation strategies of *Camellia chekiangoleosa* analysed by SSR Markers

Qiang Wen[1,2], *Li-An Xu*[2], *Wen Cai Zhou*[1], *Lin Chu Xu*[1]

1. Jiangxi Academy of Forestry

2. Nanjing Forestry University

Camellia chekiangoleosa Hu. (Theaceae), a member of genus *Camellia*, is endemic to the south of China. It has high values in ornamentals and edible oil use. This tree may be used as the important parent in hybridization for breeding ornamental camellias and important edible oil varieties in the hill country. Although the resource distribution of *C. chekiangoleosa* is mostly in the wild or semi-wild states, but its natural habitat has been increasingly fragmented due to human activity and other reasons. It was found out that the distributions, areas and resource quantities of *C. chekiangoleosa* are reducing rapidly, close to be in imminent danger through field surveys of germplasm resources and literature comparative analysis. Therefore it is urgent and necessary to do a systematic research on its current status quo and the genetic diversity so as to propose conservative strategies to protect it. In this study, based on 454 sequences of Camellias, microsatellite was mined to develop a batch of SSR markers. Meanwhile, the low-copy nuclear genes were obtained by RACE for the phylogenesis of Sect. *Camellia* and the molecular screening of *C. chekiangoleosa* and its close relative species was carried out using the introns 12-16 and 23 of the RPB2 gene of nuclear DNA. Subsequently, combining with the results of phylogenesis and screening the authenticity of the population of *C. chekiangoleosa*, the population genetic structure of 12 natural populations of *C. chekiangoleosa* was studied by 16 polymorphic SSR markers to reveal the genetic diversity and to investigate the effects of natural and artificial selection, genetic drift and other factors on the genetic structure. Compared with other species, the low genetic diversity of *C. chekiangoleosa* was found (h=0.596). The characteristic of diversity distribution is that: the populations located in the range of Huaiyu and Wuyi mountain chain had highest diversity level (such as Damao, Sanqing and Gutian) and it may be the diversity distribution center of *C. chekiangoleosa*, in complete contrast to the population located Hengshan mountains chain, where the diversity level was lowest, and the population located east to the center (such as Wuyalin and Xiapu) had medium level of diversity. The population had obviously genetic structure and had a high level of genetic differentiation. The variation among populations was about 24% of the total (FST=0.234, GST=0.239). It may be caused by habitat fragmentation and isolation, small population scale, bottleneck experienced (such as Hengshan and Lingshan), the inbreeding and little gene flow (Nm=0.796), *etc.* It is urgent to make the reasonable protection strategy. For the higher diversity populations such as the damao, shanqing and gutian, in situ conservation of forest genetic resources may be realized by establishing nature reserves to cut down the damage. And for the lower genetic diversity populations such as Hengshan and Fuan, the species recovery should be promoted by reintroduction. In addition, for lower diversity distribution area, such as Midwest of Jiangxi to Hunan, it is suggested to rich resources by the introduction method.

T1

A069

Genetic diversity of *Ophiocordyceps sinensis* and its host moths on the Qinghai-Tibetan Plateau as rewealed by ITS and *COI* sequences

Hai-Jun Wu, *Yi-Jian Yao*

State Key Laboratory of Mycology, Institute of Microbiology, Chinese Academy of Sciences, Beijing, China

Ophicordyceps sinensis, one of the most valuable medicinal fungi in China, is a native fungal parasite of the ghost moth larvae in Hepialidae (Lepidopterra). It is restricted to the Qinghai-Tibetan Plateau from 3,000 meter altitude upwards to the snow line. The annual yield of the fungus has gradually declined in recent decades and it has been listed as a state protected species in China. In this study, a total of 221 individuals of the fungus collected from 43 counties on the Plateau, covering the known distribution area of the species, were used for analyses of the genetic diversity of both the fungus and its moth hosts by sequencing the gene fragments of nrDNA ITS from the former and mitochondrial *COI* from the latter. A total of 18 ITS haplotypes and 81 *COI* haplotypes were detected. Analyses of ITS haplotypes revealed that high nucleotide diversity was located in the Aba and Garzê area in Sichuan and the Nyingchi area in Tibet. The same pattern was followed by the analyses of *COI* haplotypes. Based at 97% sequence similarity cut-off, a total of 15 distinct *COI* OTUs were identified. Among them, 11 OTUs could be named at the species level by the Blast search in Gene Bank, while there was no match sequence for the other four OTUs. The congruence of both ITS and *COI* analyses indicted the true diversity of the fungus and its hosts and also the close link between both fungus and moth populations. The results presented here provided useful information for understanding the genetic diversity of *O. sinensis* and also provide valuable information for protection and sustainable use of this treasure fungal resource.

T1

A070

Identifying *Camellia oleifera* Abel. germplasm accessions with inter-simple sequence repeat (ISSR) markers

Hua Cheng[1,2], ***Jun Xiang***[1,2], *Linling Li*[1,2]

1. Economic Forest Germplasm Improvement and Comprehensive Utilization of Resources of Hubei Key Laboratories

2. College of Life Sciences, Huanggang Normal University

The diversity and genetic relationship among and within major populations of *Camellia oleifera* 109 strains were analyzed using ISSR markers. The result showed, 23 ISSR primer, which selected from 49 primers, yield about 487 legible bands of which 335 were polymorphic markers, and the ratio of polymorphism was 68.86%; Base on the bands, the genetic similarity coefficient ranged from 0.61 to 0.93 using NTSYS2.10e software. When coefficient was 0.75,109 cultivars were divided into 11 categories and categories I contain 79 varietys by UPGMA cluster analysis. The test variety divided into 7 subgroup when categories is 0.75, which show a close genetic relationship. The results wil provide a theoretical foundation for the identification of *Camellia oleifera* cultivars, and for the efficient of their germplasm resources.

T1

A071

The conservation biology of *Iris ensata* in East Asia

Yue-E Xiao, *Shu-Cheng Feng, Guang-Qi Zhao*

1. Shanghai Botanical Garden Shanghai, China;

2. Shanghai Engineering Research Center of Sustainable Plant Innovation, Shanghai 200231, China

Iris L., with about 280-300 species, is the largest genus in the Iridaceae and is believed to have originated in the temperate northern hemisphere. *Iris ensata* Thunb., a member of series *Laevigatae* (Diels) G. H. M. Lawr. (subgenus *Limniris* (Tausch) Spach section *Limniris* Tausch), is native to China, Japan, Korea, and Eastern Russia. This species is one of the most widely cultivated, hybridized, and important horticultural species and has more than 5000 cultivars known as 'Japanese Iris' (English name) or 'Hanashobu' (Japanese name). However, excessive agricultural reclamation has resulted in habitat destruction and habitat fragmentation of natural populations. Thus, it makes sense to precede the conservation biology of *Iris ensata* in East Asia. Firstly, we investigated morphological variation of wild *I. ensata* in East Asia in the past ten years. There are plentiful phenotypic variations in floral size, type and color in East Asia. The floral size differs from 5 to 12 cm. The vertical, horizontal and intermediate floral types between vertical and horizontal type of *I. ensata* are common in the field, while it was reported that pentamerous flowers were only found in North Honshu. In contrast, there are abundant flower color forms intra-species *I. ensata*, including dark-purple, fuchsia, tawny, pink and white except common different shades of purple and bluish violet. Among these forms, tawny, the pink and white forms are very rare, which can be found in North East of China, central Honshu and north of Honshu. Furthermore, the genetic from 48 natural populations in East Asia were tested by the applicability of nine nuclear microsatellites (nSSR) loci. Moderate genetic diversity ($H_E = 0.618$, $Ho = 0.618$) and high genetic divergence ($F_{ST} = 0.245$, $P < 0.001$) were detected at the species level based on nSSR datasets. The genetic diversity level of South Korea populations was significant higher than those of other regions ($P < 0.05$). Remarkable genetic sub-structures were detected in Tianmu Mts., Japan, and South Korea. Finally, the conservation strategy of *I. ensata* in East Asia was proposed based on information of phenotypic variation and genetic diversity.

T1

A072

Ethnobotanical study on medicinal plants used by the Buyi people in eastern Yunnan

Yong Xiong[1], *Xueyi Sui*[2], *Selena Ahmed*[3], ***Chunlin Long***[1,3]

1. College of Life and Environmental Sciences, Minzu University of China, Beijing, China.

2. Sustainable Food and Bioenergy Systems Program, Department of Health and Human Development, Montana State University, Bozeman, MT, USA.

3. Kunming Institute of Botany, Chinese Academy of Sciences, Kunming, China.

Background: The Buyi is an ethnic group in southwest China. They have developed their own medicinal knowledge over a long history. The medicinal plants used traditional by one branch of them located in Luoping county (Yunnan province) serve as an inseparable part of the whole Buyi medicinal cultures. Previously, no detailed literature, concerning its traditions and using of plants has been recorded in this area. The objective of this paper is to document the medicinal plants and associated traditional knowledge.Methods: The investigations were conducted in 4 different villages in Lubuge Township of Luoping County, through ethnobotanical approaches: participatory observation, semi-structure, ranking exercise, key information interview, focus group discussion, participatory rural appraisals (PRA). Results: A total of 122 ethnomedicinal plant species belonging to 64 families were used by the local healer to cure different diseases. Among these plants, 56 species were documented to have medicinal values for the first time, compared with all existed Buyi medicinal literaures. The most frequently used medicinal part was leaf (24.9%) and the most common preparation was decoction. Medicinal plants were mainly used to treat rheumatism, trauma and injuries. Medicinal plants were used other purposes which were edible, fodder, fence and ornamental, *etc.* Conclusion: Due to the similarities in terms of living habits and environmental factors, these Buyi marginal regions may contribute to the integrity of Buyi medicinal knowledge, which should not be ignored. Buyi medicines have multiple uses, and some have poisonous too. The paper results showed that the traditional Buyi culture played a positive role the utilization of Buyi medicine. However, the ethnomedicinal knowledge and experiences are facing the danger of extinction due to several factors, which indicates the requirement for extensive investigation and relative compulsory policies to ensure their conservation and sustainable development.

T1

A073

Reactive oxygen species induced by cold stratification promote germination of *Hedysarum scoparium* seeds

Liqiang Su[1], *Qinying Lan*[2], *Hugh W. Pritchard*[3], *Hua Xue**[1], *Xiaofeng Wang**[1]

1. National Engineering Laboratory for Tree Breeding, College of Biological Sciences and Biotechnology, Beijing Forestry University, No.35, Tsinghua East road, Beijing, 100083, PR China

2. Key Laboratory of Tropical Forest Ecology, Xishuangbanna Tropical Botanical Garden, Chinese Academy of Sciences, Germplasm Bank, Mengla, 666303 Yunnan, PR China

3. Royal Botanic Gardens, Kew, Wakehurst Place, West Sussex, RH176TN, UK

Seed germination is comprehensively regulated by multiple intrinsic and extrinsic factors, and reactive oxygen species (ROS) are relatively new among these factors. However, the role and underlying mechanisms of ROS in germination regulation remain largely unknown. In this study, we initially found that cold stratification could promote germination and respiration of Hedysarum scoparium seeds, especially at low temperature. We then noted that a ROS environment change induced by hydrogen peroxide (H_2O_2) or methylviologen (MV) could similarly promote seed germination. On the other hand, the ROS scavenger N-acetyl-L-cysteine (NAC) suppressed germination of cold-stratified H. scoparium seeds, indicating a stimulatory role of ROS upon seed germination. An increased accumulation of O2- was detected in embryonic axes of cold-stratified seeds, and stratification-induced ROS generation as well as progressive accumulation of ROS during germination was further confirmed at the cellular level by confocal microscopy. Moreover, protein carbonylation in cold-stratified seeds was enhanced during germination, which was reversed by NAC treatment. Finally, the relationship between ROS and abscisic acid (ABA) or gibberellin (GA) in germination regulation was investigated. ABA treatment significantly inhibited germination and reduced the H_2O_2 content in both cold-stratified and non-cold-stratified seeds. Furthermore, we found that cold stratification mediates the down-regulation of the ABA content and increase of GA, suggesting an interaction between ROS and ABA/GA. These results in H. scoparium shed new light on the positive role of ROS and their cross-talk between plant hormones in seed germination.

T1

A074

Polycyclic natural products and their anti-AD activities of two medicinally important plants in the genus *Hypericum*

Yongbo Xue*, *Yi Guo*, *Na Zhang*, *Shuangshuang Xie*, *Yonghui Zhang**

School of Pharmacy, Tongji Medical College, Huazhong University of Science and Technology

Hypericum is one of the largest genus in family Guttiferae, many species of which have been applied as folk medicines throughout the world. For instances, *Hypericum perforatum*, also known as "St John's Wort", is utilized externally for wound healing and internally to relieve the symptoms of neurological disorders. The extract of St John's Wort (ESJW) can alleviate mild to moderate depression that is mainly attributed to the inhibition of various neurotransmitter receptors and exerts versatile biological properties including anti-inflammatory, cytotoxic and neuroprotective properties. Polycyclic prenylated acylphloroglucinols (PPAPs) are usually characterized by a highly oxygenated and densely substituted bicyclo[3.3.1]nonane-2,4, 9-trione or other related core structures that were decorated with isoprenyl, geranyl, and acyl groups. Over the past decades, almost 300 structurally complex PPAPs endowed with different skeletons have been reported. PPAPs are mainly isolated from the plants of the Guttiferae family and exhibit a wide variety of bioactivities, such as antitumor, antibacterial, antioxidant, anti-HIV, and anti-neurodegenerative. In light of their complexstructures and potential biological activities, PPAPs have attracted great interest from the scientific community due to their intriguing molecular structures and interesting pharmacological effects. Alzheimer's disease (AD) is one of the most common disease found in people live with dementia. Most recently, the World Alzheimer Report 2015 demonstrated that at present there are over 46 million people live with dementia worldwide, and this number is estimated to increase to 131.5 million by 2050. Aimed to discover more bioactive polycyclic natural products from the genus of *Hypericum* collected in central China regions, our group previously reported the isolation of a large number of PPAPs from *H. sampsonii*, *H. attenuatum*, *H. japonicum* and *H. ascyron*. Currently, the chemical study by our research group on two species of *Hypericum*, *H. longistylum* and *H. perforatum*, led to the isolation of 40 new PPAPs with diverse carbon scaffolds, along with 20 known ones. Some of the new isolates exhibited significant inhibitory activities against protein phosphatase 2A (PP2A) and beta-secretase (BACE1) enzymes in Alzheimer's disease. The finding will provide new insights into the skeletal diversity of PPAPs and natural PP2A and BACE1 inhibitors.

T1

A075

Dynamic characteristics of *Rhododendron maculiferum* community in Shennongjia

Linsen Yang

Administration of Shennongjia National Park, Shennongjia, China

A study was conducted on the sub-alpine *Rhododendron maculiferum* community in Shennongjia via ten 400 m^2 quadrats. There were 84 plant species of 58 genera and 32 families in ten quadrats. Rosaceae is the most in woody layer with 19 species. *Fargesia* had the highest importance value in the shrub layer. *Fragaria orientalis*, *Rubus fockeanus* and *Asteropyrum peltatum* were common in the herb layer. On average, there were 6 species in the tree layer, 15 species in the shrub layer, and 20 species in the herb layer, and the total was 38 species. The species richness of tree layer and shrub layer decreased with altitude, while the species richness of the herb layer slightly changed. In the tree layer most individuals were in the 5-8 cm diameter class, accounting for 61.3% of the total, and 80.9% of the individuals were in the 4-7 m height class. The ratio of the dead standing trees and the living is 7.1%, and the dead standing trees of *Rhododendron* accounted for 86% of the total number of dead standing trees. 70.8% of the dead standing trees of *Rhododendron* were in 3-4 cm diameter class, indicating that under high crown closure the objects of the self-thinning process in the community is mainly saplings. With the rise of altitude, the population density decreased and the community levels tend to be simple. The anophanerophyte accounting for 31% and hemicryptophyte for 27.4% are the main elements of the community. This life form structure adapted to the local conditions of long cold winter and humid climate. The north temperate genera is dominant in the flora, which is of obvious temperate attribute. Upper *Abies fargesii* layer disappeared after disafforestation decades ago, so the *Rhododendron maculiferum* at under layer developed into the single dominant species. Thence the species diversity of the tree layer, shrub layer and herb layer decreased, and the average tree height became smaller. The *Rhododendron* community at prime period is relatively stable, while the invasion of trees like *Abies* and *Betula utilis* in the community or on the edge promotes the succession. The gaps caused by the dead wood is the main impetus for the regeneration. There is an evolution tendency that *Abies fargesii* will replace *Rhododendron*. *Fargesia murielae* inhibited the renewal of *Rhododendron*. However, the death and regeneration of *Fargesia* since the 1990s have promoted the regeneration of *Rhododendron* to a certain extent.

T1

A076

Dye plant resources in Liangshan Yi autonomous prefecture

Bei-Xi Zhang[1], Bin-Sheng Luo[1], Qu-Ye Diri[1], Chun-Lin Long[1,2]

1. College of Life and Environmental Sciences, Minzu University of China, Beijing 100081, China

2. Kunming Institute of Botany, Chinese Academy of Sciences, Kunming 650201, China

Natural dyes have attracted more and more people's attention. Yi, an ethnic group in China, is remarkable for their traditional knowledge of dye plants. Based on literatures review and ethnobotanical investigation on traditional dye plants and their dyeing processing knowledge in Liangshan Yi Autonomous Prefecture, Sichuan Province of Southwest China, we systematically collected and identified seven species of commonly-used local dyeing plants. They are *Juglans regia, Coriaria nepalensis, Toxicodendron vernicifluum, Rubia cordifolia, Coptis chinensis, Rumex dentatus*, and *Polygonum tinctorium*. The process of traditional dyeing was recorded as well. By interviewing with indigenous Yi people, it was found that local farmers are the main user of dye plants and holders of associated dye knowledge. The current situation of local dye plants was analyzed. Results showed that traditional knowledge associated with dye plants was becoming endangered. This study can not only help saving the traditional knowledge related to dye plants, but also provide informative knowledge for future sustainable development of local dye plants.

T1

A077

Alpinia officinarum rhizomes of different varietas can be applied personalizedly

Chunrong Zhang, Xiaomin Tang, Xuanxuan Cheng, Liming Pan, Quan Yang

Guangdong Pharmaceutical University

Alpinia officinarum, known as lesser galangal, is a herbaceous plant in the Zingiberaceae family. It originated in China and is cultivated in Southeast Asia. The rhizomes, valued for their sweet spicy flavor and aromatic scent, have been used as spice and perfumes in Asia and Europe for more than 2000 years. In China, the rhizomes are also used as a traditional Chinese medicine in curing epigastric pain, vomiting, dyspepsia, gastric and duodenal ulcer, chronic gastritis, acute gastroenteritis, *etc.* Some new usages such as additives for feed and cosmetics, fresh keeper for fruit, vegetable and meat, are also found out. The long history of huge market demand has nearly exhausted the natural resources. Resources available on the market are all domestic. There are four *A. officinarum* varietas which are named Niu Jiang, Fengwo Jiang, Ji Jiang and Zhutou Jiang in China. In most places, Niu Jiang or Fengwo Jiang, which is in higher yield, is vegetatively propagated for years and years. This is disadvantageous to population evolution and new varieties formation. In our present work, macroscopic characteristics of the rhizomes were identified among the four varietas. Niu Jiang is in the highest yield and tastes sweet spicy with a little bitterness. Fengwo Jiang smells the strongest aromatic and tastes sweet spicy with no bitterness. Zhotou Jiang tastes less spicy. Ji Jiang is in the lowest yield and has the lest water content. These scent and flavor differences indicate that the *A. officinarum* rhizomes of different varietas can be applied personalizedly. For example, Niu Jiang and Fengwo Jiang can be used as medicines and perfumes. Fengwo Jiang and Ji Jiang are good for food additives. Zhutou Jiang is better for spice. To reveal their differences in chemical compositions, metabolome analysis will be in our later research.

T1

A078

The exploration of the water use pattern of populus euphratica in different ages at the Tarim River downstrean

Jiang Zhang

Shihezi University

Water is one of the factors that affect the growth and reproduction of vegetation in terrestrial ecosystem, and it is also a key factor to control the structure and dynamics of the vegetation community. Especially in the arid area, it will be the key factor to limit growth of piant.And it is important for explain the spatial and temporal distribution of vegetation under the global climate change that expound the utilization pattern of plant's water.The stable hydrogen and oxygen isotope technique used in the research of water source determination in plants mainly focus on two aspects, one is the gradient of soil water by absorption of plants; the other is the main source of water absorbed by the plants and their contribution degree.Compare the water of the plant xylem with the water in the growth environment, using two or more isotope mixing models to determine the water utilization of plants of different water sources and the relative contribution of plant water, can also study it's changes in time and space. $\delta^{13}C$ can not only reflect the carbon isotope ratio of atmospheric CO_2, also reflect the value of Ci/Ca.According to the definition of water use efficiency, water use efficiency of plant relate to Ci and Ca, so ^{13}C can indirectly reveal the long period of plant water use efficiency.A large number of studies have confirmed that the $\delta^{13}C$ is highly related to

WUE, which can be used as an effective indicator to measure the water use efficiency of plants in the long term.Therefore, we measure and analyze soil moisture and δD, $\delta^{18}O$ of the soil water, groundwater and xylem and $\delta^{13}C$ of leaf, which analyze the mean depth of water uptake of different ages of Populus euphratica, and discuss the water use patterns and he adaptation mechanism in the arid environment, which provide scientific basis for the formulation of ecological water in lower reaches of Tarim River and restoration of Populus euphratica forest.

T1

A079

Altitudinal variation of the reproductive strategies and the future distribution prediction of *Primula florindae*

Jie Zhang[1], *Xianfeng Jiang*[1], *Lingling Chen*[1], *Qingjun Li*[2]

1. Xishuangbanna tropical botanical garden

2. Yunnan University

The alpine ecosystem harbours extremely high biodiversity, however, the harsh environment factors seriously impact the growth and reproduction of plants, such as low temperature, strong wind, intense radiation, frequent rain and snow. In particular, the reproductive season shows striking sensitivity to surrounding conditions, at the same time, plants can also integrate various strategies for acclimatization. The flower, as a reproductive organ, exhibits floral traits variations in different altitudinal conditions, so it is of great importance in reproductive strategies research. *Primula florindae*, which distributes from 3100 m to the treeline (4350 m) in the Sygera Mountains, as an endemic species in the Qinghai-Tibet Plateau, is an ideal material for exploring the relationship between elevation and reproductive strategies. In this research, we select flowering phenology, floral display traits, pollinators, sex allocation, and natural seed set as characters, by comparing these characters among five populations along altitudinal gradients, we highlight three conclusions: (1) Initial blooming time of *P. florindae* did not show a correlation with elevation: the middle elevation population bloomed earliest and the highest population bloomed last. Nevertheless, with the increase of elevation, flower longevity and floral display area of both morphs of *P. florindae* increased significantly, but varieties of pollinators decreased, and both morphs tend to allocate more resources to assure female sexual function. Overall, reproductive strategies of *P. florindae* were affected by biotic and abiotic factors generated by the altitudinal gradients, but varying responses among the traits studied resulted in the intraspecies variation of reproductive strategies. (2) With the increase of elevation, flower number of *P. florindae* significantly diminished but both herbivory number and rate also reduced, so effective fruit number of all populations showed no distinct disparity, and there were also no significant differences in seed number, showing that by shaping corresponding reproductive strategies to different environment, *P. florindae* guaranteed breeding success, and then stablized population in all elevations. (3) Climate change will have profound and lasting impacts on the distribution of *P. florindae*, the lower altitudinal limit may shift up, but sandy soils at higher elevations are unfavorable to its growth, while the temporal-spatial distribution of pollinators under climate change are still unknown. We predict the lower altitudinal limit of *P. florindae* will move up to middle elevation and result in a decrease in its distribution over the next 100 years.

T1

A080

EST-SSR marker based DNA fingerprinting for cultivar Identification and genetic diversity analysis of *Elymus sibiricus* germplasm in China

Junchao Zhang, *Wengang Xie, Xuhong Zhao, Zongyu Zhang, Yongqiang Zhao, Yanrong Wang*

The State Key Laboratory of Grassland Agro-ecosystems, College of Pastoral Agriculture Science and Technology, Lanzhou University

In this study, EST-SSR markers were used to analyze the genetic diversity of 52 *Elymus sibiricus* accessions, and construct the DNA fingerprinting profiles of 7 *E. sibiricus* cultivated accessions and cultivars in China. A total of 204 bands were generated from 20 EST-SSR markers with an average of 10.2 bands. 176 (86.27%) bands were polymorphic, indicating a high level of polymorphism. The primers Elw404 and Elw195 could easily distinguish 7 cultivated accessions and cultivars from other materials. The molecular variance analysis (AMOVA) showed that genetic variation was greater within geographical regions (73.94%) than between them (26.06%) in *E. sibiricus*. The structure analysis suggested that the 52 accessions were clustered into five groups, similar to result of principal coordinate analysis (PCoA). Wild accessions (NPB=174, PPB=85.29%, I=0.2966, H=0.1798, Na=1.8529) possessed higher genetic diversity when compared with cultivars and cultivated accessions, suggesting wild accessions could be used as important genetic resources for future breeding programs. Furthermore, ex situ and in situ conservation strategies were suggested for *E. sibiricus* germplasm. The results of the present study showed that EST-SSR markers were efficient in assessing the genetic diversity and identifying cultivars.

T1

A081

The first flowering plant on Mars

Huien Zhao

Beijing Forestry University

Mars is the most Earth-like planet in the solar system. Modern-day Mars is a harsh place for life where frigid temperatures, intense radiation, highly oxidizing soil and extreme aridity combine to present significant obstacles to any potential extant life forms. Lichens were regarded as the survivors on Mars since some lichen species are able to live under harsh conditions, characterized by low temperatures, aridity and high UV radiation fluxes and several lichens are even able to survive simulated and real space conditions. Redesigning living organisms to survive on Mars were proposed recently. Here we think that amphiploid somatic transgenic multigeneric hybrids of Anthemideae and *Saussurea* within Asteraceae might be the first flowering plant colonized on Mars without any protection. There are many elite genetic resources distributed in Polar Regions, high mountains, deserts or cracksn within Ajania group. We can get Multi-generic hybrids with embryo rescue using the related genera distributed in harsh environments within Asteraceae-Anthemideae-Artemisiinae. Further more, Somatic transgenic multigeneric hybrids of Anthemideae and Saussurea could be produced using transgenic multigeneric hybrids with diverse genes from extremophiles. At last, Amphiploid somatic transgenic multigeneric hybrids of Anthemideae and Saussurea could be bred which might be suitable on Mars.

T1

A082

A brief report of pilot province in the Fourth National Survey of Chinese Materia Medica Resources: Hebei Province

Lin Pei[1], ***Lin Li***[2], *Pei He*[1], *Hong-Zhu Liang*[2], *Xu-Hong Duan*[1,2], *Shuo Shi*[2], *Jian-Cheng Zhao*[2]

1. Institute of Traditional Chinese Medicine

2. Hebei Normal University

Traditional Chinese medicine resources are one of the important strategic resources of the country. With the rapid development of Chinese medicine industry in recent years, great changes in the medicinal resources of China in terms of types, distribution, reserves and quality, have taken place. The national level's census of traditional Chinese medicine had carried out for three times. And the lasted census was carried out in 1983-1987, nearly 30 years ago. Problems such as Chinese traditional medicine resource status were not clear and lack of exchange of resources information still exists. Based on this situation the State Administration of Traditional Chinese Medicine organized the Fourth National Survey of Chinese Materia Medica Resources. The preparatory work started from 2009, and formal work began in 2011. Hebei Province and other 15 provinces, as the second batch of pilot provinces, carried out the program in 2012. 280 professional fellows from different research units, including Hebei University of Chinese Medicine, Hebei Normal University, Hebei Academy of Agriculture and Forestry Science and Chengde Medical University, constituted 16 survey teams arranged by Administration of Traditional Chinese Medicine of Hebei Province. The survey was carried out since 2013. According to the technical regulations, the survey teams carried out field investigations and got plentiful data. In the program, 1,317 quadrat were investigated, 1,975 species medicinal plants were collected, of which 265 were important traditional medicine. More than 20,000 specimens were made and 9,500 specimens were submitted to the National Resource Center for traditional Chinese medicine. 807 medicinal materials of 120 species and 2,335 collections of seeds of 788 species were also submitted. The traditional Chinese medicine raw materials quality monitoring and technical service center was established. The monitoring stations in Anguo City and Julu County have been put into use. In the circumstance of the survey, the Chinese Medicine Resources Center of Hebei Province was established; specimens of medicinal plants and crude drug materials would be well stored in it. More than 10 papers related to the survey were published. Through this survey, a group of technical backbone was growing up. A foundation for the following work of the survey was made.

Our work was supported by Special Fund on Public Health from the Central Finance "Resource survey and monitoring on original plant of essential medicine", award No. Cai She[2011]76 and the Special Fund of the National Traditional Chinese Medicine, award No. 201207002.

T1

A083

Identification and reserve gap analysis of the key biodiversity areas in china

Li-Na Zhao[1,2], *Jin-Ya Li*[3], *Hui-Yuan Liu*[1], *Yu-Chang Yang*[1,2], *Hai-Ning Qin*[1]

1. Institute of Botany, Chinese Academy of Sciences

2. University of Chinese Academy of Sciences

3. Research Center for Eco-Environmental Sciences, Chinese Academy of Sciences

China's biodiversity loss is serious, but the available protection resources are very limited. It is a more realistic and efficient way to concentrate the limited resources on the important biodiversity regions. And the construction of a key biodiversity areas evaluation index system which is operational and consistent with the specific situation of our country is the prerequisite and basis for setting the protection priority area. Based on IUCN's Key Biodiversity Areas (KBA) evaluation criteria and relevant studies, we build the China's KBA evaluation index system by considering the scientificity, operability and availability of data. Then, the key biodiversity areas of China were identified and GAP analysis was also carried out, which would be helpful to the formulation of effective policies and the rational allocation of resources.

T1

A084

Canopy seed bank of *Sophora alopecuroides* L. in Ili River Valley

Ying Liu[1,2], *Yu Zhao*[1,2]

1. Biology and Geography School, Yili Normal University

2. University and College Key Lab of Natural Product Chemistry and Application in Xinjiang

Sophora alopecuroides is a tipical perennial clone and dominant species in the Ili River Valley, which is a familiar kind of poisonous and harmful plant in this area. Just because of its typical canopy seed bank, this project take *S. alopecuroides* as the research object. By using the the field study method in desert steppe of Huocheng county, an experiment was conducted to study the size and composition, seed vigor, Supplementary effect on soil seed bank, and the retention ratio dynamics of *S.alopecuroidesopy*'s canopy seed bank. The result showed that canopy seed bank possess 271.50±124.43 beans/m^2, the maximum was 489.00 beans/m^2. *S.alopecuroidesopy*'s canopy seed bank was composed of two parts: the whole seeds, moldy and rotten seeds. The number of whole seeds reached 144.07±137.93 beans/m^2, and its percentage was 53.06%. Retention ratio reached 25.01±16.92, the maximum was 51.46, and the minimum was 4.29. After mechanical damage, germination rate reached 76.98±11.89%, seed vigor index was 72.44±39.01, which indicated that canopy seed bank could keep higher seed vigor. Canopy seed bank lasted for nine months, the retention rate was about 25 percent at the beginning of March of the next year. Even in May or June, some certain degree of seeds still stained in the canopy. As a result, canopy seed bank could be an effective supplement to the soil seed bank, and both of them performed effective supply bank of propagules during the population regeneration and community succession. The results would be helpful to systematically understand the reproductive characteristics, reveal the bio-ecological mechanism of its malignant invasion, and provide theoretical support for scientific management of grassland ecosystem in the Ili River Valley.

T1

A085

Jujube tea processing and its antioxidant activity

Xiaomin Lu, Xu Yan, Xu Liu, Yuemeng Cao, Yacui Wang, Xiaoyue Ma, ***Zhihui Zhao***
Horticulture College, Agricultural University of Hebei

Chinese jujube (*Ziziphus jujuba* Mill.), a native plant ofChina, belongs to the Rhamnaceae family and has been cultivated for over 3000 years. InChina, its fruits are used in traditional medicine exhibiting tonic activity. Chinese jujube plant tree has abundant flowers in its fruit development stage. The flowers of Chinese jujube have mild but amazing smell. We collected fresh jujube flowers at June 2015, fresh red jujube fruits at September 2015 and bought fresh fruits of apple, mango, pawpaw, pineapple, kiwi, longan, Hami melon and flowers of rose. And then some of jujube fruits were dried by oven, the others and other fresh fruits, flowers were lyophilized. We mixed different fruits and flowers with jujube fruits, flowers as fruit teas. Eight different products of jujube teas we processed were used for the evaluation. These processed products were evaluated on the basis of their appearance, taste, components and antioxidant activities. The results showed the eight products contained titratable acid of 1.98%-11.33%, reducing sugar of 2.93%-5.6%, soluble sugar of 6.22%-51.33%, polysaccharides of 0.37%-0.77% and vitamin C of 9.81mg/100mg -38.60 mg/100mg based on the dried weight. Two products with more kiwi fruits mixed had good taste and higher antioxidant activities and they had higher contents of titratable acid and soluble sugar. So, we made two jujube tea products with good taste and higher antioxidant activities.

T1

A505

The aasis of city for research and conservation of endangered species in East China—Hangzhou Botanical Garden

Chuan Chen*, Jinliang Yu, Ting Wang, Yahong Gao*
Hangzhou Botanical Garden, Hangzhou, China

Hangzhou Botanical Garden founded in 1956 is located within the World Cultural Heritage Site—Hangzhou West Lake Scenic Area, covering an area of 248.46 hectares. Besides 14 specialized gardens, such as classification garden, economic plants garden, bamboo garden, medicinal herb garden, endangered trees garden, *Acer* and *Rhododendron* garden, *Osmanthus* garden, 4,836 species of living plants are *ex-situ* conserved in Hangzhou Botanical Garden, of which more than hundreds tree species are endangered and endemic to China. One of our main missions is research and conservation for endangered species of East China on which many strategies, managements and studies have been conducted. Having scientific content, park appearance and cultural connotation, the garden is not only a comprehensive botanical garden, but also the oasis of East China for endangered species.

T1

A512

Chemical constituents from leaves of *Taxus chinensis*

Aiping Meng, Juan Li, ***Sheban Pu***
China Pharmaceutical University

A new compound, (*E*)-1-methoxy-3-*O*- (*p*-coumaroyl)-*myo*-inositol (1), and 13 known compounds, *p*-hydroxybenzaldehyde (2), *p*-hydroxybenzoic acid (3), palmitic acid (4), protocatechuic acid (5), sciadopitysin (6), ginkgetin (7), sequoiaflavonoids (8), taxacin (9), 10-deacetylbaccatin III (10), 5-deacetyltaxachitriene B (11), makisterone C (12), 7-β-xylosyl-10-decetyltaxol (13), taxiphyllin (14) were isolated from leaves of *Taxus chinensis* (Pilger) Rehd. The structure of the new compound was elucidated by analysis of spectroscopic data. Compoud 12 was isolated from the genus *Taxus* for the first time, and compounds 2,3, 8 were isolated from *Taxus chinensis* (Pilger) Rehd for the first time.

T1

A519

The genetic diversity research of the national protect plant *Bhesa sinica*

Zehua Ma
Guangzhou University

In this paper, the ISSR molecular marker technology is applied to the study on genetic diversity of 2 wild *Bhesa Sinica* populations. Using the optimized reaction system, 5 ISSR primers were screened out to amplify on 2 natural populations of a total of 18 samples in the distribution area, and amplified 69 bands in total. Using PopGene 32 for data analysis, the results show that: the genetic diversity of *Bhesa Sinica* ranks in the middle level at the species level, the percentage of polymorphic loci (*PPB*) is 69.57%, Nei's gene diversity (*He*) is 0.1310, and Shannon's information index (*Hsppo*p) is 0.2438; however the genetic diversity at population level is relatively low, (*PPB*) is 49.28%, and Shannon's information index (*Hpop*) is 0.2071. The total population gene diversity of *Bhesa Sinica* (*Ht*) is 0.1512, gene diversity within populations (*Hs*) is 0.1310, the coefficient of genetic differentiation (*Gst*) is 0.1333, that is, the variation within population accounted for 86.67% while variation among populations accounted for 13.33%. Gene flow among populations of *Bhesa Sinica* (*Nm*) is 1.6259, which has prevented the population differentiation caused by genetic drift. By comparing the coefficient of genetic differentiation (*Gst*), gene flow among populations (*Nm*) and the percentage of polymorphic loci (*P*), low genetic diversity is one of the causes of *Bhesa Sinica* on the verge of extinction. Based on the relationship between genetic identity and genetic distance, NTSYS software, version 2.01 is applied to 2 *Bheas Sinica* populations of 18 samples to make UPGMA cluster analysis, the results show that: the individuals of 2 *Bhesa Sinica* populations basically clusters according to geographical distance . The results show that, as a result of the genetic variation mainly existed among individuals within populations, a large number of samples should be taken within different populations by the *ex situ* conservation in order to achieve the preservation of genetic diversity of *Bhesa Sinica* in maximum.

T1
A520
Local habitat disturbance best predicts biological indices of mosses and vascular plants in Ohio (USA) Wetlands
Martin Stapanian[1], *Brian Gara*[2], *William Schumacher*[2], *Jean Adams*[1]
1. U.S. Geolgical Survey
2. Ohio Environmental Protection Agency

Loss of wetland habitats and their associated biological communities is a major environmental concern. Quality assessment indices (QAIs) and indices of biological integrity (IBIs) are useful for assessing the responses of taxa to wetland habitat quality and land use in the surrounding landscape. Mosses and vascular plants have been shown to be reliable indicators of wetland habitat delineation and environmental quality. Knowledge of the best ecological predictors of the quality of wetland moss and vascular plant communities may determine if similar management practices would simultaneously enhance both populations. We used Akaike's Information Criterion to identify models predicting a moss quality assessment index (MQAI) and a vascular plant index of biological integrity based on floristic quality (VIBI-FQ) from 27 emergent and 13 forested wetlands in Ohio, USA. The set of predictors included the six metrics from a wetlands disturbance index (ORAM) and two landscape development intensity indices (LDIs). The best single predictor of MQAI and one of the predictors of VIBI-FQ was an ORAM metric that assesses habitat alteration and disturbance within the wetland, such as mowing, grazing, and agricultural practices. However, the best single predictor of VIBI-FQ was an ORAM metric that assessed wetland vascular plant communities, interspersion, and microtopography. LDIs better predicted MQAI than VIBI-FQ, suggesting that mosses may either respond more rapidly to, or recover more slowly from, anthropogenic disturbance in the surrounding landscape than vascular plants. These results supported previous predictive studies on amphibian indices and metrics and a separate vegetation index, indicating that similar wetland management practices may enhance three vastly different wetland biological communities (amphibians, vascular plants, and mosses). This may lead to more efficient use of available resources by wetland management agencies.

T1
A525
In vitro root induction of *Gyrinops walla* (Gaertn.) using micropropagated multiple shoots
Arosha Nirmal Buddhapriya Welle Hewa Kankanamge, *Sandun Kumari Senarath Wathuge Thamara Priyanthie*
University of Sri Jayewardenepura

Gyrinops walla (Thymelaeaceae) is one of the major sources which provides agarwood resin for the manufacture of valuable perfumes and also for medicinal purposes. This species is naturally grown in the Southeast Asia and the Indian Subcontinent regions. In Sri Lanka, *G. walla* distributed through the humid lowland forests as evergreen trees with tall, strait and slender trunk. The reddish brown latosolic soil is more favourable for the growth of *G. walla* and their roots were adapted for that. Development of plant tissue culture protocol for mass propagation of *G. walla* has been started as a solution for scarcity of natural population due to slow natural growth and illegal felling. Multiple shoots of *G. walla* were induced using *in vitro* germinated seedling parts with optimized Murashige and Skoog (MS) medium supplemented with plant growth regulators. Eight weeks old *G. walla* multiple shoots were used for the *in vitro* root induction. Healthily grown multiple shoots were carefully separated into single shoots. Then the shoots were transferred to half strength MS medium supplemented with different concentrations of indole-3-acetic acid (IAA) (0.0-2.0 mg/L), indole-3-butyric acid (IBA) (0.0-2.0 mg/L) and 1-naphthaleneacetic acid (NAA) (0.0-2.0 mg/L) separately. There were fifteen replicates in each treatment for root induction. Cultures were incubated for 18 hour photoperiod at 25 ± 1°C and observed weekly over a period of ten weeks. Mean number of roots and mean root length were measured by taking rooted plantlets out of the culture vessels before being transferred for acclimatization, after ten week period. Like naturally grown *G. walla* plants, micropropagated plantlets show tap root system which was characterized by having one main root (the taproot) from which smaller branch roots emerge. All tested treatments including plant growth regulator free medium (control) induced roots. Half strength MS medium supplemented with 1.0 mg/L IBA was found to be the most effective treatment in stimulating roots with the highest mean number of roots per shoot (5.2 ± 0.5) and highest mean root length (8.6 ± 0.2 cm). Conclusively, micropropagated *G. walla* plantlets were managed in acclimatization process.

T1
A528
Diversity of Arecaceae in Amapá: The beginning
Tonny Medeiros
Instituto de Pesquisas Científicas e Tecnológicas do Estado do Amapá

The palm trees (Arecaceae) form a group with high species richness in the Amazon Forest. They are widely distributed, occurring mainly in the tropics and subtropics, in diverse habitats and at different reliefs, including the maritime borders to interior regions, emphasizing that there is a gradient of density and decreasing wealth from the Andes to the Atlantic Coast. In many neotropical forests, palm trees stand out for the abundance and richness of species, both in the sub-forest and in the upper strata. The great abundance and diversity of the palm trees in these forests indicate that they are elements of great importance in the structure and functioning of ecosystems. In addition to its ecological importance, the number of basic products for human subsistence is higher in the Arecaceae family than in any other family of plants. Studies in the Amazon have shown that these plants account for between 6 and 7% of the total aerial biomass of the forest, but their fruits rich in lipids and carbohydrates contribute about 60% of the energy content of all fruits produced. The number of species is considered floating and controversial, since specialized literature estimates quantities ranging from 2,500 to 3,500, with approximately 230 genera. In general, the Brazil is considered the third richest country in native palm diversity, with around 37 genera and 288 species. The Amazon flora has a significant number of palms, the surveys already recorded about 232 species and 32 genera, considering that other species are to be discovered and described as new regions are explored. Although research on Arecaceae has advanced in the last decades, most of the phytosociological studies carried out in the Amazon use criteria for inclusion in samplings of individuals that exclude palm trees. When these are included, few species are mentioned due to the minimum diameter and the difficulties in the collection and herborization procedures, as a consequence, the available data on floristic composition and palm dynamics are still very scarce. This is the case of the state of Amapá that concentrates one of the largest biodiversity in natural environments, since it is part of two geographical domains, the Amazonian and the oceanic. But little is known about the diversity and richness of palm trees. The Amapá presents most of its territory converted into protected areas, but it is the only Brazilian state in which there is no official list of plants, due to several factors, but mainly due to the incipience in botanical collections in certain places, as pointed out, in recent studies that in Brazil among the priority regions for botanical inventories is included the Tumucumaque Mountains National Park, located in the State of Amapá. Records point to about 36 species and 11 genera: *Astrocaryum*, *Bactris*, *Desmoncus*, *Geonoma*, *Hyospathe*, *Manicaria*, *Mauritia*, *Mauritiella*, *Oenocarpus*, *Socratea*, *Syagrus*, from palm trees of the Amapá. Facing the diversity of Arecaceae in the Amazon, this number denotes the need for studies with the family on diversity, wealth and biotic associations.

T1

A533

Ethnobotany in the Amapá: Situation and perspective

Raullyan Silva

Instituto de Pesquisas Científicas e Tecnológicas do Estado do Amapá

The ethnobotanical and ethnopharmacological studies are important to provide information on the extent and intensity of traditional use of plant resources by the population and in recent years there has been an increase in this type of study in Amapá state providing information about medicinal plants, ornamental, toxic and food through various sources such as academic work in its various levels, as well as through scientific papers published in scientific journals. This whole arsenal of scientific information that could (and should) serve to support more accurate studies for the development of natural products phytochemical in the discovery of new bioactive substances and production of new drugs, the improvement of existing drugs, the production of insecticides, repellents and phytocosmetics, but unfortunately this wide range of information, though important, are not systematized, leaving to be effectively useful for their applicability purposes. Therefore, this research aimed to make the state the rescue of art ethnobotany and ethnopharmacology produced in Amapá. How to collect and record data instrument was used a form to register to published works as well as a portable scanner was used to make the digital recording information in a practical, direct and safe, and then transported for processing database. They recorded 231 species included in 76 families and 184 genera. The families with the highest number of species were Fabaceae, Lamiaceae, Asteraceae, Solanaceae, Arecaceae and Rutaceae. The genera that had the highest number of species were: *Citrus, Piper, Bauhinia, Capsicum, Ocimum* and *Solanum*. Regarding the growth habit of the plant species recorded, there are the woody plants, followed by those of herbaceous, shrub, the habit of scandent and sub-shrubs. It was demonstrated the wealth of plant diversity known and used by the residents of communities in the state of Amapá, making a motivating factor to broaden and deepen research in agricultural sciences, forestry, ethnobotanical and pharmacological.

T1

A539

Flavonoid-containing plants of flora of Azerbaijan and prospects of their use

Eldar Novruzov [1], *Elnura Jafarova*[2], *Mehriban Zulfugarova*[1], *Aydan Zeynalova*[1]

1. Institute of Botany Azerbaijan National Academy of Sciences

2. Genetic Resources Institute of the Azerbaijan National Academy of Sciences

Investigation of biologically active substances (BAS) in plant sources has always been a topical issue of plants resources. Among the BAS of plant origin, flavonoids occupy a special place. Flavonoids, including numerous and varied in structure phenyl-chromone derivatives have strong anti-allergic, anti-carcinogenic, anti-inflammatory, antiviral and other properties. This is due to the presence in their molecules reactive hydroxyl and carbonyl groups. Flavonoids unlike phenolic antioxidants (tocopherols), except direct antiradical action, are capable of binding to a metal ion with a variable valence and form chelate complexes, which leads to the inhibition of free radical processes. Besides inhibiting monoamine oxidase, inhibit reactive oxygen species, which is very promising for the treatment of certain neurodegenerative diseases, inhibit lipid peroxidation, thus preventing cardiovascular diseases, promote the stabilization of production collagen, inhibit platelet aggregation and stimulate the endothelium of prostaglandins, reduce the rate of cancer cell division, can inhibit carcinogenesis. In the analysis of official sources of drugs found that there are 140 trade names of drugs based on flavonoids. According to pharmotherapetic actions flavonoid drugs grouped as angioprotectors, gepatoprotektory, hypnotics, sedatives, cardiac, improve of cerebral and peripheral circulation, and others.

In the study of flavonoid-containing plants of flora of Azerbaijan revealed that 364 species belonging to 110 genera and 46 families contain flavonoids. Of these, 314 species of flavonoid is installed for the first time. Content of flavonoids depending on characteristics of plant species in fruit varies from 9.4 to 2710.5 mg%, in the flowers of 51.9 to 5213.4 mg%. Most are rich in flavonoids species of the family Rosaceae, Eleagnaceae, Asteraceae, Apiaceae, Hypericaceae, Polygonaceae, Fabaceae, Berberidaceae, Salicaceae, Sambucaceae, Punicaceae. Most rich content of flavonoids in the flowers are considered the following: *Salix caprea* L. – 1011.3 mg%, *Populus transcaucasica* A. Jarmol. - 2115.5 mg%, *Berberis iberica* Stev.et Fisch. ex. DC. - 1018.6 mg%, *Crataegus pentagyna* Waldst.et Kit – 1117.9 mg%, *Crataegus caucasica* C.Koch – 1025.8 mg%, *Hippophae rhamnoides* L. – 1513.1 mg%, *Helichrysum plicatum* DC. - 5213.4 mg%, *Helichrysum callichrysum* DC. - 3758.9 mg%, *Helichrysum azerbaidzhanicum* D. Sosn. - 3570.4 mg%, *Tragopodon gramnifolius* DC. - 1034.5 mg%, *Spartium junceum* L. – 2211.6 mg% . It was found that the content of flavonoids depends on soil and climatic conditions and the members of the phytocenoses and qualitative composition is a biological feature of the type acquired in the course of evolution, and not subject to change. The regularities flavonoid-containing types of vertical zonation and botanical and geographical areas. It was found that with increasing altitude above sea level increases the number of flavonoid-containing species. By saturation different lower and middle belts. Flavonoid-containing plants are concentrated in the Greater Caucasus, Nakhchivan Autonomous Republic and Talysh. From the researched types more than 90 types as on content, high-quality structure, and also on raw material stocks it is possible to carry to perspective for receipt of medicines, dietary supplements, and also for increase in biological value of food and cosmetic products.

T1

A542

Genetic diversity assessment of under-exploited African plants for genetic improvement and food security

***Conrad A. Omonhinmin**, Jacob O. Popoola, Fisayo Daramola, Samuel A. Ejoh, Omolola E. Omotosho, Raphael Mordi, Ayoola Ayodeji, Olugbenga Taiwo*

Covenant University, Ota, Ogun State,

Climate change, loss of natural habitats and environmental degradation due to developmental drives are major factors accelerating the alarming rate of plant and animal diversity loss. Changes in consumer demand and the use of only a few species, varieties and breeds have also exacerbated genetic erosion. Above all, a North-South dichotomy in research focuses on global production of food, medicine etc, have relegated African plants to the background, creating the "Orphan plants or crops" of today. The pressure on forest lands, forests and the effects of unsustainable use of forest resources, pose a grave concern on the loss of potentially important plants species; and with only an estimated one percent (1%) of the over 80 000 tree species studied so far, efforts to carry out proper genetic diversity and conserve valuable food, spices, beverage and medicinal resources of African plants that are endemic, indigenously cultivated and an integral part of the way-of-life of Africans stand to offer the only true food security antidote for Africa. The present study focuses on a solution-based research effort on improved genomic information generation on indigenous but under-exploited African plant species. In our present research effort, we have commenced the genetic diversity study of 30 indigenous plants across Northern and Southern Nigeria. The survey spanned 192 communities, in 101 Local Government Areas of 35 States and the Federal Capital; generating a total of 682 accessions of 30 plant species. The research proposes to generate reliable molecular information on the plants applying rbcL sequencing and DNA barcoding tools. The results will impact immensely on biodiversity conservation of the species, draw the attention of global researchers and developmental organizations to these plants as well as directly affect other core SDG targets like; Poverty, Hunger and Food security, Health, Energy, Economic growth, Inequality, Gender equality and Women empowerment, Sustainable consumption and production, Climate change, Forest desertification and partnership.

T1

A544

Tropical tree genomics reveals fragmentation of the African rainforest during the ice ages

Rosalia Piñeiro

Royal Botanic Gardens, Kew

The fossil record in tropical Africa suggests that dry conditions during the Ice Ages caused the contraction of rainforests in favour of savannahs. This seems to explain why the African rainforest harbours less species than rainforests in South America and South East Asia. Based on the rate of endemic species, the location of forest refugia has been proposed. However, to what extent the forest was fragemented, where it survived, and the timing of demographic events remain unknown. In order to address these questions, genomic data from 250 trees of six rainforest tree species were generated using nuclear microsatellites and Genotyping by Sequencing (GBS). Phylogenomic reconstructions, Bayesian clustering, and GIS tools showed congruence of the intraspecific genetic discontinuities across species in Central Africa, pointing at common barriers to gene flow. Demographic model tests revealed the intraspecific gene pools diverged during the Pleistocene (<2Myr ago), so that intraspecific differentiation is the appropriate scale to test the effect of Pleistocene dry periods on tree populations. Demographic tests revealed clear genetic signals of population expansion in both taxa, possibly following bottleneck events after forest fragmentation.

T2

A086

The genus *Lejeunea* in Eocene amber from Baltic in the west coast of Russia

***Wei Cao**, Yuanxin Xiong, Xianzhen Luo, Hongying Tan, Degang Zhao*

College of Life Science, Guizhou University

A sterile liverwort inclusion in a piece of Eocene amber which collected from Baltic and deposited in in Guizhou University Museum, is described and assigned to the extant genus *Lejeunea*. Cenozoic ambers are a valuable source of bryophyte inclusion and provide evidence for the presence of numerous extant genera of leafy liverworts in the Eocene and Oligocene. Most Eocene amber inclusion of liverworts come from Baltic. Some of these inclusions do not match the morphology characteristics due to morphology stability of liverworts over a period of 15-20 million years. The inclusion resembles a complex of extent Russia *Lejeunea* species (*L.cavifolia*, *L. japonia*, *L. otiana* and *L. alaskana*) that share a small size, ovate to obovate leaf lobes with rounded apices, leaf lobule not reduced, first tooth spreading towards leaf apex. It is impossible that the sample can be definited to one of these species due to lack of several diagnostically relevant charaters such as perianths and oil bodies. The inclusion is clearly different from the extinct species *Lejeunea* palaeomexicana from Mexican amber, so the *Lejeunea* species first occurred in the Baltic.

T2

A087

The new discovery in the pattern on vegetative propagation

of *Cistanche lanzhouensis*

***Xue-Lin Chen**, Jin-Yuan Chen*
College of Life Science, Northwest Normal University

Cistanche lanzhouensis come into being the developed "vegetative propagate stem", this was not appeared in the research before. Vegetative propagate stem present dark yellow, its texture like the stem of *C.lanzhouensis*, easy been broken, could extend around under the earth, and could build parasitic relationship with more host root.There are many new buds grow out from the multiple parts on the surface of vegetative propagate stem.

T2

A088

Molecular systematics and biogeography of genus *Saussurea* DC. (Asteraceae)

***Yousheng Chen**, Qian Yuan*
Institute of Botany, Chinese Adademy of Sciences

Saussurea DC., one of the largest genera of Asteraceae tribe Cardueae, includes approximately 460 species, and is a typical group occurs in the Northern Hemisphere. Previous molecular systematics studies showed that the subgenera and sections in genus *Saussurea* are mostly not monophyletic, and genus *Saussurea* needs a more nature classification system. Besides, the origin and evolution, the present distribution pattern of *Saussurea* is not clear. In the present study, *Saussurea* species have been sampled in a large scale, and their phylogenetic relationships were reconstructed using molecular phylogenetic analyses and ancestral character reconstruction methods. Meanwhile, the origin and evolution, and the mechanisms of shaping the present distribution pattern in this genus were also discussed based on our phylogenetic framework. 1) Molecular phylogeny. We sampled 441 accessions belong to 263 species in *Saussurea* most representing the species diversity and geographical distributions. Molecular phylogenetic analyses based on two nuclear genes (ITS and ETS) and seven plastid genomic regions (*ycf1, rpl32-ndhF, psbA-trnH, trnL-F, rbcL, trnY-rpoB, trnL-rpl32*) provided a relatively better phylogenetic framework: (1) the species of *Saussurea* formed a highly supported monophyly, and the sister group is *Polytaxis*. (2) traditional taxonomy system of genus *Saussurea* was broken completely; (3) and *Saussurea* should be divided into five clades. 2) The evolution history of several morphological characters. Seven morphological characters (habitat, stem obviously or not, stem has wing or not, leaf shape, inflorescence, involucre shape, the density of hairs on involucral, phyllary reflexed or not) have been used to reconstruct their evolutionary pattern. The results show that these morphological characters which were usually used to classify and identificate *Saussurea* species originated many times, and have experienced convergence evolution. 3) Biogeographical history of *Saussurea*. Based on the phylogenetic tree, molecular dating analysis was done using BEAST, and ancestral area reconstruction was done by three methods: S-DIVA, DEC and S-DEC. The results indicated that (1) *Saussurea* might originate from Hengduan mountains during late-Miocene; (2) The first divergence of *Saussurea* fell at ca. 7.66 Mya, affected by the third major Qinghai-Tibetan Plateau (QTP) uplifts; (3) the forth QTP rapid uplift since 3.6 Mya might driven the radiation and diversification of *Saussurea*. (4) the global temperatures drop significantly from late Miocene might result in dispersal to other mountains the QTP and Hengduan mountains; (5) *Saussurea* might arrive at North America during Pleistocene.

T2

A089

Holly leaf epidermal cell classification by the LeNet CNN model

*Yuan Liu[2], Jingyao Tu2, Miao Ma[1,2], **Yuli Chen**[1,2]*
1. Key Laboratory of Modern Teaching Technology, Ministry of Education
2. Shaanxi Normal University

The worldwide genus *Ilex* Linnaeus is of very important economic, ornamental and medicinal value, and encompasses more than 500 species of dioecious shrubs and small trees distributed in tropical to warm temperate regions. However, extant *Ilex* species have extremely high intraspecific variability in leaf morphology, which resulted in a heated controversy on the intrageneric taxonomy and systematic relationships. The leaf epidermal characteristics of hollies (such as stomatal apparatus, anticlinal wall and epidermal hairs) can be proved to objectively circumscribe the phylogenetic relationships of different taxa and hence has great potential for understanding the classical taxonomy, paleobotany and phylogenetics. Understanding the variable, complex cuticle characters of holly leaves is one of the most challenging problems in phytotomy. Machine learning offers opportunities to deal with large numbers of microscopic images, to discover novel cuticular features of the intrageneric taxa that may have phylogenetic significance, and to use those characters to classify unknowns. Here, we tested whether a LeNet Convolutional Neural Networks (CNN) model (LeCun *et al.*, 1998) could use a database of 196,211 *Ilex* leaf anatomical images (including upper and lower epidermis) from 101 species of the subgenus *Ilex* (sections *Lioprinus*, *Paltoria*, *Ilex* and *Pseudoaquifolium*) and subgenus *Prinos* to learn features of botanical subgenera, sections and species, then classify novel images. Convolutional Neural Networks (CNN) is a kind of multi-layer neural networks trained with back-propagation algorithm. In this paper, we employ the classical deep LeNet CNN model to identify the database of holly leaf cuticle images (Li *et al.*, 2010) as shown in Table 1. The LeNet CNN model consists of convolution layers and subsampling (pooling) layers alternately, and ends with a fully connected layer. We conducted five experiments for the four sections and one subgenus, respectively. For each experiment, we further performed three sub-experiments for analyzing upper, lower and mixed epidermal cells, respectively. All of the images in the database are separated as the training and validation sets. Our experiments are carried out under the Matconvnet toolbox (http://www.vlfeat.org/matconvnet/). The input of the LeNet model is a holly leaf cuticle image. The output of the model is a vector of elements whose amount is identical to the species number. The element values indicate the probabilities of the input image belonging to the corresponding species. The species with the highest probability is regarded as the predicted result. The accuracy rates of the five experiments showed that the LeNet CNN model has a good performance on the classification of holly leaf cuticle images. Among the four sections and one subgenus of holly, the best accuracy rate reaches 95% and the lowest one is 79% as shown in Table 2. Therefore, the LeNet CNN model is likely to prompt recognition and interpretation of a wealth of cuticle characters. With assistance from computer vision, cuticles are poised to make numerous new contributions to systematic and paleobotanical studies.

T2

A090

Oligocene *Acer* fruits from the Ningming Basin in Guangxi, South China

Yunfa Chen[1], *Qi Wang*[2], *Qian Hu*[1], *Yongqing Liufu*[1], *Zhiming Xie*[1]
1. Guangxi Museum of Natural History
2. Institute of Botany, Chinese Academy of Sciences

Acer ningmingensis sp.nov. is described for asymmetric samara from the Oligocene Ningming Formation in Guangxi Zhuang Autonomous Region, South China. Fruits samaroid; nutlet at basal end; nutlets 0.7 to 0.8 cm long, 0.4 to 0.5cm wide, outline elliptic; nutlet flattened or very thin; attachment scar 0.4-0.5cm long, attachment angle 30°to 45°, nutlet angle 20°to 25°; veins on nutlet obscure, nutlet surface nearly smooth; proximal margin of nutlet either not or only slightly expanded beyond proximal margin of wing; wing 3.0-3.4cm long, 0.9-1cm wide; wing apex narrowly rounded and conspicuously inclining to proximal side; proximal margin broadly concave, distal margin broadly rounded - broadly convex, forming a broadly shallowly V-shaped sulcus with nutlet; proximal margin with 5 prominent veins coalesced along the wing proximal margin, extending from apical margin of nutlet; wing veins diverging from proximal margin at angles of 20° to 40°· nearly straightly extending toward distal margin, the angles increasing from the apex to the base, dichotomizing 2 to 4 times, few anastomoses. The winged fruits described in this paper are similar to the fossils of *Acer stewarti* in the North America. However, the nutlet outline of *Acer stewarti* is circular, concave-convex, and wing proximal margin straight, veins in wing body dichotomizing more (5 to 6 times). The present fossil differs from modern *Acer laxiflorum* fruit in Guangxi, South China which is characterized by straight proximal margin and much bigger attachment angle (>70°). The fossil is the lowest latitude Oligocene *Acer* documented in China, even in Asia, and the only recorded Oligocene *Acer* samaras from a tropical zone today. It suggests that, plant *Acer* had been widely distributed in the Oligocene in Northeast Asia and East Asia, and the climate of the Ningming area in the Oligocene would be cooler than at present.

T2

A091

Species richness and distribution of *Plagiochila* (Plagiochilaceae) in Mt. Apo and Mt. Limbawon, Mindanao, Philippines

Fulgent Coritico, *Victor Amoroso*
Central Mindanao University

There is a limited study regarding the hepatic flora in the Philippines. One of the largest genera of the liverworts (hepatics) is *Plagiochila*. It is one of the main components of bryophyte communities in tropical mossy forests. This research was conducted to determine the species richness and distribution of *Plagiochila* (Plagiochilaceae) based on field surveys from the two mountain ecosystems in Mindanao, viz., Mt. Apo and Mt. Limbawon and available herbarium data. A total of fourteen species are known from these mountains, *viz*, *Plagiochila arbuscula*, *P. blepharophora*, *P. dendroides*, *P. frondescens*, *P. fusca*, *P. javanica*, *P. massalongoana*, *P. philippinensis*, *P. propinqua*, *P. renitens*, *P. salacensis*, *P. sciophila*, *P. trabeculata* and *P. sp*. This is about 28% of the total number of species found in the Philippines. The richness of *Plagiochila* is comparatively similar in Mt. Banahao, Luzon and comparatively higher in Mt. Makiling, Mt. Pulog, and Mt. Data in Luzon Island. Majority of the species observed are closely related to the species found in the neighboring island of Mindanao such as Java, Sumatra and Borneo. Assessment of the species revealed one endemic species, *P. philippinensis*. The study obtained the first records of *P. arbuscula, P. frondescens, P. javanica, P. massalongoana, P. philippinensis, P. salacensis, P. sciophila, P. trabeculata* for the island of Mindanao and one possible new species.

T2

A092

Identification of 220 strains of *Alternaria* in China

Hui Deng, *Yan Min Hou, Yong Chun Niu*
Institute of Agricultural Resources and Regional Planning, Chinese Academy of Agricultural Sciences

Alternaria is a genus with abundant saprophytic, endophytic and pathogenic species associated with various plant substrates. According to DNA-based studies during recent years, *Alternaria* is currently divided into 26 sections. To identify 220 strains of *Alternaria* preserved in Agricultural Culture Collection of China (ACCC), morphological and molecular phylogenetic analysis were studied on them. All of the 220 strains were sequenced by the glyceraldehyde-3-phosphate dehydrogenase (GPDH) gene, and some strains were sequenced by the translation elongation factor1-a (EF-1a) gene, RNA polymerase second largest subunit (rpb2) or *Alternaria* major allergen gene (Alt a 1) regions. The data show that the 220 strains are identified as species in six sections, sect. *Alternaria*, sect. *Brassicicola*, sect. *Panax*, sect. *Porri*, sect. *Pseudoulocladium*, sect. *Ulocladium*. There are six species in sect. *Alternaria*, *Alternaria alternata*, *A. burnsii*, *A. gossypina*, *A. jacinthicola*, *A. longipes*, and one possible new species of six strains isolated from various plants in Hainan and Guangdong provinces.

T2

A093

Taxonomic and systematic studies on Acanthaceae from China

Yunfei Deng
Key Laboratory of Plant Resources Conservation & Sustainable Utilization, South China Botanical Garden, Chinese Academy of Sciences

Acanthaceae is a tropical family of about 230 genera and 4500 species. In mostly recent publication of Volume 19 of *Flora of China*, 35 genera and 304 species were recognized. After completing the treatment for *Flora of China*, the relationships among Chiense Acanthaceae were studied, some genera were re-delimited, and the thorough taxonomic revision was carried out. Six chloroplast DNA regions and one nuclear region were used in the phylogenetic study, the result indicated that Acanthaceae from China can be divided into four subfamilies, viz. Nelsonioideae, Avicennioideae, Thunbergioideae and Acanthoideae. Acanthoideae can be divided into four tribes, viz. Acantheae, Ruellieae, Andrographideae and Justiceae. The subdivision at subtribal level within Ruellieae and Justiceae are also discussed. The generic circumscription of *Ruellia* and *Strobilanthes* in broad sence is supported, but *Justicia* is a polyphyletic and the treatment in broad sense is not supported and further studies need be done. *Leptosiphonium venustum* might be treated to be a new genus *Psendosiphonium* Y.F. Deng *et al.*, ; the species of *Echinacanthus* from China and Vietnam are quite different from the type of the genus and should be described as a new genus *Sinoacanthus* Y.F. Deng *et al.*, *Justicia grossa* and *J. microdonta*, two species previously placed in the genus *Justicia* might be moved from Jusciciinae to Isoglossinae and *Tetramerium*-lineage, and they present two un-described new genera, *Hanseniacanthus* Y.F. Deng and *Wuacanthus* Y.F Deng *et al.*, respectively. Two genera *Asystasiella* and *Haplanthus* are resurrected. *Cystacanthus* is merged with *Phlogacanthus*. After the treatment of the family Acanthaceae for *Flora of China*, 10 new species were discovered, 6 names were reduced to the synonymy. Consequently, 40 genera and 308 species are currently recognized.

T2

A094

Molecular phylogeny of *Cladrastis* clade and resurrection of *Platyosprion* (Leguminosae: Papilionoideae)

Lei Duan[1], *Jun Wen*[2]
1. South China Botanical Garden
2. Smithsonian Institution

Based on sequence data of nuclear ITS and plastid *matK*, *trnL-F* and *psbA-trnH* markers, the phylogeny of *Cladrastis*, *Pickeringia* and *Styphnolobium* in the tribe Sophoreae was inferred. The results showed these three genera form a clade, phylogenetically away from the rest of Sophoreae. *Cladrastis* is polyphyletic, which is redelimited to include *C. delavayi*, *C. kentukea*, *C. shikokiana* and *C. wilsonii*. *C. chingii*, *C. parvifolia* and *C. scandens* are treated as synonyms of *C. platycarpa*. The genus *Platyosprion* is resurrected on the basis of *C. platycarpa*. The monophyly of *Pickeringia* and *Styphnolobium* are supported, and two subgenera were recognized within *Styphnolobium*.

T2

A095

Genomic SSR in *Gentiana lawrencei* var. *farreri*: Characterization, marker identification and transferability in its congeneric species

Peng-Cheng Fu[1], *Shan-Shan Sun*[1], *Yan-Wei Cheng*[1], *Hui-Yuan Ya*[1], *Shi-Long Chen*[2]
1. Luoyang Normal University
2. Northwest Institute of Plateau Biology, Chinese Academy of Sciences

Gentiana lawrencei var. *farreri* T. N. Ho is a popular perennial herb in the Qinghai-Tibetan Plateau (QTP) alpine meadow and has been used in traditional Chinese and Tibetan medicine. Reliable and highly polymorphic molecular markers are required for genetic diversity and conservation studies in this species. In this study, enriched DNA libraries of *G. lawrencei* var. *farreri* were used for high-throughput microsatellite development by 454 FLX+ platform. We obtained 87,097 reads with an average length of 809 bp. The total size of examined sequences was 23,848,926 bp, from which 6,381 SSRs were identified. Of these sequences, 4,831 contained SSRs, with 24.07% featuring more than one SSR. Only 23.37% SSRs were present in compound formation. SSRs with repeat motifs of 2–3 bp (di- and tri-nucleotide repeats) accounted for 97.05% of SSRs. Tri-nucleotide repeats were the most abundant. The predominant motif was (AAC/GTT)n and (AG/CT)n. A total of 3031 primer pairs were designed and 96 were selected for validation. After screening using a series of criteria, a set of 20 fluorescently labeled primer pairs was further selected and screened for polymorphisms in two populations of *G. lawrencei* var. *farreri* and two populations of *G. veitchiorum*. The average number of alleles per locus was 15.2 in the two taxa. Among the four populations, mean observed and expected heterozygosities ranged from 0.222 to 1 and 0.221 to 0.914, respectively. Of the 20 loci, 12 showed significant deviation from Hardy-Weinberg equilibrium (HWE) in one or more populations and only one locus showed significant deviation from HWE in all the four populations. Finally, removing linked loci, a set of 17 unlinked loci was in HWE. The genetic clustering showed that the four populations from two taxa should be grouped into two clusters, indicated the 20 SSRs could be reliable for species delimitation of the two taxa. In conclusion, the newly identified SSRs should be useful for study about genetic diversity, gene flow and evolution in *Gentiana*.

T2

A096

The complete chloroplast genome sequence of *Gentiana lawrencei* var. *farreri* (Gentianaceae) and comparative analysis with its congeneric species

Peng-Cheng Fu[1], *Shan-Shan Sun*[1], *Yan-Zhao Zhang*[1], *Shi-Long Chen*[2]
1. Luoyang Normal University
2. Northwest Institute of Plateau Biology, Chinese Academy of Sciences

Background. The chloroplast (cp) genome is useful in plant systematics, genetic diversity analysis, molecular identification and divergence dating. The genus *Gentiana* contains 362 species, but there are only two valuable complete cp genomes. We report the characterization of complete cp genome of *G. lawrencei* var. *farreri*, which is endemic to the Qinghai-Tibetan Plateau (QTP). Methods. Using high throughput sequencing technology, we got the complete nucleotide sequence of the *G. lawrencei* var. *farreri* cp genome. The comparison analysis including genome difference and gene divergence was performed with its congeneric species *G. straminea*. Results. The cp genome of *G. lawrencei* var. *farreri* is a circular molecule of 138,750 bp, containing a pair of 24,653 bp inverted repeats which are separated by small and large single-copy regions of 11,365 and 78,082 bp, respectively. The cp genome contains 130 known genes, including 85 protein coding genes (PCGs), eight ribosomal RNA genes and 37 tRNA genes. Comparative analyses indicated that *G. lawrencei* var. *farreri* is 10,241 bp shorter than its congeneric species *G. straminea*. Four large gaps were detected that are responsible for 85% of the total sequence loss. Further detailed analyses revealed that 10 PCGs were included in the four gaps that encode nine NADH dehydrogenase subunits. The cp gene content, order and orientation are similar to those of its congeneric species, but with some variation among the PCGs. Three genes, *ndhB*, *ndhF* and *clpP*, have high non-synonymous to synonymous values. There are 34 SSRs in the *G. lawrencei* var. *farreri* cp genome, of which 25 are mononucleotide repeats: no dinucleotide repeats were detected. Comparison with the *G. straminea* cp genome indicated that five SSRs have length polymorphisms and 23 SSRs are species-specific. The phylogenetic analysis of 48 PCGs from 12 Gentianales taxa cp genomes clearly identified three clades, which indicated the potential of cp genomes in phylogenetics. Discussion. The "missing" sequence of *G. lawrencei* var. *farreri* mainly consistent of *ndh* genes which could be dispensable under chilling-stressed conditions in the QTP. The complete cp genome sequence of *G. lawrencei var. farreri* provides intragenic information that will contribute to genetic and phylogenetic research in the Gentianaceae.

T2
A097
Gene flow results in high genetic similarity between *Sibiraea* (Rosaceae) species in the Qinghai-Tibetan Plateau

Peng-Cheng Fu[2], *Qing-Bo Gao*[1], *Fa-Qi Zhang*[1], *Rui Xing*[1], *Jiu-Li Wang*[1,3], *Hai-Rui Liu*[1,3], *Shi-Long Chen*[1]
1. Northwest Institute of Plateau Biology, Chinese Academy of Sciences
2. Luoyang Normal University
3. University of Chinese Academy of Sciences

Studying closely related species and divergent populations provides insight into the process of speciation. Previous studies showed that the *Sibiraea* complex's evolutionary history on the Qinghai-Tibetan Plateau (QTP) was confusing and could not be distinguishable on the molecular level. In this study, the genetic structure and gene flow of *S. laevigata* and *S. angustata* on the QTP was examined across 45 populations using 8 microsatellite loci. Microsatellites revealed high genetic diversity in *Sibiraea* populations. Most of the variance was detected within populations (87.45%) rather than between species (4.39%). We found no significant correlations between genetic and geographical distances among populations. Bayesian cluster analysis grouped all individuals in the sympatric area of *Sibiraea* into one cluster and other individuals of *S. angustata* into another. Divergence history analysis based on the approximate Bayesian computation method indicated that the populations of *S. angustata* at the sympatric area derived from the admixture of the 2 species. The assignment test assigned all individuals to populations of their own species rather than its congeneric species. Consistently, intraspecies were detected rather than interspecies first-generation migrants. The bidirectional gene flow in long-term patterns between the 2 species was asymmetric, with more from *S. angustata* to *S. laevigata*. In conclusion, the *Sibiraea* complex was distinguishable on the molecular level using microsatellite loci. We found that the high genetic similarity of the complex resulted from huge bidirectional gene flow, especially on the sympatric area where population admixtures occurred. This study sheds light on speciation with gene flow in the QTP.

T2
A098
Systemetics and taxonomy of the genus *Sinoacanthus* (Acanthaceae)

Chunming Gao[1], *Yunfei Deng*[2]
1. Binzhou university
2. South China Botanical Garden

Sinoacanthus, a new genus of Ruellinae in Acanthaceae restircted to Sino-Vietnamese karst flora, and including three species, S. *longipes*, *S. longzhouensis* and *S.lofouensis,* is segregated from *Echinacanthus.* There are some differences between *Sinoacanthus* and *Echinacanthus* based on the morphological, pollen and molecular characters. Firstly, the inflorescence of *Echinacanthus* is thyrsus, but, the *Sinoacanthus* is cyme. Then, *Sinoacanthus* has appendaged anthers with two appendages per thecae (*S. lofouensis* has one two-in-one appendaged per thecae based on our study). But, *Echinacanthus* species have appendaged anthers with one appendaged per thecae. Furthermore, pollen aperture of *E. attenuatus* arranged on the band and reticulate exine, pollen aperture of Chinese pants are located in pseudocolpi and perforate surface. Finally, *Echinacanthus* is included in subtribe Petalidiinae and *Sinoacanthus* is a member of subtribe Erantheminae based on our phylogenetic relationship. *Sinoacanthus* in Subtribe Erantheminae of Ruellinae. The new genus, *Sinoacanthus,* is a monophyletic group based on our molecular study. Furthermore, *Sinoacanthus longzhouensis* formed sister group with the clade of *Sinoacanthus longipes and Sinoacanthus lofouensis. Sinoacanthus* is a typical genus of Sino-Vietnamese karst flora. The there species have Relatively independent niche. *Sinoacanthus longzhouensis* is distributed endemic to Guangxi, Guangdong of China. It grows in limestone hills in forests (300-400m). *Sinoacanthus longipes* is distributed Guangxi, Yunnan of China and Vietnam. It grows in of limestone hills in forests (500-2000m). *Sinoacanthus lofouensis* is a species endemic to Guangxi, Guizhou in China and grows in limestone hills in forests about 500-1000m. It may be an examples in the development of Sino-Vietnamese karst.

T2
A099
Population genetic differentiation and taxonomic suggestions of two highly fragmented *Saxifraga* species endemic to the Qinghai-Tibetan Plateau

Qingbo Gao[1], *Yan Li*[1,2], *Zhuoma Gengji*[1,2], *Shilong Chen*[1]
1. Key Laboratory of Adaptation and Evolution of Plateau Biota, Northwest Institute of Plateau Biology, Chinese Academy of Sciences, Xining 810001, Qinghai, China
2. University of Chinese Academy of Sciences, Beijing 100039, China

Saxifraga pasumensis and *S. umbellulata* are two narrowly distributed and well isolated species endemic to the Qinghai-Tibetan Plateau (QTP). Taxonomy of *S. pasumensis* is still in controversy. In this study, three chloroplast DNA fragments (*rbcL*, *trnL-F*, *trnS-G*) and nuclear ribosomal DNA internal transcribed spacer (ITS) were employed to study genetic structure across 104 individuals from 12 populations of *S. pasumensis* and *S. umbellulata*, including one population which was recognized as a novel species, *S. banmaensis*. Chloroplast DNA (cpDNA) phylogenies revealed two well supported clades, corresponding exactly to the two species as identified based on morphology. Topology of ITS phylogenies was largely congruent with that generated from cpDNA haplotypes, but with minor conflicts which might be caused by incomplete lineage sorting. Analyses of molecular variance based on both cpDNA and ITS datasets revealed that most variance was detected between species (92.31% for cpDNA; 69.78% for ITS), suggesting high differentiation between *S. pasumensis* and *S. umbellulata*. Analyses of molecular clock and S-DIVA suggested a vacariant allopatric speciation of the two species at ca. 4.37 Ma, coinciding with the most recent uplift event of the QTP. Vacariance associated with QTP uplifts should also be responsible for intraspecific divergence in *S. pasumensis*. However, intraspecific divergence of *S. umbellulata* was most likely triggered by Quaternary glaciations. The extremely low gene diversity within populations of *S. pasumensis* could have resulted from both fragmentation and human collection due to its medical utilization. Combining morphological characteristics, biogeographic distribution and our molecular results, *S. pasumensis* is better to be treated as a species instead of a variety, and *S. banmaensis* should be treated as a synonym of *S. pasumensis*.

T2

A100

Liliaceae Juss. in the Azerbaijan flora

Parvana Garakhani[1], *Tariyel Talibov*[2]

1. Institute of Botany of Azerbaijan National Academy of Sciences

2. The Institute of Bioresources of Nakhchivan Division of Azerbaijan National Academy of Sciences

Some researchers of the recent years studying Liliaceae Juss. stated that extent of this family is disputable and consists of genera of different origins in the heterogeneous content. Modern systematists determined several evolution branches within the family and accepted them as independent families. Representatives of the family are encountered in many areas of world. Liliaceae Juss. species are distributed in the forests, lowlands, even in the deserts. Lilies is not only grass plants, it has shrub and/or tree representatives in Africa. Species from the family are decorative plants. Liliaceae Juss. of Azerbaijan flora is divided into several genera: *Tulipa* L., *Gagea* Salisb., *Lilium* L., *Fritillaria* L., *Liloydia* Reichenb. Among these genera *Tulipa L.* is are clearly distinguished with its decorative capasities. Tulipa comprises up to 100 species in the world and 7 in Azerbaijan. In recent years, based on the detailed results, 2 new species recorded and the taxon number of this family increased up to 9 species. *Fritillaria* L. is represented with 8 species, 1 sub-species in the Azerbaijan flora. *Lilium* L. comprises 2 species, 2 sub-species. *Liloydia* Reichenb. is represented with 1 sub-species, and *Gagea* Salisb. with 18 species in the Azerbaijan flora. Many of the species included to the family Liliaceae are endemic and rare species. These species are included to the 2nd editon of the Red Book of the Republic of Azerbaijan (2013). These species are as followings: *Fritillaria caucasica* Adams – ENA1c:B1ab (iii). Rare species of Azerbaijan; *Fritillaria gibbosa* Boiss. (Rinopetalum gibbosum Boiss.) Losinsk et Vved. – VUA2c+3c. Rare species of Azerbaijan; *Fritillaria grandiflora* Grossh. – CRB1ab (v). Endemic species of Azerbaijan; *Fritillaria kotchyana* Herb. – ENA2c. Rare species of Azerbaijan; *Gagea glacialis* C.Koch.– NT. Rare species of Azerbaijan; *Lilium ledebourii* (Baker.) Boiss. – ENA2c. Rare species of Azerbaijan, endemic species of Caucasus; *Tulipa biebersteiniana* Schult.et Schult. – VUA2c+3c. Rare species of Azerbaijan; *Tulipa biflora* Pall.– VUA2c+3c. Rare species of Azerbaijan; *Tulipa echleri* Regel. – VUA2c+3c. Rare species of Azerbaijan, endemic of Caucasus; *Tulipa florenskyi* Woronov. – VUA2c+3c. Rare species of Azerbaijan, endemic of Caucasus; *Tulipa gesneriana* L. – VUC2a (ii). Rare species of Azerbaijan; *Tulipa julia* C.Koch. – VUA2c+3c. Rare species of Azerbaijan, endemic of Caucasus; *Tulipa karabachensis* Grossh. – DD. Rare species of Azerbaijan; *Tulipa schmidtii* Fomin. – NT. Rare species of Azerbaijan.

T2

A101

Floral ontogeny in *Idesia Polycarpa* (Flacourtiaceae)

Li-Ping Ge

Shanxi Agricultural University

Idesia traditionally belongs to the family Flacourtiaceae. *Idesia Polycarpa*, only species of the genus *Idesia*, is recently concerned and utilized as a kind of biomass energy tree because its seeds can produce more oil. However, there are still many disagreement of classification on the family Flacourtiaceae among different taxonomist, and more phylogenetic analyses show that the family is not a monophyletic group, consequently some former members of Flacourtiaceae are moved into other different family according to new research results. Floral organogenesis has been shown to be valuable in elucidating phylogenetic position for generic and supra-generic taxa, but the floral ontogeny in *Idesia* has not been reported so far. The present work aims to provide data of floral ontogeny for the *Idesia Polycarpa*. The dynamic process on the floral organogenesis both female and male flowers of *Idesia Polycarpa* were examined respectively using scanning electron microscopy (SEM). Both female and male flowers, sepals are initiated successively (not simultaneously) on the periphery of the floral apical meristem; Usually, the adaxial whorl stamens initiate first, followed by the second stamens, and then the outermost stamens. The sequence of the stamen initiation in the polystemonous androecium is somewhat irregular but is whorled (three to four whorl) and centrifugal overall. The pistil is initiated after initiation of the first or second whorl stamen. Significantly, early stamen development in the male flower is resembles that of female flower, nevertheless stamens are turn into staminode during the later development period of the female flower; likewise, pistil also develop in the early male flower, but pistil are turn eventually into pistillode in the male flower. These observation result will be provide further information into the circumscription between female and male flowers.

This research was supported by Natural Science Foundation of Shanxi (2010011039–1)

T2

A102

Phylogenetic relationships in *Anaphalis* DC. (Asteraceae),

inferred from chloroplast genomes

Junyu Gu, *Zhen Wei, Xuemin Xu, Shixin Zhu*
Zhengzhou University

Anaphalis DC. belongs to the tribe Gnaphalinae, family Asteraceae. The circumscription of the *Anaphalis* is ambiguous due to its unstable and variable characters. Therefore, it is one of the most problematic groups in Asteraceae. DNA sequence analysis is a powerful tool to detect the variation of genetic material, and has been proven to be robust in revealing the evolution and phylogenetic relationships of plants. As sequencing technologies developing, chloroplast genomes have been used to construct phylogenetics on species, genus, tribe and family levels. In this study, we sequenced chloroplast genome sequences of thirty *Anaphalis* species using Illumina Hiseq-PE150 platform. The cp genomes were then annotated and coding/non-coding DNA and RNA sequences were extracted respectively. The phylogenetic relationships in *Anaphalis* were constructed and compared based on different datasets.

T2

A103

Ferns' homologs of leaf polarity gene *ARF4* in *Arabidopsis*

Gui-Sheng Li
Jishou University

Because the evidence at the aspect of gene function on the hypothesis of multiple origin of leaves is scarce, this project studied the evolution of leaf polarity gene *ARF4*. Firstly, this project aiming at the development of fern leaves conducted transcriptome sequencing to two samples of *Ceratopteris pteridoides* and each one sample of *Cyrtomium guizhouense* and *Parathelypteris nipponica*, producing 11.6,12.3,13.3 and 13.3 Gb clean data respectively, and with the combination of RT-PCR obtained the complete coding region of *ARF4* from these three species. It is possible that the three clades of the *ARF* family, *ARF5/6*, *ARF10* and *ARF2/4* are involved in leaf development in ferns. The *ARF2* and the *ARF4* sub-clade have diverged in ferns, with the former being of multiple copy while the latter single copy. *ARF4* has undergone a gene duplication in the common ancestor of seed plants and the common ancestor of angiosperms. Fern *ARF4* may possess a single tasiR-ARF binding site, while for gymnosperm *ARF4* some have two but some have only one such site, meanwhile angiosperm *ARF4* always have two such sites. It is similar to seed plants that ferns contain two types of *TAS3* genes which can produce tasiR-ARF. Small RNA sequencing of two samples of *Ceratopteris pteridoides* found 284 known miRNA and 173 brandnew miRNA, especially mining out miR390 responsible for *TAS3* processing and small RNA matching tasiR-ARF. In *Ceratopteris pteridoides* *ARF2* may have been sliced by tasiR-ARF, however whether *ARF4* is cut by tasiR-ARF remained unclear. *ARF4* is expressed in sporangia which are located at the back of leaves. Therefore, the sequence and expression pattern of fern *ARF4* is inclined to the opinion of independent origin of lycophyte microphylls, fern leaves and leaves of seed plants respectively.

T2

A104

Taxonomic studies on species of *Macromitrium* (Orthotrichaceae, Musci) with irregular lamina stratose

Dandan Li[1], *Si He*[2], ***Shuiliang Guo***[1], *Tong Cao*[1]
1. Shanghai Normal University
2. Missouri Botanical Garden

Eight species with irregular lamina stratose in the genus *Macromitrium* (Orthotrichaceae) are morphologically illustrated. Among them, *M. diaphanum* Müll. Hal., *M. nepalense* (Hook. & Grev.) Schwägr., *M. serpens* (Hook. & Grev.) Schwägr. and *M. tongense* Sull. have the character of irregular 1-3 (4)-stratose proliferation with pluripapillose cells over both leaf upper surfaces. The leaf upper parts of *M. st.-johnii* E.B. Bartram and *M. amaniense* P. de la Varde are also special, similar to those of *M. longipes* (Hook. & Grev.) Schwägr. and *M. longirostre* (Hook. & Grev.) Schwägr., partially and variably bistratose, but occasionally arranged jaggedly. Their morphological characteristics are further revealed graphically by compiling figures of digitalized transverse sections of branch leaves. Based on the overall similarities of their gametophytic features, *M. tongense* is reduced to synonymy of *M. serpense*. A key is given to separate the species of the genus *Macromitrium* with irregular lamina stratose stratose.

T2

A105

New synonyms and a quantitative study of the patterns of morphological variation within *Macromitrium microstomum* complex (Bryophyta, Orthotrichaceae)

Jing Yu[1], *Si He*[2], ***Shuiliang Guo***[1], *Tong Cao*[1]
1. Shanghai Normal University
2. Missouri Botanical Garden

Macromitrium microstomum (Hook.& Grev.)Schwägr. is the most widely distributed species of the genus in the world and exhibits a great deal of morphological variations as evident by the inclusion of 28 synonyms. A quantitative study of 98 relevant specimens of *M. microstomum*, including 17 types from its synonyms, was carried out to test and understand the patterns of morphological variation within the species. Data of 63 morphological traits were used to construct a dendroogram. A two-dimensional scatter plot and the variation patterns were evaluated by Cluster Analysis and Detrended Correspondence Analysis methods. The morphological variations were further revealed graphically by compiling figures of digitalized branch leaves and perichaetila leaves. Our quantitative analyses show that variation coefficients differ enormously ranging from 0.187 to 0.647among morphological traits, especially from the branch leaves in terms of leaf length, length/width ratios at median and upper parts as well as the angles of leaf apex. The average variation coefficient of the morphological indices is larger in perichaetial leaves than in branch leaves. Other variations from sporophytic traits, such as the width at capsule mouth and seta length of *M. microstomum* were also analysed. Our study shows that variations, particularly in morphological traits obtained from leaf angles and leaf length do exist in the specimens from different geographical regions, although the overall morphological variations are continuous among 98 specimens. *Microstomum filicaule* Müll. Hal., *M. pacificum* var. *brevisetum* Thér., and *M. pacificum* var. *longisetum* Thér., and *Macromitrium subnitidum* Müll. Hal are four new synonyms of *M. microstomum*.

T2

A106

De novo genome assembly of the mahogany mangrove *Xylocarpus granatum*

***Zixiao Guo**, Zhang Zhang, Suhua Shi*

State Key Laboratory of Biocontrol, School of Life Sciences, Sun Yat-sen University

Plants of Meliaceae family are well known for their high-quality timber as "mahogany" in furniture trade. Being the only mangrove genus in mahogany family, *Xylocarpus* is further interested for its ecological adaptation into intertidal zones as well as medicinal usage of natural products isolated from all parts of the plants such as various limonoids. Hence, we first sequenced and assembled a draft genome of *Xylocarpus granatum,* which is the most widely distributed species in the genus. A total of 96.63 short reads are generated on Hiseq 2000 platform with inserts from 0.3 to 8 Kb to *de novo* assemble the genome. The assembled genome contains 44,231 scaffolds and accumulates to 293.91 Mb, covering 85.22% of the genome which was estimated as 344.87 Mb by K-mer estimation. The longest, N50 and N90 scaffolds are 6.54,2.50 and 0.063 Mb respectively. The assembly is of high quality reflecting by 98.08% of the transcriptome sequences generated from leaf, flower and root being covered, 93.55% of the 248 ultra-conserved CEGs presenting in the assembly. We estimated that 42% of the genome consists of repetitive sequences, 88% of which (or 37% of the genome) are Long Terminal Repeats (LTR). In total, 30455 protein coding genes are predicted, among which 28492 genes could be matched to homolog in NR, SWISSPROT; 22,311 could be assigned at least one Gene Ontology term and 8659 could assigned at least one KEGG ortholog term. The genome experienced at least one Whole Genome Duplication (WGD) event, indicating by that 614 collinear blocks covering 22,682 genes or 68.49% of the genome were identified. The draft mahogany genome will provide valuable resource for investigating novel genes involved in bioactive compounds biosynthesis in this plant as well as its evolutionary mechanism of transition from terrestrial plants to mangroves.

T2

A107

Evolutionary analysis of the *LAFL* genes involved in the land plant seed maturation program

***Jing-Dan Han**, Guang-Yuan Rao*

Peking University

Seeds are one of the most significant innovations in the land plant lineage, critical to the diversification and adaptation of plants to terrestrial environments. From perspective of seed evo-devo, the most crucial developmental stage in this innovation is seed maturation, which includes accumulation of storage reserves, acquisition of desiccation tolerance, and induction of dormancy. Based on previous studies of seed development in the model plant *Arabidopsis thaliana*, seed maturation is mainly controlled by the LAFL regulatory network, which includes *LEAFY COTYLEDON1* (*LEC1*) and *LEC1-LIKE* (*L1L*) of the *NF-YB* gene family, and *ABSCISIC ACID INSENSITIVE3* (*ABI3*), *FUSCA3* (*FUS3*), and *LEC2* (*LEAFY COTYLEDON2*) of the B3-*AFL* gene family. In the present study, molecular evolution of these *LAFL* genes was analyzed, using representative species from across the major plant lineages. Additionally, to elucidate the molecular mechanisms of the seed maturation program, co-expression pattern analyses of *LAFL* genes were conducted across vascular plants. The results show that the origin of *AFL* gene family dates back to a common ancestor of bryophytes and vascular plants, while *LEC1*-type genes are only found in vascular plants. *LAFL* genes of vascular plants likely specify their co-expression in two different developmental phrases, spore and seed maturation, respectively, and expression patterns vary slightly across the major vascular plants lineages. All the information presented in this study will provide insights into the origin and diversification of seed plants.

T2

A108

Taxonomic notes on *Eritrichium* (Boraginaceae: Cynoglosseae) in the Pan-Himalaya

***Jia-Chen Hao**, Quan-Ru Liu*

College of Life Sciences, Beijing Normal University

In order to compile the *Flora of Pan-Himalayas* (*Eritrichium*), we have consulted many reports regarding *Eritrichium*, checked all relevant specimens and verified 46 scientific names which belong to 26 species, 1 subspecies and 5 varieties. Based on above works, we carried out comprehensive field investigations, population observations, combined with macro- and micro-morphology studies of nutlets. As a result, we found three taxa need to be revised: initially, *E. heterocarpum* Y. S. Lian et J. Q. Wang is conspecific with *E. echinocaryum* (I. M. Johnst.) Y. S. Lian et J. Q. Wang. There are no differences in external morphology between these two taxa. And we found that some individuals produced heteromorphic nutlets just like *E. heterocarpum* and others developed homomorphic nutlets like *E. echinocaryum* in same populations. Also the distribution areas of *E. heterocarpum* overlap with *E. echinocaryum*'s. Furthermore, *E. longifolium* Decne. is shown to be significantly different from *E. aktonense* Y. S. Lian & J. Q. Wang, but to be conspecific with *E. petiolare* var. *villosum* W. T. Wang. Additionally, *E. lasiocarpum* W. T. Wang is identical with *E. minimum* (Brand) H. Hara and *E. lasiocarpum* is treated as a synonym of the latter. Finally, it is concluded that there are 23 species, 1 subspecies and 5 varieties in total of the genus *Eritrichium* in the Pan-Himalaya region.

T2

A109

Some new recorded plants from Hebei

Tiran Huang

Beijing University of Agriculture

Hongsongwa Nature Reserve (N42°10'~42°20', E117°18'~117°35') is located in the northern part of Weichang manzu Mongolian autonomous county, Hebei province, and seated in the eastern part of the Daxing'anling Mountains and the eastern section of the Yanshan Mountains. It is a plateau platform, which is located in the southern edge of the Mongolian plateau, and its main body is the mountains. The average elevation is 1750m, which changes from 1400 to 2067m. The vegetation types is tessellated meadow which is the typical lot of the Saihanba forest edge landscape. From the division of flora, The Nature Reserve is the transition area from North China flora to the Mongolian flora[1], and there are some components of the Northeast flora. In recent years, in the process of the investigation of completing the plant background resources and the effective management of biodiversity in the Nature Reserve, we carried out a detailed field investigation on the Nature Reserve and its surrounding areas, collected a large number of specimens, after the specimen sorting, identification and literature research, eight species of plants were identified, which had not been recorded in the Flora of Hebei. The specimens cited in this text are collected in the Plant Specimen Room of College of Biological Sciences China Agricultal University: *Galatella dahurica* DC. In Prodr. 5: 256. 1836; FOC Vol. 20-21: 561,562. Heibei: Hongsongwa, Meadows, grasslands on slopes. Heilongjiang, Jilin, Liaoning, Nei Mongol, Xinjiang [Kazakhstan, Mongolia, Russia (Far East, Siberia), Uzbekistan]; *Bromus pumpellianus* Scribn. In Bull. Torrey Bot. Club. 15: 9. 1888, FOC Vol. 22: 372,375. Heibei: Hongsongwa, Mid-mountain meadows, thickets. Heilongjiang, Nei Mongol, Shanxi [Russia; W North America]. Huang Tiran, Liu Haifeng, KY009; *Agropyron mongolicum* Keng In J. Wash. Acad. Sci. 28: 305. 1938, FOC Vol. 22 : 438. Heibei: Hongsongwa, Dry steppes, sandy places. Gansu, Nei Mongol, Ningxia, Shaanxi, Shanxi, Xinjiang; *Koeleria altaica* (Domin) Kryl. In Fl. Sibir. Occid. 2: 261. 1928, FOC Vol. 22 Page 330,331. Heibei: Hongsongwa, Dry stony grassy hills, Huang Tiran, Liu Haifeng, KY152, Li Lianfang, 0008112.Nei Mongol, Xinjiang [Kazakhstan (Tarbagatai Mountains), Mongolia, Russia (S Siberia)]; *Carex tenuiformis* Levl. In Léveillé & Vaniot Bull. Acad. Int. Géogr. Bot. 11: 104. 1902, FOC Vol. 23 Page 363,364. Heibei: Hongsongwa, grasslands on forest margins or in thickets, mountain slopes, Huang Tiran, Liu Haifeng, KY073.Heilongjiang, Nei Mongol [Japan, Korea, Russia (Far East, E Siberia)]; *Carex schmidtii* Meinsh. In Beitr. Kenntn. Russ. Reiches. 26: 224. 1871, FOC Vol. 23:421,423. Heibei: Hongsongwa, Marshes, streamsides, forest margins, Huang Tiran, Liu Haifeng, KY072. Heilongjiang, Jilin, Nei Mongol [Japan, Korea, Mongolia, Russia (Far East, E Siberia)]; *Carex ovatispiculata* Y. L. Chang ex S. Y. Liang In Acta Phytotax. Sin. 28: 155. 1990, FOC Vol. 23:456. Heibei: Hongsongwa, Streambanks, wet places, Huang Tiran, Liu Haifeng, KY154. Hunan, Shaanxi, Sichuan, Xizang, Yunnan; *Kobresia capillifolia* (Decne.) C. B. Clarke In J. Linn. Soc., Bot. 20: 378. 1883, FOC Vol. 23:269,271,275,279,280. Heibei: Hongsongwa, Dry and open grassy and rocky slopes, Huang Tiran, Liu Haifeng, KY060. Gansu, Qinghai, Sichuan, Xinjiang, Xizang [Afghanistan, Bhutan, India (Himachal Pradesh, Sikkim), Kashmir, Kazakhstan, Kyrgyzstan, Mongolia, Nepal, Pakistan, Tajikistan].

T2

A110

Polyploid speciation and genome evolution of *Panax* (Araliaceae)

Peng Jiang[1], *Fengxue Shi*[1], *Yaling Li*[1], *Mingrui Li*[1], *Bao Liu*[1], *Linfeng Li*[2]

1. Northeast Normal University

2. South China Botanical Garden, Chinese Academy of Sciences

Panax L. (Araliaceae) is a medicinally important genus in the East Asia and almost every species within the genus has cultural significance for traditional Chinese medicine. The taxonomy of *Panax* has been controversial due to the circumscription of *P. pseudoginseng* and *P. japonicus*. Recent studies based on species geographic distribution, chromosome number and phylogenetic position suggested that the genus *Panax* includes three tetraploid, four diploid species and one species complex. Further biogeographic and phylogenetic analyses revealed two independent origins of the East Asia and North America disjunct distribution of this genus and multiple rounds of whole genome duplications (WGDs) might have occurred during the evolutionary process. In our recent studies, we employed multiple chloroplast and nuclear markers to investigate the evolution and diversification of *Panax*. Our phylogenetic analyses confirmed previous observations of the independent origins of disjunct distribution and both ancient and recent WGDs have occurred within *Panax*. The estimations of divergence time implied that the ancient WGD might have occurred before the establishment of *Panax*. Thereafter, at least two independent recent WGD events have occurred within *Panax*, one of which has led to the formation of three geographically isolated tetraploid species *P. ginseng*, *P. japonicus* and *P. quinquefolius*. Population genetic analyses showed that the diploid species *P. notoginseng* harbored significantly lower nucleotide diversity than those of the two tetraploid species *P. ginseng* and *P. quinquefolius* and the three species showed distinct nucleotide variation patterns at exon regions. Our findings based on the phylogenetic and population genetic analyses, coupled with the species distribution patterns of *Panax*, suggested that the two rounds of WGD along with the geographic and ecological isolations might have together contributed to the evolution and diversification of this genus.

T2

A111

Chasing ghosts: Allopolyploid origin of *Oxyria sinensis* (Polygonaceae) from its only diploid congener and an unknown ancestor

Jianquan Liu

Sichuan University

Reconstructing the origin of a polyploid species is particularly challenging when an ancestor has become extinct. Under such circumstances the extinct donor of a genome found in the polyploid may be treated as a ‘ghost’ species in that its prior existence is recognised through the presence of its genome in the polyploid. In this study, we aimed to determine the polyploid origin of *Oxyria sinensis* (2n =40) for which only one congeneric species is known, i.e. diploid *O. digyna* (2n=14). Transcriptome, genomic *in situ* hybridization (GISH) population genetics analyses and ecological niche modeling, were conducted for this purpose. Results indicate that following divergence from *O. digyna* (2n=14), involving genome duplication around 12 million years ago (Ma), a second genome duplication occurred approximately 6 Ma to give rise to *O. sinensis*. *Oxyria sinensis* was shown to contain homologous gene sequences divergent from those present in *O. digyna* as well as a set that clustered with those in *O. digyna*. In addition, GISH revealed that *O. sinensis* comprised 14 chromosomes from *O. digyna* and 26 chromosomes from an unknown ancestor. Coalescent simulations indicated that *O. sinensis* expanded its distribution approximately 6-7 Ma, possibly following the second polyploidization event, whereas *O. digyna* expanded its range much later. It was also indicated that the distributions of both species contracted and re-expanded during the Pleistocene climatic oscillations. The ecological niche modeling similarly suggested that both species experienced changes of the distributional ranges in response to the recent Quaternary climatic changes. The extinction of the unknown 'ghost' species implicated in the origin of *O. sinensis* could have resulted from superior adaptation of *O. sinensis* to repeated climatic changes in the region where it now occurs.

T2

A112

Morphological and karyological changes in multiple lineages of *Artemisia* L. (Asteraceae) from Qinghai-Tibetan Plateau

Bohan Jiao[1,2], *Tiangang Gao*[1], *Guojin Zhang*[1,2], *Bao Nie*[1,2], *Li Song*[1,2], *Hong Yan*[1,2], *Jiye Zheng*[1,2], *Guohao Niu*[1,2]
1. Institute of Botany, Chinese Academy of Sciences
2. University of Chinese Academy of Sciences

Early studies noticed some trends of morphological and karyological changes in flowering plants when they moved to high mountains. However, we know little about the trends in the "third pole"—Qinghai-Tibetan Plateau. Here we surveyed morphological and karyological differences between closely related but independent evolved lineages of *Artemisia*. We first reconstructed the phylogenetic relationship of 297 species of Artemisia. Based on the phylogenetic result and geographical distribution of *Artemisia*, we defined the Qinghai-Tibetan Plateau lineages and their sister groups. We compared the morphological traits, such as maximum height, cauline leave blade size and capitula diameters and tested whether traits were significantly different between Qinghai-Tibetan Plateau lineages and their sister groups. At the same time, we compared percentage of polyploidy between Qinghai-Tibetan Plateau taxa and other taxa. Our results indicated: (1) *Artemisia* has independently adapted to Qinghai-Tibetan Plateau more than 10 times; (2)Plant height and blade size in Qinghai-Tibetan Plateau lineages were definitely reduced; (3)Capitula diameters in Qinghai-Tibetan Plateau lineages was definitely increased; (4)Percentage of polyploidy in Qinghai-Tibetan Plateau taxa was notably lower than other taxa. Our result is contrary to early conclusion that polyploidy increases with environmental severity. But the effect of polyploidy can be replaced by enlarged head diameters. Both of them can create high heterozygosity in the offspring. Our result provided new insights for understanding the responses and adaptations of plants when they moved to high mountains.

T2

A113

Morphobiological features of the nomenclatural type of *Allium nigrum* L., nom. cons.

Lyudmila Krassovskaya, ***Igor Levichev***
Komarov Botanical Institute of the Russian Academy of Sciences

The absence of an authentic sample of *Allium nigrum* L. in Carl Linnaeus' herbarium and partial compliance of characteristics of this taxon to *A. magicum* L. has forced to conserve the first binomen with a typical sample of the second. Conservation has formally excluded application of the name *A. magicum*. The fusion of two linnaeons resulted in nomenclatural doubts concerning a dozen of other taxa with glossy, black ovaries. We observed that *A. magicum* always reproduces a uniquely original set of properties and signs: absence of black ovaries, compulsory formation of vegetative bulbils at leaf tips, active deepening of the cotyledon basis into the soil, and other constant ontogenetic features. Ovaries and capsules of this taxon are always green with a scabrous surface that contrasts to the black ovaries in other related species. The formation of a vegetative bulbil on the tip of the innermost leaf lamina is an unique biological feature of *A. magicum* specified by Linnaeus after many predecessors and images in the studies of L'Obel, Gerard, Clusius, Sweert, *etc.* This bulbil is regularly formed starting from the second year of life of a seedling, and from the first year in vegetative reproduction. At the start of ontogenesis (about 10 years long), the gemmiferous leaf is short and the bulbil is below the soil surface. Every year this leaf becomes longer. The bulbil emerges over the soil surface only in the mature reproductive age. In older plants, this leaf reaches 20-40 cm length, and every year bifurcation of its lamina tip is extended, reminiscent of a swallow tail. The width of other leaves increases to 10-11 cm. During ontogenesis, the phyllome in this taxon starts with the unifacial type of the cotyledon and bulbous scales. Later leaves are flat, but own quasibifacial nature: two rows of vascular bundles are ring-like positioned. Tepals are bifacial. The seed germination of *A. magicum* demonstrates a phenomenal functionality: the root rudiment and the hypocotyl deepen to 10-20 cm where the first bulb is also formed. There is a paradoxical situation – the cotyledon actively grows above the hypocotyl, actually it grows upwards but pushes the root node downwards. The short (less than 1 mm) root primordium starts extending only after the formation of a bulb on the 4th metamer position. For this reason, the bulb is protected by the cotyledon sheath and rudiments of the 2 first true leaves. Along with the bulb, under the soil surface, the diameter of the cotyledon increases and marks the starting point of its digeotropic growth (see e-poster). The specified unique traits do not correspond to characteristics of *A. nigrum* sensu Linnaei. Therefore, *A. magicum* cannot serve as type of this name. It is necessary to select the epitype among the *A. nigrum* forms which are still cultivated in botanical gardens of European countries.

T2

A114

Phylogeny and biogeography of the Neotropical fern genus *Campyloneurum* (Polypodiaceae)

Paulo Labiak[1], *Robbin Moran*[2]
1. Universidade Federal do Parana
2. The New York Botanical Garden

Campyloneurum belongs to the Polypodiaceae and has about 60 species, all restricted to the Neotropics. It is characterized mainly by its simple and entire leaves (three species are 1-pinnate), prominent lateral veins that run from the midrib to the margin that are cross-connected by finer veins to create a series of areoles. Within each areole there are 1–5 free veinlets, and the sori are born at the apices of these veinlets. Because of its simple morphology, *Campyloneurum* is one of the most difficult genera taxonomically. Several species complexes need study. Previous phylogenetic studies have suggested that *Campyloneurum* is closely related to *Niphidium* and *Microgramma*, but a comprehensive phylogeny for the genus is still lacking. We investigated the phylogenetic relationships of *Campyloneurum* with other Polypodiaceae and of the species within the genus. Phylogenetic relationships were recovered under Maximum Likelihood and Bayesian methods, using a dataset that includes about 85% of the known species, and are based on four plastid markers (*rbcL, rps4-trnS, trnG-trnR, trnL-trnF*). *Campyloneurum* was recovered as monophyletic, and more closely related to *Niphidium*. Our results suggest that convergence frequently occurs in *Campyloneurum*, both in habitat specialization and morphology, and several characters that have been used to delimit species need to be re-evaluated. However, some morphological characters can be used to diagnose some clades, such as lamina division, and rhizome pruinosity. The molecular phylogeny also suggests that several species currently recognized are paraphyletic, and many new species have yet to be described. In terms of geography, the Andes are the most important area of phylogenetic diversity for the genus. Other important areas are the mountains of Central America and the Atlantic Rain Forest of Brazil, where some endemic clades can be found.

T2

A115

Evolutionary modes and trends during origination of angiosperms

Igor Levichev

Komarov Botanical Institute of the Russian Academy of Sciences

The author of telome theory Walter Zimmermann specified six elementary processes that formed cormophytes: *overtopping, planation, coalescence to leaf, anastomoses of axis, reduction, incurvation*. The presence of telomes at vascular plants ancestors coming on the land determined the fundamental evolutionary mode - metamerism. Dichotomous branching caused formation of telomes, thus turning the original telomes into mesomes. In fact, the body of ancestral vascular plants consisted of mesomes and sterile phylloids that carried out photosynthesis. Roots were missing. Individual stopped pairwise branching, and historically later individuals became a single sporangium bearing telomes. Solitary terminal and lateral formations of the anisotomic (monopodial) axis resulted from indisputable predominance of dichopodial development in unsupporting air environment. The single, unitary organism of the metameric-colonial body was united by the bipolar association of vascular bundles. The telome was originally capable of branching on both ends: ageotropic and geotropic. In later evolutionary phases, one of the geotropic bifurcations became the root, the second one – into the reserve sympodial branch. On land it was not telome that evolved but its derivate, the bipolar mesome. New divisions, classes, *etc.* arose on earth by quantitatively and qualitatively different mesome associations. For example, the angiosperms arose by union of monopodium and sympodium into a single metamer, which was accompanied by the debut of the true (dimeric) leaf of sporangial nature. This concept is detailed below (see e-poster). A-modus. Metamerism; Aa-trend. Modularity. Ab. Multilayer shoot. Ac. Specialization of vegetative and reproductive mesomes; B. Bipolarity; Ba. Ageotropic phyllome. Bb. Ageotropic continuation of axis ("monopodium"). Bc. Geotropic (root) branching and stele (*anastomoses of axix*). Bd. Side (geotropic) bud and reserve sympodium; C. Rotation (right, left) or 7-dimensional capture of habitat (above and below the soil surface + lifespan); Ca. Vertical balance of soma. CB. Orientation of branches and phylloms, phyllotaxis; D. Growth; Da. From isodichopodial to dichopodial, (*overtopping*). Dd. From herbs of surf line to wood giants of dry land. DC. Intercalary growth and grandization. DD. Cyclicism, *reduction*, (long-livers, annuals); E. Association; Ea. Convergence of gematophyte and sporophyte, vegetative and reproductive propagation. Eb. *Planation*, lateral webbing of vaia (*coalescence to leaf*). Ec Supraposition of metamers (ageotropic and geotropic). Ed. Juxtaposition. Ee. Monosympodia (dimery of axes and phyllome). Ef. Gorgonoid, podium; F. Protection of bud and ovule; Fa. Phyllome and sheath. Fb. Phyllome morphogenesis of sporangium, strobilus, cones, flower. Fc. Integuments, *incurvation*, obturators; G. Heredity; Ga. Symbiogenesis and unification of genome. Gb. Hybridization, ploidy, recombination, gene transfer, *etc.*; H. Phytospreading; Ha. Migration. Hb. Selection. Hc. Symbiosis with animals and man (pollination, dispersion, selection); I. Neoteny (main biogenetic law); Ia. Recapitulation (neotenic divergence). Ib. Oligomerization, namism, herbaceous habit. It is particularly indicative that any degree of evolutionary rollback at neotenic recapitulations never results in alien formations like "horns", "hooves", root instead of phyllome, *etc.*, but always reproduces embryological ancestral form, up to "telome" with mandatory increase of the protection degree of each plant body element during embryogenesis and histogenesis.

T2

A116

A synopsis of the genus *Premna* L. (Lamiaceae) in China

Bo Li

Jiangxi Agricultural University

The genus *Premna* L. belongs to the family Lamiaceae with ca. 200 species worldwide and is distributed chiefly in Old World tropics and subtropics. *Premna* ranks among the most taxonomically difficult and complicated genera of Lamiaceae and is considered one of the fifth largest genera with 46 species in the mint family of the Flora of China. In this paper, a taxonomic revision of the genus in China is presented. Totally 35 species and four varieties are recognized. Keys, descriptions, notes on ecology, distribution, local uses; vernacular names and conservation status, and some illustrations are also provided. Three names are lectotypified in this paper, five new names are adopted and 13 names are placed into synonymy for the first time. Two new species from Yunnan Province were described and illustrated.

T2

A117

A numerical taxonomy of *Epimedium* L.

Chao Li[1], *Mingli Zhang*[2]

1. North China University of Science and Technology

2. Xinjiang Institute of Ecology and Geography, Chinese Academy of Sciences

Based on 30 qualitative and 15 quantitative characters of 55 species of *Epimedium*, a numerical taxonomy is presented by using the cluster analysis and principal components analysis. Results of cluster analysis shown that the genus can be divided into macro-flower taxa and micro-flower taxa, treatment of Section *Macroceras*, Section *Polyphyllon* and Subgenus *Rhizophyllum* is justified which is established by Stearn, taxonomic status of Sect. *Epimedium* is doubted. In principal components analysis, cumulative ratio of the first three principal components seems lower only 51.86%, this is probably related to the diversity variation and taxonomic complication of *Epimedium* evolution. however, result of principal components analysis also indicates the generic division of macro-flower and micro-flower taxa. These important taxonomic characters in *Epimedium* resulted from principal component analysis, such as ratio of petal length to inner sepal length, spur, whorls of sepal, are valuable for taxonomy of this genus.

T2

A118

Global versus Chinese perspectives on the phylogeny of the N-fixing clade

Hong-Lei Li[1,2,3], *Wei Wang*[2], *Shou-Zhou Zhang*[1], *Hong Wu*[3], *Zhi-Duan Chen*[2], *An-Ming Lu*[2]

1. Fairy Lake Botanical Garden, Shenzhen & Chinese Academy of Sciences

2. State Key Laboratory of Systematic and Evolutionary Botany, Institute of Botany, Chinese Academy of Sciences

3. College of Life Sciences, South China Agricultural University, Guangzhou

There has been increasing interest in integrating a regional tree of life with community assembly rules in the ecological research. This raises questions regarding the impacts of taxon sampling strategies at the regional versus global scales on the topology. To address this concern, we constructed two trees for the nitrogen-fixing clade: (i) a genus-level global tree including 1023 genera; and (ii) a regional tree comprising 303 genera, with taxon sampling limited to China. We used the supermatrix approach and performed maximum likelihood analyses on combined matK, rbcL, and trnL-F plastid sequences. We found that the topology of the global and the regional tree of the N-fixing clade were generally congruent. However, whereas relationships among the four orders obtained with the global tree agreed with the accepted topology obtained in focused analyses with more genes, the regional topology obtained different relationships, albeit weakly supported. At a finer scale, the phylogenetic position of the family Myricaceae was found to be sensitive to sampling density. We expect that internal support throughout the phylogeny could be improved with denser taxon sampling. The taxon sampling approach (global vs. regional) did not have a major impact on fine-level branching patterns of the N-fixing clade. Thus, a well-resolved phylogeny with relatively dense taxon sampling strategy at the regional scale appears, in this case, to be a good representation of the overall phylogenetic pattern and could be used in ecological research. Otherwise, the regional tree should be adjusted according to the correspondingly reliable global tree.

T2

A119

Molecular phylogenetics and historical biogeography of *Sorbus* sensu stricto (Rosaceae)

Meng Li[1,2,3], *Ohi-Toma Tetsuo*[4], *Yun-Dong Gao*[1], *Bo Xu*[1], *Zhang-Ming Zhu*[1], *Wen-Bin Ju*[1], *Xin-Fen Gao*[1]

1. Chengdu Institute of Biology, Chinese Academy of Sciences

2. University of Chinese Academy of Sciences

3. Arnold Arboretum of Harvard University

4. Botanical Gardens, Graduate School of Science, The University of Tokyo

Explaining how plants from eastern Asia migrated to other Northern Hemisphere regions is still challenging. Here, genus *Sorbus* sensu stricto (including *c.* 88 species) is considered as a good example to illuminate this scenario due to wide occurrence in the temperate zone and diversity in the Himalayas and H-D Mountains. Based on four nuclear markers (*LEAFY*-2, *GBSSI*-1, *SBEI* and *WD*) and one chloroplast marker (*rps16-trnK*), we reconstructed the phylogenetics relationship of *Sorbus* using 54 taxa (60% of the genus) representing all subgenera, sections or series and geographical areas in the previous classifications, and estimated divergence time and historical biogeography of the genus. Phylogenetic analyses supported that the subgenera *Sorbus* and *Albo-carmesinae* defined by traditional morphological characteristics are suitable, and associated with the two different kinds (orange-red & crimson-white) of fruit color. But, suggested that the genus *Sorbus* comprises five well-defined clades that are not consistent with the traditional section or series. Molecular dating and biogeographic reconstruction showed that the age of the most recent common ancestor was estimated at 41 mya (95% HPD: 35–49 mya) in the eastern Asia. Four dispersal events are assumed to explain the wide distribution of *Sorbus* in the temperate zone and diversification in the edges of Qinghai-Tibet Plateau (QTP). Some species dispersed to eastern Asia, Tianshan Mountain, North America and Europe in Oligocene and Miocene. We found that polyploidization occurred mainly in the subgenus *Albo-carmesinae*, in the area of Tianshan Mountains, Himalayas and H-D Mountains. Lastly, the striking species diversity in Himalayas and H-D Mountains, when contrasted to other regions may be caused by polyploidization and hybridization in response to the multi-stage uplifting and the subsequent increases in geological, ecological and climatological diversity.

T2

A120

An east–west phylogeographic break in the Qinghai-Tibet Plateau (QTP) across the 500 mm isohyet, revealed by the genetic structures of six *Allium* species (Amaryllidaceae)

Minjie Li

Sichuan University

The Qinghai-Tibet Plateau (QTP) has been assumed to be fractionized into the east monsoonal and the west continental climatic zones along the 500 mm isohyet marking the boundary between humid–semihumid and arid–semiarid in the present time. And simultaneously it is hypothesized that the Hengduan Mountains Region (HMR) uplift around 4–3 Ma and Pleistocene climatic oscillations could have contributed to the environmental changes in the QTP. However, these biogeographic hypotheses have rarely been tested using a phylogeographic approach, let alone the effect of the actual equilibrium line altitudes (ELAs) to the local survival from the Quaternary glacial crisis. Three chloroplast DNA regions, one nuclear ITS sequence and 9 nuclear microsatellite loci, were screened for genetic variation for 71 populations represented by 577 individuals across the distributional ranges for *Allium* subgenus *Cyathophora* and *A. fasciculatum*, supplemented with the principal component ordination (PCA) of bioclimatic variables, the ecological niche model (ENM) and the seed weight. Our genetic data consistently identified the east–west break through the distributional ranges of *Cyathophora* and *A. fasciculatum* along the 500 mm isohyet, which is also strongly supported by the different environmental space. The causative factor for this divide was suggested to be the climatic conditions differentiation due to the development of South Asian monsoon in the HMR and the strengthening of aridification in the west QTP, until to the boundary between humid–semihumid and arid–semiarid are clearly existent marked by the 500 mm isohyet, but other processes, particularly divergent selection for the morphological adaptation within *Cyathophora*, may also play a role in maintenance and deepening this split, while the postglacial expansion from east to west within *A. fasciculatum*, blurred this split. Moreover, the montane isolation resulted from the uplift QTP could be responsible for the divergence in the ancestral areas during Pliocene-Pleistocene, and the Pleistocene climatic fluctuations for the intraspecific isolation by contraction-expansion patterns. Additionally, the Pleistocene climatic fluctuations could have created new open micro-habitat for the self-sustaining allotetraploid species occurring at the current contact zone of *A. farreri* and *A. cyathophorum* during the postglacial. The postglacial expansion of this stable hybrid zone could have made introgression to *A. cyathophorum*. As thus, the range displacement due to the moving hybrid zone could have promoted the divergence of *A. cyathophorum*. The genetic structures revealed that the glacial refugia in the QTP mostly located in the montane regions with lower ELAs. These findings highlight the importance of orogeny and climatic changes in the QTP in shaping species diversification during the Pliocene-Pleistocene.

T2

A121

B-class MADS-box genes and gibberellin in response to variations in photoperiod are associated with chasmogamous and cleistogamous flower development in *Viola philippica*

***Kun Sun**, Qiaoxia Li*

Northwest Normal University

Some plants develop a breeding system that produces both chasmogamous (CH) and cleistogamous (CL) flowers. However, its underlying molecular mechanism remains elusive. *Viola philippica* develops CH flowers with short daylight, whereas an extended photoperiod induces the formation of intermediate CL and CL flowers. In response to long daylight, the respective number and size of petals and stamens was lower and smaller than those of normally developed CH flowers, and the use of ≥14h light induced complete CL flowers that had no petals but developed two reduced fertile stamens. The floral ABC model indicated that B-class MADS-box genes largely influenced the development of the affected two-whorl floral organs. Moreover, gibberellins also affected the development of stamens and petals, and regulated the expression of B-class MADS-box genes. Therefore, we focused on characterizing the B-class MADS-box genes and *GA3oxidase* in *V. philippica* to understand this particular developmental transition. Three B-class MADS-box genes were isolated, and respectively designated as *VpTM6-1*, *VpTM6-2*, *VpPI*. In line with no distinguishable differences during early flower development and no observed floral homeosis, these genes shared a relatively similar expression pattern during floral organogenesis of CH, CL, and inCL flowers, indicating that these genes might play a conserved role in establishing floral organ identity. However, *VpTM6-1*, *VpTM6-2*, and *VpPI* were differentially expressed in the later flower development stage (particularly in petals and stamens) and the highest level of expression was observed in CH flowers; significantly low levels were detected in intermediate CL flowers, and the lowest level in CL flowers. The observed variations in the levels of expression after floral induction and organogenesis apparently occurred in response to variations in photoperiod. We assumed that the differential expression of these B-class MADS-box genes might be influenced by the gibberellin content in flowers. Therefore, *GA3oxidase* was isolated, and designated as *VpGA3ox*. VpGA3ox was the key enzyme during the synthesis process of active GAs in the last step. With the extension of photoperiod, the expression of *VpGA3Oox* in flower buds was decreased, and the highest level of *VpGA3Oox* expression was observed in CH flowers, significantly low level was detected in intermediate CL flowers, and the lowest level in CL flowers. So the expression of *VpGA3ox* was also in response to variations of photoperiod, and the level of expression was accorded with the three B-class MADS-box genes in flowers. Therefore, inhibition of the development of petals and stamens might be due to the downregulation of B-class MADS-box gene expression dosage by long daylight or by lower content of gibberellin, thereby inducing the generation of CL flowers. Our work contributes to the understanding of the formation of dimorphic flowers in plants during adaptive evolution.

T2

A122

Four new synonyms of *Jackiella javanica* Schiffn. (Jackiellaceae, Hepaticae)

Wei Li

Key Laboratory of Forest Ecology and Management, Institute of Applied Ecology, Chinese Academy of Sciences

Jackiellaceae R.M. Schust. with ten species (including varieties) worldwide, is a monogeneric family. The monogeneric status of *Jackiella* is supported by a molecular study based on three loci by Hendry *et al.*, (2007). When I checked the type specimens located in herbarium of Edinburgh Royal Botanical Garden (E), and plants distributed in China, I found *Jackiella angustifolia* Herzog, *Jackiella javanica* var. *cavifolia* Schiffn., *Jackiella sinensis* (W.E. Nicholson) Grolle, and *Jackiella renifolia* Schiffn. as four new synonyms of *Jackiella javanica* Schiffn. I thank the National Natural Science Foundation of China (No. 31370003, 31000100) for supporting this project. Thanks are also given to the herbarium of Edinburgh Royal Botanical Garden (E) for loan of type specimens.

T2

A123

The Miocene biota of Zeku County, Qinghai Province and its taphonomic environment

Xiangchuan Li[1,2], *Liang Xiao*[1], *Yunzhi Yao*[3], *Dong Ren*[3], *Wenlong He*[4], *Zhengchuang Hui*[5]

1. Chang'an University

2. Nanjing Institute of Geology and Palaeontology, Chinese Academy of Sciences,

3. Capital Normal University

4. Xian University of Science and Technology

5. Lanzhou University

Palaeobotanical studies on the Cenozoic fossil floras in eastern China have been exhaustively reported in the past dozen years. But contrarily, the similar achievements on the coeval floras from northwestern China can be hardly found. Recently, we collected abundant fossils from the Miocene of Zeku, Qinghai. The fossiliferous horizons here are confined to the Early-Middle Miocene based on the magnetostratigraphic and palynological data (Li *et al.*, 2016). This palaeobiota yields numerous exquisitely-preserved animal and plant fossils, including insects, spiders, fruits, seeds, woods or stems, pollen grains, leaves and bird feathers (Li *et al.*, 2016). First, concerning the plant macrofossils, Li *et al.*, (2016) established a fossil species *Koelreuteria quasipaniculata* based on several exquisitely preserved fossil membranous capsules. This is a characteristic anemochorous tree. Furthermore, most co-occurring plant fossils are numerous winged fruits or seeds which belong to *Acer*, *Ulmus*, *Fraxinus* and Pinaceae *etc.* They are also typical of wind dispersal and predominately show temperate affinities. Whereas, there are many fossil legumes discovered here and recognized as *Podocarpium podocarpum* (Li *et al.*, 2013) and *Cercis*. It is amazing that the anemochorous arboreal species exhibits a high abundance relative to the overall paleofloral composition. The large diversity of anemochorously dispersed tree plants here probably indicates an actual original characteristic of the vegetation rather than taphonomic bias. It is well known that oil shale can only be deposited in the swamp or deep lacustrine environment. Whereas, the fossil stems are always discovered in the shallow lake area because the decreasing water power towards the deep lake area can hardly carry bigger items forwards, and insects usually live on the grass near the shallow water. Therefore, the co-occurrence of fossil stems, insects or spiders with fine hairs and bird feathers support the conclusion that their taphonomic condition of the fossiliferous horizons should be the swamp not deep lacustrine facies. The predominance of winged fruits and seeds suggests that wind may gradually become an essential factor for plant dispersal and fossil burial, probably an ecological response to the Tibetan Plateau uplift and potential regional environmental change. Secondarily, concerning the fossil insects, Li *et al.*, (2017) described a cicada species *Cixius discretus*, which is the first record of Cixiidae from the Miocene of the Tibetan Plateau. The diet of extant *Cixius* adults consist entirely of various shrubs and trees, which demonstrates the view point that the temperate forest occurred here in the past. According to the preliminary identification results, the fossil insect taxa also include *Mycetophila*, *Dryomyza*, *Dasysyrphus*, Tipulidae, Notonectidae, Ichneumon, Corixidae and Odonata *etc.* Among them, Corixidae dwell in slow rivers and bodies of still water, e.g. ponds, and most of them feed on aquatic plants and algae. Odonata juveniles inhabit in various bodies of shallow water for over two years and then clime up out of the water along the aquatic weeds and become the adults. Therefore, they all support the view point that the fossil biota was buried in a swamp not in the central area of a large lake.

T2

A124

Diversification of *Aristolochia* subgenus *Siphisia* (Aristolochiaceae) in East and South Asia

Xiao-Qin Li[1], *Shuai Liao*[1], *Yun-Juan Zuo*[1], *Xin-Xin Zhu*[2], ***Jin-Shuang Ma***[1]

1. Shanghai Chenshan Plant Science Research Center, Chinese Academy of Sciences

2. College of Life Sciences, Xinyang Normal University

Aristolochia subgenus *Siphisia* (Raf.) Duch comprises about 70 species, 50 of which occur in East and South Asia, while the remaining 20 species are restricted to Central and North America. Taxonomic treatments of the subgenus *Siphisia* are controversial, especially for those distributed in the region of East and South Asia. Furthermore, some new species were discovered from Asia recent years. A robust phylogeny with extensive and intensive sampling should be urgently needed to clarify the phylogenetic relationships among species and it will also provide assistance with the taxonomy of this genus. We constructed phylogenic trees of about 120 accessions based on maximum likelihood, maximum parsimony and Bayesian phylogenetic analyses using five chloroplast regions (*rps16-trnK, petB-petD, trnH-psbA, atpB-rbcL* and *rpl16* region) and the nuclear gene *phyA*. Divergence time was estimated with relaxed molecular clock approach using two fossil calibrations. Reconstruction of the ancestral distribution areas within *Siphisia* are based on nuclear *phyA* tree from the Beast implemented in the software RASP. The combined trees resolved the relationships of major clades of *Aristolochia* and reflected distinguished geographic patterns. Strong genetic differentiation was also detected between the Sino-Himalaya and Sino-Japan regions. While the topologies of nrDNA and plastid phylogenies are not significantly congruent, suggesting reticulate evolution or insufficiency of informative sites. The genetic structure of populations within *Siphisia* reflected the interactions of various evolutionary processes in the long-term evolutionary history. Based on the analysis of molecular dating and ancestral area reconstructions, favored a major diversification of the Sino-Himalaya within *Siphisia* occurred around the time during the Miocene to Pliocene/Pleistocene and involved vicariance and dispersal events. Paleoclimatic oscillations may be contributed to differential rates diversification and successive vicariance events. Divergence within the subgenus might be closely related with habitat fragmentation, gene duplication/polyploidization, specialization of pollinator and characteristics of floral organ.

T2

A125

Superiortrapa weichangensis gen. et sp. nov. (Lythraceae s.l.)

from the Miocene of North China

Ya Li[1,2], *Xiao-Qing Liang*[3], *Yong-Dong Wang*[1], *Cheng-Sen Li*[2]

1. Nanjing Institute of Geology and Palaeontology, Chinese Academy of Sciences

2. Institute of Botany, Chinese Academy of Sciences

3. Yuxi Normal University

Superiortrapa weichangensis gen. et sp. nov. (Lythraceae s.l.) is described on the basis of receptacle, calyx and fruit remains, collected from the early Miocene of Weichang County, Hebei Province, North China. *Superiortrapa* has actinomorphic, 4-merous, hypogynous flower, four decussate antennae-shaped sepals with recurved barbs, four inter-sepal appendages (or epicalyx segments), and a superior ovary. Fruits are fusiform with persistent sepals and approximately eight slender ribs on the fruit body. Detailed morphological analyses and comparison of the genus with related fossil and extant genera allowed the placement into the subfamily Trapoideae of the Lythraceae s.l. based on a suite of characters. *Superiortrapa* can be distinguished from all the other genera of the family Lythraceae s.l. by having hypogynous flower rather than perigynous flowers in Lythraceae s.str., or hemi-epigynous to epigynous flowers in *Duabanga*, *Punica*, *Sonneratia* and *Trapa*. It might represent an extinct linage, probably secondarily derived from the extinct genus *Hemitrapa* of Trapoideae, Lythraceae s.l.

T2

A126

DNA barcoding evaluation and its taxonomic implications in the recently evolved genus *Oberonia* Lindl. (Orchidaceae) in China

Yuling Li[1,2], *Yi Tong*[1], *Fuwu Xing*[1]

1. Key Laboratory of Plant Resources Conservation and Sustainable Utilization, South China Botanical Garden, Chinese Academy of Sciences

2. University of the Chinese Academy of Sciences

The orchid genus *Oberonia* Lindl., is a taxonomically complex genus characterized by recent species radiations and many closely related species. All *Oberonia* species are under conservation as listed in the CITES and the IUCN Red List Categories and Criteria. Given its difficulties in taxonomy and conservation status, *Oberonia* is an excellent model for developing DNA barcodes. Three analytical methods and five DNA barcoding regions (*rbcL*, *matK*, *trnH-psbA*, ITS and ITS2) were evaluated on 127 individuals representing 40 species and 1 variety of *Oberonia* from China. All the three plastid candidates tested (*rbcL*, *matK* and *trnH-psbA*) have a lower discriminatory power than the nuclear regions (ITS and ITS2), and ITS had the highest resolution rate (82.14%). Two to four combinations of these gene sets were not better than the ITS alone, but when considering modes of inheritance, *rbcL*+ITS and *matK*+ITS were the best barcodes for identifying *Oberonia* species. Furthermore, the present barcoding system has many new insights in the current *Oberonia* taxonomy, such as correcting species identification, resolving taxonomic uncertainties, and the underlying presence of new or cryptic species in a genus with a complex speciation history.

T2

A127

Organogenesis of reproductive organs in *Loranthus* Jacquin (Loranthaceae): The origin and identity of the bract and the calyculus

Ruozhu Lin[1], *Bei Cui*[2], *Wenxia Zhao*[1]

1. Research Institute of Forest Ecology, Environment and Protection, Chinese Academy of Forestry

2. Research Institute of Forestry, Chinese Academy of Forestry

The parasitic family Loranthaceae are characterized by monochlamydeous bisexual flowers, which are usually subtended by an outermost bract (some with several bracteoles) and a circular rim, denominated calyculus, positioned outside the perianth (petals). Various hypotheses have been proposed to explain the origin of the calyculus: an outgrowth of the axis without relation to the perianth, equivalent to an epicalyx, a bracteolar (prophyllar) origin, or a reduced calyx. To find out the origin and identity of the calyculus, we investigated the flowers of *Loranthus tanakae* Franchet & Savatier using scanning electron microscopy. The results show that the pedicel is extremely short at the beginning and slightly elongate thereafter to become concave at the base of the mature flower where a "bract" is thought to be located on. The calyculus is initiated as a semicircular primordium outside the flower primordium on the abaxial side. It then develops towards the axis and both sides are adnate to each other on the adaxial side to form a circular structure, encircling the young flower. The calyculus keep growing upwards and eventually adnate to the receptacle to become a cylinder. Inside of the calyculus, the hemisphere of flower primordium cleaves into five or six (occasional four) undifferentiated perianth from its centre, before or after the calyculus closed. Isomerous stamens are opposite to petals. The ovary is inferior and adnate to the cylinder of receptacle and calyculus at maturity. Evidence based on our observations suggests that the concave structure previously known as a "bract" is only a part of the pedicel rather than a real organ. There is no "bract" primordium is initiated outside the calyculus. Our results support the bracteolar origin of the calyculus, and further consider it equivalent to a bract for two reasons. First, the calyculus is initiated as a single primordium which shows the same morphology and the same position as a bract does. Secondly, the developmental processes of the calyculus is relatively independent of other floral organs: there is no quantitative, spatial and temporal relations between the calyculus and the petals, which means the calyculus is not likely a reduced calyx. This study led up to the conclusion that *Loranthus* has one circular bract, namely calyculus, and three whorls of floral organs: petals, stamens, and carpels.

T2

A128

Taxonomic studies on *Rungia* (Acanthaceae) in China

Zheli Lin[1,2], *Yunfei Deng*[1]

1. Key Laboratory of Plant Resources Conservation & Sustainable Utilization, South China Botanical Garden, Chinese Academy of Sciences, 510650 Guangzhou, People's Republic of China

2. University of Chinese Academy of Sciences, 100049 Beijing, People's Republic

Rungia Nees (Acanthaceae: Justicieae) is a genus of about 50 species distributed in tropical and subtropical regions of the Old World, and sixteen species with twelve endemic species are recognized in recently published volume 19 of *Flora of China.* It is generally accepted to be closely related to *Justicia* Linn. nowadays, but differs in the rising placenta in ripe capsule and generally with one sterile bract and one fertile bract at each node of the spike. Some researchers suggested to merge *Rungia* with *Justicia*, however, as the genus *Justicia* itself is paraphyletic and the morphological studies indicated some differences between two genera, we prefer to treat *Rungia* as an independent genus for the present before there is sufficient evidence to demonstrate that *Rungia* should be merged into *Justicia*. This study was based on examinations on literatures and specimens, field investigations, pollen morphology, seed morphology and cytological analyses, the main results are as follows. 1. Pollen morphology of sixteen species were observed, the result indicated that the pollen morphology of *Rungia* is of great diversity and can be divided into 7 types; seed morphology of thirteen species were observed and can be divided into 4 types; chromosome number and karyotype of fifteen species were analysed, the basic chromosome number are: x=10,11,13,14 and 25. The results mostly match the macromorphology and are significant in taxonomic studies of *Rungia*. 2. According to the macromorphology and the results above, we revised the *Rungia* species in China. Sixteen species of *Rungia* in China were recognized. Six names were reduced to the synonymy. *R. monetaria* turned out to be not ocuured in China, four species, *R. angustifolia*, *R. burmanica*, *R. eberhardtii* and *R. khasiana*, are newly recorded from China and three new species are discovered, viz. *R. flava* Z. L. Lin & Y. F. Deng, *R. sinothailandica* Z. L. Lin & Y. F. Deng and *R. glabra* Y. H. Tan, Z. L. Lin & Y. F. Deng.

T2

A129

Three new species of *Ulnaria* (Bacillariophyta) from the Wuling Mountains Area, China

***Bing Liu**, Qi-Jian Tian, Gong-Xi Chen, Jing Li*

Key Laboratory of Plant Resources Conservation and Utilization, College of Hunan Province; College of Biology and Environmental Science, Jishou University

Three new species of *Ulnaria* from the Wuling Mountains Area, China, have been studied by light and scanning electron microscopy. All possess at least six closed girdle bands. The pattern of valve striation is not uniform as some do not have a corresponding opposite structure. The cells of *Ulnaria sinensis sp. nov.* are linked and form colonies with frustules connected by interlocking linking spines. *U. sinensis* has very long valves (296–512 μm) and uniseriate striae. Both *Ulnaria ulnabiseriata sp. nov.* and *Ulnaria gaowangjiensis sp. nov.* have biseriate striae. *U. ulnabiseriata* differs from *U. gaowangjiensis* in LM by its more lanceolate valve shape as the valves of *U. ulnabiseriata* taper towards the pole from one third of the length of the valve, whist *U. gaowangjiensis* tapers towards the pole from two thirds of the length of the valve, and by its larger valve length (105–229 μm vs 61–108 μm). The three new species are all epilithic and live in freshwater habitats in or near nature reserves.

T2

A130

Diversification and historical biogeography of the parasitic family Loranthaceae: Synchronous with the radiation of tropical forests and songbirds

***Bing Liu**, Zhiduan Chen*

Institute of Botany, Chinese Academy of Sciences

The mistletoe family Loranthaceae originated in Australasia in Mid Cretaceous, and then spread to South America via Antarctic but was absent in Africa. Some genera of Elytrantheae and Lorantheae dispersed from Australasia to Asia and back to Australasia for several times via the island chains in Southeast Asia since the Late Eocene. The lineages distributed in Africa and Europe are young members migrated from Asia after Late Eocene. Loranthaceae rapidly diversified along with the formation of tropical forests in most of the tropical and subtropical regions during Early Eocene, and simultaneously converted the trophic mode to aerial parasitic and made a special adaptation to bird-dispersal. Trees in tropical forests as hosts and songbirds as dispersers are both significant factors during the diversification and historical biogeography of Loranthaceae.

T2

A131

Transcriptomic analysis of heteromorphic stamens in *Cassia biscapsularis* L.

***Zhong-Lai Luo**[1], Jin Hu[2], Zhong-Tao Zhao[1], Dian-Xiang Zhang[1]*

1. South China Botanical Garden, the Chinese Academy of Sciences

2. Shenzhen Park Service

Hermaphroditic flowers have evolved primarily under the selection on male function. Evolutionary modification often leads to stamen differentiation within flowers, or "heteranthery", a phenomenon intrigued scientists since the 18th century until recently. However, the genetic basis and molecular regulation mechanism has barely been touched. Here we conducted comparative transcriptome profiling in *Cassia biscapsularis* L., a heterantherous species with representative patterns of stamen differentiation. Numerous differentially expressed genes (DEGs) were detected between the staminodes (the degenerated stamens) and fertile stamens, while much fewer genes differentially expressed among the three sets of fertile stamens. GO term enrichment and KEGG pathway analysis characterized functional properties of DEGs in different stamen types. Transcripts showing close correlation between expression pattern and stamen types were identified. Transcription factors from the bHLH family were suggested to have taken crucial part in the formation of staminodes. This first global transcriptomic analysis focusing on stamen differentiation opens the door toward a more comprehensive understanding on the molecular regulation of floral organ evolution. Especially, the generated unigene resource would be valuable for developing male sterile lines in agronomy.

T2

A132

Reproductive isolation between sympatric sister species,

Mussaenda kwangtungensis and *M. pubescens* var. *alba*

Zhong-Lai Luo[1], Ting-Ting Duan[1], Shuai Yuan[1], Shi Chen[2], Xiufeng Bai[3], Dian-Xiang Zhang[1]

1. South China Botanical Garden, the Chinese Academy of Sciences

2. Beneficial Insects Institute, Fujian Agriculture and Forestry University

3. Department of Cell Biology, the Third Military Medical University

Reproductive isolation defines the biological species concept and plays a key role in the formation and maintenance of species. The relative contributions of different isolating stages have been suggested to be closely associated with phylogenetic relatedness. Few studies have focused on the relative contributions of pre- versus postzygotic mechanisms, and even fewer have been conducted under strict phylogenetic frameworks. Pre- and post-zygotic reproductive isolation stages have been investigated in the sister species *Mussaenda kwangtungensis* and *M. pubescens* var. *alba*. The two species have partly overlapping distribution ranges and flowering times, while the principal pollinators differed strikingly for them, demonstrating strong pre-zygotic isolations. Natural hybrids were detected by simple sequence repeat markers and their maternal parents were identified based on chloroplast gene sequences.Five out of 81 individuals were suggested to be hybrids that fall into the categories F2, BC1, and BC2 by the NewHybrids analysis. Interspecific crossings resulted in significantly reduced fruit set and seed germination rates. Phylogenetic analysis revealed short Kimura-2-parameter distance between *M. kwangtungensis* and *M. pubescens* var. *alba*. These findings strongly supported the hypothesis that for species with a closer phylogenetic relationship, prezygotic isolation plays an important part in limiting gene exchange in sympatric areas.

T2

A133

Phylogenetics, reticulate evolution and taxonomic evaluation of *Miscanthus* and related genera (Andropogoneae, Poaceae)

Hong-Zheng Ma[1], Wen-Li Chen[2], Zhe Cai[2], Si-Lan Dai[3], **Song Ge**[2]

1. College of Horticulture and Landscape, Tianjin Agricultural University

2. State Key Laboratory of Systematic and Evolutionary Botany, Institute of Botany, Chinese Academy of Sciences

3. College of Landscape Architecture, Beijing Forestry University

Miscanthus species, disjunctly distributed in Asia and Africa, are important germplasm resource for bioenergy crops with excellent tolerance to biotic and abiotic stresses. Despite many studies, taxonomic delimitation and evolutionary history of this genus remain dispute. We herein collected 59 samples representing 15 species of *Miscanthus* and 18 species from its related genera through their entire distribution area. Based on sequences of 14 plastid fragments and 4 nuclear gene (*ITS*, *FLO/LFY*-like and *Ahd1* gene), we reconstructed the phylogeny of the genus and its relatives. Although discordance between nuclear and chloroplast phylogenies exists, the inferred species tree confirmed three main clades: first one including Eastern and Southeastern Asian *Miscanthus* species (ESA clade), the second consisting of species in subgen. *Rubimons* (*Rubimons* clade) and the last representing '*Saccharum* complex' members (complex clade). The results support the speculation that *Miscanthus s.l.* and related genera experienced extensive reticulate evolution. Biogeography analysis indicated that *Miscanthus* and its related genera originated in Southeast China and diverged within recent three million years (My), and all species in ESA clade occurred in the last one My when the fluctuation of sea level gradually slowed down. Moreover, the current geographic distribution of *Miscanthus* species in Himalaya, Africa and America is the consequence of both the vicariance and the multiple long-distance dispersals. On these bases, we suggest that ESA clade consisting of two sections should be delimited as *Miscanthus* and support the treatment of independent genera for *Rubimons* and *Diandranthus*.

T2

A134

Study on taxonomy and flora of mosses in Altun Mountains National Nature Reserve, Xinjiang, China

Sulayman Mamtimin, Shuayib Yusup, Winira Ilghar, Sabiram Esrayi, Rabiye Gupur

College of Life Science and Technology, Xinjiang University

The Altun Mountains National Nature Reserve is located in Qakilik County, Xinjiang Uygur Autonomous Region, northwestern China. The reserve covers 46,800 square kilometers and has an average altitude of 4,500 meters. The reserve was regarded as one of the rare gene banks of plateau species by International Union for Conservation of Nature (IUCN) and World Wildlife Fund (WWF). Branching off from the Kunlun Mountains, the mountain range runs for more than 650 kilometers from southwest to northeast to form the boundary between the Tarim Basin to the north and the minor basin of Lake Ayakekumu and the Qaidam Basin areas of interior drainage to the south (https://global.britannica.com/place/Altun-Mountains). Unique geographic position and natural conditions of the Altun Mountains cause bryophyte of the area can bear bitter cold on highland. Bryophyte is one of the important components of montane vegetation in this area. Considering the bryophyte flora of the Altun Mountains remains poorly studied, recently, we carried out field expedition and specimen collection. Basing on the systematic field investigation of 22 sampling sides, identification of specimens and data analysis, the taxonomy and flora of mosses in Altun Mountains National Nature Reserve have been conducted, which can provide basic data and scientific evidence for mosses flora of Xinjiang and plant conservation and sustainable utilization of Altun Mountains National Reserve. The main results are as follows:1. One hundred and thirteen species belonging to 38 genera and 12 families have been identified. Among those species, *Grimmia orbicularis* Bruch ex Wilson and *Didymodon validus* Limpr. are newly recorded in China, and 14 species are new to Xinjiang: *Brachymenium pendulum* Meut., *Bryoerythrophyllum neimonggolicum* X. -L. Bai & C. Feng, *Bryum clavatum* (Schimp.) Mull. Hal., *B. leptocaulon* Cardot, *B. pachytheca* Mull. Hal., *B. riparium* I. Hagen, *Didymodon mongolicus* D. P. Zhao & T . R. Zhang, *D. acutus* (Brid.) Saito, *D. gaochenii* B. C. Tan & Y. Jia, *Grimmia crassiuscula* H. C. Greven & C. Feng, *Gymnostomum rividulum* Brid., *Stegonia latifolia* var. *pilifera* (Dicks.) Broth., *Tortella alpicola* Dix., and *Tortulalanceola* R. H. Zander (=*Pottia lanceolata*). 2. Pottiaceae Schimp., Bryaceae Schwägr., Amblystegiaceae G. Roth, Encalyptaceae Schimp. and Ditrichaceae Limpr. are five dominant families (≥ 5 species), totally including 29 genera and 97 species (account for 85.84% of the total number of species). *Bryum* Dill., *Didymodon* Hedw., *Tortula* Hedw., *Encalypta* Hedw., *Bryoerythrophyllum* P.C.Chen, *Syntrichia* Brid., and *Grimmia* Hedw. are seven dominant genera (≥ 4 species), totally including 60 species (account for 51.72% of the total number of species). 3. Mosses flora in Altun Mountains includes 9 types of floristic elements. Among those elements, North temperate element (account for 61.05%) plays a dominant role, followed by East Asia element (8.42%), Old World Temperate element (6.32%), Endemic to China element (6.32%), Temperate Asia element (6.32%), East Asia and North America disjuncted element (4.21%), Pantropic element (4.21%) and tropical Asia element (3.16%).

Acknowledgements: The National Natural Science Foundation of China (No. 31460048,31660052) are thanked for their financial support.

T2

A135

Pollen morphology of *Eritrichium* (Boraginaceae) in Pan-Himalayas and its taxonomic significance

Paras Mazari, *Jiachen Hao, Quanru Liu*

College of Life Sciences, Beijing Normal university, China

The genus *Eritrichium* comprises about 50 species, mainly distributed in Central Asia and Pan-Himalaya, with a few species in Europe and North America.There are about 43 species in China and Pan-Himalaya. The Infrageneric classification of *Eritrichium* has long been controversial and pollen morphology of *Eritrichium* species in the Pan-Himalayan region is still largely unknown. Furthermore, Boraginaceae is a eurapalynous family and its complexity is rare in angiosperms especially in regard of pollen apertures. The aim of our study is to analyze the pollen morphology of *Eritrichium* and to evaluate its taxonomic significance. In the present work pollen morphology of 38 samples representing 33 species of *Eritrichium* were examined by using scanning electron microscope. The result showed great diversity in shape and aperture characters of the pollen grains of the *Eritrichium*. The common type of the pollen grains were cocoon-shaped, 6-heterocolpate and smooth surfaced with minute pores. Within species the shape, number and position of apertures were also significantly different. In some species like *E. sessilifructum* pollen grains were oblong and without constriction in the equator part. Pollen grains of *E. thymifolium* subsp. *latialatum* had only one aperture located in the center of the pollen grains while in other species like *E. longipes* single aperture was located on one terminal pole. However *E. hemisphaericum* had two apertures located on each pole of the pollen grain.The pollen morphology is helpful for identification of species, although pollen characters alone are insufficient to reconstruct the taxonomic relationships within *Eritrichium* but palynological data can provide some useful information for correct identification and the species-level revisions. Furthermore, Cluster analysis (CA) and Principal component analysis (PCA) were conducted based on four quantitative and three qualitative pollen characters, including pollen shapes, relative size of polar/equatorial, number and location of apertures characters.

T2

A136

Embryonic development pattern of Leptosporangiatae: A case study of *Dryopteris goeringiana*

Gaowa Morigen, Xin Wang, Huining Jia, Longyun Tan, ***Baodong Liu****

Harbin Normal University, Key Laboratory of Plant biology, College of Heilongjiang Province, Harbin

Relative to other plants, the embryonic development of the Leptosporangiatae was extremely weak. As early as 1992, the embryonic development process of families from angiosperms had been compared in detail. While the latest report in 2014 only showed 8 diagrams of *Bolbitis angustipinna*. *Dryopteris goeringiana* was chosen as material. The artificial cultivation of gametophyte generation was taken in simulation natural conditions. With paraffin method, the whole process of embryo development was recorded. The results showed that the primitive cell of organ formed in the 16-cell embryo period. Meanwhile, the formation of the first leaf primitive cell was earlier than the first root primitive cell. In the hippocampal embryo period, the first leaf primordium and the first root primordium broke through cap-shaped parts, and then formed obvious vascular bundles of the first leaf and the first root respectively. Stem was a complex which was composed of residue of roots and petioles. Free nuclei existed during embryonic development. Firstly, this paper made 8 diagram enriched 13 points line graph, including the process from zygote formation to the first leaf and the first root maturity, the division mode of early embryonic cells, the characteristic of the cell nuclei and nucleoli, organogenesis order and their mutual relations. Secondly, the formation of roots and leaves was concluded from three aspects, which was the primitive cell, the tissue and the organ. It was the first time to summarize the embryonic development mode of leptosporangiatae in detail which supported Nayar's Phyllorhize Theory. This paper couldn't agree with the theory that the vegetative organ of seed plants evolved from the Leptosporangiatae, but advocated the stems from Leptosporangiatae and seed plants were different in essence, but with a common ancestor, were two groups of the parallel development.

T2

A137

Plant radiations in hotspots of biodiversity surrounding the Qinghai-Tibet Plateau: An integrative approach

Alexandra Muellner-Riehl[1], *Jana Ebersbach*[1], *Hang Sun*[2], *Adrien Favre*[1]

1. Leipzig University

2. Kunming Institute of Botany

Many studies invoke orogenesis as the main driving force for the radiation of plants in mountains. The role of climate oscillations and key opportunities as well as key innovations as contributors to the establishment of high levels of biodiversity around the Qinghai-Tibet Plateau (QTP) is often neglected. To understand the contribution of orogenesis to plant radiations and the importance of the QTP for the diversification of temperate plants, we investigated the evolutionary history of two taxa with high species numbers in the QTP region, Gentianinae and *Saxifraga* using an integrative approach. We combined extensive field work campaigns and herbaria sampling to obtain well-resolved phylogenies covering our target groups and extensive outgroups. As a key element to our study, we performed divergence time estimation in BEAST using uncorrelated log-normal molecular clock models and multiple, carefully chosen fossil calibration points. In order to understand the role of the QTP region as a cradle for northern temperate plants, we used the resulting dated phylogenies for ancestral range estimation using BioGeoBEARS. Contrary to the "out of Tibet" hypothesis, not all lineages that radiated in the QTP region had originated there, and immigration contributed to the high species richness of this region. Next, we performed diversification rate analyses in BAMM and BayesRate to investigate the temporal association between the uplift of the QTP itself and the adjacent mountain ranges and species-rich radiations in both Gentininae and *Saxifraga*. Similar to several other studies, we found a significant delay between the start of orogenesis and radiations. In addition, we investigated the role of key innovations with BiSSE during diversification and found several traits that were of importance. The evolution of berry-like fruits (as a means of efficient dispersal) were associated with higher speciation rates in *Tripterospermum* (subtropical Gentianinae) while the evolution of the cushion life form (as an adaptation to alpine conditions) played an important role during the diversification process in *Saxifraga*. Finally, we used species distribution models (Maxent) and stochastic character mapping to shed light on the niche evolution of mountain plants and found that niche divergence, as exemplified by larger niche breadth and shifts to below-treeline habitats, dominated in species rich clades. We conclude that the underlying causes for plant radiations in mountains are likely to be multi-faceted. Rather than geomorphological processes themselves, it emerges that the interaction of lineage-specific traits and complex geological settings and/or climatic modifications providing key opportunities are drivers of mountain biodiversity.

T2

A138

No evidence of S-RNase-based gametophytic self-incompatibility in Orchidaceae, similar to that known for Rosaceae, Solanaceae, and Plantaginaceae

Shan-Ce Niu

Institute Of Botany, Chinese Academy Of Sciences

Self-incompatibility (SI) is found in approximately 40% of flowering plant species and in at least 100 families. Although orchids belong to the largest angiosperm family, only 10% of the orchid species present SI, which might be gametophytic self-incompatibility (GSI), while a majority (72%) of the *Dendrobium* species, one of the largest Orchidaceae genera, show SI. Molecular characterization of S-determinants of GSI is well advanced in Solanaceae, Rosaceae, and Plantaginaceae, which are S-ribonuclease (S-RNase) lineages. In the present study, two SI species (*Dendrobium longicornu* and *D. chrysanthum*) were identified using fluorescence microscopy. The S-RNase and SSK1-like genes present in their transcriptomes and in the genomes of *Phalaenopsis equestris*, *D. catenatum*, *Vanilla shenzhenica*, and *Apostasia shenzhenica* were identified. Sequence, phylogenetic, and tissue-specific expression analyses revealed that none of the genes identified was an S-determinant, suggesting Orchidaceae might have novel SI mechanism. The results also suggested that RNase-based GSI might have evolved after the split of monocots and dicots but before the split of Asteridae and Rosidae. However, studies on gene identification, differential expression, and segregation analyses in controlled crosses are needed to further evaluate the genes showing high expression levels in GSI tissues.

T2

A140

Postglacial contraction of a cold-adapted herb, *Triplostegia* (Caprifoliaceae): Insights into Hengduan Mountain-Taiwan disjunctions

Yanting Niu[1,2], *Jianfei Ye*[1,2,3], *Limin Lu*[1], *Tuo Yang*[1], *Xiaoxin Wei*[1], *Jianhua Li*[4], *Zhiduan Chen*[1]

1. State Key Laboratory of Systematic and Evolutionary Botany, Institute of Botany, Chinese Academy of Sciences,

2. Graduate University of Chinese Academy of Sciences

3. Beijing Botanical Garden, Institute of Botany, Chinese Academy of Sciences

4. Biology Department, Hope College, Holland

Current disjunction patterns can be caused by long-distance dispersal or distribution contraction due to climate fluctuations. We herein investigate the evolutionary history of *Triplostegia*, an ecologically sensitive herb, to gain insights into the formation of its disjunction distribution between the Hengduan Mountain region (HM) and Taiwan. Five chloroplast DNA regions (*psb*K–*psb*I, *rpl*20–*rps*12, *trn*H–*psb*A, *trn*L–F and *trn*S–*trn*G) and the ribosomal nuclear internal transcribed spacer (ITS) for 397 individuals from 48 populations were used to investigate the genetic diversity and structure of *Triplostegia*. Divergence time estimates, ancestral area reconstruction and ecological niche modeling (ENM) were conducted to examine the demographic history of the *Triplostegia* populations. Both the chloroplast and ITS data supported three major lineages within *Triplostegia* corresponding to their distribution in Southwest China (SW), Central China (CC) and Taiwan (TW). Great genetic differentiations among populations from SW, CC and TW were detected. *Triplostegia* was estimated to have originated in SW at *c.* 6.79 million years ago (Ma), migrated to CC at *c.* 4.03 Ma, and finally arrived TW at *c.* 2.15 Ma. The ENM suggested distribution expansion during the Last Glacial Maximum (LGM) whereas range retraction during the Last Inter Glacial (LIG) in *Triplostegia*. *Triplostegia* shares similar adaptive patterns as other cold-adapted species. The disjunct distribution of *Triplostegia* between HM and Taiwan is most likely the consequence of range contraction since the LGM. The predicted distribution of *Triplostegia* in 2070 shows a shrinking and northward scenario. With continuing global warming and human-induced deforestation, extinction risk may increase for cold-adapted species and appropriate strategies should be made for the conservation of ecosystem.

T2

A141

Preliminary study on abnormal morphological characters of *Cycas revoluta*

Aifang Pan

Fuzhou Botanical Garden

Two kinds of morphological anomalies in *Cycas revoluta* were observed in Fuzhou, Fujian Province. The first was that a sucking bud seedling had grown 5 pieces of abnormal leaflets, one piece with a bifurcated midrib was longer than normal leaflets, and another had a double-pieced base. And the second was that a male cone had a rosette tip, as if there were a female cone on its top. These kinds of anomaly would provide some new basic data for cycad systematics and evolution.

T2

A144

Demographic and spatial determinants of plant hybrization and consequences for speciation

Remy Petit, *Etienne Klein*
INRA

Hybridization is a key evolutionary process with major consequences for speciation. Its study in a spatial context is a new research line with considerable potential. It offers a way to evaluate in situ the strength of sexual barriers. In turn, the strength of sexual barriers, in conjunction with pollen dispersal and abundance, determine the degree of intermixing that result in hybridization rates sufficiently low for species to coexist. We explore this idea experimentally in oaks in a long-term mixed species plot as well as analytically and through simulations. To obtain baseline predictions on hybridization rate between two species differing in abundance and to evaluate the effect of pollen limitation, we first used a mean-field model. We then explored a spatially explicit individual-based mating model relying on pollen dispersal kernels to predict hybridization rates in a mixed-species stand, under different scenarios for (i) the strength of sexual barriers; (ii) the spatial distribution of individuals within the site; (iii) the variation in individual fecundity; and (iv) the magnitude, shape and asymmetry of pollen dispersal kernels. Pollen limitation was shown to have the potential to greatly increase hybridization rates. Similarly, fine-scale variation in species composition can result in elevated hybridization rates compared to mean-field predictions, especially with strong sexual barriers. However, species clustering in combination with reduced pollen dispersal has the opposite effect, protecting from hybridization. Our simulation results show that variation of the pollen pool composition at the scale of individuals or stigmas due to spatial configurations or pollen limitation can substantially modify hybridization rates. We explain this by the disproportionate effect of some pollen environments on average hybridization rates. This suggests thoroughly evaluating individual behaviour in terms of hybridization, especially for rare or patchy species. The analyses also support the idea that asymmetric barriers could evolve in response to asymmetric contact between species, for instance during succession, as proposed in oaks.

T2

A145

Studies on the micromorphology of *Cynanchum* (Asclepiadaceae) in China

Xinsheng Qin
South China Agricultural University

The microphology of leaf epidermis of 42 species, 1 subspecies and 1 variety in *Cynanchum* in China were examined under environmental scanning electron microsope (ESEM). The results showed that it had high diversity in the micromorphology of leaf epidermis in *Cynanchum*. A few species are distinctively different from others in their size of stomata and ornamentation of cuticular membrane. The similar features of the leaf epidermis in *Cynanchum* mainly displayed in the shapes of leaf epidermal cells were usually polygonal or irregular. All the abaxial epidermis had stomata, except *C. stenophyllum*, and most of the shapes of the stomata were not sunken. Many species of *Cynanchum* had glabrous leaves and thick cuticular membrane on leaves. A few species are distinctively different from others in their size of stomata and ornamentation of cuticular membrane. It is supported that *C. muliense* and *C. steppicolum* were treated as synonyms of *C. forrestii* in *Flora of China* (Vol.16), and it is not supported that *C. alatum* was treated as a misidentification of *C. callialatum, C. cathayense* as a synonym of *C. acutum* subsp. *sibiricum*, and *C. komarovii* as a synonym of *C. mongolicum*. The feature of the pollinarium morphology of 29 species and 1 subspecies of *Cynanchum* in China were observed under light microscope (LM). Three types of pollinia (spheroidal, prolate and perprolate) and six types of retinaculum (oblong, near roundness, square, near square, regular triangle and near triangle) were observed. This had important significance for the classification of *Cynanchum otophyllum* and supported the treatment of *Cynanchum* Sect. *Vincetoxicum* in *Flora of China* (Vol.63). The caudicle of *C.alatum* was the least developed, while *C. decipiens* was most developed. The feature of the pollinarium morphology of *C. atratum* and *C. officinale*, *C.corymbosum* and *C. verticillatum* were similar. It was considered that *C. alatum* was the false identification of *C. callialatum* in *Flora of China* (Vol.16), but the results showed their pollinarium morphology were different.As the pollinarium in *Cynanchum* was sagging while horizontal in *Tylophora*, it support that *Tylophora* should be a unique group. Seed morphology characters of 9 species of *Cynanchum* were observed under ESEM. The results showed that the seeds of *Cynanchum* had some differences in shape, size and epidermal ornamentation among species. Seeds of *Cynanchum* Sect.*Vincetoxicum* were relative consistency in appearance, size and epidermal ornamentation. Seed ornamentation characteristics in *Cynanchum* Sect. *Endotropis* were significantly different. The seed surface of *Cynanchum auriculatum* presents immersed rumination, irregular mesh and that of *Cynanchum giraldii* immersed reticulation with more regular mesh. These results supported the treatment of *Cynanchum* Sect. *Vincetoxicum* in *Flora Reipublicae Popularis Sinicae* (Vol.63).

T2

A146

Genus *Scutellaria* L. (Lamiaceae) in Azerbaijan

Rashad Salimov[1], *Parvana Garakhani*[2], *Hajiaga Safarov*[3]
1. Institute of Botany of Azerbaijan National Academy of Sciences
2. Institute of Botany of Azerbaijan National Academy of Sciences
3. Hirkan National Park, Azerbaijan

Azerbaijan demonstrates an extremely diverse native flora, consisting of about 4,500 species of vascular plants of which around 950 are Caucasian and 200 national endemics. Azerbaijan flora represents 64% of the total species known for the Caucasus region, one of the biodiversity and endemism Hotspots of the world, certifies its unique feature and justifies the existence of global importance of its conservation. *Scuttelaria L.* (Lamiaceae) has a specific place within the Azerbaijan flora according to its composition of rare and endemic species (Solomon *et al.*, 2013). *Scuttelaria L.* is a large and taxonomically complex genus comprising 350 to 400 species distributed all continents except Antarctica. The main reason for the complexity of taxonomy of the genus is the ambiguous treatment of species due to their high polymorphism. *Scutellaria* L. is insufficiently studied in the flora of Azerbaijan (Flora Azerbaijan, 1957). When *Scutellaria* L. was compared with the number of the taxa and endemism ratio in the adjacent countries' flora; it is represented by 3 sections of 15 species determined in Azerbaijan flora. Five species (33.3%) of them are Azerbaijan and six (40%) are Caucasian endemics (Solomon & *et al.*, 2014). And In the Flora Iranica (Rechinger, 1982) there are 40 species belonging to the genus, and the number of endemic taxa is 39 (72%). Thirteen species (41%) of the genus represented by 34 taxa in Turkey are endemic (Edmondson, 1982). A total of 7 taxa were included in the Armenian flora (Flora Armenia, 1987). In Georgian flora (Flora Georgia, 1987), only 2 species (15.5%) of 13 species are national endemic. The rarity status of all endemic species of *Scutellaria* L. were assessed for the Red List of the Caucasus (V. Ali-zade et al, 2013. *In*: Solomon *et al.*,), two of them were listed in the 2nd edition of the Red Book Azerbaijan (2013): *Scutllaria grossheimiana* Juz.-NT and *S. prilipkoana* Grossh.- VU D2. Distribution and ecology of the species, and causes of extinction of their populations are considered. The phylogenic study using both morphological and molecular data for treatments *Scuttelaria L.* (Lamiaceae) in Azerbaijan flora are planned.

T2

A147

Phylogeographic Study of *Chrysanthemum lavandulifolium* complex

*Yina Xu, **Chuze Shen**, Guangyuan Rao*
School of Life Sciences, Peking University

Chrysanthemum lavandulifolium complex consists of several closely related species, i.e., *C. lavandulifolium, C. potentilloides, C. dichrum* and some types of *C. indicum*. This species complex distributes in East Asia, but its diversity center locates in the Taihang-Lvliang Region of China, where *C. potentilloides* and *C. dichrum* are endemic. For a better understanding of evolutionary trajectory of this complex, especially of endemic *C. potentilloides* and *C. dichrum*, a total of 658 individuals was sampled in this study from 28 populations in the Taihang-Lvliang Region. Leaf morphological analysis was conducted by the data extracted by landmark and LAMINA, and ploidy level was identified for every studied individual by flow cytometry or chromosome counting. Population genetic structure and relationships were inferred from phylogeographic and phylogenetic analyses of three chloroplast DNA fragments and nine nuclear DNA SSR loci. The results show that polyploidization is very popular in *C. lavandulifolium*, even within populations, for polyploids exist in each population while diploids not. No IBD patterns are detected for this complex populations. The highest morphological and genetic diversity are found in the populations of southern Taihang Mountains based on leaf, cpDNA and SSR analyses. Meanwhile, phylogeographic analysis show that ancestral lineages of this complex are also located this region. This suggests that southern Taihang Mountain region is a critical area for the evolution of *C. lavandulifolium* complex. In addition, *C. dichrum* is an extreme type of *C. lavandulifolium* complex, and it is not yet a good species taxonomically according to our morphologic and population genetic analyses. Strong gene flow is found from Taihang Mountains populations to ones in Lvliang Mountains. Niche modeling analysis shows that there are two refuges in the northern and southern Taihang mountains during glacial periods, respectively. Survivors from these refuges spread along Taihang moutains during interglacial time, also expand to Lvliang Mountains. Polyploidization occurs by auto-polyploidy and hybridization-polyploidy in this process.

T2

A148

Phylogenetic analysis of *Pohlia* Hedw. using chloroplast and ITS sequences

Jing-Yuan Niu[1,2]*, Shuo Shi*[1]*, Lin Li*[1]*, Hui Li*[1]*, Xiao-Rui Wang*[1]*, Yong-Ying Liu*[3]*, Yu-Lu Niu*[4]*, Xu-Hong Duan*[1]*, Jian-Cheng Zhao*[1]
1. Hebei Normal University
2. Xinxiang Medical University
3. Jiaozuo Teachers College
4. Hengshui University

The genus *Pohlia* Hedw. have become one of the largest moss genus in the past years. It has been placed in the Bryaceae as traditionally circumscribed and has been traditionally divided into three subgenera or sections, *Pohlia, Mniobryum and Cacodon*. However, the molecular phylogenetic analysis indicated that the genus *Pohlia* is more closely allied with genera traditionally classified in the Mniaceae or Mielichhoferiaceae. In this study, we performed a molecular phylogenetic analysis, using evidence from four chloroplast regions (*atp*B-*rbc*L, *rps4*, *trn*G, *trn*L-*trn*F) and one nuclear ribosomal intragenic spacer regions (ITS), with different analyses (BI, MP, ML). Two large clades, outgroup clade and ingroup clade, were well supported in all datasets. In the ingroup clade, the traditionally defined species of Bryaceae, including *Anomobryum gemmigerum*, *Brachymenium pendulum*, *B. caespiticium*, *B. capillare* and *Rhodobryum ontariense*, formed a monophyletic clade (Clade A) with high support. Clade B corresponds with the Mielichhoferiaceae-Mniaceae complex with a high support. All samples of *Pohlia*, *Mielichhoferia* and Mniaceae were included in the clade. The most noticeable result of the phylogenetic analysis is that the genus *Pohlia* is non-monophyletic. *Pohlia* were dividing into three parts: Clade C, Clade D and *P. wahlenbergii*. Clade C was divided into two clades, Clade C1 and Clade C2. Clade C1 including *P. minor*, *P. crudoides*, *P. elongata*, *P. gedeana*, *P. nutans* and two samples of *Mielichhoferia*. *Mielichhoferia mielichhoferiana* and *M. elongata* formed a monophyletic clade with high support. Clade C2 included *P. bolanderi*, *P. cruda*, *P. hyaloperistoma*, *P. longicollis*, *P. laticuspes* and *P. lutescens*. And *Pohlia annotina*, *P. flexuosa* and *P. proligera* formed a monophyletic clade (Clade D) with high support. The results represents that the subgenra/sections of *Pohlia* are more closed to the infrageneric classification opinion of Hill and all species included in Clade C1 can be treated as the new circumscribed genus *Pohlia.* The analysis is identical with the conclusion that *Pohlia* are more close to the genera of Mielichhoferiaceae and Miniaceae. Comparing morphological characters, the species of *Mielichhoferia* and *Minum* are distinct from one another. *Pohlia* is morphologically very different from most genera of Mniaceae. We concur with the opinion that Mielichhoferiaceae and Minaceae can be divided into independent families and *Pohlia* is included in Mielichhoferiaceae. Our work was funded by the National Natural Science Foundation of China (No. 31370237 & No. 31070184), Specialized Research Fund for the Doctoral Program of Higher Education of China (No. 20131303l2005) and Research Foundation of Hebei Normal University (No. L2015B13).

T2

A149

Re-establishment of species from synonymies based on DNA barcoding and phylogenetic analysis using *Diplopterygium simulans* (Gleicheniaceae) as an example

Jiang-Ping Shu

Shanghai Chenshan Plant Science Research Center, Chinese Academy of Sciences

Because synonymy treatment traditionally relies on morphological judgments, it usually causes many problems in species delimitation and in the biodiversity catalogue. For example, *Diplopterygium simulans*, which belongs to the Gleicheniaceae family, has been considered to be synonymous with *D. glaucum* or *D. giganteum* based mainly on the morphology of its pinna rachis and blade. In the absence of molecular evidence, these revisions remain doubtful. DNA barcoding, which is considered to be a powerful method for species-level identification, was employed to assess the genetic distance among 9 members of the *Diplopterygium* genus. The results indicate that *D. simulans* is an independent species rather than a synonymy of *D. glaucum* or *D. giganteum*. Moreover, phylogenetic analysis uncovered the sisterhood of *D. simulans* and *D. cantonense*, which is supported by their geographical distributions and morphological traits. Incorrect synonymy treatment is prevalent in the characterization of biological diversity, and our study proposes a convenient and effective method for validating synonym treatments and discovering cryptic species.

T2

A150

Species delimitation and speciation of *Orinus* (Poaceae)

Xu Su[1,2], *Yuping Liu*[1,2], ***Ting Lv***[1,2]

1. School of Geography and Life Science, Qinghai Normal University

2. Key Laboratory of Medicinal Plant and Animal Resources in the Qinghai-Tibetan Plateau, Qinghai Normal University

Orinus is an alpine endemic genus to the Qinghai-Tibet Plateau (QTP) and its adjacent regions in the grass family. This genus is the key or unique taxa growing under extremely sandy habitats in these regions. *Orinus* is of the great importance in the ecological restoration and potential usages in breeding new genetic resources. Since *Orinus* was established, six species were described for this genus. However, the boundary and taxonomic treatments of these species remain disputed and their genetic variations between and within species have never been reported. Moreover, the speciation of *Orinus* is still blank for the present. So, it provides a good model to study species delimitation and speciation of new species under the extremely sandy and alpine habitats. Hereby, in this study, we conduct an extensive field exploration for the total genus and then delimitate species boundary based on studies in morphological statistics and DNA barcoding markers at the population level to answer these unsolved questions. Simultaneously, the speciation history of the re-delimited *Orinus* species was studied together with using 10 SSR marker loci and 10 sets of nuclear gene sequences. We collected 786 individuals from 102 populations of six previously-recognized species for niche, morphological and genetic analyses. We sequenced four DNA fragments, three chloroplast DNAs (*mat*K, *rbc*L, *trn*H-*psb*A) and one nuclear ITS. In contrast to the six species previously recognized, statistical analyses based on character variation, molecular data and niche differentiation identified only two well-delimited clusters, together with a third possibly originating from relatively recent hybridization between, or historical introgression from, the other two. Based on the principle of the integrative species delimitation to reconcile different sources of data, our results together provide compelling evidence that the six previously recognized species of the genus should be reduced to two, with new circumscriptions, while the third, identified in this study, should be described as a new species. Meanwhile, we reported this new species and a complete taxonomic revision of the genus. Besides, the results of speciation history of three redefined *Orinus* species showed that *Orinus* species had a high interspecific divergence based on the AMOVA analysis. The genealogical relationships based on the haplotypes identified from 11 nuclear loci suggested that *Orinus intermedia* and *O. anomala* were more closely related to each other than to *O. thoroldii.* Models test of 10 SSR loci based on approximate Bayesian computation analyses supported the hybrid origin of *O. intermedia.* Divergence time of the speciation event between *O. thoroldii* and *O. anomala* was estimated to be around 2.51 Ma, while the other speciation event of the origin of *O. intermedia* occurred around 0.689 Ma. Furthermore, the above time was slightly later than the divergence time of 2.58 and 1.01 Ma based on the nuclear ITS sequence variations. The sustained uplifts and Quaternary climatic oscillations of the QTP, may have together derived the allopatric divergence in *Orinus* and introgression after divergence between them. The habitat differentiation due to the Pleistocene climatic changes may have leaded to the speciation of *Orinus*.

T2

A151

Seed characters and their taxonomic implications in Molluginaceae (Caryophyllales)

Alexander P. Sukhorukov[1], *Ming-Li Zhang*[2], *Maria Kushunina*[1], *Maya V. Nilova*[1], *Anastasiya A. Krinitsyna*[1], *Maxim A. Zaika*[1], *Yuri Mazei*[1]

1. Moscow M.V. Lomonosov State University

2. Xinjiang Institute of Ecology and Geography, Chinese Academy of Sciences, Urumqi

The seeds of 75 species belonging to all 11 genera of Molluginaceae were studied. Several fruit and seed characters were found to characterize certain genera or major clades within the family. 'Polpoda – Adenogramma + Psammotropha' subclade is the only lineage where the one-seeded fruits have evolved: *Polpoda* and *Psammotropha* produce both several- and one-seeded fruits, but only the one-seeded fruits are found in *Adenogramma*. The genus *Pharnaceum* in its recent circumscription is carpologically heterogeneous, and all seed characters of *Pharnaceum namaquense* and *P. pusillum*, which were transferred into this genus based on the morphological similarities, indicate their position within *Hypertelis*. *Mollugo spathulata*, a species with uncertain generic status, may also belong to *Hypertelis*. The generic placement of *Pharnaceum serpyllifolium* is also ambiguous, since this taxon is characterized by creeping stems and reticulate seed ornamentation unusual for *Pharnaceum*. The genus *Kewa* (Kewaceae, previously considered within *Hypertelis*) shares many carpological homoplasies with Molluginaceae. The most important characters (seed number, ornamentation and presence of stalactites in the outer walls of the testa cells) were reconstructed using molecular phylogeny. The reconstruction of evolutionary history of the seed number suggests that dry multi-seeded fruit (probably capsule) was an ancestral character state in all Molluginaceae, and one-seeded fruit is an autapomorphic character state in the subclade 'Polpoda – Psammotropha + Adenogramma' within Molluginaceae. The presence of 'stalactites' in the outer walls of the testa cells seems to be an ancestral character state for the whole Molluginaceae, which is retained in the large 'Trigastrotheca + Mollugo – Glinus' clade. It is also observed in the 'Paramollugo' clade. The loss of the 'stalactites' in 'Hypertelis' and 'South African' clades is presumably connected with the reduction in the testa thickness.Although Molluginaceae is not a species-rich family, a number of distinctive morphological seed characters is found compared to other core Caryophyllales. However, the anatomical seed characters, albeit diversified in the major lineages of Molluginaceae, appear to be uniform among the entire core Caryophyllales. Taxonomically significant fruit and seed characters are either genus-specific, applicable for the groups of genera (major clades), or to the species with uncertain taxonomic status.

T2
A152
Leaf shape and classification in *Rosaceae*

Xuanting Sun
Huanggang Normal University

Leaf shape is one of the most significant traits of plants. Rosaceae, a diverse and big family in anthophyta of higher organisms, has variable leaves shapes. Traditional classfication divided it into 4 parts: Maloideae, Prunoideae, Rosoideae, Spiraeoideae by distinguish similar morphology. However, after APGIV in 2016, morphological classification in Rosoideae by using these traits face challenges. This new classification by using molecular data based on geography and chromosome number, divided it into 3 parts: Dryadoideae, Rosoideae, Amygdaloideae. Some species in Dryadoideae are N-fixers, but *Dryas integrifolia* seems to be unable to fix nitrogen (Markham, 2009). The genus *Rubus* is very complicated as well, particularly in subgenus, the blackberry with polyploidy, hybridization, leading systematic botany still has to cope with grand challenges. So I chose Rosoideae and its leaf shape as materials in order to comfirm and clarify the relationship between APGIV and traditional morphological classfication in Rosaceae. With the Angiosperm Phylogeny Website, eFlora and FNA, *etc*, I collect data of leaf shape in more than 750 species in Rosoideae and paint line charts in all thirty-three genus relying on APGIV. After rank all charts relying on new classification and also, Wikipedia. I found some genus has much more variable leaf shape than several genus. For instance, in *Sibbaldia*, leaflets are obovate-oblong or oblong, base rounded to broadly cuneate. Leaves pinnate or palmately 3- 5- foliolate. *Rubus*, is a large and diverse genus of flowering plants in the rose family around the world, also, has diverse leaf shape. They has pinnately compound leaves, dentate, oval leaves. Moreover, their leaflets 0 or 3, 5, 7, or 9, terminal ovate to elliptic to obovate, roughly crenate, dentate to doubly dentate, or serrate to doubly serrate. *Chamaerhodos*, even though in Rosoideae, have less traits. Their leaf blade 1- 3 times 3- parted into narrow segments and pinnately compound leaves or simply. In this way, statistics and plot in more than 750 data is a reliable tool to figure out the leaves of Rosoideae show highly various patterns. These diverse leaves shapes indicate following morphological data, such as the shape of the leaves, do not appear to produce a same phylogenetic classification based on APGIV and only using one fundamental feature of plants can not make sure how they evolve and class them.

T2
A153
Evolutionary dynamics of *S*-alleles in *Malus sieversii*, the wild ancestor of domesticated apple

Xiang Ma[1], Zhe Cai[2], Wei Liu[2], Song Ge[2], ***Liang Tang[1,2]***
1. Institue of Tropical Agriculture and Forestry, Hainan University
2. State Key Laboratory of Systematic and Evolutionary Botany, Institute of Botany, Chinese Academy of Science

Self-incompatibility present in angiosperm can prevent self-fertilization, and it is generally controlled by a single genetic locus which encodes for the determinants of both pistil (*S*-allele) and pollen specificity. In this study, we surveyed the diversity of *S*-alleles in *Malus sieversii*, the primary wild ancestor of domesticated apple, via PCR amplification of four consensus primers. The evolution of *S*-alleles was investigated by phylogenetic inference and comparative population genetic analyses between the *S*-allele and previous reported microsatellite data. We identified 14 distinct *S*-alleles in *M. sieversii*, compared to nearly 40 *S*-alleles present in its close relative, *M. sylvestris*. Approximately 15% of amino acid sites from the Maleae self-incompatibility ribonucleases were detected under selection. In addition, this study shows that population differentiation at the *S*-locus in *M. sieversii* is substantially lower than the microsatellite loci, suggesting the action of balancing selection. In summary, our study not only elucidates the constituents and number of *S*-alleles in *M. sieversii*, which will facilitate future germplasm conservation and apple breeding, but also illustrated the potential utility of *S*-locus polymorphism in demographic inference for plant species with self-incompatibility.

T2

A154

Phylogenetic study of *Lindera* based on single copy nuclear gene

***Xiangyu Tian**, Hongfang Wang*, Lei Bao, Jianping Ge*

Ministry of Education Key Laboratory for Biodiversity Science and Engineering, School of Life Sciences, Beijing Normal University, Beijing, China

Morphology-based classification is possible confused by convergent evolution. The problem appears serious in *Lindera* and *Litsea* complex. Traditionally, *Lindera* was identified as different genera from *Litsea* only by cell numbers in anthers. *Lindera* was further divided into eight sections based on shortened brachlets, inflorescenses, and venation. However, previous phylogenetic study based on chloroplast revealed very poor support between genus or between sections within genus. Chloroplast is haploid and evolved much slower than nuclear genome, which might be the reason for the poor resolution.Therefore, the true phylogenetic relationship of genus *Lindera* is largely uncertain. In this study, to betterunderstand the phylogenetic relationship between *Lindera* species as well as related genus, we plan to employ more than 20 nuclear single copy genes. We have collected 62 *Lindera* species, covered all eight sections of the genus. To reflect the relationship between *Lindera* and other related genus, we also collected 10 *Litsea* species and 25 other Lauraceae species. To be more representative, more than one individuals from different population are collected for many species. Species tree based on *Beast or BPP is planned to be applied. Currently, we have successfully sequenced about 100 individuals for each of the 20 nuclear loci. Only few loci have been aligned and thus we only reported phylogenetic results based on single locus (*GPN*) in this abstract. Our results indicated that except few individuals of some species, most species of Sect. *Daphnidium*, Sect. *Aperula*, Sect. *Polyadeni*, and Sect. *Cupuliformes* are grouped in four separate supported clades. All sampled individuals from Sect. *Uniumbellae* are within the same supported group with Sect. *Daphnidium*. Species of Sect. *Lindera* are found in at least four supported clades and mixed with species from another section or even another genus. Two species of Sect. *Lindera* located in the basal of all *Lindera* species. Species from *Litsea* or *Parasassafras* can be found in some of the above supported clades. Collectively, our preliminary results support the messy status of the genus *Lindera*. Nonetheless, supported clades found in many sections support the phylogenetic signal of those traditional morphology traits. Carefully check will be conducted in the next step for different individuals of the same species with divergent phylogenetic position.

T2

A155

Hoya carandangiana, Hoya bicolensis, and *Hoya camphorifolia* (Apocynaceae) species delineation: Insights from leaf architecture

***Jay Torrefiel**, Inocencio Jr. Buot*

Institute of Biological Sciences, University of The Philippines, Los Baños, College

Three species of *Hoya* namely, *H. carandangiana*, *H. bicolensis*, and *H. camphorifolia* under the family Apocynaceae, which are superficially similar and consequently, controversial were studied and delineated. For the past decades there were numerous studies that were able to utilize leaf architecture such as in systematic studies aimed at resolving certain taxonomic issues and in fossil (paleobotanical) studies. In this present study, leaf architectural characters were utilized to examine the similarities and differences between the three *Hoya* species. There were unifying leaf architectural characters exhibited by the three *Hoya* species however, characters such as blade shape and class, base shape, apex shape, secondary vein category, and tertiary vein category and course were the characters separating them. Cluster analysis was also done to illustrate relationships between the three *Hoya* species by using unweighted pair-group average (UPGMA) and Single linkage. The results showed that *Hoya carandangiana, Hoya bicolensis* and *Hoya camphorifolia* are three distinct species and hence, leaf architectural characters has the capacity to serve as good taxonomic markers.

T2

A156

The obtainment of complete chloroplast genome sequence of *Gentiana veitchiorum* base on transcriptome

***Jiuli Wang**[1,2], Faqi Zhang[1], Shilong Chen[1]*

1. Northwest Institute of Plateau Biology, Chinese Academy of Sciences

2. University of Chinese Academy of Sciences

The structure and sequences of chloroplast (cp) genomes have precious information to investigate biodiversity, evolution, and inter or intraspecific phylogeny of plants. It is found that the entire cp genome is transcribed in higher plants. *Gentiana veitchiorum* Hemsl. is a perennial herb with greatly ornamental and medicinal value. To obtain the complete cp genome of *G. veitchiorum*, we mined its chloroplast DNA sequences from its transcriptome gained previously and assembled them; coherently, gaps were filled by PCR and sequencing. The total genome of *G. veitchiorum* length was 140,020 bp (42,883 A, 26,853 C, 26,158 G, and 44126 T) with a pair of inverted repeat regions of 24,794 bp each that separate an large single-copy region of 78,942 bp and an a small single-copy (SSC) region of 11,490 bp. The genome consisted of 134 annotation genes, with 89 protein coding genes, 37 transfer RNA genes, and eight ribosomal RNA genes. The phylogenetic analysis of cp genome from *G. veitchiorum* and other six tribe Gentianeae taxa (*Swertia mussotii* Franch., *G. tibetica* King ex Hook. F., *G. robusta* King ex Hook. F., *G. straminea* Maxim., *G. lawrencei var. farreri* (Balf.f.) T.N.Ho, and *G. crassicaulis* Duthie ex Burkill) indicated *G. veitchiorum* has a close relationship to *G. lawrencei* var. farreri.

T2

A157

Genome constitution and taxonomy of *Elytrigia lolioides* inferred from *Acc1*, ITS, *trn*L-F sequences and molecular cytogenetics

***Long Wang**[1], Qinghua Shi[2], Xing Fan[1], Yi Wang[1], Lina Sha[1], Houyang Kang[1], Haiqin Zhang[1], Yonghong Zhou[1,3]*

1. Triticeae Research Institute, Sichuan Agricultural University

2. State Key Laboratory of Plant Cell and Chromosome Engineering, Institute of Genetics and Developmental Biology, Chinese Academy of Science
3. Key Laboratory of Genetic Resources and Crop Improvement, Ministry of Education, Sichuan Agricultural University

Elytrigia lolioides, a hexaploid (2n=6x=42) wheatgrass distributing in Russia and Kazakhstan, was classified into *Elytrigia* and *Lophopyrum* by taxonomists based on different taxonomic principles. However, genome constitution of *Elytrigia lolioides* is still unknown. In present study, single copy nuclear gene *Acc1*, multicopy nuclear spacer DNA ITS, chloroplast gene *trn*L-F in concert with fluorescence and genomic in situ hybridization were employed to investigate genome constitution and taxonomy of *E. lolioides*. Data from cpDNA showed that *Pseudoroegneria* is maternal donor of *E. lolioides*. Despite widespread homogenization of ITS sequences, two distinct lineages (*Pseudorogneria* and *Hordeum*) were identified by the results of ITS data. *Acc1* data suggested that besides *Pseudorogneria* and *Hordeum*, *Lophopyrum* was the third potential donor of *E. lolioides*. However, data from special FISH marker for St genome displayed that *E. lolioides* has two sets of St genomes. The data from GISH were congruent with the results of FISH marker and confirmed that the third genome has been contributed by *Hordeum*. When E genome was used as the probe, there were no any signals on the forty-two chromosomes. The E-like copy of *Acc1* sequences were detected in *E. lolioides* which might due to introgression from E genome species. Accession W6-26586 from Kazakhstan of *E. lolioides* has one set of H genome that one of chromosomes has St genome signals in centromeric and pericentromeric regions. Combining with results from phylogenetic and in-situ hybridization, genome formula of *E. lolioides* was designed as StStStStHH. This species should be classified into the genus *Elymus* L. and renamed as *Elymus lolioides* (Kar. er Kir.) Meld.

T2

A158

Systematic observations on the development of gametophyte and embryo of one species of *Microlepia*

Xin Wang, *Baodong Liu**
Harbin Normal University, Key Laboratory of Plant biology, College of Heilongjiang Province, Harbin 150025, the People's Republic of China.

The embryo development was the important process of plant reproduction. It was not only necessary for classification, but the important basis of discussing the plant origin and evolution. Pteridophyte was an important group in plant evolution, but the researches on embryo were lack. One species of *Microlepia* was chosen for the experiment. The spores were cultivated with modified Knop's solid medium; the cultivation of gametophyte was taken in simulation natural conditions. With paraffin method, the whole process of embryo development was recorded. Spore of the species of *Microlepia* was homosporous with perisporium. The spore grew into the protonema in 20d. The gamephytes generally developed into the heart shape. The division mode of the embryos belonged to the shape of "XZY". Meanwhile, six-cell embryo was observed. During sixteen-cell embryo, the apical cell of embryos came out. With the division of the apical cell, the first leaf primordium formed and then developed into the first leaf. The root apical cell began to split during the development of the first leaf primordium, and then grew into the first root primordium. The apical cell of the second leaf was developing into the second leaf primordium when the first young root formed. This article summarized the initial split mode and the characteristics of the zygote splitting and predicted the process of embryo in Leptosporangiopsida. It emphasized the appearances order of the leaf and the root, vascular bundle characteristics. It was not only significant for accumulating more detailed information of fern embryology research, but also the solution of system classification.

T2

A159

Species diversity of *Trichoderma* in China

Kai Chen, ***Wen-Ying Zhuang***, *Zhao-Xiang Zhu, Wen-Tao Qin*
State Key Laboratory of Mycology, Institute of Microbiology, Chinese Academy of Sciences

Trichoderma Pers. is a wide-spread fungal genus that can be easily found in soil, on decaying wood or on other fungi. Species of this genus play an important role in agriculture and industry, such as biocontrol of plant diseases and production of enzymes. Some species were reported as causal agents of green mold disease of mushroom in cultivation and opportunistic human pathogens. Progress in taxonomy of the genus *Trichoderma* has been achieved in recent years along with introduction of molecular techniques. Species number of the group increases rapidly. Up to now, more than 270 species are known in the genus, of which about 120 species are known in China. In the five-year investigations to assess species diversity of *Trichoderma* in China, we collected more than 500 specimens on woody substrates and 750 soil samples from 29 provinces of the country. A total of 150 species have been identified by our group based on integrated studies of morphology, culture characteristics and phylogeny. Among them, 66 new species have been discovered, which are distributed in eight major clades of the genus plus several scattered terminal branches in the phylogenetic tree. Discovery of these new taxa updated our knowledge of the group, and provided rich natural resources for consideration of future use of these fungi in biocontrol and enzyme production. Our results also indicate that China is a biodiversity hotspot of *Trichoderma*. We expect to find more species in the unexplored areas.

T2

A160

Taxonomy and phylogeography of *Bruguiera* species using both chloroplast and nuclear markers

Zhengzhen Wang[1], *Zixiao Guo*[1], *Norman Duke*[2], *Zhang Zhang*[1], *Shaohua Xu*[1], *Suhua Shi*[1]
1. Sun Yat-sen University
2. James Cook University

The typical mangrove genus *Bruguiera* consists of five pure species and two hybrid species. *B. gymnorhiza* distributes widely in the Indo-West Pacific (IWP) region, while *B. sexangula*, *B. cylindrica* and *B. parviflora* are more restricted in the eastern IWP region including Indo-Malesian, Papua and northern Australia.

Notably, *B. exaristata* is only found in some places of Northern Australia. *B.* × *rhynchopetala* is a fertile hybrid between *B. gymnorhiza* and *B. sexangula*, capable of backcrossing with both parents, and *B.* × *hainesii* was recently reported as the hybrid between *B. gymnorhiza* and *B. cylindrica*. This study aimed to examine the hybridization events in the *Bruguiera* genus and compare the genetic composition of populations from different geographical sites for each species. Samples of *Bruguiera* were collected in 13 places from Thailand to Australia where at least one species was found. The sequencing of a nuclear gene *23715* and the chloroplast gene *rbc*L advocated the established taxonomic relationships in the genus. Both nuclear and chloroplast genes showed clear boundaries between the five species. Moreover, the chloroplast gene indicated that all the individuals of *B.* × *hainesii* were F_1 hybrids between *B. gymnorhiza* and *B. cylindrica*, with *B. cylindrica* being the maternal parent, while all the individuals of *B.* × *rhynchopetala* were F_1 hybrids between *B. gymnorhiza* and *B. sexangula,* both of which could be the maternal parent. Within each species, population structure was observed between populations from Australia and Malayan Peninsula, between population from east and west side of Malayan Peninsula or both. In the species that had smaller propagule including *B. cylindrica* and *B. parviflora*, the barrier of Malayan Peninsula affected greater than the geographical distance between Thailand and Australia. Yet in the species that had larger propagules including *B. gymnorhiza* and *B. sexangula*, the geographical distance seemed to contribute more to population differentiation. Geographical isolation such as the Malayan Peninsula played an important role in shaping the extent population structure of these species. The clarity of taxonomic and population structure problems in this genus provides valuable knowledge for reconstructing its evolutionary history and crucial guidance of genetic diversity conservation.

T2

A161

Notes on gemmae diversity of bryophytes from karst mountain area, Guizhou Province, China

Zhihui Wang, *Zhaohui Zhang*
Guizhou Normal University

Bryophytes have a characteristic feature which is their tendency towards extensive vegetative reproduction such as the production of gemmae. Supported by the National Natural Science Foundation of China (NSFC. No. 31360043;31360035), a sires of survey were carried out at Lu Chonggan Karst Mountain Area, Guizhou province, China. These experiment focused on the diversity of bryophyte gemmae, which was in the form of asexual reproduction of the bryophytes in the bare land, and its relationship with the environment under the karst mountainous area to explore the connection between the output of bryophytes' gemmae and microhabitat. The shape, the location, the color and the output were recorded for bryophytes with gemmae. The experiment showed that gemmae will appear in the shape of bud, round ball, rod-shaped, torch and disc; the gemmae mainly locate in the axilla and gemma cups of bryophytes, and the specific locations of gemmae in axilla have little differences; the main color of gemmae is green; the output of gemmae of bryophytes has large span in quanlity. The *Marchantia paleacea* Bertol. subsp. *dipteera* (Nees et Mont) Inoue which has the maximum density reaches hundred million in 1 m^2, which is more than four hundred times than the *Barbula rigidula* (Hedw) Mild that has the the minimum density. The research also found that the bryophytes that grow under the environment which is full of moisture and has weaker illumination intensity are easy to form gemmae.

T2

A162

Some gemmiferous species of bryophytes and their ecological response to soil metals in the karst bauxites mine in southern China

Zhihui Wang[1], *Zhaohui Zhang*[1,2]
1. Guizhou Normal University
2. Guizhou Key Laboratory for Mountainous Environmental Information and Ecological Protection

The production of gemmae is a widespread means of asexual reproduction in both liverworts and mosses in some industry ores areas. For example, there are a rich bryophyte flora with gemmiferous taxa at some karst bauxites mines in southern China. Supported by the National Natural Science Foundation of China (NSFC. No. 31360043;31360035), 780 specimens were collected from karst bauxites in Guangxi and Guizhou. Among them, 24 gemmiferous species of bryophytes belong to12 genera and 3 families were recorded. These genera found more the production of gemmae, including *Marchantia, Barbula* and *Hydrogonium* (4 species).We observed that there are more gemmiferous species in human- active environments of bauxites areas than that in the natural environments. The number of gemmae produced inconsistent, *Molendoa japonica* and *Bryum bornholmense* generated a minimum number of gemmae while *Barbula rigidula* and *Barbula ditrichoides* generated a maximum number of gemmae. By the method of the CCA, it was found that Al played a little rule for affected on bryophytes with gemmae in two province, Ca and Mg were the main factors which affected the distribution of the gemmiferous species in Guangxi while Pb, Cr and Cu were the main factors which affected the distribution of the gemmiferous species in Guizhou. On the soil environment, there was higher adaptation of Ti in *Bryum dichotomum* of Guangxi and Guizhou, and it can grow in the soil environment with high concentration of Ti. In karst bauxites areas of Guangxi, there were higher adaptation of Ca and Mn in *Weissia planifolia* and *Hyophila javanica*. It was hinted that the distribution of bryophytes with gemmae were affected by soil metal elements.

T2

A163

Study on the diversity of succulent plants from Kenya, East Africa

Neng Wei[1,2,3], *Guangwan Hu*[1,3], *Qingfeng Wang*[1,3]
1. Wuhan Botanical Garden, Chinese Academy of Sciences
2. University of Chinese Academy of Sciences
3. Sino-Africa Joint Research Center, Chinese Academy of Sciences

Succulent plants, also known as succulents, are plants with fleshy and thickened vegetative organs retaining water to adapt to dry environment and intermittent rainfall patterns. Owing to their striking and unusual appearance and life habits, they are widely used in horticulture. Based on former research and relatively narrow definition of succulents (horticultural definition exactly), Kenya shows a high diversity of succulents (428 taxa in total, among which nearly 30 percent are endemic in Kenya) and 428 taxa is an under-estimate of the possible total. During the last decade, some new succulent taxa have been found and named. In addition, the welcomed APG system, at some parts, considerably differs from the previous classification system, even at order level. Therefore, a new version of check-list of Kenyan succulents is needed. Now, we are carrying on studies on succulents' diversity and evolution in Kenya adopting an evolutionary definition. Initially, to fully understand the composition of Kenyan succulents, we made schedules to effectively conduct field investigations and paid more attention to neglected groups and potential high species-richness areas. We are going to integrate the information of specimens into BRAHMS database, which is formerly built by East Africa Herbarium. Secondly, we will sample and sequence a specific group of taxa at genera level or families level, trying to elaborate the evolutionary mechanism of character evolution about succulents by dating the origins of succulents lineages and inferring the times of succulence occurrence. Finally, we will collect and sequence all the succulents occuring in a particular area, evaluating their species diversity (SD) and phylogenetic diversity (PD). By means of this way, we will provide scientific and proper measures and suggestions to protect those threatened succulents by the ways of *in situ* conservation and *ex situ* conservation.

T2

A164

Systematics and historical biogeography of *Diplazium* (Athyriaceae)

Ran Wei

Institute of Botany, Chinese Academy of Sciences

The twin-sorus fern genus *Diplazium* (Athyriaceae), with a pantropical distribution, represents an important component of the tropical and subtropical rainforests in the world. We reconstructed the phylogenetic framework and biogeographic history of *Diplazium* to explore the phylogenetic relationships and mechanisms of its tropical disjunctions based on a world-scale species sampling and multilocus phylogeny. The phylogenetic results indicate a paraphyly of *Diplazium* together with *Allantodia*, *Callipteris* and *Monomelangium*, which calls for a broadly defined genus (*Diplazium s.l.*) to accommodate this group of ferns based on morphological synapomorphies, such as continuous grooves on rachis and costae and 'diplazioid' sori. In light of molecular and morphological evidence, an infrageneric classification with four subgenera (i.e. subgenera *Diplazium*, *Pseudallantodia*, *Sibirica* and *Callipteris*) of *Diplazium* is proposed. Molecular dating analysis and ancestral area reconstruction were run using a Bayesian Markov chain Monte Carlo (MCMC) strategy and likelihood dispersal–extinction–cladogenesis method, respectively. Our results indicate a Eurasian origin of the crown group of *Diplazium* during the Eocene. The evolutionary history of *Diplazium* involves a variety of biogeographical scenarios. Subgenus *Diplazium* (with subgenera *Pseudallantodia* and *Sibirica*) originated from the Eocene and probably migrated from Eurasia through the boreotropics to tropical America. The following amphi-Pacific disjunctions within subgenus *Diplazium* correlated to the disruption of the boreotropical belt because of the drastic global cooling from the late Eocene. In contrast, subgenus *Callipteris* diversified during the late Oligocene and middle Miocene, and island-hopping and transoceanic dispersals could be invoked for the pantropical disjunctions. The subsequent diversification of the Neotropical clade of subgenus *Callipteris* may be triggered by the major uplift of the Andes and global temperature optimum during the middle Miocene.

T2

A165

Comments on the APG's classification of angiosperms

***Wei Wang**, Xiao-Xia Zhang, Zhi-Duan Chen, An-Ming Lu*

Institute of Botany, Chinese Academy of Sciences

With the rise of plant molecular systematics, tremendous progress has been made in understanding phylogenetic relationships within angiosperms. With the basic phylogenetic framework of angiosperms established, a DNA phylogeny-based angiosperm classification system at the order and familial levels was proposed by Angiosperm Phylogeny Group (APG) in 1998 and has been updated for three times. In this paper, we summarize the major achievements of the APG system: (1) proving the repeatability of APG system and the predictability of interrelationships of angiosperms; (2) clarifying the systematic positions of some segregate taxa which were not placed based on morphological characters; (3) demonstrating the non-rationality of the dividision of dicots and monocots; (4) demonstrating the importance of tricolpate/tricolporate pollen and derivatives for angiosperm classification; (5) proving that the centrifugal development of stamens in polyandrous groups have evolved independently many times and should not be used to define classes or subclasses of angiosperms; (6) proving that most of the families delimited by broad morphological characters are natural; and (7) falling apart some families which are traditionally regarded as natural. We then point out some problems that need to be resolved in the future: (1) how to harmonize the APG system and the system integrating morphological and molecular data; (2) new morphological evolution theories need to be established on the basis of the APG system; (3) whether it is enough only using "monophyly"as a criterion to circumscribe orders and families; (4) what morphological synapormorphies of those orders in the APG system are; and (5) how to compile a key to distinguish the orders and families of the APG system and to list their diagnostic characters for the orders and families. In addition, we propose the suggestions for the phylogenetic relationships and taxonomic status of some taxa mainly distributed in Asia, especially East Asia, including each of Illiciaceae, Acanthochlamydaceae, Tetracentraceae, Leeaceae, Rhoipteiaceae, Hippcastenaceae, Aceraceae, Bretschneideraceae as familial status, and that Cornaceae *sensu lato* should be divided into Cornaceae *sensu stricto* and Nyssaceae *sensu lato*.

T2

A166

Leaf architecture of genus *Prunus* L. *sensu lato* (s. l.) and its taxonomic significance

Wenxin Huang, *Baohuan Wu, Xuming Zhao, Dafang Cui, Wei Li*
South China Agricultural University

Prunus L. *sensu lato* (s. l.) is the largest genus of stone fruit, including *Amygdalus*, *Armeniaca*, *Prunus*, *Cerasus*, *Padus*, *Laurocerasus*, *Pygeum*. It comprises roughly 200 species, and more than 100 species can found in China. The taxonomy of *Prunus* is still controversial. In this paper, leaves of 78 species from *Prunus* were studied. 41 leaf architectural characters were examined and conducted clustering analysis research with UPGMA method (unweighted pair-group method with arithmetic means) to discuss the intergeneric classification and definition of species. Our results are as follows: 1) *Prunus* had a common set of traits: simple leaf with glands; leave margins mostly with teeth; pinnate and major secondary vein mostly reaching the margin; areolation present; freely ending veinlets presented. There were also distinctive characteristics which could be used to identify subgenus and species. The genus can be divided into 7 subgenus: *Amygdalus, Prunus, Cerasus, Padus, Laurocerasus, Maddenia and Pygeum.* 2) The leaf architecture was significant for evolution of *Prunus.* Leave tooth, glands, major secondary vein framework, areolation and freely ending veinlets were different in complicated degrees and showed the evolvement characters of the *Prunus*. In the process of migration from north temperate to subtropical and tropical, leaves of *Prunus* became untoothed, at the same time the raised glands on the leaf teeth became flat glands on the abaxial surface, vein density decreased and freely ending veinlets with more branches. Meanwhile, the temperate taxa continued to evolve, the number of orders of teeth was increased, the glands on teeth showed various types, the types of the quaternary and quinternary and freely ending vein were diverse. 3) Considering the clustering analysis of leaf architecture, other morphological and molecular evidences, a revision of the *Prunus* was provided here. *P. triloba* was supported be a member of subgenus *Prunus*; Excluded the section *Microcerasus* from the subgenus *Cerasus*; *Maddenia hypoxantha* was as conspecific with *P. wilsonii*, because a wide range of variation was found on the lower leaf pubescence which is the differences between two species; *P. gongshanensis* differs from *P. himalayana* only in its almost glabrous lower leaf surfaces, it'd better to be treated as a variety of *P. himalayana*; *P. pananensis* was highly similar to *P.dielsiana* in morphology, so it should be new distribution of *P. dielsiana* in Zhejiang; *Cerasus xueluoensis* also was highly similar to *P. serrulata* in morphology, it may be a dwarf form of *P. serrulata* growing at high altitude.

T2

A167

Numerical taxonomy study on *Prunus* L. subgenus *Cerasus* (Mill.) A. Gray in East Asia

Baohuan Wu, *Wenxin Huang, Wenting Shi, Haijun Yang, Dafang Cui*
South China Agricultural University

Prunus L. subgenus *Cerasus* (Mill.) A. Gray is a group wildly grow in the temperate regions of the northern hemisphere, especially in east Asia. It's also a controversial group in taxonomy, plant taxonomists havn't reached a consensus on the classification system and the definition of some species yet. In this paper, 30 morphological characters were analyzed, using numerical taxonomic method, to discuss the infra-subgeneric and species classification. The result of UPGMA (unweighted pair-group method with arithmetic means) clustering analysis shows that 41 species are divided into 11 groups, 6 of them only contains 1 species, and others includes 2 or more species. *Prunus conradinae* Koehne, *Prunus glabra* (Pamp.) Koehne, *Prunus henryi* (C.K. Schneid.) Koehne, *Prunus duclouxii* Koehne, *Prunus yunnanensis* Franch., *Prunus pseudocerasus* Lindl. and *Prunus pusilliflora* Cardot are 7 controversial species. Most of the time, they are separated into different sections or series, but sometimes species in different sections or series were combined as 1 species. These 7 species are in 1 group in the clustering analysis, indicating their related relationship; It would be appropriate to put them in 1 section or serie. *Prunus pleiocerasus* Koehne used to be combined into *Prunus conadenia* Koehne, however, the clustering analysis shows that their distance on the morphology is not close enough to treat them as 1 species. *Prunus clarofolia* C.K. Schneid., *Prunus polytricha* Koehne, *Prunus tatsienensis* Batalin are in a group with close distance in clustering analysis. The disciform glands in the margin of bracts is the distinguishing characters of *Prunus tatsienensis* to other 2 species, but it's not found in the original literature and type specimen. And the densely villous of the *Prunus polytricha* is indistinguishable to other 2 species. The 3 species are proposed to be combined to 1 species.

T2

A168

Recircumscription of *Delphinium* subg. *Delphinium* (Ranunculaceae) and implications for its biogeography

Kun-Li Xiang[1,2], *Zeki Aytaç*[3], *Yang Liu*[4,5], *Felipe Espinosa*[6,7], *Florian Jabbour*[6], *James W. Byng*[8,9], *Cai-Fei Zhang*[10], *Andrey S. Erst*[11,12], *Wei Wang*[1,2]
1. State Key Laboratory of Systematic and Evolutionary Botany, Institute of Botany, Chinese Academy of Sciences
2. University of Chinese Academy of Sciences
3. Department of Biology, Faculty of Science, Gazi University
4. Key Laboratory of Southern Subtropical Plant Diversity, Shenzhen Fairy Lake Botanical Garden
5. Department of Ecology and Evolutionary Biology, University of Connecticut
6. Muséum national d'Histoire naturelle, Institut de Systématique, Evolution, Biodiversité, UMR 7205 ISYEB MNHN/CNRS/UPMC/EPHE, Sorbonne Universités
7. Laboratoire Ecologie, Systématique, Evolution, Université Paris Sud
8. Plant Gateway
9. Naturalis Biodiversity Center
10. Institute of Plant Biology, School of Life Sciences
11. Central Siberian Botanical Garden, Russian Academy of Sciences, Zolotodolinskaya str. 101
12. Laboratory of Systematics and Phylogeny of Plants (TK), Tomsk State University

The monophyly of *Delphinium* subg. *Delphinium* remains unresolved, owing to the controversial systematic position of the monotypic, eastern Asian endemic *D.* sect. *Anthriscifolium*. The other sect. of the subgenus, *D.* sect. *Delphinium*, is distributed in the Irano-Turanian (IT) region extending westward to the

Mediterranean Basin. Recently, a new genus endemic toTurkey, *Pseudodelphinium*, was described and considered related with *Delphinium* (Delphinieae) or *Garidella* (Nigelleae). In this study, we first conducted a broad phylogenetic analysis within Ranunculaceae using *mat*K sequences and placed *Pseudodelphinium* in Delphinieae. We then performed a series of analyses using four molecular markers (*trnK-mat*K, *trn*S-G, *trn*L-F, and ITS) focused on the tribe. The phylogenetic analyses based on the four-marker dataset indicate that *Pseudodelphinium* is embedded within sect. *Delphinium*. The Swofford-Olsen-Waddell-Hillis tests also reject the exclusion of *Pseudodelphinium* from *Delphinium* and sect. *Delphinium*. The monophyly of subg. *Delphinium* is not recognized because sect. *Anthriscifolium* unites with *D.* subg. *Delphinastrum* and *Consolida*. Based on molecular, morphological, and karyological data, we raise sect. *Anthriscifolium* to subgeneric status, whereas sect. *Delphinium* composed subg. *Delphinium* s.str. An integration of phylogenetic, molecular dating, and biogeographical methods indicates that the subgenus s.str. originated in East Asia during the latest Oligocene and began to diversify in the IT region at 8.45 Ma. Subsequently, the westward colonization events occurred at 7.06 Ma from the IT region to the Italian and Balkan Peninsulas and at 5.4 Ma fromItalyto North Africa. A dispersal from North Africa to the Iberian Peninsula was inferred in the late Pliocene, supporting a hypothesis of trans-sea dispersal. Within the Mediterranean Basin, climate aridification and eustatic sea-level changes could have initiated the westward stepwise expansion of *Delphinium*.

T2
A169
Phylogeography and allopolyploidization of *Magnolia* section *Gynopodium* (Magnoliaceae) in subtropical China

Long-Qian Xiao[1], *Qing-Qing Li*[2]

1. College of Biological and Food Engineering, Huaihua University

2. Life Science College, Southwest Forestry University

Magnolia section *Gynopodium* is an evergreen tree aggregate which includes five species exhibiting three different ploidy levels: diploid, tetraploid and hexaploid. These species have a broad distribution throughout subtropical China and extend to adjacent regions of Indochina. The chloroplast *psb*A-*trn*H sequences were used to examine the genetic structure and phylogeographical patterns for elucidating the glacial refugia and migration history of species within section *Gynopodium*, and the single copy nuclear gene *LEAFY* sequences were cloned to retrieve the different distinct allelic sequences for inferring the polyploid origins. Seven chloroplast distinct haplotypes were identified for all the individuals within section *Gynopodium*. The ancestral haplotypes were confined to the eastern edge of Yun-Gui Plateau, Emei Mountain and the northwest Yunnan. All examined nuclear *LEAFY* allelic sequences were clustered into three distinct clades. Those of two diploid species were recovered as monophyletic and nested within the same clade; whereas those of the tetraploid and hexaploid species were placed into two and three different clades, respectively. The clear-cut geographical distributions of ancestral chloroplast haplotypes indicated multiple potential refugia for species whithin the section *Gynopodium*; the chloroplast haplotype distribution patterns suggested that these species experienced different migratory histories. The nuclear *LEAFY* phylogenetic pattern, combined with evidence from chloroplast data, showed that both allopatric differentiation and allopolyploid origin involving hybridization likely took part in the rapid evolution of the section *Gynopodium*.

T2
A170
Analysis on the morphological variation in the flowers and fruits of *Sinojackia rehderiana*

Guowen Xie, *Leyi Weng, Jinwei Gao, Yisheng Zheng, Xueying Guo, Yunong Tian, Xiaodong Jin*

School of Life Sciences, Guangzhou University

Sinojackia rehderiana belongs to Styracaceae. It is an endangered species, endemic to the subtropical area of China. It mainly distributes in northern Guangdong, Hunan, and Jiangxi province. In the article, *S. rehderiana* population distributed in Yongxiu county was studied, and some fruit variation was founded. Using numerical classification analysis of fruits and flowers on morphology, the fruit variation in the population was explored. The flowers phenotypic data analysis showed that significant differences were existed on the corolla longitudinal diameter, the number of stamens, the anther length, and the diameter of the ovary. The order of variation coefficient of phenotypic characters was calyx length (17.58%) > bud vertical diameter (12.87%) > calyx wide (12.56%) > ovary length (11.53%) > bud lateral diameter (10.05%) > anther length (9.61%) > ovary diameter (9.49%) > stamen number (8.72%). The CV order among 4 fruit types was big tilt type (10.85%) < big long type (11.42%) < big short type and narrow type (11.61%). Anther length and bud vertical diameter, bud vertical diameter and bud lateral diameter, bud vertical diametere and ovary length, bud lateral diameter and ovary length were significantly correlated. Cluster analysis divided the population into three groups, they were big short type, narrow type and big tilt type, and bit long type. Principal component analysis showed that big short type subpopultion was separated from big long type subpopultion at principal component. Big tilt type subpopulation obviously separated from big long type subpopulation and big short subpopulation at principal component, only narrow type and big tilt type mixed together. The fruits phenotypic data analysis showed that all fruit phenotypic traits of *S. rehderiana* were significantly different between subpopulations. The order of variation coefficient of phenotypic characters was beak length (30.68%) > fruit weight (25.85%) > length-width ratio (13.69%) > fruit width (11.58%) > fruit length (11.55%) > bending (3.30%). The CV order among 4 subpopulations was big short type (14.25%) < big long type (16.34%) < big tilt type (16.82%) < narrow type (17.03%). Fruit width and fruit weight, length-width ratio and fruit width, fruit length and beak length, fruit length and length-width ratio were significantly correlated. Cluster analysis divided the population into five groups, they were big short type, big long type, big tilt type, narrow type and mixed type of big tilt type and narrow type. Principal component analysis showed that narrow type subpopulation was obviously separated from other three subpopultions. Big short type subpopulation separated from big long type subpopulation and big tilt subpopulation, only big long type and big tilt type mixed together. The variation of fruits and clustering results were in accordance with their geographical distribution. The narrow type subpopulation and big short type subpopulation are located respectively on both edges of the sample area, big tilt type and big long type are located in the middle transition zone. *S. rehderiana* has good adaptability and wide range of natural distribution. Some phenotypic variation was resulted from genomic hybridization because of high gene flow existed in the breeding.

T2

A171

Lowest latitude fossil record of *Asplenium* from the early Eocene of South China

Kewang Xu, *Xiaoyan Liu, Wenbo Liao, Jianhua Jin*
Sun Yat-sen university

Extant Aspleniaceae Newman is one of the most species-rich groups among leptosporangiate ferns, with about 700 species, and one of the most widespread fern groups, principally distributed in temperate and tropical regions, some species even extending into Arctic and arid areas. *Asplenium* L. is one of the largest fern genera, but the megafossil record of *Asplenium* is limited. Up to now, identifiable megafossils of *Asplenium* have been reported from the Early Cretaceous to the late Eocene of Northeast China, East and Central Russia, Japan, Greenland and Seymour Island and King George Island of Antarctica. South China is one of the *Asplenium* distribution centers, but no fossil records of this genus have yet been reported from this region. In this study, we described some new *Asplenium* fossil specimens recovered from the early Eocene Huachong Formation of Sanshui Basin, Guangdong Province, South China. They were assigned to *Asplenium* mainly based on its bipinnate frond shapes. Fronds are 14 - 8 cm long, ovate-oblong?preserved parts exhibit a maximum length of 15 cm. The main rachis is flat or shallowly sulcate with a raised supravascular ridge on the adaxial side, abaxially flat. The pinna rachis is more slender than the main rachis. Pinnules arranged alternately at angle of about 70° on the main rachis and about 0.8 - 1.5 cm apart, 5 - 9 pairs, linear, 3 - 4 cm long and about 1 cm wide, widest in the middle-lower part, tapering to the apices. The apex of the ultimate pinnule is acute. There is only one vein in the ultimate pinnule with no lateral veins. It is raised abaxially and does not reach the margin. This is the lowest modern latitude fossil record for the genus and the first fossil record of *Asplenium* reported from South China. These fossil specimens demonstrate that *Asplenium* had already spread into South China by the early Eocene. Furthermore, these specimens combined with previous fossil spore-pollen records indicate a warm and humid climate in the Sanshui Basin during the early Eocene.

T2

A172

A preliminary study on postmating reproductive isolation between *Oryza rufipogon* and *O. nivara* based on crossing experiments

Xun Xu, *Qing-Lin Meng, Mu-Fan Geng, Fu-Min Zhang, Song Ge*
Institute of Botany, Chinese Academy of Sciences

Reproductive isolation (RI) is a unique character that reduces gene exchange between populations and is thus the prerequisite for origin of new species. To determine the forms and underlying mechanisms of reproductive isolation, therefore, is critical to the understanding of speciation. *Oryza rufipogon and O. nivara* are two closely related species that are evolutionary young and distinct in morphology and ecology, representing incipient species at the early stage of speciation. Although previous studies found that the two species differed significantly in flowering time, suggesting strong premating RI, the postmating RI between the two species remains unclear despite a few of attempts. To uncover the potential postmating RI and its impact on gene flow between species, we evaluated seed setting rate of different combinations of crosses based on three pairs of *O. rufipogon* and *O. nivara* populations that were sampled from Nepal, Laos, and Cambodia, respectively. We conducted three categories of crosses: intra-population, inter-population within-species, and inter-species. First, based on the crosses involved 4,560 spikelets on 266 panicles, we found that seed set of inter-species did not differ significantly from the other two categories, implying that no obvious postmating RI is present between species. Second, we observed that inter-species crosses using *O. nivara* as the maternal parent generated higher rate of seed sets than that of the crosses using *O. rufipogon* as the maternal donor. In another word, *O. rufipogon* was less fertile than *O. nivara* in terms of female productivity, while the male fertility did not differ significantly between species. This result may arise from the decrease of pistil fertility of *O. rufipogon*, although the tradeoff between vegetative and seed propagations need further investigation. Finally, we found that the seed setting rates were significantly different between the panicles with and without castration, which may explains the inconsistent results arising from previous studies. Investigation on F1 viability and fertility of the crosses is still underway, which may provide additional insights into the RI of the speciation in this group of species.

T2

A173

The functional characterization of KCS genes involved in the synthesis of erucic acid using evo-devo strategy

Jia-Yu Xue, *Guang-Can Zhou, Yue-Yu Hang*
Institute of Botany, Jiangsu Province and Chinese Academy of Sciences

The 3-ketoacyl-CoA synthase (KCS) is the rate-limiting enzyme in the synthesis of fatty acid and genes encoding this enzyme are thereby named as KCS genes. The erucic acid is a sort of very-long-chain fatty acid (VLCFA) and a characteristic component in synthesized Brassicaceae plants. The KCS gene family consists of 21 members in Brassicaceae and different members have evolved functional specificity towards the synthesis of different VLCFAs. However, to date only one member of this gene family—*FAE1* has been confirmed to be involved in the synthesis of erucic acid. Other KCS genes have also been functionally characterized by heterologous expression experiments in yeast and *Arabidopsis*, but none participate in the synthesis of erucic acid. However, it is noteworthy that all those genes are evolutionarily distant to *FAE1*, and have diverged as early as the landing of plants. They share less sequence identity with FAE1, and are more likely to acquire distinct function. Our previous evolutionary study found that six

KCS gene were close relatives to *FAE1* in Brassicaceae, and they all emerged (including *FAE1*) after the split of Brassicaceae by duplication and have not been functionally characterized. Regarding this, we transferred to yeast to express them and erucic acid was detected in the yeast products of two genes, indicating that the two KCS genes (other than *FAE1*) were also functionally involved in the synthesis of erucic acid.

T2

A174

A multispecies phylogeography of the Himalayan blue poppy in sky-islands: Implications for speciation and species delimitation

Fu-Sheng Yang[1], *Yu-Cheng Liu*[1], *Zheng-Xiang Wang*[2], *Xiao-Quan Wang*[1]

1. State Key Laboratory of Systematic and Evolutionary Botany, Institute of Botany, Chinese Academy of Sciences

2. Faculty of Resources and Environmental Science, Hubei University

The Eastern Himalayas (EH), also called "the Earth's third pole", belongs to one of the global biodiversity hotspots. It barbors a high concentration of endemic species, especailly on sky-islands inaccessible to human. The Himalayan blue poppy (*Meconopsis*) is a flagship genus for the EH; however, species delimitation within the genus is often controversial. We carried out an extensive field investigation of *M. betonicifolia* s.l. and *M. simplicifolia* s.l., two closely related species complexes distributed on the EH's sky-islands. And then, a multilocus phylogeographical analysis was conducted to analyze the genetic diversity and retrieve the evolutionary history of the congeneric species. Interestingly, all haplotypes of *M. simplicifolia* s.l. are population-specific, demonstrating that seed-mediated gene flow is blocked completely. Similarly, pollen-mediated gene flow is blocked among the three narrow areas of *M. betonicifolia* s.l. Phylogeographical analyses of plastid and nuclear genes reveal extensive biogeographic structure among populations examined, and reciprocal monophyly of *M. simplicifolia* s.l. and *M. betonicifolia* s.l. is largely supported although some incongruence resulted from hybridization/introgression was detected. These results indicate that lower environment surrounding the EH's sky-islands acting as barriers could have greatly affected the evolutionary history of local plants, and geographic isolation and repeat range shifts by climatic oscillation have promoted population genetic differentiation, allopatric speciation, and hybrid speciation in the Himalayas. Our results do not support a taxonomic separation of *M. betonicifolia* s.l. into more species based on genetic structure, and *M. nyingchiensis* should be treated as a synonym of *M. simplicifolia*. A taxonomic treatment of the closely related species of *Meconopsis* is proposed.

T2

A175

Construction of database of Higher Plant Specimens at National Herbarium (PE), Institute of Botany, Chinese Academy of Sciences

Qi Lin[1], *Zhirong Yang*[1], *Bojian Bao*[1], *Yun Lin*[2], *Xiaobing Zhang*[3]

1. Institute of Botany, Chinese Academy of Sciences

2. Hunan Food and Drug Vocational College

3. Taiyuan Normal University

National Herbarium (PE), Institute of Botany, Chinese Academy of Sciences was established in 1928, and it is one of the oldest herbaria and the largest in Asia. In the beginning of its establishment, there were only 335,000 specimens, part of the collection dates back to the early 18th century. With the growth of the collections, curation and collection management practices evolved, up to now, PE Herbarium contains more than 2.67 million specimens, ranging from bryophytes, pteridophytes and seed plants, and herein 77% have been digitized in database. This study is based on a database of PE Herbarium records and information system, and makes more detailed introduction to the construction of the database of collection, and provide valuable and useful information regarding collections of vascular plant of PE. In additions, through the statistical analysis to the database, it is indicated that there are 2,056,722 collections recorded at PE, including bryophytes 158,218, pteridophyta 168,231, gymnosperms 29,530 and angiosperms 1,698,035, representing 363 families, 8,368 genera, 84,158 species, 16,466 subspecies, varieties and forms. The specimens were collected from 34 provinces, autonomous regions, municipalities and special administrative regions ofChinaand 147 other countries and regions. There are 18,885 type specimens, representing 195 families, 1,156 genera, 8,711 species (including subspecies, varieties and forms) recorded. The construction of database of PE Herbarium will contribute to realizing the sharing of some specimen, improving the utilization efficiency of the resources, and providing basic platform for the scientific research and social development.

T2

A176

Floral nectaries in *Euonymus* and the origin of echinate fruit surface

Chi-Yuan Yao, ***Jin-Shuang Ma***, *Cheng Du*, *Yun-Juan Zuo*

Shanghai Chenshan Plant Science Research Center, Chinese Academy of Sciences

Euonymus is a morphologically diverse and cosmopolitan genus in Celastraceae. It was argued to modify the most recent classification diagrams in previous work. Studies focused on floral nectaries of *Euonymus* were rather rare in spite of the availability of comprehensive publications dealing with anatomical and morphological aspects. The present study was undertaken to describe the variation of nectaries among the sections of *Euonymus* for the first time. In addition, we also provided a detailed analysis of the origin of echinate fruit surface.The morphology and development of floral nectaries in 21 taxa belonging to the genus *Euonymus* (19 species) and *Glyptopetalum* (2 species) were investigated using light microscopy and scanning electron microscopy. A conspicuous nectary disk is common in *Euonymus*. Organ initiation of the nectary occurred at a primordial floral stage when the flower bud was still quite small. Two types of receptacular nectaries, the intrastaminal type and a combination of inter- and extrastaminal type, were distinguished on the basis of their location. The nectary disk of the first type may form a congenitally united collar together with the stamen bases (such as *E. acanthoxanthus*, *E. pseudovagans*, *E. theifolius*, *E. fortunei*, *E. centidens*, *E. tingens* and

E. microcarpus). However, the disk of the second type extends between the corolla and the stigma, and thus also encompasses the stamen bases (such as *E. sanguineus*, *E. schensianus*, *E. verrucocarpus*, *E. wilsonii*, *E. actinocarpus*, *E. balansae*, *E. alatus*, *E. hukuangensis*. *E. yunnanensis*, *E. wui*, *E. nitidus*, *E. oblongifolius*, *G. continentale* and *G. longipedicellatum*). Variations of modified stomata were found on the apex of the nectary. The nectarostomata are sunken in pits of *E. sanguineus*, *E. centidens*, *E. alatus*, *E. hukuangensis*. *E. nitidus* and *E. oblongifolius*; and located on distinct convexities forming characteristic like chimneys or volcanoes in all species of *E.* sect. *Echinococcus* and *E.* sect. *Ilicifolia*, as well as *E. tingens* (*E.* sect. *Euonymus*); and level with adjacent epidermal cells in *G. continentale*, *G. longipedicellatum*, *E. schensianus*, *E. yunnanensis*, *E. wui*, and *E. microcarpus*. Species in *E.* sect. *Echinococcus* and *E.* sect. *Ilicifolia* exhibit a number of similarities, of which the nectarostomata are raised like chimneys or volcanoes. The nectaries in *E.* sect. *Echinococcus* continue to grow and change to appendages after anthesis, hence the difference emerges in fruit surface which is smooth in *E.* sect. *Ilicifolia* but echinate, prickly or tuberculate in *E.* sect. *Echinococcus*. What is more, it can be observed clearly that each protrusion has a nectarostomata on the top. The great diversity of floral nectaries originates in the numerous adaptations which are the consequence of the co-evolution between plants and their pollinators. This study of floral nectaries undoubtedly contributes to a better understanding of the relationships among the sections of *Euonymus*.

T2

A177

Sharp genetic discontinuity in an arid-sensitive species *Lindera obtusiloba* (Lauraceae): Solid evidence supporting the Tertiary floral subdivision in East Asia

Jun-Wei Ye

Natural History Research Center of Shanghai Natural History Museum, Shanghai Science & Technology Museum

Aim Unravelling phytogeographic divides is essential to a better understanding of flora evolution. Tertiary relict floras in East Asia have been subdivided into two northern (NEA) and southern (SEA) regions separated by an eastward arid belt. The division has been studied infrequently since it was first proposed in 2002. To test the hypothesis, we investigated the evolutionary history of a Tertiary relict species, *Lindera obtusiloba*. Location East Asia. Methods Chloroplast DNA and nuclear microsatellites (nSSRs) were genotyped among 34 populations. Regional genetic diversity was compared, and genetic differentiation between and within regions was calculated. Bayesian phylogeny and Bayesian clustering were employed for phylogeographic reconstruction. Both the most recent common ancestor (MRCA) and population divergence time of the two regions were estimated. Historical gene flow was calculated using the maximum-likelihood method. Potential habitat at present, the Last Glacial Maximum (LGM) and Last Interglacial (LIG) were predicted by ecological niche modelling (ENM). Results Both markers revealed sharp genetic discontinuity and limited migrations between regions. The MRCA and population divergence time probably occurred in the Pliocene. A clear genetic structure was found within each region. Genetic diversity of both markers and the age of chlorotypes decreased as the longitude increased within SEA. The Japanese population harboured the highest genetic diversity, with abundant private and ancestral chlorotypes within NEA. NEA had higher genetic diversity than SEA in nSSRs, while the opposite was true for cpDNA. ENM in LGM and LIG probably supports long-term isolation in multiple locations in both regions. Main conclusions Our study provides solid evidence for the proposed floral subdivision along the arid belt within East Asia. Populations in both regions remained stable over a long-term period and sustained multiple glacial refugia. Westernmost SEA and easternmost NEA sustained ancestral populations. A clear eastward expansion within SEA was detected.

T2

A178

UPLC-QTOF MS characterization of sesquiterpene pyridine alkaloids of *Monimopetalum chinense*

***Min Yin*[1]*, Ming Wang*[1]*, Xu Feng*[1]*, Qizhi Wang*[1]*, Yu Shan*[1]*, Yu Chen*[1]**

1. Institute of Botany, Jiangsu Province and Chinese Academy of Sciences

2. Nanjing Botanical Garden, Mem. Sun Yat-Sen

Monimopetalum chinense (Celastraceae) is listed as one of the Chinese national grade 2 protective rare and endangered species. Little knowledge of its chemical composition is available. 11 new compounds of β-dihydroagarofuran sesquiterpenes from *M. chinense* were first identified by UPLC-QTOF MS in this paper. These compounds are characterized by a 5-carboxy-N-methyl-2-pyridone (CNMP) ring attached to the sesquiterpene nucleus, although only three members of this class of alkaloids have been isolated from *Maytenus emarginata* (Celastraceae), *Loeseneriella merrilliana* (Hippocrateaceae), *Hippocratea excels* (Hippocrateaceae), respectively. The taxonomic position of the family Hippocrateaceae has been debated. According to Robson, the Hippocrateaceae genera are not a natural group, being derived in two separate lines from the Celastraceae. The connection between these families has been reinforced by the presence of some common chemotaxonomic markers such as the quinonemethide triterpenoids, dulcitol, and sesquiterpene pyridine alkaloids. This gives support for the botanical classification in which the two families are grouped as only one family, is united with Celastraceae, the name Hippocrateaceae is rejected in favor of Celastraceae.

T2

A179

Morphological study on flowers and fruits of genus *Schoenoplectus* (Cyperaceae)

Hui–Ying Yu

School of Nature Conservation, Beijing Forestry University

Three morphological characters on flowers and fruits of seven species of the genus *Schoenoplectus* from Shandong were observed and studied using anatomical lens. According to the color, size and shape, two types of scales were divided, i.e. I. elliptico-ovate scales of *S. ehrenbergii*, *S. triqueter* and *S. tabernaemontani*, II. rhombi-ovate of *S. jingmenensis*, *S. juncoides*, *S. hotarui* and *S. jiuxianshanensis*. Bristles of *S. ehrenbergii* were found to be longer than nutlets with statistical significance, but those of *S. triqueter*, *S. tabernaemontani*, *S. jingmenensis*, *S. juncoides*, *S. hotarui* and

S. jiuxmianshanensis were approximately equal to nutlets. For the fruit shape, obvious differences were found between the trigonous nutlets of *S. ehrenbergii* and plano-convex obovate ones of *S. triqueter*, *S. tabernaemontani*, *S. jingmenensis*, *S. juncoides*, *S. hotarui* and *S. jiuxianshanensis*. In conclusion the bristles length and fruits shape are important to the species limitation among the genus *Schoenoplectus*, while the scales morphology has failed to distinguishing those species. The results added description to these 7 kinds of plants, while also provided new evidences for taxonomic revision of the genus *Schoenoplectus*.

T2

A180

Repeated fluctuating climate drives plant evolution: Enlightenment from distribution of fleshy-fruited species in China

Shunli Yu

State Key Laboratory of Vegetation and Environmental Change, Institute of Botany, Chinese Academy of Sciences

In the evolutionary history of vascular plants, plant phyla with multiple anatomical and morphological innovations are thought to have emerged during cool-dry climate episodes of the earth's history. However, it is not fully understood (i) why plants continually evolved novel resource acquisition systems (e.g., from protostele to eustele) and why propagule evolved from spores to naked seeds and then to seeds enclosed by fruit pericarp, and (ii) why there is a synchronicity in the emergence of novel traits and evolution of plant phyla, non-synchronicity in origin and radiation of plant phyla. A hypothesis was provided that cool-dry climate episodes stimulated the evolution of more efficient resource acquisition systems, e.g. for water uptake. Clades with such innovative resource acquisition systems could diversify and radiate faster and more effectively owing to increasing niches once climate ameliorated, thus forming the genetic foundations for the evolution of future clades when cool-dry conditions reappeared. This hypothesis is proposed on basis of enlightenment from a macro-ecological analysis of the distribution of fleshy-fruited species in the flora of China. A high proportion and richness of species with high-water content fleshy fruits are associated with the high resource (e.g. water) availability of warm-humid climates and with the strong resource acquisition systems of woody plant growth forms. Thus, the evolution of propagules may be the result of the co-evolution of efficient resource acquisition (or utilization) systems and other vegetative apparatus such as leave. Consequently radiation of angiosperms and diversification of their propagules were stimulated by increased ability to occupy vacant ecological niches and attract frugivores as seed dispersers. Repeated fluctuating climate may drive plant evolution process. Our study offers further evidence with which to resolve Darwin's second "abominable mystery" about angiosperm origins and rapid dominance in the earth's terrestrial flora and vegetation.

T2

A181

Taxonomic combination and a new species of *Epimedium* from Wuling Moutians of central China

Cheng Zhang, *Ze-Long Nie, Ke-Gang Li*

Jishou University

Epimedium Linnaeus is the largest herbaceous genus of Berberidaceae, well known for its ornamental and medicinal value. The genus is comprised of 60 species disjunct distributed between the Mediterranean region and eastern Asia. China is the most important diversity center of *Epimedium*, with ca. 80% of the total species found in central-southeastern China. Moreover, all are endemicin China except for *E. koreanum* Nakai. Chinese species (sect. *Diphyllon*) was classified into four series according to flower morphology. However, phylogenetic studies could not confirm the classification based on pollen types, flavonoid types, karyotypes, and molecular marker. Although the AFLP evidence suggested that five well-supported clades are recognized in sect. *Diphyllon*, the backbone phylogeny of the whole section is still unclear. Species delimitation in *Epimedium* has been problematic because of high levels of morphological variation, especially for the Chinese species with wide variations in petal type, the form and relative size of the inner sepals and petals, and flower dimension. The Wuling Mountainsis one of the most important diversity center of *Epimedium* with 26 species found there, 16 of which are regional endemic to this region. However, it is extremely controversial for the taxonomy of the species in this region due to the poor quality of type specimen, intermediate characters of morphological variation, inadequate investigation for some species. To clarify some species taxonomy in Wuling mountains, we provide a few taxonomic combinations for some species complex group based our field investigations and herbarium works together with references of molecular evidences. (1) *E. baojingense* Q. L. Chen et B. M. Yang and *E. franchetii* Stearn are treated as synonyms of *E. hunanense* (Handel-Mazzetti) Handel-Mazzetti. Both *E. baojingense* and *E. franchetii* share similar overall morphology with *E. hunanense*. *E. hunanense* differs from *E. baojingense* with 2 trifoliolate opposite leaves while the latter with simple leaves. However, extensive field examinations in Jishou and Zhangjiajie from Hunan suggested that *E. baojingense* has predominantly trifoliolate and only occasionally unifoliolate leaves. *E. franchetii* differs from *E. hunanense* in that petals of *E. franchetii* without a distinct petaloid. However, the difference between these two species is not clear. (2) *E. shennongjiaensis*is Yan J. Zhang & J. Q. Li treated as a synonym of *E. leptorrhizum* Stearn. We find that *E. shennongjiaensis* is special ecotype with its flower similar as *E. leptorrhizum* but frozen-like. (3) *E. jingzhouense* G. H. Xia & G. Y. Li is a synonym of *E. coactum*. We have carefully examined its type specimen and plants in type locality and find it has no difference with *E. coactum*. A distinctive and undescribed species, *E. zhangii* C. Zhang, Z. L. Nie & K. G. Li sp. nov., is found from Wuling Mountainsin Hunan province. Molecular evidence indicates that the new species is closely related to *E. parvifolium* and *E. sagittatum*, but differs from them by its long stout rhizomes, flowering stem with 2 opposite leaves, trifoliolate, paniculate inflorescence, flowers purple, and petals with short spur.

T2

A182

Phylogeny and generic circumscriptions of Chinese cheilanthoid ferns

Gangmin Zhang[2], *Wenli Yang*[2]

1. Beijing Forestry University

2. Hebei Agricultural University

Cheilanthoid ferns, which constitute the subfamily Cheilanthoideae of Pteridaceae, have been one of the most contentious fern groups in terms of phylogeny and classification due to morphological convergence driven by adaptation to xeric environments. Using five DNA sequences of chloroplast makers (*rbc*L/ *trn*L–*trn*F/ *rps4*/ *rps4–trn*S/ *trn*G–*trn*R) and two nuclear DNA markers (*CRY2*/*IBR3*), combined with the studies of spore morphology and cytology, we addressed the phylogeny of Chinese cheilanthoid ferns. The results were as follows: (1) Pellaea connectens *C. Christensen*, a rare and Chinese endemic species, was robustly embedded in the newly segregated genus Argyrochosma. In addition, it is a sexual tetraploid species with 64 spores per sporangium and with a base chromosome number of x = 27, which is consistent with Argyrochosma and different from other cheilanthoid ferns. (2) *Cheilanthes tenuifolia* (N. L. Burman) Swartz and *C. nudiuscula* (R. Brown) T. Moore formed a fully supported clade (BP=100%) and distantly related to other species of Asian "Cheilanthes" group, so the two species may have an Oceania origin and represent separate evolutionary lines. (3) Except the above two species, all taxa previously placed in Leptolepidium, Aeuritopteris, Cheilosoria, Sinopteris, and the Asian species of Mildella, were intermingled together in the phylogenetic trees and constitute the Asian "Cheilanthes" group. This implies that the above small genera do not merit generic recognition and should be combined into one.

T2

A183

Parasitic angiosperms in China: Diversity, distribution and ecology

Guangfu Zhang

Nanjing Normal University

Parasitic plants play a significant role in influencing ecosystem structure and function, and thus they are generally known as key stone species and ecosystem engineers. Due to vast territory and rich climate-zones, China harbors approximately 31,000 vascular plant species, only second to Brazil in the world. However, little is known concerning the floristic composition, geographical distribution and ecological features of parasitic plants in China. Here, based on reviewing literature reports and checking herbarium records, we provide a nationwide list of parasitic plant species native to China for the first time. The list comprised a total of 678 parasitic plant species belonging to 12 families and 50 genera. The parasites accounted for 2.28% of total angiosperms (29,716 species) in China, 15.07% of the total parasites (ca 4,500 species) worldwide. The most dominant family was Scrophulariaceae with 486 species, accounting for 71.68% of parasitic plants; the most dominant genus was *Pedicularis* with 441 species, accounting for 65.04% of parasitic plants. In addition, 428 of the whole flora were endemic to China, accounting for 63.13%. Most of the 678 species identified were herbs (having 573 species, 84.51%), followed by shrubs (85, 12.54%), then subshrubs (11, 1.62%), and lianas (9, 1.33%) in order. The number of parasitic plant species in 29 province floras ranged from one (In Chongqing) to 283 (in Yunnan), indicating that the parasites were distributed unevenly in China. Additionally, no parasites were recorded in five of 34 provinces. The ecotypes of these parasites can be divided into four categories: root hemiparasites (having 516 species, making up 76.11%), root holoparasites (62, 9.14%), stem hemiparasites (87, 12.83%), stem holoparasites (13, 1.92%) in light of parasitic position and host dependency. Therefore, the flora of parasitic plants is characterized by a great diversity, a predominant family or genus, various life forms and ecotypes. More attention should be focused on distribution and ecology of parasitic plants in China, especially concerning species of dominant families or genera.

T2

A184

Inaccessible biodiversity on limestone cliffs: *Aster tianmenshanensis* (Asteraceae), a new critically endangered species from China

Guo-Jin Zhang, *Tian-Gang Gao*

Institute of Botany, Chinese Academy of Sciences

Aster tianmenshanensis G. J. Zhang & T. G. Gao, a new species of Asteraceae from southern China is described and illustrated based on evidence from morphology, micromorphology and molecular phylogeny. The new species is superficially similar to *Aster salwinensis* Onno in having rosettes of spatulate leaves and a solitary, terminal capitulum, but it differs by its glabrous leaf margins, unequal disc floret lobes and 1-seriate pappus. The molecular phylogenetic analysis, based on nuclear sequences ITS, ETS and chloroplast sequence *trn*L-F, showed that the new species was nested within the genus *Aster* and formed a well supported clade with *Aster verticillatus* (Reinw.) Brouillet *et al.,* The new species differs from the latter in having unbranched stems, much larger capitula, unequal disc floret lobes, beakless achenes and persistent pappus. In particular, *A. tianmenshanensis* has very short stigmatic lines, only ca. 0.18 mm long and less than 1/3 of the length of sterile style tip appendages, remarkably different from its congeners. This type of stigmatic line, as far as we know, has not been found in any other species of *Aster*. The very short stigmatic lines plus the unequal disc floret lobes imply that the new species may have a very specialized pollination system, which may be a consequence of habitat specialization. The new species grows only on the limestone cliffs of Mt. Tianmen, Hunan Province, at the elevation of 1,400 m. It could only be accessed when a plank walkway was built across the cliffs for tourists. As it is known only from an area estimated at less than 10 km^2 and a walkway passes through this location, its habitat could be easily disturbed. This species should best be treated as Critically Endangered based on the International Union for Conservation of Nature Red List Categories and Criteria B2a.

T2

A185

Plastid phylogenomics and adaptive evolution of *Gaultheria* series *Trichophyllae* (Ericaceae), a clade from Sky Islands of the Himalaya-Hengduan Mountains

Ming-Ying Zhang[1], *Peter W Fritsch*[2], *Peng-Fei Ma*[1], *Hong Wang*[1], *Lu Lu*[1], *De-Zhu Li*[1]

1. Kunming Institute of Botany

2. Botanical Research Institute of Texas

Gaultheria series *Trichophyllae* Airy Shaw is an angiosperm clade of high-alpine shrublets endemic to the Himalaya-Hengduan Mountains and characterized by recent species divergence and convergent character evolution that has until recently caused much confusion in species circumscription. Phylogenetic relationships among species in the group remained largely unresolved, although multiple DNA sequence regions have been employed previously. Here we examined the effectiveness of plastid genome for improving phylogenetic resolution within the *G.* series *Trichophyllae* clade. Plastid genomes of 31 samples representing all the 19 recognized species of the series and three outgroup species were sequenced with Illumina Sequencing. Maximum likelihood (ML), Maximum parsimony (MP) and Bayesian inference (BI) phylogenetic analyses were performed with various datasets, i.e., that from the whole plastid genome, coding regions, noncoding regions, large single-copy region (LSC) and inverted-repeat region a (IRa). The partitioned whole plastid genome with inverted-repeat region b (IRb) excluded was also analyzed with ML and BI. Tree topologies based on the whole plastid genome, noncoding regions, and LSC region datasets across all analyses, and that based on the partitioned dataset with ML and BI analyses, are identical and generally strongly supported. *Gaultheria* series *Trichophyllae* diverged into a clade with three species and one variety, and its sister comprising the remaining 16 species which subsequently diverged into seven main subclades. Interspecific relationships within the series are strongly supported except for those based on the coding-region and IRa-region datasets. Eight divergence hotspot regions, each possessing > 5% percent variable sites, were screened across the whole plastid genome of the 28 individuals sampled in the series. Morphological character evolution reconstruction supports several clades, and for two of these, i.e., plant habit and number of leaf marginal teeth, hypotheses of adaptive evolution are postulated.

T2

A186

Phylogenetics of Papilionoideae (Fabaceae) using pultiple genome-scale sequence

Rong Zhang[1,2], *Jian-Jun Jin*[1,2], *Ting-Shuang Yi*[1]

1. Kunming Institute of Botany, Chinese Academy of Sciences

2. University of Chinese Academy of Sciences

Papilionoideae contains Ca. 500 genera and 14,000 specie, which is the most species-rich subfamily of Fabaceae. The papilionoid legumes have global distribution in a wide range of habitats, many species are economically valued. Papilionoideae is strongly supported as monophyletic group. However, phylogenetic relationships among basal nodes ("early branching papilionoids") and some key lineages within the Papilionoideae remained poorly resolved. Therefore, it is important to reconstruct a backbone phylogeny of the subfamily of Papilionoideae. Including 253 species from 232 genera, Maximum Likelihood and Bayesian phylogenetic analyses were carried out on the complete plastome, most mitochondrial genome and nuclear 18S-5.8S-26S rDNA produced by genome skimming. Most deep relationships were fully resolved with high supports. The Swartzioid clade was strongly supported as a monophyletic group, which was sister to the remaining taxa. Then, the ADA clade, including the *Angylocalyx* clade, the Dipterygeae clade and the Aldinoid clade, was monophyletic with robust support, which was resolved as sister to the remaining taxa. Within the ADA clade, each subclade was strongly supported as a monophyletic group, and the *Angylocalyx* clade was sister to the other two subclades. Next, The *Cladrastis* clade, the *Andira* clade and the Lecointeoid clade were strongly supported to be monophyletic. The Lecointeoid clade was strongly supported as the sister to the large 50-kb inversion clade. Within the 50-kb inversion clade, a few nodes (the genus *Dermatophyllum*, the Vataireoid clade, the Dalbergioid clade and the Genistoid clade) were not well resolved, but some large clades were well supported as the Mirbelioid clade, the Millettioid/Phaseoloid clade and the Hologalegina clade, and some unplaced genera were primely resolved as *Ormosia*, *Bowdichia*. We plan to build a fully resolved and strongly supported backbone including most genera. The robust phylogenetic backbone we provided establishes a framework for future comparative studies on classification and evolution of the Papilionoideae.

T2

A187

Floral scent shapes pollinator-mediated selection in Orchidaceae pollination systems

Wufan Zhang

Institute of Botany, CAS

Insects process floral scent through their olfactory system and generate decision making in insect pollination system, which induces the attraction behavior to host plants. Different plant species releases volatile organic compounds (VOCs) varies in different components and concentration, which raises attraction to different insect pollinators and formats the pollination system. We studied the Orchidaceae plants in Huanglong Nature Reserve in Sichuan Province, analyzed the variety of orchid odor components and their pollinator system. In this study we revealed the differences in odor responses between pollination systems composed by bees and flies, and indicated the selection of pollinators by odor signals from the ecological perspective.

T2

A188

Historical geography and palaeoenvironment of *Liriodendron*

Shuang-Xing Guo[1], *Min Deng*[2]

1. Department of Palaeobotany and Palynology, Nanjing Institute of Geology and Palaeontology, Chinese Academy of Sciences

2. Shanghai Chenshan Plant Science Research Center, Chinese Academy of Sciences

Liriodendron is a small genus of Magnoliaceae adapting to humid and warm temperate to subtropical climates. This genus is only composed by two extant species, of which *L. chinense* is native to eastern China, and *L.tulipifera* is native to eastern the United States. There are 32 fossil species of genus *Liriodendron* recorded in Northern Hemisphere. The earliest fossil records were found in early Late Cretaceous in North America. At that period, *Liriodendron* maintained high species diversity level and most of the species grew under a warm temperate to subtropical climate. The species diversity of the genus kept dropping prominently since Palaeogene. The distribution range of *Liriodendron* in Palaeogene shrank significantly. The fossils of Palaeogene and Miocene of the genus existed in subtropical to warm temperate coastal humid area, with large number of the co-occurring subtropical evergreen elements. Although, in Miocene, *Liriodendron* had most abundant fossil localities and widest distribution in Cenozoic (between Lat.43-58°N in Europe and reaching to its southern most boundary as Lat.36°N in coastal regions of Asia and North American), but fewer fossil species were found than in Palaeogene. In Pliocene, *Liriodendron* species were still found in northern subtropical and warm temperate climates, but the distribution range and the abundance of evergreen elements in the fossil assembles showed significant reductions. During the Quaternary climate fluctuation, the global cooling at the glacial period forced *Liriodendron* species migrated to the southern refuge, and also caused massive local extinction, which eventually led to current relic disjunctive distribution pattern of the genus surviving in warm temperate to subtropical regions in present-day East North American and East Asia.

T2

A189

Miocene fossil leaves of *Homonoia* (Euphorbiaceae) from Tiandong County in Guangxi Province, South China

Shuang-Xing Guo[1], *Yi Zhang*[2]

1. Department of Palaeobotany and Palynology, Nanjing Institute of Geology and Palaeontology, Chinese Academy of Sciences

2. College of paleontology, Shenyang Normal University; Key Laboratory for Evolution of Past Life in Northeast Asia, MLRC

Leaf fossil specimens were collected at Yan-e village from Tiandong County in Guangxi province of South China in Miocene period. Their foliar physiognomies show petiole 5–10 mm, leaves simple, symmetrical, blade linear or narrowly lanceolate, 6–13 cm long and 1–2.2 cm wide, base cuneate or obtuse; apex acuminate to acute, margin entire or sparsely glandular-serrulate, texture papery to thin coriaceous. Venation pinnate, camptodrome, midvein stout and stright; secondary veins 11–14 pairs, alternate, base stout, sparseal, thinning upward, diverging at angles of from 35° to 45°, curving up to leaf margin, looped and closed near the margin; tertiary veins thin, reticulate network, quaternary veins fuzzy. The above characters of fossil specimens are closely similar to living *Homonoia riparia* Loureiro of Euphorbiaceae. However, the leaf blades of living species are bigger in size, number of lateral veins more and diverging angles between the lateral veins and midvein is greater. These features are distinct from leaf fossils. Therefore, these leaf fossils can be identified as a new species, *Homonoia guangxiensis* sp. nov. The living *Homonoia riparia* Lour. as a rheophyte, grows on riverbanks, rocky and gravelly stream beds, sandbars; below 1,000 m lowlands in Provinces of Guangxi, Guangdong, Guizhou, Hainan, Sichuan, Taiwan, Yunnan of South China and Counties of Cambodia, India, Indonesia, Laos, Malaysia, Myanmar, Philippines, Thailand, Vietnam of Southeast Asia. The new species of fossil plants, recorded in Tiandong County, Guangxi, probably also selected similar lowlands for living.

T2

A190

Transcriptome analyses provide insights into the phylogeny and adaptive evolution of the mangrove fern genus *Acrostichum*

Zhang Zhang[1], *Ziwen He*[1], *Shaohua Xu*[1], *Xinnian Li*[1], *Wuxia Guo*[1], *Yuchen Yang*[1], *Cairong Zhong*[2], *Renchao Zhou*[1], *Suhua Shi*[1]

1. Sun Yat-sen University

2. Hainan Dongzhai Harbor National Nature Reserve

Mangroves are a group of plants which can survive and adapt to the extremely unstable marine intertidal zone under harsh conditions, such as high salt concentrations, tidal rhythms and long-term climate changes. The genus *Acrostichum* L. (Pteridaceae) is the only fern genus growing in this environment, including three species: *A. danaeifolium*, *A. aureum* and *A. speciosum*. *A. danaeifolium* and *A. speciosum* are restricted within the Atlantic East-Pacific (AEP) area and Indo West-Pacific (IWP) area, respectively, whereas *A. aureum* is the only species of mangroves that is widely distributed in both areas. In the IWP area, *A. aureum* and *A. speciosum* often occur sympatrically but occupy different habitats with respect to light and salinity. To explore the phylogenetic relationships, origin and molecular mechanisms underlying adaptations on the genomic level, we sequenced the transcriptomes of *A. aureum* and *A. speciosum*, as well as a species in the sister genus, *Ceratopteris thalictroides*. A total of 47,517, 36,420 and 60,823 unigenes of the three ferns have been obtained, and 24.39% – 45.63% of unigenes were annotated using public databases. We reconstructed the phylogenetic tree and estimated divergence times based on the orthologous genes and chloroplast genes among six species, including published genomic data and sequences in NCBI. The calculations showed that the *Acrostichum* genus diverged from its closest relatives approximately 88.1 million years ago. However, the sister species *A. aureum* and *A. speciosum* suggested a very close relationship and diverged approximately 5.1 million years ago. The earliest fossil record was dated to the Maastrichtian in the late Cretaceous (66.0 – 72.1 million years ago), in which the aerenchyma tissue was found, and fossils of other coastal species had also been found in this region or nearby, indicating that *Acrostichum* originated in the late Cretaceous and might be one of the earliest members in mangrove ecosystem. As the two *Acrostichum* species have different preferences for intertidal habitats, it is interesting to explore the molecular mechanisms and detect whether some positively selected genes contribute their adaptations. By using two methods, the modified branch-site model and the K_h method, we identified 15-31 positively selected genes. Some of them, such as SKIP gene, NPH3 family protein, PBCP gene *etc.*, were related to stress resistance and responses to light stimulus, which may participate to their differential adaptations. Our study provides abundant transcriptome data and new insights into the evolution and adaptations of mangrove ferns in the inhospitable intertidal zone.

T2

A191

Biodiversity of bryophytes and their monitoring for metal pollution of karst bauxite area in the Guangxi, China

Zhaohui Zhang[1,2], *Zhihui Wang*[1]

1. Guizhou Normal University

2. Guizhou Key Laboratory for Mountainous Environmental Information and Ecological Protection

Karst bauxite was one of the most important aluminum mines

in China, which covered with limestone and dolomite. For the studying the biodiversity of bryophytes and their monitoring for metal pollution of karst bauxite in the southern China, a series of bryological surveys were carried out at karts area in Guangxi Autonomous Regions from 2014 to 2016. Following are some results: (1) About 753 samples and specimens of bryophytes were collected from five karst bauxite mines in Pingguo, Jingxi and Debao counties in Guangxi Autonomous Regions, China; (2) We found that the bryoflora consists of 112 species belong to 49 genera and 23 families. Among them, there are 6 species are liverworts belonging to 6 genera of 6 families, 106 species are mosses belonging to 43 genera of 17 families while no hornworts. (3) There were four life-forms of bryophytes in karst bauxite mines. They are turfs, wefts, mats and pendants. The turfs accounts for 72.56% of the species whereas the pendants are only 2.65%. (4) There were 10 phytogeogrphical elements of bryophytes in research areas. Among them, East Asia elements (20.59%) is the most, and East Asian and North Ameircan disjuncted elements elements (0.98%) is the least. (5) By means of applying ICP and Atomic Absorption Spectrometry and Atomic Fluorescence Spectrometry, the metal content of Al, Fe, Mn, Mg, Ca and Ti of mosses and their substrates were analyzed . Relationship between distribution bryophytes and bauxite pollution were assessed. It was concluded that the bryophytes can be used as indicator species in karst bauxite mines we studied area. This research was supported by the National Natural Science Foundation of China (NSFC. No.31160042;31360035).

T2

A192

Studies on biodiversity of mosses and their biogeochemistry from Carlin gold mines in China

Run Liu[1], *Zhaohui Zhang*[1,2], *Zhihui Wang*[1], *Dengfu Wang*[1], *Qimei Wu*[1]

1. Guizhou Normal University

2. Guizhou Key Laboratory for Mountainous Environmental Information and Ecological Protection

The Carlin type gold deposit is one of the most important industry ores in the karst area of Guizhou province, China.The Carlin-type Mines are characterized by very fine-grained gold and are associated with mercury, arsenic, antimony and thallium. Supported by the National Natural Science Foundation of China, the Science Fund for Distinguished Overseas Chinese Scholars of MHRSS and the Foundation of the Department of Science and Technology of Guizhou, following works were carried out. Based on a series of bryological investigation in our fieldwork, about 1026 specimens and samples were collected from nine carlin gold deposits in Guizhou. The results are showed as follow: 241 species in 80 genera of 25 families were found from the carlin gold mines. Some families such as Pottiaceae, Bryaceae, Hypnaceae, Brachytheciaceae and Thuidiaceae were the dominate group while *Barbula*, *Bryum*, *Brachythecium*, *Hydrogonium*, *Trichostomum* and *Hyophila* were the dominate genera. There were only four types life-forms of mosses in this areas, which were Short turfs, Wefts, Cushions and Pendants. Short turfs was the dominate moss life-form.There were fourteen moss phytogeogrphical elements were found in study region. Among them, North Temperate Zone elements, Tropical Asia elements, Eastern Asia elements and Endemic to China were the dominate elements. By using Atomic Absorption Spectrophotometer (AA-800) and Atomic Fluorescence Spectroscopy (AF-640), eight species of mosses with their soils were tested on the content of Au, Tl, Cu, Pb, Zn, Ca, Mg, As, Hg in them. The relationship between the distribution and biochemistry of mosses and the carlin gold mines were discussed. The results may have some certain directive function on prospecting for some valuable heavy metal deposits those are very difficult to be prospected by general methods.

T2

A193

Floristic characteristics of bryophytes and their biomonitoring for metal pollution environment in the karst bauxite mines in Guizhou province, China

Zhaohui Zhang[1,2], *Dengfu Wang*[1,2], *Qimei Wu*[1,2]

1. Guizhou Normal University

2. Guizhou Key Laboratory for Mountainous Environmental Information and Ecological Protection

Guizhou karst bauxites is one of the largest distribution regions of karst bauxites ores and bauxite mining industry in China. With over half of one century years history of karst bauxite mining activities and thousands of abandoned sites, bauxite pollution are bringing more and more serious environmental problems for local environment and society. Supported by the National Natural Science Foundation of China (NSFC. No.31160042;31360035), we carried a series of study about the bryological flora and its biomonitoring for metal pollution in karst bauxite of Guizhou. (1) From December 2013 to December 2016, we made a lot of comprehensive survey of the bryophytes in karst bauxite mines area of central Guizhou, China. Totally, 1,340 specimens and samples were collected: (2) 182 species belong to 39 genera and 21 families, of which 8 species are Hepaticae species belonging to 6 genera of 5 families, 1 species in 1 genus is Anthocerotae, 173 species are Musci belonging to 32 genera of 15 families. Among families, Pottiaceae, Brachytheciaceae, Hypnaceae, Polytrichaceae and Bryaceae were the dominant families. the dominant genera are *Polytrichum, Brachythecium, Bryum* and *Barbula* in the karst bauxite in Guizhou. (3)For reproductive systems, more than 21 gemmiferous species of bryophytes were recorded . (4) There were four types of life-forms of bryophytes in research area, they were Short turfs, Tall turfs, Wefts and Mats. The short turfs was the majority typed. (5) Through the Shannon-Wiener index, richness index and Simpson index, obtained trends in each karst bauxite. The results showed that, three index in different areas of bauxite were different changes. High canopy density, small human activities, habitat air was relatively humid, the bryophytes were relatively more, the biomass was large and grew well; and human activities, low vegetation coverage around, were the mainly reasons to reduced diversity of bryophytes. (6) With the method of ICP and AAS, eight metal elements (including Al, Ca, Fe, Ti, Mg, Cr, Cu, Ni, and Pb) of the bryophytes and their soil metal elements were studied. The correlation between the bryophyte and their environmental factors were Canonical correspondence analyzed. We suggest that the floristic characteristic, distribution patterns of bryophyte communities together with their biochemistry can be used as effectively tool to monitor some metal pollution around karst bauxite area in the Guizhou karst area.

T2

A194

Phylogeny of the climbing figs (*Ficus* subgen. *Synoecia*)

Zhen Zhang, Xiao-Mei Wang, Jian-Hang Zhang, ***Hong-Qing Li***
School of Life sciences, East China Normal University

With species diversification, reproductive system complexity and long-term coevolutionary background with fig wasps, *Ficus* (Moraceae) remains an unclear lineage in terms of its internal phylogenetic history. Berg restated subgen. *Synoecia* (including sect. *Rhizocladus* and sect. *Kissosycea*), which were split from Corner's subgen. *Ficus* according to their climbing habit mainly. To inspect the phylogenetic nature of these climbing fig trees in subgen. *Synoecia*, 68 taxa out of subgen. *Synoecia* and its allies (over 1/2 species in this group, and representing all the subsections), and additional 26 species from other subgenera to reconstruct phylogenetic relationship based on two nuclear sequences (internal transcribed spacer and external transcribed spacer). Phylogenetic analysis showed that neither subgen. *Synoecia* nor subgen. *Ficus* is monophyletic. However, the two subgenera together (except subsect. *Ficus*) compose a well-supported monophyly. In accordance with present infrageneric names, subdivision on the level of subsections is reorganized. The monophyly consists of two clades, i.e., 1) subsect. *Frutescentiae* (sect. *Ficus* subgen. *Ficus*) + subsect. *Plagiostigma* (sect. *Rhizocladus* subgen. *Synoecia*), and 2) sect. *Eriosycea* (subgen. *Ficus*) + the remains of subgen. *Synoecia*. In the first clade, subsect. *Plagiostigma* (as two separate lineages) is embedded in subsect. *Frutescentiae*. And in the second clade, a high support subclade consists of subsect. *Puntulifoliae*, subsect. *Trichocarpeae* and sect. *Kissosycea*; the other subclade includes sect. *Eriosycea* with subsect. *Pogonotrophe* (namely *F. laevis*) inserted. These results revealed that climbing habit has occurred at least 4 times and couldn't be regarded as a plesiomorphy, and the subgeneric system of *Ficus* needs further revision.

T2

A195

Molecular phylogeny of the *Ficus auriculata* complex (Moraceae)

Li-Fang Zhang, Zhen Zhang, Xiao-Mei Wang, Hai-Yan Gao, Huai-Zhen Tian, ***Hong-Qing Li***
School of Life sciences, East China Normal University

The closely related species in *Ficus* with continuous variation have been making taxonomists, co-evolutionary researchers and other related scientists confused. The boundary between species in the complex of *F. auriculata*, *F. oligodon*, *F. hainanensis*, *F. beipeiensis* and *F. variegata* is unclear for a long time. To clarify this problem, the nuclear loci ITS and *G3pdh* and chloroplast loci *trn*H-*psb*A, *trn*L-*trn*F, *trn*S-*trn*G and *psb*K-*psb*I as well as 15 pairs of SSR markers were applied to restructure the phylogeny and review their genetic background. Our results indicated that *F. variegata* should be an independent species of subsect. *Neomorphe*, sect. *Sycomorus*, which, together with *F. auriculata*, *F. oligodon*, *F. hainanensis* and *F. beipeiensis*, composes a monophyly. The last four species of this complex are of small genetic distances with indistinct differentiation, shared haplotypes and mainly overlapped geographic distribution, and should be treated as one single species.

T2

A196

Generic recircumscription of *Ptilagrostis* (Poaceae, Stipeae) as inferred by phylogenetic and morphological data

Zhongshuai Zhang[1]*, Shanwen Jiang*[1,2]*, Linglu Li*[1]*, Wenli Chen*[1]
1. State Key Laboratory of Systematic and Evolutionary Botany, Institute of Botany, The Chinese Academy of Sciences
2. University of the Chinese Academy of Sciences

The species of *Ptilagrostis* (Stipeae, Poaceae) are key elements in alpine grassland ecosystem and formed a diversification centre in Qinghai-Tibet Plateau. But generic delimitation and morphological evolution within it remain highly disputed. In this study we present a comprehensive phylogenetic study on *Ptilagrostis* and its most related genus in Stipeae. Eight chloroplast DNA fragments (representing more than 5,000 bps of the sequence) and nuclear ITS region of 46 species and 2 varieties from these genera were sequenced. The cytology, embryology, lemma epidermal micromorphology and macromorphology of these genera were studied extensively. Evidence from different disciplines coincided with each other. *Ptilagrostis* is recognized as a distinct genus which close to *Stipa* s.s. Its generic boundaries with *Stipa*, *Achnatherum*, *Piptatheropsis*, *Orthoraphium* and *Trikeraia* are redefined and the synapomorphies of each group have been given and discussed. Our study highlights the importance of molecular evidence and detailed re-examination of cytology, micromorphology and macroscopical traits that are previously neglected in defining monophyletic and natural genera.

T2

A197

Taxonomic revision of the genus *Didymodon* Hedw. (Pottiaceae, Bryophyta) in Mongolian Plateau.

Dong Ping Zhao[1]*, Tong Rui Zhang*[1]*, Jian Nan Wang*[1]*, Xue Liang Bai*[1]*, Tsogiin Tsegmed*[2]
1. School of Life Science, Inner Mongolia University, Hohhot 010021, China.
2. Institute of Botany, Academy of Science of Mongolia, Ulaanbaatar-51, Mongolia.

Mongolian Plateau is composed mainly of Inner Mongolia in China and Mongolia. The moss genus *Didymodon* Hedw. is revised for Mongolian Plateau and 28 taxa were found: *Didymodon acutus*, *D. asperifolius*, *D. australasiae*, *D. baii*, *D. constrictus* var. *constrictus*, *D. constrictus* var. *flexicuspis*, *D. cordatus*, *D. ditrichoides*, *D. fallax*, *D. ferrugineus*, *D. gaochenii*, *D. giganteus*, *D. glaucus*, *D. hedysariformis*, *D. icmadophilus*, *D. johansenii*, *D. leskeoides*, *D. mongolicus*, *D. perobtusus*, *D. rigidulus* var. *rigidulus*, *D. rigidulus* var. *subulatus*, *D. rivicola*, *D. rufidulus*, *D. subandreaeoides*, *D. tectorum*, *D. tophaceus*, *D. vinealis*, *D. zanderi*. Among these, *Didymodon baii* and *D. mongolicus* are described as a new species, *D. rigidulus* var. *subulatus* and *D. rufidulus* are new to Mongolia. *Didymodon cordatus* is new to China on the basis of comparison of Chinese material with that of Europe, previous reports being erroneous. *Didymodon anserinocapitatus*, *D. luridus*, *D. michiganensis*, *D. nigrescens* and *D. validus* are excluded from the bryoflora of Mongolian plateau. In the flora area, *Didymodon tectorum* is the most common species, and *D. australasiae*, *D. glaucus*, *D. leskeoides*, *D. rigidulus* var. *subulatus* and *D. subandreaeoides* are know only from Mongolia, while *D. baii*, *D. ditrichoides*, *D. fallax*, *D. giganteus* and *D. rivicola* are know only from Inner Mongolia of China. The key, descriptions, illustrations, distribution data, LM and SEM photographs are provided.

T2

A198

Taxonomy of Mielichhoferiaceae (Bryopsida) in China

Xiaorui Wang, *Shuo Shi, Min Li, Lin Li, Jiancheng Zhao*
Hebei Normal University

Mielichhoferiaceae Schimp. belonging to Bryales of Bryopsida, is a archaic group of bryophyte. Mielichhoferiaceae, with six genera (*Mielichhoferia* Nees & Hornsch, *Pohlia* Hedw., *Pseudopohlia* Williams, *Schizymenium* Harv., *Synthetodontium* Cardon, and *Epipterygium* Lindberg) and ca. 290 species, is a cosmopolitan family, and most species grow in mountain, plain, marsh and arid desert. Mielichhoferiaceae group were traditionally classified in the family Bryaceae. However, the recent phylogenetic researches clearly indicated that some species of Mielichhoferiaceae are closer to the family Mniaceae than Bryaceae. In this paper, morphological characters of more than 4,500 specimens of Mielichhoferiaceae and its related groups were analyzed. Phylogenetic studies using Maximum parsimony (MP), Maximum likehood (ML) and Bayesian inference (BI) analyses based on chloroplast (*rps4*, *trn*L-F, *atp*B-*rbc*L, *trn*G) and nuclear (ITS) sequences of 58 specimens for 44 species were presented. The molecular trees based on chloroplast sequences showed that the Mielichhoferiaceae group was nested within the Mniaceae, whereas the molecular trees based on nuclear sequences showed that the Mielichhoferiaceae group, including genera *Pohlia* and *Mielichhoferia*, composed a distinct lineage which was separated from Bryaceae and Mniaceae. In addition, the morphological features of Mielichhoferiaceae, such as plant size, leaf size and shape, type of leaf lamina cells, peristome configuration, *etc.*, are different from the Bryaceae and Mniaceae. Combining morphological and molecular data in the analyses, we drew a conclusion that Mielichhoferiaceae is a natural monophyletic group, and we transferred two Chinese species of *Mielichhoferia*, *M. himalayana* Mitt. and *M. sinensis* Dix. to genus *Haplodontium* (Bryaceae) as *H. himalayanum* (Mitt.) X. R. Wang et J. C. Zhao and *H. sinensis* (Dix.) X. R. Wang et J. C. Zhao separately. Currently, there are a total of five genera, *Mielichhoferia* (one species), *Pohlia* (28 species), *Pseudopohlia* (one species), *Synthetodontium* (two species) and *Epipterygium* (two species), 34 species of Mielichhoferiaceae in China. This project was financially supported by the National Natural Science Foundation of China (No. 31370237 & No. 31070184).

T2

A199

Plastid phylogeny and biogeography of the Paleotropical woody bamboos (Poaceae: Bambusoideae)

Meng-Yuan Zhou[1,3], *De-Zhu Li*[1,3], *Thomas Haevermans*[2]
1. Kunming Institute of Botany, Chinese Academy of Sciences
2. Muséum National d'Histoire Naturelle
3. University of Chinese Academy of Sciences

Paleotropical woody bamboos (PWB) are widely distributed in tropical and subtropical areas of Asia, Africa and Oceania, and they are phylogenetically and taxonomically intractable. Because of deficient sampling or lacking informative characters, phylogenetic relationships among the PWB remain incompletely resolved. Thus there are little biogeographical estimations about this group of bamboos because of ambiguous phylogeny and fossil records. This study presents the most extensively sampled phylogeny of the PWB with 18 plastid regions. Plastomes were sequenced and assembled for representative taxa in each major clade to further investigate unresolved relationships in the 18-region phylogeny. In addition, divergence time and ancestral ranges of the PWB were estimated. Results confirmed Melocanninae as the earliest diverging lineage from the rest of the group. Malagasy Hickeliinae (include *Nastus* s.str.) diverged after Melocanninae. Subtribe Bambusinae was phylogenetically heterogeneous and consisted of the *Dinochloa-Greslania-Mullerochloa-Neololeba-Sphaerobambos* (DGMNS) assemblage, *Temburongia* and the core Bambusinae clade. Malesian and Oceanian endemic *Nastus* s.l., DGMNS and Racemobambosinae formed a well-supported clade which was sister to *Temburongia*. The core Bambusinae might be redefined to include a basal grade, which contained *Kinabaluchloa*, *Holttumochloa*, *Bonia*, *Neomicrocalamus*, *Temochloa* and *Soejatmia*, and the *Bambusa-Dendrocalamus-Gigantochloa* (BDG) complex. The BDG complex was extremely diverse in morphology and was subdivided into six subclades according to topology, taxa composition, geographical distribution and network analysis. Within the Melocanninae clade, *Ochlandra*, *Davidsea* and *Neohouzeaua* were closely related to *Schizostachyum*, and the only two sympodial bamboos with long-necked rhizome, i.e. *Pseudostachyum* and *Melocanna*, had closer affinity in phylogeny. The PWB were estimated to have been originated during the late Oligocene, and their common ancestors were widely distributed across the Old World except for Africa. The extant geographical distribution patterns of major lineages were mostly evolutionary results of vicariance, within-area subset speciation and subsequent dispersal events. Diversification rate of the PWB was constant through time until ca. 5-7 Ma since when diversification rate increased rapidly. The rise of diversification rate is suspected to be due to the diversification rate shifts estimated to have occurred in the BDG complex. The intractable phylogenetic relationships within the BDG complex were potentially caused by a recently rapid radiation and frequent hybridizations among them as inferred from divergence time and diversification rate estimations of the PWB and previous studies.

T2

A200

Taxonomic status of genus *Elytrigia* Desv. (Poaceae: Triticeae) and biosystematic study on its species

Yonghong Zhou[1], *Haiqin Zhang*[1], *Yan Yang*[2], *Long Wang*[1]
1. Sichuan Agricultural University
2. China West Normal University

Elytrigia Desv. is one of the important perennial genera in the tribe Triticeae (Poaceae). Desvaux first established it based on *Triticum repense* L. in 1810 because of the strongly rhizimaous chracter, and the type species is *Elytrigia repens* (L.) Nevski. Since then, the strongly rhizimatous, long-anthered, and cross-pollinating species were included in this genus. Noway, many taxonomists consider *Elytrigia* as an independent genus. It includes about 40 species in the world, and is distributed in the subtropical and warm temperate countries of both hemispheres. Species in *Elytrigia* are excelent fodder plants, and some species are used in wheat breeding program through distant hybridization as the genetic germplasm resources for disease and drought resistance, and salt tolerance. *Elytrigia* is historically a very difficult and controversial genus in taxonomy. The definition of the genus and its precise taxonomic treatment are still under discussion. We carried out the taxonomic status of *Elytrigia* and biosystematic study on its species through morphological characters, geographic distribution, intergeneric and interspecific hybridization, fertility, cytogenetic characterization, genomic in situ hybridization (GISH) and molecular systmatic studies in nearly ten years. The main results are as follows: (1) *Elytrigia repens*, the type species, contains StStH genomes, and should be treated as *Elymus repens*. Therefore, *Elytrigia* is a invalid genus. (2) In *Elytrigia* sensu lato, the species with St genome should be included in *Pseudoroegneria*, the species with StH genomes should be trated as *Elymus*, StE species should be included in *Trichopyrum*, and EE genome species should be combined as *Lophopyrum*. (3) *Elytrigia repens* and *E. lolioides* contain StH genome, and should be treated as *Elymus*. *E. pungens* and *E. pycnantha* are with StEP genomes and should be included in *Pasmmopyrum*. *E. intermedia*, *E. geniculata* subsp. *pruinifera* and *E. geniculata* subsp. *scythica* contain StE genomes and should be treated as *Trichopyrum*. *E. pontica* contains EEEEE genomes and should be included in *Lophopyrum*. (4) *Elytrigia* s.l. contains St, E^e, E^b, H, and P basic genome. E^e genome comes from *Lophopyrum elongatum*, E^b from *Thinopyrum bessarabicum*, St from *Pseudoroegneria*, H from *Hordeum*, and P from *Agropyron*. Therefore, *Elytrigia* is an invalid genus. The species in it should be treated as species in *Elymus*, *Lophopyrum*, *Trichopyrum* and *Pasmmopyrum* respectively according their genomic constitutions.

T2
A201
Genome-wide analysis of the cellulose synthase gene superfamily in *Gossypium* cotton

Xianyan Zou

1. State Key Laboratory of Cotton Biology

2. Key Laboratory of biological and genetic breeding of cotton, The Ministry of Agriculture; Institute of Cotton Research, Chinese Academy of Agricultural Sciences

The cellulose synthase superfamily including the cellulose synthase (Ces) and cellulose synthase-like (Csl) families is involved in the synthesis of cellulose and hemicellulose. In *Gossypium* cotton this family plays a decisive role in fibre development. Using a series of bioinformatic methods, 210 *Gossypium* cotton *Ces/Csl* genes were identified and grouped into 10 subfamilies by phylogenetic analysis. Subsequently, gene structure, chromosomal location, and syntenic analyses were performed. Selection pressure ranged from 0.039 to 0.195 among the gene subfamilies indicating that the *Ces/Csl* genes had undergone purifying selection. The expression patterns of the tetraploid *CesA* genes in transcriptomic and qRT-PCR analyses revealed different expression among genes in different *CesA8* subclades. Further analysis showed differences in promoter *cis*-elements and cellulose_synt domains (PF03552) among these genes that may explain their differential expression. This study provides novel insights into the evolutionary relationships, domain structures, and expression patterns of the *Ces/Csl* gene family of *Gossypium* cotton.

T2
A202
Flow cytometric measurements reveal endopolyploidization phenomenon of *Phalaenopsis*

Zuo Li[1,2], *Zhongjian Chen*[3], *Fubing Lv*[1,2], *Wenfang Xiao*[1,2], *Heming Chen*[1,2]

1. Environmental Horticulture Research Institute of Guangdong Academy of Agricultural Sciences

2. Guangdong Provincial Key Laboratory of Ornamental Plant Germplasm Innovation and Utilization

3. Agro-biological Gene Research Center of Guangdong Academy of Agricultu

Orchidaceae species are well-known ornamental plants all over the world because of their charming flowers and various flower characteristics. Inter-specific and inter-generic hybridization within and between species of the *Orchidaceae* has increase the genetic diversity substantially. In orchid family, as popular potted flower plants, *Phalaenopsis* species have attracted a great deal of attention worldwide, regarding their long-lived and vivid flowers and various polyploidy patterns. Polyploidy has been recognized as one of the most important factors promoting speciation in the plant species, numerous genera including polyploid species complexes are well known for their high morphological variability. The endopolyploidization as a universal phenomenon existing in the higher plant, which on plant cells maintained a faster growth rate, is an important way of increasing copies of function genes. However, the understanding of polyploidy and endopolyploidization in *Phalaenopsis* genus remains limited and few research have been done. In this study, *Phalaenopsis* species were used to determine the endopolyploidization level in five kinds of tissues, including leaf tip, root tip, flower bud, blossom flower and peduncle, respectively, using the technique of flow cytometry (FCM). Meanwhile by using conventionl tabletting method, the chromosomes number of those *Phalaenopsis* samples were analyzed to help verify the FCM investigated results. The results showed that in *Phalaenopsis* the feature of endopolyploidization was specific for tissues, moreover the endopolyploidization level varied in different *Phalaenopsis* species of the same genus, even those samples owed the same chromosomes numbers. The research on the feature of plant endopolyploidization phenomenon would be helpful for regulating efficiently differentiation of organs and accelerate the formation of valuable organs. Although our research was limited to *Phalaenopsis*, the need to find a suitable tissue for FCM detection to all endopolyploid species.

T2

A506

Molecular phylogeny of eastern Sumatran species of *Mangifera* L. (Anacardiaceae): Taxonomic implication

Ibna Hayati, *Fitmawati Fitmawati, Nery Sofiyanti*
Universitas Riau

The phylogenetic relationships of the genus *Mangifera* with emphasis on eastern Sumatran species were estimated using sequence data from internal transcribed spacer (ITS) regions of nrDNA. The separation of the two traditionally accepted subgenera was not well supported. The monophyly of each section of *Deciduae, Perennes, Marchandora, Euantherae, Rawa,* and *Mangifera* within genus remained unclear. To clarify the phylogenetic relationship, we examined four different methods of phylogenetic tree reconstruction with the addition of outgroups and sequences of *Mangifera* from the NCBI GenBank as representative species of each section. The resulting tree topologies were mostly consistent with one another. Judging from the phylogenetic tree some of section should be fused. We also provide the molecular description for the common mango species found in Sumatra and create comparison among them. The result was support the resurrection of species *Mangifera sumatrana* Miq. Nucleotide variation among *Mangifera* species should be interpreted as modification of environment. Molecular data support placement of species within the genus.

T2

A510

New Taxonomic Evidence for *Mangifera laurina* Bl. (Anacardiaceae)

Fitmawati Fitmawati[1], *Alex Hartana*[2], *Mien Achmad Rifai*[4], *Bambang S Purwoko*[3], *Ibna Hayati*[1]
1. Biology Department, Universitas Riau
2. Department of Biology, Bogor Agricultural University
3. Departement of agronomy, Bogor Agricultural University
4. Lembaga Biologi Nasional

The relationship between *M. laurina* Bl. and *M. indica* L. were remain unclear. *M. laurina* Bl. is closely related to *M. indica* L. They shared similar leaves. In flower, *M. indica* could be distinguished by its flower in glomerulous and being densely pubescent while in *M. laurina* it has lax, non-glomerulate, almost glabrous inflorescence. Previous study also suggested *M. laurina* as synonymy of *M. indica* whereas *M. laurina* is an accepted name recently. Therefore we provide new evidence to clear *M. laurina* based on morphology, anatomy and molecular data using cpDNA such as *trn*L-F Intergenic Spacer and *rbc*L sequence and also nuclear ribosomal DNA of ITS Sequence and E-RAPD marker. The phylogenetic tree were also obtained from the combination between morphology and E-RAPD marker and as well as combination of cpDNA and nrDNA sequences. The results showed *M. laurina* Bl. as independent species. The result also suggested *M. laurina* as the common ancestor of *M. indica* L. Here we supported the idea about strict separation of *M. laurina* Bl. from *M. indica* L. based on these new taxonomic evidence.

T2

A511

A Jurassic wood providing insights into the earliest step in *Ginkgo* wood evolution

Zikun Jiang[1], *Yongdong Wang*[2], *Marc Philippe*[3], *Wu Zhang*[4], *Ning Tian*[5], *Shaolin Zheng*[4]
1. Chinese Academy of Geological Sciences
2. Nanjing Institute of Geology and Palaeontology, Chinese Academy of Sciences
3. UMR 5023 of the CNRS and Université Lyon 1
4. Shenyang Institute of Geology and Mineral Resources
5. College of Palaeontology, Shenyang Normal University

The fossil record of Ginkgo leaf and reproductive organs has been well dated to the Mid-Jurassic (170 Myr). However, the fossil wood record that can safely be assigned to Ginkgoales has not yet been reported from strata predating the late Early Cretaceous (ca. 100 Myr). Here, we report a new fossil wood from the Mid-Late Jurassic transition deposit (153–165 Myr) of northeastern China. The new fossil wood specimen displays several Ginkgo features, including inflated axial parenchyma and intrusive tracheid tips. The transition from early wood to late wood is gradual, with an intermediate type occupying most of the ring. The axial cells in cross-section are irregular in size and shape, and the tracheids are mostly quadrate to elliptical in the early wood, and more regularly narrowly rectangular in the late wood. Intercellular spaces are often distinct, and the tracheid wall is thick, even in the early wood, suggesting a compression wood type. The tracheid walls are destroyed in some fungi infected areas. In a tangential section, xylem rays are homogeneous, relatively low, (1) 2–4 (15) cells high and are often associated with inflated axial parenchyma. The tracheid tips are often contiguous to the ray margin or associated to inflated axial parenchyma. In the radial section, some tracheid bunches are present, with storied tips bent alongside wood rays, and the tips sometimes overlap one another. The tracheid radial pits are mostly uniseriate, round and distant, sometimes contiguous and somewhat flattened, and locally biseriate. The pattern of biseriate pits is variable, from alternate crowded to sub-opposite crowded or opposite distant, sometimes with Sanio's rims. Ray cells in each crossing field yield 4–6 tracheids in the early wood, and ray cell transverse walls are thin and unpitted, often resiniferous; cross-fields with (1) 2–4 cupressoid oculipores, with an oblique aperture. Oculipores are ordered in columns and lines. The axial parenchyma are abundant and well distributed within the ring, with possible crystalliferous chambers. Trabeculae are locally preserved, whereas resin canals are not observed. Because it is only slightly younger than the oldest recorded Ginkgo reproductive organs (the Yima Formation, 170 Myr), this fossil wood very probably represents the oldest bona fide fossil Ginkgo wood and the missing ancestral form of Ginkgo wood evolution.

T2

A523

An evolutionary continuity principle for evolutionary system of organism divisions

Da-Li Fu[1], *Hao Fu*[2]
1. Chinese Academy of Forestry
2. Central South University of Forestry and Technology

Evolutionary research is guided by the scientific basis of organism evolution. To overcome the limitations of being partial and subjective in the tree of life or phylogenetic system, the authors proposed a basic principle of organism evolution, evolutionary continuity principle. Based on this principle, the evolutions can be divided into two basic forms: vertical evolution and horizontal evolution. The vertical evolution is that evolution of the structures and features of organisms from non-existent to entirety, from simple to complex, or from primitive to advanced. The evolutionary course of vertical evolution can be divided into two or three significantly different phases, such as plants: the non-vascular → the vascular, non-seed → Gymnospermophyta D. L. Fu & H. Fu, phyl. nov. → Fructophyta D. L. Fu & H. Fu, phyl. nov., and the animals: Proenteratozoa D. L. Fu & H. Fu, phyl. nov. → Coelenteratozoa → Euenterata, non-chordata and non-gnatha → Chorda-to-gnathozoa D. L. Fu & H. Fu, phyl. nov. → Vertebrognathozoa D. L. Fu & H. Fu, phyl. nov. The horizontal evolution is that evolution of structures and features from loose to compact, from inefficient to efficient, or from primitive to advanced, which can also cause significant organism changes, such as Nudembryophyta D. L. Fu (Plantae) → Proenteratozoa D. L. Fu & H. Fu, phyl. nov. (Animalia), Bacterophyla D. L. Fu & H. Fu, phyl. nov. → Acytophyla D. L. Fu. Based on the continuity principle, the new evolutionary taxonomical system of divisions was established and the evolutionary diagram was drawn. The new system includes 20 divisions in three kingdoms, and 11 of those divisions are new. These are: I. Regnum Microbia D. L. Fu: 1.Cyanoalgophyla D. L. Fu & H. Fu, phyl. nov., 2. Bacterophyla D. L. Fu & H. Fu, phyl. nov., 3. Acytophyla D. L. Fu, 4. Monoalgophyla D. L. Fu & H. Fu, phyl. nov., 5. Monomycophyla D. L. Fu & H. Fu, phyl. nov., 6. Eualgophyla D. L. Fu & H. Fu, phyl. nov., 7. Fungophyla D. L. Fu & H. Fu, phyl. nov.; II. Regnum Plantae Haeck.: 8. Nudembryophyta D. L. Fu, 9. Bryophyta, 10. Pteridophyta, 11. Gymnospermophyta D. L. Fu & H. Fu, phyl. nov., 12. Fructophyta D. L. Fu & H. Fu, phyl. nov.; III. Regnum Animalia L.: 13. Proenteratozoa D. L. Fu & H. Fu, phyl. nov., 14. Coelenteratozoa, 15. Nematodozoa, 16. Annelidozoa, 17. Arthropodozoa, 18. Echinodermatozoa, 19. Chorda-to-gnathozoa D. L. Fu & H. Fu, phyl. nov., 20. Vertebrognathozoa. D. L. Fu & H. Fu, phyl. nov. The new evolutionary theory, new evolutionary taxa and new evolutionary system can provide scientific bases for Evolutionomy, a new science of organism evolution.

T2

A527

Study on pollen morphology of Asteraceae family at Universitas Indonesia, Depok

Rachmi Luthfikasari, *Salamah Andi*
Universitas Indonesia

Research about study on pollen morphology of Asteraceae family at Campus Universitas Indonesia, Depok conducted during October until December 2015. Eighteen species from 18 genus and 9 tribes was found and used in the research. Pollen of 14 species from 8 tribe has successfully isolated by acetolysis method. Pollen was categorized by some characters such as polarity, symmetry, pollen grains compound, shape, surface, aperture, pore shape, and size. The result from pollen analysis shows that Asteraceae has pollen with isopolar polarity, radial isopolar symmetry, monad grain, spheroidal shape on polar and equator, echinate surface, tricolporate, tetracolporate, and fenestrate aperture, circular and rectangular pore, and has medium until big size pollen. The polarity, symmetry, pollen grains compound, shape, and pollen surface characters can be used as differentiating features between Asteraceae and other families. Aperture character can be a differentiating feature between tribe in one family. Pore shape and size of pollen could not be a character to differentiate between tribes, but maybe can be used to differentiate the taxa below the tribes. Results of this study can become a reference on the regrouping of taxa especially in Asteraceae family.

T2

A529

Impact of horizontal gene transfer in the early evolution of plants.

Qia Wang[1], *Hang Sun*[1], *Jinling Huang*[1,2]
1. Kunming Institute of Botany, Chinese Academy of Sciences
2. East Carolina University

Horizontal gene transfer (HGT) refers to the movement of genetic information across normal mating barriers between more or less distantly related organisms. HGT allows recipient organisms to circumvent the slow process of ab initio gene creation and thereby accelerates genome innovation and evolution. In order to further understand the scale and evolutionary significance of HGT in multicellular eukaryotes, particularly in Plantae, we detected horizontally acquired genes in the genome of *Selaginella moellendorffii* and *Klebsormidium flaccidum* based on genome screening and phylogenetic analyses. Our study indicate that multiple gene families in *Selaginella moellendorffii* and *Klebsormidium flaccidum* were horizontally acquired from bacteria, archaea, viruses or fungi. Some of these acquired genes are involved in important processes of plant development and adaptation, such as Carbohydrate metabolism, hormone signaling, and response to stresses. These findings suggest that HGT might have facilitated the adaptation of plants to terrestrial environments and the expansion of their ecological niches.

T2

A530

Progress in study on early seed plants from China

Deming Wang
Peking University

The earliest seed plants have been mainly recorded from the Upper Devonian of Euramerica. South China is one of the centers for evolutionary radiation of early vascular plants but has lacked evidence in Devonian seeds. Our recent years' work in South China results in some knowledge of earliest seed plants: 1) We found the earliest seed in China and East Asia-*Cosmosperma*, thus expanding the geographic distribution of early seed plants. 2) *Latisemenia* bears cupulate ovules that are terminal as well as opposite on the fertile axis. It is the earliest plant with lateral ovules and may foreshadow the diverse ovule arrangements found among younger seed plant lineages. Following the telome theory, *Latisemenia* demonstrates derived features in both ovules and cupules. 3) Famennian synangia, especially *Placotheca*, do not support the current understanding that the earliest pollen organs closely resembled the fructifications of the ancestral progymnosperms. *Placotheca* is highly derived and indicates earlier diversification of pollen organs than previously expected. In contrast, the prepollen remained primitively spore-like. 4) Beyond Euramerica, we found the well preserved anatomy of earliest seed plants-*Yiduxylon*. This taxon suggests the transitional features between the early seed plants and their ancestral aneurophyte progymnosperms.

T2

A535

Hypoxis L. (Hypoxidaceae): Molecular systematics of the genus and related genera

Sanele Shiba, *Michelle Van Der Bank*
University of Johannesburg

The family Hypoxidaceae within the 'asteloid' clade of the monophyletic order Asparagales is represented by 10 genera. Of these, four are endemic to southern Africa and remain poorly studied, lacking molecular data. This is especially true for *Hypoxis* the largest genus in the family with about 90 species. The popularity of *Hypoxis* as an alternative remedy has resulted in unsustainable harvesting practices of rhizomes from the wild. This exploitation of target and related species has accelerated the need for correct species names. However, species delimitation in *Hypoxis* is problematic and despite several attempts, the systematics of the genus remains largely unresolved. This is mainly due to the lack of distinct morphological boundaries separating species. Therefore, in this study phylogenetic relationships within Hypoxidaceae, with an emphasis on *Hypoxis*, were reconstructed using four plastid DNA regions (*rbc*L*a*, *trn*L-F, *ycf1*, and *trn*S-G) for 90 taxa. Data were analysed using Maximum Parsimony and Bayesian methods. Findings from our study indicate that *Hypoxis* is not monophyletic and is represented by at least three lineages scattered throughout the family. Preliminary results of the phylogenetic analyses will be presented.

T2

A538

Comparative analysis of transcriptomes in Rhizophoraceae provides insights into the origin and adaptive evolution of mangrove plants in intertidal environments

Wuxia Guo[1], *Haidan Wu*[1], *Zhang Zhang*[1], *Chao Yang*[2], *Ling Hu*[1], *Xianggang Shi*[1], *Shuguang Jian*[3], *Suhua Shi*[1], *Yelin Huang*[1]
1. School of Life Sciences, Sun Yat-sen University
2. Department of Neurosurgery, First Affiliated Hospital of Sun Yat-sen University
3. South China Botanical Garden, Chinese Academy of Sciences

Mangroves are woody plants that grow in tropical and subtropical intertidal zones, where they exist in extreme environments of high salinity, extreme tides, strong winds, high temperatures, and muddy, anaerobic soils. As a key mangrove family, Rhizophoraceae has developed several morphological and physiological adaptations to these extreme conditions, representing an ideal system for the study of the origin and adaptive evolution of mangrove plants. In this study, we characterized and comprehensively compared the transcriptomes of four mangrove species, from all four mangrove genera, and a terrestrial relative in Rhizophoraceae, using RNA-Seq. We obtained 41,936–48,845 unigenes with 61.42–69.48% annotated for the five species in Rhizophoraceae. Orthology annotations of Gene Ontology, Kyoto Encyclopedia of Genes and Genomes, and Clusters of Orthologous Groups revealed overall similarities in the transcriptome profiles among the five species, whereas enrichment analysis identified remarkable genomic characteristics (e.g., overrepresentation of genes related to stress response and embryo development) that are conserved across the four mangrove species compared with their terrestrial relative, probably reflecting mangrove-specific gene components that are essential for their adaptation. Based on 1,816 identified orthologs, phylogeny analysis and divergence time estimation revealed a single origin for mangrove species in Rhizophoraceae, which diverged from the terrestrial lineage approximately 56.4 million years ago (Mya), suggesting that the transgression during the Paleocene–Eocene Thermal Maximum may have been responsible for the entry of the mangrove lineage of Rhizophoraceae into intertidal environments. Evidence showed that the ancestor of Rhizophoraceae may have experienced a whole genome duplication event approximately 74.6 Mya, which may have increased the adaptability and survival chances of Rhizophoraceae during the Cretaceous–Tertiary extinction through providing extra genomic materials for selection. The analysis of positive selection identified 10 positively selected genes from the ancestor branch of Rhizophoraceae mangroves, which were mainly associated with stress response, embryo development and regulation of gene expression. Positive selection of these genes may be crucial for increasing the capability of stress tolerance (i.e., defense against salt and oxidative stress) and development of adaptive traits (i.e., vivipary) of Rhizophoraceae mangroves, and thus plays an important role in their adaptation to the stressful intertidal environments.

T2

A543

Hybridisation and polyploid evolution in whitebeams

Rosalia Piñeiro, *Tim Rich, Mike Fay*
Royal Botanic Gardens, Kew

In this project we investigate how polyploid evolution occurs in nature. This topic is understudied due to the limitations of Sanger technology. However, the emergence of Next Generation Sequencing (NGS) opens exciting opportunities to study complex genomes in non-model plants. British whitebeams (*Sorbus aria* aggr., Rosaceae) are an emblematic case study of ongoing polyploid evolution in natural tree populations. SW Britain is a worldwide 'polyploid hotspot' for the genus. Only in this area, thirty highly endemic polyploid taxa occur. Three widespread species (the sexual diploid *S. aria* and the polyploid species *S. rupicola*, and *S. porrigentiformis*) have been postulated as the potential parents of most of these high endemics. Using cutting-edge DNA technology (RADseq makers and 25 NGS-sequenced nuclear microsatellites) and ploidy level measurements we have estimated the genetic diversity, compared populations sympatric and allopatric to the polyploid zone, and traced the origin of 500 trees across Britain. Conservation policies have traditionally focused on the preservation of the local endemic polyploids. However, a deep understanding of the evolution of the complex may allow preserving the polyploidisation process by protecting the parental species, even if they are widely distributed.

T2

A545

Evolutionary history of *Vitis* based on genotyping-by-sequencing data

Laura Klein

Saint Louis University

As concerns mount about sustainable agriculture in a changing climate, attention is focusing on understanding diversity and diversification in crop wild relatives. The domesticated European grapevine (*Vitis vinifera*) is among the most economically important berry crops in the world and is becoming an increasingly valuable component of rural agricultural development. In many areas of North America and around the world, grapevine cultivation includes not only *V. vinifera* but also abiotic and biotic stress resistant native North American *Vitis* species, which, aside from being grown outright, contribute to hybrid scions and rootstocks. Despite this, relatively little is known about geographic patterns of variation among natural *Vitis* populations or the evolutionary processes shaping them. The evolutionary history of the grapevine genus has been notoriously challenging, in part due to likely rampant hybridization of sympatric species. Within subgenus *Vitis*, a large North Temperate group that includes ~50 *Vitis* taxa (~18 in North America), it appears that most species are interfertile. In North America it is common for two, three, or up to five *Vitis* species to grow in sympatry, further increasing the likelihood of interspecific hybridization. In this study, we leveraged living germplasm collections maintained by the United States Department of Agriculture Agricultural Research Service (USDA-ARS) Grape Genetics Research Unit (Geneva, NY) and National Clonal Germplasm Repository (Davis, CA) to investigate relationships among *Vitis* species. Single nucleotide polymorphisms (SNP) were identified using genotyping-by-sequencing. After sequencing and filtering we generated a dataset of 358 individuals representing 17 *Vitis* species (12 from North America, the European *V. vinifera*, and four Asian species) and three *Ampelopsis* species. Individuals were genotyped at approximately 11,000 sites with a minimum mean depth of coverage of 10. To infer phylogeny, we used the Geneious Tree Builder software and the RaxML plugin with the GTRCAT model and rapid bootstrapping. Preliminary results provide support for previously recognized groups including 1) monophyletic *Vitis*; 2) monophyletic subgenus *Vitis*; 3) within subgenus *Vitis*, a monophyletic group of Eurasian species. In addition, these data suggest close relationships between *V. acerifolia, V. arizonica, V. riparia* and *V. rupestris*, and close relationships between *V. cinerea, V. palmata,* and *V. vulpina*. In some cases, sampled individuals from a single species did not cluster together in these analyses. This could represent misidentification within the USDA system; alternatively (or in addition), it is likely that interspecific hybridization is contributing to reticulate patterns of evolution within the genus *Vitis*. Future areas of work should include detailed population level sampling, and morphological and genetic analyses of *Vitis* populations where multiple species occur in sympatry as well as in regions where each of the species occur in isolation.

T3

A203

Effect of habitat heterogeneity on reproductive characters in distylous species *Primula nivalis*

Aysajan Abdusalam[1,2], *Qingjun Li*[1]

1. Key Laboratory of Tropical Forest Ecology, Xishuangbanna Tropical Botanical Garden, Chinese Academy of Sciences, Menglun Town, China

2. Key Laboratory of Ecology and Biological Resources in Yarkand Oasis, College of Biology and Geography Sciences, Kashgar University, Kasghar, China

Distylous plant is a special group of angiosperm which has reciprocal stigma and anthers high within a species. Comparative investigations of breeding system of distylous species on different elevation can be revealing the relationship between distylous plant evolution and their environments. *Primula nivalis* is a native primrose in Xinjiang, NW China, it has significant differences in plant size and seed set ratio in the populations at different elevation. In order to investigate the comparative effect of habitat heterogeneity on frequency of morphs in different populations, morphological characters of flowers, pollination characteristics, mating patterns, offspring fitness and inbreeding depression of plants were quantified at different elevation habitats in this research. The study results will clarify the reproductive strategies and adaptation mechanisms of *P. nivalis* at different elevational environments. The investigations on effect of heterogeneous habitats on *P. nivals* floral phenotype suggested that frequency of LS flower in high elevation grassland population was higher than SS flower. The floral morphological characters of LS flower among populations in different ecological conditions were significant different. The length of corolla tube and stamen, and stamen-stigma spatial distance of LS flowers were different in high elevation grassland population. This revealed that variation of LS floral characters at different population decreased the relative herkogamy level of both morph flowers sexual organs in high elevation grassland populations. Floral traits (corolla tube length, anther-stigma separation) and pollinator activities were significantly correlated with altitude both in the LS and SS individuals. Efficiency of long tongue pollinators was higher than that of short tongue pollinators, and pollination efficiency on the SS flower was lower than that on the LS. The selfing rate of the LS flower was higher than the SS flower, but pollen limitation of the LS flower was lower than that of the SS flower. Selfing and pollen limitation were correlated with anther-stigma separation, suggesting that elevation changes in stigma-anther separation evolved before self-compatibility of LS flower. Fruit set of SS flower was lower than that of LS morph at different elevation populations. Seeds from cross-pollinated flowers had higher quality than those from self-pollinated and natural pollinated flowers. *P. nivalis* seeds have a certain degree of physiological dormancy. Temperature and light conditions significantly influence the germination of seeds. Seeds from LS have higher germination rates than those of SS morphs on three different treatments. Fitness of seeds (seed germination) from cross-pollinated flowers was higher than that of seeds from natural and self-pollinated flowers of both morphs. Fitness for seeds from LS morphs was higher than that of seeds from SS in each elevation. Seed germination level inbreeding depression was increased for SS flower with the increase of elevation, but it did not differ significantly for LS flowers with increased elevation.

T3

A204

Structural and dynamic characteristics of *Eremurus anisopterus* populations in areas with different levels of disturbance

Jing An, *Ling Wu, Shaoming Wang, Xia Zhang, Chaoliang Jiang, Xiang Zhao, Zhiyong Zhang*
Shihezi University

Eremurus anisopterus (Liliaceae) is a perennial ephemeral geophyte herb. In China, *E. anisopterus* grows on fixed and semi-fixed sand dunes of the Gurbantunggut Desert of Xinjiang Uyghur Autonomous Region. This kind of ephemeral plant contributes greatly to the stabilization of sand. But increasing human activity has fragmented the primary habitat of the *E. anisopterus* population in the southern Gurbantunggut Desert, resulting in numerous separate population patches. And the frequent human activities, including engineering, reclamation of waste land, industrial discharge and over grazing, severely influence the survival and reproduction of ephemeral plants, more and more ephemeral plants are dying out. The goal of our study was to determine the current survival rates of *E. anisopterus* populations in different patches. Therefore, we surveyed 19 plots and analyzed age structure, constructed a time-specific life table, produced survivorship and mortality curves, as well as survival analysis function curves. Our results indicated that because of the various levels of human disturbance and habitat fragmentation, population dynamics of *E. anisopterus* in the 19 plots exhibited different structural features and trends. The age structure of *E. anisopterus* populations differed among the plots. The age classes within the populations were incomplete, and this phenomenon occurred more frequently in the plots in highly fragmented areas. The age structure of the b-type and c-type patches, based on the degree of fragmentation and human disturbance, exhibited a stable to declining growth trend. Population structure of the a-type patches in the most disturbed areas exhibited greater instability, with the population at higher risk of decline. The a-type population exhibited stability in the early and populations, declines in the early phase and stability in the intermediate and later phases occurred, whereas the c-type populations were stable. Therefore, it is critical to decrease or halt external disturbances in areas subjected to the greatest levels of human intermediate phases according to the survivorship curve, and in the early, intermediate, and later phases according to the survival analysis function curve. The plant population structure is the common result of the survivability of individuals and the impact of environmental conditions, and may also provide important information on the past and present regeneration of species. So the decline in *E. Anisopterus* populations likely resulted from habitat loss in the later phase. For the b-type disturbance. Studies should examine the effects of different combinations of disturbance factors on population status in different habitats to formulate scientific and practical protection and recovery strategies.

T3

A205

On-line determination of ammonia concentration and deposition at three research platforms with contrasting landscapes

Yunhua Chang[1], *Jianlin Shen*[2], *Yanlin Zhang*[1]
1. Nanjing University of Information Science and Technology
2. Institute of Subtropical Agriculture, Chinese Academy of Sciences

Ammonia (NH_3) is among the most important species in terms of forming secondary inorganic aerosols and enhancing environmental nitrogen input. High time-resolution monitoring of ambient ammonia and landscape-specified consideration of deposition parameters are in urgent need of capturing the sources and processes governing short-term fluctuations of ammonia concentrations, and accurately assessing the magnitude of ammonia deposition and its detrimental effects on various ecosystems. To this end, hourly continuous observation of ammonia and a full suit of auxiliary factors were simultaneously performed at three regional research platforms, covering urban (Pudong or PD site), rural (Qingpu site), and natural (Dongping National Forest Park in Chongming island or CM site) settings in Eastern China since 2014. This ongoing project is part of the Networking Evaluation of Ammonia in China (NEAChina) program, aiming to systematically elucidate the sources, transport, and fate of atmospheric ammonia across China, with a particular interest in non-agricultural ammonia emissions over the city clusters of the Yangtze River delta region. Here we reported our on-line measurement of ammonia concentrations at these three locations with contrasting landscapes between April 2014 and May 2016 and calculated ammonia deposition fluxes using a well-tested bi-directional ammonia exchange model with empirical parameters.

T3

A207

A new gene *B1L* regulates cold tolerance and delays flowering time in *Arabidopsis*

Tao Chen
Lanzhou University, Lanzhou, China

The process from vegetative growth to reproductive growth is pretty crucial for plants normal development and progeny reproduction. Instead of moving away like animals, plants suffer fluctuant ambient stimuli and change their development process to adapt different environmental conditions. In *Arabidopsis*, we identified *B1L*, a novel gene, was involved in cold stress pathway and flowering determination. *BIL* T-DNA inserted mutant (*b1l*) was sensitive to freezing treatment, and B1L-overexpressing transgenic plants could increase the tolerance. *b1l* plants also had an early flowering phenotype, which was restored by introducing into *B1L* sequence with its own promoter, and *B1L*-overexpressing transgenic lines delayed flowering time. It means that *B1L* positively regulates freezing tolerance and delays flowering time. To elucidate the mechanism underlying the cold response of *B1L*, we measured the expression pattern of *B1L* under low temperature. The expression of *B1L* was initially (after 1 hour) induced after treated with cold stress (4°C). GUS staining of *B1L::GUS* transgenic plants were turned more deepened after cold treatment

(4°C, 6 h), which is consistent with Q-PCR result. We tested the expression of C-repeat (CRT)-binding factors (CBFs) pathway genes which play important roles in cold stress, and found that the expression of these genes were reduced in *B1L* mutants, but increased in *B1L*-overexpressing transgenic plants. To explore how *B1L* functions in the cold resistance and flowering determination, we used yeast two hybrid system (Y2H) to screen for *B1L*-interacting proteins, and used BiFC to confirm it. GRF is one of the proteins that interacts with *B1L*, and belongs to a protein family related with many biological processes, such as plant development, biotic stress and abiotic stress. As GRF proteins usually bind to phosphorylated proteins, we then tested whether *B1L* had any phosphorylated sites that affected the interaction. A potential phosphorylated site in serine 213 was identified by bioinformatics analysis and was mutated to alanine ($B1L^{S213A}$) and phenylalanine ($B1L^{S213D}$), respectively, to mimic the nonphorylation and autophosphorylation of *B1L*. $B1L^{S213A}$ did not abolish GRF binding, but the interaction of $B1L^{S213D}$ and GRF was much stronger than any other groups. Also, we found that N-terminal of *B1L* which contains serine 213 was enough for the interaction with GRF. It reveals that Serine 213 in N-terminal of *B1L* may plays an important role for the interaction between *B1L* and GRF. *GRF/B1L* double mutations restored early-flowering of *B1L*, but not cold sensitivity, which demonstrates that the function of *B1L* in regulating flowering time requires GRF. All these data indicate that *B1L* participates in plants freezing resistance by regulates the expression of cold-regulated genes in the CBF pathway and also represses flowering in *Arabidopsis thaliana*, and GRF plays an important role for *B1L* function.

T3

A208

Community structure and species diversity of *Loropetalum subcordatum* in Maolan National Nature Reserve

Xiaoxi Chen, *Jiuxiang Huang*
South China Agricultural University

In order to provide a scientific basis to protection, ex-situ conservation and near natural-forest construction for the natural forest community of *Loropetalum subcordatum* which was rare and endangered. Community structure and species diversity of *Loropetalum subcordatum* in Maolan National Nature Reserve, Guizhou were studied by plot method. There are 142 species belongs to 60 families and 104 genera in the 4 plots with an area of 400 m^2. The dominant families were Lauraceae, Theaceae, Moraceae, Euphorbiaceae and Rubiaceae. Among them, the tropical distribution elements at genus level for 72%, and then 27% for temperate distribution elements. Species diversity results showed that there was the highest species richness in shrub layer, and the lowest one in herbal layer; Species diversity and evenness were highest in shrub layer, and the lowest one in tree layer; *Loropetalum subcordatum* population in Maolan National Nature Reserve was lack of seedling and the natural regeneration was difficult, indicating that it was a declining population. The age structure states clearly that the amount of *Loropetalum subcordatum* seedling is scarce, *Loropetalum subcordatum* may be replace by other species and became extinct; Which is urgently required for the consercation and utilization by strengthening in situ conservation, population reconstruction, propaganda and education.

T3

A209

Preference for ammonium and nitrate by four plant species during different growth stages

Jihan Cui, *Shunguo Li, Xueyan Xia, Yu Zhao, Meng Liu, Shijia Song*
Institude of Millet Crops, Hebei Academy of Agricultrual and Forestry Sciences

Uptake of ammonium and nitrate by plants at different growth stages was investigated by using four crop species (i.e., forage oat, alfalfa, common vetch and highland barley), which were commonly cultivated in the Tibetan farmlands. Plant uptake of $^{15}NH_4^+$ and $^{15}NO_3^-$ were studied by injecting dilute (1.66 mmol ^{15}N in 10 ml distilled water) solutions of respectively labeled N source into the pots and harvesting plants 24 h later. Measurements of ^{15}N in plants showed that the investigated plant species or functional groups differ in their N uptake patterns. The legumes plants showed a preference for NO_3^- at all three growth stages. By comparison, the grasses plants showed a preference for NH_4^+ at the early growth stages, and shifted its preference for NO_3^- at the later growth stages. It suggests that plant functional groups affect N uptake by plants. In spite of the controlling of two plant functional groups on N preference, it is particularly worth mentioning that N preference by both of two plant functional groups lie on the dominant N source in soil to meet their N demands, which was derived by the close relations between ratios of NH_4^+:NO_3^- uptake by plant and ratios of NH_4^+:NO_3^- in soil in the current study. Nitrogen utilization efficiency (NUE) had a close relation with ratios of NH_4^+:NO_3^- uptake which strongly depending on plant functional groups: the NUE were increased linearly with increasing ratios of NH_4^+:NO_3^- uptake in grasses plants, however, the NUE were decreased linearly with increasing ratios of NH_4^+:NO_3^- uptake in legumes plants. This confirms that uptake of N form (NH_4^+ vs. NO_3^-) has effects on plant NUE. The relations were varied by different plant functional groups because plant functional types can modify soil N cycle by absorbing and utilizing the N sources in air and soil. Study and explanation of these relations among preference for N, ratios of NH_4^+:NO_3^- and plant NUE are much ecological significance for plant adaptation mechanism of the deficiency of N in the environment, and the inherent mechanism still needs further research.

T3

A211

Reproductive biology of *Trichocereus bridgesii* (Cactaceae) an endemic cacti from dry valleys (La Paz-Bolivia)

Mariela Escobar-Torrez
Herbario Nacional de Bolivia

In recent years it has been suggested that exist a geographic dichotomy in pollination system for columnar cacti. In extratropical regions the columnar cacti show more open and generalized pollination system, meanwhile, in tropical regions columnar cacti tend to have specialized pollination system. As pollination data for South America are scarce, the aim of this study was to analyze the reproductive biology of *Trichocereus bridgesii* an endemic cactus in Ananta a dry valley located in La Paz (Bolivia). An exclusion experiment was conducted to measure the breeding systems, and the relative contribution of nocturnal and diurnal floral visitors to fruit set and seed set, also, I observe the floral visitors, their frequencies and behaviors. In addition I evaluated if *T. bridgesii* has a specialized pollination system or generalized pollination system. The experiment revealed that *Trichocereus bridgesii* exhibits a complementary general pollination system demonstrating that both pollination groups (diurnal and nocturnal) were important for plant reproduction. The importance of floral visitors on the breeding system indicates an obligate xenogamous system, self-incompatible. Seeds produced in the experimental treatments: control, xenogamy and obligate xenogamy were viable, except for the treatment of geitonogamy. During the day native bees (*Ischnomelissa* sp., *Habralictus* sp.) and the introduced *Apis mellifera* were the most frequent visitors and during the night Nitidulidae beetles, moths or bats were not registered. *T. bridgesii* shows more affinity with columnar cacti from extratropical regions than with those from tropical regions, these results demonstrated that as latitude increases in tropical regions of South America, cactaceae have a generalist pollination system. Therefore, I recommend to carry out comparative studies from cacti species located within and outside the tropics according to variations in their geographic distribution. It is expected that in very isolated fragments of vegetation, two possible consequences are generated: 1) existence of specialized pollination systems when the main pollinator is abundant and 2) existence of generalized pollination systems when the main pollinator is scarce, in this case comparative studies in disturbed and conserved areas are recommended.

T3

A212

Ecological stoichiometric in litter and soil under forest gap of *Castanopsis kawakamii* natural forests

Xueping Feng[1,2,3], *Jinfu Liu*[1,2,3], *Buajan Supaporn*[1,2,3], *Zhongsheng He*[1,2,3]

1. Fujian Agriculture and Forestry University

2. Cross-strait Nature Research Center

3. .Key Laboratory of Fujian Universities for Ecology and Resource Statistics

Ecological stoichiometry of the carbon (C), nitrogen (N), phosphorus (P), and potassium (K) in litter and soil are great significance to research the rule which was nutrient cycling of the forest ecosystem. The aims of this study is to study the ecological stoichiometry of C, N, P, and K ratios in litter and soil of different forest gap sizes using the Redundancry Analysis method (RDA). The results showed that the contents and stoichiometric ratios of C, N, P, and K in litter were significant difference between gap sizes and relevance in *Castanopsis kawakamii* natural forest. Values of the N:P ratio of 16 usually indicated the growth of plant was limited by N and P. Values of the N:P ratio was higher than 16 in small and non-gaps, indicating that the growth of plants was limited by P. Values of the N:P ratio is between the 14 and 16 in medium gaps, indicating that the growth of plants was limited by N and P. Values of the N:P ratio is lower than 14 in large gaps, indicating that the growth of plants was limited by N. We can fertilize some N and P that could promote the plant growth in gaps. The results from this study provide the knowledge of the nutrient status, which an evidence for sustainable management in *C. kawakamii* forest.

T3

A213

A primary inquiry on the microRNA of *Caragana korshinskii* along a precipitation gradient on the Loess Plateau, China

Chunmei Gong, *Sun Yingying, Zhou Wenfei*

College of Life Science, Northwest A&F University

There are some adverse effects on the survival and growth of plants due to drought. *Caragana korshinskii* is widely distributed on the Loess Plateau, China. *C. korshinskii* has excellent soil and water conservation capacity because of its drought resistance. However, it is not clear about the stress response mechanisms of *C. korshinskii* in waterless environments. MicroRNAs have been implicated in the regulation of plant responses to several types of biotic and abiotic stress. Here, we describe the miRNAs of wild *C. korshinskii* from Huangling, Yulin, and Dalad Banner, which distributed across a precipitation gradient on the Loess Plateau. Using RNA-sequencing technology and BLAST analysis, nearly 500 conserved miRNAs and about 100 novel miRNAs were characterized from the sRNAome data, with key functions determined using Gene Ontology and Kyoto Encyclopedia of Genes and Genomes pathway analyses. Then we designed stem-loop RT-qPCR to validate the expression patterns of several conserved miRNAs that obviously responded to water stress in plants grown under natural conditions with drought aggravated and found that the expression levels of miRNA were negatively correlated with their predicted target genes. This study supplied a view of miRNA may mediated gene expression to strengthen the stress responses in wild *C. korshinskii* with decreasing precipitation.

T3

A214

High phenotypic plasticity does not imply high salt stress tolerance: A case study of two pioneer species from the Yellow River Delta

Xiao Guo[1,2], *Xiao-Huang Ren*[2], *Franziska Eller*[3], *Ming-Yan Li*[2], *Ren-Qing Wang*[2], *Ning Du*[2], *Wei-Hua Guo*[2]

1. Qingdao Agricultural University

2. Shandong University

3. Aarhus University

A greenhouse experiment was conducted in which two leguminous species commonly used in the Yellow River Delta for vegetation restoration, *Robinia pseudoacacia* and *Amorpha fruticosa*, were subjected to the five salt treatments: 0, 50, 100, 150 and 200 mmol L^{-1}. We aimed at investigating which of the two species would be better suited for growth in a saline environment, and whether the acclimation capacity to salinity resulted from an inherently higher phenotypic plasticity of the species. The results showed that salinity affected most growth parameters and biomass parameters but had no effects on most leaf traits and physiological parameters of the two species. Height, relative growth rate of crown area, root biomass and leaf mass ratio of *R. pseudoacacia* were reduced by higher salinity while *A. fruticosa* was not affected. Chlorophyll a to chlorophyll b ratio and total antioxidative capacity of *A. fruti-*

cosa increased with higher salinity while those of *R. pseudoacacia* remained unchanged. Root mass ratio and vitamin C concentration of both species were not affected by salinity, whereas vitamin C concentration of *A. fruticosa* overall was higher than that of *R. pseudoacacia*. Methane dicarboxylic aldehyde concentration of *R. pseudoacacia* increased with the increase of salinity while that of *A. fruticosa* was not affected. Root to shoot ratio of *A. fruticosa* was higher than of *R. pseudoacacia* at most salt treatments. Of all leaf traits, only leaf area differed among the different salinities. *Robinia pseudoacacia* generally exhibited a greater plasticity than *A. fruticosa* in response to salinity, but *A. fruticosa* was more resistant to the higher salinities, compared to *R. pseudoacacia*, and is thus a better candidate for vegetation restoration in saline areas.

T3

A215

Evaluation of ten Chenopodiaceae species from the Tarim Basin for tolerance to salt and drought stress during seed germination

*Zhan-Jiang Han, Long Cheng, **Zhijun Li***

College of Life Science, Xinjiang Production and Construction Corps Key Laboratory of Protection and Utilization of Biological Resources in Tarim Basin, Tarim University

Seeds of ten species with wide distribution and of feed and medicinal value were subjected to salt and drought stress by germinating them on a substrate enriched with NaCl, $NaHCO_3$ or PEG6000. Final germination (G_F) and radicle length were reduced for all species, and mean germination time (MGT) was extended to varying degrees as concentrations of NaCl and $NaHCO_3$ increased. Once stress was withdrawn, seeds of *Suaeda rigida*, *Halocnemum strobilaceum*, *Halostachys caspica* and *Salicornia europaea* germinated irrespective of whether the source of stress had been NaCl or $NaHCO_3$ whereas those of *Suaeda glauca* and *S. heterophylla* were able to germinate when the source of stress had been NaCl. The G_F and radicle length of seeds of were inhibited while MGT extended continuously in various degrees with osmotic potentials descending and drought intensifying in ten species. Seeds of *Suaeda glaeda*, *S. rigida*, *S. heterophylla*, *Halocnermum strobilaceum*, *Halostachys caspica*, *Salicornia europaea* and *Halogeton arachnoideus* obtained their germination recoveries after relieving drought stress. Degree of salt or drought tolerance during germination was evaluated comprehensively through subordinate function analysis, G_F, MGT, radicle length and recovery. Overall ranking of species in descending order of tolerance to NaCl was as follows: *Halocnemum strobilaceum*, *Suaeda rigida*, *Halostachys caspica*, *S. arcuata*, *Halogeton arachnoides*, *Salicornia europaea*, *Suaeda heterophylla*, *S. glaeda*, *Chenopodium aristatum* and *S. stellatiflora*; the ranking for tolerance to $NaHCO_3$ was *Halocnemum strobilaceum*, *Halogeton arachnoides*, *Halostachys caspica*, *Suaeda rigida*, *S. arcuata*, *Salicornia europaea*, *Suaeda heterophylla*, *S. glauca*, *Chenopodium aristatum* and *S. stellatiflora*; the ranking for tolerance to drought was *Halostachys caspica*, *Halogeton arachnoideus*, *Halocnermum strobilaceum*, *Suaeda rigida*, *S. heterophylla*, *S. glaeda*, *Salicornia europaea*, *Suaeda stellatiflora*, *Chenopodium aristatum* and *S. arcuata*. The top four plants can be used for soil improvement or gene mining.

T3

A216

Germination response of four alien congeneric *Amaranthus* species to environmental factors

Jian-Hua Hao

Changshu Institute of Technology

Seed germination is the key step for successful establishment, growth and further expansion of population especially for alien plants with annual life cycle. Traits like better adaptability and germination response were thought to be associated with plant invasion. However, there are not enough empirical studies correlating adaptation to environmental factors with germination response of alien invasive plants. In this study, we conducted congeneric comparisons of germination response to different environmental factors such as light, pH, NaCl, osmotic and soil burials among four alien amaranths that differ in invasiveness and have sympatric distribution in Jiangsu Province, China. The data were used to create three-parameter sigmoid and exponential decay models, which were fitted to cumulative germination and emergence curves. The results showed higher maximum Germination (*Gmax*), shorter time for 50% germination (*G50*) and the rapid slope (*Grate*) for *Amaranthus blitum* (low-invasive) and *A.retroflexus* (high-invasive) compare to intermediately invasive *A. spinosus* and *A. viridis* in all experimental regimes. It indicated that germination potential does not necessarily constitute a trait that can efficiently distinguish highly invasive and low invasive congeners in four *Amaranthus* species. However, it was showed that the germination performances of four amaranth species were more or less correlated with their worldwide distribution area. Therefore, the germination performance can be used as a reference indicator, but not an absolute trait for invasiveness. Our results also confirmed that superior germination performance in wide environmental conditions supplementing high seed productivity in highly invasive *A. retroflexus* might be one of the reasons for its prolific growth and wide distribution. These findings lay the foundation to develop more efficient weed management practice like deep burial of seeds by turning over soil and use of tillage agriculture to control these invasive weed species.

T3

A217

Sex-related differences in adaptability to manganese stress in *Populus cathayana* when supplied with different nitrogen forms

Hongxia Zhao

Institute of Mountain Hazards and Environment, Chinese Academy of Sciences

Populus cathayana was employed as a model species to detect sexual differences in plant growth, physiological, biochemical, and ultrastructural responses to manganese (Mn) stress with two different nitrogen sources (NH_4^+ and NO_3^-). Results showed that both NH_4^+-N and NO_3^--N supply forms significantly increased plant growth, gas exchange capacity, SOD and POD enzyme activities in the leaves of both male and female *P. cathayana*. On the other hand, we found that NH_4^+-N supply could improve tolerance to Mn stress in both male and female *P. cathayana* with higher photosynthetic capabilities, biomass accumulation, antioxidant enzyme activities, superoxide radicals (O_2^-) and thiobarbituric acid-reactive substances (TBARS) contents in leaves than the individuals of both sexes under Mn stress. Conversely, female individuals grown under NO_3^- supply exhibited significant lower biomass accumulation, photosynthetic capabilities, antioxidant enzyme activities, which were mainly due to higher Mn accumulation in both above and below ground organs, and also because of higher ROS releasing rate in leaves than did males under the same condition. Moreover, serve damage to cellular ultrastructure was also observed in the leaves of males and females exposed to Mn stress, but more strongly in females with more visible degradation of mitochondrion in plant leave cells. Therefore, the different preferences of male and female *P. cathayana* to Mn stress with different N sources can serve as an important fertilizing strategy for minimizing intraspecific competition between the two genders.

T3

A218

The relationships between above- and belowground biodiversity and ecosystem multifunctionality, and its drivers in the drylands of China

Jianming Deng, *Qiajun Du, Weigang Hu*
Lanzhou University

Biodiversity is the foundation for the maintenance of ecosystems, which are valued for their ability to provide multiple functions and services simultaneously (ecosystem multifunctionality; EMF). However, we have limited knowledge of the links between biodiversity and multifunctionality as well as drivers of these relationships in large-scale natural ecosystems. This gap in our understanding is particularly true for drylands, areas covering 41% of Earth's land surface and supporting over 38% of the global human population, which are among the most sensitive ecosystems to climate change. Here, we present a regional-scale field study conducted along natural aridity gradients across 157 dryland sites from northwestern China. We surveyed plant community by measuring 30 m × 30 m plots, characterized bacterial, archaeal and fungal communities in the soil surface (top 20 cm) by using Illumina Miseq profiling of ribosomal genes and internal transcribed spacer (ITS) markers, gathered a range of key abiotic factors (i.e. climate, elevation, and soil texture), and assessed 14 ecosystem function indicators (aboveground biomass, plant carbon, plant nitrogen, plant phosphorus, soil organic carbon, soil nitrogen, soil phosphorus, nitrate, ammonium, soil available phosphorus, β-glucosidase, urease, phosphatase, and soil microbial biomass) related to soil biogeochemical cycles, soil fertility, biological productivity, and the build-up of nutrient pools. In this study, we aim to elucidate the relationships between above- and belowground biodiversity and multifunctionality, and to illustrate the interactions among abiotic factors and biotic attributes (i.e. above- and belowground biodiversity) in driving multifunctionality. We also evaluate how increases in aridity, as defined by the aridity index, affect biotic attributes and multifunctionality, and further emphasize the role of biological feedbacks on the responses of ecosystem multifunctionality to increasing aridity. This knowledge is essential to predict responses of dryland ecosystems to ongoing global change, and guide conservation and sustainable management efforts in drylands of China.

T3

A219

Combined of multidimension scale - mininum area to study the function of water source conservation of the drinking water area, in Nanning City, China

Liangmei Huang
Guangxi Environmental Monitoring Center

With the multistatistical method, as well as the combined techniques of remote sensing interpretation and ecological landscape analysis, and using field work, water quality monitoring and some research outcomes, the function index of water source conservation, landscape patterns, and the mininum area of water maintainting quantities for the two types of drinking water area, namely river and reservoir, were calculated. Then the relationship among these indices were analyzed, and their data values extracted, so the mechanism between the structure and function of the drinking water area were discussed and discovered clearly. Results showed that: (1) the function of water source conservation of river ecosystem was genernallylow, and increased from 500 m buffer area to 5 km buffer area. The function of water source conservation of reservoir ecosystem was stronger than that of river; (2) the function index of water source conservation shown canonical quantitative relationship with the water resource quantitation, the pollution index, the ration of landscape area and patches numbers, buffer area, catchment area, forest area and pollution discharges; (3) for the current state, only can mininum area of forest ecosystem dominated by soft broad-leaved forest, oak forest, sclerophyllous evergreen broad-leaved forest, bamboo, maintaint the quantitative balance of protective area, catchment area, because of its effective function of water source conservation; (4)issues such as stability of ecological system, security of water quality, and risk of water resources on the drinking water area in Nanning city should be taken into account in the course of the environmental protection and management. Ecological recovery, environmental comprehensive regulation were nacessary to carried out in the following years.

T3

A220

Population genomic study of an introduced mangrove species *Sonneratia apetala* in China using high-throughput sequencing

Haidan Wu, *Hongxian Xie, Weixi Li, Sitan Qiao, Yubin Yan, Wuxia Guo, Yelin Huang*
State Key Laboratory of Biocontrol and Guangdong Provincial Key Laboratory of Plant Resources, School of Life Sciences, Sun Yat-sen University

Sonneratia apetala, a mangrove species, is naturally distributed in Bangladesh, Myanmar, and India, including the Andaman and Nicobar islands. The species was introduced to China in 1985 for restoration purposes from Sundarban in Bangladesh. However, the further use of the species is becoming increasingly controversial as there are emerging signs that it may become invasive in certain locations. Several attempts have been made to characterize genetic variation of *S. apetala* in China. But there are no studies comparing levels of genetic diversity in native vs. introduced populations. In this study, we tried to detect the dynamics of evolution in this species by comparing the genetic diversity of five *S. apetala* populations along the coast of China, and three populations from Sundarban using deep sequencing. The 40 nuclear loci were random chosen from an EST library which were assembled de novo based on 13.8 million paired-end sequence reads for the transcriptome of *S. apetala*. The pooled DNA samples from each of eight *S. apetala* populations were sequenced respectively. The results showed that values of θ in native populations were two times higher than those in introduced populations. The result of coalescent simulation also revealed a sever bottleneck effect during the introduction of *S. apetala*. However, the values of π, which is non-sensitive to low-frequency SNP, were slightly higher in native populations than those in introduced populations. Meanwhile, we founded the extreme high heterozygosity in six founder individuals which displayed similar genetic diversity with the five *S. apetala* populations along the coast of China. No sign of population expansion or contraction after the introduction were detected based on the 40 nuclear loci. These results indicated that the genetic diversity of *S. apetala* remains constant regardless of the nearly 30 years' introduction history in China. The long generation time, high heterozygosity in founder individuals and artificial planting may contribute to slowing down the loss of genetic diversity in introduced populations after a strong genetic bottleneck. We suggested the invasion potential of *S. apetala* should not be underestimated based on its population genetic characteristics, and the long-term ecological monitoring is still needed in order to avoid ecological disaster.

T3
A221
Responds of Songnen grassland soil respiration to doubling CO_2 concentration

Li Jiang, *Xuedun Niu*
Jilin Agricultural University

Long-term doubling CO_2 concentration affect grassland is crucial to predict the future impact of global climate change on terrestrial ecosystems. How the effects of global CO_2 concentration change on grassland productivity are going to be altered by changing the soil respiration. The response of soil respiration and soil organic matter to global change is one of the main factors that affect the carbon budget of terrestrial ecosystem. The sensitivity of soil respiration to temperature (Q_{10}), to a large extent, determines the feedback relationship between global climate change and carbon cycle. This paper simulated CO_2 concentration to 700 ppm. The influence of doubling CO_2 concentration of grassland to soil respiration and sensitivity of soil respiration were measured by LI-6400-09 processing system. Soil carbon content, soil nitrogen content and soil carbon and nitrogen ratio were determinate during 2006 to 2008.The results show that the doubling CO_2 concentration increased the soil respiration rate by 19.31% in three years, compared to the control. Doubling CO_2 concentration treatment increased the Songnen grassland temperature sensitivity by 30.1%, 40.3% and 36.3% in 2006, 2007 and 2008 respectively. The active soil C pool decomposed faster with doubling CO_2 concentration in 2007, while slowly cycling C pools had longer turnover times.

T3
A222
Research of the chemotaxis response of *Plutella xylostella* to the volatile components of *Lepidium sativum* L.

Shicui Jiang, *Jinhua Zhao, Donghao Li*
Key Laboratory of Natural Resources of Changbai Mountain & Functional Molecules (Yanbian University), Ministry of Education

Lepidium sativum L. is a kind of medicinal and edible plant that commonly found in Yanbian area. It is a cruciferous vegetable, yet rarely affected by pests. Previous studies have shown that *Lepidium sativum* L. exerts biological inhibitory effect on *Culex* species. In order to research deeper into the repellent effect of *Lepidium sativum* L. towards insects, a study on the chemotaxis response of *Plutella xylostella* to the volatile components of *Lepidium sativum* L. was conducted. In this study, the sterile samples of *Lepidium sativum* L. were obtained through tissue culture technique and then the real-time chemotaxis response of *Plutella xylostella* under different regulated circumstances (inoculum density, day-age and the induction time) were rapidly and accurately being analysed by GP-MSE/GC-MS. Different inoculum density had different chemotoxis response to *Lepidium sativum* L., and the varied induction time influenced the chemorepellent effect of volatile compounds in *Lepidium sativum* L. This study provides the basic data to evaluate the utilisation of *Lepidium sativum* L. as a form of biopesticide to repel *Plutella xylostella.*

T3
A223
The interactive effect of temperature and light on seed germination of 6 Zygophyllaceae species

Peipei Jiao[1], ***Zhijun Li***[1,2], *Xiao Zhang*[1]
1. Key Laboratory of Biological Resource Protection and Utilization of Tarim Basin, Xinjiang Production and Construction Group, Xinjiang, People's Republic of China
2. College of Life Sciences, Tarim University, Xinjiang, People's Republic of China

The interactive effect of temperature and light on seed germination of 6 Zygophyllaceae species, *Sarcozygiumspecies kaschgaricum* Boriss., *Sarcozygiumspecies xanthoxylon* Bunge, *Zygophyllum brachypterum, Zygophyllum rosovii, Zygophyllum obliquum, and Zygophyllum fabago* were studied under controlled environment conditions. Treatments included two light levels (8:16 h light:dark period and 24 h dark environment), two temperature regimes (fluctuating day : night temperature regimes of 15:5°C, 20:10°C, 25:15°C, 30:20°C and 35:25°C, and constant temperature regimes of 10°C to 30°C with 5°C interval), using a completely randomized block design. There was a higher final germination percentage, shorter mean time of germination and faster germination rate for 6 species in darkness condition than in 8 h photoperiod. 6 species germinated in a wide range of temperature, most of them germinated in all designed temperatures from 10 to 35°C, fluctuating and constant. The most optimal constant temperature regime for all species' germination was 20°C in 8 h:16 h light:dark period and 25°C in darkness, The most optimal fluctuating temperature regime for germination for all species was 30:20 ℃ both in 8 h light and full darkness. The different seed germination syndromes that we found for these Zygophyllaceae species likely contribute to variable seed bank formation and emergence patterns, and species coexistence. The results are consistent with 6 Zygophyllaceae species being a tropical xerophyte species capable of establishing itself in a wide range of environment conditions.

T3

A224

Influence of elevated CO_2 on rhizosphere microbial communities associated with mangrove plant *Kandelia obovata*

Zhou Jin, *Yin Mengqing, Cai Zhonghua, Lin Guanghui*
Division of Ocean Science and Technology, Graduate School at Shenzhen, Tsinghua University

Rhizosphere microbial communities mediate the decomposition of soil organic matter. The amount of carbon that is respired leaves the soil as CO_2 and causes one of the greatest fluxes in the global carbon cycle. How rhizosphere microbial communities will respond to global warming, however, is not well understood. Here, we characterized the effect of elevated CO_2 on soil microbial communities associated with a mangrove plant Kandelia obovata. Elevated CO_2 increased the soil C:N ratio and microbial biomass. 16s rRNA analysis revealed that elevated CO_2 did not affect the soil bacteria community structure. The results of principal component analysis (PCA) and Redundancy analysis (RDA) indicate that elevated CO_2 had no strict specificity with bacterial community diversity. Organic carbon, nitrate nitrogen (NO-N), C/N, total nitrogen (N) and pH were the primary impact factors. The elevated CO_2 increased microbial activity and carbon use, probably in response to increased root exudation. Elevated CO_2 did not affect microbial structure, nor did warming affect the abundances of most microbial groups. In summary, short-term high-CO_2 had no influence on microbial community composition (or rhizosphere Microbial Communities have buffer ability to response CO_2 increase) but increased microbial metabolic activity and hence reduced carbon use efficiency.

T3

A225

Liquid hybridization and solid phase detection: A highly sensitive and accurate strategy for microRNA detection in plants and animals

Fosheng Li[1], *Lanju Mei*[1], *Cheng Zhan*[1], *Qiang Mao*[2], *Min Yao*[1], *Shenghua Wang*[1], *Lin Tang*[1], *Fang Chen*[1,3]
1. Key Laboratory of Bio-resources and Eco-environment, Ministry of Education, College of Life Science, Sichuan University, Chengdu, Sichuan, China
2. Chengdu Botanical Garden, Chengdu, Sichuan, China
3. National and Local Joint Engineering Laboratory for Energy Plant Bio-Oil Production and Application, Chengdu, Sichuan, China

MicroRNAs (miRNAs) play important roles in nearly every aspect of biology, includingphysiological, biochemical, developmental and pathological processes. Therefore, a highly sensitiveand accurate method of detection of miRNAs has great potential in research on theory andapplication, such as the clinical approach to medicine, animal and plant production, as well as stress response. Here, we report a strategic method to detect miRNAs from multicellular organisms, which mainly includes liquid hybridization and solid phase detection (LHSPD); it has been verified in various species and is much more sensitive than traditional biotin-labeled Northern blots. By using this strategy and chemiluminescent detection with digoxigenin (DIG)-labeled or biotin-labeled oligonucleotide probes, as low as 0.01–0.25 fmol [for DIG-CDP Star (disodium2-chloro-5- (4-methoxyspiro{1,2-dioxetane-3,20-(50-chloro)tricyclo[3.3.1.13,7]decan}-4-yl)phenyl phosphate) system], 0.005–0.1 fmol (for biotin-CDP Star system), or 0.05–0.5 fmol (for biotin-luminol system) of miRNA can be detected and one-base difference can be distinguished between miRNA sequences. Moreover, LHSPD performed very well in the quantitative analysis of miRNAs, and the whole process can be completed within about 9 h. The strategy of LHSPD provides an effective solution for rapid, accurate, and sensitive detection and quantitative analysis of miRNAs in plants and animals.

T3

A226

Study on soil biological character of the degradation of *Populus simonii* Plantation in the Loess Plateau

Xiang Wang, ***Jinchuan Li***, *Jianying Yue, Chunyan Guo, Ning Lu, Juan Dong*
Biology Institute of Shanxi

In order to elucidate the soil related factors of the formation of "dwarf and aged" tree, *Populus simonii* on the Loess Plateau and contrast of forest (*Robinia pseudoacacia*) were chosen as the test objects to study their soil nutrient, soil microbial characteristic and soil biological fertility. The results showed that: 1. Wither and fall layer thickness of *Populus simonii* plantation is less than CK, while field capacity, organic matter, total nitrogen, total phosphorus and available phosphorus were significantly lower than that of control; 2. Soil invertase, polyphenol oxidase, alkaline phosphatase of *Populus simonii* plantation is lower than that of the control, while urease and catalase activity were higher than that of control group; 3. The number of *Populus simonii* plantation of soil microorganism were bacteria > actinomycetes > azotobacter> denitrifying bacteria > fungi, the number of bacteria, fungi and denitrifying bacteria were higher than that of control. Furthermore, soil bacterial and fungal diversity of Pupulus simonii plantation were lower than that of control; 4. Principal component analysis and correlation analysis showed that soil nutrient and soil biological fertility of *Populus simonii* plantation were less than that of control, so the soil nutrient and soil biological fertility loss may lead to one of the important reasons about formation of "dwarf and aged" tree in the Loess Plateau.

T3

A227

Responses of productivity and species composition to elevated N deposition in alpine grassland of the Tianshan Mountains

Kaihui Li
Xinjiang Institute of Ecology and Geography, Chinese Academy of

Sciences,

Global climate change and increased nitrogen deposition are significantly altering plant primary productivity and species composition in terrestrial biomes. A 8-year field experiment with four N addition treatments was conducted in alpine grassland of the mid Tianshan mountains in China. Nitrogen addition significantly increased the aboveground biomass of dominant perennial grasses but decreased aboveground biomass of forbs. Low N addition could increase total primary under increasing temperature and growth season precipitation in alpine grassland, while mid and high N addition may suppress it. We found no significant change for species richness. These results suggest that increasing N addition will significantly effect on plant biomass of dominant herbaceous and forbs, while species composition will be not change. Whether current trends continue can only be answered by long-term monitoring. More studies need to be conducted to explain the relationships among primary productivity, climate factors and nitrogen deposition in different grassland ecosystem type.

T3

A228

Effects of foliar trichomes of *Tillandsia brachycaulos* on the retention and resuspension of atmospheric particulates

Ruiwen Zhang, Wei Zhang, Mintian Gu, Guiling Zheng, ***Peng Li***
Qingdao Agricultural University

The surface characteristics of plant leaves have great influence on the retention of atmospheric particulates, but this conclusion is mainly based on the comparison among different plants, not on the direct study of the leaf surface characteristics. In addition, atmospheric particulate matter temporarily stranded on the surface of the plants, and deposition of particulate matter will be back into the atmosphere in the certain weather conditions by resuspension, but resuspension is rare studied. Therefore, *Tillandsia brachycaulos* with typical leaf surface characteristics was selected, and its foliar trichomes were removed artificially to compare the dust retention and resuspension contents under different wind strengths and different time. The results showed that leave dust retention with all the foliar trichomes (13.942 ± 0.197 g/m^2), is significantly higher than those without foliar trichomes (10.122 ± 0.133 g/m^2). More than 94.0% of the atmospheric particulates would be resuspended under the function of wind for those with foliar trichomes, while only 51.2% for those without foliar trichomes. Resuspension ratio was increased significantly with the increase of wind speed or function time. Therefore, effects of foliar trichomes of *Tillandsia brachycaulos* on the retention and resuspension of atmospheric particulates is reflected not only in the total amount of atmospheric particulates, but also in the process of atmospheric particulate resuspension. Resuspension must be taken into account in the future study in order to accurately assess the relationship between vegetation and atmospheric particulates deposition.

T3

A229

Unique proline-benzoquinone pigment from the colored nectar of *Leucosceptrum canum* functions in bird attractions

Shi-Hong Luo[2]*, Yan Liu*[1]*,* ***Sheng-Hong Li***[1]
1. Kunming Institute of Botany, Chinese Academy of Sciences
2. Shenyang Agricultural University

Floral nectar is the bridge in mediating interactions between plants and flower visitors. In nature, at least 68 taxa have been documented to possess unusual colored nectar, but the pigmented compounds of most plants and their ecological functions are still unknown. *Leucosceptrum canum*, a Himalayan tree (up to 10 m high) with an unusual dark brown nectar, is the only colored nectar plant so far discovered in the Lamiaceae family. It is a favorite species of over 40 kinds of birds, which thus is called "bird's Coca cola tree" by ornithologists. We have been interested in the specialized metabolites of *L. canum* and their ecological functions, and found that the glandular trichomes harbor defensive sesterterpenoids with a novel C_{25} carbon framework (Luo *et al*, *Angew Chem Int Ed*, 2010, 49: 4471-4475). Our large-scale HPLC analysis of 109 nectar samples with a detection wavelength of 400 nm indicated that all samples contained a predominant peak with maximum UV absorptions at 215, 369, and 525 nm, which was accordingly targeted as the responsible pigmented compound of the dark brown nectar of *L. canum*. However, the compound was unstable which made its isolation and identification very difficult. After two years of effort, we were finally able to obtain 27.5 mg sample of the compound with sufficient purity for identification as a dark brown solid from 645 mL of *L. canum* nectar. Its chemical structure was identified as a unique symmetric proline-quinone conjugate, 2,5-di- (N- (-)-prolyl)-*para*-benzoquinone ((-)-DPBQ) by comprehensive spectroscopic studies including 1D and 2D NMR and HRMS analyses, and the absolute configuration was determined through comparison of the CD curve and optical rotation with those of (-)-DPBQ and (+)-DPBQ that were synthesized in this study. Color comparison of (-)-DPBQ in water at different concentrations with *L. canum* nectar confirmed that (-)-DPBQ, rather than the so-called 5-hydroxyflavylium characterized only by UV and mass spectral methods (Zhang *et al*, *New Phytol*, 2012, 193: 188-195), is the major cause of the dark brown color of *L. canum* nectar. The concentration of (-)-DPBQ in the nectar of *L. canum* was found to be extremely high (up to 1.39 mg/mL). A bird visitor, Japanese White-eye (*Zosterops japonicus*), was used for testing its behavioral responses to the nectar, nectar sugar, (-)-DPBQ and (+)-DPBQ. The results clearly indicated that (-)-DPBQ mainly functioned as a color attractant to bird pollinators in *L. canum* nectar (Luo *et al*, *Org Lett*, 2012, 14: 4146-4149). In addition, (-)-DPBQ exhibits selective cytotoxic, antioxidant, and NO inhibitory activities, suggesting it to be a highly functional natural pigment (ZL201210109715.1).This research was financially supported by the National Natural Science Foundation of China (31070320).

T3

A230

Stoichiometric characteristics of four submersed macrophytes in three plateau lakes with contrasting trophic statuses

Wei Li[1]*, Te Cao*[2]*, Jiayou Zhong*[3]*, Leyi Ni*[2]*, Houbao Fan*[1]
1. Nanchang Institute of Technology
2. Institute of Hydrobiology, The Chinese Academy of Sciences
3. Jiangxi Institute of Water Sciences

Plant nutrient concentrations and their stoichiometric characteristics are often used to predict nutrient availability and nutrient limitation. However, there is still debate because plants, especially submersed macrophytes, can store excess nutrients and exhibit different survival strategies. To determine the extent to which the trophic status of lakes affect nutrient concentrations of submersed macrophytes and to estimate the relative contribution of lake and interspecific differences to variation of plant nutrient concentrations, we investigated four common submersed macrophytes, *Potamogeton maackianus*, *Potamogeton pectinatus*, *Potamogeton malaianus* and *Myriophyllum spicatum*, in three plateau lakes with contrasting trophic statuses (i.e. oligotrophic: Fuxian Lake; mesotrophic: Erhai Lake; eutrophic: Dianchi Lake) in Yungui Plateau in southwest China. The results indicated that the average C:N ratios were not different among the three lakes. However, the C:P and N:P ratios in Erhai Lake and Dianchi Lake were significantly lower than those in Fuxian Lake. The N:P ratios of *P. maackianus* and *P. pectinatus* did not significantly differ among the lakes. By contrast, the N:P ratios of *P. malaianus* and *M. spicatum* indicated different levels of interspecific stoichiometric homeostasis. The C and N concentrations and C:N ratios were mainly determined by the species, while the P concentrations and C:P ratios were mainly determined by the trophic status of lakes. Our study suggested that the C:N:P stoichiometric characteristics of submersed macrophytes exhibit high interspecific differences but substantial homoeostasis in response to different trophic conditions in plateau lakes.

T3

A231

Effects of simulated acid rain on physiological characteristics and active ingredient content of *Asparagus cochinchinensis* (Lour.). Merr.

Juan Liang

College of Biological and Food Engineering, Huaihua University

The effects of simulated acid rain at pH 2.0, 3.0, 4.0, 5.6 and 6.8 (control), respectively, on the agronomic traits, chlorophyll content, resistance physiology and active ingredient content of *A. cochinchinensis* cultured in pots were studied to explore the sensitivity and tolerability of *A. cochinchinensis* to acid rain with the hope to provide the theoretical basis for the cultivation of *A. cochinchinensis*. The results showed that 1) compared to the control with pH 6.8, acid rain with pH 5.6 did not significantly affect the shape and color of leaves as well as the contents of amino acid and soluble sugar in tuber roots of *A. cochinchinensis*, but significantly enhanced the activities of superoxide dismutase (SOD) and peroxidase (POD) and 2) with the increase of acid rain stress (pH$\leq$3), the damage degree of leaves and the content of malondialdehyde (MDA) in leaves gradually increased, but the content of photosynthetic pigment, the increment of stem length and fresh weight, the activities of POD and SOD and the content of amino acid, soluble sugar and saponin in the tuber roots all decreased. The experiments indicated that 1) acid rain significantly affected the physiological characteristics and active ingredients accumulation of *A. cochinchinensis*, 2) *A. cochinchinensis* showed a certain resistance to acid rain, 3) pH $\leq$ 3 is the critical point (threshold) for acid rain to damage *A. cochinchinensis*, and 4) artificial cultivation of *A. cochinchinensis* should avoid medium and strong acid rain pollution.

T3

A232

Spatial variation of absorptive root traits and nutrient foraging strategies of Chinese fir (*Cunninghamia lanceolata*) from different locations across Subtropical China

Yingchun Liao[1], *Houbao Fan*[1], *Huimin Wang*[2]

1. Research Institute of Ecology & Environmental Science, Nanchang Institute of Technology

2. Qianyanzhou Ecological Research Station, Key Laboratory of Ecosystem Network Observation and Modeling, Institute of Geographic Sciences and Natural Resources Research

Absorptive roots and their related mycorrhizal fungi are mainly responsible for soil resource acquisition. Understanding of the spatial variation of absorptive root traits is crucial to predict the adaptation of tree absorptive roots and their nutrient foraging strategies in different environmental conditions. A study was conducted to measure the absorptive root traits related to nutrient foraging (root morphology, architecture and arbuscular mycorrhizal (AM) colonization) of Chinese fir (*Cunninghamia lanceolata*) across six areas in subtropicalChinathat differed markedly in their environmental conditions. The six Chinese fir populations were selected from the central (Shaxian, Taihe and Huitong) to the marginal portions (Tongling, Nayong and Pingxiang) of the geographic range in subtropical China. Absorptive root diameter differed significantly with root order and study sites ($P < 0.05$), while special root length (SRL) and special surface area (SSA) only showed significant difference with root order. Higher root C/N was found in marginal distribution areas except Tongling, but generally kept at lower levels in central distribution areas. Absorptive root AM colonization varied with root orders and study sites, with the 1st order roots significantly higher than the 2nd order roots ($P < 0.05$). The AM colonization of absorptive roots tended to be higher in areas with high annual mean precipitation (MAP), but to be suppressed by low soil pH. Branch ratios and AM colonization were negatively correlated ($P < 0.05$) and were at high levels in areas with low soil pH. The absorptive roots within species varied across geographic range with different climates were supposed to follow different economic strategies for nutrient acquisition. Our findings suggest root proliferation and AM colonization are the main strategies and can make complementarities in nutrient foraging.

T3

A233

Complementarity in nutrient foraging strategies of absorptive fine roots and arbuscular mycorrhizal fungi across 14 coexisting subtropical tree species

Bitao Liu

Institute of Geographic Sciences and Natural Resources Research, Chinese Academy of Sciences

In most cases, both roots and mycorrhizal fungi are needed for plant nutrient foraging. Frequently, colonization of roots by arbuscular mycorrhizal (AM) fungi seems to be higher in species with thick and sparsely branched roots than in species with thin and densely branched roots. Yet whether a complementarity exists between roots and mycorrhizal fungi across these two types of root systems remains unclear. We measured traits related to nutrient foraging (root morphology, architecture and proliferation, AM colonization and extramatrical hyphal length) across 14 coexisting AM subtropical tree species following root pruning and nutrient addition treatments. After root pruning, species with thinner roots showed more root growth but lower mycorrhizal colonization than species with thicker roots. Under multi-nutrient (NPK) addition, root growth increased but mycorrhizal colonization decreased significantly whereas no significant changes were found under nitrogen or phosphate additions. Moreover, root length proliferation was mainly achieved by altering root architecture but not root morphology. Thin-root species seem to forage nutrients mainly via roots whereas thick-root species rely more on mycorrhizal fungi. Additionally, the reliance on mycorrhizal fungi was reduced by nutrient additions across all species. These findings highlight complementary strategies for nutrient foraging across coexisting species with contrasting root traits.

T3

A234

Ecophysiological responses of *Camellia japonica* (Naidong) to different water and light conditions

***Cuiju Liu**, Xiao Guo, Qinghua Liu*
Qingdao Agricultural University

Camellia japonica (Naidong) living in Qingdao, China, is a special ecotype of *Camellia japonica* and was selected as research materials in present study. The *Camellia japonica* (Naidong) saplings were subjected to two light regimes L1, L2 (65%, 15% of full sunlight), and three water supply regimes W1, W2, W3 [75%, 50%, 25% of field capacity (FC)]. The ecophysiological and morphological responses of *Camellia japonica* (Naidong) were studied. The results showed that height (H), basal diameter (BD) and crown area (CA) of *Camellia japonica* (Naidong) were increased with the treatment time under L1 and L2, the morphological index was higher with the L1 treatment as compared to the L2, and the relative growth rate of each index increased rapidly. But with the extending of stress time, the relative growth rate increased slowly; SLA and LWC under L2 was significantly higher than L1. With the decreasing of water regimes, the Photosynthetic Rate (A), Ranspiration Rate (E), Stomatal Conductance (Gs) and Water Use Efficiency (WUE) decreased, Intercellular CO_2 Concentration (Ci) increased gradually, and Vapour Pressure Deficit (VPD) increased firstly and then decreased. The value of Maximal Fluorescence in the dark (F_m) Variable Fluorescence (F_v) and Maximal Quantum Yield (F_v/F_m) under the L1 conditions were higher than L2 conditions, and the Initial Fluorescence in the dark (F_0) showed the opposite trend. With the decreasing of moisture gradient, the F_0 increased gradually under L1, while showed the opposite trend under L2. Significant differences were found in photosynthetic pigment contents of leaves among various light and water treatments. Under high irradiation, chlorophyll a, chlorophyll b and total chlorophyll were higher than low irradiation, and the Chla/Chlb value showed the opposite trend. The contents of Superoxide Dismutase (SOD), Peroxidase (POD), Catalase (CAT), Malondialdehyde (MDA) and Proline (Pro) were significantly difference at every level of light and water treatment. And the contents of POD, SOD increased firstly and decreased later with the water gradient decreasing. The content of CAT decreased with the decreasing of water content under L2 conditions, and the contents of MDA and Pro increased gradually with the water gradient decreasing. Drought or shade has some restrictive effect to the growth of *Camellia japonica* (Naidong), certain moisture and light were favorable for plant growing. *Camellia japonica* (Naidong) could tolerate a certain degree of drought resistance and adapt the larger amplitude of light intensity. At the same time, it can change the enzyme activity and osmotic adjustment substances to against damage.

T3

A235

Latitudinal defense pattern of woody species is mainly dominated by heritability

***Jian-Feng Liu**[1], Xiao-Fei Wang[1], Yan-Yan Ni[1], Ze-Ping Jiang[1], Feng-Feng Kang[2], Yi-Hua Xiao[3], Yuan-Wen Kuang[4], Yong-Biao Lin[4], Zhi-Feng Guo[5]*
1. Research Institute of Forestry, Chinese Academy of Forestry
2. College of Forestry, Beijing Forestry University
3. Research Institute of Tropical Forestry, Chinese Academy of Forestry
4. South China Botanical Garden, Chinese Academy of Sciences
5. Heshan Institute of

Recent studies have called the latitudinal herbivory-defense hypothesis (LHDH) into question (e.g., Moles *et al.*, 2011). However, our previous field investigation conducted on natural Chinese cork oak (*Quercus variabilis* Blume) stands supported this hypothesis (Wang *et al.*, 2016), and found that climate (e.g., annual precipitation, minimum temperature of the coldest month) and herbivory, rather than edaphic factors, contributed more to the species' latitudinal defense pattern. To further test whether climate or soil or genetic factors dominate latitudinal defense pattern of oaks, two common garden experiments were designed. The seeds of five oak provenances across the north-south transect with spanning ~14° latitude (~40.2 ° N, ~37.5 ° N, ~33.5 ° N, ~30.7 ° N, ~26.3 ° N) were collected in September, 2014. Then one part of the seeds were germinated and cultivated beyond the south distribution edge of the species (~22° N, Heshan, Guangdong Provenance, South China). The other part of the seeds for seedlings were cultivated with sandy substrate in the greenhouse of CAF (~40° N, Beijing), and were treated with four soil nutrient conditions: low nitrogen with low phosphorus (LNLP), low nitrogen with optimum phosphorus (LNOP), optimum nitrogen with low phosphorus (ONLP), optimum nitrogen and phosphorus (ONOP). After one growing season (September, 2015), chemical defense substances across the seedlings' tissues (the concentrations of total phenolics, total condensed tannins and flavonoids), as well as herbivory rates in the south common garden experiment, were examined and analyzed. Our results showed that even the seedlings were grown beyond the species' south distribution edge, the latitudinal pattern of chemical defensive traits was still consistent with our previous field results, namely, compared to the north provenances, the south provenances had higher concentrations of chemical defense substances.

Consequently, the south provenances suffered less leaf loss though all provenances were under the same herbivory pressure. Under nutrient treatments, the abundance degree of N or P did not alter the geographical patterns of chemical defense substances across latitudes, as the concentrations of chemical substances decreased with the increase of the latitudes of the five provenances. Our results suggest that edaphic or climatic factors may contribute, to some extent, to latitudinal defense pattern of woody plants, but the genetic factor is paramount, which reflects long-term adaptation and coevolution of wood plants with natural enemies. Our results are enormously important to elucidate the controversies of latitudinal defensive hypothesizes, and also have practical significances on human-assisted migration of valuable woody species in the context of climate change. This work was funded by the National Natural Science Foundation of China (No. 41371075).

T3

A236

Species abundance distribution of woody plants on islands in a fragmented landscape

Jinliang Liu[1], *Fangliang He*[2,5], *Thomas J. Matthews*[3,4], *Mingjian Yu*[1]

1. College of Life Sciences, Zhejiang University

2. Sun Yat-sen University-Alberta Joint Lab for Biodiversity Conservation, State Key Laboratory of Biocontrol, School of Life Sciences, Sun Yat-sen University

3. GEES (School of Geography, Earth and Environmental Sciences), The University of Birmingham

4. Conservation Biogeography and Macroecology Programme, School of Geography and the Environment, University of Oxford

5. Department of Renewable Resources, University of Alberta

Species abundance distribution (SAD), which describes the commonness and rarity of species sampled within a community, has important applications in community assembly and biodiversity conservation. Habitat fragmentation is a main threat to the biodiversity. However, the effects of habitat fragmentation on the change of SAD have little been developed. Generally, the shapes of SAD, such as logseries model and lognormal model, are affected by the trade-off between successful colonization and extinction in fragmented ecological communities. In this study, we used the Gambin's alpha index, which reflects the shape of SAD, and nestedness analysis to study the forest community assembly in the Thousand Island Lake (TIL), eastern China. TIL is a large man-made hydroelectric reservoir formed by the construction of Xin'anjiang Dam on the Xin'an River in 1959, with more than 1,000 land-bridge islands. On 24 study islands with secondary Masson pine forests, all woody plants with diameter at breast height (DBH) $\geq$ 1 cm were tagged, measured, identified and stem-mapped on the entire island of small islands (area < 1 ha) and 0.5-1 ha plots on large islands (> 1 ha) during 2009-2010. We found that the SADs of 22 islands in all 24 islands were fitted for logseries model better than lognormal model, and Gambin's alpha index significantly decreased when islands area increased, which indicate that the shape of SAD changed from lognormal to logseries with the increase of island area. All islands were dominated by similarly common species, and the species spatial turnover of different islands mainly occurred among rare species. Moreover, with the increase of difference between the area of two islands, the nestedness (i.e. species richness loss) became higher and species spatial turnover became lower. These results suggested that island area determined the shape of SAD, and the successful colonization on large islands with the selective extinction of rare species on small islands caused the change of SADs shape in this fragmented landscape. The study results have substantial implications for conservation: a logseries distribution implies that a large fraction of species is committee to extinction and the rare species showed easier to extinction on small islands, we should focus on the rare species in the continuous habitat before fragmentation.

T3

A237

Role of indehiscent pericarp in formation of soil seed bank in five cold desert Brassicaceae species

Juan J. Lu[1], *Dun Y. Tan*[1], *Carol C. Baskin*[1,2], *Jerry M. Baskin*[1,2]

1. Xinjiang Agricultural University

2. University of Kentucky

The dispersal and germination unit of some Brassicaceae species is the fruit, but few studies have tested effects of the pericarp on formation of a soil seed bank in this family. We determined the effects of the indehiscent pericarp on germination and longevity of buried seeds of five Brassicaceae species native to cold deserts of central Asia. Germination phenology (seedling emergence) was monitored for intact dispersal units and isolated seeds of *Chorispora sibirica*, *Goldbachia laevigata*, *Spirorrhynchus sabulosus*, *Tauscheria lasiocarpa* (annuals) and *Sterigmostemum fuhaiense* (perennial) at natural temperatures in watered and nonwatered (natural precipitation) soil. Intact dispersal units and isolated seeds were buried under natural conditions and exhumed at regular intervals for 35 months to monitor germination, viability and moisture content of isolated seeds, seeds in dispersal units and seeds removed from dispersal units after burial. Isolated seeds of *Goldbachia*, *Spirorrhynchus* and *Tauscheria* germinated only the first autumn and those of *Chorispora* and *Sterigmostemum* the first autumn and first spring, with higher germination percentages in all species in watered than in nonwatered soil. A high percentage of seeds in buried dispersal units of *Chorispora*, *Goldbachia* and *Sterigmostemum* was viable after 35 months, and seeds exhibited a 6-month dormancy cycle, being nondormant only in autumn and spring. Seeds in buried dispersal units of *Spirorrhynchus* and *Tauscheria* germinated when exhumed the first spring, but all non-germinated seeds were dead after 1 year. Presence of the pericarp allows *Chorispora*, *Goldbachia* and *Sterigmostemum* to form a persistent seed bank but not *Spirorrhynchus* and *Tauscheria.*

T3

A238

Mutualism between star anise (*Illicium verum* Hook. f., Schisandraceae) and its pollinator: Larvae and adult gall midges exploit rotten floral organs for food

Zhonghui Ma[1,2], *Shixiao Luo*[1], *Dianxiang Zhang*[1]

1. South China Botanical Garden, CAS

2. Guangxi Uniersity

An understanding of the reproductive ecology of tree species is crucial for their effective conservation and management, yet such information is often lacking. In this study, the pollination system involving *Illicium verum* (star anise) was studied in southern China. Gall midges (*Anabremia* sp., Cecidomyiidae), frequently visit and oviposit in the flowers of *I. verum*, and their eggs stay dormant in the flowers untill intensively hatch into larvae after the end of the sexual phase. The fungus *Choanephora* sp. (Choanephoraceae) and *Rhizopus* sp. (Mucoraceae) infect the flowers during the post-sexual phase as the floral organs gradually becoming rotten and succulent. Adult midges feed on the mycelia and their larvae suck secretions of rotten tissues to obtain additional nutrition. Different from post-sexual phase heating found in other Asian *Illicium* species, stable lower than ambient temperature in the flower chambers during the sexual phase, and abundant mycelia of fungi growing in the post-sexual phase are among the factors maintaining the pollination mutualism between *I. verum* and the gall midges in the summer season. A similar tripartite mutualistic mode occurring in unrelated tropical species provided vivid demonstration of plasticity and parallel evolution in pollination mechanisms. The difference of frequency of adult midges visiting and ovipositing in flowers of *I. verum* reasonably explained the significant difference in the natural fruit set ratio and seed abortion rate between the origin and introduced populations. To resolve the high incidence of pollination limitations due to a pollinator-shortage in the introduced site, a series of management strategies are suggested.

T3

A239

Plant community structure responses to addition of water extracts of litter in the alpine meadow of Qinghai-Tibetan Plateau

*Zhouwen Ma, **Fujiang Hou**, Zhaofeng Wang, Saman Bowatte*
College of Pastoral Agriculture Science and Technology, Lanzhou University, Lanzhou, China.

Plant litter is a key component of grassland ecosystems that link plant productivity and soil functions. In this study, we determined the effects of litter quality on plant growth and community characteristics in alpine meadow grassland at Qinghai-Tibetan Plateau, China. We added four concentrations of water extracts of litter (0, 5, 10 and 20%) of *Kobresia setchwanensis*, *Elymus nutans* and *Saussurea hieracioides* and measured their effects on plant community species composition, height, diversity and above-ground biomass. We found addition of water extracts of litter significantly affected the plant community height, above-ground biomass plant community species diversity and plant functional groups (PFG) important value ($P < 0.05$), while plant richness and density were not affected ($P > 0.05$). The responses to addition of water extracts of litter were different depending on the litter type. Both plant community height and above-ground biomass showed a negative linear relationship with litter concentration. In contrast, plant community species diversity indices (Simpon dominance index, Shannon-Wiener Diversity index and Pielou evenness index) were positively related to litter concentration. The status and functional of Gramineae and Sedge PFGs could be decreased, but Legumes and Forbs PFGs could be increased with litter concentration. Our results indicate that litter is an important control of plant growth and community structure in alpine meadow grasslands.

T3

A240

Some biomorphological features of the northern hemiparasites (Orobanchaceae) from Eastern Europe

Svetlana Petrova
Lomonosov Moscow State University

We try to work out the problem whether the transition to parasitism in hemiparasitic Orobanchaceae occurs on the basis of different life forms, or there is a certain trend, leading to the formation of only few specific biomorphotypes. We studied the life forms and the early stages of development in 5 northern hemiparasites from Eastern Europe: *Bartsia alpina*, *Castilleja lapponica*, *Pedicularis lapponica*, *P. sceptrum-carolinum*, *Rhinanthus minor* subsp. *groenlandicus*. Three biomorphological types was distinguished: hemicryptophytes – short-rhizomatous (*P. sceptrum-carolinum*, *C. lapponica*, *B. alpina*) and long-rhizomatous (*P. lapponica*) herbaceous perennial polycarpics with adventive roots and assimilating shoots, and terophyte – an annual monocarpic with a tap root and assimilating shoots (*R. minor*). All analyzed species form lateral haustoria. We studied the early stages of development to reveal the type of hemiparasitic dependence on the host plant. Our experiments showed that all the studied perennial hemiparasites were able to germinate without a host, but their further development depended on the presence of a suitable host plant. Different experiments to stimulate a germination of *R. minor* subsp. *groenlandicus* seeds, failed. Nevertheless, some authors considered *R. minor* as the facultative hemiparasites that does not require any special factors for germination. The seedlings of studied species have much in common in their structure: they have an embryonic root with a ring of absorption hairs at the root neck, a relatively long hypocotyl, a terminal bud and two green cotyledons. The root hairs at the absorption zone occur only in *C. lapponica*; *B. alpina* and both *Pedicularis* species are lack of normal root hairs. However, in *B. alpina* we observed long haustorial hairs at the central area of the embryonic root that can be regarded as an indicator of the earliest attempt to establish a contact with a host, though without special external signals. If we assume that the main features marked the process of structural transformation to parasitism are the reduction of root hairs and increasing of specialization to chemical factors that trigger the formation of haustoria, it can be concluded that the studied species are at different stages of transition to the parasitism. The least specialized species is *C. lapponica*, more specialized taxa are *B. alpina* and then *Pedicularis* species. However, in any case, all the perennial species are obligate hemiparasites. It is believed that annuals are the final step in the process of reductional evolution of life forms, and it can be supposed that for annual hemiparasites it is easier to keep a contact with the host plants, because this contact is needed for a short time of no more than one vegetative season. Perennial hemiparasites, especially obligate, have to form more complex structures and physiological mechanisms to maintain the normal development for years. In reality, as show our results and some literature data the hemiparasites of different taxa may have different biomorphological types, and the transition to a parasitic lifestyle may occur on the basis of different life forms.

T3

A241

Overexpression of *Chorispora Bungeana ABF1* Gene in tobacco enhances cold and drought tolerance by modulating ROS scavenging

Xiaohong Pu, *Guoyan Zhang, Lizhe An, Xiule Yue*
School of Life Sciences, Lanzhou University

Plants show sensitively response to the changing external environment and biotic stress during plant life history. Among these environmental factors, drought and low temperature obviously affect the growth, reproduction and survival of plants. Our previous study demonstrated that *CbABF1*, a bZIP transcription factor from an alpine subnival plant *Chorispora bungeana*, can greatly increase the survival rate of transgenic tobacco (*Nicotiana benthamiana*) under cold and drought stresses. To further reveal the adaptive mechanisms of *CbABF1* transgenic tobacco, physiological characteristics were analyzed. In contrast with WT, *CbABF1* transgenic tobacco showed decreased relative conductivity, H_2O_2 and MDA content, but significantly increased soluble sugar content under cold and drought stress. In addition, it was found that the activities of superoxide dismutase (SOD), catalase (CAT) and peroxidase (POD) of transgenic tobacco were significantly elevated under cold and drought stresses. Together, these findings indicated that CbABF1 enhanced cold and drought tolerance through maintaining the integrity of the membrane system by weakening ROS toxic effects on cells in tobacco, which will provide theoretical support for the further study on the function of *CbABF1* gene.

T3

A242

Necromass stocks in Brazilian Atlantic Forests

Luis Carlos Quimbayo Guzmán, *Simone Aparecida Vieira*
University of Campinas

Forests have a capacity of storing carbon (C) in the form of vegetal tissues, mainly as wood. When a tree dies, an important portion of its C biomass gets stored as dead wood (DW) and litter. Eventually, most of that C is released to the atmosphere, as DW and litter are decomposed. Understanding the dynamic of death and decomposition allows to quantify the role of necromass (DW + litter) in C cycle in tropical forests and its influence in their function as C sink or source. To understand these dynamics, litter stocks and DW stocks, production and decomposition are being measured in a Brazilian Atlantic Forest along two different gradients: (1) altitudinal and (2) successional. 18 forestry inventory plots (1 ha each) are being monitored, all snags (ø > 4.8 cm) are being measured, litter stocks are being sampled with circular collectors (0.32 m^2), line intersect method is being used to measure fallen DW (ø > 2 cm) stocks and production, DW heterotrophic respiration is being measured as a proxy of decomposition with an infrared gas analyzer (EGM-4). All plots are distributed in nucleus Picinguaba, Santa Virginia and Cunha of "Serra do Mar State Park" (São Paulo, Brazil) in altitudes that range from 10 to 1,200 m. The 6 successional gradient forest plots were located between 1,000 and 1,200 m.a.s.l. and were classified as Old-growth, Selectively Logged, Late Succession and Early Succession forests. DW stocks showed a positive relation with altitude, which was determined mainly by the decrease in soil temperature with altitude and its effect in wood decomposition. Litter presented high water content in all places (> 60%), these areas have low forests fire risk. Early successional forest presented a 65% increase in litter stock after a freeze (2016 winter), other successional stages did not present such pattern, which suggests higher susceptibility to extreme weather events in early stages. Along the successional gradient, the early stage presented the lowest DW stocks because it still is a growing forest, with low DW production. DW stocks were similar among other successional stages, so in these tropical Atlantic Forests the DW component might recover its original dynamics only a few decades after disturbance.

T3

A243

Study on antioxidant activity of the extracts from *Solidago canadensis* L.

Xiao Shen[1,2], *Zhenxing Wang*[1,3], *Zhengrong Zou*[1,2]
1. College of life science, Jiangxi Normal University
2. Key Laboratory of Protection and Utilization of Subtropic Plant Resources of Jiangxi Province
3. Southwest Forestry University

Solidago canadensis L. is one of the major invasive species, which has caused the significant environmental and economic influence around in China. In order to utilize and prevent the *Solidago canadensis* L., the total flavonoids and phenolic contents in its ethanol extract (70%), petroleum ether, ethyl acetate, *n*-butanol, water extracts were investigated by $AlCl_3$ assay and Folin-Ciocalteu assay, respectively, and the in vitro antioxidant activities of them were also studied by several methods including ABTS (agreement of basic telecommunications services) assay, DPPH (1,1-diphenyl-2-picrylhydrazyl) assay, FRAP (ferric reduction ability) assay and ORAC (oxygen radical absorbance capacity) assay, with Vc (vitamin c) and BHT (butylated hydroxytoluene) as positive control. The results showed as follows: (1) The ethyl acetate extract possessed the highest amount of total flavonoids (10.12 mg/g) and phenolic (814.77 mg/g) than other extracts. (2) The ethyl acetate extract had the strongest antioxidant activity among other extracts. (3) The antioxidant activity of ethyl acetate extract was higher than Vc ($P < 0.05$). Thus, the ethyl acetate extract of *Solidago canadensis* L. can be regarded as a kind of potential resource of high-efficiency and natural antioxidants. This will provide some foundations and references for further exploiting and utilizing the *Solidago canadensis* L.

T3

A244

Biofertilizer as an ecological pathway to decrease chemical fertilizer in Maize

Safar Nasrolahzadeh[2], ***Ali Shirkhani***[1], *Saeed Zehtab Salmasi*[2]
1. Seed and Plant Improvement Institute (SPII), Kermanshah Agricultural and Natural Resources Research and Education Center. AREEO, Kermanshah, Iran
2. Department of Eco-physiology, Faculty of Agriculture, University of Tabriz, Tabriz, Iran

Iran is a major importer of Maize in the world. Maize is one of the most important crops in the western Iran. Maize production in this region has two major problems: water shortage and low percentage of organic matter in soil. In this research, effects of vermicompost and azotobacter as a boifertilizers and chemical fertilizers levels on yield and some traits of leaves of maize under normal and deficit irrigation was investigated in two years (2014 and 2015). Results showed that grain yield, leaf area index, leaf chlorophyll contents, the normalized difference vegetation index (NDVI) and leaf relative water content (RWC) were decreased by deficient irrigation. Use of azotobacter and vermicompost significantly increased these traits in normal and deficit irrigation. Results also showed that using 6 ton/ha of vermicompost and azotobacter, 50% of the maize fertilizer supplied. The results showed that combined use of biofertilizers with chemical fertilizers increased the yield and other traits. Therefore the uses of biological fertilizers significantly reduce the consumption of chemical fertilizers and reduce the adverse environmental effects. So biofertilizers could be considered as a suitable substitute for chemical fertilizer in organic agricultural systems. On the other hand from this experiment, application of vermicompost in combination with chemical fertilizers showed better performance than only chemical fertilizers, even in 100% recommendation based on soil test treatments. As a general conclusion these results suggest that integrating organic sources with 50% of recommended NPK fertilizers are appropriate for sustainable crop production in normal and deficient irrigation.

T3

A245

Quantitative classification of forest vegetation on Hainan Island

Lin Siliang[1], *Kong Fanmao*[2], *He Jiekun*[1], *Yu Jiehua*[1], *Zhu Hua*[3], ***Jiang Haisheng***[1]

1. School of Life Science, South China Normal University, Guangzhou, Guangdong

2. Guangzhou Qimao Ecological Technology Co., Ltd., Guangzhou, Guangdong,

3. Xishuangbanna Tropical Botanical Garden, Chinese Academy of Sciences, Mengla, Yunnan

Vegetation classification is the foundation and a complex problem of vegetation research. Due to the rich plant diversity and complex community struction, making it more difficult to explore and study the vegetation classification in tropical forest. As a typical tropical island of China, vegetation classification on Hainan Island has been the concern of researchers. Although most previous studies of forest vegetation classifications on Hainan Island had followed the classification system of "*Vegetation of China*", but the disputation are still exists. This paper bases on 870 forest plant plot data, adopting the quantitative classification approach, combining the environmental factors (including climate, topography and soil) to classify, analyze and explore the forest vegetation on Hainan Island. We found that forest on Hainan Island could divide into 2 Vegetation Type Groups (Evergreen forest and Deciduous forest), 6 Vegetation Types (Tropical rainforest, Tropical seasonal moist forest, Tropical montane evergreen broad-leaved forest, Tropical coniferous forest and Other evergreen forest, which are belong to Evergreen forest; Tropical monsoon forest are belongs to Deciduous forest) and 4 Vegetation Sub-Types (Tropical seasonal rainforest and Tropical montane rainforest that are belong to Tropical rainforest, Deciduous monsoon forest and Semi-deciduous monsoon forest that are belong to Tropical monsoon forest), and 32 main Formations. Our result show that forest of Hainan split into Deciduous forest group and Evergreen forest group in the high classification level. Although this result is unlike previous classifications studies, but it reflect the actual precipitation gradiant between eastern and western part of Hainan Island is the key cause in forest vegetation differentiation. The differentia of this two vegetation types is the dominance of evergreen and deciduous tree species in community. The Evergreen forest districts on the easten Hainan Island, whose climate is obvious wet; and the Deciduous forest districts on the westen Hainan Island, whose climate is obvious drought. This conclusion is similar to the other study in the tropical region. We also identified a Vegetation Types: tropical seasonal moist forest by cluster and ordination, which is an interesting vegetation community. It sharing the same climatic conditions with Tropical rainforest, and also maintaining some identification species of Tropical rainforest. But due to the limestone and sandy soil, the soil moisture content is low, which making more deciduous species growth in those community than the Tropical forest. Furthermore, the ecological physiognomy and floristic composition of community show significant different with Tropical rainforest. In addtion, it is improper to classify those community into Tropical monsoon forest only according to the ecological physiognomy. It also show significant differences in floristic composition and climatic condition with Tropical monsoon forest.

T3

A246

Nuclear AFLP loci drive adaptive population genetic differentiation in the old rare conifer *Pseudotaxus chienii* (Taxaceae): Effects of climate, soil, and their interactions

Qi Deng[1], *Zhen Wang*[2], *Ting Wang*[3], ***Yingjuan Su***[1]

1. Sun Yat-sen University

2. Nanjing Agricultural University

3. South China Agricultural University

Respective roles of climate, soil, and their interactions on the adaptive genetic differentiation in plants remain to be elucidated. In this study, we selected an old rare conifer *Pseudotaxus chienii* as a model species to explore the association between adaptive genetic variation and climate and soil variables. We measured soil and extracted climate data. We performed genome scans for 11 natural populations of *P. chienii* across its entire distributional range to detect adaptive genetic loci by using amplified fragment length polymorphism (AFLP), nuclear simple sequence repeat (nSSR), and chloroplast simple sequence repeat (cpSSR) markers. Then the spatial analysis method (SAM) was applied to evaluate the correlation between adaptive genetic loci and environmental variables. Moreover, the neutral genetic variation pattern, spatial genetic structure, linkage disequilibrium (LD), and the potential geographic distribution based on ecological niche modellings were also examined. We identified 26 AFLP but no nSSR and cpSSR adaptive loci in *P. chienii*. We found 22 of the AFLP adaptive loci significantly associated with climate (precipitation and temperature) and/or soil (K, Ca, Mg, Zn, pH, and electrical conductivity) variables, of which 17 solely linked to climate, three related to both climate and soil, whereas only two specific to soil. The ongoing global warming was predicted to induce population shrinkage in southern China and shift to Central China. *Pseudotaxus chienii* was found to possess a moderate to high level of genetic variation and differentiation, a significant fine-scale spatial genetic structure within the distance of 2 km, and an extremely low proportion of LD between genetic loci. The adaptive response to environmental changes of *P. chienii* might be predominantly driven by nuclear rather than chloroplast genome. Climate is of relatively greater importance than soil in shaping the adaptive genetic differentiation, wherein the synergistic effects of climate and climate-soil variables were also validated. The identified climate and soil variables should be taken into account when applying ex-situ conservation. This work was supported by the National Natural Science Foundation of China (31570652).

T3

A247

Species diversity and population dynamics of endangered plant *Disanthus cercidifolius* var. *longipes* in China

Yunong Tian, *Xiaodong Jin, Guowen Xie, Xueyin Guo*
School of Life science, University of Guangzhou

Disanthus cercidifolius var. *longipes* is a rare and endangered species of China, and only distributs in a few counties in Guangdong, Hunan, Jiangxi and Zhejiang provinces. the community characteristics and population dynamic of this species were studied by using typical sample method, including species diversity, age structure and population distribution pattern *etc.* The results are as follows: 1). There are rich plant resources in *D. cercidifolius* var. *longipes* community, including 331 species of vascular plants, belonging to 204 genera, 93 families. The type of the pantropical distribution is the most common type in the areal types of families, accounting for 47.56% in total, which shows this community originated from tropics. In the areal types of genera, the proportion of the tropical distribution is similar to that of the temperate one, indicating the transitional floristic characteristics from tropic to temperate. Based on cluster analysis, the type of the temperate distribution is the highest in the areal types of non-Chinese endemic species, while the type of the tropical distribution is the highest in the areal types of species endemic to China. 2). The species diversity of *D. cercidifolius* var. *longipes* community. The variation tendency in species richness index, Simpson index, Shannon-Wiener index and Pielou evenness index in different sample plots is basically consistent. The species richness index and Shannon-Wiener index of the shrub layer are higher than those of the tree layer, while Simpson index and evenness index of the tree layer are higher than those of the shrub layer. 3). The population distribution pattern of *D. cercidifolius* var. *longipes* population belongs to cluster type. The size structure is rendered as the typical pyramid, indicating that this population is in a growth period. The static life table analysis showed that the life expectancy of the population decreased with the increase of size structure. The survival curve of this population is approximately linear, indicating its growth is relatively stable. 4). The endangered causes of *D. cercidifolius* var. *longipes* were analyzed and the results showed that the internal causes include: some barriers of its seeds germination in natural conditions; forced dormancy and physiological dormancy; poor pollen competitiveness, shortage of pollinators and low pollination efficiency. The external causes include: the high seedling mortality rates due to the illumination conditions; geographical isolation and less communication between populations; the human disturbance. In order to better protect the *D. cercidifolius* var. *longipes*, in situ and ex situ conservation should be combined.

T3

A248

Higher plant as biomonitors of metal contamination in abandoned karst bauxite, Southwestern China

Dengfu Wang[1,2], *Zhaohui Zhang*[2]
1. Institute of Karst Science, Guizhou Normal University
2. Key Laboratory for information System of Mountainous Area and Protection of Ecological Environment of Guizhou Province, Guizhou Normal University

Bauxite mining on karst results in several ecological and environmental issues including metal pollution, water and soil erosion and the destruction of vegetation. However, untreated abandoned karst bauxite may result in hazards to ecosystems and human health. In general, higher plants are commonly used as biomonitors to estimate the pollution of metals in the soil. The plant specimens and soils were collected on Shangdong abandoned karst bauxite near the County of Xiuwen in the Guizhou Province of southwestern China. Higher plants containing *Pteris vittata*, *Miscanthus floridulus*, *Coriaria nepalensis*, *Pinus massoniana*, *Artemisia argyi and Senecio scandens*. The levels of six metal elements (Al, Mg, Ca, Fe, K, Ti) in the plants and soils were determined. The metal contents in soil substrates: aluminum, 9,827-57,000 mg/kg; magnesium, 612.1-8,343 mg/kg; calcium, 12,650-84,290 mg/kg; iron, 3,274-63,900 mg/kg; kalium, 5,673-19,550 mg/kg; and titanium, 158.9-8,840 mg/kg. The metal concentrations in plants tissues: root, 4.304-24,750 mg/kg; stem, 10.04-7,758 mg/kg; and leaf, 6.545-31,310 mg/kg. Compared and analyzed the concentrations of metals in plants and soils. The results showed that *Pteris vittata* plant root content of aluminum the highest, *Artemisia argyi* the lowest; *Senecio scandens* plant stem content of kalium the highest, *Artemisia argyi* the lowest, *Coriaria nepalensis* plant leaf content of calcium the highest, *Miscanthus floridulus* the lowest and different plants contained varied concentrations of metals, the same plant contained varied concentrations of metals in plants tissues, thereinto, the levels of metals in *Pteris vittata* plant and *Miscanthus floridulus* plant, root > leaf > stem, the indicated that the metal elements may have transport and transformation process in plants tissues. Correlation analysis showed that Mg, Ca and K contents were extremely significant positive correlation in soils ($p < 0.01$); Mg, Ca and Al contents were extremely significant positive correlation in plant roots ($p < 0.01$); Mg and K contents were significant positive correlation in plant stems ($p < 0.05$); Al and Ti contents were extremely significant positive correlation in plant leaves ($p < 0.01$). Through the research, we found that *Pteris vittata* plant contents of metals (Al, Ca, Fe, K) were highest in the soils, *Pteris vittata* plant root contents of six metal elements were highest and *Pteris vittata* > *Miscanthus floridulus* > *Coriaria nep-*

alensis. Therefore, *Pteris vittata* plant had strong accumulation ability to the metals contents, which its can be used as biomonitor indicator of metals contamination and provided materials for ecological monitoring and restoration in abandoned karst bauxite.

T3
A249
Effects of exclosure on community interspecific relationships in Hulunbuir Meadow Steppe

Guoqing Wang
Heilongjiang Bayi Agricultural University

We selected 17 species according to important values in each plot, using variance ratio to analyze the overall association. The χ^2 test with 2 × 2 contingency table, Ochiai index and PC were applied to analyze the inter-specific associations of dominant and companion species. The results indicated that the community was insignificantly negative correlation inside exclosure, the community was insignificantly negative correlation outside exclosure. The χ^2 test that the proportion of positive and negative correlation in the inside exclosure is below the outside exclosure, and the community is not stable inside exclosure. The Ochiai index is consistent with the PC trend in the inside exclosure and outside exclosure. The maximum value of Ochiai index and PC appears in the *Cleistogenes squarrosa* and *Leymus chinensis, Cleistogenes squarrosa* and *Poa sphondylodes, Leymus chinensis and Poa sphondylodes.* The advantage of dominant species is obvious in the inside exclosure. The Ochiai index and PC of dominant species inside exclosure is lager outside exclosure.

T3
A250
Study on plant niche of Hulunbuir meadow steppe under enclosure

Guoqing Wang
Heilongjiang Bayi Agricultural University

To study the niche characteristics of meadow steppe, The meadow steppe were monitored using the method of sampling in field investigation. The plant height, coverage, density and aboveground biomass were measured. The calculation of the niche breadth of Levins and Pianka's niche overlap is used. The results showed that the niche breadth of *Leymus chinensis, Poa sphondylodes* Trin, *Cleistogenes squarrosa* (Trin.) Keng were the largest in enclosure. The maximum of niche overlap was between *Cleistogenes squarrosa* (Trin.) Keng and *Poa sphondylodes* Trin. The niche overlap of *Cleistogenes squarrosa* (Trin.) Keng, *Poa sphondylodes* Trin, *Bupleurum chinense* DC. were the largest outside enclosure. The maximum of niche overlap was between *Galium verum* L. and *Potentilla acaulis* L. The forage of *Gramineae* was the adequatest in enclosure.

T3
A251
Niche dynamics of dominant species during restoration process of meadow steppe in enclosure

Guoqing Wang, *Guangming Du*
College of Animal Science & Veterinary Medicaine, Heilongjiang Bayi Agricultural University

The niche dynamics during restoration process of dominant in enclosure meadow steppe of Hunlunbuir were analyzed by means of niche breadth index and niche overlap index. The results showed that the total breadth of niche of *Leymus chinensis, Poa sphondylodes, Cleistogenes squarrosa* were higher other species. *Poa sphondylodes, Leymus chinensis, Allium neriniflorum, Leymus chinensis* were the species with the largest niche breadth for the restoration stage of 1, 3, 5 and 7 years respectively. The niche breadth of *Leymus chinensis* has been higher. The niche breadth of *Poa sphondylodes, Cleistogenes squarrosa* was gradually decreased. There was a trend that the niche breadth of *Allium neriniflorum, Serratula centauroides, Caragana microphylla* increasing with increasing degree of community restoration. The dominant position of Gramineae decreased. There were linear relationships of between niche breadth and niche overlap. The interspecific competition was increased with enclosure. Then, the interspecific generated niche differentiation and competition was decreased. The develop of community towards stable state.

T3
A252
Anatomy stem of species of moss from the source area of Urumqi river

Hong Wang, *Abuduaini Aifeire, Zhen Yan Yuan, Yong Man Lu*
Xinjiang University

The moss plants belong to 13 families, 19 genera and 37 species were gathered from the source area of Urumqi river. The stem structure treated paraffin sectioning was observed by scanning electron microscopy (SEM). It exits the obvious differences among the different families and genera, The stem of most species has a cellular layer with different length epidemic cell, but there are a little significant differences among the same genera different species, For example, *Ditrichum flexicaule., Distichium capillaceum* and *Distichium inclinatum* belong to the same family, although there are many similarities on their stems. But, there are a lot of differences of their stems, the epidermal stem cell wall of *Ditrichum flexicaule* is smooth and the different levels of depression, which has a variety of irregular decorations on surface, the cell walls of outside and inside cortex are thickened strongly, the small parenchyma cells which crowded into a small ball form the axis. For *Distichium capillaceum*, the outer wall of the epidermis is thin and inward concave strongly, there are a large number of particles and the small pores on the surface, the endothelial cell wall is relatively thin. Compared with the former two species, *Distichium inclinatum* has different characters on stem: epidermal stem cells studded a small amount of grain arrange irregularly, axis cells are much bigger, arrange mosaic. Thus, these micromorphological structure as one of the classification basis, and characteristics of cold and arid resistant. For example, stem epidermis cell walls are strong, the cell wall is rough, When the weather was dry the cuticle layer wrinkled, cortex cell is large, axis the cell wall corner is thickened *etc.*, and it is better to understand how these mosses to adaption the extreme environment, it can provide some reference to study the structure of the stem cells in the extreme cold environment. Therefore, it is helpful to the further research on the cell genetic and physiological ecology.

T3

A253

Leucospora multifida—A new invasion species in China

Ying Wang, *Ruifang Gao, Daojian Yu, Dongmei Chen, Caiyu Xiang, Heqing Huang, Jun Wang*
ShenZhen Entry-Exit Inspection and Quarantine Burea

In 2014, a new invasion species that had never been recorded in China, *Leucospora multifida* (Michx.) Nutt. (Narrow-leaf paleseed, cleft conobea), was found in the gap on wharf hardened cement ground used for the import of bulk raw grains in Shenzhen. *Leucospora multifida* (Scrophulariaceae) is the only specie of Genus *Leucospora*. The probability of entry is high. The plant of *Leucospora multifida* is native to southwest of United States. It is now found in the central and eastern states of US, including Florida. It was also been recorded in southeast of Canada. This plant is suitable for the shade, sufficient sunlight and the moist environment. The muddy, sandy, gravel and rock soil are also prefer to growing. The plant is annual herbaceous, short, erect, with the height of 5.0~21 cm and much branches. There are full of sticky glandular pubescent on stem, leaf, flower stalk and sepal. Terete, green or purple tinged with green of the stems. Although the plant is short, the root system is flourishing with as far as over 30 roots of a well growing individual. Each root plunged into the ground and then a number of radicular grow. The roots range cover an area of more than 1 m^2, which are firmly into the soil or cement cracks in the ground. A plant could produce lots of seeds. Each blade (opposite or juniors) has two flower leaf axils or two capsules, each capsule containing 60~90 seeds. The probability of establishment is high. This plant reproduce mainly by seeds, the seeds are as small as dust. The surface of the seed is sticky so as to be attached to waterfowl and poultry (such as duck, goose) wet feathers, which could be also mixed in bulk grain or with soil during long-distance communication of plantlet. The seed could spread by wind and rain, or in the mud with the movement of person, vehicles and animals. The probability of spead is high. According to the pest risk assessment, the pest introduction risk of the invasive species is concluded as a high level.

T3

A254

A unique method to trace the dynamic expansions of *Spartina alterniflora*

Yaqian Wu[1], *Xiangming Xiao*[1,2], *Bangqian Chen*[1]
1. Institute of Biodiversity Science, Fudan University, Shanghai, China
2. Department of Microbiology and Plant Biology, Center for Spatial Analysis, University of Oklahoma, Norman, USA

Biological invasion seriously affects the local ecosystem balance, causes a certain threat to ecological security. Three types of Landsat images were embedded in MODIS pixels which were selected by three observation points. Combined with history and field survey data, there were all beaches in each MODIS pixel in 1984 until *Spartina alterniflora* was naturally expanded. *Phragmites australis* and *Suaeda salsa* also live in this wetland. Comparing with spectral indices, *Spartina alterniflora* has a longer growth cycle and higher productivity than *Suaeda* and Reed. This is one reason why it has spread so fast here. From 1990-2000, *Spartina* showed a quick invasion trend, the rate is ranging from 15.5% to 44.4%. After that, two inland MODIS pixels have changed to pure *Spartina* pixels, while the other one which is near the coast completely was expanded in 6 year ago.

T3

A256

Phylogeography and evolution of *Quercus delavayi* from subtropical evergreen broad-leaved forest in Yunnan Plateau

Jin Xu, *Min Deng*
Shanghai Chenshan Botanical Garden/Shanghai Chenshan Plant Science Research Center, Chinese Academy of Sciences

Quercus delavayi is a character/dominant element of subtropical evergreen forests in Yunnan Plateau. This species provides an ideal model system to examine Late Tertiary/Quaternary phylogeography in relation to ancient geological and climatic events in Yunnan Plateau. In this study, we selected 33 populations of *Q. delavayi*, 314 individuals, using three chloroplast DNA (cpDNA) intergenic spacer regions (*psb*A-*trn*H, *pet*G-*trn*P and *trn*T-*trn*L) to do phylogeographic analyses, discussing the population genetic structure and genetic diversity level of this species. The results show that *Q. delavayi* has high genetic diversity ($H_T = 0.907$) and high genetic differentiation ($G_{ST} = 0.782$, $N_{ST} = 0.912$, $P<0.05$). The phylogenetic analysis revealed two major haplotype lineages (North-central Yunnan vs. Northwest Yunnan and South Yunnan). The fossil-constrained BEAST analysis revealed that haplotype divergence began at late Miocene. The topographic changes related to the uplift of Yunnan Plateau might play a significant role on intraspecific divergence of *Q. delavayi*. This species occupied multiple refugia during the Pleistocene glacial periods. The combined effects of paleogeomorphological and climatic changes of the Cenozoic have had profound impacts that shaped the extant distribution of *Q. delavayi*.

T3

A257

Super absorbent polymer seed coatings promote seed germination and seedling growth of *Caragana korshinskii* in drought

Liqiang Su, Jiaguo Li, ***Hua Xue****, *Xiaofeng Wang**
National Engineering Laboratory for Tree Breeding, College of Biological Sciences and Biotechnology, Beijing Forestry University, Beijing, PR China

Coating the seeds with water absorbent materials could improve their survival, especially for those planted in drought or barren area. In this study, effects of five kinds of super absorbent polymers (SAPs) on seed germination and seedling growth of *Caragana korshinskii* under drought condition were investigated. Our results showed that SAP coatings could significantly improve the percentage and energy of seed germination, as well as reduce relative electrical conductivity (REC), proline, MDA, H_2O_2 content and POD activity during germination. These results implied that seeds could uptake moisture from SAP coatings to alleviate drought-induced oxidative stress and membrane damages, thus exhibit a better vigor and germination performance. Meanwhile, after coating *C. korshinskii* seeds with SAPs, more seedlings emerged and grew better. Under the combined influence of water absorption capacity of SAP and other factors, the efficiency of five SAP coatings are in the sequence of D > E > B > A > C. Moreover, the function of SAP coating on promoting seedling survival was confirmed in Mu Us Sandy Land. The average seedling number of SAP D coated seeds increased to two folds of that of naked seeds. Our results are expected to be helpful on understanding and utilizing SAP seed coatings in improving plant survival under drought conditions.

T3

A258

Effect of mulching on root-zone soil environment and physiological traits of young apple trees

*Ping Huang, Ruixue Zhang, Hui Cao, **Hongqiang Yang***

College of Horticulture Science and Engineering, Shangdong Agricultural University

Soil surface mulching was one of important method to promote sustainable production in orchard system by stabilizing the soil environment. The experiment explored the variation of the moisture, temperature and available nutrients in root-zone, root architecture and nitrate metabolism transformation, leaf photosynthesis and transpiration, leaf water use efficiency (WUE) and plant growth using three-year old potted apple trees (*Malus domestica* cv. *Starkrimson*) mulched with straw mat, agricultural carpet, transparent-plastic film and horticultural fabric. The results showed that soil moisture, temperature, new root volume, root fractal dimension and root activity increased significantly by mulching the above 4 materials. The increment of relative water content in the soil mulched with straw mat and horticultural fabric was largest. Straw mat and agricultural carpet made the soil temperature more stabilized. Four treatments all enhanced the stability of soil pH and increased the content of soil organic matter, alkali-hydrolyzable N, available P and available K. The root fractal dimension was the highest and root structure was more complex after mulching straw mat, the root diameter, root volume and surface-area were the largest and the roots were thick and strong after mulching agricultural carpet; the total root length and root activity the highest after mulching transparent-plastic film; the number of root tips was the most after mulching horticultural fabric. There were significant seasonal differences in the effects of four treatments on root nitrate metabolism transformation. The activity of root nitrate reductase (NR) and nitric oxide synthase (NOS) and the production rate of nitric oxide (NO) were significantly increased in the late summer and the early winter; and the effect of straw mat was the most significant on them, while the transparent plastic film decreased them significantly in late summer. In the Mid-autumn, all the four mulching treatments increased the activity of root NR and NOS, while horticultural fabric increased root NO production. During the whole treatment, the four mulching treatments promoted the new shoots growth and stem diameter enlargement, in which the straw mat and the horticultural fabric were more effective in promoting plant growth, while transparent plastic film with minimal effect. For the comprehensive effect of the 4 mulching materials on soil, root and leaf, the order of beneficial degree was straw mat, agricultural carpet, horticultural fabric and transparent-plastic film from high to low.

T3

A259

Camellia japonica (Naidong) under different illumination and nitrogen deposition of physiological ecology response

Jikai Yang

Qingdao Agricultural University

Global climate change, including the change of illumination and nitrogen deposition, *etc.*, are growing, this paper studies different illumination and nitrogen deposition on *Camellia japonica* (Naidong) of the morphological index, chlorophyll content and chlorophyll fluorescence characteristic, the influence of the gas exchange characteristics and physiological ecology characteristics. Will help to understand and predict the impact of climate change on winter resistance camellia, to better adapt to the environment and provide a theoretical basis for protection. The research results show that the plant height (H), base diameter (BD), the crown (CA), specific leaf area (SLA) and water content (LWC) is not affected by nitrogen and nitrogen and light interaction effects, under the light and the interaction of the nitrogen morphological indicators of winter *Camellia* resistance with the increase of number of days on the rise, the same with fewer illumination, under the condition of nitrogen morphological indicators of winter camellia resistance drops, also showed the low light to its growth has certain inhibitory effect, but the L1 morphological indexes were higher than in L2. Part of the gas exchange parameters affected by illumination, nitrogen deposition, and the interaction was significant. Winter resistance of camellia vapor pressure deficit (VPD) by nitrogen deposition, illumination and the effects of the interaction of the two significant, net photosynthetic rate (A), transpiration rate (E) significant difference under nitrogen deposition and lighting effects, but no significant difference under the two interaction, water use efficiency (WUE) in significant difference under the influence of nitrogen deposition, no significant difference under the light and the interaction between the two. L2 winter resistance under low illumination camellia chlorophyll than L1 illumination condition, but Chla/Chlb above L2, chlorophyll, the only significant difference under the light conditions, under nitrogen deposition and the interaction between the difference was not significant. L2 winter camellia resistance under variable fluorescence (Fv), maximum fluorescence (Fm), PS II photochemical efficiency (Fv/Fm) and PS II potential active (Fv/F0) values are higher than that of L1, and initial fluorescence (F0) value, on the other hand, all the chlorophyll fluorescence parameters of the plants were very significant differences only under the effect of light, and under the light and the interaction between the two was not significant.

T3

A260

Differential effects of plant identities on soil nematode community in a coal gangue mountain

***Zaichao Yang**, Fei Xie, Jinghui Zuo, Hong Xiang, Xuying Wang, Wen Liao, Shiming Han*

Liupanshui Institute of Biology, Department of Life Science, Liupanshui Normal University

Coal gangue is a low combustion value and commercially worthless rock during mining. Some different plant mosaics have grown in surface soil of the accumulative coal gangue, however, few studies about how soil community respond to these plant mosaics. Here, we investigated if plant mosaics effect on soil nematode community are mediated by plant or other soil factors. We showed that nematode diversity was maximal at both legume (*Trifolium repens*) and grass (*Setaria viridis*), and nematode diversity was minimal at ruderals (*Plantago asiatica* and *Taraxacum mongolicum*). In addition, nematode diversity was positively related to litter biomass and soil moisture and was negatively related to pH. In functional groups of nematode community, the amount of bacterial-feeder, carnivores and omnivores was most abundant at legume, and that of plant-feeder was most abundant at grass, and that of fungal-feeding nematodes was most abundant at ruderals. Using a structural equation model, we explained 62% of the variance in soil nematode community, identified direct and indirect effect paths, and showed that the different groups of plant accounted for approximately three-fifths of the explained variance. We conclude that plant identity may be a key mediator in regulating soil community diversity.

T3

A261

Re-analysis of transcriptome data reveals effects of salt and drought stress on regulatory gene expression in xerophyte *Zygophyllum xanthoxylum*

Meng-Zhan Li, Qing Ma, Wei Li, Suo-Min Wang, ***Hong-Ju Yin***
Lanzhou University

Zygophyllum xanthoxylum is a succulent xerophyte species with remarkable tolerance to diverse abiotic stresses. Previous studies have preliminarily revealed physiological mechanisms of *Z. xanthoxylum* in response to abiotic stresses. Furthermore, our transcriptomics study has identified many potential functional genes which may directly involve in salt and drought tolerance in *Z. xanthoxylum*. However, the upstream regulators of these functional genes and the molecular mechanisms are still elusive. In the current study, we re-analyzed the transcriptome and digital gene expression profiling data of *Z. xanthoxylum* roots during 50 mM NaCl (salt) and -0.5 MPa osmotic (drought) treatments. Among the total 6,258 and 6,063 differentially transcribed genes, a lot of regulatory genes were up- or down-regulated in response to salt and drought respectively, including those encoding transcription factors (TFs) in various classes (Such as NAC, WRKY, AR2/ERF, MYB and bHLH), protein kinases and phosphatases, and genes associated with protein turnover (such as ubiquitination and sumoylation enzymes encoding genes). To further extend functional characterization, we identified the orthologs of differentially expressed genes in the model species *Arabidopsis thaliana*, which then allowed us to perform in silico functional inference, including gene network analysis for protein function, protein subcellular localization and gene co-expression. The data reported here indicated the extensive potential regulatory events of salt and drought tolerance in *Z. xanthoxylum*, and will shed new light on the molecular mechanisms of stress tolerance of the xerophyte plant species.

T3

A262

Salt accumulation of vegetative organs and ecological stoichiomery characteristics in typical halophytes in Xinjing, China

Yanhua Yong*, Xia Zhang, Shaoming Wang, Meilin Ren, Meilin Yang, Yuan Liu, Zhiyong Zhang*
Shihezi University

Aims: Study on the salt accumulation in vegetative organs and C, N, P stoichiometry in halophytes contributes to understanding the adaptive strategy to the saline environment and the distribution mechanism of nutrients and salinity of the halophytes. *Methods:* Four typical dominant species, including three perennial shrubs (*Halocnemum strobilaceum*, *Suaeda microphylla* and *Kalidium foliatum* and an annual herb (*Salicornia europaea*) were selected for analyzing the salt accumulation and C, N, P stoichiometry, as well as their relationships. *Important findings:* 1) The salt accumulation appeared "salt island" effect at vigorous stage, because Na^+, Cl^- and EC showed significant positive trends, when salt ions transported from the root to the top of 4 halophytes. 2) P was growth restrictive element of perennial shrubs during their growth and development, while both N and P were restrictive factors for annual herb. 3) Organs, species and their interactions affected C, N, P stoichiometry and salt icons except Mg^{2+}. 4) C was negatively correlated with N and P, but there was a significant positive correlation between N and P. 5) EC, Na^+ and K^+ were positively correlated with N, N:P and negatively correlated with C, C:N and C:P, and yet K^+, CO_3^{2-} had significant positive correlation with P. These results implied that there was some kind of mutual promoting relationship between nutrients and salinity in halophytes dealing with the saline environment, and fertilizing with N can promotes sodium chloride saline soil desalted.

T3

A263

De novo assembled and characterization of the transcriptome response to cadmium in Cosmos (*Cosmos bipinnatus* cav.) roots

Xiaofang Yu*, Yi Wang, Yujing Liu, Jian Zeng, Xing Fan, Lina Sha, Houyang Kang, Haiqin Zhang, Yonghong Zhou*
Sichuan Agricultural University

Cadmium (Cd) causes damages to various plants, while some plants have unique detoxification and metal transport mechanism to reduce Cd toxicity. In present study, *Cosmos bipinnatus* cav. could be regard as "Cd accumulation plant" and have potential for phytoextraction. Therefore, physiological and transcriptome analyses were carried out. In this study, *C. bipinnatus* accumulated 60.36 ± 1.67 $\mu g\ g^{-1}$ Cd in leaves, 321.15 ± 16.04 $\mu g\ g^{-1}$ Cd in stems, and 576 ± 41.48 $\mu g\ g^{-1}$ Cd in roots under 40 μM Cd treatment, but showed no obvious symptom response to Cd. Subcellular distribution analysis indicated that vacuole is the predominant sink for Cd sequestration in roots while cell wall can be another large part of Cd accumulation at high concentration of Cd exposure (120 μM). In addition, the result of transcriptome showed that 66,407 unigenes was assembled and 41,674 (62.75%) of the unigenes were annotated in at least one database, 40.16%

of total annotated unigenes were longer than 1,000 bp. Moreover, compared with CK, the total of 3,814 unigenes including 2,067 up-regulated unigenes and 1,747 regulated unigenes were regulated by Cd. RT-qPCR showed that the expression profiles of differentially expressed genes (DEGs) were in consistent with results from RNA-Seq analysis. Among these DEGs, we found several candidate genes encoding *ZIPs*, *ABC transporters*, *HMA2*, *HMA3*, *NRAMP2*, *NARMP3*, *GST*, *Grx* and peroxidase were response for Cd influx, accumulation, transportation and detoxification in *C. bipinnatus*. These results provide a novel insight into characterization of the transcriptome response to Cd in *C. bipinnatus*.

T3
A264
Methane uptake dependent on biotic factors rather than precipitation, warming and nitrogen deposition in a temperate desert

Ping Yue[1], *Xiaoqing Cui*[2], *Fengzhan Geng*[1], *Wenchao Wu*[1], *Kaihui Li*[1], *Yanming Gong*[1], *Xuejun Liu*[2]
1. Xinjiang Institute of Ecology and Geography, Chinese Academy of Sciences
2. China Agricultural University

Climate change and elevated nitrogen (N) deposition play vital roles in the balance of the atmospheric methane (CH_4). However, information on the response of CH_4 uptake to elevated precipitation, temperature and N deposition in temperate deserts is still scarce. An *in situ* field experiment was conducted to investigate the effect of simulated precipitation, warming and N deposition on CH_4 uptake in the Gurbantunggut Desert, northwest China, from September 2014 to October 2016. The results showed that the Gurbantunggut Desert was a weak sink for CH_4 (0.92 kg C ha^{-1} yr^{-1}) and non-growing season accounted for 28.04% of annual CH_4 uptake. Our research was different from some previous studies which reported a strong sink for CH_4 in desert soil. Treatments with increased precipitation (30% or 60 mm yr^{-1}) increase and low N deposition elevation (30 kg N ha^{-1} yr^{-1}) significantly enhanced soil CH_4 uptake, while no significant impact on CH_4 uptake by warming and higher N deposition (60 kg N ha^{-1} yr^{-1}). Therefore, the sink of CH_4 in temperate desert was not significantly impacted by warming, increased precipitation and N deposition. However, we found that biotic factors could explain CH_4 uptake by 48.8%, while only 10.4% contribution was by environmental factors. The results indicate that CH_4 uptake was mainly controlled by biotic factors (such as plant biomass, microbial biomass carbon and microbial biomass N) and was not sensitive to elevated N deposition and precipitation and warming in this temperate desert.

T3
A265
Energy flows and biodiversity in a model ecosystem

Fanqi Zeng, *Wenxu Wang*, *Jiang Zhang*
School of Systems Science, Beijing Normal University

Climate change has been caught great attention. Ecological systems coevolve with climate systems. Thus, understanding the intrinsic dynamics of ecological systems is of paramount importance for exploring and predicting climate change. Food chains, as a systematic and simplistic model, have been a paradigm for investigating the long term evolution of ecological systems in virtue of the dynamical analysis on the inter-play between the topology and the dynamical processes occurring on a food chain. We build a food chain model by considering energy exchanges among species, which is characterized by using population dynamical equations. We will compare our model with existent food chain models based on sufficiently empirical data. Our work is expected to be able to quantitatively capture and describe the interactions among species with respect to energy exchange, then deepen the understanding of ecological systems and their coevolution with climate change.

T3
A266
The elevational gradient of mountain deciduous forests, patterns and reasons

Chao Zhao[1,2], *Baiping Zhang*[1,2], *Jing Wang*[1], *Jing Wang*[1], *Yonghui Yao*[1]
1. State Key Laboratory of Resource and Environmental Information System, Institute of Geographic Sciences and Natural Resources Research, Chinese Academy of Sciences, Beijing
2. Jiangsu Center for Collaborative Innovation in Geographical Information Resource Development and Application, Nanjing

The thermal gradient along elevational are more compact than the latitudinal, together with the variety of topography, inclination, aspect, and the diverse precipitation conditions, making the mountain vegetation patterns more complex. Vegetation zonation is an easy way to assess the pattern of vegetation along elevational gradient. Most of the previous studies took a zonal belt as homogeneous one, neglecting the gradient inside. Some researchers found that the distribution range along elevation of one species can be divided into the upper zone, core zone and lower zone through detailed investigation, the upper and lower zone are pioneer communities, while the core zone is climax communities. A vegetation belt is zoned by the dominant species, thus it can be divided into three zones likewise. Typical ecological parameters (dominant species, tree heights, diameters at breast height (DBHs), coverage) of deciduous forests (800-1,600 m a.s.l) on Mt. Baotianman (1,820 m a.s.l, 33°25'~33°33'N, 111°53'~112°E), Henan Province, China was studied to understand the gradient inside of deciduous broad leaf forest. Both the tree heights and species richness showed a hump-shaped cure as altitude increases, which reached the highest height of 15-16 m at the elevation of around 1400 m. The core zone of mountain deciduous forests turn out to be the elevation between 1300-1400 m. Mass effect (or Source–Sink effect), Mid-Domain Effect (MDE), may be applied to explain the distribution pattern inside a zonal belt. This research promoted our understanding of the distribution patterns of vegetation along elevational gradient.

T3
A267
The photosynthetic characteristics and growth response of *Photinia frasery* 'Rubens' seedlings to municipal sewage sludge application

Guangqizhao Zhao

Shanghai Botanical Garden

A potted experiment was conducted to study the effects of municipal sewage sludge application on *Photinia frasery* 'Rubens' growth and photosynthetic characteristic. The results showed plant height, stem basal diameter, and biomass of the *Ph. frasery* seeding significantly increased with sewage sludge application, with 5.0% sludge treatment the maximum. All of the contents of chlorophyll a and chlorophyll b in leaves of *Ph. frasery* 'Rubens' tended to increase with the increase of sewage sludge dosage, and the contents of chl a increased faster than that of chl b. The chl a/b showed an increase tendency, whereas the net photosynthesis rate (*Pn*) and transpiration rate (*Tr*) in leaves showed a decrease. Compared with control, the 7.5% sludge treatment led to a decrease in *Pn* by 49.13%, and the *Tr* treatments of 2.5%, 5.0% and 7.5% decreased by 15.67%, 5.07%, and 21.20%, respectively. The stomatic conductance (*Gs*) reached an increase by 17.98% at 5.0% treatment compared with control. The intercellular CO_2 concentration (*Ci*) came to a climax at 7.5% treatment, increased by 28.96% compared with control and by 11.28% compared with 5.0% treatment. Basically intercellular CO_2 concentration (*Ci*) increased with the increase of sewage sludge dosage. The average and maximal value of net photosynthesis rate (*Pn*) in August increased to the maximum in different seasons after sewage sludge land application. The maximal value of stomatic conductance (*Gs*) and transpiration rate (*Tr*) decreased with the season.

T3

A268

Biomass increment and mortality losses in tropical secondary forests of Hainan, China

Jun-Fu Zhao
Hainan University

Secondary forests after severe logging are important part of China's forests. We investigated the forest biomass and its accumulation rate of a tropical secondary forest through 38 plots in Hainan Ialand, China. These secondary forests are moderate carbon sink with averaging on 1.96~2.17 tC ha^{-1} yr^{-1}. Biomass increment is majorly contributed by medium size (10~35 m) trees which differ from selective logged forest in Amazon. Tree morality account near 30% of the biomass increment and plays an negligible role in biomass accumulation estimation. Tree mortality rate is highly dependent on tree size. High mortality rate in small trees and seedlings might be related to enforced competition due to elevated irradiance after logging. In the perspective biomass and its accumulation, tropical forest is fragile to logging campaign and need careful forest management.

T3

A269

Separate and combined effect of UV-B radiation and drought on the growth and physiology in *Artemisia sphaerocephala*

***Yuanyuan Zhao**, Lijing Zhang, Xiumei Miao, Xiaolong Chen, Xiaowei Hu, Shuzhen Nan, Pengfang Yang, Guoqin Duan, Hua Fu*
College of Pastoral Agriculture Science and Technology, Lanzhou University

In natural world, plants are often affected by synthetic effect of various abiotic stresses. In this study, using *Artemisia sphaerocephala* as material, the effect of compound treatment of UV-B radiation and different level of drought stress was investigated. Results showed that plant height and biomass decreased, relative conductivity (REC) and malondialdehyde (MDA) increased when compared with CK in all treatments, except for the compound treatment of light drought stress and UV-B stress (LD + U). Under individual treatment, severe drought stress (SD) possessed less value on plant height and biomass as compared with light drought stress (LD) and UV-B stress (U), while value of REC and MDA exceeded LD and U. These results suggested that SD had the most remarkable effect of inhibiting growth in *A. sphaerocephala*. Under compound stress, plant height and biomass were significantly increased, whereas the value of REC and MDA were obviously decreased in both LD + U and the compound treatment of severe drought stress and UV-B stress (SD + U). This result indicated that additive UV-B would remit the damage which caused by drought in *A. sphaerocephala*. In addition, U had the significantly relaxative effects whether in LD or SD, and could lay the foundation of further study in the interaction mechanisms of these two environment factors.

T3

A270

Response of biological soil crust coverage to wind-blown sand burial during the succession of artificial sand-fixing vegetation in the Tengger Desert, Northern China

***Yun Zhao**[1,2], Rongliang Jia[1], Jialing Teng[1,2], Wenxiong Jia[3]*
1. Shapotou Desert Research and Experiment Station, Northwest Institute of Eco-Environment and Resources, Chinese Academy of Sciences
2. University of Chinese Academy of Sciences
3. Northwest Normal University

In artificially constructed desert ecosystems, the formation and development of a biological soil crust (hereafter crust) is one of the key indicators of successful construction of artificial sand-fixing vegetation. Importantly, with the successional progression of sand-fixing vegetation, the crust coverage undergoes substantial changes related to surface stabilities. Furthermore, surface stabilities crucially influence both the succession of sand-fixing vegetation and changes in crust coverage. Therefore, crust coverage is commonly used as a visual indicator to assess surface stability in a desert ecosystem. However, crust coverage is susceptible to sand burial (one of the common natural disturbances in desert areas) due to both its soil surface inhabitation and small stature. In order to study the characteristics and mechanisms of the response of crust coverage to sand burial during the succession of artificial sand-fixing vegetation, we used a space-replacing-time method to measure the roughness, heights, and coverage of four dominant crust types following sand burial. Furthermore, based on this, we computed the D1 value [the initial sand burial depth at which crust coverage (from 99.99%) begins to decline] and the D2 value (the threshold sand burial depth at which crust coverage declines to 0%). These crusts consist of *Bryum argenteum* crust, *Didymodon vinealis* crust, *Syntrichia caninervis* crust, and cyanobacteria-lichen-moss mixed crust in revegetated areas established in 1956,

1973, 1981 and 1987 at Shapotou, on the southeastern edge of the Tengger Desert. Our results show the following patterns. (1) With increasing sand depth, crust coverage declined along a logistic curve. (2) The mixed crust showed the lowest D1 value and the highest D2 value. Among the three moss-dominated crusts, the lowest D1 and D2 values were found in *B. argenteum* crust, whereas the D1 and D2 values of *D. vinealis* crust and *S. caninervis* crust differed depending on the history of the artificial revegetation: *D. vinealis* crust < *S. caninervis* crust at revegetated area established in 1956, and *S. caninervis* crust < *D. vinealis* crust at revegetated areas established in 1973 and 1981. (3) With successional progression of the artificial sand-fixing vegetation, the lowering rate of crust coverage gradually declined with increasing sand depth, whereas D1 and D2 values gradually increased. (4) With the development of the artificial sand-fixing vegetation, the lowering rate of crust total coverage gradually declined with increasing burial depth, indicating the increasing surface stabilities and ability to resist sand burial. (5) With successional progression of the artificial sand-fixing vegetation, the crust heights and roughness increased, significantly influencing the lowering rates of coverage and D1 and D2 values. In conclusion, our findings should provide strong support for future evaluation of the influence of sand burial on the structure and functions of crust, as well as the stability of artificial sand-fixing vegetation.

T3

A271

Impact of different vegetation on the emission of soil greenhouse gas and soil microbial activity

Jihai Zhou[1], *Jiyong Zheng*[2], *Fengpeng Han*[2], *Houbao Fan*[1]
1. Nanchang Institute of Technology
2. North West Agriculture and Forestry University

Different vegetation types change the soil environment to different extents, influencing the biogeochemical cycles. At present, many researchers are investigating the relationship between vegetation and microbial activity in natural ecological systems to provide a theoretical basis for ecological restoration in the area. Here, the effect of vegetation types on soil greenhouse gas emission and soil microbial activity was investigated in Guyuan, Ninxia Hui Autonomous Region. Soil samples were collected from different layers of soil (0-10, 10-20, 20-30 and 30-40 cm) for each of five typical vegetation types, including four artificial vegetations: *Medicago sativa*, *Prunus sibirica*, *Hongmei apricot*, *Caragana microphylla*, and one natural grassland. The results showed that soil greenhouse gas emission and microbial activity were varied in the five types of vegetations. Overall, the emission of soil CO_2, N_2O and the adsorption of CH_4 all decreased with the depths of soil. The maximal CO_2 emission occurred in natural grassland, while the minimum in *Caragana microphylla* land; the maximal N_2O emission occurred in *Hongmei apricot* land, while the minimum in *Medicago sativa* land; the maximal CH_4 adsorption occurred in natural grassland, while the minimum in *Medicago sativa* land. Compared with the natural grassland, the soil CO_2, N_2O emission in the four artificial vegetations were about 43.2%-74.9% and 82.5%-138.6%, the soil CH_4 adsorption in the four artificial vegetations decreased by 42.5%-71.1%, and the activity of soil flourescein diacetate (FDA) hydrolysis, alkaline phosphatase, urease and invertase in the four artificial plantation recovered by 50.0%-67.1%, 50.9%-91.2%, 61.9%-80.2% and 35.0%-92.6%. Further research is still required to explore soil greenhouse gas emission and its microbial mechanism in the processes of vegetation change.

T3

A272

Modeling the geographical distribution pattern and dominant climatic factors of *Cerasus schneideriana*

Hong Zhu, *Yong-Fu Li, Hua-Chen Wang, Yi-Fan Duan, Xian-Rong Wang*
Nanjing Forestry University

Cerasus schneideriana is an endemic species of cherry tree in East China, whereas its natural habitats were threatened and fragmented under climate change scenario as well as many other reasons. To understand the relationship between geographical distribution pattern of *C. schneideriana* and its dominant climatic factors on the regional scale, the DIVA-GIS software and the BIOCLIM model were jointly applied to predict the present and future potential spatial areas on the basis of actual distribution points. The results showed that *C. schneideriana*'s actual distributions covering 6 provinces of East China, including Zhejiang, Fujian, Anhui, Jiangxi, Guangxi and Hunan. The natural distribution center was located in mountainous areas of Zhejiang-Fujian border, and the north boundary was arrived at Zhejiang-Anhui border. Under future climate change situation (CCM3), the probability of potential areas of *C. schneideriana* were increased, and had a tendency to the north expansion. The principal component analysis (PCA) reveled the annual precipitation (bio12), precipitation of the driest quarter (bio6), precipitation of the warmest quarter (bio18) and temperature seasonality (bio4) were the dominant factors for geographical distribution of *C. schneideriana*. The frequency histograms further showed the factor's optimum range were 1,503~2,003 mm, 604~951 mm, 528~791 mm and 601~872 (SD*100) respectively. Evaluation by the ROC curve proved the BIOCLIM model predicted the distribution accurately (AUC = 0.998) and reliably. Therefore, our research revealed that *C. schneideriana* were suited to live in north subtropical region of China, with warm and humid condition, water and heat were the two key climatic factors to the distribution pattern.

T3

A273

Decomposition of phenolics in *Polygonum viviparum* litter on an alpine meadow

Wenzhen Zong, ***Jing Wang***, *Jing Gao, Yongsheng He*
College of Pastoral Agriculture Science and Technology, Lanzhou University

Phenolic compounds are widely distributed in higher plants as secondary metabolites, and they play an important role in nutrient cycling through litter decomposition. Most of the studies on phenolic decompositions in litter were conducted in forest ecosystem, whereas how the characteristics of alpine ecosystems affect phenolic decomposition is barely known. In addition, the report on the structural change of condensed tannins during litter decomposition is very limited. In this study, a phenolic-rich plant,

Polygonum Viviparum, on an alpine meadow in the Qinghai-Tibetan Plateau in China was selected as the research object. Phenolics and condensed tannins in litter during different decomposition stages were extracted, purified and identified to answer the two major questions: (1) Is the decomposition pattern of phenolics in litter in alpine ecosystems different from that in forest ecosystems? (2) Do low-molecular-weight phenolics and condensed tannins show dynamic structural changes during decomposition? A one-year litter bag experiment was conducted to investigate the decomposition dynamics of phenolics in *P.viviparum* litter. The content of total phenolics and condensed tannins in litter at different decomposition stages was analyzed. Low-molecular-weight phenolics and condensed tannins in litter were characterized with high performance liquid chromatography and mass spectrometer. The results showed that the content of total phenolics and condensed tannins displayed an overall decreasing trend during the decomposition process. During the early stage of decomposition, the loss of phenolic compounds could be attributed to degradation by soil microbes, which were protected by the snow cover in winter. During the middle stage of decomposition, however, the rapid loss of phenolics was mainly due to leaching, as the temperature and precipitation was elevated in the alpine growing season. This decomposition pattern was different from that in forest ecosystems, where leaching usually accounts for the loss of phenolics in the initial stage. The low-molecular-weight phenolics identified in litter included gallic acid, protocatechuic acid, chlorogenic acid, trans-ferulic acid and (+)–rutin. The monomer units of condensed tannins identified in litter included (–)–epicatechingallate, (–)–epicatechin and (–)–epigallocatechin as the extender units, and (+)–catechin, (–)–gallocatechin and (–)–epigallocatechin gallate as the terminal units. All of these phenolic compounds and monomer units of condensed tannins disappeared gradually from the litter and the emergence of new compounds was not observed. The decomposition pattern of litter phenolics in the alpine ecosystem was different from that in forest ecosystem, and the loss of phenolics during decomposition was mainly due to microbial degradation and leaching instead of chemical oxidation and polymerization. The results could help us to understand the role of phenolics in nutrient cycling in alpine meadows.

T3

A508

The modular nature of intraspecific trait variability: An experimental demonstration using a broad-niched herbaceous perennial species

Lauri Laanisto, *Tiit Hallikma*

Estonian University of Life Sciences

Although it's commonly assumed that interspecific trait variation (ITV) contributes more to functional variability in plant communities than intraspecific variation, recent studies have shown that the latter can be as high or even higher under some circumstances. Several factors related to ITV are known to have significant effects on community assembly and species interactions. Yet, despite the significance of ITV having been emphasized in reviews, and its magnitude comprehensively analyzed in studies, general patterns of ITV are yet to emerge. Further research is necessary to systematically search for the mechanisms how ITV affects community and ecosystem processes. Majority of ITV-studies with plants have so far focused on measuring key functional traits for many species in conjunction, and in just one or few locations (typically communities). Our study examined ITV from a hitherto unexplored angle, proposing that ITV dynamics in plants could be explained by modularity concept. We propose that phenotypic plasticity in plants, which is the main source of ITV is usually expressed in subindividual level, which creates complex interplay between local responses and systemic control. In order to test the modular nature of ITV, our study focused only on one species: *Phacelia secunda*–widely-distributed common herbaceous perennial plant endemic to Chile. We measured a large number of traits throughout the whole bioclimatic range of its distribution. Both generative and vegetative traits were measured, and both abiotic and biotic environmental factors potentially governing the ITV of different traits were recorded. The species studied has a wide climatic niche, and occurs in habitats ranging from sandy beaches to mountain forests. It's also known to be morphologically variable along climatic gradients. ITV responses of *Phacelia secunda* to abiotic and biotic factors showed differentiating patterns. Whereas abiotic factors (climatic factors) predominantly influenced vegetative traits of the structures that are responsible for resource gathering, biotic factors (diversity and abundance of surrounding vegetation) influenced generative organs, which translate acquired resources into fitness. These results strongly indicate that ITV has modular nature and different abiotic and biotic variables affect different plant organs differently. In our case climatic factors affected predominately the vegetative traits of *Phacelia secunda*, but biotic interactions from surrounding vegetation affected the generative traits. It remains to be tested whether this contrasting pattern is to be found with other species as well. Our study widens the perspective of ITV studies in several ways: Firstly, it's based on measurements taken from populations from the entire distribution range of one broad-niched species, which gives the opportunity to assess ITV dynamics throughout a species' full bioclimatic amplitude. Secondly, our approach was to measure as many functionally meaningful traits as possible. Thus, instead of trying to relate a specific functional trait and its variability to a specific environmental gradient, we were able to seek more general patterns in ITV. Thirdly, we measured both generative and vegetative traits, and tested ITV separately against abiotic and biotic factors.

T3

A514

A novel cotton WRKY gene, *GhWRKY6-like* improves salt tolerance by activating ABA signaling pathway and scavenging of reactive oxygen species

Abid Ullah, *Chen Lin, Heng Sun, Yang Xiyan, Zhang Xianlong*

National Key Laboratory of Crop Genetic Improvement, Huazhong Agricultural University

WRKY transcription factors are transcriptional regulators of signaling pathways involved in response to biotic and abiotic stresses. Hereby, we reported that ectopic expression of *GhWRKY6-like* gene significantly improved salt and osmotic stress tolerance in *Arabidopsis* while, silencing of *GhWRKY6-like* increased sensitivity to abiotic stresses in cotton. Expression of *GhWRKY6-like* was remarkably induced by abscisic acid (ABA), poly ethylene glycol

(PEG) and salt stresses. The *GhWRKY6-like* gene was cloned and transformed into *Arabidopsis* for further characterization. Our results showed that germination rate and root length was significantly improved in overexpressing *GhWRKY6-like* plants as compared to wild type (WT) plants under salt, mannitol and ABA treatments. And overexpressing plants showed higher salt tolerance than WT plants in soil. In addition, overexpressing plants accumulated less H_2O_2 and malondialdehyde (MDA), while higher proline content, superoxide dismutase (SOD) and peroxide dismutase (POD) activities were detected under drought and salt stresses. In contrast, virus-induced gene silencing (VIGS) of *GhWRKY6-like* in cotton showed enhanced sensitivity as compared to WT plants during salt and drought stresses. Moreover, expression analysis revealed that *GhWRKY6-like* enhance drought and salt tolerance by activating ABA signaling pathway genes. Our results revealed that *GhWRKY6-like* enhanced osmotic, drought and salt tolerance in *Arabidopsis* by scavenging of reactive oxygen species and regulating ABA signaling pathway. We suggest that overexpression of *GhWRKY6-like* gene in cotton will enhance tolerance against salt and osmotic stresses.

T3

A522

Floral nectary morphology and proteomic analysis of nectar of *Liriodendron tulipifera* Linn.

Yanwei Zhou[1,2*], *Meiping Li*[1,2], *Fangfang Zhao*[1,2], *Hongguang Zha*[3], *Liming Yang*[4], *Ye Lu*[1,2], *Guangping Wang*[1,2], *Jisen Shi*[1,2], *Jinhui Chen*[1,2]

1. Key Laboratory of Forest Genetics and Biotechnology, Ministry of Education, Nanjing Forestry University, Nanjing, China

2. Co-Innovation Center of Sustainable Forestry in Southern China, College of Forestry, Nanjing Forestry University, Nanjing, China

3. College of life and environment science, Huangshan University, Huangshan, Anhui, China

4. School of life science, Huaiyin Normal University, Huai'an, Jiangsu, China

Nectar is a primary nutrient reward for a variety of pollinators. Recent studies have demonstrated that nectar also has defensive functions against microbial invasion. In this study, the *Liriodendron tulipifera* nectary was first examined by scanning electron microscopy, and then the nectar was analysed by two-dimensional gel electrophoresis and liquid chromatography–tandem mass spectrometry, which led to identification of 42 nectar proteins involved in various biological functions. Bioinformatic analysis was then performed on an identified novel rubber elongation factor (REF) protein in *L. tulipifera* nectar. The protein was particularly abundant, representing ~60% of the major bands of 31 to 43 kDa, and showed high, stage-specific expression in nectary tissue. The REF family proteins are the major allergens in latex. We propose that REF in *L. tulipifera* nectar has defensive characteristics against microorganisms.

T3

A531

Selfing species exhibit diminished niche breadth over time despite larger geographic ranges

Daniel Park, *Charles Davis*

Harvard University

A recent study demonstrated that self-pollinating plants ("selfers") have larger geographic ranges and inhabit higher latitudes than their outcrossing relatives. The authors hypothesized that for selfers, the increased likelihood of successful colonization in new areas and initial purging of deleterious mutations offset their associated inability to adapt to new environments due to low heterozygosity. They also raised the possibility that selfers may exhibit broader environmental tolerances. Here we demonstrate that selfers do not have wider climatic niche breadths than their outcrossing sisters. Selfing sister pairs exhibited greater niche overlap than outcrossing sisters as well, implying their niches are slower to diverge. Furthermore, the niche breadth of selfers is predicted to decrease significantly faster than closely-related outcrossers even as their range sizes increase. Finally, selfers displayed higher rates of mutation accumulation than their outcrossing sisters. These results collectively indicate the long-term cost of selfing despite their initial geographic range expansion.

T3

A534

Physiological responses of *Stevia rebaudiana* to PEG-induced drought stress in greenhouse culture

Shokoofeh Hajihashemi[1], *Jan Geuns*[2]

1. Plant Biology Department, Faculty of Science, Behbahan Khatam Alanbia University of Technology, Khuzestan

2. Laboratory of Functional Biology, KULeuven, Heverlee-Leuven, Belgium

Plants frequently encounter unfavorable growth conditions. Drought is one of the major abiotic stresses that can disrupt cellular structures and impair key physiological functions. This study evaluated the physiological, antioxidant and growth responses of sweet medical herb of *Stevia rebaudiana* to drought stress-induced by polyethylene glycol (PEG 6,000, 0% to 15% w/v) treatment in greenhouse cultivation. Drought stress significantly decreased plants growth which was the consequence of a reduction in water content, photosynthesis pigments and carbohydrates biosynthesis. H_2O_2 generation increased in response to drought stress which subsequently resulted in lipid peroxidation and increase in malondialdehyde and electrolyte leakage. The ROS accumulation leaded to an increase in some non-enzymatic antioxidants (e.g. phenols, flavonoids, total amino acids, proline, α-tocopherol, glycine betaine, ascorbate pool) and the activity of some key antioxidant enzyme (e.g., catalase, ascorbate peroxidase, polyphenol oxidase, superoxide dismutase and peroxidase) to scavenge free radicals. The reduction of DPPH, $NO^\bullet$, $O_2^{\bullet-}$ and $OH^\bullet$ in the leaf extract of drought-stressed plants confirms the strong ROS scavenging properties of *Stevia*. Also, a significant increase in total antioxidant capacity (FRAP) of drought-stressed plants provide a hint at its antioxidant capacity. Collectively, our results demonstrated that reduction of *Stevia* growth and yield is the consequence of drought-induced oxidative stress, reduction of photosynthesis and carbohydrates biosynthesis, and membranes damage which lead to cell death. Hence more leaf biomass production with higher steviol glycosides is the major criterion for *Stevia* crop, sufficient irrigation is recommended to obtain higher yield.

T3
A549
Non-botanical approaches can fail to predict and classify wetland biological condition

Suneeti Jog[1], *Jason Bried*[2], *Xiao Feng*[3]
1. Northeastern State University
2. University of Arkansas
3. Oklahoma State University

Wetland biological integrity is continuously threatened by human activity. Impacts to biological health may be inferred directly (i.e. bioassessment), or estimated indirectly using rapid assessment or computer-based data layer analysis. Rapid buffer assessment is often used in lieu of intensive vegetation surveys which require considerable time and expertise. This study tests if rapid buffer assessment and surrounding land use can predict and classify wetland biological health as represented by floristic quality (FQ) and taxonomic distinctness (TD). Floristic quality is computed based on sensitivity of plants to human disturbance, while taxonomic distinctness takes species relatedness into account. From surveys of 106 non-forested wetlands in Oklahoma, we found a limited ability for land use and buffer quality to estimate site-specific FQ values. Similarly, land use was not a reliable predictor of site-specific TD values. However, limited land use had potential to identify higher levels of FQ, and more intensive land use showed potential to identify lower average TD. This suggests that both measures (FQ and average TD) are needed for a more complete classification of wetland biological response to anthropogenic disturbance. Despite growing interest in relatively rapid or low-cost methods of wetland assessment, our results do not support using such methods in lieu of direct bioassessment.

T4
A274
The effect of different water treatments on the photosynthetic characters and chromosomal control in wheat (*Triticum aestivum* L) seeding

Zhiying Bai[1], *Bin Lu*[1], *Cundong Li*[1], *Chunji Liu*[2]
1. Hebei Agricultural University
2. CSIRO Plant Industry

Wheat is one of the most important food crops in the world and water shortage is the most significant factor affecting its yield. Drought is becoming more prevalent and how to breed varieties for water-stressed environments is becoming an urgent task for all scientific disciplines related to food production. As photosynthesis is the foundation of plant production, chromosomal controls of characteristics related to photosynthesis and biomass of wheat seedlings were investigated using a set of inter-varietal chromosome substitution lines between Chinese Spring and Synthetic 6X (Syn) under a well-watered and a water-stressed treatment. The experiments were conducted in growth rooms with the controlled environment facilities (CEF) at Queensland Bioscience Precinct, Brisbane. The relative water content of pots were maintained at 75%±5% (well-watered, control) and 45%.±5% (water-stressed) respectively by weighting method. Results from this study showed that, as expected, water stress resulted in reduction in net photosynthesis rate, chlorophyll content, stomatal conductance, seedling dry mass, but promoted water use efficiency. As substitution lines with the best performance for each of the characteristics assessed were often different, different genes may have to be targeted in breeding varieties for water-stressed and non-water-stressed environments. Under well-watered condition, compared to CS, two substitution lines (5A and 5B) showed significantly increased photosynthesis rate, substitution 1A showed significant positive effect on stomatal conductance, three substitution lines (5A, 7A and 5B) were found to be significantly increased chlorophyll content, substitution 5B showed the highest water-use efficiency (WUE) among all substitution lines. It suggested that main gene responsible for regulating high photosynthesis and chlorophyll content might be located on chromosomes 5A, 5B and 5A, 5B, 7A of Syn respectively. It also suggested that chromosomes 5B and 1A of Syn might regulate high WUE and stomatal conductance respectively. Under water-stressed condition, five substitution lines (1A, 3A, 7B, 4D and 6D) showed significantly higher net photosynthetic rate than CS, four substitution lines (1A, 2A, 3A and 4D) showed higher stomatal conductance than CS, substitution 2D showed significantly higher dry mass than CS, five substitution lines (1A, 3A, 2D, 3D and 4D) showed higher WUE than CS, three substitution lines (1A, 2A and 7A) showed significantly higher dry mass. It indicated that main gene regulating high photosynthesis might be located on chromosomes 1A, 3A, 7B, 4D and 6D of Syn. The main gene regulating high stomatal conductance might be located on the chromosomes 1A, 2A, 3A and 4D of Syn. The main gene regulating high WUE and dry mass might be located on chromosomes 1A, 3A, 2D, 3D, 4D and 1A, 2A, 7A of Syn respectively. It is concluded that substitution line 1A was outstanding in that it contributed positively to each of the characteristics measured under the water-stressed treatment. Thus targeting gene (s) on this Syn chromosome in breeding varieties for water-stressed environments could be productive.

T4
A275
Diversity in wood functional traits in the hyper-diverse mangrove communities in new guinea

Lawong Balun
University of Papua New Guinea

Mangrove is a taxonomically diverse interface vegetation community between terrestrial and marine ecosystem with species of wide phylogenetic affinities exhibiting morphological and functional convergence. It is speculated that wood traits among the hyper-diverse mangrove community display narrow functional performance. This study set out to test the hypothesis that the hyper-diverse mangrove communities in New Guinea exhibit a narrow functional performance despite their wide phylogenetic affinities. I investigated the variation in wood and hydraulic functional traits of 31 mangrove species across zonations and among different growth forms from six hyper-diverse mangrove communities in New Guinea. Wood anatomical traits (wood density, vessel diameter, vessel density, hydraulic mean vessel diameter) of 31 co-existing mangrove species in six hyper-diverse mangrove communities in New Guinea varied markedly across different zonation bands. Wood density (P</span>$_{wood}$) varied significantly among different zones ($F_3,2_1$ = 40.96, $P < 0.0001$). P_{wood} varied by over 30% among different growth forms, but the differences were not significant ($P > 0.05$). Maximum P_{wood} (0.94 ± 0.03 g m-3) was observed in the mean low tide (MLT) zone and decreased in the landward direction. P_{wood} varied significantly among different rooting system types ($F_4,2_7 = 6.33$, $P < 0.001$). Vessel density (V_d) varied significantly among different zones ($F_3,1_6 = 5.01$, $P < 0.01$) but variation was not significant among different growth forms ($F_2,1_5 = 1.66$, $P > 0.22$). Vessel diameter (d_V) varied significantly among different growth forms ($F_2,1_5 = 4.57$, $P < 0.03$) but did not significantly differ among the zones ($F_2,1_7 = 0.33$, $P > 0.72$). A linear correlation of $P_W$ood and leaf mass area showed no significant correlation ($r = 0.10$, $P > 0.08$). A linear correlation of vessel density and vessel diameter showed an inverse relationship ($r = 0.18$, $P < 0.02$). Vessel density was also inversely correlated to weighted hydraulic mean diameter ($r = 0.22$, $P < 0.01$). A linear correlation of potential hydraulic conductivity (K_P) and vein density showed no significant correlation ($r = 0.0005$, $P > 0.05$). K_P was inversely correlated to: V_d ($r^2 = 0.21$, $P < 0.01$), strongly correlated to d_V ($r^2 = 0.95$, $P < 0.0001$), and strongly correlated to d_h (mean hydraulic diameter, pm) ($r = 0.78$, $P < 0.0001$). A linear correlation of calculated leaf area specific hydraulic conductivity (K_L) and vein density showed a significant relationship ($r = 0.47$, $P < 0.05$). K_L was correlated to A_{max} ($r^2 = 0.56$, $P < 0.05$).

T4

A276

Study on the regulation mechanism of AM fungi on the content of effective components in *Clematis intricate* with different levels of nitrogen and phosphorus

*Ting Liu, Hongjiao Li, Xingjiao Sun, Xiao Li, **Tebuqin Borjigin***
Inner Mongolia medical university

Clematis intricata (*Clematis intricata* Bunge) is a kind of Mongolian medicinal plants, and one of its effective components are flavonoids. The aim of this study was to investigate the effect of AM fungi on the content of effective components in clematis. The experiment was set up with 0N, 10%N, 20%N, 50%N, 100%N, 200%N six levels (KNO_3 and NH_4NO_3) and 0P, 5%P, 10%P, 20%P, 50%P, 100%P 6 P level (KH_2PO_4), and N, P levels were inoculated with AMF (*Glomus mosseae, G.m*) and without inoculation (CK) 2 treatments, each treatment was repeated 3 times, and 6 month old seedlings were used. The content of total flavonoids in plants was significantly increased by inoculation with *G.m* at different N levels. Under the condition of 0 N -20% N, the content of total flavonoids was significantly higher than that of non-inoculated ones. Under the condition of 50% N-200% N, the content of total flavonoids has no significant difference between inoculated ones. The content of total flavonoids was highest at 10% N, and the content was about 29.4 mg/g. The expression of key enzymes of flavonoid biosynthesis genes (*CiPAL, CiCHS, CiCHI, CiF3H, CiFLS* and *CiDFR*) were determined by qRT-PCR. Under different nitrogen levels, inoculated the expression of *CiPAL* in 10% N and 20% N level increased, others down regulated; *CiCHS* and *CiCHI* expression was down regulated at 50% N, others were increased, whereas the expression of *CiFLS* in 20% N increased, others were down regulated. The content of total flavonoids in plants was significantly increased by inoculation with *G.m* at different P levels. With the increase of P concentration, the content of total flavonoids increased first and then decreased in the range of 0 P to 50% P, and reached the highest at the level of 5% P and 10% P, which was 17.4 mg/g and 17.3 mg/g, respectively. Under different phosphorus levels, the content of total flavonoids was highest at 100% P, and about 20.2 mg/g, and was significantly higher than that of 5% P and 10% P. The expression of key enzymes of flavonoid biosynthesis genes (*CiPAL, CiCHS, CiCHI, CiF3H, CiFLS* and *CiDFR*) were determined by qRT-PCR and found that under different P levels, the expression of *CiPAL* was significantly up-regulated in inoculated ones except 50%P level, the expression of *CiCHS* significantly up-regulated in all P levels, while the expression of *CiFLS* in 20% P up-regulated, in other P levels were down regulated. At different nutritional conditions of nitrogen and phosphorus under the inoculation of AM fungi can significantly improve the total flavonoids content of *Clematis intricate*. The nitrogen and phosphorus may have different and complex mechanisms to influence the content of effective components.

T4

A277

Research on the dynamics of plant carbon content and nitrogen content at different successional stages in Deglaciated Terrain on Mt. Yulong

Li Chang
1. State Key Laboratory Breeding Base of Desertification and Aeolian Sand Disaster Combating, Gansu Desert Control Research Institute, Lanzhou, China
2. State Key Laboratory of Cryosphere Sciences, Cold and Arid Regions Environmental and Engineering Research Institute, Lanzhou 730000, China
3. College of Earth and Environmental Sciences, Lanzhou University, Lanzhou 730000, China

With global climate warming, deglaciated terrain, which is 1250m in length, surrounding Baishui No. 1 Glacier on Mt.Yulong was formed since the late 19th century. It provides an ideal place for plant succession in nature. In view of frequent fires and the need to restore ecosystem in Mt. Yulong, as well as fewer reported studies, it has very important theoretic and practical significance in researching ecological succession in this region. We selecting Mt. Yulong in China's alpine-monsoonal temperate glacier region as the study area, a chronosequence method (substituting space for time) was used to study the chang of physiological characteristics of dominant species. According to plant nutrient content, its overall trend of plant C or N content was Arenaria delavayi (SUC1) < Polygonum macrophyllum (SUC2) and Arenaria delavayi (SUC3) < Carex capilliformis (SUC4) and Kobresia fragilis (SUC5). While the overall trend of plant C:N was Arenaria delavayi (SUC1), Polygonum macrophyllum (SUC2), Arenaria delavayi (SUC3) and Carex capilliformis (SUC4) < Kobresia fragilis (SUC5). According to the results of regression analysis, among the 4 dominant plants, the nutrition content of Arenaria delavayi had not remarkable response to soil nutrition, while Polygonum macrophyllum has the remarkable response to soil nutrition. In the 4 soil nutrition indicators, many dominant plants have remarkable response to soil Total nitrogen content, soil Total carbon content comes second. With the succession went on, the overall trend of plant C or N content and plant C:N, which have been influenced most by the soil total nitrogen content, increased significantly.

T4

A278

Study on traits segregation of F_1 generation of Phalaenopsis-type *Dendrobium*

He-Ming Chen[1,2], *Fu-Bing Lv*[1,2], *Wen-Fang Xiao*[1,2], *Zuo Li*[1,2]
1. Environmental Horticulture Research Institute of Guangdong Academy of Agricultural Sciences
2. Guangdong Provincial Key Laboratory of Ornamental Plant Germplasm Innovation and Utilization

In recent years, the study of Phalaenopsis-type *Dendrobium* mainly focused on tissue culture, cultivation management and factory nursery, but there was almost no reported about traits separation of hybrids and genetic performance. F_1 generation and its parents were compared in this experiment and researched the rules of its genetic performance for superior individual selection of breeding progeny and matching of the parents. In October 2006, we selected the varieties of D104, D105, D106, D107, D108 for the six artificial crosses in Guangdong Ornamental Plant Germplasm nursery, and hybridization seeds were obtained after four months and sowed sterilely, one year later F_1 hybrid plants had 2100 and planted. F_1 generation was first flowering in 2009, and 100 plants were taked randomly from every combinations to investigate the following nine ornamental traits: plant height, plant width, leaf length, leaf width, pedicel length, flower diameter, flower number, flower color and the beginning of flowering. The results showed that: (1) The averages of the four quantitative traits of plant width, leaf width, pedicel length, flower diameter were relatively decreased with mid-parent value and showed the recession phenomenon in a certain degree which had a similar phenomenon such as gerbera, azaleas and lily, the reason might be that advantage was reduced after the sexual process and the average of the traits of hybrid population declined, but the variation ranges in combination were larger, the individuals of breeding larger plants and flowers larger or pedicel longer were possible mainly by improving the cultivation conditions and the level of management. (2) There were super-parent advantages in plant height, leaf length, flower number, flower color and the beginning of flowering, especially flower number and flower color, the average of number of flowers was 106.22% which was more than 37.08% with mid-parent value, and the colors of calyx and petal segregated the other 7 kinds by the 4 kinds of parents, the lip colors segregated the other 5 kinds by the 4 kinds of parents. (3) F_1 generation showed red series and segregated widely in the flower color whether with white flowers as maternal or red series, that was to say, the genetic ability of red series was stronger.

T4

A279

Study on hybridization compatibility of Nobile-type *Dendrobium* and phalaenopsis-type *Dendrobium*

He-Ming Chen[1,2], *Fu-Bing Lv*[1,2], *Zuo Li*[1,2], *Wen-Fang Xiao*[1,2]
1. Environmental Horticulture Research Institute of Guangdong Academy of Agricultural Sciences
2. Guangdong Provincial Key Laboratory of Ornamental Plant Germplasm Innovation and Utilization

In horticulture, *Dendrobium* is divided into Nobile-type *Dendrobium* and Phalaenopsis -type *Dendrobium*. Among them, Nobile-type *Dendrobium* is also known as deciduous *Dendrobium*, natural flowering is commonly March to May in the spring, Phalaenopsis-type *Dendrobium* is called evergreen *Dendrobium*, the flowering is longer, usually blossoming is August to December in autumn and winter. But the research on hybridization compatibility of Nobile-type *Dendrobium* and Phalaenopsis-type *Dendrobium* is very little. The hybridization compatibility of Nobile-type *Dendrobium* and Phalaenopsis-type *Dendrobium* has been reseached since 2008 in Environmental Horticulture Research Institute of Guangdong Academy of Agricultural Sciences. 11 varieties of Nobile-type *Dendrobium* and 7 varieties of Phalaenopsis-type *Dendrobium* were positive and negative hybridization, 12 capsules were gained from 154 hybrid combinations, and the success rate of cross was 7.79% which Phalaenopsis-type *Dendrobium* was successful as female parent, but Nobile-type *Dendrobium* was not to succeed as female parent. The result of the reseach was consistent with Wilfret and Kamenoto thinking the cross compatibility between 5% to 24% of Nobile-type *Dendrobium* and Phalaenopsis-type *Dendrobium*. But the seeds of most capsules showed no sign of germination after sterile sow, only the seeds of 2 capsules germinated and turned green after sterile sow, one was transplanted to seedlings, the other was germinating. Therefore, the histologic study will carryed out to reseach about pollen morphology, germination features on the stigma after pollination, in order to determine the stages and locations for hybrid incompatibility, and provide the basis to overcome the hybrid incompatibility in the next step.

T4

A280

Virus-induced gene sliencing as a tool for *TT12* gene functional study in strawberry

Siyu Chen, Yueming Tang, Yueyang Hu, Xiaowei Liu, Xunju Liu, Haoru Tang, ***Qing Chen***
Sichuan Agricultural University

In this study, cv. 'toyonoka' strawberry was used as materials. On the basis of cloning strawberry *transparent testa 12* (*TT12*) gene, we carried out bioinformatic and temporal expression analysis of strawberry *TT12* gene. By constructing the carrier of virus-induced gene silencing (VIGS), we silenced the *TT12* gene in the strawberry fruit and then we tested the changes of anthocyanin and proanthocyanidin content. In addition, we studied the subcellular localization expression of *TT12* gene by transient expression in Arabidopsis protoplast. Strawberry *TT12* gene contained 12 transmembrane domain structures, and *TT12* gene mainly expressed on the vacuole membrane. *TT12* gene down-regulated-expression by VIGS could significantly decrease proanthocyanidin content but no obvious change of anthocyanin content in fruit. These results revealed that *TT12* gene's role in flavonoids compounds metabolic pathways mainly by imparting in the proanthocyanidin branch and could highlight the special transportation mechanisms of different secondary metabolites.

T4

A281

Identification of 46 *MYB* transcription factor genes and analysis of their expression during abiotic stress in maize (*Zea mays*)

Yanhong Chen
Nantong University

The *MYB* transcription factor superfamily is a large gene family that plays central roles in developmental processes and defense responses in plants. Unlike in *Arabidopsis*, few members of the *R2R3-MYB* gene family have been functionally well characterized in maize, especially in abiotic stress-response pathways. Subgroup-specific conserved motifs outside the MYB domain may reflect functional conservation. A comparative genomics study using *Arabidopsis* abiotic stress-responsive MYB protein sequences identified 46 *ZmMYB* genes that may be involve in abiotic stress responses. An expression pattern analysis of the 46 *ZmMYB* genes under abiotic stress treatments was used to identify 22 *MYB* genes that were induced by one or more of the stress treatments. *ZmMYB30* was highly upregulated under the four stress treatments. The overexpression of *ZmMYB30* in transgenic *Arabidopsis* promoted salt-stress tolerance and also increased the expression levels of a number of abiotic stress-related genes, allowing the plants to overcome adverse conditions. This is the first comprehensive study of the roles of the *MYB* gene family in abiotic stress responses in maize.

T4

A282

Effects of drought stress on the growth, physiology and secondary metabolite production of *Pinellia ternata* Thunb.

Yuhang Chen[1], *Yun Chen*[2], *Qiaosheng Guo*[2], *Guosheng Zhu*[3], *Changlin Wang*[2], *Zuoyi Liu*[3]
1. Key Laboratory of Small Molecule Special Structure Drugs, Sichuan Institution of Higher Education&Chengdu Medical College, Chengdu, 610500, China
2. Institute of Chinese Medicinal Materials, Nanjing Agricultural University, Nanjing, 210095, China
3. Institute of Morden Chinese Medical Materials, Guizhou Provincial Academy of Agricaltural Sciences, Guiyang, 550006, China

Dried tuber of *Pinellia ternata* is one of the most important ingredients in many traditional medicinal preparations. Drought stress can affect the growth, physiology and secondary metabolites of plants. This study was conducted to determine the changes in terms of these variables over 35 days in *P. ternata* under four water regimes: 75% of field capacity (FC) (Control), slight drought stress (60% FC; LD), moderate drought stress (45% FC; MD), and severe drought stress (30% FC; SD). The results showed that the fresh and dry tuber weights, plant height, leaf area, number of tubers, tuber propagation index, and photosynthetic pigment content were highest in *P. ternata* plants grown under the well-watered treatment (75% field capacity). The peroxidase (POD) and catalase (CAT) activities and malondialdehyde (MDA) content were significantly higher in leaves of *P. ternata* under drought stress. The soluble protein content of *P. ternata* tubers was significantly higher under moderate drought stress, but the soluble sugar content of *P. ternata* tubers was significantly lower under severe and moderate drought stress. Additionally, the total alkaloid content of *P. ternata* tubers showed the highest increase under severe drought stress, but the guanosine and succinic acid contents of *P. ternata* tubers were significantly lower under drought stress. The maximum yield of total alkaloid, guanosine and succinic acid of *P. ternata* tubers were obtained under the well-watered regime. The results suggest that a well-watered regime (75% FC) can be an effective means of promoting tuber production and improving secondary metabolites production in *P. ternata*.

T4

A283

Proteomic analysis of alfalfa seeds response to heat stress during storage

Lingling Chen[1,2], ***Hang Cheng***[1], *Peisheng Mao*[1]
1. China Agricultural University
2. Chifeng Academy of Agricultural and Animal Sciences

Seed germination is a critical process in the life cycle of higher plants. The climate changes and global warming have affected all aspects of both our life and the biodiversity around us. The seeds also face the threat of a sharp decline in seed vigor resistance to heat stress during storage. However, knowledge about the molecular and physiological mechanisms underlying the heat stress of seeds has been lacking. The experiment was designed for alfalfa (*Medicago sativa* L.) seeds with 10% moisture content stored at 35,40,45 and 50°C for 8 d to compare the changing of seed vigour and proteomic processes. The objective was to determine the difference of protein expression level under heat stress in alfalfa seeds and to identify proteins associated with heat stress. 24 differentially expressed proteins were successfully identified by 2-DE coupled with LC-MS/MS. Among these expressed proteins, 1 were down-regulated at 40°C, 4 were down-regulated at 45°C, 17 were down-regulated at 50°C, and 2 were up-regulated at 50°C. The identified proteins were grouped into different categories of biological process according to GO and KEGG pathway enrichment analyses. Most of them were functionally categorized as cellular process, metabolic process and stress tolerance. 17.6 kDa class I heat shock protein, glyceraldehyde-3-phosphate dehydrogenase, and ubiquitin-conjugating enzyme E2 were mapped to two statistically remarkable categories in KEGG ($P<0.05$). In order to investigate the transcript level of the identified proteins in response to heat stress, quantitative RT-PCR assays were conducted. Among all the identified proteins, eight gene products were selected to investigate the expression patterns of their transcriptional levels and 75% of them had detectable expression at the mRNA level. The proteins identified in the present study provide a basis to elucidate further the molecular mechanisms underlying the adaptation of alfalfa seeds to high temperature stress. We will further study the metabolic pathways, and confirm the association of correlated proteins, which may provide a better understanding of the molecular mechanism for heat tolerance.

T4

A284

Effect of two nitrogen sources on the morphology and bio-

mass of foxtail millet

Jihan Cui, *Shunguo Li, Xueyan Xia, Meng Liu, Yu Zhao, Shijia Song, Jing Zhao*
National Foxtail Millet Improvement Center Institude of Millet Crops, Hebei Academy of Agricultrual and Forestry Sciences Minor Cereal Crops Laboratory of Hebei Province

Nitrogen (N) is one of the essential elements of plant, which plays an important role in plant growth and development. The main source of N absorbed by plant is nitrate N and ammonium N. Some plant tends to absorb ammonium N, such as rice. Some plant tends to absorb nitrate N, such as tomato. Some plant tends to absorb a mixture of ammonium N and nitrate N, such as licorice. Because there is a great difference in absorption, transportation, storage and assimilation of the two N sources, which will inevitably affect the growth and development of plant, as well as other physiological processes. At present, most of the studies on foxtail millet are focused on relationship of N application rate (or period) and yield, the effect of N forms on the morphology and biomass of foxtail millet have not been reported. In order to study ammonium N and nitrate N on morphology and biomass of foxtail millet (*Setaria italic* L.) and to give some reference evidences for preferential absorption N source of foxtail miller, vermiculite cultivation method was using with nutrient solution including two nitrogen forms. The result showed that the two N sources significantly affected the morphology and biomass of foxtail millet, and there was no significant difference in root morphology and spike length. There were significant differences in plant height, biomass, leaf area and leaf chlorophyll content which effected by two N sources: compared with ammonium N, nitrate N increased root biomass by 17%, stem biomass by 32%, leaf biomass by 39%, total biomass by 40%, height by 38%, leaf area by 40%. Compared with nitrate N, ammonium N increased chlorophyll content by 173%, and spike biomass by 12%. There were significant differences in root/shoot ratio and spike proportion: compared with nitrate N, ammonium N increased root/shoot ratio 8% and 44% spike proportion. This study showed that nitrate N played an important role in body construction by promoting plant height, leaf area and biomass, while, ammonium N played an important role in the functional structure by increasing the chlorophyll content and sex organ construction.

T4

A285

The *Reaumuria trigyna* transcription factor RtWRKY1 confers tolerance to salt stress in transgenic *Arabidopsis*

Chao Du, Lei Song, Xue Wang, ***Yingchun Wang***
Key Laboratory of Herbage & Endemic Crop Biotechnology, College of Life Sciences, Inner Mongolia University, Hohhot

Reaumuria trigyna (*Reaumuria* Linn genus, family *Tamaricaceae*) is an endangered dicotyledonous shrub with the features of a recretohalophyte, and it is endemic to the Eastern Alxa-Western Ordos area of China. It is regarded as a living fossil owing to its Tethys Ocean origin and is considered a key protected plant in Inner Mongolia. Recent studies demonstrated that *R. trigyna* shows outstanding performance because of its highly efficient antioxidant system, salt glands with secretion function, and strong osmoregulatory ability. *R.trigyna* transcriptome data analysis showed that the expression levels of 67 WRKY genes changed significantly under salt stress. A total of 171 differentially expressed genes (DEGs), and 155 DEGs were enriched in the "oxidoreductase activity" and "response to stress" GO categories, respectively. Thus, WRKY TFs may play a positive role in the response of *R.trigyna* to salt stress by regulating the expression of genes related to plant growth and homeostasis. Here, to explore the molecular mechanism underlying the response of *R.trigyna* to salt stress, the Group I WRKY transcription factor gene *RtWRKY1* was cloned from *R.trigyna* based on *R.trigyna* transcriptome data, and the expression pattern and function of *RtWRKY1* had been studied. The full-length *RtWRKY1* gene was 2100 bp, including a 1261 bp open reading frame (ORF) encoding 573 amino acids. *RtWRKY1* was mainly expressed in the stem and was induced by salt, cold stress, and ABA treatment. Overexpression of *RtWRKY1* in *Arabidopsis* significantly enhanced the chlorophyll content, root length, and fresh weight of the transgenic lines. *RtWRKY1* transgenic *Arabidopsis* exhibited higher proline content, GSH-PX, POD, SOD, and CAT activities, and lower MDA content, Na^+ content, and Na^+/K^+ ratio than wild-type *Arabidopsis* under salt stress conditions. Salt stress induced the expression of ion transport, proline biosynthesis, and antioxidant related genes, including *AtAPX1*, *AtCAT1*, *AtSOD1*, *AtP5CS1*, *AtP5CS2*, and *AtSOS1* in transgenic lines. These results suggest that *RtWRKY1* confers tolerance to salt stress in transgenic *Arabidopsis* by regulating plant growth, osmotic balance, Na^+/K^+ homeostasis, and the antioxidant system.

T4

A286

Cloning and bioinformatic analysis of three genes encoding terpenoid metabolic enzymes in tobacco

Jian Du, Yingying Huang, Yanjie Zhang, Jinyong Huang, Jia Guo, Weixin Tian, Hui Wang, ***Caipeng Yue***
School of Life Sciences, Zhengzhou University

Solanesol is a 45-carbon, unsaturated, all-trans-nonaprenol isoprenoid, widely found in higher plants, mammals and microbial organisms as natural product, and first isolated from flue-cured tobacco in 1956 by Rowland *et al.,* In the tobacco, solanesol content ranged from 0.3% to 3%. At the process of solanesol synthesis, there are three key enzymes, including 1-deoxy-xylulose 5-phosphate synthase (DXS), Farnesyl diphosphate synthase (FPS) and Solanesyl diphosphate synthase (SPS). In this study, we explore the genes function and verified them via PCR, transgenic, tissue culture, subcellular localization, phylogenetic analysis and so on. According to bioinformatics analysis, K326-*dxs* contained a 2142 bp open reading frame and encoded a hydrophilic protein which contained 713 amino acids. K326-*dxs* located in the chloroplast, do not contain signal peptide and transmembrane structure, but has transit peptides. The full length of K326-*fps* gene is 1029 bp and the gene encodes 342 amino acids. The FPS protein is a hydrophilic protein and did not contain signal peptide and transmembrane structure. The *solanesyl diphosphate synthase* gene *(sps)* has been obtained, the further research will be put into practice.

T4

A287

Effects of shift part of nitrogen fertilizer from basal to storage

root formation on growth and yield of sweetpotato (*Ipomoea batatas* L.)

Xiangbei Du, *Min Xi, Xiaoping Liu*
Anhui Academy of Agricultural Sciences

Appropriate nitrogen (N) fertilizer management is beneficial for improving growth, yield and yield components of field crops. Field experiments were conducted to determine the effects of application methods [basal application and split application levels (35 d after transplanting)] and N rates (80 and 100 kg ha^{-1}) on the sweetpotao (*Ipomoea batatas* L.) yield during summer 2015 and 2016. The experiments were carried out at the Agronomy Research Farm of Agricultural Sciences of Anhui Academy in Hefei City, China. It was concluded from the results that split application of N were more beneficial in terms of better growth, higher yield and yield components of sweetpotato under reduced N condition. More N application at later growth stage resulted in better growth and higher yield. The results further demonstrated that reducing the rate of soil applied N up to 80 kg N ha^{-1} in two splits (25% at basal and 75% at 35 d after transplanting) improve growth and sweetpotato productivity. Outcomes of this research will help producers schedule fertilization to maximize yield and nutrition use efficiency.

T4

A288

An application of hemispherical photography to analysis maize canopy structural parameters under natural light

Jiangchuan Fan, *Xinyu Guo*
Beijing Research Center for Information Technology in Agriculture/National Engineering Research Center for Information Technology in Agriculture/Beijing Key Lab of Digital Plant

Hemispherical photography (HP) has already proven to be a powerful indirect method for measuring various components of canopy structure. One of the main problems of hemispherical photography for determination of canopy structure such as leaf area index (LAI) and mean leaf inclination angle (MLA), is the selection of the optimal brightness threshold in order to distinguish leaf area from sky area thus producing a binary image. In this process, one of the challenges is how to overcome various natural light conditions which sometimes strongly affect the profile of the crop images taken outdoor. In this paper an illumination invariant multiple exposure images fusion and mapping method was proposed in order to squash negative impact of variant illumination. Firstly, a series of multiple exposure maize canopy hemispherical images was captured under natural light condition. The camera (Canon EOS 5D Mark III) with aperture (sigma 8mm f3.5 ex DG FISHEYE) was placed in the bottom of canopy towards to the sky provided vertical 360 degree horizontal 180 degree canopy images. Which were captured at different exposure times such as 1/800s, 1/400s, 1/200s, 1/125s, 1/30s, 1s. Secondly, the multiple photographs fused into a single radiance map whose pixel truncated in shadowed and lighted parts of original images expended to higher range. We were able to determine the irradiance value at each pixel, we could plot it against the measured pixel value discretized according to the 256 pixel values commonly observed in 8-bit images for each exposure time. The pixel values are proportional to the true irradiance values in the scene. The pixel values, exposure times, and irradiance values form a least squares problem. Finally, we also employed a histogram equalisation method to map irradiance values to RGB color space. After mapping processing, brightness of image have a more properly distribution, dark regions were lit more brightly and saturated regions were depressed to normal brightness condition, moreover, histograms of images share the similar distribution, which make the brightness and color of image more or less have the same values. The comparison results show that canopy gaps fraction of HP acquired at 14:00 and 17:00 with threshold value 180 has difference of 15.4% percent, and our method reduces the difference up to 2.8% percent. In canopy structure parameters analysis phase, canopy gaps were extracted from those photos, and then leaf area index (LAI) and mean leaf inclination angle (MLA) could be inversed by Beer-Lambert theory, based on the quantitative relationship between the radiation condition and the canopy structure. The experiment was conducted in 2013, density was 60000 plants/ha, normal water and fertilizer management. The hemispherical images were obtained in 8/06,8/13,8/19,8/22,8/26,9/12, the distribution of LAI and MLA was consistent with rules of growth and development of maize canopy. Moreover, a performance comparison of direct surveying method and our method was carried out, LAI and MLA values of 13 samples were collected with two methods, results of regression analysis shows that our method have a high consistency with canopy structure parameter direct surveying method, the correlation coefficient between two methods hit 0.94. The line slope was 1.463, our method measurement values were lower than direct surveying method. Our method expands the HP canopy structure parameters acquire timing, provides an automatic monitoring solution.

T4

A289

Cloning of cDNAs of flowering time genes from different flowering periods of *Camellia* plants

Shuguo Fan
School of Chemistry and Life Sciences, Chuxiong Normal University

The optimal timing of transition from vegetative to reproductive growth, known as the floral transition, is crucial for successful sexual reproduction of flowering plants. This developmental transition is precisely regulated by various environmental stimuli and endogenous signals, such as light, temperature, nutrients, and plant age. In plants, the transition from vegetative growth to flowering is regulated by a complex combination of environmental and internal signals. This developmental transition is controlled by environmental cues, such as seasonal changes in day length (photoperiod) or winter cold (vernalization) as well as ambient conditions including light intensity and spectral quality. It has been reported that floral integrator gene expression is repressed by the MADS box transcription factor SHORT VEGETATIVE PHASE (SVP), an inhibitor of flowering. Mutations in *SVP* cause early flowering under noninductive short days (SD) and under long days (LDs). SVP plays a similar role in response to vernalization where it forms a heterodimer with the MADS box transcription factor FLOWERING LOCUS C (FLC) to strongly repress flowering before exposure to cold. In *Arabidopsis*, the floral transition and floral organ identity are controlled by a subset of MADS-box transcription factors such as AGL24 and SHORT VEGETATIVE PHASE (SVP). These two closely related MADS-box genes have been shown to be important for various stages of reproductive development. No related results have been reported in *Camellia japonica*. In order to elucidate the function of *SVP*-, *FLC*- and *AG*-like genes in camellia, related primers have been designed for *CjSVP* (F: 5′- ATGGCGAGAGAAAAGATTCAG-3′; R: 5′- CTAACCACCATACGGTAAGCC-3′), *CjFLC* (F: 5′- TGACATAGAAAGCTACCACC-3′; R: 5′-CTCCCTTCACTATTCTTCATC-3′) and *CjAG* (F: 5′- CACCGGATCGGTGGCAGAAATTAATGC-3′; R: 5′-GGATCGGATTCGGGTAATACTTCTCTC-3′), respectively. RNAs were extracted from the leaves and flowers from different camellia varieties with different flowering periods (17 Camellia varieties were used, flowering from December to March), followed by measuring the purity and quality of RNAs, respectively. The total RNA was used as a template for reverse transcription to obtain cDNA first strand, and then for gene cloning. The results showed that the total RNA could be extracted from flowers than that in leaves of *Camellia japonica*. RT-PCR was performed and cDNA was cloned by using primers mentioned above. RT-PCR results showed that some specific bands could be clearly visualized, which were to be sequenced, and sequence alignments were to be performed.

T4

A290

Preliminary observations on plant morphology and morphological and anatomical characteristics of the flower bud differentiation on *Camellia japonica*

Shuguo Fan

School of Chemistry and Life Sciences, Chuxiong Normal University

The morphological characteristics of different flowering periods of *Camellia japonica* varieties were primarily investigated, which are trees, up to 13 m, or shrubs; young branchlets glabrous; leaves leathery, elliptic, 5~10 cm long, apex blunt or sharp mucronate, base broadly cuneate, glabrous on both surfaces, lateral veins 7~8 pairs, obtusely toothed; petiole 0.8~1.5 cm long; single flowers terminal and axillary, red; flowers sessile; bracts and sepals 10, semi orbicular or orbicular, 0.4~2 cm long, sericeous, glabrescent; petals 6~7, outer 2 suborbicular, free, 2 cm long, more than 5 pieces of hair, obovate, 3~4.5 cm long, connate 8 mm, glabrous; stamens 3 round; outer filament tube 2.5~3 cm long, 1.5 cm long; ovary glabrous, style 2.5 cm long, apex 3 crack; capsule globose, 3~5 cm in diameter, March cracked, woody mericarps, thickness of 6~8 mm, 1~2 seeds per locule; seed glabrous; flowering in December to March of the following year. Morphological and anatomical characteristics and external morphology of *C. japonica* flower bud during differentiation were investigated for nearly four years from January 2013 to October 2016. The results show that (1) Solitary flower bud was axillary or terminal; (2) The differentiation of flower bud started at the beginning of May and ended at the middle and later September; (3) The process of flower bud differentiation can be divided into six periods: pre-differentiation of flower bud, sepal formation, petal formation, stamen and pistil formation, anther and ovule formation, as well as stamen and pistil maturity, stamen and pistil primordium differentiated almost simultaneously, and stamen and pistil nearly synchronized development during stamen and pistil formation. Development of pistil was slightly earlier than that of stamen during anther and ovule formation in almost all the camellia varieties, firstly the lower part of pistil enlarged, then extended upward and closure up, forming ovary, locule, ovule and stigma. Subsequently stamen elongated quickly and formed anther. Meanwhile ovule was also elongated, when stamen and pistil maturity. At the period of stamen and pistil maturity, all the organs were becoming mature; (4) Correspondingly, external morphology changed at different differentiation stages. The above results lay a solid foundation on the research of flowering mechanism of Camellia.

T4

A291

Separation of cDNA of the *FT* flowering gene from *Paeonia rockii*

Shuguo Fan

School of Chemistry and Life Sciences, Chuxiong Normal University

Plants integrate day length and ambient temperature to determine the optimal timing for developmental transitions. In *Arabidopsis* (*Arabidopsis thaliana*), the floral integrator FLOWERING LOCUS T (FT) and its closest homolog TWIN SISTER OF FT promote flowering in response to their activator CONSTANS under long-day inductive conditions. Low ambient temperature (16°C) delays flowering, even under inductive photoperiods, through repression of FT, revealing the importance of floral repressors acting at low temperatures. *FT* gene is an important factor of integration into flowering; it plays an important role in the regulation of flowering time aspect. DNA and RNA were extracted from the leaves and flowers from ornamental plants *Paeonia rockii*, followed by measuring the purity and quality of DNA and RNA, respectively. The total RNA was used as a template for reverse transcription to obtain cDNA first strand, and then for *FT* gene cloning. The results showed that the total RNA was easier to be extracted from flowers than that in leaves of *P. rockii*. RT-PCR was performed and *FT* cDNA was cloned by using primers (F:5'-ATGCCTAGAAATAGGGATCCACTA-3', R:5'- TTATCTTCTTCCGCCTGACC-3'), a 1000-bp single band was clearly visualized, which was to be sequenced, and sequence alignment was to be performed.

T4

A292

De Novo transcriptome analysis of polysacchides biosynthesis and metabolic pathways involving nucleotide-sugar interconversion enzymes in *Codonopsis pilosula*

Jianping Gao[1], *Lingya Cao*[2], *Jing Li*[1], *Xiaolin Wang*[1], *Xiaoxiao Li*[1], *Zhong Cheng*[2]

1. College of Pharmacy, Shanxi Medical University

2. College of Basic Medicine, Shanxi Medical University

Codonopsis genus has been extensively used in China as a therapeutic agent, and *Codonopsis Radix* one of the valuable herbs among the 42 species of the *Codonopsis* genus used for replenishing qi deficiency, strengthening the immune system, improving poor gastrointestinal function, *etc.* Codonopsis polysaccharides (CPPs) are of the key pharmacologically active components in *C. pilosula*. However, the regulation of synthetic pathways of CPPs has not been fully investigated. To explore the genes of cluster related with CPPs biosynthesis, we performed *de novo* transcriptome analysis of *C. pilosula*, and established the plant culture system in vitro for the purpose of understanding the regulation of metabolic pathways involving nucleotide-sugar interconversion enzymes. An EST collection was created from a transcriptome library prepared from different parts of *C. pilosula* (Franch.) Nannf. We constructed the metabolic pathways related to biosynthesis of CPPs, which is divided into three main steps. The first step is the formation of UDP-Glu and GDP-Man from Glu-6-P and Fru-6-P, which are then converted in the second step into other NDP-sugars. Finally, the active monosaccharide units, NDP-sugars, are added to the sugar residues of various polysaccharides and glycoconjugates by the action of various glycosyltransferases (GTs). The large number of genes were found to encode enzymes that catalyze the formation of the CPPs, a total of 723 GT unique sequences containing 170 UGTs were obtained. The stable prokaryotic expression systems of four genes, such as UDP-glucose pyrophosphorylase (*UGPase*), sucrose transporter (*SUC*), phosphomannose isomerase (*PMI*) and phosphomannomutase (*PMM*), were constructed and confirmed by PCR, TA cloning and sequencing, sequence analysis were also performed. UGPase is a key enzyme that functions upstream of the biosynthesis pathway of Astragalus polysaccharides, and is responsible for the conversion of glucose-1-phosphate to UDP-Glu. PMI, on one hand, allows plants carrying it to convert mannose-6-phosphate, on the other hand, PMI-based selection system is harmless and highly versatile in plants, which means that PMI could be used to monitor transgenic events and visually separate transgenic material from non-transformed material. The cloning and expression of genes involved in polysaccharides biosynthesis would discover the regulation of polysaccharides metabolism and understand the related pathways. To achieve the purpose of exploring the functions of genes, the adventitious root and cluster bud induction and culture system of *C. pilosula* (Franch.) Nannf. were established. Seeds from HuGuan were treated with 400 mg/L gibberellin for 8 h, 25℃ light culture. Then the adventitious roots and cluster buds were obtained through tissue culture by manipulation of NAA and 6-BA concentration, respectively. Our results provide understanding of the biosynthesis of CPPs at the molecular level in *C. pilosula* inferred by RT-qPCR analysis of candidate genes functioning at key separate branches of biosynthesis pathway of CPPs. Therefore, the plant signaling molecules were discovered to enhance the efficient and sustainable production of CPPs and related bioactive natural products. The tissue culture system would be applied to understand the metabolic mechanism related with the production of CPPs *in vitro*.

T4

A293

Sucrose plays a crucial role in response to heat and chilling stress by reducing oxidative damage in potato

Qiao-Qiao Chen, Zai-Ping Feng, Hui-Jun Yuan, ***Hui-Ling Gong***
School of Life Science and Engineering, Lanzhou University of Technology

Abiotic stress is strongly detrimental to plant growth and development. Heat and chilling stress is the primary abiotic stresses that limit plant biomass production and productivity. Sucrose represents the major transport form of photosynthetically assimilated carbohydrates and plays versatile roles in plants. On the other hand, sugars function directly as protective agents at higher concentrations, and they serve as substrates or signals for stress-induced alterations at low concentrations, while the mechanism of the protective function seems to be unclear.To determine the alleviative effects of exogenous sucrose on chilling and heart stress, potato seedlings were pretreated with sucrose for 1 day and exposed to heat and chilling, respectively. The results showed that: 1) Pretreatment with sucrose led to a significantly increase in survival rate in potato seedlings suffering from heat and chilling stress. 2) Pretreatment with sucrose increased the activities of superoxide dismutase, ascorbate peroxidase, hydrogen peroxidase and peroxidase under heat and chilling stress. 3) Sucrose pretreatment reduced the production of reactive oxygen species, the superoxide anion (O_2^-) and the hydrogen peroxide (H_2O_2) were at a lower level than control. At the same time, the content of malonaldehyde was also reduced. In summary, we propose pretreatment with sucrose enhances chilling and heat tolerance in potato by activating antioxidant enzymes, and sucrose may act as a signal molecule interacting with antioxidant pathway.

T4

A294

High phenotypic plasticity does not imply high salt stress tolerance: A case study of two pioneer species from the Yellow River Delta

Xiao Guo[1,2]*, Xiao-Huang Ren*[2]*, Franziska Eller*[3]*, Ming-Yan Li*[2]*, Ren-Qing Wang*[2]*, Ning Du*[2]*, Wei-Hua Guo*[2]
1. Qingdao Agricultural University
2. Shandong University
3. Aarhus University

A greenhouse experiment was conducted in which two leguminous species commonly used in the Yellow River Delta for vegetation restoration, Robinia pseudoacacia and Amorpha fruticosa, were subjected to the five salt treatments: 0, 50, 100, 150 and 200 mmol L-1. We aimed at investigating which of the two species would be better suited for growth in a saline environment, and whether the acclimation capacity to salinity resulted from an inherently higher phenotypic plasticity of the species. The results showed that salinity affected most growth parameters and biomass parameters but had no effects on most leaf traits and physiological parameters of the two species. Height, relative growth rate of crown area, root biomass and leaf mass ratio of R. pseudoacacia were reduced by higher salinity while A. fruticosa was not affected. Chlorophyll a to chlorophyll b ratio and total antioxidative capacity of A. fruticosa increased with higher salinity while those of R. pseudoacacia remained unchanged. Root mass ratio and vitamin C concentration of both species were not affected by salinity, whereas vitamin C concentration of A. fruticosa overall was higher than that of R. pseudoacacia. Methane dicarboxylic aldehyde concentration of R. pseudoacacia increased with the increase of salinity while that of A. fruticosa was not affected. Root to shoot ratio of A. fruticosa was higher than of R. pseudoacacia at most salt treatments. Of all leaf traits, only leaf area differed among the different salinities. Robinia pseudoacacia generally exhibited a greater plasticity than A. fruticosa in response to salinity, but A. fruticosa was more resistant to the higher salinities, compared to R. pseudoacacia, and is thus a better candidate for vegetation restoration in saline areas.

T4

A295

Differences of photosynthetic characters among the typical geographic populations in the *Houttuynia* of China

Anna He, *Xianjin Wu*

Department of Life Science, Huaihua University

Houttuynia cordata is a perennial herb in genus *Houttuynia* of family Saururaceae. It is a functional food with enriched nutrition and a herbal medicine with long history for treatment of pneumonia, acute and chronic bronchitis, urinary tract infections and other diseases. In this paper, morphology of six typical geographical *Houttuynia* taxa and their differences in photosynthesis at normal temperature and adverse conditions such as high and low temperatures and drought were studied. The results showed that the morphological differences of the six taxa mainly manifested as plant color, main stem height, internode length of aerial stems, leaf length and leaf width. Under normal conditions, EM and HH (from Hunan Huaihua) had similar values and diurnal variation trends of Pn, Tr, Gs and Ci. Moreover, Pn was not significantly different among the six taxa and Tr, Gs and WUE differ significantly only among a few taxa. By contrast, adverse treatment such as heat (40°C), cold (5°C) and drought (70% soil water holding capacity) could substantially decrease Pn, Tr and Gs, but increase WUE. The differences in Pn, Tr and Gs were significant only among multiple taxa, while the difference in WUE was significant among all taxa. In addition, the alteration trends of chlorophyll fluorescence parameters in EM and HH after adverse treatments were very similar. Comprehensive analysis of the changes in gas exchange and chlorophyll fluorescence parameters after adverse treatments found besides *Houttuynia* taxon from high latitude Rugao, Jiangsu had high tolerance to cold, all other *Houttuynia* taxa with strong tolerance to heat and cold were from Southwest and South-central China. Photosynthetic stress resistance of all *Houttuynia* taxa was inheritable and closely related to the eco-climatic conditions of their provenance. EM may be considered a taxonomical variant of *Houttuynia cordata*.

T4

A296

The role of H_2O_2 in the regulation of cyclic electron flow in the salt tolerance of soybean

Yi He, *Junliang Fu, Dean Jiang*

State Key Laboratory of Plant Physiology and Biochemistry, College of Life Sciences, Zhejiang University, Hangzhou, China

NDH (NAD (P)H dehydrogenase complex)-dependent cyclic electron flow (CEF) plays an important role when plant was exposed to stress environments. Meanwhile, stress environments induced the production of reaction oxygen species (ROS). H_2O_2 as one of ROS might play a critical role in the regulation of CEF. However, under stress conditions, the relationship between H_2O_2 and CEF is not yet clear. In this study, to further elucidate how H_2O_2 as a signal molecule induced the activation of CEF, we monitored the localization and content of H_2O_2 in soybean's leaf, petiole and stem, as well as related CEF activity in three different soybean varieties. The results showed that the CEF of the 3rd compound leaves could be upregulated in about 5-10 min by exogenous H_2O_2 treated in the 2nd leaves, especially in the salt-tolerant S111-9. In addition, the H_2O_2 signal could transfer from leaves to petioles, stems and adjacent leaves via apoplastic pathways and this signal could initiate the CEF of the 3rd compound leaf and regulate the activation of its CEF. Meanwhile, the speed of transmission was related to the salt tolerance of soybean varieties. Furthermore, Salt stress and H_2O_2 could increase the CEF and ATP content of leaves, but no longer increased the H_2O_2 signal. Therefore, these results suggested that the H_2O_2 signal was also strongly interconnected with CEF in the response to salt stress. This study was not only helpful to our understanding for the defense signal transduction under stress conditions, but also had an important theoretical significance and application value for the improvement of crop salt tolerance.

T4

A297

Embryonic factor 10 encodes a DEAD-box RNA helicase that regulates embryo development and meristem homeostasis in *Arabidopsis*

Xiu-Li Hou[1,2], *Hua-Qin Gong*[1], ***Chun-Ming Liu***[1]

1. Institute of Botany, Chinese Academy of Sciences

2. University of Chinese Academy of Sciences

Embryogenesis is a critical process for the life cycle of flowering plants. Shoot and root meristems formed during embryogenesis establish the above- and below-ground tissues, respectively, after seed germination. Factors linking embryogenesis and post-embryonic development are still unknown. By screening of ethyl methanesulfonate (EMS)-mutagenized populations of *Arabidopsis,* we identified the *embryonic factor 10-1* (*fac10-1*) mutant that showed zygote-lethal phenotype. *FAC10* gene was cloned through map-based cloning and confirmed by a genetic complementation test against a T-DNA insertion allele *fac10-2*. *FAC10* encodes a putative DEAD-box RNA helicase highly conserved in prokaryotes and eukaryotes. Expression studies using *pFAC10:FAC10-sYFP* and *pFAC10:FAC10-GUS* transgenic plants revealed that *FAC10* is highly expressed in actively dividing cells in root and shoot meristems. Homozygous *fac10-2* displayed short root with reduced meristematic cells in roots, variable sized and multiple shoot apical meristems, delayed germination and slow growth phenotypes. Expression patterns of RAM- and SAM-specific marker genes such as *pWOX5:erGFP*, *pSCR:erGFP*, *pSHR:SHR-GPF*, *pWUS:-GUS* and *pCLV3:GUS* were altered greatly in 5-day old *fac10-2* seedlings. Quantitative RT-PCR and northern blot results showed miR164 reduced significantly in *fac10-2*. These results suggest that the DEAD-box RNA helicase *FAC10* plays a critical role in embryo development and meristem homeostasis in *Arabidopsis*.

T4

A298

Cytosolic and nucleosolic calcium signaling are mutually ex-

cluded in *Arabidopsis*

Feifei Huang
Beijing Key Laboratory of Gene Resource and Molecular Development, College of Life Sciences, Beijing Normal University, Beijing.

Calcium acts as a universal second messenger to be involved in responses to environmental stresses and developmental processes. Previous reseaches showed that a number of stimuli induced the $[Ca^{2+}]$ increases both in the cytoplasm and nucleus in plants. However, the relationship between cytosolic and nucleosolic calcium signaling is still obscure. Here, we have generated the transgenic plants containing the fusion protein of human parvabumin (PV) with nuclear export signal sequence (NES-PV) or nuclear localized signal sequence (NLS-PV) in order to selectively block cytoplasmic calcium or nuclear calcium. Firstly, we find that osmotic-stress induced $[Ca^{2+}]_{cyt}$ increase ($OICI_{cyt}$) and salt-stress induced $[Ca^{2+}]_{cyt}$ increase ($SICI_{cyt}$) were impaired in *NES-PV* lines as compared with *Arabidopsis* wild type (WT). Similarly, osmotic-stress induced $[Ca^{2+}]_{nuc}$ increase ($OICI_{nuc}$) and salt-stress induced $[Ca^{2+}]_{nuc}$ increase ($SICI_{nuc}$) were also disrupted in *NLS-PV* lines. These results indicated that PV can effectively block the increase of $[Ca^{2+}]$ in response to various stimuli in plants. However, $OICI_{cyt}$ and $SICI_{cyt}$ were similar in *NLS-PV* plants as WT, and vice versa. Furthermore, we found that osmotic-stress and salt-stress induced stomatal closure were reduced dramatically both in *NES-PV* and *NLS-PV* lines. Interestingly, the phytohormone abscisic acid and Methyl jasmonate induced stomatal closure were abolished only in NES-PV lines, but not in NLS-PV lines as compared with WT. These results implied that cytosolic and nucleosolic calcium signaling are independent and coexistent to play the pivotal roles in plant growth and development.

T4

A299

GsCHX19.3, a member of cation/H$^+$ exchanger superfamily from wild soybean contributes to high salinity and carbonate alkaline tolerance

Bowei Jia[1,2], *Mingzhe Sun*[1,2], *Junkai Yang*[1], *Jianwei Li*[1], ***Xiaoli Sun***[1]
1. Heilongjiang Bayi Agricultural University
2. Northeast Agricultural University

Cation/H$^+$ exchangers (CHX) are characterized to be involved in plant growth, development and stress responses. Although soybean genome sequencing has been completed, the GmCHX family hasn't yet been systematically analyzed. Here, a total of 40 *GmCHXs* were identified and phylogenetically clustered into five groups. Members within each group showed high conservation in the exon-intron organization and motif architecture. Interestingly, only Group IVa members exhibited high expression in different tissues, as well as induced expression under carbonate stress. Among them, *GsCHX19.3* displayed the greatest up-regulation in response to carbonate alkaline stress, which was further confirmed by quantitative real-time PCR analysis. We also observed the ubiquitous expression of *GsCHX19.3* in different tissues and its localization on plasma membrane. Moreover, we found that *GsCHX19.3* expression in AXT4K, a yeast mutant lacking four ion transporters conferred resistance to low K^+ at alkali pH, as well as carbonate stress. In *Arabidopsis*, *GsCHX19.3* overexpression increased plant tolerance to high salt and carbonate alkaline stresses. Furthermore, *GsCHX19.3* transgenic lines showed lower Na^+ concentration and higher K^+/Na^+ values under salt-alkaline stress. Whereas *atchx19* knockout mutant displayed decreased stress tolerance, higher Na^+ concentration and lower K^+/Na^+ values. Taken together, our findings indicated that *GsCHX19.3* contributed to high salinity and carbonate alkaline tolerance.

T4

A300

Nutrient status of nipa (*Nypa fruticans* Wurmb.) from semi-wild stands in selected areas of Mindanao, the Philippines

Junaldo Mantiquilla, *Reynaldo Abad, Sandy Faye Salmasan*
University of the Philippines Mindanao

Nipa (*Nypa fruticans* Wurmb.) is a potential source of biofuel from its sap. Increasing sap yield requires understanding nutrient requirements of this palm necessary for cultural management practices. Hence, the nutrient status of nipa in its natural habitats in southern Philippines (Western, Northern, Eastern and Southern Mindanao) were determined to establish baseline data of essential elements using soil and leaf analyses. Results revealed that nipa palms thrived in a wide variety of soils from clayey, loamy to sandy clay loam types. Elemental variations were observed across sampling sites owing to their locations in the estuaries, human activities and soil pH. Exchangeable bases Na, Ca, Mg and K varied considerably across areas affecting leaf levels of nipa. However, leaf B levels were shown to be regulated even at toxicity levels of B in the soil. These growing conditions indicate that nipa palm can tolerate highly variable physico-chemical factors that exist in the estuaries. The essential elements had the following levels in the leaf tissue: 1.30±0.30% N, 0.13±0.03% P, 0.82±0.14% K, 0.12±0.29% Ca, 0.03±0.03% Mg, 0.16±0.13% Na, 0.53±0.10% S, 0.70±0.10 mg/kg Cu, 4.30±0.90 mg/kg Zn, 20.00±2.30 mg/kg Fe, 64.20±31.70 mg/kg Mn and 4.40±1.70% B. These leaf nutrient levels will serve as benchmark information for fertilizer research in the future.

T4

A301

Morpho-physiological indices for drought tolerance in pigeonpea (*Cajanus cajan* L. Millsp.) genotypes under rainfed and irrigated conditions

Satish Khadia, *Suresh Kanuri*
S. D. Agricultural University, S.K.Nagar

A field experiment was carried at Centre of Excellence for Research on Pulses, Sardar Krushinagar Dantiwada Agricultural University (S.D.A.U.), Sardarkrushinagar, Gujarat during Kharif 2013 to study the morpho-physiological indices for drought tolerance of different pigeonpea genotypes under rainfed and irrigated conditions. Significant differences exhibited amongst the genotypes for initial and final plant stand, days to flowering, days to maturity, 100 seed weight, harvest index. The genotypes SKNP 1004, SKNP 1005 and SKNP 0805 recorded minimum percent reduction in yield due to moisture stress. Relative Water Content (RWC), Drought Tolerance Efficiency (DTE) were found to be the most useful parameters while selecting genotypes for drought tolerance. The genotypes SKNP 1004, SKNP 1005, SKNP 0805 and BAN-AS were promising for yield and yield contributing characters in both rainfed and irrigated conditions.

T4

A302

Arabidopsis acyl-CoA-binding protein ACBP3 participates inplant response to hypoxia by modulating very-long-chain-fatty acid metabolism

***Lijuan Xie**, Yong Xia Lai, Shi Xiao*
Sun Yat-sen University

In *Arabidopsis thaliana*, acyl-CoA-binding proteins (ACBPs) are encoded by a family of six genes (*ACBP1* to *ACBP6*), and are essential for diverse cellular activities. Recent investigations suggest that the membranean chored ACBPs are involved in oxygen sensing by sequestration of group VII ethylene-responsive factor sunder normoxia. Here, we demonstrate the involvement of *Arabidopsis* ACBP3 in hypoxic tolerance. *ACBP3* transcription was remarkably induced following submergence under both dark (DS) and light (LS) conditions. *ACBP3*-overexpressors (*ACBP3*-OEs) showed hypersensitivity to DS, LS and ethanolic stresses, with reduced transcription of hypoxia-responsive genes as well as accumulation of hydrogen peroxide in the rosettes. In contrast, suppression of ACBP3 in ACBP3-KOs enhanced plant tolerance to DS, LS and ethanol treatments. Byanalyses of double combinations of OE-1 with *npr1-5*, *coi1-2*, *ein3-1* as well as *ctr1-1* mutants, we observed that the attenuated hypoxic tolerance in *ACBP3*-OEs was dependent on NPR1- and CTR1-mediated signalling pathways. Lipid profiling revealed that both the total amounts and very-long-chain species of phosphatidylserine (C42:2- and C42:3-PS) and glucosylinositol phosphorylceramides (C22:0-, C22:1-, C24:0-, C24:1-, and C26:1-GIPC) were significantly lower in ACBP3-OEs but increased in ACBP3-KOs upon LS exposure. By microscalethermophoresis analysis, the recombinant ACBP3 protein bound VLC acyl-CoA esters with high affinitiesin vitro. Further, a knockout mutant of *MYB30*, a master regulator of very-long-chain fatty acid (VLCFA) biosynthesis, exhibited enhanced sensitivities to LS and ethanolic stresses, phenotypes that were ameliorated by ACBP3-RNAi. Taken together, these findings suggest that *Arabidopsis* ACBP3 participates in plantresponse to hypoxia by modulating VLCFA metabolism.

T4

A303

Developing anti-stress sweet potato by over-expressing stress-tolerant genes

Ge Li
Jiangsu Normal University

Higher plants are usually more sensitive to environmental stresses, such as high salt, drought, high/low temperature and so on. Various environmental stresses are the main limiting factors for plant growth and distribution. And with the growing population and demand for food and energy, therefore, the cultivation of stress-tolerant crops is also increasingly important. To solve the problem, we need to develop new environmentally friendly industrial crop varieties to be grown on marginal land and desertification areas for sustainable development. My research is to improve the yield of sweetpotato, which is an attractive crop for coping with world food and energy problems, by transgenic. Screening the stress-tolerant genes based on the functional expression of studying genes (*IbMYB2*, *IBMYB3* and *IbMYB4*) and over-expressing them in the sweet potato via construction of over-expressing vector, and the gene with higher resistance have been screened out.

T4

A304

Expression of *cadR* enhances its specific activity for Cd detoxification and accumulation in *Arabidopsis*

Jingrui Li
Institute of Botany, the Chinese Academy of Sciences

Cadmium (Cd) is a transition metal that is highly toxic in biological systems. Anthropogenic emissions of Cd have increased biogeochemical cycling and the amount of Cd in the biosphere. Here we studied the utility of a bacterial Cd-binding protein CadR for the remediation of Cd contamination. CadR was successfully targeted to chloroplasts using a constitutive *CaMV35S* promoter or a shoot-specific chlorophyll a/b-binding protein 2 gene (*CAB2*) promoter and an RbcS (small subunit of the Rubisco complex) transit peptide. Under short-term (2-d) exposure to Cd, the *cadR* transgenic plants with showed up to 2.9-fold Cd accumulation in roots compared with the untransformed plants. Under mid-term (7-d) exposure to Cd, the concentrations of Cd in leaves began to increase but there were no differences between the WT and the *cadR* transgenic plants. Under long-term (16-d) exposure to Cd, the *cadR* transgenic plants accumulated greater amounts of Cd in leaves than the untransformed plants. Total Cd accumulation (μg per plant) in shoots and roots of the plants expressing *cadR* were significantly higher (up to 3.5-fold in shoots and 5.2-fold in roots) than those of the untransformed plants. We also found that targeting CadR to chloroplasts facilitated chloroplastic metal homeostasis and chlorophyll *b* accumulation. Our results demonstrate that manipulating chelating capacity in chloroplasts or in cytoplasm may be effective in modifying both the accumulation of and resistance to Cd.

T4

A305

Light affects salt stress-induced transcriptional memory of *P5CS1* in *Arabidopsis*

*Xuanjun Feng, **Jingrui Li***
Institute of Botany, the Chinese Academy of Sciences

To cope with environmental stresses, plants often adopt a memory response upon primary stress exposure to facilitate a quicker and stronger reaction to recurring stresses. However, it remains unknown whether light is involved in the manifestation of stress memory. Proline accumulation is a striking metabolic adaptation of higher plants during various environmental stresses. Here we show that salinity-induced proline accumulation is memorable and HY5-dependent light signaling is required for such a memory response. Primary salt stress induced the expression of Δ1-pyrroline-5-carboxylate synthetase 1 (*P5CS1*), encoding a proline biosynthetic enzyme and proline accumulation, which were reduced to basal level during the recovery stage. Reoccurring salt stress-induced stronger *P5CS1* expression and proline accumulation were dependent upon light exposure during the recovery stage. Further studies demonstrated that salt-induced transcriptional memory of *P5CS1* is associated with the retention of increased H3K4me3 level at *P5CS1* during the recovery stage. HY5 binds directly to light-responsive element, C/A-box, in the *P5CS1* promoter. Deletion of the C/A-box or *hy5* mutations caused rapid reduction of H3K4me3 level at *P5CS1* during the recovery stage, resulting in impairment of the stress memory response. These results unveil a previously unrecognized mechanism whereby light regulates salt-induced transcriptional memory via the function of HY5 in maintaining H3K4me3 level at the memory gene.

T4

A306

Identification of a regulatory element responsible for salt induction of rice *OsRAV2* through *ex situ* and *in situ* promoter analysis

Juan Li[1], *Yong-Bo Duan*[1,2], *Rong-Fang Xu*[1], *Rui-Ying Qin*[1], *Hao Li*[1], *Ya-Chun Yang*[1], *Hui Ma*[1], *Li Li*[1], *Peng-Cheng Wei*[1], *Jian-Bo Yang*[1]

1. Rice Research Institute, Anhui Academy of Agricultural Sciences

2. College of Life Sciences, Huaibei Normal University

Salt is a major environmental stress factor that can affect rice growth and yields. Recent studies suggested that members of the AP2/ERF domain-containing RAV (related to ABI3/VP1) TF family are involved in abiotic stress adaptation. However, the transcriptional response of rice *RAV* genes (*OsRAVs*) to salt has not yet been fully characterized. In this study, the expression patterns of all five *OsRAVs* were examined under salt stress. Only one gene, *OsRAV2*, was stably induced by high-salinity treatment. Further expression profile analyses indicated that *OsRAV2* is transcriptionally regulated by salt, but not KCl, osmotic stress, cold or ABA (abscisic acid) treatment. To elucidate the regulatory mechanism of the stress response at the transcriptional level, we isolated and characterized the promoter region of *OsRAV2* (P<em>$_{OsRAV2}$). Transgenic analysis indicated that P_{OsRAV2} is induced by salt stress but not osmotic stress or ABA treatment. Serial 5' deletions and site-specific mutations in P_{OsRAV2} revealed that a GT-1 element located at position -664 relative to the putative translation start site is essential for the salt induction of P_{OsRAV2}. The regulatory function of the GT-1 element in the salt induction of *OsRAV2* was verified *in situ* in plants with targeted mutations generated using the CRISPR/Cas9 (clustered regularly interspaced short palindromic repeats/CRISPR-associated protein 9) system. Taken together, our results indicate that the GT-1 element directly controls the salt response of *OsRAV2*. This study provides a better understanding of the putative functions of *OsRAVs* and the molecular regulatory mechanisms of plant genes under salt stress.

T4

A307

Expression of a Na^+/H^+ antiporter *RtNHX1* from a recretohalophyte *Reaumuria trigyna* improved salt tolerance of transgenic *Arabidopsis thaliana*

Ningning Li, *Xue Wang, Linlin Zheng, Yingchun Wang*

Key Laboratory of Herbage and Endemic Crop Biotechnology, and College of Life Sciences, Inner Mongolia University, Hohhot, China

Reaumuria trigyna is an endangered recretohalophyte endemic to the Eastern Alxa-Western Ordos area of Inner Mongolia in china. Based on transcriptome data, a Na^+/H^+ antiporter gene, referred to as *RtNHX1*, was isolated from *R. trigyna*, which included a 1,662-bp open reading frame encoding a protein of 553 amino acids. *RtNHX1* was rapidly induced by NaCl and exogenous ABA treatment, and showed different tissue expression pattern before and after salt stress. Expression of the *RtNHX1* gene in yeast strain *AXT3* accumulated more Na^+ and K^+ of vacuolar, maintained relatively constant Na^+ concentration of the cytoplasm and lower Na^+/K^+ ratio of yeast cell, revealed that *R.trigyna* RtNHX1 are a functional antiporter capable of sequestering redundant Na^+ and K^+ into the vacuole. Transgenic *Arabidopsis* overexpressing *RtNHX1* accumulated more K^+ and less Na^+ in leaves, and exhibited lower Na^+/K^+ ratio than WT under salt stress, accordingly, the ions transport related genes, such as *AtAVP1*, *AtKUP6* and *AtKUP8*, were also up-regulated, indicating that over-expression of *RtNHX1* confers salt tolerance of transgenic *Arabidopsis* involving in maintaining Na^+/K^+ homeostasis. In addition, over-expression of *RtNHX1* significantly enhanced seed germination, biomass accumulation, contents of chlorophyll and root elongation of transgenic *Arabidopsis* under salinity stress, and dramatically rescued the salt-sensitive defects of the *nhx1* mutant. Transgenic *Arabidopsis* exhibited higher activities of POD and CAT, and contents of proline, relative water, and lower contents of MDA than that of WT under salt stress, moreover, the antioxidant and proline biosynthesis related genes, including *AtPOD1*, *AtCAT1*, *AtP5CS1*, *AtP5CS2*, were corresponding up-regulated, indicating that expressing *RtNHX1* confers salt tolerance of transgenic *Arabidopsis* through efficient capacity of osmotic adjustment and reactive oxygen scavenging, suggesting that *RtNHX1* may be a good candidate gene for improvement of salt tolerance in plants.

T4

A308

Terpenoids diversity in plant glandular trichomes and sesterterpenoid biosynthesis

Yan Liu[1], *Shi-Hong Luo*[2], *Chun-Huan Li*[1], *Ying Wang*[1], ***Sheng-Hong Li***[1]

1. Kunming Institute of Botany, Chinese Academy of Sciences

2. Shenyang Agricultural University

Plants glandular trichomes (GTs) are the adaptive defensive structures on the surfaces of aerial organs of most terrestrial plants. Plant GTs could produce, store and secrete a wide variety of secondary metabolites with defensive functions and important pharmaceutical value, and thus are regarded as phytochemical factories. To precisely investigate the bioactive compounds and their defensive functions in plant GTs, we have developed a new and precise approach, laser microdissection (LMD) coupled with NMR and UPLC-MS/MS. Two unique classes of C_{25} sesterterpenoids (named leucosceptroids and colquhounoids) with defensive functions against insect herbivores and pathogenic fungi

were discovered in the GTs of the Himalayan Lamiaceae plants *Leucosceptrum canum* and *Colquhounia coccinea* var. *mollis*, respectively (Luo *et al*, 2010; Li *et al*, 2013). Leucosceptroids and colquhounoids are similar by their core structures, but are clearly distinct by their partially adverse stereochemistries and different post-modifications. Moreover, three new defensive clerodane diterpenoids (named seguiniilactones) with significant insect antifeedant activity were found in the peltate GTs of *C. seguinii* (Li *et al*, 2014). In addition, four new phytotoxic labdane diterpenoids (named paraguhenryisins) were identified in the capitate GTs of *Paragutzlaffia henryi* (Acanthaceae) (Wang *et al*, 2015). The biosynthesis of leucosceptroids has been investigated. A plant geranylfarnesyl diphosphate synthase (GFDPS), the enzyme producing the C_{25} prenyl diphosphate precursor to all sesterterpenoids, was cloned and functionally characterized from the GTs of *L. canum*. GFDPS is localized to the plastids, and inhibitor studies indicated its use of isoprenyl diphosphate substrates supplied by 2-C-methyl-D-erythritol 4-phosphate pathway. Phylogenetic analysis suggested that GFDPS probably evolved from plant geranylgeranyl diphosphate synthase through gene duplication and neofunctionalization under the influence of positive selection (Liu *et al*, 2016).
This research was financially supported by the National Natural Science Foundation of China (31525005, U1202263,31070320).

T4

A309

Effects of different water and nitrogen levels on starch synthesis enzymes in wheat grains during grain filling stage

Shuang Li, *Yang Gao, Xiaojun Shen, Zhuanyun Si, Aiwang Duan**
Ministry of Agriculture Farmland Irrigation Research Institute

Under the condition treated with three levels of irrigation and four levels of nitrogen fertilizer, the winter wheat cultivars of Aikang58 was planted in the field to investigate the effects of different water and nitrogen treatments on sucrose metabolism, starch content, the activities of enzymes of starch synthesis during the grain filling stage. Results showed that suitable irrigation and nitrogen fertilizer management could improve the ability of photosynthate supply and transformation, increase the activity of starch synthesis enzymes, promote grain filling proceeding and increase grain yield. Water stress could accelerate the rate of sucrose accumulation, shorten the peak time and accelerate wheat maturation process, while decrease the activities of ADPGPPase, SSS and GBSS in grains, and correspondingly reduce the starch content and amylopectin content in wheat grains. Nitrogen fertilization would mainly affect grain physiological metabolism at the late filling stage, while the impact was small at the early stage. Excessive nitrogen application did not improve the starch synthesis activity and grain yield, so reasonable management of water and nitrogen could generate positive interaction effects. Our experiment also indicated that under the field environment, the activity of SSS was mainly controlled by ambient temperature. Low temperature could decrease the activity of SSS and slow down the rate of starch accumulation.

T4

A310

The NAC transcription factor SIN1 influences the salinity tolerance of soybean by directly regulating ABA and ROS generation genes

Shuo Li, *Nan Wang, Dandan Ji, Fengning Xiang*
Shandong University

Soil salinity compromises the productivity of soybean, but the full nature of the plant's response and of its adaptation to the stresses still not understood. Here, the over-expression in soybean of a gene encoding GmSIN1, a salinity-induced NAC family transcription factor, was shown to enhance its productivity by up to 35% under non-stressed conditions and by 19% under conditions of salinity stress. The effect of the up-regulation was to maintain root elongation when the plants experienced salinity stress, effected by a simultaneous elevation of the level of both abscisic acid (ABA) and reactive oxygen species (ROS) to a more optimal level. GmSIN1 interacted with both *GmNCED3s* (ABA synthesis) and the *GmRbohBs* (ROS generation) genes and up-regulated their transcription by binding to their promoters at a novel site which has not been described previously. *GmSIN1*, *GmNCED3s* and *GmRbohBs* appear to establish a positive feedback system, which allowed for the rapid accumulation of ABA and H_2O_2, in turn amplifying the early salinity stress signal and improving the salinity tolerance. The suggestion is that the combined manipulation of ABA and ROS content should improve the salinity tolerance of soybean.

T4

A311

Functional analysis of maize silk specific gene *ZmSS1*

Wanying Li, *Jingjuan Yu, Qian Zhao*
China Agricultural University

Basic leucine zipper (bZIP) transcription factors regulate many processes including pathogen defence, seed maturation, flower development and stress signaling. The bZIP family in maize can be divided into 11 groups, Group D genes participate in two different processes: defence against pathogens and development. *ZmSS1*, which belongs to D group of bZIP transcription factor family, was cloned from maize silks. RNA-seq analysis revealed that the expression of *ZmSS1* was tissue-specific in maize silks, which was confirmed by qRT-PCR. *In situ* hybridization showed a specific accumulation of *ZmSS1* transcript in the xylem of silks. *ZmSS1* was not induced by the treatment of PEG, ABA, NaCl and SA in maize leaves. However, the expression of *ZmSS1* was increased in silk after treated with 100mM methyl jasmonate. The subcellular localization showed that the ZmSS1 protein is located in the nucleus of maize protoplasts. However, ZmSS1 don't have the ability of transcriptional activation in yeast and yeast-one hybrid assay showed that it could not bind to C-box or G-box, the predicted common binding core elements of bZIP D group. Heterologous expression of *ZmSS1* in *Arabidopsis thaliana* gave rise to a dwarf phenotype and the expression level of pathogenesis-related (PR) genes was up-regulated. The similar dramatic dwarfism phenotype was also shown in *ZmSS1* over-expression transgenic maize, but no obviously abnormality was observed in *ZmSS1* RNAi plants. Based on present results, we speculated that the *ZmSS1* is likely to involved in responses to pathogen challenge.

T4

A312

The role of *MSR1* in glutathione redox-dependent root development

Xin Yu[2,3], *Klaus Palme*[1,2], ***Xugang Li***[1,2]
1. State Key Laboratory of Crop Biology, College of Life Sciences, Shandong Agricultural University
2. Institute of Biology II/Molecular Plant Physiology, Faculty of Biology, the University of Freiburg
3. Citrus Research Institute, Chinese Academy of Agricultural Science/Southwest University

Glutathione is involved in thiol redox signaling and acts as a major redox buffer against reactive oxygen species, helping to maintain a reducing environment in planta. The crucial role of glutathione redox status in root growth has been revealed. We have previously shown that root apical meristem (RAM) and root stem cell niche were maintained by glutathione redox status in plastids, which was controlled by glutathione reductase 2 (GR2). The loss-of-function GR2 mutant *miao*, exhibited remarkable defects in root developmental. However, the downstream factors regulated by glutathione redox status in root development remains elusive. Here, we screened miao suppressors (*msr*) from a ethyl methanesulfonate -mutagenized miao population and isolated *msr1*, further identified *MSR1* gene. The aim of this study is to uncover the molecular mechanism how *MSR1* is involved in glutathione redox-dependent root development.

T4

A313

Dual phosphorylation of an ERF transcription factor by MPK3/MPK6 and CPK11 regulates pathogen resistance in *Arabidopsis*

Yihao Li
Beijing Key Laboratory of Gene Resource and Molecular Development, College of Life Sciences, Beijing Normal University

Plants have evolved an innate immune system to resist the pathogens during their lifespan. Recent studies have shown that pathogen-associated molecular patterns (PAMPs) can be recognized by the pattern recognition receptors (PPRs), which were located on the cell membrane, then activate the mitogen-activated protein kinase (MAPK) cascade and cytosolic calcium-dependent protein kinases (CPKs) signal transduction pathway to participate in pathogen resistance in plants. However, the cross-talk mechanism between these two signaling pathways remains unclear. Here, we found an *Arabidopsis* Ethylene Responsive Factor 72 (ERF72) can be phosphorylated in vitro by the kinases MPK3/6 and CPK11 which were all involved in plant immunity. Three CPK11 phosphorylation sites of ERF72 were identified by mass spectrometry. Overexpression of ERF72 enhanced the resistance to *Pseudomonas syringae* pv. tomato DC3000. Transcriptome analysis of transgenic plants which overexpressed non-degrading form of ERF72 (MA-ERF72) have found that many genes associated with disease resistance are up-regulated. By analyzing the GCC box elements in these gene promoters, transcriptional activation assay, ChIP-qPCR and EMSA, we proved that ERF72 directly binds to the GCC box on the *WRKY33* promoter to activate its transcription. Phosphorylation of ERF72 by MPK3/6 and CPK11 stimulates its transcription activity. In summary, ERF72 is co-regulated by MAPK and CDPK signaling pathway and regulates the transcription of *WRKY33*, which is involved in the immune response in *Arabidopsis*.

T4

A314

The relationship between morphology of leaves and branches and hormone during individual development stage transition of *Populus euphratica*

Jiahao Li, Peipei Jiao, ***Zhijun Li***
College of life science, Tarim University

We choose *Populus euphratica* with different diameters on the same site as the research object, through the field investigation and laboratory analysis, the morphology characteristics and endogenous hormone contents variation of *Populus euphratica* branches and leaves, and the relationship among them were investigated. The results showed: (1) As the diameter class increased, the morphology of branches and hormone contents on the crown from base to top showed a stage, hierarchy variation, all the morphology parameter of branches and leaves tested, endogenous hormone contents, such as GA_3, IAA, and ZR, hormones ratio of IAA/ABA, ZR/IAA, GA_3/ABA, and ZR/ABA in leaves, IAA/ABA, GA_3/ABA ratio in branches, showed a significant or very significant positive or negative relationship with diameter classes and crown height. (2) Leaves number per branch, ZR/IAA of branches and leaves, ZR contents in leaves, and ZR/IAA ratio in branches can be used as an important indicator for *Populus euphratica* transition from juvenile to adult. leaf shape index, petiole length, diameter of branches, the flower bud numbers per branch and leaf IAA content, and IAA/ABA ratio can be used as important index for flower bud number increasing in adulthood trees; (3) The content of GA3, IAA and ZR in leaf, and the ratio of IAA/ABA, GA3/ABA, ZR/ABA showed a significant or highly significant positive or negative correlation with all leaf morphological indexes, and ZR/IAA ratio of leaves were very significant positively or negatively correlated with leaf index and petiole length, GA3/ABA ratio in branches was significant or very significant positively correlated with branch length and leaves number per branch. Preliminary results suggested that the leaf morphological variations of *Populus euphratica* leaves during individual development may influenced by the content of GA3, IAA, ZR and hormone ratio of IAA/ABA, GA3/ABA, and ZR/ABA. Leaf shape index and petiole length changes affected by leaf ZR/IAA ratio, while, the changes of GA3/ABA ratio on branches may result in the changes of shoot length and leaf numbers on per branch.

T4

A315

Relationship between morphology and nutrition characteristics of branches and leaves of *Populus euphratica* during transition of the individual developmental stages

Jiahao Li, Shuaifei Liu, ***Zhijun Li****, Peipei Jiao*
College of life science, Tarim University

Endangered tree *Populus euphratica* with different diameters on the same site was chosen as the research object, through the field investigation and laboratory analysis, variation of morphology and nutrition characteristics of *Populus euphratica* branches and leaves, and the relationship among them was investigated as the diameter increased. The results showed, as the diameter classes increased, leaf morphology and nutrition changes on the crown from the base to top showed a stage, hierarchy variation. All the morphology parameter tested, content of total nitrogen, total phosphorus, organic carbon of branches and leaves, and carbon nitrogen ratio of leaves were significantly or very significantly positive or negative correlated with diameter class and crown height; a comprehensive analysis showed that the shoot length, number of leaves per branch, the total nitrogen content of leaves and carbon nitrogen ratio of branches and leaves can be used as an important indicator for *Populus euphratica* transition from juvenile to adult, and leaf index, petiole length, diameter of branches, the number of flower buds per branch, organic carbon content can be used as important index for flower bud number increasing turning point in adulthood trees; leaf index can be used as an important index for transition *of Populus euphratica* from juvenile region to adult region on the crown, and leaf area, petiole length, organic carbon content of branches and leaves, total nitrogen content of leaves can be used as morphological index for flower buds increasing turning point in adulthood tress; meanwhile, total nitrogen, total phosphorus, total potassium and the content of organic carbon and carbon nitrogen ratio may play an important role in the regulation of leaf morphological changes during the stage transition.

T4

A316

Effects of *OsIAAGLU* on IAA content and plant architecture in rice

Xiao-Lu Yu[1], *Hai-Yan Wang*[1], *David W.M. Leung*[2], *Xin-Xiang Peng*[1], ***E-E Liu***[1]

1. South China Agricultural University

2. University of Canterbury

Plant architecture is an important agronomic trait that is closely related to yields. The distribution and levels of auxin or indole-3-acetic acid (IAA) within plant tissues are well known to be associated with plant architecture determination. An important mechanism of regulating free IAA levels in monocots is formation of IAA ester conjugates, which account for approximately 62-70% of the total IAA in rice seeds. In this study, the gene *OsIAAGLU* in the rice genome database was found to have 67% sequence similarity with IAA-glucose synthase of maize (ZmIAAGLU), which is known to catalyze the reaction of free IAA with glucose and generate IAA-glucose. Expression of *OsIAAGLU* was mainly in the leaf sheath of rice seedlings, and could be induced by IAA and NAA. OsIAAGLU was localized in the cytoplasm. The number of tillers and leaf angle were significantly increased in the transgenic rice lines over-expressing *OsIAAGLU* compared to the wild-type plants (WT), while plant height and panicle length as well as seed set percentage decreased. *OsIAAGLU* mutants constructed by CRISPR/Cas9 system exhibited increased height and a smaller leaf angle compared to WT. Overexpression of *OsIAAGLU* resulted in reduced sensitivity to NAA, while *OsIAAGLU* mutants were more sensitive to NAA than WT. Free IAA contents in the leaves of *OsIAAGLU*-overexpressing transgenic lines were lower and higher in *OsIAAGLU* mutants than that of WT. These results suggested that *OsIAAGLU* could play a regulatory role in IAA homeostasis and rice plant architecture.

T4

A317

Involvement of HbMC1-mediated programmed cell death in tapping panel dryness of rubber tree (*Hevea brasiliensis*)

Hui Liu, *Zhi Deng, Hong Yang, Longjun Dai, Dejun Li*

Key Laboratory of Biology and Genetic Resources of Rubber Tree, Ministry of Agriculture, Rubber Research Institute, Chinese Academy of Tropical Agricultural Sciences, Danzhou, China

Tapping panel dryness (TPD) in rubber tree (*Hevea brasiliensis*) is the key factor limiting the yield of natural rubber. Developing therapeutic targets and drugs to prevent and treat TPD has become a prominent problem required to be solved urgently for the sustainable and healthy development of natural rubber industry. It was reported that TPD was a programmed cell death (PCD) process, but lack further systematic study. Metacaspases, a family of cysteine proteases, are key regulators of PCD in plants. To reveal the role of metacaspase in TPD, we carried out genome-scale identification and expression analysis of metacaspase gene family in rubber tree. A total of nine *HbMC* genes were identified in rubber tree genome. Multiple sequence alignment and phylogenetic analyses suggested that these genes were divided into two types: type I (*HbMC1–HbMC7*) and type II (*HbMC8* and *HbMC9*). Gene structure analysis demonstrated that type I and type II *HbMCs* separately contained four and two introns, indicating the conserved exon–intron organization of *HbMCs*. Quantitative real-time PCR analysis revealed that *HbMCs* showed distinct expression patterns in different tissues, suggesting the functional diversity of *HbMCs* in various tissues during development. Most of the *HbMCs* were regulated by drought, cold, and salt stress, implying their possible functions in regulating abiotic stress-induced cell death. Of the nine *HbMCs*, *HbMC1* expression was found to positively correlate with TPD. In addition, it was up-regulated by TPD-inducing factors, such as wounding, ethephon, and H_2O_2 treatments. Ectopic expression of *HbMC1* in yeast and tobacco enhanced oxidative stress-induced PCD. We surmised that the up-regulation of *HbMC1* would trigger PCD, and then lead to TPD. *HbMC1* may act as a potential therapeutic target for TPD.

T4

A318

Mechansims of a DnaJ protein from the resurrection plant *Boea hygrometrica* to protect chloroplasts under severe drought stress

Jie Liu[1,2], *Chunying Shen*[1], *Zhenna Zhang*[1]

1. Insititue of Botany, the Chinese Academy of Sciences

2. Weifang University of Science and Technology

Plant development and growth are severely affected by drought stress. Photosynthesis, as one of the most sensitive physiological processes to such impact, results in damage to ultrastructure in chlorophylls, degradation of chloroplast pigment, enzymatic activity loss in photosystem and accumulation of reactive oxygen species. DnaJ is a heat shock protein considered involving in chloroplast development and stress protection. In this study, a gene of DnaJ protein localized in chloroplast from the resurrection plant *Boea hygrometrica* (BhDnaJ) was inducible expressed by both abiotic stress (drought, heat, salt, alkaline and high calcium concentration) and ABA. The transgenic *Arabidopsis* overexpressing *BhDnaJ* confer improved tolerance to drought and drought-heat stresses. The fluorescence parameters compared to the wild type (WT) indicates less photodamage in PSII while maintain higher PSII activity and electron transport rate. No significant difference in the photoprotection parameters between *Arabidopsis* overexpressing *BhDnaJ* and WT was found. By yeast two-hybrid system, three proteins interacted with BhDanJ were screened for association with electron transport and photosynthetic phosphorylation. The results showed that BhDnaJ might interact with the chloroplast proteins directly to protect the structure and quantity, thereby participating in the protection of chloroplasts under stress.

T4

A319

CmCADs and three lignin biosynthesis related enzymes play diverse roles in response to abiotic stresses in oriental melon seedlings

Wei Liu, ***Hongyan Qi***

Shenyang Agricultural University

Cinnamyl alcohol dehydrogenase (CAD) is a key enzyme in lignin monomer synthesis pathway, which is responsible for the conversion of cinnamyl aldehydes to cinnamyl alcohols. In some species, CAD could not only take part in plant development through lignin construction but also play a role in resisting biotic and/or abiotic stresses mainly by activated the lignin synthesis pathway. Our previous research identified five *CAD* genes in the melon genome and named *CmCAD1* to *CmCAD5*, respectively. Three abiotic stresses were set as drought (8% and 15% PEG6000), salt (50 mM and 100 mM NaCl) and wound (leaf and stem wound) according to *CmCADs* promoter analysis that revealed motifs involved in stress response and signal molecule regulation. The results indicated that lignin contents accumulated significantly higher in stems and roots than the control. Lignin staining with phloroglucinol-HCl of stem sections obtained consistent results with lignin analysis. Phenylalanine ammonia-lyase (PAL) and peroxidase (POD) were generally induced to significant levels except for PAL activity in roots under drought stress. CAD activity also increased greatly under the abiotic stresses. *CmCAD1* and *CmCAD2* were generally up-regulated by drought and wound stresses in different organs. But under 100 mM NaCl treatment of salt stress, the expression of *CmCAD1* and *CmCAD2* were sharply elevated only in roots and stems, respectively. *CmCAD3*, *CmCAD4* and *CmCAD5* presented differential expression patterns in different organs under the three abiotic stress conditions. Collectively, under the three abiotic stresses, the accumulation of lignin, induction of lignin synthesis related enzymes (PAL, POD and CAD) and differential responses of *CmCADs* in oriental melon seedlings revealed that *CmCADs* might play diverse roles in resisting abiotic stresses through activating the lignin biosynthesis pathway.

T4

A320

AHP2, *AHP3*, and *AHP5* act downstream of C[illegible] in *Arabidopsis* female gametophyte development

Zhenning Liu[1,2,3], *Li Yuan*[1], *Xiaoya Song*[1], *Xiaolin Yu*[2], *Venkatesan Sundaresan*[1]

1. Department of Plant Biology, University of California

2. Institute of Vegetable sciences, Zhejiang University

3. College of Agriculture and Forestry science, Linyi University

Histidine phosphotransfer proteins (HPs) are key elements of the two-component signaling system, which act as a shuttle to transfer phosphorylation signals from His kinases (HKs) to response regulators (RRs). *CYTOKININ INDEPENDENT 1* (*CKI1*) is a master regulator of *Arabidopsis* female gametophytic central cell specification in a linear pathway with the *Arabidopsis* His phosphotransfer proteins (AHPs). There are five *HP* genes in *Arabidopsis*, *AHP1-AHP5*, but it remains unknown which *AHPs* act downstream of *CKI1* in *Arabidopsis* female gametophyte development. Promoter activity analysis of *AHP1–AHP5* in embryo sacs revealed *AHP1*, *AHP2*, *AHP3*, and *AHP5* expression in the central cell. Various combinations of *ahp* mutants were further constructed, seeds development in siliques were checked, cleared ovules were observed and results showed that mutations in *AHP2*, *AHP3* and *AHP5* caused seed abortion phenotype due to the unfused polar nuclei and degenerated embryo sacs. Besides, female gametophyte cell-specific single and double markers were introduced into the *ahp2-2 ahp3 ahp5-2/+* triple mutant and their expression patterns were observed, and results showed that the central cell and antipodal cell fates were lost and transformed into the egg cell or synergid cell fates in *ahp2-2 ahp3 ahp5-2/+* embryo sacs, which resembled the *cki1-9/+* phenotypes. These data indicate that *AHP2*, *AHP3* and *AHP5* act downstream of *CKI1* in *Arabidopsis* female gametophyte development, which could contribute to clarify the two-component system signaling network.

T4

A321

Study on biomass-based maize leaf morphological structure model

Xianju Lu, *Xinyu Guo*, *Qi Wu*, *Zetao Yu*

Beijing Research Center for Information Technology in Agriculture/National Engineering Research Center for Information Technology in Agriculture/Beijing Key Lab of Digital Plant

Maize plant morphology directly influences its eco-physiological function and the aboveground morphogenesis is an important basis of plant morphological construction and visualization. It will be benificial to plant type designing and cultivation management. Leaf is an important aboveground organ, its morphology determines maize photosynthetic area and influences photosynthetic production. For quantitatively analyzing the relationship between the leaf architectural parameters and biomass, field experiments with different varieties (Xianyu 335, Jingke 968 and Zhengdan 958) were carried out in 2015 and 2016 seasons. Leaf morphology parameters such as leaf length, maximum width, leaf sheath length were measured in different growth stages. After the morphology parameters measurement, each leaf's fresh weight and dry weight were measured respectively. Simulation models for leaf architectural of maize were built with the 2016 season dataset of leaf architectural parameters and leaf biomass and validated by the 2015 season dataset. The results showed that the models exhibited satisfactory predictions for leaf blade length, maximum width. The model's coefficient of determination were 0.9023 and 0.9827 for leaf blade length and maximum width respectively. Therefore, the models built in this study are suitable to simulate leaf architecture of mazie and they can be used in maize functional-structural model study.

T4

A322

A putative *miR172*-targeted *PhAPETALA2*-like gene is involved in regulating chasmogamous and cleistogamous floral patterning of *Pseudostellaria heterophylla* (Caryophyllaceae)

Yan Luo, *Peng-Fei Wang, Yong-Mei Xia*

Xishuangbanna tropical Botanical Garden, Chinese Academy of Sciences

Individuals of cleistogamous species can produce both open chasmogamous flowers (CH) that are available for outcrossing and closed cleistogamous flowers (CL) that obligately self-fertilize. There are structural differences and functional differences exit in two floral forms. The genetic basis of production of two floral forms is poorly known. The interaction between microRNA *miR172* and its *APETALA2* (*AP2*) target gene is deeply conserved and has an ancient role in flower developmental regulation in angiosperms. *AP2* is a floral organ identity gene, and negatively regulated by *miR172*. Recent studies in *Arabidopsis*, rice and other flowering plants have shown that *miR172* and its *AP2* target play important roles in the determination of floral organ identity. *Pseudostellaria heterophylla* belongs to *Caryophyllaceae*, which produces both chasmogamous and cleistogamous flowers on the same plant. To further understand flowering and flower organs formation in *P. heterophylla*, we cloned an *APETALA2* cDNA sequence of the AP2/ERF transcription factor family designated *PhAP2*. The length of *PhAP2* is 2,294 bp and encoding 593 amino acids. Bioinformatics analysis with putative orthologous sequences from various plant species suggested that *PhAP2* cDNA sequences encode for AP2-like proteins with the AP2 characteristic motifs and amino acids. Phylogenetically, the isolated sequences were closest to the *APETALA2* gene from *Beta vulgaris* and *Spinacia oleracea* which belong to the Order Caryophyllales. Splice variants were identified, including a variant that potentially produces a shorter transcript without the miRNA172 targeting site. Expression of *PhAP2* was compared in floral organs and developmental stages in and CH flowers and CL flowers. The expression pattern in various developmental stages and reproductive organs was interrogated by RT-PCR. In CH flowers, *PhAP2* was ubiquitously expressed, with highest levels of expression detected in the earlier developmental stages; in mature flowers, highest *AP2* transcript accumulation was detected in petal tissue. The accumulation of *AP2* transcript during flower differentiation from the visible floral bud showed minor variation in CL flower. In situ hybridization demonstrated *PhAP2* accumulation in central floral meristems, sepal primordia both in CH and CL flowers. *PhAP2* accumulation was detected in the stamen primordia in CH flowers but not in CL flowers. A stem-loop RT-PCR was performed to detect relative levels of miRNA in various developmental stages and floral organs of CH flowers and CL flowers. In CH flowers, *miR172* was higher expression in later developmental stages and in stamens and pistils. In contrast, *miR172* was only detected in the later developmental stage. In situ hybridization demonstrated *miR172* was mostly absent from all the developing floral organs. High expression in CH flowers and CL flowers, particularly petal tissue and increased accumulation in the earlier flower developmental stages, implied that PhAP2 might have a role in the control of perianth identity in *P. heterophylla*. *MiR172* was detected in the later flower developmental stages of CH and CL flower.The increase in AP2 transcript accumulation was probably the result of decreased negative regulation exterted by *miR172*, resulting from decreased accumulation of *miR172* itself, consistent with the role of *miR172* in flower development in Arabdopsis. This study provides a new insight into the genetic basis of production of dimorphic flower in cleistogamous species and has great significance for flower morphological evolution in flowering plants.

T4

A323

Ultrastructure, antioxidant enzymes, and photosynthetic performance of a rice leaf photo-oxidation mutant *812hs*

Jing Ma[1], Guoxiang Chen[1], Zhiping Gao[1], ***Chuangen Lv[2]***

1. Nanjing Normal University

2. Jiangsu Academy of Agricultural Sciences

To explore the physiological mechanism of a rice leaf photo-oxidation mutant *812hs*, effects of high light intensity on morphological traits, photosynthetic performance, protein and activities of antioxidative enzymes of 812HS were investigated using its wild type 812S as control in field growth conditions. The chloroplast ultrastructure was observed as well. Results showed that (1) Morphological traits, chlorophyll content, chloroplast ultrastructure, photophosphorylation, ATP levels, activities of Ca^{2+}-ATPase, Mg^{2+}-ATPase, proteins, reactive oxygen species (ROS) and antioxidant enzymes activities in 812HS were similar to those in its wild type before exposure to a period of high light intensity. (2) After a period of high light intensity, all the above indices of 812HS were affected and shown significant difference in its wild type with the loss of green. Transmission electron microscopy showed that photo-oxidation resulted in significant chloroplast damage. These results were confirmed by inhibited photophosphorylation, reduced ATP content and its coupling factor Ca^{2+}-ATPase, and Mg^{2+}-ATPase activities. Further, overtly increased activities of antioxidative enzymes were observed under photo-oxidation. Malondialdehyde, hydrogen peroxide and superoxide generation rates also increased. The findings indicate that excessive energy triggers the production of toxic ROS and promotes lipid peroxidation in rice plant, causing severe damage to cell membranes, degradation of photosynthetic pigments, and inhibition of photosynthesis.

T4

A324

Shading fruit with opaque paper bag affecting the expression of peach (*Prunus persica L. Batsch*) allergen-encoding genes

***Yingtao Ma**, Jianming Han, Xuejiao Zhao*
Luoyang Normal University

Freshly consumed peach can cause allergic reactions in the worldwide population because of the presence of four classes of allergens (Pru p 1, Pru p 2, Pru p 3, and Pru p 4), and their cross-reactivity with sensitizing allergens in other species. Fruit bagging has been developed into a common field practice in peach cultivation, which not only protects fruit against damage from pests, birds, diseases, and mechanical scratches, but also reduces pesticide residues. Knowledge of influence of fruit bagging on expression of peach (*Prunus persica L. Batsch*) allergen-encoding genes would provide important information to peach growers and consumers for the adoption of agronomic practices to decrease the allergenic potential, and for the consumption of fruits with reduced levels of allergens. In the present research, the expression profiles of some selected genes encoding peach allergen isoforms, including Pru p 1.01, Pru p 1.06B, Pru p 2.01B, Pru p 2.02, Pru p 3.01, Pru p 4.01 and Pru p 4.02 were studied by means of real-time PCR. Peach samples were collected from fruits which were shaded with opaque paper bag, set up to assess the effect of shadowing on the expression of allergen genes. The results indicate that shading significantly affected the transcription of allergen-encoding genes. For Pru p 1 family members were concerned, a decreasing trend in the transcript accumulation was observed in epicarp of bagged fruit, whereas the opposite effect was observed in the mesocarp. As far as Pru p 2.01B expression was concerned, fruit bagging was shown to significantly decrease the transcriptional rate in epicarp, a slightly increasing trend was observed in the mesocarp. With regard to Pru p 2.02, the expression level were both down regulated in the epicarp and mesocarp .In the case of Pru p 3.01 gene, a strong down-regulation were observed both in the mesocarp and the epicarp. As far as the two profilin-encoding genes are concerned, the same trend were found. Fruit bagging was shown to significantly affect Pru p 4.01 and Pru p 4.02 transcription, reversing its expression pattern in the mesocarp. According to these data, new growing practices could be set up to obtain hypoallergenic peach fruits and eventually combined with the cultivation of hypoallergenic genotypes to obtain a significant reduction of the allergenic potential.

T4

A325

Preparation and identification of monoclonal antibody against Pru p 3.01 and development of a double monoclonal antibody-based sandwich ELISA

***Yingtao Ma**[1], Zhongshan Gao[2], Hongxia Wu[3], Pengcheng Fu[1], Yanwei Cheng[1], Ling Zheng[1], Xiaojun Zhou[1]*
1. Life Science College, Luoyang Normal University, Luoyang
2. College of Agriculture and Biotechnology, Zhejiang University, Hangzhou
3. South Subtropical Crops Research Institute, Chinese Academy of Tropical Agricultural Sciences, Zhanjiang

Peach lipid transfer protein (LTP1), Pru p 3.01, is a major allergen causing severe systemic reactions in peach allergic patients in Mediterranean countries and China. In this study, recombinant Pru p 3 was used to immunize Bal b/c mice (experimental animal use permission, SCXK (HU)2012-002, China) and five hybrid cell lines 4#, A7, 6#, 11# and B10 excreting monoclonal antibody against Pru p 3 were obtained by fusing mouse SP2/O myeloma cells with the spleen cells from the immunized mouses. Antibody-producing hybridoma cells were screened by ELISA and Western blotted with rPru p 3 and peach peel extract. Hybridoma cells were then injected into the abdomen of Bal b/c mice, and the anti-Pru p 3.01 mcAbs were harvested from ascites and purified on HiTrap Protein A. Seven monoclonal antibodies (4-1, 4-6, A7-1, A7-2, 6-F7, 11-B and B10-5) were produced from the immunization, the titer of which were 1:20000, 1:10000, 1:60000, 1:20000, 1:20000, 1:2000 and 1:20000, respectively. And the isotype of 4-1 and 4-6 are IgG2a (κ chain) and the rest five is IgG1 (κ chain). The western blotting analysis showed all of these antibodies were highly specific to Pru p 3.01. A monoclonal antibodies-based sandwich ELISA was established with mcAb B10-5 as capture mcAb (bound to the solid phase) and biotinylated mcAb 4-1 as the detector, which shows high accuracy, resolution, reproducity and stability. The specificity of this measurement was confirmed using peach and other fruit nsLTPs. Therefore, these obtained anti-Pru p 3.01 can be used to detect Pru p 3.01 in peach fruit and related samples.

T4

A326

The effect of nano potassium chelate and nitrogen biological fertilizers on qualitative performance of sesame seeds (*Sesamum indicum* L.) in irrigation regimes

***Akram Mahdavi**, Jafar Sinaki, Amini Dehaghy Majid*
Azad university, Damghan branch

Nanofertilizer achieved delivery of nutrients to the plant roots through improving the efficiency of fertilizers dramatic gains such as increased crop yields, lower production costs, and environmental protection. The effect of drought last season is disturbance of balance nutrition in the plant. With the completion of the consumption nutrients, can be plant growth and finally improve product quality in terms of stress. Sesame (*Sesamum indicum* L.) is one of the oldest oil plants of tropical and subtropical region and seeds due to the quantity and quality of protein and oil have high nutritional value. Therefore, the experiment was conducted split plot factorial experiment in which different irrigation regimes (full irrigation, irrigation up to 50% flowering, irrigation up to 50% seed ripening) in the main plots and various combinations of nitrogen (Nitrogen, 100% urea based on soil testing, integrated of 50% urea based on soil testing and nitrogen) and various combinations of potassium (no fertilizer, foliar application of nano potassium chelate, foliar application of potassium dioxide and soil application of nano-chelate potassium) were the subplots. This study was conducted with three replications in Agricultural Research Station, Shahed University of Iran in 2016 on the Darabl genotype of sesame in which were measured grain yield, percentage and yield of protein content and percentage and yield of oil. The results showed that the effect of different combinations of nitrogen fertilizers, the interaction of different irrigation regimes and nitrogen compounds and triple interaction (nitrogen compounds in potassium compounds in different irrigation regimes) were significant on grain yield. Among the different combinations of nitrogen was achieved the highest yield in the use of urea fertilizer based on soil test (100%), and integrated systems had yield higher than in comparison with biological fertilizer of nitrogen. Effects of different combinations of nitrogen and potassium were significant on protein content and interaction of nitrogen in potassium compound in different irrigation regimes was significant on protein yield. Protein and oil content had the highest mean in the integrated of nitrogen fertilizer. Among the combination of potassium fertilizers, the use of nano-chelated potassium for foliar and foliar application potassium dioxide compared with the other treatments increased the mean grain yield, protein and oil yield. Generally, used foliar application of potassium fertilizer and nitrogen fertilizer integrated system (according to the environmental impact caused by the indiscriminate use of industrial inputs such as chemical fertilizers) in full irrigation system and providing 100% require plant based on soil test from urea in terms of irrigation up to 50% seed ripening increased the quality and quantity of Sesame seed (in terms of protein and oil).

T4

A327

Root and shot allometric relation of seven forage crops under abiotic stresses

Amrakh Mamedov[1], *Khatira Gasimova*[1], *Emin Husiyev*[1], *Vahid Farzaliyev*[2], *Kristina Toderich*[3], *Valida Ali-Zada*[1]

1. Institute of Botany, ANAS

2. Central Botanical Garden, ANAS

3. International Center for Biosaline Agriculture (ICBA)

Knowledge of root and shoot biomass dynamics is vital to improving our understanding of the contribution of abiotic stress on plant growth and yield, transportation of photosynthate, and nutrient cycling, carbon distribution and land development. Allometric relationships between plant organs allows to link plant structural development and primary physiological processes, and hence established relationship in ontogeny can be used as components of crop models, and to estimate plant parameters that are difficult to measure. The objective of the study was (i) to investigate morphological, physiological, and biomolecular mechanisms of the forage crops salinity tolerance, (ii) using allometric relationships to estimate crop performance and variability under current and future climates, and hence (iii) to identify prime adaptation strategies to reduce susceptibility of agro-ecosystems to climate change. During three year we have studied a change of allometry between root and shoot of seven forage crops (Alfalfa, Amaranthus, Maize, Pearl Millet, Quinoa, Rapeseeds, Sorghum) affected by a salt and drought stress in the typical agricultural fields of Central Azerbaijan, characterized by clay soil texture, and different soil salinity level (EC = 4-12 dS/m) and shortage of water resources. During vegetation growth measurements at various phenological stages of crops were conducted, root, shoot and soil samples were taken for appropriate chemical and biochemical analysis. Allometric relationships between height and stem mass or stem diameter, developmental rates of different organs, and between growth rate and cell biomass was established. Results showed that (i) plants growth performance depend on the integrated functioning of both roots and aboveground leaves, and their structure and function are interrelated affecting transportation of macro- and microelements in the soil-root-plant system; (ii) root: shout ratio of the forage crops increased with decrease in root weight of the plants; quinoa and sorghum had one of lowest root: shout ratio, while having highest root weight, showing their better adaptability to combined abiotic stresses, than other crops; (iii) allometry can be used to study the ability of plants to adapt to the environment and abiotic stress condition, and assist to evaluate the contribution of crop varieties and management; (iv) allometric relationship between stem diameter and plant parameters (e.g. root and shoot weight) was established; the power or exponential function described well such relationship and may allow understanding contribution of growth period or crop root, shoot or leaf performance; allometric differences was found (exponent or coefficient of equations) between the crops or cultivars of the same crops and or between control and fertilizer treatments; (v) using easily measurable plant parameters can provide data required to estimate C accumulation rates and water and nutrient requirements for salt tolerant forage crops. The improved knowledge of the mechanisms of resistance of the crops to adverse abiotic factors will help selecting and or breeding cultivars adapted to the most diverse abiotic stresses, and develop techniques for overcoming the constraints imposed by severe environments existing in the region, and promoting an improvement in agricultural sustainability.

T4

A328

Differential expression of wound-induced genes mediating suberization in postharvest tomato fruit

Linchun Mao, *Xueyuan Han*, *Wenjing Lu*

Zhejiang University

Fruit wounding occurred at harvest, handling and transportation requires rapid suberization as a major part of the healing process to prevent infection and desiccation. The induction and expression of responding genes are essential for wound-induced suberization. The objective of this research was to determine expression profiles of specific genes involved in wound suberization in tomato fruit. The results showed that obvious suberization process with the accumulation of suberin polyphenolics (SPP) and suberin polyaliphatics (SPA) was observed through confocal autofluorescence microscopy and Sudan IV staining. Certain genes that were specifically involved in the process did not remain uniformly up or down-regulated. Expressions of *SlPAL* encoding phenylalnine ammonialyase and *Sl4CL* encoding 4-coumarate ligase in synthesis of SPP reached the highest at 4 d and 8 d after wounding, respectively. The genes regulating SPA synthesis pathway showed different variation patterns. The expression of *SlLACS1* encoding long chain acyl-CoA synthetases was induced to up-regulation, especially at 8 d after wounding in abscisic acid (ABA) treated group. Instead, *SlLACS2* showed the most enhanced expression at 6 d. Gene expressions including *SlKCS6* and *SlKCS19* encoding β-ketoacyl-CoA synthase, *SlCYP86B1* encoding fatty acyl ω-hydroxylase, *SlFAR3* encoding fatty acyl-CoA reductases, *SlGPAT4* and *SlGPAT6* encoding glycerol-3-phosphate acyltransferase were dramatically up-regulated at 2 d after wounding in ABA treated group, but then slightly declined near completion of the healing at 6–8 d. This differential in the expression profile may be the demonstration of measurable changes in the biological flux of wound-induced synthesis of SPP and SPA. In addition, both polyphenol oxidase (PPO) and lipoxygenase (LOX) activities and gene expressions increased rapidly upon wounding, and then decreased simultaneously with suberization process. The temporary oxidation upon wounding may be important for the initiation of wound suberization.

T4

A329

Two novel MADS-box transcription factor, *FaMADS1a* and *FaMADS2a*, act as negative regulators in the ripening of strawberry fruit

Linchun Mao, *Wenjing Lu, Jingxin Chen*

Zhejiang University

MADS-box genes encode highly conserved gene family of transcriptional factors that regulate abundant developmental processes in plants. In this study, two octaploid strawberry (*Fragaria* × *ananassa*) MADS-box genes, *FaMADS1a* and *FaMADS2a*, were cloned and their tissue-specific expression profiles were analyzed. *FaMADS1a* and *FaMADS2a* were expressed in petal and fruit, and their expression levels decreased significantly in accordance with fruit ripening. Multiple sequence alignments suggested that FaMADS1a protein exhibited conserved MADS_MEF2_like motif, K-box region, MADS domain, SRF-TF motif and ARG80 domain. FaMADS2a protein possesses conserved K-box region. Phylogenetic analysis revealed that FaMADS1a and FaMADS2a belonged to the SEP1/2 and SEP3 clade, respectively. Phytohormone abscisic acid (ABA) suppressed the expressions of *FaMADS1a* and *FaMADS2a* and accelerated the ripening process, while indole-3-acetic acid (IAA) induced their expressions with delayed ripening phenotype. The depression of *FaMADS1a* and *FaMADS2a* might be the essential prerequisite for the expression of *FaDFR* and *FaANS*. *FaMADS1a* and *FaMADS2a* might play a negative role in fruit ripening via the downstream of phytohormone signal transduction pathway connected to related metabolic process.

T4

A330

L-DOPA triggers the early response of *Arabidopsis thaliana* to iron deficiency

Xiangxiang Meng, *Ping Lan**

State Key Laboratory of Soil and Sustainable Agriculture, Institute of Soil Science, Chinese Academy of Sciences

L-DOPA (L-3,4-dihydroxyphenylalanine), an uncoded amino acid in some mammals and plants, plays important roles in the maintenance of internal environment of lots of living organisms. It has received increasing attention as one of allelochemicals in plants owing to its potential function to be a bioherbicide without pollution, and besides, as a medicine to alleviate the Parkinson's disease and dopamine-responsive dystonia in hominine. In the current work, the effects of L-DOPA on the responses of plants to iron deficiency were investigated by treating the two-week-old Arabidopsis seedlings with or without 100 mg/L L-DOPA . Time course studies showed that the ferric reductase activity (FCR) was gradually increased along with the processing time, i.e. by 1.5 times ($P < 0.05$), 2 times ($P < 0.01$), 4 times ($P < 0.01$) and 6 times ($P < 0.01$) for 6 hours, 9 hours, 12 hours and 24 hours respectively. Perls staining showed that the Fe^{3+} was immobilized in the whole lateral roots and vascular after 24 hours. In addition, the expression of iron acquisition-related genes, including *FIT*, *FRO2* and *IRT1*, and genes involved in iron homeostasis, such as *FRD3*, *NAS4* and *IRT3*, was significantly up-regulated by L-DOPA treatment. Melanin was largely concentrated in the roots after treatment for 12 hours. These results suggest that L-DOPA might disturb the iron homeostasis by depositing iron in apoplast through the increase of melanin concentrations.

T4

A331

Molecular and biochemical analysis of chalcone synthase from *Freesia hybrid* in flavonoid biosynthetic pathway

*Xiangyu Meng, Ling Li, Tongtong Zhou, Wei Sun, Yueqing Li, Li Wang**, ***Xiang Gao****

Key Laboratory of Molecular Epigenetics of MOE and Institute of Genetics & Cytology, Northeast Normal University, Changchun

Flavonoids are the most abundant and widely distributed pigments in the plant kingdom, and have a wide variety of biological functions, including protection of cells against UV radiation, resistances against phytopathogens and herbivores, signal molecules in plants and microbes interaction, *etc.* Chalcone synthase (CHS) catalyzes the first committed step in the flavonoid biosynthetic pathway, with the condensation of one molecule of p-coumaroyl-CoA with three molecules of malonyl-CoA, to yeild naringenin chalcone. Presently, a whelming proportion of *CHS* gene family have been well elucidated in core eudicot plants, whereas they were studied randomly in monocot plants, which are of crucial importance to decipher their duplication and functional divergences in monocot plants during evolution. *Freesia hybrida*, a monocotyledonous ornamental species, widely distributes in the world and is becoming increasingly popular as a cut flower. The range of flower colours available in this species includes red, pink, yellow, white, blue, lavender, purple and bicolors. Specialized metabolites analysis showed that five kinds of anthocyanin aglycons, i. e. delphinidin, petunidin, malvinidin, peonidin, and cyanidin, as well as flavonols and PAs were detected in the *Freesia* flowers. In this study, a strategy combining transcriptomic data, expression profiles, enzyme assays and ectopic expression was undertaken to investigate the functional divergences of chalcone synthase family. In total, seven candidate genes (*FhCHS1~7*) were isolated from *F. hybrida* and then analyzed. Multiple sequence alignments showed that both the conserved CHS active site residues and CHS signature sequences were found in the deduced amino acid sequence of FhCHSs. Meanwhile, crystallographic analysis revealed that protein structure of FhCHS1 is highly similar to that of alfalfa CHS2. However, the *in vitro* biochemical analysis indicated that only FhCHS1 and FhCHS6 have the capacity to catalyze the formation of naringenin chalcone. Quantitative Real-time PCR was performed to detect the transcript levels of *FhCHSs* in flowers, different tissues, and flowers exposed under UV-B irradiation, and the results demonstrated that the expression patterns of *FhCHS1* showed significant correlation to the accumulation patterns of anthocyanin during flower development, whereas *FhCHS6* was obviously responded to UV-B treatment. This phenomenon illustrated the fact that *FhCHS1* is the key enzyme involved in the anthocyanin biosynthetic pathway and the up-regulation of *FhCHS6* could be one of the strategies to enhance tolerance against UV-B stress in *F. hybrida*. To further characterize the functionality of *FhCHSs*, their ectopic expression in *Arabidopsis thaliana tt4* mutants and *Petunia hybrida* was performed. The results showed that the over-expression of *FhCHS1* and *FhCHS6* could fully restore the pigmentation phenotype of the mutant plants. In addition, transgenic petunia plants expressing *FhCHS1* also showed flower color alteration from white to pink. In summary, these results suggested that *FhCHS1* and *FhCHS6* might play differential roles in the biosynthesis of flavonoid in *F. hybrida*, and the finds will probably contribute to the investigation of the origination and evolution of CHS gene family in angiosperm.

T4

A333

Gibberellin inhibits floral initiation in the biofuel plant *Jatropha curcas*

*Jialong Li, Bz Pan, Longjian Niu, **Zengfu Xu***
Xishuangbanna Tropical Botanical Garden, Chinese Academy of Science

Jatropha curcas L. has great potential as renewable energy plant due to the high oil content in the seed. However, poor yield is an obstacle for the industrialization of this plant. *J. curcas* has vigorous vegetative growth but relatively poor reproductive growth. A feasible way to improve the seed yield is to inhibit the vegetative growth and increase the reproductive growth. Gibberellin (GA) is an important phytohormone and has been reported to control the transition from vegetative growth to reproductive growth. This study was undertaken to determine the effects of GA on the floral transition in *J. curcas*. The exogenous application of gibberellic acid 3 (GA_3) inhibited flowering, while GA biosynthesis inhibitors promoted flowering. The promotive effect of the GA biosynthesis inhibitor paclobutrazol (PAC) on floral initiation could be diminished by GA_3 treatment. The expression levels of GA biosynthesis genes *Jatropha GA 3-oxidase 3* (*JcGA3ox3*) and the GA receptor genes, *Jatropha GA-INSENSITIVE DWARF1A* (*JcGID1C*) were decreased, while a GA turnover gene, *Jatropha GA 2-oxidase 8* (*JcGA2ox8*), was elevated during floral transition. These results indicated that GA inhibits floral initiation in *J. curcas*, and the transition from the vegetative to the reproductive phase may be related to a change in the GA content of the shoot apex.

T4

A334

Modulating role for ROS in re-establishing desiccation tolerance in germinating seeds of *Caragana korshinskii* Kom

Long Peng
National Engineering Laboratory for Tree Breeding, College of Biological Sciences and Biotechnology, Beijing Forestry University

In close coincidence with visible germination, orthodox seeds lose desiccation tolerance (DT). This trait can be regained under osmotic stress, but the mechanisms are poorly understood. In this study, germinating seeds of *Caragana korshinskii* Kom was investigated focusing on the potential modulating roles of reactive oxygen species (ROS) in DT re-establishment. Seeds with emerged radicles measuring 2 mm long can be rendered tolerant to desiccation by incubation in a polyethylene glycol (PEG) solution (-1.7 MPa). Upon PEG incubation, ROS accumulation was detected in the tip of radicles by nitroblue tetrazolium chloride (NBT) staining and further confirmed by confocal microscopy. The PEG-induced re-establishment of DT was repressed when ROS scavenger was added to the PEG solution. Moreover, ROS act downstream of ABA to modulate PEG-mediated DT re-establishment, and serve as a new inducer to re-establish DT independently. Transcriptomic analysis revealed that DT re-establishment by ROS involved the up-regulation of key genes in phenylpropanoid-flavonoid pathway, and total flavonoids content and key enzyme activity increased after ROS treatment. Furthermore, DT was repressed by phenylalanine ammonia lyase (PAL) inhibitor. Our data suggest ROS play a key role in DT re-establishment by regulating stress-related genes and phenylpropanoid-flavonoid pathway.

T4

A335

The TRAF-family proteins TRAF1a and TRAF1b regulate autophagy dynamics by modulating ATG6 stability in *Arabidopsis*

Hua Qi
Sun Yat-Sen University

Eukaryotic cells use autophagy to recycle cellular components. In autophagy, autophagosomes deliver cytoplasmic contents to the vacuole or lysosome for breakdown. Mammalian cells could regulate the dynamics of autophagy by ubiquitin-mediated proteolysis of autophagy proteins. Here, we show that the Arabidopsis Tumor necrosis factor Receptor-Associated Factor (TRAF)-family proteins, TRAF1a and TRAF1b (previously named MUSE14 and MUSE13, respectively), help regulate autophagy via ubiquitination. Upon starvation, cytoplasmic TRAF1a and TRAF1b translocated to the autophagosomes. Knockout *traf1a/b* lines showed reduced tolerance to nutrient deficiency, increased salicylic acid and reactive oxygen species levels, and constitutive cell death in rosettes, resembling the phenotypes of autophagy-defective mutants. Starvation-activated autophagosome accumulation decreased in *traf1a/b* root cells, indicating that TRAF1a and TRAF1b function redundantly in regulating autophagosome formation. TRAF1a and TRAF1b interacted *in planta* with ATG6, the ring finger E3 ligases SINAT1 and SINAT2, and the RING-finger-truncated SINAT6. SINAT1 and SINAT2 require the presence of TRAF1a and TRAF1b to ubiquitinate and destabilize ATG6 *in vivo*. Conversely, starvation-induced SINAT6 reduced SINAT1- and SINAT2-mediated ubiquitination and degradation of ATG6. Consistent with this, *SINAT1/SINAT2* and *SINAT6* knockout mutants exhibited increased tolerance and sensitivity, respectively, to nutrient starvation. These findings demonstrate that TRAF1a and TRAF1b function as molecular adaptors that help regulate autophagy by modulating ATG6 stability in *Arabidopsis*.

T4

A336

A new method for fast extraction and determination of chlorophylls in natural water

Nianwei Qiu
Qufu Normal University

Spectrophotometry is the most classical and common method for determination of chlorophylls in natural water. Although this method had been improved several times in recent years, many problems still remain, such as a long collection time for the algae sample preparation, high cost in filters, low chlorophyll extraction rate from algae, and tedious operation steps. To solve the above problems, a new method for fast extraction was proposed in this study. First, the high-speed refrigerated centrifugation method (HSRC) was used for algae collection from water samples instead of filtration. Second, chlorophyll was extracted with dimethyl sulfoxide (DMSO) at a high temperature to accelerate the extraction rate. Third, the chlorophyll extract was diluted with 90% acetone for accurate determination of chlorophyll concentration. The optimum centrifugation condition for collecting algae was determined as: 5,000 g for 15 min at 4°C and the optimal dilution ratio of DMSO: 90% acetone (v/v) was determined as 1:4. This new method saved the cost of filter, simplified the extraction process, improved the algae acquisition efficiency and increased the chlorophyll extraction rate, thereby improving the experimental efficiency. The specific steps were as follows: The algae in water samples was collected by centrifugation at 5000 g at 4°C for 15 min. The precipitated algae was suspended with 2 ml DMSO. Then the sample was transferred to a 15 mL screw-cap glass centrifuge tube, and the tube was incubated at 65°C for 1-2 h in the dark until the sample turned white. After cooling, the chlorophyll extract was added with 8 mL 90% acetone and well mixed, before centrifugation at 5,000 g for 5 min. The absorbance of the supernatants at 750 nm, 664 nm, 647 nm and 630 nm were used for the calculation of chlorophyll concentrations by the trichromatic equations.

T4
A337

Cytology observation of anther development of the male sterile plant in *Psophocarpus teragonolobus*

Miaohua Quan
College of Biological and Food Engineering, Huaihua University

The anther development process of the male sterile and normal plant in *Psophocarpus teragonolobus* were studied in this paper. The results showed that the microspore abortion of the male sterile plants first took place at microsporocyte stage. Mainly, because its tapetum cells became vacuolated ahead of schedule during the micospore development, which made the tapetum cells couldn't provide enough nutrient for the micorspore development. Following, the tapetum cells were highly vacuolated and expanded in radialdirection at the tetrad and the pollen formation stage, microsporocytes were squeezed, which made grow abnormally and resulted in the abortion finally. In this paper, compared with the fertile pollen development, the abortion process and characteristics of the male sterile anthers in *Psophocarpus teragonolobus* were disclosed, providing a foundation for the greater insight into the abortion mechanism of male sterile plant.

T4
A339

Response of *Stevia rebaudiana Bertoni* to foliar application with yeast or seaweed extracts on growth, anatomical and chemical structure

***Azza Salama**, Ahmed Attia, Mohamed Negm*
Cairo University

The current study was carried out during the two successive seasons of 2013-2014 and 2014-2015 at Giza Agricultural Research Station, Agricultural Research Center, Giza, Egypt, to investigate the effect of foliar spray with yeast or seaweed extracts and their interaction on some morphological, anatomical as well as chemical characters of stevia (cv. China-1). Yeast extract was sprayed at concentrations of 2 and 4 g/L whereas; seaweed extract at 0.5 and 1.0 ml/L and their interaction, whereas control plants were sprayed with tap water. The results showed a significant increase in plant height, stem diameter, number of branches/plant, fresh and dry weight of leaves/plant when plants spraying with different concentration of yeast or seaweed extract as compared with the untreated plants. The maximum significant increase was recorded when plant sprayed with a mixture of 2 g yeast extract/L and 1.0 ml seaweed extract/L. Also, it was clear from the anatomical characters of stem and leaves that, such treatment increased stem diameter due to the increase in the thickness of cortex, phloem tissue, xylem tissue and parenchymatous area of the pith more than those of the control. As well as, the thickness of midvein, lamina and the dimension of the main midvein bundle increase. The foliar application with mixture of 2 g yeast extract/L+1.0 ml seaweed extract/L resulted in the highest rebaudioside-A and stevioside percentage being, 7.0 and 3.94%, respectively.

T4
A340

Population densities and functional diversity of Rhizobacteria found in root biofilms of agriculture crops in saline soils

Meenu Saraf
Gujarat University

The rhizobacteria play an intrinsic role in enhancing the plant growth by an array of activities leading to biofertilization and biocontrol. In any rhizosphere, the metabolic versatility of a bacterial population is based on its genetic variability and its putative interactive ability with other rhizosphere population of other kind including inter and intra species population with functional diversity. Evaluating the relative strengths of bacterial species richness and estimation on diversity found in rhizosphere of chick pea is of vital importance to the understanding the different function carried out by bacteria which could be specifically suited for saline agriculture. The success of microbes is attributed to effective colonization of plant roots and subsequent growth to form biofilms . To achieve this objective a study was designed targeting the rhizosphere of cicer arietinum, which were planted in various soil conditions on five distinctive sites of Gujarat state (India). These sites were constantly monitored and studied for the species richness and evenness ("heterogeneity") The isolates were checked for their Plant Growth Promoting potentials like Phosphate solubilisation, Siderophore production, Indole acetic acid production,

ACC deaminase production, HCN production, EPS production and Ammonia production .The data obtained were used to calculate richness, evenness and diversity indices. The combinatorial estimates provide the information on their distribution and roles in the habitat. The rhizobacteria in their biofilms mode perform well in inhibiting competing organisms, nutrient uptake, quick responses, and adaptation to environmental conditions and communicate to each other via quorum sensing, disease reduction by PGPR also occurs by production of antimicrobial substances and indirectly through induction of systemic resistance. The present study demonstrated the ability of native strains of saline areas to tolerate salinity and show plant growth promoting potential. From total of 176 bacterial isolates, 62 isolates have potential to tolerate salinity up to 1 M NaCl concentration. The five bacterial isolates were screened out that not only can tolerate the high salinity but play an essential role in helping plants to establish and grow in salinity conditions. Isolates were identified by 16sRNA as belonging to *Pseudomonas stutzeri* and *P.mosselii*. Isolate MSC1 increase the root length by 71% and shoot length by 34% compared to untreated plants. MSC4 was another effective growth promoter which increases the root length by 76%. Isolates MSC4 and MSC1 also shows maximum increase in number of flowers by 63% and 54% respectively. Plants inoculated with MSC4 and MSC1 showed increase in number of fruits by 112% and 105% in comparison with uninoculated control. The sequences obtained in this study were deposited in the NCBI GenBank nucleotide sequence database under the accession numbers KC669688 (*Pseudomonas stutzeri*- MBCU1) and KC669689 (*Pseudomons mosselii* – MBCU3). However, this has to be tested with many other crops to confirm the ability of this PGPR to confer salt tolerance under field conditions. Therefore, it is plausible to mention that PGPR are an effective approach for improving the growth of salt-sensitive plants and this strategy could be applied for sustainable agriculture.

T4

A341

Comparison of the light response models of photosynthesis in leaves of *Magnolia sinostellata* under different light intensities and moisture conditions

Qin Yu
Zhejiang A&F University

Our objectives were to find the optimal models for the light response of photosynthesis in leaves of *Magnolia sinostellata* seedlings by comparing the light response curves under different intensities of light and moisture conditions and define the adaptability of *Magnolia sinostellata* to low light. We measured light response of photosynthesis in leaves of *Magnolia sinostellata* under different intensities of light and moisture conditions using Li-6400 portable photosynthesis system, and fitted and analyzed the light response curves by five models: exponential, rectangular hyperbolic, non-rectangular hyperbolic, quadratic function, and modified rectangular hyperbola. (1) Within the set up illumination range, *Magnolia sinostellata* never show obvious the phenomenon of photo-inhibition, the net photosynthetic rate (Pn) is lower than CK under situation W_1, compared to situation W_2: when Photosynthetically Active Radiation (PAR) < 800 $\mu mol \cdot m^{-2} \cdot s^{-1}$, Pn is the biggest under situation L_3, and when PAR >800 $\mu mol \cdot m^{-2} \cdot s^{-1}$, Pn will be the biggest under situation L_2. (2) The sequence of fitting effect of the five light response models was in descending order: exponential model > modified rectangular hyperbola model > rectangular hyperbola model > non-rectangular hyperbola model > quadratic function model. (3) the light response parameters in fitting effect indicates: exponential model can figure out more accurate maximum net photosynthetic rate (Pnmax) and light saturation point (LSP), the rectangular hyperbolic model is the best at figuring out light compensation point (LCP), and it is the most reasonable to figure out dark respiration rate (R_d) by non-rectangular hyperbolic model. Among them, *Magnolia sinostellata* seedlings have the characteristics of heliophilic plant and can adapt to the certain environment of low light.

T4

A342

The effect of light quality on sex determination in gametophytes of the fern *Drynaria roosii*

Meiyu Sun[1], Jingyi Li[1,2], Di Wang[1], Dong Li[1], Hui Li[1], Quan Xing[1], ***Lei Shi***[1]
1. Key Laboratory of Plant Resources and Beijing Botanical Garden, Institute of Botany, Chinese Academy of Sciences
2. University of Chinese Academy of Sciences

Sex determination of fern gametophytes is influenced by the environmental conditions, especially light. Still, the research about the function of light quality on development of fern sexual organs and sex ratios has been incomplete. *Drynaria roosii* (Nakaike) is homosporous, a source of Traditional Chinese Medicine commonly known as "Gu-Sui-Bu", but the molecular mechanism of gametophytes responding to different light signals in *Drynaria roosii* have remained elusive. Here, we used four different LED-lights (red, blue, green, white; PAR = 30-40 mol $m^{-2} s^{-1}$) and fluorescent light (control, PAR = 30-40 mol $m^{-2} s^{-1}$) to treat gametophytes in different developmetal stages (filament stage, prothallial plate stage, cordate prothalli stage, sex differentiation stage) and found that light quality had a significant and reproducible effect on sex determination in *Drynaria roosii* grown in vitro. Our results confirmed the male-promoting effect of red light and female-promoting effect of blue light in *Drynaria roosii*. However, white light and green light had less impact on the sex determination compared with control light. The data also showed that the number of sexual organs (antheridia and archegonia) and sex ratios were induced and altered more dramatically when cordate prothalli stage or sex differentiation stage treated with red or blue light. Furthermore, researches of the gametophyte transcriptome in *Drynaria roosii* have never been reported before. Thus, we applied short-read Next-Generation Sequencing (NGS) to various stages of gametophytes in *Drynaria roosii* treated with different light quality (red, blue, control light) and constructed gene network using Weighted Gene Co-expression Network Analysis (WGCNA) to get further insight into the interactions among related signal pathways during the sex determination. This study provided an initial network-based framework for the function of light qulity on sex determination in gametophytes of the fern *Drynaria roosii*.

T4
A343
Evaluating *SlPIN1* expression distribution in tomato pedicel during abscission using *in situ* hybridization

Zihang Shi, *Tianlai Li*
Shenyang Agricultural University

Auxin transport and distribution is of fundamental importance in organ abscission, much information is still unknown about its function and molecular basis. To address these issues, the temporal and spatial expression of *SlPIN1* -- the auxin efflux carrier -- during tomato (*Solanum lycopersicum*) pedicel abscission were investigated Evaluating *SlPIN1* expression distribution in tomato (*Solanum lycopersicum*) pedicel during abscission by *in situ* hybridization in our research. The expression of *SlPIN1* were down-regulated after flower removal, and differential distribution become apparent among abscission zone (AZ), distal and proximal part of pedicel, indicating different auxin response ability among these tissues. The abscission layer (AL) remained relatively higher expression of *SlPIN1* in early stage, indicating an active auxin flux through AL. Our study revealed a possible role for *SlPIN1* in the regulation of pedicel abscission.

T4
A344
Functional characterization of a *Glycine soja* Ca^{2+}ATPase in salt-alkaline stress responses

Mingzhe Sun[1,2], *Bowei Jia*[1,2], *Junkai Yang*[1], *Jianwei Li*[1], ***Xiaoli Sun***[1]
1. Heilongjiang Bayi Agricultural University
2. Northeast Agricultural University

It is widely accepted that Ca^{2+}ATPase family proteins play important roles in plant environmental stress responses. However, up to now, most researches are limited in the reference plants Arabidopsis and rice. The function of Ca^{2+}ATPases from non-reference plants was rarely reported, especially its regulatory role in carbonate alkaline stress responses. Hence, in this study, we identified the P-type II Ca^{2+}ATPase family genes in soybean genome, determined their chromosomal location and gene architecture, and analyzed their amino acid sequence and evolutionary relationship. Based on above results, we pointed out the existence of gene duplication for soybean Ca^{2+}ATPases. Then, we investigated the expression profiles of the ACA subfamily genes in wild soybean (*Glycine soja*) under carbonate alkaline stress, and functionally characterized one representative gene *GsACA1* by using transgenic alfalfa. Our results suggested that *GsACA1* overexpression in alfalfa obviously increased plant tolerance to both carbonate alkaline and neutral salt stresses, as evidenced by lower levels of membrane permeability and MDA content, and higher levels of SOD activity, proline concentration and chlorophyll content under stress conditions. Taken together, for the first time, we reported a P-type II Ca^{2+}ATPase from wild soybean, *GsACA1*, which could positively regulate plant tolerance to both carbonate alkaline and neutral salt stresses.

T4
A345
Characterization of *TpSnRK2.2* involved in abiotic stress responses

Chao Wang, *Fan Peng, Yiran Cheng, Songyue Chai, Wendi Shuai, Yonghong Zhou, Yi Wang*
Triticeae Research Institute, Sichuan Agricultural University

To date, more than 50 *SnRK2s* were cloned and characterized in various plants. Increasing studies indicated that SnRK2s play a key role in abiotic stress signaling transduction pathway. In wheat, several SnRK2s were also cloned, but only four SnRK2, TaSnRK2.3,2.4,2.7 and 2.8 were functionally characterized in *Arabidopsis*. In the present study, *TpSnRK2.2*, cloned from dwarfing polish wheat, was characterized in terms of metal transport activity and cellular localization. We found that TpSnRK2.2 was a functional cadmium-efflux transporter in yeast, because it reduced the Cd concentration in yeast. Overexpression of *TpSnRK2.2* in *Arabidopsis thaliana* increased cadmium concentration in root, but there was no difference in shoot when compared with control plants. Subcellular localization using wheat protoplast showed TpSnRK2.2 expressed in the cell membrane, endoplasmic reticulum, and nucleus. Additionally, overexpression of *TpSnRK2.2* also enhanced the tolerance to drought in Arabidopsis. Thus, *TpSnRK2.2* may act as a regulatory factor involved drought stress pathway and heavy metal cadmium transporting.

T4
A346
Holliday junction resolvases mediates DNA repair by homologous recombination in Rice

Chong Wang[1], *James D Higgins*[2], *Yi He*[1], *Pingli Lu*[3], *Dabing Zhang*[1,4], *Wanqi Liang*[1]
1. School of Life Sciences and Biotechnology, Shanghai Jiao Tong University, Shanghai 20040, China
2. Department of Genetics, University of Leicester, University Road, Leicester, LE1 7RH, United Kingdom
3. Institute of Plant Biology, School of Life Sciences, Fudan University, Shanghai 200433, China
4. School of Agriculture, Food and Wine, University of Adelaide, Waite Campus, Urrbrae, South Australia 5064, Australia

Yen1/GEN1 are canonical Holliday Junction (HJ) resolvases that belong to the RAD2/XPG family. In eukaryotes, such as budding yeast, mice, worms, and humans, Yen1/GEN1 work together with Mus81-Mms4/MUS81-EME1 and Slx1-Slx4/SLX1-SLX4 in DNA repair by homologous recombination to maintain genome stability. However in plants, the biological function of Yen1/GEN1 remains largely unclear. We characterized the loss of function mutants of *OsGEN1* and *OsSEND1*, a pair of paralogs of Yen1/GEN1 in rice. We first investigated the role of *OsGEN1* during meiosis and found a reduction in chiasma frequency by ~6% in *osgen1* mutants, compared to wild type, suggesting a possible involvement of OsGEN1 in the formation of crossovers. Post-meiosis, OsGEN1 foci were detected in wild-type microspore nuclei, but not in the *Osgen1* mutant concomitant with an increase in double strand breaks. Persistent double strand breaks led to programmed cell death of the male gametes and complete male sterility. In contrast, depletion of *OsSEND1* had no effects on plant development and did not enhance *osgen1* defects. Our results indicate that OsGEN1 is essential for homologous recombinational DNA repair at two stages of microsporogenesis in rice. It was reported that MUS81 is involved in interference-insensitive CO formation. In order to understand OsMUS81 biological function and its relationship with OsGEN1, we used CRISPR to knock out *OsMUS81* in rice cultivar 9522 and got *osmus81* homozygotes. We found seed setting percentage decreased in *osmus81* mutant, which indicated that OsMUS81 also functions in rice reproductive process. *Osmus81 Osgen1* double mutant displayed more severe meiotic defects, suggesting that OsGEN1 and OsMUS81 may be involved in different pathways during meiosis in rice.

T4

A347

Physiological and molecular responses of poplar seedlings to phosphate deficiency

Fei Wang[1,2], *Ping Lan*[2], *Li Wenfeng*[1,2]

1. Collaborative Innovation Center of Sustainable Forestry in Southern China of Jiangsu Province, College of Biology and the Environment, Nanjing Forestry University

2. State Key Laboratory of Soil and Sustainable Agriculture, Institute of Soil Science, Chinese

Phosphorus is one of the essential nutrients, whose supply is often under the demand for plant development and growth in natural and agricultural ecosystem. While the molecular mechanisms of phosphate (Pi) deficiency responses is well studied in herbaceous plants, limited information has been available on the responses of the woody plants including poplar to Pi deficiency. As an important woody plant, especially on marginal soils, poplar is getting an important model woody plant, with the whole genome sequencing of *Populus* being completed; and with an increasing demand for woody biomass, an extension of poplar plantation is frequently expected. To explore the physiological and molecular responses of poplar upon Pi deficiency, we established a tissue culture system using the poplar cultivar 'soma clone' as the material and the Hoagland solution as the culture media. Based on this system, we found that poplar's growth was severely stunted upon Pi starvation, with the shoot length being reduced to 2/5 of the ones grown under normal conditions; while the primary root length and the excreted purple acid phosphatases activity were increased dramatically. Besides, an obviously early leaf abscission was observed upon Pi deficiency. Transcriptome sequencing is being conducted to further investigate the molecular mechanism of poplar responding to Pi deficiency. This research will greatly facilitate the breeding and cultivation of nutritional-efficient poplars.

T4

A348

Arabidopsis SAURs are critical for differential light regulation of the development of various organs

Ning Sun[1], ***Jiajun Wang***[1], *Zhaoxu Gao*[1], *Jie Dong*[1], *Hang He*[1], *William Terzaghi*[2], *Ning Wei*[3], *Xing Wang Deng*[1], *Haodong Chen*[1]

1. Peking University

2. Wilkes University

3. Yale University

During deetiolation of *Arabidopsis* seedlings, light promotes the expansion of cotyledons but inhibits the elongation of hypocotyls. The mechanism of this differential regulation of cell enlargement is unclear. Our organ-specific transcriptomic analysis identified 32 Small Auxin Up RNA (SAUR) genes whose transcripts were light-induced in cotyledons and/or repressed in hypocotyls. We therefore named these SAURs as lirSAURs. Both overexpression and mutation analyses demonstrated that lirSAURs could promote cotyledon expansion and opening and enhance hypocotyl elongation, possibly by inhibiting phosphatase activity of D-clade type 2C protein phosphatases (PP2CDs). Light reduced auxin levels to down-regulate the expression of lirSAURs in hypocotyls. Further, phytochrome-interacting factors (PIFs) were shown to directly bind the genes encoding these SAURs and differentially regulate their expression in cotyledons and hypocotyls. Together, our study demonstrates that light mediates auxin levels and PIF stability to differentially regulate the expression of lirSAURs in cotyledons and hypocotyls, and these lirSAURs further mediate the differential growth of these two organs.

T4

A349

The photosynthetic response in *Scenedesmus quadricauda* after carbon-ion irradiation

Jie Wang, *Xin Li, Jufang Wang, Wenjian Li, Yan Du, Wei Wei, Dong Lu, Liang Ma*

Gansu Key Laboratory of Microbial Resources and Application, Institute of Modern Physics, Chinese Academy of Sciences

A large proportion of mutants with altered pigment features have been obtained via exposure to heavy-ion beams, a technique that is efficient for trait improvement in the breeding of plants and algae. However, little is known about the changes of the photosynthetic response of microalgae after exposure. In our group, six progenies of *Scenedesmus quadricauda* deficient in chlorophyll a were isolated after carbon-ion exposure that were provided by the heavy ion research facility in Lanzhou (HIRFL), China. Two progenies were picked up because their photosynthetic efficiency and the photoprotection ability were markedly different from the wild type. What is more, the proteomics studies of the two progenies were analyzed. The most differential proteins in the two progenies were from light harvesting complexes. In the other aspect, our group analyzed the chlorophyll fluorescent parameters (Fv/Fm, φPSII, and NPQ), the photoprotective pigment lutein, and the transcriptional expression of Lhcb1 and Lhcb2 in *Scenedesmus quadricauda* after exposure to $^{12}C^{6+}$ ions. Exposure to 20 Gy of carbon ions improved the photosynthetic efficiency of *Scenedesmus quadricauda* more quickly than exposure to 60 or 120 Gy during 48 h of culture after irradiation. The thermal dissipation by *Scenedesmus quadricauda* was initiated more quickly after exposure to 20 Gy than exposure to 60 or, 120 Gy. The transcriptional expression of Lhcb1 and Lhcb2 was up-regulated within 4 h of culture after exposure to 20 Gy of carbon ions. Thus, from the mutant strains we found that light harvesting proteins expressed differently, and the photosynthetic characteristics of the two progenies changed markedly after carbon-ion irradiation. From group mutagenic effects we found that the low dose of carbon ion irradiation caused hormetic effects on the photosynthetic efficiency, thermal dissipation ability and transcriptional regulation of the light harvesting complex II antenna proteins in *Scenedesmus quadricauda*.

T4

A350

RNAi-mediated suppression of hexokinase3 leads to inhibited growth and development, inhanced photosynthesis in *Brassica napus* L.

Xiaomin Wang, *Jingxue Wang*
School of Life Sciences, Shanxi University

Oilseed (*Brassica napus* L.) is an important oil crop, which has great economic development and utilization value. Carbohydrate metabolism plays an important role in plant growth and development. Glucose (Glc) from plant photosynthesis is phosphorylated by hexokinase (HXK) to generate Glc-6-phosphate (Glc-6-P) and enters the glycolysis pathway and converted to pyruvic acid and ATP. HXK is the key and rate-limiting enzyme involved in the glycolysis pathway and plays an important role in plant growth and development. In this study, the plasmid pCaMRNAi containing interferenced *HXK3* gene were introduced into oilseed 7B plants by pollen-mediated method. Using GUS markers, the transgenic plants were identification. Total 97 transgenic plants were identified by PCR amplification. The results of quantitative real-time PCR showed that the expression of *HXK3* gene in transgenic plants was 15.41% ~ 75.43% lower than wild-type plants. Transgenic seedlings grew euphylla later than wild-type plants. During plant growth, the transgenic plants were dwarfer than the wild-type plants, and the leaves of some transgenic plants became much curly. Moreover, the flowering time of transgenic plants was earlier, and some floral organs were abnormal and unable to harvest normal seeds. In addition, the chlorophyll content of the transgenic plants was 10.95% ~ 71.07% higher than that of the wild-type plants, and the photosynthetic rate was also increased. Quantitative real-time PCR showed that the expression of *POA* gene related chlorophyll synthesis in transgenic plants was significantly increased. Further observation, we found that some visible liquid secretion on the leaf surface of transgenic plants, which is proved to be sugar and pests are also more serious. Stained with toluidine blue, the blue color area is larger in wild-type plants than in transgenic plants. Expression of *GPAT6* gene related fatty acid metabolism in transgenic plants was significantly decreased by quantitative real-time PCR amplified analysis. The function of *HXK3* gene in oilseed was improved by this study.

T4

A351

Identification of cold stress responsive proteins in *Anabasis aphylla* seedlings by iTRAQ proteomics technique

Tingting Wang[1], *Chunxiu Ye*[2], *Ping Jiang*[1], *Guangming Chu*[1], *Mei Wang*[1]
1. Shihezi University
2. Xinjiang Academy of Agricultural and Reclamation Sciences
3. Shihezi University
4. Shihezi University
5. Shihezi University

Cold stress is one of the major environmental factors that affect plant growth, development and geographical distribution. In order to study the protein molecular mechanism of *Anabasis aphylla* seedlings in response to cold stress, we used the iTRAQ technique to identify the differentially protein expression under cold stress. In total, 211 differentially proteins are identified, including 109 up-accumulated proteins and 102 down-accumulated proteins, these cold stress response proteins mainly involved carbohydrate and energy metabolism, protein metabolism and translation, stress response, transcription related, amino acid metabolism, signal transduction and membrane and transport. In our results, glyceraldehyde-3-phosphate dehydrogenase and pyruvate kinase were up-accumulated in differentially proteins expression analysis, that means *A. aphylla* seedling adapted to cold stress throughout changed its energy metabolism pathway. We identified several proteins which involved in protein translation, refolding and degradation in these iTRAQ data, and it suggested that different elongation factors had different functions in response to cold stress. Some ROS scavenging systems enzymes were also identified to be associated with cold stress response such as SOD, POD, MDH and GSH-Px. Protein TOPLESS and mitogen-activated protein kinase homolog NTF6 were found to relate to transcription and signal transduction respectively. 5-methyltetrahydropteroyltriglutamate-homocysteine methyltransferase had an important role in L-methionine biosynthesis. Membrane steroid-binding protein and mitochondrial dicarboxylate transporter which involved in transmembrane transport had been investigated successfully. Such data were discussed in the context of our current understandings of *A. aphylla* seedlings to indicated the molecule mechanisms in the cold response. Meanwhile, the correspondence between mRNA transcript level and proteins abundance were identified by qRT-PCR.

T4

A352

Proteomic characterization of hormone-responsive, differential abundance proteins, which are involved in the regulation of coleoptile growth in maize

Chenhui Xu, *Xiaolin Wu*, *Wei Wang*
Henan Agricultural University

Coleoptile is a model organ for studying cell elongation since Charles Darwin first described this organ in 1880. The coleoptile is a unique structure in monocot crops, such as maize, wheat and rice. It protects and leads the upward growth of the enclosed primary leaves during seed germination in the soil, which is a key step for the successful seedling establishment. Moreover, the coleoptile contains a small quantity of chloroplasts, which can provide additional nutrients for the seedling through photosynthesis.

In the present study, maize coleoptile was sampled at four growth stages, i.e., initial, slow, rapid, and stable growth stage, based on the growth curve, and used for proteomic, physiological and microscopic studies, so as to identify and characterize differential abundance proteins (DAPs) during coleoptile growth in maize. Total proteins were extracted using phenol extraction and separated using 2-DE. Endogenous hormone change was analyzed by HPLC-MS/MS. Moreover, microsection was made to observe cell morphological change during coleoptile elongation.

As a result, 69 DAPs were identified by MALTI-TOF/TOF analysis. These DAPs are mainly involved in hormone signaling pathway, stress response, heat shock protein, storage protein, carbohydrate metabolism, cell structure, S-adenosylmethionine metabolism, glutathione metabolism and others. Especially, beta-glucosidase involving in cytokinin signaling pathway, proteins involving S-adenosylmethionine metabolism and glutathione metabolism, accumulated in high amount during coleoptile elongation process, implying their importance for coleoptile elongation. Meanwhile, UDP-glucose dehydrogenase, a key enzyme to formation cell wall precursor, alpha-tubulin and beta-tubulin were needed for coleoptile elongation. The cessation of coleoptile growth was found to be associated with the accumulation of abscisic acid (ABA) responsive proteins. Moreover, the expression of the mRNAs of the DAPs, endogenous hormone and cell morphological changes in coleoptiles were also analyzed. This study will be helpful for further understanding of the elongation mechanism of the coleoptile in maize.

T4

A353

KcNHX1 from *Karelinia caspia* confers drought, salt, and heat stress tolerance in *Arabidopsis*

Yanqin Wang
Tarim University

Environmental stresses such as drought, salinity and heat affect plant growth and development. *Karelinia caspia* (*K. caspia*) is a unique perennial herb with remarkable tolerance to environmental stresses because it grows in desert area for long-term. The Na^+/H^+ antiporter gene was cloned from eremophyte *K. caspia* (*KcNHX1*), to investigate the possible roles of *KcNHX1* in the abiotic stress response of *K. caspia* and the underlying regulatory mechanisms. The overexpression vector was constructed and transformed into *A. thaliana*. The physiological results showed that the overexpression of *KcNHX1* in *A. thaliana* enhanced tolerance not only to drought and salt stress, but also to heat stress in the seedling stage. The overexpression of *KcNHX1* maintained higher seed yield and IAA content in the floral organs under high temperature than wild type plants during florescence. Meanwhile, application of low concentration of IAA led to higher fertility of *KcNHX1* overexpression plants under high temperature. The results indicate that *KcNHX1* might activate the auxin biosynthesis and signal pathway in *A. thaliana* to improve the tolerance to heat. *KcNHX1* from an eremophyte could be applied as a candidate gene for enhancing multiple stress tolerance in other plant species.

T4

A354

Plant physiological and nutritional responses to deficit irrigation

Yaosheng Wang[1], *Weiping Hao*[1], *Changrong Yan*[1], *Yanqing Zhang*[1], *Daozhi Gong*[1], *Fulai Liu*[2]
1. Institute of Environment and Sustainable Development, Chinese Academy of Agricultural Sciences
2. Department of Plant and Environmental Sciences, University of Copenhagen

Increasing shortage of freshwater for agriculture and more frequent and severe drought due to climate change have stimulated research into high efficient strategies to produce more crops per drop water. Deficit irrigation (DI) is a traditionally recommended irrigation strategy intensively studied on a wide range of crops in many regions of the world. In DI the entire root zone is watered with an amount of water less than the potential evapotranspiration during the non-sensitive growth stages and the mild stress has minimal effects on the yield. Therefore, the water use efficiency (WUE) is improved. Partial root-zone drying irrigation (PRI) is a further refinement of DI developed based on the drought-induced root-to-shoot chemical signaling in the drying roots. It causes soil moisture heterogeneity in the two root zones, in which only half part of the root zone is watered in order to keep the leaves hydrated while the other half is allowed to dry to a predetermined level before alternating irrigation between the drying and rewetting zone to sustain the root-to-shoot ABA signaling for controlling stomatal opening. PRI has been demonstrated to remarkably reduce water use, but increase WUE and partitioning of dry mass into fruits or carbon remobilization into grain thereby maintain the yield and increase harvest index. Both PRI and DI can induce the ABA-based root-to-shoot chemical signaling regulating stomatal conductance and leaf expansion growth. Nonetheless, at a similar degree of water saving, PRI can intensify ABA signaling relative to DI, resulting in a better control of plant transpiration causing further improvement of WUE. Moreover, PRI has been shown to improve product quality. Repeated soil drying and rewetting (DRW) cycles occur in rainfed and irrigated agriculture. The intensity and frequency of DRW cycles regulate both microbial physiology and soil physical processes, hereby affecting the mineralization and immobilization of soil nutrients and their bioavailability. PRI induces soil DRW cycles and more soil water dynamics in the root zone, which enhance soil nutrient mineralization process and nutrient substrate availability, thus increase the bioavailability of soil nutrients, resulting in improved nitrogen (N) and phosphorus (P) uptake, in which soil microbes paly an essential role in the mineralization and immobilization turnover process. Practically, PRI can be applied in an easy way that the crops are irrigated every other row in furrow irrigation or by repeatedly using single irrigation pipeline. The application of low-cost drip irrigation pipes together with fertigation can make it promising as a feasible, costless and profitable water- and fertilizer-saving strategy for agricultural production in the regions especially arid regions, where rainfall is scarce. However, there are still some uncertainties regarding effect of drought and deficit irrigation techniques in terms of climate change and intensity and frequency of DRW cycles in the field, which merit further studies.

T4

A355

Transcriptomic and metabolomic studies disclose key metabolism pathways contributing to well-maintained photosynthesis in the drought and the consequent drought-tolerance in rice

Xiaosong Ma, Hui Xia, Haibin Wei, Liang Chen, Hongyan Liu, ***Lijun Luo***
Shanghai Agrobiological Gene Center

In contrast to wild species, drought-tolerance in crops requires a fully functional metabolism during drought (particularly photosynthetic processes) to maintain relatively higher productivity. Photosynthesis is therefore a key metabolic process under drought.

However, the link between drought-tolerance, photosynthetic regulation during drought, and the associated transcript and metabolic foundation, remains largely unknown. We used two rice cultivars with contrasting drought-tolerance (the drought-intolerant cultivar IRAT109 and the drought-tolerant cultivar IAC1246) to explore transcript and metabolic responses to long-term drought. The drought-tolerant cultivar represented higher osmotic adjustment and antioxidant capacity, as well as higher relative photosynthesis rate under a progressive drought stress occurred in a modified field with shallow soil-layers. A total of 4,059 and 2,677 differentially expressed genes (DEGs) were identified in IRAT109 and IAC1246 between the drought and well-watered conditions, respectively. A total of 69 and 47 differential metabolites (DMs) were identified between the two treatments in IRAT109 and IAC1246, respectively. Compared to IRAT109, the DEGs of IAC1246 displayed enhanced regulatory amplitude during drought. We found significant correlations between DEGs and the osmolality and total antioxidant capacity (AOC) of both cultivars. During the early stages of drought, we detected up-regulation of DEGs in IAC1246 related to photosynthesis, in accordance with its higher relative photosynthesis rate. The contents of six differential metabolites were correlated with the osmotic potential and AOC. Moreover, they were differently regulated between the two cultivars. Particularly, great up-regulations of 4-hydroxycinnamic acid and ferulic acid were consistent with the performance of photosynthesis-related DEGs at the early stages of drought in IAC1246. Metabolic pathways of these two differently regulated metabolites were also enriched by DEGs detected in IAC1246. It was therefore 4-hydroxycinnamic and ferulic acid are considered as key metabolites for rice drought-tolerance. They could be applied in rice drought-tolerance improving. In addition, DEGs involved in pathways of these two metabolites were correlated with osmolality and AOC. They are consequently expected to be good candidate genes to improve drought-tolerance. In conclusion, well-maintained photosynthesis under drought should contribute to improved drought-tolerance in rice. Metabolites, such as 4-hydroxycinnamic and ferulic acid, play vital roles in photosynthesis protection under dehydration *via* osmotic adjustments and/or antioxidant mechanisms. A metabolite-based method was thus an effective way to explore drought candidate genes. Metabolic accompanied by transcript responses to drought stress should be further studied to find more useful metabolites, pathways, and genes.

T4

A356

Promoter methylation affects plant regeneration from *in vitro* cultures of *Arabidopsis thaliana*

Fengning Xiang
Shandong University

Methylation within the promoter is a critical regulator of gene expression. A set of plant regeneration related genes showing differential expression between regenerable (RC) and non-regenerable (NRC) *Arabidopsis thaliana* callus were identified by microarray analysis, and of these, 20 were up-regulated in the calli of DNA methyltransferase loss-of-function mutants. The promoters of *FLOWERING WAGENINGEN (FWA)*, *ACETYL-COENZYME A CARBOXYLASE 1 (ACC1)*, *TERMINAL FLOWER 1 (TFL1)*, *MORE AXILLARY BRANCHING (MAX3)* and *At2g05520* were methylated and the genes' expression was down-regulated in one or other of RC and NRC, while the promoter of *At5g55450* was methylated in both RC and NRC, but the gene's expression was only down-regulated in RC. Promoter methylation of *FWA*, *ACC1*, *TFL1* and *MAX3* was associated with histone methylation and deacetylation. The two demethylated mutants *fwa-1* and *fwa-2* were less regenerable than the wild type. The shoot meristem associated *WUSCHEL-RELATED HOMEOBOX* (*WOX*) gene *WOX9* was directly, while *WUSCHEL (WUS)*, *WOX2*, *WOX8* and the auxin efflux protein *PINFORMED1* (*PIN1*) were indirectly repressed. Thus, methylation of the *FWA* promoter appears to be important for the regulation of *FWA* expression and plant regeneration from *in vitro* culture.

T4

A357

Observation on petal epidermis of different color *Phalaenopsis* flowers

Wenfang Xiao[1,2], *Zuo Li*[1,2], *Heming Chen*[1,2], *Fubing Lyu*[1,2]
1. Environmental Horticulture Research Institute of Guangdong Academy of Agricultural Sciences
2. Guangdong Provincial Key Laboratory of Ornamental Plant Germplasm Innovation and Utilization

Phalaenopsis species are popular ornamental plants worldwide because of their long-lived flowers with a wind range of colors and various pigmentation patterns, such as evenly colors, shade, edges, stripes, nets, spots, irregular blotches and combinations of these. Inter-specific and inter-generic hybridization within and between members of the *Orchidaceae* has further widened the range of the patterns. This study designed to explore the relationship between flower color and petal epidermal cells shape in *Phalaenopsis*. Scanning electron microscopy (SEM) was used to explore the petal microstructure differences of *Phalaenopsis*. According to the Royal Horticultural Society Color Chart (RHSCC), flower color of 8 evenly colored flowers was grouped into four classes: white, green-yellow, purple and purple-violet. In addition, transverse section from the middle of the petal of the junction of the ground color and the over color in each spotted flowers were taken using a sharp razor blade, mounted in 0.25% PEG solution which had an osmotic potential close to that of petal cells, and examined using a Zeiss microscope. The two white flowers had papillate epidermal cells; all the other 10 flowers had dome-shaped epidermal cells. In addition, there was a small amount of pores in the petal epidermis of some samples. The upper epidermis cell's shape of the same color from the different varieties of *Phalaenopsis* petals may be different (papillate or dome), but the cell shape in different color regions of the same petal is the same. These finding prove that flower color, no matter ground color or over color, is not the determinant effect of the petal epidermal cell shape in *Phalaenopsis*. The above comparisons study between different color *Phalaenopsis* will provide some reference for the future research about flower color in *Phalaenopsis*.

T4

A358

Ectopic expression of a *SOC1* homolog from *Phyllostachys*

violascens* alters flowering time and identity of floral organs in *Arabidopsis thaliana

Shinan Liu[1,2,3], *Tiantian Qi*[1], *Jingjing Ma*[1], *Tengfei Ma*[1,4], *Luyi Ma*[3], ***Xinchun Lin***[1]
1. Zhejiang Agriculture and Forestry University
2. Guangxi University
3. Beijing Forestry University
4. The Forestry Bureau of Liandu District

As a floral activator, *SUPPRESSOR OF OVEREXPRESSION OF CO1/AGAMOUS-LIKE 20* (*SOC1/AGL20*) gene plays a key role in the flowering pathway of *Arabidopsis*. Bamboo MADS box gene *PvMADS56*, a homolog of *SOC1/AGL20*, was cloned from *Phyllostachys violascens*. Sequence comparison and phylogenetic analysis showed that PvMADS56 was closely related to MADS56-like proteins, which are the members of SOC1-like family. *PvMADS56* was widely expressed in all the tested tissues of flowering and non-flowering bamboo plants, and its function was investigated by ectopic expression in transgenic *Arabidopsis* plants. The results showed that the overexpression promoted flowering in wild-type *Arabidopsis* and complemented the delayed flowering phenotype of *soc1 Arabidopsis*. Meanwhile the transgenic plants displayed abnormal floral organs and leaves, low fertility and dwarfism. Overexpression of *PvMADS56* in the wild-type *Arabidopsis* not only caused early flowering by upregulating *Flowering Locus T* (*FT*) and downregulating *Flowering Locus C* (*FLC*) expression, but also abnormal floral organs by downregulating *APETALA1* (*AP1*), *APETALA3* (*AP3*), *PISTILLATA* (*PI*), and *AGAMOUS* (*AG*). Furthermore, PvMADS56 might be a nuclear protein, and interacted with PvAP1 and PvSEP3 from *Ph. violascens* in the yeast two-hybrid assay. In addition, the activity of *PvMADS56* promoter was enhanced by exogenous abscisic acid (ABA) and methyl jasmonate (MeJA). Taken together, *PvMADS56* may be a multifunctional gene that not only regulates the flowering time but also involves in the identity of floral organs in response to ABA and MeJA.

T4

A359

Callus induction and regeneration via shoot tips of *Dendrocalamus hamiltonii*

Zang Qiaolu, Zhou Ling, Zhuge Fei, Yang Haiyun, Wang Xiaoqin, ***Lin Xinchun***
Zhejiang Agriculture and Forestry University

Bamboos are the members of the grass family, including more than 88 genera and 1,400 species worldwide. Because of rapid growth, high output, highly maintaining soil and water, and other values, they are economically, socially and ecologically important in China. *Dendrocalamus hamiltonii* is one of the three most important bamboo species with sweet shoots in the world and is cultivated for its shoots and timber in Xishuangbanna and Puer, Yunnan province, in China. The inflorescences, embryos and seeds of the bamboos are good resources for explants, but they are difficult to obtain because bamboos rarely blossom and bear fruit. However, shoot tips are available at any time and easy to obtain. By using shoot tips as explants, various media and culture conditions for callus induction and proliferation, shoot differentiation, root induction and plantlet transplantation to develop an efficient and reliable regeneration system with *Dendrocalamus hamiltonii* were tested. Murashige and Skoog (MS) medium supplemented with 3 mg/l 2,4-dichlorophenoxyacetic acid (2,4-D), 1 mg/l benzyladenine (BA), 500 mg/l glutamine (Gln), 500 mg/l proline (Pro), and 500 mg/l casein hydrolysate (CH) yielded the best rates of callus induction and granular-compact callus induction. MS medium supplemented with 1 mg/l BA, 0.3 mg/l kinetin (KT) and 0.3 mg/l naphthaleneacetic acid (NAA) conferred the highest differentiation rate of calli. The maximum rooting rate was obtained in 1/2 MS medium supplemented with 3 mg/l indole-3-butyric acid (IBA), and the roots were long and thick. All hardened plantlets survived after transfer to an equal ratio mixture of peat, vermiculite and perlite. The regeneration system of *D. hamiltonii* developed is efficient and provides a useful tool for genetic transformation in bamboo species.

T4

A360

Photosynthetic characteristics of 12 species epiphytic ferns of polypodiaceae from Ailao Mt. in Yunnan Province

Chengdong Xu[1], *Jianmeng Feng*[2], *Fenliang Kuang*[1]
1. Institute of Bio-resources Research and Utilization in Central Yunnan, Chuxiong Normal University
2. College of Life Science and Agronomy, Dali University

Epiphytic ferns play an important role in construction and maintain of biodiversity in forest ecosystem. To present the adaptability of physiological ecology on epiphytic ferns, the photosynthetic parameter of 12 species epiphytic ferns in Polypodiaceae were measured by LI-6400XT in Ailaoshan National Nature Reserve. Photoresponse curve was fitted by a new model and the interrelation between photosynthetic characteristics and ecological adaptability was analyzed. The result showed that light saturation of the 12 species epiphytic ferns is presented in the lower photo flux density. *Phymatopsis integrifolia* have a mild photoinhibition phenomenon and *Polypodiastrum mengtzeense* have a intense photoinhibition phenomenon, when photo flux density exceed the light saturation point, however, the rest of the epiphytic ferns have a light sleep phenomenon. 12 species of epiphytic ferns are divided into four ecological types which include type of heliophilous and intolerance of shade, type of heliophilous and shade-tolerance, type of shade-requiring and intolerance of shade, and the type of shade-requiring and shade-tolerance, according to light saturation point and light compensation point. There are different characteristics of distribution in the four ecological types.

T4

A361

Rice peroxisome sulfite oxidase 1 is required for chloroplast integrity, photosynthesis, plant development and sulfite response

Dawei Xu[1], *Muhammad Uzair*[1], *Ronghui Pan*[2], *Linda Savage*[2], *Sheng Quan*[1], *Zhu Lu*[1], *Gynheung An*[3], *David Kramer*[2], *Dabing Zhang*[1,4], *Wanqi Liang*[1], *Jianping Hu*[2]
1. Shanghai Jiao Tong University
2. Michigan State University
3. Kyung Hee University

4. *University of Adelaide*

Sulfur is an essential macronutrient for animals, plants and microorganisms. In plants, sulfite is an intermediate in sulfate assimilation in chloroplasts, and it can also be oxidized into sulfate in the peroxisome by sulfite oxidase (SO). Unlike in animals where SO is mitochondrial, plant SO is exclusively localized to the peroxisome. SO has been shown to play key roles in protecting plants against toxic levels of sulfite in *Arabidopsis*, tomato and poplar plants. However, whether SO is required for plant development under non-stressed conditions was unclear. To determine the role of SO in monocot species such as cereal crops, we characterized rice *Oryza sativa* sulfite oxidase 1 (OsSO1) and its loss-of-function mutant *Osso1*. Here we show that in addition to enhanced susceptibility to Na_2SO_3 treatment and SO_2 fumigation, *Osso1* displays a series of defects under normal growth conditions, including impaired chloroplast structure, loss of chloroplasts, reduced chlorophyll levels and photosynthetic capabilities, and deficiencies in key agronomical traits such as plant height, heading time, and grain weight. Lack of SO leads to accumulation of amino acids and aberrant expression of many genes involved in chloroplast assembly and function, and sulfite metabolism in young leaves. Our study suggests that rice SO not only serves as a "safety valve" for the plant during response to excess sulfite, but also plays a housekeeping role under optimal growth conditions.

T4

A362

Two bHLH transcription factors EAT1 and TDR cooperatively regulate the anther development and pollen wall formation in rice

Jie Xu

School of Life Sciences and Biotechnology, Shanghai Jiao Tong University,

The pollen wall is a specialized extracellular cell wall matrix that protects male gametophytes and plays an essential role in flowering plant reproduction. Elucidation of the regulatory mechanisms underlying pollen wall formation will not only expand our understanding of the functions of transcription factors in anther development, but will also facilitate efforts to manipulate this process for crop improvement. Here, we show that a bHLH transcription factor, ETERNAL TAPETUM1 (EAT1), is crucial for pollen exine morphogenesis. Using evidence generated from various analytical approaches we demonstrate that EAT1 affects pollen wall formation via directly activating 7 lipid metabolism-associated target genes that are putatively involved in pollen wall formation. The role of which were further confirmed by genetic and morphological analyses on one target gene, *OsLTPL64*, in which RNAi knockdown of expression impaired pollen wall architecture. Chromatin Immunoprecipitation (ChIP) - PCR analysis demonstrated that target regulation was determined by the formation of heterodimers with TDR, or homodimer alone, to regulate both overlapping and distinct sets of genes involved in the anther and pollen development. Transient assays in tobacco revealed that EAT1 and TDR can both act directly on two downstream lipid transport related genes, *OsLTPL68/OsC6* and *OsLTPL64*. Taken together, we have shown that EAT1 and TDR function collaboratively as important regulators of rice anther and pollen development, and that EAT1 and TDR have both distinct and shared regulatory pathways responsible for rice male reproductive development.

T4

A363

Identification of chilling-responsive microRNAs and their targets in vegetable soybean (*Glycine max* L.)

Shengchun Xu[1], *Na Liu*[1], *Weihua Mao*[2], *Qizan Hu*[1], *Guofu Wang*[3], *Yaming Gong*[1]

1. *Institute of Vegetables, Zhejiang Academy of Agricultural Sciences*

2. *Center of Analysis and Measurement, Zhejiang University*

3. *Department of Life Science, Yuanpei College, Shaoxing University*

Chilling stress is a major factor limiting the yield and quality of vegetable soybean (*Glycine max* L.) on a global scale. In the present study, systematic identification and functional analysis of miRNAs under chilling stress were carried out to clarify the molecular mechanism of chilling resistance. Two independent small RNA libraries from leaves of soybean were constructed and sequenced with the high-throughput Illumina Solexa system. A total of 434 known miRNAs and 3 novel miRNAs were identified. Thirty-five miRNAs were verified by qRT-PCR analysis. Furthermore, their gene targets were identified via high-throughput degradome sequencing. A total of 898 transcripts were targeted by 54 miRNA families attributed to five categories. More importantly, we identified 51 miRNAs differentially expressed between chilling stress and control conditions. The targets of these miRNAs were enriched in oxidation-reduction, signal transduction, and metabolic process functional categories. Our qRT-PCR analysis confirmed a negative relationship among the miRNAs and their targets under chilling stress. Our work thus provides comprehensive molecular evidence supporting the involvement of miRNAs in chilling-stress responses in vegetable soybean.

T4

A364

Over-expression of *RHR1* enhances root hair elongation under phosphate deficiency in *Arabidopsis*

Caiwen Xue[1,2], *Jiexue Huang*[1,2], *Lan Ping*[1]

1. *State Key Laboratory of Soil and Sustainable Agriculture, Institute of Soil Science, Chinese Academy of Sciences*

2. *University of Chinese Academy of Sciences*

Phosphate (Pi) is a limited crucial macro-nutrient for plants due to its poor mobility in soil. Crops growth and productivity are severely impaired by Pi deficiency. To cope with Pi shortage, plants have evolved various adaptations, including an increase of root hair density and elongation and higher acid phosphatase activity. RHR1 (Root Hair Related 1), a cell-wall associated receptor kinase, has been reported to be a negative modulator of Arabidopsis root hair growth. In this research, we constructed over-expressed transgenic lines of RHR1 driven by CaMV 35S promoter and interrogated its function under Pi starvation conditions. Compared with the wild type plants, the over-expression lines showed significantly longer root hairs and higher acid phosphatase activity under phosphate deficiency condition (0 μM Pi), although they are comparable under Pi replete condition (250 μM Pi). The results indicate that the over expression of RHR1 could improve the adaptations of palnts to Pi deficiency with longer root hairs and higher phosphatase activity, which would contribute to the improvements of crop yields in agricultural system via molecular design.

T4

A365

A phosphate starvation induced gene *AtLIT1* positively regulates iron deficiency response in *Arabidopsis*

Mingke Yan[1,2], *Ping Lan*[1]

1. Institute of Soil Science, Chinese Academy Sciences

2. University of Chinese Academy of Sciences

Phosphorus and iron (Fe) are two essential nutrient elements of both plants and human, and the antagonistic interaction between them has been widely noted in plants. Here we report *Low Iron Tolerant 1* (*AtLIT1*), whose expression is significantly induced by phosphate (Pi) starvation but shows no notable change in phenotype, could markedly alleviate Fe deficiency symptoms in *Arabidopsis thaliana*. Overexpression of *AtLIT1* resulted in approximately two times longer of the primary root compared with the wild type, and increased the overall biomass as well as the proportion of green leaves of plants under Fe deficiency conditions. Real time qPCR analysis showed no significant change for Fe marker genes, such as *IRT1* and *FRO2*, in both Fe deficiency and sufficiency conditions; while, the double mutation of *AtLIT1* and one of its homologous gene resulted in down-regulation of various Fe marker genes. These results suggest that *AtLIT1* is transcriptionally mediated by Pi status, but involves in Fe limitation responses, possibly in a post transcriptional manner, which further expanded the understanding of Pi and Fe interaction and homeostasis in plants.

T4

A366

Genome-wide analysis of PHD-finger protein family in *Glycine max*

Junkai Yang, Jianwei Li, Mingzhe Sun, Bowei Jia, ***Xiaoli Sun***

Heilongjiang Bayi Agricultural University

PHD (Plant homeodomain)-finger proteins are found universally in eukaryotes, and play major roles in regulating transcription and chromatin state. The structure and function of PHD-finger proteins have been well studied in animals. On the contrary, only a few PHD-finger proteins have been characterized to engage in plant growth and development. In this study, we conducted an extensive analysis of soybean PHD family. By using Phytozome and SMART searches, a total of 99 PHD-finger genes (termed as *GmPHD1* to *GmPHD99*) were identified in soybean genome. All these GmPHD proteins contained one or several PHD-finger domains that were comprised of approximately 60 amino acids and a characteristic Cys4-His-Cys3 motif. And further phylogenetic analysis revealed that these *GmPHDs* were clustered into twenty groups (A-T). Members in each group displayed high similarity in terms of intron-exon organization pattern. Among them, 98 genes were unevenly distributed on all 20 chromosomes and contained multiple duplication events. The Ka/Ks ratios analysis suggested that the duplicated genes of the PHD-finger family mainly underwent purifying selection with restrictive functional divergence after the duplication events. Expression profiles analysis indicated that 99 GmPHDs were differentially expressed in various tissues. These results will provide useful information for the further structural and functional studies of PHD-finger genes.

T4

A367

Identification of a novel (-)-5-Epieremophilene synthase from *Salvia miltiorrhiza* via transcriptome mining

Xin Fang[2], *Yue Yang*[1], *Chen-Yi Li*[2], *Xiao-Ya Chen*[1,2], ***Lei Yang***[1]

1. Shanghai Chenshan Plant Science Research Center, Chinese Academy of Sciences, Shanghai Chenshan Botanical Garden

2. Shanghai Institute of Plant Physiology and Ecology, Chinese Academy of Sciences

Terpenes are the largest class of plant metabolites, with diverse functions in plant growth, development, bio-interactions and stress resistance. Aroma is characteristic of *Lamiaceae* family, and the volatiles with biological activities are often the major interests of the resources. Terpenes are often the important constituents of plant essential oils and have a variety of biological functions, among which sesquiterpenes form the largest group and are greatly valuable for their aromatic and pharmaceutical properties. *Salvia miltiorrhiza,* a medicinal plant in the family of *Lamiaceae*, has been used in China for thousands of years to treat coronary heart diseases. Although biosynthesis of tanshinones, a group of diterpenoids in *S. miltiorrhiza*, has been extensively investigated, to date we know little about the formation of monoterpenes and sesquiterpenes in this medicinal plant. Our previous analysis of *S. miltiorrhiza* transcriptome has annotated 24 unigenes as TPSs, most of them have not been functionally characterized. Here, we report the characterization of three sesquiterpene synthases, named SmSTPS1, SmSTPS2 and SmSTPS3, which catalyzed the formation of a new compound, (-)-5-epieremophilene. Additionally, the (-)-5-epieremophilene biosynthesis activity of SmSTPS1 was confirmed by transient expression in *Nicotiana benthamiana.* Despite the similar enzyme activities of SmSTPS1, SmSTPS2 and SmSTPS3, the three (-)-5-epieremophilene synthase genes displayed different spatial expression patterns and responded differently to hormone treatments, implicating their specific roles in plant-environment interactions. Our results provide valuable data to understanding the biosynthesis and composition of terpenes in plant.

T4

A368

Understanding dark metabolism and its regulation to photosynthesis in *Chlamydomonas*

Wenqiang Yang[1], *Tyler Wittkopp*[2], *Matthew Posewitz*[3], *Arthur Grossman*[2]

1. Photosynthesis Research Center, Key Laboratory of Photobiology, Institute of Botany, Chinese Academy of Sciences, Beijing, China

2. Department of Plant Biology, Carnegie Institution for Science, Stanford, CA, USA

3. Department of Chemistry and Geochemistry, Colorado School of Mines, Golden, CO, USA

Half of entire life of photosynthetic organisms is in the dark, and dark metabolism (heterotrophic and anoxic/hypoxic) is critical and necessary for photosynthetic organisms to recover photosynthesis when receiving light again. Dark metabolism can provide necessary materials and energy, re-oxidize NADH to NAD^+, store metabolites from photosynthesis, keep the correct carbon flux and the normal redox balance. The overall carbon budget of photosynthetic cells requires an understanding of metabolic features that accompany the movement of the cells between light and dark conditions, and specific electron carriers, such as ferredoxins, are tailored for light or dark metabolism. *Chlamydomonas reinhardtii* is a unicellular, soil-dwelling (and aquatic) green alga that has significant metabolic flexibility for balancing redox equivalents and generating ATP when it experiences hypoxic/anoxic conditions. The diversity of pathways available to ferment sugars is often revealed in mutants in which the activities of specific branches of fermentative metabolism have been eliminated; compensatory pathways having little activity in parental strains under standard laboratory fermentative conditions are often activated. The ways in which these pathways are regulated and integrated are little explored. We use Chlamydomonas as model to study dark metabolism, dark anoxic/hypoxic fermentation and photosynthesis as well, which will not only advance our understanding of the dark metabolism and photosynthesis, it should also be able to provide new evidences and insights into genetic improvement of photosynthesis efficiency.

T4

A369

Cardioprotective effect of *Dioscorea* saponins induced by biotransformation via intestinal microflora

Tao Yi, *Hubiao Chen*

School of Chinese Medicine, Hong Kong Baptist University

Dioscorea nipponica Makino (DN), *D. panthaica* Prain et Burkill (DP), and *D. zingiberensis* C. H. Wright (DZ) are three plants with significant medicinal values, which have been successfully developed as effective herbal medicines for treating ischemic heart disease (IHD) in China for decades. In our previous study, we found that the major constitutes of the three herbs are saponins. However, it is still unknown that whether these three herbs act via similar active constituents. As diosgenin has been reported for the protective action in isoprenaline-induced myocardial infarction, we hypothesized that diosgenin is one of the bioactive sapogenin related to the anti-myocardial ischemia (MI) activity of these three herbs. Recently, the effect of intestinal microflora on the drug metabolism has received increasing attention. Anaerobic bacteria presenting in small intestines are quite diverse in species, and different species produce enzymes different in functions; it is these enzymes are also responsible for the drug biotransformation in organism. Incubation of test drugs, particular phytochemicals, with fresh fecal specimen is a common means for investigating this kind of biotransformation. To verify our hypothesis, three total *Dioscorea* saponins were dissolved individually in appropriate amount of DMSO to produce a test solution. A 30 g GAM broth (General anaerobic medium broth) was dissolved in 1000 mL water, filtrated while hot, after treated with anti-bacteria process with high pressure (0.15 MPa) and temperature (121 ℃) for 20 min, cooled to 45 ℃ . The GAM broth solution was then transferred to an anaerobic chamber (37 ℃ , anaerobic condition), and 1 mg vitamin K1 and 6 mg hematin chloride were dissolved in the solution. Then biotransformation vessels (50 mL volume; one vessel per experiment group) were sterilized and filled with 30 mL of GAM broth solution. Vessels were inoculated with 3 mL of fecal suspensions (1:10, w/v) and then 1 mL of each sterile total *Dioscorea* saponins was added. *In vitro* biotransformation was run under anaerobic conditions for a period of 48 h. The the biotransformation mixtures were extracted by 50 mL ethyl acetate for three times. The remaining residues were re-extracted three times with 50 mL water-saturated n-butanol. The combined n-butanol layers were washed with water three times. Then the ethyl acetate and n-butanol layers were mixed homogeneously and concentrated under vacuum, and then diluted to the desired volume with methanol. All the solutions were centrifuged before being injected for LC–Q-TOF/MS analysis. The results demonstrated that diosgenin, as one of the main metabolites, was commonly found in feces from all groups receiving different *Dioscorea* total saponins. The finding of the present study provided a metabolic basis for understanding of the active components and their action modes of *Dioscorea* herbs, and supported their clinical use in IHD management.

This work was supported by the Faculty Research Grant of Hong Kong Baptist University (FRG2/15-16/022) and the Guangdong Natural Science Foundation (2016A030313008).

T4

A370

Cadmium stress inhibits the growth of roots and interferes auxin homeostasis in *Sorghum bicolor* seedlings

Chenliang Yu[1], *Yihua Zhan*[2], *Chenghao Zhang*[1]

1. Zhejiang Academy of Agricultural Sciences

2. Zhejiang University

Phytotoxic effects of cadmium (Cd), a heavy metal pollutant, on plants have been extensively examined. Auxin plays important roles in various aspects of plant growth and development. The relationship between root growth and auxin signaling in Cd-stressed *Sorghum bicolor* was analyzed in our study. Root and shoot length in *S. bicolor* seedlings were significantly reduced after exposure to Cd stress. Inhibition of the root cap growth, lipid peroxidation and cell death by Cd were observed at the root tips. Cd was found to be mainly localized in the meristematic zone using a Cd-specific probe. Cd stress remarkably influenced the cell cycle progression at the root tip by EdU (ethynyl deoxyuridine) assay. IAA content was significantly reduced in the roots of Cd-stressed *S. bicolor*, and this decrease was accompanied by an increase in the activity of IAA oxidase. Auxin transport inhibitors, 1-naphthylphthalamic acid (NPA) and 1-naphthoxyacetic acid (1-NOA), significantly reduced plant tolerance to Cd stress, whereas exogenous supply of 1-naphthaleneacetic acid (NAA) improved Cd tolerance of *S. bicolor* seedlings. Cd stress altered the expression of several putative auxin biosynthetic and metabolic genes as well as auxin transport genes. In addition, NAA interfered with the homeostasis of Cd-induced reactive oxygen species (ROS). These results indicated that Cd stress disturbed the growth of *S. bicolor* seedlings by affecting the homeostasis of auxin and ROS.

T4

A371

A Rice Ca^{2+} binding protein is required for tapetum function and pollen formation

Jing Yu
Shanghai Jiaotong University

In flowering plants, successful male reproduction requires the sophisticated interaction between somatic anther wall layers and reproductive cells. Timely degradation of the innermost tissue of the anther wall layer, the tapetal layer, is critical for pollen development. Ca^{2+} is a well-known stimulus for plant development, but whether it plays a role in affecting male reproduction remains elusive. Here we report a role of Defective in Exine Formation 1 (OsDEX1) in rice (*Oryza sativa*), a Ca^{2+} binding protein, in regulating rice tapetal cell degradation and pollen formation. In *Osdex1* anthers, tapetal cell degeneration is delayed and degradation of the callose wall surrounding the microspores is compromised, leading to aborted pollen formation and complete male sterility. *OsDEX1* is expressed in tapetal cells and microspores during early anther development. Recombinant OsDEX1 is able to bind Ca^{2+} and regulate Ca^{2+} homeostasis *in vitro*, and *osdex1* exhibited disturbed Ca^{2+} homeostasis in tapetal cells. Phylogenetic analysis suggested that OsDEX1 may have a conserved function in binding Ca^{2+} in flowering plants, and genetic complementation of pollen wall defects of an *Arabidopsis (Arabidopsis thaliana) dex1* mutant confirmed its evolutionary conservation in pollen development. A series of Ca^{2+} binding and Ca^{2+}-dependent kinase genes, protease related genes as well as sporopollenin synthesis genes are down regulated in *osdex1* mutant which further suggests the important role of Ca^{2+} on tapetal cell death process and pollen wall formation. Collectively, these findings suggest that OsDEX1 plays a fundamental role in the development of tapetal cells and pollen formation, possibly via modulating the Ca^{2+} homeostasis during pollen development.

T4

A372

Jasmonate regulates plant responses to post-submergence reoxygenation through transcriptional activation of antioxidant synthesis

***Libing Yuan**, Yang-Shuo Dai, Li-Juan Xie, Shi Xiao*
State Key Laboratory of Biocontrol, Guangdong Provincial Key Laboratory of Plant Resources, School of Life Sciences, Sun Yat-sen University

Submergence induces hypoxia in plants; exposure to oxygen following submergence, termed reoxygenation, produces a burst of reactive oxygen species. The mechanisms of hypoxia sensing and signaling in plants have been well studied, but how plants respond to reoxygenation remains unclear. Here, we show that reoxygenation in *Arabidopsis thaliana* involves rapid accumulation of jasmonates (JAs) and increased transcript levels of JA biosynthesis genes. Application of exogenous methyl jasmonate improved tolerance to reoxygenation in wild-type *Arabidopsis*; also, mutants deficient in JA biosynthesis and signaling were very sensitive to reoxygenation. Moreover, overexpression of the transcription factor gene *MYC2* enhanced tolerance to post-hypoxic stress and *myc2* knockout mutants showed increased sensitivity to reoxygenation, indicating that MYC2 functions as a key regulator in the JA-mediated reoxygenation response. MYC2 transcriptionally activates members of the *VITAMIN C DEFECTIVE* (*VTC*) and *GLUTATHIONE SYNTHETASE* (*GSH*) gene families, which encode rate-limiting enzymes in the ascorbate and glutathione synthesis pathways. Overexpression of *VTC1* and *GSH1* in the *myc2-2* mutant suppressed the post-hypoxic hypersensitive phenotype. The JA-inducible accumulation of antioxidants may alleviate oxidative damage caused by reoxygenation, improving plant survival after submergence. Taken together, our findings demonstrate that JA signaling interacts with the antioxidant pathway to regulate reoxygenation responses in *Arabidopsis*.

T4

A373

Stage-specific transcriptomic ananlysis of maize germinal cells in developing anther provides insights into mitosis-meiosis transition

***Tinglu Yuan**, Weijie Huang, Weihua Tang*
Institute of Plant Physiology & Ecology, SIBS, CAS

In flowering plants, germ lines are induced from somatic meristems within reproductive organs. Within anthers, germinal cell initials first undergo several rounds of mitotic proliferation before synchronously entering meiosis. Current understanding of the transition from mitosis to meiosis is still limited. Using laser microdissection-based expression profiling, we studied the transcriptome dynamics of germinal cells at three sequential stages, i.e. mitotic archesporial cells (AR), enlarging pollen mother cells at the pre-meiosis interphase (ePMC) and pollen mother cells at the early prophase of meiosis (PMC). These cells possess similar transcriptomes that are very different from those of mature pollen, stem pith parenchyma cells andseedlings. Nevertheless, cells undergoing the mitosis-meiosis switch exhibit robust transcriptional changes. The three stages are featured by the expression of genes encoding transcription factor subsets, meiotic chromosome recombination proteins and distinct E3 ubiquitin ligases, respectively. Transcription of genes encoding protein turnover machinery is higher in these three stages of germinal cells than in mature pollen, parenchyma cells and seedlings. Many meiotic genes are transcribed in germinal cells prior to meiosis, and translated in pre-meiosis interphase ePMC. We suggest that the ePMC stage represents a crucial turning point from mitosis to meiosis for developing germinal cells.

T4

A374

Elucidating the role of microRNAs during cotton response to drought and salinity stress

***Baohong Zhang**[1], Qinglian Wang[2], Fuliang Xie[1], Runrun Sun[2]*
1. East Carolina University
2. Henan Institute of Sciences and Technology

Drought and salinity are two major environmental factors adversely affecting plant growth and productivity. However, the regulatory mechanism is unknown. In this study, the potential roles of small regulatory microRNAs (miRNAs) in cotton response to those stresses were investigated. Using next-generation deep sequencing, a total of 337 miRNAs with precursors were identified, comprising 289 known miRNAs and 48 novel miRNAs. Of these miRNAs, 155 miRNAs were expressed differentially. Target prediction, Gene Ontology (GO)-based functional classification, and Kyoto Encyclopedia of Genes and Genomes (KEGG)-based functional enrichment show that these miRNAs might play roles in response to salinity and drought stresses through targeting a series of stress-related genes. Degradome sequencing analysis showed that at least 55 predicted target genes were further validated to be regulated by 60 miRNAs. CitationRank-based literature mining was employed to determinhe the importance of genes related to drought and salinity stress. The *NAC*, *MYB* and *MAPK* families were ranked top under the context of drought and salinity, indicating their important roles for the plant to combat drought and salinity stress. According to target prediction, a series of cotton miRNAs are associated with these top-ranked genes, including *miR164*, *miR172*, *miR396*, *miR1520*, *miR6158*, *ghr-n24*, *ghr-n56*, and *ghr-n59*. Interestingly, 163 cotton miRNAs were also identified to target 210 genes that are important in fibre development. These results will contribute to cotton stress-resistant breeding as well as understanding fibre development.

T4

A375

Identification of two 1-Deoxy-D-xylulose-5-phosphate synthase cDNAs expressed differentially in *Alpinia officinarum*

Chunrong Zhang, *Xiaomin Tang, Xuanxuan Cheng, Liming Pan, Quan Yang*
Guangdong Pharmaceutical University

Alpinia officinarum Hance (Zingiberaceae) is a perennial herbaceous flowering plant native to southeastern Asia. The dried rhizome of *A. officinarum*, named lesser galangal or smaller galangal, is a traditional Ayurvedic, Chinese and Thai medicine with antimicrobial, anti-inflammatory, antioxidant, antiemetic, anti-obesity, antihyperlipidemic, anti-psychiatric, analgesic activities and has been used for relieving stomachache, vomiting, hiccups and diarrhea, treating colds, edema, ascites and cancer, invigorating the circulatory system. With the wide and increasing application, wild resources is on the verge of extinction, and the market products are all cultivated. However, the quality of the cultivated *A. officinarum* varied because of different origins and cultivars. Pathway engineering and molecular breeding are emerging strategies for increasing productivity and quality of medicinal plants. To realize these strategies, numerous key biosynthetic genes for bioactive plant constitutes have been cloned and analyzed. With a pungent, aromatic and ginger-like odor, essential oil makes the characteristic constituent of *A. officinarum*, and has been used for quality evaluation. The main constituents of the *A. officinarum* essential oil are monoterpenes. In plants, monoterpenes are synthesized through the 2-*C*-methyl-D-erythritol 4-phosphate (MEP) pathway in plastids. The 1-deoxy-D-xylulose 5-phosphate synthase (DXS, EC 2.2.1.7) enzyme catalyses the first step of MEP pathway, and has been recognized as a rate limiting enzyme and a regulation point in terpene biosynthasis. Overexpression of plant *DXS* genes can enhance levels of specific isoprenoids in many plants. In our present work, two full-length cDNAs encoding DXS were cloned from the rhizome of *A. officinarum* by homology-based RT-PCR and RACE methods. The new cDNAs were designated as *AoDXS1* and *AoDXS2* with a 2,148 bp and a 2,136 bp open reading frame respectively, encoding polypeptides of 715 and 711 amino acids with calculated molecular mass of 77.23 and 51.48 KDa respectively. Bioinformatic analyses revealed that AoDXS1 and AoDXS2 showed extensive homology with DXSs from other plant species and belonged to class I and class II DXS respectively. The expression of *AoDXS1* and *AoDXS2* was analyzed in rhizomes, roots, stems and leaves of 6-month-old plant by quantitative real-time PCR. Among the detected tissues, *AoDXS1* expressed mostly in stems, while *AoDXS2* expressed mostly in leaves. In all detected tissues, the expression levels of *AoDXS2* were much higher than those of *AoDXS1*. These findings may provide valuable information for future experiments on the molecular mechanisms underlying the isoprenoid biosynthesis via the MEP pathway, and on the pathway engineering and molecular breeding of *A. officinarum*.

T4

A376

Exogenous SA increased the glycolysis in winter wheat (*Triticum aestivum* L.) under low temperature etress

Deyi Meng, Hu Zhao, Bing Wu, Jing Cang, Qinghua Xu, ***Da Zhang***
Northeast Agricultural University

Salicylic acid (SA) plays a key role in plant growth, development and the responses of biotic or abiotic stress. As nutrients, osmotic regulation substances and metabolic signals, soluble sugar, can improve plant stress resistance. After exogenous SA treatment winter wheat (*Triticum aestivum* L.), our previous research results showed that: the enzyme activities and gene expression levels of *sucrose synthase* (*TaSuS*), *sucrose phosphate synthase* (*TaSPS*) were increased; the enzyme activities and gene expression levels of *sucrose soluble acid invertase* (*TaSAInv*), *alkaline invertase* (*TaA/N-Inv*) were decreased. For further exploring the effect of exogenous SA on the glycolysis, the strong cold-resistance variety (DN1) and the weak cold-resistance variety (JM22) were selected as experiment materials; That 1 mM SA was sprayed on wheat leaves at trefoil stage was experimental treatment and 5% ethanol was control treatment. Leaves and rhizome were sampled at 5℃ , 0℃ , -10℃ and -25℃ under the condition of natural cooling. We measured the content of fructose and pyruvic acid, and screened the enzyme activities and gene expression levels of *hexokinase* (*TaHxK*), *phosphofructokinase* (*TaPFK*), *pyruvate kinase* (*TaPK*). The results showed, after exogenous SA treatment, the content of fructose and pyruvic acid in leaf and rhizome in DN1 were more definitely increased than in JM22; the enzyme activities and gene expression levels of *HxK* (*TaHxk*), *PFK* (*TaPFK*) and *PK* (*TaPK*) were up-regulated in DN1. So soluble sugar in glycolysis increased, osmotic adjustment ability enhanced, and then the cold resistance of winter wheat was improved.

T4

A377

The abundance of certain metabolites responds to drought

stress in the highly drought tolerant plant *Caragana korshinskii*

Jiwei Zhang
Key Laboratory of Stress Physiology and Ecology in Cold and Arid Regions, Gansu Province, Northwest Institute of Eco-Environment and Resources, Chinese Academy of Sciences

Metabolomics offers opportunities for studying the systematic response of an organism to a genetic and/or an environmental change. Here, the metabolic consequences of drought stress were characterized in the highly drought tolerant plant Caragana korshinskii. The time-of-flight mass spectrometry platform employed identified several hundred metabolites in extracts of the leaf, stem, root collar and root of plants which had been either subjected to drought stress or were well-watered. Each of the four organs harbored a number of potential metabolite markers for the drought response. An increased abundance of various small carbohydrates and soluble amino acids in each of the four organs was induced by the stress; these compounds may act as compatible solutes or antioxidants. Across the whole plant there was a fall in the content of several Krebs cycle and glycolysis intermediates, as well as in that of the amino acids glutamic acid and aspartic acid. Pathway analysis suggested that most of the potential metabolite markers were involved in energy metabolism and amino acid metabolism. The implication was that energy metabolism and photosynthesis are compromised during the adaptation of *C. korshinskii* to drought stress. Given the different spectrum of metabolites associated with the drought response in the four organs, it was concluded that each organ employs a distinct strategy to cope with drought stress.

T4

A378

Overexpression of *Actinidia deliciosa pyruvate decarboxylase 2* gene enhances waterlogging and heat stress in transgene *Arabidopsis*

Ji-Yu Zhang[1,2], *Sheng-Nan Huang*[1,2], *Tao Wang*[1,2], *Gang Wang*[1,2], *Zhong-Ren Guo*[1,2]
1. Institute of Botany, Jiangsu Province and Chinese Academy of Sciences
2. Jiangsu Key Laboratory for the Research and Utilization of Plant Resources

Previous report showed that both *PDC* genes were significantly up-regulated in kiwifruit after treatment with waterlogging using Illumina sequencing technology, and kiwifruit *AdPDC1* gene is required during waterlogging but might not be required during other environmental stresses. Here, the function of another PDC gene named *AdPDC2* was analyzed. The expression of *AdPDC2* gene were determined using qRT-PCR, and the results showed that the expression levels of *AdPDC2* in reproductive organs were much higher than those in nutritive organs, waterlogging, NaCl and Heat could induce the expression of *AdPDC2*. Overexpression of kiwifruit *AdPDC2* in transgenic *Arabidopsis* enhanced the resistance to waterlogging and heat stresses at five weeks old seedlings, but could not enhance resistance to NaCl and mannitol stresses at the stage of seed germination and in early seedlings. These results suggested that the kiwifruit *AdPDC2* gene play key role in resistance to waterlogging and heat stresses in kiwifruit.

T4

A379

OsATG8b-mediated autophagy is involved in nitrogen remobilization to seeds in rice

Mingyong Zhang, *Tian Fan*
South China Botanical Garden, Chinese Academy of Sciences

Improving nutrient use efficiency is recognized as an effective measure to reduce excessive input of fertilizers and increase crop yield. Autophagy is an evolutionarily conserved degradation system in eukaryotic cells and plays an important role in N remobilization. Central to autophagy are two conjugation pathways that attach ATG8 to PE and ATG5 to ATG12, which then help with vesicle elongation and enclosure. We identified the rice gene *OsATG8b* and characterized its role in N remobilization by generating transgenic plants with its over-expression and knockdown. Our study showed that *OsATG8b* restored autophagosome formation in yeast *scatg8* mutant and that OsATG8b is localized to autophagosomes in yeast and rice. ^{15}N pulse-chase analysis revealed that *OsATG8b* over-expression transgenic rice plants conferred higher N remobilization efficiency to seeds, while *OsATG8b* knockdown transgenic plants with lower N remobilization efficiency and poorer seed quality. We conclude that *OsATG8b*-mediated autophagy is involved in nitrogen remobilization to seeds in rice. This result may provide strategic guidance for N application in rice molecular breeding and cultivation.

T4

A380

Fine-mapping a key gene controlling rice leaf angle

Qijun Zhang, *Shijian Xia*, *Yanda Zeng*, *Chuangen Lu*
Institute of food crops, Jiangsu Province Academy of Agricultural Sciences

Rice (*Oryza Sativa* L.) is a staple food for China, as well as the entire world. Consistently and sustainably high yield is essential for an important guarantee to maintain food security in our nation. And good plant architecture is the key factor of rice high yield. In order to make full use of the light energy and increase the yield, all reported rice ideal plant architectural models are required to leaves drooping and faster growing at seedling stage, while post functional leaves relatively erect and smaller leaf angle. This architecture can help plant improve photosynthetic efficiency, reduce pest damage and increase productivities. So leaf angle is an important factor for rice morphology and largely affects its yield. Therefore, studying and cloning key gene responsible for rice good plant architectural breeding is of great significance. In this study, we used map-based cloning method to fine map a new recessive nuclear candidate gene for a control of leaves drooping at seedling stage, while smaller leaf angle, shorter and more erect leaf blade after 5 leaves and 1 hearts period, larger grain in a mutant generated from Thai aromatic rice "Khao Dawk Mali 105" by radiation mutagenesis. This gene was fine mapped between RM3437 and RM237 on rice chromosome 5, co-segregation with RM18532. Re-sequencing results showed that there was only one candidate gene existed protein sequence difference between wild-type and mutant. The candidate gene encodes a putative E3 ubiquitin ligase predicated by RiceDB, it is similar to the *TUD1* gene, but the mutant phenotype is very different from *tud1*. The results of hormone sensitivity analysis showed that the mutant had a response to Gibberellin, but no response to brassinolide. This study will set up a foundation to clone this mutant gene and use this genetic resource for breeding new rice varieties with ideal plant architecture and high yield.

T4

A381

Involvement of salicylic acid in antioxidative action and proline synthesis in *Brassica napus*

Qian Zhang, *Wei-Feng Xu, Ying-Jiao Zhang*

College of Life Science and Center for Plant Water-use and Nutrition Regulation, Fujian Agriculture and Forestry University

To address the influence of salicylic acid (SA) on antioxidative action and proline synthesis, the content of H_2O_2, proline, the activity of antioxidant enzymes and proline biosynthesis-related genes were analyzed in leaves of *Brassica napus*. The reactive oxygen species H_2O_2 content was considerably accumulated in the plants exposed to SA. Accompany with the increase in H_2O_2, antioxidant enzymes including catalase, ascorbate peroxidase and guaiacol peroxidase were activated, showing a dramatic increase in SA-applied plants. Exogenous SA application highly increased proline in the presence of sulphur (S). Transcript levels of pyrroline-5-carboxylate synthetase 1 (*P5CS1*) and proline dehydrogenase (*PDH*) were up-regulated by SA. The results suggested that SA might enhance oxidative stress resistance by activating proline synthesis and antioxidant enzymes.

T4

A382

Control of seed size by SOD7 and its interacting protein (SIP1) in *Arabidopsis thaliana*

Yueying Zhang[1,2], *Liang Du*[1], *Ran Xu*[1], *Rongfeng Cui*[1], *Yunhai Li*[1]

1. State Key Laboratory of Plant Cell and Chromosome Engineering, Institute of Genetics and Developmental Biology

2. University of Chinese Academy of Sciences

Although the size of seeds is one of the most important agronomic traits in plants, the genetic and molecular mechanisms that set the final size of seeds are largely unknown. We have previously demonstrated that the Ubiquitin receptor DA1 controls seed size through restricting cell proliferation. Through a genetic screening for *da1-1* modifiers, we have identified a suppressor of *da1-1 (sod7)*. *SOD7* encodes a B3 domain transcriptional repressor NGAL2, acts maternally to control seed size by restricting cell proliferation in the integuments of ovules and developing seeds. Overexpression of *SOD7* significantly decreases seed size of wild-type plants, while the simultaneous disruption of *SOD7* and its closest homolog *DPA4/NGAL3* increases seed size. Genetic analyses indicate that SOD7 and DPA4 act in a common pathway with the seed size regulator KLU to control seed growth, but do so independently of DA1. SOD7 directly binds to the promoter of *KLU in vitro* and *in vivo* and represses expression of *KLU*. We further identify two SOD7-Interacting Proteins (SIPs), SIP1 and SIP2. Genetic analyses indicate that *SOD7* functions in the same pathway with *SIP1* to control seed and organ size. Therefore, our findings reveal the genetic and molecular mechanisms of SOD7, DPA4, KLU and SIP1 in seed and organ size control, suggesting that these pathways are promising targets for crop yield improvement.

T4

A384

Contribution of the alternative respiratory pathway to PSII photoprotection in C_3 and C_4 plants

Zishan Zhang[1], *Meijun Liu*[2], *Huiyuan Gao*[1]

1. Shandong Agricultural University

2. Xinjiang Agricultural University

The mechanism by which the mitochondrial alternative oxidase (AOX) pathway contributes to photosystem II (PSII) photoprotection is in dispute. It was generally thought that the AOX pathway protects photosystems by dissipating excess reducing equivalents exported from chloroplasts through the malate/oxaloacetate (Mal/OAA) shuttle and thus preventing the over-reduction of chloroplasts. In this study, using the aox1a Arabidopsis mutant and nine other C_3 and C_4 plant species, we revealed an additional action model of the AOX pathway in PSII photoprotection. Although the AOX pathway contributes to PSII photoprotection in C_3 leaves treated with high light, this contribution was observed to disappear when photorespiration was suppressed. Disruption or inhibition of the AOX pathway significantly decreased the photorespiration in C_3 leaves. Moreover, the AOX pathway did not respond to high light and contributed little to PSII photoprotection in C_4 leaves possessing a highly active Mal/OAA shuttle but with little photorespiration. These results demonstrate that the AOX pathway contributes to PSII photoprotection in C_3 plants by maintaining photorespiration to detoxify glycolate and via the indirect export of excess reducing equivalents from chloroplasts by the Mal/OAA shuttle. This new action model explains why the AOX pathway does not contribute to PSII photoprotection in C_4 plants.

T4

A385

Protonema development and gametophyte regeneration of two species of *Bryum* Hedw.

Hong-Zhu Liang[1,2], *Jian-Cheng Zhao*[1], *Lin Li*[1], *Yan-Yan Shang*[1]

1. Hebei Normal University

2. University of Chinese Academy of Sciences

Bryophytes are commonly known as the representative transitional plants from aquatic to terrestrial habitats, which have only simple formation and have not true root and vascular bundle, preferring to aquatic circumstance. Bryophytes, one of the largest groups of the earth plants, contain more than 23,000 species throughout the world. But hitherto, the bryophyte flora is verified continually impoverished throughout the earth. Generally, axenic culturing is regarded as a common propagation phase for the ex situ conservation of bryophytes. Over the recent years, in vitro propagation, one of the typical systems to study morphogenetic models of bryophytes has attracted increasing attention, because of their biological traits and characteristic haploid life style.

In this research, we selected the mature capsules of *Bryum coronatum* and *B. recurvulum* as the research materials, which were collected from the Xiaowutaishan National Nature Reserve of Hebei Province and Fugong County of Yunnan Province, and now they are conserved in the Herbarium of Hebei Normal University (HBNU). After collecting of the materials, firstly they were identified for taxonomic evaluation, then they would be washed and dried by airing, and lastly they would be conserved in the refrigerator by 4°C and -40°C for spores germinating and protonema development purpose. This paper described the protonema development features of *B. coronatum* and *B. recurvulum*, from spore germination to young gametophyte, by solid and liquid cultivating in laboratory. Spores samples for gametophyte cultivation were taken from various mature sporophytes then sterilized in 75% alcohol solution, washed by distilled water, and lastly sown on Knob's medium. Spores were sowed both in the culture dishes (solid medium) and in the conical flasks (liquid medium). Spore germination of *B. coronatum* and *B. recurvulum* are both of the Bryum-type. Spore germination positions of the two species are both locating inside the cell wall. Gametophyte produces both on chloronema and on caulonema, without rhizoids occurring. Hollow cells occurs only on chloronema, where one filiform protonema cracking into two. Gametophytes occurred about 30 days for *B. coronatum* and 40 days for *B. recurvulum* after sowing. The protonemata system occurred on the 16th day after sowing in the liquid culture of *B. coronatum*, the spore germination was 1-2 polar germinating. In comparison with the solid culture, hollow cells and browning cells are more common in the liquid culture. Gametophyte occurred about 80 days after the sowing. In brief, it made it possible to large scale of rapid propagation of bryophyte by the protonema development and gametophyte cultivation research. The taxonomic significances of the protonemata system and gametophyte morphology were discussed in this paper.

T4

A386

Product from programmed cell death involved in rubber granule synthesis in laticiferous canal of *Decaisnea fargesii*

Yafu Zhou[1], *Shaoli Mao*[1], *Xinwei Shi*[1], *Yuchao Wang*[1], *Wenzhe Liu*[2]
1. Institute of Botany of Shaanxi Province
2. Key Laboratory of Resource Biology and Biotechnology in Western China, Ministry of Education, School of Life Science, Northwest University

Controversies have long been dedicated to the manner of rubber granule synthesis. *Decaisnea fargesii* is a latex-producing plant that is characterized by the presence of a special type of secretory structure in the pericarp, the laticiferous canal, which has been confirmed to be a typical programmed cell death process, during which the rubber granules occur and accumulate. In this paper, a combination of light microscopy, scanning electron microscope, transmission electron microscopy and histochemistry are used to detect the relationship between rubber granule synthesis and programmed cell death in laticiferous canal of *Decaisnea fargesii*. During the early developmental stage of the laticiferous canal, little osmiophilic flocculent material or membranous mass structures have been detected in the secretory cells, which expand by combining with the free material from cytoplasm. In the degeneration stage (especially in sunken stage and expanding stage), the precursors of rubber granules are characterized by the osmiophilic flocculent material or membranous mass structures which are close to degenerated plastids, mitochondria, or on the end of swelling endoplasmic reticulum. Sequentially degenerated cellular components integrate into the rubber precursors, and the volume of the rubber granules expands in this process. The rubber precursors or rubber granules are also accumulated within the central vacuole by autophagy, and integrate with the vesicular, flocculent and mass structures continually, even the organelle enveloped by the rubber granule was also observed, finally shapely mature rubber granules form. It could be concluded that synthesis of the rubber precursor or rubber granule in laticiferous canal of *Decaisnea fargesii* is a random event and are not associated with a specific organelle or stage. And components from cellular disintegration are mainly used to synthesize rubber granules during the PCD process, which provides new evidences for rubber granule synthesis in plant.

T4

A387

Unsaturation of very-long-chain ceramides protects plant from hypoxia-induced damages by modulating ethylene signaling in *Arabidopsis*

Lijuan Xie, *Ying Zhou*, *Shi Xiao*
Sun Yat-Sen University

Lipid remodeling is crucial for hypoxic tolerance in animals, whilst little is known about the hypoxia-induced lipid dynamics in plants. Here we performed a mass spectrometry-based analysis to survey the lipid profiles of *Arabidopsis* rosettes under various hypoxic conditions. We observed that hypoxia caused a significant increase in total amounts of phosphatidylserine, phosphatidic acid and oxidized lipids, but a decrease in phosphatidylcholine (PC) and phosphatidylethanolamine (PE). Particularly, significant gains in the polyunsaturated species of PC, PE and phosphatidylinositol, and losses in their saturated and mono-unsaturated species were evident during hypoxia. Moreover, hypoxia led to a remarkable elevation of ceramides and hydroxyceramides. Disruption of ceramide synthases LOH1, LOH2 and LOH3 enhanced plant sensitivity to dark submergence, but displayed more resistance to submergence under light than wild type. Consistently, levels of unsaturated very-long-chain (VLC) ceramide species (22:1,24:1 and 26:1) predominantly declined in the *loh1*, *loh2* and *loh3* mutants under dark submergence. In contrast, significant reduction of VLC ceramides in the *loh1-1 loh3-1* knockdown double mutant and lacking of VLC unsaturated ceramides in the *ads2* mutants impaired plant tolerance to both dark and light submergences. Evidence that C24:1-ceramide interacted with recombinant CTR1 protein and inhibited its kinase activity *in vitro*, enhanced ER-to-nucleus translocation of EIN2-GFP and stabilization of EIN3-GFP *in vivo*, suggests a role of ceramides in modulating CTR1-mediated ethylene signaling. The dark submergence-sensitive phenotypes of *loh* mutants were rescued by a *ctr1-1* mutation. Thus, our findings demonstrate that unsaturation of VLC ceramides is a protective strategy for hypoxic tolerance in *Arabidopsis*.

T4

A388

Genome-wide association study identifies main factors con-

trolling the anthocyanidin accumulation in rice leaves

Jie Zheng, *Zaihui Zhou, Hao Wu, Huabing Zhu, Zijin Weng, Hao Chen*
Huazhong Agricultural University

Genome-wide association study (GWAS) has been widely used to detect associations between genes and crop traits. Anthocyanidin is a natural pigment for protection from damage by biotic and abiotic stresses and for reproduction. The synthesis of anthocyanidin in plants mainly involves 7 to 8 structural genes and several regulatory transcription factors. The anthocyanidin contents in leaves of 49 wild rice species were analyzed using high performance liquid chromatography (HPLC). The result showed that anthocyanidin accumulation was detected more or less in most wild rice species, indicating that the anthocyanidin accumulation is a common phenotype in wild rice. However, purple leaf is a rare phenotype in cultivated rice, demonstrating it is a trait with artificial selection during rice demonstration. The anthocyanidin in leaves of a worldwide germplasm collection consisting of 533 rice landraces were measured by HPLC, and then GWAS was conducted. The result would help illustrate the key factors under artificial selection for non-anthocyanidin phenotype during rice demonstration.

T4

A389

A nucleocytoplasmic trafficking WD40 protein regulates ABA signaling through affecting ABI5 stability

Guobin Jiang, Wang Wan, Xuezhong Xu, ***Guohui Zhu***
College of Life Sciences, South China Agricultural University

Nucleocytoplasmic trafficking of proteins is being recognized as an important step in plant signaling pathways, including responses to hormone, abiotic and biotic stresses and environmental changes. Here, we report a nucleocytoplasmic trafficking WD40 protein directly interacts with ABI5 and affects ABI5 stability, and therefore regulating ABA signaling. Quantitative RT-PCR analysis and GUS staining analysis showed that WD40 protein was ubiquitously expressed, with relative higher expressions in the seedlings and inflorescence of mature plants. Expression level of *WD40* was continuously decreased during seed germination and seedling development, and its transcripts were induced upon exposure to exogenous ABA. Compared with Col-0, *wd40* mutant seemed less sensitive to ABA treatment during seed germination and seedling development. Moreover, WD40 affects ABA-induced stomatal closing and thus positively regulates drought stress tolerance. We also determined the expressions of several typical ABA responsive genes in the Col-0 and wd40 mutant. Expression levels of *ABI3, ABI5, EM1* and *EM6* were less increased in the wd40 mutant compared with Col-0 in response to 0.5μM ABA. Similarity, expressions of *RD29A, RD29B* and *ABI5* were less up-regulated in the wd40 seedlings when treated by 50μM ABA, these results indicated WD40 protein positively regulates ABA signaling. Subcellular localization analysis showed that WD40 protein is a nucleocytoplasmic localized protein. Yeast two-hybrid, BiFC and Co-IP analyses showed that WD40 directly interacts with XPO1A, an importin β-like nuclear transport receptor. WD40 protein was obviously accumulated in nucleus in tobacco cells after leptomycine B (LMB) treatment, suggesting WD40 protein directly interacts with and nuclear exported by XPO1A. Since WD40 is a nucleocytoplasmic localization protein and is involved in ABA response, we have interests to know if the nucleocytoplasmic participation of WD40 is affected by ABA. GFP signals were significantly enhanced in the nucleus of WD40-GFP transgenic plant cells in response to ABA treatment. Western blot analysis confirmed that ABA treatment increased GFP abundance in the nucleus. These date suggested that ABA regulates the nucleocytoplasmic participation of WD40. Interestingly, environmental stresses, i.e. salt and osmotic stress, also enhanced GFP intensities in the nucleus, implying the nuclear retention of WD40 may be a universal mechanism for plants response to environmental stress. Yeast two-hybrid analysis showed that WD40 directly interacts with ABI5, an important transcript factor positively regulating ABA signaling. Western blot showed that *wd40* mutants had less ABI5 protein in response to ABA compared with Col-0 seeds. Moreover, ABI5 proteins were degraded much faster in *wd40* than in Col-0 after removal of ABA treatment. These data indicated WD40 directly interacts with ABI5 and affects ABI5 stability, and therefore regulating ABA signaling.

T4

A390

SCF^{COI1}-mediated jasmonate signaling represses COP1 for inhibiting skotomorphogenesis in *Arabidopsis*

Ziqiang Zhu
Nanjing Normal University

A germinating seedling undergoes skotomorphogenesis to emerge from soil by reaching for light. During this phase, the cotyledons are closed and the hypocotyl elongates. Upon exposure to light, the seedling rapidly switches to photomorphogenesis by opening its cotyledons and suppressing hypocotyl elongation. The RING type E3 ubiquitin ligase CONSTITUTIVE PHOTOMORPHOGENIC 1 (COP1) is critical for maintaining skotomorphogenesis. Here we show that jasmonate (JA) suppresses skotomorphogenesis (JSS) through the SCF^{COI1}-mediated JA signaling. Arabidopsis seedlings treated with JA phenocopy *cop1* mutants in the dark. Then we check the stabilities of several COP1-targetted transcription factors, including HY5 and PAR1/PAR2, and find that JA stabilizes them in a COP1-dependent manner. Seedlings impaired in the expression of *HY5* or overexpression of *PAR1/PAR2* are hyposensitive or hypersensitive to JA respectively, which suggests that these transcription factors contribute to JSS. We further demonstrate that JA suppresses COP1 activity probably through at least two distinct mechanisms: decreasing COP1 protein accumulation in the nucleus and reducing the physical interaction between COP1 and its activator, SUPPRESSOR OF PHYTOCHROME A 1 (SPA1). Our work reveals that a defense hormone triggers the switch from skotomorphogenesis to photomorphogenesis through the tandem actions of two evolutionally unrelated E3 ubiquitin ligases (SCF^{COI1} and COP1).

T4

A391

The distribution of characteristic events of programmed cell death (PCD) and its main regulators during the petal senescence of *Gardenia jasminoides*

Jingjing Zou[1], *Xuan Cai*[2], *Xiangling Zeng*[1], *Hongguo Chen*[1], *Caiyun*

Wang[2]
1. School of Nuclear Technology and Chemistry & Biology, Hubei University of Science and Technology, Xianning, China
2. Key Laboratory of Horticultural Plant Biology, Ministry of Education, Huazhong Agricultural University, Wuhan, Hubei province, China

Programmed cell death (PCD) is a genetically controlled program involving an ordered set of events, which provides a model system to predict exactly when and how the flowers will senesce. In the present study, six developmental stages (stage 1-6) of *G. jasminoides* var radicans were selected to discuss the spatial distribution of PCD across the petal development and senescence. The results show that the DNA degradation, reactive oxygen species (ROS) generation and lipid peroxidation all started late at the veraison stage. However, the rapid increase of ethylene production, decrease of adenosine triphosphate (ATP) levels and the content of soluble sugar were found to occurred since the flower opening, which implied an early regulatory role of ethylene and source of energy during the petal senescence.of *G. jasminoides* var radicans. Thus, cut flowers of *G. jasminoides* var radicans were treated with ethephon, silver thiosulphate (STS) and sugar to further investigate the role of ethylene and energy loss in flower senescence. The results showed that the vase life of cut flowers was significantly prolonged 28.8 h and 16.8 h by 0.2 mM STS and 1% sugar treatments respectively, but reduced 28.8 h and 14.4 h by 50 mg L^{-1} ethephon and 5% sugar treatments, compared to distilled water. Furthermore, the endogenous ethylene production was increased by ethephon and 5% sugar treatments but reduced by STS and 1% sugar treatments throughout vase life. It is concluded that *G. jasminoides* is an ethylene sensitive flower and the petal senescence is regulated by the endogenous ethylene production. Both ethylene and sugar are important regulators during the petal senescence of cut *Gardenia* flowers, and sugar play the regulatory role by increasing the endogenous ethylene production. This work would offer new theoretical basis for postponing the senescence of woody flowers, and provide the academic and practical basis for selecting the long-blooming resources of *Gardenia* flowers.

T4

A392

Effect of light and temperature on the characteristic color compounds and aroma compounds in *Osmanthus fragrans*

Jingjing Zou[1], *Xuan Cai*[2], *Xiangling Zeng*[1], *Hongguo Chen*[1], *Caiyun Wang*[2]
1. School of Nuclear Technology and Chemistry and Biology, Hubei University of Science and Technology, Xianning, China
2. Key Laboratory of Horticultural Plant Biology of Ministry of Education, College of Horticulture and Forestry Sciences, Huazhong Agricultural University, Wuhan, China

Sweet osmanthus (*Osmanthus fragrans*), belonging to the family Oleacea, is an important ornamental and economical woody plant that has been commercially cultivated in China for over 2,500 years. As one of the top ten traditional flowers in China, it is not only widely applied as avenue tree and shade tree in landscape, but also a famous aromatic plant to produce commercial ingrediants for tea, food, spice, wine and soft beverages industries. Flower opening of *O. fragrans* after flower bud differentiation is usually conducted by a period of cold accumulated temperature in. During the flower opening, environment factors, such as the temperature and light conditions, that have great fluctuations will significantly affect flower scent and flower color. However, how and which of color and aroma compounds have been changed was not clear. In our previous study, the characteristic color and aroma compounds, which have contributed most effectively to the flower aroma and flower color have been identified by GC-MS and HPLC-MS during the flower opening. In the present study, changes of these characteristic flower color and aroma compounds treated in different light and temperature were further analyzed, using *O. fragrans* 'Houban Yingui' as the plant material. The results showed that, under light treatment, the quercetin-O-glycoside content was higher than in dark treatment and the content of naringenin-O-glycoside was lower, resulting in paler color in light and deeper color in dark. Under dark treatment, linalool and its oxides and trans-β-ocimene contents were decreased; cis-β-ocimene and neo-allo-ocimene were not dectected. Under the low temperature treatment, the contents of quercetin-O-glycoside and naringenin-O-glycoside were higher than in high temperature treatment, resulting in brighter color in low temperature and paler color in high temperature. High temperature accelerated the senescence of corolla lobes, and promoted the release of cis-3-hexenyl butanoate and hexyl butanoate. Low temperature increased the contents of trans-β-ionone, α-ionone and many monoterpenes and their oxides.

T4

A393

Comparative analysis of miRNAs and their targets during blueberry fruit maturation using small RNA and degradome sequencing

Yanming Hou, Xin Xie, Lulu Zhai, Xuyan Li, ***Shaomin Bian***
College of Plant Science, Jilin University

MicroRNAs (miRNAs) are a class of endogenous, non-coding, small RNAs, which regulate gene expression mainly via cleavage-induced degradation of mRNAs in plants. Increasing evidences indicate that miRNAs play vital roles in the regulation of fruit development and ripening. Blueberry is an important small berry fruit crop with economical and nutritional value. However, it is totally unknown about the miRNAs and their targets involved in blueberry fruit development and maturation. In this study, using high-throughput sequencing of small RNAs and degradome fragments, 84 known miRNAs belonging to 28 families and 16 novel miRNAs were identified in white fruit (WF) and blue fruit (BF) libraries, which represent fruit ripening onset and in-progress, respectively. Among them, 41 miRNAs were shown to be differentially expressed during fruit maturation; and 16 miRNAs representing 16 families were further chosen to validate the sRNA sequencing data by stem-loop qRT-PCR. Meanwhile, 178 targets were identified for 41 known and 7 novel miRNAs in WF and BF libraries, and targets of miR160 were validated using RLM-RACE approach. Moreover, the expression patterns of 6 miRNAs and their targets were examined during fruit development and ripening. Finally, integrative analysis of miRNAs and their targets revealed a complex miRNA-mediated regulation network involving a wide variety of biological processes such as auxin signal pathway and hormonal crosstalk, cell enlargement, and fruit coloration during blueberry fruit ripening. The findings will greatly facilitate future investigation of the mechanisms that regulate blueberry fruit development and ripening.

T4

A394

Genome-wide analysis of *DUF221* gene family in *Glycine soja*

Jianwei Li, *Junkai Yang, Bowei Jia, Mingzhe Sun, Xiaoli Sun*
Heilongjiang Bayi Agricultural University

DUF221 belongs to the transmembrane calcium-permeable cation channel proteins. It plays vital roles in hyperosmotic stress responses. However, up to now, little studies were reported about the *DUF221* gene family, especially in legume crops. Hence, we used bioinformatic method to screen *DUF221* family genes in soybean genome, and identified a total of 21 *GmDUF221* genes. Chromosomal location and synteny analysis revealed that 21 *GmDUF221s* located on 15 chromosomes, and possessed 8 gene duplication events. Analysis of the Ka (non-synonymous substitution rate)/Ks (synonymous substitution rate) ratios suggested that the duplication of soybean *DUF221*genes mainly underwent purifying selection. We further constructed a phylogenetic tree, and found that they were clustered into four subfamilies (I-IV), and members within each subfamily displayed similar intron-exon organization pattern. Furthermore, we also suggested that soybean DUF221 family protein sequences were highly conserved, including three conserved functional domains (Late exocytosis, Cytosolic domain of 10TM putative phosphate transporter and Calcium-dependent channel), 25 conserved motifs and at least 6 transmenbrane domains. Expression patterns analysis indicated that soybean *DUF221* family genes were differentially expressed in various tissues. Among them, *GmDUF221j-2* was highly expressed in flower, leaf and root. These results could provide new valuable information and will facilitate further research.

T4

A509

Phosphoproteomics unveils stable energy supply as key to flooding tolerance in *Kandelia candel*

Dezhuo Pan[1], Lingxia Wang[1,2], Fanlin Tan[3], Shipin Chen[4], Chi-Lien Cheng[5], ***Wei Chen***[1]
1. College of LIfe Sciences, Fujian Agriculture andForestry University, Fuzhou 350002, P. R. China
2. College of Life Sciences, Ningxia University, Yinchuan 750021, P. R. China
3. Fujian Academy of Forestry, Fuzhou 350012, P .R. China
4. College of Forestry, Fujian Agriculture and Forestry University, Fuzhou 350002, P. R. China
5. Department of Biology, The University of Iowa, 210 Biology Building, Iowa City, Iowa 52242, United States of America

The woody mangrove *Kandelia candel* experiences daily flooding cycles in its natural habitat. To explore molecular mechanisms for physiological adaptation to flooding, the role of protein phosphorylation that may play in *K. candel*'s flooding response was investigated by a large-scale quantitative phosphoproteomic analysis using isobaric tag for relative and absolute quantitatiion (iTRAQ). A total of 2,141 unique phosphopeptides and 2,603 non-redundant phosphorylation sites were identified from 1,516 phosphoproteins in *K. candel* leaves. In addition to known phosphorylation motifs, three new motifs [GSP], [GxxSP], and [RSxS] were discovered. Ninety-six differentially expressed phosphoproteins including those involved in pyruvate metabolism and energy production altered their phosphorylation levels in response to flooding. The physiological parameters relevant to flooding response including photosynthesis, pyruvate metabolism, and ROS production were investigated and all found to be robust under flooding conditions. The consistent results between the physiological analysis and the phosphoproteomics reinforce each other to demonstrate that the physiology of *K. candel* is adapted to flooding and that protein phosphorylation plays a role in this adaptation.

T4

A517

A putative MYB-like transcription factor involved in secondary cell wall formation in *Arabidopsis*

Chunming Wang, *Meihui Yu, Junxian He*
School of Life Sciences and State Key Laboratory of Agribiotechnology, The Chinese University of Hong Kong

Secondary cell walls (SCWs) are important for plants to support their growth and adapt to changing environments. SCWs are composed mainly of three biopolymers, namely lignin, cellulose and hemicellulose. Studies have suggested that the biosynthesis of these biopolymers is regulated by a complex transcriptional network. However, the details of this transcriptional network have not been fully uncovered. From an activation tagging based genetic mutant screen in *Arabidopsis*, we identified a new mutant, *rif1* (*Respressor of Interfascicular Fibers 1*), which has severe defects in secondary wall formation in stems, making the plant cannot stand erectly. Microscopy study indicated that the SCW deposition in interfascicular fibers of *rif1* is completely missing, and SCW components assay showed that the levels of the three biopolymers are all significantly decreased in the *rif1* mutant. *RIF1* encodes a MYB like transcription factor in *Arabidopsis* which has not been characterized before. RIF1 is localized in nucleus and has putative transcriptional regulatory activity. From a protoplast transcription transient assay, we demonstrate that RIF1 acts as a transcriptional repressor and negatively regulate several important SCW biosynthesis genes. Our results suggest that RIF1 may define a novel negative transcription regulator in SCW formation in plants. This work was supported by the GRF grants (14121915 and 14148916) and the AoE grants (AoE/M-05/12, AoE/M-403/16) from the Research Grant Council of Hong Kong, the Shenzhen Science & Technology Research & Development Funding - Peacock Scheme, and the Direct Grants from The Chinese University of Hong Kong.

T4

A524

Drought stress release increased growth rate but did not affect levels of storage carbohydrates in scots pine trees

Leonie Schönbeck, *Arthur Gessler, Andreas Rigling, Marcus Schaub, Mai-He Li*
Swiss Federal Institute for Forest, Snow and Landscape Research

(WSL)

For trees, energy storage in the form of non-structural carbohydrates (NSCs) plays an important role for survival and growth, especially during stress events such as drought. It is hypothesized, that tree individuals that experience long-term drought stress use up larger amounts of NSCs than trees that do not experience drought. Consequently, such drought-induced depletion might lead to a decrease in tree vigor and carbon starvation, a mechanism that is subject of intensive debates in recent literature. Hence, if carbon starvation is occurring during drought, drought stress release should again increase NSC concentrations. A long-term (13 years) irrigation experiment is being conducted in the Pfyn forest, the largest *Pinus sylvestris* dominated forest in Switzerland, located in the dry inner-Alpine Swiss Rhone valley (average precipitation ~600 mm/year, with frequent dry spells). Water addition (~600 mm/year) is executed every year during the growing season between April and October. Tree height, stem diameter and crown transparency are being measured since 2003. In February, July and October 2015, roots, stem sapwood and needles were harvested from 30 irrigated and 30 control trees and 5 different crown transparency classes. Shoot length, needle morphology, soluble sugars, starch concentrations, needle $\delta^{13}C$ and $\delta^{15}N$ were measured. Shoot and stem growth were higher in irrigated trees than in control trees. Growth decreased with increasing crown transparency in both treatments. Only in July, needle starch levels were higher in irrigated trees than in control trees but there was no treatment effect for wood and root starch concentrations. Tissue starch and sugar levels were negatively correlated with crown transparency, particularly in the roots ($p<0.001$), independent of the treatment. Needle $\delta^{13}C$ values were higher in the control trees than in the irrigated trees, where needle $\delta^{13}C$ values were positively correlated with increasing transparency ($p<0.01$). Annual shoot growth was positively correlated with starch levels in the roots. The results show that 13 years of irrigation did lead to increased growth but not to increased NSC levels hence not confirming our initial hypothesis. $\delta^{13}C$ levels indicate that control trees experienced more drought stress than irrigated trees. However, we found irrigated trees from high crown transparency classes with similar $\delta^{13}C$ levels as for non-irrigated control trees. The release of drought stress has benefited the initially vital trees, whereas the initially inferior trees still show signs of drought stress. The results point to a 'winner takes it all principle', where differences between individuals increase when environment conditions improve. This caused the irrigation treatment not being effective in generally releasing drought stress and NSC depletion in all trees. As increasing crown transparency over both treatments is correlated with decreasing growth and decreasing NSC levels, there are still indications that reduced NSC is related to reduced tree vigor under drought.

T4
A526
A journey of evolution: Phytohormones responses in wound-induced regeneration of land plants

Regina Yeo
National Institute of Education

The evolutionary transition of plants from the aquatic environment to the terrestrial environment is one of the major events in the history of life. On land, plants are highly vulnerable to attack and wounding by herbivores. Thankfully, plants have remarkable regenerative abilities due to their high degrees of developmental plasticity. As such, plants have long exploited the mechanism of regeneration as a response to injuries via regenerating wounded tissues or lost structures. The replacement of a wounded or lost structure in plants can occur via the development of an intermediate mass of specialised tissue, called callus. Phytohormones are known to play vital roles in modulating plant regeneration where optimal plant fitness relies on the regulation of these phytohormones that link damage-associated signals to appropriate changes in growth and development for regeneration to ensue. Although extensive studies have been carried out on the relations of phytohormones to plant development and growth, the evolutionary origin of phytohormone signalling in plant regeneration remains elusive. Furthermore, it is important to note that there are marked differences in basal plants and the higher plants with regards to their initiation, processes and capabilities of regeneration. Through evolution, plants have evolved sophisticated and highly regulated regeneration systems to defend against attackers or sustained injuries. Here, we investigate selected phytohormone signalling pathways to determine their engagement across the plant lineages, from the evolutionary lower to higher plants, in the process of wound-induced regeneration.

T4
A532
Enhancement of *Vicia faba* drought tolerance by gamma irradiation

Mona G. Dawood[1], *H. Kamel* [2], *Magdi T. Abdelhamid* [1]
1. National Research Centre, Cairo, Egypt
2. Nuclear Resrarch Center, Cairo, Egypt

For ameliorating the negative effects of moderate and severe drought stress on faba bean plants, two pot experiments were conducted to examine the effect of different doses of gamma radiation (10,25 and 50 Gy). Data revealed that 10 and 25 Gy enhanced germination percentage, fresh and dry weight of seedling, meanwhile 50 Gy showed inhibition effects. Drought decreased plant height, branches number/plant, leaflet length, specific leaf area (SLA), leaf weight ratio (LWR), net assimilation rate (NAR) and relative growth rate (RGR) accompanied by significant increases in soluble protein, phenolic content, proline and free amino acid in dry faba bean leaves. In well watered plants, 10 and 25 Gy caused marked increases in plant height, leaflet lengths, SLA, NAR and RGR relative to control, meanwhile, 50 Gy showed opposite trend. In drought stressed plants, 10 and 25 Gy caused gradual increases in plant height, whereas, 50 Gy decreased plant height relative to corresponding control. Otherwise, leaflet lengths of plants undergo medium or severe drought showed variable responses under the effect of different doses of gamma radiation. Plants undergo severe drought showed that all doses of gamma radiation increased SLA and LWR and decreased NAR and RGR. All doses of gamma radiation caused significant increases in phenolic content, proline and free amino acids, accompanied by significant decreases in soluble protein percentage in drought stressed plants. It could be concluded that 25 Gy was the optimum dose in ameliorating harmful effects of drought stress on faba bean plants.

T4

A546

Two SUMO proteases SUMO PROTEASE RELATED TO FERTILITY 1 and -2 are required for fertility in *Arabidopsis*

Yong-Fu Fu
Institute of Crop Sciences, Chinese Academy of Agricultural Sciences

In *planta*, SUMO is proved to be implicated in extensive cellular processes, including flowering-time control and responses to biotic and abiotic stresses. Here, we report that two proteases, SUMO PROTEASE RELATED TO FERTILITY (SPF) 1 and 2, regulate the fertility in *Arabidopsis thaliana*. The *spf1* single mutant exhibits abnormalities in flower structure and embryos, while the *spf2* single mutant exhibits largely wild type. However, *spf2* enhances the phenotype of the *spf1* mutant, so that the *spf1 spf2* double mutant exhibits severe abnormalities in micro- and mega-gametophytes and embryos, suggesting the two genes function redundantly. *SPF1* and *SPF2* genes are expressed highly in reproductive tissues, and impairment of them results in misexpression of generative- and embryo-specific genes. The SPF1 and SPF2 proteins display *in vitro* the activity of a SUMO-processing protease, while SPF1 displays desumoylating activity *in vivo* on its potential substrate EDA9 (EMBRYO SAC DEVELOPMENT ARREST9). The SUMO-conjugates accumulate at higher levels in the inflorescences, but not in seedlings, in the *spf1 spf2* double mutant than in the wild type. SPF1 and SPF2 also display their own specificity on substrates and interacting proteins. Our results indicate that SPF1 and SPF2 are two SUMO proteases contributed to fertility in *Arabidopsis*.

T5

A395

Diversity and seasonality of arbuscular mycorrhizal Fungi of *Artemisia halodendron* community in Horqin sandy land

Tebuqin Borjigin[1,2], ***Yuying Bao***[1]
1. Inner monglia university
2. Inner mongolia medical university

Artemisia halodendron is an important plant in Horqin sandy land and plays an important role in the stability and development of local plant communities. But its dominance shows a declining trend in the succession process of shifting sandy land to the fixed sandy land. In the 6-9 months of 2006, in the Naiman dune belongs to Horqin sandy land, we studied the shifting sandy land, semi fixed sandy land and fixed sandy land, and collect the rhizosphere soil samples of *Artemisia halodendron* in 0-5 cm, 5-15 cm, 15-30 cm layers, and study the mycorrhizal infection rate and AM fungi diversity. The results showed that the roots of *Artemisia halodendron* were I- type (Intermediate type), the highest colonization rate and the maximum spore density appeared in the 0-5 cm soil layer, and decreased with soil depth. We identified 5 fungal genera and 21 species of AM fungi in the rhizosphere of *Artemisia halodendron*, of which 12 species belonged to Glomus, 3 species belonged to Diverspora, 3 species belonged to Gigaspora, 2 species belonged to Acaulospora, 1 species belonged to Entrophospora. There were 11 kinds of species appeared in all kinds of sandy land, they were *Acaulospora rehmii, Acau. scrobiculata, Diversispora etunicatum, Dive. spurcum, Glomus claroideum, G. constrictum, G. microaggregatum, G. tortuosum, Scutellospora calospora, Scut. pellucida, Scut. persica. Entrospora infrequens,* and *Glomus intraradices* only appeared in fixed sandy land, *Glomus fasciculatum and G. luteum* only appeared in the semi fixed sandy land, *Glomus deserticola, G. geosporum, G. hyderabadensis, G. manihotis* and *G. mosseae* don't appear in shifting sandy land. The dominant genera were the *Gigaspora*. From the sandy land type, AM fungi species were 11 in shifting sandy land, 19 in semi fixed sandy land, 19 in fixed sandy land. For one month the AM fungi species number were, fixed sandy land>semi fixed sand> shifting sandy land. It implied a close relationship between plant degeneration and AM fungal community changes. With the fixed degree of sandy land the AM fungal communities increased, and reduced the dominance of *Artemisia halodendron* population. The results provide a basis for studying the relationship between plant community and ecological distribution of AM fungi.

T5

A396

Metabolomic analysis of alfalfa (*Medicago sativa* L.) root with symbiotic rhizobia under alkali stress

Hua Cai*, Tingting Song, Huihui Xu, Liu Jiang*
Northeast Agricultural University

With respect to salinity stress, alkaline salt causes more severe damage to plants than neutral salt, due to the additional effect of pH stress. However, the mechanism by which plants respond to alkali stress is not fully understood. Here, we studied the effects of root nodule symbiosis on alkali stress in alfalfa using comparative metabolomics. *Rhizobium*-nodulized (RI group) and non-nodulized (NI group) alfalfa were treated with 200 mmol/L $NaHCO_3$ and root samples were subjected to malondialdehydyde (MDA), proline, glutathione (GSH), superoxide dismutase (SOD), and peroxidase (POD) content analysis, as well as metabolite profiling with gas chromatography combined with time-of-flight mass spectrometry (GC/TOF-MS). Phenotypically, the RI alfalfa exhibited a greater resistance to alkali stress than NI plants. Physiological analysis and metabolic profiling revealed that RI plants accumulated more antioxidants (SOD, POD, GSH), osmolytes (sugar, glycols, proline), organic acids (succinic acid, fumaric acid, and alpha-ketoglutaric acid), and metabolites that are involved in nitrogen fixation. Our pairwise metabolomics comparisons revealed that RI alfalfa exhibited a distinct metabolic profile in conferring alkali resistance relative to NI alfalfa. Our study provides new information about the relationship between non-nodulized, rhizobium-nodulized alfalfa and alkali resistance, giving valuable insight into the mechanisms mediating alkali resistance in legume crops.

T5

A397

Molecular regulation mechanism of cold resistance response of MiR398 interacts with cold-resistant lncRNA in winter wheat

Jing Cang*, Qiuwei Lu*
Northeast Agricultural University

Dongnongdongmai 1 (dn 1) was the first freezing-tolerant (resistant to low temperature of -30 ℃) winter wheat that can live through the winter in extreme frigid region in Heilongjiang Province. miRNAs targets was regulated by miRNA, then initiated the cold resistant response. lncRNA (long noncoding RNA) competitivly binded miRNA and coregulated the expression level of miRNA targets. Our previous study indicated that during the natural cooling process the expression of miR398 and its target gene CSD in winter wheat was significantly different.The expression level of miRNA was also affected by the expression level of some noncoding RNA, such as lncRNA. So the cold resistance of winter wheat was affected, but the mechanism was not cleared. dn1 was used as materials, the rhizomes were sampled under natural cooling condition in fields. Using high throughput sequencing, molecular biology and bioinformatics technology combined with cell biology and physiological and biochemical analysis method, MiR398 and the interaction of lncRNA in the cold function, expression regulation model were studied, the members of ceRNA (lncRNA-miRNA-target mRNA) were determined and their function was identified.It is revealed that the complex biological regulation network between miR398 and lncRNA and target mRNA provides a new idea for the study of regulating mechanism of winter wheat miRNA, At the same time, miRNA-lncRNA can affect the morphological characteristics and growth and development of plants, and can also provide theoretical guidance for the cultivation of crops.

T5

A398

Genetic and pathogenic diversity of *Ralstonia solanacearum* causing potato brown rot in China

Li Wang, Bingsen Wang, Guozhen Zhao, Xingkui Cai, ***Huilan Chen***
Key Laboratory of Potato Biology and Biotechnology, Ministry of Agriculture /College of Horticulture and Forestry Sciences, Huazhong Agricultural University

Causing potato brown rot, *Ralstonia solanacearum* (*R. solanacearum*) strains are reported as one of the most destructive bacteria to potato (*Solanum tuberosum* L.) in China. In this study, 113 strains were isolated from potato, collected in the four major agroecological zones in China. The study showed that 102 strains belonged to the phylotype IIB sequevar 1 (race 3 biovar 2). The 11 remaining strains belonged to the phylotype I, sequevar 13,17 18,16 or 14 M, a new sequevar closely related to sequevar 14. Thirty-four strains were further characterized according to their virulence at low temperature on three wild potato species. IIB-1 strains all belonged to high and moderate virulence, while others belonged to the low virulence group, which had limited pathogenicity.

T5

A399

Functional characterization of the *Ginkgo biloba GbDXR* gene promoter in transgenic tobacco

Hua Cheng[1,2], *Jun Xiang*[1,2], *Honghui Yuan*[1,2]
1. College of Life Sciences, Huanggang Normal University
2. Economic Forest Germplasm Improvement and Comprehensive Utilization of Resources of Hubei Key Laboratories

1-Deoxy-D-xylulose 5-phosphate reductoisomerase (DXR) is one of the key enzyme that catalyze the first critical step of the MEP pathway for terpene trilactones biosynthesis. By using genomic walking method, a 3111 bp 5′ flanking region (promoter sequence) of the *GbDXR* gene was obtained. The *GbDXR* gene promoter (named *GbDXRP*) was analyzed in the plant cis-acting regulatory element databases and various of cis-elements were found in the sequence. Except the large number of basal TATA box and CAAT box, *GbDXRP* include a lot of specific elements that can response to light regulated, plant hormones and most adversity stress. To study the function of *GbDXRP*, the fusion vectors PCAMBIA1304+ (DXRP1~P5) and PCAMBIA1304-*35SP* (negative control) were constructed and infiltrated into tobacco epidermal cells by LBA4404. According to the epifluorescence results, all the five promoter fragments (DXRP1~P5) could promote the expression of green fluorescent proteins (GFP) in where? Deletion analysis of the *GbDXRP* in transgenic tobacco indicated that the promoter have an efficient promote function, even the promoter fragment is less than 1000 bp. The significant positive relationship between the *GbDXR* gene expression level and the content of terpene trilactones under Uv-B, Drought and Ethylene treatments just reflect the regulatory function involved in *GbDXRP* expression and lactone terpene accumulation in *G. biloba.*

T5

A400

Proteomic profiling of the interactions of Cd/Zn in the roots of dwarf polish wheat (*Ttiticum polonicum* L.)

Yiran Cheng[1], *Xiaolu Wang*[1], *Chao Wang*[1], *Rujiao Wang*[1], *Fan Peng*[1], *Xue Xiao*[1], *Jian Zeng*[2], *Xing Fan*[1], *Houyang Kang*[1], *Lina Sha*[1], *Haiqin Zhang*[1], *Yi Wang*[1]
1. Triticeae Research Institute, Sichuan Agricultural University
2. College of Resources, Sichuan Agricultural University

Cd and Zn have been shown to interact antagonistically or synergistically in various plants. In the present study of dwarf polish wheat (DPW)roots, Cd uptake was inhibited by Zn, and Zn uptake was inhibited by Cd, suggesting that Cd and Zn interact antagonistically in this plant. A study of proteomic changes showed that Cd, Zn, and Cd+Zn stresses altered the expression of 206,303, and 190 proteins respectively. Among these, 53 proteins were altered significantly in response to all these stresses (Cd, Zn, and Cd+Zn), whereas 58,131, and 47 proteins were altered in response to individual stresses (Cd, Zn, and Cd+Zn, respectively). Sixty-one differentially expressed proteins (DEPs) were induced in response to both Cd and Zn stresses; 33 proteins were induced in response to both Cd and Cd+Zn stresses; and 57 proteins were induced in response to both Zn and Cd+Zn stresses. These results indicate that Cd and Zn induce differential molecular responses, which result in differing interactions of Cd/Zn. A number of proteins that mainly participate in oxidation-reduction and GSH, SAM, and sucrose metabolisms were induced in response to Cd stress, but not Cd+Zn stress. This result indicates that these proteins participate in Zn inhibition of Cd uptake and ultimately cause Zn detoxification of Cd. Meanwhile, a number of proteins that mainly participate in sucrose and organic acid metabolisms and oxidation-reduction were induced in response to Zn stress but not Cd+Zn stress. This result indicates that these proteins participate in Cd inhibition of Zn uptake and ultimately cause the Cd detoxification of Zn.

Other proteins induced in response to Cd, Zn, or Cd+Zn stress, participate in ribosome biogenesis, DNA metabolism, and protein folding/modification and may also participate in the differential defense mechanisms.

T5

A402

Whole genome re-sequencing of a dwarf mutant induced by carbon ion beam irradiation in *Arabidopsis thaliana*

Yan Du, *Libin Zhou, Lixia Yu, Shanwei Luo, Tao Cui*

Institute of Modern Physics, Chinese Academy of Sciences

In the post genomics era, mutants play an important role for gene function exploring. Therefore, acquiring various variants and studying on them are useful for understanding the mechanism of plant development and molecular design for plant breeding. In recent decades, due to its unique advantages in Physics and Biology, heavy ion beams, with high mutation rate and wide mutation spectrum, have been widely used in plant breeding as a novel and efficient physical mutagen. In present work, an *Arabidopsis thaliana* variant numbered #541 displaying dwarfism phenotype was induced by carbon ion beams accelerated by the Heavy Ion Research Facility in Lanzhou (HIRFL). In fact, dwarf is one of essential agronomic features and the important research field in life science as well as the research focus in plant breeding. The height of #541 adult plants was only 20.73% that of the wild type plants, however there was no difference in the final number of siliques in single plant between #541 and WT, even though the length of the penultimate and antepenult silique of #541 was shorter than that of WT. As the ever-increasing release of whole genome of various plant species, a direct and effective way to identify the mutations that underlying variation phenotypes of interest is the whole genome re-sequencing. To mine the genes that are responsible for the mutant phenotypes quickly, young leaves of Lab-WT and #541 M3 lines were collected for DNA extraction. And then the genomic re-sequencing was performed by IIlumina HiSeq TM 2500 system. The variations shared by WT and #541 were filtered in order to detect the real mutations that induced by irradiation. Based on the above filter criteria, finally there were 22 SNP (Single Nucleotide Polymorphism), 4 small InDel (insertion-deletion) were detected out through bioinformatics analysis. Among them there were 6 mutations happened in the exon (1 was synonymous, 2 were missense, 1 was nonsense, 2 were frame shift), 19 located in the upstream or downstream regions, while 1 in untranslated regions. The mutations which occurred in intron or up and down stream regions had less probability to change the gene function or expression level, therefore, only the sites whose effects resulted in non-synonymous and frame-shift mutation were considered to be the candidate mutations which related to #541 phenotypes. Thus AT1G49210.1, AT3G15410.1, AT1G64900.1, AT1G62800.2 were screened out according to the SnpEff annotation. Gene Ontology (GO) indicated that the above candidate genes involved in zinc ion binding, leucine-rich repeat (LRR) family protein, cytochrome P450 (CYP89A2), and aspartate aminotransferase, respectively. According to the algorithms of VarScan, AT1G49210.1 was a homozygous mutation, while others were heterozygous. Considering that the dwarfism phenotype of #541 exhibit stable inheritance, since AT1G49210.1 was the most possible candidate gene that might be responsible for the dwarfism of #541.

T5

A403

The complete chloroplast genome of *Primulina* and two novel strategies for development of high polymorphic loci for population genetic and phylogenetic studies

Chao Feng[1], Meizhen Xu[1,2], Chen Feng[1,2], ***Ming Kang***[1]

1. Key Laboratory of Plant Resources Conservation and Sustainable Utilization

2. University of Chinese Academy of Sciences

Chloroplast genomes can provide important information for phylogenetic and population genetic studies. Recent advances in next-generation sequencing (NGS) techniques greatly facilitate the sequencing whole chloroplast genomes for multiple individuals. In this study, three complete *Primulina* chloroplast genomes were assembled from genome survey data, and then two novel strategies were developed to yield highly polymorphic markers. The first protocol was to develop lineage-specific highly variable marker from the true high variation regions (Con_Seas) across the whole cp genomes, instead of traditional noncoding regions. The experimental evaluation results based on a set of *Primulina* species showed that the newly developed markers are more variable than traditional ones. While the second method, which were developed to assemble fragments (poTs) and sub-super-marker (Cp-Contigs) through our SACRing pipeline, fundamentally altered the strategies of phylogenomic and population genetic studies based on cp markers, from traditional Sanger sequencing to RAD-Seq. This method was firstly proposed here, and successfully applied in population genetic studies from three natural populations of *Primulina*. We believe that these two novel strategies may provide a reference to similar research in other non-model species, and the newly developed high polymorphic loci in this study would promote the phylogenetic and population genetic studies in *Primulina* and other genera of the family Gesneriaceae.

T5

A404

Populations of *Arabidopsis thaliana* from Tibet, a unique ecotype at high elevation

Tianshu Sun, Yihao Shi, Chenqi Zhu, Xuefei Wang, ***Hongya Gu***

State Key Laboratory for Protein and Plant Gene Research, School of Life Sciences, Peking University

Arabidopsis thaliana is found in different geographic regions with different climatic conditions. Six populations were collected from Tibet, the upper reaches of the Yangtze River, China. They are found in habitats at 3,800-4400 meters above sea level, and they formed a monophyletic clade based on chloroplast DNA. These populations have distinct morphological and physiological characters such as very short, sometimes multi-branched, low germination rate, very sensitive to freezing temperature, and *etc.* The low tolerance to freezing temperature is partially due to the malfunctioning of a group of C-repeat binding factor (*CBF*) genes which have been found playing an important role in responding to low temperature in plants. The expression level of *CBF1~3* is lower in all 6 populations than that of Col ecotype at 4°C, and a loss-of-function mutation was found in *CBF2* and/or *CBF3* of these populations. An RNA-seq analysis was performed on the plants treated at 4°C for 3 h and 12h. It was found that at 3 h, the 6 populations shared 98 exclusive up-regulated genes which were mainly involved in responses to various phytohormones such as ABA, auxin, and JA, and to cold and salt stress; while Col had 198 exclusive up-regulated genes involving in regulation of gene expression, protein autophosphorylation, ethylene signaling pathway, and systemic acquired resistance. At 12 h, the 6 populations had a totally different set of up-regulated genes from Col. We will discuss about the adaptation significance of these variations, together with the genetic and molecular characterizations of some other interesting traits in this unique ecotype from Tibet.

T5

A405

Functional analysis of *Rphq4*, a locus for quantitative resistance to leaf rust (*Puccinia hordei*) in barley

*Yajun Wang, Hongjiang Han, **Xiaoquan Qi***
Institute of Botany, Chinese Academy of Sciences

Barley is encountered with numerous diseases such as leaf rust, powdery mildew, sheath blight. The leaf rust resistance controlled by quantitative trait loci (QTLs) has been characterized by the combined components, such as longer latent period, fewer and smaller uredia, lower spore production and slower disease progress, and is usually considered to be more durable and broad-spectrum. *Rphq4*, an adult plant resistant QTL, was mapped on the short arm of 5H chromosome in barley. Map-based cloning showed that two genes are underline *Rphq4* locus. Virus induced gene silencing (VIGS) indicated that both *Rphq4-G1* and *Rphq4-G2* were required in resistance to *Puccinia hordei*. The full-length candidate genes driven by ubiquitin promoter have been cloned into a binary vector to transform into susceptible lines, Golden Promise. The T1 transformed plants have been obtained and will be used for functional validation of *Rphq4-G1* and *Rphq4-G2*.

T5

A406

Arabidopsis glutamate:glyoxylate aminotransferase 1 (Ler) mutants generated by CRISPR/Cas9 and their characteristics

*Yaping Liang, Xiuying Zeng, Xinxiang Peng, **Xuewen Hou***
South-China Agricultural University

Genome editing technology provides an easy tool to edit genomes of plant, animal and microorganism. CRISPR/Cas9 is the most successful and broadly applied genome editing technology nowadays. In this paper, CRISPR/Cas9 was used to edit photorespiration key enzyme gene *GGAT1* (*GLUTAMATE:GLYOXYLATE AMINOTRANSFERASE 1*) of *Arabidopsis* (Ler). Two *GGAT1* target sgRNAs were constructed into pYLCRISPR/Cas9-DN, and flowering *Arabidopsis* (Ler) plants were transformed through *Agrobacterium tumafacien*-mediated floral dip method. Eleven chimeric and two heterozygous *GGAT1* edited T1 lines of target 1 were screened out from positive transgenic lines, respectively. A TCC-deletion and T-deletion homozygous T2 lines were generated from above 13 T1 lines. The edited mutation sites were stable through generations no matter the T-DNA was existed or not; the genetic separation of mutation sites obeyed the Mendel single gene separation rule; no mutations were detected at the possible off-target site; the two independent *ggat1* mutants had similar photorespiration phenotypes and down-regulated GGAT enzyme activity. All these results indicated that genetically stable *ggat1* (Ler) mutants were generated by CRISPR/Cas9 genome editing technology, and these mutants will promote the dissection of how *AtGGAT1* is regulated.

T5

A407

Development and characterization of 18 polymorphic SSR markers for *Barthea barthei* (Melastomataceae)

Guilian Huang
Sun Yat-sen University

Premise of the study: To examine genetic diversity and population differentiation, SSR markers were developed and characterized in *Barthea barthei*, an evergreen shrub with the southern mainland China-Taiwan disjunct distribution.
Methods and Results: We used Illumina HiSeq to develop a genomic library for SSR identification. 18 SSR loci were polymorphic in 60 individuals from three populations of *B. barthei*. The number of alleles per SSR locus ranged from two to nine at the species level and at the population level, the observed heterozygosity and expected heterozygosity per locus varied from 0 to 0.850 and from 0 to 0.809, respectively. Higher genetic differentiation between the two populations of *B. barthei* var. *barthei* (Fst = 0.474) was observed, relative to that between the two varieties (Fst = 0.387 and 0.418, respectively).
Conclusions: These polymorphic genomic SSR markers may be useful for understanding phytogeographic history of *B. barthei*. Lower genetic differentiation between the two varieties than the populations suggests that the taxonomic treatment may not hold.

T5

A408

The effect of $12C^{6+}$ heavy ion irradiation on proliferation, differentiation and mutation of rhizomes of *Cymbidium* 'JunHao'

Jenny Lee
South China Agriculture University

The objective of this study is to breed the new variety of *Cymbidium* interspecific hybrids that has beautiful flower color, dark leaf color, variegation leaf and easy flowering instead of grey orange flower color, late florescence not within Chinese New Year, light leaf color and no variegation leaf. Investigations were carried out on mutagenic effects of different does of 0,30,50,70,90,120 Gy of $12C^{6+}$ heavy ion radiation on proliferation, differentiation and mutation of rhizomes of *Cymbidium* 'JunHao' and preliminary selection of variants form by observing the phenotypic variation. Heavy ion irradiation played the role of inhibition in proliferation and differentiation of rhizomes, the higher the irradiation was, the stronger the inhibition effect was. Proliferation and differentiation of rhizomes irradiated with 50 Gy heavy ion came to be normal in SC_1 generation, while 70 Gy did in SC_2, and differentiation recovered in SC_3 generation. 50 Gy radiation had an obvious positive effect on differentiation of posterity rhizomes of *Cymbidium* 'JunHao'. 50 Gy was the most appropriate irradiation does. The phenotypic variation rates of plants regenerated from rhizomes irradiated by 50,70 and 90 Gy were 67.35%, 68.97% and 100.00%. Regenerated plants were found mutant in sturdy or weak, erect, dark or light, more leaves, wide leaves, variegation, deformity and so on. The above genetic stability needs to be confirmed further.

T5

A409

Characterization of a PVY isolate — PVY-CJ breaking down the resistance of *va* genotype tobacco in China

Ruo Li[1], *Xiuqing Wan*[1], *Xia Liu*[2], *Chan Qiao*[1], *Zunqiang Li*[1], *Xiao Cui*[1,3], *Yuehan Zhu*[1,3], *Zhaokui Guo*[1]

1. Tobacco Science Research Institute of Heilongjiang Province, China Tobacco Corporation

2. School of Basic Medical Sciences, Mudanjiang Medical University

3. School of Life Science and Technology, Mudanjiang Normal University

Potato virus Y (PVY) which is able to infect many Solanacae crops including potato, pepper, tomato, tobacco is a plant virus with very important economical and agronomical impacts on agricultural production. PVY is also one of the main diseases of tobacco production and causes huge economic loss globally every year. Cultivated tobacco plants infected by PVY showed dwarf, vein and stem necrosis, leaf mottling symptoms resulting in dramatic reduce of the yield and quality of tobacco leaves. PVY is the type member of *Potyviruses* genus. The virus has a single-stranded positive-sense RNA genome of about 10 kb in length. A recessive resistant gene *va* had been introduced into cultivated tobacco since a X-ray treated tobacco variety Virgin A Mutant (VAM) was released in 1958 (Koelle *et al.,*). Since then *va* had been deployed in PVY resistant breeding for many years. The selection pressure of *va* on PVY inevitably led to the emergence of mutated PVY population which could overcome the resistance of *va* and this breaking down of *va* resistance has been reported in *va* genotype tobacco growing field globally according to a field survey of Cooperation Centre for Scientific Research Relative to Tobacco (CORESTA) on a worldwide basis (Verrier *et al.,*). Here we report a PVY isolate collected in China which breaks down the resistance of *va* genotype tobacco for the first time. The PVY isolate we named as PVY-CJ was collected from a VAM tobacco plant on a tobacco growing field in Ning An of Hei Longjiang Province of China in 2016. The infected tobacco plant showed typical vein necrosis, leaf mottling symptoms caused by PVY. Using multiplex PCR assay, PVY-CJ was characterized as PVY^{NTN} strain (Ali *et al.,*). The whole genome of PVY-CJ was then obtained by using RT-PCR method (Zhang *et al.,*). Sequence analysis showed that PVY-CJ had a 9,614 nt ORF encoding a polyprotein of 3,061 amino acid residues. The full-length genome sequence of PVY-CJ was subjected to phylogenetic analysis against 60 full-length PVY sequences deposited to the GenBank (Hu *et al.,*). Result showed that PVY-CJ clustered with PVY^{NTN} strains. How PVY-CJ breaks down the resistance of *va*? We then did some primary analysis on the genome sequence of PVY-CJ. Some previous investigations indicated that the Virus Genome-linked Protein (VPg) was responsible for the *va* adaption of PVY to resistant tobacco cultivars (Masuta *et al.,*), while virulence gain in PVY occurred rapidly and preferentially by substitution at position AA105 in the VPg (Janzac *et al.,*). The amino acid sequence of VPg of PVY-CJ was aligned with several other PVY^{NTN} strains. The sequence alignment result showed that there was a substitution of lysine (K) to glutamic acid (E) at position AA105 of VPg of PVY-CJ compared with those of other PVY^{NTN} strains. This result is in accordance with previous investigations. More work may be needed to confirm the role of this single amino acid substitution on the resistance breaking down of PVY-CY on *va* genotype tobacco.

T5

A410

Rice gene co-expression network during anther development

Hong Lin[1,2], *Jing Yu*[1], *Simon Pearce*[3], *Zoe Wilson*[2], *Dabing Zhang*[1]

1. School of Life Sciences and Biotechnology, Shanghai Jiao Tong University

2. Division of Plant & Crop Sciences, School of Biosciences, University of Nottingham

3. Faculty of Biology, University of Manchester

In plants, anther and pollen development includes many important biological events. Complex regulatory relationships occur between genes during this process. Better understanding of the relationships between gene interactions and biological events will help crop breeding and increase yields. In rice, many genes have been confirmed as being responsible for anther development in different stages. However, large numbers of genes in the regulatory network remain unknown. Gene co-expression has been revealed as a good way to characterize global gene interactions because genes in the same pathway or with similar functions usually show very similar expression patterns over different biological processes and experimental conditions. In this work, a rice gene co-expression network during anther development named RiceAntherNet is presented. It was generated from 57 Affymetrix Rice Genome Arrays that specifically focus on anther tissue during whole anther development processes in wild-type rice samples. A website is provided to present expression profile in rice anthers during different stages for users' reference. Microarray datasets of 9 male sterile mutants, including *msp1, ostdl1a, gamyb-2, tip2, udt1-1, tdr, eat1, ptc1* and *mads3-4* mutants were used to focus on example clusters related to anther development. 3 clades with distinguish expression patterns were further constructed to highlights genes associated with particular biological process. Especially for clade1, all 7 reported genes in it have been confirmed to participate in pollen wall formation in rice, furthermore, 16 orthologs in Arabidopsis have been reported as key genes during pollen development in Arabidopsis, which suggest that the rest unknown genes may be involved in the same developmental process. The RiceAntherNet provide a useful library for novel gene identification during anther and pollen development in rice.

T5

A411

Identification of miRNAs involved in light-induced anthocyanin biosynthesis in bagging-treated red Chinese sand pear, 'Mantianhong'

***Hainan Liu**, Jun Wu*, Yangfan Zhu, Gaifang Yao, Meiling Ming, Yaojun Chang*
Nanjing Agricultural University

Bagged fruits of red Chinese sand pear cultivar 'Mantianhong (*Pyrus pyrifolia* Nakai cv. Mantianhong)' have been found to turn red after debagging during ripening stage. miRNAs play very important roles in many developmental processes, such as flower bud formation, flowering, leaf, root and other morphological formation, fruit ripening, and developmental phase transition (juvenile-adult-reproductive transition), as well as stress, hormonal responses and disease resistance in plants. We investigated the roles of miRNAs in response to debagging to explore whether miRNAs paticipate in this phenomenon of 'Mantianhong'. Six small RNA libraries and seven transcriptome libraries were primarily constructed from peels of 'Mantianhong' pears subjected to bagging followed by sunlight re-exposure treatments (0 h, 4,8 and 10 day) (debagging group) in ripening stage, and taking peels without bagging removal (bagging group) as control. A total of 279 known miRNAs belonging to 40 miRNA families and 416 novel miRNAs were identified via high-throughput sequencing, and 50,452,493 clean reads at average were generated by Illumina sequencing of debagging and control groups after removing low-quality sequences. Some miRNAs and their corresponding target genes were found to be differentially expressed compared debagging group with control, which indicated that miRNAs affected anthocyanin accumulation through their targets in pear. To further explore the effect of debagging on miRNAs regulating the expression of anthocyanin related genes, the expression of four differentially expressed miRNAs and their target genes, which expressed differentially in transcriptome libraries, were identified by qRT-PCR, miRNA-mediated mRNA cleavage sites of two target genes were detected by 5' RLM-RACE (RNA Ligase Mediated Rapid Amplification of cDNA Ends). Results showed that miRNAs play different roles in the whole stage of anthocyanin accumulation. Additional analysis of gene ontology for the differentially expressed miRNAs after debagging treatments and their predicted target genes showed that they were involved in light response and play important roles in light-induced anthocyanins biosynthesis in red-skinned Chinese sand pear.

T5

A412

Cloning and characterization of a novel dehydrin gene, sidhn, from *Saussurea involucrata* Kar. et Kir. that enhances cold and drought tolerance in transgenic tobacco

Hongling Liu
College of life science, Shihezi University

Saussurea involucrata Kar. et Kir. is a hardy, dicotyledonous plant capable of tolerating severe abiotic stress conditions. In a previous study, we created a cDNA library to determine what factors are associated with the cold acclimation response of S. involucrata. A full-length cDNA of a dehydrin-like gene (SiDHN) was obtained by RT-PCR. The SiDHN gene was characterized in this study. The full-length SiDHN cDNA consisted of 703 bp containing an open reading frame of 333 bp and specified a protein of 111 amino acids. An alignment of the deduced amino acid sequence showed that SiDHN shared 36% identity with Helianthus annuus Agrobacterium tumefaciens was used to transform pBIN438-SiDHN constructs into tobacco. The germination rates of the transgenic plants were measured as well as their resistance to freezing and drought stress. The transgenic plants showed greater resistance to freezing and drought stress than the wild-type. This study demonstrates that SiDHN confers cold tolerance and drought resistance. We suggest that SiDHN has potential to be used to genetically improve the resistance of plants to abiotic stress.

T5

A414

De novo sequencing and characterization of juvenile sporophyte transcriptome of a fern, *Dicranopteris dichotoma*

***Liting Liu**, Quanquan Jia*
Jiangxi Academy of Forestry

The sporophyte transcriptome of *Dicranopteris dichotoma* was sequenced by Illumina MiSeq 250 to provide molecular information of its growth, development, metabolism, and the micro evolutionary mechanism. The functional annotations, metabolic pathways and microsatellite analysis of some Unigenes were conducted using bioinformatics methods. A total of 18 463 296 reads containing 4. 62 Gbp of sequence information were generated. A total of 63 169 unigenes were formed by initial sequence splicing, with an average read length of 863 bp and N50 value of 1 587 bp. 26 826 unigenes were annotated using BLASTX searches against the Nr, Nt and SwissProt databases. The unigenes of the transcriptome of *D. dichotoma* were roughly divided into cellular components, molecule function and biological processes categories of 47 branches by gene ontology, of which related with cellular process cell, binding, metabolism processes and catalytic activities. Further annotated based on COG category, Unigenes could be grouped into 26 functional categories. KEGG pathway analysis showed that Unigenes could be divided into 276 classes based on their metabolic function. Meanwhile, 13 286 SSRs (simple sequence repeats) were mined with repeat motif of 2 to 6 bp by MISA. The trinucleotide repeats were most dominant, accounting for a total of 40. 41%. AG/CT (14. 45%) and AAG/CTT (12. 39%) were the most common repeat motifs. Polymorphic SSR markers were developed from repeat motifs, which could be used for genotyping of different individuals of *D. dichotoma*. A higher quality of transcriptome database was obtained in this study, which could reveal the general characteristics of gene expression in the process of growth and development, and lay the foundation for further gene function mining and the large-scale development of molecular markers of *D. dichotoma*. In this study, 23 Unigene sequences which closely related to heavy metal-binding proteins were obtained, furthermore, their functional predictions showed that they were mainly composed of copper, zinc and cadmium. The results proved that *D. dichotoma* has a strong ability to absorb heavy metals at the molecular level. It is great value in the restoration of heavy metal pollution wastelands.

T5

A415

Transcriptome-wide identification and expression analyses of *MATE* genes in response to aluminum toxicity in *Medicago sativa* L.

Xueyang Min, *Zhengshe Zhang, Xingyi Wei, Yanrong Wang, Wenxian Liu*
The State Key Laboratory of Grassland Agro-ecosystems, College of Pastoral Agriculture Science and Technology, Lanzhou University, Lanzhou, 730020, P.R. China.

Crop yields are significantly reduced by aluminum toxicity on acidic soils. The Multidrug and toxic compound extrusion (MATE) transporters contribute to multidrug resistance by extruding a wide variety of drugs across cell membranes, including aluminum translocation. Alfalfa (*Medicago sativa* L.) is the most extensively cultivated forage legume, yet most alfalfa cultivars are not aluminum tolerant, and the analysis of MATE family has not been reported in this species. In this study, a total of 88 *MsMATE* genes were identified at the whole transcriptome level. The predicted protein products ranged from 106 to 595 amino acids, with molecular weights from 13.22 to 65.29 kDa. The number of transmembrane helices of these proteins ranged from 0 to 12, and their predicted isoelectric point varied from 4.76 to 9.79. In addition, the subcellular location of these proteins are mainly in the plasma membrane. Phylogenetic analysis of *MsMATE* could classify them into four subfamilies comprising ten smaller subgroups, with diverse potential functions such as aluminum detoxification/iron translocation, flavonoids accumulation, disease resistance and response to abiotic stresses. Ten MsMATE transporters could be clustered together with the previously reported aluminum detoxification and iron translocation related MATE proteins, suggesting their similar potential functions in alfalfa. We further explored the expression pattern of all putative *MsMATE* genes in different tissues by using RNA-Seq data and qRT-PCR method. The most of MsMATE transcripts were found to be expressed in various tissues and showed differentially expression patterns, suggesting that they might be involved in regulating growth and development in alfalfa. Considering the limited functional understanding of MATE transcripts in this important forage legume, our findings will be valuable for the functional investigation and application of this gene family in alfalfa and other related legume species.

T5

A416

Characterization of the fatty acid desaturase genes in *Medicago truncatula*: Structure, phylogeny, and expression patterns

Zhengshe Zhang, *Xueyang Min, Xingyi Wei, Yanrong Wang, Wenxian Liu*
The State Key Laboratory of Grassland Agro-ecosystems, College of Pastoral Agriculture Science and Technology, Lanzhou University, Lanzhou, 730020, P.R. China.

Fatty acid desaturases (FADs) introduce double bonds into the hydrocarbon chains of fatty acids to produce unsaturated fatty acids, and play critical roles in plant development and acclimation to environmental stresses. However, no reports about characterization of the *FAD* genes have been reported in the model dicotyledonous grass species *Medicago truncatula*. In the present study, 20 full-length *FAD* genes in *M. truncatula* were identified in a genome-wide level, including ten *MtFAB2* genes, three *MtFAD2* genes, one *MtFAD5* gene, and two genes each for *MtFAD3*, *MtFAD6*, *MtFAD7*. These *MtFAD* genes were distributed on seven chromosomes except the chromosome 8. Based on a phylogenetic analysis, the *M. truncatula* FAD proteins were clustered into six subfamilies with their counterparts from other plants, and the gene structures and protein sequences were considerably conserved in each subfamily. For the extron-intron gene structures, all the *MtFAD3* and *MtFAD7* clusters (encoding the ω-3 *FADs*) have the same structure, while for the genes encoding the ω-6 *FADs* (*MtFAD2.1*, *MtFAD2.2*, *MtFAD2.3*, *MtFAD6.1* and *MtFAD6.2*) were quite different. Ten MtFAB2 proteins shared conserved structural with the known plant soluble FAD proteins. The other MtFADs belonged to the membrane-bound FADs and contained three highly conserved histidine boxes. Additionally, the putative endoplasmic reticulum retention signal was found at the C-termini of the MtFAD2 and one MtFAD3 proteins, while the N-termini of MtFAD5, MtFAD6, MtFAD7 and two MtFAB2 contained a predicted chloroplast signal peptide, which was consistent with their associated metabolic pathways. Furthermore, the gene expression analysis showed that *MtFAD2.2*, *MtFAD2.3*, *MtFAB2.9*, *MtFAB2.10* and *MtFAD3.1* were universally expressed in all tested tissues, whereas the other *MtFAD* genes were preferentially expressed in the seeds. This study would provide an important foundation for future cloning and functional studies of *FAD* genes in *M. truncatula* and other related legume species.

T5

A417

Genetic diversity and population structure of flower lotus (*Nelumbo Adans.*) cultivars assessed by SSR markers

Xiaojing Liu, ***Fengfeng Du***, *Yajun Chang, Naiwei Li, Dongrui Yao*
Institute of botany, Jiangsu province and Chinese academy of sciences

Lotus (*Nelumbo* Adans.) is an important aquatic crops in Asia, Oceania and North America. Flower lotus is widely used in aquatic gardens for its high ornamental value. To determine its genetic diversity and population structure, 36 flower lotus accessions were genotyped with 17 pair SSR primers. A total of 76 alleles were generated and 87.5% were polymorphic. The polymorphism information content (PIC) value ranged from 0.20 to 0.81. Analysis of unweighted pair-group method with arithmetic mean (UPGMA) clustering and principal coordinate analysis (PCoA) distinctly grouped flower lotus accessions from Chinese lotus, American lotus and Sino-American hybrids into different clusters. However, the hybrids dispersed among clusters. Population structure analysis further confirmed the clustering of lotus accessions, and it also indicated that there were several subgroups within Chinese lotus population. The genetic diversity and population structure of flower lotus accessions herein are beneficial for germplasm utilization and genetic improvements of flower lotus in the future.

T5

A418

Genome-wide identification and expression analysis of the MATH-domain proteins in rice (*Oryza sativa*)

Lujun Yu, *Chen Qinfang, Xiao Shi*
State Key Laboratory of Biocontrol, Guangdong Key Laboratory of Plant Resources, Collaborative Innovation Center of Genetics and Development, School of Life Sciences, Sun Yat-sen University

The MATH (meprin and TRAF homology) domain contained a conserved 180 amino acids region and played functions in the protein and protein interaction. *In silico* investigation revealed 81 MATH domain proteins in the full complement of sequenced rice genome. MATH domain genes were categorized into diverse classes according to the phylogenetic analysis and domain organizations, indicating their relation among the various subfamilies. Chromosomal localization and duplication analysis indicated their unevenly distribution on 11 chromosomes without in chromosome 9, with 57 genes being present on rice tandemly duplicated chromosome regions. The rice MATH detailed expression analysis revealed that 29 *OsMATH* genes were differentially expressed during panicle development, with 35 *OsMATH* during seed development, and 43 *OsMATH* genes were differentially expressed under salinity, drought, or cold abiotic stresses. These data provide the valuable information for the further functional genomics study of rice MATH domain-containing genes.

T5

A419

Comparative transcriptomics study reveals two distinct phases of C_4 evolution

Ming-Ju Lyu

Partner Institute of Computational Biology, Shanghai Institutes for Biological Sciences, CAS

C_4 photosynthesis is a remarkable complex trait, not only because species using this biochemistry have higher light, nitrogen and water use efficiencies than their C_3 ancestors, but also because the trait has emerged independently in more than 60 evolutionary lineages. Determining the evolutionary trajectory of C_4 photosynthesis from the ancestral C_3 pathway would elucidate the basic principles of complex trait evolution and help guide engineering of C_3 crops for higher yields. Here, we use the genus *Flaveria*, an evolutionarily significant group that contains C_3, C_3-C_4 intermediate, C_4-like and C_4 species, to systematically characterize transcript abundance and predicted protein sequence changes at different stages of C_4 evolution. This study revealed the following: a) gene transcript abundance, protein amino acid sequence and leaf anatomy changed in a coordinated manner during C_4 evolution; b) formation of a C_4 cycle and augmentation of C_4 photosynthesis represent two distinct phases of C_4 evolution, where cyclic electron transport (CET) plays two distinct roles; c) novel links to metabolic or regulatory processes in addition to CET for C_4 evolution were present, including starch metabolism, protein degradation, cellulose metabolism, and amino acid transport; moreover potential functional sites in the proteins associated with these processes were identified. Overall, this study demonstrates the power of using a comparative transcriptomics approach to gain insight into the evolutionary history of a complex trait.

T5

A420

Multi-scale modelling scheme of maize venation networks

Liming Ma

National Engineering Research Center for Information Technology in Agriculture (NERCITA)

Leaf venation networks is built from lignified xylem and phloem tissue and serves multiple functions, including transport of water and sugars, mechanical support, and herbivory resistance . Because of this diversity in function, these networks are of interest to a wide range of fields, including plant physiology, systematics, and physics. However, the study of these networks is limited by slow and destructive imaging methods. Here, we use micro-CT and confocal microscope which are rapid, of high resolution, and nondestructive to build a three dimensional model of maize venation networks. We acquired serials of cross-sectional images using micro-CT, which included vein features in a wide section of maize leaf. Both major and minor veins could be consistently visualized with the use of iodine contrast agent. Internal feature of the vein was obtained by confocal microscope. After fluorescent dyeing, we acquired microscopic characteristics of xylem and phloem tissue through Z-Stack scanning method. Dataset from micro-CT and confocal microscope were combined to build maize venation model. The generated models are easy to compute the structural and topological properties of leaf veins and their components (the number of connected components, adjacency relationship, volume *etc.*). What is more, they can be used, for example, to simulate the water transports in leaf veins and investigate hydraulic resistance.

T5

A422

Effect of aspergillus niger and trichoderma viride on the increase in plant biomass of *Casuarina equsetifolia*

Shalini Tiwari, *Arun Arya*

Maharaja Sayajirao University of Baroda

Casuarina is a tree crop that provides fuel wood, land reclamation, dune stabilization, shelter belts and pulp production. Roots of this plant are associated with a number of microbes in rhizospheric zone. They have the basic physiology for survival in diverse situation, and they often serve to colonize the plant which grow in poor soils. Many species of *Casuarina* grow on soil with low fertility, some thriving on light soil, others on heavy soil, *C. equsetifolia* thrives in dunes or near the seaside. The outstanding ability of various *Casuarinas* to grow vigorously on poor soils is due partly to their unusual symbiosis with an actinomycetes *Frankia*, that enables them to use nitrogen directly from the atmosphere. *Frankia* which plays a key role on the growth of *Casuarina*, there are other bioinoculants present around the root, which are very important and beneficial for the plant. Mycorrhizal fungi may enter the root and facilitate the uptake of minerals, notably phosphorus and some trace elements. Furthermore, a range of unidentified microorganism interact with the plant to produce dense mats of "Proteoid roots". The role of various fungi and phosphate solubilizing fungi on the growth of the plants was done in various ways like root shoot ratio, fresh and dry weight, photosynthetic capacity, nutrient utilization and seedling quality. *Frankia* nodules were also collected and observed from these treated plants. Seedlings of *C.equsetifolia* were grown in the botanical garden of MS University of Baroda, Vadodara. These were inoculated with fungi like *A. Niger and T.viride,* either individually or in combinations. Growth was recorded at different time intervals. Inoculation of bioinocunlats affected seedling growth, improved seedling quality and nutrient uptake. Nodules were also collected after harvesting and different characteristic were recorded. Nodules recorded from *A.niger* were more in number and bigger in size than *T.viride*. *A.niger* and *T.viride* showed the better growth as compared to other fungi.

T5

A423

Three-dimensional reconstruction and anatomical trait analysis of maize root

Xiaodi Pan[1,2], *Ying Zhang*[1,2], *Meng Shao*[1,2], *Liming Ma*[1,2], *Jinglu Wang*[1,2], *Jianjun Du*[1,2], *Xinyu Guo*[1,2]
1. Beijing Key Lab for Digital Plant
2. National Engineering Research Center for Information Technology in Agriculture

Roots play an essential role in the acquisition of water and nutrients from soils, and are an important component in plant stress tolerance and adaptation. Root anatomical traits have significant effects on root functions, such as acquisition of resources from the soil, water transport within the plant, and metabolic cost of soil exploration. However, two-dimentional microscopic images are often used to acquire useful information on anatomical features, which are not capable of displaying the spatial organization of microstructures and their interactions. In recent years, X-ray computed tomography (CT) provides an attractive alternative for imaging and analysis on the internal anatomy of plant tissues, which is a non-destructive and non-invasive 3-D imaging technique that enable plant tissues to be studied in their natural state. This technique has been widely applied to 3-D imaging of plant leaves, stem vasculature, and crop seeds. However, there have been few reports about its application on root anatomy. In this work, we plan to establish a rapid and non-destructive method for 3-D reconstruction of maize root depending on X-ray micro-CT and develop a 3-D image analysis software for micro-CT data, which is available for high-throughput 3-D information extraction and analysis of root anatomical traits. The above all is in favor of the further understanding of root 3-D anatomy and modelling of hydromechanics.

T5

A424

Genome-wide identification, phylogenetic analysis of the *UXS* gene family in cotton

Yuxin Pan
North China Univrsity of Science and Technology

UDP-Xylose Synthase (UXS) is an enzyme in plants and participate in cell wall noncellulose. Phylogenetic analysis suggested that *UXS* genes could be divided into four subgroups and members within the same subgroup shared similar exon-intron structures and subcellular location. Notably, gene colinearity information indicates 50% cases of aberrant tree topology of gene subgroups, and helps obtain corrected phylogeny, *UXS* genes were subjected to significant positive selection on key evolutionary nodes.

T5

A425

TpNRAMP3 and *5* from dwarf polish wheat (*Triticum polonicum* L.) transport Cd and Co

Fan Peng, *Chao Wang, Yiran Cheng, Songyue Chai, Wendi Shuai, Yonghong Zhou, Yi Wang*
Triticeae Research Institute, Sichuan Agricultural University

Many heavy metal transporter families, such as P_{1B}-ATPase-HMA (P<sub>1B-ATPase-Heavy metal associated proteins), YSL (Yellow-stripe 1-like proteins), ZIP (Zinc-regulated transporters-like proteins) and NRMAP (Natural resistance-associated macrophage proteins), were identified in plant. In rice and Arabidopsis, seven and six NRAMPs were cloned, respectively. Most of them were functionally identified. However, no information of this family was reported in wheat. In the present study, two *NRAMPs*, *TpNRAMP3* and *TpNRAMP5*, were cloned from dwarf polish wheat. Their lengths of open reading frame were 1647 bp and 1623 bp, individually encoded 548 and 540 amino acids. They were mapped on 7BL and 4AS, respectively. *TpNRAMP3* contains 13 exons and 12 introns; *TpNRAMP5* contains 12 exons and 11 introns. At the seeding stage, *TpNRAMP3* was mainly expressed in the leaves; it was markedly up-regulated in the roots by Cd and Fe. Conversely, *TpNRAMP5* was mainly expressed in the roots; its expression in the leaves was slightly up-regulated by Cd and Zn, but was not by Pb, Fe or Cu. Expression of these two genes in yeast indicated that TpNRAMP3 and TpNRAMP5 transported Cd and Co, but did not transport Zn, Fe or Al. Meanwhile, TpNRAMP3 expressed in cell membrane and endoplasmic reticulum using wheat protoplast. These findings suggested that these two genes play an important role in the translocation of Cd and Co in dwarf polish wheat.

T5

A426

Non-destructive and fast visualization and quantification of maize kernel 3D structure using X-ray micro-computed tomography in standardized laboratory

Meng Shao, *Xinyu Guo*
Beijing Key Lab of Digital Plant, Beijing Research Center for Information Technology in Agriculture, Beijing Academy of Agriculture and Forestry Sciences

Observing the internal structure of plant organs or tissues is one of an important content of plant developmental biology and plant functional genetics. However, traditional optical microscopic observation methods have some disadvantages such as sample preparation process is time consuming, destructive, and constrained to two dimensions or limited in optical depth. X-ray micro-CT which is a non-destructive imaging technology don't need sample preparation process and can obtain non-destructive three-dimensional image of plant organs. Exploring the application of this technology in 3D anatomy of maize seeds, is seeking to establish an accurate and fast method for identifying plant phenotypic traits and studying the influence on morphological structure by environmental or genetic factors. Using X-ray micro-CT scan and take 3D reconstruction of maize seeds which are Nongda 178 harvested from drought treatment environment. Graphic image analysis software (Simpleware) was used to realize image segmentation process of 3D anatomy of maize seeds. The 3D images of maize seeds were obtained by X-ray micro-CT. Different structures such as pericarp, embryo, endosperm and cavity on different sections were observed by doing different levels virtual slices on 3D images of maize seeds. Among these structures, the cavity structure can reflect the endosperm enrichment degree of maize plants which were affected by environmental factors such as drought in the process of growth and development. Segmentation of the different structures of maize seeds were obtained by using "floodfill", "region grow", "dilate" and "erode" and the morphometric parameters were acquired by using Simpleware software. This article is the first time to perform the global anatomical 3D segmentation of the compartments of maize seed. It demonstrate the X-ray micro-CT can overcome the traditional optical sectioning complicated sample preparation process and adapt to operate in standardized laboratory to obtain three dimensional structure internal plant tissues. Meanwhile, we achieved the image segmentation of different structures of maize on the three dimensional, and calculate morphological parameters of these different compartments.

T5

A427

Comparative molecular cytogenetic analysis of five wild *Vigna* species (Fabaceae)

Chaowen She[1,2,3], *Ying Mao*[1,2,3]

1. Key Laboratory of Research and Utilisation of Ethnomedicinal Plant Resources of Hunan Province, Huaihua University

2. Key Laboratory of Xiangxi Medicinal Plant and Ethnobotany of Hunan Higher Education, Huaihua University

3. College of Biological and Food Engineering, Huaihua University

The genus *Vigna* Savi (Fabaceae) includes over 100 species distributed throughout the Old and New Worlds. However, chromosome characterization of *Vigna* species has been minimal to date. In this study, the genome organization of five wild *Vigna* species; *V. luteola*, *V. vexillata*, *V. minima*, *V. caracalla* and *V. trilobata,* was determined by sequential combined PI and DAPI (CPD) staining and dual-colour fluorescence *in situ* hybridization (FISH) with 5S and 45S rDNA probes. For phylogenetic analysis, genomic *in situ* hybridization (cGISH) with *Vigna umbellata* genomic DNA probe onto somatic chromosomes was also conducted. Quantitative karyotypes were established using chromosome measurements, fluorochrome bands, and rDNA FISH signals. All species have chromosome number 2n = 2x = 22, and symmetrical karyotypes composed of metacentric and submetacentric chromosomes. CPD staining revealed the 45S rDNA sites in all species, (peri)centromeric GC-rich heterochromatin in *V. luteola*, *V. caracalla* and *V. trilobata*, and proximal AT-rich heterochromatin in *V. caracalla*. rDNA FISH revealed one 45S locus in *V. luteola* and *V. caracalla*, two 45S loci in *V. vexillata* and *V. trilobata*, and five 45S loci in *V. minima*; two 5S loci in *V. caracalla* and one 5S locus in the other four species. cGISH of *V. umbellata* genomic DNA to the chromosomes of all species produced strong or weak signals in all pericentromeric regions of *V. luteola* and *V. trilobata*. The combined data indicated that *V. luteola* and *V. trilobata* are closely related to each other, and relatively distant from the other three species.

T5

A428

GmHMA3, a P1B-ATPase, transports Cd from cell wall to organelles in soybean

Wendi Shuai[1], *Chao Wang*[1], *Yujing Liu*[1], *Yan Huang*[2], *Kangfu Yu*[3], *Yonghong Zhou*[1], *Yi Wang*[1]

1. Triticeae Research Institute, Sichuan Agricultural University

2. College of Life Science, Sichuan Agricultural University

3. Harrow Research and Development Centre, Agriculture and Agri-Food Canada

A single point mutation in GmHMA3 (Glycine Max heavy metal-associated) was tightly associated with the seed Cd concentration. The GmHMA3w (wild type) functions as a Cd/Zn transporter in yeast. Therefore, it was recommended as a target gene responsible for limiting the Cd concentration in soybean seeds. In the present study, knocking down of GmHMA3w using virus-induced gene silencing (VIGS) reduced the Cd concentration in roots and leaves and enhanced the Cd tolerance. Overexpression of GmHMA3w significantly increased the Cd concentration in roots, but significantly reduced Cd concentration in stems, which indicated that GmHMA3w inhibited root-to-shoot Cd translocation. Subcellular localization of GmHMA3w and 3a indicated that GmHMA3w localized at the plasma membrane and the organelles (except of chloroplasts). Therefore, GmHMA3w transported Cd from cell wall to organelles, so that decreasing the Cd tolerance and reducing the biomass production, finally limiting the Cd transport from root to stem and controlling the Cd accumulation in whole plants and tissues. Meanwhile, GmHMA3w did not transport Z, Fe, Co and Pb. Therefore, it is a Cd specific transporter.

T5

A429

In vivo tracking of genetically encoded FRET based isoleucine nanosensors

Shruti Singh

Jamia Hamdard

With the advancement of the technology, microorganisms are metabolically engineered *in vivo* for the production of biological molecules. To investigate the cellular and subcellular levels of metabolites, genetically encoded fluorescent nanosensors are used. The fluorescence resonance energy transfer (FRET) phenomenon is employed to sense the presence of small molecules (substrate, products and cofactor) in *E. coli* and eukaryotic cells. Fluorescent protein attached to an additional protein sequence that makes it sensitive. The isoleucine periplasmic binding protein (LivJ) was sandwiched between the cyan fluorescent proteins (CFP) and yellow fluorescent protein (YFP) for real time monitoring of isoleucine levels, within live cells as it is an essential amino acid that has stimulant activity. These genetically encoded nanosensors provide a new tool for fluxomics or changes in metabolites levels during cell migration. FRET nanosensors are used in bio imaging, labelling, separation, disease diagnosis and analysis of proteins for deep tissue imaging in mammals.

T5

A430

Large-scale analyses of *BES1* gene family reveal the distinct evolutionary patterns between Brassicaceae and Poaceae species

Xiaoming Song
North China University of Science and Technology

The BES1 gene family, an important class of plant-specific transcription factors, play key roles in the BR signaling pathway in plants, regulating various development processes, such as leaf development, stem elongation, pollen tube growth, xylem cell differentiation, senescence, and photomorphogenesis. Until now, there has been no large-scale pan-genomis analysis of *BES1* gene family between Brassicaceae and Poaceae species. Here, we identified 132 *BES1* genes in 19 representative angiosperms, and performed a comprehensive analysis by checking their evolutionary pattern, gene synteny, gene duplication, positive selection, and interaction network. The evolutionary pattern of *BES1* genes was determined through analysis of phylogeny and conserved motif. We detected an additional domain contained M_4 to M_8 in groups F_1 and F_2, which did not exist in all other groups. The WGD/T or segmental duplication contributed the most to expansion of this gene family among all 6 Brassicaceae species, while dispersed duplication contributed the most to expansion in 3 Poaceae species. The percentage of *BES1* genes located in collinear blocks was significant larger than genome-wide level among nearly all examined species, and this percentage in Brassicaceae was higher than Poaceae. There were more positive selection nodes in Brassicaceae than Poaceae species. In conclusion, this study provides useful resources for future studies on the function of *BES1* genes in other related species.

T5

A431

Phylogeography of *Pseudotaxus chienii* (Taxaceae), an old rare species endemic to China

Ning Li[1], *Zhen Wang*[2], *Ting Wang*[3], ***Yingjuan Su***[1]
1. Sun Yat-sen University
2. Nanjing Agricultural University
3. South China Agricultural University

Pseudotaxus chienii is the sole species in monotypic genus *Pseudotaxus* of Taxaceae endemic to China. The species is a dioecious shrub to 4 m tall with distinguished features including white stomatal bands and aril. *P. chienii* is widely but fragmentally distributed in China including montane regions of southern Zhejiang, southwestern Jiangxi, northwestern Hunan, and central Guangxi. The plant grows in warm temperate regions, where is temperate, cool, foggy, and humid with annual precipitation around 1,800-2,000 mm. As an old rare species, *P. chienii* well adapts to cold and arid environment. However, it is unknown how *P. chienii* responded to the Quaternary climatic cycles, and whether phylogeography of the plant conforms to the expansion–contraction (EC) model or exists *in situ* survival. Here, we sequenced three chloroplast DNA (cpDNA) intergenic spacer fragments (*trn*L-*trn*F, *trn*D-*trn*T, and *psb*C-*trn*S) and two nuclear DNA (nDNA) regions (ITS and *NEEDLY3*) in populations of *P. chienii*, respectively, which covered its entire distribution range. We determined 18 chloroplast haplotypes based on the concatenated cpDNA fragments. The haplotype diversity and nucleotide diversity at the species level were 0.751 ± 0.020 and 0.00059 ± 0.00005, respectively. In addition, nine ITS ribotypes and six *NEEDLY3* haplotypes were revealed. Compare to *NEEDLY3* (0.647 ± 0.033), the ITS (0.805 ± 0.014) had a higher level of haplotype diversity, and lower nucleotide diversity (ITS: 0.00471 ± 0.00016; *NEEDLY3*: 0.00593 ± 0.00031). Private haplotypes were identified in several populations of Zhejiang and Guangxi, which possessed high chloroplast and nuclear haplotype diversity and nucleotide diversity. Neutrality tests (Tajima's *D* and Fu's *FS*), the Bayesian skyline plots analyses, and approximate Bayesian computation (ABC) approach consistently showed that no population expansion occurred across the entire geographic distribution. The results of AMOVA indicated that the majority of genetic variation were partitioned among populations. Mantel test revealed that the pattern of isolation by distance existed in cpDNA (r = 0.440014; P = 0.0091) and *NEEDLY* (r = 0.639079; P = 0. 0012). We identified three groups through the BARRIER analyses, which were further conducted to isolation-with-migration (IM) models analyses with two pairs. The results showed that gene flow in both directions was close to zero for each pairwise comparison. Ecological niche model indicated that no significant range shift for the potential distribution of *P. chienii* during the last glacial maximum (LGM). We speculated that *P. chienii* likely survived *in situ* and existed multiple refugia during the LGM, which was further confirmed by ancestral area reconstructions. This work was supported by the National Natural Science Foundation of China (31570652), and the National Natural Science Foundation of Guangdong Province (2016A030313320).

T5

A432

Preliminary investigation on the genetic variation of *Amentotaxus argotaenia* (Taxaceae) based on gSSR and EST-SSR markers

Xiaoxian Ruan[1], *Qiqi Huang*[1], *Zhen Wang*[2], *Ting Wang*[3], ***Yingjuan Su***[1]
1. Sun Yat-sen University
2. Nanjing Agricultural University
3. South China Agricultural University

Amentotaxus argotaenia (Taxaceae) is a vulnerable species endemic to China. It has a wide distribution range with small isolated populations. The species has well adapted to habitat fragment and ecological heterogeneity across 11 provinces of South and Central China such as Fujian, Guangdong, Guangxi, Guizhou, and Zhejiang. The ISSR fingerprinting has indicated that *A. argotaenia* complex have low genetic diversity. However, the low level of genetic diversity is unable to provide raw material for adaptive evolution. Hence, it is necessary to accurately evaluate genetic variation of *A. argotaenia* using different molecular markers. First, we developed gSSR and EST-SSR through FIASCO method and transcriptome, respectively. Totally, 22 EST-SSRs and 23 gSSRs with polymorphism were isolated. For EST-SSRs, the number of observed alleles and the effective number of alleles were ranged from 3 to 10 and 1.097 to 3.506 per locus, respectively. For gSSRs, the number of observed alleles and the effective number of alleles was 1-13 and 1.00-7.86 per locus, respectively. These markers also successfully cross-amplified in *A. yunnanensis*. Then, we separately evaluated genetic variation of *A. argotaenia* from four natural populations using these developed EST-SSRs and gSSRs. The results showed that *A. argotaenia* possessed mediate genetic variation and high genetic differentiation. However, compared to EST-SSRs, the genetic diversity based gSSRs was higher, whereas genetic differentiation was lower. AMOVA analysis indicated that majority of genetic variation was within populations. The Mantel test based on the two markers solidly showed that *A. argotaenia* lacked pattern of isolation by distance. The STRUCTURE results of EST-SSRs and gSSRs were inconsistent. In addition, no adaptive loci was detected using Bayescan and Dfdist. MAXENT predicted that the future suitable habitat of *A. argotaenia* was decreasing in comparison with the current one. The results will lay a solid foundation for further investigating adaptive evolution and conservation strategies of *A. argotaenia*. This work was supported by the National Natural Science Foundation of China (31570652), and the National Natural Science Foundation of Guangdong Province (2016A030313320).

T5

A433

A comparative study on chloroplast genome between the invasive species *Mikania micrantha* and its indigenous congener *M. cordata* in south China: Structure variation, identification of highly divergent noncoding regions, divergence time estimation, and phylogenetic analysis

***Yingjuan Su*[1], *Lu Huang*[1], *Zhen Wang*[2], *Ting Wang*[3]
1. Sun Yat-sen University
2. Nanjing Agricultural University
3. South China Agricultural University

Mikania micrantha and *M. cordata* are only two species in genus *Mikania* (Asteraceae) in China. They exhibit high similarity in morphology, life history, and growth characteristics. However, their impacts on the environment are totally different. In south China, *M. cordata* is no harmful to native plants or habitats, while *M. micrantha* is a noxious exotic invasive weed. Comparison of differences between invasive species with its indigenous close congener can provide new insights to profoundly comprehend invasive characteristics and mechanisms. Hence, we obtained complete chloroplast genome sequence of *M. cordata* through the Illumina sequencing, and compared it to *M. micrantha* chloroplast genome. The *M. cordata* chloroplast genome was 151,984 bp in length, and possessed typical quadripartite structure. Numbers and distributions of protein coding genes, tRNA genes, and rRNA genes of *M. cordata* were identical to *M. micrantha* chloroplast genome. The main difference of two *Mikania* species laid in genome length and *ndhF*. Fourteen highly divergent regions, 235 SNPs, and 58 InDels between two *Mikania* species were identified. Phylogenetic inferences and age estimations separately demonstrated a sister relationship between *M. micrantha* and *M. cordata* and divergence of early Pleistocene. Twelve cpSSR loci were detected to be polymorphic, and adopted to preliminary survey genetic diversity and adaptation of *M. micrantha* populations. The results will lay a solid foundation to further investigate invasive mechanism of *M. micrantha*. This work was supported by the National Natural Science Foundation of China (31570652), and the National Natural Science Foundation of Guangdong Province (2016A030313320).

T5

A434

The complete chloroplast genome sequence of *Neolepisorus ovatus* (Polypodiaceae), an important medicinal and ornamental fern

Liu Shanshan[1], *Zhen Wang*[2], *Ting Wang*[3], ***Yingjuan Su***[1]
1. Sun Yat-sen University
2. Nanjing Agricultural University
3. South China Agricultural University

It is the first report on complete plastid genome of *Neolepisorus ovatus*, which is an important medicinal and ornamental fern. The complete chloroplast genome of the fern is 151,936 bp in length with 41.34% GC content. The chloroplast genome exhibits typical quadripartite structure possessing a pair of two inverted repeats (24,609 bp), a small single copy (SSC, 21,601 bp), and a large single copy (LSC, 81,117 bp). The *N. ovatus* genome encodes 132 genes, including 89 protein-coding genes, 35 tRNA genes, and eight rRNA genes. Among of them, 16 genes including six protein-coding genes, six tRNA, and four rRNA genes are duplicated within the IR regions. In total, 15 genes (*rps*16, *atp*F, *rpo*C1, *pet*B, *pet*D, *rpl*2, *ndh*B, *ndh*A, *rpl*16, *trn*G-UUC, *trn*L-UAA, *trn*V-UAC, *trn*A-UGC, *trn*I-GAU, and *trn*T-UGU) contain one intron, whereas three genes (*rps*12, *ycf*3, and *clp*P) occur two introns. The ML phylogenetic analysis based on 18 complete cpDNA from ferns showed that *N. ovatus* is has a close sister relationship with *Cyrtomium devexiscapulae* and *Cyrtomium falcatum*. The complete cpDNA of *Neolepisorus ovatus* genome will be useful for phylogeny and comparative chloroplast genomics of ferns.
This work was supported by the National Natural Science Foundation of China (31570652,31670200)

T5

A435

Expression regulation of *FATTY ACID ELONGATION1* expression by swapping: sequence variation, epigenetic modification, or spatial-temporal localization?

Xiaoqin Sun, *Zhao Lei*, *Jiayu Xue*, *Yueyu Hang*
Institute of Botany, Jiangsu Province and Chinese Academy of Sciences

FAE1 is a key gene in control of erucic acid synthesis in plant seeds. The expression of the *FAE1* genes in *Brassica napus* and *Capsella rubella*, representives of high and low erucic acid species, was characterized to provide insight into the regulation of very long chain fatty acid (VLCFA) biosynthesis in seeds. Almost none methylation was detected either in *B. oleracea* or *C. rubella*, suggesting that the modification of promoter methylation might not be a predominant mechanism. The swapping constructs were specifically designed so that the contribution of the variations in the coding regions or promoters to the changes in the *FAE1* expression levels could be quantified and compared in parallel. Significant higher fold change in erucic acid content was observed by swapping coding regions than by swapping promoters, indicating that the coding region should be a major determinant of catalytic power of KCS proteins. It was proposed that the common motifs might be essential to the preservation of the basic gene expression patterns such as the seed specific expression; by contrast, the occurrence variation of common *cis*-elements or the presence of species-specific *cis*-elements might be a plausible mechanism for changes of the expression levels in different organisms. Meanwhile, the conflicted observation in previous reports associated with the *FAE1* expression is discussed here, and we suggest that caution should be taken in selecting a plant transformation vector and in the interpretation of the results obtained from vectors carrying the CaMV 35S promoter within their regions.

T5

A436

SSR primers of *M. Truncatula* were tested for their transferability in fababean and genetic and analysis on segregation distortion of molecular markers in f2 population

***Yingying Tian**, Wanwei Hou, Yujiao Liu*

Qinghai University, Xining Qinghai Academy of Agricultural and Forestry Sciences

In this study, 216 SSR primers of *M. Truncatula* were tested for their transferability and polymorphism by PCR amplification with the genomic DNA of four fababean varietys. F_2 materials derived from a cross between Yun122×TF42 were used as mapping population in cotton. The results indicated that the transferability rate of *M. Truncatula* SSR in fababean was 76.39%, and the ratio of polymorphism SSR primers in these crops was 9.4%. Among the 84 polymorphic markers analyzed 19 markers (76%) showed the genetic distortion ($p<0.05$) in the F2 population. Among these segregation distortion SSR markers, 12 SSR marker alleles were distorted to the male parent TF42 (63.2%), 3 SSR markers allels were distorted to the female parent Yun122 (15.8%), and 4 SSR marker alleles were distorted to the heterozygote (21%).

T5

A437

The reproductive pathways of floating *Ulva prolifera* and the response mechanisms to stress

Guangce Wang

Institute of Oceanology, Chinese Academy of Sciences

Macroalgae are the important component of the marine ecosystem, which are important to the offshore environmental protection and ecological balance. *Ulva prolifera* is multicellular types of algae, artificial breeding in the South China Sea area such as Zhejiang province, used as food or feed, *etc.* The green tide has been developed in the Yellow Sea, China since the year 2008, of which free-floating *U. prolifera* is one of the causative species. In the late spring of 2008, without prior warning, the actual coverage area of *U. prolifera* was about 3800 km^2 and the thickness was about 0.5 m in less than two weeks. The influencing area was tens of thousands square kilometers, which directly affected the 2008 Beijing Olympic Games sailing competition in Qingdao. The key research of the project aims to study the photosynthetic carbon fixation in the rapid accumulation and floating state of the thalli and the molecular mechanism of upper floating thalli to stress.

(1) Rapid reproductive means and mechanism of sporogenesis in floating *Ulva prolifera*

In nature environments intact thallus of floating *Ulva prolifera* is easily cut into fragments by seawater flow, gnawing of animal, and ship propeller. Those thallus fragments are an important origin that results in rapid biomass accumulation of *Ulva prolifera.* During fragments excised from intact thallus were cultured in laboratory, the cell in thallus fragments changed greatly. Those cells developed into sporangium after 48 h culture. Subsequently zoospores were released form those sporangia.

The cells in fragments of floating *Ulva prolifera* were vegetative during 0-48 h culture. Meanwhile photochemical quantum yield of PS I Y (I) and effective PS II quantum yield Y (II) were relatively stable and the variation tendencies between them were similar. The formation of sporangia and spore release occurred after 48 h and spore release reached maximum degree at 60 h in culture. At 48 h Y (II) dropped to its lowest level, while Y (I) rose to its highest level. Those changes suggested that linear electron flow (LEF) declined while cyclic electron flow around PS I (CEF) was enhanced markedly at sporulation onset. These results indicated that both CEF enhancement and the oxidation of the plastoquinone pool are essential for sporulation onset. The further research suggested carbon fixation in photosynthesis has a relationship with sporulation. The decline of LEF can decrease inorganic carbon fixation.

(2) The pigment composition of floating *U. prolifera* thalli and the carbon source for photosynthetic carbon fixation process

U. prolifera thalli in free-floating mats could be separated into three groups according to color: light green, intermediate green, and dark green. The effective photosystem II quantum yield (YII) of the light green thalli cells had the highest value. These thalli, which are on the surface of the mats, usually are exposed to the air for long periods of time, during which it is difficult for them to acquire dissolved inorganic carbon from the seawater. In addition, although the thalli in the lower layers of the mat are in contact with seawater, photosynthesis is limited because of limited exposure to sunlight. So, why the growth rate of free-floating *U. prolifera* is extremly high when green tides occur? Our results found that *U. prolifera* thalli can utilize CO_2 in the air directly. CO_2 assimilation was enhanced as CO_2 increased in the air, even when under moderate stress, such as dehydration and salt stress.

(3) The tolerant mechanism of *Ulva prolifera* to desiccation

During green tides, massive mats of the floating *U. prolifera* could be accumulated rapidly and the thickness of the floating mats were closed to 0.5 m. Moreover, the thalli on the surface of the mats usually undergo desiccation stress lasting several hours in the daytime. We found that during desiccation, the PSI-driven cyclic electron flow in *U. prolifera* increased significantly. Although the activity of cyclic electron flow is lower than that of linear electron flow which dominates under optimal conditions, the cyclic electron flow might provide some protection of the photosynthetic apparatus and extra ATP under desiccated conditions. In addition, the activity of PSI was restored faster than that of PSII during re-hydration. Based on these results, we propose that the physiological tolerance and stability of photochemical systems, especially PSI-driven cyclic electron flow, might be one of the most important factors that make *U. prolifera* well suited to withstand repeated cycles of desiccation and re-hydration during daily low and high tides. These results will increase our understanding of the tolerant mechanism of the intertidal macro-algae to desiccation and will lay the foundation for the solution of green tides.

T5

A438

Integrated gene expression analysis of maize (*Zea mays L.*) leaves under biological stress reveals several sensitive loci related to pathological mechanism

Jinglu Wang, *Xinyu Guo*

Beijing Key Lab of Digital Plant, National Engineering Research Center for Information Technology in Agriculture

Large-scale expression profiling has become a favorable method for understanding the molecular mechanism of plant expression, and some progress has been made in botany research. Among these, microarray technology was used as a practical method to analyze the gene dysfunction contributed to maize leaves under different stress, in particular, the abiotic stresses. Recent years, plenty of gene expression profiling studies of the changes of maize leaves under biological stress have been published. But the heterogeneity of each study is apparent because of various chip platforms and data processing procedures. As a consequence, an integrated analysis for the genome-wide expression data sets of maize leaves should be conducted. In order to make the results comparable, the same pre-processing and quality control procedures are necessarily used to analyze each data set separately. Notably, a standard linear mixed-effects model would be used to compute the differentially expressed genes. And pathway enrichment analysis of the DEGs would conducted by DAVID to explore their functions. Through the analysis of the differences in gene expression between healthy leaves and leaves infected by pathogens, we can not only identify some candidate loci, but also can help us understand the phenotypic differences between different maize species.

T5

A439

Hierarchically aligning plant genomes establishes genomics research platform and contributes to understanding polyploid evolution

Jinpeng Wang[1,2], *Pengchuan Sun*[1,2], *Yinzhe Liu*[1,2], *Yuxian Li*[1,2], *Ruiyan Xia*[1], *Xuelian Ma*[1], *Jigao Yu*[1,2], *Nanshan Yang*[1,2], *Sangrong Sun*[1,2], *Tianyu Lei*[1,2], *Xiaojian Liu*[1], *Beibei Jiao*[2], *Yue Xing*[2], *Weina Ge*[1,2], *Li Wang*[1,2], *Zhenyi Wang*[1,2], *Lan Zhang*[1,2], *Xiaoming Song*[1,2], *Di Guo*[1,2], *Zhang Jiaqi*[1,2], *Dianchuan Jin*[2], *Yuxin Pan*[1,2], *Tao Liu*[2], *Ling Jin*[2], *Jinshuai Sun*[1], *Jiaxiang Yu*[1,2], *Rui Cheng*[1], *Xueqian Duan*[1], *Shaoqi Shen*[1], *Xiyin Wang*[1,2]

1. School of Life Sciences, North China University of Science and Technology, Tangshan, Hebei 063000, China

2. Center for Genomics and Computational Biology, North China University of Science and Technology, Tangshan, Hebei 063000, China

Polyploidy contributes to the divergence and domestication of land plants. However, for the sophisticated complexity of plant genomes due to recursive polyploidizations, often it is much difficult to deconvolute their structures . By performing comparative genomics analysis, we managed to make a hierarchical and event-related alignment of the 8 Poaceae, Pineapple, Cotton, Cacao, and Grape genomes, deconvoluted layers of the aligned genomic regions correspond to ancestral polyploidizations and speciations, and deciphered gene-colinearity-supported paralogs derived from these events. Similarly, we showed multiple cotton genome structures affected by decaploidy in their common ancestors. With these data sets, we explored how polyploidization contributed divergently to copy number variation of gene families. These present efforts laid a solid foundation further genomics exploration in plant research community and beyond.

T5

A440

Subcellular locations of potential cell wall proteins in plants: Predictors, databases and cross-referencing

Qinbin Zhang, *Xiaolin Wu*, ***Wei Wang***

Henan Agricultural University

The cell wall is the most striking feature that distinguishes plant cells from animal cells. It plays an essential role in cell shape, stability and growth as well as protection against environmental stresses. Despite being present in small amounts, cell wall proteins (CWPs) continue to emerge as crucial components of the plant cell wall. In general, the cell wall proteome consists of components such as *sensu stricto* CWPs, apoplast proteins, and extracellular secreted proteins. Compared with proteins in intracellular compartments, the bioinformatic prediction of the subcellular location of possible CWPs is challenging because many aspects of these proteins have not been experimentally characterized. Furthermore, there are no specific, accurate predictors. Currently, there is a need for the bioinformatic analysis of a tremendous number of protein sequences (including possible CWPs) that have been generated from genomic, transcriptomic and proteomic research. By introducing the biological relevance of the cell wall and CWPs, we critically evaluate state-of-the-art predictors for the subcellular location of CWPs based on a large, independent database of Arabidopsis and rice proteins. We also propose promising predictors that are suitable for CWPs. Furthermore, we outline the publicly available databases that can be used for cross-referencing the subcellular location of CWPs. Lastly, we propose a flowchart of subcellular location predictions and a cross-reference of possible CWPs. We also discuss challenges and perspectives in the bioinformatic analysis of CWPs. This work provides practical guidance regarding CWPs for non-specialists in bioinformatics.

T5

A441

Mitochondrial phylogenomics of Nectriaceae (Ascomycota, Fungi)

Xin-Cun Wang, *Zhao-Qing Zeng*, ***Wen-Ying Zhuang***

State Key Laboratory of Mycology, Institute of Microbiology, Chinese Academy of Sciences

Nectriaceae includes a group of fungi with agricultural, industrial and medicinal importance. Phylogeny of the family inferred from multiple nuclear genes have been thoroughly studied, however, phylogenetic inference based on mitochondrial genes is hitherto lacking. In this study, 26 mitochondrial genomes representing 21 genera of the family were determined using the next-generation sequencing technology. The genomes are all circular molecules. The genome size ranges from 27,129 bp to 149,875 bp, and the GC contents are from 25.95% to 32.61%. Gene prediction revealed at least 42 genes encoding 15 conserved proteins, no less than 25 tRNAs, and large and small ribosomal RNAs. All genes are located on the same strand. The number of introns in the protein-coding genes reaches up to 40, while no intron was found in ten mitochondrial genomes. The mitochondrial genome size was correlated with numbers of intron and tRNA, but had no effect on phylogeny of the family. The same gene order of 42 genes was shared by eight species. Genome rearrangements include gene lost, tRNA gene translocation and duplication. GC content and some variations in mitochondrial genome structure carry phylogenetic information. The gene order of 15 protein-coding genes was conserved in Nectriaceae, which were also found in other four families of Hypocreales and the related orders of the class Sordariomycetes. It represents an ancestral state in the class. Mitochondrial phylogeny of Nectriaceae was reconstructed using the concatenated amino acid sequences of the 15 protein-coding genes derived from 42 mitochondrial genomes. The tree topology was very similar to that obtained from the multiple nuclear genes. Potential changes among genera of the family were roughly investigated. Mitochondrial genomes of Nectriaceae will give us a better understanding of phylogeny and evolution of the group and its related fungi.

T5

A442

Ancestral karyotype reconstruction unravels independent genomic repatterning between Triticeae and *Brachypodium*

Zhenyi Wang
North China University of Science and Technology

Based on the theory of our latest research results, we found that *Brachypodium distachyon* was not of enormous relevance for comparative genomics, evolutionary studies of large genomes of the Triticeae, wheat and barley have their unique evolution process, rather than the other researchers concluded that, the common ancestor of *Brachypodium* and the core Pooideae had a genome structure quite similar to that of rice, genome rearrangements between *B. distachyon* and Triticeae are independent of rearrangements between *B. distachyon* and rice, consistent with the phylogeny. The results of our study is different from their researches, the common ancestor of Triticeae and rice had a genome structure quite similar to that of *Brachypodium distachyon*, the genome evolution of Triticeae is distantly related to *Brachypodium distachyon*, while related to rice. Thus rice could serve as an anchor genome to help locate important chromosomal locations in Triticeae crops.

T5

A443

Preliminary functional characterization of a rice mitogen-activated protein kinase (MAPKK-426)

Jian Wei, Jianshuo Zhang, Yalu Guo, Jianan Shi, Shijuan Dou, Liyun Li, Lijuan Liu, ***Guozhen Liu***
College of Life Sciences, Hebei Agricultural University

Rice is one of the most important food crops and also an important model plant in the world. Most of the mitogen-activated protein kinase (MAPKs) in plant is serine/threonine protein kinase, which stimulated by a variety of extracellular stimuli and transmit signals to the nucleus. In this research, a MAPKK was chosen as target (Namely MAPKK-426). Over expression and RNA interference vectors were constructed rice TP309 was transformed by Agrobacterium-Mediated protocol In T0 generation, 20 over expression Lines was obtained, PMI protein detection by western blot analysis verified 17 of them are positive and 3 are negative ones. Fifty-one lines were obtained via two different RNAi constructs and detected by NPT-II antibody. It showed that 22 positive and 29 negative. In T1 generation, genomic DNA were extracted and PCR analysis were carried out. Three homozygous lines and 2 heterozygous lines were identified in 5 over expression lines tested. Ten homozygous lines and 16 heterozygous lines were identified among 26 lines derived from two RNAi constructs. In T2 generation cultivated from positive seeds, 4 homozygous lines and 4 heterozygous lines were identified in 8 over expression lines detected by PCR, 10 homozygous lines and 10 heterozygous lines were identified from RNAi constructs. To monitor the expression of target protein, western blot analysis was carried out by specific antibody, it showed that the expression of target protein were up-regulated in 6 over expression lines, while it down-regulated in 3 homozygous RNAi Lines. It also found that RNAi plants showed dwarf and etiolation compared with the control in tillering stage. Taken together, stable transgenic rice with over or down-regulated MAPKK-426 Lines were identified with phenotypic changes, which will helpful for the functional characterization of the gene.

T5

A444

Genome-wide analysis of DREBs subfamily in foxtail millet

***Songchang Shi**, Zhang Ruye, Zhao Zhenting, Yang Lei, Ge Weina*
College of life sciences, North China University of Science and Technology

DREBs (dehydration responsive element binding) are unique transcription factors in plants, and play important roles in drought, cold, salt and other environmental stresses. Bioinformatic analysis of the DREB subfamily was performed at the genome-wide level in foxtail millet (*Setaria italic*). 65 genes encoding DREB were found in the millet genome, which were distributed on 9 chromosomes. According to the sequence characteristics, they were divided into six groups (A1 to A6). Each group has their own characteristic in the gene structure and the composition of protein motifs. Segmental duplication and tandem duplication play important roles for the expansion of the *SiDREB* subfamily. *SiDREB* genes were strongly positive selected during the evolution of millet, and that was also the reason for the rapid evolution of *DREB* genes. These results will lay a foundation for further study on the function of *DREB* genes, and provide some clues for revealing the mechanism of the resistance of millet.

T5

A445

Distinct transcriptome profiles reveal gene expression patterns during fruit development and maturation in five main cultivated species of pear (*Pyrus L.*)

Jun Wu, *Ming-Yue Zhang, Cheng Xue*
Centre of Pear Engineering Technology Research, State Key Laboratory of Crop Genetics and Germplasm Enhancement, Nanjing Agricultural University

The transcriptomes of five pear cultivars, 'Hosui' (*P. pyrifolia*), 'Yali' (*P. bretschneideri*), 'Kuerlexiangli' (*P. sinkiangensis)*, 'Nanguoli' (*P. ussuriensis*), and 'Starkrimson' (*P. communis*) were sequenced at seven key fruit developmental stages, from fruit setting to maturation and fruit senescence after harvesting. In total, 33,136 genes that could be mapped by reads, were analyzed. Most gene expression cluster models showed a steadily decreasing trend. Gene expression patterns had obvious differences according to maturity type, that is, post-ripening cultivars were still vigorous at maturity, and showed a higher proportion of up-regulated genes; non post-ripening cultivars had a gradually decreasing tendency during fruit maturation. Meanwhile, differentially expressed genes related to fruit quality and development, such as stone cells, sugar, acid and hormones, were identified. Co-expression analysis revealed that several ethylene synthesis genes and polyphenoloxidase-related genes interacted with each other directly, and an indirect relationship was reflected between ethylene synthesis genes and ethylene response genes. In addition, the highly diverse SNPs represented the great differences between oriental and occidental pears. Understanding how RNA-seq based gene-expression patterns and differential gene expression contribute to fruit quality allows us to build models for gene-expression for fruit development of *Pyrus* species.

T5

A446

Improving eating and cooking qualities of rice grain by molecular marker-assisted breeding

Mingwei Wu[1,2], *Chunming Liu*[1]
1. Key Laboratory of Plant Molecular Physiology, Institute of Botany, Chinese Academy of Sciences
2. University of Chinese Academy of Sciences

Rice (*Oryza sativa* L.), one of the important cereal crops, is the major stable food for more than 50% population in the world. Increasing rice grain yield is a crucial challenge, and improving its eating and cooking quality (ECQ) is also an important breeding goal. Starch is the major storage material in rice seeds, which comprises two main types of glucan homopolymer: amylose and amylopectin. The eating and cooking qualities of rice grain are closely related to the ratio of amylose to amylopectin and the architectural features of amylopectin. For improving the ECQ of cultivated species in Northeast of China, the apparent amylose content (AAC) and Rapid Visco Analyser (RVA) profile of 130 rice varieties were measured; nine rice varieties showed highly starch quality were used as donor candidates, and hybridized with the cultivated specie Longjing26 in Northeast of China expecting ultimately to obtain the superior rice variety. Meanwhile, we hope to find some quantitative trait loci (QTLs) related to the starch quality by constructing Recombinant Inbred Line (RIL). Now, some AAC related QTLs have been found by mapping the Recombinant Inbred Line of RCC147xLongjing26; next, we hope to use those QTLs to improve ECQ of Longjing26

T5

A447

Batch annotation tool, a novel automatic annotation tool for chloroplast genome

Ping Wu, *Shiliang Zhou*
Institution of Botany, Chinese Academy of Science

Chloroplast genomes is widely used in plant phylogenetic studies, especially for angiosperms due to widespread maternal inheritance. Previously the application of chloroplast genome sequence was limited to several genes or fragments, such as rbcL, matK, trnH-psbA, *etc.* In recent year, the price of high throughput sequencing was dramatically decreased, and the sequencing technique for chloroplast became mature. All of above lead to an explosion for sequencing chloroplast. In spite of this, the lack of automatic and accurate annotation tools slowed the pace of utilizing these data. The annotation of chloroplast still stay on the manual phases that researchers have to waste lots of time on software or website. Moreover, the correction of errors caused by those tools merely depend on their experience, which may bring possible mistakes that was neglected before. Herein, we developed a native software Batch Annotation Tool (BAT) to solve this problem. Based on existing chloroplast genomes in Genbank integrated into BAT, there are two strategies for choice. First, BAT could use high resolution fragments such as ycf1 and trnH-psbA to identify given input sequence to find the most similar reference genome and then generate annotation information according to reference sequence. This method is suitable for numerous similar samples, for example, that chloroplasts from same species but comes from different habitats. Second, BAT splits given sequence into discontinuous parts and search them in the database of BAT by similarity. To reduce errors, BAT repeats this process with different parameters to get multiple possible annotations and keep the most common one. This method could be used to annotate chloroplasts without allied species in database or various chloroplasts comes from distant species at one time. BAT could batch annotate chloroplast genomes quickly, robotically and only require assembled FASTA format sequence of genomes as input. With little manual intervention, accurate annotation file was generated as proper format that is suitable for submitting to Genbank. We hope this tool could reduce the work of data process and accelerate the utilization of chloroplast genome in plant phylogenetics.

T5

A448

Composition and function of the epiphytic microbial community of the red alga *Gracilaria lemaneiformis*

Xinfei Xie
Institute of Hydrobiology, Department of Ecology, Jinan University

The red alga, *Gracilaria lemaneiformis*, is cultivated widely along the coast of China and has proved to be an effective bioremediation in coastal ecosystems. Even though the important role of *G. lemaneiformis* in improving water quality has been acknowledged, little is known about the composition and function of the epiphytic microbial communities living on its surface. In this study, we use 16S rRNA gene amplicon and ITS sequencing, and GeoChip 5.0 to determine the composition, function and variability of the epiphytic community on the surface of *G. lemaneiformis* as well as the free-living microbial community in the surrounding water. Significant divergences in composition and function of the microbial community were observed between epiphytic groups and free-living groups according to a dissimilarity test. Some epiphytic genera, such as *Croceitalea*, *Formosa*, *Granulosicoccus*, *Leucothrix*, *Polaribacter*, *Aureobasidium*, *Penicillium* in high relative abundances specialize in an alga-associated lifestyle with agarolytic or algal polysaccharides degrading potential. Genes associated with adaptation or response to the seaweed surface is overrepresented in the epiphytic community. Most genes mediate attenuation of the biogeochemical cycle, while assimilatory N reduction, sulfide/sulfur oxidation, chlorinated solvents and herbicide related compounds are accelerated in the epiphytic community compared with the free-living community. Our results demonstrated that the special microbial community on the surface of *Gracilaria lemaneiformis* plays an indispensable role in its life cycle. Understanding the composition and function of the epiphytic microbial community of *G. lemaneiformis* helps define the basic nature of the interaction between epiphytic microbes and the host seaweed.

T5

A449

Functional analysis of PmNF-YB transcription factors in *Prunus mume*

Jie Yang[1,2], *Xueli Wan*[1], *Junwei Zhang*[1], *Manzhu Bao*[1]

1. Key Laboratory of Horticultural Plant Biology, Ministry of Education, College of Horticulture and Forestry Sciences, Huazhong Agricultural University

2. School of Nuclear Technology and Chemistry and Biology, Hubei University of Science and Technology

Mei flower (*Prunus mume* Sieb. et. Zucc.) is one of the tradition famous flowers and has very long cultivation history in China. It has high economic value and outstanding ornamental features, such as trees' pose, flower's fragrance and color. In this study, a total of 6 *PmNF-YA*, 13 *PmNF-YB*, and 8 *PmNF-YC* genes in the *Prunus mume* genome were identified and annotated by bioinformatics procedures. Based on Real-time PCR analyses, we investigated the gene expression patterns of all *PmNF-Y* genes in ABA, osmotic, salt, ROS stresses. *PmNF-YA1/2/4/5/6*, *PmNF-YB3/4/8/10/11/13*, and *PmNF-YC1/2/4/5/6/8* were found to be up-regulated expression under ABA and osmotic treatment. The expression levels of *PmNF-YA1/2/3/4/5/6*, *PmNF-YB1/3/8/10/11/13*, and *PmNF-YC1/2/5/6/8* were obviously induced by ROS stress. In addition, only 2 *PmNF-Ys* (*PmNF-YA2* and *PmNF-YB3*) also enhanced expression in salt stress. These findings suggested that *PmNF-Y* genes may be involved in multiple abiotic stresses response. Based on the gene expression patterns analysis of *PmNF-YB* genes under abiotic stresses, three *PmNF-YB* genes (*PmNF-YB3*, *PmNF-YB8* and *PmNF-YB13*), whose expression levels were increased significantly under abiotic stresses, were cloned from *P. mume* 'Xue Mei' and overexpression vectors were constructed. Then these genes were transformed to *Arabidopsis* to investigate their functions in plant stress response. We found that transformed *Arabidopsis* with over-expressing *PmNF-YB3* or *PmNF-YB8* exhibited higher seed germination and faster root elongation under salt and osmotic treatment compared to wild type. So, the result showed that transgenic *Arabidopsis* overexpressing *PmNF-YB3* or *PmNF-YB8* improved tolerance to salt and osmotic, while overexpression of *PmNF-YB8* or *PmNF-YB13* in *Arabidopsis* showed higher survival rate than wild type under drought stress, which hinted that these *PmNF-YB* genes were involved in response to abiotic stresses.

T5

A450

A new origin discovery of hybridization and introgression in *Nicotiana tobaccum*: An universality of the complex process of allotetraploid evolution

Wang Yi[2], *Li Huayang*[1], ***Yang Long***[1]

1. Agricultural Big-Data Research Center, College of Plant Protection, Shandong Agricultural University

2. Beijing Key Laboratory of Grape Science and Enology and Key Laboratory of Plant Resource, Institute of Botany, Chinese Academy of Sciences

Nicotiana tabacum is the most important model plant in the plant science, it plays a vital role in the molecular science and plant protection studies. Previous studies reveal *N. sylvestris* (S genome) is the maternal donor and *N. tomentosiformis* (T genome) is the paternal donor of *N. tabacum*. Although *N. otophora* (O genome) is a sister species to *N. tomentosiformis*, limited evidence suggests that *N. otophora* is an alternative paternal donor, too. In this study, we used 6 draft genomes of the *Nicotiana*, including three wild types (*N. sylvestris, N. tomentosiformis* and *N. otophora*), three cultivation types (*N. tabacum* var. K326, *N. tabacum* var. TN90 and *N. tabacum* var. BX). Using the comparative genomics and bioinformatics methods, all wild types were clipped into 100-mer sequences and mapped to the genome of the cultivation, and then the mapped region were surveyed according to the depth of the mapping result. The results show that the modern cultivation type contains the genetics information of all three wild types, more than 2.15 G bases genome are found high conversed between the wild and modern cultivation types. Among these conserved regions, the distribution frequency is very consistent across each cultivation type. The average frequency of *N. tabacum* is 54.10% from S genome, 40.04% from T genome and 5.86% from O genome. That means that *N. otophora* also is one of the donor of *N.tabacum*. The genome of modern cultivation types are mainly come from the S, T and O genome. It is supposed that there are three steps of the origin of the modern cultivation types. Firstly, the genome of O and T combine into a middle species. Secondly, the middle species hybrid with the genome S. Thirdly, an allotetraploid comes from natural doubling of chromosomes of hybrids. Besides these, there may be some other wild types also take part in the origin of *N.tabacum*. In the comparison and concluding, a new origin model of *N. tabacum* is conjectured as follow: S × {[

(O × T) × T] × T}. This model shows the main hybrid generation process of *N. tabacum*. In this model, S genome (50%) as the maternal donor crosses with the hybrids from O and T genome at the nearest generations. And an interspecific hybrid from T genome (43.75%) backcross three times with O genome (6.25%) is the paternal donor of *N. Tabacum*. This finding provides a valuable evidence for the presumed origin of hybridization and introgression in *N. tabacum*, and also provides an universality of the complex process of allotetraploid evolution.

T5

A451

ESCRT-I component VPS23A affects ABA signaling by recognizing ABA receptors for endosomal degradation

Feifei Yu[1,2], *Lijuan Lou*[1], *Miaomiao Tian*[1], *Qingliang Li*[1], *Yanglin Ding*[3], *Xiaoqiang Cao*[1], *Yaorong Wu*[1], *Borja Belda-Palazon*[4], *Pedro L. Rodriguez*[4], *Shuhua Yang*[3], *Qi Xie*[1]
1. Institute of Genetics and Developmental Biology, Chinese Academy of Sciences
2. University of Chinese Academy of Sciences
3. China Agricultural University
4. Consejo Superior de Investigaciones Científicas-Universidad Politécnica de Valencia

The recent discovery of PYR/PYL/RCAR-type ABA receptors has been one of most significant advances in plant science. In mammals, endosome sorting acts as an important pathway to down-regulate different types of receptors, but this kind of regulation of hormone signaling is poorly understood in plants. Here, we report that an ubiquitin E2-like protein, VPS23A, also a key component of ESCRT-I, negatively regulates ABA signaling. *VPS23A* has epistatic relation with *PYR/PYL/RCAR* type ABA receptors and deletion of *VPS23A* enhances the activity of key kinase OST1 in ABA signaling pathway under ABA treatment. Moreover, VPS23A interacts with PYR1/PYLs and K63-linked diubiquitin, and PYL4 possesses K63-linked ubiquitinated modification *in vivo*. VPS23A affects the subcellular localization of PYR1 and stability of the PYL4 by *vps23a* mutant analysis. These findings supports that VPS23A affects PYR1/PYL4 via vacuole-mediated degradation besides 26S proteasome system, further strengthen our understanding of both the turnover of ABA receptors and ESCRTs in plant hormone signaling.

T5

A452

A brine shrimp (*Artemia salina* L.) larvae drug toxicity screening model based on microfluidics

Qiyuan Chen, Ying Liu, Yuhuan Wu, ***Zhiming Yu***
College of Life and Environmental Sciences, Hangzhou Normal University

The vast majority of plants can't avoid or evade danger and adversity like animals. They have to tolerate the environmental, biological and other adverse effects in their situation. To cope with harmful and unfavorable conditions, plants have to rely on their own chemical diversity during their metabolic pathways. Our smart and industrious human ancestors have used the plants' secondary metabolites to treat many serious diseases. So far, humans have found a lot of drugs from the plants. For example, Artemisinin, extracted from plant *Artemisia annua* L. (Asteraceae), is one of the most effective malaria treatment today. Terpene lactones, extracted from *Ginkgo biloba*, have been clinically proven and widely used in clinic as inhibitors of platelet activating factor. Plants' secondary metabolites are a treasure house for mankind, but the current research results are only the tip of the iceberg. How to track active ingredients in a simple and effective way is a prerequisite for discovering more valuable compounds from medicinal plants. As a toxicology and pharmacological pre-screening model materials, Brine shrimp (*Artemia salina L.*) lethality assay is low-cost and user friendly. Early use of this model only concerned its IC50 (half maximal inhibitory concentration), so some of the target analytes' potential biological characteristics are not given sufficient attention. For this reason, smarter model should be proposed. Brine shrimp, a genus of aquatic crustaceans, has changed little externally since the Triassic period. While the phototaxis of brine shrimp larvae has been studied and widely identified, there are few published articles using this feature to develop a screening model. We have subtly constructed a new screening model by using brine shrimp larvae's phototaxis feature. To assess the test subjects' pharmacological and toxicological activity precisely as well as scientifically, we have introduced microfluidic technology to our brine shrimp model. The brine shrimp larvae exposed to the testing samples are cultured in darkness. After treatment, a beam of light passes through microfluidic chip which is connected to the culture pond. Due to the desire to the light, the more energetic larvae will swim into the tubular space. The space only allows one larva entering once a time and also has no enough space to make the larvae turn around. According to the swimming distance of a larva, we can get quantitative data about the sample' physiological activity to the brine shrimp larvae. Using this new evaluation system, we can give standards for chemical compounds or crude extracts from medical plants. From these standards, we may understand the potential properties and the possible biological value of the compounds.

The project was supported by NSFC (Grant No. 31200913 and No. 41571049)

T5

A453

CPK11 phosphorylates the transcription factor FD in *Arabidopsis thaliana*

Min Yuan[1,2], *Kyle Bender*[1], *Steve Huber*[1], *Yoshie Hanzawa*[1]
1. University of Illinois at Urbana and Champaign
2. North China University of Science and Technology

Calcium-dependent protein kinases (CPKs), a class of Ser/Thr protein kinases constituting a large multigene family, are known as a sensor of Ca^{2+} flux and play important roles in various physiological processes, including cell growth, development and stress responses in plants. Recently, CPK6 and CPK33 are implicated in the regulation of flowering time control through phosphorylation of a bZIP transcription factor, FD. Here we report characterization of CPK11 in phosphorylation of FD. CPK11 is distantly related to CPK6 and CPK33. *CPK11* gene was cloned into the PET28a expression vector and transformed into *E.coli* strain BL21 to obtain the His_6-CPK11 recombinant protein. The His_6-CPK11 recombinant protein was induced by 0.5mmol/L IPTG at room temperature. The Ni-NTA system was used to purify the His_6-CPK11 recombinant protein and the purified protein was specifically detected by anti-His antibody in the western blot assay. Our *in vitro* kinase assays demonstrated that His_6-CPK11 phosphorylated GST-tagged FD protein. Our result provides further evidence to support the role of CPKs in FD phosphorylation in flowering control.

T5

A454

MicroRNA roles in cotton fiber initiation and early development

Baohong Zhang[1], *Qinglian Wang*[2]

1. East Carolina University

2. Henan Institute of Science and Technology

Cotton fiber development is a fundamental biological process; investigating cotton fiber initiation and development provides a unique window into the regulation of cell differentiation, cellulose biosynthesis, and further increasing cotton fiber quality and yield. For several decades, a great deal of research has been aimed at elucidating the underlying molecular pathways. Nevertheless, the mechanisms by which cotton fiber differentiates and develops remains unclear. In our previous studies, we found that microRNAs (miRNAs) are differentially expressed among cotton fiber and other tissues. Comparing the miRNA expression profiles based on deep sequencing technology, we found that both conserved and novel miRNAs have unique pattern in cotton developing ovule. To further test the potential role of miRNAs during cotton fiber development, we selected 58 miRNAs and profiled their expression in fiber with different development stages (from -2 DPA to 20 DPA) in both wild type and fibreless mutants. Using transgenic, RNAi and genome editing technologies show that overexpression and knockout/knockdown of an individual miRNAs affected cotton fiber development and further affect cotton fiber length and quality.

T5

A455

The complete chloroplast genome of dendrobium (Orchidaceae): Phylogenetics and a 'cp-DNA-Barcode'

*Zhimin Zhong, **Guifang Zhang**, Xiaoping Lai, Xiaoqiao Yan, Yifan Li*

Guangzhou University of Chinese Medicine

DNA barcoding is currently a widely used and promising tool for rapid and accurate identification of plant species. However, because of a much slower mutation rate, a universal barcode in plants has not found yet. *Dendrobium*, one of the largest and most diverse genera of Ochidaceae, is generally under severe threat of extinction mainly due to over-collection and habitat destruction. Methods for accurate classification and identification of *Dendrobium* species are critical in the fields of biological conservation and sustainable utilization of *Denbrobium* as plant resources. Here, the complete chloroplast genome (cp-genome) of 50 samples from 41 *Dendrobium* species was sequenced by using Illumina and PacBio sequencing platform. The sequences were assembled using the Hierarchical Genome Assembly Process 2 (HGAP2) protocol from SMRT Analysis version 2.0 package (18), resulting in the complete circular 152,185 bp cp-genome. The cp-genome of *Dendrobium officinale 'zhong ke* Ⅳ *hao'* was assembled using the SOAPdenovo (v2.04) and Celera Assembler (v8.0) and annotated with Dual Organellar GenoMe Annotator (DOGMA). Genetic distances and neighbor-joining analysis was performed with Mega 6.06. The success rate of PCR amplification and sequencing was 100%. Many single-nucleotide polymorphisms (SNPs) demonstrated a significant divergence between the inter- and intra-specific genetic distance of cp-genomes. Our results showed that the cp-genome could be function as a the 'cp-DNA-barcode' and the traditional Chinese medicine regional marker sequences, which could be used to infer phylogenetic relationships of *Dendrobium* species and to identify the same *Dendrobium* species from different origins. In addition, results of a phylogenetic tree and principal component analysis showed that the cp-genome could be effectively used to classify and identify this family and could provide a reference for study of the molecular systematics of *Dendrobium*. Additionally, these new potential 'cp-DNA-barcodes' would contribute to the phylogenetic analysis of plants more generally.

T5

A456

Preliminary functional characterization of a rice mitogen-activated protein kinase (MAPK-520) in the resistance response mediated by *Xa21*

*Jianshuo Zhang, Yalu Guo, Jian Wei, Jinjiao Ma, Jianan Shi, Shijuan Dou, Liyun Li, Lijuan Liu, **Guozhen Liu***

College of Life Sciences, Hebei Agricultural University

Rice is one of the most important food crops in the world, and it is also an important model organism in plant molecular biology. Bacterial blight is one of the most serious diseases in rice causing dramatic yield losses. *Xa21* is the first cloned resistance genes in rice against *Xanthomonas oryzae* pv.*oryzae* (*Xoo*), which encoding a receptor-like kinase. It is proposed that MAPK cascade play important role in the *Xa21*- mediated signal transduction pathway. Most of the mitogen-activated protein kinase (MAPKs) in plant belongs to serine/threonine protein kinase, which stimulated by a variety of extracellular stimuli and transmit signals to the nucleus. In this research, a MAPK was choosen as target (Namely MAPK-520). Over expression (OX) and RNA interference vectors were constructed, rice TP309 confers *Xa21* gene was transformed by agrobacterium-mediated protocol. In T0 generation, 25 OX lines were obtained, PMI protein detection by western blot analysis verified 18 of them are positive and 7 are negative ones. Thirty-eight lines were obtained via two different RNAi constructs, and detected by NPT- II antibody. It showed that 28 positive and 10 negative. In T1 generation, genomic DNA was extracted and PCR analyses were carried out. Eight positive OX lines and 20 positive RNAi lines were found. In T2 generation, 3 homozygous lines were identified by PCR in OX lines, 9 homozygous lines were identified from RNAi constructs by PCR. To monitor the expression of target protein, western blot analysis was carried out by specific antibody, it showed that the expression of target protein of 3 PCR positive OX lines were up-regulated, while it down-regulated in 6 RNAi lines. Innoculation of *Xoo* at tillering stage revealed alteration of resistance mediated by *Xa21*. Taken together, stable transgenic rice with over or down-regulated MAPK-520 lines were identified, which will be helpful for the functional characterization of the gene in the process of *Xa21*-mediated resistance.

T5

A457

Comparative genomics research of *Amborella* and several flowering plants

Lan Zhang, *Min Yuan*
North China University of Science and Technology

Angiosperm, as the most important plants in the worldwide, has experienced a complicated evolution. In order to illustrate the evolution process of angiosperm, the genomes of *Amborella, Arabidopsis*, rice, grape, peach and cacao were taken as the study objects in this study. The genomes of *Arabidopsis* have went through another two times diplodization event, while the genomes of *Arabidopsis*, grape, peach and cacao have experienced the same eudicot-common triplication. Rice has went through three times diplodization event. *Amborella*, the single living species of the sister lineage to all other extant flowering plants, provides a unique reference of ancestral gene content and genome structure in the most recent common ancestral of all living angiosperms. We performed comparative genomics analysis of *Amborella* and these flowing plants, and found that the more complicated genome ployploidy, the less number of homologous blocks and genes between *Amborella a*nd other flowering plants. According to the length of homologous blocks was greater than 20, there were 62 homologous blocks and 1,776 homologous genes between *Amborella* and grape, which was the longest blocks in the study. The least number of homologous blocks and genes existed between *Amborella* and *Arabidopsis*, including 1 homologous block and 20 homologous genes. Finally, this study would provide a basis for studying the angiosperm evolution.

T5

A459

Complete pear mitochondrion genome assembly revealing 129 pear samples classification and maternal inheritance

Ming-Yue Zhang, *Jun Wu*
Nanjing Agriculture University

Mitochondrion is an important organelle mainly responding for energy metabolism. We assembled the complete pear mitochondrial genome, and genome size is 458,897 bp, with G+C contents of 45.2%. A total of twenty-eight open reading frames in the assembled mitochondrial genome were annotated, including fifteen electron transport complex, one intron maturase, one transport membrane, five cytochrome C biogenesis. What is more, twenty tRNA, three rRNA genes and eight ribosomal subunits were also presented. There were 129 pear samples distributed around the world to reveal the pear classification and wild origin in our study. Occidental pears manifested great differences from the oriental pears in the genetic level, and the same species varied because of geographic distribution. Maternal inheritance should be the main inheriting method for pear mitochondrion and chloroplast, but ‘Chikusui’ is an exceptant. Furthermore, we suppose pear mitochondrial genome is less diverse than the chloroplast genome based on mutation rates and tandom repeats. The co-linearity of mitochondrion from distinct species varied a lot, but the gene region had a relatively high co-linearity. Comparative mitochondrial gene expression revealed a potential tRNA gene perhaps responsible for pear post mature character.

T5

A460

Field-based phenomics platform: Image-based methods for analyzing morphological traits and agronomic performance under different nitrogen treatments for sweet potato

Yu Zhang
The University of Tokyo

Spatial variability in a crop field creates a need for precision agriculture. Early detection and management of problems associated with crop yield indicators can help increase yield and subsequent profit. Remote sensing can be used to assess spatial variability in crop yield. While highly predictable spatial variability may be amenable to multiple approaches of site-specific N management, strong temporal variability is much more difficult to manage. Remote sensing techniques that use the crop to indicate its N status show considerable promise for improving N management. Infrared and near-infrared (NIR) imaging systems are advantageous in their ability to visualize plant grown in field, but are currently limited by their small analysis volume and often low resolution and precision, as well as their cost, accessibility, and low-throughput. With further advancements, IR/NIR technologies have the greatest scale-up potential for the eventual non-destructive imaging and phenotyping of field-grown plant root systems. Although these current root growth systems and imaging technologies are still unable to accurately visualize and quantify complex, mature plant root systems grown under field conditions, they have contributed greatly to increase the precision and efficiency of 2D and 3D spatial. Management of fertilizer and watering in agriculture is very difficult. A part of N is lost in the atmosphere or leached into groundwater, lakes, and rivers, which causes severe environmental pollution. Despite that major crops can utilize only 30 – 40% of the applied N. In addition, plant response to each N source show stage specific difference and very sensitive. To solve this case of problem, development of remote sensing technology is useful to evaluate agronomic performance. In this study, we set up three different N application fields to make different yield performance with diverse genotypes of sweet potato as one simulation of different agronomic performance.

T5

A461

3D modeling and estimating of leaf and stem shape of nursery Paprika plants by a novel multi-camera photography system

Yu Zhang
The University of Tokyo

In plant growth and phenotyping, accurate measurements of plant structure parameters are very crucial. Over the past three decades, 2D imaging has been applied to a variety of aspects, ranging from structural analysis, stomata movements and photosynthetic diagnosis, herbicide control, disease detection and yield estimation. However, as 2D imaging is not robust against the occlusion of plant organs, such as overlapping leaves and stems and changing shapes during the measurement, most of these applications are either for large-scale analysis or for simple plant variables' measurements at the early growth stage. Recently, 3D imaging technologies for the robust measurement of plant growth and phenotype parameters have emerged and been applied for solving this problem owing to the advance of sensing technology. Among the most noticeable applications of 3D imaging technology in agriculture are the 3D model construction for plant and the properties estimation for plant growth. In this research, a novel solution was proposed to improve surface representation and construct 3D plant models through automated feature extraction and key point matching across images based on the Structure from Motion (SfM) algorithm. Our method could solve some problems or limitations of other methods, such as high cost, self-occlusion, low accuracy, information missing, noises, and so on. This study aimed at estimating and comparing plant parameters with different cameras and lenses for 3D imaging. We have, therefore, developed a high-efficiency multi-camera photography (MCP) system combining Multi View Stereovision (MVS) with Structure from Motion (SfM) algorithm. In this study, we measured six variables of nursery paprika plants and investigated the accuracy of 3D models reconstructed from photos taken by four lens types at four different positions. The results demonstrated that error between the estimated and measured values was small and the root mean square error (RMSE) for leaf width/length and stem height/diameter was 1.65mm (R^2=0.98) and 0.57mm (R^2=0.99), respectively. The accuracies of 3D model reconstruction of leaf and stem by 28 mm lens at the first and third camera positions were the highest and the percentages of the number of leaf and stem whose 3D model shape surface area reconstructed over 95% to real shape were about 47% and 100%, respectively. The results confirmed the feasibility of the proposed new methodology for the reconstruction of fine-scale plant model and accurate estimation of the plant parameters. It also displayed that our system is an ideal system for capturing high-resolution 3D images of nursery plants with high-efficiency.

T5

A463

Construction of a high-density genetic map and its application to QTL identification for fiber strength in upland cotton

Zhen Zhang

Institute of Cotton Research of Chinese Academy of Agricultural Science

Cotton (Gossypium sp.) is an important worldwide cash crop that provides a competitive renewable natural fiber supply for the demands of textile industry. The development of new textile technologies and the improvement of living standards increase the demands for both fiber quantity and fiber quality. '0–153' is an upland cotton cultivar with excellent fiber quality derived from Asiatic cotton sources, especially with regards to fiber strength. To identify quantitative trait loci (QTLs) for fiber strength in this line, a recombinant inbred line population consisting of 196 lines was developed from a cross between it and 'sGK9708'. A genetic linkage map consisting of 2393 loci was constructed using this recombinant inbred line population, with single nucleotide polymorphism (SNP) markers from the IntlCottonSNPConsortium_70k chip. Quantitative trait loci for fiber strength were detected across 11 environments using both single-environment and combined multipleenvironment models. A total of 63 QTLs controlling fiber strength were detected by the single-environment model. Sixteen QTLs were identified by the combined multipleenvironment model. These QTLs could make a contribution to the improvement of fiber quality via marker-assisted selection and provide useful information for QTL fine mapping and functional gene research activities as well.

T5

A464

De novo assembly and characterization of the transcriptome and development of microsatellite markers in a Chinese endemic *Euphorbia kansui*

Xueyan Zhao[2], *Xia Cai*[1]

1. Northwest University

2. Shaanxi Institute of International Trade and Commerce

Euphorbia kansui Liou is an endemic and traditional medicinal plant in China. Despite the value, the genomic and transcriptomic information of *E. kansui* is limited. Next generation sequencing (NGS) technology has been widely applied to transcriptomics and the development of molecular markers, and functional gene studies of non-model species. In the present study, a total of 43,211,690 high quality reads were generated using Illumina paired-end sequencing technology. After assembly, 58,362 unigenes were recovered with an N50 length of 1,683 bp. 36,396 (62.36%) of these unigenes had at least one blast hit against the multiple public databases. Based on Kyoto Encyclopedia of Genes and Genomes (KEGG) pathway analyses, unigenes were discovered to encode known enzymes that catalyze the formation of the terpenoid backone and diterpenoid. In addition, the unigenes related to casbene synthase which catalyzes the formation of the major active constituents (diterpenoids) in *E. kansui* were identified. For functional marker development, 7,016 candidate simple sequence repeats (SSR, microsatellites) were identified from 6,150 unigenes, and 40 SSR loci were selected randomly and amplified in two populations of *E. kansui*. A total of 28 loci were amplified successfully and 23 loci exhibited polymorphisms. The number of alleles of each locus ranged from 2 to 8 alleles and the mean number of alleles per locus was 3.391. The observed heterozygosity and expected heterozygosity varied from 0.100 to 1.000 and 0.099 to 0.809, respectively. These microsatellite markers developed here will be useful for studies on population genetics of this endemic species. Moreover, the information of transcriptom will be provided useful resource for the future researches on metabolism, regulation of the terpenoids, genetic studies of *E. kansui*.

T5

A465

Application of capillary electrophoresis to genetic diversity evaluation of *Glycine soja*

Yong Zhao[1,2], *Hongkun Zhao*[2], *Yingshan Dong*[1,2]

1. Jilin Agricultural University

2. Jilin Academy of Agricultural Science

The 60 pairs of SSR primers were selected and the genetic diversity of 102 northeast *Glycine soja* accessions were evaluated by means of simple sequence repeats (SSR) and capillary electrophoresis (CE). The results showed that: 385 alleles were detected by 40 pairs of SSR markers, with an average of 9.6 alleles for each locus. Shannon-Weaver diversity indexes averaged 2.0560, *Glycine soja* has a rich genetic variation in northeast China; SSR technique had such characteristics as intervarietal recognizability, environmental stability and accuracy, and could be applied to evaluating the genetic diversity of *Glycine soja* accessions; CE is a kind of effective separation technology, which has characteristics as high efficiency, celerity, sensitivity and less sample consumption.

T5

A466

Screening for salinity tolerance in soybean germplasm

Yong Zhao[1,2], *Hongkun Zhao*[1], *Guangxun Qi*[1]

1. Jilin Academy of Agricultural Science

2. Jilin Agricultural University

Salt and alkali stress are two of the main environmental factors, which can induce a series of complex physiological and biochemical reactions and then result in a dramatic decline of production in soybean. Planting salt-tolerant soybean is the main strategy to use saline-alkaline soils effectively. The aim of our research were to select salt tolerant and high yield soybean varieties and to develop a genetic map of major salt tolerance QTLs for soybean improvement for the long term. A total of 134 soybean germplasm accessions, including 2 parental lines (ZYD00659 and Jikedou 1) and 132 Progeny populations, were evaluated for the salt injury index under 140Mm NaCl treatment at the seeding stage over three biological replications. Salt-tolerant ZYD00659 and salt-sensitive Jikedou 1 were used as the controls. As a result, the salt injury index for ZYD00659 and jikedou 1 were 89% and 65%, respectively. The salt injury index for 25 accessions were higher than 89%. The salt injury index for 73 accessions ranged from 65% to 89%. All these accessions will be used in genetic mapping in the near future. Previous studies have evaluated the salt tolerance in soybean using germination and seedling stage physiological index; however, the tolerance at the adult stage as indicated by yields are usually ignored. In this study, all of the 134 accessions were cultivated in a aline-alkaline soils, and the salt tolerance at the adult stage will be evaluated by investigating yield before long.

T5

A467

Genetic diversity of half-sib offspring of sour jujubes from Huanghua Ancient Chenier

Xingjuan Zheng[1], *Jin Zhao*[2], *Mengjun Liu* [3], *Jiurui Wang* [1]

1. College of Forestry, Agricultural University of Hebei

2. College of Life Science, Agricultural University of Hebei

3. Research Center of Chinese Jujube, Agricultural University of Hebei

Exploration of genetic variation of half-sib families is a basis for their utilization in plant genetic improvement. In addition to phenotypic, physiological and biochemical variation, variation revealed by molecular markers can provide direct insight into the genetic variation of half-sib families. Sour jujube (*Ziziphus acidojujuba* C.Y. Cheng et M. J. Liu), one of the most popular wild fruits and important medicinal plants in China, belongs to the genus of *Ziziphus* Mill. of family Rhamnaceae and regarded as the ancestor species of cultivated Chinese jujube (*Z. jujuba* Mill.). Huanghua Ancient Chenier is close to the Bohai Bay, where the germplasms of sour jujube are potential resources for root-stocks with resistance to saline-alkali. In order to evaluate, preserve and utilize the endangered sour jujube germplasms, genetic diversity of two half-sib families of sour jujube sampled at Huanghua Ancient Chenier were investigated by SSR markers in this study. 155 offspring from two half-sib families consisted of 96 and 59 progenies of sour jujube B07 and B16 respectively and genomic DNA of leaf from them was extracted by modified CTAB method. 29 SSR primers pairs which produce repetitive and clear bands were used and PCR reaction products were detected by LabChip GX Touch. The date was analyzed by the POPGENE software, which had been transformed into the format of software identification. A total of 146 alleles were obtained and each locus had an average of 5.034 alleles. The number of alleles varied from 3 to 7. The observed and expected levels of heterozygosity ranged from 0.2500 to 0.8431 (the mean 0.5193) and from 0.1990 to 0.7932 (the mean 0.5721), respectively. Shannon's average index was 1.0718 and Nei's average information index was 0.5702. The coefficient of gene differentiation in different sites was slight different and the maximum and the minimum was 0.3189 and 0.0069, separately, with an average of 0.0852. It indicated that 8.52% of the total genetic variation was from the different groups and most of them from offspring individuals. The genetic diversity of progeny of sour jujube B16 (expected heterozygosity = 0.5468, Shannon's average index = 0.9944, Nei's average information index = 0.5421) was more abundant than that of sour jujube B07 (expected heterozygosity = 0.5134, Shannon's average index = 0.9475, Nei's average information index = 0.5106), The results showed that the genetic diversity of the two half-sib families was rich, and the half-sib offspring has potential in breeding.

T5

A468

Application and evaluation of fluorescent SSR markers detection technique with capillary electrophoresis in *Phalaenopsis* genotypes

Zuo Li[1,2], *Wenfang Xiao*[1,2], *Heming Chen*[1,2], *Fubing Lv*[1,2]

1. Environmental Horticulture Research Institute of Guangdong Academy of Agricultural Sciences

2. Guangdong Provincial Key Laboratory of Ornamental Plant Germplasm Innovation and Utilization

Phalaenopsis, one of the most popular potted flower plants, this species have attracted a great deal of attention worldwide, regarding their long-lived and gorgeous flowers and various pigmentation patterns. However, the understanding of this ornamental genus remains limited and little genomic research has been done. This research used fluorescence simple sequence repeat (SSR) markers detection technique with capillary electrophoresis (CE) system to analyze the genetic relationships among *Phalaenopsis* cultivated germplasm collections. In total, 10 out of 55 SSR primers including genomic-SSR and EST-SSR, were screened and selected for detection in this study. Screening all the collected accessions using those 10 selected SSR primers yielded a total of 145 amplified bands in the range of 130 to 532 bp, of which 144 were polymorphic (99.31%). The number of polymorphic bands per primer ranged from 4 to 21, with an average of 14.4. The averages of PIC and MI were 0.66 and 27.42, respectively. Although the polymorphism revealed by genomic-SSR is lower than by EST-SSR, it can effectively differentiate diverse genotypes as well. The results suggest that the *Phalaenopsis* germplasm collection in this study has a high level of genetic diversity. This should facilitate subsequent studies on genetic mapping and molecular breeding. Compare with traditional agarose gel electrophoresis, capillary electrophoresis with fluorescence detection has the advantages of higher accuracy, sensitiveness, reliability and effectiveness. And a comprehensive research on the genetic diversity of *Phalaenopsis* make significant sence, would both protect the diversity of germplasm resources and make sustainable utilization of those resources. This research was financially supported by National Natural Science Foundation of P. R. China (Grant No. 31201650), Guangdong Provincial Science and Technology Project (Grant No. 2014A030304046, Grant No. 2016B070701014, Grant No. 2015B020231005, Grant No. 2014A020208063, Grant No. 2013B060400032).

T5

A469

Analysis of inter-specific SNPs linked to petal texture traits using SLAF-Seq technology in *Phalaenopsis*

Zuo Li[1,2], *Wenfang Xiao*[1,2], *Heming Chen*[1,2], *Fubing Lv*[1,2]

1. Environmental Horticulture Research Institute of Guangdong Academy of Agricultural Sciences

2. Guangdong Provincial Key Laboratory of Ornamental Plant Germplasm Innovation and Utilization

Orchidaceae species are famous ornamental plants all over the world because of their gorgeous flowers and various flower characteristics. Inter-specific and inter-generic hybridization within and between species of the *Orchidaceae* has further enriched the genetic diversity. In Orchid family, as popular potted flower plants, *Phalaenopsis* flowers owned two kinds of special natural flower texture characteristics, namely waxy petal texture and velvety petal texture. This study was concerned with the genes which associated with the petal texture in *Phalaenopsis*. The genetic control of plant architecture is a promising approach to breed desirable varieties, especially in ornamental flowers. In this study, the F_1 offspring characteristics separation on petal texture derived from a cross-bred combination of *Phalaenopsis* 'Ruey Lih Beauty' (female parent, velvety petal texture) with *Phalaenopsis.* SH127 (male parent, waxy petal texture) were studied. Twenty velvety and waxy petal texture seedlings were pooled and subjected to a specific-locus amplified fragment sequencing (SLAF-seq) method. In this study, the sequencing reads couldn't map to the reference genomes *P. equestris* and *Vanilla planifolia,* in the orchid family (Orchidaceae), thus *Oryza sativa* was used as control to assess the effectiveness of enzyme digestion. The results coming from the alignment between the sequences of CK *O. sativa* and its reference genome indicated that the construction of SLAF library fitted well to the standard, with its paired-end mapped reads reaching 84.37%, normal digestion ratio reached 91.03%. With the help of sequencing alignment as well as other bioinformatics technologies, a total of 79.75 Mb reads were obtained. The average values of Q30 was 87.72%, and the average value of GC content was 37.51%, which was low enough to process the sequencing. After SLAF library construction and high-throughput sequencing, a total of 362,381 high quality SLAF tags were obtained and 101,551 polymorphic SLAF tags were identified, with a polymorphism rate of 28.02%. And then 16,639 SNPs selected to do association analysis, by one genotype deriving from male parent and the other from female parent. Finally, 121 SLAF markers with SNP index of > 0.375, including 186 SNPs, were found. However, annotation of those SNPs were useless for now, only some microsatellite sequences of *Phalaenopsis* with undiscovered function. In next step of this study, a direct comparison of the two progeny pools, abandoning the parents, will be done to find more useful markers related candidate regions. This paper provided results that SLAF-seq technology can be further used for the analysis of population evolution and developing specific SNP markers. This research was financially supported by Guangdong Provincial Science and Technology Project (Grant No. 2014A030304046, Grant No. 2016B070701014, Grant No. 2015B020231005, Grant No. 2014A020208063, Grant No. 2013B060400032).

T5

A470

Conservation and diversification of the miR166 family in soybean and potential roles of newly identified miR166s

Xin Xie, Lulu Zhai, Yanming Hou, Shaomin Bian, ***Xuyan Li***

College of Plant Science, Jilin University

MicroR166 (miR166) is a highly conserved family of miRNAs implicated in a wide range of cellular and physiological processes in plants. miR166 family generally comprises multiple miR166 members in plants, which might exhibit functional redundancy and specificity. The soybean miR166 family consists of 21 members according to the miRBase database. However, the evolutionary conservation and functional diversification of miR166 family members in soybean remain poorly understood. In this study, we identified 5 novel miR166s in soybean by data mining approach, thus enlarging the size of miR166 family from 21 to 26 members. Phylogenetic analyses of the 26 miR166s and their precursors indicated that soybean miR166 family exhibited both evolutionary conservation and diversification, and 10 pairs of miR166 precursors with high sequence identity were individually grouped into a discrete clade in the phylogenetic tree. The analysis of genomic organization and evolution of *MIR166* gene family revealed that 8 segmental duplications and 4 tandem duplications occurred during evolution of the miR166 family in soybean. The cis-elements in promoters of *MIR166* family genes and their putative targets pointed to their possible contributions to the functional conservation and diversification. Further, the expression patterns of the 5 newly identified *MIR166*s were examined during seed development and in response to abiotic stresses, which provided important clues for dissecting their functions and isoform specificity. These findings have laid a foundation for elucidating functional conservation and diversification of miR166 family members, especially during seed development or under abiotic stresses.

T5

A471

Transcriptome sequencing and analysis of male sterile flower buds in *Catalpa bungei*

Hongwei Liang, *Weibing Mao, Faju Chen, Changlan Wang*
Three Gorges University

Objective: In order to revealed molecular mechanism of *Catalpa bungei* male sterility from the gene expression levels, we research the regulation of gene expression patterns for *C. bungei* male sterility in the natural state. So this research can provide new and meaningful reference on male sterility for *C. bungei* and other Woody plants. Method: The transcriptome sequencing was carried out on the flower bud of male sterile and fertile flower bud. We use the method of comparative analysis to deal with the fertile bud and sterility bud transcript through bioinformatics, and predict and filter genes about *C. bungei* male sterility. Results: Transcriptome sequencing produced a total of 27.18 Gb data. Finally we obtained 86 076 Unigene after assembling and removing redundant. Then we make use of the seven function database ((NR, NT, GO, COG, KEGG, Swissprot and Interpro)) annotating the all Unigene, eventually Unigene that can be annotated at any one database in a database of seven is 64 600 (75.05%). We analyze the expression levels of Unigene between the experimental group (male sterility flower buds) and control group (fertile flower buds). Different expressed genes which are different expression and high reliability were screened out. In the noise distribution (Multiple differentially expressed in 2 above, Credibility in 0.8 above), 6 915 up-regulated genes and 3 504 down-regulated genes were selected from the experimental group. Poisson distribution (Multiple different expressed in 2 above, The incidence of errors under 0.001), in three biological repeat SL-1-VS-FL-1, SL-2-VS-FL-2, SL-3-VS-FL-3 obtained different up-regulated expressed gene respectively 13979,13513,13055 and different down-regulated expressed genes respectively 12 170,13 807,10 411. By GO functional analysis for different expressed gene showed that the terms which are significantly enriched and higher cluster frequency reproductive process, during the development of the reproductive, reproductive system development, development of reproductive structures in the biological processes. In molecular function the Auxin efflux transmembrane transporter activity is significantly enriched. By the KEGG pathway analysis for different expressed genes, different expressed genes are mapped to 127 different biological pathways. Those which are significantly enriched and larger number of different expressed genes in metabolic pathways mainly contain metabolic pathways, biosynthesis of secondary metabolites, spliceosome, RNA transport and Metabolism of Glycerophospholipid and Starch and sucrose. By comparing different expressed genes with these genes that are related to male sterility that have been reported. 246 High homology Unigene were distinguished. The COG function classification most gathered in RNA processing and modification, cell cycle control, cell division and chromosome partitions, transcription, *etc.* The different expressed genes which are high homology with the male sterility genes are mapped to the pyruvate metabolism, the plant hormone signal transduction pathway. Conclusion: Male sterility of *C. bungei* possibly due to the involvement of pyruvate metabolism process appeared abnormal, leading to brassinosteroid synthesis abnormal so that the tapetum appear dysplasia and further affect the meiosis of microspore. Eventually *C. bungei* pollen become abortive.

T5

A507

Eneumeration of begomovirus particles by Real Time PCR (RT-PCR) and proteomic analysis for agroinnoculated plants

Bhavin Bhatt[1], *Achuit Singh*[2]
1. Shree Ramkrishna Institute of Computer Education and Applied Sciences
2. Senior Scientist, Crop Improvement Division, ICAR - Indian Institute of Vegetable Research

Begomoviruses have emerged as major constraint for vegetable production. French bean is one of the most important vegetable crops of India which is consumed widely. Dwarf mosaic disease (DMD) and Yellow mosaic disease (YMD) are two premier challenges for the quantitative production of, not only French bean, but major legume plants. We have cloned and sequenced begomoviral genomic components associated with disease manifested leaf samples from Varanasi and Bengaluru, India. Sequence analyses confirmed bipartite nature of begomovirus. *Mungbean yellow mosaic India virus* (MYMIV) is causal agent of DMD in Varanasi and *Horsegram Yellow Mosaic Virus* (HgYMV) is c causal agent of DMD in Bengaluru. Compatible pathogen incidence on plants changes spatial and temporal expression of genome of host plants. Partial tandem repeats (PTRs) were generated for each viral clone in plant binary vector pCAMBIA 2300 and agroinnoculated on *N. benthamiana* and *Phaseolus vulgaris*. Viral component accumulation in agro-inoculated plants was carried out by real time PCR (RT-PCR). During the exponential phase of PCR, the amplified products emit fluorescence signal which are directly proportional to the amplicons generated. Thus, RT-PCR enables quantification of the viral copy numbers into inoculated plants. Absolute quantification of virus DNA was carried out through SyBr Green based RT-PCR. Quantitative polymerase chain reaction results suggest that associated DNA - B molecule is absolutely essential for systemic movement and maintenance of helper virus inside plants. Twenty protein spots were identified and in response to begomovirus infection on French bean by Two-dimensional gel electrophoresis (2-DE). The relative expression levels of most of identified proteins were upregulated. These proteins were mainly involved in defense, signal transduction, energy and metabolic regulation. In our studies, molecular characterization in conjugation with proteomics approach was implemented to get information for molecular events occurring during plant virus interaction. Moreover, our analysis revealed that mock plants had no significant protein profile alteration as compared to virus inoculated plants demonstrating that virus infection initiates modulation of proteomic profile.

T5

A513

Fern genomes on the horizon

Fay-Wei Li
1. Boyce Thompson Institute

2. Cornell University

Ferns are the final frontier in plant comparative genomics. Although ferns are known to have notoriously large genome sizes (average 1C>10 Gb), our ongoing work has identified some small-genome species. As a community effort, we have assembled and annotated two complete heterosporous fern genomes: *Salvinia cucullata* and *Azolla filiculoides*, and here I will highlight some of our key findings. I will also present our progress on three additional fern genome projects, which together with *Ceratopteris*, will provide a broader picture of fern genome evolution.

T5

A516

OsbZIP23 mediated transcriptional regulation work together with H3K4me3 modification to regulate gene expression under drought stress in rice

Wei Zong[2], ***Jun Yang***[1], *Lizhong Xiong*[1]
1. *Huazhong Agricultural University*
2. *Southern University of Science and Technology*

Under drought stress conditions, thousands of genes were differentially expressed (DEGs) in rice. However, the regulatory mechanism of these DEGs remains largely unclear. Here, we revealed an interplay between histone H3K4me3 modification and bZIP transcription factor OsbZIP23 regulation of four clustered dehydrin genes under drought stress in rice. Four dehydrin genes could be rapidly and highly induced by drought stresses, whereas the transposon gene within the cluster could not. The whole chromatin status of this region except the transposon gene was significantly affected after drought stresses. By modulating H3K4me3 modification levels in rice, the expression levels of the dehydrin genes were positively affected and the binding levels of OsbZIP23 to the promoter of the dehydrin genes were enhanced under drought stress. Moreover, H3K4me3 modification and gene expression levels of the dehydrin genes under drought stress were down regulated in *osbzip23* mutants. Taken together, our study uncovers a connection between transcriptional and chromatin regulation under drought stress in rice, which will help us to further understand the gene regulation mechanism under stress conditions in plants.

T5

A521

Genetic control of phenotypic variability: An often overlooked dimension of genetic variation

Simon Forsberg, *Örjan Carlborg*
Dep. of Medical Biochemistry and Microbiology, Uppsala University

Genome Wide Association Studies (GWAS) have been widely used to identify loci contributing to quantitative traits. Such studies typically look for genetic effects on the trait mean, ignoring potential effects of alternative alleles on trait variability. This traditional approach might thereby overlook this important dimension of genetic variation. By searching for loci where alternative alleles are associated with a difference in trait variability (vGWAS), interesting genetic mechanisms can be unraveled that might otherwise have gone unnoticed. Through a vGWAS in *Arabidopsis Thaliana*, we identified multiple functional alleles in the vicinity of the *MOT1* gene, affecting molybdenum accumulation. Such multi-allelic loci are problematic in a GWAS setting, which typically uses bi-allelic genetic markers. They might, however, give rise to an effect on the trait variability that can be detected even with such markers. Here, we identified two structural polymorphisms in the promoter region of *MOT1* that displayed strong effects on the molybdenum levels in the plants. We also revealed two additional associations to molybdenum levels, and by evaluating T-DNA insertion alleles, we were able to suggest two novel candidate genes (*COPT6* and *AT2G27030*) involved in molybdenum homeostasis. Collectively, these alleles gave rise to the observed effect on trait variability that we detected in our vGWA analysis. Searching for genetic effects on variability can also reveal signatures of selection. We found one such case on the gene *CMT2* in *A. Thaliana*. Plants carrying a disruptive mutation in this gene were present across a wide range of climates, possibly indicating environmental plasticity. Plants with the alternative allele were confined to a narrower climate range. This disruptive mutation was further found to affect the genome wide methylation pattern and conveyed increased heat tolerance. Finally, a difference in trait variability between alternative alleles might also be caused by interactions (gene-by-gene or gene-by-environment). If we observe such a locus, it might thus be an indication that it is interacting with some other unknown factor. vGWAS might therefore also be used as a shortcut to detect epistatic loci. We demonstrated this in a large yeast cross, where we characterized extensive networks of interacting loci. We further studied higher order interactions involving up to six loci in these networks, revealing how their phenotypic effects where mutually dependent on each other. In particular we found many examples of genetic capacitors, with the ability reduce or completely negate the phenotypic effects of many other QTLs. In summary, since traditional mean focused GWASs ignore genetic effects on trait variability, they might miss important links between genotype and phenotype. We demonstrate how vGWAS is a valuable tool to detect novel associations of biological importance in natural populations.

T5

A536

H2A monoubiquitination in *Arabidopsis* is generally independent of LHP1 and PRC2 activity

Yue Zhou[1], *Francisco Romero-Campero*[2], *Ángeles Gómez-Zambrano*[3], *Myriam Calonje*[3], ***Franziska Turck***[1]
1. *Max Planck Institute for Plant Breeding Research, Department Plant Developmental Biology*
2. *University of Sevilla, Department of Computer Science and Artificial Intelligence*
3. *University of Sevilla, Institute of Plant Biochemistry and Photosynthesis*

Polycomb group (PcG) complexes PRC1 and PRC2 repress gene expression at the chromatin level in eukaryotes. The classical recruitment model of PcG complexes in which PRC2-mediated H3K27 trimethylation recruits PRC1 for H2A monoubiquitination was recently challenged by data showing that PRC1 activity can also recruit PRC2. However, the prevalence of these two mechanisms is unknown, especially in plants as H2AK121ub marks were examined at only a handful of PcG targets. By using genome-wide analyses, we show that H2AK121ub marks are surprisingly widespread in Arabidopsis, often co-localizing with H3K27me3 but also occupying a set of transcriptionally active genes devoid of H3K27me3. Furthermore, by profiling H2AK121ub and H3K27me3 marks in *atbmi1a/b/c*, *clf28/swn7* and *lhp1* mutants we found that PRC2 activity is not required for H2AK121ub marking at most genes. In contrast, loss of AtBMI1 function impacts the incorporation of H3K27me3 marks at many PcG targets. Our findings show the relationship between H2AK121ub and H3K27me3 marks across the *Arabidopsis* genome and unveil that ubiquitination by PRC1 is largely independent of PRC2 activity in plants, while the inverse is true for H3K27 trimethylation.

T5

A537

Isolation and characterization of anthocyanidin synthase (ANS) gene from callus cultures of roselle, *Hibiscus sabdariffa* L. (Family Malvaceae)

Therese Julienne Medina, *Lourdes Cardenas*
Institute of Biological Sciences, University of the Philippines Los Baños

Characterization of anthocyanidin synthase (ANS) gene of *Hibiscus sabdariffa* was done based on the isolated partial sequence derived from its callus cultures. The partial ANS gene isolated from roselle was made up of 345 bp that is translated into 115 amino acids (GenBank accession number KX818221). The partial sequence showed significant similarities in ANS gene nucleotide and amino acid sequences to *Gossypium hirsutum* (84%) and *Theobroma cacao* (84%). Based on the alignment of amino acid sequences, the partial ANS gene showed homology to the members of the family of 2-oxoglutarate-dependent oxygenases. The information gathered in this study can be used in the improvement of anthocyanin production of *H. sabdariffa in vitro* and possible molecular modification of the ANS gene that would target on the promotion of its expression.

T5

A540

De Novo transcriptome assembly and profiling of antioxidant transcripts in the mature seed embryos of *Moringa oleifera* Lam.

Vivian Panes, *Divina Discar*
Department of Biology, School of Science and Engineering, Ateneo de Manila University

Moringa oleifera Lam. is a widely used crop that produces seeds with a plethora of benefits encompassing nutrition, medicine and bioremediation. These benefits are attributed to the compounds present in the seeds including secondary metabolites. Exploring the transcriptome of the seeds paves the way for the discovery of the expressed secondary metabolites. Although various phytochemical researches reported the presence of the secondary metabolites in *M. oleifera* Lam. seeds, there is a lack of information on the expression of genes that encode for enzymes leading to the synthesis of secondary metabolites through transcriptome analysis. In the present study, RNA sequencing was used to analyze the transcriptome of the mature embryo of *M. oleifera* Lam. More than 41 million sequencing reads were generated and assembled *de novo* using Trinity and SOAP assemblers. Trinity assembly produced 177,417 contigs, of which 52.73% (93,546 transcripts) were annotated from NCBI non-redundant database. The SOAPdenovo assembly produced 49,170 contigs of which 57.83% (28,433 transcripts) were annotated from the NCBI non-redundant database. Biological pathway analysis revealed genes encoding 18 enzymes involved in the catalytic steps in the phenylpropanoid pathway and 11 enzymes involved in the catalytic steps in the flavonoid pathway. Furthermore, there are 19 genes involved in the catalytic steps of the alkaloid pathway. This study provides the transcriptome profile and the candidate set of genes for the biosynthesis of major secondary metabolites such as the phenylpropanoids, flavonoids and alkaloids in the *M. oleifera* seeds. This may serve as a baseline information on the secondary metabolites expressed in the mature *M. oleifera* seeds that may pave the way for improving the yield and quality of the secondary metabolites in the seeds of *M. oleifera*.

T5

A547

Enhanced photosynthetic-H_2 production regulated by non-coding RNAs in eukaryotic microalga *Chlamydomonas reinhardtii*

Hui Li[1], *Zhiyong Zeng*[2], *Yuting Wang*[1], *Sulin Lou*[3], *Xiulan Zhu*[1], *Xiaoqi Cai*[2], *Jiangxin Wang*[3], *Chaogang Wang*[2], *Zhangli Hu*[1]
1. College of Life Sciences & Oceanography, Shenzhen University
2. Guangdong Technology Research Center for Marine Algal Bioengineering, Shenzhen University
3. Shenzhen Key Laboratory of Marine Bioresource & Eco-environmental Science, Shenzhen University

Hydrogen (H_2) is considered as the most attractive alternative fuel and photosynthetic-H_2 produced by green microalgae is regarded as the most promising biofuel. However, incompatibility in the simultaneous O_2 and H_2 evolution from green microalgae prevented the H_2 production applications. How to make the normally cultured algae produce H_2 in a continuous and sustainable way is still the biggest barrier for this issue. Recently, a new strategy on regulation of H_2 production by a model eukaryotic microalga *Chlamydomonas reinhardtii* was investigated in our lab by understanding the functions of non-coding RNAs (ncRNAs). For the first time, microRNAs (miRNAs) and long non-coding RNAs (lncRNAs) have been successfully identified and characterized in *C. reinhardtii*. Several heat-inducible expression vectors, which contain either artificial miRNAs or endogenous miRNAs targeting photosystem II related genes have been constructed and transformed into *C. reinhardtii*. With a more rapidly O_2 consumption, some transgenic algae showed that miRNA gene knock down was correlated with the higher H_2 production. Without imposing any nutrient deprived stress, this novel strategy provided a convenient and efficient way for regulation of photosynthetic-H_2 production in green microalgae by simply "turn-on" the expression of a designed ncRNA.

TT5

A550

Genomic variation associated with local adaptation of weedy

rice during de-domestication

*Jie Qiu[1], Yongjun Zhou[2], **Lingfeng Mao**[1], Chuyu Ye[1], Weidi Wang[1], Yongyi Yu[1], Michael P. Timko[3], Song Ge[4], Longjiang Fan[1], Yongliang Lu[2]*

1. Institutue of Crop Science & Institute of Bioinformatics,Zhejiang University, Hangzhou 310058

2. China National Rice Research Institute, Chinese Academy of Agricultural Sciences, Hangzhou 310006

3. Department of Biology, University of Virginia, Charlottesville, VA 22904

4. State Key Laboratory of Systematic and Evolutionary Botany, Institute of Botany, Chinese Academy of Sciences, Beijing 100093

De-domestication is a unique evolutionary process by which domesticated crops are converted into 'wild predecessor like' forms. Weedy rice (Oryza sativa f. spontanea) is an excellent model to dissect the molecular processes underlying de-domestication. We collected 155 weedy and 76 locally-cultivated rice accessions from four representative regions in China and sequenced their genomes to an average 18.2× coverage. Phylogenetic and demographic analyses indicate that Chinese weedy rice were de-domesticated independently from cultivated rice and experienced a strong genetic bottleneck. Although evolving from multiple origins, critical genes underlying convergent evolution of different weedy types can be found. Allele frequency analyses suggest that standing variations and new mutations contribute differently to japonica and indica weedy rice. A Mb-scale genomic region present in weedy rice but not cultivated rice genomes was identified with evidence of balancing selection suggesting that there might be more complexity inherent to the process of de-domestication.

T5

A551

Origin and evolution of the *MORC* gene family in green plants

***Wei Dong**, Fei Chen, Liangsheng Zhang*

Center for Genomics and Biotechnology, Fujian Agriculture and Forestry University

Microrchidia (MORC) protein is a kind of novel epigenetic regulator that contains a typical GHKL (Gyrase, Hsp90, Histidine kinase, MutL) ATPase domain. Here, we conducted a comprehensive evolutionary analysis of *MORC* genes in green plant lineage. The MORC sequences revealed a successive highly conserved motif composition in addition to the ATPase domain. We defined these conserved motif composition as the MORC domain.This complete MORC domain distributed widely among all examined green plants and its origin could be traced back to the common ancestor of Streptophyta. The MORC domain was present in the Charophyte algae but not in the earlier Chlorophyte algae, showed that it originate from the Charophyte. Phylogenetic analysis showed that the *MORC* gene family in green plant lineage could be divided into two major groups designated as Group I and Group II. Moreover, Group II could be further subdivided into Group II-1 and Group II-2. Group II-2 only contained seed plants that revealed that they divided at least before seed plants. Among them, Group I and Group II-1 had undergone several gene duplication events that resulted in the expansion of *MORC* gene family in angiosperms. Synteny analysis showed that there were both eudicot-shared ancient whole genome duplication and subsequently several family-specificwhole genome duplications occurred in Group I and Group II-1. In Group II-2, only one copy was found in most species, suggesting the Group II-2 was relatively conserved compared with the Group I and Group II-1. Expression of *MORC* genes in Arabidopsis, soybean, and rice revealed that they exhibited relatively higher expression level in reproductive tissues and several paralogous gene pairs showed divergent expression patterns. Overall, our results systematically elucidated the evolutionary relationships of the *MORC* gene family in the green plant lineage, which would help to further reveal their functions in plant epigenetic regulation process.

T5

A552

Development and Characterization of Polymorphic EST-SSR of *Pteroceltis tatarinowii*, an endangered plant endemic to China

Yunyan Zhang

Nanjing University

With wide distribution in the genome, the microsatellite markers relied on expressed sequence tags (EST) can be linked to genetic traits and applied broadly in population genetics, molecular marker assisted breeding and molecular marker screening of target traits. However, the lack of the *P. tatarinowii* genomic and transcriptional dataset limits the application of molecular markers in genetic studies of its populations. In this paper, we used Illumina high-throughput sequencing technique to sequence the *P. tatarinowii* leaf transcriptome, based on which polymorphic microsatellite loci is developed. A total of 42477 unigenes with an average length of 815bp were obtained. After aligning with NR, SwissProt, GO, COG and KEGG databases, 23688 unigenes were assigned. These data provided the foundation for gene expression and function analysis of *P. tatarinowii*. Among the unigenes, a total of 6543 EST-SSRs were identified and130 EST-SSRs were selected for validation as EST-SSR markers by PCR amplification. Of these, 48 EST-SSRs were amplified successfully and 32 EST-SSRs were polymorphic among 47 *P. tatarinowii*individuals. Additionally, cross-amplifications of EST-SSR were detected in *Ulmus gaussenii* and*Ulmus chenmoui*, and the versatility and polymorphism were 25% and 87.5%, respectively. The development of the *P. tatarinowii* microsatellite markers provided the molecular basis for the study of the *P. tatarinowii* and other plants in family Ulmaceae.

T6

A472

Effects of endophytic fungi on perennial ryegrass growth and physiological characteristics under different PH latosol extraction

***Zhenjiang Chen**, Xuekai Wei, Chunjie Li*

(State Key Laboratory of Grassland Agro-ecosystems, College of Pastoral Agriculture Science and Technology, Lanzhou, China

The paper was conducted to compared the endophyte-infected (E+) and endophyte-free (E-) of ryegrass (*Lolium perenne*) on morphological and physiological changes under different pH laterite leaching solution, to analyze the effect of the endophytic fungi on the seeding growth and physiological charactristics of perennial ryegrass. The result showed that: 1) E+ and E– morphological indicators and free proline content (Pro), malondialdehyde content (MDA), superoxide dismutase activity (SOD) and peroxidase activity (POD) reached peak value under pH 6.12 (CK), and were significantly ($P<0.05$) higher than the other five treatments, but malondialdehyde were significantly ($P<0.05$) lower than soil/water treatments. 2) Endophyte-infected significantly ($P<0.05$) improved germination rate, germination energy, germination index, vigor index, shoot length, root length and dry weight and Pro content, SOD and POD activity, and decreased the content of MDA under pH value at 4.47 ~ 5.01, the root length of perennial ryegrassthatendophyte-infected (E+) had no significant differences ($P>0.05$) with endophyte-free (E-) at pH 4.47 ~ 6.12. The showed that endophytic fungi increased the growth of host, and plants had a wide application range of leaching solution pH, the perennial ryegrass can be used in acid soil improvement.

T6

A473

Effects of *Epichloë* on seed germination and seedling growth of *Lolium Perenne* under aluminum stress

Zhenjiang Chen, *Chunjie Li*

State Key Laboratory of Grassland Agro-ecosystems, College of Pastoral Agriculture Science and Technology, Lanzhou, China

Endophytic fungi is a kind of fungi that its partial or hole life history in plant tissue and doesn't cause obvious disease to the plant. According to incomplete statistics, as many as 170 species of endophytic fungi have been found, which involved multiple fungal taxa such as ascomycete, basidiomycete, zygomycete, mitosporic fungi, mycelia sterilia and so on. Grass endophytes mainly focus in ascomycete class Clavicipitaceae family *Epichloë* and *Neotyphodium* genus. Grass endophytes mainly exist in the shoots of plants, which can promote the healthy growth of host plants through providing mineral nutrition for grass. Therefore the grasses infected by endophytic fungi play a important role in phytoremediation of heavy metal contaminated soil with the help of the heavy metal accumulation of grass endophytes hyphae, higher grass biomass and stronger grass root activity. In recent years, scientists have payed a close attention to grass endophytes can promote heavy metal tolerance of grass plant but the related reports are less. These research currently focused on heavy metal element such as Zn, Cd, Pb and Cu, but the study is rare that ryegrass endophytic fungi resistant to Al stress, especially the infection of endophytic fungi on the seed germination of perennial ryegrass under aluminum stress was less. This is the first study of aluminum tolerance effect on perennial ryegrass endophyte. An effort was made in this study to compareperennial ryegrass (*Lolium Perenne* L.) that was lendophyte-infected (E+) and endophyte-free (E-) in term of seed germination under different aluminum stress conditions through germination test on filter paper. Results showed that either control or stress conditions, the germination rate, germinative force, vitality index, embryo length, root tolerance indexes and root activity all decreased with the increase of aluminum concentration both endophyte-infected (E+) and endophyte-free (E-). This may be because that Al is not the essential element of plant mineral nutrition and trace Al can't promote normal plant growth and physiological metabolism, while high concentration of aluminium ions will seriously affect the grass plant growth, physiological and biochemical metabolic pathways. All measurement indexes of seeds that lendophyte-infected (E+) seed were significantly higher than those seeds that endophyte-infected (E+) and endophyte-free (E-) ($P<0.05$). This implied that *Neotyphodium* endophyte can improve the germination rate, germination speed as well as growth of seedling after germination under aluminum stress. But we can see from the weight of embryo, the weight of radicle and tolerance index of seminal root that the decreased level endophyte-infected (E+) seed of seeds wes higher than endophyte-free (E-) seed, and with the increasing of aluminum ion concentration, the differences were gradually narrowing between lendophyte-infected (E+) and endophyte-free (E-) seed of seeds. When aluminum ion concentration reached a certain value there may be no difference (P > *0.05*). In this study, Al stress inhibited the seed germination of perennial ryegrass, while endophyte infection was good for seed germination under Al stress.

T6

A474

Chinese names of plant taxonomists from traditional to modern system

Ling-Li Deng[1], ***Jin-Shuang Ma***[2]

1. Shanghai Chenshan Plant Science Research Center, Chinese Academy of Sciences

2. Shanghai Chenshan Plant Science Research Center, Chinese Academy of Sciences

The authors of the Chinese plant taxonomists recorded in the earlier literature were mainly based on the Wade–Giles System, but nowadays, except for Taiwan, there are a few people who know and use the Wade–Giles system. At the same time, before the official use of Chinese phonetic alphabet, some scholars have also used Postal Spelling System for their names. In addition, the abbreviated names of the Chinese authors in the historical records abroad were also used, which often fall into the confused state. In all, it causes intangible and real problems in the current scholar's reading and checking. In order to not only improve the various types of information and scattered information around international authoritative websites about Chinese names of modern plant taxonomists, but also to provide a clearer author name spelling method of historical background for the researchers who work on history of plant taxonomy, this paper mainly explores the causes of the phonetic transcription of the names of plant Chinese taxonomists, and summarizes the transformation rules of Wade–Giles system and Chinese phonetic alphabet. We also list some earlier records of the author's name in the book of errata, eventually developed a more detailed modern taxonomic description of Chinese name information of plant taxonomists.

T6

A475

The involvement of serine/arginine-rich (SR) proteins in the response of *Arabidopsis thaliana* to iron deficiency

Junjiao Feng[1,2], *Ping Lan*[1], *Wengfeng Li*[1,2]

1. Collaborative Innovation Center of Sustainable Forestry in Southern China of Jiangsu Province, College of Biology and the

Environment, Nanjing Forestry University, Nanjing, P. R. China
2. State Key Laboratory of Soil and Sustainable Agriculture, Institute of Soil Science, Chinese Academy of Sciences, Nanjing, P. R. China

Iron (Fe) is essential micronutrients for plant growth and development, and the deficiency of Fe will reduce the yield and quality of crops. Alternative splicing (AS) is a main source of transcriptome and proteome diversity in metazoans and plants, and AS patterns will be dramatically changed in response to various biotic and abiotic stresses, including Fe deficiency. Serine/arginine-rich (SR) proteins, the significant modulators of constitutive and alternative splicing have been involved in variety of abiotic stresses. However, whether the SR proteins participate in plant iron deficiency response and its mechanism remain unknown. In this study, homozygotes of SR T-DNA insertion mutants, confirmed by DNA sequencing were used to test their roles in iron deficiency. Results showed mutation in one of the SR proteins resulted in more severely yellowing in the leaves than that of wild type plants, which is consistent with the lower content of total chlorophyll in the mutant. Moreover, the mutant plants were showed smaller in size and shorter in the root length than that of wild type plants under iron deficiency. Underlying mechanisms of this SR protein associated with iron deficiency are under the way.

T6
A476
Identification and analysis chemical composition of essential oils from the peel of 6 species of mandarins

***Lulu Feng**, Yunzhi Mao, Hui Ran, Shiyao Liu*
College of Horticulture and Landscape Architecture, Southwest University

Peel essential oil is one of the important characteristic traits of citrus, different species of citrus there are significant differences in the composition and content of essential oil components, between different varieties of the same species of essential oil components is also different, Therefore, citrus essential oil can be used as the basis for the study of citrus origin, genetic relationship and classification. in this study, the peel essential oils from six species of mandarins:Nanfeng tangrine (*Citrus kinokuni* Hort. ex Tan), Satsuma mandarins (*Citru-s unshiu* Marc), No 3 Ponkan (*Citrus reticulata* Blanco), Mangshanyegan (*Citrus nobilis* Lauriro), Cheokan (*Citrus tankan* Tanaka), tangerine (*Citrus tangerina* Tanaka), cultivated under the same pedoclimatic and cultural conditions. Their chemical composition was investigated by solid phase micro-extraction headspace GC-MS. The results showed that there were significant differences among the cultivars in composition and relatively content of volatile and content of essential oils, Cheokan peel essential oil content was the highest in 1.56%, No 3 Ponkan peel essential oil content was the lowest of 0.77%; Totally 64 volatiles belonging to various chemical classes were detected, and there are 8 kinds of common components, for unique components, Nanfeng tangerine contained 1, Satsuma mandarins and No 3 Ponkan both contained 5, Mangshanyegan contained 7, Cheokan and tangerine contained 2 and 6 components respectively;For the identified components, D-limonene (66.91-91.51%), γ-terpinene (tr-14.73%), β-myrcene (2.54-24.77%), linalool (0.1-3.23%), 1R-α-Pinene (0.54-2.43%) were characteristic elements of six kinds of mandarins peel essential oils. According to a cluster analysis, the 6 species of mandarins were divided into two categories corresponding two major chemotypes, limonene and limonene/ γ-terpinene. Nanfeng tangerine, Satsuma mandarins, Mangshanyegan and Cheokan belonged to limonene chemotype with a very high amount of D-limonene (70.04-91.51%), Which Satsuma mandarins and Cheokan closest relative, Mangshanyegan followed, Nanfeng tangerine far;No 3 Ponkan and tangerine contained limonene (72.54%, 66.91%) and γ-terpinene (14.22%, 14.73%) as the main components, so they belonged to limonene/ γ-terpinene chemotype.This study enriched the methods of classification and genetic relationship of mandarins plants, also revealed that the composition of the volatile constituents of the essential oil of the six species of original seed mandarins trees and to provide the basis for exploring the evolution of mandarins fruit flavor.

T6
A477
New ornamental plants in anthropogenically transformed environment in conditions of the steppe zone

*Alexandr Glukhov, **Ludmila Kharkhota***
Donetsk Botanic Garden

The Donetsk region is situated in the steppe zone of southeastern Europe. Tree plantations of this region exist in specific natural conditions of the southern steppe zone. The climate is continental with distinctly appearing droughts and hot winds, uneven annual distribution and high variations of rainfall from 223 to 655 mm. The climate is characterized by significant daily, annual and absolute air temperature fluctuations (from -35 ℃ to +39 ℃). Limiting factors for plant growth are the low temperature in winter and the high temperature in summer period, early autumn and late spring frosts, lack of moisture, soil and air dryness. Donbass is a region with a high concentration of modern industrial production resulting in the increased level of air, soil and groundwater pollution. Large energetic, coal mining, coking, metallurgical and engineering enterprises are located in towns of the region. On almost all indicators, the level of environmental pollution exceeds the current allowable concentrations in 3-5 times. Natural climatic conditions along with the intensive anthropogenic load lead to the decrease in longevity of tree plantations. The lifespan of woody plants is reduced in 3-4 times and rarely exceeds 50-60 years (only 4% of trees achieve this age). In such conditions it is important to enlarge the range of tree plantations for optimization of environment. Natural flora of woody plants in Donbass includes 107 species, not more than 40 of them are in culture, the basis of dendrocenosis is formed by 10-15 species. An important stage in solving the problems of phytooptimization of the environment in the region is connected with works of the Donetsk Botanic Garden (DBG) established more than 50 years ago. The collection fund comprising more than 1,000 species and sorts of woody plants has been created during this period. Nowadays a bit more than 200 species of trees and shrubs are used in landscaping of industrial towns of Donbass. The majority of them are trees and shrubs involved from other floristic regions and brought in culture as a result of intensive introduction activity of DBG concerning the use of valuable species of the world flora. From East Asian floristic region alone 272 species of plants have been introduced in DBG.

Such species as *Acer ginnala* Maxim., *Ailanthus altissima* (Mill.) Swingle, *Ampelopsis aconitifolia* Bunge, *Cerasus tomentosa* (Thunb.) Wall., *Chaenomeles japonica* (Thunb.) Spach, *Chamaecyparis pisifera* (Siebold & Zucc.) Endl., *Deutzia scabra* Thunb., *Larix leptolepis* (Siebold & Zucc.) Gordon, *Picea asperata* Mast., *Platycladus orientalis* (L.) Franco, *Sorbaria sorbifolia* (L.) A. Br., *Ulmus parvifolia* Jacq. have adapted well to local conditions and are widely used in ornamental plantations of the region. Accumulation of the collection fund of woody plants is continuing. Among the species which have passed through introduction, are *Prinsepia sinensis* (Oliver) Bean, *Stephanandra tanakae* (Franch. et Sav.), *Viburnum carlesii* Hemsl., species and cultivars *Berberis thunbergii* DC., *Juniperus chinensis* L., *Spiraea japonica* L. f. The effective methods for their propagation have been developed. During the adaptation to our climatic conditions, some of these species change the growth form. For example, *Buddleia davidii* Franch., being a high shrub from mountainous regions of the subtropical zone of China, every year frosts up to the level of root crown. However, in the next vegetative period it grows again up to the height of 1.5-2.0 m, gets an effective form, blooms abundantly, becoming a perfect decoration for exquisite landscape compositions. The use of dendrologic resources of the world flora and the enrichment of the species composition of plantations are of prime importance in their optimization and hardening in conditions of the technogenic environment.

T6

A478

A hexaploid triticale 4D (4B) substitution line confers superior stripe rust resistance

Daiyan Li[1], *Yujie Wang*[1], *Yi Wang*[1], *Xing Fan*[1], *Jian Zeng*[2], *Lina Sha*[1], *Haiqin Zhang*[1], *Yonghong Zhou*[1], *Houyang Kang*[1]

1. Triticeae Research Institute, Sichuan Agricultural University

2. College of Resources, Sichuan Agricultural University

Triticale lines tend to become less resistant to stripe rust and other fungal diseases over time, and exhibit relatively limited genetic diversity. Therefore, it is important that new triticale varieties with superior agronomic traits are continually produced to enrich the available genetic pool. In this study, a new hexaploid triticale line (K14-827-1), which was derived from the progenies of a wheat–rye–Psathyrostachys huashanica trigeneric hybrid, was identified and analyzed using genomic and fluorescence in situ hybridizations, seed protein profiling, and molecular markers. Meiotic pairing studies suggested that the mean chromosomal configuration of K14-827-1 was 2n = 42 = 0.24 I + 18.23 II (ring) + 2.65 II (rod). The in situ hybridization karyotyping results indicated that K14-827-1 was a 4D (4B) substitution line, consisting of complete R and A genomes and chromosomes 4D, 1B–3B, and 5B–7B. Simple sequence repeat analysis of K14-827-1 confirmed that wheat chromosome 4B had been substituted by chromosome 4D. The seed protein profiling results uncovered polymorphic 75K γ-secalin and low-molecular-weight glutenin subunits between K14-827-1 and its recurrent triticale parent (Zhongsi828). Furthermore, the K14-827-1 plants were highly resistant to the stripe rust pathogen (*Puccinia striiformis* f. sp. tritici) prevalent in China, including race V26/Gui22, than Zhongsi828 plants at the seedling and adult stages. This new hexaploid triticale line may be useful for diversifying triticale germplasms and breeding new varieties with improved forage grass traits.

T6

A479

The effects of the nitrogen form ratios on photosynthetic efficiency of Populus × euramericana 'Neva' under condition of phenolic acids

Hui Li

Linyi university

The objective of this study was to investigate the effects of nitrogen form ratio on photosynthetic productivity of poplar under the environment of phenolic acid. Poplar (Populus × euramericana 'Neva') seedlings were treated with 9 nutrient solutions composed by phenolic acid with three concentrations (T0, T1, and T2) and nitrate salts with three nitrogen form ratios (the ratios of ammonium nitroge to nitrate nitrogen (NO3--N) were 1:3,1:7, and 1:14). The photosynthetic physiology parameters, leaf nitrogen content, and total biomass of poplar were measured. Results showed that photosynthetic physiology parameters were influenced by both phenolic acid concentration and nitrogen form ratio. The nitrogen form ratio was the major factor on net photosyntheticrate (PN). When poplar were treated by solutions with NH4+-N to NO3--N ratio of 1:14, the major limitation factor of photosynthesis was non stomatal factor.When poplar were treated by solutions with NH4+-N to NO3--N ratio of 1:3, the major limitation factor of photosynthesis changed to stomatal factor. The leaf nitrogen content and total biomass were obviously positively related with PN ($P<0.05$). Phenolic acid inhibited photosynthesis of poplar in a major way and this effect decreased with increase of the content of NH4+-N. The order of PN values obtained from 9 treatments was T1-1:3, T0-1:3, T2-1:3, T0-1:7, T1-1:7, T0-1:14, T2-1:7, T1-1:14, and T2-1:14 (from high to low). The photosynthesis productivity and biomass of poplar reached the highest values when the ratio of NH4+-N to NO3--N was 1:3 in each treatment group, respectively. The study results could provide guidance on handling the continuously cropped poplar obstacle issue faced nowadays.

T6

A480

Synthesis and antifungal activity of bis (indolyl)methanes

Fei Liu[1], *Qizhi Wang*[1], *Shu Xu*[1], *Min Yin*[1], *Xiancui Guo*[2], *Jinyue Luo*[2], *Xu Feng*[1]

1. Institute of Botany, Jiangsu Province and Chinese Academy of Sciences, Nanjing Botanical Garden Mem. Sun Yat-Sen

2. Nanjing Forestry University

Bis (indolyl)methanes are a kind of alkaloids with important biological activity, and many bis (indolyl)methanes have been isolated from a variety of terrestrial and marine natural sources. We found 3,3'- (phenylmethylene)bis (1*H*-indole) can be gained in the asymmetric Friedel-Crafts Henry Reaction (F-C-H reaction) in the presence of $NiCl_2$. Therefore, we report the use of $NiCl_2$ as a cheap catalyst for the synthesis of bis (indolyl)methanes based on indole and benzaldehyde. By using the catalyst $NiCl_2$, the reaction was optimized in terms of the solvent, reaction temperature, reaction time, mole ratio of indole and benzaldehyde, mass fraction of catalyst. The suitable conditions for the synthesis of 3,3'-(phenylmethylene)bis (1*H*-indole) are as follows: when indole is 2.0 mmol, *n* (indole):*n* (benzaldehyde) = 2:1.1; the amount of catalyst ($NiCl_2$) is 5% of the amount of benzaldehyde; reaction time is 5 h and the reaction temperature is 40 ℃. The reaction was performed under the suitable conditions three times and the average yield of 3,3'- (phenylmethylene)bis (1*H*-indole) is 82.6% (the yields are 82.3%, 83.1% and 82.5%, respectively). The catalyst ($NiCl_2$) is easy preparation from hydrate ($NiCl_2 \cdot 6H_2O$) and cheap in the reaction system. Meanwhile, the reaction condition is mild and the post-processing is simple in this process. Then 5 bis (indolyl)methanes were synthesized by the synthetic process and high yield is obtained (more than 80%). The structure of the bis (indolyl)methanes purified by column chromatography was identified by infrared spectra, HRMS (high resolution mass spectrometer), ^{1}H NMR and ^{13}C NMR. Then the antifungal activity of bis (indolyl) methanes were studied by using the cross method. Compared with the control group, the inhibition ratio of the 3,3'- (phenylmethylene)bis (1*H*-indole) is 61.2% (*Sclerotiniasclerotiorum* (Lib.) de Bary) and 60.7% (*Botrytis cinerea* Pers. ex Fr.), respectively. The study of antifungal activity of bis (indolyl)methanes expands the application fields of bis (indolyl)methanes and it suggests that bis (indolyl)methanes maybe play a more important role in plant disease control.

T6

A481

Ethnic resources of medicinal plant in Guizhou province -- medicinal plants categories for treating nephropathy

Sizhao Liu[1], *Chunlin Long*[1,2]

1. Minzu university of China

2. Kunming Institute of Botany

Guizhou is particularly rich in biological resources which contains a plenty of medicinal plants. It has formed a variety of unique ethnic medicine after long-term practice. Based on the related literature research and modern chemical technology, the study has collated the main medicinal plants which can treat Nephropathy. It includes the following categories: *Dioscorea opposite Thunb* : the medicinal parts are its block and stem; *Atractylodes macrocephala* Koidz. : the medicinal parts are its tubers; *Plantago major* L. : the medicinal parts are the seeds and entire plants; *Eucommia ulmoides* Oliv. : the medicinal parts are barks. It and its tree farm have been planted a lot in Zunyi. *Cibotium barometz* (L.) J. Simth : the medicinal parts are its tubers. It is widely distributed throughout Guizhou. *Ficus pumila* L. : the medicinal parts are steams and branches, leaves and its fruits. It located in southwestern, central, southern, southeastern of limestone slopes in Guizhou. *Drnaria fortune* (Kunze) J. Smith : the medicinal parts are its tubers. *Psoralca coryliflia* L. : the medicinal parts are its fruits. *Lysimachia christinae* Hance : its entire plants can be used as medicine. *Allium tuberosum* Rottl. ex Spreng : the medicinal parts are its seeds, roots and leaves. *Cuscuta chinensis* Lam. : its seeds can be used as medicine. *Epimedium accuminatum* Franch. : the entire plants can be used as medicine. These plants not only play a prominent role in the treatment of nephropathy, but also play an important role in the treatment of other diseases. This paper hopes to provide the information of medicinal plant resource with reference values for the treatment of nephropathy in the pharmaceutical industry through this research, and promote the healthy development of Guizhou botanical medicine industry.

T6

A482

Origin of plant life unity of meteorite strike energy model and bioenergetic model

Ling Lu

Cological Society of Guangdong Province

Meteorite strike is a common and extensive geological process in the solar system, the Earth is no exception. Based on the causes of meteorite impact on the earth's crust, the cognition to celestial spiral motion and particle structure in the galaxy, as well as the comparison between meteorite strike energy model and various biomorph on the earth, matching between meteorite strike structure on ground rock stratum and plant and biota, it is inferred that the morphological characteristics of the life on the Earth, including fungi, plants and animals is highly consistent with the development of meteorite strike energy model. This study suggests that the said fact provides the key clue proving the mechanism of the origin of life, including plants, on the Earth. In the course of celestial bodies hitting the earth's surface, the superior potential energy converted into high-energy physics process such as kinetic energy, thermal energy and nuclear energy, causing the fission and fusion of the particles, the basic material objects of falling celestial bodies, when the structure of these particles is swelling, and the energy information exchanges, the space-time order of energy movement during the meteorite strike process can be "memorized". As the germ of life on the Earth and under certain conditions, the information they have memorized can express in sequence, such as "gene models". Plant energy model corresponds to the phase before the main clue of meteorite strike energy model ruptured, while the animal energy model extended to a later phase. The life embryo created during the meteorite strike process continued to develop and evolve under the specific ecological environment on the Earth.

T6

A483

Comparison and evaluation on nut quality of Chinese-grown pecan [*Carya illinoensis* (Wangenh.) K. Koch] cultivars

Xiaodong Jia[1,2], *Huiting Luo*[1], *Min Zhai*[1], *Yongrong Li*[3], *Jiping Xuan*[1], *Jiyu Zhang*[1], *Gang Wang*[1], *Tao Wang*[1], *Zhongren Guo*[1]

1. Institute of Botany, Jiangsu Province and Chinese Academy of Sciences

2. Institute of Botany, Jiangsu Province and Chinese Academy of Sciences

3. Nanjing Green Universe Pecan Science & Technology Co., Ltd.

14 Chinese-grown pecan [*Carya illinoensis* (Wangenh.) K. Koch] cultivars, including "Caddo", "Chickasaw", "Cheyenne", "Choctaw", "Desirable", "Kanza", "Mahan", "Pawnee", "Stuart", "Western", "Wichita", "Jinhua", "Lvzhou 1", "Shaoxing" were investigated. Their phenotypic characters, fatty acid, tocopherol, total phenolic (TP), condensed tannin (CT), total flavonoids (TF) content, and antioxidant capacity (AC) were evaluated for each

cultivar. The results of phenotypic characters analysis showed that among all pecan cultivars, "Desirable" had the highest mean nut weight, kernel weight and oil content (11.45 g, 6.60 g, and 79.73%, respectively), "Mahan" had the highest mean nut shape index (2.24), "Wichita" had the highest mean kernel rate (61.57%) and the lowest mean shell thickness (0.70 mm). Fatty acid composition was analyzed by Gas Chromatograph (GC). The results showed that the mean relative content of palmitic, stearic, oleic, linoleic, and linolenic acids were 6.03%, 2.09%, 68.93%, 21.93%, and 1.02% respectively. Unsaturated fatty acid (UFA) was the major fatty acid present and "Lvzhou 1" cultivated in China had the highest UFA content (93.72%) among all the cultivars. The total tocopherols content varied from 225.61 $mg \cdot kg^{-1}$ ("Kanza") to 466.95 $mg \cdot kg^{-1}$ ("Pawnee") oil, while the predominant gamma-tocopherol was determined to be 211.99 $mg \cdot kg^{-1}$ ("Kanza") to 448.92 $mg \cdot kg^{-1}$ ("Pawnee") oil. TP, CT and TF were quantified in acetonic extracts. Fresh kernel contained 1.14 ("Chickasaw") to 2.14 ("Mahan") g ellagic acid equivalents/100 g. A strong correlation was found between the content of TP and the antioxidant activity. The antioxidant ability of pecan kernels varied from 28.57 ("Choctaw") to 134.38 ("Mahan")μmol Trolox equivalent/g dry weight (ABTSradical-scavenging activity) and 57.64 ("Stuart") to 126.97 ("Mahan") mg Trolox equivalent/g dry weight (DPPH radical-scavenging activity). The results indicated that "Mahan" had stronger antioxidantabilitythan other cultivars and pecan can be considered as an important source of antioxidants.

T6

A484

Identification and application analysis of volatile components of the essential oils from leaves of *Lindera* Thunb. growing in Jinyun Mountain

Yunzhi Mao, *Lulu Feng, Hui Ran, Shiyao Liu*

College of Horticulture and Landscape Architecture, Southwest University

Essential oils were kinds of volatile aromatic compounds, which contained components had a varietits of biological activities. *Lauraceae* was an important resource of the plant essential oils, common Lauraceae essential oils were included camphor, aromatic alcohol, safrole, eucalyptol (e), cineole and so on. *Lindera* Thunb. was rich in volatile components of plants, a large number of studies have indicated that the volatile components of essential oil in the genus *Lindera* Thunb. were observably different. In this paper, the essential oils obtained from the leaves of three species of *Lindera* Thunb. in Jinyun Mountain of Chongqing, were analyzed for volatile constituents and the application value by hydrodistillation, GC-MS, and physical and chemical properties. Results shown that the extraction rates of the essential oil from leaves of *Lindera glauca* (Sieb. et Zucc.) Bl, *Lindera megaphylla* Hemsl., *Lindera kwangtungensis* (Liou) Allen were 0.1738%, 0.0802% and 0.1510%, respectively; a total of 148 volatile compounds (59 kinds of alcohols, 43 kinds of alcohols, 14 kinds of ketones, 12 kinds of esters, 7 kinds of aldehydes, 6 kinds of organic acids, 1 kind of pyran and 6 kinds of others) were isolated from three kinds of leaf essential oils, among them, 16 kinds of common components (9 kinds of hydrocarbons, 6 kinds of alcohols, 1 kinds of others); *L. glauca*'s main ingredients were α- cadinol (9.41%) and (+)-δ-cadinene (8.32%), α-humulene (7.37%) and γ-muurolene (6.23%) can be used for the deployment of a variety of soap and cosmetics fragrance; *L. megaphylla*'s main ingredients were phytol (14.9%), α-cadinol (6.58%) and hexadecanoic acid (6.09%), dextrose-limonene (4.67%) had an effect on refreshing and promoting metabolism; *L. kwangtungensis*'s main ingredients were trans-nerolidol (48.44%), farnesyl alcohol (6.17%) and caryophyllene oxide (5.21%), among them, trans-nerolidol was important synthetic precursor of isophytol, medicine, cosmetics, spices and chemical synthesis intermediates. This paper revealed the volatile constituents' regular pattern of essential oils of three kinds of native *Lindera* Thunb. plants leaves of Jinyun Mountain in Chongqing, analyzed the physical and chemical properties and application potential of the main volatile components, and provided a scientific basis for the development and utilization of native plant resources of *Lauraceae* in Chongqing.

T6

A485

Leaf venation and its taxonomical significance in Sect. *Denticulatae* (*Salix*) of China

Xiangyu Meng, Huiying Yu, ***Zhixiang Zhang***

College of Nature Conservation, Beijing Forestry University

Comparative observation on leaf venation of 17 species and 1 variety from *Salix* Sect. *Denticulatae* were made performed in order to provide new morphological evidence for the classification of *Salix* by the means of this section. The results showed that the venation of all taxa was found to be craspedodromously pinnate. Most species have intersecondary veins. Tertiary and higher level veins were divided into three types. The areoles are regular or irregular. Veinlets are branched or absent. The results also indicated that the areoles and veinlets were stable characters which showed significant differences bwteeen species. Based on the UPGMA analysis for venation pattern of Sect. *Denticulatae* C.K. Schneid., *S. praticola* Hand.-Mazz. ex Enander types and *S. delavayana* Hand.-Mazz. types were founed. Leaf venation characters provides new characters. They could be used, therefore, as important diagnostic features for identifying these species.

T6

A486

Determination of volatile components from *Trib. litseeae Mez* leaves by GC-MS in Jinyun Mountain Chongqing

Hui Ran, *Lulu Feng, Yunzhi Mao, Shiyao Liu*

Southwest University College of Horticulture and Landscape Architecture

Lauraceae plants are important timber species, spices, aromatic oils and medicinal resources, with great economic value. Lauraceae resources are rich in Jinyun Mountain Chongqing, having identified the local Lauraceae plants 10 genus 22 species 3 variants. In this study, four kinds of plants were used to study the extraction of essential oil from leaves and the identification of volatile components by using distillation and GC-MS. The extraction rate of essential oil and the composition of volatile components were obtained. The extraction rates of essential oil from the *Neolitsea levinei Merr.*, *Neolitsea aurata var. glauca Yang*, *Litsea elongate var. subverticillate (Yang) Yang et P.H.Huang* and *Litsea mollis Hemsl.* were 0.2116%, 0.6982%, 0.2584% and 0.0634% respectively. By GC-MS qualitative analysis, 165 volatile components were identified from the 4 species, and *Neolitsea levinei Merr.*, *Neolitsea aurata var. glauca Yang*, *Litsea elongata var. subverticillate (Yang) Yang et P.H.Huang* and *Litsea mollis Hemsl.* contained 78,66,68 and 67 components respectively. There are 16 common components among the 4 species. For the identified components, the volatile components of the leaves were mainly composed of hydrocarbons and alcohols. *Neolitsea levinei Merr.* mainly contained volatile components such Borneol acetate, β-Eudesmol, 4-Terpineol; *Neolitsea aurata var. glauca Yang* mainly contained 4-Terpineol, Longifolene, β-Caryophyllene; *Litsea elongata var.subverticillate (Yang) Yang et P.H.Huang* mainly contained α-Bergamotene, α-Pinene, α-Caryophyllene; *Litsea mollis Hemsl.* mainly contained D-Limonene, β-Elemene, trans-Geraniol. The differences of composition of the volatile components among the 4 species were great, so the use value of them shuold be discriminatory. This study provides a theoretical basis for the development and utilization of essential oil in the leaves of Trib. litseeae plant resources in Chongqing area.

T6

A487

Analysis of nutritional components and evaluation of nutritional value for different hay of *Foxtail Millet* germplasm resources

Xiaoli Ren[1], *Shijia Song*[1], *Meng Liu*[1], *Zhigang Shi*[1], *Chunmei Nan*[1], *Fei Liu*[1], *Jihan Cui*[1], *Juan Wang*[2], *Xueyan Xia*[1], *Shunguo Li*[1]

1. Institute of Millet Crops, Hebei Academy of Agriculture and Forestry Sciences, National Foxtail Millet Improvement Center, Minor Cereal Crops Laboratory of Hebei

2. Shijiazhuang engineering vocational college

Foxtail Millet has strong adaptability to abiotic stresses, especially drought and poor soil. It also has high nutritional value. The objective of this study was to improve the selection and breeding efficiency of forage foxtail millet germplasm resources. The contents of 11 nutritional indexs in hay of 82 foxtail millet germplasm resources were detected, which were moisture (MC), crude ash (Ash), crude protein (CP), ether extract (EE), crude fibre (CF), carbohydrate (CA), phosphorus (P), calcium (Ca), ferrum (Fe), zinc (Zn), selenium (Se), then the analysis of correlation and clustering for nutrient quality characters were carried out by the SPSS statistics software 19.0. The results indicated that: (1) There were significant differences in the contents of 11 nutrient components in different millet germplasm resources; (2) Crude protein was significantly positively related to carbohydrate, phosphorus, calcium and zinc, but significantly negatively related to crude fibre; ether extract (EE) was not significantly related to other nutritional indexs; there was a significantly negative correlation between moisture and carbohydrate; crude ash was significantly positively related to calcium and ferrum; crude fiber was significantly negatively related to carbohydrate, phosphorus, zinc and selenium; carbohydrate was significantly positively related to phosphorus and selenium; ferrum was not significantly related to phosphorus; there was a significantly positive correlation between ferrum, calcium, zinc and selenium; zine was significantly positively related to phosphorus, calcium, ferrum and selenium; selenium was significantly positively related to phosphorus, calcium, ferrum and zine. (3) The 82 varieties of millet germplasm resources were classified into six types by hierarchical cluster analysis (squared euclidean distance). 25 varieties of the first type, such as Jigu 25, contain higher crude protein, carbohydrate and mineral, the content of crude fibre was lower, these were high quality forage varieties. The second type India 90-39 and other five varieties, carbohydrate and ether extract were relatively higher. The third type of Jigu 24, the contents of ether extract, crude ash, phosphorus and selenium were relatively higher. The founth type of Jigu 22 and other 46 varieties, rich in minerals. The fifth type of changsheng 06 and lvsuigu, ether extract was relatively higher, but the contents of ferrum and selenium were very low.The sixth type was Z281, the highest moisture and higher ferrum, the other indexes were lower. In conclusion, *Foxtail Millet* was a high quality grain and forage millet. The crude protein of the millet stalk after the harvest of grains was 4.29–9.56%, equal to sorghum, superior to wheat. The content of phosphorus was superior to the normal range of forage grass, the content of zine and selenium were on the normal range of forage grass. The study also showed that the Ca/P ratios of 64 varieties were on the optimal range of ruminants. It had great significance for the development and growth of livestock. This research provided an important foundation for breeding and reasonable utilization of high quality forage millet germplasm resources.

T6

A488

Studies on chemical constituents in the roots of *Suaeda glauca*

Yu Shan, *Qizhi Wang, Ming Wang, Fuqin Guan, Min Yin, Fei Liu, Yu Chen, Shu Xu, Xu Feng*

Institute of Botany, Jiangsu Province and Chinese Academy of Sciences, Nanjing Botanical Garden Mem. Sun Yat-Sen

Suaeda glauca (Bge.) Bge. belongs to the family Chenopodiaceae, which is one of the wild resources in saline soil. It is also a dominant species of halophyte in developing the seawater-irrigation agriculture. As a high-quality vegetable and plant resource with the value of health care, there is a broad prospect for its development, which is lack of system research. The chemical constituents and its activities in the roots of *S. glauca* were systematically studied in this paper. In our study, 21 compounds were isolated from the ethanol extract of *S. glauca* by a variety of modern chromatography and spectroscopic methods. Their structures were elucidated by spectroscopic methods, including 1D-, 2D-nuclear magnetic resonance (NMR) and high-resolution time of flight electrospray ionization mass spectrometry (HR-TOF-ESI-MS), including Fourteen flavones: 6,7, 2′-trimethoxy- 5-hydroxyisoflavanone (SG-2), Rutin (SG-8), 5,8-dimethoxy-7,2′-dihydroxy isoflavanone (SG-9), 5,7, 8,2′-tetrahydroxyisoflavanone (SG-10), 5,8, 2′-trimethoxy- 7-hydroxyisoflavanone (SG-11), 5,6-dimethoxy-5,2′-dihydroxyisoflavanone (SG-12), 5,6-dimethoxy-7,2′-dihydroxyisoflavanone (SG-13), Quercetin 3-O-β-D-glucoside (SG-14), 7,8, 2′-trimethoxy-5-dihydroxyisoflavanone (SG-15), 2′-hydroxy-6,7-methylenedioxy-S-methoxyisoflavone (SG-16), Kaempferol (SG-17), 5,6, 7,2′-tetrahydroxyisoflavanone (SG-18), 8,2′-dimethoxy-5,7-dihydroxyisoflavanone (SG-19), Quercetin (SG-23);

Five sterols: Ergosterol (SG-1), β-daucosterol (SG-3), β-sitosterol (SG-7-a), Stigmasterol (SG-7-b), Spinasterol (SG-7-c); One triterpenses: β-Amyrin (SG-4); One lignan: (-)-syringaresinol (SG-21). Among which, SG-11 and SG-13 are two new compounds, SG-2, SG-10, SG-15, SG-18 are four new natrual products, SG-9, SG-12, SG-16, SG-19 are isolated from the genus *Suaeda* for the first time. The pharmacological activities (cytotoxic effects, antioxidant effects) of isoflavones were evaluated in vitro. SG-2 and SG-12 show antiproliferation activity on human breast cancer cell lines (MDA-MB-231), with an IC50 value of 27.28 μM and 38.31 μM, respectively. The study provides essential chemical information for further development of *S. glauca*.

T6

A489

Function analysis of a glyceraldehyde-3-phosphate dehydrogenase gene from *Spartina alterniflora* under salt stress

Sheng Xu, *Lili Cai, Huiyan Yuan, Bin Sun, Bing Xia, Ren Wang*
Institute of Botany, Jiangsu Province and Chinese Academy of Sciences, Nanjing, China

Besides its role in energy metabolism, glyceraldehyde-3-phosphate dehydrogenase (GAPDH) has been demonstrated to be a multifunctional protein involved in various cellular processes in plants. In this study, a putative *GAPDH* gene from *S. alterniflora* (*SaGAPDH*) was identified based on the full-length cDNA library, and further its potential role in response to salt stress was addressed. The results showed that *SaGAPDH* was highly conserved from plant *GAPDH*s, and localized in cytosol. The expression of *SaGAPDH* gene is induced by salt stress. Expression of the recombinant *SaGAPDH* in *E. coli* BL21 (DE3) pLysS showed an increasing salt stress resistance. By using *Agrobacterium*-mediated gene transfer, we generated transgenic *Arabidopsis* plants ectopically antisense-expressing *SaGAPDH* under the control of cauliflower mosaic virus (CaMV) 35S promoter. Polymerase chain reaction (PCR) results showed that *SaGAPDH* was ectopically expressed in *Arabidopsis*, and a specific down-regulation of *AtGAPC1* and the GAPDH enzyme activity was observed. Compared to wild-type plants, transgenic *Arabidopsis* plants antisense-overexpressing *SaGAPDH* showed decreased tolerance to salt stress and down-regulation of some antioxidant defense genes including *superoxide dismutase* (*SOD*), *catalase* (*CAT*) and *guaiacol peroxidase* (*POD*). Furthermore, these results were confirmed by the aggravation of oxidative damage, as indicated by the enhancement of reactive oxygen species (ROS) such as superoxide anion ($O_2^{\cdot-}$) and hydrogen peroxide (H_2O_2) accumulation. Taken together, we proposed that *SaGAPDH* may play important roles in response to salt stress by the regulation of redox homeostasis and relative genes expression.

T6

A490

Research on genetic difference of *Desmodium styracifolium* by chemical mutagen on phenotypic traits and SRAP

Xiaomin Tang
Guangdong Pharmaceutical University

The seeds of *Desmodium styracifolium* were treated with chemical mutagens including sodium azide (NaN_3), ethyl methyl sulfone (EMS) and colchicine. As a result, a lot of macroscopic variations such as plant height, basal diameter and branch numbers occured. indexes including 13 plants with phenotypic variation were selected as materials. SRAP markers were used to conduct DNA molecular identification of the variants of *D. styracifolium* and the control plants. The results showed that a total of 3,285 DNA bands were amplified by 10 pairs of primers. and the percentage of polymorphic bands was 96.45%. The genetic similarity coefficients of the variants ranged from 0.326 to 0.476, which were relatively low. Experiments proved that genetic differences between mutants and the control materials of *D. styracifolium* were true.

T6

A491

Incidence and severity of common diseases of carrot and their pathogenic agents in Northern Nigeria

Nasir Hassan Wagini, *Aliyu Musa Yusuf, Asma'U Ibrahim Safana, Sama'Ila Abubakar*
Umaru Musa Yar'adua University Katsina

The incidence and severity of common fungal and viral diseases affecting cropping and dry season carrot were determined on the carrot growing field of the Sakijiki irrigation site, Batsari Local Government Area, 46 km western Katsina State located on latitude 12°45′05″ N, longitude 7°14′40″ E and 475 m above sea level in the Northern semi-arid ecological zone of Nigeria. The growth period was partitioned into three stages viz: seedling/vegetative stage (1), pre-flowering stage (2), and flowering/fruiting stage (3), during which the incidence of fungal, viral or mixed fungal and viral diseases were investigated. The aim was to identify the relative prevalence of the disease categories for each growth stages. Leaf blight and root knot diseases are the common infections observed during the investigation. The study indicates that *Meloidygene incognita, Cercospora carota* and *Alternaria dauci* were found to be the disease causing organisms in the samples analysed. *M. incognita* was isolated from samples infected with root knot disease, while *C. carotae* and *A. dauci* were isolated from samples infected with leaf blight disease. However, the results showed that plants in the earliest stage of growth with the lowest rate of infection. Nevertheless, viral disease incidence was the significantly highest (22.7%) at this stage. The incidence of double infections increased gradually from the second growth stage reaching 29% in the third stage. Comparatively, infection with virus alone in the third stage was 28.5%, while fungus alone was 0. 59%. The highest disease severity was in plants that were mixed infected by the two categories of pathogens. Such plants had significant retardation of growth (height and mean number of leaves/leaflets) and yield attributes such as the number of fruits produced per plant compared to singly infected and apparently healthy plants. Conclusively, this research work can be used as initial basis for research on carrot.

T6

A492

Sharing the routines at botanical gardens to satisfy demand of the public

Kang Wang
Beijing Botanical Garden

The routines, such as seedling, pruning, cutting, rooting, labeling, *etc.*, at botanical gardens seem very mysterious for the public. When learning more knowledge on botany and horticulture, the public will be more interested in everything about plants. It might be possible for the routine jobs to be a great resource for the education programs. In order to collect more admission fee, most botanical gardens in China are trying to accept more visitors as regular tourists. The public can enjoy the beautiful plants and flowers with nice landscapes as well, but the basic education program with stereotyped exhibitions and interpretation are very poor and not enough far away from the real demand. Beijing Botanical Garden started to open the steps and corners of daily routines to the registered visitors and friends of the garden as more as possible. The first success was made by professional horticulturists, who lead the special tours when the spectacular flower blossoms. Most participants enjoyed the face-to-face and interactive talks, even if everything they talked used to appear on the mounted interpretation or the printed brochures. According the survey, most local visitors want to learn more technical skill on horticulture and botany, which are the regular routines for the botanists and horticulturists at botanical garden. With benefits from internet, the education program made for the public needs locally will open a new gate, which is more widely than the one for the regular admission only. This might be a key step for Chinese botanical garden community to realize the importance of the localization.

T6

A493

The rediscovery of extinct *Lepisanthes unilocularis* Leenhouts (subgenera *Otophora*) endemic to Hainan, China

***Qing-Long Wang**, Zhu-Nian Wang, Qing Yang, Mao-Yuan Wang*

Tropical Crops Genetic Resources Institute, Chinese Academy of Tropical Agricultural Sciences

Lepisanthes unilocularis Leenhouts is an extinct plant species endemic to Hainan, China. In 2013, Nearly 80 years after it disappeared, the plant was found again in Ledong Hainan. The species belongs to *Lepisanthes* subgenus *Otophora*. It grows in the bushes near of the coast, rare. Whether the reproductive process of endangered plants exists weaknesses is the key factor affecting its extinction; while to the extinction fact of the critically endangered plant *L. unilocularis*, we will study the mechanism of this endangered species from the perspective of conservation biology, seeking practical technology to prevent its extinction. Through wild populations and environmental factors surveys, intends to understand the current distribution, habitat conditions, and wild populations. To analyze their weaknesses and limiting factors in the reproductive process, we focus on observing flowering phenology under various specific conditions, estimating outcrossing index, ovule ratio and detecting the quantity and viability of pollens, fertilization ability of stigma, advancing bagging experiment, ascertaining the pollinators and fruit set, pollination efficiency, researching seed germination characteristics under different pH, temperature and humidity. Explaining its propagation characteristics, ecological adaptability and the limiting factors in the reproductive process, revealing some scientific issues in the *L. unilocularis* endangered mechanism, propagation characteristics and ecological adaptability. This study can provide a theoretical basis for developing measures to rational conservation and management measures to prevent its extinction.

T6

A494

The power of technology: the practices of using mobile internet technologies to make botanical gardens more influential in the general public of China

Jimmy Wu

Beijing Lotosky Information Technology Co., Ltd.

Many botanical gardens in China have been very active in utilizing online technologies, since The Internet Plus concept was first presented by Premier Li Keqiang in March of 2015 when delivering the government work report, and unveils Internet Plus action plan to fuel growth of the country. A large proportion of botanical gardens have launched their official sites on the Social Network of Wechat, which is Chinese Facebook. Besides, the most influential botanical gardens in China published their official mobile applications to improve the access of the general public to the useful information of botanical gardens and their plants one after another, the gardens include Beijing Botanical Garden, Wuhan Botanical Garden, Lushan Botanical Garden, Chongqing Nanshan Botanical Garden, Hunan Forest Botanical Garden, Hangzhou Botanical Garden. There are also a number of very proactive cases of applications of IoT (Internet of Things) in public programs of Chinese botanical gardens and showed how IoT can facilitate the public programs and collect data of participants at the meantime, and how the data are processed and analyzed, such as Xishuangbanna Tropical Botanical Garden Chinese Academy of Science.

T6

A495

Landscape and gardening application of Zingiberaceae plants

***Zhi Wu**[1], Nian Liu[2]*

1. ShenZhen Techand Ecology&Environment CO., LTD.

2. Zhongkai University of Agriculture and Engineering

There are 1,500 species Zingiberaceae plants around the worth, and 210 species in China. As perennial herb, Zingiberaceae plants are the important component of tropical and subtropical plant community. Zingiberaceae plants always grow at understory, forest edge, and near stream. Some grow in limestone area, high altitude area, or on trees and stones. Some grow better under shade, or some under sun, or some in water. Some are evergreen or deciduous. Plant type, leave, flower and fruit of most Zingiberaceae plants appear to be abundant and various. Under special seasons and microenvironment, Zingiberaceae plants have irreplaceable role on allocation effect and ornamental value in garden, especially in health garden. According to local conditions, Zingiberaceae plants are used as garden flowers or ground covers in flower beds and flower borders, and as flowers for potted plant and combined potted plant. At present, Zingiberaceae plants *Hedychium coronarium*, *Alpinia zerumbet* and *Alpinia zerumbet* cv Variegata are used widely in landscape and gardening application. By field survey, introduction, domestication, and hybridization breeding, a series of Zingiberaceous plants were applied in landscape and gardening in Southern China. *Zingiber zerumbet*, *Costus speciosus* and *Alpinia polyantha* are suitable as cutting flowers, and have unique flower shape, fresh color, long rachis and long vase time. *Hedychium flavum*, *Hedychium cocineum* and *Costus speciosus* are suitable as understory flowers, and have better shade tolerance, outstretched leaves, tall plant, and upright rachis. *Alpinia pumila*, *Curcuma kwangsiensis* and *Kaempferia rotunda* are suitable as potted flowers, and have low plant, fresh leaves, unique flower shape and better stress resistance.

T6

A496

Antifungal activities of different solvent extracts and chemical components of *Tetradium ruticarpum*

Shu Xu, *Qizhi Wang, Bi Wang, Fei Liu, Yu Shan, Ming Wang, Xu Feng*

Institute of Botany, Jiangsu Province and Chinese Academy of Sciences/ Nanjing Botanical Garden Mem. Sun Yat-Sen

Tetradium ruticarpum is a medicinal plant commonly used in China. In order to investigate the antifungal activities of *T. ruticarpum*, *T. ruticarpum* fruits were successively extracted with petroleum ether, dichloromethane, ethyl acetate, and methanol by using a sonication method. Antifungal activities of these extracts against five plant-pathogenic fungi (*Sclerotinia sclerotiorum*, *Fusarium graminearum*, *Botrytis cinerea*, *Colletotrichum gloeosporioides*, and *Fusarium moniliforme*) were determined with mycelial growth rate method. Eleven compounds (six indole quinoline alkaloids, three quinolone alkaloids, and two limonins) were subsequently isolated from the dichloromethane extract by multi-column chromatography method and recrystallization method, and their antifungal activities were also determined. The results showed that the dichloromethane extract was most effective against these plant-pathogenic fungi, while other extracts had less or none antifungal activities, indicating that the effective components mainly distributed in the dichloromethane extract. Further investigation showed that evodiamine isolated from the dichloromethane extract accounted for most of the antifungal activities, while other compounds showed little effect against tested fungi. In conclusion, the dichloromethane extract of *Tetradium ruticarpum* showed the best antifungal activities among all four extracts, and evodiamine isolated from the dichloromethane extract was the dominant active component in it.

T6

A497

"iFlowers" ——an image-based immediate plant identification system

Min Li[1], *Jing Xuan*[1], *Zhi Liang*[2]

1. Institute of Botany, CAS

2. Robusoft (Beijing) Co., Ltd.

"iFlowers" is an APP for image-based plant identification developed by Institute of Botany, CAS and Robusoft Company. It is built by taking advantage of deep learning techniques and based on about 3 million classified plant photographs of over 25 thousand plant species from the database "Plant Photo Bank of China" (PPBC). Through comprehensive use of deep convolution network, image semantic analysis, fine-grained identification and other fields of technology to achieve the image recognition training and detection, the program has realized the immediate identification of plant species. The program's image recognition training has been completed for China's common wild or cultivated plant species, and the real-time shooting recognition has been achieved for more than 11,000 species that covering most of the common species. Not like other APPs, the images used for identification in "iFlowers" are not limited to specific parts of a plant, and the program have realized the identification of all types of vascular plants, such as ferns, gymnosperms and angiosperms. This program can provide service through APP on android and iOS, WeChat public platform (cn-flora) and the website (http://stu.iplant.cn).

T6

A498

The compilation of *Plants of China*—an illustrated toolbook for 20 thousand species

Min Li, *Jing Xuan*

Institute of Botany, CAS

The Plant Photo Bank of China (PPBC) is a full-time agency for plant picture management in Herbarium (PE) of Institute of Botany, CAS. Since founded in 2008, PPBC has collected about 3 million photographs of over 2,5000 species from 4,146 genera in 538 families. Based on the photograph data, we've started to compile a set of books named *Plants of China*, which is divided into 65 volumes. The set of books is expected to cover all genera and families of native, invasive or naturalized and introduced plant species in China and the species involved are based on the latest classification system. Besides species photographs, their information including the common name, scientific name, distribution map, morphological characters is described in the book. The compilation work is conducted through online editing system which can realize online image selection, text editing and digital typesetting, *etc.* All books are expected to be completed by 2018.

T6

A499

Effects of season on chemical composition and nutrient digestibility of herbage on the Qinghai-Tibetan Plateau

Chuntao Yang[1], *Peng Gao*[1], *Fujiang Hou*[1], *Tianhai Yan*[2], *Shenghua Chang*[1], *Zhaofeng Wang*[1]

1. College of Pastoral Agriculture Science and Technology, State Key Laboratory of Grassland Agro-ecosystems, Lanzhou University

2. Sustainable Agri-Food Sciences Division, Agriculture Branch, Agri-Food and Biosciences Institute

Livestock production relying on pastures has traditionally been the common practice in many areas of the world such as parts of China, New Zealand, and UK. The growth dynamics and productivity of grasslands in China are strongly influenced and restricted by the region's climate, especially in the continental cold/humid type with the rainy season in summer and dry season in winter. The response of grasslands to changes in climate influences the yield and quality of herbage, which affects livestock productivity, but limited information exists concerning herbage nutrient value and livestock performance and the relationship between them in grazing systems of the Qinghai-Tibetan Plateau (QTP). We therefore evaluated the impact of three grass growth seasons on the chemical composition and nutrient digestibility of herbage on the QTP for Tibetan sheep. Herbage was harvested on 14 consecutive days during each of the three annual grass growth seasons (primary-growth period: mid June, bloom phase: mid August, grass-withering period: mid December) from 2011 to 2013. At the same time, we carried out a metabolism trial with six male Tibetan sheep (body weight: 25.9 ± 3.1 kg (mean ± SD)) fed fresh herbage *ad libitum* for 14 days. Grass, feces, and urine samples were collected during the trial period to analyze the chemical composition and to calculate nutrient digestibility. The results shown that growth season affected both the content of all nutrients and digestibility of herbage. The crude protein (CP) content decreased significantly over the three growth seasons ($P < 0.05$). In contrast, the content of both neutral detergent fiber (NDF) and acid detergent fiber (ADF) increased significantly over the three growth seasons ($P < 0.05$). All of the following were highest during the primary-growth period, intermediate during the bloom phase, and lowest during the grass-withering period ($P < 0.05$): the digestibility of dry matter (DM), organic matter, CP, NDF, ADF, digestible energy, and the ratios of digestible energy/gross energy and metabolizable energy/gross energy. Nutrient digestibility and energy parameters correlated positively with the content of both herbage CP and gross energy but correlated negatively with the content of NDF, ADF, organic matter, and ether extract of herbage on the QTP. We found that the nutrient quality and digestibility of herbage on the QTP decreased over the three growth seasons, which may be one of the reasons suppressed ADG and DMI of Tibetan sheep. The parameters of herbage CP, NDF, and ADF content were most closely related to nutrient digestibility on the QTP. The results provided basic dataset better grazing utilization on the QTP.

T6

A500

Some medicinal plants used as folk medicine for rheumatism diseases in European Turkey

Gulden Yilmaz

Trakya University, Faculty of Science, Department of Biology

This paper includes some traditional medicinal plants which are used for rheumatism diseases as folk medicine in European Turkey. In this presentation the photographs of some of the plants, locations of them in European Turkey and information about how they could be used for rheumatism diseases as folk medicines, plant parts which were used for treatment, preparations, therapeutic effects, using dosages were given. The evaluation of relevant data in literature revealed that, roots of *Cynodon dactylon* (L.) Pers. var. *dactylon* Regel, branches and leaves of *Rapistrum rugosum* (L.) All., *Salix* sp., *Urtica urens* L., leaves of *Sambucus nigra* L., *Urtica dioica* L., *Tribulus terrestris* L., *Malva neglecta* Wallr., leaves and flowers of *Malva sylvestris* L. were used for rheumatism diseases as folk medicine in European Turkey. In spite of the well-developed medicinal facilities, the local people still use folk medicines for the treatment of rheumatism diseases. It is important to document the valuable information because the transmission of knowledge from the old to the new generation.

T6

A501

Taxonomic re-evaluation of some *Gelidocalamus* (Poaceae: Bambusoideae) taxa from Southeast China

***Wengen Zhang**, Weijian Li, Ming Tang, Yuguang Liu, Guangyao Yang*

Jiangxi Provincial Key Laboratory for Bamboo Germplasm Resources and Utilization, Forestry College, Jiangxi Agricultural University

In the present study, we employed a classical taxonomic approach associated with information on micromorphological characters of the leaf epidermis to reassess the taxonomy of four *Gelidocalamus* taxa, a woody bamboo genus endemic to China. We propose that *G. subsolidus* is distinct from *G. tessellatus*, and *G. albopubescens* should be placed in synonymy under *G. subsolidus*. A formal taxonomic treatment is presented, including description, comments, illustrations, a distribution map, and SEM images of the abaxial leaf epidermis.

T6

A502

Genetic diversity of a range of *Taxodium distichum* genotypes and cultivars based on ISSR marker

***Yuhong Zheng** [1,3], Bea Clack[2], David Creech[1], Yunlong Yin[3]*

1. Arthur Temple College of Forestry and Agriculture, Stephen F. Austin State University

2. Department of Biology, Stephen F. Austin State University

3. Institute of Botany, Jiangsu Province & Chinese Academy of Sciences

Taxodium is in the cypress family, Cupressaceae, one of several ancient genera in the family commonly known as cypresses. Once three separate species under the genus Taxodium, we are currently accepting Taxodium as one species with three botanical varieties: *Taxodium distichum* (L.) Rich.var. distichum (Baldcypress, BC), 2) *T. distichum* var. imbricarium (Nutt.) Croom (Pondcypress, PC), and 3) *T. distichum* var. mexicana Gordon (Montezuma cypress, MC). In recent years, Taxodium has found favor in coastal windbreak forests, tidal lands, and bottomland restoration projects and is appreciated for tolerance to flood, salinity, alkalinity and hurricanes. In this research project, the genetic diversity of the three botanical varieties, purported hybrids from China, were evaluated by Inter-Simple Sequence Repeat (ISSR) markers. A total of 135 individuals produced 108 polymorphic bands when amplified using nine simple sequence repeat ISSR primers. The Nei's gene diversity (h), the Shannon's Information index (I) and the Percentage of Polymorphic Loci (PPL) between different genotypes of baldcypress were 0.2581,0.3931 and 87.96%. For PC, they were 0.2278,0.3445 and 72.22%, respectively. For MC, the values were 0.2068,0.3120 and 61.11%. The results suggested rich genetic diversity between the genotypes of Taxodium, and less within the genotypes. Gene differentiation analysis showed that the gene flow (Nm) and coefficients of gene differentiation (Gst) of the three taxa were 0.7189 and 0.1955 for BC, 0.2696 and 0.2942 for

PC, 0.9861 and 0.0071 for MC. The results indicated that there was less gene exchange between genotypes in Taxodium that will lead to the population differentiation.

T6

A503

Selection of medium for cultivation of *Idesia polycarpa* Maxim. var. *vestita* Diels

Jingbo Zhou[1], *Zirui Zhao*[2], *Hong Chen*[1], *Yuan Jiang*[3]
1. Anhui Vocational & Technical College of Forestry
2. Shanghai Bohao Biotechnology Corporation
3. Anhui Agricultural University

Idesia polycarpa Maxim. var. *vestita* Diels is a kind of important energy plant that has the very high utilization value. In order to improve the seed germination rate and cultivate the high quality seedlings, there is need to find an optimal cultivation medium for seedling growth of *Idesia polycarpa* in a plastic cave tray (plug seedling). In this study, four types of media including M1 (soil), M2 (peat : perlite = 2:1), M3 (peat), and M4 (peat : perlite : vermiculite = 2:1:1) were tested in industrialized plug seedling. The seedling growth of *Idesia polycarpa* including plant height, crown diameter, leaf counts, ground diameter, the relative chlorophyll contents of leaves, *etc.* was measured. The results indicated that medium M2 was the best one for the increases in seedling height, crown diameter, leaf counts, ground diameter and the relative chlorophyll contents of leaves. Therefore, medium M2 is recommended in the industrialized nursery process for the seedling growth of *Idesia polycarpa.*

T6

A504

Exogenous salicylic acid and abscisic acid affect physiology *Epichloë-Achnatherum inebrians* symbiosis

Jing Liu, *Chunjie Li*
State Key Laboratory of Grassland Agro-ecosystems, College of Pastoral Agriculture Science and Technology

Drunken horse grass (*Achnatherum inebrians*) is an intoxicating grass that grows in the native grasslands of Northwest China. It forms a symbiotic relationship with *Achnatherum inebrians*, *Epichloë-Achnatherum inebrians* symbiosis, which can promote plants growth, and improve the biotic and abiotic resistance. To test physiological change of *Achnatherum inebrians* under exogenous salicylic acid (SA) and abscisic acid (ABA) treatment, endophyte-infected (E+) and endophyte-free (E-). *A. inebrians* were set experience and control group, respectively. The result showed that total chlorophyll content, proline acid content, superoxide dismutase activity, peroxidase activity and soluble sugar content increased, but malondialdehyde content decreased under both SA and ABA treatments in E+plants, comparing to E-plants. Under ABA treatment, chlorophyll content, proline acid content, superoxide dismutase activity and peroxidase activity significantly increased, and proline acid content and superoxide dismutase activity significantly increased under SA treatment in E+plants, comparing with E-plants. These results showed that *Epichloë* endophyte could alleviate the damage from exogenous ABA and SA stress in *Epichloë-Achnatherum inebrians* symbiosis.

T6

A515

Triterpenoid saponins from *Salicornia europaea* and their α-glucosidase inhibition activites

Xin Ma, Hui Lv, Fuqin Guan, Xu Feng, ***Yu Chen***
Jiangsu Key Laboratory for the Research and Utilization of Plant Resources, Institute of Botany, Jiangsu Province and Chinese Academy of Sciences

Salicornia. (Chenopodiaceae) is a genus of annual apparently leafless halophytic herbs that have articulated and succulent stems. *S. europaea* is used in traditional medicine for treating hypertension, cephalalgia and scurvy in China. Although the economic value of this plant has already been discovered and developed, its medicinal value has been ignored for a long time. The present study was carried out to examine the chemical constituents of *S. europaea* and their α-glucosidase inhibition activities to support further pharmaceutical research. Previously two classes of principle ingredients have been reported from this species, including flavones, and triterpenoid saponins. In the current research, a new triterpenoid saponin, 3-O-*β*-D-glucopyranosyl- (1→2)-[*β*-D-xylopyranosyl- (1→3)]-*β*-D-glucuronopyranosyl 30-noroleanolic acid 28-O-β-D-glucopyranosyl ester (Salieuropaea A), was isolated from the plant together with thirteen known triterpenoid saponins. Their structures were established by spectral analysis. The α-glucosidase inhibition activities of the isolated triterpenoid saponins were also investigated. Experimental results show that aglycones of the saponins give potent α-glucosidase inhibition activities, which indicated that the α-glucosidase inhibition activity of aglycone may precede that of saponin.

T6

A541

Antimicrobial *in vitro* activities of selected species of medicinal plants of Kaimosi in the Lake Victoria Basin

Leonard Sida Malweyi[1], *Donald F Otieno*[2], *Elizabeth W Njenga*[2]
1. University of Nairobi
2. University of Eldoret

The efficacy claims made by herbal practitioners in Kaimosi area on use of 11 medicinal plants for antimicrobial treatment in the area was investigated in 2009. The disc diffusion method was used to ascertain the efficacy of crude plant extracts against thirteen test microbes obtained using solvents of increasing polarity (petroleum ether, chloroform, methanol and water) and subsequently the minimum inhibitory concentrations of extracts with good results determined using the micro-dilution method. The extracts from *Lantana trifolia* were found to be most active against bacteria and those of *Fuerstia africana* against fungi with activity ranging from 1 mm to 7 mm for the former and from 1 mm to 2 mm for the latter. The isolates most susceptible to the extracts were Shigella sp., *Bacillus subtilis* and *Staphylococcus aureus* each out of 44 extracts. Chloroform and methanol extracts of *L. trifolia* had the largest inhibition zone of 7 mm diameter against *Sh. flexneri* and 6.5 mm against *S. aureus* respectively. *Fuerstia africana* produced the most promising results on both fungal and bacterial isolates, giving a MIC value of 0.051 mg/mL against *Shigella flexneri* and 0.102 mg/mL against *Staphylococcus aureus*. The results of this study supported the use of medicinal plants in management of infectious diseases in the study area.

T6

A548

Assessment physiological effects of gamma rays on trigonella (*In vivo* and *In vitro*)

Mahmoud Khater
National Research Centre

This investigation was carried out during two successive seasons 2013/2014, 2014/2015 at the Greenhouse and Tissue culture laboratory, Botany Department, National Research Centre, Giza, Egypt. Molecular studies were carried out at Genetics Department, Faculty of Agriculture, Zagazig University, Egypt. To study the physiological and molecular changes in fenugreek (*Trigonella foenum graecum* L.) *In vivo* and *In vitro* as a response to gamma rays with doses (0, 100, 150, 200, 250 and 300 Gy). Results revealed that callus induction frequency increased in response to the increase of 2,4-D concentrations and varied from one gamma-ray dose to another. The best interaction between gammaray and 2,4-D concentrations was between 2.0 mg/L 2,4-D and all doses of γ-rays especially 200 Gy which recorded the highest percentage of callus (above 90%). On the other hand, in RAPD analysis there were 44 bands (total bands) divided to polymorphic bands (25), monomorphic bands (2) and unique bands (17) with an average percentage 88.00% polymorphism and numerous molecular sizes that ranging from 265,122 to 1,911,225 bp, Moreover, results of ISSR analysis illustrated that there were 22 bands (total bands) divided to polymorphic bands (3), monomorphic bands (11) and unique bands (8) with an average percentage 76.00% polymorphism and different molecular weights that ranging from 260,582 to 1,333,384 bp.

Author Index

G

Q

R